Climate Change 2013
The Physical Science Basis

Working Group I Contribution to the Fifth Assessment Report of the Intergovernmental Panel on Climate Change

Edited by

Thomas F. Stocker
Working Group I Co-Chair
University of Bern

Dahe Qin
Working Group I Co-Chair
China Meteorological Administration

Gian-Kasper Plattner
Director of Science

Melinda M.B. Tignor
Director of Operations

Simon K. Allen
Senior Science Officer

Judith Boschung
Administrative Assistant

Alexander Nauels
Science Assistant

Yu Xia
Science Officer

Vincent Bex
IT Officer

Pauline M. Midgley
Head

Working Group I Technical Support Unit

CAMBRIDGE
UNIVERSITY PRESS
www.cambridge.org

CAMBRIDGE
UNIVERSITY PRESS

32 Avenue of the Americas, New York NY 10013-2473, USA

Cambridge University Press is part of the University of Cambridge.

It furthers the University's mission by disseminating knowledge in the pursuit of education, learning, and research at the highest international levels of excellence.

www.cambridge.org
Information on this title: www.cambridge.org/9781107661820

First published 2014

Printed in the United States of America

A catalog record for this publication is available from the British Library.

ISBN 978-1-107-05799-9 Hardback
ISBN 978-1-107-66182-0 Paperback

Please use the following reference to the whole report:
IPCC, 2013: *Climate Change 2013: The Physical Science Basis. Contribution of Working Group I to the Fifth Assessment Report of the Intergovernmental Panel on Climate Change* [Stocker, T.F., D. Qin, G.-K. Plattner, M. Tignor, S.K. Allen, J. Boschung, A. Nauels, Y. Xia, V. Bex and P.M. Midgley (eds.)]. Cambridge University Press, Cambridge, United Kingdom and New York, NY, USA, 1535 pp.

Cover photo:
Folgefonna glacier on the high plateaus of Sørfjorden, Norway (60°03′ N - 6°20′ E) © Yann Arthus-Bertrand / Altitude.

Foreword, Preface and Dedication

Foreword

"Climate Change 2013: The Physical Science Basis" presents clear and robust conclusions in a global assessment of climate change science—not the least of which is that the science now shows with 95 percent certainty that human activity is the dominant cause of observed warming since the mid-20th century. The report confirms that warming in the climate system is unequivocal, with many of the observed changes unprecedented over decades to millennia: warming of the atmosphere and the ocean, diminishing snow and ice, rising sea levels and increasing concentrations of greenhouse gases. Each of the last three decades has been successively warmer at the Earth's surface than any preceding decade since 1850.

These and other findings confirm and enhance our scientific understanding of the climate system and the role of greenhouse gas emissions; as such, the report demands the urgent attention of both policymakers and the general public.

As an intergovernmental body jointly established in 1988 by the World Meteorological Organization (WMO) and the United Nations Environment Programme (UNEP), the Intergovernmental Panel on Climate Change (IPCC) has provided policymakers with the most authoritative and objective scientific and technical assessments. Beginning in 1990, this series of IPCC Assessment Reports, Special Reports, Technical Papers, Methodology Reports and other products have become standard works of reference.

This Working Group I contribution to the IPCC's Fifth Assessment Report contains important new scientific knowledge that can be used to produce climate information and services for assisting society to act to address the challenges of climate change. The timing is particularly significant, as this information provides a new impetus, through clear and indisputable physical science, to those negotiators responsible for concluding a new agreement under the United Nations Framework Convention on Climate Change in 2015.

Climate change is a long-term challenge, but one that requires urgent action given the pace and the scale by which greenhouse gases are accumulating in the atmosphere and the risks of a more than 2 degree Celsius temperature rise. Today we need to focus on the fundamentals and on the actions otherwise the risks we run will get higher with every year.

This Working Group I assessment was made possible thanks to the commitment and dedication of many hundreds of experts worldwide, representing a wide range of disciplines. WMO and UNEP are proud that so many of the experts belong to their communities and networks. We express our deep gratitude to all authors, review editors and expert reviewers for devoting their knowledge, expertise and time. We would like to thank the staff of the Working Group I Technical Support Unit and the IPCC Secretariat for their dedication.

We are also grateful to the governments that supported their scientists' participation in developing this report and that contributed to the IPCC Trust Fund to provide for the essential participation of experts from developing countries and countries with economies in transition. We would like to express our appreciation to the government of Italy for hosting the scoping meeting for the IPCC's Fifth Assessment Report, to the governments of China, France, Morocco and Australia for hosting drafting sessions of the Working Group I contribution and to the government of Sweden for hosting the Twelfth Session of Working Group I in Stockholm for approval of the Working Group I Report. The generous financial support by the government of Switzerland, and the logistical support by the University of Bern (Switzerland), enabled the smooth operation of the Working Group I Technical Support Unit. This is gratefully acknowledged.

We would particularly like to thank Dr. Rajendra Pachauri, Chairman of the IPCC, for his direction and guidance of the IPCC and we express our deep gratitude to Professor Qin Dahe and Professor Thomas Stocker, the Co-Chairs of Working Group I for their tireless leadership throughout the development and production of this report.

M. Jarraud
Secretary-General
World Meteorological Organization

A. Steiner
Executive Director
United Nations Environment Programme

Preface

The Working Group I contribution to the Fifth Assessment Report of the Intergovernmental Panel on Climate Change (IPCC) provides a comprehensive assessment of the physical science basis of climate change. It builds upon the Working Group I contribution to the IPCC's Fourth Assessment Report in 2007 and incorporates subsequent new findings from the Special Report on Managing the Risks of Extreme Events and Disasters to Advance Climate Change Adaptation, as well as from research published in the extensive scientific and technical literature. The assessment considers new evidence of past, present and projected future climate change based on many independent scientific analyses from observations of the climate system, paleoclimate archives, theoretical studies of climate processes and simulations using climate models.

Scope of the Report

During the process of scoping and approving the outline of its Fifth Assessment Report, the IPCC focussed on those aspects of the current understanding of the science of climate change that were judged to be most relevant to policymakers.

In this report, Working Group I has extended coverage of future climate change compared to earlier reports by assessing near-term projections and predictability as well as long-term projections and irreversibility in two separate chapters. Following the decisions made by the Panel during the scoping and outline approval, a set of new scenarios, the Representative Concentration Pathways, are used across all three Working Groups for projections of climate change over the 21st century. The coverage of regional information in the Working Group I report is expanded by specifically assessing climate phenomena such as monsoon systems and their relevance to future climate change in the regions.

The Working Group I Report is an assessment, not a review or a text book of climate science, and is based on the published scientific and technical literature available up to 15 March 2013. Underlying all aspects of the report is a strong commitment to assessing the science comprehensively, without bias and in a way that is relevant to policy but not policy prescriptive.

Structure of the Report

This report consists of a short Summary for Policymakers, a longer Technical Summary and fourteen thematic chapters plus annexes. An innovation in this Working Group I assessment is the Atlas of Global and Regional Climate Projections (Annex I) containing time series and maps of temperature and precipitation projections for 35 regions of the world, which enhances accessibility for stakeholders and users.

The Summary for Policymakers and Technical Summary of this report follow a parallel structure and each includes cross-references to the chapter and section where the material being summarised can be found in the underlying report. In this way, these summary components of the report provide a road-map to the contents of the entire report and a traceable account of every major finding.

In order to facilitate the accessibility of the findings of the Working Group I assessment for a wide readership and to enhance their usability for stakeholders, each section of the Summary for Policymakers has a highlighted headline statement. Taken together, these 19 headline statements provide an overarching summary in simple and quotable language that is supported by the scientists and approved by the member governments of the IPCC. Another innovative feature of this report is the presentation of Thematic Focus Elements in the Technical Summary that provide end to end assessments of important cross-cutting issues in the physical science basis of climate change.

Introduction (Chapter 1): This chapter provides information on the progress in climate change science since the First Assessment Report of the IPCC in 1990 and gives an overview of key concepts, indicators of climate change, the treatment of uncertainties and advances in measurement and modelling capabilities. This includes a description of the future scenarios and in particular the Representative Concentration Pathway scenarios used across all Working Groups for the IPCC's Fifth Assessment Report.

Observations and Paleoclimate Information (Chapters 2, 3, 4, 5): These chapters assess information from all climate system components on climate variability and change as obtained from instrumental records and climate archives. They cover all relevant aspects of the atmosphere including the stratosphere, the land surface, the oceans and the cryosphere. Timescales from days to decades (Chapters 2, 3 and 4) and from centuries to many millennia (Chapter 5) are considered.

Process Understanding (Chapters 6 and 7): These chapters cover all relevant aspects from observations and process understanding to projections from global to regional scales for two key topics. Chapter 6 covers the carbon cycle and its interactions with other biogeochemical cycles, in particular the nitrogen cycle, as well as feedbacks on the climate system. For the first time, there is a chapter dedicated to the assessment of the physical science basis of clouds and aerosols, their interactions and chemistry, and the role of water vapour, as well as their role in feedbacks on the climate system (Chapter 7).

From Forcing to Attribution of Climate Change (Chapters 8, 9, 10): All the information on the different drivers (natural and anthropogenic) of climate change is collected, expressed in terms of Radiative Forcing and assessed in Chapter 8. In Chapter 9, the hierarchy of climate models used in simulating past and present climate change is assessed and evaluated against observations and paleoclimate reconstructions.

Information regarding detection of changes on global to regional scales and their attribution to the increase in anthropogenic greenhouse gases is assessed in Chapter 10.

Future Climate Change, Predictability and Irreversibility (Chapters 11 and 12): These chapters assess projections of future climate change derived from climate models on time scales from decades to centuries at both global and regional scales, including mean changes, variability and extremes. Fundamental questions related to the predictability of climate as well as long term climate change, climate change commitments and inertia in the climate system are addressed. Knowledge on irreversible changes and surprises in the climate system is also assessed.

Integration (Chapters 13 and 14): These chapters synthesise all relevant information for two key topics of this assessment: sea level change (Chapter 13) and climate phenomena across the regions (Chapter 14). Chapter 13 presents an end to end assessment of information on sea level change based on paleoclimate reconstructions, observations and process understanding, and provides projections from global to regional scales. Chapter 14 assesses the most important modes of variability in the climate system, such as El Niño-Southern Oscillation, monsoon and many others, as well as extreme events. Furthermore, this chapter deals with interconnections between the climate phenomena, their regional expressions and their relevance for future regional climate change.

Maps assessed in Chapter 14, together with Chapters 11 and 12, form the basis of the Atlas of Global and Regional Climate Projections in Annex I, which is also available in digital format. Radiative forcings and estimates of future atmospheric concentrations from Chapters 7, 8, 11 and 12 form the basis of the Climate System Scenario Tables presented in Annex II. All material including high-resolution versions of the figures, underlying data and Supplementary Material to the chapters is also available online: www.climatechange2013.org.

The scientific community and the climate modelling centres around the world brought together their activities in the Coordinated Modelling Intercomparison Project Phase 5 (CMIP5), providing the basis for most of the assessment of future climate change in this report. Their efforts enable Working Group I to deliver comprehensive scientific information for the policymakers and the users of this report, as well as for the specific assessments of impacts carried out by IPCC Working Group II, and of costs and mitigation strategies, carried out by IPCC Working Group III.

Following the successful introduction in the previous Working Group I assessment in 2007, all chapters contain Frequently Asked Questions. In these the authors provide scientific answers to a range of general questions in a form that will be accessible to a broad readership and serves as a resource for teaching purposes. Finally, the report is accompanied by extensive Supplementary Material which is made available in the online versions of the report to provide an additional level of detail, such as description of datasets, models, or methodologies used in chapter analyses, as well as material supporting the figures in the Summary for Policymakers.

The Process

This Working Group I Assessment Report represents the combined efforts of hundreds of leading experts in the field of climate science and has been prepared in accordance with rules and procedures established by the IPCC. A scoping meeting for the Fifth Assessment Report was held in July 2009 and the outlines for the contributions of the three Working Groups were approved at the 31st Session of the Panel in November 2009. Governments and IPCC observer organisations nominated experts for the author team. The team of 209 Coordinating Lead Authors and Lead Authors plus 50 Review Editors selected by the Working Group I Bureau was accepted at the 41st Session of the IPCC Bureau in May 2010. In addition, more than 600 Contributing Authors provided draft text and information to the author teams at their request. Drafts prepared by the authors were subject to two rounds of formal review and revision followed by a final round of government comments on the Summary for Policymakers. A total of 54,677 written review comments were submitted by 1089 individual expert reviewers and 38 governments. The Review Editors for each chapter monitored the review process to ensure that all substantive review comments received appropriate consideration. The Summary for Policymakers was approved line-by-line and the underlying chapters were then accepted at the 12th Session of IPCC Working Group I from 23–27 September 2007.

Acknowledgements

We are very grateful for the expertise, hard work, commitment to excellence and integrity shown throughout by the Coordinating Lead Authors and Lead Authors with important help by the many Contributing Authors. The Review Editors have played a critical role in assisting the author teams and ensuring the integrity of the review process. We express our sincere appreciation to all the expert and government reviewers. We would also like to thank the members of the Bureau of Working Group I: Jean Jouzel, Abdalah Mokssit, Fatemeh Rahimizadeh, Fredolin Tangang, David Wratt and Francis Zwiers, for their thoughtful advice and support throughout the preparation of the report.

We gratefully acknowledge the long-term efforts of the scientific community, organized and facilitated through the World Climate Research Programme, in particular CMIP5. In this effort by climate modelling centres around the world, more than 2 million gigabytes of numerical data have been produced, which were archived and distributed under the stewardship of the Program for Climate Model Diagnosis and Intercomparison. This represents an unprecedented concerted effort by the scientific community and their funding institutions.

Our sincere thanks go to the hosts and organizers of the four Working Group I Lead Author Meetings and the 12th Session of Working Group I. We gratefully acknowledge the support from the host countries: China, France, Morocco, Australia and Sweden. The support for their scientists provided by many governments as well as through the IPCC Trust Fund is much appreciated. The efficient operation of the Working Group I Technical Support Unit was made possible by the generous financial support provided by the government of Switzerland and logistical support from the University of Bern (Switzerland).

We would also like to thank Renate Christ, Secretary of the IPCC, and the staff of the IPCC Secretariat: Gaetano Leone, Jonathan Lynn, Mary Jean Burer, Sophie Schlingemann, Judith Ewa, Jesbin Baidya, Werani Zabula, Joelle Fernandez, Annie Courtin, Laura Biagioni and Amy Smith. Thanks are due to Francis Hayes who served as the conference officer for the Working Group I Approval Session.

Finally our particular appreciation goes to the Working Group I Technical Support Unit: Gian-Kasper Plattner, Melinda Tignor, Simon Allen, Judith Boschung, Alexander Nauels, Yu Xia, Vincent Bex and Pauline Midgley for their professionalism, creativity and dedication. Their tireless efforts to coordinate the Working Group I Report ensured a final product of high quality. They were assisted in this by Adrien Michel and Flavio Lehner with further support from Zhou Botao and Sun Ying. In addition, the following contributions are gratefully acknowledged: David Hansford (editorial assistance with the Frequently Asked Questions), UNEP/GRID-Geneva and University of Geneva (graphics assistance with the Frequently Asked Questions), Theresa Kornak (copyedit), Marilyn Anderson (index) and Michael Shibao (design and layout).

Rajendra K. Pachauri
IPCC Chair

Qin Dahe
IPCC WGI Co-Chair

Thomas F. Stocker
IPCC WGI Co-Chair

Dedication

Bert Bolin
(15 May 1925 – 30 December 2007)

The Working Group I contribution to the Fifth Assessment Report of the Intergovernmental Panel on Climate Change (IPCC) *Climate Change 2013: The Physical Science Basis* is dedicated to the memory of Bert Bolin, the first Chair of the IPCC.

As an accomplished scientist who published on both atmospheric dynamics and the carbon cycle, including processes in the atmosphere, oceans and biosphere, Bert Bolin realised the complexity of the climate system and its sensitivity to anthropogenic perturbation. He made a fundamental contribution to the organisation of international cooperation in climate research, being involved in the establishment of a number of global programmes.

Bert Bolin played a key role in the creation of the IPCC and its assessments, which are carried out in a unique and formalized process in order to provide a robust scientific basis for informed decisions regarding one of the greatest challenges of our time. His vision and leadership of the Panel as the founding Chair from 1988 to 1997 laid the basis for subsequent assessments including this one and are remembered with deep appreciation.

Contents

Summary for Policymakers

Summary for Policymakers

Drafting Authors:

Lisa V. Alexander (Australia), Simon K. Allen (Switzerland/New Zealand), Nathaniel L. Bindoff (Australia), François-Marie Bréon (France), John A. Church (Australia), Ulrich Cubasch (Germany), Seita Emori (Japan), Piers Forster (UK), Pierre Friedlingstein (UK/Belgium), Nathan Gillett (Canada), Jonathan M. Gregory (UK), Dennis L. Hartmann (USA), Øystein Jansen (Norway), Ben Kirtman (USA), Reto Knutti (Switzerland), Krishna Kumar Kanikicharla (India), Peter Lemke (Germany), Jochem Marotzke (Germany), Valérie Masson-Delmotte (France), Gerald A. Meehl (USA), Igor I. Mokhov (Russian Federation), Shilong Piao (China), Gian-Kasper Plattner (Switzerland), Qin Dahe (China), Venkatachalam Ramaswamy (USA), David Randall (USA), Monika Rhein (Germany), Maisa Rojas (Chile), Christopher Sabine (USA), Drew Shindell (USA), Thomas F. Stocker (Switzerland), Lynne D. Talley (USA), David G. Vaughan (UK), Shang-Ping Xie (USA)

Draft Contributing Authors:

Myles R. Allen (UK), Olivier Boucher (France), Don Chambers (USA), Jens Hesselbjerg Christensen (Denmark), Philippe Ciais (France), Peter U. Clark (USA), Matthew Collins (UK), Josefino C. Comiso (USA), Viviane Vasconcellos de Menezes (Australia/Brazil), Richard A. Feely (USA), Thierry Fichefet (Belgium), Arlene M. Fiore (USA), Gregory Flato (Canada), Jan Fuglestvedt (Norway), Gabriele Hegerl (UK/Germany), Paul J. Hezel (Belgium/USA), Gregory C. Johnson (USA), Georg Kaser (Austria/Italy), Vladimir Kattsov (Russian Federation), John Kennedy (UK), Albert M. G. Klein Tank (Netherlands), Corinne Le Quéré (UK), Gunnar Myhre (Norway), Timothy Osborn (UK), Antony J. Payne (UK), Judith Perlwitz (USA), Scott Power (Australia), Michael Prather (USA), Stephen R. Rintoul (Australia), Joeri Rogelj (Switzerland/Belgium), Matilde Rusticucci (Argentina), Michael Schulz (Germany), Jan Sedláček (Switzerland), Peter A. Stott (UK), Rowan Sutton (UK), Peter W. Thorne (USA/Norway/UK), Donald Wuebbles (USA)

This Summary for Policymakers should be cited as:

IPCC, 2013: Summary for Policymakers. In: *Climate Change 2013: The Physical Science Basis. Contribution of Working Group I to the Fifth Assessment Report of the Intergovernmental Panel on Climate Change* [Stocker, T.F., D. Qin, G.-K. Plattner, M. Tignor, S.K. Allen, J. Boschung, A. Nauels, Y. Xia, V. Bex and P.M. Midgley (eds.)]. Cambridge University Press, Cambridge, United Kingdom and New York, NY, USA.

A. Introduction

The Working Group I contribution to the IPCC's Fifth Assessment Report (AR5) considers new evidence of climate change based on many independent scientific analyses from observations of the climate system, paleoclimate archives, theoretical studies of climate processes and simulations using climate models. It builds upon the Working Group I contribution to the IPCC's Fourth Assessment Report (AR4), and incorporates subsequent new findings of research. As a component of the fifth assessment cycle, the IPCC Special Report on Managing the Risks of Extreme Events and Disasters to Advance Climate Change Adaptation (SREX) is an important basis for information on changing weather and climate extremes.

This Summary for Policymakers (SPM) follows the structure of the Working Group I report. The narrative is supported by a series of overarching highlighted conclusions which, taken together, provide a concise summary. Main sections are introduced with a brief paragraph in italics which outlines the methodological basis of the assessment.

The degree of certainty in key findings in this assessment is based on the author teams' evaluations of underlying scientific understanding and is expressed as a qualitative level of confidence (from *very low* to *very high*) and, when possible, probabilistically with a quantified likelihood (from *exceptionally unlikely* to *virtually certain*). Confidence in the validity of a finding is based on the type, amount, quality, and consistency of evidence (e.g., data, mechanistic understanding, theory, models, expert judgment) and the degree of agreement[1]. Probabilistic estimates of quantified measures of uncertainty in a finding are based on statistical analysis of observations or model results, or both, and expert judgment[2]. Where appropriate, findings are also formulated as statements of fact without using uncertainty qualifiers. (See Chapter 1 and Box TS.1 for more details about the specific language the IPCC uses to communicate uncertainty).

The basis for substantive paragraphs in this Summary for Policymakers can be found in the chapter sections of the underlying report and in the Technical Summary. These references are given in curly brackets.

B. Observed Changes in the Climate System

Observations of the climate system are based on direct measurements and remote sensing from satellites and other platforms. Global-scale observations from the instrumental era began in the mid-19th century for temperature and other variables, with more comprehensive and diverse sets of observations available for the period 1950 onwards. Paleoclimate reconstructions extend some records back hundreds to millions of years. Together, they provide a comprehensive view of the variability and long-term changes in the atmosphere, the ocean, the cryosphere, and the land surface.

Warming of the climate system is unequivocal, and since the 1950s, many of the observed changes are unprecedented over decades to millennia. The atmosphere and ocean have warmed, the amounts of snow and ice have diminished, sea level has risen, and the concentrations of greenhouse gases have increased (see Figures SPM.1, SPM.2, SPM.3 and SPM.4). {2.2, 2.4, 3.2, 3.7, 4.2–4.7, 5.2, 5.3, 5.5–5.6, 6.2, 13.2}

---

1 In this Summary for Policymakers, the following summary terms are used to describe the available evidence: limited, medium, or robust; and for the degree of agreement: low, medium, or high. A level of confidence is expressed using five qualifiers: very low, low, medium, high, and very high, and typeset in italics, e.g., *medium confidence*. For a given evidence and agreement statement, different confidence levels can be assigned, but increasing levels of evidence and degrees of agreement are correlated with increasing confidence (see Chapter 1 and Box TS.1 for more details).

2 In this Summary for Policymakers, the following terms have been used to indicate the assessed likelihood of an outcome or a result: virtually certain 99–100% probability, very likely 90–100%, likely 66–100%, about as likely as not 33–66%, unlikely 0–33%, very unlikely 0–10%, exceptionally unlikely 0–1%. Additional terms (extremely likely: 95–100%, more likely than not >50–100%, and extremely unlikely 0–5%) may also be used when appropriate. Assessed likelihood is typeset in italics, e.g., *very likely* (see Chapter 1 and Box TS.1 for more details).

SPM

B.1 Atmosphere

Each of the last three decades has been successively warmer at the Earth's surface than any preceding decade since 1850 (see Figure SPM.1). In the Northern Hemisphere, 1983–2012 was *likely* the warmest 30-year period of the last 1400 years (*medium confidence*). {2.4, 5.3}

- The globally averaged combined land and ocean surface temperature data as calculated by a linear trend, show a warming of 0.85 [0.65 to 1.06] °C[3], over the period 1880 to 2012, when multiple independently produced datasets exist. The total increase between the average of the 1850–1900 period and the 2003–2012 period is 0.78 [0.72 to 0.85] °C, based on the single longest dataset available[4] (see Figure SPM.1). {2.4}

- For the longest period when calculation of regional trends is sufficiently complete (1901 to 2012), almost the entire globe has experienced surface warming (see Figure SPM.1). {2.4}

- In addition to robust multi-decadal warming, global mean surface temperature exhibits substantial decadal and interannual variability (see Figure SPM.1). Due to natural variability, trends based on short records are very sensitive to the beginning and end dates and do not in general reflect long-term climate trends. As one example, the rate of warming over the past 15 years (1998–2012; 0.05 [–0.05 to 0.15] °C per decade), which begins with a strong El Niño, is smaller than the rate calculated since 1951 (1951–2012; 0.12 [0.08 to 0.14] °C per decade)[5]. {2.4}

- Continental-scale surface temperature reconstructions show, with *high confidence*, multi-decadal periods during the Medieval Climate Anomaly (year 950 to 1250) that were in some regions as warm as in the late 20th century. These regional warm periods did not occur as coherently across regions as the warming in the late 20th century (*high confidence*). {5.5}

- It is *virtually certain* that globally the troposphere has warmed since the mid-20th century. More complete observations allow greater confidence in estimates of tropospheric temperature changes in the extratropical Northern Hemisphere than elsewhere. There is *medium confidence* in the rate of warming and its vertical structure in the Northern Hemisphere extra-tropical troposphere and *low confidence* elsewhere. {2.4}

- *Confidence* in precipitation change averaged over global land areas since 1901 is *low* prior to 1951 and *medium* afterwards. Averaged over the mid-latitude land areas of the Northern Hemisphere, precipitation has increased since 1901 (*medium confidence* before and *high confidence* after 1951). For other latitudes area-averaged long-term positive or negative trends have *low confidence* (see Figure SPM.2). {TS TFE.1, Figure 2; 2.5}

- Changes in many extreme weather and climate events have been observed since about 1950 (see Table SPM.1 for details). It is *very likely* that the number of cold days and nights has decreased and the number of warm days and nights has increased on the global scale[6]. It is *likely* that the frequency of heat waves has increased in large parts of Europe, Asia and Australia. There are *likely* more land regions where the number of heavy precipitation events has increased than where it has decreased. The frequency or intensity of heavy precipitation events has *likely* increased in North America and Europe. In other continents, *confidence* in changes in heavy precipitation events is at most *medium*. {2.6}

---

3 In the WGI contribution to the AR5, uncertainty is quantified using 90% uncertainty intervals unless otherwise stated. The 90% uncertainty interval, reported in square brackets, is expected to have a 90% likelihood of covering the value that is being estimated. Uncertainty intervals are not necessarily symmetric about the corresponding best estimate. A best estimate of that value is also given where available.

4 Both methods presented in this bullet were also used in AR4. The first calculates the difference using a best fit linear trend of all points between 1880 and 2012. The second calculates the difference between averages for the two periods 1850–1900 and 2003–2012. Therefore, the resulting values and their 90% uncertainty intervals are not directly comparable. {2.4}

5 Trends for 15-year periods starting in 1995, 1996, and 1997 are 0.13 [0.02 to 0.24] °C per decade, 0.14 [0.03 to 0.24] °C per decade, and, 0.07 [–0.02 to 0.18] °C per decade, respectively.

6 See the Glossary for the definition of these terms: cold days/cold nights, warm days/warm nights, heat waves.

SPM

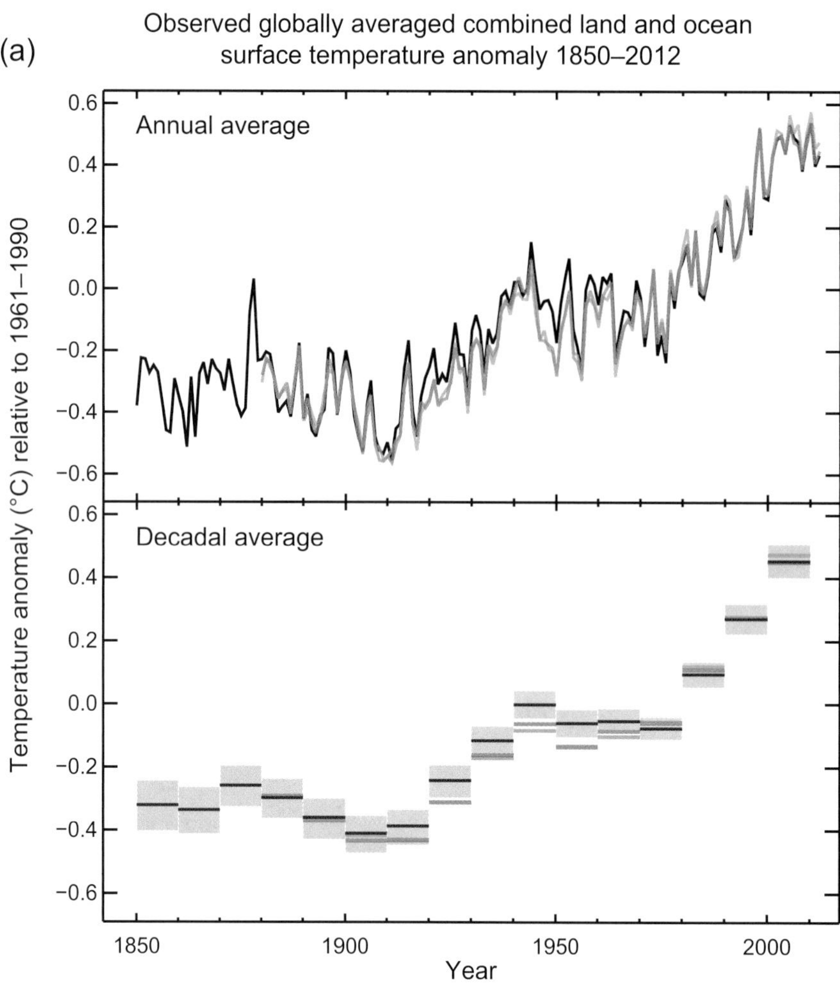

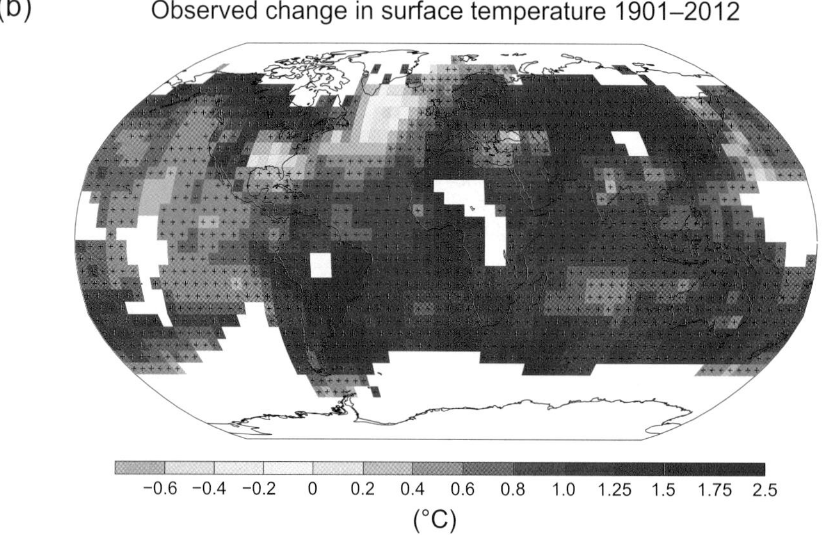

Figure SPM.1 | (a) Observed global mean combined land and ocean surface temperature anomalies, from 1850 to 2012 from three data sets. Top panel: annual mean values. Bottom panel: decadal mean values including the estimate of uncertainty for one dataset (black). Anomalies are relative to the mean of 1961−1990. (b) Map of the observed surface temperature change from 1901 to 2012 derived from temperature trends determined by linear regression from one dataset (orange line in panel a). Trends have been calculated where data availability permits a robust estimate (i.e., only for grid boxes with greater than 70% complete records and more than 20% data availability in the first and last 10% of the time period). Other areas are white. Grid boxes where the trend is significant at the 10% level are indicated by a + sign. For a listing of the datasets and further technical details see the Technical Summary Supplementary Material. {Figures 2.19–2.21; Figure TS.2}

Table SPM.1 | Extreme weather and climate events: Global-scale assessment of recent observed changes, human contribution to the changes, and projected further changes for the early (2016–2035) and late (2081–2100) 21st century. Bold indicates where the AR5 (black) provides a revised* global-scale assessment from the SREX (blue) or AR4 (red). Projections for early 21st century were not provided in previous assessment reports. Projections in the AR5 are relative to the reference period of 1986–2005, and use the new Representative Concentration Pathway (RCP) scenarios (see Box SPM.1) unless otherwise specified. See the Glossary for definitions of extreme weather and climate events.

Phenomenon and direction of trend	Assessment that changes occurred (typically since 1950 unless otherwise indicated)	Assessment of a human contribution to observed changes	Likelihood of further changes	
			Early 21st century	Late 21st century
Warmer and/or fewer cold days and nights over most land areas	*Very likely* {2.6} *Very likely* *Very likely*	***Very likely*** {10.6} *Likely* *Likely*	*Likely* {11.3}	*Virtually certain* {12.4} *Virtually certain* *Virtually certain*
Warmer and/or more frequent hot days and nights over most land areas	*Very likely* {2.6} *Very likely* *Very likely*	***Very likely*** {10.6} *Likely* *Likely* (nights only)	*Likely* {11.3}	*Virtually certain* {12.4} *Virtually certain* *Virtually certain*
Warm spells/heat waves. Frequency and/or duration increases over most land areas	***Medium confidence*** on a global scale *Likely* in large parts of Europe, Asia and Australia {2.6} *Medium confidence* in many (but not all) regions *Likely*	***Likely***[a] {10.6} Not formally assessed *More likely than not*	Not formally assessed[b] {11.3}	*Very likely* {12.4} *Very likely* *Very likely*
Heavy precipitation events. Increase in the frequency, intensity, and/or amount of heavy precipitation	*Likely* more land areas with increases than decreases[c] {2.6} *Likely* more land areas with increases than decreases *Likely* over most land areas	***Medium confidence*** {7.6, 10.6} *Medium confidence* *More likely than not*	*Likely* over many land areas {11.3}	***Very likely*** over most of the mid-latitude land masses and over wet tropical regions {12.4} *Likely* over many areas *Very likely* over most land areas
Increases in intensity and/or duration of drought	***Low confidence*** on a global scale *Likely* changes in some regions[d] {2.6} *Medium confidence* in some regions *Likely* in many regions, since 1970[e]	***Low confidence*** {10.6} *Medium confidence*[f] *More likely than not*	*Low confidence*[g] {11.3}	***Likely (medium confidence)*** on a regional to global scale[h] {12.4} *Medium confidence* in some regions *Likely*[e]
Increases in intense tropical cyclone activity	***Low confidence*** in long term (centennial) changes *Virtually certain* in North Atlantic since 1970 {2.6} *Low confidence* *Likely* in some regions, since 1970	***Low confidence***[i] {10.6} *Low confidence* *More likely than not*	*Low confidence* {11.3}	***More likely than not*** in the Western North Pacific and North Atlantic[j] {14.6} *More likely than not* in some basins *Likely*
Increased incidence and/or magnitude of extreme high sea level	*Likely* (since 1970) {3.7} *Likely* (late 20th century) *Likely*	***Likely***[k] {3.7} *Likely*[k] *More likely than not*[k]	*Likely*[l] {13.7}	***Very likely***[l] {13.7} *Very likely*[m] *Likely*

* The direct comparison of assessment findings between reports is difficult. For some climate variables, different aspects have been assessed, and the revised guidance note on uncertainties has been used for the SREX and AR5. The availability of new information, improved scientific understanding, continued analyses of data and models, and specific differences in methodologies applied in the assessed studies, all contribute to revised assessment findings.

Notes:

[a] Attribution is based on available case studies. It is *likely* that human influence has more than doubled the probability of occurrence of some observed heat waves in some locations.

[b] Models project near-term increases in the duration, intensity and spatial extent of heat waves and warm spells.

[c] In most continents, *confidence* in trends is not higher than *medium* except in North America and Europe where there have been *likely* increases in either the frequency or intensity of heavy precipitation with some seasonal and/or regional variation. It is *very likely* that there have been increases in central North America.

[d] The frequency and intensity of drought has *likely* increased in the Mediterranean and West Africa, and *likely* decreased in central North America and north-west Australia.

[e] AR4 assessed the area affected by drought.

[f] SREX assessed *medium confidence* that anthropogenic influence had contributed to some changes in the drought patterns observed in the second half of the 20th century, based on its attributed impact on precipitation and temperature changes. SREX assessed *low confidence* in the attribution of changes in droughts at the level of single regions.

[g] There is *low confidence* in projected changes in soil moisture.

[h] Regional to global-scale projected decreases in soil moisture and increased agricultural drought are *likely (medium confidence)* in presently dry regions by the end of this century under the RCP8.5 scenario. Soil moisture drying in the Mediterranean, Southwest US and southern African regions is consistent with projected changes in Hadley circulation and increased surface temperatures, so there is *high confidence* in *likely* surface drying in these regions by the end of this century under the RCP8.5 scenario.

[i] There is *medium confidence* that a reduction in aerosol forcing over the North Atlantic has contributed at least in part to the observed increase in tropical cyclone activity since the 1970s in this region.

[j] Based on expert judgment and assessment of projections which use an SRES A1B (or similar) scenario.

[k] Attribution is based on the close relationship between observed changes in extreme and mean sea level.

[l] There is *high confidence* that this increase in extreme high sea level will primarily be the result of an increase in mean sea level. There is *low confidence* in region-specific projections of storminess and associated storm surges.

[m] SREX assessed it to be *very likely* that mean sea level rise will contribute to future upward trends in extreme coastal high water levels.

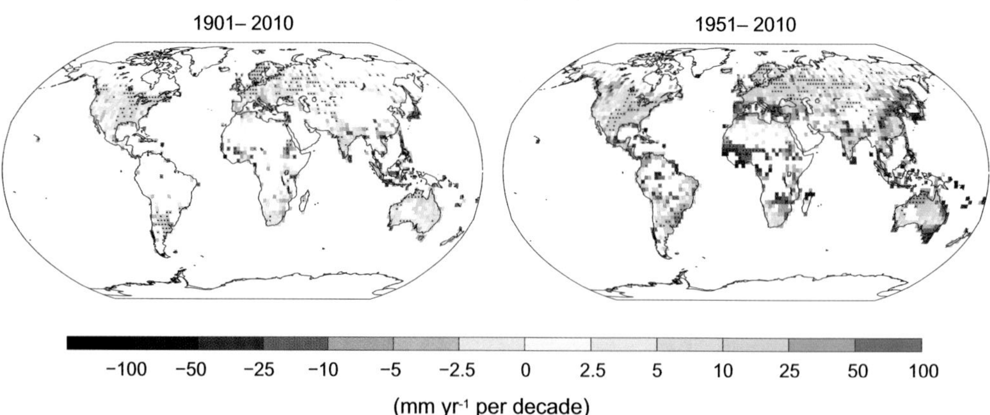

Figure SPM.2 | Maps of observed precipitation change from 1901 to 2010 and from 1951 to 2010 (trends in annual accumulation calculated using the same criteria as in Figure SPM.1) from one data set. For further technical details see the Technical Summary Supplementary Material. {TS TFE.1, Figure 2; Figure 2.29}

B.2 Ocean

Ocean warming dominates the increase in energy stored in the climate system, accounting for more than 90% of the energy accumulated between 1971 and 2010 (*high confidence*). It is *virtually certain* that the upper ocean (0–700 m) warmed from 1971 to 2010 (see Figure SPM.3), and it *likely* warmed between the 1870s and 1971. {3.2, Box 3.1}

- On a global scale, the ocean warming is largest near the surface, and the upper 75 m warmed by 0.11 [0.09 to 0.13] °C per decade over the period 1971 to 2010. Since AR4, instrumental biases in upper-ocean temperature records have been identified and reduced, enhancing confidence in the assessment of change. {3.2}

- It is *likely* that the ocean warmed between 700 and 2000 m from 1957 to 2009. Sufficient observations are available for the period 1992 to 2005 for a global assessment of temperature change below 2000 m. There were *likely* no significant observed temperature trends between 2000 and 3000 m for this period. It is *likely* that the ocean warmed from 3000 m to the bottom for this period, with the largest warming observed in the Southern Ocean. {3.2}

- More than 60% of the net energy increase in the climate system is stored in the upper ocean (0–700 m) during the relatively well-sampled 40-year period from 1971 to 2010, and about 30% is stored in the ocean below 700 m. The increase in upper ocean heat content during this time period estimated from a linear trend is *likely* 17 [15 to 19] × 10^{22} J [7] (see Figure SPM.3). {3.2, Box 3.1}

- It is *about as likely as not* that ocean heat content from 0–700 m increased more slowly during 2003 to 2010 than during 1993 to 2002 (see Figure SPM.3). Ocean heat uptake from 700–2000 m, where interannual variability is smaller, *likely* continued unabated from 1993 to 2009. {3.2, Box 9.2}

- It is *very likely* that regions of high salinity where evaporation dominates have become more saline, while regions of low salinity where precipitation dominates have become fresher since the 1950s. These regional trends in ocean salinity provide indirect evidence that evaporation and precipitation over the oceans have changed (*medium confidence*). {2.5, 3.3, 3.5}

- There is no observational evidence of a trend in the Atlantic Meridional Overturning Circulation (AMOC), based on the decade-long record of the complete AMOC and longer records of individual AMOC components. {3.6}

[7] A constant supply of heat through the ocean surface at the rate of 1 W m^{-2} for 1 year would increase the ocean heat content by 1.1 × 10^{22} J.

B.3 Cryosphere

Over the last two decades, the Greenland and Antarctic ice sheets have been losing mass, glaciers have continued to shrink almost worldwide, and Arctic sea ice and Northern Hemisphere spring snow cover have continued to decrease in extent (*high confidence*) (see Figure SPM.3). {4.2–4.7}

- The average rate of ice loss[8] from glaciers around the world, excluding glaciers on the periphery of the ice sheets[9], was *very likely* 226 [91 to 361] Gt yr^{-1} over the period 1971 to 2009, and *very likely* 275 [140 to 410] Gt yr^{-1} over the period 1993 to 2009[10]. {4.3}

- The average rate of ice loss from the Greenland ice sheet has *very likely* substantially increased from 34 [–6 to 74] Gt yr^{-1} over the period 1992 to 2001 to 215 [157 to 274] Gt yr^{-1} over the period 2002 to 2011. {4.4}

- The average rate of ice loss from the Antarctic ice sheet has *likely* increased from 30 [–37 to 97] Gt yr^{-1} over the period 1992–2001 to 147 [72 to 221] Gt yr^{-1} over the period 2002 to 2011. There is *very high confidence* that these losses are mainly from the northern Antarctic Peninsula and the Amundsen Sea sector of West Antarctica. {4.4}

- The annual mean Arctic sea ice extent decreased over the period 1979 to 2012 with a rate that was *very likely* in the range 3.5 to 4.1% per decade (range of 0.45 to 0.51 million km^2 per decade), and *very likely* in the range 9.4 to 13.6% per decade (range of 0.73 to 1.07 million km^2 per decade) for the summer sea ice minimum (perennial sea ice). The average decrease in decadal mean extent of Arctic sea ice has been most rapid in summer (*high confidence*); the spatial extent has decreased in every season, and in every successive decade since 1979 (*high confidence*) (see Figure SPM.3). There is *medium confidence* from reconstructions that over the past three decades, Arctic summer sea ice retreat was unprecedented and sea surface temperatures were anomalously high in at least the last 1,450 years. {4.2, 5.5}

- It is *very likely* that the annual mean Antarctic sea ice extent increased at a rate in the range of 1.2 to 1.8% per decade (range of 0.13 to 0.20 million km^2 per decade) between 1979 and 2012. There is *high confidence* that there are strong regional differences in this annual rate, with extent increasing in some regions and decreasing in others. {4.2}

- There is *very high confidence* that the extent of Northern Hemisphere snow cover has decreased since the mid-20th century (see Figure SPM.3). Northern Hemisphere snow cover extent decreased 1.6 [0.8 to 2.4] % per decade for March and April, and 11.7 [8.8 to 14.6] % per decade for June, over the 1967 to 2012 period. During this period, snow cover extent in the Northern Hemisphere did not show a statistically significant increase in any month. {4.5}

- There is *high confidence* that permafrost temperatures have increased in most regions since the early 1980s. Observed warming was up to 3°C in parts of Northern Alaska (early 1980s to mid-2000s) and up to 2°C in parts of the Russian European North (1971 to 2010). In the latter region, a considerable reduction in permafrost thickness and areal extent has been observed over the period 1975 to 2005 (*medium confidence*). {4.7}

- Multiple lines of evidence support very substantial Arctic warming since the mid-20th century. {Box 5.1, 10.3}

---

8 All references to 'ice loss' or 'mass loss' refer to net ice loss, i.e., accumulation minus melt and iceberg calving.

9 For methodological reasons, this assessment of ice loss from the Antarctic and Greenland ice sheets includes change in the glaciers on the periphery. These peripheral glaciers are thus excluded from the values given for glaciers.

10 100 Gt yr^{-1} of ice loss is equivalent to about 0.28 mm yr^{-1} of global mean sea level rise.

SPM

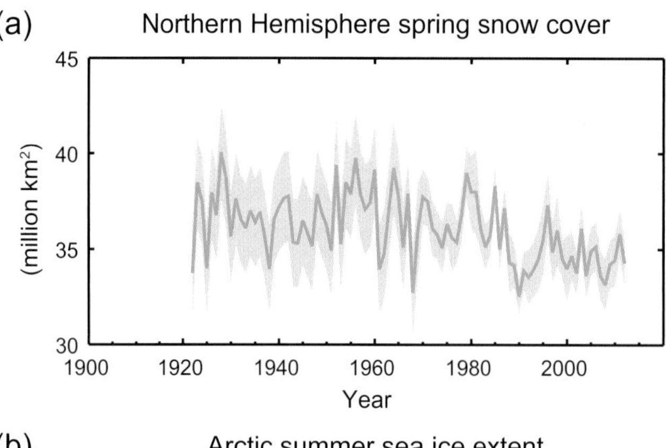

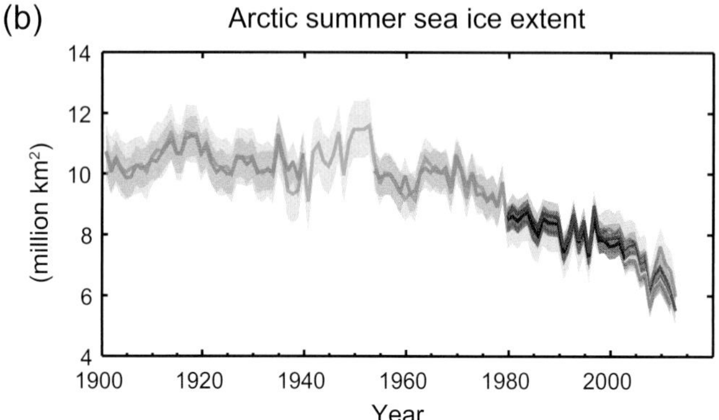

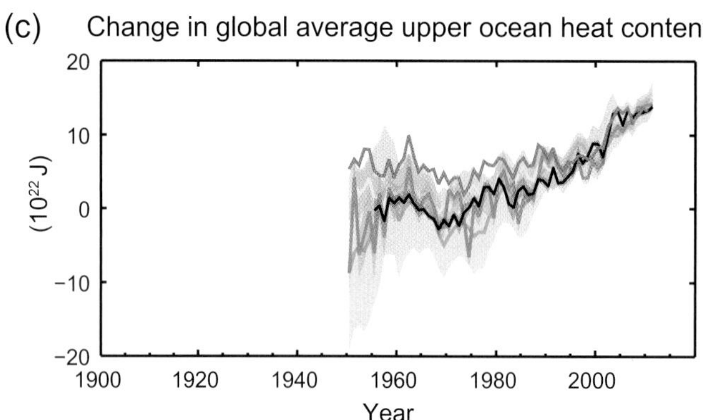

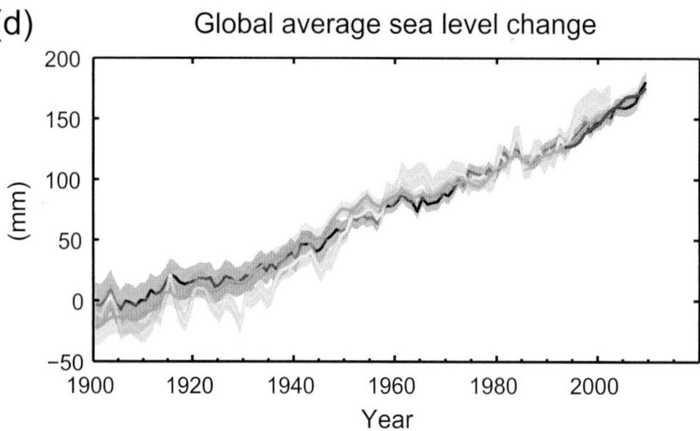

Figure SPM.3 | Multiple observed indicators of a changing global climate: (a) Extent of Northern Hemisphere March-April (spring) average snow cover; (b) extent of Arctic July-August-September (summer) average sea ice; (c) change in global mean upper ocean (0–700 m) heat content aligned to 2006–2010, and relative to the mean of all datasets for 1970; (d) global mean sea level relative to the 1900–1905 mean of the longest running dataset, and with all datasets aligned to have the same value in 1993, the first year of satellite altimetry data. All time-series (coloured lines indicating different data sets) show annual values, and where assessed, uncertainties are indicated by coloured shading. See Technical Summary Supplementary Material for a listing of the datasets. {Figures 3.2, 3.13, 4.19, and 4.3; FAQ 2.1, Figure 2; Figure TS.1}

B.4 Sea Level

The rate of sea level rise since the mid-19th century has been larger than the mean rate during the previous two millennia (*high confidence*). Over the period 1901 to 2010, global mean sea level rose by 0.19 [0.17 to 0.21] m (see Figure SPM.3). {3.7, 5.6, 13.2}

- Proxy and instrumental sea level data indicate a transition in the late 19th to the early 20th century from relatively low mean rates of rise over the previous two millennia to higher rates of rise (*high confidence*). It is *likely* that the rate of global mean sea level rise has continued to increase since the early 20th century. {3.7, 5.6, 13.2}

- It is *very likely* that the mean rate of global averaged sea level rise was 1.7 [1.5 to 1.9] mm yr^{-1} between 1901 and 2010, 2.0 [1.7 to 2.3] mm yr^{-1} between 1971 and 2010, and 3.2 [2.8 to 3.6] mm yr^{-1} between 1993 and 2010. Tide-gauge and satellite altimeter data are consistent regarding the higher rate of the latter period. It is *likely* that similarly high rates occurred between 1920 and 1950. {3.7}

- Since the early 1970s, glacier mass loss and ocean thermal expansion from warming together explain about 75% of the observed global mean sea level rise (*high confidence*). Over the period 1993 to 2010, global mean sea level rise is, with *high confidence*, consistent with the sum of the observed contributions from ocean thermal expansion due to warming (1.1 [0.8 to 1.4] mm yr^{-1}), from changes in glaciers (0.76 [0.39 to 1.13] mm yr^{-1}), Greenland ice sheet (0.33 [0.25 to 0.41] mm yr^{-1}), Antarctic ice sheet (0.27 [0.16 to 0.38] mm yr^{-1}), and land water storage (0.38 [0.26 to 0.49] mm yr^{-1}). The sum of these contributions is 2.8 [2.3 to 3.4] mm yr^{-1}. {13.3}

- There is *very high confidence* that maximum global mean sea level during the last interglacial period (129,000 to 116,000 years ago) was, for several thousand years, at least 5 m higher than present, and *high confidence* that it did not exceed 10 m above present. During the last interglacial period, the Greenland ice sheet *very likely* contributed between 1.4 and 4.3 m to the higher global mean sea level, implying with *medium confidence* an additional contribution from the Antarctic ice sheet. This change in sea level occurred in the context of different orbital forcing and with high-latitude surface temperature, averaged over several thousand years, at least 2°C warmer than present (*high confidence*). {5.3, 5.6}

B.5 Carbon and Other Biogeochemical Cycles

The atmospheric concentrations of carbon dioxide, methane, and nitrous oxide have increased to levels unprecedented in at least the last 800,000 years. Carbon dioxide concentrations have increased by 40% since pre-industrial times, primarily from fossil fuel emissions and secondarily from net land use change emissions. The ocean has absorbed about 30% of the emitted anthropogenic carbon dioxide, causing ocean acidification (see Figure SPM.4). {2.2, 3.8, 5.2, 6.2, 6.3}

- The atmospheric concentrations of the greenhouse gases carbon dioxide (CO_2), methane (CH_4), and nitrous oxide (N_2O) have all increased since 1750 due to human activity. In 2011 the concentrations of these greenhouse gases were 391 ppm[11], 1803 ppb, and 324 ppb, and exceeded the pre-industrial levels by about 40%, 150%, and 20%, respectively. {2.2, 5.2, 6.1, 6.2}

- Concentrations of CO_2, CH_4, and N_2O now substantially exceed the highest concentrations recorded in ice cores during the past 800,000 years. The mean rates of increase in atmospheric concentrations over the past century are, with *very high confidence*, unprecedented in the last 22,000 years. {5.2, 6.1, 6.2}

[11] ppm (parts per million) or ppb (parts per billion, 1 billion = 1,000 million) is the ratio of the number of gas molecules to the total number of molecules of dry air. For example, 300 ppm means 300 molecules of a gas per million molecules of dry air.

- Annual CO_2 emissions from fossil fuel combustion and cement production were 8.3 [7.6 to 9.0] GtC[12] yr^{-1} averaged over 2002–2011 (*high confidence*) and were 9.5 [8.7 to 10.3] GtC yr^{-1} in 2011, 54% above the 1990 level. Annual net CO_2 emissions from anthropogenic land use change were 0.9 [0.1 to 1.7] GtC yr^{-1} on average during 2002 to 2011 (*medium confidence*). {6.3}

- From 1750 to 2011, CO_2 emissions from fossil fuel combustion and cement production have released 375 [345 to 405] GtC to the atmosphere, while deforestation and other land use change are estimated to have released 180 [100 to 260] GtC. This results in cumulative anthropogenic emissions of 555 [470 to 640] GtC. {6.3}

- Of these cumulative anthropogenic CO_2 emissions, 240 [230 to 250] GtC have accumulated in the atmosphere, 155 [125 to 185] GtC have been taken up by the ocean and 160 [70 to 250] GtC have accumulated in natural terrestrial ecosystems (i.e., the cumulative residual land sink). {Figure TS.4, 3.8, 6.3}

- Ocean acidification is quantified by decreases in pH[13]. The pH of ocean surface water has decreased by 0.1 since the beginning of the industrial era (*high confidence*), corresponding to a 26% increase in hydrogen ion concentration (see Figure SPM.4). {3.8, Box 3.2}

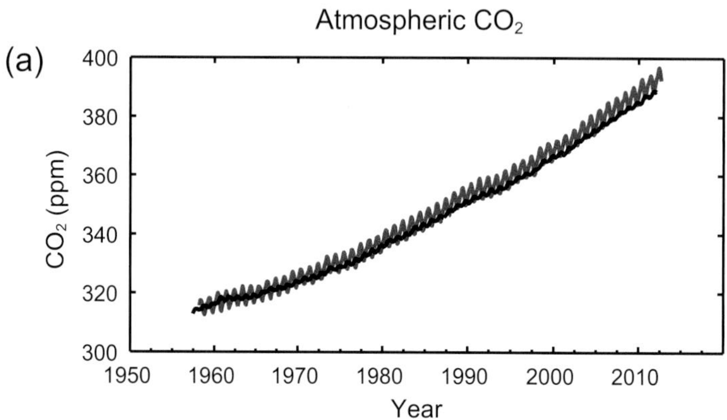

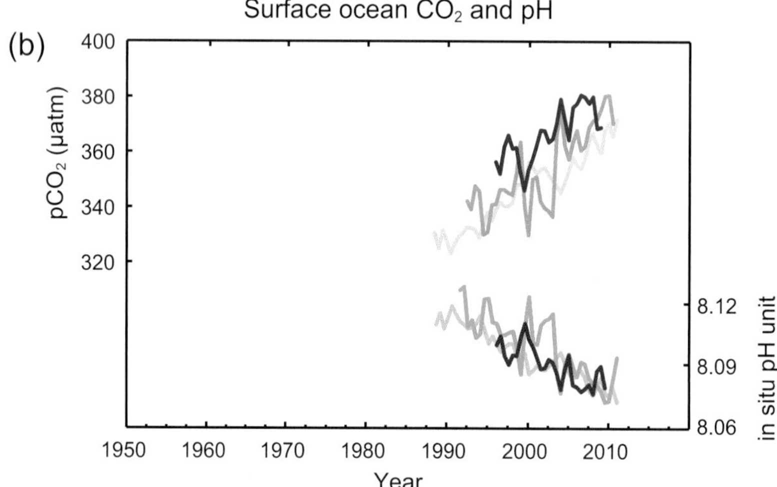

Figure SPM.4 | Multiple observed indicators of a changing global carbon cycle: (a) atmospheric concentrations of carbon dioxide (CO_2) from Mauna Loa (19°32'N, 155°34'W – red) and South Pole (89°59'S, 24°48'W – black) since 1958; (b) partial pressure of dissolved CO_2 at the ocean surface (blue curves) and in situ pH (green curves), a measure of the acidity of ocean water. Measurements are from three stations from the Atlantic (29°10'N, 15°30'W – dark blue/dark green; 31°40'N, 64°10'W – blue/green) and the Pacific Oceans (22°45'N, 158°00'W – light blue/light green). Full details of the datasets shown here are provided in the underlying report and the Technical Summary Supplementary Material. {Figures 2.1 and 3.18; Figure TS.5}

---

12 1 Gigatonne of carbon = 1 GtC = 10^{15} grams of carbon. This corresponds to 3.667 $GtCO_2$.

13 pH is a measure of acidity using a logarithmic scale: a pH decrease of 1 unit corresponds to a 10-fold increase in hydrogen ion concentration, or acidity.

SPM

C. Drivers of Climate Change

Natural and anthropogenic substances and processes that alter the Earth's energy budget are drivers of climate change. Radiative forcing[14] (RF) quantifies the change in energy fluxes caused by changes in these drivers for 2011 relative to 1750, unless otherwise indicated. Positive RF leads to surface warming, negative RF leads to surface cooling. RF is estimated based on in-situ and remote observations, properties of greenhouse gases and aerosols, and calculations using numerical models representing observed processes. Some emitted compounds affect the atmospheric concentration of other substances. The RF can be reported based on the concentration changes of each substance[15]. Alternatively, the emission-based RF of a compound can be reported, which provides a more direct link to human activities. It includes contributions from all substances affected by that emission. The total anthropogenic RF of the two approaches are identical when considering all drivers. Though both approaches are used in this Summary for Policymakers, emission-based RFs are emphasized.

Total radiative forcing is positive, and has led to an uptake of energy by the climate system. The largest contribution to total radiative forcing is caused by the increase in the atmospheric concentration of CO_2 since 1750 (see Figure SPM.5). {3.2, Box 3.1, 8.3, 8.5}

- The total anthropogenic RF for 2011 relative to 1750 is 2.29 [1.13 to 3.33] W m^{-2} (see Figure SPM.5), and it has increased more rapidly since 1970 than during prior decades. The total anthropogenic RF best estimate for 2011 is 43% higher than that reported in AR4 for the year 2005. This is caused by a combination of continued growth in most greenhouse gas concentrations and improved estimates of RF by aerosols indicating a weaker net cooling effect (negative RF). {8.5}

- The RF from emissions of well-mixed greenhouse gases (CO_2, CH_4, N_2O, and Halocarbons) for 2011 relative to 1750 is 3.00 [2.22 to 3.78] W m^{-2} (see Figure SPM.5). The RF from changes in concentrations in these gases is 2.83 [2.26 to 3.40] W m^{-2}. {8.5}

- Emissions of CO_2 alone have caused an RF of 1.68 [1.33 to 2.03] W m^{-2} (see Figure SPM.5). Including emissions of other carbon-containing gases, which also contributed to the increase in CO_2 concentrations, the RF of CO_2 is 1.82 [1.46 to 2.18] W m^{-2}. {8.3, 8.5}

- Emissions of CH_4 alone have caused an RF of 0.97 [0.74 to 1.20] W m^{-2} (see Figure SPM.5). This is much larger than the concentration-based estimate of 0.48 [0.38 to 0.58] W m^{-2} (unchanged from AR4). This difference in estimates is caused by concentration changes in ozone and stratospheric water vapour due to CH_4 emissions and other emissions indirectly affecting CH_4. {8.3, 8.5}

- Emissions of stratospheric ozone-depleting halocarbons have caused a net positive RF of 0.18 [0.01 to 0.35] W m^{-2} (see Figure SPM.5). Their own positive RF has outweighed the negative RF from the ozone depletion that they have induced. The positive RF from all halocarbons is similar to the value in AR4, with a reduced RF from CFCs but increases from many of their substitutes. {8.3, 8.5}

- Emissions of short-lived gases contribute to the total anthropogenic RF. Emissions of carbon monoxide (CO) are *virtually certain* to have induced a positive RF, while emissions of nitrogen oxides (NO_x) are *likely* to have induced a net negative RF (see Figure SPM.5). {8.3, 8.5}

- The RF of the total aerosol effect in the atmosphere, which includes cloud adjustments due to aerosols, is –0.9 [–1.9 to –0.1] W m^{-2} (*medium confidence*), and results from a negative forcing from most aerosols and a positive contribution

[14] The strength of drivers is quantified as Radiative Forcing (RF) in units watts per square metre (W m^{-2}) as in previous IPCC assessments. RF is the change in energy flux caused by a driver, and is calculated at the tropopause or at the top of the atmosphere. In the traditional RF concept employed in previous IPCC reports all surface and tropospheric conditions are kept fixed. In calculations of RF for well-mixed greenhouse gases and aerosols in this report, physical variables, except for the ocean and sea ice, are allowed to respond to perturbations with rapid adjustments. The resulting forcing is called Effective Radiative Forcing (ERF) in the underlying report. This change reflects the scientific progress from previous assessments and results in a better indication of the eventual temperature response for these drivers. For all drivers other than well-mixed greenhouse gases and aerosols, rapid adjustments are less well characterized and assumed to be small, and thus the traditional RF is used. {8.1}

[15] This approach was used to report RF in the AR4 Summary for Policymakers.

SPM

from black carbon absorption of solar radiation. There is *high confidence* that aerosols and their interactions with clouds have offset a substantial portion of global mean forcing from well-mixed greenhouse gases. They continue to contribute the largest uncertainty to the total RF estimate. {7.5, 8.3, 8.5}

- The forcing from stratospheric volcanic aerosols can have a large impact on the climate for some years after volcanic eruptions. Several small eruptions have caused an RF of –0.11 [–0.15 to –0.08] W m^{-2} for the years 2008 to 2011, which is approximately twice as strong as during the years 1999 to 2002. {8.4}

- The RF due to changes in solar irradiance is estimated as 0.05 [0.00 to 0.10] W m^{-2} (see Figure SPM.5). Satellite observations of total solar irradiance changes from 1978 to 2011 indicate that the last solar minimum was lower than the previous two. This results in an RF of –0.04 [–0.08 to 0.00] W m^{-2} between the most recent minimum in 2008 and the 1986 minimum. {8.4}

- The total natural RF from solar irradiance changes and stratospheric volcanic aerosols made only a small contribution to the net radiative forcing throughout the last century, except for brief periods after large volcanic eruptions. {8.5}

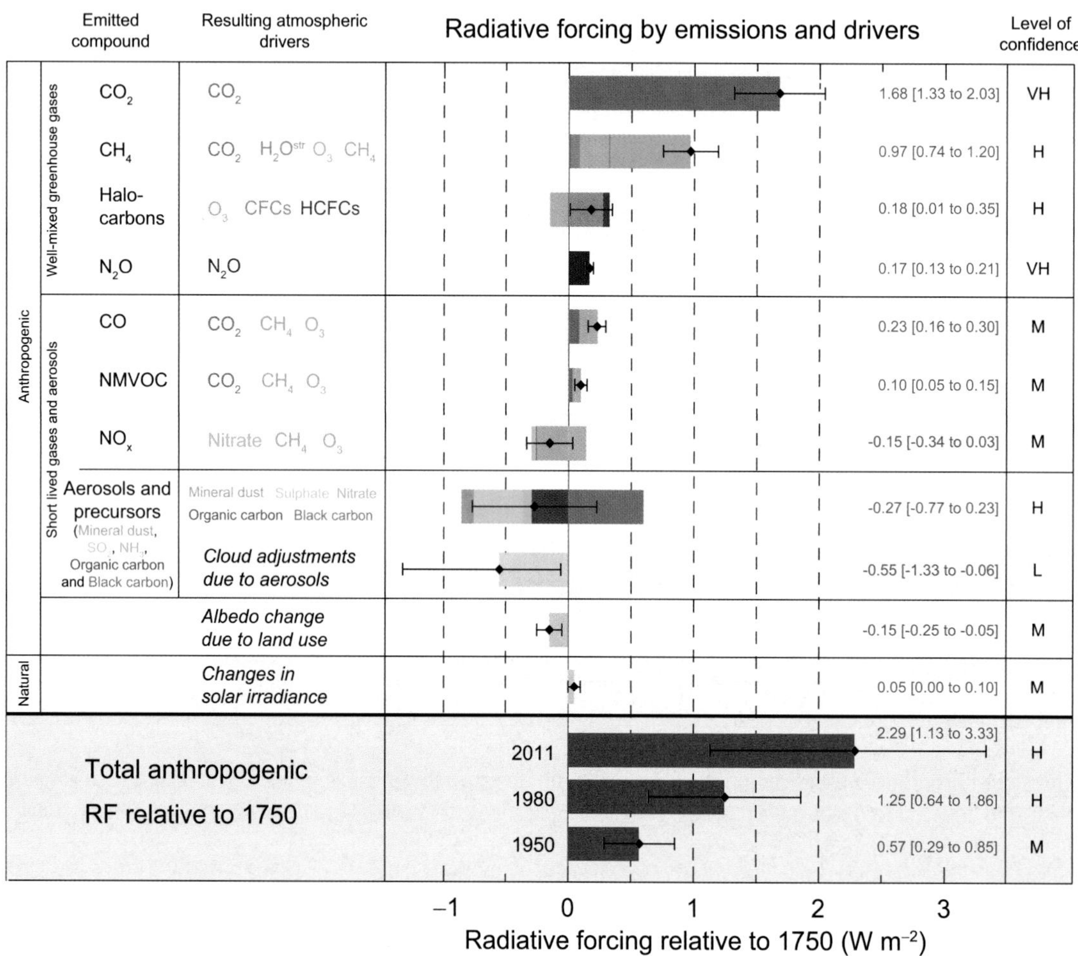

Figure SPM.5 | Radiative forcing estimates in 2011 relative to 1750 and aggregated uncertainties for the main drivers of climate change. Values are global average radiative forcing (RF[14]), partitioned according to the emitted compounds or processes that result in a combination of drivers. The best estimates of the net radiative forcing are shown as black diamonds with corresponding uncertainty intervals; the numerical values are provided on the right of the figure, together with the confidence level in the net forcing (VH – *very high*, H – *high*, M – *medium*, L – *low*, VL – *very low*). Albedo forcing due to black carbon on snow and ice is included in the black carbon aerosol bar. Small forcings due to contrails (0.05 W m^{-2}, including contrail induced cirrus), and HFCs, PFCs and SF_6 (total 0.03 W m^{-2}) are not shown. Concentration-based RFs for gases can be obtained by summing the like-coloured bars. Volcanic forcing is not included as its episodic nature makes is difficult to compare to other forcing mechanisms. Total anthropogenic radiative forcing is provided for three different years relative to 1750. For further technical details, including uncertainty ranges associated with individual components and processes, see the Technical Summary Supplementary Material. {8.5; Figures 8.14–8.18; Figures TS.6 and TS.7}

D. Understanding the Climate System and its Recent Changes

Understanding recent changes in the climate system results from combining observations, studies of feedback processes, and model simulations. Evaluation of the ability of climate models to simulate recent changes requires consideration of the state of all modelled climate system components at the start of the simulation and the natural and anthropogenic forcing used to drive the models. Compared to AR4, more detailed and longer observations and improved climate models now enable the attribution of a human contribution to detected changes in more climate system components.

Human influence on the climate system is clear. This is evident from the increasing greenhouse gas concentrations in the atmosphere, positive radiative forcing, observed warming, and understanding of the climate system. {2–14}

D.1 Evaluation of Climate Models

Climate models have improved since the AR4. Models reproduce observed continental-scale surface temperature patterns and trends over many decades, including the more rapid warming since the mid-20th century and the cooling immediately following large volcanic eruptions (*very high confidence*). {9.4, 9.6, 9.8}

- The long-term climate model simulations show a trend in global-mean surface temperature from 1951 to 2012 that agrees with the observed trend (*very high confidence*). There are, however, differences between simulated and observed trends over periods as short as 10 to 15 years (e.g., 1998 to 2012). {9.4, Box 9.2}

- The observed reduction in surface warming trend over the period 1998 to 2012 as compared to the period 1951 to 2012, is due in roughly equal measure to a reduced trend in radiative forcing and a cooling contribution from natural internal variability, which includes a possible redistribution of heat within the ocean (*medium confidence*). The reduced trend in radiative forcing is primarily due to volcanic eruptions and the timing of the downward phase of the 11-year solar cycle. However, there is *low confidence* in quantifying the role of changes in radiative forcing in causing the reduced warming trend. There is *medium confidence* that natural internal decadal variability causes to a substantial degree the difference between observations and the simulations; the latter are not expected to reproduce the timing of natural internal variability. There may also be a contribution from forcing inadequacies and, in some models, an overestimate of the response to increasing greenhouse gas and other anthropogenic forcing (dominated by the effects of aerosols). {9.4, Box 9.2, 10.3, Box 10.2, 11.3}

- On regional scales, the confidence in model capability to simulate surface temperature is less than for the larger scales. However, there is *high confidence* that regional-scale surface temperature is better simulated than at the time of the AR4. {9.4, 9.6}

- There has been substantial progress in the assessment of extreme weather and climate events since AR4. Simulated global-mean trends in the frequency of extreme warm and cold days and nights over the second half of the 20th century are generally consistent with observations. {9.5}

- There has been some improvement in the simulation of continental-scale patterns of precipitation since the AR4. At regional scales, precipitation is not simulated as well, and the assessment is hampered by observational uncertainties. {9.4, 9.6}

- Some important climate phenomena are now better reproduced by models. There is *high confidence* that the statistics of monsoon and El Niño-Southern Oscillation (ENSO) based on multi-model simulations have improved since AR4. {9.5}

SPM

- Climate models now include more cloud and aerosol processes, and their interactions, than at the time of the AR4, but there remains *low confidence* in the representation and quantification of these processes in models. {7.3, 7.6, 9.4, 9.7}

- There is robust evidence that the downward trend in Arctic summer sea ice extent since 1979 is now reproduced by more models than at the time of the AR4, with about one-quarter of the models showing a trend as large as, or larger than, the trend in the observations. Most models simulate a small downward trend in Antarctic sea ice extent, albeit with large inter-model spread, in contrast to the small upward trend in observations. {9.4}

- Many models reproduce the observed changes in upper-ocean heat content (0–700 m) from 1961 to 2005 (*high confidence*), with the multi-model mean time series falling within the range of the available observational estimates for most of the period. {9.4}

- Climate models that include the carbon cycle (Earth System Models) simulate the global pattern of ocean-atmosphere CO_2 fluxes, with outgassing in the tropics and uptake in the mid and high latitudes. In the majority of these models the sizes of the simulated global land and ocean carbon sinks over the latter part of the 20th century are within the range of observational estimates. {9.4}

D.2 Quantification of Climate System Responses

Observational and model studies of temperature change, climate feedbacks and changes in the Earth's energy budget together provide confidence in the magnitude of global warming in response to past and future forcing. {Box 12.2, Box 13.1}

- The net feedback from the combined effect of changes in water vapour, and differences between atmospheric and surface warming is *extremely likely* positive and therefore amplifies changes in climate. The net radiative feedback due to all cloud types combined is *likely* positive. Uncertainty in the sign and magnitude of the cloud feedback is due primarily to continuing uncertainty in the impact of warming on low clouds. {7.2}

- The equilibrium climate sensitivity quantifies the response of the climate system to constant radiative forcing on multi-century time scales. It is defined as the change in global mean surface temperature at equilibrium that is caused by a doubling of the atmospheric CO_2 concentration. Equilibrium climate sensitivity is *likely* in the range 1.5°C to 4.5°C (*high confidence*), *extremely unlikely* less than 1°C (*high confidence*), and *very unlikely* greater than 6°C (*medium confidence*)[16]. The lower temperature limit of the assessed *likely* range is thus less than the 2°C in the AR4, but the upper limit is the same. This assessment reflects improved understanding, the extended temperature record in the atmosphere and ocean, and new estimates of radiative forcing. {TS TFE.6, Figure 1; Box 12.2}

- The rate and magnitude of global climate change is determined by radiative forcing, climate feedbacks and the storage of energy by the climate system. Estimates of these quantities for recent decades are consistent with the assessed *likely* range of the equilibrium climate sensitivity to within assessed uncertainties, providing strong evidence for our understanding of anthropogenic climate change. {Box 12.2, Box 13.1}

- The transient climate response quantifies the response of the climate system to an increasing radiative forcing on a decadal to century timescale. It is defined as the change in global mean surface temperature at the time when the atmospheric CO_2 concentration has doubled in a scenario of concentration increasing at 1% per year. The transient climate response is *likely* in the range of 1.0°C to 2.5°C (*high confidence*) and *extremely unlikely* greater than 3°C. {Box 12.2}

- A related quantity is the transient climate response to cumulative carbon emissions (TCRE). It quantifies the transient response of the climate system to cumulative carbon emissions (see Section E.8). TCRE is defined as the global mean

[16] No best estimate for equilibrium climate sensitivity can now be given because of a lack of agreement on values across assessed lines of evidence and studies.

surface temperature change per 1000 GtC emitted to the atmosphere. TCRE is *likely* in the range of 0.8°C to 2.5°C per 1000 GtC and applies for cumulative emissions up to about 2000 GtC until the time temperatures peak (see Figure SPM.10). {12.5, Box 12.2}

- Various metrics can be used to compare the contributions to climate change of emissions of different substances. The most appropriate metric and time horizon will depend on which aspects of climate change are considered most important to a particular application. No single metric can accurately compare all consequences of different emissions, and all have limitations and uncertainties. The Global Warming Potential is based on the cumulative radiative forcing over a particular time horizon, and the Global Temperature Change Potential is based on the change in global mean surface temperature at a chosen point in time. Updated values are provided in the underlying Report. {8.7}

D.3 Detection and Attribution of Climate Change

Human influence has been detected in warming of the atmosphere and the ocean, in changes in the global water cycle, in reductions in snow and ice, in global mean sea level rise, and in changes in some climate extremes (see Figure SPM.6 and Table SPM.1). This evidence for human influence has grown since AR4. It is *extremely likely* that human influence has been the dominant cause of the observed warming since the mid-20th century. {10.3–10.6, 10.9}

- It is *extremely likely* that more than half of the observed increase in global average surface temperature from 1951 to 2010 was caused by the anthropogenic increase in greenhouse gas concentrations and other anthropogenic forcings together. The best estimate of the human-induced contribution to warming is similar to the observed warming over this period. {10.3}

- Greenhouse gases contributed a global mean surface warming *likely* to be in the range of 0.5°C to 1.3°C over the period 1951 to 2010, with the contributions from other anthropogenic forcings, including the cooling effect of aerosols, *likely* to be in the range of –0.6°C to 0.1°C. The contribution from natural forcings is *likely* to be in the range of –0.1°C to 0.1°C, and from natural internal variability is *likely* to be in the range of –0.1°C to 0.1°C. Together these assessed contributions are consistent with the observed warming of approximately 0.6°C to 0.7°C over this period. {10.3}

- Over every continental region except Antarctica, anthropogenic forcings have *likely* made a substantial contribution to surface temperature increases since the mid-20th century (see Figure SPM.6). For Antarctica, large observational uncertainties result in *low confidence* that anthropogenic forcings have contributed to the observed warming averaged over available stations. It is *likely* that there has been an anthropogenic contribution to the very substantial Arctic warming since the mid-20th century. {2.4, 10.3}

- It is *very likely* that anthropogenic influence, particularly greenhouse gases and stratospheric ozone depletion, has led to a detectable observed pattern of tropospheric warming and a corresponding cooling in the lower stratosphere since 1961. {2.4, 9.4, 10.3}

- It is *very likely* that anthropogenic forcings have made a substantial contribution to increases in global upper ocean heat content (0–700 m) observed since the 1970s (see Figure SPM.6). There is evidence for human influence in some individual ocean basins. {3.2, 10.4}

- It is *likely* that anthropogenic influences have affected the global water cycle since 1960. Anthropogenic influences have contributed to observed increases in atmospheric moisture content in the atmosphere (*medium confidence*), to global-scale changes in precipitation patterns over land (*medium confidence*), to intensification of heavy precipitation over land regions where data are sufficient (*medium confidence*), and to changes in surface and sub-surface ocean salinity (*very likely*). {2.5, 2.6, 3.3, 7.6, 10.3, 10.4}

SPM

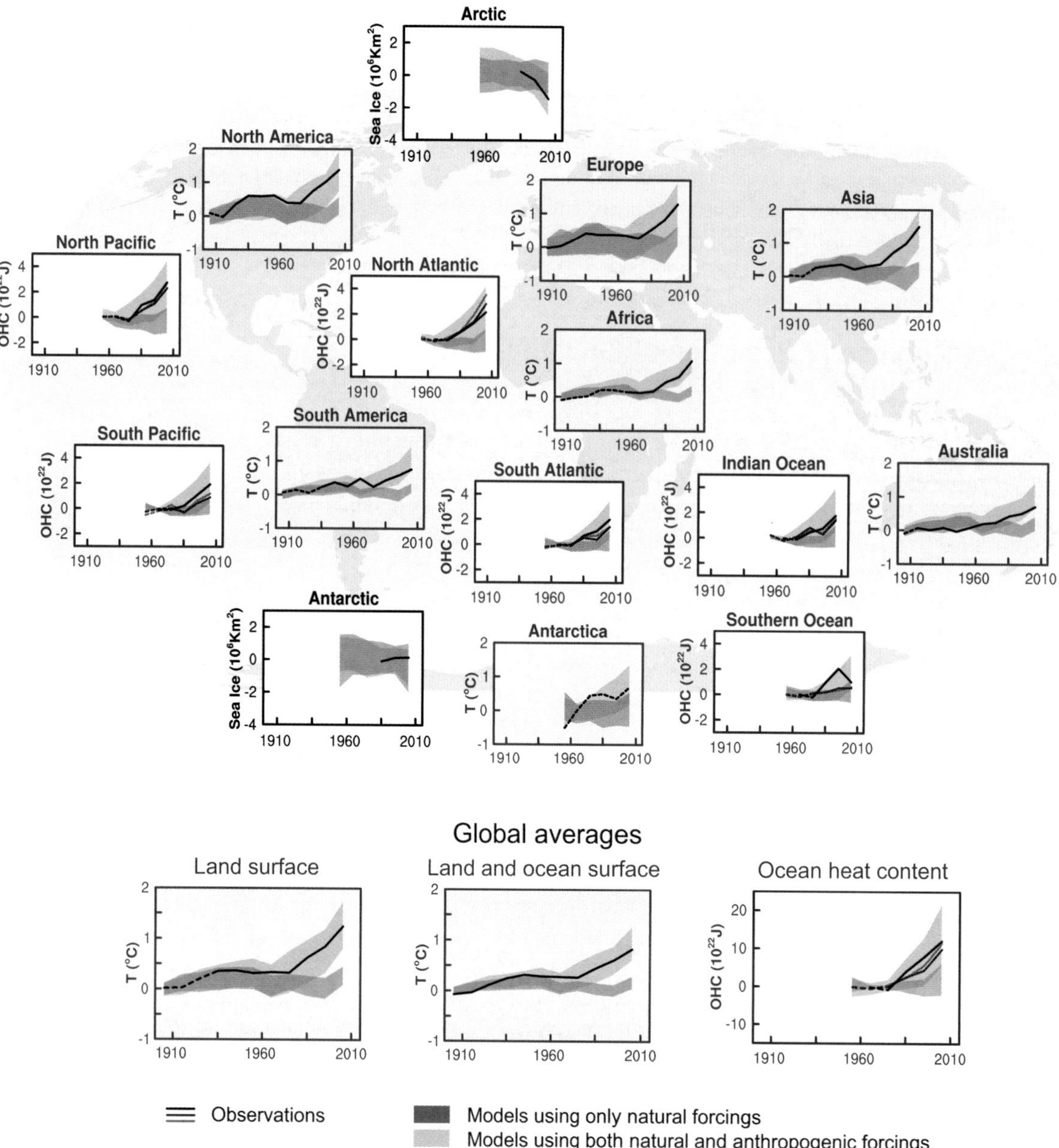

Figure SPM.6 | Comparison of observed and simulated climate change based on three large-scale indicators in the atmosphere, the cryosphere and the ocean: change in continental land surface air temperatures (yellow panels), Arctic and Antarctic September sea ice extent (white panels), and upper ocean heat content in the major ocean basins (blue panels). Global average changes are also given. Anomalies are given relative to 1880–1919 for surface temperatures, 1960–1980 for ocean heat content and 1979–1999 for sea ice. All time-series are decadal averages, plotted at the centre of the decade. For temperature panels, observations are dashed lines if the spatial coverage of areas being examined is below 50%. For ocean heat content and sea ice panels the solid line is where the coverage of data is good and higher in quality, and the dashed line is where the data coverage is only adequate, and thus, uncertainty is larger. Model results shown are Coupled Model Intercomparison Project Phase 5 (CMIP5) multi-model ensemble ranges, with shaded bands indicating the 5 to 95% confidence intervals. For further technical details, including region definitions see the Technical Summary Supplementary Material. {Figure 10.21; Figure TS.12}

- There has been further strengthening of the evidence for human influence on temperature extremes since the SREX. It is now *very likely* that human influence has contributed to observed global scale changes in the frequency and intensity of daily temperature extremes since the mid-20th century, and *likely* that human influence has more than doubled the probability of occurrence of heat waves in some locations (see Table SPM.1). {10.6}

- Anthropogenic influences have *very likely* contributed to Arctic sea ice loss since 1979. There is *low confidence* in the scientific understanding of the small observed increase in Antarctic sea ice extent due to the incomplete and competing scientific explanations for the causes of change and *low confidence* in estimates of natural internal variability in that region (see Figure SPM.6). {10.5}

- Anthropogenic influences *likely* contributed to the retreat of glaciers since the 1960s and to the increased surface mass loss of the Greenland ice sheet since 1993. Due to a low level of scientific understanding there is *low confidence* in attributing the causes of the observed loss of mass from the Antarctic ice sheet over the past two decades. {4.3, 10.5}

- It is *likely* that there has been an anthropogenic contribution to observed reductions in Northern Hemisphere spring snow cover since 1970. {10.5}

- It is *very likely* that there is a substantial anthropogenic contribution to the global mean sea level rise since the 1970s. This is based on the *high confidence* in an anthropogenic influence on the two largest contributions to sea level rise, that is thermal expansion and glacier mass loss. {10.4, 10.5, 13.3}

- There is *high confidence* that changes in total solar irradiance have not contributed to the increase in global mean surface temperature over the period 1986 to 2008, based on direct satellite measurements of total solar irradiance. There is *medium confidence* that the 11-year cycle of solar variability influences decadal climate fluctuations in some regions. No robust association between changes in cosmic rays and cloudiness has been identified. {7.4, 10.3, Box 10.2}

E. Future Global and Regional Climate Change

Projections of changes in the climate system are made using a hierarchy of climate models ranging from simple climate models, to models of intermediate complexity, to comprehensive climate models, and Earth System Models. These models simulate changes based on a set of scenarios of anthropogenic forcings. A new set of scenarios, the Representative Concentration Pathways (RCPs), was used for the new climate model simulations carried out under the framework of the Coupled Model Intercomparison Project Phase 5 (CMIP5) of the World Climate Research Programme. In all RCPs, atmospheric CO_2 concentrations are higher in 2100 relative to present day as a result of a further increase of cumulative emissions of CO_2 to the atmosphere during the 21st century (see Box SPM.1). Projections in this Summary for Policymakers are for the end of the 21st century (2081–2100) given relative to 1986–2005, unless otherwise stated. To place such projections in historical context, it is necessary to consider observed changes between different periods. Based on the longest global surface temperature dataset available, the observed change between the average of the period 1850–1900 and of the AR5 reference period is 0.61 [0.55 to 0.67] °C. However, warming has occurred beyond the average of the AR5 reference period. Hence this is not an estimate of historical warming to present (see Chapter 2) .

Continued emissions of greenhouse gases will cause further warming and changes in all components of the climate system. Limiting climate change will require substantial and sustained reductions of greenhouse gas emissions. {6, 11–14}

- Projections for the next few decades show spatial patterns of climate change similar to those projected for the later 21st century but with smaller magnitude. Natural internal variability will continue to be a major influence on climate, particularly in the near-term and at the regional scale. By the mid-21st century the magnitudes of the projected changes are substantially affected by the choice of emissions scenario (Box SPM.1). {11.3, Box 11.1, Annex I}

SPM

- Projected climate change based on RCPs is similar to AR4 in both patterns and magnitude, after accounting for scenario differences. The overall spread of projections for the high RCPs is narrower than for comparable scenarios used in AR4 because in contrast to the SRES emission scenarios used in AR4, the RCPs used in AR5 are defined as concentration pathways and thus carbon cycle uncertainties affecting atmospheric CO_2 concentrations are not considered in the concentration-driven CMIP5 simulations. Projections of sea level rise are larger than in the AR4, primarily because of improved modelling of land-ice contributions.{11.3, 12.3, 12.4, 13.4, 13.5}

E.1 Atmosphere: Temperature

Global surface temperature change for the end of the 21st century is *likely* to exceed 1.5°C relative to 1850 to 1900 for all RCP scenarios except RCP2.6. It is *likely* to exceed 2°C for RCP6.0 and RCP8.5, and *more likely than not* to exceed 2°C for RCP4.5. Warming will continue beyond 2100 under all RCP scenarios except RCP2.6. Warming will continue to exhibit interannual-to-decadal variability and will not be regionally uniform (see Figures SPM.7 and SPM.8). {11.3, 12.3, 12.4, 14.8}

- The global mean surface temperature change for the period 2016–2035 relative to 1986–2005 will *likely* be in the range of 0.3°C to 0.7°C (*medium confidence*). This assessment is based on multiple lines of evidence and assumes there will be no major volcanic eruptions or secular changes in total solar irradiance. Relative to natural internal variability, near-term increases in seasonal mean and annual mean temperatures are expected to be larger in the tropics and subtropics than in mid-latitudes (*high confidence*). {11.3}

- Increase of global mean surface temperatures for 2081–2100 relative to 1986–2005 is projected to *likely* be in the ranges derived from the concentration-driven CMIP5 model simulations, that is, 0.3°C to 1.7°C (RCP2.6), 1.1°C to 2.6°C (RCP4.5), 1.4°C to 3.1°C (RCP6.0), 2.6°C to 4.8°C (RCP8.5). The Arctic region will warm more rapidly than the global mean, and mean warming over land will be larger than over the ocean (*very high confidence*) (see Figures SPM.7 and SPM.8, and Table SPM.2). {12.4, 14.8}

- Relative to the average from year 1850 to 1900, global surface temperature change by the end of the 21st century is projected to *likely* exceed 1.5°C for RCP4.5, RCP6.0 and RCP8.5 (*high confidence*). Warming is *likely* to exceed 2°C for RCP6.0 and RCP8.5 (*high confidence*), *more likely than not* to exceed 2°C for RCP4.5 (*high confidence*), but *unlikely* to exceed 2°C for RCP2.6 (*medium confidence*). Warming is *unlikely* to exceed 4°C for RCP2.6, RCP4.5 and RCP6.0 (*high confidence*) and is *about as likely as not* to exceed 4°C for RCP8.5 (*medium confidence*). {12.4}

- It is *virtually certain* that there will be more frequent hot and fewer cold temperature extremes over most land areas on daily and seasonal timescales as global mean temperatures increase. It is *very likely* that heat waves will occur with a higher frequency and duration. Occasional cold winter extremes will continue to occur (see Table SPM.1). {12.4}

E.2 Atmosphere: Water Cycle

Changes in the global water cycle in response to the warming over the 21st century will not be uniform. The contrast in precipitation between wet and dry regions and between wet and dry seasons will increase, although there may be regional exceptions (see Figure SPM.8). {12.4, 14.3}

- Projected changes in the water cycle over the next few decades show similar large-scale patterns to those towards the end of the century, but with smaller magnitude. Changes in the near-term, and at the regional scale will be strongly influenced by natural internal variability and may be affected by anthropogenic aerosol emissions. {11.3}

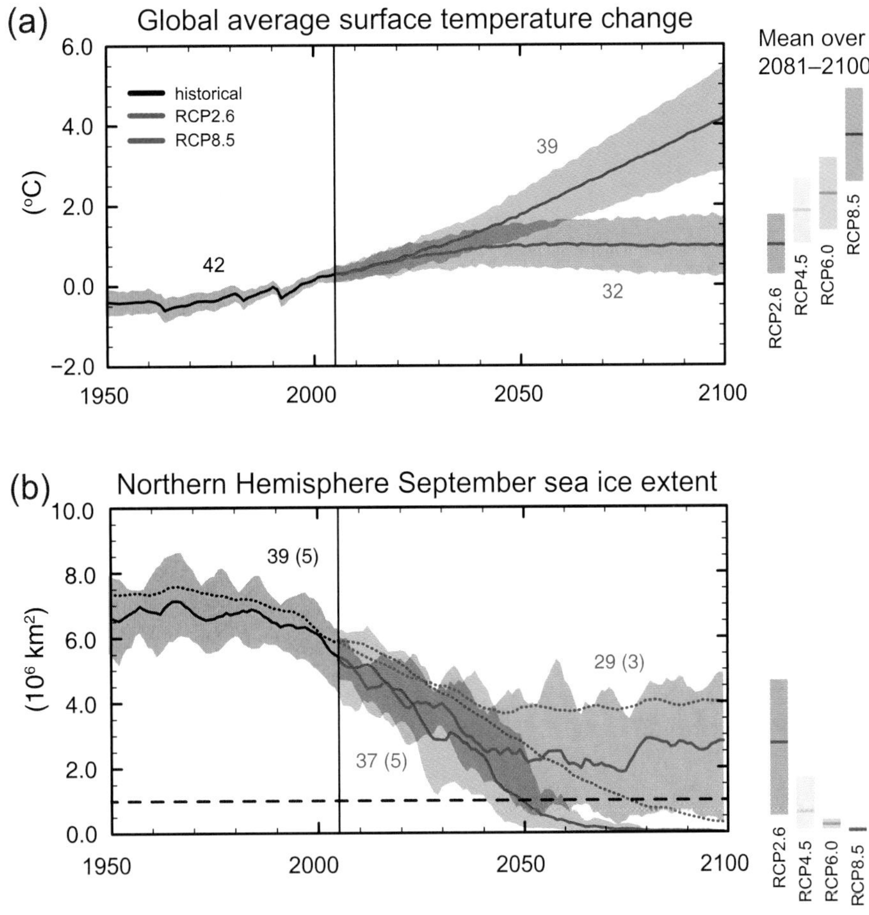

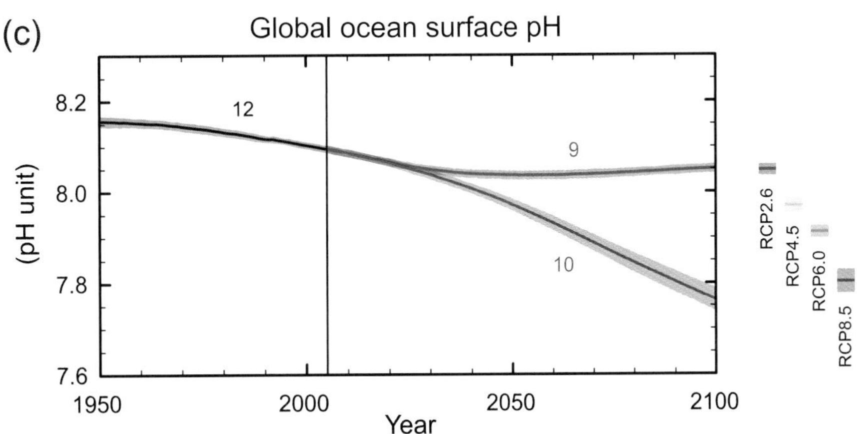

Figure SPM.7 | CMIP5 multi-model simulated time series from 1950 to 2100 for (a) change in global annual mean surface temperature relative to 1986–2005, (b) Northern Hemisphere September sea ice extent (5-year running mean), and (c) global mean ocean surface pH. Time series of projections and a measure of uncertainty (shading) are shown for scenarios RCP2.6 (blue) and RCP8.5 (red). Black (grey shading) is the modelled historical evolution using historical reconstructed forcings. The mean and associated uncertainties averaged over 2081−2100 are given for all RCP scenarios as colored vertical bars. The numbers of CMIP5 models used to calculate the multi-model mean is indicated. For sea ice extent (b), the projected mean and uncertainty (minimum-maximum range) of the subset of models that most closely reproduce the climatological mean state and 1979 to 2012 trend of the Arctic sea ice is given (number of models given in brackets). For completeness, the CMIP5 multi-model mean is also indicated with dotted lines. The dashed line represents nearly ice-free conditions (i.e., when sea ice extent is less than 10^6 km² for at least five consecutive years). For further technical details see the Technical Summary Supplementary Material {Figures 6.28, 12.5, and 12.28–12.31; Figures TS.15, TS.17, and TS.20}

SPM

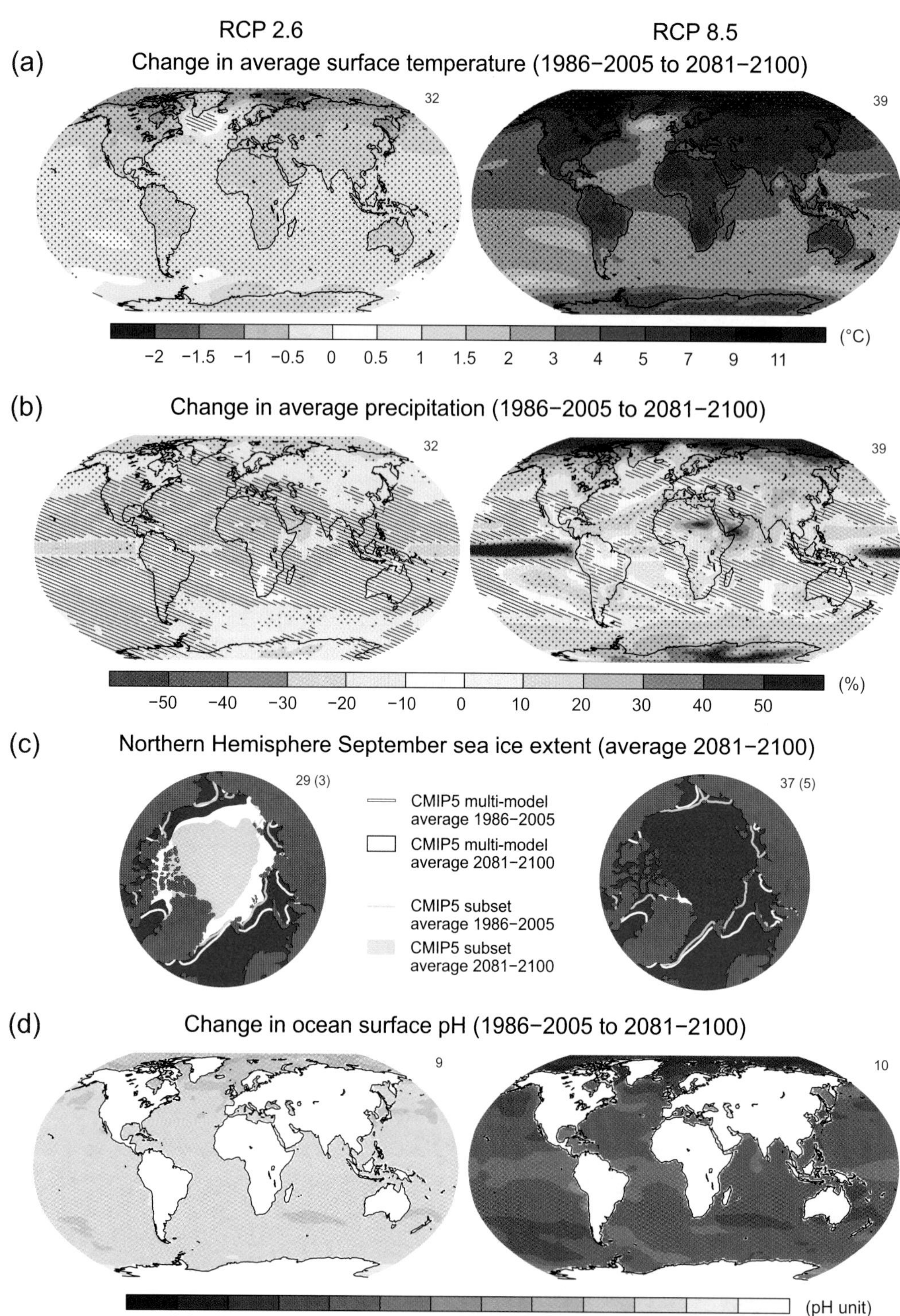

Figure SPM.8 | Maps of CMIP5 multi-model mean results for the scenarios RCP2.6 and RCP8.5 in 2081–2100 of (a) annual mean surface temperature change, (b) average percent change in annual mean precipitation, (c) Northern Hemisphere September sea ice extent, and (d) change in ocean surface pH. Changes in panels (a), (b) and (d) are shown relative to 1986–2005. The number of CMIP5 models used to calculate the multi-model mean is indicated in the upper right corner of each panel. For panels (a) and (b), hatching indicates regions where the multi-model mean is small compared to natural internal variability (i.e., less than one standard deviation of natural internal variability in 20-year means). Stippling indicates regions where the multi-model mean is large compared to natural internal variability (i.e., greater than two standard deviations of natural internal variability in 20-year means) and where at least 90% of models agree on the sign of change (see Box 12.1). In panel (c), the lines are the modelled means for 1986–2005; the filled areas are for the end of the century. The CMIP5 multi-model mean is given in white colour, the projected mean sea ice extent of a subset of models (number of models given in brackets) that most closely reproduce the climatological mean state and 1979 to 2012 trend of the Arctic sea ice extent is given in light blue colour. For further technical details see the Technical Summary Supplementary Material. {Figures 6.28, 12.11, 12.22, and 12.29; Figures TS.15, TS.16, TS.17, and TS.20}

- The high latitudes and the equatorial Pacific Ocean are *likely* to experience an increase in annual mean precipitation by the end of this century under the RCP8.5 scenario. In many mid-latitude and subtropical dry regions, mean precipitation will *likely* decrease, while in many mid-latitude wet regions, mean precipitation will *likely* increase by the end of this century under the RCP8.5 scenario (see Figure SPM.8). {7.6, 12.4, 14.3}

- Extreme precipitation events over most of the mid-latitude land masses and over wet tropical regions will *very likely* become more intense and more frequent by the end of this century, as global mean surface temperature increases (see Table SPM.1). {7.6, 12.4}

- Globally, it is *likely* that the area encompassed by monsoon systems will increase over the 21st century. While monsoon winds are *likely* to weaken, monsoon precipitation is *likely* to intensify due to the increase in atmospheric moisture. Monsoon onset dates are *likely* to become earlier or not to change much. Monsoon retreat dates will *likely* be delayed, resulting in lengthening of the monsoon season in many regions. {14.2}

- There is *high confidence* that the El Niño-Southern Oscillation (ENSO) will remain the dominant mode of interannual variability in the tropical Pacific, with global effects in the 21st century. Due to the increase in moisture availability, ENSO-related precipitation variability on regional scales will *likely* intensify. Natural variations of the amplitude and spatial pattern of ENSO are large and thus *confidence* in any specific projected change in ENSO and related regional phenomena for the 21st century remains *low*. {5.4, 14.4}

Table SPM.2 | Projected change in global mean surface air temperature and global mean sea level rise for the mid- and late 21st century relative to the reference period of 1986–2005. {12.4; Table 12.2, Table 13.5}

		2046–2065		**2081–2100**	
	Scenario	**Mean**	***Likely* range[c]**	**Mean**	***Likely* range[c]**
Global Mean Surface Temperature Change (°C)[a]	RCP2.6	1.0	0.4 to 1.6	1.0	0.3 to 1.7
	RCP4.5	1.4	0.9 to 2.0	1.8	1.1 to 2.6
	RCP6.0	1.3	0.8 to 1.8	2.2	1.4 to 3.1
	RCP8.5	2.0	1.4 to 2.6	3.7	2.6 to 4.8
	Scenario	**Mean**	***Likely* range[d]**	**Mean**	***Likely* range[d]**
Global Mean Sea Level Rise (m)[b]	RCP2.6	0.24	0.17 to 0.32	0.40	0.26 to 0.55
	RCP4.5	0.26	0.19 to 0.33	0.47	0.32 to 0.63
	RCP6.0	0.25	0.18 to 0.32	0.48	0.33 to 0.63
	RCP8.5	0.30	0.22 to 0.38	0.63	0.45 to 0.82

Notes:

[a] Based on the CMIP5 ensemble; anomalies calculated with respect to 1986–2005. Using HadCRUT4 and its uncertainty estimate (5–95% confidence interval), the observed warming to the reference period 1986–2005 is 0.61 [0.55 to 0.67] °C from 1850–1900, and 0.11 [0.09 to 0.13] °C from 1980–1999, the reference period for projections used in AR4. *Likely* ranges have not been assessed here with respect to earlier reference periods because methods are not generally available in the literature for combining the uncertainties in models and observations. Adding projected and observed changes does not account for potential effects of model biases compared to observations, and for natural internal variability during the observational reference period {2.4; 11.2; Tables 12.2 and 12.3}

[b] Based on 21 CMIP5 models; anomalies calculated with respect to 1986–2005. Where CMIP5 results were not available for a particular AOGCM and scenario, they were estimated as explained in Chapter 13, Table 13.5. The contributions from ice sheet rapid dynamical change and anthropogenic land water storage are treated as having uniform probability distributions, and as largely independent of scenario. This treatment does not imply that the contributions concerned will not depend on the scenario followed, only that the current state of knowledge does not permit a quantitative assessment of the dependence. Based on current understanding, only the collapse of marine-based sectors of the Antarctic ice sheet, if initiated, could cause global mean sea level to rise substantially above the *likely* range during the 21st century. There is *medium confidence* that this additional contribution would not exceed several tenths of a meter of sea level rise during the 21st century.

[c] Calculated from projections as 5–95% model ranges. These ranges are then assessed to be *likely* ranges after accounting for additional uncertainties or different levels of confidence in models. For projections of global mean surface temperature change in 2046–2065 *confidence* is *medium*, because the relative importance of natural internal variability, and uncertainty in non-greenhouse gas forcing and response, are larger than for 2081–2100. The *likely* ranges for 2046–2065 do not take into account the possible influence of factors that lead to the assessed range for near-term (2016–2035) global mean surface temperature change that is lower than the 5–95% model range, because the influence of these factors on longer term projections has not been quantified due to insufficient scientific understanding. {11.3}

[d] Calculated from projections as 5–95% model ranges. These ranges are then assessed to be *likely* ranges after accounting for additional uncertainties or different levels of confidence in models. For projections of global mean sea level rise *confidence* is *medium* for both time horizons.

E.3 Atmosphere: Air Quality

- The range in projections of air quality (ozone and PM2.5[17] in near-surface air) is driven primarily by emissions (including CH_4), rather than by physical climate change (*medium confidence*). There is *high confidence* that globally, warming decreases background surface ozone. High CH_4 levels (as in RCP8.5) can offset this decrease, raising background surface ozone by year 2100 on average by about 8 ppb (25% of current levels) relative to scenarios with small CH_4 changes (as in RCP4.5 and RCP6.0) (*high confidence*). {11.3}

- Observational and modelling evidence indicates that, all else being equal, locally higher surface temperatures in polluted regions will trigger regional feedbacks in chemistry and local emissions that will increase peak levels of ozone and PM2.5 (*medium confidence*). For PM2.5, climate change may alter natural aerosol sources as well as removal by precipitation, but no confidence level is attached to the overall impact of climate change on PM2.5 distributions. {11.3}

E.4 Ocean

The global ocean will continue to warm during the 21st century. Heat will penetrate from the surface to the deep ocean and affect ocean circulation. {11.3, 12.4}

- The strongest ocean warming is projected for the surface in tropical and Northern Hemisphere subtropical regions. At greater depth the warming will be most pronounced in the Southern Ocean (*high confidence*). Best estimates of ocean warming in the top one hundred meters are about 0.6°C (RCP2.6) to 2.0°C (RCP8.5), and about 0.3°C (RCP2.6) to 0.6°C (RCP8.5) at a depth of about 1000 m by the end of the 21st century. {12.4, 14.3}

- It is *very likely* that the Atlantic Meridional Overturning Circulation (AMOC) will weaken over the 21st century. Best estimates and ranges[18] for the reduction are 11% (1 to 24%) in RCP2.6 and 34% (12 to 54%) in RCP8.5. It is *likely* that there will be some decline in the AMOC by about 2050, but there may be some decades when the AMOC increases due to large natural internal variability. {11.3, 12.4}

- It is *very unlikely* that the AMOC will undergo an abrupt transition or collapse in the 21st century for the scenarios considered. There is *low confidence* in assessing the evolution of the AMOC beyond the 21st century because of the limited number of analyses and equivocal results. However, a collapse beyond the 21st century for large sustained warming cannot be excluded. {12.5}

E.5 Cryosphere

It is *very likely* that the Arctic sea ice cover will continue to shrink and thin and that Northern Hemisphere spring snow cover will decrease during the 21st century as global mean surface temperature rises. Global glacier volume will further decrease. {12.4, 13.4}

- Year-round reductions in Arctic sea ice extent are projected by the end of the 21st century from multi-model averages. These reductions range from 43% for RCP2.6 to 94% for RCP8.5 in September and from 8% for RCP2.6 to 34% for RCP8.5 in February (*medium confidence*) (see Figures SPM.7 and SPM.8). {12.4}

---

17 PM2.5 refers to particulate matter with a diameter of less than 2.5 micrometres, a measure of atmospheric aerosol concentration.

18 The ranges in this paragraph indicate a CMIP5 model spread.

- Based on an assessment of the subset of models that most closely reproduce the climatological mean state and 1979 to 2012 trend of the Arctic sea ice extent, a nearly ice-free Arctic Ocean[19] in September before mid-century is *likely* for RCP8.5 (*medium confidence*) (see Figures SPM.7 and SPM.8). A projection of when the Arctic might become nearly ice-free in September in the 21st century cannot be made with confidence for the other scenarios. {11.3, 12.4, 12.5}

- In the Antarctic, a decrease in sea ice extent and volume is projected with *low confidence* for the end of the 21st century as global mean surface temperature rises. {12.4}

- By the end of the 21st century, the global glacier volume, excluding glaciers on the periphery of Antarctica, is projected to decrease by 15 to 55% for RCP2.6, and by 35 to 85% for RCP8.5 (*medium confidence*). {13.4, 13.5}

- The area of Northern Hemisphere spring snow cover is projected to decrease by 7% for RCP2.6 and by 25% in RCP8.5 by the end of the 21st century for the model average (*medium confidence*). {12.4}

- It is *virtually certain* that near-surface permafrost extent at high northern latitudes will be reduced as global mean surface temperature increases. By the end of the 21st century, the area of permafrost near the surface (upper 3.5 m) is projected to decrease by between 37% (RCP2.6) to 81% (RCP8.5) for the model average (*medium confidence*). {12.4}

E.6 Sea Level

Global mean sea level will continue to rise during the 21st century (see Figure SPM.9). Under all RCP scenarios, the rate of sea level rise will *very likely* exceed that observed during 1971 to 2010 due to increased ocean warming and increased loss of mass from glaciers and ice sheets. {13.3–13.5}

- Confidence in projections of global mean sea level rise has increased since the AR4 because of the improved physical understanding of the components of sea level, the improved agreement of process-based models with observations, and the inclusion of ice-sheet dynamical changes. {13.3–13.5}

- Global mean sea level rise for 2081–2100 relative to 1986–2005 will *likely* be in the ranges of 0.26 to 0.55 m for RCP2.6, 0.32 to 0.63 m for RCP4.5, 0.33 to 0.63 m for RCP6.0, and 0.45 to 0.82 m for RCP8.5 (*medium confidence*). For RCP8.5, the rise by the year 2100 is 0.52 to 0.98 m, with a rate during 2081 to 2100 of 8 to 16 mm yr^{-1} (*medium confidence*). These ranges are derived from CMIP5 climate projections in combination with process-based models and literature assessment of glacier and ice sheet contributions (see Figure SPM.9, Table SPM.2). {13.5}

- In the RCP projections, thermal expansion accounts for 30 to 55% of 21st century global mean sea level rise, and glaciers for 15 to 35%. The increase in surface melting of the Greenland ice sheet will exceed the increase in snowfall, leading to a positive contribution from changes in surface mass balance to future sea level (*high confidence*). While surface melting will remain small, an increase in snowfall on the Antarctic ice sheet is expected (*medium confidence*), resulting in a negative contribution to future sea level from changes in surface mass balance. Changes in outflow from both ice sheets combined will *likely* make a contribution in the range of 0.03 to 0.20 m by 2081–2100 (*medium confidence*). {13.3–13.5}

- Based on current understanding, only the collapse of marine-based sectors of the Antarctic ice sheet, if initiated, could cause global mean sea level to rise substantially above the *likely* range during the 21st century. However, there is *medium confidence* that this additional contribution would not exceed several tenths of a meter of sea level rise during the 21st century. {13.4, 13.5}

---

[19] Conditions in the Arctic Ocean are referred to as nearly ice-free when the sea ice extent is less than 10^6 km^2 for at least five consecutive years.

SPM

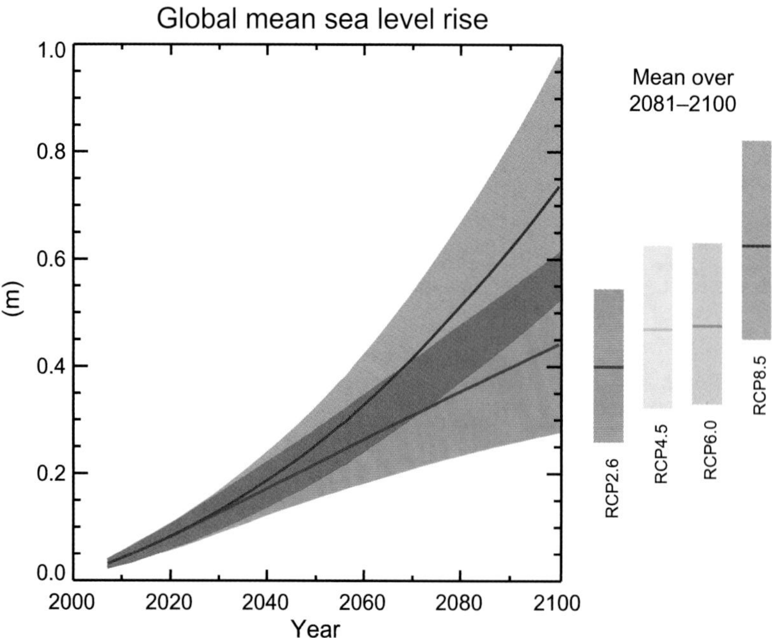

Figure SPM.9 | Projections of global mean sea level rise over the 21st century relative to 1986–2005 from the combination of the CMIP5 ensemble with process-based models, for RCP2.6 and RCP8.5. The assessed *likely* range is shown as a shaded band. The assessed *likely* ranges for the mean over the period 2081–2100 for all RCP scenarios are given as coloured vertical bars, with the corresponding median value given as a horizontal line. For further technical details see the Technical Summary Supplementary Material {Table 13.5, Figures 13.10 and 13.11; Figures TS.21 and TS.22}

- The basis for higher projections of global mean sea level rise in the 21st century has been considered and it has been concluded that there is currently insufficient evidence to evaluate the probability of specific levels above the assessed *likely* range. Many semi-empirical model projections of global mean sea level rise are higher than process-based model projections (up to about twice as large), but there is no consensus in the scientific community about their reliability and there is thus *low confidence* in their projections. {13.5}

- Sea level rise will not be uniform. By the end of the 21st century, it is *very likely* that sea level will rise in more than about 95% of the ocean area. About 70% of the coastlines worldwide are projected to experience sea level change within 20% of the global mean sea level change. {13.1, 13.6}

E.7 Carbon and Other Biogeochemical Cycles

Climate change will affect carbon cycle processes in a way that will exacerbate the increase of CO_2 in the atmosphere (*high confidence*). Further uptake of carbon by the ocean will increase ocean acidification. {6.4}

- Ocean uptake of anthropogenic CO_2 will continue under all four RCPs through to 2100, with higher uptake for higher concentration pathways (*very high confidence*). The future evolution of the land carbon uptake is less certain. A majority of models projects a continued land carbon uptake under all RCPs, but some models simulate a land carbon loss due to the combined effect of climate change and land use change. {6.4}

- Based on Earth System Models, there is *high confidence* that the feedback between climate and the carbon cycle is positive in the 21st century; that is, climate change will partially offset increases in land and ocean carbon sinks caused by rising atmospheric CO_2. As a result more of the emitted anthropogenic CO_2 will remain in the atmosphere. A positive feedback between climate and the carbon cycle on century to millennial time scales is supported by paleoclimate observations and modelling. {6.2, 6.4}

Table SPM.3 | Cumulative CO_2 emissions for the 2012 to 2100 period compatible with the RCP atmospheric concentrations simulated by the CMIP5 Earth System Models. {6.4, Table 6.12, Figure TS.19}

Scenario	Cumulative CO_2 Emissions 2012 to 2100[a]			
	GtC		$GtCO_2$	
	Mean	Range	Mean	Range
RCP2.6	270	140 to 410	990	510 to 1505
RCP4.5	780	595 to 1005	2860	2180 to 3690
RCP6.0	1060	840 to 1250	3885	3080 to 4585
RCP8.5	1685	1415 to 1910	6180	5185 to 7005

Notes:
[a] 1 Gigatonne of carbon = 1 GtC = 10^{15} grams of carbon. This corresponds to 3.667 $GtCO_2$.

- Earth System Models project a global increase in ocean acidification for all RCP scenarios. The corresponding decrease in surface ocean pH by the end of 21st century is in the range[18] of 0.06 to 0.07 for RCP2.6, 0.14 to 0.15 for RCP4.5, 0.20 to 0.21 for RCP6.0, and 0.30 to 0.32 for RCP8.5 (see Figures SPM.7 and SPM.8). {6.4}

- Cumulative CO_2 emissions[20] for the 2012 to 2100 period compatible with the RCP atmospheric CO_2 concentrations, as derived from 15 Earth System Models, range[18] from 140 to 410 GtC for RCP2.6, 595 to 1005 GtC for RCP4.5, 840 to 1250 GtC for RCP6.0, and 1415 to 1910 GtC for RCP8.5 (see Table SPM.3). {6.4}

- By 2050, annual CO_2 emissions derived from Earth System Models following RCP2.6 are smaller than 1990 emissions (by 14 to 96%). By the end of the 21st century, about half of the models infer emissions slightly above zero, while the other half infer a net removal of CO_2 from the atmosphere. {6.4, Figure TS.19}

- The release of CO_2 or CH_4 to the atmosphere from thawing permafrost carbon stocks over the 21st century is assessed to be in the range of 50 to 250 GtC for RCP8.5 (*low confidence*). {6.4}

E.8 Climate Stabilization, Climate Change Commitment and Irreversibility

Cumulative emissions of CO_2 largely determine global mean surface warming by the late 21st century and beyond (see Figure SPM.10). Most aspects of climate change will persist for many centuries even if emissions of CO_2 are stopped. This represents a substantial multi-century climate change commitment created by past, present and future emissions of CO_2. {12.5}

- Cumulative total emissions of CO_2 and global mean surface temperature response are approximately linearly related (see Figure SPM.10). Any given level of warming is associated with a range of cumulative CO_2 emissions[21], and therefore, e.g., higher emissions in earlier decades imply lower emissions later. {12.5}

- Limiting the warming caused by anthropogenic CO_2 emissions alone with a probability of >33%, >50%, and >66% to less than 2°C since the period 1861–1880[22], will require cumulative CO_2 emissions from all anthropogenic sources to stay between 0 and about 1570 GtC (5760 $GtCO_2$), 0 and about 1210 GtC (4440 $GtCO_2$), and 0 and about 1000 GtC (3670 $GtCO_2$) since that period, respectively[23]. These upper amounts are reduced to about 900 GtC (3300 $GtCO_2$), 820 GtC (3010 $GtCO_2$), and 790 GtC (2900 $GtCO_2$), respectively, when accounting for non-CO_2 forcings as in RCP2.6. An amount of 515 [445 to 585] GtC (1890 [1630 to 2150] $GtCO_2$), was already emitted by 2011. {12.5}

[20] From fossil fuel, cement, industry, and waste sectors.

[21] Quantification of this range of CO_2 emissions requires taking into account non-CO_2 drivers.

[22] The first 20-year period available from the models.

[23] This is based on the assessment of the transient climate response to cumulative carbon emissions (TCRE, see Section D.2).

SPM

- A lower warming target, or a higher likelihood of remaining below a specific warming target, will require lower cumulative CO_2 emissions. Accounting for warming effects of increases in non-CO_2 greenhouse gases, reductions in aerosols, or the release of greenhouse gases from permafrost will also lower the cumulative CO_2 emissions for a specific warming target (see Figure SPM.10). {12.5}

- A large fraction of anthropogenic climate change resulting from CO_2 emissions is irreversible on a multi-century to millennial time scale, except in the case of a large net removal of CO_2 from the atmosphere over a sustained period. Surface temperatures will remain approximately constant at elevated levels for many centuries after a complete cessation of net anthropogenic CO_2 emissions. Due to the long time scales of heat transfer from the ocean surface to depth, ocean warming will continue for centuries. Depending on the scenario, about 15 to 40% of emitted CO_2 will remain in the atmosphere longer than 1,000 years. {Box 6.1, 12.4, 12.5}

- It is *virtually certain* that global mean sea level rise will continue beyond 2100, with sea level rise due to thermal expansion to continue for many centuries. The few available model results that go beyond 2100 indicate global mean sea level rise above the pre-industrial level by 2300 to be less than 1 m for a radiative forcing that corresponds to CO_2 concentrations that peak and decline and remain below 500 ppm, as in the scenario RCP2.6. For a radiative forcing that corresponds to a CO_2 concentration that is above 700 ppm but below 1500 ppm, as in the scenario RCP8.5, the projected rise is 1 m to more than 3 m (*medium confidence*). {13.5}

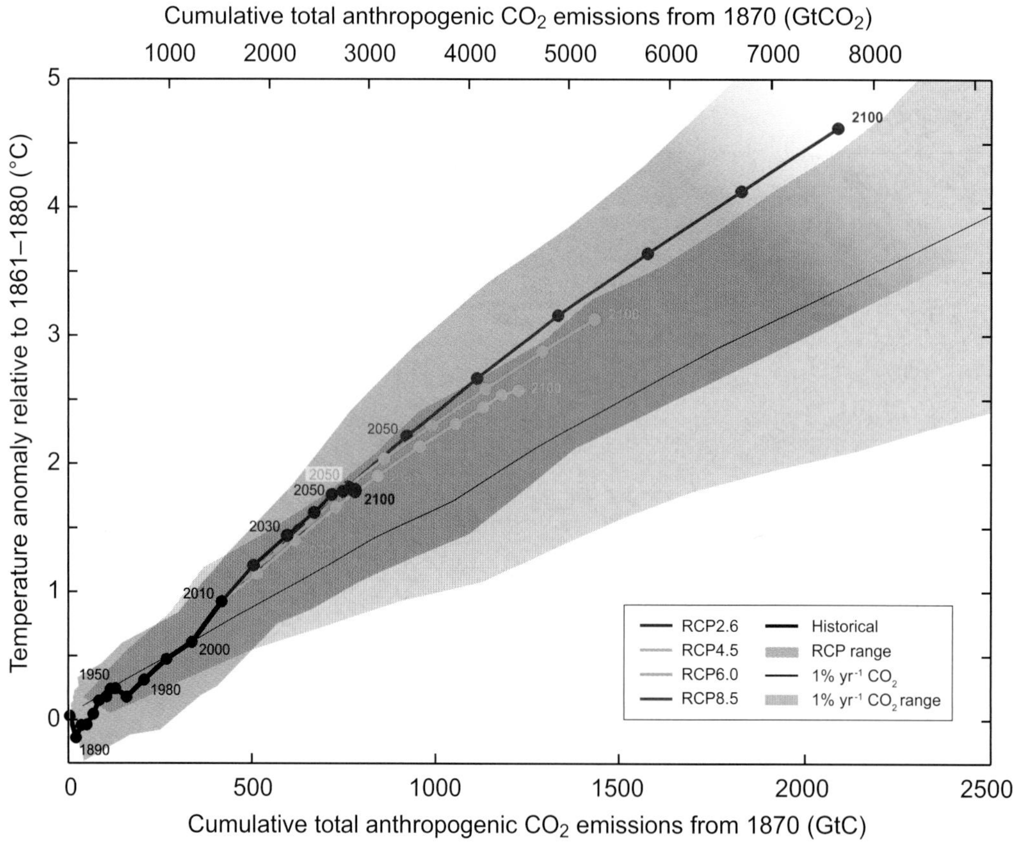

Figure SPM.10 | Global mean surface temperature increase as a function of cumulative total global CO_2 emissions from various lines of evidence. Multi-model results from a hierarchy of climate-carbon cycle models for each RCP until 2100 are shown with coloured lines and decadal means (dots). Some decadal means are labeled for clarity (e.g., 2050 indicating the decade 2040–2049). Model results over the historical period (1860 to 2010) are indicated in black. The coloured plume illustrates the multi-model spread over the four RCP scenarios and fades with the decreasing number of available models in RCP8.5. The multi-model mean and range simulated by CMIP5 models, forced by a CO_2 increase of 1% per year (1% yr^{-1} CO_2 simulations), is given by the thin black line and grey area. For a specific amount of cumulative CO_2 emissions, the 1% per year CO_2 simulations exhibit lower warming than those driven by RCPs, which include additional non-CO_2 forcings. Temperature values are given relative to the 1861–1880 base period, emissions relative to 1870. Decadal averages are connected by straight lines. For further technical details see the Technical Summary Supplementary Material. {Figure 12.45; TS TFE.8, Figure 1}

- Sustained mass loss by ice sheets would cause larger sea level rise, and some part of the mass loss might be irreversible. There is *high confidence* that sustained warming greater than some threshold would lead to the near-complete loss of the Greenland ice sheet over a millennium or more, causing a global mean sea level rise of up to 7 m. Current estimates indicate that the threshold is greater than about 1°C (*low confidence*) but less than about 4°C (*medium confidence*) global mean warming with respect to pre-industrial. Abrupt and irreversible ice loss from a potential instability of marine-based sectors of the Antarctic ice sheet in response to climate forcing is possible, but current evidence and understanding is insufficient to make a quantitative assessment. {5.8, 13.4, 13.5}

- Methods that aim to deliberately alter the climate system to counter climate change, termed geoengineering, have been proposed. Limited evidence precludes a comprehensive quantitative assessment of both Solar Radiation Management (SRM) and Carbon D ioxide Removal (CDR) and their impact on the climate system. CDR methods have biogeochemical and technological limitations to their potential on a global scale. There is insufficient knowledge to quantify how much CO_2 emissions could be partially offset by CDR on a century timescale. Modelling indicates that SRM methods, if realizable, have the potential to substantially offset a global temperature rise, but they would also modify the global water cycle, and would not reduce ocean acidification. If SRM were terminated for any reason, there is *high confidence* that global surface temperatures would rise very rapidly to values consistent with the greenhouse gas forcing. CDR and SRM methods carry side effects and long-term consequences on a global scale. {6.5, 7.7}

Box SPM.1: Representative Concentration Pathways (RCPs)

Climate change projections in IPCC Working Group I require information about future emissions or concentrations of greenhouse gases, aerosols and other climate drivers. This information is often expressed as a scenario of human activities, which are not assessed in this report. Scenarios used in Working Group I have focused on anthropogenic emissions and do not include changes in natural drivers such as solar or volcanic forcing or natural emissions, for example, of CH_4 and N_2O.

For the Fifth Assessment Report of IPCC, the scientific community has defined a set of four new scenarios, denoted Representative Concentration Pathways (RCPs, see Glossary). They are identified by their approximate total radiative forcing in year 2100 relative to 1750: 2.6 W m^{-2} for RCP2.6, 4.5 W m^{-2} for RCP4.5, 6.0 W m^{-2} for RCP6.0, and 8.5 W m^{-2} for RCP8.5. For the Coupled Model Intercomparison Project Phase 5 (CMIP5) results, these values should be understood as indicative only, as the climate forcing resulting from all drivers varies between models due to specific model characteristics and treatment of short-lived climate forcers. These four RCPs include one mitigation scenario leading to a very low forcing level (RCP2.6), two stabilization scenarios (RCP4.5 and RCP6), and one scenario with very high greenhouse gas emissions (RCP8.5). The RCPs can thus represent a range of 21st century climate policies, as compared with the no-climate policy of the Special Report on Emissions Scenarios (SRES) used in the Third Assessment Report and the Fourth Assessment Report. For RCP6.0 and RCP8.5, radiative forcing does not peak by year 2100; for RCP2.6 it peaks and declines; and for RCP4.5 it stabilizes by 2100. Each RCP provides spatially resolved data sets of land use change and sector-based emissions of air pollutants, and it specifies annual greenhouse gas concentrations and anthropogenic emissions up to 2100. RCPs are based on a combination of integrated assessment models, simple climate models, atmospheric chemistry and global carbon cycle models. While the RCPs span a wide range of total forcing values, they do not cover the full range of emissions in the literature, particularly for aerosols.

Most of the CMIP5 and Earth System Model simulations were performed with prescribed CO_2 concentrations reaching 421 ppm (RCP2.6), 538 ppm (RCP4.5), 670 ppm (RCP6.0), and 936 ppm (RCP 8.5) by the year 2100. Including also the prescribed concentrations of CH_4 and N_2O, the combined CO_2-equivalent concentrations are 475 ppm (RCP2.6), 630 ppm (RCP4.5), 800 ppm (RCP6.0), and 1313 ppm (RCP8.5). For RCP8.5, additional CMIP5 Earth System Model simulations are performed with prescribed CO_2 emissions as provided by the integrated assessment models. For all RCPs, additional calculations were made with updated atmospheric chemistry data and models (including the Atmospheric Chemistry and Climate component of CMIP5) using the RCP prescribed emissions of the chemically reactive gases (CH_4, N_2O, HFCs, NO_x, CO, NMVOC). These simulations enable investigation of uncertainties related to carbon cycle feedbacks and atmospheric chemistry.

Technical Summary

TS Technical Summary

Coordinating Lead Authors:
Thomas F. Stocker (Switzerland), Qin Dahe (China), Gian-Kasper Plattner (Switzerland)

Lead Authors:
Lisa V. Alexander (Australia), Simon K. Allen (Switzerland/New Zealand), Nathaniel L. Bindoff (Australia), François-Marie Bréon (France), John A. Church (Australia), Ulrich Cubasch (Germany), Seita Emori (Japan), Piers Forster (UK), Pierre Friedlingstein (UK/Belgium), Nathan Gillett (Canada), Jonathan M. Gregory (UK), Dennis L. Hartmann (USA), Eystein Jansen (Norway), Ben Kirtman (USA), Reto Knutti (Switzerland), Krishna Kumar Kanikicharla (India), Peter Lemke (Germany), Jochem Marotzke (Germany), Valérie Masson-Delmotte (France), Gerald A. Meehl (USA), Igor I. Mokhov (Russian Federation), Shilong Piao (China), Venkatachalam Ramaswamy (USA), David Randall (USA), Monika Rhein (Germany), Maisa Rojas (Chile), Christopher Sabine (USA), Drew Shindell (USA), Lynne D. Talley (USA), David G. Vaughan (UK), Shang-Ping Xie (USA)

Contributing Authors:
Myles R. Allen (UK), Olivier Boucher (France), Don Chambers (USA), Jens Hesselbjerg Christensen (Denmark), Philippe Ciais (France), Peter U. Clark (USA), Matthew Collins (UK), Josefino C. Comiso (USA), Viviane Vasconcellos de Menezes (Australia/Brazil), Richard A. Feely (USA), Thierry Fichefet (Belgium), Gregory Flato (Canada), Jesús Fidel González Rouco (Spain), Ed Hawkins (UK), Paul J. Hezel (Belgium/USA), Gregory C. Johnson (USA), Simon A. Josey (UK), Georg Kaser (Austria/Italy), Albert M.G. Klein Tank (Netherlands), Janina Körper (Germany), Gunnar Myhre (Norway), Timothy Osborn (UK), Scott B. Power (Australia), Stephen R. Rintoul (Australia), Joeri Rogelj (Switzerland/Belgium), Matilde Rusticucci (Argentina), Michael Schulz (Germany), Jan Sedláček (Switzerland), Peter A. Stott (UK), Rowan Sutton (UK), Peter W. Thorne (USA/Norway/UK), Donald Wuebbles (USA)

Review Editors:
Sylvie Joussaume (France), Joyce Penner (USA), Fredolin Tangang (Malaysia)

This Technical Summary should be cited as:
Stocker, T.F., D. Qin, G.-K. Plattner, L.V. Alexander, S.K. Allen, N.L. Bindoff, F.-M. Bréon, J.A. Church, U. Cubasch, S. Emori, P. Forster, P. Friedlingstein, N. Gillett, J.M. Gregory, D.L. Hartmann, E. Jansen, B. Kirtman, R. Knutti, K. Krishna Kumar, P. Lemke, J. Marotzke, V. Masson-Delmotte, G.A. Meehl, I.I. Mokhov, S. Piao, V. Ramaswamy, D. Randall, M. Rhein, M. Rojas, C. Sabine, D. Shindell, L.D. Talley, D.G. Vaughan and S.-P. Xie, 2013: Technical Summary. In: *Climate Change 2013: The Physical Science Basis. Contribution of Working Group I to the Fifth Assessment Report of the Intergovernmental Panel on Climate Change* [Stocker, T.F., D. Qin, G.-K. Plattner, M. Tignor, S.K. Allen, J. Boschung, A. Nauels, Y. Xia, V. Bex and P.M. Midgley (eds.)]. Cambridge University Press, Cambridge, United Kingdom and New York, NY, USA.

Table of Contents

Supplementary Material is available in online versions of the report.

TS.1 Introduction

Climate Change 2013: The Physical Science Basis is the contribution of Working Group I (WGI) to the Fifth Assessment Report (AR5) of the Intergovernmental Panel on Climate Change (IPCC). This comprehensive assessment of the physical aspects of climate change puts a focus on those elements that are relevant to understand past, document current and project future climate change. The assessment builds on the IPCC Fourth Assessment Report (AR4)[1] and the recent Special Report on Managing the Risk of Extreme Events and Disasters to Advance Climate Change Adaptation (SREX)[2] and is presented in 14 chapters and 3 annexes. The chapters cover direct and proxy observations of changes in all components of the climate system; assess the current knowledge of various processes within, and interactions among, climate system components, which determine the sensitivity and response of the system to changes in forcing; and quantify the link between the changes in atmospheric constituents, and hence radiative forcing (RF)[3], and the consequent detection and attribution of climate change. Projections of changes in all climate system components are based on model simulations forced by a new set of scenarios. The Report also provides a comprehensive assessment of past and future sea level change in a dedicated chapter. Regional climate change information is presented in the form of an Atlas of Global and Regional Climate Projections (Annex I). This is complemented by Annex II: Climate System Scenario Tables and Annex III: Glossary.

The primary purpose of this Technical Summary (TS) is to provide the link between the complete assessment of the multiple lines of independent evidence presented in the 14 chapters of the main report and the highly condensed summary prepared as the WGI Summary for Policymakers (SPM). The Technical Summary thus serves as a starting point for those readers who seek the full information on more specific topics covered by this assessment. This purpose is facilitated by including pointers to the chapters and sections where the full assessment can be found. Policy-relevant topics, which cut across many chapters and involve many interlinked processes in the climate system, are presented here as Thematic Focus Elements (TFEs), allowing rapid access to this information.

An integral element of this report is the use of uncertainty language that permits a traceable account of the assessment (Box TS.1). The degree of certainty in key findings in this assessment is based on the author teams' evaluations of underlying scientific understanding and is expressed as a level of confidence that results from the type, amount, quality and consistency of evidence and the degree of agreement in the scientific studies considered[4]. Confidence is expressed qualitatively. Quantified measures of uncertainty in a finding are expressed probabilistically and are based on a combination of statistical analyses of observations or model results, or both, and expert judgement. Where appropriate, findings are also formulated as statements of fact without using uncertainty qualifiers (see Chapter 1 and Box TS.1 for more details).

The Technical Summary is structured into four main sections presenting the assessment results following the storyline of the WGI contribution to AR5: Section TS.2 covers the assessment of observations of changes in the climate system; Section TS.3 summarizes the information on the different drivers, natural and anthropogenic, expressed in terms of RF; Section TS.4 presents the assessment of the quantitative understanding of observed climate change; and Section TS.5 summarizes the assessment results for projections of future climate change over the 21st century and beyond from regional to global scale. Section TS.6 combines and lists key uncertainties from the WGI assessment from Sections TS.2 to TS.5. The overall nine TFEs, cutting across the various components of the WGI AR5, are dispersed throughout the four main TS sections, are visually distinct from the main text and should allow stand-alone reading.

The basis for substantive paragraphs in this Technical Summary can be found in the chapter sections of the underlying report. These references are given in curly brackets.

---

1 IPCC, 2007: *Climate Change 2007: The Physical Science Basis.* Contribution of Working Group I to the Fourth Assessment Report of the Intergovernmental Panel on Climate Change [Solomon, S., D. Qin, M. Manning, Z. Chen, M. Marquis, K.B. Averyt, M. Tignor and H.L. Miller (eds.)]. Cambridge University Press, Cambridge, United Kingdom and New York, NY, USA, 996 pp.

2 IPCC, 2012: *Managing the Risks of Extreme Events and Disasters to Advance Climate Change Adaptation*. A Special Report of Working Groups I and II of the Intergovernmental Panel on Climate Change [Field, C.B., V. Barros, T.F. Stocker, D. Qin, D.J. Dokken, K.L. Ebi, M.D. Mastrandrea, K.J. Mach, G.-K. Plattner, S.K. Allen, M. Tignor and P. M. Midgley (eds.)]. Cambridge University Press, Cambridge, UK, and New York, NY, USA, 582 pp.

3 Radiative forcing (RF) is a measure of the net change in the energy balance of the Earth system in response to some external perturbation. It is expressed in watts per square metre (W m^{-2}); see Box TS.2.

4 Mastrandrea, M.D., C.B. Field, T.F. Stocker, O. Edenhofer, K.L. Ebi, D.J. Frame, H. Held, E. Kriegler, K.J. Mach, P.R. Matschoss, G.-K. Plattner, G.W. Yohe, and F.W. Zwiers, 2010: *Guidance Note for Lead Authors of the IPCC Fifth Assessment Report on Consistent Treatment of Uncertainties*. Intergovernmental Panel on Climate Change (IPCC).

Box TS.1 | Treatment of Uncertainty

Based on the Guidance Note for Lead Authors of the IPCC Fifth Assessment Report on Consistent Treatment of Uncertainties, this WGI Technical Summary and the WGI Summary for Policymakers rely on two metrics for communicating the degree of certainty in key findings, which is based on author teams' evaluations of underlying scientific understanding:

- Confidence in the validity of a finding, based on the type, amount, quality and consistency of evidence (e.g., mechanistic understanding, theory, data, models, expert judgement) and the degree of agreement. Confidence is expressed qualitatively.
- Quantified measures of uncertainty in a finding expressed probabilistically (based on statistical analysis of observations or model results, or expert judgement).

The AR5 Guidance Note refines the guidance provided to support the IPCC Third and Fourth Assessment Reports. Direct comparisons between assessment of uncertainties in findings in this Report and those in the AR4 and the SREX are difficult, because of the application of the revised guidance note on uncertainties, as well as the availability of new information, improved scientific understanding, continued analyses of data and models and specific differences in methodologies applied in the assessed studies. For some climate variables, different aspects have been assessed and therefore a direct comparison would be inappropriate.

Each key finding is based on an author team's evaluation of associated evidence and agreement. The confidence metric provides a qualitative synthesis of an author team's judgement about the validity of a finding, as determined through evaluation of evidence and agreement. If uncertainties can be quantified probabilistically, an author team can characterize a finding using the calibrated likelihood language or a more precise presentation of probability. Unless otherwise indicated, high or very high confidence is associated with findings for which an author team has assigned a likelihood term.

The following summary terms are used to describe the available evidence: limited, medium, or robust; and for the degree of agreement: low, medium, or high. A level of confidence is expressed using five qualifiers very low, low, medium, high, and very high, and typeset in italics, e.g., *medium confidence.* Box TS.1, Figure 1 depicts summary statements for evidence and agreement and their relationship to confidence. There is flexibility in this relationship; for a given evidence and agreement statement, different confidence levels can be assigned, but increasing levels of evidence and degrees of agreement correlate with increasing confidence.

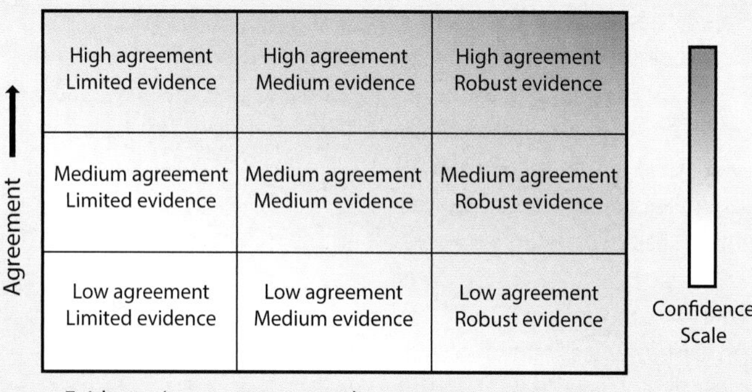

Box TS.1, Figure 1 | A depiction of evidence and agreement statements and their relationship to confidence. Confidence increases toward the top right corner as suggested by the increasing strength of shading. Generally, evidence is most robust when there are multiple, consistent independent lines of high quality. {Figure 1.11}

The following terms have been used to indicate the assessed likelihood, and typeset in italics:

Term*	**Likelihood of the outcome**
Virtually certain	99–100% probability
Very likely	90–100% probability
Likely	66–100% probability
About as likely as not	33–66% probability
Unlikely	0–33% probability
Very unlikely	0–10% probability
Exceptionally unlikely	0–1% probability

* Additional terms (*extremely likely*: 95–100% probability, *more likely than not*: >50–100% probability, and *extremely unlikely*: 0–5% probability) may also be used when appropriate.

TS.2 Observation of Changes in the Climate System

TS.2.1 Introduction

Observations of the climate system are based on direct physical and biogeochemical measurements, and remote sensing from ground stations and satellites; information derived from paleoclimate archives provides a long-term context. Global-scale observations from the instrumental era began in the mid-19th century, and paleoclimate reconstructions extend the record of some quantities back hundreds to millions of years. Together, they provide a comprehensive view of the variability and long-term changes in the atmosphere, the ocean, the cryosphere and at the land surface.

The assessment of observational evidence for climate change is summarized in this section. Substantial advancements in the availability, acquisition, quality and analysis of observational data sets for the atmosphere, land surface, ocean and cryosphere have occurred since the AR4. Many aspects of the climate system are showing evidence of a changing climate. {2, 3, 4, 5, 6, 13}

TS.2.2 Changes in Temperature

TS.2.2.1 Surface

It is certain that global mean surface temperature (GMST) has increased since the late 19th century (Figures TS.1 and TS.2). Each of the past three decades has been successively warmer at the Earth's surface than any the previous decades in the instrumental record, and the decade of the 2000's has been the warmest. The globally averaged combined land and ocean temperature data as calculated by a linear trend[5], show a warming of 0.85 [0.65 to 1.06] °C[6], over the period 1880–2012, when multiple independently produced datasets exist, about 0.89 [0.69 to 1.08] °C over the period 1901–2012, and about 0.72 [0.49 to 0.89] °C over the period 1951–2012 when based on three independently-produced data sets. The total increase between the average of the 1850–1900 period and the 2003–2012 period is 0.78 [0.72 to 0.85] °C, based on the Hadley Centre/Climatic Research Unit gridded surface temperature data set 4 (HadCRUT4), the global mean surface temperature dataset with the longest record of the three independently-produced data sets. The warming from 1850–1900 to 1986–2005 (reference period for the modelling chapters and the Atlas in Annex I) is 0.61 [0.55 to 0.67] °C, when calculated using HadCRUT4 and its uncertainty estimates. It is also *virtually certain* that maximum and minimum temperatures over land have increased on a global scale since 1950.[7] {2.4.1, 2.4.3; Chapter 2 Supplementary Material Section 2.SM.3}

Despite the robust multi-decadal warming, there exists substantial interannual to decadal variability in the rate of warming, with several periods exhibiting weaker trends (including the warming hiatus since 1998) (Figure TS.1). The rate of warming over the past 15 years (1998–2012; 0.05 [–0.05 to +0.15] °C per decade) is smaller than the trend since 1951 (1951–2012; 0.12[0.08 to 0.14] °C per decade). Trends for short periods are uncertain and very sensitive to the start and end years. For example, trends for 15-year periods starting in 1995, 1996, and 1997 are 0.13 [0.02 to 0.24] °C per decade, 0.14 [0.03 to 0.24] °C per decade and 0.07 [–0.02 to 0.18] °C per decade, respectively. Several independently analysed data records of global and regional land surface air temperature obtained from station observations are in broad agreement that land surface air temperatures have increased. Sea surface temperatures (SSTs) have also increased. Intercomparisons of new SST data records obtained by different measurement methods, including satellite data, have resulted in better understanding of errors and biases in the records. {2.4.1–2.4.3; Box 9.2}

It is *unlikely* that any uncorrected urban heat island effects and land use change effects have raised the estimated centennial globally averaged land surface air temperature trends by more than 10% of the reported trend. This is an average value; in some regions that have rapidly developed urban heat island and land use change impacts on regional trends may be substantially larger. {2.4.1}

There is *high confidence* that annual mean surface warming since the 20th century has reversed long-term cooling trends of the past 5000 years in mid-to-high latitudes of the Northern Hemisphere (NH). For average annual NH temperatures, the period 1983–2012 was *very likely* the warmest 30-year period of the last 800 years (*high confidence*) and *likely* the warmest 30-year period of the last 1400 years (*medium confidence*). This is supported by comparison of instrumental temperatures with multiple reconstructions from a variety of proxy data and statistical methods, and is consistent with AR4. Continental-scale surface temperature reconstructions show, with *high confidence*, multi-decadal periods during the Medieval Climate Anomaly (950–1250) that were in some regions as warm as in the mid-20th century and in others as warm as in the late 20th century. With *high confidence*, these regional warm periods were not as synchronous across regions as the warming since the mid-20th century. Based on the comparison between reconstructions and simulations, there is *high confidence* that not only external orbital, solar and volcanic forcing, but also internal

[5] The warming is reported as an unweighted average based on linear trend estimates calculated from Hadley Centre/Climatic Research Unit gridded surface temperature data set 4 (HadCRUT4), Merged Land–Ocean Surface Temperature Analysis (MLOST) and Goddard Institute for Space Studies Surface Temperature Analysis (GISTEMP) data sets (see Figure TS.2; Section 2.4.3).

[6] In the WGI contribution to the AR5, uncertainty is quantified using 90% uncertainty intervals unless otherwise stated. The 90% uncertainty interval, reported in square brackets, is expected to have a 90% likelihood of covering the value that is being estimated. The upper endpoint of the uncertainty interval has a 95% likelihood of exceeding the value that is being estimated and the lower endpoint has a 95% likelihood of being less than that value. A best estimate of that value is also given where available. Uncertainty intervals are not necessarily symmetric about the corresponding best estimate.

[7] Both methods presented in this paragraph to calculate temperature change were also used in AR4. The first calculates the difference using a best fit linear trend of all points between two years, e.g., 1880 and 2012. The second calculates the difference between averages for the two periods, e.g., 1850 to 1900 and 2003 to 2012. Therefore, the resulting values and their 90% uncertainty intervals are not directly comparable.

variability, contributed substantially to the spatial pattern and timing of surface temperature changes between the Medieval Climate Anomaly and the Little Ice Age (1450–1850). {5.3.5, 5.5.1}

TS.2.2.2 Troposphere and Stratosphere

Based on multiple independent analyses of measurements from radiosondes and satellite sensors, it is *virtually certain* that globally the troposphere has warmed and the stratosphere has cooled since the mid-20th century (Figure TS.1). Despite unanimous agreement on the sign of the trends, substantial disagreement exists between available estimates as to the rate of temperature changes, particularly outside the NH extratropical troposphere, which has been well sampled by radiosondes. Hence there is only *medium confidence* in the rate of change and its vertical structure in the NH extratropical troposphere and *low confidence* elsewhere. {2.4.4}

TS.2.2.3 Ocean

It is *virtually certain* that the upper ocean (above 700 m) has warmed from 1971 to 2010, and *likely* that it has warmed from the 1870s to 1971 (Figure TS.1). There is less certainty in changes prior to 1971 because of relatively sparse sampling in earlier time periods. Instrumental biases in historical upper ocean temperature measurements have been identified and reduced since AR4, diminishing artificial decadal variation in temperature and upper ocean heat content, most prominent during the 1970s and 1980s. {3.2.1–3.2.3, 3.5.3}

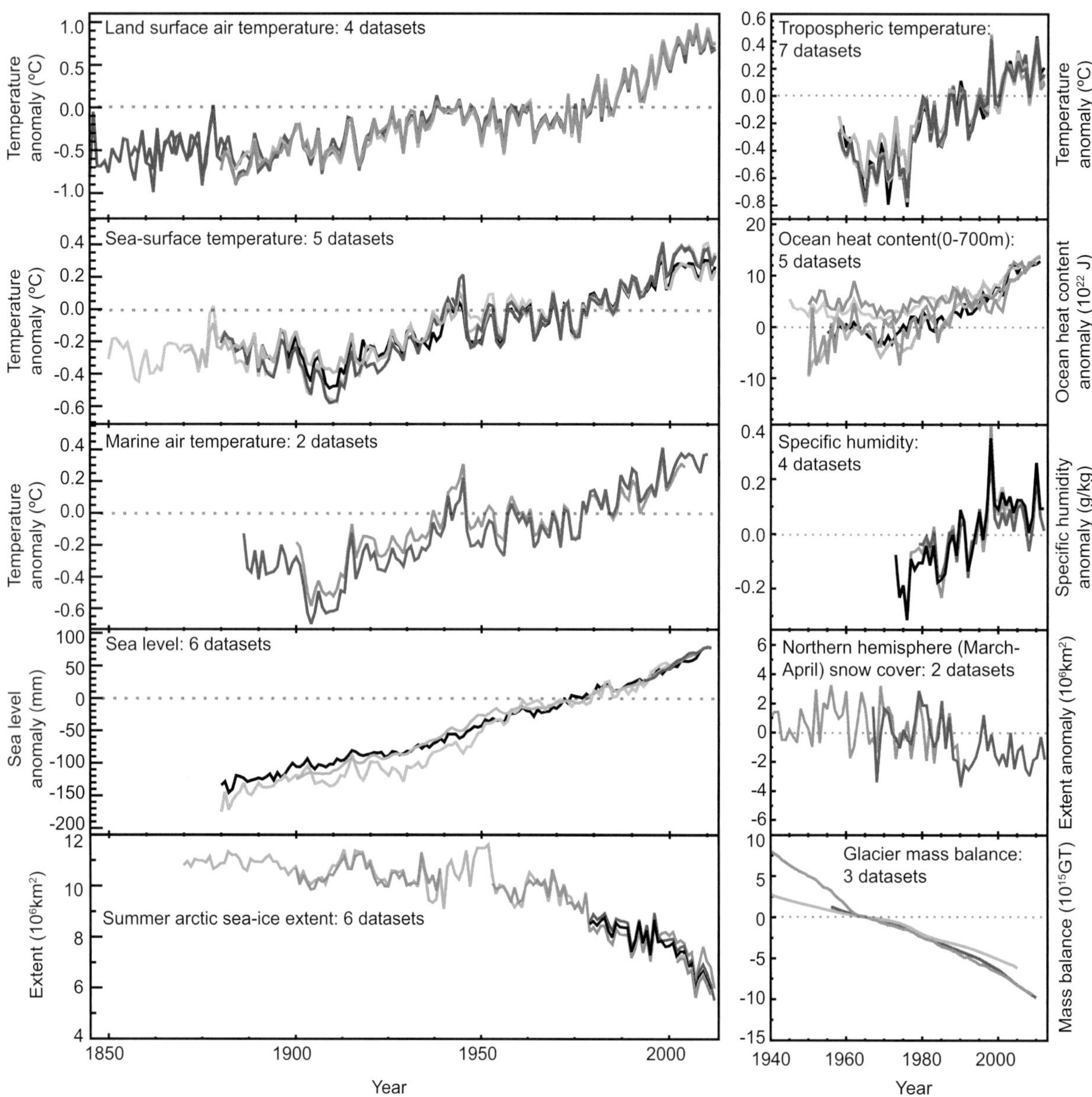

Figure TS.1 | Multiple complementary indicators of a changing global climate. Each line represents an independently derived estimate of change in the climate element. The times series presented are assessed in Chapters 2, 3 and 4. In each panel all data sets have been normalized to a common period of record. A full detailing of which source data sets go into which panel is given in Chapter 2 Supplementary Material Section 2.SM.5 and in the respective chapters. Further detail regarding the related Figure SPM.3 is given in the TS Supplementary Material. {FAQ 2.1, Figure 1; 2.4, 2.5, 3.2, 3.7, 4.5.2, 4.5.3}

It is *likely* that the ocean warmed between 700-2000 m from 1957 to 2009, based on 5-year averages. It is *likely* that the ocean warmed from 3000 m to the bottom from 1992 to 2005, while no significant trends in global average temperature were observed between 2000 and 3000 m depth from circa 1992 to 2005. Below 3000 m depth, the largest warming is observed in the Southern Ocean. {3.2.4, 3.5.1; Figures 3.2b, 3.3; FAQ 3.1}

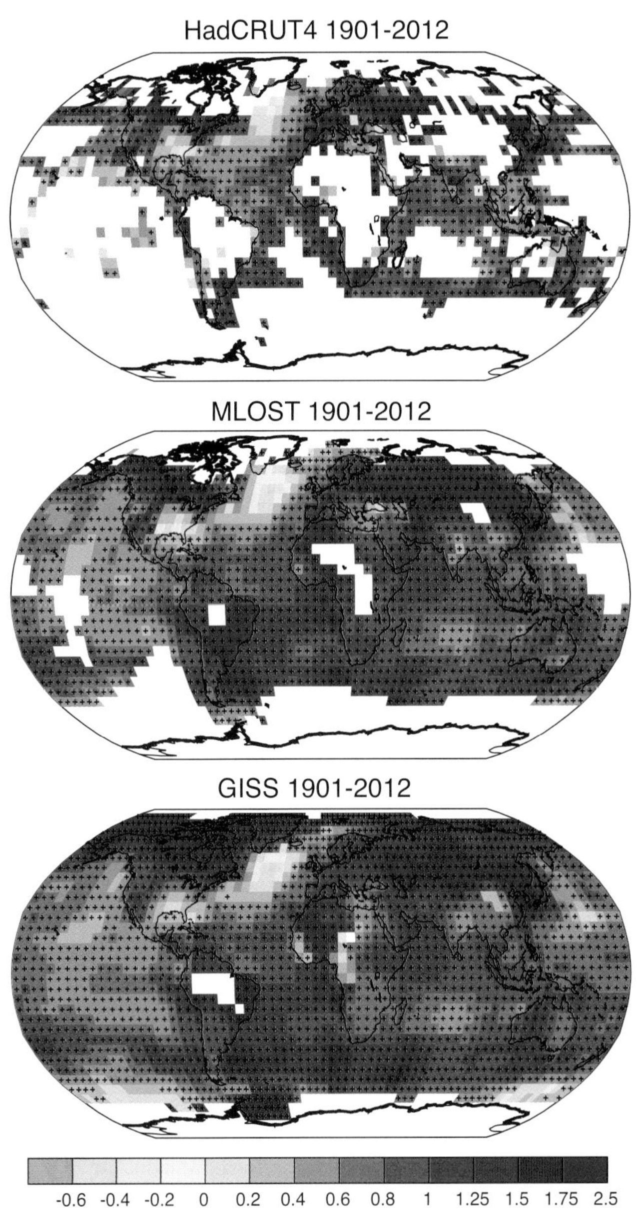

Figure TS.2 | Change in surface temperature over 1901–2012 as determined by linear trend for three data sets. White areas indicate incomplete or missing data. Trends have been calculated only for those grid boxes with greater than 70% complete records and more than 20% data availability in the first and last 10% of the time period. Black plus signs (+) indicate grid boxes where trends are significant (i.e., a trend of zero lies outside the 90% confidence interval). Differences in coverage primarily reflect the degree of interpolation to account for data void regions undertaken by the data set providers ranging from none beyond grid box averaging (Hadley Centre/Climatic Research Unit gridded surface temperature data set 4 (HadCRUT4)) to substantial (Goddard Institute for Space Studies Surface Temperature Analysis (GISTEMP)). Further detail regarding the related Figure SPM.1 is given in the TS Supplementary Material. {Figure 2.21}

TS.2.3 Changes in Energy Budget and Heat Content

The Earth has been in radiative imbalance, with more energy from the Sun entering than exiting the top of the atmosphere, since at least about 1970. It is *virtually certain* that the Earth has gained substantial energy from 1971 to 2010. The estimated increase in energy inventory between 1971 and 2010 is 274 [196 to 351] $\times 10^{21}$ J (*high confidence*), with a heating rate of 213 $\times 10^{12}$ W from a linear fit to the annual values over that time period (see also TFE.4). {Boxes 3.1, 13.1}

Ocean warming dominates that total heating rate, with full ocean depth warming accounting for about 93% (*high confidence*), and warming of the upper (0 to 700 m) ocean accounting for about 64%. Melting ice (including Arctic sea ice, ice sheets and glaciers) and warming of the continents each account for 3% of the total. Warming of the atmosphere makes up the remaining 1%. The 1971–2010 estimated rate of ocean energy gain is 199 $\times 10^{12}$ W from a linear fit to data over that time period, equivalent to 0.42 W m^{-2} heating applied continuously over the Earth's entire surface, and 0.55 W m^{-2} for the portion owing to ocean warming applied over the ocean's entire surface area. The Earth's estimated energy increase from 1993 to 2010 is 163 [127 to 201] $\times 10^{21}$ J with a trend estimate of 275 $\times 10^{15}$ W. The ocean portion of the trend for 1993–2010 is 257 $\times 10^{12}$ W, equivalent to a mean heat flux into the ocean of 0.71 W m^{-2}. {3.2.3, 3.2.4; Box 3.1}

It is *about as likely as not* that ocean heat content from 0–700 m increased more slowly during 2003 to 2010 than during 1993 to 2002 (Figure TS.1). Ocean heat uptake from 700–2000 m, where interannual variability is smaller, *likely* continued unabated from 1993 to 2009. {3.2.3, 3.2.4; Box 9.2}

TS.2.4 Changes in Circulation and Modes of Variability

Large variability on interannual to decadal time scales hampers robust conclusions on long-term changes in atmospheric circulation in many instances. *Confidence* is *high* that the increase of the northern mid-latitude westerly winds and the North Atlantic Oscillation (NAO) index from the 1950s to the 1990s, and the weakening of the Pacific Walker Circulation from the late 19th century to the 1990s, have been largely offset by recent changes. With *high confidence*, decadal and multi-decadal changes in the winter NAO index observed since the 20th century are not unprecedented in the context of the past 500 years. {2.7.2, 2.7.5, 2.7.8, 5.4.2; Box 2.5; Table 2.14}

It is *likely* that circulation features have moved poleward since the 1970s, involving a widening of the tropical belt, a poleward shift of storm tracks and jet streams and a contraction of the northern polar vortex. Evidence is more robust for the NH. It is *likely* that the Southern Annular Mode (SAM) has become more positive since the 1950s. The increase in the strength of the observed summer SAM since 1950 has been anomalous, with *medium confidence*, in the context of the past 400 years. {2.7.5, 2.7.6, 2.7.8, 5.4.2; Box 2.5; Table 2.14}

New results from high-resolution coral records document with *high confidence* that the El Niño-Southern Oscillation (ENSO) system has remained highly variable throughout the past 7000 years, showing no discernible evidence for an orbital modulation of ENSO. {5.4.1}

Recent observations have strengthened evidence for variability in major ocean circulation systems on time scales from years to decades. It is *very likely* that the subtropical gyres in the North Pacific and South Pacific have expanded and strengthened since 1993. Based on measurements of the full Atlantic Meridional Overturning Circulation (AMOC) and its individual components at various latitudes and different time periods, there is no evidence of a long-term trend. There is also no evidence for trends in the transports of the Indonesian Throughflow, the Antarctic Circumpolar Current (ACC) or in the transports between the Atlantic Ocean and Nordic Seas. However, a southward shift of the ACC by about 1° of latitude is observed in data spanning the time period 1950–2010 with *medium confidence*. {3.6}

TS.2.5 Changes in the Water Cycle and Cryosphere

TS.2.5.1 Atmosphere

Confidence in precipitation change averaged over global land areas is *low* prior to 1951 and *medium* afterwards because of insufficient data, particularly in the earlier part of the record (for an overview of observed and projected changes in the global water cycle see TFE.1). Further, when virtually all the land area is filled in using a reconstruction method, the resulting time series shows little change in land-based precipitation since 1901. NH mid-latitude land areas do show a *likely* overall increase in precipitation (*medium confidence* prior to 1951, but *high confidence* afterwards). For other latitudes area-averaged long-term positive or negative trends have *low confidence* (TFE.1, Figure 1). {2.5.1}

It is *very likely* that global near surface and tropospheric air specific humidity have increased since the 1970s. However, during recent years the near-surface moistening trend over land has abated (*medium confidence*) (Figure TS.1). As a result, fairly widespread decreases in relative humidity near the surface are observed over the land in recent years. {2.4.4, 2.5.5, 2.5.6}

Although trends of cloud cover are consistent between independent data sets in certain regions, substantial ambiguity and therefore *low confidence* remains in the observations of global-scale cloud variability and trends. {2.5.7}

TS.2.5.2 Ocean and Surface Fluxes

It is *very likely* that regional trends have enhanced the mean geographical contrasts in sea surface salinity since the 1950s: saline surface waters in the evaporation-dominated mid-latitudes have become more saline, while relatively fresh surface waters in rainfall-dominated tropical and polar regions have become fresher. The mean contrast between high- and low-salinity regions increased by 0.13 [0.08 to 0.17] from 1950 to 2008. It is *very likely* that the inter-basin contrast in freshwater content has increased: the Atlantic has become saltier and the Pacific and Southern Oceans have freshened. Although similar conclusions were reached in AR4, recent studies based on expanded data sets and new analysis approaches provide *high confidence* in this assessment. {3.3.2, 3.3.3, 3.9; FAQ 3.2}

The spatial patterns of the salinity trends, mean salinity and the mean distribution of evaporation minus precipitation are all similar (TFE.1, Figure 1). These similarities provide indirect evidence that the pattern of evaporation minus precipitation over the oceans has been enhanced since the 1950s (*medium confidence*). Uncertainties in currently available surface fluxes prevent the flux products from being reliably used to identify trends in the regional or global distribution of evaporation or precipitation over the oceans on the time scale of the observed salinity changes since the 1950s. {3.3.2–3.3.4, 3.4.2, 3.4.3, 3.9; FAQ 3.2}

TS.2.5.3 Sea Ice

Continuing the trends reported in AR4, there is *very high confidence* that the Arctic sea ice extent (annual, multi-year and perennial) decreased over the period 1979–2012 (Figure TS.1). The rate of the annual decrease was *very likely* between 3.5 and 4.1% per decade (range of 0.45 to 0.51 million km^2 per decade). The average decrease in decadal extent of annual Arctic sea ice has been most rapid in summer and autumn (*high confidence*), but the extent has decreased in every season, and in every successive decade since 1979 (*high confidence*). The extent of Arctic perennial and multi-year ice decreased between 1979 and 2012 (*very high confidence*). The rates are *very likely* 11.5 [9.4 to 13.6]% per decade (0.73 to 1.07 million km^2 per decade) for the sea ice extent at summer minimum (perennial ice) and *very likely* 13.5 [11 to 16] % per decade for multi-year ice. There is *medium confidence* from reconstructions that the current (1980–2012) Arctic summer sea ice retreat was unprecedented and SSTs were anomalously high in the perspective of at least the last 1,450 years. {4.2.2, 5.5.2}

It is *likely* that the annual period of surface melt on Arctic perennial sea ice lengthened by 5.7 [4.8 to 6.6] days per decade over the period 1979–2012. Over this period, in the region between the East Siberian Sea and the western Beaufort Sea, the duration of ice-free conditions increased by nearly 3 months. {4.2.2}

There is *high confidence* that the average winter sea ice thickness within the Arctic Basin decreased between 1980 and 2008. The average decrease was *likely* between 1.3 m and 2.3 m. *High confidence* in this assessment is based on observations from multiple sources: submarine, electromagnetic probes and satellite altimetry; and is consistent with the decline in multi-year and perennial ice extent. Satellite measurements made in the period 2010–2012 show a decrease in sea ice volume compared to those made over the period 2003–2008 (*medium confidence*). There is *high confidence* that in the Arctic, where the sea ice thickness has decreased, the sea ice drift speed has increased. {4.2.2}

It is *very likely* that the annual Antarctic sea ice extent increased at a rate of between 1.2 and 1.8% per decade (0.13 to 0.20 million km^2 per decade) between 1979 and 2012 (*very high confidence*). There was a greater increase in sea ice area, due to a decrease in the percentage of open water within the ice pack. There is *high confidence* that there are strong regional differences in this annual rate, with some regions increasing in extent/area and some decreasing. There are also contrasting regions around the Antarctic where the ice-free season has lengthened, and others where it has decreased over the satellite period (*high confidence*). {4.2.3}

TS.2.5.4 Glaciers and Ice Sheets

There is *very high confidence* that glaciers world-wide are persistently shrinking as revealed by the time series of measured changes in glacier length, area, volume and mass (Figures TS.1 and TS.3). The few exceptions are regionally and temporally limited. Measurements of glacier change have increased substantially in number since AR4. Most of the new data sets, along with a globally complete glacier inventory, have been derived from satellite remote sensing {4.3.1, 4.3.3}

There is *very high confidence* that, during the last decade, the largest contributions to global glacier ice loss were from glaciers in Alaska, the Canadian Arctic, the periphery of the Greenland ice sheet, the Southern Andes and the Asian mountains. Together these areas account for more than 80% of the total ice loss. Total mass loss from all glaciers in the world, excluding those on the periphery of the ice sheets, was *very likely* 226 [91 to 361] Gt yr^{-1} (sea level equivalent, 0.62 [0.25 to 0.99] mm yr^{-1}) in the period 1971–2009, 275 [140 to 410] Gt yr^{-1} (0.76 [0.39 to 1.13] mm yr^{-1}) in the period 1993–2009 and 301 [166 to 436] Gt yr^{-1} (0.83 [0.46 to 1.20] mm yr^{-1}) between 2005 and 2009[8]. {4.3.3; Tables 4.4, 4.5}

There is *high confidence* that current glacier extents are out of balance with current climatic conditions, indicating that glaciers will continue to shrink in the future even without further temperature increase. {4.3.3}

There is *very high confidence* that the Greenland ice sheet has lost ice during the last two decades. Combinations of satellite and airborne remote sensing together with field data indicate with *high confidence* that the ice loss has occurred in several sectors and that large rates of mass loss have spread to wider regions than reported in AR4 (Figure TS.3). There is *high confidence* that the mass loss of the Greenland ice sheet has accelerated since 1992: the average rate has *very likely* increased from 34 [–6 to 74] Gt yr^{-1} over the period 1992–2001 (sea level equivalent, 0.09 [–0.02 to 0.20] mm yr^{-1}), to 215 [157 to 274] Gt yr^{-1} over the period 2002–2011 (0.59 [0.43 to 0.76] mm yr^{-1}). There is *high confidence* that ice loss from Greenland resulted from increased surface melt and runoff and increased outlet glacier discharge, and these occurred in similar amounts. There is *high confidence* that the area subject to summer melt has increased over the last two decades. {4.4.2, 4.4.3}

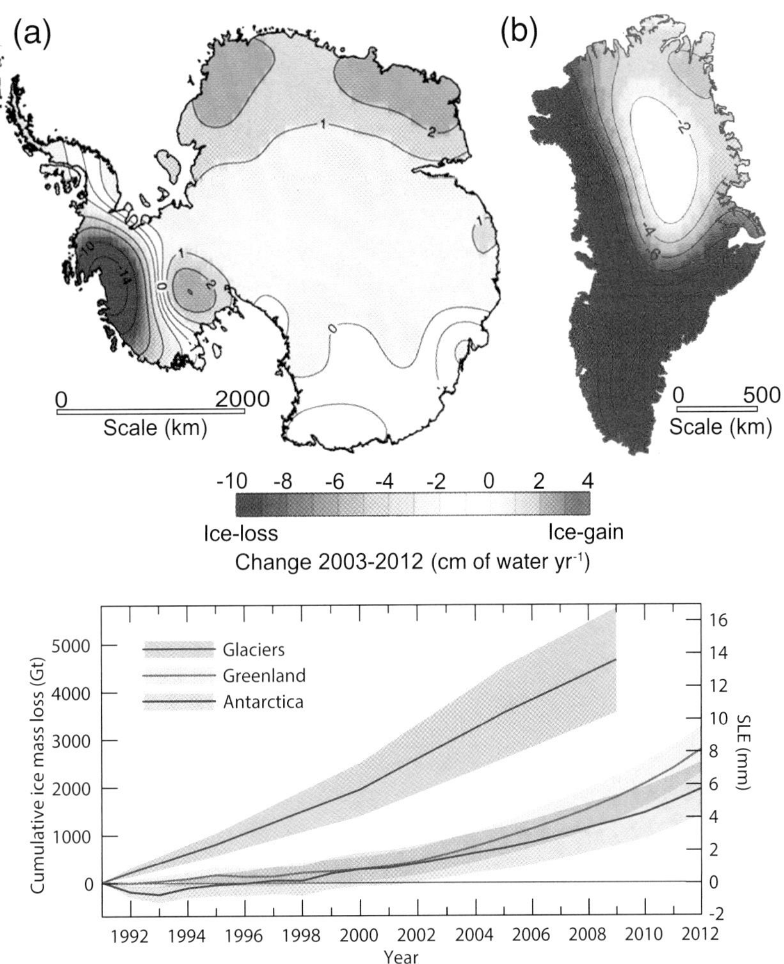

Figure TS.3 | (Upper) Distribution of ice loss determined from Gravity Recovery and Climate Experiment (GRACE) time-variable gravity for (a) Antarctica and (b) Greenland, shown in centimetres of water per year (cm of water yr^{-1}) for the period 2003–2012. (Lower) The assessment of the total loss of ice from glaciers and ice sheets in terms of mass (Gt) and sea level equivalent (mm). The contribution from glaciers excludes those on the periphery of the ice sheets. {4.3.4; Figures 4.12–4.14, 4.16, 4.17, 4.25}

[8] 100 Gt yr^{-1} of ice loss corresponds to about 0.28 mm yr^{-1} of sea level equivalent.

Thematic Focus Elements

TFE.1 | Water Cycle Change

The water cycle describes the continuous movement of water through the climate system in its liquid, solid and vapour forms, and storage in the reservoirs of ocean, cryosphere, land surface and atmosphere. In the atmosphere, water occurs primarily as a gas, water vapour, but it also occurs as ice and liquid water in clouds. The ocean is primarily liquid water, but the ocean is partly covered by ice in polar regions. Terrestrial water in liquid form appears as surface water (lakes, rivers), soil moisture and groundwater. Solid terrestrial water occurs in ice sheets, glaciers, snow and ice on the surface and permafrost. The movement of water in the climate system is essential to life on land, as much of the water that falls on land as precipitation and supplies the soil moisture and river flow has been evaporated from the ocean and transported to land by the atmosphere. Water that falls as snow in winter can provide soil moisture in springtime and river flow in summer and is essential to both natural and human systems. The movement of fresh water between the atmosphere and the ocean can also influence oceanic salinity, which is an important driver of the density and circulation of the ocean. The latent heat contained in water vapour in the atmosphere is critical to driving the circulation of the atmosphere on scales ranging from individual thunderstorms to the global circulation of the atmosphere. {12.4.5; FAQ 3.2, FAQ 12.2}

Observations of Water Cycle Change

Because the saturation vapour pressure of air increases with temperature, it is expected that the amount of water vapour in air will increase with a warming climate. Observations from surface stations, radiosondes, global positioning systems and satellite measurements indicate increases in tropospheric water vapour at large spatial scales (TFE.1, Figure 1). It is *very likely* that tropospheric specific humidity has increased since the 1970s. The magnitude of the observed global change in tropospheric water vapour of about 3.5% in the past 40 years is consistent with the observed temperature change of about 0.5°C during the same period, and the relative humidity has stayed approximately constant. The water vapour change can be attributed to human influence with *medium confidence*. {2.5.4, 10.3.2}

Changes in precipitation are harder to measure with the existing records, both because of the greater difficulty in sampling precipitation and also because it is expected that precipitation will have a smaller fractional change than the water vapour content of air as the climate warms. Some regional precipitation trends appear to be robust (TFE.1, Figure 2), but when virtually all the land area is filled in using a reconstruction method, the resulting time series of global mean land precipitation shows little change since 1900. At present there is *medium confidence* that there has been a significant human influence on global scale changes in precipitation patterns, including increases in Northern Hemisphere (NH) mid-to-high latitudes. Changes in the extremes of precipitation, and other climate extremes related to the water cycle are comprehensively discussed in TFE.9. {2.5.1, 10.3.2}

Although direct trends in precipitation and evaporation are difficult to measure with the available records, the observed oceanic surface salinity, which is strongly dependent on the difference between evaporation and precipitation, shows significant trends (TFE.1, Figure 1). The spatial patterns of the salinity trends since 1950 are very similar to the mean salinity and the mean distribution of evaporation minus precipitation: regions of high salinity where evaporation dominates have become more saline, while regions of low salinity where rainfall dominates have become fresher (TFE.1, Figure 1). This provides indirect evidence that the pattern of evaporation minus precipitation over the oceans has been enhanced since the 1950s (*medium confidence*). The inferred changes in evaporation minus precipitation are consistent with the observed increased water vapour content of the warmer air. It is *very likely* that observed changes in surface and subsurface salinity are due in part to anthropogenic climate forcings. {2.5, 3.3.2–3.3.4, 3.4, 3.9, 10.4.2; FAQ 3.2}

In most regions analysed, it is *likely* that decreasing numbers of snowfall events are occurring where increased winter temperatures have been observed. Both satellite and *in situ* observations show significant reductions in the NH snow cover extent over the past 90 years, with most of the reduction occurring in the 1980s. Snow cover decreased most in June when the average extent decreased *very likely* by 53% (40 to 66%) over the period 1967 to 2012. From 1922 to 2012 only data from March and April are available and show *very likely* a 7% (4.5 to 9.5%) decline. Because of earlier spring snowmelt, the duration of the NH snow season has declined by 5.3 days per decade since the 1972/1973 winter. It is *likely* that there has been an anthropogenic component to these observed reductions in snow cover since the 1970s. {4.5.2, 10.5.1, 10.5.3}

(continued on next page)

TFE.1 (continued)

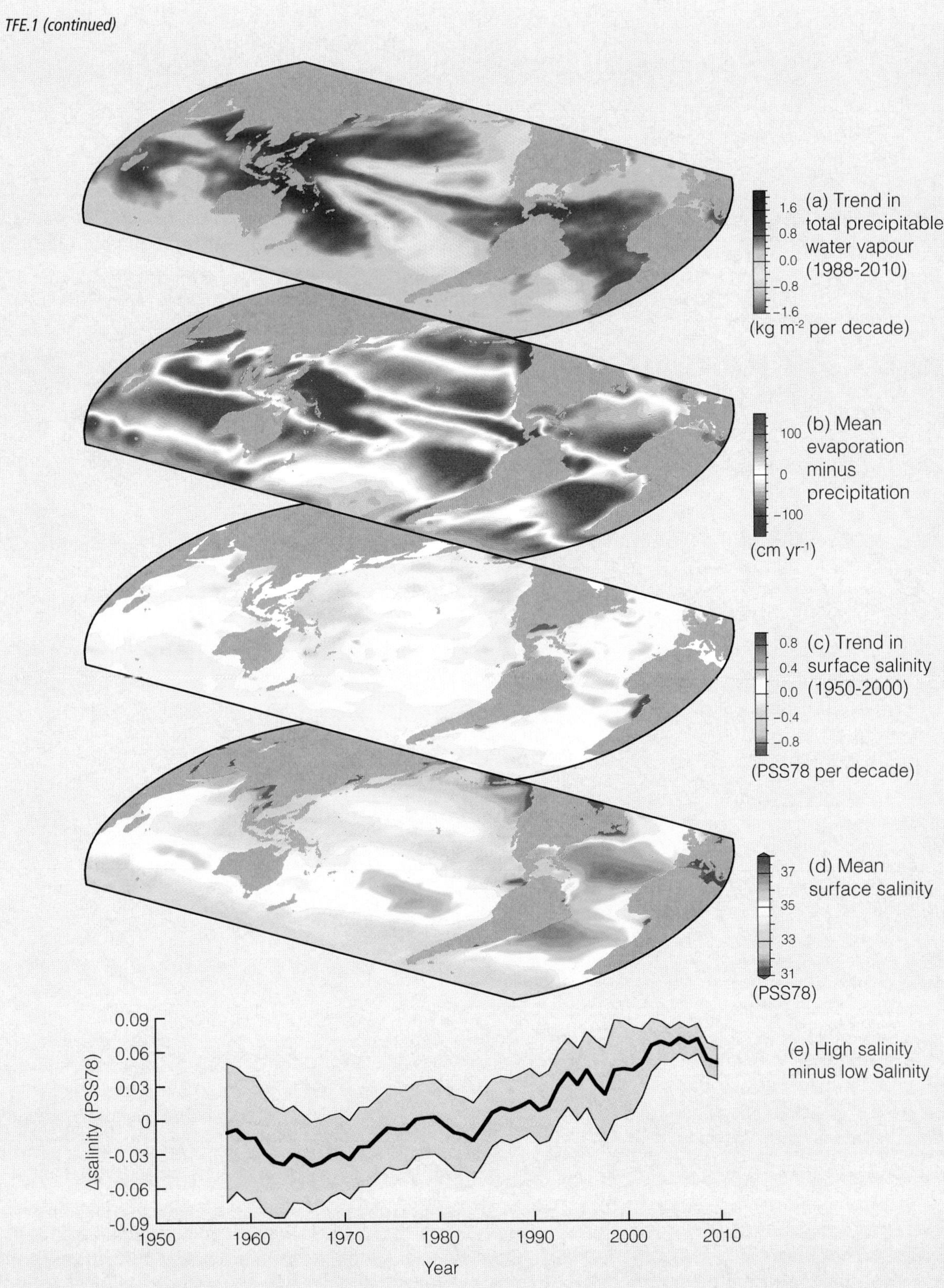

TFE.1, Figure 1 | Changes in sea surface salinity are related to the atmospheric patterns of evaporation minus precipitation (*E* – *P*) and trends in total precipitable water: (a) Linear trend (1988 to 2010) in total precipitable water (water vapour integrated from the Earth's surface up through the entire atmosphere) (kg m^{-2} per decade) from satellite observations. (b) The 1979–2005 climatological mean net evaporation minus precipitation (cm yr^{-1}) from meteorological reanalysis data. (c) Trend (1950–2000) in surface salinity (Practical Salinity Scale 78 (PSS78) per 50 years). (d) The climatological mean surface salinity (PSS78) (blues <35; yellows-reds >35). (e) Global difference between salinity averaged over regions where the sea surface salinity is greater than the global mean sea surface salinity ("High Salinity") and salinity averaged over regions with values below the global mean ('Low Salinity'). For details of data sources see Figure 3.21 and FAQ 3.2, Figure 1. {3.9}

TFE.1 (continued)

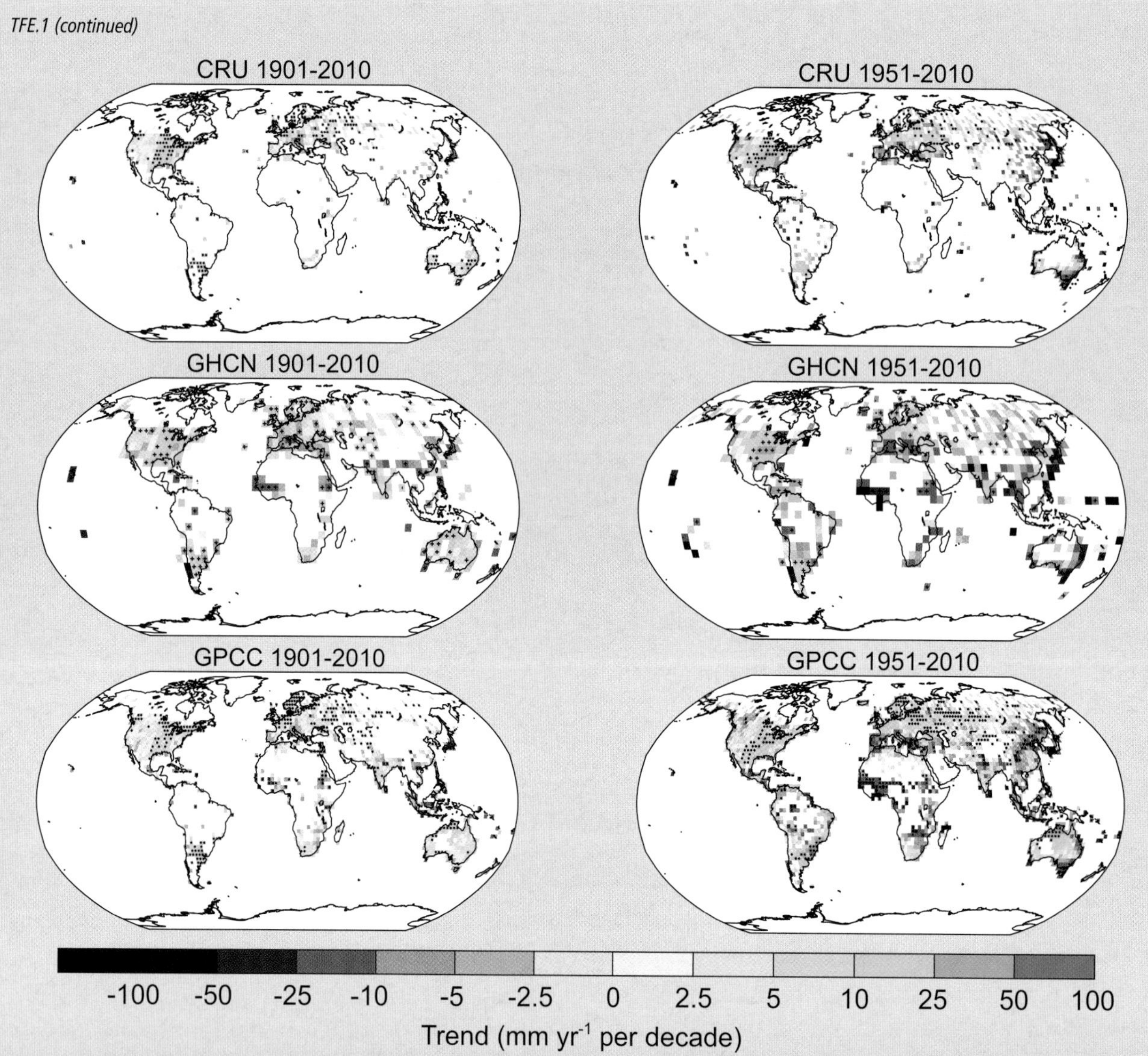

TFE.1, Figure 2 | Maps of observed precipitation change over land from 1901 to 2010 (left-hand panels) and 1951 to 2010 (right-hand panels) from the Climatic Research Unit (CRU), Global Historical Climatology Network (GHCN) and Global Precipitation Climatology Centre (GPCC) data sets. Trends in annual accumulation have been calculated only for those grid boxes with greater than 70% complete records and more than 20% data availability in first and last decile of the period. White areas indicate incomplete or missing data. Black plus signs (+) indicate grid boxes where trends are significant (i.e., a trend of zero lies outside the 90% confidence interval). Further detail regarding the related Figure SPM.2 is given in the TS Supplementary Material. {Figure 2.29; 2.5.1}

The most recent and most comprehensive analyses of river runoff do not support the IPCC Fourth Assessment Report (AR4) conclusion that global runoff has increased during the 20th century. New results also indicate that the AR4 conclusions regarding global increasing trends in droughts since the 1970s are no longer supported. {2.5.2, 2.6.2}

Projections of Future Changes

Changes in the water cycle are projected to occur in a warming climate (TFE.1, Figure 3, see also TS 4.6, TS 5.6, Annex I). Global-scale precipitation is projected to gradually increase in the 21st century. The precipitation increase is projected to be much smaller (about 2% K^{-1}) than the rate of lower tropospheric water vapour increase (about 7% K^{-1}), due to global energetic constraints. Changes of average precipitation in a much warmer world will not be uniform, with some regions experiencing increases, and others with decreases or not much change at all. The high latitude land masses are *likely* to experience greater amounts of precipitation due to the additional water carrying capacity of the warmer troposphere. Many mid-latitude and subtropical arid and semi-arid regions will *likely* experience less precipitation. The largest precipitation changes over northern Eurasia and North America are projected to occur during the winter. {12.4.5, Annex I}

(continued on next page)

TFE.1 (continued)

Regional to global-scale projections of soil moisture and drought remain relatively uncertain compared to other aspects of the water cycle. Nonetheless, drying in the Mediterranean, southwestern USA and southern African regions are consistent with projected changes in the Hadley Circulation, so drying in these regions as global temperatures increase is *likely* for several degrees of warming under the Representative Concentration Pathway RCP8.5. Decreases in runoff are *likely* in southern Europe and the Middle East. Increased runoff is *likely* in high northern latitudes, and consistent with the projected precipitation increases there. {12.4.5}

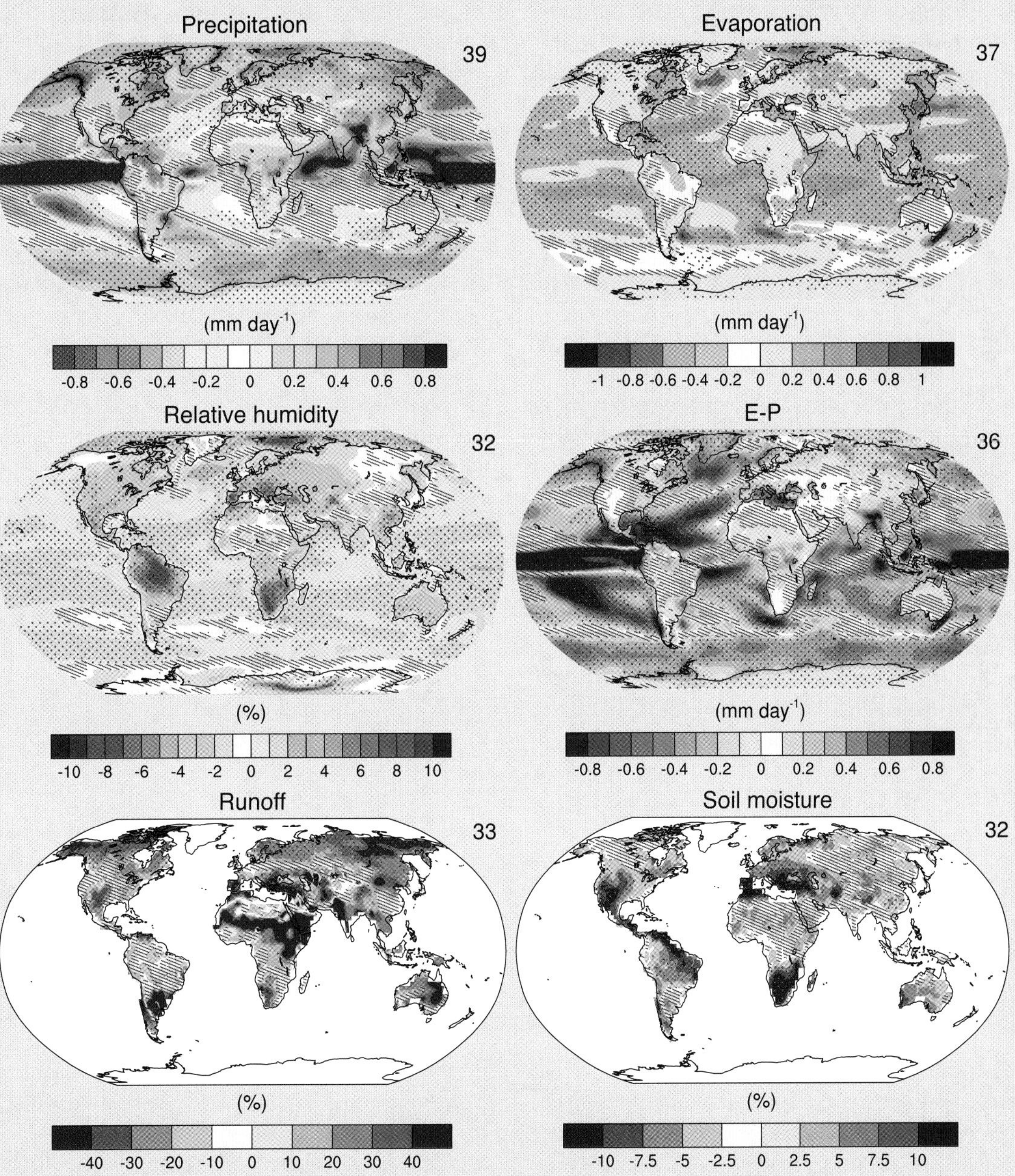

TFE.1, Figure 3 | Annual mean changes in precipitation (*P*), evaporation (*E*), relative humidity, *E* – *P*, runoff and soil moisture for 2081–2100 relative to 1986–2005 under the Representative Concentration Pathway RCP8.5 (see Box TS.6). The number of Coupled Model Intercomparison Project Phase 5 (CMIP5) models to calculate the multi-model mean is indicated in the upper right corner of each panel. Hatching indicates regions where the multi-model mean change is less than one standard deviation of internal variability. Stippling indicates regions where the multi-model mean change is greater than two standard deviations of internal variability and where 90% of models agree on the sign of change (see Box 12.1). {Figures 12.25–12.27}

There is *high confidence* that the Antarctic ice sheet has been losing ice during the last two decades (Figure TS.3). There is *very high confidence* that these losses are mainly from the northern Antarctic Peninsula and the Amundsen Sea sector of West Antarctica and *high confidence* that they result from the acceleration of outlet glaciers. The average rate of ice loss from Antarctica *likely* increased from 30 [–37 to 97] Gt yr^{-1} (sea level equivalent, 0.08 [–0.10 to 0.27] mm yr^{-1}) over the period 1992–2001, to 147 [72 to 221] Gt yr^{-1} over the period 2002–2011 (0.40 [0.20 to 0.61] mm yr^{-1}). {4.4.2, 4.4.3}

There is *high confidence* that in parts of Antarctica floating ice shelves are undergoing substantial changes. There is *medium confidence* that ice shelves are thinning in the Amundsen Sea region of West Antarctica, and *low confidence* that this is due to high ocean heat flux. There is *high confidence* that ice shelves around the Antarctic Peninsula continue a long-term trend of retreat and partial collapse that began decades ago. {4.4.2, 4.4.5}

TS.2.5.5 Snow Cover, Freshwater Ice and Frozen Ground

There is *very high confidence* that snow cover extent has decreased in the NH, especially in spring (Figure TS.1). Satellite records indicate that over the period 1967–2012, snow cover extent *very likely* decreased; the largest change, –53% [–40 to –66%], occurred in June. No month had statistically significant increases. Over the longer period, 1922–2012, data are available only for March and April, but these show *very likely* a 7% [4.5 to 9.5%] decline and a negative correlation (–0.76) with March to April 40°N to 60°N land temperature. In the Southern Hemisphere (SH), evidence is too limited to conclude whether changes have occurred. {4.5.2, 4.5.3}

Permafrost temperatures have increased in most regions around the world since the early 1980s (*high confidence*). These increases were in response to increased air temperature and to changes in the timing and thickness of snow cover (*high confidence*). The temperature increase for colder permafrost was generally greater than for warmer permafrost (*high confidence*). {4.7.2; Table 4.8}

TS.2.6 Changes in Sea Level

The primary contributions to changes in the volume of water in the ocean are the expansion of the ocean water as it warms and the transfer to the ocean of water currently stored on land, particularly from glaciers and ice sheets. Water impoundment in reservoirs and ground water depletion (and its subsequent runoff to the ocean) also affect sea level. Change in sea level relative to the land (relative sea level) can be significantly different from the global mean sea level (GMSL) change because of changes in the distribution of water in the ocean, vertical movement of the land and changes in the Earth's gravitational field. For an overview on the scientific understanding and uncertainties associated with recent (and projected) sea level change see TFE.2. {3.7.3, 13.1}

During warm intervals of the mid Pliocene (3.3 to 3.0 Ma), when there is *medium confidence* that GMSTs were 1.9°C to 3.6°C warmer than for pre-industrial climate and carbon dioxide (CO_2) levels were between 350 and 450 ppm, there is *high confidence* that GMSL was above present, implying reduced volume of polar ice sheets. The best estimates from various methods imply with *high confidence* that sea level has not exceeded +20 m during the warmest periods of the Pliocene, due to deglaciation of the Greenland and West Antarctic ice sheets and areas of the East Antarctic ice sheet. {5.6.1, 13.2}

There is *very high confidence* that maximum GMSL during the last interglacial period (129 to 116 ka) was, for several thousand years, at least 5 m higher than present and *high confidence* that it did not exceed 10 m above present, implying substantial contributions from the Greenland and Antarctic ice sheets. This change in sea level occurred in the context of different orbital forcing and with high-latitude surface temperature, averaged over several thousand years, at least 2°C warmer than present (*high confidence*). Based on ice sheet model simulations consistent with elevation changes derived from a new Greenland ice core, the Greenland ice sheet *very likely* contributed between 1.4 m and 4.3 m sea level equivalent, implying with *medium confidence* a contribution from the Antarctic ice sheet to the GMSL during the Last Interglacial Period. {5.3.4, 5.6.2, 13.2.1}

Proxy and instrumental sea level data indicate a transition in the late 19th to the early 20th century from relatively low mean rates of rise over the previous two millennia to higher rates of rise (*high confidence*) {3.7, 3.7.4, 5.6.3, 13.2}

GMSL has risen by 0.19 [0.17 to 0.21] m, estimated from a linear trend over the period 1901–2010, based on tide gauge records and additionally on satellite data since 1993. It is *very likely* that the mean rate of sea level rise was 1.7 [1.5 to 1.9] mm yr^{-1} between 1901 and 2010. Between 1993 and 2010, the rate was *very likely* higher at 3.2 [2.8 to 3.6] mm yr^{-1}; similarly high rates *likely* occurred between 1920 and 1950. The rate of GMSL rise has *likely* increased since the early 1900s, with estimates ranging from 0.000 [–0.002 to 0.002] to 0.013 [–0.007 to 0.019] mm yr^{-2}. {3.7, 5.6.3, 13.2}

TS.2.7 Changes in Extremes

TS.2.7.1 Atmosphere

Recent analyses of extreme events generally support the AR4 and SREX conclusions (see TFE.9 and in particular TFE.9, Table 1, for a synthesis). It is *very likely* that the number of cold days and nights has decreased and the number of warm days and nights has increased on the global scale between 1951 and 2010. Globally, there is *medium confidence* that the length and frequency of warm spells, including heat waves, has increased since the middle of the 20th century, mostly owing to lack of data or studies in Africa and South America. However, it is *likely* that heat wave frequency has increased over this period in large parts of Europe, Asia and Australia. {2.6.1; Tables 2.12, 2.13}

It is *likely* that since about 1950 the number of heavy precipitation events over land has increased in more regions than it has decreased. Confidence is highest for North America and Europe where there have been *likely* increases in either the frequency or intensity of heavy precipitation with some seasonal and regional variations. It is *very likely* that there have been trends towards heavier precipitation events in central North America. {2.6.2; Table 2.13}

Thematic Focus Elements

TFE.2 | Sea Level Change: Scientific Understanding and Uncertainties

After the Last Glacial Maximum, global mean sea levels (GMSLs) reached close to present-day values several thousand years ago. Since then, it is *virtually certain* that the rate of sea level rise has increased from low rates of sea level change during the late Holocene (order tenths of mm yr^{-1}) to 20th century rates (order mm yr^{-1}, Figure TS1). {3.7, 5.6, 13.2}

Ocean thermal expansion and glacier mass loss are the dominant contributors to GMSL rise during the 20th century (*high confidence*). It is *very likely* that warming of the ocean has contributed 0.8 [0.5 to 1.1] mm yr^{-1} of sea level change during 1971–2010, with the majority of the contribution coming from the upper 700 m. The model mean rate of ocean thermal expansion for 1971–2010 is close to observations. {3.7, 13.3}

Observations, combined with improved methods of analysis, indicate that the global glacier contribution (excluding the peripheral glaciers around Greenland and Antarctica) to sea level was 0.25 to 0.99 mm yr^{-1} sea level equivalent during 1971–2010. *Medium confidence* in global glacier mass balance models used for projections of glacier changes arises from the process-based understanding of glacier surface mass balance, the consistency of observations and models of glacier changes, and the evidence that Atmosphere–Ocean General Circulation Model (AOGCM) climate simulations can provide realitistic climate input. A simulation using observed climate data shows a larger rate of glacier mass loss during the 1930s than the simulations using AOGCM input, possibly a result of an episode of warming in Greenland associated with unforced regional climate variability. {4.3, 13.3}

Observations indicate that the Greenland ice sheet has *very likely* experienced a net loss of mass due to both increased surface melting and runoff, and increased ice discharge over the last two decades (Figure TS.3). Regional climate models indicate that Greenland ice sheet surface mass balance showed no significant trend from the 1960s to the 1980s, but melting and consequent runoff has increased since the early 1990s. This tendency is related to pronounced regional warming, which may be attributed to a combination of anomalous regional variability in recent years and anthropogenic climate change. *High confidence* in projections of future warming in Greenland and increased surface melting is based on the qualitative agreements of models in projecting amplified warming at high northern latitudes for well-understood physical reasons. {4.4, 13.3}

There is *high confidence* that the Antarctic ice sheet is in a state of net mass loss and its contribution to sea level is also *likely* to have increased over the last two decades. Acceleration in ice outflow has been observed since the 1990s, especially in the Amundsen Sea sector of West Antarctica. Interannual variability in accumulation is large and as a result no significant trend is present in accumulation since 1979 in either models or observations. Surface melting is currently negligible in Antarctica. {4.4, 13.3}

Model-based estimates of climate-related changes in water storage on land (as snow cover, surface water, soil moisture and ground water) do not show significant long-term contributions to sea level change for recent decades. However, human-induced changes (reservoir impoundment and groundwater depletion) have each contributed at least several tenths of mm yr^{-1} to sea level change. Reservoir impoundment exceeded groundwater depletion for the majority of the 20th century but the rate of groundwater depletion has increased and now exceeds the rate of impoundment. Their combined net contribution for the 20th century is estimated to be small. {13.3}

The observed GMSL rise for 1993–2010 is consistent with the sum of the observationally estimated contributions (TFE.2, Figure 1e). The closure of the observational budget for recent periods within uncertainties represents a significant advance since the IPCC Fourth Assessment Report in physical understanding of the causes of past GMSL change, and provides an improved basis for critical evaluation of models of these contributions in order to assess their reliability for making projections. {13.3}

The sum of modelled ocean thermal expansion and glacier contributions and the estimated change in land water storage (which is relatively small) accounts for about 65% of the observed GMSL rise for 1901–1990, and 90% for 1971–2010 and 1993–2010 (TFE.2, Figure 1). After inclusion of small long-term contributions from ice sheets and the possible greater mass loss from glaciers during the 1930s due to unforced climate variability, the sum of the modelled contribution is close to the observed rise. The addition of the observed ice sheet contribution since 1993 improves the agreement further between the observed and modelled sea level rise (TFE.2, Figure 1). The evidence now available gives a clearer account than in previous IPCC assessments of 20th century sea level change. {13.3}

(continued on next page)

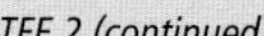

TFE.2 (continued)

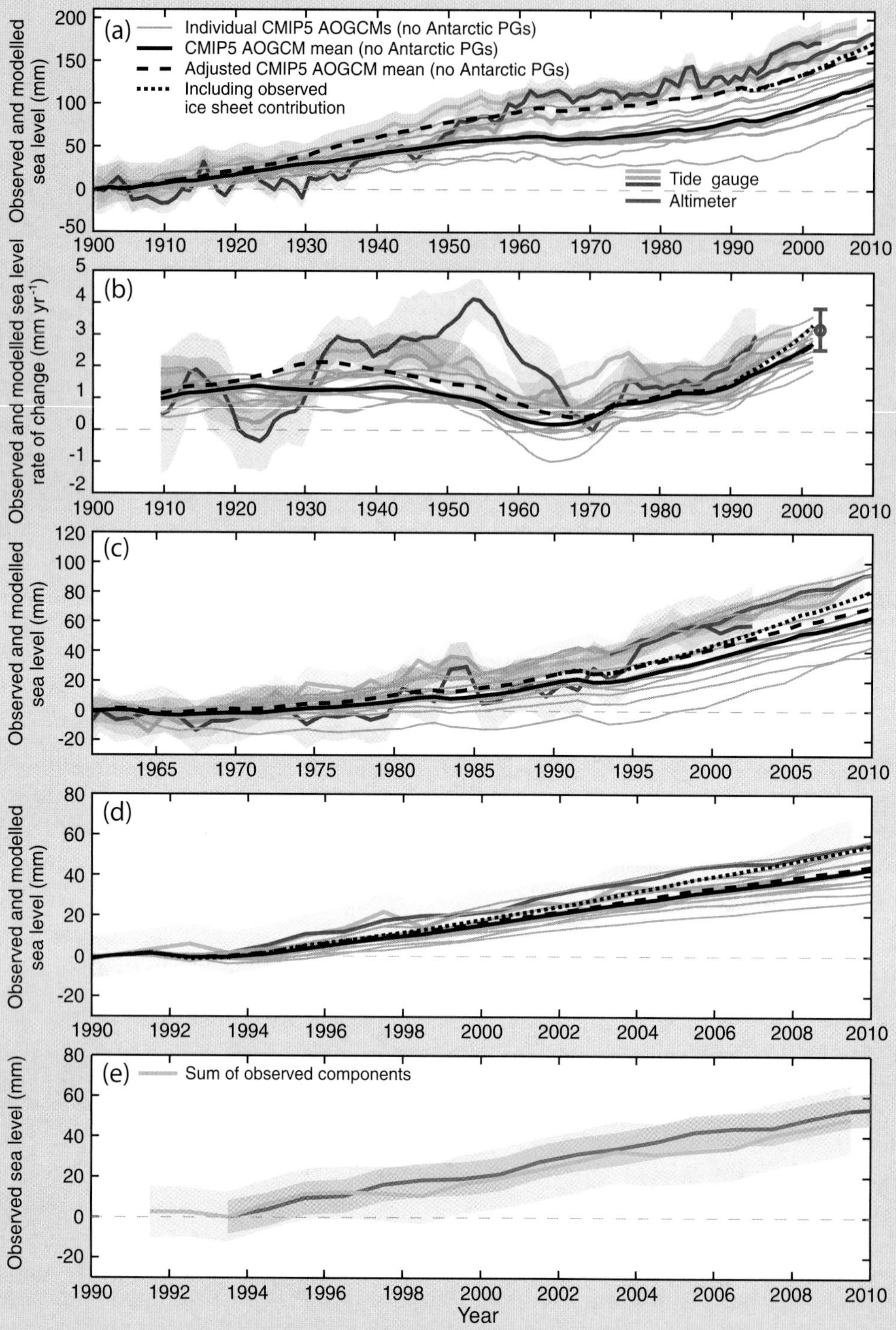

TFE.2, Figure 1 | (a) The observed and modelled sea level for 1900 to 2010. (b) The rates of sea level change for the same period, with the satellite altimeter data shown as a red dot for the rate. (c) The observed and modelled sea level for 1961 to 2010. (d) The observed and modelled sea level for 1990 to 2010. Panel (e) compares the sum of the observed contributions (orange) and the observed sea level from the satellite altimeter data (red). Estimates of GMSL from different sources are given, with the shading indicating the uncertainty estimates (two standard deviations). The satellite altimeter data since 1993 are shown in red. The grey lines in panels (a)-(d) are the sums of the contributions from modelled ocean thermal expansion and glaciers (excluding glaciers peripheral to the Antarctic ice sheet), plus changes in land-water storage (see Figure 13.4). The black line is the mean of the grey lines plus a correction of thermal expansion for the omission of volcanic forcing in the Atmosphere–Ocean General Circulation Model (AOGCM) control experiments (see Section 13.3.1). The dashed black line (adjusted model mean) is the sum of the corrected model mean thermal expansion, the change in land water storage, the glacier estimate using observed (rather than modelled) climate (see Figure 13.4), and an illustrative long-term ice-sheet contribution (of 0.1 mm yr^{-1}). The dotted black line is the adjusted model mean but now including the observed ice-sheet contributions, which begin in 1993. Because the observational ice-sheet estimates include the glaciers peripheral to the Greenland and Antarctic ice sheets (from Section 4.4), the contribution from glaciers to the adjusted model mean excludes the peripheral glaciers (PGs) to avoid double counting. {13.3; Figure 13.7}

TFE.2 (continued)

When calibrated appropriately, recently improved dynamical ice sheet models can reproduce the observed rapid changes in ice sheet outflow for individual glacier systems (e.g., Pine Island Glacier in Antarctica; *medium confidence*). However, models of ice sheet response to global warming and particularly ice sheet–ocean interactions are incomplete and the omission of ice sheet models, especially of dynamics, from the model budget of the past means that they have not been as critically evaluated as other contributions. {13.3, 13.4}

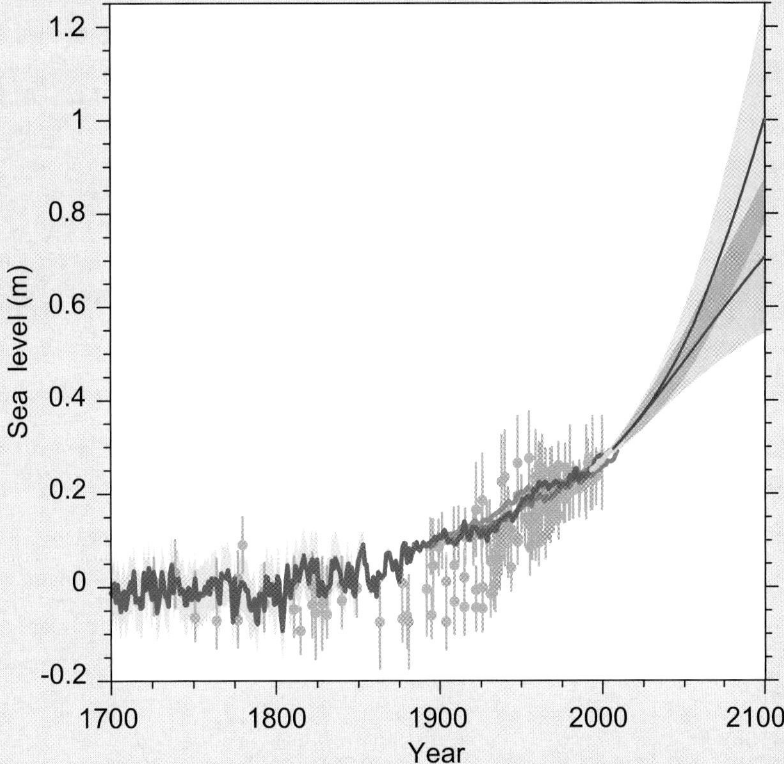

TFE.2, Figure 2 | Compilation of paleo sealevel data (purple), tide gauge data (blue, red and green), altimeter data (light blue) and central estimates and *likely* ranges for projections of global mean sea level rise from the combination of CMIP5 and process-based models for RCP2.6 (blue) and RCP8.5 (red) scenarios, all relative to pre-industrial values. {Figures 13.3, 13.11, 13.27}

GMSL rise for 2081–2100 (relative to 1986–2005) for the Representative Concentration Pathways (RCPs) will *likely* be in the 5 to 95% ranges derived from Coupled Model Intercomparison Project Phase 5 (CMIP5) climate projections in combination with process-based models of other contributions (*medium confidence*), that is, 0.26 to 0.55 m (RCP2.6), 0.32 to 0.63 m (RCP4.5), 0.33 to 0.63 m (RCP6.0), 0.45 to 0.82 (RCP8.5) m (see Table TS.1 and Figure TS.15 for RCP forcing). For RCP8.5 the range at 2100 is 0.52 to 0.98 m. Confidence in the projected *likely* ranges comes from the consistency of process-based models with observations and physical understanding. It is assessed that there is currently insufficient evidence to evaluate the probability of specific levels above the *likely* range. Based on current understanding, only the collapse of marine-based sectors of the Antarctic ice sheet, if initiated, could cause GMSL to rise substantially above the *likely* range during the 21st century. There is a lack of consensus on the probability for such a collapse, and the potential additional contribution to GMSL rise cannot be precisely quantified, but there is *medium confidence* that it would not exceed several tenths of a metre of sea level rise during the 21st century. It is *virtually certain* that GMSL rise will continue beyond 2100. {13.5.1, 13.5.3}

Many semi-empirical models projections of GMSL rise are higher than process-based model projections, but there is no consensus in the scientific community about their reliability and there is thus *low confidence* in their projections. {13.5.2, 13.5.3}

TFE.2, Figure 2 combines the paleo, tide gauge and altimeter observations of sea level rise from 1700 with the projected GMSL change to 2100. {13.5, 13.7, 13.8}

There is *low confidence* in a global-scale observed trend in drought or dryness (lack of rainfall), owing to lack of direct observations, dependencies of inferred trends on the index choice and geographical inconsistencies in the trends. However, this masks important regional changes and, for example, the frequency and intensity of drought have *likely* increased in the Mediterranean and West Africa and *likely* decreased in central North America and northwest Australia since 1950. {2.6.2; Table 2.13}

There is *high confidence* for droughts during the last millennium of greater magnitude and longer duration than those observed since the beginning of the 20th century in many regions. There is *medium confidence* that more megadroughts occurred in monsoon Asia and wetter conditions prevailed in arid Central Asia and the South American monsoon region during the Little Ice Age (1450–1850) compared to the Medieval Climate Anomaly (950–1250). {5.5.4, 5.5.5}

Confidence remains *low* for long-term (centennial) changes in tropical cyclone activity, after accounting for past changes in observing capabilities. However, for the years since the 1970s, it is *virtually certain* that the frequency and intensity of storms in the North Atlantic have increased although the reasons for this increase are debated (see TFE.9). There is *low confidence* of large-scale trends in storminess over the last century and there is still insufficient evidence to determine whether robust trends exist in small-scale severe weather events such as hail or thunderstorms. {2.6.2–2.6.4}

With *high confidence*, floods larger than recorded since the 20th century occurred during the past five centuries in northern and central Europe, the western Mediterranean region and eastern Asia. There is *medium confidence* that in the Near East, India and central North America, modern large floods are comparable or surpass historical floods in magnitude and/or frequency. {5.5.5}

TS.2.7.2 Oceans

It is *likely* that the magnitude of extreme high sea level events has increased since 1970 (see TFE.9, Table 1). Most of the increase in extreme sea level can be explained by the mean sea level rise: changes in extreme high sea levels are reduced to less than 5 mm yr^{-1} at 94% of tide gauges once the rise in mean sea level is accounted for. There is *medium confidence* based on reanalysis forced model hindcasts and ship observations that mean significant wave height has increased since the 1950s over much of the North Atlantic north of 45°N, with typical winter season trends of up to 20 cm per decade. {3.4.5, 3.7.5}

TS.2.8 Changes in Carbon and Other Biogeochemical Cycles

Concentrations of the atmospheric greenhouse gases (GHGs) carbon dioxide (CO_2), methane (CH_4) and nitrous oxide (N_2O) in 2011 exceed the range of concentrations recorded in ice cores during the past 800 kyr. Past changes in atmospheric GHG concentrations are determined with *very high confidence* from polar ice cores. Since AR4 these records have been extended from 650 ka to 800 ka. {5.2.2}

With *very high confidence*, the current rates of CO_2, CH_4 and N_2O rise in atmospheric concentrations and the associated increases in RF are unprecedented with respect to the 'highest resolution' ice core records of the last 22 kyr. There is *medium confidence* that the rate of change of the observed GHG rise is also unprecedented compared with the lower resolution records of the past 800 kyr. {2.2.1, 5.2.2}

In several periods characterized by high atmospheric CO_2 concentrations, there is *medium confidence* that global mean temperature was significantly above pre-industrial level. During the mid-Pliocene (3.3 to 3.0 Ma), atmospheric CO_2 concentration between 350 ppm and 450 ppm (*medium confidence*) occurred when GMST was 1.9°C to 3.6°C warmer (*medium confidence*) than for pre-industrial climate. During the Early Eocene (52 to 48 Ma), atmospheric CO_2 concentration exceeded about 1000 ppm when GMST was 9°C to 14°C higher (*medium confidence*) than for pre-industrial conditions. {5.3.1}

TS.2.8.1 Carbon Dioxide

Between 1750 and 2011, CO_2 emissions from fossil fuel combustion and cement production are estimated from energy and fuel use statistics to have released 375 [345 to 405] PgC[9]. In 2002–2011, average fossil fuel and cement manufacturing emissions were 8.3 [7.6 to 9.0] PgC yr^{-1} (*high confidence*), with an average growth rate of 3.2% yr^{-1} (Figure TS.4). This rate of increase of fossil fuel emissions is higher than during the 1990s (1.0% yr^{-1}). In 2011, fossil fuel emissions were 9.5 [8.7 to 10.3] PgC. {2.2.1, 6.3.1; Table 6.1}

Between 1750 and 2011, land use change (mainly deforestation), derived from land cover data and modelling, is estimated to have released 180 [100 to 260] PgC. Land use change emissions between 2002 and 2011 are dominated by tropical deforestation, and are estimated at 0.9 [0.1 to 1.7] PgC yr^{-1} (*medium confidence*), with possibly a small decrease from the 1990s due to lower reported forest loss during this decade. This estimate includes gross deforestation emissions of around 3 PgC yr^{-1} compensated by around 2 PgC yr^{-1} of forest regrowth in some regions, mainly abandoned agricultural land. {6.3.2; Table 6.2}

Of the 555 [470 to 640] PgC released to the atmosphere from fossil fuel and land use emissions from 1750 to 2011, 240 [230 to 250] PgC accumulated in the atmosphere, as estimated with very high accuracy from the observed increase of atmospheric CO_2 concentration from 278 [273 to 283] ppm[10] in 1750 to 390.5 [390.4 to 390.6] ppm in 2011. The amount of CO_2 in the atmosphere grew by 4.0 [3.8 to 4.2] PgC yr^{-1} in the first decade of the 21st century. The distribution of observed atmospheric CO_2 increases with latitude clearly shows that the increases are driven by anthropogenic emissions that occur primarily in the industrialized countries north of the equator. Based on annual average concentrations, stations in the NH show slightly higher concentrations than stations in the SH. An independent line of evidence

---

[9] 1 Petagram of carbon = 1 PgC = 10^{15} grams of carbon = 1 Gigatonne of carbon = 1 GtC. This corresponds to 3.667 $GtCO_2$.

[10] ppm (parts per million) or ppb (parts per billion, 1 billion = 1000 million) is the ratio of the number of greenhouse gas molecules to the total number of molecules of dry air. For example, 300 ppm means 300 molecules of a greenhouse gas per million molecules of dry air.

for the anthropogenic origin of the observed atmospheric CO_2 increase comes from the observed consistent decrease in atmospheric oxygen (O_2) content and a decrease in the stable isotopic ratio of CO_2 ($^{13}C/^{12}C$) in the atmosphere (Figure TS.5). {2.2.1, 6.1.3}

The remaining amount of carbon released by fossil fuel and land use emissions has been re-absorbed by the ocean and terrestrial ecosystems. Based on high agreement between independent estimates using different methods and data sets (e.g., oceanic carbon, oxygen and transient tracer data), it is *very likely* that the global ocean inventory of anthropogenic carbon increased from 1994 to 2010. In 2011, it is estimated to be 155 [125 to 185] PgC. The annual global oceanic uptake rates calculated from independent data sets (from changes in the oceanic inventory of anthropogenic carbon, from measurements of the atmospheric oxygen to nitrogen ratio (O_2/N_2) or from CO_2 partial pressure (pCO_2) data) and for different time periods agree with each other within their uncertainties, and *very likely* are in the range of 1.0 to 3.2 PgC yr^{-1}. Regional observations of the storage rate of anthropogenic carbon in the ocean are in broad agreement with the expected rate resulting from the increase in atmospheric CO_2

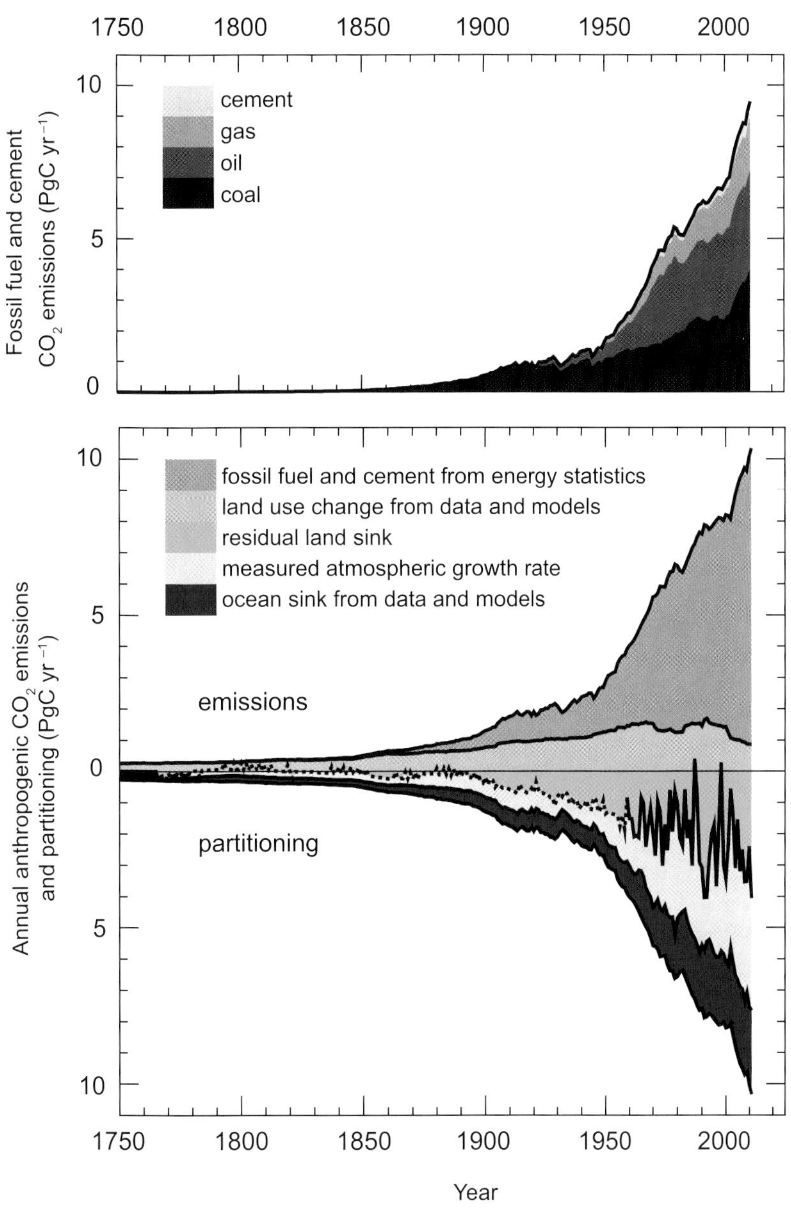

Figure TS.4 | Annual anthropogenic CO_2 emissions and their partitioning among the atmosphere, land and ocean (PgC yr^{-1}) from 1750 to 2011. (Top) Fossil fuel and cement CO_2 emissions by category, estimated by the Carbon Dioxide Information Analysis Center (CDIAC). (Bottom) Fossil fuel and cement CO_2 emissions as above. CO_2 emissions from net land use change, mainly deforestation, are based on land cover change data (see Table 6.2). The atmospheric CO_2 growth rate prior to 1959 is based on a spline fit to ice core observations and a synthesis of atmospheric measurements from 1959. The fit to ice core observations does not capture the large interannual variability in atmospheric CO_2 and is represented with a dashed line. The ocean CO_2 sink is from a combination of models and observations. The residual land sink (term in green in the figure) is computed from the residual of the other terms. The emissions and their partitioning include only the fluxes that have changed since 1750, and not the natural CO_2 fluxes (e.g., atmospheric CO_2 uptake from weathering, outgassing of CO_2 from lakes and rivers and outgassing of CO_2 by the ocean from carbon delivered by rivers; see Figure 6.1) between the atmosphere, land and ocean reservoirs that existed before that time and still exist today. The uncertainties in the various terms are discussed in Chapter 6 and reported in Table 6.1 for decadal mean values. {Figure 6.8}

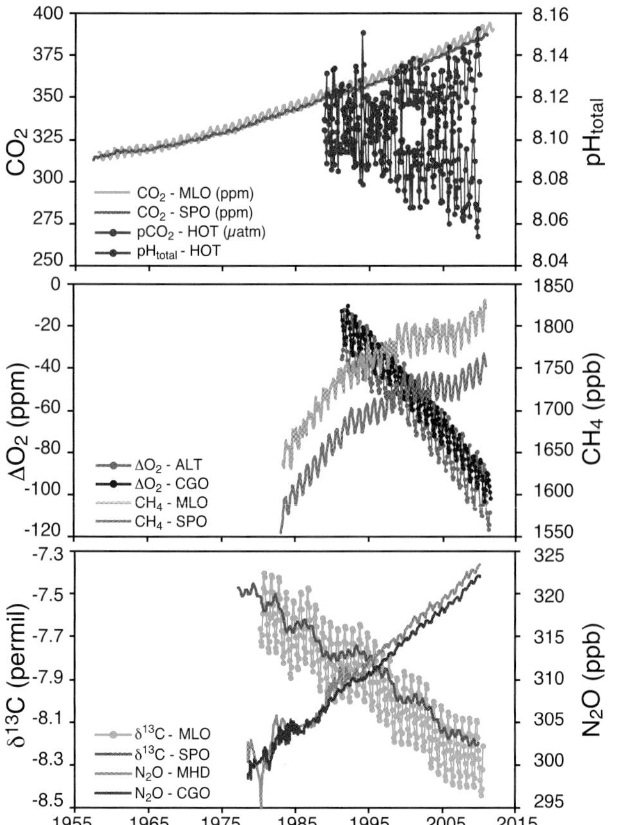

Figure TS.5 | Atmospheric concentration of CO_2, oxygen, $^{13}C/^{12}C$ stable isotope ratio in CO_2, as well as CH_4 and N_2O atmospheric concentrations and oceanic surface observations of CO_2 partial pressure (pCO_2) and pH, recorded at representative time series stations in the Northern and the Southern Hemispheres. MLO: Mauna Loa Observatory, Hawaii; SPO: South Pole; HOT: Hawaii Ocean Time-Series station; MHD: Mace Head, Ireland; CGO: Cape Grim, Tasmania; ALT: Alert, Northwest Territories, Canada. Further detail regarding the related Figure SPM.4 is given in the TS Supplementary Material. {Figures 3.18, 6.3; FAQ 3.3, Figure 1}

concentrations, but with significant spatial and temporal variations. {3.8.1, 6.3}

Natural terrestrial ecosystems (those not affected by land use change) are estimated by difference from changes in other reservoirs to have accumulated 160 [70 to 250] PgC between 1750 and 2011. The gain of carbon by natural terrestrial ecosystems is estimated to take place mainly through the uptake of CO_2 by enhanced photosynthesis at higher CO_2 levels and nitrogen deposition and longer growing seasons in mid and high latitudes. Natural carbon sinks vary regionally owing to physical, biological and chemical processes acting on different time scales. An excess of atmospheric CO_2 absorbed by land ecosystems gets stored as organic matter in diverse carbon pools, from short-lived (leaves, fine roots) to long-lived (stems, soil carbon). {6.3; Table 6.1}

TS.2.8.2 Carbon and Ocean Acidification

Oceanic uptake of anthropogenic CO_2 results in gradual acidification of the ocean. The pH[11] of ocean surface water has decreased by 0.1 since the beginning of the industrial era (*high confidence*), corresponding to a 26% increase in hydrogen ion concentration. The observed pH trends range between –0.0014 and –0.0024 per year in surface waters. In the ocean interior, natural physical and biological processes, as well as uptake of anthropogenic CO_2, can cause changes in pH over decadal and longer time scales. {3.8.2; Box 3.2; Table 3.2; FAQ 3.3}

TS.2.8.3 Methane

The concentration of CH_4 has increased by a factor of 2.5 since pre-industrial times, from 722 [697 to 747] ppb in 1750 to 1803 [1799 to 1807] ppb in 2011 (Figure TS.5). There is *very high confidence* that the atmospheric CH_4 increase during the Industrial Era is caused by anthropogenic activities. The massive increase in the number of ruminants, the emissions from fossil fuel extraction and use, the expansion of rice paddy agriculture and the emissions from landfills and waste are the dominant anthropogenic CH_4 sources. Anthropogenic emissions account for 50 to 65% of total emissions. By including natural geological CH_4 emissions that were not accounted for in previous budgets, the fossil component of the total CH_4 emissions (i.e., anthropogenic emissions related to leaks in the fossil fuel industry and natural geological leaks) is now estimated to amount to about 30% of the total CH_4 emissions (*medium confidence*). {2.2.1, 6.1, 6.3.3}

In recent decades, CH_4 growth in the atmosphere has been variable. CH_4 concentrations were relatively stable for about a decade in the 1990s, but then started growing again starting in 2007. The exact drivers of this renewed growth are still debated. Climate-driven fluctuations of CH_4 emissions from natural wetlands (177 to 284 $\times 10^{12}$ g (CH_4) yr^{-1} for 2000–2009 based on bottom-up estimates) are the main drivers of the global interannual variability of CH_4 emissions (*high confidence*), with a smaller contribution from biomass burning emissions during high fire years {2.2.1, 6.3.3; Table 6.8}.

TS.2.8.4 Nitrous Oxide

Since pre-industrial times, the concentration of N_2O in the atmosphere has increased by a factor of 1.2 (Figure TS.5). Changes in the nitrogen cycle, in addition to interactions with CO_2 sources and sinks, affect emissions of N_2O both on land and from the ocean. {2.2.1, 6.4.6}

TS.2.8.5 Oceanic Oxygen

High agreement among analyses provides *medium confidence* that oxygen concentrations have decreased in the open ocean thermocline in many ocean regions since the 1960s. The general decline is consistent with the expectation that warming-induced stratification leads to a decrease in the supply of oxygen to the thermocline from near surface waters, that warmer waters can hold less oxygen and that changes in wind-driven circulation affect oxygen concentrations. It is *likely* that the tropical oxygen minimum zones have expanded in recent decades. {3.8.3}

---

[11] pH is a measure of acidity: a decrease in pH value means an increase in acidity, that is, acidification.

TS.3 Drivers of Climate Change

TS.3.1 Introduction

Human activities have changed and continue to change the Earth's surface and atmospheric composition. Some of these changes have a direct or indirect impact on the energy balance of the Earth and are thus drivers of climate change. Radiative forcing (RF) is a measure of the net change in the energy balance of the Earth system in response to some external perturbation (see Box TS.2), with positive RF leading to a warming and negative RF to a cooling. The RF concept is valuable for comparing the influence on GMST of most individual agents affecting the Earth's radiation balance. The quantitative values provided in AR5 are consistent with those in previous IPCC reports, though there have been some important revisions (Figure TS.6). Effective radiative forcing (ERF) is now used to quantify the impact of some forcing agents that involve rapid adjustments of components of the atmosphere and surface that are assumed constant in the RF concept (see Box TS.2). RF and ERF are estimated from the change between 1750 and 2011, referred to as 'Industrial Era', if other time periods are not explicitly stated. Uncertainties are given associated with the best estimates of RF and ERF, with values representing the 5 to 95% (90%) confidence range. {8.1, 7.1}

In addition to the global mean RF or ERF, the spatial distribution and temporal evolution of forcing, as well as climate feedbacks, play a role in determining the eventual impact of various drivers on climate. Land surface changes may also impact the local and regional climate through processes that are not radiative in nature. {8.1, 8.3.5, 8.6}

TS.3.2 Radiative Forcing from Greenhouse Gases

Human activity leads to change in the atmospheric composition either directly (via emissions of gases or particles) or indirectly (via atmospheric chemistry). Anthropogenic emissions have driven the changes in well-mixed greenhouse gas (WMGHG) concentrations during the Industrial Era (see Section TS.2.8 and TFE.7). As historical WMGHG concentrations since the pre-industrial are well known based on direct measurements and ice core records, and WMGHG radiative properties are also well known, the computation of RF due to concentration changes provides tightly constrained values (Figure TS.6). There has not been significant change in our understanding of WMGHG radiative impact, so that the changes in RF estimates relative to AR4 are due essentially to concentration increases. The best estimate for WMGHG ERF is the same as RF, but the uncertainty range is twice as large due to the poorly constrained cloud responses. Owing to high-quality observations, it is certain that increasing atmospheric burdens of most WMGHGs, especially CO_2, resulted in a further increase in their RF from 2005 to 2011. Based on concentration changes, the RF of all WMGHGs in 2011 is 2.83 [2.54 to 3.12] W m^{-2} (*very high confidence*). This is an increase since AR4 of 0.20 [0.18 to 0.22] W m^{-2}, with nearly all of the increase due to the increase in the abundance of CO_2 since 2005. The Industrial Era RF for CO_2 alone is 1.82 [1.63 to 2.01] W m^{-2}. Over the last 15 years, CO_2 has been the dominant contributor to the increase in RF from the WMGHGs, with RF of CO_2 having an average growth rate slightly less than 0.3 W m^{-2} per decade. The uncertainty in the WMGHG RF is due in part to its radiative properties but mostly to the full accounting of atmospheric radiative transfer including clouds. {2.2.1, 5.2, 6.3, 8.3, 8.3.2; Table 6.1}

After a decade of near stability, the recent increase of CH_4 concentration led to an enhanced RF compared to AR4 by 2% to 0.48 [0.43 to 0.53] W m^{-2}. It is *very likely* that the RF from CH_4 is now larger than that of all halocarbons combined. {2.2.1, 8.3.2}

Atmospheric N_2O has increased by 6% since AR4, causing an RF of 0.17 [0.14 to 0.20] W m^{-2}. N_2O concentrations continue to rise while those of dichlorodifluoromethane (CF_2Cl_2, CFC-12), the third largest WMGHG contributor to RF for several decades, are decreasing due to phase-out of emissions of this chemical under the Montreal Protocol. Since

Box TS.2 | Radiative Forcing and Effective Radiative Forcing

RF and ERF are used to quantify the change in the Earth's energy balance that occurs as a result of an externally imposed change. They are expressed in watts per square metre (W m^{-2}). RF is defined in AR5, as in previous IPCC assessments, as the change in net downward flux (shortwave + longwave) at the tropopause after allowing for stratospheric temperatures to readjust to radiative equilibrium, while holding other state variables such as tropospheric temperatures, water vapour and cloud cover fixed at the unperturbed values (see Glossary). {8.1.1}

Although the RF concept has proved very valuable, improved understanding has shown that including rapid adjustments of the Earth's surface and troposphere can provide a better metric for quantifying the climate response. These rapid adjustments occur over a variety of time scales, but are relatively distinct from responses to GMST change. Aerosols in particular impact the atmosphere temperature profile and cloud properties on a time scale much shorter than adjustments of the ocean (even the upper layer) to forcings. The ERF concept defined in AR5 allows rapid adjustments to perturbations, for all variables except for GMST or ocean temperature and sea ice cover. The ERF and RF values are significantly different for the anthropogenic aerosols, owing to their influence on clouds and on snow or ice cover. For other components that drive the Earth's energy balance, such as GHGs, ERF and RF are fairly similar, and RF may have comparable utility given that it requires fewer computational resources to calculate and is not affected by meteorological variability and hence can better isolate small forcings. In cases where RF and ERF differ substantially, ERF has been shown to be a better indicator of the GMST response and is therefore emphasized in AR5. {7.1, 8.1; Box 8.1}

AR4, N_2O has overtaken CFC-12 to become the third largest WMGHG contributor to RF. The RF from halocarbons is very similar to the value in AR4, with a reduced RF from CFCs but increases in many of their replacements. Four of the halocarbons (trichlorofluoromethane ($CFCl_3$, CFC-11), CFC-12, trichlorotrifluoroethane ($CF_2ClCFCl_2$, CFC-113) and chlorodifluoromethane (CHF_2Cl, HCFC-22) account for 85% of the total halocarbon RF. The former three compounds have declining RF over the last 5 years but are more than compensated for by the increased RF from HCFC-22. There is *high confidence* that the growth rate in RF from all WMGHG is weaker over the last decade than in the 1970s and 1980s owing to a slower increase in the non-CO_2 RF. {2.2.1, 8.3.2}

The short-lived GHGs ozone (O_3) and stratospheric water vapour also contribute to anthropogenic forcing. Observations indicate that O_3 *likely* increased at many undisturbed (background) locations through the 1990s. These increases have continued mainly over Asia (though

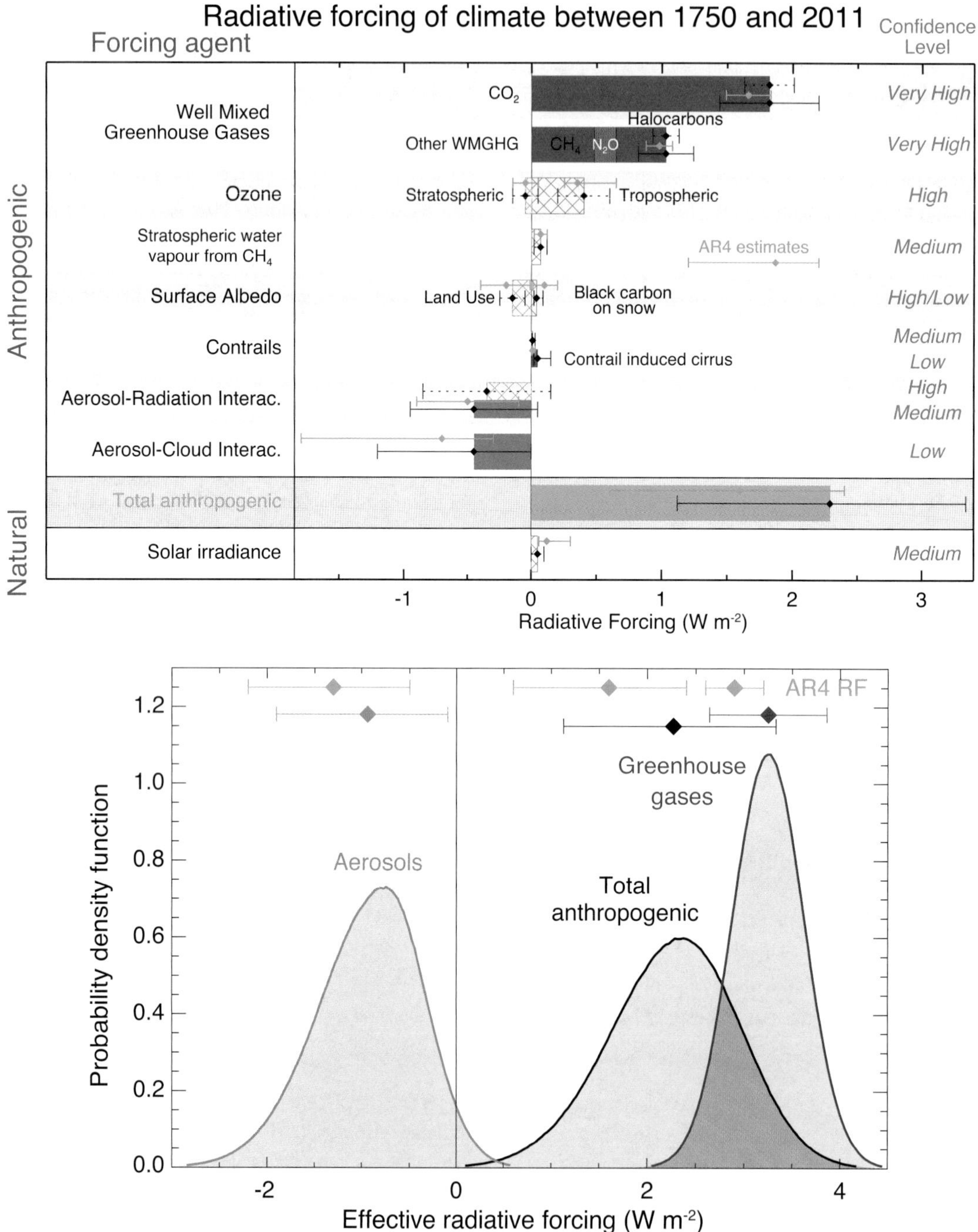

Figure TS.6 | Radiative forcing (RF) and Effective radiative forcing (ERF) of climate change during the Industrial Era. (Top) Forcing by concentration change between 1750 and 2011 with associated uncertainty range (solid bars are ERF, hatched bars are RF, green diamonds and associated uncertainties are for RF assessed in AR4). (Bottom) Probability density functions (PDFs) for the ERF, for the aerosol, greenhouse gas (GHG) and total. The green lines show the AR4 RF 90% confidence intervals and can be compared with the red, blue and black lines which show the AR5 ERF 90% confidence intervals (although RF and ERF differ, especially for aerosols). The ERF from surface albedo changes and combined contrails and contrail-induced cirrus is included in the total anthropogenic forcing, but not shown as a separate PDF. For some forcing mechanisms (ozone, land use, solar) the RF is assumed to be representative of the ERF but an additional uncertainty of 17% is added in quadrature to the RF uncertainty. {Figures 8.15, 8.16}

observations cover a limited area) and flattened over Europe during the last decade. The total RF due to changes in O_3 is 0.35 [0.15 to 0.55] W m^{-2} (*high confidence*), with RF due to tropospheric O_3 of 0.40 [0.20 to 0.60] W m^{-2} (*high confidence*) and due to stratospheric O_3 of –0.05 [–0.15 to +0.05] W m^{-2} (*high confidence*). O_3 is not emitted directly into the atmosphere; instead it is formed by photochemical reactions. In the troposphere these reactions involve precursor compounds that are emitted into the atmosphere from a variety of natural and anthropogenic sources. Tropospheric O_3 RF is largely attributed to increases in emissions of CH_4, carbon monoxide, volatile organics and nitrogen oxides, while stratospheric RF results primarily from O_3 depletion by anthropogenic halocarbons. However, there is now strong evidence for substantial links between the changes in tropospheric and stratospheric O_3 and a total O_3 RF of 0.50 [0.30 to 0.70] W m^{-2} is attributed to tropospheric O_3 precursor emissions and –0.15 [–0.30 to 0.00] W m^{-2} to O_3 depletion by halocarbons. There is strong evidence that tropospheric O_3 also has a detrimental impact on vegetation physiology, and therefore on its CO_2 uptake. This reduced uptake leads to an indirect increase in the atmospheric CO_2 concentration. Thus a fraction of the CO_2 RF should be attributed to ozone or its precursors rather than direct emission of CO_2, but there is a *low confidence* on the quantitative estimates. RF for stratospheric water vapour produced from CH_4 oxidation is 0.07 [0.02 to 0.12] W m^{-2}. Other changes in stratospheric water vapour, and all changes in water vapour in the troposphere, are regarded as a feedback rather than a forcing. {2.2.2, 8.1–8.3; FAQ 8.1}

TS.3.3 Radiative Forcing from Anthropogenic Aerosols

Anthropogenic aerosols are responsible for an RF of climate through multiple processes which can be grouped into two types: aerosol–radiation interactions (ari) and aerosol–cloud interactions (aci). There has been progress since AR4 on observing and modelling climate-relevant aerosol properties (including their size distribution, hygroscopicity, chemical composition, mixing state, optical and cloud nucleation properties) and their atmospheric distribution. Nevertheless, substantial uncertainties remain in assessments of long-term trends of global aerosol optical depth and other global properties of aerosols due to difficulties in measurement and lack of observations of some relevant parameters, high spatial and temporal variability and the relatively short observational records that exist. The anthropogenic RFari is given a best estimate of –0.35 [–0.85 to +0.15] W m^{-2} (*high confidence*) using evidence from aerosol models and some constraints from observations. The RFari is caused by multiple aerosol types (see Section TS3.6). The rapid adjustment to RFari leads to further negative forcing, in particular through cloud adjustments, and is attributable primarily to black carbon. As a consequence, the ERFari is more negative than the RFari (*low confidence*) and given a best estimate of –0.45 [–0.95 to +0.05] W m^{-2}. The assessment for RFari is less negative than reported in AR4 because of a re-evaluation of aerosol absorption. The uncertainty estimate is wider but more robust. {2.2.3, 7.3, 7.5.2}

Improved understanding of aerosol–cloud interactions has led to a reduction in the magnitude of many global aerosol–cloud forcings estimates. The total ERF due to aerosols (ERFari+aci, excluding the effect of absorbing aerosol on snow and ice) is assessed to be –0.9 [–1.9 to –0.1] W m^{-2} (*medium confidence*). This estimate encompasses all rapid adjustments, including changes to the cloud lifetime and aerosol microphysical effects on mixed-phase, ice and convective clouds. This range was obtained by giving equal weight to satellite-based studies and estimates from climate models. It is consistent with multiple lines of evidence suggesting less negative estimates for aerosol–cloud interactions than those discussed in AR4. {7.4, 7.5, 8.5}

The RF from black carbon (BC) on snow and ice is assessed to be 0.04 [0.02 to 0.09] W m^{-2} (*low confidence*). Unlike in the previous IPCC assessment, this estimate includes the effects on sea ice, accounts for more physical processes and incorporates evidence from both models and observations. This RF causes a two to four times larger GMST change per unit forcing than CO_2 primarily because all of the forcing energy is deposited directly into the cryosphere, whose evolution drives a positive albedo feedback on climate. This effect thus can represent a significant forcing mechanism in the Arctic and other snow- or ice-covered regions. {7.3, 7.5.2, 8.3.4, 8.5}

Despite the large uncertainty ranges on aerosol forcing, there is a *high confidence* that aerosols have offset a substantial portion of GHG forcing. Aerosol–cloud interactions can influence the character of individual storms, but evidence for a systematic aerosol effect on storm or precipitation intensity is more limited and ambiguous. {7.4, 7.6, 8.5}

TS.3.4 Radiative Forcing from Land Surface Changes and Contrails

There is robust evidence that anthropogenic land use changes such as deforestation have increased the land surface albedo, which leads to an RF of –0.15 [–0.25 to –0.05] W m^{-2}. There is still a large spread of quantitative estimates owing to different assumptions for the albedo of natural and managed surfaces (e.g., croplands, pastures). In addition, the time evolution of the land use change, and in particular how much was already completed in the reference year 1750, are still debated. Furthermore, land use change causes other modifications that are not radiative but impact the surface temperature, including modifications in the surface roughness, latent heat flux, river runoff and irrigation. These are more uncertain and they are difficult to quantify, but they tend to offset the impact of albedo changes at the global scale. As a consequence, there is low agreement on the sign of the net change in global mean temperature as a result of land use change. Land use change, and in particular deforestation, also has significant impacts on WMGHG concentrations. It contributes to the corresponding RF associated with CO_2 emissions or concentration changes. {8.3.5}

Persistent contrails from aviation contribute a positive RF of 0.01 [0.005 to 0.03] W m^{-2} (*medium confidence*) for year 2011, and the combined contrail and contrail-cirrus ERF from aviation is assessed to be 0.05 [0.02 to 0.15] W m^{-2} (*low confidence*). This forcing can be much larger regionally but there is now *medium confidence* that it does not produce observable regional effects on either the mean or diurnal range of surface temperature. {7.2.7}

TS.3.5 Radiative Forcing from Natural Drivers of Climate Change

Solar and volcanic forcings are the two dominant natural contributors to global climate change during the Industrial Era. Satellite observations of total solar irradiance (TSI) changes since 1978 show quasi-periodic cyclical variation with a period of roughly 11 years. Longer term forcing is typically estimated by comparison of solar minima (during which variability is least). This gives an RF change of –0.04 [–0.08 to 0.00] W m^{-2} between the most recent (2008) minimum and the 1986 minimum. There is some diversity in the estimated trends of the composites of various satellite data, however. Secular trends of TSI before the start of satellite observations rely on a number of indirect proxies. The best estimate of RF from TSI changes over the industrial era is 0.05 [0.00 to 0.10] W m^{-2} (*medium confidence*), which includes greater RF up to around 1980 and then a small downward trend. This RF estimate is substantially smaller than the AR4 estimate due to the addition of the latest solar cycle and inconsistencies in how solar RF was estimated in earlier IPCC assessments. The recent solar minimum appears to have been unusually low and long-lasting and several projections indicate lower TSI for the forthcoming decades. However, current abilities to project solar irradiance are extremely limited so that there is *very low confidence* concerning future solar forcing. Nonetheless, there is a *high confidence* that 21st century solar forcing will be much smaller than the projected increased forcing due to WMGHGs. {5.2.1, 8.4.1; FAQ 5.1}

Changes in solar activity affect the cosmic ray flux impinging upon the Earth's atmosphere, which has been hypothesized to affect climate through changes in cloudiness. Cosmic rays enhance aerosol nucleation and thus may affect cloud condensation nuclei production in the free troposphere, but the effect is too weak to have any climatic influence during a solar cycle or over the last century (medium evidence, high agreement). No robust association between changes in cosmic rays and cloudiness has been identified. In the event that such an association existed, a mechanism other than cosmic ray–induced nucleation of new aerosol particles would be needed to explain it. {7.3, 7.4.6}

The RF of stratospheric volcanic aerosols is now well understood and there is a large RF for a few years after major volcanic eruptions (Box TS.5, Figure 1). Although volcanic eruptions inject both mineral particles and sulphate aerosol precursors into the atmosphere, it is the latter, because of their small size and long lifetimes, that are responsible for RF important for climate. The emissions of CO_2 from volcanic eruptions are at least 100 times smaller than anthropogenic emissions, and inconsequential for climate on century time scales. Large tropical volcanic eruptions have played an important role in driving annual to decadal scale climate change during the Industrial Era owing to their sometimes very large negative RF. There has not been any major volcanic eruption since Mt Pinatubo in 1991, which caused a 1-year RF of about –3.0 W m^{-2}, but several smaller eruptions have caused an RF averaged over the years 2008–2011 of –0.11 [–0.15 to –0.08] W m^{-2} (*high confidence*), twice as strong in magnitude compared to the 1999–2002 average. The smaller eruptions have led to better understanding of the dependence of RF on the amount of material from high-latitude injections as well as the time of the year when they take place. {5.2.1, 5.3.5, 8.4.2; Annex II}

TS.3.6 Synthesis of Forcings; Spatial and Temporal Evolution

A synthesis of the Industrial Era forcing finds that among the forcing agents, there is a *very high confidence* only for the WMGHG RF. Relative to AR4, the confidence level has been elevated for seven forcing agents owing to improved evidence and understanding. {8.5; Figure 8.14}

The time evolution of the total anthropogenic RF shows a nearly continuous increase from 1750, primarily since about 1860. The total anthropogenic RF increase rate since 1960 has been much greater than during earlier Industrial Era periods, driven primarily by the continuous increase in most WMGHG concentrations. There is still low agreement on the time evolution of the total aerosol ERF, which is the primary factor for the uncertainty in the total anthropogenic forcing. The fractional uncertainty in the total anthropogenic forcing decreases gradually after 1950 owing to the smaller offset of positive WMGHG forcing by negative aerosol forcing. There is robust evidence and high agreement that natural forcing is a small fraction of the WMGHG forcing. Natural forcing changes over the last 15 years have *likely* offset a substantial fraction (at least 30%) of the anthropogenic forcing increase during this period (Box TS.3). Forcing by CO_2 is the largest single contributor to the total forcing during the Industrial Era and from 1980–2011. Compared to the entire Industrial Era, the dominance of CO_2 forcing is larger for the 1980–2011 change with respect to other WMGHGs, and there is *high confidence* that the offset from aerosol forcing to WMGHG forcing during this period was much smaller than over the 1950–1980 period. {8.5.2}

Forcing can also be attributed to emissions rather than to the resulting concentration changes (Figure TS.7). Carbon dioxide is the largest single contributor to historical RF from either the perspective of changes in the atmospheric concentration of CO_2 or the impact of changes in net emissions of CO_2. The relative importance of other forcing agents can vary markedly with the perspective chosen, however. In particular, CH_4 emissions have a much larger forcing (about 1.0 W m^{-2} over the Industrial Era) than CH_4 concentration increases (about 0.5 W m^{-2}) due to several indirect effects through atmospheric chemistry. In addition, carbon monoxide emissions are *virtually certain* to cause a positive forcing, while emissions of reactive nitrogen oxides *likely* cause a net negative forcing but uncertainties are large. Emissions of ozone-depleting halocarbons *very likely* cause a net positive forcing as their direct radiative effect is larger than the impact of the stratospheric ozone depletion that they induce. Emissions of SO_2, organic carbon and ammonia cause a negative forcing, while emissions of black carbon lead to positive forcing via aerosol–radiation interactions. Note that mineral dust forcing may include a natural component or a climate feedback effect. {7.3, 7.5.2, 8.5.1}

Although the WMGHGs show a spatially fairly homogeneous forcing, other agents such as aerosols, ozone and land use changes are highly heterogeneous spatially. RFari showed maximum negative values over eastern North America and Europe during the early 20th century, with large negative values extending to East and Southeast Asia, South America and central Africa by 1980. Since then, however, the magnitude has decreased over eastern North America and Europe due to pollution control, and the peak negative forcing has shifted to South and East Asia primarily as a result of economic growth and the resulting increase in emissions in those areas. Total aerosol ERF shows similar behaviour for locations with maximum negative forcing, but also shows substantial positive forcing over some deserts and the Arctic. In contrast, the global mean whole atmosphere ozone forcing increased throughout the 20th century, and has peak positive amplitudes around 15°N to 30°N but negative values over Antarctica. Negative land use forcing by albedo changes has been strongest in industrialized and biomass burning regions. The inhomogeneous nature of these forcings can cause them to have a substantially larger influence on the hydrologic cycle than an equivalent global mean homogeneous forcing. {8.3.5, 8.6}

Over the 21st century, anthropogenic RF is projected to increase under the Representative Concentration Pathways (RCPs; see Box TS.6). Simple model estimates of the RF resulting from the RCPs, which include WMGHG emissions spanning a broad range of possible futures, show anthropogenic RF relative to 1750 increasing to 3.0 to 4.8 W m^{-2} in 2050, and 2.7 to 8.4 W m^{-2} at 2100. In the near term, the RCPs are quite similar to one another (and emissions of near-term climate forcers do not span the literature range of possible futures), with RF at 2030 ranging only from 2.9 to 3.3 W m^{-2} (additional 2010 to 2030 RF of 0.7 to 1.1 W m^{-2}), but they show highly diverging values for the second half of the 21st century driven largely by CO_2. Results based on the RCP scenarios suggest only small changes in aerosol ERF between 2000 and 2030, followed by a strong reduction in the aerosols and a substantial weakening of the negative total aerosol ERF. Nitrate aerosols are an exception to this reduction, with a substantially increased negative forcing which is a robust feature among the few available models. The divergence across the RCPs indicates that, although a certain amount of future climate change is already 'in the system' due to the current radiative imbalance caused by historical emissions and the long lifetime of some atmospheric forcing agents, societal choices can still have a very large effect on future RF, and hence on climate change. {8.2, 8.5.3, 12.3; Figures 8.22, 12.4}

TS.3.7 Climate Feedbacks

Feedbacks will also play an important role in determining future climate change. Indeed, climate change may induce modification in the water, carbon and other biogeochemical cycles which may reinforce (positive feedback) or dampen (negative feedback) the expected

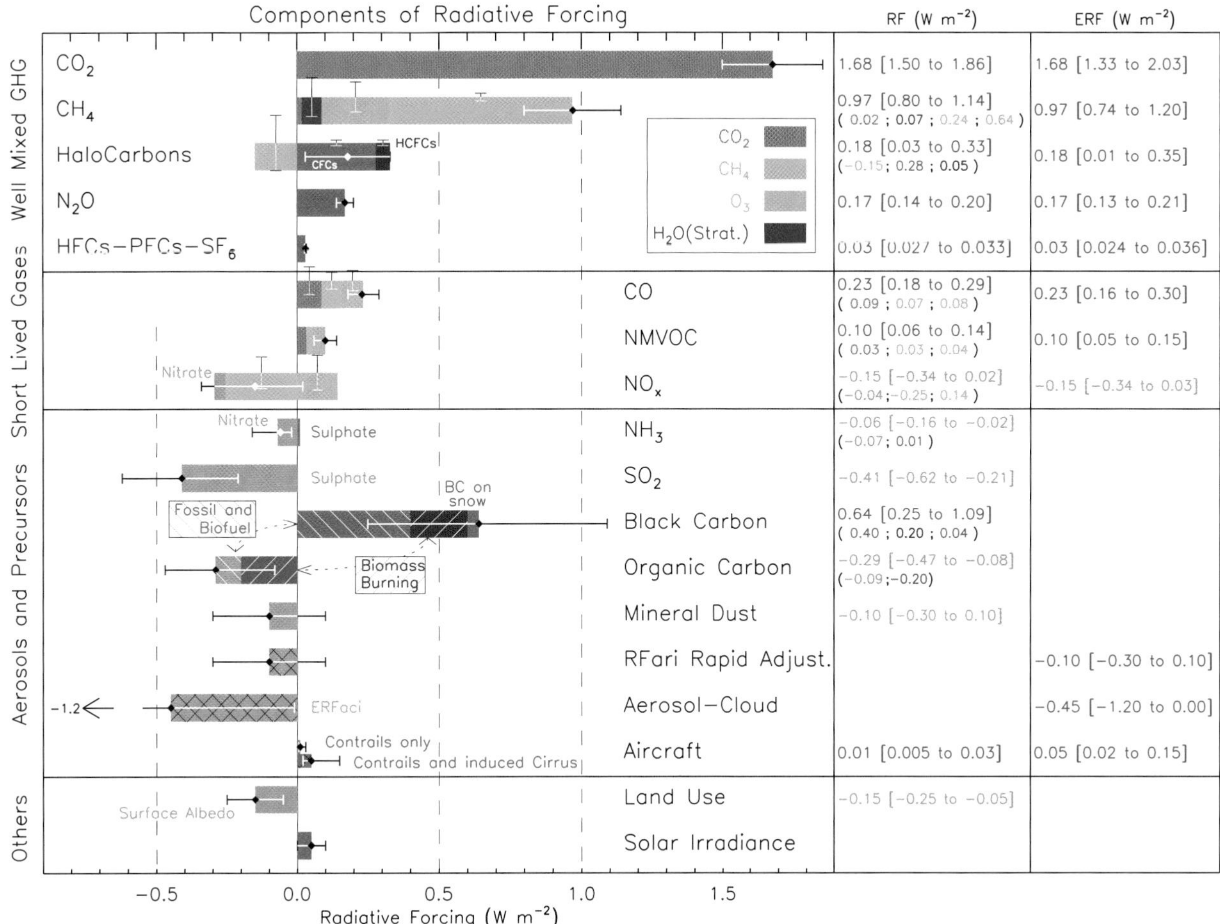

Figure TS.7 | Radiative forcing (RF) of climate change during the Industrial Era shown by emitted components from 1750 to 2011. The horizontal bars indicate the overall uncertainty, while the vertical bars are for the individual components (vertical bar lengths proportional to the relative uncertainty, with a total length equal to the bar width for a ±50% uncertainty). Best estimates for the totals and individual components (from left to right) of the response are given in the right column. Values are RF except for the effective radiative forcing (ERF) due to aerosol–cloud interactions (ERFaci) and rapid adjustment associated with the RF due to aerosol-radiation interaction (RFari Rapid Adjust.). Note that the total RF due to aerosol-radiation interaction (–0.35 Wm^{-2}) is slightly different from the sum of the RF of the individual components (–0.33 Wm^{-2}). The total RF due to aerosol-radiation interaction is the basis for Figure SPM.5. Secondary organic aerosol has not been included since the formation depends on a variety of factors not currently sufficiently quantified. The ERF of contrails includes contrail induced cirrus. Combining ERFaci –0.45 [–1.2 to 0.0] Wm^{-2} and rapid adjustment of ari –0.1 [–0.3 to +0.1] Wm^{-2} results in an integrated component of adjustment due to aerosols of –0.55 [–1.33 to –0.06] Wm^{-2}. CFCs = chlorofluorocarbons, HCFCs = hydrochlorofluorocarbons, HFCs = hydrofluorocarbons, PFCs = perfluorocarbons, NMVOC = Non-Methane Volatile Organic Compounds, BC = black carbon. Further detail regarding the related Figure SPM.5 is given in the TS Supplementary Material. {Figure 8.17}

temperature increase. Snow and ice albedo feedbacks are known to be positive. The combined water vapour and lapse rate feedback is *extremely likely* to be positive and now fairly well quantified, while cloud feedbacks continue to have larger uncertainties (see TFE.6). In addition, the new Coupled Model Intercomparison Project Phase 5 (CMIP5) models consistently estimate a positive carbon-cycle feedback, that is, reduced natural CO_2 sinks in response to future climate change. In particular, carbon-cycle feedbacks in the oceans are positive in the models. Carbon sinks in tropical land ecosystems are less consistent, and may be susceptible to climate change via processes such as drought and fire that are sometimes not yet fully represented. A key update since AR4 is the introduction of nutrient dynamics in some of the CMIP5 land carbon models, in particular the limitations on plant growth imposed by nitrogen availability. The net effect of accounting for the nitrogen cycle is a smaller projected land sink for a given trajectory of anthropogenic CO_2 emissions (see TFE.7). {6.4, Box 6.1, 7.2}

Models and ecosystem warming experiments show high agreement that wetland CH_4 emissions will increase per unit area in a warmer climate, but wetland areal extent may increase or decrease depending on regional changes in temperature and precipitation affecting wetland hydrology, so that there is *low confidence* in quantitative projections of wetland CH_4 emissions. Reservoirs of carbon in hydrates and permafrost are very large, and thus could potentially act as very powerful feedbacks. Although poorly constrained, the 21st century global release of CH_4 from hydrates to the atmosphere is *likely* to be low due to the under-saturated state of the ocean, long ventilation time of the ocean and slow propagation of warming through the seafloor. There is *high confidence* that release of carbon from thawing permafrost provides a positive feedback, but there is *low confidence* in quantitative projections of its strength. {6.4.7}

Aerosol-climate feedbacks occur mainly through changes in the source strength of natural aerosols or changes in the sink efficiency of natural and anthropogenic aerosols; a limited number of modelling studies have assessed the magnitude of this feedback to be small with a *low confidence*. There is *medium confidence* for a weak feedback (of uncertain sign) involving dimethylsulphide, cloud condensation nuclei and cloud albedo due to a weak sensitivity of cloud condensation nuclei population to changes in dimethylsulphide emissions. {7.3.5}

TS.3.8 Emission Metrics

Different metrics can be used to quantify and communicate the relative and absolute contributions to climate change of emissions of different substances, and of emissions from regions/countries or sources/sectors. Up to AR4, the most common metric has been the Global Warming Potential (GWP) that integrates RF out to a particular time horizon. This metric thus accounts for the radiative efficiencies of the various substances, and their lifetimes in the atmosphere, and gives values relative to those for the reference gas CO_2. There is now increasing focus on the Global Temperature change Potential (GTP), which is based on the change in GMST at a chosen point in time, again relative to that caused by the reference gas CO_2, and thus accounts for climate response along with radiative efficiencies and atmospheric lifetimes. Both the GWP and the GTP use a time horizon (Figure TS.8 top), the choice of which is subjective and context dependent. In general, GWPs for near-term climate forcers are higher than GTPs due to the equal time weighting in the integrated forcing used in the GWP. Hence the choice of metric can greatly affect the relative importance of near-term climate forcers and WMGHGs, as can the choice of time horizon. Analysis of the impact of current emissions (1-year pulse of emissions) shows that near-term climate forcers, such as black carbon, sulphur dioxide or CH_4, can have contributions comparable to that of CO_2 for short time horizons (of either the same or opposite sign), but their impacts become progressively less for longer time horizons over which emissions of CO_2 dominate (Figure TS.8 top). {8.7}

A large number of other metrics may be defined down the driver–response–impact chain. No single metric can accurately compare all consequences (i.e., responses in climate parameters over time) of different emissions, and a metric that establishes equivalence with regard to one effect will not give equivalence with regard to other effects. The choice of metric therefore depends strongly on the particular consequence one wants to evaluate. It is important to note that the metrics do not define policies or goals, but facilitate analysis and implementation of multi-component policies to meet particular goals. All choices of metric contain implicit value-related judgements such as type of effect considered and weighting of effects over time. Whereas GWP integrates the effects up to a chosen time horizon (i.e., giving equal weight to all times up to the horizon and zero weight thereafter), the GTP gives the temperature just for one chosen year with no weight on years before or after. {8.7}

The GWP and GTP have limitations and suffer from inconsistencies related to the treatment of indirect effects and feedbacks, for instance, if climate–carbon feedbacks are included for the reference gas CO_2 but not for the non-CO_2 gases. The uncertainty in the GWP increases with time horizon, and for the 100-year GWP of WMGHGs the uncertainty can be as large as ±40%. Several studies also point out that this metric is not well suited for policies with a maximum temperature target. Uncertainties in GTP also increase with time as they arise from the same factors contributing to GWP uncertainties along with additional contributions from it being further down the driver–response–impact chain and including climate response. The GTP metric is better suited to target-based policies, but is again not appropriate for every goal. Updated metric values accounting for changes in knowledge of lifetimes and radiative efficiencies and for climate–carbon feedbacks are now available. {8.7, Table 8.7, Table 8.A.1, Chapter 8 Supplementary Material Table 8.SM.16}

With these emission metrics, the climate impact of past or current emissions attributable to various activities can be assessed. Such activity-based accounting can provide additional policy-relevant information, as these activities are more directly affected by particular societal choices than overall emissions. A single year's worth of emissions (a pulse) is often used to quantify the impact on future climate. From this perspective and with the absolute GTP metric used to illustrate the results, energy and industry have the largest contributions to warming over the next 50 to 100 years (Figure TS.8, bottom). Household fossil and biofuel, biomass burning and on-road transportation are also relatively large contributors to warming over these time scales, while current emissions from sectors that emit large amounts of CH_4 (animal husbandry, waste/landfills and agriculture) are also important over shorter time horizons (up to about 20 years). Another useful perspective is to examine the effect of sustained current emissions. Because emitted substances are removed according to their residence time, short-lived species remain at nearly constant values while long-lived gases accumulate in this analysis. In both cases, the sectors that have the greatest long-term warming impacts (energy and industry) lead to cooling in the near term (primarily due to SO_2 emissions), and thus emissions from those sectors can lead to opposite global mean temperature responses at short and long time scales. The relative importance of the other sectors depends on the time and perspective chosen. As with RF or ERF, uncertainties in aerosol impacts are large, and in particular attribution of aerosol–cloud interactions to individual components is poorly constrained. {8.7; Chapter 8 Supplementary Material Figures 8.SM.9, 8.SM.10}

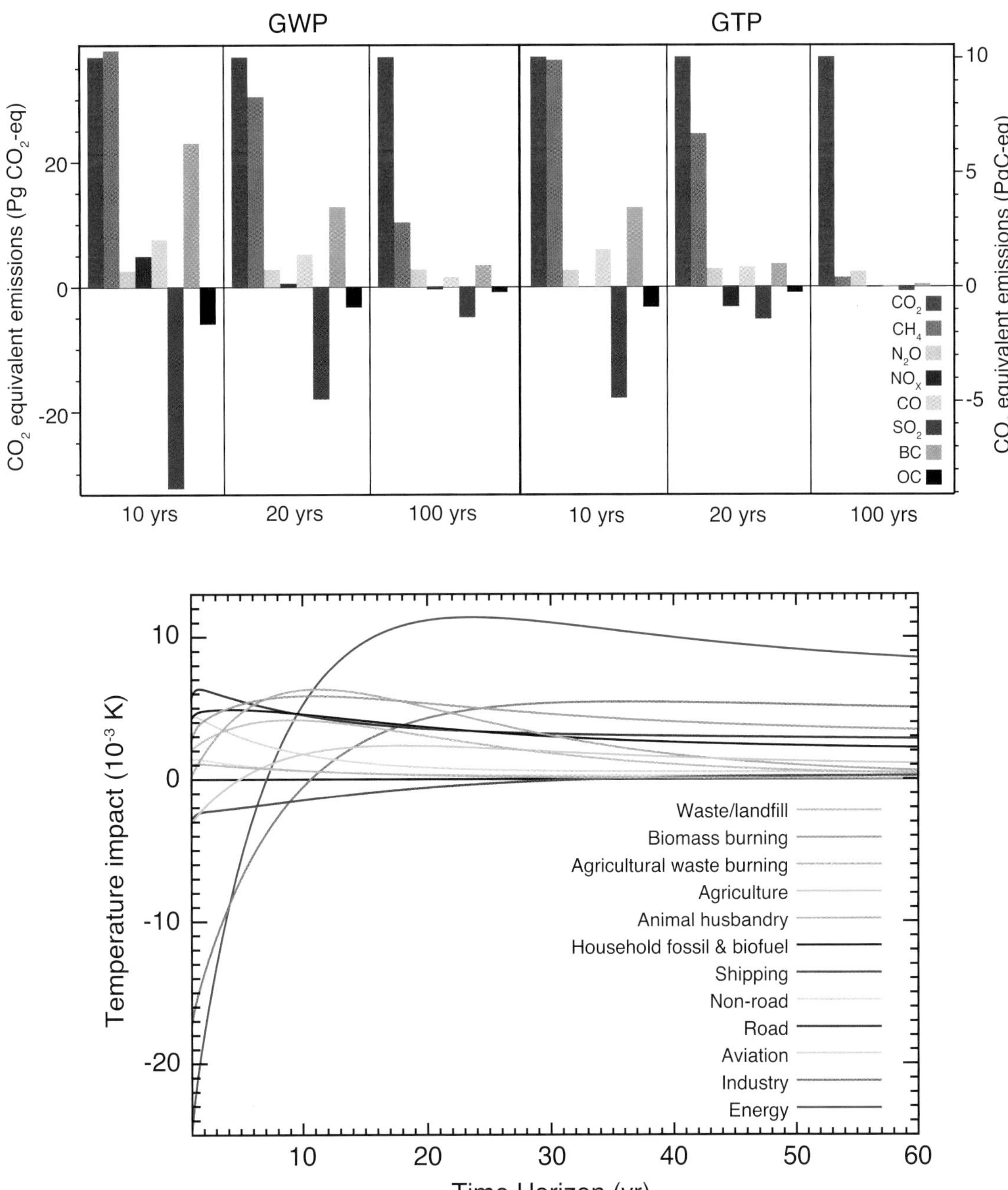

Figure TS.8 | (Upper) Global anthropogenic present-day emissions weighted by the Global Warming Potential (GWP) and the Global Temperature change Potential (GTP) for the chosen time horizons. Year 2008 (single-year pulse) emissions weighted by GWP, which is the global mean radiative forcing (RF) per unit mass emitted integrated over the indicated number of years relative to the forcing from CO_2 emissions, and GTP which estimates the impact on global mean temperature based on the temporal evolution of both RF and climate response per unit mass emitted relative to the impact of CO_2 emissions. The units are 'CO_2 equivalents', which reflects equivalence only in the impact parameter of the chosen metric (integrated RF over the chosen time horizon for GWP; temperature change at the chosen point in time for GTP), given as Pg(CO_2)eq (left axis) and PgCeq (right axis). (Bottom) The Absolute GTP (AGTP) as a function of time multiplied by the present-day emissions of all compounds from the indicated sectors is used to estimate global mean temperature response (AGTP is the same as GTP, except is not normalized by the impact of CO_2 emissions). There is little change in the relative values for the sectors over the 60 to 100-year time horizon. The effects of aerosol–cloud interactions and contrail-induced cirrus are not included in the upper panel. {Figures 8.32, 8.33}

TS.4 Understanding the Climate System and Its Recent Changes

TS.4.1 Introduction

Understanding of the climate system results from combining observations, theoretical studies of feedback processes and model simulations. Compared to AR4, more detailed observations and improved climate models (see Box TS.4) now enable the attribution of detected changes to human influences in more climate system components. The consistency of observed and modelled changes across the climate system, including in regional temperatures, the water cycle, global energy budget, cryosphere and oceans (including ocean acidification), points to global climate change resulting primarily from anthropogenic increases in WMGHG concentrations. {10}

TS.4.2 Surface Temperature

Several advances since the AR4 have allowed a more robust quantification of human influence on surface temperature changes. Observational uncertainty has been explored much more thoroughly than previously and the assessment now considers observations from the first decade of the 21st century and simulations from a new generation of climate models whose ability to simulate historical climate has improved in many respects relative to the previous generation of models considered in AR4. Observed GMST anomalies relative to 1880–1919 in recent years lie well outside the range of GMST anomalies in CMIP5 simulations with natural forcing only, but are consistent with the ensemble of CMIP5 simulations including both anthropogenic and natural forcing (Figure TS.9) even though some individual models overestimate the warming trend, while others underestimate it. Simulations with WMGHG changes only, and no aerosol changes, generally exhibit stronger warming than has been observed (Figure TS.9). Observed temperature trends over the period 1951–2010, which are characterized by warming over most of the globe with the most intense warming over the NH continents, are, at most observed locations, consistent with the temperature trends in CMIP5 simulations including anthropogenic and natural forcings and inconsistent with the temperature trends in CMIP5 simulations including natural forcings only. A number of studies have investigated the effects of the Atlantic Multi-decadal Oscillation (AMO) on GMST. Although some studies find a significant role for the AMO in driving multi-decadal variability in GMST, the AMO exhibited little trend over the period 1951–2010 on which the current assessments are based, and the AMO is assessed with *high confidence* to have made little contribution to the GMST trend between 1951 and 2010 (considerably less than 0.1°C). {2.4, 9.8.1, 10.3; FAQ 9.1}

It is *extremely likely* that human activities caused more than half of the observed increase in global average surface temperature from 1951 to 2010. This assessment is supported by robust evidence from multiple studies using different methods. In particular, the temperature trend attributable to all anthropogenic forcings combined can be more closely constrained in multi-signal detection and attribution analyses. Uncertainties in forcings and in climate models' responses to those forcings, together with difficulty in distinguishing the patterns of temperature response due to WMGHGs and other anthropogenic forcings, prevent as precise a quantification of the temperature changes attributable to

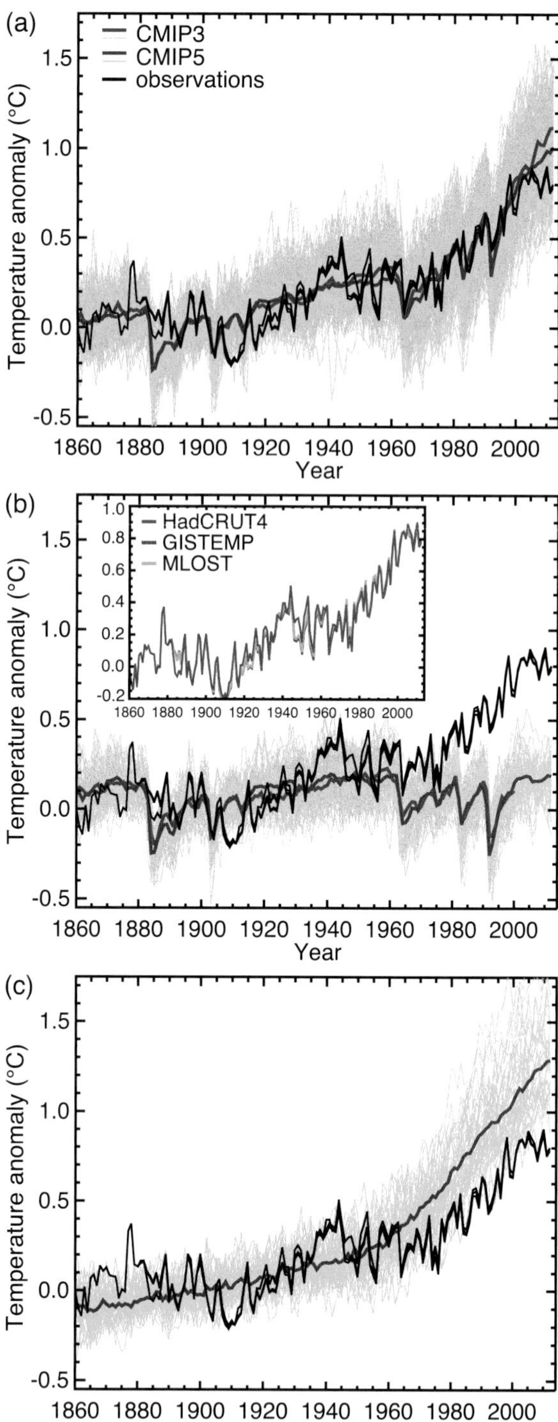

Figure TS.9 | Three observational estimates of global mean surface temperature (black lines) from the Hadley Centre/Climatic Research Unit gridded surface temperature data set 4 (HadCRUT4), Goddard Institute for Space Studies Surface Temperature Analysis (GISTEMP), and Merged Land–Ocean Surface Temperature Analysis (MLOST), compared to model simulations (CMIP3 models— thin blue lines and CMIP5 models—thin yellow lines) with anthropogenic and natural forcings (a), natural forcings only (b) and greenhouse gas forcing only (c). Thick red and blue lines are averages across all available CMIP5 and CMIP3 simulations respectively. All simulated and observed data were masked using the HadCRUT4 coverage (as this data set has the most restricted spatial coverage), and global average anomalies are shown with respect to 1880–1919, where all data are first calculated as anomalies relative to 1961–1990 in each grid box. Inset to (b) shows the three observational data sets distinguished by different colours. {Figure 10.1}

Box TS.3 | Climate Models and the Hiatus in Global Mean Surface Warming of the Past 15 Years

The observed GMST has shown a much smaller increasing linear trend over the past 15 years than over the past 30 to 60 years (Box TS.3, Figure 1a, c). Depending on the observational data set, the GMST trend over 1998–2012 is estimated to be around one third to one half of the trend over 1951–2012. For example, in HadCRUT4 the trend is 0.04°C per decade over 1998–2012, compared to 0.11°C per decade over 1951–2012. The reduction in observed GMST trend is most marked in NH winter. Even with this 'hiatus' in GMST trend, the decade of the 2000s has been the warmest in the instrumental record of GMST. Nevertheless, the occurrence of the hiatus in GMST trend during the past 15 years raises the two related questions of what has caused it and whether climate models are able to reproduce it. {2.4.3, 9.4.1; Box 9.2; Table 2.7}

Fifteen-year-long hiatus periods are common in both the observed and CMIP5 historical GMST time series. However, an analysis of the full suite of CMIP5 historical simulations (augmented for the period 2006–2012 by RCP4.5 simulations) reveals that 111 out of 114 realizations show a GMST trend over 1998–2012 that is higher than the entire HadCRUT4 trend ensemble (Box TS.3, Figure 1a; CMIP5 ensemble mean trend is 0.21°C per decade). This difference between simulated and observed trends could be caused by some combination of (a) internal climate variability, (b) missing or incorrect RF, and (c) model response error. These potential sources of the difference, which are not mutually exclusive, are assessed below, as is the cause of the observed GMST trend hiatus. {2.4.3, 9.3.2, 9.4.1; Box 9.2}

TS

Internal Climate Variability

Hiatus periods of 10 to 15 years can arise as a manifestation of internal decadal climate variability, which sometimes enhances and sometimes counteracts the long-term externally forced trend. Internal variability thus diminishes the relevance of trends over periods as short as 10 to 15 years for long-term climate change. Furthermore, the timing of internal decadal climate variability is not expected to be matched by the CMIP5 historical simulations, owing to the predictability horizon of at most 10 to 20 years (CMIP5 historical simulations are typically started around nominally 1850 from a control run). However, climate models exhibit individual decades of GMST trend hiatus even during a prolonged phase of energy uptake of the climate system, in which case the energy budget would be balanced by increasing subsurface–ocean heat uptake. {2.4.3, 9.3.2, 11.2.2; Boxes 2.2, 9.2}

Owing to sampling limitations, it is uncertain whether an increase in the rate of subsurface–ocean heat uptake occurred during the past 15 years. However, it is *very likely* that the climate system, including the ocean below 700 m depth, has continued to accumulate energy over the period 1998–2010. Consistent with this energy accumulation, GMSL has continued to rise during 1998–2012, at a rate only slightly and insignificantly lower than during 1993–2012. The consistency between observed heat content and sea level changes yields *high confidence* in the assessment of continued ocean energy accumulation, which is in turn consistent with the positive radiative imbalance of the climate system. By contrast, there is limited evidence that the hiatus in GMST trend has been accompanied by a slower rate of increase in ocean heat content over the depth range 0 to 700 m, when comparing the period 2003–2010 against 1971–2010. There is low agreement on this slowdown, as three of five analyses show a slowdown in the rate of increase while the other two show the increase continuing unabated. {3.2.3, 3.2.4, 3.7, 8.5.1, 13.3; Boxes 3.1, 13.1}

During the 15-year period beginning in 1998, the ensemble of HadCRUT4 GMST trends lies below almost all model-simulated trends (Box TS.3, Figure 1a), whereas during the 15-year period ending in 1998, it lies above 93 out of 114 modelled trends (Box TS.3, Figure 1b; HadCRUT4 ensemble mean trend 0.26°C per decade, CMIP5 ensemble mean trend 0.16°C per decade). Over the 62-year period 1951–2012, observed and CMIP5 ensemble mean trend agree to within 0.02°C per decade (Box TS.3, Figure 1c; CMIP5 ensemble mean trend 0.13°C per decade). There is hence *very high confidence* that the CMIP5 models show long-term GMST trends consistent with observations, despite the disagreement over the most recent 15-year period. Due to internal climate variability, in any given 15-year period the observed GMST trend sometimes lies near one end of a model ensemble, an effect that is pronounced in Box TS.3, Figure 1a, b as GMST was influenced by a very strong El Niño event in 1998. {Box 9.2}

Unlike the CMIP5 historical simulations referred to above, some CMIP5 predictions were initialized from the observed climate state during the late 1990s and the early 21st century. There is medium evidence that these initialized predictions show a GMST lower by about 0.05°C to 0.1°C compared to the historical (uninitialized) simulations and maintain this lower GMST during the first few years of the simulation. In some initialized models this lower GMST occurs in part because they correctly simulate a shift, around 2000, from a positive to a negative phase of the Inter-decadal Pacific Oscillation (IPO). However, the improvement of this phasing of the IPO through initialization is not universal across the CMIP5 predictions. Moreover, although part of the GMST reduction through initialization indeed results from initializing at the correct phase of internal variability, another part may result from correcting a model bias that was caused by incorrect past forcing or incorrect model response to past forcing, especially in the ocean. The relative magnitudes of these effects are at present unknown; moreover, the quality of a forecasting system cannot be evaluated from a single prediction (here, a 10-year prediction within

(continued on next page)

Box TS.3 (continued)

the period 1998–2012). Overall, there is *medium confidence* that initialization leads to simulations of GMST during 1998–2012 that are more consistent with the observed trend hiatus than are the uninitialized CMIP5 historical simulations, and that the hiatus is in part a consequence of internal variability that is predictable on the multi-year time scale. {11.1, 11.2.3; Boxes 2.5, 9.2, 11.1, 11.2}

Radiative Forcing

On decadal to interdecadal time scales and under continually increasing ERF, the forced component of the GMST trend responds to the ERF trend relatively rapidly and almost linearly (*medium confidence*). The expected forced-response GMST trend is related to the ERF trend by a factor that has been estimated for the 1% per year CO_2 increases in the CMIP5 ensemble as 2.0 [1.3 to 2.7] W m^{-2} °C^{-1} (90% uncertainty range). Hence, an ERF trend can be approximately converted to a forced-response GMST trend, permitting an assessment of how much of the change in the GMST trends shown in Box TS.3, Figure 1 is due to a change in ERF trend. {Box 9.2}

The AR5 best-estimate ERF trend over 1998–2011 is 0.22 [0.10 to 0.34] W m^{-2} per decade (90% uncertainty range), which is substantially lower than the trend over 1984–1998 (0.32 [0.22 to 0.42] W m^{-2} per decade; note that there was a strong volcanic eruption in 1982) and the trend over 1951–2011 (0.31 [0.19 to 0.40] W m^{-2} per decade; Box TS.3, Figure 1d–f; the end year 2011 is chosen because data availability is more limited than for GMST). The resulting forced-response GMST trend would approximately be 0.12 [0.05 to 0.29] °C per decade, 0.19 [0.09 to 0.39] °C per decade, and 0.18 [0.08 to 0.37] °C per decade for the periods 1998–2011, 1984–1998, and 1951–2011, respectively (the uncertainty ranges assume that the range of the conversion factor to GMST trend and the range of ERF trend itself are independent). The AR5 best-estimate ERF forcing trend difference between 1998–2011 and 1951–2011 thus might explain about one-half (0.05 °C per decade) of the observed GMST trend difference between these periods (0.06 to 0.08 °C per decade, depending on observational data set). {8.5.2}

The reduction in AR5 best-estimate ERF trend over 1998–2011 compared to both 1984–1998 and 1951–2011 is mostly due to decreasing trends in the natural forcings, –0.16 [–0.27 to –0.06] W m^{-2} per decade over 1998–2011 compared to 0.01 [–0.00 to +0.01] W m^{-2} per decade over 1951–2011. Solar forcing went from a relative maximum in 2000 to a relative minimum in 2009, with a peak-to-peak difference of around 0.15 W m^{-2} and a linear trend over 1998–2011 of around –0.10 W m^{-2} per decade. Furthermore, a series of small volcanic eruptions has increased the observed stratospheric aerosol loading after 2000, leading to an additional negative ERF linear-trend contribution of around –0.06 W m^{-2} per decade over 1998–2011 (Box TS.3, Figure 1d, f). By contrast, satellite-derived estimates of tropospheric aerosol optical depth suggests little overall trend in global mean aerosol optical depth over the last 10 years, implying little change in ERF due to aerosol–radiative interaction (*low confidence* because of *low confidence* in aerosol optical depth trend itself). Moreover, because there is only *low confidence* in estimates of ERF due to aerosol–cloud interaction, there is likewise *low confidence* in its trend over the last 15 years. {2.2.3, 8.4.2, 8.5.1, 8.5.2, 10.3.1; Box 10.2; Table 8.5}

For the periods 1984–1998 and 1951–2011, the CMIP5 ensemble mean ERF trend deviates from the AR5 best-estimate ERF trend by only 0.01 W m^{-2} per decade (Box TS.3, Figure 1e, f). After 1998, however, some contributions to a decreasing ERF trend are missing in the CMIP5 models, such as the increasing stratospheric aerosol loading after 2000 and the unusually low solar minimum in 2009. Nonetheless, over 1998–2011 the CMIP5 ensemble mean ERF trend is lower than the AR5 best-estimate ERF trend by 0.03 W m^{-2} per decade (Box TS.3, Figure 1d). Furthermore, global mean aerosol optical depth in the CMIP5 models shows little trend over 1998–2012, similar to the observations. Although the forcing uncertainties are substantial, there are no apparent incorrect or missing global mean forcings in the CMIP5 models over the last 15 years that could explain the model–observations difference during the warming hiatus. {9.4.6}

Model Response Error

The discrepancy between simulated and observed GMST trends during 1998–2012 could be explained in part by a tendency for some CMIP5 models to simulate stronger warming in response to increases in greenhouse-gas concentration than is consistent with observations. Averaged over the ensembles of models assessed in Section 10.3.1, the best-estimate GHG and other anthropogenic scaling factors are less than one (though not significantly so, Figure 10.4), indicating that the model-mean GHG and other anthropogenic responses should be scaled down to best match observations. This finding provides evidence that some CMIP5 models show a larger response to GHGs and other anthropogenic factors (dominated by the effects of aerosols) than the real world (*medium confidence*). As a consequence, it is argued in Chapter 11 that near-term model projections of GMST increase should be scaled down by about 10%. This downward scaling is, however, not sufficient to explain the model mean overestimate of GMST trend over the hiatus period. {10.3.1, 11.3.6}

Another possible source of model error is the poor representation of water vapour in the upper atmosphere. It has been suggested that a reduction in stratospheric water vapour after 2000 caused a reduction in downward longwave radiation and hence a surface-cooling contribution, possibly missed by the models. However, this effect is assessed here to be small, because there was a recovery in stratospheric water vapour after 2005. {2.2.2, 9.4.1; Box 9.2} *(continued on next page)*

Box TS.3 (continued)

In summary, the observed recent warming hiatus, defined as the reduction in GMST trend during 1998–2012 as compared to the trend during 1951–2012, is attributable in roughly equal measure to a cooling contribution from internal variability and a reduced trend in external forcing (expert judgement, *medium confidence*). The forcing trend reduction is due primarily to a negative forcing trend from both volcanic eruptions and the downward phase of the solar cycle. However, there is *low confidence* in quantifying the role of forcing trend in causing the hiatus, because of uncertainty in the magnitude of the volcanic forcing trend and *low confidence* in the aerosol forcing trend. {Box 9.2}

Almost all CMIP5 historical simulations do not reproduce the observed recent warming hiatus. There is *medium confidence* that the GMST trend difference between models and observations during 1998–2012 is to a substantial degree caused by internal variability, with possible contributions from forcing error and some CMIP5 models overestimating the response to increasing GHG forcing. The CMIP5 model trend in ERF shows no apparent bias against the AR5 best estimate over 1998–2012. However, *confidence* in this assessment of CMIP5 ERF trend is *low*, primarily because of the uncertainties in model aerosol forcing and processes, which through spatial heterogeneity might well cause an undetected global mean ERF trend error even in the absence of a trend in the global mean aerosol loading. {Box 9.2}

The causes of both the observed GMST trend hiatus and of the model–observation GMST trend difference during 1998–2012 imply that, barring a major volcanic eruption, most 15-year GMST trends in the near-term future will be larger than during 1998–2012 (*high confidence*; see Section 11.3.6 for a full assessment of near-term projections of GMST). The reasons for this implication are fourfold: first, anthropogenic GHG concentrations are expected to rise further in all RCP scenarios; second, anthropogenic aerosol concentration is expected to decline in all RCP scenarios, and so is the resulting cooling effect; third, the trend in solar forcing is expected to be larger over most near-term 15-year periods than over 1998–2012 (*medium confidence*), because 1998–2012 contained the full downward phase of the solar cycle; and fourth, it is *more likely than not* that internal climate variability in the near term will enhance and not counteract the surface warming expected to arise from the increasing anthropogenic forcing. {Box 9.2}

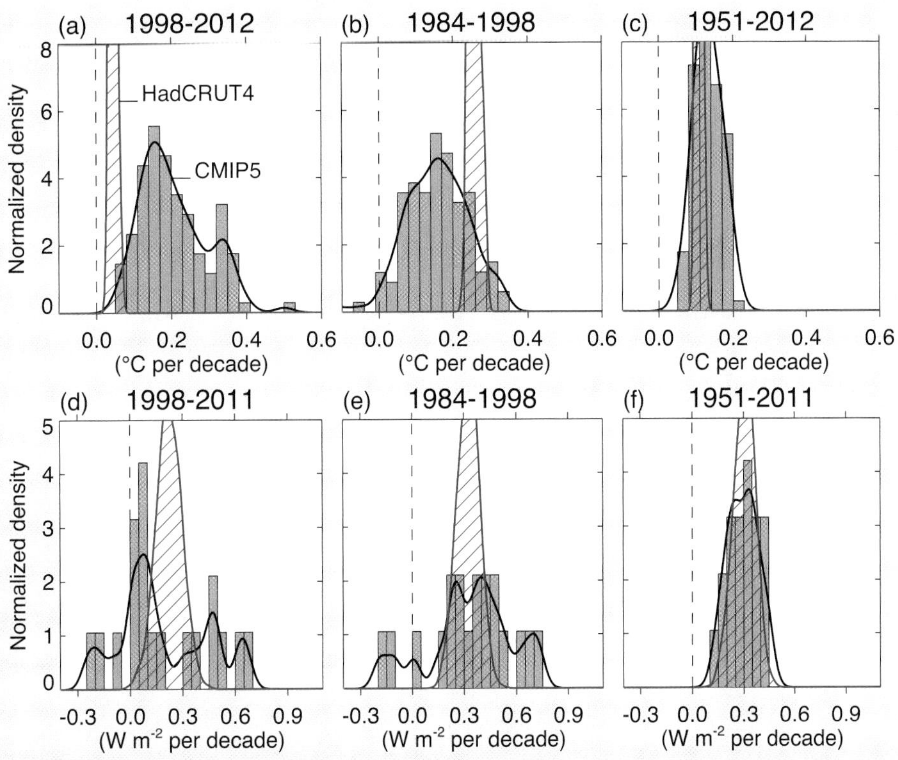

Box TS.3, Figure 1 | (Top) Observed and simulated GMST trends in °C per decade, over the periods 1998–2012 (a), 1984–1998 (b), and 1951–2012 (c). For the observations, 100 realizations of the Hadley Centre/Climatic Research Unit gridded surface temperature data set 4 (HadCRUT4) ensemble are shown (red, hatched). The uncertainty displayed by the ensemble width is that of the statistical construction of the global average only, in contrast to the trend uncertainties quoted in Section 2.4.3, which include an estimate of internal climate variability. Here, by contrast, internal variability is characterized through the width of the model ensemble. For the models, all 114 available CMIP5 historical realizations are shown, extended after 2005 with the RCP4.5 scenario and through 2012 (grey, shaded). (Bottom) Trends in effective radiative forcing (ERF, in W m^{-2} per decade) over the periods 1998–2011 (d), 1984–1998 (e), and 1951–2011 (f). The figure shows AR5 best-estimate ERF trends (red, hatched) and CMIP5 ERF (grey, shaded). Black lines are smoothed versions of the histograms. Each histogram is normalized so that its area sums up to one. {2.4.3, 8.5.2; Box 9.2; Figure 8.18; Box 9.2, Figure 1}

Thematic Focus Elements

TFE.3 | Comparing Projections from Previous IPCC Assessments with Observations

Verification of projections is arguably the most convincing way of establishing the credibility of climate change science. Results of projected changes in carbon dioxide (CO_2), global mean surface temperature (GMST) and global mean sea level (GMSL) from previous IPCC assessment reports are quantitatively compared with the best available observational estimates. The comparison between the four previous reports highlights the evolution in our understanding of how the climate system responds to changes in both natural and anthropogenic forcing and provides an assessment of how the projections compare with observational estimates. TFE.3, Figure 1, for example, shows the projected and observed estimates of: (1) CO_2 changes (top row), (2) GMST anomaly relative to 1961–1990 (middle row) and (3) GMSL relative to 1961–1990 (bottom row). Results from previous assessment reports are in the left-hand column, and for completeness results from current assessment are given in the right-hand column. {2.4, 3.7, 6.3, 11.3, 13.3} *(continued on next page)*

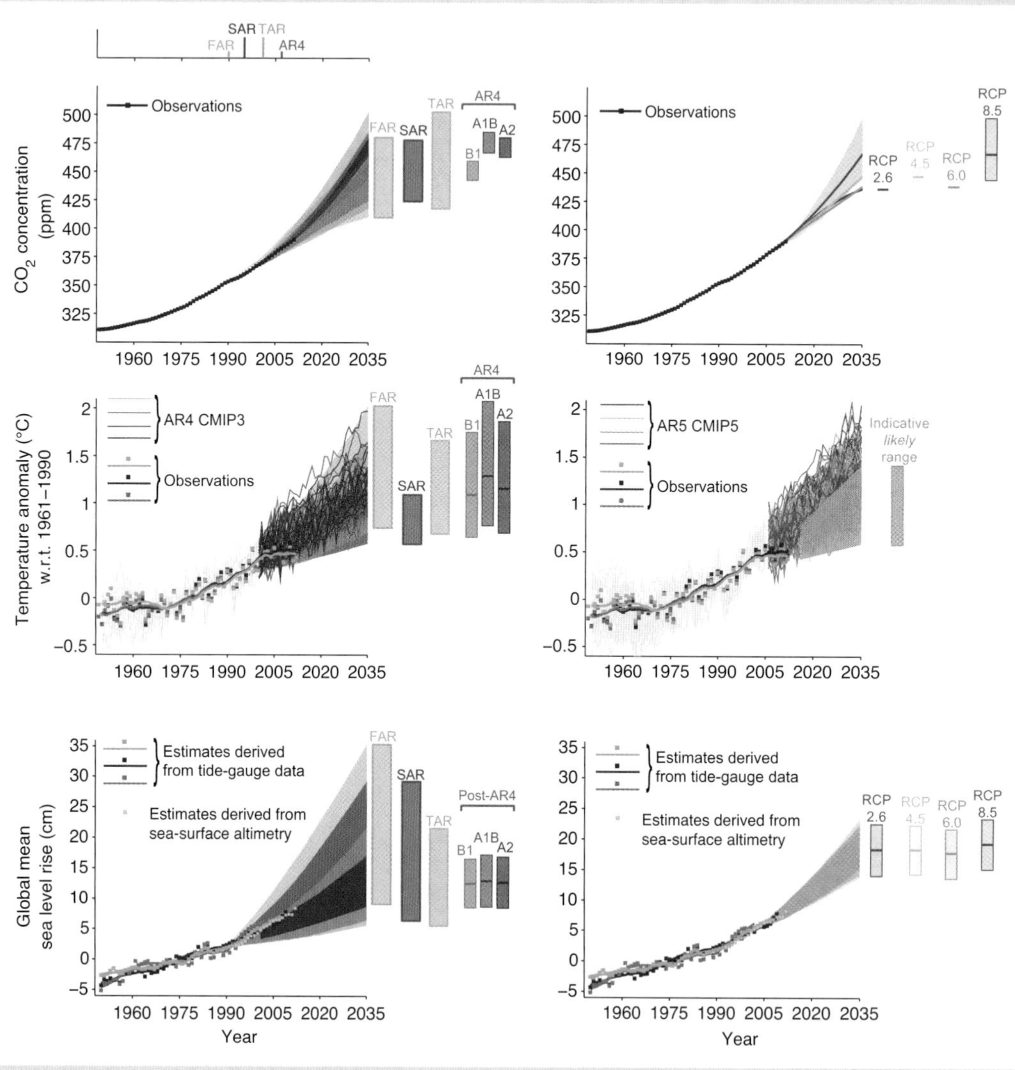

TFE.3, Figure 1 | (Top left) Observed globally and annually averaged CO_2 concentrations in parts per million (ppm) since 1950 compared with projections from the previous IPCC assessments. Observed global annual CO_2 concentrations are shown in dark blue. The shading shows the largest model projected range of global annual CO_2 concentrations from 1950 to 2035 from FAR (First Assessment Report; Figure A.3 in the Summary for Policymakers (SPM) of IPCC 1990), SAR (Second Assessment Report; Figure 5b in the TS of IPCC 1996), TAR (Third Assessment Report; Appendix II of IPCC 2001), and for the IPCC Special Report on Emission Scenarios (SRES) A2, A1B and B1 scenarios presented in the AR4 (Fourth Assessment Report; Figure 10.26). The publication years of the assessment reports are shown. (Top right) Same observed globally averaged CO_2 concentrations and the projections from this report. Only RCP8.5 has a range of values because the emission-driven senarios were carried out only for this RCP. For the other RCPs the best estimate is given. (Middle left) Estimated changes in the observed globally and annually averaged surface temperature anomaly relative to 1961–1990 (in °C) since 1950 compared with the range of projections from the previous IPCC assessments. Values are harmonized

TFE.3 (continued)

to start from the same value at 1990. Observed global annual temperature anomaly, relative to 1961–1990, from three data sets is shown as squares and smoothed time series as solid lines from the Hadley Centre/Climatic Research Unit gridded surface temperature data set 4 (HadCRUT4; bright green), Merged Land–Ocean Surface Temperature Analysis (MLOST; warm mustard) and Goddard Institute for Space Studies Surface Temperature Analysis (GISTEMP; dark blue) data sets. The coloured shading shows the projected range of global annual mean near surface temperature change from 1990 to 2035 for models used in FAR (Figure 6.11), SAR (Figure 19 in the TS of IPCC 1996), TAR (full range of TAR, Figure 9.13(b)). TAR results are based on the simple climate model analyses presented in this assessment and not on the individual full three-dimensional climate model simulations. For the AR4 results are presented as single model runs of the CMIP3 ensemble for the historical period from 1950 to 2000 (light grey lines) and for three SRES scenarios (A2, A1B and B1) from 2001 to 2035. For the three SRES scenarios the bars show the CMIP3 ensemble mean and the *likely* range given by –40 % to +60% of the mean as assessed in Chapter 10 of AR4. (Middle right) Projections of annual mean global mean surface air temperature (GMST) for 1950–2035 (anomalies relative to 1961–1990) under different RCPs from CMIP5 models (light grey and coloured lines, one ensemble member per model), and observational estimates the same as the middle left panel. The grey shaded region shows the indicative *likely* range for annual mean GMST during the period 2016–2035 for all RCPs (see Figure TS.14 for more details). The grey bar shows this same indicative *likely* range for the year 2035. (Bottom left) Estimated changes in the observed global annual mean sea level (GMSL) since 1950. Different estimates of changes in global annual sea level anomalies from tide gauge data (dark blue, warm mustard, dark green) and based on annual averages of altimeter data (light blue) starting in 1993 (the values have been aligned to fit the 1993 value of the tide gauge data). Squares indicate annual mean values, solid lines smoothed values. The shading shows the largest model projected range of global annual sea level rise from 1950 to 2035 for FAR (Figures 9.6 and 9.7), SAR (Figure 21 in TS of IPCC, 1996), TAR (Appendix II of IPCC, 2001) and based on the CMIP3 model results available at the time of AR4 using the SRES A1B scenario. Note that in the AR4 no full range was given for the sea level projections for this period. Therefore, the figure shows results that have been published subsequent to the AR4. The bars at the right hand side of each graph show the full range given for 2035 for each assessment report. (Bottom right) Same observational estimate as bottom left. The bars are the *likely* ranges (*medium confidence*) for global mean sea level rise at 2035 with respect to 1961–1990 following the four RCPs. Appendix 1.A provides details on the data and calculations used to create these figures. See Chapters 1, 11 and 13 for more details. {Figures 1.4, 1.5, 1.10, 11.9, 11.19, 11.25, 13.11}

Carbon Dioxide Changes

From 1950 to 2011 the observed concentrations of atmospheric CO_2 have steadily increased. Considering the period 1990–2011, the observed CO_2 concentration changes lie within the envelope of the scenarios used in the four assessment reports. As the most recent assessment prior to the current, the IPCC Fourth Assessment Report (AR4) (TFE.3.Figure 1; top left) has the narrowest scenario range and the observed concentration follows this range. The results from the IPCC Fifth Assessment Report (AR5) (TFE.3, Figure 1; top right) are consistent with AR4, and during 2002–2011, atmospheric CO_2 concentrations increased at a rate of 1.9 to 2.1 ppm yr^{-1}. {2.2.1, 6.3; Table 6.1}

Global Mean Temperature Anomaly

Relative to the 1961–1990 mean, the GMST anomaly has been positive and larger than 0.25°C since 2001. Observations are generally well within the range of the extent of the earlier IPCC projections (TFE.3, Figure1, middle left) This is also true for the Coupled Model Intercomparison Project Phase 5 (CMIP5) results (TFE.3, Figure 1; middle right) in the sense that the observed record lies within the range of the model projections, but on the lower end of the plume. Mt Pinatubo erupted in 1991 (see FAQ 11.2 for discussion of how volcanoes impact the climate system), leading to a brief period of relative global mean cooling during the early 1990s. The IPCC First, Second and Third Assessment Reports (FAR, SAR and TAR) did not include the effects of volcanic eruptions and thus failed to include the cooling associated with the Pinatubo eruption. AR4 and AR5, however, did include the effects from volcanoes and did simulate successfully the associated cooling. During 1995–2000 the global mean temperature anomaly was quite variable—a significant fraction of this variability was due to the large El Niño in 1997–1998 and the strong back-to-back La Niñas in 1999–2001. The projections associated with these assessment reports do not attempt to capture the actual evolution of these El Niño and La Niña events, but include them as a source of uncertainty due to natural variability as encompassed by, for example, the range given by the individual CMIP3 and CMIP5 simulations and projection (TFE.3, Figure 1). The grey wedge in TFE.3, Figure 1 (middle right) corresponds to the indicative *likely* range for annual temperatures, which is determined from the Representative Concentration Pathways (RCPs) assessed value for the 20-year mean 2016–2035 (see discussion of Figure TS.14 and Section 11.3.6 for details). From 1998 to 2012 the observational estimates have largely been on the low end of the range given by the scenarios alone in previous assessment reports and CMIP3 and CMIP5 projections. {2.4; Box 9.2}

Global Mean Sea Level

Based on both tide gauge and satellite altimetry data, relative to 1961–1990, the GMSL has continued to rise. While the increase is fairly steady, both observational records show short periods of either no change or a slight decrease. The observed estimates lie within the envelope of all the projections except perhaps in the very early 1990s. The sea level rise uncertainty due to scenario-related uncertainty is smallest for the most recent assessments (AR4 and AR5) and observed estimates lie well within this scenario-related uncertainty. It is *virtually certain* that over the 20th century sea level rose. The mean rate of sea level increase was 1.7 mm yr^{-1} with a *very likely* range between 1.5 to 1.9 between 1901 and 2010 and this rate increased to 3.2 with a *likely* range of 2.8 to 3.6 mm yr^{-1} between 1993 and 2010 (see TFE.2). {3.7.2, 3.7.4}

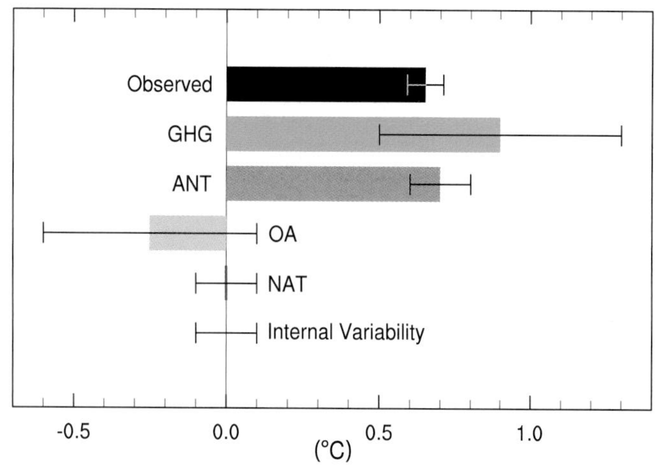

Figure TS.10 | Assessed *likely* ranges (whiskers) and their midpoints (bars) for warming trends over the 1951–2010 period due to well-mixed greenhouse gases (GHG), anthropogenic forcings (ANT) anthropogenic forcings other than well-mixed greenhouse gases (OA), natural forcings (NAT) and internal variability. The trend in the Hadley Centre/Climatic Research Unit gridded surface temperature data set 4 (HadCRUT4) observations is shown in black with its 5 to 95% uncertainty range due only to observational uncertainty in this record. {Figure 10.5}

WMGHGs and other anthropogenic forcings individually. Consistent with AR4, it is assessed that more than half of the observed increase in global average surface temperature from 1951 to 2010 is *very likely* due to the observed anthropogenic increase in WMGHG concentrations. WMGHGs contributed a global mean surface warming *likely* to be between 0.5°C and 1.3°C over the period between 1951 and 2010, with the contributions from other anthropogenic forcings *likely* to be between –0.6°C and 0.1°C and from natural forcings *likely* to be between –0.1°C and 0.1°C. Together these assessed contributions are consistent with the observed warming of approximately 0.6°C over this period (Figure TS.10). {10.3}

Solar forcing is the only known natural forcing acting to warm the climate over the 1951–2010 period but it has increased much less than WMGHG forcing, and the observed pattern of long-term tropospheric warming and stratospheric cooling is not consistent with the expected response to solar irradiance variations. Considering this evidence together with the assessed contribution of natural forcings to observed trends over this period, it is assessed that the contribution from solar forcing to the observed global warming since 1951 is *extremely unlikely* to be larger than that from WMGHGs. Because solar forcing has *very likely* decreased over a period with direct satellite measurements of solar output from 1986 to 2008, there is *high confidence* that changes in total solar irradiance have not contributed to global warming during that period. However, there is *medium confidence* that the 11-year cycle of solar variability influences decadal climate fluctuations in some regions through amplifying mechanisms. {8.4, 10.3; Box 10.2}

Observed warming over the past 60 years is far outside the range of internal climate variability estimated from pre-instrumental data, and it is also far outside the range of internal variability simulated in climate models. Model-based simulations of internal variability are assessed to be adequate to make this assessment. Further, the spatial pattern of observed warming differs from those associated with internal variability. Based on this evidence, the contribution of internal variability to the 1951–2010 GMST trend was assessed to be *likely* between –0.1°C and 0.1°C, and it is *virtually certain* that warming since 1951 cannot be explained by internal variability alone. {9.5, 10.3, 10.7}

The instrumental record shows a pronounced warming during the first half of the 20th century. Consistent with AR4, it is assessed that the early 20th century warming is *very unlikely* to be due to internal variability alone. It remains difficult to quantify the contributions to this early century warming from internal variability, natural forcing and anthropogenic forcing, due to forcing and response uncertainties and incomplete observational coverage. {10.3}

TS.4.3 Atmospheric Temperature

A number of studies since the AR4 have investigated the consistency of simulated and observed trends in free tropospheric temperatures (see section TS.2). Most, though not all, CMIP3 and CMIP5 models overestimate the observed warming trend in the tropical troposphere during the satellite period 1979–2012. Roughly one half to two thirds of this difference from the observed trend is due to an overestimate of the SST trend, which is propagated upward because models attempt to maintain static stability. There is *low confidence* in these assessments, however, owing to the *low confidence* in observed tropical tropospheric trend rates and vertical structure. Outside the tropics, and over the period of the radiosonde record beginning in 1961, the discrepancy between simulated and observed trends is smaller. {2.4.4, 9.4, 10.3}

Analysis of both radiosonde and satellite data sets, combined with CMIP5 and CMIP3 simulations, continues to find that observed tropospheric warming is inconsistent with internal variability and simulations of the response to natural forcings alone. Over the period 1961–2010 CMIP5 models simulate tropospheric warming driven by WMGHG changes, with only a small offsetting cooling due to the combined effects of changes in reflecting and absorbing aerosols and tropospheric ozone. Taking this evidence together with the results of multi-signal detection and attribution analyses, it is *likely* that anthropogenic forcings, dominated by WMGHGs, have contributed to the warming of the troposphere since 1961. Uncertainties in radiosonde and satellite records makes assessment of causes of observed trends in the upper troposphere less confident than an assessment of the overall atmospheric temperature changes. {2.4.4, 9.4, 10.3}

CMIP5 simulations including WMGHGs, ozone and natural forcing changes broadly reproduce the observed evolution of lower stratospheric temperature, with some tendency to underestimate the observed cooling trend over the satellite era (see Section TS.2). New studies of stratospheric temperature, considering the responses to natural forcings, WMGHGs and ozone-depleting substances, demonstrate that it is *very likely* that anthropogenic forcings, dominated by the depletion of the ozone layer due to ozone depleting substances have contributed to the cooling of the lower stratosphere since 1979. CMIP5 models simulate only a very weak cooling of the lower stratosphere in response to historical WMGHG changes, and the influence of WMGHGs on lower stratospheric temperature has not been formally detected. Considering both regions together, it is *very likely* that anthropogenic

Thematic Focus Elements

TFE.4 | The Changing Energy Budget of the Global Climate System

The global energy budget is a fundamental aspect of the Earth's climate system and depends on many phenomena within it. The ocean has stored about 93% of the increase in energy in the climate system over recent decades, resulting in ocean thermal expansion and hence sea level rise. The rate of storage of energy in the Earth system must be equal to the net downward radiative flux at the top of the atmosphere, which is the difference between effective radiative forcing (ERF) due to changes imposed on the system and the radiative response of the system. There are also significant transfers of energy between components of the climate system and from one location to another. The focus here is on the Earth's global energy budget since 1970, when better global observational data coverage is available. {3.7, 9.4, 13.4; Box 3.1}

The ERF of the climate system has been positive as a result of increases in well-mixed (long-lived) greenhouse gas (GHG) concentrations, changes in short-lived GHGs (tropospheric and stratospheric ozone and stratospheric water vapour), and an increase in solar irradiance (TFE.4, Figure 1a). This has been partly compensated by a negative contribution to the ERF of the climate system as a result of changes in tropospheric aerosol, which predominantly reflect sunlight and furthermore enhance the brightness of clouds, although black carbon produces positive forcing. Explosive volcanic eruptions (such as El Chichón in Mexico in 1982 and Mt Pinatubo in the Philippines in 1991)

(continued on next page)

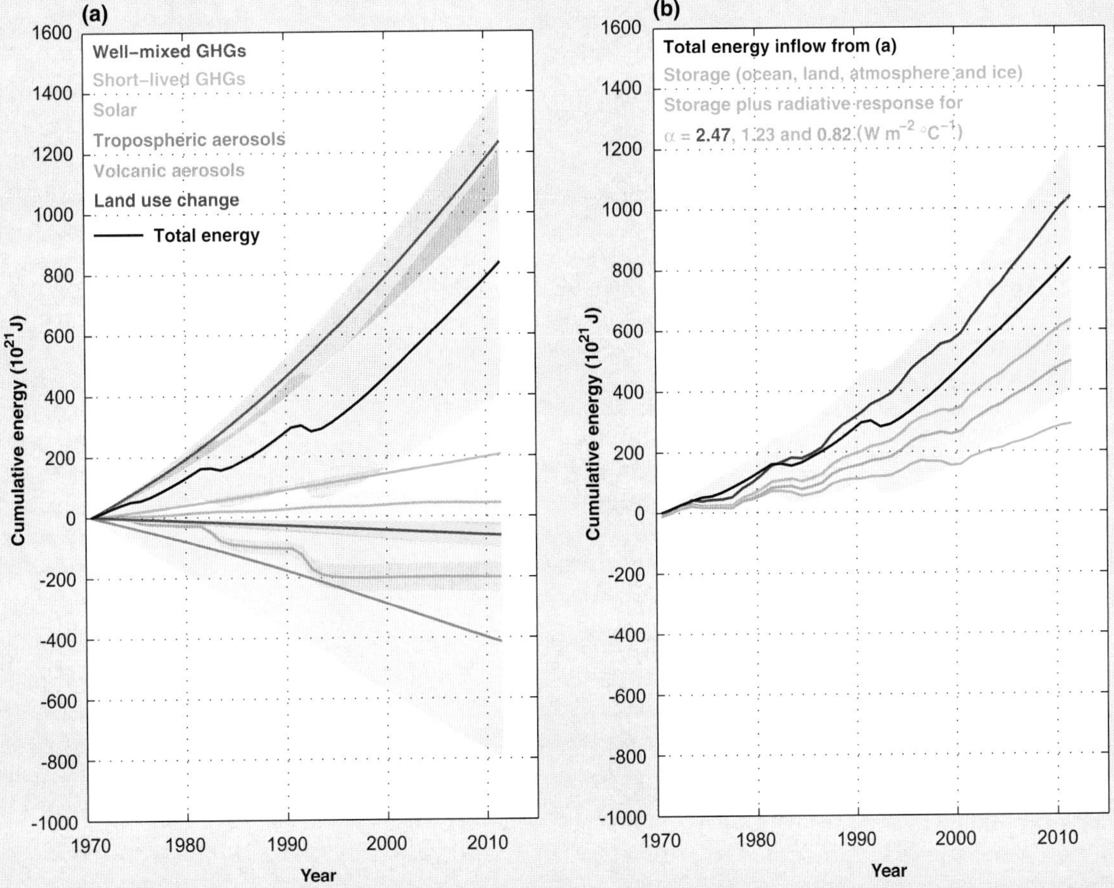

TFE.4, Figure 1 | The Earth's energy budget from 1970 through 2011. (a) The cumulative energy inflow into the Earth system from changes in well-mixed and short-lived greenhouse gases, solar forcing, tropospheric aerosol forcing, volcanic forcing and changes in surface albedo due to land use change (all relative to 1860–1879) are shown by the coloured lines; these contributions are added to give the total energy inflow (black; contributions from black carbon on snow and contrails as well as contrail-induced cirrus are included but not shown separately). (b) The cumulative total energy inflow from (a, black) is balanced by the sum of the energy uptake of the Earth system (blue; energy absorbed in warming the ocean, the atmosphere and the land, as well as in the melting of ice) and an increase in outgoing radiation inferred from changes in the global mean surface temperature. The sum of these two terms is given for a climate feedback parameter α of 2.47, 1.23 and 0.82 W m^{-2} °C^{-1}, corresponding to an equilibrium climate sensitivity of 1.5°C, 3.0°C and 4.5°C, respectively; 1.5°C to 4.5°C is assessed to be the *likely* range of equilibrium climate sensitivity. The energy budget would be closed for a particular value of α if the corresponding line coincided with the total energy inflow. For clarity, all uncertainties (shading) shown are *likely* ranges. {Box 12.2; Box 13.1, Figure 1}

TS

TFE.4 (continued)

can inject sulphur dioxide into the stratosphere, giving rise to stratospheric aerosol, which persists for several years. Stratospheric aerosol reflects some of the incoming solar radiation and thus gives a negative forcing. Changes in surface albedo from land use change have also led to a greater reflection of shortwave radiation back to space and hence a negative forcing. Since 1970, the net ERF of the climate system has increased, and the integrated impact of these forcings is an energy inflow over this period (TFE.4, Figure 1a). {2.3, 8.5; Box 13.1}

As the climate system warms, energy is lost to space through increased outgoing radiation. This radiative response by the system is due predominantly to increased thermal radiation, but it is modified by climate feedbacks such as changes in water vapour, clouds and surface albedo, which affect both outgoing longwave and reflected shortwave radiation. The top of the atmosphere fluxes have been measured by the Earth Radiation Budget Experiment (ERBE) satellites from 1985 to 1999 and the Cloud and the Earth's Radiant Energy System (CERES) satellites from March 2000 to the present. The top of the atmosphere radiative flux measurements are highly precise, allowing identification of changes in the Earth's net energy budget from year to year within the ERBE and CERES missions, but the absolute calibration of the instruments is not sufficiently accurate to allow determination of the absolute top of the atmosphere energy flux or to provide continuity across missions. TFE.4, Figure 1b relates the cumulative total energy change of the Earth system to the change in energy storage and the cumulative outgoing radiation. Calculation of the latter is based on the observed global mean surface temperature multiplied by the climate feedback parameter α, which in turn is related to the equilibrium climate sensitivity. The mid-range value for α, 1.23 W m^{-2} °C^{-1}, corresponds to an ERF for a doubled carbon dioxide (CO_2) concentration of 3.7 [2.96 to 4.44] W m^{-2} combined with an equilibrium climate sensitivity of 3.0°C. The climate feedback parameter α is *likely* to be in the range from 0.82 to 2.47 W m^{-2} °C^{-1} (corresponding to the *likely* range in equilibrium climate sensitivity of 1.5°C to 4.5°C). {9.7.1; Box 12.2}

If ERF were fixed, the climate system would eventually warm sufficiently that the radiative response would balance the ERF, and there would be no further change in energy storage in the climate system. However, the forcing is increasing, and the ocean's large heat capacity means that the climate system is not in radiative equilibrium and its energy content is increasing (TFE.4, Figure 1b). This storage provides strong evidence of a changing climate. The majority of this additional heat is in the upper 700 m of the ocean, but there is also warming in the deep and abyssal ocean. The associated thermal expansion of the ocean has contributed about 40% of the observed sea level rise since 1970. A small amount of additional heat has been used to warm the continents, warm and melt glacial and sea ice and warm the atmosphere. {13.4.2; Boxes 3.1, 13.1}

In addition to these forced variations in the Earth's energy budget, there is also internal variability on decadal time scales. Observations and models indicate that, because of the comparatively small heat capacity of the atmosphere, a decade of steady or even decreasing surface temperature can occur in a warming world. Climate model simulations suggest that these periods are associated with a transfer of heat from the upper to the deeper ocean, of the order 0.1 W m^{-2}, with a near-steady or an increased radiation to space, again of the order 0.1 W m^{-2}. Although these natural fluctuations represent a large amount of heat, they are significantly smaller than the anthropogenic forcing of the Earth's energy budget, particularly on time scales of several decades or longer. {9.4; Boxes 9.2, 13.1}

The available independent estimates of ERF, of observed heat storage, and of surface warming combine to give an energy budget for the Earth that is consistent with the assessed *likely* range of equilibrium climate sensitivity to within estimated uncertainties (*high confidence*). Quantification of the terms in the Earth's energy budget and verification that these terms balance over recent decades provides strong evidence for our understanding of anthropogenic climate change. {Box 13.1}

forcing, particularly WMGHGs and stratospheric ozone depletion, has led to a detectable observed pattern of tropospheric warming and lower stratospheric cooling since 1961. {2.4, 9.4, 10.3}

TS.4.4 Oceans

The observed upper-ocean warming during the late 20th and early 21st centuries and its causes have been assessed more completely since AR4 using updated observations and more simulations (see Section TS.2.2). The long term trends and variability in the observations are most consistent with simulations of the response to both anthropogenic forcing and volcanic forcing. The anthropogenic fingerprint in observed upper-ocean warming, consisting of global mean and basin-scale pattern changes, has also been detected. This result is robust to a number of observational, model and methodological or structural uncertainties. It is *very likely* that anthropogenic forcings have made a substantial contribution to upper ocean warming (above 700 m) observed since the 1970s. This anthropogenic ocean warming has contributed to global sea level rise over this period through thermal expansion. {3.2.2, 3.2.3, 3.7.2, 10.4.1, 10.4.3; Box 3.1}

Observed surface salinity changes also suggest a change in the global water cycle has occurred (see TFE.1). The long-term trends show that there is a strong positive correlation between the mean climate of the surface salinity and the temporal changes of surface salinity from 1950 to 2000. This correlation shows an enhancement of the climatological salinity pattern—so fresh areas have become fresher and salty areas saltier. The strongest anthropogenic signals are in the tropics (30°S to 30°N) and the Western Pacific. The salinity contrast between the Pacific and Atlantic Oceans has also increased with significant contributions from anthropogenic forcing. {3.3, 10.3.2, 10.4.2; FAQ 3.2}

On a global scale, surface and subsurface salinity changes (1955–2004) over the upper 250 m of the water column do not match changes expected from natural variability but do match the modelled distribution of forced changes (WMGHGs and tropospheric aerosols). Natural external variability taken from the simulations with just the variations in solar and volcanic forcing does not match the observations at all, thus excluding the hypothesis that observed trends can be explained by just solar or volcanic variations. These lines of evidence and our understanding of the physical processes leads to the conclusion that it is *very likely* that anthropogenic forcings have made a discernible contribution to surface and subsurface oceanic salinity changes since the 1960s. {10.4.2; Table 10.1}

Oxygen is an important physical and biological tracer in the ocean. Global analyses of oxygen data from the 1960s to 1990s extend the spatial coverage from local to global scales and have been used in attribution studies with output from a limited range of Earth System Models (ESMs). It is concluded that there is *medium confidence* that the observed global pattern of decrease in dissolved oxygen in the oceans can be attributed in part to human influences. {3.8.3, 10.4.4; Table 10.1}

The observations show distinct trends for ocean acidification (which is observed to be between –0.0014 and –0.0024 pH units per year). There is *high confidence* that the pH of ocean surface seawater decreased by about 0.1 since the beginning of the industrial era as a consequence of the oceanic uptake of anthropogenic CO_2. {3.8.2, 10.4.4; Box 3.2; Table 10.1}

TS.4.5 Cryosphere

The reductions in Arctic sea ice extent and NH snow cover extent and widespread glacier retreat and increased surface melt of Greenland are all evidence of systematic changes in the cryosphere. All of these changes in the cryosphere have been linked to anthropogenic forcings. {4.2.2, 4.4–4.6, 10.5.1, 10.5.3; Table 10.1}

Attribution studies, comparing the seasonal evolution of Arctic sea ice extent from observations from the 1950s with that simulated by coupled model simulations, demonstrate that human influence on the sea ice extent changes can be robustly detected since the early 1990s. The anthropogenic signal is also detectable for individual months from May to December, suggesting that human influence, strongest in late summer, now also extends into colder seasons. From these simulations of sea ice and observed sea ice extent from the instrumental record with high agreement between studies, it is concluded that anthropogenic forcings are *very likely* to have contributed to Arctic sea ice loss since 1979 (Figure TS.12). {10.5.1}

For Antarctic sea ice extent, the shortness of the observed record and differences in simulated and observed variability preclude an assessment of whether or not the observed increase since 1979 is inconsistent with internal variability. Untangling the processes involved with trends and variability in Antarctica and surrounding waters remains complex and several studies are contradictory. In conclusion, there is *low confidence* in the scientific understanding of the observed increase in Antarctic sea ice extent since 1979, due to the large differences between sea ice simulations from CMIP5 models and to the incomplete and competing scientific explanations for the causes of change and *low confidence* in estimates of internal variability (Figure TS.12). {9.4.3, 10.5.1; Table 10.1}

The Greenland ice sheet shows recent major melting episodes in response to record temperatures relative to the 20th century associated with persistent shifts in early summer atmospheric circulation, and these shifts have become more pronounced since 2007. Although many Greenland instrumental records are relatively short (two decades), regional modelling and observations tell a consistent story of the response of Greenland temperatures and ice sheet runoff to shifts in regional atmospheric circulation associated with larger scale flow patterns and global temperature increases. Mass loss and melt is also occurring in Greenland through the intrusion of warm water into the major fjords containing glaciers such as Jacobshaven Glacier. It is *likely* that anthropogenic forcing has contributed to surface melting of the Greenland ice sheet since 1993. {10.5.2; Table 10.1}

Estimates of ice mass in Antarctica since 2000 show that the greatest losses are at the edges. An analysis of observations underneath a floating ice shelf off West Antarctica leads to the conclusion that ocean warming in this region and increased transport of heat by ocean circulation are largely responsible for accelerating melt rates. The observational record of Antarctic mass loss is short and the internal variability of the ice sheet is poorly understood. Due to a low level of scientific understanding there is *low confidence* in attributing the causes of the observed loss of mass from the Antarctic ice sheet since 1993. {3.2, 4.2, 4.4.3, 10.5.2}

The evidence for the retreat of glaciers due to warming and moisture change is now more complete than at the time of AR4. There is *high confidence* in the estimates of observed mass loss and the estimates of natural variations and internal variability from long-term glacier records. Based on these factors and our understanding of glacier response to climatic drivers there is *high confidence* that a substantial part of the mass loss of glaciers is *likely* due to human influence. It is *likely* that there has been an anthropogenic component to observed reductions in NH snow cover since 1970. {4.3.3, 10.5.2, 10.5.3; Table 10.1}

Thematic Focus Elements

TFE.5 | Irreversibility and Abrupt Change

A number of components or phenomena within the climate system have been proposed as potentially exhibiting threshold behaviour. Crossing such thresholds can lead to an abrupt or irreversible transition into a different state of the climate system or some of its components.

Abrupt climate change is defined in this IPCC Fifth Assessment Report (AR5) as a large-scale change in the climate system that takes place over a few decades or less, persists (or is anticipated to persist) for at least a few decades and causes substantial disruptions in human and natural systems. There is information on potential consequences of some abrupt changes, but in general there is *low confidence* and little consensus on the likelihood of such events over the 21st century. Examples of components susceptible to such abrupt change are the strength of the Atlantic Meridional Overturning Circulation (AMOC), clathrate methane release, tropical and boreal forest dieback, disappearance of summer sea ice in the Arctic Ocean, long-term drought and monsoonal circulation. {5.7, 6.4.7, 12.5.5; Table 12.4}

TS

A change is said to be *irreversible* if the recovery time scale from this state due to natural processes is significantly longer than the time it takes for the system to reach this perturbed state. Such behaviour may arise because the time scales for perturbations and recovery processes are different, or because climate change may persist due to the long residence time of a carbon dioxide (CO_2) perturbation in the atmosphere (see TFE.8). Whereas changes in Arctic Ocean summer sea ice extent, long-term droughts and monsoonal circulation are assessed to be reversible within years to decades, tropical or boreal forest dieback may be reversible only within centuries. Changes in clathrate methane and permafrost carbon release, Greenland and Antarctic ice sheet collapse may be irreversible during millennia after the causal perturbation. {5.8, 6.4.7, 12.5.5, 13.4.3, 13.4.4; Table 12.4}

Abrupt Climate Change Linked with AMOC

New transient climate model simulations have confirmed with *high confidence* that strong changes in the strength of the AMOC produce abrupt climate changes at global scale with magnitude and pattern resembling past glacial Dansgaard–Oeschger events and Heinrich stadials. Confidence in the link between changes in North Atlantic climate and low-latitude precipitation has increased since the IPCC Fourth Assessment Report (AR4). From new paleoclimate reconstructions and modelling studies, there is *very high confidence* that a reduced strength of the AMOC and the associated surface cooling in the North Atlantic region caused southward shifts of the Atlantic Intertropical Convergence Zone and affected the American (north and south), African and Asian monsoons. {5.7}

The interglacial mode of the AMOC can recover (*high confidence*) from a short-lived freshwater input into the subpolar North Atlantic. Approximately 8.2 ka, a sudden freshwater release occurred during the final stages of North America ice sheet melting. Paleoclimate observations and model results indicate, with *high confidence*, a marked reduction in the strength of the AMOC followed by a rapid recovery, within approximately 200 years after the perturbation. {5.8.2}

Although many more model simulations have been conducted since AR4 under a wide range of future forcing scenarios, projections of the AMOC behaviour have not changed. It remains *very likely* that the AMOC will weaken over the 21st century relative to 1850-1900 values. Best estimates and ranges for the reduction from the Coupled Model Intercomparison Project Phase 5 (CMIP5) are 11% (1 to 24%) for the Representative Concentration Pathway RCP2.6 and 34% (12 to 54%) for RCP8.5, but there is *low confidence* on the magnitude of weakening. It also remains *very unlikely* that the AMOC will undergo an abrupt transition or collapse in the 21st century for the scenarios considered (*high confidence*) (TFE.5, Figure 1). For an abrupt transition of the AMOC to occur, the sensitivity of the AMOC to forcing would have to be far greater than seen in current models, or would require meltwater flux from the Greenland ice sheet greatly exceeding even the highest of current projections. Although neither possibility can be excluded entirely, it is *unlikely* that the AMOC will collapse beyond the end of the 21st century for the scenarios considered, but a collapse beyond the 21st century for large sustained warming cannot be excluded. There is *low confidence* in assessing the evolution of AMOC beyond the 21st century because of limited number of analyses and equivocal results. {12.4.7, 12.5.5}

Potential Irreversibility of Changes in Permafrost, Methane Clathrates and Forests

In a warming climate, permafrost thawing may induce decomposition of carbon accumulated in frozen soils which could persist for hundreds to thousands of years, leading to an increase of atmospheric CO_2 and/or methane (CH_4)

(continued on next page)

TFE.5 (continued)

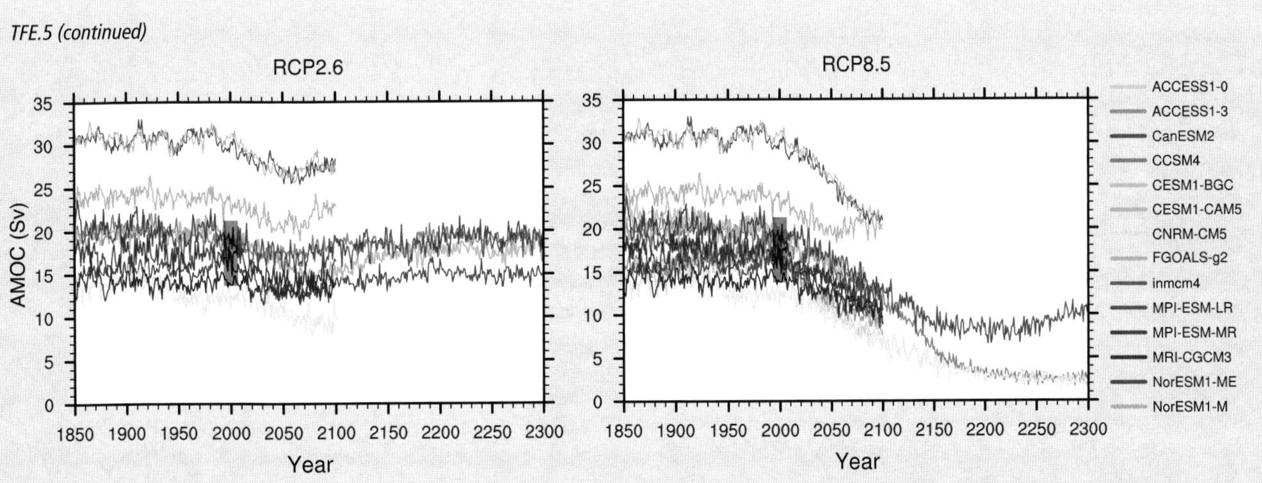

TFE.5, Figure 1 | Atlantic Meridional Overturning Circulation (AMOC) strength at 30°N (Sv) as a function of year, from 1850 to 2300 as simulated by different Atmosphere–Ocean General Circulation Models in response to scenario RCP2.6 (left) and RCP8.5 (right). The vertical black bar shows the range of AMOC strength measured at 26°N, from 2004 to 2011 {Figures 3.11, 12.35}

concentrations. The existing modelling studies of permafrost carbon balance under future warming that take into account at least some of the essential permafrost-related processes do not yield consistent results, beyond the fact that present-day permafrost will become a net emitter of carbon during the 21st century under plausible future warming scenarios (*low confidence*). This also reflects an insufficient understanding of the relevant soil processes during and after permafrost thaw, including processes leading to stabilization of unfrozen soil carbon, and precludes any quantitative assessment of the amplitude of irreversible changes in the climate system potentially related to permafrost degassing and associated feedbacks. {6.4.7, 12.5.5}

Anthropogenic warming will *very likely* lead to enhanced CH_4 emissions from both terrestrial and oceanic clathrates. Deposits of CH_4 clathrates below the sea floor are susceptible to destabilization via ocean warming. However, sea level rise due to changes in ocean mass enhances clathrate stability in the ocean. While difficult to formally assess, initial estimates of the 21st century feedback from CH_4 clathrate destabilization are small but not insignificant. It is *very unlikely* that CH_4 from clathrates will undergo catastrophic release during the 21st century (*high confidence*). On multi-millennial time scales, such CH_4 emissions may provide a positive feedback to anthropogenic warming and may be irreversible, due to the diffference between release and accumulation time scales. {6.4.7, 12.5.5}

The existence of critical climate change driven dieback thresholds in the Amazonian and other tropical rainforests purely driven by climate change remains highly uncertain. The possibility of a critical threshold being crossed in precipitation volume and duration of dry seasons cannot be ruled out. The response of boreal forest to projected climate change is also highly uncertain, and the existence of critical thresholds cannot at present be ruled out. There is *low confidence* in projections of the collapse of large areas of tropical and/or boreal forests. {12.5.5}

Potential Irreversibility of Changes in the Cryosphere

The reversibility of sea ice loss has been directly assessed in sensitivity studies to CO_2 increase and decrease with Atmosphere–Ocean General Circulation Models (AOGCMs) or Earth System Models (ESMs). None of them show evidence of an irreversible change in Arctic sea ice at any point. By contrast, as a result of the strong coupling between surface and deep waters in the Southern Ocean, the Antarctic sea ice in some models integrated with ramp-up and ramp-down atmospheric CO_2 concentration exhibits some hysteresis behaviour. {12.5.5}

At present, both the Greenland and Antarctic ice sheets have a positive surface mass balance (snowfall exceeds melting), although both are losing mass because ice outflow into the sea exceeds the net surface mass balance. A positive feedback operates to reduce ice sheet volume and extent when a decrease of the surface elevation of the ice sheet induces a decreased surface mass balance. This arises generally through increased surface melting, and therefore applies in the 21st century to Greenland, but not to Antarctica, where surface melting is currently very small. Surface melting in Antarctica is projected to become important after several centuries under high well-mixed greenhouse gas radiative forcing scenarios. {4.4, 13.4.4; Boxes 5.2, 13.2}

Abrupt change in ice sheet outflow to the sea may be caused by unstable retreat of the grounding line in regions where the bedrock is below sea level and slopes downwards towards the interior of the ice sheet. This mainly

(continued on next page)

TFE.5 (continued)

applies to West Antarctica, but also to parts of East Antarctica and Greenland. Grounding line retreat can be triggered by ice shelf decay, due to warmer ocean water under ice shelves enhancing submarine ice shelf melt, or melt water ponds on the surface of the ice shelf promoting ice shelf fracture. Because ice sheet growth is a slow process, such changes would be irreversible in the definition adopted here. {4.4.5; Box 13.2}

There is *high confidence* that the volumes of the Greenland and West Antarctic ice sheets were reduced during periods of the past few million years that were globally warmer than present. Ice sheet model simulations and geological data suggest that the West Antarctic ice sheet is very sensitive to subsurface ocean warming and imply with *medium confidence* a West Antarctic ice sheet retreat if atmospheric CO_2 concentration stays within, or above, the range of 350–450 ppm for several millennia. {5.8.1, 13.4.4; Box 13.2}

The available evidence indicates that global warming beyond a threshold would lead to the near-complete loss of the Greenland ice sheet over a millennium or longer, causing a global mean sea level rise of approximately 7 m. Studies with fixed present-day ice sheet topography indicate that the threshold is greater than 2°C but less than 4°C (*medium confidence*) of global mean surface temperature rise above pre-industrial. The one study with a dynamical ice sheet suggests the threshold is greater than about 1°C (*low confidence*) global mean warming with respect to pre-industrial. Considering the present state of scientific uncertainty, a *likely* range cannot be quantified. The complete loss of the Greenland ice sheet is not inevitable because this would take a millennium or more; if temperatures decline before the ice sheet has completely vanished, the ice sheet might regrow. However, some part of the mass loss might be irreversible, depending on the duration and degree of exceedance of the threshold, because the ice sheet may have multiple steady states, due to its interaction with regional climate. {13.4.3, 13.4.4}

TS.4.6 Water Cycle

Since the AR4, new evidence has emerged of a detectable human influence on several aspects of the water cycle. There is *medium confidence* that observed changes in near-surface specific humidity since 1973 contain a detectable anthropogenic component. The anthropogenic water vapour fingerprint simulated by an ensemble of climate models has been detected in lower tropospheric moisture content estimates derived from Special Sensor Microwave/Imager (SSM/I) data covering the period 1988–2006. An anthropogenic contribution to increases in tropospheric specific humidity is found with *medium confidence*. {2.5, 10.3}

Attribution studies of global zonal mean terrestrial precipitation and Arctic precipitation both find a detectable anthropogenic influence. Overall there is *medium confidence* in a significant human influence on global scale changes in precipitation patterns, including increases in NH mid-to-high latitudes. Remaining observational and modelling uncertainties and the large effect of internal variability on observed precipitation preclude a more confident assessment. {2.5, 7.6, 10.3}

Based on the collected evidence for attributable changes (with varying levels of confidence and likelihood) in specific humidity, terrestrial precipitation and ocean surface salinity through its connection to precipitation and evaporation, and from physical understanding of the water cycle, it is *likely* that human influence has affected the global water cycle since 1960. This is a major advance since AR4. {2.4, 2.5, 3.3, 9.4.1, 10.3, 10.4.2; Table 10.1; FAQ 3.2}

TS.4.7 Climate Extremes

Several new attribution studies have found a detectable anthropogenic influence in the observed increased frequency of warm days and nights and decreased frequency of cold days and nights. Since the AR4 and SREX, there is new evidence for detection of human influence on extremely warm daytime temperature and there is new evidence that the influence of anthropogenic forcing may be detected separately from the influence of natural forcing at global scales and in some continental and sub-continental regions. This strengthens the conclusions from both AR4 and SREX, and it is now *very likely* that anthropogenic forcing has contributed to the observed changes in the frequency and intensity of daily temperature extremes on the global scale since the mid-20th century. It is *likely* that human influence has significantly increased the probability of occurrence of heat waves in some locations. See TFE.9 and TFE.9, Table 1 for a summary of the assessment of extreme weather and climate events. {10.6}

Since the AR4, there is some new limited direct evidence for an anthropogenic influence on extreme precipitation, including a formal detection and attribution study and indirect evidence that extreme precipitation would be expected to have increased given the evidence of anthropogenic influence on various aspects of the global hydrological cycle and *high confidence* that the intensity of extreme precipitation events will increase with warming, at a rate well exceeding that of the mean precipitation. In land regions where observational coverage is sufficient for assessment, there is *medium confidence* that anthropogenic forcing has contributed to a global-scale intensification of heavy precipitation over the second half of the 20th century. {7.6, 10.6}

Globally, there is *low confidence* in attribution of changes in tropical cyclone activity to human influence. This is due to insufficient observational evidence, lack of physical understanding of the links between anthropogenic drivers of climate and tropical cyclone activity, and the low level of agreement between studies as to the relative importance of internal variability, and anthropogenic and natural forcings. In the North Atlantic region there is *medium confidence* that a reduction in aerosol forcing over the North Atlantic has contributed at least in part to the observed increase in tropical cyclone activity there since the 1970s. There remains substantial disagreement on the relative importance of internal variability, WMGHG forcing and aerosols for this observed trend. {2.6, 10.6, 14.6}

Although the AR4 concluded that it is *more likely than not* that anthropogenic influence has contributed to an increased risk of drought in the second half of the 20th century, an updated assessment of the observational evidence indicates that the AR4 conclusions regarding global increasing trends in hydrological droughts since the 1970s are no longer supported. Owing to the *low confidence* in observed large-scale trends in dryness combined with difficulties in distinguishing decadal-scale variability in drought from long-term climate change, there is now *low confidence* in the attribution of changes in drought over global land since the mid-20th century to human influence. {2.6, 10.6}

TS.4.8 From Global to Regional

Taking a longer term perspective shows the substantial role played by external forcings in driving climate variability on hemispheric scales in pre-industrial times (Box TS.5). It is *very unlikely* that NH temperature variations from 1400 to 1850 can be explained by internal variability alone. There is *medium confidence* that external forcing contributed to NH temperature variability from 850 to 1400 and that external forcing contributed to European temperature variations over the last 5 centuries. {5.3.3, 5.5.1, 10.7.2, 10.7.5; Table 10.1}

Changes in atmospheric circulation are important for local climate change because they could lead to greater or smaller changes in climate in a particular region than elsewhere. It is *likely* that human influence has altered sea level pressure patterns globally. There is *medium confidence* that stratospheric ozone depletion has contributed to the observed poleward shift of the southern Hadley Cell border during austral summer. It is *likely* that stratospheric ozone depletion has contributed to the positive trend in the SAM seen in austral summer since the mid-20th century which corresponds to sea level pressure reductions over the high latitudes and increase in the subtropics (Figure TS.11). {10.3}

The evidence is stronger that observed changes in the climate system can now be attributed to human activities on global and regional scales in many components (Figure TS.12). Observational uncertainty has been explored much more thoroughly than previously, and fingerprints of human influence have been deduced from a new generation of climate models. There is improved understanding of ocean changes, including salinity changes, that are consistent with large scale intensification of the water cycle predicted by climate models. The changes in near surface temperatures, free atmosphere temperatures, ocean temperatures and NH snow cover and sea ice extent, when taken together, show not just global mean changes, but also distinctive regional patterns consistent with the expected fingerprints of change from anthropogenic forcings and the expected responses from volcanic eruptions (Figure TS.12). {10.3–10.6, 10.9}

Human influence has been detected in nearly all of the major assessed components of the climate system (Figure TS.12). Taken together, the combined evidence increases the overall level of confidence in the attribution of observed climate change, and reduces the uncertainties associated with assessment based on a single climate variable. From this combined evidence it is *virtually certain* that human influence has warmed the global climate system. Anthropogenic influence has been identified in changes in temperature near the surface of the Earth, in the atmosphere and in the oceans, as well as in changes in the cryosphere, the water cycle and some extremes. There is strong evidence that excludes solar forcing, volcanoes and internal variability as the strongest drivers of warming since 1950. {10.9; Table 10.1; FAQ 5.1}

Over every continent except Antarctica, anthropogenic influence has *likely* made a substantial contribution to surface temperature increases since the mid-20th century (Figure TS.12). It is *likely* that there has been a significant anthropogenic contribution to the very substantial warming in Arctic land surface temperatures over the past 50 years. For Antarctica large observational uncertainties result in *low confidence* that anthropogenic influence has contributed to observed warming averaged over available stations. Detection and attribution at regional

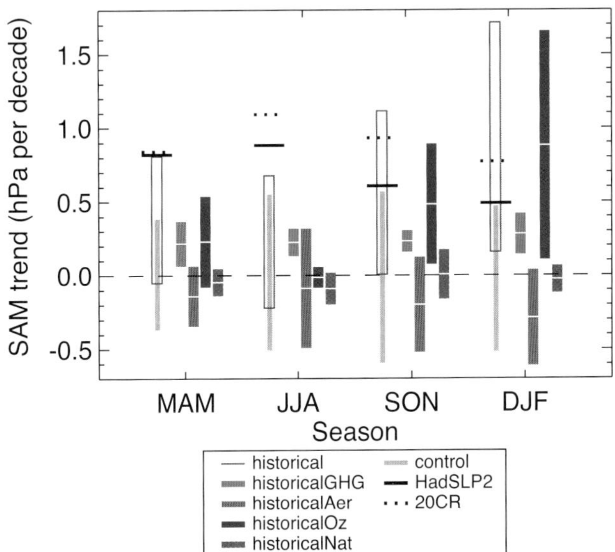

Figure TS.11 | Simulated and observed 1951–2011 trends in the Southern Annular Mode (SAM) index by season. The SAM index is a difference between zonal mean sea level pressure (SLP) at 40°S and 65°S. The SAM index is defined without normalization, so that the magnitudes of simulated and observed trends can be compared. Black lines show observed trends from the Hadley Centre Sea Level Pressure 2r (HadSLP2r) data set (solid), and the 20th Century Reanalysis (dotted). Grey bars show 5th to 95th percentile ranges of control trends, and red boxes show the 5th to 95th percentile range of trends in historical simulations including anthropogenic and natural forcings. Coloured bars show ensemble mean trends and their associated 5 to 95% confidence ranges simulated in response to well-mixed greenhouse gas (light green), aerosol (dark green), ozone (magenta) and natural forcing changes (blue) in CMIP5 individual-forcing simulations. {Figure 10.13b}

scales is complicated by the greater role played by dynamical factors (circulation changes), a greater range of forcings that may be regionally important, and the greater difficulty of modelling relevant processes at regional scales. Nevertheless, human influence has *likely* contributed to temperature increases in many sub-continental regions. {10.3; Box 5.1}

The coherence of observed changes with simulations of anthropogenic and natural forcing in the physical system is remarkable (Figure TS.12), particularly for temperature-related variables. Surface temperature and ocean heat content show emerging anthropogenic and natural signals in both records, and a clear separation from the alternative hypothesis of just natural variations. These signals do not appear just in the global means, but also appear at regional scales on continents and in ocean basins in each of these variables. Sea ice extent emerges clearly from the range of internal variability for the Arctic. At sub-continental scales human influence is *likely* to have substantially increased the probability of occurrence of heat waves in some locations. {Table 10.1}

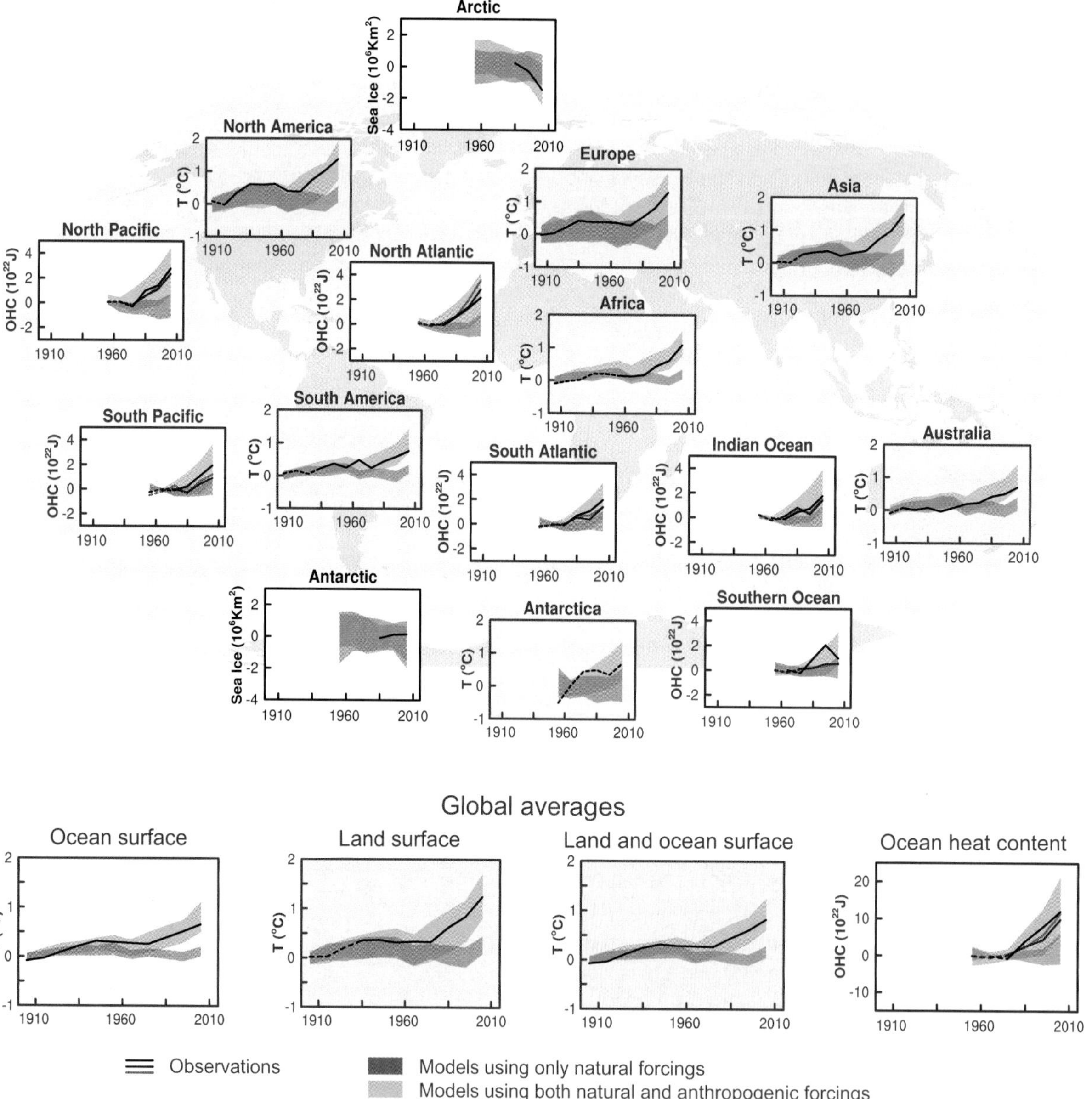

Figure TS.12 | Comparison of observed and simulated change in the climate system, at regional scales (top panels) and global scales (bottom four panels). Brown panels are land surface temperature time series, blue panels are ocean heat content time series and white panels are sea ice time series (decadal averages). Each panel shows observations (black or black and shades of grey), and the 5 to 95% range of the simulated response to natural forcings (blue shading) and natural and anthropogenic forcings (pink shading), together with the corresponding ensemble means (dark blue and dark red respectively). The observed surface temperature is from the Hadley Centre/Climatic Research Unit gridded surface temperature data set 4 (HadCRUT4). Three observed records of ocean heat content (OHC) are shown. Sea ice anomalies (rather than absolute values) are plotted and based on models in Figure 10.16. The observations lines are either solid or dashed and indicate the quality of the observations and estimates. For land and ocean surface temperatures panels and precipitation panels, solid observation lines indicate where spatial coverage of areas being examined is above 50% coverage and dashed observation lines where coverage is below 50%. For example, data coverage of Antarctica never goes above 50% of the land area of the continent. For ocean heat content and sea ice panels the solid observations line is where the coverage of data is good and higher in quality, and the dashed line is where the data coverage is only adequate. This figure is based on Figure 10.21 except presented as decadal averages rather than yearly averages. Further detail regarding the related Figure SPM.6 is given in the TS Supplementary Material. {Figure 10.21}

Box TS.4 | Model Evaluation

Climate models have continued to be improved since the AR4, and many models have been extended into Earth System Models (ESMs) by including the representation of biogeochemical cycles important to climate change. Box TS.4, Figure 1 provides a partial overview of model capabilities as assessed in this report, including improvements or lack thereof relative to models that were assessed in the AR4 or that were available at the time of the AR4. {9.1, 9.8.1; Box 9.1}

The ability of climate models to simulate surface temperature has improved in many, though not all, important aspects relative to the generation of models assessed in the AR4. There continues to be *very high confidence* that models reproduce the observed large-scale time-mean surface temperature patterns (pattern correlation of about 0.99), although systematic errors of several degrees Celsius are found in some regions. There is *high confidence* that on the regional scale (sub-continental and smaller), time-mean surface temperature is better simulated than at the time of the AR4; however, confidence in model capability is lower than for the large scale. Models are able to reproduce the magnitude of the observed global mean or northern-hemisphere-mean temperature variability on interannual to centennial time scales. Models are also able to reproduce the large-scale patterns of temperature during the Last Glacial Maximum indicating an ability to simulate a climate state much different from the present (see also Box TS.5). {9.4.1, 9.6.1}

There is *very high confidence* that models reproduce the general features of the global and annual mean surface temperature changes over the historical period, including the warming in the second half of the 20th century and the cooling immediately following large volcanic eruptions. Most simulations of the historical period do not reproduce the observed reduction in global mean surface warming trend over the last 10 to 15 years (see Box TS.3). There is *medium confidence* that the trend difference between models and observations during 1998–2012 is to a substantial degree caused by internal variability, with possible contributions from forcing inadequacies in models and some models overestimating the response to increasing greenhouse gas forcing. Most, though not all, models overestimate the observed warming trend in the tropical troposphere over the last 30 years, and tend to underestimate the long-term lower-stratospheric cooling trend. {9.4.1; Box 9.2}

The simulation of large-scale patterns of precipitation has improved somewhat since the AR4, although models continue to perform less well for precipitation than for surface temperature. The spatial pattern correlation between modelled and observed annual mean precipitation has increased from 0.77 for models available at the time of the AR4 to 0.82 for current models. At regional scales, precipitation is not simulated as well, and the assessment remains difficult owing to observational uncertainties. {9.4.1, 9.6.1}

Many models are able to reproduce the observed changes in upper-ocean heat content from 1961 to 2005. The time series of the multi-model mean falls within the range of the available observational estimates for most of the period. {9.4.2}

There is robust evidence that the downward trend in Arctic summer sea ice extent is better simulated than at the time of the AR4. About one quarter of the models show a trend as strong as, or stronger, than the trend in observations over the satellite era 1979–2012. Most models simulate a small decreasing trend in Antarctic sea ice extent, albeit with large inter-model spread, in contrast to the small increasing trend in observations. {9.4.3}

There has been substantial progress since the AR4 in the assessment of model simulations of extreme events. Changes in the frequency of extreme warm and cold days and nights over the second half of the 20th century are consistent between models and observations, with the ensemble mean global mean time series generally falling within the range of observational estimates. The majority of models underestimate the sensitivity of extreme precipitation to temperature variability or trends, especially in the tropics. {9.5.4}

In the majority of the models that include an interactive carbon cycle, the simulated global land and ocean carbon sinks over the latter part of the 20th century fall within the range of observational estimates. However, models systematically underestimate the NH land sink implied by atmospheric inversion techniques. {9.4.5}

Regional downscaling methods provide climate information at the smaller scales needed for many climate impact studies. There is *high confidence* that downscaling adds value both in regions with highly variable topography and for various small-scale phenomena. {9.6.4}

The model spread in equilibrium climate sensitivity ranges from 2.1°C to 4.7°C and is very similar to the assessment in the AR4. There is *very high confidence* that the primary factor contributing to the spread in equilibrium climate sensitivity continues to be the cloud feedback. This applies to both the modern climate and the last glacial maximum. There is likewise *very high confidence* that, consistent with observations, models show a strong positive correlation between tropospheric temperature and water vapour on regional to global scales, implying a positive water vapour feedback in both models and observations. {5.3.3, 9.4.1, 9.7} *(continued on next page)*

Box TS.4 (continued)

Climate models are based on physical principles, and they reproduce many important elements of observed climate. Both aspects contribute to our confidence in the models' suitability for their application in detection and attribution studies (see Chapter 10) and for quantitative future predictions and projections (see Chapters 11 to 14). There is increasing evidence that some elements of observed variability or trends are well correlated with inter-model differences in model projections for quantities such as Arctic summer sea ice trends, the snow–albedo feedback, and the carbon loss from tropical land. However, there is still no universal strategy for transferring a model's past performance to a relative weight of this model in a multi-model-ensemble mean of climate projections. {9.8.3}

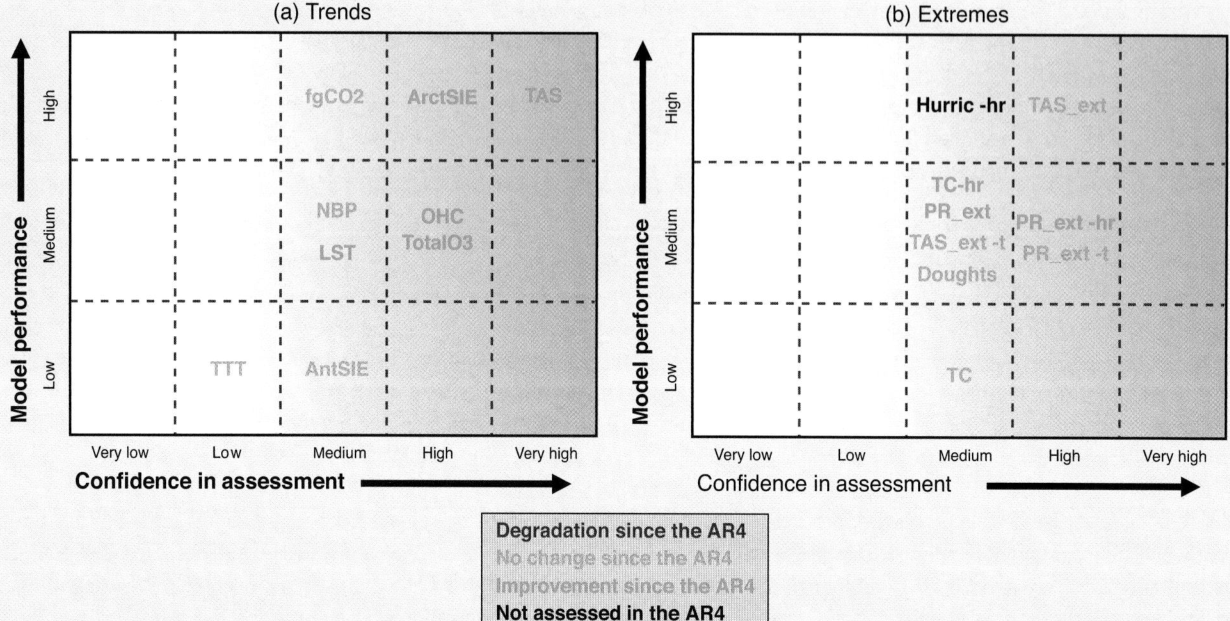

Box TS.4, Figure1 | Summary of how well the current-generation climate models simulate important features of the climate of the 20th century. Confidence in the assessment increases towards the right as suggested by the increasing strength of shading. Model quality increases from bottom to top. The colour coding indicates improvements from the models available at the time of the AR4 to the current assessment. There have been a number of improvements since the AR4, and some some modelled quantities are not better simulated. The major climate quantities are listed in this summary and none shows degradation. The assessment is based mostly on the multi-model mean, not excluding that deviations for individual models could exist. Assessed model quality is simplified for representation in this figure; details of each assessment are found in Chapter 9. {9.8.1; Figure 9.44}

The figure highlights the following key features, with the sections that back up the assessment added in brackets:

(a) Trends in:

AntSIE	Antarctic sea ice extent {9.4.3}
ArctSIE	Arctic sea ice extent {9.4.3}
fgCO2	Global ocean carbon sink {9.4.5}
LST	Lower-stratospheric temperature {9.4.1.}
NBP	Global land carbon sink {9.4.5}
OHC	Global ocean heat content {9.4.2}
TotalO3	Total-column ozone {9.4.1}
TAS	Surface air temperature {9.4.1}
TTT	Tropical tropospheric temperature {9.4.1}

(b) Extremes:

Droughts	Droughts {9.5.4}
Hurric-hr	Year-to-year count of Atlantic hurricanes in high-resolution AGCMs {9.5.4}
PR_ext	Global distribution of precipitation extremes {9.5.4}
PR_ext-hr	Global distribution of precipitation extremes in high-resolution AGCMs {9.5.4}
PR_ext-t	Global trends in precipitation extremes {9.5.4}
TAS_ext	Global distributions of surface air temperature extremes {9.5.4}
TAS_ext-t	Global trends in surface air temperature extremes {9.5.4}
TC	Tropical cyclone tracks and intensity {9.5.4}
TC-hr	Tropical cyclone tracks and intensity in high-resolution AGCMs {9.5.4}

Box TS.5 | Paleoclimate

Reconstructions from paleoclimate archives allow current changes in atmospheric composition, sea level and climate (including extreme events such as droughts and floods), as well as future projections, to be placed in a broader perspective of past climate variability (see Section TS.2). {5.2–5.6, 6.2, 10.7}

Past climate information also documents the behaviour of slow components of the climate system including the carbon cycle, ice sheets and the deep ocean for which instrumental records are short compared to their characteristic time scales of responses to perturbations, thus informing on mechanisms of abrupt and irreversible changes. Together with the knowledge of past external climate forcings, syntheses of paleoclimate data have documented polar amplification, characterized by enhanced temperature changes in the Arctic compared to the global mean, in response to high or low CO_2 concentrations. {5.2.1, 5.2.2, 5.6, 5.7, 5.8, 6.2, 8.4.2, 13.2.1, 13.4; Boxes 5.1, 5.2}

Since AR4, the inclusion of paleoclimate simulations in the PMIP3 (Paleoclimate Modelling Intercomparison Project)/CMIP5 framework has enabled paleoclimate information to be more closely linked with future climate projections. Paleoclimate information for the mid-Holocene (6 ka), the Last Glacial Maximum (approximately 21 ka), and last millennium has been used to test the ability of models to simulate realistically the magnitude and large-scale patterns of past changes. Combining information from paleoclimate simulations and reconstructions enables to quantify the response of the climate system to radiative perturbations, constraints to be placed on the range of equilibrium climate sensitivity, and past patterns of internal climate variability to be documented on inter-annual to multi-centennial scales. {5.3.1–5.3.5, 5.4, 5.5.1, 9.4.1, 9.4.2, 9.5.3, 9.7.2, 10.7.2, 14.1.2}

Box TS.5, Figure 1 illustrates the comparison between the last millennium Paleoclimate Modelling Intercomparison Project Phase 3 (PMIP3)/CMIP5 simulations and reconstructions, together with the associated solar, volcanic and WMGHG RFs. For average annual NH temperatures, the period 1983–2012 was *very likely* the warmest 30-year period of the last 800 years (*high confidence*) and *likely* the warmest 30-year period of the last 1400 years (*medium confidence*). This is supported by comparison of instrumental temperatures with multiple reconstructions from a variety of proxy data and statistical methods, and is consistent with AR4. In response to solar, volcanic and anthropogenic radiative changes, climate models simulate multi-decadal temperature changes in the last 1200 years in the NH that are generally consistent in magnitude and timing with reconstructions, within their uncertainty ranges. Continental-scale temperature reconstructions show, with *high confidence*, multi-decadal periods during the Medieval Climate Anomaly (about 950 to 1250) that were in some regions as warm as the mid-20th century and in others as warm as in the late 20th century. With *high confidence*, these regional warm periods were not as synchronous across regions as the warming since the mid-20th century. Based on the comparison between reconstructions and simulations, there is *high confidence* that not only external orbital, solar and volcanic forcing but also internal variability contributed substantially to the spatial pattern and timing of surface temperature changes between the Medieval Climate Anomaly and the Little Ice Age (about 1450 to 1850). However, there is only *very low confidence* in quantitative estimates of their relative contributions. It is *very unlikely* that NH temperature variations from 1400 to 1850 can be explained by internal variability alone. There is *medium confidence* that external forcing contributed to Northern Hemispheric temperature variability from 850 to 1400 and that external forcing contributed to European temperature variations over the last 5 centuries. {5.3.5, 5.5.1, 10.7.2, 10.7.5; Table 10.1} *(continued on next page)*

Box TS.5 (continued)

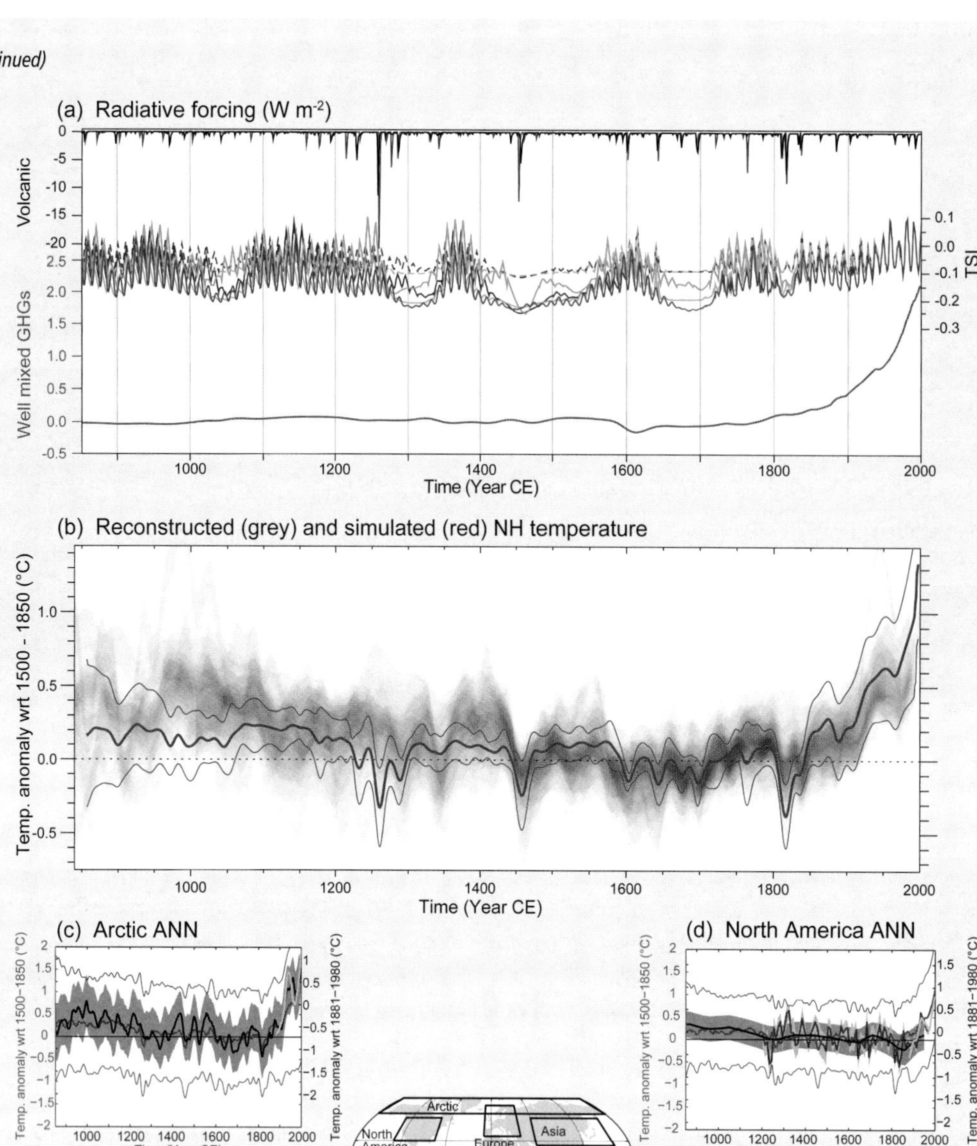

Box TS.5, Figure 1 | Last-millennium simulations and reconstructions. (a) 850–2000 PMIP3/CMIP5 radiative forcing due to volcanic, solar and well-mixed greenhouse gases. Different colours illustrate the two existing data sets for volcanic forcing and four estimates of solar forcing. For solar forcing, solid (dashed) lines stand for reconstruction variants in which background changes in irradiance are (not) considered; (b) 850–2000 PMIP3/CMIP5 simulated (red) and reconstructed (shading) Northern Hemisphere (NH) temperature changes. The thick red line depicts the multi-model mean while the thin red lines show the multi-model 90% range. The overlap of reconstructed temperatures is shown by grey shading; all data are expressed as anomalies from their 1500–1850 mean and smoothed with a 30-year filter. Note that some reconstructions represent a smaller spatial domain than the full NH or a specific season, while annual temperatures for the full NH mean are shown for the simulations. (c), (d), (e) and (f) Arctic and North America annual mean temperature, and Europe and Asia June, July and August (JJA) temperature, from 950 to 2000 from reconstructions (black line), and PMIP3/CMIP5 simulations (thick red, multi-model mean; thin red, 90% multi-model range). All red curves are expressed as anomalies from their 1500–1850 mean and smoothed with a 30-year filter. The shaded envelope depicts the uncertainties from each reconstruction (Arctic: 90% confidence bands, North American: ±2 standard deviation. Asia: ±2 root mean square error. Europe: 95% confidence bands). For comparison with instrumental record, the Climatic Research Unit land station Temperature (CRUTEM4) data set is shown (yellow line). These instrumental data are not necessarily those used in calibration of the reconstrctions, and thus may show greater or lesser correspondence with the reconstructions than the instrumental data actually used for calibration; cutoff timing may also lead to end effects for smoothed data shown. All lines are smoothed by applying a 30-year moving average. Map shows the individual regions for each reconstruction. {5.3.5; Table 5.A.1; Figures 5.1, 5.8, 5.12}

TS.5 Projections of Global and Regional Climate Change

TS.5.1 Introduction

Projections of changes in the climate system are made using a hierarchy of climate models ranging from simple climate models, to models of intermediate complexity, to comprehensive climate models, and Earth System Models (ESMs). These models simulate changes based on a set of scenarios of anthropogenic forcings. A new set of scenarios, the Representative Concentration Pathways (RCPs), was used for the new climate model simulations carried out under the framework of the Coupled Model Intercomparison Project Phase 5 (CMIP5) of the World Climate Research Programme. A large number of comprehensive climate models and ESMs have participated in CMIP5, whose results form the core of the climate system projections.

This section summarizes the assessment of these climate change projections. First, future forcing and scenarios are presented. The following subsections then address various aspects of projections of global and regional climate change, including near-term (up to about mid-century) and long-term (end of the 21st century) projections in the atmosphere, ocean and cryosphere; projections of carbon and other biogeochemical cycles; projections in sea level change; and finally changes to climate phenomena and other aspects of regional climate over the 21st century. Projected changes are given relative to the 1986–2005 average unless indicated otherwise. Projections of climate change on longer term and information on climate stabilization and targets are provided in TFE.8. Methods to counter climate change, termed geoengineering, have been proposed and an overview is provided in Box TS.7. {11.3, 12.3–12.5, 13.5–13.7, 14.1–14.6, Annex I}

TS.5.2 Future Forcing and Scenarios

In this assessment report a series of new RCPs are used that largely replace the IPCC Special Report on Emission Scenarios (SRES) scenarios (see Box TS.6 and Annex II for Climate System Scenario Tables). They produce a range of responses from ongoing warming, to approximately stabilized forcing, to a stringent mitigation scenario (RCP2.6) that stabilizes and then slowly reduces the RF after mid-21st century. In contrast to the AR4, the climate change from the RCP scenarios in the AR5 is framed as a combination of adaptation and mitigation. Mitigation actions starting now in the various RCP scenarios do not produce discernibly different climate change outcomes for the next 30 years or so, whereas long-term climate change after mid-century is appreciably different across the RCPs. {Box 1.1}

TS

Box TS.6 | The New Representative Concentration Pathway Scenarios and Coupled Model Intercomparison Project Phase 5 Models

Future anthropogenic emissions of GHGs, aerosol particles and other forcing agents such as land use change are dependent on socioeconomic factors, and may be affected by global geopolitical agreements to control those emissions to achieve mitigation. AR4 made extensive use of the SRES scenarios that do not include additional climate initiatives, which means that no scenarios were available that explicitly assume implementation of the United Nations Framework Convention on Climate Change (UNFCCC) or the emissions targets of the Kyoto Protocol. However, GHG emissions are directly affected by non-climate change policies designed for a wide range of other purposes. The SRES scenarios were developed using a sequential approach, that is, socioeconomic factors fed into emissions scenarios, which were then used in simple climate models to determine concentrations of GHGs, and other agents required to drive the more complex AOGCMs. In this report, outcomes of climate simulations that use new scenarios (some of which include implied policy actions to achieve mitigation) referred to as RCPs are assessed. These RCPs represent a larger set of mitigation scenarios and were selected to have different targets in terms of radiative forcing at 2100 (about 2.6, 4.5, 6.0 and 8.5 W m^{-2}; Figure TS.15). The scenarios should be considered plausible and illustrative, and do not have probabilities attached to them. {12.3.1; Box 1.1}

The RCPs were developed using Integrated Assessment Models (IAMs) that typically include economic, demographic, energy, and simple climate components. The emission scenarios they produce are then run through a simple model to produce time series of GHG concentrations that can be run in AOGCMs. The emission time series from the RCPs can then be used directly in ESMs that include interactive biogeochemistry (at least a land and ocean carbon cycle). {12.3.1; Box 1.1}

The CMIP5 multi-model experiment (coordinated through the World Climate Research Programme) presents an unprecedented level of information on which to base assessments of climate variability and change. CMIP5 includes new ESMs in addition to AOGCMs, new model experiments and more diagnostic output. CMIP5 is much more comprehensive than the preceding CMIP3 multi-model experiment that was available at the time of the IPCC AR4. CMIP5 has more than twice as many models, many more experiments (that also include experiments to address understanding of the responses in the future climate change scenario runs), and nearly 2×10^{15} bytes of data (as compared to over 30×10^{12} bytes of data in CMIP3). A larger number of forcing agents are treated more completely in the CMIP5 models, with respect to aerosols and land use particularly. Black carbon aerosol is now a commonly included forcing agent. Considering CO_2, both 'concentrations-driven' projections and 'emissions-driven' projections are assessed from CMIP5. These allow quantification of the physical response uncertainties as well as climate–carbon cycle interactions. {1.5.2}

(continued on next page)

Box TS.6 (continued)

The assessment of the mean values and ranges of global mean temperature changes in AR4 would not have been substantially different if the CMIP5 models had been used in that report. The differences in global temperature projections can largely be attributed to the different scenarios. The global mean temperature response simulated by CMIP3 and CMIP5 models is very similar, both in the mean and the model range, transiently and in equilibrium. The range of temperature change across all scenarios is wider because the RCPs include a strong mitigation scenario (RCP2.6) that had no equivalent among the SRES scenarios used in CMIP3. For each scenario, the 5 to 95% range of the CMIP5 projections is obtained by approximating the CMIP5 distributions by a normal distribution with same mean and standard deviation and assessed as being *likely* for projections of global temperature change for the end of the 21st century. Probabilistic projections with simpler models calibrated to span the range of equilibrium climate sensitivity assessed by the AR4 provide uncertainty ranges that are consistent with those from CMIP5. In AR4 the uncertainties in global temperature projections were found to be approximately constant when expressed as a fraction of the model mean warming (constant fractional uncertainty). For the higher RCPs, the uncertainty is now estimated to be smaller than with the AR4 method for long-term climate change, because the carbon cycle–climate feedbacks are not relevant for the concentration-driven RCP projections (in contrast, the assessed projection uncertainties of global temperature in AR4 did account of carbon cycle–climate feedbacks, even though these were not part of the CMIP3 models). When forced with RCP8.5, CO_2 emissions, as opposed to the RCP8.5 CO_2 concentrations, CMIP5 ESMs with interactive carbon cycle simulate, on average, a 50 (–140 to +210) ppm (CMIP5 model spread) larger atmospheric CO_2 concentration and 0.2°C larger global surface temperature increase by 2100. For the low RCPs the fractional uncertainty is larger because internal variability and non-CO_2 forcings make a larger relative contribution to the total uncertainty. {12.4.1, 12.4.8, 12.4.9} *(continued on next page)*

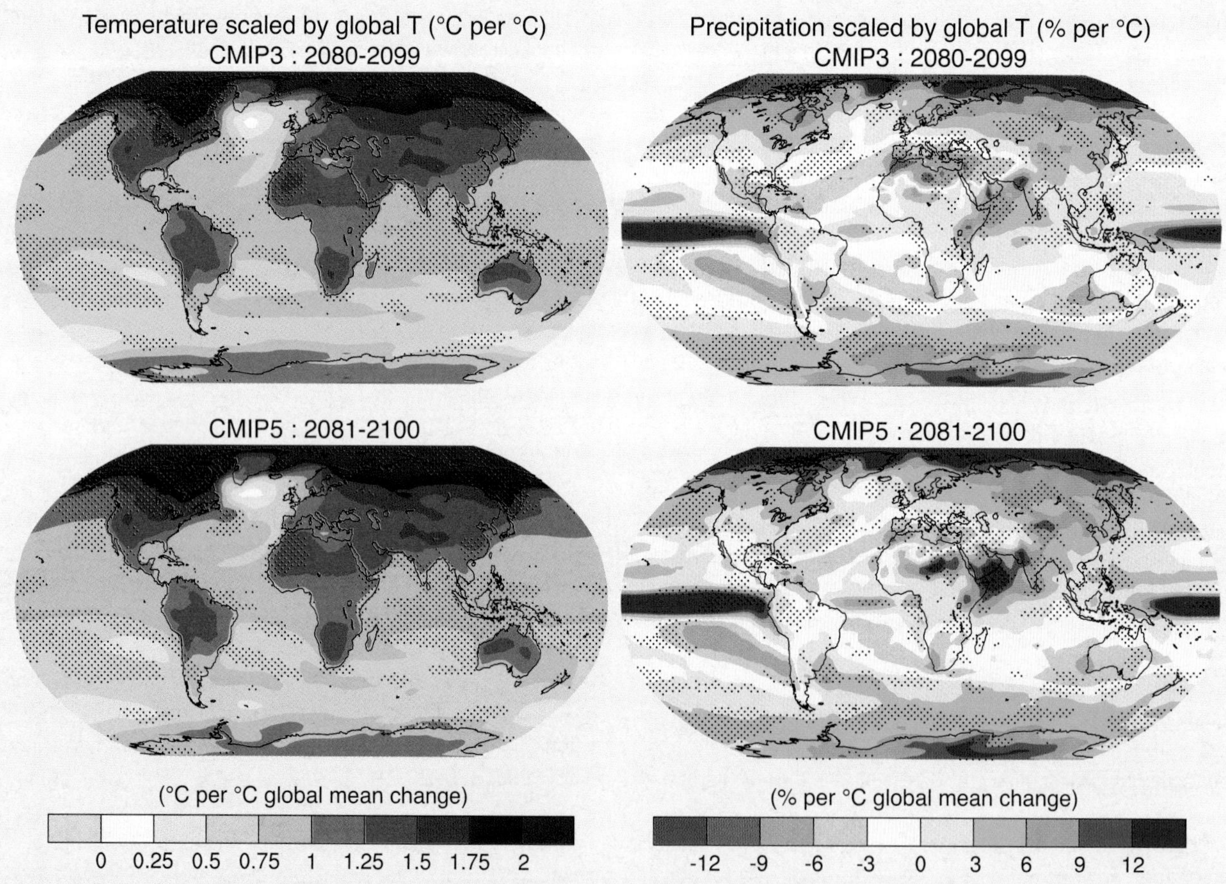

Box TS.6, Figure 1 | Patterns of temperature (left column) and percent precipitation change (right column) for the CMIP3 models average (first row) and CMIP5 models average (second row), scaled by the corresponding global average temperature changes. The patterns are computed in both cases by taking the difference between the averages over the last 20 years of the 21st century experiments (2080–2099 for CMIP3 and 2081–2100 for CMIP5) and the last 20 years of the historic experiments (1980–1999 for CMIP3, 1986–2005 for CMIP5) and rescaling each difference by the corresponding change in global average temperature. This is done first for each individual model, then the results are averaged across models. Stippling indicates a measure of significance of the difference between the two corresponding patterns obtained by a bootstrap exercise. Two subsets of the pooled set of CMIP3 and CMIP5 ensemble members of the same size as the original ensembles, but without distinguishing CMIP3 from CMIP5 members, were randomly sampled 500 times. For each random sample the corresponding patterns and their difference are computed, then the true difference is compared, grid-point by grid-point, to the distribution of the bootstrapped differences, and only grid-points at which the value of the difference falls in the tails of the bootstrapped distribution (less than the 2.5th percentiles or the 97.5th percentiles) are stippled. {Figure 12.41}

Box TS.6 (continued)

There is overall consistency between the projections of temperature and precipitation based on CMIP3 and CMIP5, both for large-scale patterns and magnitudes of change (Box TS.6, Figure 1). Model agreement and confidence in projections depends on the variable and on spatial and temporal averaging, with better agreement for larger scales. Confidence is higher for temperature than for those quantities related to the water cycle or atmospheric circulation. Improved methods to quantify and display model robustness have been developed to indicate where lack of agreement across models on local trends is a result of internal variability, rather than models actually disagreeing on their forced response. Understanding of the sources and means of characterizing uncertainties in long-term large scale projections of climate change has not changed significantly since AR4, but new experiments and studies have continued to work towards a more complete and rigorous characterization. {9.7.3, 12.2, 12.4.1, 12.4.4, 12.4.5, 12.4.9; Box 12.1}

The well-established stability of geographical patterns of temperature and precipitation change during a transient experiment remains valid in the CMIP5 models (Box TS.6, Figure 1). Patterns are similar over time and across scenarios and to first order can be scaled by the global mean temperature change. There remain limitations to the validity of this technique when it is applied to strong mitigation scenarios, to scenarios where localized forcings (e.g., aerosols) are significant and vary in time and for variables other than average seasonal mean temperature and precipitation. {12.4.2}

The range in anthropogenic aerosol emissions across all scenarios has a larger impact on near-term climate projections than the corresponding range in long-lived GHGs, particularly on regional scales and for hydrological cycle variables. The RCP scenarios do not span the range of future aerosol emissions found in the SRES and alternative scenarios (Box TS.6). {11.3.1, 11.3.6}

If rapid reductions in sulphate aerosol are undertaken for improving air quality or as part of decreasing fossil-fuel CO_2 emissions, then there is *medium confidence* that this could lead to rapid near-term warming. There is evidence that accompanying controls on CH_4 emissions would offset some of this sulphate-induced warming, although the cooling from CH_4 mitigation will emerge more slowly than the warming from sulphate mitigation due to the different time scales over which atmospheric concentrations of these substances decrease in response to decreases in emissions. Although removal of black carbon aerosol could also counter warming associated with sulphate removal, uncertainties are too large to constrain the net sign of the global temperature response to black carbon emission reductions, which depends on reduction of co-emitted (reflective) aerosols and on aerosol indirect effects. {11.3.6}

Including uncertainties in projecting the chemically reactive GHGs CH_4 and N_2O from RCP emissions gives a range in abundance pathways that is *likely* 30% larger than the range in RCP concentrations used to force the CMIP5 climate models. Including uncertainties in emission estimates from agricultural, forest and land use sources, in atmospheric lifetimes, and in chemical feedbacks, results in a much wider range of abundances for N_2O, CH_4 and HFCs and their RF. In the case of CH_4, by year 2100 the *likely* range of RCP8.5 CH_4 abundance extends 520 ppb above the single-valued RCP8.5 CH_4 abundance, and RCP2.6 CH_4 extends 230 ppb below RCP2.6 CH_4. {11.3.5}

There is *very low confidence* in projections of natural forcing. Major volcanic eruptions cause a negative RF up to several watts per square metre, with a typical lifetime of one year, but the possible occurrence and timing of future eruptions is unknown. Except for the 11-year solar cycle, changes in the total solar irradiance are uncertain. Except where explicitly indicated, future volcanic eruptions and changes in total solar irradiance additional to a repeating 11-year solar cycle are not included in the projections of near- and long-term climate assessed. {8, 11.3.6}

TS.5.3 Quantification of Climate System Response

Estimates of the equilibrium climate sensitivity (ECS) based on observed climate change, climate models and feedback analysis, as well as paleoclimate evidence indicate that ECS is positive, *likely* in the range 1.5°C to 4.5°C with *high confidence*, *extremely unlikely* less than 1°C (*high confidence*) and *very unlikely* greater than 6°C (*medium confidence*). Earth system sensitivity over millennia time scales including long-term feedbacks not typically included in models could be significantly higher than ECS (see TFE.6 for further details). {5.3.1, 10.8; Box 12.2}

With *high confidence* the transient climate response (TCR) is positive, *likely* in the range 1°C to 2.5°C and *extremely unlikely* greater than 3°C, based on observed climate change and climate models (see TFE.6 for further details). {10.8; Box 12.2}

The ratio of GMST change to total cumulative anthropogenic carbon emissions is relatively constant and independent of the scenario, but is model dependent, as it is a function of the model cumulative airborne fraction of carbon and the transient climate response. For any given temperature target, higher emissions in earlier decades therefore imply lower emissions by about the same amount later on. The transient climate response to cumulative carbon emission (TCRE) is *likely* between 0.8°C to 2.5°C per 1000 PgC (*high confidence*), for cumulative carbon emissions less than about 2000 PgC until the time at which temperatures peak (see TFE.8 for further details). {10.8, 12.5.4; Box 12.2}

TS

Thematic Focus Elements

TFE.6 | Climate Sensitivity and Feedbacks

The description of climate change as a response to a forcing that is amplified by feedbacks goes back many decades. The concepts of radiative forcing (RF) and climate feedbacks continue to be refined, and limitations are now better understood; for instance, feedbacks may be much faster than the surface warming, feedbacks depend on the type of forcing agent (e.g., greenhouse gas (GHG) vs. solar forcing), or may have intrinsic time scales (associated mainly with vegetation change and ice sheets) of several centuries to millennia. The analysis of physical feedbacks in models and from observations remains a powerful framework that provides constraints on transient future warming for different scenarios, on climate sensitivity and, combined with estimates of carbon cycle feedbacks (see TFE.5), determines the GHG emissions that are compatible with climate stabilization or targets (see TFE.8). {7.1, 9.7.2, 12.5.3; Box 12.2}

The water vapour/lapse rate, albedo and cloud feedbacks are the principal determinants of equilibrium climate sensitivity. All of these feedbacks are assessed to be positive, but with different levels of likelihood assigned ranging from *likely* to *extremely likely*. Therefore, there is *high confidence* that the net feedback is positive and the black body response of the climate to a forcing will therefore be amplified. Cloud feedbacks continue to be the largest uncertainty. The net feedback from water vapour and lapse rate changes together is *extremely likely* positive and approximately doubles the black body response. The mean value and spread of these two processes in climate models are essentially unchanged from the IPCC Fourth Assessment Report (AR4), but are now supported by stronger observational evidence and better process understanding of what determines relative humidity distributions. Clouds respond to climate forcing mechanisms in multiple ways and individual cloud feedbacks can be positive or negative. Key issues include the representation of both deep and shallow cumulus convection, microphysical processes in ice clouds and partial cloudiness that results from small-scale variations of cloud-producing and cloud-dissipating processes. New approaches to diagnosing cloud feedback in General Circulation Models (GCMs) have clarified robust cloud responses, while continuing to implicate low cloud cover as the most important source of intermodel spread in simulated cloud feedbacks. The net radiative feedback due to all cloud types is *likely* positive. This conclusion is reached by considering a plausible range for unknown contributions by processes yet to be accounted for, in addition to those occurring in current climate models. Observations alone do not currently provide a robust, direct constraint, but multiple lines of evidence now indicate positive feedback contributions from changes in both the height of high clouds and the horizontal distribution of clouds. The additional feedback from low cloud amount is also positive in most climate models, but that result is not well understood, nor effectively constrained by observations, so *confidence* in it is *low*. {7.2.4–7.2.6, 9.7.2}

The representation of aerosol–cloud processes in climate models continues to be a challenge. Aerosol and cloud variability at scales significantly smaller than those resolved in climate models, and the subtle responses of clouds to aerosol at those scales, mean that, for the foreseeable future, climate models will continue to rely on parameterizations of aerosol–cloud interactions or other methods that represent subgrid variability. This implies large uncertainties for estimates of the forcings associated with aerosol–cloud interactions. {7.4, 7.5.3, 7.5.4}

Equilibrium climate sensitivity (ECS) and transient climate response (TCR) are useful metrics summarising the global climate system's temperature response to an externally imposed RF. ECS is defined as the equilibrium change in annual mean global mean surface temperature (GMST) following a doubling of the atmospheric carbon dioxide (CO_2) concentration, while TCR is defined as the annual mean GMST change at the time of CO_2 doubling following a linear increase in CO_2 forcing over a period of 70 years (see Glossary). Both metrics have a broader application than these definitions imply: ECS determines the eventual warming in response to stabilisation of atmospheric composition on multi-century time scales, while TCR determines the warming expected at a given time following any steady increase in forcing over a 50- to 100-year time scale. {Box 12.2; 12.5.3}

ECS and TCR can be estimated from various lines of evidence (TFE.6, Figures 1 and 2). The estimates can be based on the values of ECS and TCR diagnosed from climate models, or they can be constrained by analysis of feedbacks in climate models, patterns of mean climate and variability in models compared to observations, temperature fluctuations as reconstructed from paleoclimate archives, observed and modelled short term perturbations of the energy balance like those caused by volcanic eruptions, and the observed surface and ocean temperature trends since pre-industrial. For many applications, the limitations of the forcing-feedback analysis framework and the dependence of feedbacks on time scales and the climate state must be kept in mind. {5.3.1, 5.3.3, 9.7.1–9.7.3, 10.8.1, 10.8.2, 12.5.3; Box 5.2; Table 9.5} *(continued on next page)*

TFE.6 (continued)

Newer studies of constraints on ECS are based on the observed warming since pre-industrial, analysed using simple and intermediate complexity models, improved statistical methods and several different and newer data sets. Together with paleoclimate constraints but without considering the CMIP based evidence these studies show ECS is *likely* between 1.5°C to 4.5°C (*medium confidence*) and *extremely unlikely* less than 1.0°C. {5.3.1, 5.3.3, 10.8.2; Boxes 5.2, 12.2}

Estimates based on Atmosphere–Ocean General Circulation Models (AOGCMs) and feedback analysis indicate a range of 2°C to 4.5°C, with the Coupled Model Intercomparison Project Phase 5 (CMIP5) model mean at 3.2°C, similar to CMIP3. High climate sensitivities are found in some perturbed parameter ensembles models, but recent comparisons of perturbed-physics ensembles against the observed climate find that models with ECS values in the range 3°C to 4°C show the smallest errors for many fields. Relationships between climatological quantities and climate sensitivity are often found within a specific perturbed parameter ensemble model but in many cases the relationship is not robust across perturbed parameter ensembles models from different models or in CMIP3 and CMIP5. The assessed literature suggests that the range of climate sensitivities and transient responses covered by CMIP3 and CMIP5 cannot be narrowed significantly by constraining the models with observations of the mean climate and variability. Studies based on perturbed parameter ensembles models and CMIP3 support the conclusion that a credible representation of the mean climate and variability is very difficult to achieve with ECSs below 2°C. {9.2.2, 9.7.3; Box 12.2}

New estimates of ECS based on reconstructions and simulations of the Last Glacial Maximum (21 ka to 19 ka) show that values below 1°C as well as above 6°C are *very unlikely*. In some models climate sensitivity differs between warm and cold climates because of differences in the representation of cloud feedbacks. Estimates of an Earth system sensitivity including slow feedbacks (e.g., ice sheets or vegetation) are even more difficult to relate to climate sensitivity of the current climate state. The main limitations of ECS estimates from paleoclimate states are uncertainties in proxy data, spatial coverage of the data, uncertainties in some forcings, and structural limitations in models used in model–data comparisons. {5.3, 10.8.2, 12.5.3}

Bayesian methods to constrain ECS or TCR are sensitive to the assumed prior distributions. They can in principle yield narrower estimates by combining constraints from the observed warming trend, volcanic eruptions, model climatology and paleoclimate, and that has been done in some studies, but there is no consensus on how this should be done robustly. This approach is sensitive to the assumptions regarding the independence of the various lines of evidence, the possibility of shared biases in models or feedback estimates and the assumption that each individual line of evidence is unbiased. The combination of different estimates in this assessment is based on expert judgement. {10.8.2; Box 12.2}

Based on the combined evidence from observed climate change including the observed 20th century warming, climate models, feedback analysis and paleoclimate, as discussed above, ECS is *likely* in the range 1.5°C to 4.5°C with *high confidence*. ECS is positive, *extremely unlikely*

(continued on next page)

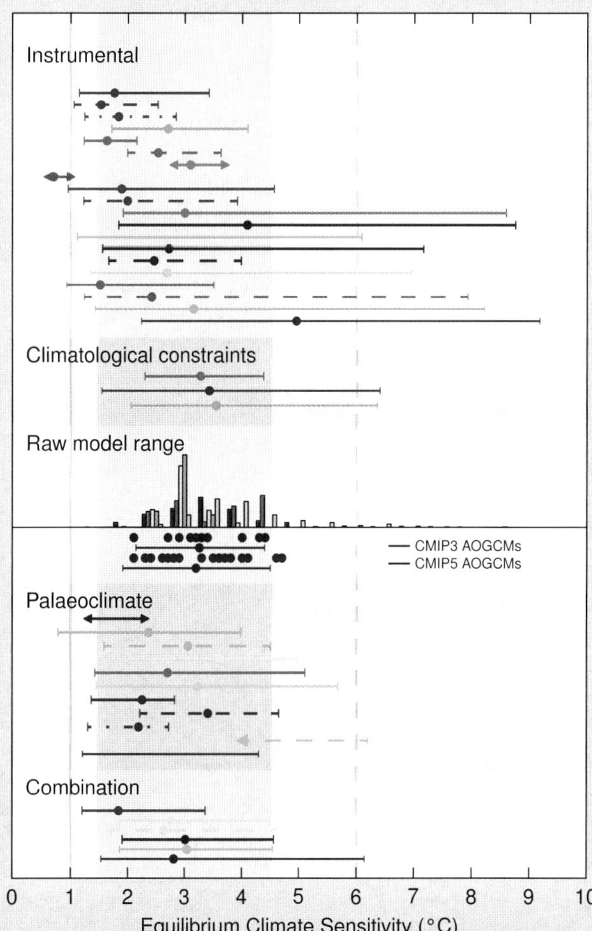

TFE.6, Figure 1 | Probability density functions, distributions and ranges for equilibrium climate sensitivity, based on Figure 10.20b plus climatological constraints shown in IPCC AR4 (Box AR4 10.2 Figure 1), and results from CMIP5 (Table 9.5). The grey shaded range marks the *likely* 1.5°C to 4.5°C range, grey solid line the *extremely unlikely* less than 1°C, the grey dashed line the *very unlikely* greater than 6°C. See Figure 10.20b and Chapter 10 Supplementary Material for full caption and details. {Box 12.2, Figure 1}

TFE.6 (continued)

less than 1°C (*high confidence*), and *very unlikely* greater than 6°C (*medium confidence*). The tails of the ECS distribution are now better understood. Multiple lines of evidence provide *high confidence* that an ECS value less than 1°C is *extremely unlikely*. The upper limit of the *likely* range is unchanged compared to AR4. The lower limit of the *likely* range of 1.5°C is less than the lower limit of 2°C in AR4. This change reflects the evidence from new studies of observed temperature change, using the extended records in atmosphere and ocean. These studies suggest a best fit to the observed surface and ocean warming for ECS values in the lower part of the *likely* range. Note that these studies are not purely observational, because they require an estimate of the response to RF from models. In addition, the uncertainty in ocean heat uptake remains substantial. Accounting for short-term variability in simple models remains challenging, and it is important not to give undue weight to any short time period which might be strongly affected by internal variability. On the other hand, AOGCMs with observed climatology with ECS values in the upper part of the 1.5 to 4.5°C range show very good agreement with observed climatology, but the simulation of key feedbacks like clouds remains challenging in those models. The estimates from the observed warming, paleoclimate, and from climate models are consistent within their uncertainties, each is supported by many studies and multiple data sets, and in combination they provide *high confidence* for the assessed *likely* range. Even though this assessed range is similar to previous reports, confidence today is much higher as a result of high quality and longer observational records with a clearer anthropogenic signal, better process understanding, more and better understood evidence from paleoclimate reconstructions, and better climate models with higher resolution that capture many more processes more realistically. All these lines of evidence individually support the assessed *likely* range of 1.5°C to 4.5°C. {3.2, 9.7.3, 10.8; Boxes 9.2, 13.1}

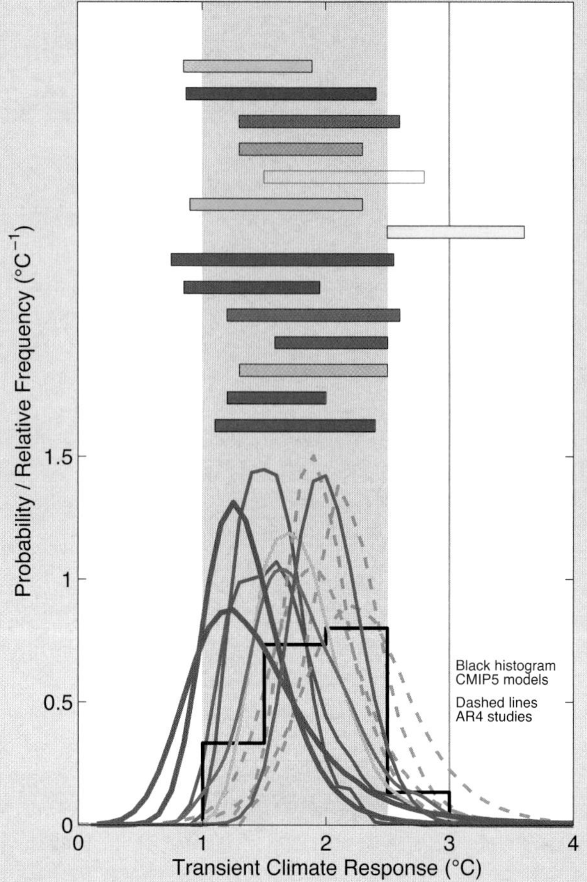

TFE.6, Figure 2 | Probability density functions, distributions and ranges (5 to 95%) for the transient climate response from different studies, based on Figure 10.20a, and results from CMIP5 (black histogram, Table 9.5). The grey shaded range marks the *likely* 1°C to 2.5°C range, the grey solid line marks the *extremely unlikely* greater than 3°C. See Figure 10.20a and Chapter 10 Supplementary Material for full caption and details. {Box 12.2, Figure 2}

On time scales of many centuries and longer, additional feedbacks with their own intrinsic time scales (e.g., vegetation, ice sheets) may become important but are not usually modelled in AOGCMs. The resulting equilibrium temperature response to a doubling of CO_2 on millennial time scales or Earth system sensitivity is less well constrained but *likely* to be larger than ECS, implying that lower atmospheric CO_2 concentrations are compatible with limiting warming to below a given temperature level. These slow feedbacks are less likely to be proportional to global mean temperature change, implying that Earth system sensitivity changes over time. Estimates of Earth system sensitivity are also difficult to relate to climate sensitivity of the current climate state. {5.3.3, 10.8.2, 12.5.3}

For scenarios of increasing RF, TCR is a more informative indicator of future climate change than ECS. This assessment concludes with *high confidence* that the TCR is *likely* in the range 1°C to 2.5°C, close to the estimated 5 to 95% range of CMIP5 (1.2°C to 2.4°C), is positive and *extremely unlikely* greater than 3°C. As with the ECS, this is an expert-assessed range, supported by several different and partly independent lines of evidence, each based on multiple studies, models and data sets. TCR is estimated from the observed global changes in surface temperature, ocean heat uptake and RF including detection/attribution studies identifying the response patterns to increasing GHG concentrations, and the results of CMIP3 and CMIP5. Estimating TCR suffers from fewer difficulties in terms of state- or time-dependent feedbacks, and is less affected by uncertainty as to how much energy is taken up by the

(continued on next page)

TFE.6 (continued)

ocean. Unlike ECS, the ranges of TCR estimated from the observed warming and from AOGCMs agree well, increasing our confidence in the assessment of uncertainties in projections over the 21st century.

The assessed ranges of ECS and TCR are largely consistent with the observed warming, the estimated forcing and the projected future warming. In contrast to AR4, no best estimate for ECS is given because of a lack of agreement on the best estimate across lines of evidence and studies and an improved understanding of the uncertainties in estimates based on the observed warming. Climate models with ECS values in the upper part of the *likely* range show very good agreement with observed climatology, whereas estimates derived from observed climate change tend to best fit the observed surface and ocean warming for ECS values in the lower part of the *likely* range. In estimates based on the observed warming the most likely value is sensitive to observational and model uncertainties, internal climate variability and to assumptions about the prior distribution of ECS. In addition, "best estimate" and "most likely value" are defined in various ways in different studies. {9.7.1, 10.8.1, 12.5.3; Table 9.5}

TS.5.4 Near-term Climate Change

Near-term decadal climate prediction provides information not available from existing seasonal to interannual (months to a year or two) predictions or from long-term (mid 21st century and beyond) climate change projections. Prediction efforts on seasonal to interannual time scales require accurate estimates of the initial climate state with less focus extended to changes in external forcing[12], whereas long-term climate projections rely more heavily on estimations of external forcing with little reliance on the initial state of internal variability. Estimates of near-term climate depend on the committed warming (caused by the inertia of the oceans as they respond to historical external forcing) the time evolution of internally generated climate variability, and the future path of external forcing. Near-term predictions out to about a decade (Figure TS.13) depend more heavily on an accurate depiction of the internally generated climate variability. {11.1, 12, 14}

Further near-term warming from past emissions is unavoidable owing to thermal inertia of the oceans. This warming will be increased by ongoing emissions of GHGs over the near term, and the climate observed in the near term will also be strongly influenced by the internally generated variability of the climate system. Previous IPCC Assessments only described climate-change projections wherein the externally forced component of future climate was included but no attempt was made to initialize the internally generated climate variability. Decadal climate predictions, on the other hand, are intended to predict both the externally forced component of future climate change, and the internally generated component. Near-term predictions do not provide detailed information of the evolution of weather. Instead they can provide estimated changes in the time evolution of the statistics of near-term climate. {11.1, 11.2.2; Box 11.1; FAQ 11.1}

Retrospective prediction experiments have been used to assess forecast quality. There is *high confidence* that the retrospective prediction experiments for forecast periods of up to 10 years exhibit positive skill when verified against observations over large regions of the planet and of the global mean. Observation-based initialization of the forecasts contributes to the skill of predictions of annual mean temperature for the first couple of years and to the skill of predictions of the GMST and the temperature over the North Atlantic, regions of the South Pacific and the tropical Indian Ocean up to 10 years (*high confidence*) partly due to a correction of the forced response. Probabilistic temperature predictions are statistically reliable (see Section 11.2.3 for definition of reliability) owing to the correct representation of global trends, but still unreliable at the regional scale when probabilities are computed from the multi-model ensemble. Predictions initialized over 2000–2005 improve estimates of the recent global mean temperature hiatus. Predictions of precipitation over continental areas with large forced trends also exhibit positive skill. {11.2.2, 11.2.3; Box 9.2}

TS.5.4.1 Projected Near-term Changes in Climate

Projections of near-term climate show small sensitivity to GHG scenarios compared to model spread, but substantial sensitivity to uncertainties in aerosol emissions, especially on regional scales and for hydrological cycle variables. In some regions, the local and regional responses in precipitation and in mean and extreme temperature to land use change will be larger than those due to large-scale GHGs and aerosol forcing. These scenarios presume that there are no major volcanic eruptions and that anthropogenic aerosol emissions are rapidly reduced during the near term. {11.3.1, 11.3.2, 11.3.6}

TS.5.4.2 Projected Near-term Changes in Temperature

In the absence of major volcanic eruptions—which would cause significant but temporary cooling—and, assuming no significant future long-term changes in solar irradiance, it is *likely* that the GMST anomaly for the period 2016–2035, relative to the reference period of 1986–2005 will be in the range 0.3°C to 0.7°C (*medium confidence*). This is based on multiple lines of evidence. This range is consistent

[12] Seasonal-to-interannual predictions typically include the impact of external forcing.

with the range obtained by using CMIP5 5 to 95% model trends for 2012–2035. It is also consistent with the CMIP5 5 to 95% range for all four RCP scenarios of 0.36°C to 0.79°C, using the 2006–2012 reference period, after the upper and lower bounds are reduced by 10% to take into account the evidence that some models may be too sensitive to anthropogenic forcing (see Table TS.1 and Figure TS.14). {11.3.6}

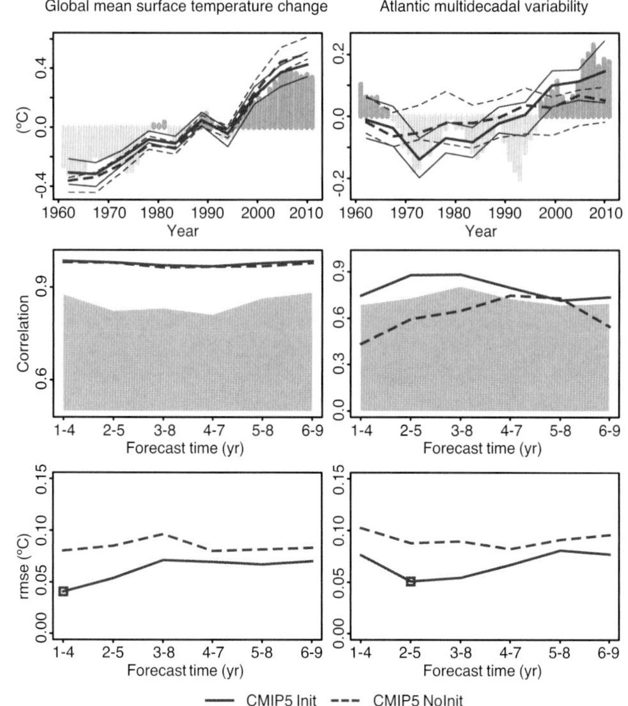

Figure TS.13 | Decadal prediction forecast quality of several climate indices. (Top row) Time series of the 2- to 5-year average ensemble mean initialized hindcast anomalies and the corresponding non-initialized experiments for three climate indices: global mean surface temperature (GMST, left) and the Atlantic Multi-decadal Variability (AMV, right). The observational time series, Goddard Institute of Space Studies Goddard Institute for Space Studies Surface Temperature Analysis (GISTEMP) global mean temperature and Extended Reconstructed Sea Surface Tempearture (ERSST) for the AMV, are represented with dark grey (positive anomalies) and light grey (negative anomalies) vertical bars, where a 4-year running mean has been applied for consistency with the time averaging of the predictions. Predicted time series are shown for the CMIP5 Init (solid) and NoInit (dotted) simulations with hindcasts started every 5 years over the period 1960–2005. The lower and upper quartile of the multi-model ensemble are plotted using thin lines. The AMV index was computed as the sea surface temperature (SST) anomalies averaged over the region Equator to 60°N and 80°W to 0°W minus the SST anomalies averaged over 60°S to 60°N. Note that the vertical axes are different for each time series. (Middle row) Correlation of the ensemble mean prediction with the observational reference along the forecast time for 4-year averages of the three sets of CMIP5 hindcasts for Init (solid) and NoInit (dashed). The one-sided 95% confidence level with a *t* distribution is represented in grey. The effective sample size has been computed taking into account the autocorrelation of the observational time series. A two-sided *t* test (where the effective sample size has been computed taking into account the autocorrelation of the observational time series) has been used to test the differences between the correlation of the initialized and non-initialized experiments, but no differences were found statistically significant with a confidence equal or higher than 90%. (Bottom row) Root mean square error (RMSE) of the ensemble mean prediction along the forecast time for 4-year averages of the CMIP5 hindcasts for Init (solid) and NoInit (dashed). A two-sided *F* test (where the effective sample size has been computed taking into account the autocorrelation of the observational time series) has been used to test the ratio between the RMSE of the Init and NoInit, and those forecast times with differences statistically significant with a confidence equal or higher than 90% are indicated with an open square. {Figure 11.3}

Higher concentrations of GHGs and lower amounts of sulphate aerosol lead to greater warming. In the near-term, differences in global mean surface air temperature across RCP scenarios for a single climate model are typically smaller than across climate models for a single RCP scenario. In 2030, the CMIP5 ensemble median values for global mean temperature differ by at most 0.2°C between the RCP scenarios, whereas the model spread (defined as the 17 to 83% range) for each RCP is around 0.4°C. The inter-scenario spread increases in time and by 2050 is comparable to the model spread. Regionally, the largest differences in surface air temperature between RCP scenarios are found in the Arctic. {11.3.2. 11.3.6}

The projected warming of global mean temperatures implies *high confidence* that new levels of warming relative to 1850-1900 mean climate will be crossed, particularly under higher GHG emissions scenarios. Relative to a reference period of 1850–1900, under RCP4.5 or RCP6.0, it is *more likely than not* that the mean GMST for the period 2016–2035 will be more than 1°C above the mean for 1850–1900, and *very unlikely* that it will be more than 1.5°C above the 1850–1900 mean (*medium confidence*). {11.3.6}

A future volcanic eruption similar in size to the 1991 eruption of Mt Pinatubo would cause a rapid drop in global mean surface air temperature of about 0.5°C in the following year, with recovery over the next few years. Larger eruptions, or several eruptions occurring close together in time, would lead to larger and more persistent effects. {11.3.6}

Possible future changes in solar irradiance could influence the rate at which GMST increases, but there is *high confidence* that this influence will be small in comparison to the influence of increasing concentrations of GHGs in the atmosphere. {11.3.6}

The spatial patterns of near-term warming projected by the CMIP5 models following the RCP scenarios (Figure TS.15) are broadly consistent with the AR4. It is *very likely* that anthropogenic warming of surface air temperature over the next few decades will proceed more rapidly over land areas than over oceans, and it is *very likely* that the anthropogenic warming over the Arctic in winter will be greater than the global mean warming, consistent with the AR4. Relative to background levels of internally generated variability there is *high confidence* that the anthropogenic warming relative to the reference period is expected to be larger in the tropics and subtropics than in mid-latitudes. {11.3.2}

It is *likely* that in the next decades the frequency of warm days and warm nights will increase in most land regions, while the frequency of cold days and cold nights will decrease. Models also project increases in the duration, intensity and spatial extent of heat waves and warm spells for the near term. These changes may proceed at a different rate than the mean warming. For example, several studies project that European high-percentile summer temperatures are projected to warm faster than mean temperatures (see also TFE.9). {11.3.2}

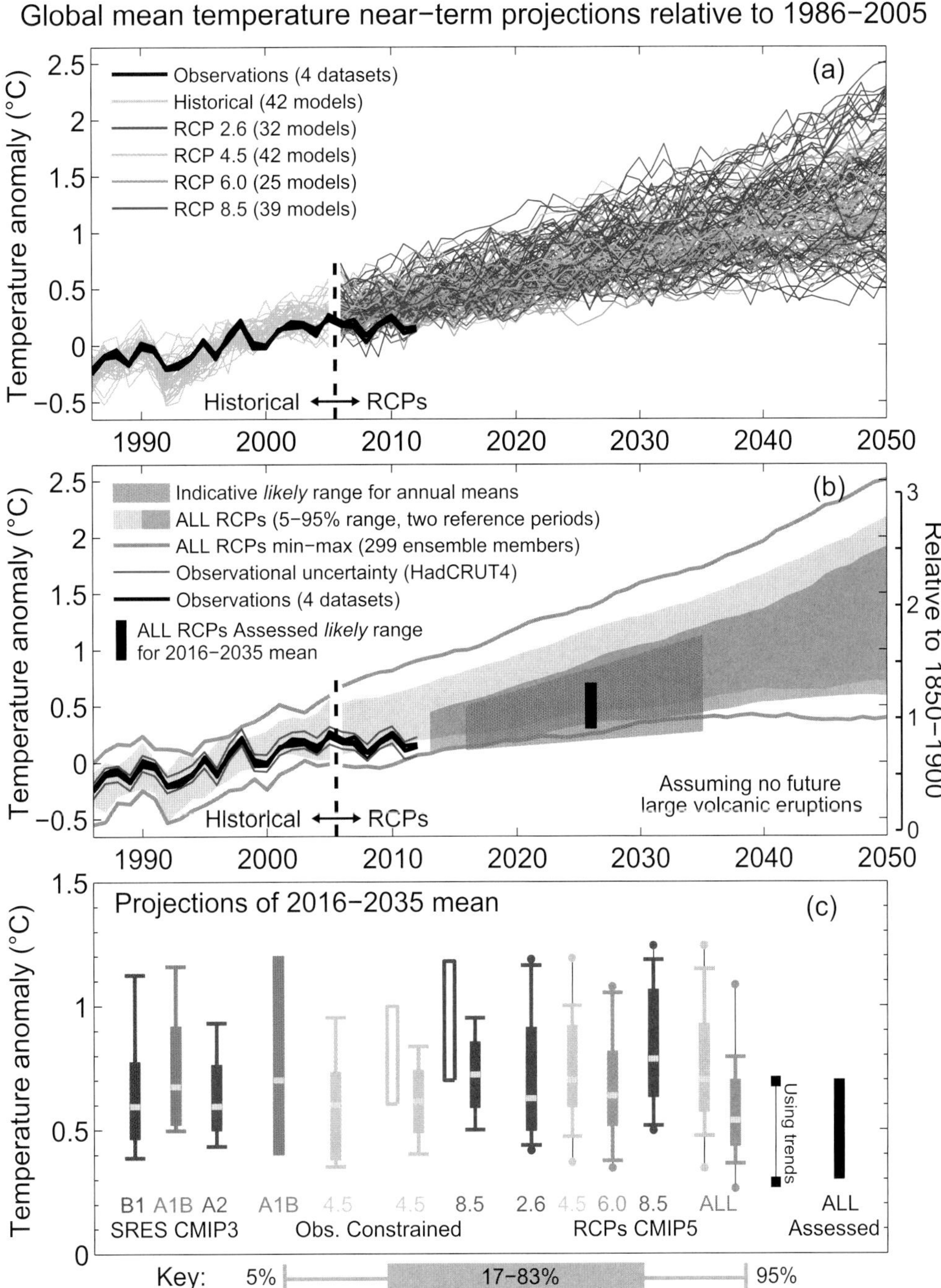

Figure TS.14 | Synthesis of near-term projections of global mean surface air temperature (GMST). (a) Projections of annual mean GMST 1986–2050 (anomalies relative to 1986–2005) under all RCPs from CMIP5 models (grey and coloured lines, one ensemble member per model), with four observational estimates (Hadley Centre/Climatic Research Unit gridded surface temperature data set 4 (HadCRUT4), European Centre for Medium Range Weather Forecasts (ECMWF) interim re-analysis of the global atmosphere and surface conditions (ERA-Interim), Goddard Institute for Space Studies Surface Temperature Analysis (GISTEMP), National Oceanic and Atmospheric Administration (NOAA)) for the period 1986–2012 (black lines). (b) As (a) but showing the 5 to 95% range of annual mean CMIP5 projections (using one ensemble member per model) for all RCPs using a reference period of 1986–2005 (light grey shade) and all RCPs using a reference period of 2006–2012, together with the observed anomaly for (2006–2012) minus (1986–2005) of 0.16°C (dark grey shade). The percentiles for 2006 onwards have been smoothed with a 5-year running mean for clarity. The maximum and minimum values from CMIP5 using all ensemble members and the 1986–2005 reference period are shown by the grey lines (also smoothed). Black lines show annual mean observational estimates. The red shaded region shows the indicative *likely* range for annual mean GMST during the period 2016–2035 based on the 'ALL RCPs Assessed' *likely* range for the 20-year mean GMST anomaly for 2016–2035, which is shown as a black bar in both (b) and (c) (see text for details). The temperature scale relative to 1850-1900 mean climate on the right-hand side assumes a warming of GMST prior to 1986–2005 of 0.61°C estimated from HadCRUT4. (c) A synthesis of projections for the mean GMST anomaly for 2016–2035 relative to 1986–2005. The box and whiskers represent the 66% and 90% ranges. Shown are: unconstrained SRES CMIP3 and RCP CMIP5 projections; observationally constrained projections for the SRES A1B and, the RCP4.5 and 8.5 scenarios; unconstrained projections for all four RCP scenarios using two reference periods as in (b) (light grey and dark grey shades), consistent with (b); 90% range estimated using CMIP5 trends for the period 2012–2035 and the observed GMST anomaly for 2012; an overall *likely* (>66%) assessed range for all RCP scenarios. The dots for the CMIP5 estimates show the maximum and minimum values using all ensemble members. The medians (or maximum likelihood estimate; green filled bar) are indicated by a grey band. (Adapted from Figure 11.25.) See Section 11.3.6 for details. {Figure 11.25}

TS.5.4.3 Projected Near-term Changes in the Water Cycle

Zonal mean precipitation will *very likely* increase in high and some of the mid latitudes, and will *more likely than not* decrease in the subtropics. At more regional scales precipitation changes may be dominated by a combination of natural internal variability, volcanic forcing and anthropogenic aerosol effects. {11.3.2}

Over the next few decades increases in near-surface specific humidity are *very likely*. It is *likely* that there will be increases in evaporation in many regions. There is *low confidence* in projected changes in soil moisture and surface runoff. {11.3.2}

In the near term, it is *likely* that the frequency and intensity of heavy precipitation events will increase over land. These changes are primarily driven by increases in atmospheric water vapour content, but also affected by changes in atmospheric circulation. The impact of anthropogenic forcing at regional scales is less obvious, as regional-scale changes are strongly affected by natural variability and also depend on the course of future aerosol emissions, volcanic forcing and land use changes (see also TFE.9). {11.3.2}

TS

TS.5.4.4 Projected Near-term Changes in Atmospheric Circulation

Internally generated climate variability and multiple RF agents (e.g., volcanoes, GHGs, ozone and anthropogenic aerosols) will all contribute to near-term changes in the atmospheric circulation. For example, it is *likely* that the annual mean Hadley Circulation and the SH mid-latitude westerlies will shift poleward, while it is *likely* that the projected recovery of stratospheric ozone and increases in GHG concentrations will have counteracting impacts on the width of the Hadley Circulation and the meridional position of the SH storm track. Therefore it is *unlikely* that they will continue to expand poleward as rapidly as in recent decades. {11.3.2}

There is *low confidence* in near-term projections of the position and strength of NH storm tracks. Natural variations are larger than the projected impact of GHGs in the near term. {11.3.2}

There is *low confidence* in basin-scale projections of changes in intensity and frequency of tropical cyclones in all basins to the mid-21st century. This *low confidence* reflects the small number of studies exploring near-term tropical cyclone activity, the differences across published projections of tropical cyclone activity, and the large role for natural variability. There is *low confidence* in near-term projections for increased tropical cyclone intensity in the Atlantic; this projection is in part due to projected reductions in aerosol loading. {11.3.2}

TS.5.4.5 Projected Near-term Changes in the Ocean

It is *very likely* that globally averaged surface and vertically averaged ocean temperatures will increase in the near-term. In the absence of multiple major volcanic eruptions, it is *very likely* that globally averaged surface and depth-averaged temperatures averaged for 2016–2035 will be warmer than those averaged over 1986–2005. {11.3.3}

It is *likely* that salinity will increase in the tropical and (especially) subtropical Atlantic, and decrease in the western tropical Pacific over the next few decades. Overall, it is *likely* that there will be some decline in the Atlantic Meridional Overturning Circulation by 2050 (*medium confidence*). However, the rate and magnitude of weakening is very uncertain and decades when this circulation increases are also to be expected. {11.3.3}

TS.5.4.6 Projected Near-term Changes in the Cryosphere

A nearly ice-free Arctic Ocean (sea ice extent less than 10^6 km^2 for at least five consecutive years) in September is *likely* before mid-century under RCP8.5 (*medium confidence*). This assessment is based on a subset of models that most closely reproduce the climatological mean state and 1979 to 2012 trend of Arctic sea ice cover. It is *very likely* that there will be further shrinking and thinning of Arctic sea ice cover, and decreases of northern high-latitude spring time snow cover and near surface permafrost as GMST rises (Figures TS.17 and TS.18). There is *low confidence* in projected near-term decreases in the Antarctic sea ice extent and volume. {11.3.4}

TS.5.4.7 Possibility of Near-term Abrupt Changes in Climate

There are various mechanisms that could lead to changes in global or regional climate that are abrupt by comparison with rates experienced in recent decades. The likelihood of such changes is generally lower for the near term than for the long term. For this reason the relevant mechanisms are primarily assessed in the TS.5 sections on long-term changes and in TFE.5. {11.3.4}

TS.5.4.8 Projected Near-term Changes in Air Quality

The range in projections of air quality (O_3 and $PM_{2.5}$ in surface air) is driven primarily by emissions (including CH_4), rather than by physical climate change (*medium confidence*). The response of air quality to climate-driven changes is more uncertain than the response to emission-driven changes (*high confidence*). Globally, warming decreases background surface O_3 (*high confidence*). High CH_4 levels (such as RCP8.5 and SRES A2) can offset this decrease, raising 2100 background surface O_3 on average by about 8 ppb (25% of current levels) relative to scenarios with small CH_4 changes (such as RCP4.5 and RCP6.0) (*high confidence*). On a continental scale, projected air pollution levels are lower under the new RCP scenarios than under the SRES scenarios because the SRES did not incorporate air quality legislation (*high confidence*). {11.3.5, 11.3.5.2; Figures 11.22 and 11.23ab, AII.4.2, AII.7.1–AII.7.4}

Observational and modelling evidence indicates that, all else being equal, locally higher surface temperatures in polluted regions will trigger regional feedbacks in chemistry and local emissions that will increase peak levels of O_3 and $PM_{2.5}$ (*medium confidence*). Local emissions combined with background levels and with meteorological conditions conducive to the formation and accumulation of pollution are known to produce extreme pollution episodes on local and regional scales. There is *low confidence* in projecting changes in meteorological blocking associated with these extreme episodes. For $PM_{2.5}$, climate change may alter natural aerosol sources (wildfires, wind-lofted dust, biogenic precursors) as well as precipitation scavenging, but no confidence level is attached to the overall impact of climate change on $PM_{2.5}$ distributions. {11.3.5, 11.3.5.2; Box 14.2}

TS.5.5 Long-term Climate Change

TS.5.5.1 Projected Long-term Changes in Global Temperature

Global mean temperatures will continue to rise over the 21st century under all of the RCPs. From around the mid-21st century, the rate of global warming begins to be more strongly dependent on the scenario (Figure TS.15). {12.4.1}

Under the assumptions of the concentration-driven RCPs, GMSTs for 2081–2100, relative to 1986–2005 will *likely* be in the 5 to 95% range of the CMIP5 models; 0.3°C to 1.7°C (RCP2.6), 1.1 to 2.6°C (RCP4.5), 1.4°C to 3.1°C (RCP6.0), 2.6°C to 4.8°C (RCP8.5) (see Table TS.1). With *high confidence*, the 5 to 95% range of CMIP5 is assessed as *likely* rather than *very likely* based on the assessment of TCR (see TFE.6). The 5 to 95% range of CMIP5 for global mean temperature change is also assessed as *likely* for mid-21st century, but only with *medium confidence*. With respect to 1850–1900 mean conditions, global temperatures averaged in the period 2081–2100 are projected to *likely* exceed 1.5°C above 1850–1900 values for RCP4.5, RCP6.0 and RCP8.5 (*high confidence*) and are *likely* to exceed 2°C above 1850–1900 values for RCP6.0 and RCP8.5 (*high confidence*). Temperature change above 2°C relative to 1850–1900 under RCP2.6 is *unlikely* (*medium confidence*). Warming above 4°C by 2081–2100 is *unlikely* in all RCPs (*high confidence*) except for RCP8.5, where it is *about as likely as not* (*medium confidence*). {12.4.1; Tables 12.2, 12.3}

TS.5.5.2 Projected Long-term Changes in Regional Temperature

There is *very high confidence* that globally averaged changes over land will exceed changes over the ocean at the end of the 21st century by a factor that is *likely* in the range 1.4 to 1.7. In the absence of a strong reduction in the Atlantic Meridional Overturning, the Arctic region is projected to warm most (*very high confidence*) (Figure TS.15). As

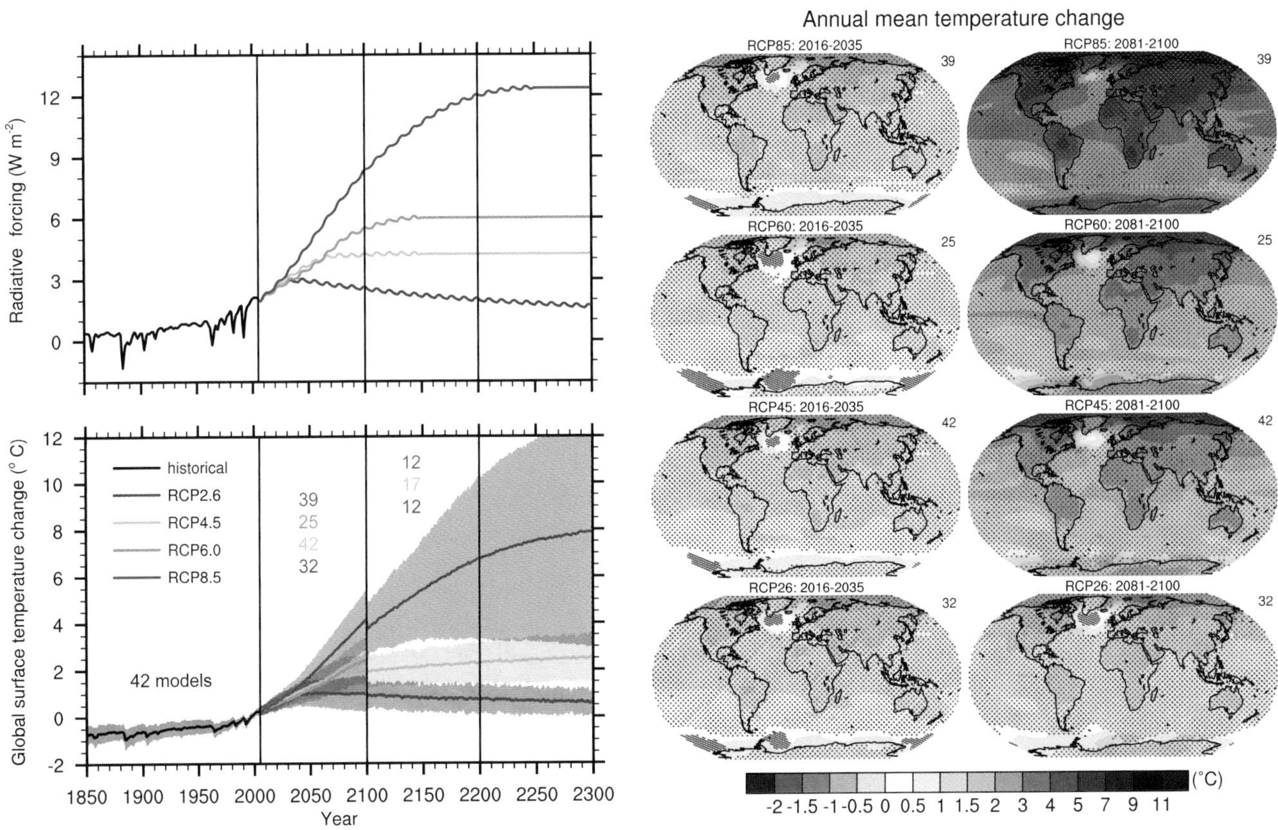

Figure TS.15 | (Top left) Total global mean radiative forcing for the four RCP scenarios based on the Model for the Assessment of Greenhouse-gas Induced Climate Change (MAGICC) energy balance model. Note that the actual forcing simulated by the CMIP5 models differs slightly between models. (Bottom left) Time series of global annual mean surface air temperature anomalies (relative to 1986–2005) from CMIP5 concentration-driven experiments. Projections are shown for each RCP for the multi-model mean (solid lines) and ±1.64 standard deviation (5 to 95%) across the distribution of individual models (shading), based on annual means. The 1.64 standard deviation range based on the 20 yr averages 2081–2100, relative to 1986–2005, are interpreted as *likely* changes for the end of the 21st century. Discontinuities at 2100 are due to different numbers of models performing the extension runs beyond the 21st century and have no physical meaning. Numbers in the same colours as the lines indicate the number of different models contributing to the different time periods. Maps: Multi-model ensemble average of annual mean surface air temperature change (compared to 1986–2005 base period) for 2016–2035 and 2081–2100, for RCP2.6, 4.5, 6.0 and 8.5. Hatching indicates regions where the multi-model mean signal is less than one standard deviation of internal variability. Stippling indicates regions where the multi-model mean signal is greater than two standard deviations of internal variability and where 90% of the models agree on the sign of change. The number of CMIP5 models used is indicated in the upper right corner of each panel. Further detail regarding the related Figures SPM.7a and SPM.8.a is given in the TS Supplementary Material. {Box 12.1; Figures 12.4, 12.5, 12.11; Annex I}

Table TS.1 | Projected change in global mean surface air temperature and global mean sea level rise for the mid- and late 21st century relative to the reference period of 1986–2005. {12.4.1; Tables 12.2, 13.5}

		2046–2065		2081–2100	
	Scenario	**Mean**	***Likely* range[c]**	**Mean**	***Likely* range[c]**
Global Mean Surface Temperature Change (°C)[a]	RCP2.6	1.0	0.4 to 1.6	1.0	0.3 to 1.7
	RCP4.5	1.4	0.9 to 2.0	1.8	1.1 to 2.6
	RCP6.0	1.3	0.8 to 1.8	2.2	1.4 to 3.1
	RCP8.5	2.0	1.4 to 2.6	3.7	2.6 to 4.8
	Scenario	**Mean**	***Likely* range[d]**	**Mean**	***Likely* range[d]**
Global Mean Sea Level Rise (m)[b]	RCP2.6	0.24	0.17 to 0.32	0.40	0.26 to 0.55
	RCP4.5	0.26	0.19 to 0.33	0.47	0.32 to 0.63
	RCP6.0	0.25	0.18 to 0.32	0.48	0.33 to 0.63
	RCP8.5	0.30	0.22 to 0.38	0.63	0.45 to 0.82

Notes:

[a] Based on the CMIP5 ensemble; anomalies calculated with respect to 1986–2005. Using HadCRUT4 and its uncertainty estimate (5–95% confidence interval), the observed warming to the reference period 1986–2005 is 0.61 [0.55 to 0.67] °C from 1850–1900, and 0.11 [0.09 to 0.13] °C from 1980–1999, the reference period for projections used in AR4. *Likely* ranges have not been assessed here with respect to earlier reference periods because methods are not generally available in the literature for combining the uncertainties in models and observations. Adding projected and observed changes does not account for potential effects of model biases compared to observations, and for natural internal variability during the observational reference period. {2.4; 11.2; Tables 12.2 and 12.3}

[b] Based on 21 CMIP5 models; anomalies calculated with respect to 1986–2005. Where CMIP5 results were not available for a particular AOGCM and scenario, they were estimated as explained in Chapter 13, Table 13.5. The contributions from ice sheet rapid dynamical change and anthropogenic land water storage are treated as having uniform probability distributions, and as largely independent of scenario. This treatment does not imply that the contributions concerned will not depend on the scenario followed, only that the current state of knowledge does not permit a quantitative assessment of the dependence. Based on current understanding, only the collapse of marine-based sectors of the Antarctic ice sheet, if initiated, could cause global mean sea level to rise substantially above the *likely* range during the 21st century. There is *medium confidence* that this additional contribution would not exceed several tenths of a metre of sea level rise during the 21st century.

[c] Calculated from projections as 5–95% model ranges. These ranges are then assessed to be *likely* ranges after accounting for additional uncertainties or different levels of confidence in models. For projections of global mean surface temperature change in 2046–2065 *confidence* is *medium*, because the relative importance of natural internal variability, and uncertainty in non-greenhouse gas forcing and response, are larger than for 2081–2100. The *likely* ranges for 2046–2065 do not take into account the possible influence of factors that lead to the assessed range for near-term (2016–2035) global mean surface temperature change that is lower than the 5–95% model range, because the influence of these factors on longer term projections has not been quantified due to insufficient scientific understanding. {11.3}

[d] Calculated from projections as 5–95% model ranges. These ranges are then assessed to be *likely* ranges after accounting for additional uncertainties or different levels of confidence in models. For projections of global mean sea level rise *confidence* is *medium* for both time horizons.

GMST rises, the pattern of atmospheric zonal mean temperatures show warming throughout the troposphere and cooling in the stratosphere, consistent with previous assessments. The consistency is especially clear in the tropical upper troposphere and the northern high latitudes. {12.4.3; Box 5.1}

It is *virtually certain* that, in most places, there will be more hot and fewer cold temperature extremes as global mean temperatures increase. These changes are expected for events defined as extremes on both daily and seasonal time scales. Increases in the frequency, duration and magnitude of hot extremes along with heat stress are expected; however, occasional cold winter extremes will continue to occur. Twenty-year return values of low-temperature events are projected to increase at a rate greater than winter mean temperatures in most regions, with the largest changes in the return values of low temperatures at high latitudes. Twenty-year return values for high-temperature events are projected to increase at a rate similar to or greater than the rate of increase of summer mean temperatures in most regions. Under RCP8.5 it is *likely* that, in most land regions, a current 20-year high-temperature event will occur more frequently by the end of the 21st century (at least doubling its frequency, but in many regions becoming an annual or 2-year event) and a current 20-year low-temperature event will become exceedingly rare (See also TFE.9). {12.4.3}

Models simulate a decrease in cloud amount in the future over most of the tropics and mid-latitudes, due mostly to reductions in low clouds. Changes in marine boundary layer clouds are most uncertain. Increases in cloud fraction and cloud optical depth and therefore cloud reflection are simulated in high latitudes, poleward of 50°. {12.4.3}

TS.5.5.3 Projected Long-term Changes in Atmospheric Circulation

Mean sea level pressure is projected to decrease in high latitudes and increase in the mid-latitudes as global temperatures rise. In the tropics, the Hadley and Walker Circulations are *likely* to slow down. Poleward shifts in the mid-latitude jets of about 1 to 2 degrees latitude are *likely* at the end of the 21st century under RCP8.5 in both hemispheres (*medium confidence*), with weaker shifts in the NH. In austral summer, the additional influence of stratospheric ozone recovery in the SH opposes changes due to GHGs there, though the net response varies strongly across models and scenarios. Substantial uncertainty and thus *low confidence* remains in projecting changes in NH storm tracks, especially for the North Atlantic basin. The Hadley Cell is *likely* to widen, which translates to broader tropical regions and a poleward encroachment of subtropical dry zones. In the stratosphere, the Brewer–Dobson circulation is *likely* to strengthen. {12.4.4}

TS.5.5.4 Projected Long-term Changes in the Water Cycle

On the planetary scale, relative humidity is projected to remain roughly constant, but specific humidity to increase in a warming climate. The projected differential warming of land and ocean promotes changes in atmospheric moistening that lead to small decreases in near-surface relative humidity over most land areas with the notable exception of parts of tropical Africa (*medium confidence*) (see TFE.1, Figure 1). {12.4.5}

It is *virtually certain* that, in the long term, global precipitation will increase with increased GMST. Global mean precipitation will increase at a rate per °C smaller than that of atmospheric water vapour. It will *likely* increase by 1 to 3% °C^{-1} for scenarios other than RCP2.6. For RCP2.6 the range of sensitivities in the CMIP5 models is 0.5 to 4% °C^{-1} at the end of the 21st century. {7.6.2, 7.6.3, 12.4.1}

Changes in average precipitation in a warmer world will exhibit substantial spatial variation under RCP8.5. Some regions will experience increases, other regions will experience decreases and yet others will not experience significant changes at all (see Figure TS.16). There is *high confidence* that the contrast of annual mean precipitation between dry and wet regions and that the contrast between wet and dry seasons will increase over most of the globe as temperatures increase. The general pattern of change indicates that high latitudes are *very likely* to experience greater amounts of precipitation due to the increased specific humidity of the warmer troposphere as well as increased transport of water vapour from the tropics by the end of this century under the RCP8.5 scenario. Many mid-latitude and subtropical arid and semi-arid regions will *likely* experience less precipitation and many moist mid-latitude regions will *likely* experience more precipitation by the end of this century under the RCP8.5 scenario. Maps of precipitation change for the four RCP scenarios are shown in Figure TS.16. {12.4.2, 12.4.5}

Globally, for short-duration precipitation events, a shift to more intense individual storms and fewer weak storms is *likely* as temperatures increase. Over most of the mid-latitude land masses and over wet tropical regions, extreme precipitation events will *very likely* be more intense and more frequent in a warmer world. The global average sensitivity of the 20-year return value of the annual maximum daily precipitation ranges from 4% °C^{-1} of local temperature increase (average of CMIP3 models) to 5.3% °C^{-1} of local temperature increase (average of CMIP5 models), but regionally there are wide variations. {12.4.2, 12.4.5}

Annual surface evaporation is projected to increase as global temperatures rise over most of the ocean and is projected to change over land following a similar pattern as precipitation. Decreases in annual runoff are *likely* in parts of southern Europe, the Middle East and southern Africa by the end of this century under the RCP8.5 scenario. Increases in annual runoff are *likely* in the high northern latitudes corresponding to large increases in winter and spring precipitation by the end of the 21st century under the RCP8.5 scenario. Regional to global-scale projected decreases in soil moisture and increased risk of agricultural drought are *likely* in presently dry regions and are projected with *medium confidence* by the end of this century under the RCP8.5 scenario. Prominent

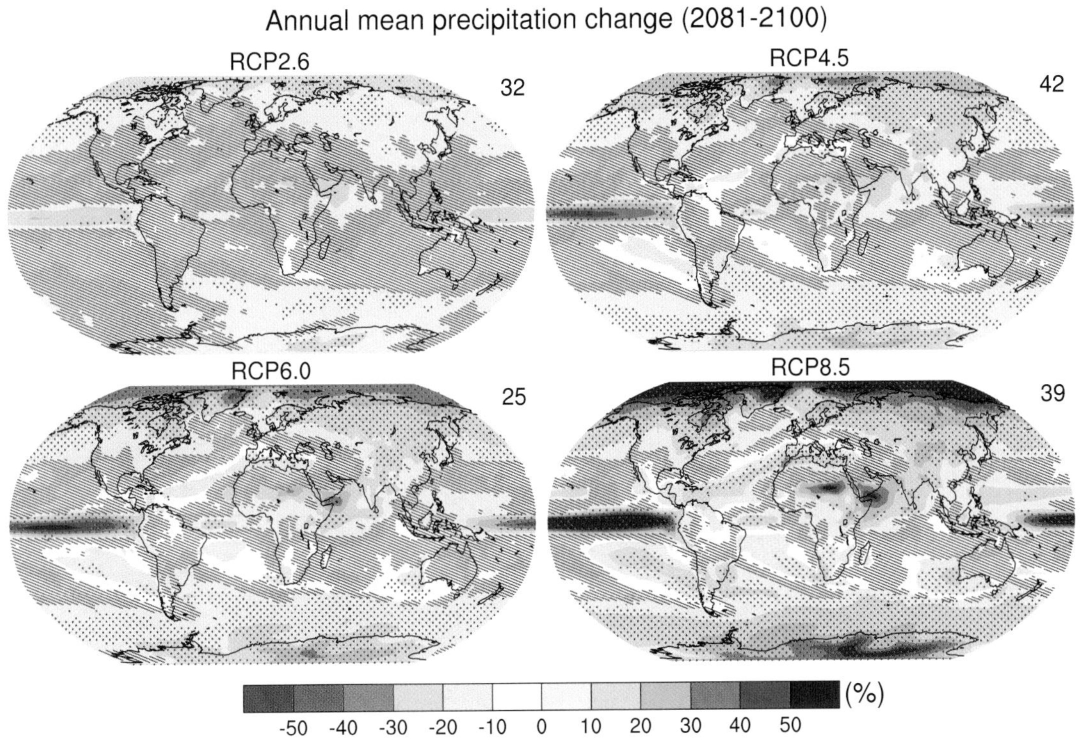

Figure TS.16 | Maps of multi-model results for the scenarios RCP2.6, RCP4.5, RCP6.0 and RCP8.5 in 2081–2100 of average percent change in mean precipitation. Changes are shown relative to 1986–2005. The number of CMIP5 models to calculate the multi-model mean is indicated in the upper right corner of each panel. Hatching indicates regions where the multi- model mean signal is less than 1 standard deviation of internal variability. Stippling indicates regions where the multi- model mean signal is greater than 2 standard deviations of internal variability and where 90% of models agree on the sign of change (see Box 12.1). Further detail regarding the related Figure SPM.8b is given in the TS Supplementary Material. {Figure 12.22; Annex I}

areas of projected decreases in evaporation include southern Africa and northwestern Africa along the Mediterranean. Soil moisture drying in the Mediterranean and southern African regions is consistent with projected changes in Hadley Circulation and increased surface temperatures, so surface drying in these regions as global temperatures increase is *likely* with *high confidence* by the end of this century under the RCP8.5 scenario. In regions where surface moistening is projected, changes are generally smaller than natural variability on the 20-year time scale. A summary of the projected changes in the water cycle from the CMIP5 models is shown in TFE.1, Figure 1. {12.4.5; Box 12.1}

TS.5.5.5 Projected Long-term Changes in the Cryosphere

It is *very likely* that the Arctic sea ice cover will continue shrinking and thinning year-round in the course of the 21st century as GMST rises. At the same time, in the Antarctic, a decrease in sea ice extent and volume is expected, but with *low confidence*. The CMIP5 multi-model projections give average reductions in Arctic sea ice extent for 2081–2100 compared to 1986–2005 ranging from 8% for RCP2.6 to 34% for RCP8.5 in February and from 43% for RCP2.6 to 94% for RCP8.5 in September (*medium confidence*) (Figure TS.17). A nearly ice-free Arctic Ocean (sea ice extent less than 10^6 km^2 for at least five consecutive years) in September before mid-century is *likely* under RCP8.5 (*medium confidence*), based on an assessment of a subset of models that most closely reproduce the climatological mean state and 1979–2012 trend of the Arctic sea ice cover. Some climate projections exhibit 5- to 10-year periods of sharp summer Arctic sea ice decline—even steeper than observed over the last decade—and it is *likely* that such instances of rapid ice loss will occur in the future. There is little evidence in global climate models of a tipping point (or critical threshold) in the transition from a perennially ice-covered to a seasonally ice-free Arctic Ocean beyond which further sea ice loss is unstoppable and irreversible. In the Antarctic, the CMIP5 multi-model mean projects a decrease in sea ice extent that ranges from 16% for RCP2.6 to 67% for RCP8.5 in February and from 8% for RCP2.6 to 30% for RCP8.5 in September for 2081–2100 compared to 1986–2005. There is, however, *low confidence* in those projections because of the wide inter-model spread and the inability of almost all of the available models to reproduce the overall increase of the Antarctic sea ice areal coverage observed during the satellite era. {12.4.6, 12.5.5}

It is *very likely* that NH snow cover will reduce as global temperatures rise over the coming century. A retreat of permafrost extent with rising global temperatures is *virtually certain*. Snow cover changes result from precipitation and ablation changes, which are sometimes opposite. Projections of the NH spring snow covered area by the end of the 21st century vary between a decrease of 7 [3 to 10] % (RCP2.6) and 25 [18 to 32] % (RCP8.5) (Figure TS.18), but *confidence* is those numbers is only *medium* because snow processes in global climate models are strongly simplified. The projected changes in permafrost are a response not only to warming, but also to changes in snow cover, which exerts a control on the underlying soil. By the end of the 21st century, diagnosed near-surface permafrost area is projected to decrease by between 37% (RCP2.6) to 81% (RCP8.5) (*medium confidence*). {12.4.6}

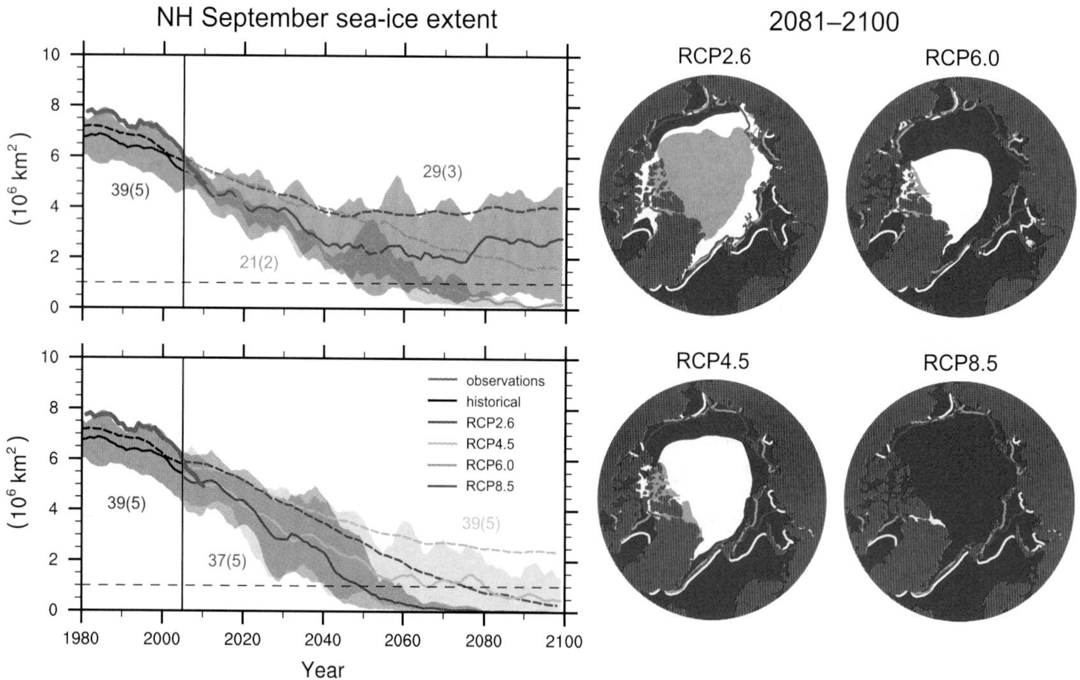

Figure TS.17 | Northern Hemisphere (NH) sea ice extent in September over the late 20th century and the whole 21st century for the scenarios RCP2.6, RCP4.5, RCP6.0 and RCP8.5 in the CMIP5 models, and corresponding maps of multi-model results in 2081–2100 of NH September sea ice extent. In the time series, the number of CMIP5 models to calculate the multi-model mean is indicated (subset in brackets). Time series are given as 5-year running means. The projected mean sea ice extent of a subset of models that most closely reproduce the climatological mean state and 1979–2012 trend of the Arctic sea ice is given (solid lines), with the minimum to maximum range of the subset indicated with shading. Black (grey shading) is the modelled historical evolution using historical reconstructed forcings. The CMIP5 multi-model mean is indicated with dashed lines. In the maps, the CMIP5 multi-model mean is given in white and the results for the subset in grey. Filled areas mark the averages over the 2081–2100 period, lines mark the sea ice extent averaged over the 1986–2005 period. The observed sea ice extent is given in pink as a time series and averaged over 1986–2005 as a pink line in the map. Further detail regarding the related Figures SPM.7b and SPM.8c is given in the TS Supplementary Material. {Figures 12.18, 12.29, 12.31}

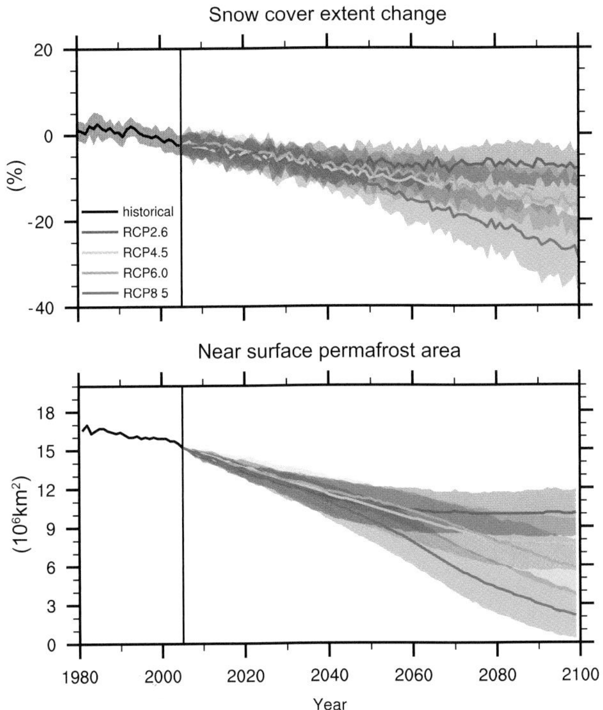

Figure TS.18 | (Top) Northern Hemisphere (NH) spring (March to April average) relative snow-covered area (RSCA) in CMIP5, obtained by dividing the simulated 5-year box smoothed spring snow-covered area (SCA) by the simulated average spring SCA of 1986–2005 reference period. (Bottom) NH diagnosed near-surface permafrost area in CMIP5, using 20-year average monthly surface air temperatures and snow depths. Lines indicate the multi model average, shading indicates the inter-model spread (one standard deviation). {Figures 12.32, 12.33}

TS.5.5.6 Projected Long-term Changes in the Ocean

Over the course of the 21st century, the global ocean will warm in all RCP scenarios. The strongest ocean warming is projected for the surface in subtropical and tropical regions. At greater depth the warming is projected to be most pronounced in the Southern Ocean. Best estimates of ocean warming in the top one hundred metres are about 0.6°C (RCP2.6) to 2.0°C (RCP8.5), and 0.3°C (RCP2.6) to 0.6°C (RCP8.5) at a depth of about 1 km by the end of the 21st century. For RCP4.5 by the end of the 21st century, half of the energy taken up by the ocean is in the uppermost 700 m, and 85% is in the uppermost 2000 m. Due to the long time scales of this heat transfer from the surface to depth, ocean warming will continue for centuries, even if GHG emissions are decreased or concentrations kept constant, and will result in a continued contribution to sea level rise (see Section TS5.7). {12.4.3, 12.4.7}

TS.5.6 Long-term Projections of Carbon and Other Biogeochemical Cycles

Projections of the global carbon cycle to 2100 using the CMIP5 ESMs represent a wider range of complex interactions between the carbon cycle and the physical climate system. {6}

With *very high confidence*, ocean carbon uptake of anthropogenic CO_2 will continue under all four RCPs through to 2100, with higher uptake in higher concentration pathways. The future evolution of the land carbon uptake is much more uncertain. A majority of CMIP5 ESMs project a continued net carbon uptake by land ecosystems through 2100. Yet, a minority of models simulate a net CO_2 source to the atmosphere by 2100 due to the combined effect of climate change and land use change. In view of the large spread of model results and incomplete process representation, there is *low confidence* on the magnitude of modelled future land carbon changes. {6.4.3}

There is *high confidence* that climate change will partially offset increases in global land and ocean carbon sinks caused by rising atmospheric CO_2. Yet, there are regional differences among CMIP5 ESMs in the response of ocean and land CO_2 fluxes to climate. There is high agreement between models that tropical ecosystems will store less carbon in a warmer climate. There is medium agreement between the CMIP5 ESMs that at high latitudes warming will increase land carbon storage, although none of these models accounts for decomposition of carbon in permafrost which may offset increased land carbon storage. There is *high confidence* that reductions in permafrost extent due to warming will cause thawing of some currently frozen carbon. However, there is *low confidence* on the magnitude of carbon losses through CO_2 and CH_4 emissions to the atmosphere with a range from 50 to 250 PgC between 2000 and 2100 for RCP8.5. {6.4.2, 6.4.3}

The loss of carbon from frozen soils constitutes a positive radiative feedback that is missing in current coupled ESM projections. There is high agreement between CMIP5 ESMs that ocean warming and circulation changes will reduce the rate of ocean carbon uptake in the Southern Ocean and North Atlantic, but that carbon uptake will nevertheless persist in those regions. {6.4.2}

It is *very likely*, based on new experimental results and modelling, that nutrient shortage will limit the effect of rising atmospheric CO_2 on future land carbon sinks for the four RCP scenarios. There is *high confidence* that low nitrogen availability will limit carbon storage on land even when considering anthropogenic nitrogen deposition. The role of phosphorus limitation is more uncertain. {6.4.6}

For the ESMs simulations driven by CO_2 concentrations, representation of the land and ocean carbon cycle allows quantification of the fossil fuel emissions compatible with the RCP scenarios. Between 2012 and 2100, ESM results imply cumulative compatible fossil fuel emissions of 270 [140 to 410] PgC for RCP2.6, 780 [595 to 1005] PgC for RCP4.5, 1060 [840 to 1250] PgC for RCP6.0 and 1685 [1415 to 1910] PgC for RCP8.5 (values quoted to nearest 5 PgC, range ±1 standard deviation derived from CMIP5 model results) (Figure TS.19). For RCP2.6, the models project an average 50% (range 14 to 96%) emission reduction by 2050 relative to 1990 levels. By the end of the 21st century, about half of the models infer emissions slightly above zero, while the other half infer a net removal of CO_2 from the atmosphere (see also Box TS.7). {6.4.3; Table 6.12}

When forced with RCP8.5 CO_2 emissions, as opposed to the RCP8.5 CO_2 concentrations, CMIP5 ESMs with interactive carbon cycles simulate, on average, a 50 (–140 to +210) ppm larger atmospheric CO_2 concentration and a 0.2 (–0.4 to +0.9) °C larger global surface temperature increase by 2100 (CMIP5 model spread). {12.4.8}

It is *virtually certain* that the increased storage of carbon by the ocean will increase acidification in the future, continuing the observed trends of the past decades. Ocean acidification in the surface ocean will follow atmospheric CO_2 and it will also increase in the deep ocean as CO_2 continues to penetrate the abyss. The CMIP5 models consistently project worldwide increased ocean acidification to 2100 under all RCPs. The corresponding decrease in surface ocean pH by the end of 21st century is 0.065 (0.06 to 0.07) for RCP2.6, 0.145 (0.14 to 0.15) for RCP4.5, 0.203 (0.20 to 0.21) for RCP6.0 and 0.31 (0.30 to 0.32) for RCP8.5 (CMIP5 model spread) (Figure TS.20). Surface waters are projected to become seasonally corrosive to aragonite in parts of the Arctic and in some coastal upwelling systems within a decade, and

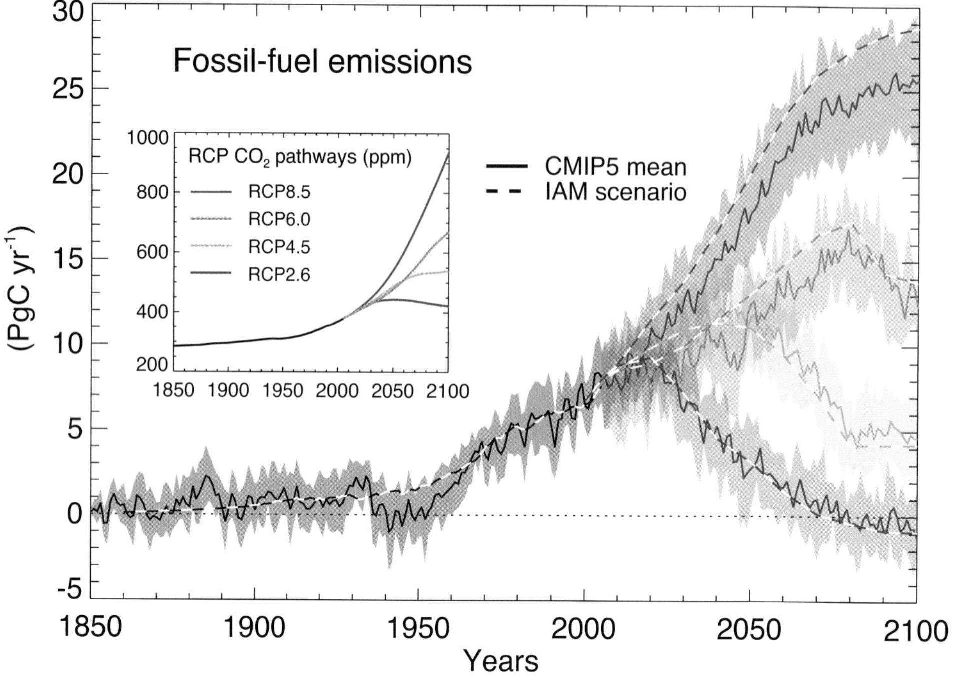

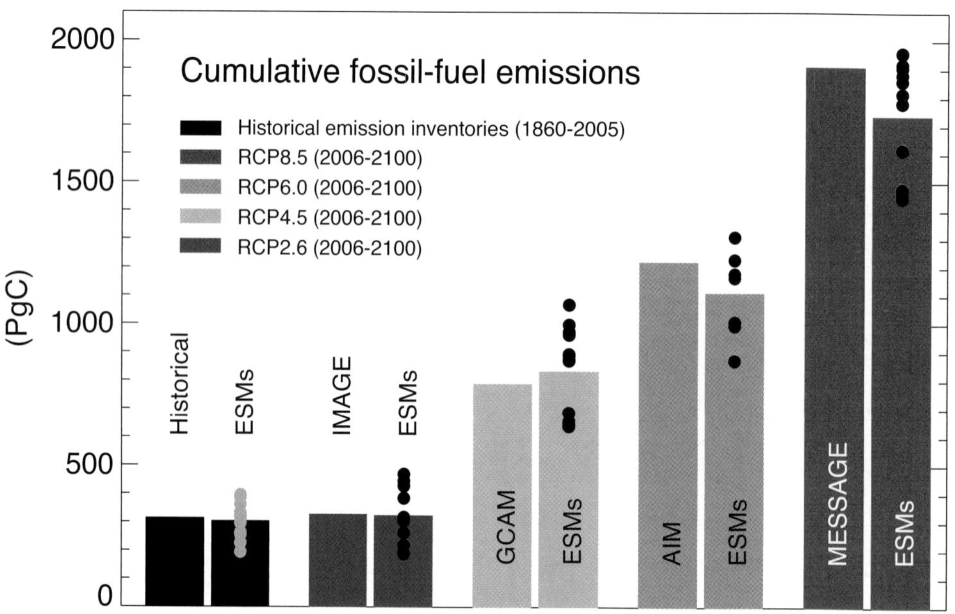

Figure TS.19 | Compatible fossil fuel emissions simulated by the CMIP5 models for the four RCP scenarios. (Top) Time series of annual emission (PgC yr^{-1}). Dashed lines represent the historical estimates and RCP emissions calculated by the Integrated Assessment Models (IAMs) used to define the RCP scenarios, solid lines and plumes show results from CMIP5 Earth System Models (ESMs, model mean, with one standard deviation shaded). (Bottom) Cumulative emissions for the historical period (1860–2005) and 21st century (defined in CMIP5 as 2006–2100) for historical estimates and RCP scenarios. Left bars are cumulative emissions from the IAMs, right bars are the CMIP5 ESMs multi-model mean estimate and dots denote individual ESM results. From the CMIP5 ESMs results, total carbon in the land-atmosphere–ocean system can be tracked and changes in this total must equal fossil fuel emissions to the system. Hence the compatible emissions are given by cumulative emissions = $\Delta C_A + \Delta C_L + \Delta C_O$, while emission rate = $d/dt\ [C_A + C_L + C_O]$, where C_A, C_L, C_O are carbon stored in atmosphere, land and ocean respectively. Other sources and sinks of CO_2 such as from volcanism, sedimentation or rock weathering, which are very small on centennial time scales are not considered here. {Box 6.4; Figure 6.25}

in parts of the Southern Ocean within one to three decades in most scenarios. Aragonite, a less stable form of calcium carbonate, undersaturation becomes widespread in these regions at atmospheric CO_2 levels of 500 to 600 ppm. {6.4.4}

It is *very likely* that the dissolved oxygen content of the ocean will decrease by a few percent during the 21st century in response to surface warming. CMIP5 models suggest that this decrease in dissolved oxygen will predominantly occur in the subsurface mid-latitude oceans, caused by enhanced stratification, reduced ventilation and warming. However, there is no consensus on the future development of the volume of hypoxic and suboxic waters in the open ocean because of large uncertainties in potential biogeochemical effects and in the evolution of tropical ocean dynamics. {6.4.5}

With *very high confidence*, the carbon cycle in the ocean and on land will continue to respond to climate change and atmospheric CO_2 increases that arise during the 21st century (see TFE.7 and TFE 8). {6.4}

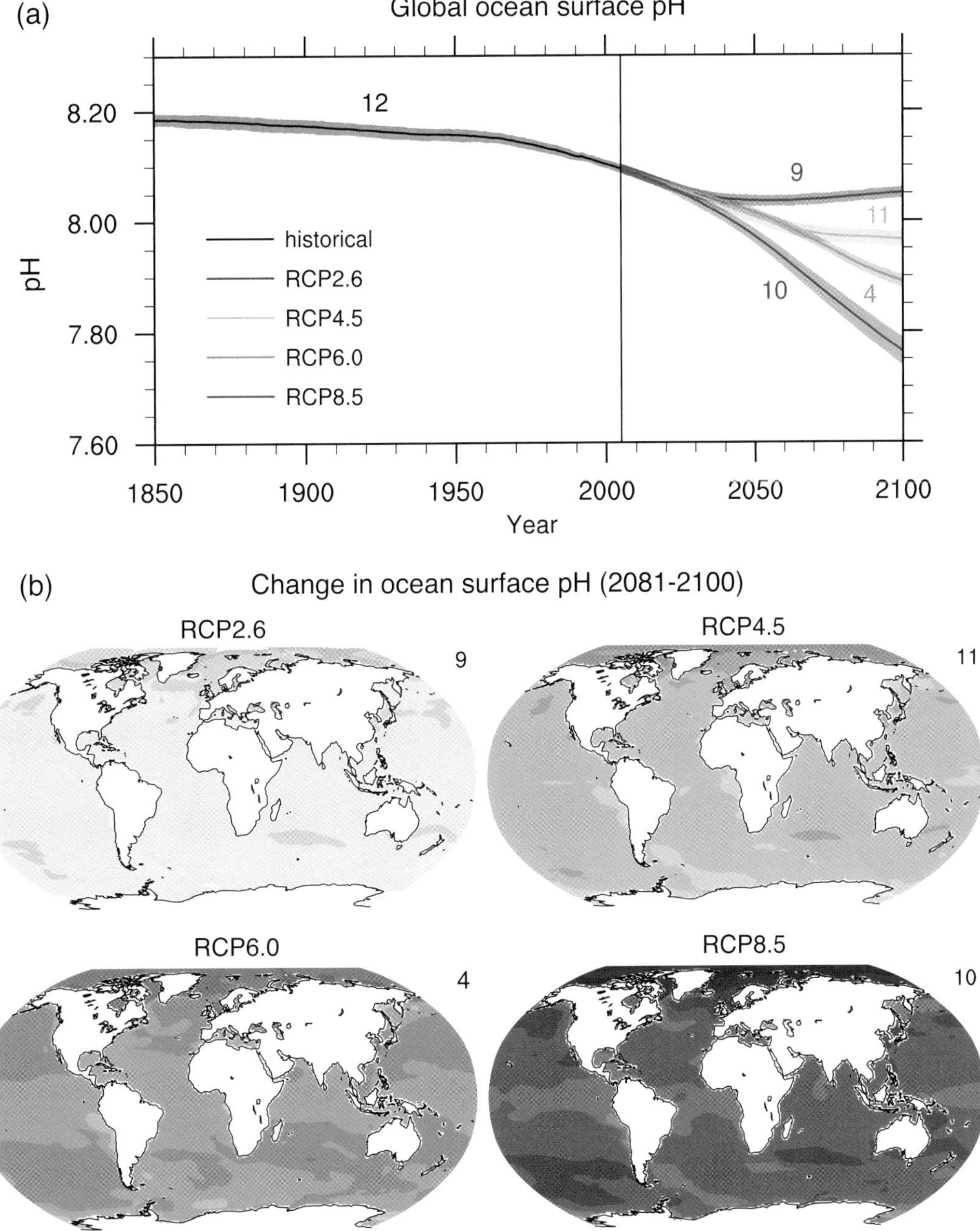

Figure TS.20 | (a) Time series (model averages and minimum to maximum ranges) and (b) maps of multi-model surface ocean pH for the scenarios RCP2.6, RCP4.5, RCP6.0 and RCP8.5 in 2081–2100. The maps in (b) show change in global ocean surface pH in 2081–2100 relative to 1986–2005. The number of CMIP5 models to calculate the multi-model mean is indicated in the upper right corner of each panel. Further detail regarding the related Figures SPM.7c and SPM.8.d is given in the TS Supplementary Material. {Figure 6.28}

Thematic Focus Elements

TFE.7 | Carbon Cycle Perturbation and Uncertainties

The natural carbon cycle has been perturbed since the beginning of the Industrial Revolution (about 1750) by the anthropogenic release of carbon dioxide (CO_2) to the atmosphere, virtually all from fossil fuel combustion and land use change, with a small contribution from cement production. Fossil fuel burning is a process related to energy production. Fossil fuel carbon comes from geological deposits of coal, oil and gas that were buried in the Earth crust for millions of years. Land use change CO_2 emissions are related to the conversion of natural ecosystems into managed ecosystems for food, feed and timber production with CO_2 being emitted from the burning of plant material or from the decomposition of dead plants and soil organic carbon. For instance when a forest is cleared, the plant material may be released to the atmosphere quickly through burning or over many years as the dead biomass and soil carbon decay on their own. {6.1, 6.3; Table 6.1}

The human caused excess of CO_2 in the atmosphere is partly removed from the atmosphere by carbon sinks in land ecosystems and in the ocean, currently leaving less than half of the CO_2 emissions in the atmosphere. Natural carbon sinks are due to physical, biological and chemical processes acting on different time scales. An excess of atmospheric CO_2 supports photosynthetic CO_2 fixation by plants that is stored as plant biomass or in the soil. The residence times of stored carbon on land depends on the compartments (plant/soil) and composition of the organic carbon, with time horizons varying from days to centuries. The increased storage in terrestrial ecosystems not affected by land use change is *likely* to be caused by enhanced photosynthesis at higher CO_2 levels and nitrogen deposition, and changes in climate favoring carbon sinks such as longer growing seasons in mid-to-high latitudes. {6.3, 6.3.1}

The uptake of anthropogenic CO_2 by the ocean is primarily a response to increasing CO_2 in the atmosphere. Excess atmospheric CO_2 absorbed by the surface ocean or transported to the ocean through aquatic systems (e.g., rivers, groundwaters) gets buried in coastal sediments or transported to deep waters where it is stored for decades to centuries. The deep ocean carbon can dissolve ocean carbonate sediments to store excess CO_2 on time scales of centuries to millennia. Within a 1 kyr, the remaining atmospheric fraction of the CO_2 emissions will be between 15 and 40%, depending on the amount of carbon released (TFE.7, Figure 1). On geological time scales of 10 kyr or longer, additional CO_2 is removed very slowly from the atmosphere by rock weathering, pulling the remaining atmospheric CO_2 fraction down to 10 to 25% after 10 kyr. {Box 6.1}

The carbon cycle response to future climate and CO_2 changes can be viewed as two strong and opposing feedbacks. The concentration–carbon feedback determines changes in storage due to elevated CO_2, and the climate–carbon feedback determines changes in carbon storage due to changes in climate. There is *high confidence* that increased atmospheric CO_2 will lead to increased land and ocean carbon uptake but by an uncertain amount. Models agree on the positive sign of land and ocean response to rising CO_2 but show only medium and low agreement for the magnitude of ocean and land carbon uptake respectively (TFE.7, Figure 2). Future climate change will decrease land and ocean carbon uptake compared to the case with constant climate (*medium confidence*). This is further supported by paleoclimate observations and modelling indicating that there is a positive feedback between climate and the carbon cycle on century to millennial time scales. Models agree on the sign, globally negative, of land and ocean response to climate change but show low agreement on the magnitude of this response, especially for the land (TFE.7, Figure 2). A key update since the IPCC Fourth Assessment Report (AR4) is the introduction of nutrient dynamics in some land carbon models, in particular the limitations on plant growth imposed by nitrogen availability. There is *high confidence* that, at the global scale, relative to the Coupled Model Intercomparison Project Phase 5 (CMIP5) carbon-only Earth System

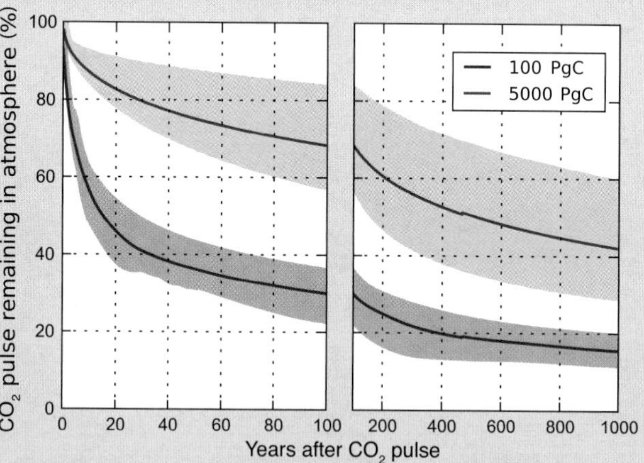

TFE.7, Figure 1 | Percentage of initial atmospheric CO_2 perturbation remaining in the atmosphere in response to an idealized instantaneous CO_2 emission pulse in year 0 as calculated by a range of coupled climate–carbon cycle models. Multi-model mean (line) and the uncertainty interval (maximum model range, shading) simulated during 100 years (left) and 1 kyr (right) following the instantaneous emission pulse of 100 PgC (blue) and 5,000 PgC (red). {Box 6.1, Figure 1}

(continued on next page)

TS

TFE.7 (continued)

Models (ESMs), CMIP5 ESMs including a land nitrogen cycle will reduce the strength of both the concentration–carbon feedback and the climate–carbon feedback of land ecosystems (TFE.7, Figure 2). Inclusion of nitrogen-cycle processes increases the spread across the CMIP5 ensemble. The CMIP5 spread in ocean sensitivity to CO_2 and climate appears reduced compared to AR4 (TFE.7, Figure 2). {6.2.3, 6.4.2}

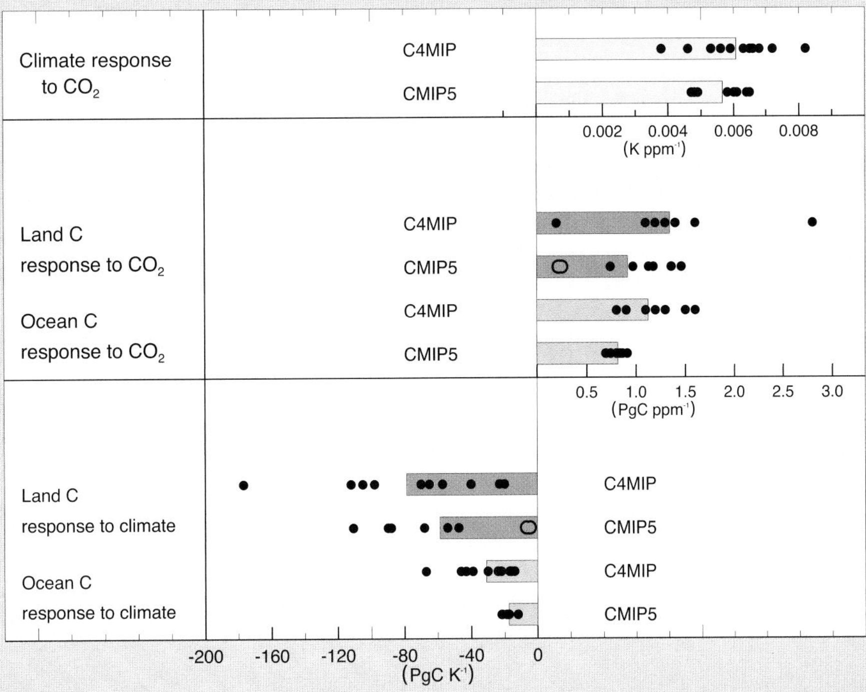

TFE.7, Figure 2 | Comparison of carbon cycle feedback metrics between the ensemble of seven General Circulation Models (GCMs) and four Earth System Models of Intermediate Complexity (EMICs) at the time of AR4 (Coupled Carbon Cycle Climate Model Intercomparison Project (C⁴MIP)) under the SRES A2 scenario and the eight CMIP5 models under the 140-year 1% CO_2 increase per year scenario. Black dots represent a single model simulation and coloured bars the mean of the multi-model results, grey dots are used for models with a coupled terrestrial nitrogen cycle. The comparison with C⁴MIP models is for context, but these metrics are known to be variable across different scenarios and rates of change (see Section 6.4.2). The SRES A2 scenario is closer in rate of change to a 0.5% CO_2 increase per year scenario and as such it should be expected that the CMIP5 climate–carbon sensitivity terms are comparable, but the concentration–carbon sensitivity terms are *likely* to be around 20% smaller for CMIP5 than for C⁴MIP due to lags in the ability of the land and ocean to respond to higher rates of CO_2 increase. This dependence on scenario reduces confidence in any quantitative statements of how CMIP5 carbon cycle feedbacks differ from C⁴MIP. {Figure 6.21}

With *very high confidence*, ocean carbon uptake of anthropogenic CO_2 emissions will continue under all four Representative Concentration Pathways (RCPs) through to 2100, with higher uptake corresponding to higher concentration pathways. The future evolution of the land carbon uptake is much more uncertain, with a majority of models projecting a continued net carbon uptake under all RCPs, but with some models simulating a net loss of carbon by the land due to the combined effect of climate change and land use change. In view of the large spread of model results and incomplete process representation, there is *low confidence* on the magnitude of modelled future land carbon changes. {6.4.3; Figure 6.24}

Biogeochemical cycles and feedbacks other than the carbon cycle play an important role in the future of the climate system, although the carbon cycle represents the strongest of these. Changes in the nitrogen cycle, in addition to interactions with CO_2 sources and sinks, affect emissions of nitrous oxide (N_2O) both on land and from the ocean. The human-caused creation of reactive nitrogen has increased steadily over the last two decades and is dominated by the production of ammonia for fertilizer and industry, with important contributions from legume cultivation and combustion of fossil fuels. {6.3}

Many processes, however, are not yet represented in coupled climate-biogeochemistry models (e.g., other processes involving other biogenic elements such as phosphorus, silicon and iron) so their magnitudes have to be estimated in offline or simpler models, which make their quantitative assessment difficult. It is *likely* that there will be nonlinear interactions between many of these processes, but these are not yet well quantified. Therefore any assessment of the future feedbacks between climate and biogeochemical cycles still contains large uncertainty. {6.4}

Box TS.7 | Climate Geoengineering Methods

Geoengineering is defined as the deliberate large-scale intervention in the Earth system to counter undesirable impacts of climate change on the planet. Carbon Dioxide Reduction (CDR) aims to slow or perhaps reverse projected increases in the future atmospheric CO_2 concentrations, accelerating the natural removal of atmospheric CO_2 and increasing the storage of carbon in land, ocean and geological reservoirs. Solar Radiation Management (SRM) aims to counter the warming associated with increasing GHG concentrations by reducing the amount of sunlight absorbed by the climate system. A related technique seeks to deliberately decrease the greenhouse effect in the climate system by altering high-level cloudiness. {6.5, 7.7; FAQ 7.3}

CDR methods could provide mitigation of climate change if CO_2 can be reduced, but there are uncertainties, side effects and risks, and implementation would depend on technological maturity along with economic, political and ethical considerations. CDR would *likely* need to be deployed at large-scale and over at least one century to be able to significantly reduce CO_2 concentrations. There are biogeochemical, and currently technical limitations that make it difficult to provide quantitative estimates of the potential for CDR. It is *virtually certain* that CO_2 removals from the atmosphere by CDR would be partially offset by outgassing of CO_2 previously stored in ocean and terrestrial carbon reservoirs. Some of the climatic and environmental side effects of CDR methods are associated with altered surface albedo from afforestation, ocean de-oxygenation from ocean fertilization, and enhanced N_2O emissions. Land-based CDR methods would probably face competing demands for land. The level of *confidence* on the effectiveness of CDR methods and their side effects on carbon and other biogeochemical cycles is *low*. {6.5; Box 6.2; FAQ 7.3}

SRM remains unimplemented and untested but, if realizable, could offset a global temperature rise and some of its effects. There is *medium confidence* that SRM through stratospheric aerosol injection is scalable to counter the RF and some of the climate effects expected from a twofold increase in CO_2 concentration. There is no consensus on whether a similarly large RF could be achieved from cloud brightening SRM due to insufficient understanding of aerosol–cloud interactions. It does not appear that land albedo change SRM could produce a large RF. Limited literature on other SRM methods precludes their assessment. {7.7.2, 7.7.3}

Numerous side effects, risks and shortcomings from SRM have been identified. SRM would produce an inexact compensation for the RF by GHGs. Several lines of evidence indicate that SRM would produce a small but significant decrease in global precipitation (with larger differences on regional scales) if the global surface temperature were maintained. Another side effect that is relatively well characterized is the likelihood of modest polar stratospheric ozone depletion associated with stratospheric aerosol SRM. There could also be other as yet unanticipated consequences. {7.6.3, 7.7.3, 7.7.4}

As long as GHG concentrations continued to increase, the SRM would require commensurate increase, exacerbating side effects. In addition, scaling SRM to substantial levels would carry the risk that if the SRM were terminated for any reason, there is *high confidence* that surface temperatures would increase rapidly (within a decade or two) to values consistent with the GHG forcing, which would stress systems sensitive to the rate of climate change. Finally, SRM would not compensate for ocean acidification from increasing CO_2. {7.7.3, 7.7.4}

TS.5.7 Long-term Projections of Sea Level Change

TS.5.7.1 Projections of Global Mean Sea Level Change for the 21st Century

GMSL rise for 2081–2100 (relative to 1986–2005) for the RCPs will *likely* be in the 5 to 95% ranges derived from CMIP5 climate projections in combination with process-based models of glacier and ice sheet surface mass balance, with possible ice sheet dynamical changes assessed from the published literature. These *likely* ranges are 0.26 to 0.55 m (RCP2.6), 0.32 to 0.63 m (RCP4.5), 0.33 to 0.63 m (RCP6.0) and 0.45 to 0.82 m (RCP8.5) (*medium confidence*) (Table TS.1, Figure TS.21). For RCP8.5 the range at 2100 is 0.52 to 0.98 m. The central projections for GMSL rise in all scenarios lie within a range of 0.05 m until the middle of the century, when they begin to diverge; by the late 21st century, they have a spread of 0.25 m. Although RCP4.5 and RCP6.0 are very similar at the end of the century, RCP4.5 has a greater rate of rise earlier in the century than RCP6.0. GMSL rise depends on the pathway of CO_2 emissions, not only on the cumulative total; reducing emissions earlier rather than later, for the same cumulative total, leads to a larger mitigation of sea level rise. {12.4.1, 13.4.1, 13.5.1; Table 13.5}

Confidence in the projected *likely* ranges comes from the consistency of process-based models with observations and physical understanding. The basis for higher projections has been considered and it has been concluded that there is currently insufficient evidence to evaluate the probability of specific levels above the *likely* range. Based on current understanding, only the collapse of marine-based sectors of the Antarctic ice sheet, if initiated, could cause GMSL to rise substantially above the *likely* range during the 21st century. There is a lack of consensus on the probability for such a collapse, and the potential additional contribution to GMSL rise cannot be precisely quantified, but there is *medium confidence* that it would not exceed several tenths of a metre of sea level rise during the 21st century. {13.5.1, 13.5.3}

Under all the RCP scenarios, the time-mean rate of GMSL rise during the 21st century is *very likely* to exceed the rate observed during 1971–2010. In the projections, the rate of rise initially increases. In RCP2.6 it becomes roughly constant (central projection about 4.5 mm yr^{-1}) before the middle of the century, and subsequently declines slightly. The rate of rise becomes roughly constant in RCP4.5 and RCP6.0 by the end of the 21st century, whereas acceleration continues throughout the century in RCP8.5 (reaching 11 [8 to 16] mm yr^{-1} during 2081–2100). {13.5.1; Table 13.5}

In all RCP scenarios, thermal expansion is the largest contribution, accounting for about 30 to 55% of the total. Glaciers are the next largest, accounting for 15-35%. By 2100, 15 to 55% of the present glacier volume is projected to be eliminated under RCP2.6, and 35 to 85% under RCP8.5 (*medium confidence*). The increase in surface melting in Greenland is projected to exceed the increase in accumulation, and there is *high confidence* that the surface mass balance changes on the Greenland ice sheet will make a positive contribution to sea level rise over the 21st century. On the Antarctic ice sheet, surface melting is projected to remain small, while there is *medium confidence* that snowfall will increase (Figure TS.21). {13.3.3, 13.4.3, 13.4.4, 13.5.1; Table 13.5}

There is *medium confidence* in the ability to model future rapid changes in ice sheet dynamics on decadal time scales. At the time of the AR4, scientific understanding was not sufficient to allow an assessment of the possibility of such changes. Since the publication of the AR4, there has been substantial progress in understanding the relevant processes as well as in developing new ice sheet models that are capable of simulating them. However, the published literature as yet provides only a partially sufficient basis for making projections related to particular scenarios. In our projections of GMSL rise by 2081–2100, the *likely* range from rapid changes in ice outflow is 0.03 to 0.20 m from the two ice sheets combined, and its inclusion is the most important reason why the projections are greater than those given in the AR4. {13.1.5, 13.5.1, 13.5.3}

Semi-empirical models are designed to reproduce the observed sea level record over their period of calibration, but do not attribute sea level rise to its individual physical components. For RCPs, some semi-empirical models project a range that overlaps the process-based *likely* range while others project a median and 95-percentile that are about twice as large as the process-based models. In nearly every case, the semi-empirical model 95th percentile is higher than the process-based *likely* range. For 2081–2100 (relative to 1986–2005) under RCP4.5, semi-empirical models give median projections in the range 0.56 to 0.97 m, and their 95th percentiles extend to about 1.2 m. This difference implies either that there is some contribution which is presently

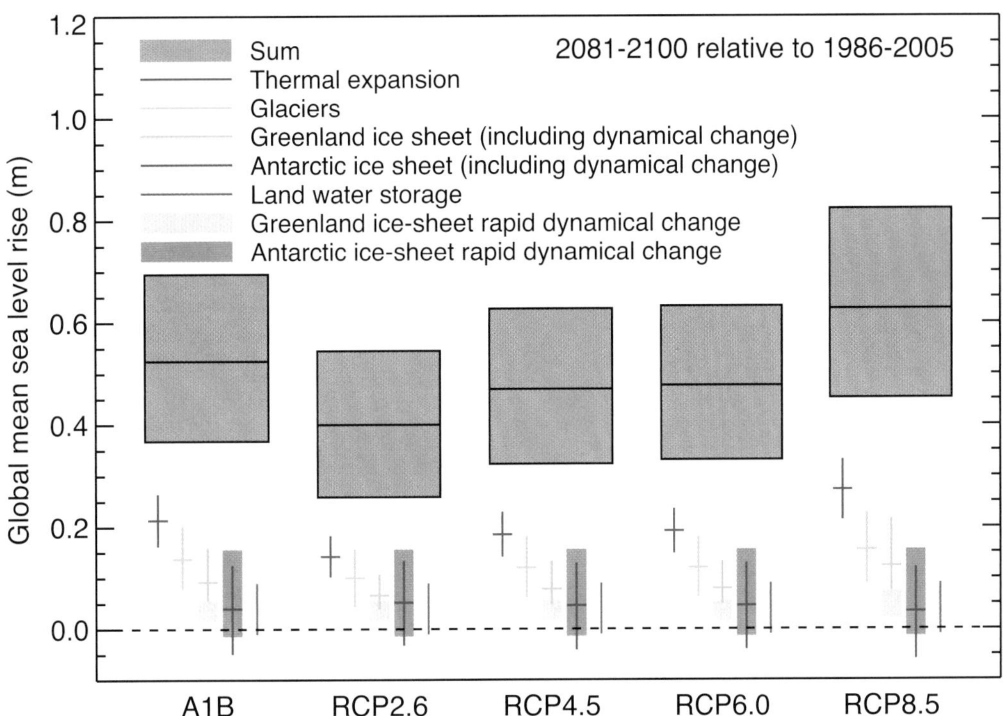

Figure TS.21 | Projections from process-based models with *likely* ranges and median values for global mean sea level (GMSL) rise and its contributions in 2081–2100 relative to 1986–2005 for the four RCP scenarios and scenario SRES A1B used in the AR4. The contributions from ice sheets include the contributions from ice sheet rapid dynamical change, which are also shown separately. The contributions from ice sheet rapid dynamics and anthropogenic land water storage are treated as having uniform probability distributions, and as independent of scenario (except that a higher rate of change is used for Greenland ice sheet outflow under RCP8.5). This treatment does not imply that the contributions concerned will not depend on the scenario followed, only that the current state of knowledge does not permit a quantitative assessment of the dependence. See discussion in Sections 13.5.1 and 13.5.3 and Supplementary Material for methods. Based on current understanding, only the collapse of the marine-based sectors of the Antarctic ice sheet, if initiated, could cause GMSL to rise substantially above the *likely* range during the 21st century. This potential additional contribution cannot be precisely quantified but there is *medium confidence* that it would not exceed several tenths of a metre during the 21st century. {Figure 13.10}

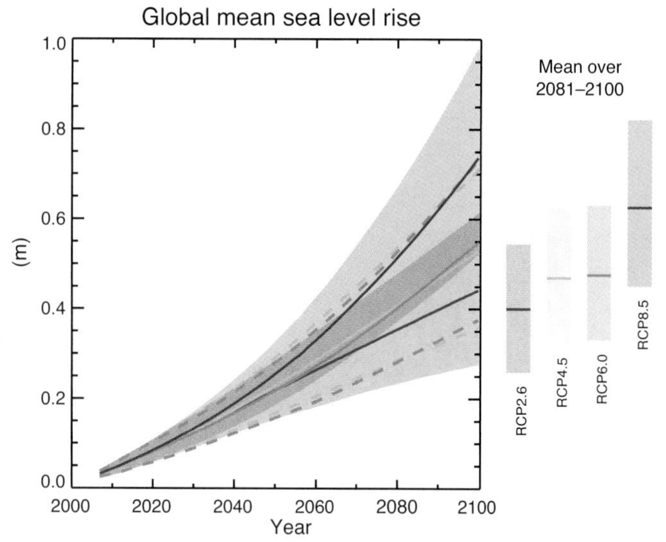

Global mean sea level rise

Figure TS.22 | Projections from process-based models of global mean sea level (GMSL) rise relative to 1986–2005 for the four RCP scenarios. The solid lines show the median projections, the dashed lines show the *likely* ranges for RCP4.5 and RCP6.0, and the shading the *likely* ranges for RCP2.6 and RCP8.5. The time means for 2081–2100 are shown as coloured vertical bars. See Sections 13.5.1 and 13.5.3 and Supplementary Material for methods. Based on current understanding, only the collapse of the marine-based sectors of the Antarctic ice sheet, if initiated, could cause GMSL to rise substantially above the *likely* range during the 21st century. This potential additional contribution cannot be precisely quantified but there is *medium confidence* that it would not exceed several tenths of a metre during the 21st century. Further detail regarding the related Figure SPM.9 is given in the TS Supplementary Material. {Table 13.5; Figures 13.10, 13.11}

unidentified or underestimated by process-based models, or that the projections of semi-empirical models are overestimates. Making projections with a semi-empirical model assumes that sea level change in the future will have the same relationship as it has had in the past to RF or global mean temperature change. This may not hold if potentially nonlinear physical processes do not scale in the future in ways which can be calibrated from the past. There is no consensus in the scientific community about the reliability of semi-empirical model projections, and *confidence* in them is assessed to be *low*. {13.5.2, 13.5.3}

TS.5.7.2 Projections of Global Mean Sea Level Change Beyond 2100

It is *virtually certain* that GMSL rise will continue beyond 2100. The few available model results that go beyond 2100 indicate global mean sea level rise above the pre-industrial level (defined here as an equilibrium 280 ppm atmospheric CO_2 concentration) by 2300 to be less than 1 m for a RF that corresponds to CO_2 concentrations that peak and decline and remain below 500 ppm, as in the scenario RCP2.6. For a RF that corresponds to a CO_2 concentration that is above 700 ppm but below 1500 ppm, as in the scenario RCP8.5, the projected rise is 1 m to more than 3 m (*medium confidence*). {13.5.4}

Sea level rise due to ocean thermal expansion will continue for centuries to millennia. The amount of ocean thermal expansion increases with global warming (models give a range of 0.2 to 0.6 m °C^{-1}). The glacier contribution decreases over time as their volume (currently about 0.43 m sea level equivalent) decreases. In Antarctica, beyond 2100 and with higher GHG scenarios, the increase in surface melting could exceed the increase in accumulation. {13.5.2, 13.5.4}

The available evidence indicates that global warming greater than a certain threshold would lead to the near-complete loss of the Greenland ice sheet over a millennium or more, causing a GMSL rise of about 7 m. Studies with fixed present-day ice sheet topography indicate the threshold is greater than 2°C but less than 4°C of GMST rise with respect to pre-industrial (*medium confidence*). The one study with a dynamical ice sheet suggests the threshold is greater than about 1°C (*low confidence*) global mean warming with respect to pre-industrial. Considering the present state of scientific uncertainty, a *likely* range cannot be quantified. The complete loss of the ice sheet is not inevitable because this would take a millennium or more; if temperatures decline before the ice sheet is eliminated, the ice sheet might regrow. However, some part of the mass loss might be irreversible, depending on the duration and degree of exceedance of the threshold, because the ice sheet may have multiple steady states, due to its interaction with its regional climate. {13.4.3, 13.5.4}

Currently available information indicates that the dynamical contribution of the ice sheets will continue beyond 2100, but *confidence* in projections is *low*. In Greenland, ice outflow induced from interaction with the ocean is self-limiting as the ice sheet margin retreats inland from the coast. By contrast, the bedrock topography of Antarctica is such that there may be enhanced rates of mass loss as the ice retreats. About 3.3 m of equivalent global sea level of the West Antarctic ice sheet is grounded on areas with downward sloping bedrock, which may be subject to potential ice loss via the marine ice sheet instability. Abrupt and irreversible ice loss from a potential instability of marine-based sectors of the Antarctic Ice Sheet in response to climate forcing is possible, but current evidence and understanding is insufficient to make a quantitative assessment. Due to relatively weak snowfall on Antarctica and the slow ice motion in its interior, it can be expected that the West Antarctic ice sheet would take at least several thousand years to regrow if it was eliminated by dynamic ice discharge. Consequently any significant ice loss from West Antarctic that occurs within the next century will be irreversible on a multi-centennial to millennial time scale. {5.8, 13.4.3, 13.4.4, 13.5.4}

TS.5.7.3 Projections of Regional Sea Level Change

Regional sea level will change due to dynamical ocean circulation changes, changes in the heat content of the ocean, mass redistribution in the entire Earth system and changes in atmospheric pressure. Ocean dynamical change results from changes in wind and buoyancy forcing (heat and freshwater), associated changes in the circulation, and redistribution of heat and freshwater. Over time scales longer than a few days, regional sea level also adjusts nearly isostatically to regional changes in sea level atmospheric pressure relative to its mean over the ocean. Ice sheet mass loss (both contemporary and past), glacier mass loss and changes in terrestrial hydrology cause water mass redistribution among the cryosphere, the land and the oceans, giving rise to distinctive regional changes in the solid Earth, Earth rotation and the gravity field. In some coastal locations, changes in the hydrological cycle, ground subsidence associated with anthropogenic activity, tectonic processes and coastal processes can dominate the relative sea level change, that is, the change in sea surface height relative to the land. {13.1.3, 13.6.2, 13.6.3, 13.6.4}

By the end of the 21st century, sea level change will have a strong regional pattern, which will dominate over variability, with many regions *likely* experiencing substantial deviations from the global mean change (Figure TS.23). It is *very likely* that over about 95% of the ocean will experience regional relative sea level rise, while most regions experiencing a sea level fall are located near current and former glaciers and ice sheets. Local sea level changes deviate more than 10% and 25% from the global mean projection for as much as 30% and 9% of the ocean area, respectively, indicating that spatial variations can be large. Regional changes in sea level reach values of up to 30% above the global mean value in the Southern Ocean and around North America, between 10% and 20% in equatorial regions, and up to 50% below the global mean in the Arctic region and some regions near Antarctica. About 70% of the coastlines worldwide are projected to experience a relative sea level change within 20% of the GMSL change. Over decadal periods, the rates of regional relative sea level change as a result of climate variability can differ from the global average rate by more than 100%. {13.6.5}

TS.5.7.4 Projections of Change in Sea Level Extremes and Waves During the 21st Century

It is *very likely* that there will be a significant increase in the occurrence of future sea level extremes by the end of the 21st century, with a *likely* increase in the early 21st century (see TFE.9, Table 1). This increase will primarily be the result of an increase in mean sea level (*high confidence*), with extreme return periods decreasing by at least an order of magnitude in some regions by the end of the 21st century. There is *low confidence* in region-specific projections of storminess and associated storm surges. {13.7.2}

It is *likely* (*medium confidence*) that annual mean significant wave heights will increase in the Southern Ocean as a result of enhanced wind speeds. Southern Ocean–generated swells are *likely* to affect heights, periods and directions of waves in adjacent basins. It is *very likely* that wave heights and the duration of the wave season will increase in the Arctic Ocean as a result of reduced sea ice extent. In general, there is *low confidence* in region-specific projections due to the *low confidence* in tropical and extratropical storm projections, and to the challenge of down-scaling future wind states from coarse resolution climate models. {13.7.3}

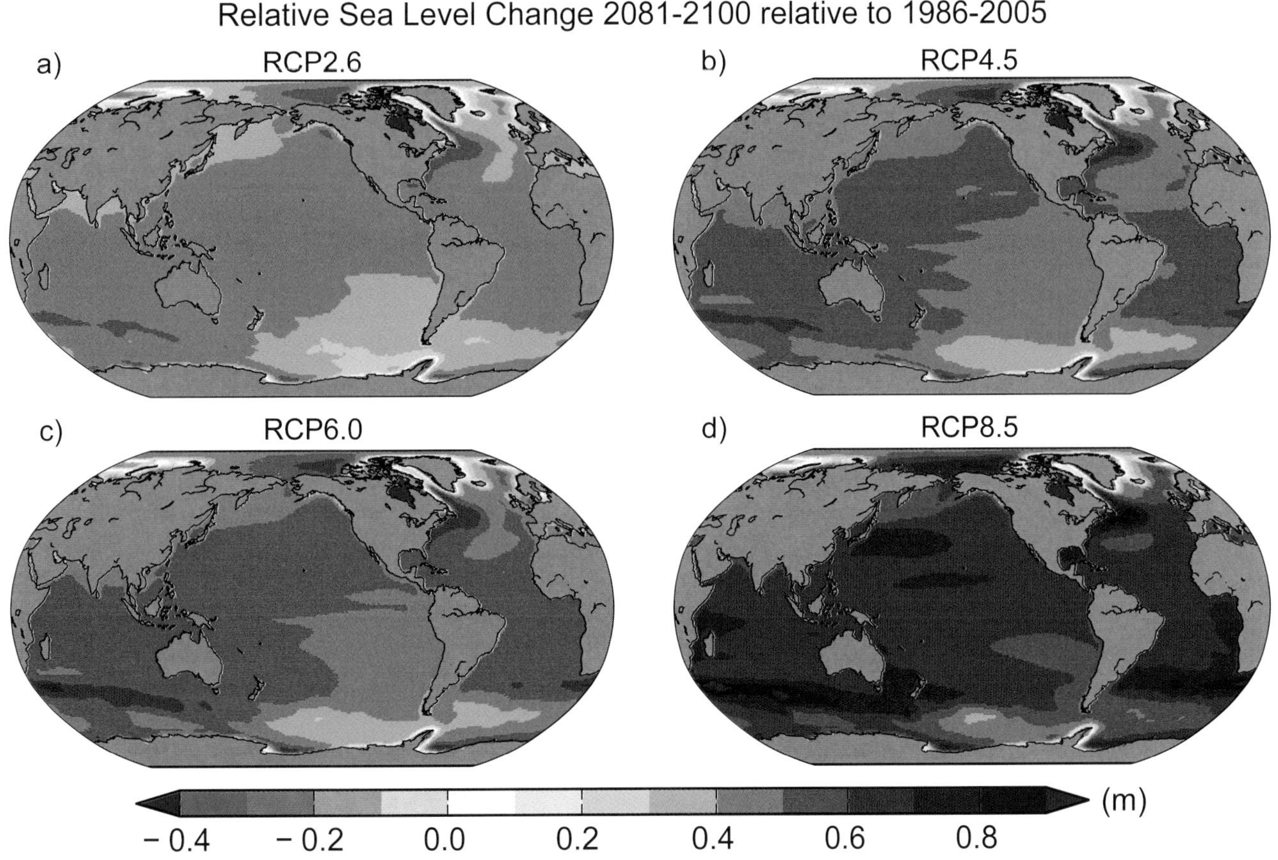

Figure TS.23 | Ensemble mean net regional relative sea level change (metres) evaluated from 21 CMIP5 models for the RCP scenarios (a) 2.6, (b) 4.5, (c) 6.0 and (d) 8.5 between 1986–2005 and 2081–2100. Each map includes effects of atmospheric loading, plus land-ice, glacial isostatic adjustment (GIA) and terrestrial water sources. {Figure 13.20}

TS

Thematic Focus Elements

TFE.8 | Climate Targets and Stabilization

The concept of stabilization is strongly linked to the ultimate objective of the United Nations Framework Convention on Climate Change (UNFCCC), which is 'to achieve [...] stabilization of greenhouse gas concentrations in the atmosphere at a level that would prevent dangerous anthropogenic interference with the climate system'. Recent policy discussions focused on limits to a global temperature increase, rather than to greenhouse gas (GHG) concentrations, as climate targets in the context of the UNFCCC objectives. The most widely discussed is that of 2°C, that is, to limit global temperature increase relative to pre-industrial times to below 2°C, but targets other than 2°C have been proposed (e.g., returning warming to well below 1.5°C global warming relative to pre-industrial, or returning below an atmospheric carbon dioxide (CO_2) concentration of 350 ppm). Climate targets generally mean avoiding a warming beyond a predefined threshold. Climate impacts, however, are geographically diverse and sector specific, and no objective threshold defines when dangerous interference is reached. Some changes may be delayed or irreversible, and some impacts could be beneficial. It is thus not possible to define a single critical objective threshold without value judgements and without assumptions on how to aggregate current and future costs and benefits. This TFE does not advocate or defend any threshold or objective, nor does it judge the economic or political feasibility of such goals, but assesses, based on the current understanding of climate and carbon cycle feedbacks, the climate projections following the Representative Concentration Pathways (RCPs) in the context of climate targets, and the implications of different long-term temperature stabilization objectives on allowed carbon emissions. Further below it is highlighted that temperature stabilization does not necessarily imply stabilization of the entire Earth system. {12.5.4}

Temperature targets imply an upper limit on the total radiative forcing (RF). Differences in RF between the four RCP scenarios are relatively small up to 2030, but become very large by the end of the 21st century and dominated by CO_2 forcing. Consequently, in the near term, global mean surface temperatures (GMSTs) are projected to continue to rise at a similar rate for the four RCP scenarios. Around the mid-21st century, the rate of global warming begins to be more strongly dependent on the scenario. By the end of the 21st century, global mean temperatures will be warmer than present day under all the RCPs, global temperature change being largest (>0.3°C per decade) in the highest RCP8.5 and significantly lower in RCP2.6, particularly after about 2050 when global surface temperature response stabilizes (and declines thereafter) (see Figure TS.15). {11.3.1, 12.3.3, 12.4.1}

In the near term (2016–2035), global mean surface warming is *more likely than not* to exceed 1°C and *very unlikely* to be more than 1.5°C relative to the average from year 1850 to 1900 (assuming 0.61°C warming from 1850-1900 to 1986–2005) (*medium confidence*). By the end of the 21st century (2081–2100), global mean surface warming, relative to 1850-1900, is *likely* to exceed 1.5°C for RCP4.5, RCP6.0 and RCP8.5 (*high confidence*) and is *likely* to exceed 2°C for RCP6.0 and RCP8.5 (*high confidence*). It is *more likely than not* to exceed 2°C for RCP4.5 (*medium confidence*). Global mean surface warming above 2°C under RCP2.6 is *unlikely* (*medium confidence*). Global mean surface warming above 4°C by 2081–2100 is *unlikely* in all RCPs (*high confidence*) except for RCP8.5 where it is *about as likely as not* (*medium confidence*). {11.3.6, 12.4.1; Table 12.3}

Continuing GHG emissions beyond 2100 as in the RCP8.5 extension induces a total RF above 12 W m^{-2} by 2300, with global warming reaching 7.8 [3.0 to 12.6] °C for 2281–2300 relative to 1986–2005. Under the RCP4.5 extension, where radiative forcing is kept constant (around 4.5 W m^{-2}) beyond 2100, global warming reaches 2.5 [1.5 to 3.5] °C. Global warming reaches 0.6 [0.0 to 1.2] °C under the RCP2.6 extension where sustained negative emissions lead to a further decrease in RF, reaching values below present-day RF by 2300. See also Box TS.7. {12.3.1, 12.4.1, 12.5.1}

The total amount of anthropogenic CO_2 released in the atmosphere since pre-industrial (often termed cumulative carbon emission, although it applies only to CO_2 emissions) is a good indicator of the atmospheric CO_2 concentration and hence of the global warming response. The ratio of GMST change to total cumulative anthropogenic CO_2 emissions is relatively constant over time and independent of the scenario. This near-linear relationship between total CO_2 emissions and global temperature change makes it possible to define a new quantity, the transient climate response to cumulative carbon emission (TCRE), as the transient GMST change for a given amount of cumulated anthropogenic CO_2 emissions, usually 1000 PgC (TFE.8, Figure 1). TCRE is model dependent, as it is a function of the cumulative CO_2 airborne fraction and the transient climate response, both quantities varying significantly across models. Taking into account the available information from multiple lines of evidence (observations, models and process understanding), the near linear relationship between cumulative CO_2 emissions and peak global mean temperature is

(continued on next page)

TFE.8 (continued)

well established in the literature and robust for cumulative total CO_2 emissions up to about 2000 PgC. It is consistent with the relationship inferred from past cumulative CO_2 emissions and observed warming, is supported by process understanding of the carbon cycle and global energy balance, and emerges as a robust result from the entire hierarchy of models. Expert judgment based on the available evidence suggests that TCRE is *likely* between 0.8°C and 2.5°C per 1000 PgC, for cumulative emissions less than about 2000 PgC until the time at which temperature peaks (TFE.8, Figure 1a). {6.4.3, 12.5.4; Box 12.2}

CO_2-induced warming is projected to remain approximately constant for many centuries following a complete cessation of emissions. A large fraction of climate change is thus irreversible on a human time scale, except if net anthropogenic CO_2 emissions were strongly negative over a sustained period. Based on the assessment of TCRE (assuming a normal distribution with a ±1 standard deviation range of 0.8 to 2.5°C per 1000 PgC), limiting the warming caused by anthropogenic CO_2 emissions alone (i.e., ignoring other radiative forcings) to less than 2°C since the period 1861–1880 with a probability of >33%, >50% and >66%, total CO_2 emissions from all anthropogenic sources would need to be below a cumulative budget of about 1570 PgC, 1210 PgC and 1000 PgC since 1870, respectively. An amount of 515 [445 to 585] PgC was emitted between 1870 and 2011 (TFE.8, Figure 1a,b). Higher emissions in earlier decades therefore imply lower or even negative emissions later on. Accounting for non-CO_2 forcings contributing to peak warming implies lower cumulated CO_2 emissions. Non-CO_2 forcing constituents are important, requiring either assumptions on how CO_2 emission reductions are linked to changes in other forcings, or separate emission budgets and climate modelling for short-lived and long-lived gases. So far, not many studies have considered non-CO_2 forcings. Those that do consider them found significant effects, in particular warming of several tenths of a degree for abrupt reductions in emissions of short-lived species, like aerosols. Accounting for an unanticipated release of GHGs from permafrost or methane hydrates, not included in studies assessed here, would also reduce the anthropogenic CO_2 emissions compatible with a given temperature target. Requiring a higher likelihood of temperatures remaining below a given temperature target would further reduce the compatible emissions (TFE.8, Figure 1c). When accounting for the non-CO_2 forcings as in the RCP scenarios, compatible carbon emissions since 1870 are reduced to about 900 PgC, 820 PgC and 790 PgC to limit warming to less than 2°C since the period 1861–1880 with a probability of >33%, >50%, and >66%, respectively. These estimates were derived by computing the fraction of the Coupled Model Intercomparison Project Phase 5 (CMIP5) Earth System Models (ESMs) and Earth System Models of Intermediate Complexity (EMICs) that stay below 2°C for given cumulative emissions following RCP8.5, as shown in TFE.8 Fig. 1c. The non-CO_2 forcing in RCP8.5 is higher than in RCP2.6. Because all likelihood statements in calibrated IPCC language are open intervals, the estimates provided are thus both conservative and consistent choices valid for non-CO_2 forcings across all RCP scenarios. There is no RCP scenario which limits warming to 2°C with probabilities of >33% or >50%, and which could be used to directly infer compatible cumulative emissions. For a probability of >66% RCP2.6 can be used as a comparison. Combining the average back-calculated fossil fuel carbon emissions for RCP2.6 between 2012 and 2100 (270 PgC) with the average historical estimate of 515 PgC gives a total of 785 PgC, i.e., 790 PgC when rounded to 10 PgC. As the 785 PgC estimate excludes an explicit assessment of future land-use change emissions, the 790 PgC value also remains a conservative estimate consistent with the overall likelihood assessment. The ranges of emissions for these three likelihoods based on the RCP scenarios are rather narrow, as they are based on a single scenario and on the limited sample of models available (TFE.8 Fig. 1c). In contrast to TCRE they do not include observational constraints or account for sources of uncertainty not sampled by the models. The concept of a fixed cumulative CO_2 budget holds not just for 2°C, but for any temperature level explored with models so far (up to about 5°C, see Figures 12.44 to 12.46). Higher temperature targets would allow larger cumulative budgets, while lower temperature target would require lower cumulative budgets (TFE.8, Figure 1). {6.3.1, 12.5.2, 12.5.4}

The climate system has multiple time scales, ranging from annual to multi-millennial, associated with different thermal and carbon reservoirs. These long time scales induce a commitment warming 'already in the pipe-line'. Stabilization of the forcing would not lead to an instantaneous stabilization of the warming. For the RCP scenarios and their extensions to 2300, the fraction of realized warming, at that time when RF stabilizes, would be about 75 to 85% of the equilibrium warming. For a 1% yr^{-1} CO_2 increase to 2 × CO_2 or 4 × CO_2 and constant forcing thereafter, the fraction of realized warming would be much smaller, about 40 to 70% at the time when the forcing is kept constant. Owing to the long time scales in the deep ocean, full equilibrium is reached only after hundreds to thousands of years. {12.5.4}

(continued on next page)

TS

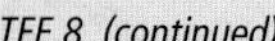

TFE.8 (continued)

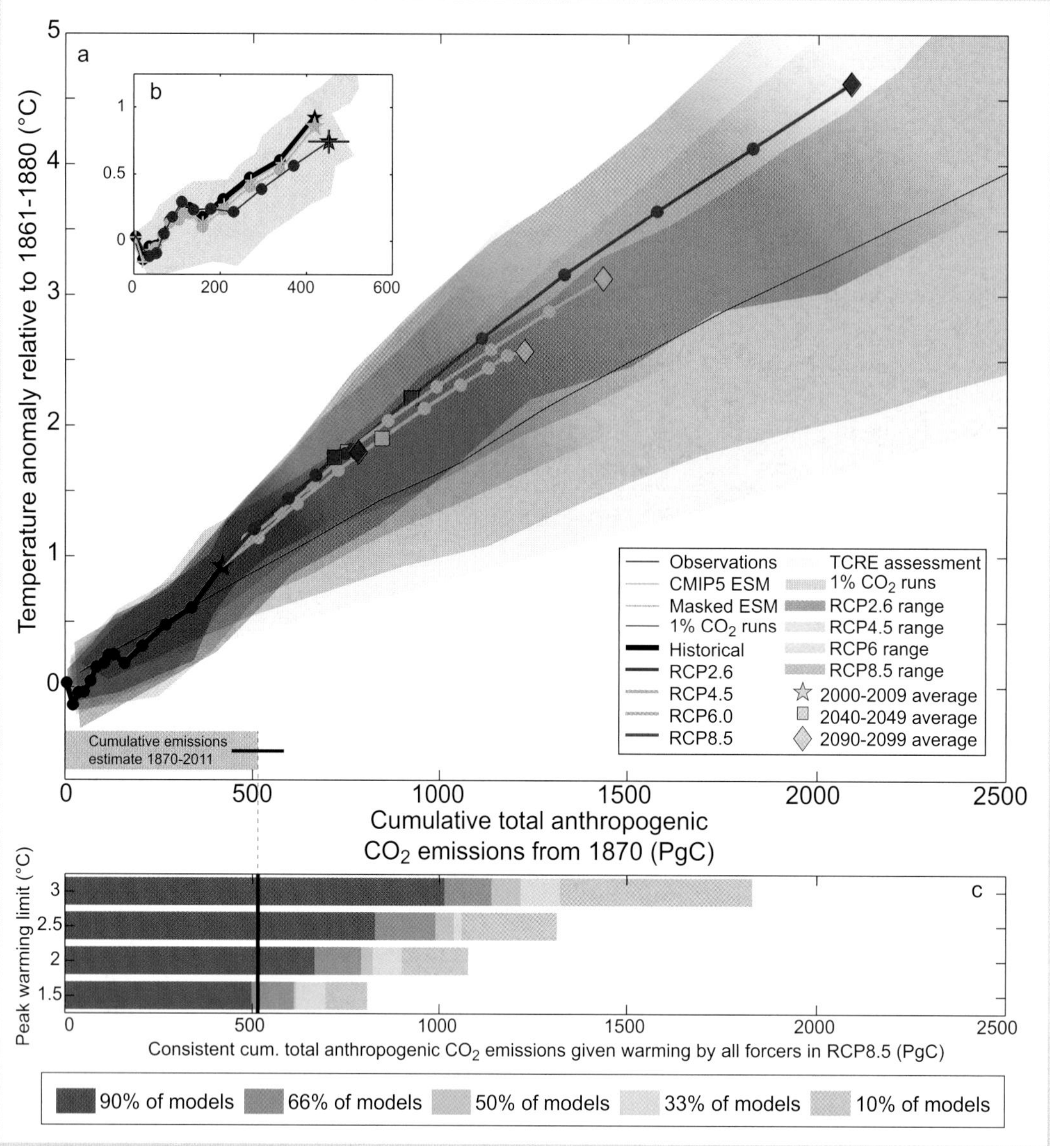

TFE.8, Figure 1 | Global mean temperature increase since 1861–1880 as a function of cumulative total global CO_2 emissions from various lines of evidence. (a) Decadal average results are shown over all CMIP5 Earth System Model of Intermediate Complexity (EMICs) and Earth System Models (ESMs) for each RCP respectively, with coloured lines (multi-model average), decadal markers (dots) and with three decades (2000–2009, 2040–2049 and 2090–2099) highlighted with a star, square and diamond, respectively. The historical time period up to decade 2000–2009 is taken from the CMIP5 historical runs prolonged by RCP8.5 for 2006–2010 and is indicated with a black thick line and black symbols. Coloured ranges illustrate the model spread (90% range) over all CMIP5 ESMs and EMICs and do not represent a formal uncertainty assessment. Ranges are filled as long as data of all models is available and until peak temperature. They are faded out for illustrative purposes afterward. CMIP5 simulations with 1% yr^{-1} CO_2 increase only are illustrated by the dark grey area (range definition similar to RCPs above) and the black thin line (multi-model average). The light grey cone represents this Report's assessment of the transient climate response to emissions (TCRE) from CO_2 only. Estimated cumulative historical CO_2 emissions from 1870 to 2011 with associated uncertainties are illustrated by the grey bar at the bottom of (a). (b) Comparison of historical model results with observations. The magenta line and uncertainty ranges are based on observed emissions from Carbon Dioxide Information Analysis Center (CDIAC) extended by values of the Global Carbon project until 2010 and observed temperature estimates of the Hadley Centre/Climatic Research Unit gridded surface temperature data set 4 (HadCRUT4). The uncertainties in the last decade of observations are based on the assessment in this report. The black thick line is identical to the one in (a). The thin green line with crosses is as the black line but for ESMs only. The yellow-brown line and range show these ESM results until 2010, when corrected for HadCRUT4's incomplete geographical coverage over time. All values are given relative to the 1861–1880 base period. All time-series are derived from decadal averages to illustrate the long-term trends. Note that observations are in addition subject to internal climate variability, adding an uncertainty of about 0.1°C. (c) Cumulative CO_2 emissions over the entire industrial era, consistent with four illustrative peak global temperature limits (1.5°C, 2°C, 2.5°C and 3°C, respectively) when taking into account warming by all forcers. Horizontal bars indicate consistent cumulative emission budgets as a function of the fraction of models (CMIP5 ESMs and EMICs) that at least hold warming below a given temperature limit. Note that the fraction of models cannot be interpreted as a probability. The budgets are derived from the RCP8.5 runs, with relative high non-CO_2 forcing over the 21st century. If non-CO_2 are significantly reduced, the CO_2 emissions compatible with a specific temperature limit might be slightly higher, but only to a very limited degree, as illustrated by the other coloured lines in (a), which assume significantly lower non-CO_2 forcing. Further detail regarding the related Figure SPM.10 is given in the TS Supplementary Material. {Figure 12.45}

TFE.8 (continued)

The commitment to past emissions is a persistent warming for hundreds of years, continuing at about the level of warming that has been realized when emissions were ceased. The persistence of this CO_2-induced warming after emission have ceased results from a compensation between the delayed commitment warming described above and the slow reduction in atmospheric CO_2 resulting from ocean and land carbon uptake. This persistence of warming also results from the nonlinear dependence of RF on atmospheric CO_2, that is, the relative decrease in forcing being smaller than the relative decrease in CO_2 concentration. For high climate sensitivities, and in particular if sulphate aerosol emissions are eliminated at the same time as GHG emissions, the commitment from past emission can be strongly positive, and is a superposition of a fast response to reduced aerosols emissions and a slow response to reduced CO_2. {12.5.4}

Stabilization of global temperature does not imply stabilization for all aspects of the climate system. Processes related to vegetation change, changes in the ice sheets, deep ocean warming and associated sea level rise and potential feedbacks linking, for example, ocean and the ice sheets have their own intrinsic long time scales. Ocean acidification will *very likely* continue in the future as long as the oceans will continue to take up atmospheric CO_2. Committed land ecosystem carbon cycle changes will manifest themselves further beyond the end of the 21st century. It is *virtually certain* that global mean sea level rise will continue beyond 2100, with sea level rise due to thermal expansion to continue for centuries to millennia. Global mean sea level rise depends on the pathway of CO_2 emissions, not only on the cumulative total; reducing emissions earlier rather than later, for the same cumulative total, leads to a larger mitigation of sea level rise. {6.4.4, 12.5.4, 13.5.4}

TS.5.8 Climate Phenomena and Regional Climate Change

This section assesses projected changes over the 21st century in large-scale climate phenomena that affect regional climate (Table TS.2). Some of these phenomena are defined by climatology (e.g., monsoons), and some by interannual variability (e.g., El Niño), the latter affecting climate extremes such as floods, droughts and heat waves. Changes in statistics of weather phenomena such as tropical cyclones and extratropical storms are also summarized here. {14.8}

TS.5.8.1 Monsoon Systems

Global measures of monsoon by the area and summer precipitation are *likely* to increase in the 21st century, while the monsoon circulation weakens. Monsoon onset dates are *likely* to become earlier or not to change much while monsoon withdrawal dates are *likely* to delay, resulting in a lengthening of the monsoon season in many regions (Figure TS.24). The increase in seasonal mean precipitation is pronounced in the East and South Asian summer monsoons while the change in other monsoon regions is subject to larger uncertainties. {14.2.1}

There is *medium confidence* that monsoon-related interannual rainfall variability will increase in the future. Future increase in precipitation extremes related to the monsoon is *very likely* in South America, Africa, East Asia, South Asia, Southeast Asia and Australia. {14.2.1, 14.8.5, 14.8.7, 14.8.9, 14.8.11–14.8.13}

There is *medium confidence* that overall precipitation associated with the Asian-Australian monsoon will increase but with a north–south asymmetry: Indian monsoon rainfall is projected to increase, while projected changes in the Australian summer monsoon rainfall are small. There is *medium confidence* in that the Indian summer monsoon circulation weakens, but this is compensated by increased atmospheric moisture content, leading to more rainfall. For the East Asian summer monsoon, both monsoon circulation and rainfall are projected to increase. {14.2.2, 14.8.9, 14.8.11, 14.8.13}

There is *low confidence* in projections of the North American and South American monsoon precipitation changes, but *medium confidence* that the North American monsoon will arrive and persist later in the annual cycle, and *high confidence* in expansion of South American Monsoon area. {14.2.3, 14.8.3–14.8.5}

There is *low confidence* in projections of a small delay in the West African rainy season, with an intensification of late-season rains. The limited skills of model simulations for the region suggest *low confidence* in the projections. {14.2.4, 14.8.7}

TS.5.8.2 Tropical Phenomena

Precipitation change varies in space, increasing in some regions and decreasing in some others. The spatial distribution of tropical rainfall changes is *likely* shaped by the current climatology and ocean warming pattern. The first effect is to increase rainfall near the currently rainy regions, and the second effect increases rainfall where the ocean warming exceeds the tropical mean. There is *medium confidence* that tropical rainfall projections are more reliable for the seasonal than annual mean changes. {7.6.2, 12.4.5, 14.3.1}

There is *medium confidence* in future increase in seasonal mean precipitation on the equatorial flank of the Intertropical Convergence Zone and a decrease in precipitation in the subtropics including parts

Table TS.2 | Overview of projected regional changes and their relation to major climate phenomena. A phenomenon is considered relevant when there is both sufficient confidence that it has an influence on the given region, and when there is sufficient confidence that the phenomenon will change, particularly under the RCP4.5 or higher end scenarios. See Section 14.8 and Tables 14.2 and 14.3 for full assessment of the confidence in these changes, and their relevance for regional climate. {14.8; Tables 14.2, 14.3}

Regions	Projected Major Changes in Relation to Phenomena
Arctic {14.8.2}	Wintertime changes in temperature and precipitation resulting from the small projected increase in North Atlantic Oscillation (NAO); enhanced warming and sea ice melting; significant increase in precipitation by mid-century due mostly to enhanced precipitation in extratropical cyclones.
North America {14.8.3}	Monsoon precipitation will shift later in the annual cycle; increased precipitation in extratropical cyclones will lead to large increases in wintertime precipitation over the northern third of the continent; extreme precipitation increases in tropical cyclones making landfall along the western coast of USA and Mexico, the Gulf Mexico, and the eastern coast of USA and Canada.
Central America and Caribbean {14.8.4}	Projected reduction in mean precipitation and increase in extreme precipitation; more extreme precipitation in tropical cyclones making landfall along the eastern and western coasts.
South America {14.8.5}	A southward displaced South Atlantic Convergence Zone increases precipitation in the southeast; positive trend in the Southern Annular Mode displaces the extratropical storm track southward, decreasing precipitation in central Chile and increasing it at the southern tip of South America.
Europe and Mediterranean {14.8.6}	Enhanced extremes of storm-related precipitation and decreased frequency of storm-related precipitation over the eastern Mediterranean.
Africa {14.8.7}	Enhanced summer monsoon precipitation in West Africa; increased short rain in East Africa due to the pattern of Indian Ocean warming; increased rainfall extremes of landfall cyclones on the east coast (including Madagascar).
Central and North Asia {14.8.8}	Enhanced summer precipitation; enhanced winter warming over North Asia.
East Asia {14.8.9}	Enhanced summer monsoon precipitation; increased rainfall extremes of landfall typhoons on the coast; reduction in the midwinter suppression of extratropical cyclones.
West Asia {14.8.10}	Increased rainfall extremes of landfall cyclones on the Arabian Peninsula; decreased precipitation in northwest Asia due to a northward shift of extra-tropical storm tracks.
South Asia {14.8.11}	Enhanced summer monsoon precipitation; increased rainfall extremes of landfall cyclones on the coasts of the Bay of Bengal and Arabian Sea.
Southeast Asia {14.8.12}	Reduced precipitation in Indonesia during July to October due to the pattern of Indian Ocean warming; increased rainfall extremes of landfall cyclones on the coasts of the South China Sea, Gulf of Thailand and Andaman Sea.
Australia and New Zealand {14.8.13}	Summer monsoon precipitation may increase over northern Australia; more frequent episodes of the zonal South Pacific Convergence Zone may reduce precipitation in northeastern Australia; increased warming and reduced precipitation in New Zealand and southern Australia due to projected positive trend in the Southern Annular Mode; increased extreme precipitation associated with tropical and extratropical storms
Pacific Islands {14.8.14}	Tropical convergence zone changes affect rainfall and its extremes; more extreme precipitation associated with tropical cyclones
Antarctica {14.8.15}	Increased warming over Antarctic Peninsula and West Antarctic related to the positive trend in the Southern Annular Mode; increased precipitation in coastal areas due to a poleward shift of storm track.

of North and Central Americas, the Caribbean, South America, Africa and West Asia. There is *medium confidence* that the interannual occurrence of zonally oriented South Pacific Convergence Zone events will increase, leading possibly to more frequent droughts in the southwest Pacific. There is *medium confidence* that the South Atlantic Convergence Zone will shift southwards, leading to a precipitation increase over southeastern South America and a reduction immediately north of the convergence zone. {14.3.1, 14.8.3–14.8.5, 14.8.7, 14.8.11, 14.8.14}

The tropical Indian Ocean is *likely* to feature a zonal pattern with reduced warming and decreased rainfall in the east (including Indonesia), and enhanced warming and increased rainfall in the west (including East Africa). The Indian Ocean dipole mode of interannual variability is *very likely* to remain active, affecting climate extremes in East Africa, Indonesia and Australia. {14.3.3, 14.8.7, 14.8.12}

There is *low confidence* in the projections for the tropical Atlantic—both for the mean and interannual modes, because of large errors in model simulations in the region. Future projections in Atlantic hurricanes and tropical South American and West African precipitation are therefore of *low confidence*. {14.3.4, 14.6.1, 14.8.5,14.8.7}

It is currently not possible to assess how the Madden–Julian Oscillation will change owing to the poor skill in model simulations of this intraseasonal phenomenon and the sensitivity to ocean warming patterns. Future projections of regional climate extremes in West Asia, Southeast Asia and Australia are therefore of *low confidence*. {9.5.2, 14.3.4, 14.8.10, 14.8.12, 14.8.13}

TS.5.8.3 El Niño-Southern Oscillation

There is *high confidence* that the El Niño-Southern Oscillation (ENSO) will remain the dominant mode of natural climate variability in the 21st century with global influences in the 21st century, and that regional rainfall variability it induces *likely* intensifies. Natural variations of the amplitude and spatial pattern of ENSO are so large that *confidence* in any projected change for the 21st century remains *low*. The projected change in El Niño amplitude is small for both RCP4.5 and RCP8.5 compared to the spread of the change among models (Figure TS.25). Over the North Pacific and North America, patterns of temperature and precipitation anomalies related to El Niño and La Niña (teleconnections) are *likely* to move eastwards in the future (*medium confidence*), while *confidence* is *low* in changes in climate impacts on other regions including Central and South Americas, the Caribbean, Africa, most of Asia, Australia and most Pacific Islands. In a warmer climate, the increase in atmospheric moisture intensifies temporal variability

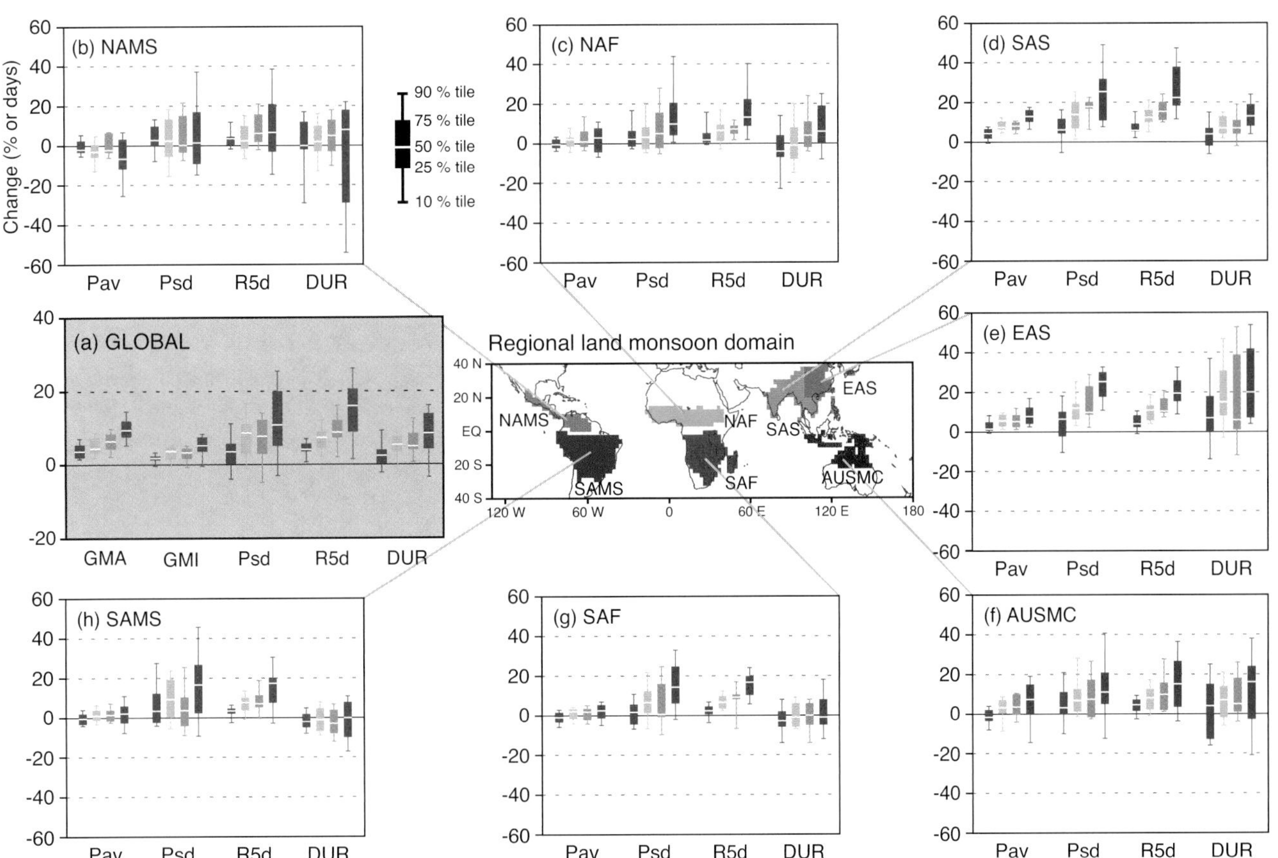

Figure TS.24 | Future change in monsoon statistics between the present-day (1986–2005) and the future (2080–2099) based on CMIP5 ensemble from RCP2.6 (dark blue; 18 models), RCP4.5 (blue; 24), RCP6.0 (yellow; 14), and RCP8.5 (red; 26) simulations. (a) GLOBAL: Global monsoon area (GMA), global monsoon intensity (GMI), standard deviation of inter-annual variability in seasonal precipitation (Psd), seasonal maximum 5-day precipitation total (R5d) and monsoon season duration (DUR). Regional land monsoon domains determined by 24 multi-model mean precipitation in the present-day. (b)–(h) Future change in regional land monsoon statistics: seasonal average precipitation (Pav), Psd, R5d, and DUR in (b) North America (NAMS), (c) North Africa (NAF), (d) South Asia (SAS), (e) East Asia (EAS), (f) Australia-Maritime continent (AUSMC), (g) South Africa (SAF) and (h) South America (SAMS). Units are % except for DUR (days). Box-and-whisker plots show the 10th, 25th, 50th, 75th and 90th percentiles. All the indices are calculated for the summer season (May to September for the Northern, and November to March for the Southern Hemisphere) over each model's monsoon domains. {Figures 14.3, 14.4, 14.6, 14.7}

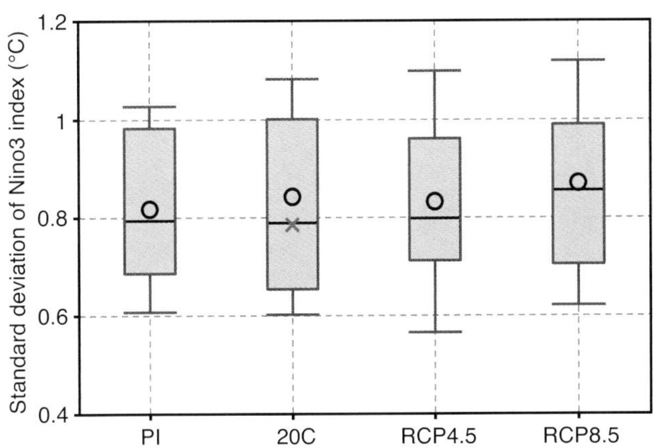

Figure TS.25 | Standard deviation in CMIP5 multi-model ensembles of sea surface temperature variability over the eastern equatorial Pacific Ocean (Nino3 region: 5°S to 5°N, 150°W to 90°W), a measure of El Niño amplitude, for the pre-industrial (PI) control and 20th century (20C) simulations, and 21st century projections using RCP4.5 and RCP8.5. Open circles indicate multi-model ensemble means, and the red cross symbol is the observed standard deviation for the 20th century. Box-and-whisker plots show the 16th, 25th, 50th, 75th and 84th percentiles. {Figure 14.14}

of precipitation even if atmospheric circulation variability remains the same. This applies to ENSO-induced precipitation variability but the possibility of changes in ENSO teleconnections complicates this general conclusion, making it somewhat regional-dependent. {12.4.5, 14.4, 14.8.3–14.8.5, 14.8.7, 14.8.9, 14.8.11–14.8.14}

TS.5.8.4 Cyclones

Projections for the 21st century indicate that it is *likely* that the global frequency of tropical cyclones will either decrease or remain essentially unchanged, concurrent with a *likely* increase in both global mean tropical cyclone maximum wind speed and rain rates (Figure TS.26). The influence of future climate change on tropical cyclones is *likely* to vary by region, but there is *low confidence* in region-specific projections. The frequency of the most intense storms will *more likely than not* increase in some basins. More extreme precipitation near the centers of tropical cyclones making landfall is projected in North and Central America, East Africa, West, East, South and Southeast Asia as well as in Australia and many Pacific islands (*medium confidence*). {14.6.1, 14.8.3, 14.8.4, 14.8.7, 14.8.9–14.8.14}

The global number of extratropical cyclones is *unlikely* to decrease by more than a few percent and future changes in storms are *likely* to be small compared to natural interannual variability and substantial variations between models. A small poleward shift is *likely* in the SH storm track but the magnitude of this change is model dependent. It is *unlikely* that the response of the North Atlantic storm track in climate projections is a simple poleward shift. There is *medium confidence* in a projected poleward shift in the North Pacific storm track. There is *low confidence* in the impact of storm track changes on regional climate at the surface. More precipitation in extratropical cyclones leads to a winter precipitation increase in Arctic, Northern Europe, North America and the mid-to-high-latitude SH. {11.3.2, 12.4.4, 14.6.2, 14.8.2, 14.8.3, 14.8.5, 14.8.6, 14.8.13, 14.8.15}

TS.5.8.5 Annular and Dipolar Modes of Variability

Future boreal wintertime North Atlantic Oscillation (NAO is *very likely* to exhibit large natural variations as observed in the past. The NAO is *likely* to become slightly more positive (on average), with some, but not very well documented implications for winter conditions in the Arctic, North America and Eurasia. The austral summer/autumn positive trend in Southern Annular Mode (SAM) is *likely* to weaken considerably as stratospheric ozone recovers through the mid-21st century with some, but not very well documented, implications for South America, Africa, Australia, New Zealand and Antarctica. {11.3.2, 14.5.2,14.8.5, 14.8.7, 14.8.13, 14.8.15}

TS.5.8.6 Additional Phenomena

It is *unlikely* that the Atlantic Multi-decadal Oscillation (AMO will change its behaviour as the mean climate changes. However, natural fluctuations in the AMO over the coming few decades are *likely* to influence regional climates at least as strongly as will human-induced changes with implications for Atlantic major hurricane frequency, the West African monsoon and North American and European summer conditions. {14.2.4, 14.5.1, 14.6.1, 14.7.6, 14.8.2, 14.8.3, 14.8.6, 14.8.8}

There is *medium confidence* that the frequency of NH and SH blocking will not increase, while the trends in blocking intensity and persistence remain uncertain. {Box 14.2}

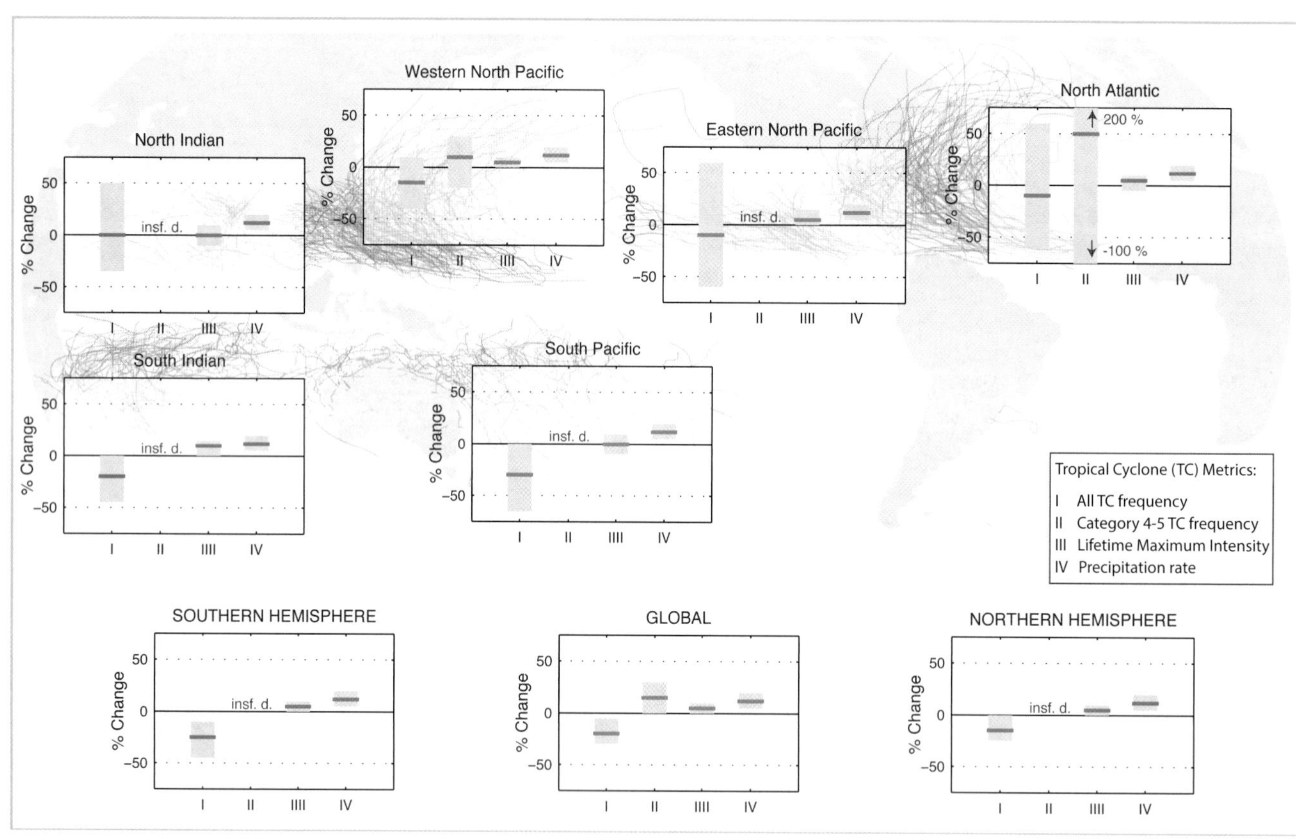

Figure TS.26 | Projected changes in tropical cyclone statistics. All values represent expected percent change in the average over period 2081–2100 relative to 2000–2019, under an A1B-like scenario, based on expert judgement after subjective normalization of the model projections. Four metrics were considered: the percent change in I) the total annual frequency of tropical storms, II) the annual frequency of Category 4 and 5 storms, III) the mean Lifetime Maximum Intensity (LMI; the maximum intensity achieved during a storm's lifetime) and IV) the precipitation rate within 200 km of storm center at the time of LMI. For each metric plotted, the solid blue line is the best guess of the expected percent change, and the coloured bar provides the 67% (*likely*) confidence interval for this value (note that this interval ranges across –100% to +200% for the annual frequency of Category 4 and 5 storms in the North Atlantic). Where a metric is not plotted, there are insufficient data (denoted X) available to complete an assessment. A randomly drawn (and coloured) selection of historical storm tracks are underlaid to identify regions of tropical cyclone activity. See Section 14.6.1 for details. {14.6.1}

Thematic Focus Elements

TFE.9 | Climate Extremes

Assessing changes in climate extremes poses unique challenges, not just because of the intrinsically rare nature of these events, but because they invariably happen in conjunction with disruptive conditions. They are strongly influenced by both small- and large-scale weather patterns, modes of variability, thermodynamic processes, land–atmosphere feedbacks and antecedent conditions. Much progress has been made since the IPCC Fourth Assessment Report (AR4) including the comprehensive assessment of extremes undertaken by the IPCC Special Report on Managing the Risk of Extreme Events and Disasters to Advance Climate Change Adaptation (SREX) but also because of the amount of observational evidence available, improvements in our understanding and the ability of models to simulate extremes. {1.3.3, 2.6, 7.6, 9.5.4}

For some climate extremes such as droughts, floods and heat waves, several factors need to be combined to produce an extreme event. Analyses of rarer extremes such as 1-in-20- to 1-in-100-year events using Extreme Value Theory are making their way into a growing body of literature. Other recent advances concern the notion of 'fraction of attributable risk' that aims to link a particular extreme event to specific causal relationships. {1.3.3, 2.6.1, 2.6.2, 10.6.2, 12.4.3; Box 2.4}

TFE.9, Table 1 indicates the changes that have been observed in a range of weather and climate extremes over the last 50 years, the assessment of the human contribution to those changes, and how those extremes are expected to change in the future. The table also compares the current assessment with that of the AR4 and the SREX where applicable. {2.6, 3.7, 10.6, 11.3, 12.4, 14.6}

Temperature Extremes, Heat Waves and Warm Spells

It is *very likely* that both maximum and minimum temperature extremes have warmed over most land areas since the mid-20th century. These changes are well simulated by current climate models, and it is *very likely* that anthropogenic forcing has affected the frequency of these extremes and *virtually certain* that further changes will occur. This supports AR4 and SREX conclusions although with greater confidence in the anthropogenic forcing component. {2.6.1, 9.5.4, 10.6.1, 12.4.3}

For land areas with sufficient data there has been an overall increase in the number of warm days and nights. Similar decreases are seen in the number of cold days and nights. It is *very likely* that increases in unusually warm days and nights and/or reductions in unusually cold days and nights including frosts have occurred over this period across most continents. Warm spells or heat waves containing consecutive extremely hot days or nights are often associated with quasi-stationary anticyclonic circulation anomalies and are also affected by pre-existing soil conditions and the persistence of soil moisture anomalies that can amplify or dampen heat waves particularly in moisture-limited regions. Most global land areas, with a few exceptions, have experienced more heat waves since the middle of the 20th century. Several studies suggest that increases in mean temperature account for most of the changes in heat wave frequency, however, heat wave intensity/amplitude is highly sensitive to changes in temperature variability and the shape of the temperature distribution and heat wave definition also plays a role. Although in some regions instrumental periods prior to the 1950s had more heat waves (e.g., USA), for other regions such as Europe, an increase in heat wave frequency in the period since the 1950s stands out in long historical temperature series. {2.6, 2.6.1, 5.5.1; Box 2.4; Tables 2.12, 2.13; FAQ 2.2}

The observed features of temperature extremes and heat waves are well simulated by climate models and are similar to the spread among observationally based estimates in most regions. Regional downscaling now offers credible information on the spatial scales required for assessing extremes and improvements in the simulation of the El Niño-Southern Oscillation from Coupled Model Intercomparison Project Phase 3 (CMIP3) to Phase 5 (CMIP5) and other large-scale phenomena is crucial. However simulated changes in frequency and intensity of extreme events is limited by observed data availability and quality issues and by the ability of models to reliably simulate certain feedbacks and mean changes in key features of circulation such as blocking. {2.6, 2.7, 9.4, 9.5.3, 9.5.4, 9.6, 9.6.1, 10.3, 10.6, 14.4; Box 14.2}

Since AR4, the understanding of mechanisms and feedbacks leading to changes in extremes has improved. There continues to be strengthening evidence for a human influence on the observed frequency of extreme temperatures and heat waves in some regions. Near-term (decadal) projections suggest *likely* increases in temperature extremes but with little distinguishable separation between emissions scenarios (TFE.9, Figure 1). Changes may proceed at

(continued on next page)

TS

TFE.9, Table 1 | Extreme weather and climate events: Global-scale assessment of recent observed changes, human contribution to the changes and projected further changes for the early (2016–2035) and late (2081–2100) 21st century. Bold indicates where the AR5 (black) provides a revised* global-scale assessment from the Special Report on Managing the Risk of Extreme Events and Disasters to Advance Climate Change Adaptation (SREX, blue) or AR4 (red). Projections for early 21st century were not provided in previous assessment reports. Projections in the AR5 are relative to the reference period of 1986–2005, and use the new RCP scenarios unless otherwise specified. See the Glossary for definitions of extreme weather and climate events.

Phenomenon and direction of trend	Assessment that changes occurred (typically since 1950 unless otherwise indicated)	Assessment of a human contribution to observed changes	Likelihood of further changes: Early 21st century	Likelihood of further changes: Late 21st century
Warmer and/or fewer cold days and nights over most land areas	*Very likely* {2.6} *Very likely* *Very likely*	***Very likely*** {10.6} *Likely* *Likely*	*Likely* {11.3}	*Virtually certain* {12.4} *Virtually certain* *Virtually certain*
Warmer and/or more frequent hot days and nights over most land areas	*Very likely* {2.6} *Very likely* *Very likely*	***Very likely*** {10.6} *Likely* *Likely* (nights only)	*Likely* {11.3}	*Virtually certain* {12.4} *Virtually certain* *Virtually certain*
Warm spells/heat waves. Frequency and/or duration increases over most land areas	***Medium confidence*** on a global scale *Likely* in large parts of Europe, Asia and Australia {2.6} *Medium confidence* in many (but not all) regions *Likely*	***Likely***[a] {10.6} Not formally assessed *More likely than not*	Not formally assessed[b] {11.3}	*Very likely* {12.4} *Very likely* *Very likely*
Heavy precipitation events. Increase in the frequency, intensity, and/or amount of heavy precipitation	*Likely* more land areas with increases than decreases[c] {2.6} *Likely* more land areas with increases than decreases *Likely* over most land areas	***Medium confidence*** {7.6, 10.6} *Medium confidence* *More likely than not*	*Likely* over many land areas {11.3}	***Very likely*** **over most of the mid-latitude land masses and over wet tropical regions** {12.4} *Likely* over many areas *Very likely* over most land areas
Increases in intensity and/or duration of drought	***Low confidence*** **on a global scale** ***Likely*** **changes in some regions**[d] {2.6} *Medium confidence* in some regions *Likely* in many regions, since 1970[e]	***Low confidence*** {10.6} *Medium confidence*[f] *More likely than not*	*Low confidence*[g] {11.3}	***Likely*** **(*medium confidence*) on a regional to global scale**[h] {12.4} *Medium confidence* in some regions *Likely*[e]
Increases in intense tropical cyclone activity	***Low confidence*** **in long term (centennial) changes** ***Virtually certain*** **in North Atlantic since 1970** {2.6} *Low confidence* *Likely* in some regions, since 1970	***Low confidence***[i] {10.6} *Low confidence* *More likely than not*	*Low confidence* {11.3}	***More likely than not*** **in the Western North Pacific and North Atlantic**[j] {14.6} *More likely than not* in some basins *Likely*
Increased incidence and/or magnitude of extreme high sea level	*Likely* (since 1970) {3.7} *Likely* (late 20th century) *Likely*	***Likely***[k] {3.7} *Likely*[k] *More likely than not*[k]	*Likely*[l] {13.7}	***Very likely***[l] {13.7} *Very likely*[m] *Likely*

* The direct comparison of assessment findings between reports is difficult. For some climate variables, different aspects have been assessed, and the revised guidance note on uncertainties has been used for the SREX and AR5. The availability of new information, improved scientific understanding, continued analyses of data and models, and specific differences in methodologies applied in the assessed studies, all contribute to revised assessment findings.

Notes:

[a] Attribution is based on available case studies. It is *likely* that human influence has more than doubled the probability of occurrence of some observed heat waves in some locations.

[b] Models project near-term increases in the duration, intensity and spatial extent of heat waves and warm spells.

[c] In most continents, *confidence* in trends is not higher than *medium* except in North America and Europe where there have been *likely* increases in either the frequency or intensity of heavy precipitation with some seasonal and/or regional variation. It is *very likely* that there have been increases in central North America.

[d] The frequency and intensity of drought has *likely* increased in the Mediterranean and West Africa and *likely* decreased in central North America and north-west Australia.

[e] AR4 assessed the area affected by drought.

[f] SREX assessed *medium confidence* that anthropogenic influence had contributed to some changes in the drought patterns observed in the second half of the 20th century, based on its attributed impact on precipitation and temperature changes. SREX assessed *low confidence* in the attribution of changes in droughts at the level of single regions.

[g] There is *low confidence* in projected changes in soil moisture.

[h] Regional to global-scale projected decreases in soil moisture and increased agricultural drought are *likely* (*medium confidence*) in presently dry regions by the end of this century under the RCP8.5 scenario. Soil moisture drying in the Mediterranean, Southwest USA and southern African regions is consistent with projected changes in Hadley circulation and increased surface temperatures, so there is *high confidence* in *likely* surface drying in these regions by the end of this century under the RCP8.5 scenario.

[i] There is *medium confidence* that a reduction in aerosol forcing over the North Atlantic has contributed at least in part to the observed increase in tropical cyclone activity since the 1970s in this region.

[j] Based on expert judgment and assessment of projections which use an SRES A1B (or similar) scenario.

[k] Attribution is based on the close relationship between observed changes in extreme and mean sea level.

[l] There is *high confidence* that this increase in extreme high sea level will primarily be the result of an increase in mean sea level. There is *low confidence* in region-specific projections of storminess and associated storm surges.

[m] SREX assessed it to be *very likely* that mean sea level rise will contribute to future upward trends in extreme coastal high water levels.

TFE.9 (continued)

a different rate than the mean warming however, with several studies showing that projected European high-percentile summer temperatures will warm faster than mean temperatures. Future changes associated with the warming of temperature extremes in the long-term are *virtually certain* and scale with the strength of emissions scenario, that is, greater anthropogenic emissions correspond to greater warming of extremes (TFE.9, Figure 1). For high-emissions scenarios, it is *likely* that, in most land regions, a current 1-in-20-year maximum temperature event

(continued on next page)

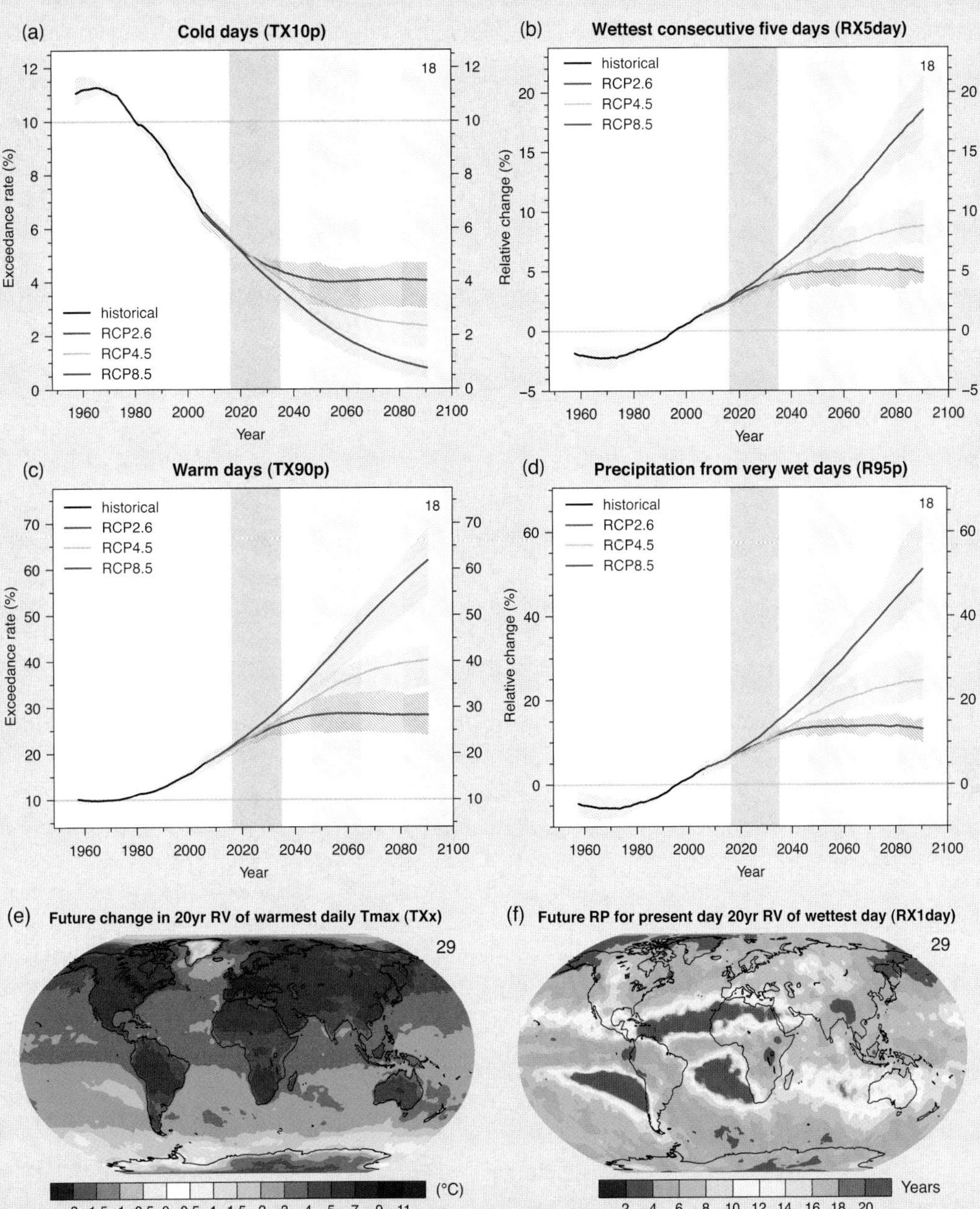

TFE.9, Figure 1 | Global projections of the occurrence of (a) cold days (TX10p)- percentage of days annually with daily maximum surface air temperature (Tmax) below the 10th percentile of Tmax for 1961 to 1990, (b) wettest consecutive 5 days (RX5day) —percentage change relative to 1986–2005 in annual maximum consecutive 5-day precipitation totals, (c) warm days (TX90p)—percentage of days annually with daily maximum surface air temperature (Tmax) exceeding the 90th percentile of Tmax for 1961 to 1990 and (d) very wet day precipitation (R95p)—percentage change relative to 1986–2005 of annual precipitation from days >95th percentile. Results are shown from CMIP5 for the RCP2.6, RCP4.5 and RCP8.5 scenarios. Solid lines indicate the ensemble median and shading indicates the interquartile spread between individual projections (25th and 75th percentiles). Maps show (e) the change from 1986–2005 to 2081–2100 in 20-year return values (RV) of daily maximum temperatures, TXx, and (f) the 2081–2100 return period (RP) for rare daily precipitation values, RX1day, that have a 20-year return period during 1986–2005. Both maps are based on the CMIP5 RCP8.5 scenario. The number of models used to calculate the multi-model mean is indicated in each panel. See Box 2.4, Table 1 for index definitions. {Figures 11.17, 12.14, 12.26, 12.27}

TFE.9 (continued)

will at least double in frequency but in many regions will become an annual or a 1-in-2-year event by the end of the 21st century. The magnitude of both high and low temperature extremes is expected to increase at least at the same rate as the mean, but with 20-year return values for low temperature events projected to increase at a rate greater than winter mean temperatures in most regions. {10.6.1, 11.3.2, 12.4.3}

Precipitation Extremes
It is *likely* that the number of heavy precipitation events over land has increased in more regions than it has decreased in since the mid-20th century, and there is *medium confidence* that anthropogenic forcing has contributed to this increase. {2.6.2, 10.6.1}

There has been substantial progress between CMIP3 and CMIP5 in the ability of models to simulate more realistic precipitation extremes. However, evidence suggests that the majority of models underestimate the sensitivity of extreme precipitation to temperature variability or trends especially in the tropics, which implies that models may underestimate the projected increase in extreme precipitation in the future. While progress has been made in understanding the processes that drive extreme precipitation, challenges remain in quantifying cloud and convective effects in models for example. The complexity of land surface and atmospheric processes limits confidence in regional projections of precipitation change, especially over land, although there is a component of a 'wet-get-wetter' and 'dry-get-drier' response over oceans at the large scale. Even so, there is *high confidence* that, as the climate warms, extreme precipitation rates (e.g., on daily time scales) will increase faster than the time average. Changes in local extremes on daily and sub-daily time scales are expected to increase by roughly 5 to 10% per °C of warming (*medium confidence*). {7.6, 9.5.4}

For the near and long term, CMIP5 projections confirm a clear tendency for increases in heavy precipitation events in the global mean seen in the AR4, but there are substantial variations across regions (TFE.9, Figure 1). Over most of the mid-latitude land masses and over wet tropical regions, extreme precipitation will *very likely* be more intense and more frequent in a warmer world. {11.3.2, 12.4.5}

Floods and Droughts
There continues to be a lack of evidence and thus *low confidence* regarding the sign of trend in the magnitude and/or frequency of floods on a global scale over the instrumental record. There is *high confidence* that past floods larger than those recorded since 1900 have occurred during the past five centuries in northern and central Europe, western Mediterranean region, and eastern Asia. There is *medium confidence* that modern large floods are comparable to or surpass historical floods in magnitude and/or frequency in the Near East, India and central North America. {2.6.2, 5.5.5}

Compelling arguments both for and against significant increases in the land area affected by drought and/or dryness since the mid-20th century have resulted in a *low confidence* assessment of observed and attributable large-scale trends. This is due primarily to a lack and quality of direct observations, dependencies of inferred trends on the index choice, geographical inconsistencies in the trends and difficulties in distinguishing decadal scale variability from long term trends. On millennial time scales, there is *high confidence* that proxy information provides evidence of droughts of greater magnitude and longer duration than observed during the 20th century in many regions. There is *medium confidence* that more megadroughts occurred in monsoon Asia and wetter conditions prevailed in arid Central Asia and the South American monsoon region during the Little Ice Age (1450 to 1850) compared to the Medieval Climate Anomaly (950 to 1250). {2.6.2, 5.5.4, 5.5.5, 10.6.1}

Under the Representative Concentration Pathway RCP8.5, projections by the end of the century indicate an increased risk of drought is *likely* (*medium confidence*) in presently dry regions linked to regional to global-scale projected decreases in soil moisture. Soil moisture drying is most prominent in the Mediterranean, Southwest USA, and southern Africa, consistent with projected changes in the Hadley Circulation and increased surface temperatures, and surface drying in these regions is *likely* (*high confidence*) by the end of the century under RCP8.5. {12.4.5}

Extreme Sea Level
It is *likely* that the magnitude of extreme high sea level events has increased since 1970 and that most of this rise can be explained by increases in mean sea level. When mean sea level changes is taken into account, changes in extreme high sea levels are reduced to less than 5 mm yr^{-1} at 94% of tide gauges. In the future it is *very likely* that there will be a significant increase in the occurrence of sea level extremes and similarly to past observations, this increase will primarily be the result of an increase in mean sea level. {3.7.5, 13.7.2}

(continued on next page)

TFE.9 (continued)

Tropical and Extratropical Cyclones

There is *low confidence* in long-term (centennial) changes in tropical cyclone activity, after accounting for past changes in observing capabilities. However over the satellite era, increases in the frequency and intensity of the strongest storms in the North Atlantic are robust (*very high confidence*). However, the cause of this increase is debated and there is *low confidence* in attribution of changes in tropical cyclone activity to human influence owing to insufficient observational evidence, lack of physical understanding of the links between anthropogenic drivers of climate and tropical cyclone activity and the low level of agreement between studies as to the relative importance of internal variability, and anthropogenic and natural forcings. {2.6.3, 10.6.1, 14.6.1}

Some high-resolution atmospheric models have realistically simulated tracks and counts of tropical cyclones and models generally are able to capture the general characteristics of storm tracks and extratropical cyclones with evidence of improvement since the AR4. Storm track biases in the North Atlantic have improved slightly, but models still produce a storm track that is too zonal and underestimate cyclone intensity. {9.4.1, 9.5.4}

While projections indicate that it is *likely* that the global frequency of tropical cyclones will either decrease or remain essentially unchanged, concurrent with a *likely* increase in both global mean tropical cyclone maximum wind speed and rainfall rates, there is lower confidence in region-specific projections of frequency and intensity. However, due to improvements in model resolution and downscaling techniques, it is *more likely than not* that the frequency of the most intense storms will increase substantially in some basins under projected 21st century warming (see Figure TS.26). {11.3.2, 14.6.1}

Research subsequent to the AR4 and SREX continues to support a *likely* poleward shift of storm tracks since the 1950s. However over the last century there is *low confidence* of a clear trend in storminess due to inconsistencies between studies or lack of long-term data in some parts of the world (particularly in the Southern Hemisphere (SH)). {2.6.4, 2.7.6}

Despite systematic biases in simulating storm tracks, most models and studies are in agreement that the global number of extratropical cyclones is *unlikely* to decrease by more than a few per cent. A small poleward shift is *likely* in the SH storm track. It is *more likely than not* (*medium confidence*) for a projected poleward shift in the North Pacific storm track but it is *unlikely* that the response of the North Atlantic storm track is a simple poleward shift. There is *low confidence* in the magnitude of regional storm track changes, and the impact of such changes on regional surface climate. {14.6.2}

TS.6 Key Uncertainties

This final section of the Technical Summary provides readers with a short overview of key uncertainties in the understanding of the climate system and the ability to project changes in response to anthropogenic influences. The overview is not comprehensive and does not describe in detail the basis for these findings. These are found in the main body of this Technical Summary and in the underlying chapters to which each bullet points in the curly brackets.

TS.6.1 Key Uncertainties in Observation of Changes in the Climate System

- There is only *medium* to *low confidence* in the rate of change of tropospheric warming and its vertical structure. Estimates of tropospheric warming rates encompass surface temperature warming rate estimates. There is *low confidence* in the rate and vertical structure of the stratospheric cooling. {2.4.4}

- *Confidence* in global precipitation change over land is *low* prior to 1951 and *medium* afterwards because of data incompleteness. {2.5.1}

- Substantial ambiguity and therefore *low confidence* remains in the observations of global-scale cloud variability and trends. {2.5.6}

- There is *low confidence* in an observed global-scale trend in drought or dryness (lack of rainfall), due to lack of direct observations, methodological uncertainties and choice and geographical inconsistencies in the trends. {2.6.2}

- There is *low confidence* that any reported long-term (centennial) changes in tropical cyclone characteristics are robust, after accounting for past changes in observing capabilities. {2.6.3}

- Robust conclusions on long-term changes in large-scale atmospheric circulation are presently not possible because of large variability on interannual to decadal time scales and remaining differences between data sets. {2.7}

- Different global estimates of sub-surface ocean temperatures have variations at different times and for different periods, suggesting that sub-decadal variability in the temperature and upper heat content (0 to to 700 m) is still poorly characterized in the historical record. {3.2}

- Below ocean depths of 700 m the sampling in space and time is too sparse to produce annual global ocean temperature and heat content estimates prior to 2005. {3.2.4}

- Observational coverage of the ocean deeper than 2000 m is still limited and hampers more robust estimates of changes in global ocean heat content and carbon content. This also limits the quantification of the contribution of deep ocean warming to sea level rise. {3.2, 3.7, 3.8; Box 3.1}

- The number of continuous observational time series measuring the strength of climate relevant ocean circulation features (e.g., the meridional overturning circulation) is limited and the existing time series are still too short to assess decadal and longer trends. {3.6}.

- In Antarctica, available data are inadequate to assess the status of change of many characteristics of sea ice (e.g., thickness and volume). {4.2.3}

- On a global scale the mass loss from melting at calving fronts and iceberg calving are not yet comprehensively assessed. The largest uncertainty in estimated mass loss from glaciers comes from the Antarctic, and the observational record of ice–ocean interactions around both ice sheets remains poor. {4.3.3, 4.4}

TS.6.2 Key Uncertainties in Drivers of Climate Change

- Uncertainties in aerosol–cloud interactions and the associated radiative forcing remain large. As a result, uncertainties in aerosol forcing remain the dominant contributor to the overall uncertainty in net anthropogenic forcing, despite a better understanding of some of the relevant atmospheric processes and the availability of global satellite monitoring. {2.2, 7.3–7.5, 8.5}

- The cloud feedback is *likely* positive but its quantification remains difficult. {7.2}

- Paleoclimate reconstructions and Earth System Models indicate that there is a positive feedback between climate and the carbon cycle, but *confidence* remains *low* in the strength of this feedback, particularly for the land. {6.4}

TS.6.3 Key Uncertainties in Understanding the Climate System and Its Recent Changes

- The simulation of clouds in AOGCMs has shown modest improvement since AR4; however, it remains challenging. {7.2, 9.2.1, 9.4.1, 9.7.2}

- Observational uncertainties for climate variables other than temperature, uncertainties in forcings such as aerosols, and limits in process understanding continue to hamper attribution of changes in many aspects of the climate system. {10.1, 10.3, 10.7}

- Changes in the water cycle remain less reliably modelled in both their changes and their internal variability, limiting confidence in attribution assessments. Observational uncertainties and the large effect of internal variability on observed precipitation also precludes a more confident assessment of the causes of precipitation changes. {2.5.1, 2.5.4, 10.3.2}

- Modelling uncertainties related to model resolution and incorporation of relevant processes become more important at regional scales, and the effects of internal variability become more significant. Therefore, challenges persist in attributing observed change to external forcing at regional scales. {2.4.1, 10.3.1}

- The ability to simulate changes in frequency and intensity of extreme events is limited by the ability of models to reliably simulate mean changes in key features. {10.6.1}

- In some aspects of the climate system, including changes in drought, changes in tropical cyclone activity, Antarctic warming, Antarctic sea ice extent, and Antarctic mass balance, *confidence* in attribution to human influence remains *low* due to modelling uncertainties and low agreement between scientific studies. {10.3.1, 10.5.2, 10.6.1}

TS.6.4 Key Uncertainties in Projections of Global and Regional Climate Change

- Based on model results there is limited confidence in the predictability of yearly to decadal averages of temperature both for the global average and for some geographical regions. Multi-model results for precipitation indicate a generally low predictability. Short-term climate projection is also limited by the uncertainty in projections of natural forcing. {11.1, 11.2, 11.3.1, 11.3.6; Box 11.1}

- There is *medium confidence* in near-term projections of a northward shift of NH storm track and westerlies. {11.3.2}

- There is generally *low confidence* in basin-scale projections of significant trends in tropical cyclone frequency and intensity in the 21st century. {11.3.2, 14.6.1}

- Projected changes in soil moisture and surface run off are not robust in many regions. {11.3.2, 12.4.5}

- Several components or phenomena in the climate system could potentially exhibit abrupt or nonlinear changes, but for many phenomena there is *low confidence* and little consensus on the likelihood of such events over the 21st century. {12.5.5}

- There is *low confidence* on magnitude of carbon losses through CO_2 or CH_4 emissions to the atmosphere from thawing permafrost. There is *low confidence* in projected future CH_4 emissions from natural sources due to changes in wetlands and gas hydrate release from the sea floor. {6.4.3, 6.4.7}

- There is *medium confidence* in the projected contributions to sea level rise by models of ice sheet dynamics for the 21st century, and *low confidence* in their projections beyond 2100. {13.3.3}

- There is *low confidence* in semi-empirical model projections of global mean sea level rise, and no consensus in the scientific community about their reliability. {13.5.2, 13.5.3}

- There is *low confidence* in projections of many aspects of climate phenomena that influence regional climate change, including changes in amplitude and spatial pattern of modes of climate variability. {9.5.3, 14.2–14.7}

Chapters

1 Introduction

Coordinating Lead Authors:
Ulrich Cubasch (Germany), Donald Wuebbles (USA)

Lead Authors:
Deliang Chen (Sweden), Maria Cristina Facchini (Italy), David Frame (UK/New Zealand), Natalie Mahowald (USA), Jan-Gunnar Winther (Norway)

Contributing Authors:
Achim Brauer (Germany), Lydia Gates (Germany), Emily Janssen (USA), Frank Kaspar (Germany), Janina Körper (Germany), Valérie Masson-Delmotte (France), Malte Meinshausen (Australia/Germany), Matthew Menne (USA), Carolin Richter (Switzerland), Michael Schulz (Germany), Uwe Schulzweida (Germany), Bjorn Stevens (Germany/USA), Rowan Sutton (UK), Kevin Trenberth (USA), Murat Türkeş (Turkey), Daniel S. Ward (USA)

Review Editors:
Yihui Ding (China), Linda Mearns (USA), Peter Wadhams (UK)

This chapter should be cited as:
Cubasch, U., D. Wuebbles, D. Chen, M.C. Facchini, D. Frame, N. Mahowald, and J.-G. Winther, 2013: Introduction. In: *Climate Change 2013: The Physical Science Basis. Contribution of Working Group I to the Fifth Assessment Report of the Intergovernmental Panel on Climate Change* [Stocker, T.F., D. Qin, G.-K. Plattner, M. Tignor, S.K. Allen, J. Boschung, A. Nauels, Y. Xia, V. Bex and P.M. Midgley (eds.)]. Cambridge University Press, Cambridge, United Kingdom and New York, NY, USA.

1

Table of Contents

Executive Summary

Human Effects on Climate

Human activities are continuing to affect the Earth's energy budget by changing the emissions and resulting atmospheric concentrations of radiatively important gases and aerosols and by changing land surface properties. Previous assessments have already shown through multiple lines of evidence that the climate is changing across our planet, largely as a result of human activities. The most compelling evidence of climate change derives from observations of the atmosphere, land, oceans and cryosphere. Unequivocal evidence from *in situ* observations and ice core records shows that the atmospheric concentrations of important greenhouse gases such as carbon dioxide (CO_2), methane (CH_4), and nitrous oxide (N_2O) have increased over the last few centuries. {1.2.2, 1.2.3}

The processes affecting climate can exhibit considerable natural variability. Even in the absence of external forcing, periodic and chaotic variations on a vast range of spatial and temporal scales are observed. Much of this variability can be represented by simple (e.g., unimodal or power law) distributions, but many components of the climate system also exhibit multiple states—for instance, the glacial–interglacial cycles and certain modes of internal variability such as El Niño-Southern Oscillation (ENSO). Movement between states can occur as a result of natural variability, or in response to external forcing. The relationship among variability, forcing and response reveals the complexity of the dynamics of the climate system: the relationship between forcing and response for some parts of the system seems reasonably linear; in other cases this relationship is much more complex. {1.2.2}

Multiple Lines of Evidence for Climate Change

Global mean surface air temperatures over land and oceans have increased over the last 100 years. Temperature measurements in the oceans show a continuing increase in the heat content of the oceans. Analyses based on measurements of the Earth's radiative budget suggest a small positive energy imbalance that serves to increase the global heat content of the Earth system. Observations from satellites and *in situ* measurements show a trend of significant reductions in the mass balance of most land ice masses and in Arctic sea ice. The oceans' uptake of CO_2 is having a significant effect on the chemistry of sea water. Paleoclimatic reconstructions have helped place ongoing climate change in the perspective of natural climate variability. {1.2.3; Figure 1.3}

Observations of CO_2 concentrations, globally averaged temperature and sea level rise are generally well within the range of the extent of the earlier IPCC projections. The recently observed increases in CH_4 and N_2O concentrations are smaller than those assumed in the scenarios in the previous assessments. Each IPCC assessment has used new projections of future climate change that have become more detailed as the models have become more advanced. Similarly, the scenarios used in the IPCC assessments have themselves changed over time to reflect the state of knowledge. The range of climate projections from model results provided and assessed in the first IPCC assessment in 1990 to those in the 2007 AR4 provides an opportunity to compare the projections with the actually observed changes, thereby examining the deviations of the projections from the observations over time. {1.3.1, 1.3.2, 1.3.4; Figures 1.4, 1.5, 1.6, 1.7, 1.10}

Climate change, whether driven by natural or human forcing, can lead to changes in the likelihood of the occurrence or strength of extreme weather and climate events or both. Since the AR4, the observational basis has increased substantially, so that some extremes are now examined over most land areas. Furthermore, more models with higher resolution and a greater number of regional models have been used in the simulations and projections of extremes. {1.3.3; Figure 1.9}

Treatment of Uncertainties

For AR5, the three IPCC Working Groups use two metrics to communicate the degree of certainty in key findings: (1) Confidence is a qualitative measure of the validity of a finding, based on the type, amount, quality and consistency of evidence (e.g., data, mechanistic understanding, theory, models, expert judgment) and the degree of agreement[1]; and (2) Likelihood provides a quantified measure of uncertainty in a finding expressed probabilistically (e.g., based on statistical analysis of observations or model results, or both, and expert judgement)[2]. {1.4; Figure 1.11}

Advances in Measurement and Modelling Capabilities

Over the last few decades, new observational systems, especially satellite-based systems, have increased the number of observations of the Earth's climate by orders of magnitude. Tools to analyse and process these data have been developed or enhanced to cope with this large increase in information, and more climate proxy data have been acquired to improve our knowledge of past changes in climate. Because the Earth's climate system is characterized on multiple spatial and temporal scales, new observations may reduce the uncertainties surrounding the understanding of short timescale

---

1 In this Report, the following summary terms are used to describe the available evidence: limited, medium, or robust; and for the degree of agreement: low, medium, or high. A level of confidence is expressed using five qualifiers: very low, low, medium, high, and very high, and typeset in italics, e.g., *medium confidence*. For a given evidence and agreement statement, different confidence levels can be assigned, but increasing levels of evidence and degrees of agreement are correlated with increasing confidence (see Section 1.4 and Box TS.1 for more details).

2 In this Report, the following terms have been used to indicate the assessed likelihood of an outcome or a result: Virtually certain 99–100% probability, Very likely 90–100%, Likely 66–100%, About as likely as not 33–66%, Unlikely 0–33%, Very unlikely 0–10%, Exceptionally unlikely 0–1%. Additional terms (Extremely likely: 95–100%, More likely than not >50–100%, and Extremely unlikely 0–5%) may also be used when appropriate. Assessed likelihood is typeset in italics, e.g., *very likely* (see Section 1.4 and Box TS.1 for more details).

1

processes quite rapidly. However, processes that occur over longer timescales may require very long observational baselines before much progress can be made. {1.5.1; Figure 1.12}

Increases in computing speed and memory have led to the development of more sophisticated models that describe physical, chemical and biological processes in greater detail. Modelling strategies have been extended to provide better estimates of the uncertainty in climate change projections. The model comparisons with observations have pushed the analysis and development of the models. The inclusion of 'long-term' simulations has allowed incorporation of information from paleoclimate data to inform projections. Within uncertainties associated with reconstructions of past climate variables from proxy record and forcings, paleoclimate information from the Mid Holocene, Last Glacial Maximum, and Last Millennium have been used to test the ability of models to simulate realistically the magnitude and large-scale patterns of past changes. {1.5.2; Figures 1.13, 1.14}

As part of the process of getting model analyses for a range of alternative images of how the future may unfold, four new scenarios for future emissions of important gases and aerosols have been developed for the AR5, referred to as Representative Concentration Pathways (RCPs). {Box 1.1}

1.1 Chapter Preview

This introductory chapter serves as a lead-in to the science presented in the Working Group I (WGI) contribution to the Intergovernmental Panel on Climate Change (IPCC) Fifth Assessment Report (AR5). Chapter 1 in the IPCC Fourth Assessment Report (AR4) (Le Treut et al., 2007) provided a historical perspective on the understanding of climate science and the evidence regarding human influence on the Earth's climate system. Since the last assessment, the scientific knowledge gained through observations, theoretical analyses, and modelling studies has continued to increase and to strengthen further the evidence linking human activities to the ongoing climate change. In AR5, Chapter 1 focuses on the concepts and definitions applied in the discussions of new findings in the other chapters. It also examines several of the key indicators for a changing climate and shows how the current knowledge of those indicators compares with the projections made in previous assessments. The new scenarios for projected human-related emissions used in this assessment are also introduced. Finally, the chapter discusses the directions and capabilities of current climate science, while the detailed discussion of new findings is covered in the remainder of the WGI contribution to the AR5.

1.2 Rationale and Key Concepts of the WGI Contribution

1.2.1 Setting the Stage for the Assessment

The IPCC was set up in 1988 by the World Meteorological Organization and the United Nations Environment Programme to provide governments with a clear view of the current state of knowledge about the science of climate change, potential impacts, and options for adaptation and mitigation through regular assessments of the most recent information published in the scientific, technical and socio-economic literature worldwide. The WGI contribution to the IPCC AR5 assesses the current state of the physical sciences with respect to climate change. This report presents an assessment of the current state of research results and is not a discussion of all relevant papers as would be included in a review. It thus seeks to make sure that the range of scientific views, as represented in the peer-reviewed literature, is considered and evaluated in the assessment, and that the state of the science is concisely and accurately presented. A transparent review process ensures that disparate views are included (IPCC, 2012a).

As an overview, Table 1.1 shows a selection of key findings from earlier IPCC assessments. This table provides a non-comprehensive selection of key assessment statements from previous assessment reports—IPCC First Assessment Report (FAR, IPCC, 1990), IPCC Second Assessment Report (SAR, IPCC, 1996), IPCC Third Assessment Report (TAR, IPCC, 2001) and IPCC Fourth Assessment Report (AR4, IPCC, 2007)—with a focus on policy-relevant quantities that have been evaluated in each of the IPCC assessments.

Scientific hypotheses are contingent and always open to revision in light of new evidence and theory. In this sense the distinguishing features of scientific enquiry are the search for truth and the willingness to subject itself to critical re-examination. Modern research science conducts this critical revision through processes such as the peer review. At conferences and in the procedures that surround publication in peer-reviewed journals, scientific claims about environmental processes are analysed and held up to scrutiny. Even after publication, findings are further analysed and evaluated. That is the self-correcting nature of the scientific process (more details are given in AR4 Chapter 1 and Le Treut et al., 2007).

Science strives for objectivity but inevitably also involves choices and judgements. Scientists make choices regarding data and models, which processes to include and which to leave out. Usually these choices are uncontroversial and play only a minor role in the production of research. Sometimes, however, the choices scientists make are sources of disagreement and uncertainty. These are usually resolved by further scientific enquiry into the sources of disagreement. In some cases, experts cannot reach a consensus view. Examples in climate science include how best to evaluate climate models relative to observations, how best to evaluate potential sea level rise and how to evaluate probabilistic projections of climate change. In many cases there may be no definitive solution to these questions. The IPCC process is aimed at assessing the literature as it stands and attempts to reflect the level of reasonable scientific consensus as well as disagreement.

To assess areas of scientific controversy, the peer-reviewed literature is considered and evaluated. Not all papers on a controversial point can be discussed individually in an assessment, but every effort has been made here to ensure that all views represented in the peer-reviewed literature are considered in the assessment process. A list of topical issues is given in Table 1.3.

The Earth sciences study the multitude of processes that shape our environment. Some of these processes can be understood through idealized laboratory experiments, by altering a single element and then tracing through the effects of that controlled change. However, as in other natural and the social sciences, the openness of environmental systems, in terms of our lack of control of the boundaries of the system, their spatially and temporally multi-scale character and the complexity of interactions, often hamper scientists' ability to definitively isolate causal links. This in turn places important limits on the understanding of many of the inferences in the Earth sciences (e.g., Oreskes et al., 1994). There are many cases where scientists are able to make inferences using statistical tools with considerable evidential support and with high degrees of confidence, and conceptual and numerical modelling can assist in forming understanding and intuition about the interaction of dynamic processes.

1.2.2 Key Concepts in Climate Science

Here, some of the key concepts in climate science are briefly described; many of these were summarized more comprehensively in earlier IPCC assessments (Baede et al., 2001). We focus only on a certain number of them to facilitate discussions in this assessment.

First, it is important to distinguish the meaning of weather from climate. Weather describes the conditions of the atmosphere at a certain place and time with reference to temperature, pressure, humidity, wind, and other key parameters (meteorological elements); the

Table 1.1 | Historical overview of major conclusions of previous IPCC assessment reports. The table provides a non-comprehensive selection of key statements from previous assessment reports—IPCC First Assessment Report (FAR; IPCC, 1990), IPCC Second Assessment Report (SAR; IPCC, 1996), IPCC Third Assessment Report (TAR; IPCC, 2001) and IPCC Fourth Assessment Report (AR4; IPCC, 2007)—with a focus on global mean surface air temperature and sea level change as two policy relevant quantities that have been covered in IPCC since the first assessment report.

Topic		FAR SPM Statement	SAR SPM Statement	TAR SPM Statement	AR4 SPM Statement
Human and Natural Drivers of Climate Change		There is a natural greenhouse effect which already keeps the Earth warmer than it would otherwise be. Emissions resulting from human activities are substantially increasing the atmospheric concentrations of the greenhouse gases carbon dioxide, methane, chlorofluorocarbons and nitrous oxide. These increases will enhance the greenhouse effect, resulting on average in an additional warming of the Earth's surface. Continued emissions of these gases at present rates would commit us to increased concentrations for centuries ahead.	Greenhouse gas concentrations have continued to increase. These trends can be attributed largely to human activities, mostly fossil fuel use, land use change and agriculture. Anthropogenic aerosols are short-lived and tend to produce negative radiative forcing.	Emissions of greenhouse gases and aerosols due to human activities continue to alter the atmosphere in ways that are expected to affect the climate. The atmospheric concentration of CO_2 has increased by 31% since 1750 and that of methane by 151%. Anthropogenic aerosols are short-lived and mostly produce negative radiative forcing by their direct effect. There is more evidence for their indirect effect, which is negative, although of very uncertain magnitude. Natural factors have made small contributions to radiative forcing over the past century.	Global atmospheric concentrations of carbon dioxide, methane and nitrous oxide have increased markedly as a result of human activities since 1750 and now far exceed pre-industrial values determined from ice cores spanning many thousands of years. The global increases in carbon dioxide concentration are due primarily to fossil fuel use and land use change, while those of methane and nitrous oxide are primarily due to agriculture. *Very high confidence* that the global average net effect of human activities since 1750 has been one of warming, with a radiative forcing of +1.6 [+0.6 to +2.4] W m^{-2}.
Direct Observations of Recent Climate Change	**Temperature**	Global mean surface air temperature has increased by 0.3°C to 0.6°C over the last 100 years, with the five global-average warmest years being in the 1980s.	Climate has changed over the past century. Global mean surface temperature has increased by between about 0.3 and 0.6°C since the late 19th century. Recent years have been among the warmest since 1860, despite the cooling effect of the 1991 Mt. Pinatubo volcanic eruption.	An increasing body of observations gives a collective picture of a warming world and other changes in the climate system. The global average temperature has increased since 1861. Over the 20th century the increase has been 0.6°C. Some important aspects of climate appear not to have changed.	Warming of the climate system is unequivocal, as is now evident from observations of increases in global average air and ocean temperatures, widespread melting of snow and ice, and rising global average sea level. Eleven of the last twelve years (1995–2006) rank among the 12 warmest years in the instrumental record of global surface temperature (since 1850). The updated 100-year linear trend (1906 to 2005) of 0.74°C [0.56°C to 0.92°C] is therefore larger than the corresponding trend for 1901 to 2000 given in the TAR of 0.6°C [0.4°C to 0.8°C]. Some aspects of climate have not been observed to change.
	Sea Level	Over the same period global sea level has increased by 10 to 20 cm These increases have not been smooth with time, nor uniform over the globe.	Global sea level has risen by between 10 and 25 cm over the past 100 years and much of the rise may be related to the increase in global mean temperature.	Tide gauge data show that global average sea level rose between 0.1 and 0.2 m during the 20th century.	Global average sea level rose at an average rate of 1.8 [1.3 to 2.3] mm per year over 1961 to 2003. The rate was faster over 1993 to 2003: about 3.1 [2.4 to 3.8] mm per year. The total 20th century rise is estimated to be 0.17 [0.12 to 0.22] m.
A Palaeoclimatic Perspective		Climate varies naturally on all timescales from hundreds of millions of years down to the year-to-year. Prominent in the Earth's history have been the 100,000 year glacial–interglacial cycles when climate was mostly cooler than at present. Global surface temperatures have typically varied by 5°C to 7°C through these cycles, with large changes in ice volume and sea level, and temperature changes as great as 10°C to 15°C in some middle and high latitude regions of the Northern Hemisphere. Since the end of the last ice age, about 10,000 years ago, global surface temperatures have probably fluctuated by little more than 1°C. Some fluctuations have lasted several centuries, including the Little Ice Age which ended in the nineteenth century and which appears to have been global in extent.	The limited available evidence from proxy climate indicators suggests that the 20th century global mean temperature is at least as warm as any other century since at least 1400 AD. Data prior to 1400 are too sparse to allow the reliable estimation of global mean temperature.	New analyses of proxy data for the Northern Hemisphere indicate that the increase in temperature in the 20th century is *likely* to have been the largest of any century during the past 1,000 years. It is also *likely* that, in the Northern Hemisphere, the 1990s was the warmest decade and 1998 the warmest year. Because less data are available, less is known about annual averages prior to 1,000 years before present and for conditions prevailing in most of the Southern Hemisphere prior to 1861.	Palaeoclimatic information supports the interpretation that the warmth of the last half century is unusual in at least the previous 1,300 years. The last time the polar regions were significantly warmer than present for an extended period (about 125,000 years ago), reductions in polar ice volume led to 4 to 6 m of sea level rise.

(continued on next page)

(Table 1.1 continued)

Topic		FAR SPM Statement	SAR SPM Statement	TAR SPM Statement	AR4 SPM Statement
Understanding and Attributing Climate Change		The size of this warming is broadly consistent with predictions of climate models, but it is also of the same magnitude as natural climate variability. Thus the observed increase could be largely due to this natural variability; alternatively this variability and other human factors could have offset a still larger human-induced greenhouse warming. The unequivocal detection of the enhanced greenhouse effect from observations is not likely for a decade or more.	The balance of evidence suggests a discernible human influence on global climate. Simulations with coupled atmosphere–ocean models have provided important information about decade to century timescale natural internal climate variability.	There is new and stronger evidence that most of the warming observed over the last 50 years is attributable to human activities. There is a longer and more scrutinized temperature record and new model estimates of variability. Reconstructions of climate data for the past 1,000 years indicate this warming was unusual and is *unlikely* to be entirely natural in origin.	Most of the observed increase in global average temperatures since the mid-20th century is *very likely* due to the observed increase in anthropogenic greenhouse gas concentrations. Discernible human influences now extend to other aspects of climate, including ocean warming, continental-average temperatures, temperature extremes and wind patterns.
Projections of Future Changes in Climate	**Temperature**	Under the IPCC Business-as-Usual-emissions of greenhouse gases, a rate of increase of global mean temperature during the next century of about 0.3°C per decade (with an uncertainty range of 0.2°C to 0.5°C per decade); this is greater than that seen over the past 10,000 years.	Climate is expected to continue to change in the future. For the mid-range IPCC emission scenario, IS92a, assuming the 'best estimate' value of climate sensitivity and including the effects of future increases in aerosols, models project an increase in global mean surface air temperature relative to 1990 of about 2°C by 2100.	Global average temperature and sea level are projected to rise under all IPCC SRES scenarios. The globally averaged surface temperature is projected to increase by 1.4°C to 5.8°C over the period 1990 to 2100. Confidence in the ability of models to project future climate has increased. Anthropogenic climate change will persist for many centuries.	For the next two decades, a warming of about 0.2°C per decade is projected for a range of SRES emission scenarios. Even if the concentrations of all greenhouse gases and aerosols had been kept constant at year 2000 levels, a further warming of about 0.1°C per decade would be expected. There is now higher confidence in projected patterns of warming and other regional-scale features, including changes in wind patterns, precipitation and some aspects of extremes and of ice. Anthropogenic warming and sea level rise would continue for centuries, even if greenhouse gas concentrations were to be stabilised.
	Sea Level	An average rate of global mean sea level rise of about 6 cm per decade over the next century (with an uncertainty range of 3 to 10 cm per decade) is projected.	Models project a sea level rise of 50 cm from the present to 2100.	Global mean sea level is projected to rise by 0.09 to 0.88 m between 1990 and 2100.	Global sea level rise for the range of scenarios is projected as 0.18 to 0.59 m by the end of the 21st century.

1

1

presence of clouds, precipitation; and the occurrence of special phenomena, such as thunderstorms, dust storms, tornados and others. Climate in a narrow sense is usually defined as the average weather, or more rigorously, as the statistical description in terms of the mean and variability of relevant quantities over a period of time ranging from months to thousands or millions of years. The relevant quantities are most often surface variables such as temperature, precipitation and wind. Classically the period for averaging these variables is 30 years, as defined by the World Meteorological Organization. Climate in a wider sense also includes not just the mean conditions, but also the associated statistics (frequency, magnitude, persistence, trends, etc.), often combining parameters to describe phenomena such as droughts. Climate change refers to a change in the state of the climate that can be identified (e.g., by using statistical tests) by changes in the mean and/or the variability of its properties, and that persists for an extended period, typically decades or longer.

The Earth's climate system is powered by solar radiation (Figure 1.1). Approximately half of the energy from the Sun is supplied in the visible part of the electromagnetic spectrum. As the Earth's temperature has been relatively constant over many centuries, the incoming solar energy must be nearly in balance with outgoing radiation. Of the incoming solar shortwave radiation (SWR), about half is absorbed by the Earth's surface. The fraction of SWR reflected back to space by gases and aerosols, clouds and by the Earth's surface (albedo) is approximately 30%, and about 20% is absorbed in the atmosphere. Based on the temperature of the Earth's surface the majority of the outgoing energy flux from the Earth is in the infrared part of the spectrum. The longwave radiation (LWR, also referred to as infrared radiation) emitted from the Earth's surface is largely absorbed by certain atmospheric constituents—water vapour, carbon dioxide (CO_2), methane (CH_4), nitrous oxide (N_2O) and other greenhouse gases (GHGs); see Annex III for Glossary—and clouds, which themselves emit LWR into all directions. The downward directed component of this LWR adds heat to the lower layers of the atmosphere and to the Earth's surface (greenhouse effect). The dominant energy loss of the infrared radiation from the Earth is from higher layers of the troposphere. The Sun provides its energy to the Earth primarily in the tropics and the subtropics; this energy is then partially redistributed to middle and high latitudes by atmospheric and oceanic transport processes.

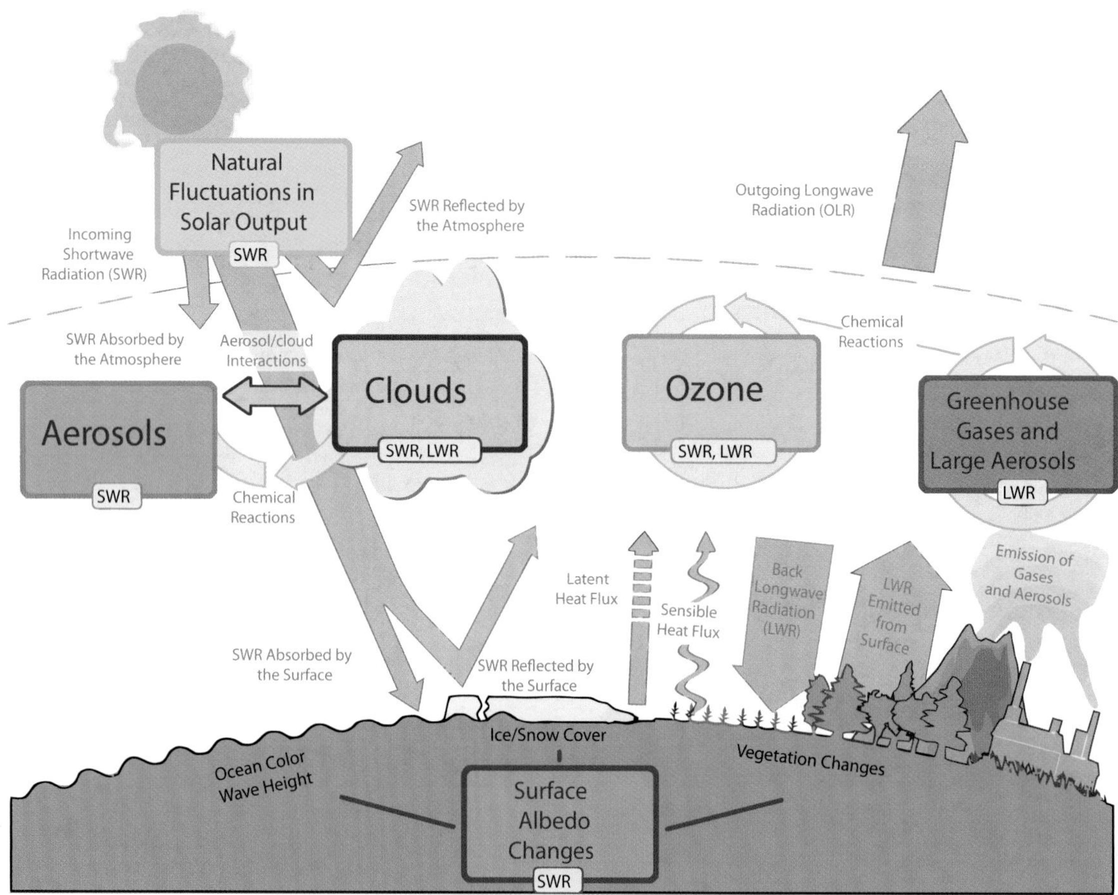

Figure 1.1 | Main drivers of climate change. The radiative balance between incoming solar shortwave radiation (SWR) and outgoing longwave radiation (OLR) is influenced by global climate 'drivers'. Natural fluctuations in solar output (solar cycles) can cause changes in the energy balance (through fluctuations in the amount of incoming SWR) (Section 2.3). Human activity changes the emissions of gases and aerosols, which are involved in atmospheric chemical reactions, resulting in modified O_3 and aerosol amounts (Section 2.2). O_3 and aerosol particles absorb, scatter and reflect SWR, changing the energy balance. Some aerosols act as cloud condensation nuclei modifying the properties of cloud droplets and possibly affecting precipitation (Section 7.4). Because cloud interactions with SWR and LWR are large, small changes in the properties of clouds have important implications for the radiative budget (Section 7.4). Anthropogenic changes in GHGs (e.g., CO_2, CH_4, N_2O, O_3, CFCs) and large aerosols (>2.5 μm in size) modify the amount of outgoing LWR by absorbing outgoing LWR and re-emitting less energy at a lower temperature (Section 2.2). Surface albedo is changed by changes in vegetation or land surface properties, snow or ice cover and ocean colour (Section 2.3). These changes are driven by natural seasonal and diurnal changes (e.g., snow cover), as well as human influence (e.g., changes in vegetation types) (Forster et al., 2007).

Changes in the global energy budget derive from either changes in the net incoming solar radiation or changes in the outgoing longwave radiation (OLR). Changes in the net incoming solar radiation derive from changes in the Sun's output of energy or changes in the Earth's albedo. Reliable measurements of total solar irradiance (TSI) can be made only from space, and the precise record extends back only to 1978. The generally accepted mean value of the TSI is about 1361 W m^{-2} (Kopp and Lean, 2011; see Chapter 8 for a detailed discussion on the TSI); this is lower than the previous value of 1365 W m^{-2} used in the earlier assessments. Short-term variations of a few tenths of a percent are common during the approximately 11-year sunspot solar cycle (see Sections 5.2 and 8.4 for further details). Changes in the outgoing LWR can result from changes in the temperature of the Earth's surface or atmosphere or changes in the emissivity (measure of emission efficiency) of LWR from either the atmosphere or the Earth's surface. For the atmosphere, these changes in emissivity are due predominantly to changes in cloud cover and cloud properties, in GHGs and in aerosol concentrations. The radiative energy budget of the Earth is almost in balance (Figure 1.1), but ocean heat content and satellite measurements indicate a small positive imbalance (Murphy et al., 2009; Trenberth et al., 2009; Hansen et al., 2011) that is consistent with the rapid changes in the atmospheric composition.

In addition, some aerosols increase atmospheric reflectivity, whereas others (e.g., particulate black carbon) are strong absorbers and also modify SWR (see Section 7.2 for a detailed assessment). Indirectly, aerosols also affect cloud albedo, because many aerosols serve as cloud condensation nuclei or ice nuclei. This means that changes in aerosol types and distribution can result in small but important changes in cloud albedo and lifetime (Section 7.4). Clouds play a critical role in climate because they not only can increase albedo, thereby cooling the planet, but also because of their warming effects through infrared radiative transfer. Whether the net radiative effect of a cloud is one of cooling or of warming depends on its physical properties (level of occurrence, vertical extent, water path and effective cloud particle size) as well as on the nature of the cloud condensation nuclei population (Section 7.3). Humans enhance the greenhouse effect directly by emitting GHGs such as CO_2, CH_4, N_2O and chlorofluorocarbons (CFCs) (Figure 1.1). In addition, pollutants such as carbon monoxide (CO), volatile organic compounds (VOC), nitrogen oxides (NO_x) and sulphur dioxide (SO_2), which by themselves are negligible GHGs, have an indirect effect on the greenhouse effect by altering, through atmospheric chemical reactions, the abundance of important gases to the amount of outgoing LWR such as CH_4 and ozone (O_3), and/or by acting as precursors of secondary aerosols. Because anthropogenic emission sources simultaneously can emit some chemicals that affect climate and others that affect air pollution, including some that affect both, atmospheric chemistry and climate science are intrinsically linked.

In addition to changing the atmospheric concentrations of gases and aerosols, humans are affecting both the energy and water budget of the planet by changing the land surface, including redistributing the balance between latent and sensible heat fluxes (Sections 2.5, 7.2, 7.6 and 8.2). Land use changes, such as the conversion of forests to cultivated land, change the characteristics of vegetation, including its colour, seasonal growth and carbon content (Houghton, 2003; Foley et al., 2005). For example, clearing and burning a forest to prepare agricultural land reduces carbon storage in the vegetation, adds CO_2 to the atmosphere, and changes the reflectivity of the land (surface albedo), rates of evapotranspiration and longwave emissions (Figure 1.1).

Changes in the atmosphere, land, ocean, biosphere and cryosphere—both natural and anthropogenic—can perturb the Earth's radiation budget, producing a radiative forcing (RF) that affects climate. RF is a measure of the net change in the energy balance in response to an external perturbation. The drivers of changes in climate can include, for example, changes in the solar irradiance and changes in atmospheric trace gas and aerosol concentrations (Figure 1.1). The concept of RF cannot capture the interactions of anthropogenic aerosols and clouds, for example, and thus in addition to the RF as used in previous assessments, Sections 7.4 and 8.1 introduce a new concept, effective radiative forcing (ERF), that accounts for rapid response in the climate system. ERF is defined as the change in net downward flux at the top of the atmosphere after allowing for atmospheric temperatures, water vapour, clouds and land albedo to adjust, but with either sea surface temperatures (SSTs) and sea ice cover unchanged or with global mean surface temperature unchanged.

Once a forcing is applied, complex internal feedbacks determine the eventual response of the climate system, and will in general cause this response to differ from a simple linear one (IPCC, 2001, 2007). There are many feedback mechanisms in the climate system that can either amplify ('positive feedback') or diminish ('negative feedback') the effects of a change in climate forcing (Le Treut et al., 2007) (see Figure 1.2 for a representation of some of the key feedbacks). An example of a positive feedback is the water vapour feedback whereby an increase in surface temperature enhances the amount of water vapour present in the atmosphere. Water vapour is a powerful GHG: increasing its atmospheric concentration enhances the greenhouse effect and leads to further surface warming. Another example is the ice albedo feedback, in which the albedo decreases as highly reflective ice and snow surfaces melt, exposing the darker and more absorbing surfaces below. The dominant negative feedback is the increased emission of energy through LWR as surface temperature increases (sometimes also referred to as blackbody radiation feedback). Some feedbacks operate quickly (hours), while others develop over decades to centuries; in order to understand the full impact of a feedback mechanism, its timescale needs to be considered. Melting of land ice sheets can take days to millennia.

A spectrum of models is used to project quantitatively the climate response to forcings. The simplest energy balance models use one box to represent the Earth system and solve the global energy balance to deduce globally averaged surface air temperature. At the other extreme, full complexity three-dimensional climate models include the explicit solution of energy, momentum and mass conservation equations at millions of points on the Earth in the atmosphere, land, ocean and cryosphere. More recently, capabilities for the explicit simulation of the biosphere, the carbon cycle and atmospheric chemistry have been added to the full complexity models, and these models are called Earth System Models (ESMs). Earth System Models of Intermediate Complexity include the same processes as ESMs, but at reduced resolution, and thus can be simulated for longer periods (see Annex III for Glossary and Section 9.1).

An equilibrium climate experiment is an experiment in which a climate model is allowed to adjust fully to a specified change in RF. Such experiments provide information on the difference between the initial and final states of the model simulated climate, but not on the time-dependent response. The equilibrium response in global mean surface air temperature to a doubling of atmospheric concentration of CO_2 above pre-industrial levels (e.g., Arrhenius, 1896; see Le Treut et al., 2007 for a comprehensive list) has often been used as the basis for the concept of equilibrium climate sensitivity (e.g., Hansen et al., 1981; see Meehl et al., 2007 for a comprehensive list). For more realistic simulations of climate, changes in RF are applied gradually over time, for example, using historical reconstructions of the CO_2, and these simulations are called transient simulations. The temperature response in these transient simulations is different than in an equilibrium simulation. The transient climate response is defined as the change in global surface temperature at the time of atmospheric CO_2 doubling in a global coupled ocean–atmosphere climate model simulation where concentrations of CO_2 were increased by 1% yr^{-1}. The transient climate response is a measure of the strength and rapidity of the surface temperature response to GHG forcing. It can be more meaningful for some problems as well as easier to derive from observations (see Figure 10.20; Section 10.8; Chapter 12; Knutti et al., 2005; Frame et al., 2006; Forest et al., 2008), but such experiments are not intended to replace the more realistic scenario evaluations.

Climate change commitment is defined as the future change to which the climate system is committed by virtue of past or current forcings. The components of the climate system respond on a large range of timescales, from the essentially rapid responses that characterise some radiative feedbacks to millennial scale responses such as those associated with the behaviour of the carbon cycle (Section 6.1) and ice sheets (see Figure 1.2 and Box 5.1). Even if anthropogenic emissions were immediately ceased (Matthews and Weaver, 2010) or if climate forcings were fixed at current values (Wigley, 2005), the climate system would continue to change until it came into equilibrium with those forcings (Section 12.5). Because of the slow response time of some components

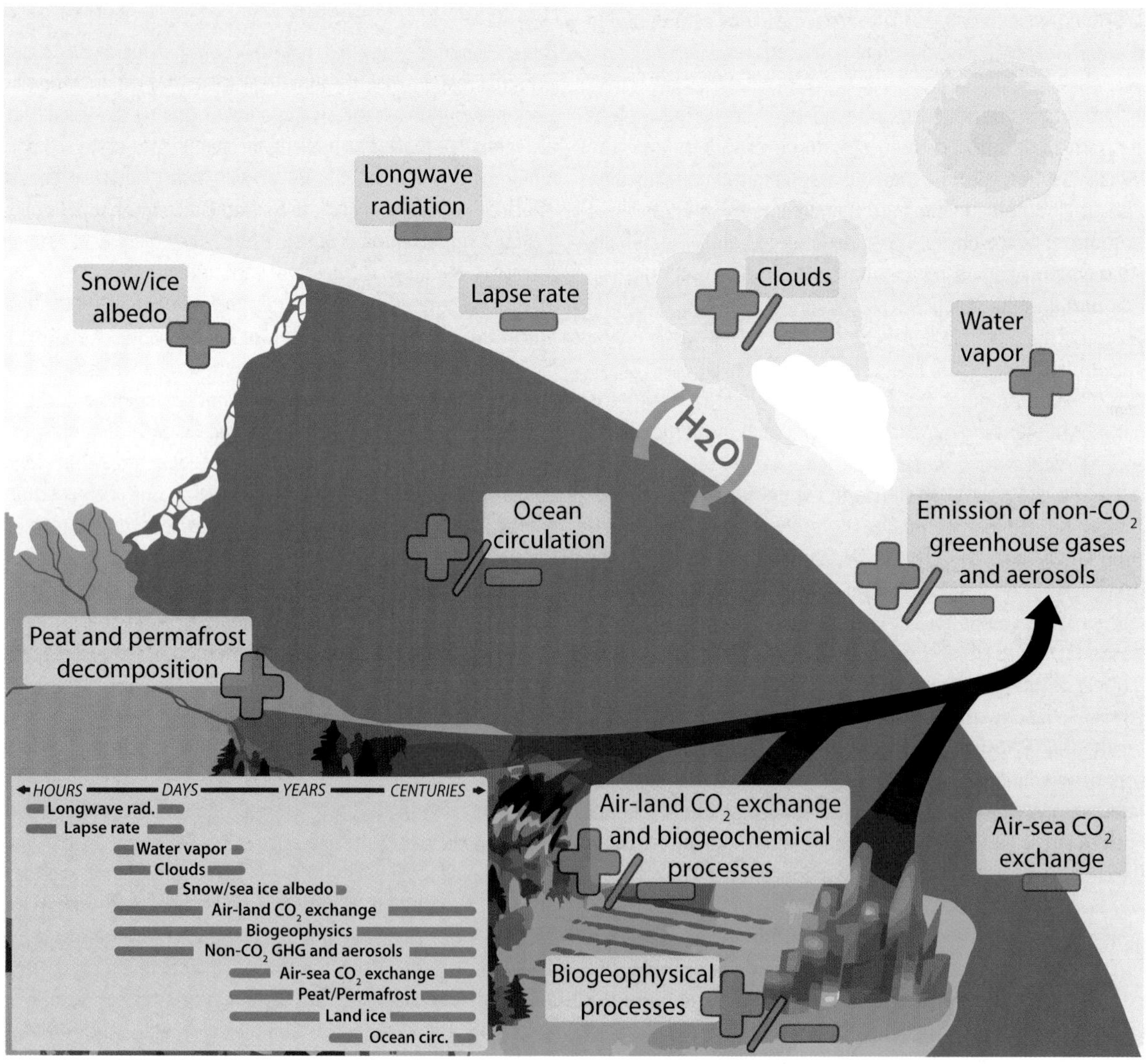

Figure 1.2 | Climate feedbacks and timescales. The climate feedbacks related to increasing CO_2 and rising temperature include negative feedbacks (–) such as LWR, lapse rate (see Glossary in Annex III), and air–sea carbon exchange and positive feedbacks (+) such as water vapour and snow/ice albedo feedbacks. Some feedbacks may be positive or negative (±): clouds, ocean circulation changes, air–land CO_2 exchange, and emissions of non-GHGs and aerosols from natural systems. In the smaller box, the large difference in timescales for the various feedbacks is highlighted.

of the climate system, equilibrium conditions will not be reached for many centuries. Slow processes can sometimes be constrained only by data collected over long periods, giving a particular salience to paleoclimate data for understanding equilibrium processes. Climate change commitment is indicative of aspects of inertia in the climate system because it captures the ongoing nature of some aspects of change.

A summary of perturbations to the forcing of the climate system from changes in solar radiation, GHGs, surface albedo and aerosols is presented in Box 13.1. The energy fluxes from these perturbations are balanced by increased radiation to space from a warming Earth, reflection of solar radiation and storage of energy in the Earth system, principally the oceans (Box 3.1, Box 13.1).

The processes affecting climate can exhibit considerable natural variability. Even in the absence of external forcing, periodic and chaotic variations on a vast range of spatial and temporal scales are observed. Much of this variability can be represented by simple (e.g., unimodal or power law) distributions, but many components of the climate system also exhibit multiple states—for instance, the glacial-interglacial cycles and certain modes of internal variability such as El Niño-Southern Oscillation (ENSO) (see Box 2.5 for details on patterns and indices of climate variability). Movement between states can occur as a result of natural variability, or in response to external forcing. The relationship between variability, forcing and response reveals the complexity of the dynamics of the climate system: the relationship between forcing and response for some parts of the system seems reasonably linear; in other cases this relationship is much more complex, characterised by hysteresis (the dependence on past states) and a non-additive combination of feedbacks.

Related to multiple climate states, and hysteresis, is the concept of irreversibility in the climate system. In some cases where multiple states and irreversibility combine, bifurcations or 'tipping points' can been reached (see Section 12.5). In these situations, it is difficult if not impossible for the climate system to revert to its previous state, and the change is termed irreversible over some timescale and forcing range. A small number of studies using simplified models find evidence for global-scale 'tipping points' (e.g., Lenton et al., 2008); however, there is no evidence for global-scale tipping points in any of the most comprehensive models evaluated to date in studies of climate evolution in the 21st century. There is evidence for threshold behaviour in certain aspects of the climate system, such as ocean circulation (see Section 12.5) and ice sheets (see Box 5.1), on multi-centennial-to-millennial timescales. There are also arguments for the existence of regional tipping points, most notably in the Arctic (e.g., Lenton et al., 2008; Duarte et al., 2012; Wadhams, 2012), although aspects of this are contested (Armour et al., 2011; Tietsche et al., 2011).

1.2.3 Multiple Lines of Evidence for Climate Change

While the first IPCC assessment depended primarily on observed changes in surface temperature and climate model analyses, more recent assessments include multiple lines of evidence for climate change. The first line of evidence in assessing climate change is based on careful analysis of observational records of the atmosphere, land, ocean and cryosphere systems (Figure 1.3). There is incontrovertible evidence from *in situ* observations and ice core records that the atmospheric concentrations of GHGs such as CO_2, CH_4, and N_2O have increased substantially over the last 200 years (Sections 6.3 and 8.3). In addition, instrumental observations show that land and sea surface temperatures have increased over the last 100 years (Chapter 2). Satellites allow a much broader spatial distribution of measurements, especially over the last 30 years. For the upper ocean temperature the observations indicate that the temperature has increased since at least 1950 (Willis et al., 2010; Section 3.2). Observations from satellites and *in situ* measurements suggest reductions in glaciers, Arctic sea ice and ice sheets (Sections 4.2, 4.3 and 4.4). In addition, analyses based on measurements of the radiative budget and ocean heat content suggest a small imbalance (Section 2.3). These observations, all published in peer-reviewed journals, made by diverse measurement groups in multiple countries using different technologies, investigating various climate-relevant types of data, uncertainties and processes, offer a wide range of evidence on the broad extent of the changing climate throughout our planet.

Conceptual and numerical models of the Earth's climate system offer another line of evidence on climate change (discussions in Chapters 5 and 9 provide relevant analyses of this evidence from paleoclimatic to recent periods). These use our basic understanding of the climate system to provide self-consistent methodologies for calculating impacts of processes and changes. Numerical models include the current knowledge about the laws of physics, chemistry and biology, as well as hypotheses about how complicated processes such as cloud formation can occur. Because these models can represent only the existing state of knowledge and technology, they are not perfect; they are, however, important tools for analysing uncertainties or unknowns, for testing different hypotheses for causation relative to observations, and for making projections of possible future changes.

One of the most powerful methods for assessing changes occurring in climate involves the use of statistical tools to test the analyses from models relative to observations. This methodology is generally called detection and attribution in the climate change community (Section 10.2). For example, climate models indicate that the temperature response to GHG increases is expected to be different than the effects from aerosols or from solar variability. Radiosonde measurements and satellite retrievals of atmospheric temperature show increases in tropospheric temperature and decreases in stratospheric temperatures, consistent with the increases in GHG effects found in climate model simulations (e.g., increases in CO_2, changes in O_3), but if the Sun was the main driver of current climate change, stratospheric and tropospheric temperatures would respond with the same sign (Hegerl et al., 2007).

Resources available prior to the instrumental period—historical sources, natural archives, and proxies for key climate variables (e.g., tree rings, marine sediment cores, ice cores)—can provide quantitative information on past regional to global climate and atmospheric composition variability and these data contribute another line of evidence. Reconstructions of key climate variables based on these data sets have provided important information on the responses of the Earth system to a variety of external forcings and its internal variability over a wide range of timescales (Hansen et al., 2006; Mann et al.,

1

2008). Paleoclimatic reconstructions thus offer a means for placing the current changes in climate in the perspective of natural climate variability (Section 5.1). AR5 includes new information on external RFs caused by variations in volcanic and solar activity (e.g., Steinhilber et al., 2009; see Section 8.4). Extended data sets on past changes in atmospheric concentrations and distributions of atmospheric GHG concentrations (e.g., Lüthi et al., 2008; Beerling and Royer, 2011) and mineral aerosols (Lambert et al., 2008) have also been used to attribute reconstructed paleoclimate temperatures to past variations in external forcings (Section 5.2).

1.3 Indicators of Climate Change

There are many indicators of climate change. These include physical responses such as changes in the following: surface temperature, atmospheric water vapour, precipitation, severe events, glaciers, ocean and land ice, and sea level. Some key examples of such changes in important climate parameters are discussed in this section and all are assessed in much more detail in other chapters.

As was done to a more limited extent in AR4 (Le Treut et al., 2007), this section provides a test of the planetary-scale hypotheses of climate change against observations. In other words, how well do the projections used in the past assessments compare with observations to date? Seven additional years of observations are now available to evaluate earlier model projections. The projected range that was given in each assessment is compared to observations. The largest possible range of scenarios available for a specific variable for each of the previous assessment reports is shown in the figures.

Based on the assessment of AR4, a number of the key climate and associated environmental parameters are presented in Figure 1.3, which updates the similar figure in the Technical Summary (TS) of IPCC (2001). This section discusses the recent changes in several indicators, while more thorough assessments for each of these indicators are

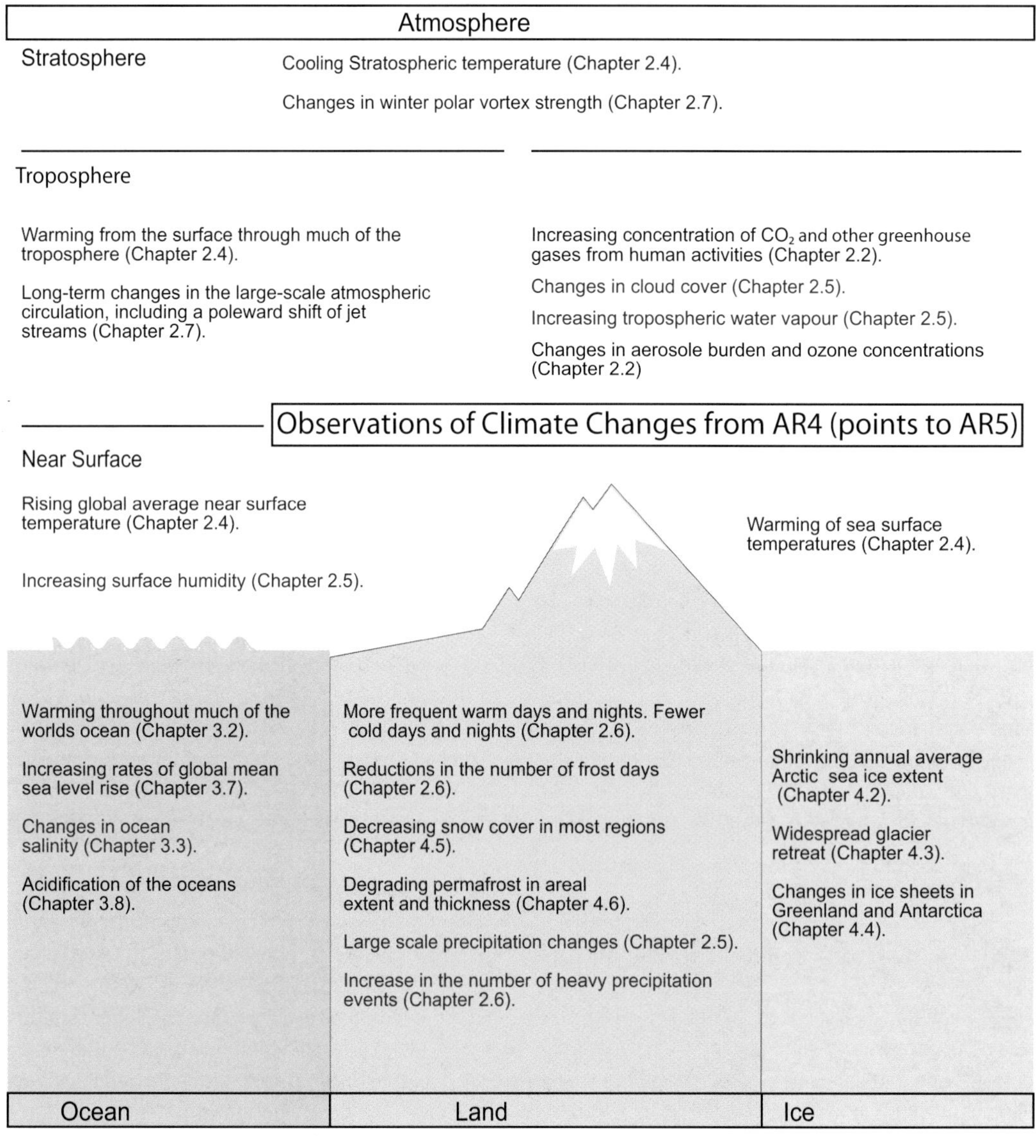

Figure 1.3 | Overview of observed climate change indicators as listed in AR4. Chapter numbers indicate where detailed discussions for these indicators are found in AR5 (temperature: red; hydrological: blue; others: black).

provided in other chapters. Also shown in parentheses in Figure 1.3 are the chapter and section where those indicators of change are assessed in AR5.

Note that projections presented in the IPCC assessments are not predictions (see the Glossary in Annex III); the analyses in the discussion below only examine the short-term plausibility of the projections up to AR4, including the scenarios for future emissions and the models used to simulate these scenarios in the earlier assessments. Model results from the Coupled Model Intercomparison Project Phase 5 (CMIP5) (Taylor et al., 2012) used in AR5 are therefore not included in this section; Chapters 11 and 12 describe the projections from the new modelling studies. Note that none of the scenarios examined in the IPCC assessments were ever intended to be short-term predictors of change.

1.3.1 Global and Regional Surface Temperatures

Observed changes in global mean surface air temperature since 1950 (from three major databases, as anomalies relative to 1961–1990) are shown in Figure 1.4. As in the prior assessments, global climate models generally simulate global temperatures that compare well with observations over climate timescales (Section 9.4). Even though the projections from the models were never intended to be predictions over such a short timescale, the observations through 2012 generally fall within the projections made in all past assessments. The 1990–2012 data have been shown to be consistent with the FAR projections (IPCC, 1990), and not consistent with zero trend from 1990, even in the presence of substantial natural variability (Frame and Stone, 2013).

The scenarios were designed to span a broad range of plausible futures, but are not aimed at predicting the most likely outcome. The scenarios considered for the projections from the earlier reports (FAR, SAR) had a much simpler basis than those of the Special Report on Emission Scenarios (SRES) (IPCC, 2000) used in the later assessments. For example, the FAR scenarios did not specify future aerosol distributions. AR4 presented a multiple set of projections that were simulated using comprehensive ocean–atmosphere models provided by CMIP3 and these projections are continuations of transient simulations of the 20th century climate. These projections of temperature provide in addition a measure of the natural variability that could not be obtained

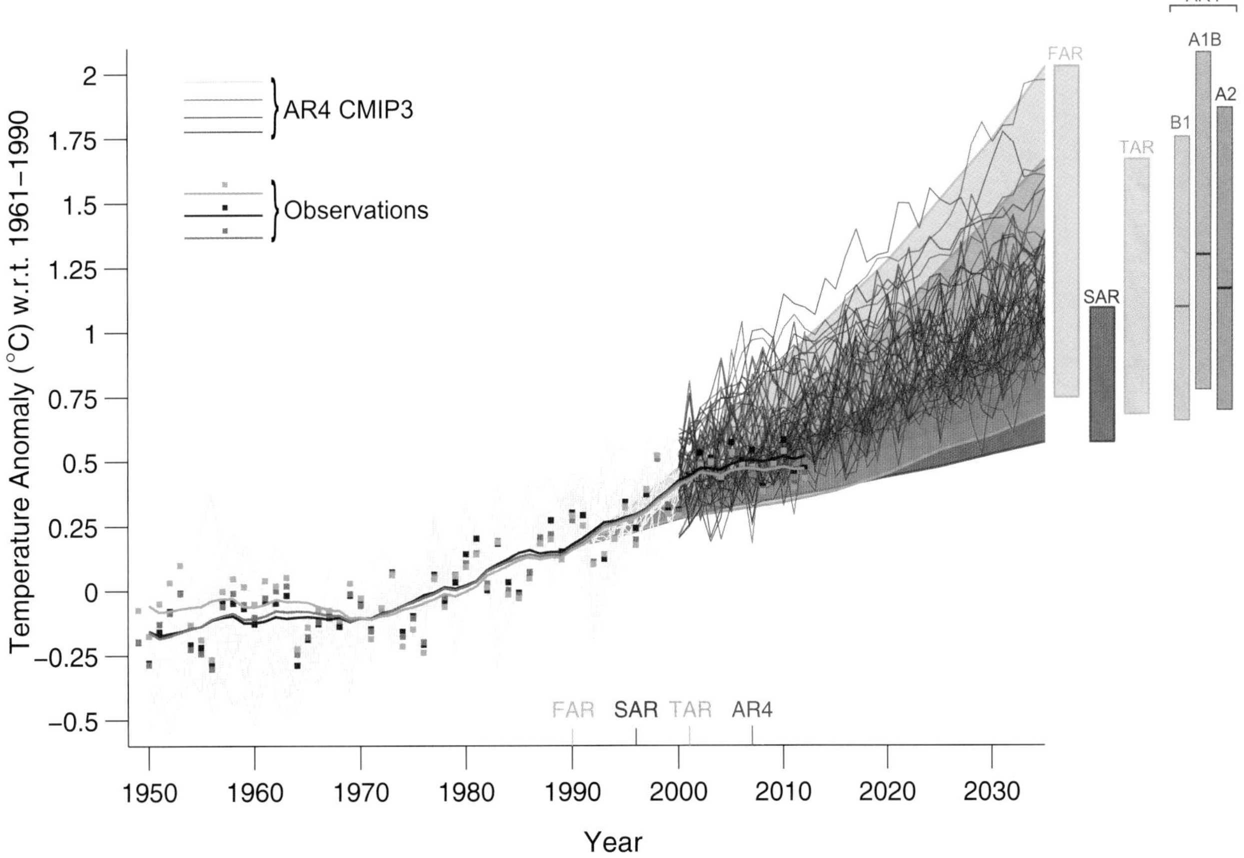

Figure 1.4 | Estimated changes in the observed globally and annually averaged surface temperature anomaly relative to 1961–1990 (in °C) since 1950 compared with the range of projections from the previous IPCC assessments. Values are harmonized to start from the same value in 1990. Observed global annual mean surface air temperature anomaly, relative to 1961–1990, is shown as squares and smoothed time series as solid lines (NASA (dark blue), NOAA (warm mustard), and the UK Hadley Centre (bright green) reanalyses). The coloured shading shows the projected range of global annual mean surface air temperature change from 1990 to 2035 for models used in FAR (Figure 6.11 in Bretherton et al., 1990), SAR (Figure 19 in the TS of IPCC, 1996), TAR (full range of TAR Figure 9.13(b) in Cubasch et al., 2001). TAR results are based on the simple climate model analyses presented and not on the individual full three-dimensional climate model simulations. For the AR4 results are presented as single model runs of the CMIP3 ensemble for the historical period from 1950 to 2000 (light grey lines) and for three scenarios (A2, A1B and B1) from 2001 to 2035. The bars at the right-hand side of the graph show the full range given for 2035 for each assessment report. For the three SRES scenarios the bars show the CMIP3 ensemble mean and the *likely* range given by –40% to +60% of the mean as assessed in Meehl et al. (2007). The publication years of the assessment reports are shown. See Appendix 1.A for details on the data and calculations used to create this figure.

1

from the earlier projections based on models of intermediate complexity (Cubasch et al., 2001).

Note that before TAR the climate models did not include natural forcing (such as volcanic activity and solar variability). Even in AR4 not all models included natural forcing and some also did not include aerosols. Those models that allowed for aerosol effects presented in the AR4 simulated, for example, the cooling effects of the 1991 Mt Pinatubo eruption and agree better with the observed temperatures than the previous assessments that did not include those effects.

The bars on the side for FAR, SAR and TAR represent the range of results for the scenarios at the end of the time period and are not error bars. In contrast to the previous reports, the AR4 gave an assessment of the individual scenarios with a mean estimate (cross bar; ensemble mean of the CMIP3 simulations) and a *likely* range (full bar; –40% to +60% of the mean estimate) (Meehl et al., 2007).

In summary, the trend in globally averaged surface temperatures falls within the range of the previous IPCC projections. During the last decade the trend in the observations is smaller than the mean of the projections of AR4 (see Section 9.4.1, Box 9.2 for a detailed assessment of the hiatus in global mean surface warming in the last 15 years). As shown by Hawkins and Sutton (2009), trends in the observations during short-timescale periods (decades) can be dominated by natural variability in the Earth's climate system. Similar episodes are also seen in climate model experiments (Easterling and Wehner, 2009). Due to their experimental design these episodes cannot be duplicated with the same timing as the observed episodes in most of the model simulations; this affects the interpretation of recent trends in the scenario evaluations (Section 11.2). Notwithstanding these points, there is evidence that early forecasts that carried formal estimates of uncertainty have proved highly consistent with subsequent observations (Allen et al., 2013). If the contributions of solar variability, volcanic activity and ENSO are removed from the observations the remaining trend of surface air temperature agree better with the modelling studies (Rahmstorf et al., 2012).

1.3.2 Greenhouse Gas Concentrations

Key indicators of global climate change also include the changing concentrations of the radiatively important GHGs that are significant drivers for this change (e.g., Denman et al., 2007; Forster et al., 2007). Figures 1.5 through 1.7 show the recent globally and annually averaged observed concentrations for the gases of most concern, CO_2, CH_4, and N_2O (see Sections 2.2, 6.3 and 8.3 for more detailed discussion of these and other key gases). As discussed in the later chapters, accurate measurements of these long-lived gases come from a number of monitoring stations throughout the world. The observations in these figures are compared with the projections from the previous IPCC assessments.

The model simulations begin with historical emissions up to 1990. The further evolution of these gases was described by scenario projections. TAR and AR4 model concentrations after 1990 are based on the SRES

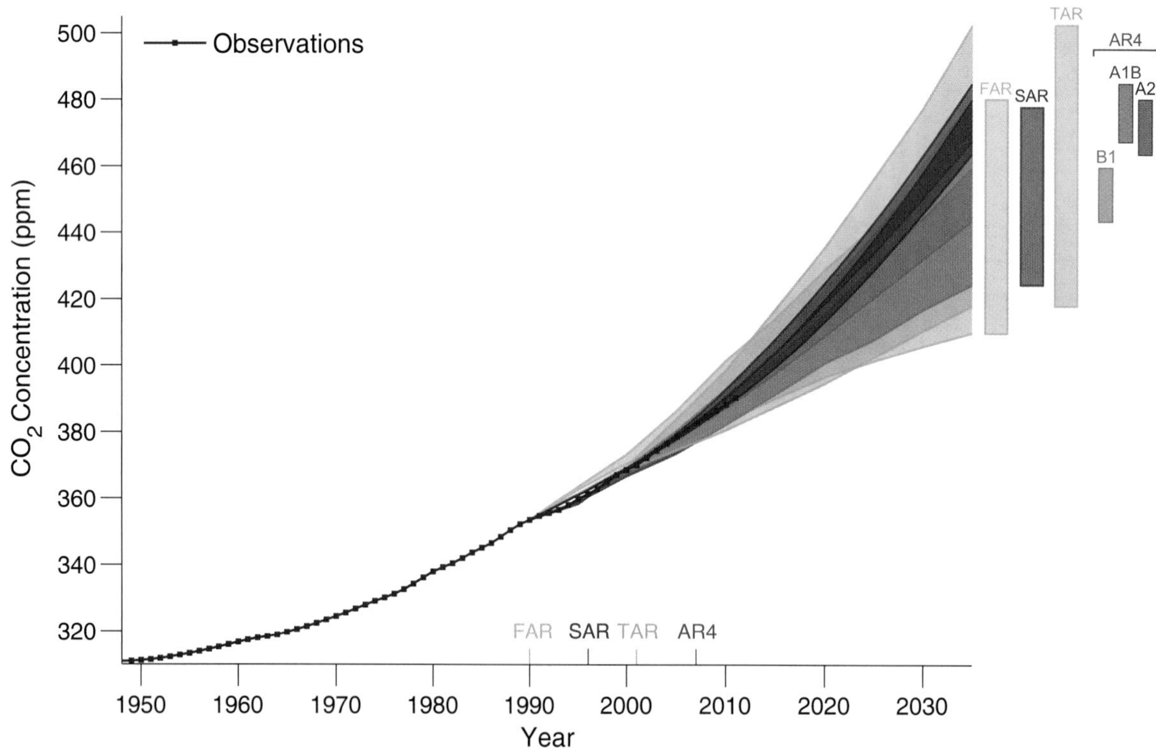

Figure 1.5 | Observed globally and annually averaged CO_2 concentrations in parts per million (ppm) since 1950 compared with projections from the previous IPCC assessments. Observed global annual CO_2 concentrations are shown in dark blue. The shading shows the largest model projected range of global annual CO_2 concentrations from 1950 to 2035 from FAR (Figure A.3 in the Summary for Policymakers of IPCC, 1990); SAR (Figure 5b in the Technical Summary of IPCC, 1996); TAR (Appendix II of IPCC, 2001); and from the A2, A1B and B1 scenarios presented in the AR4 (Figure 10.26 in Meehl et al., 2007). The bars at the right-hand side of the graph show the full range given for 2035 for each assessment report. The publication years of the assessment reports are shown. See Appendix 1.A for details on the data and calculations used to create this figure.

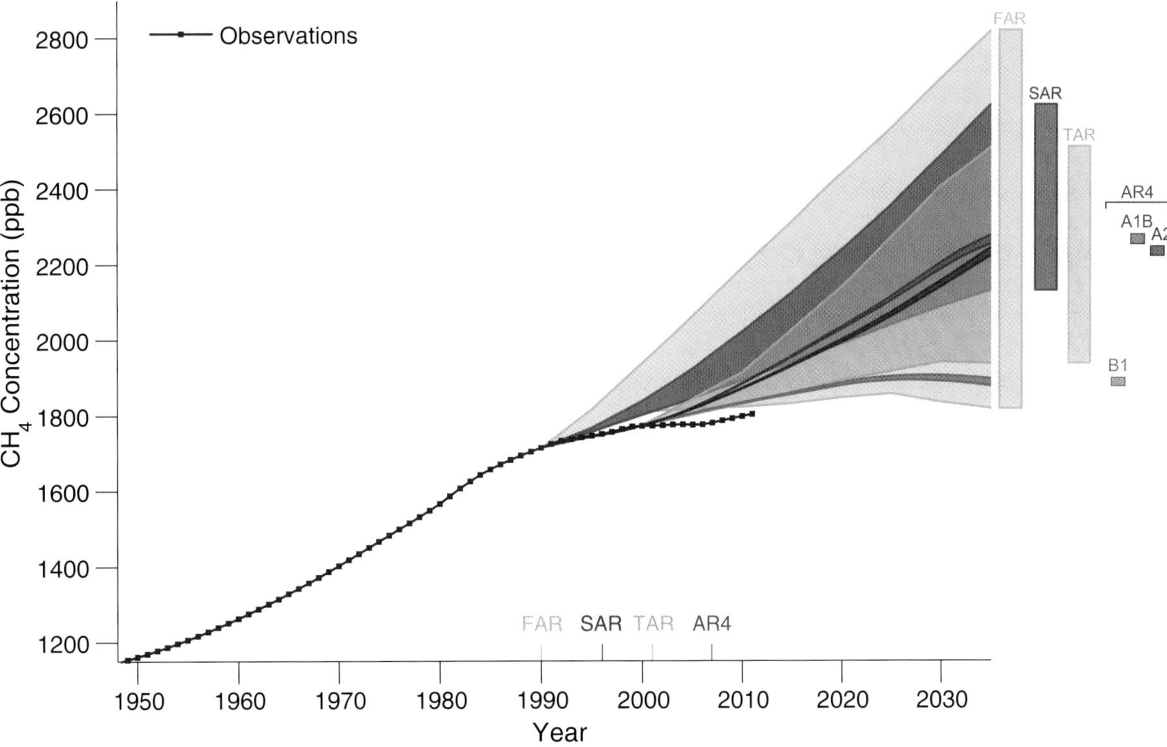

Figure 1.6 | Observed globally and annually averaged CH_4 concentrations in parts per billion (ppb) since 1950 compared with projections from the previous IPCC assessments. Estimated observed global annual CH_4 concentrations are shown in dark blue. The shading shows the largest model projected range of global annual CH_4 concentrations from 1950 to 2035 from FAR (Figure A.3 of the Annex of IPCC, 1990); SAR (Table 2.5a in Schimel et al., 1996); TAR (Appendix II of IPCC, 2001); and from the A2, A1B and B1 scenarios presented in the AR4 (Figure 10.26 in Meehl et al., 2007). The bars at the right-hand side of the graph show the full range given for 2035 for each assessment report. The publication years of the assessment reports are shown. See Appendix 1.A for details on the data and calculations used to create this figure.

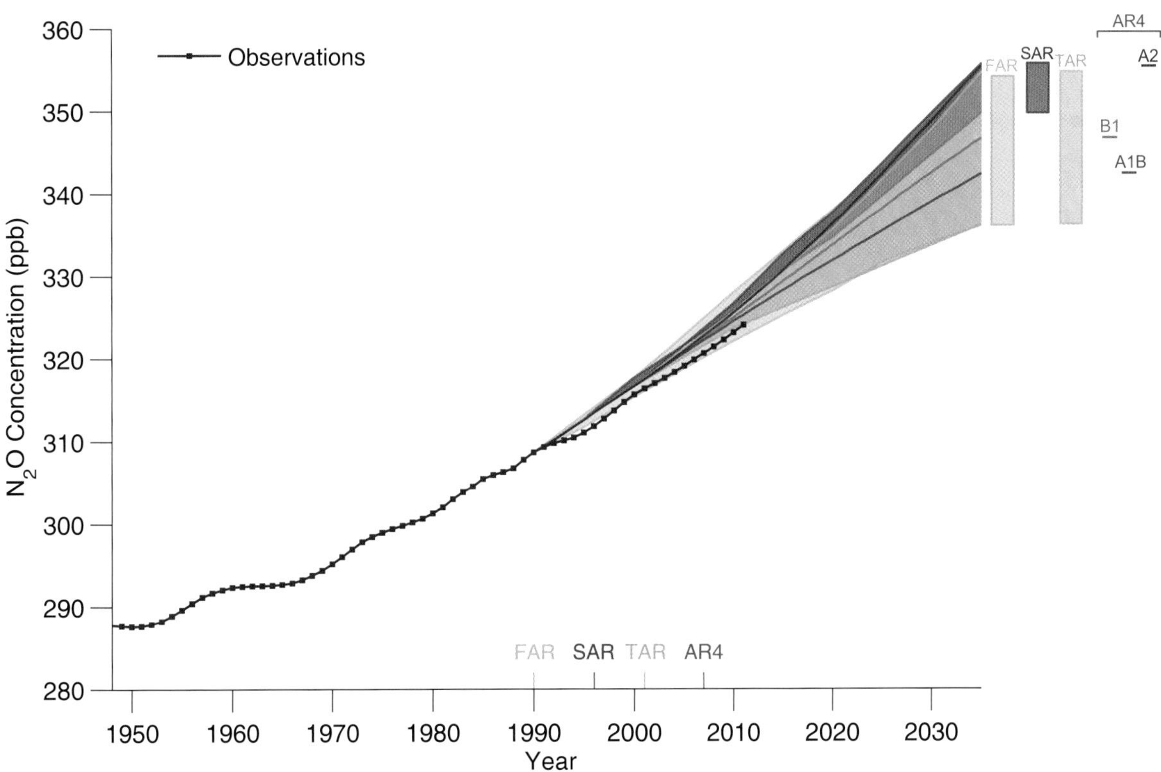

Figure 1.7 | Observed globally and annually averaged N_2O concentrations in parts per billion (ppb) since 1950 compared with projections from the previous IPCC assessments. Observed global annual N_2O concentrations are shown in dark blue. The shading shows the largest model projected range of global annual N_2O concentrations from 1950 to 2035 from FAR (Figure A3 in the Annex of IPCC, 1990), SAR (Table 2.5b in Schimel et al., 1996), TAR (Appendix II of IPCC, 2001), and from the A2, A1B and B1 scenarios presented in the AR4 (Figure 10.26 in Meehl et al., 2007). The bars at the right hand side of the graph show the full range given for 2035 for each assessment report. The publication years of the assessment reports are shown. See Appendix 1.A for details on the data and calculations used to create this figure.

scenarios but those model results may also account for historical emissions analyses. The recent observed trends in CO_2 concentrations tend to be in the middle of the scenarios used for the projections (Figure 1.5).

As discussed in Dlugokencky et al. (2009), trends in CH_4 showed a stabilization from 1999 to 2006, but CH_4 concentrations have been increasing again starting in 2007 (see Sections 2.2 and 6.3 for more discussion on the budget and changing concentration trends for CH_4). Because at the time the scenarios were developed (e.g., the SRES scenarios were developed in 2000), it was thought that past trends would continue, the scenarios used and the resulting model projections assumed in FAR through AR4 all show larger increases than those observed (Figure 1.6).

Concentrations of N_2O have continued to increase at a nearly constant rate (Elkins and Dutton, 2010) since about 1970 as shown in Figure 1.7. The observed trends tend to be in the lower part of the projections for the previous assessments.

1.3.3 Extreme Events

Climate change, whether driven by natural or human forcings, can lead to changes in the likelihood of the occurrence or strength of extreme weather and climate events such as extreme precipitation events or warm spells (see Chapter 3 of the IPCC Special Report on Managing the Risks of Extreme Events and Disasters to Advance Climate Change Adaptation (SREX); Seneviratne et al., 2012). An extreme weather event is one that is rare at a particular place and/or time of year. Definitions of 'rare' vary, but an extreme weather event would normally be as rare as or rarer than the 10th or 90th percentile of a probability density function estimated from observations (see also Glossary in Annex III and FAQ 2.2). By definition, the characteristics of what is called extreme weather may vary from place to place in an absolute sense. At present, single extreme events cannot generally be directly attributed to anthropogenic influence, although the change in likelihood for the event to occur has been determined for some events by accounting for observed changes in climate (see Section 10.6). When a pattern of extreme weather persists for some time, such as a season, it may be classified as an extreme climate event, especially if it yields an average or total that is itself extreme (e.g., drought or heavy rainfall over a season). For some climate extremes such as drought, floods and heat waves, several factors such as duration and intensity need to be combined to produce an extreme event (Seneviratne et al., 2012).

The probability of occurrence of values of a climate or weather variable can be described by a probability density function (PDF) that for some variables (e.g., temperature) is shaped similar to a Gaussian curve. A PDF is a function that indicates the relative chances of occurrence of different outcomes of a variable. Simple statistical reasoning indicates that substantial changes in the frequency of extreme events (e.g., the maximum possible 24-hour rainfall at a specific location) can result from a relatively small shift in the distribution of a weather or climate variable. Figure 1.8a shows a schematic of such a PDF and illustrates the effect of a small shift in the mean of a variable on the frequency of extremes at either end of the distribution. An increase in the frequency of one extreme (e.g., the number of hot days) can be accompanied by

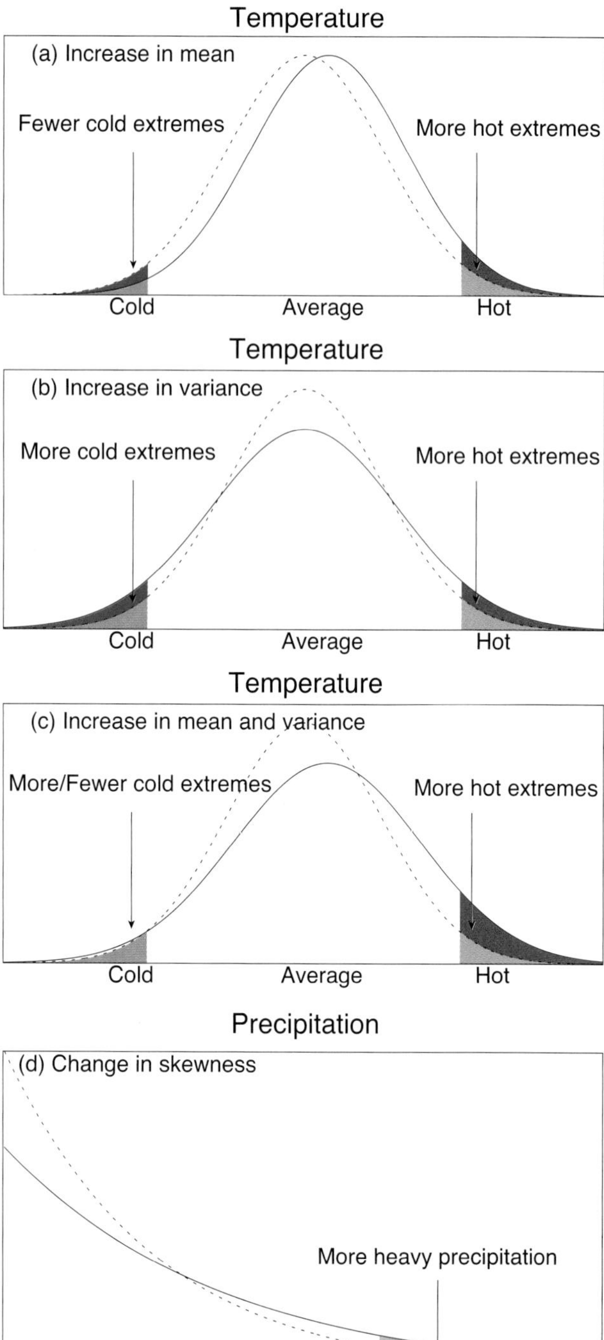

Figure 1.8 | Schematic representations of the probability density function of daily temperature, which tends to be approximately Gaussian, and daily precipitation, which has a skewed distribution. Dashed lines represent a previous distribution and solid lines a changed distribution. The probability of occurrence, or frequency, of extremes is denoted by the shaded areas. In the case of temperature, changes in the frequencies of extremes are affected by changes (a) in the mean, (b) in the variance or shape, and (c) in both the mean and the variance. (d) In a skewed distribution such as that of precipitation, a change in the mean of the distribution generally affects its variability or spread, and thus an increase in mean precipitation would also imply an increase in heavy precipitation extremes, and vice-versa. In addition, the shape of the right-hand tail could also change, affecting extremes. Furthermore, climate change may alter the frequency of precipitation and the duration of dry spells between precipitation events. (Parts a–c modified from Folland et al., 2001, and d modified from Peterson et al., 2008, as in Zhang and Zwiers, 2012.)

a decline in the opposite extreme (in this case the number of cold days such as frost days). Changes in the variability, skewness or the shape of the distribution can complicate this simple picture (Figure 1.8b, c and d).

While the SAR found that data and analyses of extremes related to climate change were sparse, improved monitoring and data for changes in extremes were available for the TAR, and climate models were being analysed to provide projections of extremes. In AR4, the observational basis of analyses of extremes had increased substantially, so that some extremes were now examined over most land areas (e.g., rainfall extremes). More models with higher resolution, and a larger number of regional models have been used in the simulation and projection of extremes, and ensemble integrations now provide information about PDFs and extremes.

Since the TAR, climate change studies have especially focused on changes in the global statistics of extremes, and observed and projected changes in extremes have been compiled in the so-called 'Extremes'-Table (Figure 1.9). This table has been modified further to account for the SREX assessment. For some extremes ('higher maximum temperature', 'higher minimum temperature', 'precipitation extremes', 'droughts or dryness'), all of these assessments found an increasing trend in the observations and in the projections. In the observations for

Changes in Phenomenon	Uncertainty in observed changes (since about the mid-20th century)			Uncertainty in projected changes (up to 2100)		
IPCC Assessment Report	TAR	AR4	SREX	TAR	AR4	SREX
Higher maximum temperatures and more hot days	*Likely* over nearly all land areas	*Very Likely* over most land areas	*Very Likely* at a global scale	*Very Likely* over nearly all land areas	*Virtually Certain* over most land areas	*Virtually Certain* at a global scale
Higher minimum temperatures, fewer cold days	*Very Likely* over nearly all land areas	*Very Likely* over most land areas	*Very Likely* at a global scale	*Very Likely* over nearly all land areas	*Virtually Certain* over most land areas	*Virtually Certain* at a global scale
Warm spells/heat waves. frequency, length or intensity increases	-	*Likely* over most land areas	*Medium Confidence* in many regions	-	*Very Likely* over most land areas	*Very Likely* over most land areas
Precipitation extremes	*Likely*[1], over many Northern Hemisphere mid- to high latitude land areas	*Likely*[2] over most areas	*Likely*[3]	*Very Likely*[1] over many areas	*Very Likely*[2]	*Likely*[2,4] in many land areas of the globe
Droughts or dryness	*Likely*[5], in a few areas	*Likely*[6], in many regions since 1970s	*Medium Confidence* in more intense and longer droughts in some regions , but some opposite trend exists	*Likely*[5], over most mid-latitude continental interiors (Lack of consistent projections in other areas)	*Likely*[6]	*Medium Confidence*[7] that droughts will intensify in some seasons and areas; Overall *low confidence* elsewhere
Changes in tropical cyclone activity (i.e. intensity, frequency, duration)	Not Observed[8], in the few analyses available	*Likely*[9], in some regions since 1970	*Low confidence*[10]	*Likely*[8], over some areas	*Likely*[9]	*Likely*[11]
Increase in extreme sea level (excludes tsunamis)	-	*Likely*	*Likely*[12]	-	*Likely*	*Very Likely*[13]

1 More intense precipitation events
2 Heavy precipitation events. Frequency (or proportion of total rainfall from heavy falls) increases
3 Statistically significant trends in the number of heavy precipitation events in some regions. It is *likely* that more of these regions have experienced increases than decreases.
4 See SREX Table 3-3 for details on precipitation extremes for the different regions.
5 Increased summer continental drying and associated risk of drought
6 Area affected by droughts increases
7 Some areas include southern Europe and the Mediterranean region, central Europe, central North America and Mexico, northeast Brazil and southern Africa
8 Increase in tropical cyclone peak wind intensities
9 Increase in intense tropical cyclone activity
10 In any observed long-term (i.e., 40 years or more) after accounting for past changes in observing capabilities (see SREX, section 3.4.4)
11 Increase in average tropical cyclone maximum wind speed is, although not in all ocean basins; either decrease or no change in the global frequency of tropical cyclones
12 Increase in extreme coastal high water worldwide related to increases in mean sea level in the late 20th century
13 Mean sea level rise will contribute to upward trends in extreme coastal high water levels

Figure 1.9 | Change in the confidence levels for extreme events based on prior IPCC assessments: TAR, AR4 and SREX. Types of extreme events discussed in all three reports are highlighted in green. Confidence levels are defined in Section 1.4. Similar analyses for AR5 are discussed in later chapters. Please note that the nomenclature for confidence level changed from AR4 to SREX and AR5.

1

the 'higher maximum temperature' the likelihood level was raised from *likely* in the TAR to *very likely* in SREX. While the diurnal temperature range was assessed in the Extremes-Table of the TAR, it was no longer included in the Extremes-Table of AR4, since it is not considered a climate extreme in a narrow sense. Diurnal temperature range was, however, reported to decrease for 21st century projections in AR4 (Meehl et al., 2007). In projections for precipitation extremes, the spatial relevance has been improved from *very likely* 'over many Northern Hemisphere mid-latitudes to high latitudes land areas' from the TAR to *very likely* for all regions in AR4 (these 'uncertainty labels' are discussed in Section 1.4). However, likelihood in trends in projected precipitation extremes was downscaled to *likely* in the SREX as a result of a perception of biases and a fairly large spread in the precipitation projections in some regions. SREX also had less confidence than TAR and AR4 in the trends for droughts and dryness, 'due to lack of direct observations, some geographical inconsistencies in the trends, and some dependencies of inferred trends on the index choice' (IPCC, 2012b).

For some extremes (e.g., 'changes in tropical cyclone activity') the definition changed between the TAR and the AR4. Whereas the TAR only made a statement about the peak wind speed of tropical cyclones, the AR4 also stressed the overall increase in intense tropical cyclone activity. The '*low confidence*' for any long term trend (>40 years) in the observed changes of the tropical cyclone activities is due to uncertainties in past observational capabilities (IPCC, 2012b). The 'increase in extreme sea level' has been added in the AR4. Such an increase is *likely* according to the AR4 and the SREX for observed trends, and *very likely* for the climate projections reported in the SREX.

The assessed likelihood of anthropogenic contributions to trends is lower for variables where the assessment is based on indirect evidence. Especially for extremes that are the result of a combination of factors such as droughts, linking a particular extreme event to specific causal relationships is difficult to determine (e.g., difficult to establish the clear role of climate change in the event) (see Section 10.6 and Peterson et al., 2012). In some cases (e.g., precipitation extremes), however, it may be possible to estimate the human-related contribution to such changes in the probability of occurrence of extremes (Pall et al., 2011; Seneviratne et al., 2012).

1.3.4 Climate Change Indicators

Climate change can lead to other effects on the Earth's physical system that are also indicators of climate change. Such integrative indicators include changes in sea level (ocean warming + land ice melt), in ocean acidification (ocean uptake of CO_2) and in the amount of ice on ocean and land (temperature and hydrological changes). See Chapters 3, 4 and 13 for detailed assessment.

1.3.4.1 Sea Level

Global mean sea level is an important indicator of climate change (Section 3.7 and Chapter 13). The previous assessments have all shown that observations indicate that the globally averaged sea level is rising. Direct observations of sea level change have been made for more than 150 years with tide gauges, and for more than 20 years with satellite radar altimeters. Although there is regional variability from non-uniform density change, circulation changes, and deformation of ocean basins, the evidence indicates that the global mean sea level is rising, and that this is *likely* (according to AR4 and SREX) resulting from global climate change (ocean warming plus land ice melt; see Chapter 13 for AR5 findings). The historical tide gauge record shows that the average rate of global mean sea level rise over the 20th century was 1.7 ± 0.2 mm yr^{-1} (e.g., Church and White, 2011). This rate increased to 3.2 ± 0.4 mm yr^{-1} since 1990, mostly because of increased thermal expansion and land ice contributions (Church and White, 2011; IPCC, 2012b). Although the long-term sea level record shows decadal and multi-decadal oscillations, there is evidence that the rate of global mean sea level rise during the 20th century was greater than during the 19th century.

All of the previous IPCC assessments have projected that global sea level will continue to rise throughout this century for the scenarios examined. Figure 1.10 compares the observed sea level rise since 1950 with the projections from the prior IPCC assessments. Earlier models had greater uncertainties in modelling the contributions, because of limited observational evidence and deficiencies in theoretical understanding of relevant processes. Also, projections for sea level change in the prior assessments are scenarios for the response to anthropogenic forcing only; they do not include unforced or natural interannual variability. Nonetheless, the results show that the actual change is in the middle of projected changes from the prior assessments, and towards the higher end of the studies from TAR and AR4.

1.3.4.2 Ocean Acidification

The observed decrease in ocean pH resulting from increasing concentrations of CO_2 is another indicator of global change. As discussed in AR4, the ocean's uptake of CO_2 is having a significant impact on the chemistry of sea water. The average pH of ocean surface waters has fallen by about 0.1 units, from about 8.2 to 8.1 (total scale) since 1765 (Section 3.8). Long time series from several ocean sites show ongoing declines in pH, consistent with results from repeated pH measurements on ship transects spanning much of the globe (Sections 3.8 and 6.4; Byrne et al., 2010; Midorikawa et al., 2010). Ocean time-series in the North Atlantic and North Pacific record a decrease in pH ranging between –0.0015 and –0.0024 per year (Section 3.8). Due to the increased storage of carbon by the ocean, ocean acidification will increase in the future (Chapter 6). In addition to other impacts of global climate change, ocean acidification poses potentially serious threats to the health of the world's oceans ecosystems (see AR5 WGII assessment).

1.3.4.3 Ice

Rapid sea ice loss is one of the most prominent indicators of Arctic climate change (Section 4.2). There has been a trend of decreasing Northern Hemisphere sea ice extent since 1978, with the summer of 2012 being the lowest in recorded history (see Section 4.2 for details). The 2012 minimum sea ice extent was 49% below the 1979 to 2000 average and 18% below the previous record from 2007. The amount of multi-year sea ice has been reduced, i.e., the sea ice has been thinning and thus the ice volume is reduced (Haas et al., 2008; Kwok et al., 2009). These changes make the sea ice less resistant to wind forcing.

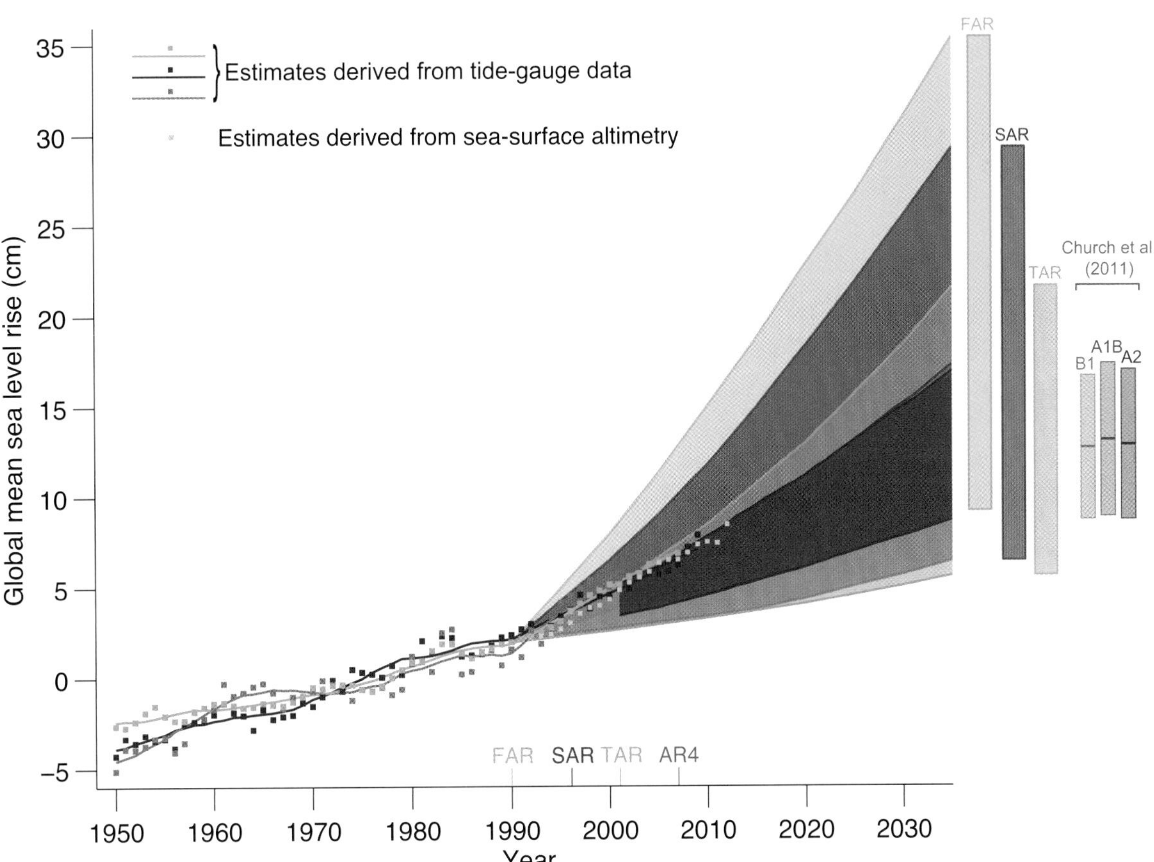

Figure 1.10 | Estimated changes in the observed global annual mean sea level (GMSL) since 1950 relative to 1961–1990. Estimated changes in global annual sea level anomalies are presented based on tide gauge data (warm mustard: Jevrejeva et al., 2008; dark blue: Church and White, 2011; dark green: Ray and Douglas, 2011) and based on sea surface altimetry (light blue). The altimetry data start in 1993 and are harmonized to start from the mean 1993 value of the tide gauge data. Squares indicate annual mean values and solid lines smoothed values. The shading shows the largest model projected range of global annual sea level rise from 1950 to 2035 for FAR (Figures 9.6 and 9.7 in Warrick and Oerlemans, 1990), SAR (Figure 21 in TS of IPCC, 1996), TAR (Appendix II of IPCC, 2001) and for Church et al. (2011) based on the Coupled Model Intercomparison Project Phase 3 (CMIP3) model results not assessed at the time of AR4 using the SRES B1, A1B and A2 scenarios. Note that in the AR4 no full range was given for the sea level projections for this period. Therefore, the figure shows results that have been published subsequent to the AR4. The bars at the right-hand side of the graph show the full range given for 2035 for each assessment report. For Church et al. (2011) the mean sea level rise is indicated in addition to the full range. See Appendix 1.A for details on the data and calculations used to create this figure.

Sea ice extent has been diminishing significantly faster than projected by most of the AR4 climate models (SWIPA, 2011). While AR4 found no consistent trends in Antarctica sea ice, more recent studies indicate a small increase (Section 4.2). Various studies since AR4 suggest that this has resulted in a deepening of the low-pressure systems in West Antarctica that in turn caused stronger winds and enhanced ice production in the Ross Sea (Goosse et al., 2009; Turner and Overland, 2009).

AR4 concluded that taken together, the ice sheets in Greenland and Antarctica have *very likely* been contributing to sea level rise. The Greenland Ice Sheet has lost mass since the early 1990s and the rate of loss has increased (see Section 4.4). The interior, high-altitude areas are thickening due to increased snow accumulation, but this is more than counterbalanced by the ice loss due to melt and ice discharge (AMAP, 2009; Ettema et al., 2009). Since 1979, the area experiencing surface melting has increased significantly (Tedesco, 2007; Mernild et al., 2009), with 2010 breaking the record for surface melt area, runoff, and mass loss, and the unprecedented areal extent of surface melt of the Greenland Ice Sheet in 2012 (Nghiem et al., 2012). Overall, the Antarctic continent now experiences a net loss of ice (Section 4.4). Significant mass loss has been occurring in the Amundsen Sea sector of West Antarctica and the northern Antarctic Peninsula. The ice sheet on the rest of the continent is relatively stable or thickening slightly (Lemke et al., 2007; Scott et al., 2009; Turner et al., 2009). Since AR4, there have been improvements in techniques of measurement, such as gravity, altimetry and mass balance, and understanding of the change (Section 4.4).

As discussed in the earlier assessments, most glaciers around the globe have been shrinking since the end of the Little Ice Age, with increasing rates of ice loss since the early 1980s (Section 4.3). The vertical profiles of temperature measured through the entire thickness of mountain glaciers, or through ice sheets, provide clear evidence of a warming climate over recent decades (e.g., Lüthi and Funk, 2001; Hoelzle et al., 2011). As noted in AR4, the greatest mass losses per unit area in the last four decades have been observed in Patagonia, Alaska, northwest USA, southwest Canada, the European Alps, and the Arctic. Alaska and the Arctic are especially important regions as contributors to sea level rise (Zemp et al., 2008, 2009).

1.4 Treatment of Uncertainties

1.4.1 Uncertainty in Environmental Science

Science always involves uncertainties. These arise at each step of the scientific method: in the development of models or hypotheses, in measurements and in analyses and interpretation of scientific assumptions. Climate science is not different in this regard from other areas of science. The complexity of the climate system and the large range of processes involved bring particular challenges because, for example, gaps in direct measurements of the past can be filled only by reconstructions using proxy data.

Because the Earth's climate system is characterized by multiple spatial and temporal scales, uncertainties do not usually reduce at a single, predictable rate: for example, new observations may reduce the uncertainties surrounding short-timescale processes quite rapidly, while longer timescale processes may require very long observational baselines before much progress can be made. Characterization of the interaction between processes, as quantified by models, can be improved by model development, or can shed light on new areas in which uncertainty is greater than previously thought. The fact that there is only a single realization of the climate, rather than a range of different climates from which to draw, can matter significantly for certain lines of enquiry, most notably for the detection and attribution of causes of climate change and for the evaluation of projections of future states.

1.4.2 Characterizing Uncertainty

'Uncertainty' is a complex and multifaceted property, sometimes originating in a lack of information, and at other times from quite fundamental disagreements about what is known or even knowable (Moss and Schneider, 2000). Furthermore, scientists often disagree about the best or most appropriate way to characterize these uncertainties: some can be quantified easily while others cannot. Moreover, appropriate characterization is dependent on the intended use of the information and the particular needs of that user community.

Scientific uncertainty can be partitioned in various ways, in which the details of the partitioning usually depend on the context. For instance, the process and classifications used for evaluating observational uncertainty in climate science is not the same as that employed to evaluate projections of future change. Uncertainty in measured quantities can arise from a range of sources, such as statistical variation, variability, inherent randomness, inhomogeneity, approximation, subjective judgement, and linguistic imprecision (Morgan et al., 1990), or from calibration methodologies, instrumental bias or instrumental limitations (JCGM, 2008).

In the modelling studies that underpin projections of future climate change, it is common to partition uncertainty into four main categories: scenario uncertainty, due to uncertainty of future emissions of GHGs and other forcing agents; 'model uncertainty' associated with climate models; internal variability and initial condition uncertainty; and forcing and boundary condition uncertainty for the assessment of historical and paleoclimate simulations (e.g., Collins and Allen, 2002; Yip et al., 2011).

Model uncertainty is an important contributor to uncertainty in climate predictions and projections. It includes, but is not restricted to, the uncertainties introduced by errors in the model's representation of dynamical and physical and bio-geochemical aspects of the climate system as well as in the model's response to external forcing. The phrase 'model uncertainty' is a common term in the climate change literature, but different studies use the phrase in different senses: some use it to represent the range of behaviours observed in ensembles of climate model (model spread), while others use it in more comprehensive senses (see Sections 9.2, 11.2 and 12.2). Model spread is often used as a measure of climate response uncertainty, but such a measure is crude as it takes no account of factors such as model quality (Chapter 9) or model independence (e.g., Masson and Knutti, 2011; Pennell and Reichler, 2011), and not all variables of interest are adequately simulated by global climate models.

To maintain a degree of terminological clarity this report distinguishes between 'model spread' for this narrower representation of climate model responses and 'model uncertainty' which describes uncertainty about the extent to which any particular climate model provides an accurate representation of the real climate system. This uncertainty arises from approximations required in the development of models. Such approximations affect the representation of all aspects of the climate including the response to external forcings.

Model uncertainty is sometimes decomposed further into parametric and structural uncertainty, comprising, respectively, uncertainty in the values of model parameters and uncertainty in the underlying model structure (see Section 12.2). Some scientific research areas, such as detection and attribution and observationally-constrained model projections of future climate, incorporate significant elements of both observational and model-based science, and in these instances both sets of relevant uncertainties need to be incorporated.

Scenario uncertainty refers to the uncertainties that arise due to limitations in our understanding of future emissions, concentration or forcing trajectories. Scenarios help in the assessment of future developments in complex systems that are either inherently unpredictable, or that have high scientific uncertainties (IPCC, 2000). The societal choices defining future climate drivers are surrounded by considerable uncertainty, and these are explored by examining the climate response to a wide range of possible futures. In past reports, emissions scenarios from the SRES (IPCC, 2000) were used as the main way of exploring uncertainty in future anthropogenic climate drivers. Recent research has made use of Representative Concentration Pathways (RCP) (van Vuuren et al., 2011a, 2011b).

Internal or natural variability, the natural fluctuations in climate, occur in the absence of any RF of the Earth's climate (Hawkins and Sutton, 2009). Climate varies naturally on nearly all time and space scales, and quantifying precisely the nature of this variability is challenging, and is characterized by considerable uncertainty. The analysis of internal and forced contributions to recent climate is discussed in Chapter 10. The fractional contribution of internal variability compared with other forms of uncertainty varies in time and in space, but usually diminishes with time as other sources of uncertainty become more significant (Hawkins and Sutton, 2009; see also Chapter 11 and FAQ 1.1).

In the WGI contribution to the AR5, uncertainty is quantified using 90% uncertainty intervals unless otherwise stated. The 90% uncertainty interval, reported in square brackets, is expected to have a 90% likelihood of covering the value that is being estimated. The value that is being estimated has a 5% likelihood of exceeding the upper endpoint of the uncertainty interval, and the value has a 5% likelihood of being less than that the lower endpoint of the uncertainty interval. A best estimate of that value is also given where available. Uncertainty intervals are not necessarily symmetric about the corresponding best estimate.

In a subject as complex and diverse as climate change, the information available as well as the way it is expressed, and often the interpretation of that material, varies considerably with the scientific context. In some cases, two studies examining similar material may take different approaches even to the quantification of uncertainty. The interpretation of similar numerical ranges for similar variables can differ from study to study. Readers are advised to pay close attention to the caveats and conditions that surround the results presented in peer-reviewed studies, as well as those presented in this assessment. To help readers in this complex and subtle task, the IPCC draws on specific, calibrated language scales to express uncertainty (Mastrandrea et al., 2010), as well as specific procedures for the expression of uncertainty (see Table 1.2). The aim of these structures is to provide tools through which chapter teams might consistently express uncertainty in key results.

1.4.3 Treatment of Uncertainty in IPCC

In the course of the IPCC assessment procedure, chapter teams review the published research literature, document the findings (including uncertainties), assess the scientific merit of this information, identify the key findings, and attempt to express an appropriate measure of the uncertainty that accompanies these findings using a shared guidance procedure. This process has changed over time. The early Assessment Reports (FAR and SAR) were largely qualitative. As the field has grown and matured, uncertainty is being treated more explicitly, with a greater emphasis on the expression, where possible and appropriate, of quantified measures of uncertainty.

Although IPCC's treatment of uncertainty has become more sophisticated since the early reports, the rapid growth and considerable diversity of climate research literature presents ongoing challenges. In the wake of the TAR the IPCC formed a Cross-Working Group team charged with identifying the issues and compiling a set of Uncertainty Guidance Notes that could provide a structure for consistent treatment of uncertainty across the IPCC's remit (Manning et al., 2004). These expanded on the procedural elements of Moss and Schneider (2000) and introduced calibrated language scales designed to enable chapter teams to use the appropriate level of precision to describe findings. These notes were revised between the TAR and AR4 and again between AR4 and AR5 (Mastrandrea et al., 2010).

Recently, increased engagement of social scientists (e.g., Patt and Schrag, 2003; Kandlikar et al., 2005; Risbey and Kandlikar, 2007; Broomell and Budescu, 2009; Budescu et al., 2009; CCSP, 2009) and expert advisory panels (CCSP, 2009; InterAcademy Council, 2010) in the area of uncertainty and climate change has helped clarify issues and procedures to improve presentation of uncertainty. Many of the recommendations of these groups are addressed in the revised Guidance Notes. One key revision relates to clarification of the relationship between the 'confidence' and 'likelihood' language, and pertains to demarcation between qualitative descriptions of 'confidence' and the numerical representations of uncertainty that are expressed by the likelihood scale. In addition, a finding that includes a probabilistic measure of uncertainty does not require explicit mention of the level of confidence associated with that finding if the level of *confidence* is *high* or *very high*. This is a concession to stylistic clarity and readability: if something is described as having a high likelihood, then in the absence of additional qualifiers it should be inferred that it also has *high* or *very high confidence*.

1.4.4 Uncertainty Treatment in This Assessment

All three IPCC Working Groups in the AR5 have agreed to use two metrics for communicating the degree of certainty in key findings (Mastrandrea et al., 2010):

- Confidence in the validity of a finding, based on the type, amount, quality, and consistency of evidence (e.g., data, mechanistic understanding, theory, models, expert judgment) and the degree of agreement. Confidence is expressed qualitatively.
- Quantified measures of uncertainty in a finding expressed probabilistically (based on statistical analysis of observations or model results, or expert judgement).

A level of confidence synthesizes the Chapter teams' judgements about the validity of findings as determined through evaluation of the available evidence and the degree of scientific agreement. The evidence and agreement scale underpins the assessment, as it is on the basis of evidence and agreement that statements can be made with scientific confidence (in this sense, the evidence and agreement scale replaces the 'level of scientific understanding' scale used in previous WGI assessments). There is flexibility in this relationship; for a given evidence and agreement statement, different confidence levels could be assigned, but increasing levels of evidence and degrees of agreement are correlated with increasing confidence. Confidence cannot necessarily be assigned for all combinations of evidence and agreement, but where key variables are highly uncertain, the available evidence and scientific agreement regarding that variable are presented and discussed. Confidence should not be interpreted probabilistically, and it is distinct from 'statistical confidence'.

The confidence level is based on the evidence (robust, medium and limited) and the agreement (high, medium and low). A combination of different methods, e.g., observations and modelling, is important for evaluating the confidence level. Figure 1.11 shows how the combined evidence and agreement results in five levels for the confidence level used in this assessment.

The qualifier 'likelihood' provides calibrated language for describing quantified uncertainty. It can be used to express a probabilistic estimate of the occurrence of a single event or of an outcome, for example, a climate parameter, observed trend, or projected change

1

Frequently Asked Questions

FAQ 1.1 | If Understanding of the Climate System Has Increased, Why Hasn't the Range of Temperature Projections Been Reduced?

The models used to calculate the IPCC's temperature projections agree on the direction of future global change, but the projected size of those changes cannot be precisely predicted. Future greenhouse gas (GHG) emission rates could take any one of many possible trajectories, and some underlying physical processes are not yet completely understood, making them difficult to model. Those uncertainties, combined with natural year-to-year climate variability, produce an 'uncertainty range' in temperature projections.

The uncertainty range around projected GHG and aerosol precursor emissions (which depend on projections of future social and economic conditions) cannot be materially reduced. Nevertheless, improved understanding and climate models—along with observational constraints—may reduce the uncertainty range around some factors that influence the climate's response to those emission changes. The complexity of the climate system, however, makes this a slow process. (FAQ1.1, Figure 1)

Climate science has made many important advances since the last IPCC assessment report, thanks to improvements in measurements and data analysis in the cryosphere, atmosphere, land, biosphere and ocean systems. Scientists also have better understanding and tools to model the role of clouds, sea ice, aerosols, small-scale ocean mixing, the carbon cycle and other processes. More observations mean that models can now be evaluated more thoroughly, and projections can be better constrained. For example, as models and observational analysis have improved, projections of sea level rise have become more accurate, balancing the current sea level rise budget.

Despite these advances, there is still a range in plausible projections for future global and regional climate—what scientists call an 'uncertainty range'. These uncertainty ranges are specific to the variable being considered (precipitation vs. temperature, for instance) and the spatial and temporal extent (such as regional vs. global averages). Uncertainties in climate projections arise from natural variability and uncertainty around the rate of future emissions and the climate's response to them. They can also occur because representations of some known processes are as yet unrefined, and because some processes are not included in the models.

There are fundamental limits to just how precisely annual temperatures can be projected, because of the chaotic nature of the climate system. Furthermore, decadal-scale projections are sensitive to prevailing conditions—such as the temperature of the deep ocean—that are less well known. Some natural variability over decades arises from interactions between the ocean, atmosphere, land, biosphere and cryosphere, and is also linked to phenomena such as the El Niño-Southern Oscillation (ENSO) and the North Atlantic Oscillation (see Box 2.5 for details on patterns and indices of climate variability).

Volcanic eruptions and variations in the sun's output also contribute to natural variability, although they are externally forced and explainable. This natural variability can be viewed as part of the 'noise' in the climate record, which provides the backdrop against which the 'signal' of anthropogenic climate change is detected.

Natural variability has a greater influence on uncertainty at regional and local scales than it does over continental or global scales. It is inherent in the Earth system, and more knowledge will not eliminate the uncertainties it brings. However, some progress is possible—particularly for projections up to a few years ahead—which exploit advances in knowledge of, for instance, the cryosphere or ocean state and processes. This is an area of active research. When climate variables are averaged over decadal timescales or longer, the relative importance of internal variability diminishes, making the long-term signals more evident (FAQ1.1, Figure 1). This long-term perspective is consistent with a common definition of climate as an average over 30 years.

A second source of uncertainty stems from the many possible trajectories that future emission rates of GHGs and aerosol precursors might take, and from future trends in land use. Nevertheless, climate projections rely on input from these variables. So to obtain these estimates, scientists consider a number of alternative scenarios for future human society, in terms of population, economic and technological change, and political choices. They then estimate the likely emissions under each scenario. The IPCC informs policymaking, therefore climate projections for different emissions scenarios can be useful as they show the possible climatic consequences of different policy choices. These scenarios are intended to be compatible with the full range of emissions scenarios described in the current scientific literature, with or without climate policy. As such, they are designed to sample uncertainty in future scenarios. *(continued on next page)*

FAQ 1.1 (continued)

Projections for the next few years and decades are sensitive to emissions of short-lived compounds such as aerosols and methane. More distant projections, however, are more sensitive to alternative scenarios around long-lived GHG emissions. These scenario-dependent uncertainties will not be reduced by improvements in climate science, and will become the dominant uncertainty in projections over longer timescales (e.g., 2100) (FAQ 1.1, Figure 1).

The final contribution to the uncertainty range comes from our imperfect knowledge of how the climate will respond to future anthropogenic emissions and land use change. Scientists principally use computer-based global climate models to estimate this response. A few dozen global climate models have been developed by different groups of scientists around the world. All models are built on the same physical principles, but some approximations are needed because the climate system is so complex. Different groups choose slightly different approximations to represent specific processes in the atmosphere, such as clouds. These choices produce differences in climate projections from different models. This contribution to the uncertainty range is described as 'response uncertainty' or 'model uncertainty'.

The complexity of the Earth system means that future climate could follow many different scenarios, yet still be consistent with current understanding and models. As observational records lengthen and models improve, researchers should be able, within the limitations of the range of natural variability, to narrow that range in probable temperature in the next few decades (FAQ 1.1, Figure 1). It is also possible to use information about the current state of the oceans and cryosphere to produce better projections up to a few years ahead.

As science improves, new geophysical processes can be added to climate models, and representations of those already included can be improved. These developments can appear to increase model-derived estimates of climate response uncertainty, but such increases merely reflect the quantification of previously unmeasured sources of uncertainty (FAQ1.1, Figure 1). As more and more important processes are added, the influence of unquantified processes lessens, and there can be more confidence in the projections.

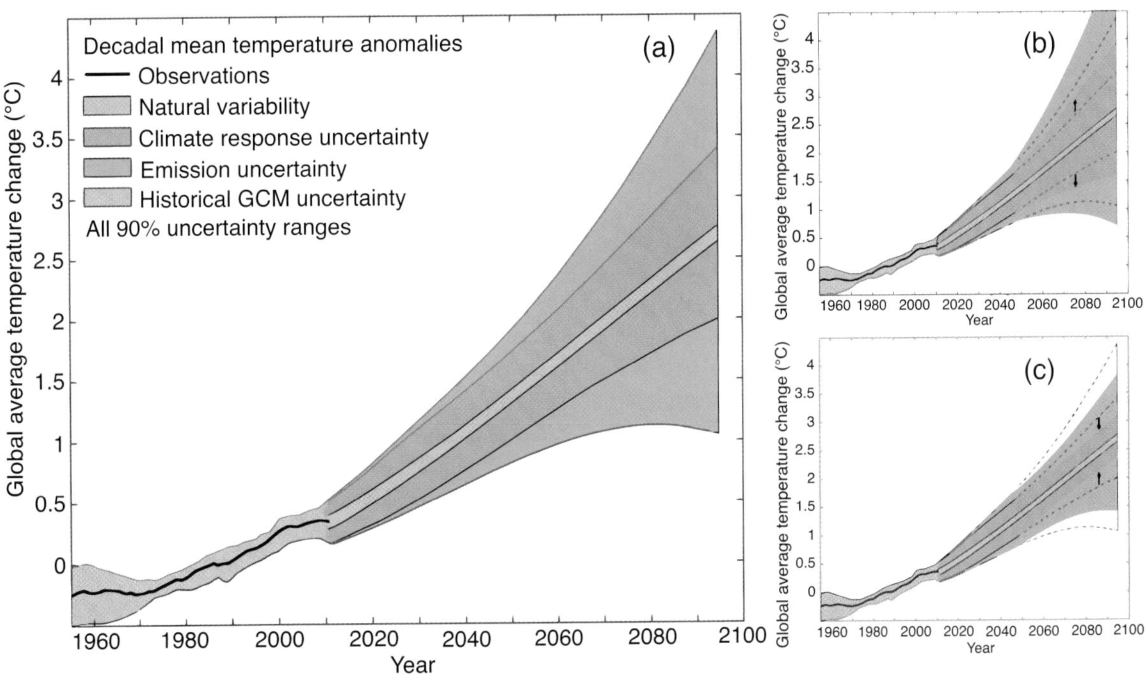

FAQ 1.1, Figure 1 | Schematic diagram showing the relative importance of different uncertainties, and their evolution in time. (a) Decadal mean surface temperature change (°C) from the historical record (black line), with climate model estimates of uncertainty for historical period (grey), along with future climate projections and uncertainty. Values are normalised by means from 1961 to 1980. Natural variability (orange) derives from model interannual variability, and is assumed constant with time. Emission uncertainty (green) is estimated as the model mean difference in projections from different scenarios. Climate response uncertainty (blue-solid) is based on climate model spread, along with added uncertainties from the carbon cycle, as well as rough estimates of additional uncertainty from poorly modelled processes. Based on Hawkins and Sutton (2011) and Huntingford et al. (2009). (b) Climate response uncertainty can appear to increase when a new process is discovered to be relevant, but such increases reflect a quantification of previously unmeasured uncertainty, or (c) can decrease with additional model improvements and observational constraints. The given uncertainty range of 90% means that the temperature is estimated to be in that range, with a probability of 90%.

1

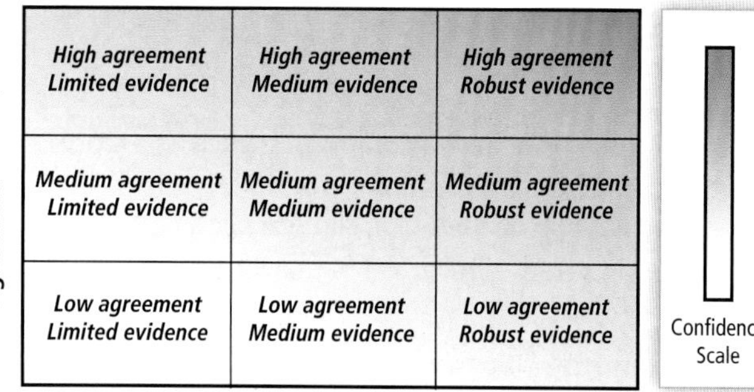

Figure 1.11 | The basis for the confidence level is given as a combination of evidence (limited, medium, robust) and agreement (low, medium and high) (Mastrandrea et al., 2010).

lying in a given range. Statements made using the likelihood scale may be based on statistical or modelling analyses, elicitation of expert views, or other quantitative analyses. Where sufficient information is available it is preferable to eschew the likelihood qualifier in favour of the full probability distribution or the appropriate probability range. See Table 1.2 for the list of 'likelihood' qualifiers to be used in AR5.

Many social sciences studies have found that the interpretation of uncertainty is contingent on the presentation of information, the context within which statements are placed and the interpreter's own lexical preferences. Readers often adjust their interpretation of probabilistic language according to the magnitude of perceived potential consequences (Patt and Schrag, 2003; Patt and Dessai, 2005). Furthermore, the framing of a probabilistic statement impinges on how it is interpreted (Kahneman and Tversky, 1979): for example, a 10% chance of dying is interpreted more negatively than a 90% chance of surviving.

In addition, work examining expert judgement and decision making shows that people—including scientific experts—are prone to a range of heuristics and biases that affect their judgement (e.g., Kahneman et al., 1982). For example, in the case of expert judgements there is a tendency towards overconfidence both at the individual level (Morgan et al., 1990) and at the group level as people converge on a view and draw confidence in its reliability from each other. However, in an assessment of the state of scientific knowledge across a field such as climate change—characterized by complexity of process and heterogeneity of data constraints—some degree of expert judgement is inevitable (Mastrandrea et al., 2010).

Table 1.2 | Likelihood terms associated with outcomes used in the AR5.

Term	Likelihood of the Outcome
Virtually certain	99–100% probability
Very likely	90–100% probability
Likely	66–100% probability
About as likely as not	33–66% probability
Unlikely	0–33% probability
Very unlikely	0–10% probability
Exceptionally unlikely	0–1% probability

Notes:
Additional terms that were used in limited circumstances in the AR4 (*extremely likely* = 95–100% probability, *more likely than not* = >50–100% probability, and *extremely unlikely* = 0–5% probability) may also be used in the AR5 when appropriate.

These issues were brought to the attention of chapter teams so that contributors to the AR5 might be sensitized to the ways presentation, framing, context and potential biases might affect their own assessments and might contribute to readers' understanding of the information presented in this assessment. There will always be room for debate about how to summarize such a large and growing literature. The uncertainty guidance is aimed at providing a consistent, calibrated set of words through which to communicate the uncertainty, confidence and degree of consensus prevailing in the scientific literature. In this sense the guidance notes and practices adopted by IPCC for the presentation of uncertainties should be regarded as an interdisciplinary work in progress, rather than as a finalized, comprehensive approach. Moreover, one precaution that should be considered is that translation of this assessment from English to other languages may lead to a loss of precision.

1.5 Advances in Measurement and Modelling Capabilities

Since AR4, measurement capabilities have continued to advance. The models have been improved following the progress in the understanding of physical processes within the climate system. This section illustrates some of those developments.

1.5.1 Capabilities of Observations

Improved understanding and systematic monitoring of Earth's climate requires observations of various atmospheric, oceanic and terrestrial parameters and therefore has to rely on various technologies (ranging from ground-based instruments to ships, buoys, ocean profilers, balloons, aircraft, satellite-borne sensors, etc.). The Global Climate Observing System (GCOS, 2009) defined a list of so-called Essential Climate Variables, that are technically and economically feasible to observe, but some of the associated observing systems are not yet operated in a systematic manner. However, during recent years, new observational systems have increased the number of observations by orders of magnitude and observations have been made at places where there have been no data before (see Chapters 2, 3 and 4 for an assessment of changes in observations). Parallel to this, tools to analyse and process the data have been developed and enhanced to cope with the increase of information and to provide a more comprehensive picture of the Earth's climate. At the same time, it should be kept in mind that there has been some limited progress in developing countries in filling gaps in their *in situ* observing networks, but developed countries have made little progress in ensuring long-term continuity for several important observing systems (GCOS, 2009). In addition, more proxy (non-instrumental) data have been acquired to provide a more comprehensive picture of climate changes in the past (see Chapter 5). Efforts are also occurring to digitize historic observations, mainly of ground-station data from periods prior to the second half of the 20th century (Brunet and Jones, 2011).

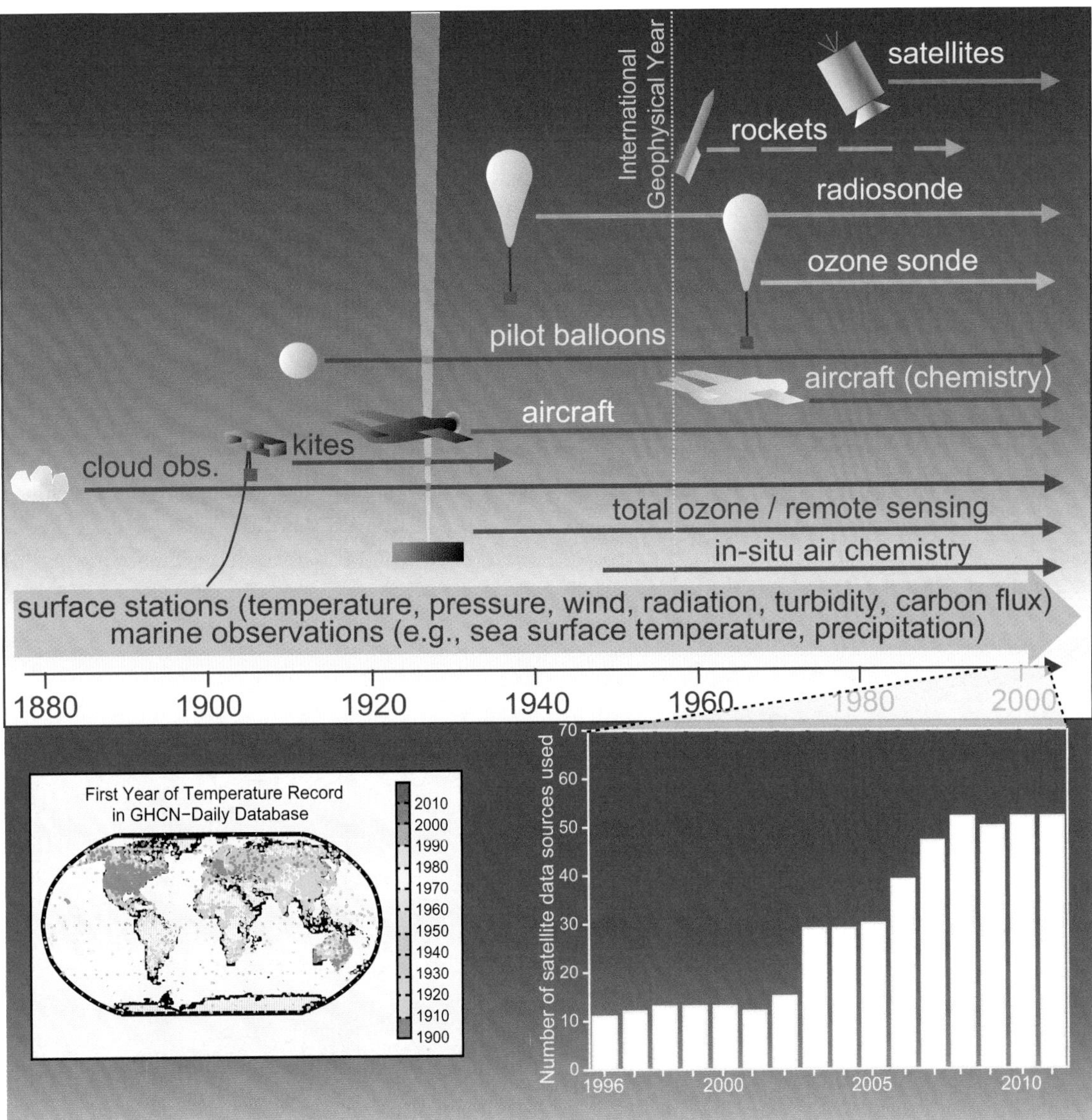

Figure 1.12 | Development of capabilities of observations. Top: Changes in the mix and increasing diversity of observations over time create challenges for a consistent climate record (adapted from Brönnimann et al., 2008). Bottom left: First year of temperature data in Global Historical Climatology Network (GHCN) daily database (available at http://www.ncdc.noaa.gov/oa/climate/ghcn-daily/; Menne et al., 2012). Bottom right: Number of satellite instruments from which data have been assimilated in the European Centre for Medium-Range Weather Forecasts production streams for each year from 1996 to 2010. This figure is used as an example to demonstrate the fivefold increase in the usage of satellite data over this time period.

Reanalysis is a systematic approach to produce gridded dynamically consistent data sets for climate monitoring and research by assimilating all available observations with help of a climate model (Box 2.3). Model-based reanalysis products play an important role in obtaining a consistent picture of the climate system. However, their usefulness in detecting long-term climate trends is currently limited by changes over time in observational coverage and biases, linked to the presence of biases in the assimilating model (see also Box 2.3 in Chapter 2). Because AR4 both the quantity and quality of the observations that are assimilated through reanalysis have increased (GCOS, 2009). As an example, there has been some overall increase in mostly atmospheric observations assimilated in European Centre for Medium-Range Weather Forecasts Interim Reanalysis since 2007 (Dee et al., 2011). The overwhelming majority of the data, and most of the increase over recent years, come from satellites (Figure 1.12) (GCOS, 2011). For example, information from Global Positioning System radio occultation measurements has increased significantly since 2007. The increases in data from fixed stations are often associated with an increased frequency of reporting, rather than an increase in the number of stations. Increases in data quality come from improved instrument design or from more accurate correction in the ground-station processing that is applied before the data are transmitted to users and data centres. As an example for *in situ* data, temperature biases of radiosonde measurements from radiation effects have been reduced over recent years. The new generation of satellite sensors such as the high spectral resolution infrared sounders (such as the Atmospheric Infrared Sounder and the Infrared Atmospheric Sounding Interferometer) are instrumental to achieving a better temporal stability for recalibrating sensors such as the High-Resolution Infrared Radiation Sounder. Few instruments (e.g., the Advanced Very High Resolution Radiometer) have now been in orbit for about three decades, but these were not originally designed for climate applications and therefore require careful re-calibration.

A major achievement in ocean observation is due to the implementation of the Argo global array of profiling floats system (GCOS, 2009). Deployment of Argo floats began in 2000, but it took until 2007 for numbers to reach the design target of 3000 floats. Since 2000 the ice-free upper 2000 m of the ocean have been observed systematically for temperature and salinity for the first time in history, because both the Argo profiling float and surface drifting buoy arrays have reached global coverage at their target numbers (in January 2009, there were 3291 floats operating). Biases in historical ocean data have been identified and reduced, and new analytical approaches have been applied (e.g., Willis et al., 2009). One major consequence has been the reduction of an artificial decadal variation in upper ocean temperature and heat content that was apparent in the observational assessment for AR4 (see Section 3.2). The spatial and temporal coverage of biogeochemical measurements in the ocean has also expanded. Satellite observations for sea level (Sections 3.7 and 13.2), sea surface salinity (Section 3.3), sea ice (Section 4.2) and ocean colour have also been further developed over the past few years.

Progress has also been made with regard to observation of terrestrial Essential Climate Variables. Major advances have been achieved in remote sensing of soil moisture due to the launch of the Soil Moisture and Oceanic Salinity mission in 2009 but also due to new retrieval techniques that have been applied to data from earlier and ongoing missions (see Seneviratne et al., 2010 for a detailed review). However, these measurements have limitations. For example, the methods fail under dense vegetation and they are restricted to the surface soil. Updated Advanced Very High Resolution Radiometer-based Normalized Differenced Vegetation Index data provide new information on the change in vegetation. During the International Polar Year 2007–2009 the number of borehole sites was significantly increased and therefore allows a better monitoring of the large-scale permafrost features (see Section 4.7).

1.5.2 Capabilities in Global Climate Modelling

Several developments have especially pushed the capabilities in modelling forward over recent years (see Figure 1.13 and a more detailed discussion in Chapters 6, 7 and 9).

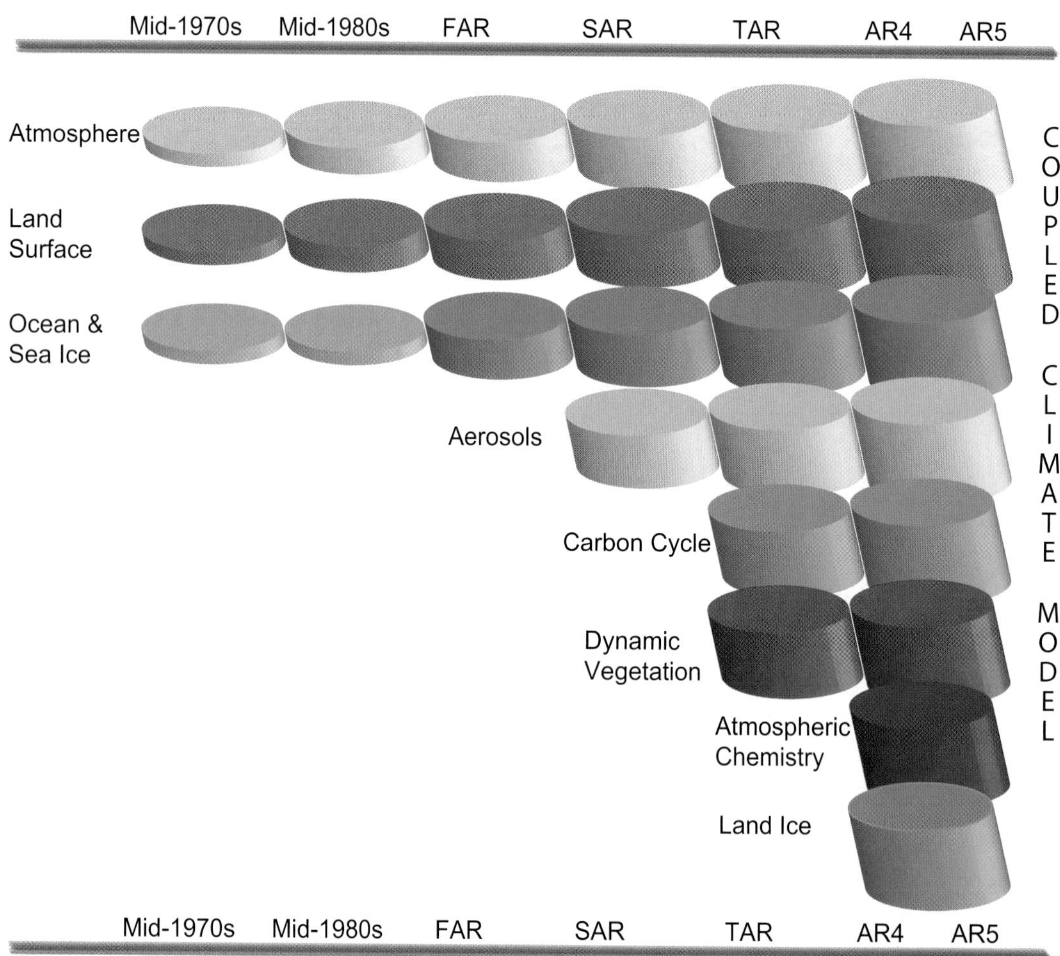

Figure 1.13 | The development of climate models over the last 35 years showing how the different components were coupled into comprehensive climate models over time. In each aspect (e.g., the atmosphere, which comprises a wide range of atmospheric processes) the complexity and range of processes has increased over time (illustrated by growing cylinders). Note that during the same time the horizontal and vertical resolution has increased considerably e.g., for spectral models from T21L9 (roughly 500 km horizontal resolution and 9 vertical levels) in the 1970s to T95L95 (roughly 100 km horizontal resolution and 95 vertical levels) at present, and that now ensembles with at least three independent experiments can be considered as standard.

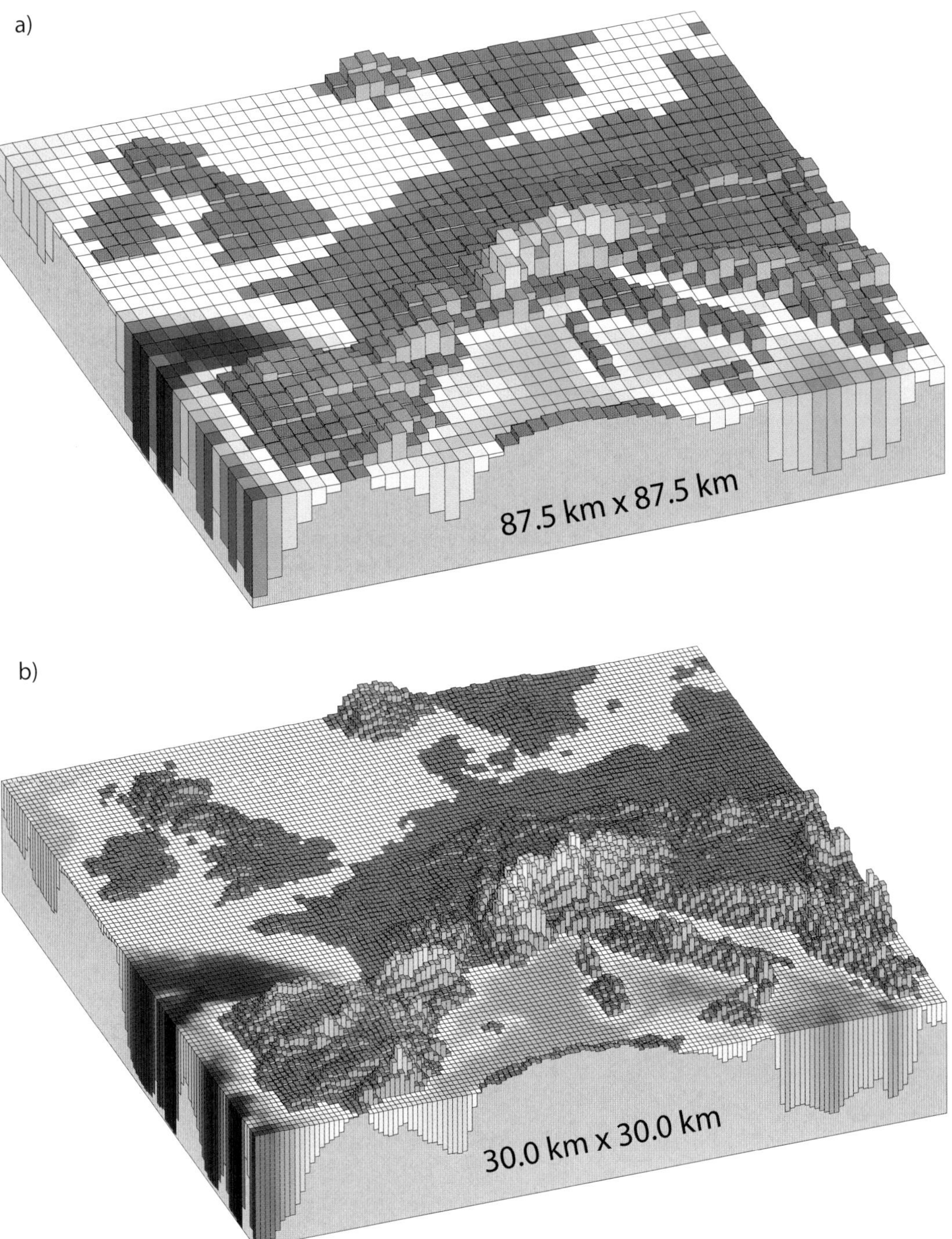

Figure 1.14 | Horizontal resolutions considered in today's higher resolution models and in the very high resolution models now being tested: (a) Illustration of the European topography at a resolution of 87.5 × 87.5 km; (b) same as (a) but for a resolution of 30.0 × 30.0 km.

There has been a continuing increase in horizontal and vertical resolution. This is especially seen in how the ocean grids have been refined, and sophisticated grids are now used in the ocean and atmosphere models making optimal use of parallel computer architectures. More models with higher resolution are available for more regions. Figure 1.14a and 1.14b show the large effect on surface representation from a horizontal grid spacing of 87.5 km (higher resolution than most current global models and similar to that used in today's highly resolved models) to a grid spacing of 30.0 km (similar to the current regional climate models).

Representations of Earth system processes are much more extensive and improved, particularly for the radiation and the aerosol cloud interactions and for the treatment of the cryosphere. The representation of the carbon cycle was added to a larger number of models and has been improved since AR4. A high-resolution stratosphere is now included in many models. Other ongoing process development in climate models includes the enhanced representation of nitrogen effects on the carbon cycle. As new processes or treatments are added to the models, they are also evaluated and tested relative to available observations (see Chapter 9 for more detailed discussion).

1 Ensemble techniques (multiple calculations to increase the statistical sample, to account for natural variability, and to account for uncertainty in model formulations) are being used more frequently, with larger samples and with different methods to generate the samples (different models, different physics, different initial conditions). Coordinated projects have been set up to generate and distribute large samples (ENSEMBLES, climateprediction.net, Program for Climate Model Diagnosis and Intercomparison).

The model comparisons with observations have pushed the analysis and development of the models. CMIP5, an important input to the AR5, has produced a multi-model data set that is designed to advance our understanding of climate variability and climate change. Building on previous CMIP efforts, such as the CMIP3 model analysis reported in AR4, CMIP5 includes 'long-term' simulations of 20th century climate and projections for the 21st century and beyond. See Chapters 9, 10, 11 and 12 for more details on the results derived from the CMIP5 archive.

Since AR4, the incorporation of 'long-term' paleoclimate simulations in the CMIP5 framework has allowed incorporation of information from paleoclimate data to inform projections. Within uncertainties associated with reconstructions of past climate variables from proxy records and forcings, paleoclimate information from the Mid Holocene, Last Glacial Maximum and Last Millennium have been used to test the ability of models to simulate realistically the magnitude and large-scale patterns of past changes (Section 5.3, Box 5.1 and 9.4).

The capabilities of ESMs continue to be enhanced. For example, there are currently extensive efforts towards developing advanced treatments for the processes affecting ice sheet dynamics. Other enhancements are being aimed at land surface hydrology, and the effects of agriculture and urban environments.

As part of the process of getting model analyses for a range of alternative assumptions about how the future may unfold, scenarios for future emissions of important gases and aerosols have been generated for the IPCC assessments (e.g., see the SRES scenarios used in TAR and AR4). The emissions scenarios represent various development pathways based on well-defined assumptions. The scenarios are used to calculate future changes in climate, and are then archived in the Climate Model Intercomparison Project (e.g., CMIP3 for AR4; CMIP5 for AR5). For CMIP5, four new scenarios, referred to as Representative Concentration Pathways (RCPs) were developed (Section 12.3; Moss et al., 2010). See Box 1.1 for a more thorough discussion of the RCP scenarios. Because results from both CMIP3 and CMIP5 will be presented in the later chapters (e.g., Chapters 8, 9, 11 and 12), it is worthwhile considering the differences and similarities between the SRES and the RCP scenarios. Figure 1.15, acting as a prelude to the discussion in Box 1.1, shows that the RF for several of the SRES and RCP scenarios are similar over time and thus should provide results that can be used to compare climate modelling studies.

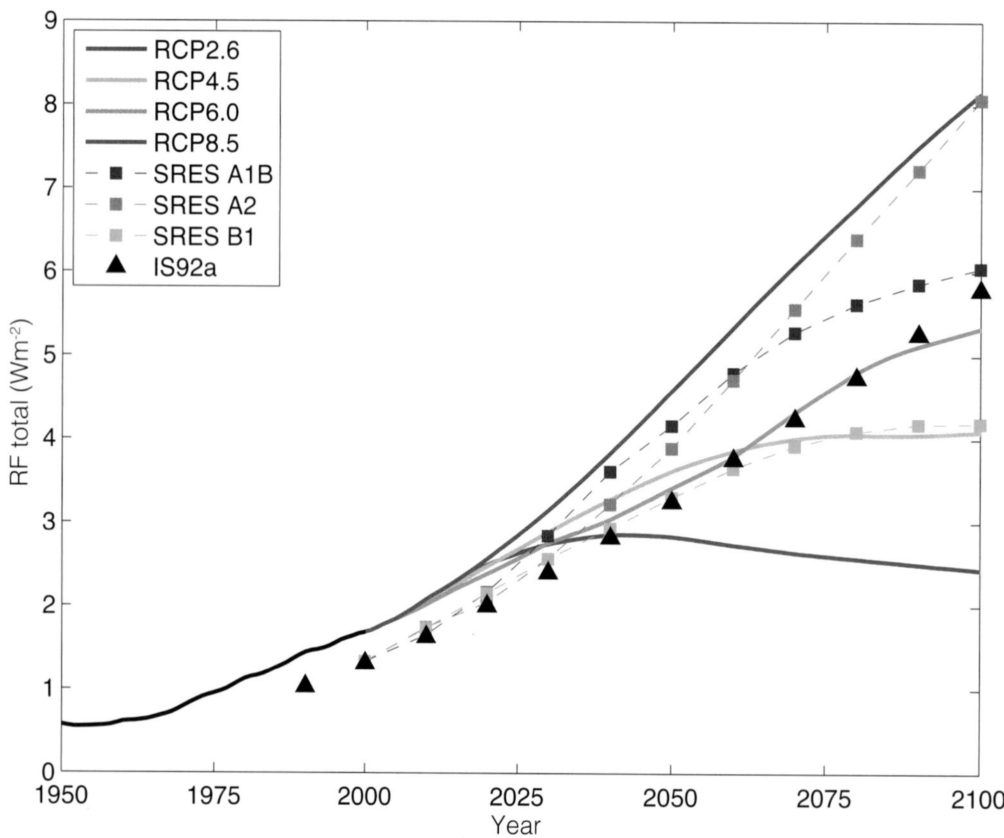

Figure 1.15 | Historical and projected total anthropogenic RF (W m^{-2}) relative to preindustrial (about 1765) between 1950 and 2100. Previous IPCC assessments (SAR IS92a, TAR/AR4 SRES A1B, A2 and B1) are compared with representative concentration pathway (RCP) scenarios (see Chapter 12 and Box 1.1 for their extensions until 2300 and Annex II for the values shown here). The total RF of the three families of scenarios, IS92, SRES and RCP, differ for example, for the year 2000, resulting from the knowledge about the emissions assumed having changed since the TAR and AR4.

Box 1.1 | Description of Future Scenarios

Long-term climate change projections require assumptions on human activities or natural effects that could alter the climate over decades and centuries. Defined scenarios are useful for a variety of reasons, e.g., assuming specific time series of emissions, land use, atmospheric concentrations or RF across multiple models allows for coherent climate model intercomparisons and synthesis. Scenarios can be formed in a range of ways, from simple, idealized structures to inform process understanding, through to comprehensive scenarios produced by Integrated Assessment Models (IAMs) as internally consistent sets of assumptions on emissions and socio-economic drivers (e.g., regarding population and socio-economic development).

Idealized Concentration Scenarios

As one example of an idealized concentration scenario, a 1% yr^{-1} compound increase of atmospheric CO_2 concentration until a doubling or a quadrupling of its initial value has been widely used in the past (Covey et al., 2003). An exponential increase of CO_2 concentrations induces an essentially linear increase in RF (Myhre et al., 1998) due to a 'saturation effect' of the strong absorbing bands. Such a linear ramp function is highly useful for comparative diagnostics of models' climate feedbacks and inertia. The CMIP5 intercomparison project again includes such a stylized pathway up to a quadrupling of CO_2 concentrations, in addition to an instantaneous quadrupling case.

The Socio-Economic Driven SRES Scenarios

The SRES suite of scenarios were developed using IAMs and resulted from specific socio-economic scenarios from storylines about future demographic and economic development, regionalization, energy production and use, technology, agriculture, forestry and land use (IPCC, 2000). The climate change projections undertaken as part of CMIP3 and discussed in AR4 were based primarily on the SRES A2, A1B and B1 scenarios. However, given the diversity in models' carbon cycle and chemistry schemes, this approach implied differences in models' long lived GHG and aerosol concentrations for the same emissions scenario. As a result of this and other shortcomings, revised scenarios were developed for AR5 to allow atmosphere-ocean general circulation model (AOGCM) (using concentrations) simulations to be compared with those ESM simulations that use emissions to calculate concentrations.

Representative Concentration Pathway Scenarios and Their Extensions

Representative Concentration Pathway (RCP) scenarios (see Section 12.3 for a detailed description of the scenarios; Moss et al., 2008; Moss et al., 2010; van Vuuren et al., 2011b) are new scenarios that specify concentrations and corresponding emissions, but are not directly based on socio-economic storylines like the SRES scenarios. The RCP scenarios are based on a different approach and include more consistent short-lived gases and land use changes. They are not necessarily more capable of representing future developments than the SRES scenarios. Four RCP scenarios were selected from the published literature (Fujino et al., 2006; Smith and Wigley, 2006; Riahi et al., 2007; van Vuuren et al., 2007; Hijioka et al., 2008; Wise et al., 2009) and updated for use within CMIP5 (Masui et al., 2011; Riahi et al., 2011; Thomson et al., 2011; van Vuuren et al., 2011a). The four scenarios are identified by the 21st century peak or stabilization value of the RF derived by the reference model (in W m^{-2}) (Box 1.1, Figure 1): the lowest RCP, RCP2.6 (also referred to as

(continued on next page)

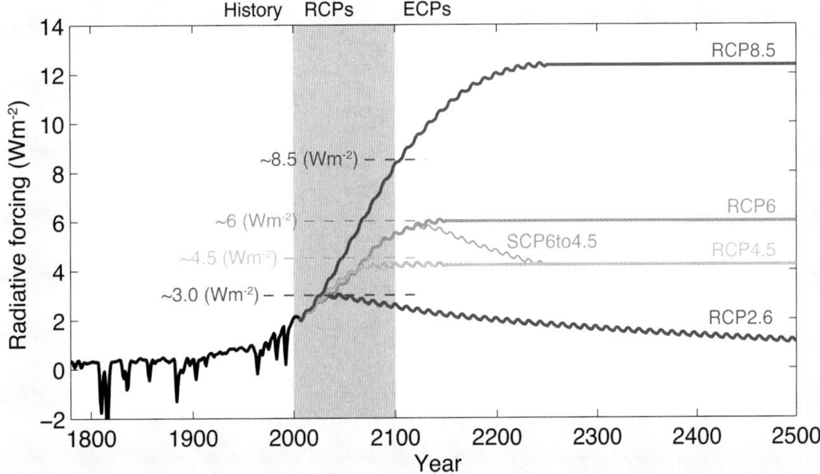

Box 1.1, Figure 1 | Total RF (anthropogenic plus natural) for RCPs and extended concentration pathways (ECP)—for RCP2.6, RCP4.5, and RCP6, RCP8.5, as well as a supplementary extension RCP6 to 4.5 with an adjustment of emissions after 2100 to reach RCP4.5 concentration levels in 2250 and thereafter. Note that the stated RF levels refer to the illustrative default median estimates only. There is substantial uncertainty in current and future RF levels for any given scenario. Short-term variations in RF are due to both volcanic forcings in the past (1800–2000) and cyclical solar forcing assuming a constant 11-year solar cycle (following the CMIP5 recommendation), except at times of stabilization. (Reproduced from Figure 4 in Meinshausen et al., 2011.)

Box 1.1 (continued)

RCP3-PD) which peaks at 3 W m^{-2} and then declines to approximately 2.6 W m^{-2} by 2100; the medium-low RCP4.5 and the medium-high RCP6 aiming for stabilization at 4.5 and 6 W m^{-2}, respectively around 2100; and the highest one, RCP8.5, which implies a RF of 8.5 W m^{-2} by 2100, but implies rising RF beyond that date (Moss et al., 2010). In addition there is a supplementary extension SCP6to4.5 with an adjustment of emissions after 2100 to reach RCP 4.5 concentration levels in 2250 and thereafter. The RCPs span the full range of RF associated with emission scenarios published in the peer-reviewed literature at the time of the development of the RCPs, and the two middle scenarios where chosen to be roughly equally spaced between the two extremes (2.6 and 8.5 W m^{-2}). These forcing values should be understood as comparative labels representative of the forcing associated with each scenario, which will vary somewhat from model to model. This is because concentrations or emissions (rather than the RF) are prescribed in the CMIP5 climate model runs.

Various steps were necessary to turn the selected 'raw' RCPs into emission scenarios from IAMs and to turn these into data sets usable by the climate modelling community, including the extension with historical emissions (Granier et al., 2011; Meinshausen et al., 2011), the harmonization (smoothly connected historical reconstruction) and gridding of land use data sets (Hurtt et al., 2011), the provision of atmospheric chemistry modelling studies, particularly for tropospheric ozone (Lamarque et al., 2011), analyses of 2000–2005 GHG emission levels, and extension of GHG concentrations with historical GHG concentrations and harmonization with analyses of 2000–2005 GHG concentrations levels (Meinshausen et al., 2011). The final RCP data sets comprise land use data, harmonized GHG emissions and concentrations, gridded reactive gas and aerosol emissions, as well as ozone and aerosol abundance fields (Figures 2, 3, and 4 in Box 1.1). *(continued on next page)*

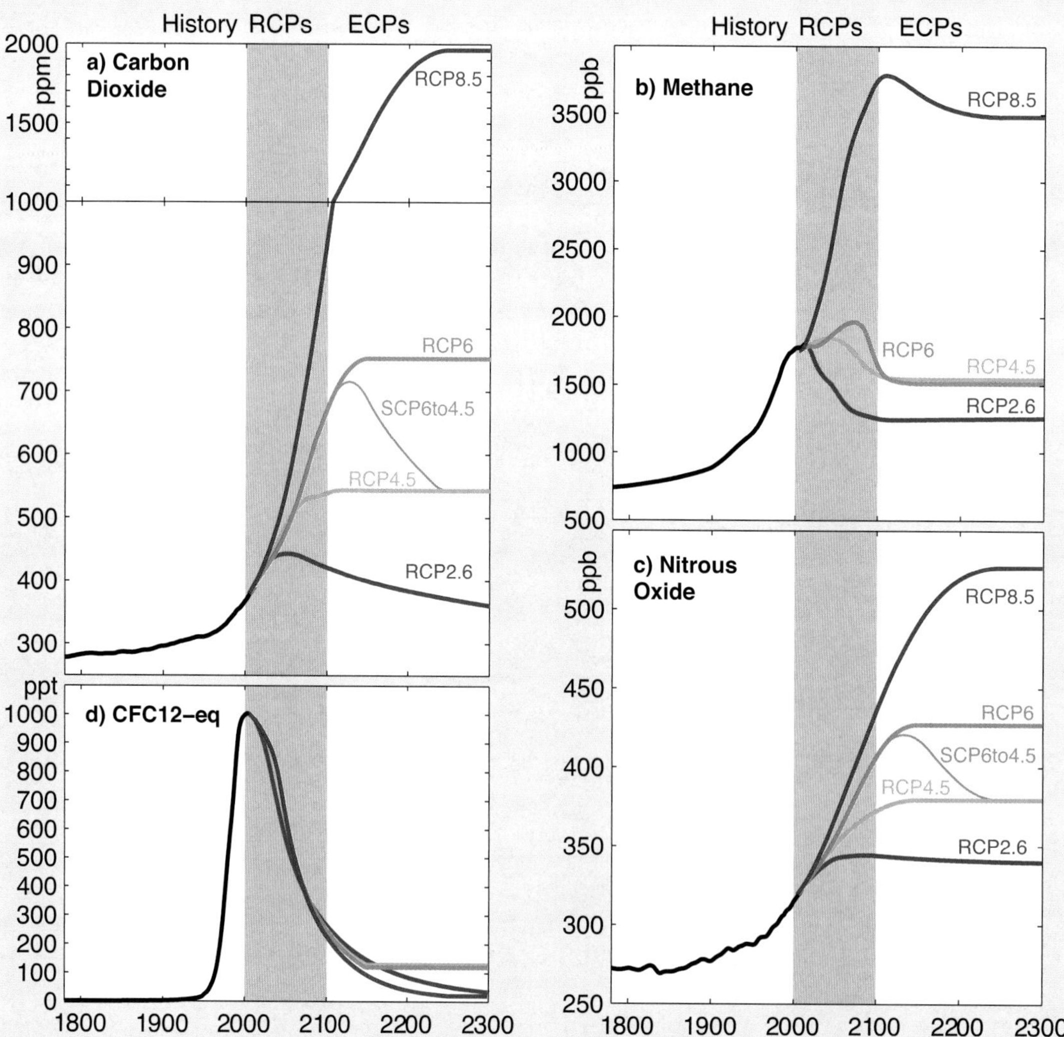

Box 1.1, Figure 2 | Concentrations of GHG following the 4 RCPs and their extensions (ECP) to 2300. (Reproduced from Figure 5 in Meinshausen et al., 2011.) Also see Annex II Table AII.4.1 for CO_2, Table AII.4.2 for CH_4, Table AII.4.3 for N_2O.

Box 1.1 (continued)

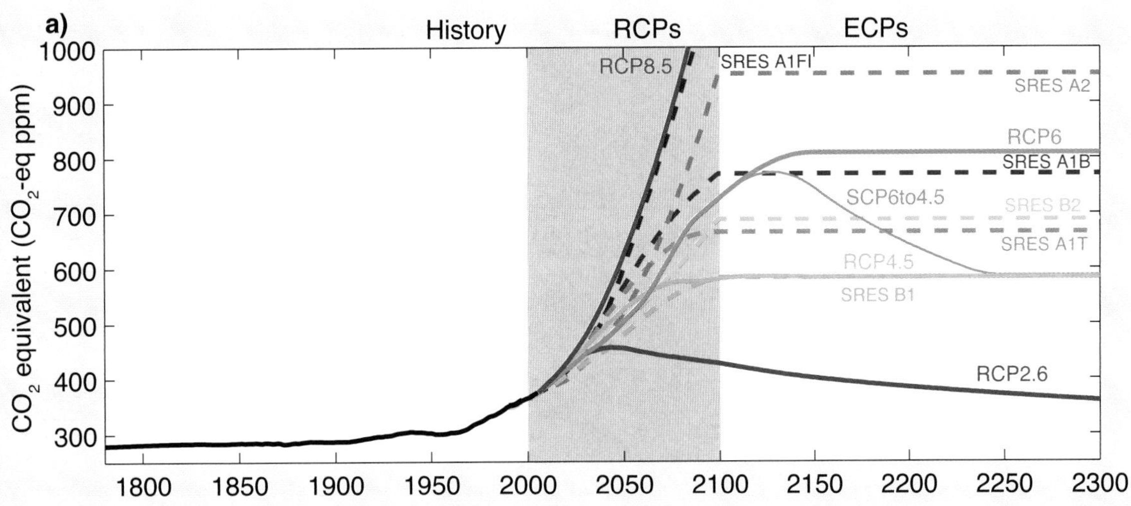

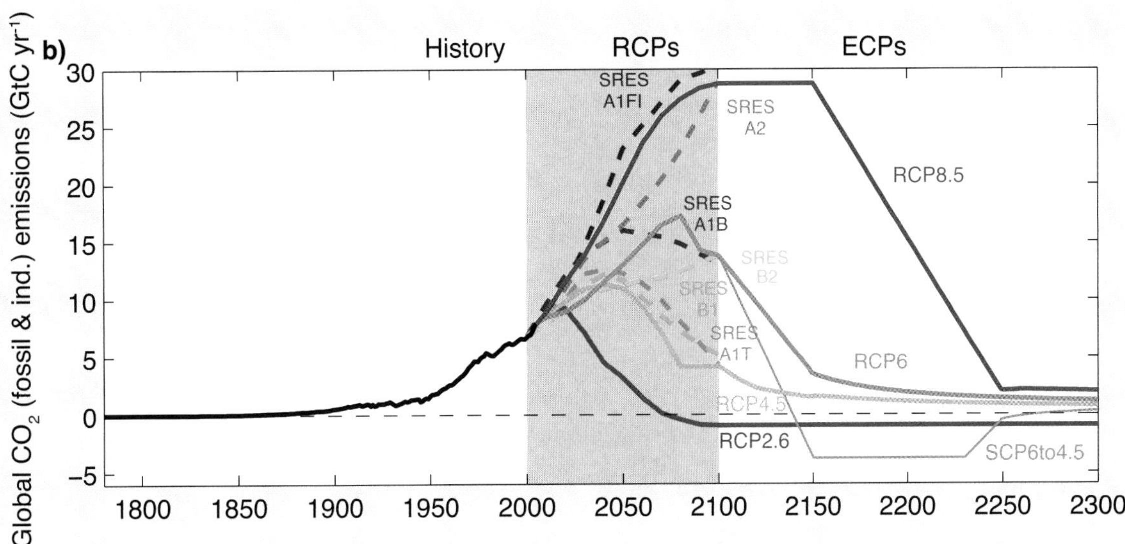

Box 1.1, Figure 3 | (a) Equivalent CO_2 concentration and (b) CO_2 emissions (except land use emissions) for the four RCPs and their ECPs as well as some SRES scenarios.

To aid model understanding of longer-term climate change implications, these RCPs were extended until 2300 (Meinshausen et al., 2011) under reasonably simple and somewhat arbitrary assumptions regarding post-2100 GHG emissions and concentrations. In order to continue to investigate a broad range of possible climate futures, the two outer RCPs, RCP2.6 and RCP8.5 assume constant emissions after 2100, while the two middle RCPs aim for a smooth stabilization of concentrations by 2150. RCP8.5 stabilizes concentrations only by 2250, with CO_2 concentrations of approximately 2000 ppm, nearly seven times the pre-industrial levels. As the RCP2.6 implies netnegative CO_2 emissions after around 2070 and throughout the extension, CO_2 concentrations are slowly reduced towards 360 ppm by 2300.

Comparison of SRES and RCP Scenarios

The four RCP scenarios used in CMIP5 lead to RF values that span a range larger than that of the three SRES scenarios used in CMIP3 (Figure 12.3). RCP4.5 is close to SRES B1, RCP6 is close to SRES A1B (more after 2100 than during the 21st century) and RCP8.5 is somewhat higher than A2 in 2100 and close to the SRES A1FI scenario (Figure 3 in Box 1.1). RCP2.6 is lower than any of the SRES scenarios (see also Figure 1.15). *(continued on next page)*

1

Box 1.1 (continued)

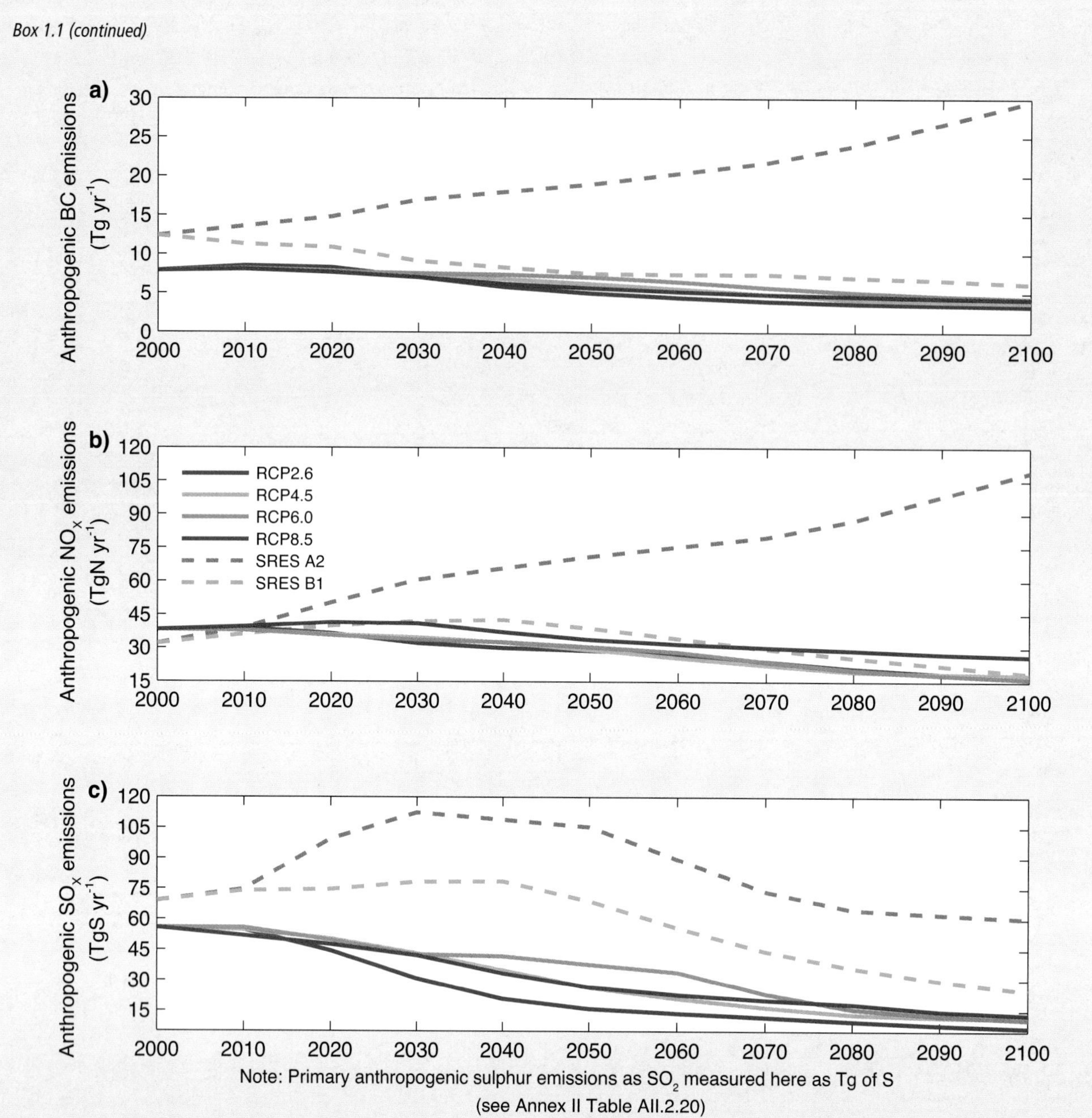

Box 1.1, Figure 4 | (a) Anthropogenic BC emissions (Annex II Table AII.2.22), (b) anthropogenic NO_x emissions (Annex II Table AII.2.18), and (c) anthropogenic SO_x emissions (Annex II Table II.2.20).

1.6 Overview and Road Map to the Rest of the Report

As this chapter has shown, understanding of the climate system and the changes occurring in it continue to advance. The notable scientific advances and associated peer-reviewed publications since AR4 provide the basis for the assessment of the science as found in Chapters 2 to 14. Below a quick summary of these chapters and their objectives is provided.

Observations and Paleoclimate Information (Chapters 2, 3, 4 and 5): These chapters assess information from all climate system components on climate variability and change as obtained from instrumental records and climate archives. This group of chapters covers all relevant aspects of the atmosphere including the stratosphere, the land surface, the oceans and the cryosphere. Information on the water cycle, including evaporation, precipitation, runoff, soil moisture, floods, drought, etc. is assessed. Timescales from daily to decades (Chapters 2, 3 and 4) and from centuries to many millennia (Chapter 5) are considered.

Process Understanding (Chapters 6 and 7): These chapters cover all relevant aspects from observations and process understanding, to projections from global to regional scale. Chapter 6 covers the carbon cycle and its interactions with other biogeochemical cycles, in particular the nitrogen cycle, as well as feedbacks on the climate system. Chapter 7 treats in detail clouds and aerosols, their interactions and chemistry, the role of water vapour, as well as their role in feedbacks on the climate system.

From Forcing to Attribution of Climate Change (Chapters 8, 9 and 10): In these chapters, all the information on the different drivers (natural and anthropogenic) of climate change is collected, expressed in terms of RF, and assessed (Chapter 8). As part of this, the science of metrics commonly used in the literature to compare radiative effects from a range of agents (Global Warming Potential, Global Temperature Change Potential and others) is covered. In Chapter 9, the hierarchy of climate models used in simulating past and present climate change is assessed. Information regarding detection and attribution of changes on global to regional scales is assessed in Chapter 10.

Future Climate Change and Predictability (Chapters 11 and 12): These chapters assess projections of future climate change derived from climate models on timescales from decades to centuries at both global and regional scales, including mean changes, variability and extremes. Fundamental questions related to the predictability of climate as well as long-term climate change, climate change commitments and inertia in the climate system are addressed.

Integration (Chapters 13 and 14): These chapters integrate all relevant information for two key topics in WGI AR5: sea level change (Chapter 13) and climate phenomena across the regions (Chapter 14). Chapter 13 assesses information on sea level change ranging from observations and process understanding to projections from global to regional scales. Chapter 14 assesses the most important modes of variability in the climate system and extreme events. Furthermore, this chapter deals with interconnections between the climate phenomena, their regional expressions, and their relevance for future regional climate change. Maps produced and assessed in Chapter 14, together with Chapters 11 and 12, form the basis of the Atlas of Global and Regional Climate Projections in Annex I. RFs and estimates of future atmospheric concentrations from Chapters 7, 8, 11 and 12 form the basis of the Climate System Scenario Tables in Annex II.

1.6.1 Topical Issues

A number of topical issues are discussed throughout the assessment. These issues include those of areas where there is contention in the peer-reviewed literature and where questions have been raised that are being addressed through ongoing research. Table 1.3 provides a non-comprehensive list of many of these and the chapters where they are discussed.

Table 1.3 | Key topical issues discussed in the assessment.

Topic	Section
Abrupt change and irreversibility	5.7, 12.5, 13.4
Aerosols	6.4, 7.3, 7.4, 7.5, 7.6, 8.3, 11.3, 14.1
Antarctic climate change	5.8, 9.4, 10.3, 13.3
Arctic sea ice change	4.2, 5.5, 9.4, 10.3, 11.3, 12.4
Hydrological cycle changes	2.5, 2.6, 3.3, 3.4, 3.5, 7.6, 10.3, 12.4
Carbon-climate feedbacks	6.4, 12.4
Climate sensitivity	5.3, 9.7, 10.8, 12.5
Climate stabilization	6.3, 6.4, 12.5
Cloud feedbacks	5.3, 7.2, 9.7, 11.3, 12.4
Cosmic ray effects on clouds	7.4
Decadal climate variability	5.3, 9.5, 10.3
Earth's Energy (trends, distribution and budget)	2.3, 3.2, 13.3
El Niño-Southern Oscillation	2.7, 5.4, 9.4, 9.5, 14.4
Geo-engineering	6.4, 7.7
Glacier change	4.3, 5.5, 10.5, 13.3
Ice sheet dynamics and mass balance assessment	4.4, 5.3, 5.6, 10.5, 13.3
Monsoons	2.7, 5.5, 9.5, 14.2
Ocean acidification	3.8, 6.4
Permafrost change	4.7, 6.3, 10.5
Solar effects on climate change	5.2, 8.4
Sea level change, including regional effects	3.7, 5.6, 13.1
Temperature trends since 1998	2.4, 3.2, 9.4
Tropical cyclones	2.6, 10.6, 14.6
Upper troposphere temperature trends	2.4, 9.4

1

References

Allen, M. R., J. F. B. Mitchell, and P. A. Stott, 2013: Test of a decadal climate forecast. *Nature Geosci.*, **6**, 243–244.

AMAP, 2009: Summary – The Greenland Ice Sheet in a Changing Climate: Snow, Water, Ice and Permafrost in the Arctic (SWIPA). Arctic Monitoring and Assessment Programme (AMAP), 22 pp.

Armour, K. C., I. Eisenman, E. Blanchard-Wrigglesworth, K. E. McCusker, and C. M. Bitz, 2011: The reversibility of sea ice loss in a state-of-the-art climate model. *Geophys. Res. Lett.*, **38**.

Arrhenius, S., 1896: On the influence of carbonic acid in the air upon the temperature of the ground. *Philos. Mag.*, **41**, 237–276.

Baede, A. P. M., E. Ahlonsou, Y. Ding, and D. Schimel, 2001: The climate system: An overview. In: *Climate Change 2001: The Scientific Basis. Contribution of Working Group I to the Third Assessment Report of the Intergovernmental Panel on Climate Change* [J. T. Houghton, Y. Ding, D. J. Griggs, M. Noquer, P. J. van der Linden, X. Dai, K. Maskell and C. A. Johnson (eds.)]. Cambridge University Press, Cambridge, United Kingdom and New York, NY, USA.

Beerling, D. J., and D. L. Royer, 2011: Convergent Cenozoic CO_2 history. *Nature Geosci.*, **4**, 418–420.

Bretherton, F. P., K. Bryan, and J. D. Woodes, 1990: Time-dependent greenhouse-gas-induced climate change. In: *Climate Change: The IPCC Scientific Assessment* [J. T. Houghton, G. J. Jenkins and J. J. Ephraums (eds.)]. Cambridge University Press, Cambridge, United Kingdom and New York, NY, USA, 177–193.

Brönnimann, S., T. Ewen, J. Luterbacher, H. F. Diaz, R. S. Stolarski, and U. Neu, 2008: A focus on climate during the past 100 years. In: *Climate Variability and Extremes during the Past 100 Years* [S. Brönnimann, J. Luterbacher, T. Ewen, H. F. Diaz, R. S. Stolarski and U. Neu (eds.)]. Springer Science+Business Media, Heidelberg, Germany and New York, NY, USA, pp. 1–25.

Broomell, S., and D. Budescu, 2009: Why are experts correlated? Decomposing correlations between judges. *Psychometrika*, **74**, 531–553.

Brunet, M., and P. Jones, 2011: Data rescue initiatives: Bringing historical climate data into the 21st century. *Clim. Res.*, **47**, 29–40.

Budescu, D., S. Broomell, and H.-H. Por, 2009: Improving communication of uncertainty in the reports of the Intergovernmental Panel on Climate Change. *Psychol. Sci.*, **20**, 299–308.

Byrne, R., S. Mecking, R. Feely, and X. Liu, 2010: Direct observations of basin-wide acidification of the North Pacific Ocean. *Geophys. Res. Lett.*, **37**.

CCSP, 2009: *Best Practice Approaches for Characterizing, Communicating, and Incorporating Scientific Uncertainty in Climate Decision Making.* U.S. Climate Change Science Program, Washington, DC, USA, 96 pp.

Church, J. A., and N. J. White, 2011: Sea-level rise from the late 19th to the early 21st century. *Surv. Geophys.*, **32**, 585–602.

Church, J. A., J. M. Gregory, N. J. White, S. M. Platten, and J. X. Mitrovica, 2011: Understanding and projecting sea level change. *Oceanography*, **24**, 130–143.

Cleveland, W. S., 1979: Robust locally weighted regression and smoothing scatterplots. *J. Am. Stat. Assoc.*, **74**, 829–836.

Collins, M., and M. R. Allen, 2002: Assessing the relative roles of initial and boundary conditions in interannual to decadal climate predictability. *J. Clim.*, **15**, 3104–3109.

Covey, C., et al., 2003: An overview of results from the Coupled Model Intercomparison Project. *Global Planet. Change*, **37**, 103–133.

Cubasch, U., et al., 2001: Projections of future climate change. In: *Climate Change 2001: The Scientific Basis. Contribution of Working Group I to the Third Assessment Report of the Intergovernmental Panel on Climate Change* [J. T. Houghton, Y. Ding, D. J. Griggs, M. Noquer, P. J. van der Linden, X. Dai, K. Maskell and C. A. Johnson (eds.)]. Cambridge University Press, Cambridge, United Kingdom and New York, NY, USA, 527–582.

Dee, D. P., et al., 2011: The ERA-Interim reanalysis: Configuration and performance of the data assimilation system. *Q. J. R. Meteorol. Soc.*, **137**, 553–597.

Denman, K. L., et al., 2007: Couplings between changes in the climate system and biogeochemistry. In: *Climate Change 2007: The Physical Science Basis. Contribution of Working Group I to the Fourth Assessment Report of the Intergovernmental Panel on Climate Change* [Solomon, S., D. Qin, M. Manning, Z. Chen, M. Marquis, K. B. Averyt, M. Tignor and H. L. Miller (eds.)]. Cambridge University Press, Cambridge, United Kingdom and New York, NY, USA, 501–587.

Dlugokencky, E. J., et al., 2009: Observational constraints on recent increases in the atmospheric CH_4 burden. *Geophys. Res. Lett.*, **36**, L18803.

Duarte, C. M., T. M. Lenton, P. Wadhams, and P. Wassmann, 2012: Commentary: Abrupt climate change in the Arctic. *Nature Clim. Change*, **2**, 60–62.

Easterling, D. R., and M. F. Wehner, 2009: Is the climate warming or cooling? *Geophys. Res. Lett.*, **36**, L08706.

Elkins, J., and G. Dutton, 2010: Nitrous oxide and sulfur hexaflouride. Section in State of the Climate in 2009. *Bull. Am. Meteorol. Soc.*, **91**, 44–45.

Ettema, J., M. R. van den Broeke, E. van Meijgaard, W. J. van de Berg, J. L. Bamber, J. E. Box, and R. C. Bales, 2009: Higher surface mass balance of the Greenland ice sheet revealed by high-resolution climate modeling. *Geophys. Res. Lett.*, **36**, L12501.

Foley, J., et al., 2005: Global consequences of land use. *Science*, **309**, 570–574.

Folland, C. K., et al., 2001: Observed climate variability and change. In: *Climate Change 2001: The Scientific Basis. Contribution of Working Group I to the Third Assessment Report of the Intergovernmental Panel on Climate Change* [J. T. Houghton, Y. Ding, D. J. Griggs, M. Noquer, P. J. van der Linden, X. Dai, K. Maskell and C. A. Johnson (eds.)]. Cambridge University Press, Cambridge, United Kingdom and New York, NY, USA, 101–181.

Forest, C. E., P. H. Stone, and A. P. Sokolov, 2008: Constraining climate model parameters from observed 20th century changes. *Tellus A*, **60**, 911–920.

Forster, P., et al., 2007: Changes in atmospheric constituents and in radiative forcing. In: *Climate Change 2007: The Physical Science Basis. Contribution of Working Group I to the Fourth Assessment Report of the Intergovernmental Panel on Climate Change* [Solomon, S., D. Qin, M. Manning, Z. Chen, M. Marquis, K. B. Averyt, M. Tignor and H. L. Miller (eds.)]. Cambridge University Press, Cambridge, United Kingdom and New York, NY, USA, 131–234.

Frame, D. J., and D. A. Stone, 2013: Assessment of the first consensus prediction on climate change. *Nature Clim. Change*, **3**, 357–359.

Frame, D. J., D. A. Stone, P. A. Stott, and M. R. Allen, 2006: Alternatives to stabilization scenarios. *Geophys. Res. Lett.*, **33**.

Fujino, J., R. Nair, M. Kainuma, T. Masui, and Y. Matsuoka, 2006: Multi-gas mitigation analysis on stabilization scenarios using aim global model. *Energy J.*, **0**, 343–353.

GCOS, 2009: Progress Report on the Implementation of the Global Observing System for Climate in Support of the UNFCCC 2004–2008, GCOS-129 (WMO/TD-No. 1489; GOOS-173; GTOS-70) , Geneva, Switzerland.

GCOS , 2011: Systematic Observation Requirements for Satellite-based Products for Climate Supplemental details to the satellite-based component of the Implementation Plan for the Global Observing System for Climate in Support of the UNFCCC – 2011 Update, (GCOS-154) – December 2011, Geneva, Switzerland.

Goosse, H., W. Lefebvre, A. de Montety, E. Crespin, and A. H. Orsi, 2009: Consistent past half-century trends in the atmosphere, the sea ice and the ocean at high southern latitudes. *Clim. Dyn.*, **33**, 999–1016.

Granier, C., et al., 2011: Evolution of anthropogenic and biomass burning emissions of air pollutants at global and regional scales during the 1980–2010 period. *Clim. Change*, **109**, 163–190.

Haas, C., A. Pfaffling, S. Hendricks, L. Rabenstein, J. L. Etienne, and I. Rigor, 2008: Reduced ice thickness in Arctic Transpolar Drift favors rapid ice retreat. *Geophys. Res. Lett.*, **35**, L17501.

Hansen, J., D. Johnson, A. Lacis, S. Lebedeff, P. Lee, D. Rind, and G. Russell, 1981: Climate impact of increasing atmospheric carbon dioxide. *Science*, **213**, 957–966.

Hansen, J., M. Sato, R. Ruedy, K. Lo, D. W. Lea, and M. Medina-Elizade, 2006: Global temperature change. *Proc. Natl. Acad. Sci. U.S.A.*, **103**, 14288–14293.

Hansen, J., R. Ruedy, M. Sato, and K. Lo, 2010: Global surface temperature change. *Rev. Geophys.*, **48**, RG4004.

Hansen, J., M. Sato, P. Kharecha, and K. von Schuckmann, 2011: Earth's energy imbalance and implications. *Atmos. Chem. Phys.*, **11**, 13421–13449.

Hawkins, E., and R. Sutton, 2009: The potential to narrow uncertainty in regional climate predictions. *Bull. Am. Meteorol. Soc.*, **90**, 1095–1107.

Hawkins, E., and R. Sutton, 2011: The potential to narrow uncertainty in projections of regional precipitation change. *Clim. Dyn.*, **37**, 407–418.

Hegerl, G. C., et al., 2007: Understanding and attributing climate change. In: *Climate Change 2007: The Physical Science Basis. Contribution of Working Group I to the Fourth Assessment Report of the Intergovernmental Panel on Climate Change* [Solomon, S., D. Qin, M. Manning, Z. Chen, M. Marquis, K. B. Averyt, M. Tignor and H. L. Miller (eds.)]. Cambridge University Press, Cambridge, United Kingdom and New York, NY, USA, 665–745.

Hijioka, Y., Y. Matsuoka, H. Nishomoto, M. Masui, and M. Kainuma, 2008: Global GHG emission scenarios under GHG concentration stabilization targets. *JGEE*, **13**, 97–108.

Hoelzle, M., G. Darms, M. P. Lüthi, and S. Suter, 2011: Evidence of accelerated englacial warming in the Monte Rosa area, Switzerland/Italy. *Cryosphere*, **5**, 231–243.

Houghton, R., 2003: Revised estimates of the annual net flux of carbon to the atmosphere from changes in land use and land management 1850–2000. *Tellus B*, **55**, 378–390.

Huntingford, C., J. Lowe, B. Booth, C. Jones, G. Harris, L. Gohar, and P. Meir, 2009: Contributions of carbon cycle uncertainty to future climate projection spread. *Tellus B*, doi:10.1111/j.1600–0889.2009.00414.x, 355–360.

Hurtt, G. C., et al., 2011: Harmonization of land-use scenarios for the period 1500–2100: 600 years of global gridded annual land-use transitions, wood harvest, and resulting secondary lands. *Clim. Change*, **109**, 117–161.

InterAcademy Council, 2010: Climate change assessments. In: *Review of the Processes and Procedures of the IPCC*, Amsterdam, The Netherlands.

IPCC, 1990: *Climate Change: The IPCC Scientific Assessment* [J. T. Houghton, G. J. Jenkins and J. J. Ephraums (eds.)]. Cambridge University Press, Cambridge, United Kingdom and New York, NY, USA, 212 pp.

IPCC , 1996: *Climate Change 1995: The Science of Climate Change. Contribution of Working Group I to the Second Assessment Report of the Intergovernmental Panel on Climate Change*. Cambridge University Press, Cambridge, United Kingdom and New York, NY, USA, 584 pp.

IPCC , 2000: IPCC Special Report on Emissions Scenarios. Prepared by Working Group III of the Intergovernmental Panel on Climate Change, Cambridge University Press, Cambrudge, United Kingdom, pp 570.

IPCC , 2001: *Climate Change 2001: The Scientific Basis. Contribution of Working Group I to the Third Assessment Report of the Intergovernmental Panel on Climate Change* [J. T. Houghton, Y. Ding, D. J. Griggs, M. Noquer, P. J. van der Linden, X. Dai, K. Maskell and C. A. Johnson (eds.)]. Cambridge University Press, Cambridge, United Kingdom and New York, NY, USA, 881 pp.

IPCC , 2007: *Climate Change 2007: The Physical Science Basis. Contribution of Working Group I to the Fourth Assessment Report of the Intergovernmental Panel on Climate Change (IPCC)* [Solomon, S., D. Qin, M. Manning, Z. Chen, M. Marquis, K. B. Averyt, M. Tignor and H. L. Miller (eds.)]. Cambridge University Press, 996 pp.

IPCC , 2012a: Procedures for the preparation, review, acceptance, adoption, approval and publication of IPCC reports. Appendix A to the Principles Governing IPCC Work, Geneva, Switzerland, 6-9 June 2012, 29 pp.

IPCC , 2012b: *Managing the Risks of Extreme Events and Disasters to Advance Climate Change Adaptation. Special Report of the Intergovernmental Panel on Climate Change*. [Field, C. B., V. Barros, T. F. Stocker, D. Qin, D. J. Dokken, K. L. Ebi, M. D. Mastrandrea, K. J. Mach, G.-K. Plattner, S. K. Allen, M. Tignor, and P. M. Midgley (Eds.)]. Cambridge University Press, Cambridge, United Kingdom, 582 pp.

JCGM, 2008: JCGM 100: 2008. GUM 1995 with minor corrections. Evaluation of measurement data—Guide to the expression of uncertainty in measurement. Joint Committee for Guides in Metrology.

Jevrejeva, S., J. C. Moore, A. Grinsted, and P. L. Woodworth, 2008: Recent global sea level acceleration started over 200 years ago? *Geophys. Res. Lett.*, **35**, L08715.

Kahneman, D., and A. Tversky, 1979: Prospect theory: An analysis of decision under risk. *Econometrica*, **47**, 263–291.

Kahneman, D., P. Slovic, and A. Tversky, Eds., 1982: *Judgment under Uncertainty: Heuristics and Biases*. Cambridge University Press, Cambridge, United Kingdom and New York, NY, USA, 544 pp.

Kandlikar, M., J. Risbey, and S. Dessai, 2005: Representing and communicating deep uncertainty in climate-change assessments. *C. R. Geosci.*, **337**, 443–455.

Knutti, R., F. Joos, S. A. Müller, G. K. Plattner, and T. F. Stocker, 2005: Probabilistic climate change projections for CO_2 stabilization profiles. *Geophys. Res. Lett.*, **32**, L20707.

Knutti, R., et al., 2008: A review of uncertainties in global temperature projections over the twenty-first century. *J. Clim.*, **21**, 2651–2663.

Kopp, G., and J. L. Lean, 2011: A new, lower value of total solar irradiance: Evidence and climate significance. *Geophys. Res. Lett.*, **38**, L01706.

Kwok, R., G. F. Cunningham, M. Wensnahan, I. Rigor, H. J. Zwally, and D. Yi, 2009: Thinning and volume loss of the Arctic Ocean sea ice cover: 2003–2008. *J. Geophys. Res. Oceans*, **114**, C07005.

Lamarque, J. F., et al., 2011: Global and regional evolution of short-lived radiatively-active gases and aerosols in the Representative Concentration Pathways. *Clim. Change*, **109**, 191–212.

Lambert, F., et al., 2008: Dust-climate couplings over the past 800,000 years from the EPICA Dome C ice core. *Nature*, **452**, 616–619.

Lemke, P., et al., 2007: Observations: Changes in snow, ice and frozen ground. In: *Climate Change 2007: The Physical Science Basis. Contribution of Working Group I to the Fourth Assessment Report of the Intergovernmental Panel on Climate Change* [Solomon, S., D. Qin, M. Manning, Z. Chen, M. Marquis, K. B. Averyt, M. Tignor and H. L. Miller (eds.)]. Cambridge University Press, Cambridge, United Kingdom and New York, NY, USA, 339–383.

Lenton, T., H. Held, E. Kriegler, J. Hall, W. Lucht, S. Rahmstorf, and H. Schellnhuber, 2008: Tipping elements in the Earth's climate system. *Proc. Natl. Acad. Sci.U.S.A.*, **105**, 1786–1793.

Le Treut, H., et al., 2007: Historical Overview of Climate Change. In: *Climate Change 2007: The Physical Science Basis. Contribution of Working Group I to the Fourth Assessment Report of the Intergovernmental Panel on Climate Change* [Solomon, S., D. Qin, M. Manning, Z. Chen, M. Marquis, K. B. Averyt, M. Tignor and H. L. Miller (eds.)]. Cambridge University Press, Cambridge, United Kingdom and New York, NY, USA, pp. 94–127.

Lüthi, M., and M. Funk, 2001: Modelling heat flow in a cold, high-altitude glacier: Interpretation of measurements from Colle Gnifetti, Swiss Alps. *J. Glaciol.*, **47**, 314–324.

Lüthi, D., et al., 2008: High-resolution carbon dioxide concentration record 650,000–800,000 years before present. *Nature*, **453**, 379–382.

Mann, M., Z. Zhang, M. Hughes, R. Bradley, S. Miller, S. Rutherford, and F. Ni, 2008: Proxy-based reconstructions of hemispheric and global surface temperature variations over the past two millennia. *Proc. Natl. Acad. Sci. U.S.A.*, **105**, 13252–13257.

Manning, M., et al., 2004: *IPCC workshop Report: Describing scientific uncertainties in climate change to support analysis of risk and of options* [IPCC IPCC Working Group I Technical Support Unit (ed.)]. Available at http://www.ipcc.ch/ (accessed 07-10-2013), 138.

Masson, D., and R. Knutti, 2011: Climate model genealogy. *Geophys. Res. Lett.*, **38**, L08703.

Mastrandrea, M. D., et al., 2010: Guidance notes for lead authors of the IPCC Fifth Assessment Report on Consistent Treatment of Uncertainties. Available at http://www.ipcc.ch (accessed 07-10-2013).

Masui, T., et al., 2011: An emission pathway for stabilization at 6 Wm^{-2} radiative forcing. *Clim. Change*, **109**, 59–76.

Matthews, H. D., and A. J. Weaver, 2010: Committed climate warming. *Nature Geosci.*, **3**, 142–143.

Meehl, G. A., et al., 2007: Global climate projections. In: *Climate Change 2007: The Physical Science Basis. Contribution of Working Group I to the Fourth Assessment Report of the Intergovernmental Panel on Climate Change* [Solomon, S., D. Qin, M. Manning, Z. Chen, M. Marquis, K. B. Averyt, M. Tignor and H. L. Miller (eds.)]. Cambridge University Press, Cambridge, United Kingdom and New York, NY, USA, 749–845.

Meinshausen, M., et al., 2011: The RCP greenhouse gas concentrations and their extensions from 1765 to 2300. *Clim. Change*, **109**, 213–241.

Menne, M. J., I. Durre, R. S. Vose, B. E. Gleason, and T. G. Houston, 2012: An overview of the Global Historical Climatology Network-Daily Database. *J. Atmos. Ocean. Technol.*, **29**, 897–910.

Mernild, S. H., G. E. Liston, C. A. Hiemstra, K. Steffen, E. Hanna, and J. H. Christensen, 2009: Greenland ice sheet surface mass-balance modelling and freshwater flux for 2007, and in a 1995–2007 perspective. *Hydrol. Proc,*, **23**, 2470–2484.

Midorikawa, T., et al., 2010: Decreasing pH trend estimated from 25-yr time series of carbonate parameters in the western North Pacific. *Tellus B*, **62**, 649–659.

Morgan, M. G., M. Henrion, and M. Small, 1990: *Uncertainty: A Guide to Dealing with Uncertainty in Quantitative Risk and Policy Analysis.* Cambridge University Press, Cambridge, United Kingdom and New York, NY, USA, 332 pp.

Morice, C. P., J. J. Kennedy, N. A. Rayner, and P. D. Jones, 2012: Quantifying uncertainties in global and regional temperature change using an ensemble of observational estimates: The HadCRUT4 data set. *J. Geophys. Res. Atmos.*, **117**, D08101.

Moss, R. H., and S. H. Schneider, 2000: Uncertainties in the IPCC TAR: Recommendations to lead authors for more consistent assessment and reporting. In: *Guidance Papers on the Cross Cutting Issues of the Third Assessment Report of the IPCC*. World Meteorological Organization, Geneva, pp. 33–51.

Moss, R., et al., 2008: *Towards New Scenarios for Analysis of Emissions, Climate Change, Impacts, and Response Strategies*. Geneva, Intergovernmental Panel on Climate Change, 132 pp.

Moss, R., et al., 2010: The next generation of scenarios for climate change research and assessment. *Nature*, **463**, 747–756.

Murphy, D., S. Solomon, R. Portmann, K. Rosenlof, P. Forster, and T. Wong, 2009: An observationally based energy balance for the Earth since 1950. *J. Geophys. Res. Atmos.*, **114**, D17107.

Myhre, G., E. Highwood, K. Shine, and F. Stordal, 1998: New estimates of radiative forcing due to well mixed greenhouse gases. *Geophys. Res. Lett.*, **25**, 2715–2718.

Nghiem, S. V., et al., 2012: The extreme melt across the Greenland ice sheet in 2012. *Geophys. Res. Lett.*, **39**, L20502.

Oreskes, N., K. Shrader-Frechette, and K. Belitz, 1994: Verification, validation, and confirmation of numerical models in the earth sciences. *Science*, 263, 641–646.

Pall, P., et al., 2011: Anthropogenic greenhouse gas contribution to flood risk in England and Wales in autumn 2000. *Nature*, **470**, 382–385.

Patt, A. G., and D. P. Schrag, 2003: Using specific language to describe risk and probability. *Clim. Change*, 61, 17–30.

Patt, A. G., and S. Dessai, 2005: Communicating uncertainty: Lessons learned and suggestions for climate change assessment. *C. R. Geosci.*, **337**, 425–441.

Pennell, C., and T. Reichler, 2011: On the effective number of climate models. *J. Clim.*, **24**, 2358–2367.

Peterson, T. C., P. A. Stott, and S. Herring, 2012: Explaining extreme events of 2011 from a climate perspective. *Bull. Am. Meteorol. Soc.*, **93**, 1041–1067.

Peterson, T. C., et al., 2008: Why weather and climate extremes matter. In: *Weather and Climate Extremes in a Changing Climate. Regions of Focus: North America, Hawaii, Caribbean, and U.S. Pacific Islands*, [Karl, T. R., G. A. Meehl, C. D. Miller, S. J. Hassol, A. M. Waple, and W. L. Murray (eds.)]. A Report by the U.S.Climate Change Science Program and the Subcommittee on Global Change Research, Washington, DC., USA, 11–33.

Rahmstorf, S., G. Foster, and A. Cazenave, 2012: Comparing climate projections to observations up to 2011. *Environ. Res. Lett.*, **7**, 044035.

Ray, R. D., and B. C. Douglas, 2011: Experiments in reconstructing twentieth-century sea levels. *Prog. Oceanogr.*, **91**, 496–515.

Riahi, K., A. Grübler, and N. Nakicenovic, 2007: Scenarios of long-term socio-economic and environmental development under climate stabilization. *Technol. ForecastSoc Change*, **74**, 887–935.

Riahi, K., et al., 2011: RCP 8.5–A scenario of comparatively high greenhouse gas emissions. *Clim. Change*, **109**, 33–57.

Risbey, J. S., and M. Kandlikar, 2007: Expressions of likelihood and confidence in the IPCC uncertainty assessment process. *Clim. Change*, **85**, 19–31.

Schimel, D., et al., 1996: Radiative forcing of climate change. In: *Climate Change 1995: The Science of Climate Change, Contribution of Working Group I to the Second Assessment Report of the Intergovernmental Panel on Climate Change* [J. T. Houghton, L. G. Meiro Filho, B. A. Callander, N. Harris, A. Kattenburg and K. Maskell (eds.)]. Cambridge University Press, Cambridge, United Kingdom and New York, NY, USA, 69–131.

Scott, J. T. B., G. H. Gudmundsson, A. M. Smith, R. G. Bingham, H. D. Pritchard, and D. G. Vaughan, 2009: Increased rate of acceleration on Pine Island Glacier strongly coupled to changes in gravitational driving stress. *The Cryosphere*, **3**, 125–131.

Seneviratne, S., et al., 2010: Investigating soil moisture-climate interactions in a changing climate: A review. *Earth-Sci. Rev.*, **99**, 125–161.

Seneviratne, S. I., et al., 2012: Chapter 3: Changes in climate extremes and their Impacts on the Natural Physical Environment. In: *SREX: Special Report on Managing the Risks of Extreme Events and Disasters to Advance Climate Change Adaptation* [C. B. Field, et al. (eds.]. Cambridge University Press, Cambridge, United Kingdom and New York, NY, USA, pp.109–230.

Smith, S., and T. Wigley, 2006: Multi-gas forcing stabilization with Minicam. *Energy J.*, 373–391.

Smith, T. M., R. W. Reynolds, T. C. Peterson, and J. Lawrimore, 2008: Improvements to NOAA's historical merged land-ocean surface temperature analysis (1880–2005). *J. Clim.*, **21**, 2283–2296.

Steinhilber, F., J. Beer, and C. Fröhlich, 2009: Total solar irradiance during the Holocene. *Geophys. Res. Lett.*, **36**, L19704.

SWIPA, 2011: Snow, water, ice and permafrost in the Arctic. SWIPA 2011 Executive Summary. AMAP, Oslo, Norway, 16 pp.

Taylor, K. E., R. J. Stouffer, and G. A. Meehl, 2012: An overview of CMIP5 and the experiment design. *Bull. Am. Meteorol. Soc.*, **93**, 485–498.

Tedesco, M., 2007: A new record in 2007 for melting in Greenland. *EOS, Trans. Am. Geophys. Union*, **88**, 383.

Thomson, A. M., et al., 2011: RCP4.5: A pathway for stabilization of radiative forcing by 2100. *Clim. Change*, **109**, 77–94.

Tietsche, S., D. Notz, J. H. Jungclaus, and J. Marotzke, 2011: Recovery mechanisms of Arctic summer sea ice. *Geophys. Res. Lett.*, **38**, L02707.

Trenberth, K. E., J. T. Fasullo, and J. Kiehl, 2009: Earth's global energy budget. *Bull. Am. Meteorol. Soc.*, **90**, 311–323.

Turner, J., and J. E. Overland, 2009: Contrasting climate change in the two polar regions. *Polar Res.*, **28**, 146–164.

Turner, J., et al., 2009: Antarctic Climate Change and the environment. Scientific Committee on Antarctic Research, Cambridge, United Kingdom, 526 pp.

van Vuuren, D., et al., 2007: Stabilizing greenhouse gas concentrations at low levels: An assessment of reduction strategies and costs. *Clim. Change*, **81**, 119–159.

van Vuuren, D. P., et al., 2011a: RCP2.6: Exploring the possibility to keep global mean temperature increase below 2°C. *Clim. Change*, **109**, 95–116.

van Vuuren, D. P., et al., 2011b: The representative concentration pathways: An overview. *Clim. Change*, **109**, 5–31.

Wadhams, P., 2012: Arctic ice cover, ice thickness and tipping points. *Ambio*, **41**, 23–33.

Warrick, R., and J. Oerlemans, 1990: Sea level rise. In: *Climate Change: The IPCC Scientific Assessment* [J. T. Houghton, G. J. Jenkins and J. J. Ephraums (eds.)]. Cambridge University Press, Cambridge, United Kingdom and New York, NY, USA, 261–281.

Wigley, T. M. L., 2005: The climate change commitment. *Science*, 307, 1766–1769.

Willis, J. K., J. M. Lyman, G. C. Johnson, and J. Gilson, 2009: In situ data biases and recent ocean heat content variability. *J. Atmos. Ocean. Technol.*, **26**, 846–852.

Willis, J., D. Chambers, C. Kuo, and C. Shum, 2010: Global sea level rise recent progress and challenges for the decade to come. *Oceanography*, **23**, 26–35.

Wise, M., et al., 2009: Implications of limiting CO_2 concentrations for land use and energy. *Science*, **324**, 1183–1186.

Yip, S., C. A. T. Ferro, D. B. Stephenson, and E. Hawkins, 2011: A simple, coherent framework for partitioning uncertainty in climate predictions. *J. Climate*, **24**, 4634–4643.

Zemp, M., I. Roer, A. Kääb, M. Hoelzle, F. Paul, and W. Haeberli, 2008: Global glacier changes: Facts and figures. United Nations Environment Programme and World Glacier Monitoring Service, 88 pp.

Zemp, M., M. Hoelzle, and W. Haeberli, 2009: Six decades of glacier mass-balance observations: A review of the worldwide monitoring network. *Ann. Glaciol.*, **50**, 101–111.

Zhang, X., and F. Zwiers, 2012: Statistical indices for the diagnosing and detecting changes in extremes. In: *Extremes in a Changing Climate: Detection, Analysis and Uncertainty* [A. AghaKouchak, D. Easterling, K. Hsu, S. Schubert, and S. Sorooshian (eds.)]. Springer Science+Business Media, Heidelberg, Germany and New York, NY, USA, 1–14.

Appendix 1.A: Notes and Technical Details on Figures Displayed in Chapter 1

Figure 1.4: Documentation of Data Sources

Observed Temperature

NASA GISS evaluation of the observations: Hansen et al. (2010) updated: The data were downloaded from http://data.giss.nasa.gov/gistemp/tabledata_v3/GLB.Ts+dSST.txt. Annual means are used (January to December) and anomalies are calculated relative to 1961–1990.

NOAA NCDC evaluation of the observations: Smith et al. (2008) updated: The data were downloaded from ftp://ftp.ncdc.noaa.gov/pub/data/anomalies/annual.land_ocean.90S.90N.df_1901–2000mean.dat. Annual mean anomalies are calculated relative to 1961–1990.

Hadley Centre evaluation of the observations: Morice et al. (2012): The data were downloaded from http://www.metoffice.gov.uk/hadobs/hadcrut4/data/current/download.html#regional_series. Annual mean anomalies are calculated relative to 1961–1990 based on the ensemble median.

IPCC Range of Projections

Table 1.A.1 | FAR: The data have been digitized using a graphics tool from FAR Chapter 6, Figure 6.11 (Bretherton et al., 1990) in 5-year increments as anomalies relative to 1990 (°C).

Year	Lower Bound (Scenario D)	Upper Bound (Business as Usual)
1990	0.00	0.00
1995	0.09	0.14
2000	0.15	0.30
2005	0.23	0.53
2010	0.28	0.72
2015	0.33	0.91
2020	0.39	1.11
2025	0.45	1.34
2030	0.52	1.58
2035	0.58	1.86

Table 1.A.2 | SAR: The data have been digitized using a graphics tool from Figure 19 of the TS (IPCC, 1996) in 5-year increments as anomalies relative to 1990. The scenarios include changes in aerosols beyond 1990 (°C).

Year	Lower Bound (IS92c/1.5)	Upper Bound (IS92e/4.5)
1990	0.00	0.00
1995	0.05	0.09
2000	0.11	0.17
2005	0.16	0.28
2010	0.19	0.38
2015	0.23	0.47
2020	0.27	0.57
2025	0.31	0.67
2030	0.36	0.79
2035	0.41	0.92

Table 1.A.3 | TAR: The data have been digitized using a graphics tool from Figure 9.13(b) (Cubasch et al., 2001) in 5-year increments based on the GFDL_R15_a and DOE PCM parameter settings (°C).

Year	Lower Bound	Upper Bound
1990	0.00	0.00
1995	0.05	0.09
2000	0.11	0.20
2005	0.14	0.34
2010	0.17	0.52
2015	0.22	0.70
2020	0.28	0.87
2025	0.37	1.08
2030	0.43	1.28
2035	0.52	1.50

AR4: The temperature projections of the AR4 are presented for three SRES scenarios: B1, A1B and A2. Annual mean anomalies relative to 1961–1990 of the individual CMIP3 ensemble simulations (as used in AR4 SPM Figure SPM5) are shown. One outlier has been eliminated based on the advice of the model developers because of the model drift that leads to an unrealistic temperature evolution. As assessed by Meehl et al. (2007), the *likely* range for the temperature change is given by the ensemble mean temperature change +60% and –40% of the ensemble mean temperature change. Note that in the AR4 the uncertainty range was explicitly estimated for the end of the 21st century results. Here, it is shown for 2035. The time dependence of this range has been assessed in Knutti et al. (2008). The relative uncertainty is approximately constant over time in all estimates from different sources, except for the very early decades when natural variability is being considered (see Figure 3 in Knutti et al., 2008).

Data Processing

Observations

The observations are shown from 1950 to 2012 as annual mean anomaly relative to 1961–1990 (squares). For smoothing, first, the trend of each of the observational data sets was calculated by locally weighted scatter plot smoothing (Cleveland, 1979; f = 1/3). Then, the 11-year running means of the residuals were determined with reflected ends for the last 5 years. Finally, the trend was added back to the 11-year running means of the residuals.

Projections

For FAR, SAR and TAR, the projections have been harmonized to match the average of the three smoothed observational data sets at 1990.

1

Figure 1.5: Documentation of Data Sources

Observed CO_2 Concentrations

Global annual mean CO_2 concentrations are presented as annual mean values from Annex II Table AII.1.1a.

IPCC Range of Projections

Table 1.A.4 | FAR: The data have been digitized using a graphics tool from Figure A.3 (Annex, IPCC, 1990) as anomalies compared to 1990 in 5-year increments (ppm) and the observed 1990 value (353.6) has been added.

Year	Lower Bound (Scenario D)	Upper Bound (Business as Usual)
1990	353.6	353.6
1995	362.8	363.7
2000	370.6	373.3
2005	376.5	386.5
2010	383.2	401.5
2015	390.2	414.3
2020	396.6	428.8
2025	401.5	442.0
2030	406.0	460.7
2035	410.0	480.3

Table 1.A.5 | SAR: The data have been digitized using a graphics tool from Figure 5b in the TS (IPCC, 1996) in 5-year increments (ppm) as anomalies compared to 1990 and the observed 1990 value (353.6) has been added.

Year	Lower Bound (IS92c)	Upper Bound (IS92e)
1990	353.6	353.6
1995	358.4	359.0
2000	366.8	369.2
2005	373.7	380.4
2010	382.3	392.9
2015	391.4	408.0
2020	400.7	423.0
2025	408.0	439.6
2030	416.9	457.7
2035	424.5	477.7

TAR: The data were taken in 10-year increments from table Appendix II (IPCC, 2001) SRES Data Tables Table II.2.1 (ISAM model high and low setting). The scenarios that give the upper bound or lower bound respectively vary over time.

AR4: The data used was obtained from Figure 10.26 in Chapter 10 of AR4 (Meehl et al., 2007, provided by Malte Meinshausen). Annual means are used.

Data Processing

The projections have been harmonized to start from the observed value in 1990.

Figure 1.6: Documentation of Data Sources

Observed CH_4 Concentrations

Global annual mean CH_4 concentrations are presented as annual mean values from Annex II Table AII.1.1a.

IPCC Range of Projections

Table 1.A.6 | FAR: The data have been digitized using a graphics tool from FAR SPM Figure 5 (IPCC, 1990) in 5-year increments (ppb) as anomalies compared to 1990 the observed 1990 value (1714.4) has been added.

Year	Lower Bound (Scenario D)	Upper Bound (Business as Usual)
1990	1714.4	1714.4
1995	1775.7	1816.7
2000	1809.7	1938.7
2005	1819.0	2063.8
2010	1823.1	2191.1
2015	1832.3	2314.1
2020	1847.7	2441.3
2025	1857.9	2562.3
2030	1835.3	2691.6
2035	1819.0	2818.8

SAR: The data were taken in 5-year increments from Table 2.5a (Schimel et al., 1996). The scenarios that give the upper bound or lower bound respectively vary over time.

TAR: The data were taken in 10-year increments from Appendix II SRES Data Tables Table II.2.2 (IPCC, 2001). The upper bound is given by the A1p scenario, the lower bound by the B1p scenario.

AR4: The data used was obtained from Figure 10.26 in Chapter 10 of AR4 (Meehl et al., 2007, provided by Malte Meinshausen). Annual means are used.

Data Processing

The observations are shown as annual means. The projections have been harmonized to start from the same value in 1990.

Figure 1.7: Documentation of Data Sources

Observed N_2O Concentrations

Global annual mean N_2O concentrations are presented as annual mean values from Annex II Table AII.1.1a.

IPCC Range of Projections

Table 1.A.7: FAR | The data have been digitized using a graphics tool from FAR A.3 (Annex, IPCC, 1990) in 5-year increments (ppb) as anomalies compared to 1990 and the observed 1990 value (308.7) has been added.

Year	Lower Bound (Scenario D)	Upper Bound (Business as Usual)
1990	308.7	308.7
1995	311.7	313.2
2000	315.4	317.7
2005	318.8	322.9
2010	322.1	328.0
2015	325.2	333.0
2020	328.2	337.9
2025	331.7	343.0
2030	334.0	348.9
2035	336.1	354.1

SAR: The data were taken in 5-year increments from Table 2.5b (Schimel et al., 1996). The upper bound is given by the IS92e and IS92f scenario, the lower bound by the IS92d scenario.

TAR: The data were taken in 10-year increments from Appendix II SRES Data Tables Table II.2.3 (IPCC, 2001). The upper bound is given by the A1FI scenario, the lower bound by the B2 and A1T scenario.

AR4: The data used was obtained from Figure 10.26 in Chapter 10 of AR4 (Meehl et al., 2007, provided by Malte Meinshausen). Annual means are used.

Data Processing

The observations are shown as annual means. No smoothing is applied. The projections have been harmonized to start from the same value in 1990.

Figure 1.10: Documentation of Data Sources

Observed Global Mean Sea Level Rise

Three data sets based on tide gauge measurements are presented: Church and White (2011), Jevrejeva et al. (2008), and Ray and Douglas (2011). Annual mean anomalies are calculated relative to 1961–1990.

Estimates based on sea surface altimetry are presented as the ensemble mean of five different data sets (Section 3.7, Figure 3.13, Section 13.2, Figure 13.3) from 1993 to 2012. Annual means have been calculated. The data are harmonized to start from the mean of the three tide gauge based estimates (see above) at 1993.

IPCC Range of Projections

Table 1.A.8 | FAR: The data have been digitized using a graphics tool from Chapter 9, Figure 9.6 for the upper bound and Figure 9.7 for the lower bound (Warrick and Oerlemans, 1990) in 5-year increments as anomalies relative to 1990 (cm) and the observed anomaly relative to 1961–1990 (2.0 cm) has been added.

Year	Lower Bound (Scenario D)	Upper Bound (Business as Usual)
1990	2.0	2.0
1995	2.7	5.0
2000	3.7	7.9
2005	4.6	11.3
2010	5.5	15.0
2015	6.3	18.7
2020	6.9	22.8
2025	7.7	26.7
2030	8.4	30.9
2035	9.2	35.4

Table 1.A.9 | SAR: The data have been digitized using a graphics tool from Figure 21 (TS, IPCC, 1996) in 5-year increments as anomalies relative to 1990 (cm) and the observed anomaly relative to 1961–1990 (2.0 cm) has been added.

Year	Lower Bound (IS92c/1.5)	Upper Bound (IS92e/4.5)
1990	2.0	2.0
1995	2.4	4.3
2000	2.7	6.5
2005	3.1	9.0
2010	3.4	11.7
2015	3.8	14.9
2020	4.4	18.3
2025	5.1	21.8
2030	5.7	25.4
2035	6.4	29.2

TAR: The data are given in Table II.5.1 in 10-year increments. They are harmonized to start from mean of the observed anomaly relative to 1961–1990 at 1990 (2.0 cm).

1

AR4: The AR4 did not give a time-dependent estimate of sea level rise. These analyses have been conducted post AR4 by Church et al. (2011) based on the CMIP3 model results that were available at the time of AR4. Here, the SRES B1, A1B and A2 scenarios are shown from Church et al. (2011). The data start in 2001 and are given as anomalies with respect to 1990. They are displayed from 2001 to 2035, but the anomalies are harmonized to start from mean of the observed anomaly relative to 1961–1990 at 1990 (2.0 cm).

Data Processing

The observations are shown from 1950 to 2012 as the annual mean anomaly relative to 1961–1990 (squares) and smoothed (solid lines). For smoothing, first, the trend of each of the observational data sets was calculated by locally weighted scatterplot smoothing (Cleveland, 1979; f = 1/3). Then, the 11-year running means of the residuals were determined with reflected ends for the last 5 years. Finally, the trend was added back to the 11-year running means of the residuals.

2 Observations: Atmosphere and Surface

Coordinating Lead Authors:
Dennis L. Hartmann (USA), Albert M.G. Klein Tank (Netherlands), Matilde Rusticucci (Argentina)

Lead Authors:
Lisa V. Alexander (Australia), Stefan Brönnimann (Switzerland), Yassine Abdul-Rahman Charabi (Oman), Frank J. Dentener (EU/Netherlands), Edward J. Dlugokencky (USA), David R. Easterling (USA), Alexey Kaplan (USA), Brian J. Soden (USA), Peter W. Thorne (USA/Norway/UK), Martin Wild (Switzerland), Panmao Zhai (China)

Contributing Authors:
Robert Adler (USA), Richard Allan (UK), Robert Allan (UK), Donald Blake (USA), Owen Cooper (USA), Aiguo Dai (USA), Robert Davis (USA), Sean Davis (USA), Markus Donat (Australia), Vitali Fioletov (Canada), Erich Fischer (Switzerland), Leopold Haimberger (Austria), Ben Ho (USA), John Kennedy (UK), Elizabeth Kent (UK), Stefan Kinne (Germany), James Kossin (USA), Norman Loeb (USA), Carl Mears (USA), Christopher Merchant (UK), Steve Montzka (USA), Colin Morice (UK), Cathrine Lund Myhre (Norway), Joel Norris (USA), David Parker (UK), Bill Randel (USA), Andreas Richter (Germany), Matthew Rigby (UK), Ben Santer (USA), Dian Seidel (USA), Tom Smith (USA), David Stephenson (UK), Ryan Teuling (Netherlands), Junhong Wang (USA), Xiaolan Wang (Canada), Ray Weiss (USA), Kate Willett (UK), Simon Wood (UK)

Review Editors:
Jim Hurrell (USA), Jose Marengo (Brazil), Fredolin Tangang (Malaysia), Pedro Viterbo (Portugal)

This chapter should be cited as:
Hartmann, D.L., A.M.G. Klein Tank, M. Rusticucci, L.V. Alexander, S. Brönnimann, Y. Charabi, F.J. Dentener, E.J. Dlugokencky, D.R. Easterling, A. Kaplan, B.J. Soden, P.W. Thorne, M. Wild and P.M. Zhai, 2013: Observations: Atmosphere and Surface. In: *Climate Change 2013: The Physical Science Basis. Contribution of Working Group I to the Fifth Assessment Report of the Intergovernmental Panel on Climate Change* [Stocker, T.F., D. Qin, G.-K. Plattner, M. Tignor, S.K. Allen, J. Boschung, A. Nauels, Y. Xia, V. Bex and P.M. Midgley (eds.)]. Cambridge University Press, Cambridge, United Kingdom and New York, NY, USA.

Table of Contents

Supplementary Material is available in online versions of the report.

Executive Summary

The evidence of climate change from observations of the atmosphere and surface has grown significantly during recent years. At the same time new improved ways of characterizing and quantifying uncertainty have highlighted the challenges that remain for developing long-term global and regional climate quality data records. Currently, the observations of the atmosphere and surface indicate the following changes:

Atmospheric Composition

It is certain that atmospheric burdens of the well-mixed greenhouse gases (GHGs) targeted by the Kyoto Protocol increased from 2005 to 2011. The atmospheric abundance of carbon dioxide (CO_2) was 390.5 ppm (390.3 to 390.7)[1] in 2011; this is 40% greater than in 1750. Atmospheric nitrous oxide (N_2O) was 324.2 ppb (324.0 to 324.4) in 2011 and has increased by 20% since 1750. Average annual increases in CO_2 and N_2O from 2005 to 2011 are comparable to those observed from 1996 to 2005. Atmospheric methane (CH_4) was 1803.2 ppb (1801.2 to 1805.2) in 2011; this is 150% greater than before 1750. CH_4 began increasing in 2007 after remaining nearly constant from 1999 to 2006. Hydrofluorocarbons (HFCs), perfluorocarbons (PFCs), and sulphur hexafluoride (SF_6) all continue to increase relatively rapidly, but their contributions to radiative forcing are less than 1% of the total by well-mixed GHGs. {2.2.1.1}

For ozone-depleting substances (Montreal Protocol gases), it is certain that the global mean abundances of major chlorofluorocarbons (CFCs) are decreasing and HCFCs are increasing. Atmospheric burdens of major CFCs and some halons have decreased since 2005. HCFCs, which are transitional substitutes for CFCs, continue to increase, but the spatial distribution of their emissions is changing. {2.2.1.2}

Because of large variability and relatively short data records, *confidence*[2] in stratospheric H_2O vapour trends is *low*. Near-global satellite measurements of stratospheric water vapour show substantial variability but small net changes for 1992–2011. {2.2.2.1}

It is certain that global stratospheric ozone has declined from pre-1980 values. Most of the decline occurred prior to the mid 1990s; since then ozone has remained nearly constant at about 3.5% below the 1964–1980 level. {2.2.2.2}

***Confidence* is *medium* in large-scale increases of tropospheric ozone across the Northern Hemisphere (NH) since the 1970s. *Confidence* is *low* in ozone changes across the Southern Hemisphere (SH) owing to limited measurements.** It is *likely*[3] that surface ozone trends in eastern North America and Western Europe since 2000 have levelled off or decreased and that surface ozone strongly increased in East Asia since the 1990s. Satellite and surface observations of ozone precursor gases NO_x, CO, and non-methane volatile organic carbons indicate strong regional differences in trends. Most notably NO_2 has *likely* decreased by 30 to 50% in Europe and North America and increased by more than a factor of 2 in Asia since the mid-1990s. {2.2.2.3, 2.2.2.4}

It is *very likely* that aerosol column amounts have declined over Europe and the eastern USA since the mid 1990s and increased over eastern and southern Asia since 2000. These shifting aerosol regional patterns have been observed by remote sensing of aerosol optical depth (AOD), a measure of total atmospheric aerosol load. Declining aerosol loads over Europe and North America are consistent with ground-based *in situ* monitoring of particulate mass. *Confidence* in satellite based global average AOD trends is *low*. {2.2.3}

Radiation Budgets

Satellite records of top of the atmosphere radiation fluxes have been substantially extended since AR4, and it is *unlikely* that significant trends exist in global and tropical radiation budgets since 2000. Interannual variability in the Earth's energy imbalance related to El Niño-Southern Oscillation is consistent with ocean heat content records within observational uncertainty. {2.3.2}

Surface solar radiation *likely* underwent widespread decadal changes after 1950, with decreases ('dimming') until the 1980s and subsequent increases ('brightening') observed at many land-based sites. There is *medium confidence* for increasing downward thermal and net radiation at land-based observation sites since the early 1990s. {2.3.3}

Temperature

It is certain that Global Mean Surface Temperature has increased since the late 19th century. Each of the past three decades has been successively warmer at the Earth's surface than all the previous decades in the instrumental record, and the first decade of the 21st century has been the warmest. The globally averaged combined land and ocean surface temperature data as calculated by a linear trend, show a warming of 0.85 [0.65 to 1.06] °C, over the period 1880–2012, when multiple independently produced datasets exist, and

---

1 Values in parentheses are 90% confidence intervals. Elsewhere in this chapter usually the half-widths of the 90% confidence intervals are provided for the estimated change from the trend method.

2 In this Report, the following summary terms are used to describe the available evidence: limited, medium, or robust; and for the degree of agreement: low, medium, or high. A level of confidence is expressed using five qualifiers: very low, low, medium, high, and very high, and typeset in italics, e.g., *medium confidence*. For a given evidence and agreement statement, different confidence levels can be assigned, but increasing levels of evidence and degrees of agreement are correlated with increasing confidence (see Section 1.4 and Box TS.1 for more details).

3 In this Report, the following terms have been used to indicate the assessed likelihood of an outcome or a result: Virtually certain 99–100% probability, Very likely 90–100%, Likely 66–100%, About as likely as not 33–66%, Unlikely 0–33%, Very unlikely 0–10%, Exceptionally unlikely 0–1%. Additional terms (Extremely likely: 95–100%, More likely than not >50–100%, and Extremely unlikely 0–5%) may also be used when appropriate. Assessed likelihood is typeset in italics, e.g., *very likely* (see Section 1.4 and Box TS.1 for more details).

about 0.72°C [0.49°C to 0.89°C] over the period 1951–2012. The total increase between the average of the 1850–1900 period and the 2003–2012 period is 0.78 [0.72 to 0.85] °C and the total increase between the average of the 1850–1900 period and the reference period for projections, 1986–2005, is 0.61 [0.55 to 0.67] °C, based on the single longest dataset available. For the longest period when calculation of regional trends is sufficiently complete (1901–2012), almost the entire globe has experienced surface warming. In addition to robust multi-decadal warming, global mean surface temperature exhibits substantial decadal and interannual variability. Owing to natural variability, trends based on short records are very sensitive to the beginning and end dates and do not in general reflect long-term climate trends. As one example, the rate of warming over the past 15 years (1998–2012; 0.05 [–0.05 to +0.15] °C per decade), which begins with a strong El Niño, is smaller than the rate calculated since 1951 (1951–2012; 0.12 [0.08 to 0.14] °C per decade). Trends for 15-year periods starting in 1995, 1996, and 1997 are 0.13 [0.02 to 0.24], 0.14 [0.03 to 0.24] and 0.07 [–0.02 to 0.18], respectively. Several independently analyzed data records of global and regional land-surface air temperature (LSAT) obtained from station observations are in broad agreement that LSAT has increased. Sea surface temperatures (SSTs) have also increased. Intercomparisons of new SST data records obtained by different measurement methods, including satellite data, have resulted in better understanding of uncertainties and biases in the records. {2.4.1, 2.4.2, 2.4.3; Box 9.2}

It is *unlikely* that any uncorrected urban heat-island effects and land use change effects have raised the estimated centennial globally averaged LSAT trends by more than 10% of the reported trend. This is an average value; in some regions with rapid development, urban heat island and land use change impacts on regional trends may be substantially larger. {2.4.1.3}

***Confidence* is *medium* in reported decreases in observed global diurnal temperature range (DTR), noted as a key uncertainty in the AR4.** Several recent analyses of the raw data on which many previous analyses were based point to the potential for biases that differently affect maximum and minimum average temperatures. However, apparent changes in DTR are much smaller than reported changes in average temperatures and therefore it is *virtually certain* that maximum and minimum temperatures have increased since 1950. {2.4.1.2}

Based on multiple independent analyses of measurements from radiosondes and satellite sensors it is *virtually certain* that globally the troposphere has warmed and the stratosphere has cooled since the mid-20th century. Despite unanimous agreement on the sign of the trends, substantial disagreement exists among available estimates as to the rate of temperature changes, particularly outside the NH extratropical troposphere, which has been well sampled by radiosondes. Hence there is only *medium confidence* in the rate of change and its vertical structure in the NH extratropical troposphere and *low confidence* elsewhere. {2.4.4}

Hydrological Cycle

***Confidence* in precipitation change averaged over global land areas since 1901 is *low* for years prior to 1951 and *medium* afterwards. Averaged over the mid-latitude land areas of the Northern Hemisphere, precipitation has *likely* increased since 1901 (*medium confidence* before and *high confidence* after 1951).** For other latitudinal zones area-averaged long-term positive or negative trends have *low confidence* due to data quality, data completeness or disagreement amongst available estimates. {2.5.1.1, 2.5.1.2}

It is *very likely* that global near surface and tropospheric air specific humidity have increased since the 1970s. However, during recent years the near surface moistening over land has abated (*medium confidence*). As a result, fairly widespread decreases in relative humidity near the surface are observed over the land in recent years. {2.4.4, 2.5.4, 2.5.5}

While trends of cloud cover are consistent between independent data sets in certain regions, substantial ambiguity and therefore *low confidence* remains in the observations of global-scale cloud variability and trends. {2.5.6}

Extreme Events

It is *very likely* that the numbers of cold days and nights have decreased and the numbers of warm days and nights have increased globally since about 1950. There is only *medium confidence* that the length and frequency of warm spells, including heat waves, has increased since the middle of the 20th century mostly owing to lack of data or of studies in Africa and South America. However, it is *likely* that heatwave frequency has increased during this period in large parts of Europe, Asia and Australia. {2.6.1}

It is *likely* that since about 1950 the number of heavy precipitation events over land has increased in more regions than it has decreased. *Confidence* is *highest* for North America and Europe where there have been *likely* increases in either the frequency or intensity of heavy precipitation with some seasonal and/or regional variation. It is *very likely* that there have been trends towards heavier precipitation events in central North America. {2.6.2.1}

***Confidence* is *low* for a global-scale observed trend in drought or dryness (lack of rainfall) since the middle of the 20th century, owing to lack of direct observations, methodological uncertainties and geographical inconsistencies in the trends.** Based on updated studies, AR4 conclusions regarding global increasing trends in drought since the 1970s were probably overstated. However, this masks important regional changes: the frequency and intensity of drought have *likely* increased in the Mediterranean and West Africa and *likely* decreased in central North America and north-west Australia since 1950. {2.6.2.2}

***Confidence* remains *low* for long-term (centennial) changes in tropical cyclone activity, after accounting for past changes in observing capabilities.** However, it is *virtually certain* that the frequency and intensity of the strongest tropical cyclones in the North Atlantic has increased since the 1970s. {2.6.3}

***Confidence* in large-scale trends in storminess or storminess proxies over the last century is *low* owing to inconsistencies**

between studies or lack of long-term data in some parts of the world (particularly in the SH). {2.6.4}

Because of insufficient studies and data quality issues *confidence* is also *low* for trends in small-scale severe weather events such as hail or thunderstorms. {2.6.2.4}

Atmospheric Circulation and Indices of Variability

It is *likely* that circulation features have moved poleward since the 1970s, involving a widening of the tropical belt, a poleward shift of storm tracks and jet streams, and a contraction of the northern polar vortex. Evidence is more robust for the NH. It is *likely* that the Southern Annular Mode has become more positive since the 1950s. {2.7.5, 2.7.6, 2.7.8; Box 2.5}

Large variability on interannual to decadal time scales hampers robust conclusions on long-term changes in atmospheric circulation in many instances. *Confidence* is *high* that the increase in the northern mid-latitude westerly winds and the North Atlantic Oscillation (NAO) index from the 1950s to the 1990s and the weakening of the Pacific Walker circulation from the late 19th century to the 1990s have been largely offset by recent changes. {2.7.5, 2.7.8, Box 2.5}

***Confidence* in the existence of long-term changes in remaining aspects of the global circulation is *low* owing to observational limitations or limited understanding.** These include surface winds over land, the East Asian summer monsoon circulation, the tropical cold-point tropopause temperature and the strength of the Brewer Dobson circulation. {2.7.2, 2.7.4, 2.7.5, 2.7.7}

2.1 Introduction

This chapter assesses the scientific literature on atmospheric and surface observations since AR4 (IPCC, 2007). The most likely changes in physical climate variables or climate forcing agents are identified based on current knowledge, following the IPCC AR5 uncertainty guidance (Mastrandrea et al., 2011).

As described in AR4 (Trenberth et al., 2007), the climate comprises a variety of space- and timescales: from the diurnal cycle, to interannual variability such as the El Niño-Southern Oscillation (ENSO), to multi-decadal variations. 'Climate change' refers to a change in the state of the climate that can be identified by changes in the mean and/or the variability of its properties and that persists for an extended period of time (Annex III: Glossary). In this chapter, climate change is examined for the period with instrumental observations, since about 1850. Change prior to this date is assessed in Chapter 5. The word 'trend' is used to designate a long-term movement in a time series that may be regarded, together with the oscillation and random component, as composing the observed values (Annex III: Glossary). Where numerical values are given, they are equivalent linear changes (Box 2.2), though more complex nonlinear changes in the variable will often be clear from the description and plots of the time series.

In recent decades, advances in the global climate observing system have contributed to improved monitoring capabilities. In particular, satellites provide additional observations of climate change, which have been assessed in this and subsequent chapters together with more traditional ground-based and radiosonde observations. Since AR4, substantial developments have occurred including the production of revised data sets, more digital data records, and new data set efforts. New dynamical reanalysis data sets of the global atmosphere have been published (Box 2.3). These various innovations have improved understanding of data issues and uncertainties (Box 2.1).

Developing homogeneous long-term records from these different sources remains a challenge. The longest observational series are land surface air temperatures (LSATs) and sea surface temperatures (SSTs). Like all physical climate system measurements, they suffer from non-climatic artefacts that must be taken into account (Box 2.1). The global combined LSAT and SST remains an important climate change measure for several reasons. Climate sensitivity is typically assessed in the context of global mean surface temperature (GMST) responses to a doubling of CO_2 (Chapter 8) and GMST is thus a key metric in the climate change policy framework. Also, because it extends back in time farther than any other global instrumental series, GMST is key to understanding both the causes of change and the patterns, role and magnitude of natural variability (Chapter 10). Starting at various points in the 20th century, additional observations, including balloon-borne measurements and satellite measurements, and reanalysis products allow analyses of indicators such as atmospheric composition, radiation budgets, hydrological cycle changes, extreme event characterizations and circulation indices. A full understanding of the climate system characteristics and changes requires analyses of all such variables as well as ocean (Chapter 3) and cryosphere (Chapter 4) indicators. Through such a holistic analysis, a clearer and more robust assessment of the changing climate system emerges (FAQ 2.1).

This chapter starts with an assessment of the observations of the abundances of greenhouse gases (GHGs) and of aerosols, the main drivers of climate change (Section 2.2). Global trends in GHGs are indicative of the imbalance between sources and sinks in GHG budgets, and play an important role in emissions verification on a global scale. The radiative forcing (RF) effects of GHGs and aerosols are assessed in Chapter 8. The observed changes in radiation budgets are discussed in Section 2.3. Aerosol–cloud interactions are assessed in Chapter 7. Section 2.4 provides an assessment of observed changes in surface and atmospheric temperature. Observed change in the hydrological cycle, including precipitation and clouds, is assessed in Section 2.5. Changes in variability and extremes (such as cold spells, heat waves, droughts and tropical cyclones) are assessed in Section 2.6. Section 2.7 assesses observed changes in the circulation of the atmosphere and its modes of variability, which help determine seasonal and longer-term anomalies at regional scales (Chapter 14).

Trends have been assessed where possible for multi-decadal periods starting in 1880, 1901 (referred to as long-term trends) and in 1951, 1979 (referred to as short-term trends). The time elapsed since AR4 extends the period for trend calculation from 2005 to 2012 for many variables. The GMST trend since 1998 has also been considered (see also Box 9.2) as well as the trends for sequential 30-year segments of the time series. For many variables derived from satellite data, information is available for 1979–2012 only. In general, trend estimates are more reliable for longer time intervals, and trends computed on short intervals have a large uncertainty. Trends for short intervals are very sensitive to the start and end years. An exception to this is trends in GHGs, whose accurate measurement and long lifetimes make them well-mixed and less susceptible to year-to-year variability, so that trends computed on relatively short intervals are very meaningful for these variables. Where possible, the time interval 1961–1990 has been chosen as the climatological reference period (or normal period) for averaging. This choice enables direct comparisons with AR4, but is different from the present-day climate period (1986–2005) used as a reference in the modelling chapters of AR5 and Annex I: Atlas of Global and Regional Climate Projections.

It is important to note that the question of whether the observed changes are outside the possible range of natural internal climate variability and consistent with the climate effects from changes in atmospheric composition is not addressed in this chapter, but rather in Chapter 10. No attempt has been undertaken to further describe and interpret the observed changes in terms of multi-decadal oscillatory (or low-frequency) variations, (long-term) persistence and/or secular trends (e.g., as in Cohn and Lins, 2005; Koutsoyiannis and Montanari, 2007; Zorita et al., 2008; Lennartz and Bunde, 2009; Mills, 2010; Mann, 2011; Wu et al., 2011; Zhou and Tung, 2012; Tung and Zhou, 2013). In this chapter, the robustness of the observed changes is assessed in relation to various sources of observational uncertainty (Box 2.1). In addition, the reported trend significance and statistical confidence intervals provide an indication of how large the observed trend is compared to the range of observed variability in a given aspect of the climate system (see Box 2.2 for a description of the statistical trend model applied). Unless otherwise stated, 90% confidence intervals are given. The chapter also examines the physical consistency across different observations, which helps to provide additional confidence in the reported changes. Additional information about data sources and methods is described in the Supplementary Material to Chapter 2.

2.2 Changes in Atmospheric Composition

2.2.1 Well-Mixed Greenhouse Gases

AR4 (Forster et al., 2007; IPCC, 2007) concluded that increasing atmospheric burdens of well-mixed GHGs resulted in a 9% increase in their RF from 1998 to 2005. Since 2005, the atmospheric abundances of many well-mixed GHG increased further, but the burdens of some ozone-depleting substances (ODS) whose production and use were controlled by the Montreal Protocol on Substances that Deplete the Ozone Layer (1987; hereinafter, 'Montreal Protocol') decreased.

Based on updated *in situ* observations, this assessment concludes that these trends resulted in a 7.5% increase in RF from GHGs from 2005 to 2011, with carbon dioxide (CO_2) contributing 80%. Of note is an increase in the average growth rate of atmospheric methane (CH_4) from ~0.5 ppb yr^{-1} during 1999–2006 to ~6 ppb yr^{-1} from 2007 through 2011. Current observation networks are sufficient to quantify global annual mean burdens used to calculate RF and to constrain global emission rates (with knowledge of loss rates), but they are not sufficient for accurately estimating regional scale emissions and how they are changing with time.

The globally, annually averaged well-mixed GHG mole fractions reported here are used in Chapter 8 to calculate RF. A direct, inseparable connection exists between observed changes in atmospheric composition and well-mixed GHG emissions and losses (discussed in Chapter 6 for CO_2, CH_4, and N_2O). A global GHG budget consists of the total atmospheric burden, total global rate of production or emission (i.e., sources), and the total global rate of destruction or removal (i.e., sinks). Precise, accurate systematic observations from independent globally distributed measurement networks are used to estimate global annual mean well-mixed GHG mole fractions at the Earth's surface, and these allow estimates of global burdens. Emissions are predominantly from surface sources, which are described in Chapter 6 for CO_2, CH_4, and N_2O. Direct

Box 2.1 | Uncertainty in Observational Records

The vast majority of historical (and modern) weather observations were not made explicitly for climate monitoring purposes. Measurements have changed in nature as demands on the data, observing practices and technologies have evolved. These changes almost always alter the characteristics of observational records, changing their mean, their variability or both, such that it is necessary to process the raw measurements before they can be considered useful for assessing the true climate evolution. This is true of all observing techniques that measure physical atmospheric quantities. The uncertainty in observational records encompasses instrumental/recording errors, effects of representation (e.g., exposure, observing frequency or timing), as well as effects due to physical changes in the instrumentation (such as station relocations or new satellites). All further processing steps (transmission, storage, gridding, interpolating, averaging) also have their own particular uncertainties. Because there is no unique, unambiguous, way to identify and account for non-climatic artefacts in the vast majority of records, there must be a degree of uncertainty as to how the climate system has changed. The only exceptions are certain atmospheric composition and flux measurements whose measurements and uncertainties are rigorously tied through an unbroken chain to internationally recognized absolute measurement standards (e.g., the CO_2 record at Mauna Loa; Keeling et al., 1976a).

Uncertainty in data set production can result either from the choice of parameters within a particular analytical framework—parametric uncertainty, or from the choice of overall analytical framework— structural uncertainty. Structural uncertainty is best estimated by having multiple independent groups assess the same data using distinct approaches. More analyses assessed now than in AR4 include published estimates of parametric or structural uncertainty. It is important to note that the literature includes a very broad range of approaches. Great care has been taken in comparing the published uncertainty ranges as they almost always do not constitute a like-for-like comparison. In general, studies that account for multiple potential error sources in a rigorous manner yield larger uncertainty ranges. This yields an apparent paradox in interpretation as one might think that smaller uncertainty ranges should indicate a better product. However, in many cases this would be an incorrect inference as the smaller uncertainty range may instead reflect that the published estimate considered only a subset of the plausible sources of uncertainty. Within the timeseries figures, where this issue would be most acute, such parametric uncertainty estimates are therefore not generally included. Consistent with AR4 HadCRUT4 uncertainties in GMST are included in Figure 2.19, which in addition includes structural uncertainties in GMST.

To conclude, the vast majority of the raw observations used to monitor the state of the climate contain residual non-climatic influences. Removal of these influences cannot be done definitively and neither can the uncertainties be unambiguously assessed. Therefore, care is required in interpreting both data products and their stated uncertainty estimates. Confidence can be built from: redundancy in efforts to create products; data set heritage; and cross-comparisons of variables that would be expected to co-vary for physical reasons, such as LSATs and SSTs around coastlines. Finally, trends are often quoted as a way to synthesize the data into a single number. Uncertainties that arise from such a process and the choice of technique used within this chapter are described in more detail in Box 2.2.

use of observations of well-mixed GHG to model their regional budgets can also play an important role in verifying inventory estimates of emissions (Nisbet and Weiss, 2010).

Systematic measurements of well-mixed GHG in ambient air began at various times during the last six decades, with earlier atmospheric histories being reconstructed from measurements of air stored in air archives and trapped in polar ice cores or in firn. In contrast to the physical meteorological parameters discussed elsewhere in this chapter, measurements of well-mixed GHG are reported relative to standards developed from fundamental SI base units (SI = International System of Units) as dry-air mole fractions, a unit that is conserved with changes in temperature and pressure (Box 2.1). This eliminates dilution by H_2O vapour, which can reach 4% of total atmospheric composition. Here, the following abbreviations are used: ppm = µmol mol^{-1}; ppb = nmol mol^{-1}; and ppt = pmol mol^{-1}. Unless noted otherwise, averages of National Oceanic and Atmospheric Administration (NOAA) and Advanced Global Atmospheric Gases Experiment (AGAGE) annually averaged surface global mean mole fractions is described in Section 2.2.1 (see Supplementary Material 2.SM.2 for further species not listed here).

Table 2.1 summarizes globally, annually averaged well-mixed GHG mole fractions from four independent measurement programs. Sampling strategies and techniques for estimating global means and their uncertainties vary among programs. Differences among measurement programs are relatively small and will not add significantly to uncertainty in RF. Time series of the well-mixed GHG are plotted in Figures 2.1 (CO_2), 2.2 (CH_4), 2.3 (N_2O), and 2.4 (halogen-containing compounds).

2.2.1.1 Kyoto Protocol Gases (Carbon Dioxide, Methane, Nitrous Oxide, Hydrofluorocarbons, Perfluorocarbons and Sulphur Hexafluoride)

2.2.1.1.1 Carbon Dioxide

Precise, accurate systematic measurements of atmospheric CO_2 at Mauna Loa, Hawaii and South Pole were started by C. D. Keeling from Scripps Institution of Oceanography in the late 1950s (Keeling et al., 1976a; Keeling et al., 1976b). The 1750 globally averaged abundance of atmospheric CO_2 based on measurements of air extracted from ice cores and from firn is 278 ± 2 ppm (Etheridge et al., 1996). Globally averaged CO_2 mole fractions since the start of the instrumental record

Table 2.1 | Global annual mean surface dry-air mole fractions and their change since 2005 for well-mixed greenhouse gases (GHGs) from four measurement networks. Units are ppt except where noted. Uncertainties are 90% confidence intervals[a]. REs (radiative efficiency) and lifetimes (except CH_4 and N_2O, which are from Prather et al., 2012) are from Chapter 8.

			2011 Global Annual Mean			Global Increase from 2005 to 2011		
Species	**Lifetime (yr)**	**RE (W m^{-2} ppb^{-1})**	**UCI**	**SIO[b]/AGAGE**	**NOAA**	**UCI**	**SIO[b]/AGAGE**	**NOAA**
CO_2 (ppm)		1.37 × 10^{-5}		390.48 ± 0.28	390.44 ± 0.16		11.67 ± 0.37	11.66 ± 0.13
CH_4 (ppb)	9.1	3.63 × 10^{-4}	1798.1 ± 0.6	1803.1 ± 4.8	1803.2 ± 1.2	26.6 ± 0.9	28.9 ± 6.8	28.6 ± 0.9
N_2O (ppb)	131	3.03 × 10^{-3}		324.0 ± 0.1	324.3 ± 0.1		4.7 ± 0.2	5.24 ± 0.14
SF_6	3200	0.575		7.26 ± 0.02	7.31 ± 0.02		1.65 ± 0.03	1.64 ±0.01
CF_4	50,000	0.1		79.0 ± 0.1			4.0 ± 0.2	
C_2F_6	10,000	0.26		4.16 ± 0.02			0.50 ± 0.03	
HFC-125	28.2	0.219		9.58 ± 0.04			5.89 ± 0.07	
HFC-134a	13.4	0.159	63.4 ± 0.9	62.4 ± 0.3	63.0 ± 0.6	27.7 ± 1.4	28.2 ± 0.4	28.2 ± 0.1
HFC-143a	47.1	0.159		12.04 ± 0.07			6.39 ± 0.10	
HFC-152a	1.5	0.094		6.4 ± 0.1			3.0 ± 0.2	
HFC-23	222	0.176		24.0 ± 0.3			5.2 ± 0.6	
CFC-11	45	0.263	237.9 ± 0.8	236.9 ± 0.1	238.5 ± 0.2	–13.2 ± 0.8	–12.7 ± 0.2	–13.0 ± 0.1
CFC-12	100	0.32	525.3 ± 0.8	529.5 ± 0.2	527.4 ± 0.4	–12.8 ± 0.8	–13.4 ± 0.3	–14.1 ± 0.1
CFC-113	85	0.3	74.9 ± 0.6	74.29 ± 0.06	74.40 ± 0.04	–4.6 ± 0.8	–4.25 ± 0.08	–4.35 ±0.02
HCFC-22	11.9	0.2	209.0 ± 1.2	213.4 ± 0.8	213.2 ± 1.2	41.5 ± 1.4	44.6 ± 1.1	44.3 ± 0.2
HCFC-141b	9.2	0.152	20.8 ± 0.5	21.38 ± 0.09	21.4 ± 0.2	3.7 ± 0.5	3.70 ± 0.1	3.76 ± 0.03
HCFC-142b	17.2	0.186	21.0 ± 0.5	21.35 ± 0.06	21.0 ± 0.1	4.9 ± 0.5	5.72 ± 0.09	5.73 ± 0.04
CCl_4	26	0.175	87.8 ± 0.6	85.0 ± 0.1	86.5 ± 0.3	–6.4 ± 0.5	–6.9 ± 0.2	–7.8 ± 0.1
CH_3CCl_3	5	0.069	6.8 ± 0.6	6.3 ± 0.1	6.35 ± 0.07	–14.8 ± 0.5	–11.9 ± 0.2	–12.1 ± 0.1

Notes:

AGAGE = Advanced Global Atmospheric Gases Experiment; NOAA = National Oceanic and Atmospheric Administration, Earth System Research Laboratory, Global Monitoring Division; SIO = Scripps Institution of Oceanography, University of California, San Diego; UCI = University of California, Irvine, Department of Chemistry. HFC-125 = CHF_2CF_3; HFC-134a = CH_2FCF_3; HFC-143a = CH_3CF_3; HFC-152a = CH_3CHF_2; HFC-23 = CHF_3; CFC-11 = CCl_3F; CFC-12 = CCl_2F_2; CFC-113 = $CClF_2CCl_2F$; HCFC-22 = $CHClF_2$; HCFC-141b = CH_3CCl_2F; HCFC-142b = CH_3CClF_2.

[a] Each program uses different methods to estimate uncertainties.

[b] SIO reports only CO_2; all other values reported in these columns are from AGAGE. SIO CO_2 program and AGAGE are not affiliated with each other.

Budget lifetimes are shown; for CH_4 and N_2O, perturbation lifetimes (12.4 years for CH_4 and 121 years for N_2O) are used to estimate global warming potentials (Chapter 8).

Year 1750 values determined from air extracted from ice cores are below detection limits for all species except CO_2 (278 ± 2 ppm), CH_4 (722 ± 25 ppb), N_2O (270 ± 7 ppb) and CF_4 (34.7 ± 0.2 ppt). Centennial variations up to 10 ppm CO_2, 40 ppb CH_4, and 10 ppb occur throughout the late-Holocene (Chapter 6).

are plotted in Figure 2.1. The main features in the contemporary CO_2 record are the long-term increase and the seasonal cycle resulting from photosynthesis and respiration by the terrestrial biosphere, mostly in the Northern Hemisphere (NH). The main contributors to increasing atmospheric CO_2 abundance are fossil fuel combustion and land use change (Section 6.3). Multiple lines of observational evidence indicate that during the past few decades, most of the increasing atmospheric burden of CO_2 is from fossil fuel combustion (Tans, 2009). Since the last year for which the AR4 reported (2005), CO_2 has increased by 11.7 ppm to 390.5 ppm in 2011 (Table 2.1). From 1980 to 2011, the average annual increase in globally averaged CO_2 (from 1 January in one year to 1 January in the next year) was 1.7 ppm yr^{-1} (1 standard deviation = 0.5 ppm yr^{-1}; 1 ppm globally corresponds to 2.1 PgC increase in the atmospheric burden). Since 2001, CO_2 has increased at 2.0 ppm yr^{-1} (1 standard deviation = 0.3 ppm yr^{-1}). The CO_2 growth rate varies from year to year; since 1980 the range in annual increase is 0.7 ± 0.1 ppm in 1992 to 2.9 ± 0.1 ppm in 1998. Most of this interannual variability in growth rate is driven by small changes in the balance between photosynthesis and respiration on land, each having global fluxes of ~120 PgC yr^{-1} (Chapter 6).

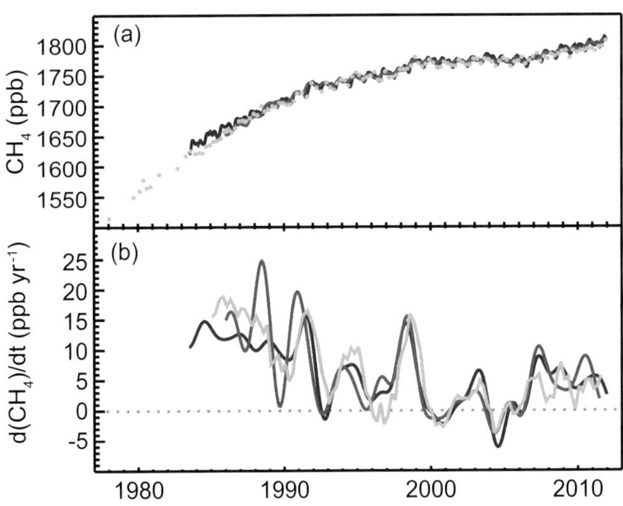

Figure 2.2 | (a) Globally averaged CH_4 dry-air mole fractions from UCI (green; four values per year, except prior to 1984, when they are of lower and varying frequency), AGAGE (red; monthly), and NOAA/ESRL/GMD (blue; quasi-weekly). (b) Instantaneous growth rate for globally averaged atmospheric CH_4 using the same colour code as in (a). Growth rates were calculated as in Figure 2.1.

2.2.1.1.2 Methane

Globally averaged CH_4 in 1750 was 722 ± 25 ppb (after correction to the NOAA-2004 CH_4 standard scale) (Etheridge et al., 1998; Dlugokencky et al., 2005), although human influences on the global CH_4 budget may have begun thousands of years earlier than this time that is normally considered 'pre-industrial' (Ruddiman, 2003; Ferretti et al., 2005; Ruddiman, 2007). In 2011, the global annual mean was 1803 ± 2 ppb. Direct atmospheric measurements of CH_4 of sufficient spatial coverage to calculate global annual means began in 1978 and are plotted through 2011 in Figure 2.2a. This time period is characterized by a decreasing growth rate (Figure 2.2b) from the early 1980s until 1998, stabilization from 1999 to 2006, and an increasing atmospheric burden from 2007 to 2011 (Rigby et al., 2008; Dlugokencky et al., 2009). Assuming no long-term trend in hydroxyl radical (OH) concentration, the observed decrease in CH_4 growth rate from the early 1980s through 2006 indicates an approach to steady state where total global emissions have been approximately constant at ~550 Tg (CH_4) yr^{-1}. Superimposed on the long-term pattern is significant interannual variability; studies of this variability are used to improve understanding of the global CH_4 budget (Chapter 6). The most likely drivers of increased atmospheric CH_4 were anomalously high temperatures in the Arctic in 2007 and greater than average precipitation in the tropics during 2007 and 2008 (Dlugokencky et al., 2009; Bousquet, 2011). Observations of the difference in CH_4 between zonal averages for northern and southern polar regions (53° to 90°) (Dlugokencky et al., 2009, 2011) suggest that, so far, it is unlikely that there has been a permanent measureable increase in Arctic CH_4 emissions from wetlands and shallow sub-sea CH_4 clathrates.

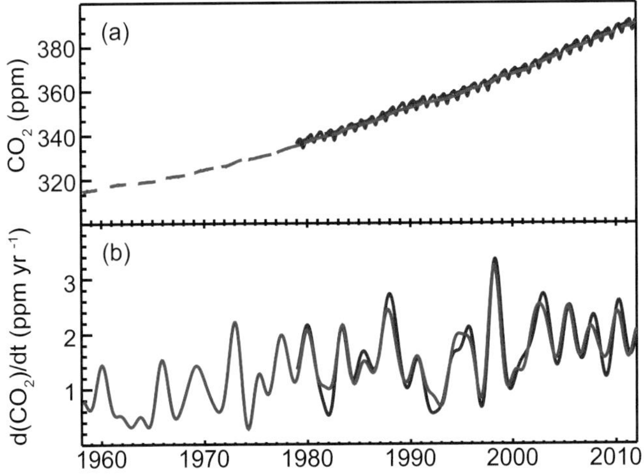

Figure 2.1 | (a) Globally averaged CO_2 dry-air mole fractions from Scripps Institution of Oceanography (SIO) at monthly time resolution based on measurements from Mauna Loa, Hawaii and South Pole (red) and NOAA/ESRL/GMD at quasi-weekly time resolution (blue). SIO values are deseasonalized. (b) Instantaneous growth rates for globally averaged atmospheric CO_2 using the same colour code as in (a). Growth rates are calculated as the time derivative of the deseasonalized global averages (Dlugokencky et al., 1994).

Reaction with the hydroxyl radical (OH) is the main loss process for CH_4 (and for hydrofluorocarbons (HFCs) and hydrochlorofluorocarbons (HCFCs)), and it is the largest term in the global CH_4 budget. Therefore, trends and interannual variability in OH concentration significantly impact our understanding of changes in CH_4 emissions. Methyl chloroform (CH_3CCl_3; Section 2.2.1.2) has been used extensively to estimate globally averaged OH concentrations (e.g., Prinn et al., 2005). AR4 reported no trend in OH from 1979 to 2004, and there is no evidence from this assessment to change that conclusion for 2005 to 2011. Montzka et al. (2011a) exploited the exponential decrease and small emissions in CH_3CCl_3 to show that interannual variations in OH concentration from 1998 to 2007 are 2.3 ± 1.5%, which is consistent with estimates based on CH_4, tetrachloroethene (C_2Cl_4), dichloromethane (CH_2Cl_2), chloromethane (CH_3Cl) and bromomethane (CH_3Br).

2.2.1.1.3 Nitrous Oxide

Globally averaged N_2O in 2011 was 324.2 ppb, an increase of 5.0 ppb over the value reported for 2005 in AR4 (Table 2.1). This is an increase

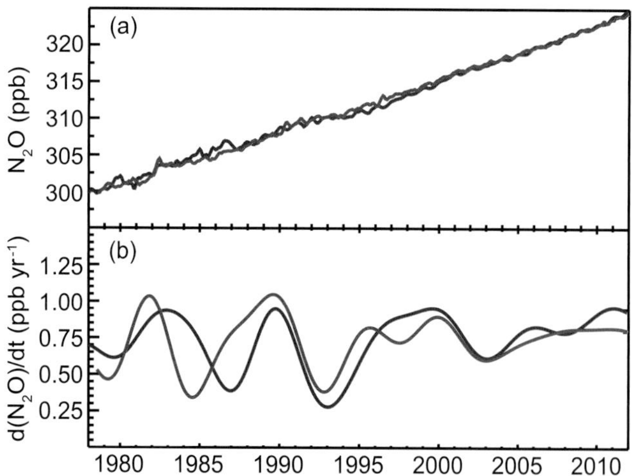

Figure 2.3 | (a) Globally averaged N_2O dry-air mole fractions from AGAGE (red) and NOAA/ESRL/GMD (blue) at monthly resolution. (b) Instantaneous growth rates for globally averaged atmospheric N_2O. Growth rates were calculated as in Figure 2.1.

of 20% over the estimate for 1750 from ice cores, 270 ± 7 ppb (Prather et al., 2012). Measurements of N_2O and its isotopic composition in firn air suggest the increase, at least since the early 1950s, is dominated by emissions from soils treated with synthetic and organic (manure) nitrogen fertilizer (Rockmann and Levin, 2005; Ishijima et al., 2007; Davidson, 2009; Syakila and Kroeze, 2011). Since systematic measurements began in the late 1970s, N_2O has increased at an average rate of ~0.75 ppb yr^{-1} (Figure 2.3). Because the atmospheric burden of CFC-12 is decreasing, N_2O has replaced CFC-12 as the third most important well-mixed GHG contributing to RF (Elkins and Dutton, 2011).

Persistent latitudinal gradients in annually averaged N_2O are observed at background surface sites, with maxima in the northern subtropics, values about 1.7 ppb lower in the Antarctic, and values about 0.4 ppb lower in the Arctic (Huang et al., 2008). These persistent gradients contain information about anthropogenic emissions from fertilizer use at northern tropical to mid-latitudes and natural emissions from soils and ocean upwelling regions of the tropics. N_2O time series also contain seasonal variations with peak-to-peak amplitudes of about 1 ppb in high latitudes of the NH and about 0.4 ppb at high southern and tropical latitudes. In the NH, exchange of air between the stratosphere (where N_2O is destroyed by photochemical processes) and troposphere is the dominant contributor to observed seasonal cycles, not seasonality in emissions (Jiang et al., 2007). Nevison et al. (2011) found correlations between the magnitude of detrended N_2O seasonal minima and lower stratospheric temperature, providing evidence for a stratospheric influence on the timing and amplitude of the seasonal cycle at surface monitoring sites. In the Southern Hemisphere (SH), observed seasonal cycles are also affected by stratospheric influx, and by ventilation and thermal out-gassing of N_2O from the oceans.

2.2.1.1.4 Hydrofluorocarbons, Perfluorocarbons, Sulphur Hexafluoride and Nitrogen Trifluoride

The budgets of HFCs, PFCs and SF_6 were recently reviewed in Chapter 1 of the Scientific Assessment of Ozone Depletion: 2010 (Montzka et al., 2011b), so only a brief description is given here. The current atmospheric abundances of these species are summarized in Table 2.1 and plotted in Figure 2.4.

Atmospheric HFC abundances are low and their contribution to RF is small relative to that of the CFCs and HCFCs they replace (less than 1% of the total by well-mixed GHGs; Chapter 8). As they replace CFCs and HCFCs phased out by the Montreal Protocol, however, their contribution to future climate forcing is projected to grow considerably in the absence of controls on global production (Velders et al., 2009).

HFC-134a is a replacement for CFC-12 in automobile air conditioners and is also used in foam blowing applications. In 2011, it reached 62.7 ppt, an increase of 28.2 ppt since 2005. Based on analysis of high-frequency measurements, the largest emissions occur in North America, Europe and East Asia (Stohl et al., 2009).

HFC-23 is a by-product of HCFC-22 production. Direct measurements of HFC-23 in ambient air at five sites began in 2007. The 2005 global annual mean used to calculate the increase since AR4 in Table 2.1, 5.2 ppt, is based on an archive of air collected at Cape Grim, Tasmania (Miller et al., 2010). In 2011, atmospheric HFC-23 was at 24.0 ppt. Its growth rate peaked in 2006 as emissions from developing countries

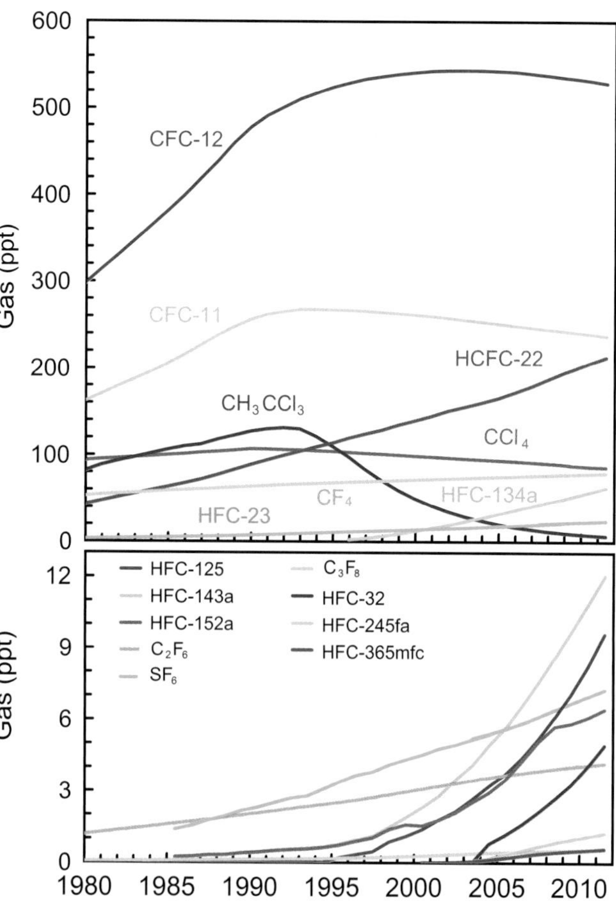

Figure 2.4 | Globally averaged dry-air mole fractions at the Earth's surface of the major halogen-containing well-mixed GHG. These are derived mainly using monthly mean measurements from the AGAGE and NOAA/ESRL/GMD networks. For clarity, only the most abundant chemicals are shown in different compound classes and results from different networks have been combined when both are available.

increased, then declined as emissions were reduced through abatement efforts under the Clean Development Mechanism (CDM) of the UNFCCC. Estimates of total global emissions based on atmospheric observations and bottom-up inventories agree within uncertainties (Miller et al., 2010; Montzka et al., 2010). Currently, the largest emissions of HFC-23 are from East Asia (Yokouchi et al., 2006; Kim et al., 2010; Stohl et al., 2010); developed countries emit less than 20% of the global total. Keller et al. (2011) found that emissions from developed countries may be larger than those reported to the UNFCCC, but their contribution is small. The lifetime of HFC-23 was revised from 270 to 222 years since AR4.

After HFC-134a and HFC-23, the next most abundant HFCs are HFC-143a at 12.04 ppt in 2011, 6.39 ppt greater than in 2005; HFC-125 (O'Doherty et al., 2009) at 9.58 ppt, increasing by 5.89 ppt since 2005; HFC-152a (Greally et al., 2007) at 6.4 ppt with a 3.0 ppt increase since 2005; and HFC-32 at 4.92 ppt in 2011, 3.77 ppt greater than in 2005. Since 2005, all of these were increasing exponentially except for HFC-152a, whose growth rate slowed considerably in about 2007 (Figure 2.4). HFC-152a has a relatively short atmospheric lifetime of 1.5 years, so its growth rate will respond quickly to changes in emissions. Its major uses are as a foam blowing agent and aerosol spray propellant while HFC-143a, HFC-125, and HFC-32 are mainly used in refrigerant blends. The reasons for slower growth in HFC-152a since about 2007 are unclear. Total global emissions of HFC-125 estimated from the observations are within about 20% of emissions reported to the UNFCCC, after accounting for estimates of unreported emissions from East Asia (O'Doherty et al., 2009).

CF_4 and C_2F_6 (PFCs) have lifetimes of 50 kyr and 10 kyr, respectively, and they are emitted as by-products of aluminium production and used in plasma etching of electronics. CF_4 has a natural lithospheric source (Deeds et al., 2008) with a 1750 level determined from Greenland and Antarctic firn air of 34.7 ± 0.2 ppt (Worton et al., 2007; Muhle et al., 2010). In 2011, atmospheric abundances were 79.0 ppt for CF_4, increasing by 4.0 ppt since 2005, and 4.16 ppt for C_2F_6, increasing by 0.50 ppt. The sum of emissions of CF_4 reported by aluminium producers and for non-aluminium production in EDGAR (Emission Database for Global Atmospheric Research) v4.0 accounts for only about half of global emissions inferred from atmospheric observations (Muhle et al., 2010). For C_2F_6, emissions reported to the UNFCCC are also substantially lower than those estimated from atmospheric observations (Muhle et al., 2010).

The main sources of atmospheric SF_6 emissions are electricity distribution systems, magnesium production, and semi-conductor manufacturing. Global annual mean SF_6 in 2011 was 7.29 ppt, increasing by 1.65 ppt since 2005. SF_6 has a lifetime of 3200 years, so its emissions accumulate in the atmosphere and can be estimated directly from its observed rate of increase. Levin et al. (2010) and Rigby et al. (2010) showed that SF_6 emissions decreased after 1995, most likely because of emissions reductions in developed countries, but then increased after 1998. During the past decade, they found that actual SF_6 emissions from developed countries are at least twice the reported values.

NF_3 was added to the list of GHG in the Kyoto Protocol with the Doha Amendment, December, 2012. Arnold et al. (2013) determined 0.59 ppt for its global annual mean mole fraction in 2008, growing from almost zero in 1978. In 2011, NF_3 was 0.86 ppt, increasing by 0.49 ppt since 2005. These abundances were updated from the first work to quantify NF_3 by Weiss et al. (2008). Initial bottom-up inventories underestimated its emissions; based on the atmospheric observations, NF_3 emissions were 1.18 ± 0.21Gg in 2011 (Arnold et al., 2013).

In summary, it is certain that atmospheric burdens of well-mixed GHGs targeted by the Kyoto Protocol increased from 2005 to 2011. The atmospheric abundance of CO_2 was 390.5 ± 0.2 ppm in 2011; this is 40% greater than before 1750. Atmospheric N_2O was 324.2 ± 0.2 ppb in 2011 and has increased by 20% since 1750. Average annual increases in CO_2 and N_2O from 2005 to 2011 are comparable to those observed from 1996 to 2005. Atmospheric CH_4 was 1803.2 ± 2.0 ppb in 2011; this is 150% greater than before 1750. CH_4 began increasing in 2007 after remaining nearly constant from 1999 to 2006. HFCs, PFCs, and SF_6 all continue to increase relatively rapidly, but their contributions to RF are less than 1% of the total by well-mixed GHGs (Chapter 8).

2.2.1.2 Ozone-Depleting Substances (Chlorofluorocarbons, Chlorinated Solvents, and Hydrochlorofluorocarbons)

CFC atmospheric abundances are decreasing (Figure 2.4) because of the successful reduction in emissions resulting from the Montreal Protocol. By 2010, emissions from ODSs had been reduced by ~11 Pg CO_2-eq yr^{-1}, which is five to six times the reduction target of the first commitment period (2008–2012) of the Kyoto Protocol (2 PgCO_2-eq yr^{-1}) (Velders et al., 2007). These avoided equivalent-CO_2 emissions account for the offsets to RF by stratospheric O_3 depletion caused by ODSs and the use of HFCs as substitutes for them. Recent observations in Arctic and Antarctic firn air further confirm that emissions of CFCs are entirely anthropogenic (Martinerie et al., 2009; Montzka et al., 2011b). CFC-12 has the largest atmospheric abundance and GWP-weighted emissions (which are based on a 100-year time horizon) of the CFCs. Its tropospheric abundance peaked during 2000–2004. Since AR4, its global annual mean mole fraction declined by 13.8 ppt to 528.5 ppt in 2011. CFC-11 continued the decrease that started in the mid-1990s, by 12.9 ppt since 2005. In 2011, CFC-11 was 237.7 ppt. CFC-113 decreased by 4.3 ppt since 2005 to 74.3 ppt in 2011. A discrepancy exists between top-down and bottom-up methods for calculating CFC-11 emissions (Montzka et al., 2011b). Emissions calculated using top-down methods come into agreement with bottom-up estimates when a lifetime of 64 years is used for CFC-11 in place of the accepted value of 45 years; this longer lifetime (64 years) is at the upper end of the range estimated by Douglass et al. (2008) with models that more accurately simulate stratospheric circulation. Future emissions of CFCs will largely come from 'banks' (i.e., material residing in existing equipment or stores) rather than current production.

The mean decrease in globally, annually averaged carbon tetrachloride (CCl_4) based on NOAA and AGAGE measurements since 2005 was 7.4 ppt, with an atmospheric abundance of 85.8 ppt in 2011 (Table 2.1). The observed rate of decrease and inter-hemispheric difference of CCl_4 suggest that emissions determined from the observations are on average greater and less variable than bottom-up emission estimates, although large uncertainties in the CCl_4 lifetime result in large uncertainties in the top-down estimates of emissions (Xiao et al., 2010;

Montzka et al., 2011b). CH_3CCl_3 has declined exponentially for about a decade, decreasing by 12.0 ppt since 2005 to 6.3 ppt in 2011.

HCFCs are classified as 'transitional substitutes' by the Montreal Protocol. Their global production and use will ultimately be phased out, but their global production is not currently capped and, based on changes in observed spatial gradients, there has likely been a shift in emissions within the NH from regions north of about 30°N to regions south of 30°N (Montzka et al., 2009). Global levels of the three most abundant HCFCs in the atmosphere continue to increase. HCFC-22 increased by 44.5 ppt since 2005 to 213.3 ppt in 2011. Developed country emissions of HCFC-22 are decreasing, and the trend in total global emissions is driven by large increases from south and Southeast Asia (Saikawa et al., 2012). HCFC-141b increased by 3.7 ppt since 2005 to 21.4 ppt in 2011, and for HCFC-142b, the increase was 5.73 ppt to 21.1 ppt in 2011. The rates of increase in these three HCFCs increased since 2004, but the change in HCFC-141b growth rate was smaller and less persistent than for the other two, which approximately doubled from 2004 to 2007 (Montzka et al., 2009).

In summary, for ODS, whose production and consumption are controlled by the Montreal Protocol, it is certain that the global mean abundances of major CFCs are decreasing and HCFCs are increasing. Atmospheric burdens of CFC-11, CFC-12, CFC-113, CCl_4, CH_3CCl_3 and some halons have decreased since 2005. HCFCs, which are transitional substitutes for CFCs, continue to increase, but the spatial distribution of their emissions is changing.

2.2.2 Near-Term Climate Forcers

This section covers observed trends in stratospheric water vapour; stratospheric and tropospheric ozone (O_3); the O_3 precursor gases, nitrogen dioxide (NO_2) and carbon monoxide (CO); and column and surface aerosol. Since trend estimates from the cited literature are used here, issues such as data records of different length, potential lack of comparability among measurement methods and different trend calculation methods, add to the uncertainty in assessing trends.

2.2.2.1 Stratospheric Water Vapour

Stratospheric H_2O vapour has an important role in the Earth's radiative balance and in stratospheric chemistry. Increased stratospheric H_2O vapour causes the troposphere to warm and the stratosphere to cool (Manabe and Strickler, 1964; Solomon et al., 2010), and also causes increased rates of stratospheric O_3 loss (Stenke and Grewe, 2005). Water vapour enters the stratosphere through the cold tropical tropopause. As moisture-rich air masses are transported through this region, most water vapour condenses resulting in extremely dry lower stratospheric air. Because tropopause temperature varies seasonally, so does H_2O abundance there. Other contributions include oxidation of methane within the stratosphere, and possibly direct injection of H_2O vapour in overshooting deep convection (Schiller et al., 2009). AR4 reported that stratospheric H_2O vapour showed significant long-term variability and an upward trend over the last half of the 20th century, but no net increase since 1996. This updated assessment finds large interannual variations that have been observed by independent measurement techniques, but no significant net changes since 1996.

The longest continuous time series of stratospheric water vapour abundance is from *in situ* measurements made with frost point hygrometers starting in 1980 over Boulder, USA (40°N, 105°W) (Scherer et al., 2008), with values ranging from 3.5 to 5.5 ppm, depending on altitude. These observations have been complemented by long-term global satellite observations from SAGE II (1984–2005; Stratospheric Aerosol and Gas Experiment II (Chu et al., 1989)), HALOE (1991–2005; HALogen Occultation Experiment (Russell et al., 1993)), Aura MLS (2004–present; Microwave Limb Sounder (Read et al., 2007)) and Envisat MIPAS (2002-2012; Michelson Interferometer for Passive Atmospheric Sounding (Milz et al., 2005; von Clarmann et al., 2009)). Discrepancies in water vapour mixing ratios from these different instruments can be attributed to differences in the vertical resolution of measurements, along with other factors. For example, offsets of up to 0.5 ppm in lower stratospheric water vapour mixing ratios exist between the most current versions of HALOE (v19) and Aura MLS (v3.3) retrievals during their 16-month period of overlap (2004 to 2005), although such biases can be removed to generate long-term records. Since AR4, new studies characterize the uncertainties in measurements from individual types of *in situ* H_2O sensors (Vömel et al., 2007b; Vömel et al., 2007a; Weinstock et al., 2009), but discrepancies between different instruments (50 to 100% at H_2O mixing ratios less than 10 ppm), particularly for high-altitude measurements from aircraft, remain largely unexplained.

Observed anomalies in stratospheric H_2O from the near-global combined HALOE+MLS record (1992–2011) (Figure 2.5) include effects linked to the stratospheric quasi-biennial oscillation (QBO) influence on tropopause temperatures, plus a step-like drop after 2001 (noted in AR4), and an increasing trend since 2005. Variability during 2001–2011 was large yet there was only a small net change from 1992 through 2011. These interannual water vapour variations for the satellite record are closely linked to observed changes in tropical tropopause temperatures (Fueglistaler and Haynes, 2005; Randel et al., 2006; Rosenlof and Reid, 2008; Randel, 2010), providing reasonable understanding of observed changes. The longer record of Boulder balloon measurements (since 1980) has been updated and reanalyzed (Scherer et al., 2008; Hurst et al., 2011), showing deca dal-scale variability and a long-term stratospheric (16 to 26 km) increase of 1.0 ± 0.2 ppm for 1980–2010. Agreement between interannual changes inferred from the Boulder and HALOE+MLS data is good for the period since 1998 but was poor during 1992–1996. About 30% of the positive trend during 1980–2010 determined from frost point hygrometer data (Fujiwara et al., 2010; Hurst et al., 2011) can be explained by increased production of H_2O from CH_4 oxidation (Rohs et al., 2006), but the remainder cannot be explained by changes in tropical tropopause temperatures (Fueglistaler and Haynes, 2005) or other known factors.

In summary, near-global satellite measurements of stratospheric H_2O show substantial variability for 1992–2011, with a step-like decrease after 2000 and increases since 2005. Because of this large variability and relatively short time series, *confidence* in long-term stratospheric H_2O trends is *low*. There is good understanding of the relationship between the satellite-derived H_2O variations and tropical tropopause temperature changes. Stratospheric H_2O changes from temporally sparse balloon-borne observations at one location (Boulder, Colorado) are in good agreement with satellite observations from 1998 to the present, but discrepancies exist for changes during 1992–1996. Long-

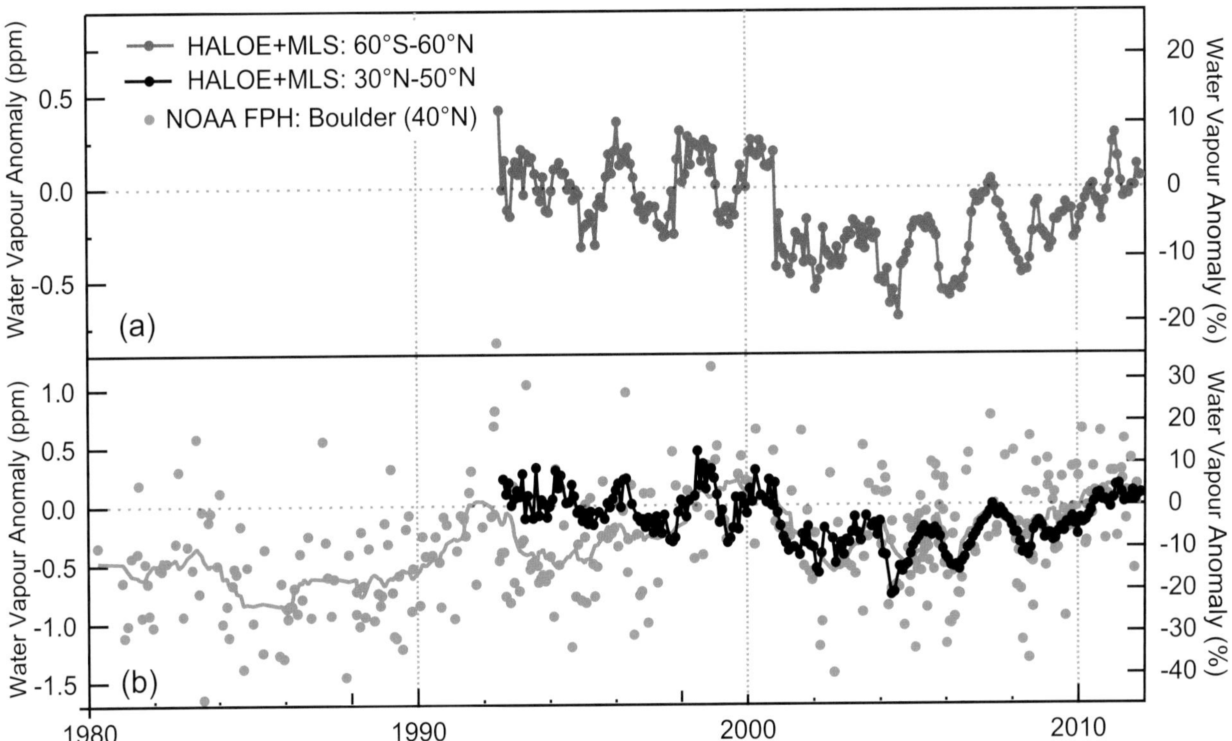

Figure 2.5 | Water vapour anomalies in the lower stratosphere (~16 to 19 km) from satellite sensors and *in situ* measurements normalized to 2000–2011. (a) Monthly mean water vapour anomalies at 83 hPa for 60°S to 60°N (blue) determined from HALOE and MLS satellite sensors. (b) Approximately monthly balloon-borne measurements of stratospheric water vapour from Boulder, Colorado at 40°N (green dots; green curve is 15-point running mean) averaged over 16 to 18 km and monthly means as in (a), but averaged over 30°N to 50°N (black)

term balloon measurements from Boulder indicate a net increase of 1.0 ± 0.2 ppm over 16 to 26 km for 1980–2010, but these long-term increases cannot be fully explained by changes in tropical tropopause temperatures, methane oxidation or other known factors.

2.2.2.2 Stratospheric Ozone

AR4 did not explicitly discuss measured stratospheric ozone trends. For the current assessment report such trends are relevant because they are the basis for revising the RF from –0.05 ± 0.10 W m^{-2} in 1750 to –0.10 ± 0.15 W m^{-2} in 2005 (Section 8.3.3.2). These values strongly depend on the vertical distribution of the stratospheric ozone changes.

Total ozone is a good proxy for stratospheric ozone because tropospheric ozone accounts for only about 10% of the total ozone column. Long-term total ozone changes over various latitudinal belts, derived from Weber et al. (2012), are illustrated in Figure 2.6 (a–d). Annually averaged total column ozone declined during the 1980s and early 1990s and has remained constant for the past decade, about 3.5 and 2.5% below the 1964–1980 average for the entire globe (not shown) and 60°S to 60°N, respectively, with changes occurring mostly outside the tropics, particularly the SH, where the current extratropical (30°S to 60°S) mean values are 6% below the 1964–1980 average, compared to 3.5% for the NH extratropics (Douglass et al., 2011). In the NH, the 1993 minimum of about –6% was caused primarily by ozone loss through heterogeneous reactions on volcanic aerosols from Mt. Pinatubo.

Two altitude regions are mainly responsible for long-term changes in total column ozone (Douglass et al., 2011). In the upper stratosphere (35 to 45 km), there was a strong and statistically significant decline (about 10%) up to the mid-1990s and little change or a slight increase since. The lower stratosphere, between 20 and 25 km over mid-latitudes, also experienced a statistically significant decline (7 to 8%) between 1979 and the mid-1990s, followed by stabilization or a slight (2 to 3%) ozone increase.

Springtime averages of total ozone poleward of 60° latitude in the Arctic and Antarctic are shown in Figure 2.6e. By far the strongest ozone loss in the stratosphere occurs in austral spring over Antarctica (ozone hole) and its impact on SH climate is discussed in Chapters 11, 12 and 14. Interannual variability in polar stratospheric ozone abundance and chemistry is driven by variability in temperature and transport due to year-to-year differences in dynamics. This variability is particularly large in the Arctic, where the most recent large depletion occurred in 2011, when chemical ozone destruction was, for the first time in the observational record, comparable to that in the Antarctic (Manney et al., 2011).

In summary, it is certain that global stratospheric ozone has declined from pre-1980 values. Most of the decline occurred prior to the mid-1990s; since then there has been little net change and ozone has remained nearly constant at about 3.5% below the 1964–1980 level.

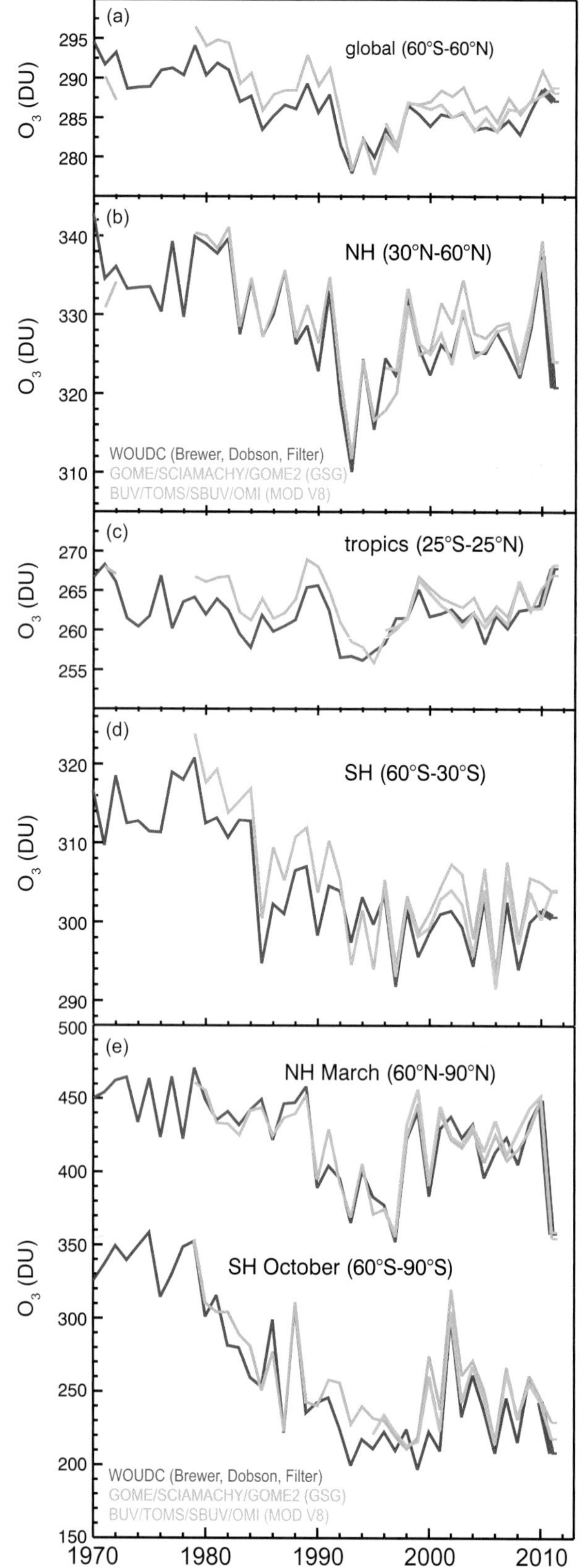

Figure 2.6 | Zonally averaged, annual mean total column ozone in Dobson Units (DU; 1 DU = 2.69×10^{16} O_3/cm^2) from ground-based measurements combining Brewer, Dobson, and filter spectrometer data WOUDC (red), GOME/SCIAMACHY/GOME-2 GSG (green) and merged satellite BUV/TOMS/SBUV/OMI MOD V8 (blue) for (a) Non-Polar Global (60°S to 60°N), (b) NH (30°N to 60°N), (c) Tropics (25°S to 25°N), (d) SH (30°S to 60°S) and (e) March NH Polar (60°N to 90°N) and October SH Polar. (Adapted from Weber et al., 2012; see also for abbreviations.)

2.2.2.3 Tropospheric Ozone

Tropospheric ozone is a short-lived trace gas that either originates in the stratosphere or is produced *in situ* by precursor gases and sunlight (e.g., Monks et al., 2009). An important GHG with an estimated RF of 0.40 ± 0.20 W m^{-2} (Chapter 8), tropospheric ozone also impacts human health and vegetation at the surface. Its average atmospheric lifetime of a few weeks produces a global distribution highly variable by season, altitude and location. These characteristics and the paucity of long-term measurements make the assessment of long-term global ozone trends challenging. However, new studies since AR4 provide greater understanding of surface and free tropospheric ozone trends from the 1950s through 2010. An extensive compilation of measured ozone trends is presented in the Supplementary Material, Figure 2.SM.1 and Table 2.SM.2.

The earliest (1876–1910) quantitative ozone observations are limited to Montsouris near Paris where ozone averaged 11 ppb (Volz and Kley, 1988). Semiquantitative ozone measurements from more than 40 locations around the world in the late 1800s and early 1900s range from 5 to 32 ppb with large uncertainty (Pavelin et al., 1999). The low 19th century ozone values cannot be reproduced by most models (Section 8.2.3.1), and this discrepancy is an important factor contributing to uncertainty in RF calculations (Section 8.3.3.1). Limited quantitative measurements from the 1870s to 1950s indicate that surface ozone in Europe increased by more than a factor of 2 compared to observations made at the end of the 20th century (Marenco et al., 1994; Parrish et al., 2012).

Satellite-based tropospheric column ozone retrievals across the tropics and mid-latitudes reveal a greater burden in the NH than in the SH (Ziemke et al., 2011). Tropospheric column ozone trend analyses are few. An analysis by Ziemke et al. (2005) found no trend over the tropical Pacific Ocean but significant positive trends (5 to 9% per decade) in the mid-latitude Pacific of both hemispheres during 1979–2003. Significant positive trends (2 to 9% per decade) were found across broad regions of the tropical South Atlantic, India, southern China, southeast Asia, Indonesia and the tropical regions downwind of China (Beig and Singh, 2007).

Long-term ozone trends at the surface and in the free troposphere (of importance for calculating RF, Chapter 8) can be assessed only from *in situ* measurements at a limited number of sites, leaving large areas such as the tropics and SH sparsely sampled (Table 2.SM.2, Figure 2.7). Nineteen predominantly rural surface sites or regions around the globe have long-term records that stretch back to the 1970s, and in two cases the 1950s (Lelieveld et al., 2004; Parrish et al., 2012; Oltmans et al., 2013). Thirteen of these sites are in the NH, and 11 sites have statistically significant positive trends of 1 to 5 ppb per decade, corresponding to >100% ozone increases since the 1950s and 9 to 55% ozone increases since the 1970s. In the SH, three of six sites have significant trends of approximately 2 ppb per decade and three have insignificant trends. Free tropospheric monitoring since the 1970s is more limited. Significant positive trends since 1971 have been observed using ozone sondes above Western Europe, Japan and coastal Antarctica (rates of increase range from 1 to 3 ppb per decade), but not at all levels (Oltmans et al., 2013). In addition, aircraft have measured significant upper tropospheric trends in one or more seasons above the north-eastern USA, the North Atlantic Ocean, Europe, the Middle East, northern India, southern China and Japan (Schnadt Poberaj et al., 2009). Insignificant free tropospheric trends were found above the Mid-Atlantic USA (1971–2010) (Oltmans et al., 2013) and in the upper troposphere above the western USA (1975–2001) (Schnadt Poberaj et al., 2009). No site or region showed a significant negative trend.

In recent decades ozone precursor emissions have decreased in Europe and North America and increased in Asia (Granier et al., 2011), impacting ozone production on regional and hemispheric scales (Skeie et al., 2011). Accordingly, 1990–2010 surface ozone trends vary regionally. In Europe ozone generally increased through much of the 1990s but since 2000 ozone has either levelled off or decreased at rural and mountaintop sites, as well as for baseline ozone coming ashore at Mace Head, Ireland (Tarasova et al., 2009; Logan et al., 2012; Parrish et al., 2012; Oltmans et al., 2013). In North America surface ozone has increased in eastern and Arctic Canada, but is unchanged in central and western Canada (Oltmans et al., 2013). Surface ozone has increased in baseline air masses coming ashore along the west coast of the USA (Parrish et al., 2012) and at half of the rural sites in the western USA during spring (Cooper et al., 2012). In the eastern USA surface ozone has decreased strongly in summer, is largely unchanged in spring and has increased in winter (Lefohn et al., 2010; Cooper et al., 2012). East Asian surface ozone is generally increasing (Table 2.SM.2) and at downwind sites ozone is increasing at Mauna Loa, Hawaii but decreasing at Minami Tori Shima in the subtropical western North Pacific (Oltmans et al., 2013). In the SH ozone has increased at the eight available sites, although trends are insignificant at four sites (Helmig et al., 2007; Oltmans et al., 2013).

Owing to methodological changes, free tropospheric ozone observations are most reliable since the mid-1990s. Ozone has decreased above Europe since 1998 (Logan et al., 2012) and is largely unchanged above Japan (Oltmans et al., 2013). Otherwise the remaining regions with measurements (North America, North Pacific Ocean, SH) show a range of positive trends (both significant and insignificant) depending on altitude, with no site having a negative trend at any altitude (Table 2.SM.2).

In summary, there is *medium confidence* from limited measurements in the late 19th through mid-20th century that European surface ozone more than doubled by the end of the 20th century. There is *medium confidence* from more widespread measurements beginning in the 1970s that surface ozone has increased at most (non-urban) sites in the NH (1 to 5 ppb per decade), while there is *low confidence* for ozone increases (2 ppb per decade) in the SH. Since 1990 surface ozone has *likely* increased in East Asia, while surface ozone in the eastern USA and Western Europe has levelled off or is decreasing. Ozone monitoring in the free troposphere since the 1970s is very limited and indicates a weaker rate of increase than at the surface. Satellite instruments can now quantify the present-day tropospheric ozone burden on a near-global basis; significant tropospheric ozone column increases were observed over extended tropical regions of southern Asia, as well as mid-latitude regions of the South and North Pacific Ocean since 1979.

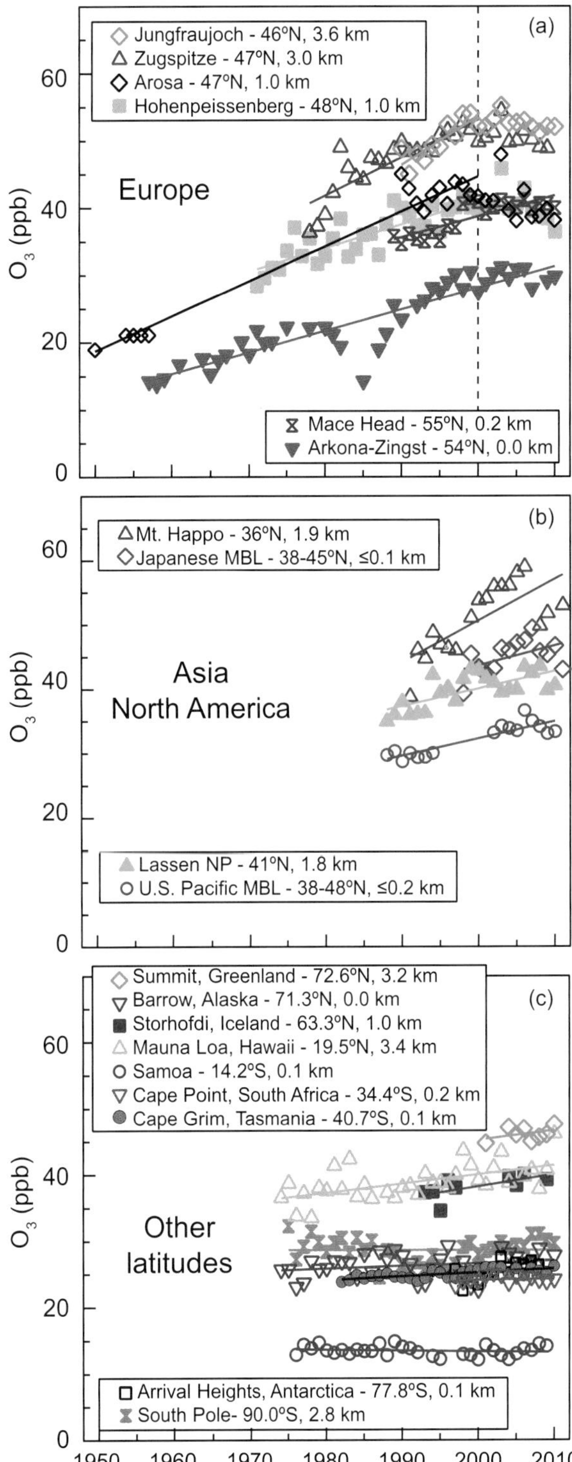

Figure 2.7 | Annual average surface ozone concentrations from regionally representative ozone monitoring sites around the world. (a) Europe. (b) Asia and North America. (c) Remote sites in the Northern and Southern Hemispheres. The station name in the legend is followed by its latitude and elevation. Time series include data from all times of day and trend lines are linear regressions following the method of Parrish et al. (2012). Trend lines are fit through the full time series at each location, except for Jungfraujoch, Zugspitze, Arosa and Hohenpeissenberg where the linear trends end in 2000 (indicated by the dashed vertical line in (a)). Twelve of these 19 sites have significant positive ozone trends (i.e., a trend of zero lies outside the 95% confidence interval); the seven sites with non-significant trends are: Japanese MBL (marine boundary layer), Summit (Greenland), Barrow (Alaska), Storhofdi (Iceland), Samoa (tropical South Pacific Ocean), Cape Point (South Africa) and South Pole (Antarctica).

2

2.2.2.4 Carbon Monoxide, Non-Methane Volatile Organic Compounds and Nitrogen Dioxide

Emissions of carbon monoxide (CO), non-methane volatile organic compounds (NMVOCs) and NO_x ($NO + NO_2$) do not have a direct effect on RF, but affect climate indirectly as precursors to tropospheric O_3 and aerosol formation, and their impacts on OH concentrations and CH_4 lifetime. NMVOCs include aliphatic, aromatic and oxygenated hydrocarbons (e.g., aldehydes, alcohols and organic acids), and have atmospheric lifetimes ranging from hours to months. Global coverage of NMVOC measurements is poor, except for a few compounds. Reports on trends generally indicate declines in a range of NMVOCs in urban and rural regions of North America and Europe on the order of a few percent to more than 10% yr^{-1}. Global ethane levels reported by Simpson et al. (2012) declined by about 21% from 1986 to 2010. Measurements of air extracted from firn suggest that NMVOC concentrations were growing until 1980 and declined afterwards (Aydin et al., 2011; Worton et al., 2012). Satellite retrievals of formaldehyde column abundances from 1997 to 2007 show significant positive trends over northeastern China (4% yr^{-1}) and India (1.6% yr^{-1}), possibly related to strong increases in anthropogenic NMVOC emissions, whereas negative trends of about –3% yr^{-1} are observed over Tokyo, Japan and the northeast USA urban corridor as a result of pollution regulation (De Smedt et al., 2010).

The major sources of atmospheric CO are *in situ* production by oxidation of hydrocarbons (mostly CH_4 and isoprene) and direct emission resulting from incomplete combustion of biomass and fossil fuels. An analysis of MOPITT (Measurements of Pollutants in the Troposphere) and AIRS (Atmospheric Infrared Sounder) satellite data suggest a clear and consistent decline of CO columns for 2002–2010 over a number of polluted regions in Europe, North America and Asia with a global trend of about –1% yr^{-1} (Yurganov et al., 2010; Fortems-Cheiney et al., 2011; Worden et al., 2013). Analysis of satellite data using two more instruments for recent overlapping years shows qualitatively similar decreasing trends (Worden et al., 2013), but the magnitude of trends remains uncertain owing to the presence of instrument biases. Small CO decreases observed in the NOAA and AGAGE networks are consistent with slight declines in global anthropogenic CO emissions over the same time (Supplementary Material 2.SM.2).

Due to its short atmospheric lifetime (approximately hours), NO_x concentrations are highly variable in time and space. AR4 described the potential of satellite observations of NO_2 to verify and improve NO_x emission inventories and their trends and reported strong NO_2 increases by 50% over the industrial areas of China from 1996 to 2004. An extension of this analysis reveals increases between a factor of 1.7 and 3.2 over parts of China, while over Europe and the USA NO_2 has decreased by 30 to 50% between 1996 and 2010 (Hilboll et al., 2013).

Figure 2.8 shows the changes relative to 1996 in satellite-derived tropospheric NO_2 columns, with a strong upward trend over central eastern China and an overall downward trend in Japan, Europe and the USA. NO_2 reductions in the USA are very pronounced after 2004, related to differences in effectiveness of NO_x emission abatements in the USA and also to changes in atmospheric chemistry of NO_x (Russell et al., 2010). Increasingly, satellite data are used to derive trends in anthropogenic

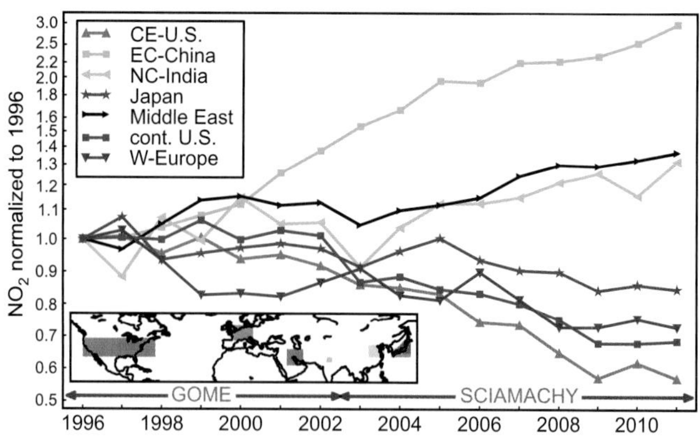

Figure 2.8 | Relative changes in tropospheric NO_2 column amounts (logarithmic scale) in seven selected world regions dominated by high NO_x emissions. Values are normalized for 1996 and derived from the GOME (Global Ozone Monitoring Experiment) instrument from 1996 to 2002 and SCIAMACHY (Scanning Imaging Spectrometer for Atmospheric Cartography) from 2003 to 2011 (Hilboll et al., 2013). The regions are indicated in the map inset.

NO_x emissions, with Castellanos and Boersma (2012) reporting overall increases in global emissions, driven by Asian emission increases of up to 29% yr^{-1} (1996–2006), while moderate decreases up to 7% yr^{-1} (1996–2006) are reported for North America and Europe.

In summary, satellite and surface observations of ozone precursor gases NO_x, CO, and non-methane volatile organic carbons indicate strong regional differences in trends. Most notably, NO_2 has *likely* decreased by 30 to 50% in Europe and North America and increased by more than a factor of 2 in Asia since the mid-1990s.

2.2.3 Aerosols

This section assesses trends in aerosol resulting from both anthropogenic and natural sources. The significance of aerosol changes for global dimming and brightening is discussed in Section 2.3. Chapter 7 provides additional discussion of aerosol properties, while Chapter 8 discusses future RF and the ice-core records that contain information on aerosol changes prior to the 1980s. Chapter 11 assesses air quality–climate change interactions. Because of the short lifetime (days to weeks) of tropospheric aerosol, trends have a strong regional signature. Aerosol from anthropogenic sources (i.e., fossil and biofuel burning) are confined mainly to populated regions in the NH, whereas aerosol from natural sources, such as desert dust, sea salt, volcanoes and the biosphere, are important in both hemispheres and likely dependent on climate and land use change (Carslaw et al., 2010). Owing to interannual variability, long-term trends in aerosols from natural sources are more difficult to identify (Mahowald et al., 2010).

2.2.3.1 Aerosol Optical Depth from Remote Sensing

AOD is a measure of the integrated columnar aerosols load and is an important parameter for evaluating aerosol–radiation interactions. AR4 described early attempts to retrieve AOD from satellites but did not provide estimates of temporal changes in tropospheric aerosol. Little high-accuracy information on AOD changes exists prior to 1995. Better satellite sensors and ground-based sun-photometer networks, along with improved retrieval methods and methodological intercomparisons, allow assessment of regional AOD trends since about 1995.

AOD sun photometer measurements at two stations in northern Germany, with limited regional representativity, suggest a long-term decline of AOD in Europe since 1986 (Ruckstuhl et al., 2008). Ground-based, cloud-screened solar broadband radiometer measurements provide longer time-records than spectrally selective sun-photometer data, but are less specific for aerosol retrieval. Multi-decadal records over Japan (Kudo et al., 2011) indicate an AOD increase until the mid-1980s, followed by an AOD decrease until the late 1990s and almost constant AOD in the 2000s. Similar broad-band solar radiative flux multi-decadal trends have been observed for urban–industrial regions of Europe and North America (Wild et al., 2005), and were linked to successful measures to reduce sulphate (precursor) emissions since the mid-1980s (Section 2.3). An indirect method to estimate AOD is offered by ground-based visibility observations. These data are more ambiguous to interpret, but records go further back in time than broadband, sun photometer and satellite data. A multi-regional analysis for 1973–2007 (Wang et al., 2009a) shows that prior to the 1990s visibility-derived AOD was relatively constant in most regions analysed (except for positive trends in southern Asia), but after 1990 positive AOD trends were observed over Asia, and parts of South America, Australia and Africa, and mostly negative AOD trends were found over Europe. In North America, a small stepwise decrease of visibility after 1993 was likely related to methodological changes (Wang et al., 2012f).

AOD can be determined most accurately with sun photometers that measure direct solar intensity in the absence of cloud interferences with an absolute uncertainty of single measurements of ± 0.01% (Holben et al., 1998). AERONET (AErosol RObotic NETwork) is a global sun photometer network (Holben et al., 1998), with densest coverage over Europe and North America. AERONET AOD temporal trends were examined in independent studies (de Meij et al., 2012; Hsu et al., 2012; Yoon et al., 2012), using different data selection and statistical methods. Hsu et al. (2012) investigated AOD trends at 12 AERONET sites with data coverage of at least 10 years between 1997 and 2010. Yoon et al. (2012) investigated AOD and size trends at 14 AERONET sites with data coverage varying between 4 and 12 years between 1997 and 2009. DeMeij et al. (2012) investigated AOD trends between 2000 and 2009 (550 nm; monthly data) at 62 AERONET sites mostly located in USA and Europe. Each of these studies noted an increase in AOD over East Asia and reductions in North America and Europe. The only dense sun photometer network over southern Asia, ARFINET (Aerosol Radiative Forcing over India NETwork), shows an increase in AOD of about 2% yr^{-1} during the last one to two decades (Krishna Moorthy et al., 2013), with an absolute uncertainty of ± 0.02 at 500 nm (Krishna Moorthy et al., 2007). In contrast, negative AOD trends are identified at more than 80% of examined European and North American AERONET sites (de Meij et al., 2012). Decreasing AOD is also observed near the west coast of northern Africa, where aerosol loads are dominated by Saharan dust outflow. Positive AOD trends are found over the Arabian Peninsula, where aerosol is dominated by dust. Inconsistent AOD trends reported for stations in central Africa result from the use of relatively short time series with respect to the large interannual variability caused by wildfires and dust emissions.

Aerosol products from dedicated satellite sensors complement surface-based AOD with better spatial coverage. The quality of the satellite-derived AOD strongly depends on the retrieval's ability to remove scenes contaminated by clouds and to accurately account for reflectivity at the Earth's surface. Due to relatively weak reflectance of incoming sunlight by the sea surface, the typical accuracy of retrieved AOD over oceans (uncertainty of 0.03 +0.05*AOD; Kahn et al. (2007)) is usually better than over continents (uncertainty of 0.05 +0.15*AOD, Levy et al. (2010)).

2

Satellite-based AOD trends at 550 nm over oceans from conservatively cloud-screened MODIS data (Zhang and Reid, 2010) for 2000–2009 are presented in Figure 2.9. Strongly positive AOD trends were observed over the oceans adjacent to southern and eastern Asia. Positive AOD trends are also observed over most tropical oceans. The negative MODIS AOD trends observed over coastal regions of Europe and near the east coast of the USA are in agreement with sun photometer observations and *in situ* measurements (Section 2.2.3.2) of aerosol mass in these regions. These regional changes over oceans are consistent with analyses of AVHRR (Advanced Very High Resolution Radiometer) trends for 1981–2005 (Mishchenko et al., 2007; Cermak et al., 2010; Zhao et al., 2011), except over the Southern Ocean (45°S to 60°S), where negative AOD trends of AVHRR retrievals are neither confirmed by MODIS after 2001 (Zhang and Reid, 2010) nor by ATSR-2 (Along Track Scanning Radiometer) for 1995–2001 (Thomas et al., 2010).

Satellite-based AOD changes for both land and oceans (Figure 2.9b) were examined with re-processed SeaWiFS (Sea-viewing Wide Field-of-view Sensor) AOD data for 1998–2010 (Hsu et al., 2012). A small positive global average AOD trend is reported, which is likely influenced by interannual natural aerosol emissions variability (e.g., related to ENSO or North Atlantic Oscillation (NAO); Box 2.5), and compensating larger positive and negative regional AOD trends. In addition, temporal changes in aerosol composition are ignored in the retrieval algorithms, giving more uncertain trends than suggested by statistical analysis alone (Mishchenko et al., 2012). Thus, *confidence* is *low* for global satellite derived AOD trends over these relatively short time periods.

The sign and magnitude of SeaWiFS regional AOD trends over continents are in agreement with most AOD trends by ground-based sun photometer data (see above) and with MODIS trends (Figure 2.9). The strong positive AOD trend over the Arabian Peninsula occurs mainly during spring (MAM) and summer (JJA), during times of dust transport, and is also visible in MODIS data (Figure 2.9). The positive AOD trend over southern and eastern Asia is strongest during the dry seasons (DJF, MAM), when reduced wet deposition allows anthropogenic aerosol to accumulate in the troposphere. AOD over the Saharan outflow region off western Africa displays the strongest seasonal AOD trend differences, with AOD increases only in spring, but strong AOD decreases during the other seasons. SeaWifs AOD decreases over Europe and the USA and increases over southern and eastern Asia (especially during the dry season) are in agreement with reported temporal trends in anthropogenic emissions, and surface observations (Section 2.2.3.2).

In summary, based on satellite- and surface-based remote sensing it is *very likely* that AOD has decreased over Europe and the eastern

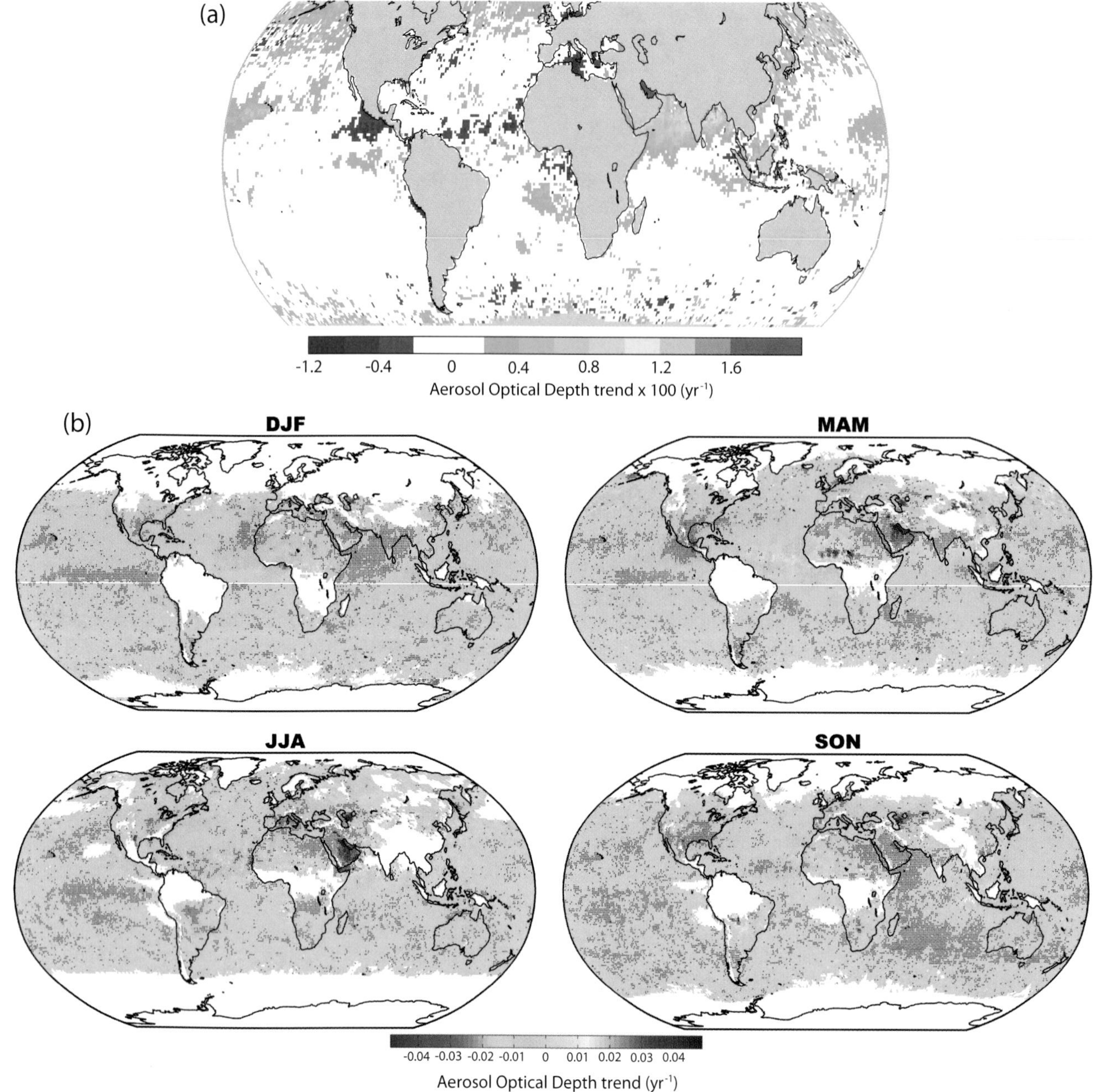

Figure 2.9 | (a) Annual average aerosol optical depth (AOD) trends at 0.55 µm for 2000–2009, based on de-seasonalized, conservatively cloud-screened MODIS aerosol data over oceans (Zhang and Reid, 2010). Negative AOD trends off Mexico are due to enhanced volcanic activity at the beginning of the record. Most non-zero trends are significant (i.e., a trend of zero lies outside the 95% confidence interval). (b) Seasonal average AOD trends at 0.55 µm for 1998–2010 using SeaWiFS data (Hsu et al., 2012). White areas indicate incomplete or missing data. Black dots indicate significant trends (i.e., a trend of zero lies outside the 95% confidence interval).

USA since the mid 1990s and increased over eastern and southern Asia since 2000. In the 2000s dust-related AOD has been increasing over the Arabian Peninsula and decreasing over the North Atlantic Ocean. Aerosol trends over other regions are less strong or not significant during this period owing to relative strong interannual variability. Overall, *confidence* in satellite-based global average AOD trends is *low*.

2.2.3.2 *In Situ* Surface Aerosol Measurements

AR4 did not report trends in long-term surface-based *in situ* measurements of particulate matter, its components or its properties. This section summarizes reported trends of PM_{10}, $PM_{2.5}$ (particulate matter with aerodynamic diameters <10 and <2.5 µm, respectively), sulphate and equivalent black carbon/elemental carbon, from regionally representative measurement networks. An overview of current networks and definitions pertinent to aerosol measurements is given in Supplementary Material 2.SM.2.3. Studies reporting trends representative of regional changes are presented in Table 2.2. Long-term data are almost entirely from North America and Europe, whereas a few individual studies on aerosol trends in India and China are reported in Supplementary Material 2.SM.2.3. Figure 2.10 gives an overview of observed PM_{10}, $PM_{2.5}$, and sulphate trends in North America and Europe for 1990–2009 and 2000–2009.

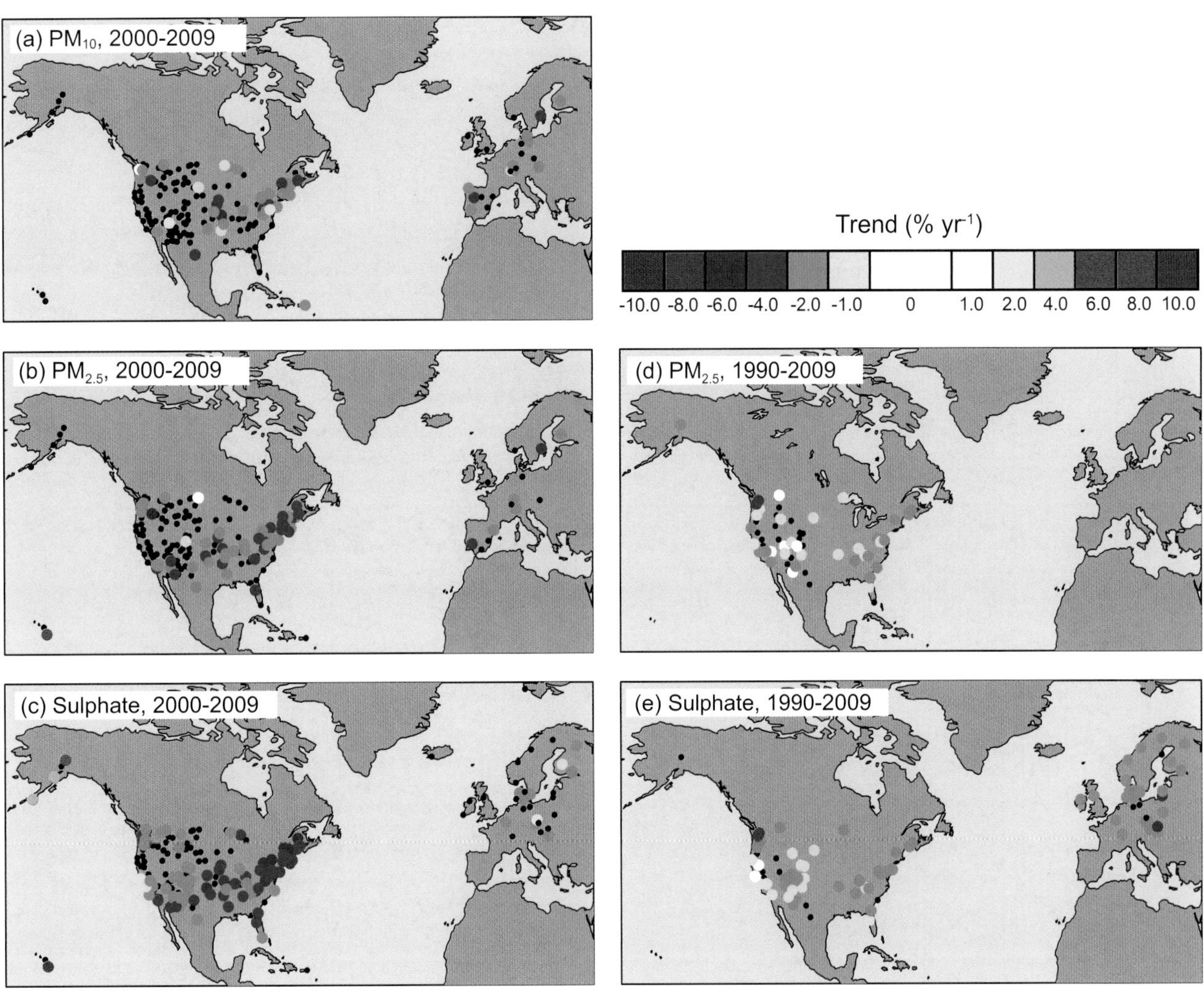

Figure 2.10 | Trends in particulate matter (PM_{10} and $PM_{2.5}$ with aerodynamic diameters <10 and <2.5 µm, respectively) and sulphate in Europe and USA for two overlapping periods 2000–2009 (a, b, c) and 1990–2009 (d, e). The trends are based on measurements from the EMEP (Torseth et al., 2012) and IMPROVE (Hand et al., 2011) networks in Europe and USA, respectively. Sites with significant trends (i.e., a trend of zero lies outside the 95% confidence interval) are shown in colour; black dots indicate sites with non-significant trends.

In Europe, strong downward trends are observed for PM_{10}, $PM_{2.5}$ and sulphate from the rural stations in the EMEP (European Monitoring and Evaluation Programme) network. For 2000–2009, $PM_{2.5}$ shows an average reduction of 3.9% yr^{-1} for the six stations with significant trends, while trends are not significant at seven other stations. Over 2000–2009, PM_{10} at 12 (out of 24) sites shows significant downward trend of on average 2.6% yr^{-1}. Similarly sulphate strongly decreased at 3.1% yr^{-1} from 1990 to 2009 with 26 of 30 sites having significant reductions. The largest decrease occurred before 2000, while for 2000–2009, the trends were weaker and less robust. This is consistent with reported emission reductions of 65% from 1990 to 2000 and 28% from 2001 to 2009 (Yttri et al., 2011; Torseth et al., 2012). Model analysis (Pozzoli et al., 2011) attributed the trends in large part to emission changes.

In the USA, the largest reductions in PM and sulphate are observed in the 2000s, rather than the 1990s as in Europe. IMPROVE (U.S. Interagency Monitoring of Protected Visual Environments Network) $PM_{2.5}$ measurements (Hand et al., 2011) show significant downward trends averaging 4.0% yr^{-1} for 2000–2009 at sites with significant trends, and 2.1% yr^{-1} at all sites, and PM_{10} decreases of 3.1% yr^{-1} for 2000–2009. Declines of $PM_{2.5}$ and SO_4^{2-} in Canada are very similar (Hidy and Pennell, 2010), with annual mean $PM_{2.5}$ at urban measurement sites decreasing by 3.6% yr^{-1} during 1985–2006 (Canada, 2012).

In the eastern and southwestern USA, IMPROVE data show strong sulphate declines, which range from 2 to 6% yr^{-1}, with an average of 2.3% yr^{-1} for the sites with significant negative trends for 1990–2009. However, four IMPROVE sites show strong SO_4^{2-} increases from 2000 to 2009, amounting to 11.9% yr^{-1}, at Hawaii (1225 m above sea level), and 4 to 7% yr^{-1} at three sites in southwest Alaska.

A recent study on long-term trends in aerosol optical properties from 24 globally distributed background sites (Collaud Coen et al., 2013) reported statistically significant trends at 16 locations, but the sign and magnitude of the trends varied largely with the aerosol property considered and geographical region (Table 2.3). Among the sites, this study reported strong increases in absorption and scattering coefficients in the free troposphere at Mauna Loa, Hawaii (3400 m above sea level), which is a regional feature also evident in the satellite-based AOD

Table 2.2 | Trend estimates for various aerosol variables reported in the literature, using data sets with at least 10 years of measurements. Unless otherwise noted, trends of individual stations were reported in % yr^{-1}, and 95% confidence intervals. The standard deviation (in parentheses) is determined from the individual trends of a set of regional stations.

Aerosol variable	Trend, % yr^{-1} (1σ, standard deviation)	Period	Reference	Comments
Europe				
$PM_{2.5}$	−2.9 (1.31) −3.9 (0.87)[b]	2000–2009	(Adapted from Torseth et al., 2012) Regional background sites	13 sites available, 6 sites show statistically significant results. Average change was −0.37 and −0.52[b] mg m^{-3} yr^{-1}.
PM_{10}	−1.9 (1.43) −2.6 (1.19)[b]	2000–2009		24 sites available, 12 sites show statistically significant results. Average change was −0.29 and −0.40[b] mg m^{-3} yr^{-1} .
SO_4^{2-}	−3.0 (0.82) −3.1 (0.72)[b]	1990–2009		30 sites available, 26 sites show statistically significant results. Average change was −0.04 and −0.04[b] mg m^{-3} yr^{-1}.
SO_4^{2-}	− 1.5 (1.41) −2.0 (1.8)[b]	2000–2009		30 sites available, 10 sites show statistically significant results. Average change was −0.01 and −0.01[b] mg m^{-3} yr^{-1}.
PM_{10}	−1.9	1991–2008	(Barmpadimos et al., 2012) Rural and urban sites	10 sites in Switzerland. The trend is adjusted for change in meteorology—unadjusted data did not differ strongly. The average change was −0.51 mg m^{-3} yr^{-1}.
USA				
$PM_{2.5}$	−2.1 (2.08) −4.0 (1.01)[b]	2000–2009	Adapted from (Hand et al., 2011) Regional background sites	153 sites available, 52 sites show statistically significant negative results. Only 1 site shows statisticallly positive trend.
$PM_{2.5}$	−1.5 (1.25) −2.1 (0.97)[b]	1990–2009		153 sites available, 39 sites show statistically significant results.
PM_{10}	−1.7 (2.00) −3.1 (1.65)[b]	2000–2009		154 sites available, 37 sites show statistically significant results.
SO_4^{2-}	−3.0 (2.86) −3.0 (0.62)[b]	2000–2009		154 sites available, 83 sites show statistically significant negative results. 4 sites showed statistical positive trend.
SO_4^{2-}	−2.0 (1.07) −2.3 (0.85)[b]	1990–2009		103 sites available, 41 sites show statistically significant results.
Total Carbon	−2.5 to −7.5	1989–2008	(Hand et al., 2011) Regional background sites	The trend interval includes about 50 sites mainly located along the East and West Coasts of the USA; fewer sites were situated in the central part of the continent.
Arctic				
EBC[a]	−3.8 (0.7)[c]	1989–2008	(Hirdman et al., 2010)	Alert, Canada 62.3°W 82.5°N
SO_4^{2-}	−3.0 (0.6)[c]	1985–2006		
EBC[a]	Not sig.[c]	1998–2008		Barrow, Alaska, 156.6°W 71.3° N
SO_4^{2-}	Not sig.[c]	1997–2008		
EBC[a]	−9.0 (5.0)[c]	2002–2009		Zeppelin, Svalbard, 11.9°E 78.9° N
SO_4^{2-}	−1.9 (1.7)[c]	1990–2008		

Notes:

[a] Equivalent black carbon.

[b] Trend numbers indicated refer to the subset of stations with significant changes over time—generally in regions strongly influenced by anthropogenic emissions (Figure 2.10).

[c] Trend values significant at 1% level.

trends (illustrated in Figure 2.9). Possible explanations for these changes include the influence of increasing Asian emissions and changes in clouds and removal processes. More and longer Asian time series, coupled with transport analyses, are needed to corroborate these findings. Aerosol number concentrations (Asmi et al., 2013) are declining significantly at most sites in Europe, North America, the Pacific and the Caribbean, but increasing at South Pole based on a study of 17 globally distributed remote sites.

Total carbon (= light absorbing carbon + organic carbon) measurements indicate highly significant downward trends between 2.5 and 7.5% yr^{-1} along the east and west coasts of the USA, and smaller and less significant trends in other regions of the USA from 1989 to 2008 (Hand et al., 2011; Murphy et al., 2011).

In Europe, Torseth et al. (2012) suggest a slight reduction in elemental carbon concentrations at two stations from 2001 to 2009, subject to large interannual variability. Collaud Coen et al. (2013) reported consistent negative trends in the aerosol absorption coefficient at stations in the continental USA, Arctic and Antarctica, but mostly insignificant trends in Europe over the last decade.

In the Arctic, changes in aerosol impact the atmosphere's radiative balance as well as snow and ice albedo. Similar to Europe and the USA, Hirdman et al. (2010) reported downward trends in equivalent black carbon and SO_4^{2-} for two out of total three Arctic stations and attributed them to emission changes.

In summary, declining AOD in Europe and North America is corroborated by *very likely* downward trends in ground-based *in situ* particulate matter measurements since the mid-1980s. Robust evidence from around 200 regional background sites with *in situ* ground based aerosol measurements indicate downward trends in the last two decades of $PM_{2.5}$ in parts of Europe (2 to 6% yr^{-1}) and the USA (1 to 2.5% yr^{-1}),

Table 2.3 | Summary table of aerosol optical property trends reported in the literature, using data sets with at least 10 years of measurements. Otherwise as in Table 2.2.

Region	Trend, % yr^{-1}(1σ, standard deviation)	Period	Reference	Comments
Scattering coefficient				
Europe (4/1)	+0.6 (1.9) +2.7[a]	2001–2010	Adapted from (Collaud Coen et al., 2013) Regional background sites	Trend study including 24 regional background sites with more than 10 years of observations. Regional averages for last 10 years are included here. Values in parenthesis show total number of sites/number of sites with significant trend.
USA (14/10)	−2.0 (2.5) −2.9 (2.4)[a]			
Mauna Loa (1/1)	+2.7			
Arctic (1/0)	+2.4			
Antarctica (1/0)	+2.5			
Absorption coefficient				
Europe (3/0)	+0.3 (0.4)	2001–2010	Adapted from (Collaud Coen et al., 2013) Regional background sites	Trend study of aerosol optical properties including 24 regional background sites with more than 10 years of observations. Regional averages for last 10 years are included here. Values in parenthesis show total number of sites and number of sites with significant trend.
USA (1/1)	−2.0			
Mauna Loa (1/1)	+9.0			
Arctic (1/1)	−6.5			
Antarctica (1/1)	−0.1			
Particle number concentration				
Europe (4/2)	−0.9 (1.8) −2.3 (1.0)[a]	2001–2010	Adapted from (Asmi et al., 2013) Regional background sites	Trend study of particle number concentration (N) and size distribution including 17 regional background sites. Regional averages of particle number concentration for last 10 years are included here. Values in parentheses show total number of sites and number of sites with significant trend.
North America and Caribbean (3/3)	−5.3 (2.8) −6.6 (1.1)[a]			
Mauna Loa (1/1)	−3.5			
Arctic (1/0)	−1.3			
Antarctica (2/2)	+2.7 (1.4)			

Notes:

[a] Trend numbers indicated refer to the subset of stations with significant changes over time—generally in regions strongly influenced by anthropogenic emissions (Figure 2.10).

Box 2.2 | Quantifying Changes in the Mean: Trend Models and Estimation

Many statistical methods exist for estimating trends in environmental time series (see Chandler and Scott, 2011 for a review). The assessment of long-term changes in historical climate data requires trend models that are transparent and robust, and that can provide credible uncertainty estimates.

Linear Trends

Historical climate trends are frequently described and quantified by estimating the linear component of the change over time (e.g., AR4). Such linear trend modelling has broad acceptance and understanding based on its frequent and widespread use in the published research assessed in this report, and its strengths and weaknesses are well known (von Storch and Zwiers, 1999; Wilks, 2006). Challenges exist in assessing the uncertainty in the trend and its dependence on the assumptions about the sampling distribution (Gaussian or otherwise), uncertainty in the data, dependency models for the residuals about the trend line, and treating their serial correlation (Von Storch, 1999; Santer et al., 2008).

The quantification and visualization of temporal changes are assessed in this chapter using a linear trend model that allows for first-order autocorrelation in the residuals (Santer et al., 2008; Supplementary Material 2.SM.3). Trend slopes in such a model are the same as ordinary least squares trends; uncertainties are computed using an approximate method. The 90% confidence interval quoted is solely that arising from sampling uncertainty in estimating the trend. Structural uncertainties, to the extent sampled, are apparent from the range of estimates from different data sets. Parametric and other remaining uncertainties (Box 2.1), for which estimates are provided with some data sets, are not included in the trend estimates shown here, so that the same method can be applied to all data sets considered.

Nonlinear Trends

There is no *a priori* physical reason why the long-term trend in climate variables should be linear in time. Climatic time series often have trends for which a straight line is not a good approximation (e.g., Seidel and Lanzante, 2004). The residuals from a linear fit in time often do not follow a simple autoregressive or moving average process, and linear trend estimates can easily change when recalculated

(continued on next page)

2

Box 2.2 (continued)

for shorter or longer time periods or when new data are added. When linear trends for two parts of a longer time series are calculated separately, the trends calculated for two shorter periods may be very different (even in sign) from the trend in the full period, if the time series exhibits significant nonlinear behavior in time (Box 2.2, Table 1).

Many methods have been developed for estimating the long-term change in a time series without assuming that the change is linear in time (e.g., Wu et al., 2007; Craigmile and Guttorp, 2011). Box 2.2, Figure 1 shows the linear least squares and a nonlinear trend fit to the GMST values from the HadCRUT4 data set (Section 2.4.3). The nonlinear trend is obtained by fitting a smoothing spline trend (Wood, 2006; Scinocca et al., 2010) while allowing for first-order autocorrelation in the residuals (Supplementary Material 2.SM.3). The results indicate that there are significant departures from linearity in the trend estimated this way.

Box 2.2, Table 1 shows estimates of the change in the GMST from the two methods. The methods give similar estimates with 90% confidence intervals that overlap one another. Smoothing methods that do not assume the trend is linear can provide useful information on the structure of change that is not as well treated with linear fits. The linear trend fit is used in this chapter because it can be applied consistently to all the data sets, is relatively simple, transparent and easily comprehended, and is frequently used in the published research assessed here.

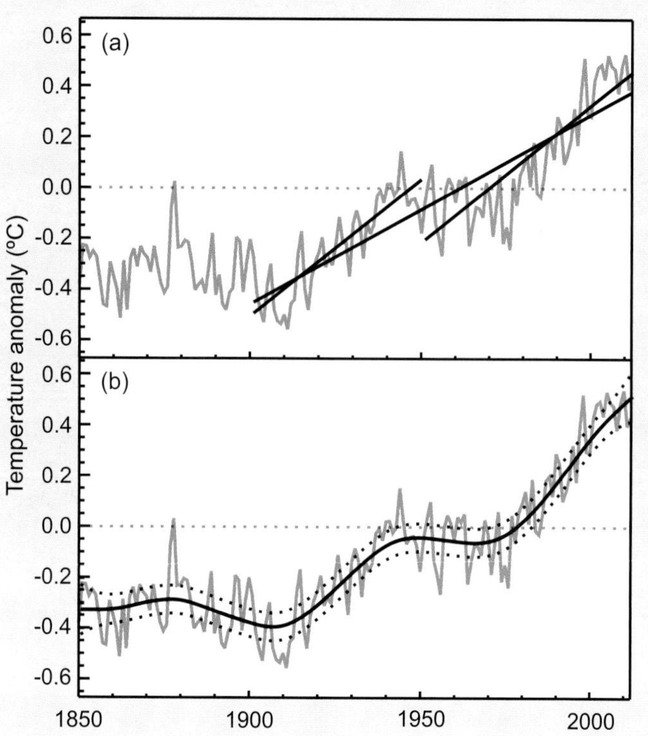

Box 2.2, Figure 1 | (a) Global mean surface temperature (GMST) anomalies relative to a 1961–1990 climatology based on HadCRUT4 annual data. The straight black lines are least squares trends for 1901–2012, 1901–1950 and 1951–2012. (b) Same data as in (a), with smoothing spline (solid curve) and the 90% confidence interval on the smooth curve (dashed lines). Note that the (strongly overlapping) 90% confidence intervals for the least square lines in (a) are omitted for clarity. See Figure 2.20 for the other two GMST data products.

Box 2.2, Table 1 | Estimates of the mean change in global mean surface temperature (GMST) between 1901 and 2012, 1901 and 1950, and 1951 and 2012, obtained from the linear (least squares) and nonlinear (smoothing spline) trend models. Half-widths of the 90% confidence intervals are also provided for the estimated changes from the two trend methods.

	Trends in °C per decade		
Method	**1901–2012**	**1901–1950**	**1951–2012**
Least squares	0.075 ± 0.013	0.107 ± 0.026	0.106 ± 0.027
Smoothing spline	0.081 ± 0.010	0.070 ± 0.016	0.090 ± 0.018

and also for SO_4^{2-} (2 to 5% yr^{-1}). The strongest decreases were in the 1990s in Europe and in the 2000s in the USA. There is robust evidence for downward trends of light absorbing aerosol in the USA and the Arctic, while elsewhere in the world *in situ* time series are lacking or not long enough to reach statistical significance.

2.3 Changes in Radiation Budgets

The radiation budget of the Earth is a central element of the climate system. On average, radiative processes warm the surface and cool the atmosphere, which is balanced by the hydrological cycle and sensible heating. Spatial and temporal energy imbalances due to radiation and latent heating produce the general circulation of the atmosphere and oceans. Anthropogenic influence on climate occurs primarily through perturbations of the components of the Earth radiation budget.

The radiation budget at the top of the atmosphere (TOA) includes the absorption of solar radiation by the Earth, determined as the difference between the incident and reflected solar radiation at the TOA, as well as the thermal outgoing radiation emitted to space. The surface radiation budget takes into account the solar fluxes absorbed at the Earth's surface, as well as the upward and downward thermal radiative fluxes emitted by the surface and atmosphere, respectively. In view of new observational evidence since AR4, the mean state as well as multi-decadal changes of the surface and TOA radiation budgets are assessed in the following.

2.3.1 Global Mean Radiation Budget

Since AR4, knowledge on the magnitude of the radiative energy fluxes in the climate system has improved, requiring an update of the global annual mean energy balance diagram (Figure 2.11). Energy exchanges between Sun, Earth and Space are observed from space-borne platforms such as the Clouds and the Earth's Radiant Energy System (CERES, Wielicki et al., 1996) and the Solar Radiation and Climate Experiment (SORCE, Kopp and Lawrence, 2005) which began data collection in 2000 and 2003, respectively. The total solar irradiance (TSI) incident at the TOA is now much better known, with the SORCE Total Irradiance Monitor (TIM) instrument reporting uncertainties as low as 0.035%, compared to 0.1% for other TSI instruments (Kopp et al., 2005). During the 2008 solar minimum, SORCE/TIM observed a solar irradiance of 1360.8 ± 0.5 W m^{-2} compared to 1365.5 ± 1.3 W m^{-2} for instruments launched prior to SORCE and still operating in 2008 (Section 8.4.1.1). Kopp and Lean (2011) conclude that the SORCE/TIM value of TSI is the most credible value because it is validated by a National Institute of Standards and Technology calibrated cryogenic radiometer. This revised TSI estimate corresponds to a solar irradiance close to 340 W m^{-2} globally averaged over the Earth's sphere (Figure 2.11).

The estimate for the reflected solar radiation at the TOA in Figure 2.11, 100 W m^{-2}, is a rounded value based on the CERES Energy Balanced and Filled (EBAF) satellite data product (Loeb et al., 2009, 2012b) for the period 2001–2010. This data set adjusts the solar and thermal TOA fluxes within their range of uncertainty to be consistent with independent estimates of the global heating rate based on *in situ* ocean observations (Loeb et al., 2012b). This leaves 240 W m^{-2} of solar radiation absorbed by the Earth, which is nearly balanced by thermal emission to space of about 239 W m^{-2} (based on CERES EBAF), considering a global heat storage of 0.6 W m^{-2} (imbalance term in Figure 2.11) based on Argo data from 2005 to 2010 (Hansen et al., 2011; Loeb et al., 2012b; Box 3.1). The stated uncertainty in the solar reflected TOA fluxes from CERES due to uncertainty in absolute calibration alone is about 2% (2-sigma), or equivalently 2 W m^{-2} (Loeb et al., 2009). The uncertainty of the outgoing thermal flux at the TOA as measured by CERES due to calibration is ~3.7 W m^{-2} (2σ). In addition to this, there is uncertainty in removing the influence of instrument spectral response on measured radiance, in radiance-to-flux conversion, and in time–space averaging, which adds up to another 1 W m^{-2} (Loeb et al., 2009).

2

The components of the radiation budget at the surface are generally more uncertain than their counterparts at the TOA because they cannot be directly measured by passive satellite sensors and surface measurements are not always regionally or globally representative. Since AR4, new estimates for the downward thermal infrared (IR) radiation at

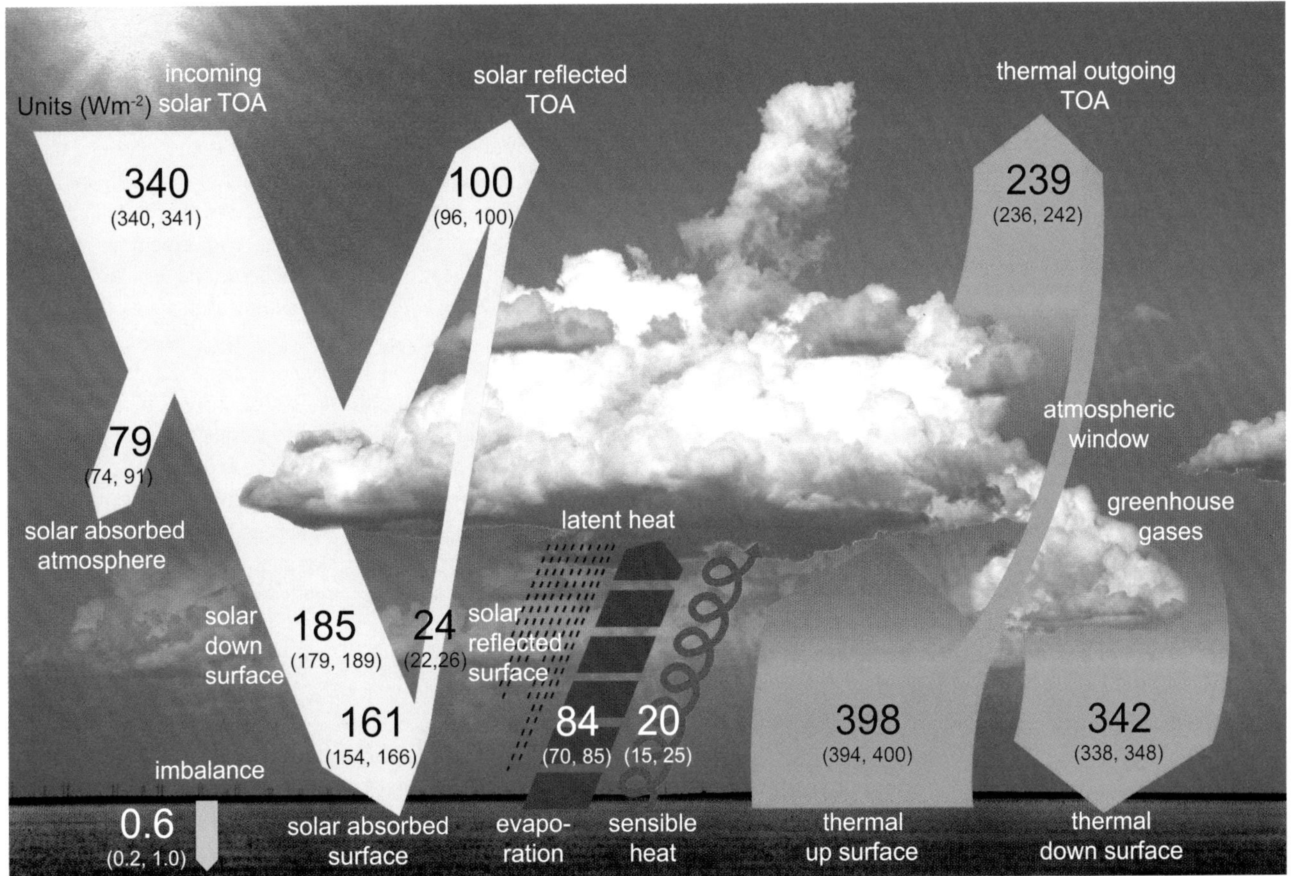

Figure 2.11: | Global mean energy budget under present-day climate conditions. Numbers state magnitudes of the individual energy fluxes in W m^{-2}, adjusted within their uncertainty ranges to close the energy budgets. Numbers in parentheses attached to the energy fluxes cover the range of values in line with observational constraints. (Adapted from Wild et al., 2013.)

the surface have been established that incorporate critical information on cloud base heights from space-borne radar and lidar instruments (L'Ecuyer et al., 2008; Stephens et al., 2012a; Kato et al., 2013). In line with studies based on direct surface radiation measurements (Wild et al., 1998, 2013) these studies propose higher values of global mean downward thermal radiation than presented in previous IPCC assessments and typically found in climate models, exceeding 340 W m^{-2} (Figure 2.11). This aligns with the downward thermal radiation in the ERA-Interim and ERA-40 reanalyses (Box 2.3), of 341 and 344 W m^{-2}, respectively (Berrisford et al., 2011). Estimates of global mean downward thermal radiation computed as a residual of the other terms of the surface energy budget (Kiehl and Trenberth, 1997; Trenberth et al., 2009) are lower (324 to 333 W m^{-2}), highlighting remaining uncertainties in estimates of both radiative and non-radiative components of the surface energy budget.

2

Estimates of absorbed solar radiation at the Earth's surface include considerable uncertainty. Published global mean values inferred from satellite retrievals, reanalyses and climate models range from below 160 W m^{-2} to above 170 W m^{-2}. Recent studies taking into account surface observations as well as updated spectroscopic parameters and continuum absorption for water vapor favour values towards the lower bound of this range, near 160 W m^{-2}, and an atmospheric solar absorption around 80 W m^{-2} (Figure 2.11) (Kim and Ramanathan, 2008; Trenberth et al., 2009; Kim and Ramanathan, 2012; Trenberth and Fasullo, 2012b; Wild et al., 2013). The ERA-Interim and ERA-40 reanalyses further support an atmospheric solar absorption of this magnitude (Berrisford et al., 2011). Latest satellite-derived estimates constrained by CERES now also come close to these values (Kato et al., in press). Recent independently derived surface radiation estimates favour therefore a global mean surface absorbed solar flux near 160 W m^{-2} and a downward thermal flux slightly above 340 W m^{-2}, respectively (Figure 2.11).

The global mean latent heat flux is required to exceed 80 W m^{-2} to close the surface energy balance in Figure 2.11, and comes close to the 85 W m^{-2} considered as upper limit by Trenberth and Fasullo (2012b) in view of current uncertainties in precipitation retrieval in the Global Precipitation Climatology Project (GPCP, Adler et al., 2012) (the latent heat flux corresponds to the energy equivalent of evaporation, which globally equals precipitation; thus its magnitude may be constrained by global precipitation estimates). This upper limit has recently been challenged by Stephens et al. (2012b). The emerging debate reflects potential remaining deficiencies in the quantification of the radiative and non-radiative energy balance components and associated uncertainty ranges, as well as in the consistent representation of the global mean energy and water budgets (Stephens et al., 2012b; Trenberth and Fasullo, 2012b; Wild et al., 2013). Relative uncertainty in the globally averaged sensible heat flux estimate remains high owing to the very limited direct observational constraints (Trenberth et al., 2009; Stephens et al., 2012b).

In summary, newly available observations from both space-borne and surface-based platforms allow a better quantification of the Global Energy Budget, even though notable uncertainties remain, particularly in the estimation of the non-radiative surface energy balance components.

2.3.2 Changes in Top of the Atmosphere Radiation Budget

While the previous section emphasized the temporally-averaged state of the radiation budget, the focus in the following is on the temporal (multi-decadal) changes of its components. Variations in TSI are discussed in Section 8.4.1. AR4 reported large changes in tropical TOA radiation between the 1980s and 1990s based on observations from the Earth Radiation Budget Satellite (ERBS) (Wielicki et al., 2002; Wong et al., 2006). Although the robust nature of the large decadal changes in tropical radiation remains to be established, several studies have suggested links to changes in atmospheric circulation (Allan and Slingo, 2002; Chen et al., 2002; Clement and Soden, 2005; Merrifield, 2011) (Section 2.7).

Since AR4, CERES enabled the extension of satellite records of TOA fluxes into the 2000s (Loeb et al., 2012b). The extended records from CERES suggest no noticeable trends in either the tropical or global radiation budget during the first decade of the 21st century (e.g., Andronova et al., 2009; Harries and Belotti, 2010; Loeb et al., 2012a, 2012b). Comparisons between ERBS/CERES thermal radiation and that derived from the NOAA High Resolution Infrared Radiation Sounder (HIRS) (Lee et al., 2007) show good agreement until approximately 1998, corroborating the rise of 0.7 W m^{-2} between the 1980s and 1990s reported in AR4. Thereafter the HIRS thermal fluxes show much higher values, likely due to changes in the channels used for HIRS/3 instruments launched after October 1998 compared to earlier HIRS instruments (Lee et al., 2007).

On a global scale, interannual variations in net TOA radiation and ocean heating rate (OHR) should correspond, as oceans have a much larger effective heat capacity than land and atmosphere, and therefore serve as the main reservoir for heat added to the Earth–atmosphere system (Box 3.1). Wong et al. (2006) showed that interannual variations in these two data sources are in good agreement for 1992–2003. In the ensuing 5 years, however, Trenberth and Fasullo (2010) note that the two diverge with ocean *in situ* measurements (Levitus et al., 2009), indicating a decline in OHR, in contrast to expectations from the observed net TOA radiation. The divergence after 2004 is referred to as "missing energy" by Trenberth and Fasullo (2012b), who further argue that the main sink of the missing energy likely occurs at ocean depths below 275 m. Loeb et al. (2012b) compared interannual variations in CERES net radiation with OHRs derived from three independent ocean heat content anomaly analyses and included an error analysis of both CERES and the OHRs. They conclude that the apparent decline in OHR is not statistically robust and that differences between interannual variations in OHR and satellite net TOA flux are within the uncertainty of the measurements (Figure 2.12). They further note that between January 2001 and December 2012, the Earth has been steadily accumulating energy at a rate of 0.50 ± 0.43 W m^{-2} (90% CI). Hansen et al. (2011) obtained a similar value for 2005–2010 using an independent analysis of the ocean heat content anomaly data (von Schuckmann and Le Traon, 2011). The variability in the Earth's energy imbalance is strongly influenced by ocean circulation changes relating to the ENSO (Box 2.5); during cooler La Niña years (e.g., 2009) less thermal radiation is emitted and the climate system gains heat while the reverse is true for warmer El Niño years (e.g., 2010) (Figure 2.12).

In summary, satellite records of TOA radiation fluxes have been substantially extended since AR4. It is *unlikely* that significant trends exist in global and tropical radiation budgets since 2000. Interannual variability in the Earth's energy imbalance related to ENSO is consistent with ocean heat content records within observational uncertainty.

2.3.3 Changes in Surface Radiation Budget

2.3.3.1 Surface Solar Radiation

Changes in radiative fluxes at the surface can be traced further back in time than the satellite-based TOA fluxes, although only at selected terrestrial locations where long-term records exist. Monitoring of radiative fluxes from land-based stations began on a widespread basis in the mid-20th century, predominantly measuring the downward solar component, also known as global radiation or surface solar radiation (SSR).

AR4 reported on the first indications for substantial decadal changes in observational records of SSR. Specifically, a decline of SSR from the beginning of widespread measurements in the 1950s until the mid-1980s has been observed at many land-based sites (popularly known as 'global dimming'; Stanhill and Cohen, 2001; Liepert, 2002), as well as a partial recovery from the 1980s onward ('brightening'; Wild et al., 2005) (see the longest available SSR series of Stockholm, Sweden, in Figure 2.13 as an illustrative example).

Since AR4, numerous studies have substantiated the findings of significant decadal SSR changes observed both at worldwide distributed terrestrial sites (Dutton et al., 2006; Wild et al., 2008; Gilgen et al., 2009; Ohmura, 2009; Wild, 2009 and references therein) as well as in specific regions. In Europe, Norris and Wild (2007) noted a dimming between 1971 and 1986 of 2.0 to 3.1 W m^{-2} per decade and subsequent brightening of 1.1 to 1.4 W m^{-2} per decade from 1987 to 2002 in a pan-European time series comprising 75 sites. Similar tendencies were found at sites in northern Europe (Stjern et al., 2009), Estonia (Russak, 2009) and Moscow (Abakumova et al., 2008). Chiacchio and Wild (2010) pointed out that dimming and subsequent brightening in Europe is seen mainly in spring and summer. Brightening in Europe from the 1980s onward was further documented at sites in Switzerland, Germany, France, the Benelux, Greece, Eastern Europe and the Iberian Peninsula (Ruckstuhl et al., 2008; Wild et al., 2009; Zerefos et al., 2009; Sanchez-Lorenzo et al., 2013). Concurrent brightening of 2 to 8 W m^{-2} per decade was also noted at continental sites in the USA (Long et al., 2009; Riihimaki et al., 2009; Augustine and Dutton, 2013). The general pattern of dimming and consecutive brightening was further found at numerous sites in Japan (Norris and Wild, 2009; Ohmura, 2009; Kudo et al., 2011) and in the SH in New Zealand (Liley, 2009). Analyses of observations from sites in China confirmed strong declines in SSR from the 1960s to 1980s on the order of 2 to 8 W m^{-2} per decade, which also did not persist in the 1990s (Che et al., 2005; Liang and Xia, 2005; Qian et al., 2006; Shi et al., 2008; Norris and Wild, 2009; Xia, 2010a). On the other hand, persistent dimming since the mid-20th

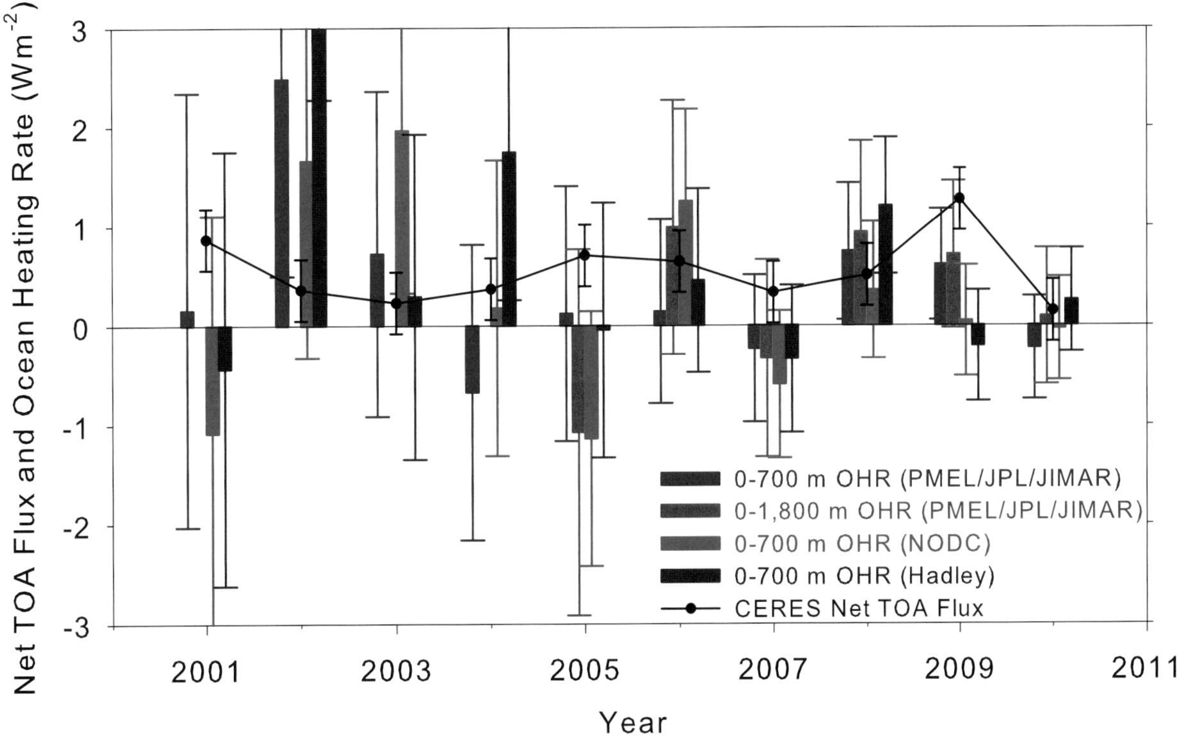

Figure 2.12 | Comparison of net top of the atmosphere (TOA) flux and upper ocean heating rates (OHRs). Global annual average (July to June) net TOA flux from CERES observations (based on the EBAF-TOA_Ed2.6r product) (black line) and 0–700 (blue) and 0–1800 m (red) OHR from the Pacific Marine Environmental Laboratory/Jet Propulsion Laboratory/Joint Institute for Marine and Atmospheric Research (PMEL/JPL/JIMAR), 0–700 m OHR from the National Oceanic Data Center (NODC) (green; Levitus et al., 2009), and 0–700 m OHR from the Hadley Center (brown; Palmer et al., 2007). The length of the coloured bars corresponds to the magnitude of OHR. Thin vertical lines are error bars, corresponding to the magnitude of uncertainties. Uncertainties for all annual OHR are given at one standard error derived from ocean heat content anomaly uncertainties (Lyman et al., 2010). CERES net TOA flux uncertainties are given at the 90% confidence level determined following Loeb et al. (2012b). (Adapted from Loeb et al., 2012b.)

century with no evidence for a trend reversal was noted at sites in India (Wild et al., 2005; Kumari et al., 2007; Kumari and Goswami, 2010; Soni et al., 2012) and in the Canadian Prairie (Cutforth and Judiesch, 2007). Updates on latest SSR changes observed since 2000 provide a less coherent picture (Wild, 2012). They suggest a continuation of brightening at sites in Europe, USA, and parts of Asia, a levelling off at sites in Japan and Antarctica, and indications for a renewed dimming in parts of China (Wild et al., 2009; Xia, 2010a).

The longest observational SSR records, extending back to the 1920s and 1930s at a few sites in Europe, further indicate some brightening during the first half of the 20th century, known as 'early brightening' (cf. Figure 2.13) (Ohmura, 2009; Wild, 2009). This suggests that the decline in SSR, at least in Europe, was confined to a period between the 1950s and 1980s.

A number of issues remain, such as the quality and representativeness of some of the SSR data as well as the large-scale significance of the phenomenon (Wild, 2012). The historic radiation records are of variable quality and rigorous quality control is necessary to avoid spurious trends (Dutton et al., 2006; Shi et al., 2008; Gilgen et al., 2009; Tang et al., 2011; Wang et al., 2012e; Sanchez-Lorenzo et al., 2013). Since the mid-1990s, high-quality data are becoming increasingly available from new sites of the Baseline Surface Radiation Network (BSRN) and Atmospheric Radiation Measurement (ARM) Program, which allow the determination of SSR variations with unprecedented accuracy (Ohmura et al., 1998). Alpert et al. (2005) and Alpert and Kishcha (2008) argued that the observed SSR decline between 1960 and 1990 was larger in densely populated than in rural areas. The magnitude of this 'urbanization effect' in the radiation data is not yet well quantified. Dimming and brightening is, however, also notable at remote and rural sites (Dutton et al., 2006; Karnieli et al., 2009; Liley, 2009; Russak, 2009; Wild, 2009; Wang et al., 2012d).

Globally complete satellite estimates have been available since the early 1980s (Hatzianastassiou et al., 2005; Pinker et al., 2005; Hinkelman et al., 2009). Because satellites do not directly measure the surface fluxes, they have to be inferred from measurable TOA signals using empirical or physical models to remove atmospheric perturbations. Available satellite-derived products qualitatively agree on a brightening from the mid-1980s to 2000 averaged globally as well as over oceans, on the order of 2 to 3 W m^{-2} per decade (Hatzianastassiou et al., 2005; Pinker et al., 2005; Hinkelman et al., 2009). Averaged over land, however, trends are positive or negative depending on the respective satellite product (Wild, 2009). Knowledge of the decadal variation of aerosol burdens and optical properties, required in satellite retrievals of SSR and considered relevant for dimming/brightening particularly over land, is very limited (Section 2.2.3). Extensions of satellite-derived SSR beyond 2000 indicate tendencies towards a renewed dimming at the beginning of the new millennium (Hinkelman et al., 2009; Hatzianastassiou et al., 2012).

Reconstructions of SSR changes from more widely measured meteorological variables can help to increase their spatial and temporal coverage. Multi-decadal SSR changes have been related to observed changes in sunshine duration, atmospheric visibility, diurnal temperature range (DTR; Section 2.4.1.2) and pan evaporation (Section 2.5.3).

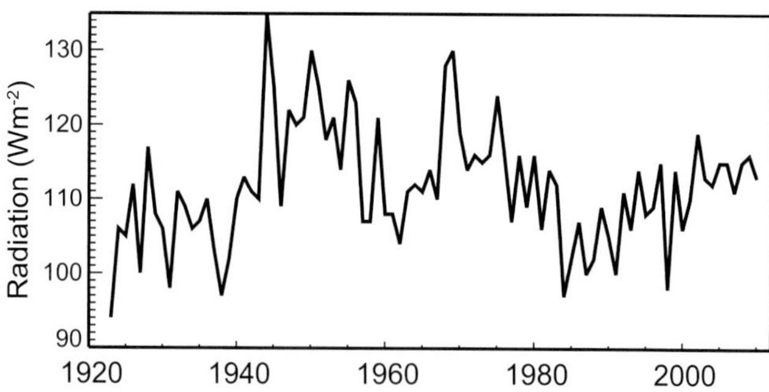

Figure 2.13 | Annual mean Surface Solar Radiation (SSR) as observed at Stockholm, Sweden, from 1923 to 2010. Stockholm has the longest SSR record available worldwide. (Updated from Wild (2009) and Ohmura (2009).)

Overall, these proxies provide independent evidence for the existence of large-scale multi-decadal variations in SSR. Specifically, widespread observations of declines in pan evaporation from the 1950s to the 1980s were related to SSR dimming amongst other factors (Roderick and Farquhar, 2002). The observed decline in DTR over global land surfaces from the 1950s to the 1980s (Section 2.4.1.2), and its stabilisation thereafter fits to a large-scale dimming and subsequent brightening, respectively (Wild et al., 2007). Widespread brightening after 1980 is further supported by reconstructions from sunshine duration records (Wang et al., 2012e). Over Europe, SSR dimming and subsequent brightening is consistent with concurrent declines and increases in sunshine duration (Sanchez-Lorenzo et al., 2008), evaporation in energy limited environments (Teuling et al., 2009), visibility records (Vautard et al., 2009; Wang et al., 2009b) and DTR (Makowski et al., 2009). The early brightening in the 1930s and 1940s seen in a few European SSR records is in line with corresponding changes in sunshine duration and DTR (Sanchez-Lorenzo et al., 2008; Wild, 2009; Sanchez-Lorenzo and Wild, 2012). In China, the levelling off in SSR in the 1990s after decades of decline coincides with similar tendencies in the pan evaporation records, sunshine duration and DTR (Liu et al., 2004a; Liu et al., 2004b; Qian et al., 2006; Ding et al., 2007; Wang et al., 2012d). Dimming up to the 1980s and subsequent brightening is also indicated in a set of 237 sunshine duration records in South America (Raichijk, 2011).

2.3.3.2 Surface Thermal and Net Radiation

Thermal radiation, also known as longwave, terrestrial or far-IR radiation is sensitive to changes in atmospheric GHGs, temperature and humidity. Long-term measurements of the thermal surface components as well as surface net radiation are available at far fewer sites than SSR. Downward thermal radiation observations started to become available during the early 1990s at a limited number of globally distributed terrestrial sites. From these records, Wild et al. (2008) determined an overall increase of 2.6 W m^{-2} per decade over the 1990s, in line with model projections and the expectations of an increasing greenhouse effect. Wang and Liang (2009) inferred an increase in downward thermal radiation of 2.2 W m^{-2} per decade over the period 1973–2008 from globally available terrestrial observations of temperature, humidity and cloud fraction. Prata (2008) estimated a slightly lower increase of 1.7 W m^{-2} per decade for clear sky conditions over the earlier period 1964–1990, based on observed temperature and humidity profiles from globally distributed land-based radiosonde stations and radiative transfer calculations. Philipona et al. (2004; 2005) and Wacker et al. (2011) noted increasing downward thermal fluxes recorded in the Swiss Alpine Surface Radiation Budget (ASRB) network since the mid-1990s, corroborating an increasing greenhouse effect. For mainland Europe, Philipona et al. (2009) estimated an increase of downward thermal radiation of 2.4 to 2.7 W m^{-2} per decade for the period 1981–2005.

There is limited observational information on changes in surface net radiation, in large part because measurements of upward fluxes at the surface are made at only a few sites and are not spatially representative. Wild et al. (2004, 2008) inferred a decline in land surface net radiation on the order of 2 W m^{-2} per decade from the 1960s to the 1980s, and an increase at a similar rate from the 1980s to 2000, based on estimated changes of the individual radiative components that constitute the surface net radiation. Philipona et al. (2009) estimated an increase in surface net radiation of 1.3 to 2 W m^{-2} per decade for central Europe and the Alps between 1981 and 2005.

2.3.3.3 Implications from Observed Changes in Related Climate Elements

The observed multi-decadal SSR variations cannot be explained by changes in TSI, which are an order of magnitude smaller (Willson and Mordvinov, 2003). They therefore have to originate from alterations in the transparency of the atmosphere, which depends on the presence of clouds, aerosols and radiatively active gases (Kvalevag and Myhre, 2007; Kim and Ramanathan, 2008). Cloud cover changes (Section 2.5.7) effectively modulate SSR on an interannual basis, but their contribution to the longer-term SSR trends is ambiguous. Although cloud cover changes were found to explain the trends in some areas (e.g., Liley, 2009), this is not always the case, particularly in relatively polluted regions (Qian et al., 2006; Norris and Wild, 2007, 2009; Wild, 2009; Kudo et al., 2012). SSR dimming and brightening has also been observed under cloudless atmospheres at various locations, pointing to a prominent role of atmospheric aerosols (Wild et al., 2005; Qian et al., 2007; Ruckstuhl et al., 2008; Sanchez-Lorenzo et al., 2009; Wang et al., 2009b; Zerefos et al., 2009).

Box 2.3 | Global Atmospheric Reanalyses

Dynamical reanalyses are increasingly used for assessing weather and climate phenomena. Given their more frequent use in this assessment compared to AR4, their characteristics are described in more detail here.

Reanalyses are distinct from, but complement, more 'traditional' statistical approaches to assessing the raw observations. They aim to produce continuous reconstructions of past atmospheric states that are consistent with all observations as well as with atmospheric physics as represented in a numerical weather prediction model, a process termed data assimilation. Unlike real-world observations, reanalyses are uniform in space and time and provide non-observable variables (e.g., potential vorticity).

Several groups are actively pursuing reanalysis development at the global scale, and many of these have produced several generations of reanalyses products (Box 2.3, Table 1). Since the first generation of reanalyses produced in the 1990s, substantial development has taken place. The NASA Modern-Era Retrospective Analysis for Research and Applications (MERRA) and ERA-Interim reanalyses show improved tropical precipitation and hence better represent the global hydrological cycle (Dee et al., 2011b). The NCEP/CFSR reanalysis

(continued on next page)

Box 2.3, Table 1 | Overview of global dynamical reanalysis data sets (ranked by start year; the period extends to present if no end year is provided). A further description of reanalyses and their technical derivation is given in pp. S33–35 of Blunden et al. (2011). Approximate resolution is calculated as 1000 km * *20/N* (with N denoting the spectral truncation, Laprise, 1992).

Institution	Reanalysis	Period	Approximate Resolution at Equator	Reference
Cooperative Institute for Research in Environmental Sciences (CIRES), National Oceanic and Atmospheric Administration (NOAA), USA	20th Century Reanalysis, Vers. 2 (20CR)	1871–2010	320 km	Compo et al. (2011)
National Centers for Environmental Prediction (NCEP) and National Center for Atmospheric Research (NCAR), USA	NCEP/NCAR R1 (NNR)	1948–	320 km	Kistler et al. (2001)
European Centre for Medium-Range Weather Forecasts (ECMWF)	ERA-40	1957–2002	125 km	Uppala et al. (2005)
Japan Meteorological Agency (JMA)	JRA-55	1958–	60 km	Ebita et al. (2011)
National Centers for Environmental Prediction (NCEP), US Department of Energy, USA	NCEP/DOE R2	1979–	320 km	Kanamitsu et al. (2002)
Japan Meteorological Agency (JMA)	JRA-25	1979–	190 km	Onogi et al. (2007)
National Aeronautics and Space Administration (NASA), USA	MERRA	1979–	75 km	Rienecker et al. (2011)
European Centre for Medium-Range Weather Forecasts (ECMWF)	ERA-Interim	1979–	80 km	Dee et al. (2011b)
National Centers for Environmental Prediction (NCEP), USA	CFSR	1979–	50 km	Saha et al. (2010)

Box 2.3 (continued)

uses a coupled ocean–atmosphere–land-sea–ice model (Saha et al., 2010). The 20th Century Reanalyses (20CR, Compo et al., 2011) is a 56-member ensemble and covers 140 years by assimilating only surface and sea level pressure (SLP) information. This variety of groups and approaches provides some indication of the robustness of reanalyses when compared. In addition to the global reanalyses, several regional reanalyses exist or are currently being produced.

Reanalyses products provide invaluable information on time scales ranging from daily to interannual variability. However, they may often be unable to characterize long-term trends (Trenberth et al., 2011). Although reanalyses projects by definition use a 'frozen' assimilation system, there are many other sources of potential errors. In addition to model biases, changes in the observational systems (e.g., coverage, introduction of satellite data) and time-dependent errors in the underlying observations or in the boundary conditions lead to step changes in time, even in latest generation reanalyses (Bosilovich et al., 2011).

2

Errors of this sort were ubiquitous in early generation reanalyses and rendered them of limited value for trend characterization (Thorne and Vose, 2010). Subsequent products have improved and uncertainties are better understood (Dee et al., 2011a), but artefacts are still present. As a consequence, trend adequacy depends on the variable under consideration, the time period and the region of interest. For example, surface air temperature and humidity trends over land in the ERA-Interim reanalysis compare well with observations (Simmons et al., 2010), but polar tropospheric temperature trends in ERA-40 disagree with trends derived from radiosonde and satellite observations (Bitz and Fu, 2008; Grant et al., 2008; Graversen et al., 2008; Thorne, 2008; Screen and Simmonds, 2011) owing to problems that were resolved in ERA-Interim (Dee et al., 2011a).

Studies based on reanalyses are used cautiously in AR5 and known inadequacies are pointed out and referenced. Later generation reanalyses are preferred where possible; however, literature based on these new products is still sparse.

Aerosols can directly attenuate SSR by scattering and absorbing solar radiation, or indirectly, through their ability to act as cloud condensation nuclei, thereby changing cloud reflectivity and lifetime (Chapter 7). SSR dimming and brightening is often reconcilable with trends in anthropogenic emission histories and atmospheric aerosol loadings (Stern, 2006; Streets et al., 2006; Mishchenko et al., 2007; Ruckstuhl et al., 2008; Ohvril et al., 2009; Russak, 2009; Streets et al., 2009; Cermak et al., 2010; Wild, 2012). Recent trends in aerosol optical depth derived from satellites indicate a decline in Europe since 2000 (Section 2.2.3), in line with evidence from SSR observations. However, direct aerosol effects alone may not be able to account for the full extent of the observed SSR changes in remote regions with low pollution levels (Dutton and Bodhaine, 2001; Schwartz, 2005). Aerosol indirect effects have not yet been well quantified, but have the potential to amplify aerosol-induced SSR trends, particularly in relatively pristine environments, such as over oceans (Wild, 2012).

SSR trends are also qualitatively in line with observed multi-decadal surface warming trends (Chapter 10), with generally smaller warming rates during phases of declining SSR, and larger warming rates in phases of increasing SSR (Wild et al., 2007). This is seen more pronounced for the relatively polluted NH than the more pristine SH (Wild, 2012). For Europe, Vautard et al. (2009) found that a decline in the frequency of low-visibility conditions such as fog, mist and haze over the past 30 years and associated SSR increase may be responsible for 10 to 20% of Europe's recent daytime warming, and 50% of Eastern European warming. Philipona (2012) noted that both warming and brightening are weaker in the European Alps compared to the surrounding lowlands with stronger aerosol declines since 1981.

Reanalyses and observationally based methods have been used to show that increased atmospheric moisture with warming (Willett et al., 2008; Section 2.5) enhances thermal radiative emission of the atmosphere to the surface, leading to reduced net thermal cooling of the surface (Prata, 2008; Allan, 2009; Philipona et al., 2009; Wang and Liang, 2009).

In summary, the evidence for widespread multi-decadal variations in solar radiation incident on land surfaces has been substantiated since AR4, with many of the observational records showing a decline from the 1950s to the 1980s ('dimming'), and a partial recovery thereafter ('brightening'). *Confidence* in these changes is *high* in regions with high station densities such as over Europe and parts of Asia. These *likely* changes are generally supported by observed changes in related, but more widely measured variables, such as sunshine duration, DTR and hydrological quantities, and are often in line with aerosol emission patterns. Over some remote land areas and over the oceans, *confidence* is *low* owing to the lack of direct observations, which hamper a truly global assessment. Satellite-derived SSR fluxes support the existence of brightening also over oceans, but are less consistent over land surface where direct aerosol effects become more important. There are also indications for increasing downward thermal and net radiation at terrestrial stations since the early 1990s with *medium confidence*.

2.4 Changes in Temperature

2.4.1 Land Surface Air Temperature

2.4.1.1 Large-Scale Records and Their Uncertainties

AR4 concluded global land-surface air temperature (LSAT) had increased over the instrumental period of record, with the warming rate approximately double that reported over the oceans since 1979. Since AR4, substantial developments have occurred including the production of revised data sets, more digital data records, and new data set efforts. These innovations have improved understanding of data issues and uncertainties, allowing better quantification of regional changes. This reinforces confidence in the reported globally averaged LSAT time series behaviour.

Global Historical Climatology Network Version 3 (GHCNv3) incorporates many improvements (Lawrimore et al., 2011) but was found to be virtually indistinguishable at the global mean from version 2 (used in AR4). Goddard Institute of Space Studies (GISS) continues to provide an estimate based upon primarily GHCN, accounting for urban impacts through nightlights adjustments (Hansen et al., 2010). CRUTEM4 (Jones et al., 2012) incorporates additional station series and also newly homogenized versions of many individual station records. A new data product from a group based predominantly at Berkeley (Rohde et al., 2013a) uses a method that is substantially distinct from earlier efforts (further details on all the data sets and data availability are given in Supplementary Material 2.SM.4). Despite the range of approaches, the long-term variations and trends broadly agree among these various LSAT estimates, particularly after 1900. Global LSAT has increased (Figure 2.14, Table 2.4).

Since AR4, various theoretical challenges have been raised over the verity of global LSAT records (Pielke et al., 2007). Globally, sampling and methodological independence has been assessed through sub-sampling (Parker et al., 2009; Jones et al., 2012), creation of an entirely new and structurally distinct product (Rohde et al., 2013b) and a complete reprocessing of GHCN (Lawrimore et al., 2011). None of these yielded more than minor perturbations to the global LSAT records since 1900. Willett et al. (2008) and Peterson et al. (2011) explicitly showed that changes in specific and relative humidity (Section 2.5.5) were physically consistent with reported temperature trends, a result replicated in the ERA reanalyses (Simmons et al., 2010). Various investigators (Onogi et al., 2007; Simmons et al., 2010; Parker, 2011; Vose et al., 2012a) showed that LSAT estimates from modern reanalyses were in quantitative agreement with observed products.

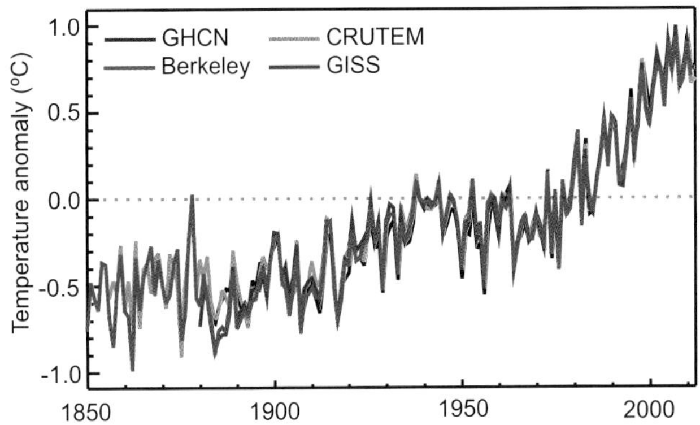

Figure 2.14 | Global annual average land-surface air temperature (LSAT) anomalies relative to a 1961–1990 climatology from the latest versions of four different data sets (Berkeley, CRUTEM, GHCN and GISS).

Particular controversy since AR4 has surrounded the LSAT record over the United States, focussed on siting quality of stations in the US Historical Climatology Network (USHCN) and implications for long-term trends. Most sites exhibit poor current siting as assessed against official WMO siting guidance, and may be expected to suffer potentially large siting-induced absolute biases (Fall et al., 2011). However, overall biases for the network since the 1980s are *likely* dominated by instrument type (owing to replacement of Stevenson screens with maximum minimum temperature systems (MMTS) in the 1980s at the majority of sites), rather than siting biases (Menne et al., 2010; Williams et al., 2012). A new automated homogeneity assessment approach (also used in GHCNv3, Menne and Williams, 2009) was developed that has been shown to perform as well or better than other contemporary approaches (Venema et al., 2012). This homogenization procedure *likely* removes much of the bias related to the network-wide changes in the 1980s (Menne et al., 2010; Fall et al., 2011; Williams et al., 2012). Williams et al. (2012) produced an ensemble of data set realizations using perturbed settings of this procedure and concluded through assessment against plausible test cases that there existed a propensity to under-estimate adjustments. This propensity is critically dependent upon the (unknown) nature of the inhomogeneities in the raw data records. Their homogenization increases both minimum temperature and maximum temperature centennial-time-scale USA average LSAT trends. Since 1979 these adjusted data agree with a range of reanalysis products whereas the raw records do not (Fall et al., 2010; Vose et al., 2012a).

Regional analyses of LSAT have not been limited to the United States. Various national and regional studies have undertaken assessments for Europe (Winkler, 2009; Bohm et al., 2010; Tietavainen et al., 2010; van

Table 2.4: | Trend estimates and 90% confidence intervals (Box 2.2) for LSAT global average values over five common periods.

Data Set	Trends in °C per decade				
	1880–2012	1901–2012	1901–1950	1951–2012	1979–2012
CRUTEM4.1.1.0 (Jones et al., 2012)	0.086 ± 0.015	0.095 ± 0.020	0.097 ± 0.029	0.175 ± 0.037	0.254 ± 0.050
GHCNv3.2.0 (Lawrimore et al., 2011)	0.094 ± 0.016	0.107 ± 0.020	0.100 ± 0.033	0.197 ± 0.031	0.273 ± 0.047
GISS (Hansen et al., 2010)	0.095 ± 0.015	0.099 ± 0.020	0.098 ± 0.032	0.188 ± 0.032	0.267 ± 0.054
Berkeley (Rohde et al., 2013)	0.094 ± 0.013	0.101 ± 0.017	0.111 ± 0.034	0.175 ± 0.029	0.254 ± 0.049

der Schrier et al., 2011), China (Li et al., 2009; Zhen and Zhong-Wei, 2009; Li et al., 2010a; Tang et al., 2010), India (Jain and Kumar, 2012), Australia (Trewin, 2012), Canada (Vincent et al., 2012), South America, (Falvey and Garreaud, 2009) and East Africa (Christy et al., 2009). These analyses have used a range of methodologies and, in many cases, more data and metadata than available to the global analyses. Despite the range of analysis techniques they are generally in broad agreement with the global products in characterizing the long-term changes in mean temperatures. This includes some regions, such as the Pacific coast of South America, that have exhibited recent cooling (Falvey and Garreaud, 2009). Of specific importance for the early global records, large (>1°C) summer time warm bias adjustments for many European 19th century and early 20th century records were revisited and broadly confirmed by a range of approaches (Bohm et al., 2010; Brunet et al., 2011).

Since AR4 efforts have also been made to interpolate Antarctic records from the sparse, predominantly coastal ground-based network (Chapman and Walsh, 2007; Monaghan et al., 2008; Steig et al., 2009; O'Donnell et al., 2011). Although these agree that Antarctica as a whole has warmed since the late 1950s, substantial multi-annual to multi-decadal variability and uncertainties in reconstructed magnitude and spatial trend structure yield only *low confidence* in the details of pan-Antarctic regional LSAT changes.

In summary, it is certain that globally averaged LSAT has risen since the late 19th century and that this warming has been particularly marked since the 1970s. Several independently analyzed global and regional LSAT data products support this conclusion. There is *low confidence* in changes prior to 1880 owing to the reduced number of estimates, non-standardized measurement techniques, the greater spread among the estimates and particularly the greatly reduced observational sampling. *Confidence* is also *low* in the spatial detail and magnitude of LSAT trends in sparsely sampled regions such as Antarctica. Since AR4 significant efforts have been undertaken to identify and adjust for data issues and new estimates have been produced. These innovations have further strengthened overall understanding of the global LSAT records.

2.4.1.2 Diurnal Temperature Range

In AR4 diurnal temperature range (DTR) was found, globally, to have narrowed since 1950, with minimum daily temperatures increasing faster than maximum daily temperatures. However, significant multi-decadal variability was highlighted including a recent period from 1997 to 2004 of no change, as both maximum and minimum temperatures rose at similar rates. The Technical Summary of AR4 highlighted changes in DTR and their causes as a key uncertainty. Since AR4, uncertainties in DTR and its physical interpretation have become even more apparent.

No dedicated global analysis of DTR has been undertaken subsequent to Vose et al. (2005a), although global behaviour has been discussed in two broader ranging analyses. Rohde et al. (2012) and Wild et al. (2007) note an apparent reversal since the mid-1980s; with DTR subsequently increasing. This decline and subsequent increase in DTR over global land surfaces is qualitatively consistent with the dimming and subsequent brightening noted in Section 2.3.3.1. Donat et al. (2013c) using HadEX2 (Section 2.6) find significant decreasing DTR trends in more than half of the land areas assessed but less than 10% of land with significant increases since 1951. Available trend estimates (–0.04 ± 0.01°C per decade over 1950–2011 (Rohde et al., 2013b) and –0.066°C per decade over 1950–2004 (Vose et al., 2005a)) are much smaller than global mean LSAT average temperature trends over 1951–2012 (Table 2.4). It therefore logically follows that globally averaged maximum and minimum temperatures over land have both increased by in excess of 0.1°C per decade since 1950.

Regionally, Makowski et al. (2008) found that DTR behaviour in Europe over 1950 to 2005 changed from a decrease to an increase in the 1970s in Western Europe and in the 1980s in Eastern Europe. Sen Roy and Balling (2005) found significant increases in both maximum and minimum temperatures for India, but little change in DTR over 1931–2002. Christy et al. (2009) reported that for East Africa there has been no pause in the narrowing of DTR in recent decades. Zhou and Ren (2011) reported a significant decrease in DTR over mainland China of –0.15°C per decade during 1961–2008.

Various investigators (e.g., Christy et al. (2009), Pielke and Matsui (2005), Zhou and Ren (2011)) have raised doubts about the physical interpretation of minimum temperature trends, hypothesizing that microclimate and local atmospheric composition impacts are more apparent because the dynamical mixing at night is much reduced. Parker (2006) investigated this issue arguing that if data were affected in this way, then a trend difference would be expected between calm and windy nights. However, he found no such minimum temperature differences on a global average basis. Using more complex boundary layer modelling techniques, Steeneveld et al. (2011) and McNider et al. (2012) showed much lower sensitivity to windspeed variations than posited by Pielke and Matsui but both concluded that boundary layer understanding was key to understanding the minimum temperature changes. Data analysis and long-term side-by-side instrumentation field studies show that real non-climatic data artefacts certainly affect maximum and minimum differently in the raw records for both recent (Fall et al., 2011; Williams et al., 2012) and older (Bohm et al., 2010; Brunet et al., 2011) records. Hence there could be issues over interpretation of apparent DTR trends and variability in many regions (Christy et al., 2006, 2009; Fall et al., 2011; Zhou and Ren, 2011; Williams et al., 2012), particularly when accompanied by regional-scale land-use/land-cover (LULC) changes (Christy et al., 2006).

In summary, *confidence* is *medium* in reported decreases in observed global DTR, noted as a key uncertainty in AR4. Several recent analyses of the raw data on which many previous analyses were based point to the potential for biases that differently affect maximum and minimum average temperatures. However, apparent changes in DTR are much smaller than reported changes in average temperatures and therefore it is *virtually certain* that maximum and minimum temperatures have increased since 1950.

2.4.1.3 Land Use Change and Urban Heat Island Effects

In AR4 Urban Heat Island (UHI) effects were concluded to be real local phenomena with negligible impact on large-scale trends. UHI and land-use land-cover change (LULC) effects arise mainly because the modified surface affects the storage and transfer of heat, water and airflow. For single discrete locations these impacts may dominate all other factors.

Regionally, most attention has focused on China. A variety of investigations have used methods as diverse as SST comparisons (e.g., Jones et al., 2008), urban minus rural (e.g., Ren et al., 2008; Yang et al., 2011), satellite observations (Ren and Ren, 2011) and observations minus reanalysis (e.g., Hu et al., 2010; Yang et al., 2011). Interpretation is complicated because often studies have used distinct versions of station series. For example, the effect in Beijing is estimated at 80% (Ren et al., 2007) or 40% (Yan et al., 2010) of the observed trend depending on data corrections applied. A representative sample of these studies suggest the effect of UHI and LULC is approximately 20% of the trend in Eastern China as a whole and of the order 0.1°C per decade nationally (Table 1 in Yang et al., 2011) over the last 30 years, but with very substantial uncertainties. These effects have *likely* been partially or completely accounted for in many homogenized series (e.g., Li et al., 2010b; Yan et al., 2010). Fujibe (2009) ascribes about 25% of Japanese warming trends in 1979–2006 to UHI effects. Das et al. (2011) confirmed that many Japanese sites have experienced UHI warming but that rural stations show unaffected behaviour when compared to nearby SSTs.

There is an important distinction to be made between UHI trend effects in regions underseeing rapid development and those that have been developed for a long time. Jones and Lister (2009) and Wilby et al. (2011) using data from London (UK) concluded that some sites that have always been urban and where the UHI has not grown in magnitude will exhibit regionally indicative trends that agree with nearby rural locations and that in such cases the time series may exhibit multi-decadal trends driven primarily by synoptic variations. A lack of obvious time-varying UHI influences was also noted for Sydney, Melbourne and Hobart in Australia by Trewin (2012). The impacts of urbanization also will be dependent on the natural LULC characteristics that they replace. Zhang et al. (2010) found no evidence for urban influences in the desert North West region of China despite rapid urbanization.

Global adjusted data sets *likely* account for much of the UHI effect present in the raw data. For the US network, Hausfather et al. (2013) showed that the adjustments method used in GHCNv3 removed much of an apparent systematic difference between urban and rural locations, concluding that this arose from adjustment of biased urban location data. Globally, Hansen et al. (2010) used satellite-based nightlight radiances to estimate the worldwide influence on LSAT of local urban development. Adjustments reduced the global 1900–2009 temperature change (averaged over land and ocean) only from 0.71°C to 0.70°C. Wickham et al. (2013) also used satellite data and found that urban locations in the Berkeley data set exhibited even less warming than rural stations, although not statistically significantly so, over 1950 to 2010.

Studies of the broader effects of LULC since AR4 have tended to focus on the effects of irrigation on temperatures, with a large number of studies in the Californian central belt (Christy et al., 2006; Kueppers et al., 2007; Bonfils et al., 2008; Lo and Famiglietti, 2013). They find cooler average temperatures and a marked reduction in DTR in areas of active irrigation and ascribe this to increased humidity; effectively a repartitioning of moist and dry energy terms. Reanalyses have also been used to estimate the LULC signature in LSAT trends. Fall et al. (2010) found that the North American Regional Reanalysis generated overall surface air temperature trends for 1979–2003 similar to observed records. Observations-minus-reanalysis trends were most positive for barren and urban areas, in accord with the results of Lim et al. (2008) using the NCEP/NCAR and ERA-40 reanalyses, and negative in agricultural areas.

McKitrick and Michaels (2004) and de Laat and Maurellis (2006) assessed regression of trends with national socioeconomic and geographical indicators, concluding that UHI and related LULC have caused much of the observed LSAT warming. AR4 concluded that this correlation ceases to be statistically significant if one takes into account the fact that the locations of greatest socioeconomic development are also those that have been most warmed by atmospheric circulation changes but provided no explicit evidence for this overall assessment result. Subsequently McKitrick and Michaels (2007) concluded that about half the reported warming trend in global-average land surface air temperature in 1980–2002 resulted from local land surface changes and faults in the observations. Schmidt (2009) undertook a quantitative analysis that supported AR4 conclusions that much of the reported correlation largely arose due to naturally occurring climate variability and model over-fitting and was not robust. Taking these factors into account, modified analyses by McKitrick (2010) and McKitrick and Nierenberg (2010) still yielded significant evidence for such contamination of the record.

In marked contrast to regression based studies, several studies have shown the methodologically diverse set of modern reanalysis products and the various LSAT records at global and regional levels to be similar since at least the mid-20th century (Simmons et al., 2010; Parker, 2011; Ferguson and Villarini, 2012; Jones et al., 2012; Vose et al., 2012a). These reanalyses do not directly assimilate the LSAT measurements but rather infer LSAT estimates from an observational constraint provided by much of the rest of the global observing system, thus representing an independent estimate. A hypothesized residual significant warming artefact argued for by regression-based analyses is therefore physically inconsistent with many other components of the global observing system according to a broad range of state-of-the-art data assimilation models (Box 2.3). Further, Efthymiadis and Jones (2010) estimated an absolute upper limit on urban influence globally of 0.02°C per decade, or about 15% of the total LSAT trends, in 1951–2009 from trends of coastal land and SST.

In summary, it is indisputable that UHI and LULC are real influences on raw temperature measurements. At question is the extent to which they remain in the global products (as residual biases in broader regionally representative change estimates). Based primarily on the range of urban minus rural adjusted data set comparisons and the degree of agreement of these products with a broad range of reanalysis products, it is *unlikely* that any uncorrected urban heat-island effects and LULC change effects have raised the estimated centennial globally averaged LSAT trends by more than 10% of the reported trend (*high confidence,* based on robust evidence and high agreement). This is an average value; in some regions with rapid development, UHI and LULC change impacts on regional trends may be substantially larger.

2.4.2 Sea Surface Temperature and Marine Air Temperature

AR4 concluded that 'recent' warming (since the 1950s) is strongly evident at all latitudes in SST over each ocean. Prominent spatio-temporal structures including the ENSO and decadal variability patterns in the Pacific Ocean (Box 2.5) and a hemispheric asymmetry in the Atlantic Ocean were highlighted as contributors to the regional differences in surface warming rates, which in turn affect atmospheric circulation. Since AR4 the availability of metadata has increased, data completeness has improved and a number of new SST products have been produced. Intercomparisons of data obtained by different measurement methods, including satellite data, have resulted in better understanding of errors and biases in the record.

2.4.2.1 Advances in Assembling Data Sets and in Understanding Data Errors

2.4.2.1.1 *In situ* data records

Historically, most SST observations were obtained from moving ships. Buoy measurements comprise a significant and increasing fraction of *in situ* SST measurements from the 1980s onward (Figure 2.15). Improvements in the understanding of uncertainty have been expedited by the use of metadata (Kent et al., 2007) and the recovery of observer instructions and other related documents. Early data were systematically cold biased because they were made using canvas or wooden buckets that, on average, lost heat to the air before the measurements were taken. This effect has long been recognized (Brooks, 1926), and prior to AR4 represented the only artefact adjusted in gridded SST products, such as HadSST2 (Rayner et al., 2006) and ERSST (Smith et al., 2005, 2008), which were based on 'bucket correction' methods by Folland and Parker (1995) and Smith and Reynolds (2002), respectively. The adjustments, made using ship observations of Night Marine Air Temperature (NMAT) and other sources, had a striking effect on the SST global mean estimates: note the difference in 1850–1941 between HadSST2 and International Comprehensive Ocean-Atmosphere Data Set (ICOADS) curves in Figure 2.16 (a brief description of SST and NMAT data sets and their methods is given in Supplementary Material 2.SM.4.3).

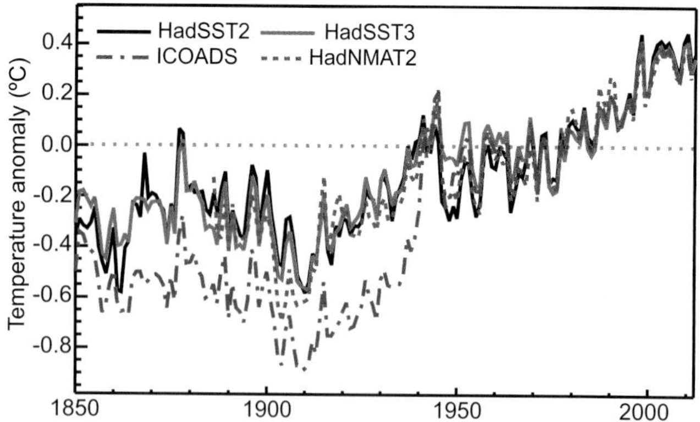

Figure 2.16 | Global annual average sea surface temperature (SST) and Night Marine Air Temperature (NMAT) relative to a 1961–1990 climatology from gridded data sets of SST observations (HadSST2 and its successor HadSST3), the raw SST measurement archive (ICOADS, v2.5) and night marine air temperatures data set HadNMAT2 (Kent et al., 2013). HadSST2 and HadSST3 both are based on SST observations from versions of the ICOADS data set, but differ in degree of measurement bias correction.

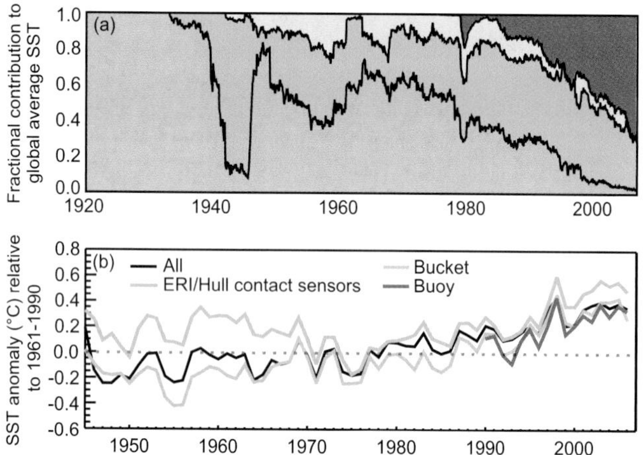

Figure 2.15 | Temporal changes in the prevalence of different measurement methods in the International Comprehensive Ocean-Atmosphere Data Set (ICOADS). (a) Fractional contributions of observations made by different measurement methods: bucket observations (blue), engine room intake (ERI) and hull contact sensor observations (green), moored and drifting buoys (red), and unknown (yellow). (b) Global annual average sea surface temperature (SST) anomalies based on different kinds of data: ERI and hull contact sensor (green), bucket (blue), buoy (red), and all (black). Averages are computed over all 5° × 5° grid boxes where both ERI/hull and bucket measurements, but not necessarily buoy data, were available. (Adapted from Kennedy et al., 2011a.)

Buckets of improved design and measurement methods with smaller, on average, biases came into use after 1941 (Figure 2.15, top); average biases were reduced further in recent decades, but not eliminated (Figure 2.15, bottom). Increasing density of SST observations made possible the identification (Reynolds et al., 2002, 2010; Kennedy et al., 2012) and partial correction of more recent period biases (Kennedy et al., 2011a). In particular, it is hypothesized that the proximity of the hot engine often biases engine room intake (ERI) measurements warm (Kent et al., 2010). Because of the prevalence of the ERI measurements among SST data from ships, the ship SSTs are biased warm by 0.12°C to 0.18°C on average compared to the buoy data (Reynolds et al., 2010; Kennedy et al., 2011a, 2012). An assessment of the potential impact of modern biases can be ascertained by considering the difference

Table 2.5 | Trend estimates and 90% confidence intervals (Box 2.2) for two subsequent versions of the HadSST data set over five common periods. HadSST2 has been used in AR4; HadSST3 is used in this chapter.

Data Set	Trends in °C per decade				
	1880–2012	1901–2012	1901–1950	1951–2012	1979–2012
HadSST3 (Kennedy et al., 2011a)	0.054 ± 0.012	0.067 ± 0.013	0.117 ± 0.028	0.074 ± 0.027	0.124 ± 0.030
HadSST2 (Rayner et al., 2006)	0.051 ± 0.015	0.069 ± 0.012	0.084 ± 0.055	0.098 ± 0.017	0.121 ± 0.033

between HadSST3 (bias corrections applied throughout) and HadSST2 (bucket corrections only) global means (Figure 2.16): it is particularly prominent in 1945–1970 period, when rapid changes in prevalence of ERI and bucket measurements during and after the World War II affect HadSST2 owing to the uncorrected measurement biases (Thompson et al., 2008), while these are corrected in HadSST3. Nevertheless, for periods longer than a century the effect of HadSST3-HadSST2 differences on linear trend slopes is small relative to the trend uncertainty (Table 2.5). Some degree of independent check on the validity of HadSST3 adjustments comes from a comparison to sub-surface temperature data (Gouretski et al., 2012) (see Section 3.2).

The traditional approach to modeling random error of *in situ* SST data assumed the independence of individual measurements. Kent and Berry (2008) identified the need to account for error correlation for measurements from the same "platform" (i.e., an individual ship or buoy), while measurement errors from different platforms remain independent.. Kennedy et al. (2011b) achieved that by introducing platform-dependent biases, which are constant within the same platform, but change randomly from one platform to another. Accounting for such correlated errors in HadSST3 resulted in estimated error for global and hemispheric monthly means that are more than twice the estimates given by HadSST2. The uncertainty in many, but not all, components of the HadSST3 product is represented by the ensemble of its realizations (Figure 2.17).

Data sets of marine air temperatures (MATs) have traditionally been restricted to nighttime series only (NMAT data sets) due to the direct solar heating effect on the daytime measurements, although corrected daytime MAT records for 1973–present are already available (Berry and Kent, 2009). Other major biases, affecting both nighttime and daytime MAT are due to increasing deck height with the general increase in the size of ships over time and non-standard measurement practices. Recently these biases were re-examined and explicit uncertainty calculation undertaken for NMAT by Kent et al. (2013), resulting in the HadNMAT2 data set.

2.4.2.1.2 Satellite SST data records

Satellite SST data sets are based on measuring electromagnetic radiation that left the ocean surface and got transmitted through the atmosphere. Because of the complexity of processes involved, the majority of such data has to be calibrated on the basis of *in situ* observations. The resulting data sets, however, provide a description of global SST fields with a level of spatial detail unachievable by *in situ* data only. The principal IR sensor is the Advanced Very High Resolution Radiometer (AVHRR). Since AR4, the AVHRR time series has been reprocessed consistently back to March 1981 (Casey et al., 2010) to create the AVHRR Pathfinder v5.2 data set. Passive microwave data sets of SST are available since 1997 equatorward of 40° and near-globally since 2002 (Wentz et al., 2000; Gentemann et al., 2004). They are generally less accurate than IR-based SST data sets, but their superior coverage in areas of persistent cloudiness provides SST estimates where the IR record has none (Reynolds et al., 2010).

The (Advanced) Along Track Scanning Radiometer (A)ATSR) series of three sensors was designed for climate monitoring of SST; their com-

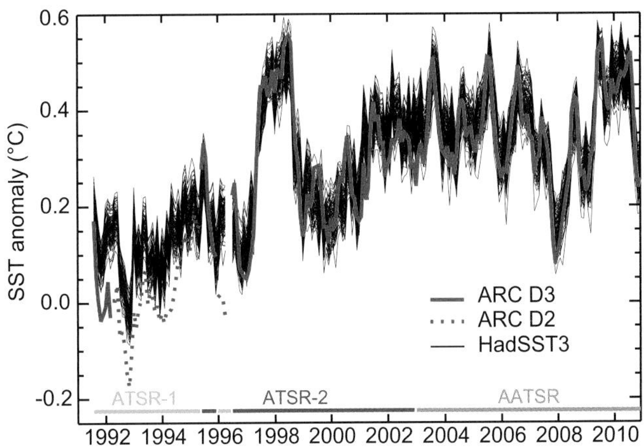

Figure 2.17 | Global monthly mean sea surface temperature (SST) anomalies relative to a 1961–1990 climatology from satellites (ATSRs) and *in situ* records (HadSST3). Black lines: the 100-member HadSST3 ensemble. Red lines: ATSR-based nighttime subsurface temperature at 0.2 m depth ($SST_{0.2m}$) estimates from the ATSR Reprocessing for Climate (ARC) project. Retrievals based on three spectral channels (D3, solid line) are more accurate than retrievals based on only two (D2, dotted line). Contributions of the three different ATSR missions to the curve shown are indicated at the bottom. The *in situ* and satellite records were co-located within 5° × 5° monthly grid boxes: only those where both data sets had data for the same month were used in the comparison. (Adapted from Merchant et al. 2012.)

bined record starts in August 1991 and exceeds two decades (it stopped with the demise of the ENVISAT platform in 2012). The (A) ATSRs are 'dual-view' IR radiometers intended to allow atmospheric effects removal without the use of *in situ* observations. Since AR4, (A)ATSR observations have been reprocessed with new estimation techniques (Embury and Merchant, 2011). The resulting SST products seem to be more accurate than many *in situ* observations (Embury et al., 2011). In terms of monthly global means, the agreement is illustrated in Figure 2.17. By analyzing (A)ATSR and *in situ* data together, Kennedy at al. (2012) verified and extended existing models for biases and random errors of *in situ* data.

2.4.2.2 Interpolated SST Products and Trends

SST data sets form a major part of global surface temperature analyses considered in this assessment report. To use an SST data set as a boundary condition for atmospheric reanalyses products (Box 2.3) or in atmosphere-only climate simulations (considered in Chapter 9 onwards), gridded data sets with complete coverage over the global ocean are typically needed. These are usually based on a special form of kriging (optimal interpolation) procedure that retains large-scale correlation structures and can accommodate very sparse data coverage. For the pre-satellite era (generally, before October 1981) only *in situ* data are used; for the latter period some products also use AVHRR data. Figure 2.18 compares interpolated SST data sets that extend back to the 19th century with the uninterpolated HadSST3 and HadNMAT2 products. Linear trend estimates for global mean SSTs from those products updated through 2012 are presented in Table 2.6. Differences between the trends from different data sets are larger when the calculation period is shorter (1979–2012) or has lower quality data (1901–1950); these are due mainly to different data coverage of underlying observational data sets and bias correction methods used in these products.

2

Table 2.6 | Trend estimates and 90% confidence intervals (Box 2.2) for interpolated SST data sets (uninterpolated state-of-the-art HadSST3 data set is included for comparison). Dash indicates not enough data available for trend calculation.

Data Set	Trends in °C per decade				
	1880–2012	1901–2012	1901–1950	1951–2012	1979–2012
HadISST (Rayner et al., 2003)	0.042 ± 0.007	0.052 ± 0.007	0.067 ± 0.024	0.064 ± 0.015	0.072 ± 0.024
COBE-SST (Ishii et al., 2005)	–	0.058 ± 0.007	0.066 ± 0.032	0.071 ± 0.014	0.073 ± 0.020
ERSSTv3b (Smith et al., 2008)	0.054 ± 0.015	0.071 ± 0.011	0.097 ± 0.050	0.088 ± 0.017	0.105 ± 0.031
HadSST3 (Kennedy et al., 2011a)	0.054 ± 0.012	0.067 ± 0.013	0.117 ± 0.028	0.074 ± 0.027	0.124 ± 0.030

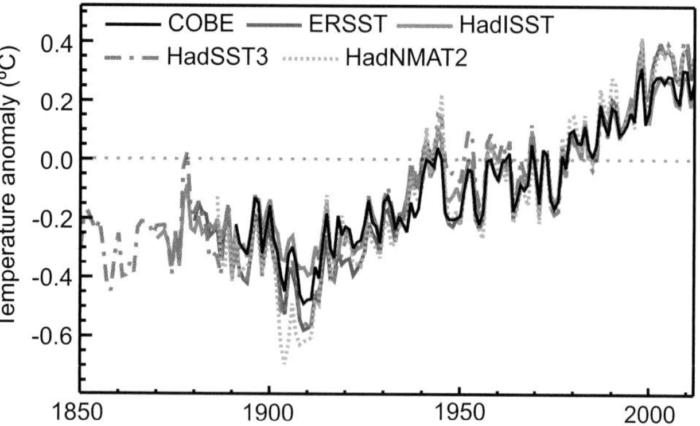

Figure 2.18 | Global annual average sea surface temperature (SST) and Night Marine Air Temperature (NMAT) relative to a 1961–1990 climatology from state of the art data sets. Spatially interpolated products are shown by solid lines; non-interpolated products by dashed lines.

In summary, it is certain that global average sea surface temperatures (SSTs) have increased since the beginning of the 20th century. Since AR4, major improvements in availability of metadata and data completeness have been made, and a number of new global SST records have been produced. Intercomparisons of new SST data records obtained by different measurement methods, including satellite data, have resulted in better understanding of uncertainties and biases in the records. Although these innovations have helped highlight and quantify uncertainties and affect our understanding of the character of changes since the mid-20th century, they do not alter the conclusion that global SSTs have increased both since the 1950s and since the late 19th century.

2.4.3 Global Combined Land and Sea Surface Temperature

AR4 concluded that the GMST had increased, with the last 50 years increasing at almost double the rate of the last 100 years. Subsequent developments in LSAT and SST have led to better understanding of the data and their uncertainties as discussed in preceding sections. This improved understanding has led to revised global products.

Changes have been made to all three GMST data sets that were used in AR4 (Hansen et al., 2010; Morice et al., 2012; Vose et al., 2012b). These are now in somewhat better agreement with each other over recent years, in large part because HadCRUT4 now better samples the NH high latitude land regions (Jones et al., 2012; Morice et al., 2012) which comparisons to reanalyses had shown led to a propensity for HadCRUT3 to underestimate recent warming (Simmons et al., 2010). Starting in the 1980s each decade has been significantly warmer at the Earth's surface than any preceding decade since the 1850s in HadCRUT4, a data set that explicitly quantifies a large number of sources of uncertainty (Figure 2.19). Each of the last three decades is also the warmest in the other two GMST data sets, but these have substantially less mature and complete uncertainty estimates, precluding such an assessment of significance of their decadal differences. The GISS and MLOST data sets fall outside the 90% CI of HadCRUT4 for several decades in the 20th century (Figure 2.19). These decadal differences could reflect residual biases in one or more data set, an incomplete treatment of uncertainties in HadCRUT4.1 or a combination of these effects (Box 2.1). The data sets utilize different LSAT (Section 2.4.1) and SST (Section 2.4.2) component records (Supplementary Material 2.SM.4.3.4) that in the case of SST differ somewhat in their multi-decadal trend behaviour (Table 2.6 compare HadSST3 and ERSSTv3b).

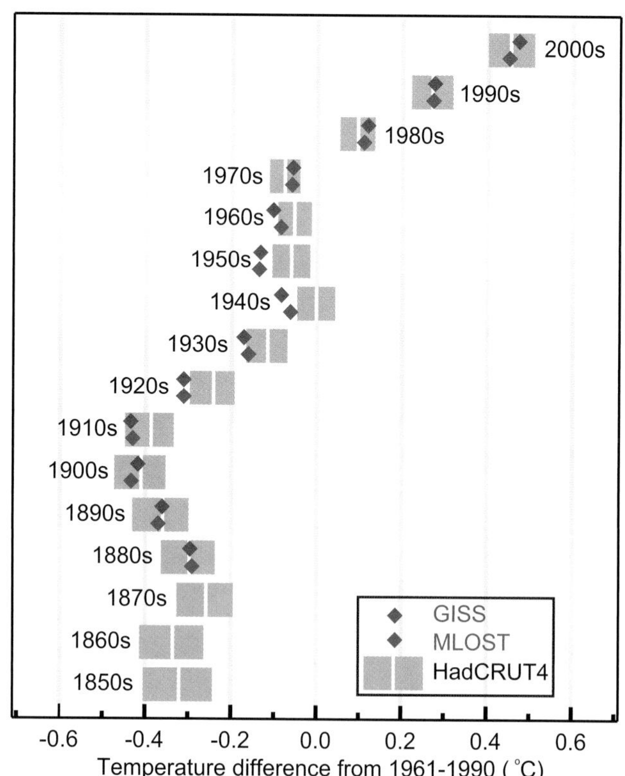

Figure 2.19 | Decadal global mean surface temperature (GMST) anomalies (white vertical lines in grey blocks) and their uncertainties (90% confidence intervals as grey blocks) based upon the land-surface air temperature (LSAT) and sea surface temperature (SST) combined HadCRUT4 (v4.1.1.0) ensemble (Morice et al., 2012). Anomalies are relative to a 1961–1990 climatology. 1850s indicates the period 1850-1859, and so on. NCDC MLOST and GISS data set best-estimates are also shown.

All ten of the warmest years have occurred since 1997, with 2010 and 2005 effectively tied for the warmest year on record in all three products. However, uncertainties on individual annual values are sufficiently large that the ten warmest years are statistically indistinguishable from one another. The global-mean trends are significant for all data sets and multi-decadal periods considered in Table 2.7. Using HadCRUT4 and its uncertainty estimates, the warming from 1850–1900 to 1986–2005 (reference period for the modelling chapters and Annex I) is 0.61 [0.55 to 0.67] °C (90% confidence interval), and the warming from 1850–1900 to 2003–2012 (the most recent decade) is 0.78 [0.72 to 0.85] °C (Supplementary Material 2.SM.4.3.3).

Differences between data sets are much smaller than both interannual variability and the long-term trend (Figure 2.20). Since 1901 almost the whole globe has experienced surface warming (Figure 2.21). Warming has not been linear; most warming occurred in two periods: around 1900 to around 1940 and around 1970 onwards (Figure 2.22. Shorter periods are noisier and so proportionately less of the sampled globe exhibits statistically significant trends at the grid box level (Figure 2.22). The two periods of global mean warming exhibit very distinct spatial signatures. The early 20th century warming was largely a NH mid- to high-latitude phenomenon, whereas the more recent warming is more global in nature. These distinctions may yield important information as to causes (Chapter 10). Differences between data sets are larger in earlier periods (Figures 2.19, 2.20), particularly prior to the 1950s when observational sampling is much more geographically incomplete (and many of the well sampled areas may have been globally unrepresentative (Brönnimann, 2009)), data errors and subsequent methodological impacts are larger (Thompson et al., 2008), and different ways of accounting for data void regions are more important (Vose et al., 2005b).

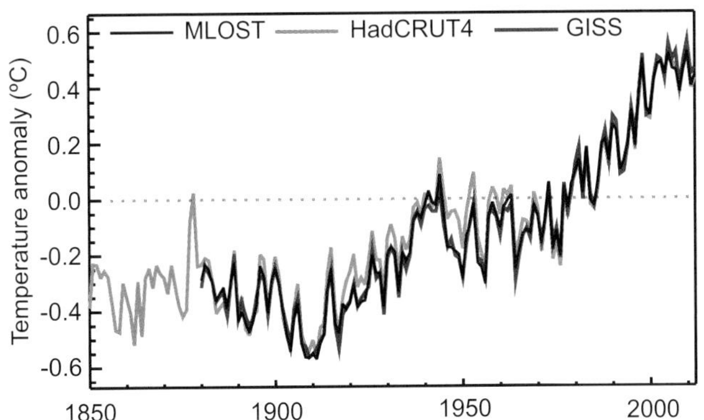

Figure 2.20 | Annual global mean surface temperature (GMST) anomalies relative to a 1961–1990 climatology from the latest version of the three combined land-surface air temperature (LSAT) and sea surface temperature (SST) data sets (HadCRUT4, GISS and NCDC MLOST). Published data set uncertainties are not included for reasons discussed in Box 2.1.

Much interest has focussed on the period since 1998 and an observed reduction in warming trend, most marked in NH winter (Cohen et al., 2012). Various investigators have pointed out the limitations of such short-term trend analysis in the presence of auto-correlated series variability and that several other similar length phases of no warming exist in all the observational records and in climate model simulations

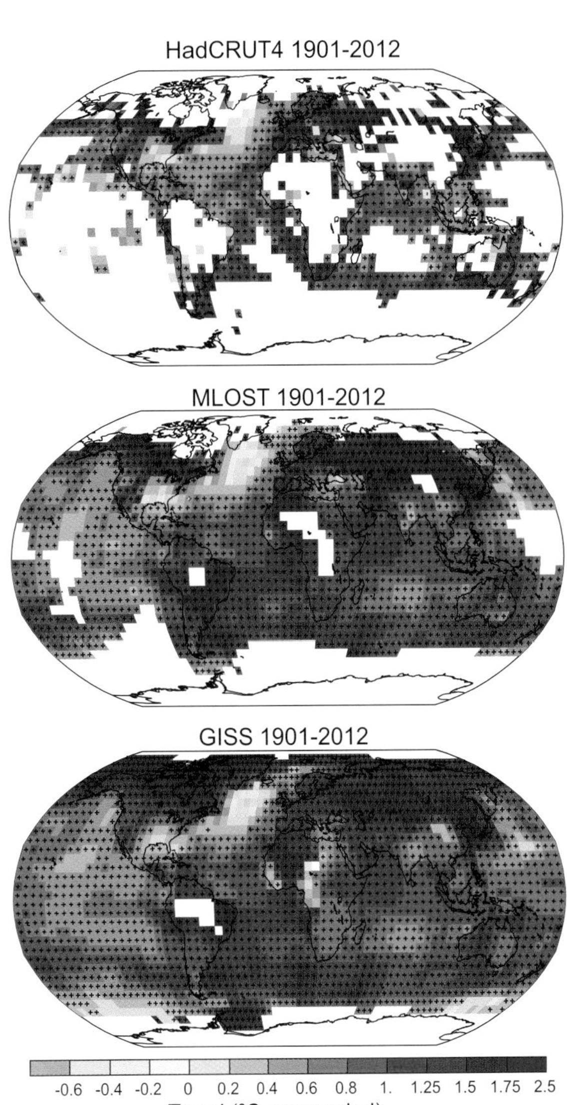

Figure 2.21 | Trends in surface temperature from the three data sets of Figure 2.20 for 1901–2012. White areas indicate incomplete or missing data. Trends have been calculated only for those grid boxes with greater than 70% complete records and more than 20% data availability in first and last decile of the period. Black plus signs (+) indicate grid boxes where trends are significant (i.e., a trend of zero lies outside the 90% confidence interval). Differences in coverage primarily reflect the degree of interpolation to account for data void regions undertaken by the data set providers ranging from none beyond grid box averaging (HadCRUT4) to substantial (GISS).

Table 2.7 | Same as Table 2.4, but for global mean surface temperature (GMST) over five common periods.

Data Set	Trends in °C per decade				
	1880–2012	1901–2012	1901–1950	1951–2012	1979–2012
HadCRUT4 (Morice et al., 2012)	0.062 ± 0.012	0.075 ± 0.013	0.107 ± 0.026	0.106 ± 0.027	0.155 ± 0.033
NCDC MLOST (Vose et al., 2012b)	0.064 ± 0.015	0.081 ± 0.013	0.097 ± 0.040	0.118 ± 0.021	0.151 ± 0.037
GISS (Hansen et al., 2010)	0.065 ± 0.015	0.083 ± 0.013	0.090 ± 0.034	0.124 ± 0.020	0.161 ± 0.033

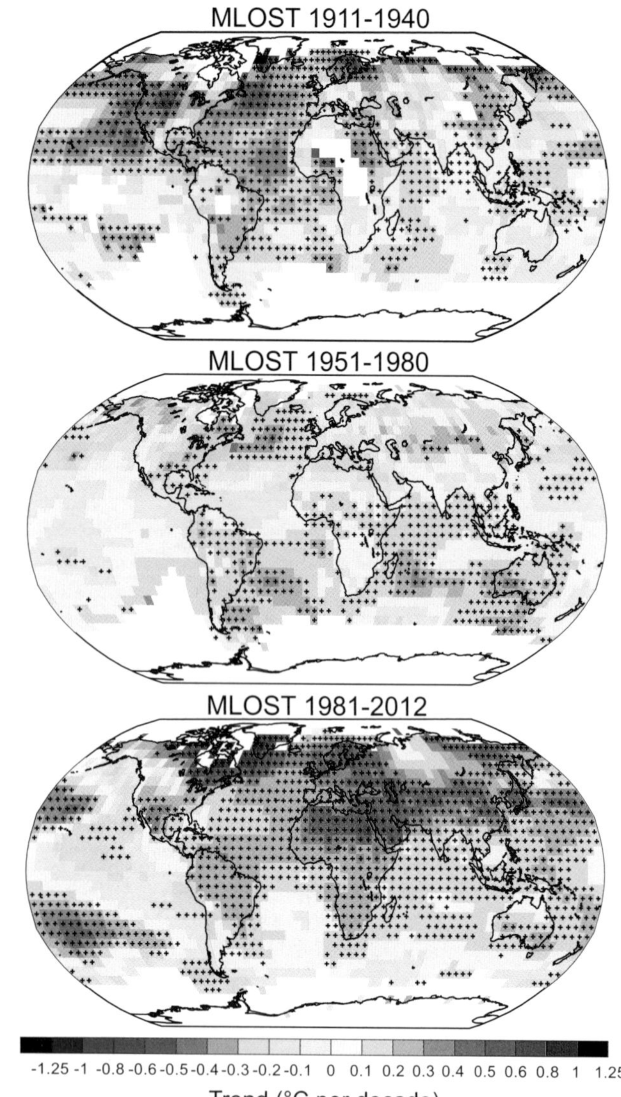

Figure 2.22 | Trends in surface temperature from NCDC MLOST for three non-consecutive shorter periods (1911–1940; 1951–1980; 1981–2012). White areas indicate incomplete or missing data. Trends and significance have been calculated as in Figure 2.21.

(Easterling and Wehner, 2009; Peterson et al., 2009; Liebmann et al., 2010; Foster and Rahmstorf, 2011; Santer et al., 2011). This issue is discussed in the context of model behaviour, forcings and natural variability in Box 9.2 and Section 10.3.1. Regardless, all global combined LSAT and SST data sets exhibit a statistically non-significant warming trend over 1998–2012 (0.042°C ± 0.093°C per decade (HadCRUT4); 0.037°C ± 0.085°C per decade (NCDC MLOST); 0.069°C ± 0.082°C per decade (GISS)). An average of the trends from these three data sets yields an estimated change for the 1998–2012 period of 0.05 [–0.05 to +0.15] °C per decade. Trends of this short length are very sensitive to the precise period selection with trends calculated in the same manner for the 15-year periods starting in 1995, 1996, and 1997 being 0.13 [0.02 to 0.24], 0.14 [0.03 to 0.24] and 0.07 [–0.02 to 0.18] (all °C per decade), respectively.

In summary, it is certain that globally averaged near surface temperatures have increased since the late 19th century. Each of the past three decades has been warmer than all the previous decades in the instrumental record, and the decade of the 2000s has been the warmest. The globally averaged combined land and ocean surface temperature data as calculated by a linear trend, show a warming of 0.85 [0.65 to 1.06] °C, over the period 1880–2012, when multiple independently produced datasets exist, about 0.89°C [0.69 to 1.08] °C over the period 1901–2012, and about 0.72 [0.49° to 0.89] °C over the period 1951–2012. The total increase between the average of the 1850–1900 period and the 2003–2012 period is 0.78 [0.72 to 0.85] °C and the total increase between the average of the 1850–1900 period and the reference period for projections 1986–2005 is 0.61 [0.55 to 0.67] °C, based on the single longest dataset available. For the longest period when calculation of regional trends is sufficiently complete (1901–2012), almost the entire globe has experienced surface warming. In addition to robust multi-decadal warming, global mean surface temperature exhibits substantial decadal and interannual variability. Owing to natural variability, trends based on short records are very sensitive to the beginning and end dates and do not in general reflect long-term climate trends. As one example, the rate of warming over the past 15 years (1998–2012; 0.05 [–0.05 to +0.15] °C per decade), which begins with a strong El Niño, is smaller than the rate calculated since 1951 (1951–2012; 0.12 [0.08 to 0.14] °C per decade)Trends for 15-year periods starting in 1995, 1996, and 1997 are 0.13 [0.02 to 0.24], 0.14 [0.03 to 0.24] and 0.07 [–0.02 to 0.18], respectively..

2.4.4 Upper Air Temperature

AR4 summarized that globally the troposphere had warmed at a rate greater than the GMST over the radiosonde record, while over the shorter satellite era the GMST and tropospheric warming rates were indistinguishable. Trends in the tropics were more uncertain than global trends although even this region was concluded to be warming. Globally, the stratosphere was reported to be cooling over the satellite era starting in 1979. New advances since AR4 have highlighted the substantial degree of uncertainty in both satellite and balloon-borne radiosonde records and led to some revisions and improvements in existing products and the creation of a number of new data products.

2.4.4.1 Advances in Multi-Decadal Observational Records

The major global radiosonde records extend back to 1958, with temperatures, measured as the balloon ascends, reported at mandatory pressure levels. Satellites have monitored tropospheric and lower stratospheric temperature trends since late 1978 through the Microwave Sounding Unit (MSU) and its follow-on Advanced Microwave Sounding Unit (AMSU) since 1998. These measures of upwelling radiation represent bulk (volume averaged) atmospheric temperature (Figure 2.23). The 'Mid-Tropospheric' (MT) MSU channel that most directly corresponds to the troposphere has 10 to 15% of its signal from both the skin temperature of the Earth's surface and the stratosphere. Two alternative approaches have been suggested for removing the stratospheric component based on differencing of view angles (LT) and statistical recombination (*G) with the 'Lower Stratosphere' (LS) channel (Spencer and Christy, 1992; Fu et al., 2004). The MSU satellite series also included a Stratospheric Sounding Unit (SSU) that measured at higher altitudes (Seidel et al., 2011).

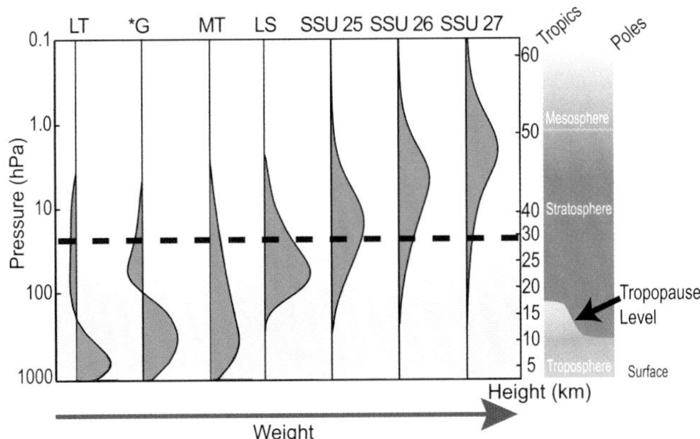

Figure 2.23 | Vertical weighting functions for those satellite temperature retrievals discussed in this chapter (modified from Seidel et al. (2011)). The dashed line indicates the typical maximum altitude achieved in the historical radiosonde record. The three SSU channels are denoted by the designated names 25, 26 and 27. LS (Lower Stratosphere) and MT (Mid Troposphere) are two direct MSU measures and LT (Lower Troposphere) and *G (Global Troposphere) are derived quantities from one or more of these that attempt to remove the stratospheric component from MT.

At the time of AR4 there were only two 'global' radiosonde data sets that included treatment of homogeneity issues: RATPAC (Free et al., 2005) and HadAT (Thorne et al., 2005). Three additional estimates have appeared since AR4 based on novel and distinct approaches. A group at the University of Vienna have produced RAOBCORE and RICH (Haimberger, 2007; Haimberger et al., 2008, 2012) using ERA reanalysis products (Box 2.3). Sherwood and colleagues developed an iterative universal kriging approach for radiosonde data to create IUK (Sherwood et al., 2008) and concluded that non-climatic data issues leading to spurious cooling remained in the deep tropics even after homogenization. The HadAT group created an automated version, undertook systematic experimentation and concluded that the parametric uncertainty (Box 2.1) was of the same order of magnitude as the apparent climate signal (McCarthy et al., 2008; Titchner et al., 2009; Thorne et al., 2011). A similar ensemble approach has also been applied to the RICH product (Haimberger et al., 2012). These various ensembles and new products exhibit more tropospheric warming / less stratospheric cooling than pre-existing products at all levels. Globally the radiosonde records all imply the troposphere has warmed and the stratosphere cooled since 1958 but with uncertainty that grows with height and is much greater outside the better-sampled NH extra-tropics (Thorne et al., 2011; Haimberger et al., 2012), where it is of the order 0.1°C per decade.

For MSU, AR4 considered estimates produced from three groups: UAH (University of Alabama in Huntsville); RSS (Remote Sensing Systems) and VG2 (now no longer updated). A new product has been created by NOAA labelled STAR, using a fundamentally distinct approach for the critical inter-satellite warm target calibration step (Zou et al., 2006a). STAR exhibits more warming/less cooling at all levels than UAH and RSS. For MT and LS, Zou and Wang (2010) concluded that this does not relate primarily to use of their inter-satellite calibration technique but rather differences in other processing steps. RSS also produced a parametric uncertainty ensemble (Box 2.1) employing a Monte Carlo approach allowing methodological inter-dependencies to be fully expressed (Mears et al., 2011). For large-scale trends dominant effects were inter-satellite offset determinations and, for tropospheric channels, diurnal drift. Uncertainties were concluded to be of the order 0.1°C per decade at the global mean for both tropospheric channels (where it is of comparable magnitude to the long-term trends) and the stratospheric channel.

SSU provides the only long-term near-global temperature data above the lower stratosphere, with the series terminating in 2006. Some AMSU-A channels have replaced this capability and efforts to understand the effect of changed measurement properties have been undertaken (Kobayashi et al., 2009). Until recently only one SSU data set existed (Nash and Edge, 1989), updated by Randel et al. (2009). Liu and Weng (2009) have produced an intermediate analysis for Channels 25 and 26 (but not Channel 27). Wang et al. (2012g), building on insights from several of these recent studies, have produced a more complete analysis. Differences between the independent estimates are much larger than differences between MSU records or radiosonde records at lower levels, with substantial inter-decadal time series behaviour departures, zonal trend structure, and global trend differences of the order 0.5°C per decade (Seidel et al., 2011; Thompson et al., 2012; Wang et al., 2012g). Although all SSU data sets agree that the stratosphere is cooling, there is therefore *low confidence* in the details above the lower stratosphere.

In summary, many new data sets have been produced since AR4 from radiosondes and satellites with renewed interest in satellite measurements above the lower stratosphere. Several studies have attempted to quantify the parametric uncertainty (Box 2.1) more rigorously. These various data sets and analyses have served to highlight the degree of uncertainty in the data and derived products.

2.4.4.2 Intercomparisons of Various Long-Term Radiosonde and MSU Products

Since AR4 there have been a large number of intercomparisons between radiosonde and MSU data sets. Interpretation is complicated, as most studies considered data set versions that have since been superseded. Several studies compared UAH and RSS products to local, regional or global raw/homogenized radiosonde data (Christy and Norris, 2006, 2009; Christy et al., 2007, 2010, 2011; Randall and Herman, 2008; Mears et al., 2012; Po-Chedley and Fu, 2012). Early studies focussed on the time of transition from NOAA-11 to NOAA-12 (early 1990s) which indicated an apparent issue in RSS. Christy et al. (2007) noted that this coincided with the Mt Pinatubo eruption and that RSS was the only product, either surface or tropospheric, that exhibited tropical warming immediately after the eruption when cooling would be expected. Using reanalysis data Bengtsson and Hodges (2011) also found evidence of a potential jump in RSS in 1993 over the tropical oceans. Mears et al. (2012) cautioned that an El Niño event quasi-simultaneous with Pinatubo complicates interpretation. They also highlighted several other periods of disagreement between radiosonde records and MSU records. All MSU records were most uncertain when satellite orbits are drifting rapidly (Christy and Norris, 2006, 2009). Mears et al. (2011) found that trend differences between RSS and other data sets could not be explained in many cases by parametric uncertainties in RSS alone. It was repeatedly cautioned that there were potential common biases (of varying magnitude) between the different MSU records or between the different radiosonde records which complicate intercomparisons (Christy and Norris, 2006, 2009; Mears et al., 2012).

In summary, assessment of the large body of studies comparing various long-term radiosonde and MSU products since AR4 is hampered by data set version changes, and inherent data uncertainties. These factors substantially limit the ability to draw robust and consistent inferences from such studies about the true long-term trends or the value of different data products.

2.4.4.3 Additional Evidence from Other Technologies and Approaches

Global Positioning System (GPS) radio occultation (RO) currently represents the only self-calibrated SI traceable raw satellite measurements (Anthes et al., 2008; Anthes, 2011). The fundamental observation is time delay of the occulted signal's phase traversing the atmosphere. The time delay is a function of several atmospheric physical state variables. Subsequent analysis converts the time delay to temperature and other parameters, which inevitably adds some degree of uncertainty to the derived temperature data. Intercomparisons of GPS-RO products show that differences are largest for derived geophysical parameters (including temperature), but are still small relative to other observing technologies (Ho et al., 2012). Comparisons to MSU and radiosondes (Kuo et al., 2005; Ho et al., 2007, 2009a, 2009b; He et al., 2009; Baringer et al., 2010; Sun et al., 2010; Ladstadter et al., 2011) show substantive agreement in interannual behaviour, but also some multi-year drifts that require further examination before this additional data source can usefully arbitrate between different MSU and radiosonde trend estimates.

Atmospheric winds are driven by thermal gradients. Radiosonde winds are far less affected by time-varying biases than their temperatures (Gruber and Haimberger, 2008; Sherwood et al., 2008; Section 2.7.3). Allen and Sherwood (2007) initially used radiosonde wind to infer temperatures within the Tropical West Pacific warm pool region, then extended this to a global analysis (Allen and Sherwood, 2008) yielding a distinct tropical upper tropospheric warming trend maximum within the vertical profile, but with large uncertainty. Winds can only quantify relative changes and require an initialization (location and trend at that location) (Allen and Sherwood, 2008). The large uncertainty range was predominantly driven by this initialization choice, a finding later confirmed by Christy et al. (2010), who in addition questioned the stability given the sparse geographical sampling, particularly in the tropics, and possible systematic sampling effects amongst other potential issues. Initializing closer to the tropics tended to reduce or remove the appearance of a tropical upper tropospheric warming trend maximum (Allen and Sherwood, 2008; Christy et al., 2010). There is only *low confidence* in trends inferred from 'thermal winds' given the relative immaturity of the analyses and their large uncertainties.

In summary, new technologies and approaches have emerged since AR4. However, these new technologies and approaches either constitute too short a record or are too immature to inform assessments of long-term trends at the present time.

2.4.4.4 Synthesis of Free Atmosphere Temperature Estimates

Global-mean lower tropospheric temperatures have increased since the mid-20th century (Figure 2.24, bottom). Structural uncertainties (Box 2.1) are larger than at the surface but it can still be concluded that globally the troposphere has warmed (Table 2.8). On top of this long-term trend are superimposed short-term variations that are highly correlated with those at the surface but of somewhat greater amplitude. Global mean lower stratospheric temperatures have decreased since the mid-20th century punctuated by short-lived warming events associated with explosive volcanic activity (Figure 2.24a). However, since the mid-1990s little net change has occurred. Cooling rates are on average greater from radiosonde data sets than MSU data sets. This *very likely* relates to widely recognized cooling biases in radiosondes (Mears et al., 2006) which all data set producers explicitly caution are *likely* to remain to some extent in their final products (Free and Seidel, 2007; Haimberger et al., 2008; Sherwood et al., 2008; Thorne et al., 2011).

In comparison to the surface (Figure 2.22), tropospheric layers exhibit smoother geographic trends (Figure 2.25) with warming dominating cooling north of approximately 45°S and greatest warming in high northern latitudes. The lower stratosphere cooled almost everywhere but this cooling exhibits substantial large-scale structure. Cooling is greatest in the highest southern latitudes and smallest in high northern latitudes. There are also secondary stratospheric cooling maxima in the mid-latitude regions of each hemisphere.

Available global and regional trends from radiosondes since 1958 (Figure 2.26) show agreement that the troposphere has warmed and the stratosphere cooled. While there is little ambiguity in the sign of the changes, the rate and vertical structure of change are distinctly data set dependent, particularly in the stratosphere. Differences are greatest in the tropics and SH extra-tropics where the historical radiosonde data coverage is poorest. Not shown in the figure for clarity are estimates of parametric data set uncertainties or trend-fit uncertainties—both of which are of the order of at least 0.1°C per decade (Section 2.4.4.1).

Differences in trends between available radiosonde data sets are greater during the satellite era than for the full radiosonde period of record in all regions and at most levels (Figure 2.27; cf. Figure 2.26). The RAOBCORE product exhibits greater vertical trend gradients than other data sets and it has been posited that this relates to its dependency on reanalysis fields (Sakamoto and Christy, 2009; Christy et al., 2010). MSU trend estimates in the troposphere are generally bracketed by the radiosonde range. In the stratosphere MSU deep layer estimates tend to show slightly less cooling. Over both 1958–2011 and 1979–2011 there is some evidence in the radiosonde products taken as a whole that the tropical tropospheric trends increase with height. But the magnitude and the structure is highly data set dependent.

In summary, based on multiple independent analyses of measurements from radiosondes and satellite sensors it is *virtually certain* that globally the troposphere has warmed and the stratosphere has cooled since the mid-20th century. Despite unanimous agreement on the sign of the trends, substantial disagreement exists among available estimates as to the rate of temperature changes, particularly outside the NH extratropical troposphere, which has been well sampled by radiosondes.

Table 2.8 | Trend estimates and 90% confidence intervals (Box 2.2) for radiosonde and MSU data set global average values over the radiosonde (1958–2012) and satellite periods (1979–2012). LT indicates Lower Troposphere, MT indicates Mid Troposphere and LS indicates Lower Stratosphere (Figure 2.23. Satellite records start only in 1979 and STAR do not produce an LT product.

Data Set	Trends in °C per decade					
	1958–2012			1979–2012		
	LT	MT	LS	LT	MT	LS
HadAT2 (Thorne et al., 2005)	0.159 ± 0.038	0.095 ± 0.034	–0.339 ± 0.086	0.162 ± 0.047	0.079 ± 0.057	–0.436 ± 0.204
RAOBCORE 1.5 (Haimberger et al., 2012)	0.156 ± 0.031	0.109 ± 0.029	–0.186 ± 0.087	0.139 ± 0.049	0.079 ± 0.054	–0.266 ± 0.227
RICH-obs (Haimberger et al., 2012)	0.162 ± 0.031	0.102 ± 0.029	–0.285 ± 0.087	0.158 ± 0.046	0.081 ± 0.052	–0.331 ± 0.241
RICH-tau (Haimberger et al., 2012)	0.168 ± 0.032	0.111 ± 0.030	–0.280 ± 0.085	0.160 ± 0.046	0.083 ± 0.052	–0.345 ± 0.238
RATPAC (Free et al., 2005)	0.136 ± 0.028	0.076 ± 0.028	–0.338 ± 0.092	0.128 ± 0.044	0.039 ± 0.051	–0.468 ± 0.225
UAH (Christy et al., 2003)				0.138 ± 0.043	0.043 ± 0.042	–0.372 ± 0.201
RSS (Mears and Wentz, 2009a, 2009b)				0.131 ± 0.045	0.079 ± 0.043	–0.268 ± 0.177
STAR (Zou and Wang, 2011)					0.123 ± 0.047	–0.320 ± 0.175

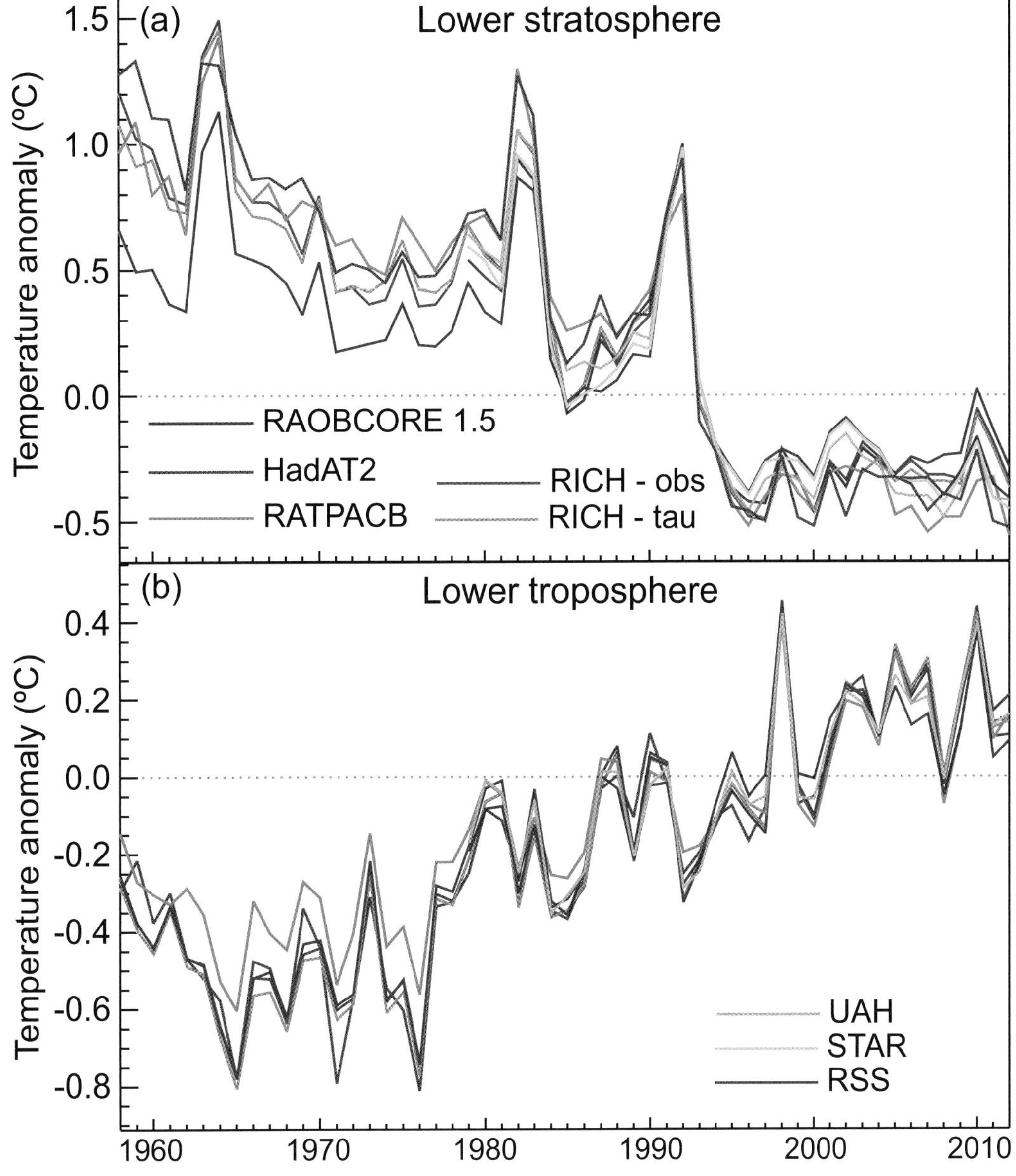

Figure 2.24 | Global annual average lower stratospheric (top) and lower tropospheric (bottom) temperature anomalies relative to a 1981–2010 climatology from different data sets. STAR does not produce a lower tropospheric temperature product. Note that the *y*-axis resolution differs between the two panels.

Frequently Asked Questions

FAQ 2.1 | How Do We Know the World Has Warmed?

Evidence for a warming world comes from multiple independent climate indicators, from high up in the atmosphere to the depths of the oceans. They include changes in surface, atmospheric and oceanic temperatures; glaciers; snow cover; sea ice; sea level and atmospheric water vapour. Scientists from all over the world have independently verified this evidence many times. That the world has warmed since the 19th century is unequivocal.

Discussion about climate warming often centres on potential residual biases in temperature records from land-based weather stations. These records are very important, but they only represent one indicator of changes in the climate system. Broader evidence for a warming world comes from a wide range of independent physically consistent measurements of many other, strongly interlinked, elements of the climate system (FAQ 2.1, Figure 1).

A rise in global average surface temperatures is the best-known indicator of climate change. Although each year and even decade is not always warmer than the last, global surface temperatures have warmed substantially since 1900.

Warming land temperatures correspond closely with the observed warming trend over the oceans. Warming oceanic air temperatures, measured from aboard ships, and temperatures of the sea surface itself also coincide, as borne out by many independent analyses.

The atmosphere and ocean are both fluid bodies, so warming at the surface should also be seen in the lower atmosphere, and deeper down into the upper oceans, and observations confirm that this is indeed the case. Analyses of measurements made by weather balloon radiosondes and satellites consistently show warming of the troposphere, the active weather layer of the atmosphere. More than 90% of the excess energy absorbed by the climate system since at least the 1970s has been stored in the oceans as can be seen from global records of ocean heat content going back to the 1950s. *(continued on next page)*

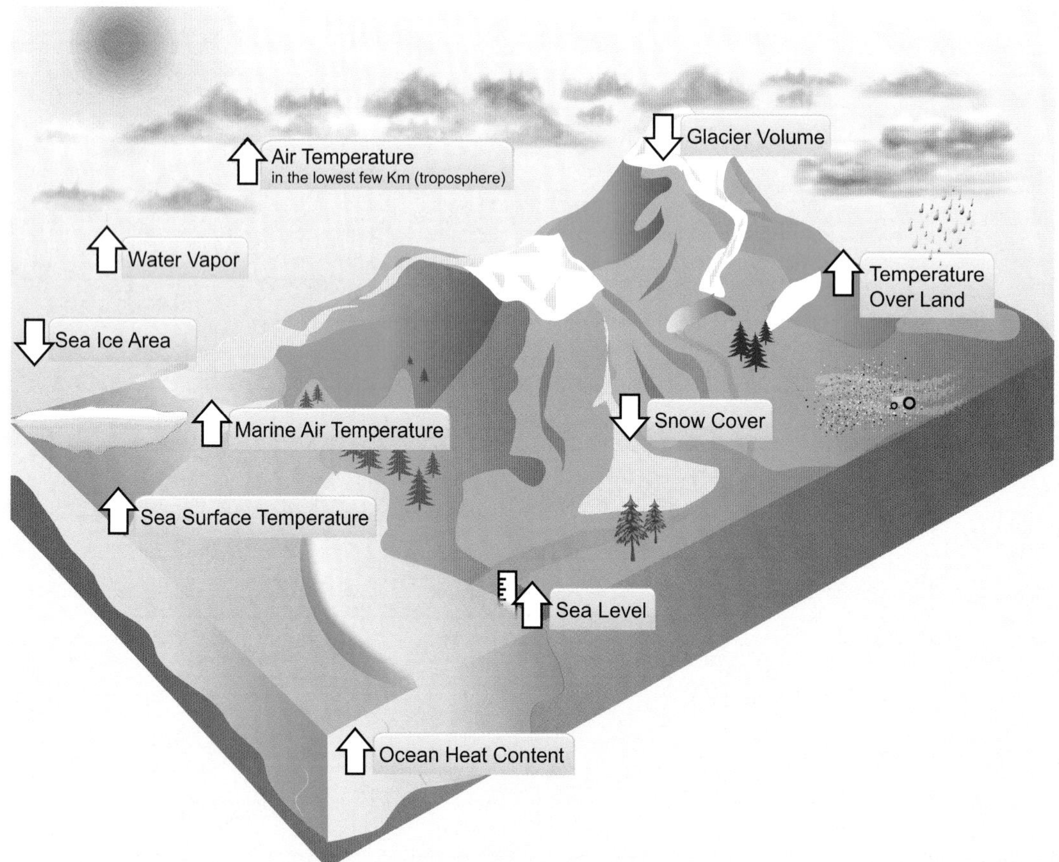

FAQ 2.1, Figure 1 | Independent analyses of many components of the climate system that would be expected to change in a warming world exhibit trends consistent with warming (arrow direction denotes the sign of the change), as shown in FAQ 2.1, Figure 2.

FAQ 2.1 (continued)

As the oceans warm, the water itself expands. This expansion is one of the main drivers of the independently observed rise in sea levels over the past century. Melting of glaciers and ice sheets also contribute, as do changes in storage and usage of water on land.

A warmer world is also a moister one, because warmer air can hold more water vapour. Global analyses show that specific humidity, which measures the amount of water vapour in the atmosphere, has increased over both the land and the oceans.

The frozen parts of the planet—known collectively as the cryosphere—affect, and are affected by, local changes in temperature. The amount of ice contained in glaciers globally has been declining every year for more than 20 years, and the lost mass contributes, in part, to the observed rise in sea level. Snow cover is sensitive to changes in temperature, particularly during the spring, when snow starts to melt. Spring snow cover has shrunk across the NH since the 1950s. Substantial losses in Arctic sea ice have been observed since satellite records began, particularly at the time of the mimimum extent, which occurs in September at the end of the annual melt season. By contrast, the increase in Antarctic sea ice has been smaller.

Individually, any single analysis might be unconvincing, but analysis of these different indicators and independent data sets has led many independent research groups to *all* reach the same conclusion. From the deep oceans to the top of the troposphere, the evidence of warmer air and oceans, of melting ice and rising seas all points unequivocally to one thing: the world has warmed since the late 19th century (FAQ 2.1, Figure 2).

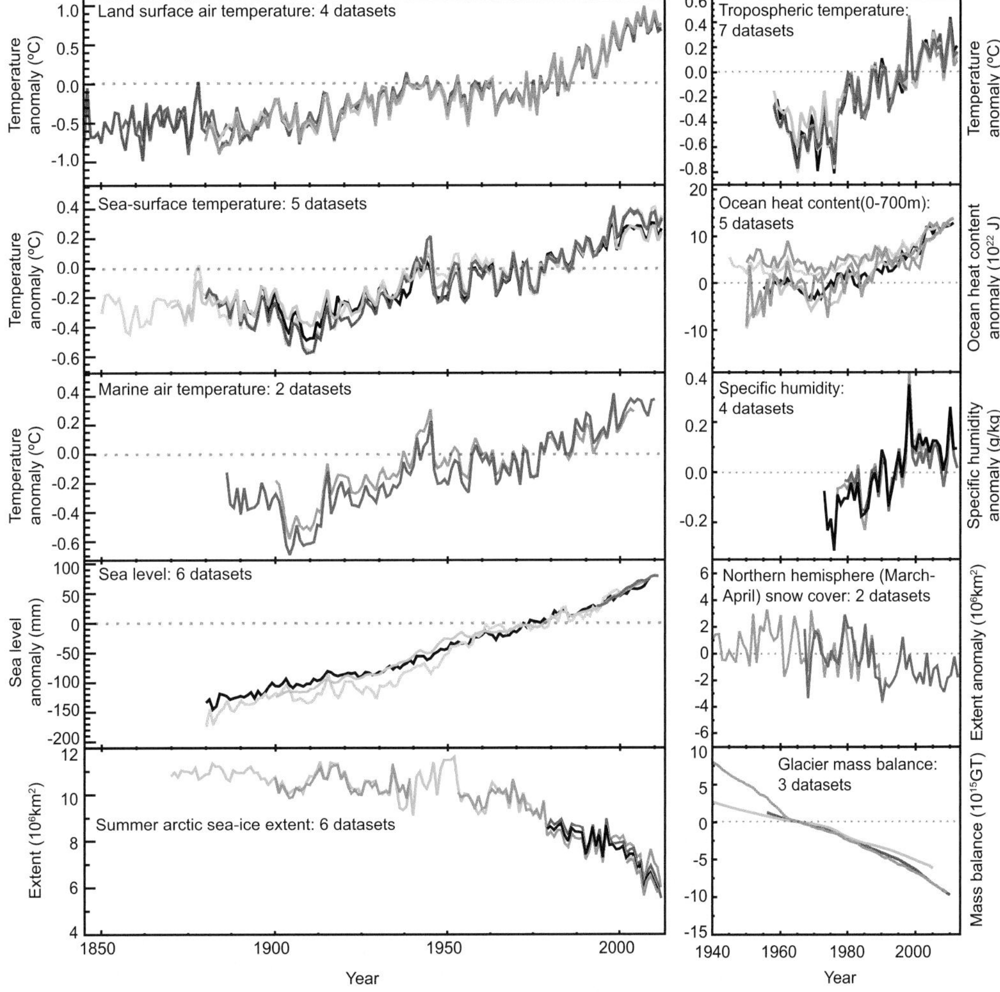

FAQ 2.1, Figure 2 | Multiple independent indicators of a changing global climate. Each line represents an independently derived estimate of change in the climate element. In each panel all data sets have been normalized to a common period of record. A full detailing of which source data sets go into which panel is given in the Supplementary Material 2.SM.5.

2

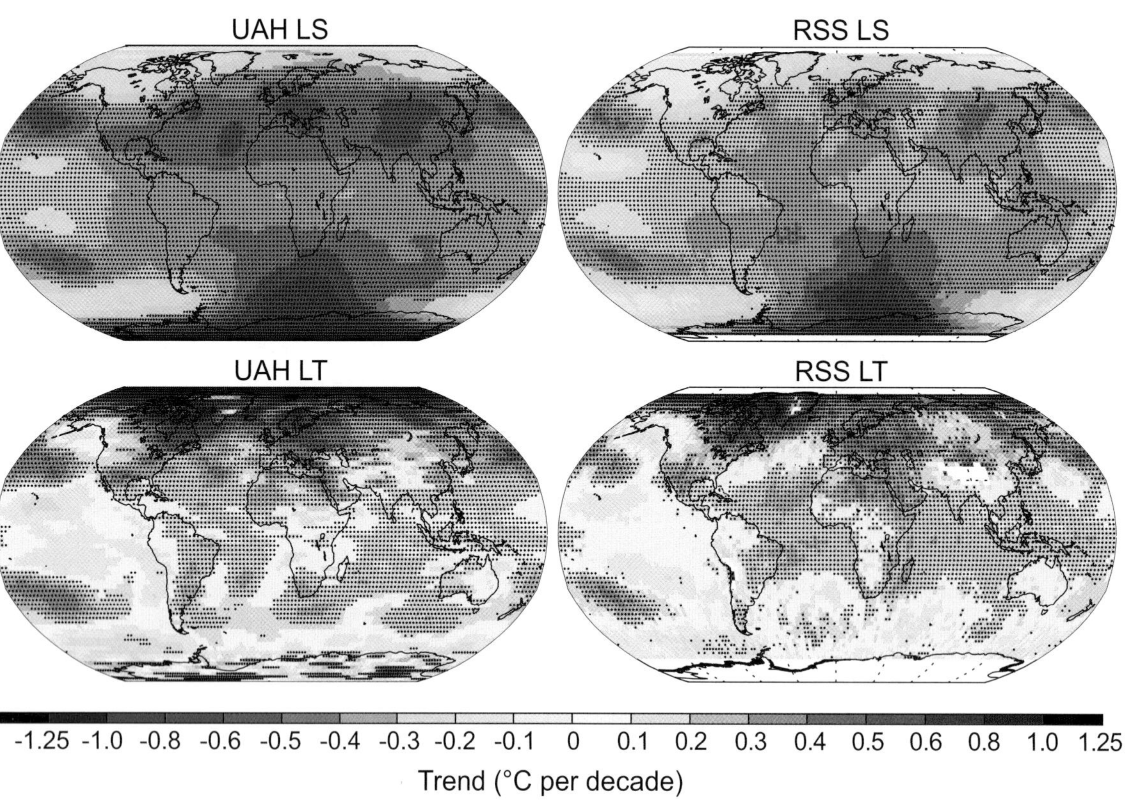

Figure 2.25 | Trends in MSU upper air temperature over 1979–2012 from UAH (left-hand panels) and RSS (right-hand panels) and for LS (top row) and LT (bottom row). Data are temporally complete within the sampled domains for each data set. White areas indicate incomplete or missing data. Black plus signs (+) indicate grid boxes where trends are significant (i.e., a trend of zero lies outside the 90% confidence interval).

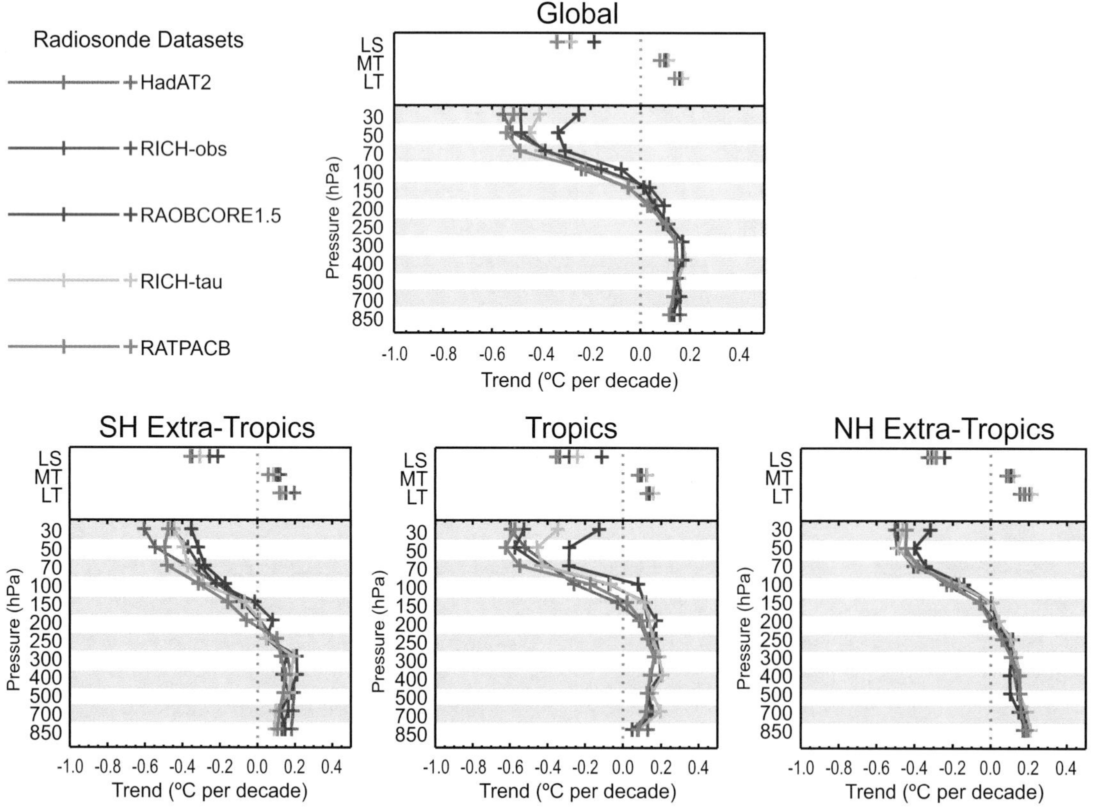

Figure 2.26 | Trends in upper air temperature for all available radiosonde data products that contain records for 1958–2012 for the globe (top) and tropics (20°N to 20°S) and extra-tropics (bottom). The bottom panel trace in each case is for trends on distinct pressure levels. Note that the pressure axis is not linear. The top panel points show MSU layer equivalent measure trends. MSU layer equivalents have been processed using the method of Thorne et al. (2005). No attempts have been made to sub-sample to a common data mask.

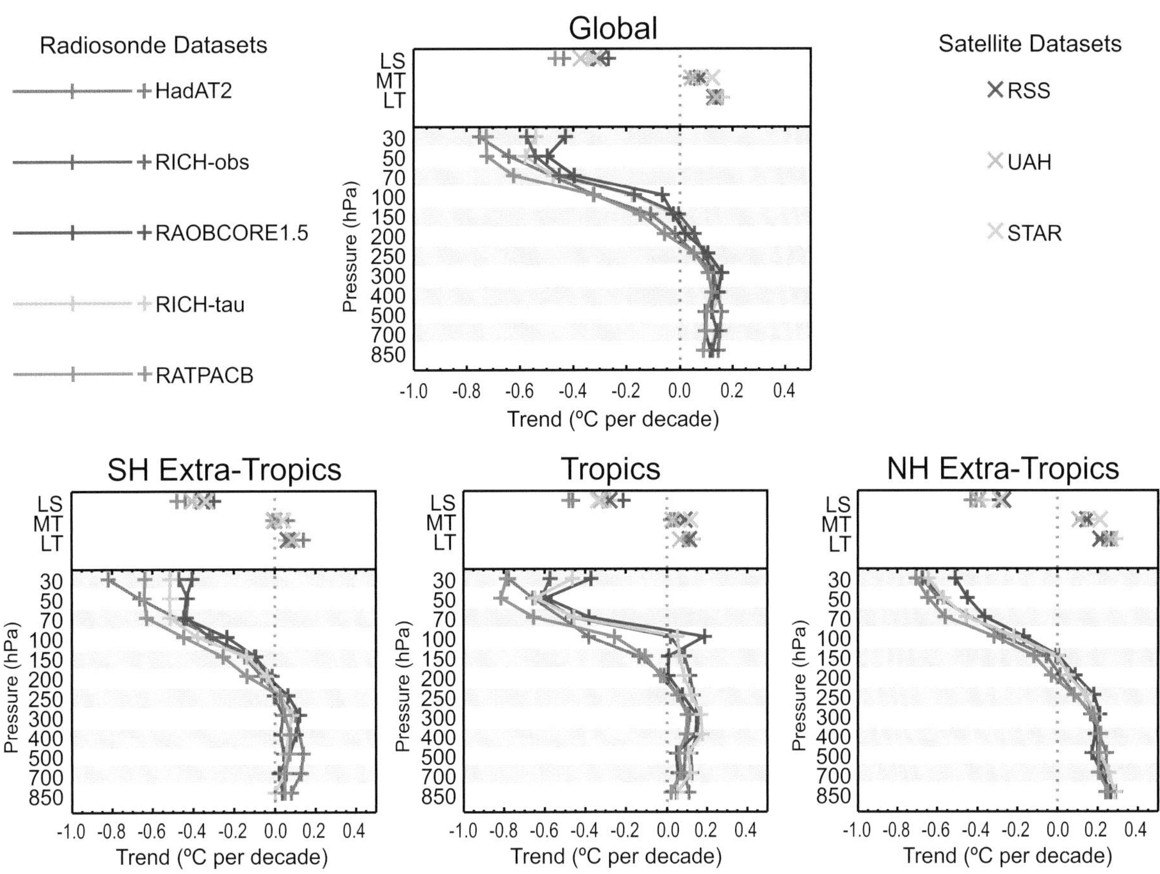

Figure 2.27 | As Figure 2.26 except for the satellite era 1979–2012 period and including MSU products (RSS, STAR and UAH).

Hence there is only *medium confidence* in the rate of change and its vertical structure in the NH extratropical troposphere and *low confidence* elsewhere.

2.5 Changes in Hydrological Cycle

This section covers the main aspects of the hydrological cycle, including large-scale average precipitation, stream flow and runoff, soil moisture, atmospheric water vapour, and clouds. Meteorological drought is assessed in Section 2.6. Ocean precipitation changes are assessed in Section 3.4.3 and changes in the area covered by snow in Section 4.5.

2.5.1 Large-Scale Changes in Precipitation

2.5.1.1 Global Land Areas

AR4 concluded that precipitation has generally increased over land north of 30°N over the period 1900–2005 but downward trends dominate the tropics since the 1970s. AR4 included analysis of both the GHCN (Vose et al., 1992) and CRU (Mitchell and Jones, 2005) gauge-based precipitation data sets for the globally averaged annual precipitation over land. For both data sets the overall linear trend from 1900 to 2005 (1901–2002 for CRU) was positive but not statistically significant (Table 3.4 from AR4). Other periods covered in AR4 (1951–2005 and 1979–2005) showed a mix of negative and positive trends depending on the data set.

Since AR4, existing data sets have been updated and a new data set developed. Figure 2.28 shows the century-scale variations and trends on globally and zonally averaged annual precipitation using five data sets: GHCN V2 (updated through 2011; Vose et al., 1992), Global Precipitation Climatology Project V2.2 (GPCP) combined raingauge–satellite product (Adler et al., 2003), CRU TS 3.10.01 (updated from Mitchell and Jones, 2005), Global Precipitation Climatology Centre V6 (GPCC) data set (Becker et al., 2013) and a reconstructed data set by Smith et al. (2012). Each data product incorporates a different number of station series for each region. The Smith et al. product is a statistical reconstruction using Empirical Orthogonal Functions, similar to the NCDC MLOST global temperature product (Section 2.4.3) that does provide coverage for most of the global surface area although only land is included here. The data sets based on *in situ* observations only start in 1901, but the Smith et al. data set ends in 2008, while the other three data sets contain data until at least 2010.

For the longest common period of record (1901–2008) all datasets exhibit increases in globally averaged precipitation, with three of the four showing statistically significant changes (Table 2.9). However, there is a factor of almost three spread in the magnitude of the change which serves to create *low confidence*. Global trends for the shorter period (1951–2008) show a mix of statistically non-significant positive and negative trends amongst the four data sets with the infilled Smith et al. (2012) analysis showing increases and the remainder decreases. These differences among data sets indicate that long-term increases

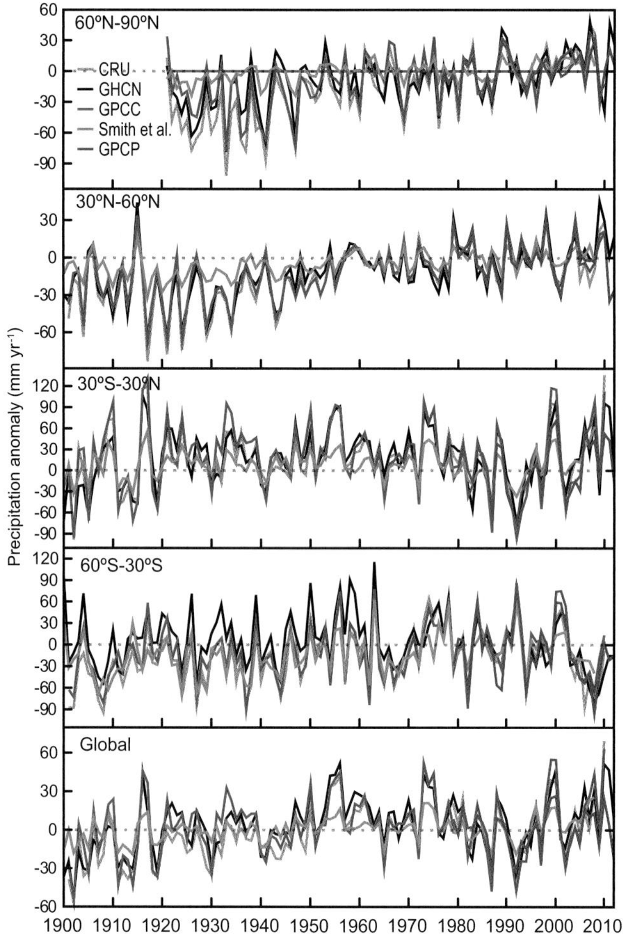

Figure 2.28 | Annual precipitation anomalies averaged over land areas for four latitudinal bands and the globe from five global precipitation data sets relative to a 1981–2000 climatology.

in global precipitation discussed in AR4 are uncertain, owing in part to issues in data coverage in the early part of the 20th century (Wan et al., 2013).

In summary, *confidence* in precipitation change averaged over global land areas is *low* for the years prior to 1950 and *medium* afterwards because of insufficient data, particularly in the earlier part of the record. Available globally incomplete records show mixed and non-significant long-term trends in reported global mean changes. Further, when virtually all the land area is filled in using a reconstruction method, the resulting time series shows less change in land-based precipitation since 1900.

2.5.1.2 Spatial Variability of Observed Trends

The latitude band plots in Figure 2.28 suggest that precipitation over tropical land areas (30°S to 30°N) has increased over the last decade reversing the drying trend that occurred from the mid-1970s to mid-1990s. As a result the period 1951–2008 shows no significant overall trend in tropical land precipitation in any of the datasets (Table 2.10). Longer term trends (1901–2008) in the tropics, shown in Table 2.10, are also non-significant for each of the four data sets. The mid-latitudes of the NH (30°N to 60°N) show an overall increase in precipitation from 1901 to 2008 with statistically significant trends for each data set. For the shorter period (1951–2008) the trends are also positive but non-significant for three of the four data sets. For the high latitudes of the NH (60°N to 90°N) where data completeness permits trend calculations solely for the 1951–2008 period, all datasets show increases but there is a wide range of magnitudes and the infilled Smith et al. series shows small and insignificant trends (Table 2.10). Fewer data from high latitude stations make these trends less certain and yield *low confidence* in resulting zonal band average estimates. In the mid-latitudes of the SH (60°S to 30°S) there is limited evidence of long-term increases with three data sets showing significant trends for the 1901–2008 period but GHCN having negative trends that are not significant. For the 1951–2008 period changes in SH mid-latitude precipitation are less certain, with one data set showing a significant trend towards drying, two showing non-significant drying trends and the final dataset suggesting increases in precipitation. All data sets show an abrupt decline in SH mid-latitude precipitation in the early 2000s (Figure 2.28) consistent with enhanced drying that has very recently recovered. These results for latitudinal changes are broadly consistent with the global satellite observations for the 1979–2008 period (Allan et al., 2010) and land-based gauge measurements for the 1950–1999 period (Zhang et al., 2007a).

In AR4, maps of observed trends of annual precipitation for 1901–2005 were calculated using GHCN interpolated to a 5° × 5° latitude/longitude grid. Trends (in percent per decade) were calculated for each grid box and showed statistically significant changes, particularly increases in eastern and northwestern North America, parts of Europe and Russia, southern South America and Australia, declines in the Sahel region of Africa, and a few scattered declines elsewhere.

Figure 2.29 shows the spatial variability of long-term trends (1901–2010) and more recent trends (1951–2010) over land in annual precipitation using the CRU, GHCN and GPCC data sets. The trends are computed from land-only grid box time series using each native data set grid resolution. The patterns of these absolute trends (in mm yr^{-1} per decade) are broadly similar to the trends (in percent per decade) relative

Table 2.9 | Trend estimates and 90% confidence intervals (Box 2.2) for annual precipitation for each time series in Figure 2.28 over two common periods of record.

Data Set	Area	Trends in mm yr^{-1} per decade	
		1901–2008	1951–2008
CRU TS 3.10.01 (updated from Mitchell and Jones, 2005)	Global	2.77 ± 1.46	–2.12 ± 3.52
GHCN V2 (updated through 2011; Vose et al., 1992)	Global	2.08 ± 1.66	–2.77 ± 3.92
GPCC V6 (Becker et al., 2013)	Global	1.48 ± 1.65	–1.54 ± 4.50
Smith et al. (2012)	Global	1.01 ± 0.64	0.68 ± 2.07

Table 2.10 | Trend estimates and 90% confidence intervals (Box 2.2) for annual precipitation for each time series in Figure 2.28 over two periods. Dashes indicate not enough data available for trend calculation. For the latitudinal band 90°S to 60°S not enough data exist for each product in either period.

Data Set	Area	Trends in mm yr^{-1} per decade	
		1901–2008	1951–2008
CRU TS 3.10.01 (updated from Mitchell and Jones, 2005)	60°N–90°N	–	5.82 ± 2.72
	30°N–60°N	3.82 ± 1.14	1.13 ± 2.01
	30°S–30°N	0.89 ± 2.89	–4.22 ± 8.27
	60°S–30°S	3.88 ± 2.28	–3.73 ± 5.94
GHCN V2 (updated through 2011; Vose et al., 1992)	60°N–90°N	–	4.52 ± 2.64
	30°N–60°N	3.23 ± 1.10	1.39 ± 1.98
	30°S–30°N	1.01 ± 3.00	–5.15 ± 7.28
	60°S–30°S	–0.57 ± 2.27	–8.01 ± 5.63
GPCC V6 (Becker et al., 2013)	60°N–90°N	–	2.69 ± 2.54
	30°N–60°N	3.14 ± 1.05	1.50 ± 1.93
	30°S–30°N	–0.48 ± 3.35	–4.16 ± 9.65
	60°S–30°S	2.40 ± 2.01	–0.51 ± 5.45
Smith et al. (2012)	60°N–90°N	–	0.63 ± 1.27
	30°N–60°N	1.44 ± 0.50	0.97 ± 0.88
	30°S–30°N	0.43 ± 1.48	0.67 ± 4.75
	60°S–30°S	2.94 ± 1.40	0.78 ± 3.31

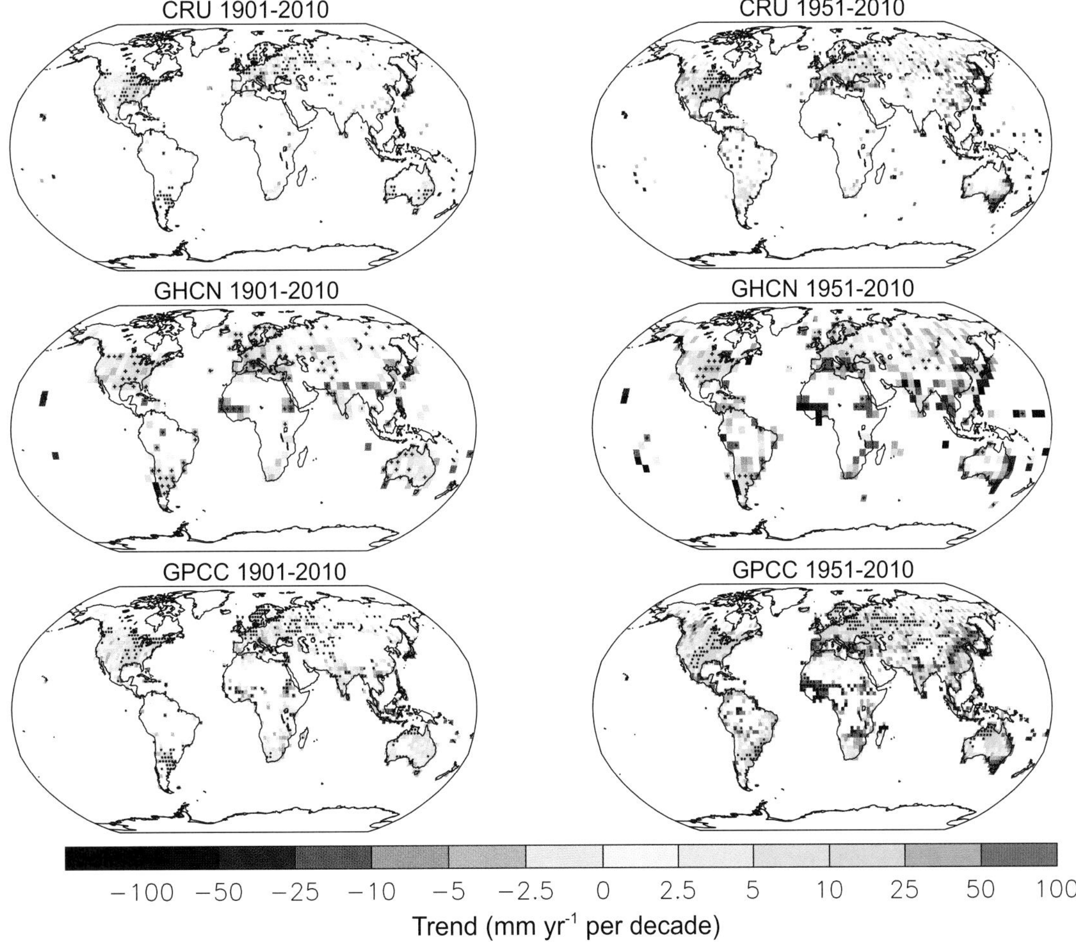

Figure 2.29 | Trends in annual precipitation over land from the CRU, GHCN and GPCC data sets for 1901–2010 (left-hand panels) and 1951–2010 (right-hand panels). Trends have been calculated only for those grid boxes with greater than 70% complete records and more than 20% data availability in first and last decile of the period. White areas indicate incomplete or missing data. Black plus signs (+) indicate grid boxes where trends are significant (i.e., a trend of zero lies outside the 90% confidence interval).

to local climatology (Supplementary Material 2.SM.6.1). Increases for the period 1901–2010 are seen in the mid- and higher-latitudes of both the NH and SH consistent with the reported changes for latitudinal bands. At the grid box scale, statistically significant trends occur in most of the same areas, in each data set but are far more limited than for temperature over a similar length period (cf. Figure 2.21). The GPCC map shows the most areas with significant trends. Comparing the maps in Figure 2.29, most areas for which trends can be calculated for both periods show similar trends between the 1901–2010 period and the 1951–2010 period with few exceptions (e.g., South Eastern Australia,). Trends over shorter periods can differ from those implied for the longest periods. For example, since the late 1980s trends in the Sahel region have been significantly positive (not shown).

In summary, when averaged over the land areas of the mid-latitudes of the NH, all datasets show a *likely* overall increase in precipitation (*medium confidence* since 1901, but *high confidence* after 1951). For all other zones one or more of data sparsity, quality, or a lack of quantitative agreement amongst available estimates yields *low confidence* in characterisation of such long-term trends in zonally averaged precipitation. Nevertheless, changes in some more regional or shorter-term recent changes can be quantified. It is *likely* there was an abrupt decline in SH mid-latitude precipitation in the early 2000s consistent with enhanced drying that has very recently recovered. Precipitation in the tropical land areas has increased (*medium confidence*) over the last decade, reversing the drying trend that occurred from the mid-1970s to mid-1990s reported in AR4.

2.5.1.3 Changes in Snowfall

AR4 draws no conclusion on global changes in snowfall. Changes in snowfall are discussed on a region-by-region basis, but focussed mainly on North America and Eurasia. Statistically significant increases were found in most of Canada, parts of northern Europe and Russia. A number of areas showed a decline in the number of snowfall events, especially those where climatological averaged temperatures were close to 0°C and where warming led to earlier onset of spring. Also, an increase in lake-effect snowfall was found for areas near the North American Great Lakes.

Since AR4, most published literature has considered again changes in snowfall in North America. These studies have confirmed that more winter-time precipitation is falling as rain rather than snow in the western USA (Knowles et al., 2006), the Pacific Northwest and Central USA (Feng and Hu, 2007). Kunkel et al. (2009) analyzed trends using a specially quality-controlled data set of snowfall observations over the contiguous USA and found that snowfall has been declining in the western USA, northeastern USA and southern margins of the seasonal snow region, but increasing in the western Great Plains and Great Lakes regions. Snowfall in Canada has increased mainly in the north while a significant decrease was observed in the southwestern part of the country for 1950–2009 (Mekis and Vincent, 2011).

Other regions that have been analyzed include Japan (Takeuchi et al., 2008), where warmer winters in the heavy snowfall areas on Honshu are associated with decreases in snowfall and precipitation in general. Shekar et al. (2010) found declines in total seasonal snowfall along with increases in maximum and minimum temperatures in the western Himalaya. Serquet et al. (2011) analyzed snowfall and rainfall days since 1961 and found the proportion of snowfall days to rainfall days in Switzerland was declining in association with increasing temperatures. Scherrer and Appenzeller (2006) found a trend in a pattern of variability of snowfall in the Swiss Alps that indicated decreasing snow at low altitudes relative to high altitudes, but with large decadal variability in key snow indicators (Scherrer et al., 2013). Van Ommen and Morgan (2010) draw a link between increased snowfall in coastal East Antarctica and increased southwest Western Australia drought. However, Monaghan and Bromwich (2008) found an increase in snow accumulation over all Antarctica from the late 1950s to 1990, then a decline to 2004. Thus snowfall changes in Antarctica remain uncertain.

In summary, in most regions analyzed, it is *likely* that decreasing numbers of snowfall events are occurring where increased winter temperatures have been observed (North America, Europe, Southern and East Asia). *Confidence* is *low* for the changes in snowfall over Antarctica.

2.5.2 Streamflow and Runoff

AR4 concluded that runoff and river discharge generally increased at high latitudes, with some exceptions. No consistent long-term trend in discharge was reported for the world's major rivers on a global scale.

River discharge is unique among water cycle components in that it both spatially and temporally integrates surplus waters upstream within a catchment (Shiklomanov et al., 2010) which makes it well suited for *in situ* monitoring (Arndt et al., 2010). The most recent comprehensive analyses (Milliman et al., 2008; Dai et al., 2009) do not support earlier work (Labat et al., 2004) that reported an increasing trend in global river discharge associated with global warming during the 20th century. It must be noted that many if not most large rivers, especially those for which a long-term streamflow record exists, have been impacted by human influences such as dam construction or land use, so results must be interpreted with caution. Dai et al. (2009) assembled a data set of 925 most downstream stations on the largest rivers monitoring 80% of the global ocean draining land areas and capturing 73% of the continental runoff. They found that discharges in about one-third of the 200 largest rivers (including the Congo, Mississippi, Yenisey, Paraná, Ganges, Colombia, Uruguay and Niger) show statistically significant trends during 1948–2004, with the rivers having downward trends (45) outnumbering those with upward trends (19). Decreases in streamflow were found over many low and mid-latitude river basins such as the Yellow River in northern China since 1960s (Piao et al., 2010) where precipitation has decreased. Increases in streamflow during the latter half of the 20th century also have been reported over regions with increased precipitation, such as parts of the USA (Groisman et al., 2004), and in the Yangtze River in southern China (Piao et al., 2010). In the Amazon basin an increase of discharge extremes is observed over recent decades (Espinoza Villar et al., 2009). For France, Giuntoli et al. (2013) found that the sign of the temporal trends in natural streamflows varies with period studied. In that case study, significant correlations between median to low flows and the Atlantic Multidecadal Oscillation (AMO; Section 2.7.8) result in long quasi-periodic oscillations.

At high latitudes, increasing winter base flow and mean annual stream flow resulting from possible permafrost thawing were reported in northwest Canada (St. Jacques and Sauchyn, 2009). Rising minimum daily flows also have been observed in northern Eurasian rivers (Smith et al., 2007). For ocean basins other than the Arctic, and for the global ocean as a whole, the data for continental discharge show small or downward trends, which are statistically significant for the Pacific (–9.4 $km^3 yr^{-1}$). Precipitation is a major driver for the discharge trends and for the large interannual-to-decadal variations (Dai et al., 2009). However, for the Arctic drainage areas, Adam and Lettenmaier (2008) found that upward trends in streamflow are not accompanied by increasing precipitation, especially over Siberia, based on available observations. Zhang et al. (2012a) argued that precipitation measurements are sparse and exhibit large cold-season biases in the Arctic drainage areas and hence there would be large uncertianties using these data to investigate their influence on streamflow.

Recently, Stahl et al. (2010) and Stahl and Tallaksen (2012) investigated streamflow trends based on a data set of near-natural streamflow records from more than 400 small catchments in 15 countries across Europe for 1962–2004. A regional coherent pattern of annual streamflow trends was revealed with negative trends in southern and eastern regions, and generally positive trends elsewhere. Subtle regional differences in the subannual changes in various streamflow metrics also can be captured in regional studies such as by Monk et al. (2011) for Canadian rivers.

In summary, the most recent comprehensive analyses lead to the conclusion that *confidence* is *low* for an increasing trend in global river discharge during the 20th century.

2.5.3 Evapotranspiration Including Pan Evaporation

AR4 concluded that decreasing trends were found in records of pan evaporation over recent decades over the USA, India, Australia, New Zealand, China and Thailand and speculated on the causes including decreased surface solar radiation, sunshine duration, increased specific humidity and increased clouds. However, AR4 also reported that direct measurements of evapotranspiration over global land areas are scarce, and concluded that reanalysis evaporation fields are not reliable because they are not well constrained by precipitation and radiation.

Since AR4 gridded data sets have been developed that estimate actual evapotranspiration from either atmospheric forcing and thermal remote sensing, sometimes in combination with direct measurements (e.g., from FLUXNET, a global network of flux towers), or interpolation of FLUXNET data using regression techniques, providing an unprecedented look at global evapotranspiration (Mueller et al., 2011). On a global scale, evapotranspiration over land increased from the early 1980s up to the late 1990s (Wild et al., 2008; Jung et al., 2010; Wang et al., 2010) and Wang et al. (2010) found that global evapotranspiration increased at a rate of 0.6 W m^{-2} per decade for the period 1982–2002. After 1998, a lack of moisture availability in SH land areas, particularly decreasing soil moisture, has acted as a constraint to further increase of global evapotranspiration (Jung et al., 2010).

Zhang et al. (2007b) found decreasing pan evaporation at stations across the Tibetan Plateau, even with increasing air temperature. Similarly, decreases in pan evaporation were also found for northeastern India (Jhajharia et al., 2009) and the Canadian Prairies (Burn and Hesch, 2007). A continuous decrease in reference and pan evaporation for the period 1960–2000 was reported by Xu et al. (2006a) for a humid region in China, consistent with reported continuous increase in aerosol levels over China (Qian et al., 2006; Section 2.2.4). Roderick et al. (2007) examined the relationship between pan evaporation changes and many of the possible causes listed above using a physical model and conclude that many of the decreases (USA, China, Tibetan Plateau, Australia) cited previously are related to declining wind speeds and to a lesser extent decreasing solar radiation. Fu et al. (2009) provided an overview of pan evaporation trends and concluded the major possible causes, changes in wind speed, humidity and solar radiation, have been occurring, but that the importance of each is regionally dependent.

The recent increase in incoming shortwave radiation in regions with decreasing aerosol concentrations (Section 2.2.3) can explain positive evapotranspiration trends only in the humid part of Europe. In semiarid and arid regions, trends in evapotranspiration largely follow trends in precipitation (Jung et al., 2010). Trends in surface winds (Section 2.7.2) and CO_2 (Section 2.2.1.1.1) also alter the partitioning of available energy into evapotranspiration and sensible heat. While surface wind trends may explain pan evaporation trends over Australia (Rayner, 2007; Roderick et al., 2007), their impact on actual evapotranspiration is limited due to the compensating effect of boundary-layer feedbacks (van Heerwaarden et al., 2010). In vegetated regions, where a large part of evapotranspiration comes from transpiration through plants' stomata, rising CO_2 concentrations can lead to reduced stomatal opening and evapotranspiration (Idso and Brazel, 1984; Leakey et al., 2006). Additional regional effects that impact evapotranspiration trends are lengthening of the growing season and land use change.

In summary, there is *medium confidence* that pan evaporation continued to decline in most regions studied since AR4 related to changes in wind speed, solar radiation and humidity. On a global scale, evapotranspiration over land increased (*medium confidence*) from the early 1980s up to the late 1990s. After 1998, a lack of moisture availability in SH land areas, particularly decreasing soil moisture, has acted as a constraint to further increase of global evapotranspiration.

2.5.4 Surface Humidity

AR4 reported widespread increases in surface air moisture content since 1976, along with near-constant relative humidity over large scales though with some significant changes specific to region, time of day or season.

In good agreement with previous analysis from Dai (2006), Willett et al. (2008) show widespread increasing specific humidity across the globe from the homogenized gridded monthly mean anomaly product HadCRUH (1973–2003). Both Dai and HadCRUH products that are blended land and ocean data products end in 2003 but HadISDH (1973–2012) (Willett et al., 2013) and the NOCS product (Berry and Kent, 2009) are available over the land and ocean respectively through 2012. There

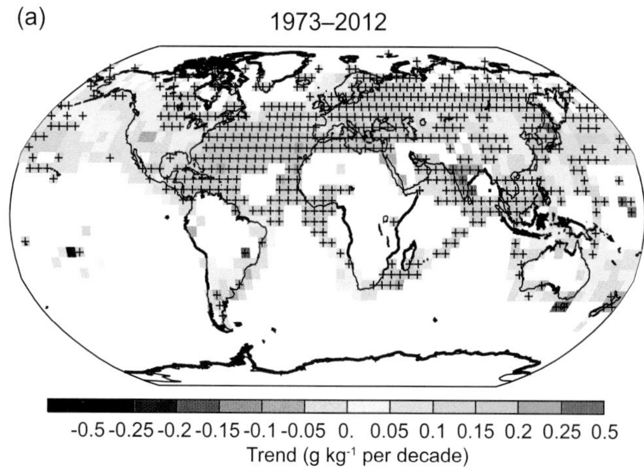

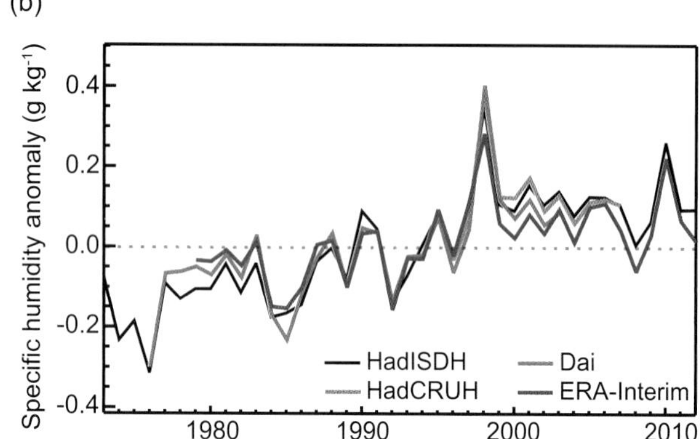

Figure 2.30 | (a) Trends in surface specific humidity from HadISDH and NOCS over 1973–2012. Trends have been calculated only for those grid boxes with greater than 70% complete records and more than 20% data availability in first and last decile of the period. White areas indicate incomplete or missing data. Black plus signs (+) indicate grid boxes where trends are significant (i.e., a trend of zero lies outside the 90% confidence interval). (b) Global annual average anomalies in land surface specific humidity from Dai (2006; red), HadCRUH (Willett et al., 2013; orange), HadISDH (Willett et al., 2013; black), and ERA-Interim (Simmons et al., 2010; blue). Anomalies are relative to the 1979–2003 climatology.

are some small isolated but coherent areas of drying over some of the more arid land regions (Figure 2.30a). Moistening is largest in the tropics and in the extratropics during summer over both land and ocean. Large uncertainty remains over the SH where data are sparse. Global specific humidity is sensitive to large-scale phenomena such as ENSO (Figure 2.30b; Box 2.5). It is strongly correlated with land surface temperature averages over the 23 Giorgi and Francisco (2000) regions for the period 1973–1999 and exhibits increases mostly at or above the increase expected from the Clausius–Clapeyron relation (about 7% $°C^{-1}$; Annex III: Glossary) with *high confidence* (Willett et al., 2010). Land surface humidity trends are similar in ERA-Interim to observed estimates of homogeneity-adjusted data sets (Simmons et al., 2010; Figure 2.30b).

Since 2000 surface specific humidity over land has remained largely unchanged (Figure 2.30) whereas land areas have on average warmed slightly (Figure 2.14), implying a reduction in land region relative humidity. This may be linked to the greater warming of the land surface relative to the ocean surface (Joshi et al., 2008). The marine specific humidity (Berry and Kent, 2009), like that over land, shows widespread increases that correlate strongly with SST. However, there is a marked decline in marine relative humidity around 1982. This is reported in Willett et al. (2008) where its origin is concluded to be a non-climatic data issue owing to a change in reporting practice for dewpoint temperature.

In summary, it is *very likely* that global near surface air specific humidity has increased since the 1970s. However, during recent years the near surface moistening over land has abated (*medium confidence*). As a result, fairly widespread decreases in relative humidity near the surface are observed over the land in recent years.

2.5.5 Tropospheric Humidity

As reported in AR4, observations from radiosonde and GPS measurements over land, and satellite measurements over ocean indicate increases in tropospheric water vapour at near-global spatial scales which are consistent with the observed increase in atmospheric temperature over the last several decades. Tropospheric water vapour plays an important role in regulating the energy balance of the surface and TOA, provides a key feedback mechanism and is essential to the formation of clouds and precipitation.

2.5.5.1 Radiosonde

Radiosonde humidity data for the troposphere were used sparingly in AR4, noting a renewed appreciation for biases with the operational radiosonde data that had been highlighted by several major field campaigns and intercomparisons. Since AR4 there have been three distinct efforts to homogenize the tropospheric humidity records from operational radiosonde measurements (Durre et al., 2009; McCarthy et al., 2009; Dai et al., 2011) (Supplementary Material 2.SM.6.1, Table 2.SM.9). Over the common period of record from 1973 onwards, the resulting estimates are in substantive agreement regarding specific

Table 2.11 | Trend estimates and 90% confidence intervals (Box 2.2) for surface humidity over two periods.

	Data Set	Trends in % per decade	
		1976–2003	1973–2012
Land	HadISDH (Willett et al., 2008)	0.127 ± 0.037	0.091 ± 0.023
	HadCRUH_land (Willett et al., 2008)	0.128 ± 0.043	
	Dai_land (Dai, 2006)	0.099 ± 0.046	
Ocean	NOCS (Berry and Kent, 2009)	0.114 ± 0.064	0.090 ± 0.033
	HadCRUH_marine (Willett et al., 2008)	0.065 ± 0.049	
	Dai_marine (Dai, 2006)	0.058 ± 0.044	

humidity trends at the largest geographical scales. On average, the impact of the correction procedures is to remove an artificial temporal trend towards drying in the raw data and indicate a positive trend in free tropospheric specific humidity over the period of record. In each analysis, the rate of increase in the free troposphere is concluded to be largely consistent with that expected from the Clausius–Clapeyron relation (about 7% per degree Celsius). There is no evidence for a significant change in free tropospheric relative humidity, although a decrease in relative humidity at lower levels is observed (Section 2.5.5). Indeed, McCarthy et al. (2009) show close agreement between their radiosonde product at the lowest levels and HadCRUH (Willett et al., 2008).

2.5.5.2 Global Positioning System

Since the early 1990s, estimates of column integrated water vapour have been obtained from ground-based Global Positioning System (GPS) receivers. An international network started with about 100 stations in 1997 and has currently been expanded to more than 500 (primarily land-based) stations. Several studies have compiled GPS water vapour data sets for climate studies (Jin et al., 2007; Wang et al., 2007; Wang and Zhang, 2008, 2009). Using such data, Mears et al. (2010) demonstrated general agreement of the interannual anomalies between ocean-based satellite and land-based GPS column integrated water vapour data. The interannual water vapour anomalies are closely tied to the atmospheric temperature changes in a manner consistent with that expected from the Clausius–Clapeyron relation. Jin et al. (2007) found an average column integrated water vapour trend of about 2 kg m^{-2} per decade during 1994–2006 for 150 (primarily land-based) stations over the globe, with positive trends at most NH stations and negative trends in the SH. However, given the short length (about 10 years) of the GPS records, the estimated trends are very sensitive to the start and end years and the analyzed time period (Box 2.2).

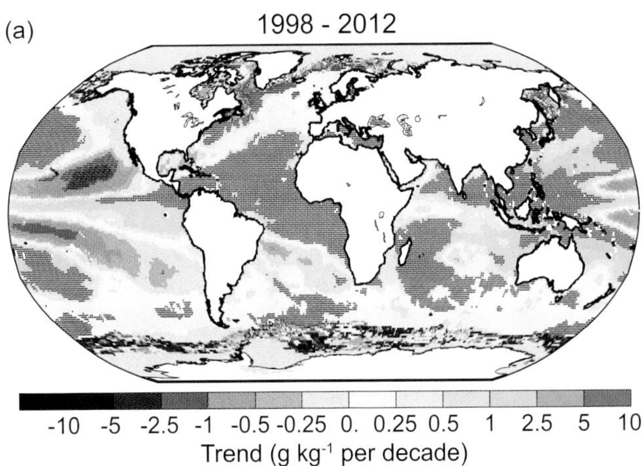

Figure 2.31 | (a) Trends in column integrated water vapour over ocean surfaces from Special Sensor Microwave Imager (Wentz et al., 2007) for the period 1988–2010. Trends have been calculated only for those grid boxes with greater than 70% complete records and more than 20% data availability in first and last decile of the period. Black plus signs (+) indicate grid boxes where trends are significant (i.e., a trend of zero lies outside the 90% confidence interval). (b) Global annual average anomalies in column integrated water vapour averaged over ocean surfaces. Anomalies are relative to the 1988–2007 average.

2.5.5.3 Satellite

AR4 reported positive decadal trends in lower and upper tropospheric water vapour based on satellite observations for the period 1988–2004. Since AR4, there has been continued evidence for increases in lower tropospheric water vapour from microwave satellite measurements of column integrated water vapour over oceans (Santer et al., 2007; Wentz et al., 2007) and globally from satellite measurements of spectrally resolved reflected solar radiation (Mieruch et al., 2008). The interannual variability and longer-term trends in column-integrated water vapour over oceans are closely tied to changes in SST at the global scale and interannual anomalies show remarkable agreement with low-level specific humidity anomalies from HadCRUH (O'Gorman et al., 2012). The rate of moistening at large spatial scales over oceans is close to that expected from the Clausius–Clapeyron relation (about 7% per degree Celsius) with invariant relative humidity (Figure 2.31). Satellite measurements also indicate that the globally averaged upper tropospheric relative humidity has changed little over the period 1979–2010 while the troposphere has warmed, implying an increase in the mean water vapour mass in the upper troposphere (Shi and Bates, 2011).

Interannual variations in temperature and upper tropospheric water vapour from IR satellite data are consistent with a constant RH behavior at large spatial scales (Dessler et al., 2008; Gettelman and Fu, 2008; Chung et al., 2010). On decadal time-scales, increased GHG concentrations reduce clear-sky outgoing long-wave radiation (Allan, 2009; Chung and Soden, 2010), thereby influencing inferred relationships between moisture and temperature. Using Meteosat IR radiances, Brogniez et al. (2009) demonstrated that interannual variations in free tropospheric humidity over subtropical dry regions are heavily influenced by meridional mixing between the deep tropics and the extra tropics. Regionally, upper tropospheric humidity changes in the tropics were shown to relate strongly to the movement of the ITCZ based upon microwave satellite data (Xavier et al., 2010). Shi and Bates (2011) found an increase in upper tropospheric humidity over the equatorial tropics from 1979 to 2008. However there was no significant trend found in tropical-mean or global-mean averages, indicating that on these time and space scales the upper troposphere has seen little change in relative humidity over the past 30 years. While microwave satellite measurements have become increasingly relied upon for studies of upper tropospheric humidity, the absence of a homogenized data set across multiple satellite platforms presents some difficulty in documenting coherent trends from these records (John et al., 2011).

2.5.5.4 Reanalyses

Using NCEP reanalyses for the period 1973–2007, Paltridge et al. (2009) found negative trends in specific humidity above 850 hPa over both the tropics and southern mid-latitudes, and above 600 hPa in the NH mid-latitudes. However, as noted in AR4, reanalysis products suffer from time dependent biases and have been shown to simulate unrealistic trends and variability over the ocean (Mears et al., 2007; John et al., 2009) (Box 2.3). Some reanalysis products do reproduce observed variability in low level humidity over land (Simmons et al., 2010), more complete assesments of multiple reanalysis products yield substantially different and even opposing trends in free tropospheric specific humidity (Chen et al., 2008; Dessler and Davis, 2010). Consequently, reanalysis products are still considered to be unsuitable for the analysis of tropospheric water vapour trends (Sherwood et al., 2010).

2

In summary, radiosonde, GPS and satellite observations of tropospheric water vapour indicate *very likely* increases at near global scales since the 1970s occurring at a rate that is generally consistent with the Clausius-Clapeyron relation (about 7% per degree Celsius) and the observed increase in atmospheric temperature. Significant trends in tropospheric relative humidity at large spatial scales have not been observed, with the exception of near-surface air over land where relative humidity has decreased in recent years (Section 2.5.5).

2.5.6 Clouds

2.5.6.1 Surface Observations

AR4 reported that surface-observed total cloud cover may have increased over many land areas since the middle of the 20th century, including the USA, the former USSR, Western Europe, mid-latitude Canada and Australia. A few regions exhibited decreases, including China and central Europe. Trends were less globally consistent since the early 1970s, with regional reductions in cloud cover reported for western Asia and Europe but increases over the USA.

Analyses since AR4 have indicated decreases in cloud occurrence/cover in recent decades over Poland (Wibig, 2008), China and the Tibetan Plateau (Duan and Wu, 2006; Endo and Yasunari, 2006; Xia, 2010b), in particular for upper level clouds (Warren et al., 2007) and also over Africa, Eurasia and in particular South America (Warren et al., 2007). Increased frequency of overcast conditions has been reported for some regions, such as Canada, from 1953 to 2002 (Milewska, 2004), with no statistically significant trends evident over Australia (Jovanovic et al., 2011) and North America (Warren et al., 2007). A global analysis of surface observations spanning the period 1971–2009 (Eastman and Warren, 2012) indicates a small decline in total cloud cover of about 0.4% per decade which is largely attributed to declining mid- and high-level cloud cover and is most prominent in the middle latitudes.

Regional variability in surface-observed cloudiness over the ocean appeared more credible than zonal and global mean variations in AR4. Multidecadal changes in upper-level cloud cover and total cloud cover over particular areas of the tropical Indo-Pacific Ocean were consistent with island precipitation records and SST variability. This has been extended more recently by Deser et al. (2010a), who found that an eastward shift in tropical convection and total cloud cover from the western to central equatorial Pacific occurred over the 20th century and attributed it to a long-term weakening of the Walker circulation (Section 2.7.5). Eastman et al. (2011) report that, after the removal of apparently spurious globally coherent variability, cloud cover decreased in all subtropical stratocumulus regions from 1954 to 2008.

2.5.6.2 Satellite Observations

Satellite cloud observations offer the advantage of much better spatial and temporal coverage compared to surface observations. However they require careful efforts to identify and correct for temporal discontinuities in the data sets associated with orbital drift, sensor degradation, and inter-satellite calibration differences. AR4 noted that there were substantial uncertainties in decadal trends of cloud cover in all satellite data sets available at the time and concluded that there was no clear consensus regarding the decadal changes in total cloud cover. Since AR4 there has been continued effort to assess the quality of and develop improvements to multi-decadal cloud products from operational satellite platforms (Evan et al., 2007; O'Dell et al., 2008; Heidinger and Pavolonis, 2009).

Several satellite data sets offer multi-decadal records of cloud cover (Stubenrauch et al., 2013). AR4 noted that there were discrepancies in global cloud cover trends between ISCCP and other satellite data products, notably a large downward trend of global cloudiness in ISCCP since the late 1980s which is inconsistent with PATMOS-x and surface observations (Baringer et al., 2010). Recent work has confirmed the conclusion of AR4, that much of the downward trend in ISCCP is spurious and an artefact of changes in satellite viewing geometry (Evan et al., 2007). An assesment of long-term variations in global-mean cloud amount from nine different satellite data sets by Stubenrauch et al. (2013) found differences between data sets were comparable in magnitude to the interannual variability (2.5 to 3.5%). Such inconsistencies result from differnces in sampling as well as changes in instrument calibration and inhibit an accurate assessment of global-scale cloud cover trends.

Satellite observations of low-level marine clouds suggest no long-term trends in cloud liquid water path or optical properties (O'Dell et al., 2008; Rausch et al., 2010). On regional scales, trends in cloud properties over China have been linked to changes in aerosol concentrations (Qian et al., 2009; Bennartz et al., 2011) (Section 2.2.3).

In summary, surface-based observations show region- and height-specific variations and trends in cloudiness but there remains substantial ambiguity regarding global-scale cloud variations and trends, especially from satellite observations. Although trends of cloud cover are consistent between independent data sets in certain regions, substantial ambiguity and therefore *low confidence* remains in the observations of global-scale cloud variability and trends.

2.6 Changes in Extreme Events

AR4 highlighted the importance of understanding changes in extreme climate events (Annex III: Glossary) because of their disproportionate impact on society and ecosystems compared to changes in mean climate (see also IPCC Working Group II). More recently a comprehensive assessment of observed changes in extreme events was undertaken by the IPCC Special Report on Managing the Risks of Extreme Events and Disasters to Advance Climate Change Adaptation (SREX) (Seneviratne et al., 2012; Section 1.3.3).

Data availability, quality and consistency especially affect the statistics of extremes and some variables are particularly sensitive to changing measurement practices over time. For example, historical tropical cyclone records are known to be heterogeneous owing to changing observing technology and reporting protocols (Section 14.6.1) and when records from multiple ocean basins are combined to explore global trends, because data quality and reporting protocols vary substantially between regions (Knapp and Kruk, 2010). Similar problems have been discovered when analysing wind extremes, because of the sensitivity of measurements to changing instrumentation and observing practice (e.g., Smits et al., 2005; Wan et al., 2010).

Numerous regional studies indicate that changes observed in the frequency of extremes can be explained or inferred by shifts in the overall probability distribution of the climate variable (Griffiths et al., 2005; Ballester et al., 2010; Simolo et al., 2011). However, it should be noted that these studies refer to counts of threshold exceedance—frequency, duration—which closely follow mean changes. Departures from high percentiles/return periods (intensity, severity, magnitude) are highly sensitive to changes in the shape and scale parameters of the distribution (Schär et al., 2004; Clark et al., 2006; Della-Marta et al., 2007a, 2007b; Fischer and Schär, 2010) and geographical location. Debate continues over whether variance as well as mean changes are affecting global temperature extremes (Hansen et al., 2012; Rhines and Huybers, 2013) as illustrated in Figure 1.8 and FAQ 2.2, Figure 1. In the following sections the conclusions from both AR4 and SREX are reviewed along with studies subsequent to those assessments.

2.6.1 Temperature Extremes

AR4 concluded that it was *very likely* that a large majority of global land areas had experienced decreases in indices of cold extremes and increases in indices of warm extremes, since the middle of the 20th century, consistent with warming of the climate. In addition, globally averaged multi-day heat events had *likely* exhibited increases over a similar period. SREX updated AR4 but came to similar conclusions while using the revised AR5 uncertainty guidance (Seneviratne et al., 2012). Further evidence since then indicates that the level of *confidence* that the majority of warm and cool extremes show warming remains *high*.

A large amount of evidence continues to support the conclusion that most global land areas analysed have experienced significant warming of both maximum and minimum temperature extremes since about 1950 (Donat et al., 2013c). Changes in the occurrence of cold and warm days (based on daily maximum temperatures) are generally less marked (Figure 2.32). ENSO (Box 2.5) influences both maximum and minimum temperature variability especially around the Pacific Rim (e.g., Kenyon and Hegerl, 2008; Alexander et al., 2009) but often affecting cold and warm extremes differently. Different data sets using different gridding methods and/or input data (Supplementary Material 2.SM.7) indicate large coherent trends in temperature extremes globally, associated with warming (Figure 2.32). The level of quality control varies between these data sets. For example, HadEX2 (Donat et al., 2013c) uses more rigorous quality control which leads to a reduced station sample compared to GHCNDEX (Donat et al., 2013a) or HadGHCND (Caesar et al., 2006). However, despite these issues data sets compare remarkably consistently even though the station networks vary through time (Figure 2.32; Table 2.12). Other data sets that have assessed these indices, but cover a shorter period, also agree very well over the period of overlapping data, e.g., HadEX (Alexander et al., 2006) and Duke (Morak et al., 2011, 2013).

The shift in the distribution of nighttime temperatures appears greater than daytime temperatures although whether distribution changes are simply linked to increases in the mean or other moments is an active area of research (Ballester et al., 2010; Simolo et al., 2011; Donat and Alexander, 2012; Hansen et al., 2012). Indeed, all data sets examined (Duke, GHCNDEX, HadEX, HadEX2 and HadGHCND), indicate a faster increase in minimum temperature extremes than maximum temperature extremes. While DTR declines have only been assessed with *medium confidence* (Section 2.4.1.2), *confidence* of accelerated increases in minimum temperature extremes compared to maximum temperature extremes is *high* due to the more consistent patterns of warming in minimum temperature extremes globally.

Regional changes in a range of climate indices are assessed in Table 2.13. These indicate *likely* increases across most continents in unusually warm days and nights and/or reductions in unusually cold days and nights including frosts. Some regions have experienced close to a doubling of the occurrence of warm and a halving of the occurrence of cold nights, for example, parts of the Asia-Pacific region (Choi et al., 2009) and parts of Eurasia (Klein Tank et al., 2006; Donat et al., 2013a, 2013c) since the mid-20th century. Changes in both local and global SST patterns (Section 2.4.2) and large scale circulation patterns (Section 2.7) have been shown to be associated with regional changes in temperature extremes (Barrucand et al., 2008; Scaife et al., 2008;

Table 2.12 | Trend estimates and 90% confidence intervals (Box 2.2) for global values of cold nights (TN10p), cold days (TX10p), warm nights (TN90p) and warm days (TX90p) over the periods 1951–2010 and 1979–2010 (see Box 2.4, Table 1 for more information on indices).

Data Set	Trends in % per decade							
	TN10p		TX10p		TN90p		TX90p	
	1951–2010	1979–2010	1951–2010	1979–2010	1951–2010	1979–2010	1951–2010	1979–2010
HadEX2 (Donat et al., 2013c)	–3.9 ± 0.6	–4.2 ± 1.2	–2.5 ± 0.7	–4.1 ± 1.4	4.5 ± 0.9	6.8 ± 1.8	2.9 ± 1.2	6.3 ± 2.2
HadGHCND (Caesar et al., 2006)	–4.5 ± 0.7	–4.0 ± 1.5	–3.3 ± 0.8	–5.0 ± 1.6	5.8 ± 1.3	8.6 ± 2.3	4.2 ± 1.8	9.4 ± 2.7
GHCNDEX (Donat et al., 2013a)	–3.9 ± 0.6	–3.9 ± 1.3	–2.6 ± 0.7	–3.9 ± 1.4	4.3 ± 0.9	6.3 ± 1.8	2.9 ± 1.2	6.1 ± 2.2

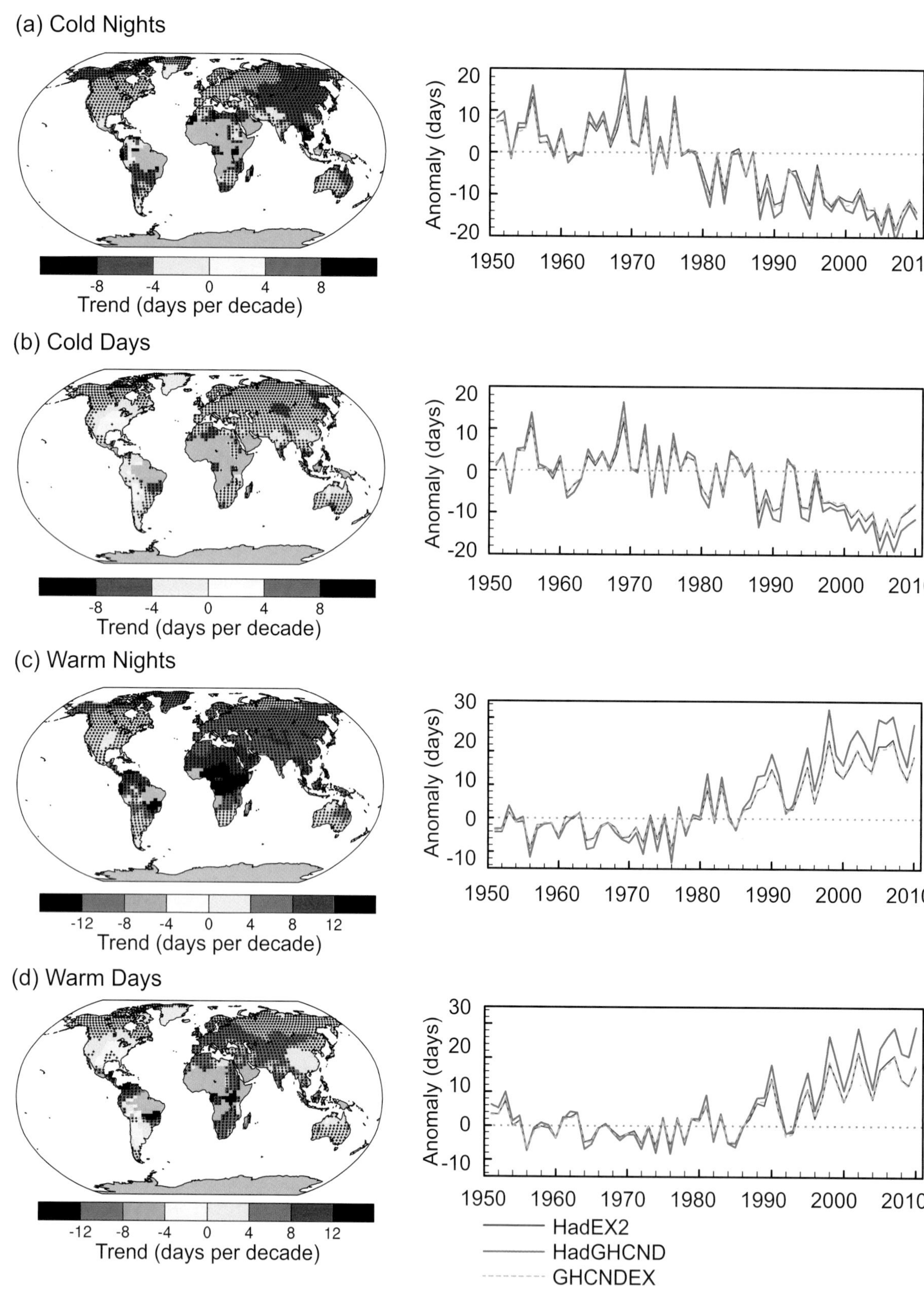

Figure 2.32 | Trends in annual frequency of extreme temperatures over the period 1951–2010, for (a) cold nights (TN10p), (b) cold days (TX10p), (c) warm nights (TN90p) and (d) warm days (TX90p) (Box 2.4, Table 1). Trends were calculated only for grid boxes that had at least 40 years of data during this period and where data ended no earlier than 2003. Grey areas indicate incomplete or missing data. Black plus signs (+) indicate grid boxes where trends are significant (i.e., a trend of zero lies outside the 90% confidence interval). The data source for trend maps is HadEX2 (Donat et al., 2013c) updated to include the latest version of the European Climate Assessment data set (Klok and Tank, 2009). Beside each map are the near-global time series of annual anomalies of these indices with respect to 1961–1990 for three global indices data sets: HadEX2 (red); HadGHCND (Caesar et al., 2006; blue) and updated to 2010 and GHCNDEX (Donat et al., 2013a; green). Global averages are only calculated using grid boxes where all three data sets have at least 90% of data over the time period. Trends are significant (i.e., a trend of zero lies outside the 90% confidence interval) for all the global indices shown.

Table 2.13 | Regional observed changes in a range of climate indices since the middle of the 20th century. Assessments are based on a range of 'global' studies and assessments (Groisman et al., 2005; Alexander et al., 2006; Caesar et al., 2006; Sheffield and Wood, 2008; Dai, 2011a, 2011b, 2013; Seneviratne et al., 2012; Sheffield et al., 2012; Donat et al., 2013a, 2013c; van der Schrier et al., 2013) and selected regional studies as indicated. Bold text indicates where the assessment is somewhat different to SREX Table 3-2. In each such case a footnote explains why the assessment is different. See also Figures 2.32 and 2.33.

Region	Warm Days (e.g., TX90p[a])	Cold Days (e.g., TX10p[a])	Warm Nights (e.g., TN90p[a], TR[a])	Cold Nights/Frosts (e.g., TN10p[a], FD[a])	Heat Waves / Warm Spells[g]	Extreme Precipitation (e.g., RX1day[a], R95p[a], R99p[a])	Dryness (e.g,. CDD[a]) / Drought[h]
North America and Central America	*High confidence*: *Likely* overall increase but spatially varying trends[1,2]	*High confidence*: *Likely* overall decrease but with spatially varying trends[1,2]	*High confidence*: *Likely* overall increase[1,2]	*High confidence*: *Likely* overall decrease[1,2]	*Medium confidence*: increases in more regions than decreases[1,3] but 1930s dominates longer term trends in the USA[4]	*High confidence*: *Likely* overall increase[1,2,5] but some spatial variation *High confidence*: *Very likely* increase central North America[6,7]	*Medium confidence*: decrease[1] but spatially varying trends ***High confidence***[b]: ***Likely* decrease central North America**[4]
South America	***Medium confidence***[b]: **Overall increase**[8]	***Medium confidence***[b]: **Overall decrease**[8]	***Medium confidence***[b]: **Overall increase**[8]	***Medium confidence***[b]: **Overall decrease**[8]	*Low confidence*: insufficient evidence (lack of literature) and spatially varying trends but some evidence of increases in more areas than decreases[8]	***Medium confidence***[b]: **Increases in more regions than decreases**[8,9] **but spatially varying trends**	*Low confidence*: limited literature and spatially varying trends[8]
Europe and Mediterranean	*High confidence*: *Likely* overall increase[10,11,12]	*High confidence*: *Likely* overall decrease[11,12]	*High confidence*: *Likely* overall increase[11,12]	*High confidence*: *Likely* overall decrease[10,11,12]	***High confidence***[b]: ***Likely* increases in most regions**[3,13]	***High confidence***[b,c]: ***Likely* increases in more regions than decreases**[5,15,16] **but regional and seasonal variation**	*Medium confidence*: spatially varying trends ***High confidence***[b]: ***Likely* increase in Mediterranean**[17,18]
Africa and Middle East	***Low to medium confidence***[b,d]: **limited data in many regions but increases in most regions assessed** ***Medium confidence***[b]: **increase North Africa and Middle East**[19,20] ***High confidence***[b]: ***Likely* increase southern Africa**[21,22,23]	***Low to medium confidence***[b,d]: **limited data in many regions but decreases in most regions assessed** ***Medium confidence***[b]: **decrease North Africa and Middle East**[19,20] ***High confidence***[b]: ***Likely* decrease southern Africa**[21,22,23]	***Medium confidence***[b,d]: **limited data in many regions but increases in most regions assessed** ***Medium confidence***[b]: **increase North Africa and Middle East**[19,20] ***High confidence***[b]: ***Likely* increase southern Africa**[21,22,23]	***Medium confidence***[b,d]: **limited data in many regions but decreases in most regions assessed** ***Medium confidence***[b]: **decrease North Africa and Middle East**[19,20] ***High confidence***[b]: ***Likely* decrease southern Africa**[21,22,23]	*Low confidence*[d]: insufficient evidence (lack of literature) *Medium confidence*: increase in North Africa and Middle East and southern Africa[3,19,21,22]	*Low confidence*[d]: insufficient evidence and spatially varying trends ***Medium confidence***[b]: **increases in more regions than decreases in southern Africa but spatially varying trends depending on index**[5,21,22]	*Medium confidence*[d]: increase[19,22,24] ***High confidence***[b]: ***Likely* increase in West Africa**[25,26] **although 1970s Sahel drought dominates the trend**
Asia (excluding South-east Asia)	***High confidence***[b,e]: ***Likely* overall increase**[27,28,29,30,31,32]	***High confidence***[b,e]: ***Likely* overall decrease**[27,28,29,30,31,32]	***High confidence***[b,e]: ***Likely* overall increase**[27,28,29,30,31,32]	***High confidence***[b,e]: ***Likely* overall increase**[27,28,29,30,31,32]	***Medium confidence***[b,e]: **Spatially varying trends and insufficient data in some regions** ***High confidence***[b,c]: ***Likely* more areas of increases than decreases**[3,28,33]	***Low to medium confidence***[b,e]: ***Low confidence* due to insufficient evidence or spatially varying trends.** ***Medium confidence*: increases in more regions than decreases**[5,34,35,36]	***Low to medium confidence***[b,e] ***Medium confidence*: Increase in eastern Asia**[36,37]
South-east Asia and Oceania	***High confidence***[b,f]: ***Likely* overall increase**[27,38,39,40]	***High confidence***[b,f]: ***Likely* overall decrease**[27,38,39]	***High confidence***[b,f]: ***Likely* overall increase**[27,38,39,40]	***High confidence***[b,f]: ***Likely* overall decrease**[27,38,39]	***Low confidence* (due lack of literature) to *high confidence***[b,f] **depending on region** ***High confidence***[2]: ***Likely* overall increase in Australia**[3,14,41]	***Low confidence* (lack of literature) to *high confidence***[b,f] *High confidence*: *Likely* decrease in southern Australia[42,43] but index and season dependent	***Low to medium confidence***[b,f]: **inconsistent trends between studies in SE Asia. Overall increase in dryness in southern and eastern Australia** ***High confidence***[b]: ***Likely* decrease northwest Australia**[25,26,44]

(continued on next page)

(Table 2.13 continued)

Notes:

[a] See Table 1 in Box 2.4, for definitions.

[b] More recent literature updates the assessment from SREX Table 3-2 (including 'global' studies).

[c] This represents a measure of the area affected which is different from what was assessed in SREX Table 3-2.

[d] This represents a slightly different region than that assessed in SREX Table 3-2 as it includes the Middle East.

[e] This represents a slightly different region than that assessed in SREX Table 3-2 as it excludes Southeast Asia.

[f] This represents a slightly different region than that assessed in SREX Table 3-2 as it combines SE Asia and Oceania.

[g] Definitions for warm spells and heat waves vary (Perkins and Alexander, 2012) but here we are commonly assessing the Warm Spell Duration Index (WSDI; Zhang et al., 2011) or other heat wave indices (e.g., HWF, HWM; (Fischer and Schär, 2010; Perkins et al., 2012) that have defined multi-day heat extremes relative to either daily maximum or minimum temperatures (or both) above a high (commonly 90th) percentile relative to a late-20th century reference period.

[h] See Box 2.4 and Section 2.6.1 for definitions.

[1] Kunkel et al. (2008), [2] Peterson et al. (2008), [3] Perkins et al. (2012), [4] Peterson et al. (2013), [5] Westra et al. (2013), [6] Groisman et al. (2012), [7] Villarini et al. (2013), [8] Skansi et al. (2013), [9] Haylock et al. (2006), [10] Andrade et al. (2012), [11] Efthymiadis et al. (2011), [12] Moberg et al. (2006), [13] Della-Marta et al. (2007a), [14] Perkins and Alexander (2012), [15] Van den Besselaar et al. (2012), [16] Zolina et al. (2009), [17] Sousa et al. (2011), [18] Hoerling et al. (2012), [19] Donat et al. (2013b), [20] Zhang et al. (2005), [21] Kruger and Sekele (2013), [22] New et al. (2006), [23] Vincent et al. (2011), [24] Aguilar et al. (2009), [25] Dai (2013), [26] Sheffield et al. (2012), [27] Choi et al. (2009), [28] Rahimzadeh et al. (2009), [29] Revadekar et al. (2012), [30] Tank et al. (2006), [31] You et al. (2010), [32] Zhou and Ren (2011), [33] Ding et al. (2010), [34] Krishna Moorthy et al. (2009), [35] Pattanaik and Rajeevan (2010), [36] Wang et al. (2012b), [37] Fischer et al. (2011), [38] Caesar et al. (2011), [39] Chambers and Griffiths (2008), [40] Wang et al. (2013), [41] Tryhorn and Risbey (2006), [42] Gallant et al. (2007), [43] King et al. (2013), [44] Jones et al. (2009).

Alexander et al., 2009; Li et al., 2012), particularly in regions around the Pacific Rim (Kenyon and Hegerl, 2008). Globally, there is evidence of large-scale warming trends in the extremes of temperature, especially minimum temperature, since the beginning of the 20th century (Donat et al., 2013c).

There are some exceptions to this large-scale warming of temperature extremes including central North America, eastern USA (Alexander et al., 2006; Kunkel et al., 2008; Peterson et al., 2008) and some parts of South America (Alexander et al., 2006; Rusticucci and Renom, 2008; Skansi et al., 2013) which indicate changes consistent with cooling in these locations. However, these exceptions appear to be mostly associated with changes in maximum temperatures (Donat et al., 2013c). The so-called 'warming hole' in central North America and eastern USA, where temperatures have cooled relative to the significant warming elsewhere in the region, is associated with observed changes in the hydrological cycle and land–atmosphere interaction (Pan et al., 2004; Portmann et al., 2009a; Portmann et al., 2009b; Misra et al., 2012) and decadal and multi-decadal variability linked with the Atlantic and Pacific Oceans (Meehl et al., 2012; Weaver, 2012).

Since AR4 many studies have analysed local to regional changes in multi-day temperature extremes in more detail, specifically addressing different heat wave aspects such as frequency, intensity, duration and spatial extent (Box 2.4, FAQ 2.2). Several high-profile heat waves have occurred in recent years (e.g., in Europe in 2003 (Beniston, 2004), Australia in 2009 (Pezza et al., 2012), Russia in 2010 (Barriopedro et al., 2011; Dole et al., 2011; Trenberth and Fasullo, 2012a) and USA in 2010/2011 (Hoerling et al., 2012) (Section 10.6.2) which have had severe impacts (see WGII). Heat waves are often associated with quasi-stationary anticyclonic circulation anomalies that produce prolonged hot conditions at the surface (Black and Sutton, 2007; Garcia-Herrera et al., 2010), but long-term changes in the persistence of these anomalies are still relatively poorly understood (Section 2.7). Heat waves can also be amplified by pre-existing dry soil conditions in transitional climate zones (Ferranti and Viterbo, 2006; Fischer et al., 2007; Seneviratne et al., 2010; Mueller and Seneviratne, 2012) and the persistence of those soil-mositure anomalies (Lorenz et al., 2010). Dry soil-moisture conditions are either induced by precipitation deficits (Della-Marta et al., 2007b; Vautard et al., 2007), or evapotranspiration excesses (Black and Sutton, 2007; Fischer et al., 2007), or a combination of both (Seneviratne et al., 2010). This amplification of soil moisture–temperature feedbacks is suggested to have partly enhanced the duration of extreme summer heat waves in southeastern Europe during the latter part of the 20th century (Hirschi et al., 2011), with evidence emerging of a signature in other moisture-limited regions (Mueller and Seneviratne, 2012).

Table 2.13 shows that there has been a *likely* increasing trend in the frequency of heatwaves since the middle of the 20th century in Europe and Australia and across much of Asia where there are sufficient data. However, *confidence* on a global scale is *medium* owing to lack of studies over Africa and South America but also in part owing to differences in trends depending on how heatwaves are defined (Perkins et al., 2012). Using monthly means as a proxy for heatwaves Coumou et al. (2013) and Hansen et al. (2012) indicate that record-breaking temperatures in recent decades substantially exceed what would be expected by chance but caution is required when making inferences between these studies and those that deal with multi-day events and/or use more complex definitions for heatwave events. There is also evidence in some regions that periods prior to the 1950s had more heatwaves (e.g., over the USA, the decade of the 1930s stands out and is also associated with extreme drought conditions (Peterson et al., 2013) whereas conversely in other regions heatwave trends may have been underestimated owing to poor quality and/or consistency of data (e.g., Della-Marta et al. (2007a) over Western Europe; Kuglitsch et al. (2009, 2010) over the Mediterranean). Recent available studies also suggest that the number of cold spells has reduced significantly since the 1950s (Donat et al., 2013a, 2013c).

In summary, new analyses continue to support the AR4 and SREX conclusions that since about 1950 it is *very likely* that the numbers of cold days and nights have decreased and the numbers of warm days and nights have increased overall on the global scale, that is, for land areas with sufficient data. It is *likely* that such changes have also occurred across most of North America, Europe, Asia and Australia. There is *low* to *medium confidence* in historical trends in daily temperature extremes in Africa and South America as there is either insufficient data or trends vary across these regions. This, combined with issues with defining events, leads to the assessment that there is *medium confidence* that globally the length and frequency of warm spells, including heat waves, has increased since the middle of the 20th century although it is *likely* that heatwave frequency has increased during this period in large parts of Europe, Asia and Australia.

2.6.2 Extremes of the Hydrological Cycle

In Section 2.5 mean state changes in different aspects of the hydrological cycle are discussed. In this section we focus on the more extreme aspects of the cycle including extreme rainfall, severe local weather events like hail, flooding and droughts. Extreme events associated with tropical and extratropical storms are discussed in Sections 2.6.3 and 2.6.4 respectively.

2.6.2.1 Precipitation Extremes

AR4 concluded that substantial increases are found in heavy precipitation events. It was *likely* that annual heavy precipitation events had disproportionately increased compared to mean changes between 1951 and 2003 over many mid-latitude regions, even where there had been a reduction in annual total precipitation. Rare precipitation (such as the highest annual daily precipitation total) events were *likely* to have increased over regions with sufficient data since the late 19th century. SREX supported this view, as have subsequent analyses, but noted large spatial variability within and between regions (Table 3.2 of Seneviratne et al., 2012).

Given the diverse climates across the globe, it has been difficult to provide a universally valid definition of 'extreme precipitation'. However, Box 2.4 Table 1 indicates some of the common definitions that are used in the scientific literature. In general, statistical tests indicate changes in precipitation extremes are consistent with a wetter climate (Section 7.6.5), although with a less spatially coherent pattern of change than temperature, in that there are large areas that show increasing trends and large areas that show decreasing trends and a lower level of statistical significance than for temperature change (Alexander et al., 2006; Donat et al., 2013a, 2013c). Using R95p and SDII indices (Box 2.4), Figures 2.33a and 2.33b show these areas for heavy precipitation amounts and precipitation intensity where sufficient data are available in the HadEX2 data set (Donat et al., 2013c) although there are more areas showing significant increases than decreases. Although changes in large-scale circulation patterns have a substantial influence on precipitation extremes globally (Alexander et al., 2009; Kenyon and Hegerl, 2010), Westra et al. (2013) showed, using *in situ* data over land, that trends in the wettest day of the year indicate more increases than would be expected by chance. Over the tropical oceans satellite measurements show an increase in the frequency of the heaviest rainfall during warmer (El Niño) years (Allan and Soden, 2008).

Regional trends in precipitation extremes since the middle of the 20th century are varied (Table 2.13). In most continents *confidence* in trends is not higher than *medium* except in North America and Europe where there have been *likely* increases in either the frequency or intensity of heavy precipitation. This assessment increases to *very likely* for central North America. For North America it is also *likely* that increases have occurred during the whole of the 20th century (Pryor et al., 2009; Donat et al., 2013c; Villarini et al., 2013). For South America the most recent integrative studies indicate heavy rain events are increasing in frequency and intensity over the contient as a whole (Donat et al., 2013c; Skansi et al., 2013). For Europe and the Mediterranean, the assessment masks some regional and seasonal variation. For example, much of the increase reported in Table 2.13 is found in winter although with decreasing trends in some other regions such as northern Italy, Poland and some Mediterranean coastal sites (Pavan et al., 2008; Lupikasza, 2010; Toreti et al., 2010). There are mixed regional trends across Asia and Oceania but with some indication that increases are being observed in more regions than decreases while recent studies focused on Africa, in general, have not found significant trends in extreme precipitation (see Chapter 14 for more on regional variations and trends).

The above studies generally use indices which reflect 'moderate' extremes, for example, events occurring as often as 5% or 10% of the time (Box 2.4). Only a few regions have sufficient data to assess trends in rarer precipitation events reliably, for example, events occurring on average once in several decades. Using Extreme Value Theory, DeGaetano (2009) showed a 20% reduction in the return period for extreme precipitation events over large parts of the contiguous USA from 1950 to 2007. For Europe from 1951 to 2010, Van den Besselaar et al. (2012) reported a median reduction in 5- to 20-year return periods of 21%, with a range between 2% and 58% depending on the subregion and season. This decrease in return times for rare extremes is qualitatively similar to the increase in moderate extremes for these regions reported above, and also consistent with earlier local results for the extreme tail of the distribution reported in AR4.

The aforementioned studies refer to daily precipitation extremes, although rainfall will often be limited to part of the day only. The literature on sub-daily scales is too limited for a global assessment although it is clear that analysis and framing of questions regarding sub-daily precipitation extremes is becoming more critical (Trenberth, 2011). Available regional studies have shown results that are even more complex than for daily precipitation and with variations in the spatial patterns of trends depending on event formulation and duration. However, regional studies show indications of more increasing than decreasing trends (Sen Roy, 2009; for India) (Sen Roy and Rouault, 2013; for South Africa) (Westra and Sisson, 2011; for Australia). Some studies present evidence of scaling of sub-daily precipitation with temperature that is outside that expected from the Clausius–Clapeyron relation (about 7% per degree Celsius) (Lenderink and Van Meijgaard, 2008; Haerter et al., 2010; Jones et al., 2010; Lenderink et al., 2011; Utsumi et al., 2011), but scaling beyond that expected from thermodynamic theories is controversial (Section 7.6.5).

In summary, further analyses continue to support the AR4 and SREX conclusions that it is *likely* that since 1951 there have been statistically significant increases in the number of heavy precipitation events (e.g., above the 95th percentile) in more regions than there have been statistically significant decreases, but there are strong regional and subregional variations in the trends. In particular, many regions present statistically non-significant or negative trends, and, where seasonal changes have been assessed, there are also variations between seasons (e.g., more consistent trends in winter than in summer in Europe). The overall most consistent trends towards heavier precipitation events are found in central North America (*very likely* increase) but assessment for Europe shows *likely* increases in more regions than decreases.

2.6.2.2 Floods

AR4 WGI Chapter 3 (Trenberth et al., 2007) did not assess changes in floods but AR4 WGII concluded that there was not a general global trend in the incidence of floods (Kundzewicz et al., 2007). SREX went further to suggest that there was low agreement and thus *low confidence* at the global scale regarding changes in the magnitude or frequency of floods or even the sign of changes.

AR5 WGII assesses floods in regional detail accounting for the fact that trends in floods are strongly influenced by changes in river management (see also Section 2.5.2). Although the most evident flood trends appear to be in northern high latitudes, where observed warming trends have been largest, in some regions no evidence of a trend in extreme flooding has been found, for example, over Russia based on daily river discharge (Shiklomanov et al., 2007). Other studies for Europe (Hannaford and Marsh, 2008; Renard et al., 2008; Petrow and Merz, 2009; Stahl et al., 2010) and Asia (Jiang et al., 2008; Delgado et al., 2010) show evidence for upward, downward or no trend in the magnitude and frequency of floods, so that there is currently no clear and widespread evidence for observed changes in flooding except for the earlier spring flow in snow-dominated regions (Seneviratne et al., 2012).

In summary, there continues to be a lack of evidence and thus *low confidence* regarding the sign of trend in the magnitude and/or frequency of floods on a global scale.

2.6.2.3 Droughts

AR4 concluded that droughts had become more common, especially in the tropics and sub-tropics since about 1970. SREX provided a comprehensive assessment of changes in observed droughts (Section 3.5.1 and Box 3.3 of SREX), updated the conclusions provided by AR4 and stated that the type of drought considered and the complexities in defining drought (Annex III: Glossary) can substantially affect the conclusions regarding trends on a global scale (Chapter 10). Based on evidence since AR4, SREX concluded that there were not enough direct observations of dryness to suggest *high confidence* in observed trends globally, although there was *medium confidence* that since the 1950s some regions of the world have experienced more intense and longer droughts. The differences between AR4 and SREX are due primarily to analyses post-AR4, differences in how both assessments considered drought and updated IPCC uncertainty guidance.

There are very few direct measurements of drought related variables, such as soil moisture (Robock et al., 2000), so drought proxies (e.g., PDSI, SPI, SPEI; Box 2.4) and hydrological drought proxies (e.g., Vidal et al., 2010; Dai, 2011b) are often used to assess drought. The chosen proxy (e.g., precipitation, evapotranspiration, soil moisture or streamflow) and time scale can strongly affect the ranking of drought events (Sheffield et al., 2009; Vidal et al., 2010). Analyses of these indirect indices come with substantial uncertainties. For example, PDSI may not be comparable across climate zones. A self-calibrating (sc-) PDSI can replace the fixed empirical constants in PDSI with values representative of the local climate (Wells et al., 2004). Furthermore, for studies using simulated soil moisture, the type of potential evapotranspiration model used can lead to significant differences in the estimation of the regions affected and the areal extent of drought (Sheffield et al., 2012), but the overall effect of a more physically realistic parameterisation is debated (van der Schrier et al., 2013).

Because drought is a complex variable and can at best be incompletely represented by commonly used drought indices, discrepancies in the interpretation of changes can result. For example, Sheffield and Wood (2008) found decreasing trends in the duration, intensity and severity of drought globally. Conversely, Dai (2011a,b) found a general global increase in drought, although with substantial regional variation and individual events dominating trend signatures in some regions (e.g., the 1970s prolonged Sahel drought and the 1930s drought in the USA and Canadian Prairies). Studies subsequent to these continue to provide somewhat different conclusions on trends in global droughts and/or dryness since the middle of the 20th century (Sheffield et al., 2012; Dai, 2013; Donat et al., 2013c; van der Schrier et al., 2013).

Van der Schrier et al. (2013), using monthly sc-PDSI, found no strong case either for notable drying or moisture increase on a global scale over the periods 1901–2009 or 1950–2009, and this largely agrees with the results of Sheffield et al. (2012) over the latter period. A comparison between the sc-PDSI calculated by van der Schrier et al. (2013) and that of Dai (2011a) shows that the dominant mode of variability is very similar, with a temporal evolution suggesting a trend toward drying. However, the same analysis for the 1950–2009 period shows an initial increase in drying in the Van der Schrier et al. data set, followed by a decrease from the mid-1980s onwards, while the Dai data show a continuing increase until 2000. The difference in trends between the sc-PDSI data set of Van der Schrier et al. and Dai appears to be due to the different calibration periods used, the shorter 1950–1979 period in the latter study resulting in higher index values from 1980 onwards, although the associated spatial patterns are similar. In addition, the observed precipitation forcing data set differs between studies, with van der Schrier et al. (2013) and Sheffield et al. (2012) using CRU TS 3.10.01 (updated from Mitchell and Jones, 2005). This data set uses fewer stations and has been wetter than some other precipitation products in the last couple of decades (Figure 2.29, Table 2.9), although the best data set to use is still an open question. Despite this, a measure of sc-PDSI with potential evapotranspiration estimated using the Penman–Montieth equation shows an increase in the percentage of land area in drought since 1950 (Sheffield et al., 2012; Dai, 2013), while van der Schrier et al. (2013) also finds a slight increase in the percentage of land area in severe drought using the same measure. This is qualitatively consistent with the trends in surface soil moisture found for the shorter period 1988–2010 by Dorigo et al. (2012) using a new multi-satellite data set and changes in observed streamflow (Dai, 2011b). However all these studies draw somewhat different conclusions and the compelling arguments both for (Dai, 2011b, 2013) and against (Sheffield et al., 2012; van der Schrier et al., 2013) a significant increase in the land area experiencing drought has hampered global assessment.

Studies that support an increasing trend towards the land area affected by drought seem to be at odds with studies that look at trends in dryness (i.e., lack of rainfall). For example, Donat et al. (2013c) found that the annual maximum number of consecutive dry days has declined since the 1950s in more regions than it has increased (Figure 2.33c). However, only regions in Russia and the USA indicate significant changes and there is a lack of information for this index over large regions, especially Africa. Most other studies focussing on global dryness find similar results, with decadal variability dominating longer-term trends (Frich et al., 2002; Alexander et al., 2006; Donat et al., 2013a). However, Giorgi et al. (2011) indicate that 'hydroclimatic intensity' (Box 2.4, Chapter 7), a measure which combines both dry spell length and precipitation intensity, has increased over the latter part of the 20th century in response to a warming climate. They show that positive trends (reflecting an increase in the length of drought and/or extreme precipitation events) are most marked in Europe, India, parts of South America and East Asia although trends appear to have decreased (reflecting a decrease in the length of drought and/or extreme precipitation events) in Australia and northern South America (Figure 2.33c). Data availability, quality and length of record remain issues in drawing conclusions on a global scale, however.

Despite differences between the conclusions drawn by global studies, there are some areas in which they agree. Table 2.13 indicates that there is *medium confidence* of an increase in dryness or drought in East Asia with *high confidence* that this is the case in the Mediterannean and West Africa. There is also *high confidence* of decreases in dryness or drought in central North America and north-west Australia.

In summary, the current assessment concludes that there is not enough evidence at present to suggest more than *low confidence* in a global-scale observed trend in drought or dryness (lack of rainfall) since the middle of the 20th century, owing to lack of direct observations, geographical inconsistencies in the trends, and dependencies of inferred trends on the index choice. Based on updated studies, AR4 conclusions regarding global increasing trends in drought since the 1970s were probably overstated. However, it is *likely* that the frequency and intensity of drought has increased in the Mediterranean and West Africa and decreased in central North America and north-west Australia since 1950.

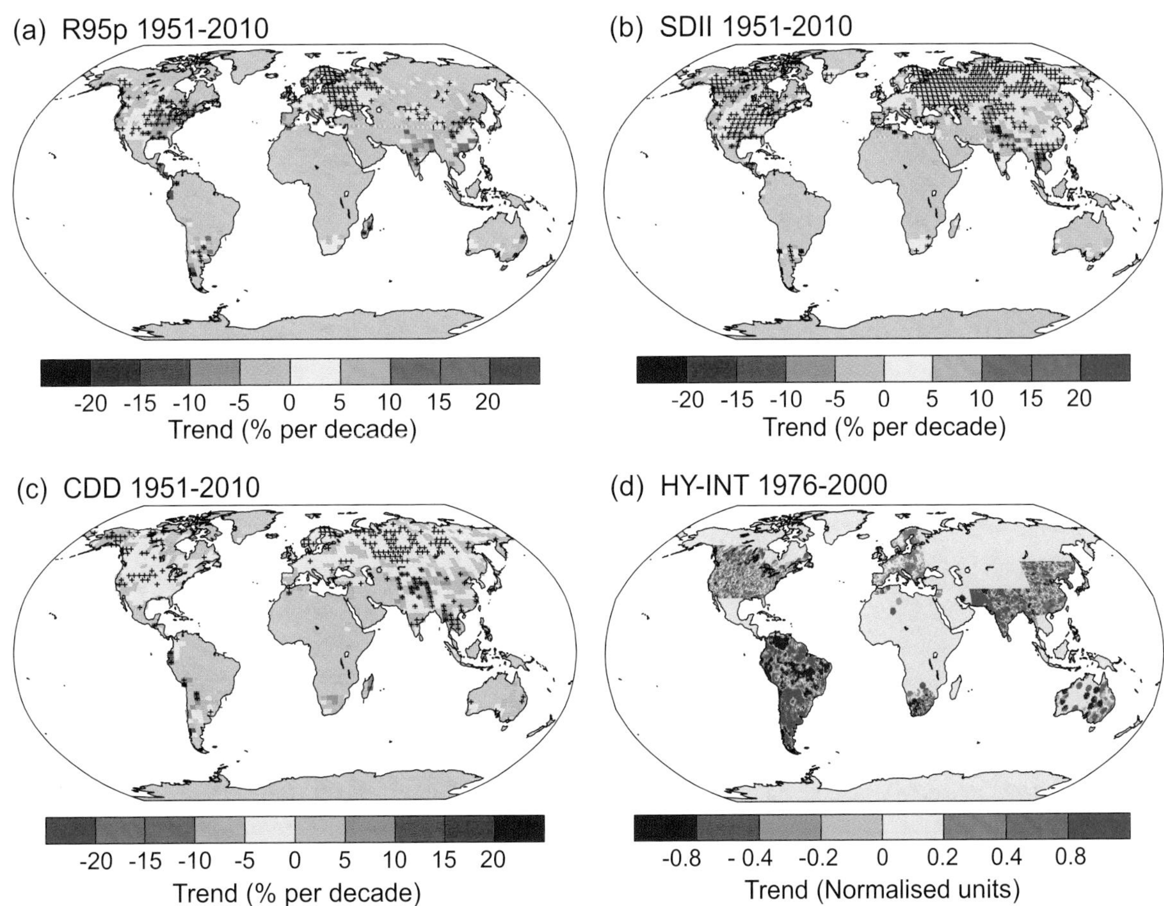

Figure 2.33 | Trends in (a) annual amount of precipitation from days >95th percentile (R95p), (b) daily precipitation intensity (SDII) and (c) frequency of the annual maximum number of consecutive dry days (CDD) (Box 2.4, Table 1). Trends are shown as relative values for better comparison across different climatic regions. Trends were calculated only for grid boxes that had at least 40 years of data during this period and where data ended no earlier than 2003. Grey areas indicate incomplete or missing data. Black plus signs (+) indicate grid boxes where trends are significant (i.e., a trend of zero lies outside the 90% confidence interval). The data source for trend maps is HadEX2 (Donat et al., 2013a) updated to include the latest version of the European Climate Assessment data set (Klok and Tank, 2009). (d) Trends (normalized units) in hydroclimatic intensity (HY-INT: a multiplicative measure of length of dry spell and precipitation intensity) over the period 1976–2000 (adapted from Giorgi et al., 2011). An increase (decrease) in HY-INT reflects an increase (decrease) in the length of drought and /or extreme precipitation events.

2.6.2.4 Severe Local Weather Events

Another extreme aspect of the hydrological cycle is severe local weather phenomena such as hail or thunder storms. These are not well observed in many parts of the world because the density of surface meteorological observing stations is too coarse to measure all such events. Moreover, homogeneity of existing reporting is questionable (Verbout et al., 2006; Doswell et al., 2009). Alternatively, measures of severe thunderstorms or hailstorms can be derived by assessing the environmental conditions that are favourable for their formation but this method is very uncertain (Seneviratne et al., 2012). SREX highlighted studies such as those of Brooks and Dotzek (2008), who found significant variability but no clear trend in the past 50 years in severe thunderstorms in a region east of the Rocky Mountains in the USA, Cao (2008), who found an increasing frequency of severe hail events in Ontario, Canada during the period 1979–2002 and Kunz et al. (2009), who found that hail days significantly increased during the period 1974–2003 in southwest Germany. Hailpad studies from Italy (Eccel et al., 2012) and France (Berthet et al., 2011) suggest slight increases in larger hail sizes and a correlation between the fraction of precipitation falling as hail with average summer temperature while in Argentina between 1960 and 2008 the annual number of hail events was found to be increasing in some regions and decreasing in others (Mezher et al., 2012). In China between 1961 and 2005, the number of hail days has been found to generally decrease, with the highest occurrence between 1960 and 1980 but with a sharp drop since the mid-1980s (CMA, 2007; Xie et al., 2008). However, there is little consistency in hail size changes in different regions of China since 1980 (Xie et al., 2010). Remote sensing offers a potential alterative to surface-based meteorological networks for detecting changes in small scale severe weather phenomenon such as proxy measurements of lightning from satellites (Zipser et al., 2006) but there remains little convincing evidence that changes in severe thunderstorms or hail have occurred since the middle of the 20th century (Brooks, 2012).

In summary, there is *low confidence* in observed trends in small-scale severe weather phenomena such as hail and thunderstorms because of historical data inhomogeneities and inadequacies in monitoring systems.

2.6.3 Tropical Storms

AR4 concluded that it was *likely* that an increasing trend had occurred in intense tropical cyclone activity since 1970 in some regions but that there was no clear trend in the annual numbers of tropical cyclones. Subsequent assessments, including SREX and more recent literature indicate that it is difficult to draw firm conclusions with respect to the confidence levels associated with observed trends prior to the satellite era and in ocean basins outside of the North Atlantic.

Section 14.6.1 discusses changes in tropical storms in detail. Current data sets indicate no significant observed trends in global tropical cyclone frequency over the past century and it remains uncertain whether any reported long-term increases in tropical cyclone frequency are robust, after accounting for past changes in observing capabilities (Knutson et al., 2010). Regional trends in tropical cyclone frequency and the frequency of very intense tropical cyclones have been identified in the North Atlantic and these appear robust since the 1970s (Kossin et al. 2007) (*very high confidence*). However, argument reigns over the cause of the increase and on longer time scales the fidelity of these trends is debated (Landsea et al., 2006; Holland and Webster, 2007; Landsea, 2007; Mann et al., 2007b) with different methods for estimating undercounts in the earlier part of the record providing mixed conclusions (Chang and Guo, 2007; Mann et al., 2007a; Kunkel et al., 2008; Vecchi and Knutson, 2008, 2011). No robust trends in annual numbers of tropical storms, hurricanes and major hurricanes counts have been identified over the past 100 years in the North Atlantic basin. Measures of land-falling tropical cyclone frequency (Figure 2.34) are generally considered to be more reliable than counts of all storms which tend to be strongly influenced by those that are weak and/or short lived. Callaghan and Power (2011) find a statistically significant decrease in Eastern Australia land-falling tropical cyclones since the late 19th century although including 2010/2011 season data this trend becomes non-significant (i.e., a trend of zero lies just inside the 90% confidence interval). Significant trends are not found in other oceans on shorter time scales (Chan and Xu, 2009; Kubota and Chan, 2009; Mohapatra et al., 2011; Weinkle et al., 2012), although Grinsted et al. (2012) find a significant positive trend in eastern USA using tide-guage data from 1923–2008 as a proxy for storm surges associated with land-falling hurricanes. Differences between tropical cyclone studies highlight the challenges that still lie ahead in assessing long-term trends.

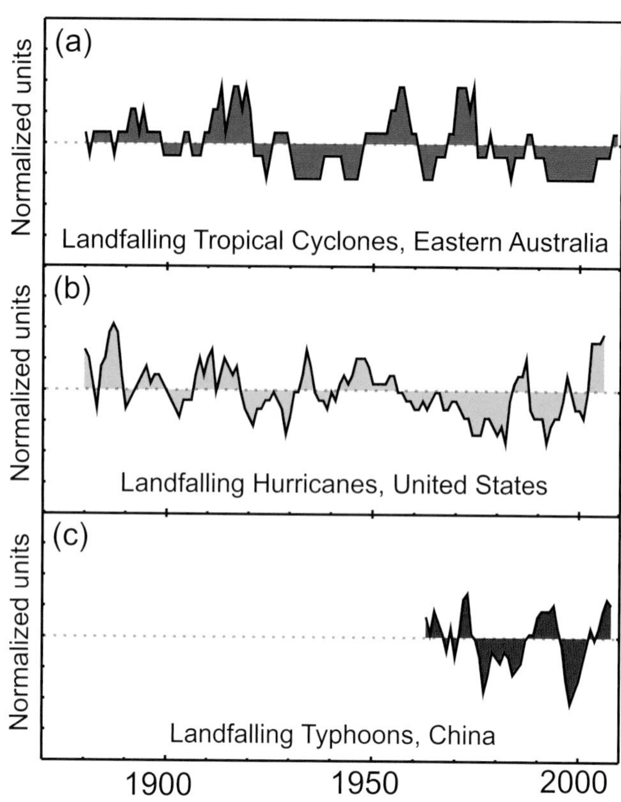

Figure 2.34 | Normalized 5-year running means of the number of (a) adjusted land falling eastern Australian tropical cyclones (adapted from Callaghan and Power (2011) and updated to include 2010//2011 season) and (b) unadjusted land falling U.S. hurricanes (adapted from Vecchi and Knutson (2011) and (c) land-falling typhoons in China (adapted from CMA, 2011). Vertical axis ticks represent one standard deviation, with all series normalized to unit standard deviation after a 5-year running mean was applied.

Arguably, storm frequency is of limited usefulness if not considered in tandem with intensity and duration measures. Intensity measures in historical records are especially sensitive to changing technology and improving methodology. However, over the satellite era, increases in the intensity of the strongest storms in the Atlantic appear robust (Kossin et al., 2007; Elsner et al., 2008) but there is limited evidence for other regions and the globe. Time series of cyclone indices such as power dissipation, an aggregate compound of tropical cyclone frequency, duration and intensity that measures total wind energy by tropical cyclones, show upward trends in the North Atlantic and weaker upward trends in the western North Pacific since the late 1970s (Emanuel, 2007), but interpretation of longer-term trends is again constrained by data quality concerns (Landsea et al., 2011).

In summary, this assessment does not revise the SREX conclusion of *low confidence* that any reported long-term (centennial) increases in tropical cyclone activity are robust, after accounting for past changes in observing capabilities. More recent assessments indicate that it is *unlikely* that annual numbers of tropical storms, hurricanes and major hurricanes counts have increased over the past 100 years in the North Atlantic basin. Evidence, however, is for a *virtually certain* increase in the frequency and intensity of the strongest tropical cyclones since the 1970s in that region.

2.6.4 Extratropical Storms

AR4 noted a *likely* net increase in frequency/intensity of NH extreme extratropical cyclones and a poleward shift in storm tracks since the 1950s. SREX further consolidated the AR4 assessment of poleward shifting storm tracks, but revised the assessment of the confidence levels associated with regional trends in the intensity of extreme extratropical cyclones.

Studies using reanalyses continue to support a northward and eastward shift in the Atlantic cyclone activity during the last 60 years with both more frequent and more intense wintertime cyclones in the high-latitude Atlantic (Schneidereit et al., 2007; Raible et al., 2008; Vilibic and Sepic, 2010) and fewer in the mid-latitude Atlantic (Wang et al., 2006b; Raible et al., 2008). Some studies show an increase in intensity and number of extreme Atlantic cyclones (Paciorek et al., 2002; Lehmann et al., 2011) while others show opposite trends in eastern Pacific and North America (Gulev et al., 2001). Comparisons between studies are hampered because of the sensitivities in identification schemes and/or different definitions for extreme cyclones (Ulbrich et al., 2009; Neu et al., 2012). The fidelity of research findings also rests largely with the underlying reanalyses products that are used (Box 2.3). See also Section 14.6.2.

Over longer periods studies of severe storms or storminess have been performed for Europe where long running *in situ* pressure and wind observations exist. Direct wind speed measurements, however, either have short records or are hampered by inconsistencies due to changing instrumentation and observing practice over time (Smits et al., 2005; Wan et al., 2010). In most cases, therefore wind speed or storminess proxies are derived from *in situ* pressure measurements or reanalyses data, the quality and consistency of which vary. *In situ* observations indicate no clear trends over the past century or longer (Hanna et al., 2008; Matulla et al., 2008; Allan et al., 2009; Barring and Fortuniak, 2009), with substantial decadal and longer fluctuations but with some regional and seasonal trends (Wang et al., 2009c, 2011). Figure 2.35 shows some of these changes for boreal winter using geostrophic wind speeds indicating that decreasing trends outnumber increasing trends (Wang et al., 2011), although with few that are statistically significant. Although Donat et al. (2011) and Wang et al. (2012h) find significant increases in both the strength and frequency of wintertime storms for large parts of Europe using the 20CR (Compo et al., 2011), there is debate over whether this is an artefact of the changing number of assimilated observations over time (Cornes and Jones, 2011; Krueger et al., 2013) even though Wang et al. (2012h) find good agreement between the 20CR trends and those derived from geostropic wind extremes in the North Sea region.

SREX noted that available studies using reanalyses indicate a decrease in extratropical cyclone activity (Zhang et al., 2004) and intensity (Zhang et al., 2004; Wang et al., 2009d) over the last 50 years has been reported for northern Eurasia (60°N to 40°N) linked to a possible northward shift with increased cyclone frequency in the higher latitudes and decrease in the lower latitudes. The decrease at lower latitudes was also found in East Asia (Wang et al., 2012h) and is also supported by a study of severe storms by Zou et al. (2006b) who used sub-daily *in situ* pressure data from a number of stations across China.

SREX also notes that, based on reanalyses, North American cyclone numbers have increased over the last 50 years, with no statistically significant change in cyclone intensity (Zhang et al., 2004). Hourly SLP data from Canadian stations showed that winter cyclones have become significantly more frequent, longer lasting, and stronger in the lower Canadian Arctic over the last 50 years (1953–2002), but less frequent and weaker in the south, especially along the southeast and southwest Canadian coasts (Wang et al., 2006a). Further south, a tendency toward weaker low-pressure systems over the past few decades was found for U.S. east coast winter cyclones using reanalyses, but no statistically significant trends in the frequency of occurrence of systems (Hirsch et al., 2001).

Using the 20CR (Compo et al., 2011), Wang et al. (2012h) found substantial increases in extratropical cyclone activity in the SH (20°S to 90°S). However, for southeast Australia, a decrease in activity is found and this agrees well with geostrophic wind extremes derived from *in situ* surface pressure observations (Alexander et al., 2011). This strengthens the evidence of a southward shift in storm tracks previously noted using older reanalyses products (Fyfe, 2003; Hope et al., 2006). Frederiksen and Frederiksen (2007) linked the reduction in cyclogenesis at 30°S and southward shift to a decrease in the vertical mean meridional temperature gradient. There is some inconsistency among reanalysis products for the SH regarding trends in the frequency of intense extratropical cyclones (Lim and Simmonds, 2007; Pezza et al., 2007; Lim and Simmonds, 2009) although studies tend to agree on a trend towards more intense systems, even when inhomogeneities associated with changing numbers of observations have been taken into account (Wang et al., 2012h). However, further undetected contamination of these trends owing to issues with the reanalyses products cannot be ruled out (Box 2.3) and this lowers our confidence in long-term trends. Links between extratropical cyclone activity and large-scale variability are discussed in Sections 2.7 and 14.6.2.

Frequently Asked Questions

FAQ 2.2 | Have There Been Any Changes in Climate Extremes?

There is strong evidence that warming has lead to changes in temperature extremes—including heat waves—since the mid-20th century. Increases in heavy precipitation have probably also occurred over this time, but vary by region. However, for other extremes, such as tropical cyclone frequency, we are less certain, except in some limited regions, that there have been discernable changes over the observed record.

From heat waves to cold snaps or droughts to flooding rains, recording and analysing climate extremes poses unique challenges, not just because these events are rare, but also because they invariably happen in conjunction with disruptive conditions. Furthermore, there is no consistent definition in the scientific literature of what constitutes an extreme climatic event, and this complicates comparative global assessments.

Although, in an absolute sense, an extreme climate event will vary from place to place—a hot day in the tropics, for instance, may be a different temperature to a hot day in the mid-latitudes—international efforts to monitor extremes have highlighted some significant global changes.

For example, using consistent definitions for cold (<10th percentile) and warm (>90th percentile) days and nights it is found that warm days and nights have increased and cold days and nights have decreased for most regions of the globe; a few exceptions being central and eastern North America, and southern South America but mostly only related to daytime temperatures. Those changes are generally most apparent in minimum temperature extremes, for example, warm nights. Data limitations make it difficult to establish a causal link to increases in average temperatures, but FAQ 2.2, Figure 1 indicates that daily global temperature extremes have indeed changed. Whether these changes are simply associated with the average of daily temperatures increasing (the dashed lines in FAQ 2.2, Figure 1) or whether other changes in the distribution of daytime and nighttime temperatures have occurred is still under debate.

Warm spells or heat waves, that is, periods containing consecutive extremely hot days or nights, have also been assessed, but there are fewer studies of heat wave characteristics than those that compare changes in merely warm days or nights. Most global land areas with available data have experienced more heat waves since the middle of the 20th century. One exception is the south-eastern USA, where heat wave frequency and duration measures generally show decreases. This has been associated with a so-called 'warming hole' in this region, where precipitation has also increased and may be related to interactions between the land and the atmosphere and long-term variations in the Atlantic and Pacific Oceans. However, for large regions, particularly in Africa and South America, information on changes in heatwaves is limited.

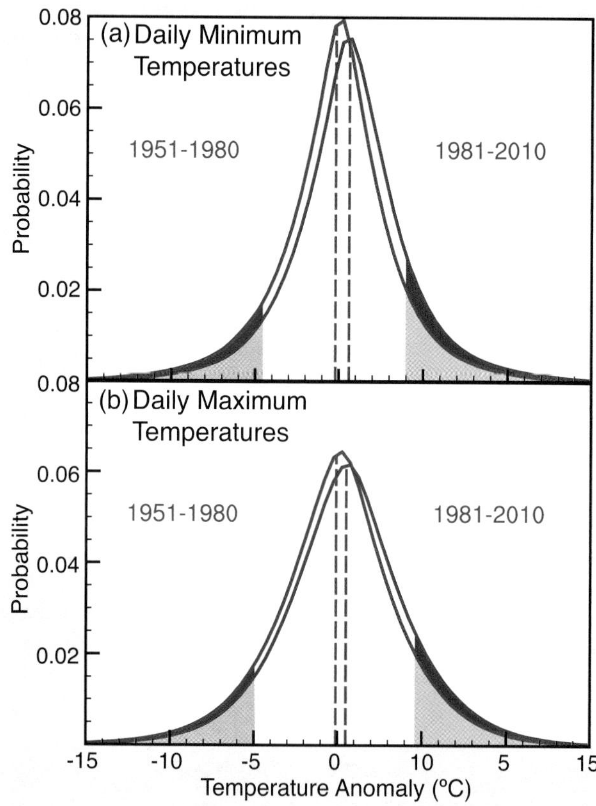

FAQ 2.2, Figure 1 | Distribution of (a) daily minimum and (b) daily maximum temperature anomalies relative to a 1961–1990 climatology for two periods: 1951–1980 (blue) and 1981–2010 (red) using the HadGHCND data set. The shaded blue and red areas represent the coldest 10% and warmest 10% respectively of (a) nights and (b) days during the 1951–1980 period. The darker shading indicates by how much the number of the coldest days and nights has reduced (dark blue) and by how much the number of the warmest days and nights has increased (dark red) during the 1981–2010 period compared to the 1951–1980 period.

For regions such as Europe, where historical temperature reconstructions exist going back several hundreds of years, indications are that some areas have experienced a disproportionate number of extreme heat waves in recent decades. *(continued on next page)*

FAQ 2.2 (continued)

Changes in extremes for other climate variables are generally less coherent than those observed for temperature, owing to data limitations and inconsistencies between studies, regions and/or seasons. However, increases in precipitation extremes, for example, are consistent with a warmer climate. Analyses of land areas with sufficient data indicate increases in the frequency and intensity of extreme precipitation events in recent decades, but results vary strongly between regions and seasons. For instance, evidence is most compelling for increases in heavy precipitation in North America, Central America and Europe, but in some other regions—such as southern Australia and western Asia—there is evidence of decreases. Likewise, drought studies do not agree on the sign of the global trend, with regional inconsistencies in trends also dependent on how droughts are defined. However, indications exist that droughts have increased in some regions (e.g., the Mediterranean) and decreased in others (e.g., central North America) since the middle of the 20th century.

Considering other extremes, such as tropical cyclones, the latest assessments show that due to problems with past observing capabilities, it is difficult to make conclusive statements about long-term trends. There is very strong evidence, however, that storm activity has increased in the North Atlantic since the 1970s.

Over periods of a century or more, evidence suggests slight decreases in the frequency of tropical cyclones making landfall in the North Atlantic and the South Pacific, once uncertainties in observing methods have been considered. Little evidence exists of any longer-term trend in other ocean basins. For extratropical cyclones, a poleward shift is evident in both hemispheres over the past 50 years, with further but limited evidence of a decrease in wind storm frequency at mid-latitudes. Several studies suggest an increase in intensity, but data sampling issues hamper these assessments.

FAQ 2.2, Figure 2 summarizes some of the observed changes in climate extremes. Overall, the most robust global changes in climate extremes are seen in measures of daily temperature, including to some extent, heat waves. Precipitation extremes also appear to be increasing, but there is large spatial variability, and observed trends in droughts are still uncertain except in a few regions. While robust increases have been seen in tropical cyclone frequency and activity in the North Atlantic since the 1970s, the reasons for this are still being debated. There is limited evidence of changes in extremes associated with other climate variables since the mid-20th century.

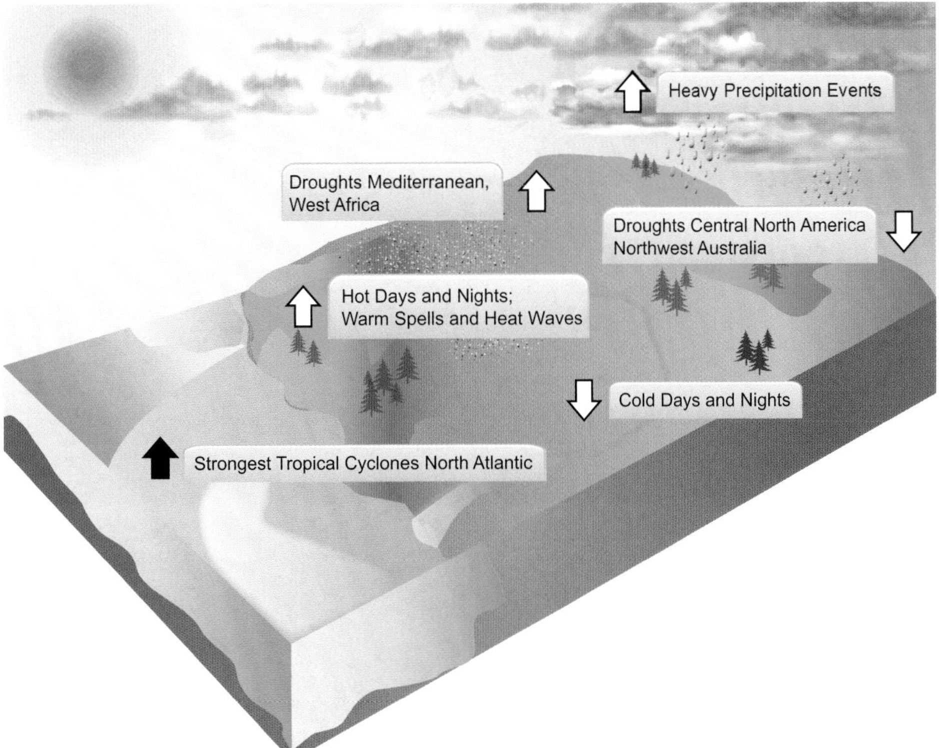

FAQ 2.2, Figure 2 | Trends in the frequency (or intensity) of various climate extremes (arrow direction denotes the sign of the change) since the middle of the 20th century (except for North Atlantic storms where the period covered is from the 1970s).

Studies that have examined trends in wind extremes from observations or regional reanalysis products tend to point to declining trends in extremes in mid-latitudes (Pirazzoli and Tomasin, 2003; Smits et al., 2005; Pryor et al., 2007; Zhang et al., 2007b) and increasing trends in high latitudes (Lynch et al., 2004; Turner et al., 2005; Hundecha et al., 2008; Stegall and Zhang, 2012). Other studies have compared the trends from observations with reanalysis data and reported differing or even opposite trends in the reanalysis products (Smits et al., 2005; McVicar et al., 2008). On the other hand, declining trends reported by Xu et al. (2006b) over China between 1969 and 2000 were generally consistent with trends in NCEP reanalysis. Trends extracted from reanalysis products must be treated with caution however, although usually with later generation products providing improvements over older products (Box 2.3).

In summary, *confidence* in large scale changes in the intensity of extreme extratropical cyclones since 1900 is *low*. There is also *low confidence* for a clear trend in storminess proxies over the last century due to inconsistencies between studies or lack of long-term data in some parts of the world (particularly in the SH). Likewise, *confidence* in trends in extreme winds is *low*, owing to quality and consistency issues with analysed data.

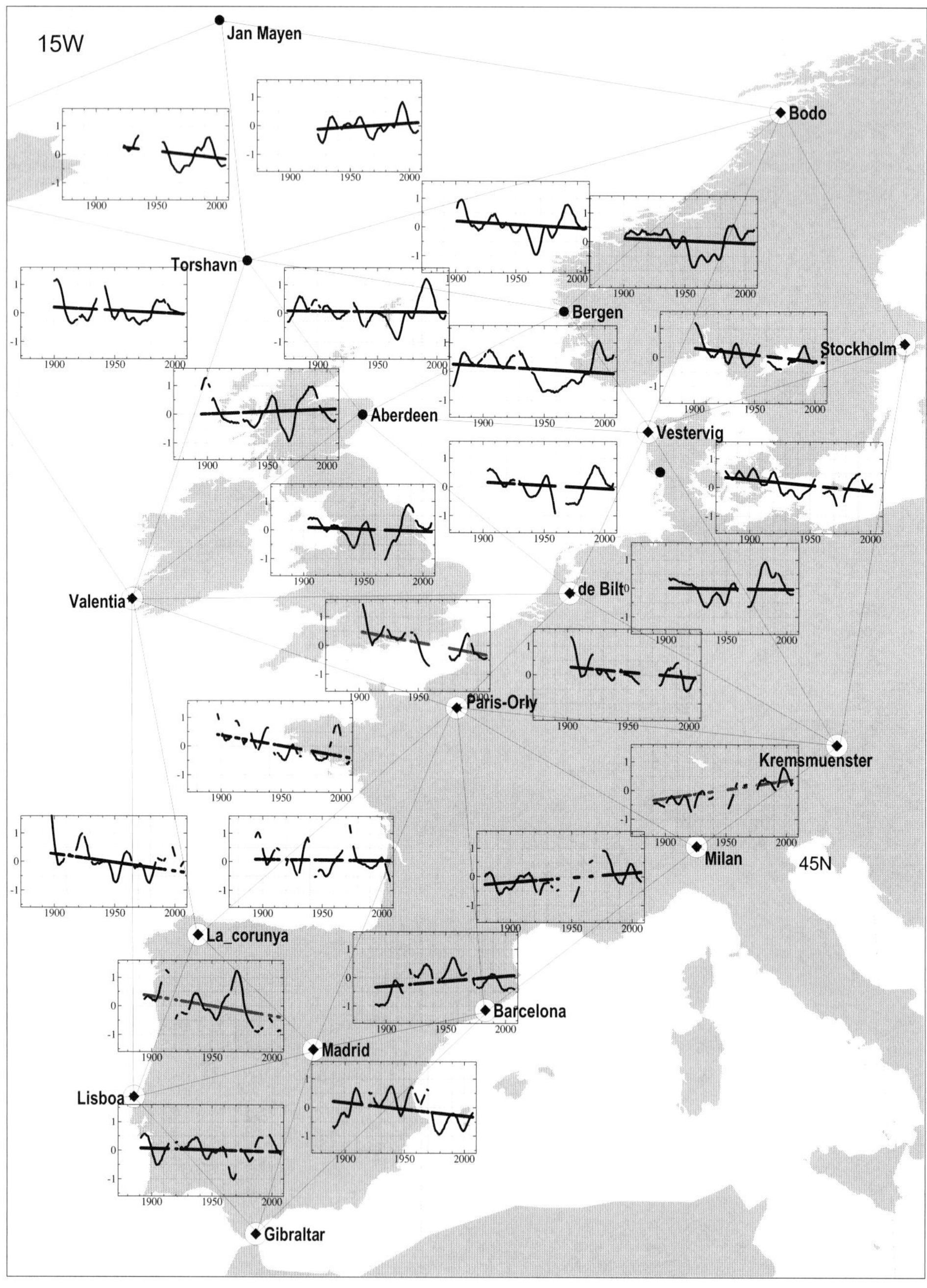

Figure 2.35 | 99th percentiles of geostrophic wind speeds for winter (DJF). Triangles show regions where geostrophic wind speeds have been calculated from *in situ* surface pressure observations. Within each pressure triangle, Gaussian low-pass filtered curves and estimated linear trends of the 99th percentile of these geostrophic wind speeds for winter are shown. The ticks of the time (horizontal) axis range from 1875 to 2005, with an interval of 10 years. Disconnections in lines show periods of missing data. Red (blue) trend lines indicate upward (downward) significant trends (i.e., a trend of zero lies outside the 95% confidence interval). (From Wang et al., 2011.)

Box 2.4 | Extremes Indices

As SREX highlighted, there is no unique definition of what constitutes a climate extreme in the scientific literature given variations in regions and sectors affected (Stephenson et al., 2008). Much of the available research is based on the use of so-called 'extremes indices' (Zhang et al., 2011). These indices can either be based on the probability of occurrence of given quantities or on absolute or percentage threshold exceedances (relative to a fixed climatological period) but also include more complex definitions related to duration, intensity and persistence of extreme events. For example, the term 'heat wave' can mean very different things depending on the index formulation for the application for which it is required (Perkins and Alexander, 2012).

Box 2.4, Table 1 lists a number of specific indices that appear widely in the literature and have been chosen to provide some consistency across multiple chapters in AR5 (along with the location of associated figures and text). These indices have been generally chosen for their robust statistical properties and their applicability across a wide range of climates. Another important criterion is that data for these indices are broadly available over both space and time. The existing near-global land-based data sets cover at least the post-1950 period but for regions such as Europe, North America, parts of Asia and Australia much longer analyses are available. The same indices used in observational studies (this chapter) are also used to diagnose climate model output (Chapters 9, 10, 11 and 12).

The types of indices discussed here do not include indices such as NIÑO3 representing positive and negative phases of ENSO (Box 2.5), nor do they include extremes such as 1 in 100 year events. Typically extreme indices assessed here reflect more 'moderate' extremes, for example, events occurring as often as 5% or 10% of the time (Box 2.4, Table 1). Predefined extreme indices are usually easier to obtain than the underlying daily climate data, which are not always freely exchanged by meteorological services. However, some of these indices do represent rarer events, for example, annual maxima or minima. Analyses of these and rarer extremes (e.g., with longer

(continued on next page)

Box 2.4, Table 1 | Definitions of extreme temperature and precipitation indices used in IPCC (after Zhang et al., 2011). The most common units are shown but these may be shown as normalized or relative depending on application in different chapters.

Index	Descriptive name	Definition	Units	Figures/Tables	Section
TXx	Warmest daily Tmax	Seasonal/annual maximum value of daily maximum temperature	°C	Box 2.4, Figure 1, Figures 9.37, 10.17, 12.13	Box 2.4, 9.5.4.1, 10.6.1.1, 12.4.3.3
TNx	Warmest daily Tmin	Seasonal/annual maximum value of daily minimum temperature	°C	Figures 9.37, 10.17	9.5.4.1, 10.6.1.1
TXn	Coldest daily Tmax	Seasonal/annual minimum value of daily maximum temperature	°C	Figures 9.37, 10.17, 12.13	9.5.4.1, 10.6.1.1, 12.4.3.3
TNn	Coldest daily Tmin	Seasonal/annual minimum value of daily minimum temperature	°C	Figures 9.37, 10.17, 12.13	9.5.4.1, 10.6.1.1
TN10p	Cold nights	Days (or fraction of time) when daily minimum temperature <10th percentile	Days (%)	Figures 2.32, 9.37, 10.17 Tables 2.11, 2.12	2.6.1, 9.5.4.1, 10.6.1.1, 11.3.2.5.1
TX10p	Cold days	Days (or fraction of time) when daily maximum temperature <10th percentile	Days (%)	Figures 2.32, 9.37, 10.17, 11.17	2.6.1, 9.5.4.1, 10.6.1.1, 11.3.2.5.1,
TN90p	Warm nights	Days (or fraction of time) when daily minimum temperature >90th percentile	Days (%)	Figures 2.32, 9.37, 10.17 Tables 2.11, 2.12	2.6.1, 9.5.4.1, 10.6.1.1, 11.3.2.5.1
TX90p	Warm days	Days (or fraction of time) when daily maximum temperature >90th percentile	Days (%)	Figures 2.32, 9.37, 10.17, 11.17 Tables 2.11, 2.12	2.6.1, 9.5.4.1, 10.6.1.1, 11.3.2.5.1,
FD	Frost days	Frequency of daily minimum temperature <0°C	Days	Figures 9.37, 12.13 Table 2.12	2.6.1, 9.5.4.1, 10.6.1.1, 12.4.3.3
TR	Tropical nights	Frequency of daily minimum temperature >20°C	Days	Figures 9.37, 12.13	9.5.4.1, 12.4.3.3
RX1day	Wettest day	Maximum 1-day precipitation	mm	Figures 9.37, 10.10 Table 2.12, 12.27	2.6.2.1, 9.5.4.1, 10.6.1.2, 12.4.5.5
RX5day	Wettest consecutive five days	Maximum of consecutive 5-day precipitation	mm	Figures 9.37, 12.26, 14.1	9.5.4.1, 10.6.1.2, 12.4.5.5, 14.2.1
SDII	Simple daily intensity index	Ratio of annual total precipitation to the number of wet days (≥1 mm)	mm day^{-1}	Figures 2.33, 9.37, 14.1	2.6.2.1, 9.5.4.1, 14.2.1
R95p	Precipitation from very wet days	Amount of precipitation from days >95th percentile	mm	Figures 2.33, 9.37, 11.17 Table 2.12	2.6.2.1, 9.5.4.1, 11.3.2.5.1
CDD	Consecutive dry days	Maximum number of consecutive days when precipitation <1 mm	Days	Figures 2.33, 9.37, 12.26, 14.1	2.6.2.3, 9.5.4.1, 12.4.5.5, 14.2.1

Box 2.4 (continued)

return period thresholds) are making their way into a growing body of literature which, for example, are using Extreme Value Theory (Coles, 2001) to study climate extremes (Zwiers and Kharin, 1998; Brown et al., 2008; Sillmann et al., 2011; Zhang et al., 2011; Kharin et al., 2013).

Extreme indices are more generally defined for daily temperature and precipitation characteristics (Zhang et al., 2011) although research is developing on the analysis of sub-daily events but mostly only on regional scales (Sen Roy, 2009; Shiu et al., 2009; Jones et al., 2010; Jakob et al., 2011; Lenderink et al., 2011; Shaw et al., 2011). Temperature and precipitation indices are sometimes combined to investigate 'extremeness' (e.g., hydroclimatic intensity, HY-INT; Giorgi et al., 2011) and/or the areal extent of extremes (e.g., the Climate Extremes Index (CEI) and its variants (Gleason et al., 2008; Gallant and Karoly, 2010; Ren et al., 2011). Indices rarely include other weather and climate variables, such as wind speed, humidity or physical impacts (e.g., streamflow) and phenomena. Some examples are available in the literature for wind-based (Della-Marta et al., 2009) and pressure-based (Beniston, 2009) indices, for health-relevant indices combining temperature and relative humidity characteristics (Diffenbaugh et al., 2007; Fischer and Schär, 2010) and for a range of dryness or drought indices (e.g., Palmer Drought Severity Index (PDSI) Palmer, 1965; Standardised Precipitation Index (SPI), Standardised Precipitation Evapotranspiration Index (SPEI) Vicente-Serrano et al., 2010) and wetness indices (e.g., Standardized Soil Wetness Index (SSWI); Vidal et al., 2010). *(continued on next page)*

In addition to the complication of defining an index, the results depend also on the way in which indices are calculated (to create global averages, for example). This is due to the fact that different algorithms may be employed to create grid box averages from station data, or that extremes indices may be calculated from gridded daily data or at station locations and then gridded. All of these factors add uncertainty to the calculation of an extreme. For example, the spatial patterns of trends in the hottest day of the year differ slightly between data sets, although when globally averaged, trends are similar over the second half of the 20th century (Box 2.4, Figure 1). Further discussion of the parametric and structural uncertainties in data sets is given in Box 2.1.

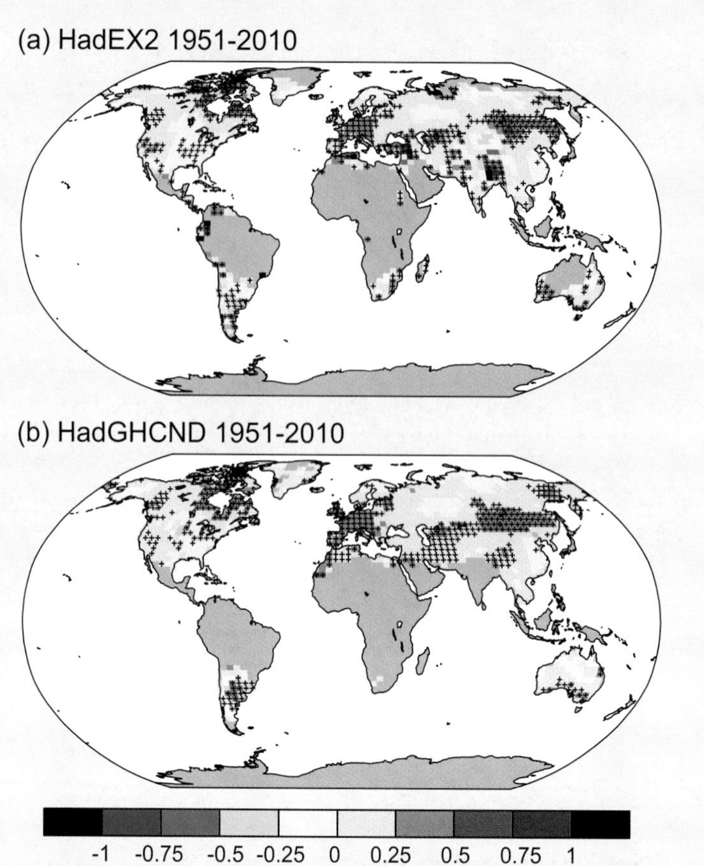

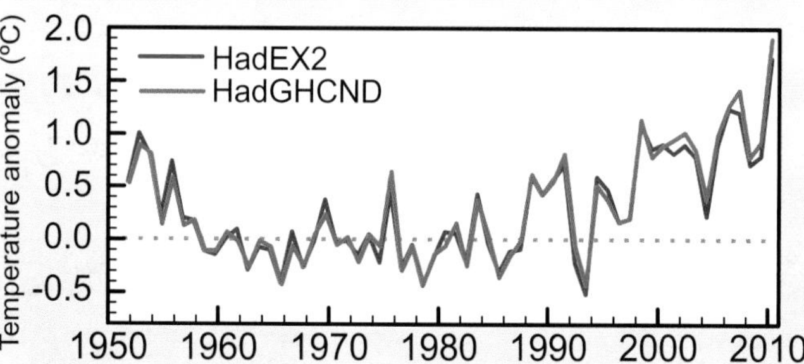

Box 2.4, Figure 1 | Trends in the warmest day of the year using different data sets for the period 1951–2010. The data sets are (a) HadEX2 (Donat et al., 2013c) updated to include the latest version of the European Climate Assessment data set (Klok and Tank, 2009), (b) HadGHCND (Caesar et al., 2006) using data updated to 2010 (Donat et al., 2013a) and (c) Globally averaged annual warmest day anomalies for each data set. Trends were calculated only for grid boxes that had at least 40 years of data during this period and where data ended no earlier than 2003. Grey areas indicate incomplete or missing data. Black plus signs (+) indicate grid boxes where trends are significant (i.e., a trend of zero lies outside the 90% confidence interval). Anomalies are calculated using grid boxes only where both data sets have data and where 90% of data are available.

2.7 Changes in Atmospheric Circulation and Patterns of Variability

Changes in atmospheric circulation and indices of climate variability, as expressed in sea level pressure (SLP), wind, geopotential height (GPH), and other variables were assessed in AR4. Substantial multi-decadal variability was reported in the large-scale atmospheric circulation over the Atlantic and the Pacific. With respect to trends, a decrease was found in tropospheric GPH over high latitudes of both hemispheres and an increase over the mid-latitudes in boreal winter for the period 1979–2001. These changes were found to be associated with an intensification and poleward displacement of Atlantic and southern mid-latitude jet streams and enhanced storm track activity in the NH from the 1960s to at least the 1990s. Changes in the North Atlantic Oscillation (NAO) and the Southern Annular Mode (SAM) towards their positive phases were observed, but it was noted that the NAO returned to its long-term mean state from the mid-1990s to the early 2000s.

Since AR4, more and improved observational data sets and reanalysis data sets (Box 2.3) have been published. Uncertainties and inaccuracies in all data sets are better understood (Box 2.1). The studies since AR4 assessed in this section support the poleward movement of circulation features since the 1970s and the change in the SAM. At the same time, large decadal-to-multidecadal variability in atmospheric circulation is found that partially offsets previous trends in other circulation features such as the NAO or the Pacific Walker circulation.

This section assesses observational evidence for changes in atmospheric circulation in fields of SLP, GPH, and wind, in circulation features (such as the Hadley and Walker circulation, monsoons, or jet streams; Annex III: Glossary), as well as in circulation variability modes. Regional climate effects of the circulation changes are discussed in Chapter 14.

2.7.1 Sea Level Pressure

AR4 concluded that SLP in December to February decreased between 1948 and 2005 in the Arctic, Antarctic and North Pacific. More recent studies using updated data for the period 1949–2009 (Gillett and Stott, 2009) also find decreases in SLP in the high latitudes of both hemispheres in all seasons and increasing SLP in the tropics and subtropics most of the year. However, due to decadal variability SLP trends are sensitive to the choice of the time period (Box 2.2), and they depend on the data set.

The spatial distribution of SLP represents the distribution of atmospheric mass, which is the surface imprint of the atmospheric circulation. Barometric measurements are made in weather stations or on board ships. Fields are produced from the observations by interpolation or using data assimilation into weather models. One of the most widely used observational data sets is HadSLP2 (Allan and Ansell, 2006), which integrates 2228 historical global terrestrial stations with marine observations from the ICOADS on a 5° × 5°grid. Other observation products (e.g., Trenberth and Paolino, 1980; for the extratropical NH) or reanalyses are also widely used to address changes in SLP. Although the quality of SLP data is considered good, there are discrepancies between gridded SLP data sets in regions with sparse observations, e.g., over Antarctica (Jones and Lister, 2007).

Van Haren et al. (2012) found a strong SLP decrease over the Mediterranean in January to March from 1961 to 2000. For the more recent period (1979–2012) trends in SLP, consistent across different data sets (shown in Figure 2.36 for ERA-Interim), are negative in the tropical and northern subtropical Atlantic during most of the year as well as, in May to October, in northern Siberia. Positive trends are found year-round over the North and South Pacific and South Atlantic. Trends in

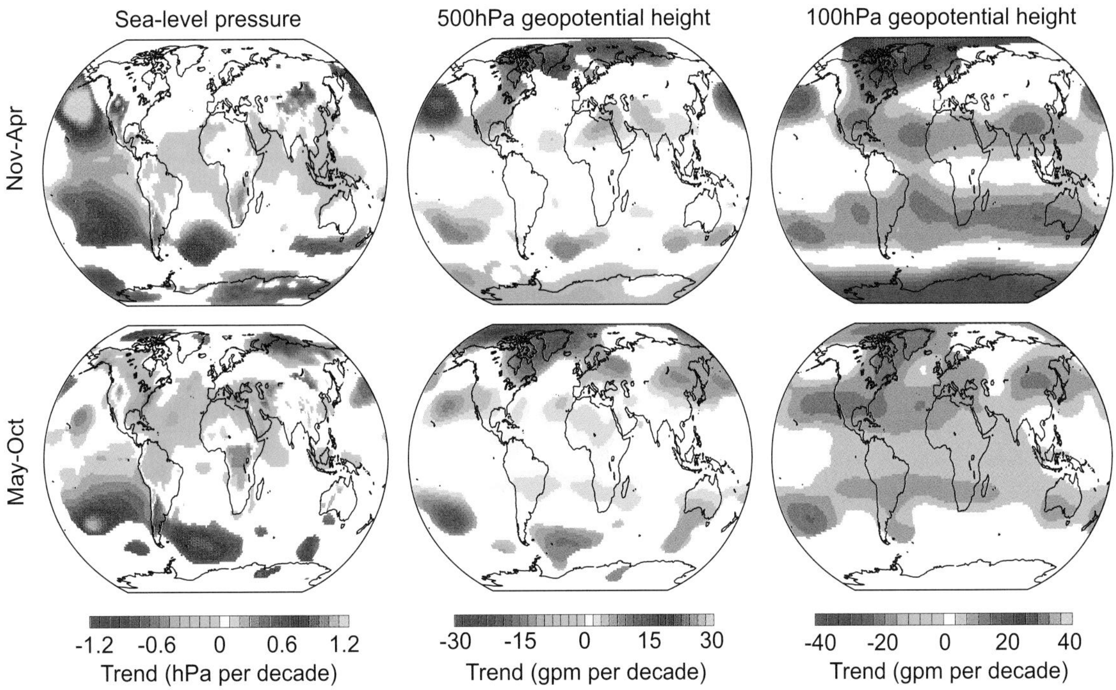

Figure 2.36 | Trends in (left) sea level pressure (SLP), (middle) 500 hPa geopotential height (GPH) and (right) 100 hPa GPH in (top) November to April 1979/1980 to 2011/2012 and (bottom) May to October 1979 to 2011 from ERA-Interim data. Trends are shown only if significant (i.e., a trend of zero lies outside the 90% confidence interval).

2

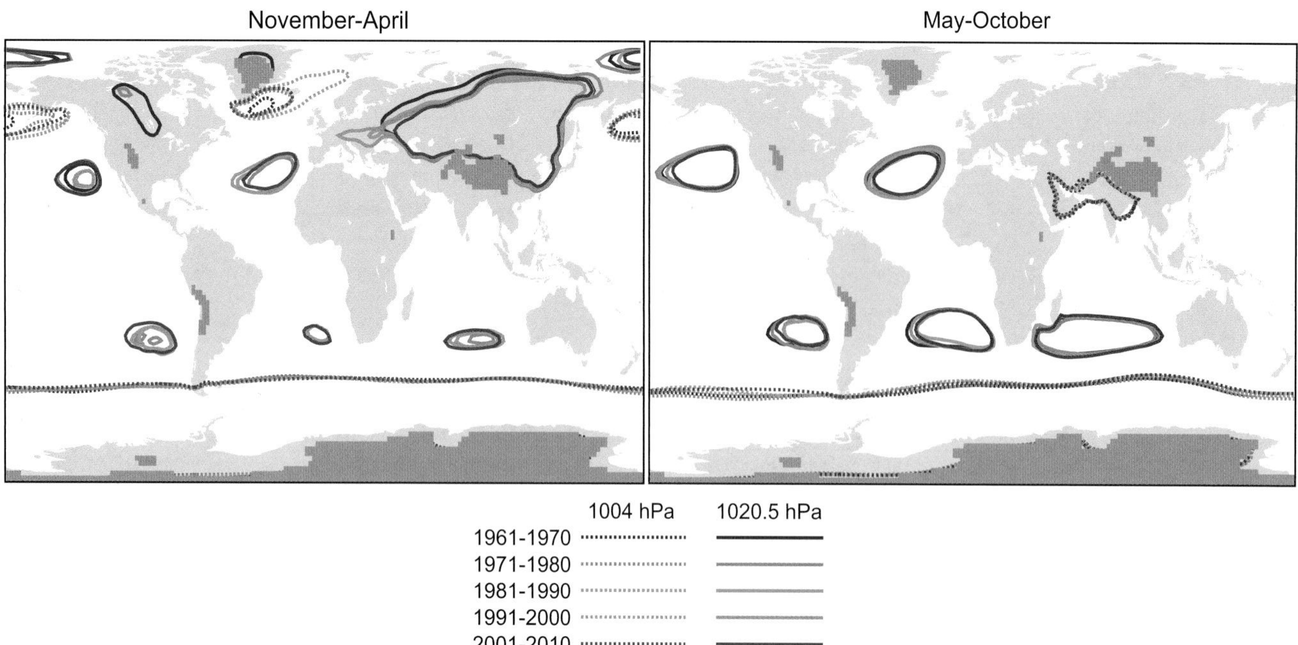

Figure 2.37 | Decadal averages of sea level pressure (SLP) from the 20th Century Reanalysis (20CR) for (left) November of previous year to April and (right) May to October shown by two selected contours: 1004 hPa (dashed lines) and 1020.5 hPa (solid lines). Topography above 2 km above mean sea level in 20CR is shaded in dark grey.

the equatorial Pacific zonal SLP gradient during the 20th century (e.g., Vecchi et al., 2006; Power and Kociuba, 2011a, 2011b) are discussed in Section 2.7.5.

The position and strength of semi-permanent pressure centres show no clear evidence for trends since 1951. However, prominent variability is found on decadal time scales (Figure 2.37). Consistent across different data sets, the Azores high and the Icelandic low in boreal winter, as captured by the high and low SLP contours, were both small in the 1960s and 1970s, large in the 1980s and 1990s, and again smaller in the 2000s. Favre and Gershunov (2006) find an eastward shift of the Aleutian low from the mid-1970s to 2001, which persisted during the 2000s (Figure 2.37). The Siberian High exhibits pronounced decadal-to-multidecadal variability (Panagiotopoulos et al., 2005; Huang et al., 2010), with a recent (1998 to 2012) strengthening and northwestward expansion (Zhang et al., 2012b). In boreal summer, the Atlantic and Pacific high-pressure systems extended more westward in the 1960s and 1970s than later. On interannual time scales, variations in pressure centres are related to modes of climate variability. Trends in the indices that capture the strength of these modes are reported in Section 2.7.8, their characteristics and impacts are discussed in Chapter 14.

In summary, sea level pressure has *likely* decreased from 1979 to 2012 over the tropical Atlantic and increased over large regions of the Pacific and South Atlantic, but trends are sensitive to the time period analysed owing to large decadal variability.

2.7.2 Surface Wind Speed

AR4 concluded that mid-latitude westerly winds have generally increased in both hemispheres. Because of shortcomings in the observations, SREX stated that *confidence* in surface wind trends is *low*. Further studies assessed here confirm this assessment.

Surface wind measurements over land and ocean are based on largely separate observing systems. Early marine observations were based on ship speed through the water or sails carried or on visual estimates of sea state converted to the wind speed using the Beaufort scale. Anemometer measurements were introduced starting in the 1950s. The transition from Beaufort to measured winds introduced a spurious trend, compounded by an increase in mean anemometer height over time (Kent et al., 2007; Thomas et al., 2008). ICOADS release 2.5 (Woodruff et al., 2011) contains information on measurement methods and wind measurement heights, permitting adjustment for these effects. The ICOADS-based data set WASWind (1950–2010; Tokinaga and Xie, 2011a) and the interpolated product NOCS v.2.0 (1973–present; Berry and Kent, 2011) include such corrections, among other improvements.

Marine surface winds are also measured from space using various microwave range instruments: scatterometers and synthetic aperture radars retrieve wind vectors, while altimeters and passive radiometers measure wind speed only (Bourassa et al., 2010). The latter type provides the longest continuous record, starting in July 1987. Satellite-based interpolated marine surface wind data sets use objective analysis methods to blend together data from different satellites and atmospheric reanalyses. The latter provide wind directions as in Blended Sea Winds (BSW; Zhang et al., 2006), or background fields as in Cross-Calibrated Multi-Platform winds (CCMP; Atlas et al., 2011) and OAFlux (Yu and Weller, 2007). CCMP uses additional dynamical constraints, *in situ* data and a recently homogenized data set of SSM/I observations (Wentz et al., 2007), among other satellite sources.

Figure 2.38 compares 1988–2010 linear trends in surface wind speeds from interpolated data sets based on satellite data, from interpolated and non-interpolated data sets based on *in situ* data, and from atmospheric reanalyses. Note that these trends over a 23-year-long period primarily reflect decadal variability in winds, rather than long-term climate change (Box 2.2). Kent et al. (2012) recently intercompared several of these data sets and found large differences. The differences in trend patterns in Figure 2.38 are large as well. Nevertheless, some statistically significant features are present in most data sets, including a pattern of positive and negative trend bands across the North Atlantic Ocean (Section 2.7.6.2.) and positive trends along the west coast of North America. Strengthening of the Southern Ocean winds, consistent with the increasing trend in the SAM (Section 2.7.8) and with the observed changes in wind stress fields described in Section 3.4.4, can be seen in satellite-based analyses and atmospheric reanalyses in Figure 2.38. Alternating Southern Ocean trend signs in the NOCS v.2.0 panel are due to interpolation of very sparse *in situ* data (cf. the panel for the uninterpolated WASWind product).

Surface winds over land have been measured with anemometers on a global scale for decades, but until recently the data have been rarely used for trend analysis. Global data sets lack important meta information on instrumentation and siting (McVicar et al., 2012). Long, homogenized instrumental records are rare (e.g., Usbeck et al., 2010; Wan et al., 2010). Moreover, wind speed trends are sensitive to the anemometer height (Troccoli et al., 2012). Winds near the surface can be derived from reanalysis products (Box 2.3), but discrepancies are found when comparing trends therein with trends for land stations (Smits et al., 2005; McVicar et al., 2008).

Over land, a weakening of seasonal and annual mean as well as maximum winds is reported for many regions from around the 1960s or 1970s to the early 2000s (a detailed review is given in McVicar et al. (2012)), including China and the Tibetan Plateau (Xu et al., 2006b; Guo et al., 2010) (but levelling off since 2000; Lin et al., 2012), Western and southern Europe (e.g., Earl et al., 2013), much of the USA (Pryor et al., 2007), Australia (McVicar et al., 2008) and southern and western

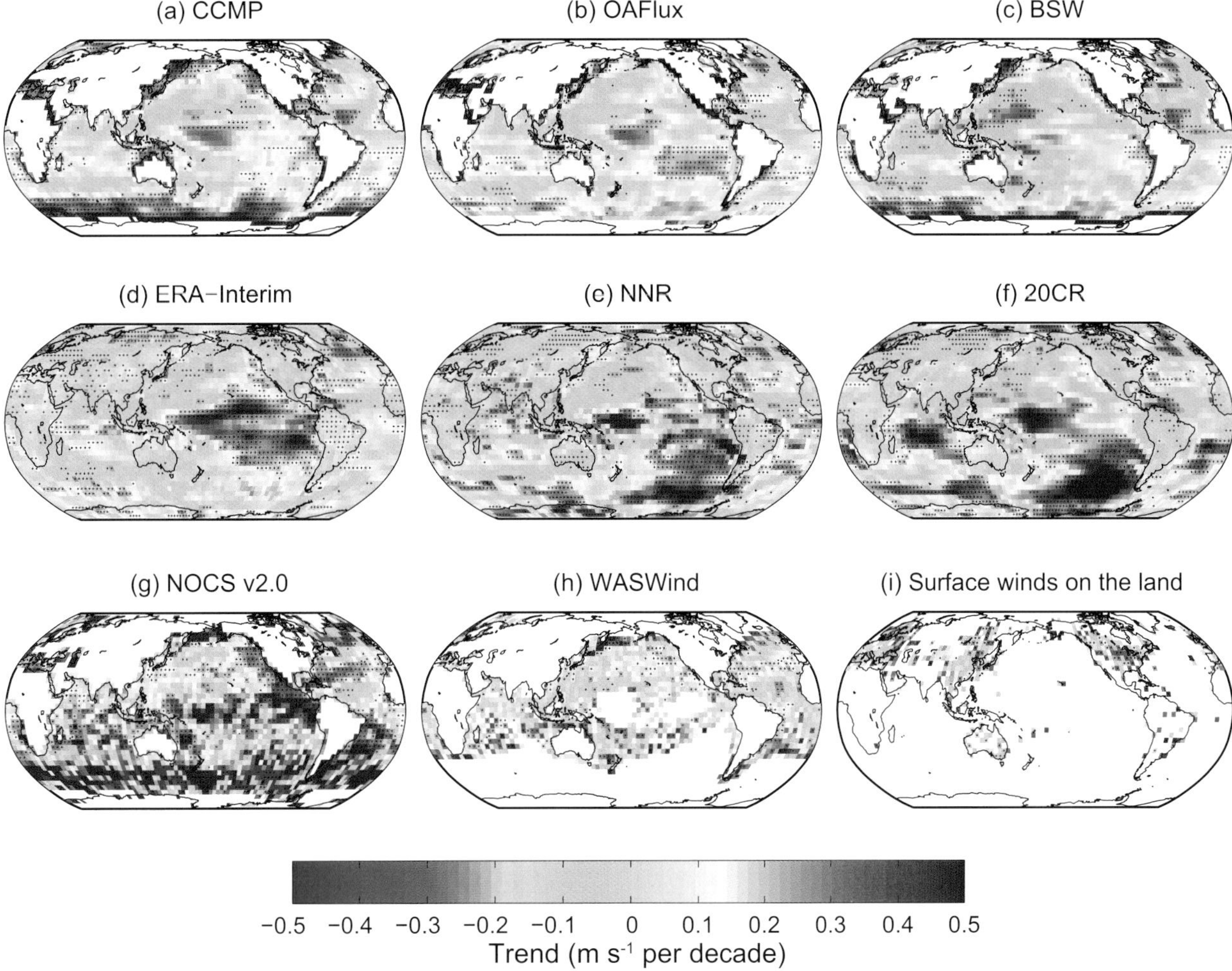

Figure 2.38 | Trends in surface wind speed for 1988–2010. Shown in the top row are data sets based on the satellite wind observations: (a) Cross-Calibrated Multi-Platform wind product (CCMP; Atlas et al., 2011); (b) wind speed from the Objectively Analyzed Air-Sea Heat Fluxes data set, release 3 (OAFlux); (c) Blended Sea Winds (BSW; Zhang et al., 2006); in the middle row are data sets based on surface observations: (d) ERA-Interim; (e) NCEP-NCAR, v.1 (NNR); (f) 20th Century Reanalysis (20CR, Compo et al., 2011), and in the bottom row are surface wind speeds from atmospheric reanalyses: (g) wind speed from the Surface Flux Data set, v.2, from NOC, Southampton, UK (Berry and Kent, 2009); (h) Wave- and Anemometer-based Sea Surface Wind (WASWind; Tokinaga and Xie, 2011a)); and (i) Surface Winds on the Land (Vautard et al., 2010). Wind speeds correspond to 10 m heights in all products. Land station winds (panel f) are also for 10 m (but anemometer height is not always reported) except for the Australian data where they correspond to 2 m height. To improve readability of plots, all data sets (including land station data) were averaged to the 4° × 4° uniform longitude-latitude grid. Trends were computed for the annually averaged timeseries of 4° × 4° cells. For all data sets except land station data, an annual mean was considered available only if monthly means for no less than eight months were available in that calendar year. Trend values were computed only if no less than 17 years had values and at least 1 year was available among the first and last 3 years of the period. White areas indicate incomplete or missing data. Black plus signs (+) indicate grid boxes where trends are significant (i.e., a trend of zero lies outside the 90% confidence interval).

Canada (Wan et al., 2010). Increasing wind speeds were found at high latitudes in both hemispheres, namely in Alaska from 1921 to 2001 (Lynch et al., 2004), in the central Canadian Arctic and Yukon from the 1950 to the 2000s (Wan et al., 2010) and in coastal Antarctica over the second half of the 20th century (Turner et al., 2005). A global review of 148 studies showed that near-surface terrestrial wind speeds are declining in the Tropics and the mid-latitudes of both hemispheres at a rate of –0.14 m s^{-1} per decade (McVicar et al., 2012). Vautard et al. (2010), analysing a global land surface wind data set from 1979 to 2008, found negative trends on the order of –0.1 m s^{-1} per decade over large portions of NH land areas. The wind speed trend pattern over land inferred from their data (1988–2010, Figure 2.38) has many points with magnitudes much larger than those in the reanalysis products, which appear to underestimate systematically the wind speed over land, as well as in coastal regions (Kent et al., 2012).

2

In summary, *confidence* is *low* in changes in surface wind speed over the land and over the oceans owing to remaining uncertainties in data sets and measures used.

2.7.3 Upper-Air Winds

In contrast to surface winds, winds above the planetary boundary layer have received little attention in AR4. Radiosondes and pilot balloon observations are available from around the 1930s (Stickler et al., 2010). Temporal inhomogeneities in radiosonde wind records are less common, but also less studied, than those in radiosonde temperature records (Gruber and Haimberger, 2008; Section 2.4.4.3). Upper air winds can also be derived from tracking clouds or water vapour in satellite imagery (Menzel, 2001) or from measurements using wind profilers, aircraft or thermal observations, all of which serve as an input to reanalyses (Box 2.3).

In the past few years, interest in an accurate depiction of upper air winds has grown, as they are essential for estimating the state and changes of the general atmospheric circulation and for explaining changes in the surface winds (Vautard et al., 2010). Allen and Sherwood (2008), analysing wind shear from radiosonde data, found significant positive zonal mean zonal wind trends in the northern extratropics in the upper troposphere and stratosphere and negative trends in the tropical upper troposphere for the period 1979–2005. Vautard et al. (2010) find increasing wind speed in radiosonde observations in the lower and middle troposphere from 1979 to 2008 over Europe and North America and decreasing wind speeds over Central and East Asia. However, systematic global trend analyses of radiosonde winds are rare, prohibiting an assessment of upper-air wind trends (specific features such as monsoons, jet streams and storms are discussed in Sections 2.7.5, 2.7.6 and 2.6, respectively).

In summary, upper-air winds are less studied than other aspects of the circulation, and less is known about the quality of data products, hence *confidence* in upper-air wind trends is *low*.

2.7.4 Tropospheric Geopotential Height and Tropopause

AR4 concluded that over the NH between 1960 and 2000, boreal winter and annual means of tropospheric GPH decreased over high latitudes and increased over the mid-latitudes. AR4 also reported an increase in tropical tropopause height and a slight cooling of the tropical cold-point tropopause.

Changes in GPH, which can be addressed using radiosonde data or reanalysis data (Box 2.3), reflect SLP and temperature changes in the atmospheric levels below. The spatial gradients of the trend indicate changes in the upper-level circulation. As for SLP, tropopsheric GPH trends strongly depend on the period analysed due to pronounced decadal variability. For the 1979–2012 period, trends for 500 hPa GPH from the ERA-Interim reanalysis (Figure 2.36) as well as for other reanalyses show a significant decrease only at southern high latitudes in November to April, but significant positive GPH trends in the subtropics and northern high latitudes. Hence the change in the time period leads to a different trend pattern as compared to AR4. The seasonality and spatial dependence of 500 hPa GPH trends over Antarctica was highlighted by Neff et al. (2008), based upon radiosonde data over the period 1957–2007.

Minimum temperatures near the tropical tropopause (and therefore tropical tropopause height) are important as they affect the water vapour input into the stratosphere (Section 2.2.2.1). Studies since AR4 confirm the increase in tropopause height (Wang et al., 2012c). For tropical tropopause temperatures, studies based on radiosonde data and reanalyses partly support a cooling between the 1990s and the early 2000s (Randel et al., 2006; Randel and Jensen, 2013), but uncertainties in long-term trends of the tropical cold-point tropopause temperature from radiosondes (Wang et al., 2012c; Randel and Jensen, 2013) and reanalyses (Gettelman et al., 2010) are large and *confidence* is therefore *low*.

In summary, tropospheric geopotential height *likely* decreased from 1979 to 2012 at SH high latitudes in austral summer and increased in the subtropics and NH high latitudes. *Confidence* in trends of the tropical cold-point tropopause is *low* owing to remaining uncertainties in the data.

2.7.5 Tropical Circulation

In AR4, large interannual variability of the Hadley and Walker circulation was highlighted, as well as the difficulty in addressing changes in these features in the light of discrepancies between data sets. AR4 also found that rainfall in many monsoon systems exhibits decadal changes, but that data uncertainties restrict confidence in trends. SREX also attributed *low confidence* to observed trends in monsoons.

Observational evidence for trends and variability in the strength of the Hadley and Walker circulations (Annex III: Glossary), the monsoons, and the width of the tropical belt is based on radiosonde and reanalyses data (Box 2.3). In addition, changes in the tropical circulation imprint on other fields that are observed from space (e.g., total ozone, outgoing longwave radiation). Changes in the average state of the tropical circulation are constrained to some extent by changes in the water cycle (Held and Soden, 2006; Schneider et al., 2010). Changes in the monsoon systems are expressed through altered circulation, moisture transport and convergence, and precipitation. Only a few monsoon studies address circulation changes, while most work focuses on precipitation.

Several studies report a weakening of the global monsoon circulations as well as a decrease of global land monsoon rainfall or of the number of precipitation days over the past 40 to 50 years (Zhou et al., 2008, see also SREX; Liu et al., 2011). Concerning the East Asian Monsoon, a year-round decrease is reported for wind speeds over China at the surface and in the lower troposphere based on surface observations and radiosonde data (Guo et al., 2010; Jiang et al., 2010; Vautard et al., 2010; Xu et al., 2010). The changes in wind speed are concomitant with changes in pressure centres such as a westward extension of the Western Pacific Subtropical High (Gong and Ho, 2002; Zhou et al., 2009b). A weakening of the East Asian summer monsoon since the 1920s is also found in SLP gradients (Zhou et al., 2009a). However, trends derived from wind observations and circulation trends from reanalysis data carry large uncertainties (Figure 2.38), and monsoon rainfall trends depend, for example, on the definition of the monsoon area (Hsu et al., 2011). For instance, using a new definition of monsoon area, an increase in northern hemispheric and global summer monsoon (land and ocean) precipitation is reported from 1979 to 2008 (Hsu et al., 2011; Wang et al., 2012a).

The additional data sets that became available since AR4 confirm the large interannual variability of the Hadley and Walker circulation. The strength of the northern Hadley circulation (Figure 2.39) in boreal winter and of the Pacific Walker circulation in boreal fall and winter is largely related to the ENSO (Box 2.5). This association dominates interannual variability and affects trends. Data sets do not agree well with respect to trends in the Hadley circulation (Figure 2.39). Two widely used reanalysis data sets, NNR and ERA-40, both have demonstrated

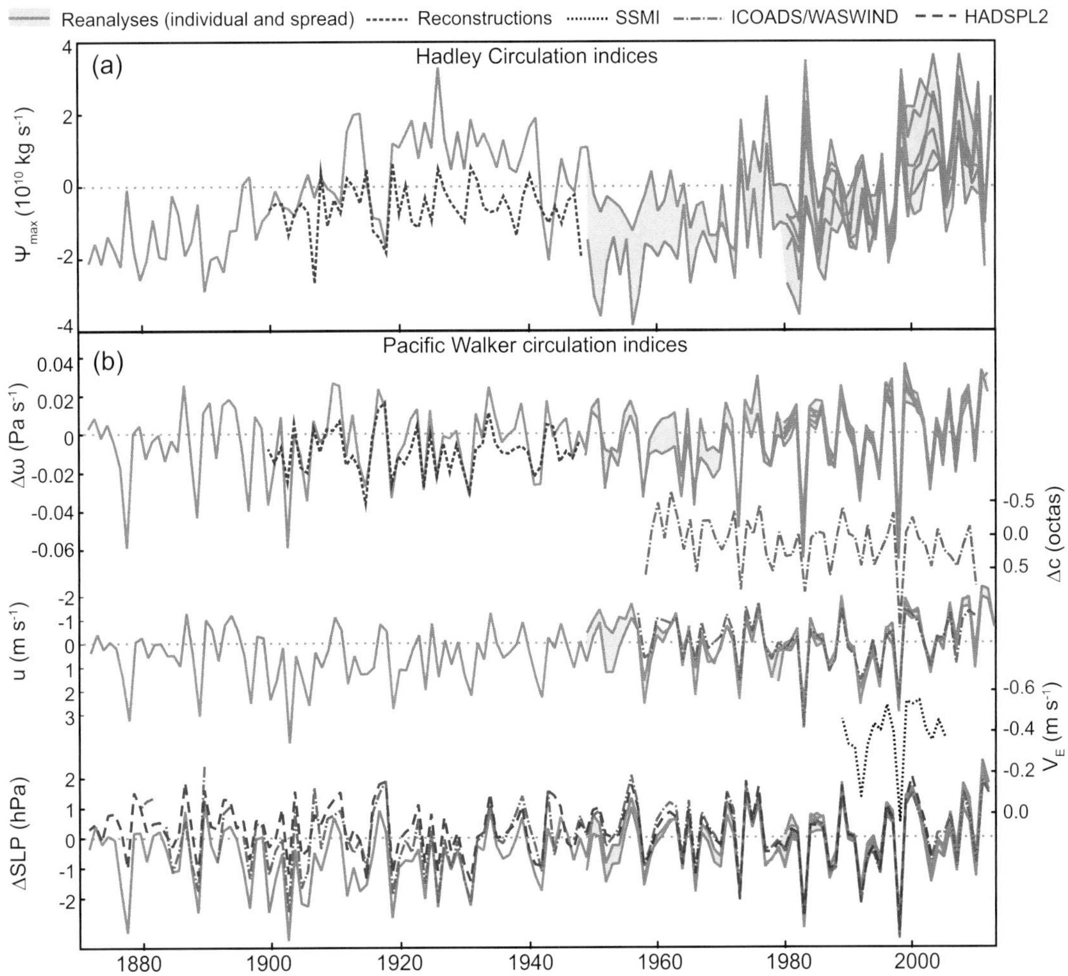

Figure 2.39 | (a) Indices of the strength of the northern Hadley circulation in December to March (Ψ_{max} is the maximum of the meridional mass stream function at 500 hPa between the equator and 40°N). (b) Indices of the strength of the Pacific Walker circulation in September to January ($\Delta\omega$ is the difference in the vertical velocity between [10°S to 10°N, 180°W to 100°W] and [10°S to 10°N, 100°E to 150°E] as in Oort and Yienger (1996), Δc is the difference in cloud cover between [6°N to 12°S, 165°E to 149°W] and [18°N to 6°N, 165°E to 149°W] as in Deser et al. (2010a), v_E is the effective wind index from SSM/I satellite data, updated from Sohn and Park (2010), u is the zonal wind at 10 m averaged in the region [10°S to 10°N, 160°E to 160°W], ΔSLP is the SLP difference between [5°S to 5°N, 160°W to 80°W] and [5°S to 5°N, 80°E to 160°E] as in Vecchi et al. (2006)). Reanalysis data sets include 20CR, NCEP/NCAR, ERA-Interim, JRA-25, MERRA, and CFSR, except for the zonal wind at 10 m (20CR, NCEP/NCAR, ERA-Interim), where available until January 2013. ERA-40 and NCEP2 are not shown as they are outliers with respect to the strength trend of the northern Hadley circulation (Mitas and Clement, 2005; Song and Zhang, 2007; Hu et al., 2011; Stachnik and Schumacher, 2011). Observation data sets include HadSLP2 (Section 2.7.1), ICOADS (Section 2.7.2; only 1957–2009 data are shown) and WASWIND (Section 2.7.2), reconstructions are from Brönnimann et al. (2009). Where more than one time series was available, anomalies from the 1980/1981 to 2009/2010 mean values of each series are shown.

shortcomings with respect to tropical circulation; hence their increases in the Hadley circulation strength since the 1970s might be artificial (Mitas and Clement, 2005; Song and Zhang, 2007; Hu et al., 2011; Stachnik and Schumacher, 2011). Later generation reanalysis data sets including ERA-Interim (Brönnimann et al., 2009; Nguyen et al., 2013) as well as satellite humidity data (Sohn and Park, 2010) also suggest a strengthening from the mid 1970s to present, but the magnitude is strongly data set dependent.

Consistent changes in different observed variables suggest a weakening of the Pacific Walker circulation during much of the 20th century that has been largely offset by a recent strengthening. A weakening is indicated by trends in the zonal SLP gradient across the equatorial Pacific (Section 2.7.1, Table 2.14) from 1861 to 1992 (Vecchi et al., 2006), or from 1901 to 2004 (Power and Kociuba, 2011b). Boreal spring and summer contribute most strongly to the centennial trend (Nicholls, 2008; Karnauskas et al., 2009), as well as to the trend in the second half of the 20th century (Tokinaga et al., 2012). For boreal fall and winter, when the circulation is strongest, no trend is found in the Pacific Walker circulation based on the vertical velocity at 500 hPa from reanalyses (Compo et al., 2011), equatorial Pacific 10 m zonal winds, or SLP in Darwin (Nicholls, 2008; Figure 2.39). However, there are inconsistencies between ERA-40 and NNR (Chen et al., 2008). Deser et al. (2010a) find changes in marine air temperature and cloud cover over the Pacific that are consistent with a weakening of the Walker circulation during most of the 20th century (Section 2.5.7.1 and Yu and Zwiers, 2010). Tokinaga et al. (2012) find robust evidence for a weakening of the Walker circulation (most notably over the Indian Ocean) from 1950 to 2008 based on observations of cloud cover, surface wind, and SLP. Since the 1980s or 1990s, however, trends in the Pacific Walker circulation have reversed (Figure 2.39; Luo et al., 2012). This is evident from changes in SLP (see equatorial Southern Oscillation Index (SOI) trends in Table 2.14 and Box 2.5, Figure 1), vertical velocity (Compo et al., 2011), water vapour flux from satellite and reanalysis data (Sohn and Park, 2010), or sea level height (Merrifield, 2011). It is also consistent with the SST trend pattern since 1979 (Meng et al., 2012; see also Figure 2.22).

Observed changes in several atmospheric parameters suggest that the width of the tropical belt has increased at least since 1979 (Seidel et al., 2008; Forster et al., 2011; Hu et al., 2011). Since AR4, wind, temperature, radiation, and ozone information from radiosondes, satellites, and reanalyses had been used to diagnose the tropical belt width and estimate their trends. Annual mean time series of the tropical belt width from various sources are shown in Figure 2.40.

Since 1979 the region of low column ozone values typical of the tropics has expanded in the NH (Hudson et al., 2006; Hudson, 2012). Based on radiosonde observations and reanalyses, the region of the high tropical tropopause has expanded since 1979, and possibly since 1960 (Seidel and Randel, 2007; Birner, 2010; Lucas et al., 2012), although widening estimates from different reanalyses and using different methodologies show a range of magnitudes (Seidel and Randel, 2007; Birner, 2010).

Several lines of evidence indicate that climate features at the edges of the Hadley cell have also moved poleward since 1979. Subtropical jet metrics from reanalysis zonal winds (Strong and Davis, 2007, 2008; Archer and Caldeira, 2008b, 2008a) and layer-average satellite temperatures (Fu et al., 2006; Fu and Lin, 2011) also indicate widening, although 1979–2009 wind-based trends (Davis and Rosenlof, 2011)

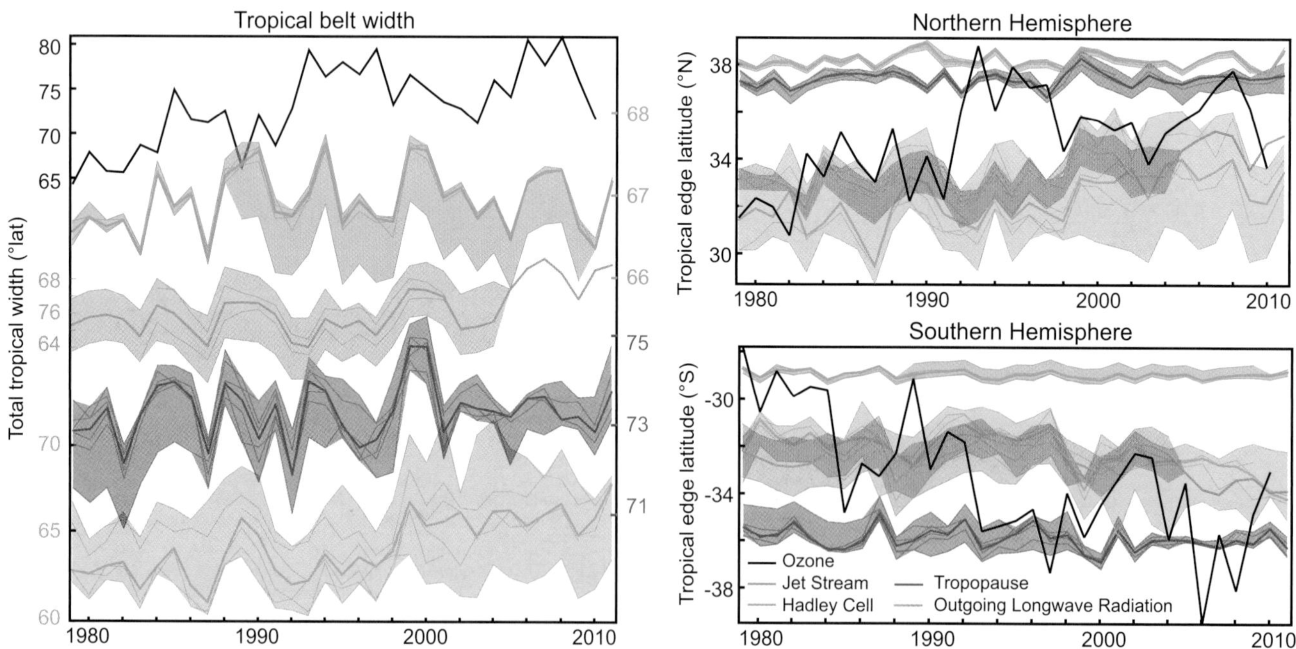

Figure 2.40 | Annual average tropical belt width (left) and tropical edge latitudes in each hemisphere (right). The tropopause (red), Hadley cell (blue), and jet stream (green) metrics are based on reanalyses (NCEP/NCAR, ERA-40, JRA25, ERA-Interim, CFSR, and MERRA, see Box 2.3); outgoing longwave radiation (orange) and ozone (black) metrics are based on satellite measurements. The ozone metric refers to equivalent latitude (Hudson et al., 2006; Hudson, 2012). Adapted and updated from Seidel et al. (2008) using data presented in Davis and Rosenlof (2011) and Hudson (2012). Where multiple data sets are available for a particular metric, all are shown as light solid lines, with shading showing their range and a heavy solid line showing their median.

are not statistically significant. Changes in subtropical outgoing longwave radiation, a surrogate for high cloud, also suggest widening (Hu and Fu, 2007), but the methodology and results are disputed (Davis and Rosenlof, 2011). Widening of the tropical belt is also found in precipitation patterns (Hu and Fu, 2007; Davis and Rosenlof, 2011; Hu et al., 2011; Kang et al., 2011; Zhou et al., 2011), including in SH regions (Cai et al., 2012).

The qualitative consistency of these observed changes in independent data sets suggests a widening of the tropical belt between at least 1979 and 2005 (Seidel et al., 2008), and possibly longer. Widening estimates range between around 0° and 3° latitude per decade, but their uncertainties have been only partially explored (Birner, 2010; Davis and Rosenlof, 2011).

In summary, large interannual-to-decadal variability is found in the strength of the Hadley and Walker circulation. The *confidence* in trends in the strength of the Hadley circulation is *low* due to uncertainties in reanalysis data sets. Recent strengthening of the Pacific Walker circulation has largely offset the weakening trend from the 19th century to the 1990s (*high confidence*). Several lines of independent evidence indicate a widening of the tropical belt since the 1970s. The suggested weakening of the East Asian monsoon has *low confidence*, given the nature and quality of the evidence.

2.7.6 Jets, Storm Tracks and Weather Types

2.7.6.1 Mid-latitude and Subtropical Jets and Storm Track Position

AR4 reported a poleward displacement of Atlantic and southern polar front jet streams from the 1960s to at least the mid-1990s and a poleward shift of the northern hemispheric storm tracks. However, it was also noted that uncertainties are large and that NNR and ERA-40 disagree in important aspects. SREX also reported a poleward shift of NH and SH storm tracks. Studies since AR4 confirm that in the NH, the jet core has been migrating towards the pole since the 1970s, but trends in the jet speed are uncertain. Additional studies assessed here further support the poleward shift of the North Atlantic storm track from the 1950s to the early 2000s.

Subtropical and mid-latitude jet streams are three-dimensional entities that vary meridionally, zonally, and vertically. The position of the mid-latitude jet streams is related to the position of the mid-latitude storm tracks; regions of enhanced synoptic activity due to the passage of cyclones (Section 2.6). Jet stream winds can be determined from radiosonde measurements of GPH using quasi-geostrophic flow assumptions. Using reanalysis data sets (Box 2.3), it is possible to track three-dimensional jet variations by identifying a surface of maximum wind (SMW), although a high vertical resolution is required for identification of jets.

Various new analyses based on NCEP/NCAR and ERA-40 reanalyses as well as MSU/AMSU lower stratospheric temperatures (Section 2.4.4) confirm that the jet streams (mid-latitude and subtropical) have been moving poleward in most regions in the NH over the last three decades (Fu et al., 2006; Hu and Fu, 2007; Strong and Davis, 2007; Archer and Caldeira, 2008a; Fu and Lin, 2011) but no clear trend is found in the SH (Swart and Fyfe, 2012). There is inconsistency with respect to jet speed trends based upon whether one uses an SMW-based or isobaric-based approach (Strong and Davis, 2007, 2008; Archer and Caldeira, 2008b, 2008a) and the choice of analysis periods due to inhomogeneities in reanalyses (Archer and Caldeira, 2008a). In general, jets have become more common (and jet speeds have increased) over the western and central Pacific, eastern Canada, the North Atlantic and Europe (Strong and Davis, 2007; Barton and Ellis, 2009), trends that are concomitant with regional increases in GPH gradients and circumpolar vortex contraction (Frauenfeld and Davis, 2003; Angell, 2006). From a climate dynamics perspective, these trends are driven by regional patterns of tropospheric and lower stratospheric warming or cooling and thus are coupled to large-scale circulation variability.

The North Atlantic storm track is closely associated with the NAO (Schneidereit et al., 2007). Studies based on ERA-40 reanalysis (Schneidereit et al., 2007), SLP measurements from ships (Chang, 2007), sea level time series (Vilibic and Sepic, 2010), and cloud analyses (Bender et al., 2012) support a poleward shift and intensification of the North Atlantic cyclone tracks from the 1950s to the early 2000s (Sorteberg and Walsh, 2008; Cornes and Jones, 2011).

2.7.6.2 Weather Types and Blocking

In AR4, weather types were not assessed as such, but an increase in blocking frequency in the Western Pacific and a decrease in North Atlantic were noted.

Changes in the frequency of weather types are of interest since weather extremes are often associated with specific weather types. For instance, persistent blocking of the westerly flow was essential in the development of the 2010 heat wave in Russia (Dole et al., 2011) (Section 9.5.2.2 and Box 14.2). Synoptic classifications or statistical clustering (Philipp et al., 2007) are commonly used to classify the weather on a given day. Feature-based methods are also used (Croci-Maspoli et al., 2007a). All these methods require daily SLP or upper-level fields.

Trends in synoptic weather types have been best analysed for central Europe since the mid-20th century, where several studies describe an increase in westerly or cyclonic weather types in winter but an increase of anticyclonic, dry weather types in summer (Philipp et al., 2007; Werner et al., 2008; Trnka et al., 2009). An eastward shift of blocking events over the North Atlantic (fewer cases of blocking over Greenland and more frequent blocking over the eartern North Atlantic) and the North Pacific was found by Davini et al. (2012) using NCEP/NCAR reanalysis since 1951 and by Croci-Maspoli et al. (2007a) in ERA-40 reanalysis during the period 1957–2001. Mokhov et al. (2013) find an increase in blocking duration over the NH year-round since about 1990 in a study based on NCEP/NCAR reanalysis data from 1969–2011. For the SH, Dong et al. (2008) found a decrease in number of blocking days but increase in intensity of blocking over the period 1948–1999. Differences in blocking index definitions, the sensitivity of some indices to changes in the mean field, and strong interannual variability in all seasons (Kreienkamp et al., 2010), partly related to circulation variability modes (Croci-Maspoli et al., 2007b), complicate a global assessment of blocking trends.

In summary, there is evidence for a poleward shift of storm tracks and jet streams since the 1970s. Based on the consistency of these trends with the widening of the tropical belt (Section 2.7.5), trends that are based on many different data sets, variables, and approaches, it is *likely* that circulation features have moved poleward since the 1970s. Methodological differences between studies mean there is *low confidence* in characterizing the global nature of any change in blocking.

2.7.7 Stratospheric Circulation

Changes in the polar vortices were assessed in AR4. A significant decrease in lower-stratospheric GPH in summer over Antarctica since 1969 was found, whereas trends in the Northern Polar Vortex were considered uncertain owing to its large variability.

The most important characteristics of the stratospheric circulation for climate and for trace gas distribution are the winter and spring polar vortices and Sudden Stratospheric Warmings (rapid warmings of the middle stratosphere that may lead to a collapse of the Polar Vortex), the Quasi-Biennial Oscillation (an oscillation of equatorial zonal winds with a downward phase propagation) and the Brewer-Dobson circulation (BDC, the meridional overturning circulation transporting air upward in the tropics, poleward to the winter hemisphere, and downward at polar and subpolar latitudes; Annex III: Glossary). Radiosonde observations, reanalysis data sets and space-borne temperature or trace gas observations are used to address changes in the stratospheric circulation, but all of these sources of information carry large trend uncertainties.

The AR4 assessment was corroborated further in Forster et al. (2011) and in updated 100 hPa GPH trends from ERA-Interim reanalysis (Box 2.3, Figure 2.36). There is *high confidence* that lower stratospheric GPH over Antarctica has decreased in spring and summer at least since 1979. Cohen et al. (2009) reported an increase in the number of Arctic sudden stratospheric warmings during the last two decades. However, interannual variability in the Arctic Polar Vortex is large, uncertainties in reanalysis products are high (Tegtmeier et al., 2008), and trends depend strongly on the time period analysed (Langematz and Kunze, 2008).

The BDC is only indirectly observable via wave activity diagnostics (which represent the main driving mechanism of the BDC), via temperatures or via the distribution of trace gases which may allow the determination of the 'age of air' (i.e., the time an air parcel has resided in the stratosphere after its entry from the troposphere). Randel et al. (2006), found a sudden decrease in global lower stratospheric water vapour and ozone around 2001 that is consistent with an increase in the mean tropical upwelling, that is, the tropical branch of the BDC (Rosenlof and Reid, 2008; Section 2.2.2.1; Lanzante, 2009; Randel and Jensen, 2013). On the other hand, Engel et al. (2009) found no statistically significant change in the age of air in the 24-35 km layer over the NH mid-latitudes from measurements of chemically inert trace gases from 1975 to 2005. However, this does not rule out trends in the lower stratospheric branch of the BDC or trends in mid to low latitude mixing (Bonisch et al., 2009; Ray et al., 2010). All of these methods are subject to considerable uncertainties, and they might shed light only on some aspects of the BDC. *Confidence* in trends in the BDC is therefore *low*.

In summary, it is *likely* that lower-stratopheric geopotential height over Antarctica has decreased in spring and summer at least since 1979. Owing to uncertainties in the data and approaches used, *confidence* in trends in the Brewer–Dobson circulation is *low*.

2.7.8 Changes in Indices of Climate Variability

AR4 assessed changes in indices of climate variability. The NAO and SAM were found to exhibit positive trends (strengthened mid-latitude westerlies) from the 1960s to 1990s, but the NAO has returned to its long-term mean state since then.

Indices of climate variability describe the state of the climate system with regards to individual modes of climate variability. Together with corresponding spatial patterns, they summarize large fractions of spatio-temporal climate variability. Inferences about significant trends in indices are generally hampered by relative shortness of climate records, their uncertainties and the presence of large variability on decadal and multidecadal time scales.

Table 2.14 summarizes observed changes in well-known indices of climate variability (see Box 2.5, Table 1 for precise definitions). Even the indices that explicitly include detrending of the entire record (e.g., Deser et al., 2010b), can exhibit statistically significant trends over shorter sub-periods. Confidence intervals in Table 2.14 that do not contain zero indicate trend significance at 10% level; however, the trends significant at 5% and 1% levels are emphasized in the discussion that follows. Chapter 14 discusses the main features and physical meaning of individual climate modes.

The NAO index reached very low values in the winter of 2010 (Osborn, 2011). As a result, with the exception of the principal component (PC) -based NAO index, which still shows a 5% significant positive trend from 1951 to present, other NAO or North Annular Mode (NAM) indices do not show significant trends of either sign for the periods presented in Table 2.14. In contrast, the SAM maintained the upward trend (Table 2.14). Fogt et al. (2009) found a positive trend in the SAM index from 1957 to 2005. Visbeck (2009), in a station-based index, found an increase in recent decades (1970s to 2000s).

The observed detrended multidecadal SST anomaly averaged over the North Atlantic Ocean area is often called Atlantic Multi-decadal Oscillation Index (AMO; see Box 2.5, Table 1, Figure 1). The warming trend in the "revised" AMO index since 1979 is significant at 1% level (Table 2.14) but cannot be readily interpreted because of the difficulty with reliable removal of the SST warming trend from it (Deser et al., 2010b).

On decadal and inter-decadal time scales the Pacific climate shows an irregular oscillation with long periods of persistence in individual stages and prominent shifts between them. Pacific Decadal Oscillation (PDO), Inter-decadal Pacific Oscillation (IPO) and North Pacific Index (NPI) indices characterize this variability for both hemispheres and agree well with each other (Box 2.5, Figure 1). While AR4 noted climate impacts of the 1976–1977 PDO phase transition, the shift in the opposite direction, both in PDO and IPO, may have occurred at the end of 1990s (Cai and van Rensch, 2012; Dai, 2012). Significance of 1979–2012 trends in PDO and NPI then would be an artefact of this

Table 2.14 | Trends for selected indices listed in Box 2.5, Table 1. Each index was standardized for its longest available period contained within the 1870–2012 interval. Standardization was done on the December-to-March (DJFM) means for the NAO, NAM and Pacific-North American pattern (PNA), on seasonal anomalies for Pacific-South American patterns (PSA1,PSA2) and on monthly anomalies for all other indices. Standardized monthly and seasonal anomalies were further averaged to annual means. Trend values computed for annual or DJFM means are given in standard deviation per decade with their 90% confidence intervals. Index records where the source is not explicitly indicated were computed from either HadISST1 (for SST-based indices), or HadSLP2r (for SLP-based indices) or NNR fields of 500 hPa or 850 hPa geopotential height. CoA stands for 'Centers of Action' index definitions. Linear trends for 1870–2012 were removed from ATL3, BMI and DMI.

Index Name	Trends in standard deviation units per decade		
	1901–2012	1951–2012	1979–2012
(−1)*SOI from CPC		0.004 ± 0.103	−0.243 ± 0.233
(−1)*SOI Troup from BOM records	0.012 ± 0.039	0.018 ± 0.104	−0.247 ± 0.236
SOI Darwin from BOM records	0.028 ± 0.036	0.082 ± 0.085	−0.116 ± 0.195
(−1)*EQSOI	0.001 ± 0.051	−0.076 ± 0.143	−0.558[b] ± 0.297
NIÑO3.4	−0.003 ± 0.042	0.012 ± 0.105	−0.156 ± 0.274
NIÑO3.4 (ERSST v.3b)	0.067[a] ± 0.045	0.054 ± 0.103	−0.085 ± 0.259
NIÑO3.4 (COBE SST)	0.024 ± 0.041	0.008 ± 0.107	−0.154 ± 0.289
NIÑO3	0.007 ± 0.039	0.043 ± 0.095	−0.143 ± 0.256
NIÑO3 (ERSST v.3b)	0.069 ± 0.039	0.098 ± 0.092	−0.073 ± 0.236
NIÑO3 (COBE SST)	0.034 ± 0.036	0.054 ± 0.096	−0.113 ± 0.258
NIÑO4	0.026 ± 0.054	0.068 ± 0.145	−0.102 ± 0.380
EMI	−0.059 ± 0.061	−0.119 ± 0.189	−0.131 ± 0.580
(−1)*TNI	0.019 ± 0.052	0.066 ± 0.167	0.030 ± 0.550
PDO from Mantua et al. (1997)	−0.017 ± 0.071	0.112 ± 0.189	−0.460[a] ± 0.284
(−1)*NPI	−0.026[a] ± 0.022	0.010 ± 0.046	−0.169[a] ± 0.105
AMO revised	−0.001 ± 0.111	−0.012 ± 0.341	0.779[b] ± 0.291
NAO stations from Jones et al. (1997)	−0.044 ± 0.056	0.095 ± 0.149	−0.136 ± 0.394
NAO stations from Hurrell (1995)	−0.001 ± 0.066	0.171 ± 0.179	−0.214 ± 0.400
NAO PC from Hurrell (1995)	0.012 ± 0.059	0.198[a] ± 0.148	−0.037 ± 0.401
NAM PC	0.003 ± 0.048	0.141 ± 0.123	0.029 ± 0.360
SAM Z850 PC		0.268[b] ± 0.063	0.100 ± 0.109
SAM SLP grid 40°S to 70°S	0.139[b] ± 0.026	0.198[b] ± 0.052	0.294[b] ± 0.131
SAM SLP stations from Marshall (2003)			0.128[a] ± 0.097
PNA CoA		0.113 ± 0.114	−0.103 ± 0.298
PNA RPC from CPC		0.202[b] ± 0.111	0.019 ± 0.271
PSA Karoly (1989) CoA definition		−0.267[b] ± 0.079	−0.233[a] ± 0.174
(−1)*PSA Yuan and Li (2008) CoA definition		−0.211[b] ± 0.069	−0.208 ± 0.189
PSA1 PC		−0.163[a] ± 0.103	−0.368[a] ± 0.245
PSA2 PC		0.200[b] ± 0.066	0.036 ± 0.156
ATL3	0.035 ± 0.043	0.125[a] ± 0.088	0.186 ± 0.193
AONM PC	0.064[a] ± 0.051	0.138[a] ± 0.109	0.327[a] ± 0.230
AMM PC	0.019 ± 0.058	−0.015 ± 0.155	0.309 ± 0.324
IOBM PC	0.075[a] ± 0.051	0.314[b] ± 0.082	0.201 ± 0.206
BMI	0.072[a] ± 0.050	0.294[b] ± 0.083	0.189 ± 0.206
IODM PC	−0.016 ± 0.034	−0.031 ± 0.093	−0.052 ± 0.203
DMI	0.030 ± 0.033	0.080 ± 0.090	0.211 ± 0.210

Notes:

[a] Trend values significant at the 5% level.

[b] Trend values significant at the 1% level.

2

change; incidentally, no significant trends in these indices were seen for longer periods (Table 2.14). Nevertheless, Pacific changes since the 1980s (positive for NPI and negative for PDO and IPO) are consistent with the observed SLP changes (Section 2.7.1) and with reversing trends in the Walker Circulation (Section 2.7.5), which was reported to be slowing down during much of the 20th century but sped up again since the 1990s. Equatorial SOI shows an increasing trend since 1979 at 1% significance; more traditionally defined SOI indices do not show significant trends (Table 2.14).

NIÑO3.4 and NIÑO3 show a century-scale warming trend significant at 5% level, if computed from the ERSSTv3b data set (Section 2.4.2) but not if calculated from other data sets (Table 2.14). Furthermore, the sign (and significance) of the trend in east–west SST gradient across the Pacific remains ambiguous (Vecchi and Soden, 2007; Bunge and Clarke, 2009; Karnauskas et al., 2009; Deser et al., 2010a) (Section 14.4.1).

In addition to changes in the mean values of climate indices, changes in the associated spatial patterns are also possible. In particular, the diversity of detail of different ENSO events and possible distinction between their "flavors" have received significant attention (Section 14.4.2). These efforts also intensified the discussion of useful ENSO indices in the literature. Starting from the work of Trenberth and Stepaniak (2001), who proposed to characterize the evolution of ENSO events with the Trans-Niño Index (TNI), which is virtually uncorrelated with the standard ENSO index NIÑO3.4, other alternative ENSO indices have been introduced and proposals were made for classifying ENSO events according to the indices they primarily maximize. While a traditional, 'canonical' El Niño event type (Rasmusson and Carpenter, 1982) is viewed as the 'eastern Pacific' type, some of the alternative indices purport to identify events that have central Pacific maxima and are called dateline El Niño (Larkin and Harrison, 2005), Modoki (Ashok et al., 2007), or Central Pacific El Niño (Kao and Yu, 2009). However, no consensus has been reached regarding the appropriate classification of ENSO events. Takahashi et al. (2011) and Ren and Jin (2011) have presented many of the popular ENSO indices as elements in a two-dimensional linear space spanned by a pair of such indices. ENSO indices that involve central and western Pacific SST (NIÑO4, EMI, TNI) show no significant trends.

Significant positive PNA trends and negative and positive trends in the first and second PSA modes respectively are observed over the last 60 years (Table 2.14). However, the level of significance of these trends depends on the index definition and on the data set used. The positive trend in the Atlantic Ocean 'Niño' mode (AONM) index and in ATL3 are due to the intensified warming in the eastern Tropical Atlantic that causes the the weakening of the Atlantic equatorial cold tongue: these changes were noticed by Tokinaga and Xie (2011b) with regards to the last 60-year period. The Indian Ocean Basin Mode (IOBM) has a strong warming trend (significant at 1% since the middle of the 20th century). This phenomenon is well-known (Du and Xie, 2008) and its consequences for the regional climate is a subject of active research (Du et al., 2009; Xie et al., 2009).

In summary, large variability on interannual to decadal time scales and remaining differences between data sets precludes robust conclusions on long-term changes in indices of climate variability. *Confidence* is *high* that the increase in the NAO index from the 1950s to the 1990s has been largely offset by recent changes. It is *likely* that the SAM index has become more positive since the 1950s.

Box 2.5 | Patterns and Indices of Climate Variability

Much of the spatial structure of climate variability can be described as a combination of 'preferred' patterns. The most prominent of these are known as modes of climate variability and they impact weather and climate on many spatial and temporal scales (Chapter 14). Individual climate modes historically have been identified through spatial teleconnections: correlations between regional climate variations at widely separated, geographically fixed spatial locations. An index describing temporal variations of the climate mode in question can be formed, for example, by adding climate anomalies calculated from meteorological records at stations exhibiting the strongest correlation with the mode and subtracting anomalies at stations exhibiting anticorrelation. By regressing climate records from other places on this index, one derives a spatial climate pattern characterizing this mode. Patterns of climate variability have also been derived using a variety of mathematical techniques such as principal component analysis (PCA). These patterns and their indices are useful both because they efficiently describe climate variability in terms of a few preferred modes and also because they can provide clues about how the variablility is sustained (Box 14.1 provides formal definitions of these terms).

Box 2.5, Table 1 lists some prominent modes of large-scale climate variability and indices used for defining them. Changes in these indices are associated with large-scale climate variations on interannual and longer time scales. With some exceptions, indices shown have been used by a variety of authors. They are defined relatively simply from raw or statistically analyzed observations of a single climate variable, which has a history of surface observations. For most of these indices at least a century-long record is available for climate research. *(continued on next page)*

Box 2.5, Table 1 | Established indices of climate variability with global or regional influence. Z500, Z700 and Z850 denote geopotential height at the 500, 700 and 850 hPa levels, respectively. The subscripts s and a denote "standardized" and "anomalies", respectively. Further information is given in Supplementary Material 2.SM.8. Climate impacts of these modes are listed in Box 14.1. *(continued on next page)*

Climate Phenomenon		Index Name	Index Definition	Primary Reference(s)
El Niño – Southern Oscillation (ENSO)	Traditional indices of ENSO-related Tropical Pacific variability	NIÑO1+2	SSTa averaged over [10°S–0°, 90°W–80°W]	Rasmusson and Wallace (1983), Cane (1986)
		NIÑO3	Same as above but for [5°S–5°N, 150°W–90°W]	
		NIÑO4	Same as above but for [5°S–5°N, 160°E–150°W]	
		NIÑO3.4	Same as above but for [5°S–5°N, 170°W–120°W]	Trenberth (1997)
		Troup Southern Oscillation Index (SOI)	Standardized for each calendar month SLP_a difference: Tahiti minus Darwin, ×10	Troup (1965)
		SOI	Standardized difference of SLP_{sa}: Tahiti minus Darwin	Trenberth (1984); Ropelewski and Jones (1987)
		Darwin SOI	Darwin SLP_{sa}	Trenberth and Hoar (1996)
		Equatorial SOI (EQSOI)	Standardized difference of standardized averages of SLP_a over equatorial [5°S–5°N] Pacific Ocean areas: [130°W–80°W] minus [90°E–140°E]	Bell and Halpert (1998)
	Indices of ENSO events evolution and type	Trans-Niño Index (TNI)	$NIÑO1+2_s$ minus $NIÑO4_s$	Trenberth and Stepaniak (2001)
		El Niño Modoki Index (EMI)	SST_a [165°E–140°W, 10°S–10°N] minus ½*[110°W–70°W, 15°S–5°N] minus ½*[125°E–145°E, 10°S–20°N]	Ashok et al. (2007)
Pacific Decadal and Interdecadal Variability		Pacific Decadal Oscillation (PDO)	1st PC of monthly N. Pacific SST_a field [20°N–70°N] with subtracted global mean	Mantua et al. (1997); Zhang et al. (1997)
		Inter-decadal Pacific Oscillation (IPO)	Projection of a global SST_a onto the IPO pattern, which is found as one of the leading Empirical Orthogonal Functions of a low-pass filtered global SST_a field	Folland et al. (1999); Power et al. (1999); Parker et al. (2007)
		North Pacific Index (NPI)	SLP_a averaged over [30°N–65°N; 160°E–140°W]	Trenberth and Hurrell (1994)
North Atlantic Oscillation (NAO)		Azores-Iceland NAO Index	SLP_{sa} difference: Lisbon/Ponta Delgada minus Stykkisholmur/ Reykjavik	Hurrell (1995)
		PC-based NAO Index	Leading PC of SLP_a over the Atlantic sector	Hurrell (1995)
		Gibraltar – South-west Iceland NAO Index	Standardized for each calendar month SLP_a difference: Gibraltar minus SW Iceland / Reykjavik	Jones et al. (1997)
Annular modes	Northern Annular Mode (NAM)	PC-based NAM or Arctic Oscillation (AO) index	1st PC of the monthly mean SLP_a poleward of 20°N	Thompson and Wallace (1998, 2000)
	Southern Annular Mode (SAM)	PC-based SAM or Antarctic Oscillation (AAO) index	1st PC of $Z850_a$ or $Z700_a$ south of 20°S	Thompson and Wallace (2000)
		Grid-based SAM index: 40°S–70°S difference	Difference between standardized zonally averaged SLP_a at 40°S and 70°S, using gridded SLP fields	Nan and Li (2003)
		Station-based SAM index: 40°S–65°S	Difference in standardized zonal mean SLP_a at 40°S and 65°S, using station data	Marshall (2003)
Pacific/North America (PNA) atmospheric teleconnection		PNA index based on centers of action	¼[(20°N, 160°W) – (45°N, 165°W) + (55°N, 115°W) – (30°N, 85°W)] in the $Z500_{sa}$ field	Wallace and Gutzler (1981)
		PNA from rotated PCA	Rotated PC (RPC) from the analysis of the NH $Z500_a$ field	Barnston and Livezey (1987)
Pacific/South America (PSA) atmospheric teleconnection		PSA1 and PSA2 mode indices (PC-based)	2nd and 3rd PCs respectively of SH seasonal $Z500_a$	Mo and Paegle (2001)
		PSA index (centers of action)	[–(35°S, 150°W) + (60°S, 120°W) – (45°S, 60°W)] in the $Z500_a$ field	Karoly (1989)
			[(45°S, 170°W) – (67.5°S, 120°W) + (50°S, 45°W)]/3 in the $Z500_a$ field	Yuan and Li (2008)
Atlantic Ocean Multidecadal Variability		Atlantic Multi-decadal Oscillation (AMO) index	10-year running mean of linearly detrended Atlantic mean SST_a [0°–70°N]	Enfield et al. (2001)
		Revised AMO index	As above, but detrended by subtracting SST_a [60°S–60°N] mean	Trenberth and Shea (2006)
Tropical Atlantic Ocean Variability	Atlantic Ocean Niño Mode (AONM)	ATL3	SST_a averaged over [3°S–3°N, 20°W–0°]	Zebiak (1993)
		PC-based AONM	1st PC of the detrended tropical Atlantic monthly SST_a (20°S–20°N)	Deser et al. (2010b)
	Tropical Atlantic Meridional Mode (AMM)	PC-based AMM Index	2nd PC of the detrended tropical Atlantic monthly SST_a (20°S–20°N)	
Tropical Indian Ocean Variability	Indian Ocean Basin Mode (IOBM)	Basin mean index (BMI)	SST_a averaged over [40°–110°E, 20°S–20°N]	Yang et al. (2007)
		IOBM, PC-based Index	The first PC of the IO detrended SST_a (40°E–110° E, 20°S–20°N)	Deser et al. (2010b)
	Indian Ocean Dipole Mode (IODM)	PC-based IODM index	The second PC of the IO detrended SST_a (40°E–110° E, 20°S–20°N)	
		Dipole Mode Index (DMI)	SST_a difference: [50°E–70°E, 10°S–10°N] minus [90°E–110°E, 10°S–0°]	Saji et al. (1999)

Box 2.5 (continued)

Most climate modes are illustrated by several indices (Box 2.5, Figure 1), which often behave similarly to each other. Spatial patterns of SST or SLP associated with these climate modes are illustrated in Box 2.5, Figure 2. They can be interpreted as a change in the SST or SLP field associated with one standard deviation change in the index. *(continued on next page)*

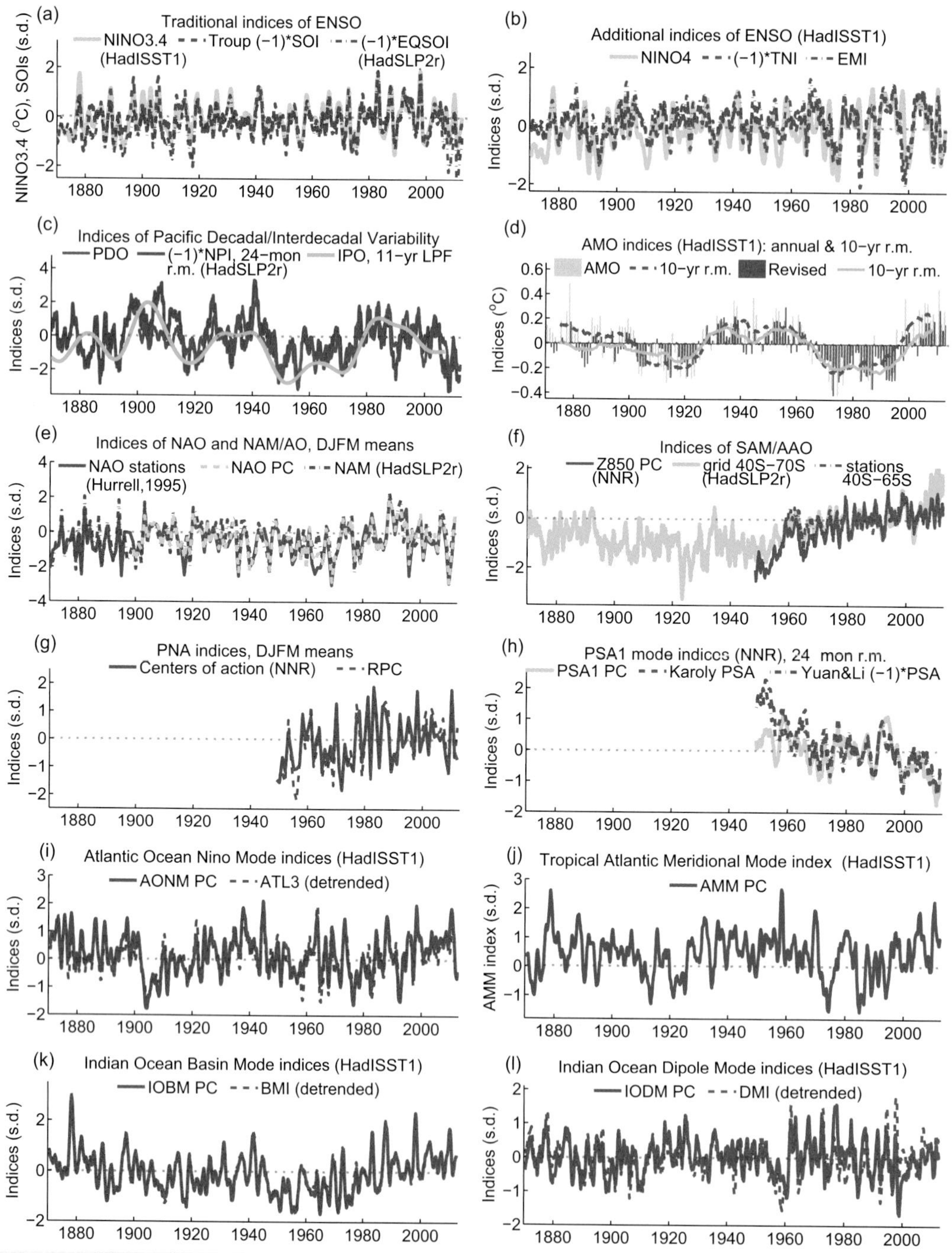

Box 2.5, Figure 1 | Some indices of climate variability, as defined in Box 2.5, Table 1, plotted in the 1870–2012 interval. Where 'HadISST1', 'HadSLP2r', or 'NNR' are indicated, the indices were computed from the sea surface temperature (SST) or sea level pressure (SLP) values of the former two data sets or from 500 or 850 hPa geopotential height fields from the NNR. Data set references given in the panel titles apply to all indices shown in that panel. Where no data set is specified, a publicly available version of an index from the authors of a primary reference given in Box 2.5, Table 1 was used. All indices were standardized with regard to 1971–2000 period except for NIÑO3.4 (centralized for 1971–2000) and AMO indices (centralized for 1901–1970). Indices marked as "detrended" had their linear trend for 1870–2012 removed. All indices are shown as 12-month running means except when the temporal resolution is explicitly indicated (e.g., 'DJFM' for December-to-March averages) or smoothing level (e.g., 11-year LPF for a low-pass filter with half-power at 11 years).

Box 2.5 (continued)

The difficulty of identifying a universally 'best' index for any particular climate mode is due to the fact that no simply defined indicator can achieve a perfect separation of the target phenomenon from all other effects occurring in the climate system. As a result, each index is affected by many climate phenomena whose relative contributions may change with the time period and the data set used. Limited length and quality of the observational record further compound this problem. Thus the choice of index is always application specific.

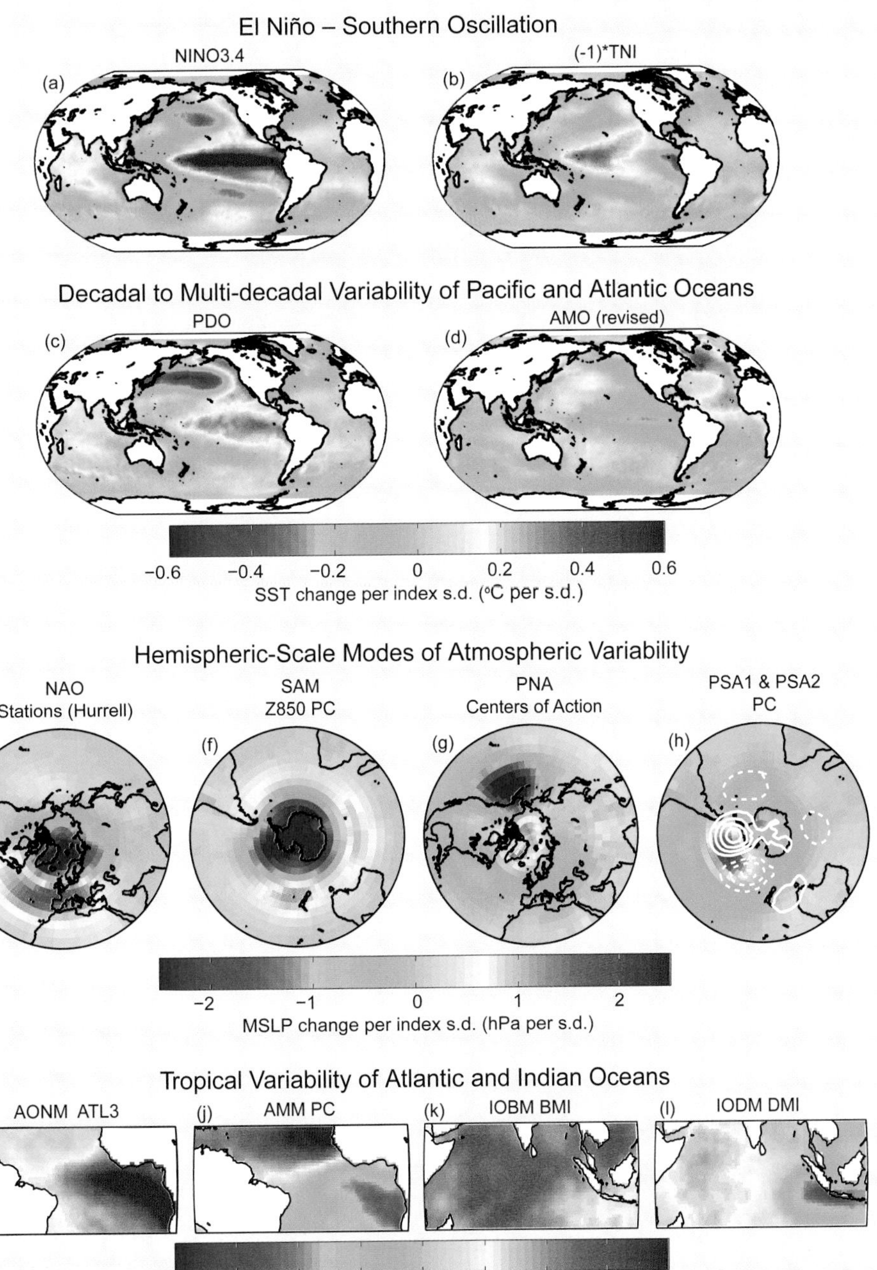

Box 2.5, Figure 2 | Spatial patterns of climate modes listed in Box 2.5, Table 1. All patterns shown here are obtained by regression of either sea surface temperature (SST) or sea level pressure (SLP) fields on the standardized index of the climate mode. For each climate mode one of the specific indices shown in Box 2.5, Figure 1 was used, as identified in the panel subtitles. SST and SLP fields are from HadISST1 and HadSLP2r data sets (interpolated gridded products based on data sets of historical observations). Regressions were done on monthly means for all patterns except for NAO and PNA, which were done with the DJFM means, and for PSA1 and PSA2, where seasonal means were used. Each regression was done for the longest period within the 1870-2012 interval when the index was available. For each pattern the time series was linearly de-trended over the entire regression interval. All patterns are shown by color plots, except for PSA2, which is shown by white contours over the PSA1 color plot (contour steps are 0.5 hPa, zero contour is skipped, negative values are indicated by dash).

Acknowledgements

The authors of Chapter 2 wish to thank Wenche Aas (NILU, Kjeller), Erika Coppola (ICTP, Trieste), Ritesh Gautam (NASA GSFC, Greenbelt), Jenny Hand (CIRA, Fort Collins), Andreas Hilboll (U. Bremen, Bremen), Glenn Hyatt (NOAA NCDC, Asheville), David Parrish (NOAA ESRL-CSD, Boulder), Deborah Misch (LMI, Inc, Asheville), Jared Rennie (CICS-NC, Asheville), Deborah Riddle (NOAA NCDC, Asheville), Sara Veasey (NOAA NCDC, Asheville), Mark Weber (U. Bremen, Bremen), Yin Xungang (STG Inc., Asheville), Teresa Young (STG, Asheville) and Jianglong Zhang (U. North Dakota, Grand Forks) for their critical contributions to the production of figures in this work.

References

Abakumova, G. M., E. V. Gorbarenko, E. I. Nezval, and O. A. Shilovtseva, 2008: Fifty years of actinometrical measurements in Moscow. *Int. J. Remote Sens.*, **29**, 2629–2665.

Adam, J. C., and D. P. Lettenmaier, 2008: Application of new precipitation and reconstructed streamflow products to streamflow trend attribution in northern Eurasia. *J. Clim.*, **21**, 1807–1828.

Adler, R. F., G. J. Gu, and G. J. Huffman, 2012: Estimating climatological bias errors for the global Precipitation Climatology Project (GPCP). *J. Appl. Meteor. Climatol.*, **51**, 84–99.

Adler, R. F., et al., 2003: The version-2 global precipitation climatology project (GPCP) monthly precipitation analysis (1979–present). *J. Hydrometeor.*, **4**, 1147–1167.

Aguilar, E., et al., 2009: Changes in temperature and precipitation extremes in western central Africa, Guinea Conakry, and Zimbabwe, 1955–2006. *J. Geophys. Res. Atmos.*, **114**, D02115.

Alexander, L. V., P. Uotila, and N. Nicholls, 2009: Influence of sea surface temperature variability on global temperature and precipitation extremes. *J. Geophys. Res. Atmos.*, **114**, D18116.

Alexander, L. V., X. L. L. Wang, H. Wan, and B. Trewin, 2011: Significant decline in storminess over southeast Australia since the late 19th century. *Aust. Meteor. Ocean. J.*, **61**, 23–30.

Alexander, L. V., et al., 2006: Global observed changes in daily climate extremes of temperature and precipitation. *J. Geophys. Res. Atmos.*, **111**, D05109.

Allan, R., and T. Ansell, 2006: A new globally complete monthly historical gridded mean sea level pressure dataset (HadSLP2): 1850–2004. *J. Clim.*, **19**, 5816–5842.

Allan, R., S. Tett, and L. Alexander, 2009: Fluctuations in autumn-winter severe storms over the British Isles: 1920 to present. *Int. J. Climatol.*, **29**, 357–371.

Allan, R. P., 2009: Examination of relationships between clear-sky longwave radiation and aspects of the atmospheric hydrological cycle in climate models, reanalyses, and observations. *J. Clim.*, **22**, 3127–3145.

Allan, R. P., and A. Slingo, 2002: Can current climate model forcings explain the spatial and temporal signatures of decadal OLR variations? *Geophys. Res. Lett.*, **29**, 1141.

Allan, R. P., and B. J. Soden, 2008: Atmospheric warming and the amplification of precipitation extremes. *Science*, **321**, 1481–1484.

Allan, R. P., B. J. Soden, V. O. John, W. Ingram, and P. Good, 2010: Current changes in tropical precipitation. *Environ. Res. Lett.*, **5**, 025205.

Allen, R. J., and S. C. Sherwood, 2007: Utility of radiosonde wind data in representing climatological variations of tropospheric temperature and baroclinicity in the western tropical Pacific. *J. Clim.*, **20**, 5229–5243.

Allen, R. J., and S. C. Sherwood, 2008: Warming maximum in the tropical upper troposphere deduced from thermal winds. *Nature Geosci.*, **1**, 399–403.

Alpert, P., and P. Kishcha, 2008: Quantification of the effect of urbanization on solar dimming. *Geophys. Res. Lett.*, **35**, L08801.

Alpert, P., P. Kishcha, Y. J. Kaufman, and R. Schwarzbard, 2005: Global dimming or local dimming? Effect of urbanization on sunlight availability. *Geophys. Res. Lett.*, **32**, L17802.

Andrade, C., S. Leite, and J. Santos, 2012: Temperature extremes in Europe: Overview of their driving atmospheric patterns. *Nat. Hazards Earth Syst. Sci.*, **12**, 1671–1691.

Andronova, N., J. E. Penner, and T. Wong, 2009: Observed and modeled evolution of the tropical mean radiation budget at the top of the atmosphere since 1985. *J. Geophys. Res. Atmos.*, **114**, D14106.

Angell, J. K., 2006: Changes in the 300-mb North Circumpolar Vortex, 1963–2001. *J. Clim.*, **19**, 2984–2994.

Anthes, R. A., 2011: Exploring Earth's atmosphere with radio occultation: contributions to weather, climate and space weather. *Atmos. Meas. Tech.*, **4**, 1077–1103.

Anthes, R. A., et al., 2008: The COSMOC/FORMOSAT-3 Mission—Early results. *Bull. Am. Meteor. Soc.*, **89**, 313.

Archer, C. L., and K. Caldeira, 2008a: Reply to comment by Courtenay Strong and Robert E. Davis on "Historical trends in the jet streams". *Geophys. Res. Lett.*, **35**. L24807.

Archer, C. L., and K. Caldeira, 2008b: Historical trends in the jet streams. *Geophys. Res. Lett.*, **35**, L08803.

Arndt, D. S., M. O. Baringer, and M. R. Johnson, 2010: State of the Climate in 2009. *Bull. Am. Meteor. Soc.*, **91**, S1–.

Arnold, T., et al., 2013: Nitrogen trifluoride global emissions estimated from updated atmospheric measurements. *Proc. Natl. Acad. Sci. U.S.A.*, **110**, 2029–2034.

Ashok, K., S. K. Behera, S. A. Rao, H. Y. Weng, and T. Yamagata, 2007: El Niño Modoki and its possible teleconnection. *J. Geophys. Res. Oceans*, **112**, C11007.

Asmi, A., et al., 2013: Aerosol decadal trends – Part 2: In-situ aerosol particle number concentrations at GAW and ACTRIS stations. *Atmos. Chem. Phys.*, **13**, 895–916.

Atlas, R., R. Hoffman, J. Ardizzone, S. Leidner, J. Jusem, D. Smith, and D. Gombos, 2011: A cross-calibrated mutiplatform ocean wind velocity product for meteorlogical and oceanographic applications. *Bull. Am. Meteor. Soc.*, **92**, 157–.

Augustine, J. A., and E. G. Dutton, 2013: Variability of the surface radiation budget over United States from 1996 through 2011 from high-quality measurements. *J. Geophys. Res.*, **118**, 43-53,

Aydin, M., et al., 2011: Recent decreases in fossil-fuel emissions of ethane and methane derived from firn air. *Nature*, **476**, 198–201.

Ballester, J., F. Giorgi, and X. Rodo, 2010: Changes in European temperature extremes can be predicted from changes in PDF central statistics. *Clim. Change*, **98**, 277–284.

Baringer, M. O., D. S. Arndt, and M. R. Johnson, 2010: State of the Climate in 2009. *Bull. Am. Meteor. Soc.*, **91**, S1–.

Barmpadimos, I., J. Keller, D. Oderbolz, C. Hueglin, and A. S. H. Prévôt, 2012: One decade of parallel fine (PM2.5) and coarse (PM10–PM2.5) particulate matter measurements in Europe: trends and variability. *Atmos. Chem. Phys.*, **12**, 3189–3203.

Barnston, A. G., and R. E. Livezey, 1987: Classification, seasonality and persistence of low-frequency atmospheric circulation patterns. *Mon. Weather Rev.*, **115**, 1083–1126.

Barring, L., and K. Fortuniak, 2009: Multi-indices analysis of southern Scandinavian storminess 1780–2005 and links to interdecadal variations in the NW Europe-North Sea region. *Int. J. Climatol.*, **29**, 373–384.

Barriopedro, D., E. M. Fischer, J. Luterbacher, R. Trigo, and R. Garcia-Herrera, 2011: The hot summer of 2010: Redrawing the temperature record map of Europe. *Science*, **332**, 220–224.

Barrucand, M., M. Rusticucci, and W. Vargas, 2008: Temperature extremes in the south of South America in relation to Atlantic Ocean surface temperature and Southern Hemisphere circulation. *J. Geophys. Res. Atmos.*, **113**, D20111.

Barton, N. P., and A. W. Ellis, 2009: Variability in wintertime position and strength of the North Pacific jet stream as represented by re-analysis data. *Int. J. Climatol.*, **29**, 851–862.

Becker, A., et al., 2013: A description of the global land-surface precipitation data products of the Global Precipitation Climatology Centre with sample applications including centennial (trend) analysis from 1901–present. *Earth Syst. Sci. Data*, **5**, 71–99.

Beig, G., and V. Singh, 2007: Trends in tropical tropospheric column ozone from satellite data and MOZART model. *Geophys. Res. Lett.*, **34**, L17801.

Bell, G. D., and M. S. Halpert, 1998: Climate assessment for 1997. *Bull. Am. Meteor. Soc.*, **79**, S1–S50.

Bender, F. A. M., V. Ramanathan, and G. Tselioudis, 2012: Changes in extratropical storm track cloudiness 1983–2008: Observational support for a poleward shift. *Clim. Dyn.*, **38**, 2037–2053.

Bengtsson, L., and K. I. Hodges, 2011: On the evaluation of temperature trends in the tropical troposphere. *Clim. Dyn.*, **36**, 419–430.

Beniston, M., 2004: The 2003 heat wave in Europe: A shape of things to come? An analysis based on Swiss climatological data and model simulations. *Geophys. Res. Lett.*, **31**, L02202.

Beniston, M., 2009: Decadal-scale changes in the tails of probability distribution functions of climate variables in Switzerland. *Int. J. Climatol.*, **29**, 1362–1368.

Bennartz, R., J. Fan, J. Rausch, L. Leung, and A. Heidinger, 2011: Pollution from China increases cloud droplet number, suppresses rain over the East China Sea. *Geophys. Res. Lett.*, **38**, L09704.

Berrisford, P., et al., 2011: Atmospheric conservation properties in ERA-Interim. *Q. J. R. Meteor. Soc.*, **137**, 1381–1399.

Berry, D., and E. Kent, 2011: Air-Sea fluxes from ICOADS: The construction of a new gridded dataset with uncertainty estimates. *Int. J. Climatol.*, **31**, 987–1001.

Berry, D. I., and E. C. Kent, 2009: A new air-sea interaction gridded dataset from ICOADS with uncertainty estimates. *Bull. Am. Meteor. Soc.*, **90**, 645-656.

Berthet, C., J. Dessens, and J. Sanchez, 2011: Regional and yearly variations of hail frequency and intensity in France. *Atmos. Res.*, **100**, 391–400.

Birner, T., 2010: Recent widening of the tropical belt from global tropopause statistics: Sensitivities. *J. Geophys. Res. Atmos.*, **115,** D23109.

Bitz, C. M., and Q. Fu, 2008: Arctic warming aloft is data set dependent. *Nature*, **455**, E3–E4.

Black, E., and R. Sutton, 2007: The influence of oceanic conditions on the hot European summer of 2003. *Clim. Dyn.*, **28**, 53–66.

Blunden, J., D. S. Arndt, and M. O. Baringer, 2011: State of the Climate in 2010. *Bull. Am. Meteor. Soc.*, **92**, S17–.

Bohm, R., P. D. Jones, J. Hiebl, D. Frank, M. Brunetti, and M. Maugeri, 2010: The early instrumental warm-bias: A solution for long central European temperature series 1760–2007. *Clim. Change*, **101**, 41–67.

Bonfils, C., P. B. Duffy, B. D. Santer, T. M. L. Wigley, D. B. Lobell, T. J. Phillips, and C. Doutriaux, 2008: Identification of external influences on temperatures in California. *Clim. Change*, **87**, S43–S55.

Bonisch, H., A. Engel, J. Curtius, T. Birner, and P. Hoor, 2009: Quantifying transport into the lowermost stratosphere using simultaneous in-situ measurements of SF_6 and CO_2. *Atmos. Chem. Phys.*, **9**, 5905–5919.

Bosilovich, M. G., F. R. Robertson, and J. Chen, 2011: Global energy and water budgets in MERRA. *J. Clim.*, **24**, 5721-5739..

Bourassa, M. A., S. T. Gille, D. L. Jackson, J. B. Roberts, and G. A. Wick, 2010: Ocean winds and turbulent air-sea fluxes inferred from remote sensing. *Oceanography*, **23**, 36–51.

Bousquet, P., 2011: Source attribution of the changes in atmospheric methane for 2006–2008. *Atmos. Chem. Phys. Discuss.*, **10**, 27603–27630.

Brogniez, H., R. Roca, and L. Picon, 2009: Study of the free tropospheric humidity interannual variability using meteosat data and an advection-condensation transport model. *J. Clim.*, **22**, 6773–6787.

Brönnimann, S., 2009: Early twentieth-century warming. *Nature Geosci.*, **2**, 735–736.

Brönnimann, S., et al., 2009: Variability of large-scale atmospheric circulation indices for the Northern Hemisphere during the past 100 years. *Meteorol. Z.*, **18**, 379–396.

Brooks, C. F., 1926: Observing water-surface temperatures at sea. *Mon. Wea. Rev.*, **54,** 241–253.

Brooks, H., 2012: Severe thunderstorms and climate change. *Atmos. Res.*, **123**, SI, 129-138.

Brooks, H. E., and N. Dotzek, 2008: The spatial distribution of severe convective storms and an analysis of their secular changes. In: *Climate Extremes and Society* [H. F. Diaz and R. J. Murnane (eds.] Cambridge University Press, pp. 35–53.

Brown, S. J., J. Caesar, and C. A. T. Ferro, 2008: Global changes in extreme daily temperature since 1950. *J. Geophys. Res. Atmos.*, **113,** D05115.

Brunet, M., et al., 2011: The minimization of the screen bias from ancient Western Mediterranean air temperature records: an exploratory statistical analysis. *Int. J. Climatol.*, **31**, 1879–1895.

Bunge, L., and A. J. Clarke, 2009: A verified estimation of the El Niño Index Niño-3.4 since 1877. *J. Clim.*, **22**, 3979–3992.

Burn, D. H., and N. M. Hesch, 2007: Trends in evaporation for the Canadian prairies. *J. Hydrol.*, **336**, 61–73.

Caesar, J., L. Alexander, and R. Vose, 2006: Large-scale changes in observed daily maximum and minimum temperatures: Creation and analysis of a new gridded data set. *J. Geophys. Res. Atmos.*, **111,** D05101.

Caesar, J., et al., 2011: Changes in temperature and precipitation extremes over the Indo-Pacific region from 1971 to 2005. *Int. J. Climatol.*, **31**, 791–801.

Cai, W., and P. van Rensch, 2012: The 2011 southeast Queensland extreme summer rainfall: A confirmation of a negative Pacific Decadal Oscillation phase? *Geophys. Res. Lett.*, **39**, L08702.

Cai, W., T. Cowan, and M. Thatcher, 2012: Rainfall reductions over Southern Hemisphere semi-arid regions: The role of subtropical dry zone expansion. *Sci. Rep.*, **2,** 702.

Callaghan, J., and S. B. Power, 2011: Variability and decline in the number of severe tropical cyclones making land-fall over eastern Australia since the late nineteenth century. *Clim. Dyn.*, **37**, 647–662.

Canada, 2012: Canadian Smog Science Assessment – Highlights and Key Messages. Environment Canada and Health Canada, 64 pp.

Cane, M. A., 1986: El-Niño. *Annu. Rev. Earth Planet. Sci.*, **14**, 43–70.

Cao, Z. H., 2008: Severe hail frequency over Ontario, Canada: Recent trend and variability. *Geophys. Res. Lett.*, **35.** L14803.

Carslaw, K. S., O. Boucher, D. Spracklen, G. Mann, J. G. Rae, S. Woodward, and M. Kumala, 2010: A review of natural aerosol interactions and feedbacks within the Earth system. *Atmos. Chem. Phys.*, **10**, 1701–1737.

Casey, K. S., T. B. Brandon, P. Cornillon, and R. Evans, 2010: The past, present and future of the AVHRR Pathfinder SST Program. In: *Oceanography from Space: Revisited* [V. Barale, J. F. R. Gower, and L. Alberotanza (eds.)]. Springer Science+Business Media, New York, 323-341.

Castellanos, P., and K. F. Boersma, 2012: Reductions in nitrogen oxides over Europe driven by environmental policy and economic recession. *Sci. Rep.*, **2.** 265.

Cermak, J., M. Wild, R. Knutti, M. I. Mishchenko, and A. K. Heidinger, 2010: Consistency of global satellite-derived aerosol and cloud data sets with recent brightening observations. *Geophys. Res. Lett.*, **37**, L21704.

Chambers, L., and G. Griffiths, 2008: The changing nature of temperature extremes in Australia and New Zealand. *Aust. Meteorol. Mag.*, **57**, 13–35.

Chan, J. C. L., and M. Xu, 2009: Inter-annual and inter-decadal variations of landfalling tropical cyclones in East Asia. Part I: time series analysis. *Int. J. Climatol.*, **29**, 1285–1293.

Chandler, R. E., and E. M. Scott, 2011: *Statistical Methods for Trend Detection and Analysis in the Environmental Sciences.* John Wiley & Sons, Hoboken, NJ.

Chang, E. K. M., 2007: Assessing the increasing trend in Northern Hemisphere winter storm track activity using surface ship observations and a statistical storm track model. *J. Clim.*, **20**, 5607–5628.

Chang, E. K. M., and Y. J. Guo, 2007: Is the number of North Atlantic tropical cyclones significantly underestimated prior to the availability of satellite observations? *Geophys. Res. Lett.*, **34.** L14801.

Chapman, W. L., and J. E. Walsh, 2007: A synthesis of Antarctic temperatures. *J. Clim.*, **20**, 4096–4117.

Che, H. Z., et al., 2005: Analysis of 40 years of solar radiation data from China, 1961–2000. *Geophys. Res. Lett.*, **32**, L06803.

Chen, J. Y., B. E. Carlson, and A. D. Del Genio, 2002: Evidence for strengthening of the tropical general circulation in the 1990s. *Science*, **295**, 838–841.

Chen, J. Y., A. D. Del Genio, B. E. Carlson, and M. G. Bosilovich, 2008: The spatiotemporal structure of twentieth-century climate variations in observations and reanalyses. Part I: Long-term trend. *J. Clim.*, **21**, 2611–2633.

Chiacchio, M., and M. Wild, 2010: Influence of NAO and clouds on long-term seasonal variations of surface solar radiation in Europe. *J. Geophys. Res. Atmos.*, **115**, D00d22.

Choi, G., et al., 2009: Changes in means and extreme events of temperature and precipitation in the Asia Pacific Network region, 1955–2007. *Int. J. Climatol.*, **29**, 1906–1925.

Christy, J. R., and W. B. Norris, 2006: Satellite and VIZ-radiosonde intercomparisons for diagnosis of nonclimatic influences. *J. Atmos. Ocean. Technol.*, **23**, 1181–1194.

Christy, J. R., and W. B. Norris, 2009: Discontinuity issues with radiosonde and satellite temperatures in the Australian Region 1979–2006. *J. Atmos. Ocean Technol.*, **26**, 508–522.

Christy, J. R., W. B. Norris, and R. T. McNider, 2009: Surface temperature variations in East Africa and possible causes. *J. Clim.*, **22**, 3342–3356.

Christy, J. R., R. W. Spencer, and W. B. Norris, 2011: The role of remote sensing in monitoring global bulk tropospheric temperatures. *Int. J. Remote Sens.*, **32**, 671–685.

Christy, J. R., W. B. Norris, K. Redmond, and K. P. Gallo, 2006: Methodology and results of calculating central california surface temperature trends: Evidence of human-induced climate change? *J. Clim.*, **19**, 548–563.

Christy, J. R., W. B. Norris, R. W. Spencer, and J. J. Hnilo, 2007: Tropospheric temperature change since 1979 from ,tropical radiosonde and satellite measurements. *J. Geophys. Res. Atmos.*, **112,** D06102.

Christy, J. R., R. W. Spencer, W. B. Norris, W. D. Braswell, and D. E. Parker, 2003: Error estimates of version 5.0 of MSU-AMSU bulk atmospheric temperatures. *J. Atmos. Ocean Technol.*, **20**, 613–629.

Christy, J. R., et al., 2010: What do observational datasets say about modeled tropospheric temperature trends since 1979? , 2148–2169.

Chu, W. P., M. P. McCormick, J. Lenoble, C. Brogniez, and P. Pruvost, 1989: SAGE II inversion algorithm. *J. Geophys. Res. Atmos.*, **94**, 8339–8351.

Chung, E. S., and B. J. Soden, 2010: Investigating the influence of carbon dioxide and the stratosphere on the long-term tropospheric temperature monitoring from HIRS. *J. Appl. Meteor. Climatol.*, **49**, 1927–1937.

Chung, E. S., D. Yeomans, and B. J. Soden, 2010: An assessment of climate feedback processes using satellite observations of clear-sky OLR. *Geophys. Res. Lett.*, **37**, L02702.

Clark, R. T., S. J. Brown, and J. M. Murphy, 2006: Modeling Northern Hemisphere summer heat extreme changes and their uncertainties using a physics ensemble of climate sensitivity experiments. *J. Clim.*, **19**, 4418–4435.

Clement, A. C., and B. Soden, 2005: The sensitivity of the tropical-mean radiation budget. *J. Clim.*, **18**, 3189–3203.

CMA, 2007: *Atlas of China Disastrous Weather and Climate*. Chinese Meteorological Administration. Beijing, China.

CMA, 2011: China Climate Change Bulletin. Chinese Meteorological Administration. Beijing, China.

Cohen, J., M. Barlow, and K. Saito, 2009: Decadal fluctuations in planetary wave forcing modulate global warming in late boreal winter. *J. Clim.*, **22**, 4418–4426.

Cohen, J. L., J. C. Furtado, M. Barlow, V. A. Alexeev, and J. E. Cherry, 2012: Asymmetric seasonal temperature trends. *Geophys. Res. Lett.*, **39**, L04705.

Cohn, T. A., and H. F. Lins, 2005: Nature's style: Naturally trendy. *Geophys. Res. Lett.*, **32**, L23402.

Coles, S., 2001: *An Introduction to Statistical Modeling of Extreme Values.* Springer Science+Business Media, New York, 208 pp.

Collaud Coen, M., et al., 2013: Aerosol decadal trends – Part 1: In-situ optical measurements at GAW and IMPROVE stations. *Atmos. Chem. Phys.*, **13**, 869–894.

Compo, G. P., et al., 2011: The twentieth century reanalysis project. *Q. J. Roy. Meteorol. Soc.*, **137**, 1–28.

Cooper, O. R., R. S. Gao, D. Tarasick, T. Leblanc, and C. Sweeney, 2012: Long-term ozone trends at rural ozone monitoring sites across the United States, 1990–2010. *J. Geophys. Res.*, **117**, D22307.

Cornes, R. C., and P. D. Jones, 2011: An examination of storm activity in the northeast Atlantic region over the 1851–2003 period using the EMULATE gridded MSLP data series. *J. Geophys. Res. Atmos.*, **116**, D16110.

Coumou, D., A. Robinson, and S. Rahmstorf, 2013: Global increase in record-breaking monthly-mean temperatures. *Clim. Change*, 118, 771-782.

Craigmile, P. F., and P. Guttorp, 2011: Space-time modelling of trends in temperature series. *J. Time Ser. Anal.*, **32**, 378–395.

Croci-Maspoli, M., C. Schwierz, and H. C. Davies, 2007a: A multifaceted climatology of atmospheric blocking and its recent linear trend. *J. Clim.*, **20**, 633–649.

Croci-Maspoli, M., C. Schwierz, and H. C. Davies, 2007b: Atmospheric blocking: Space-time links to the NAO and PNA. *Clim. Dyn.*, **29**, 713–725.

Cutforth, H. W., and D. Judiesch, 2007: Long-term changes to incoming solar energy on the Canadian Prairie. *Agr. Forest Meteor.*, **145**, 167–175.

Dai, A., 2006: Recent climatology, variability, and trends in global surface humidity. *J. Clim.*, **19**, 3589–3606.

Dai, A., 2011a: Characteristics and trends in various forms of the Palmer Drought Severity Index during 1900–2008. *J. Geophys. Res. Atmos.*, **116**, D12115.

Dai, A., 2012: The influence of the inter-decadal Pacific oscillation on US precipitation during 1923–2010. *Clim. Dyn.*, 41, 633-646.

Dai, A., 2013: Increasing drought under global warming in observations and models. *Nature Clim. Change*, **3**, 52–58.

Dai, A., T. T. Qian, K. E. Trenberth, and J. D. Milliman, 2009: Changes in continental freshwater diSchärge from 1948 to 2004. *J. Clim.*, **22**, 2773–2792.

Dai, A. G., 2011b: Drought under global warming: A review. *Clim. Change*, **2**, 45–65.

Dai, A. G., J. H. Wang, P. W. Thorne, D. E. Parker, L. Haimberger, and X. L. L. Wang, 2011: A new approach to homogenize daily radiosonde humidity data. *J. Clim.*, **24**, 965–991.

Das, L., J. D. Annan, J. C. Hargreaves, and S. Emori, 2011: Centennial scale warming over Japan: Are the rural stations really rural? *Atmos. Sci. Lett.*, **12**, 362-367.

Davidson, E., 2009: The contribution of manure and fertilizer nitrogen to atmospheric nitrous oxide since 1860. *Nature Geosci.*, **2**, 659–662.

Davini, P., C. Cagnazzo, S. Gualdi, and A. Navarra, 2012: Bidimensional diagnostics, variability, and trends of Northern Hemisphere blocking. *J. Clim.*, **25**, 6496–6509.

Davis, S. M., and K. H. Rosenlof, 2011: A multidiagnostic intercomparison of tropical width time series using reanalyses and satellite observations. *J. Clim.*, **25**, 1061-1078.

De Laat, A. T. J., and A. N. Maurellis, 2006: Evidence for influence of anthropogenic surface processes on lower tropospheric and surface temperature trends. *Int. J. Climatol.*, **26**, 897–913.

de Meij, A., A. Pozzer, and J. Lelieveld, 2012: Trend analysis in aerosol optical depths and pollutant emission estimates between 2000 and 2009. *Atmos. Environ.*, **51**, 75–85.

De Smedt, I., T. Stavrakou, J. F. Müller, R. J. van der A, and M. Van Roozendael, 2010: Trend detection in satellite observations of formaldehyde tropospheric columns. *Geophys. Res. Lett.*, **37**, L18808.

Dee, D. P., E. Kallen, A. J. Simmons, and L. Haimberger, 2011a: Comments on "Reanalyses suitable for characterizing long-term trends". *Bull. Am. Meteor. Soc.*, **92**, 65–70.

Dee, D. P., et al., 2011b: The ERA-Interim reanalysis: Configuration and performance of the data assimilation system. *Q. J. R. Meteor. Soc.*, **137**, 553–597.

Deeds, D., et al., 2008: Evidence for crustal degassing of CF4 and SF6 in Mojave Desert groundwaters. *Geochim. Cosmochim. Acta*, **72**, 999–1013.

DeGaetano, A. T., 2009: Time-dependent changes in extreme-precipitation return-period amounts in the continental United States. *J. Appl. Meteor. Climatol.*, **48**, 2086–2099.

Delgado, J. M., H. Apel, and B. Merz, 2010: Flood trends and variability in the Mekong River. *Hydrol. Earth Syst. Sci.*, **14**, 407–418.

Della-Marta, P. M., M. R. Haylock, J. Luterbacher, and H. Wanner, 2007a: Doubled length of western European summer heat waves since 1880. *J. Geophys. Res. Atmos.*, **112**, D15103.

Della-Marta, P. M., J. Luterbacher, H. von Weissenfluh, E. Xoplaki, M. Brunet, and H. Wanner, 2007b: Summer heat waves over western Europe 1880–2003, their relationship to large-scale forcings and predictability. *Clim. Dyn.*, **29**, 251–275.

Della-Marta, P. M., H. Mathis, C. Frei, M. A. Liniger, J. Kleinn, and C. Appenzeller, 2009: The return period of wind storms over Europe. *Int. J. Climatol.*, **29**, 437–459.

Deser, C., A. S. Phillips, and M. A. Alexander, 2010a: Twentieth century tropical sea surface temperature trends revisited. *Geophys. Res. Lett.*, **37**, L10701.

Deser, C., M. A. Alexander, S. P. Xie, and A. S. Phillips, 2010b: Sea surface temperature variability: Patterns and mechanisms. *Annu. Rev. Mar. Sci.*, **2**, 115–143.

Dessler, A. E., and S. M. Davis, 2010: Trends in tropospheric humidity from reanalysis systems. *J. Geophys. Res. Atmos.*, **115**, D19127.

Dessler, A. E., Z. Zhang, and P. Yang, 2008: Water-vapor climate feedback inferred from climate fluctuations, 2003–2008. *Geophys. Res. Lett.*, **35**, L20704.

Diffenbaugh, N. S., J. S. Pal, F. Giorgi, and X. J. Gao, 2007: Heat stress intensification in the Mediterranean climate change hotspot. *Geophys. Res. Lett.*, **34**, L11706.

Ding, T., W. H. Qian, and Z. W. Yan, 2010: Changes in hot days and heat waves in China during 1961–2007. *Int. J. Climatol.*, **30**, 1452–1462.

Ding, Y. H., G. Y. Ren, Z. C. Zhao, Y. Xu, Y. Luo, Q. P. Li, and J. Zhang, 2007: Detection, causes and projection of climate change over China: An overview of recent progress. *Adv. Atmos. Sci.*, **24**, 954–971.

Dlugokencky, E., E. Nisbet, R. Fisher, and D. Lowry, 2011: Global atmospheric methane: Budget, changes and dangers. *Philos. Trans. R. Soc. London Ser. A*, **369**, 2058–2072.

Dlugokencky, E., et al., 2005: Conversion of NOAA atmospheric dry air CH4 mole fractions to a gravimetrically prepared standard scale. *J. Geophys. Res. Atmos.*, **110**, D18306.

Dlugokencky, E., et al., 2009: Observational constraints on recent increases in the atmospheric CH4 burden. *Geophys. Res. Lett.*, L18803.

Dlugokencky, E. J., K. A. Masaire, P. M. Lang, P. P. Steele, and E. G. Nisbet, 1994: A dramatic decrease in the growth rate of atmospheric methane in the Northern Hemisphere during 1992. *Geophys. Res. Lett.*, **21**, 45–48.

Dole, R., et al., 2011: Was there a basis for anticipating the 2010 Russian heat wave? *Geophys. Res. Lett.*, **38**, L06702.

Donat, M. G., and L. V. Alexander, 2012: The shifting probability distribution of global daytime and night-time temperatures. *Geophys. Res. Lett.*, **39**, L14707.

Donat, M. G., D. Renggli, S. Wild, L. V. Alexander, G. C. Leckebusch, and U. Ulbrich, 2011: Reanalysis suggests long-term upward trends in European storminess since 1871. *Geophys. Res. Lett.*, **38**, L14703.

Donat, M. G., L. V. Alexander, H. Yang, I. Durre, R. Vose, and J. Caesar, 2013a: Global land-based datasets for monitoring climatic extremes. *Bull. Am. Meteor. Soc.*, **94**, 997-1006.

Donat, M. G., et al., 2013b: Changes in extreme temperature and precipitation in the Arab region: Long-term trends and variability related to ENSO and NAO. *Int. J. Climatol.*, doi:10.1002/joc.3707.

Donat, M. G., et al., 2013c: Updated analyses of temperature and precipitation extreme indices since the beginning of the twentieth century: The HadEX2 data-set. *J. Geophys. Res. Atmos.*, **118**, 2098-2118.

Dong, L., T. J. Vogelsang, and S. J. Colucci, 2008: Interdecadal trend and ENSO-related interannual variability in Southern Hemisphere blocking. *J. Clim.*, **21**, 3068–3077.

Dorigo, W., R. de Jeu, D. Chung, R. Parinussa, Y. Liu, W. Wagner, and D. Fernández-Prieto, 2012: Evaluating global trends (1988–2010) in harmonized multi-satellite surface soil moisture. *Geophys. Res. Lett.*, **39**, L18405.

Doswell, C., H. Brooks, and N. Dotzek, 2009: On the implementation of the enhanced Fujita scale in the USA. *Atmos. Res.*, **93**, 554–563.

Douglass, A., et al., 2008: Relationship of loss, mean age of air and the distribution of CFCs to stratospheric circulation and implications for atmospheric lifetimes. *J. Geophys. Res. Atmos.*, **113**, D14309.

Douglass, A., et al., 2011: WMO/UNEP scientific assessment of ozone depletion: 2010. In: *Stratospheric Ozone and Surface Ultraviolet Radiation*. World Meteorological Organisation, Geneva, Switzerland.

Du, Y., and S. Xie, 2008: Role of atmospheric adjustments in the tropical Indian Ocean warming during the 20th century in climate models. *Geophys. Res. Lett.*, **35,** L08712.

Du, Y., S. Xie, G. Huang, and K. Hu, 2009: Role of air-sea interaction in the long persistence of El Niño-induced North Indian Ocean warming. *J. Clim.*, **22**, 2023–2038.

Duan, A. M., and G. X. Wu, 2006: Change of cloud amount and the climate warming on the Tibetan Plateau. *Geophys. Res. Lett.*, **33**, L22704.

Durre, I., C. N. Williams, X. G. Yin, and R. S. Vose, 2009: Radiosonde-based trends in precipitable water over the Northern Hemisphere: An update. *J. Geophys. Res. Atmos.*, **114,** D05112.

Dutton, E. G., and B. A. Bodhaine, 2001: Solar irradiance anomalies caused by clear-sky transmission variations above Mauna Loa: 1958–99. *J. Clim.*, **14**, 3255–3262.

Dutton, E. G., D. W. Nelson, R. S. Stone, D. Longenecker, G. Carbaugh, J. M. Harris, and J. Wendell, 2006: Decadal variations in surface solar irradiance as observed in a globally remote network. *J. Geophys. Res. Atmos.*, **111**, D19101.

Earl, N., S. Dorling, R. Hewston, and R. von Glasow, 2013: 1980–2010 Variability in U.K. surface wind climate. *J. Climate*, **26**, 1172–1191.

Easterling, D., and M. Wehner, 2009: Is the climate warming or cooling? *Geophys. Res. Lett.*, **36**, L08706.

Eastman, R., and S. G. Warren, 2012: A 39-yr survey of cloud changes from land stations worldwide 1971–2009: Long-term trends, relation to aerosols, and expansion of the Tropical Belt. *J. Clim.*, **26**, 1286–1303.

Eastman, R., S. G. Warren, and C. J. Hahn, 2011: Variations in cloud cover and cloud types over the ocean from surface observations, 1954–2008. *J. Clim.*, **24**, 5914–5934.

Ebita, A., et al., 2011: The Japanese 55-year reanalysis "JRA-55": An interim report. *Sola*, **7**, 149–152.

Eccel, E., P. Cau, K. Riemann-Campe, and F. Biasioli, 2012: Quantitative hail monitoring in an alpine area: 35-year climatology and links with atmospheric variables. *Int. J. Climatol.*, **32**, 503–517.

Efthymiadis, D., C. M. Goodess, and P. D. Jones, 2011: Trends in Mediterranean gridded temperature extremes and large-scale circulation influences. *Nat. Hazards Earth Syst. Sci.*, **11**, 2199–2214.

Efthymiadis, D. A., and P. D. Jones, 2010: Assessment of maximum possible urbanization influences on land temperature data by comparison of land and marine data around coasts. *Atmosphere*, **1**, 51–61.

Elkins, J. W., and G. S. Dutton, 2011: Nitrous oxide and sulfur hexaflouride. *Bull. Am. Meteor. Soc.*, **92**, 2.

Elsner, J. B., J. P. Kossin, and T. H. Jagger, 2008: The increasing intensity of the strongest tropical cyclones. *Nature*, **455**, 92–95.

Emanuel, K., 2007: Environmental factors affecting tropical cyclone power dissipation. *J. Clim.*, **20**, 5497–5509.

Embury, O., and C. J. Merchant, 2011: Reprocessing for climate of sea surface temperature from the along-track scanning radiometers: A new retrieval scheme. *Remote Sens. Environ.*, **116**, 47-61.

Embury, O., C. J. Merchant, and G. K. Corlett, 2011: A reprocessing for climate of sea surface temperature from the along-track scanning radiometers: Preliminary validation, accounting for skin and diurnal variability. *Remote Sens. Environ.*, **116**, 62-78.

Endo, N., and T. Yasunari, 2006: Changes in low cloudiness over China between 1971 and 1996. *J. Clim.*, **19**, 1204–1213.

Enfield, D. B., A. M. Mestas-Nunez, and P. J. Trimble, 2001: The Atlantic multidecadal oscillation and its relation to rainfall and river flows in the continental US. *Geophys. Res. Lett.*, **28**, 2077–2080.

Engel, A., et al., 2009: Age of stratospheric air unchanged within uncertainties over the past 30 years. *Nature Geosci.*, **2**, 28–31.

Espinoza Villar, J. C., et al., 2009: Contrasting regional diSchärge evolutions in the Amazon basin (1974–2004). *J. Hydrol.*, **375**, 297–311.

Etheridge, D., L. Steele, R. Francey, and R. Langenfelds, 1998: Atmospheric methane between 1000 AD and present: Evidence of anthropogenic emissions and climatic variability. *J. Geophys. Res. Atmos.*, 15979–15993.

Etheridge, D. M., L. P. Steele, R. L. Langenfelds, R. J. Francey, J. M. Barnola, and V. I. Morgan, 1996: Natural and anthropogenic changes in atmospheric CO_2 over the last 1000 years from air in Antarctic ice and firn. *J. Geophys. Res. Atmos.*, 4115–4128.

Evan, A. T., A. K. Heidinger, and D. J. Vimont, 2007: Arguments against a physical long-term trend in global ISCCP cloud amounts. *Geophys. Res. Lett.*, **34**, L04701.

Fall, S., D. Niyogi, A. Gluhovsky, R. A. Pielke, E. Kalnay, and G. Rochon, 2010: Impacts of land use land cover on temperature trends over the continental United States: Assessment using the North American regional reanalysis. *Int. J. Climatol.*, **30**, 1980–1993.

Fall, S., A. Watts, J. Nielsen-Gammon, E. Jones, D. Niyogi, J. R. Christy, and R. A. Pielke, 2011: Analysis of the impacts of station exposure on the US Historical Climatology Network temperatures and temperature trends. *J. Geophys. Res. Atmos.*, **116,** D14120.

Falvey, M., and R. D. Garreaud, 2009: Regional cooling in a warming world: Recent temperature trends in the southeast Pacific and along the west coast of subtropical South America (1979–2006). *J. Geophys. Res. Atmos.*, **114,** D04102.

Favre, A., and A. Gershunov, 2006: Extra-tropical cyclonic/anticyclonic activity in north-eastern Pacific and air temperature extremes in western North America. *Clim. Dyn.*, **26**, 617–629.

Feng, S., and Q. Hu, 2007: Changes in winter snowfall/precipitation ratio in the contiguous United States. *J. Geophys. Res. Atmos.*, **112,** D15109.

Ferguson, C. R., and G. Villarini, 2012: Detecting inhomogeneities in the twentieth century reanalysis over the central United States. *J. Geophys. Res. Atmos.*, **117,** D05123.

Ferranti, L., and P. Viterbo, 2006: The European summer of 2003: Sensitivity to soil water initial conditions. *J. Clim.*, **19**, 3659–3680.

Ferretti, D., et al., 2005: Unexpected changes to the global methane budget over the past 2000 years. *Science*, **309**, 1714–1717.

Fischer, E. M., and C. Schär, 2010: Consistent geographical patterns of changes in high-impact European heatwaves. *Nature Geosci.*, **3**, 398–403.

Fischer, E. M., S. I. Seneviratne, P. L. Vidale, D. Luthi, and C. Schär, 2007: Soil moisture–atmosphere interactions during the 2003 European summer heat wave. *J. Clim.*, **20**, 5081–5099.

Fischer, T., M. Gemmer, L. Liu, and B. Su, 2011: Temperature and precipitation trends and dryness/wetness pattern in the Zhujiang River Basin, South China, 1961–2007. *Quatern. Int.*, **244**, 138–148.

Fogt, R. L., J. Perlwitz, A. J. Monaghan, D. H. Bromwich, J. M. Jones, and G. J. Marshall, 2009: Historical SAM variability. Part II: Twentieth-century variability and trends from reconstructions, observations, and the IPCC AR4 models. *J. Clim.*, **22**, 5346–5365.

Folland, C. K., and D. E. Parker, 1995: Correction of instrumental biases in historical sea-surface temperature data. *Q. J. R. Meteor. Soc.*, **121**, 319–367.

Folland, C. K., D. E. Parker, A. Colman, and W. R., 1999: Large scale modes of ocean surface temperature since the late nineteenth century. In: *Beyond El Niño: Decadal and Interdecadal Climate Variability* [A. Navarra (ed.)] Springer-Verlag, New York, pp. 73–102.

Forster, P., et al., 2007: Changes in atmospheric constituents and in radiative forcing. In: *Climate Change 2007: The Physical Science Basis. Contribution of Working Group I to the Fourth Assessment Report of the Intergovernmental Panel on Climate Change* [Solomon, S., D. Qin, M. Manning, Z. Chen, M. Marquis, K. B. Averyt, M. Tignor and H. L. Miller (eds.)] Cambridge University Press, Cambridge, United Kingdom and New York, NY, USA, 129-234.

Forster, P. M., et al., 2011: Stratospheric changes and climate. Scientific Assessment of Ozone Depletion: 2010. Global Ozone Research and Monitoring Project–Report No. 52. World Meteorological Organization, Geneva, Switzerland, 1-60.

Fortems-Cheiney, A., F. Chevallier, I. Pison, P. Bousquet, S. Szopa, M. N. Deeter, and C. Clerbaux, 2011: Ten years of CO emissions as seen from Measurements of Pollution in the Troposphere (MOPITT). *J. Geophys. Res.*, **116**, D05304.

Foster, G., and S. Rahmstorf, 2011: Global temperature evolution 1979–2010. *Environ. Res. Lett.*, **6**, 044022.

Frauenfeld, O. W., and R. E. Davis, 2003: Northern Hemisphere circumpolar vortex trends and climate change implications. *J. Geophys. Res. Atmos.*, **108,** 4423.

Frederiksen, J. S., and C. S. Frederiksen, 2007: Interdecadal changes in Southern Hemisphere winter storm track modes. *Tellus A*, **59**, 599–617.

Free, M., and D. J. Seidel, 2007: Comments on "biases in stratospheric and tropospheric temperature trends derived from historical radiosonde data". *J. Clim.*, **20**, 3704–3709.

Free, M., D. J. Seidel, J. K. Angell, J. Lanzante, I. Durre, and T. C. Peterson, 2005: Radiosonde Atmospheric Temperature Products for Assessing Climate (RATPAC): A new data set of large-area anomaly time series. *J. Geophys. Res. Atmos.*, **110**, D22101.

Frich, P., L. V. Alexander, P. Della-Marta, B. Gleason, M. Haylock, A. Tank, and T. Peterson, 2002: Observed coherent changes in climatic extremes during the second half of the twentieth century. *Clim. Res.*, **19**, 193–212.

Fu, G. B., S. P. Charles, and J. J. Yu, 2009: A critical overview of pan evaporation trends over the last 50 years. *Clim. Change*, **97**, 193–214.

Fu, Q., and P. Lin, 2011: Poleward shift of subtropical jets inferred from satellite-observed lower stratospheric temperatures. *J. Clim.*, **24**, 5597–5603.

Fu, Q., C. M. Johanson, S. G. Warren, and D. J. Seidel, 2004: Contribution of stratospheric cooling to satellite-inferred tropospheric temperature trends. *Nature*, **429**, 55–58.

Fu, Q., C. M. Johanson, J. M. Wallace, and T. Reichler, 2006: Enhanced mid-latitude tropospheric warming in satellite measurements. *Science*, **312**, 1179–1179.

Fueglistaler, S., and P. H. Haynes, 2005: Control of interannual and longer-term variability of stratospheric water vapor. *J. Geophys. Res. Atmos.*, **110**, D24108.

Fujibe, F., 2009: Detection of urban warming in recent temperature trends in Japan. *Int. J. Climatol.*, **29**, 1811–1822.

Fujiwara, M., et al., 2010: Seasonal to decadal variations of water vapor in the tropical lower stratosphere observed with balloon-borne cryogenic frost point hygrometers. *J. Geophys. Res. Atmos.*, **115**, D18304.

Fyfe, J. C., 2003: Extratropical southern hemisphere cyclones: Harbingers of climate change? *J. Clim.*, **16**, 2802–2805.

Gallant, A., K. Hennessy, and J. Risbey, 2007: Trends in rainfall indices for six Australian regions: 1910–2005. *Aust. Meteor. Mag.*, **56**, 223–239.

Gallant, A. J. E., and D. J. Karoly, 2010: A Combined Climate Extremes Index for the Australian Region. *J. Clim.*, **23**, 6153–6165.

Garcia-Herrera, R., J. Diaz, R. M. Trigo, J. Luterbacher, and E. M. Fischer, 2010: A review of the European summer heat wave of 2003. *Crit. Rev. Environ. Sci. Technol.*, **40**, 267–306.

Gentemann, C., F. Wentz, C. Mears, and D. Smith, 2004: In situ validation of Tropical Rainfall Measuring Mission microwave sea surface temperatures. *J. Geophys. Res. Oceans*, **109**, C04021.

Gettelman, A., and Q. Fu, 2008: Observed and simulated upper-tropospheric water vapor feedback. *J. Clim.*, **21**, 3282–3289.

Gettelman, A., et al., 2010: Multimodel assessment of the upper troposphere and lower stratosphere: Tropics and global trends. *J. Geophys. Res. Atmos.*, **115**, D00M08.

Gilgen, H., A. Roesch, M. Wild, and A. Ohmura, 2009: Decadal changes in shortwave irradiance at the surface in the period from 1960 to 2000 estimated from Global Energy Balance Archive Data. *J. Geophys. Res. Atmos.*, **114**, D00d08.

Gillett, N. P., and P. A. Stott, 2009: Attribution of anthropogenic influence on seasonal sea level pressure. *Geophys. Res. Lett.*, **36**, L23709.

Giorgi, F., and R. Francisco, 2000: Evaluating uncertainties in the prediction of regional climate change. *Geophys. Res. Lett.*, **27**, 1295–1298.

Giorgi, F., E. S. Im, E. Coppola, N. S. Diffenbaugh, X. J. Gao, L. Mariotti, and Y. Shi, 2011: Higher hydroclimatic intensity with global warming. *J. Clim.*, **24**, 5309–5324.

Giuntoli, I., B. Renard, J. P. Vidal, and A. Bard, 2013: Low flows in France and their relationship to large-scale climate indices. *J. Hydrol.*, **482**, 105-118.

Gleason, K. L., J. H. Lawrimore, D. H. Levinson, T. R. Karl, and D. J. Karoly, 2008: A revised US Climate Extremes Index. *J. Clim.*, **21**, 2124–2137.

Gong, D. Y., and C. H. Ho, 2002: The Siberian High and climate change over middle to high latitude Asia. *Theor. Appl. Climatol.*, **72**, 1–9.

Gouretski, V., J. Kennedy, T. Boyer, and A. Kohl, 2012: Consistent near-surface ocean warming since 1900 in two largely independent observing networks. *Geophys. Res. Lett.*, **39**, L19606.

Granier, C., et al., 2011: Evolution of anthropogenic and biomass burning emissions of air pollutants at global and regional scales during the 1980–2010 period. *Clim. Change*, **109**, 163–190.

Grant, A. N., S. Brönnimann, and L. Haimberger, 2008: Recent Arctic warming vertical structure contested. *Nature*, **455**, E2–E3.

Graversen, R. G., T. Mauritsen, M. Tjernstrom, E. Kallen, and G. Svensson, 2008: Vertical structure of recent Arctic warming. *Nature*, **451**, 53–U54.

Greally, B., et al., 2007: Observations of 1,1-difluoroethane (HFC-152a) at AGAGE and SOGE monitoring stations in 1994–2004 and derived global and regional emission estimates. *J. Geophys. Res. Atmos.*, **112**, D06308.

Griffiths, G. M., et al., 2005: Change in mean temperature as a predictor of extreme temperature change in the Asia-Pacific region. *Int. J. Climatol.*, **25**, 1301–1330.

Grinsted, A., J. C. Moore, and S. Jevrejeva, 2012: Homogeneous record of Atlantic hurricane surge threat since 1923. *Proc. Natl. Acad. Sci. U.S.A.* **109**, 19601-19605.

Groisman, P., R. Knight, and T. Karl, 2012: Changes in intense precipitation over the central United States. *J. Hydrometeor.*, **13**, 47–66.

Groisman, P., R. Knight, T. R. Karl, D. Easterling, B. M. Sun, and J. Lawrimore, 2004: Contemporary changes of the hydrological cycle over the contiguous United States: Trends derived from in situ observations. *J. Hydrometeor.*, **5**, 64–85.

Groisman, P. Y., R. W. Knight, D. R. Easterling, T. R. Karl, G. C. Hegerl, and V. A. N. Razuvaev, 2005: Trends in intense precipitation in the climate record. *J. Clim.*, **18**, 1326–1350.

Gruber, C., and L. Haimberger, 2008: On the homogeneity of radiosonde wind time series. *Meteorol. Z.*, **17**, 631–643.

Gulev, S. K., O. Zolina, and S. Grigoriev, 2001: Extratropical cyclone variability in the Northern Hemisphere winter from the NCEP/NCAR reanalysis data. *Clim. Dyn.*, **17**, 795–809.

Guo, H., M. Xu, and Q. Hub, 2010: Changes in near-surface wind speed in China: 1969–2005. *Int. J. Climatol.*, **31**, 349-358.

Haerter, J., P. Berg, and S. Hagemann, 2010: Heavy rain intensity distributions on varying time scales and at different temperatures. *J. Geophys. Res. Atmos.*, **115**, D17102.

Haimberger, L., 2007: Homogenization of radiosonde temperature time series using innovation statistics. *J. Clim.*, **20**, 1377–1403.

Haimberger, L., C. Tavolato, and S. Sperka, 2008: Toward elimination of the warm bias in historic radiosonde temperature records—Some new results from a comprehensive intercomparison of upper-air data. *J. Clim.*, **21**, 4587–4606.

Haimberger, L., C. Tavolato, and S. Sperka, 2012: Homogenization of the global radiosonde temperature dataset through combined comparison with reanalysis background series and neighboring stations. *J. Clim.*, **25**, 8108–8131.

Hand, J. L., et al., 2011: IMPROVE, spatial and seasonal patterns and temporal variability of haze and its constituents in the United States. Cooperative Institute for Research in the Atmosphere and Colorado University.

Hanna, E., J. Cappelen, R. Allan, T. Jonsson, F. Le Blancq, T. Lillington, and K. Hickey, 2008: New insights into North European and North Atlantic surface pressure variability, storminess, and related climatic change since 1830. *J. Clim.*, **21**, 6739–6766.

Hannaford, J., and T. Marsh, 2008: High-flow and flood trends in a network of undisturbed catchments in the UK. *Int. J. Climatol.*, **28**, 1325–1338.

Hansen, J., M. Sato, and R. Ruedy, 2012: Perception of climate change. *Proc. Natl. Acad. Sci. U.S.A.*, **109**, E2415–E2423.

Hansen, J., R. Ruedy, M. Sato, and K. Lo, 2010: Global surface temperature change. *Rev. Geophys.*, **48**, RG4004.

Hansen, J., M. Sato, P. Kharecha, and K. von Schuckmann, 2011: Earth's energy imbalance and implications. *Atmos. Chem. Phys.*, **11**, 13421–13449.

Harries, J. E., and C. Belotti, 2010: On the variability of the global net radiative energy balance of the nonequilibrium Earth. *J. Clim.*, **23**, 1277–1290.

Hatzianastassiou, N., C. Matsoukas, A. Fotiadi, K. G. Pavlakis, E. Drakakis, D. Hatzidimitriou, and I. Vardavas, 2005: Global distribution of Earth's surface shortwave radiation budget. *Atmos. Chem. Phys.*, **5**, 2847–2867.

Hatzianastassiou, N., C. D. Papadimas, C. Matsoukas, K. Pavlakis, A. Fotiadi, M. Wild, and I. Vardavas, 2012: Recent regional surface solar radiation dimming and brightening patterns: inter-hemispherical asymmetry and a dimming in the Southern Hemisphere. *Atmos. Sci. Lett.*, **13**, 43–48.

Hausfather, Z., M. J. Menne, C. N. Williams, T. Masters, R. Broberg, and D. Jones, 2013: Quantifying the effect of urbanization on U.S. Historical Climatology Network temperature records. *J. Geophys. Res. Atmos.*, **118**, 481-494.

Haylock, M. R., et al., 2006: Trends in total and extreme South American rainfall in 1960–2000 and links with sea surface temperature. *J. Clim.*, **19**, 1490–1512.

He, W. Y., S. P. Ho, H. B. Chen, X. J. Zhou, D. Hunt, and Y. H. Kuo, 2009: Assessment of radiosonde temperature measurements in the upper troposphere and lower stratosphere using COSMIC radio occultation data. *Geophys. Res. Lett.*, **36**, L17807.

Heidinger, A. K., and M. J. Pavolonis, 2009: Gazing at cirrus clouds for 25 years through a split window. Part I: Methodology. *J. Appl. Meteor. Climatol.*, **48**, 1100–1116.

Held, I. M., and B. J. Soden, 2006: Robust responses of the hydrological cycle to global warming. *J. Clim.*, **19**, 5686–5699.

Helmig, D., et al., 2007: A review of surface ozone in the polar regions. *Atmos. Environ.*, **41**, 5138–5161.

Hidy, G. M., and G. T. Pennell, 2010: Multipollutant air quality management: 2010 critical review. *J. Air Waste Manage. Assoc.*, **60**, 645–674.

Hilboll, A., A. Richter, and J. P. Burrows, 2013: Long-term changes of tropospheric NO_2 over megacities derived from multiple satellite instruments. *Atmos. Chem. Phys.*, **13**, 4145-4169.

Hinkelman, L. M., P. W. Stackhouse, B. A. Wielicki, T. P. Zhang, and S. R. Wilson, 2009: Surface insolation trends from satellite and ground measurements: Comparisons and challenges. *J. Geophys. Res. Atmos.*, **114**, D00d20.

Hirdman, D., et al., 2010: Long-term trends of black carbon and sulphate aerosol in the Arctic: Changes in atmospheric transport and source region emissions. *Atmos. Chem. Phys.*, **10**, 9351–9368.

Hirsch, M. E., A. T. DeGaetano, and S. J. Colucci, 2001: An East Coast winter storm climatology. *J. Clim.*, **14**, 882–899.

Hirschi, M., et al., 2011: Observational evidence for soil-moisture impact on hot extremes in southeastern Europe. *Nature Geosci.*, **4**, 17–21.

Ho, S. P., W. He, and Y. H. Kuo, 2009a: Construction of consistent temperature records in the lower stratosphere using Global Positioning System Radio Occultation Data and Microwave Sounding measurements. New Horizons in Occultation Research, Springer-Verlag Berlin, 207–217.

Ho, S. P., Y. H. Kuo, Z. Zeng, and T. C. Peterson, 2007: A comparison of lower stratosphere temperature from microwave measurements with CHAMP GPS RO data. *Geophys. Res. Lett.*, **34**, L15701.

Ho, S. P., M. Goldberg, Y. H. Kuo, C. Z. Zou, and W. Schreiner, 2009b: Calibration of temperature in the lower stratosphere from microwave measurements using COSMIC radio occultation data: Preliminary results. *Terr. Atmos. Ocean. Sci.*, **20**, 87–100.

Ho, S. P., et al., 2012: Reproducibility of GPS radio occultation data for climate monitoring: Profile-to-profile inter-comparison of CHAMP climate records 2002 to 2008 from six data centers. *J. Geophys. Res. Atmos.*, **117**, D18111.

Hoerling, M., et al., 2012: Anatomy of an extreme event. *J. Clim.*, **26**, 2811–2832.

Holben, B. N., et al., 1998: AERONET—A federated instrument network and data archive for aerosol characterization. *Remote Sens. Environ.*, **66**, 1–16.

Holland, G. J., and P. J. Webster, 2007: Heightened tropical cyclone activity in the North Atlantic: Natural variability or climate trend? *Philos. Trans. R. Soc. London Ser. A*, **365**, 2695–2716.

Hope, P. K., W. Drosdowsky, and N. Nicholls, 2006: Shifts in the synoptic systems influencing southwest Western Australia. *Clim. Dyn.*, **26**, 751–764.

Hsu, N. C., et al., 2012: Global and regional trends of aerosol optical depth over land and ocean using SeaWiFS measurements from 1997 to 2010. *Atmos. Chem. Phys. Discuss.*, **12**, 8465–8501.

Hsu, P. C., T. Li, and B. Wang, 2011: Trends in global monsoon area and precipitation over the past 30 years. *Geophys. Res. Lett.*, **38**, L08701.

Hu, Y., and Q. Fu, 2007: Observed poleward expansion of the Hadley circulation since 1979. *Atmos. Chem. Phys.*, **7**, 5229–5236.

Hu, Y. C., W. J. Dong, and Y. He, 2010: Impact of land surface forcings on mean and extreme temperature in eastern China. *J. Geophys. Res. Atmos.*, **115**, 11.

Hu, Y. Y., C. Zhou, and J. P. Liu, 2011: Observational evidence for the poleward expansion of the Hadley circulation. *Adv. Atmos. Sci.*, **28**, 33–44.

Huang, J., et al., 2008: Estimation of regional emissions of nitrous oxide from 1997 to 2005 using multinetwork measurements, a chemical transport model, and an inverse method. *J. Geophys. Res. Atmos.*, **113**, D17313.

Huang, W.-R., S.-Y. Wang, and J. C. L. Chan, 2010: Discrepancies between global reanalyses and observations in the interdecadal variations of Southeast Asian cold surge. *Int. J. Climatol.*, **31**, 2272-2280..

Hudson, R. D., 2012: Measurements of the movement of the jet streams at mid-latitudes, in the Northern and Southern Hemispheres, 1979 to 2010. *Atmos. Chem. Phys.*, **12**, 7797–7808.

Hudson, R. D., M. F. Andrade, M. B. Follette, and A. D. Frolov, 2006: The total ozone field separated into meteorological regimes—Part II: Northern Hemisphere mid-latitude total ozone trends. *Atmos. Chem. Phys.*, **6**, 5183–5191.

Hundecha, Y., A. St-Hilaire, T. Ouarda, S. El Adlouni, and P. Gachon, 2008: A nonstationary extreme value analysis for the assessment of changes in extreme annual wind speed over the Gulf of St. Lawrence, Canada. *J. Appl. Meteor. Climatol.*, **47**, 2745–2759.

Hurrell, J. W., 1995: Decadal trends in the North Atlantic Oscillation: Regional temperatures and precipitation. *Science*, **269**, 676–679.

Hurst, D., 2011: Stratospheric water vapor trends over Boulder, Colorado: Analysis of the 30 year Boulder record. *J. Geophys. Res.*, **116**, D02306.

Idso, S. B., and A. J. Brazel, 1984: Rising atmospheric carbon-dioxide concentrations may increase streamflow. *Nature*, **312**, 51–53.

IPCC, 2007: *Clim. Change 2007: The Physical Science Basis. Contribution of Working Group I to the Fourth Assessment Report of the Intergovernmental Panel on Climate Change (IPCC)* [Solomon, S., D. Qin, M. Manning, Z. Chen, M. Marquis, K. B. Averyt, M. Tignor and H. L. Miller (eds.)]. Cambridge University Press, Cambridge, United Kingdom and New York, NY, USA, 996 pp.

Ishii, M., A. Shouji, S. Sugimoto, and T. Matsumoto, 2005: Objective analyses of sea-surface temperature and marine meteorological variables for the 20th century using icoads and the Kobe collection. *Int. J. Climatol.*, **25**, 865–879.

Ishijima, K., et al., 2007: Temporal variations of the atmospheric nitrous oxide concentration and its delta N-15 and delta O-18 for the latter half of the 20th century reconstructed from firn air analyses. *J. Geophys. Res. Atmos.*, **112**, D03305 .

Jain, S. K., and V. Kumar, 2012: Trend analysis of rainfall and temperature data for India. *Curr. Sci.*, **102**, 37–49.

Jakob, D., D. Karoly, and A. Seed, 2011: Non-stationarity in daily and sub-daily intense rainfall—Part 2: Regional assessment for sites in south-east Australia. *Nat. Hazards Earth Syst. Sci.*, **11**, 2273–2284.

Jhajharia, D., S. Shrivastava, D. Sarkar, and S. Sarkar, 2009: Temporal characteristics of pan evaporation trends under humid conditions of northeast India. *Agr. Forest Meteorol.*, **336**, 61–73.

Jiang, T., Z. W. Kundzewicz, and B. Su, 2008: Changes in monthly precipitation and flood hazard in the Yangtze River Basin, China. *Int. J. Climatol.*, **28**, 1471–1481.

Jiang, X., W. Ku, R. Shia, Q. Li, J. Elkins, R. Prinn, and Y. Yung, 2007: Seasonal cycle of N_2O: Analysis of data. *Global Biogeochem. Cycles*, **21**, GB1006.

Jiang, Y., Y. Luo, Z. C. Zhao, and S. W. Tao, 2010: Changes in wind speed over China during 1956–2004. *Theor. Appl. Climatol.*, **99**, 421–430.

Jin, S. G., J. U. Park, J. H. Cho, and P. H. Park, 2007: Seasonal variability of GPS-derived zenith tropospheric delay (1994–2006) and climate implications. *J. Geophys. Res. Atmos.*, **112**, D09110.

John, V. O., R. P. Allan, and B. J. Soden, 2009: How robust are observed and simulated precipitation responses to tropical ocean warming? *Geophys. Res. Lett.*, **36**, L14702.

John, V. O., G. Holl, R. P. Allan, S. A. Buehler, D. E. Parker, and B. J. Soden, 2011: Clear-sky biases in satellite infrared estimates of upper tropospheric humidity and its trends. *J. Geophys. Res. Atmos.*, **116**, D14108.

Jones, D. A., W. Wang, and R. Fawcett, 2009: High-quality spatial climate data-sets for Australia. *Australian, Meteor. Ocean. J.*, **58**, 233–248.

Jones, P. D., and D. H. Lister, 2007: Intercomparison of four different Southern Hemisphere sea level pressure datasets. *Geophys. Res. Lett.*, **34**, L10704.

Jones, P. D., and D. H. Lister, 2009: The urban heat island in Central London and urban-related warming trends in Central London since 1900. *Weather*, **64**, 323–327.

Jones, P. D., T. Jonsson, and D. Wheeler, 1997: Extension to the North Atlantic Oscillation using early instrumental pressure observations from Gibraltar and southwest Iceland. *Int. J. Climatol.*, **17**, 1433–1450.

Jones, P. D., D. H. Lister, and Q. Li, 2008: Urbanization effects in large-scale temperature records, with an emphasis on China. *J. Geophys. Res. Atmos.*, **113**, D16122.

Jones, P. D., D. H. Lister, T. J. Osborn, C. Harpham, M. Salmon, and C. P. Morice, 2012: Hemispheric and large-scale land-surface air temperature variations: An extensive revision and an update to 2010. *J. Geophys. Res. Atmos.*, **117**, D05127.

Jones, R., S. Westra, and A. Sharma, 2010: Observed relationships between extreme sub-daily precipitation, surface temperature, and relative humidity. *Geophys. Res. Lett.*, **37**, L22805.

Joshi, M. M., J. M. Gregory, M. J. Webb, D. M. H. Sexton, and T. C. Johns, 2008: Mechanisms for the land/sea warming contrast exhibited by simulations of climate change. *Clim. Dyn.*, **30**, 455–465.

Jovanovic, B., D. Collins, K. Braganza, D. Jakob, and D. A. Jones, 2011: A high-quality monthly total cloud amount dataset for Australia. *Clim. Change*, **108**, 485-517.

Jung, M., et al., 2010: Recent decline in the global land evapotranspiration trend due to limited moisture supply. *Nature*, **467**, 951–954.

Kahn, R. A., et al., 2007: Satellite-derived aerosol optical depth over dark water from MISR and MODIS: Comparisons with AERONET and implications for climatological studies. *J. Geophys. Res. Atmos.*, **112**, D18205.

Kanamitsu, M., W. Ebisuzaki, J. Woollen, S. K. Yang, J. J. Hnilo, M. Fiorino, and G. L. Potter, 2002: NCEP-DOE AMIP-II reanalysis (R-2). *Bull. Am. Meteor. Soc.*, **83**, 1631–1643.

Kang, S. M., L. M. Polvani, J. C. Fyfe, and M. Sigmond, 2011: Impact of polar ozone depletion on subtropical Precipitation. *Science*, **332**, 951–954.

Kao, H. Y., and J. Y. Yu, 2009: Contrasting Eastern-Pacific and Central-Pacific types of ENSO. *J. Clim.*, **22**, 615–632.

Karnauskas, K. B., R. Seager, A. Kaplan, Y. Kushnir, and M. A. Cane, 2009: Observed strengthening of the zonal sea surface temperature gradient across the equatorial Pacific Ocean. *J. Clim.*, **22**, 4316–4321.

Karnieli, A., et al., 2009: Temporal trend in anthropogenic sulfur aerosol transport from central and eastern Europe to Israel. *J. Geophys. Res. Atmos.*, **114**, D00d19.

Karoly, D., 1989: Southern-Hemisphere circulation features associated with El Niño-Southern Oscillation. *J. Clim.*, **2**, 1239–1252.

Kato, S., et al.: Surface irradiances consistent with CERES-derived top-of-atmosphere shortwave and longwave irradiances. *J. Clim*. **26**, 2719-2740

Keeling, C., R. Bacastow, A. Bainbridge, C. Ekdahl, P. Guenther, L. Waterman, and J. Chin, 1976a: Atmospheric Carbon-Dioxide Variations at Mauna-Loa Observatory, Hawaii. *Tellus*, **28**, 538–551.

Keeling, C. D., J. A. Adams, and C. A. Ekdahl, 1976b: Atmospheric carbo-dioxide variations at South Pole. *Tellus*, **28**, 553–564.

Keller, C., D. Brunner, S. Henne, M. Vollmer, S. O'Doherty, and S. Reimann, 2011: Evidence for under-reported western European emissions of the potent greenhouse gas HFC-23. *Geophys. Res. Lett.*, **38**, L15808.

Kennedy, J. J., N. A. Rayner, and R. O. Smith, 2012: Using AATSR data to assess the quality of in situ sea surface temperature observations for climate studies. *Remote Sens. Environ.*, **116**, 79–92.

Kennedy, J. J., N. A. Rayner, R. O. Smith, D. E. Parker, and M. Saunby, 2011a: Reassessing biases and other uncertainties in sea surface temperature observations measured in situ since 1850: 2. Biases and homogenization. *J. Geophys. Res. Atmos.*, **116**, D14104.

Kennedy, J. J., N. A. Rayner, R. O. Smith, M. Saunby, and D. E. Parker, 2011b: Reassessing biases and other uncertainties in sea surface temperature observations since 1850, part 1: Measurement and sampling uncertainties. *J. Geophys. Res.*, **116**, D14103.

Kent, E. C., and D. I. Berry, 2008: Assessment of the Marine Observing System (ASMOS): Final report. *National Oceanography Centre Southampton Research and Consultancy Report*, 55 pp.

Kent, E. C., S. D. Woodruff, and D. I. Berry, 2007: Metadata from WMO publication no. 47 and an assessment of voluntary observing ship observation heights in ICOADS. *J. Atmos. Ocean Technol.*, **24**, 214–234.

Kent, E. C., S. Fangohr, and D. I. Berry, 2012: A comparative assessment of monthly mean wind speed products over the global ocean. *Int. J. Climatol.*, 33, 2530-2541.

Kent, E. C., J. J. Kennedy, D. I. Berry, and R. O. Smith, 2010: Effects of instrumentation changes on sea surface temperature measured in situ. *Clim. Change*, **1**, 718–728.

Kent, E. C., N. A. Rayner, D. I. Berry, M. Saunby, B. I. Moat, J. J. Kennedy, and D. E. Parker, 2013: Global analysis of night marine air temperature and its uncertainty since 1880, the HadNMAT2 Dataset, *J. Geophys. Res.*, **118**, 1281-1298.

Kenyon, J., and G. C. Hegerl, 2008: Influence of modes of climate variability on global temperature extremes. *J. Clim.*, **21**, 3872–3889.

Kenyon, J., and G. C. Hegerl, 2010: Influence of modes of climate variability on global precipitation extremes. *J. Clim.*, **23**, 6248–6262.

Kharin, V., F. Zwiers, X. Zhang, and M. Wehner, 2013: Changes in temperature and precipitation extremes in the CMIP5 ensemble. *Climatic Change*, **119**, 345-357.

Kiehl, J. T., and K. E. Trenberth, 1997: Earth's annual global mean energy budget. *Bull. Am. Meteor. Soc.*, **78**, 197–208.

Kim, D., and V. Ramanathan, 2012: Improved estimates and understanding of global albedo and atmospheric solar absorption. *Geophys. Res. Lett.*, **39**, L24704.

Kim, D. Y., and V. Ramanathan, 2008: Solar radiation budget and radiative forcing due to aerosols and clouds. *J. Geophys. Res. Atmos.*, **113**, D02203.

Kim, J., et al., 2010: Regional atmospheric emissions determined from measurements at Jeju Island, Korea: Halogenated compounds from China. *Geophys. Res. Lett.*, **37**, L12801.

King, A., L. Alexander, and M. Donat, 2013: The efficacy of using gridded data to examine extreme rainfall characteristics: A case study for Australia. *Inter. J. Climatol.*, **33**, 2376-2387.

Kistler, R., et al., 2001: The NCEP-NCAR 50-year reanalysis: Monthly means CD-ROM and documentation. *Bull. Am. Meteor. Soc.*, **82**, 247–267.

Klein Tank, A. M. G., et al., 2006: Changes in daily temperature and precipitation extremes in central and south Asia. *J. Geophys. Res. Atmos.*, **111**, D16105.

Klok, E. J., and A. Tank, 2009: Updated and extended European dataset of daily climate observations. *Int. J. Climatol.*, **29**, 1182–1191.

Knapp, K. R., and M. C. Kruk, 2010: Quantifying interagency differences in tropical cyclone best-track wind speed estimates. *Mon. Weather Rev.*, **138**, 1459–1473.

Knowles, N., M. D. Dettinger, and D. R. Cayan, 2006: Trends in snowfall versus rainfall in the western United States. *J. Clim.*, **19**, 4545–4559.

Knutson, T. R., et al., 2010: Tropical cyclones and climate change. *Nature Geosci.*, **3**, 157–163.

Kobayashi, S., M. Matricardi, D. Dee, and S. Uppala, 2009: Toward a consistent reanalysis of the upper stratosphere based on radiance measurements from SSU and AMSU-A. *Q. J. R. Meteorol. Soc.*, **135**, 2086–2099.

Kopp, G., and G. Lawrence, 2005: The Total Irradiance Monitor (TIM): Instrument design. *Solar Phys.*, **230**, 91–109.

Kopp, G., and J. L. Lean, 2011: A new, lower value of total solar irradiance: Evidence and climate significance. *Geophys. Res. Lett.*, **38**, L01706.

Kopp, G., G. Lawrence, and G. Rottman, 2005: The Total Irradiance Monitor (TIM): Science results. *Solar Phys.*, **230**, 129–139.

Kossin, J. P., K. R. Knapp, D. J. Vimont, R. J. Murnane, and B. A. Harper, 2007: A globally consistent reanalysis of hurricane variability and trends. *Geophys. Res. Lett.*, **34**, L04815.

Koutsoyiannis, D., and A. Montanari, 2007: Statistical analysis of hydroclimatic time series: Uncertainty and insights. *Water Resour. Res.*, **43**, W05429.

Kreienkamp, F., A. Spekat, and W. Enke, 2010: Stationarity of atmospheric waves and blocking over Europe-based on a reanalysis dataset and two climate scenarios. *Theor. Appl. Climatol.*, **102**, 205–212.

Krishna Moorthy, K., S. Suresh Babu, and S. K. Satheesh, 2007: Temporal heterogeneity in aerosol characteristics and the resulting radiative impact at a tropical coastal station—Part 1: Microphysical and optical properties. *Ann. Geophys.*, **25**, 2293–2308.

Krishna Moorthy, K., S. Suresh Babu, M. R. Manoj, and S. K. Satheesh, 2013: Buildup of Aerosols over the Indian Region. *Geophys. Res. Lett.*, **40**, 1011-1014

Krishna Moorthy, K., S. S. Babu, S. K. Satheesh, S. Lal, M. M. Sarin, and S. Ramachandran, 2009: Climate implications of atmospheric aerosols and trace gases: Indian Scenario, Climate Sense. World Meteorological Organisation, Geneva, Switzerland, pp. 157–160.

Krueger, O., F. Schenk, F. Feser, and R. Weisse, 2013: Inconsistencies between long-term trends in storminess derived from the 20CR reanalysis and observations. *J. Clim.*, **26**, 868–874.

Kruger, A., and S. Sekele, 2013: Trends in extreme temperature indices in South Africa: 1962–2009. *Int. J. Climatol.*, **33**, 661-676.

Kubota, H., and J. C. L. Chan, 2009: Interdecadal variability of tropical cyclone landfall in the Philippines from 1902 to 2005. *Geophys. Res. Lett.*, **36**, L12802.

Kudo, R., A. Uchiyama, A. Yamazaki, T. Sakami, and O. Ijima, 2011: Decadal changes in aerosol optical thickness and single scattering albedo estimated from ground-based broadband radiometers: A case study in Japan. *J. Geophys. Res.*, **116**, D03207.

Kudo, R., A. Uchiyama, O. Ijima, N. Ohkawara, and S. Ohta, 2012: Aerosol impact on the brightening in Japan. *J. Geophys. Res. Atmos.*, **117**, 11.

Kueppers, L. M., M. A. Snyder, and L. C. Sloan, 2007: Irrigation cooling effect: Regional climate forcing by land-use change. *Geophys. Res. Lett.*, **34**, L03703.

Kuglitsch, F. G., A. Toreti, E. Xoplaki, P. M. Della-Marta, J. Luterbacher, and H. Wanner, 2009: Homogenization of daily maximum temperature series in the Mediterranean. *J. Geophys. Res. Atmos.*, **114**, D15108.

Kuglitsch, F. G., A. Toreti, E. Xoplaki, P. M. Della-Marta, C. S. Zerefos, M. Turkes, and J. Luterbacher, 2010: Heat wave changes in the eastern Mediterranean since 1960. *Geophys. Res. Lett.*, **37**, L04802.

Kumari, B. P., and B. N. Goswami, 2010: Seminal role of clouds on solar dimming over the Indian monsoon region. *Geophys. Res. Lett.*, **37**, L06703.

Kumari, B. P., A. L. Londhe, S. Daniel, and D. B. Jadhav, 2007: Observational evidence of solar dimming: Offsetting surface warming over India. *Geophys. Res. Lett.*, **34**, L21810.

Kundzewicz, Z. W., et al., 2007: Freshwater resources and their management. *Climate Change 2007: Impacts, Adaptation and Vulnerability. Contribution of Working Group II to the Fourth Assessment Report of the Intergovernmental Panel on Climate Change* [Solomon, S., D. Qin, M. Manning, Z. Chen, M. Marquis, K. B. Averyt, M. Tignor and H. L. Miller (eds.)].Cambridge University Press, Cambridge, United Kingdom and New York, NY, USA,172-210.

Kunkel, K. E., M. A. Palecki, L. Ensor, D. Easterling, K. G. Hubbard, D. Robinson, and K. Redmond, 2009: Trends in twentieth-century US extreme snowfall seasons. *J. Clim.*, **22**, 6204–6216.

Kunkel, K. E., et al., 2008: Observed changes in weather and climate extremes. In: *Weather and Climate Extremes in a Changing Climate. Regions of Focus: North America, Hawaii, Caribbean, and U.S. Pacific Islands* [T. R. Karl, G. A. Meehl, D. M. Christopher, S. J. Hassol, A. M. Waple, and W. L. Murray (eds.)]. A Report by the U.S. Climate Change Science Program and the Subcommittee on Global Change Research.

Kunz, M., J. Sander, and C. Kottmeier, 2009: Recent trends of thunderstorm and hailstorm frequency and their relation to atmospheric characteristics in southwest Germany. *Int. J. Climatol.*, **29**, 2283–2297.

2

Kuo, Y. H., W. S. Schreiner, J. Wang, D. L. Rossiter, and Y. Zhang, 2005: Comparison of GPS radio occultation soundings with radiosondes. *Geophys. Res. Lett.*, **32**, L05817.

Kvalevag, M. M., and G. Myhre, 2007: Human impact on direct and diffuse solar radiation during the industrial era. *J. Clim.*, **20**, 4874–4883.

L'Ecuyer, T. S., N. B. Wood, T. Haladay, G. L. Stephens, and P. W. Stackhouse, 2008: Impact of clouds on atmospheric heating based on the R04 CloudSat fluxes and heating rates data set. *J. Geophys. Res. Atmos.*, **113**, 15.

Labat, D., Y. Godderis, J. L. Probst, and J. L. Guyot, 2004: Evidence for global runoff increase related to climate warming. *Adv. Water Resour.*, **27**, 631–642.

Ladstadter, F., A. K. Steiner, U. Foelsche, L. Haimberger, C. Tavolato, and G. Kirchnebngast, 2011: An assessment of differences in lower stratospheric temperature records from (A)MSU, radiosondes and GPS radio occultation. *Atmos. Meas. Tech.*, **4**, 1965–1977.

Landsea, C. W., 2007: Counting Atlantic tropical cyclones back to 1900. *EOS Trans. (AGU)*, **88**, 197–202.

Landsea, C. W., B. A. Harper, K. Hoarau, and J. A. Knaff, 2006: Can we detect trends in extreme tropical cyclones? *Science*, **313**, 452–454.

Landsea, C. W., et al., 2011: A reanalysis of the 1921–30 Atlantic Hurricane Database. *J. Clim.*, **25**, 865–885.

Langematz, U., and M. Kunze, 2008: Dynamical changes in the Arctic and Antarctic stratosphere during spring. In: *Climate Variability and Extremes during the Past 100 Years. Advances in Global Change Research* [S. Brönnimann, J. Luterbacher, T. Ewen, H. F. Diaz, R. S. Stolarski, and U. Neu (eds.)], Springer, pp. 293–301.

Lanzante, J. R., 2009: Comment on "Trends in the temperature and water vapor content of the tropical lower stratosphere: Sea surface connection" by Karen H. Rosenlof and George C. Reid. *J. Geophys. Res. Atmos.*, **114**, D12104.

Laprise, R., 1992: The resolution of global spectroal models. *Bull. Am. Meteor. Soc.*, **73**, 1453–1454.

Larkin, N. K., and D. E. Harrison, 2005: On the definition of El Niño and associated seasonal average US weather anomalies. *Geophys. Res. Lett.*, **32**, L13705.

Lawrimore, J. H., M. J. Menne, B. E. Gleason, C. N. Williams, D. B. Wuertz, R. S. Vose, and J. Rennie, 2011: An overview of the Global Historical Climatology Network monthly mean temperature data set, version 3. *J. Geophys. Res. Atmos.*, **116**, D19121.

Leakey, A. D. B., M. Uribelarrea, E. A. Ainsworth, S. L. Naidu, A. Rogers, D. R. Ort, and S. P. Long, 2006: Photosynthesis, productivity, and yield of maize are not affected by open-air elevation of CO_2 concentration in the absence of drought. *Plant Physiol.*, **140**, 779–790.

Lee, H. T., A. Gruber, R. G. Ellingson, and I. Laszlo, 2007: Development of the HIRS outgoing longwave radiation climate dataset. *J. Atmos. Ocean Technol.*, **24**, 2029–2047.

Lefohn, A. S., D. Shadwick, and S. J. Oltmans, 2010: Characterizing changes in surface ozone levels in metropolitan and rural areas in the United States for 1980–2008 and 1994–2008. *Atmos. Environ.*, **44**, 5199–5210.

Lehmann, A., K. Getzlaff, and J. Harlass, 2011: Detailed assessment of climate variability in the Baltic Sea area for the period 1958 to 2009. *Clim. Res.*, **46**, 185–196.

Lelieveld, J., J. van Aardenne, H. Fischer, M. de Reus, J. Williams, and P. Winkler, 2004: Increasing ozone over the Atlantic Ocean. *Science*, **304**, 1483–1487.

Lenderink, G., and E. Van Meijgaard, 2008: Increase in hourly precipitation extremes beyond expectations from temperature changes. *Nature Geosci.*, **1**, 511–514.

Lenderink, G., H. Y. Mok, T. C. Lee, and G. J. van Oldenborgh, 2011: Scaling and trends of hourly precipitation extremes in two different climate zones – Hong Kong and the Netherlands. *Hydrol. Earth Syst. Sci. Discuss.*, **8**, 4701–4719.

Lennartz, S., and A. Bunde, 2009: Trend evaluation in records with long-term memory: Application to global warming. *Geophys. Res. Lett.*, **36**, L16706.

Levin, I., et al., 2010: The global SF6 source inferred from long-term high precision atmospheric measurements and its comparison with emission inventories. *Atmos. Chem. Phys.*, **10**, 2655–2662.

Levitus, S., J. I. Antonov, T. P. Boyer, R. A. Locarnini, H. E. Garcia, and A. V. Mishonov, 2009: Global ocean heat content 1955–2008 in light of recently revealed instrumentation problems. *Geophys. Res. Lett.*, **36**, 5.

Levy, R. C., L. A. Remer, R. G. Kleidman, S. Mattoo, C. Ichoku, R. Kahn, and T. F. Eck, 2010: Global evaluation of the Collection 5 MODIS dark-target aerosol products over land. *Atmos. Chem. Phys.*, **10**, 10399–10420.

Li, Q., H. Zhang, X. Liu, J. Chen, W. Li, and P. Jones, 2009: A mainland China homogenized historical temperature dataset of 1951–2004. *Bull. Am. Meteor. Soc.*, **90**, 1062–1065.

Li, Q., W. Dong, W. Li, X. Gao, P. Jones, J. Kennedy, and D. Parker, 2010a: Assessment of the uncertainties in temperature change in China during the last century. *Chin. Sci. Bull.*, **55**, 1974–1982.

Li, Q. X., et al., 2010b: Assessment of surface air warming in northeast China, with emphasis on the impacts of urbanization. *Theor. Appl. Climatol.*, **99**, 469–478.

Li, Z., et al., 2012: Changes of daily climate extremes in southwestern China during 1961–2008. *Global Planet. Change*, **80–81**, 255–272.

Liang, F., and X. A. Xia, 2005: Long-term trends in solar radiation and the associated climatic factors over China for 1961–2000. *Ann. Geophys.*, **23**, 2425–2432.

Liebmann, B., R. M. Dole, C. Jones, I. Blade, and D. Allured, 2010: Influence of choice of time period on global surface temperature trend estimates. *Bull. Am. Meteor. Soc.*, **91**, 1485–U1471.

Liepert, B. G., 2002: Observed reductions of surface solar radiation at sites in the United States and worldwide from 1961 to 1990. *Geophys. Res. Lett.*, **29**, 1421.

Liley, J. B., 2009: New Zealand dimming and brightening. *J. Geophys. Res. Atmos.*, **114**, D00d10.

Lim, E. P., and I. Simmonds, 2007: Southern Hemisphere winter extratropical cyclone characteristics and vertical organization observed with the ERA-40 data in 1979–2001. *J. Clim.*, **20**, 2675–2690.

Lim, E. P., and I. Simmonds, 2009: Effect of tropospheric temperature change on the zonal mean circulation and SH winter extratropical cyclones. *Clim. Dyn.*, **33**, 19–32.

Lim, Y. K., M. Cai, E. Kalnay, and L. Zhou, 2008: Impact of vegetation types on surface temperature change. *J. Appl. Meteor. Climatol.*, **47**, 411–424.

Lin, C., K. Yang, J. Qin, and R. Fu, 2012: Observed coherent trends of surface and upper-air wind speed over China since 1960. *J. Clim.*,**26**, 2891-2903..

Liu, B., M. Xu, and M. Henderson, 2011: Where have all the showers gone? Regional declines in light precipitation events in China, 1960–2000. *Int. J. Climatol.*, **31**, 1177–1191.

Liu, B. H., M. Xu, M. Henderson, and W. G. Gong, 2004a: A spatial analysis of pan evaporation trends in China, 1955–2000. *J. Geophys. Res. Atmos.*, **109**, D15102.

Liu, B. H., M. Xu, M. Henderson, Y. Qi, and Y. Q. Li, 2004b: Taking China's temperature: Daily range, warming trends, and regional variations, 1955–2000. *J. Clim.*, **17**, 4453–4462.

Liu, Q. H., and F. Z. Weng, 2009: Recent stratospheric temperature observed from satellite measurements. *Sola*, **5**, 53–56.

Lo, M.-H., and J. S. Famiglietti, 2013: Irrigation in California's Central Valley strengthens the southwestern U.S. water cycle. *Geophys. Res. Lett.*, **40**, 301-306.

Loeb, N. G., et al., 2009: Toward optimal closure of the Earth's top-of-atmosphere radiation budget. *J. Clim.*, **22**, 748–766.

Loeb, N. G., et al., 2012a: Advances in understanding top-of-atmosphere radiation variability from satellite observations. *Surv. Geophys.*, **33**, 359–385.

Loeb, N. G., et al., 2012b: Observed changes in top-of-the-atmosphere radiation and upper-ocean heating consistent within uncertainty. *Nature Geosci.*, **5**, 110–113.

Logan, J. A., et al., 2012: Changes in ozone over Europe since 1990: Analysis of ozone measurements from sondes, regular aircraft (MOZAIC), and alpine surface sites. *J. Geophys. Res.*, **117**, D09301.

Long, C. N., E. G. Dutton, J. A. Augustine, W. Wiscombe, M. Wild, S. A. McFarlane, and C. J. Flynn, 2009: Significant decadal brightening of downwelling shortwave in the continental United States. *J. Geophys. Res. Atmos.*, **114**, D00d06.

Lorenz, R., E. B. Jaeger, and S. I. Seneviratne, 2010: Persistence of heat waves and its link to soil moisture memory. *Geophys. Res. Lett.*, **37**, L09703.

Lucas, C., H. Nguyen, and B. Timbal, 2012: An observational analysis of Southern Hemisphere tropical expansion. *J. Geohys. Res.*, **117**, D17112.

Luo, J. J., W. Sasaki, and Y. Masumoto, 2012: Indian Ocean warming modulates Pacific climate change. *Proc. Natl. Acad. Sci. U.S.A.*, **109**, 18701–18706.

Lupikasza, E., 2010: Spatial and temporal variability of extreme precipitation in Poland in the period 1951–2006. *Int. J. Climatol.*, **30**, 991–1007.

Lyman, J. M., et al., 2010: Robust warming of the global upper ocean. *Nature*, **465**, 334–337.

Lynch, A. H., J. A. Curry, R. D. Brunner, and J. A. Maslanik, 2004: Toward an integrated assessment of the impacts of extreme wind events on Barrow, Alaska. *Bull. Am. Meteor. Soc.*, **85**, 209–.

Mahowald, N., et al., 2010: Observed 20th century desert dust variability: Impact on climate and biogeochemistry. *Atmos. Chem. Phys.*, **10**, 10875–10893.

Makowski, K., M. Wild, and A. Ohmura, 2008: Diurnal temperature range over Europe between 1950 and 2005. *Atmos. Chem. Phys.*, **8**, 6483–6498.

Makowski, K., E. B. Jaeger, M. Chiacchio, M. Wild, T. Ewen, and A. Ohmura, 2009: On the relationship between diurnal temperature range and surface solar radiation in Europe. *J. Geophys. Res. Atmos.*, **114**, D00d07.

Manabe, S., and R. F. Strickler, 1964: Thermal equilibrium of the atmosphere with a convective adjustment. *Journal of the Atmospheric Sciences*, **21**, 361–385.

Mann, M. E., 2011: On long range dependence in global surface temperature series. *Clim. Change*, **107**, 267–276.

Mann, M. E., T. A. Sabbatelli, and U. Neu, 2007a: Evidence for a modest undercount bias in early historical Atlantic tropical cyclone counts. *Geophys. Res. Lett.*, **34**, L22707.

Mann, M. E., K. A. Emanual, G. J. Holland, and P. J. Webster, 2007b: Atlantic tropical cyclones revisited. *EOS Transactions (AGU)*, **88**, 349–350.

Manney, G. L., et al., 2011: Unprecedented Arctic ozone loss in 2011. *Nature*, **478**, 469–475.

Mantua, N. J., S. R. Hare, Y. Zhang, J. M. Wallace, and R. C. Francis, 1997: A Pacific interdecadal climate oscillation with impacts on salmon production. *Bull. Am. Meteor. Soc.*, **78**, 1069–1079.

Marenco, A., H. Gouget, P. Nédélec, and J. P. Pagés, 1994: Evidence of a long-term increase in tropospheric ozone from Pic du Midi series: Consequences: positive radiative forcing. *J. Geophys. Res.*, **99**, 16,617–616, 632.

Marshall, G. J., 2003: Trends in the southern annular mode from observations and reanalyses. *J. Clim.*, **16**, 4134–4143.

Martinerie, P., et al., 2009: Long-lived halocarbon trends and budgets from atmospheric chemistry modelling constrained with measurements in polar firn. *Atmos. Chem. Phys.*, 3911–3934.

Mastrandrea, M., et al., 2011: The IPCC AR5 guidance note on consistent treatment of uncertainties: A common approach across the working groups. *Clim. Change*, **108**, 675–691.

Matulla, C., W. Schoner, H. Alexandersson, H. von Storch, and X. L. Wang, 2008: European storminess: Late nineteenth century to present. *Clim. Dyn.*, **31**, 125–130.

McCarthy, M. P., P. W. Thorne, and H. A. Titchner, 2009: An analysis of tropospheric humidity trends from radiosondes. *J. Clim.*, **22**, 5820–5838.

McCarthy, M. P., H. A. Titchner, P. W. Thorne, S. F. B. Tett, L. Haimberger, and D. E. Parker, 2008: Assessing bias and uncertainty in the HadAT-adjusted radiosonde climate record. *J. Clim.*, **21**, 817–832.

McKitrick, R., 2010: Atmospheric circulations do not explain the temperature-industrialization correlation. *Stat. Politics Policy*, **1**, issue 1 .

McKitrick, R., and P. J. Michaels, 2004: A test of corrections for extraneous signals in gridded surface temperature data. *Clim. Res.*, **26**, 159–173.

McKitrick, R., and N. Nierenberg, 2010: Socioeconomic patterns in climate data. *J. Econ. Soc. Meas*, **35**, 149–175.

McKitrick, R. R., and P. J. Michaels, 2007: Quantifying the influence of anthropogenic surface processes and inhomogeneities on gridded global climate data. *J. Geophys. Res. Atmos.*, **112**, D24S09.

McNider, R. T., et al., 2012: Response and sensitivity of the nocturnal boundary layer over land to added longwave radiative forcing. *J. Geophys. Res.*, **117**, D14106.

McVicar, T. R., T. G. Van Niel, L. T. Li, M. L. Roderick, D. P. Rayner, L. Ricciardulli, and R. J. Donohue, 2008: Wind speed climatology and trends for Australia, 1975–2006: Capturing the stilling phenomenon and comparison with near-surface reanalysis output. *Geophys. Res. Lett.*, **35**, L20403.

McVicar, T. R., et al., 2012: Global review and synthesis of trends in observed terrestrial near-surface wind speeds: Implications for evaporation. *J. Hydrol.*, **416**, 182–205.

Mears, C., J. Wang, S. Ho, L. Zhang, and X. Zhou, 2010: Total column water vapor, in State of the Climate in 2009. *Bull. Am. Meteor. Soc.* [D. S. Arndt, M. O. Baringer, and M. R. Johnson (eds.)].

Mears, C. A., and F. J. Wentz, 2009a: Construction of the remote sensing systems V3.2 atmospheric temperature records from the MSU and AMSU microwave sounders. *J. Atmos. Ocean Technol.*, **26**, 1040–1056.

Mears, C. A., and F. J. Wentz, 2009b: Construction of the RSS V3.2 lower-tropospheric temperature dataset from the MSU and AMSU microwave sounders. *J. Atmos. Ocean Technol.*, **26**, 1493–1509.

Mears, C. A., F. J. Wentz, and P. W. Thorne, 2012: Assessing the value of Microwave Sounding Unit-radiosonde comparisons in ascertaining errors in climate data records of tropospheric temperatures. *J. Geophys. Res. Atmos.*, **117**, D19103.

Mears, C. A., F. J. Wentz, P. Thorne, and D. Bernie, 2011: Assessing uncertainty in estimates of atmospheric temperature changes from MSU and AMSU using a Monte-Carlo estimation technique. *J. Geophys. Res. Atmos.*, **116**.

Mears, C. A., C. E. Forest, R. W. Spencer, R. S. Vose, and R. W. Reynolds, 2006: What is our understanding of the contribution made by observational or methodological uncertainties to the previously reported vertical differences in temperature trends? In: *Temperature Trends in the Lower Tmosphere: Steps for Understanding and Reconciling Differences* [T. R. Karl, S. J. Hassol, C. D. Miller, and W. L. Murray (eds.)], 71-88.

Mears, C. A., B. D. Santer, F. J. Wentz, K. E. Taylor, and M. F. Wehner, 2007: Relationship between temperature and precipitable water changes over tropical oceans. *Geophys. Res. Lett.*, **34**, L24709.

Meehl, G. A., J. M. Arblaster, and G. Branstator, 2012: Mechanisms contributing to the warming hole and the consequent U.S. East–West differential of heat extremes. *J. Clim.*, **25**, 6394–6408.

Mekis, É., and L. A. Vincent, 2011: An overview of the second generation adjusted daily precipitation dataset for trend analysis in Canada. *Atmosphere-Ocean*, **49**, 163–177.

Meng, Q. J., M. Latif, W. Park, N. S. Keenlyside, V. A. Semenov, and T. Martin, 2012: Twentieth century Walker Circulation change: Data analysis and model experiments. *Clim. Dyn.*, **38**, 1757–1773.

Menne, M. J., and C. N. Williams, 2009: Homogenization of temperature series via pairwise comparisons. *J. Clim.*, **22**, 1700–1717.

Menne, M. J., C. N. Williams, and M. A. Palecki, 2010: On the reliability of the US surface temperature record. *J. Geophys. Res. Atmos.*, **115**, D11108.

Menzel, W. P., 2001: Cloud tracking with satellite imagery: From the pioneering work of Ted Fujita to the present. *Bull. Am. Meteor. Soc.*, **82**, 33–47.

Merchant, C. J., et al., 2012: A 20 year independent record of sea surface temperature for climate from Along Track Scanning Radiometer. *J. Geophys. Res.*, **117**. C12013.

Merrifield, M. A., 2011: A shift in western tropical Pacific sea level trends during the 1990s. *J. Clim.*, **24**, 4126–4138.

Mezher, R. N., M. Doyle, and V. Barros, 2012: Climatology of hail in Argentina. *Atmos. Res.*, **114–115**, 70–82.

Mieruch, S., S. Noel, H. Bovensmann, and J. P. Burrows, 2008: Analysis of global water vapour trends from satellite measurements in the visible spectral range. *Atmos. Chem. Phys.*, **8**, 491–504.

Milewska, E. J., 2004: Baseline cloudiness trends in Canada 1953–2002. *Atmos. Ocean*, **42**, 267–280.

Miller, B., et al., 2010: HFC-23 (CHF3) emission trend response to HCFC-22 (CHClF2) production and recent HFC-23 emission abatement measures. *Atmos. Chem. Phys.*, **10**, 7875–7890.

Milliman, J. D., K. L. Farnsworth, P. D. Jones, K. H. Xu, and L. C. Smith, 2008: Climatic and anthropogenic factors affecting river diSchärge to the global ocean, 1951–2000. *Global Planet. Change*, **62**, 187–194.

Mills, T. C., 2010: 'Skinning a cat': Alternative models of representing temperature trends. *Clim. Change*, **101**, 415–426.

Milz, M., et al., 2005: Water vapor distributions measured with the Michelson Interferometer for passive atmospheric sounding on board Envisat (MIPAS/Envisat). *J. Geophys. Res.*, **110**, D24307.

Mishchenko, M. I., et al., 2007: Long-term satellite record reveals likely recent aerosol trend. *Science*, **315**, 1543–1543.

Mishchenko, M. I., et al., 2012: Aerosol retrievals from channel-1 and -2 AVHRR radiances: Long-term trends updated and revisited. *J. Quant. Spectr. Radiat. Trans.*, **113**, 1974–1980.

Misra, V., J. P. Michael, R. Boyles, E. P. Chassignet, M. Griffin, and J. J. O'Brien, 2012: Reconciling the spatial distribution of the surface temperature trends in the southeastern United States. *J. Clim.*, **25**, 3610–3618.

Mitas, C. M., and A. Clement, 2005: Has the Hadley cell been strengthening in recent decades? *Geophys. Res. Lett.*, **32**, L030809.

Mitchell, T. D., and P. D. Jones, 2005: An improved method of constructing a database of monthly climate observations and associated high-resolution grids. *Int. J. Climatol.*, **25**, 693–712.

Mo, K., and J. Paegle, 2001: The Pacific-South American modes and their downstream effects. *Int. J. Climatol.*, **21**, 1211–1229.

Moberg, A., et al., 2006: Indices for daily temperature and precipitation extremes in Europe analyzed for the period 1901–2000. *J. Geophys. Res. Atmos.*, **111**, D22106.

Mohapatra, M., B. K. Mandyopadhyay, and A. Tyagi, 2011: Best track parameters of tropical cyclones over the North Indian Ocean: A review. *Natural Hazards,* **63**, 1285-1317.

Mokhov, I. I., M. G. Akperov, M. A. Prokofyeva, A. V. Timazhev, A. R. Lupo, and H. Le Treut, 2013: Blockings in the Northern Hemisphere and Euro-Atlantic region: Estimates of changes from reanalyses data and model simulations. *Doklady, Earth Sci.*, 449, 430-433.

Monaghan, A. J., and D. H. Bromwich, 2008: Advances describing recent Antarctic climate variability. *Bull. Am. Meteorol. Soc.*, **89**, 1295–1306.

Monaghan, A. J., D. H. Bromwich, W. Chapman, and J. C. Comiso, 2008: Recent variability and trends of Antarctic near-surface temperature. *J. Geophys. Res. Atmos.*, **113**, D04105.

Monk, W., D. L. Peters, D. J. Baird, and R. A. Curry, 2011: Trends in indicator hydrological variables for Canadian rivers. *Hydrol. Proc.*, **25**, 3086–3100.

Monks, P. S., et al., 2009: Atmospheric composition change – global and regional air quality. *Atmos. Environ.*, **43**, 5268–5350.

Montzka, S., B. Hall, and J. Elkins, 2009: Accelerated increases observed for hydrochlorofluorocarbons since 2004 in the global atmosphere. *Geophys. Res. Lett.*, **36**, L03804 .

Montzka, S., M. Krol, E. Dlugokencky, B. Hall, P. Jockel, and J. Lelieveld, 2011: Small interannual variability of global atmospheric hydroxyl. *Science*, **331**, 67-69.

Montzka, S., L. Kuijpers, M. Battle, M. Aydin, K. Verhulst, E. Saltzman, and D. Fahey, 2010: Recent increases in global HFC-23 emissions. *Geophys. Res. Lett.*, **37**, L02808.

Montzka, S. A., et al., 2011b: Ozone-depleting substances (ODSs) and related chemicals. In Scientific Assessment of Ozone Depletion: 2010, Global Ozone Research and Monitoring Project—Report No. 52. World Meteorological Organization, Geneva, Switzerland, 516 pp.

Morak, S., G. C. Hegerl, and J. Kenyon, 2011: Detectable regional changes in the number of warm nights. *Geophys. Res. Lett.*, **38**, 5.

Morak, S., G. C. Hegerl, and N. Christidis, 2013: Detectable changes in the frequency of temperature extremes. *J. Clim.*, **26**, 1561–1574.

Morice, C. P., J. J. Kennedy, N. A. Rayner, and P. D. Jones, 2012: Quantifying uncertainties in global and regional temperature change using an ensemble of observational estimates: The HadCRUT4 data set. *J. Geophys. Res. Atmos.*, **117**, 22.

Mueller, B., and S. Seneviratne, 2012: Hot days induced by precipitation deficits at the global scale. *Proc. Natl. Acad. Sci. U.S.A.*, 109, 12398-12403.

Mueller, B., et al., 2011: Evaluation of global observations-based evapotranspiration datasets and IPCC AR4 simulations. *Geophys. Res. Lett.*, **38**, L06402.

Muhle, J., et al., 2010: Perfluorocarbons in the global atmosphere: tetrafluoromethane, hexafluoroethane, and octafluoropropane. *Atmos. Chem. Phys.*, 10, 5145-5164.

Murphy, D. M., et al., 2011: Decreases in elemental carbon and fine particle mass in the United States. *Atmos. Chem. Phys.*, **11**, 4679–4686.

Nan, S., and J. P. Li, 2003: The relationship between the summer precipitation in the Yangtze River Valley and the boreal spring Southern Hemisphere annular mode. *Geophys. Res. Lett.*, **30**, 2266.

Nash, J., and P. R. Edge, 1989: Temperature changes in the stratosphere and lower mesosphere 197–1988 inferred from TOVS radiance observations. *Adv. Space Res.*, **9**, 333–341.

Neff, W., J. Perlwitz, and M. Hoerling, 2008: Observational evidence for asymmetric changes in tropospheric heights over Antarctica on decadal time scales. *Geophys. Res. Lett.*, **35**, L18703.

Neu, U., et al., 2012: IMILAST: A community effort to intercompare extratropical cyclone detection and tracking algorithms. *Bull. Am. Meteor. Soc.*, **94**, 529–547.

Nevison, C., et al., 2011: Exploring causes of interannual variability in the seasonal cycles of tropospheric nitrous oxide. *Atmos. Chem. Phys.*, **11**, 3713–3730.

New, M., et al., 2006: Evidence of trends in daily climate extremes over southern and west Africa. *J. Geophys. Res. Atmos.*, **111**, D14102.

Nguyen, H., B. Timbal, I. Smith, A. Evans, and C. Lucas, 2013: The Hadley circulation in reanalyses: Climatology, variability and change. *J. Clim.*, **26**, 3357–3376.

Nicholls, N., 2008: Recent trends in the seasonal and temporal behaviour of the El Niño-Southern Oscillation. *Geophys. Res. Lett.*, **35**, L19703.

Nisbet, E., and R. Weiss, 2010: Top-down versus bottom-up. *Science*, **328**, 1241-1243.

Norris, J. R., and M. Wild, 2007: Trends in aerosol radiative effects over Europe inferred from observed cloud cover, solar "dimming" and solar "brightening". *J. Geophys. Res. Atmos.*, **112**, D08214.

Norris, J. R., and M. Wild, 2009: Trends in aerosol radiative effects over China and Japan inferred from observed cloud cover, solar dimming, and solar brightening. *J. Geophys. Res. Atmos.*, **114**, D00d15.

O'Dell, C. W., F. J. Wentz, and R. Bennartz, 2008: Cloud liquid water path from satellite-based passive microwave observations: A new climatology over the global oceans. *J. Clim.*, **21**, 1721–1739.

O'Doherty, S., et al., 2009: Global and regional emissions of HFC-125 (CHF2CF3) from in situ and air archive atmospheric observations at AGAGE and SOGE observatories. *J. Geophys. Res. Atmos.*, **109**, D06310.

O'Donnell, R., N. Lewis, S. McIntyre, and J. Condon, 2011: Improved methods for PCA-based reconstructions: Case study using the Steig et al. (2009) Antarctic Temperature Reconstruction. *J. Clim.*, **24**, 2099–2115.

O'Gorman, P., R. P. Allan, M. P. Byrne, and M. Previdi, 2012: Energentic constraints on precipitation under climate change. *Surv. Geophys.*, **33**, 585–608.

Ohmura, A., 2009: Observed decadal variations in surface solar radiation and their causes. *J. Geophys. Res. Atmos.*, **114**, D00d05.

Ohmura, A., et al., 1998: Baseline Surface Radiation Network (BSRN/WCRP): New precision radiometry for climate research. *Bull. Am. Meteor. Soc.*, **79**, 2115–2136.

Ohvril, H., et al., 2009: Global dimming and brightening versus atmospheric column transparency, Europe, 1906–2007. *J. Geophys. Res. Atmos.*, **114**, D00d12.

Oltmans, S. J., et al., 2013: Recent tropospheric ozone changes – A pattern dominated by slow or no growth. *Atmos. Environ.*, **67**, 331–351.

Onogi, K., et al., 2007: The JRA-25 reanalysis. *J. Meteorol. Soc. Jpn.*, **85**, 369–432.

Oort, A. H., and J. J. Yienger, 1996: Observed interannual variability in the Hadley circulation and its connection to ENSO. *J. Clim.*, **9**, 2751–2767.

Osborn, T. J., 2011: Winter 2009/2010 temperatures and a record breaking North Atlantic Oscillation index. *Weather*, **66**, 19–21.

Paciorek, C. J., J. S. Risbey, V. Ventura, and R. D. Rosen, 2002: Multiple indices of Northern Hemisphere cyclone activity, winters 1949–99. *J. Clim.*, **15**, 1573–1590.

Palmer, M. D., K. Haines, S. F. B. Tett, and T. J. Ansell, 2007: Isolating the signal of ocean global warming. *Geophys. Res. Lett.*, **34**, 6.

Palmer, W. C., 1965: Meteorological drought. *US Weather Bureau Research Paper*, 45, 58 pages.

Paltridge, G., A. Arking, and M. Pook, 2009: Trends in middle- and upper-level tropospheric humidity from NCEP reanalysis data. *Theor. Appl. Climatol.*, **98**, 351–359.

Pan, Z. T., R. W. Arritt, E. S. Takle, W. J. Gutowski, C. J. Anderson, and M. Segal, 2004: Altered hydrologic feedback in a warming climate introduces a "warming hole". *Geophys. Res. Lett.*, **31**, L17109.

Panagiotopoulos, F., M. Shahgedanova, A. Hannachi, and D. B. Stephenson, 2005: Observed trends and teleconnections of the Siberian high: A recently declining center of action. *J. Clim.*, **18**, 1411–1422.

Parker, D., C. Folland, A. Scaife, J. Knight, A. Colman, P. Baines, and B. Dong, 2007: Decadal to multidecadal variability and the climate change background. *J. Geophys. Res. Atmos.*, **112**. D18115.

Parker, D. E., 2006: A demonstration that large-scale warming is not urban. *J. Clim.*, **19**, 2882–2895.

Parker, D. E., 2011: Recent land surface air temperature trends assessed using the 20th century reanalysis. *J. Geophys. Res. Atmos.*, **116**, D20125.

Parker, D. E., P. Jones, T. C. Peterson, and J. Kennedy, 2009: Comment on "Unresolved issues with the assessment of multidecadal global land surface temperature trends" by Roger A. Pielke Sr. et al. *J. Geophys. Res. Atmos.*, **114**. D05104.

Parrish, D. D., et al., 2012: Long-term changes in lower tropospheric baseline ozone concentrations at northern mid-latitudes. *Atmos. Chem. Phys.*, **12**, 11485–11504.

Pattanaik, D. R., and M. Rajeevan, 2010: Variability of extreme rainfall events over India during southwest monsoon season. *Meteorol. Appl.*, **17**, 88–104.

Pavan, V., R. Tomozeiu, C. Cacciamani, and M. Di Lorenzo, 2008: Daily precipitation observations over Emilia-Romagna: Mean values and extremes. *Int. J. Climatol.*, **28**, 2065–2079.

Pavelin, E. G., C. E. Johnson, S. Rughooputh, and R. Toumi, 1999: Evaluation of pre-industrial surface ozone measurements made using Sch\onbein's method. *Atmos. Environ.*, **33**, 919–929.

Perkins, S. E., and L. V. Alexander, 2012: On the measurement of heat waves. *J. Clim.*, **26**, 4500-4517 .

Perkins, S. E., L. V. Alexander, and J. R. Nairn, 2012: Increasing frequency, intensity and duration of observed global heatwaves and warm spells. *Geophys. Res. Lett.*, **39**. L20714.

Peterson, T. C., K. M. Willett, and P. W. Thorne, 2011: Observed changes in surface atmospheric energy over land. *Geophys. Res. Lett.*, **38,** L16707.

Peterson, T. C., X. B. Zhang, M. Brunet-India, and J. L. Vazquez-Aguirre, 2008: Changes in North American extremes derived from daily weather data. *J. Geophys. Res. Atmos.*, **113,** D07113.

Peterson, T. C., et al., 2009: State of the Climate in 2008. *Bull. Am. Meteor. Soc.*, **90**, S13-.

Peterson, T. C., et al., 2013: Monitoring and understanding changes in heat waves, cold waves, floods and droughts in the United States: State of knowledge. *Bull. Am. Meteor. Soc.*, 94, 821-834.

Petrow, T., and B. Merz, 2009: Trends in flood magnitude, frequency and seasonality in Germany in the period 1951–2002. *J. Hydrol.*, **371**, 129–141.

Pezza, A., P. van Rensch, and W. Cai, 2012: Severe heat waves in Southern Australia: Synoptic climatology and large scale connections. *Clim. Dyn.*, **38**, 209–224.

Pezza, A. B., I. Simmonds, and J. A. Renwick, 2007: Southern Hemisphere cyclones and anticyclones: Recent trends and links with decadal variability in the Pacific Ocean. *Int. J. Climatol.*, **27**, 1403–1419.

Philipona, R., 2012: Greenhouse warming and solar brightening in and around the Alps. *Int. J. Climatol.*, **33**, 1530-1537.

Philipona, R., K. Behrens, and C. Ruckstuhl, 2009: How declining aerosols and rising greenhouse gases forced rapid warming in Europe since the 1980s. *Geophys. Res. Lett.*, **36**, L02806.

Philipona, R., B. Dürr, A. Ohmura, and C. Ruckstuhl, 2005: Anthropogenic greenhouse forcing and strong water vapor feedback increase temperature in Europe. *Geophys. Res. Lett.*, **32,** L19809.

Philipona, R., B. Dürr, C. Marty, A. Ohmura, and M. Wild, 2004: Radiative forcing-measured at Earth's surface—corroborate the increasing greenhouse effect. *Geophys. Res. Lett.*, **31**, L03202.

Philipp, A., P. M. Della-Marta, J. Jacobeit, D. R. Fereday, P. D. Jones, A. Moberg, and H. Wanner, 2007: Long-term variability of daily North Atlantic-European pressure patterns since 1850 classified by simulated annealing clustering. *J. Clim.*, **20**, 4065–4095.

Piao, S., et al., 2010: The impacts of climate change on water resources and agriculture in China. *Nature*, **467**, 43–51.

Pielke, R. A., and T. Matsui, 2005: Should light wind and windy nights have the same temperature trends at individual levels even if the boundary layer averaged heat content change is the same? *Geophys. Res. Lett.*, **32,** L21813.

Pielke, R. A., Sr., et al., 2007: Unresolved issues with the assessment of multidecadal global land surface temperature trends. *J. Geophys. Res. Atmos.*, **112,** D24S08.

Pinker, R. T., B. Zhang, and E. G. Dutton, 2005: Do satellites detect trends in surface solar radiation? *Science*, **308**, 850–854.

Pirazzoli, P. A., and A. Tomasin, 2003: Recent near-surface wind changes in the central Mediterranean and Adriatic areas. *Int. J. Climatol.*, **23**, 963–973.

Po-Chedley, S., and Q. Fu, 2012: A bias in the Midtropospheric Channel Warm Target Factor on the NOAA-9 Microwave Sounding Unit. *J. Atmos. Ocean Technol.*, **29**, 646–652.

Portmann, R., S. Solomon, and G. Hegerl, 2009a: Spatial and seasonal patterns in climate change, temperatures, and precipitation across the United States. *Proc. Natl. Acad. Sci. U.S.A.*, **106**, 7324–7329.

Portmann, R. W., S. Solomon, and G. C. Hegerl, 2009b: Linkages between climate change, extreme temperature and precipitation across the United States. *Proc. Natl. Acad. Sci. U.S.A.*, **106**, 7324–7329.

Power, S., T. Casey, C. Folland, A. Colman, and V. Mehta, 1999: Inter-decadal modulation of the impact of ENSO on Australia. *Clim. Dyn.*, **15**, 319–324.

Power, S. B., and G. Kociuba, 2011a: The impact of global warming on the Southern Oscillation Index. *Clim. Dynamics*, **37**, 1745–1754.

Power, S. B., and G. Kociuba, 2011b: What caused the observed twentieth-century weakening of the Walker Circulation? *J. Clim.*, **24**, 6501–6514.

Pozzoli, L., et al., 2011: Reanalysis of tropospheric sulfate aerosol and ozone for the period 1980–2005 using the aerosol-chemistry-climate model ECHAM5–HAMMOZ. *Atmos. Chem. Phys.*, **11**, 9563–9594.

Prata, F., 2008: The climatological record of clear-sky longwave radiation at the Earth's surface: Evidence for water vapour feedback? *Int. J. Remote Sens.*, **29**, 5247–5263.

Prather, M., C. Holmes, and J. Hsu, 2012: Reactive greenhouse gas scenarios: Systematic exploration of uncertainties and the role of atmospheric chemistry. *Geophys. Res. Lett.*, **39**, L09803.

Prinn, R., et al., 2005: Evidence for variability of atmospheric hydroxyl radicals over the past quarter century. *Geophys. Res. Lett.*, L07809.

Pryor, S. C., R. J. Barthelmie, and E. S. Riley, 2007: Historical evolution of wind climates in the USA - art. no. 012065. In: *Science of Making Torque from Wind* [M. O. L. Hansen and K. S. Hansen (eds.)], **75**, 12065–12065.

Pryor, S. C., J. A. Howe, and K. E. Kunkel, 2009: How spatially coherent and statistically robust are temporal changes in extreme precipitation in the contiguous USA? *Int. J. Climatol.*, **29**, 31–45.

Qian, Y., D. P. Kaiser, L. R. Leung, and M. Xu, 2006: More frequent cloud-free sky and less surface solar radiation in China from 1955 to 2000. *Geophys. Res. Lett.*, **33**, L01812.

Qian, Y., W. G. Wang, L. R. Leung, and D. P. Kaiser, 2007: Variability of solar radiation under cloud-free skies in China: The role of aerosols. *Geophys. Res. Lett.*, **34**, L12804.

Qian, Y., D. Gong, J. Fan, L. Leung, R. Bennartz, D. Chen, and W. Wang, 2009: Heavy pollution suppresses light rain in China: Observations and modeling. *J. Geophys. Res. Atmos.*, **114**, D00K02.

Rahimzadeh, F., A. Asgari, and E. Fattahi, 2009: Variability of extreme temperature and precipitation in Iran during recent decades. *Int. J. Climatol.*, **29**, 329–343.

Raible, C. C., P. M. Della-Marta, C. Schwierz, H. Wernli, and R. Blender, 2008: Northern hemisphere extratropical cyclones: A comparison of detection and tracking methods and different reanalyses. *Mon. Weather Rev.*, **136**, 880–897.

Raichijk, C., 2011: Observed trends in sunshine duration over South America. *Int. J. Climatol.*, **32**, 669-680.

Randall, R. M., and B. M. Herman, 2008: Using limited time period trends as a means to determine attribution of discrepancies in microwave sounding unit-derived tropospheric temperature time series. *J. Geophys. Res. Atmos.*, **113**, D05105.

Randel, W. J., 2010: Variability and trends in stratospheric temperature and water vapor. *The Stratosphere: Dynamics, Transport and Chemistry*, S. Polvani, and Waugh, Ed., American Geophysical Union, 123–135.

Randel, W. J., and E. J. Jensen, 2013: Physical processes in the tropical tropopause layer and their roles in a changing climate. *Nature Geosci.*, 169–176.

Randel, W. J., F. Wu, H. Vömel, G. E. Nedoluha, and P. Forster, 2006: Decreases in stratospheric water vapor after 2001: Links to changes in the tropical tropopause and the Brewer-Dobson circulation. *J. Geophys. Res. Atmos.*, **111**, D12312.

Randel, W. J., et al., 2009: An update of observed stratospheric temperature trends. *J. Geophys. Res. Atmos.*, **114**, D02107.

Rasmusson, E. M., and T. H. Carpenter, 1982: Variations in tropical sea surface temperature and surface wind fields associated with the Southern Oscillation/El Niño. *Mon. Weather Rev.*, **110**, 354–384.

Rasmusson, E. M., and J. M. Wallace, 1983: Meteorological aspects of the El Niño-Southern Oscillation. *Science*, **222**, 1195–1202.

Rausch, J., A. Heidinger, and R. Bennartz, 2010: Regional assessment of microphysical properties of marine boundary layer cloud using the PATMOS-x dataset. *J. Geophys. Res. Atmos.*, **115**, D23212.

Ray, E. A., et al., 2010: Evidence for changes in stratospheric transport and mixing over the past three decades based on multiple data sets and tropical leaky pipe analysis. *J. Geophys. Res. Atmos.*, **115,** D21304.

Rayner, D. P., 2007: Wind run changes: The dominant factor affecting pan evaporation trends in Australia. *J. Clim.*, **20**, 3379–3394.

Rayner, N. A., et al., 2003: Global analyses of sea surface temperature, sea ice, and night marine air temperature since the late nineteenth century. *J. Geophys. Res. Atmos.*, **108**, 37.

Rayner, N. A., et al., 2006: Improved analyses of changes and uncertainties in sea surface temperature measured in situ sice the mid-nineteenth century: The HadSST2 dataset. *J. Clim.*, **19**, 446–469.

Read, W. G., et al., 2007: Aura Microwave Limb Sounder upper tropospheric and lower stratospheric H_2O and relative humidity with respect to ice validation. *J. Geophys. Res. Atmos.*, **112**, D24S35.

Ren, G., et al., 2011: Change in climatic extremes over mainland China based on an integrated extreme climate index. *Clim. Res.*, **50**, 113–124.

Ren, G. Y., Z. Y. Chu, Z. H. Chen, and Y. Y. Ren, 2007: Implications of temporal change in urban heat island intensity observed at Beijing and Wuhan stations. *Geophys. Res. Lett.*, **34**, L05711.

Ren, G. Y., Y. Q. Zhou, Z. Y. Chu, J. X. Zhou, A. Y. Zhang, J. Guo, and X. F. Liu, 2008: Urbanization effects on observed surface air temperature trends in north China. *J. Clim.*, **21**, 1333–1348.

Ren, H.-L., and F.-F. Jin, 2011: Niño indices for two types of ENSO. *Geophys. Res. Lett.*, **38**, L04704.

Ren, Y. Y., and G. Y. Ren, 2011: A remote-sensing method of selecting reference stations for evaluating urbanization effect on surface air temperature trends. *J. Clim.*, **24**, 3179–3189.

Renard, B., et al., 2008: Regional methods for trend detection: Assessing field significance and regional consistency. *Water Resourc. Res.*, **44**, W08419.

Revadekar, J., D. Kothawale, S. Patwardhan, G. Pant, and K. Kumar, 2012: About the observed and future changes in temperature extremes over India. *Nat. Hazards*, **60**, 1133–1155.

Reynolds, R., N. Rayner, T. Smith, D. Stokes, and W. Wang, 2002: An improved in situ and satellite SST analysis for climate. *J. Clim.*, **15**, 1609–1625.

Reynolds, R. W., C. L. Gentemann, and G. K. Corlett, 2010: Evaluation of AATSR and TMI Satellite SST Data. *J. Clim.*, **23**, 152–165.

Rhines, A., and P. Huybers, 2013: Frequent summer temperature extremes reflect changes in the mean, not the variance. *Proc. Natl. Acad. Sci. U.S.A.*, **110**, E546–E546.

Rienecker, M. M., Suarez, M.J., Gelaro, R., Todling, R., Bacmeister, J., Liu, E., Bosilovich, M. G., Schubert, S. D., Takacs, L., Kim, G.-K., Bloom, S., Chen, J., Collins, D., Conaty, A., da Silva, A., Gu, W., Joiner, J., Koster, R. D., Lucchesi, R., Molod, A., Owens, T., Pawson, S., Pegion, P., Redder, C. R., Reichle, R., Robertson, F. R., Ruddick, A. G., Sienkiewicz, M., and Woollen, J., 2011: MERRA: NASA's modern-era retrospective analysis for research and applications. *J. Clim.*, 24, 3624-3648.

Rigby, M., et al., 2008: Renewed growth of atmospheric methane. *Geophys. Res. Lett.*, **35**, L22805.

Rigby, M., et al., 2010: History of atmospheric SF6 from 1973 to 2008. *Atmos. Chem. Phys.*, **10**, 10305-10320.

Riihimaki, L. D., F. E. Vignola, and C. N. Long, 2009: Analyzing the contribution of aerosols to an observed increase in direct normal irradiance in Oregon. *J. Geophys. Res. Atmos.*, **114**, D00d02.

Robock, A., et al., 2000: The Global Soil Moisture Data Bank. *Bull. Am. Meteor. Soc.*, **81**, 1281–1299.

Rockmann, T., and I. Levin, 2005: High-precision determination of the changing isotopic composition of atmospheric N2O from 1990 to 2002. *J. Geophys. Res. Atmos.*, **110**, D21304.

Roderick, M. L., and G. D. Farquhar, 2002: The cause of decreased pan evaporation over the past 50 years. *Science*, **298**, 1410–1411.

Roderick, M. L., L. D. Rotstayn, G. D. Farquhar, and M. T. Hobbins, 2007: On the attribution of changing pan evaporation. *Geophys. Res. Lett.*, **34**, L17403.

Rohde, R., et al., 2013a: A new estimate of the average Earth surface land temperature spanning 1753 to 2011. *Geoinfor. Geostat.: An Overview*, **1**, doi:10.4172/gigs.1000101.

Rohde, R., et al., 2013b: Berkeley Earth temperature averaging process. *Geoinfor Geostat: An Overview*, **1**, doi:10.4172/gigs.1000103.

Rohs, S., et al., 2006: Long-term changes of methane and hydrogen in the stratosphere in the period 1978–2003 and their impact on the abundance of stratospheric water vapor. *J. Geophys. Res. Atmos.*, **111**, D14315.

Ropelewski, C. F., and P. D. Jones, 1987: An extension of the Tahiti–Darwin Southern Oscillation Index. *Mon. Weather Rev.*, **115**, 2161–2165.

Rosenlof, K. H., and G. C. Reid, 2008: Trends in the temperature and water vapor content of the tropical lower stratosphere: Sea surface connection. *J. Geophys. Res. Atmos.*, **113**, D06107.

Ruckstuhl, C., et al., 2008: Aerosol and cloud effects on solar brightening and the recent rapid warming. *Geophys. Res. Lett.*, **35**, L12708.

Ruddiman, W., 2003: The anthropogenic greenhouse era began thousands of years ago. *Clim. Change*, 261–293.

Ruddiman, W., 2007: The early anthropogenic hypothesis: Challenges and responses. *Rev. Geophys.*, **45**, RG4001.

Russak, V., 2009: Changes in solar radiation and their influence on temperature trend in Estonia (1955–2007). *J. Geophys. Res. Atmos.*, **114**, D00d01.

Russell, A. R., L. C. Valin, E. J. Bucsela, M. O. Wenig, and R. C. Cohen, 2010: Space-based constraints on spatial and temporal patterns of NOx emissions in California, 2005–2008. *Environ. Sci. Technol.*, **44**, 3608–3615.

Russell, J., et al., 1993: The halogen occultation experiment. *J. Geophys. Res. Atmos.*, **98**, 10777–10797.

Rusticucci, M., and M. Renom, 2008: Variability and trends in indices of quality-controlled daily temperature extremes in Uruguay. *Int. J. Climatol.*, **28**, 1083–1095.

Saha, S., et al., 2010: The NCEP climate forecaset system reanalysis. *Bull. Am. Meteor. Soc.*, **91**, 1015–1057.

Saikawa, E., et al., 2012: Global and regional emissions estimates for HCFC-22. *Atmos. Chem. Phys. Discuss.*, **12**, 18423–18285.

Saji, N. H., B. N. Goswami, P. N. Vinayachandran, and T. Yamagata, 1999: A dipole mode in the tropical Indian Ocean. *Nature*, **401**, 360–363.

Sakamoto, M., and J. R. Christy, 2009: The influences of TOVS radiance assimilation on temperature and moisture tendencies in JRA-25 and ERA-40. *J. Atmos. Ocean Technol.*, **26**, 1435–1455.

Sanchez-Lorenzo, A., and M. Wild, 2012: Decadal variations in estimated surface solar radiation over Switzerland since the late 19th century. *Atmos. Chem. Phys. Discussion*, **12**, 10815–10843.

Sanchez-Lorenzo, A., J. Calbo, and J. Martin-Vide, 2008: Spatial and temporal trends in sunshine duration over Western Europe (1938–2004). *J. Clim.*, **21**, 6089–6098.

Sanchez-Lorenzo, A., J. Calbo, and M. Wild, 2013: Global and diffuse solar radiation in Spain: Building a homogeneous dataset and assessing their trends. *Global Planet. Change*, **100**, 343–352.

Sanchez-Lorenzo, A., J. Calbo, M. Brunetti, and C. Deser, 2009: Dimming/brightening over the Iberian Peninsula: Trends in sunshine duration and cloud cover and their relations with atmospheric circulation. *J. Geophys. Res. Atmos.*, **114**, D00d09.

Santer, B., et al., 2008: Consistency of modelled and observed temperature trends in the tropical troposphere. *Int. J. Climatol.*, **28**, 1703–1722.

Santer, B. D., et al., 2007: Identification of human-induced changes in atmospheric moisture content. *Proc. Natl. Acad. Sci. U.S.A.*, **104**, 15248–15253.

Santer, B. D., et al., 2011: Separating signal and noise in atmospheric temperature changes: The importance of timescale. *J. Geophys. Res. Atmos.*, **116**, D22105.

Scaife, A., C. Folland, L. Alexander, A. Moberg, and J. Knight, 2008: European climate extremes and the North Atlantic Oscillation. *J. Clim.*, **21**, 72–83.

Schär, C., P. L. Vidale, D. Luthi, C. Frei, C. Haberli, M. A. Liniger, and C. Appenzeller, 2004: The role of increasing temperature variability in European summer heatwaves. *Nature*, **427**, 332–336.

Scherer, M., H. Vömel, S. Fueglistaler, S. J. Oltmans, and J. Staehelin, 2008: Trends and variability of midlatitude stratospheric water vapour deduced from the re-evaluated Boulder balloon series and HALOE. *Atmos. Chem. Phys.*, **8**, 1391–1402.

Scherrer, S. C., and C. Appenzeller, 2006: Swiss Alpine snow pack variability: Major patterns and links to local climate and large-scale flow. *Clim. Res.*, **32**, 187–199.

Scherrer, S. C., C. Wüthrich, M. Croci-Maspoli, R. Weingartner, and C. Appenzeller, 2013: Snow variability in the Swiss Alps 1864–2009. *Int. J. Climatol.*, doi: 10.1002/joc.3653.

Schiller, C., J. U. Grooss, P. Konopka, F. Plager, F. H. Silva dos Santos, and N. Spelten, 2009: Hydration and dehydration at the tropical tropopause. *Atmos. Chem. Phys.*, **9**, 9647–9660.

Schmidt, G. A., 2009: Spurious correlations between recent warming and indices of local economic activity. *Int. J. Climatol.*, **29**, 2041–2048.

Schnadt Poberaj, C., J. Staehelin, D. Brunner, V. Thouret, H. De Backer, and R. Stübi, 2009: Long-term changes in UT/LS ozone between the late 1970s and the 1990s deduced from the GASP and MOZAIC aircraft programs and from ozonesondes. *Atmos. Chem. Phys.*, **9**, 5343–5369.

Schneider, T., P. A. O'Gorman, and X. J. Levine, 2010: Water vapour and the dynamics of climate changes. *Rev. Geophys*., RG3001.

Schneidereit, A., R. Blender, K. Fraedrich, and F. Lunkeit, 2007: Icelandic climate and north Atlantic cyclones in ERA-40 reanalyses. *Meteorol. Z.*, **16**, 17–23.

Schwartz, R. D., 2005: Global dimming: Clear-sky atmospheric transmission from astronomical extinction measurements. *J. Geophys. Res. Atmos.*, **110**.

Scinocca, J. F., D. B. Stephenson, T. C. Bailey, and J. Austin, 2010: Estimates of past and future ozone trends from multimodel simulations using a flexible smoothing spline methodology. *J. Geophys. Res. Atmos.*, **115**.

Screen, J. A., and I. Simmonds, 2011: Erroneous Arctic temperature trends in the ERA-40 reanalysis: A closer look. *J. Clim.*, **24**, 2620–2627.

Seidel, D. J., and J. R. Lanzante, 2004: An assessment of three alternatives to linear trends for characterizing global atmospheric temperature changes. *J. Geophys. Res. Atmos.*, **109,** D14108.

Seidel, D. J., and W. J. Randel, 2007: Recent widening of the tropical belt: Evidence from tropopause observations. *J. Geophys. Res. Atmos.*, **112,** D20113.

Seidel, D. J., Q. Fu, W. J. Randel, and T. J. Reichler, 2008: Widening of the tropical belt in a changing climate. *Nature Geosci.*, **1**, 21–24.

Seidel, D. J., N. P. Gillett, J. R. Lanzante, K. P. Shine, and P. W. Thorne, 2011: Stratospheric temperature trends: Our evolving understanding. *Clim. Change*, **2**, 592–616.

Sen Roy, S., 2009: A spatial analysis of extreme hourly precipitation patterns in India. *Int. J. Climatol.*, **29**, 345–355.

Sen Roy, S., and R. C. Balling, 2005: Analysis of trends in maximum and minimum temperature, diurnal temperature range, and cloud cover over India. *Geophys. Res. Lett.*, **32,** L12702.

Sen Roy, S., and M. Rouault, 2013: Spatial patterns of seasonal scale trends in extreme hourly precipitation in South Africa. *Appl. Geogr.*, **39**, 151–157.

Seneviratne, S. I., et al., 2010: Investigating soil moisture-climate interactions in a changing climate: A review. *Earth Sci. Rev.*, **99**, 125–161.

Seneviratne, S. I., et al., 2012: Changes in climate extremes and their impacts on the natural physical environment. In: *IPCC Special Report on Extremes*, 109-230.

Serquet, G., C. Marty, J. P. Dulex, and M. Rebetez, 2011: Seasonal trends and temperature dependence of the snowfall/precipitation-day ratio in Switzerland. *Geophys. Res. Lett.*, **38**, L07703.

Shaw, S. B., A. A. Royem, and S. J. Riha, 2011: The relationship between extreme hourly precipitation and surface temperature in different hydroclimatic regions of the United States. *J. Hydrometeor.*, **12**, 319–325.

Sheffield, J., and E. F. Wood, 2008: Global trends and variability in soil moisture and drought characteristics, 1950–2000, from observation-driven simulations of the terrestrial hydrologic cycle. *J. Clim.*, **21**, 432–458.

Sheffield, J., E. Wood, and M. Roderick, 2012: Little change in global drought over the past 60 years. *Nature*, **491**, 435–.

Sheffield, J., K. Andreadis, E. Wood, and D. Lettenmaier, 2009: Global and continental drought in the second half of the twentieth century: Severity-area-duration analysis and temporal variability of large-scale events. *J. Clim.*, **22**, 1962–1981.

Shekar, M., H. Chand, S. Kumar, K. Srinivasan, and A. Ganju, 2010: Climate change studies in the western Himalaya. *Ann. Glaciol.*, **51**, 105-112.

Sherwood, S. C., R. Roca, and T. M. Weckwerth, 2010: Tropospheric water vapor, convection, and climate. *Rev. Geophys.*, **48,** RG2001.

Sherwood, S. C., C. L. Meyer, R. J. Allen, and H. A. Titchner, 2008: Robust tropospheric warming revealed by iteratively homogenized radiosonde data. *J. Clim.*, **21**, 5336–5350.

Shi, G. Y., et al., 2008: Data quality assessment and the long-term trend of ground solar radiation in China. *J. Appl. Meteor. Climatol.*, **47**, 1006–1016.

Shi, L., and J. J. Bates, 2011: Three decades of intersatellite-calibrated High-Resolution Infrared Radiation Sounder upper tropospheric water vapor. *J. Geophys. Res. Atmos.*, **116**, D04108.

Shiklomanov, A. I., R. B. Lammers, M. A. Rawlins, L. C. Smith, and T. M. Pavelsky, 2007: Temporal and spatial variations in maximum river diSchärge from a new Russian data set. *J. Geophys. Res. Biogeosci.*, **112.**

Shiklomanov, I. A., V. Y. Georgievskii, V. I. Babkin, and Z. A. Balonishnikova, 2010: Research problems of formation and estimation of water resources and water availability changes of the Russian Federation. *Russ. Meteorol, Hydrol.*, **35**, 13–19.

Shiu, C. J., S. C. Liu, and J. P. Chen, 2009: Diurnally asymmetric trends of temperature, humidity, and precipitation in Taiwan. *J. Clim.*, **22**, 5635–5649.

Sillmann, J., M. Croci-Maspoli, M. Kallache, and R. W. Katz, 2011: Extreme cold winter temperatures in Europe under the influence of North Atlantic atmospheric blocking. *J. Clim.*, **24**, 5899–5913.

Simmons, A. J., K. M. Willett, P. D. Jones, P. W. Thorne, and D. P. Dee, 2010: Low-frequency variations in surface atmospheric humidity, temperature, and precipitation: Inferences from reanalyses and monthly gridded observational data sets. *J. Geophys. Res. Atmos.*, **115,** D01110.

Simolo, C., M. Brunetti, M. Maugeri, and T. Nanni, 2011: Evolution of extreme temperatures in a warming climate. *Geophys. Res. Lett.*, **38**, 6.

Simpson, I. J., et al., 2012: Long-term decline of global atmospheric ethane concentrations and implications for methane. *Nature*, **488**, 490–494.

Skansi, M., et al., 2013: Warming and wetting signals emerging from analysis of changes in climate extreme indices over South America. *Global Planet. Change*, **100**, 295–307.

Skeie, R. B., T. K. Berntsen, G. Myhre, K. Tanaka, M. M. Kvalevåg, and C. R. Hoyle, 2011: Anthropogenic radiative forcing time series from pre-industrial times until 2010. *Atmos. Chem. Phys.*, **11**, 11827–11857.

Smith, L. C., T. Pavelsky, G. MacDonald, I. A. Shiklomanov, and R. Lammers, 2007: Rising minimum daily flows in northern Eurasian rivers suggest a growing influence of groundwater in the high-latitude water cycle. *J. Geophys. Res.*, **112**, G04S47.

Smith, T. M., and R. W. Reynolds, 2002: Bias corrections for historical sea surface temperatures based on marine air temperatures. *J. Clim.*, **15**, 73–87.

Smith, T. M., T. C. Peterson, J. H. Lawrimore, and R. W. Reynolds, 2005: New surface temperature analyses for climate monitoring. *Geophys. Res. Lett.*, **32,** L14712.

Smith, T. M., R. W. Reynolds, T. C. Peterson, and J. Lawrimore, 2008: Improvements to NOAA's historical merged land-ocean surface temperature analysis (1880–2006). *J. Clim.*, **21**, 2283–2296.

Smith, T. M., P. A. Arkin, L. Ren, and S. S. P. Shen, 2012: Improved reconstruction of global precipitation since 1900. *J. Atmos. Ocean. Technol.*, **29**, 1505–1517.

Smits, A., A. Tank, and G. P. Konnen, 2005: Trends in storminess over the Netherlands, 1962–2002. *Int. J. Climatol.*, **25**, 1331–1344.

Sohn, B. J., and S. C. Park, 2010: Strengthened tropical circulations in past three decades inferred from water vapor transport. *J. Geophys. Res. Atmos.*, **115**, D15112.

Solomon, S., K. Rosenlof, R. Portmann, J. Daniel, S. Davis, T. Sanford, and G. Plattner, 2010: Contributions of stratospheric water vapor to decadal changes in the rate of global warming. *Science*, **327**, 1219-1223.

Song, H., and M. H. Zhang, 2007: Changes of the boreal winter Hadley circulation in the NCEP-NCAR and ECMWF reanalyses: A comparative study. *J. Clim.*, **20**, 5191–5200.

Soni, V. K., G. Pandithurai, and D. S. Pai, 2012: Evaluation of long-term changes of solar radiation in India. *Int. J. Climatol.*, **32**, 540–551.

Sorteberg, A., and J. E. Walsh, 2008: Seasonal cyclone variability at 70 degrees N and its impact on moisture transport into the Arctic. *Tellus A*, **60**, 570–586.

Sousa, P., R. Trigo, P. Aizpurua, R. Nieto, L. Gimeno, and R. Garcia-Herrera, 2011: Trends and extremes of drought indices throughout the 20th century in the Mediterranean. *Nat. Hazards Earth Syst. Sci.*, **11**, 33–51.

Spencer, R. W., and J. R. Christy, 1992: Precision and radiosonde validation of satellite gridpoint temperature anomalies. 2. A troposhperic retrieval and trends during 1979–90. *J. Clim.*, **5**, 858–866.

St. Jacques, J.-M., and D. Sauchyn, 2009: Increasing winter baseflow and mean annual streamflow from possible permafrost thawing in the Northwest Territories, Canada. *Geophys. Res. Lett.*, **36,** L01401.

Stachnik, J. P., and C. Schumacher, 2011: A comparison of the Hadley circulation in modern reanalyses. *J. Geophys. Res. Atmos.*, **116,** D22102.

Stahl, K., and L. M. Tallaksen, 2012: Filling the white space on maps of European runoff trends: Estimates from a multi-model ensemble. *Hydrol. Earth Syst. Sci. Discuss.*, **9**, 2005–2032.

Stahl, K., et al., 2010: Streamflow trends in Europe: Evidence from a dataset of near-natural catchments. *Hydrol. Earth Syst. Sci.*, **14**, 2367–2382.

Stanhill, G., and S. Cohen, 2001: Global dimming: A review of the evidence for a widespread and significant reduction in global radiation with discussion of its probable causes and possible agricultural consequences. *Agr. Forest Meteorol.*, **107**, 255–278.

Steeneveld, G. J., A. A. M. Holtslag, R. T. McNider, and R. A. Pielke, 2011: Screen level temperature increase due to higher atmospheric carbon dioxide in calm and windy nights revisited. *J. Geophys. Res. Atmos.*, **116**.

Stegall, S., and J. Zhang, 2012: Wind field climatology, changes, and extremes in the Chukchi-Beaufort Seas and Alaska North Slope during 1979–2009. *J. Clim.*, **25**, 8075–8089.

Steig, E. J., D. P. Schneider, S. D. Rutherford, M. E. Mann, J. C. Comiso, and D. T. Shindell, 2009: Warming of the Antarctic ice-sheet surface since the 1957 International Geophysical Year. *Nature*, **460**, 766–766.

Stenke, A., and V. Grewe, 2005: Simulation of stratospheric water vapor trends: Impact on stratospheric ozone chemistry. *Atmos. Chem. Phys.*, **5**, 1257–1272.

Stephens, G. L., M. Wild, P. W. Stackhouse, T. L'Ecuyer, S. Kato, and D. S. Henderson, 2012a: The global character of the flux of downward longwave radiation. *J. Clim.*, **25**, 2329–2340.

Stephens, G. L., et al., 2012b: An update on Earth's energy balance in light of the latest global observations. *Nature Geosci.*, **5**, 691–696.

Stephenson, D. B., H. F. Diaz, and R. J. Murnane, 2008: Definition, diagnosis and origin of extreme weather and climate events. In: *Climate Extremes and Society* [R. J. Murnane, and H. F. Diaz (eds.)] Cambridge University Press, Cambridge, United Kingdom and New York, NY, USA, pp. 11–23.

Stern, D. I., 2006: Reversal of the trend in global anthropogenic sulfur emissions. *Global Environ. Change Hum. Policy Dimens.*, **16**, 207–220.

Stickler, A., et al., 2010: The Comprehensive Historical Upper-Air Network. *Bull. Am. Meteor. Soc.*, **91**, 741–751.

Stjern, C. W., J. E. Kristjansson, and A. W. Hansen, 2009: Global dimming and global brightening - an analysis of surface radiation and cloud cover data in northern Europe. *Int. J. Climatol.*, **29**, 643–653.

Stohl, A., et al., 2009: An analytical inversion method for determining regional and global emissions of greenhouse gases: Sensitivity studies and application to halocarbons. *Atmos. Chem. Phys.*, 1597–1620.

Stohl, A., et al., 2010: Hydrochlorofluorocarbon and hydrofluorocarbon emissions in East Asia determined by inverse modeling. *Atmos. Chem. Phys.*, 3545–3560.

Streets, D. G., Y. Wu, and M. Chin, 2006: Two-decadal aerosol trends as a likely explanation of the global dimming/brightening transition. *Geophys. Res. Lett.*, **33**, L15806.

Streets, D. G., et al., 2009: Anthropogenic and natural contributions to regional trends in aerosol optical depth, 1980–2006. *J. Geophys. Res. Atmos.*, **114**, D00D18.

Strong, C., and R. E. Davis, 2007: Winter jet stream trends over the Northern Hemisphere. *Q. J. R. Meteorol. Soc.*, **133**, 2109–2115.

Strong, C., and R. E. Davis, 2008: Comment on "Historical trends in the jet streams" by Cristina L. Archer and Ken Caldeira. *Geophys. Res. Lett.*, L24806.

Stubenrauch, C. J., et al., 2013: Assessment of global cloud datasets from satellite: Project and database initiated by the GEWEX radiation panel. *Bull. Am. Meteorol. Soc.*, **94**, 1031-1049.

Sun, B. M., A. Reale, D. J. Seidel, and D. C. Hunt, 2010: Comparing radiosonde and COSMIC atmospheric profile data to quantify differences among radiosonde types and the effects of imperfect collocation on comparison statistics. *J. Geophys. Res. Atmos.*, **115**, D23104.

Swart, N. C., and J. C. Fyfe, 2012: Observed and simulated changes in the Southern Hemisphere surface westerly wind-stress. *Geophys. Res. Lett.*, 39, L16711.

Syakila, A., and C. Kroeze, 2011: The global nitrous oxide budget revisited. *Greenhouse Gas Meas. Management*, **1**, 17–26.

Takahashi, K., A. Montecinos, K. Goubanova, and B. Dewitte, 2011: ENSO regimes: Reinterpreting the canonical and Modoki El Niño. *Geophys. Res. Lett.*, **38**, L10704.

Takeuchi, Y., Y. Endo, and S. Murakami, 2008: High correlation between winter precipitation and air temperature in heavy-snowfall areas in Japan. *Ann. Glaciol.*, **49**, 7–10.

Tang, G., Y. Ding, S. Wang, G. Ren, H. Liu, and L. Zhang, 2010: Comparative analysis of China surface air temperature series for the past 100 years. *Adv. Climate Change Res.*, **1**, 11–19.

Tang, W. J., K. Yang, J. Qin, C. C. K. Cheng, and J. He, 2011: Solar radiation trend across China in recent decades: a revisit with quality-controlled data. *Atmos. Chem. Phys.*, **11**, 393–406.

Tank, A., et al., 2006: Changes in daily temperature and precipitation extremes in central and south Asia. *J. Geophys. Res. Atmos.*, **111**, D16105.

Tans, P., 2009: An accounting of the observed increase in oceanic and atmospheric CO_2 and an outlook for the future. *Oceanography*, 26–35.

Tarasova, O. A., I. A. Senik, M. G. Sosonkin, J. Cui, J. Staehelin, and A. S. H. PrÃ©vÃ´t, 2009: Surface ozone at the Caucasian site Kislovodsk High Mountain Station and the Swiss Alpine site Jungfraujoch: Data analysis and trends (1990–2006). *Atmos. Chem. Phys.*, **9**, 4157–4175.

Tegtmeier, S., K. Kruger, I. Wohltmann, K. Schoellhammer, and M. Rex, 2008: Variations of the residual circulation in the Northern Hemispheric winter. *J. Geophys. Res. Atmos.*, **113**, D16109.

Teuling, A. J., et al., 2009: A regional perspective on trends in continental evaporation. *Geophys. Res. Lett.*, **36**, L02404.

Thomas, B., E. Kent, V. Swail, and D. Berry, 2008: Trends in ship wind speeds adjusted for observation method and height. *Int. J. Climatol.*, **28**, 747–763.

Thomas, G. E., et al., 2010: Validation of the GRAPE single view aerosol retrieval for ATSR-2 and insights into the long term global AOD trend over the ocean. *Atmos. Chem. Phys.*, **10**, 4849–4866.

Thompson, D. W. J., and J. M. Wallace, 1998: The Arctic Oscillation signature in the wintertime geopotential height and temperature fields. *Geophys. Res. Lett.*, **25**, 1297–1300.

Thompson, D. W. J., and J. M. Wallace, 2000: Annular modes in the extratropical circulation. Part I: Month-to-month variability. *J. Clim.*, **13**, 1000–1016.

Thompson, D. W. J., J. J. Kennedy, J. M. Wallace, and P. D. Jones, 2008: A large discontinuity in the mid-twentieth century in observed global-mean surface temperature. *Nature*, **453**, 646–649.

Thompson, D. W. J., et al., 2012: The mystery of recent stratospheric temperature trends. *Nature*, **491**, 692–697.

Thorne, P. W., 2008: Arctic tropospheric warming amplification? *Nature*, **455**, E1–E2.

Thorne, P. W., and R. S. Vose, 2010: Reanalyses suitable for characterizing long-term trends: Are They Really Achievable? *Bull. Am. Meteor. Soc.*, **91**, 353–.

Thorne, P. W., D. E. Parker, S. F. B. Tett, P. D. Jones, M. McCarthy, H. Coleman, and P. Brohan, 2005: Revisiting radiosonde upper air temperatures from 1958 to 2002. *J. Geophys. Res. Atmos.*, **110**.

Thorne, P. W., et al., 2011: A quantification of uncertainties in historical tropical tropospheric temperature trends from radiosondes. *J. Geophys. Res. Atmos.*, **116**, D12116.

Tietavainen, H., H. Tuomenvirta, and A. Venalainen, 2010: Annual and seasonal mean temperatures in Finland during the last 160 years based on gridded temperature data. *Int. J. Climatol.*, **30**, 2247–2256.

Titchner, H. A., P. W. Thorne, M. P. McCarthy, S. F. B. Tett, L. Haimberger, and D. E. Parker, 2009: Critically reassessing tropospheric temperature trends from radiosondes using realistic validation experiments. *J. Clim.*, **22**, 465–485.

Tokinaga, H., and S.-P. Xie, 2011a: Wave and anemometer-based sea surface wind (WASWind) for climate change analysis. *J. Clim.*, 267-285.

Tokinaga, H., and S. P. Xie, 2011b: Weakening of the equatorial Atlantic cold tongue over the past six decades. *Nature Geosci.*, **4**, 222-226.

Tokinaga, H., S. P. Xie, A. Timmermann, S. McGregor, T. Ogata, H. Kubota, and Y. M. Okumura, 2012: Regional patterns of tropical Indo-Pacific climate change: Evidence of the Walker circulation weakening. *J. Clim.*, **25**, 1689–1710.

Toreti, A., E. Xoplaki, D. Maraun, F. G. Kuglitsch, H. Wanner, and J. Luterbacher, 2010: Characterisation of extreme winter precipitation in Mediterranean coastal sites and associated anomalous atmospheric circulation patterns. *Nat. Hazards Earth Syst. Sci.*, **10**, 1037–1050.

Torseth, K., et al., 2012: Introduction to the European Monitoring and Evaluation Programme (EMEP) and observed atmospheric composition change during 1972–2009. *Atmos. Chem. Phys. Discuss.*, **12**, 1733–1820.

Trenberth, K., 2011: Changes in precipitation with climate change. *Clim. Res.*, **47**, 123–138.

Trenberth, K., and J. Fasullo, 2012a: Climate extremes and climate change: The Russian heat wave and other climate extremes of 2010. *J. Geophys. Res. Atmos.*, **117**, D17103.

Trenberth, K. E., 1984: Signal versus noise in the Southern Oscillation. *Mon. Weather Rev.*, **112**, 326–332.

Trenberth, K. E., 1997: The definition of El Niño. *Bull. Am. Meteor. Soc.*, **78**, 2771–2777.

Trenberth, K. E., and D. A. Paolino, 1980: The Northern Hemisphere Sea-Level Pressure Data Set – Trends, errors, and discontinuities. *Mon. Weather Rev.*, **108**, 855–872.

Trenberth, K. E., and J. W. Hurrell, 1994: Decadal atmosphere-ocean variations in the Pacific. *Clim. Dyn.*, **9**, 303–319.

Trenberth, K. E., and T. J. Hoar, 1996: The 1990–1995 El Niño Southern Oscillation event: Longest on record. *Geophys. Res. Lett.*, **23**, 57–60.

Trenberth, K. E., and D. P. Stepaniak, 2001: Indices of El Niño evolution. *J. Clim.*, **14**, 1697–1701.

Trenberth, K. E., and D. J. Shea, 2006: Atlantic hurricanes and natural variability in 2005. *Geophys. Res. Lett.*, **33**, L12704.

Trenberth, K. E., and J. T. Fasullo, 2010: Climate change tracking Earth's energy. *Science*, **328**, 316–317.

Trenberth, K. E., and J. T. Fasullo, 2012b: Tracking Earth's energy: From El Niño to global warming. *Surv. Geophys.*, **33**, 413–426.

Trenberth, K. E., J. T. Fasullo, and J. Kiehl, 2009: Earth's Global energy budget. *Bull. Am. Meteor. Soc.*, **90**, 311.

Trenberth, K. E., J. T. Fasullo, and J. Mackaro, 2011: Atmospheric moisture transports from ocean to land and global energy flows in reanalyses. *J. Clim.*, **24**, 4907–4924.

Trenberth, K. E., et al., 2007: Observations: Surface and atmospheric climate change. In: *Climate Change 2007: The Physical Science Basis. Contribution of Working Group I to the Fourth Assessment Report of the Intergovernmental Panel on Climate Change* [Solomon, S., D. Qin, M. Manning, Z. Chen, M. Marquis, K. B. Averyt, M. Tignor and H. L. Miller (eds.)]. Cambridge University Press, Cambridge, United Kingdom and New York, NY, USA.

Trewin, B., 2012: A daily homogenized temperature data set for Australia. *Int. J. Climatol.*, 33, 1510-1529.

Trnka, M., J. Kysely, M. Mozny, and M. Dubrovsky, 2009: Changes in Central-European soil-moisture availability and circulation patterns in 1881–2005. *Int. J. Climatol.*, **29**, 655–672.

Troccoli, A., K. Muller, P. Coppin, R. Davy, C. Russell, and A. L. Hirsch, 2012: Long-term wind speed trends over Australia. *J. Climate*, **25**, 170–183.

Troup, A. J., 1965: Southern Oscillation. *Q. J. R. Meteorol. Soc.*, **91**, 490–.

Tryhorn, L., and J. Risbey, 2006: On the distribution of heat waves over the Australian region. *Aust. Meteorol. Mag.*, **55**, 169–182.

Tung, K.-K., and J. Zhou, 2013: Using data to attribute episodes of warming and cooling in instrumental records. *Proc. Natl. Acad. Sci. U.S.A.*, 110, 2058-2063.

Turner, J., et al., 2005: Antarctic climate change during the last 50 years. *Int. J. Climatol.*, **25**, 279–294.

Ulbrich, U., G. C. Leckebusch, and J. G. Pinto, 2009: Extra-tropical cyclones in the present and future climate: A review. *Theor. Appl. Climatol.*, **96**, 117–131.

Uppala, S. M., et al., 2005: The ERA-40 re-analysis. *Q. J. R. Meteorol. Soc.*, **131**, 2961–3012.

Usbeck, T., T. Wohlgemuth, C. Pfister, R. Volz, M. Beniston, and M. Dobbertin, 2010: Wind speed measurements and forest damage in Canton Zurich (Central Europe) from 1891 to winter 2007. *Int. J. Climatol.*, **30**, 347–358.

Utsumi, N., S. Seto, S. Kanae, E. Maeda, and T. Oki, 2011: Does higher surface temperature intensify extreme precipitation? *Geophys. Res. Lett.*, **38,** L16708.

van den Besselaar, E. J. M., A. M. G. Klein Tank, and T. A. Buishand, 2012: Trends in European precipitation extremes over 1951–2010. *Int. J. Climatol.*, **33**, 2682–2689.

van der Schrier, G., A. van Ulden, and G. J. van Oldenborgh, 2011: The construction of a Central Netherlands temperature. *Clim. Past*, **7**, 527–542.

van der Schrier, G., J. Barichivich, K. R. Briffa, and P. D. Jones, 2013: A scPDSI-based global dataset of dry and wet spells for 1901–2009. *J. Geophys. Res. Atmos.*, **118**, 4025-4048.

van Haren, R., G. J. van Oldenborgh, G. Lenderink, M. Collins, and W. Hazeleger, 2012: SST and circulation trend biases cause an underestimation of European precipitation trends. *Clim. Dyn.*, **40**, 1-20.

van Heerwaarden, C. C., J. V. G. de Arellano, and A. J. Teuling, 2010: Land-atmosphere coupling explains the link between pan evaporation and actual evapotranspiration trends in a changing climate. *Geophys. Res. Lett.*, **37**, L21401.

van Ommen, T. D., and V. Morgan, 2010: Snowfall increase in coastal East Antarctica linked with southwest Western Australian drought. *Nature Geosci.*, **3**, 267–272.

Vautard, R., P. Yiou, and G. J. van Oldenborgh, 2009: Decline of fog, mist and haze in Europe over the past 30 years. *Nature Geosci.*, **2**, 115–119.

Vautard, R., J. Cattiaux, P. Yiou, J. N. The paut, and P. Ciais, 2010: Northern Hemisphere atmospheric stilling partly attributed to an increase in surface roughness. *Nature Geosci.*, **3**, 756-761.

Vautard, R., et al., 2007: Summertime European heat and drought waves induced by wintertime Mediterranean rainfall deficit. *Geophys. Res. Lett.*, **34.**, L07711

Vecchi, G. A., and B. J. Soden, 2007: Global warming and the weakening of the tropical circulation. *J. Clim.*, **20**, 4316–4340.

Vecchi, G. A., and T. R. Knutson, 2008: On estimates of historical north Atlantic tropical cyclone activity. *J. Clim.*, **21**, 3580–3600.

Vecchi, G. A., and T. R. Knutson, 2011: Estimating annual numbers of Atlantic hurricanes missing from the HURDAT database (1878–1965) using ship track density. *J. Clim.*, **24**, 1736–1746.

Vecchi, G. A., B. J. Soden, A. T. Wittenberg, I. M. Held, A. Leetmaa, and M. J. Harrison, 2006: Weakening of tropical Pacific atmospheric circulation due to anthropogenic forcing. *Nature*, **441**, 73–76.

Velders, G., S. Andersen, J. Daniel, D. Fahey, and M. McFarland, 2007: The importance of the Montreal Protocol in protecting climate. *Proc. Natl. Acad. Sci. U.S.A.*, **104**, 4814-4819.

Velders, G., D. Fahey, J. Daniel, M. McFarland, and S. Andersen, 2009: The large contribution of projected HFC emissions to future climate forcing. *Proc. Natl. Acad. Sci. U.S.A.*, **106**, 10949-10954.

Venema, V. K. C., et al., 2012: Benchmarking homogenization algorithms for monthly data. *Clim. Past*, **8**, 89–115.

Verbout, S., H. Brooks, L. Leslie, and D. Schultz, 2006: Evolution of the US tornado database: 1954–2003. *Weather Forecast.*, **21**, 86–93.

Vicente-Serrano, S. M., S. Begueria, and J. I. Lopez-Moreno, 2010: A multiscalar drought index sensitive to global warming: The Standardized Precipitation Evapotranspiration Index. *J. Clim.*, **23**, 1696–1718.

Vidal, J., E. Martin, L. Franchisteguy, F. Habets, J. Soubeyroux, M. Blanchard, and M. Baillon, 2010: Multilevel and multiscale drought reanalysis over France with the Safran-Isba-Modcou hydrometeorological suite. *Hydrol. Earth Syst. Sci.*, **14**, 459–478.

Vilibic, I., and J. Sepic, 2010: Long-term variability and trends of sea level storminess and extremes in European Seas. *Global Planet. Change*, **71**, 1–12.

Villarini, G., J. Smith, and G. Vecchi, 2013: Changing frequency of heavy rainfall over the central United States. *J. Clim.*, **26**, 351–357.

Vincent, L., et al., 2011: Observed trends in indices of daily and extreme temperature and precipitation for the countries of the western Indian Ocean, 1961–2008. *J. Geophys. Res. Atmos.*, **116**, D10108.

Vincent, L. A., X. L. L. Wang, E. J. Milewska, H. Wan, F. Yang, and V. Swail, 2012: A second generation of homogenized Canadian monthly surface air temperature for climate trend analysis. *J. Geophys. Res. Atmos.*, **117**, D18110.

Visbeck, M., 2009: A station-based Southern Annular Mode Index from 1884 to 2005. *J. Clim.*, **22**, 940–950.

Volz, A., and D. Kley, 1988: Evaluation of the Montsouris series of ozone measurements made in the 19th century. *Nature*, **332**, 240–242.

Vömel, H., D. E. David, and K. Smith, 2007a: Accuracy of tropospheric and stratospheric water vapor measurements by the cryogenic frost point hygrometer: Instrumental details and observations. *J. Geophys. Res. Atmos.*, **112,** D08305.

Vömel, H., et al., 2007b: Validation of Aura Microwave Limb Sounder water vapor by balloon-borne cryogenic frost point hygrometer measurements. *J. Geophys. Res. Atmos.*, **112,** D24S37.

von Clarmann, T., et al., 2009: Retrieval of temperature, H_2O, O_3, HNO_3, CH4, N_2O, $ClONO_2$ and ClO from MIPAS reduced resolution nominal mode limb emission measurements. *Atmos. Meas. Tech.*, **2**, 159–175.

von Schuckmann, K., and P.-Y. Le Traon, 2011: How well can we derive global ocean indicators from Argo data? *Ocean Sci.*, **7**, 783–791.

Von Storch, H., 1999: Misuses of statistical analysis in climate research. *Analysis of Climate Variability: Applications of Statistical Techniques*, 2nd edition [H. Von Storch and A. Navarra (eds.)]. Springer-Verlag, New York, and Heidelberg, Germany, pp. 11–26.

von Storch, H., and F. W. Zwiers, 1999: *Statistical Analysis in Climate Research.* Cambridge University Press, Cambridge, United Kingdom and New York, NY, USA, 484 pp.

Vose, R. S., D. R. Easterling, and B. Gleason, 2005a: Maximum and minimum temperature trends for the globe: An update through 2004. *Geophys. Res. Lett.*, **32,** L23822.

Vose, R. S., D. Wuertz, T. C. Peterson, and P. D. Jones, 2005b: An intercomparison of trends in surface air temperature analyses at the global, hemispheric, and grid-box scale. *Geophys. Res. Lett.*, **32**, L18718.

Vose, R. S., S. Applequist, M. J. Menne, C. N. Williams, Jr., and P. Thorne, 2012a: An intercomparison of temperature trends in the US Historical Climatology Network and recent atmospheric reanalyses. *Geophys. Res. Lett.*, **39**, L10703.

Vose, R. S., Oak Ridge National Laboratory. Environmental Sciences Division., U.S. Global Change Research Program, United States. Dept. of Energy. Office of Health and Environmental Research., Carbon Dioxide Information Analysis Center (U.S.), and Martin Marietta Energy Systems Inc., 1992: *The Global Historical Climatology Network: Long-Term Monthly Temperature, Precipitation, Sea Level Pressure, and Station Pressure Data.* Carbon Dioxide Information Analysis Center. Available to the public from N.T.I.S., 1 v. (various pagings)

Vose, R. S., et al., 2012b: NOAA's Merged Land-Ocean Surface Temperature Analysis. *Bull. Am. Meteor. Soc.*, **93**, 1677–1685.

Wacker, S., J. Grobner, K. Hocke, N. Kampfer, and L. Vuilleumier, 2011: Trend analysis of surface cloud-free downwelling long-wave radiation from four Swiss sites. *J. Geophys. Res. Atmos.*, **116**, 13.

Wallace, J. M., and D. S. Gutzler, 1981: Teleconnections in the geopotential height field during the Northern Hemisphere winter. *Mon. Weather Rev.*, **109**, 784–812.

Wan, H., X. L. Wang, and V. R. Swail, 2010: Homogenization and trend analysis of Canadian near-surface wind speeds. *J. Clim.*, **23**, 1209–1225.

Wan, H., X. Zhang, F. Zwiers, S. Emori, and H. Shiogama, 2013: Effect of data coverage on the estimation of mean and variability of precipitation at global and regional scales. *J. Geophys. Res.*, **118**, 534–546.

Wang, B., J. Liu, H. J. Kim, P. J. Webster, and S. Y. Yim, 2012a: Recent change of the global monsoon precipitation (1979–2008). *Clim. Dyn.*, **39**, 1123–1135.

Wang, H., et al., 2012b: Extreme climate in China: Facts, simulation and projection. *Meteorol. Z.*, **21**, 279–304.

Wang, J. H., and L. Y. Zhang, 2008: Systematic errors in global radiosonde precipitable water data from comparisons with ground-based GPS measurements. *J. Clim.*, **21**, 2218–2238.

Wang, J. H., and L. Y. Zhang, 2009: Climate applications of a global, 2-hourly atmospheric precipitable water dataset derived from IGS tropospheric products. *J. Geodes.*, **83**, 209–217.

Wang, J. H., L. Y. Zhang, A. Dai, T. Van Hove, and J. Van Baelen, 2007: A near-global, 2-hourly data set of atmospheric precipitable water from ground-based GPS measurements. *J. Geophys. Res. Atmos.*, **112**, D11107.

Wang, J. S., D. J. Seidel, and M. Free, 2012c: How well do we know recent climate trends at the tropical tropopause? *J. Geophys. Res. Atmos.*, **117**, D09118.

Wang, K., R. E. Dickinson, and S. Liang, 2009a: Clear sky visibility has decreased over land globally from 1973 to 2007. *Science*, **323**, 1468–1470.

Wang, K., H. Ye, F. Chen, Y. Z. Xiong, and C. P. Wang, 2012d: Urbanization effect on the diurnal temperature range: Different roles under solar dimming and brightening. *J. Clim.*, **25**, 1022–1027.

Wang, K. C., and S. L. Liang, 2009: Global atmospheric downward longwave radiation over land surface under all-sky conditions from 1973 to 2008. *J. Geophys. Res. Atmos.*, **114**, D19101.

Wang, K. C., R. E. Dickinson, and S. L. Liang, 2009b: Clear sky visibility has decreased over land globally from 1973 to 2007. *Science*, **323**, 1468–1470.

Wang, K. C., R. E. Dickinson, M. Wild, and S. L. Liang, 2010: Evidence for decadal variation in global terrestrial evapotranspiration between 1982 and 2002: 2. Results. *J. Geophys. Res. Atmos.*, **115**, D20113.

Wang, K. C., R. E. Dickinson, M. Wild, and S. Liang, 2012e: Atmospheric impacts on climatic variability of surface incident solar radiation. *Atmos. Chem. Phys.*, **12**, 9581–9592.

Wang, K. C., R. E. Dickinson, L. Su, and K. E. Trenberth, 2012f: Contrasting trends of mass and optical properties of aerosols over the Northern Hemisphere from 1992 to 2011. *Atmos. Chem. Phys.*, **12**, 9387–9398.

Wang, L. K., C. Z. Zou, and H. F. Qian, 2012g: Construction of stratospheric temperature data records from Stratospheric Sounding Units. *J. Clim.*, **25**, 2931–2946.

Wang, X., H. Wan, and V. Swail, 2006a: Observed changes in cyclone activity in Canada and their relationships to major circulation regimes. *J. Clim.*, **19**, 896–915.

Wang, X., B. Trewin, Y. Feng, and D. Jones, 2013: Historical changes in Australian temperature extremes as inferred from extreme value distribution analysis. *Geophys. Res. Lett.*, 40, 573-578.

Wang, X., Y. Feng, G. P. Compo, V. R. Swail, F. W. Zwiers, R. J. Allan, and P. D. Sardeshmukh, 2012: Trends and low frequency variability of extra-tropical cyclone activity in the ensemble of twentieth century reanalysis. *Clim. Dyn.*, **40**, 2775-2800.

Wang, X., et al., 2011: Trends and low-frequency variability of storminess over western Europe, 1878–2007. *Clim. Dyn.*, **37**, 2355-2371.

Wang, X. L. L., V. R. Swail, and F. W. Zwiers, 2006b: Climatology and changes of extratropical cyclone activity: Comparison of ERA-40 with NCEP-NCAR reanalysis for 1958–2001. *J. Clim.*, **19**, 3145–3166.

Wang, X. L. L., F. W. Zwiers, V. R. Swail, and Y. Feng, 2009c: Trends and variability of storminess in the Northeast Atlantic region, 1874–2007. *Clim. Dyn.*, **33**, 1179–1195.

Wang, X. M., P. M. Zhai, and C. C. Wang, 2009d: Variations in extratropical cyclone activity in northern East Asia. *Adv. Atmos. Sci.*, **26**, 471–479.

Warren, S. G., R. M. Eastman, and C. J. Hahn, 2007: A survey of changes in cloud cover and cloud types over land from surface observations, 1971–96. *J. Clim.*, **20**, 717–738.

Weaver, S. J., 2012: Factors associated with decadal variability in Great Plains summertime surface temperatures. *J. Clim.*, **26**, 343–350.

Weber, M., W. Steinbrecht, C. Long, V. E. Fioletov, S. H. Frith, R. Stolarski, and P. A. Newman, 2012: Stratospheric ozone [in "State of the Climate in 2011"]. *Bull. Am. Met. Soc.*, **93**, S46–S44.

Weinkle, J., R. Maue, and R. Pielke, 2012: Historical global tropical cyclone landfalls. *J. Clim.*, **25**, 4729–4735.

Weinstock, E. M., et al., 2009: Validation of the Harvard Lyman-alpha in situ water vapor instrument: Implications for the mechanisms that control stratospheric water vapor. *J. Geophys. Res. Atmos.*, **114**.

Weiss, R., J. Muhle, P. Salameh, and C. Harth, 2008: Nitrogen trifluoride in the global atmosphere. *Geophys. Res. Lett.*, **35**, L20821.

Wells, N., S. Goddard, and M. J. Hayes, 2004: A self-calibrating Palmer Drought Severity Index. *J. Clim.*, **17**, 2335–2351.

Wentz, F., C. Gentemann, D. Smith, and D. Chelton, 2000: Satellite measurements of sea surface temperature through clouds. *Science*, **288**, 847–850.

Wentz, F. J., L. Ricciardulli, K. Hilburn, and C. Mears, 2007: How much more rain will global warming bring? *Science*, **317**, 233–235.

Werner, P. C., F. W. Gerstengarbe, and F. Wechsung, 2008: Grosswetterlagen and precipitation trends in the Elbe River catchment. *Meteorol. Z.*, **17**, 61–66.

Westra, S., and S. Sisson, 2011: Detection of non-stationarity in precipitation extremes using a max-stable process model. *J. Hydrol.*, **406**, 119–128.

Westra, S., L. Alexander, and F. Zwiers, 2013: Global increasing trends in annual maximum daily precipitation. *J. Clim.*, **26**, 3904-3918.

Wibig, J., 2008: Cloudiness variations in Lodz in the second half of the 20th century. *Int. J. Climatol.*, 28, 479–491.

Wickham, C., et al., 2013: Influence of urban heating on the global temperature land average using rural sites identified from MODIS classifications. *Geoinfor Geostat: An Overview*, **1**, 1:2. doi:10.4172/gigs.1000104.

Wielicki, B. A., B. R. Barkstrom, E. F. Harrison, R. B. Lee, G. L. Smith, and J. E. Cooper, 1996: Clouds and the Earth's radiant energy system (CERES): An Earth observing system experiment. *Bull. Am. Meteor. Soc.*, **77**, 853–868.

Wielicki, B. A., et al., 2002: Evidence for large decadal variability in the tropical mean radiative energy budget. *Science*, **295**, 841–844.

Wilby, R. L., P. D. Jones, and D. H. Lister, 2011: Decadal variations in the nocturnal heat island of London. *Weather*, **66**, 59–64.

Wild, M., 2009: Global dimming and brightening: A review. *J. Geophys. Res. Atmos.*, **114**, D00D16.

Wild, M., 2012: Enlightening global dimming and brightening. *Bull. Am. Meteor. Soc.*, **93**, 27–37.

Wild, M., A. Ohmura, and K. Makowski, 2007: Impact of global dimming and brightening on global warming. *Geophys. Res. Lett.*, **34**, L04702.

Wild, M., J. Grieser, and C. Schaer, 2008: Combined surface solar brightening and increasing greenhouse effect support recent intensification of the global land-based hydrological cycle. *Geophys. Res. Lett.*, **35**, L17706.

Wild, M., A. Ohmura, H. Gilgen, and D. Rosenfeld, 2004: On the consistency of trends in radiation and temperature records and implications for the global hydrological cycle. *Geophys. Res. Lett.*, **31**, L11201.

Wild, M., A. Ohmura, H. Gilgen, E. Roeckner, M. Giorgetta, and J. J. Morcrette, 1998: The disposition of radiative energy in the global climate system: GCM-calculated versus observational estimates. *Clim. Dyn.*, **14**, 853–869.

Wild, M., D. Folini, C. Schär, N. Loeb, E. G. Dutton, and G. König-Langlo, 2013: The global energy balance from a surface perspective. *Clim. Dyn.*, **40**, 3107-3134.

Wild, M., B. Truessel, A. Ohmura, C. N. Long, G. Konig-Langlo, E. G. Dutton, and A. Tsvetkov, 2009: Global dimming and brightening: An update beyond 2000. *J. Geophys. Res. Atmos.*, **114**, D00d13.

Wild, M., et al., 2005: From dimming to brightening: Decadal changes in solar radiation at Earth's surface. *Science*, **308**, 847–850.

Wilks, D. S., 2006: *Statistical Methods in the Atmospheric Sciences,* 2nd edition. Elsevier, Philadelphia, 627 pp.

Willett, K. M., P. D. Jones, N. P. Gillett, and P. W. Thorne, 2008: Recent Changes in Surface Humidity: Development of the HadCRUH dataset. *J. Clim.*, **21**, 5364–5383.

Willett, K. M., P. D. Jones, P. W. Thorne, and N. P. Gillett, 2010: A comparison of large scale changes in surface humidity over land in observations and CMIP3 general circulation models. *Environ. Res. Lett.*, **5**.

Willett, K. M., et al., 2013: HadISDH: an updateable land surface specific humidity product for climate monitoring. *Clim. Past*, **9**, 657–677.

Williams, C. N., M. J. Menne, and P. W. Thorne, 2012: Benchmarking the performance of pairwise homogenization of surface temperatures in the United States. *J. Geophys. Res. Atmos.*, **117**.

Willson, R. C., and A. V. Mordvinov, 2003: Secular total solar irradiance trend during solar cycles 21–23. *Geophys. Res. Lett.*, **30**, 1199.

Winkler, P., 2009: Revision and necessary correction of the long-term temperature series of Hohenpeissenberg, 1781–2006. *Theor. Appl. Climatol.*, **98**, 259–268.

Wong, T., B. A. Wielicki, R. B. Lee, ., G. L. Smith, K. A. Bush, and J. K. Willis, 2006: Reexamination of the observed decadal variability of the earth radiation budget using altitude-corrected ERBE/ERBS nonscanner WFOV data. *J. Clim.*, **19**, 4028–4040.

Wood, S. N., 2006: *Generalized Additive Models: An Introduction with R*. CRC/Chapman & Hall, Boca Raton, FL, USA.

Woodruff, S. D., et al., 2011: ICOADS Release 2.5: Extensions and enhancements to the Surface Marine Meteorological Archive. *Int. J. Climatol.*, **31**, 951–967.

Worden, H. M., et al., 2013: Decadal record of satellite carbon monoxide observations. *Atmos. Chem. Phys.*, **13**, 837–850.

Worton, D., et al., 2007: Atmospheric trends and radiative forcings of CF4 and C2F6 inferred from firn air. *Environ. Sci. Technol.*, **41**, 2184-2189.

Worton, D. R., et al., 2012: Evidence from firn air for recent decreases in non-methane hydrocarbons and a 20th century increase in nitrogen oxides in the northern hemisphere. *Atmos. Environ.*, **54**, 592–602.

Wu, Z., N. E. Huang, S. R. Long, and C.-K. Peng, 2007: On the trend, detrending, and variability of nonlinear and nonstationary time series. *Proc. Natl. Acad. Sci. U.S.A.*, **104**, 14889–14894.

Wu, Z., N. E. Huang, J. M. Wallace, B. V. Smoliak, and X. Chen, 2011: On the time-varying trend in global-mean surface temperature. *Clim. Dyn.*, **37**, 759–773.

Xavier, P. K., V. O. John, S. A. Buehler, R. S. Ajayamohan, and S. Sijikumar, 2010: Variability of Indian summer monsoon in a new upper tropospheric humidity data set. *Geophys. Res. Lett.*, **37**, L05705.

Xia, X., 2010a: A closer looking at dimming and brightening in China during 1961–2005. *Ann. Geophys.*, **28**, 1121–1132.

Xia, X. G., 2010b: Spatiotemporal changes in sunshine duration and cloud amount as well as their relationship in China during 1954–2005. *J. Geophys. Res. Atmos.*, **115**, D00K06.

Xiao, X., et al., 2010: Atmospheric three-dimensional inverse modeling of regional industrial emissions and global oceanic uptake of carbon tetrachloride. *Atmos. Chem. Phys.*, **10**, 10421–10434.

Xie, B., Q. Zhang, and Y. Wang, 2010: Observed characteristics of hail size in four regions in China during 1980–2005. *J. Clim.*, **23**, 4973–4982.

Xie, B. G., Q. H. Zhang, and Y. Q. Wang, 2008: Trends in hail in China during 1960–2005. *Geophys. Res. Lett.*, **35**, L13801.

Xie, S., K. Hu, J. Hafner, H. Tokinaga, Y. Du, G. Huang, and T. Sampe, 2009: Indian Ocean capacitor effect on Indo-Western Pacific climate during the summer following El Niño. *J. Clim.*, **22**, 730–747.

Xu, C. Y., L. B. Gong, J. Tong, and D. L. Chen, 2006a: Decreasing reference evapotranspiration in a warming climate – A case of Changjiang (Yangtze) River catchment during 1970–2000. *Adv. Atmos. Sci.*, **23**, 513–520.

Xu, K. H., J. D. Milliman, and H. Xu, 2010: Temporal trend of precipitation and runoff in major Chinese Rivers since 1951. *Global Planet. Change*, **73**, 219–232.

Xu, M., C. P. Chang, C. B. Fu, Y. Qi, A. Robock, D. Robinson, and H. M. Zhang, 2006b: Steady decline of east Asian monsoon winds, 1969–2000: Evidence from direct ground measurements of wind speed. *J. Geophys. Res. Atmos.*, **111**.

Yan, Z. W., Z. Li, Q. X. Li, and P. Jones, 2010: Effects of site change and urbanisation in the Beijing temperature series 1977–2006. *Int. J. Climatol.*, **30**, 1226–1234.

Yang, J., Q. Liu, S.-P. Xie, Z. Liu, and L. Wu, 2007: Impact of the Indian Ocean SST basin mode on the Asian summer monsoon. *Geophys. Res. Lett.*, **34**, L02708.

Yang, X. C., Y. L. Hou, and B. D. Chen, 2011: Observed surface warming induced by urbanization in east China. *J. Geophys. Res. Atmos.*, **116**, 12.

Yokouchi, Y., S. Taguchi, T. Saito, Y. Tohjima, H. Tanimoto, and H. Mukai, 2006: High frequency measurements of HFCs at a remote site in east Asia and their implications for Chinese emissions. *Geophys. Res. Lett.*, **33**, L21814.

Yoon, J., W. von Hoyningen-Huene, A. A. Kokhanovsky, M. Vountas, and J. P. Burrows, 2012: Trend analysis of aerosol optical thickness and Angstrom exponent derived from the global AERONET spectral observations. *Atmos. Meas. Tech.*, **5**, 1271–1299.

You, Q., et al., 2010: Changes in daily climate extremes in China and their connection to the large scale atmospheric circulation during 1961–2003. *Clim. Dyn.*, **36**, 2399-2417.

Yttri, K. E., et al., 2011: Transboundary particulate matter in Europe, Status Report 2011. In: *Co-operative Programme for Monitoring and Evaluation of the Long Range Transmission of Air Pollutants (Joint CCC, MSC-W, CEIP and CIAM report 2011)*. NILU - Chemical Coordinating Centre - CCC. http://emep.int/publ/common_publications.html

Yu, B., and F. W. Zwiers, 2010: Changes in equatorial atmospheric zonal circulations in recent decades. *Geophys. Res. Lett.*, **37**, L05701.

Yu, L., and R. Weller, 2007: Objectively analyzed air-sea heat fluxes for the global ice-free oceans (1981–2005). *Bull. Am. Meteor. Soc.*, **88**, 527-539.

Yuan, X., and C. Li, 2008: Climate modes in southern high latitudes and their impacts on Antarctic sea ice. *J. Geophys. Res. Oceans*, **113**, C06S91.

Yurganov, L., W. McMillan, E. Grechko, and A. Dzhola, 2010: Analysis of global and regional CO burdens measured from space between 2000 and 2009 and validated by ground-based solar tracking spectrometers. *Atmos. Chem. Phys.*, **10**, 3479–3494.

Zebiak, S. E., 1993: Air-sea interaction in the equatorial Atlantic region. *J. Clim.*, **6**, 1567–1568.

Zerefos, C. S., et al., 2009: Solar dimming and brightening over Thessaloniki, Greece, and Beijing, China. *Tellus B*, **61**, 657–665.

Zhang, A. Y., G. Y. Ren, J. X. Zhou, Z. Y. Chu, Y. Y. Ren, and G. L. Tang, 2010: On the urbanization effect on surface air temperature trends over China. *Acta Meteorol. Sin.*, **68**, 957–966.

Zhang, H., J. Bates, and R. Reynolds, 2006: Assessment of composite global sampling: Sea srface wind speed. *Geophys. Res. Lett.*, **33**, L17714.

Zhang, J., and J. S. Reid, 2010: A decadal regional and global trend analysis of the aerosol optical depth using a data-assimilation grade over-water MODIS and Level 2 MISR aerosol products. *Atmos. Chem. Phys.*, **10**, 10949–10963.

Zhang, X., J. He, J. Zhang, I. Polaykov, R. Gerdes, J. Inoue, and P. Wu, 2012a: Enhanced poleward moisture transport and amplified northern high-latitude wetting trend. *Nature Clim. Change*, **3**, 47-51.

Zhang, X., et al., 2007a: Detection of human influence on twentieth-century precipitation trends. *Nature*, **448**, 461–U464.

Zhang, X., et al., 2011: Indices for monitoring changes in extremes based on daily temperature and precipitation data. *Wiley Interdis. Rev. Clim. Change*, **2**, 851-870.

Zhang, X. B., et al., 2005: Trends in Middle East climate extreme indices from 1950 to 2003. *J. Geophys. Res. Atmos.*, **110**, D22104.

Zhang, X. D., C. H. Lu, and Z. Y. Guan, 2012b: Weakened cyclones, intensified anticyclones and recent extreme cold winter weather events in Eurasia. *Environ. Res. Lett.*, **7**, 044044.

Zhang, X. D., J. E. Walsh, J. Zhang, U. S. Bhatt, and M. Ikeda, 2004: Climatology and interannual variability of Arctic cyclone activity: 1948–2002. *J. Clim.*, **17**, 2300–2317.

Zhang, Y., J. M. Wallace, and D. S. Battisti, 1997: ENSO-like interdecadal variability: 1900–93. *J. Clim.*, **10**, 1004–1020.

Zhang, Y. Q., C. M. Liu, Y. H. Tang, and Y. H. Yang, 2007b: Trends in pan evaporation and reference and actual evapotranspiration across the Tibetan Plateau. *J. Geophys. Res. Atmos.*, **112**, D12110.

Zhao, X. P. T., A. K. Heidinger, and K. R. Knapp, 2011: Long-term trends of zonally averaged aerosol optical thickness observed from operational satellite AVHRR instrument. *Meteorol. Appl.*, **18**, 440–445.

Zhen, L., and Y. Zhong-Wei, 2009: Homogenized daily mean.maximum/minimum temperature series for China from 1960–2008. 237–243.

Zhou, J., and K.-K. Tung, 2012: Deducing multidecadal anthropogenic global warming trends using multiple regression Analysis. *J. Atmos. Sci.*, **70**, 3–8.

Zhou, T., L. Zhang, and H. Li, 2008: Changes in global land monsoon area and total rainfall accumulation over the last half century. *Geophys. Res. Lett.*, **35**, L16707.

Zhou, T. J., D. Y. Gong, J. Li, and B. Li, 2009a: Detecting and understanding the multi-decadal variability of the East Asian summer monsoon – Recent progress and state of affairs. *Meteorol. Z.*, **18**, 455–467.

Zhou, T. J., et al., 2009b: Why the Western Pacific subtropical high has extended westward since the late 1970s. *J. Clim.*, **22**, 2199–2215.

Zhou, Y. P., K. M. Xu, Y. C. Sud, and A. K. Betts, 2011: Recent trends of the tropical hydrological cycle inferred from Global Precipitation Climatology Project and International Satellite Cloud Climatology Project data. *J. Geophys. Res. Atmos.*, **116**, D09101

Zhou, Y. Q., and G. Y. Ren, 2011: Change in extreme temperature event frequency over mainland China, 1961–2008. *Clim. Res.*, **50**, 125–139.

Ziemke, J. R., S. Chandra, and P. K. Bhartia, 2005: A 25-year data record of atmospheric ozone in the Pacific from Total Ozone Mapping Spectrometer (TOMS) cloud slicing: Implications for ozone trends in the stratosphere and troposphere. *J. Geophys. Res.*, **110**, D15105.

Ziemke, J. R., S. Chandra, G. J. Labow, P. K. Bhartia, L. Froidevaux, and J. C. Witte, 2011: A global climatology of tropospheric and stratospheric ozone derived from Aura OMI and MLS measurements. *Atmos. Chem. Phys.*, **11**, 9237–9251.

Zipser, E. J., C. Liu, D. J. Cecil, S. W. Nesbitt, and D. P. Yorty, 2006: Where are the most intense thunderstorms on Earth? *Bull. Am. Meteor. Soc.*, **87**, 1057–1071.

Zolina, O., C. Simmer, K. Belyaev, A. Kapala, and S. Gulev, 2009: Improving estimates of heavy and extreme precipitation using daily records from European rain gauges. *J. Hydrometeor.*, **10**, 701–716.

Zorita, E., T. F. Stocker, and H. von Storch, 2008: How unusual is the recent series of warm years? *Geophys. Res. Lett.*, **35**, L24706.

Zou, C. Z., and W. H. Wang, 2010: Stability of the MSU-derived atmospheric temperature trend. *J. Atmos. Ocean Technol.*, **27**, 1960–1971.

Zou, C. Z., and W. H. Wang, 2011: Intersatellite calibration of AMSU-A observations for weather and climate applications. *J. Geophys. Res. Atmos.*, **116,** D23113.

Zou, C. Z., M. D. Goldberg, Z. H. Cheng, N. C. Grody, J. T. Sullivan, C. Y. Cao, and D. Tarpley, 2006a: Recalibration of microwave sounding unit for climate studies using simultaneous nadir overpasses. *J. Geophys. Res. Atmos.*, **111,** L17701.

Zou, X., L. V. Alexander, D. Parker, and J. Caesar, 2006b: Variations in severe storms over China. *Geophys. Res. Lett.*, **33**.

Zwiers, F. W., and V. V. Kharin, 1998: Changes in the extremes of the climate simulated by CCC GCM2 under CO_2 doubling. *J. Clim.*, **11**, 2200–2222.

3 Observations: Ocean

Coordinating Lead Authors:
Monika Rhein (Germany), Stephen R. Rintoul (Australia)

Lead Authors:
Shigeru Aoki (Japan), Edmo Campos (Brazil), Don Chambers (USA), Richard A. Feely (USA), Sergey Gulev (Russian Federation), Gregory C. Johnson (USA), Simon A. Josey (UK), Andrey Kostianoy (Russian Federation), Cecilie Mauritzen (Norway), Dean Roemmich (USA), Lynne D. Talley (USA), Fan Wang (China)

Contributing Authors:
Ian Allison (Australia), Michio Aoyama (Japan), Molly Baringer (USA), Nicholas R. Bates (Bermuda), Timothy Boyer (USA), Robert H. Byrne (USA), Sarah Cooley (USA), Stuart Cunningham (UK), Thierry Delcroix (France), Catia M. Domingues (Australia), Scott Doney (USA), John Dore (USA), Paul. J. Durack (USA/Australia), Rana Fine (USA), Melchor González-Dávila (Spain), Simon Good (UK), Nicolas Gruber (Switzerland), Mark Hemer (Australia), David Hydes (UK), Masayoshi Ishii (Japan), Stanley Jacobs (USA), Torsten Kanzow (Germany), David Karl (USA), Georg Kaser (Austria/Italy), Alexander Kazmin (Russian Federation), Robert Key (USA), Samar Khatiwala (USA), Joan Kleypas (USA), Ronald Kwok (USA), Kitack Lee (Republic of Korea), Eric Leuliette (USA), Melisa Menéndez (Spain), Calvin Mordy (USA), Jon Olafsson (Iceland), James Orr (France), Alejandro Orsi (USA), Geun-Ha Park (Republic of Korea), Igor Polyakov (USA), Sarah G. Purkey (USA), Bo Qiu (USA), Gilles Reverdin (France), Anastasia Romanou (USA), Sunke Schmidtko (UK), Raymond Schmitt (USA), Koji Shimada (Japan), Doug Smith (UK), Thomas M. Smith (USA), Uwe Stöber (Germany), Lothar Stramma (Germany), Toshio Suga (Japan), Neil Swart (Canada/South Africa), Taro Takahashi (USA), Toste Tanhua (Germany), Karina von Schuckmann (France), Hans von Storch (Germany), Xiaolan Wang (Canada), Rik Wanninkhof (USA), Susan Wijffels (Australia), Philip Woodworth (UK), Igor Yashayaev (Canada), Lisan Yu (USA)

Review Editors:
Howard Freeland (Canada), Silvia Garzoli (USA), Yukihiro Nojiri (Japan)

This chapter should be cited as:
Rhein, M., S.R. Rintoul, S. Aoki, E. Campos, D. Chambers, R.A. Feely, S. Gulev, G.C. Johnson, S.A. Josey, A. Kostianoy, C. Mauritzen, D. Roemmich, L.D. Talley and F. Wang, 2013: Observations: Ocean. In: *Climate Change 2013: The Physical Science Basis. Contribution of Working Group I to the Fifth Assessment Report of the Intergovernmental Panel on Climate Change* [Stocker, T.F., D. Qin, G.-K. Plattner, M. Tignor, S.K. Allen, J. Boschung, A. Nauels, Y. Xia, V. Bex and P.M. Midgley (eds.)]. Cambridge University Press, Cambridge, United Kingdom and New York, NY, USA.

Table of Contents

Executive Summary

Temperature and Heat Content Changes

It is *virtually certain*[1] that the upper ocean (above 700 m) has warmed from 1971 to 2010, and *likely* that it has warmed from the 1870s to 1971. *Confidence* in the assessment for the time period since 1971 is *high*[2] based on increased data coverage after this date and on a high level of agreement among independent observations of subsurface temperature [3.2], sea surface temperature [2.4.2], and sea level rise, which is known to include a substantial component due to thermal expansion [3.7, Chapter 13]. There is less certainty in changes prior to 1971 because of relatively sparse sampling in earlier time periods. The strongest warming is found near the sea surface (0.11 [0.09 to 0.13] °C per decade in the upper 75 m between 1971 and 2010), decreasing to about 0.015°C per decade at 700 m. It is *very likely* that the surface intensification of this warming signal increased the thermal stratification of the upper ocean by about 4% between 0 and 200 m depth. Instrumental biases in historical upper ocean temperature measurements have been identified and reduced since AR4, diminishing artificial decadal variation in temperature and upper ocean heat content, most prominent during the 1970s and 1980s. {3.2.1–3.2.3, Figures 3.1, 3.2 and 3.9}

It is *likely* that the ocean warmed between 700 and 2000 m from 1957 to 2009, based on 5-year averages. It is *likely* that the ocean warmed from 3000 m to the bottom from 1992 to 2005, while no significant trends in global average temperature were observed between 2000 and 3000 m depth during this period. Warming below 3000 m is largest in the Southern Ocean {3.2.4, 3.5.1, Figures 3.2b and 3.3, FAQ 3.1}

It is *virtually certain* that upper ocean (0 to 700 m) heat content increased during the relatively well-sampled 40-year period from 1971 to 2010. Published rates for that time period range from 74 TW to 137 TW, with generally smaller trends for estimates that assume zero anomalies in regions with sparse data. Using a statistical analysis of ocean variability to estimate change in sparsely sampled areas and to estimate uncertainties results in a rate of increase of global upper ocean heat content of 137 [120–154] TW (*medium confidence*). Although not all trends agree within their statistical uncertainties, all are positive, and all are statistically different from zero. {3.2.3, Figure 3.2}

Warming of the ocean between 700 and 2000 m *likely* contributed about 30% of the total increase in global ocean heat content (0 to 2000 m) between 1957 and 2009. Although globally integrated ocean heat content in some of the 0 to 700 m estimates increased more slowly from 2003 to 2010 than over the previous decade, ocean heat uptake from 700 to 2000 m *likely* continued unabated during this period. {3.2.4, Figure 3.2, Box 9.2}

Ocean warming dominates the global energy change inventory. Warming of the ocean accounts for about 93% of the increase in the Earth's energy inventory between 1971 and 2010 (*high confidence*), with warming of the upper (0 to 700 m) ocean accounting for about 64% of the total. Melting ice (including Arctic sea ice, ice sheets and glaciers) and warming of the continents and atmosphere account for the remainder of the change in energy. The estimated net increase in the Earth's energy storage between 1971 and 2010 is 274 [196 to 351] ZJ (1 ZJ = 10^{21} Joules), with a heating rate of 213 TW from a linear fit to annual inventories over that time period, equivalent to 0.42 W m^{-2} heating applied continuously over the Earth's entire surface, and 0.55 W m^{-2} for the portion due to ocean warming applied over the ocean surface area. {Section 3.2.3, Figure 3.2, Box 3.1}

Salinity and Freshwater Content Changes

It is *very likely* that regional trends have enhanced the mean geographical contrasts in sea surface salinity since the 1950s: saline surface waters in the evaporation-dominated mid-latitudes have become more saline, while relatively fresh surface waters in rainfall-dominated tropical and polar regions have become fresher. The mean contrast between high- and low-salinity regions increased by 0.13 [0.08 to 0.17] from 1950 to 2008. It is *very likely* that the interbasin contrast in freshwater content has increased: the Atlantic has become saltier and the Pacific and Southern oceans have freshened. Although similar conclusions were reached in AR4, recent studies based on expanded data sets and new analysis approaches provide *high confidence* in the assessment of trends in ocean salinity. {3.3.2, 3.3.3, 3.3.5, Figures 3.4, 3.5 and 3.21d, FAQ 3.2}

It is *very likely* that large-scale trends in salinity have also occurred in the ocean interior. It is *likely* that both the subduction of surface water anomalies formed by changes in evaporation – precipitation (E – P) and the movement of density surfaces due to warming have contributed to the observed changes in subsurface salinity. {3.3.2–3.3.4, Figures 3.5 and 3.9}

The spatial patterns of the salinity trends, mean salinity and the mean distribution of E – P are all similar. This provides, with *medium confidence*, indirect evidence that the pattern of E – P over the oceans has been enhanced since the 1950s. {3.3.2–3.3.4, Figures 3.4, 3.5 and 3.20d, FAQ 3.2}.

---

1 In this Report, the following terms have been used to indicate the assessed likelihood of an outcome or a result: Virtually certain 99–100% probability, Very likely 90–100%, Likely 66–100%, About as likely as not 33–66%, Unlikely 0–33%, Very unlikely 0-10%, Exceptionally unlikely 0–1%. Additional terms (Extremely likely: 95–100%, More likely than not >50–100%, and Extremely unlikely 0–5%) may also be used when appropriate. Assessed likelihood is typeset in italics, e.g., *very likely* (see Section 1.4 and Box TS.1 for more details).

2 In this Report, the following summary terms are used to describe the available evidence: limited, medium, or robust; and for the degree of agreement: low, medium, or high. A level of confidence is expressed using five qualifiers: very low, low, medium, high, and very high, and typeset in italics, e.g., *medium confidence*. For a given evidence and agreement statement, different confidence levels can be assigned, but increasing levels of evidence and degrees of agreement are correlated with increasing confidence (see Section 1.4 and Box TS.1 for more details).

Air–Sea Flux and Wave Height Changes

Uncertainties in air–sea heat flux data sets are too large to allow detection of the change in global mean net air-sea heat flux, of the order of 0.5 W m^{-2} since 1971, required for consistency with the observed ocean heat content increase. The products cannot yet be reliably used to directly identify trends in the regional or global distribution of evaporation or precipitation over the oceans on the time scale of the observed salinity changes since 1950. {3.4.2, 3.4.3, Figures 3.6 and 3.7}

Basin-scale wind stress trends at decadal to centennial time scales have been observed in the North Atlantic, Tropical Pacific and Southern Ocean with *low to medium confidence*. These results are based largely on atmospheric reanalyses, in some cases a single product, and the confidence level is dependent on region and time scale considered. The evidence is strongest for the Southern Ocean, for which there is *medium confidence* that zonal mean wind stress has increased in strength since the early 1980s. {3.4.4, Figure 3.8}

There is *medium confidence* based on ship observations and reanalysis forced wave model hindcasts that mean significant wave height has increased since the 1950s over much of the North Atlantic north of 45°N, with typical winter season trends of up to 20 cm per decade. {3.4.5}

3

Changes in Water Masses and Circulation

Observed changes in water mass properties *likely* reflect the combined effect of long-term trends in surface forcing (e.g., warming of the surface ocean and changes in E – P) and inter-annual-to-multi-decadal variability related to climate modes. Most of the observed temperature and salinity changes in the ocean interior can be explained by subduction and spreading of water masses with properties that have been modified at the sea surface. From 1950 to 2000, it is *likely* that subtropical salinity maximum waters became more saline, while fresh intermediate waters formed at higher latitude have generally become fresher. For Upper North Atlantic Deep Water changes in properties and formation rates are *very likely* dominated by decadal variability. The Lower North Atlantic Deep Water has *likely* cooled from 1955 to 2005, and the freshening trend highlighted in AR4 reversed in the mid-1990s. It is *likely* that the Antarctic Bottom Water warmed and contracted globally since the 1980s and freshened in the Indian/Pacific sectors from 1970 to 2008. {3.5, FAQ 3.1}

Recent observations have strengthened evidence for variability in major ocean circulation systems on time scales from years to decades. It is *very likely* that the subtropical gyres in the North Pacific and South Pacific have expanded and strengthened since 1993. It is about *as likely as not* that this is linked to decadal variability in wind forcing rather than being part of a longer-term trend. Based on measurements of the full Atlantic Meridional Overturning Circulation and its individual components at various latitudes and different time periods, there is no evidence of a long-term trend. There is also no evidence for trends in the transports of the Indonesian Throughflow, the Antarctic Circumpolar Current (ACC), or between the Atlantic Ocean and Nordic Seas. However, there is *medium confidence* that the ACC shifted south between 1950 and 2010, at a rate equivalent to about 1° of latitude in 40 years. {3.6, Figures 3.10, 3.11}

Sea Level Change

Global mean sea level (GMSL) has risen by 0.19 [0.17 to 0.21] m over the period 1901–2010, calculated using the mean rate over these 110 years, based on tide gauge records and since 1993 additionally on satellite data. It is *very likely* that the mean rate was 1.7 [1.5 to 1.9] mm yr^{-1} between 1901 and 2010 and increased to 3.2 [2.8 to 3.6] mm yr^{-1} between 1993 and 2010. This assessment is based on high agreement among multiple studies using different methods, long tide gauge records corrected for vertical land motion and independent observing systems (tide gauges and altimetry) since 1993 (see also TFE.2, Figure 1). It is *likely* that GMSL rose between 1920 and 1950 at a rate comparable to that observed between 1993 and 2010, as individual tide gauges around the world and reconstructions of GMSL show increased rates of sea level rise during this period. Rates of sea level rise over broad regions can be several times larger or smaller than that of GMSL for periods of several decades due to fluctuations in ocean circulation. High agreement between studies with and without corrections for vertical land motion suggests that it is *very unlikely* that estimates of the global average rate of sea level change are significantly biased owing to vertical land motion that has been unaccounted for. {3.7.2, 3.7.3, Table 3.1, Figures 3.12, 3.13, 3.14}

It is *very likely* that warming of the upper 700 m has been contributing an average of 0.6 [0.4 to 0.8] mm yr^{-1} of sea level rise since 1971. It is *likely* that warming between 700 m and 2000 m has been contributing an additional 0.1 mm yr^{-1} [0 to 0.2] of sea level rise since 1971, and that warming below 2000 m has been contributing another 0.1 [0.0 to 0.2] mm yr^{-1} of sea level rise since the early 1990s. {3.7.2, Figure 3.13}

It is *likely* that the rate of sea level rise increased from the early 19th century to the early 20th century, and increased further over the 20th century. The inference of 19th century change is based on a small number of very long tide gauge records from northern Europe and North America. Multiple long tide gauge records and reconstructions of global mean sea level confirm a higher rate of rise from the late 19th century. It is *likely* that the average acceleration over the 20th century is [–0.002 to 0.019] mm yr^{-2}, as two of three reconstructions extending back to at least 1900 show an acceleration during the 20th century. {3.7.4}

It is *likely* that the magnitude of extreme high sea level events has increased since 1970. A rise in mean sea level can explain most of the increase in extreme sea levels: changes in extreme high sea levels are reduced to less than 5 mm yr^{-1} at 94% of tide gauges once the rise in mean sea level is accounted for. {3.7.5, Figure 3.15}

Changes in Ocean Biogeochemistry

Based on high agreement between independent estimates using different methods and data sets (e.g., oceanic carbon, oxygen, and transient tracer data), it is *very likely* that the global ocean inventory of anthropogenic carbon (C_{ant}) increased from 1994 to 2010. The oceanic C_{ant} inventory in 2010 is estimated to be 155 PgC with an uncertainty of ±20%. The annual global oceanic uptake rates calculated from independent data sets (from oceanic C_{ant} inventory changes, from atmospheric O_2/N_2 measurements or from partial pressure of carbon dioxide (pCO_2) data) and for different time periods agree with each other within their uncertainties and *very likely* are in the range of 1.0 to 3.2 PgC yr^{-1} {3.8.1, Figure 3.16}

Uptake of anthropogenic CO_2 results in gradual acidification of the ocean. The pH of surface seawater has decreased by 0.1 since the beginning of the industrial era, corresponding to a 26% increase in hydrogen ion concentration (*high confidence*). The observed pH trends range between –0.0014 and –0.0024 yr^{-1} in surface waters. In the ocean interior, natural physical and biological processes, as well as uptake of anthropogenic CO_2, can cause changes in pH over decadal and longer time scales. {3.8.2, Table 3.2, Box 3.2, Figures 3.18, 3.19, FAQ 3.3}

High agreement among analyses provides *medium confidence* that oxygen concentrations have decreased in the open ocean thermocline in many ocean regions since the 1960s. The general decline is consistent with the expectation that warming-induced stratification leads to a decrease in the supply of oxygen to the thermocline from near surface waters, that warmer waters can hold less oxygen, and that changes in wind-driven circulation affect oxygen concentrations. It is *likely* that the tropical oxygen minimum zones have expanded in recent decades. {3.8.3, Figure 3.20}

Synthesis

The observations summarized in this chapter provide strong evidence that ocean properties of relevance to climate have changed during the past 40 years, including temperature, salinity, sea level, carbon, pH, and oxygen. The observed patterns of change in the subsurface ocean are consistent with changes in the surface ocean in response to climate change and natural variability and with known physical and biogeochemical processes in the ocean, providing *high confidence* in this assessment. {3.9, Figures 3.21, 3.22}

3.1 Introduction

The ocean influences climate by storing and transporting large amounts of heat, freshwater, and carbon, and by exchanging these properties with the atmosphere. About 93% of the excess heat energy stored by the Earth over the last 50 years is found in the ocean (Church et al., 2011; Levitus et al., 2012). The ability of the ocean to store vast amounts of heat reflects the large mass and heat capacity of seawater relative to air and the fact that ocean circulation connects the surface and interior ocean. More than three quarters of the total exchange of water between the atmosphere and the Earth's surface through evaporation and precipitation takes place over the oceans (Schmitt, 2008). The ocean contains 50 times more carbon than the atmosphere (Sabine et al., 2004) and is at present acting to slow the rate of climate change by absorbing about 30% of human emissions of carbon dioxide (CO_2) from fossil fuel burning, cement production, deforestation and other land use change (Mikaloff-Fletcher et al., 2006; Le Quéré et al., 2010). Changes in the ocean may result in climate feedbacks that either increase or reduce the rate of climate change. Climate variability and change on time scales from seasons to millennia is therefore closely linked to the ocean and its interactions with the atmosphere and cryosphere. The large inertia of the oceans means that they naturally integrate over short-term variability and often provide a clearer signal of longer-term change than other components of the climate system. Observations of ocean change therefore provide a means to track the evolution of climate change, and a relevant benchmark for climate models.

3

The lack of long-term measurements of the global ocean and changes in the observing system over time makes documenting and understanding change in the oceans a difficult challenge (Appendix 3.A). Many of the issues raised in Box 2.1 regarding uncertainty in atmospheric climate records are common to oceanographic data. Despite the limitations of historical records, AR4 identified significant trends in a number of ocean variables relevant to climate change, including ocean heat content, sea level, regional patterns of salinity, and biogeochemical parameters (Bindoff et al., 2007). Since AR4, substantial progress has been made in improving the quality and coverage of ocean observations. Biases in historical measurements have been identified and reduced, providing a clearer record of past change. The Argo array of profiling floats has provided near-global, year-round measurements of temperature and salinity in the upper 2000 m since 2005. The satellite altimetry record is now more than 20 years in length. Longer continuous time series of important components of the meridional overturning circulation and tropical oceans have been obtained. The spatial and temporal coverage of biogeochemical measurements in the ocean has expanded. As a result of these advances, there is now stronger evidence of change in the ocean, and our understanding of the causes of ocean change is improved.

This chapter summarizes the observational evidence of change in the ocean, with an emphasis on basin- and global-scale changes relevant to climate, with a focus on studies published since the AR4. As in Chapter 2, the robustness of observed changes is assessed relative to sources of observational uncertainty. The attribution of ocean change, including the degree to which observed changes are consistent with anthropogenic climate change, is addressed in Chapter 10. The evidence for changes in subsurface ocean temperature and heat content is assessed in Section 3.2; changes in sea surface temperature (SST) are covered in Chapter 2. Changes in ocean heat content dominate changes in the global energy inventory (Box 3.1). Recent studies have strengthened the evidence for regional changes in ocean salinity and their link to changes in evaporation and precipitation over the oceans (Section 3.3), a connection already identified in AR4. Evidence for changes in the fluxes of heat, water and momentum (wind stress) across the air–sea interface is assessed in Section 3.4. Considering ocean changes from a water-mass perspective adds additional insight into the nature and causes of ocean change (Section 3.5). Although direct observations of ocean circulation are more limited than those of temperature and salinity, there is growing evidence of variability and change of ocean current patterns relevant to climate (Section 3.6). Observations of sea level change are summarized in Section 3.7; Chapter 13 builds on the evidence presented in this and other chapters to provide an overall synthesis of past and future sea level change. Biogeochemical changes in the ocean, including ocean acidification, are covered in Section 3.8. Chapter 6 combines observations with models to discuss past and present changes in the carbon cycle. Section 3.9 provides an overall synthesis of changes observed in the ocean during the instrumental period and highlights key uncertainties. Unless otherwise noted, uncertainties (in square brackets) represent 5 to 95% confidence intervals.

3.2 Changes in Ocean Temperature and Heat Content

3.2.1 Effects of Sampling on Ocean Heat Content Estimates

Temperature is the most often measured subsurface ocean variable. Historically, a variety of instruments have been used to measure temperature, with differing accuracies, precisions, and sampling depths. Both the mix of instruments and the overall sampling patterns have changed in time and space (Boyer et al., 2009), complicating efforts to determine and interpret long-term change. The evolution of the observing system for ocean temperature is summarized in Appendix 3.A. Upper ocean temperature (hence heat content) varies over multiple time scales including seasonal (e.g., Roemmich and Gilson, 2009), interannual (e.g. associated with El Niño, which has a strong influence on ocean heat uptake, Roemmich and Gilson, 2011), decadal (e.g., Carson and Harrison, 2010), and centennial (Gouretski et al., 2012; Roemmich et al., 2012). Ocean data assimilation products using these data exhibit similar significant variations (e.g., Xue et al., 2012). Sparse historical sampling coupled with large amplitude variations on shorter time and spatial scales raise challenges for estimating globally averaged upper ocean temperature changes. Uncertainty analyses indicate that the historical data set begins to be reasonably well suited for this purpose starting around 1970 (e.g., Domingues et al., 2008; Lyman and Johnson, 2008; Palmer and Brohan, 2011). UOHC uncertainty estimates shrink after 1970 with improved sampling, so this assessment focuses on changes since 1971. Estimates of UOHC have been extended back to 1950 by averaging over longer time intervals, such as 5-year running means, to compensate for sparse data distributions in earlier time periods (e.g., Levitus et al., 2012). These estimates may be most appropriate in the deeper ocean, where strong interannual variability in upper ocean temperature distributions such as that associated with El Niño (Roemmich and Gilson, 2011) is less likely to be aliased.

Since AR4 the significant impact of measurement biases in some of the widely used instruments (the expendable (XBT) and mechanical bathythermograph (MBT) as well as a subset of Argo floats) on estimates of ocean temperature and upper (0 to 700 m) ocean heat content (hereafter UOHC) changes has been recognized (Gouretski and Koltermann, 2007; Barker et al., 2011). Careful comparison of measurements from the less accurate instruments with those from the more accurate ones has allowed some of the biases to be identified and reduced (Wijffels et al., 2008; Ishii and Kimoto, 2009; Levitus et al., 2009; Gouretski and Reseghetti, 2010; Hamon et al., 2012). One major consequence of this bias reduction has been the reduction of an artificial decadal variation in upper ocean heat content that was apparent in the observational assessment for AR4, in notable contrast to climate model output (Domingues et al., 2008). Substantial time-dependent XBT and MBT biases introduced spurious warming in the 1970s and cooling in the early 1980s in the analyses assessed in AR4. Most ocean state estimates that assimilate biased data (Carton and Santorelli, 2008) also showed this artificial decadal variability while one (Stammer et al., 2010) apparently rejected these data on dynamical grounds. More recent estimates assimilating better-corrected data sets (Giese et al., 2011) also result in reduced artificial decadal variability during this time period.

Recent estimates of upper ocean temperature change also differ in their treatment of unsampled regions. Some studies (e.g., Ishii and Kimoto, 2009; Levitus et al., 2012) effectively assume a temperature anomaly of zero in these regions, while other studies (Palmer et al., 2007; Lyman and Johnson, 2008) assume that the averages of sampled regions are representative of the global mean in any given year, and yet others (Smith and Murphy, 2007; Domingues et al., 2008) use ocean statistics (from numerical model output and satellite altimeter data, respectively) to extrapolate temperature anomalies in sparsely sampled areas and estimate uncertainties. These differences in approach, coupled with choice of background climatology, can lead to significant divergence in basin-scale averages (Gleckler et al., 2012), especially in sparsely sampled regions (e.g., the extratropical Southern Hemisphere (SH) prior to Argo), and as a result can produce different global averages (Lyman et al., 2010). However, for well-sampled regions and times, the various analyses of temperature changes yield results in closer agreement, as do reanalyses (Xue et al., 2012).

3.2.2 Upper Ocean Temperature

Depth-averaged 0 to 700 m ocean temperature trends from 1971 to 2010 are positive over most of the globe (Levitus et al., 2009; Figure 3.1a). The warming is more prominent in the Northern Hemisphere (NH), especially the North Atlantic. This result holds in different analyses, using different time periods, bias corrections and data sources (e.g., with or without XBT or MBT data) (e.g., Palmer et al., 2007; Durack and Wijffels, 2010; Gleckler et al., 2012; Figures 3.1 and 3.9). However, the greater volume of the SH oceans increases the contribution of their warming to global heat content. Zonally averaged upper ocean temperature trends show warming at nearly all latitudes and depths (Levitus et al., 2009, Figure 3.1b). A maximum in warming south of 30°S appears in Figure 3.1b, but is not as strong as in other analyses (e.g., Gille, 2008), likely because the data are relatively sparse in this location so anomalies are attenuated by the objectively analyzed fields used for Figure 3.1 and because warming in the upper 1000 m of the Southern Ocean was stronger between the 1930s and the 1970s than between the 1970s and 1990s (Gille, 2008). Another warming maximum is present at 25°N to 65°N. Both warming signals extend to 700 m (Levitus et al., 2009, Figure 3.1b), and are consistent with poleward displacement of the mean temperature field. Other zonally averaged temperature changes are also consistent with poleward displacement of the mean temperatures. For example, cooling at depth between 30°S and the equator (Figure 3.1b) is consistent with a southward shift of cooler water near the equator. Poleward displacements of some subtropical and subpolar zonal currents and associated wind changes are discussed in Section 3.6.

Globally averaged ocean temperature anomalies as a function of depth and time (Figure 3.1c) relative to a 1971–2010 mean reveal warming at all depths in the upper 700 m over the relatively well-sampled 40-year period considered. Strongest warming is found closest to the

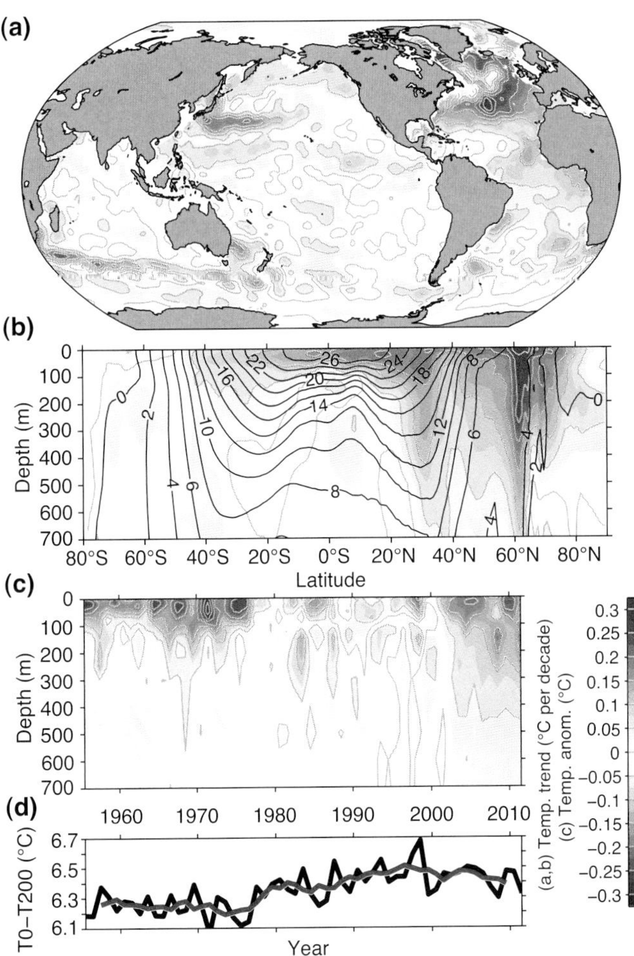

Figure 3.1 | (a) Depth-averaged 0 to 700 m temperature trend for 1971–2010 (longitude vs. latitude, colours and grey contours in degrees Celsius per decade). (b) Zonally averaged temperature trends (latitude vs. depth, colours and grey contours in degrees Celsius per decade) for 1971–2010 with zonally averaged mean temperature over-plotted (black contours in degrees Celsius). (c) Globally averaged temperature anomaly (time vs. depth, colours and grey contours in degrees Celsius) relative to the 1971–2010 mean. (d) Globally averaged temperature difference between the ocean surface and 200 m depth (black: annual values, red: 5-year running mean). All panels are constructed from an update of the annual analysis of Levitus et al. (2009).

sea surface, and the near-surface trends are consistent with independently measured SST (Chapter 2). The global average warming over this period is 0.11 [0.09 to 0.13] °C per decade in the upper 75 m, decreasing to 0.015°C per decade by 700 m (Figure 3.1c). Comparison of Argo data to *Challenger* expedition data from the 1870s suggests that warming started earlier than 1971, and was also larger in the Atlantic than in the Pacific over that longer time interval (Roemmich et al., 2012). An observational analysis of temperature in the upper 400 m of the global ocean starting in the year 1900 (Gouretski et al., 2012) finds warming between about 1900 and 1945, as well as after 1970, with some evidence of slight cooling between 1945 and 1970.

The globally averaged temperature difference between the ocean surface and 200 m (Figure 3.1d) increased by about 0.25°C from 1971 to 2010 (Levitus et al., 2009). This change, which corresponds to a 4% increase in density stratification, is widespread in all the oceans north of about 40°S.

A potentially important impact of ocean warming is the effect on sea ice, floating glacial ice and ice sheet dynamics (see Chapter 4 for a discussion of these topics). Although some of the global integrals of UOHC neglect changes poleward of ±60° (Ishii and Kimoto, 2009) or ±65° (Domingues et al., 2008) latitude, at least some parts of the Arctic have warmed: In the Arctic Ocean, subsurface pulses of relatively warm water of Atlantic origin can be traced around the Eurasian Basin, and analyses of data from 1950–2010 show a decadal warming of this water mass since the late 1970s (Polyakov et al., 2012), as well as a shoaling, by 75 to 90 m (Polyakov et al., 2010). Arctic surface waters have also warmed, at least in the Canada Basin, from 1993 to 2007 (Jackson et al., 2010).

3.2.3 Upper Ocean Heat Content

Global integrals of 0 to 700 m UOHC (Figure 3.2a) estimated from ocean temperature measurements all show a gain from 1971 to 2010 (Palmer et al., 2007; Smith and Murphy, 2007; e.g., Domingues et al., 2008; Ishii and Kimoto, 2009; Levitus et al., 2012) . These estimates usually start around 1950, although as noted in Section 3.2.1 and discussed in the Appendix, historical data coverage is sparse, so global integrals are increasingly uncertain for earlier years, especially prior to 1970. There is some convergence towards agreement in instrument bias correction algorithms since AR4 (Section 3.2.1), but other sources of uncertainty include the different assumptions regarding mapping and integrating UOHCs in sparsely sampled regions, differences in quality control of temperature data, and differences among baseline climatologies used for estimating changes in heat content (Lyman et al., 2010). Although there are still apparent interannual variations about the upward trend of global UOHC since 1970, different global estimates have variations at different times and for different periods, suggesting that sub-decadal variability in the time rate of change is still quite uncertain in the historical record. Most of the estimates in Figure 3.2a do exhibit decreases for a few years immediately following major volcanic eruptions in 1963, 1982 and 1991 (Domingues et al., 2008).

Again, all of the global integrals of UOHC in Figure 3.2a have increased between 1971 and 2010. Linear trends fit to the UOHC estimates for the relatively well-sampled 40-year period from 1971 to 2010 estimate

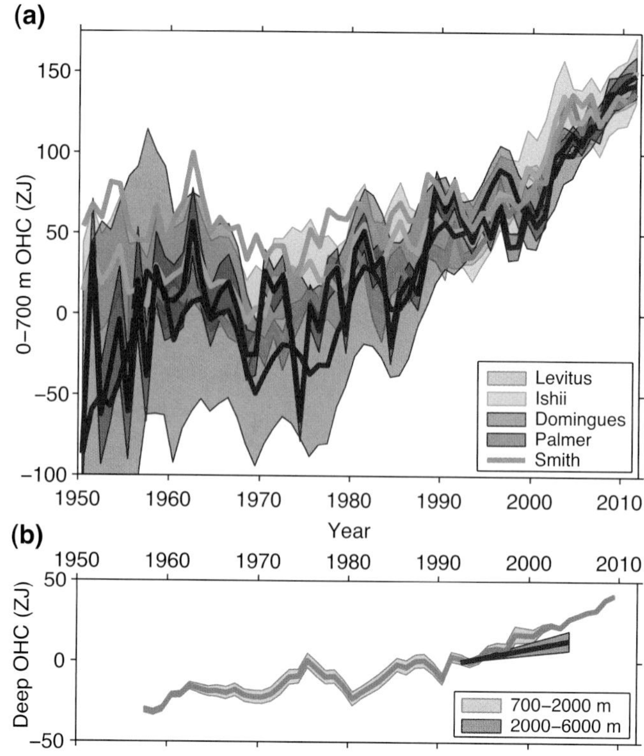

Figure 3.2: | (a) Observation-based estimates of annual global mean upper (0 to 700 m) ocean heat content in ZJ (1 ZJ = 10^{21} Joules) updated from (see legend): Levitus et al. (2012), Ishii and Kimoto (2009), Domingues et al. (2008), Palmer et al. (2007) and Smith and Murphy (2007). Uncertainties are shaded and plotted as published (at the one standard error level, except one standard deviation for Levitus, with no uncertainties provided for Smith). Estimates are shifted to align for 2006–2010, 5 years that are well measured by Argo, and then plotted relative to the resulting mean of all curves for 1971, the starting year for trend calculations. (b) Observation-based estimates of annual 5-year running mean global mean mid-depth (700 to 2000 m) ocean heat content in ZJ (Levitus et al., 2012) and the deep (2000 to 6000 m) global ocean heat content trend from 1992 to 2005 (Purkey and Johnson, 2010), both with one standard error uncertainties shaded (see legend).

the heating rate required to account for this warming: 118 [82 to 154] TW (1 TW = 10^{12} watts) for Levitus et al. (2012), 98 [67 to 130] TW for Ishii and Kimoto (2009), 137 [120 to 154] TW for Domingues et al. (2008), 108 [80 to 136] TW for Palmer et al. (2007), and 74 [43 to 105] TW for Smith and Murphy (2007). Uncertainties are calculated as 90% confidence intervals for an ordinary least squares fit, taking into account the reduction in the degrees of freedom implied by the temporal correlation of the residuals. Although these rates of energy gain do not all agree within their statistical uncertainties, all are positive, and all are statistically different from zero. Generally the smaller trends are for estimates that assume zero anomalies in areas of sparse data, as expected for that choice, which will tend to reduce trends and variability. Hence the assessment of the Earth's energy uptake (Box 3.1) employs a global UOHC estimate (Domingues et al., 2008) chosen because it fills in sparsely sampled areas and estimates uncertainties using a statistical analysis of ocean variability patterns.

Globally integrated ocean heat content in three of the five 0 to 700 m estimates appear to be increasing more slowly from 2003 to 2010 than over the previous decade (Figure 3.2a). Although this apparent change is concurrent with a slowing of the increase global mean surface temperature, as discussed in Box 9.2, this is also a time period when the ocean observing system transitioned from predominantly XBT to predominantly Argo temperature measurements (Johnson and Wijffels, 2011). Shifts in observing systems can sometimes introduce spurious signals, so this apparent recent change should be viewed with caution.

3.2.4 Deep Ocean Temperature and Heat Content

Below 700 m data coverage is too sparse to produce annual global ocean heat content estimates prior to about 2005, but from 2005 to 2010 and 0 to 1500 m the global ocean warmed (von Schuckmann and Le Traon, 2011). Five-year running mean estimates yield a 700 to 2000 m global ocean heat content trend from 1957 to 2009 (Figure 3.2b) that is about 30% of that for 0 to 2000 m over the length of the record (Levitus et al., 2012). Ocean heat uptake from 700 to 2000 m *likely* continues unabated since 2003 (Figure 3.2b); as a result, ocean heat content from 0 to 2000 m shows less slowing after 2003 than does 0 to 700 m heat content (Levitus et al., 2012).

Global sampling of the ocean below 2000 m is limited to a number of repeat oceanographic transects, many occupied only in the last few decades (Figure 3.3b), and several time-series stations, some of which extend over decades. This sparse sampling in space and time makes assessment of global deep ocean heat content variability less certain than that for the upper ocean (Ponte, 2012), especially at mid-depths, where vertical gradients are still sufficiently large for transient variations (ocean eddies, internal waves, and internal tides) to alias estimates made from sparse data sets. However, the deep North Atlantic Ocean is better sampled than the rest of the globe, making estimates of full-depth deep ocean heat content changes there feasible north of 20°N since the mid-1950s (Mauritzen et al., 2012).

Based on the limited information available, it is *likely* that the global ocean did not show a significant temperature trend between 2000 and 3000 m depth from about 1992–2005 (Figures 3.2b and 3.3a; Kouketsu et al., 2011). At these depths it has been around a millennium on average since waters in the Indian and Pacific Oceans were last exposed to air–sea interaction (Gebbie and Huybers, 2012).

Warming from 1992 to 2005 is *likely* greater than zero from 3000 m to the ocean floor (Figures 3.2b and 3.3a; Kouketsu et al., 2011), especially in recently formed Antarctic Bottom Water (AABW). South of the Sub-Antarctic Front (Figure 3.3b), much of the water column warmed between 1992 and 2005 (Purkey and Johnson, 2010). Globally, deep warming rates are highest near 4500 m (Figure 3.3a), usually near the sea floor where the AABW influence is strongest, and attenuate towards the north (Figure 3.3b), where the AABW influence weakens. Global scale abyssal warming on relatively short multi-decadal time scales is possible because of communication of signals by planetary waves originating within the Southern Ocean, reaching even such remote regions as the North Pacific (Kawano et al., 2010; Masuda et al., 2010). This AABW warming may partly reflect a recovery from cool conditions induced by the 1970s Weddell Sea Polynya (Robertson et al., 2002), but further north, in the Vema Channel of the South Atlantic, observations since 1970 suggest strong bottom water warming did not commence there until about 1991 (Zenk and Morozov, 2007).

(a)

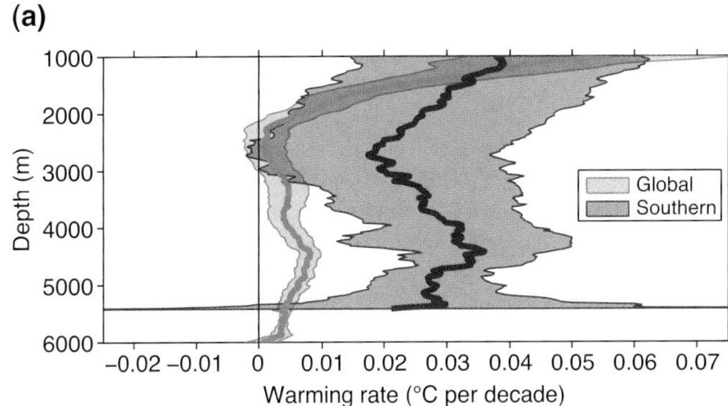

(b)

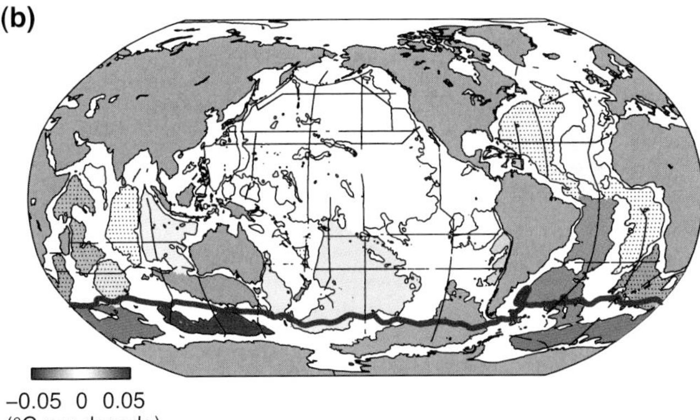

Figure 3.3 | (a) Areal mean warming rates (°C per decade) versus depth (thick lines) with 5 to 95% confidence limits (shading), both global (orange) and south of the Sub Antarctic Front (purple), centred on 1992–2005. (b) Mean warming rates (°C per decade) below 4000 m (colour bar) estimated for deep ocean basins (thin black outlines), centred on 1992–2005. Stippled basin warming rates are not significantly different from zero at 95% confidence. The positions of the Sub-Antarctic Front (purple line) and the repeat oceanographic transects from which these warming rates are estimated (thick black lines) also shown. (Data from Purkey and Johnson, 2010.)

In the North Atlantic, strong decadal variability in North Atlantic Deep Water (NADW) temperature and salinity (Wang et al., 2010), largely associated with the North Atlantic Oscillation (NAO, Box 2.5) (e.g., Yashayaev, 2007; Sarafanov et al., 2008), complicates efforts to determine long-term trends from the historical record. Heat content in the North Atlantic north of 20°N from 2000 m to the ocean floor increased slightly from 1955 to 1975, and then decreased more strongly from 1975 to 2005 (Mauritzen et al., 2012), with a net cooling trend of –4 TW from 1955–2005 estimated from a linear fit. The global trend estimate below 2000 m is +35 TW from 1992 to 2005 (Purkey and Johnson, 2010), with strong warming in the Southern Ocean.

3.2.5 Conclusions

It is *virtually certain* that the upper ocean (0 to 700 m) warmed from 1971 to 2010. This result is supported by three independent and consistent methods of observation including (1) multiple analyses of subsurface temperature measurements described here; (2) SST data (Section 2.4.2) from satellites and *in situ* measurements from surface drifters and ships; and (3) the record of sea level rise, which is known to include a substantial component owing to thermosteric expansion (Section 3.7 and Chapter 13). The warming rate is 0.11 [0.09 to 0.13]°C per decade in the upper 75 m, decreasing to about 0.015°C per decade by 700 m. It is *very likely* that surface intensification of the warming

Box 3.1 | Change in Global Energy Inventory

The Earth has been in radiative imbalance, with less energy exiting the top of the atmosphere than entering, since at least about 1970 (Murphy et al., 2009; Church et al., 2011; Levitus et al., 2012). Quantifying this energy gain is essential for understanding the response of the climate system to radiative forcing. Small amounts of this excess energy warm the atmosphere and continents, evaporate water and melt ice, but the bulk of it warms the ocean (Box 3.1, Figure 1). The ocean dominates the change in energy because of its large mass and high heat capacity compared to the atmosphere. In addition, the ocean has a very low albedo and absorbs solar radiation much more readily than ice.

The global atmospheric energy change inventory accounting for specific heating and water evaporation is estimated by combining satellite estimates for temperature anomalies in the lower troposphere (Mears and Wentz, 2009a; updated to version 3.3) from 70°S to 82.5°N and the lower stratosphere (Mears and Wentz, 2009b; updated to version 3.3) from 82.5°S to 82.5°N weighted by the ratio of the portions of atmospheric mass they sample (0.87 and 0.13, respectively). These temperature anomalies are converted to energy changes using a total atmospheric mass of 5.14×10^{18} kg, a mean total water vapor mass of 12.7×10^{15} kg (Trenberth and Smith, 2005), a heat capacity of 1 J g^{-1} °C^{-1}, a latent heat of vaporization of 2.464 J kg^{-1} and a fractional increase of integrated water vapor content of 0.075 °C^{-1} (Held and Soden, 2006). Smaller changes in potential and kinetic energy are considered negligible. Standard deviations for each year of data are used for uncertainties, and the time series starts in 1979. The warming trend from a linear fit from 1979 to 2010 amounts to 2 TW (1 TW = 10^{12} watts).

The global average rate of continental warming and its uncertainty has been estimated from borehole temperature profiles from 1500 to 2000 at 50-year intervals (Beltrami et al., 2002). The 1950–2000 estimate of land warming, 6 TW, is extended into the first decade of the 21st century, although that extrapolation is almost certainly an underestimate of the energy absorbed, as land surface air temperatures for years since 2000 are some of the warmest on record (Section 2.4.1).

All annual ice melt rates (for glaciers and ice-caps, ice sheets and sea ice from Chapter 4) are converted into energy change using a heat of fusion (334×10^3 J kg^{-1}) and density (920 kg m^{-3}) for freshwater ice. The heat of fusion and density of ice may vary, but only slightly among the different ice types, and warming the ice from sub-freezing temperatures requires much less energy than that to melt it, so these second-order contributions are neglected here. The linear trend of energy storage from 1971 to 2010 is 7 TW.

For the oceans, an estimate of global upper (0 to 700 m depth) ocean heat content change using ocean statistics to extrapolate to sparsely sampled regions and estimate uncertainties (Domingues et al., 2008) is used (see Section 3.2), with a linear trend from 1971 to 2010 of 137 TW. For the ocean from 700 to 2000 m, annual 5-year running mean estimates are used from 1970 to 2009 and annual estimates for 2010–2011 (Levitus et al., 2012). For the ocean from 2000 m to bottom, a uniform rate of energy gain of 35 [6 to 61] TW from warming rates centred on 1992–2005 (Purkey and Johnson, 2010) is applied from 1992 to 2011, with no warming below 2000 m assumed prior to 1992. Their 5 to 95% uncertainty estimate may be too small, as it

(continued on next page)

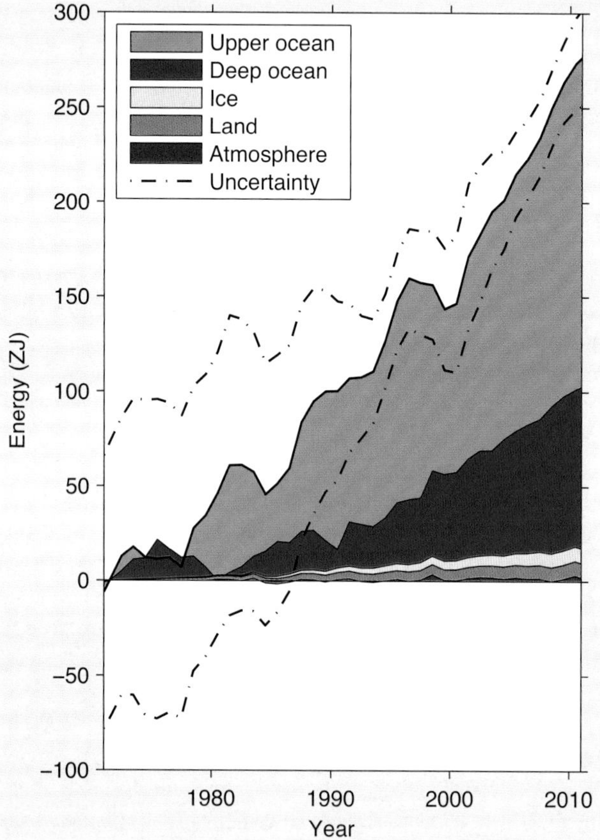

Box 3.1, Figure 1 | Plot of energy accumulation in ZJ (1 ZJ = 10^{21} J) within distinct components of the Earth's climate system relative to 1971 and from 1971 to 2010 unless otherwise indicated. See text for data sources. Ocean warming (heat content change) dominates, with the upper ocean (light blue, above 700 m) contributing more than the mid-depth and deep ocean (dark blue, below 700 m; including below 2000 m estimates starting from 1992). Ice melt (light grey; for glaciers and ice caps, Greenland and Antarctic ice sheet estimates starting from 1992, and Arctic sea ice estimate from 1979 to 2008); continental (land) warming (orange); and atmospheric warming (purple; estimate starting from 1979) make smaller contributions. Uncertainty in the ocean estimate also dominates the total uncertainty (dot-dashed lines about the error from all five components at 90% confidence intervals).

Box 3.1 (continued)

assumes the usually sparse sampling in each deep ocean basin analysed is representative of the mean trend in that basin. The linear trend for heating the ocean below 700 m is 62 TW for 1971–2010.

It is *virtually certain* that the Earth has gained substantial energy from 1971 to 2010 — the estimated increase in energy inventory between 1971 and 2010 is 274 [196 to 351] ZJ (1 ZJ = 10^{21} J), with a rate of 213 TW from a linear fit to the annual values over that time period (Box 3.1, Figure 1). An energy gain of 274 ZJ is equivalent to a heating rate of 0.42 W m^{-2} applied continuously over the surface area of the earth (5.10×10^{14} m^2). Ocean warming dominates the total energy change inventory, accounting for roughly 93% on average from 1971 to 2010 (*high confidence*). The upper ocean (0-700 m) accounts for about 64% of the total energy change inventory. Melting ice (including Arctic sea ice, ice sheets and glaciers) accounts for 3% of the total, and warming of the continents 3%. Warming of the atmosphere makes up the remaining 1%. The 1971–2010 estimated rate of oceanic energy gain is 199 TW from a linear fit to data over that time period, implying a mean heat flux of 0.55 W m^{-2} across the global ocean surface area (3.60×10^{14} m^2). The Earth's net estimated energy increase from 1993 to 2010 is 163 [127 to 201] ZJ with a trend estimate of 275 TW. The ocean portion of the trend for 1993–2010 is 257 TW, equivalent to a mean heat flux into the ocean of 0.71 W m^{-2} over the global ocean surface area.

signal increased the thermal stratification of the upper ocean by about 4% (between 0 and 200 m depth) from 1971 to 2010. It is also *likely* that the upper ocean warmed over the first half of the 20th century, based again on these same three independent and consistent, although much sparser, observations. Deeper in the ocean, it is *likely* that the waters from 700 to 2000 m have warmed on average between 1957 and 2009 and *likely* that no significant trend was observed between 2000 and 3000 m from 1992 to 2005. It is *very likely* that the deep (2000 m to bottom) North Atlantic Ocean north of 20°N warmed from 1955 to 1975, and then cooled from 1975 to 2005, with an overall cooling trend. It is *likely* that most of the water column south of the Sub-Antarctic Front warmed at a rate of about 0.03°C per decade from 1992 to 2005, and waters of Antarctic origin warmed below 3000 m at a global average rate approaching 0.01°C per decade at 4500 m over the same time period. For the deep ocean. Sparse sampling is the largest source of uncertainty below 2000 m depth.

3.3 Changes in Salinity and Freshwater Content

3.3.1 Introduction

The ocean plays a pivotal role in the global water cycle: about 85% of the evaporation and 77% of the precipitation occurs over the ocean (Schmitt, 2008). The horizontal salinity distribution of the upper ocean largely reflects this exchange of freshwater, with high surface salinity generally found in regions where evaporation exceeds precipitation, and low salinity found in regions of excess precipitation and runoff (Figure 3.4a,b). Ocean circulation also affects the regional distribution of surface salinity. The subduction (Section 3.5) of surface waters transfers the surface salinity signal into the ocean interior, so that subsurface salinity distributions are also linked to patterns of evaporation, precipitation and continental run-off at the sea surface. Melting and freezing of ice (both sea ice and glacial ice) also influence ocean salinity.

Regional patterns and amplitudes of atmospheric moisture transport could change in a warmer climate, because warm air can contain more moisture (FAQ 3.2). The water vapour content of the troposphere *likely* has increased since the 1970s, at a rate consistent with the observed warming (Sections 2.4.4, 2.5.5 and 2.5.6).

It has not been possible to detect robust trends in regional precipitation and evaporation over the ocean because observations over the ocean are sparse and uncertain (Section 3.4.2). Ocean salinity, on the other hand, naturally integrates the small difference between these two terms and has the potential to act as a rain gauge for precipitation minus evaporation over the ocean (e.g., Lewis and Fofonoff, 1979; Schmitt, 2008; Yu, 2011; Pierce et al., 2012; Terray et al., 2012; Section 10.4). Diagnosis and understanding of ocean salinity trends is also important because salinity changes, like temperature changes, affect circulation and stratification, and therefore the ocean's capacity to store heat and carbon as well as to change biological productivity. Salinity changes also contribute to regional sea level change (Steele and Ermold, 2007).

In AR4, surface and subsurface salinity changes consistent with a warmer climate were highlighted, based on linear trends for the period between 1955 and 1998 in the historical global salinity data set (Boyer et al., 2005) as well as on more regional studies. In the early few decades the salinity data distribution was good in the NH, especially the North Atlantic, but the coverage was poor in some regions such as the central South Pacific, central Indian and polar oceans (Appendix 3.A). However, Argo provides much more even spatial and temporal coverage in the 2000s. These additional observations, improvements in the availability and quality of historical data and new analysis approaches now allow a more complete assessment of changes in salinity.

'Salinity' refers to the weight of dissolved salts in a kilogram of seawater. Because the total amount of salt in the ocean does not change, the salinity of seawater can be changed only by addition or removal of fresh water. All salinity values quoted in the chapter are expressed on the Practical Salinity Scale 1978 (PSS78) (Lewis and Fofonoff, 1979).

Frequently Asked Questions
FAQ 3.1 | Is the Ocean Warming?

Yes, the ocean is warming over many regions, depth ranges and time periods, although neither everywhere nor constantly. The signature of warming emerges most clearly when considering global, or even ocean basin, averages over time spans of a decade or more.

Ocean temperature at any given location can vary greatly with the seasons. It can also fluctuate substantially from year to year—or even decade to decade—because of variations in ocean currents and the exchange of heat between ocean and atmosphere.

Ocean temperatures have been recorded for centuries, but it was not until around 1971 that measurements were sufficiently comprehensive to estimate the average global temperature of the upper several hundred meters of the ocean confidently for any given year. In fact, before the international Argo temperature/salinity profiling float array first achieved worldwide coverage in 2005, the global average upper ocean temperature for any given year was sensitive to the methodology used to estimate it.

Global mean upper ocean temperatures have increased over decadal time scales from 1971 to 2010. Despite large uncertainty in most yearly means, this warming is a robust result. In the upper 75 m of the ocean, the global average warming trend has been 0.11 [0.09 to 0.13]°C per decade over this time. That trend generally lessens from the surface to mid-depth, reducing to about 0.04°C per decade by 200 m, and to less than 0.02°C per decade by 500 m.

Temperature anomalies enter the subsurface ocean by paths in addition to mixing from above (FAQ3.1, Figure 1). Colder—hence denser—waters from high latitudes can sink from the surface, then spread toward the equator beneath warmer, lighter, waters at lower latitudes. At a few locations—the northern North Atlantic Ocean and the Southern Ocean around Antarctica—ocean water is cooled so much that it sinks to great depths, even to the sea floor. This water then spreads out to fill much of the rest of the deep ocean. As ocean surface waters warm, these sinking waters also warm with time, increasing temperatures in the ocean interior much more quickly than would downward mixing of surface heating alone.

In the North Atlantic, the temperature of these deep waters varies from decade to decade—sometimes warming, sometimes cooling—depending on prevailing winter atmospheric patterns. Around Antarctica, bottom waters have warmed detectably from about 1992–2005, perhaps due to the strengthening and southward shift of westerly winds around the Southern Ocean over the last several decades. This warming signal in the deepest coldest bottom waters of the world ocean is detectable, although it weakens northward in the Indian, Atlantic and Pacific Oceans. Deep warming rates are generally less pronounced than ocean surface rates (around 0.03°C per decade since the 1990s in the deep and bottom waters around Antarctica, and smaller in many other locations). However, they occur over a large volume, so deep ocean warming contributes significantly to the total increase in ocean heat.

Estimates of historical changes in global average ocean temperature have become more accurate over the past several years, largely thanks to the recognition, and reduction, of systematic measurement errors. By carefully comparing less accurate measurements with sparser, more accurate ones at adjacent locations and similar times, scientists have reduced some spurious instrumental biases in the historical record. These improvements revealed that the global average ocean temperature has increased much more steadily from year to year than was reported prior to 2008. Nevertheless, the global average warming rate may not be uniform in time. In some years, the ocean appears to warm faster than average; in others, the warming rate seems to slow.

The ocean's large mass and high heat capacity allow it to store huge amounts of energy—more than 1000 times that in the atmosphere for an equivalent increase in temperature. The Earth is absorbing more heat than it is emitting back into space, and nearly all this excess heat is entering the oceans and being stored there. The ocean has absorbed about 93% of the combined heat stored by warmed air, sea, and land, and melted ice between 1971 and 2010.

The ocean's huge heat capacity and slow circulation lend it significant thermal inertia. It takes about a decade for near-surface ocean temperatures to adjust in response to climate forcing (Section 12.5), such as changes in greenhouse gas concentrations. Thus, if greenhouse gas concentrations could be held at present levels into the future, increases in the Earth's surface temperature would begin to slow within about a decade. However, deep ocean temperature would continue to warm for centuries to millennia (Section 12.5), and thus sea levels would continue to rise for centuries to millennia as well (Section 13.5). *(continued on next page)*

FAQ 3.1 (continued)

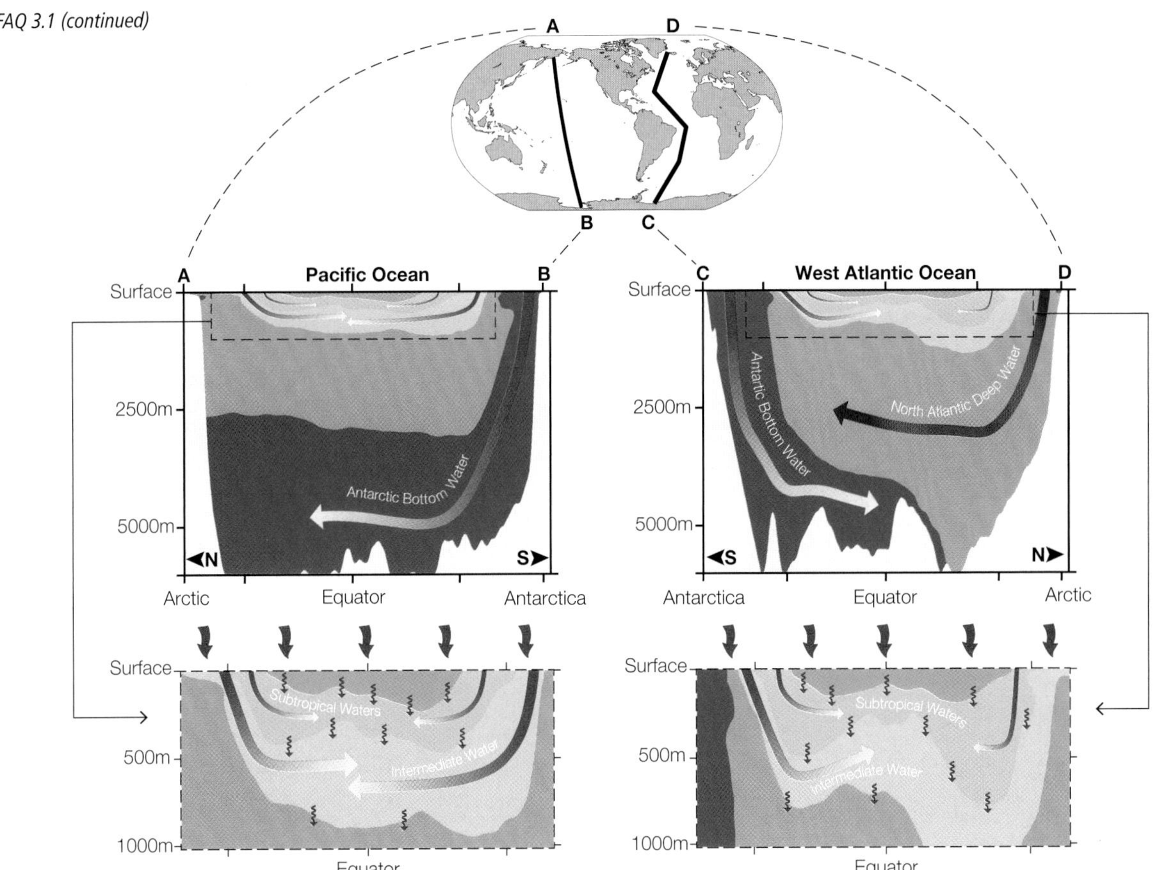

FAQ 3.1, Figure 1 | Ocean heat uptake pathways. The ocean is stratified, with the coldest, densest water in the deep ocean (upper panels: use map at top for orientation). Cold Antarctic Bottom Water (dark blue) sinks around Antarctica then spreads northward along the ocean floor into the central Pacific (upper left panel: red arrows fading to white indicate stronger warming of the bottom water most recently in contact with the ocean surface) and western Atlantic oceans (upper right panel), as well as the Indian Ocean (not shown). Less cold, hence lighter, North Atlantic Deep Water (lighter blue) sinks in the northern North Atlantic Ocean (upper right panel: red and blue arrow in the deep water indicates decadal warming and cooling), then spreads south above the Antarctic Bottom Water. Similarly, in the upper ocean (lower left panel shows Pacific Ocean detail, lower right panel the Atlantic), cool Intermediate Waters (cyan) sink in sub-polar regions (red arrows fading to white indicating warming with time), before spreading toward the equator under warmer Subtropical Waters (green), which in turn sink (red arrows fading to white indicate stronger warming of the intermediate and subtropical waters most recently in contact with the surface) and spread toward the equator under tropical waters, the warmest and lightest (orange) in all three oceans. Excess heat or cold entering at the ocean surface (top curvy red arrows) also mixes slowly downward (sub-surface wavy red arrows).

3.3.2 Global to Basin-Scale Trends

The salinity of near-surface waters is changing on global and basin scales, with an increase in the more evaporative regions and a decrease in the precipitation-dominant regions in almost all ocean basins.

3.3.2.1 Sea Surface Salinity

Multi-decadal trends in sea surface salinity have been documented in studies published since AR4 (Boyer et al., 2007; Hosoda et al., 2009; Roemmich and Gilson, 2009; Durack and Wijffels, 2010), confirming the trends reported in AR4 based mainly on Boyer et al. (2005). The spatial pattern of surface salinity change is similar to the distribution of surface salinity itself: salinity tends to increase in regions of high mean salinity, where evaporation exceeds precipitation, and tends to decrease in regions of low mean salinity, where precipitation dominates (Figure 3.4). For example, salinity generally increased in the surface salinity maxima formed in the evaporation-dominated subtropical gyres. The surface salinity minima at subpolar latitudes and the intertropical convergence zones have generally freshened. Interbasin salinity differences are also enhanced: the relatively salty Atlantic has become more saline on average, while the relatively fresh Pacific has become fresher (Figures 3.5 and 3.9). No well-defined trend is found in the subpolar North Atlantic , which is dominated by decadal variability from atmospheric modes like the North Atlantic Oscillation (NAO, Box 2.5). The 50-year salinity trends in Figure 3.4c, both positive and negative, are statistically significant at the 99% level over 43.8% of the global ocean surface (Durack and Wijffels, 2010); trends were less significant over the remainder of the surface. The patterns of salinity change in the complementary Hosoda et al. (2009) study of differences between the periods 1960–1989 and 2003–2007 (Figure 3.4d), using a different methodology, have a point-to-point correlation of 0.64 with

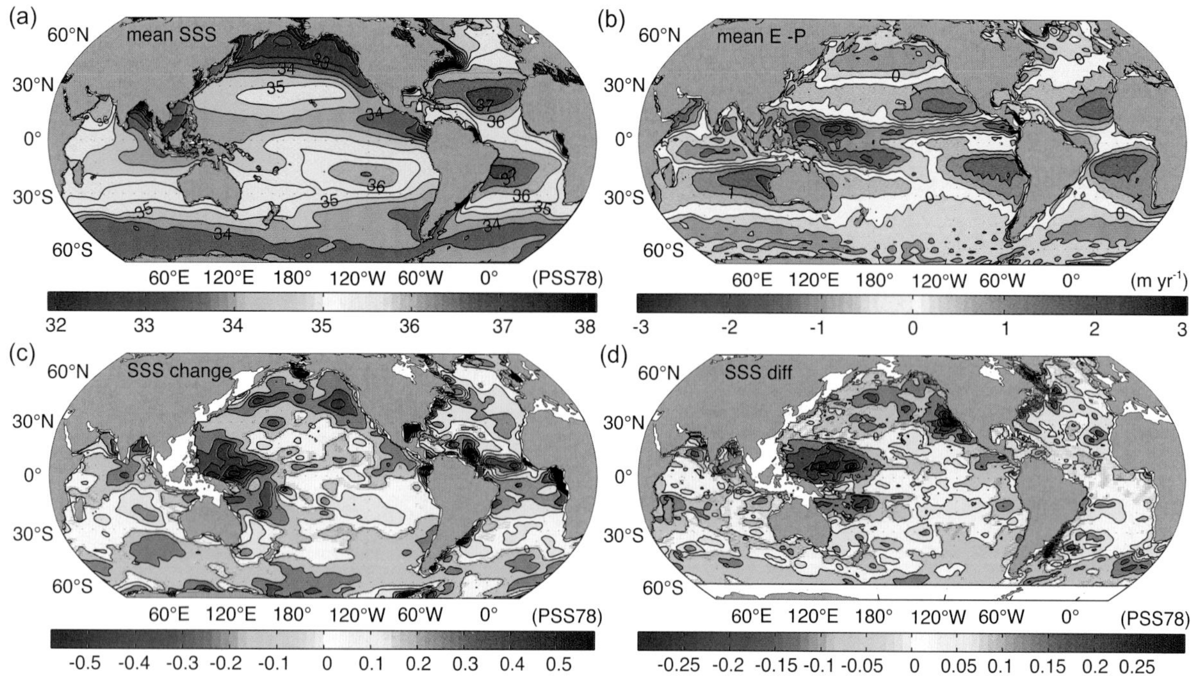

Figure 3.4 | (a) The 1955–2005 climatological-mean sea surface salinity (World Ocean Atlas 2009 of Antonov et al., 2010) colour contoured at 0.5 PSS78 intervals (black lines). (b) Annual mean evaporation–precipitation averaged over the period 1950–2000 (NCEP) colour contoured at 0.5 m yr^{-1} intervals (black lines). (c) The 58-year (2008 minus 1950) sea surface salinity change derived from the linear trend (PSS78), with seasonal and El Niño-Southern Oscillation (ENSO) signals removed (Durack and Wijffels, 2010) colour contoured at 0.116 PSS78 intervals (black lines). (d) The 30-year (2003–2007 average centred at 2005, minus the 1960–1989 average, centred at 1975) sea surface salinity difference (PSS78) (Hosoda et al., 2009) colour contoured at 0.06 PSS78 intervals (black lines). Contour intervals in (c) and (d) are chosen so that the trends can be easily compared, given the different time intervals in the two analyses. White areas in (c) to (d) are marginal seas where the calculations are not carried out. Regions where the change is not significant at the 99% confidence level are stippled in grey.

the Durack and Wijffels (2010) results, with significant differences only in limited locations such as adjacent to the West Indies, Labrador Sea, and some coastlines (Figure 3.4c and d).

It is *very likely* that the globally averaged contrast between regions of high and low salinity relative to the global mean salinity has increased. The contrast between high and low salinity regions, averaged over the ocean area south of 70°N, increased by 0.13 [0.08 to 0.17] PSS78 from 1950 to 2008 using the data set of Durack and Wijffels (2010) , and by 0.12 [0.10 to 0.15] PSS78 using the data set of Boyer et al. (2009) with the range reported in brackets signifying a 99% confidence interval (Figure 3.21d).

3.3.2.2 Upper Ocean Subsurface Salinity

Compatible with observed changes in surface salinity, robust multi-decadal trends in subsurface salinity have been detected (Boyer et al., 2005; Boyer et al., 2007; Steele and Ermold, 2007; Böning et al., 2008; Durack and Wijffels, 2010; Helm et al., 2010; Wang et al., 2010). Global, zonally averaged multi-decadal salinity trends (1950–2008) in the upper 500 m (Figures 3.4, 3.5, 3.9 and Section 3.5) show salinity increases at the salinity maxima of the subtropical gyres, freshening of the low-salinity intermediate waters sinking in the Southern Ocean (Subantarctic Mode Water and Antarctic Intermediate Water) and North Pacific (North Pacific Intermediate Water). On average, the Pacific freshened, and the Atlantic became more saline. These trends, shown in Figures 3.5 and 3.9, are significant at a 95% confidence interval.

Freshwater content in the upper 500 m *very likely* changed, based on the World Ocean Database 2009 (Boyer et al., 2009), analyzed by Durack and Wijffels (2010) and independently as an update to Boyer et al. (2005) for 1955–2010 (Figure 3.5a, b, e, f). Both show freshening in the North Pacific, salinification in the North Atlantic south of 50°N and salinification in the northern Indian Ocean (trends significant at 90% confidence). A significant freshening is observed in the circumpolar Southern Ocean south of 50S.

Density layers that are ventilated (connected to the sea surface) in precipitation-dominated regions have freshened, while those ventilated in evaporation-dominated regions have increased in salinity, compatible with an enhancement of the mean surface freshwater flux pattern (Helm et al., 2010). In addition, where warming has caused surface outcrops of density layers to move (poleward) into higher salinity surface waters, the subducted salinity in the density layers has increased; where outcrops have moved into fresher surface waters, the subducted salinity decreased (Durack and Wijffels, 2010). Vertical and lateral shifts of density surfaces, due to both changes in water mass renewal rates and wind-driven circulation, have also contributed to the observed subsurface salinity changes (Levitus, 1989; Bindoff and McDougall, 1994).

A change in total, globally integrated freshwater content and salinity requires an addition or removal of freshwater; the only significant source is land ice (ice sheets and glaciers). The estimate of change in globally averaged salinity and freshwater content remains smaller than

Frequently Asked Questions

FAQ 3.2 | Is There Evidence for Changes in the Earth's Water Cycle?

The Earth's water cycle involves evaporation and precipitation of moisture at the Earth's surface. Changes in the atmosphere's water vapour content provide strong evidence that the water cycle is already responding to a warming climate. Further evidence comes from changes in the distribution of ocean salinity, which, due to a lack of long-term observations of rain and evaporation over the global oceans, has become an important proxy rain gauge.

The water cycle is expected to intensify in a warmer climate, because warmer air can be moister: the atmosphere can hold about 7% more water vapour for each degree Celsius of warming. Observations since the 1970s show increases in surface and lower atmospheric water vapour (FAQ 3.2, Figure 1a), at a rate consistent with observed warming. Moreover, evaporation and precipitation are projected to intensify in a warmer climate.

Recorded changes in ocean salinity in the last 50 years support that projection. Seawater contains both salt and fresh water, and its salinity is a function of the weight of dissolved salts it contains. Because the total amount of salt—which comes from the weathering of rocks—does not change over human time scales, seawater's salinity can only be altered—over days or centuries—by the addition or removal of fresh water.

The atmosphere connects the ocean's regions of net fresh water loss to those of fresh water gain by moving evaporated water vapour from one place to another. The distribution of salinity at the ocean surface largely reflects the spatial pattern of evaporation minus precipitation, runoff from land, and sea ice processes. There is some shifting of the patterns relative to each other, because of the ocean's currents.

Subtropical waters are highly saline, because evaporation exceeds rainfall, whereas seawater at high latitudes and in the tropics—where more rain falls than evaporates—is less so (FAQ 3.2, Figure 1b, d). The Atlantic, the saltiest ocean basin, loses more freshwater through evaporation than it gains from precipitation, while the Pacific is nearly neutral (i.e., precipitation gain nearly balances evaporation loss), and the Southern Ocean (region around Antarctica) is dominated by precipitation.

Changes in surface salinity and in the upper ocean have reinforced the mean salinity pattern. The evaporation-dominated subtropical regions have become saltier, while the precipitation-dominated subpolar and tropical regions have become fresher. When changes over the top 500 m are considered, the evaporation-dominated Atlantic has become saltier, while the nearly neutral Pacific and precipitation-dominated Southern Ocean have become fresher (FAQ 3.2, Figure 1c).

Observing changes in precipitation and evaporation directly and globally is difficult, because most of the exchange of fresh water between the atmosphere and the surface happens over the 70% of the Earth's surface covered by ocean. Long-term precipitation records are available only from over the land, and there are no long-term measurements of evaporation.

Land-based observations show precipitation increases in some regions, and decreases in others, making it difficult to construct a globally integrated picture. Land-based observations have shown more extreme rainfall events, and more flooding associated with earlier snow melt at high northern latitudes, but there is strong regionality in the trends. Land-based observations are so far insufficient to provide evidence of changes in drought.

Ocean salinity, on the other hand, acts as a sensitive and effective rain gauge over the ocean. It naturally reflects and smoothes out the difference between water gained by the ocean from precipitation, and water lost by the ocean through evaporation, both of which are very patchy and episodic. Ocean salinity is also affected by water runoff from the continents, and by the melting and freezing of sea ice or floating glacial ice. Fresh water added by melting ice on land will change global-averaged salinity, but changes to date are too small to observe.

Data from the past 50 years show widespread salinity changes in the upper ocean, which are indicative of systematic changes in precipitation and runoff minus evaporation, as illustrated in FAQ 3.2, Figure 1.

FAQ 3.2 is based on observations reported in Chapters 2 and 3, and on model analyses in Chapters 9 and 12.

(continued on next page)

FAQ 3.2 (continued)

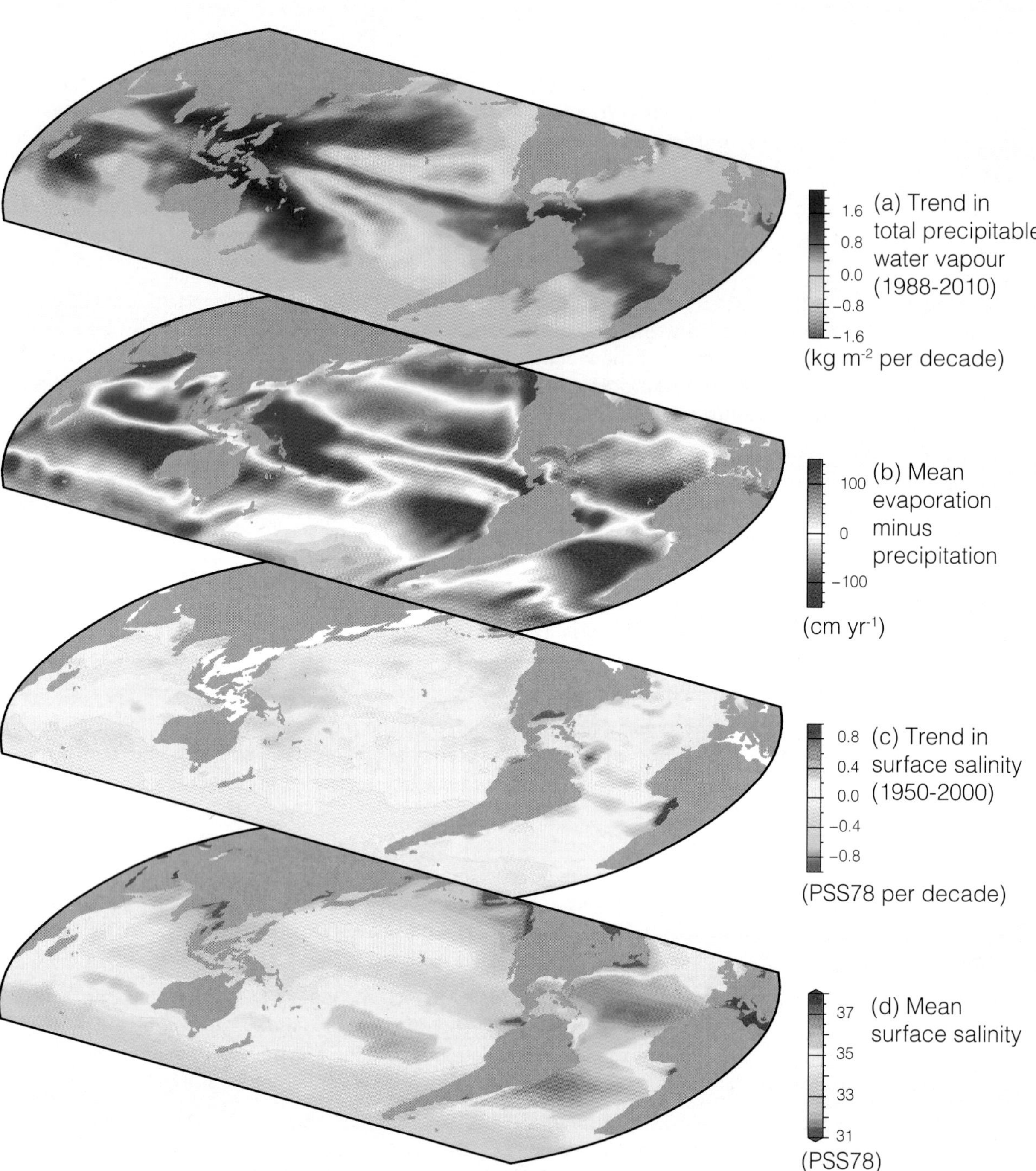

FAQ 3.2, Figure 1 | Changes in sea surface salinity are related to the atmospheric patterns of evaporation minus precipitation (E – P) and trends in total precipitable water: (a) Linear trend (1988–2010) in total precipitable water (water vapor integrated from the Earth's surface up through the entire atmosphere) (kg m^{-2} per decade) from satellite observations (Special Sensor Microwave Imager) (after Wentz et al., 2007) (blues: wetter; yellows: drier). (b) The 1979–2005 climatological mean net E –P (cm yr^{-1}) from meteorological reanalysis (National Centers for Environmental Prediction/National Center for Atmospheric Research; Kalnay et al., 1996) (reds: net evaporation; blues: net precipitation). (c) Trend (1950–2000) in surface salinity (PSS78 per 50 years) (after Durack and Wijffels, 2010) (blues freshening; yellows-reds saltier). (d) The climatological-mean surface salinity (PSS78) (blues: <35; yellows–reds: >35).

its uncertainty, as was true in the AR4 assessment. For instance, the globally averaged sea surface salinity change from 1950 to 2008 is small (+0.003 [–0.056 to 0.062]) compared to its error estimate (Durack and Wijffels, 2010). Thus a global freshening due to land ice loss has not yet been discerned in global surface salinity change even if it were assumed that all added freshwater were in the ocean's surface layer.

3.3.3 Regional Changes in Upper Ocean Salinity

Regional changes in ocean salinity are broadly consistent with the conclusion that regions of net precipitation (precipitation greater than evaporation) have *very likely* become fresher, while regions of net evaporation have become more saline. This pattern is seen in salinity trend maps (Figure 3.4); zonally averaged salinity trends and freshwater inventories for each ocean (Figure 3.5); and the globally averaged contrast between regions of high and low salinity (Figure 3.21d). In the high-latitude regions, higher runoff, increased melting of ice and changes in freshwater transport by ocean currents have *likely* also contributed to observed salinity changes (Bersch et al., 2007; Polyakov et al., 2008; Jacobs and Giulivi, 2010).

3.3.3.1 Pacific and Indian Oceans

In the tropical Pacific, surface salinity has declined by 0.1 to 0.3 over 50 years in the precipitation-dominated western equatorial regions and by up to 0.6 to 0.75 in the Intertropical Convergence Zone and the South Pacific Convergence Zone (Cravatte et al., 2009), while surface salinity has increased by up to 0.1 over the same period in the evaporation-dominated zones in the southeastern and north-central tropical Pacific (Figure 3.9). The fresh, low-density waters in the warm pool of the western equatorial Pacific expanded in area as the surface salinity front migrated eastward by 1500 to 2500 km over the period 1955–2003 (Delcroix et al., 2007; Cravatte et al., 2009). Similarly, in the Indian Ocean, the net precipitation regions in the Bay of Bengal and the warm pool contiguous with the tropical Pacific warm pool have been freshening by up to 0.1 to 0.2, while the saline Arabian Sea and south Indian Ocean have been getting saltier by up to 0.2 (Durack and Wijffels, 2010).

In the North Pacific, the subtropical thermocline has freshened by 0.1 since the early 1990s, following surface freshening that began around 1984 (Ren and Riser, 2010); the freshening extends down through the intermediate water that is formed in the northwest Pacific (Nakano et al., 2007), continuing the freshening documented by Wong et al. (1999). Warming of the surface water that subducts to supply the intermediate water is one reason for this signal, as the freshwater from the subpolar North Pacific is now entering the subtropical thermocline at lower density.

Salinity changes, together with temperature changes (Section 3.2.2), affect stratification; salinity has more impact than temperature in some regions. In the western tropical Pacific, for example, the density changes from 1970 to 2003 at a trend of –0.013 kg m^{-3} yr^{-1}, about 60% of that due to salinity (Delcroix et al., 2007). The decreasing density trend mainly occurs near the surface only, which should affect stratification across the base of the mixed layer. In the Oyashio region of the western North Pacific, salinity decrease near the surface accounts for about 60% of the density decrease of –0.004 kg m^{-3} yr^{-1} from 1968 to 1998 (Ono et al., 2001).

3.3.3.2 Atlantic Ocean

The net evaporative North Atlantic has become saltier as a whole over the past 50 years (Figure 3.9; Boyer et al., 2007). The largest increase in the upper 700 m occurred in the Gulf Stream region (0.006 per decade between 1955–1959 and 2002–2006) (Wang et al., 2010). Salinity increase is also evident following the circulation pathway of Mediterranean Outflow Water (Figure 3.9; Fusco et al., 2008). This increase can be traced back to the western basin of the Mediterranean, where salinity of the deep water increased during the period from 1943 to the mid-2000s (Smith et al., 2008; Vargas-Yáñez et al., 2010).

During the time period between 1955–1959 and 2002–2006 (using salinities averaged over the indicated 5-year ranges), the upper 700 m of the subpolar North Atlantic freshened by up to 0.002 per decade (Wang et al., 2010), while an increase in surface salinity was found between the average taken over 1960–1989 and the 5-year average over 2003–2007 (Hosoda et al., 2009). Decadal and multi-decadal variability in the subpolar gyre and Nordic Seas is vigorous and has been related to various climate modes such as the NAO, the Atlantic multi-decadal oscillation (AMO, Box 2.5), and even El Niño-Southern Oscillation (ENSO; Polyakov et al., 2005; Yashayaev and Loder, 2009), obscuring long-term trends. The 1970s to 1990s freshening of the northern North Atlantic and Nordic Seas (Dickson et al., 2002; Curry et al., 2003; Curry and Mauritzen, 2005) reversed to salinification (0 to 2000 m depth) starting in the late 1990s (Boyer et al., 2007; Holliday et al., 2008), and the propagation of this signal could be followed along the eastern boundary from south of 60°N in the Northeast Atlantic to Fram Strait at 79°N (Holliday et al., 2008). Advection has also played a role in moving higher salinity subtropical waters to the subpolar gyre (Hatun et al., 2005; Bersch et al., 2007; Lozier and Stewart, 2008; Valdimarsson et al., 2012). The variability of the cross equatorial transport contribution to this budget is highly uncertain. Reversals of North Atlantic surface salinity of similar amplitude and duration to those observed in the last 50 years are apparent in the early 20th century (Reverdin et al., 2002; Reverdin, 2010). The evaporation-dominated subtropical South Atlantic has become saltier by 0.1 to 0.3 during the period from 1950 to 2008 (Hosoda et al., 2009; Durack and Wijffels, 2010; Figure 3.4).

3.3.3.3 Arctic Ocean

Sea ice in the Arctic has declined significantly in recent decades (Section 4.2), which might be expected to reduce the surface salinity and increase freshwater content as freshwater locked in multi-year sea ice is released. Generally, strong multi-decadal variability, regional variability, and the lack of historical observations have made it difficult to assess long-term trends in ocean salinity and freshwater content for the Arctic as a whole (Rawlins et al., 2010). The signal that is now emerging, including salinity observations from 2005 to 2010, indicates increased freshwater content, with *medium confidence*.

Over the 20th century (1920–2003) the central Arctic Ocean in the upper 150 m became fresher in the 1950s and then more saline by

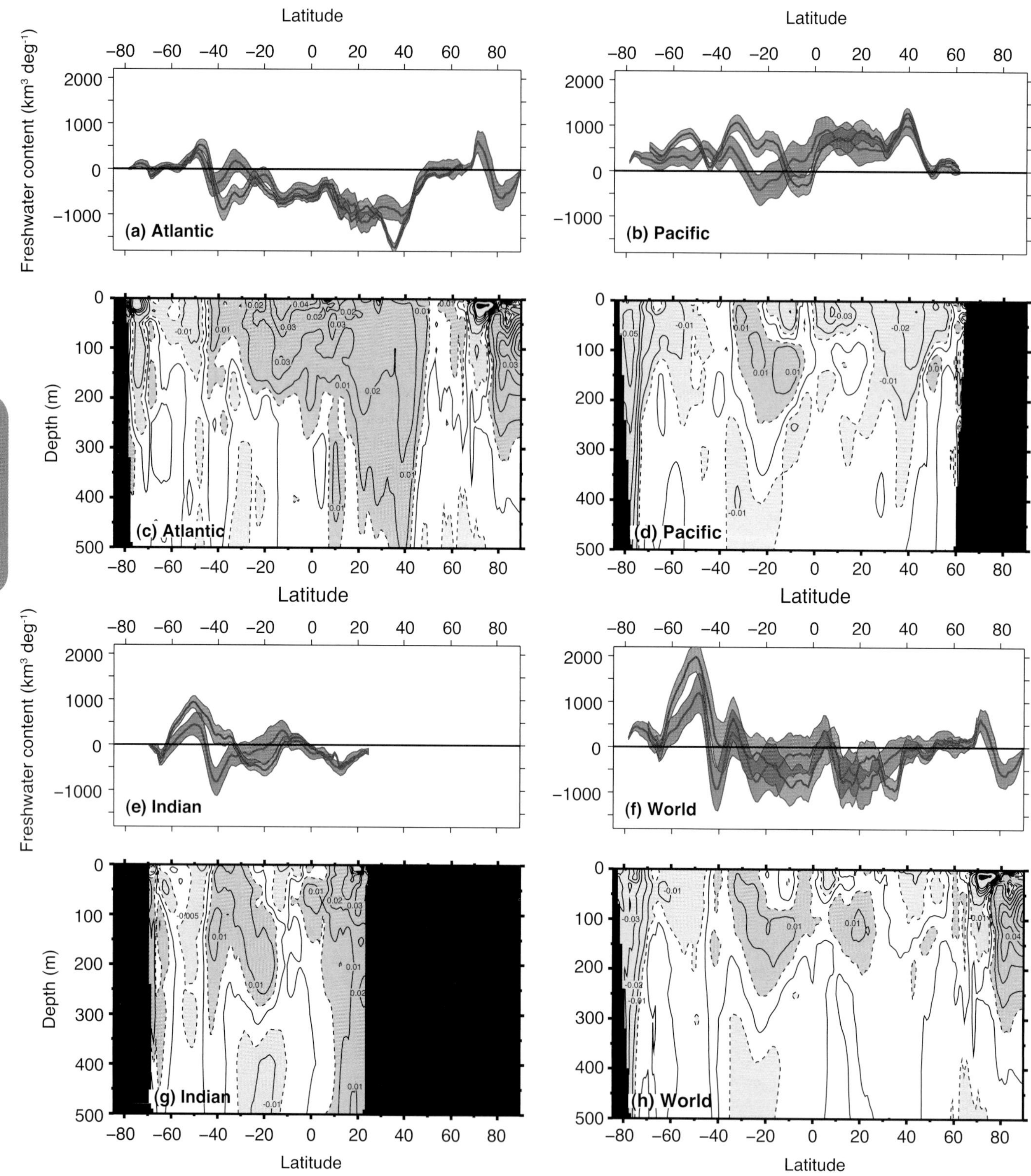

3

Figure 3.5 | Zonally integrated freshwater content changes (FWCC; km^3 per degree of latitude) in the upper 500 m over one-degree zonal bands and linear trends (1955–2010) of zonally averaged salinity (PSS78; lower panels) in the upper 500 m of the (a) and (c) Atlantic, (b) and (d) Pacific, (e) and (g) Indian and (f) and (h) World Oceans. The FWCC time period is from 1955 to 2010 (Boyer et al., 2005; blue lines) and 1950 to 2008 (Durack and Wijffels, 2010; red lines). Data are updated from Boyer et al. (2005) and calculations of FWCC are done according to the method of Boyer et al. (2007), using 5-year averages of salinity observations and fitting a linear trend to these averages. Error estimates are 95% confidence intervals. The contour interval of salinity trend in the lower panels is 0.01 PSS78 per decade and dashed contours are 0.005 PSS78 per decade. Red shading indicates values equal to or greater than 0.05 PSS78 per decade and blue shading indicates values equal to or less than –0.005 PSS78 per decade.

the early 2000s, with a net small salinification over the whole record (Polyakov et al., 2008), while at the Siberian Shelf the river discharge increased (Shiklomanov and Lammers, 2009) and the shelf waters became fresher (Polyakov et al., 2008).

Upper ocean freshening has also been observed regionally in the southern Canada basin from the period 1950–1980 to the period 1990–2000s (Proshutinsky et al., 2009; Yamamoto-Kawai et al., 2009). These are the signals reflected in the freshwater content trend from 1955 to 2010 shown in Figure 3.5a, f: salinification at the highest latitudes and a band of freshening at about 70°N to 80°N. Ice production and sustained export of freshwater from the Arctic Ocean in response to winds are suggested as key contributors to the high- latitude salinification (Polyakov et al., 2008; McPhee et al., 2009). The contrasting changes in different regions of the Arctic have been attributed to the effects of Ekman transport, sea ice formation (and melt) and a shift in the pathway of Eurasian river runoff (McPhee et al., 2009; Yamamoto-Kawai et al., 2009; Morison et al., 2012).

Between the periods 1992–1999 and 2006–2008, not only the central Arctic Ocean freshened (Rabe et al., 2011; Giles et al., 2012), but also freshening is now observed in all regions including those that were becoming more saline through the early 2000s (updated from Polyakov et al., 2008). Moreover, freshwater transport out of the Arctic has increased in that time period (McPhee et al., 2009).

3.3.3.4 Southern Ocean

Widespread freshening (trend of –0.01 per decade, significant at 95% confidence, from the 1980s to 2000s) of the upper 1000 m of the Southern Ocean was inferred by taking differences between modern data (mostly Argo) and a long-term climatology along mean streamlines (Böning et al., 2008). Decadal variability, although notable, does not overwhelm this trend (Böning et al., 2008). Both a southward shift of the Antarctic Circumpolar Current and water-mass changes contribute to the observed trends during the period 1992–2009 (Meijers et al., 2011). The zonally averaged freshwater content for each ocean and the world (Figure 3.5) shows this significant Southern Ocean freshening, which exceeds other regional trends and is present in each basin (Indian, Atlantic and Pacific, Figure 3.9).

3.3.4 Evidence for Change of the Hydrological Cycle from Salinity Changes

The similarity between the geographic distribution of significant salinity and freshwater content trends (Figures 3.4, 3.5 and 3.21) and both the mean salinity pattern and the distribution of mean evaporation – precipitation (E – P; Figure 3.4) indicates, *with medium confidence,* that the large-scale pattern of net evaporation minus precipitation over the oceans has been enhanced. Whereas the surface salinity pattern could be enhanced by increased stratification due to surface warming, the large-scale changes in column-integrated freshwater content are *very unlikely* to result from changes in stratification in the thin surface layer. Furthermore, the large spatial scale of the observed changes in freshwater content cannot be explained by changes in ocean circulation such as shifts of gyre boundaries. The observed changes in surface and subsurface salinity require additional horizontal atmospheric water transport from regions of net evaporation to regions of net precipitation. A similar conclusion was reached in AR4 (Bindoff et al., 2007). The water vapour in the troposphere has *likely* increased since the 1970s, due to warming (2.4.4, 2.5.5, 2.5.6; FAQ 3.2). The inferred enhanced pattern of net E – P can be related to water vapor increase, although the linkage is complex (Emori and Brown, 2005; Held and Soden, 2006). From 1950 to 2000, the large-scale pattern of surface salinity has amplified at a rate that is larger than model simulations for the historical 20th century and 21st century projections. The observed rate of surface salinity amplification is comparable to the rate expected from a water cycle response following the Clausius–Clapeyron relationship (Durack et al., 2012).

Studies published since AR4, based on expanded data sets and new analysis approaches, have substantially decreased the level of uncertainty in the salinity and freshwater content trends (e.g., Stott et al., 2008; Hosoda et al., 2009; Roemmich and Gilson, 2009; Durack and Wijffels, 2010; Helm et al., 2010), and thus increased confidence in the inferred changes of evaporation and precipitation over the ocean.

3.3.5 Conclusions

Both positive and negative trends in ocean salinity and freshwater content have been observed throughout much of the ocean, both at the sea surface and in the ocean interior. While similar conclusions were reached in AR4, the recent studies summarized here, based on expanded data sets and new analysis approaches, provide *high confidence* in the assessment of trends in ocean salinity. It is *virtually certain* that the salinity contrast between regions of high and low surface salinity has increased since the 1950s. It is *very likely* that since the 1950s, the mean regional pattern of upper ocean salinity has been enhanced: saline surface waters in the evaporation-dominated mid-latitudes have become more saline, while the relatively fresh surface waters in rainfall-dominated tropical and polar regions have become fresher. Similarly, it is *very likely* that the interbasin contrast between saline Atlantic and fresh Pacific surface waters has increased, and it is *very likely* that freshwater content in the Southern Ocean has increased. There is *medium confidence* that these patterns in salinity trends are caused by increased horizontal moisture transport in the atmosphere, suggesting changes in evaporation and precipitation over the ocean as the lower atmosphere has warmed.

Trends in salinity have been observed in the ocean interior as well. It is *likely* that the subduction of surface water mass anomalies and the movement of density surfaces have contributed to the observed salinity changes on depth levels. Changes in freshwater flux and the migration of surface density outcrops caused by surface warming (e.g., to regions of lower or higher surface salinity) have *likely* both contributed to the formation of salinity anomalies on density surfaces.

3.4 Changes in Ocean Surface Fluxes

3.4.1 Introduction

Exchanges of heat, water and momentum (wind stress) at the sea surface are important factors for driving the ocean circulation. Changes in the air–sea fluxes may result from variations in the driving surface meteorological state variables (air temperature and humidity, SST, wind speed, cloud cover, precipitation) and can impact both water-mass formation rates and ocean circulation. Air–sea fluxes also influence temperature and humidity in the atmosphere and, therefore, the hydrological cycle and atmospheric circulation. AR4 concluded that, at the global scale, the accuracy of the observations is insufficient to permit a direct assessment of changes in heat flux (AR4 Section 5.2.4). As described in Section 3.4.2, although substantial progress has been made since AR4, that conclusion still holds for this assessment.

The net air–sea heat flux is the sum of two turbulent (latent and sensible) and two radiative (shortwave and longwave) components. Ocean heat gain from the atmosphere is defined to be positive according to the sign convention employed here. The latent and sensible heat fluxes are computed from the state variables using bulk parameterizations; they depend primarily on the products of wind speed and the vertical near-sea-surface gradients of humidity and temperature respectively. The air–sea freshwater flux is the difference of precipitation (P) and evaporation (E). It is linked to heat flux through the relationship between evaporation and latent heat flux. Thus, when considering potential trends in the global hydrological cycle, consistency between observed heat budget and evaporation changes is required in areas where evaporation is the dominant term in hydrological cycle changes. Ocean surface shortwave and longwave radiative fluxes can be inferred from satellite measurements using radiative transfer models, or computed using empirical formulae, involving astronomical parameters, atmospheric humidity, cloud cover and SST. The wind stress is given by the product of the wind speed squared, and the drag coefficient. For detailed discussion of all terms see, for example, Gulev et al. (2010).

3

Atmospheric reanalyses, discussed in Box 2.3, are referred to frequently in the following sections and for clarity the products cited are summarised here: ECMWF 40-year Reanalysis (referred to as ERA40 hereafter, Uppala et al., 2005), ECMWF Interim Reanalysis (ERAI, Dee et al., 2011), NCEP/NCAR Reanalysis 1 (NCEP1, Kalnay et al., 1996), NCEP/DOE Reanalysis 2 (NCEP2, Kanamitsu et al., 2002), NCEP Climate Forecast System Reanalysis (CFSR, Saha et al., 2010), NASA Modern Era Reanalysis for Research and Applications (MERRA, Rienecker et al., 2011) and NOAA-CIRES 20th Century Reanalysis, version 2 (20CRv2, Compo et al., 2011).

3.4.2 Air–Sea Heat Fluxes

3.4.2.1 Turbulent Heat Fluxes and Evaporation

The latent and sensible heat fluxes have a strong regional dependence, with typical values varying in the annual mean from close to zero to –220 W m^{-2} and –70 W m^{-2} respectively over strong heat loss sites (Yu and Weller, 2007). Estimates of these terms have many potential sources of error (e.g., sampling issues, instrument biases, changing data sources, uncertainty in the flux computation algorithms). These sources may be spatially and temporally dependent, and are difficult to quantify (Gulev et al., 2007); consequently flux error estimates have a high degree of uncertainty. Spurious temporal trends may arise as a result of variations in measurement method for the driving meteorological state variables, in particular wind speed (Tokinaga and Xie, 2011). The overall uncertainty of the annually averaged global ocean mean for each term is expected to be in the range 10 to 20%. In the case of the latent heat flux term, this corresponds to an uncertainty of up to 20 W m^{-2}. In comparison, changes in global mean values of individual heat flux components expected as a result of anthropogenic climate change since 1900 are at the level of <2 W m^{-2} (Pierce et al., 2006).

Many new turbulent heat flux data sets have become available since AR4 including products based on atmospheric reanalyses, satellite and *in situ* observations, and hybrid or synthesized data sets that combine information from these three different sources. It is not possible to identify a single best product as each has its own strengths and weaknesses (Gulev et al., 2010); several data sets are summarised here to illustrate the key issues. The Hamburg Ocean-Atmosphere Parameters and Fluxes from Satellite (HOAPS) data product provides global turbulent heat fluxes (and precipitation) developed from observations at microwave and infrared wavelengths (Andersson et al., 2011). In common with other satellite data sets it provides globally complete fields, however, it spans a relatively short period (1987 onwards) and is thus of limited utility for identifying long-term changes. A significant advance in flux data set development methodology is the 1 × 1 degree grid Objectively Analysed Air–Sea heat flux (OAFlux) data set that covers 1958 onwards and for the first time synthesizes state variables (SST, air temperature and humidity, wind speed) from reanalyses and satellite observations, prior to flux calculation (Yu and Weller, 2007). OAFlux has the potential to minimize severe spatial sampling errors that limit the usefulness of data sets based on ship observations alone and provides a new resource for temporal variability studies. However, the data sources for OAFlux changed in the 1980s, with the advent of satellite data, and the consequences of this change need to be assessed. In an alternative approach, Large and Yeager (2009) modified NCEP1 reanalysis state variables prior to flux calculation using various adjustment techniques, to produce the hybrid Coordinated Ocean-ice Reference Experiments (CORE) turbulent fluxes for 1948–2007 (Griffies et al., 2009). However, as the adjustments employed to produce the CORE fluxes were based on limited periods (e.g., 2000–2004 for wind speed) it is not clear to what extent CORE can be reliably used for studies of interdecadal variability over the 60-year period that it spans.

Analysis of OAFlux suggests that global mean evaporation may vary at inter-decadal time scales, with the variability being relatively small compared to the mean (Yu, 2007; Li et al., 2011; Figure 3.6a). Changing data sources, particularly as satellite observations became available in the 1980s, may contribute to this variability (Schanze et al., 2010) and it is not yet possible to identify how much of the variability is due to changes in the observing system. The latent heat flux variations (Figure 3.6b) closely follow those in evaporation (with allowance for the sign definition which results in negative values of latent heat flux corresponding to positive values of evaporation) but do not scale exactly as there is an additional minor dependence on SST through the latent heat of evaporation. The large uncertainty ranges that are evident in each of the time series highlight the difficulty in establishing whether there is a trend in global ocean mean evaporation or latent heat flux. The uncertainty range for latent heat flux is much larger than the 0.5 W m^{-2} level of net heat flux change expected from the ocean heat content increase (Box 3.1). Thus, it is not yet possible to use such data sets to establish global ocean multi-decadal trends in evaporation or latent

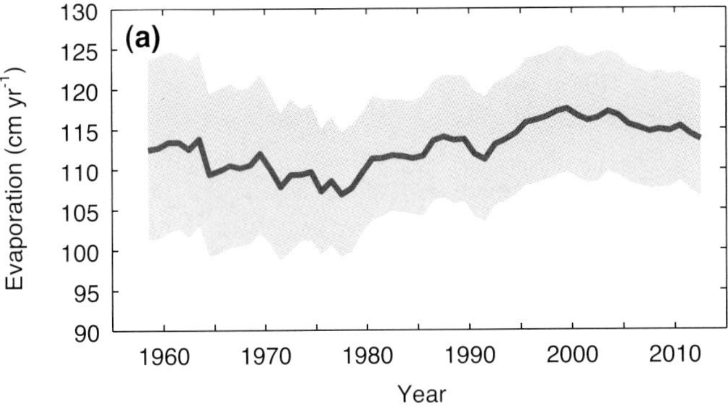

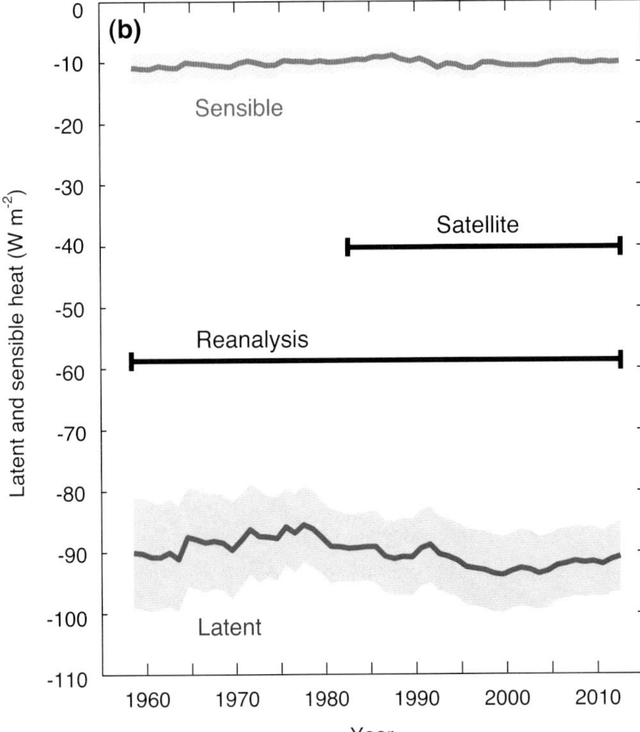

Figure 3.6 | Time series of annual mean global ocean average evaporation (red line, a), sensible heat flux (green line, b) and latent heat flux (blue line, b) from 1958 to 2012 determined by Yu from a revised and updated version of the original OAFlux data set Yu and Weller (2007). Shaded bands show uncertainty estimates and the black horizontal bars in (b) show the time periods for which reanalysis output and satellite observations were employed in the OAFlux analysis; they apply to both panels.

heat flux at this level. The globally averaged sensible heat flux is smaller in magnitude than the latent heat flux and has a smaller absolute range of uncertainty (Figure 3.6b).

3.4.2.2 Surface Fluxes of Shortwave and Longwave Radiation

The surface shortwave flux has a strong latitudinal dependence with typical annual mean values of 250 W m^{-2} in the tropics. The annual mean surface net longwave flux ranges from –30 to –70 W m^{-2}. Estimates of these terms are available from *in situ* climatologies, from atmospheric reanalyses, and, since the 1980s, from satellite observations. These data sets have many potential sources of error that include: uncertainty in the satellite retrieval algorithms and *in situ* formulae, cloud representation in reanalyses, sampling issues and changing satellite sensors (Gulev et al., 2010). As for the turbulent fluxes, the uncertainty of the annually averaged global ocean mean shortwave or longwave flux is difficult to determine and in the range 10–20%.

High accuracy *in situ* radiometer measurements are available at land sites since the 1960s (see Wild, 2009 Figure 1), allowing analysis of decadal variations in the surface shortwave flux. However, this is not the case over the oceans, where there are very few *in situ* measurements (the exception being moored buoy observations in the tropical band 15°S to 15°N since the 1990s, Pinker et al., 2009). Consequently, for global ocean shortwave analyses it is necessary to rely on satellite observations, which are less accurate (compared to *in situ* determination of radiative fluxes), restrict the period that can be considered to the mid-1980s onwards, but do provide homogeneous sampling. Detailed discussion of variations in global (land and ocean) averaged surface solar radiation is given in Section 2.3.3; *confidence* in variability of radiation averaged over the global ocean is *low* owing to the lack of direct observations.

3.4.2.3 Net Heat Flux and Ocean Heat Storage Constraints

The most reliable source of information for changes in the global mean net air–sea heat flux comes from the constraints provided by analyses of changes in ocean heat storage. The estimate of increase in global ocean heat content for 1971–2010 quantified in Box 3.1 corresponds to an increase in mean net heat flux from the atmosphere to the ocean of 0.55 W m^{-2}. In contrast, closure of the global ocean mean net surface heat flux budget to within 20 W m^{-2} from observation based surface flux data sets has still not been reliably achieved (e.g., Trenberth et al., 2009). The increase in mean net air–sea heat flux is thus small compared to the uncertainties of the global mean. Large and Yeager (2012) examined global ocean average net heat flux variability using the CORE data set over 1984–2006 and concluded that natural variability, rather than long-term climate change, dominates heat flux changes over this relatively short, recent period. Since AR4, some studies have shown consistency in regional net heat flux variability at sub-basin scale since the 1980s, notably in the Tropical Indian Ocean (Yu et al., 2007) and North Pacific (Kawai et al., 2008). However, detection of a change in air–sea fluxes responsible for the long-term ocean warming remains beyond the ability of currently available surface flux data sets.

3.4.3 Ocean Precipitation and Freshwater Flux

Assessment of changes in ocean precipitation at multi-decadal time scales is very difficult owing to the lack of reliable observation based data sets prior to the satellite era. The few studies available rely on reconstruction techniques. Remote sensing based precipitation observations from the Global Precipitation Climatology Project (GPCP) for 1979–2003 have been used by Smith et al. (2009, 2012) to reconstruct precipitation for 1900–2008 (over 75°S to 75°N) by employing statistical techniques that make use of the correlation between precipitation and both SST and sea level pressure (SLP). Each of the reconstructions shows both centennial and decadal variability in global ocean mean precipitation (Figure 3.7). The trend from 1900 to 2008 is 1.5 mm per month per century according to Smith et al. (2012). For the period of overlap, the reconstructed global ocean mean precipitation

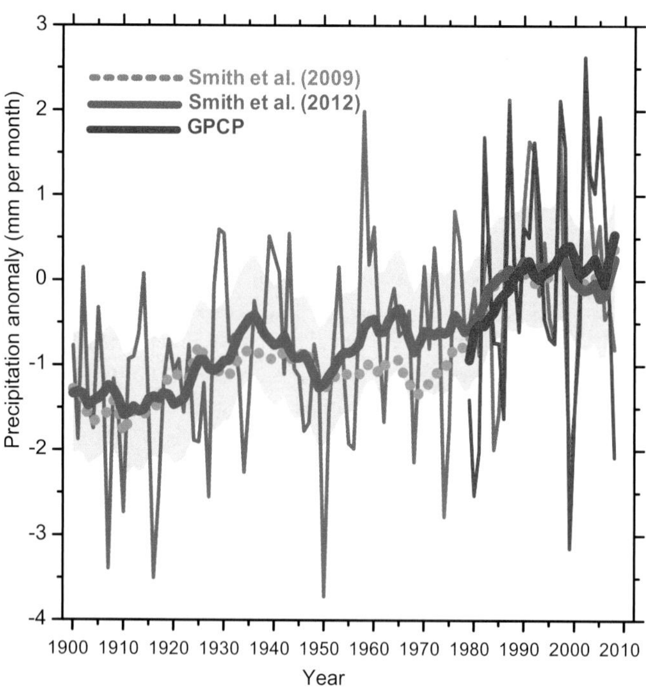

Figure 3.7 | Long-term reconstruction of ocean precipitation anomaly averaged over 75°S to 75°N from Smith et al. (2012): Annual values, thin blue line; low-pass filtered (15-year running mean) values, bold blue line with uncertainty estimates (shading). Smith et al. (2009) low-pass filtered values, dotted grey line. Also shown is the corresponding GPCPv2.2 derived ocean precipitation anomaly time series averaged over the same latitudinal range (annual values, thin magenta line; low-pass filtered values, bold magenta line); note Smith et al. (2012) employed an earlier version of the GPCP data set leading to minor differences relative to the published time series in their paper. Precipitation anomalies were taken relative to the 1979–2008 period.

time series show consistent variability with GPCP as is to be expected (Figure 3.7). Focusing on the Tropical Ocean (25°S to 25°N) for the recent period 1979–2005, Gu et al. (2007) have identified a precipitation trend of 0.06 mm day^{-1} per decade using GPCP. Concerns have been expressed in the cited studies over the need for further work both to determine the most reliable approach to precipitation reconstruction and to evaluate the remotely sensed precipitation data sets. Given these concerns, *confidence* in ocean precipitation trend results is *low*.

Evaporation and precipitation fields from atmospheric reanalyses can be tested for internal consistency of different components of the hydrological cycle. Specifically, the climatological mean value for E – P averaged over the global ocean should equal both the corresponding mean for P – E averaged over land and the moisture transport from ocean to land. Trenberth et al. (2011) find in an assessment of eight atmospheric reanalyses that this is not the case for each product considered, and they also report spurious trends due to variations in the observing system with time. Schanze et al. (2010) examine interannual variability within the OAFlux evaporation and GPCP precipitation data sets, and find that use of satellite data prior to 1987 is limited by discontinuities attributable to variations in data type. Thus, it is not yet possible to use such data sets to establish whether there are significant multi-decadal trends in mean E – P. However, regional trends in surface salinity since the 1950s do suggest trends in E – P over the same time (see Section 3.3.4).

3.4.4 Wind Stress

Wind stress fields are available from reanalyses, satellite-based data sets, and *in situ* observations. Basin scale wind stress trends at decadal to centennial time scales have been reported for the Southern Ocean, the North Atlantic and the Tropical Pacific as detailed below. However, these results are based largely on atmospheric reanalyses, in some cases a single product, and consequently the *confidence* level is *low* to *medium* depending on region and time scale considered.

In the Southern Ocean, the majority of reanalyses in the most comprehensive study available show an increase in the annual mean zonal wind stress (Swart and Fyfe, 2012; Figure 3.8). They find an increase in annual mean wind stress strength in four (NCEP1, NCEP2, ERAI and 20CRv2) of the six reanalyses considered (Figure 3.8). The mean of all reanalyses available at a given time (Figure 3.8, black line) also shows an upward trend from about 0.15 N m^{-2} in the early 1950s to 0.20 N m^{-2} in the early 2010s. An earlier study, covering 1979–2009, found a wind stress increase in two of four reanalyses considered (Xue et al., 2010). A positive trend of zonal wind stress from 1980 to 2000 was also reported by Yang et al. (2007) using a single reanalysis (ERA40) and found to be consistent with increases in wind speed observations made on Macquarie Island (54.5°S, 158.9°E) and by the SSM/I satellite (data from 1987 onwards). The wind stress strengthening is found by Yang et al. (2007) to have a seasonal dependence, with strongest trends in January, and has been linked by them to changes in the Southern Annular Mode (SAM, Box 2.5), which has continued to show an upward trend since AR4 (Section 2.7.8). Taken as a whole, these studies provide *medium confidence* that Southern Ocean wind stress has strengthened since the early 1980s. A strengthening of the related wind speed field in the Southern Ocean, consistent with the increasing trend in the SAM, has also been noted in Section 2.7.2 from satellite-based analyses and atmospheric reanalyses.

In the Tropical Pacific, a reanalysis based study found a strengthening of the trade wind associated wind stress for 1990–2009, but for the earlier period 1959–1989 there is no clear trend (Merrifield, 2011). Strengthening of the related Tropical Pacific Ocean wind speed field in recent decades is evident in reanalysis and satellite based data sets. Taken together with evidence for rates of sea level rise in the western Pacific larger than the global mean (Section 3.7.3) these studies provide *medium confidence* that Tropical Pacific wind stress has increased since 1990. This increase may be related to the Pacific Decadal Oscillation (Merrifield et al., 2012). At centennial time scales, attempts have been made to reconstruct the wind stress field in the Tropical Pacific by making use of the relationship between wind stress and SLP in combination with historic SLP data. Vecchi et al. (2006), using this approach, found a reduction of 7% in zonal mean wind stress across the Equatorial Pacific from the 1860s to the 1990s and related it to a possible weakening of the tropical Walker circulation. Observations discussed in Section 2.7.5 indicate that this weakening has largely been offset by a stronger Walker circulation since the 1990s.

Changes in winter season wind stress curl over the North Atlantic from 1950 to early 2000s from NCEP1 and ERA40 have leading modes that are highly correlated with the NAO and East Atlantic circulation patterns; each of these patterns demonstrates a trend towards more positive index values superimposed on pronounced decadal variability over the period from the early 1960s to the late 1990s (Sugimoto and Hanawa, 2010). Wu et al. (2012) find a poleward shift over the past century of the zero wind stress curl line by 2.5° [1.5° to 3.5°] in the North Atlantic and 3.0° [1.6° to 4.4°] in the North Pacific from 20CRv2. *Confidence* in these results is *low* as they are based on a single product, 20CRv2 (the only century time scale reanalysis), which may be affected by temporal inhomogeneity in the number of observations assimilated (Krueger et al., 2013).

3.4.5 Changes in Surface Waves

Surface wind waves are generated by wind forcing and are partitioned into two components, namely wind–sea (wind-forced waves propagating slower than surface wind) and swell (resulting from the wind–sea development and propagating typically faster than surface wind). Significant wave height (SWH) represents the measure of the wind wave field consisting of wind–sea and swell and is approximately equal to the highest one-third of wave heights. Local wind changes influence wind–sea properties, while changes in remote storms affect swell. Thus, patterns of wind wave and surface wind variability may differ because wind waves integrate wind properties over a larger domain. As wind waves integrate characteristics of atmospheric dynamics over a range of scales they potentially serve as an indicator of climate variability and change. Global and regional time series of wind waves characteristics are available from buoy data, Voluntary Observing Ship (VOS) reports, satellite measurements and model wave hindcasts. No source is superior, as all have their strengths and weaknesses (Sterl and Caires, 2005; Gulev and Grigorieva, 2006; Wentz and Ricciardulli, 2011).

3.4.5.1 Changes in Surface Waves from Voluntary Observing Ship and Wave Model Hindcasts Forced by Reanalyses

AR4 reported statistically significant positive SWH trends during 1900–2002 in the North Pacific (up to 8 to 10 cm per decade) and stronger trends (up to 14 cm per decade) from 1950 to 2002 for most of the mid-latitudinal North Atlantic and North Pacific, with insignificant trends, or small negative trends, in most other regions (Trenberth et al., 2007). Studies since AR4 have provided further evidence for SWH trends with more detailed quantification and regionalization.

Model hindcasts based on 20CRv2 (spanning 1871–2010) and ERA40 (spanning 1958–2001) show increases in annual and winter mean SWH in the north-east Atlantic, although the trend magnitudes depend on the reanalysis products used (Sterl and Caires, 2005; Wang et al., 2009, 2012; Semedo et al., 2011). Analysis of VOS observations for 1958–2002 reveals increases in winter mean SWH over much of the North Atlantic, north of 45°N, and the central to eastern mid-latitude North Pacific with typical trends of up to 20 cm per decade (Gulev and Grigorieva, 2006).

3.4.5.2 Changes in Surface Waves from Buoy Data

Positive regional trends in extreme wave heights have been reported at several buoy locations since the late 1970s, with some evidence for seasonal dependence, including at sites on the east and west coasts of

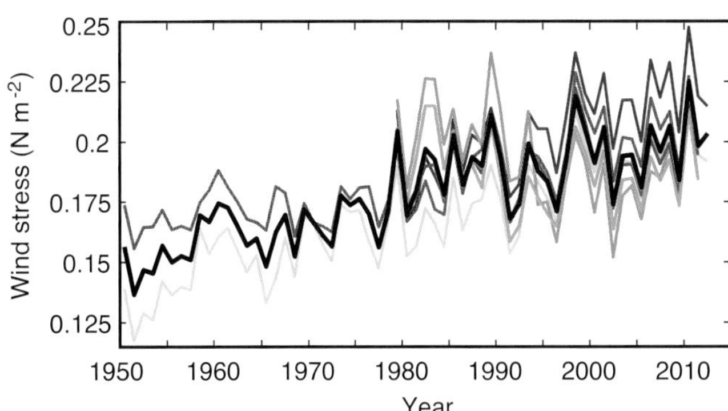

Figure 3.8 | Time series of annual average maximum zonal-mean zonal wind stress (N m^{-2}) over the Southern Ocean for various atmospheric reanalyses: CFSR (orange), NCEP1 (cyan), NCEP2 (red), ERAI (dark blue), MERRA (green), 20CR (grey), and mean of all reanalyses at a given time (thick black), see Box 2.3 for details of reanalyses. Updated version of Figure 1a in Swart and Fyfe (2012), with CFSR, MERRA and the mean of all reanalyses added.

the USA (Komar and Allan, 2008; Ruggiero et al., 2010) and the north-east Pacific coast (Menéndez et al., 2008). However, Gemmrich et al. (2011) found for the Pacific buoys that some trends may be artefacts due to step-type historical changes in the instrument types, observational practices and post-processing procedures. Analysis of data from a single buoy deployed west of Tasmania showed no significant trend in the frequency of extreme waves contrary to a significant positive trend seen in the ERA40 reanalysis (Hemer, 2010).

3.4.5.3 Changes in Surface Waves from Satellite Data

Satellite altimeter observations provide a further data source for wave height variability since the mid-1980s. Altimetry is of particular value in the southern hemisphere, and in some poorly sampled regions of the northern hemisphere, where analysis of SWH trends remains a challenge due to limited *in situ* data and temporal inhomogeneity in the data used for reanalysis products. In the Southern Ocean, altimeter-derived SWH and model output both show regions with increasing wave height although these regions cover narrower areas in the altimeter analysis than in the models and have smaller trends (Hemer et al., 2010). Young et al. (2011a) compiled global maps of mean and extreme (90th and 99th percentile) surface wind speed and SWH trends for 1985–2008 using altimeter measurements. As the length of the data set is short, it is not possible to determine whether their results reflect long-term SWH and wind speed trends, or are part of a multi-decadal oscillation. For mean SWH, their analysis shows positive linear trends of up to 10 to 15 cm per decade in some parts of the Southern Ocean (with the strongest changes between 80°E and 160°W) that may reflect the increase in strength of the wind stress since the early 1980s (see Section 3.4.4). Young et al. (2011a) note, however, that globally the level of statistical significance is generally low in the mean and 90th percentile SWH trends but increases for the 99th percentile. Small negative mean SWH trends are found in many NH ocean regions and these are of opposite sign to, and thus inconsistent with, trends in wind speed — the latter being primarily positive. Nevertheless, for the 99th SWH percentile, strong positive trends up to 50 to 60 cm per decade were identified in the Southern Ocean, North

Atlantic and North Pacific and these are consistent in sign with the extreme wind speed trends. Subsequent analysis has shown that the Young et al. (2011a) wind speed trends tend to be biased high when compared with microwave radiometer data (Wentz and Ricciardulli, 2011; Young et al., 2011b).

3.4.6 Conclusions

Uncertainties in air–sea heat flux data sets are too large to allow detection of the change in global mean net air–sea heat flux, on the order of 0.5 W m^{-2} since 1971, required for consistency with the observed ocean heat content increase. The accuracy of reanalysis and satellite observation based freshwater flux products is limited by changing data sources. Consequently, the products cannot yet be reliably used to directly identify trends in the regional or global distribution of evaporation or precipitation over the oceans on the time scale of the observed salinity changes since 1950.

Basin scale wind stress trends at decadal to centennial time scales have been observed in the North Atlantic, Tropical Pacific, and Southern Oceans with *low to medium confidence*. These results are based largely on atmospheric reanalyses, in some cases a single product, and the confidence level is dependent on region and time scale considered. The evidence is strongest for the Southern Ocean for which there is *medium confidence* that zonal mean wind stress has increased in strength since the early 1980s.

There is *medium confidence* based on ship observations and reanalysis forced wave model hindcasts that mean significant wave height has increased since the 1950s over much of the North Atlantic north of 45°N, with typical winter season trends of up to 20 cm per decade.

3.5 Changes in Water-Mass Properties

3.5.1 Introduction

To a large degree, water properties are set at the sea surface through interaction between the ocean and the overlying atmosphere (and ice, in polar regions). The water characteristics resulting from these interactions (e.g., temperature, salinity and concentrations of dissolved gases and nutrients) are transferred to various depths in the world ocean, depending on the density of the water. Warm, light water masses supply (or "ventilate") the upper ocean at low to mid-latitudes, while the colder, denser water masses formed at higher latitudes supply the intermediate and deep layers of the ocean (see schematic in FAQ 3.1, Figure 1). The formation and subduction of water masses are important for the ocean's capacity to store heat, freshwater, carbon, oxygen and other properties relevant to climate. In this section, the evidence for change in some of the major water masses of the world ocean is assessed.

The zonal-mean distributions of salinity, density, and temperature in each ocean basin (black contours in Figure 3.9) reflect the formation of water masses at the sea surface and their subsequent spreading into the ocean interior. For example, warm, salty waters formed in the regions of net evaporation between 10° and 30° latitude (Figure 3.4b) supply the subtropical salinity maximum waters found in the upper few hundred meters in each basin (Figure 3.9). Relatively fresh water masses produced at higher latitude, where precipitation exceeds evaporation, sink and spread equatorward to form salinity minimum layers at intermediate depths. Outflow of saline water from the Mediterranean Sea and Red Sea, where evaporation is very strong, accounts for the relatively high salinity observed in the upper 1000 m in the subtropical North Atlantic and North Indian basins, respectively.

Many of the observed changes in zonally averaged salinity, density and temperature are aligned with the spreading paths of the major water masses (Figure 3.9, trends from 1950 to 2000 shown in colours and white contours), illustrating how the formation and spreading of water masses transfer anomalies in surface climate to the ocean interior. The strongest anomalies in a water mass are found near its source region. For instance, bottom and deep water anomalies are strongest in the Southern Ocean and the northern North Atlantic, with lessening amplitudes along the spreading paths of these water masses. In each basin, the subtropical salinity maximum waters have become more saline, while the low-salinity intermediate waters have become fresher (Figure 3.9 a, d, g, j; see also Section 3.3). Strongest warming is observed in the upper 100 m, which has warmed almost everywhere, with reduced warming (Atlantic) or regions of cooling (Indian and Pacific) observed between 100 and 500 m depth.

Warming is observed throughout the upper 2000 m south of 40°S in each basin. Shifts in the location of ocean circulation features can also contribute to the observed trends in temperature and salinity, as discussed in Section 3.2. Density decreased throughout most of the upper 2000 m of the global ocean (middle column of Figure 3.9). The decrease in near-surface density (hence increase in stratification) is largest in the Pacific, where warming and freshening both act to reduce density, and smallest in the Atlantic where the salinity and temperature trends have opposite effects on density.

The remainder of this section focuses on evidence of change in globally relevant intermediate, deep and bottom water masses.

3.5.2 Intermediate Waters

3.5.2.1 North Pacific Intermediate Water

The North Pacific Intermediate Water (NPIW) has freshened over the last two decades (Wong et al., 1999; Nakano et al., 2007; Figure 3.9g) and has warmed since the 1950s, as reported in AR4, Chapter 5. NPIW in the northwestern North Pacific warmed by 0.5°C from 1955 to 2004 and is now entering the subtropics at lower density; oxygen concentrations in the NPIW have declined, indicating weaker ventilation (Nakanowatari et al., 2007; Kouketsu et al., 2010). The strongest trends are in the Sea of Okhotsk, where NPIW is formed, and have been tentatively linked to increased air temperature and decreased sea-ice extent in winter (Nakanowatari et al., 2007; Figure 3.9i).

3.5.2.2 Antarctic Intermediate Water

In AR4, Chapter 5, Antarctic Intermediate Water (AAIW) was reported to have warmed and freshened since the 1960s (Figure 3.9). In most recent studies, usually—but not always—a dipole pattern was found: on isopycnals denser than the AAIW salinity minimum, a warming and salinification was observed and on isopycnals lighter than the AAIW salinity minimum, a cooling and freshening trend (Böning et al., 2008; Durack and Wijffels, 2010; Helm et al., 2010; McCarthy et al., 2011). The salinity minimum core of the AAIW also underwent changes consistent with these patterns on isopycnals: In 1970–2009, south of 30°S, the AAIW salinity minimum core showed a strong, large-scale shoaling (30 to 50 dbar per decade) and warming (0.05°C to 0.15°C per decade), leading to lighter densities (up to 0.03 kg m^{-3} per decade), while the salinity trends varied regionally. A long-term freshening of the AAIW core is found in the southwest Atlantic, southeast Pacific, and south-central Indian oceans, with salinification south of Africa and Australia. All trends were strongest close to the AAIW formation latitude just north of the Antarctic Circumpolar Current (Schmidtko and Johnson, 2012).

Both an increase in precipitation—evaporation and poleward migration of density surfaces caused by warming have *likely* contributed to the observed trends (Section 3.3; Böning et al., 2008; Durack and Wijffels, 2010; Helm et al., 2010; McCarthy et al., 2011). Changes in AAIW properties in particular locations have also been linked to other processes, including exchange between the Indian and Atlantic basins (McCarthy et al., 2011) and changes in surface forcing related to modes of climate variability like ENSO and the SAM (Garabato et al., 2009). Whether these changes in properties also affected the formation rates of AAIW cannot be assessed from the available observations.

3.5.3 Deep and Bottom Waters

Deep and bottom layers of the ocean are supplied by roughly equal volumes of dense water sinking in the northern North Atlantic (Lower North Atlantic Deep Water, LNADW) and around Antarctica (Antarctic Bottom Water, AABW) (FAQ 3.1, Figure 1).

3.5.3.1 Upper North Atlantic Deep Water

Upper North Atlantic Deep Water (UNADW) is formed by deep convection in the Labrador Sea between Canada and Greenland, so is also known as Labrador Sea Water (LSW). It is the shallowest component of the NADW, located above the overflow water masses that supply the Lower North Atlantic Deep Water (LNADW). AR4 Chapter 5 assessed the variability in water mass properties of LSW from the 1950s. Recent studies have confirmed the large interannual-to-multi-decadeal variability of LSW properties and provided new information on variability in formation rates and the impact on heat and carbon (Section 3.8.1) uptake by the deep ocean.

During the 1970s and 1980s and especially the 1990s the UNADW has been cold and fresh. In Figure 3.9A it is the strong freshening signal from the 1960s to the 1990s that dominates the trend. This freshening trend reversed in the late 1990s (Boyer et al., 2007; Holliday et al., 2008; see Section 3.3.3.2). Estimates of the LSW formation rate[3] decreased from about 7.6 to 8.9 Sv in 1997–1999 (Kieke et al., 2006) to roughly 0.5 Sv in 2003–2005 (Rhein et al., 2011), and since 1997, only less dense LSW was formed compared to the high NAO years before. There is, however, evidence that formation of denser LSW occurred in 2008 (Våge et al., 2009; Yashayaev and Loder, 2009), but not in the following years (Yashayaev and Loder, 2009; Rhein et al., 2011).

The strong variability in the formation of UNADW affected significantly the heat transfer into the deep North Atlantic (Mauritzen et al., 2012). Substantial heat entered the deep North Atlantic during the low NAO years of the 1960s, when salinity was large enough to compensate for the high temperatures, and dense LSW was still formed and exported to the subtropics.

3.5.3.2 Lower North Atlantic Deep Water

Dense waters overflowing the sills between Greenland and Scotland supply the Lower North Atlantic Deep Water (LNADW). Both overflows freshened from the mid-1960s to the mid-1990s (Dickson et al., 2008). The salinity of the Faroe Bank overflow increased by 0.015 to 0.02 from 1997 to 2004, implying a density increase on the order of 0.01 kg m^{-3} (Hansen and Osterhus, 2007). The other main overflow, through Denmark Strait, shows large interannual variability in temperature and salinity, but no trends for the time period 1996–2011 (Jochumsen et al., 2012). Observations of the transport of the dense overflows are dominated by short-term variability and there is no evidence of a trend in the short time series available (see Section 3.6). As both overflow components descend into the North Atlantic, they entrain substantial amounts of ambient subpolar waters to create LNADW. As a whole, the LNADW in the North Atlantic cooled from the 1950s to 2005 (Mauritzen et al., 2012), a signal thus stemming primarily from the entrained waters, possibly an adjustment from an unusually warm period observed in the 1920s and 1930s (Drinkwater, 2006).

3.5.3.3 Antarctic Bottom Water

The Antarctic Bottom Water (AABW) has warmed since the 1980s or 1990s, most noticeably near Antarctica (Aoki et al., 2005; Rintoul, 2007; Johnson et al., 2008a; Purkey and Johnson, 2010; Kouketsu et al., 2011), but with warming detectable into the North Pacific and North Atlantic Oceans (Johnson et al., 2008b; Kawano et al., 2010). The warming of AABW between the 1990s and 2000s contributed to global ocean heat uptake (Section 3.2). The global volume of the AABW layer decreased by 8.2 [5.6 to 10.8] Sv during the last two decades (Johnson et al., 2008b; Mauritzen et al., 2012; Purkey and Johnson, 2012), making it *more likely than not* that at least the export rate of AABW from the Southern Ocean declined during this period.

The sources of AABW in the Indian and Pacific sectors of the Southern Ocean have freshened in recent decades. The strongest signal (0.03 per decade, between 1970 and 2008) is observed in the Ross Sea and has been linked to inflow of glacial melt water from the Amundsen and Bellingshausen Seas (Shepherd et al., 2004; Rignot et al., 2008; Jacobs and Giulivi, 2010). Freshening has been observed in AABW since the 1970s in the Indian sector (Rintoul, 2007) and between the

3 The formation rate of a water mass is the volume of water per year that is transformed into the density range of this water mass by surface processes (for instance cooling), eventually modified through ocean interior processes (for instance mixing). Formation rates are reported in Sverdrups (Sv). 1 Sv equals 10^6 m^3 s^{-1}.

1990s and 2000s in the Pacific sector (Swift and Orsi, 2012; Purkey and Johnson, 2013).

In the Weddell Sea (the primary source of AABW in the Atlantic), a contraction of the bottom water mass was observed between 1984 and 2008 at the Prime Meridian, accompanied by warming of about 0.015°C, and by salinity variability on a multi-annual time scale. Transient tracer observations between 1984 and 2011 confirmed that the AABW there has become less well ventilated over that time period. The changes in the AABW, however, seem to be caused by the much stronger trends observed in the Warm Deep Water, as WDW is entrained into the AABW while sinking to the bottom, and not by changes in the AABW formation rate (Huhn et al., 2008; Huhn et al., 2013).

3.5.4 Conclusions

AR4 Chapter 5 concluded that observed changes in upper ocean water masses reflect the combination of long-term trends and interannual to decadal variability related to climate modes like ENSO, NAO and SAM. The time series are still generally too short and incomplete to distinguish decadal variability from long-term trends, but understanding of the nature and causes of variability has improved in this assessment. The observed patterns of change in subsurface temperature and salinity (Sections 3.2 and 3.3) are consistent with understanding of how and where water masses form, enhancing the level of confidence in the assessment of the observed changes.

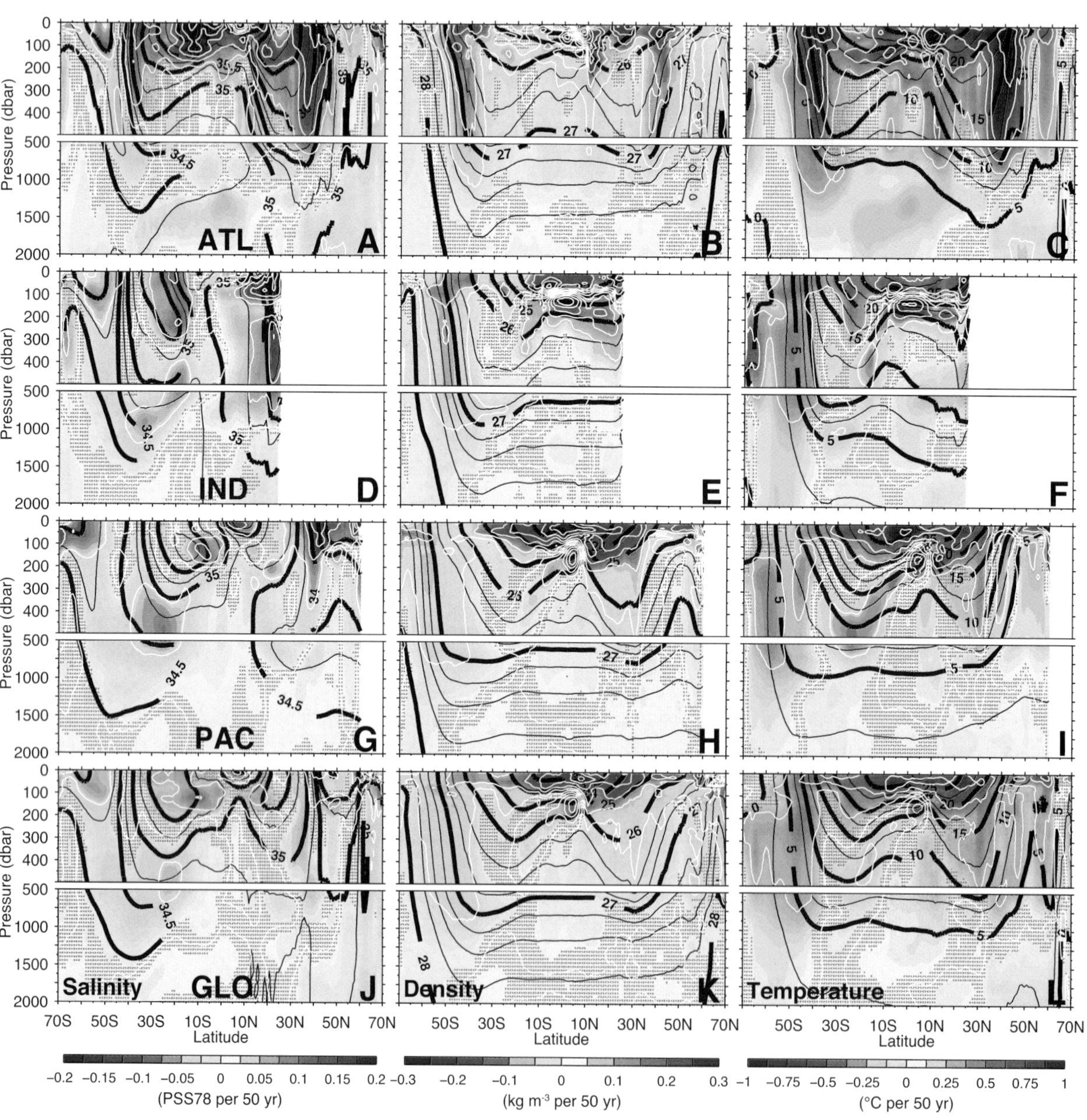

Figure 3.9 | Upper 2000 dbar zonally-averaged linear trend (1950 to 2000) (colours with white contours) of salinity changes (column 1, PSS-78 per 50 yr), neutral density changes (column 2, kg m^{-3} per 50 yr), and potential temperature changes (column 3, °C per 50 yr), for the Atlantic Ocean (ATL) in row 1, Indian Ocean (IND), row 2, Pacific Ocean (PAC), row 3, and global ocean (GLO) in row 4. Mean fields are shown as black lines (salinity: thick black contours 0.5 PSS-78, thin contours 0.25 PSS-78; neutral density: thick black contours 1.0 kg m^{-3}, thin contours 0.25 kg m^{-3}; potential temperature: thick black contours 5.0°C, thin contours 2.5°C). Trends are calculated on pressure surfaces (1 dbar pressure is approximately equal to 1 m in depth). Regions where the resolved linear trend is not significant at the 90% confidence level are stippled in grey. Salinity results are republished from Durack and Wijffels (2010) with the unpublished temperature and density results from that study also presented.

Recent studies showed that the warming of the upper ocean (Section 3.2.2) *very likely* affects properties of water masses in the interior, in direct and indirect ways. Transport of SST and SSS anomalies caused by changes in surface heat and freshwater fluxes are brought into the ocean's interior by contact with the surface ocean (Sections 3.2 and 3.3). Vertical and horizontal displacements of isopycnals due to surface warming could change salinity and temperature (Section 3.3). Circulation changes (Section 3.6) could also change salinity by shifting the outcrop area of this isopycnal in regions with higher (or lower) E – P. Properties of several deep and bottom water masses are the product of near surface processes and significant mixing or entrainment of other ambient water masses (Section 3.5). Changes in the properties of the entrained or admixed water mass could dominate the observed deep and bottom water mass changes, for instance, in the LNADW and the AABW in the Weddell Sea.

From 1950 to 2000, it is *likely* that subtropical salinity maximum waters have become more saline, while fresh intermediate waters formed at higher latitudes have generally become fresher. In the extratropical North Atlantic, it is *very likely* that the temperature, salinity, and formation rate of the UNADW is dominated by strong decadal variability related to NAO. It is *likely* that LNADW has cooled from 1955 to 2005. It is *likely* that the abyssal layer ventilated by AABW warmed over much of the globe since the 1980s or 1990s respectively, and the volume of cold AABW has been reduced over this time period.

3.6 Changes in Ocean Circulation

3.6.1 Global Observations of Ocean Circulation Variability

The present-day ocean observing system includes global observations of velocity made at the sea surface by the Global Drifter Program (Dohan et al., 2010), and at 1000 m depth by the Argo Program (Freeland et al., 2010). In addition, Argo observes the geostrophic shear between 2000 m and the sea surface. These two recently implemented observing systems, if sustained, will continue to document the large-spatial scale, long-time-scale variability of circulation in the upper ocean. The drifter program achieved its target of 1250 drifters in 2005, and Argo its target of 3000 floats in 2007.

Historically, global measurements of ocean circulation are much sparser, so estimates of decadal and longer-term changes in circulation are very limited. Since 1992, high-precision satellite altimetry has measured the time variations in sea surface height (SSH), whose horizontal gradients are proportional to the surface geostrophic velocity. In addition, a single global top-to-bottom hydrographic survey was carried out by the World Ocean Circulation Experiment (WOCE), mostly during 1991–1997, measuring geostrophic shear as well as velocity from mid-depth floats and from lowered acoustic Doppler current profilers. A subset of WOCE and pre-WOCE transects is being repeated at 5- to 10-year intervals (Hood et al., 2010).

Ocean circulation studies in relation to climate have focused on variability in the wind-driven gyres (Section 3.6.2) and changes in the meridional overturning circulations (MOCs, Sections 3.6.3 and 3.6.4) influenced by buoyancy loss and water-mass formation as well as wind forcing. The MOCs are responsible for much of the ocean's capacity to carry excess heat from the tropics to middle latitudes, and also are important in the ocean's sequestration of carbon. The connections between ocean basins (Section 3.6.5) have also been subject to study because of the significance of inter-basin exchanges in wind-driven and thermohaline variability, and also because these can be logistically advantageous regions for measurement ("chokepoints"). An assessment is now possible of the recent mean and the changes in global geostrophic circulation over the previous decade (Figure 3.10, and discussion in Section 3.6.2). In general, changes in the slope of SSH across ocean basins indicate changes in the major gyres and the interior component of MOCs. Changes occurring in high gradient regions such as the Antarctic Circumpolar Current (ACC) may indicate shifts in the location of those currents. In the following, the best-studied and most significant aspects of circulation variability and change are assessed including wind-driven circulation in the Pacific, the Atlantic and Antarctic MOCs, and selected interbasin exchanges.

3.6.2 Wind-Driven Circulation Variability in the Pacific Ocean

The Pacific covers over half of the global ocean area and its wind-driven variability is of interest both for its consistency with wind stress observations and for potential air–sea feedbacks that could influence climate. Changes in Pacific Ocean circulation since the early 1990s to the present, from the subarctic gyre to the southern ocean, observed with satellite ocean data and *in situ* ocean measurements, are in good agreement and consistent with the expected dynamical response to observed changes in wind stress forcing.

The subarctic gyre in the North Pacific poleward of 40°N consists of the Alaska Gyre to the east and the Western Subarctic Gyre (WSG). Since 1993, the cyclonic Alaska Gyre has intensified while decreasing in size. The shrinking is seen in the northward shift of the North Pacific Current

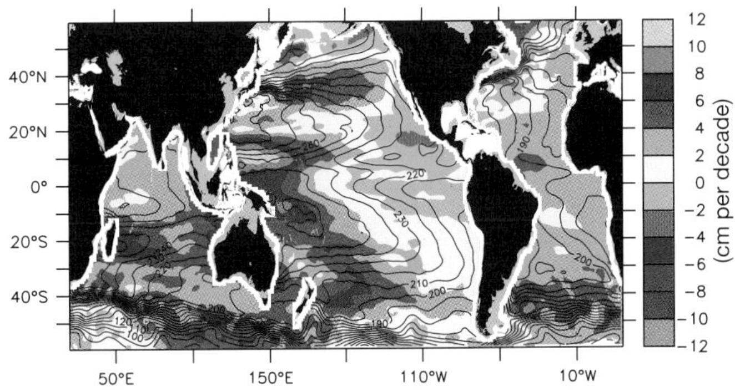

Figure 3.10 | Mean steric height of the sea surface relative to 2000 decibars (black contours at 10-cm intervals) shows the pattern of geostrophic flow for the Argo era (2004–2012) based on Argo profile data, updated from Roemmich and Gilson (2009). The sea surface height (SSH) trend (cm per decade, colour shading) for the period 1993–2011 is based on the AVISO altimetry "reference" product (Ducet et al., 2000). Spatial gradients in the SSH trend, divided by the (latitude-dependant) Coriolis parameter, are proportional to changes in surface geostrophic velocity. For display, the mean steric height contours and SSH trends are spatially smoothed over 5° longitude and 3° latitude.

(NPC, the high gradient region centred about 40°N in Figure 3.10) and has been described using the satellite altimeter, XBT/hydrography, and, more recently, Argo profiling float data (Douglass et al., 2006; Cummins and Freeland, 2007). A similar 20-year trend is detected in the WSG, with the northern WSG in the Bering Sea having intensified while the southern WSG south of the Aleutian Islands has weakened. These decadal changes are attributable to strengthening and northward expansion of the Pacific High and Aleutian Low atmospheric pressure systems over the subarctic North Pacific Ocean (Carton et al., 2005).

The subtropical gyre in the North Pacific also expanded along its southern boundary over the past two decades. The North Equatorial Current (NEC) shifted southward along the 137°E meridian (Qiu and Chen, 2012; also note the SSH increase east of the Philippines in Figure 3.10 indicating the southward shift). The NEC's bifurcation latitude along the Philippine coast migrated southward from a mean latitude of 13°N in the early 1990s to 11°N in the late 2000s (Qiu and Chen, 2010). These changes are due to a recent strengthening of the Walker circulation generating a positive wind stress curl anomaly (Tanaka et al., 2004; Mitas and Clement, 2005). The enhanced regional sea level rise, >10 mm yr^{-1} in the western tropical North Pacific Ocean (Timmermann et al., 2010, Figure 3.10), is indicative of the changes in ocean circulation. The 20-year time-scale expansion of the North Pacific subtropical gyre has *high confidence* owing to the good agreement seen in satellite altimetry, subsurface ocean data and wind stress changes. This sea level increase in the western tropical Pacific also indicates a strengthening of the equatorward geostrophic limb of the subtropical cells. However, the 20-year increase reversed a longer term weakening of the subtropical cells (Feng et al., 2010), illustrating the high difficulty of separating secular trends from multi-decadal variability.

3

Variability in the mid-latitude South Pacific over the past two decades is characterized by a broad increase in SSH in the 35°S to 50°S band and a lesser increase south of 50°S along the path of the ACC (Figure 3.10). These SSH fluctuations are induced by the intensification in the SH westerlies (i.e., the SAM; see also Section 3.4.4), generating positive and negative wind stress curl anomalies north and south of 50°S. In response, the southern limb of the South Pacific subtropical gyre has intensified in the past two decades (Cai, 2006; Qiu and Chen, 2006; Roemmich et al., 2007) along with a southward expansion of the East Australian Current (EAC) into the Tasman Sea (Hill et al., 2008). The intensification in the South Pacific gyre extends to a greater depth (>1800 m) than that in the North Pacific gyre (Roemmich and Gilson, 2009). As in the north, the 20-year changes in the South Pacific are seen with *high confidence* as they occur consistently in multiple lines of medium and high-quality data. Multiple linear regression analysis of the 20-year Pacific SSH field (Zhang and Church, 2012) indicated that interannual and decadal modes explain part of the circulation variability seen in SSH gradients, and once the aliasing by these modes is removed, the SSH trends are weaker and more spatially uniform than in a single variable trend analysis.

The strengthening of SH westerlies is a multi-decadal signal, as seen in SLP difference between middle and high southern latitudes from 1949 to 2009 (Gillett and Stott, 2009; also Section 3.4.4). The multi-decadal warming in the Southern Ocean (e.g., Figure 3.1, and Gille, 2008, for the past 50 to 70 years) is consistent with a poleward displacement of the ACC and the southern limb of the subtropical gyres, by about 1° of latitude per 40 years (Gille, 2008). The warming and corresponding sea level rise signals are not confined to the South Pacific, but are seen globally in zonal mean fields (e.g., at 40°S to 50°S in Figures 3.9 l and 3.10). Alory et al. (2007) describe the broad warming consistent with a southward shift of the ACC in the South Indian Ocean. In the Atlantic, a southward trend in the location of the Brazil-Malvinas confluence (at around 39°S) is described from surface drifters and altimetry by Lumpkin and Garzoli (2011), and in the location of the Brazil Current separation point from SST and altimetry by Goni et al. (2011). Enhanced surface warming and poleward displacement, globally, of the western boundary currents is described by Wu et al. (2012).

Changes in Pacific Ocean circulation over the past two decades since 1993, observed with *medium to high confidence*, include intensification of the North Pacific subpolar gyre, the South Pacific subtropical gyre, and the subtropical cells, plus expansion of the North Pacific subtropical gyre and a southward shift of the ACC. It is *likely* that these wind-driven changes are predominantly due to interannual-to-decadal variability, and in the case of the subtropical cells represent reversal of earlier multi-decadal change. Sustained time series of wind stress forcing and ocean circulation will permit increased skill in separating interannual and decadal variability from long-term trends (e.g., Zhang and Church, 2012).

3.6.3 The Atlantic Meridional Overturning Circulation

The Atlantic Meridional Overturning Circulation (AMOC) consists of an upper limb with net northward transport between the surface and approximately 1200 m depth, and a lower limb of denser, colder, fresher waters returning southward between 1200 m and 5000 m. The AMOC is responsible for most of the meridional transport of heat and carbon by the mid-latitude NH ocean and associated with the production of about half of the global ocean's deep waters in the northern North Atlantic. Coupled climate models find that a slowdown of the AMOC in the next decades is *very likely*, though with uncertain magnitude (Section 11.3.3.3). Observations of the AMOC are directed toward detecting possible long-term changes in its amplitude, its northward energy transport, and in the ocean's capacity to absorb excess heat and greenhouse gases, as well as characterizing short-term variability and its relationship to changes in forcing.

Presently, variability in the full AMOC and meridional heat flux are being estimated on the basis of direct observations at 26.5°N by the RAPID/MOCHA array (Cunningham et al., 2007; Kanzow et al., 2007; Johns et al., 2011). The array showed a mean AMOC magnitude of 18 ± 1.0 Sv (±1 standard deviation of annual means) between April 2004 and April 2009, with 10-day values ranging from 3 to 32 Sv (McCarthy et al., 2012). Earlier estimates of AMOC strength from five shipboard expeditions over 47 years at 24°N (Bryden et al., 2005) were in the range of variability seen by RAPID/MOCHA. For the 1-year period 1 April 2009 to 31 March 2010, the AMOC mean strength decreased to 12.8 Sv. This decrease was manifest in a shift of southward interior transport from the deep layers to the upper 1000 m. Although the AMOC weakening in 2009/2010 was large, it subsequently rebounded and with the large year-to-year changes no trend is detected in the updated time-series (Figure 3.11b).

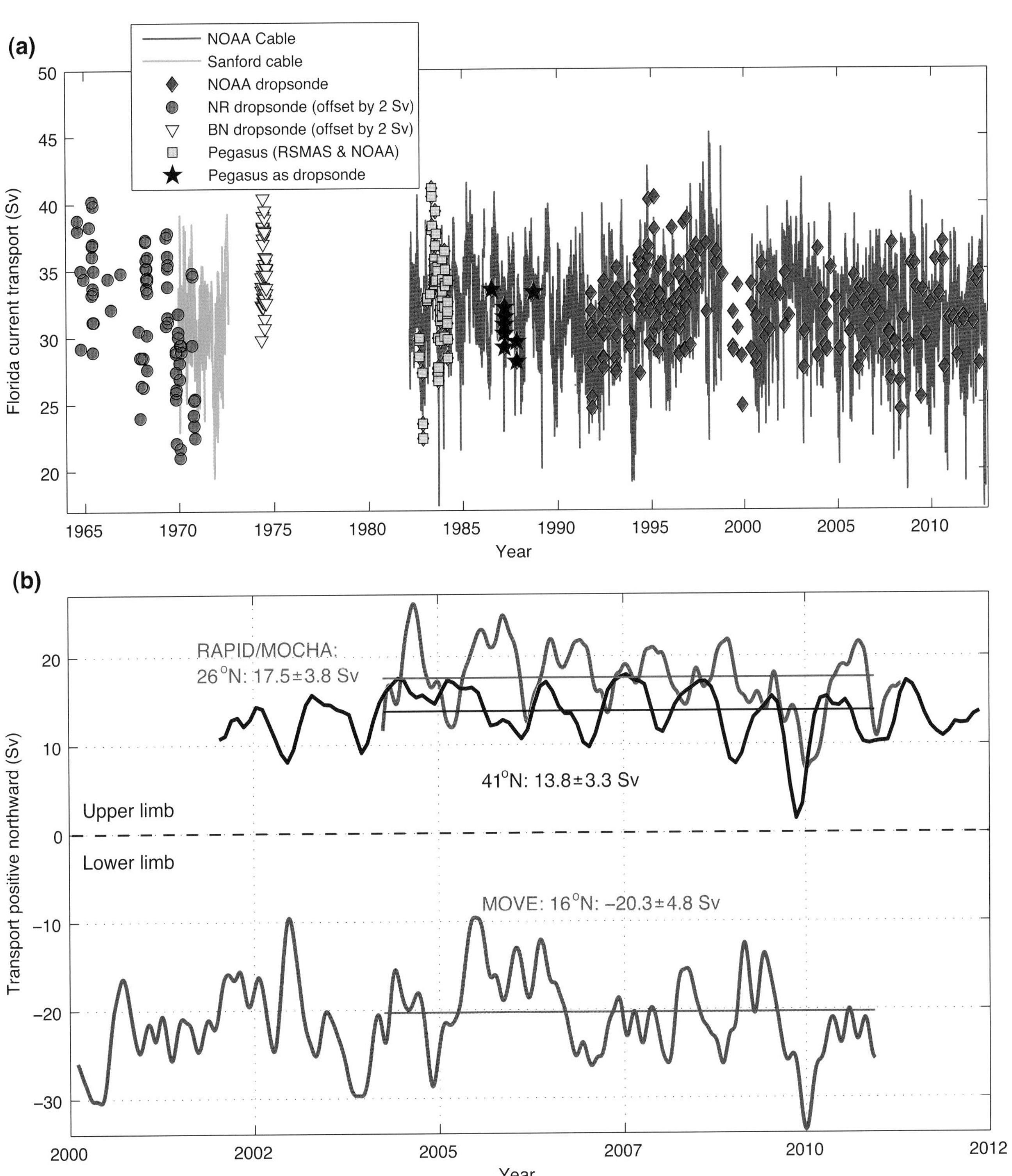

Figure 3.11 | (a) Volume transport in Sverdrups (Sv; where 1 Sv = 10^6 m^3 s^{-1}) of the Florida Current between Florida and the Bahamas, from dropsonde measurements (symbols) and cable voltages (continuous line), extending the time-series shown in Meinen et al. (2010) (b) Atlantic Meridional Overturning Circulation (AMOC) transport estimates (Sv): 1. RAPID/MOCHA (Rapid Climate Change programme / Meridional Ocean Circulation and Heatflux Array) at 26.5°N (red). The array monitors the top-to-bottom Atlantic wide circulation, ensuring a closed mass balance across the section, and hence a direct measure of the upper and lower limbs of the AMOC. 2. 41°N (black): An index of maximum AMOC strength from Argo float measurements in the upper 2000 m only, combined with satellite altimeter data. The lower limb is not measured. 3. Meridional Overturning Variability Experiment (MOVE) at 16°N (blue) measuring transport of North Atlantic Deep Water in the lower limb of the AMOC between 1100 m and 4800 m depth between the Caribbean and the mid-Atlantic Ridge. This transport is thought to be representative of maximum MOC variability based on model validation experiments. The temporal resolution of the three time series is 10 days for 16°N and 26°N and 1 month for 41°N. The data have been 3-month low-pass filtered. Means and standard deviations for the common period of 2 April 2004 to 1 April 2010 are 17.5 ± 3.8 Sv, 13.8 ± 3.3 Sv and −20.3 ± 4.8 Sv (negative indicating the southward lower limb) for 26.5°N, 41°N and 16°N respectively. The means over this period are indicated by the horizontal line on each time series.

Observations targeting one limb of the AMOC include Willis (2010) at 41°N combining velocities from Argo drift trajectories, Argo temperature/salinity profiles, and satellite altimeter data (Figure 3.11b). Here the upper limb AMOC magnitude is 15.5 Sv ± 2.4 from 2002 to 2009 (Figure 3.11b). This study suggests an increase in the AMOC strength by about 2.6 Sv from 1993 to 2010, though with *low confidence* because it is based on SSH alone in the pre-Argo interval of 1993–2001. At 16°N, geostrophic array-based estimates of the southward transport of the AMOC's lower limb, in the depth range 1100 to 4700 m, have been made continuously since 2000 (Kanzow et al., 2008). These are the longest continuous measurements of the southward flow of NADW in the western basin. Whereas the period 2000 to mid-2009 suggested a downward trend (Send et al., 2011), the updated time series (Figure 3.11b) has no apparent trend. In the South Atlantic at 35°S, estimates of the AMOC upper limb were made using 27 high-resolution XBT transects (2002–2011) and Argo float data (Garzoli et al., 2013). The upper-limb AMOC magnitude was 18.1 Sv ± 2.3 (1 standard deviation based on cruise values), consistent with the NH estimates.

3

The continuous AMOC estimates at 16°N, 26.5°N and 41°N have time series of length 11, 7, and 9 years respectively (Figure 3.11b). All show a substantial variability of ~3 to 5 Sv for 3-month low-pass time series, with a peak-to-peak interannual variability of 5 Sv. The shortness of these time series and the relatively large interannual variability emerging in them suggests that trend estimates be treated cautiously, and no trends are seen at 95% confidence in any of the time series.

Continuous time series of AMOC components, longer than those of the complete system at 26.5°N, have been obtained using moored instrumentation. These include the inflow into the Arctic through Fram Strait (since 1997, Schauer and Beszczynska-Möller, 2009) and through the Barents Sea (since 1997, Ingvaldsen et al., 2004; Mauritzen et al., 2011), dense inflows across sills between Greenland and Scotland (since 1999 and 1995 respectively, Olsen et al., 2008; Jochumsen et al., 2012) and North Atlantic Deep Water carried southward within the Deep Western Boundary Current at 53°N (since 1997, Fischer et al., 2010) and at 39°N (Line W, since 2004, Toole et al., 2011). The longest time series of observations of ocean transport in the world (dropsonde and cable voltage measurements in the Florida Straits), extend from the mid-1960s to the present (Meinen et al., 2010), with small decadal variability of about 1 Sv and no evidence of a multi-decadal trend (Figure 3.11a). Similarly, none of the other direct, continuous transport estimates of single components of the AMOC exhibit long-term trends at 95% significance.

Indirect estimates of the annual average AMOC strength and variability can be made (Grist et al., 2009; Josey et al., 2009) from diapycnal transports driven by air–sea fluxes (NCEP-NCAR reanalysis fields from 1960 to 2007) or by inverse techniques (Lumpkin and Speer, 2007). Decadal fluctuations of up to 2 Sv are seen, but no trend. Consistent with Grist et al. (2009), the sea level index of the strength of the AMOC, based on several coherent western boundary tide gauge records between 39°N and 43°N at the American coast (Bingham and Hughes, 2009) shows no long-term trend from 1960 to 2007.

In summary, measurements of the AMOC and of circulation elements contributing to it, at various latitudes and covering different time periods, agree that the range of interannual variability is 5 Sv (Figure 3.11b). These estimates do not have trends, in either the subtropical or the subpolar gyre. However, the observational record of AMOC variability is short, and there is insufficient evidence to support a finding of change in the transport of the AMOC.

3.6.4 The Antarctic Meridional Overturning Circulation

Sinking of AABW near Antarctica supplies about half of the deep and abyssal waters in the global ocean (Orsi et al., 1999). AABW spreads northward as part of the global overturning circulation and ventilates the bottom-most portions of much of the ocean. Observed widespread warming of AABW in recent decades (Section 3.5.4) implies a concomitant reduction in its northward spread. Reductions of 1 to 4 Sv in northward transports of AABW across 24°N have been estimated by geostrophic calculations using repeat oceanographic section data between 1981 and 2010 in the North Atlantic Ocean (Johnson et al., 2008b; Frajka-Williams et al., 2011) and between 1985 and 2005 in the North Pacific (Kouketsu et al., 2009). A global full-depth ocean data assimilation study shows a reduction of northward AABW flow across 35°S of >2 Sv in the South Pacific starting around 1985 and >1 Sv in the western South Atlantic since around 1975 (Kouketsu et al., 2011). This reduction is consistent with the contraction in volume of AABW (Purkey and Johnson, 2012) discussed in Section 3.5.4.

Several model studies have suggested that changes in wind stress over the Southern Ocean (Section 3.4) may drive a change in the Southern Ocean overturning circulation (e.g., Le Quéré et al., 2007). A recent analysis of changes in chlorofluorocarbon (CFC) concentrations in the Southern Ocean supports the idea that the overturning cell formed by upwelling of deep water and sinking of intermediate waters has slowed, but does not quantify the change in transport (Waugh et al., 2013).

3.6.5 Water Exchange Between Ocean Basins

3.6.5.1 The Indonesian Throughflow

The transport of water from the Pacific to the Indian Ocean via the Indonesian archipelago is the only low-latitude exchange between oceans, and is significant because it is a fluctuating sink/source for very warm tropical water in the two oceans. The Indonesian Throughflow (ITF) transport has been estimated from hydrographic and XBT transects between Australia and Indonesia, and as a synthesis of these together with satellite altimetry, wind stress, and other data (Wunsch, 2010), and from moorings in the principal Indonesian passages. The most comprehensive observations were obtained in 2004–2006 in three passages by the INSTANT mooring array (Sprintall et al., 2009), and show a westward transport of 15.0 (±4) Sv. For the main passage, Makassar Strait, Susanto et al. (2012) find 13.3 (±3.6) Sv in the period 2004–2009, with small year-to-year differences. On a longer time scale, the Wunsch (2010) estimate for 1992–2007 was 11.5 Sv (±2.4) westward, and thus consistent with INSTANT. Wainwright et al. (2008) analyzed data between Australia and Indonesia beginning in the early 1950s, and found a change in the slope of the thermocline for data before and after 1976, indicating a decrease in geostrophic transport by 23%, consistent with a weakening of the tradewinds (e.g.,

Vecchi et al. (2006), who described a downward trend in the Walker circulation since the late 19th century). Other transport estimates based on the IX1 transect show correlation with ENSO variability (Potemra and Schneider, 2007) and no significant trend for the period since 1984 having continuous sampling along IX1 (Sprintall et al., 2002). Overall, the limited evidence provides *low confidence* that a trend in ITF transport has been observed.

3.6.5.2 The Antarctic Circumpolar Current

There is *medium confidence* that the westerly winds in the Southern Ocean have increased since the early 1980s (Section 3.4.4), associated with a positive trend in the SAM (Marshall, 2003); also see Sections 3.4.4 and 3.6.3). Although a few observational studies have found evidence for correlation between SAM and ACC transport on subseasonal to interannual scales (e.g., Hughes et al., 2003; Meredith et al., 2004), there is no significant observational evidence of an increase in ACC transport associated with the multi-decadal trend in wind forcing over the Southern Ocean. Repeat hydrographic sections spread unevenly over 35 years in Drake Passage (e.g., Cunningham et al., 2003; Koshlyakov et al., 2007, 2011; Gladyshev et al., 2008), south of Africa (Swart et al., 2008) and south of Australia (Rintoul et al., 2002) reveal moderate variability but no significant trends in these sparse and discontinuous records. A comparison of recent Argo data and a long-term climatology showed that the slope of density surfaces (hence baroclinic transport) associated with the ACC had not changed in recent decades (Böning et al., 2008). Eddy-resolving models suggest the ACC transport is relatively insensitive to trends in wind forcing, consistent with the ACC being in an "eddy-saturated" state where increases in wind forcing are compensated by changes in the eddy field (Hallberg and Gnanadesikan, 2006; Farneti et al., 2010; Spence et al., 2010). While there is limited evidence for (or against) multi-decadal changes in transport of the ACC, observations of changes in temperature, salinity and SSH indicate the current system has shifted poleward (*medium confidence*) (Böning et al., 2008; Gille, 2008; Morrow et al., 2008; Sokolov and Rintoul, 2009; Kazmin, 2012).

3.6.5.3 North Atlantic/Nordic Seas Exchange

There is no observational evidence of changes during the past two decades in the flow across the Greenland–Scotland Ridge, which connects the North Atlantic with the Norwegian and Greenland Seas. Direct current measurements since the mid-1990s have not shown any significant trends in volume transport for any of the three inflow branches (Østerhus et al., 2005; Hansen et al., 2010; Mauritzen et al., 2011; Jónsson and Valdimarsson, 2012).

The two primary pathways for the deep southward overflows across the Greenland–Scotland Ridge are the Denmark Strait and Faroe Bank Channel. Moored measurements of the Denmark Strait overflow demonstrate significant interannual transport variations (Macrander et al., 2005; Jochumsen et al., 2012), but the time series is not long enough to detect a multi-decadal trend. Similarly, a 10-year time series of moored measurements in the Faroe Bank channel (Olsen et al., 2008) does not show a trend in transport.

3.6.6 Conclusions

Recent observations have greatly increased the knowledge of the amplitude of variability in major ocean circulation systems on time scales from years to decades. It is *very likely* that the subtropical gyres in the North Pacific and South Pacific have expanded and strengthened since 1993, but it is *about as likely as not* that this reflects a decadal oscillation linked to changes in wind forcing, including changes in winds associated with the modes of climate variability. There is no evidence for a long-term trend in the AMOC amplitude, based on a decade of continuous observations plus several decades of sparse hydrographic transects, or in the longer records of components of the AMOC such as the Florida Current (since 1965), although there are large interannual fluctuations. Nor is there evidence of a trend in the transports of the ITF (over about 20 years), the ACC (about 30 years sparsely sampled), or between the Atlantic and Nordic Seas (about 20 years). Given the short duration of direct measurements of ocean circulation, we have *very low confidence* that multi-decadal trends can be separated from decadal variability.

3.7 Sea Level Change, Including Extremes

3.7.1 Introduction and Overview of Sea Level Measurements

3

Sea level varies as the ocean warms or cools, as water is transferred between the ocean and continents, between the ocean and ice sheets, and as water is redistributed within the ocean due to the tides and changes in the oceanic and atmospheric circulation. Sea level can rise or fall on time scales ranging from hours to centuries, spatial scales from <1 km to global, and with height changes from a few millimeters to a meter or more (due to tides). Sea level integrates and reflects multiple climatic and dynamical signals. Measurements of sea level are the longest-running ocean observation system. This section assesses interannual and longer variations in non-tidal sea level from the instrumented period (late 18th century to the present). Sections 4.3.3 and 4.4.2 assess contributions of glaciers and ice sheets to sea level, Section 5.6 assess reconstructions of sea level from the geological record, Section 10.4.3 assesses detection and attribution of human influences on sea level change, and Chapter 13 synthesizes results and assesses projections of sea level change.

The sea level observing system has evolved over time. There are intermittent records of sea level at four sites in Northern Europe starting in the 1700s. By the late 1800s, there were more tide gauges being operated in Northern Europe, on both North American coasts, and in Australia and New Zealand in the SH (Appendix 3.A). Tide gauges began to be placed on islands far from continental coasts starting in the early 20th century, but a majority of deep-ocean islands did not have an operating tide gauge suitable for climate studies until the early 1970s.

Tide gauge records measure the combined effect of ocean volume change and vertical land motion (VLM). For detecting climate related variability of the ocean volume, the VLM signal must be removed. One component that can be accounted for to a certain extent is the VLM associated with glacial isostatic adjustment (GIA) (Peltier, 2001). In some areas, however, VLM from tectonic activity, groundwater mining, or hydrocarbon extraction is greater than GIA (e.g., Wöppelmann et al., 2009; King et al., 2012); these effects can be reduced by selecting gauges with no known tectonic or subsidence issues (e.g., Douglas, 2001) or by selecting gauges where GIA models have small differences (Spada and Galassi, 2012). More recently, Global Positioning System (GPS) receivers have been installed at tide gauge sites to measure VLM as directly as possible (e.g., Wöppelmann et al., 2009; King et al., 2012). However, these measurements of VLM are only available since the late 1990s at the earliest, and either have to be extrapolated into the past to apply to older records, or used to identify sites without extensive VLM.

Satellite radar altimeters in the 1970s and 1980s made the first nearly global observations of sea level, but these early measurements were highly uncertain and of short duration. The first precise record began with the launch of TOPEX/Poseidon (T/P) in 1992. This satellite and its successors (Jason-1, Jason-2) have provided continuous measurements of sea level variability at 10-day intervals between approximately ±66° latitude. Additional altimeters in different orbits (ERS-1, ERS-2, Envisat, Geosat Follow-on) have allowed for measurements up to ±82° latitude and at different temporal sampling (3 to 35 days), although these measurements are not as accurate as those from the T/P and Jason satellites. Unlike tide gauges, altimetry measures sea level relative to a geodetic reference frame (classically a reference ellipsoid that coincides with the mean shape of the Earth, defined within a globally realized terrestrial reference frame) and thus will not be affected by VLM, although a small correction that depends on the area covered by the satellite (~0.3 mm yr^{-1}) must be added to account for the change in location of the ocean bottom due to GIA relative to the reference frame of the satellite (Peltier, 2001; see also Section 13.1.2).

Tide gauges and satellite altimetry measure the combined effect of ocean warming and mass changes on ocean volume. Although variations in the density related to upper-ocean salinity changes cause regional changes in sea level, when globally averaged their effect on sea level rise is an order of magnitude or more smaller than thermal effects (Lowe and Gregory, 2006). The thermal contribution to sea level can be calculated from *in situ* temperature measurements (Section 3.2). It has only been possible to directly measure the mass component of sea level since the launch of the Gravity Recovery and Climate Experiment (GRACE) in 2002 (Chambers et al., 2004). Before that, estimates were based either on estimates of glacier and ice sheet mass losses or using residuals between sea level measured by altimetry or tide gauges and estimates of the thermosteric component (e.g., Willis et al., 2004; Domingues et al., 2008), which allowed for the estimation of seasonal and interannual variations as well. GIA also causes a gravitational signal in GRACE data that must be removed in order to determine present-day mass changes; this correction is of the same order of magnitude as the expected trend and is still uncertain at the 30% level (Chambers et al., 2010).

3.7.2 Trends in Global Mean Sea Level and Components

Tide gauges with the longest nearly continuous records of sea level show increasing sea level over the 20th century (Figure 3.12; Woodworth et al., 2009; Mitchum et al., 2010). There are, however, significant interannual and decadal-scale fluctuations about the average rate of sea level rise in all records. Different approaches have been used to compute the mean rate of 20th century global mean sea level (GMSL) rise from the available tide gauge data: computing average rates from only very long, nearly continuous records (Douglas, 2001; Holgate, 2007); using more numerous but shorter records and filters to separate nonlinear trends from decadal-scale quasi-periodic variability (Jevrejeva et al., 2006, 2008); neural network methods (Wenzel and Schroeter, 2010); computing regional sea level for specific basins then

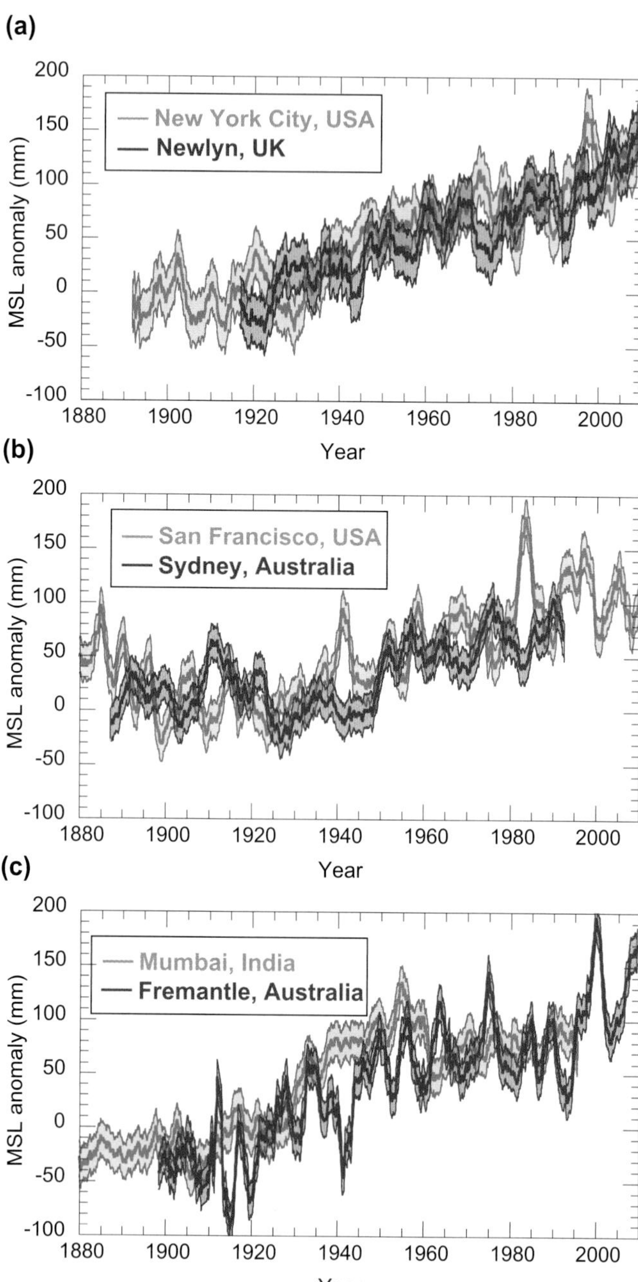

Figure 3.12 | 3-year running mean sea level anomalies (in millimeters) relative to 1900–1905 from long tide gauge records representing each ocean basin from the Permanent Service for Mean Sea Level (PSMSL) (http://www.psmsl.org), obtained May 2011. Data have been corrected for Glacial Isostatic Adjustment (GIA) (Peltier, 2004), using values available from http://www.psmsl.org/train_and_info/geo_signals/gia/peltier/. Error bars reflect the 5 to 95% confidence interval, based on the residual monthly variability about the 3-year running mean.

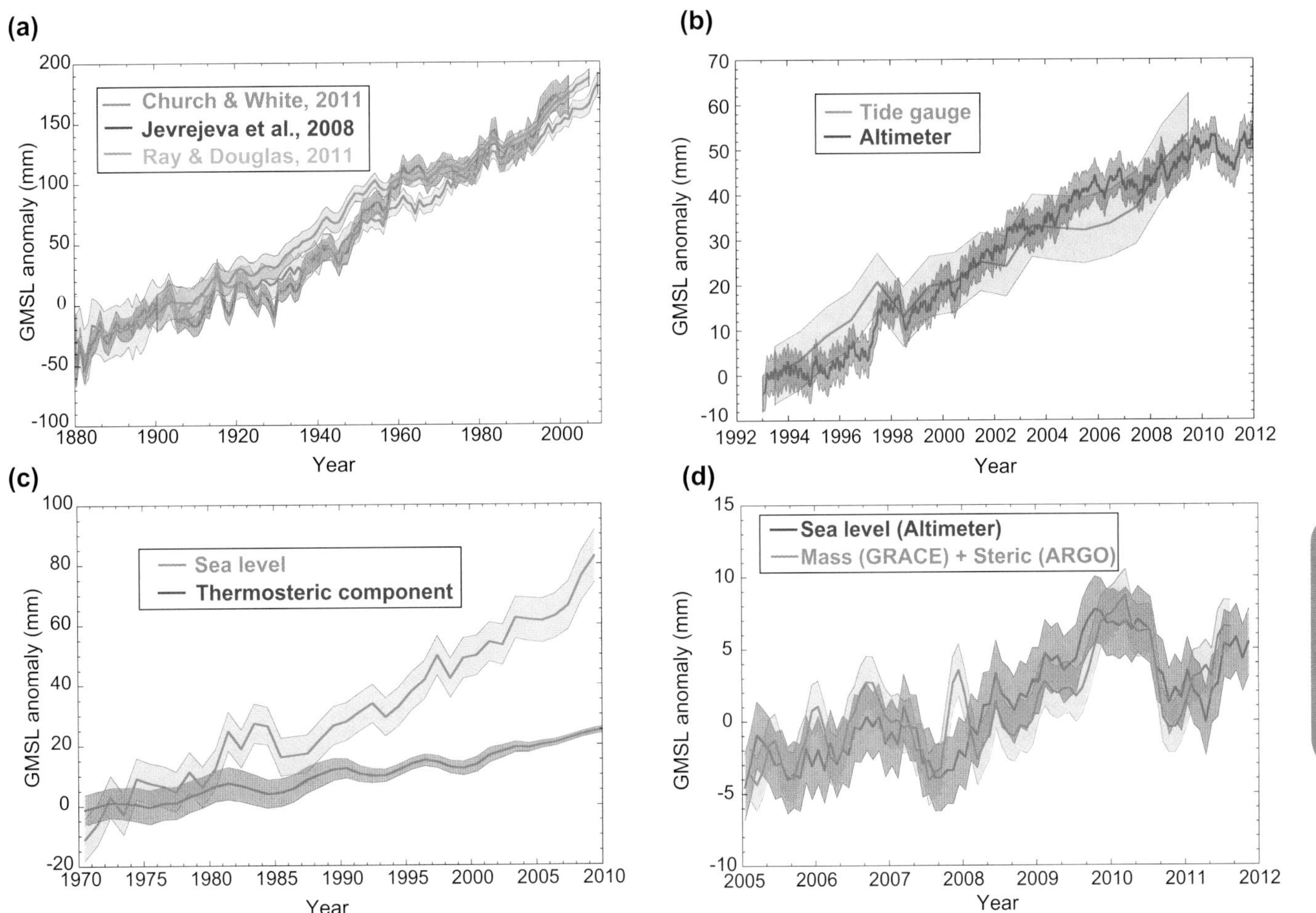

Figure 3.13 | Global mean sea level anomalies (in mm) from the different measuring systems as they have evolved in time, plotted relative to 5-year mean values that start at (a) 1900, (b) 1993, (c) 1970 and (d) 2005. (a) Yearly average GMSL reconstructed from tide gauges (1900–2010) by three different approaches (Jevrejeva et al., 2008; Church and White, 2011; Ray and Douglas, 2011). (b) GMSL (1993–2010) from tide gauges and altimetry (Nerem et al., 2010) with seasonal variations removed and smoothed with a 60-day running mean. (c) GMSL (1970–2010) from tide gauges along with the thermosteric component to 700 m (3-year running mean) estimated from in situ temperature profiles (updated from Domingues et al., 2008). (d) The GMSL (nonseasonal) from altimetry and that computed from the mass component (GRACE) and steric component (Argo) from 2005 to 2010 (Leuliette and Willis, 2011), all with a 3-month running mean filter. All uncertainty bars are one standard error as reported by the authors. The thermosteric component is just a portion of total sea level, and is not expected to agree with total sea level.

averaging (Jevrejeva et al., 2006, 2008; Merrifield et al., 2009; Wöppelmann et al., 2009); or projecting tide gauge records onto empirical orthogonal functions (EOFs) computed from modern altimetry (Church et al., 2004; Church and White, 2011; Ray and Douglas, 2011) or EOFs from ocean models (Llovel et al., 2009; Meyssignac et al., 2012). Different approaches show very similar long-term trends, but noticeably different interannual and decadal-scale variability (Figure 3.13a). Only the time series from Church and White (2011) extends to 2010, so it is used in the assessment of rates of sea level rise. The rate from 1901 to 2010 is 1.7 [1.5 to 1.9] mm yr^{-1} (Table 3.1), which is unchanged from the value in AR4. Rates computed using alternative approaches over the longest common interval (1900–2003) agree with this estimate within the uncertainty.

Since AR4, significant progress has been made in quantifying the uncertainty in GMSL associated with unknown VLM and uncertainty in GIA models. Differences between rates of GMSL rise computed with and without VLM from GPS are smaller than the estimated uncertainties (Merrifield et al., 2009; Wöppelmann et al., 2009). Use of different GIA models to correct tide gauge measurements results in differences less than 0.2 mm yr^{-1} (one standard error), and rates of GMSL rise computed from uncorrected tide gauges differ from rates computed from GIA-corrected gauges by only 0.4 mm yr^{-1} (Spada and Galassi, 2012), again within uncertainty estimates. This agreement gives increased confidence that the 20th century rate of GMSL rise is not biased high due to unmodeled VLM at the gauges.

Satellite altimetry can resolve interannual fluctuations in GMSL better than tide gauge records because less temporal smoothing is required (Figure 3.13b). It is clear that deviations from the long-term trend can exist for periods of several years, especially during El Niño (e.g., 1997–1998) and La Niña (e.g., 2011) events (Nerem et al., 1999; Boening et al., 2012; Cazenave et al., 2012). The rate of GMSL rise from 1993–2010 is 3.2 [2.8 to 3.6] mm yr^{-1} based on the average of altimeter time series published by multiple groups (Ablain et al., 2009; Beckley et al., 2010; Leuliette and Scharroo, 2010; Nerem et al., 2010; Church and White, 2011; Masters et al., 2012, Figure 3.13). As noted in AR4, this rate continues to be statistically higher than that for the 20th century

(Table 3.1). There is *high confidence* that this change is real and not an artefact of the different sampling or change in instrumentation, as the trends estimated over the same period from tide gauges and altimetry are consistent. Although the rate of GMSL rise has a slightly lower trend between 2005 and 2010 (Nerem et al., 2010), this variation is consistent with earlier interannual fluctuations in the record (e.g., 1993–1997), mostly attributable to El Niño/La Niña cycles (Box 9.2). At least 15 years of data are required to reduce the impact of interannual variations associated with El Niño or La Niña on estimated trends (Nerem et al., 1999).

Since AR4, estimates of both the thermosteric component and mass component of GMSL rise have improved, although estimates of the mass component are possible only since the start of the GRACE measurements in 2002. After correcting for biases in older XBT data [3.2], the rate of thermosteric sea level rise in the upper 700 m since 1971 is 50% higher than estimates used for AR4 (Domingues et al., 2008; Wijffels et al., 2008). Because of much sparser upper ocean measurements before 1971, we estimate the trend only since then (Section 3.2). The warming of the upper 700 m from 1971 to 2010 caused an estimated mean thermosteric rate of rise of 0.6 [0.4 to 0.8] mm yr^{-1} (90% confidence), which is 30% of the observed rate of GMSL rise for the same period (Table 3.1; Figure 3.13c). Although still a short record, more numerous, better distributed, and higher quality profile measurements from the Argo program are now being used to estimate the steric component for the upper 700 m as well as for the upper 2000 m (Domingues et al., 2008; Willis et al., 2008, 2010; Cazenave et al., 2009; Leuliette and Miller, 2009; Leuliette and Willis, 2011; Llovel et al., 2011; von Schuckmann and Le Traon, 2011; Levitus et al., 2012). However, these data have been shown to be best suited for global analyses after 2005 owing to a combination of interannual variability and large biases when using data before 2005 owing to sparser sampling (Leuliette and Miller, 2009; von Schuckmann and Le Traon, 2011). Comparison of sparse but accurate temperature measurements from the World Ocean Circulation Experiment in the 1990s with Argo data from 2006 to 2008 also indicates a significant rise in global thermosteric sea level, although the estimate is uncertain owing to relatively sparse 1990s sampling (Freeland and Gilbert, 2009).

Observations of the contribution to sea level rise from warming below 700 m are still uncertain due to limited historical data, especially in the Southern Ocean (Section 3.2). Before Argo, they are based on 5-year averages to 2000 m depth (Levitus et al., 2012). From 1971 to 2010, the estimated trend for the contribution between 700 m and 2000 m is 0.1 [0 to 0.2] mm yr^{-1} (Table 3.1; Levitus et al., 2012). To measure the contribution of warming below 2000 m, much sparser but very accurate temperature profiles along repeat hydrographic sections are utilized (Purkey and Johnson, 2010; Kouketsu et al., 2011). The studies have found a significant warming trend between 1000 and 4000 m within and south of the Sub-Antarctic Front (Figure 3.3). The estimated total contribution of warming below 2000 m to global mean sea level rise between about 1992 and 2005 is 0.1 [0.0 to 0.2] mm yr^{-1} (95% confidence as reported by authors; Purkey and Johnson, 2010).

Detection of the mass component of sea level from the GRACE mission was not assessed in AR4, as the record was too short and there was still considerable uncertainty in the measurements and corrections required. Considerable progress has been made since AR4, and the mass component of sea level measured by GRACE has been increasing at a rate between 1 and 2 mm yr^{-1} since 2002 (Willis et al., 2008, 2010; Cazenave et al., 2009; Leuliette and Miller, 2009; Chambers et al., 2010; Llovel et al., 2010; Leuliette and Willis, 2011). Differences between studies are due partially to the time periods used to compute trends, as there are significant interannual variations in the mass component of GMSL (Willis et al., 2008; Chambers et al., 2010; Llovel et al., 2010; Boening et al., 2012), but also to substantial differences in GIA corrections applied, of order 1 mm yr^{-1}. Recent evaluations of the GIA correction have found explanations for the difference (Chambers et al., 2010; Peltier et al., 2012), but uncertainty of 0.3 mm yr^{-1} is still probable. Measurements of sea level from altimetry and the sum of observed steric and mass components are also consistent at monthly scales during the time period when Argo data have global distribution (Figure 3.13d), which gives *high confidence* that the current ocean observing system is capable of resolving the rate of sea level rise and its components.

3.7.3 Regional Distribution of Sea Level Change

Large-scale spatial patterns of sea level change are known to high precision only since 1993, when satellite altimetry became available (Figure 3.10). These data have shown a persistent pattern of change since the early 1990s in the Pacific, with rates of rise in the Warm Pool of the western Pacific up to three times larger than those for GMSL, while rates over much of the eastern Pacific are near zero or negative (Beckley et al., 2010). The increasing sea level in the Warm Pool started shortly before the launch of TOPEX/Poseidon (Merrifield, 2011), and is caused by an intensification of the trade winds (Merrifield and Maltrud, 2011) since the late 1980s that may be related to the Pacific Decadal Oscillation (PDO) (Merrifield et al., 2012; Zhang and Church, 2012). The lower rate of sea level rise since 1993 along the western coast of the United States has also been attributed to changes in the wind stress curl over the North Pacific associated with the PDO (Bromirski et al., 2011). While global maps can be created using EOF analysis (e.g., Church et al., 2004; Llovel et al., 2009), pre-1993 results are still uncertain, as the method assumes that the EOFs since 1993 are capable of representing the patterns in previous decades, and results may be biased in the middle of the ocean where there are no tide gauges to constrain the estimate (Ray and Douglas, 2011). Several studies have examined individual long tide gauge records in the North Atlantic and found coherent decadal-scale fluctuations along both the USA east coast (Sturges and Hong, 1995; Hong et al., 2000; Miller and Douglas, 2007), the European coast (Woodworth et al., 2010; Sturges and Douglas, 2011; Calafat et al., 2012), and the marginal seas in the western North Pacific (Marcos et al., 2012), all related to natural climate variability.

There is still considerable uncertainty on how long large-scale patterns of regional sea level change can persist, especially in the Pacific where the majority of tide gauge records are less than 40 years long. Based on analyses of the longest records in the Atlantic, Indian and Pacific Oceans (including the available gauges in the Southern Ocean) there are significant multi-decadal variations in regional sea level (Holgate, 2007; Woodworth et al., 2009, 2011; Mitchum et al., 2010; Chambers et al., 2012). Hence local rates of sea level rise can be considerably higher or lower than the global mean rate for periods of a decade or more.

The preceding discussion of regional sea level trends has focused on effects that appear to be related to regional ocean volume change, and not those due to vertical land motion. As discussed in Section 3.7.1, vertical land motion can dramatically affect local sea level change. Some extreme examples of vertical land motion are in Neah Bay, Washington, where the signal is +3.8 mm yr^{-1} (uplift from tectonic activity); Galveston, Texas, where the value is –5.9 mm yr^{-1} (subsidence from groundwater mining); and Nedre Gavle, Sweden where the value is +7.1 mm yr^{-1} (uplift from GIA), all computed from nearby GPS receivers (Wöppelmann et al., 2009). These areas will all have long-term rates of sea level rise that are significantly higher or lower than those due to ocean volume change alone, but as these rates are not related to climate change, they are not discussed here.

3.7.4 Assessment of Evidence for Accelerations in Sea Level Rise

AR4 concluded that there was "*high confidence* that the rate of global sea level rise increased from the 19th to the 20th century" but could not be certain as to whether the higher rate since 1993 was reflective of decadal variability or a further increase in the longer-term trend. Since AR4, there has been considerable effort to quantify the level of decadal and multi-decadal variability and to detect acceleration in GMSL and mean sea level at individual tide gauges. It has been clear for some time that there was a significant increase in the rate of sea level rise in the four oldest records from Northern Europe starting in the early to mid-19th century (Ekman, 1988; Woodworth, 1990, 1999; Mitchum et al., 2010). Estimates of the change in the rate have been computed, either by comparing trends over 100-year intervals for the Stockholm site (Ekman, 1988; Woodworth, 1990), or by fitting a quadratic term to all the long records starting before 1850 (Woodworth, 1990, 1999). The results are consistent and indicate a significant acceleration that started in the early to mid-19th century (Woodworth, 1990, 1999), although some have argued it may have started in the late 1700s (Jevrejeva et al., 2008). The increase in the rate of sea level rise at Stockholm (the longest record that extends past 1900) has been based on differencing 100-year trends from 1774–1884 and 1885–1985. The estimated change is 1.0 [0.7 to 1.3] mm yr^{-1} per century (1 standard error, as calculated by Woodworth, 1990). Although sites in other ocean basins do show an increased trend after 1860 (e.g., Figure 3.12), it is impossible to detect a change in the early to mid-1800s in other parts of the ocean using tide gauge data alone, as there are no observations.

Numerous studies have attempted to quantify if a detectable acceleration has continued into the 20th century, typically by fitting a quadratic to data at individual tide gauges (Woodworth, 1990; Woodworth et al., 2009, 2011; Houston and Dean, 2011; Watson, 2011) as well as to reconstructed time series of GMSL (Church and White, 2006; Jevrejeva et al., 2008; Church and White, 2011; Rahmstorf and Vermeer, 2011), or by examining differences in long-term rates computed at different tide gauges (Sallenger et al., 2012). Woodworth et al. (2011) find significant quadratic terms at the sites that begin before 1860 (all in the NH). Other authors using more numerous but significantly shorter records have found either insignificant or small negative quadratic terms in sea level around the United States and Australia since 1920 (Houston and Dean, 2011; Watson, 2011), or large positive quadratic values since 1950 along the U.S. east coast (Sallenger et al., 2012). However, fitting a quadratic term to tide gauge data after 1920 results in highly variable, insignificant quadratic terms (Rahmstorf and Vermeer, 2011), and so only studies that use data before 1920 and that extend until 2000 or beyond are suitable for evaluating long-term acceleration of sea level.

A long time scale is needed because significant multi-decadal variability appears in numerous tide gauge records during the 20th century (Holgate, 2007; Woodworth et al., 2009, 2011; Mitchum et al., 2010; Chambers et al., 2012). The multi-decadal variability is marked by an increasing trend starting in 1910–1920, a downward trend (i.e., leveling of sea level if a long-term trend is not removed) starting around 1950, and an increasing trend starting around 1980. The pattern can be seen in New York, Mumbai and Fremantle records, for instance (Figure 3.12), as well as 14 other gauges representing all ocean basins (Chambers et al., 2012), and in all reconstructions (Figure 3.14). It is also seen in an analysis of upper 400 m temperature (Gouretski et al., 2012; Section 3.3.2). Although the calculations of 18-year rates of GMSL rise based on the different reconstruction methods disagree by as much as 2 mm yr^{-1} before 1950 and on details of the variability (Figure 3.14), all do indicate 18-year trends that were significantly higher than the 20th century average at certain times (1920–1950, 1990–present) and lower at other periods (1910–1920, 1955–1980), *likely* related to multi-decadal variability. Several studies have suggested these variations may be linked to climate fluctuations like the Atlantic Multi-decadal Oscillation (AMO) and/or Pacific Decadal Oscillation (PDO, Box 2.5) (Holgate, 2007; Jevrejeva et al., 2008; Chambers et al., 2012), but these results are not conclusive.

While technically correct that these multi-decadal changes represent acceleration/deceleration of sea level, they should not be interpreted as change in the longer-term rate of sea level rise, as a time series longer than the variability is required to detect those trends. Using data

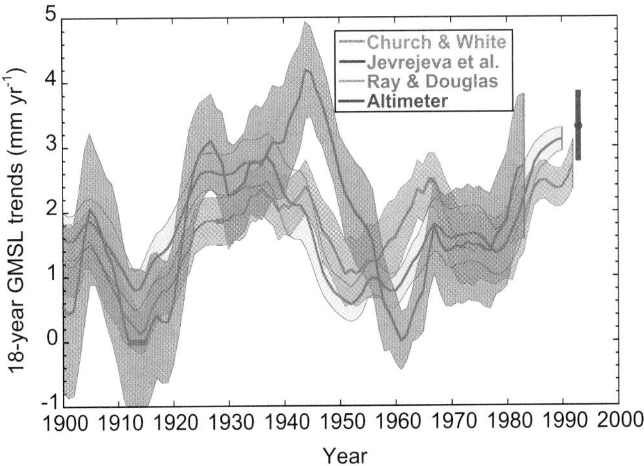

Figure 3.14 | 18-year trends of GMSL rise estimated at 1-year intervals. The time is the start date of the 18-year period, and the shading represents the 90% confidence. The estimate from satellite altimetry is also given, with the 90% confidence given as an error bar. Uncertainty is estimated by the variance of the residuals about the fit, and accounts for serial correlation in the residuals as quantified by the lag-1 autocorrelation.

extending from 1900 to after 2000, the quadratic term computed from both individual tide gauge records and GMSL reconstructions is significantly positive (Jevrejeva et al., 2008; Church and White, 2011; Rahmstorf and Vermeer, 2011; Woodworth et al., 2011). Church and White (2006) report that the estimated acceleration term in GMSL (twice the quadratic parameter) is 0.009 [0.006 to 0.012] mm yr^{-2} (1 standard deviation) from 1880 to 2009, which is consistent with the other published estimates (e.g., Jevrejeva et al., 2008; Woodworth et al., 2009) that use records longer than 100 years. Chambers et al. (2012) find that modelling a period near 60 years removes much of the multi-decadal variability of the 20th century in the tide gauge reconstruction time series. When a 60-year oscillation is modeled along with an acceleration term, the estimated acceleration in GMSL since 1900 ranges from: 0.000 [–0.002 to 0.002] mm yr^{-2} in the Ray and Douglas (2011) record, 0.013 [0.007 to 0.019] mm yr^{-2} in the Jevrejeva et al. (2008) record, and 0.012 [0.009 to 0.015] mm yr^{-2} in the Church and White (2011) record. Thus, while there is more disagreement on the value of a 20th century acceleration in GMSL when accounting for multi-decadal fluctuations, two out of three records still indicate a significant positive value. The trend in GMSL observed since 1993, however, is not significantly larger than the estimate of 18-year trends in previous decades (e.g., 1920–1950).

3.7.5 Changes in Extreme Sea Level

Aside from non-climatic events such as tsunamis, extremes in sea level (i.e., coastal flooding, storm surge, high water events, etc.) tend to be caused by large storms, especially when they occur at times of high tide. However, any low-pressure system offshore with associated high winds can cause a coastal flooding event depending on the duration and direction of the winds. Evaluation of changes in frequency and intensity of storms have been treated in Sections 2.6.3 and 2.6.4, as well as SREX Chapter 3 (Section 3.5.2). The main conclusions from both are that there is *low confidence* of any trend or long term change in tropical or extratropic storm frequency or intensity in any ocean basin, although there is robust evidence for an increase in the most intense tropical cyclones in the North Atlantic basin since the 1970s. The magnitude and frequency of extreme events can still increase without a change in storm intensity, however, if the mean water level is also increasing. AR4 concluded that the highest water levels have been increasing since the 1950s in most regions of the world, caused mainly by increasing mean sea level. Studies published since AR4 continue to support this conclusion, although higher regional extremes are also caused by large interannual and multi-decadal variations in sea level associated with climate fluctuations such as ENSO, the North Atlantic Oscillation and the Atlantic Multi-decadal Oscillation, among others (e.g., Abeysirigunawardena and Walker, 2008; Haigh et al., 2010; Menéndez and Woodworth, 2010; Park et al., 2011).

Global analyses of the changes in extreme sea level are limited, and most reports are based on analysis of regional data (see Lowe et al., 2010 for a review). Estimates of changes in extremes rely either on the analysis of local tide gauge data, or on multi-decadal hindcasts of a dynamical model (WASA-Group, 1998). Most analyses have focused on specific regions and find that extreme values have been increasing since the 1950s, using various statistical measures such as annual maximum surge, annual maximum surge-at-high-water, monthly mean high water level, changes in number of high storm surge events, or changes in 99th percentile events (e.g., Church et al., 2006; D'Onofrio et al., 2008; Marcos et al., 2009; Haigh et al., 2010; Letetrel et al., 2010; Tsimplis and Shaw, 2010; Vilibic and Sepic, 2010; Grinsted et al., 2012). A global analysis of tide gauge records has been performed for data from the 1970s onwards when the global data sampling has been robust and finds that the magnitude of extreme sea level events has increased in all regions studied since that time (Woodworth and Blackman, 2004; Menéndez and Woodworth, 2010; Woodworth et al., 2011).

The height of a 50-year flood event has increased anywhere from 2 to more than 10 cm per decade since 1970 (Figure 3.15a), although some areas have seen a negative rate because vertical land motion is much larger than the rate of mean sea level rise. However, when the annual median height at each gauge is removed to reduce the effect of local mean sea level rise, interannual and decadal fluctuations, and vertical land motion, the rate of extreme sea level change drops in 49% of the gauges to below significance (Figure 3.15b), while at 45% it fell to less than 5 mm yr^{-1}. Only 6% of tide gauge records evaluated had a change in the amplitude of more than 5 mm yr^{-1} after removing mean sea level variations, mainly in the southeast United States, the western Pacific, Southeast Asia and a few locations in Northern Europe. The higher rates in the southeastern United States have been linked to larger storm surge events unconnected to global sea level rise (Grinsted et al., 2012).

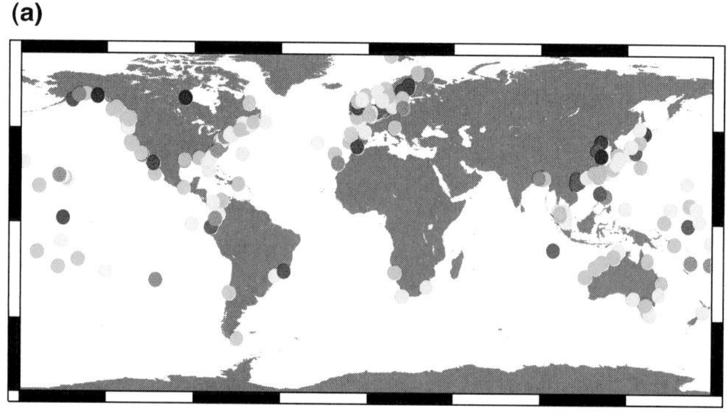

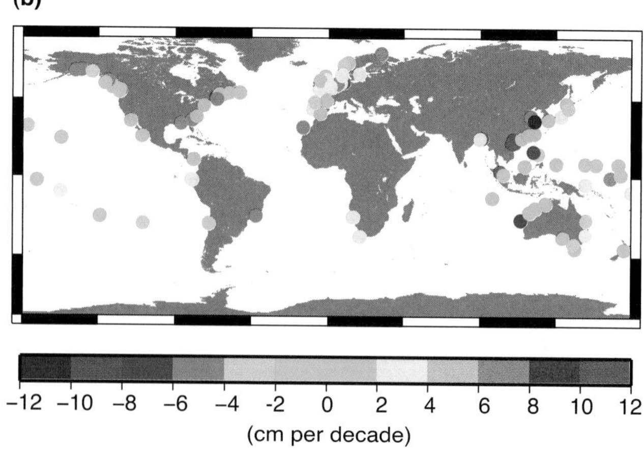

Figure 3.15 | Estimated trends (cm per decade) in the height of a 50-year event in extreme sea level from (a) total elevation and (b) total elevation after removal of annual medians. Only trends significant at the 95% confidence level are shown. (Data are from Menéndez and Woodworth, 2010.)

3.7.6 Conclusions

It is *virtually certain* that globally averaged sea level has risen over the 20th century, with a *very likely* mean rate between 1900 and 2010 of 1.7 [1.5 to 1.9] mm yr^{-1} and 3.2 [2.8 and 3.6] mm yr^{-1} between 1993 and 2010. This assessment is based on high agreement among multiple studies using different methods, and from independent observing systems (tide gauges and altimetry) since 1993. It is *likely* that a rate comparable to that since 1993 occurred between 1920 and 1950, possibly due to a multi-decadal climate variation, as individual tide gauges around the world and all reconstructions of GMSL show increased rates of sea level rise during this period. Although local vertical land motion can cause even larger rates of sea level rise (or fall) relative to the coastline, it is *very likely* that this does not affect the estimates of the global average rate, based on multiple estimations of the average with and without VLM corrections.

It is *virtually certain* that interannual and decadal changes in the large-scale winds and ocean circulation can cause significantly higher or lower rates over shorter periods at individual locations, as this has been observed in tide gauge records around the world. Warming of the upper 700 m of the ocean has *very likely* contributed an average of 0.6 [0.4 to 0.8] mm yr^{-1} of sea level change since 1971. Warming between 700 m and 2000 m has *likely* been contributing an additional 0.1 mm yr^{-1} [0 to 0.2] of sea level rise since 1971, and warming below 2000 m *likely* has been contributing another 0.1 [0.0 to 0.2] mm yr^{-1} of sea level rise since the early 1990s.

It is *very likely* that the rate of mean sea level rise along Northern European coastlines has accelerated since the early 1800s and that this has continued through the 20th century, as the increased rate since 1875 has been observed in multiple long tide gauge records and by different groups using different analysis techniques. It is *likely* that sea level rise throughout the NH has also accelerated since 1850, as this is also observed in a smaller number of gauges along the coast of North America. Two of the three time series based on reconstructing GMSL from tide gauge data back to 1900 or earlier indicate a significant positive acceleration, while one does not. The range is –0.002 to 0.019 mm yr^{-2}, so it is *likely* that GMSL has accelerated since 1900. Finally, it is *likely* that extreme sea levels have increased since 1970, largely as a result of the rise in mean sea level.

3.8 Ocean Biogeochemical Changes, Including Anthropogenic Ocean Acidification

The oceans can store large amounts of CO_2. The reservoir of inorganic carbon in the ocean is roughly 50 times that of the atmosphere (Sabine et al., 2004). Therefore even small changes in the ocean reservoir can have an impact on the atmospheric concentration of CO_2. The ocean

3

Table 3.1 | Estimated trends in GMSL and components over different periods from representative time-series. Trends and uncertainty have been estimated from a time series provided by the authors using ordinary least squares with the uncertainty representing the 90% confidence interval. The model fit for yearly averaged time series was a bias + trend; the model fit for monthly and 10-day averaged data was a bias + trend + seasonal sinusoids. Uncertainty accounts for correlations in the residuals.

Quantity	Period	Trend (mm yr^{-1})	Source	Resolution
GMSL	1901–2010	1.7 [1.5 to 1.9]	Tide Gauge Reconstruction (Church and White, 2011)	Yearly
	1901–1990	1.5 [1.3 to 1.7]	Tide Gauge Reconstruction (Church and White, 2011)	Yearly
	1971–2010	2.0 [1.7 to 2.3]	Tide Gauge Reconstruction (Church and White, 2011)	Yearly
	1993–2010	2.8 [2.3 to 3.3]	Tide Gauge Reconstruction (Church and White, 2011)	Yearly
	1993–2010	3.2 [2.8 to 3.6][a]	Altimetry (Nerem et al., 2010) time-series	10-Day
Thermosteric Component (upper 700 m)	1971–2010	0.6 [0.4 to 0.8]	XBT Reconstruction (updated from Domingues et al., 2008)	3-Year running means
	1993–2010	0.8 [0.5 to 1.1]	XBT Reconstruction (updated from Domingues et al., 2008)	3-Year running means
Thermosteric Component (700 to 2000 m)	1971–2010	0.1 [0 to 0.2]	Objective mapping of historical temperature data (Levitus et al., 2012)	5-Year averages
	1993–2010	0.2 [0.1 to 0.3]	Objective mapping of historical temperature data (Levitus et al., 2012)	5-Year averages
Thermosteric Component (below 2000 m)	1992–2005	0.11 [0.01 to 0.21][b]	Deep hydrographic sections (Purkey and Johnson, 2010)	Trend only
Thermosteric Component (whole depth)	1971–2010	0.8 [0.5 to 1.1][c]	Combination of estimates from 0 to 700 m, 700 to 2000 m, and below 2000 m[c]	Trend only
	1993–2010	1.1 [0.8 to 1.4][c]	Combination of estimates from 0–700 m, 700 to 2000 m, and below 2000 m[c]	Trend only

Notes:

[a] Uncertainty estimated from fit to Nerem et al. (2010) time series and includes potential systematic error owing to drift of altimeter, estimated to be ±0.4 mm yr^{-1} (Beckley et al., 2010; Nerem et al., 2010), applied as the root-sum-square (RSS) with the least squares error estimate. The uncertainty in drift contains uncertainty in the reference frame, orbit and instrument.

[b] Trend value taken from Purkey and Johnson (2010), Table 1. Uncertainty represents the 2.5–97.5% confidence interval.

[c] Assumes no trend below 2000 m before 1 January 1992, then value from Purkey and Johnson (2010) afterwards. Uncertainty for 0 to 700 m, 700 to 2000 m and below 2000 m is assumed to be uncorrelated, and uncertainty is calculated as RSS of the uncertainty for each layer.

also provides an important sink for carbon dioxide released by human activities, the anthropogenic CO_2 (C_{ant}). Currently, an amount of CO_2 equivalent to approximately 30% of the total human emissions of CO_2 to the atmosphere is accumulating in the ocean (Mikaloff-Fletcher et al., 2006; Le Quéré et al., 2010). In this section, observations of change in the ocean uptake of carbon, the inventory of C_{ant}, and ocean acidification are assessed, as well as changes in oxygen and nutrients. Chapter 6 provides a synthesis of the overall carbon cycle, including the ocean, atmosphere and biosphere and considering both past trends and future projections.

3.8.1 Carbon

3.8.1.1 Ocean Uptake of Carbon

The air–sea flux of CO_2 is computed from the observed difference in the partial pressure of CO_2 (pCO_2) across the air–water interface ($\Delta pCO_2 = pCO_{2,sw} - pCO_{2,air}$), the solubility of CO_2 in seawater, and the gas transfer velocity (Wanninkhof et al., 2009). However, the limited geographic and temporal coverage of the ΔpCO_2 measurement as well as uncertainties in wind forcing and transfer velocity parameterizations mean that uncertainties in global and regional fluxes calculated from measurements of ΔpCO_2 can be as larges as ±50% (Wanninkhof et al., 2013). Using ΔpCO_2 data in combination with the riverine input Gruber et al. (2009) estimated a global uptake rate of 1.9 [1.2 to 2.5] PgC yr^{-1} for the time period 1995–2000 and Takahashi et al. (2009) found 2.0 [1.0 to 3.0] PgC yr^{-1} normalized to the year 2000. Uncertainties in fluxes calculated from ΔpCO_2 are too large to detect trends in global ocean carbon uptake.

Trends in surface ocean pCO_2 are calculated from ocean time series stations and repeat hydrographic sections in the North Atlantic and North Pacific (Table 3.2). At all locations and for all time periods shown, pCO_2 in both the atmosphere and ocean has increased, while pH and $[CO_3^{2-}]$ have decreased. At some sites, oceanic surface pCO_2 increased faster than the atmospheric trend, implying a decreasing uptake of atmospheric CO_2 at those locations. The oceanic pCO_2 trend can differ from that in the atmosphere owing to changes in the intensity of biological production and changes in physical conditions, for instance between El Niño and La Niña (Keeling et al., 2004; Midorikawa et al., 2005; Yoshikawa-Inoue and Ishii, 2005; Takahashi et al., 2006, 2009; Schuster and Watson, 2007; Ishii et al., 2009; McKinley et al., 2011; Bates, 2012; Lenton et al., 2012).

Although local variations of ΔpCO_2 with time have little effect on the atmospheric CO_2 growth rate in the short term, they provide important information on the dynamics of the ocean carbon cycle and the potential for longer-term climate feedbacks. For example, El Niño and La Niña can drive large changes in the efflux of CO_2 in the Pacific. Differences in ΔpCO_2 can exceed 100 µatm in the eastern and central equatorial Pacific between El Niño and La Niña; an increase in ΔpCO_2 observed between 1998 and 2004 was attributed to wind and circulation changes associated with the Pacific Decadal Oscillation (Feely et al., 2006). CO_2 uptake in the North Atlantic decreased by 0.24 [0.19–0.29] PgC yr^{-1} between 1994 and 2003 (Schuster and Watson, 2007) and has partially recovered since then (Watson et al., 2009). Linear trends for the North Atlantic from 1995 to 2009 reveal an increased uptake (Schuster et al., 2013). Uptake of CO_2 in the Subtropical Mode Water (STMW) of the North Atlantic was enhanced during the 1990s, a predominantly positive phase of the NAO, and much reduced in the 2000s when the NAO phase was neutral or negative (Bates, 2012). Observations in the Indian and Pacific sectors of the Southern Ocean were interpreted as evidence for reduced winter-time CO_2 uptake as a result of increased winds, increased upwelling and outgassing of natural CO_2 (Metzl, 2009; Lenton et al., 2012).

3.8.1.2 Changes in the Oceanic Inventory of Anthropogenic Carbon Dioxide

Ocean carbon uptake and storage is inferred from changes in the inventory of anthropogenic carbon. C_{ant} cannot be measured directly but is calculated from observations of ocean properties (Appendix 3.A discusses the sampling on which the ocean carbon inventory is based). Two independent data-based methods to calculate anthropogenic carbon inventories exist: the ΔC^* method (Sabine et al., 2004), and the transit time distribution (TTD) method (Waugh et al., 2006). The Green's function approach that applies the maximum entropy de-convolution methodology (Khatiwala et al., 2009) is related to the latter. These approaches use different tracer data, for instance, the TTD method is based mostly on chlorofluorcarbon measurements. Changes due to variability in ocean productivity (Chavez et al., 2011) are not considered.

Estimates of the global inventory of C_{ant} (including marginal seas) calculated using these methods have a mean value of 118 PgC and a range of 93 to 137 PgC in 1994 and a mean of 160 PgC and range of 134 to 186 PgC in 2010 (Sabine et al., 2004; Waugh et al., 2006; Khatiwala et al., 2009, 2013). When combined with model results (Mikaloff-Fletcher et al., 2006; Doney et al., 2009; Gerber et al., 2009; Graven et al., 2012), Khatiwala et al. (2013) arrive at a "best" estimate of the global ocean inventory (including marginal seas) of anthropogenic carbon from 1750 to 2010 of 155 PgC with an uncertainty of ±20% (Figure 3.16). While the estimates of total inventory agree within their uncertainty, the different methods result in significant differences in the inferred spatial distribution of C_{ant}, particularly at high latitudes.

The C_{ant} inventory "best" estimate of 155 PgC (Khatiwala et al., 2013; Figure 3.16) corresponds to an uptake rate of 2.3 (range of 1.7 to 2.9) PgC yr^{-1} from 2000 to 2010, in close agreement with an independent estimate of 2.5 (range of 1.8 to 3.2) PgC yr^{-1} based on atmospheric O_2/N_2 measurements obtained for the same period (Ishidoya et al., 2012). The O_2/N_2 method resulted in 2.2 ± 0.6 PgC yr^{-1} for the time period 1990 to 2000 and 2.5 ± 0.6 for the period from 2000 to 2010 (Keeling and Manning, 2014). These estimates are also consistent with an independent estimate of 1.9 ± 0.4 PgC yr^{-1} for the period between 1970 and 1990 based on depth-integrated $\delta^{13}C$ changes (Quay et al., 2003) and with estimates inferred from ΔpCO_2.

The storage rate of anthropogenic CO_2 is assessed by calculating the change in C_{ant} concentrations between two time periods. Regional observations of the storage rate are in general agreement with that expected from the increase in atmospheric CO_2 concentrations and with the tracer-based estimates. However, there are significant spatial and temporal variations in the degree to which the inventory of C_{ant}

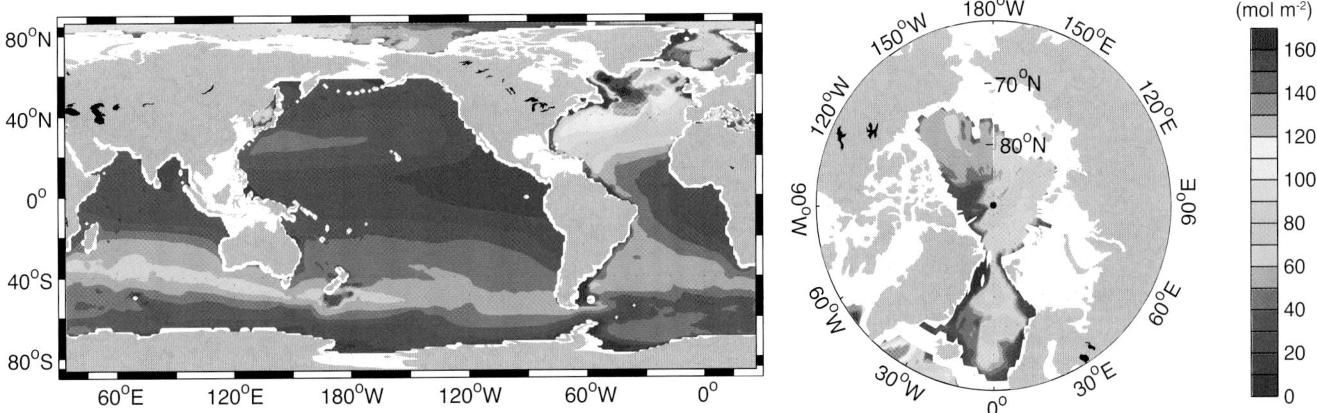

Figure 3.16 | Compilation of the 2010 column inventories (mol m^{-2}) of anthropogenic CO_2: the global Ocean excluding the marginal seas (updated from Khatiwala et al., 2009) 150 ± 26 PgC; Arctic Ocean (Tanhua et al., 2009) 2.7 to 3.5 PgC; the Nordic Seas (Olsen et al., 2010) 1.0 to 1.6 PgC; the Mediterranean Sea (Schneider et al., 2010) 1.6 to 2.5 PgC; the Sea of Japan(Park et al., 2006) 0.40 ± 0.06 PgC. From Khatiwala et al. (2013).

tracks changes in the atmosphere (Figure 3.17). The North Atlantic, in particular, is an area with high variability in circulation and deep water formation, influencing the C_{ant} inventory. As a result of the decline in Labrador Sea Water (LSW) formation since 1997 (Rhein et al., 2011), the C_{ant} increase between 1997 and 2003 was smaller in the subpolar North Atlantic than expected from the atmospheric increase, in contrast to the subtropical and equatorial Atlantic (Steinfeldt et al., 2009). Perez et al. (2010) also noted the dependence of the C_{ant} storage rate in the North Atlantic on the NAO, with high C_{ant} storage rate during phases of high NAO (i.e., high LSW formation rates) and low storage during phases of low NAO (low formation). Wanninkhof et al. (2010) found a smaller inventory increase in the North Atlantic compared to the South Atlantic between 1989 and 2005.

Ocean observations are insufficient to assess whether there has been a change in the rate of total (anthropogenic plus natural) carbon uptake by the global ocean. Evidence from regional ocean studies (often covering relatively short time periods), atmospheric observations and models is equivocal, with some studies suggesting the ocean uptake rate of total CO_2 may have declined (Le Quéré et al., 2007; Schuster and Watson, 2007; McKinley et al., 2011) while others conclude that there is little evidence for a decline (Knorr, 2009; Gloor et al., 2010; Sarmiento et al., 2010). A study based on atmospheric CO_2 observations and emission inventories concluded that global carbon uptake by land and oceans doubled from 1960 to 2010, implying that it is *unlikely* that on a global scale both land and ocean sinks decreased (Ballantyne et al., 2012).

In summary, the high agreement between multiple lines of independent evidence for increases in the ocean inventory of C_{ant} underpins the conclusion that it is *virtually certain* that the ocean is sequestering anthropogenic carbon dioxide and *very likely* that the oceanic C_{ant} inventory increased from 1994 to 2010. Oceanic carbon uptake rates calculated using different data sets and methods agree within their uncertainties and *very likely* range between 1.0 and 3.2 PgC yr^{-1}.

3.8.2 Anthropogenic Ocean Acidification

The uptake of CO_2 by the ocean changes the chemical balance of seawater through the thermodynamic equilibrium of CO_2 with seawater. Dissolved CO_2 forms a weak acid (H_2CO_3) and, as CO_2 in seawater increases, the pH, carbonate ion (CO_3^{2-}), and calcium carbonate ($CaCO_3$) saturation state of seawater decrease while bicarbonate ion (HCO_3^-) increases (FAQ 3.3). Variations in oceanic total dissolved inorganic carbon ($C_T = CO_2 + CO_3^{2-} + HCO_3^-$) and pCO_2 reflect changes in both the natural carbon cycle and the uptake of anthropogenic CO_2 from the atmosphere. The mean pH (total scale) of surface waters ranges between 7.8 and 8.4 in the open ocean, so the ocean remains mildly basic (pH > 7) at present (Orr et al., 2005a; Feely et al., 2009). Ocean uptake of CO_2 results in gradual acidification of seawater; this

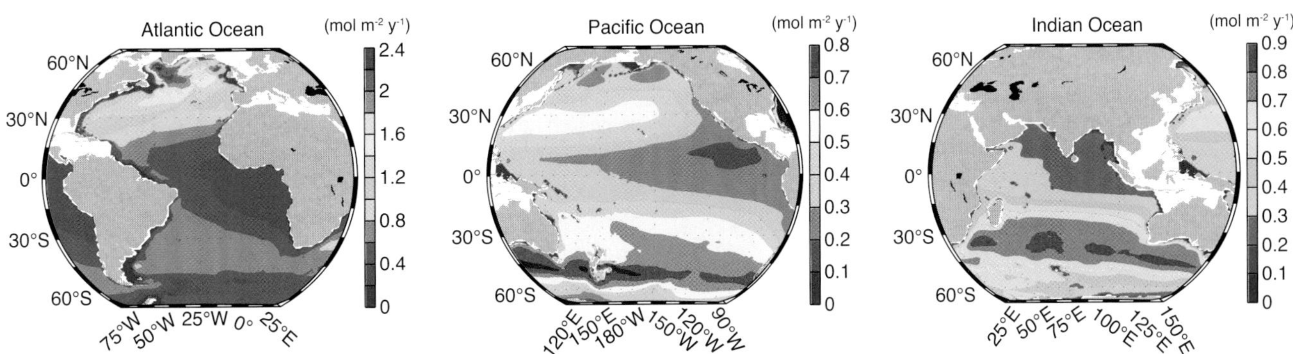

Figure 3.17 | Maps of storage rate distribution of anthropogenic carbon (mol m^{-2} yr^{-1}) for the three ocean basins (left to right: Atlantic, Pacific and Indian Ocean) averaged over 1980–2005 estimated by the Green's function approach (Khatiwala et al., 2009). Note that a different colour scale is used in each basin.

process is termed ocean acidification (Box 3.2) (Broecker and Clark, 2001; Caldeira and Wickett, 2003). The observed decrease in ocean pH of 0.1 since the beginning of the industrial era corresponds to a 26% increase in the hydrogen ion concentration [H^+] concentration of seawater (Orr et al., 2005b; Feely et al., 2009). The consequences of changes in pH, CO_3^{2-}, and the saturation state of $CaCO_3$ minerals for marine organisms and ecosystems are just beginning to be understood (see WGII Chapters 5, 6, 28 and 30).

A global mean decrease in surface water pH of 0.08 from 1765 to 1994 was calculated based on the inventory of anthropogenic CO_2 (Sabine et al., 2004), with the largest reduction (–0.10) in the northern North Atlantic and the smallest reduction (–0.05) in the subtropical South Pacific. These regional variations in the size of the pH decrease are consistent with the generally lower buffer capacities of the high latitude oceans compared to lower latitudes (Egleston et al., 2010).

3

Direct measurements on ocean time-series stations in the North Atlantic and North Pacific record decreasing pH with rates ranging between –0.0014 and –0.0024 yr^{-1} (Table 3.2, Figure 3.18; Bates, 2007, 2012; Santana-Casiano et al., 2007; Dore et al., 2009; Olafsson et al., 2009; González-Dávila et al., 2010). Directly measured pH differences in the surface mixed layer along repeat transects in the central North Pacific Ocean between Hawaii and Alaska showed a –0.0017 yr^{-1} decline in pH between 1991 and 2006, in agreement with observations at the time-series sites (Byrne et al., 2010). This rate of pH change is also consistent with repeat transects of CO_2 and pH measurements in the western North Pacific (winter: –0.0018 ± 0.0002 yr^{-1}; summer: –0.0013 ± 0.0005 yr^{-1}) (Midorikawa et al., 2010). The pH changes in southern ocean surface waters are less certain because of the paucity of long-term time-series observations there, but pCO_2 measurements collected by ships-of-opportunity indicate similar rates of pH decrease there (Takahashi et al., 2009).

Uptake of anthropogenic CO_2 is the dominant cause of observed changes in the carbonate chemistry of surface waters (Doney et al., 2009). Changes in carbonate chemistry in subsurface waters can also reflect local physical and biological variability. As an example, while pH changes in the mixed layer of the North Pacific Ocean can be explained solely by equilibration with atmospheric CO_2, declines in pH between 800 m and the mixed layer in the time period 1991–2006 were attributed in approximately equal measure to anthropogenic and natural variations (Byrne et al., 2010). Figure 3.19 shows the portion of pH changes between the surface and 1000 m that were attributed solely to the effects of anthropogenic CO_2. Seawater pH and [CO_3^{2-}] decreased by 0.0014 to 0.0024 yr^{-1} and ~0.4 to 0.9 µmol kg^{-1} yr^{-1}, respectively, between 1988 and 2009 (Table 3.2). Over longer time periods, anthropogenic changes in ocean chemistry are expected to become increasingly prominent relative to changes imparted by physical and biological variability.

The consistency of these observations demonstrates that the pH of surface waters has decreased as a result of ocean uptake of anthropogenic CO_2 from the atmosphere. There is *high confidence* that the pH decreased by 0.1 since the preindustrial era.

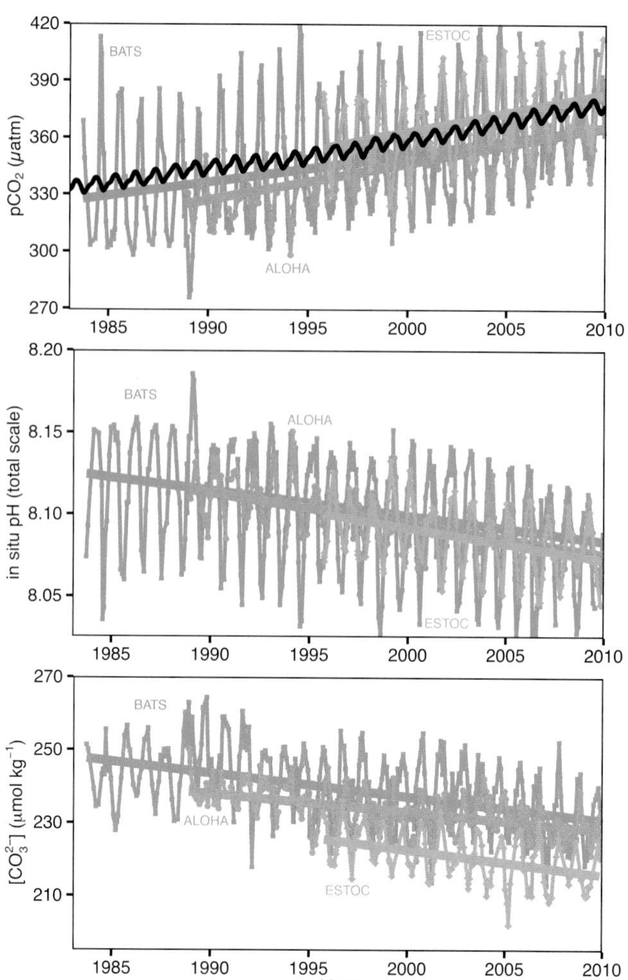

Figure 3.18 | Long-term trends of surface seawater pCO_2 (top), pH (middle) and carbonate ion (bottom) concentration at three subtropical ocean time series in the North Atlantic and North Pacific Oceans, including (a) Bermuda Atlantic Time-series Study (BATS, 31°40'N, 64°10'W; green) and Hydrostation S (32°10', 64°30'W) from 1983 to present (updated from Bates, 2007); (b) Hawaii Ocean Time-series (HOT) at Station ALOHA (A Long-term Oligotrophic Habitat Assessment; 22°45'N, 158°00'W; orange) from 1988 to present (updated from Dore et al., 2009) and (c) European Station for Time series in the Ocean (ESTOC, 29°10'N, 15°30'W; blue) from 1994 to present (updated from González-Dávila et al., 2010). Atmospheric pCO_2 (black) from the Mauna Loa Observatory Hawaii is shown in the top panel. Lines show linear fits to the data, whereas Table 3.2 give results for harmonic fits to the data (updated from Orr, 2011).

3.8.3 Oxygen

As a consequence of the early introduction of standardized methods and the relatively wide interest in the distribution of dissolved oxygen, the historical record of marine oxygen observations is generally richer than that of other biogeochemical parameters, although still sparse compared to measurements of temperature and salinity (Appendix 3.A). Dissolved oxygen changes in the ocean thermocline has generally decreased since 1960, but with strong regional variations (Keeling et al., 2010; Keeling and Manning, 2014). Oxygen concentrations at 300 dbar decreased between 50°S and 50°N at a mean rate of 0.63 µmol kg^{-1} per decade between 1960 and 2010 (Stramma et al., 2012). For the period 1970 to 1990, the mean annual global oxygen loss between 100 m and 1000 m was calculated to be 0.55 ± 0.13 × 10^{14} mol yr^{-1} (Helm et al., 2011).

Box 3.2 | Ocean Acidification

Ocean acidification refers to a reduction in pH of the ocean over an extended period, typically decades or longer, caused primarily by the uptake of carbon dioxide (CO_2) from the atmosphere. Ocean acidification can also be caused by other chemical additions or subtractions from the oceans that are natural (e.g., increased volcanic activity, methane hydrate releases, long-term changes in net respiration) or human-induced (e.g., release of nitrogen and sulphur compounds into the atmosphere). Anthropogenic ocean acidification refers to the component of pH reduction that is caused by human activity (IPCC, 2011).

Since the beginning of the industrial era, the release of CO_2 from industrial and agricultural activities has resulted in atmospheric CO_2 concentrations that have increased from approximately 280 ppm to about 392 ppm in 2012 (Chapter 6). The oceans have absorbed approximately 155 PgC from the atmosphere over the last two and a half centuries (Sabine et al., 2004; Khatiwala et al., 2013). This natural process of absorption has benefited humankind by significantly reducing the greenhouse gas levels in the atmosphere and abating some of the impacts of global warming. However, the ocean's uptake of carbon dioxide is having a significant impact on the chemistry of seawater. The average pH of ocean surface waters has already fallen by about 0.1 units, from about 8.2 to 8.1 (total scale), since the beginning of the industrial revolution (Orr et al., 2005a; Figure 1; Feely et al., 2009). Estimates of future atmospheric and oceanic carbon dioxide concentrations indicate that, by the end of this century, the average surface ocean pH could be lower than it has been for more than 50 million years (Caldeira and Wickett, 2003).

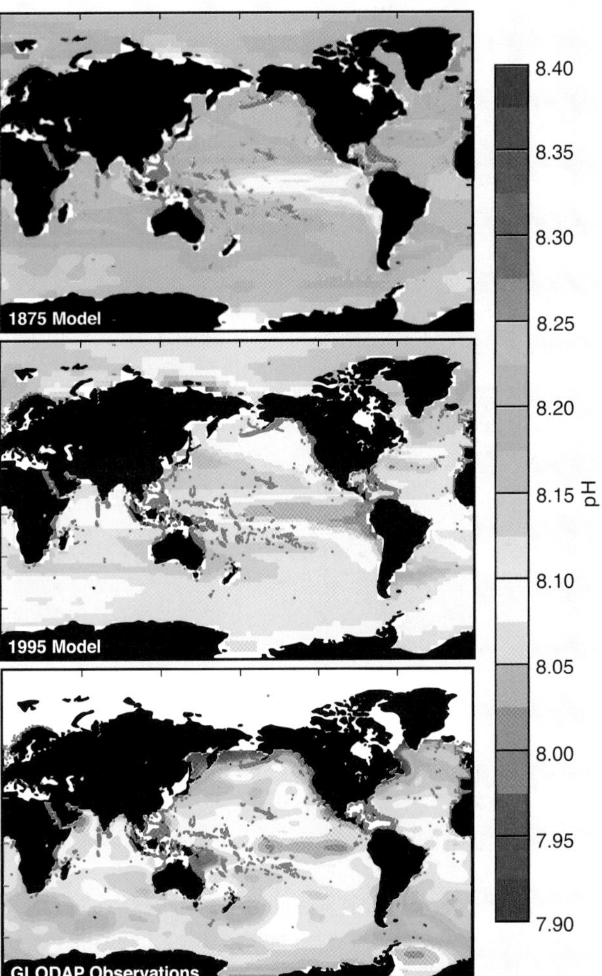

Box 3.2, Figure 1 | National Center for Atmospheric Research Community Climate System Model 3.1 (CCSM3)-modeled decadal mean pH at the sea surface centred on the years 1875 (top) and 1995 (middle). Global Ocean Data Analysis Project (GLODAP)-based pH at the sea surface, nominally for 1995 (bottom). Deep and shallow-water coral reefs are indicated with magenta dots. White areas indicate regions with no data. (After Feely et al., 2009.)

The major controls on seawater pH are atmospheric CO_2 exchange, the production and respiration of dissolved and particulate organic matter in the water column, and the formation and dissolution of calcium carbonate minerals. Oxidation of organic matter lowers dissolved oxygen concentrations, adds CO_2 to solution, reduces pH, carbonate ion (CO_3^{2-}) and calcium carbonate ($CaCO_3$) saturation states (Box 3.2, Figure 2), and lowers the pH of seawater in subsurface waters (Byrne et al., 2010). As a result of these processes, minimum pH values in the oceanic water column are generally found near the depths of the oxygen minimum layer. When CO_2 reacts with seawater it forms carbonic acid (H_2CO_3), which is highly reactive and reduces the concentration of carbonate ion (Box 3.2, Figure 2) and can affect shell formation for marine animals such as corals, plankton, and shellfish. This process could affect fundamental biological and chemical processes of the sea in coming decades (Fabry et al., 2008; Doney et al., 2009; WGII Chapters 5, 6, 28 and 30).

(continued on next page)

The long-term deoxygenation of the open ocean thermocline is consistent with the expectation that warmer waters can hold less dissolved oxygen (solubility effect), and that warming-induced stratification leads to a decrease in the transport of dissolved oxygen from surface to subsurface waters (stratification effect) (Matear and Hirst, 2003; Deutsch et al., 2005; Frölicher et al., 2009). Observations of oxygen change suggested that about 15% of the oxygen decline between 1970 and 1990 could be explained by warming and the remainder by reduced ventilation due to increased stratification (Helm et al., 2011; see Table 6.14).

Oxygen concentrations in the tropical ocean thermocline decreased in each of the ocean basins over the last 50 years (Ono et al., 2001; Stramma et al., 2008; Keeling et al., 2010; Helm et al., 2011), resulting in an expansion of the dissolved oxygen minimum zones. A comparison of data between 1960 and 1974 with those from 1990 to 2008 showed

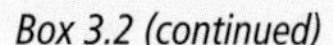

Box 3.2 (continued)

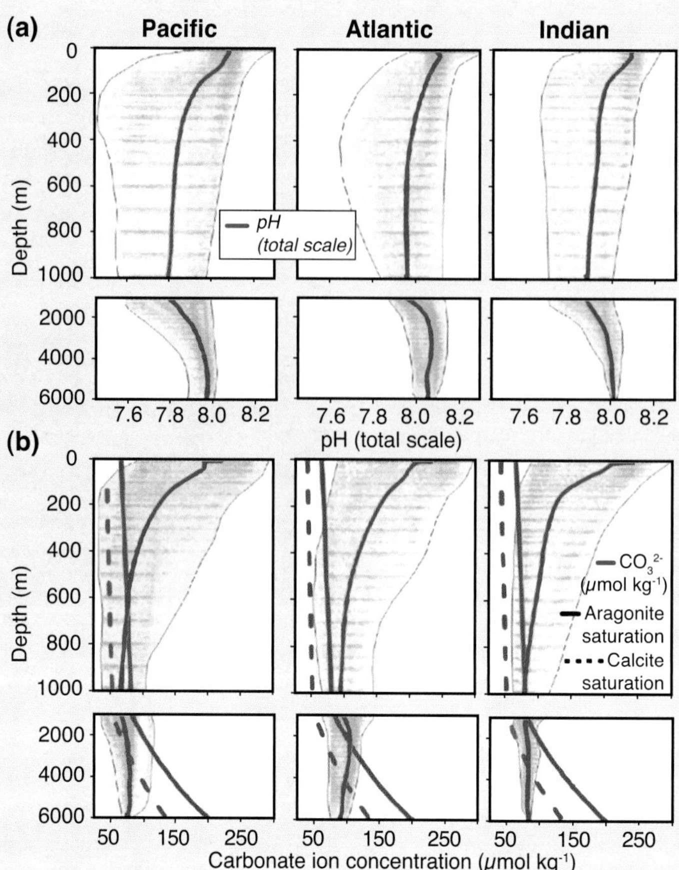

Box 3.2, Figure 2 | Distribution of (a) pH and (b) carbonate (CO_3^{2-}) ion concentration in the Pacific, Atlantic and Indian Oceans. The data are from the World Ocean Circulation Experiment/Joint Global Ocean Flux Study/Ocean Atmosphere Carbon Exchange Study global carbon dioxide (CO_2) survey (Sabine et al., 2005). The lines show the mean pH (red solid line, top panel), mean CO_3^{2-} (red solid line, bottom panel), and aragonite and calcite (black solid and dashed lines, bottom panel) saturation values for each of these basins (modified from Feely et al., 2009). The shaded areas show the range of values within the ocean basins. Dissolution of aragonite and calcite shells and skeletons occurs when CO_3^{2-} concentrations drop below the saturation level, reducing the ability of calcifying organisms to produce their shells and skeletons.

that oxygen concentrations decreased in most tropical regions at an average rate of 2 to 3 µmol kg^{-1} per decade (Figure 3.20; Stramma et al., 2010). Data from one of the longest time-series sites in the subpolar North Pacific (Station Papa, 50°N, 145°W) reveal a persistent declining oxygen trend in the thermocline over the last 50 years (Whitney et al., 2007), superimposed on oscillations with time scales of a few years to two decades. Stendardo and Gruber (2012) found dissolved oxygen decreases in upper water masses of the North Atlantic and increases in intermediate water masses. The changes were caused by changes in solubility as well as changes in ventilation and circulation over time. In contrast to the widely distributed oxygen declines, oxygen increased in the thermoclines of the Indian and South Pacific Oceans from the 1990s to the 2000s (McDonagh et al., 2005; Álvarez et al., 2011), apparently due to strengthened circulation driven by stronger winds (Cai, 2006; Roemmich et al., 2007). In the southern Indian Ocean below the thermocline, east of 75°E, oxygen decreased between 1960 and 2010 most prominently on the isopycnals σ_θ = 26.9 to 27.0 (Kobayashi et al., 2012). While some studies suggest a widespread decline of oxygen in the Southern Ocean (e.g., Helm et al., 2011), other studies show regions of alternating sign (e.g., Stramma et al., 2010), reflecting differences in data and period considered.

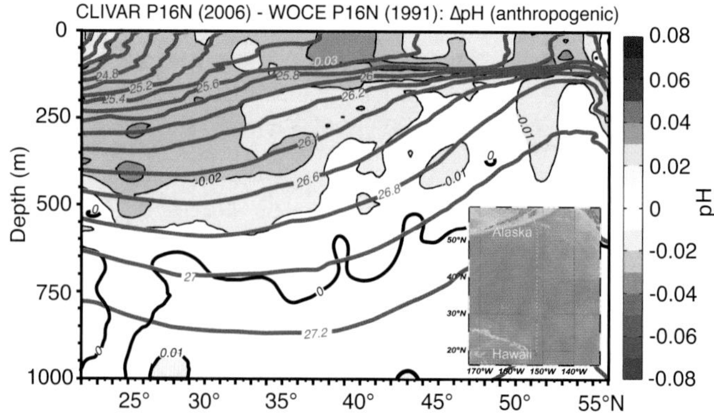

Figure 3.19 | ΔpH_{ant}: pH change attributed to the uptake of anthropogenic carbon between 1991 and 2006, at about 150°W, Pacific Ocean (from Byrne et al., 2010). The red lines show the layers of constant density.

Frequently Asked Questions

FAQ 3.3 | How Does Anthropogenic Ocean Acidification Relate to Climate Change?

Both anthropogenic climate change and anthropogenic ocean acidification are caused by increasing carbon dioxide concentrations in the atmosphere. Rising levels of carbon dioxide (CO_2), along with other greenhouse gases, indirectly alter the climate system by trapping heat as it is reflected back from the Earth's surface. Anthropogenic ocean acidification is a direct consequence of rising CO_2 concentrations as seawater currently absorbs about 30% of the anthropogenic CO_2 from the atmosphere.

Ocean acidification refers to a reduction in pH over an extended period, typically decades or longer, caused primarily by the uptake of CO_2 from the atmosphere. pH is a dimensionless measure of acidity. Ocean acidification describes the direction of pH change rather than the end point; that is, ocean pH is decreasing but is not expected to become acidic (pH < 7). Ocean acidification can also be caused by other chemical additions or subtractions from the oceans that are natural (e.g., increased volcanic activity, methane hydrate releases, long-term changes in net respiration) or human-induced (e.g., release of nitrogen and sulphur compounds into the atmosphere). Anthropogenic ocean acidification refers to the component of pH reduction that is caused by human activity.

Since about 1750, the release of CO_2 from industrial and agricultural activities has resulted in global average atmospheric CO_2 concentrations that have increased from 278 to 390.5 ppm in 2011. The atmospheric concentration of CO_2 is now higher than experienced on the Earth for at least the last 800,000 years and is expected to continue to rise because of our dependence on fossil fuels for energy. To date, the oceans have absorbed approximately 155 ± 30 PgC from the atmosphere, which corresponds to roughly one-fourth of the total amount of CO_2 emitted (555 ± 85 PgC) by human activities since preindustrial times. This natural process of absorption has significantly reduced the greenhouse gas levels in the atmosphere and minimized some of the impacts of global warming. However, the ocean's uptake of CO_2 is having a significant impact on the chemistry of seawater. The average pH of ocean surface waters has already fallen by about 0.1 units, from about 8.2 to 8.1 since the beginning of the Industrial Revolution. Estimates of projected future atmospheric and oceanic CO_2 concentrations indicate that, by the end of this century, the average surface ocean pH could be 0.2 to 0.4 lower than it is today. The pH scale is logarithmic, so a change of 1 unit corresponds to a 10-fold change in hydrogen ion concentration.

When atmospheric CO_2 exchanges across the air–sea interface it reacts with seawater through a series of four chemical reactions that increase the concentrations of the carbon species: dissolved carbon dioxide ($CO_{2(aq)}$), carbonic acid (H_2CO_3) and bicarbonate (HCO_3^-):

$$CO_{2(atmos)} \rightleftarrows CO_{2(aq)} \quad (1)$$

$$CO_{2(aq)} + H_2O \rightleftarrows H_2CO_3 \quad (2)$$

$$H_2CO_3 \rightleftarrows H^+ + HCO_3^- \quad (3)$$

$$HCO_3^- \rightleftarrows H^+ + CO_3^{2-} \quad (4)$$

Hydrogen ions (H^+) are produced by these reactions. This increase in the ocean's hydrogen ion concentration corresponds to a reduction in pH, or an increase in acidity. Under normal seawater conditions, more than 99.99% of the hydrogen ions that are produced will combine with carbonate ion (CO_3^{2-}) to produce additional HCO_3^-. Thus, the addition of anthropogenic CO_2 into the oceans lowers the pH and consumes carbonate ion. These reactions are fully reversible and the basic thermodynamics of these reactions in seawater are well known, such that at a pH of approximately 8.1 approximately 90% the carbon is in the form of bicarbonate ion, 9% in the form of carbonate ion, and only about 1% of the carbon is in the form of dissolved CO_2. Results from laboratory, field, and modeling studies, as well as evidence from the geological record, clearly indicate that marine ecosystems are highly susceptible to the increases in oceanic CO_2 and the corresponding decreases in pH and carbonate ion.

Climate change and anthropogenic ocean acidification do not act independently. Although the CO_2 that is taken up by the ocean does not contribute to greenhouse warming, ocean warming reduces the solubility of carbon dioxide in seawater; and thus reduces the amount of CO_2 the oceans can absorb from the atmosphere. For example, under doubled preindustrial CO_2 concentrations and a 2°C temperature increase, seawater absorbs about 10% less CO_2 (10% less total carbon, C_T) than it would with no temperature increase (compare columns 4 and 6 in Table 1), but the pH remains almost unchanged. Thus, a warmer ocean has less capacity to remove CO_2 from the atmosphere, yet still experiences ocean acidification. The reason for this is that bicarbonate is converted to carbonate in a warmer ocean, releasing a hydrogen ion thus stabilizing the pH. *(continued on next page)*

3

FAQ 3.3 (continued)

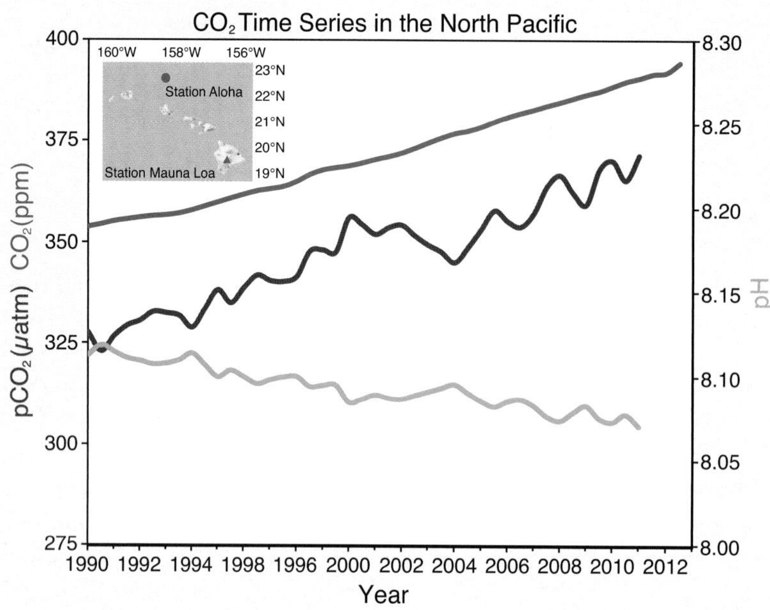

FAQ 3.3, Figure 1 | A smoothed time series of atmospheric CO_2 mole fraction (in ppm) at the atmospheric Mauna Loa Observatory (top red line), surface ocean partial pressure of CO_2 (pCO_2; middle blue line) and surface ocean pH (bottom green line) at Station ALOHA in the subtropical North Pacific north of Hawaii for the period from1990–2011 (after Doney et al., 2009; data from Dore et al., 2009). The results indicate that the surface ocean pCO_2 trend is generally consistent with the atmospheric increase but is more variable due to large-scale interannual variability of oceanic processes.

FAQ 3.3, Table 1 | Oceanic pH and carbon system parameter changes in surface water for a CO_2 doubling from the preindustrial atmosphere without and with a 2°C warming[a].

Parameter	Pre-industrial (280 ppmv) 20°C	2 × Pre-industrial (560 ppmv) 20°C	(% change relative to pre-industrial)	2 × Pre-industrial (560 ppmv) 22°C	(% change relative to pre-industrial)
pH	8.1714	7.9202	–	7.9207	–
H^+ (mol kg^{-1})	6.739e^{-9}	1.202e^{-8}	(78.4)	1.200e^{-8}	(78.1)
$CO_{2(aq)}$ (μmol kg^{-1})	9.10	18.10	(98.9)	17.2	(89.0)
HCO_3^- (μmol kg^{-1})	1723.4	1932.8	(12.15)	1910.4	(10.9)
CO_3^{2-} (μmol kg^{-1})	228.3	143.6	(-37.1)	152.9	(–33.0)
C_T (μmol kg^{-1})	1960.8	2094.5	(6.82)	2080.5	(6.10)

Notes:
[a] $CO_{2(aq)}$ = dissolved CO_2, H_2CO_3 = carbonic acid, HCO_3^- = bicarbonate, CO_3^{2-} = carbonate, C_T = total carbon = $CO_{2(aq)} + HCO_3^- + CO_3^{2-}$).

Coastal regions have also experienced long-term dissolved oxygen changes. Bograd et al. (2008) reported a substantial reduction of the thermocline oxygen content in the southern part of the California Current from 1984 to 2002, resulting in a shoaling of the hypoxic boundary (marked by oxygen concentrations of about 60 μmol kg^{-1}). Off the British Columbia coast, oxygen concentrations in the near bottom waters decreased an average of 1.1 μmol kg^{-1} yr^{-1} over a 30-year period (Chan et al., 2008). These changes along the west coast of North America appear to have been largely caused by the open ocean dissolved oxygen decrease and local processes associated with decreased vertical dissolved oxygen transport following near-surface warming and increased stratification. Gilbert et al. (2010) found evidence that for the time period 1976–2000 oxygen concentrations between 0 and 300 m depth were declining about 10 times faster in the coastal ocean than in the open ocean, and an increase in the number of hypoxic zones was observed since the 1960s (Diaz and Rosenberg, 2008).

3.8.4 Nutrients

Nutrient concentrations in the surface ocean surface are influenced by human impacts on coastal runoff and on atmospheric deposition, and by changing nutrient supply from the ocean's interior into the mixed layer (for instance due to increased stratification). Changing nutrient distributions might influence the magnitude and variability of the ocean's biological carbon pump.

Globally, the manufacture of nitrogen fertilizers has continued to increase (Galloway et al., 2008) accompanied by increasing eutrophi-

Table 3.2 | Published and updated long-term trends of atmospheric (pCO_2^{atm}) and seawater carbonate chemistry (i.e., surface-water pCO_2, and corresponding calculated pH, CO_3^{2-}, and aragonite saturation state (Ωa) at four ocean time series in the North Atlantic and North Pacific oceans: (1) Bermuda Atlantic Time-series Study (BATS, 31°40'N, 64°10'W) and Hydrostation S (32°10'N, 64°30'W) from 1983 to present (Bates, 2007); (2) Hawaii Ocean Time series (HOT) at Station ALOHA (A Long-term Oligotrophic Habitat Assessment; 22°45'N, 158°00'W) from 1988 to the present (Dore et al., 2009); (3) European Station for Time series in the Ocean (ESTOC, 29°10'N, 15°30'W) from 1994 to the present (González-Dávila et al., 2010); and (4) Iceland Sea (IS, 68.0°N, 12.67°W) from 1985 to 2006 (Olafsson et al., 2009). Trends at the first three time-series sites are from observations with the seasonal cycle removed. Also reported are the wintertime trends in the Iceland Sea as well as the pH difference trend for the North Pacific Ocean between transects in 1991 and 2006 (Byrne et al., 2010) and repeat sections in the western North Pacific between 1983 and 2008 (Midorikawa et al., 2010).

Site	Period	pCO_2^{atm} (µatm yr^{-1})	pCO_2^{sea} (µatm yr^{-1})	pH* (yr^{-1})	[CO_3^{2-}] (µmol kg^{-1} yr^{-1})	Ω_a (yr^{-1})
a. Published trends						
BATS	1983–2005[a]	1.78 ± 0.02	1.67 ± 0.28	–0.0017 ± 0.0003	–0.47 ± 0.09	–0.007 ± 0.002
	1983–2005[b]	1.80 ± 0.02	1.80 ± 0.13	–0.0017 ± 0.0001	–0.52 ± 0.02	–0.006 ± 0.001
ALOHA	1988–2007[c]	1.68 ± 0.03	1.88 ± 0.16	–0.0019 ± 0.0002	—	–0.0076 ± 0.0015
	1998–2007[d]	—	—	–0.0014 ± 0.0002	—	—
ESTOC	1995–2004[e]	—	1.55 ± 0.43	–0.0017 ± 0.0004	—	—
	1995–2004[f]	1.6 ± 0.7	1.55	–0.0015 ± 0.0007	–0.90 ± 0.08	–0.0140 ± 0.0018
IS	1985–2006[g]	1.69 ± 0.04	2.15 ± 0.16	–0.0024 ± 0.0002	—	–0.0072 ± 0.0007[g]
N. Pacific	1991–2006[h]	—	—	–0.0017	—	—
N. Pacific	1983–2008[i]	Summer 1.54 ± 0.08 Winter 1.65 ± 0.05	Summer 1.37 ± 0.33 Winter 1.58 ± 0.12	Summer –0.0013 ± 0.0005 Winter –0.0018 ± 0.0002	—	—
Coast of western N. Pacific	1994–2008[k]	1.99 ± 0.02	1.54 ± 0.33	–0.0020 ± 0.0007	—	–0.012 ± 0.005
b. Updated trends[j,l]						
BATS	1983–2009	1.66 ± 0.01	1.92 ± 0.08	–0.0019 ± 0.0001	–0.59 ± 0.04	–0.0091 ± 0.0006
	1985–2009	1.67 ± 0.01	2.02 ± 0.08	–0.0020 ± 0.0001	–0.68 ± 0.04	–0.0105 ± 0.0006
	1988–2009	1.73 ± 0.01	2.22 ± 0.11	–0.0022 ± 0.0001	–0.87 ± 0.05	–0.0135 ± 0.0008
	1995–2009	1.90 ± 0.01	2.16 ± 0.18	–0.0021 ± 0.0002	–0.80 ± 0.08	–0.0125 ± 0.0013
ALOHA	1988–2009	1.73 ± 0.01	1.82 ± 0.07	–0.0018 ± 0.0001	–0.52 ± 0.04	–0.0083 ± 0.0007
	1995–2009	1.92 ± 0.01	1.58 ± 0.13	–0.0015 ± 0.0001	–0.40 ± 0.07	–0.0061 ± 0.0028
ESTOC	1995–2009	1.88 ± 0.02	1.83 ± 0.15	–0.0017 ± 0.0001	–0.72 ± 0.05	–0.0123 ± 0.0015
IS	1985–2009[m]	1.75 ± 0.01	2.07 ± 0.15	–0.0024 ± 0.0002	–0.47 ± 0.04	–0.0071 ± 0.0006
	1988–2009[m]	1.70 ± 0.01	1.96 ± 0.22	–0.0023 ± 0.0003	–0.48 ± 0.05	–0.0073 ± 0.0008
	1995–2009[m]	1.90 ± 0.01	2.01 ± 0.37	–0.0022 ± 0.0004	–0.40 ± 0.08	–0.0062 ± 0.0012

Notes:

* pH on the total scale.

a Bates (2007, Table 1): Simple linear fit.

b Bates (2007, Table 2): Seasonally detrended (including linear term for time).

c Dore et al. (2009): Linear fit with calculated pH and pCO_2 from measured DIC and TA (full time series); corresponding Ωa from Feely et al. (2009).

d Dore et al. (2009): Linear fit with measured pH (partial time series).

e Santana-Casiano et al. (2007): Seasonal detrending (including linear terms for time and temperature).

f González-Dávila et al. (2010): Seasonal detrending (including linear terms for time, temperature and mixed-layer depth).

g Olafsson et al. (2009): Multivariable linear regression (linear terms for time and temperature) for winter data only.

h Byrne et al. (2010): Meridional section originally occupied in 1991 and repeated in 2006.

i Midorikawa et al. (2010): Winter and summer observations along 137°E.

j Trends are for linear time term in seasonal detrending with harmonic periods of 12, 6 and 4 months. Harmonic analysis made after interpolating data to regular monthly grids (except for IS, which was sampled much less frequently):

1983–2009 = September 1983 to December 2009 (BATS/Hydrostation S sampling period),
1985–2009 = February 1985 to December 2009 (IS sampling period),
1988–2009 = November 1988 to December 2009 (ALOHA/HOT sampling period), and
1995–2009 = September 1995 to December 2009 (ESTOC sampling period).

k Ishii et al. (2011) - time-series observations in the coast of western North Pacific, with the seasonal cycle removed

l Atmospheric pCO_2 trends computed from same harmonic analysis (12-, 6- and 4-month periods) on the GLOBALVIEW-CO_2 (2010) data product for the marine boundary layer referenced to the latitude of the nearest atmospheric measurement station (BME = Bermuda; MLO = ALOHA; IZO = ESTOC; ICE = Iceland).

m Winter ocean data, collected during dark period (between 19 January and 7 March), as per Olafsson et al. (2009) to reduce scatter from large interannual variations in intense short-term bloom events, undersampled in time, fit linearly ($y = at + bT + c$).

cation of coastal waters (Diaz and Rosenberg, 2008; Seitzinger et al., 2010; Kim et al., 2011), which amplifies the drawdown of CO_2 (Borges and Gypens, 2010; Provoost et al., 2010). In addition, atmospheric deposition of anthropogenic fixed nitrogen may now account for up to about 3% of oceanic new production, and this nutrient source is projected to increase (Duce et al., 2008).

Satellite observations of chlorophyll reveal that oligotrophic provinces in four of the world's major oceans expanded at average rates of 0.8 to 4.3% yr^{-1} from 1998 to 2006 (Polovina et al., 2008; Irwin and Oliver, 2009), consistent with a reduction in nutrient availability owing to increases in stratification. Model and observational studies suggest interannual and multi-decadal fluctuations in nutrients are coupled with variability of mode water and the NAO in the Atlantic Ocean (Cianca et al., 2007; Pérez et al., 2010), climate modes of variability in the Pacific Ocean (Wong et al., 2007; Di Lorenzo et al., 2009), and variability of subtropical gyre circulation in the Indian Ocean (Álvarez et al., 2011). However, there are no published studies quantifying long-term trends in ocean nutrient concentrations.

3.8.5 Conclusions

Based on high agreement between independent estimates using different methods and data sets (e.g., oceanic carbon, oxygen, and transient tracer data), it is *very likely* that the global ocean inventory of anthropogenic carbon (C_{ant}) increased from 1994 to 2010. The oceanic C_{ant} inventory in 2010 is estimated to be 155 PgC with an uncertainty of ±20%. The annual global oceanic uptake rates calculated from independent data sets (from oceanic C_{ant} inventory changes, from atmospheric O_2/N_2 measurements or from pCO_2 data) and for different time periods agree with each other within their uncertainties and *very likely* are in the range of 1.0 to 3.2 PgC yr^{-1}. (Section 3.8.1, Figures 3.16 and 3.17)

Oceanic uptake of anthropogenic CO_2 results in gradual acidification of the ocean. The pH of surface seawater has decreased by 0.1 since the beginning of the industrial era, corresponding to a 26% increase in hydrogen ion concentration. The observed pH trends range between –0.0014 and –0.0024 yr^{-1} in surface waters. In the ocean interior, natural physical and biological processes, as well as uptake of anthropogenic CO_2, can cause changes in pH over decadal and longer time scales (Section 3.8.2, Table 3.2, Box 3.2, Figures 3.18 and 3.19, FAQ 3.3).

High agreement among analyses provides *medium confidence* that oxygen concentrations have decreased in the open ocean thermocline in many ocean regions since the 1960s. The general decline is consistent with the expectation that warming-induced stratification leads to a decrease in the supply of oxygen to the thermocline from near surface

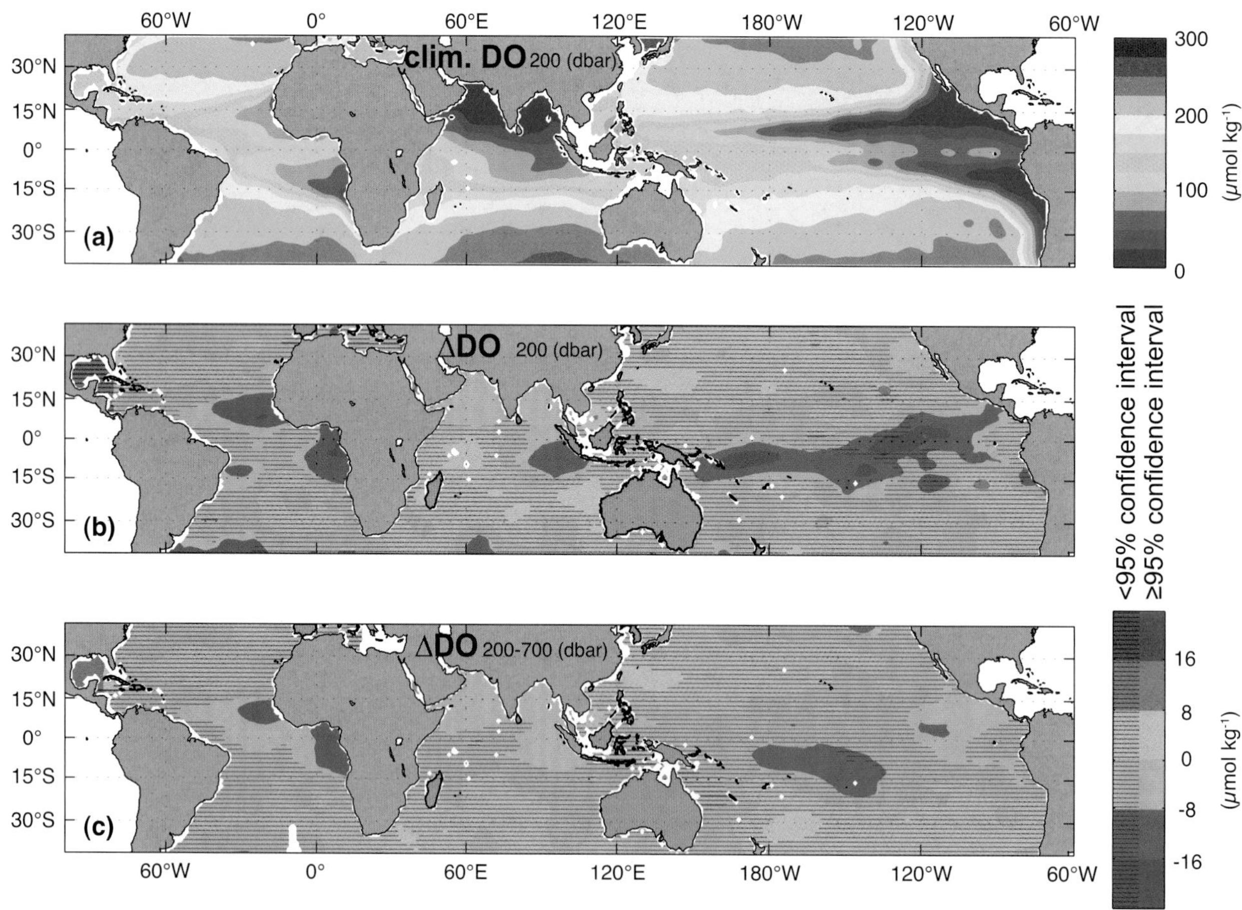

Figure 3.20 | Dissolved oxygen (DO) distributions (in µmol kg^{-1}) between 40°S and 40°N for: (a) the climatological mean (Boyer et al., 2006) at 200 dbar, as well as changes between 1960 and 1974 and 1990 and 2008 of (b) dissolved oxygen (ΔDO) at 200 dbar and (c) ΔDO vertically averaged over 200 to 700 dbar. In (b) and (c) increases are red and decreases blue, and areas with differences below the 95% confidence interval are shaded by black horizontal lines. (After Stramma et al., 2010.)

waters, that warmer waters can hold less oxygen, and that changes in wind-driven circulation affect oxygen concentrations. It is *likely* that the tropical oxygen minimum zones have expanded in recent decades (Section 3.8.3, Figure 3.20).

3.9 Synthesis

Substantial progress has been made since AR4 in documenting and understanding change in the ocean. The major findings of this chapter are largely consistent with those of AR4, but in many cases statements can now be made with greater confidence because more data are available, biases in historical data have been identified and reduced, and new analytical approaches have been applied.

Changes have been observed in a number of ocean properties of relevance to climate. It is *virtually certain* that the upper ocean (0 to 700 m) has warmed from 1971 to 2010 (Section 3.2.2, Figures 3.1 and 3.2). Warming between 700 and 2000 m *likely* contributed about 30% of the total increase in global ocean heat content between 1957 and 2009 (Section 3.2.4, Figure 3.2). Global mean sea level has risen by 0.19 [0.17 to 0.21] m over the period 1901–2010. It is *very likely* that the mean rate was 1.7 [1.5 to 1.9] mm yr^{-1} between 1901 and 2010 and increased to 3.2 [2.8 to 3.6] mm yr^{-1} between 1993 and 2010 (Section 3.7, Figure 3.13). The rise in mean sea level can explain most of the observed increase in extreme sea levels (Figure 3.15). Regional trends in sea surface salinity have *very likely* enhanced the mean geographical contrasts in sea surface salinity since the 1950s: saline surface waters in evaporation-dominated regions have become more saline, while fresh surface waters in rainfall-dominated regions have become fresher. It is *very likely* that trends in salinity have also occurred in the ocean interior. These salinity changes provide indirect evidence that the pattern of evaporation minus precipitation over the oceans has been enhanced since the 1950s (Section 3.4, Figures 3.4 and 3.5]. Observed changes in water mass properties *likely* reflect the combined effect of long-term trends in surface forcing (e.g., warming and changes in evaporation minus precipitation) and variability associated with climate modes (Section 3.5, Figure 3.9). It is *virtually certain* that the ocean is storing anthropogenic CO_2 and *very likely* that the ocean inventory of anthropogenic CO_2 increased from 1994 to 2010 (Section 3.8, Figures 3.16 and 3.17). The uptake of anthropogenic CO_2 has *very likely* caused acidification of the ocean (Section 3.8.2, Box 3.2).

For some ocean properties, the short and incomplete observational record is not sufficient to detect trends. For example, there is no observational evidence for or against a change in the strength of the AMOC (Section 3.6, Figure 3.11). However, recent observations have strengthened evidence for variability in major ocean circulation systems and water mass properties on time scales from years to decades. Much of the variability observed in ocean currents and in water masses can be linked to changes in surface forcing, including wind changes associated with the major modes of climate variability such as the NAO, SAM, ENSO, PDO and the AMO (Section 3.6, Box 2.5).

The consistency between the patterns of change in a number of independent ocean parameters enhances confidence in the assessment that the physical and biogeochemical state of the oceans has changed. This consistency is illustrated here with two simple figures (Figures 3.21 and 3.22). Four global measures of ocean change have increased since the 1950s: the inventory of anthropogenic CO_2, global mean sea level, upper ocean heat content, and the salinity contrast between regions of high and low sea surface salinity (Figure 3.21). High agreement among multiple lines of evidence based on independent data and different methods provides *high confidence* in the observed increase in these global metrics of ocean change.

The distributions of trends in subsurface water properties, summarized in a schematic zonally averaged view in Figure 3.22, are consistent both with each other and with well-understood dynamics of ocean circulation and water mass formation. The largest changes in temperature, salinity, anthropogenic CO_2, and other properties are observed along known ventilation pathways (indicated by arrows in Figure 3.22), where surface waters are transferred to the ocean interior, or in regions where changes in ocean circulation (e.g., contraction or expansion of gyres, or a southward shift of the Antarctic Circumpolar

3

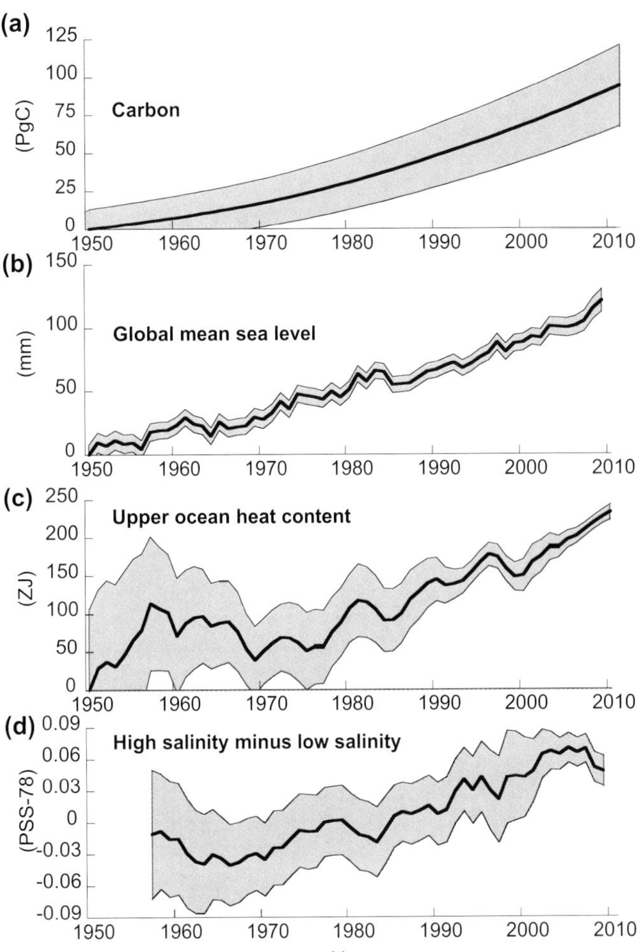

Figure 3.21 | Time series of changes in large-scale ocean climate properties. From top to bottom: global ocean inventory of anthropogenic carbon dioxide, updated from Khatiwala et al. (2009); global mean sea level (GMSL), from Church and White (2011); global upper ocean heat content anomaly, updated from Domingues et al. (2008); the difference between salinity averaged over regions where the sea surface salinity is greater than the global mean sea surface salinity ("High Salinity") and salinity averaged over regions values below the global mean ("Low Salinity"), from Boyer et al. (2009).

Current) result in large anomalies. Zonally averaged warming trends are widespread throughout the upper 2000 m, with largest warming near the sea surface. Water masses formed in the precipitation-dominated mid to high latitudes have freshened, while water masses formed in the evaporation-dominated subtropics have become saltier. Anthropogenic CO_2 has accumulated in surface waters and been transferred into the interior, primarily by water masses formed in the North Atlantic and Southern Oceans.

In summary, changes have been observed in ocean properties of relevance to climate during the past 40 years, including temperature, salinity, sea level, carbon, pH, and oxygen. The observed patterns of change are consistent with changes in the surface ocean (warming, changes in salinity and an increase in C_{ant}) in response to climate change and variability and with known physical and biogeochemical processes in the ocean, providing *high confidence* in this assessment. Chapter 10 discusses the extent to which these observed changes can be attributed to human or natural forcing.

Improvements in the quality and quantity of ocean observations has allowed for a more definitive assessment of ocean change than was possible in AR4. However, substantial uncertainties remain. In many cases, the observational record is still too short or incomplete to detect trends in the presence of energetic variability on time scales of years to decades. Recent improvements in the ocean observing system, most notably the Argo profiling float array, mean that temperature and salinity are now being sampled routinely in most of the ocean above 2000 m depth for the first time. However, sparse sampling of the deep ocean and of many biogeochemical variables continues to limit the ability to detect and understand changes in the global ocean.

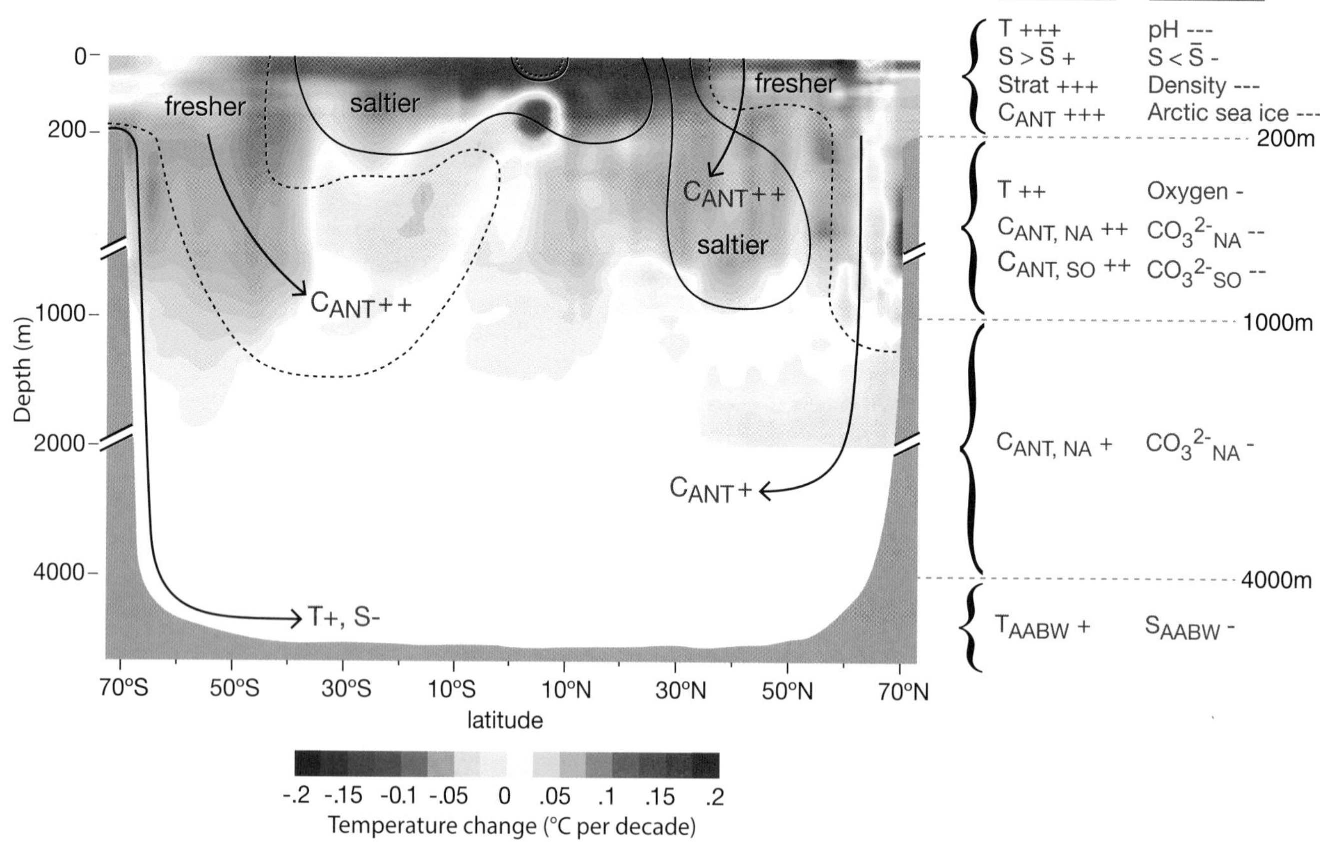

Figure 3.22 | Summary of observed changes in zonal averages of global ocean properties. Temperature trends (degrees Celsius per decade) are indicated in colour (red = warming, blue = cooling); salinity trends are indicated by contour lines (dashed = fresher; solid = saltier) for the upper 2000 m of the water column (50-year trends from data set of Durack and Wijffels (2010); trends significant at >90% confidence are shown). Arrows indicate primary ventilation pathways. Changes in other physical and chemical properties are summarised to the right of the figure, for each depth range (broken axes symbols delimit changes in vertical scale). Increases are shown in red, followed by a plus sign; decreases are shown in blue, followed by a minus sign; the number of + and – signs indicates the level of confidence associated with the observation of change (+++, *high confidence*; ++, *medium confidence*; +, *low confidence*). T = temperature, S = salinity, Strat = stratification, C_{ant} = anthropogenic carbon, CO_3^{2-} = carbonate ion, NA = North Atlantic, SO = Southern Ocean, AABW = Antarctic Bottom Water. $S > \bar{S}$ refers to the salinity averaged over regions where the sea surface salinity is greater than the global mean sea surface salinity; $S < \bar{S}$ refers to the average over regions with values below the global mean.

References

Abeysirigunawardena, D. S., and I. J. Walker, 2008: Sea level responses to climatic variability and change in Northern British Columbia. *Atmos. Ocean*, **46**, 277–296.

Ablain, M., A. Cazenave, S. Guinehut, and G. Valladeau, 2009: A new assessment of global mean sea level from altimeters highlights a reduction of global slope from 2005 to 2008 in agreement with in-situ measurements. *Ocean Sci.*, **5**, 193 - 201.

Alory, G., S. Wijffels, and G. Meyers, 2007: Observed temperature trends in the Indian Ocean over 1960–1999 and associated mechanisms. *Geophys. Res. Lett.*, **34**, L02606.

Álvarez, M., T. Tanhua, H. Brix, C. Lo Monaco, N. Metzl, E. L. McDonagh, and H. L. Bryden, 2011: Decadal biogeochemical changes in the subtropical Indian Ocean associated with Subantarctic Mode Water. *J. Geophys. Res. Oceans*, **116**, C09016.

Andersson, A., C. Klepp, K. Fennig, S. Bakan, H. Grassl, and J. Schulz, 2011: Evaluation of HOAPS-3 ocean surface freshwater flux components. *J. Appl. Meteorol. Climatol.*, **50**, 379–398.

Antonov, J. I., et al., 2010: *World Ocean Atlas 2009,* Vol. 2: *Salinity. NOAA Atlas NESDIS 68.* S. Levitus, Ed. U.S. Government Printing Office, Washington, DC, USA, 184 pp.

Aoki, S., S. R. Rintoul, S. Ushio, S. Watanabe, and N. L. Bindoff, 2005: Freshening of the Adelie Land Bottom Water near 140°E. *Geophys. Res. Lett.*, **32**, L23601.

Ballantyne, A. P., C. B. Alden, J. B. Miller, P. P. Tans, and J. W. C. White, 2012: Increase in observed net carbon dioxide uptake by land and oceans during the past 50 years. *Nature*, **488**, 70–72.

Barker, P. M., J. R. Dunn, C. M. Domingues, and S. E. Wijffels, 2011: Pressure sensor drifts in Argo and their impacts. *J. Atmos. Ocean. Technol.*, **28**, 1036–1049.

Bates, N. R., 2007: Interannual variability of the oceanic CO_2 sink in the subtropical gyre of the North Atlantic Ocean over the last 2 decades. *J. Geophys. Res. Oceans*, **112**, C09013.

Bates, N. R., 2012: Multi-decadal uptake of carbon dioxide into subtropical mode water of the North Atlantic Ocean. *Biogeosciences*, **9**, 2649–2659.

Beckley, B. D., et al., 2010: Assessment of the Jason-2 extension to the TOPEX/Poseidon, Jason-1 sea-surface height time series for global mean sea level monitoring. *Mar. Geodesy*, **33**, 447–471.

Beltrami, H., J. E. Smerdon, H. N. Pollack, and S. P. Huang, 2002: Continental heat gain in the global climate system. *Geophys. Res. Lett.*, **29**, 3.

Bersch, M., I. Yashayaev, and K. P. Koltermann, 2007: Recent changes of the thermohaline circulation in the subpolar North Atlantic. *Ocean Dyn.*, **57**, 223–235.

Bindoff, N. L., and T. J. McDougall, 1994: Diagnosing climate-change and ocean ventilation using hydrographic data. *J. Phys. Oceanogr.*, **24**, 1137–1152.

Bindoff, N. L., et al., 2007: Observations: Oceanic climate change and sea level. In: *Climate Change 2007: The Physical Science Basis. Contribution of Working Group I to the Fourth Assessment Report of the Intergovernmental Panel on Climate Change* [Solomon, S., D. Qin, M. Manning, Z. Chen, M. Marquis, K. B. Averyt, M. Tignor and H. L. Miller (eds.)] Cambridge University Press, Cambridge, United Kingdom and New York, NY, USA.

Bingham, R. J., and C. W. Hughes, 2009: Signature of the Atlantic meridional overturning circulation in sea level along the east coast of North America. *Geophys. Res. Lett.*, **36**, L02603.

Boening, C., J. K. Willis, F. W. Landerer, R. S. Nerem, and J. Fasullo, 2012: The 2011 La Niña: So strong, the oceans fell. *Geophys. Res. Lett.*, **39**, L19602.

Bograd, S. J., C. G. Castro, E. Di Lorenzo, D. M. Palacios, H. Bailey, W. Gilly, and F. P. Chavez, 2008: Oxygen declines and the shoaling of the hypoxic boundary in the California Current. *Geophys. Res. Lett.*, **35**, L12607.

Böning, C. W., A. Dispert, M. Visbeck, S. R. Rintoul, and F. U. Schwarzkopf, 2008: The response of the Antarctic Circumpolar Current to recent climate change. *Nature Geosci.*, **1**, 864–869.

Borges, A. V., and N. Gypens, 2010: Carbonate chemistry in the coastal zone responds more strongly to eutrophication than to ocean acidification. *Limnol. Oceanogr.*, **55**, 346–353.

Boyer, T., S. Levitus, J. Antonov, R. Locarnini, A. Mishonov, H. Garcia, and S. A. Josey, 2007: Changes in freshwater content in the North Atlantic Ocean 1955–2006. *Geophys. Res. Lett.*, **34**, L16603.

Boyer, T. P., S. Levitus, J. I. Antonov, R. A. Locarnini, and H. E. Garcia, 2005: Linear trends in salinity for the World Ocean, 1955–1998. *Geophys. Res. Lett.*, **32**, L01604.

Boyer, T. P., et al., 2006: Introduction. *World Ocean Database 2005 (DVD), NOAA Atlas NESDIS,* Vol. 60 [S. Levitus, (ed.)]. US Government Printing Office, Washington, DC, pp. 15–37.

Boyer, T. P., et al., 2009: Chapter 1: Introduction. *World Ocean Database 2009, NOAA Atlas NESDIS 66*, DVD ed., S. Levitus, Ed., U.S. Gov. Printing Office, Wash., D.C., USA, pp. 216.

Broecker, W., and E. Clark, 2001: A dramatic Atlantic dissolution event at the onset of the last glaciation. *Geochem. Geophys. Geosyst.*, **2**, 2001GC000185.

Bromirski, P. D., A. J. Miller, R. E. Flick, and G. Auad, 2011: Dynamical suppression of sea level rise along the Pacific coast of North America: Indications for imminent acceleration. *J. Geophys. Res. Oceans*, **116**, C07005.

Bryden, H. L., H. R. Longworth, and S. A. Cunningham, 2005: Slowing of the Atlantic meridional overturning circulation at 25°N. *Nature*, **438**, 655–657.

Byrne, R. H., S. Mecking, R. A. Feely, and X. W. Liu, 2010: Direct observations of basin-wide acidification of the North Pacific Ocean. *Geophys. Res. Lett.*, **37**, L02601.

Cai, W., 2006: Antarctic ozone depletion causes an intensification of the Southern Ocean super-gyre circulation. *Geophys. Res. Lett.*, **33**, L03712.

Calafat, F. M., D. P. Chambers, and M. N. Tsimplis, 2012: Mechanisms of decadal sea level variability in the eastern North Atlantic and the Mediterranean Sea. *J. Geophys. Res. Oceans*, **117**, C09022.

Caldeira, K., and M. E. Wickett, 2003: Anthropogenic carbon and ocean pH. *Nature*, **425**, 365–365.

Carson, M., and D. E. Harrison, 2010: Regional interdecadal variability in bias-corrected ocean temperature data. *J. Clim.*, **23**, 2847–2855.

Carton, J. A., and A. Santorelli, 2008: Global decadal upper-ocean heat content as viewed in nine analyses. *J. Clim.*, **21**, 6015–6035.

Carton, J. A., B. S. Giese, and S. A. Grodsky, 2005: Sea level rise and the warming of the oceans in the Simple Ocean Data Assimilation (SODA) ocean reanalysis. *J. Geophys. Res. Oceans*, **110**, C09006.

Cazenave, A., et al., 2009: Sea level budget over 2003–2008: A re-evaluation from GRACE space gravimetry, satellite altimetry and Argo. *Mar. Geodesy*, **65**, 447 - 471.

Cazenave, A., et al., 2012: Estimating ENSO influence on the global mean sea level, 1993–2010. *Mar. Geodesy*, **35**, 82–97.

Chambers, D. P., J. Wahr, and R. S. Nerem, 2004: Preliminary observations of global ocean mass variations with GRACE. *Geophys. Res. Lett.*, **31**, L13310.

Chambers, D. P., M. A. Merrifield, and R. S. Nerem, 2012: Is there a 60-year oscillation in global mean sea level? *Geophys. Res. Lett.*, **39**, L18607.

Chambers, D. P., J. Wahr, M. E. Tamisiea, and R. S. Nerem, 2010: Ocean mass from GRACE and glacial isostatic adjustment. *J. Geophys. Res.–Sol. Ea.*, **115**, B11415.

Chan, F., J. A. Barth, J. Lubchenco, A. Kirincich, H. Weeks, W. T. Peterson, and B. A. Menge, 2008: Emergence of anoxia in the California current large marine ecosystem. *Science*, **319**, 920–920.

Chavez, F. P., M. Messié, and J. T. Pennington, 2011: Marine primary production in relation to climate variability and change. *Annu. Rev. Mar. Sci.*, **3**, 227–260.

Church, J. A., and N. J. White, 2006: A 20th century acceleration in global sea-level rise. *Geophys. Res. Lett.*, **33**, L01602.

Church, J. A., and N. J. White, 2011: Sea-level rise from the late 19th to the early 21st century. *Surv. Geophys.*, **32**, 585–602.

Church, J. A., J. R. Hunter, K. L. McInnes, and N. J. White, 2006: Sea-level rise around the Australian coastline and the changing frequency of extreme sea-level events. *Aust. Meteorol. Mag.*, **55**, 253–260.

Church, J. A., N. J. White, R. Coleman, K. Lambeck, and J. X. Mitrovica, 2004: Estimates of the regional distribution of sea level rise over the 1950–2000 period. *J. Clim.*, **17**, 2609–2625.

Church, J. A., et al., 2011: Revisiting the Earth's sea-level and energy budgets from 1961 to 2008. *Geophys. Res. Lett.*, **38**, L18601.

Cianca, A., P. Helmke, B. Mourino, M. J. Rueda, O. Llinas, and S. Neuer, 2007: Decadal analysis of hydrography and in situ nutrient budgets in the western and eastern North Atlantic subtropical gyre. *J. Geophys. Res. Oceans*, **112**, C07025.

Compo, G. P., et al., 2011: The Twentieth Century Reanalysis Project. *Q. J. R. Meteor. Soc.*, **137**, 1–28.

Cravatte, S., T. Delcroix, D. X. Zhang, M. McPhaden, and J. Leloup, 2009: Observed freshening and warming of the western Pacific Warm Pool. *Clim. Dyn.*, **33**, 565–589.

Cummins, P. F., and H. J. Freeland, 2007: Variability of the North Pacific current and its bifurcation. *Prog. Oceanogr.*, **75**, 253–265.

Cunningham, S. A., S. G. Alderson, B. A. King, and M. A. Brandon, 2003: Transport and variability of the Antarctic Circumpolar Current in Drake Passage. *J. Geophys. Res. Oceans*, **108**, 8084.

Cunningham, S. A., et al., 2007: Temporal variability of the Atlantic meridional overturning circulation at 26.5°N. *Science*, **317**, 935–938.

Curry, R., and C. Mauritzen, 2005: Dilution of the northern North Atlantic Ocean in recent decades. *Science*, **308**, 1772–1774.

Curry, R., B. Dickson, and I. Yashayaev, 2003: A change in the freshwater balance of the Atlantic Ocean over the past four decades. *Nature*, **426**, 826–829.

D'Onofrio, E. E., M. M. E. Fiore, and J. L. Pousa, 2008: Changes in the regime of storm surges at Buenos Aires, Argentina. *J. Coast. Res.*, **24**, 260–265.

Dee, D. P., et al., 2011: The ERA-Interim reanalysis: Configuration and performance of the data assimilation system. *Q. J. R. Meteor. Soc.*, **137**, 553–597.

Delcroix, T., S. Cravatte, and M. J. McPhaden, 2007: Decadal variations and trends in tropical Pacific sea surface salinity since 1970. *J. Geophys. Res. Oceans*, **112**, C03012.

Deutsch, C., S. Emerson, and L. Thompson, 2005: Fingerprints of climate change in North Pacific oxygen. *Geophys. Res. Lett.*, **32**, L16604.

Di Lorenzo, E., et al., 2009: Nutrient and salinity decadal variations in the central and eastern North Pacific. *Geophys. Res. Lett.*, **36**, L14601.

Diaz, R. J., and R. Rosenberg, 2008: Spreading dead zones and consequences for marine ecosystems. *Science*, **321**, 926–929.

Dickson, B., I. Yashayaev, J. Meincke, B. Turrell, S. Dye, and J. Holfort, 2002: Rapid freshening of the deep North Atlantic Ocean over the past four decades. *Nature*, **416**, 832–837.

Dickson, R. R., et al., 2008: The overflow flux west of Iceland: variability, origins and forcing. In: *Arctic-Subarctic Ocean Fluxes* [R. R. Dickson, J. Meincke, and P. B. Rhines (eds.)] Springer Science+Business Media, New York, NY, USA, and Heidelberg, Germany, 443-474.

Dohan, K., et al., 2010: Measuring the global ocean surface circulation with satellite and in situ observations. In: *Proceedings of OceanObs'09: Sustained Ocean Observations and Information for Society* (Vol. 2). Venice, Italy,21-25 September 2009, Hall, J., Harrison, D.E. & Stammer, D., Eds., European Space Agency, ESA Publication WPP-306, doi:10.5270/OceanObs09.cwp.23

Domingues, C. M., J. A. Church, N. J. White, P. J. Gleckler, S. E. Wijffels, P. M. Barker, and J. R. Dunn, 2008: Improved estimates of upper-ocean warming and multi-decadal sea-level rise. *Nature*, **453**, 1090–1093.

Doney, S. C., V. J. Fabry, R. A. Feely, and J. A. Kleypas, 2009: Ocean Acidification: The other CO_2 problem. *Annu. Rev. Mar. Sci.*, **1**, 169–192.

Dore, J. E., R. Lukas, D. W. Sadler, M. J. Church, and D. M. Karl, 2009: Physical and biogeochemical modulation of ocean acidification in the central North Pacific. *Proc. Natl. Acad. Sci. U.S.A.*, **106**, 12235–12240.

Douglas, B. C., 2001: Sea level change in the era of the recording tide gauge. In: *Sea Level Rise: History and Consequences* [B. C. Douglas, M. S. Kearney, and S. P. Leatherman (eds.)]. Academic Press, San Diego,CA, USA, pp. 37–64.

Douglass, E., D. Roemmich, and D. Stammer, 2006: Interannual variability in northeast Pacific circulation. *J. Geophys. Res. Oceans*, **111**, C04001.

Drinkwater, K. F., 2006: The regime shift of the 1920s and 1930s in the North Atlantic. *Prog. Oceanogr.*, **68**, 134–151.

Duce, R. A., et al., 2008: Impacts of atmospheric anthropogenic nitrogen on the open ocean. *Science*, **320**, 893–897.

Ducet, N., P. Y. Le Traon, and G. Reverdin, 2000: Global high-resolution mapping of ocean circulation from TOPEX/Poseidon and ERS-1 and-2. *J. Geophys. Res. Oceans*, **105**, 19477–19498.

Durack, P. J., and S. E. Wijffels, 2010: Fifty-year trends in global ocean salinities and their relationship to broad-scale warming. *J. Clim.*, **23**, 4342–4362.

Durack, P. J., S. E. Wijffels, and R. J. Matear, 2012: Ocean salinities reveal strong global water cycle intensification during 1950 to 2000. *Science*, **336**, 455–458.

Egleston, E. S., C. L. Sabine, and F. M. M. Morel, 2010: Revelle revisited: Buffer factors that quantify the response of ocean chemistry to changes in DIC and alkalinity. *Global Biogeochem. Cycles*, **24**, GB1002.

Ekman, M., 1988: The world's longest continued series of sea-level observations. *Pure Appl. Geophys.*, **127**, 73–77.

Emori, S., and S. J. Brown, 2005: Dynamic and thermodynamic changes in mean and extreme precipitation under changed climate. *Geophys. Res. Lett.*, **32**, L17706.

Fabry, V. J., B. A. Seibel, R. A. Feely, and J. C. Orr, 2008: Impacts of ocean acidification on marine fauna and ecosystem processes. *Ices J. Mar. Sci.*, **65**, 414–432.

Farneti, R., T. L. Delworth, A. J. Rosati, S. M. Griffies, and F. Zeng, 2010: The role of mesoscale eddies in the rectification of the Southern Ocean response to climate change. *J. Phys. Oceanogr.*, **40**, 1539–1557.

Feely, R. A., S. C. Doney, and S. R. Cooley, 2009: Ocean acidification: Present conditions and future changes in a high-CO_2 world. *Oceanography*, **22**, 36–47.

Feely, R. A., T. Takahashi, R. Wanninkhof, M. J. McPhaden, C. E. Cosca, S. C. Sutherland, and M. E. Carr, 2006: Decadal variability of the air-sea CO_2 fluxes in the equatorial Pacific Ocean. *J. Geophys. Res. Oceans*, **111**, C08s90.

Feng, M., M. J. McPhaden, and T. Lee, 2010: Decadal variability of the Pacific subtropical cells and their influence on the southeast Indian Ocean. *Geophys. Res. Lett.*, **37**, L09606.

Fischer, J., M. Visbeck, R. Zantopp, and N. Nunes, 2010: Interannual to decadal variability of outflow from the Labrador Sea. *Geophys. Res. Lett.*, **37**, L24610.

Frajka-Williams, E., S. A. Cunningham, H. Bryden, and B. A. King, 2011: Variability of Antarctic Bottom Water at 24.5°N in the Atlantic. *J. Geophys. Res. Oceans*, **116**, C11026.

Freeland, H., et al., 2010: Argo—A decade of progress. In: *Proceedings of OceanObs'09: Sustained Ocean Observations and Information for Society* (Vol. 2). Venice, Italy,21-25 September 2009, Hall, J., Harrison, D.E. & Stammer, D., Eds., European Space Agency, ESA Publication WPP-306, doi:10.5270/OceanObs09.cwp.32

Freeland, H. J., and D. Gilbert, 2009: Estimate of the steric contribution to global sea level rise from a comparison of the WOCE one-time survey with 2006–2008 Argo observations. *Atmos. Ocean*, **47**, 292–298.

Frölicher, T. L., F. Joos, G. K. Plattner, M. Steinacher, and S. C. Doney, 2009: Natural variability and anthropogenic trends in oceanic oxygen in a coupled carbon cycle-climate model ensemble. *Global Biogeochem. Cycles*, **23**, Gb1003.

Fusco, G., V. Artale, Y. Cotroneo, and G. Sannino, 2008: Thermohaline variability of Mediterranean Water in the Gulf of Cadiz, 1948–1999. *Deep-Sea Res. Pt. I*, **55**, 1624–1638.

Galloway, J. N., et al., 2008: Transformation of the nitrogen cycle: Recent trends, questions, and potential solutions. *Science*, **320**, 889–892.

Garabato, A. C. N., L. Jullion, D. P. Stevens, K. J. Heywood, and B. A. King, 2009: Variability of Subantarctic Mode Water and Antarctic Intermediate Water in the Drake Passage during the late-twentieth and early-twenty-first centuries. *J. Clim.*, **22**, 3661–3688.

Garzoli, S. L., M. O. Baringer, S. F. Dong, R. C. Perez, and Q. Yao, 2013: South Atlantic meridional fluxes. *Deep-Sea Res. Pt. I*, **71**, 21–32.

Gebbie, G., and P. Huybers, 2012: The mean age of ocean waters inferred from radiocarbon observations: sensitivity to surface sources and accounting for mixing histories. *J. Phys. Oceanogr.*, **42**, 291–305.

Gemmrich, J., B. Thomas, and R. Bouchard, 2011: Observational changes and trends in northeast Pacific wave records. *Geophys. Res. Lett.*, **38**, L22601.

Gerber, M., F. Joos, M. Vázquez-Rodríguez, F. Touratier, and C. Goyet, 2009: Regional air-sea fluxes of anthropogenic carbon inferred with an Ensemble Kalman Filter. *Global Biogeochem. Cycles*, **23**, Gb1013.

Giese, B. S., G. A. Chepurin, J. A. Carton, T. P. Boyer, and H. F. Seidel, 2011: Impact of bathythermograph temperature bias models on an ocean reanalysis. *J. Clim.*, **24**, 84–93.

Gilbert, D., N. N. Rabalais, R. J. Diaz, and J. Zhang, 2010: Evidence for greater oxygen decline rates in the coastal ocean than in the open ocean. *Biogeosciences*, **7**, 2283–2296.

Giles, K. A., S. W. Laxon, A. L. Ridout, D. J. Wingham, and S. Bacon, 2012: Western Arctic Ocean freshwater storage increased by wind-driven spin-up of the Beaufort Gyre. *Nature Geosci.*, **5**, 194–197.

Gille, S. T., 2008: Decadal-scale temperature trends in the Southern Hemisphere ocean. *J. Clim.*, **21**, 4749–4765.

Gillett, N. P., and P. A. Stott, 2009: Attribution of anthropogenic influence on seasonal sea level pressure. *Geophys. Res. Lett.*, **36**, L23709.

Gladyshev, S. V., M. N. Koshlyakov, and R. Y. Tarakanov, 2008: Currents in the Drake Passage based on observations in 2007. *Oceanology*, **48**, 759–770.

Gleckler, P. J., et al., 2012: Human-induced global ocean warming on multidecadal timescales. *Nature Clim. Change*, **2**, 524–529.

Gloor, M., J. L. Sarmiento, and N. Gruber, 2010: What can be learned about carbon cycle climate feedbacks from the CO_2 airborne fraction? *Atmos. Chem. Phys.*, **10**, 7739–7751.

Goni, G. J., F. Bringas, and P. N. DiNezio, 2011: Observed low frequency variability of the Brazil Current front. *J. Geophys. Res. Oceans*, **116**, C10037.

González-Dávila, M., J. M. Santana-Casiano, M. J. Rueda, and O. Llinas, 2010: The water column distribution of carbonate system variables at the ESTOC site from 1995 to 2004. *Biogeosciences*, **7**, 3067–3081.

Gouretski, V., and K. P. Koltermann, 2007: How much is the ocean really warming? *Geophys. Res. Lett.*, **34**, L01610.

Gouretski, V., and F. Reseghetti, 2010: On depth and temperature biases in bathythermograph data: Development of a new correction scheme based on analysis of a global ocean database. *Deep-Sea Res. Pt. I*, **57**, 812–833.

Gouretski, V., J. Kennedy, T. Boyer, and A. Kohl, 2012: Consistent near-surface ocean warming since 1900 in two largely independent observing networks. *Geophys. Res. Lett.*, **39**, L19606.

Graven, H. D., N. Gruber, R. Key, S. Khatiwala, and X. Giraud, 2012: Changing controls on oceanic radiocarbon: New insights on shallow-to-deep ocean exchange and anthropogenic CO_2 uptake. *J. Geophys. Res. Oceans*, **117**, C10005.

Griffies, S. M., et al., 2009: Coordinated Ocean-ice Reference Experiments (COREs). *Ocean Model.*, **26**, 1–46.

Grinsted, A., J. C. Moore, and S. Jevrejeva, 2012: Homogeneous record of Atlantic hurricane surge threat since 1923. *Proc. Natl. Acad. Sci. U.S.A.*, **109**, 19601–19605.

Grist, J. P., R. Marsh, and S. A. Josey, 2009: On the relationship between the North Atlantic Meridional Overturning Circulation and the surface-forced overturning streamfunction. *J. Clim.*, **22**, 4989–5002.

Gruber, N., et al., 2009: Oceanic sources, sinks, and transport of atmospheric CO_2. *Global Biogeochem. Cycles*, **23**, Gb1005.

Gu, G. J., R. F. Adler, G. J. Huffman, and S. Curtis, 2007: Tropical rainfall variability on interannual-to-interdecadal and longer time scales derived from the GPCP monthly product. *J. Clim.*, **20**, 4033–4046.

Gulev, S., T. Jung, and E. Ruprecht, 2007: Estimation of the impact of sampling errors in the VOS observations on air-sea fluxes. Part II: Impact on trends and interannual variability. *J. Clim.*, **20**, 302–315.

Gulev, S., et al., 2010: Surface Energy and CO_2 Fluxes in the Global Ocean-Atmosphere-Ice System. In: *Proceedings of OceanObs'09: Sustained Ocean Observations and Information for Society*. Venice, Italy. 21-25 September 2009, Hall, J., Harrison, D.E. & Stammer, D., Eds., European Space Agency, ESA Publication WPP-306, doi:10.5270/OceanObs09.pp.19

Gulev, S. K., and V. Grigorieva, 2006: Variability of the winter wind waves and swell in the North Atlantic and North Pacific as revealed by the voluntary observing ship data. *J. Clim.*, **19**, 5667–5685.

Haigh, I., R. Nicholls, and N. Wells, 2010: Assessing changes in extreme sea levels: Application to the English Channel, 1900–2006. *Cont. Shelf Res.*, **30**, 1042–1055.

Hallberg, R., and A. Gnanadesikan, 2006: The role of eddies in determining the structure and response of the wind-driven Southern Hemisphere overturning: Results from the Modeling Eddies in the Southern Ocean (MESO) project. *J. Phys. Oceanogr.*, **36**, 2232–2252.

Hamon, M., G. Reverdin, and P. Y. Le Traon, 2012: Empirical correction of XBT data. *J. Atmos. Ocean. Technol.*, **29**, 960–973.

Hansen, B., and S. Osterhus, 2007: Faroe Bank Channel overflow 1995–2005. *Prog. Oceanogr.*, **75**, 817–856.

Hansen, B., H. Hatun, R. Kristiansen, S. M. Olsen, and S. Osterhus, 2010: Stability and forcing of the Iceland-Faroe inflow of water, heat, and salt to the Arctic. *Ocean Sci.*, **6**, 1013–1026.

Hatun, H., A. B. Sando, H. Drange, B. Hansen, and H. Valdimarsson, 2005: Influence of the Atlantic subpolar gyre on the thermohaline circulation. *Science*, **309**, 1841–1844.

Held, I. M., and B. J. Soden, 2006: Robust responses of the hydrological cycle to global warming. *J. Clim.*, **19**, 5686–5699.

Helm, K. P., N. L. Bindoff, and J. A. Church, 2010: Changes in the global hydrological-cycle inferred from ocean salinity. *Geophys. Res. Lett.*, **37**, L18701.

Helm, K. P., N. L. Bindoff, and J. A. Church, 2011: Observed decreases in oxygen content of the global ocean. *Geophys. Res. Lett.*, **38**, L23602.

Hemer, M. A., 2010: Historical trends in Southern Ocean storminess: Long-term variability of extreme wave heights at Cape Sorell, Tasmania. *Geophys. Res. Lett.*, **37**, L18601.

Hemer, M. A., J. A. Church, and J. R. Hunter, 2010: Variability and trends in the directional wave climate of the Southern Hemisphere. *Int. J. Climatol.*, **30**, 475–491.

Hill, K. L., S. R. Rintoul, R. Coleman, and K. R. Ridgway, 2008: Wind forced low frequency variability of the East Australia Current. *Geophys. Res. Lett.*, **35**, L08602.

Holgate, S. J., 2007: On the decadal rates of sea level change during the twentieth century. *Geophys. Res. Lett.*, **34**, L01602.

Holliday, N., et al., 2008: Reversal of the 1960s to 1990s freshening trend in the northeast North Atlantic and Nordic Seas. *Geophys. Res. Lett.*, **35**, L03614.

Hong, B. G., W. Sturges, and A. J. Clarke, 2000: Sea level on the US East Coast: Decadal variability caused by open ocean wind-curl forcing. *J. Phys. Oceanogr.*, **30**, 2088–2098.

Hood, M., et al., 2010: Ship-based repeat hydrography: A strategy for a sustained global program. In: *Proceedings of OceanObs'09: Sustained Ocean Observations and Information for Society* (Vol. 2). Venice, Italy. 21-25 September 2009, Hall, J., Harrison, D.E. & Stammer, D., Eds., European Space Agency, ESA Publication WPP-306, doi:10.5270/OceanObs09.cwp.44

Hosoda, S., T. Suga, N. Shikama, and K. Mizuno, 2009: Global surface layer salinity change detected by Argo and its implication for hydrological cycle intensification. *J. Oceanogr.*, **65**, 579–586.

Houston, J. R., and R. G. Dean, 2011: Sea-level acceleration based on US tide gauges and extensions of previous global-gauge analyses. *J. Coast. Res.*, **27**, 409–417.

Hughes, C. W., P. L. Woodworth, M. P. Meredith, V. Stepanov, T. Whitworth, and A. R. Pyne, 2003: Coherence of Antarctic sea levels, Southern Hemisphere Annular Mode, and flow through Drake Passage. *Geophys. Res. Lett.*, **30**, 1464.

Huhn, O., M. Rhein, M. Hoppema, and S. van Heuven, 2013: Decline of deep and bottom water ventilation and slowing down of anthropogenic carbon storage in the Weddell Sea, 1984–2011. *Deep-Sea Res. Pt. I*, **76**, 66–84.

Huhn, O., H. H. Hellmer, M. Rhein, C. Rodehacke, W. G. Roether, M. P. Schodlok, and M. Schröder, 2008: Evidence of deep- and bottom-water formation in the western Weddell Sea. *Deep-Sea Res. Pt. II*, **55**, 1098–1116.

Ingvaldsen, R. B., L. Asplin, and H. Loeng, 2004: Velocity field of the western entrance to the Barents Sea. *J. Geophys. Res. Oceans*, **109**, C03021.

IPCC, 2011: Workshop Report of the Intergovernmental Panel on Climate Change Workshop on Impacts of Ocean Acidification on Marine Biology and Ecosystems [C. B. Field, V. Barros, T. F. Stocker, D. Qin, K. J. Mach, G.-K. Plattner, M. D. Mastrandrea, M. Tignor and K. L. Ebi (eds.)]. IPCC Working Group II Technical Support Unit, Carnegie Institution, Stanford, CA, USA 164 pp.

Irwin, A. J., and M. J. Oliver, 2009: Are ocean deserts getting larger? *Geophys. Res. Lett.*, **36**, L18609.

Ishidoya, S., S. Aoki, D. Goto, T. Nakazawa, S. Taguchi, and P. K. Patra, 2012: Time and space variations of the O_2/N_2 ratio in the troposphere over Japan and estimation of the global CO_2 budget for the period 2000–2010. *Tellus B*, **64**, 18964.

Ishii, M., and M. Kimoto, 2009: Reevaluation of historical ocean heat content variations with time-varying XBT and MBT depth bias corrections. *J. Oceanogr.*, **65**, 287–299.

Ishii, M., N. Kosugi, D. Sasano, S. Saito, T. Midorikawa, and H. Y. Inoue, 2011: Ocean acidification off the south coast of Japan: A result from time series observations of CO_2 parameters from 1994 to 2008. *J. Geophys. Res. Oceans*, **116**, C06022.

Ishii, M., et al., 2009: Spatial variability and decadal trend of the oceanic CO_2 in the western equatorial Pacific warm/fresh water. *Deep-Sea Res. Pt. II.*, **56**, 591–606.

Jackson, J. M., E. C. Carmack, F. A. McLaughlin, S. E. Allen, and R. G. Ingram, 2010: Identification, characterization, and change of the near-surface temperature maximum in the Canada Basin, 1993–2008. *J. Geophys. Res. Oceans*, **115**, C05021.

Jacobs, S. S., and C. F. Giulivi, 2010: Large multidecadal salinity trends near the Pacific-Antarctic continental margin. *J. Clim.*, **23**, 4508–4524.

Jevrejeva, S., A. Grinsted, J. C. Moore, and S. Holgate, 2006: Nonlinear trends and multiyear cycles in sea level records. *J. Geophys. Res. Oceans*, **111**, C09012.

Jevrejeva, S., J. C. Moore, A. Grinsted, and P. L. Woodworth, 2008: Recent global sea level acceleration started over 200 years ago? *Geophys. Res. Lett.*, **35**, L08715.

Jochumsen, K., D. Quadfasel, H. Valdimarsson, and S. Jonsson, 2012: Variability of the Denmark Strait overflow: Moored time series from 1996–2011. *J. Geophys. Res. Oceans*, **117**, C12003.

Johns, W. E., et al., 2011: Continuous, array-based estimates of Atlantic Ocean heat transport at 26.5°N. *J. Clim.*, **24**, 2429–2449.

Johnson, G. C., and S. E. Wijffels, 2011: Ocean density change contributions to sea level rise. *Oceanography*, **24**, 112–121.

Johnson, G. C., S. G. Purkey, and J. L. Bullister, 2008a: Warming and freshening in the abyssal southeastern Indian Ocean. *J. Clim.*, **21**, 5351–5363.

Johnson, G. C., S. G. Purkey, and J. M. Toole, 2008b: Reduced Antarctic meridional overturning circulation reaches the North Atlantic Ocean. *Geophys. Res. Lett.*, **35**, L22601.

Jónsson, S., and H. Valdimarsson, 2012: Water mass transport variability to the North Icelandic shelf, 1994–2010. *Ices J. Mar. Sci.*, **69**, 809–815.

Josey, S. A., J. P. Grist, and R. Marsh, 2009: Estimates of meridional overturning circulation variability in the North Atlantic from surface density flux fields. *J. Geophys. Res. Oceans*, **114**, C09022.

Kalnay, E., et al., 1996: The NCEP/NCAR 40-year reanalysis project. *Bull. Am. Meteorol. Soc.*, **77**, 437–471.

Kanamitsu, M., W. Ebisuzaki, J. Woollen, S. K. Yang, J. J. Hnilo, M. Fiorino, and G. L. Potter, 2002: NCEP-DOE AMIP-II reanalysis (R-2). *Bull. Am. Meteorol. Soc.*, **83**, 1631–1643.

Kanzow, T., U. Send, and M. McCartney, 2008: On the variability of the deep meridional transports in the tropical North Atlantic. *Deep-Sea Res. Pt. I*, **55**, 1601–1623.

Kanzow, T., et al., 2007: Observed flow compensation associated with the MOC at 26.5°N in the Atlantic. *Science*, **317**, 938–941.

Kawai, Y., T. Doi, H. Tomita, and H. Sasaki, 2008: Decadal-scale changes in meridional heat transport across 24°N in the Pacific Ocean. *J. Geophys. Res. Oceans*, **113**, C08021.

Kawano, T., T. Doi, H. Uchida, S. Kouketsu, M. Fukasawa, Y. Kawai, and K. Katsumata, 2010: Heat content change in the Pacific Ocean between the 1990s and 2000s. *Deep-Sea Res. Pt. II*, **57**, 1141–1151.

Kazmin, A. S., 2012: Variability of the large-scale frontal zones: analysis of the global satellite information. *Mod. Prob. Remote Sens. Ea. Space*, **9**, 213–218 (in Russian).

Keeling, C. D., H. Brix, and N. Gruber, 2004: Seasonal and long-term dynamics of the upper ocean carbon cycle at Station ALOHA near Hawaii. *Global Biogeochem. Cycles*, **18**, GB4006.

Keeling, R.F. and A. C. Manning, 2014: Studies of Recent Changes in Atmospheric O_2 Content. In: Holland, H.D. and Turekian, K.K., eds. Treatise on Geochemistry, 2nd Edition, Volume 5, pp.385-404. Oxford: Elsevier

Keeling, R. F., A. Kortzinger, and N. Gruber, 2010: Ocean deoxygenation in a warming world. *Annu. Rev. Mar. Sci.*, **2**, 199–229.

Key, R. M., et al., 2004: A global ocean carbon climatology: Results from Global Data Analysis Project (GLODAP). *Global Biogeochem. Cycles*, **18**, Gb4031.

Khatiwala, S., F. Primeau, and T. Hall, 2009: Reconstruction of the history of anthropogenic CO_2 concentrations in the ocean. *Nature*, **462**, 346–349.

Khatiwala, S., et al., 2013: Global ocean storage of anthropogenic carbon. *Biogeosciences*, **10**, 2169–2191.

Kieke, D., M. Rhein, L. Stramma, W. M. Smethie, D. A. LeBel, and W. Zenk, 2006: Changes in the CFC inventories and formation rates of Upper Labrador Sea Water, 1997–2001. *J. Phys. Oceanogr.*, **36**, 64–86.

Kim, T. W., K. Lee, R. G. Najjar, H. D. Jeong, and H. J. Jeong, 2011: Increasing N abundance in the northwestern Pacific Ocean due to atmospheric nitrogen deposition. *Science*, **334**, 505–509.

King, M. A., M. Keshin, P. L. Whitehouse, I. D. Thomas, G. Milne, and R. E. M. Riva, 2012: Regional biases in absolute sea-level estimates from tide gauge data due to residual unmodeled vertical land movement. *Geophys. Res. Lett.*, **39**, L14604.

Knorr, W., 2009: Is the airborne fraction of anthropogenic CO_2 emissions increasing? *Geophys. Res. Lett.*, **36**, L21710.

Kobayashi, T., K. Mizuno, and T. Suga, 2012: Long-term variations of surface and intermediate waters in the southern Indian Ocean along 32°S. *J. Oceanogr.*, **68**, 243–265.

Komar, P. D., and J. C. Allan, 2008: Increasing hurricane-generated wave heights along the US East Coast and their climate controls. *J. Coast. Res.*, **24**, 479–488.

Koshlyakov, M. N., Lisina, II, E. G. Morozov, and R. Y. Tarakanov, 2007: Absolute geostrophic currents in the Drake Passage based on observations in 2003 and 2005. *Oceanology*, **47**, 451–463.

Koshlyakov, M. N., S. V. Gladyshev, R. Y. Tarakanov, and D. A. Fedorov, 2011: Currents in the western Drake Passage by the observations in January 2010. *Oceanology*, **51**, 187–198.

Kouketsu, S., M. Fukasawa, D. Sasano, Y. Kumamoto, T. Kawano, H. Uchida, and T. Doi, 2010: Changes in water properties around North Pacific intermediate water between the 1980s, 1990s and 2000s. *Deep-Sea Res. Pt. II*, **57**, 1177–1187.

Kouketsu, S., et al., 2009: Changes in water properties and transports along 24°N in the North Pacific between 1985 and 2005. *J. Geophys. Res. Oceans*, **114**, C01008.

Kouketsu, S., et al., 2011: Deep ocean heat content changes estimated from observation and reanalysis product and their influence on sea level change. *J. Geophys. Res. Oceans*, **116**, C03012.

Krueger, O., F. Schenk, F. Feser, and R. Weisse, 2013: Inconsistencies between long-term trends in storminess derived from the 20CR reanalysis and observations. *J. Clim.*, **26**, 868–874.

Large, W. G., and S. G. Yeager, 2009: The global climatology of an interannually varying air-sea flux data set. *Clim. Dyn.*, **33**, 341–364.

Large, W. G., and S. G. Yeager, 2012: On the observed trends and changes in global sea surface temperature and air–sea heat fluxes (1984–2006). *J. Clim.*, **25**, 6123–6135.

Le Quéré, C., T. Takahashi, E. T. Buitenhuis, C. Roedenbeck, and S. C. Sutherland, 2010: Impact of climate change and variability on the global oceanic sink of CO_2. *Global Biogeochem. Cycles*, **24**, Gb4007.

Le Quéré, C., et al., 2007: Saturation of the Southern Ocean CO_2 sink due to recent climate change. *Science*, **316**, 1735–8.

Lenton, A., et al., 2012: The observed evolution of oceanic pCO_2 and its drivers over the last two decades. *Global Biogeochem. Cycles*, **26**, Gb2021.

Letetrel, C., M. Marcos, B. M. Miguez, and G. Wöppelmann, 2010: Sea level extremes in Marseille (NW Mediterranean) during 1885–2008. *Cont. Shelf Res.*, **30**, 1267–1274.

Leuliette, E. W., and L. Miller, 2009: Closing the sea level rise budget with altimetry, Argo, and GRACE. *Geophys. Res. Lett.*, **36**, L04608.

Leuliette, E. W., and R. Scharroo, 2010: Integrating Jason-2 into a multiple-altimeter climate data record. *Mar. Geodesy*, **33**, 504–517.

Leuliette, E. W., and J. K. Willis, 2011: Balancing the sea level budget. *Oceanography*, **24**, 122–129.

Levitus, S., 1989: Interpentadal variability of temperature and salinity at intermediate depths of the North-Atlantic Ocean, 1970–1974 variabilityersus 1955–1959. *J. Geophys. Res. Oceans*, **94**, 6091–6131.

Levitus, S., J. I. Antonov, T. P. Boyer, R. A. Locarnini, H. E. Garcia, and A. V. Mishonov, 2009: Global ocean heat content 1955–2008 in light of recently revealed instrumentation problems. *Geophys. Res. Lett.*, **36**, L07608.

Levitus, S., et al., 2012: World ocean heat content and thermosteric sea level change (0–2000m) 1955–2010. *Geophys. Res. Lett.*, **39**, L10603.

Lewis, E. L., and N. P. Fofonoff, 1979: A practical salinity scale. *J. Phys. Oceanogr.*, **9**, 446.

Li, G., B. Ren, J. Zheng, and C. Yang, 2011: Trend singular value decomposition analysis and its application to the global ocean surface latent heat flux and SST anomalies. *J. Clim.*, **24**, 2931–2948.

Llovel, W., S. Guinehut, and A. Cazenave, 2010: Regional and interannual variability in sea level over 2002–2009 based on satellite altimetry, Argo float data and GRACE ocean mass. *Ocean Dyn.*, **60**, 1193–1204.

Llovel, W., B. Meyssignac, and A. Cazenave, 2011: Steric sea level variations over 2004–2010 as a function of region and depth: Inference on the mass component variability in the North Atlantic Ocean. *Geophys. Res. Lett.*, **38**, L15608.

Llovel, W., A. Cazenave, P. Rogel, A. Lombard, and M. B. Nguyen, 2009: Two-dimensional reconstruction of past sea level (1950–2003) from tide gauge data and an Ocean General Circulation Model. *Clim. Past*, **5**, 217–227.

Lowe, J. A., and J. M. Gregory, 2006: Understanding projections of sea level rise in a Hadley Centre coupled climate model. *J. Geophys. Res. Oceans*, **111**, C11014.

Lowe, J. A., et al., 2010: Past and future changes in extreme sea levels and waves. In: *Understanding Sea-Level Rise and Variability* [J. A. Church, P. L. Woodworth, T. Aarup, and W. S. Wilson (eds.)]. Wiley-Blackwell, New York, NY, USA, 326-375.

Lozier, M. S., and N. M. Stewart, 2008: On the temporally varying northward penetration of Mediterranean Overflow Water and eastward penetration of Labrador Sea water. *J. Phys. Oceanogr.*, **38**, 2097–2103.

Lumpkin, R., and K. Speer, 2007: Global ocean meridional overturning. *J. Phys. Oceanogr.*, **37**, 2550–2562.

Lumpkin, R., and S. Garzoli, 2011: Interannual to decadal changes in the western South Atlantic's surface circulation. *J. Geophys. Res. Oceans*, **116**, C01014.

Lyman, J. M., and G. C. Johnson, 2008: Estimating annual global upper-ocean heat content anomalies despite irregular in situ ocean sampling. *J. Clim.*, **21**, 5629–5641.

Lyman, J. M., et al., 2010: Robust warming of the global upper ocean. *Nature*, **465**, 334–337.

Macrander, A., U. Send, H. Valdimarsson, S. Jonsson, and R. H. Kase, 2005: Interannual changes in the overflow from the Nordic Seas into the Atlantic Ocean through Denmark Strait. *Geophys. Res. Lett.*, **32**, L06606.

Marcos, M., M. N. Tsimplis, and A. G. P. Shaw, 2009: Sea level extremes in southern Europe. *J. Geophys. Res. Oceans*, **114**, C01007.

Marcos, M., M. N. Tsimplis, and F. M. Calafat, 2012: Inter-annual and decadal sea level variations in the north-western Pacific marginal seas. *Prog. Oceanogr.*, **105**, 4–21.

Marshall, G. J., 2003: Trends in the southern annular mode from observations and reanalyses. *J. Clim.*, **16**, 4134–4143.

Masters, D., R. S. Nerem, C. Choe, E. Leuliette, B. Beckley, N. White, and M. Ablain, 2012: Comparison of global mean sea level time series from TOPEX/Poseidon, Jason-1, and Jason-2. *Mar. Geodesy*, **35**, 20–41.

Masuda, S., et al., 2010: Simulated rapid warming of abyssal North Pacific waters. *Science*, **329**, 319–322.

Matear, R. J., and A. C. Hirst, 2003: Long-term changes in dissolved oxygen concentrations in the ocean caused by protracted global warming. *Global Biogeochem. Cycles*, **17**, 1125.

Mauritzen, C., A. Melsom, and R. T. Sutton, 2012: Importance of density-compensated temperature change for deep North Atlantic Ocean heat uptake. *Nature Geosci.*, **5**, 905–910.

Mauritzen, C., et al., 2011: Closing the loop—Approaches to monitoring the state of the Arctic Mediterranean during the International Polar Year 2007–2008. *Prog. Oceanogr.*, **90**, 62–89.

McCarthy, G., E. McDonagh, and B. King, 2011: Decadal variability of thermocline and intermediate waters at 24°S in the South Atlantic. *J. Phys. Oceanogr.*, **41**, 157–165.

McCarthy, G., et al., 2012: Observed interannual variability of the Atlantic meridional overturning circulation at 26.5°N. *Geophys. Res. Lett.*, **39**, L19609.

McDonagh, E. L., H. L. Bryden, B. A. King, R. J. Sanders, S. A. Cunningham, and R. Marsh, 2005: Decadal changes in the south Indian Ocean thermocline. *J. Clim.*, **18**, 1575–1590.

McKinley, G. A., A. R. Fay, T. Takahashi, and N. Metzl, 2011: Convergence of atmospheric and North Atlantic carbon dioxide trends on multidecadal timescales. *Nature Geosci.*, **4**, 606–610.

McPhee, M. G., A. Proshutinsky, J. H. Morison, M. Steele, and M. B. Alkire, 2009: Rapid change in freshwater content of the Arctic Ocean. *Geophys. Res. Lett.*, **36**, L10602.

Mears, C. A., and F. J. Wentz, 2009a: Construction of the RSS V3.2 lower-tropospheric temperature dataset from the MSU and AMSU microwave sounders. *J. Atmos. Ocean. Technol.*, **26**, 1493–1509.

Mears, C. A., and F. J. Wentz, 2009b: Construction of the Remote Sensing Systems V3.2 atmospheric temperature records from the MSU and AMSU microwave sounders. *J. Atmos. Ocean. Technol.*, **26**, 1040–1056.

Meijers, A. J. S., N. L. Bindoff, and S. R. Rintoul, 2011: Frontal movements and property fluxes: Contributions to heat and freshwater trends in the Southern Ocean. *J. Geophys. Res. Oceans*, **116**, C08024.

Meinen, C. S., M. O. Baringer, and R. F. Garcia, 2010: Florida Current transport variability: An analysis of annual and longer-period signals. *Deep-Sea Res. Pt. I*, **57**, 835–846.

Menéndez, M., and P. L. Woodworth, 2010: Changes in extreme high water levels based on a quasi-global tide-gauge data set. *J. Geophys. Res. Oceans*, **115**, C10011.

Menéndez, M., F. J. Méndez, I. J. Losada, and N. E. Graham, 2008: Variability of extreme wave heights in the northeast Pacific Ocean based on buoy measurements. *Geophys. Res. Lett.*, **35**, L22607.

Meredith, M. P., P. L. Woodworth, C. W. Hughes, and V. Stepanov, 2004: Changes in the ocean transport through Drake Passage during the 1980s and 1990s, forced by changes in the Southern Annular Mode. *Geophys. Res. Lett.*, **31**, L21305.

Merrifield, M. A., 2011: A shift in western tropical Pacific sea level trends during the 1990s. *J. Clim.*, **24**, 4126–4138.

Merrifield, M. A., and M. E. Maltrud, 2011: Regional sea level trends due to a Pacific trade wind intensification. *Geophys. Res. Lett.*, **38**, L21605.

Merrifield, M. A., S. T. Merrifield, and G. T. Mitchum, 2009: An anomalous recent acceleration of global sea level rise. *J. Clim.*, **22**, 5772–5781.

Merrifield, M. A., P. R. Thompson, and M. Lander, 2012: Multidecadal sea level anomalies and trends in the western tropical Pacific. *Geophys. Res. Lett.*, **39**, L13602.

Metzl, N., 2009: Decadal increase of oceanic carbon dioxide in Southern Indian Ocean surface waters (1991–2007). *Deep-Sea Res. Pt. II*, **56**, 607–619.

Meyssignac, B., M. Becker, W. Llovel, and A. Cazenave, 2012: An assessment of two-dimensional past sea level reconstructions over 1950–2009 based on tide-gauge data and different input sea level grids. *Surv. Geophys.*, **33**, 945–972.

Midorikawa, T., K. Nemoto, H. Kamiya, M. Ishii, and H. Y. Inoue, 2005: Persistently strong oceanic CO_2 sink in the western subtropical North Pacific. *Geophys. Res. Lett.*, **32**, L05612.

Midorikawa, T., et al., 2010: Decreasing pH trend estimated from 25-yr time series of carbonate parameters in the western North Pacific. *Tellus B*, **62**, 649–659.

Mikaloff-Fletcher, S. E., et al., 2006: Inverse estimates of anthropogenic CO_2 uptake, transport, and storage by the ocean. *Global Biogeochem. Cycles*, **20**, Gb2002.

Miller, L., and B. C. Douglas, 2007: Gyre-scale atmospheric pressure variations and their relation to 19th and 20th century sea level rise. *Geophys. Res. Lett.*, **34**, L16602.

Mitas, C. M., and A. Clement, 2005: Has the Hadley cell been strengthening in recent decades? *Geophys. Res. Lett.*, **32**, L03809.

Mitchum, G. T., R. S. Nerem, M. A. Merrifield, and W. R. Gehrels, 2010: Modern sea-level-change estimates. In: *Understanding Sea-Level Rise and Variability* [J. A. Church, P. L. Woodworth, T. Aarup, and W. S. Wilson (eds.)]. Wiley-Blackwell, New York, NY, USA, 122-142.

Morison, J., R. Kwok, C. Peralta-Ferriz, M. Alkire, I. Rigor, R. Andersen, and M. Steele, 2012: Changing Arctic Ocean freshwater pathways. *Nature*, **481**, 66–70.

Morrow, R., G. Valladeau, and J. B. Sallee, 2008: Observed subsurface signature of Southern Ocean sea level rise. *Prog. Oceanogr.*, **77**, 351–366.

Murphy, D. M., S. Solomon, R. W. Portmann, K. H. Rosenlof, P. M. Forster, and T. Wong, 2009: An observationally based energy balance for the Earth since 1950. *J. Geophys. Res. Atmos.*, **114**, D17107.

Nakano, T., I. Kaneko, T. Soga, H. Tsujino, T. Yasuda, H. Ishizaki, and M. Kamachi, 2007: Mid-depth freshening in the North Pacific subtropical gyre observed along the JMA repeat and WOCE hydrographic sections. *Geophys. Res. Lett.*, **34**, L23608.

Nakanowatari, T., K. Ohshima, and M. Wakatsuchi, 2007: Warming and oxygen decrease of intermediate water in the northwestern North Pacific, originating from the Sea of Okhotsk, 1955–2004. *Geophys. Res. Lett.*, **34**, L04602.

Nerem, R. S., D. P. Chambers, C. Choe, and G. T. Mitchum, 2010: Estimating mean sea level change from the TOPEX and Jason altimeter missions. *Mar. Geodesy*, **33**, 435–446.

Nerem, R. S., D. P. Chambers, E. W. Leuliette, G. T. Mitchum, and B. S. Giese, 1999: Variations in global mean sea level associated with the 1997–1998 ENSO event: Implications for measuring long term sea level change. *Geophys. Res. Lett.*, **26**, 3005–3008.

Olafsson, J., S. R. Olafsdottir, A. Benoit-Cattin, M. Danielsen, T. S. Arnarson, and T. Takahashi, 2009: Rate of Iceland Sea acidification from time series measurements. *Biogeosciences*, **6**, 2661–2668.

Olsen, A., A. M. Omar, E. Jeansson, L. G. Anderson, and R. G. J. Bellerby, 2010: Nordic seas transit time distributions and anthropogenic CO_2. *J. Geophys. Res. Oceans*, **115**, C05005.

Olsen, S. M., B. Hansen, D. Quadfasel, and S. Osterhus, 2008: Observed and modelled stability of overflow across the Greenland-Scotland ridge. *Nature*, **455**, 519–22.

Ono, T., T. Midorikawa, Y. W. Watanabe, K. Tadokoro, and T. Saino, 2001: Temporal increases of phosphate and apparent oxygen utilization in the subsurface waters of western subarctic Pacific from 1968 to 1998. *Geophys. Res. Lett.*, **28**, 3285–3288.

Orr, J. C., 2011: Recent and future changes in ocean carbonate chemistry. In: *Ocean Acidification* [J.-P. Gattuso and L. Hansson (eds.)]. Oxford University Press, Oxford, UK, and New York, NY, USA, pp. 41–66.

Orr, J. C., S. Pantoja, and H. O. Pörtner, 2005a: Introduction to special section: The ocean in a high-CO_2 world. *J. Geophys. Res. Oceans*, **110**, C09S01.

Orr, J. C., et al., 2005b: Anthropogenic ocean acidification over the twenty-first century and its impact on calcifying organisms. *Nature*, **437**, 681–686.

Orsi, A. H., G. C. Johnson, and J. L. Bullister, 1999: Circulation, mixing, and production of Antarctic Bottom Water. *Prog. Oceanogr.*, **43**, 55–109.

Østerhus, S., W. R. Turrell, S. Jonsson, and B. Hansen, 2005: Measured volume, heat, and salt fluxes from the Atlantic to the Arctic Mediterranean. *Geophys. Res. Lett.*, **32**, L07603.

Palmer, M., and P. Brohan, 2011: Estimating sampling uncertainty in fixed-depth and fixed-isotherm estimates of ocean warming. *Int. J. Climatol.*, **31**, 980–986.

Palmer, M., K. Haines, S. Tett, and T. Ansell, 2007: Isolating the signal of ocean global warming. *Geophys. Res. Lett.*, **34**, L23610.

Park, G. H., et al., 2006: Large accumulation of anthropogenic CO_2 in the East (Japan) Sea and its significant impact on carbonate chemistry. *Global Biogeochem. Cycles*, **20**, Gb4013.

Park, J., J. Obeysekera, M. Irizarry, J. Barnes, P. Trimble, and W. Park-Said, 2011: Storm surge projections and implications for water management in South Florida. *Clim. Change*, **107**, 109–128.

Peltier, W. R., 2001: Global glacial isostatic adjustment and modern instrumental records of relative sea level history. In: *Sea Level Rise* [B. C. Douglas, M. S. Kearney, and S. P. Leatherman (eds.)]. Elsevier, Amsterdam, the Netherlands, and Philadelphia, PA, USA, pp. 65–95.

Peltier, W. R., 2004: Global glacial isostasy and the surface of the ice-age earth: The ice-5G (VM2) model and grace. *Annu. Rev. Earth Planet. Sci.*, **32**, 111–149.

Peltier, W. R., R. Drummond, and K. Roy, 2012: Comment on "Ocean mass from GRACE and glacial isostatic adjustment" by D. P. Chambers et al. *J. Geophys. Res.–Sol. Ea.*, **117**, B11403.

Pérez, F. F., M. Vázquez-Rodríguez, H. Mercier, A. Velo, P. Lherminier, and A. F. Ríos, 2010: Trends of anthropogenic CO_2 storage in North Atlantic water masses. *Biogeosciences*, **7**, 1789–1807.

Pierce, D. W., P. J. Gleckler, T. P. Barnett, B. D. Santer, and P. J. Durack, 2012: The fingerprint of human-induced changes in the ocean's salinity and temperature fields. *Geophys. Res. Lett.*, **39**, L21704.

Pierce, D. W., T. P. Barnett, K. M. AchutaRao, P. J. Gleckler, J. M. Gregory, and W. M. Washington, 2006: Anthropogenic warming of the oceans: Observations and model results. *J. Clim.*, **19**, 1873–1900.

Pinker, R. T., H. M. Wang, and S. A. Grodsky, 2009: How good are ocean buoy observations of radiative fluxes? *Geophys. Res. Lett.*, **36**, L10811.

Polovina, J. J., E. A. Howell, and M. Abecassis, 2008: Ocean's least productive waters are expanding. *Geophys. Res. Lett.*, **35**, L03618.

Polyakov, I. V., A. V. Pnyushkov, and L. A. Timokhov, 2012: Warming of the Intermediate Atlantic Water of the Arctic Ocean in the 2000s. *J. Clim.*, **25**, 8362–8370.

Polyakov, I. V., V. A. Alexeev, U. S. Bhatt, E. I. Polyakova, and X. D. Zhang, 2010: North Atlantic warming: patterns of long-term trend and multidecadal variability. *Clim. Dyn.*, **34**, 439–457.

Polyakov, I. V., U. S. Bhatt, H. L. Simmons, D. Walsh, J. E. Walsh, and X. Zhang, 2005: Multidecadal variability of North Atlantic temperature and salinity during the twentieth century. *J. Clim.*, **18**, 4562–4581.

Polyakov, I. V., et al., 2008: Arctic ocean freshwater changes over the past 100 years and their causes. *J. Clim.*, **21**, 364–384.

Ponte, R. M., 2012: An assessment of deep steric height variability over the global ocean. *Geophys. Res. Lett.*, **39**, L04601.

Potemra, J. T., and N. Schneider, 2007: Interannual variations of the Indonesian throughflow. *J. Geophys. Res. Oceans*, **112**, C05035.

Proshutinsky, A., et al., 2009: Beaufort Gyre freshwater reservoir: State and variability from observations. *J. Geophys. Res. Oceans*, **114**, C00A10.

Provoost, P., S. van Heuven, K. Soetaert, R. W. P. M. Laane, and J. J. Middelburg, 2010: Seasonal and long-term changes in pH in the Dutch coastal zone. *Biogeosciences*, **7**, 3869–3878.

Purkey, S. G., and G. C. Johnson, 2010: Warming of global abyssal and deep Southern Ocean waters between the 1990s and 2000s: Contributions to global heat and sea level rise budgets. *J. Clim.*, **23**, 6336–6351.

——, 2012: Global contraction of Antarctic Bottom Water between the 1980s and 2000s. *J. Clim.*, **25**, 5830–5844.

Purkey, S. G., and G. C. Johnson, 2013: Antarctic Bottom Water warming and freshening: Contributions to sea level rise, ocean freshwater budgets, and global heat gain. *J. Clim.*, doi:10.1175/JCLI-D-12-00834.1.

Qiu, B., and S. Chen, 2006: Decadal variability in the large-scale sea surface height field of the South Pacific Ocean: Observations and causes. *J. Phys. Oceanogr.*, **36**, 1751–1762.

Qiu, B., and S. Chen, 2010: Interannual-to-decadal variability in the bifurcation of the North Equatorial Current off the Philippines. *J. Phys. Oceanogr.*, **40**, 2525–2538.

Qiu, B., and S. Chen, 2012: Multi-decadal sea level and gyre circulation variability in the northwestern tropical Pacific Ocean. *J. Phys. Oceanogr.*, **42**, 193–206.

Quay, P., R. Sonnerup, T. Westby, J. Stutsman, and A. McNichol, 2003: Changes in the C-13/C-12 of dissolved inorganic carbon in the ocean as a tracer of anthropogenic CO_2 uptake. *Global Biogeochem. Cycles*, **17**, 1004.

Rabe, B., et al., 2011: An assessment of Arctic Ocean freshwater content changes from the 1990s to the 2006–2008 period. *Deep-Sea Res. Pt. I*, **58**, 173–185.

Rahmstorf, S., and M. Vermeer, 2011: Discussion of: Houston, J.R. and Dean, R.G., 2011. Sea-level acceleration based on U.S. tide gauges and extensions of previous global-gauge analyses. *J. Coast. Res.*, 27(3), 409–417. *J. Coast. Res.*, **27**, 784–787.

Rawlins, M. A., et al., 2010: Analysis of the Arctic System for freshwater cycle intensification: Observations and expectations. *J. Clim.*, **23**, 5715–5737.

Ray, R. D., and B. C. Douglas, 2011: Experiments in reconstructing twentieth-century sea levels. *Prog. Oceanogr.*, **91**, 496–515.

Ren, L., and S. C. Riser, 2010: Observations of decadal time scale salinity changes in the subtropical thermocline of the North Pacific Ocean. *Deep-Sea Res. Pt. II*, **57**, 1161–1170.

Reverdin, G., 2010: North Atlantic subpolar gyre surface variability (1895–2009). *J. Clim.*, **23**, 4571–4584.

Reverdin, G., F. Durand, J. Mortensen, F. Schott, H. Valdimarsson, and W. Zenk, 2002: Recent changes in the surface salinity of the North Atlantic subpolar gyre. *J. Geophys. Res. Oceans*, **107**, 8010.

Rhein, M., et al., 2011: Deep water formation, the subpolar gyre, and the meridional overturning circulation in the subpolar North Atlantic. *Deep-Sea Res. Pt. II*, **58**, 1819–1832.

Rienecker, M. M., et al., 2011: MERRA: NASA's Modern-Era Retrospective Analysis for Research and Applications. *J. Clim.*, **24**, 3624–3648.

Rignot, E., J. L. Bamber, M. R. Van Den Broeke, C. Davis, Y. H. Li, W. J. Van De Berg, and E. Van Meijgaard, 2008: Recent Antarctic ice mass loss from radar interferometry and regional climate modelling. *Nature Geosci.*, **1**, 106–110.

Rintoul, S. R., 2007: Rapid freshening of Antarctic Bottom Water formed in the Indian and Pacific oceans. *Geophys. Res. Lett.*, **34**, L06606.

Rintoul, S. R., S. Sokolov, and J. Church, 2002: A 6 year record of baroclinic transport variability of the Antarctic Circumpolar Current at 140°E derived from expendable bathythermograph and altimeter measurements. *J. Geophys. Res. Oceans*, **107**, 3155.

Robertson, R., M. Visbeck, A. L. Gordon, and E. Fahrbach, 2002: Long-term temperature trends in the deep waters of the Weddell Sea. *Deep-Sea Res. Pt. II*, **49**, 4791–4806.

Roemmich, D., and J. Gilson, 2009: The 2004–2008 mean and annual cycle of temperature, salinity, and steric height in the global ocean from the Argo Program. *Prog. Oceanogr.*, **82**, 81–100.

Roemmich, D., and J. Gilson, 2011: The global ocean imprint of ENSO. *Geophys. Res. Lett.*, **38**, L13606.

Roemmich, D., W. J. Gould, and J. Gilson, 2012: 135 years of global ocean warming between the *Challenger* expedition and the Argo Programme. *Nature Clim. Change*, **2**, 425–428.

Roemmich, D., J. Gilson, R. Davis, P. Sutton, S. Wijffels, and S. Riser, 2007: Decadal spinup of the South Pacific subtropical gyre. *J. Phys. Oceanogr.*, **37**, 162–173.

Ruggiero, P., P. D. Komar, and J. C. Allan, 2010: Increasing wave heights and extreme value projections: The wave climate of the U.S. Pacific Northwest. *Coast. Engng.*, **57**, 539–552.

Sabine, C. L., et al., 2005: Global Ocean Data Analysis Project (GLODAP): Results and data. ORNL/CDIAC-145, NDP-083. Carbon Dioxide Information Analysis Center, Oak Ridge National Laboratory, U.S. Department of Energy, 110 pp.

Sabine, C. L., et al., 2004: The oceanic sink for anthropogenic CO_2. *Science*, **305**, 367–371.

Saha, S., et al., 2010: The NCEP Climate Forecast System Reanalysis. *Bull. Am. Meteorol. Soc.*, **91**, 1015–1057.

Sallenger, A. H., K. S. Doran, and P. A. Howd, 2012: Hotspot of accelerated sea-level rise on the Atlantic coast of North America. *Nature Clim. Change*, **2**, 884–888.

Santana-Casiano, J. M., M. González-Dávila, M. J. Rueda, O. Llinas, and E. F. González-Dávila, 2007: The interannual variability of oceanic CO_2 parameters in the northeast Atlantic subtropical gyre at the ESTOC site. *Global Biogeochem. Cycles*, **21**, GB1015.

Sarafanov, A., A. Falina, A. Sokov, and A. Demidov, 2008: Intense warming and salinification of intermediate waters of southern origin in the eastern subpolar North Atlantic in the 1990s to mid-2000s. *J. Geophys. Res. Oceans*, **113**, C12022.

Sarmiento, J. L., et al., 2010: Trends and regional distributions of land and ocean carbon sinks. *Biogeosciences*, **7**, 2351–2367.

Schanze, J. J., R. W. Schmitt, and L. L. Yu, 2010: The global oceanic freshwater cycle: A state-of-the-art quantification. *J. Mar. Res.*, **68**, 569–595.

Schauer, U., and A. Beszczynska-Möller, 2009: Problems with estimation and interpretation of oceanic heat transport – conceptual remarks for the case of Fram Strait in the Arctic Ocean. *Ocean Sci.*, **5**, 487–494.

Schmidtko, S., and G. C. Johnson, 2012: Multi-decadal warming and shoaling of Antarctic Intermediate Water. *J. Clim.*, **25**, 201–221.

Schmitt, R. W., 2008: Salinity and the global water cycle. *Oceanography*, **21**, 12–19.

Schneider, T., P. A. O'Gorman, and X. J. Levine, 2010: Water vapor and the dynamics of climate changes. *Rev. Geophys.*, **48**, Rg3001.

Schuster, U., and A. J. Watson, 2007: A variable and decreasing sink for atmospheric CO_2 in the North Atlantic. *J. Geophys. Res. Oceans*, **112**, C11006.

Schuster, U., et al., 2013: An assessment of the Atlantic and Arctic sea-air CO_2 fluxes, 1990–2009. *Biogeosciences*, **10**, 607–627.

Seitzinger, S. P., et al., 2010: Global river nutrient export: A scenario analysis of past and future trends. *Global Biogeochem. Cycles*, **24**, Gb0a08.

Semedo, A., K. Suselj, A. Rutgersson, and A. Sterl, 2011: A global view on the wind sea and swell climate and variability from ERA-40. *J. Clim.*, **24**, 1461–1479.

Send, U., M. Lankhorst, and T. Kanzow, 2011: Observation of decadal change in the Atlantic Meridional Overturning Circulation using 10 years of continuous transport data. *Geophys. Res. Lett.*, **38**, L24606.

Shepherd, A., D. Wingham, and E. Rignot, 2004: Warm ocean is eroding West Antarctic Ice Sheet. *Geophys. Res. Lett.*, **31**, L23402.

Shiklomanov, A. I., and R. B. Lammers, 2009: Record Russian river discharge in 2007 and the limits of analysis. *Environ. Res. Lett.*, **4**, 045015.

Smith, D. M., and J. M. Murphy, 2007: An objective ocean temperature and salinity analysis using covariances from a global climate model. *J. Geophys. Res. Oceans*, **112**, C02022.

Smith, R. O., H. L. Bryden, and K. Stansfield, 2008: Observations of new western Mediterranean deep water formation using Argo floats 2004–2006. *Ocean Sci.*, **4**, 133–149.

Smith, T. M., P. A. Arkin, and M. R. P. Sapiano, 2009: Reconstruction of near-global annual precipitation using correlations with sea surface temperature and sea level pressure. *J. Geophys. Res. Atmos.*, **114**, D12107.

Smith, T. M., P. A. Arkin, L. Ren, and S. S. P. Shen, 2012: Improved reconstruction of global precipitation since 1900. *J. Atmos. Ocean. Technol.*, **29**, 1505–1517.

Sokolov, S., and S. R. Rintoul, 2009: Circumpolar structure and distribution of the Antarctic Circumpolar Current fronts: 2. Variability and relationship to sea surface height. *J. Geophys. Res. Oceans*, **114**, C11019.

Spada, G., and G. Galassi, 2012: New estimates of secular sea level rise from tide gauge data and GIA modelling. *Geophys. J. Int.*, **191**, 1067–1094.

Spence, P., J. C. Fyfe, A. Montenegro, and A. J. Weaver, 2010: Southern Ocean response to strengthening winds in an eddy-permitting global climate model. *J. Clim.*, **23**, 5332–5343.

Sprintall, J., S. Wijffels, T. Chereskin, and N. Bray, 2002: The JADE and WOCE I10/IR6 Throughflow sections in the southeast Indian Ocean. Part 2: velocity and transports. *Deep-Sea Res. Pt. II*, **49**, 1363–1389.

Sprintall, J., S. E. Wijffels, R. Molcard, and I. Jaya, 2009: Direct estimates of the Indonesian Throughflow entering the Indian Ocean: 2004–2006. *J. Geophys. Res. Oceans*, **114**, C07001.

Stammer, D. et al, 2010: Ocean Information Provided Through Ensemble Ocean Syntheses in *Proceedings of OceanObs'09: Sustained Ocean Observations and Information for Society (Vol. 2)*, Venice, Italy, 21-25 September 2009, Hall, J., Harrison, D.E. & Stammer, D., Eds., European Space Agency, ESA Publication WPP-306, doi:10.5270/OceanObs09.cwp.85

Steele, M., and W. Ermold, 2007: Steric sea level change in the Northern Seas. *J. Clim.*, **20**, 403–417.

Steinfeldt, R., M. Rhein, J. L. Bullister, and T. Tanhua, 2009: Inventory changes in anthropogenic carbon from 1997–2003 in the Atlantic Ocean between 20°S and 65°N. *Global Biogeochem. Cycles*, **23**, GB3010.

Stendardo, I., and N. Gruber, 2012: Oxygen trends over five decades in the North Atlantic. *J. Geophys. Res. Oceans*, **117**, C11004.

Sterl, A., and S. Caires, 2005: Climatology, variability and extrema of ocean waves: The web-based KNMI/ERA-40 wave atlas. *Int. J. Climatol.*, **25**, 963–977.

Stott, P. A., R. T. Sutton, and D. M. Smith, 2008: Detection and attribution of Atlantic salinity changes. *Geophys. Res. Lett.*, **35**, L21702.

Stramma, L., A. Oschlies, and S. Schmidtko, 2012: Mismatch between observed and modeled trends in dissolved upper-ocean oxygen over the last 50 yr. *Biogeosciences*, **9**, 4045–4057.

Stramma, L., G. C. Johnson, J. Sprintall, and V. Mohrholz, 2008: Expanding oxygen-minimum zones in the tropical oceans. *Science*, **320**, 655–658.

Stramma, L., S. Schmidtko, L. A. Levin, and G. C. Johnson, 2010: Ocean oxygen minima expansions and their biological impacts. *Deep-Sea Res. Pt. I*, **57**, 587–595.

Sturges, W., and B. G. Hong, 1995: Wind forcing of the Atlantic thermocline along 32°N at low-frequencies. *J. Phys. Oceanogr.*, **25**, 1706–1715.

Sturges, W., and B. C. Douglas, 2011: Wind effects on estimates of sea level rise. *J. Geophys. Res. Oceans*, **116**, C06008.

Sugimoto, S., and K. Hanawa, 2010: The wintertime wind stress curl field in the North Atlantic and its relation to atmospheric teleconnection patterns. *J. Atmos. Sci.*, **67**, 1687–1694.

Susanto, R. D., A. Ffield, A. L. Gordon, and T. R. Adi, 2012: Variability of Indonesian throughflow within Makassar Strait, 2004–2009. *J. Geophys. Res. Oceans*, **117**, C09013.

Swart, N. C., and J. C. Fyfe, 2012: Observed and simulated changes in the Southern Hemisphere surface westerly wind-stress. *Geophys. Res. Lett.*, **39**, L16711.

Swart, S., S. Speich, I. J. Ansorge, G. J. Goni, S. Gladyshev, and J. R. E. Lutjeharms, 2008: Transport and variability of the Antarctic Circumpolar Current South of Africa. *J. Geophys. Res. Oceans*, **113**, C09014.

Swift, J. H., and A. H. Orsi, 2012: Sixty-four days of hydrography and storms: RVIB Nathaniel B. Palmer's 2011 S04P Cruise. *Oceanography*, **25**, 54–55.

Takahashi, T., S. C. Sutherland, R. A. Feely, and R. Wanninkhof, 2006: Decadal change of the surface water pCO_2 in the North Pacific: A synthesis of 35 years of observations. *J. Geophys. Res. Oceans*, **111**, C07s05.

Takahashi, T., et al., 2009: Climatological mean and decadal change in surface ocean pCO_2, and net sea-air CO_2 flux over the global oceans (vol 56, pg 554, 2009). *Deep-Sea Res. Pt. I*, **56**, 2075–2076.

Tanaka, H. L., N. Ishizaki, and A. Kitoh, 2004: Trend and interannual variability of Walker, monsoon and Hadley circulations defined by velocity potential in the upper troposphere. *Tellus A*, **56**, 250–269.

Tanhua, T., E. P. Jones, E. Jeansson, S. Jutterstrom, W. M. Smethie, D. W. R. Wallace, and L. G. Anderson, 2009: Ventilation of the Arctic Ocean: Mean ages and inventories of anthropogenic CO_2 and CFC-11. *J. Geophys. Res. Oceans*, **114**, C01002.

Terray, L., L. Corre, S. Cravatte, T. Delcroix, G. Reverdin, and A. Ribes, 2012: Near-surface salinity as nature's rain gauge to detect human influence on the tropical water cycle. *J. Clim.*, **25**, 958–977.

Timmermann, A., S. McGregor, and F. F. Jin, 2010: Wind effects on past and future regional sea level trends in the southern Indo-Pacific. *J. Clim.*, **23**, 4429–4437.

Tokinaga, H., and S.-P. Xie, 2011: Wave- and Anemometer-based Sea surface Wind (WASWind) for climate change analysis. *J. Clim.*, **24**, 267–285.

Toole, J. M., R. G. Curry, T. M. Joyce, M. McCartney, and B. Pena-Molino, 2011: Transport of the North Atlantic Deep Western Boundary Current about 39°N, 70°W: 2004–2008. *Deep-Sea Res. Pt. II*, **58**, 1768–1780.

Trenberth, K. E., and L. Smith, 2005: The mass of the atmosphere: A constraint on global analyses. *J. Clim.*, **18**, 864–875.

Trenberth, K. E., J. T. Fasullo, and J. Kiehl, 2009: Earth's global energy budget. *Bull. Am. Meteorol. Soc.*, **90**, 311–323.

Trenberth, K. E., J. T. Fasullo, and J. Mackaro, 2011: Atmospheric moisture transports from ocean to land and global energy flows in reanalyses. *J. Clim.*, **24**, 4907–4924.

Trenberth, K. E., et al., 2007: Observations: Surface and atmospheric climate change. In: *Climate Change 2007: The Physical Science Basis. Contribution of Working Group I to the Fourth Assessment Report of the Intergovernmental Panel on Climate Change* [Solomon, S., D. Qin, M. Manning, Z. Chen, M. Marquis, K. B. Averyt, M. Tignor and H. L. Miller (eds.)] Cambridge University Press, Cambridge, United Kingdom and New York, NY, USA.

Tsimplis, M. N., and A. G. P. Shaw, 2010: Seasonal sea level extremes in the Mediterranean Sea and at the Atlantic European coasts. *Nat. Hazards Earth Syst. Sci.*, **10**, 1457–1475.

Uppala, S. M., et al., 2005: The ERA-40 re-analysis. *Q. J. R. Meteor. Soc.*, **131**, 2961–3012.

Våge, K., et al., 2009: Surprising return of deep convection to the subpolar North Atlantic Ocean in winter 2007–2008. *Nature Geosci.*, **2**, 67–72.

Valdimarsson, H., O. S. Astthorsson, and J. Palsson, 2012: Hydrographic variability in Icelandic waters during recent decades and related changes in distribution of some fish species. *Ices J. Mar. Sci.*, **69**, 816–825.

Vargas-Yáñez, M., et al., 2010: How much is the western Mediterranean really warming and salting? *J. Geophys. Res. Oceans*, **115**, C04001.

Vecchi, G. A., B. J. Soden, A. T. Wittenberg, I. M. Held, A. Leetmaa, and M. J. Harrison, 2006: Weakening of tropical Pacific atmospheric circulation due to anthropogenic forcing. *Nature*, **441**, 73–76.

Vilibic, I., and J. Sepic, 2010: Long-term variability and trends of sea level storminess and extremes in European Seas. *Global Planet. Change*, **71**, 1–12.

von Schuckmann, K., and P. Y. Le Traon, 2011: How well can we derive Global Ocean Indicators from Argo data? *Ocean Sci.*, **7**, 783–791.

Wainwright, L., G. Meyers, S. Wijffels, and L. Pigot, 2008: Change in the Indonesian Throughflow with the climatic shift of 1976/77. *Geophys. Res. Lett.*, **35**, L03604.

Wang, C. Z., S. F. Dong, and E. Munoz, 2010: Seawater density variations in the North Atlantic and the Atlantic meridional overturning circulation. *Clim. Dyn.*, **34**, 953–968.

Wang, X., Y. Feng, and V. R. Swail, 2012: North Atlantic wave height trends as reconstructed from the Twentieth Century Reanalysis. *Geophys. Res. Lett.*, **39**, L18705.

Wang, X. L. L., and V. R. Swail, 2006: Climate change signal and uncertainty in projections of ocean wave heights. *Clim. Dyn.*, **26**, 109–126.

Wang, X. L. L., V. R. Swail, F. W. Zwiers, X. B. Zhang, and Y. Feng, 2009: Detection of external influence on trends of atmospheric storminess and northern oceans wave heights. *Clim. Dyn.*, **32**, 189–203.

Wanninkhof, R., W. E. Asher, D. T. Ho, C. Sweeney, and W. R. McGillis, 2009: Advances in quantifying air-sea gas exchange and environmental forcing. *Annu. Rev. Mar. Sci.*, **1**, 213–244.

Wanninkhof, R., S. C. Doney, J. L. Bullister, N. M. Levine, M. Warner, and N. Gruber, 2010: Detecting anthropogenic CO_2 changes in the interior Atlantic Ocean between 1989 and 2005. *J. Geophys. Res.*, **115**, C11028.

Wanninkhof, R., G. H. Park, T. Takahashi, R. A. Feely, J. L. Bullister, and S. C. Doney, 2013: Changes in deep-water CO_2 concentrations over the last several decades determined from discrete pCO_2 measurements. *Deep-Sea Res. Pt. I*, **74**, 48–63.

WASA-Group, 1998: Changing waves and storm in the Northern Atlantic? *Bull. Am. Meteorol. Soc.*, **79**, 741–760.

Watson, A. J., et al., 2009: Tracking the variable North Atlantic sink for atmospheric CO_2. *Science*, **326**, 1391–1393.

Watson, P. J., 2011: Is there evidence yet of an acceleration in mean sea level rise around mainland Australia? *J. Coast. Res.*, **27**, 368–377.

Waugh, D. M., F. Primeau, T. DeVries, and M. Holzer, 2013: Recent changes in the ventilation of the Southern Oceans. *Science*, **339**, 568–570.

Waugh, D. W., T. M. Hall, B. I. McNeil, R. Key, and R. J. Matear, 2006: Anthropogenic CO_2 in the oceans estimated using transit-time distributions. *Tellus B*, **58**, 376–389.

Wentz, F. J., and L. Ricciardulli, 2011: Comment on "Global trends in wind speed and wave height." *Science*, **334**, 905–905.

Wentz, F. J., L. Ricciardulli, K. Hilburn, and C. Mears, 2007: How much more rain will global warming bring? *Science*, **317**, 233–235.

Wenzel, M., and J. Schroeter, 2010: Reconstruction of regional mean sea level anomalies from tide gauges using neural networks. *J. Geophys. Res. Oceans*, **115**, C08013.

Whitney, F. A., H. J. Freeland, and M. Robert, 2007: Persistently declining oxygen levels in the interior waters of the eastern subarctic Pacific. *Prog. Oceanogr.*, **75**, 179–199.

Wijffels, S. E., et al., 2008: Changing expendable bathythermograph fall rates and their impact on estimates of thermosteric sea level rise. *J. Clim.*, **21**, 5657–5672.

Wild, M., 2009: Global dimming and brightening: A review. *J. Geophys. Res. Atmos.*, **114**, D00D16.

Willis, J. K., 2010: Can in situ floats and satellite altimeters detect long-term changes in Atlantic Ocean overturning? *Geophys. Res. Lett.*, **37**, L06602.

Willis, J. K., D. Roemmich, and B. Cornuelle, 2004: Interannual variability in upper ocean heat content, temperature, and thermosteric expansion on global scales. *J. Geophys. Res. Oceans*, **109**, C12036.

Willis, J. K., D. P. Chambers, and R. S. Nerem, 2008: Assessing the globally averaged sea level budget on seasonal to interannual timescales. *J. Geophys. Res. Oceans*, **113**, C06015.

Willis, J. K., D. P. Chambers, C.-Y. Kuo, and C. K. Shum, 2010: Global sea level rise: Recent progress and challenges for the decade to come. *Oceanography*, **23**, 26–35.

Wong, A. P. S., N. L. Bindoff, and J. A. Church, 1999: Large-scale freshening of intermediate waters in the Pacific and Indian oceans. *Nature*, **400**, 440–443.

Wong, C. S., L. S. Xie, and W. W. Hsieh, 2007: Variations in nutrients, carbon and other hydrographic parameters related to the 1976/77 and 1988/89 regime shifts in the sub-arctic Northeast Pacific. *Prog. Oceanogr.*, **75**, 326–342.

Woodworth, P. L., 1990: A search for accelerations in records of European mean sea-level. *Int. J. Climatol.*, **10**, 129–143.

Woodworth, P. L., 1999: High waters at Liverpool since 1768: the UK's longest sea level record. *Geophys. Res. Lett.*, **26**, 1589–1592.

Woodworth, P. L., and D. L. Blackman, 2004: Evidence for systematic changes in extreme high waters since the mid-1970s. *J. Clim.*, **17**, 1190–1197.

Woodworth, P. L., N. Pouvreau, and G. Woeppelmann, 2010: The gyre-scale circulation of the North Atlantic and sea level at Brest. *Ocean Sci.*, **6**, 185–190.

Woodworth, P. L., M. Menéndez, and W. R. Gehrels, 2011: Evidence for century-timescale acceleration in mean sea levels and for recent changes in extreme sea levels. *Surv. Geophys.*, **32**, 603–618.

Woodworth, P. L., N. J. White, S. Jevrejeva, S. J. Holgate, J. A. Church, and W. R. Gehrels, 2009: Evidence for the accelerations of sea level on multi-decade and century timescales. *Int. J. Climatol.*, **29**, 777–789.

Wöppelmann, G., et al., 2009: Rates of sea-level change over the past century in a geocentric reference frame. *Geophys. Res. Lett.*, **36**, L12607.

Wu, L., et al., 2012: Enhanced warming over the global subtropical western boundary currents. *Nature Clim. Change*, **2**, 161–166.

Wunsch, C., 2010: Variability of the Indo-Pacific Ocean exchanges. *Dyn. Atmos. Oceans*, **50**, 157–173.

Xue, Y., B. Huang, Z.-Z. Hu, A. Kumar, C. Wen, D. Behringer, and S. Nadiga, 2010: An assessment of oceanic variability in the NCEP climate forecast system reanalysis. *Clim. Dyn.*, **37**, 2541–2550.

Xue, Y., et al., 2012: A comparative analysis of upper ocean heat content variability from an ensemble of operational ocean reanalyses. *J. Clim.*, **25**, 6905–6929.

Yamamoto-Kawai, M., F. A. McLaughlin, E. C. Carmack, S. Nishino, K. Shimada, and N. Kurita, 2009: Surface freshening of the Canada Basin, 2003–2007: River runoff versus sea ice meltwater. *J. Geophys. Res. Oceans*, **114**, C00A05.

Yang, X. Y., R. X. Huang, and D. X. Wang, 2007: Decadal changes of wind stress over the Southern Ocean associated with Antarctic ozone depletion. *J. Clim.*, **20**, 3395–3410.

Yashayaev, I., 2007: Hydrographic changes in the Labrador Sea, 1960–2005. *Prog. Oceanogr.*, **73**, 242–276.

Yashayaev, I., and J. W. Loder, 2009: Enhanced production of Labrador Sea Water in 2008. *Geophys. Res. Lett.*, **36**, L01606.

Yoshikawa-Inoue, H., and M. Ishii, 2005: Variations and trends of CO_2 in the surface seawater in the Southern Ocean south of Australia between 1969 and 2002. *Tellus B*, **57**, 58–69.

Young, I. R., S. Zieger, and A. V. Babanin, 2011a: Global trends in wind speed and wave height. *Science*, **332**, 451–455.

Young, I. R., A. V. Babanin, and S. Zieger, 2011b: Response to comment on "Global trends in wind speed and wave height". *Science*, **334**, 905–905.

Yu, L., 2011: A global relationship between the ocean water cycle and near-surface salinity. *J. Geophys. Res. Oceans*, **116**, C10025.

Yu, L., and R. A. Weller, 2007: Objectively analyzed air-sea flux fields for the global ice-free oceans (1981–2005). *Bull. Am. Meteorol. Soc.*, **88**, 527–539.

Yu, L. S., 2007: Global variations in oceanic evaporation (1958–2005): The role of the changing wind speed. *J. Clim.*, **20**, 5376–5390.

Yu, L. S., X. Z. Jin, and R. A. Weller, 2007: Annual, seasonal, and interannual variability of air-sea heat fluxes in the Indian Ocean. *J. Clim.*, **20**, 3190–3209.

Zenk, W., and E. Morozov, 2007: Decadal warming of the coldest Antarctic Bottom Water flow through the Vema Channel. *Geophys. Res. Lett.*, **34**, L14607.

Zhang, X. B., and J. A. Church, 2012: Sea level trends, interannual and decadal variability in the Pacific Ocean. *Geophys. Res. Lett.*, **39**, L21701.

3

Appendix 3.A: Availability of Observations for Assessment of Change in the Oceans

Sampling of the ocean has been highly heterogeneous since 1950. The coverage in space, time, depth and number of ocean variables has evolved over time, reflecting changes in technology and the contribution of major oceanographic research programs. Changes in the distribution and quality of ocean measurements over time complicate efforts to detect and interpret change in the ocean. This Appendix provides some illustrative examples of the evolution of the ocean observing system on which the assessment of ocean change in this chapter is based. A more comprehensive discussion of ocean sampling is provided in the literature cited in this chapter. Sampling of sea surface temperature is discussed in Chapter 2.

3.A.1 Subsurface Ocean Temperature and Heat Content

Temperature is the best-sampled oceanographic variable, but even for temperature sampling is far from ideal or complete. Early oceanographic expeditions included the *Challenger* voyage around the world in the 1870s, the *Meteor* survey of the Atlantic in the 1920s, and the *Discovery* investigations of the Southern Ocean starting in the 1920s. More frequent basin-scale sampling commenced in the late 1950s with the International Geophysical Year. The number of profiles available for assessment of changes in temperature and ocean heat content in the upper 700 m generally increases with time since the 1950s (Figure 3.A.1). Near-global coverage of the upper half of the ocean was not achieved until the widespread deployment of Argo profiling floats in the 2000s (Figure 3.A.2).

Early measurements of temperature were made using reversing thermometers and Nansen bottles that were lowered from ships on station (not moving). Starting in the 1960s conductivity-temperature-depth (CTD) instruments with Niskin bottles gradually gained dominance for high-quality data and deep data collected on station during oceanographic cruises. From at least 1950 through about 1970, most subsurface measurements of ocean temperature were made with mechanical bathythermographs, an advance because these instruments could be deployed from a moving ship, albeit a slowly moving one, but these casts were generally limited to depths shallower than 250 m. Expendable bathythermographs (XBTs) that could be deployed from a rapidly moving ship and sampled to 400 m came into widespread use in the late 1960s, and those that sampled to 700 m became predominant in the 1990s, greatly expanding oceanographic sampling. Starting in 2000, Argo floats began sampling the ocean to a target depth of 2000 m, building to near-global coverage by 2005. Prior to the Argo era, sampling of the ocean below 700 m was almost solely achieved from ships on station deploying Nansen bottles with reversing thermometers or later using CTDs with Niskin bottles. Today ship-based station data still dominates sampling for waters deeper than 2000 m depth (the maximum depth currently sampled by Argo floats). An illustration of the limited data available for assessment of change in the deep ocean is provided in Figure 3.3, which shows locations of full-depth oceanographic CTD sections that have been occupied more than once since about 1980. The depth coverage of the ocean observing system has changed over time (Figure 3A.2, top panel) with a hemispheric bias (Figure 3.A.2, middle and lower panels). The Northern Hemisphere (NH) has been consistently better sampled than the Southern Hemisphere (SH) prior to the Argo era.

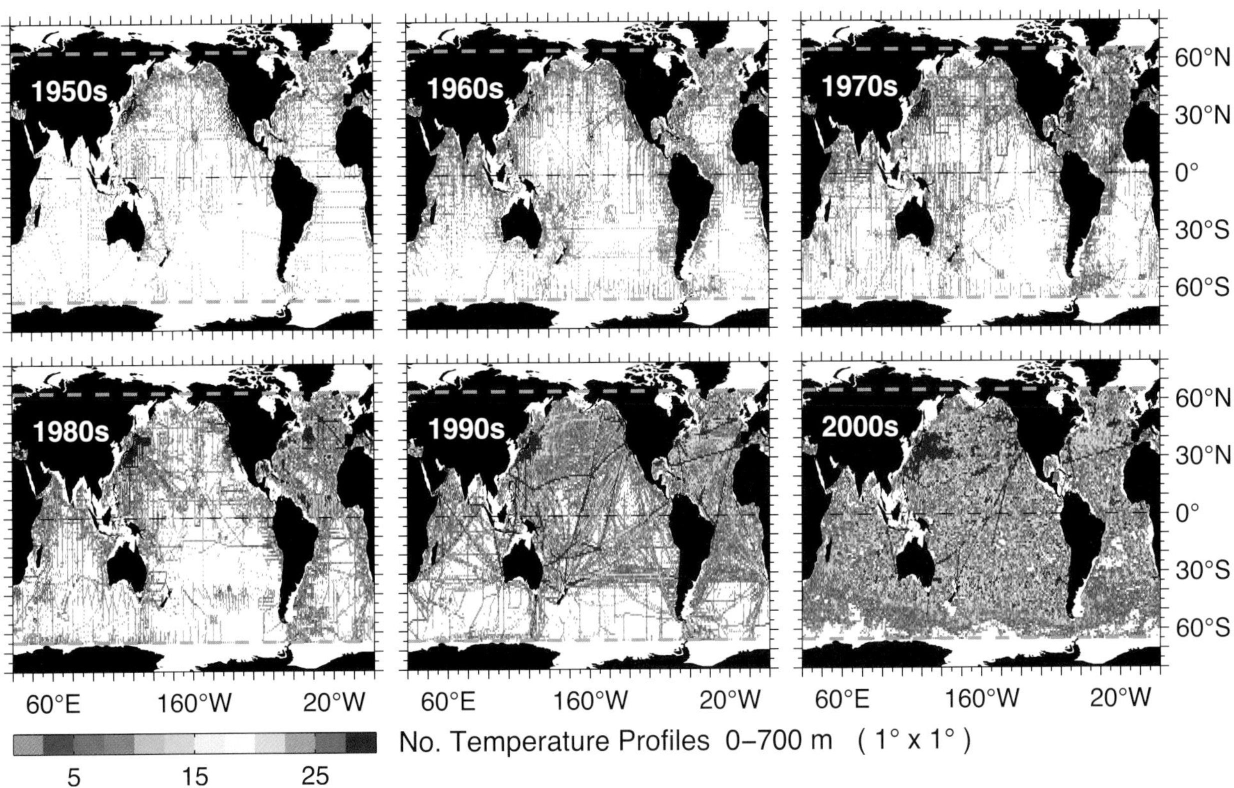

Figure 3.A.1 | Number of temperature profiles extending to 700 m depth in each 1° × 1° square, by decade, between 65°N and 65°S.

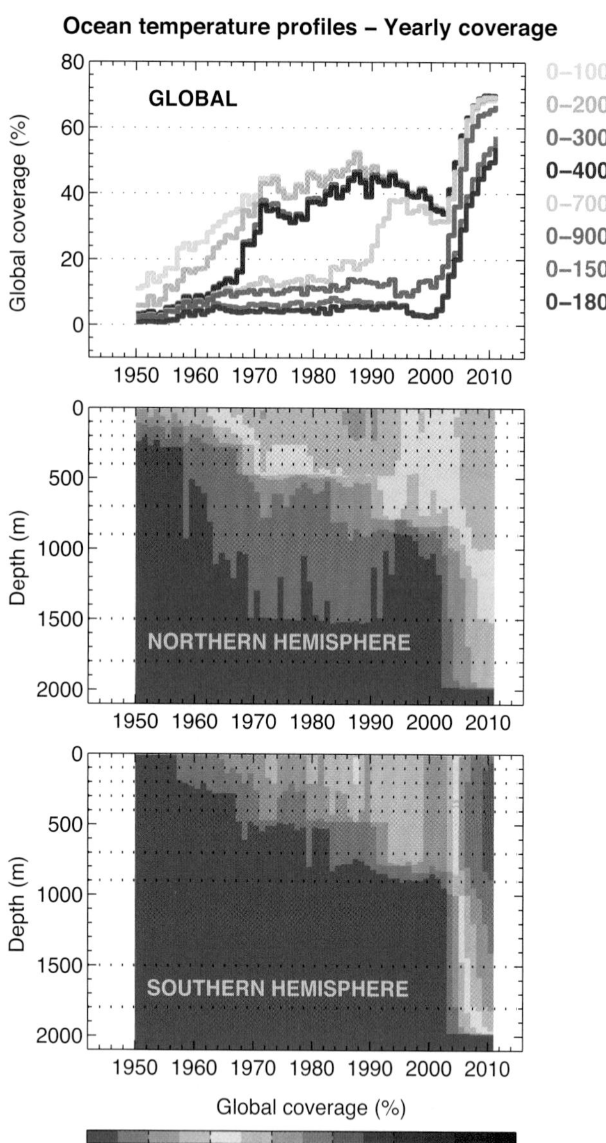

Figure 3.A.2 | (Top) Percentage of global coverage of ocean temperature profiles as a function of depth in 1° latitude by 1° longitude by 1-year bins (top panel) shown versus time. Different colours indicate profiles to different depths (middle panel). Percentage of global coverage as a function of depth and time, for the Northern Hemisphere. (Bottom panel) As above, but for the Southern Hemisphere.

3.A.2 Salinity

Measurements of subsurface salinity have relied almost solely on data collected from bottle and CTD casts from ships on station (and, more recently, using profiling floats that sample both temperature and salinity with CTDs). Hence fewer measurements of salinity are available than of temperature (by roughly a factor of 3). However, the evolution with time of subsurface salinity sampling shows a progression similar to that of temperature (Figure 3.A.3). Coverage generally improves with time, but there is a strong NH bias, particularly in the North Atlantic. A shift in focus from basin-to-basin as major field programs were carried out is evident. Near-global coverage of ocean salinity above 2000 m was not achieved until after 2005, when the Argo array approached full deployment. For depths greater than 2000 m, outside the relatively well sampled North Atlantic, information on changes in ocean salinity is largely restricted to the repeat hydrographic transects (see Figure 3.3).

3.A.3 Sea Level

Direct observations of sea level are made using tide gauges since the 1700s and high-precision satellite altimeters since 1992. Tide gauge measurements are limited to coastlines and islands. There are intermittent records of sea level at Amsterdam from 1700 and at three more sites in Northern Europe starting after 1770. By the late 1800s, more tide gauges were being operated in Northern Europe and in North America, as well as in Australia and New Zealand (Figure 3.A.4). It was not until the late 1970s to early 1980s that a majority of deep-ocean islands had operating tide gauges suitable for climate studies. Although tide gauges have continued to be deployed since 1990, they have been complemented by continuous, near-global measurements of sea level from space since 1992. Measurements are made along the satellite's ground track on the Earth surface, typically averaged over approximately 7 km to reduce noise and improve precision. The maximum latitude extent of the measurement is limited by the inclination of the orbital plane, which has been between ±66° for the TOPEX/Poseidon and Jason series of altimeters. The spacing between ground tracks is much greater than the spacing along the ground track. As an example, the groundtrack separation of the TOPEX/Poseidon-type of orbits is about 300 km at the equator, but is less than 100 km at latitudes poleward of 50° latitude. On average, the spacing is between 100 and 200 km. Satellites are limited in the temporal sampling as well due to the orbit configuration. For a specific location along a groundtrack, the return time for a TOPEX/Poseidon-type of orbit is 9.9 days. If one relaxes the requirement to a measurement within a 300 km radius, the return time can be as short as a few hours at high latitudes to about 3 days at the equator. As noted in Section 3.6, satellite altimeter observations of sea level are also an important tool for observing large-scale ocean circulation.

3.A.4 Biogeochemistry

The data available for assessing changes in biogeochemical parameters is less complete than for temperature and salinity. The global data base on which the Global Ocean Data Analysis Project (GLODAP, Key et al., 2004) ocean carbon inventory is based is illustrated in Figure 3.A.5. Changes in the ocean inventory of anthropogenic CO_2 have been estimated using measurements of carbon parameters and other tracers collected at these roughly 12,000 stations, mostly occupied since 1990. The majority of these stations extend through the full water depth. A subset of these stations have been repeated one or more times. The distribution of oxygen measurements at 300 m depth in 10-year periods since 1960 is shown in Figure 3.A.6, as used in the global study of Stramma et al. (2012). As for temperature and salinity, the sampling is heterogeneous, coverage generally improves with time but shifts between basins as major field programs come and go, and tends to be concentrated in the NH.

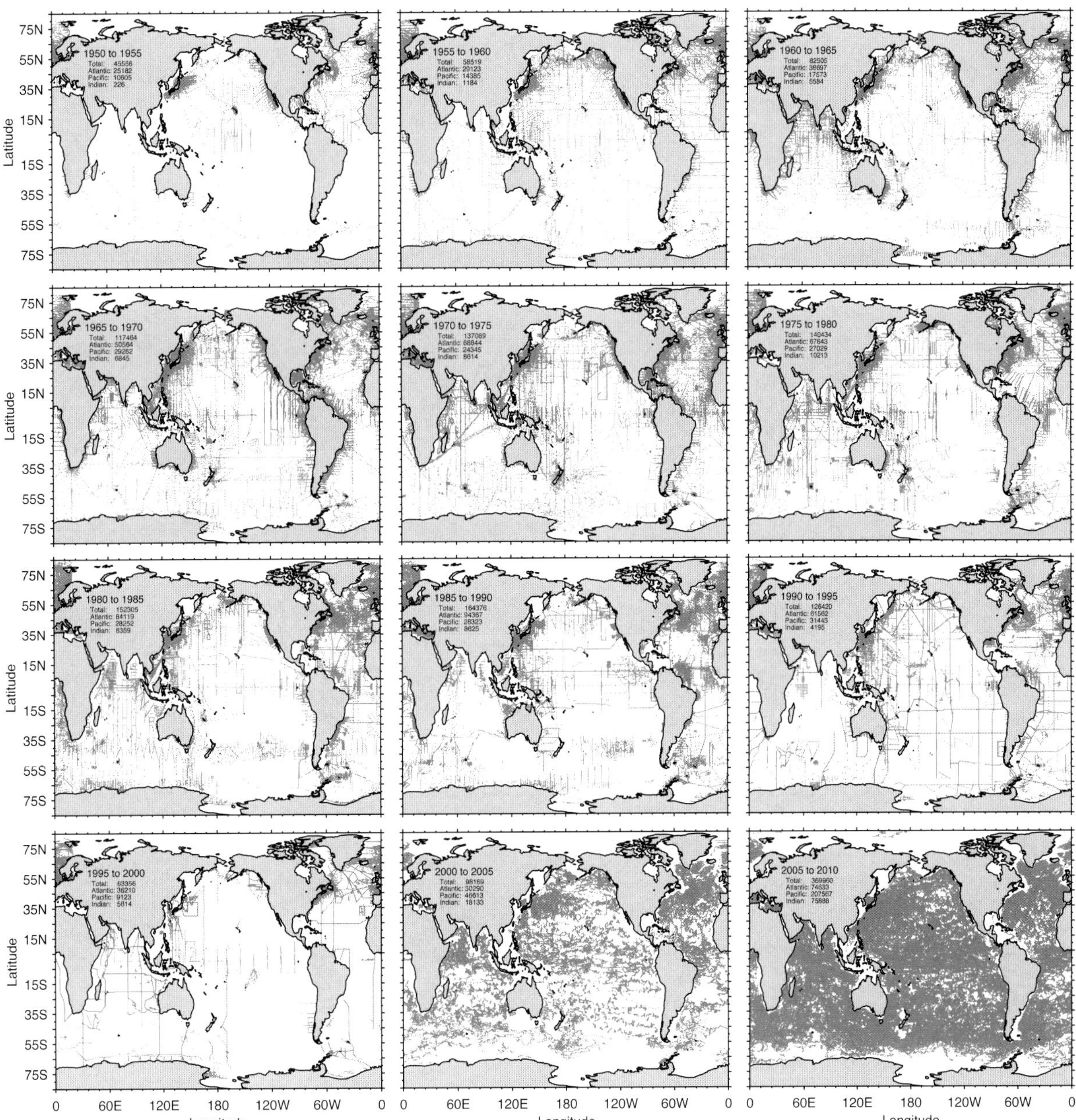

Figure 3.A.3 | Hydrographic profile data used in the Durack and Wijffels (2010) study. Station locations for 5-year temporal bins from 1950–1955 (top left) to 2005–2010 (bottom right).

3

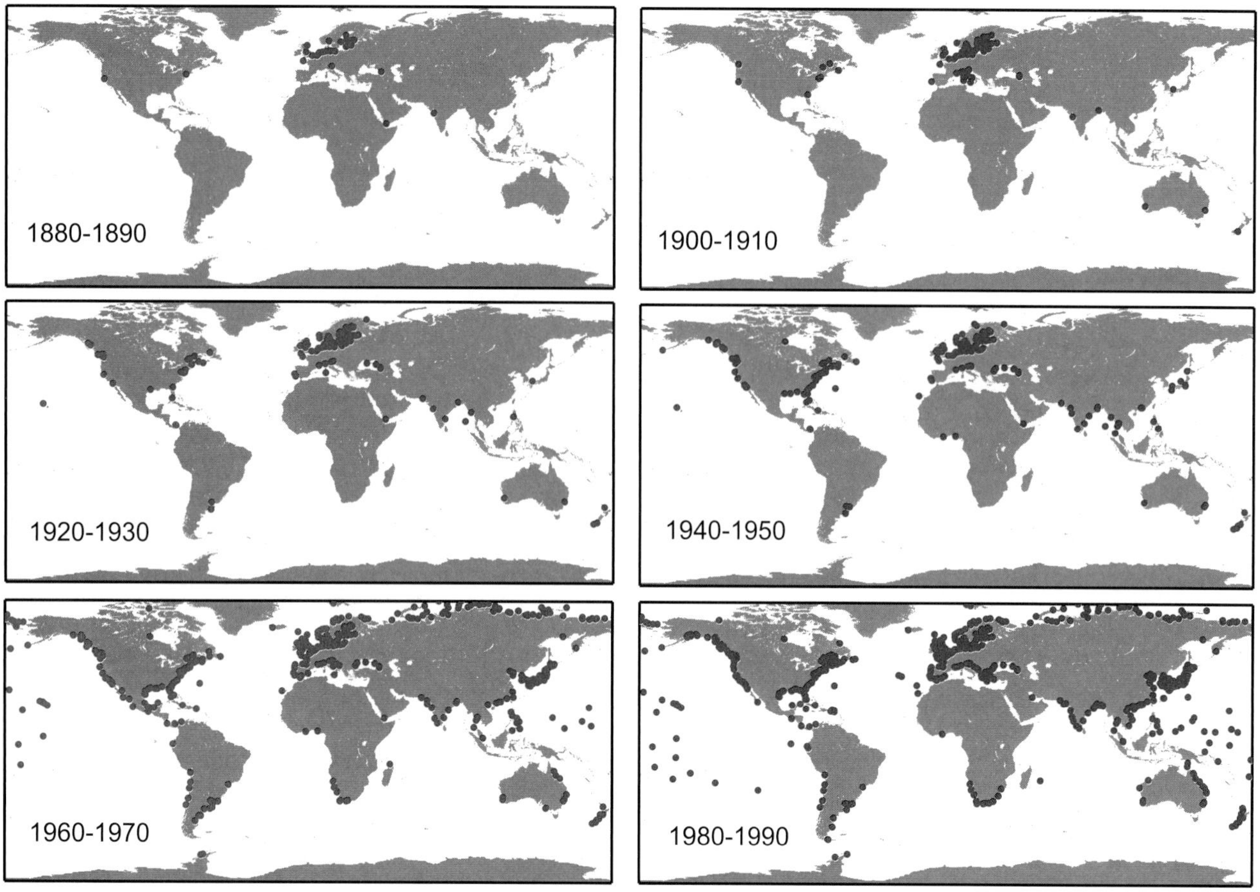

Figure 3.A.4 | Location of tide gauges (red dots) that had at least 1 year of observations within the decade indicated.

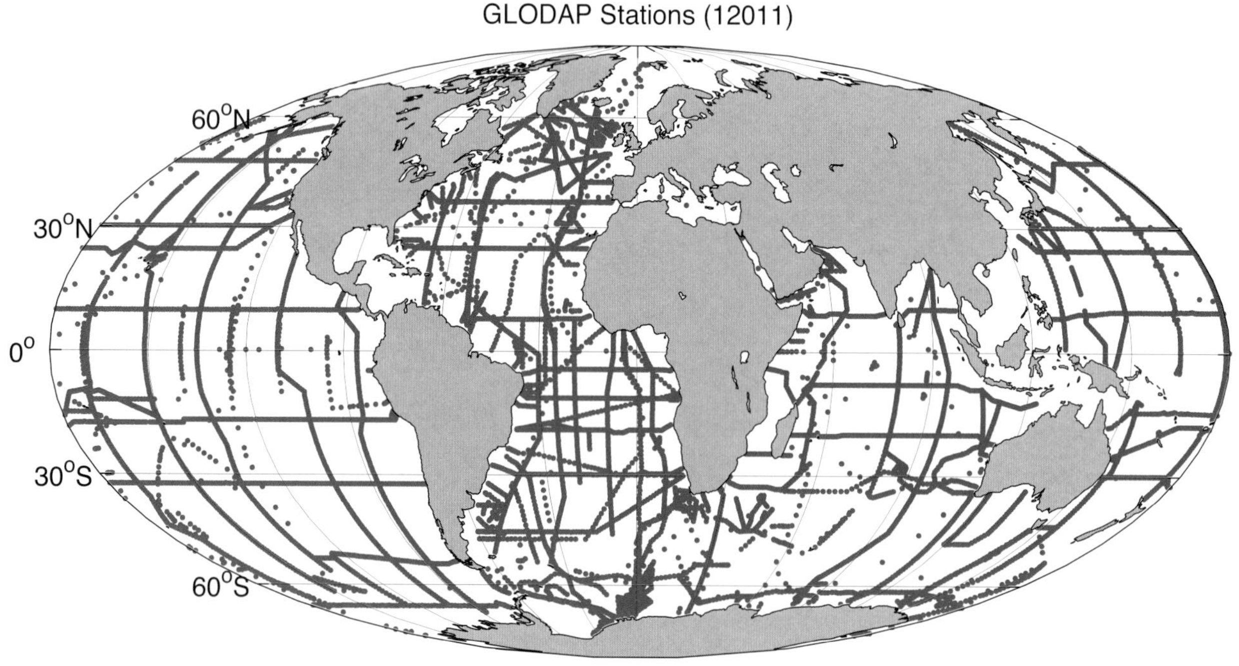

Figure 3.A.5 | Location of profiles used to construct the Global Ocean Data Analysis Project (GLODAP) ocean carbon climatology.

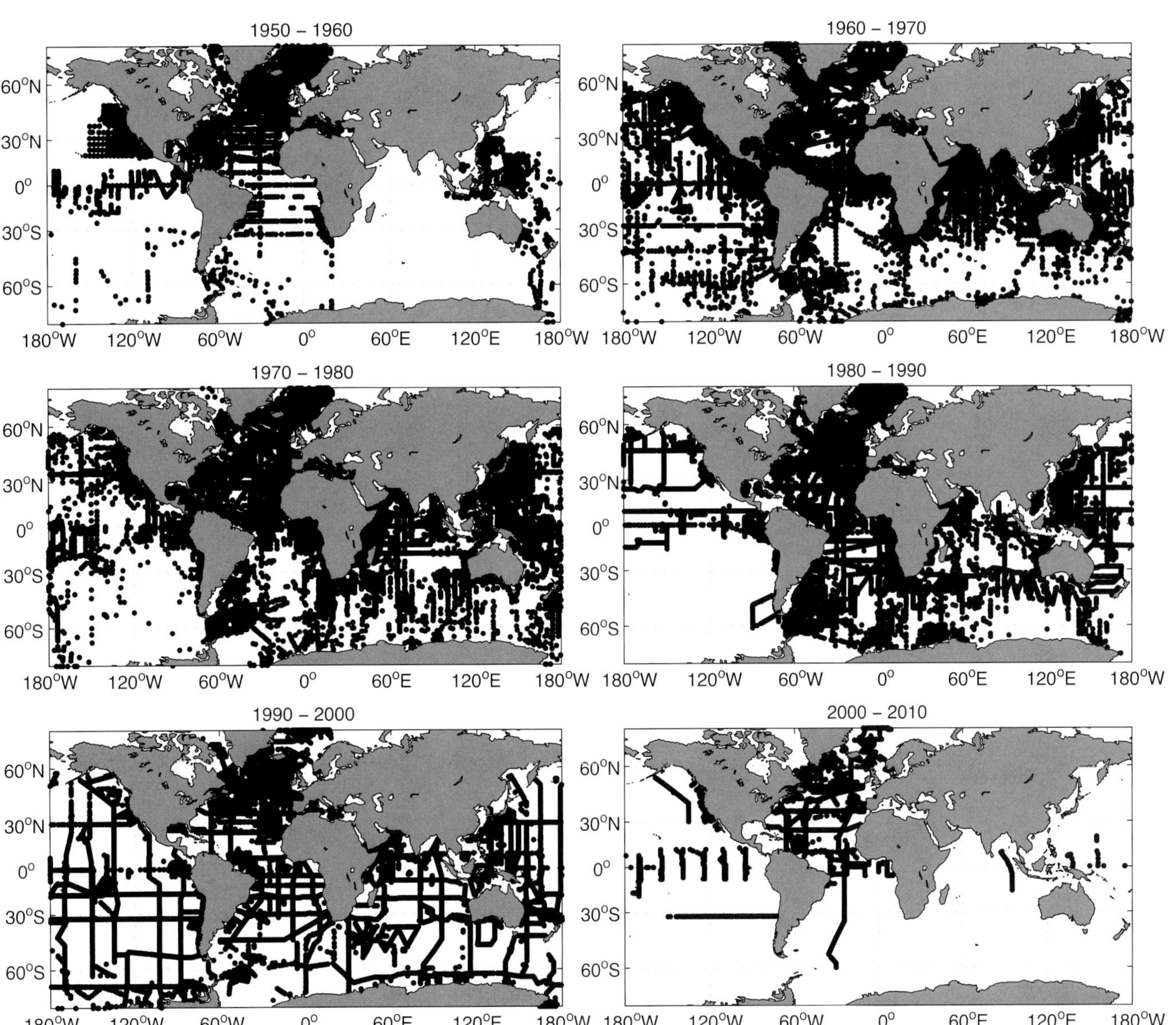

Figure 3.A.6 | Distribution of oxygen measurements at 300 dbar for the decades 1950 to 1960 (upper left) to 2000 to 2010 (lower right frame). (From Stramma et al., 2012.) [Note that additional oxygen data have become available for the 2000–2010 period since that study was completed.]

4 Observations: Cryosphere

Coordinating Lead Authors:
David G. Vaughan (UK), Josefino C. Comiso (USA)

Lead Authors:
Ian Allison (Australia), Jorge Carrasco (Chile), Georg Kaser (Austria/Italy), Ronald Kwok (USA), Philip Mote (USA), Tavi Murray (UK), Frank Paul (Switzerland/Germany), Jiawen Ren (China), Eric Rignot (USA), Olga Solomina (Russian Federation), Konrad Steffen (USA/Switzerland), Tingjun Zhang (USA/China)

Contributing Authors:
Anthony A. Arendt (USA), David B. Bahr (USA), Michiel van den Broeke (Netherlands), Ross Brown (Canada), J. Graham Cogley (Canada), Alex S. Gardner (USA), Sebastian Gerland (Norway), Stephan Gruber (Switzerland), Christian Haas (Canada), Jon Ove Hagen (Norway), Regine Hock (USA), David Holland (USA), Matthias Huss (Switzerland), Thorsten Markus (USA), Ben Marzeion (Austria), Rob Massom (Australia), Geir Moholdt (USA), Pier Paul Overduin (Germany), Antony Payne (UK), W. Tad Pfeffer (USA), Terry Prowse (Canada), Valentina Radić (Canada), David Robinson (USA), Martin Sharp (Canada), Nikolay Shiklomanov (USA), Sharon Smith (Canada), Sharon Stammerjohn (USA), Isabella Velicogna (USA), Peter Wadhams (UK), Anthony Worby (Australia), Lin Zhao (China)

Review Editors:
Jonathan Bamber (UK), Philippe Huybrechts (Belgium), Peter Lemke (Germany)

This chapter should be cited as:
Vaughan, D.G., J.C. Comiso, I. Allison, J. Carrasco, G. Kaser, R. Kwok, P. Mote, T. Murray, F. Paul, J. Ren, E. Rignot, O. Solomina, K. Steffen and T. Zhang, 2013: Observations: Cryosphere. In: *Climate Change 2013: The Physical Science Basis. Contribution of Working Group I to the Fifth Assessment Report of the Intergovernmental Panel on Climate Change* [Stocker, T.F., D. Qin, G.-K. Plattner, M. Tignor, S.K. Allen, J. Boschung, A. Nauels, Y. Xia, V. Bex and P.M. Midgley (eds.)]. Cambridge University Press, Cambridge, United Kingdom and New York, NY, USA.

Table of Contents

Supplementary Material

Supplementary Material is available in online versions of the report.

4

Executive Summary

The cryosphere, comprising snow, river and lake ice, sea ice, glaciers, ice shelves and ice sheets, and frozen ground, plays a major role in the Earth's climate system through its impact on the surface energy budget, the water cycle, primary productivity, surface gas exchange and sea level. The cryosphere is thus a fundamental control on the physical, biological and social environment over a large part of the Earth's surface. Given that all of its components are inherently sensitive to temperature change over a wide range of time scales, the cryosphere is a natural integrator of climate variability and provides some of the most visible signatures of climate change.

Since AR4, observational technology has improved and key time series of measurements have been lengthened, such that our identification and measurement of changes and trends in all components of the cryosphere has been substantially improved, and our understanding of the specific processes governing their responses has been refined. Since the AR4, observations show that there has been a continued net loss of ice from the cryosphere, although there are significant differences in the rate of loss between cryospheric components and regions. The major changes occurring to the cryosphere are as follows.

Sea Ice

Continuing the trends reported in AR4, the annual Arctic sea ice extent decreased over the period 1979–2012. The rate of this decrease was *very likely*[1] between 3.5 and 4.1% per decade (0.45 to 0.51 million km^2 per decade). The average decrease in decadal extent of Arctic sea ice has been most rapid in summer and autumn (*high confidence*[2]), but the extent has decreased in every season, and in every successive decade since 1979 (*high confidence*). {4.2.2, Figure 4.2}

The extent of Arctic perennial and multi-year sea ice decreased between 1979 and 2012 (*very high confidence*). The perennial sea ice extent (summer minimum) decreased between 1979 and 2012 at 11.5 ± 2.1% per decade (0.73 to 1.07 million km^2 per decade) (*very likely*) and the multi-year ice (that has survived two or more summers) decreased at a rate of 13.5 ± 2.5% per decade (0.66 to 0.98 million km^2 per decade) (*very likely*). {4.2.2, Figures 4.4, 4.6}

The average winter sea ice thickness within the Arctic Basin decreased between 1980 and 2008 (*high confidence*). The average decrease was *likely* between 1.3 and 2.3 m. *High confidence* in this assessment is based on observations from multiple sources: submarine, electro-magnetic (EM) probes, and satellite altimetry, and is consistent with the decline in multi-year and perennial ice extent {4.2.2, Figures 4.5, 4.6} Satellite measurements made in the period 2010–2012 show a decrease in sea ice volume compared to those made over the period 2003–2008 (*medium confidence*). There is *high confidence* that in the Arctic, where the sea ice thickness has decreased, the sea ice drift speed has increased. {4.2.2, Figure 4.6}

It is *likely* that the annual period of surface melt on Arctic perennial sea ice lengthened by 5.7 ± 0.9 days per decade over the period 1979–2012. Over this period, in the region between the East Siberian Sea and the western Beaufort Sea, the duration of ice-free conditions increased by nearly 3 months. {4.2.2, Figure 4.6}

It is *very likely* that the annual Antarctic sea ice extent increased at a rate of between 1.2 and 1.8% per decade (0.13 to 0.20 million km^2 per decade) between 1979 and 2012. There was a greater increase in sea ice area, due to a decrease in the percentage of open water within the ice pack. There is *high confidence* that there are strong regional differences in this annual rate, with some regions increasing in extent/area and some decreasing {4.2.3, Figure 4.7}

Glaciers

Since AR4, almost all glaciers worldwide have continued to shrink as revealed by the time series of measured changes in glacier length, area, volume and mass (*very high confidence*). Measurements of glacier change have increased substantially in number since AR4. Most of the new data sets, along with a globally complete glacier inventory, have been derived from satellite remote sensing. {4.3.1, 4.3.3, Figures 4.9, 4.10, 4.11}

Between 2003 and 2009, most of the ice lost was from glaciers in Alaska, the Canadian Arctic, the periphery of the Greenland ice sheet, the Southern Andes and the Asian Mountains (*very high confidence*). Together these regions account for more than 80% of the total ice loss. {4.3.3, Figure 4.11, Table 4.4}

Total mass loss from all glaciers in the world, excluding those on the periphery of the ice sheets, was *very likely* 226 ± 135 Gt yr^{-1} (sea level equivalent, 0.62 ± 0.37 mm yr^{-1}) in the period 1971–2009, 275 ± 135 Gt yr^{-1} (0.76 ± 0.37 mm yr^{-1}) in the period 1993–2009, and 301 ± 135 Gt yr^{-1} (0.83 ± 0.37 mm yr^{-1}) between 2005 and 2009. {4.3.3, Figure 4.12, Table 4.5}

Current glacier extents are out of balance with current climatic conditions, indicating that glaciers will continue to shrink in the future even without further temperature increase (*high confidence*). {4.3.3}

4

1 In this Report, the following terms have been used to indicate the assessed likelihood of an outcome or a result: Virtually certain 99–100% probability, Very likely 90–100%, Likely 66–100%, About as likely as not 33–66%, Unlikely 0–33%, Very unlikely 0–10%, Exceptionally unlikely 0–1%. Additional terms (Extremely likely: 95–100%, More likely than not >50–100%, and Extremely unlikely 0–5%) may also be used when appropriate. Assessed likelihood is typeset in italics, e.g., *very likely* (see Section 1.4 and Box TS.1 for more details).

2 In this Report, the following summary terms are used to describe the available evidence: limited, medium, or robust; and for the degree of agreement: low, medium, or high. A level of confidence is expressed using five qualifiers: very low, low, medium, high, and very high, and typeset in italics, e.g., *medium confidence*. For a given evidence and agreement statement, different confidence levels can be assigned, but increasing levels of evidence and degrees of agreement are correlated with increasing confidence (see Section 1.4 and Box TS.1 for more details).

Ice Sheets

The Greenland ice sheet has lost ice during the last two decades (*very high confidence*). Combinations of satellite and airborne remote sensing together with field data indicate with *high confidence* that the ice loss has occurred in several sectors and that large rates of mass loss have spread to wider regions than reported in AR4. {4.4.2, 4.4.3, Figures 4.13, 4.15, 4.17}

The rate of ice loss from the Greenland ice sheet has accelerated since 1992. The average rate has *very likely* increased from 34 [–6 to 74] Gt yr^{-1} over the period 1992–2001 (sea level equivalent, 0.09 [–0.02 to 0.20] mm yr^{-1}), to 215 [157 to 274] Gt yr^{-1} over the period 2002–2011 (0.59 [0.43 to 0.76] mm yr^{-1}). {4.4.3, Figures 4.15, 4.17}

Ice loss from Greenland is partitioned in approximately similar amounts between surface melt and outlet glacier discharge (*medium confidence*), and both components have increased (*high confidence*). The area subject to summer melt has increased over the last two decades (*high confidence*). {4.4.2}

The Antarctic ice sheet has been losing ice during the last two decades (*high confidence*). There is *very high confidence* that these losses are mainly from the northern Antarctic Peninsula and the Amundsen Sea sector of West Antarctica, and *high confidence* that they result from the acceleration of outlet glaciers. {4.4.2, 4.4.3, Figures 4.14, 4.16, 4.17}

The average rate of ice loss from Antarctica *likely* increased from 30 [–37 to 97] Gt yr^{-1} (sea level equivalent, 0.08 [–0.10 to 0.27] mm yr^{-1}) over the period 1992–2001, to 147 [72 to 221] Gt yr^{-1} over the period 2002–2011 (0.40 [0.20 to 0.61] mm yr^{-1}). {4.4.3, Figures 4.16, 4.17}

In parts of Antarctica, floating ice shelves are undergoing substantial changes (*high confidence*). There is *medium confidence* that ice shelves are thinning in the Amundsen Sea region of West Antarctica, and *medium confidence* that this is due to high ocean heat flux. There is *high confidence* that ice shelves round the Antarctic Peninsula continue a long-term trend of retreat and partial collapse that began decades ago. {4.4.2, 4.4.5}

Snow Cover

Snow cover extent has decreased in the Northern Hemisphere, especially in spring (*very high confidence*). Satellite records indicate that over the period 1967–2012, annual mean snow cover extent decreased with statistical significance; the largest change, –53% [*very likely*, –40% to –66%], occurred in June. No months had statistically significant increases. Over the longer period, 1922–2012, data are available only for March and April, but these show a 7% [*very likely*, 4.5% to 9.5%] decline and a strong negative [–0.76] correlation with March–April 40°N to 60°N land temperature. {4.5.2, 4.5.3}

Station observations of snow, nearly all of which are in the Northern Hemisphere, generally indicate decreases in spring, especially at warmer locations (*medium confidence*). Results depend on station elevation, period of record, and variable measured (e.g., snow depth or duration of snow season), but in almost every study surveyed, a majority of stations showed decreasing trends, and stations at lower elevation or higher average temperature were the most liable to show decreases. In the Southern Hemisphere, evidence is too limited to conclude whether changes have occurred. {4.5.2, 4.5.3, Figures 4.19, 4.20, 4.21}

Freshwater Ice

The limited evidence available for freshwater (lake and river) ice indicates that ice duration is decreasing and average seasonal ice cover shrinking (*low confidence*). For 75 Northern Hemisphere lakes, for which trends were available for 150-, 100- and 30-year periods ending in 2005, the most rapid changes were in the most recent period (*medium confidence*), with freeze-up occurring later (1.6 days per decade) and breakup earlier (1.9 days per decade). In the North American Great Lakes, the average duration of ice cover declined 71% over the period 1973–2010. {4.6}

Frozen Ground

Permafrost temperatures have increased in most regions since the early 1980s (*high confidence*) although the rate of increase has varied regionally. The temperature increase for colder permafrost was generally greater than for warmer permafrost (*high confidence*). {4.7.2, Table 4.8, Figure 4.24}

Significant permafrost degradation has occurred in the Russian European North (*medium confidence*). There is *medium confidence* that, in this area, over the period 1975–2005, warm permafrost up to 15 m thick completely thawed, the southern limit of discontinuous permafrost moved north by up to 80 km and the boundary of continuous permafrost moved north by up to 50 km. {4.7.2}

***In situ* measurements and satellite data show that surface subsidence associated with degradation of ice-rich permafrost occurred at many locations over the past two to three decades (*medium confidence*).** {4.7.4}

In many regions, the depth of seasonally frozen ground has changed in recent decades (*high confidence*). In many areas since the 1990s, active layer thicknesses increased by a few centimetres to tens of centimetres (*medium confidence*). In other areas, especially in northern North America, there were large interannual variations but few significant trends (*high confidence*). The thickness of the seasonally frozen ground in some non-permafrost parts of the Eurasian continent *likely* decreased, in places by more than 30 cm from 1930 to 2000 (*high confidence*) {4.7.4}

4.1 Introduction

The cryosphere is the collective term for the components of the Earth system that contain a substantial fraction of water in the frozen state (Table 4.1). The cryosphere comprises several components: snow, river and lake ice; sea ice; ice sheets, ice shelves, glaciers and ice caps; and frozen ground which exist, both on land and beneath the oceans (see Glossary and Figure 4.1). The lifespan of each component is very different. River and lake ice, for example, are transient features that generally do not survive from winter to summer; sea ice advances and retreats with the seasons but especially in the Arctic can survive to become multi-year ice lasting several years. The East Antarctic ice sheet, on the other hand, is believed to have become relatively stable around 14 million years ago (Barrett, 2013). Nevertheless, all components of the cryosphere are inherently sensitive to changes in air temperature and precipitation, and hence to a changing climate (see Chapter 2).

Changes in the longer-lived components of the cryosphere (e.g., glaciers) are the result of an integrated response to climate, and the cryosphere is often referred to as a 'natural thermometer'. But as our understanding of the complexity of this response has grown, it is increasingly clear that elements of the cryosphere should rather be considered as a 'natural climate-meter', responsive not only to temperature but also to other climate variables (e.g., precipitation). However, it remains the case that the conspicuous and widespread nature of changes in the cryosphere (in particular, sea ice, glaciers and ice sheets) means these changes are frequently used emblems of the impact of changing climate. It is thus imperative that we understand the context of current change within the framework of past changes and natural variability.

The cryosphere is, however, not simply a passive indicator of climate change; changes in each component of the cryosphere have a significant and lasting impact on physical, biological and social systems. Ice sheets and glaciers exert a major control on global sea level (see Chapters 5 and 13), ice loss from these systems may affect global ocean circulation and marine ecosystems, and the loss of glaciers near populated areas as well as changing seasonal snow cover may have direct impacts on water resources and tourism (see WGII Chapters 3 and 24). Similarly, reduced sea ice extent has altered, and in the future may continue to alter, ocean circulation, ocean productivity and regional climate and will have direct impacts on shipping and mineral and oil exploration (see WGII, Chapter 28). Furthermore, decline in snow cover and sea ice will tend to amplify regional warming through snow and ice-albedo feedback effects (see Glossary and Chapter 9). In addition,

Table 4.1 | Representative statistics for cryospheric components indicating their general significance.

Ice on Land	Percent of Global Land Surface[a]	Sea Level Equivalent[b] (metres)
Antarctic ice sheet[c]	8.3	58.3
Greenland ice sheet[d]	1.2	7.36
Glaciers[e]	0.5	0.41
Terrestrial permafrost[f]	9–12	0.02–0.10[g]
Seasonally frozen ground[h]	33	Not applicable
Seasonal snow cover (seasonally variable)[i]	1.3–30.6	0.001–0.01
Northern Hemisphere freshwater (lake and river) ice[j]	1.1	Not applicable
Total[k]	**52.0–55.0%**	**~66.1**
Ice in the Ocean	**Percent of Global Ocean Area[a]**	**Volume[l] (10^3 km^3)**
Antarctic ice shelves	0.45[m]	~380
Antarctic sea ice, austral summer (spring)[n]	0.8 (5.2)	3.4 (11.1)
Arctic sea ice, boreal autumn (winter/spring)[n]	1.7 (3.9)	13.0 (16.5)
Sub-sea permafrost[o]	~0.8	Not available
Total[p]	**5.3–7.3**	

Notes:

a Assuming a global land area of 147.6 Mkm^2 and ocean area of 362.5 Mkm^2.

b See Glossary. Assuming an ice density of 917 kg m^{-3}, a seawater density of 1028 kg m^{-3}, with seawater replacing ice currently below sea level.

c Area of grounded ice sheet not including ice shelves is 12.295 Mkm^2 (Fretwell et al., 2013).

d Area of ice sheet and peripheral glaciers is 1.801 Mkm^2 (Kargel et al., 2012). SLE (Bamber et al., 2013).

e Calculated from glacier outlines (Arendt et al., 2012), includes glaciers around Greenland and Antarctica. For sources of SLE see Table 4.2.

f Area of permafrost excluding permafrost beneath the ice sheets is 13.2 to 18.0 Mkm^2 (Gruber, 2012).

g Value indicates the full range of estimated excess water content of Northern Hemisphere permafrost (Zhang et al., 1999).

h Long-term average maximum of seasonally frozen ground is 48.1 Mkm^2 (Zhang et al., 2003); excludes Southern Hemisphere.

i Northern Hemisphere only (Lemke et al., 2007).

j Areas and volume of freshwater (lake and river ice) were derived from modelled estimates of maximum seasonal extent (Brooks et al., 2012).

k To allow for areas of permafrost and seasonally frozen ground that are also covered by seasonal snow, total area excludes seasonal snow cover.

l Antarctic austral autumn (spring) (Kurtz and Markus, 2012); and Arctic boreal autumn (winter) (Kwok et al., 2009). For the Arctic, volume includes only sea ice in the Arctic Basin.

m Area is 1.617 Mkm^2 (Griggs and Bamber, 2011).

n Maximum and minimum areas taken from this assessment, Sections 4.2.2 and 4.2.3.

o Few estimates of the area of sub-sea permafrost exist in the literature. The estimate shown, 2.8 Mkm^2, has significant uncertainty attached and was assembled from other publications by Gruber (2012).

p Summer and winter totals assessed separately.

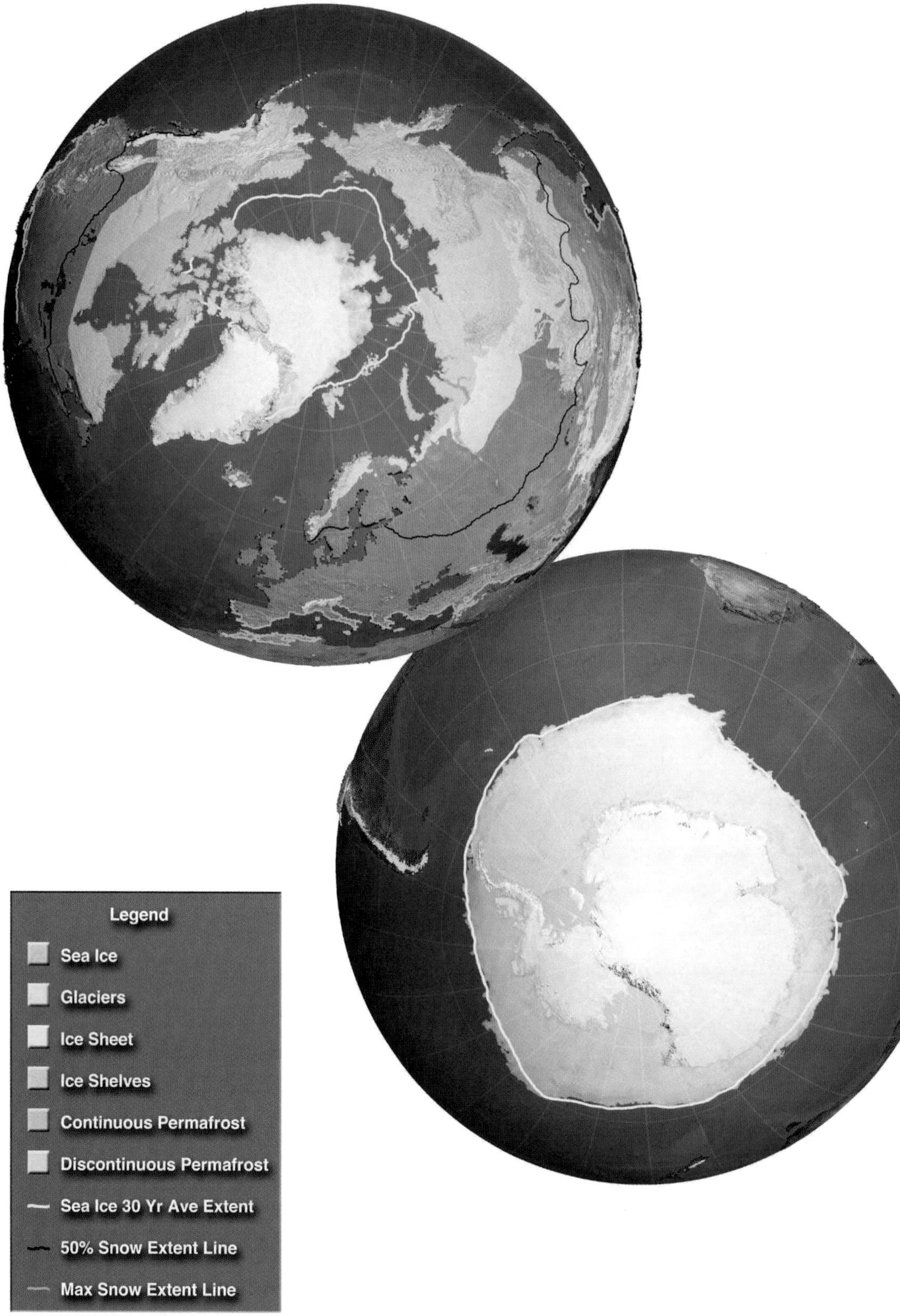

Figure 4.1 | The cryosphere in the Northern and Southern Hemispheres in polar projection. The map of the Northern Hemisphere shows the sea ice cover during minimum summer extent (13 September 2012). The yellow line is the average location of the ice edge (15% ice concentration) for the yearly minima from 1979 to 2012. Areas of continuous permafrost (see Glossary) are shown in dark pink, discontinuous permafrost in light pink. The green line along the southern border of the map shows the maximum snow extent while the black line across North America, Europe and Asia shows the 50% contour for frequency of snow occurrence. The Greenland ice sheet (blue/grey) and locations of glaciers (small gold circles) are also shown. The map of the Southern Hemisphere shows approximately the maximum sea ice cover during an austral winter (13 September 2012). The yellow line shows the average ice edge (15% ice concentration) during maximum extent of the sea ice cover from 1979 to 2012. Some of the elements (e.g., some glaciers and snow) located at low latitudes are not visible in this projection (see Figure 4.8). The source of the data for sea ice, permafrost, snow and ice sheet are data sets held at the National Snow and Ice Data Center (NSIDC), University of Colorado, on behalf of the North American Atlas, Instituto Nacional de Estadística, Geografía e Informática (Mexico), Natural Resources Canada, U.S. Geological Survey, Government of Canada, Canada Centre for Remote Sensing and The Atlas of Canada. Glacier locations were derived from the multiple data sets compiled in the Randolph Glacier Inventory (Arendt et al., 2012).

changes in frozen ground (in particular, ice-rich permafrost) will damage some vulnerable Arctic infrastructure (see WGII, Chapter 28), and could substantially alter the carbon budget through the release of methane (see Chapter 6).

Since AR4, substantial progress has been made in most types of cryospheric observations. Satellite technologies now permit estimates of regional and temporal changes in the volume and mass of the ice sheets. The longer time series now available enable more accurate assessments of trends and anomalies in sea ice cover and rapid identification of unusual events such as the dramatic decline of Arctic summer sea ice extent in 2007 and 2012. Similarly, Arctic sea ice thickness can now be estimated using satellite altimetry, allowing pan-Arctic measurements of changes in volume and mass. A new global glacier inventory includes nearly all glaciers (Arendt et al., 2012) (42% in AR4) and allows for much better estimates of the total ice volume and its past and future changes. Remote sensing measurements of regional glacier volume change are also now available widely and modelling of glacier mass change has improved considerably. Finally, fluctuations in the cryosphere in the distant and recent past have been mapped with increasing certainty, demonstrating the potential for rapid ice loss, compared to slow recovery, particularly when related to sea level rise.

This chapter describes the current state of the cryosphere and its individual components, with a focus on recent improvements in understanding of the observed variability, changes and trends. Projections of future cryospheric changes (e.g., Chapter 13) and potential drivers (Chapter 10) are discussed elsewhere. Earlier IPCC reports used cryospheric terms that have specific scientific meanings (see Cogley et al., 2011), but have rather different meanings in everyday language. To avoid confusion, this chapter uses the term 'glaciers' for what was previously termed 'glaciers and ice caps' (e.g., Lemke et al., 2007). For the two largest ice masses of continental size, those covering Greenland and Antarctica, we use the term 'ice sheets'. For simplicity, we use units such as gigatonnes (Gt, 10^9 tonnes, or 10^{12} kg). One gigatonne is approximately equal to one cubic kilometre of freshwater (1.1 km^3 of ice), and 362.5 Gt of ice removed from the land and immersed in the oceans will cause roughly 1 mm of global sea level rise (Cogley, 2012).

4.2 Sea Ice

4.2.1 Background

Sea ice (see Glossary) is an important component of the climate system. A sea ice cover on the ocean changes the surface albedo, insulates the ocean from heat loss, and provides a barrier to the exchange of momentum and gases such as water vapour and CO_2 between the ocean and atmosphere. Salt ejected by growing sea ice alters the density structure and modifies the circulation of the ocean. Regional climate changes affect the sea ice characteristics and these changes can feed back on the climate system, both regionally and globally. Sea ice is also a major component of polar ecosystems; plants and animals at all trophic levels find a habitat in, or are associated with, sea ice.

Most sea ice exists as pack ice, and wind and ocean currents drive the drift of individual pieces of ice (called floes). Divergence and shear in sea ice motion create areas of open water where, during colder months, new ice can quickly form and grow. On the other hand, convergent ice motion causes the ice cover to thicken by deformation. Two relatively thin floes colliding with each other can 'raft', stacking one on top of the other and thickening the ice. When thicker floes collide, thick ridges may be built from broken pieces, with a height above the surface (ridge sail) of 2 m or more, and a much greater thickness (~10 m) and width below the ocean surface (ridge keel).

Sea ice thickness also increases by basal freezing during winter months. But the thicker the ice becomes the more it insulates heat loss from the ocean to the atmosphere and the slower the basal growth is. There is an equilibrium thickness for basal ice growth that is dependent on the surface energy balance and heat from the deep ocean below. Snow cover lying on the surface of sea ice provides additional insulation, and also alters the surface albedo and aerodynamic roughness. But also, and particularly in the Antarctic, a heavy snow load on thin sea ice can depress the ice surface and allow seawater to flood the snow. This saturated snow layer freezes quickly to form 'snow ice' (see FAQ 4.1).

Because sea ice is formed from seawater it contains sea salt, mostly in small pockets of concentrated brine. The total salt content in newly formed sea ice is only 25 to 50% of that in the parent seawater, and the residual salt rejected as the sea ice forms alters ocean water density and stability. The salinity of the ice decreases as it ages, and particularly during the Arctic summer when melt water (including from melt ponds that form on the surface) drains through and flushes the ice. The salinity and porosity of sea ice affect its mechanical strength, its thermal properties and its electrical properties – the latter being very important for remote sensing.

Geographical constraints play a dominant but not an exclusive role in determining the quite different characteristics of sea ice in the Arctic and the Antarctic (see FAQ 4.1). This is one of the reasons why changes in sea ice extent and thickness are very different in the north and the south. We also have much more information on Arctic sea ice thickness than we do on Antarctic sea ice thickness, and so discuss Arctic and Antarctic separately in this assessment.

4.2.2 Arctic Sea Ice

Regional sea ice observations, which span more than a century, have revealed significant interannual changes in sea ice coverage (Walsh and Chapman, 2001). Since the advent of satellite multichannel passive microwave imaging systems in 1979, which now provide more than 34 years of continuous coverage, it has been possible to monitor the entire extent of sea ice with a temporal resolution of less than a day. A number of procedures have been used to convert the observed microwave brightness temperature into sea ice concentration— the fractional area of the ocean covered by ice—and thence to derive sea ice extent and area (Markus and Cavalieri, 2000; Comiso and Nishio, 2008). Sea ice extent is defined as the sum of ice covered areas with concentrations of at least 15%, while ice area is the product of the ice concentration and area of each data element within the ice extent. A brief description of the different techniques for deriving sea ice concentration is provided in the Supplementary Material. The trends in the sea ice concentration, ice extent and ice area, as inferred from data derived from the different techniques, are generally compatible. A comparison of derived ice extents from different sources is presented in the next section and in the Supplementary Material. Results presented in this assessment are based primarily on a single technique (Comiso and Nishio, 2008) but the use of data from other techniques would provide generally the same conclusions.

Arctic sea ice cover varies seasonally, with average ice extent varying between about 6×10^6 km^2 in the summer and about 15×10^6 km^2 in the winter (Comiso and Nishio, 2008; Cavalieri and Parkinson, 2012; Meier et al., 2012). The summer ice cover is confined to mainly the Arctic Ocean basin and the Canadian Arctic Archipelago, while winter sea ice reaches as far south as 44°N, into the peripheral seas. At the end of summer, the Arctic sea ice cover consists primarily of the previously thick, old and ridged ice types that survived the melt period. Interannual variability is largely determined by the extent of the ice cover in the peripheral seas in winter and by the ice cover that survives the summer melt in the Arctic Basin.

4.2.2.1 Total Arctic Sea Ice Extent and Concentration

Figure 4.2 (derived from passive microwave data) shows both the seasonality of the Arctic sea ice cover and the large decadal changes that have occurred over the last 34 years. Typically, Arctic sea ice reaches its maximum seasonal extent in February or March whereas the minimum occurs in September at the end of summer melt. Changes in decadal averages in Arctic ice extent are more pronounced in summer than in winter. The change in winter extent between 1979–1988 and 1989–1998 was negligible. Between 1989–1998 and 1999–2008, there was a decrease in winter extent of around 0.6×10^6 km^2. This can be contrasted to a decrease in ice extent at the end of the summer (September) of 0.5×10^6 km^2 between 1979–1988 and 1989–1998, followed by a further decrease of 1.2×10^6 km^2 between 1989–1998 and 1999–2008. Figure 4.2 also shows that the change in extent from 1979–1988 to 1989–1998 was statistically significant mainly in spring and summer while the change from 1989–1998 to 1999–2008 was statistically significant during winter and summer. The largest interannual changes occur during the end of summer when only the thick components of the winter ice cover survive the summer melt (Comiso et al., 2008; Comiso, 2012).

4

For comparison, the average extents during the 2009–2012 period are also presented: the extent during this period was considerably less than in earlier periods in all seasons, except spring. The summer minimum extent was at a record low in 2012 following an earlier record set in 2007 (Stroeve et al., 2007; Comiso et al., 2008). The minimum ice extent in 2012 was 3.44×10^6 km^2 while the low in 2007 was 4.22×10^6 km^2. For comparison, the record high value was 7.86×10^6 km^2 in 1980. The low extent in 2012 (which is 18.5% lower than in 2007) was probably caused in part by an unusually strong storm in the Central Arctic Basin on 4 to 8 August 2012 (Parkinson and Comiso, 2013). The extents for 2007 and 2012 were almost the same from June until the storm period in 2012, after which the extent in 2012 started to trend considerably lower than in 2007. The error bars, which represent 1 standard deviation (1σ) of samples used to estimate each data point, are smallest in the first decade and get larger with subsequent decades indicating much higher interannual variability in recent years. The error bars are also comparable in summer and winter during the first decade but become progressively larger for summer compared to winter in subsequent decades. These results indicate that the largest interannual variability has occurred in the summer and in the recent decade.

Although relatively short as a climate record, the 34-year satellite record is long enough to allow determination of significant and consistent trends of the time series of monthly anomalies (i.e., difference between the monthly and the averages over the 34-year record) of ice extent, area and concentration. The trends in ice concentration for the winter, spring, summer and autumn for the period November 1978 to December 2012 are shown in Figure 4.2 (b, c, d and e). The seasonal trends for different regions, except the Bering Sea, are negative. Ice cover changes are relatively large in the eastern Arctic Basin and most peripheral seas in winter and spring, while changes are pronounced almost everywhere in the Arctic Basin, except at greater than 82°N, in summer and autumn. In connection with a comprehensive observational research program during the International Polar Year 2007–2008, regional studies primarily on the Canadian side of the Arctic revealed very similar patterns of spatial and interannual variability of the sea ice cover (Derksen et al., 2012).

From the monthly anomaly data, the trend in sea ice extent in the Northern Hemisphere (NH) for the period from November 1978 to December 2012 is -3.8 ± 0.3% per decade (*very likely*) (see FAQ 4.1). The error quoted is calculated from the standard deviation of the slope of the regression line. The baseline for the monthly anomalies is the average of all data for each month from November 1978 to December 2012. The trends for different regions vary greatly, ranging from +7.3% per decade in the Bering Sea to −13.8% per decade in the Gulf of St. Lawrence. This large spatial variability is associated with the complexity of the atmospheric and oceanic circulation system as manifested in the Arctic Oscillation (Thompson and Wallace, 1998). The trends also differ with season (Comiso and Nishio, 2008; Comiso et al., 2011). For the entire NH, the trends in ice extent are -2.3 ± 0.5%, -1.8 ± 0.5%, -6.1 ± 0.8% and -7.0 ± 1.5% per decade (*very likely*) in winter, spring, summer and autumn, respectively. The corresponding trends in ice area are -2.8 ± 0.5%, -2.2 ± 0.5%, -7.2 ± 1.0%, and -7.8 ± 1.3% per decade (*very likely*). Similar results were obtained by (Cavalieri and Parkinson, 2012) but cannot be compared directly since their data are for the period from 1979 to 2010 (see Supplementary Material). The trends for ice extent and ice area are comparable except in the summer and autumn, when the trend in ice area is significantly more than that in ice extent. This is due in part to increasing open water areas within the pack that may be caused by more frequent storms and more divergence in the summer (Simmonds et al., 2008). The trends are larger in the summer and autumn mainly because of the rapid decline in the multi-year ice cover (Comiso, 2012), as discussed in Section 4.2.2.3. The trends in km^2 yr^{-1} were estimated as in Comiso and Nishio (2008) and Comiso (2012) but the percentage trends presented in this chapter were calculated differently. Here the percentage is calculated as a difference from the first data point on the trend line whereas the earlier estimations used the difference from the mean value. The new percentage trends are only slightly different from the previous ones and the conclusions about changes are the same.

4.2.2.2 Longer Records of Arctic Ice Extent

For climate analysis, the variability of the sea ice cover prior to the commencement of the satellite record in 1979 is also of interest. There are a number of pre-satellite records, some based on regional observations taken from ships or aerial reconnaissance (e.g., Walsh and Chapman, 2001; Polyakov et al., 2003) while others were based on terrestrial proxies (e.g., Macias Fauria et al., 2010; Kinnard et al., 2011). The records constructed by Kinnard et al. (2011) and Macias Fauria et al. (2010) suggest that the decline of sea ice over the last few decades has been unprecedented over the past 1450 years (see Section 5.5.2). In a study of the marginal seas near the Russian coastline using ice extent data from 1900 to 2000, Polyakov et al. (2003) found a low frequency multi-decadal oscillation near the Kara Sea that shifted to a dominant decadal oscillation in the Chukchi Sea.

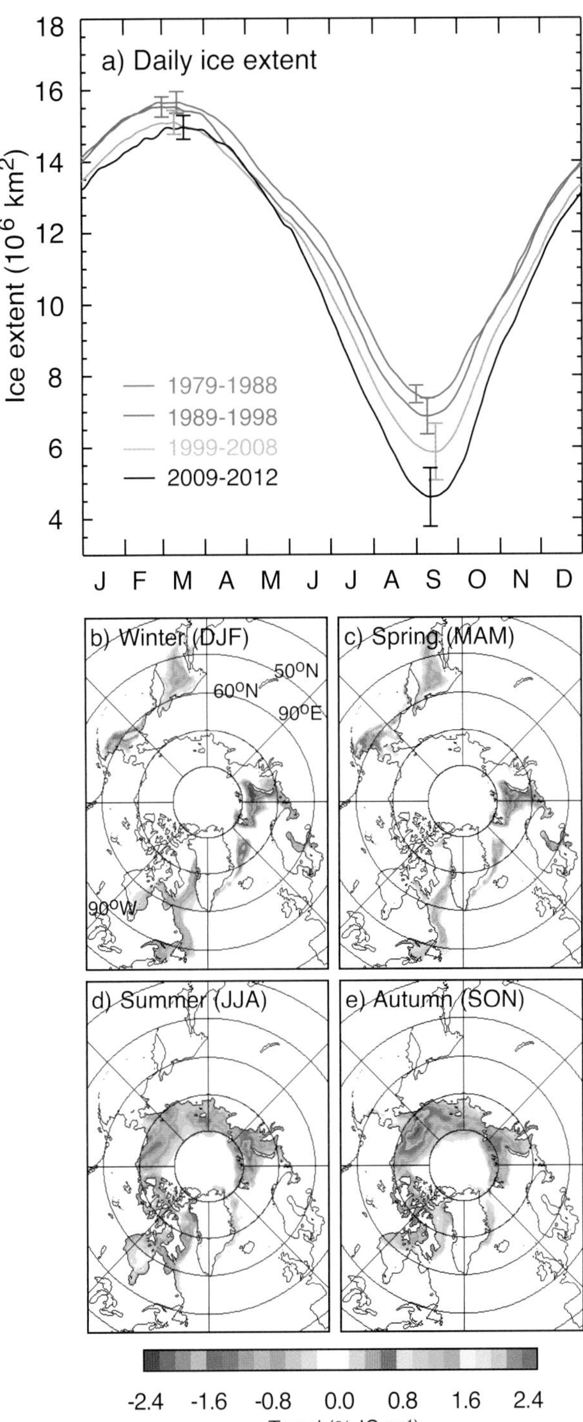

Figure 4.2 | (a) Plots of decadal averages of daily sea ice extent in the Arctic (1979 to 1988 in red, 1989 to 1998 in blue, 1999 to 2008 in gold) and a 4-year average daily ice extent from 2009 to 2012 in black. Maps indicate ice concentration trends (1979–2012) in (b) winter, (c) spring, (d) summer and (e) autumn (updated from Comiso, 2010).

A more comprehensive basin-wide record, compiled by Walsh and Chapman (2001), showed very little interannual variability until the last three to four decades. For the period 1901 to 1998, their results show a summer mode that includes an anomaly of the same sign over nearly the entire Arctic and that captures the sea-ice trend from recent satellite data. Figure 4.3 shows an updated record of the Walsh and Chapman data set with longer time coverage (1870 to 1978) that is more robust because it includes additional historical sea ice observations (e.g., from Danish meteorological stations). A comparison of this updated data set with that originally reported by Walsh and Chapman (2001) shows similar interannual variability that is dominated by a nearly constant extent of the winter (January–February–March) and autumn (October–November–December) ice cover from 1870 to the 1950s. The absence of interannual variability during that period is due to the use of climatology to fill gaps, potentially masking the natural signal. Sea ice data from 1900–2011 as compiled by Met Office Hadley Centre are also plotted for comparison. In this data set, the 1979–2011 values were derived from various sources, including satellite data, as described by Rayner et al. (2003). Since the 1950s, more *in situ* data are available and have been homogenized with the satellite record (Meier et al., 2012). These data show a consistent decline in the sea ice cover that is relatively moderate during the winter but more dramatic during the summer months. Satellite data from other sources are also plotted in Figure 4.3, including Scanning Multichannel Microwave Radiometer (SMMR) and Special Sensor Microwave/Imager (SSM/I) data using the Bootstrap Algorithm (SBA) as described by Comiso and Nishio (2008) and National Aeronautics and Space Administration (NASA) Team Algorithm (NT1) as described by Cavalieri et al. (1984) (see Supplementary Material). Data from the Advanced Microwave Scanning Radiometer - Earth Observing System (AMSR-E) using the Bootstrap Algorithm (ABA) and the NASA Team Algorithm Version 2 (NT2) are also presented. The error bars represent one standard deviation of the interannual variability during the satellite period. Because of the use of climatology to fill data gaps from 1870 to 1953, the error bars in the Walsh and Chapman data were set to twice that of the satellite period and 1.5 times higher for 1954 to 1978. The apparent reduction of the sea ice extent from 1978 to 1979 is in part due to the change from surface observations to satellite data. Generally, the temporal distributions from the various sources are consistent with some exceptions that may be attributed to possible errors in the data (e.g., Screen, 2011 and Supplementary Material). Taking this into account, the various sources provide similar basic information and conclusions about the changing extent and variability of the Arctic sea ice cover.

4.2.2.3 Multi-year/Seasonal Ice Coverage

The winter extent and area of the perennial and multi-year ice cover in the Central Arctic (i.e., excluding Greenland Sea multi-year ice) for

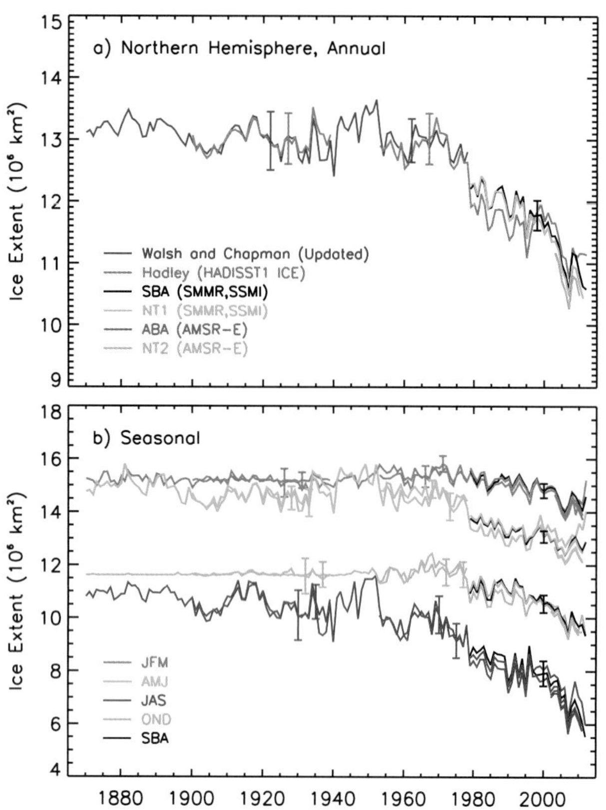

Figure 4.3 | Ice extent in the Arctic from 1870 to 2011. (a) Annual ice extent and (b) seasonal ice extent using averages of mid-month values derived from *in situ* and other sources including observations from the Danish meteorological stations from 1870 to 1978 (updated from, Walsh and Chapman, 2001). Ice extent from a joint Hadley and National Oceanic and Atmospheric Administration (NOAA) project (called HADISST1_Ice) from 1900 to 2011 is also shown. The yearly and seasonal averages for the period from 1979 to 2011 are shown as derived from Scanning Multichannel Microwave Radiometer (SMMR) and Special Sensor Microwave/Imager (SSM/I) passive microwave data using the Bootstrap Algorithm (SBA) and National Aeronautics and Space Administration (NASA) Team Algorithm, Version 1 (NT1), using procedures described in Comiso and Nishio (2008), and Cavalieri et al. (1984), respectively; and from Advanced Microwave Scanning Radiometer, Version 2 (AMSR2) using algorithms called AMSR Bootstrap Algorithm (ABA) and NASA Team Algorithm, Version 2 (NT2), described in Comiso and Nishio (2008) and Markus and Cavalieri (2000). In (b), data from the different seasons are shown in different colours to illustrate variation between seasons, with SBA data from the procedure in Comiso and Nishio (2008) shown in black.

1979–2012 are shown in Figure 4.4. Perennial ice is that which survives the summer, and the ice extent at summer minimum has been used as a measure of its coverage (Comiso, 2002). Multi-year ice (as defined by World Meteorological Organization) is ice that has survived at least two summers. Generally, multi-year ice is less saline and has a distinct microwave signature that differs from the seasonal ice, and thus can be discriminated and monitored with satellite microwave radiometers (Johannessen et al., 1999; Zwally and Gloersen, 2008; Comiso, 2012).

Figure 4.4 shows similar interannual variability and large trends for both perennial and multi-year ice for the period 1979 to 2012. The extent of the perennial ice cover, which was about 7.9 × 10^6 km^2 in 1980, decreased to as low as 3.5 × 10^6 km^2 in 2012. Similarly, the multi-year ice extent decreased from about 6.2 × 10^6 km^2 in 1981 to about 2.5 × 10^6 km^2 in 2012. The trends in perennial ice extent and ice area were strongly negative at –11.5 ± 2.1 and –12.5 ± 2.1% per decade (*very likely*) respectively. These values indicate an increased rate of decline from the –6.4% and –8.5% per decade, respectively, reported for the 1979 to 2000 period by Comiso (2002). The trends in multi-year ice extent and area are even more negative, at –13.5 ± 2.5 and –14.7 ± 3.0% per decade (*very likely*), respectively, as updated for the period 1979 to 2012 (Comiso, 2012). The more negative trend in ice area than in ice extent indicates that the average ice concentration of multi-year ice in the Central Arctic has also been declining. The rate of decline in the extent and area of multi-year ice cover is consistent with the observed decline of old ice types from the analysis of ice drift and ice age by Maslanik et al. (2007), confirming that older and thicker ice types in the Arctic have been declining significantly. The more negative trend for the thicker multi-year ice area than that for the perennial ice area implies that the average thickness of the ice, and hence the ice volume, has also been declining.

Drastic changes in the multi-year ice coverage from QuikScat (satellite radar scatterometer) data, validated using high-resolution Synthetic Aperture Radar data (Kwok, 2004; Nghiem et al., 2007), have also been reported. Some of these changes have been attributed to the near zero replenishment of the Arctic multi-year ice cover by ice that survives the summer (Kwok, 2007).

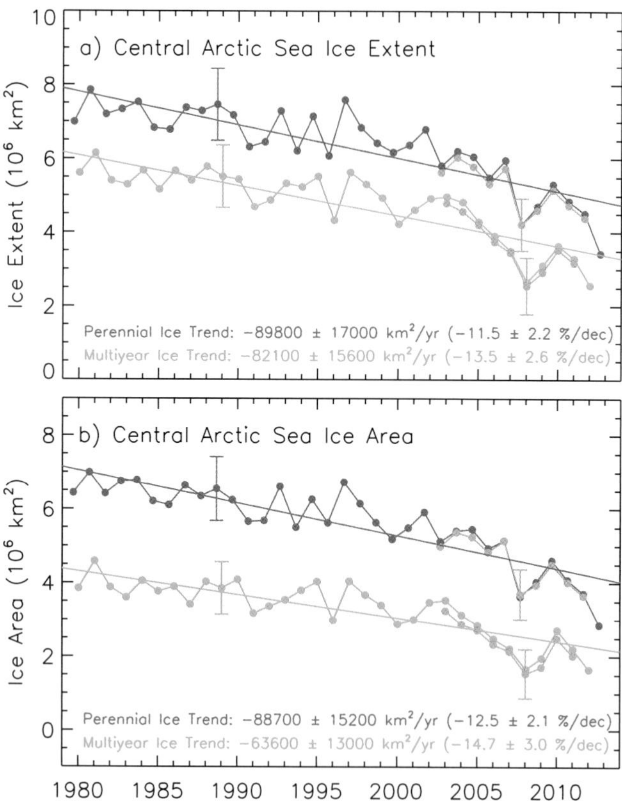

Figure 4.4 | Annual perennial (blue) and multi-year (green) sea ice extent (a) and sea ice area (b) in the Central Arctic from 1979 to 2012 as derived from satellite passive microwave data (updated from Comiso, 2012). Perennial ice values are derived from summer minimum ice extent, while the multi-year ice values are averages of those from December, January and February. The gold lines (after 2002) are from AMSR-E data. Uncertainties in the observations (*very likely* range) are indicated by representative error bars, and uncertainties in the trends are given (*very likely* range).

4.2.2.4 Ice Thickness and Volume

For the Arctic, there are several techniques available for estimating the thickness distribution of sea ice. Combined data sets of draft and thickness from submarine sonars, satellite altimetry and airborne electromagnetic sensing provide broadly consistent and strong evidence of decrease in Arctic sea ice thickness in recent years (Figure 4.6c).

Data collected by upward-looking sonar on submarines operating beneath the Arctic pack ice provided the first evidence of 'basin-wide' decreases in ice thickness (Wadhams, 1990). Sonar measurements are of average draft (the submerged portion of sea ice), which is converted to thickness by assuming an average density for the measured floe including its snow cover. With the then available submarine records, Rothrock et al. (1999) found that ice draft in the mid-1990s was less than that measured between 1958 and 1977 in each of six regions within the Arctic Basin. The change was least (–0.9 m) in the Beaufort and Chukchi seas and greatest (–1.7 m) in the Eurasian Basin. The decrease averaged about 42% of the average 1958 to 1977 thickness. This decrease matched the decline measured in the Eurasian Basin between 1976 and 1996 using UK submarine data (Wadhams and Davis, 2000), which was 43%.

A subsequent analysis of US Navy submarine ice draft (Rothrock et al., 2008) used much richer and more geographically extensive data from 34 cruises within a data release area that covered almost 38% of the area of the Arctic Ocean. These cruises were equally distributed in spring and autumn over a 25-year period between 1975 and 2000. Observational uncertainty associated with the ice draft from these is 0.5 m (Rothrock and Wensnahan, 2007). Multiple regression analysis was used to separate the interannual changes (Figure 4.6c), the annual cycle and the spatial distribution of draft in the observations. Results of that analysis show that the annual mean ice thickness declined from a peak of 3.6 m in 1980 to 2.4 m in 2000, a decrease of 1.2 m. Over the period, the most rapid change was –0.08 m yr^{-1} in 1990.

The most recent submarine record, Wadhams et al. (2011), found that tracks north of Greenland repeated between the winters of 2004 and 2007 showed a continuing shift towards less multi-year ice.

Satellite altimetry techniques are now capable of mapping sea ice freeboard to provide relatively comprehensive pictures of the distribution of Arctic sea ice thickness. Similar to the estimation of sea ice thickness from ice draft, satellite measured freeboard (the height of sea ice above the water surface) is converted to thickness, assuming an average density of ice and snow. The principal challenges to accurate thickness estimation using satellite altimetry are in the discrimination of ice and open water, and in estimating the thickness of the snow cover.

Since 1993, radar altimeters on the European Space Agency (ESA), European Remote Sensing (ERS) and Envisat satellites have provided Arctic observations south of 81.5°N. With the limited latitudinal reach of these altimeters, however, it has been difficult to infer basin-wide changes in thickness. The ERS-1 estimates of ice thickness show a downward trend but, because of the high variability and short time series (1993–2001), Laxon et al. (2003) concluded that the trend in a region of mixed seasonal and multi-year ice (i.e., below 81.5°N) cannot be considered as significant. Envisat observations showed a large decrease in thickness (0.25 m) following September 2007 when ice extent was the second lowest on record (Giles et al., 2008b). This was associated with the large retreat of the summer ice cover, with thinning regionally confined to the Beaufort and Chukchi seas, but with no significant changes in the eastern Arctic. These results are consistent with those from the NASA Ice, Cloud and land Elevation Satellite (ICESat) laser altimeter (see comment on ICESat data in Section 4.4.2.1), which show thinning in the same regions between 2007 and 2008 (Kwok, 2009) (Figure 4.5). Large decreases in thickness due to the 2007 minimum in summer ice are clearly seen in both the radar and laser altimeter thickness estimates.

The coverage of the laser altimeter on ICESat (which ceased operation in 2009) extended to 86°N and provided a more complete spatial pattern of the thickness distribution in the Arctic Basin (Figure 4.6c). Thickness estimates are consistently within 0.5 m of sonar measurements from near-coincident submarine tracks and profiles from sonar moorings in the Chukchi and Beaufort seas (Kwok, 2009). Ten ICESat campaigns between autumn 2003 and spring 2008 showed seasonal differences in thickness and thinning and volume losses of the Arctic Ocean ice cover (Kwok, 2009). Over these campaigns, the multi-year sea ice thickness in spring declined by ~0.6 m (Figure 4.5), while the average thickness of the first-year ice (~2 m) had a negligible trend. The average sea ice volume inside the Arctic Basin in spring (February/March) was ~14,000 km^3. Between 2004 and 2008, the total multi-year ice volume in spring (February/March) experienced a net loss of 6300 km^3 (>40%). Residual differences between sonar mooring and satellite thicknesses suggest basin-scale volume uncertainties of approximately 700 km^3. The rate of volume loss (–1237 km^3 yr^{-1}) during autumn (October/November), while highlighting the large changes during the short ICESat record compares with a more moderate loss rate (–280 ± 100 km^3 yr^{-1}) over a 31-year period (1979–2010) estimated from a sea ice reanalysis study using the Pan-Arctic Ice-Ocean Modelling and Assimilation system (Schweiger et al., 2011).

The CryoSat-2 radar altimeter (launched in 2010), which provides coverage up to 89°N, has provided new thickness and volume estimates of Arctic Ocean sea ice (Laxon et al., 2013). These show that the ice volume inside the Arctic Basin decreased by a total of 4291 km^3 in autumn (October/November) and 1479 km^3 in winter (February/March) between the ICESat (2003–2008) and CryoSat-2 (2010–2012) periods. Based on ice thickness estimates from sonar moorings, an inter-satellite bias between ICESat and CryoSat-2 of 700 km^3 can be expected. This is much less than the change in volume between the two periods.

Airborne electro-magnetic (EM) sounding measures the distance between an EM instrument near the surface or on an aircraft and the ice/water interface, and provides another method to measure ice thickness. Uncertainties in these thickness estimates are 0.1 m over level ice. Comparison with drill-hole measurements over a mix of level and ridged ice found differences of 0.17 m (Haas et al., 2011).

Repeat EM surveys in the Arctic, though restricted in time and space, have provided a regional view of the changing ice cover. From repeat ground-based and helicopter-borne EM surveys, Haas et al. (2008) found significant thinning in the region of the Transpolar Drift (an

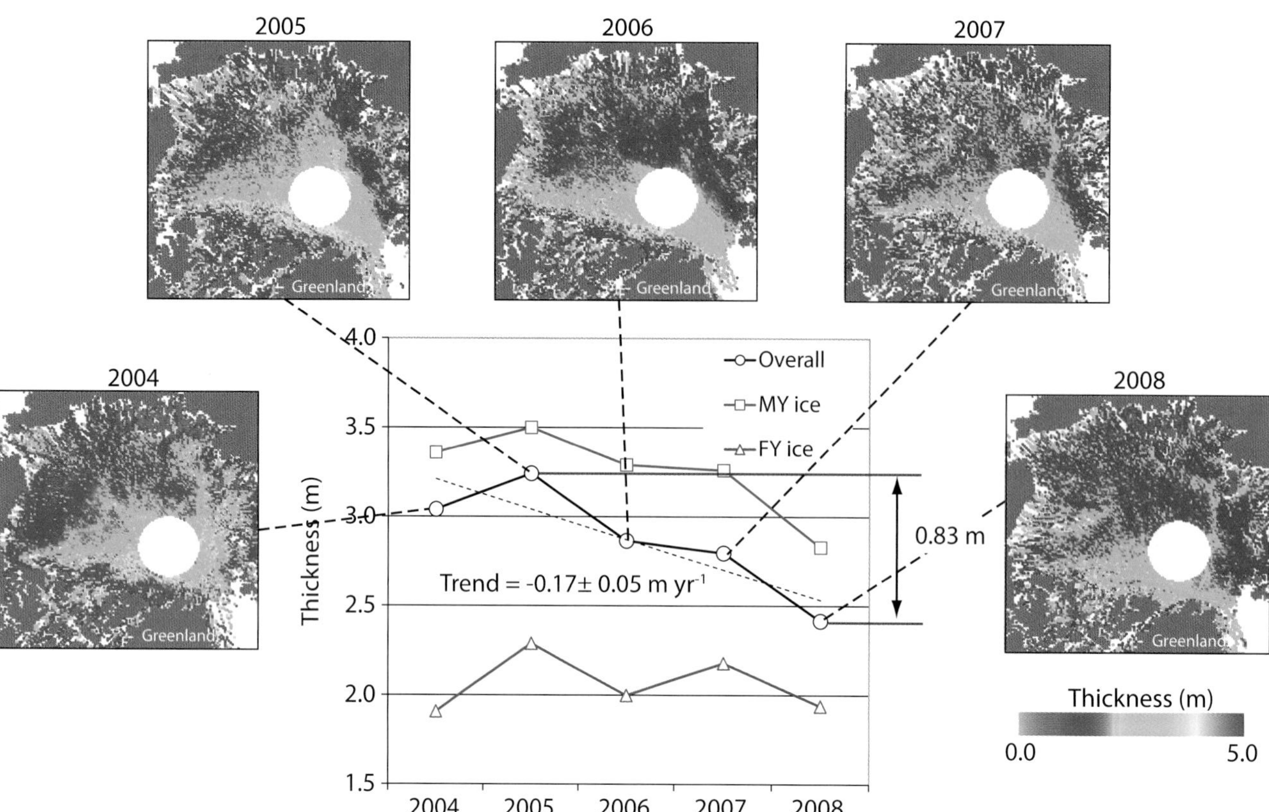

Figure 4.5 | The distribution of winter sea ice thickness in the Arctic and the trends in average, first-year (FY and multi-year (MY) ice thickness derived from ICESat data between 2004 and 2008 (Kwok, 2009).

4

average wind-driven drift pattern that transports sea ice from the Siberian coast of Russia across the Arctic Basin to Fram Strait). Between 1991 and 2004, the modal ice thickness decreased from 2.5 m to 2.2 m, with a larger decline to 0.9 m in 2007. Mean ice thicknesses also decreased strongly. This thinning was associated with reduction of the age of the ice, and replacement of second-year ice by first-year ice in 2007 (following the large decline in summer ice extent in 2007) as seen in satellite observations. Ice thickness estimates from EM surveys near the North Pole can be compared to submarine estimates (Figure 4.6c). Airborne EM measurements from the Lincoln Sea between 83°N and 84°N since 2004 (Haas et al., 2010) showed some of the thickest ice in the Arctic, with mean and modal thicknesses of more than 4.5 m and 4 m, respectively. Since 2008, the modal thickness in this region has declined to 3.5 m, which is most likely related to the narrowing of the remaining band of old ice along the northern coast of Canada.

4.2.2.5 Arctic Sea Ice Drift

Ice motion influences the distribution of sea ice thickness in the Arctic Basin: locally, through deformation and creation of open water areas; regionally, through advection of ice from one area to another; and basin-wide, through export of ice from polar seas to lower latitudes where it melts. The drift and deformation of sea ice is forced primarily by winds and surface currents, but depends also on ice strength, top and bottom surface roughness, and ice concentration. On time scales of days to weeks, winds are responsible for most of the variance in sea ice motion.

Drifting buoys have been used to measure Arctic sea ice motion since 1979. From the record of buoy drift archived by the International Arctic Buoy Programme, Rampal et al. (2009) found an increase in average drift speed between 1978 and 2007 of 17 ± 4.5% per decade in winter and 8.5 ± 2.0% per decade in summer. Using daily satellite ice motion fields, which provide a basin-wide picture of the ice drift, Spreen et al. (2011) found that, between 1992 and 2008, the spatially averaged winter ice drift speed increased by 10.6 ± 0.9% per decade, but varied regionally between –4 and +16% per decade (Figure 4.6d). Increases in drift speed are seen over much of the Arctic except in areas with thicker ice (Figure 4.6b, e.g., north of Greenland and the Canadian Archipelago). The largest increases occurred during the second half of the period (2001–2009), coinciding with the years of rapid ice thinning discussed in Section 4.2.2.4. Both Rampal et al. (2009) and Spreen et al. (2011) suggest that, since atmospheric reanalyses do not show stronger winds, the positive trend in drift speed is probably due to a weaker and thinner ice cover, especially during the period after 2003.

In addition to freezing and melting, sea ice export through Fram Strait is a major component of the Arctic Ocean ice mass balance. Approximately10% of the area of Arctic Ocean ice is exported annually. Over a 32-year satellite record (1979–2010), the mean annual outflow of ice area through Fram Strait was 699 ± 112 × 10^3 km^2 with a peak during the 1994–1995 winter (updated from , Kwok, 2009), but with no significant decadal trend. Decadal trends in ice volume export—a more definitive measure of change—is far less certain owing to the lack of an extended record of the thickness of sea ice exported through Fram

Strait. Comparison of volume outflow using ICESat thickness estimates (Spreen et al., 2009) with earlier estimates by Kwok and Rothrock (1999) and Vinje (2001) using thicknesses from moored upward looking sonars shows no discernible change.

Between 2005 and 2008, more than a third of the thicker and older sea ice loss occurred by transport of thick, multi-year ice, typically found west of the Canadian Archipelago, into the southern Beaufort Sea, where it melted in summer (Kwok and Cunningham, 2010). Uncertainties remain in the relative contributions of in-basin melt and export to observed changes in Arctic ice volume loss, and it has also been shown that export of thicker ice through Nares Strait could account for a small fraction of the loss (Kwok, 2005).

4.2.2.6 Timing of Sea Ice Advance, Retreat and Ice Season Duration; Length of Melt Season

Importantly from both physical and biological perspectives, strong regional changes have occurred in the seasonality of sea ice in both polar regions (Massom and Stammerjohn, 2010; Stammerjohn et al., 2012). However, there are distinct regional differences in when seasonally the change is strongest (Stammerjohn et al., 2012).

Seasonality collectively describes the annual time of sea ice advance and retreat, and its duration (the time between day of advance and retreat). Daily satellite ice-concentration records (1979–2012) are used to determine the day to which sea ice advanced, and the day from which it retreated, for each satellite pixel location. Maps of the timing of sea ice advance, retreat and duration are derived from these data (see Parkinson (2002) and Stammerjohn et al. (2008) for detailed methods).

Most regions in the Arctic show trends towards shorter ice season duration. One of the most rapidly changing areas (showing greater than 2 days yr^{-1} change) extends from the East Siberian Sea to the western Beaufort Sea. Here, between 1979 and 2011, sea ice advance occurred 41 ± 6 days later (or 1.3 ± 0.2 days yr^{-1}), sea ice retreat 49 ± 7 days earlier (–1.5 ± 0.2 days yr^{-1}), and duration became 90 ± 16 days shorter (–2.8 ± 0.5 days yr^{-1}) (Stammerjohn et al., 2012). This 3-month lengthening of the summer ice-free season places Arctic summer sea ice extent loss into a seasonal perspective and underscores impacts to the marine ecosystem (e.g., Grebmeier et al., 2010).

The timing of surface melt onset in spring, and freeze-up in autumn, can be derived from satellite microwave data as the emissivity of the surface changes significantly with snow melt (Smith, 1998; Drobot and Anderson, 2001; Belchansky et al., 2004). The amount of solar energy absorbed by the ice cover increases with the length of the melt season. Longer melt seasons with lower albedo surfaces (wet snow, melt ponds and open water) increase absorption of incoming shortwave radiation and ice melt (Perovich et al., 2007). Hudson (2011) estimates that the observed reduction in Arctic sea ice has contributed approximately 0.1 W m^{-2} of additional global radiative forcing, and that an ice-free summer Arctic Ocean will result in a forcing of about 0.3 W m^{-2}. The satellite record (Markus et al., 2009) shows a trend toward earlier melt and later freeze-up nearly everywhere in the Arctic (Figure 4.6e). Over the last 34 years, the mean melt season over the Arctic ice cover has increased at a rate of 5.7 ± 0.9 days per decade. The largest and most significant trends (at the 99% level) of more than 10 days per decade are seen in the coastal margins and peripheral seas: Hudson Bay, the East Greenland Sea, the Laptev/East Siberian seas, and the Chukchi/Beaufort seas.

4.2.2.7 Arctic Polynyas

High sea ice production in coastal polynyas (anomalous regions of open water or low ice concentration) over the continental shelves of the Arctic Ocean is responsible for the formation of cold saline water, which contributes to the maintenance of the Arctic Ocean halocline (see Glossary). A new passive microwave algorithm has been used to estimate thin sea ice thicknesses (<0.15 m) in the Arctic Ocean (Tamura and Ohshima, 2011), providing the first circumpolar mapping of sea ice production in coastal polynyas. High sea ice production is confined to the most persistent Arctic coastal polynyas, with the highest ice production rate being in the North Water Polynya. The mean annual sea ice production in the 10 major Arctic polynyas is estimated to be 2942 ± 373 km^3 and decreased by 462 km^3 between 1992 and 2007 (Tamura and Ohshima, 2011).

4.2.2.8 Arctic Land-Fast Ice

Shore- or land-fast ice is sea ice attached to the coast. Land-fast ice along the Arctic coast is usually grounded in shallow water, with the seaward edge typically around the 20 to 30 m isobath (Mahoney et al., 2007). In fjords and confined bays, land-fast ice extends into deeper water.

There are no reliable estimates of the total area or interannual variability of land-fast ice in the Arctic. However, both significant and non-significant trends have been observed regionally. Long-term monitoring near Hopen, Svalbard, revealed thinning of land-fast ice in the Barents Sea region by 11 cm per decade between 1966 and 2007 (Gerland et al., 2008). Between 1936 and 2000, the trends in land-fast ice thickness (in May) at four Siberian sites (Kara Sea, Laptev Sea, East Siberian Sea, Chukchi Sea) are insignificant (Polyakov et al., 2003). A more recent composite time series of land-fast ice thickness between the mid 1960s and early 2000s from 15 stations along the Siberian coast revealed an average rate of thinning of 0.33 cm yr^{-1}(Polyakov et al., 2010). End-of-winter ice thickness for three stations in the Canadian Arctic reveal a small downward trend at Eureka, a small positive trend at Resolute Bay, and a negligible trend at Cambridge Bay (updated from Brown and Coté, 1992; Melling, 2012), but these trends are small and not statistically significant. Even though the trend in the land-fast ice extent near Barrow, Alaska has not been significant (Mahoney et al., 2007), relatively recent observations by Mahoney et al. (2007) and Druckenmiller et al. (2009) found longer ice-free seasons and thinner land-fast ice compared to earlier records (Weeks and Gow, 1978; Barry et al., 1979). As freeze-up happens later, the growth season shortens and the thinner ice breaks up and melts earlier.

4.2.2.9 Decadal Trends in Arctic Sea Ice

The average decadal extent of Arctic sea ice has decreased in every season and in every successive decade since satellite observations commenced. The data set is robust with continuous and consistent global coverage on a daily basis thereby providing very reliable trend results (*very high confidence*). The annual Arctic sea ice cover *very likely* declined within the range 3.5 to 4.1% per decade (0.45 to 0.51 million km^2 per decade) during the period 1979–2012 with larger changes occurring in summer and autumn (*very high confidence*). Much larger changes apply to the perennial ice (the summer minimum extent) which *very likely* decreased in the range from 9.4 % to 13.6 % per decade (0.73 to 1.07 million km^2 per decade) and multiyear sea ice (more than 2 years old) which *very likely* declined in the range from 11.0 % to 16.0% per decade (0.66 to 0.98 million km^2 per decade) (*very high confidence*; Figure 4.4b). The rate of decrease in ice area has been greater than that in extent (Figure 4.4b) because the ice concentration has also decreased. The decline in multiyear ice cover as observed by QuikScat from 1992 to 1910 is presented in Figure 4.6b and shown to be consistent with passive microwave data (Figure 4.4b).

The decrease in perennial and multi-year ice coverage has resulted in a strong decrease in ice thickness, and hence in ice volume. Declassified submarine sonar measurements, covering ~38% of the Arctic Ocean, indicate an overall mean winter thickness of 3.64 m in 1980, which *likely* decreased by 1.8 [1.3 to 2.3] m by 2008 (*high confidence*, Figure 4.6c). Between 1975 and 2000, the steepest rate of decrease was 0.08 m yr^{-1} in 1990 compared to a slightly higher winter/summer rate of 0.10/0.20 m yr^{-1} in the 5-year ICESat record (2003–2008). This combined analysis (Figure 4.6c) shows a long-term trend of sea ice thinning that spans five decades. Satellite measurements made in the period 2010–2012 show a decrease in basin-scale sea ice volume compared to those made over the period 2003–2008 (*medium confidence*). The Arctic sea ice is becoming increasingly seasonal with thinner ice, and it will take several years for any recovery.

The decreases in both concentration and thickness reduces sea ice strength reducing its resistance to wind forcing, and drift speed has increased (Figure 4.6d) (Rampal et al., 2009; Spreen et al., 2011). Other significant changes to the Arctic Ocean sea ice include lengthening in the duration of the surface melt on perennial ice of 6 days per decade (Figure 4.6e) and a nearly 3-month lengthening of the ice-free season in the region from the East Siberian Sea to the western Beaufort Sea.

4.2.3 Antarctic Sea Ice

The Antarctic sea ice cover is largely seasonal, with average extent varying from a minimum of about 3 × 10^6 km^2 in February to a maximum of about 18 × 10^6 km^2 in September (Zwally et al., 2002a; Comiso et al., 2011). The relatively small fraction of Antarctic sea ice that survives the summer is found mostly in the Weddell Sea, but with some perennial ice also surviving on the western side of the Antarctic Peninsula and in small patches around the coast. As well as being mostly first-year ice, Antarctic sea ice is also on average thinner, warmer, more saline and more mobile than Arctic ice (Wadhams and Comiso, 1992). These characteristics, which reduce the capabilities of some remote sensing techniques, together with its more distant location from inhabited continents, result in far less being known about the properties of Antarctic sea ice than of that in the Arctic.

4.2.3.1 Total Antarctic Sea Ice Extent and Concentration

Figure 4.7a shows the seasonal variability of Antarctic sea ice extent using 34 years of satellite passive microwave data updated from Comiso and Nishio (2008). In contrast to the Arctic, decadal monthly averages almost overlap with each other, and the seasonal variability of the total Antarctic sea ice cover has not changed much over the period. In winter, the values for the 1999–2008 decade were slightly higher than those of the other decades; whereas in autumn the values for 1989–1998 and 1999–2008 decades were higher than those of 1979–1988. There was more seasonal variability in the period 2009–2012 than for earlier decadal periods, with relatively high values in late autumn, winter and spring.

Trend maps for winter, spring, summer and autumn extent are presented in Figure 4.7 (b, c, d and e respectively). The seasonal trends are significant mainly near the ice edge, with the values alternating between positive and negative around Antarctica. Such an alternating pattern is similar to that described previously as the Antarctic Circumpolar Wave (ACW) (White and Peterson, 1996) but the ACW may not be associated with the trends because the trends have been strongly positive in the Ross Sea and negative in the Bellingshausen/Amundsen seas but with almost no trend in the other regions (Comiso et al., 2011). In the winter, negative trends are evident at the tip of the Antarctic Peninsula and the western part of the Weddell Sea, while positive trends are prevalent in the Ross Sea. The patterns in spring are very similar to those of winter, whereas in summer and autumn negative trends are mainly confined to the Bellingshausen/Amundsen seas, while positive trends are dominant in the Ross Sea and the Weddell Sea.

The regression trend in the monthly anomalies of Antarctic sea ice extent from November 1978 to December 2012 (updated from Comiso and Nishio, 2008) is slightly positive, at 1.5 ± 0.3% per decade, or 0.13 to 0.20 million km^2 per decade (*very likely*) (see FAQ 4.1). The seasonal trends in ice extent are 1.2 ± 0.5%, 1.0 ± 0.5%, 2.5 ± 2.0% and 3.0 ± 2.0% per decade (*very likely*) in winter, spring, summer and autumn, respectively, as updated from Comiso et al. (2011). The corresponding trends in ice area (also updated) are 1.9 ± 0.7%, 1.6 ± 0.5%, 3.0 ± 2.1%, and 4.4 ± 2.3% per decade (*very likely*). The values are all positive, with the largest trends occurring in the autumn. The trends are consistently higher for ice area than ice extent, indicating less open water (possibly due to less storms and divergence) within the pack in later years. Trends reported by Parkinson and Cavalieri (2012) using data from 1978 to 2010 are slightly different, in part because they cover a different time period (see Supplementary Material). The overall interannual trends for various sectors around Antarctica are given in FAQ 4.1, and show large regional variability. Changes in ice drift and wind patterns as reported by Holland and Kwok (2012) may be related to this phenomenon.

4.2.3.2 Antarctic Sea Ice Thickness and Volume

Since AR4, some advances have been made in determining the thickness of Antarctic sea ice, particularly in the use of ship-based observations and satellite altimetry. However, there is still no information on large-scale Antarctic ice thickness change. Worby et al. (2008) compiled 25 years of ship-based data from 83 Antarctic voyages on

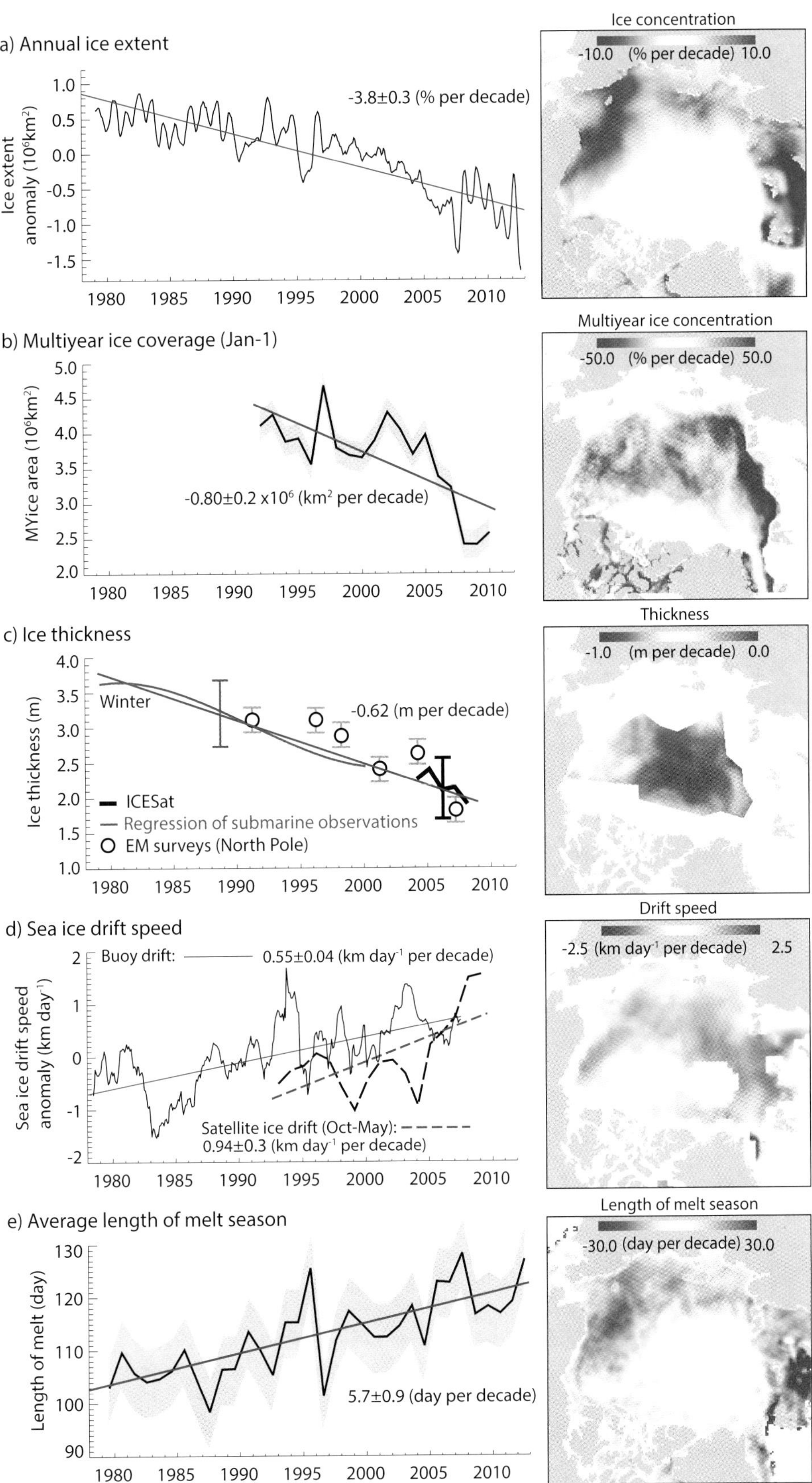

Figure 4.6 | Summary of linear decadal trends (red lines) and pattern of changes in the following: (a) Anomalies in Arctic sea ice extent from satellite passive microwave observations (Comiso and Nishio, 2008, updated to include 2012). Uncertainties are discussed in the text. (b) Multi-year sea ice coverage on January 1st from analysis of the QuikSCAT time series (Kwok, 2009); grey band shows uncertainty in the retrieval. (c) Sea ice thickness from submarine (blue), satellites (black) (Kwok and Rothrock, 2009), and *in situ*/electromagnetic (EM) surveys (circles) (Haas et al., 2008); trend in submarine ice thickness is from multiple regression of available observations within the data release area (Rothrock et al., 2008). Error bars show uncertainties in observations. (d) Anomalies in buoy (Rampal et al., 2009) and satellite-derived sea ice drift speed (Spreen et al., 2011). (e) Length of melt season (updated from Markus et al., 2009); grey band shows the basin-wide variability.

4

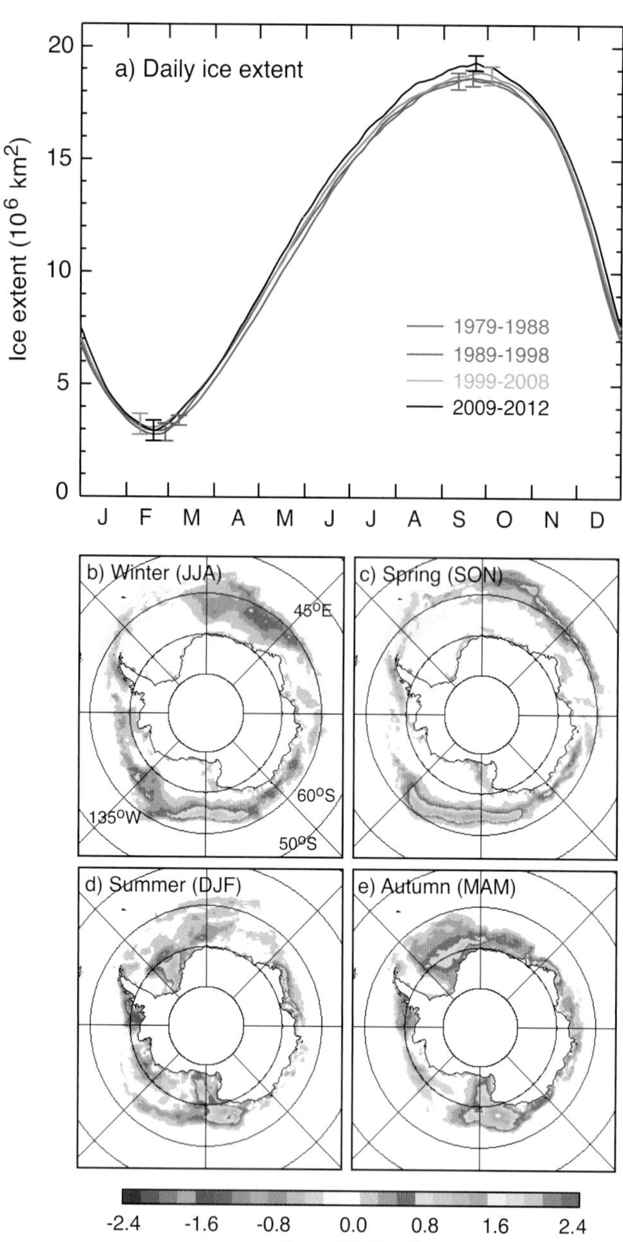

Figure 4.7 | (a) Plots of decadal averages of daily sea ice extent in the Antarctic (1979–1988 in red, 1989–1998 in blue, 1999– 2008 in gold) and a 4-year average daily ice extent from 2009 to 2012 in black. Maps indicate ice concentration trends (1979–2012) in (b) winter, (c) spring, (d) summer and (e) autumn (updated from Comiso, 2010).

which routine observations of sea ice and snow properties were made. Their compilation included a gridded data set that reflects the regional differences in sea ice thickness. A subset of these ship observations, and ice charts, was used by DeLiberty et al. (2011) to estimate the annual cycle of sea ice thickness and volume in the Ross Sea, and to investigate the relationship between ice thickness and extent. They found that maximum sea ice volume was reached later than maximum extent. While ice is advected to the northern edge and melts, the interior of the sea ice zone is supplied with ice from higher latitudes and continues to thicken by thermodynamic growth and deformation. Satellite retrievals of sea ice freeboard and thickness in the Antarctic (Mahoney et al., 2007; Zwally et al., 2008; Xie et al., 2011) are under development but progress is limited by knowledge of snow thickness and the paucity of suitable validation data sets. A recent analysis of the ICESat record by Kurtz and Markus (2012), assuming zero ice freeboard, found negligible trends in ice thickness over the 5-year record.

4.2.3.3 Antarctic Sea Ice Drift

Using a 19-year data set (1992–2010) of satellite-tracked sea ice motion, Holland and Kwok (2012) found large and statistically significant decadal trends in Antarctic ice drift that in most sectors are caused by changes in local winds. These trends suggest acceleration of the wind-driven Ross Gyre and deceleration of the Weddell Gyre. The changes in meridional ice transport affect the freshwater budget near the Antarctic coast. This is consistent with the increase of 30,000 km^2 yr^{-1} in the net area export of sea ice from the Ross Sea shelf coastal polynya region between 1992 and 2008 (Comiso et al., 2011). Assuming an annual average thickness of 0.6 m, Comiso et al. (2011) estimated an increase in volume export of 20 km^3 yr^{-1} which is similar to the rate of production in the Ross Sea coastal polynya region for the same period discussed in Section 4.2.3.5.

4.2.3.4 Timing of Sea Ice Advance, Retreat and Ice Season Duration

In the Antarctic there are regionally different patterns of strong change in ice duration (>2 days yr^{-1}). In the northeast and west Antarctic Peninsula and southern Bellingshausen Sea region, later ice advance (+61 ± 15 days), earlier retreat (–39 ± 13 days) and shorter duration (+100 ± 31 days, a trend of –3.1 ± 1.0 days yr^{-1}) occurred over the period 1979/1980–2010/2011 (Stammerjohn et al., 2012). These changes have strong impacts on the marine ecosystem (Montes-Hugo et al., 2009; Ducklow et al., 2011). The opposite is true in the adjacent western Ross Sea, where substantial lengthening of the ice season of 79 ± 12 days has occurred (+2.5 ± 0.4 days yr^{-1}) due to earlier advance (+42 ± 8 days) and later retreat (–37 ± 8 days). Patterns of change in the relatively narrow East Antarctic sector are generally of a lower magnitude and zonally complex, but in certain regions involve changes in the timing of sea ice advance and retreat of the order of ±1 to 2 days yr^{-1} (for the period 1979–2009) (Massom et al., 2013).

4.2.3.5 Antarctic Polynyas

Polynyas are commonly found along the coast of Antarctica. There are two different processes that cause a polynya. Warm water upwelling keeps the surface water near the freezing point and reduces ice production (sensible heat polynya), and wind or ocean currents move ice away and increase further ice production (latent heat polynya).

An increase in the extent of coastal polynyas in the Ross Sea caused increased ice production (latent heat effect) that is primarily responsible for the positive trend in ice extent in the Antarctic (Comiso et al., 2011). Drucker et al. (2011) show that in the Ross Sea, the net ice export equals the annual ice production in the Ross Sea polynya (approximately 400 km^3 in 1992), and that ice production increased by 20 km^3 yr^{-1} from 1992 to 2008. However, the ice production in the Weddell Sea, which is three times less, has had no statistically significant

Frequently Asked Questions

FAQ 4.1 | How Is Sea Ice Changing in the Arctic and Antarctic?

The sea ice covers on the Arctic Ocean and on the Southern Ocean around Antarctica have quite different characteristics, and are showing different changes with time. Over the past 34 years (1979–2012), there has been a downward trend of 3.8% per decade in the annual average extent of sea ice in the Arctic. The average winter thickness of Arctic Ocean sea ice has thinned by approximately 1.8 m between 1978 and 2008, and the total volume (mass) of Arctic sea ice has decreased at all times of year. The more rapid decrease in the extent of sea ice at the summer minimum is a consequence of these trends. In contrast, over the same 34-year period, the total extent of Antarctic sea ice shows a small increase of 1.5% per decade, but there are strong regional differences in the changes around the Antarctic. Measurements of Antarctic sea ice thickness are too few to be able to judge whether its total volume (mass) is decreasing, steady, or increasing.

A large part of the total Arctic sea ice cover lies above 60°N (FAQ 4.1, Figure 1) and is surrounded by land to the south with openings to the Canadian Arctic Archipelago, and the Bering, Barents and Greenland seas. Some of the ice within the Arctic Basin survives for several seasons, growing in thickness by freezing of seawater at the base and by deformation (ridging and rafting). Seasonal sea ice grows to only ~2 m in thickness but sea ice that is more than 1 year old (perennial ice) can be several metres thicker. Arctic sea ice drifts within the basin, driven by wind and ocean currents: the mean drift pattern is dominated by a clockwise circulation pattern in the western Arctic and a Transpolar Drift Stream that transports Siberian sea ice across the Arctic and exports it from the basin through the Fram Strait.

Satellites with the capability to distinguish ice and open water have provided a picture of the sea ice cover changes. Since 1979, the annual average extent of ice in the Arctic has decreased by 3.8% per decade. The decline in extent at the end of summer (in late September) has been even greater at 11% per decade, reaching a record minimum in 2012. The decadal average extent of the September minimum Arctic ice cover has decreased for each decade since satellite records began. Submarine and satellite records suggest that the thickness of Arctic ice, and hence the total volume, is also decreasing. Changes in the relative amounts of perennial and seasonal ice are contributing to the reduction in ice volume. Over the 34-year record, approximately 17% of this type of sea ice per decade has been lost to melt and export out of the basin since 1979 and 40% since 1999. Although the area of Arctic sea ice coverage can fluctuate from year to year because of variable seasonal production, the proportion of thick perennial ice, and the total sea ice volume, can recover only slowly.

Unlike the Arctic, the sea ice cover around Antarctica is constrained to latitudes north of 78°S because of the presence of the continental land mass. The Antarctic sea ice cover is largely seasonal, with an average thickness of only ~1 m at the time of maximum extent in September. Only a small fraction of the ice cover survives the summer minimum in February, and very little Antarctic sea ice is more than 2 years old. The ice edge is exposed to the open ocean and the snowfall rate over Antarctic sea ice is higher than in the Arctic. When the snow load from snowfall is sufficient to depress the ice surface below sea level, seawater infiltrates the base of the snow pack and snow-ice is formed when the resultant slush freezes. Consequently, snow-to-ice conversion (as well as basal freezing as in the Arctic) contributes to the seasonal growth in ice thickness and total ice volume in the Antarctic. Snow-ice formation is sensitive to changes in precipitation and thus changes in regional climate. The consequence of changes in precipitation on Antarctic sea ice thickness and volume remains a focus for research.

Unconstrained by land boundaries, the latitudinal extent of the Antarctic sea ice cover is highly variable. Near the Antarctic coast, sea ice drift is predominantly from east to west, but further north, it is from west to east and highly divergent. Distinct clockwise circulation patterns that transport ice northward can be found in the Weddell and Ross seas, while the circulation is more variable around East Antarctica. The northward extent of the sea ice cover is controlled in part by the divergent drift that is conducive in winter months to new ice formation in persistent open water areas (polynyas) along the coastlines. These zones of ice formation result in saltier and thus denser ocean water and become one of the primary sources of the deepest water found in the global oceans.

Over the same 34-year satellite record, the annual extent of sea ice in the Antarctic increased at about 1.5% per decade. However, there are regional differences in trends, with decreases seen in the Bellingshausen and Amundsen seas, but a larger increase in sea ice extent in the Ross Sea that dominates the overall trend. Whether the smaller overall increase in Antarctic sea ice extent is meaningful as an indicator of climate is uncertain because the extent

(continued on next page)

FAQ 4.1 (continued)

varies so much from year to year and from place to place around the continent. Results from a recent study suggest that these contrasting trends in ice coverage may be due to trends in regional wind speed and patterns. Without better ice thickness and ice volume estimates, it is difficult to characterize how Antarctic sea ice cover is responding to changing climate, or which climate parameters are most influential.

There are large differences in the physical environment and processes that affect the state of Arctic and Antarctic sea ice cover and contribute to their dissimilar responses to climate change. The long, and unbroken, record of satellite observations have provided a clear picture of the decline of the Arctic sea ice cover, but available evidence precludes us from making robust statements about overall changes in Antarctic sea ice and their causes.

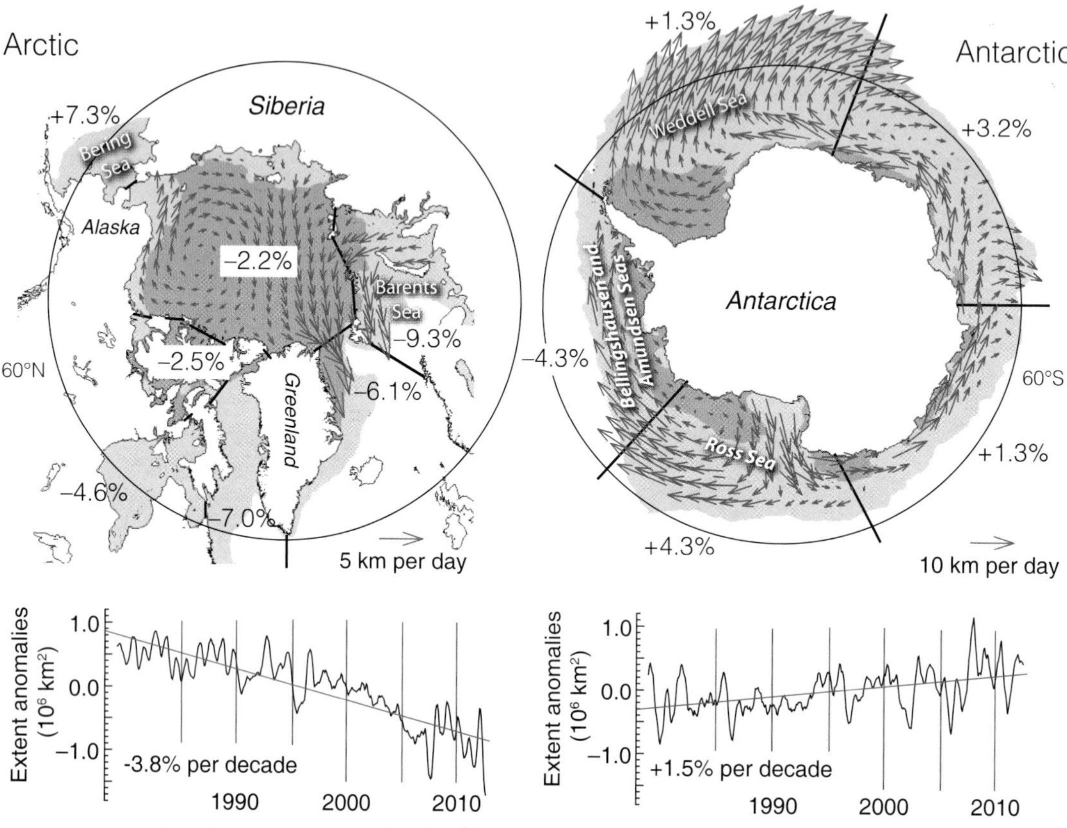

FAQ 4.1, Figure 1 | The mean circulation pattern of sea ice and the decadal trends (%) in annual anomalies in ice extent (i.e., after removal of the seasonal cycle), in different sectors of the Arctic and Antarctic. Arrows show the average direction and magnitude of ice drift. The average sea ice cover for the period 1979 through 2012, from satellite observations, at maximum (minimum) extent is shown as orange (grey) shading.

trend over the same period. Variability in the ice cover in this region is linked to changes in the Southern Annular Mode (SAM). Between 1974 and 1976, the large Weddell Sea Polynya, which is a sensible heat polynya, was created by the injection of relatively warm deep water into the surface layer due to sustained deep-ocean convection (sensible heat effect) during negative SAM, but since the late 1970s the SAM has been mainly positive, resulting in warmer and wetter condition forestalling any reoccurrence of the Weddell Sea Polynya (Gordon et al., 2007).

4.2.3.6 Antarctic Land-Fast Ice

Land-fast ice forms around the coast of Antarctica, typically in narrow coastal bands of varying width up to 150 km from the coast and in water depths of up to 400 to 500 m. Around East Antarctica, it comprises generally between 5% (winter) and 35% (summer) of the overall sea ice area (Fraser et al., 2012), and a greater fraction of ice volume (Giles et al., 2008a).

Variability in the distribution and extent of land-fast ice is sensitive to processes of ice formation and to processes such as ocean swell and waves, and strong wind events that cause the ice to break-up. Historical records of Antarctic land-fast ice extent, such as that of Kozlovsky et al. (1977) covering 0° to 160°E, were limited by sparse and sporadic sampling. Recently, using cloud-free Moderate Resolution Imaging Spectrometer (MODIS) composite images, Fraser et al. (2012) derived a high-resolution time series of land-fast sea ice extent along the East Antarctic coast, showing a statistically significant increase (1.43 ± 0.30% yr^{-1}) between March 2000 and December 2008. There is a strong increase in the Indian Ocean sector (20°E to 90°E, 4.07 ± 0.42% yr^{-1}), and a non-significant decrease in the sector from 90°E to 160°E (–0.40 ± 0.37% yr^{-1}). An apparent shift from a negative to a positive trend was noted in the Indian Ocean sector from 2004, which coincided with greater interannual variability. Although significant changes are observed, this record is only 9 years in length.

4.2.3.7 Decadal Trends in Antarctic Sea Ice

For the Antarctic, any changes in many sea ice characteristics are unknown. There has been a small but significant increase in total annual mean sea ice extent that is *very likely* in the range of 1.2 to 1.8 % per decade between 1979 and 2012 (0.13 to 0.20 million km^2 per decade) (*very high confidence*). There was also a greater increase in ice area associated with an increase in ice concentration. But there are strong regional differences within this total, with some regions increasing in extent/area and some decreasing (*high confidence*). Similarly, there are contrasting regions around the Antarctic where the ice-free season has lengthened, and others where it has decreased over the satellite period (*high confidence*). There are still inadequate data to make any assessment of changes to Antarctic sea ice thickness and volume.

4.3 Glaciers

This section considers all perennial surface land ice masses (defined in 4.1 and Glossary) outside of the Antarctic and Greenland ice sheets. Glaciers occur where climate conditions and topographic characteristics allow snow to accumulate over several years and to transform gradually into firn (snow that persists for at least one year) and finally to ice. Under the force of gravity, this ice flows downwards to elevations with higher temperatures where various processes of ablation (loss of snow and ice) dominate over accumulation (gain of snow and ice). The sum of all accumulation and ablation processes determines the mass balance of a glacier. Accumulation is in most regions due mainly to solid precipitation (in general snow), but also results from refreezing of liquid water, especially in polar regions or at high altitudes where firn remains below melting temperature. Ablation is, in most regions, mainly due to surface melting with subsequent runoff, but loss of ice by calving (on land or in water; see Glossary) or sublimation (important in dry regions) can also dominate. Re-distribution of snow by wind and avalanches can contribute to both accumulation and ablation. The energy and mass fluxes governing the surface mass balance are directly linked to atmospheric conditions and are modified by topography (e.g., due to shading). Glaciers are sensitive climate indicators because they adjust their size in response to changes in climate (e.g., temperature and precipitation) (FAQ 4.2). Glaciers are also important seasonal to long-term hydrologic reservoirs (WGII, Chapter 3) on a regional scale and a major contributor to sea level rise on a global scale (see Section 4.3.3.4 and Chapter 13). In the following, we report global glacier coverage (Section 4.3.1), how changes in length, area, volume and mass are determined (Section 4.3.2) and the observed changes in these parameters through time (Section 4.3.3).

4.3.1 Current Area and Volume of Glaciers

The total area covered by glaciers was only roughly known in AR4, resulting in large uncertainties for all related calculations (e.g., overall glacier volume or mass changes). Since AR4, the world glacier inventory (WGMS, 1989) was gradually extended by Cogley (2009a) and Radić and Hock (2010); and for AR5, a new globally complete data set of glacier outlines (Randolph Glacier Inventory (RGI)) was compiled from a wide range of data sources from the 1950s to 2010 with varying levels of detail and quality (Arendt et al., 2012). Regional glacier-covered areas for 19 regions were extracted from the RGI and supplemented with the percentage of the area covered by glaciers terminating in tidewater (Figure 4.8 and Table 4.2). The areas covered by glaciers that are in contact with freshwater lakes are only locally available. The separation of so-called peripheral glaciers from the ice sheets in Greenland and Antarctica is not easy. A new detailed inventory of the glaciers in Greenland (Rastner et al., 2012) allows for estimation of their area, volume, and mass balance separately from those of the ice sheet. This separation is still incomplete for Antarctica, and values discussed here (Figures 4.1, 4.8 to 4.11, Tables 4.2 and 4.4) refer to the glaciers on the islands in the Antarctic and Sub-Antarctic (Bliss et al., 2013) but exclude glaciers on the mainland of Antarctica that are separate from the ice sheet. Regionally variable accuracy of the glacier outlines leads to poorly quantified uncertainties. These uncertainties, along with the regional variation in the minimum size of glaciers included in the inventory, and the subdivision of contiguous ice masses, also makes the total number of glaciers uncertain; the current best estimate is around 170,000 covering a total area of about 730,000 km^2. When summed up, nearly 80% of the glacier area found in regions Antarctic and Subantarctic (region 19), Canadian Arctic (regions 3 and 4), High Mountain Asia (regions 13, 14 and 15), Alaska (region 5), and Greenland (region 17) (Table 4.2).

From the glacier areas in the new inventory, total glacier volumes and masses have been determined by applying both simple scaling relations and ice-dynamical considerations (Table 4.2, and references therein), however, both methods are calibrated with only a few hundred glacier thickness measurements. This small sample means that uncertainties are large and difficult to quantify. The range of values as derived from four global-scale studies for each of the 19 RGI regions is given in Table 4.2, suggesting a global glacier mass that is *likely* between 114,000 and 192,000 Gt (314 to 529 mm SLE). The numbers and areas of glaciers reported in Table 4.2 are directly taken from RGI 2.0 (Arendt et al., 2012), with updates for the Low Latitudes (region 16) and the Southern Andes (region 17).

4.3.2 Methods to Measure Changes in Glacier Length, Area and Volume/Mass

To measure changes in glacier length, area, mass and volume, a wide range of observational techniques has been developed. Each technique has individual benefits over specific spatial and temporal scales; their main characteristics are summarized in Table 4.3. Monitoring programs include complex climate-related observations at a few glaciers, index measurements of mass balance at about a hundred glaciers, annual length changes for a few hundred glaciers, and repeat geodetic estimates of area and volume changes at regional scales using remote sensing methods (e.g., Haeberli et al., 2007). Although *in situ* measurements of glacier changes are biased towards glaciers that are easily accessible, comparatively small and simple to interpret, a large proportion of all glaciers in the world is debris covered or tidewater calving (see Table 4.2) and changes of such glaciers are more difficult to interpret in climatic terms (Yde and Pasche, 2010). In addition, many of the remote-sensing based assessments do not discriminate these types.

4.3.2.1 Length Change Measurements

For the approximately 500 glaciers worldwide that are regularly observed, front variations (commonly called length changes) are usually obtained through annual measurements of the glacier terminus position. Globally coordinated observations were started in 1894, providing one of the longest available time series of environmental change (WGMS, 2008). More recently, particularly in regions that are difficult to access, aerial photography and satellite imaging have been used to determine glacier length changes over the past decades. For selected glaciers globally, historic terminus positions have been reconstructed from maps, photographs, satellite imagery, also paintings, dated moraines and other sources (e.g., Masiokas et al., 2009; Lopez et al., 2010; Nussbaumer et al., 2011; Davies and Glasser, 2012; Leclercq and Oerlemans, 2012; Rabatel et al., 2013). Early reconstructions are sparsely distributed in both space and time, generally at intervals of decades. The terminus fluctuations of some individual glaciers have been reconstructed for periods of more than 3000 years (Holzhauser et al., 2005), with a much larger number of records available as far back as the 16th or 17th centuries (Zemp et al., 2011, and references therein). The reconstructed glacier length records are globally well distributed and were used, for example, to determine the contribution of glaciers to global sea level rise (Leclercq et al., 2011) (Section 4.3.3.4), and for an independent temperature reconstruction at a hemispheric scale (Leclercq and Oerlemans, 2012).

Table 4.2 | The 19 regions used throughout this chapter and their respective glacier numbers and area (absolute and in percent) are derived from the RGI 2.0 (Arendt et al., 2012); the tidewater fraction is from Gardner et al. (2013). The minimum and maximum values of glacier mass are the minimum and maximum of the estimates given in four studies: Grinsted (2013), Huss and Farinotti (2012), Marzeion et al. (2012) and Radić et al. (2013). The mean sea level equivalent (SLE) of the mean glacier mass is the mean of estimates from the same four studies, using an ocean area of 362.5×10^6 km^2 for conversion. All values were derived with globally consistent methods; deviations from more precise national data sets are thus possible. Ongoing improvements may lead to revisions of these (RGI 2.0) numbers in future releases of the RGI.

Region	Region Name	Number of Glaciers	Area (km^2)	Percent of total area	Tidewater fraction (%)	Mass (minimum) (Gt)	Mass (maximum) (Gt)	Mean SLE (mm)
1	Alaska	23,112	89,267	12.3	13.7	16,168	28,021	54.7
2	Western Canada and USA	15,073	14,503.5	2.0	0	906	1148	2.8
3	Arctic Canada North	3318	103,990.2	14.3	46.5	22,366	37,555	84.2
4	Arctic Canada South	7342	40,600.7	5.6	7.3	5510	8845	19.4
5	Greenland	13,880	87,125.9	12.0	34.9	10,005	17,146	38.9
6	Iceland	290	10,988.6	1.5	0	2390	4640	9.8
7	Svalbard	1615	33,672.9	4.6	43.8	4821	8700	19.1
8	Scandinavia	1799	2833.7	0.4	0	182	290	0.6
9	Russian Arctic	331	51,160.5	7.0	64.7	11,016	21,315	41.2
10	North Asia[a]	4403	3425.6	0.4	0	109	247	0.5
11	Central Europe	3920	2058.1	0.3	0	109	125	0.3
12	Caucasus	1339	1125.6	0.2	0	61	72	0.2
13	Central Asia	30,200	64,497	8.9	0	4531	8591	16.7
14	South Asia (West)	22,822	33,862	4.7	0	2900	3444	9.1
15	South Asia (East)	14,006	21,803.2	3.0	0	1196	1623	3.9
16	Low Latitudes[a]	2601	2554.7	0.6	0	109	218	0.5
17	Southern Andes[a]	15,994	29,361.2	4.5	23.8	4241	6018	13.5
18	New Zealand	3012	1160.5	0.2	0	71	109	0.2
19	Antarctic and Sub-Antarctic	3274	13,2267.4	18.2	97.8	27,224	43,772	96.3
	Total	168,331	726,258.3		38.5	113,915	191,879	412.0

Notes:

[a] For regions 10, 16 and 17 the number and area of glaciers are corrected to allow for over-inclusion of seasonal snow in the glacierized extent of RGI 2.0 and for improved outlines (region 10) compared to RGI 2.0 (updated from, Arendt et al., 2012).

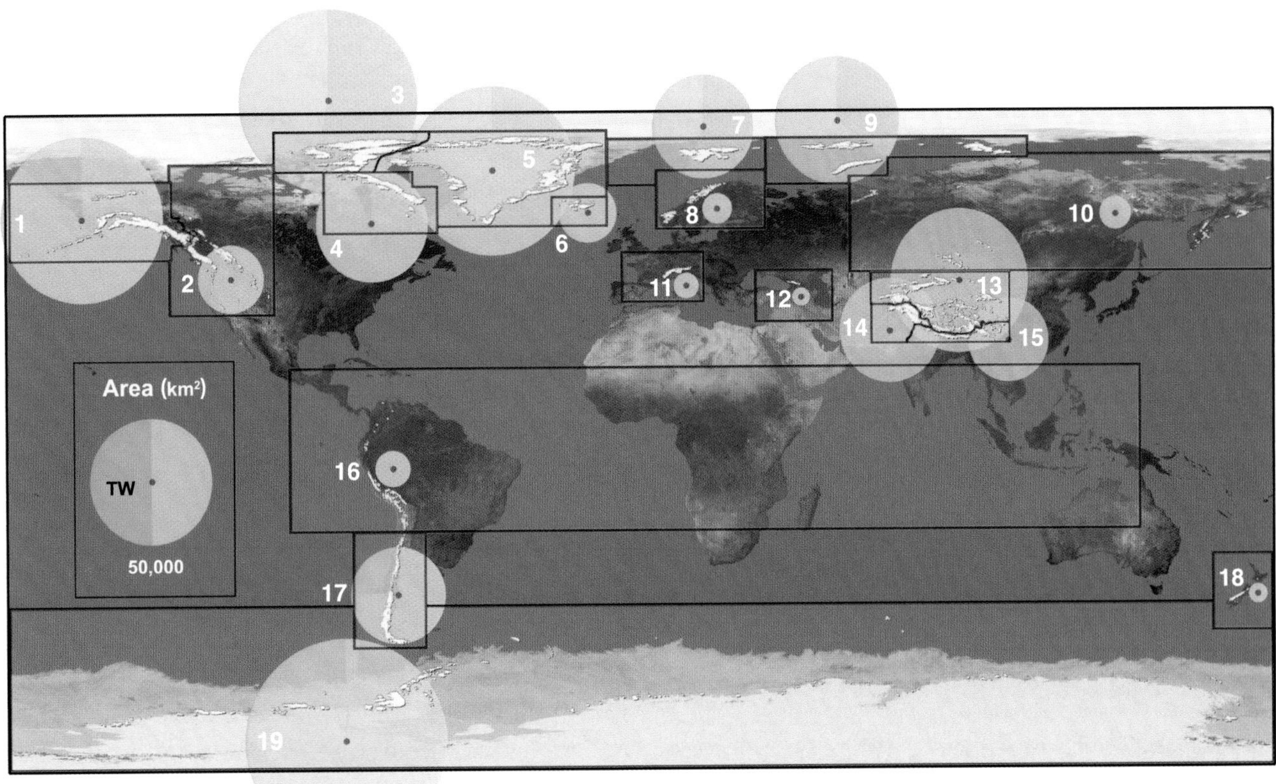

Figure 4.8 | Global distribution of glaciers (yellow, area increased for visibility) and area covered (diameter of the circle), sub-divided into the 19 RGI regions (white number) referenced in Table 4.2. The area percentage covered by tidewater (TW) glaciers in each region is shown in blue. Data from Arendt et al. (2012) and Gardner et al. (2013).

4.3.2.2 Area Change Measurements

Glacier area changes are reported in increasing number and coverage based on repeat satellite imagery (WGMS, 2008). Although satellite-based observations are available only for the past four decades, studies using aerial photography, old maps, as well as mapped and dated moraines and trim lines show glacier areas back to the end of the so-called Little Ice Age (LIA, see Glossary) about 150 years ago (cf. Figure 6 in Rabatel et al., 2008) and beyond (e.g., Citterio et al., 2009; Davies and Glasser, 2012). The observed area changes depend (in most regions) on glacier size (with smaller glaciers shrinking at faster percentage rates) and tend to vary greatly within any one mountain range. Moreover, the time spans of the measurements of change vary from study to study and regional or global-scale estimates are therefore difficult to generate. The focus here is thus on the comparison of mean annual relative area changes averaged over entire mountain regions.

4.3.2.3 Volume and Mass Change Measurements

Several methods are in use for measuring mass changes of glaciers. Traditionally, the annual surface mass balance is derived from repeated snow density and snow/ice stake readings on individual glaciers. Estimates over larger regions are obtained by extrapolating from the measured glaciers. This labour-intensive method is generally restricted to a limited number of accessible glaciers, which are unevenly distributed over glacier regions and types. Annual measurements began in the 1940s on a few glaciers, with about 100 glaciers being measured since the 1980s—and only 37 glaciers have been measured without interruption for more than 40 years (WGMS, 2009). Potential mass loss from calving or from basal ablation is not included in the surface measurements. At present, it is not possible to quantify all sources of uncertainty in mass budgets extrapolated from measurements of individual glaciers (Cogley, 2009b).

A second method determines the volume change of all glaciers in a region by measurement of surface-elevation changes (Section 4.3.3.3 and Figure 4.11). The information is derived by subtracting digital terrain models from two points in time, including those from repeat airborne or satellite altimetry (particularly suitable for larger and flatter ice surfaces). The conversion from volume to mass change can cause a major uncertainty, in particular over short periods, as density information is required, but is generally available only from field measurements (Gardner et al., 2013, and references therein).

Since 2003, a third method used to estimate overall mass change is through measurement of the changing gravity field from satellites (GRACE mission). The coarse spatial resolution (about 300 km) and the difficulties of separating different mass change signals such as hydrological storage and glacial isostatic adjustment limit this method to regions with large continuous ice extent (Gardner et al., 2013).

A fourth method calculates the mass balance of individual glaciers, or a glacier region, with models that either convert particular glacier variables such as length changes or the altitude of the equilibrium line (see Glossary) (e.g., Rabatel et al., 2005; Luethi et al., 2010; Leclercq et al., 2011) into mass changes, or use time series of atmospheric temperature and other meteorological variables to simulate glacier mass balances at different levels of complexity (e.g., Hock et al., 2009; Machguth et al., 2009; Marzeion et al., 2012; Hirabayashi et al., 2013). The models improve fidelity and physical completeness, and add value to the scarce direct measurements.

A fifth method determines glacier mass changes as residuals of the water balance for hydrological basins rather than for glacier regions. Results from all but this last method are used in the following regional and global assessment of glacier mass changes (Sections 4.3.3.3 and 4.3.3.4).

4.3.3 Observed Changes in Glacier Length, Area and Mass

4.3.3.1 Length Changes

Despite their variability due to different response times and local conditions (see FAQ 4.2), the annually measured glacier terminus fluctuations from about 500 glaciers worldwide reveal a largely homogeneous trend of retreat (WGMS, 2008). In Figure 4.9, a selection of the available long-term records of field measurements is shown for 14 out of the 19 RGI regions. Cumulative values of retreat for large, land-terminating valley glaciers typically reach a few kilometres over the 120-year period of observation. For mid-latitude mountain and valley glaciers, typical retreat rates are of the order of 5 to 20 m yr^{-1}. Rates of up to 100 m yr^{-1} (or even more) are seen to occur under special conditions, such as the complete loss of a tongue on a steep slope (see FAQ 4.2, Figure 1c), or the disintegration of a very flat tongue. A non-calving valley glacier in Chile had reported mean annual retreat rates of 125 m from 1961 to 2011 (Rivera et al., 2012). The general tendency of retreat in the 20th century was interrupted in several regions (e.g., regions 2, 8, 11 and 17) by phases of stability lasting one or two decades, or even advance, for example in the 1920s, 1970s and 1990s (regionally variable). In regions for which long-term field measurements of several glaciers of different sizes are available, the terminus fluctuations typically show a pattern with the largest (flatter) glaciers tending to retreat continuously and by large cumulative distances, medium-sized (steeper) glaciers showing decadal fluctuations, and smaller glaciers showing high variability superimposed on smaller cumulative retreats (Figure 4.9).

The exceptional terminus advances of a few individual glaciers in Scandinavia and New Zealand in the 1990s may be related to locally specific climatic conditions such as increased winter precipitation (Nesje et al., 2000; Chinn et al., 2005; Lemke et al., 2007). In other regions, such as Iceland, the Karakoram and Svalbard, observed advances were often related to dynamical instabilities (surging) of glaciers (e.g., Murray et al., 2003; Quincey et al., 2011; Bolch et al., 2012; Björnsson et al., 2013). Glaciers with calving instabilities can retreat exceptionally rapidly (Pfeffer, 2007), while those with heavily debris-covered tongues are often close to stationary (Scherler et al., 2011). More regionally-focused studies of length change over different time periods (e.g., Citterio et al., 2009; Masiokas et al., 2009; Lopez et al., 2010; Bolch et al., 2012) justify *high confidence* about the trend of glacier length variations shown in Figure 4.9.

4.3.3.2 Area Changes

From the large number of published studies on glacier area changes in all parts of the world since AR4 (see Table 4.SM.1) a selection with examples from 16 out of the 19 RGI regions is shown in Figure 4.10. The studies reveal that (1) total glacier area has decreased in all regions, (2) the rates of change cover a similar range of values in all regions, (3) there is considerable variability of the rates of change within each region, (4) highest loss rates are found in regions 2, 11 and 16, and (5) the rates of loss have a tendency to be higher over more recent time periods. The last point (5) requires studies comparing the same sample of glaciers over multiple similar time periods. For 14 out of 19 regions listed in Table 4.SM.1 (see Supplementary Material) with such an analysis, higher loss rates were found for the more recent period.

While points (1) and (2) give *high confidence* in the global-scale shrinkage in glacier area, (3) points to a considerable regional to local-scale scatter of observed change rates. The shorter the period of investigation and the smaller the sample of glaciers analysed, the more variable

Table 4.3 | Overview of methods used to determine changes in glacier length, area and volume mass along with some typical characteristics. The techniques are not exclusive. The last three columns provide only indicative values.

Parameter	Method	Technique	Typical Accuracy	Number of Glaciers	Repeat Interval	Earliest Data
Length change	Various	Reconstruction	10 m	Dozens	Decadal – centuries	Holocene
	Field	*In situ* measurement	1 m	Hundreds	Annual	19th century
	Remote sensing	Photogrammetric survey	Two image pixels (depending on resolution)	Hundreds	Annual	20th century
Area change	Maps	Cartographic	5% of the area	Hundreds	Decadal	19th century
	Remote sensing	Image processing	5% of the area	Thousands	Sub-decadal	20th century
Volume change	Remote sensing	Laser and radar profiling	0.1 m	Hundreds	Annual	21st century
	Remote sensing	DEM differencing	0.5 m	Thousands	Decadal	20th century
Mass change	Field	Direct mass balance measurement	0.2 m	Hundreds	Seasonal	20th century
	Remote sensing	Gravimetry (GRACE)	Dependent on the region	Global	Seasonal	21st century

4

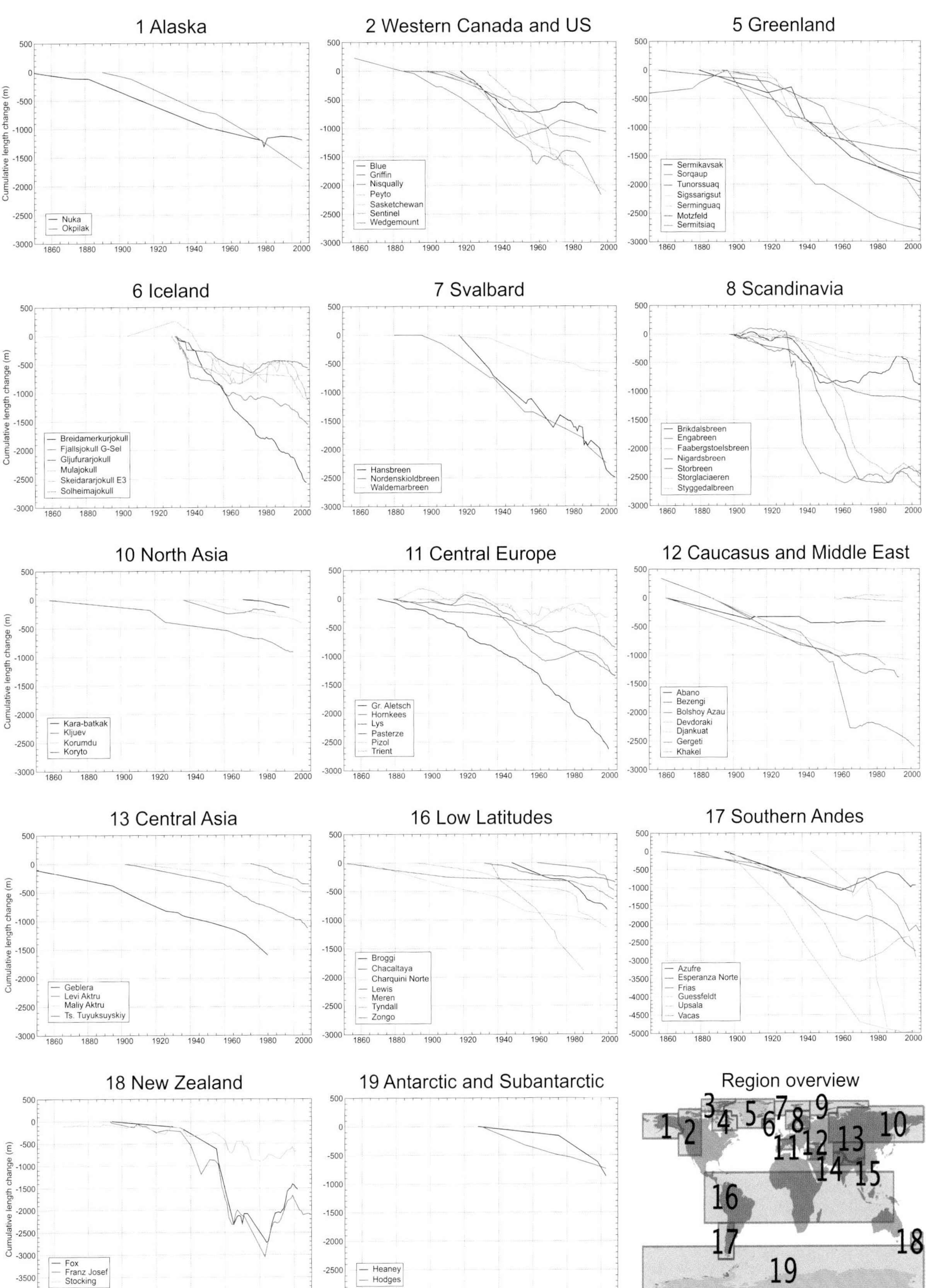

Figure 4.9 | Selection of long-term cumulative glacier length changes as compiled from *in situ* measurements (WGMS, 2008), reconstructed data points added to measured time series (region 5) from Leclercq et al. (2012), and additional time series from reconstructions (regions 1, 2, 7, 10, 12, 16, 17 and 18) from Leclercq and Oerlemans (2012). Independent of their (highly variable) temporal density, all measurement points are connected by straight lines. The glacier Mulajokull (region 6) is of surge type and some of the glaciers showing strong retreat either terminate (Guessfeldt and Upsala in region 17) or terminated (Engabreen and Nigardsbreen in region 8) in lakes. For region 11, many more time series are available (see WGMS, 2008), but are not shown for graphical reasons.

are the rates of change reported for a specific region (Table 4.SM.1). In many regions of the world, rates of area loss have increased (Table 4.SM.1), confirming that glaciers are still too large for the current climate and will continue to shrink (see FAQ 4.2 and Section 4.3.3.3).

Several studies have reported the disappearance of glaciers, among others in Arctic Canada (Thomson et al., 2011), the Rocky Mountains (Bolch et al., 2010; Tennant et al., 2012) and North Cascades (Pelto, 2006), Patagonia (Bown et al., 2008; Davies and Glasser, 2012), several tropical mountain ranges (Coudrain et al., 2005; Klein and Kincaid, 2006; Cullen et al., 2013), the European Alps (Citterio et al., 2007; Knoll and Kerschner, 2009; Diolaiuti et al., 2012), the Tien Shan (Hagg et al., 2012; Kutuzov and Shahgedanova, 2009) in Asia and on James Ross Island in Antarctica (Carrivick et al., 2012). In total, the disappearance of more than 600 glaciers has been reported, but the real number is certainly higher. Also some of the glaciers, whose annual mass balance has been measured over several years or even decades, have disappeared or started to disintegrate (Ramirez et al., 2001; Carturan and Seppi, 2007; Thibert et al., 2008). Though the number of glaciers that have disappeared is difficult to compare directly (e.g., the time periods analysed or the disappearance-criteria applied differ), glaciers that have disappeared provide *robust evidence* that the ELA (see Glossary) has risen above the highest peaks in many mountain ranges (see FAQ 4.2).

4.3.3.3 Regional Scale Glacier Volume and Mass Changes

In AR4, global and regional scale glacier mass changes were extrapolated from *in situ* measurements of mass balance on individual glaciers (Kaser et al., 2006; Lemke et al., 2007). In some regions, such as Alaska, Patagonia and the Russian Arctic, very few if any such records were available. Since AR4, geodetically derived ice volume changes have been assimilated (Cogley, 2009b), providing more consistent regional coverage and better representation of the proportion of calving glaciers. In addition, the new near complete inventory (RGI) of glacier-covered areas (Arendt et al., 2012) has improved knowledge about regional and global glacier volume and mass changes.

Figure 4.11 shows a compilation of available mean mass-balance rates for 1960–2010 for each of the 19 RGI regions. Where error estimates are reported, the 90% confidence bounds are shown. Most results shown are calculated using a single method, some merge multiple methods; those from Gardner et al. (2013) are reconciled estimates for 2003–2009 obtained by selecting the most reliable results of different observation methods, after region-by-region reanalysis and comparison.

Despite the great progress made since AR4, uncertainties inherent to specific methods, and arising from the differences between methods (cf.

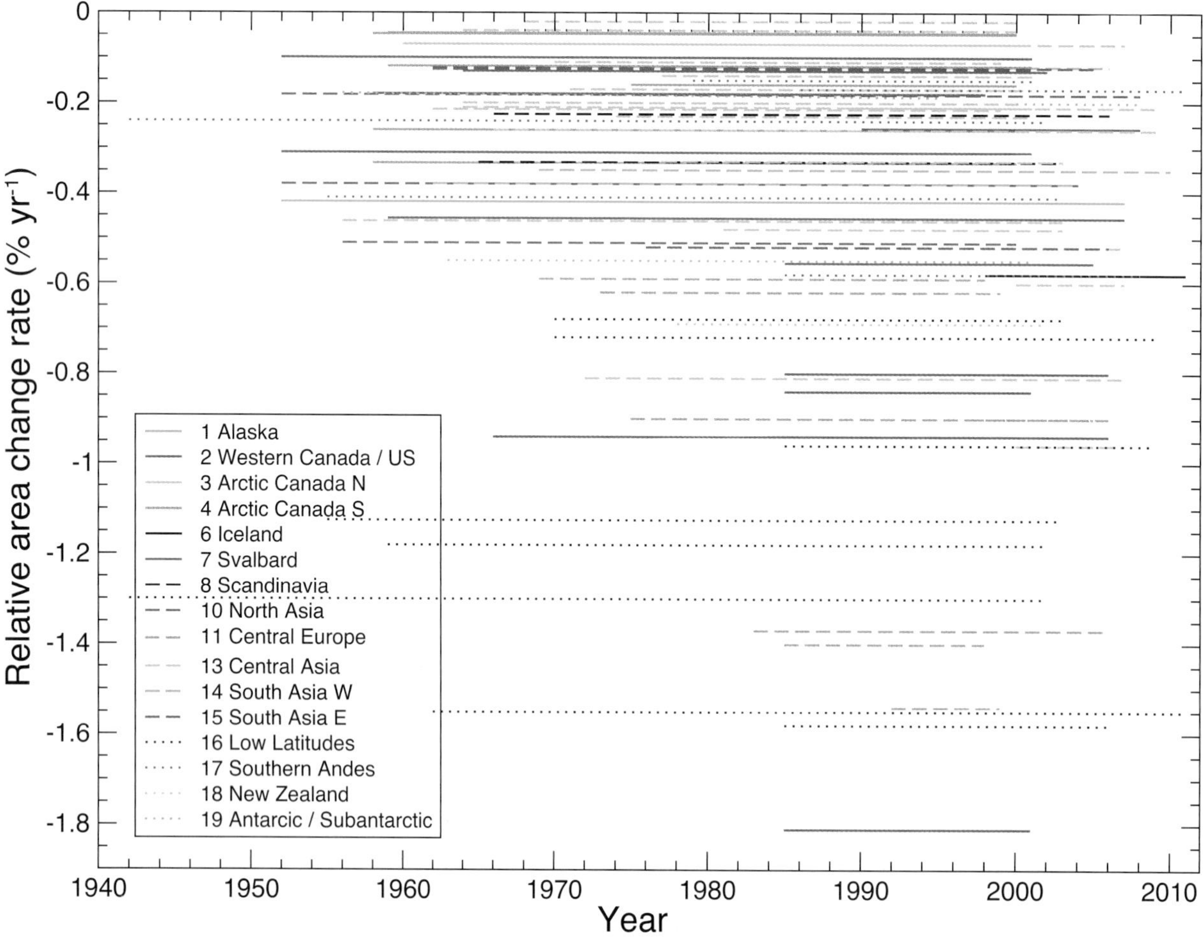

Figure 4.10 | Mean annual relative area loss rates for 16 out of the 19 RGI regions of Figure 4.8. Each line shows a measurement of the rate of percentage change in area over a mountain range from a specific publication (for sources see Table 4.SM.1), the length of the line shows the period used for averaging.

Box 2.1), remain large, and *confidence* about the absolute value of mass loss is *medium* at both regional and global scales (Figure 4.11). The highest density of measurement and best time resolution are available for Scandinavia (region 8) and Central Europe (region 11). The least coverage and some of the highest uncertainties are in the Arctic (regions 3, 4, 5, 9) and the Antarctic and Sub-Antarctic (region 19).

Gardner et al. (2013) discussed inconsistencies among, and differences between, methods and their respective results for 2003–2009. They found that results from the GRACE gravimetric mission agree well with results from ICESat laser altimetry in regions of extensive ice cover (see comment on ICESat data in Section 4.4.2.1), but show much more variable and uncertain mass changes in regions with small or scattered ice

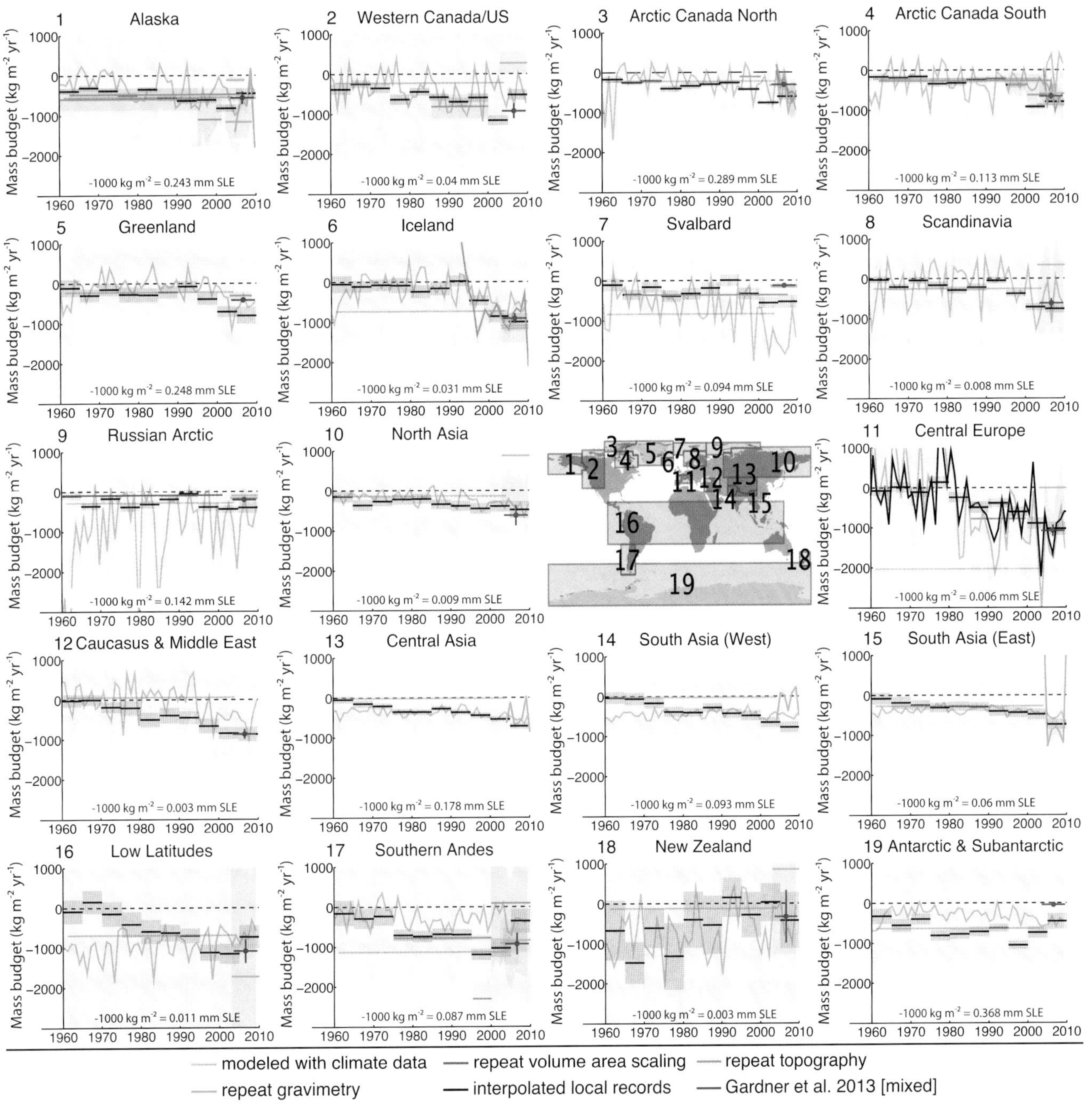

Figure 4.11 | Regional glacier mass budgets in units of kg m^{-2} yr^{-1} for the world's 19 glacierized regions (Figure 4.8 and Table 4.2). Estimates are from modelling with climate data (blue: Hock et al., 2009; Marzeion et al., 2012), repeat gravimetry (green: Chen et al., 2007; Luthcke et al., 2008; Peltier, 2009; Matsuo and Heki, 2010; Wu et al., 2010; Gardner et al., 2011; Ivins et al., 2011; Schrama and Wouters, 2011; Jacob et al., 2012, updated for RGI regions), repeat volume area scaling (magenta: Glazovsky and Macheret, 2006), interpolation of local glacier records (black: Cogley, 2009a; Huss, 2012), or airborne and/or satellite repeat topographic mapping (orange: Arendt et al., 2002; Rignot et al., 2003; Abdalati et al., 2004; Schiefer et al., 2007; Paul and Haeberli, 2008; Berthier et al., 2010; Moholdt et al., 2010, 2012; Nuth et al., 2010; Gardner et al., 2011, 2012; Willis et al., 2012; Björnsson et al., 2013; Bolch et al., 2013). Mass-budget estimates are included only for study domains that cover about 50% or more of the total regional glacier area. Mass-budget estimates include 90% confidence envelopes (not available from all studies). Conversions from specific mass budget in kg m^{-2} to mm SLE are given for each region. Gravimetric estimates are often not accompanied by estimates of glacierized area (required for conversion from Gt yr^{-1} to kg m^{-2} yr^{-1}); in such cases the RGI regional glacier areas were used.

cover such as Western Canada/USA (region 2) and the Qinghai-Xizang (Tibet) Plateau (e.g., Yao et al., 2012, and references therein). Based on ICESat measurements, Gardner et al. (2013) also found that glaciers with *in situ* measurements tend to be located in sub-regions that are thinning more rapidly than the region as a whole. Thus, extrapolation from *in situ* measurements has a negative bias in regions with sparse measurements. Based on this analysis, Gardner et al. (2013) excluded GRACE results for regions with small or scattered glacier coverage and excluded results based on extrapolation of local records for remote, sparsely sampled regions.

The 2003–2009 regionally differentiated results are given in Table 4.4. There is *very high confidence* that, between 2003 and 2009, most mass loss was from glaciers in the Canadian Arctic (regions 3 and 4), Alaska (region 1), Greenland (region 5), the Southern Andes (region 17) and the Asian Mountains (region 13 to 15), which together account for more than 80% of the global ice loss.

Despite the considerable scatter, Figure 4.11 shows mass losses in all 19 regions over the past five decades that, together with their consistency with length (Section 4.3.3.1) and area changes (Section 4.3.3.2), provide *robust evidence* and *very high confidence* in global glacier shrinkage. In many regions, ice loss has *likely* increased during the last two decades, with slightly smaller losses in some regions during the most recent years, since around 2005. In Central Europe (region 11), the increase of loss rates was earliest and strongest. In the Russian Arctic (region 9) and in the Antarctic and Sub-Antarctic (region 19), the signal is highly uncertain and trends are least clear. Gardner et al. (2013) present values close to balance for the Antarctic and Subantarctic (region 19) that result from complex regional patterns, for example, with losses on Antarctic Peninsula islands and gains on Ellsworth Land islands. The picture is also heterogeneous in High Mountain Asia (region 13 to 15) (e.g., Bolch et al., 2012; Yao et al., 2012), where glaciers in the Himalaya and the Hindu Kush have been losing mass (Kääb et al., 2012) while those in the Karakoram are close to balance (Gardelle et al., 2012).

Table 4.4 | Regional mass change rates in units of kg m^{-2} yr^{-1} and Gt y^{r-1} for the period 2003–2009 from Gardner et al. (2013). Central Asia (region 13), South Asia West (region14), and South Asia East (region 15) are merged into a single region. For the division of regions see Figure 4.8.

No.	Region Name	(kg m^{-2} yr^{-1})	(Gt yr^{-1})
1	Alaska	–570 ± 200	–50 ± 17
2	Western Canada and USA	–930 ± 230	–14 ± 3
3	Arctic Canada North	–310 ± 40	–33 ± 4
4	Arctic Canada South	–660 ± 110	–27 ± 4
5	Greenland periphery	–420 ± 70	–38 ± 7
6	Iceland	–910 ± 150	–10 ± 2
7	Svalbard	–130 ± 60	–5 ± 2
8	Scandinavia	–610 ± 140	–2 ± 0
9	Russian Arctic	–210 ± 80	–11 ± 4
10	North Asia	–630 ± 310	–2 ± 1
11	Central Europe	–1060 ± 170	–2 ± 0
12	Caucasus and Middle East	–900 ± 160	–1 ± 0
13–15	High Mountain Asia	–220 ± 100	–26 ± 12
16	Low Latitudes	–1080 ± 360	–4 ± 1
17	Southern Andes	–990 ± 360	–29 ± 10
18	New Zealand	–320 ± 780	0 ± 1
19	Antarctic and Sub-Antarctic	–50 ± 70	–6 ± 10
	Total	**–350 ± 40**	**–259 ± 28**

Several studies of recent glacier velocity change (Heid and Kääb, 2012; Azam et al., 2012) and of the worldwide present-day sizes of accumulation areas (Bahr et al., 2009) indicate that the world's glaciers are out of balance with the present climate and thus committed to losing considerable mass in the future, even without further changes in climate. Increasing ice temperatures recorded at high elevation sites in the tropical Andes (Gilbert et al., 2010) and in the European Alps (Col du Dome on Mont Blanc and Monte Rosa) (Vincent et al., 2007; Hoelzle et al., 2011), as well as the ongoing thinning of the cold surface layer on Storglaciären in northern Sweden (Gusmeroli et al., 2012), support this conclusion and give it *high confidence*.

4.3.3.4 Global Scale Glacier Mass Changes—The Contribution to Sea Level

Global time series are required to assess the continuing contribution of glacier mass changes to sea level (see Section 13.4.2 for discussion of the small proportion of ice loss from glaciers that does not contribute to sea level rise). A series of recent studies, some updated to RGI areas for this report by their respective authors, provides *very high confidence* in a considerable and continuous mass loss, despite only *medium agreement* on the specific rates (Figure 4.12 and Table 4.5). Cogley (updated from, 2009b) compiled 4,817 directly measured annual mass budgets, and 983 volume change measurements by extending the data set of WGMS (2009, and earlier issues). Global 5-year averages for 1961–2010, with uncertainties, were estimated from these using an inverse-distance-weighted interpolation. Newly available volume change measurements increased the proportion of observations from calving glaciers from 3% to 16% compared to earlier estimates reported by Lemke et al. (2007). This proportion is more realistic, but may still underestimate the relative importance of calving glaciers (Figure 4.8 and Section 4.3.3.3).

Leclercq et al. (updated from 2011) used length variations from 382 glaciers worldwide as a proxy for glacier mass loss since 1800. The length/mass change conversion was calibrated against mass balance observations for 1950–2005 from Cogley (2009b) and provide one estimate based on the arithmetic mean and another based on area-weighted extrapolation of regional averages. Uncertainty was estimated from upper and lower bounds of the calibration parameter assumptions, and cumulatively propagated backward in time. For the 19th century, the information was constrained by a limited number of observations, particularly in extensively glacierized regions that contribute most to the global mass budget.

Two global-scale time series are obtained from mass-balance modelling based on temperature and precipitation data (Marzeion et al., 2012; Hirabayashi et al., 2013). Glacier size adjustments are simulated by using area–volume power-law relations as proposed by Bahr et al. (1997) for the approximately 170,000 individual glaciers delineated in

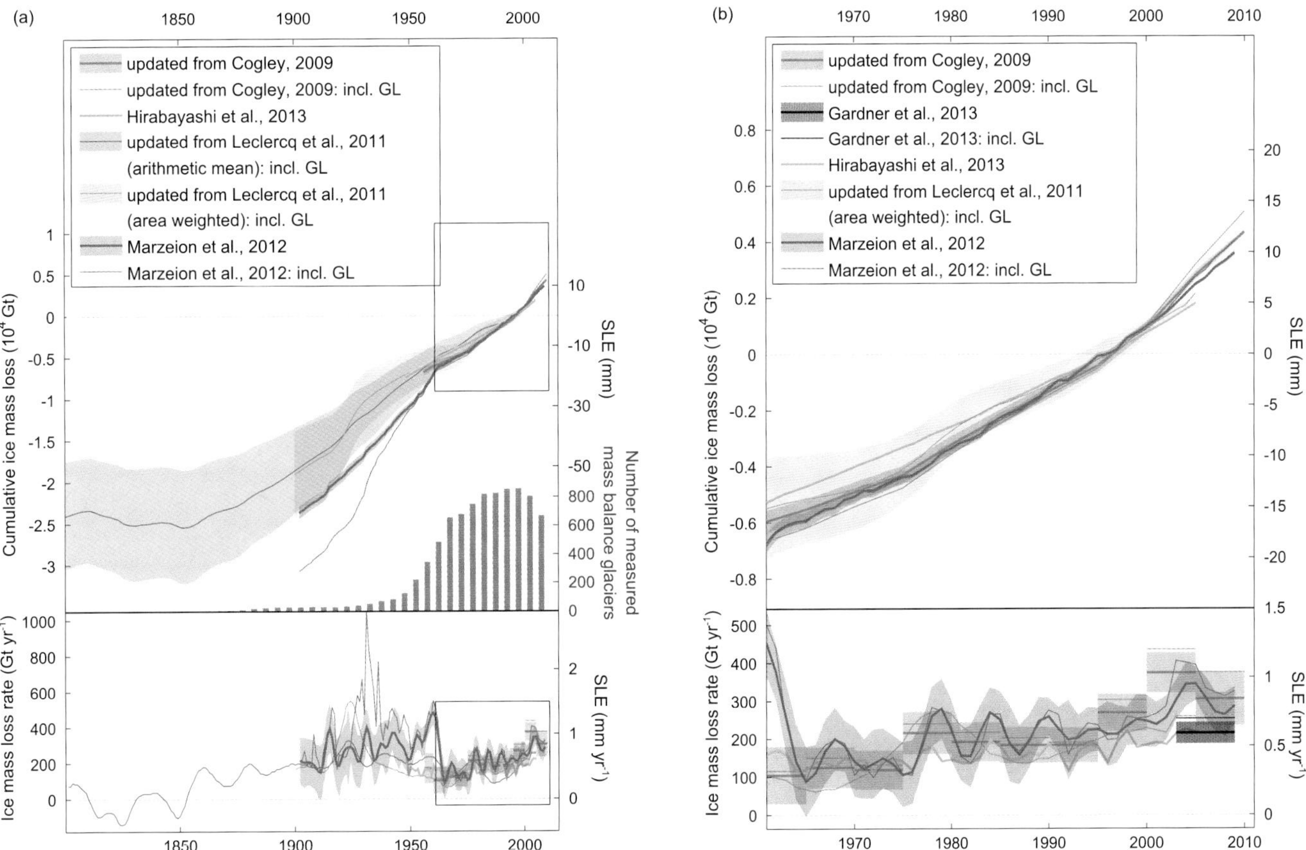

Figure 4.12 | Global cumulative (top graphs) and annual (lower graphs) glacier mass change for (a) 1801–2010 and (b) 1961–2010. The cumulative estimates are all set to zero mean over 1986–2005. Estimates are based on glacier length variations (updated from Leclercq et al., 2011), from area-weighted extrapolations of individual directly and geodetically measured glacier mass budgets (updated from Cogley, 2009b), and from modelling with atmospheric variables as input (Marzeion et al., 2012; Hirabayashi et al., 2013). Uncertainties are based on comprehensive error analyses in Cogley (2009b) and Marzeion et al. (2012) and on assumptions about the representativeness of the sampled glaciers in Leclercq et al. (2011). Hirabayashi et al. (2013) give a bulk error estimate only. For clarity in the bottom panels, uncertainties are shown only for the Cogley and Marzeion curves excluding Greenland (GL). The blue bars (a, top) show the number of measured single-glacier mass balances per pentad in the updated Cogley (2009b) time series. The mean 2003–2009 estimate of Gardner et al. (2013) is added to b, bottom.

the RGI. Marzeion et al. (2012) derive mass balances for 1902–2009 from monthly mean temperature and precipitation obtained from Mitchell and Jones (2005). The model is calibrated against measured time series and validated against independent measurements. Uncertainty estimates are obtained from comprehensive error propagation, first accumulated temporally for each glacier, and then regionally and globally. The model does not account for the subsurface mass balance or calving, but reproduces geodetically measured volume changes for land-based glaciers within the uncertainties; however, it underestimates volume loss slightly for calving glaciers. Hirabayashi et al. (2013) force an extended positive degree-day model with data from an observation-based global set of daily precipitation and near-surface temperature as updated from earlier work (Hirabayashi et al., 2008). Annual mass balance is provided for 1948–2005 with a constant root mean square error of 500 $km^3 yr^{-1}$, estimated from comparison of modelled with measured mass balances.

For the Antarctic and Sub-Antarctic (region 19) observational information is limited and difficult to incorporate into this assessment. For some studies the time spans do not match (e.g., Gardner et al., 2013 give only mean mass change for 2003–2009). Some estimates have been made simply by extrapolating the global mean to region 19 (Cogley, 2009a; Marzeion et al., 2012). One (1961–2004) mean glacier mass loss estimate based on an ECMWF 40-year reanalysis (ERA-40) driven simulation (Hock et al., 2009), is for a glacier area differing from the RGI region 19. For these reasons, the Antarctic and Sub-Antarctic (region 19) is excluded from this global glacier mass change assessment. The contribution of region 19 to sea level is assumed to be within the uncertainty bounds of the Antarctic ice sheet assessment (Section 4.4.2). Whereas Hirabayashi et al. (2013) exclude the Antarctic and Greenland from their simulations and Leclercq et al. (2011) implicitly include Greenland, both Cogley (2009a) and Marzeion et al. (2012) explicitly estimate mass changes in Greenland. In Figure 4.12, cumulative mass changes and corresponding rates are shown for global glaciers excluding regions 5 and 19 (bold lines), and also for global glaciers excluding only region 19 (thin lines). The cumulative curves are normalized such that their 1986–2005 averages are all zero.

The arithmetic-mean estimate of Leclercq et al. (2011) indicates continuous mass loss from glaciers after about 1850 (Figure 4.12a, top). During the 1920s their area-weighted extrapolation reaches considerably higher rates (Figure 4.12a, bottom) than the other estimates, but the reasons remain unclear. After 1950, mass loss rates including Greenland are all within the uncertainty bounds of those that exclude Greenland, except for the 2001–2005 period when the Greenland contribution was slightly outside the uncertainty bounds for both the Cogley and the Marzeion et al. estimates. Most notable is the rapid loss from Greenland glaciers in the Marzeion et al. simulations during the 1930s. Other studies support rapid Greenland mass loss around this time (Zeeberg and Forman, 2001; Yde and Knudsen, 2007, and references therein; Bjørk et al., 2012; Zdanowicz et al., 2012); however, the neighbouring regions in the Canadian Arctic (south and north) and Iceland have mass loss anomalies an order of magnitude lower than predicted for Greenland in the same simulation. This discrepancy may be an artefact of the uncertainties in the forcing and methods of Marzeion et al. that are considerably larger in the first than in the second half of the 20th century, so that the rates may well be overestimated. The Marzeion et al. rates are also considerably greater in the 1950s and 1960s than in the other studies; during this period, the most rapid losses are in Arctic Canada and the Russian Arctic (Marzeion et al., 2012).

Overall, there is *very high confidence* that globally, the mass loss from glaciers has increased since the 1960s, and this is evident in regional-scale estimates (Figure 4.11). For 2003–2009, Gardner et al. (2013) indicate that some regional (Section 4.3.3.3 and Figure 4.11) and also the global time series may overestimate mass loss (Figure 4.12b, bottom). That glaciers with measured mass balances are concentrated in sub-regions with higher mass losses definitely biases the estimates of Cogley (2009a) (Section 4.3.3.3), but this explanation cannot hold for the Marzeion et al. (2012) time series, for which mass changes are simulated separately for every single glacier in the inventory. It also remains unclear whether the 2003–2009 inconsistency identified by Gardner et al. (2013) applies to earlier times, and if so how it should be reconciled. Neither the evidence nor our level of understanding warrants any simple correction of the longer time series at present.

Table 4.5 summarizes global-scale glacier mass losses for different periods relevant to discussions on sea level change (Chapter 13) and the global energy budget (Chapter 3). Values are given separately for the Greenland glaciers alone (region 5) and for all glaciers excluding those in regions 5 and 19, which are included in the assessment of ice sheets (Section 4.4.2). For the more recent periods, the time series of Cogley (2009a) and Marzeion et al. (2012) are combined, while for 1901–1990 the Leclercq et al. (2011) series were separated by area-weighting and combined with the Marzeion et al. (2012) values. Each rate in Table 4.5 is thus the arithmetic mean of two series, with a confidence bound calculated from their difference, and assessed to represent the 90% likelihood range. Because differences between the two time series vary considerably, the average confidence bound of 1971–2009 is also applied uniformly to the two sub-periods 1993–2009 and 2005–2009. The 2003–2009 estimate of Gardner et al. (2013) is lower than the Cogley and Marzeion averages but those for all glaciers excluding regions 5 and 19 are within the 2005–2009 90% confidence bound. The 1991–2009 assessment is shown as a cumulative time series in Section 4.8 (see Figure 4.25). Earlier studies of the long-term contribution of glaciers to sea level change (Meier, 1984; Zuo and Oerlemans, 1997; Gregory and Oerlemans, 1998; Kaser et al., 2006; Lemke et al., 2007; Oerlemans et al., 2007; Hock et al., 2009, with removal of Antarctic glacier contribution) all give smaller estimates than those assessed here.

4.4 Ice Sheets

4.4.1 Background

Since AR4, satellite, airborne and *in situ* observations have greatly improved our ability to identify and quantify change in the vast polar ice sheets of Antarctica and Greenland. As a direct consequence, our understanding of the underlying drivers of ice-sheet change is also much improved. These observations and the insights they yield are discussed throughout Section 4.4, while the attribution of recent ice sheet change, projection of future changes in ice sheets and their future contribution to sea level rise are discussed in Chapter 10 and Chapter 13 respectively.

4.4.2 Changes in Mass of Ice Sheets

The current state of mass balance of the Greenland and Antarctic ice sheets is assessed in sections 4.4.2.2 and 4.4.2.3, but is introduced by

Table 4.5 | Average annual rates of global mass change in Gt yr^{-1} and in sea level equivalents (mm SLE yr^{-1}) for different time periods (Chapter 13) for (a) glaciers around the Greenland ice sheet (region 5 as defined by Rastner et al., 2012) and (b) all glaciers globally, excluding peripheral glaciers around the Antarctic and Greenland ice sheets (see discussion in Section 4.4.2). The values are derived by averaging the results from the references listed and uncertainty ranges give 90% confidence level. The uncertainty calculated for 1971–2009 is also applied for the sub-periods 1993–2009 and 2005–2009. The global values for 2003–2009 from Gardner et al. (2013) are within this likelihood range (italics).

	Reference	(a) Greenland glaciers (region 5)		(b) All glaciers excluding ice sheet peripheries	
		Gt yr^{-1}	mm SLE yr^{-1}	Gt yr^{-1}	mm SLE yr^{-1}
1901–1990	Marzeion et al. (2012); Leclercq et al. (2011), updated[a]	–54 ± 16	0.15 ± 0.05	–197 ± 24	0.54 ± 0.07
1971–2009	Cogley (2009a); Marzeion et al. (2012)	–21 ± 10	0.06 ± 0.03	–226 ± 135	0.62 ± 0.37
1993–2009	Cogley (2009a); Marzeion et al. (2012)	–37 ± 10	0.10 ± 0.03	–275 ± 135	0.76 ± 0.37
2005–2009	Cogley (2009a); Marzeion et al. (2012)	–56 ± 10	0.15 ± 0.03	–301 ± 135	0.83 ± 0.37
2003–2009	Gardner et al. (2013)	–38 ± 7	0.10 ± 0.02	–215 ± 26	0.59 ± 0.07

Notes:

[a] Isolation of (a) from (b) made by applying the respective annual ratios in Marzeion et al. (2012).

Frequently Asked Questions

FAQ 4.2 | Are Glaciers in Mountain Regions Disappearing?

In many mountain ranges around the world, glaciers are disappearing in response to the atmospheric temperature increases of past decades. Disappearing glaciers have been reported in the Canadian Arctic and Rocky Mountains; the Andes; Patagonia; the European Alps; the Tien Shan; tropical mountains in South America, Africa and Asia and elsewhere. In these regions, more than 600 glaciers have disappeared over the past decades. Even if there is no further warming, many more glaciers will disappear. It is also likely that some mountain ranges will lose most, if not all, of their glaciers.

In all mountain regions where glaciers exist today, glacier volume has decreased considerably over the past 150 years. Over that time, many small glaciers have disappeared. With some local exceptions, glacier shrinkage (area and volume reduction) was globally widespread already and particularly strong during the 1940s and since the 1980s. However, there were also phases of relative stability during the 1890s, 1920s and 1970s, as indicated by long-term measurements of length changes and by modelling of mass balance. Conventional *in situ* measurements—and increasingly, airborne and satellite measurements—offer robust evidence in most glacierized regions that the rate of reduction in glacier area was higher over the past two decades than previously, and that glaciers continue to shrink. In a few regions, however, individual glaciers are behaving differently and have advanced while most others were in retreat (e.g., on the coasts of New Zealand, Norway and Southern Patagonia (Chile), or in the Karakoram range in Asia). In general, these advances are the result of special topographic and/or climate conditions (e.g., increased precipitation).

It can take several decades for a glacier to adjust its extent to an instantaneous change in climate, so most glaciers are currently larger than they would be if they were in balance with current climate. Because the time required for the adjustment increases with glacier size, larger glaciers will continue to shrink over the next few decades, even if temperatures stabilise. Smaller glaciers will also continue to shrink, but they will adjust their extent faster and many will ultimately disappear entirely.

Many factors influence the future development of each glacier, and whether it will disappear: for instance, its size, slope, elevation range, distribution of area with elevation, and its surface characteristics (e.g., the amount of debris cover). These factors vary substantially from region to region, and also between neighbouring glaciers. External factors, such as the surrounding topography and the climatic regime, are also important for future glacier evolution. Over shorter time scales (one or two decades), each glacier responds to climate change individually and differently in detail.

Over periods longer than about 50 years, the response is more coherent and less dependent on local environmental details, which means that long-term trends in glacier development can be well modelled. Such models are built on an understanding of basic physical principles. For example, an increase in local mean air temperature, with no change in precipitation, will cause an upward shift of the equilibrium line altitude (ELA; see Glossary) by about 150 m for each degree Celsius of atmospheric warming. Such an upward shift and its consequences for glaciers of different size and elevation range are illustrated in FAQ 4.2, Figure 1.

Initially, all glaciers have an accumulation area (white) above and an ablation area (light blue) below the ELA (FAQ 4.2, Figure 1a). As the ELA shifts upwards, the accumulation area shrinks and the ablation area expands, thus increasing the area over which ice is lost through melt (FAQ 4.2, Figure 1b). This imbalance results in an overall loss of ice. After several years, the glacier front retreats, and the ablation area shrinks until the glacier has adjusted its extent to the new climate (FAQ 4.2, Figure 1c). Where climate change is sufficiently strong to raise the ELA permanently above the glacier's highest point (FAQ 4.2, Figure 1b, right) the glacier will eventually disappear entirely (FAQ 4.2, Figure 1c, right). Higher glaciers, which retain their accumulation areas, will shrink but not disappear (FAQ 4.2, Figure 1c, left and middle). A large valley glacier might lose much of its tongue, probably leaving a lake in its place (FAQ 4.2, Figure 1c, left). Besides air temperature, changes in the quantity and seasonality of precipitation influence the shift of the ELA as well. Glacier dynamics (e.g., flow speed) also plays a role, but is not considered in this simplified scheme.

Many observations have confirmed that different glacier types do respond differently to recent climate change. For example, the flat, low-lying tongues of large valley glaciers (such as in Alaska, Canada or the Alps) currently show the strongest mass losses, largely independent of aspect, shading or debris cover. This type of glacier is slow in

(continued on next page)

FAQ 4.2 (continued)

adjusting its extent to new climatic conditions and reacts mainly by thinning without substantial terminus retreat. In contrast, smaller mountain glaciers, with fairly constant slopes, adjust more quickly to the new climate by changing the size of their ablation area more rapidly (FAQ 4.2, Figure 1c, middle).

The long-term response of most glacier types can be determined very well with the approach illustrated in FAQ 4.2, Figure 1. However, modelling short-term glacier response, or the long-term response of more complex glacier types (e.g., those that are heavily debris-covered, fed by avalanche snow, have a disconnected accumulation area, are of surging type, or calve into water), is difficult. These cases require detailed knowledge of other glacier characteristics, such as mass balance, ice thickness distribution, and internal hydraulics. For the majority of glaciers worldwide, such data are unavailable, and their response to climate change can thus only be approximated with the simplified scheme shown in FAQ 4.2, Figure 1.

The Karakoram–Himalaya mountain range, for instance, has a large variety of glacier types and climatic conditions, and glacier characteristics are still only poorly known. This makes determining their future evolution particularly uncertain. However, gaps in knowledge are expected to decrease substantially in coming years, thanks to increased use of satellite data (e.g., to compile glacier inventories or derive flow velocities) and extension of the ground-based measurement network.

In summary, the fate of glaciers will be variable, depending on both their specific characteristics and future climate conditions. More glaciers will disappear; others will lose most of their low-lying portions and others might not change substantially. Where the ELA is already above the highest elevation on a particular glacier, that glacier is destined to disappear entirely unless climate cools. Similarly, all glaciers will disappear in those regions where the ELA rises above their highest elevation in the future.

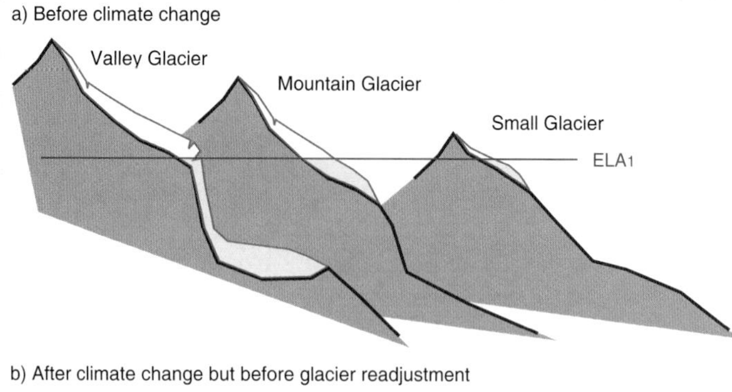

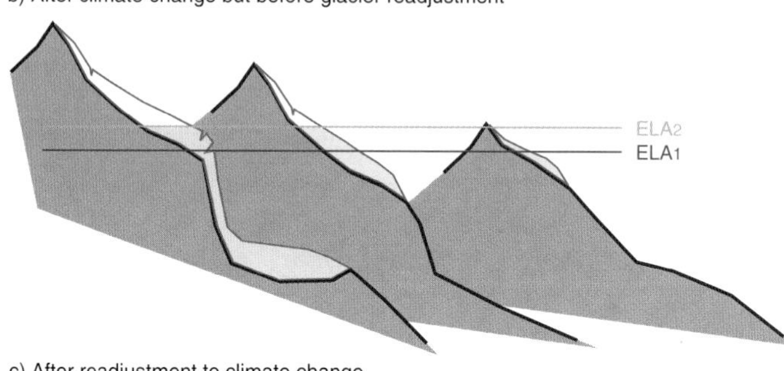

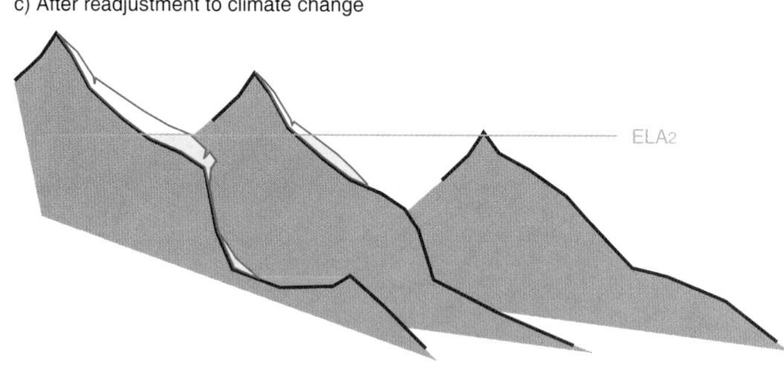

FAQ 4.2, Figure 1 | Schematic of three types of glaciers located at different elevations, and their response to an upward shift of the equilibrium line altitude (ELA). (a) For a given climate, the ELA has a specific altitude (ELA1), and all glaciers have a specific size. (b) Due to a temperature increase, the ELA shifts upwards to a new altitude (ELA2), initially resulting in reduced accumulation and larger ablation areas for all glaciers. (c) After glacier size has adjusted to the new ELA, the valley glacier (left) has lost its tongue and the small glacier (right) has disappeared entirely.

a discussion of the improvements in techniques of measurement and understanding of the change made since AR4 (e.g., Lemke et al., 2007; Cazenave et al., 2009; Chen et al., 2011).

4.4.2.1 Techniques

The three broad techniques for measuring ice-sheet mass balance are the mass budget method, repeated altimetry and measurement of temporal variations in the Earth's gravity field. Each method has been applied to both ice sheets by multiple groups, and over time scales ranging from multiple years to decades (Figures 4.13 and 4.14). The peripheral glaciers, surrounding but not strictly a part of the ice sheets, are not treated in the same manner by each technique. Peripheral glaciers are generally excluded from estimates using the mass budget method, they are sometimes, but not always, included in altimetric estimates, and they are almost always included in gravity estimates.

4.4.2.1.1 Mass budget method

The mass budget method (see Glossary) relies on estimating the difference between net surface balance over the ice sheet (input) and perimeter ice discharge flux (output). This method requires comparison of two very large numbers, and even small percentage errors in either may result in large errors in total mass balance. For ice discharge, perimeter fluxes are calculated from measurements of ice velocity and ice thickness at the grounding line. Knowledge of perimeter fluxes has improved significantly since AR4 for both ice sheets (Rignot et al., 2011b) as a result of more complete ice-thickness data (Bamber et al., 2013; Fretwell et al., 2013) and velocity data from satellite radar interferometry and other techniques (Joughin et al., 2010b; Rignot et al., 2011a). However, incomplete ice thickness mapping still causes uncertainties in ice discharge of 2 to 15% in Antarctica (Rignot et al., 2008b) and 10% in Greenland (Howat et al., 2011; Rignot et al., 2011c).

Regional atmospheric climate models (see Glossary) verified using independent *in situ* data are increasingly preferred to produce estimates of surface mass balance over models that are recalibrated or corrected with *in situ* data (Box et al., 2009), downscaling of global re-analysis data (see Glossary) (Hanna et al., 2011), or interpolation of

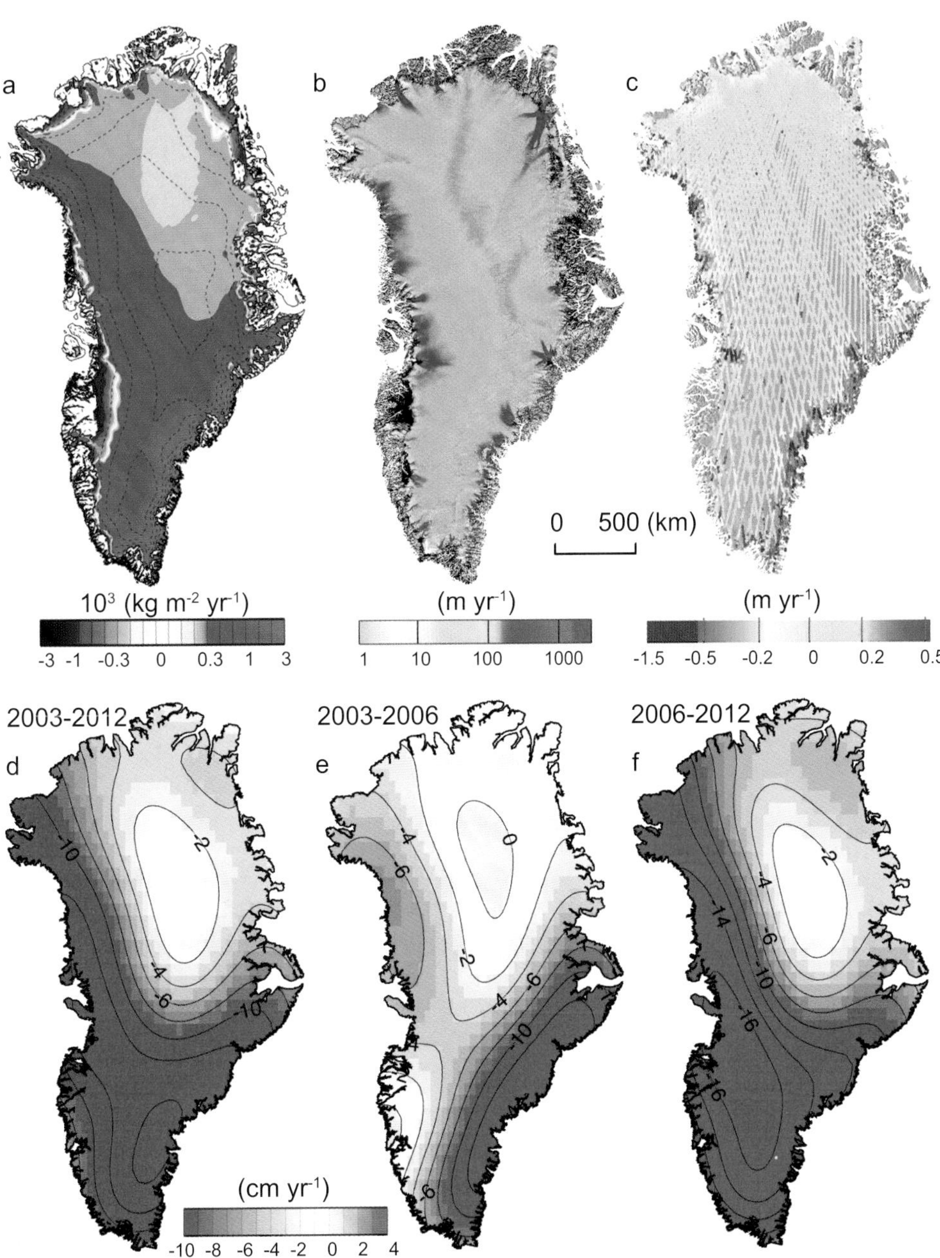

Figure 4.13 | Key variable related to the determination of the Greenland ice sheet mass changes. (a) Mean surface mass balance for 1989–2004 from regional atmospheric climate modelling (Ettema et al., 2009). (b) Ice sheet velocity for 2007–2009 determined from satellite data, showing fastest flow in red, fast flow in blue and slower flow in green and yellow (Rignot and Mouginot, 2012). (c) Changes in ice sheet surface elevation for 2003–2008 determined from ICESat altimetry, with elevation decrease in red to increase in blue (Pritchard et al., 2009). (d, e) Temporal evolution of ice loss determined from GRACE time-variable gravity, shown in centimetres of water per year for the periods (a) 2003–2012, (b) 2003–2006 and (c) 2006–2012, colour coded red (loss) to blue (gain) (Velicogna, 2009). Fields shown in (a) and (b) are used together with ice thickness (see Figure 4.18) in the mass budget method.

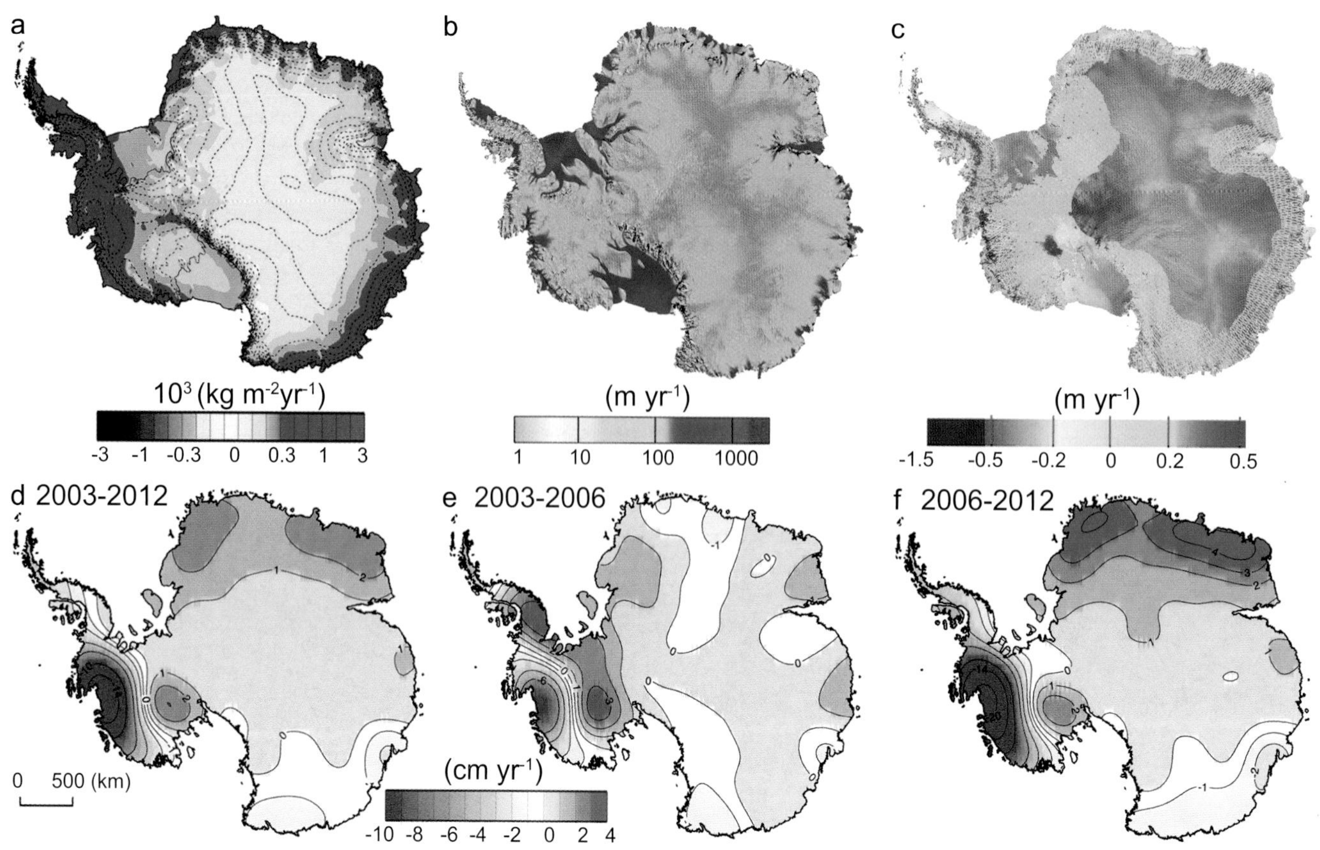

Figure 4.14 | Key fields relating to the determination of Antarctica ice sheet mass changes. (a) Mean surface mass balance for 1989–2004 from regional atmospheric climate modelling (van den Broeke et al., 2006). (b) Ice sheet velocity for 2007–2009 determined from satellite data, showing fastest flow in red, fast flow in blue, and slower flow in green and yellow (Rignot et al., 2011a). (c) Changes in ice sheet surface elevation for 2003–2008 determined from ICESat altimetry, with elevation decrease in red to increase in blue (Pritchard et al., 2009). (d, e) Temporal evolution of ice loss determined from GRACE time-variable gravity, shown in centimetres of water per year for the periods (a) 2003–2012, (b) 2003–2006 and (c) 2006–2012, colour coded red (loss) to blue (gain) (Velicogna, 2009). Fields shown in (a) and (b) are used together with ice thickness (see Figure 4.18) in the mass budget method.

in situ measurements (Arthern et al., 2006; Bales et al., 2009). In Antarctica, surface mass balance (excluding ice shelves) for 1979–2010 is estimated at 1983 ± 122 Gt yr^{-1} (van de Berg et al., 2006; Lenaerts et al., 2012) with interannual variability of 114 Gt yr^{-1} driven by snowfall variability (Figure 4.14). Comparison with 750 *in situ* observations indicates an overall uncertainty of 6% for total ice sheet mass balance, ranging from 5 to 20% for individual drainage basins (van de Berg et al., 2006; Rignot et al., 2008b; Lenaerts et al., 2012; Shepherd et al., 2012). In Greenland, total snowfall (697 Gt yr^{-1}) and rainfall (46 Gt yr^{-1}) minus runoff (248 Gt yr^{-1}) and evaporation/sublimation (26 Gt yr^{-1}) yield a surface mass balance of 469 ± 82 Gt yr^{-1} for 1958–2007 (Ettema et al., 2009). The 17% uncertainty is based on a comparison of model outputs with 350 *in situ* accumulation observations and, in the absence of runoff data, an imposed 20% uncertainty in runoff (Howat et al., 2011). Interannual variability in surface mass balance is large (107 Gt yr^{-1}) due to the out-of-phase relationship between the variability in precipitation (78 Gt yr^{-1}) and runoff (67 Gt yr^{-1}).

4.4.2.1.2 Repeated altimetry

Repeated altimetric survey allows measurement of rates of surface-elevation change, and after various corrections (for changes in snow density and bed elevation; or if the ice is floating, for tides and sea level) reveals changes in ice sheet mass. Satellite radar altimetry (SRALT) has been widely used (Thomas et al., 2008b; Wingham et al., 2009), as has laser altimetry from airplanes (Krabill et al., 2002; Thomas et al., 2009) and satellites (Pritchard et al., 2009; Abdalati et al., 2010; Sorensen et al., 2011; Zwally et al., 2011). Both radar and laser methods have significant challenges. The field-of-view of early SRALT sensors was ~20 km in diameter, and as a consequence, interpretation of the data they acquired over ice sheets with undulating surfaces or significant slopes was complex. Also, for radar altimeters, estimates are affected by penetration of the radar signal below the surface, which depends on characteristics such as snow density and wetness, and by wide orbit separation (Thomas et al., 2008b). Errors in surface-elevation change are typically determined from the internal consistency of the measurements, often after iterative removal of surface elevation-change values that exceed some multiple of the local value of their standard deviation; this results in very small error estimates (Zwally et al., 2005).

Laser altimeters have been used from aircraft for many years, but satellite laser altimetry, available for the first time from NASA's ICESat satellite launched in 2003, has provided many new results since AR4. Laser altimetry is easier to validate and interpret than radar data; the field of view is small (1 m diameter for airborne lasers, 60 m for ICESat), and there is negligible penetration below the surface. However, clouds limit data acquisition, and accuracy is affected by atmospheric conditions, laser-pointing errors, and data scarcity.

Knowledge of the density of the snow and firn in the upper layers of an ice sheet is required to convert altimetric measurements to mass change. However, snow densification rates are sensitive to snow temperature and wetness. Warm conditions favour more rapid densification (Arthern et al., 2010; Li and Zwally, 2011). Consequently, recent Greenland warming has probably caused surface lowering simply from this effect. Corrections are inferred from models that are difficult to validate and are typically less than 2 cm yr^{-1}. ICESat derived surface elevation changes supplemented with differenced ASTER (Advanced Spaceborne Thermal Emission and Reflection Radiometer) satellite digital elevation models were used for outlet glaciers in southeast Greenland (Howat et al., 2008) and for the northern Antarctic Peninsula (Shuman et al., 2011). Laser surveys from airborne platforms over Greenland yield elevation estimates accurate to 10 cm along reference targets (Krabill et al., 1999; Thomas et al., 2009) and 15 cm for ICESat using ground-based high-resolution GPS measurements (Siegfried et al., 2011). For a 5-year separation between surveys, this is an uncertainty of 2.0 cm yr^{-1} for airborne platforms and 3 cm yr^{-1} for ICESat.

Early in 2013, NASA released an elevation correction for ICESat (National Snow and Ice Data Center, 2013) that is relevant to several studies cited in this chapter, but was provided too late to be included in those studies. This correction improves shot-to-shot variability in ICESat elevations, although spatial averaging and application of inter-campaign bias corrections derived from calibration data and used in many studies already mitigates the impact of the higher variability. The correction also changes elevation trend estimates over the 2003–2009 ICESat mission period by up to –1.4 cm yr^{-1}.

To date, a thorough treatment of the impact of this finding has not been published in the peer-reviewed literature, but the overall magnitude of the effect is reported to be at the level of 1.4 cm yr^{-1}. For many studies of glaciers, ice sheets and sea ice this is substantially lower (in some cases, an order of magnitude lower) than the signal of change, but elsewhere (e.g, for elevation changes in East Antarctica) it may have an impact. However, multiple lines of evidence, of which ICESat is only one, are used to arrive at the conclusions presented in this chapter. To the degree to which it can be assessed, there is *high confidence* that the substantive conclusions this chapter will not be affected by revisions of the ICESat data products.

4.4.2.1.3 Temporal variations in Earth gravity field

Since 2002, the GRACE (Gravity Recovery and Climate Experiment) satellite mission has surveyed the Earth's time-variable gravity field. Time-variable gravity provides a direct estimate of the ice-mass change at a spatial resolution of about 300 km (Wahr, 2007). GRACE data yielded early estimates of trends in ice-mass changes over the Greenland and Antarctic ice sheets and confirmed regions of ice loss in coastal Greenland and West Antarctica (Luthcke et al., 2006; Velicogna and Wahr, 2006a, 2006b). With extended time series, now more than 10 years, estimates of ice sheet mass change from GRACE have lower uncertainties than in AR4 (e.g., Harig and Simons, 2012; King et al., 2012). The ice-loss signal from the last decade is also more distinct because the numbers have grown significantly higher (e.g., Wouters et al., 2008; Cazenave et al., 2009; Chen et al., 2009; Velicogna, 2009). The estimates of ice loss based on data from GRACE vary between published studies due to the time-variable nature of the signal, along with other factors that include (1) data-centre specific processing, (2) specific methods used to calculate the mass change, and (3) contamination by other signals within the ice sheet (e.g., glacial isostatic adjustment or GIA, see Glossary) or outside the ice sheet (continental hydrology, ocean circulation). Many of these differences have been reduced in studies published since AR4, resulting in greater agreement between GRACE estimates (Shepherd et al., 2012).

In Antarctica, the GIA signal is similar in magnitude to the ice-loss signal, with an uncertainty of ±80 Gt yr^{-1} (Velicogna and Wahr, 2006b; Riva et al., 2009; Velicogna, 2009). Correction for the GIA signal is addressed using numerical models (e.g., Ivins and James, 2005; Paulson et al., 2007; Peltier, 2009). A comparison of recent GIA models (Tarasov and Peltier, 2002; Fleming and Lambeck, 2004; Peltier, 2004; Ivins and James, 2005; Simpson et al., 2009; Whitehouse et al., 2012) with improved constraints on ice-loading history, indicate better agreement with direct observations of vertical land movements (Thomas et al., 2011a), despite a potential discrepancy between far-field sea level records and common NH deglaciation models. In Greenland, the GIA correction is less than 10% of the GRACE signal with an error of ±19 Gt yr^{-1}. However, because the GIA rate is constant over the satellite's lifetime, GIA uncertainty does not affect the estimate of any change in the rate of ice mass loss (acceleration/deceleration). In Antarctica, the adoption of new GIA models has resulted in a lowering of estimated ice-sheet mass loss (King et al., 2012; Shepherd et al., 2012).

In addition to GRACE, the elastic response of the crustal deformation shown in GPS measurements of uplift rates confirms increasing rates of ice loss in Greenland (Khan et al., 2010b; Khan et al., 2010a) and Antarctica (Thomas et al., 2011a). Analysis of a 34-year time series of the Earth's oblateness (J2) by satellite laser ranging also suggests that ice loss from Greenland and Antarctica has progressively dominated the change in oblateness trend since the 1990s (Nerem and Wahr, 2011).

4.4.2.2 Greenland

There is *very high confidence* that the Greenland ice sheet has lost ice and contributed to sea level rise over the last two decades (Ewert et al., 2012; Sasgen et al., 2012; Shepherd et al., 2012). Recent GRACE results are in better agreement than in AR4 as discussed in Section 4.4.2.1 (Baur et al., 2009; Velicogna, 2009; Pritchard et al., 2010; Wu et al., 2010; Chen et al., 2011; Schrama and Wouters, 2011). Altimetry missions report losses comparable to those from the mass budget method and from the time-variable gravity method (Thomas et al., 2006; Zwally et al., 2011) (Figure 4.13f).

Figure 4.15 shows the cumulative ice mass loss from the Greenland ice sheet over the period 1992–2012 derived from 18 recent studies made by 14 different research groups (Baur et al., 2009; Cazenave et al., 2009; Slobbe et al., 2009; Velicogna, 2009; Pritchard et al., 2010; Wu et al., 2010; Chen et al., 2011; Rignot et al., 2011c; Schrama and Wouters, 2011; Sorensen et al., 2011; Zwally et al., 2011; Ewert et al., 2012; Harig and Simons, 2012; Sasgen et al., 2012). These studies do not include earlier estimates from the same researchers when those have been updated by more recent analyses using extended data. They include estimates made from satellite gravimetry, satellite altimetry and the mass budget method. Details of the studies used for Greenland are listed in Appendix Table 4.A.1 (additional studies not selected are listed in Table 4.A.2).

The mass balance for each year is estimated as a simple average of all the selected estimates available for that particular year. Figure 4.15 shows an accumulation of these estimates since an arbitrary zero on 1 January 1992. The number of estimates available varies with time, with as few as two estimates per year in the 1990s and up to 18 per year from 2004. The cumulative uncertainty in Figure 4.15 is based on the uncertainty cited in the original studies which, when the confidence level is not specifically given, is assumed to be at the 1 standard deviation (1σ) level. However, the annual estimates from different studies often do not overlap within the original uncertainties, and hence the error limits used in this assessment are derived from the absolute maximum and minimum mass balance estimate for each year. These have been converted to the 90% confidence interval (5 to 95%, or 1.65σ). The cumulative error is weighted by $1/\sqrt{n}$, where n is the number of years accumulated.

Despite year-to-year differences between the various original analyses, this multi-study assessment yields very *high confidence* that Greenland has lost mass over the last two decades and *high confidence* that the rate of loss has increased. The increase is also shown in several individual studies (Velicogna, 2009; Chen et al., 2011; Rignot et al., 2011c; Zwally et al., 2011) (Figure 4.13a–c). The average ice mass change to Greenland from the present assessment has been –121 [–149 to –94] Gt yr^{-1} (a sea level equivalent of 0.33 [0.41 to 0.26] mm yr^{-1}) over the period 1993 to 2010, and –229 [–290 to –169] Gt yr^{-1} (0.63 [0.80 to 0.47] mm yr^{-1} sea level equivalent) over the period 2005–2010.

Greenland changes that include and exclude peripheral glaciers cannot be cleanly separated from the mixture of studies and techniques in this assessment, but for the post 2003 period there is a prevalence of gravity studies, which do include the peripheral glaciers. Hence, although the estimated mass change in Greenland peripheral glaciers of -38 ± 7 Gt yr^{-1}over the period 2003–2009 (Gardner et al., 2013) is discussed in Section 4.3.3 (Table 4.5), these changes are included within the values for ice-sheet change quoted in this section, and not as part of the total mass change for glaciers.

A reconciliation of apparent disparities between the different satellite methods was made by the Ice-sheet Mass Balance Intercomparison Experiment (IMBIE) (Shepherd et al., 2012). This intercomparison combined an ensemble of satellite altimetry, interferometry, airborne radio-echo sounding and airborne gravimetry data and regional atmospheric climate model output products, for common geographical regions and for common time intervals. Good agreement was obtained between the estimates from the different methods and, whereas the uncertainties of any method are sometimes large, the combination of methods considerably improves the overall certainty. (Note that Shepherd et al. (2012) also cannot cleanly separate estimates including or excluding peripheral glaciers). For Greenland, Shepherd et al. (2012) estimate a change in mass over the period 1992–2011, averaged

4

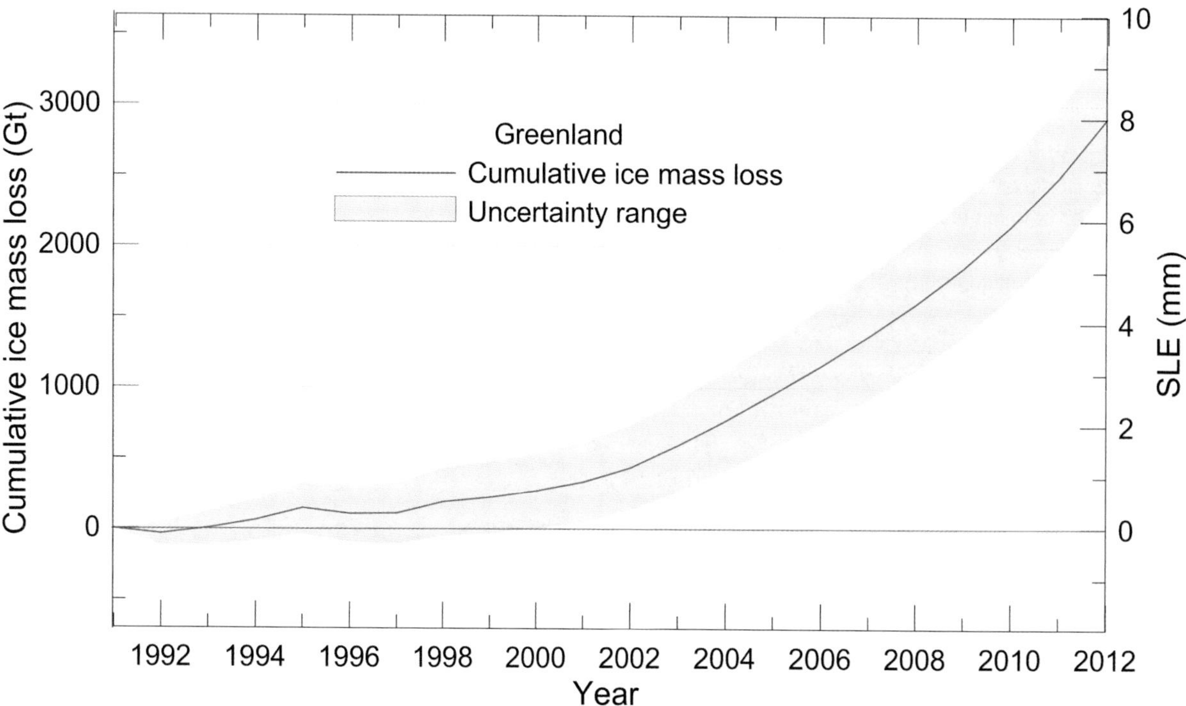

Figure 4.15 | Cumulative ice mass loss (and sea level equivalent, SLE) from Greenland derived as annual averages from 18 recent studies (see main text and Appendix 4.A for details).

across the ensemble for each method, of −142 ± 49 Gt yr^{-1} (0.39 ± 0.14 mm yr^{-1} of sea level rise). For the same period, this present assessment, which averages across individual studies, yields a slightly slower loss, with a rate of mass change of −125 ± 25 Gt yr^{-1} at the 90% confidence level (0.34 ± 0.07 mm yr^{-1} SLE). Averaging across technique ensembles in the present assessment yields a loss at a rate of −129 Gt yr^{-1} (0.36 mm). Shepherd et al. (2012) confirm an increasing mass loss from Greenland, although they also identify mass balance variations over intermediate (2- to 4-year) periods.

The mass budget method shows that ice loss from the Greenland ice sheet is partitioned in approximately similar amounts between surface mass balance (i.e., runoff) and discharge from ice flow across the grounding line (van den Broeke et al., 2009) (*medium confidence*). However, there are significant differences in the relative importance of ice discharge and surface mass balance in various regions of Greenland (Howat et al., 2007; Pritchard et al., 2009; van den Broeke et al., 2009; Sasgen et al., 2012). Dynamic losses dominate in southeast and central west regions, and also influence losses in northwest Greenland, whereas in the central north, southwest and northeast sectors, changes in surface mass balance appear to dominate.

There is *high confidence* that over the last two decades, surface mass balance has become progressively more negative as a result of an increase in surface melt and runoff, and that ice discharge across the grounding line has also been enhanced due to the increased speed of some outlet glaciers. Altimetric measurements of surface height suggest slight inland thickening in 1994–2006 (Thomas et al., 2006, 2009), but this is not confirmed by regional atmospheric climate model outputs for the period 1957–2009 (Ettema et al., 2009), nor recent ice core (see Glossary) data (Buchardt et al., 2012), hence there is *low confidence* in an increase in precipitation in Greenland in recent decades. Probable changes in accumulation are, however, exceeded by the increased runoff especially since 2006 (van den Broeke et al., 2009). The four highest runoff years over the last 140 years occurred since 1995 (Hanna et al., 2011).

The total surface melt area has continued to increase since AR4 and has accelerated in the past few years (Fettweis et al., 2011; Tedesco et al., 2011), with an extreme melt event covering more than 90% of the ice sheet for a few days in July 2012 (Nghiem et al., 2012; Tedesco et al., 2013). Annual surface mass balance in 2011–2012 was 2 standard deviations (2σ) below the 2003–2012 mean. Such extreme melt events are rare and have been observed in ice core records only twice, once in 1889, and once more, seven centuries earlier in the Medieval Warm Period (Meese et al., 1994; Alley and Anandakrishnan, 1995). Over the past decade, the surface albedo of the Greenland ice sheet has decreased by up to 18% in coastal regions, with a statistically significant increase over 87% of the ice sheet due to melting and snow metamorphism, allowing more solar energy to be absorbed for surface melting (Box et al., 2012).

GRACE results show ice loss was largest in southeast Greenland during 2005 and increased in the northwest after 2007 (Khan et al., 2010a; Chen et al., 2011; Schrama and Wouters, 2011; Harig and Simons, 2012). Subsequent to 2005, ice loss decreased in the southeast. These GRACE results agree with measurements of ice discharge from the major outlet glaciers that confirm the dominance of dynamic losses in these regions (van den Broeke et al., 2009). In particular, major outlet glacier speed-up reported in AR4 occurred in west Greenland between 1996 and 2000 (Rignot and Kanagaratnam, 2006) and in southeast Greenland from 2001 to 2006 (Rignot and Kanagaratnam, 2006; Joughin et al., 2010b). In the southeast, many outlet glaciers slowed after 2005 (Howat et al., 2007; Howat et al., 2011), with many flow speeds decreasing back towards those of the early 2000s (Murray et al., 2010; Moon et al., 2012), although most are still flowing faster and discharging more ice into the ocean than they did in 1996 (Rignot and Kanagaratnam, 2006; Howat et al., 2011).

In the northwest, the increase in the rate of ice loss from 1996–2006 to 2006–2010 was probably caused partially by a higher accumulation in the late 1990s compared to earlier and later years (Sasgen et al., 2012), but ice dynamic changes also played a role as outlet glaciers in the northwest showed an increase in speed from 2000 to 2010, with the greatest increase from 2007 to 2010 (Moon et al., 2012). Longer-term observations of surface topography in the northwest sector confirm the dynamic component of this mass loss and suggest two periods of loss in 1985–1993 and 2005–2010 separated by limited mass changes (Kjaer et al., 2012). In the southeast, an 80-year long record reveals that many land-terminating glaciers retreated more rapidly in the 1930s compared to the 2000s, but marine-terminating glaciers retreated more rapidly during the recent warming (Bjørk et al., 2012).

4.4.2.3 Antarctica

Antarctic results from the gravity method are also now more numerous and consistent than in AR4 (Figure 4.14a-c). Methods combining GPS and GRACE at the regional level indicate with *high confidence* that the Antarctic Peninsula is losing ice (Ivins et al., 2011; Thomas et al., 2011a). In other areas, large uncertainties remain in the global GRACE-GPS solutions (Wu et al., 2010).

The SMB reconstructions used in the mass budget method have improved considerably since AR4 (e.g., Rignot et al. 2008b; van den Broeke et al., 2006; Lenaerts et al., 2012; Shepherd et al., 2012). Reconstructed snowfall from regional atmospheric climate models indicates higher accumulation along the coastal sectors than in previous estimates, but little difference in total snowfall. There is *medium confidence* that there has been no long-term trend in the total accumulation over the continent over the past few decades (Monaghan et al., 2006; van den Broeke et al., 2006; Bromwich et al., 2011; Frezzotti et al., 2012; Lenaerts et al., 2012). Although anomalies in accumulation have been noted in recent decades in Eastern Wilkes Land (Boening et al., 2012; Shepherd et al., 2012) and Law Dome (Van Ommen and Morgan, 2010) in East Antarctica, their overall impact on total mass balance is not significant. Satellite laser altimetry indicates that ice volume changes are concentrated on outlet glaciers and ice streams (see Glossary), as illustrated by the strong correspondence between areas of thinning (Figure 4.14f) and areas of fast flow (Figure 4.14e) (Pritchard et al., 2009).

Figure 4.16 shows the cumulative ice mass loss from the Antarctic ice sheet over the period 1992–2012 derived from recent studies made by 10 different research groups (Cazenave et al., 2009; Chen et al., 2009; E et al., 2009; Horwath and Dietrich, 2009; Velicogna, 2009; Wu

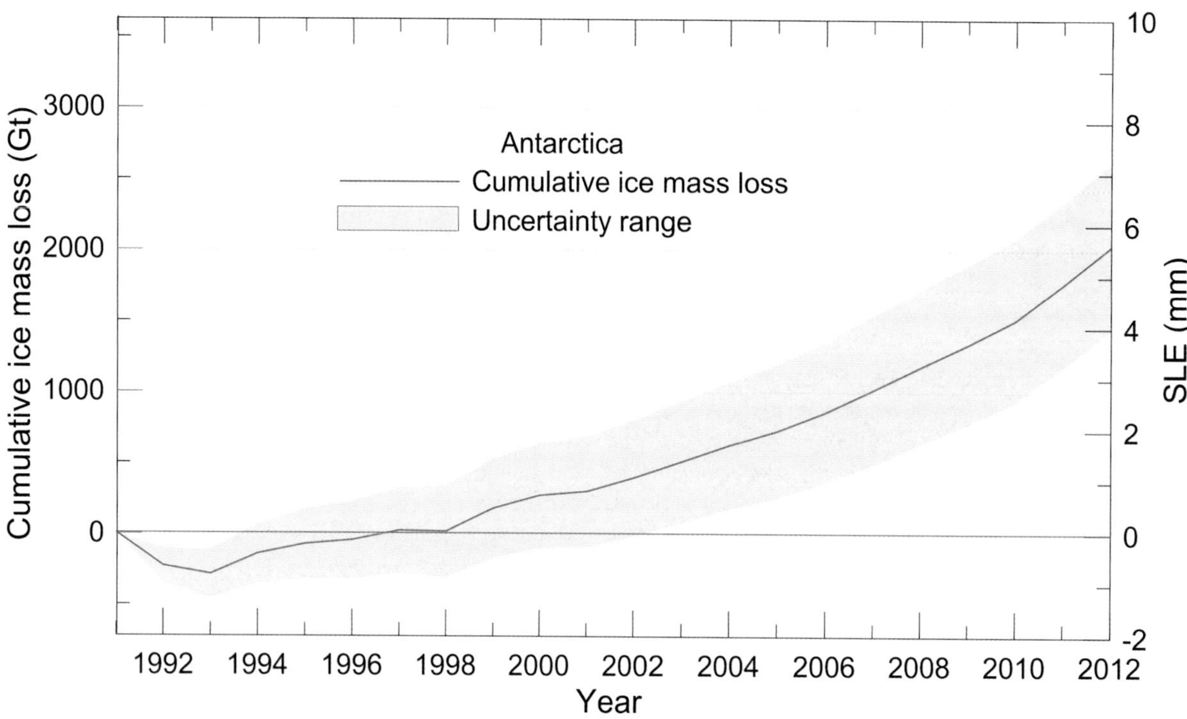

Figure 4.16 | Cumulative ice mass loss (and sea level equivalent, SLE) from Antarctica derived as annual averages from 10 recent studies (see main text and Appendix 4.A for details).

4

et al., 2010; Rignot et al., 2011c; Shi et al., 2011; King et al., 2012; Tang et al., 2012). These studies do not include earlier estimates from the same researchers when those have been updated by more recent analyses using extended data. They include estimates made from satellite gravimetry, satellite altimetry and the mass balance method. Details of the studies used for Antarctica are listed in Table 4.A.3 (additional studies not selected are listed in Table 4.A.4). The number of estimates available varies with time, with only one estimate per year in the 1990s and up to 10 per year from 2003. The cumulative curves and associated errors are derived in the same way as those for Figure 4.15 (see Section 4.4.2.2).

Overall, there is *high confidence* that the Antarctic ice sheet is currently losing mass. The average ice mass change to Antarctica from the present assessment has been –97 [–135 to –58] Gt yr^{-1} (a sea level equivalent of 0.27 mm yr^{-1} [0.37 to 0.16] mm yr^{-1}) over the period 1993–2010, and –147 [–221 to –74] Gt yr^{-1} (0.41 [0.61 to 0.20] mm yr^{-1}) over the period 2005–2010. These assessments include the Antarctic peripheral glaciers.

The recent IMBIE intercomparison (Shepherd et al., 2012) for Antarctica, where the GIA signal is less well known than in Greenland, used two new GIA models (an updated version of Ivins and James (2005), for details see Shepherd et al. (2012); and Whitehouse et al. (2012)). These new models had the effect of reducing the estimates of East Antarctic ice mass loss from GRACE data, compared with some previous estimates. For Antarctica, Shepherd et al. (2012) estimate an average change in mass for 1992–2011 of –71 ± 53 Gt yr^{-1} (0.20 ± 0.15 mm yr^{-1} of sea level equivalent). For the same period this present assessment estimates a loss of 88 ± 35 Gt yr^{-1} at the 90% confidence level (0.24 ± 0.10 mm yr^{-1} SLE). Averaging across technique ensembles in the present assessment, rather than individual estimates, yields no significant difference.

There is *low confidence* that the rate of Antarctic ice loss has increased over the last two decades (Chen et al., 2009; Velicogna, 2009; Rignot et al., 2011c; Shepherd et al., 2012); however, GRACE data gives *medium confidence* of increasing loss over the last decade (Chen et al., 2009; Velicogna, 2009) (Figure 4.16). For GRACE, this conclusion is independent of the GIA signal, which is constant over the measurement period. The mass budget method suggests that the increase in loss from the mass budget method is caused by an increase in glacier flow-speed in the eastern part of the Pacific sector of West Antarctica (Rignot, 2008; Joughin et al., 2010a) and the Antarctic Peninsula (Scambos et al., 2004; Pritchard and Vaughan, 2007; Rott et al., 2011). Comparison of GRACE and the mass budget method for 1992–2010 indicates an increase in the rate of ice loss of, on average, 14 ± 2 Gt yr^{-1} per year compared with 21 ± 2 Gt yr^{-1} per year on average for Greenland during the same time period (Rignot et al., 2011c). The recent IMBIE analysis (Shepherd et al., 2012) shows that the West Antarctic ice sheet and the Antarctic Peninsula are losing mass at an increasing rate, but that East Antarctica gained an average of 21 ± 43 Gt yr^{-1} between 1992 and 2011. Zwally and Giovinetto (2011) also estimate a mass gain for East Antarctica (+16 Gt yr^{-1} between 1992 and 2001). Their reassessment of total Antarctic change made a correction for the ice discharge estimates from regions of the ice sheet not observed in the mass budget method (see Section 4.4.2.1.1). The analysis of Shepherd et al. (2012) indicated that the missing regions contribute little to the total mass change.

In the near-absence of surface runoff and, as discussed in this section, with no evidence of multi-decadal change in total snowfall, there is *high confidence* that Antarctic multi-decadal changes in grounded ice mass must be due to increased ice discharge, although the observational record of ice dynamics extends only from the 1970s and is spatially incomplete for much of this period. Over shorter time scales, however, the interannual to decadal variability in snowfall has an important impact on ice sheet mass balance (Rignot et al., 2011c).

The three techniques are in excellent agreement as to the spatial pattern of ice loss (thinning) and gain (thickening) over Antarctica (Figure 4.14). There is *very high confidence* that the largest ice losses are located along the northern tip of the Antarctic Peninsula where the collapse of several ice shelves in the last two decades triggered the acceleration of outlet glaciers, and in the Amundsen Sea, in West Antarctica (Figure 4.14).. On the Antarctic Peninsula, there is evidence that precipitation has increased (Thomas et al., 2008a) but the resulting ice-gain is insufficient to counteract the losses (Wendt et al., 2010; Ivins et al., 2011). There is *medium confidence* that changes in the Amundsen Sea region are due to the thinning of ice shelves (Pritchard et al., 2012), and *medium confidence* that this is due to high ocean heat flux (Jacobs et al., 2011), which caused grounding line retreat (1 km yr^{-1}) (Joughin et al., 2010a) and glacier thinning (Wingham et al., 2009). Indications of dynamic change are also evident in East Antarctica, primarily around Totten Glacier, from GRACE (Chen et al., 2009), altimetry (Wingham et al., 2006; Shepherd and Wingham, 2007; Pritchard et al., 2009; Flament and Remy, 2012), and satellite radar interferometry (Rignot et al., 2008b). The contribution to the total ice loss from these areas is, however, small and not well understood.

4.4.2.4 Ice Shelves and Floating Ice Tongues

As much as 74% of the ice discharged from the grounded ice sheet in Antarctica passes through ice shelves and floating ice tongues (Bindschadler et al., 2011). Ice shelves help to buttress and restrain flow of the grounded ice (Rignot et al., 2004; Scambos et al., 2004; Hulbe et al., 2008), and so changes in thickness (Shepherd et al., 2003, 2010; Fricker and Padman, 2012), and extent (Doake and Vaughan, 1991; Scambos et al., 2004) of ice shelves influence current ice sheet change. Indeed, nearly all of the outlet glaciers and ice streams that are experiencing high rates of ice loss flow into thinning or disintegrated ice shelves (Pritchard et al., 2012). Many of the larger ice shelves however, exhibit stable conditions (King et al., 2009; Shepherd et al., 2010; Pritchard et al., 2012).

Around the Antarctic Peninsula, the reduction in ice-shelf extent has been ongoing for several decades (Cook and Vaughan, 2010; Fricker and Padman, 2012), and has continued since AR4 with substantial collapse of a section of Wilkins Ice Shelf (Humbert et al., 2010), which had been retreating since the late1990s (Scambos et al., 2000). Overall, 7 of 12 ice shelves around the Peninsula have retreated in recent decades with a total loss of 28,000 km^2, and a continuing rate of loss of around 6000 km^2 per decade (Cook and Vaughan, 2010). There is *high confidence* that this retreat of ice shelves along the Antarctic Peninsula has been related to changing atmospheric temperatures (e.g., Scambos et al., 2000; Morris and Vaughan, 2003; Marshall et al., 2006; Holland et al., 2011). There is *low confidence* that changes in the ocean have also contributed (e.g., Shepherd et al., 2003; Holland et al., 2011; Nicholls et al., 2012; Padman et al., 2012).

4.4.3 Total Ice Loss from Both Ice Sheets

The total ice loss from both ice sheets for the 20 years 1992–2011 (inclusive) has been 4260 [3060 to 5460] Gt, equivalent to 11.7 [8.4 to 15.1] mm of sea level. However, the rate of change has increased with time and most of this ice has been lost in the second decade of the 20-year period. From the data presented in Figure 4.17, the average loss in Greenland has *very likely* increased from 34 [–6 to 74] Gt yr^{-1} over the decade 1992–2001 (sea level equivalent, 0.09 [–0.02 to 0.20] mm yr^{-1}), to 215 [157 to 274] Gt yr^{-1} over the decade 2002–2011 (0.59 [0.43 to 0.76] mm yr^{-1}). In Antarctica, the loss has *likely* increased 30 [–37 to 97] Gt yr^{-1} (sea level equivalent, 0.08 [–0.10 to 0.27] mm yr^{-1}) for 1992–2001, to 147 [72 to 221] Gt yr^{-1} for 2002–2011 (0.40 [0.20 to 0.61] mm yr^{-1}). Over the last five years (2007–2011), the loss from both ice sheets combined has been equivalent to 1.2 ± 0.4 mm yr^{-1} of sea level (Figure 4.17 and Table 4.6).

4.4.4 Causes of Changes in Ice Sheets

4.4.4.1 Climatic Forcing

Changes in ice sheet mass balance are the result of an integrated response to climate, and it is imperative that we understand the context of current change within the framework of past changes and natural variability.

4.4.4.1.1 Snowfall and surface temperature

Ice sheets experience large interannual variability in snowfall, and local trends may deviate significantly from the long-term trend in integrated snowfall. However, as in AR4, the available data do not suggest any significant long-term change in accumulation in Antarctica, except for the Antarctic Peninsula (Monaghan et al., 2006; Ettema et al., 2009; van den Broeke et al., 2009; Bromwich et al., 2011).

Increasing air temperature will (when above the freezing point) increase the amount of surface melt, and can also increase the moisture bearing capacity of the air, and hence can increase snowfall. Over Greenland, temperature has risen significantly since the early 1990s, reaching values similar to those in the 1930s (Box et al., 2009). The

Table 4.6 | Average rates of ice sheet loss given as mm of sea level equivalent, derived as described for Figure 4.15 and Figure 4.16 using estimates listed in Appendix Tables 4.A.1 and 4.A.3.

Period	Ice sheet loss (mm yr^{-1} SLE)	
Greenland		
2005–2010 (6-year)	0.63	±0.17
1993–2010 (18-year)	0.33	±0.08
Antarctica		
2005–2010 (6-year)	0.41	±0.20
1993–2010 (18-year)	0.27	±0.11
Combined		
2005–2010 (6-year)	1.04	±0.37
1993–2010 (18-year)	0.60	±0.18

4

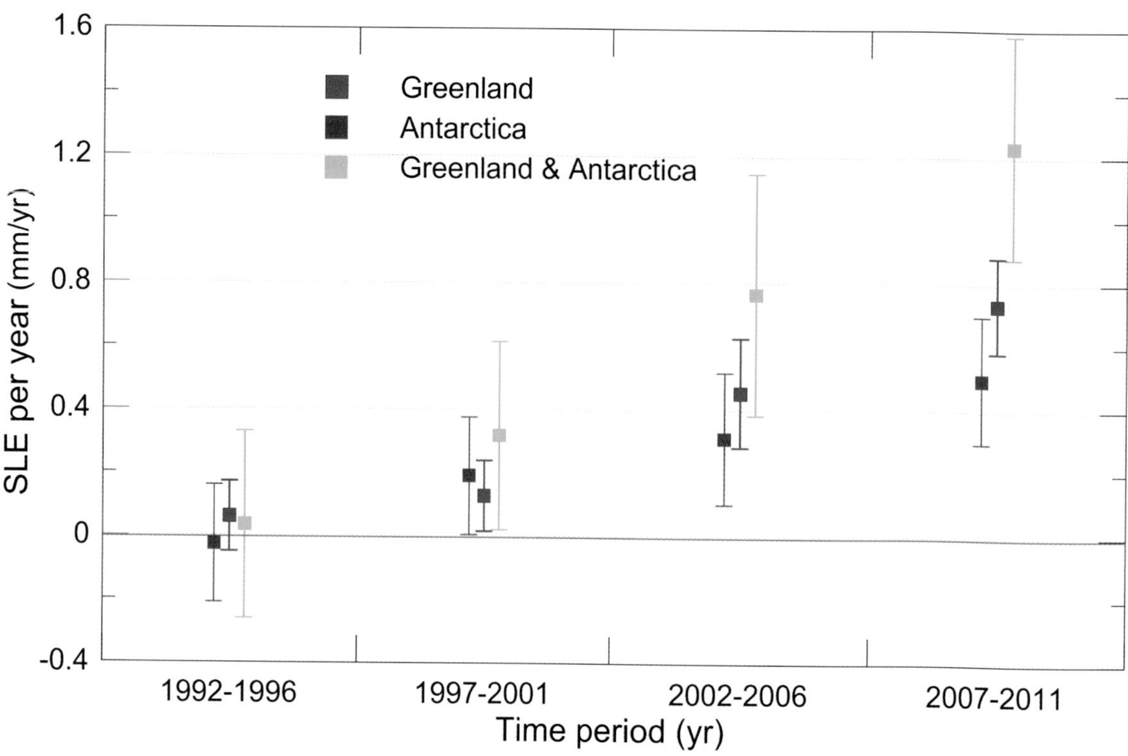

Figure 4.17 | Rate of ice sheet loss in sea level equivalent averaged over 5-year periods between 1992 and 2011. These estimates are derived from the data in Figures 4.15 and 4.16.

4

year 2010 was an exceptionally warm year in west Greenland with Nuuk having the warmest year since the start of the temperature record in 1873 (Tedesco et al., 2011). In West Antarctica, the warming since the 1950s (Steig et al., 2009; Ding et al., 2011; Schneider et al., 2012; Bromwich et al., 2013), the magnitude and seasonality of which are still debated, has not manifested itself in enhanced surface melting (Tedesco and Monaghan, 2009; Kuipers Munneke et al., 2012) nor in increased snowfall (Monaghan et al., 2006; Bromwich et al., 2011; Lenaerts et al., 2012). Statistically significant summer warming has been observed on the east coast of the northern Antarctic Peninsula (Marshall et al., 2006; Chapman and Walsh, 2007), with extension of summer melt duration (Barrand et al., 2013), while East Antarctica has showed summer cooling (Turner et al., 2005). In contrast, the significant winter warming at Faraday/Vernadsky station on the western Antarctic Peninsula is attributable to a reduction of sea ice extent (Turner et al., 2005).

4.4.4.1.2 Ocean thermal forcing

Since AR4, observational evidence has contributed to *medium confidence* that the interaction between ocean waters and the periphery of large ice sheets plays a major role in present ice sheet changes (Holland et al., 2008; Pritchard et al., 2012). Ocean waters provide the heat that can drive high melt rates beneath ice shelves (Jacobs et al., 1992; Holland and Jenkins, 1999; Rignot and Jacobs, 2002; Pritchard et al., 2012) and at marine-terminating glacier fronts (Holland et al., 2008; Rignot et al., 2010; Jacobs et al., 2011).

Ocean circulation delivers warm waters to ice sheets. Variations in wind patterns associated with the North Atlantic Oscillation (Jacobs et al., 1992; Hurrell, 1995), and tropical circulations influencing West Antarctica (Ding et al., 2011; Steig et al., 2012), are probable drivers of increasing melt at some ice-sheet margins. In some parts of Antarctica, changes in the Southern Annular Mode (Thompson and Wallace, 2000, see Glossary) may also be important. Observations have established that warm waters of subtropical origin are present within several fjords in Greenland (Holland et al., 2008; Myers et al., 2009; Straneo et al., 2010; Christoffersen et al., 2011; Daniault et al., 2011).

Satellite records and *in situ* observations indicate warming of the Southern Ocean (see Chapter 3) since the 1950s (Gille, 2002, 2008). This warming is confirmed by data from robotic ocean buoys (Argo floats) (Boening et al., 2008) but the observational record remains short and, close to Antarctica, there are only limited observations from ships (Jacobs et al., 2011), short-duration moorings and data from instrumented seals (Charrassin et al., 2008; Costa et al., 2008).

4.4.4.2 Ice Sheet Processes

4.4.4.2.1 Basal lubrication

Ice flows in part by sliding over the underlying rock and sediment, which is lubricated by water at the ice base: a process known as basal lubrication (see Glossary). In many regions close to the Greenland ice sheet margin, abundant summer meltwater on the surface of the ice sheet forms large lakes. This surface water can drain to the ice sheet bed, thus increasing basal water pressure, reducing basal friction and increasing ice flow speed (Zwally et al., 2002b).

Such drainage events are common in southwest and northeast Greenland, but rare in the most rapidly changing southeast and northwest regions (Selmes et al., 2011). The effect can be seen in diurnal flow variations of some land-terminating regions (Das et al., 2008; Shepherd et al., 2009), and after lake-drainage events, when 50 to 110% short-term speed-up of flow has been observed. However, the effect is temporally and spatially restricted (Das et al., 2008). The summer increase in speed over the annual mean is only ~10–20%, the increase is less at higher elevations (Bartholomew et al., 2011), and observations suggest most lake drainages do not affect ice sheet velocity (Hoffman et al., 2011). Theory and field studies suggest an initial increase in flow rate with increased surface meltwater supply (Bartholomew et al., 2011; Palmer et al., 2011), but if the supply of surface water continues to increase and subglacial drainage becomes more efficient, basal water pressure, and thus basal motion, is reduced (van de Wal et al., 2008; Schoof, 2010; Sundal et al., 2011; Shannon et al., 2012). Overall, there is *high confidence* that basal lubrication is important in modulating flow in some regions, especially southwest Greenland, but there is also *high confidence* that it does not explain recent dramatic regional speed-ups that have resulted in rapid increases in ice loss from calving glaciers.

4.4.4.2.2 Cryo-hydrologic warming

Percolation and refreezing of surface meltwater that drains through the ice column may alter the thermal regime of the ice sheet on decadal time scales (Phillips et al., 2010). This process is known as cryo-hydrologic warming, and it could affect ice rheology and hence ice flow.

4.4.4.2.3 Ice shelf buttressing

Recent changes in marginal regions of the Greenland and Antarctic ice sheets include some thickening and slowdown of outlet glaciers, but mostly thinning and acceleration (e.g., Pritchard et al., 2009; Sorensen et al., 2011), with some glacier speeds increasing two- to eight-fold (Joughin et al., 2004; Rignot et al., 2004; Scambos et al., 2004; Luckman and Murray, 2005; Rignot and Kanagaratnam, 2006; Howat et al., 2007). Many of the largest and fastest glacier changes appear to be at least partly a response to thinning, shrinkage or loss of ice shelves or floating ice tongues (MacGregor et al., 2012; Pritchard et al., 2012). This type of glacier response is consistent with classical models of ice shelf buttressing proposed 40 years ago (Hughes, 1973; Weertman, 1974; Mercer, 1978; Thomas and Bentley, 1978).

4.4.4.2.4 Ice–ocean interaction

Since AR4 it has become far more evident that the rates of submarine melting can be very large (e.g., Motyka et al., 2003). The rate of melting is proportional to the product of ocean thermal forcing (difference between ocean temperature and the *in situ* freezing point of seawater) and water flow speed at the ice–ocean interface (Holland and Jenkins, 1999). Melt rates along marine-terminating glacier margins are one-to-two orders of magnitude greater than for ice shelves because of the additional buoyancy forces provided by the discharge of sub-glacial melt water at the glacier base (Motyka et al., 2003; Jenkins, 2011; Straneo et al., 2012; Xu et al., 2012). In South Greenland, there is *medium confidence* that the acceleration of glaciers from the mid-1990s to mid-2000s was due to the intrusion of ocean waters of subtropical origin into glacial fjords (Holland et al., 2008; Howat et al., 2008; Murray et al., 2010; Straneo et al., 2010; Christoffersen et al., 2011; Motyka et al., 2011; Straneo et al., 2011; Rignot and Mouginot, 2012). Models suggest that the increase in ice melting by the ocean contributed to the reduction of backstress experienced by glaciers and subsequent acceleration (Payne et al., 2004; Schoof, 2007; Nick et al., 2009; Nick et al., 2013; O'Leary and Christoffersen, 2013): changes in the floating mixture of sea ice, iceberg debris and blown snow in front of the glacier may also play a part (Amundson et al., 2010).

4.4.4.2.5 Iceberg calving

Calving of icebergs from marine-terminating glaciers and ice shelves is important in their overall mass balance, but the processes that initiate calving range from seasonal melt-driven processes (Benn et al., 2007) to ocean swells and tsunamis (MacAyeal et al., 2006; Brunt et al., 2011), or the culmination of a response to gradual change (Doake et al., 1998; Scambos et al., 2000). Some of these processes show strong climate influence, while others do not. Despite arguments of rather limited progress in this area (Pfeffer, 2011), there have been some recent advances (Joughin et al., 2008a; Blaszczyk et al., 2009; Amundson et al., 2010; Nick et al., 2010, 2013), and continental-scale ice sheet models currently rely on improved parameterisations (Alley et al., 2008; Pollard and DeConto, 2009; Levermann et al., 2012). Recently more realistic models have been developed allowing the dependence of calving and climate to be explicitly investigated (e.g., Nick et al., 2013).

4.4.5 Rapid Ice Sheet Changes

The projections of sea level rise presented in AR4 explicitly excluded future rapid dynamical changes (see Glossary) in ice flow, and stated that 'understanding of these processes is limited and there is no consensus on their magnitude'. Considerable efforts have been made since AR4 to fill this knowledge gap. Chapter 13 discusses observed and likely future sea level, including model projections of changes in the volume stored in the ice sheets: in this section we summarise the processes thought to be potential causes of rapid changes in ice flow and emphasise new observational evidence that these processes are already underway.

'Rapid ice sheet changes' are defined as changes that are of sufficient speed and magnitude to impact on the mass budget and hence rate of sea level rise on time scales of several decades or shorter. A further consideration is whether and under what circumstances any such changes are 'irreversible', that is, would take several decades to centuries to reverse under a different climate forcing. For example, an effectively irreversible change might be the loss of a significant fraction of the Greenland ice sheet, because at its new lower (and therefore warmer) surface elevation, the ice sheet would be able to grow thicker only slowly even in a cooler climate (Ridley et al., 2010) (Section 13.4.3.3).

Observations suggest that some observed changes in ice shelves and glaciers on the Antarctic Peninsula are irreversible. These ice bodies continue to experience rapid and irreversible retreat, coincident with air temperatures rising at four to six times the global average rate at some stations (Vaughan et al., 2003), and with warm Circumpolar Deep Water becoming widespread on the western continental shelf (Martinson et al., 2008). Collapse of floating ice shelves on the Antarctic Peninsula, such as the 2002 collapse of the Larsen B Ice Shelf which is unprecedented in the last 10,000 years, has resulted in speed up of tributary glaciers by 300 to 800% (De Angelis and Skvarca, 2003; Rignot et al., 2004; Scambos et al., 2004; Rott et al., 2011). Even if iceberg calving was to cease entirely, regrowth of the Larsen B ice shelf to its pre-collapse state would take centuries based on the ice-shelf speed and size prior to its collapse (Rignot et al., 2004).

Surface melt that becomes runoff is a major contributor to mass loss from the Greenland ice sheet, which results in a lower (hence warmer) ice sheet surface and a lower surface albedo (allowing the surface to absorb more solar radiation); both processes further increase melt. The warm summers of the last two decades (van den Broeke et al., 2009; Hanna et al., 2011), and especially in 2012 (Hall et al., 2013), are unusual in the multi-centennial record. Exceptionally high melt events have affected even the far north of Greenland, for example, with the partial collapse of the floating ice tongues of Ostenfeld Gletscher and Zachariae Isstrom in 2000–2006 (Moon and Joughin, 2008).

The importance that subsurface warm waters play in melting the periphery of ice sheets in Greenland and Antarctica, and the evolution of these ice sheets, has become much clearer since AR4 (see Sections 4.4.3.1 and 4.4.3.2). New observations in Greenland and Antarctica, as well as advances in theoretical understanding, show that regions of ice sheets that are grounded well below sea level are most likely to experience rapid ice mass loss, especially if the supply of heat to the ice margin increases (Schoof, 2007; Holland et al., 2008; Joughin and Alley, 2011; Motyka et al., 2011; Young et al., 2011a; Joughin et al., 2012; Ross et al., 2012) (See also Figure 4.18.) Where this ice meets the

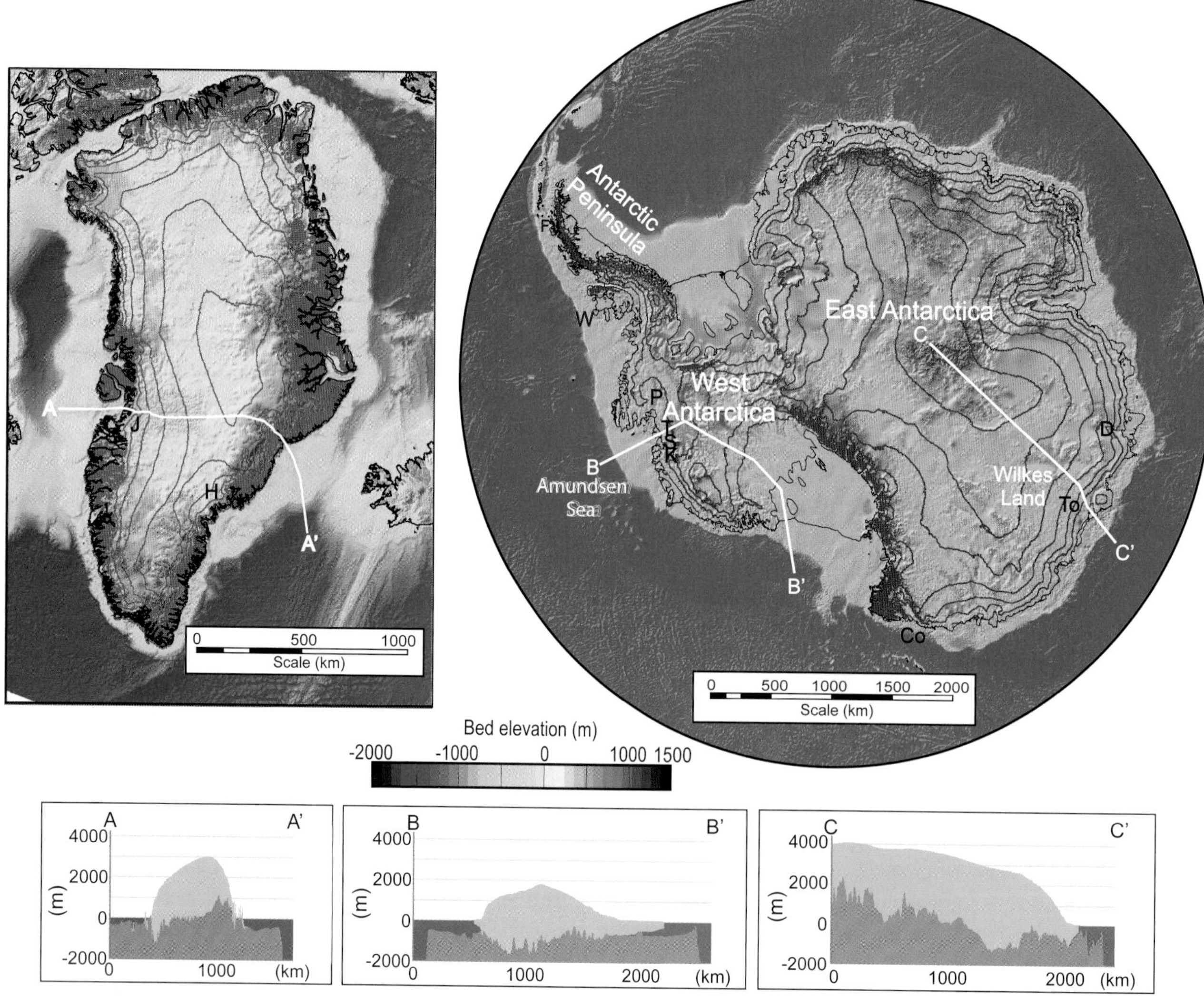

Figure 4.18 | Subglacial and seabed topography for Greenland and Antarctica derived from digital compilations (Bamber et al., 2013; Fretwell et al., 2013). Blue areas highlight the marine-based parts of the ice sheets, which are extensive in Antarctica, but in Greenland, relate to specific glacier troughs. Selected sections through the ice sheet show reverse bed gradients that exist beneath some glaciers in both ice sheets.

ocean and does not form an ice shelf, warm waters can increase melting at the ice front, causing undercutting, higher calving rates, ice-front retreat (Motyka et al., 2003; Benn et al., 2007; Thomas et al., 2011a) and consequent speed-up and thinning. Surface runoff also increases subglacial water discharge at the grounding line, which enhances ice melting at the ice front (Jenkins, 2011; Xu et al., 2012). Where an ice shelf is present, ice melt by the ocean may cause thinning of the shelf as well as migration of the grounding line further inland into deep basins, with a major impact on buttressing, flow speed and thinning rate (Thomas et al., 2011a).

The influence of the ocean on the ice sheets is controlled by the delivery of heat to the ice sheet margins, particularly to ocean cavities beneath ice shelves and to calving fronts (Jenkins and Doake, 1991; Jacobs et al., 2011). The amount of heat delivered is a function of the temperature and salinity of ocean waters; ocean circulation; and the bathymetry of continental shelves, in fjords near glacier fronts and beneath ice shelves, most of which are not known in sufficient detail (Jenkins and Jacobs, 2008; Holland et al., 2010; Dinniman et al., 2012; Galton-Fenzi et al., 2012; Padman et al., 2012). Changes in any of these parameters would have a direct and rapid impact on melt rates and potentially on calving fluxes (see Chapter 13).

Ice grounded on a reverse bed-slope, deepening towards the ice sheet interior, is potentially subject to the marine ice sheet instability (Weertman, 1974; Schoof, 2007) (see Box 13.2). Much of the bed of the West Antarctic Ice Sheet (WAIS) lies below sea level and on a reverse bed-slope, with basins extending to depths greater than 2 km (Figure 4.18). The marine parts of the WAIS contain ~3.4 m of equivalent sea level rise (Bamber et al., 2013; Fretwell et al., 2013), and a variety of evidence strongly suggests that the ice sheet volume has been much smaller than present in the last 1 million years, during periods with temperatures similar to those predicted in the next century (see also Chapter 5) (Kopp et al., 2009). Potentially unstable marine ice sheets also exist in East Antarctica, for example, in Wilkes Land (Young et al., 2011a), and these contain more ice than WAIS (9 m sea level equivalent for Wilkes Land). In northern Greenland, ice is also grounded below sea level, with reverse slopes (Figure 4.18; Joughin et al., 1999).

Observations since AR4 confirm that rapid changes are indeed occurring at the marine margins of ice sheets, and that these changes have been observed to penetrate hundreds of kilometres inland (Pritchard et al., 2009; Joughin et al., 2010b).

The Amundsen Sea sector of West Antarctica is grounded significantly below sea level and is the region of Antarctica changing most rapidly at present. Pine Island Glacier has sped up 73% since 1974 (Rignot, 2008) and has thinned throughout 1995–2008 at increasing rates (Wingham et al., 2009) due to grounding line retreat. There is *medium confidence* that retreat was caused by the intrusion of warm ocean water into the sub-ice shelf cavity (Jenkins et al., 2010; Jacobs et al., 2011; Steig et al., 2012). The neighbouring Thwaites, Smith and Kohler glaciers are also speeding-up, thinning and contributing to increasing mass loss (Figure 4.14). The present rates of thinning are more than one order of magnitude larger than millennial-scale thinning rates in this area (Johnson et al., 2008). Changes in velocity, elevation, thickness and grounding line position observed in the past two to three decades in the Pine Island/Thwaites Glacier sector are not inconsistent with the development of a marine ice sheet instability triggered by a change in climate forcing, but neither are they inconsistent solely with a response to external environmental (probably oceanic) forcing.

In Greenland, there is *medium confidence* that the recent rapid retreat of Jakobshavn Isbrae was caused by the intrusion of warm ocean water beneath the floating ice tongue (Holland et al., 2008; Motyka et al., 2011) combined with other factors, such as weakening of the floating mixture of sea ice, iceberg debris and blown snow within ice rifts (Joughin et al., 2008b; Amundson et al., 2010). There is *medium confidence* that recent variations in southeast Greenland's glaciers have been caused by intrusion of warm waters of subtropical origin into glacial fjords. Since AR4 it has become clear that the mid-2000s speed up of southeast Greenland glaciers, which caused a doubling of ice loss from the Greenland ice sheet (Luthcke et al., 2006; Rignot and Kanagaratnam, 2006; Howat et al., 2008; Wouters et al., 2008), was a pulse that was followed by a partial slow down (Howat et al., 2008; Murray et al., 2010). Although changes in elevation in the north are not as large as in the south, marine sectors were thinning in 2003–2008 (Pritchard et al., 2009; Sorensen et al., 2011).

In contrast to the rapidly changing marine margins of the ice sheets, land-terminating regions of the Greenland ice sheet are changing more slowly, and these changes are explained largely by changes in the input of snow and loss of meltwater (Sole et al., 2011). Surface meltwater, although abundant on the Greenland ice sheet, does not seem to be driving significant changes in basal lubrication that impact on ice sheet flow (Joughin et al., 2008b; Selmes et al., 2011; Sundal et al., 2011).

In Greenland, the observed changes are not all irreversible. The Helheim Glacier in southeast Greenland accelerated, retreated and increased its calving flux during the period 2002–2005 (Howat et al., 2011; Andresen et al., 2012), but its calving flux similarly increased during the late 1930s – early 1940s (Andresen et al., 2012): an episode from which the glacier subsequently recovered and re-advanced (Joughin et al., 2008b). The collapse of the floating tongue of Jakobshavn Isbrae in 2002 and consequent loss of buttressing has considerably increased ice flow speeds and discharge from the ice sheet. At present, the glacier grounding line is retreating 0.5–0.6 km yr^{-1} (Thomas et al., 2011b; Rosenau et al., 2013), with speeds in excess of 11 km yr^{-1} (Moon et al., 2012), and the glacier is retreating on a bed that deepens further inland, which could be conducive to a marine instability. However, there is evidence that Jakobshavn Isbrae has undergone significant margin changes over the last approximately 8000 years which may have been both more and less extensive than the recent ones (Young et al., 2011b).

Since AR4, many new observations indicate that changes in ice sheets can happen more rapidly than was previously recognised. Similarly, evidence presented since AR4 indicates that interactions with both the atmosphere and ocean are key drivers of decadal ice-sheet change. So, although our understanding of the detailed processes that control the evolution of ice sheets in a warming climate remains incomplete, there is no indication in observations of a slowdown in the mass loss from ice sheets; instead, recent observations suggest an ongoing increase in mass loss.

4.5 Seasonal Snow

4.5.1 Background

Snowfall is a component of total precipitation and, in that context, is discussed in Chapter 2 (See Section 2.5.1.3); here we discuss accumulated snow as a climatological indicator. Snow is measured using a variety of instruments and techniques, and reported using several metrics, including snow cover extent (SCE; see Glossary); the seasonal sum of daily snowfall; snow depth (SD); snow cover duration (SCD), that is, number of days with snow exceeding a threshold depth; or snow water equivalent (SWE; see Glossary).

Long-duration, consistent records of snow are rare owing to many challenges in making accurate and representative measurements. Although weather stations in snowy inhabited areas often report snow depth, records of snowfall are often patchy or use techniques that change over time (e.g., Kunkel et al., 2007). The density of stations and the choice of metric also varies considerably from country to country. The longest satellite-based record of SCE is the visible-wavelength weekly product of the National Oceanic and Atmospheric Administration (NOAA) dating to 1966 (Robinson et al., 1993), but this covers only the NH. Satellite mapping of snow depth and SWE has lower accuracy than SCE, especially in mountainous and heavily forested areas. Measurement challenges are particularly acute in the Southern Hemisphere (SH), where only about 11 long-duration *in situ* records continue to recent times: seven in the central Andes and four in southeast Australia. Owing to concerns about quality and duration, global satellite microwave retrievals of SWE are of less use in the data-rich NH than in the data-poor SH.

4.5.2 Hemispheric View

By blending *in situ* and satellite records, Brown and Robinson (2011) have updated a key indicator of climate change, namely the time series of NH SCE (Figure 4.19). This time series shows significant reductions over the past 90 years with most of the reductions occurring in the 1980s, and is an improvement over that presented in AR4 in several ways, not least because the uncertainty estimates are explicitly derived through the statistical analysis of multiple data sets, which leads to *very high confidence*. Snow cover decreases are largest in spring (Table 4.7), and the rate of decrease increases with latitude in response to larger albedo feedbacks (Déry and Brown, 2007). Averaged March and April NH SCE decreased 0.8% [0.5 to 1.1%] per decade over the 1922–2012 period, 1.6% [0.8 to 2.4%] per decade over the 1967–2012 period, and 2.2% [1.1 to 3.4%] per decade over the 1979–2012 period. In a new development since AR4, both absolute and relative losses in June SCE now exceed the losses in March–April SCE: 11.7% [8.8 to 14.6%] per decade or 53% [40 to 66%] total over the 1967–2012 period and 14.8% [10.3 to 19.3%] per decade over the 1979–2012 period (all ranges *very likely*). Note that these percentages differ from those given by Brown and Robinson (2011) which were calculated relative to the mean over the 1979–2000 period, rather than relative to the starting point. The loss rate of June SCE exceeds the loss rate for Coupled Model Intercomparison Project Phase 5 (CMIP5) model projections of June SCE and also exceeds the well-known loss of September sea ice extent (Derksen and Brown, 2012). Viewed another way, the NOAA SCE data indicate that, owing to earlier spring snowmelt, the duration of the snow season averaged over NH grid points declined by 5.3 days per decade since winter 1972–1973 (Choi et al., 2010).

Over Eurasia, *in situ* data show significant increases in winter snow accumulation but a shorter snowmelt season (Bulygina et al., 2009). From analysis of passive microwave satellite data since 1979, significant trends toward a shortening of the snowmelt season have been identified over much of Eurasia (Takala et al., 2009) and the pan-Arctic region (Tedesco et al., 2009), with a trend toward earlier melt of about 5 days per decade for the beginning of the melt season, and a trend of about 10 days per decade later for the end of the melt season.

The correlation between spring temperature and SCE (Figure 4.20) demonstrates that trends in spring SCE are linked to rising temperature, and for a well-understood reason: The spring snow cover-albedo feedback. This feedback contributes substantially to the hemispheric response to rising greenhouse gases and provides a useful test of global

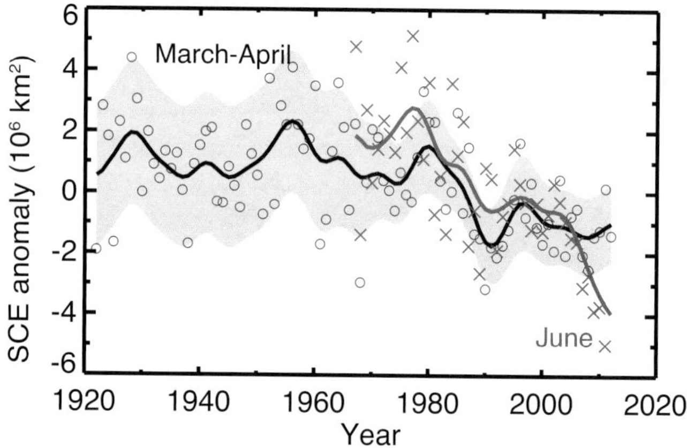

Figure 4.19 | March–April NH snow cover extent (SCE, circles) over the period of available data, filtered with a 13-term smoother and with shading indicating the 95% confidence interval; and June SCE (red crosses, from satellite data alone), also filtered with a 13-term smoother. The width of the smoothed 95% confidence interval is influenced by the interannual variability in SCE. Updated from Brown and Robinson (2011). For both time series the anomalies are calculated relative to the 1971–2000 mean.

Table 4.7 | Least-squares linear trend in Northern Hemisphere snow cover extent (SCE) in 10^6 km^2 per decade for 1967–2012. The equivalent trends for 1922–2012 (available only for March and April) are –0.19* March and –0.40* April.

Annual	Jan	Feb	March	April	May	June	July	Aug	Sep	Oct	Nov	Dec
–0.40*	0.03	–0.13	–0.50*	–0.63*	–0.90*	–1.31*	n/a	n/a	n/a	n/a	0.17	0.34

Notes:
*Denotes statistical significance at $p = 0.05$.

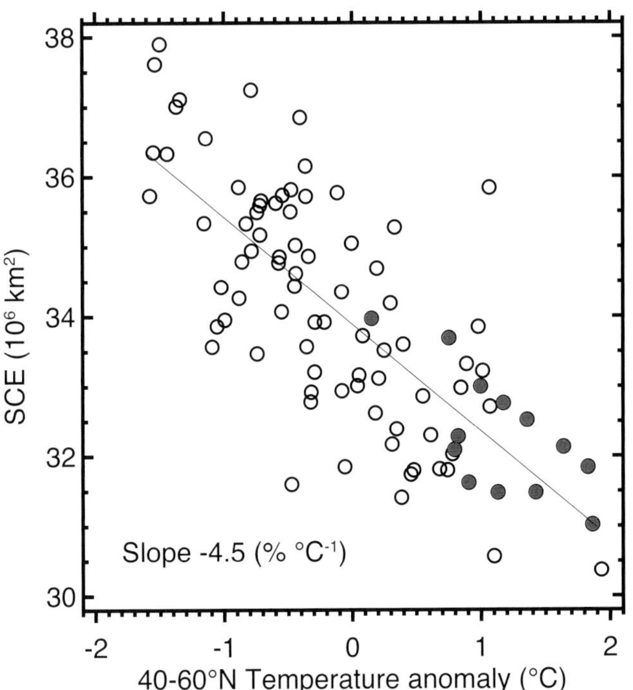

Figure 4.20 | Relationship between NH April SCE and corresponding land air temperature anomalies over 40°N to 60°N from the CRUtem4 data set (Jones et al., 2012). Red circles indicate the years 2000–2012. The correlation is 0.76. Updated from Brown and Robinson (2011).

climate models (Fernandes et al., 2009) (see also Chapter 9). Indeed, the observed declines in land snow cover and sea ice have contributed roughly the same amount to changes in the surface energy fluxes, and the albedo feedback of the NH cryosphere is likely in the range 0.3 to 1.1 W m^{-2} K^{-1} (Flanner et al., 2011). Brown et al. (2010) used satellite, reanalyses and *in situ* observations to document variability and trend in Arctic spring (May–June) SCE over the 1967–2008 period. In June, with Arctic albedo feedback at a maximum, SCE decreased 46% (as of 2012, now 53%) and air temperature explains 56% of the variability.

For the SH, as noted above (see Section 4.5.1), there are no correspondingly long visible-wavelength satellite records, but microwave data date from 1979. Foster et al. (2009) presented the first satellite study of variability and trends in any measure of snow for South America, in this case SWE from microwave data. They focused on the May-September period and noted large year-to-year variability and some lower frequency variability—the July with most extensive snow cover had almost six times as much as the July with the least extensive snow cover—but identified no trends.

4.5.3 Trends from *In Situ* Measurements

AR4 stimulated a review paper (Brown and Mote, 2009) that synthesized modelling results as well as observations from many countries. They showed that decreases in various metrics of snow are most likely to be observed in spring and at locations where air temperatures are close to the freezing point, because changes in air temperature there are most effective at reducing snow accumulation, increasing snowmelt, or both. However, unravelling the competing effects of rising temperatures and changing precipitation remains an important challenge in understanding and interpreting observed changes. Figure 4.21 shows a compilation of many published trends observed at individual locations; data were obtained either from tables in the published papers, or (when the numerical results in the figures were not tabulated) directly from the author, in some cases including updates to the published data sets. The figure shows that in most studies, a majority of sites experienced declines during the varying periods of record, and where data on site mean temperature or elevation were available, warmer/lower sites (red circles) were more likely to experience declines.

Some *in situ* studies in addition to those in Figure 4.21 deserve discussion. Ma and Qin (2012) described trends by season at 754 stations aggregated by region in China over 1951–2009; they found statistically significant trends: positive in winter SD in northwest China, and negative in SD and SWE in spring for China as a whole and spring SWE for the Qinghai-Xizang (Tibet) Plateau. Marty and Meister (2012) noted changes at six high-elevation (>2200 m) sites in the European Alps of Switzerland, Austria, and Germany, consistent with Figure 4.21: no change in SD in midwinter, shortening of SCD in spring and reduction in spring SWE and SD coincident with warming. For the Pyrenees, Lopez-Moreno and Vicente-Serrano (2007) derived proxy SD for 106 sites since 1950 from actual SD measurements since 1985 and weather measurements; they noted declines in spring SD that were related to changes in atmospheric circulation. In the SH, of seven records in the Andes, none have significant trends in maximum SWE (Masiokas et al., 2010) over their periods of record. Of four records in Australia discussed in AR4, all show decreases in spring SWE over their respective periods of record (Nicholls, 2005), and the only one that has been updated since the Nicholls (2005) paper shows a statistically significant decrease of 37% (Sanchez-Bayo and Green, 2013).

4.5.4 Changes in Snow Albedo

In addition to reductions in snow cover extent, which will reduce the mean reflectivity of particular regions, the reflectivity (albedo) of the snow itself may also be changing in response to human activities. Unfortunately, there are extremely limited data on the changes of albedo over time, and we must rely instead on analyses from ice cores, direct recent observations, and modelling. Flanner et al. (2007), using a detailed snow radiative model coupled to a global climate model and estimates of biomass burning, estimated that the human-induced radiative forcing by deposition of black carbon on snow cover is +0.054 (0.007–0.13) W m^{-2} globally, of which 80% is from fossil fuels. However, spatially comprehensive surveys of impurities in Arctic snow in the late 2000s and mid-1980s suggested that impurities decreased between those two periods (Doherty et al., 2010) and hence albedo changes have probably not made a significant contribution to recent reductions in Arctic ice and snow.

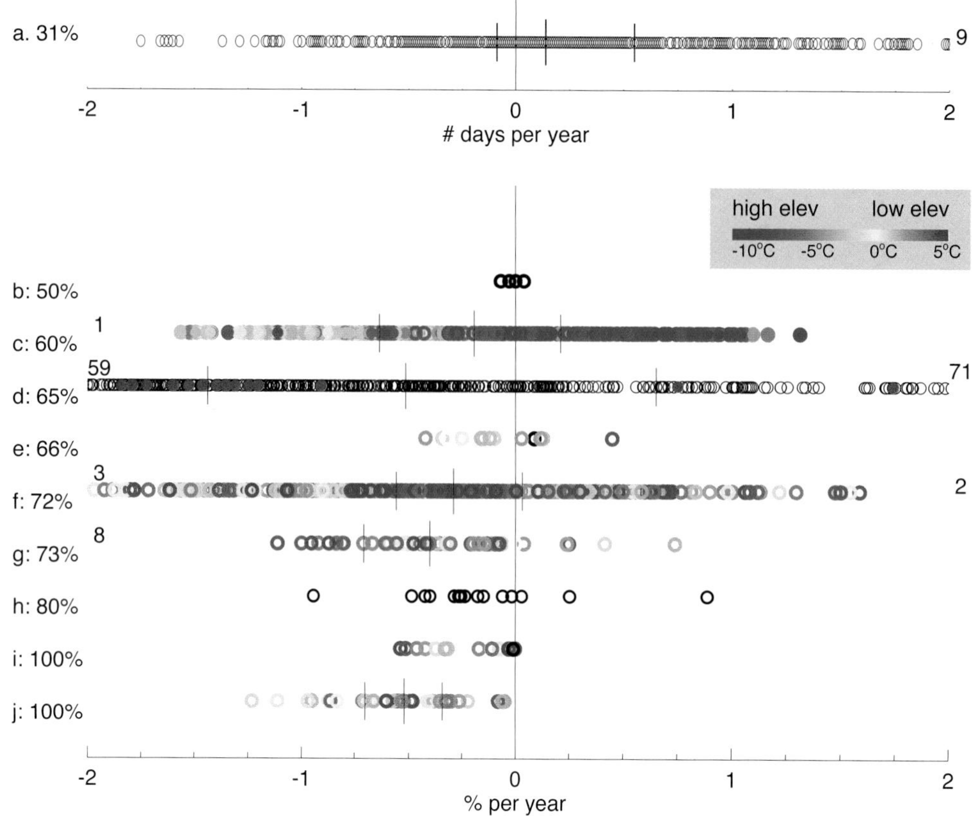

Figure 4.21 | Compilation of studies (rows) showing trends at individual stations (symbols in each row, with percentage of trends that are negative) showing that most sites studied show decreases in snow, especially at lower and/or warmer locations. For each study, if more than one quantity was presented, only the one representing spring conditions is shown. (a) Number of days per year with SD >20 cm at 675 sites in northern Eurasia, 1966–2010 (Bulygina et al., 2011). (b) March–April–May snowfall for 500 stations in California, aggregated into four regions (Christy, 2012). (c) maximum SWE at 393 sites in Norway, 1961–2009 (Skaugen et al., 2012); statistically significant trends are denoted by solid circles. (d) SD at 560 sites in China, 1957–2012 (Ma and Qin, 2012); statistically significant trends are denoted by solid red circles. (e) Snow cover duration at 15 sites in the Romanian Carpathians, 1961–2003 (Micu, 2009). (f) 1 April SWE at 799 sites, 1950–2000, in western North America (Mote, 2006). (g) Difference between 1990s and 1960s March SD at 89 sites in Japan (Ishizaka, 2004). (h) SCD at 15 sites for starting years near 1931, ending 2000 (Petkova et al., 2004). (i) SCD at 18 sites in Italy, 1950–2009 (Valt and Cianfarra, 2010). (j) SCD at 34 sites in Switzerland, 1948–2007, from Marty (2008). See text for definitions of abbreviations. For (b) through (j), the quantity plotted is the percentage change of a linear fit divided by the number of years of the fit. For studies with more than 50 sites, the median, upper and lower quartiles are shown with vertical lines. In a few cases, some trends lie beyond the edges of the graph; these are indicated by a numeral at the corresponding edge of the graph, for example, two sites >2% yr–1 in row (f). Colours indicate temperature or, for studies e) and i), elevation using the lowest and highest site in the respective data set to set the colour scale. Note the prevalence of negative trends at lower/warmer sites.

Box 4.1 | Interactions of Snow within the Cryosphere

Snow is just one component of the cryosphere, but snow also sustains ice sheets and glaciers, and has strong interactions with all the other cryospheric components, except sub-sea permafrost. For example, snow can affect the rate of sea-ice production, and can alter frozen ground through its insulating effect. Snowfall and the persistence of snow cover are strongly dependent on atmospheric temperature and precipitation, and are thus likely to change in complex ways in a changing climate (e.g, Brown and Mote, 2009).

For the Earth's climate in general, and more specifically, the cryospheric components on which snow falls, the two most important physical properties of snow are its high albedo (reflectivity of solar radiation) and its low thermal conductivity, which results because its high air content makes it an excellent thermal insulator. Both factors substantially alter the flux of energy between the atmosphere and the material beneath the snow cover. Snow also has a major impact on the total energy balance of the Earth's surface because large regions in the NH are seasonally covered by snow (e.g., Barry and Gran, 2011). When seasonal snow melts it is also an important fresh water resource.

The high albedo of snow has a strong impact on the radiative energy balance of all surfaces on which it lies, most of which (including glaciers and sea ice) are much less reflective. For example, the albedo of bare glacier or sea ice is typically only 20 to 30%, and hence

(continued on next page)

Box 4.1 (continued)

70 to 80% of solar radiation is absorbed at the surface. For ice at the melting point, this energy melts the ice. With a fresh snow cover over ice, the albedo changes to 80% or even higher and melting is greatly reduced (e.g., Oerlemans, 2001). The effect is similar for other land surfaces—bare soil, frozen ground, low-lying vegetation—but here the thermal properties of the snow cover also play an important role by insulating the ground from changes in ambient air temperature.

While an insulating snow cover can reduce the growth of sea ice, a heavy snow load, particularly in the Antarctic, often depresses the sea ice surface below sea level and this leads to faster transformation of snow to ice (see FAQ 4.1). Even without flooding, the basal snow layer on Antarctic sea ice tends to be moist and saline because brine is wicked up through the snow cover. In regions of heavily ridged and deformed sea ice, snow redistributed by wind smoothes the ice surface, reducing the drag of the air on the ice and thus slowing ice drift and reducing heat exchange (Massom et al., 2001) .

For frozen ground, the insulation characteristics of snow cover are particularly important. If the air above is colder than the material on which it lies, the presence of snow will reduce heat transfer upwards, especially for fresh snow with a low density. This could, for example, reduce the seasonal freezing of soil, slow down the freezing of the active layer (seasonally thawed layer) or protect permafrost from cooling. Alternatively, if the air is warmer than the material beneath the snow, heat transfer downwards from the air is reduced and the presence of snow cover can increase the thickness of seasonal soil freeze and protect permafrost from warming. Which process applies depends on the timing of the snowfall, its thickness, and its duration (e.g., Zhang, 2005; Smith et al., 2012).

For the preceding reasons, the timing of snowfall and the persistence of snow cover are of major importance. Whereas snow falling on glaciers and ice sheets in summer has a strongly positive (sustaining) effect on the mass budget, early snow cover can reduce radiative and conductive cooling and freezing of the active layer. During winter, snowfall is the most important source of nourishment for most glaciers, but radiative cooling of frozen ground is strongly reduced by thick snow cover (Zhang, 2005).

4.6 Lake and River Ice

The assessment of changes in lake and river ice is made more difficult by several factors. Until the satellite era, some nations collected data from numerous lakes and rivers and others none; many published studies focus on a single lake or river. Many records have been discontinued (Prowse et al., 2011), and consistency of observational methods is a challenge, especially for date of ice break-up of ice on rivers when the process of break-up can take as long as 3 months (Beltaos and Prowse, 2009).

The most comprehensive description is the analysis of 75 lakes, mostly in Scandinavia and the northern USA, but with one each in Switzerland and Russia (Benson et al., 2012). Examining 150-, 100-, and 30-year periods ending in spring 2005, they found the most rapid changes in the most recent 30-year period (*medium confidence*) with trends in freeze-up 1.6 days per decade later and breakup 1.9 days per decade earlier. Wang et al. (2012) found a total ice cover reduction on the north American Great Lakes of 71% over the 1973–2010 period of record, using weekly ice charts derived from satellite observations (*medium confidence*). Jensen et al. (2007) examined data from 65 water bodies in the Great Lakes region between Minnesota and New York (not including the Great Lakes themselves) and found trends in freeze-up 3.3 days per decade later, trends in breakup 2.1 days per decade earlier, and rates of change over 1975–2004 that were bigger than those over 1846–1995. Spatial patterns in trends are ambiguous: Latifovic and Pouliot (2007) found larger trends in higher latitudes over Canada, but Hodgkins et al. (2002) found larger trends in lower latitudes in the northeastern USA.

In the only reported study since the 1990s of ice on SH lakes, Green (2011) suggested on the basis of available evidence that break-up of ice cover on Blue Lake in the Snowy Mountains of Australia had shifted from November to October between observation periods 1970–1972 and 1998–2010.

Several studies made quantitative connections between ice cover and temperature. For instance, Benson et al. (2012) found significant correlations between mean ice duration and mean NH land air temperature in fall-winter-spring ($r^2 = 0.48$) and between spring air temperature and breakup ($r^2 = 0.36$); see also the review by Prowse et al. (2011).

Studies of changes in river ice have used both disparate data and time intervals, ranging in duration from multi-decade to more than two centuries, and most focus on a single river. Beltaos and Prowse (2009), summarizing most available information for northern rivers, noted an almost universal trend towards earlier break-up dates but considerable spatial variability in those for freeze-up, and noted too that changes were often more pronounced during the last few decades of the 20th century. They noted that the 20th century increase in mean air temperature in spring and autumn has produced in many areas a change of about 10 to 15 days toward earlier break-up and later freeze-up, although the relationship with air temperatures is complicated by the roles of snow accumulation and spring runoff.

In summary, the limited evidence available for freshwater (lake and river) ice indicates that ice duration is decreasing and average seasonal ice cover shrinking (*low confidence*), and the following general patterns (each of which has exceptions): rates of change in timing are generally, but not universally, (1) higher for spring breakup than fall freeze-up; (2) higher for more recent periods; (3) higher at higher elevations (Jensen et al., 2007) and (4) quantitatively related to temperature changes.

4.7 Frozen Ground

4.7.1 Background

Frozen ground occurs across the world at high latitudes, in mountain regions, beneath glacial ice and beneath lakes and seas. It is a product of cold weather and climate, and can be diurnal, seasonal or perennial. Wherever the ground remains at or below 0°C for at least two consecutive years, it is called permafrost (Van Everdingen, 1998), and this too can occur beneath the land surface (terrestrial permafrost) and beneath the seafloor (subsea permafrost). In this chapter, the term permafrost refers to terrestrial permafrost unless specified.

Both the temperature and extent of permafrost are highly sensitive to climate change, but the responses may be complex and highly heterogeneous (e.g., Osterkamp, 2007). Similarly, the annual freezing and thawing of seasonally frozen ground is coupled to the land surface energy and moisture fluxes, and thus to climate. Since, permafrost and seasonally frozen ground, can contain significant fractions of ice, changes in landscapes, ecosystems and hydrological processes can occur when it forms or degrades (Jorgenson et al., 2006; Gruber and Haeberli, 2007; White et al., 2007). Furthermore, frozen organic soils contain considerable quantities of carbon, more than twice the amount currently in the atmosphere (Tarnocai et al., 2009), and permafrost thawing exposes previously frozen carbon to microbial degradation and releases radiatively active gases, such as carbon dioxide (CO_2) and methane (CH_4), into the atmosphere (Zimov et al., 2006; Schuur et al., 2009; Schaefer et al., 2011) (for a detailed assessment of this issue, see Chapter 6). Similarly, recent evidence suggests that degradation of permafrost may also permit the release of nitrous oxide (N_2O), which is also radiatively active (Repo et al., 2009; Marushchak et al., 2011). Finally, permafrost degradation may directly affect the lives of people, both in northern and high-mountain areas, through impacts on the landscape, vegetation and infrastructure (WGII, Chapter 28).

4.7.2 Changes in Permafrost

4.7.2.1 Permafrost Temperature

The ice content and temperature of permafrost are the key parameters that determine its physical state. Permafrost temperature is a key parameter used to document changes to permafrost. Permafrost temperature measured at a depth where seasonal variations cease to occur is generally used as an indicator of long-term change and to represent the mean annual ground temperature (Romanovsky et al., 2010a). For most sites this depth occurs in the upper 20 m.

In the SH, permafrost temperatures as low as –23.6°C have been observed in the Antarctic (Vieira et al., 2010), but in the NH, permafrost temperatures generally range from –15°C to close to the freezing point (Figure 4.22) (Romanovsky et al., 2010a). They are usually coldest in high Arctic regions and gradually increase southwards, but substantial differences do occur at the same latitude. For example, as a result of the proximity to warm ocean currents, the southern limit of permafrost is farther north, and permafrost temperature is higher in Scandinavia and north-western Russia than it is in Arctic regions of Siberia and North America (Romanovsky et al., 2010a).

In Russia, permafrost temperature measurements reach back to the early 1930s (Romanovsky et al., 2010b), in North America to the late 1940s (Brewer, 1958) and in China to the early 1960s (Zhou et al., 2000; Zhao et al., 2010). Systematic measurements, however, began mostly in the late 1970s and early 1980s (Zhou et al., 2000; Osterkamp, 2007; Smith et al., 2010). In addition, since the AR4, considerable effort (especially during the International Polar Year) has gone into enhancing the observation network and establishing a baseline against which future changes in permafrost can be measured (Romanovsky et al., 2010a). However, it should be noted that there still exist comparatively few measurements of permafrost temperature in the SH (Vieira et al., 2010).

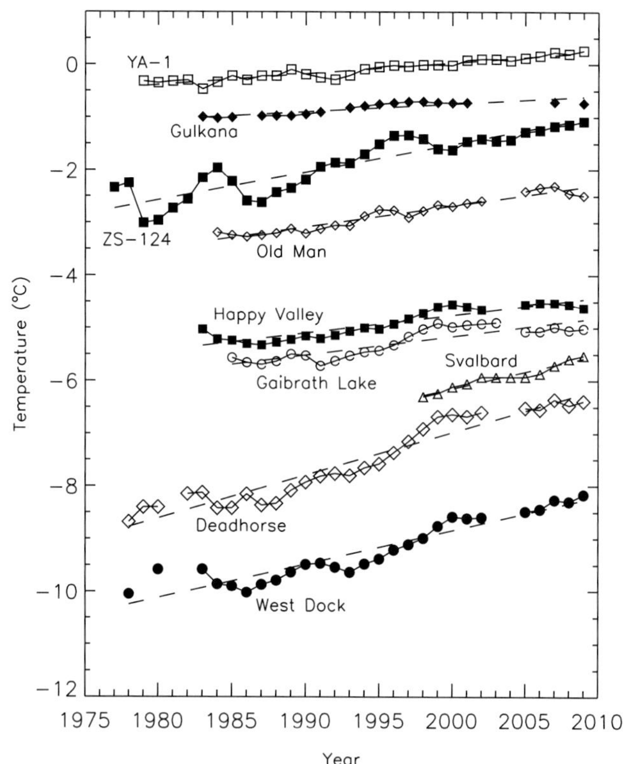

Figure 4.22 | Time series of mean annual ground temperatures at depths between 10 and 20 m for boreholes throughout the circumpolar northern permafrost regions (Romanovsky et al., 2010a). Data sources are from Romanovsky et al. (2010b) and Christiansen et al. (2010). Measurement depth is 10 m for Russian boreholes, 15 m for Gulkana and Oldman, and 20 m for all other boreholes. Borehole locations are: ZS-124, 67.48°N 063.48°E; 85-8A, 61.68°N 121.18°W; Gulkana, 62.28°N 145.58°W; YA-1, 67.58°N 648°E; Oldman, 66.48°N 150.68°W; Happy Valley, 69.18°N 148.88°W; Svalbard, 78.28°N 016.58°E; Deadhorse, 70.28°N 148.58°W and West Dock, 70.48°N 148.58°W. The rate of change (degrees Celsius per decade) in permafrost temperature over the period of each site record is: ZS-124: 0.53 ± 0.07; YA-1: 0.21 ± 0.02; West Dock: 0.64 ± 0.08; Deadhorse: 0.82 ± 0.07; Happy Valley: 0.34 ± 0.05; Gaibrath Lake: 0.35 ± 0.07; Gulkana: 0.15 ± 0.03; Old Man: 0.40 ± 0.04 and Svalvard: 0.63 ± 0.09. (The trends are *very likely* range, 90%.)

In most regions, and at most sites, permafrost temperatures have increased during the past three decades (*high confidence*): at rather fewer sites, permafrost temperatures show little change, or a slight decrease (Figure 4.22; Table 4.8). However, it is important to discriminate between *cold permafrost*, with mean annual ground temperatures below –2°C, and *warm permafrost at temperatures above* –2°C (Cheng and Wu, 2007; Smith et al., 2010; Wu and Zhang, 2010). Warm permafrost is found mostly in the discontinuous permafrost zone, while cold permafrost exists in the continuous permafrost zone and only occasionally in the discontinuous permafrost zone (Romanovsky et al., 2010a).

Overall, permafrost temperature increases are greater in cold permafrost than they are in warm permafrost (*high confidence*). This is especially true for warm ice-rich permafrost, due to heat absorbed by partial melting of interstitial ice, slowing and attenuating temperature change (Romanovsky et al., 2010a). The temperatures of cold permafrost across a range of regions have increased by up to 2°C since the 1970s (Table 4.8 and Callaghan et al., 2011); however, the timing of warming events has shown considerable spatial variability (Romanovsky et al., 2010a).

Temperatures of warm permafrost have also increased over the last three decades, but generally by less than 1°C. Warm permafrost is sometimes nearly isothermal with depth; as is observed in mountain regions such as the European Alps (Noetzli and Vonder Muehll, 2010), Scandinavia (Christiansen et al., 2010), the Western Cordillera of North America (Smith et al., 2010; Lewkowicz et al., 2011), the Qinghai-Xizang (Tibet) Plateau (Zhao et al., 2010; Wu et al., 2012) and in the northern high latitudes in the southern margins of discontinuous permafrost regions (Romanovsky et al., 2010b; Smith et al., 2010). In such areas, permafrost temperatures have shown little or no change, indicating that permafrost is thawing internally but remaining very close to the melting point (Smith et al., 2010). Cooling of permafrost due to atmospheric temperature fluctuations has been observed; for example, in the eastern Canadian Arctic until the mid-1990s (Smith et al., 2010); but some examples have been short-lived and others controlled by site-specific conditions (Marchenko et al., 2007; Wu and Zhang, 2008; Noetzli and Vonder Muehll, 2010; Zhao et al., 2010). In at least one case in the Antarctic, permafrost warming has been observed in a region with almost stable air temperatures (Guglielmin and Cannone, 2012).

Permafrost warming is mainly in response to increased air temperature and changing snow cover (see Box 4.1). In cold permafrost regions, especially in tundra regions with low ice content (such as bedrock) where permafrost warming rates have been greatest, changes in snow cover may play an important role (Zhang, 2005; Smith et al., 2010).

Table 4.8 | Permafrost temperatures during the International Polar Year (2007–2009) and their recent changes. Each line may refer to one or more measurements sites.

Region	Permafrost Temperature During IPY (°C)	Permafrost Temperature Change (°C)	Depth (m)	Period of Record	Source
North America					
Northern Alaska	–5.0 to –10.0	0.6–3	10–20	Early 1980s–2009	Osterkamp (2005, 2007); Smith et al. (2010); Romanovsky et al. (2010a)
Mackenzie Delta and Beaufort coastal region	–0.5 to –8.0	1.0–2.0	12–20	Late 1960s–2009	Burn and Kokelj (2009); Burn and Zhang (2009); Smith et al. (2010)
Canadian High Arctic	–11.8 to –14.3	1.2–1.7	12–15	1978–2008	Smith et al. (2010, 2012)
Interior of Alaska,	0.0 to –5.0	0.0–0.8	15–20	1985–2009	Osterkamp (2008); Smith et al. (2010); Romanovsky et al. (2010a)
Central and Southern Mackenzie Valley	>–2.2	0.0–0.5	10–12	1984–2008	Smith et al. (2010)
Northern Quebec	>–5.6	0.0–1.8	12–20	1993–2008	Allard et al. (1995); Smith et al. (2010)
Europe					
European Alps	>–3	0.0–0.4	15–20	1990s–2010	Haeberli et al. (2010); Noetzli and Vonder Muehll (2010); Christiansen et al. (2012)
Russian European North	–0.1 to –4.1	0.3–2.0	8–22	1971–2010	Malkova (2008); Oberman (2008); Romanovsky et al. (2010b); Oberman (2012)
Nordic Countries	–0.1 to –5.6	0.0–1.0	2–15	1999–2009	Christiansen et al. (2010); Isaksen et al. (2011)
Northern and Central Asia					
Northern Yakutia	–4.3 to –10.8	0.5–1.5	14–25	early 1950s–2009	Romanovsky et al. (2010b)
Trans-Baykal region	–4.7 to –5.1	0.5–0.8	19–20	late 1980s–2009	Romanovsky et al. (2010b)
Qinghai-Xizang Plateau	–0.2 to –3.4	0.2–0.7	6	1996–2010	Cheng and Wu (2007); Li et al. (2008); Wu and Zhang (2008); Zhao et al. (2010)
Tian Shan	–0.4 to –1.1	0.3–0.9	10–25	1974–2009	(Marchenko et al. (2007); Zhao et al. (2010)
Mongolia	0.0 to <–2.0	0.2–0.6	10–15	1970–2009	Sharkhuu et al. (2007); Zhao et al. (2010); Ishikawa et al. (2012)
Others					
Maritime Antarctica	–0.5 to –3.1	NA	20–25	2007–2009	Vieira et al. (2010)
Continental Antarctica	–13.9 to –19.1	NA	20–30	2005–2008	Vieira et al. (2010); Guglielmin et al. (2011)
East Greenland	–8.1	NA	3.25	2008–2009	Christiansen et al. (2010)

4

In forested areas, especially in warm ice-rich permafrost, changes in permafrost temperature are reduced by the effects of the surface insulation (Smith et al., 2012; Throop et al., 2012) and latent heat (Romanovsky et al., 2010a).

4.7.2.2 Permafrost Degradation

Permafrost degradation refers to a decrease in thickness and/or areal extent. In particular, the degradation can be manifested by a deepening of summer thaw, or top-down or bottom-up permafrost thawing, and a development of taliks (see Glossary). Other manifestations of degradation include geomorphologic changes such as the formation of thermokarst terrain (see Glossary and Jorgenson et al., 2006), expansion of thaw lakes (Sannel and Kuhry, 2011) active-layer detachment slides along slopes, rock falls (Ravanel et al., 2010), and destabilized rock glaciers (Kääb et al., 1997; Haeberli et al., 2006; Haeberli et al., 2010). Although most permafrost has been degrading since the Little Ice Age (Halsey et al., 1995), the trend was relatively modest until the past two decades, during which the rate of degradation has increased in some regions (Romanovsky et al., 2010b).

Significant permafrost degradation has been reported in the Russian European North (*medium confidence*). Warm permafrost with a thickness of 10 to 15 m thawed completely in the period 1975–2005 in the Vorkuta area (Oberman, 2008). And although boundaries between permafrost types are not easy to map, the southern permafrost boundary in this region is reported to have moved north by about 80 km and the boundary of continuous permafrost has moved north by 15 to 50 km (Oberman, 2008) (*medium confidence*). Taliks have also developed in relatively thick permafrost during the past several decades. In the Vorkuta region, the thickness of existing closed taliks increased by 0.6 to 6.7 m over the past 30 years (Romanovsky et al., 2010b). Permafrost thawing and talik formation has occurred in the Nadym and Urengoy regions in north- western Russian (Drozdov et al., 2010). Long-term permafrost thawing has been reported around the city of Yakutsk, but this in this case, the thawing may have been caused mainly by forest fires or human disturbance (Fedorov and Konstantinov, 2008). Permafrost degradation has also been reported on the Qinghai-Xizang (Tibet) Plateau (Cheng and Wu, 2007; Li et al., 2008).

Coastal erosion and permafrost degradation appear to be evident along many Arctic coasts in recent years, with complex interactions between them (Jones et al., 2009). In part, these interactions arise from the thermal and chemical impact of sea water on cold terrestrial permafrost (Rachold et al., 2007). Similar impacts arise for permafrost beneath new thaw lakes, which have been formed in recent years (e.g., Sannel and Kuhry, 2011). In northern Alaska, estimates of permafrost thawing under thaw lakes are in the range 0.9 to 1.7 cm a^{-1} (Ling and Zhang, 2003).

Since AR4, destabilized rock glaciers have received increased attention from researchers. A rock glacier is a mass of perennially frozen rock fragments on a slope, that contains ice in one or more forms and shows evidence of past or present movement (Van Everdingen, 1998; Haeberli et al., 2006). Time series acquired over recent decades by terrestrial surveys indicate acceleration of some rock glaciers as well as seasonal velocity changes related to ground temperatures (Bodin et al., 2009; Noetzli and Vonder Muehll, 2010; Schoeneich et al., 2010; Delaloye et al., 2011). Similarly, photo-comparison and photogrammetry have indicated collapse-like features on some rock glaciers (Roer et al., 2008). The clear relationship between mean annual air temperature at the rock glacier front and rock glacier velocity points to a likely temperature influence and a plausible causal connection to climate (Kaab et al., 2007). Strong surface lowering of rock glaciers has been reported in the Andes (Bodin et al., 2010), indicating melting of ground ice in rock glaciers and permafrost degradation.

4.7.3 Subsea Permafrost

Subsea permafrost is similar to its terrestrial counterpart, but lies beneath the coastal seas. And as with terrestrial permafrost, subsea permafrost is a substantial reservoir and/or a confining layer for gas hydrates (Koch et al., 2009). It is roughly estimated that subsea permafrost contains 2 to 65 Pg of CH_4 hydrate (McGuire et al., 2009). Observations of gas release on the East Siberian Shelf and high methane concentrations in water-column and air above (Shakhova et al., 2010a, 2010b) have led to the suggestion that permafrost thawing creates pathways for gas release.

Subsea permafrost in the Arctic is generally relict terrestrial permafrost (Vigdorchik, 1980), inundated after the last glaciation and now degrading under the overlying shelf sea. Permafrost may, however, also form when the sea is shallow, permitting sediment freezing through bottom-fast winter sea ice (Solomon et al., 2008; Stevens et al., 2010). A 76-year record of bottom water temperature in the Laptev Sea (Dmitrenko et al., 2011) showed warming of 2.1°C since 1985 in the near-shore zone (<10 m water depth), as lengthening summers reduced sea ice extent and increased solar heating. Degradation rates of the ice-bearing permafrost following inundation have been estimated to be 1 to 20 cm a^{-1} on the East Siberian Shelf (Overduin et al., 2007) and 1 to 4 cm a^{-1} in the Alaskan Beaufort Sea (Overduin et al., 2012).

4.7.4 Changes in Seasonally Frozen Ground

Seasonally frozen ground is a soil layer that freezes and thaws annually, which may or may not overlie terrestrial permafrost, and also includes some portions of the Arctic seabed that freeze in winter. A key parameter regarding seasonally frozen ground overlying permafrost is the active-layer thickness (ALT; see Glossary), which indicates the depth of the seasonal freeze–thaw cycle, and which is dependent on climate and other factors; for example, vegetation cover (Smith et al., 2009). Many observations across many regions have revealed trends in the thickness of the active laver (*high confidence*).

4.7.4.1 Changes in Active-Layer Thickness

Many observations have revealed a general positive trend in the thickness of the active layer (see Glossary) for discontinuous permafrost regions at high latitudes (*medium confidence*). Based on measurements from the International Permafrost Association (IPA) Circumpolar Active Layer Monitoring (CALM) programme, active-layer thickening has been observed since the 1970s and has accelerated since 1995 in northern Europe (Akerman and Johansson, 2008; Callaghan et al., 2010), and on Svalbard and Greenland since the late-1990s (Christiansen et al., 2010). The ALT has increased significantly in the Russian European North (Mazhitova, 2008), East Siberia (Fyodorov-Davydov et al., 2008), and Chukotka (Zamolodchikov, 2008) since the mid-1990s. Burn and Kokelj (2009) found, for a site in the Mackenzie Delta area, that ALT increased by 8 cm between 1983 and 2008, although the record does exhibit high interannual variability as has been observed at other sites in the region (Smith et al., 2009). ALT has increased since the mid-1990s in the eastern portion of the Canadian Arctic, with the largest increase occurring at bedrock sites in the discontinuous permafrost zone (Smith et al., 2010).

The interannual variations and trends of the active-layer thickness in Northern America, Northern Europe and Northern Asia from 1990 to 2012 are presented in Figure 4.23. Large regional variations in the yearly variability patterns and trends are apparent. While increases in ALT are occurring in the Eastern Canadian Region (Smith et al., 2009), a slightly declining trend is observed in the Western Canadian Region (Figure 4.23a). In Northern Europe, the trends in the study areas are similar and consistently positive (Figure 4.23b). On the other hand, in Northern Asia, trends are generally strongly positive with the exception of West Siberia, where the trend is slightly negative (Figure 4.23c). On the interior of Alaska, slightly increasing ALT from 1990 to 2010 was followed by anomalous increases in 2011 and in 2012. Overall, a general increase in ALT since the 1990s has been observed at many stations in many regions (*medium confidence*). The general increase is shown in Figure 4.23d, which shows the results of analysis of data from about 44 stations in Russia indicating a change of almost 0.2 m from 1950 to 2008.

At some measurement sites on the Qinghai-Xizang (Tibet) Plateau, ALT was reported to be increasing at 7.8 cm yr^{-1} over a period from 1995 through 2010 (Wu and Zhang, 2010). The high rates may have been the result of local disturbances since more recent studies indicate rates of 1.33 cm yr^{-1} for the period 1981–2010 and 3.6 cm yr^{-1} for the period 1998–2010 (e.g., Zhao et al., 2010; Li et al., 2012a).

During the past decade, increases in ALT up to 4.0 cm yr^{-1} were observed in Mongolian sites characterized as a warm permafrost region (Sharkhuu et al., 2007). Changes in ALT were also detected in Tian Shan (Marchenko et al., 2007; Zhao et al., 2010), and in the European Alps, where increases in ALT were largest during years of hot summers but a strong dependence on surface and subsurface characteristics was noted (Noetzli and Vonder Muehll, 2010).

In several areas, across North America and in West Siberia, large-inter annual variations obscure any trends in ALT (*high confidence*, Figure 4.23). No trend in ALT was observed on the Alaskan North Slope from 1993 to 2010 (Streletskiy et al., 2008; Shiklomanov et al., 2010) and also in the Mackenzie Valley (Smith et al., 2009) and in West Siberia (Vasiliev et al., 2008) since the mid-1990s (Figure 4.23). At some sites, such as at Western Canada (C5) and Western Siberia (R1) (Figure 4.23), the active layer thickness was actually decreasing.

The penetration of thaw into ice-rich permafrost at the base of the active layer is often accompanied by loss of volume due to consolidation. At several sites, this has been shown to cause surface subsidence (*medium confidence*). Results from ground-based measurements at

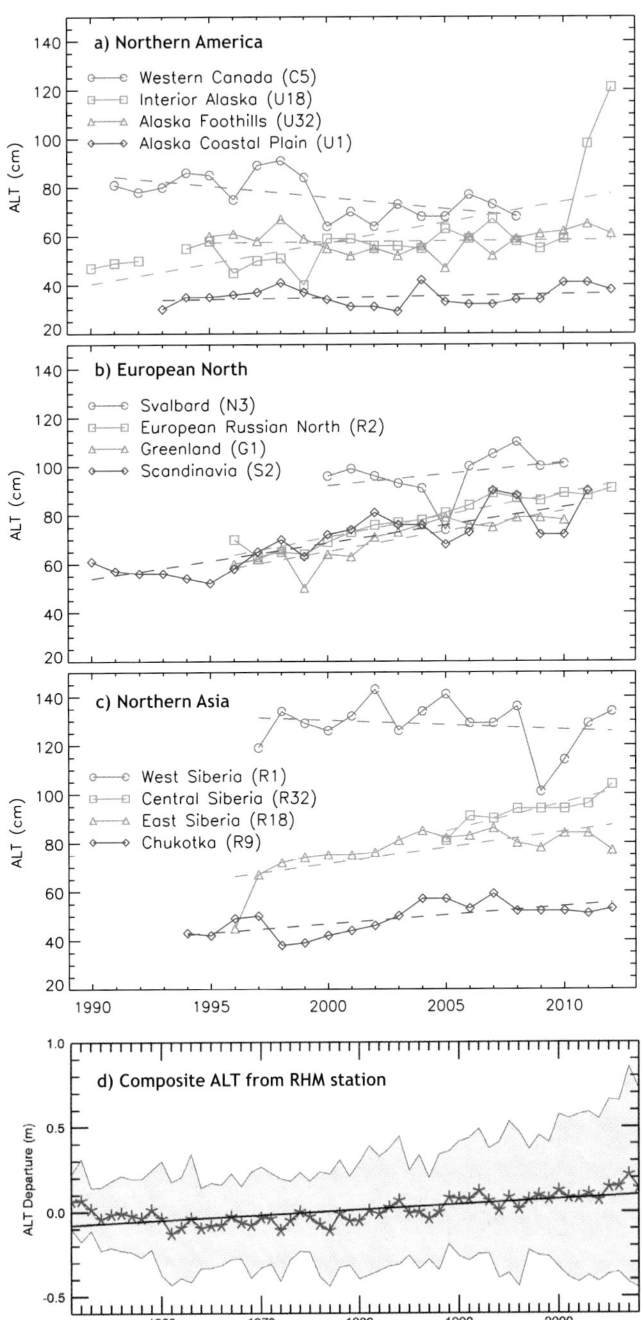

Figure 4.23 | Active layer thickness from different locations for slightly different periods between 1990 and 2012 in (a) Northern America, (b) Northern Europe, and (c) Northern Asia. The dashed lines represents linear fit to each set of data. ALT data for Northern America, Northern Asia and Northern Europe were obtained from the International Permafrost Association (IPA) CALM website (http://www.udel.edu/Geography/calm/about/permafrost.html). The number of Russian Hydrometeorological Stations (RHM) stations has expanded from 31 stations as reported from Frauenfeld et al. (2004) and Zhang et al. (2005) to 44 stations and the time series has extended from 1990 to 2008. (d) Departures from the mean of active layer thickness in Siberia from 1950 to 2008. The red asterisk represents the mean composite value, the shaded area indicates the standard deviation and the black line is the trend. Data for Siberia stations were obtained from the Russian Hydrometeorological Stations (RHM).

selected sites on the North Slope of Alaska indicate 11 to 13 cm in surface subsidence over the period 2001–2006 (Streletskiy et al., 2008), 4 to 10 cm from 2003 to 2005 in the Brooks Range (Overduin and Kane, 2006) and up to 20 cm in the Russian European North (Mazhitova and Kaverin, 2007). Subsidence has also been identified using space-borne interferometric synthetic aperture radar (InSAR) data. Surface deformation was detected using InSAR over permafrost on the North Slope of Alaska during the 1992–2000 thaw seasons and a long-term surface subsidence of 1 to 4 cm per decade (Liu et al., 2010). Such subsidence could explain why *in situ* measurements at some locations reveal negligible trends in ALT changes during the past two decades, despite the fact that atmospheric and permafrost temperatures increased during that time.

4.7.4.2 Changes in Seasonally Frozen Ground in Areas Not Underlain by Permafrost

An estimate based on monthly mean soil temperatures from 387 stations across part of the Eurasian continent suggested that the thickness of seasonally frozen ground decreased by about 0.32 m during the period 1930–2000 (*high confidence*, Figure 4.24) (Frauenfeld and Zhang, 2011). Inter-decadal variability was such that no trend could be identified until the late 1960s, after which seasonal freeze depths decreased significantly until the early 1990s. From then, until about 2008, no further change was evident. Such changes are closely linked with the freezing index, but also with mean annual air temperatures and snow depth (Frauenfeld and Zhang, 2011).

Thickness of seasonally frozen ground in western China decreased by 20 to 40 cm since the early 1960s (Li et al., 2008), whereas on the Qinghai-Xizang (Tibet) Plateau, the seasonally frozen depth decreased by up to 33 cm since the middle of 1980s (Li et al., 2009). Evidence from the satellite record indicates that the onset dates of spring thaw advanced by 14 days, whereas the autumn freeze date was delayed by 10 days on the Qinghai-Xizang (Tibet) Plateau from 1988 through 2007 (Li et al., 2012b)

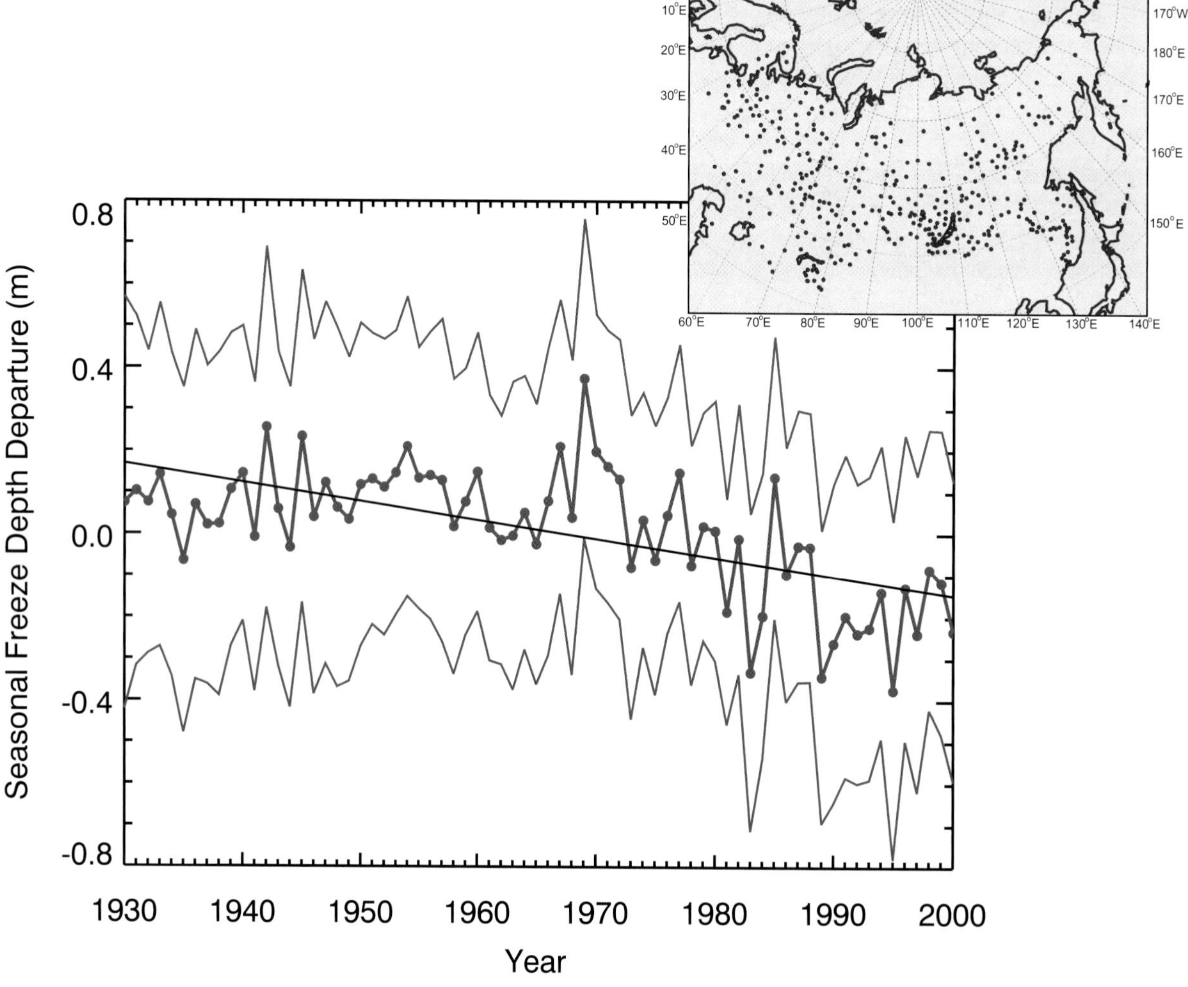

Figure 4.24 | Annual anomalies of the average thickness of seasonally frozen depth in Russia from 1930 to 2000. Each data point represents a composite from 320 stations as compiled at the Russian Hydrometeorological Stations (RHM) (upper right inset). The composite was produced by taking the sum of the thickness measurements from each station and dividing the result by the number of stations operating in that year. Although the total number of stations is 320, the number providing data may be different for each year but the minimum was 240. The yearly anomaly was calculated by subtracting the 1971–2000 mean from the composite for each year. The thin lines indicate the 1 standard deviation (1σ) (*likely*) uncertainty range. The line shows a negative trend of –4.5 cm per decade or a total decrease in the thickness of seasonally frozen ground of 31.9 cm from 1930 to 2000 (Frauenfeld and Zhang, 2011).

4.8 Synthesis

Observations show that the cryosphere has been in transition during the last few decades and that the strong and significant changes reported in AR4 have continued, and in many cases accelerated. The number of *in situ* and satellite observations of cryospheric parameters has increased considerably since AR4 and the use of the new data in trend analyses, and also in process studies, has enabled increased confidence in the quantification of most of the changes. A graphical depiction and a text summary of observed changes in the various components of the cryosphere are provided in Figure 4.25. They reveal a general decline in all components of the cryosphere, but the magnitude of the decline varies regionally and there are isolated cases where an increase is observed.

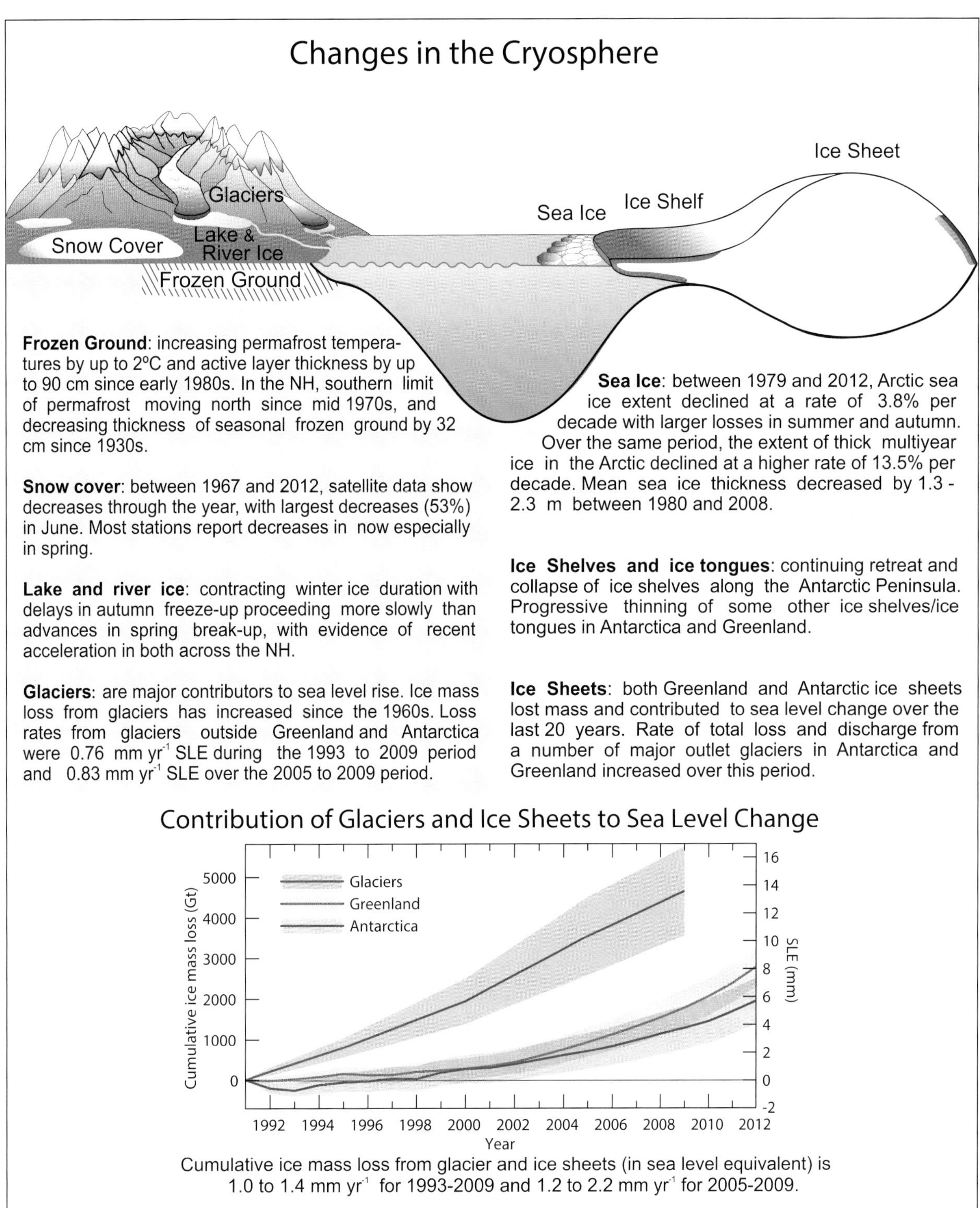

Figure 4.25 | Schematic summary of the dominant observed variations in the cryosphere. The inset figure summarises the assessment of the sea level equivalent of ice loss from the ice sheets of Greenland and Antarctica, together with the contribution from all glaciers except those in the periphery of the ice sheets (Section 4.3.3 and 4.4.2).

Some of the observed changes since AR4 have been considerable and unexpected. One of the most visible was the dramatic decline in the September minimum sea ice cover in the Arctic in 2007, which was followed by a record low value in 2012, supporting observations that the thicker components of the Arctic sea ice cover are decreasing. The trend in extent for Arctic sea ice is $-3.8 \pm 0.3\%$ per decade (*very likely*) while that for multi-year ice is $-13.5 \pm 2.5\%$ per decade (*very likely*). Observations also show marked decreases in Arctic ice thickness and volume. The pattern of melt on the surface of the Greenland ice sheet has also changed radically, with melt occurring in 2012 over almost the entire surface of the ice sheet for the first time during the satellite era. The ice mass loss in Greenland has been observed to have increased from 34 [–6 to 74] Gt yr^{-1} for the period 1992–2001 to 215 [157 to 274] Gt yr^{-1} for the period 2002–2011 while the estimates of mass loss in Antarctica have increased from 30 [–37 to 97] Gt yr^{-1} during the 1992–2001 period to 147 [72 to 221] Gt yr^{-1} during the 2002–2011 period. Observed mass loss from glaciers has also increased, with the global mass loss (excluding the glaciers peripheral to the ice sheets) estimated to be 226 [91 to 361] Gt yr^{-1} during the 1971–2009 period, 275 [140 to 410] Gt yr^{-1} over the 1993–2009 period, and 301 [166 to 436] over the 2005–2009 period. A large majority of observing stations report decreasing trends in snow depth, snow duration, or snow water equivalent, and the largest decreases are typically observed at locations with temperatures close to freezing. Most lakes and rivers with long-term records have exhibited declines in ice duration and average seasonal ice cover. Permafrost has also been degrading and retreating to the north while permafrost temperatures have increased in most regions since the 1980s.

The observed positive trend of sea ice extent in the Antarctic that was regarded as small and insignificant in AR4, has persisted, and increased slightly to about $1.5 \pm 0.2\%$ per decade. The higher-than-average Antarctic sea ice extent in recent years has been mainly due to increases in the Ross Sea region, which more than offset the declines in the Bellingshausen Sea and Amundsen Sea. Ice production in coastal polynyas (regarded as 'sea ice factories') along the Ross Sea ice shelves have been observed to be increasing. Recent work suggests strengthening of the zonal (east-west) winds and accompanying ice drift accounts for some of the increasing sea ice extent.

Satellite data have provided the ability to observe large-scale changes in the cryosphere at relatively good temporal and spatial resolution throughout the globe. Largely because of the availability of high resolution satellite data, the first near-complete global glacial inventory has been generated, leading to a more precise determination of the past, current and future contribution of glaciers to sea level rise. As more data accumulate, and as more capable sensors are launched, the data become more valuable for studies related to change assessment. The advent of new satellites and airborne missions has provided powerful tools that have enabled breakthroughs in the capability to measure some parameters and enhance our ability to interpret results. However, a longer record of measurements of the cryosphere will help increase confidence in the results, reduce uncertainties in the long-term trends, and bring more critical insights into the physical processes controlling the changes. There is thus a need for the continuation of the satellite records, and a requirement for longer and more reliable historical data from *in situ* measurements and proxies.

The sea level equivalent of mass loss from the Greenland and Antarctic ice sheets over the period 1993–2010, has been about 5.9 mm (including 1.7 mm from glaciers around Greenland) and 4.8 mm, respectively. The reliability of observations of ice loss from the ice sheets has been enhanced with the introduction of advanced satellite observation techniques. The ice loss from glaciers between 1993 and 2009 measured in terms of sea level equivalent (excluding those peripheral to the ice sheets) is estimated to be 13 mm. The inset to Figure 4.25 shows the cumulative sea level equivalent from glaciers and the ice sheets in Greenland and Antarctica. These have been contributing dominantly to sea level rise in recent decades. The contribution of the cryosphere to sea level change is discussed more fully in Chapter 13.

The overall consistency in the negative changes observed in the various components of the cryosphere (Figure 4.25), and the acceleration of these changes in recent decades, provides a strong signal of climate change. Regional differences in the magnitude and direction of the signals are apparent, but these are not unexpected considering the large variability and complexity of atmospheric and oceanic circulations. It is *very likely*, however, that the Arctic has changed substantially since 1979.

Acknowledgements

We acknowledge the kind contributions of C. Starr (NASA Visualization Group), U. Blumthaler, S. Galos (University of Innsbruck) and P. Fretwell (British Antarctic Survey), who assisted in drafting figures. M. Mahrer, R. Graber (University of Zurich) and G. Hiess (BAS) undertook valuable literature reviews, and N. E. Barrand (BAS) assisted with collation of references.

References

Abdalati, W., et al., 2004: Elevation changes of ice caps in the Canadian Arctic Archipelago. *J. Geophys. Res. Earth Surf.*, **109**, 11 (F04007).

Abdalati, W., et al., 2010: The ICESat-2 Laser Altimetry Mission. *Proc. IEEE*, **98**, 735–751.

Akerman, H. J., and M. Johansson, 2008: Thawing permafrost and thicker active layers in sub-Arctic Sweden. *Permafr. Process.*, **19**, 279–292.

Allard, M., B. L. Wang, and J. A. Pilon, 1995: Recent cooling along the southern shore of the Hudson Strait, Quebec, Canada, documented from permafrost temperatuure-measurements. *Arct. Alp. Res.*, **27**, 157–166.

Alley, R. B., and S. Anandakrishnan, 1995: Variations in melt-layer frequency in the GISP2 ice core: Implications for Holocene summer temperatures central Greenland. *Ann. Glaciol.*, **21**, 64–70.

Alley, R. B., et al., 2008: A simple law for ice-shelf calving. *Science*, **322**, 1344–1344.

Amundson, J. M., M. Fahnestock, M. Truffer, J. Brown, M. P. Luthi, and R. J. Motyka, 2010: Ice melange dynamics and implications for terminus stability, Jakobshavn Isbrae Greenland. *J. Geophys. Res. Earth Surf.*, **115**, 12

Andresen, C. S., et al., 2012: Rapid response of Helheim Glacier in Greenland to climate variability over the past century. *Nature Geosci.*, **5**, 37–41

Arendt, A. A., K. A. Echelmeyer, W. D. Harrison, C. S. Lingle, and V. B. Valentine, 2002: Rapid wastage of Alaska glaciers and their contribution to rising sea level. *Science*, **297**, 382–386.

Arendt, A., et al., 2012: Randolph Glacier Inventory [v2.0]: A Dataset of Global Glacier Outlines. Global Land Ice Measurements from Space, Boulder Colorado, USA. Digital Media 32 pp. [Available online at: http://www.glims.org/RGI/RGI_Tech_Report_V2.0.pdf]

Arthern, R. J., D. P. Winebrenner, and D. G. Vaughan, 2006: Antarctic snow accumulation mapped using polarization of 4.3-cm wavelength microwave emission. *J. Geophys. Res. Atmos.*, **111**, D06107.

Arthern, R. J., D. G. Vaughan, A. M. Rankin, R. Mulvaney, and E. R. Thomas, 2010: In-situ measurements of Antarctic snow compaction, compared with predictions of models. *J. Geophys. Res.*, **115**, F03011.

Azam, M. F., et al., 2012: From balance to imbalance: a shift in the dynamic behaviour of Chhota Shigri glacier, western Himalaya, India. *J. Glaciol.*, **58**, 315–324.

Bahr, D. B., M. F. Meier, and S. D. Peckham, 1997: The physical basis of glacier volume-area scaling. *J. Geophys. Res. Sol. Ea.*, **102**, 20355–20362.

Bahr, D. B., M. Dyurgerov, and M. F. Meier, 2009: Sea-level rise from glaciers and ice caps: A lower bound. *Geophys. Res. Lett.*, **36**, 4 (L03501).

Bales, R. C., et al., 2009: Annual accumulation for Greenland updated using ice core data developed during 2000–2006 and analysis of daily coastal meteorological data. *J. Geophys. Res. Atmos.*, **114**, D06116.

Bamber, J. L., et al., 2013: A new bed elevation dataset for Greenland. *Cryosphere*, **7**, 499–510.

Barrand, N., D. G. Vaughan, N. Steiner, M. Tedesco, P. Kuipers Munneke, M. R. van den Broeke, and J. S. Hosking, 2013: Trends in Antarctic Peninsula surface melting conditions from observations and regional climate modelling. *J. Geophys. Res.*, **118**, 1–16.

Barrett, P. J., 2013: Resolving views on Antarctic Neogene glacial history—the Sirius debate. *Earth Environ. Sci. Trans. R. Soc. Edinburgh*, **104**, 29–51.

Barry, R., and T. Y. Gran, 2011: *The Global Cryosphere: Past, Present and Future.* Cambridge University Press, Cambridge, UK, and New York, NY, USA, 498 pp.

Barry, R. G., R. E. Moritz, and J. C. Rogers, 1979: Fast ice regimes of the Beaufort and Chukchi sea coasts, Alaska. *Cold Reg. Sci. Technol.*, **1**, 129–152.

Bartholomew, I. D., P. Nienow, A. Sole, D. Mair, T. Cowton, M. A. King, and S. Palmer, 2011: Seasonal variations in Greenland Ice Sheet motion: Inland extent and behaviour at higher elevations. *Earth Planet. Sci. Lett.*, **307**, 271–278.

Baur, O., M. Kuhn, and W. E. Featherstone, 2009: GRACE-derived ice-mass variations over Greenland by accounting for leakage effects. *J. Geophys. Res. Sol. Ea.*, **114**, 13 (B06407).

Belchansky, G. I., D. C. Douglas, and N. G. Platonov, 2004: Duration of the Arctic Sea ice melt season: Regional and interannual variability, 1979–2001. *J. Clim.*, **17**, 67–80.

Beltaos, S., and T. Prowse, 2009: River-ice hydrology in a shrinking cryosphere. *Hydrol. Process.*, **23**, 122–144.

Benn, D. I., C. R. Warren, and R. H. Mottram, 2007: Calving processes and the dynamics of calving glaciers. *Earth Sci. Rev.*, **82**, 143–179.

Benson, B. J., et al., 2012: Extreme events, trends, and variability in Northern Hemisphere lake-ice phenology (1855–2005). *Clim. Change*, **112**, 299–323.

Berthier, E., E. Schiefer, G. K. C. Clarke, B. Menounos, and F. Remy, 2010: Contribution of Alaskan glaciers to sea-level rise derived from satellite imagery. *Nature Geosci.*, **3**, 92–95.

Bindschadler, R., et al., 2011: Getting around Antarctica: new high-resolution mappings of the grounded and freely-floating boundaries of the Antarctic ice sheet created for the International Polar Year. *Cryosphere*, **5**, 569–588.

Bjørk, A. A., et al., 2012: An aerial view of 80 years of climate-related glacier fluctuations in southeast Greenland. *Nature Geosci.*, **5**, 427–432.

Björnsson, H., et al., 2013: Contribution of Icelandic ice caps to sea level rise: trends and variability since the Little Ice Age. *Geophys. Res. Lett.*, **40**, 1546-1550

Blaszczyk, M., J. A. Jania, and J. O. Hagen, 2009: Tidewater glaciers of Svalbard: Recent changes and estimates of calving fluxes. *Pol. Polar Res.*, **30**, 85–142.

Bliss, A., R. Hock, and J. G. Cogley, 2013: A new inventory of mountain glaciers and ice caps for the Antarctic periphery. *Ann. Glaciol.*, **54**, 191–199.

Bodin, X., F. Rojas, and A. Brenning, 2010: Status and evolution of the cryosphere in the Andes of Santiago (Chile, 33.5 degrees S.). *Geomorphology*, **118**, 453–464.

Bodin, X., et al., 2009: Two Decades of Responses (1986–2006) to Climate by the Laurichard Rock Glacier, French Alps. *Permafr. Periglac. Process.*, **20**, 331–344.

Boening, C., M. Lebsock, F. Landerer, and G. Stephens, 2012: Snowfall-driven mass change on the East Antarctic ice sheet. *Geophys. Res. Lett.*, **39**, L21501.

Boening, C. W., A. Dispert, M. Visbeck, S. R. Rintoul, and F. U. Schwarzkopf, 2008: The response of the Antarctic Circumpolar Current to recent climate change. *Nature Geosci.*, **1**, 864–869.

Bolch, T., B. Menounos, and R. Wheate, 2010: Landsat-based inventory of glaciers in western Canada, 1985–2005. *Remote Sens. Environ.*, **114**, 127–137.

Bolch, T., L. Sandberg Sørensen, S. B. Simonsen, N. Moelg, H. Machguth, P. Rastner, and F. Paul, 2013: Mass loss of Greenland's glaciers and ice caps 2003–2008 revealed from ICESat data. *Geophys. Res. Lett.*, **40**, 875–881.

Bolch, T., et al., 2012: The state and fate of Himalayan glaciers. *Science*, **336**, 310–314.

Bown, F., A. Rivera, and C. Acuna, 2008: Recent glacier variations at the Aconcagua basin, central Chilean Andes. *Ann. Glaciol.*, **48**, 43–48.

Box, J. E., L. Yang, D. H. Bromwich, and L. S. Bai, 2009: Greenland ice sheet surface air temperature variability: 1840–2007. *J. Clim.*, **22**, 4029–4049.

Box, J. E., X. Fettweis, J. C. Stroeve, M. Tedesco, D. K. Hall, and K. Steffen, 2012: Greenland ice sheet albedo feedback: Thermodynamics and atmospheric drivers. *Cryosphere*, **6**, 821–839.

Brewer, M. C., 1958: Some results of geothermal investigations of permafrost. *Am. Geophys. Union Trans.* **39**, 19–26.

Bromwich, D. H., J. P. Nicolas, and A. J. Monaghan, 2011: An assessment of precipitation changes over Antarctica and the Southern Ocean since 1989 in contemporary global reanalyses. *J. Clim.*, **24**, 4189–4209.

Bromwich, D. H., J. P. Nicolas, A. J. Monaghan, M. A. Lazzara, L. M. Keller, G. A. Weidner, and A. B. Wilson, 2013: Central West Antarctica among the most rapidly warming regions on Earth. *Nature Geosci.*, **6**, 139–145.

Brooks, R. N., T. D. Prowse, and I. J. O'Connell, 2012: Quantifying Northern Hemisphere freshwater ice. *Geophys. Res. Lett.*, **40**, 1128–1131.

Brown, R., C. Derksen, and L. B. Wang, 2010: A multi-data set analysis of variability and change in Arctic spring snow cover extent, 1967–2008. *J. Geophys. Res. Atmos.*, **115**, D16111.

Brown, R. D., and P. Coté, 1992: Interannual variability of landfast ice thickness in the Canadian High Arctic, 1950–89. *Arctic*, **45**, 273–284.

Brown, R. D., and P. W. Mote, 2009: The response of Northern Hemisphere snow cover to a changing climate. *J. Clim.*, **22**, 2124–2145.

Brown, R. D., and D. A. Robinson, 2011: Northern Hemisphere spring snow cover variability and change over 1922–2010 including an assessment of uncertainty. *Cryosphere*, **5**, 219–229.

Brunt, K. M., E. A. Okal, and D. R. MacAyeal, 2011: Antarctic ice-shelf calving triggered by the Honshu (Japan) earthquake and tsunami, March 2011. *J. Glaciol.*, **57**, 785–788.

Buchardt, S. L., H. B. Clausen, B. M. Vinther, and D. Dahl-Jensen, 2012: Investigating the past and recent delta18O-accumulation relationship seen in Greenland ice cores. *Clim. Past*, **8**, 2053–2059.

Bulygina, O. N., V. N. Razuvaev, and N. N. Korshunova, 2009: Changes in snow cover over Northern Eurasia in the last few decades. *Environ. Res. Lett.*, **4**, 045026.

Bulygina, O. N., P. Y. Groisman, V. N. Razuvaev, and N. N. Korshunova, 2011: Changes in snow cover characteristics over Northern Eurasia since 1966. *Environ. Res. Lett.*, **6**, 045204.

Burn, C. R., and S. V. Kokelj, 2009: The environment and permafrost of the Mackenzie Delta Area. *Permafr. Periglac. Process.*, **20**, 83–105.

Burn, C. R., and Y. Zhang, 2009: Permafrost and climate change at Herschel Island (Qikiqtaruq), Yukon Territory, Canada. *J. Geophys. Res.*, **114**, F02001.

Callaghan, T. V., F. Bergholm, T. R. Christensen, C. Jonasson, U. Kokfelt, and M. Johansson, 2010: A new climate era in the sub-Arctic: Accelerating climate changes and multiple impacts. *Geophys. Res. Lett.*, **37**, L14705.

Callaghan, T. V., M. Johansson, O. Anisimov, H. H. Christiansen, A. Instanes, V. Romanovsky, and S. Smith, 2011: Changing permafrost and its impacts. In: *Snow, Water, Ice and Permafrost in the Arctic (SWIPA)*. Arctic Monitoring and Assessment Program (AMAP).

Carrivick, J. L., B. J. Davies, N. F. Glasser, D. Nyvlt, and M. J. Hambrey, 2012: Late-Holocene changes in character and behaviour of land-terminating glaciers on James Ross Island, Antarctica. *J. Glaciol.*, **58**, 1176–1190.

Carturan, L., and R. Seppi, 2007: Recent mass balance results and morphological evolution of Careser glacier (Central Alps). *Geograf. Fis. Dinam. Quat.*, **30**, 33–42.

Cavalieri, D. J., and C. L. Parkinson, 2012: Arctic sea ice variability and trends, 1979–2010. *Cryosphere*, **6**, 957–979.

Cavalieri, D. J., P. Gloersen, and W. J. Campbell, 1984: Determination of sea ice parameters with the Nimbus-7 SMMR. *J. Geophys. Res. Atmos.*, **89**, 5355–5369.

Cazenave, A., et al., 2009: Sea level budget over 2003–2008: A reevaluation from GRACE space gravimetry, satellite altimetry and Argo. *Global Planet. Change*, **65**, 83–88.

Chapman, W. L., and J. E. Walsh, 2007: A synthesis of Antarctic temperatures. *J. Clim.*, **20**, 4096–4117.

Charrassin, J. B., et al., 2008: Southern Ocean frontal structure and sea-ice formation rates revealed by elephant seals. *Proc. Natl. Acad. Sci. U.S.A.*, **105**, 11634–11639.

Chen, J. L., C. R. Wilson, and B. D. Tapley, 2006: Satellite gravity measurements confirm accelerated melting of Greenland ice sheet. *Science*, **313**, 1958–1960.

Chen, J. L., C. R. Wilson, and B. D. Tapley, 2011: Interannual variability of Greenland ice losses from satellite gravimetry. *J. Geophys. Res. Sol. Ea.*, **116**, 11(B07406).

Chen, J. L., C. R. Wilson, D. Blankenship, and B. D. Tapley, 2009: Accelerated Antarctic ice loss from satellite gravity measurements. *Nature Geosci.*, **2**, 859–862.

Chen, J. L., C. R. Wilson, B. D. Tapley, D. D. Blankenship, and E. R. Ivins, 2007: Patagonia icefield melting observed by gravity recovery and climate experiment (GRACE). *Geophys. Res. Lett.*, **34**, 6 (L22501).

Cheng, G. D., and T. H. Wu, 2007: Responses of permafrost to climate change and their environmental significance, Qinghai-Tibet Plateau. *J. Geophys. Res.*, **112**, F02S03.

Chinn, T., S. Winkler, M. J. Salinger, and N. Haakensen, 2005: Srecent glacier advances in Norway and New Zealand: A comparison of their glaciological and meteorological causes. *Geograf. Annal. A*, **87A**, 141–157.

Choi, G., D. A. Robinson, and S. Kang, 2010: Changing Northern Hemisphere snow seasons. *J. Clim.*, **23**, 5305–5310.

Christiansen, H. H., M. Guglielmin, J. Noetzli, V. Romanovsky, N. Shiklomanov, S. Smith, and L. Zhao, 2012: Cryopsphere, Permafrost thermal state. *Special Suppl. to Bull. Am. Meteorol. Soc.*, **93**(July) [J. Blunden and D. S. Arndt (eds.)], S19–S21.

Christiansen, H. H., et al., 2010: The thermal state of permafrost in the Nordic area during the International Polar Year 2007–2009. *Permafr. Periglac. Process.*, **21**, 156–181.

Christoffersen, P., et al., 2011: Warming of waters in an East Greenland fjord prior to glacier retreat: Mechanisms and connection to large-scale atmospheric conditions. *Cryosphere*, **5**, 701–714.

Christy, J. R., 2012: Searching for information in 133 years of California snowfall observations. *J. Hydrometeorol.*, **13**, 895–912.

Citterio, M., F. Paul, A. P. Ahlstrom, H. F. Jepsen, and A. Weidick, 2009: Remote sensing of glacier change in West Greenland: accounting for the occurrence of surge-type glaciers. *Ann. Glaciol.*, **50**, 70–80.

Citterio, M., G. Diolaiuti, C. Smiraglia, C. D'Agata, T. Carnielli, G. Stella, and G. B. Siletto, 2007: The fluctuations of Italian glaciers during the last century: A contribution to knowledge about Alpine glacier changes. *Geograf. Annal. A*, **89A**, 167–184.

Cogley, J. G., 2009a: A more complete version of the World Glacier Inventory. *Ann. Glaciol.*, **50**, 32–38.

Cogley, J. G., 2009b: Geodetic and direct mass-balance measurements: comparison and joint analysis. *Ann. Glaciol.*, **50**, 96–100.

Cogley, J. G., 2012: Area of the ocean. *Mar. Geodesy*, **35**, 379–388.

Cogley, J. G., et al., 2011: *Glossary of Glacier Mass Balance and Related Terms.* IHP-VII Technical Documents in Hydrology No. 86, International Association of Cryospheric Sciences, Contribution No. 2, UNESCO-IHP. 114 pp.

Comiso, J. C., 2002: A rapidly declining perennial sea ice cover in the Arctic. *Geophys. Res. Lett.*, **29**, 1956.

Comiso, J. C., 2010: *Polar Oceans from Space.* Springer Science+Business Media, New York, NY, USA and Heidelberg, Germany.

Comiso, J. C., 2012: Large decadal decline in the Arctic multiyear ice cover. *J. Clim.*, **25**, 1176–1193.

Comiso, J. C., and F. Nishio, 2008: Trends in the sea ice cover using enhanced and compatible AMSR-E, SSM/I, and SMMR data. *J. Geophys. Res. Oceans*, **113**, C02S07.

Comiso, J. C., C. L. Parkinson, R. Gersten, and L. Stock, 2008: Accelerated decline in the Arctic Sea ice cover. *Geophys. Res. Lett.*, **35**, L01703.

Comiso, J. C., R. Kwok, S. Martin, and A. L. Gordon, 2011: Variability and trends in sea ice extent and ice production in the Ross Sea. *J. Geophys. Res. Oceans*, **116**, C04021.

Cook, A. J., and D. G. Vaughan, 2010: Overview of areal changes of the ice shelves on the Antarctic Peninsula over the past 50 years. *Cryosphere*, **4**, 77–98.

Costa, D. P., J. M. Klinck, E. E. Hofmann, M. S. Dinniman, and J. M. Burns, 2008: Upper ocean variability in west Antarctic Peninsula continental shelf waters as measured using instrumented seals. *Deep-Sea Res. Pt. Ii*, **55**, 323–337.

Coudrain, A., B. Francou, and Z. W. Kundzewicz, 2005: Glacier shrinkage in the Andes and consequences for water resources. *Hydrol. Sci. J. –J. Sci. Hydrol.*, **50**, 925–932.

Cullen, N. J., P. Sirguey, T. Moelg, G. Kaser, M. Winkler, and S. J. Fitzsimons, 2013: A century of ice retreat on Kilimanjaro: The mapping reloaded. *Cryosphere*, **7**, 419–431.

Daniault, N., H. Mercier, and P. Lherminier, 2011: The 1992–2009 transport variability of the East Greenland-Irminger Current at 60 degrees N. *Geophys. Res. Lett.*, **38**, 4 (L07601).

Das, S. B., I. Joughin, M. D. Behn, I. M. Howat, M. A. King, D. Lizarralde, and M. P. Bhatia, 2008: Fracture propagation to the base of the Greenland Ice Sheet during supraglacial lake drainage. *Science*, **320**, 778–781.

Davies, B. J., and N. F. Glasser, 2012: Accelerating shrinkage of Patagonian glaciers from the "Little Ice Age" (c. AD 1870) to 2011. *J. Glaciol.*, **58**, 1063–1084.

De Angelis, H., and P. Skvarca, 2003: Glacier surge after ice shelf collapse. *Science*, **299**, 1560–1562.

Delaloye, R., et al., cited 2011: Recent interannual variations of rock glacier creep in the European Alps. [Available online at http://www.zora.uzh.ch/7031/.]

DeLiberty, T. L., C. A. Geiger, S. F. Ackley, A. P. Worby, and M. L. Van Woert, 2011: Estimating the annual cycle of sea-ice thickness and volume in the Ross Sea. *Deep-Sea Res. Pt. Ii*, **58**, 1250–1260.

Derksen, C., and R. Brown, 2012: Spring snow cover extent reductions in the 2008–2012 period exceeding climate model projections. *Geophys. Res. Lett.*, **39**, L19504.

Derksen, C., et al., 2012: Variability and change in the Canadian cryosphere. *Clim. Change*, **115**, 59–88.

Déry, S. J., and R. D. Brown, 2007: Recent Northern Hemisphere snow cover extent trends and implications for the snow-albedo feedback. *Geophys. Res. Lett.*, **34**, 6 (L22504).

Ding, Q. H., E. J. Steig, D. S. Battisti, and M. Kuttel, 2011: Winter warming in West Antarctica caused by central tropical Pacific warming. *Nature Geosci.*, **4**, 398–403.

Dinniman, M. S., J. M. Klinck, and E. E. Hofmann, 2012: Sensitivity of circumpolar deep water transport and ice shelf basal melt along the West Antarctic Peninsula to changes in the winds. *J. Clim.*, **25**, 4799–4816.

Diolaiuti, G., D. Bocchiola, C. D'Agata, and C. Smiraglia, 2012: Evidence of climate change impact upon glaciers' recession within the Italian Alps—The case of Lombardy glaciers. *Theor. Appl. Climatol.*, **109**, 429–445.

Dmitrenko, I. A., et al., 2011: Recent changes in shelf hydrography in the Siberian Arctic: Potential for subsea permafrost instability. *J. Geophys. Res. Oceans*, **116**, 10 (C10027).

Doake, C. S. M., and D. G. Vaughan, 1991: Rapid disintegration of the wordie ice shelf in response to atmospheric warming. *Nature*, **350**, 328–330.

Doake, C. S. M., H. F. J. Corr, H. Rott, P. Skvarca, and N. W. Young, 1998: Breakup and conditions for stability of the northern Larsen Ice Shelf, Antarctica. *Nature*, **391**, 778–780.

Doherty, S. J., S. G. Warren, T. C. Grenfell, A. D. Clarke, and R. E. Brandt, 2010: Light-absorbing impurities in Arctic snow. *Atmos. Chem. Phys.*, **10**, 11647–11680.

Drobot, S. D., and M. R. Anderson, 2001: An improved method for determining snowmelt onset dates over Arctic sea ice using scanning multichannel microwave radiometer and Special Sensor Microwave/Imager data. *J. Geophys. Res. Atmos.*, **106**, 24033–24049.

Drozdov, D. S., N. G. Ukraintseva, A. M. Tsarev, and S. N. Chekrygina, 2010: Changes in the temperature field and in the state of the geosystems within the territory of the Urengoy field during the last 35 years (1974–2008). *Earth Cryosphere*, **14**, 22–31.

Druckenmiller, M. L., H. Eicken, M. A. Johnson, D. J. Pringle, and C. C. Williams, 2009: Toward an integrated coastal sea-ice observatory: System components and a case study at Barrow, Alaska. *Cold Reg. Sci. Technol.*, **56**, 61–72.

Drucker, R., S. Martin, and R. Kwok, 2011: Sea ice production and export from coastal polynyas in the Weddell and Ross Seas. *Geophys. Res. Lett.*, **38**, 4 (L17502).

Ducklow, H., et al., 2011: The marine system of the Western Antarctic Peninsula In: *Antarctica: An Extreme Environment in a Changing World* [A. D. Rogers (ed.)]. New York, NY, USA, John Wiley & Sons. 121-159.

E, D.-C., Y.-D. Yang, and D.-B. Chao, 2009: The sea level change from the Antarctic ice sheet based on GRACE. *Chin. J. Geophys.–Chin. Ed.*, **52**, 2222–2228.

Ettema, J., M. R. van den Broeke, E. van Meijgaard, W. J. van de Berg, J. L. Bamber, J. E. Box, and R. C. Bales, 2009: Higher surface mass balance of the Greenland ice sheet revealed by high-resolution climate modeling. *Geophys. Res. Lett.*, **36**, L12501.

Ewert, H., A. Groh, and R. Dietrich, 2012: Volume and mass changes of the Greenland ice sheet inferred from ICESat and GRACE. *J. Geodyn.*, **59**, 111–123.

Fedorov, A. N., and P. Y. Konstantinov, 2008: Recent changes in ground temperature and the effect on permafrost landscapes in Central Yakutia. In: *Proceeding of the 9th International Conference on Permafrost, 29 June–3 July 2008, Institute of Northern Engineering, University of Alaska, Fairbanks* [D. L. Kane, and K. M. Hinkel (eds.)], pp. 433–438.

Fernandes, R., H. X. Zhao, X. J. Wang, J. Key, X. Qu, and A. Hall, 2009: Controls on Northern Hemisphere snow albedo feedback quantified using satellite Earth observations. *Geophys. Res. Lett.*, **36**, L21702.

Fettweis, X., M. Tedesco, M. R. van den Broeke, and J. Ettema, 2011: Melting trends over the Greenland ice sheet (1958–2009) from spaceborne microwave data and regional climate models. *Cryosphere*, **5**, 359–375.

Flament, T., and F. Remy, 2012: Dynamic thinning of Antarctic glaciers from along-track repeat radar altimetry. *J. Glaciol.*, **58**, 830–840.

Flanner, M. G., C. S. Zender, J. T. Randerson, and P. J. Rasch, 2007: Present-day climate forcing and response from black carbon in snow. *J. Geophys. Res.*, **112**, D11202.

Flanner, M. G., K. M. Shell, M. Barlage, D. K. Perovich, and M. A. Tschudi, 2011: Radiative forcing and albedo feedback from the Northern Hemisphere cryosphere between 1979 and 2008. *Nature Geosci.*, **4**, 151–155.

Fleming, K., and K. Lambeck, 2004: Constraints on the Greenland Ice Sheet since the Last Glacial Maximum from sea-level observations and glacial-rebound models. *Quat. Sci. Rev.*, **23**, 1053–1077.

Foster, J. L., D. K. Hall, R. E. J. Kelly, and L. Chiu, 2009: Seasonal snow extent and snow mass in South America using SMMR and SSM/I passive microwave data (1979–2006). *Remote Sens. Environ.*, **113**, 291–305.

Fraser, A. D., R. A. Massom, K. J. Michael, B. K. Galton-Fenzi, and J. L. Lieser, 2012: East Antarctic landfast sea ice distribution and variability, 2000–2008. *J. Clim.*, **25**, 1137–1156.

Frauenfeld, O. W., and T. J. Zhang, 2011: An observational 71-year history of seasonally frozen ground changes in the Eurasian high latitudes. *Environ. Res. Lett.*, **6**, 044024.

Frauenfeld, O. W., T. J. Zhang, R. G. Barry, and D. Gilichinsky, 2004: Interdecadal changes in seasonal freeze and thaw depths in Russia. *J. Geophys. Res.*, **109**, D05101.

Fretwell, P. T., et al., 2013: Bedmap2: improved ice bed, surface and thickness datasets for Antarctica. *Cryosphere*, **7**, 375–393.

Frezzotti, M., C. Scarchilli, S. Becagli, M. Proposito, and S. Urbini, 2012: A synthesis of the Antarctic Surface Mass Balance during the last eight centuries. *Cryosphere*, **7**, 303–319.

Fricker, H. A., and L. Padman, 2012: Thirty years of elevation change on Antarctic Peninsula ice shelves from multimission satellite radar altimetry. *J. Geophys. Res. Oceans*, **117**, C02026.

Fyodorov-Davydov, D. G., A. L. Kholodov, V. E. Ostroumov, G. N. Kraev, V. A. Sorokovikov, S. P. Davudov, and A. A. Merekalova, 2008: Seasonal thaw of soils in the North Yakutian ecosystems. In: *Proceedings of the 9th International Conference on Permafrost, 29 June– 3 July 2008, Institute of Northern Engineering, University of Alaska, Fairbanks* [D. L. Kane, and K. M. Hinkel (eds.)], pp. 481–486.

Galton-Fenzi, B. K., J. R. Hunter, R. Coleman, and N. Young, 2012: A decade of change in the hydraulic connection between an Antarctic epishelf lake and the ocean. *J. Glaciol.*, **58**, 223–228.

Gardelle, J., E. Berthier, and Y. Arnaud, 2012: Slight mass gain of Karakoram glaciers in the early twenty-first century. *Nature Geosci.*, **5**, 322–325.

Gardner, A., G. Moholdt, A. Arendt, and B. Wouters, 2012: Accelerated contributions of Canada's Baffin and Bylot Island glaciers to sea level rise over the past half century. *Cryosphere*, **6**, 1103–1125.

Gardner, A. S., et al., 2011: Sharply increased mass loss from glaciers and ice caps in the Canadian Arctic Archipelago. *Nature*, **473**, 357–360.

Gardner, A. S., et al., 2013: A reconciled estimate of glacier contributions to sea level rise: 2003 to 2009. *Science*, **340**, 852–857.

Gerland, S., A. H. H. Renner, F. Godtliebsen, D. Divine, and T. B. Loyning, 2008: Decrease of sea ice thickness at Hopen, Barents Sea, during 1966–2007. *Geophys. Res. Lett.*, **35**, L06501.

Gilbert, A., P. Wagnon, C. Vincent, P. Ginot, and M. Funk, 2010: Atmospheric warming at a high-elevation tropical site revealed by englacial temperatures at Illimani, Bolivia (6340 m above sea level, 16 degrees S, 67 degrees W). *J. Geophys. Res.*, **115**, D10109.

Giles, A. B., R. A. Massom, and V. I. Lytle, 2008a: Fast-ice distribution in East Antarctica during 1997 and 1999 determined using RADARSAT data. *J. Geophys. Res. Oceans*, **113**, 15 (C02S14).

Giles, K. A., S. W. Laxon, and A. L. Ridout, 2008b: Circumpolar thinning of Arctic sea ice following the 2007 record ice extent minimum. *Geophys. Res. Lett.*, **35**, L22502.

Gille, S. T., 2002: Warming of the Southern Ocean since the 1950s. *Science*, **295**, 1275–1277.

Gille, S. T., 2008: Decadal-scale temperature trends in the Southern Hemisphere Ocean. *J. Clim.*, **21**, 4749–4765.

Glazovsky, A., and Y. Macheret, 2006: Eurasian Arctic. In: *Glaciation in North and Central Eurasia in Present Time*. [V. M. Kotlyakov (ed.)]. Nauka, Saint Petersburg, Russian Federation, pp. 438–445.

Gordon, A. L., M. Visbeck, and J. C. Comiso, 2007: A possible link between the Weddell Polynya and the Southern Annular Mode. *J. Clim.*, **20**, 2558–2571.

Grebmeier, J. M., S. E. Moore, J. E. Overland, K. E. Frey, and R. Gradinger, 2010: Biological response to recent Pacific Arctic sea ice retreats. *EOS Trans. Am. Geophys. Union*, **91**, 161–163.

Green, K., 2011: Interannual and seasonal changes in the ice cover of glacial lakes in the Snowy Mountains of Australia. *J. Mount. Sci.*, **8**, 655–663.

Gregory, J. M., and J. Oerlemans, 1998: Simulated future sea-level rise due to glacier melt based on regionally and seasonally resolved temperature changes. *Nature*, **391**, 474–476.

Griggs, J., and J. L. Bamber, 2011: Antarctic ice-shelf thickness from satellite radar altimetry. *J. Glaciol.*, **57**, 485–498.

Grinsted, A., 2013: An estimate of global glacier volume. *Cryosphere*, **7**, 141–151.

Gruber, S., 2012: Derivation and analysis of a high-resolution estimate of globalpermafrost zonation. *Cryosphere*, **6**, 221–233.

Gruber, S., and W. Haeberli, 2007: Permafrost in steep bedrock slopes and its temperature-related destabilization following climate change. *J. Geophys. Res.*, **112**, 10 (F02S18).

Guglielmin, M., and N. Cannone, 2012: A permafrost warming in a cooling Antarctica? *Clim. Change*, **111**, 177–195.

Guglielmin, M., M. R. Balks, L. S. Adlam, and F. Baio, 2011: Permafrost thermal regime from two 30-m deep boreholes in Southern Victoria Land, Antarctica. *Permafr. Periglac. Process.*, **22**, 129–139.

Gunter, B., et al., 2009: A comparison of coincident GRACE and ICESat data over Antarctica. *J. Geodesy*, **83**, 1051–1060.

Gusmeroli, A., P. Jansson, R. Pettersson, and T. Murray, 2012: Twenty years of cold surface layer thinning at Storglaciären, sub-Arctic Sweden, 1989–2009. *J. Glaciol.*, **58**, 3–10.

Haas, C., S. Hendricks, H. Eicken, and A. Herber, 2010: Synoptic airborne thickness surveys reveal state of Arctic sea ice cover. *Geophys. Res. Lett.*, **37**, L09501.

Haas, C., H. Le Goff, S. Audrain, D. Perovich, and J. Haapala, 2011: Comparison of seasonal sea-ice thickness change in the Transpolar Drift observed by local ice mass-balance observations and floe-scale EM surveys. *Ann. Glaciol.*, **52**, 97–102.

Haas, C., A. Pfaffling, S. Hendricks, L. Rabenstein, J. L. Etienne, and I. Rigor, 2008: Reduced ice thickness in Arctic Transpolar Drift favors rapid ice retreat. *Geophys. Res. Lett.*, **35**, L17501.

Haeberli, W., M. Hoelzle, F. Paul, and M. Zemp, 2007: Integrated monitoring of mountain glaciers as key indicators of global climate change: the European Alps. *Ann. Glaciol.*, **46**, 150–160.

Haeberli, W., et al., 2006: Permafrost creep and rock glacier dynamics. *Permafr. Periglac. Process.*, **17**, 189–214.

Haeberli, W., et al., 2010: Mountain permafrost: development and challenges of a young research field. *J. Glaciol.*, **56**, 1043–1058.

Hagg, W., C. Mayer, A. Lambrecht, D. Kriegel, and E. Azizov, 2012: Glacier changes in the Big Naryn basin, Central Tian Shan. *Global Planet. Change*, doi:10.1016/j.gloplacha.2012.07.010.

Hall, D. K., J. C. Comiso, N. E. DiGirolamo, C. A. Shuman, J. E. Box, and L. S. Koenig, 2013: Variability in the surface temperature and melt extent of the Greenland Ice Sheet from MODIS. *Geophys. Res. Lett.*, **10**, 2114-2120.

Halsey, L. A., D. H. Vitt, and S. C. Zoltai, 1995: Disequilibrium response of permafrost in boreal continental western Canada to climate-change. *Clim. Change*, **30**, 57–73.

Hanna, E., et al., 2011: Greenland Ice Sheet surface mass balance 1870 to 2010 based on Twentieth Century Reanalysis, and links with global climate forcing. *J. Geophys. Res.*, **116**, D24121.

Harig, C., and F. J. Simons, 2012: Mapping Greenland's mass loss in space and time. *Proc. Natl. Acad. Sci. U.S.A.*, **109**, 19934–19937.

Heid, T., and A. Kääb, 2012: Repeat optical satellite images reveal widespread and long term decrease in land-terminating glacier speeds. *Cryosphere* , **6**, 467–478

Hirabayashi, Y., S. Kanae, K. Masuda, K. Motoya, and P. Doell, 2008: A 59-year (1948–2006) global near-surface meteorological data set for land surface models. Part I: Development of daily forcing and assessment of precipitation intensity. *Hydrol. Res. Lett.*, **2**, 36–40.

Hirabayashi, Y., Y. Zhang, S. Watanabe, S. Koirala, and S. Kanae, 2013: Projection of glacier mass changes under a high-emission climate scenario using the global glacier model HYOGA2. *Hydrol. Res. Lett.*, **7**, 6–11.

Hock, R., M. de Woul, V. Radic, and M. Dyurgerov, 2009: Mountain glaciers and ice caps around Antarctica make a large sea-level rise contribution. *Geophys. Res. Lett.*, **36**, L07501.

Hodgkins, A. G., I. C. James, and T. G. Huntington, 2002: Historical changes in lake ice-out dates as indicators of climate change in New England, 1850–2000. *Int. J. Clim.*, **22**, 1819–1827.

Hoelzle, M., G. Darms, M. P. Lüthi, and S. Suter, 2011: Evidence of accelerated englacial warming in the Monte Rosa area, Switzerland/Italy. *Cryosphere*, **5**, 231–243.

Hoffman, M. J., G. A. Catania, T. A. Neumann, L. C. Andrews, and J. A. Rumrill, 2011: Links between acceleration, melting, and supraglacial lake drainage of the western Greenland Ice Sheet. *J. Geophys. Res. Earth Surf.*, **116**, F04035.

Holland, D. M., and A. Jenkins, 1999: Modeling thermodynamic ice-ocean interactions at the base of an ice shelf. *J. Phys. Oceanogr.*, **29**, 1787–1800.

Holland, D. M., R. H. Thomas, B. De Young, M. H. Ribergaard, and B. Lyberth, 2008: Acceleration of Jakobshavn Isbrae triggered by warm subsurface ocean waters. *Nature Geosci.*, **1**, 659–664.

Holland, P. R., and R. Kwok, 2012: Wind-driven trends in Antarctic sea ice motion. *Nature Geosci.*, **5**, 872–875.

Holland, P. R., A. Jenkins, and D. M. Holland, 2010: Ice and ocean processes in the Bellingshausen Sea, Antarctica. *J. Geophys. Res. Oceans*, **115**, C05020.

Holland, P. R., H. F. J. Corr, H. D. Pritchard, D. G. Vaughan, R. J. Arthern, A. Jenkins, and M. Tedesco, 2011: The air content of Larsen Ice Shelf. *Geophys. Res. Lett.*, **38**, L10503.

Holzhauser, H., M. Magny, and H. J. Zumbuhl, 2005: Glacier and lake-level variations in west-central Europe over the last 3500 years. *Holocene*, **15**, 789–801.

Horwath, M., and R. Dietrich, 2009: Signal and error in mass change inferences from GRACE: The case of Antarctica. *Geophys. J. Int.*, **177**, 849–864.

Howat, I. M., I. Joughin, and T. A. Scambos, 2007: Rapid changes in ice discharge from Greenland outlet glaciers. *Science*, **315**, 1559–1561.

Howat, I. M., I. Joughin, M. Fahnestock, B. E. Smith, and T. A. Scambos, 2008: Synchronous retreat and acceleration of southeast Greenland outlet glaciers 2000–06: Ice dynamics and coupling to climate. *J. Glaciol.*, **54**, 646–660.

Howat, I. M., Y. Ahn, I. Joughin, M. R. van den Broeke, J. T. M. Lenaerts, and B. Smith, 2011: Mass balance of Greenland's three largest outlet glaciers, 2000–2010. *Geophys. Res. Lett.*, **38**, 5 (L12501).

Hudson, S. R., 2011: Estimating the global radiative impact of the sea ice-albedo feedback in the Arctic. *J. Geophys. Res. Atmos.*, **116**, D16102.

Hughes, T. J., 1973: Is the West Antarctic ice sheet disintegrating? *J. Geophys. Res.*, **78**, 7884–7910.

Hulbe, C. L., T. A. Scambos, T. Youngberg, and A. K. Lamb, 2008: Patterns of glacier response to disintegration of the Larsen B ice shelf, Antarctic Peninsula. *Global Planet. Change*, **63**, 1–8.

Humbert, A., et al., 2010: Deformation and failure of the ice bridge on the Wilkins Ice Shelf, Antarctica. *Ann. Glaciol.*, **51**, 49–55.

Hurrell, J. W., 1995: Decadal trends in the North-Atlantic oscillation—regional temperatures and precipitation. *Science*, **269**, 676–679.

Huss, M., 2012: Extrapolating glacier mass balance to the mountain-range scale: The European Alps 1900–2100. *Cryosphere*, **6**, 713–727.

Huss, M., and D. Farinotti, 2012: Distributed ice thickness and volume of 180,000 glaciers around the globe. *J. Geophys. Res.*, **117**, F04010.

Isaksen, K., et al., 2011: Degrading mountain permafrost in southern Norway: Spatial and temporal variability of mean ground temperatures, 1999–2009. *Permafr. Periglac. Process.*, **22**, 361–377.

Ishikawa, M., N. Sharkhuu, Y. Jambaljav, G. Davaa, K. Yoshikawa, and T. Ohata, 2012: Thermal state of Mongolian permafrost. In: *Proceedings of the 10th International Conference on Permafrost, June, 2012, Salekhard, Yamel-nenets Autonomous District, Russian Federation, v1.*[K. M. Hinkel (ed)]. The Northern Publisher, Salekhard, Russia, pp. 173–178.

Ishizaka, M., 2004: Climatic response of snow depth to recent warmer winter seasons in heavy-snowfall areas in Japan. *Ann. Glaciol.,* **38**, 299–304.

Ivins, E. R., and T. S. James, 2005: Antarctic glacial isostatic adjustment: A new assessment. *Antarct. Sci.*, **17**, 541–553.

Ivins, E. R., M. M. Watkins, D. N. Yuan, R. Dietrich, G. Casassa, and A. Rulke, 2011: On-land ice loss and glacial isostatic adjustment at the Drake Passage: 2003–2009. *J. Geophys. Res. Sol. Ea.*, **116**, 24 (B02403).

Jacob, T., J. Wahr, W. T. Pfeffer, and S. Swenson, 2012: Recent contributions of glaciers and ice caps to sea level rise. *Nature*, **482**, 514–518.

Jacobs, S. S., A. Jenkins, C. F. Giulivi, and P. Dutrieux, 2011: Stronger ocean circulation and increased melting under Pine Island Glacier ice shelf. *Nature Geosci.*, **4**, 519–523.

Jacobs, S. S., H. H. Helmer, C. S. M. Doake, A. Jenkins, and R. M. Frolich, 1992: Melting of the ice shelves and the mass balance of Antarctica. *J. Glaciol.*, **38**, 375–387.

Jenkins, A., 2011: Convection-driven melting near the grounding lines of ice shelves and tidewater glaciers. *J. Clim.*, **41**, 2279–2294.

Jenkins, A., and C. S. M. Doake, 1991: Ice-Ocean interaction on Ronne ice shelf, Antarctica *J. Geophys. Res. Oceans*, **96**, 791–813.

Jenkins, A., and S. Jacobs, 2008: Circulation and melting beneath George VI Ice Shelf, Antarctica. *J. Geophys. Res. Oceans*, **113**, C04013.

Jenkins, A., P. Dutrieux, S. S. Jacobs, S. D. McPhail, J. R. Perrett, A. T. Webb, and D. White, 2010: Observations beneath Pine Island Glacier in West Antarctica and implications for its retreat. *Nature Geosci.*, **3**, 468–472.

Jensen, O. P., B. J. Benson, J. J. Magnuson, V. M. Card, M. N. Futter, P. A. Soranno, and K. M. Stewart, 2007: Spatial analysis of ice phenology trends across the Laurentian Great Lakes Region during a recent warming period. *Limnol. Oceanogr.*, **52**, 2013–2026.

Jia, L. L., H. S. Wang, and L. W. Xiang, 2011: Effect of glacio-static adjustment on the estimate of ice mass balance over Antarctic and uncertainties. *Chin. J. Geophys.*, **54**, 1466–1477.

Johannessen, O. M., E. V. Shalina, and M. W. Miles, 1999: Satellite evidence for an Arctic sea ice cover in transformation. *Science*, **286**, 1937–1939.

Johnson, J. S., M. J. Bentley, and K. Gohl, 2008: First exposure ages from the Amundsen Sea embayment, West Antarctica: The late quaternary context for recent thinning of Pine Island, Smith, and Pope Glaciers. *Geology*, **36**, 223–226.

Jones, B. M., C. D. Arp, M. T. Jorgenson, K. M. Hinkel, J. A. Schmutz, and P. L. Flint, 2009: Increase in the rate and uniformity of coastline erosion in Arctic Alaska. *Geophys. Res. Lett.*, **36**, 5 (L03503).

Jones, P. D., D. H. Lister, T. J. Osborn, C. Harpham, M. Salmon, and C. P. Morice, 2012: Hemispheric and large-scale land-surface air temperature variations: An extensive revision and an update to 2010. *J. Geophys. Res. Atmos.*, **117**, D05127.

Jorgenson, M. T., Y. L. Shur, and E. R. Pullman, 2006: Abrupt increase in permafrost degradation in Arctic Alaska. *Geophys. Res. Lett.*, **33**, 4 (L02503).

Joughin, I., and R. B. Alley, 2011: Stability of the West Antarctic ice sheet in a warming world. *Nature Geosci.*, **4**, 506–513.

Joughin, I., W. Abdalati, and M. Fahnestock, 2004: Large fluctuations in speed on Greenland's Jakobshavn Isbrae glacier. *Nature*, **432**, 608–610.

Joughin, I., B. E. Smith, and D. M. Holland, 2010a: Sensitivity of 21st century sea level to ocean-induced thinning of Pine Island Glacier, Antarctica. *Geophys. Res. Lett.*, **37**, L20502.

Joughin, I., R. B. Alley, and D. M. Holland, 2012: Ice-sheet response to oceanic forcing. *Science*, **338**, 1172–1176.

Joughin, I., M. Fahnestock, R. Kwok, P. Gogineni, and C. Allen, 1999: Ice flow of Humboldt, Petermann and Ryder Gletscher, northern Greenland. *J. Glaciol.*, **45**, 231–241.

Joughin, I., B. E. Smith, I. M. Howat, T. Scambos, and T. Moon, 2010b: Greenland flow variability from ice-sheet-wide velocity mapping. *J. Glaciol.*, **56**, 415–430.

Joughin, I., S. B. Das, M. A. King, B. E. Smith, I. M. Howat, and T. Moon, 2008a: Seasonal speedup along the western flank of the Greenland ice sheet. *Science*, **320**, 781–783.

Joughin, I., et al., 2008b: Ice-front variation and tidewater behavior on Helheim and Kangerdlugssuaq Glaciers, Greenland. *J. Geophys. Res. Earth Surf.*, **113**, F01004.

Kääb, A., R. Frauenfelder, and I. Roer, 2007: On the response of rockglacier creep to surface temperature increase. *Global Planet. Change*, **56**, 172–187.

Kääb, A., W. Haeberli, and G. H. Gudmundsson, 1997: Analysing the creep of mountain permafrost using high precision aerial photogrammetry: 25 years of monitoring Gruben Rock Glacier, Swiss Alps. *Permafr. Periglac. Process.*, **8**, 409–426.

Kääb, A., E. Berthier, C. Nuth, J. Gardelle, and Y. Arnaud, 2012: Contrasting patterns of early twenty-first-century glacier mass change in the Himalayas. *Nature*, **488**, 495–498.

Kargel, J. S., et al., 2012: Greenland's shrinking ice cover: "Fast times" but not that fast. *Cryosphere*, **6**, 533–537.

Kaser, G., J. G. Cogley, M. B. Dyurgerov, M. F. Meier, and A. Ohmura, 2006: Mass balance of glaciers and ice caps: Consensus estimates for 1961–2004. *Geophys. Res. Lett.*, **33**, L19501.

Khan, S. A., J. Wahr, M. Bevis, I. Velicogna, and E. Kendrick, 2010a: Spread of ice mass loss into northwest Greenland observed by GRACE and GPS. *Geophys. Res. Lett.*, **37**, L06501.

Khan, S. A., L. Liu, J. Wahr, I. Howat, I. Joughin, T. van Dam, and K. Fleming, 2010b: GPS measurements of crustal uplift near Jakobshavn Isbrae due to glacial ice mass loss. *J. Geophys. Res. Sol. Ea.*, **115**, 13 (B09405).

King, M. A., R. J. Bingham, P. Moore, P. L. Whitehouse, M. J. Bentley, and G. A. Milne, 2012: Lower satellite-gravimetry estimates of Antarctic sea-level contribution. *Nature*, **491**, 586-589.

King, M. A., et al., 2009: A 4–decade record of elevation change of the Amery Ice Shelf, East Antarctica. *J. Geophys. Res. Earth Surf.*, **114**, F01010.

Kinnard, C., C. M. Zdanowicz, D. A. Fisher, E. Isaksson, A. De Vernal, and L. G. Thompson, 2011: Reconstructed changes in Arctic sea ice over the past 1,450 years. *Nature*, **479**, 509–U231.

Kjaer, K. H., et al., 2012: Aerial photographs reveal late-20th-century dynamic ice loss in northwestern Greenland. *Science*, **337**, 569–573.

Klein, A. G., and J. L. Kincaid, 2006: Retreat of glaciers on Puncak Jaya, Irian Jaya, determined from 2000 and 2002 IKONOS satellite images. *J. Glaciol.*, **52**, 65–79.

Knoll, C., and H. Kerschner, 2009: A glacier inventory for South Tyrol, Italy, based on airborne laser-scanner data. *Ann. Glaciol.*, **50**, 46–52.

Koch, K., C. Knoblauch, and D. Wagner, 2009: Methanogenic community composition and anaerobic carbon turnover in submarine permafrost sediments of the Siberian Laptev Sea. *Environ. Microbiol.*, **11**, 657–668.

Kopp, R. E., F. J. Simons, J. X. Mitrovica, A. C. Maloof, and M. Oppenheimer, 2009: Probabilistic assessment of sea level during the last interglacial stage. *Nature*, **462**, 863–867.

Kozlovsky, A. M., Y. L. Nazintsev, V. I. Fedotov, and N. V. Cherepanov, 1977: Fast ice of the Eastern Antarctic (in Russian). *Proc. Soviet Antarct. Expedit.*, **63**, 1–129.

Krabill, W., et al., 1999: Rapid thinning of parts of the southern Greenland ice sheet. *Science*, **283**, 1522–1524.

Krabill, W., et al., 2000: Greenland ice sheet: High-elevation balance and peripheral thinning. *Science*, **289**, 428–430.

Krabill, W. B., et al., 2002: Aircraft laser altimetry measurement of elevation changes of the Greenland ice sheet: Technique and accuracy assessment. *J. Geodyn.*, **34**, 357–376.

Kuipers Munneke, P., G. Picard, M. R. van den Broeke, J. T. M. Lenaerts, and E. Van Meijgaard, 2012: Insignificant change in Antarctic snowmelt volume since 1979. *Geophys. Res. Lett.*, **39**, (L01501).

Kunkel, K. E., M. A. Palecki, K. G. Hubbard, D. A. Robinson, K. T. Redmond, and D. R. Easterling, 2007: Trend identification in twentieth-century US snowfall: The challenges. *J. Atmos. Ocean. Technol.*, **24**, 64–73.

Kurtz, N. T., and T. Markus, 2012: Satellite observations of Antarctic sea ice thickness and volume. *J. Geophys. Res. Oceans*, **117**, C08025

Kutuzov, S., and M. Shahgedanova, 2009: Glacier retreat and climatic variability in the eastern Terskey-Alatoo, inner Tien Shan between the middle of the 19th century and beginning of the 21st century. *Global Planet. Change*, **69**, 59–70.

Kwok, R., 2004: Annual cycles of multiyear sea ice coverage of the Arctic Ocean: 1999–2003. *J. Geophys. Res.-Oceans*, **109**, C11004.

Kwok, R., 2005: Variability of Nares Strait ice flux. *Geophys. Res. Lett.*, **32**, L24502.

Kwok, R., 2007: Near zero replenishment of the Arctic multiyear sea ice cover at the end of 2005 summer. *Geophys. Res. Lett.*, **34**, L05501.

Kwok, R., 2009: Outflow of Arctic Ocean Sea Ice into the Greenland and Barents Seas: 1979–2007. *J. Clim.*, **22**, 2438–2457.

Kwok, R., and D. A. Rothrock, 1999: Variability of Fram Strait ice flux and North Atlantic Oscillation. *J. Geophys. Res. Oceans*, **104**, 5177–5189.

Kwok, R., and D. A. Rothrock, 2009: Decline in Arctic sea ice thickness from submarine and ICESat records: 1958–2008. *Geophys. Res. Lett.*, **36**, L15501.

Kwok, R., and G. F. Cunningham, 2010: Contribution of melt in the Beaufort Sea to the decline in Arctic multiyear sea ice coverage: 1993–2009. *Geophys. Res. Lett.*, **37**, L20501.

Kwok, R., G. F. Cunningham, M. Wensnahan, I. Rigor, H. J. Zwally, and D. Yi, 2009: Thinning and volume loss of the Arctic Ocean sea ice cover: 2003–2008. *J. Geophys. Res. Oceans*, **114**, C07005.

Latifovic, R., and D. Pouliot, 2007: Analysis of climate change impacts on lake ice phenology in Canada using the historical satellite data record. *Remote Sens. Environ.*, **106**, 492–507.

Laxon, S., N. Peacock, and D. Smith, 2003: High interannual variability of sea ice thickness in the Arctic region. *Nature*, **425**, 947–950.

Laxon, S. W., et al., 2013: CryoSat-2 estimates of Arctic sea ice thickness and volume. *Geophys. Res. Lett.*, **40**, 732–737.

Leclercq, P. W., and J. Oerlemans, 2012: Global and hemispheric temperature reconstruction from glacier length fluctuations. *Clim. Dyn.*, **38**, 1065–1079.

Leclercq, P. W., J. Oerlemans, and J. G. Cogley, 2011: Estimating the glacier contribution to sea-level rise for the period 1800–2005. *Surv. Geophys.*, **32**, 519–535.

Leclercq, P. W., A. Weidick, F. Paul, T. Bolch, M. Citterio, and Oerlemans.J, 2012: Brief communication—Historical glacier length changes in West Greenland. *Cryosphere*, **6**, 1339–1343.

Lemke, P., et al., 2007: Observations: Changes in snow, ice and frozen ground. In: *Climate Change 2007: The Physical Science Basis. Contribution of Working Group I to the Fourth Assessment Report of the Intergovernmental Panel on Climate Change* [Solomon, S., D. Qin, M. Manning, Z. Chen, M. Marquis, K. B. Averyt, M. Tignor and H. L. Miller (eds.)] Cambridge University Press, Cambridge, United Kingdom and New York, NY, USA, pp. 337–383.

Lenaerts, J. T. M., M. R. van den Broeke, W. J. van de Berg, E. van Meijgaard, and P. Kuipers Munneke, 2012: A new, high resolution surface mass balance map of Antarctica (1979–2010) based on regional climate modeling. *Geophys. Res. Lett.*, **39**, 1–5 (L04501).

Levermann, A., T. Albrecht, R. Winkelmann, M. A. Martin, M. Haseloff, and I. Joughin, 2012: Kinematic first-order calving law implies potential for abrupt ice-shelf retreat. *Cryosphere*, **6**, 273–286.

Lewkowicz, A. G., B. Etzelmuller, and S. L. Smith, 2011: Characteristics of discontinuous permafrost based on ground temperature measurements and electrical resistivity tomography, Southern Yukon, Canada. *Permafr. Periglac. Process.*, **22**, 320–342.

Li, J., and H. J. Zwally, 2011: Modeling of firn compaction for estimating ice-sheet mass change from observed ice-sheet elevation change. *Ann. Glaciol.*, **52**, 1–7.

Li, R., L. Zhao, and Y. Ding, 2009: The climatic characteristics of the maximum seasonal frozen depth in the Tibetan plateau. *J. Glaciol. Geocryol.*, **31**, 1050–1056.

Li, R., et al., 2012a: Temporal and spatial variations of the active layer along the Qinghai-Tibet Highway in a permafrost region. *Chin. Sci. Bull.*, **57**, 4609–4616.

Li, X., R. Jin, X. D. Pan, T. J. Zhang, and J. W. Guo, 2012b: Changes in the near-surface soil freeze-thaw cycle on the Qinghai-Tibetan Plateau. *Int. J. Appl. Earth Obs. Geoinf.*, **17**, 33–42.

Li, X., et al., 2008: Cryospheric change in China. *Global Planet. Change*, **62**, 210–218.

Ling, F., and T. Zhang, 2003: Numerical simulation of permafrost thermal regime and talik development under shallow thaw lakes on the Alaskan Arctic Coastal Plain. *J. Geophys. Res. Atmos.*, **108**, 11.

Liu, L., T. Zhang, and J. Wahr, 2010: InSAR measurements of surface deformation over permafrost on the North Slope of Alaska. *J. Geophys. Res. Earth Surf.*, **115**, F03023

Lopez, P., P. Chevallier, V. Favier, B. Pouyaud, F. Ordenes, and J. Oerlemans, 2010: A regional view of fluctuations in glacier length in southern South America. *Global Planet. Change*, **71**, 85–108.

Lopez-Moreno, J. I., and S. M. Vicente-Serrano, 2007: Atmospheric circulation influence on the interannual variability of snow pack in the Spanish Pyrenees during the second half of the 20th century. *Nordic Hydrol.*, **38**, 33–44.

Luckman, A., and T. Murray, 2005: Seasonal variation in velocity before retreat of Jakobshavn Isbrae, Greenland. *Geophys. Res. Lett.*, **32**, 4.

Luethi, M. P., A. Bauder, and M. Funk, 2010: Volume change reconstruction of Swiss glaciers from length change data. *J. Geophys. Res. Earth Surf.*, **115**, F04022.

Luthcke, S. B., A. A. Arendt, D. D. Rowlands, J. J. McCarthy, and C. F. Larsen, 2008: Recent glacier mass changes in the Gulf of Alaska region from GRACE mascon solutions. *J. Glaciol.*, **54**, 767–777.

Luthcke, S. B., et al., 2006: Recent Greenland ice mass loss by drainage system from satellite gravity observations. *Science*, **314**, 1286–1289.

Ma, L., and D. Qin, 2012: Temporal-spatial characteristics of observed key parameters of snow cover in China during 1957–2009. *Sci. Cold Arid Reg.*, **4(5)**, 384–393.

MacAyeal, D. R., et al., 2006: Transoceanic wave propagation links iceberg calving margins of Antarctica with storms in tropics and Northern Hemisphere. *Geophys. Res. Lett.*, **33**, 4 (L17502).

MacGregor, J. A., G. A. Catania, M. S. Markowski, and A. G. Andrews, 2012: Widespread rifting and retreat of ice-shelf margins in the eastern Amundsen Sea Embayment between 1972 and 2011. *J. Glaciol.*, **58**, 458–466.

Machguth, H., F. Paul, S. Kotlarski, and M. Hoelzle, 2009: Calculating distributed glacier mass balance for the Swiss Alps from regional climate model output: A methodical description and interpretation of the results. *J. Geophys. Res. Atmos.*, **114**, D19106.

Mahoney, A., H. Eicken, and L. Shapiro, 2007: How fast is landfast sea ice? A study of the attachment and detachment of nearshore ice at Barrow, Alaska. *Cold Reg. Sci. Technol.*, **47**, 233–255.

Macias Fauria, M., et al., 2010: Unprecedented low twentieth century winter sea ice extent in the Western Nordic Seas since AD 1200. *Clim. Dyn.*, **34**, 781–795.

Malkova, G. V., 2008: The last twenty-five years of changes in permafrost temperature of the European Russian Arctic. In: *Proceedings of the 9th International Conference on Permafrost, 29 June– 3 July 2008, Institute of Northern Engineering, University of Alaska, Fairbanks* [D. L. Kane, and K. M. Hinkel (eds.)], pp. 1119–1124.

Marchenko, S. S., A. P. Gorbunov, and V. E. Romanovsky, 2007: Permafrost warming in the Tien Shan Mountains, Central Asia. *Global Planet. Change*, **56**, 311–327.

Markus, T., and D. J. Cavalieri, 2000: An enhancement of the NASA Team sea ice algorithm. *IEEE Trans. Geosci. Remote Sens.*, **38**, 1387–1398.

Markus, T., J. C. Stroeve, and J. Miller, 2009: Recent changes in Arctic sea ice melt onset, freezeup, and melt season length. *J. Geophys. Res. Oceans*, **114**, C12024.

Marshall, G. J., A. Orr, N. P. M. van Lipzig, and J. C. King, 2006: The impact of a changing Southern Hemisphere Annular Mode on Antarctic Peninsula summer temperatures. *J. Clim.*, **19**, 5388–5404.

Martinson, D. G., S. E. Stammerjohn, R. A. Iannuzzi, R. C. Smith, and M. Vernet, 2008: Western Antarctic Peninsula physical oceanography and spatio-temporal variability. *Deep-Sea Res. Pt. Ii*, **55**, 1964–1987.

Marty, C., 2008: Regime shift of snow days in Switzerland. *Geophys. Res. Lett.*, **35**, L12501.

Marty, C., and R. Meister, 2012: Long-term snow and weather observations at Weissfluhjoch and its relation to other high-altitude observatories in the Alps. *Theor. Appl. Climatol.*, **110**, 573–583.

Marushchak, M. E., A. Pitkamaki, H. Koponen, C. Biasi, M. Seppala, and P. J. Martikainen, 2011: Hot spots for nitrous oxide emissions found in different types of permafrost peatlands. *Global Change Biol.*, **17**, 2601–2614.

Marzeion, B., A. H. Jarosch, and M. Hofer, 2012: Past and future sea-level change from the surface mass balance of glaciers. *Cryosphere*, **6**, 1295–1322.

Masiokas, M. H., R. Villalba, B. H. Luckman, and S. Mauget, 2010: Intra- to multidecadal variations of snowpack and streamflow records in the Andes of Chile and Argentina between 30 degrees and 37 degrees S. *J. Hydrometeorol.*, **11**, 822–831.

Masiokas, M. H., A. Rivera, L. E. Espizua, R. Villalba, S. Delgado, and J. C. Aravena, 2009: Glacier fluctuations in extratropical South America during the past 1000 years. *Palaeogeogr. Palaeoclimatol. Palaeoecol.*, **281**, 242–268.

Maslanik, J. A., C. Fowler, J. Stroeve, S. Drobot, J. Zwally, D. Yi, and W. Emery, 2007: A younger, thinner Arctic ice cover: Increased potential for rapid, extensive sea-ice loss. *Geophys. Res. Lett.*, **34**, L24501.

Massom, R. A., and S. Stammerjohn, 2010: Antarctic sea ice change and variability—Physical and ecological implications. *Polar Sci.*, 149–186.

Massom, R. A., P. Reid, B. Raymond, S. Stammerjohn, A. D. Fraser, and S. Ushio, 2013: Change and variability in East Antarctic Sea Ice Seasonality, 1979/80–2009/10. *PLoS ONE*, **8**, e64756.

Massom, R. A., et al., 2001: Snow on Antarctic Sea ice: A review of physical characteristics. *Rev. Geophys.*, **39**, 413–445.

Matsuo, K., and K. Heki, 2010: Time-variable ice loss in Asian high mountains from satellite gravimetry. *Earth Planet. Sci. Lett.*, **290**, 30–36.

Mazhitova, G. G., 2008: Soil temperature regimes in the discontinuous permafrost zone in the east European Russian Arctic. *Euras. Soil Sci.*, **41**, 48–62.

Mazhitova, G. G., and D. A. Kaverin, 2007: Thaw depth dynamics and soil surface subsidence at a Circumpolar Active Layer Monitoring (CALM) site in the East European Russian Arctic. *Kriosfera Zemli*, **XI**, **N**, 20–30.

McGuire, A. D., et al., 2009: Sensitivity of the carbon cycle in the Arctic to climate change. *Ecol. Monogr.*, **79**, 523–555.

Meese, D. A., et al., 1994: The accumulation record from the Gisp2 Core as and indicator of climate change throughout the holocene. *Science*, **266**, 1680–1682.

Meier, M. F., 1984: Contribution of small glaciers to global sea level. *Science*, **226**, 1418–1421.

Meier, W. N., J. Stroeve, A. Barrett, and F. Fetterer, 2012: A simple approach to providing a more consistent Arctic sea ice extent time series from the 1950s to present. *Cryosphere*, **6**, 1359–1368.

Melling, H., 2012: Sea-Ice Observation: Advances and challenges. In: *Arctic Climate Change: The ACSYS Decade and Beyond* [P. Lemke and H.-W. Jacobi (eds.)]. Atmospheric and Oceanographic Sciences Library. Springer Science, New York, NY, USA, and Heidelberg, Germany. 27-115.

Mercer, J. H., 1978: West Antarctic ice sheet and CO2 greenhouse effect— threat of disaster. *Nature*, **271**, 321–325.

Micu, D., 2009: Snow pack in the Romanian Carpathians under changing climatic conditions. *Meteorol. Atmos. Phys.*, **105**, 1–16.

Mitchell, T. D., and P. D. Jones, 2005: An improved method of constructing a database of monthly climate observations and associated high-resolution grids. *Int. J. Climatol.*, **25**, 693–712.

Moholdt, G., B. Wouters, and A. S. Gardner, 2012: Recent contribution to sea-level rise from glaciers and ice caps in the Russian High Arctic. *Geophys. Res. Lett.*, **39**, L10502.

Moholdt, G., C. Nuth, J. O. Hagen, and J. Kohler, 2010: Recent elevation changes of Svalbard glaciers derived from ICESat laser altimetry. *Remote Sens. Environ.*, **114**, 2756–2767.

Monaghan, A. J., D. H. Bromwich, and S. H. Wang, 2006: Recent trends in Antarctic snow accumulation from Polar MM5 simulations. *Philos. Trans. R. Soc. A*, **364**, 1683–1708.

Montes-Hugo, M., S. C. Doney, H. W. Ducklow, W. Fraser, D. Martinson, S. E. Stammerjohn, and O. Schofield, 2009: Recent changes in phytoplankton communities associated with rapid regional climate change along the western Antarctic Peninsula. *Science*, **323**, 1470–1473.

Moon, T., and I. Joughin, 2008: Changes in ice front position on Greenland's outlet glaciers from 1992 to 2007. *J. Geophys. Res. Earth Surf.*, **113**, F02022.

Moon, T., I. Joughin, B. Smith, and I. Howat, 2012: 21st-Century evolution of Greenland outlet glacier velocities. *Science*, **336**, 576–578.

Moore, P., and M. A. King, 2008: Antarctic ice mass balance estimates from GRACE: Tidal aliasing effects. *J. Geophys. Res. Earth Surf.*, **113**, F02005.

Morris, E. M., and D. G. Vaughan, 2003: Spatial and temporal variation of surface temperature on the Antarctic Peninsula and the limit of viability of ice shelves. In: *Antarctic Peninsula Climate Variability: Historical and Paleoenvironmental Perspectives* [E. Domack, A. Leventer, A. Burnett, R. Bindschadler, P. Convey, and M. Kirby (eds.)]. Antarctic Research Series, 79, American Geophysical Union, Washington, DC, pp. 61–68.

Mote, P. W., 2006: Climate-driven variability and trends in mountain snowpack in western North America. *J. Clim.*, **19**, 6209–6220.

Motyka, R. J., L. Hunter, K. A. Echelmeyer, and C. Connor, 2003: Submarine melting at the terminus of a temperate tidewater glacier, LeConte Glacier, Alaska, USA. *Ann. Glaciol.*, **36**, 57–65.

Motyka, R. J., M. Truffer, M. Fahnestock, J. Mortensen, S. Rysgaard, and I. Howat, 2011: Submarine melting of the 1985 Jakobshavn Isbrae floating tongue and the triggering of the current retreat. *J. Geophys. Res. Earth Surf.*, **116**, F01007.

Murray, T., T. Strozzi, A. Luckman, H. Jiskoot, and P. Christakos, 2003: Is there a single surge mechanism? Contrasts in dynamics between glacier surges in Svalbard and other regions. *J. Geophys. Res. Sol. Ea.*, **108**, 2237.

Murray, T., et al., 2010: Ocean regulation hypothesis for glacier dynamics in southeast Greenland and implications for ice sheet mass changes. *J. Geophys. Res. Earth Surf.*, **115**, F03026.

Myers, P. G., C. Donnelly, and M. H. Ribergaard, 2009: Structure and variability of the West Greenland Current in Summer derived from 6 repeat standard sections. *Prog. Oceanogr.*, **80**, 93–112.

National Snow and Ice Data Center, 2013: http://nsidc.org/data/icesat/correction-to-product-surface-elevations.html.

Nerem, R. S., and J. Wahr, 2011: Recent changes in the Earth's oblateness driven by Greenland and Antarctic ice mass loss. *Geophys. Res. Lett.*, **38**, 6 (L13501).

Nesje, A., O. Lie, and S. O. Dahl, 2000: Is the North Atlantic Oscillation reflected in Scandinavian glacier mass balance records? *J. Quat. Sci.*, **15**, 587–601.

Nghiem, S. V., I. G. Rigor, D. K. Perovich, P. Clemente-Colon, J. W. Weatherly, and G. Neumann, 2007: Rapid reduction of Arctic perennial sea ice. *Geophys. Res. Lett.*, **34**, 6 (L19504).

Nghiem, S. V., et al., 2012: The extreme melt across the Greenland ice sheet in 2012. *Geohys. Res. Lett.*, **39**, L20502.

Nicholls, K. W., K. Makinson, and E. J. Venables, 2012: Ocean circulation beneath Larsen C Ice Shelf, Antarctica from in situ observations. *Geophys. Res. Lett.*, **39**, L19608.

Nicholls, N., 2005: Climate variability, climate change and the Australian snow season. *Aust. Meteorol. Mag.*, **54**, 177–185.

Nick, F. M., A. Vieli, I. M. Howat, and I. Joughin, 2009: Large-scale changes in Greenland outlet glacier dynamics triggered at the terminus. *Nature Geosci.*, **2**, 110–114.

Nick, F. M., C. J. van der Veen, A. Vieli, and D. Benn, 2010: A physically based calving model applied to marine outlet glaciers and implications for their dynamics. *J. Glaciol.*, **56**, 781–794.

Nick, F. M., et al., 2013: Future sea level rise from Greenland's major outlet glaciers in a warming climate. *Nature*, **497**, 235–238.

Noetzli, J., and D. Vonder Muehll, 2010: Permafrost in Switzerland 2006/2007 and 2007/2008. Glaciological Report (Permafrost) No. 8/9 of the Cryospheric Commission of the Swiss Academy of Sciences. Cryospheric Commission of the Swiss Academy of Sciences, 68 pp.

Nussbaumer, S. U., A. Nesje, and H. J. Zumbuhl, 2011: Historical glacier fluctuations of Jostedalsbreen and Folgefonna (southern Norway) reassessed by new pictorial and written evidence. *Holocene*, **21**, 455–471.

Nuth, C., G. Moholdt, J. Kohler, J. O. Hagen, and A. Kaab, 2010: Svalbard glacier elevation changes and contribution to sea level rise. *J. Geophys. Res. Earth Surf.*, **115**, 16 (F01008).

Oberman, N. G., 2008: Contemporary permafrost degradation of Northern European Russia. In: *Proceedings of the 9th International Conference on Permafrost, 29 June– 3 July 2008, Institute of Northern Engineering, University of Alaska, Fairbanks* [D. L. Kane, and K. M. Hinkel (eds.)], pp. 1305–1310.

Oberman, N. G., 2012: Long-term temperature regime of the Northeast European permafrost region during contemporary climate warming. In: *Proceedings of the 10th International Conference on Permafrost, June, 2012, Salekhard, Yamel-Nenets Autonomous District, Russian Federation, v2.* [V. P. Melnikov, D. S. Drozdov and V. E. Romanovsky (eds)]. The Northern Publisher, Salekhard, Russia. pp. 287–291.

Oerlemans, J., 2001: *Glaciers and Climate Change.* A. A. Balkema, Lisse, the Netherlands, 160 pp.

Oerlemans, J., M. Dyurgerov, and R. De Wal, 2007: Reconstructing the glacier contribution to sea-level rise back to 1850. *Cryosphere*, **1**, 59–65.

O'Leary, M., and P. Christoffersen, 2013: Calving on tidewater glaciers amplified by submarine frontal melting. *Cryosphere*, **7**, 119–128.

Osterkamp, T.E., 2005: The recent warming of permafrost in Alaska. *Global Planet. Change*, **49**, 187–202.

Osterkamp, T. E., 2007: Characteristics of the recent warming of permafrost in Alaska. *J. Geophys. Res. Earth Surf.*, **112**, 10 (F02S02).

Osterkamp, T. E., 2008: Thermal state of permafrost in Alaska during the fourth quarter of the twentieth century. In: *Proceedings of the 9th International Conference on Permafrost, 29 June– 3 July 2008, Institute of Northern* Engineering, *University of Alaska, Fairbanks, Alaska* [D. L. Kane, and K. M. Hinkel (eds.)], pp. 1333–1338.

Overduin, P. P., and D. L. Kane, 2006: Frost boils and soil ice content: Field observations. *Permafr. Periglac. Process.*, **17**, 291–307.

Overduin, P. P., H.-W. Hubberten, V. Rachold, N. Romanovskii, M. N. Grigoriev, and M. Kasymskaya, 2007: Evolution and degradation of coastal and offshore permafrost in the Laptev and East Siberian Seas during the last climatic cycle. *GSA Special Papers*, **426**, 97–111.

Overduin, P. P., S. Westermann, K. Yoshikawa, T. Haberlau, V. Romanovsky, and S. Wetterich, 2012: Geoelectric observations of the degradation of nearshore submarine permafrost at Barrow (Alaskan Beaufort Sea). *J. Geophys. Res. Earth Surf.*, **117**, F02004.

Padman, L., et al., 2012: Oceanic controls on the mass balance of Wilkins Ice Shelf, Antarctica. *J. Geophys. Res. Oceans*, **117**, C01010

Palmer, S., A. Shepherd, P. Nienow, and I. Joughin, 2011: Seasonal speedup of the Greenland Ice Sheet linked to routing of surface water. *Earth Planet. Sci. Lett.*, **302**, 423–428.

Parkinson, C. L., 2002: Trends in the length of the Southern Ocean sea-ice season, 1979–99. *Ann. Glaciol.*, **34**, 435–440.

Parkinson, C. L., and D. J. Cavalieri, 2012: Antarctic sea ice variability and trends, 1979–2010. *Cryosphere*, **6**, 871–880.

Parkinson, C. L., and J. C. Comiso, 2013: On the 2012 record low Arctic sea ice cover: Combined impact of preconditioning and an August storm. *Geophys. Res. Lett.*, **40**, 1356–1361.

Paul, F., and W. Haeberli, 2008: Spatial variability of glacier elevation changes in the Swiss Alps obtained from two digital elevation models. *Geophys. Res. Lett.*, **35**, 5 (L21502).

Paulson, A., S. J. Zhong, and J. Wahr, 2007: Inference of mantle viscosity from GRACE and relative sea level data. *Geophys. J. Int.*, **171**, 497–508.

Payne, A. J., A. Vieli, A. P. Shepherd, D. J. Wingham, and E. Rignot, 2004: Recent dramatic thinning of largest West Antarctic ice stream triggered by oceans. *Geophys. Res. Lett.*, **31**, L23401.

Peltier, W. R., 2004: Global glacial isostasy and the surface of the ice-age earth: The ice-5G (VM2) model and grace. *Annu. Rev. Earth Planet. Sci.*, **32**, 111–149.

Peltier, W. R., 2009: Closure of the budget of global sea level rise over the GRACE era: The importance and magnitudes of the required corrections for global glacial isostatic adjustment. *Quat. Sci. Rev.*, **28**, 1658–1674.

Pelto, M. S., 2006: The current disequilibrium of North Cascade glaciers. *Hydrol. Process.*, **20**, 769–779.

Perovich, D. K., B. Light, H. Eicken, K. F. Jones, K. Runciman, and S. V. Nghiem, 2007: Increasing solar heating of the Arctic Ocean and adjacent seas, 1979–2005: Attribution and role in the ice-albedo feedback. *Geophys. Res. Lett.*, **34**, L19505.

Petkova, N., E. Koleva, and V. Alexandrov, 2004: Snow cover variability and change in mountainous regions of Bulgaria, 1931–2000. *Meteorol. Z.*, **13**, 19–23.

Pfeffer, W. T., 2007: A simple mechanism for irreversible tidewater glacier retreat. *J. Geophys. Res. Earth Surf.*, **112**, F03S25.

Pfeffer, W. T., 2011: Land ice and sea level rise: A thirty-year perspective. *Oceanography*, **24**, 94–111.

Phillips, T., H. Rajaram, and K. Steffen, 2010: Cryo-hydrologic warming: A potential mechanism for rapid thermal response of ice sheets. *Geophys. Res. Lett.*, **37**, L20503.

Pollard, D., and R. M. DeConto, 2009: Modelling West Antarctic ice sheet growth and collapse through the past five million years. *Nature*, **458**, 329–333.

Polyakov, I. V., et al., 2003: Long-term ice variability in Arctic marginal seas. *J. Clim.*, **16**, 2078–2085.

Polyakov, I. V., et al., 2010: Arctic Ocean warming contributes to reduced polar ice cap. *J. Phys. Oceanogr.*, **40**, 2743–2756

Pritchard, H. D., and D. G. Vaughan, 2007: Widespread acceleration of tidewater glaciers on the Antarctic Peninsula. *J. Geophys. Res. Earth Surf.*, **112**, F03S29.

Pritchard, H. D., S. B. Luthcke, and A. H. Fleming, 2010: Understanding ice-sheet mass balance: Progress in satellite altimetry and gravimetry. *J. Glaciol.*, **56**, 1151–1161.

Pritchard, H. D., R. J. Arthern, D. G. Vaughan, and L. A. Edwards, 2009: Extensive dynamic thinning on the margins of the Greenland and Antarctic ice sheets. *Nature*, **461**, 971–975.

Pritchard, H. D., S. R. M. Ligtenberg, H. A. Fricker, D. G. Vaughan, M. R. van den Broeke, and L. Padman, 2012: Antarctic ice loss driven by ice-shelf melt. *Nature*, **484**, 502–505.

Prowse, T., et al., 2011: Arctic freshwater ice and its climatic role. *Ambio*, **40**, 46–52.

Quincey, D. J., M. Braun, N. F. Glasser, M. P. Bishop, K. Hewitt, and A. Luckman, 2011: Karakoram glacier surge dynamics. *Geophys. Res. Lett.*, **38**, L18504.

Rabatel, A., J. P. Dedieu, and C. Vincent, 2005: Using remote-sensing data to determine equilibrium-line altitude and mass-balance time series: validation on three French glaciers, 1994–2002. *J. Glaciol.*, **51**, 539–546.

Rabatel, A., B. Francou, V. Jomelli, P. Naveau, and D. Grancher, 2008: A chronology of the Little Ice Age in the tropical Andes of Bolivia (16 degrees S) and its implications for climate reconstruction. *Q. Res.*, **70**, 198–212.

Rabatel, A., et al., 2013: Current state of glaciers in the tropical Andes: A multi-century perspective on glacier evolution and climate change. *Cryosphere*, **7**, 81–102.

Rachold, V., et al., 2007: Near-shore Arctic subsea permafrost in transition. *EOS Trans. Am. Geophys. Union*, **88**, 149–156.

Radić, V., and R. Hock, 2010: Regional and global volumes of glaciers derived from statistical upscaling of glacier inventory data. *J. Geophys. Res. Earth Surf.*, **115**, F01010.

Radić, V., A. Bliss, A. C. Beedlow, R. Hock, E. Miles, and J. G. Cogley, 2013: Regional and global projections of 21st century glacier mass changes in response to climate scenarios from global climate models. *Clim. Dyn.*, doi:10.1007/s00382-013-1719-7.

Ramillien, G., A. Lombard, A. Cazenave, E. R. Ivins, M. Llubes, F. Remy, and R. Biancale, 2006: Interannual variations of the mass balance of the Antarctica and Greenland ice sheets from GRACE. *Global Planet. Change*, **53**, 198–208.

Ramirez, E., et al., 2001: Small glaciers disappearing in the tropical Andes: a case-study in Bolivia: Glaciar Chacaltaya (16 degrees S). *J. Glaciol.*, **47**, 187–194.

Rampal, P., J. Weiss, and D. Marsan, 2009: Positive trend in the mean speed and deformation rate of Arctic sea ice, 1979–2007. *J. Geophys. Res. Oceans*, **114**, C05013.

Rastner, P., T. Bolch, N. Mölg, H. Machguth, and F. Paul, 2012: The first complete glacier inventory for entire Greenland. *Cryosphere*, **6**, 1483–1495.

Ravanel, L., F. Allignol, P. Deline, S. Gruber, and M. Ravello, 2010: Rock falls in the Mont Blanc Massif in 2007 and 2008. *Landslides*, **7**, 493–501.

Rayner, N. A., et al., 2003: Global analyses of SST, sea ice and night marine air temperature since the late nineteenth century. *J. Geophys. Res.*, **108**, 4407.

Repo, M. E., et al., 2009: Large N2O emissions from cryoturbated peat soil in tundra. *Nature Geosci.*, **2**, 189–192.

Ridley, J., J. M. Gregory, P. Huybrechts, and J. Lowe, 2010: Thresholds for irreversible decline of the Greenland ice sheet. *Clim. Dyn.*, **35**, 1065–1073.

Rignot, E., 2008: Changes in West Antarctic ice stream dynamics observed with ALOS PALSAR data. *Geophys. Res. Lett.*, **35**, L12505.

Rignot, E., and S. S. Jacobs, 2002: Rapid bottom melting widespread near Antarctic ice sheet grounding lines. *Science*, **296**, 2020–2023.

Rignot, E., and R. H. Thomas, 2002: Mass balance of polar ice sheets. *Science*, **297**, 1502–1506.

Rignot, E., and P. Kanagaratnam, 2006: Changes in the velocity structure of the Greenland ice sheet. *Science*, **311**, 986–990.

Rignot, E., and J. Mouginot, 2012: Ice flow in Greenland for the International Polar Year 2008–2009. *Geophys. Res. Lett.*, **39**, L11501.

Rignot, E., A. Rivera, and G. Casassa, 2003: Contribution of the Patagonia Icefields of South America to sea level rise. *Science*, **302**, 434–437.

Rignot, E., M. Koppes, and I. Velicogna, 2010: Rapid submarine melting of the calving faces of West Greenland glaciers. *Nature Geosci.*, **3**, 187–191.

Rignot, E., J. Mouginot, and B. Scheuchl, 2011a: Ice flow of the Antarctic ice sheet. *Science*, **333**, 1427–1430.

Rignot, E., J. Mouginot, and B. Scheuchl, 2011b: Antarctic grounding line mapping from differential satellite radar interferometry. *Geophys. Res. Lett.*, **38**, L10504.

Rignot, E., J. E. Box, E. Burgess, and E. Hanna, 2008a: Mass balance of the Greenland ice sheet from 1958 to 2007. *Geophys. Res. Lett.*, **35**, L20502.

Rignot, E., I. Velicogna, M. R. van den Broeke, A. Monaghan, and J. Lenaerts, 2011c: Acceleration of the contribution of the Greenland and Antarctic ice sheets to sea level rise. *Geophys. Res. Lett.*, **38**, 5 (L05503).

Rignot, E., G. Casassa, P. Gogineni, W. Krabill, A. Rivera, and R. Thomas, 2004: Accelerated ice discharge from the Antarctic Peninsula following the collapse of Larsen B ice shelf. *Geophys. Res. Lett.*, **31**, 4 (L18401).

Rignot, E., J. L. Bamber, M. R. van den Broeke, C. Davis, Y. H. Li, W. J. van de Berg, and E. Van Meijgaard, 2008b: Recent Antarctic ice mass loss from radar interferometry and regional climate modelling. *Nature Geosci.*, **1**, 106–110.

Riva, R. E. M., et al., 2009: Glacial Isostatic Adjustment over Antarctica from combined ICESat and GRACE satellite data. *Earth Planet. Sci. Lett.*, **288**, 516–523.

Rivera, A., F. Bown, D. Carrion, and P. Zenteno, 2012: Glacier responses to recent volcanic activity in Southern Chile. *Environ. Res. Lett.*, **7**, 014036.

Robinson, D. A., K. F. Dewey, and R. R. Heim, 1993: Global snow cover monitoring—An update. *Bull. Am. Meteorol. Soc.*, **74**, 1689–1696.

Roer, I., W. Haeberli, M. Avian, V. Kaufmann, R. Delaloye, C. Lambiel, and A. Kääb, 2008: Observations and considerations on destabilizing active rock glaciers in the European Alps. In: *Proceedings of the 9th International Conference on Permafrost, 29 June– 3 July 2008, Institute of Northern Engineering, University of Alaska, Fairbanks* [D. L. Kane, and K. M. Hinkel (eds.)], pp. 1505–1510.

Romanovsky, V. E., S. L. Smith, and H. H. Christiansen, 2010a: Permafrost thermal state in the polar Northern Hemisphere during the International Polar Year 2007–2009: A Synthesis. *Permafr. Periglac. Process.*, **21**, 106–116.

Romanovsky, V. E., et al., 2010b: Thermal state of permafrost in Russia. *Permafr. Periglac. Process.*, **21**, 136–155.

Rosenau, R., E. Schwalbe, H.-G. Maas, M. Baessler, and R. Dietrich, 2013: Grounding line migration and high resolution calving dynamics of Jakobshavn Isbræ, West Greenland. *J. Geophys. Res.*, **118**, 382-395.

Ross, N., et al., 2012: Steep reverse bed slope at the grounding line of the Weddell Sea sector in West Antarctica. *Nature Geosci*, **5**, 393–396.

Rothrock, D. A., and M. Wensnahan, 2007: The accuracy of sea ice drafts measured from US Navy submarines. *J. Atmos. Ocean. Technol.*, **24**, 1936–1949.

Rothrock, D. A., Y. Yu, and G. A. Maykut, 1999: Thinning of the Arctic sea-ice cover. *Geophys. Res. Lett.*, **26**, 3469–3472.

Rothrock, D. A., D. B. Percival, and M. Wensnahan, 2008: The decline in arctic sea-ice thickness: Separating the spatial, annual, and interannual variability in a quarter century of submarine data. *J. Geophys. Res. Oceans*, **113**, C05003.

Rott, H., F. Muller, T. Nagler, and D. Floricioiu, 2011: The imbalance of glaciers after disintegration of Larsen-B ice shelf, Antarctic Peninsula. *Cryosphere*, **5**, 125–134.

Sanchez-Bayo, F., and K. Green, 2013: Australian snowpack disappearing under the influence of global warming and solare activity. *Arct., Antarct. Alp. Res.*, **45**, 107–118.

Sannel, A. B. K., and P. Kuhry, 2011: Warming-induced destabilization of peat plateau/thermokarst lake complexes. *J. Geophys. Res. Biogeosci.*, **116**, 16 (G03035).

Sasgen, I., et al., 2012: Timing and origin of recent regional ice-mass loss in Greenland. *Earth Planet. Sci. Lett.*, **333**, 293–303.

Scambos, T. A., C. Hulbe, M. Fahnestock, and J. Bohlander, 2000: The link between climate warming and break-up of ice shelves in the Antarctic Peninsula. *J. Glaciol.*, **46**, 516–530.

Scambos, T. A., J. A. Bohlander, C. A. Shuman, and P. Skvarca, 2004: Glacier acceleration and thinning after ice shelf collapse in the Larsen B embayment, Antarctica. *Geophys. Res. Lett.*, **31**, 4.

Schaefer, K., T. J. Zhang, L. Bruhwiler, and A. P. Barrett, 2011: Amount and timing of permafrost carbon release in response to climate warming. *Tellus B*, **63**, 165–180.

Scherler, D., B. Bookhagen, and M. R. Strecker, 2011: Spatially variable response of Himalayan glaciers to climate change affected by debris cover. *Nature Geosci.*, **4**, 156–159

Schiefer, E., B. Menounos, and R. Wheate, 2007: Recent volume loss of British Columbian glaciers, Canada. *Geophys. Res. Lett.*, **34**, 6 (L16503).

Schneider, D. P., C. Deser, and Y. Okumura, 2012: An assessment and interpretation of the observed warming of West Antarctica in the austral spring. *Clim. Dyn.*, **38**, 323–347.

Schoeneich, P., X. Bodin, J. Krysiecki, P. Deline, and L. Ravanel, 2010: Permafrost in France, 1st Report. Institut de Géographie Alpine, Université Joseph Fourier, Grenoble, France. 68 pp.

Schoof, C., 2007: Ice sheet grounding line dynamics: Steady states, stability, and hysteresis. *J. Geophys. Res. Earth Surf.*, **112**, F03S28.

Schoof, C., 2010: Ice-sheet acceleration driven by melt supply variability. *Nature*, **468**, 803–806.

Schrama, E. J. O., and B. Wouters, 2011: Revisiting Greenland ice sheet mass loss observed by GRACE. *J. Geophys. Res. Sol. Ea.*, **116**, B02407.

Schuur, E. A. G., J. G. Vogel, K. G. Crummer, H. Lee, J. O. Sickman, and T. E. Osterkamp, 2009: The effect of permafrost thaw on old carbon release and net carbon exchange from tundra. *Nature*, **459**, 556–559.

Schweiger, A., R. Lindsay, J. L. Zhang, M. Steele, H. Stern, and R. Kwok, 2011: Uncertainty in modeled Arctic sea ice volume. *J. Geophys. Res. Oceans*, **116**, C00D06.

Screen, J. A., 2011: Sudden increase in Antarctic sea ice: Fact or artifact? *Geophys. Res. Lett.*, **38**, L13702.

Selmes, N., T. Murray, and T. D. James, 2011: Fast draining lakes on the Greenland Ice Sheet. *Geophys. Res. Lett.*, **38**, 5 (L15501).

Shakhova, N., I. Semiletov, A. Salyuk, V. Yusupov, D. Kosmach, and O. Gustafsson, 2010a: Extensive methane venting to the atmosphere from sediments of the East Siberian Arctic Shelf. *Science*, **327**, 1246–1250.

Shakhova, N., I. Semiletov, I. Leifer, A. Salyuk, P. Rekant, and D. Kosmach, 2010b: Geochemical and geophysical evidence of methane release over the East Siberian Arctic Shelf. *J. Geophys. Res. Oceans*, **115**, 14 (C08007).

Shannon, S., et al., 2012: Enhanced basal lubrication and the contribution of the Greenland ice sheet to future sea level rise. *Proc. Natl. Acad. Sci. U.S.A.* **110** (35), 14156-14161.

Sharkhuu, A., et al., 2007: Permafrost monitoring in the Hovsgol mountain region, Mongolia. *J. Geophys. Res. Earth Surf.*, **112**, 11 (F02S06).

Shepherd, A., and D. Wingham, 2007: Recent sea-level contributions of the Antarctic and Greenland ice sheets. *Science*, **315**, 1529–1532.

Shepherd, A., D. Wingham, T. Payne, and P. Skvarca, 2003: Larsen ice shelf has progressively thinned. *Science*, **302**, 856–859.

Shepherd, A., A. Hubbard, P. Nienow, M. King, M. McMillan, and I. Joughin, 2009: Greenland ice sheet motion coupled with daily melting in late summer. *Geophys. Res. Lett.*, **36**, L01501.

Shepherd, A., D. Wingham, D. Wallis, K. Giles, S. Laxon, and A. V. Sundal, 2010: Recent loss of floating ice and the consequent sea level contribution. *Geophys. Res. Lett.*, **37**, 5 (L13503).

Shepherd, A., et al., 2012: A reconciled estimate of ice-sheet mass balance. *Science*, **338**, 1183–1189.

Shi, H. L., Y. Lu, Z. L. Du, L. L. Jia, Z. Z. Zhang, and C. X. Zhou, 2011: Mass change detection in Antarctic ice sheet using ICESat block analysis techniques from 2003 similar to 2008. *Chin. J. Geophys.–Chin. Ed.*, **54**, 958–965.

Shiklomanov, N. I., et al., 2010: Decadal variations of active-layer thickness in moisture-controlled landscapes, Barrow, Alaska. *J. Geophys. Res. Biogeosci.*, **115**, G00I04.

Shuman, C. A., E. Berthier, and T. A. Scambos, 2011: 2001–2009 elevation and mass losses in the Larsen A and B embayments, Antarctic Peninsula. *J. Glaciol.*, **57**, 737–754.

Siegfried, M. R., R. L. Hawley, and J. F. Burkhart, 2011: High-resolution ground-based GPS measurements show intercampaign bias in ICESat Elevation data near Summit, Greenland. *IEEE Trans. Geosci. Remote Sens.*, **49**, 3393–3400.

Simmonds, I., C. Burke, and K. Keay, 2008: Arctic climate change as manifest in cyclone behavior. *J. Clim.*, **21**, 5777–5796.

Simpson, M. J. R., G. A. Milne, P. Huybrechts, and A. J. Long, 2009: Calibrating a glaciological model of the Greenland ice sheet from the Last Glacial Maximum to present-day using field observations of relative sea level and ice extent. *Quat. Sci. Rev.*, **28**, 1631–1657.

Skaugen, T., H. B. Stranden, and T. Saloranta, 2012: Trends in snow water equivalent in Norway (1931–2009). *Hydrol. Res.*, **43**, 489–499.

Slobbe, D. C., P. Ditmar, and R. C. Lindenbergh, 2009: Estimating the rates of mass change, ice volume change and snow volume change in Greenland from ICESat and GRACE data. *Geophys. J. Int.*, **176**, 95–106.

Smith, D. M., 1998: Recent increase in the length of the melt season of perennial Arctic sea ice. *Geophys. Res. Lett.*, **25**, 655–658.

Smith, S. L., J. Throop, and A. G. Lewkowicz, 2012: Recent changes in climate and permafrost temperatures at forested and polar desert sites in northern Canada. *Can. J. Earth Sci.*, **49**, 914–924.

Smith, S. L., S. A. Wolfe, D. W. Riseborough, and F. M. Nixon, 2009: Active-layer characteristics and summer climatic indices, Mackenzie Valley, Northwest Territories, Canada. *Permafr. Periglac. Process.*, **20**, 201–220.

Smith, S. L., et al., 2010: Thermal state of permafrost in North America: A contribution to the International Polar Year. *Permafr. Periglac. Process.*, **21**, 117–135.

Sole, A. J., D. W. F. Mair, P. W. Nienow, I. D. Bartholomew, M. A. King, M. J. Burke, and I. Joughin, 2011: Seasonal speedup of a Greenland marine-terminating outlet glacier forced by surface melt-induced changes in subglacial hydrology. *J. Geophys. Res. Earth Surf.*, **116**, 11 (F03014)

Solomon, S. M., A. E. Taylor, and C. W. Stevens, 2008: Nearshore ground temperatures, seasonal ice bonding and permafrost formation within the bottom-fast ice zone, Mackenzie Delta, NWT. In: *Proceedings of the 9th International Conference of Permafrost, 29 June–3 July 2008, Institute of Northern Engineering, University of Alaska, Fairbanks* [D. L. Kane, and K. M. Hinkel (eds.)], pp. 1675–1680.

Sorensen, L. S., et al., 2011: Mass balance of the Greenland ice sheet (2003–2008) from ICESat data: The impact of interpolation, sampling and firn density. *Cryosphere*, **5**, 173–186.

Spreen, G., R. Kwok, and D. Menemenlis, 2011: Trends in Arctic sea ice drift and role of wind forcing: 1992–2009. *Geophys. Res. Lett.*, **38**, 6 (L19501).

Spreen, G., S. Kern, D. Stammer, and E. Hansen, 2009: Fram Strait sea ice volume export estimated between 2003 and 2008 from satellite data. *Geophys. Res. Lett.*, **36**, L19502.

Stammerjohn, S., R. Massom, D. Rind, and D. Martinson, 2012: Regions of rapid sea ice change: An inter-hemispheric seasonal comparison. *Geophys. Res. Lett.*, **39**, L06501.

Stammerjohn, S. E., D. G. Martinson, R. C. Smith, X. Yuan, and D. Rind, 2008: Trends in Antarctic annual sea ice retreat and advance and their relation to El Niño-Southern Oscillation and Southern Annular Mode variability. *J. Geophys. Res. Oceans*, **113**, C03S90.

Steig, E. J., Q. Ding, D. S. Battisti, and A. Jenkins, 2012: Tropical forcing of Circumpolar Deep Water Inflow and outlet glacier thinning in the Amundsen Sea Embayment, West Antarctica. *Ann. Glaciol.*, **53**, 19–28.

Steig, E. J., D. P. Schneider, S. D. Rutherford, M. E. Mann, J. C. Comiso, and D. T. Shindell, 2009: Warming of the Antarctic ice-sheet surface since the 1957 International Geophysical Year. *Nature*, **457**, 459–462.

Stevens, C. W., B. J. Moorman, and S. M. Solomon, 2010: Modeling ground thermal conditions and the limit of permafrost within the nearshore zone of the Mackenzie Delta, Canada. *J. Geophys. Res. Earth Surf.*, **115**, F04027.

Straneo, F., R. G. Curry, D. A. Sutherland, G. S. Hamilton, C. Cenedese, K. Vage, and L. A. Stearns, 2011: Impact of fjord dynamics and glacial runoff on the circulation near Helheim Glacier. *Nature Geosci.*, **4**, 322–327.

Straneo, F., et al., 2010: Rapid circulation of warm subtropical waters in a major glacial fjord in East Greenland. *Nature Geosci.*, **3**, 182–186.

Straneo, F., et al., 2012: Characteristics of ocean waters reaching Greenland's glaciers. *Ann. Glaciol.*, **53**, 202–210.

Streletskiy, D. A., N. I. Shiklomanov, F. E. Nelson, and A. E. Klene, 2008: 13 Years of Observations at Alaskan CALM Sites: Long-term Active Layer and Ground Surface Temperature Trends. In: *Proceedings of the 9th International Conference on Permafrost, 29 June–3 July 2008, Institute of Northern Engineering, University of Alaska, Fairbanks* [D. L. Kane, and K. M. Hinkel (eds.)], pp. 1727–1732.

Stroeve, J., M. M. Holland, W. Meier, T. Scambos, and M. Serreze, 2007: Arctic sea ice decline: Faster than forecast. *Geophys. Res. Lett.*, **34**, L09501.

Sundal, A. V., A. Shepherd, P. Nienow, E. Hanna, S. Palmer, and P. Huybrechts, 2011: Melt-induced speed-up of Greenland ice sheet offset by efficient subglacial drainage. *Nature*, **469**, 522–U583.

Takala, M., J. Pulliainen, S. J. Metsamaki, and J. T. Koskinen, 2009: Detection of snowmelt using spaceborne microwave radiometer data in Eurasia from 1979 to 2007. *IEEE Trans. Geosci. Remote Sens.*, **47**, 2996–3007.

Tamura, T., and K. I. Ohshima, 2011: Mapping of sea ice production in the Arctic coastal polynyas. *J. Geophys. Res. Oceans*, **116**, 20 (C07030).

Tang, J. S., H. W. Cheng, and L. Liu, 2012: Using nonlinear programming to correct leakage and estimate mass change from GRACE observation and its application to Antarctica. *J. Geophys. Res. Sol. Ea.*, **117**, B11410.

Tarasov, L., and W. R. Peltier, 2002: Greenland glacial history and local geodynamic consequences. *Geophys. J. Int.*, **150**, 198–229.

Tarnocai, C., J. G. Canadell, E. A. G. Schuur, P. Kuhry, G. Mazhitova, and S. Zimov, 2009: Soil organic carbon pools in the northern circumpolar permafrost region. *Global Biogeochem. Cycles*, **23**, 11 (GB2023).

Tedesco, M., and A. J. Monaghan, 2009: An updated Antarctic melt record through 2009 and its linkages to high-latitude and tropical climate variability. *Geophys. Res. Lett.*, **36**, L18502.

Tedesco, M., M. Brodzik, R. Armstrong, M. Savoie, and J. Ramage, 2009: Pan arctic terrestrial snowmelt trends (1979–2008) from spaceborne passive microwave data and correlation with the Arctic Oscillation. *Geophys. Res. Lett.*, **36**, L21402.

Tedesco, M., X. Fettweis, T. Mote, J. Wahr, P. Alexander, J. E. Box, and B. Wouters, 2013: Evidence and analysis of 2012 Greenland records from spaceborne observations, a regional climate model and reanalysis data. *Cryosphere*, **7**, 615–630.

Tedesco, M., et al., 2011: The role of albedo and accumulation in the 2010 melting record in Greenland. *Environ. Res. Lett.*, **6**, 6 (014005).

Tennant, C., B. Menounos, R. Wheate, and J. J. Clague, 2012: Area change of glaciers in the Canadian Rocky Mountains, 1919 to 2006. *Cryosphere*, **6**, 1541–1552.

Thibert, E., R. Blanc, C. Vincent, and N. Eckert, 2008: Glaciological and volumetric mass-balance measurements: error analysis over 51 years for Glacier de Sarennes, French Alps. *J. Glaciol.*, **54**, 522–532.

Thomas, E. R., G. J. Marshall, and J. R. McConnell, 2008a: A doubling in snow accumulation in the western Antarctic Peninsula since 1850. *Geophys. Res. Lett.*, **35**, 5 (L01706)

Thomas, I. D., et al., 2011a: Widespread low rates of Antarctic glacial isostatic adjustment revealed by GPS observations. *Geophys. Res. Lett.*, **38**, L22302.

Thomas, R., E. Frederick, W. Krabill, S. Manizade, and C. Martin, 2006: Progressive increase in ice loss from Greenland. *Geophys. Res. Lett.*, **33**, 4 (L10503).

Thomas, R., E. Frederick, W. Krabill, S. Manizade, and C. Martin, 2009: Recent changes on Greenland outlet glaciers. *J. Glaciol.*, **55**, 147–162.

Thomas, R., C. Davis, E. Frederick, W. Krabill, Y. H. Li, S. Manizade, and C. Martin, 2008b: A comparison of Greenland ice-sheet volume changes derived from altimetry measurements. *J. Glaciol.*, **54**, 203–212.

Thomas, R., E. Frederick, J. Li, W. Krabill1, S. Manizade, J. Paden, J. Sonntag, R. Swift, J. Yungel., 2011b: Accelerating ice loss from the fastest Greenland and Antarctic glaciers. *Geophys. Res. Lett.*, **38**, L10502.

Thomas, R. H., and C. R. Bentley, 1978: A model for Holocene retreat of the West Antarctic Ice Sheet. *Quat. Res.*, **10**, 150–170.

Thompson, D. W. J., and J. M. Wallace, 1998: The Arctic Oscillation signature in the wintertime geopotential height and temperature fields. *Geophys. Res. Lett.*, **25**, 1297–1300.

Thompson, D. W. J., and J. M. Wallace, 2000: Annular modes in the extratropical circulation. Part I: Month-to-month variability. *J. Clim.*, **13**, 1000–1016.

Thomson, L. I., G. R. Osinski, and C. S. L. Ommanney, 2011: Glacier change on Axel Heiberg Island, Nunavut, Canada. *J. Glaciol.*, **57**, 1079–1086

Throop, J., A. G. Lewkowicz, and S. L. Smith, 2012: Climate and ground temperature relations at sites across the continuous and discontinuous permafrost zones, northern Canada. *Can. J. Earth Sci.*, **49**, 865–876.

Turner, J., et al., 2005: Antarctic climate change during the last 50 years. *Int. J. Climatol.*, **25**, 279–294.

Valt, M., and P. Cianfarra, 2010: Recent snow cover variability in the Italian Alps. *Cold Reg. Sci. Technol.*, **64**, 146–157.

van de Berg, W. J., M. R. van den Broeke, C. H. Reijmer, and E. Van Meijgaard, 2006: Reassessment of the Antarctic surface mass balance using calibrated output of a regional atmospheric climate model. *J. Geophys. Res. Atmos.*, **111**, D11104.

van de Wal, R. S. W., W. Boot, M. R. van den Broeke, C. Smeets, C. H. Reijmer, J. J. A. Donker, and J. Oerlemans, 2008: Large and rapid melt-induced velocity changes in the ablation zone of the Greenland Ice Sheet. *Science*, **321**, 111–113.

van den Broeke, M., W. J. van de Berg, and E. Van Meijgaard, 2006: Snowfall in coastal West Antarctica much greater than previously assumed. *Geophys. Res. Lett.*, **33**, L02505.

van den Broeke, M., et al., 2009: Partitioning recent Greenland mass loss. *Science*, **326**, 984–986.

Van Everdingen, R. (ed.), 1998: *Multi-language Glossary of Permafrost and Related Ground-Ice Terms*. National Snow and Ice Data Center /World Data Center for Glaciology.

Van Ommen, T. D., and V. Morgan, 2010: Snowfall increase in coastal East Antarctica linked with southwest Western Australian drought. *Nature Geosci.*, **3**, 267–272.

Vasiliev, A. A., M. O. Leibman, and N. G. Moskalenko, 2008: Active layer monitoring in West Siberia under the CALM II Program. In: *Proceedings of the 9th International Conference on Permafrost, 29 June–3 July 2008, Institute of Northern Engineering, University of Alaska, Fairbanks*, [D. L. Kane, and K. M. Hinkel (eds.)], pp. 1815–1821.

Vaughan, D. G., et al., 2003: Recent rapid regional climate warming on the Antarctic Peninsula. *Clim. Change*, **60**, 243–274.

Velicogna, I., 2009: Increasing rates of ice mass loss from the Greenland and Antarctic ice sheets revealed by GRACE. *Geophys. Res. Lett.*, **36**, L19503.

Velicogna, I., and J. Wahr, 2006a: Acceleration of Greenland ice mass loss in spring 2004. *Nature*, **443**, 329–331.

Velicogna, I., and J. Wahr, 2006b: Measurements of time-variable gravity show mass loss in Antarctica. *Science*, **311**, 1754–1756.

Vieira, G., et al., 2010: Thermal state of permafrost and active-layer monitoring in the Antarctic: Advances during the International Polar Year 2007–2009. *Permafr. Periglac. Process.*, **21**, 182–197.

Vigdorchik, M. E., 1980: *Arctic Pleistocene History and the Development of Submarine Permafrost.* Westview Press, Boulder, CO, USA, 286 pp.

Vincent, C., E. Le Meur, D. Six, P. Possenti, E. Lefebvre, and M. Funk, 2007: Climate warming revealed by englacial temperatures at Col du Dome (4250 m, Mont Blanc area). *Geophys. Res. Lett.*, **34**.

Vinje, T., 2001: Fram strait ice fluxes and atmospheric circulation: 1950–2000. *J. Clim.*, **14**, 3508–3517.

Wadhams, P., 1990: Evidence for thinning of the Arctic ice cover north of Greenland. *Nature*, **345**, 795–797.

Wadhams, P., and J. C. Comiso, 1992: The ice thickness distribution inferred using remote sensing techniques. In: *Microware Remote Sensing of Sea Ice* [F. Carsey (ed.)]. American Geophysical Union, Washington, DC, pp. 375–383.

Wadhams, P., and N. R. Davis, 2000: Further evidence of ice thinning in the Arctic Ocean. *Geophys. Res. Lett.*, **27**, 3973–3975.

Wadhams, P., N. Hughes, and J. Rodrigues, 2011: Arctic sea ice thickness characteristics in winter 2004 and 2007 from submarine sonar transects. *J. Geophys. Res. Oceans*, **116**, C00E02.

Wahr, J. M., 2007: Time-variable gravity from satellites. In: *Treatise on Geophysics* [T. A. Herring (ed.)]. Elsevier, Amsterdam, the Netherlands, and Philadelphia, PA, USA, pp. 213–237.

Walsh, J. E., and W. L. Chapman, 2001: 20th-century sea-ice variations from observational data. *Ann. Glaciol.*, **33**, 444–448.

Wang, J., X. Bai, H. Hu, A. Clites, M. Colton, and B. Lofgren, 2012: Temporal and spatial variability of Great Lakes ice cover, 1973–2010. *J. Clim.*, **25**, 13181329.

Weeks, W. F., and A. J. Gow, 1978: Preferred crystal orientations in fast ice along margins of Arctic Ocean *J. Geophys. Res. Oceans Atmos.*, **83**, 5105–5121.

Weertman, J., 1974: Stability of the junction of an ice sheet and an ice shelf. *J. Glaciol.*, **13**, 3–11.

Wendt, J., A. Rivera, A. Wendt, F. Bown, R. Zamora, G. Casassa, and C. Bravo, 2010: Recent ice-surface-elevation changes of Fleming Glacier in response to the removal of the Wordie Ice Shelf, Antarctic Peninsula. *Ann. Glaciol.*, **51**, 97–102.

WGMS, 1989: World glacier inventory—Status 1988. IAHS (ICSI)/UNEP/ UNESCO,[Haeberli, W., H. Bösch, K. Scherler, G. Østrem and C. C. Wallén (eds.)] World Glacier Monitoring Service, Zurich, Switzerland, 458 pp.

WGMS, 2008: *Global Glacier Changes: Facts and Figures.* [Zemp, M, I. Roer, A. Kääb, M. Hoelzle, F. Paul, W. G. Haeberli (eds.)] UNEP and World Glacier Monitoring Service, Zurich, Switzerland, 88 pp.

WGMS, 2009: Glacier Mass Balance Bulletin No. 10 (2006–2007). ICSU (WDS)/IUGG (IACS) /UNEP/UNESCO/WMO. [Haeberli, W., I. Gärtner-Roer, M. Hoelzle, F. Paul, M.I Zemp (eds.)] World Glacier Monitoring Service, Zurich, Switzerland. 96 pp.

White, D., et al., 2007: The arctic freshwater system: Changes and impacts. *J. Geophys. Res. Biogeosci.*, **112**, G04S54.

White, W. B., and R. G. Peterson, 1996: An Antarctic circumpolar wave in surface pressure, wind, temperature and sea-ice extent. *Nature*, **380**, 699–702.

Whitehouse, P. L., M. J. Bentley, G. A. Milne, M. A. King, and I. D. Thomas, 2012: A new glacial isostatic adjustment model for Antarctica: Calibrated and tested using observations of relative sea-level change and present-day uplift rates. *Geophys. J. Int.*, **190**, 1464–1482.

Willis, M. J., A. K. Melkonian, M. E. Pritchard, and A. Rivera, 2012: Ice loss from the Southern Patagonian Icefield. *Geophys. Res. Lett.*, **39**, L17501.

Wingham, D. J., D. W. Wallis, and A. Shepherd, 2009: Spatial and temporal evolution of Pine Island Glacier thinning, 1995–2006. *Geophys. Res. Lett.*, **36**, L17501.

Wingham, D. J., A. Shepherd, A. Muir, and G. J. Marshall, 2006: Mass balance of the Antarctic ice sheet. *Philos. Trans. R. Soc. A*, **364**, 1627–1635.

Wingham, D. J., A. J. Ridout, R. Scharroo, R. J. Arthern, and C. K. Shum, 1998: Antarctic elevation change from 1992 to 1996. *Science*, **282**, 456–458.

Worby, A. P., C. A. Geiger, M. J. Paget, M. L. Van Woert, S. F. Ackley, and T. L. DeLiberty, 2008: Thickness distribution of Antarctic sea ice. *J. Geophys. Res. Oceans*, **113**, C05S92.

Wouters, B., D. Chambers, and E. J. O. Schrama, 2008: GRACE observes small-scale mass loss in Greenland. *Geophys. Res. Lett.*, **35**, L20501.

Wu, Q., T. Zhang, and Y. Liu, 2012: Thermal state of the active layer and permafrost along the Qinghai-Xizang (Tibet) Railway from 2006 to 2010. *Cryosphere*, **6**, 607–612.

Wu, Q. B., and T. J. Zhang, 2008: Recent permafrost warming on the Qinghai-Tibetan plateau. *J. Geophys. Res. Atmos.*, **113**, D13108.

Wu, Q. B., and T. J. Zhang, 2010: Changes in active layer thickness over the Qinghai-Tibetan Plateau from 1995 to 2007. *J. Geophys. Res. Atmos.*, **115**, D09107.

Wu, X. P., et al., 2010: Simultaneous estimation of global present-day water transport and glacial isostatic adjustment. *Nature Geosci.*, **3**, 642–646.

Xie, H., et al., 2011: Sea-ice thickness distribution of the Bellingshausen Sea from surface measurements and ICESat altimetry. *Deep-Sea Res. Pt. Ii*, **58**, 1039–1051.

Xu, Y., E. Rignot, D. Menemenlis, and M. Koppes, 2012: Numerical experiments on subaqueous melting of Greenland tidewater glaciers in response to ocean warming and enhanced subglacial discharge. *Ann. Glaciol.*, **53**, 229–234.

Yao, T., et al., 2012: Different glacier status with atmospheric circulations in Tibetan Plateau and surroundings. *Nature Clim. Change*, **2**, 663–667

Yde, J. C., and N. T. Knudsen, 2007: 20th-century glacier fluctuations on Disko Island (Qeqertarsuaq), Greenland. *Ann. Glaciol.*, **46**, 209–214.

Yde, Y. C., and O. Pasche, 2010: Reconstructing climate change: Not all glaciers suitable. *EOS*, **91**, 189–190.

Young, D. A., et al., 2011a: A dynamic early East Antarctic Ice Sheet suggested by ice-covered fjord landscapes. *Nature*, **474**, 72–75.

Young, N. E., J. P. Briner, Y. Axford, B. Csatho, G. S. Babonis, D. H. Rood, and R. C. Finkel, 2011b: Response of a marine-terminating Greenland outlet glacier to abrupt cooling 8200 and 9300 years ago. *Geophys. Res. Lett.*, **38**, L24701.

Zamolodchikov, D., 2008: Recent climate and active layer changes in northeast Russia: Regional output of Circumpolar Active Layer Monitoring (CALM). In: *Proceedings of the 9th International Conference on Permafrost, 29 June–3 July 2008, Institute of Northern Engineering, University of Alaska, Fairbanks* [D. L. Kane, and K. M. Hinkel (eds.)], pp. 2021–2027.

Zdanowicz, C., A. Smetny-Sowa, D. Fisher, N. Schaffer, L. Copland, J. Eley, and F. Dupont, 2012: Summer melt rates on Penny Ice Cap, Baffin Island: Past and recent trends and implications for regional climate. *J. Geophys. Res.*, **117**, F02006.

Zeeberg, J., and S. L. Forman, 2001: Changes in glacier extent on north Novaya Zemlya in the twentieth century. *Holocene*, **11**, 161–175.

Zemp, M., H. J. Zumbhul, S. U. Nussbaumer, M. H. Masiokas, L. E. Espizua, and P. Pitte, 2011: Extending glacier monitoring into the Little Ice Age and beyond. *PAGES News*, **19**, 67–69.

Zhang, T., R. G. Barry, K. Knowles, J. A. Heginbottom, and J. Brown, 1999: Statistics and characteristics of permafrost and ground ice distribution in the Northern Hemisphere. *Polar Geogr.*, **23**, 147–169.

Zhang, T., R. G. Barry, K. Knowles, F. Ling, and R. L. Armstrong, 2003: Distribution of seasonally and perennially frozen ground in the Northern Hemisphere. In: *Proceedings of the 8th International Conference on Permafrost, 21–25 July 2003, Zurich, Switerland* [Phillips, M., S.M. Springman, and L.U. Arenson (eds)]. A. A. Balkema, Lisse, the Netherlands, pp. 1289–1294.

Zhang, T. J., 2005: Influence of the seasonal snow cover on the ground thermal regime: An overview. *Rev. Geophys.*, **43**, RG4002

Zhang, T. J., et al., 2005: Spatial and temporal variability in active layer thickness over the Russian Arctic drainage basin. *J. Geophys. Res. Atmos.*, **110**, D16101.

Zhao, L., Q. B. Wu, S. S. Marchenko, and N. Sharkhuu, 2010: Thermal state of permafrost and active layer in central Asia during the International Polar Year. *Permafr. Periglac. Process.*, **21**, 198–207.

Zhou, Y., D. Guo, G. Qiu, G. Cheng, and S. Li, 2000: *Geocryology in China.* Science Press, Beijing, China, 450 pp.

Zimov, S. A., E. A. G. Schuur, and F. S. Chapin, 2006: Permafrost and the global carbon budget. *Science*, **312**, 1612–1613.

Zuo, Z., and J. Oerlemans, 1997: Contribution of glacier melt to sea-level rise since AD 1865: A regionally differentiated calculation. *Clim. Dyn.*, **13**, 835–845.

Zwally, H. J., and P. Gloersen, 2008: Arctic sea ice surviving the summer melt: interannual variability and decreasing trend. *J. Glaciol.*, **54**, 279–296.

Zwally, H. J., and M. B. Giovinetto, 2011: Overview and assessment of Antarctic Ice-Sheet mass balance estimates: 1992–2009. *Surv. Geophys.*, **32**, 351–376.

Zwally, H. J., D. H. Yi, R. Kwok, and Y. H. Zhao, 2008: ICESat measurements of sea ice freeboard and estimates of sea ice thickness in the Weddell Sea. *J. Geophys. Res. Oceans*, **113**, C02S15.

Zwally, H. J., J. C. Comiso, C. L. Parkinson, D. J. Cavalieri, and P. Gloersen, 2002a: Variability of Antarctic sea ice 1979–1998. *J. Geophys. Res.*, **107**, 1029–1047.

Zwally, H. J., W. Abdalati, T. Herring, K. Larson, J. Saba, and K. Steffen, 2002b: Surface melt-induced acceleration of Greenland ice-sheet flow. *Science*, **297**, 218–222.

Zwally, H. J., et al., 2005: Mass changes of the Greenland and Antarctic ice sheets and shelves and contributions to sea-level rise: 1992–2002. *J. Glaciol.*, **51**, 509–527.

Zwally, H. J., et al., 2011: Greenland ice sheet mass balance: Distribution of increased mass loss with climate warming; 2003–07 versus 1992–2002. *J. Glaciol.*, **57**, 88–102.

4

Appendix 4.A: Details of Available and Selected Ice Sheet Mass Balance Estimates from 1992 to 2012

All comprehensive mass balance estimates available for Greenland, and the subset of those selected for this assessment (Section 4.4.2) are listed in Tables 4.A.1 and 4.A.2. Those available for Antarctica are shown in Tables 4.A.3 and 4.A.4. These studies include estimates made from satellite gravimetry (GRACE), satellite altimetry (radar and laser) and the mass balance (flux) method. The studies selected for this assessment are the latest made by different research groups, for each of Greenland and Antarctica. The tables indicate whether smaller glaciers peripheral to the ice sheet are included, or excluded, in the estimate, and explain why some studies were not selected (e.g., earlier estimates from the same researchers have been updated by more recent analyses using extended data).

Table 4.A.1 | Sources used for calculation of ice loss from Greenland.

Source	Method	Start	End	Gt yr^{-1}	Uncertainty	Peripheral Glaciers	Comment
Ewert et al. (2012)	Laser alt.	2003.8	2008.2	–185	28	Excluded	
	GRACE	2002.7	2009.5	–191	21	Included	
Harig and Simons (2012)	GRACE	2003.0	2011.0	–200	6	Included	Yearly estimates used in compilation. GIA uncertainty not provided.
Sasgen et al. (2012)	GRACE	2002.7	2011.7	–240	18	Included	Yearly estimates used in compilation.
	Flux	2002.7	2011.7	–244	53	Excluded	Yearly estimates used in compilation.
	Laser alt.	2003.8	2009.8	–245	28	Included	
Chen et al. (2011)	GRACE	2002.3	2005.3	–144	25	Included	
	GRACE	2005.3	2009.9	–248	43	Included	
Rignot et al. (2011c)	Flux	1992.0	2010.0	–154	51	Excluded	Yearly estimates used in compilation.
Schrama and Wouters (2011)	GRACE	2003.2	2010.1	–201	19	Included	2 standard deviation (2σ) uncertainty
Sorensen et al. (2011)	Laser alt.	2003.8	2008.2	–221	28	Included	
Zwally et al. (2011)	Radar alt.	1992.3	2002.8	–7	3	Excluded	
	Laser alt.	2003.8	2007.8	–171	4	Excluded	
Pritchard et al. (2010)	GRACE	2003.6	2009.6	–195	30	Included	
Wu et al. (2010)	GRACE +GPS	2002.4	2009.0	–104	23	Included	Global inversion technique.
Baur et al. (2009)	GRACE	2002.6	2008.6	–159	11	Included	No GIA correction.
Cazenave et al. (2009)	GRACE	2003.0	2008.0	–136	18	Included	
Slobbe et al. (2009)	GRACE	2002.5	2007.5	–178	78	Included	
	Laser alt.	2003.1	2007.3	–139	68	Included	
Velicogna (2009)	GRACE	2002.3	2009.1	–269	33	Included	Time series extended to 2012 using new data and published method. Yearly estimates derived from cited trend.

Table 4.A.2 | Sources NOT used for calculation of ice loss from Greenland.

Source	Method	Start	End	Gt yr^{-1}	Uncertainty	Peripheral Glaciers	Comment
Shepherd et al. (2012)	Flux	1992.0	2009.9	–154	51	Excluded	This comprehensive inter-comparison reconciles estimates from different techniques. The "reconciled" value is the best estimate from all techniques. This source is discussed separately and not included within the average assessment presented here.
	GRACE	2002.2	2012.0	–212	27	Included	
	Laser alt.	2004.5	2007.4	–198	23	Excluded	
	Reconciled	1992.0	2011.0	–142	49	-	
van den Broeke (2009)	Flux	2003.0	2009.0	–237	20	Excluded	Superseded by Rignot et al. (2011c).
Rignot et al. (2008a)	Flux	1996.0	1997.0	–97	47	Excluded	Superseded by Rignot et al. (2011c).
	Flux	2000.0	2001.0	–156	44	Excluded	
	Flux	2004.0	2008.0	–264	39	Excluded	
Wouters et al. (2008)	GRACE	2003.2	2008.1	–179	25	Included	Superseded by Schrama and Wouters (2011).
Chen et al. (2006)	GRACE	2002.3	2005.9	–219	21	Included	Superseded by Chen et al. (2011).
Luthcke et al. (2006)	GRACE	2003.5	2005.5	–101	16	Included	Superseded by Pritchard et al. (2010).
Ramillien et al. (2006)	GRACE	2002.5	2005.2	–129	15	Included	Superseded by Cazenave et al. (2009).
Rignot and Kanagaratnam (2006)	Flux	1996.0	1997.0	–83	28	Excluded	Superseded by Rignot et al. (2011c).
	Flux	2000.0	2001.0	–127	28	Excluded	
	Flux	2005.0	2006.0	–205	38	Excluded	
Thomas et al. (2006)	Radar alt.	1994.0	1999.0	–27	23	Excluded	Includes only half the ice sheet and fills in the rest with a melt model.
	Radar alt.	1999.0	2005.0	–81	24	Excluded	
Velicogna and Wahr (2006a)	GRACE	2002.3	2004.3	–95	49	Included	Superseded by Velicogna (2009).
	GRACE	2004.3	2006.3	–313	60	Included	
Zwally et al. (2005)	Radar alt.	1992.3	2002.8	11	3	Not known	Superseded by Zwally et al. (2011).
Krabill et al. (2000)	Laser alt. (aircraft)	1993.5	1999.5	–47		Excluded	Includes only half the ice sheet and fills in the rest with a melt model.

Table 4.A.3 | Sources used for calculation of ice loss from Antarctica.

Source	Method	Start	End	Gt yr^{-1}	Uncertainty	Peripheral Glaciers	Comment
King et al. (2012)	GRACE	2002.6	2011.0	–69	18	Included	2 standard deviation (2σ) uncertainty. This study treats systematic uncertainty as bounds not random error as in other GRACE studies.
Tang et al. (2012)	GRACE	2006.0	2011.4	–211	75	Included	
Rignot et al. (2011c)	Flux	1992.0	2010.0	–83	91	Excluded	Yearly estimates used in compilation.
Shi et al. (2011)	Laser alt.	2003.1	2008.2	–78	5	Not known	Methodology and error budget incompletely described.
Wu et al. (2010)	GRACE +GPS	2002.4	2009.0	–87	43	Included	Global inversion technique.
Cazenave et al. (2009)	GRACE	2003.0	2008.0	–198	22	Included	
Chen et al. (2009)	GRACE	2002.3	2006.0	–144	58	Included	
	GRACE	2006.0	2009.1	–220	89	Included	
E et al. (2009)	GRACE	2002.5	2007.7	–78	37	Included	Error budget incompletely explained.
Horwath and Dietrich (2009)	GRACE	2002.6	2008.1	–109	48	Included	
Velicogna (2009)	GRACE	2002.3	2013.0	–184	73	Included	Time series extended to 2012 using new data and published method. Yearly estimates derived from cited trend.

Table 4.A.4 | Sources NOT used for calculation of ice loss from Antarctica.

Source	Method	Start	End	Gt yr^{-1}	Uncertainty	Peripheral Glaciers	Comment
Shepherd et al. (2012)	Flux	1992.0	2010.0	−110	89	Excluded	This comprehensive inter-comparison reconciles estimates from different techniques. Estimates are made separately for East Antarctica, West Antarctica and the Antarctic Peninsula. The "reconciled" value is the best estimate from all techniques. The results from this study are discussed separately and not included within the average assessment presented here.
	GRACE	2003.0	2011.0	−90	44	Included	
	Laser alt.	2003.8	2008.7	+21	76	Excluded	
	Reconciled	1992.0	2011.0	-71	53	-	
Jia et al. (2011)	GRACE	2002.6	2010.0	−82	29	Included	No consideration of gravity signal leakage.
Zwally and Giovinetto (2011)	Radar alt.	1992.3	2001.3	−31	12	Excluded	Same data analysis as Zwally et al. (2005). Excludes Antarctic Peninsula.
Gunter et al. (2009)	Laser alt.	2003.1	2007.1	−100	?	Not known	No error bar and no final estimate.
Moore and King (2008)	GRACE	2002.3	2006.0	−150	73	Included	Superseded by King et al. (2012)
Rignot et al. (2008b)	Flux	1996.0	1997.0	−112	91	Excluded	Superseded by Rignot et al. (2011c).
	Flux	2006.0	2007.0	−196	92	Excluded	
Ramillien et al. (2006)	GRACE	2002.5	2005.2	−40	36	Included	Superseded by Cazenave et al. (2009).
Velicogna and Wahr (2006b)	GRACE	2002.3	2005.8	−139	73	Included	Superseded by Velicogna (2009).
Zwally et al. (2005)	Radar alt.	1992.3	2001.3	−31	52	Excluded	Antarctic Peninsula excluded.
Wingham et al. (2006)	Radar alt.	1992.8	2003.1	27	29	Not known	No data in Antarctic Peninsula; series truncated within 100 km of coast.
Rignot and Thomas (2002)	Flux	Not specific	Not specific	−26	37	Excluded	Not an ice-sheet wide estimate.
Wingham et al. (1998)	Radar alt.	1992.3	1997.0	−60	76	Not known	Superseded by Wingham et al. (2006).

4

5 Information from Paleoclimate Archives

Coordinating Lead Authors:
Valérie Masson-Delmotte (France), Michael Schulz (Germany)

Lead Authors:
Ayako Abe-Ouchi (Japan), Jürg Beer (Switzerland), Andrey Ganopolski (Germany), Jesus Fidel González Rouco (Spain), Eystein Jansen (Norway), Kurt Lambeck (Australia), Jürg Luterbacher (Germany), Tim Naish (New Zealand), Timothy Osborn (UK), Bette Otto-Bliesner (USA), Terrence Quinn (USA), Rengaswamy Ramesh (India), Maisa Rojas (Chile), XueMei Shao (China), Axel Timmermann (USA)

Contributing Authors:
Kevin Anchukaitis (USA), Julie Arblaster (Australia), Patrick J. Bartlein (USA), Gerardo Benito (Spain), Peter Clark (USA), Josefino C. Comiso (USA), Thomas Crowley (UK), Patrick De Deckker (Australia), Anne de Vernal (Canada), Barbara Delmonte (Italy), Pedro DiNezio (USA), Trond Dokken (Norway), Harry J. Dowsett (USA), R. Lawrence Edwards (USA), Hubertus Fischer (Switzerland), Dominik Fleitmann (UK), Gavin Foster (UK), Claus Fröhlich (Switzerland), Aline Govin (Germany), Alex Hall (USA), Julia Hargreaves (Japan), Alan Haywood (UK), Chris Hollis (New Zealand), Ben Horton (USA), Masa Kageyama (France), Reto Knutti (Switzerland), Robert Kopp (USA), Gerhard Krinner (France), Amaelle Landais (France), Camille Li (Norway/Canada), Dan Lunt (UK), Natalie Mahowald (USA), Shayne McGregor (Australia), Gerald Meehl (USA), Jerry X. Mitrovica (USA/Canada), Anders Moberg (Sweden), Manfred Mudelsee (Germany), Daniel R. Muhs (USA), Stefan Mulitza (Germany), Stefanie Müller (Germany), James Overland (USA), Frédéric Parrenin (France), Paul Pearson (UK), Alan Robock (USA), Eelco Rohling (Australia), Ulrich Salzmann (UK), Joel Savarino (France), Jan Sedláček (Switzerland), Jeremy Shakun (USA), Drew Shindell (USA), Jason Smerdon (USA), Olga Solomina (Russian Federation), Pavel Tarasov (Germany), Bo Vinther (Denmark), Claire Waelbroeck (France), Dieter Wolf-Gladrow (Germany), Yusuke Yokoyama (Japan), Masakazu Yoshimori (Japan), James Zachos (USA), Dan Zwartz (New Zealand)

Review Editors:
Anil K. Gupta (India), Fatemeh Rahimzadeh (Iran), Dominique Raynaud (France), Heinz Wanner (Switzerland)

This chapter should be cited as:
Masson-Delmotte, V., M. Schulz, A. Abe-Ouchi, J. Beer, A. Ganopolski, J.F. González Rouco, E. Jansen, K. Lambeck, J. Luterbacher, T. Naish, T. Osborn, B. Otto-Bliesner, T. Quinn, R. Ramesh, M. Rojas, X. Shao and A. Timmermann, 2013: Information from Paleoclimate Archives. In: *Climate Change 2013: The Physical Science Basis. Contribution of Working Group I to the Fifth Assessment Report of the Intergovernmental Panel on Climate Change* [Stocker, T.F., D. Qin, G.-K. Plattner, M. Tignor, S.K. Allen, J. Boschung, A. Nauels, Y. Xia, V. Bex and P.M. Midgley (eds.)]. Cambridge University Press, Cambridge, United Kingdom and New York, NY, USA.

Table of Contents

5

Executive Summary

Greenhouse-Gas Variations and Past Climate Responses

It is a fact that present-day (2011) concentrations of the atmospheric greenhouse gases (GHGs) carbon dioxide (CO_2), methane (CH_4) and nitrous oxide (N_2O) exceed the range of concentrations recorded in ice cores during the past 800,000 years. Past changes in atmospheric GHG concentrations can be determined with *very high confidence*[1] from polar ice cores. Since AR4 these records have been extended from 650,000 years to 800,000 years ago. {5.2.2}

With *very high confidence*, the current rates of CO_2, CH_4 and N_2O rise in atmospheric concentrations and the associated radiative forcing are unprecedented with respect to the highest resolution ice core records of the last 22,000 years. There is *medium confidence* that the rate of change of the observed GHG rise is also unprecedented compared with the lower resolution records of the past 800,000 years. {5.2.2}

There is *high confidence* that changes in atmospheric CO_2 concentration play an important role in glacial–interglacial cycles. Although the primary driver of glacial–interglacial cycles lies in the seasonal and latitudinal distribution of incoming solar energy driven by changes in the geometry of the Earth's orbit around the Sun ("orbital forcing"), reconstructions and simulations together show that the full magnitude of glacial–interglacial temperature and ice volume changes cannot be explained without accounting for changes in atmospheric CO_2 content and the associated climate feedbacks. During the last deglaciation, it is *very likely*[2] that global mean temperature increased by 3°C to 8°C. While the mean rate of global warming was *very likely* 0.3°C to 0.8°C per thousand years, two periods were marked by faster warming rates, *likely* between 1°C and 1.5°C per thousand years, although regionally and on shorter time scales higher rates may have occurred. {5.3.2}

New estimates of the equilibrium climate sensitivity based on reconstructions and simulations of the Last Glacial Maximum (21,000 years to 19,000 years ago) show that values below 1°C as well as above 6°C for a doubling of atmospheric CO_2 concentration are *very unlikely*. In some models climate sensitivity differs between warm and cold climates because of differences in the representation of cloud feedbacks. {5.3.3}

With *medium confidence*, global mean surface temperature was significantly above pre-industrial levels during several past periods characterised by high atmospheric CO_2 concentrations. During the mid-Pliocene (3.3 to 3.0 million years ago), atmospheric CO_2 concentrations between 350 ppm and 450 ppm (*medium confidence*) occurred when global mean surface temperatures were 1.9°C to 3.6°C (*medium confidence*) higher than for pre-industrial climate {5.3.1}. During the Early Eocene (52 to 48 million years ago), atmospheric CO_2 concentrations exceeded ~1000 ppm (*medium confidence*) when global mean surface temperatures were 9°C to 14°C (*medium confidence*) higher than for pre-industrial conditions. {5.3.1}

New temperature reconstructions and simulations of past climates show with *high confidence* polar amplification in response to changes in atmospheric CO_2 concentration. For high CO_2 climates such as the Early Eocene (52 to 48 million years ago) or mid-Pliocene (3.3 to 3.0 million years ago), and low CO_2 climates such as the Last Glacial Maximum (21,000 to 19,000 years ago), sea surface and land surface air temperature reconstructions and simulations show a stronger response to changes in atmospheric GHG concentrations at high latitudes as compared to the global average. {Box 5.1, 5.3.1, 5.3.3}

Global Sea Level Changes During Past Warm Periods

The current rate of global mean sea level change, starting in the late 19th-early 20th century, is, with *medium confidence*, unusually high in the context of centennial-scale variations of the last two millennia. The magnitude of centennial-scale global mean sea level variations did not exceed 25 cm over the past few millennia (*medium confidence*). {5.6.3}

There is *very high confidence* that the maximum global mean sea level during the last interglacial period (129,000 to 116,000 years ago) was, for several thousand years, at least 5 m higher than present and *high confidence* that it did not exceed 10 m above present. The best estimate is 6 m higher than present. Based on ice sheet model simulations consistent with elevation changes derived from a new Greenland ice core, the Greenland ice sheet *very likely* contributed between 1.4 and 4.3 m sea level equivalent, implying with *medium confidence* a contribution from the Antarctic ice sheet to the global mean sea level during the last interglacial period. {5.6.2}

There is *high confidence* that global mean sea level was above present during some warm intervals of the mid-Pliocene (3.3 to 3.0 million years ago), implying reduced volume of polar ice sheets. The best estimates from various methods imply with *high confidence* that sea level has not exceeded +20 m during the warmest periods of the Pliocene, due to deglaciation of the Greenland and West Antarctic ice sheets and areas of the East Antarctic ice sheet. {5.6.1}

5

1 In this Report, the following summary terms are used to describe the available evidence: limited, medium, or robust; and for the degree of agreement: low, medium, or high. A level of confidence is expressed using five qualifiers: very low, low, medium, high, and very high, and typeset in italics, e.g., *medium confidence*. For a given evidence and agreement statement, different confidence levels can be assigned, but increasing levels of evidence and degrees of agreement are correlated with increasing confidence (see Section 1.4 and Box TS.1 for more details).

2 In this Report, the following terms have been used to indicate the assessed likelihood of an outcome or a result: Virtually certain 99–100% probability, Very likely 90–100%, Likely 66–100%, About as likely as not 33–66%, Unlikely 0–33%, Very unlikely 0–10%, Exceptionally unlikely 0–1%. Additional terms (Extremely likely: 95–100%, More likely than not >50–100%, and Extremely unlikely 0–5%) may also be used when appropriate. Assessed likelihood is typeset in italics, e.g., *very likely* (see Section 1.4 and Box TS.1 for more details).

Observed Recent Climate Change in the Context of Interglacial Climate Variability

New temperature reconstructions and simulations of the warmest millennia of the last interglacial period (129,000 to 116,000 years ago) show with *medium confidence* that global mean annual surface temperatures were never more than 2°C higher than pre-industrial. High latitude surface temperature, averaged over several thousand years, was at least 2°C warmer than present (*high confidence*). Greater warming at high latitudes, seasonally and annually, confirm the importance of cryosphere feedbacks to the seasonal orbital forcing. During these periods, atmospheric GHG concentrations were close to the pre-industrial level. {5.3.4, Box 5.1}

There is *high confidence* that annual mean surface warming since the 20th century has reversed long-term cooling trends of the past 5000 years in mid-to-high latitudes of the Northern Hemisphere (NH). New continental- and hemispheric-scale annual surface temperature reconstructions reveal multi-millennial cooling trends throughout the past 5000 years. The last mid-to-high latitude cooling trend persisted until the 19th century, and can be attributed with *high confidence* to orbital forcing, according to climate model simulations. {5.5.1}

There is *medium confidence* from reconstructions that the current (1980–2012) summer sea ice retreat was unprecedented and sea surface temperatures in the Arctic were anomalously high in the perspective of at least the last 1450 years. Lower than late 20th century summer Arctic sea ice cover is reconstructed and simulated for the period between 8000 and 6500 years ago in response to orbital forcing. {5.5.2}

There is *high confidence* that minima in NH extratropical glacier extent between 8000 and 6000 years ago were primarily due to high summer insolation (orbital forcing). The current glacier retreat occurs within a context of orbital forcing that would be favourable for NH glacier growth. If glaciers continue to reduce at current rates, most extratropical NH glaciers will shrink to their minimum extent, which existed between 8000 and 6000 years ago, within this century (*medium confidence*). {5.5.3}

For average annual NH temperatures, the period 1983–2012 was *very likely* the warmest 30-year period of the last 800 years (*high confidence*) and *likely* the warmest 30-year period of the last 1400 years (*medium confidence*). This is supported by comparison of instrumental temperatures with multiple reconstructions from a variety of proxy data and statistical methods, and is consistent with AR4. In response to solar, volcanic and anthropogenic radiative changes, climate models simulate multi-decadal temperature changes over the last 1200 years in the NH, that are generally consistent in magnitude and timing with reconstructions, within their uncertainty ranges. {5.3.5}

Continental-scale surface temperature reconstructions show, with *high confidence*, multi-decadal periods during the Medieval Climate Anomaly (950 to 1250) that were in some regions as warm as in the mid-20th century and in others as warm as in the late 20th century. With *high confidence*, these regional warm periods were not as synchronous across regions as the warming since the mid-20th century. Based on the comparison between reconstructions and simulations, there is *high confidence* that not only external orbital, solar and volcanic forcing, but also internal variability, contributed substantially to the spatial pattern and timing of surface temperature changes between the Medieval Climate Anomaly and the Little Ice Age (1450 to 1850). {5.3.5.3, 5.5.1}

There is *high confidence* for droughts during the last millennium of greater magnitude and longer duration than those observed since the beginning of the 20th century in many regions. There is *medium confidence* that more megadroughts occurred in monsoon Asia and wetter conditions prevailed in arid Central Asia and the South American monsoon region during the Little Ice Age (1450 to 1850) compared to the Medieval Climate Anomaly (950 to 1250). {5.5.4 and 5.5.5}

With *high confidence*, floods larger than those recorded since 1900 occurred during the past five centuries in northern and central Europe, western Mediterranean region and eastern Asia. There is *medium confidence* that modern large floods are comparable to or surpass historical floods in magnitude and/or frequency in the Near East, India and central North America. {5.5.5}

Past Changes in Climate Modes

New results from high-resolution coral records document with *high confidence* that the El Niño-Southern Oscillation (ENSO) system has remained highly variable throughout the past 7000 years, showing no discernible evidence for an orbital modulation of ENSO. This is consistent with the weak reduction in mid-Holocene ENSO amplitude of only 10% simulated by the majority of climate models, but contrasts with reconstructions reported in AR4 that showed a reduction in ENSO variance during the first half of the Holocene. {5.4.1}

With *high confidence*, decadal and multi-decadal changes in the winter North Atlantic Oscillation index (NAO) observed since the 20th century are not unprecedented in the context of the past 500 years. Periods of persistent negative or positive winter NAO phases, similar to those observed in the 1960s and 1990 to 2000s, respectively, are not unusual in the context of NAO reconstructions during at least the past 500 years. {5.4.2}

The increase in the strength of the observed summer Southern Annular Mode since 1950 has been anomalous, with *medium confidence*, in the context of the past 400 years. No similar spatially coherent multi-decadal trend can be detected in tree-ring indices from New Zealand, Tasmania and South America. {5.4.2}

Abrupt Climate Change and Irreversibility

With *high confidence*, the interglacial mode of the Atlantic Ocean meridional overturning circulation (AMOC) can recover from a short-term freshwater input into the subpolar North Atlantic. Approximately 8200 years ago, a sudden freshwater release occurred during the final stages of North America ice sheet melting. Paleoclimate observations and model results indicate, with *high confidence*, a marked reduction in the strength of the AMOC followed by a rapid recovery, within approximately 200 years after the perturbation. {5.8.2}

Confidence in the link between changes in North Atlantic climate and low-latitude precipitation patterns has increased since AR4. From new paleoclimate reconstructions and modelling studies, there is *very high confidence* that reduced AMOC and the associated surface cooling in the North Atlantic region caused southward shifts of the Atlantic Intertropical Convergence Zone, and also affected the American (North and South), African and Asian monsoon systems. {5.7}

It is *virtually certain* that orbital forcing will be unable to trigger widespread glaciation during the next 1000 years. Paleoclimate records indicate that, for orbital configurations close to the present one, glacial inceptions only occurred for atmospheric CO_2 concentrations significantly lower than pre-industrial levels. Climate models simulate no glacial inception during the next 50,000 years if CO_2 concentrations remain above 300 ppm. {5.8.3, Box 6.2}

There is *high confidence* that the volumes of the Greenland and West Antarctic ice sheets were reduced during periods of the past few million years that were globally warmer than present. Ice sheet model simulations and geological data suggest that the West Antarctic ice sheet is very sensitive to subsurface Southern Ocean warming and imply with *medium confidence* a West Antarctic ice sheet retreat if atmospheric CO_2 concentration stays within or above the range of 350 ppm to 450 ppm for several millennia. {5.3.1, 5.6.1, 5.8.1}

5.1 Introduction

This chapter assesses the information on past climate obtained prior to the instrumental period. The information is based on data from various paleoclimatic archives and on modelling of past climate, and updates Chapter 6 of AR4 of IPCC Working Group I (Jansen et al., 2007).

The Earth system has responded and will continue to respond to various external forcings (solar, volcanic and orbital) and to changes in atmospheric composition. Paleoclimate data and modelling provide quantitative information on the Earth system response to these forcings. Paleoclimate information facilitates understanding of Earth system feedbacks on time scales longer than a few centuries, which cannot be evaluated from short instrumental records. Past climate changes also document transitions between different climate states, including abrupt events, which occurred on time scales of decades to a few centuries. They inform about multi-centennial to millennial baseline variability, against which the recent changes can be compared to assess whether or not they are unusual.

Major progress since AR4 includes the acquisition of new and more precise information from paleoclimate archives, the synthesis of regional information, and Paleoclimate Modelling Intercomparison Project Phase III (PMIP3) and Coupled Model Intercomparison Project Phase 5 (CMIP5) simulations using the same models as for projections (see Chapter 1). This chapter assesses the understanding of past climate variations, using paleoclimate reconstructions as well as climate models of varying complexity, while the model evaluation based on paleoclimate information is covered in Chapter 9. Additional paleoclimate perspectives are included in Chapters 6, 10 and 13 (see Table 5.1).

The content of this chapter is largely restricted to topics for which substantial new information has emerged since AR4. Examples include proxy-based estimates of the atmospheric carbon dioxide (CO_2) content during the past ~65 million years (Section 5.2.2) and magnitude of sea level variations during interglacial periods (Section 5.6.2). Information from glacial climates has been included only if the underlying processes are of direct relevance for an assessment of projected climate change. The impacts of past climate changes on biological systems and past civilizations are not covered, as these topics are beyond the scope of Working Group I.

The chapter proceeds from evidence for pre-industrial changes in atmospheric composition and external solar and volcanic forcings (Section 5.2, FAQ 5.1), to global and hemispheric responses (Section 5.3). After evaluating the evidence for past changes in climate modes of variability (Section 5.4), a specific focus is given to regional changes in temperature, cryosphere and hydroclimate during the current interglacial period (Section 5.5). Sections on sea level change (Section 5.6, FAQ 5.2), abrupt climate changes (Section 5.7) and illustrations of irreversibility and recovery time scales (Section 5.8) conclude the chapter. While polar amplification of temperature changes is addressed in Box 5.1, the relationships between ice sheets, sea level, atmospheric CO_2 concentration and climate are addressed in several sections (Box 5.2, Sections 5.3.1, 5.5, and 5.8.1).

Additional information to this chapter is available in the Appendix. Processed data underlying the figures are stored in the PANGAEA database (www.pangaea.de), while model output from PMIP3 is available from pmip3.lsce.ipsl.fr. In all sections, information is structured by time, going from past to present. Table 5.1 summarizes the past periods assessed in the subsections.

5.2 Pre-Industrial Perspective on Radiative Forcing Factors

5.2.1 External Forcings

5.2.1.1 Orbital Forcing

The term 'orbital forcing' is used to denote the incoming solar radiation changes originating from variations in the Earth's orbital parameters as well as changes in its axial tilt. Orbital forcing is well known from precise astronomical calculations for the past and future (Laskar et al., 2004). Changes in eccentricity, longitude of perihelion (related to precession) and axial tilt (obliquity) (Berger and Loutre, 1991) predominantly affect the seasonal and latitudinal distribution and magnitude of solar energy received at the top of the atmosphere (AR4, Box 6.1; Jansen et al., 2007), and the durations and intensities of local seasons. Obliquity also modulates the annual mean insolation at any given latitude, with opposite effects at high and low latitudes. Orbital forcing is considered the pacemaker of transitions between glacials and interglacials (*high confidence*), although there is still no consensus on exactly how the different physical processes influenced by insolation changes interact to influence ice sheet volume (Box 5.2; Section 5.3.2). The different orbital configurations make each glacial and interglacial period unique (Yin and Berger, 2010; Tzedakis et al., 2012a). Multi-millennial trends of temperature, Arctic sea ice and glaciers during the current interglacial period, and specifically the last 2000 years, have been related to orbital forcing (Section 5.5).

5.2.1.2 Solar Forcing

Solar irradiance models (e.g., Wenzler et al., 2005) have been improved to explain better the instrumental measurements of total solar irradiance (TSI) and spectral (wavelength dependent) solar irradiance (SSI). Typical changes measured over an 11-year solar cycle are 0.1% for TSI and up to several percent for the ultraviolet (UV) part of SSI (see Section 8.4). Changes in TSI directly impact the Earth's surface (see solar Box 10.2), whereas changes in UV primarily affect the stratosphere, but can influence the tropospheric circulation through dynamical coupling (Haigh, 1996). Most models attribute all TSI and SSI changes exclusively to magnetic phenomena at the solar surface (sunspots, faculae, magnetic network), neglecting any potential internal phenomena such as changes in energy transport (see also Section 8.4). The basic concept in solar models is to divide the solar surface into different magnetic features each with a specific radiative flux. The balance of contrasting dark sunspots and bright faculae and magnetic network leads to a higher TSI value during solar cycle maxima and at most wavelengths, but some wavelengths may be out of phase with the solar cycle (Harder et al., 2009; Cahalan et al., 2010; Haigh et al., 2010). TSI and SSI are calculated by adding the radiative fluxes of all features plus the contribution from

Table 5.1 | Summary of past periods for which climate information is assessed in the various sections of this chapter and other chapters of AR5. Calendar ages are expressed in Common Era (CE), geological ages are expressed in thousand years (ka) or million years (Ma) before present (BP), with present defined as 1950. Radiocarbon-based ages are quoted as the published calibrated ages.

Time Period	Age	Chapter 5 Sections							Other Chapters
		5.2	5.3	5.4	5.5	5.6	5.7	5.8	
Holocene[a]	11.65 ka[g] to present	✓	✓	✓	✓	✓			6, 9, 10
Pre-industrial period	refers to times before 1850 or 1850 values[h]								
Little Ice Age (LIA)	1450–1850[i]	✓	✓	✓	✓	✓			10
Medieval Climate Anomaly (MCA)[b]	950–1250[i]	✓	✓	✓	✓	✓			10
Last Millennium	1000–1999[j]	✓	✓	✓	✓	✓			9, 10
Mid-Holocene (MH)	~6 ka				✓				9, 13
8.2-ka event	~8.2 ka[g]							✓	
Last Glacial Termination[c]			✓			✓			6
Younger Dryas[d]	12.85–11.65 ka[g]						✓		6
Bølling-Allerød[e]	14.64–12.85 ka[g]						✓		6
Meltwater Pulse 1A (MWP-1A)	14.65–14.31 ka[k]					✓			
Heinrich stadial 1 (HS1)	~19–14.64 ka[l]						✓		
Last Glacial Maximum (LGM)	~21–19 ka[m]	✓	✓	✓					6, 9
Last Interglacial (LIG)[f]	~129–116 ka[n]	✓	✓			✓		✓	13
Mid-Pliocene Warm Period (MPWP)	~3.3–3.0 Ma[o]	✓	✓			✓		✓	13
Early Eocene Climatic Optimum (EECO)	~52–50 Ma[p]	✓	✓						
Paleocene-Eocene Thermal Maximum (PETM)	~55.5–55.3 Ma[q]	✓	✓					✓	

Notes:

[a] Also known as Marine Isotopic Stage (MIS) 1 or current interglacial.

[b] Also known as Medieval Climate Optimum or the Medieval Warm Period.

[c] Also known as Termination I or the Last Deglaciation. Based on sea level, Last Glacial Termination occurred between ~19 and ~6 ka.

[d] Also known as Greenland Stadial GS-1.

[e] Also known as Greenland Interstadial GI-a-c-e.

[f] Also known as MIS5e, which overlaps with the Eemian (Shackleton et al., 2003).

[g] As estimated from the Greenland ice core GICC05 chronology (Rasmussen et al., 2006; Thomas et al., 2007).

[h] In this chapter, when referring to comparison of radiative forcing or climate variables, pre-industrial refers to 1850 values in accordance with Taylor et al. (2012). Otherwise it refers to an extended period of time before 1850 as stated in the text. Note that Chapter 7 uses 1750 as the reference pre-industrial period.

[i] Different durations are reported in the literature. In Section 5.3.5, time intervals 950–1250 and 1450–1850 are used to calculate Northern Hemisphere temperature anomalies representative of the MCA and LIA, respectively.

[j] Note that CMIP5 "Last Millennium simulations" have been performed for the period 850–1850 (Taylor et al., 2012).

[k] As dated on Tahiti corals (Deschamps et al., 2012).

[l] The duration of Heinrich stadial 1 (e.g., Stanford et al., 2011) is longer than the associated Heinrich event, which is indicated by ice-rafted debris in deep sea sediment cores from the North Atlantic Ocean (Hemming, 2004).

[m] Period based on MARGO Project Members (2009). LGM simulations are performed for 21 ka. Note that maximum continental ice extent had already occurred at 26.5 ka (Clark et al., 2009).

[n] Ages are maximum date for the onset and minimum age for the end from tectonically stable sites (cf. Section 5.6.2).

[o] Dowsett et al. (2012).

[p] Zachos et al. (2008).

[q] Westerhold et al. (2007).

the magnetically inactive surface. These models can successfully reproduce the measured TSI changes between 1978 and 2003 (Balmaceda et al., 2007; Crouch et al., 2008), but not necessarily the last minimum of 2008 (Krivova et al., 2011). This approach requires detailed information of all the magnetic features and their temporal changes (Wenzler et al., 2006; Krivova and Solanki, 2008) (see Section 8.4).

The extension of TSI and SSI into the pre-satellite period poses two main challenges. First, the satellite period (since 1978) used to calibrate the solar irradiance models does not show any significant long-term trend. Second, information about the various magnetic features at the solar surface decreases back in time and must be deduced from proxies such as sunspot counts for the last 400 years and cosmogenic radionuclides (^{10}Be and ^{14}C) for the past millennium (Muscheler et al., 2007; Delaygue and Bard, 2011) and the Holocene (Table 5.1) (Steinhilber et al., 2009; Vieira et al., 2011). ^{10}Be and ^{14}C records reflect not only solar activity, but also the geomagnetic field intensity and effects of their respective geochemical cycles and transport pathways (Pedro et al., 2011; Steinhilber et al., 2012). The corrections for these non-solar components, which are difficult to quantify, contribute to the overall error of the reconstructions (grey band in Figure 5.1c).

TSI reconstructions are characterized by distinct grand solar minima lasting 50 to 100 years (e.g., the Maunder Minimum, 1645–1715) that are superimposed upon long-term changes. Spectral analysis of TSI records reveals periodicities of 87, 104, 150, 208, 350, 510, ~980 and ~2200 years (Figure 5.1d) (Stuiver and Braziunas, 1993), but with time-varying amplitudes (Steinhilber et al., 2009; Vieira et al., 2011). All reconstructions rely ultimately on the same data (sunspots and cosmogenic radionuclides), but differ in the details of the methodologies. As a result the reconstructions agree rather well in their shape, but differ in their amplitude (Figure 5.1b) (Wang et al., 2005; Krivova et al., 2011; Lean et al., 2011; Schrijver et al., 2011) (see Section 8.4.1).

Since AR4, most recent reconstructions show a considerably smaller difference (<0.1%) in TSI between the late 20th century and the Late Maunder Minimum (1675–1715) when the sun was very quiet, compared to the often used reconstruction of Lean et al. (1995b) (0.24%) and Shapiro et al. (2011) (~0.4%). The Lean et al. (1995a) reconstruction has been used to scale solar forcing in simulations of the last millennium prior to PMIP3/CMIP5 (Table 5.A.1). PMIP3/CMIP5 last

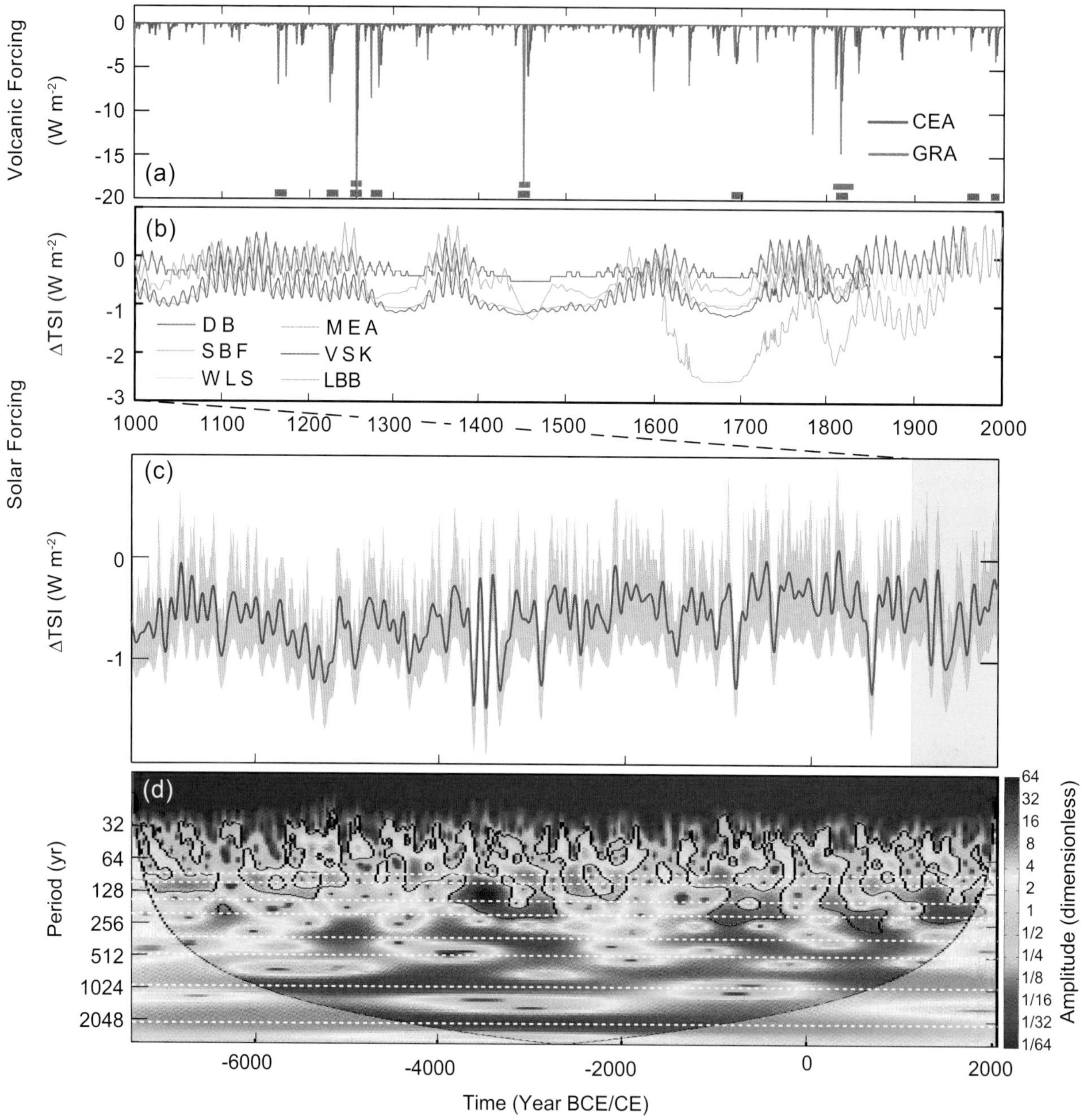

Figure 5.1 | (a) Two reconstructions of volcanic forcing for the past 1000 years derived from ice core sulphate and used for Paleoclimate Modelling Intercomparison Project Phase III (PMIP3) and Coupled Model Intercomparison Project Phase 5 (CMIP5) simulations (Schmidt et al., 2011). GRA: Gao et al. (2012); CEA: Crowley and Unterman (2013). Volcanic sulphate peaks identified from their isotopic composition as originating from the stratosphere are indicated by squares (green: Greenland; brown: Antarctica) (Baroni et al., 2008; Cole-Dai et al., 2009). (b) Reconstructed total solar irradiance (TSI) anomalies back to the year 1000. Proxies of solar activity (e.g., sunspots, ^{10}Be) are used to estimate the parameters of the models or directly TSI. All records except LBB (Lean et al., 1995b) have been used for PMIP3/CMIP5 simulations (Schmidt et al., 2011). DB: Delaygue and Bard (2011); MEA: Muscheler et al. (2007); SBF: Steinhilber et al. (2009); WLS: Wang et al. (2005); VSK: Vieira et al. (2011). For the years prior to 1600, the 11-year cycle has been added artificially to the original data with an amplitude proportional to the mean level of TSI. (c) Reconstructed TSI anomalies (100-year low-pass filtered; grey shading: 1 standard deviation uncertainty range) for the past 9300 years (Steinhilber et al., 2009). The reconstruction is based on ^{10}Be and calibrated using the relationship between instrumental data of the open magnetic field, which modulates the production of ^{10}Be, and TSI for the past four solar minima. The yellow band indicates the past 1000 years shown in more details in (a) and (b). Anomalies are relative to the 1976–2006 mean value (1366.14 W m^{-2}) of Wang et al. (2005). (d) Wavelet analysis (Torrence and Compo, 1998) of TSI anomalies from (c) with dashed white lines highlighting significant periodicities (Stuiver and Braziunas, 1993).

millennium simulations have used the weak solar forcing of recent reconstructions of TSI (Schmidt et al., 2011, 2012b) calibrated (Muscheler et al., 2007; Delaygue and Bard, 2011) or spliced (Steinhilber et al., 2009; Vieira and Solanki, 2010) to Wang et al. (2005). The larger range of past TSI variability in Shapiro et al. (2011) is not supported by studies of magnetic field indicators that suggest smaller changes over the 19th and 20th centuries (Svalgaard and Cliver, 2010; Lockwood and Owens, 2011).

Note that: (1) the recent new measurement of the absolute value of TSI and TSI changes during the past decades are assessed in Section 8.4.1.1; (2) the current state of understanding the effects of galactic cosmic rays on clouds is assessed in Sections 7.4.6 and 8.4.1.5 and (3) the use of solar forcing in simulations of the last millennium is discussed in Section 5.3.5.

5.2.1.3 Volcanic Forcing

Volcanic activity affects global climate through the radiative impacts of atmospheric sulphate aerosols injected by volcanic eruptions (see Sections 8.4.2 and 10.3.1). Quantifying volcanic forcing in the pre-satellite period is important for historical and last millennium climate simulations, climate sensitivity estimates and detection and attribution studies. Reconstructions of past volcanic forcing are based on sulphate deposition from multiple ice cores from Greenland and Antarctica, combined with atmospheric modelling of aerosol distribution and optical depth.

Since AR4, two new reconstructions of the spatial distribution of volcanic aerosol optical depth have been generated using polar ice cores, spanning the last 1500 years (Gao et al., 2008, 2012) and 1200 years (Crowley and Unterman, 2013) (Figure 5.1a). Although the relative size of eruptions for the past 700 years is generally consistent among these and earlier studies (Jansen et al., 2007), they differ in the absolute amplitude of peaks. There are also differences in the reconstructions of Icelandic eruptions, with an ongoing debate on the magnitude of stratospheric inputs for the 1783 Laki eruption (Thordarson and Self, 2003; Wei et al., 2008; Lanciki et al., 2012; Schmidt et al., 2012a). The recurrence time of past large volcanic aerosol injections (eruptions changing the radiative forcing (RF) by more than 1 W m^{-2}) varies from 3 to 121 years, with long-term mean value of 35 years (Gao et al., 2012) and 39 years (Crowley and Unterman, 2013), and only two or three periods of 100 years without such eruptions since 850.

Hegerl et al. (2006) estimated the uncertainty of the RF for a given volcanic event to be approximately 50%. Differences between reconstructions (Figure 5.1a) arise from different proxy data, identification of the type of injection, methodologies to estimate particle distribution and optical depth (Kravitz and Robock, 2011), and parameterization of scavenging for large events (Timmreck et al., 2009). Key limitations are associated with ice core chronologies (Plummer et al., 2012; Sigl et al., 2013), and deposition patterns (Moore et al., 2012).

A new independent methodology has recently been developed to distinguish between tropospheric and stratospheric volcanic aerosol deposits (Baroni et al., 2007). The stratospheric character of several large eruptions has started to be assessed from Greenland and/or Antarctic ice core sulphur isotope data (Baroni et al., 2008; Cole-Dai et al., 2009; Schmidt et al., 2012b).

The use of different volcanic forcing reconstructions in pre-PMIP3/CMIP5 (see AR4 Chapter 6) and PMIP3/CMIP5 last millennium simulations (Schmidt et al., 2011) (Table 5.A.1), together with the methods used to implement these volcanic indices with different representations of aerosols in climate models, is a source of uncertainty in model intercomparisons. The impact of volcanic forcing on climate variations of the last millennium climate is assessed in Sections 5.3.5, 5.4, 5.5.1 and 10.7.1.

5.2.2 Radiative Perturbations from Greenhouse Gases and Dust

5.2.2.1 Atmospheric Concentrations of Carbon Dioxide, Methane and Nitrous Oxide from Ice Cores

Complementing instrumental data, air enclosed in polar ice provides a direct record of past atmospheric well-mixed greenhouse gas (WMGHG) concentrations albeit smoothed by firn diffusion (Joos and Spahni, 2008; Köhler et al., 2011). Since AR4, the temporal resolution of ice core records has been enhanced (MacFarling Meure et al., 2006; Ahn and Brook, 2008; Loulergue et al., 2008; Lüthi et al., 2008; Mischler et al., 2009; Schilt et al., 2010; Ahn et al., 2012; Bereiter et al., 2012). During the pre-industrial part of the last 7000 years, millennial (20 ppm CO_2, 125 ppb CH_4) and centennial variations (up to 10 ppm CO_2, 40 ppb CH_4 and 10 ppb N_2O) are recorded (see Section 6.2.2 and Figure 6.6). Significant centennial variations in CH_4 during the last glacial occur in phase with Northern Hemisphere (NH) rapid climate changes, while millennial CO_2 changes coincide with their Southern Hemisphere (SH) bipolar seesaw counterpart (Ahn and Brook, 2008; Loulergue et al., 2008; Lüthi et al., 2008; Grachev et al., 2009; Capron et al., 2010b; Schilt et al., 2010; Bereiter et al., 2012).

Long-term records have been extended from 650 ka in AR4 to 800 ka (Figures 5.2 and 5.3) (Loulergue et al., 2008; Lüthi et al., 2008; Schilt et al., 2010). During the last 800 ka, the pre-industrial ice core WMGHG concentrations stay within well-defined natural limits with maximum interglacial concentrations of approximately 300 ppm, 800 ppb and 300 ppb for CO_2, CH_4 and N_2O, respectively, and minimum glacial concentrations of approximately 180 ppm, 350 ppb, and 200 ppb. The new data show lower than pre-industrial (280 ppm) CO_2 concentrations during interglacial periods from 800 to 430 ka (MIS19 to MIS13) (Figure 5.3). It is a fact that present-day (2011) concentrations of CO_2 (390.5 ppm), CH_4 (1803 ppb) and N_2O (324 ppm) (Annex II) exceed the range of concentrations recorded in the ice core records during the past 800 ka. With *very high confidence*, the rate of change of the observed anthropogenic WMGHG rise and its RF is unprecedented with respect to the highest resolution ice core record back to 22 ka for CO_2, CH_4 and N_2O, accounting for the smoothing due to ice core enclosure processes (Joos and Spahni, 2008; Schilt et al., 2010). There is *medium confidence* that the rate of change of the observed anthropogenic WMGHG rise is also unprecedented with respect to the lower resolution records of the past 800 ka.

Progress in understanding the causes of past WMGHG variations is reported in Section 6.2.

Frequently Asked Questions

FAQ 5.1 | Is the Sun a Major Driver of Recent Changes in Climate?

Total solar irradiance (TSI, Chapter 8) is a measure of the total energy received from the sun at the top of the atmosphere. It varies over a wide range of time scales, from billions of years to just a few days, though variations have been relatively small over the past 140 years. Changes in solar irradiance are an important driver of climate variability (Chapter 1; Figure 1.1) along with volcanic emissions and anthropogenic factors. As such, they help explain the observed change in global surface temperatures during the instrumental period (FAQ 5.1, Figure 1; Chapter 10) and over the last millennium. While solar variability may have had a discernible contribution to changes in global surface temperature in the early 20th century, it cannot explain the observed increase since TSI started to be measured directly by satellites in the late 1970s (Chapters 8, 10).

The Sun's core is a massive nuclear fusion reactor that converts hydrogen into helium. This process produces energy that radiates throughout the solar system as electromagnetic radiation. The amount of energy striking the top of Earth's atmosphere varies depending on the generation and emission of electromagnetic energy by the Sun and on the Earth's orbital path around the Sun.

Satellite-based instruments have directly measured TSI since 1978, and indicate that on average, ~1361 W m^{-2} reaches the top of the Earth's atmosphere. Parts of the Earth's surface and air pollution and clouds in the atmosphere act as a mirror and reflect about 30% of this power back into space. Higher levels of TSI are recorded when the Sun is more active. Irradiance variations follow the roughly 11-year sunspot cycle: during the last cycles, TSI values fluctuated by an average of around 0.1%.

For pre-satellite times, TSI variations have to be estimated from sunspot numbers (back to 1610), or from radioisotopes that are formed in the atmosphere, and archived in polar ice and tree rings. Distinct 50- to 100-year periods of very low solar activity—such as the Maunder Minimum between 1645 and 1715—are commonly referred to as grand solar minima. Most estimates of TSI changes between the Maunder Minimum and the present day are in the order of 0.1%, similar to the amplitude of the 11-year variability.

How can solar variability help explain the observed global surface temperature record back to 1870? To answer this question, it is important to understand that other climate drivers are involved, each producing characteristic patterns of regional climate responses. However, it is the combination of them all that causes the observed climate change. Solar variability and volcanic eruptions are natural factors. Anthropogenic (human-produced) factors, on the other hand, include changes in the concentrations of greenhouse gases, and emissions of visible air pollution (aerosols) and other substances from human activities. 'Internal variability' refers to fluctuations within the climate system, for example, due to weather variability or phenomena like the El Niño-Southern Oscillation.

The relative contributions of these natural and anthropogenic factors change with time. FAQ 5.1, Figure 1 illustrates those contributions based on a very simple calculation, in which the mean global surface temperature variation represents the sum of four components linearly related to solar, volcanic, and anthropogenic forcing, and to internal variability. Global surface temperature has increased by approximately 0.8°C from 1870 to 2010 (FAQ 5.1, Figure 1a). However, this increase has not been uniform: at times, factors that cool the Earth's surface—volcanic eruptions, reduced solar activity, most anthropogenic aerosol emissions—have outweighed those factors that warm it, such as greenhouse gases, and the variability generated within the climate system has caused further fluctuations unrelated to external influences.

The solar contribution to the record of global surface temperature change is dominated by the 11-year solar cycle, which can explain global temperature fluctuations up to approximately 0.1°C between minima and maxima (FAQ 5.1, Figure 1b). A long-term increasing trend in solar activity in the early 20th century may have augmented the warming recorded during this interval, together with internal variability, greenhouse gas increases and a hiatus in volcanism. However, it cannot explain the observed increase since the late 1970s, and there was even a slight decreasing trend of TSI from 1986 to 2008 (Chapters 8 and 10).

Volcanic eruptions contribute to global surface temperature change by episodically injecting aerosols into the atmosphere, which cool the Earth's surface (FAQ 5.1, Figure 1c). Large volcanic eruptions, such as the eruption of Mt. Pinatubo in 1991, can cool the surface by around 0.1°C to 0.3°C for up to three years. *(continued on next page)*

FAQ 5.1 (continued)

The most important component of internal climate variability is the El Niño Southern Oscillation, which has a major effect on year-to-year variations of tropical and global mean temperature (FAQ 5.1, Figure 1d). Relatively high annual temperatures have been encountered during El Niño events, such as in 1997–1998.

The variability of observed global surface temperatures from 1870 to 2010 (Figure 1a) reflects the combined influences of natural (solar, volcanic, internal; FAQ 5.1, Figure 1b–d) factors, superimposed on the multi-decadal warming trend from anthropogenic factors (FAQ 5.1, Figure 1e).

Prior to 1870, when anthropogenic emissions of greenhouse gases and aerosols were smaller, changes in solar and volcanic activity and internal variability played a more important role, although the specific contributions of these individual factors to global surface temperatures are less certain. Solar minima lasting several decades have often been associated with cold conditions. However, these periods are often also affected by volcanic eruptions, making it difficult to quantify the solar contribution.

At the regional scale, changes in solar activity have been related to changes in surface climate and atmospheric circulation in the Indo-Pacific, Northern Asia and North Atlantic areas. The mechanisms that amplify the regional effects of the relatively small fluctuations of TSI in the roughly 11-year solar cycle involve dynamical interactions between the upper and the lower atmosphere, or between the ocean sea surface temperature and atmosphere, and have little effect on global mean temperatures (see Box 10.2).

Finally, a decrease in solar activity during the past solar minimum a few years ago (FAQ 5.1, Figure 1b) raises the question of its future influence on climate. Despite uncertainties in future solar activity, there is *high confidence* that the effects of solar activity within the range of grand solar maxima and minima will be much smaller than the changes due to anthropogenic effects.

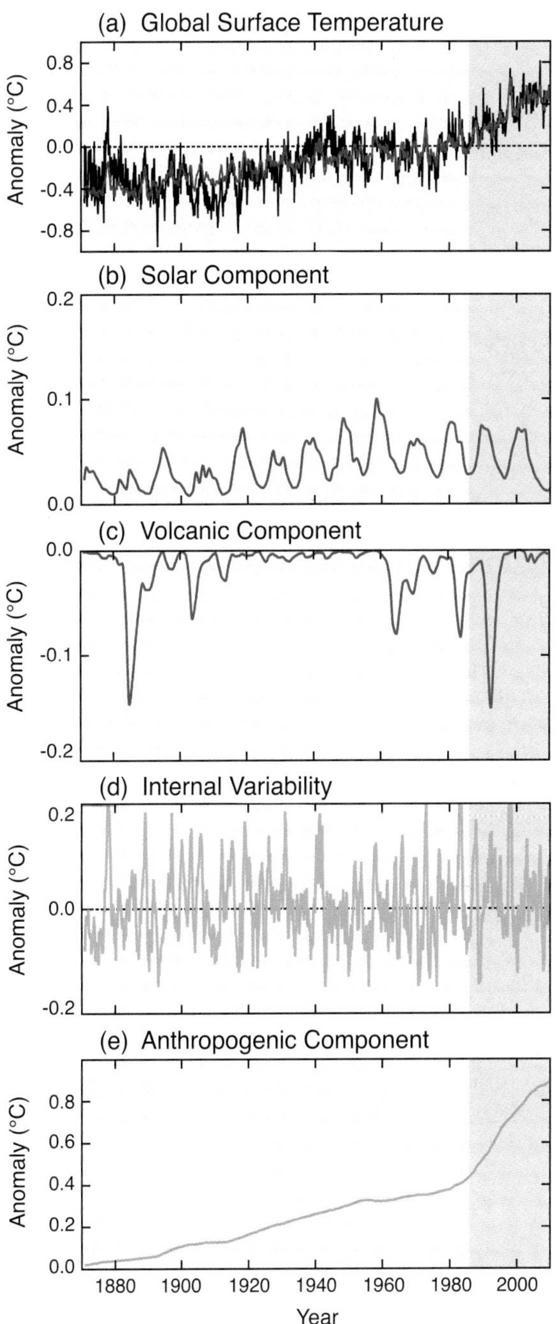

FAQ 5.1, Figure 1 | Global surface temperature anomalies from 1870 to 2010, and the natural (solar, volcanic, and internal) and anthropogenic factors that influence them. (a) Global surface temperature record (1870–2010) relative to the average global surface temperature for 1961–1990 (black line). A model of global surface temperature change (a: red line) produced using the sum of the impacts on temperature of natural (b, c, d) and anthropogenic factors (e). (b) Estimated temperature response to solar forcing. (c) Estimated temperature response to volcanic eruptions. (d) Estimated temperature variability due to internal variability, here related to the El Niño-Southern Oscillation. (e) Estimated temperature response to anthropogenic forcing, consisting of a warming component from greenhouse gases, and a cooling component from most aerosols.

5

5.2.2.2 Atmospheric Carbon Dioxide Concentrations from Geological Proxy Data

Geological proxies provide indirect information on atmospheric CO_2 concentrations for time intervals older than those covered by ice core records (see Section 5.2.2.1). Since AR4, the four primary proxy CO_2 methods have undergone further development (Table 5.A.2). A reassessment of biological respiration and carbonate formation has reduced CO_2 reconstructions based on fossil soils by approximately 50% (Breecker et al., 2010). Bayesian statistical techniques for calibrating leaf stomatal density reconstructions produce consistently higher CO_2 estimates than previously assessed (Beerling et al., 2009), resulting in more convergence between estimates from these two terrestrial proxies. Recent CO_2 reconstructions using the boron isotope proxy provide an improved understanding of foraminifer species effects and evolution of seawater alkalinity (Hönisch and Hemming, 2005) and seawater boron isotopic composition (Foster et al., 2012). Quantification of the phytoplankton cell-size effects on carbon isotope fractionation has also improved the consistency of the alkenone method (Henderiks and Pagani, 2007). These proxies have also been applied more widely and at higher temporal resolution to a range of geological archives, resulting in an increased number of atmospheric CO_2 estimates since 65 Ma (Beerling and Royer, 2011). Although there is improved consensus between the proxy CO_2 estimates, especially the marine proxy estimates, a significant degree of variation among the different techniques remains. All four techniques have been included in the assessment, as there is insufficient knowledge to discriminate between different proxy estimates on the basis of confidence (assessed in Table 5.A.2).

In the time interval between 65 and 23 Ma, all proxy estimates of CO_2 concentration span a range of 300 ppm to 1500 ppm (Figure 5.2). An independent constraint on Early Eocene atmospheric CO_2 concentration is provided by the occurrence of the sodium carbonate mineral nahcolite, in about 50 Ma lake sediments, which precipitates in association with halite at the sediment–water interface only at CO_2 levels >1125 ppm (Lowenstein and Demicco, 2006), and thus provides a potential lower bound for atmospheric concentration (*medium confidence*) during the warmest period of the last 65 Ma, the Early Eocene Climatic Optimum (EECO; 52 to 50 Ma; Table 5.1), which is inconsistent with lower estimates from stomata and paleosoils. Although the reconstructions indicate a general decrease in CO_2 concentrations since about 50 Ma (Figure 5.2), the large scatter of proxy data precludes a robust assessment of the second-order variation around this overall trend.

5

Since 23 Ma, CO_2 proxy estimates are at pre-industrial levels with exception of the Middle Miocene climatic optimum (17 to 15 Ma) and the Pliocene (5.3 to 2.6 Ma), which have higher concentrations. Although new CO_2 reconstructions for the Pliocene based on marine proxies have produced consistent estimates mostly in the range 350 ppm to 450 ppm (Pagani et al., 2010; Seki et al., 2010; Bartoli et al., 2011), the uncertainties associated with these marine estimates remain difficult to quantify. Several boron-derived data sets agree within error (±25 ppm) with the ice core records (Foster, 2008; Hönisch et al., 2009), but alkenone data for the ice core period are outside the error limits (Figure 5.2). We conclude that there is *medium confidence* that CO_2 levels were above pre-industrial interglacial concentration (~280 ppm) and did not exceed ~450 ppm during the Pliocene, with interglacial values in the upper part of that range between 350 and 450 ppm.

5.2.2.3 Past Changes in Mineral Dust Aerosol Concentrations

Past changes in mineral dust aerosol (MDA) are important for estimates of climate sensitivity (see Section 5.3.3) and for its supply of nutrients, especially iron to the Southern Ocean (see Section 6.2). MDA concentration is controlled by variations in dust sources, and by changes in atmospheric circulation patterns acting on its transport and lifetime.

Since AR4, new records of past MDA flux have been obtained from deep-sea sediment and ice cores. A 4 million-year MDA-flux reconstruction from the Southern Ocean (Figure 5.2) implies reduced dust generation and transport during the Pliocene compared to Holocene levels, followed by a significant rise around 2.7 Ma when NH ice volume increased (Martinez-Garcia et al., 2011). Central Antarctic ice core records show that local MDA deposition fluxes are ~20 times higher during glacial compared to interglacial periods (Fischer et al., 2007; Lambert et al., 2008; Petit and Delmonte, 2009). This is due to enhanced dust production in southern South America and perhaps Australia (Gaiero, 2007; De Deckker et al., 2010; Gabrielli et al., 2010; Martinez-Garcia et al., 2011; Wegner et al., 2012). The impact of changes in MDA lifetime (Petit and Delmonte, 2009) on dust fluxes in Antarctica remains uncertain (Fischer et al., 2007; Wolff et al., 2010). Equatorial Pacific glacial–interglacial MDA fluxes co-vary with Antarctic records, but with a glacial–interglacial ratio in the range of approximately three to four (Winckler et al., 2008), attributed to enhanced dust production from Asian and northern South American sources in glacial times (Maher et al., 2010). The dominant dust source regions (e.g., North Africa, Arabia and Central Asia) show complex patterns of variability (Roberts et al., 2011). A glacial increase of MDA source strength by a factor of 3 to 4 requires low vegetation cover, seasonal aridity, and high wind speeds (Fischer et al., 2007; McGee et al., 2010). In Greenland ice cores, MDA ice concentrations are higher by a factor of 100 and deposition fluxes by a factor 20 during glacial periods (Ruth et al., 2007). This is due mainly to changes in the dust sources for Greenland (Asian desert areas), increased gustiness (McGee et al., 2010) and atmospheric lifetime and transport of MDA (Fischer et al., 2007). A strong coherence is observed between dust in Greenland ice cores and aeolian deposition in European loess formations (Antoine et al., 2009).

Global data synthesis shows two to four times more dust deposition at the Last Glacial Maximum (LGM; Table 5.1) than today (Derbyshire, 2003; Maher et al., 2010). Based on data–model comparisons, estimates of global mean LGM dust RF vary from -3 W m^{-2} to $+0.1$ W m^{-2}, due to uncertainties in radiative properties. The best estimate value remains at -1 W m^{-2} as in AR4 (Claquin et al., 2003; Mahowald et al., 2006, 2011; Patadia et al., 2009; Takemura et al., 2009; Yue et al., 2010). Models may underestimate the MDA RF at high latitudes (Lambert et al., 2013).

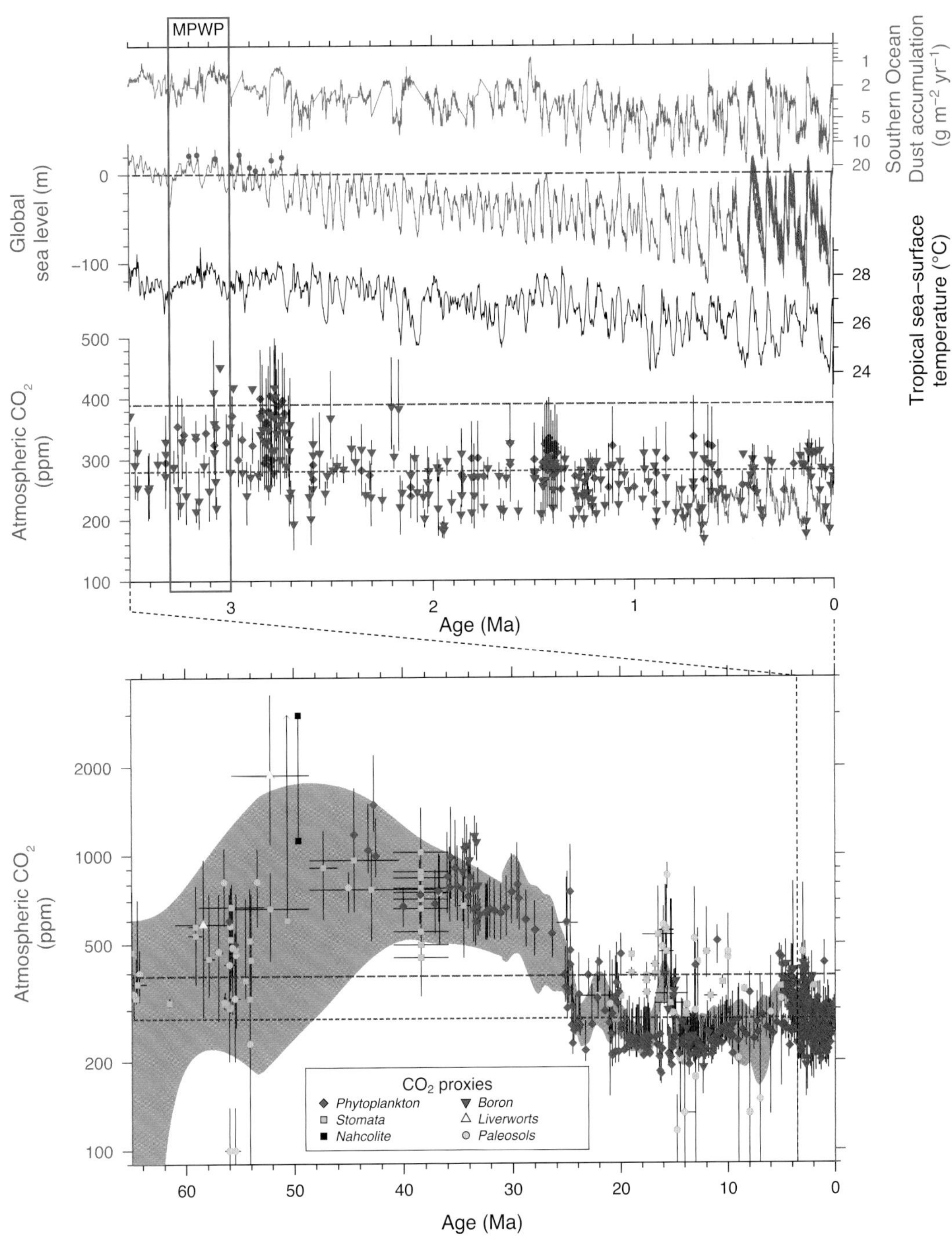

Figure 5.2 | (Top) Orbital-scale Earth system responses to radiative forcings and perturbations from 3.5 Ma to present. Reconstructed dust mass accumulation rate is from the Atlantic sector of the Southern Ocean (red) (Martinez-Garcia et al., 2011). Sea level curve (blue) is the stacked $\delta^{18}O$ proxy for ice volume and ocean temperature (Lisiecki and Raymo, 2005) calibrated to global average eustatic sea level (Naish and Wilson, 2009; Miller et al., 2012a). Also shown are global eustatic sea level reconstructions for the last 500 kyr based on sea level calibration of the $\delta^{18}O$ curve using dated coral shorelines (green line; Waelbroeck et al., 2002) and the Red Sea isotopic reconstruction (red line; Rohling et al., 2009). Weighted mean estimates (2 standard deviation uncertainty) for far-field reconstructions of eustatic peaks are shown for mid-Pliocene interglacials (red dots; Miller et al., 2012a). The dashed horizontal line represents present-day sea level. Tropical sea surface temperature (black line) based on a stack of four alkenone-based sea surface temperature reconstructions (Herbert et al., 2010). Atmospheric carbon dioxide (CO_2) measured from Antarctic ice cores (green line, Petit et al., 1999; Siegenthaler et al., 2005; Lüthi et al., 2008), and estimates of CO_2 from boron isotopes ($\delta^{11}B$) in foraminifera in marine sediments (blue triangles; Hönisch et al., 2009; Seki et al., 2010; Bartoli et al., 2011), and phytoplankton alkenone-derived carbon isotope proxies (red diamonds; Pagani et al., 2010; Seki et al., 2010), plotted with 2 standard deviation uncertainty. Present (2012) and pre-industrial CO_2 concentrations are indicated with long-dashed and short-dashed grey lines, respectively. (Bottom) Concentration of atmospheric CO_2 for the last 65 Ma is reconstructed from marine and terrestrial proxies (Cerling, 1992; Freeman and Hayes, 1992; Koch et al., 1992; Stott, 1992; van der Burgh et al., 1993; Sinha and Stott, 1994; Kürschner, 1996; McElwain, 1998; Ekart et al., 1999; Pagani et al., 1999a, 1999b, 2005a, 2005b, 2010, 2011; Kürschner et al., 2001, 2008; Royer et al., 2001a, 2001b; Beerling et al., 2002, 2009; Beerling and Royer, 2002; Nordt et al., 2002; Greenwood et al., 2003; Royer, 2003; Lowenstein and Demicco, 2006; Fletcher et al., 2008; Pearson et al., 2009; Retallack, 2009b, 2009a; Tripati et al., 2009;Seki et al., 2010; Smith et al., 2010; Bartoli et al., 2011; Doria et al., 2011; Foster et al., 2012). Individual proxy methods are colour-coded (see also Table A5.1). The light blue shading is a 1-standard deviation uncertainty band constructed using block bootstrap resampling (Mudelsee et al., 2012) for a kernel regression through all the data points with a bandwidth of 8 Myr prior to 30 Ma, and 1 Myr from 30 Ma to present. Most of the data points for CO_2 proxies are based on duplicate and multiple analyses. The red box labelled MPWP represents the mid-Pliocene Warm Period (3.3 to 3.0 Ma; Table 5.1).

Box 5.1 | Polar Amplification

Polar amplification occurs if the magnitude of zonally averaged surface temperature change at high latitudes exceeds the globally averaged temperature change, in response to climate forcings and on time scales greater than the annual cycle. Polar amplification is of global concern due to the potential effects of future warming on ice sheet stability and, therefore, global sea level (see Sections 5.6.1, 5.8.1 and Chapter 13) and carbon cycle feedbacks such as those linked with permafrost melting (see Chapter 6).

Some external climate forcings have an enhanced radiative impact at high latitudes, such as orbital forcing (Section 5.2.1.1), or black carbon (Section 8.3.4). Here, we focus on the latitudinal response of surface temperature to CO_2 perturbations. The magnitude of polar amplification depends on the relative strength and duration of different climate feedbacks, which determine the transient and equilibrium response to external forcings. This box first describes the different feedbacks operating in both polar regions, and then contrasts polar amplification depicted for past high CO_2 and low CO_2 climates with projected temperature patterns for the RCP8.5 future greenhouse gas (WMGHG) emission scenario.

In the Arctic, the sea ice/ocean surface albedo feedback plays an important role (Curry et al., 1995; Serreze and Barry, 2011). With retreating sea ice, surface albedo decreases, air temperatures increase and the ocean can absorb more heat. The resulting ocean warming contributes to further sea ice melting. The sea ice/ocean surface albedo feedback can exhibit threshold behaviour when temperatures exceed the freezing point of sea ice. This may also translate into a strong seasonality of the response characteristics. Other feedbacks, including water vapour and cloud feedbacks have been suggested as important amplifiers of Arctic climate change (Vavrus, 2004; Abbot and Tziperman, 2008, 2009; Graversen and Wang, 2009; Lu and Cai, 2009; Screen and Simmonds, 2010; Bintanja et al., 2011). In continental Arctic regions with seasonal snow cover, changes in radiative forcing (RF) can heavily influence snow cover (Ghatak et al., 2010), and thus surface albedo. Other positive feedbacks operating on time scales of decades-to-centuries in continental high-latitude regions are associated with surface vegetation changes (Bhatt et al., 2010) and thawing permafrost (e.g., Walter et al., 2006). On glacial-to-interglacial time scales, the very slow ice sheet–albedo response to external forcings (see Box 5.2) is a major contributor to polar amplification in the Northern Hemisphere.

An amplified response of Southern Ocean sea surface temperature (SST) to radiative perturbations also emerges from the sea ice–albedo feedback. However, in contrast to the Arctic Ocean, which in parts is highly stratified, mixed-layer depths in the Southern Ocean typically exceed several hundreds of meters, which allows the ocean to take up vast amounts of heat (Böning et al., 2008; Gille, 2008; Sokolov and Rintoul, 2009) and damp the SST response to external forcing. This process, and the presence of the ozone hole over the Antarctic ice sheet (Thompson and Solomon, 2002, 2009), can affect the transient response of surface warming of the Southern Ocean and Antarctica, and lead to different patterns of future polar amplification on multi-decadal to multi-centennial time scales. In response to rapid atmospheric CO_2 changes, climate models indeed project an asymmetric warming between the Arctic and Southern Oceans, with an earlier response in the Arctic and a delayed response in the Southern Ocean (Section 12.4.3). Above the Antarctic ice sheet, however, surface air temperature can respond quickly to radiative perturbations owing to the limited role of latent heat flux in the surface energy budget of Antarctica.

These differences in transient and equilibrium responses of surface temperatures on Antarctica, the Southern Ocean and over continents and oceans in the Arctic domain can explain differences in the latitudinal temperature patterns depicted in Box 5.1, Figure 1 for past periods (equilibrium response) and future projections (transient response).

Box 5.1, Figure 1 illustrates the polar amplification phenomenon for three different periods of the Earth's climate history using temperature reconstructions from natural archives and climate model simulations for: (i) the Early Eocene Climatic Optimum (EECO, 54 to 48 Ma) characterised by CO_2 concentrations of 1000 to 2000 ppm (Section 5.2.2.2) and the absence of continental ice sheets; (ii) the mid-Pliocene Warm Period (MPWP, 3.3 to 3.0 Ma), characterized by CO_2 concentrations in the range of 350 to 450 ppm (Section 5.2.2.2) and reduced Greenland and Antarctic ice sheets compared to today (see Section 5.6.1), (iii) the Last Glacial Maximum (LGM, 21 to 19 ka), characterized by CO_2 concentrations around 200 ppm and large continental ice sheets covering northern Europe and North America.

Throughout all three time periods, reconstructions and simulations reveal Arctic and Antarctic surface air temperature amplification of up to two times the global mean (Box 5.1, Figure 1c, d), and this bipolar amplification appears to be a robust feature of the equilibrium Earth system response to changes of CO_2 concentration, irrespective of climate state. The absence (EECO), or expansion (LGM) of continental ice sheets has the potential to affect the zonally averaged surface temperatures due to the lapse-rate effect (see Box 5.2), hence contributing to polar amplification. However, polar amplification is also suppressed in zonally averaged gradients of SST compared with terrestrial surface air temperature (Box 5.1, Figure 1), owing to the presence of high-latitude sea ice in the pre-industrial control,

(continued on next page)

Box 5.1 (continued)

which places a lower limit on SST. Global mean temperature estimates for these three past climates also imply an Earth system climate sensitivity to radiative perturbations up to two times higher than the equilibrium climate sensitivity (Lunt et al., 2010; Haywood et al., 2013) (see Section 5.3.1 and Box 12.2).

Polar amplification explains in part why Greenland Ice Sheet (GIS) and the West Antarctic Ice Sheet (WAIS) appear to be highly sensitive to relatively small increases in CO_2 concentration and global mean temperature. For example, global sea level during MPWP may have been up to +20m higher than present day when atmospheric CO_2 concentrations were ~350 to 450 ppm and global mean surface temperature was 2°C to 3°C above pre-industrial levels (see Sections 5.6.1 and 5.8.1). *(continued on next page)*

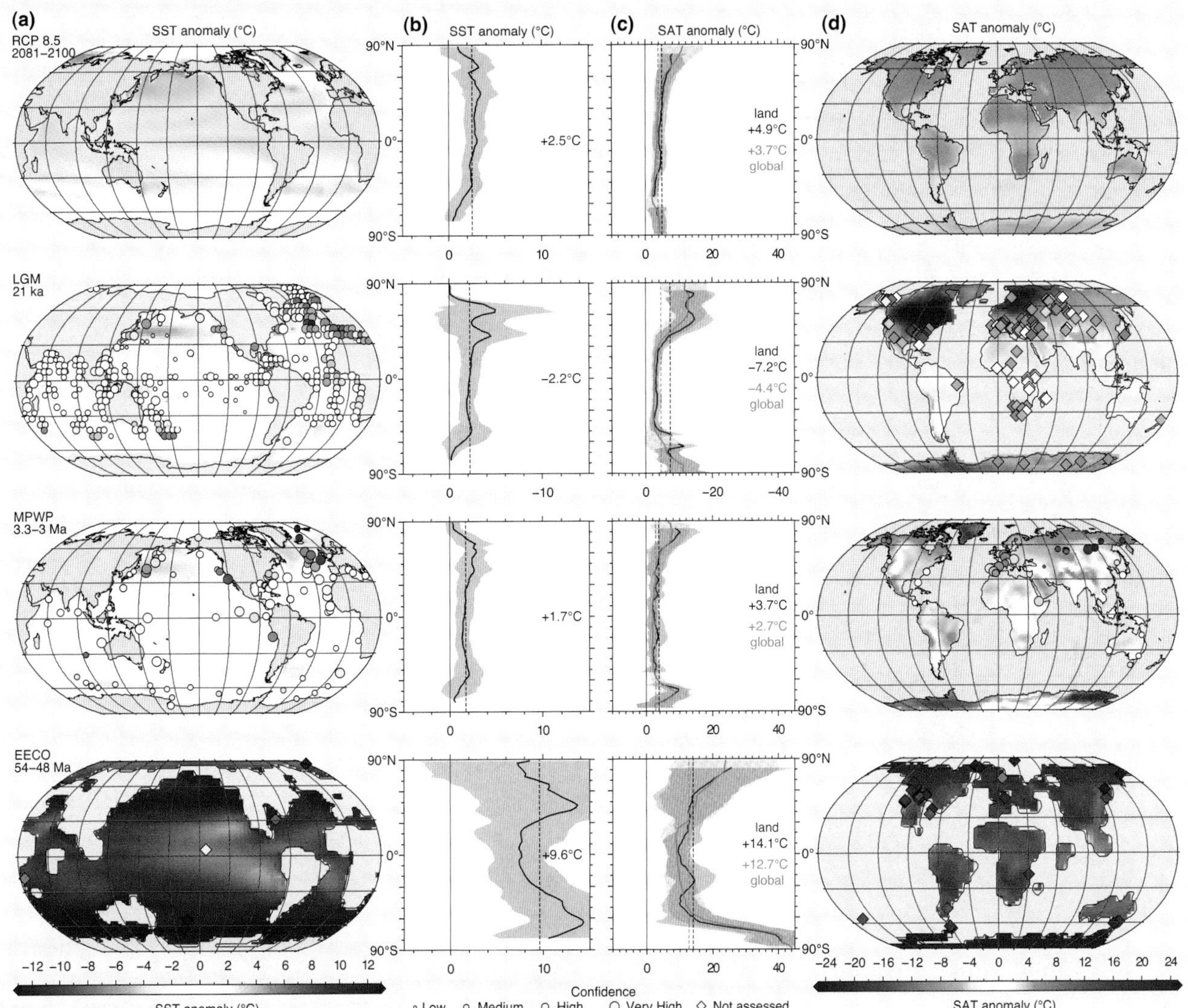

Box 5.1, Figure 1 | Comparison of data and multi-model mean (MMM) simulations, for four periods of time, showing (a) sea surface temperature (SST) anomalies, (b) zonally averaged SST anomalies, (c) zonally averaged global (green) and land (grey) surface air temperature (SAT) anomalies and (d) land SAT anomalies. The time periods are 2081–2100 for the Representative Concentration Pathway (RCP) 8.5 (top row), Last Glacial Maximum (LGM, second row), mid-Pliocene Warm Period (MPWP, third row) and Early Eocene Climatic Optimum (EECO, bottom row). Model temperature anomalies are calculated relative to the pre-industrial value of each model in the ensemble prior to calculating the MMM anomaly (a, d; colour shading). Zonal MMM gradients (b, c) are plotted with a shaded band indicating 2 standard deviations. Site specific temperature anomalies estimated from proxy data are calculated relative to present site temperatures and are plotted (a, d) using the same colour scale as the model data, and a circle-size scaled to estimates of confidence. Proxy data compilations for the LGM are from Multiproxy Approach for the Reconstruction of the Glacial Ocean surface (MARGO) Project Members (2009) and Bartlein et al. (2011), for the MPWP are from Dowsett et al. (2012), Salzmann et al. (2008) and Haywood et al. (2013) and for the EECO are from Hollis et al. (2012) and Lunt et al. (2012). Model ensemble simulations for 2081–2100 are from the CMIP5 ensemble using RCP 8.5, for the LGM are seven Paleoclimate Modelling Intercomparison Project Phase III (PMIP3) and Coupled Model Intercomparison Project Phase 5 (CMIP5) models, for the Pliocene are from Haywood et al., (2013), and for the EECO are after Lunt et al. (2012).

Box 5.1 (continued)

Based on earlier climate data–model comparisons, it has been claimed (summarised in Huber and Caballero, 2011), that models underestimated the strength of polar amplification for high CO_2 climates by 30 to 50%. While recent simulations of the EECO and the MPWP exhibit a wide inter-model variability, there is generally good agreement between new simulations and data, particularly if seasonal biases in some of the marine SST proxies from high-latitude sites are considered (Hollis et al., 2012; Lunt et al., 2012; Haywood et al., 2013).

Transient polar amplification as recorded in historical instrumental data and as projected by coupled climate models for the 21st century involves a different balance of feedbacks than for the "equilibrium" past states featured in Box 5.1, Figure 1. Since 1875, the Arctic north of 60°N latitude has warmed at a rate of 1.36°C per century, approximately twice as fast as the global average (Bekryaev et al., 2010), and since 1979, Arctic land surface has warmed at an even higher rate of 0.5°C per decade (e.g., Climatic Research Unit (CRU) Gridded Dataset of Global Historical Near-Surface Air TEMperature Anomalies Over Land version 4 (CRUTEM4), Jones et al., 2012; Hadley Centre/CRU gridded surface temperature data set version 4 (HadCRUT4), Morice et al., 2012) (see Section 2.4). This recent warming appears unusual in the context of reconstructions spanning the past 2000 years (Section 5.5) and has been attributed primarily to anthropogenic factors (Gillett et al., 2008) (see Section 10.3.1.1.4). The fact that the strongest warming occurs in autumn and early winter (Chylek et al., 2009; Serreze et al., 2009; Polyakov et al., 2010; Screen and Simmonds, 2010; Semenov et al., 2010; Spielhagen et al., 2011) strongly links Arctic amplification to feedbacks associated with the seasonal reduction in sea ice extent and duration, as well as the insulating effect of sea ice in winter (e.g., Soden et al., 2008; Serreze et al., 2009; Serreze and Barry, 2011). For future model projections (Box 5.1, Figure 1), following the RCP8.5 scenarios, annual mean Arctic (68°N to 90°N) warming is expected to exceed the global average by 2.2 to 2.4 times for the period 2081–2100 compared to 1986–2005 (see Section 12.4.3.1), which corresponds to the higher end of polar amplification implied by paleo-reconstructions.

The transient response of Antarctic and Southern Ocean temperatures to the anthropogenic perturbation appears more complex, than for the Arctic region. Zonal mean Antarctic surface warming has been modest at 0.1°C per decade over the past 50 years (Steig et al., 2009; O'Donnell et al., 2010). The Antarctic Peninsula is experiencing one of the strongest regional warming trends (0.5°C per decade over the past 50 years), more than twice that of the global mean temperature. Central West Antarctica may have also experienced a similar strong warming trend, as depicted by the only continuous meteorological station during the last 50 years (Bromwich et al., 2013), and borehole measurements spanning the same period (Orsi et al., 2012). Ice core records show enhanced summer melting in the Antarctic Peninsula since the 1950, which is unprecedented over the past 1000 years (Abram et al., 2013), and warming in West Antarctica that cannot be distinguished from natural variability over the last 2000 years (Steig et al., 2013) (see also Section 10.3.1.1.4, and Section 5.5). Polar amplification in the Southern Ocean and Antarctica is virtually absent in the transient CMIP5 RCP4.5 future simulations (2081–2100 versus 1986–2005) (see Section 12.4.3.1), although CMIP5 RCP8.5 exhibits an amplified warming in the Southern Ocean (Box 5.1, Figure 1), much smaller in magnitude than the equilibrium response implied from paleo-reconstructions for a high-CO_2 world.

In summary, *high confidence* exists for polar amplification in either one or both hemispheres, based on robust and consistent evidence from temperature reconstructions of past climates, recent instrumental temperature records and climate model simulations of past, present and future climate changes.

5.3 Earth System Responses and Feedbacks at Global and Hemispheric Scales

This section updates the information available since AR4 on changes in surface temperature on million-year to orbital time scales and for the last 2000 years. New information on changes of the monsoon systems on glacial–interglacial time scales is also assessed.

5.3.1 High-Carbon Dioxide Worlds and Temperature

Cenozoic (last 65 Ma) geological archives provide examples of natural climate states globally warmer than the present, which are associated with atmospheric CO_2 concentrations above pre-industrial levels. This relationship between global warmth and high CO_2 is complicated by factors such as tectonics and the evolution of biological systems, which play an important role in the carbon cycle (e.g., Zachos et al., 2008). Although new reconstructions of deep-ocean temperatures have been compiled since AR4 (e.g., Cramer et al., 2011), *low confidence* remains in the precise relationship between CO_2 and deep-ocean temperature (Beerling and Royer, 2011).

Since AR4 new proxy and model data have become available from three Cenozoic warm periods to enable an assessment of forcing, feedbacks and the surface temperature response (e.g., Dowsett et al., 2012; Lunt et al., 2012; Haywood et al., 2013). These are the Paleocene–Eocene Thermal Maximum (PETM; Table 5.1), the Early Eocene Climatic Optimum (EECO; Table 5.1) and the mid-Pliocene Warm Period (MPWP; Table 5.1). Reconstructions of surface temperatures based on proxy data remain challenged by (i) the limited number and uneven geographical distribution of sites, (ii) seasonal biases and (iii) the validity of assumptions required by each proxy method (assessed in Table 5.A.3). There is also a lack of consistency in the way uncertainties are reported for proxy climate estimates. In most cases error bars represent the analytical and calibration error. In some compilations qualitative confidence assessments are reported to account for the quality of the age control, number of samples, fossil preservation and abundance, performance of the proxy method utilized and agreement of multiple proxy estimates (e.g., Multiproxy Approach for the Reconstruction of the Glacial Ocean surface (MARGO) Project Members, 2009; Dowsett et al., 2012).

The PETM was marked by a massive carbon release and corresponding global ocean acidification (Zachos et al., 2005; Ridgwell and Schmidt, 2010) and, with *low confidence*, global warming of 4°C to 7°C relative to pre-PETM mean climate (Sluijs et al., 2007; McInerney and Wing, 2011). The carbon release of 4500 to 6800 PgC over 5 to 20 kyr translates into a rate of emissions of ~0.5 to 1.0 PgC yr^{-1} (Panchuk et al., 2008; Zeebe et al., 2009). GHG emissions from marine methane hydrate and terrestrial permafrost may have acted as positive feedbacks (DeConto et al., 2012).

The EECO represents the last time atmospheric CO_2 concentrations may have reached a level of ~1000 ppm (Section 5.2.2.2). There were no substantial polar ice sheets, and oceanic and continental configurations, vegetation type and distribution were significantly different from today. Whereas simulated SAT are in reasonable agreement with reconstructions (Huber and Caballero, 2011; Lunt et al., 2012) (Box 5.1, Figure 1d), there are still significant discrepancies between simulated and reconstructed mean annual SST, which are reduced if seasonal biases in some of the marine proxies are considered for the high-latitude sites (Hollis et al., 2012; Lunt et al., 2012). *Medium confidence* is placed on the reconstructed global mean surface temperature anomaly estimate of 9°C to 14°C.

The Pliocene is characterized by a long-term increase in global ice volume and decrease in temperature from ~3.3–2.6 Ma (Lisiecki and Raymo, 2005; Mudelsee and Raymo, 2005; Fedorov et al., 2013), which marks the onset of continental-scale glaciations in the NH. Superimposed on this trend, benthic $\delta^{18}O$ (Lisiecki and Raymo, 2005) and an ice proximal geological archive (Lisiecki and Raymo, 2005; Naish et al., 2009a) imply moderate fluctuations in global ice volume paced by the 41 kyr obliquity cycle. This orbital variability is also evident in far-field sea level reconstructions (Miller et al., 2012a), tropical Pacific SST (Herbert et al., 2010) and Southern Ocean MDA records (Martinez-Garcia et al., 2011), and indicate a close coupling between temperature, atmospheric circulation and ice volume/sea level (Figure 5.2). The MPWP and the following 300 kyr represent the last time atmospheric CO_2 concentrations were in the range 350 to 450 ppm (Section 5.2.2.2, Figure 5.2). Model–data comparisons (Box 5.1, Figure 1) provide *high confidence* that mean surface temperature was warmer than pre-industrial for the average interglacial climate state during the MPWP (Dowsett et al., 2012; Haywood et al., 2013). Global mean SST is estimated at +1.7°C (without uncertainty) above the 1901–1920 mean based on large data syntheses (Lunt et al., 2010; Dowsett et al., 2012). General circulation model (GCM) results agree with this SST anomaly (to within ±0.5°C), and produce a range of global mean SAT of +1.9°C and +3.6°C relative to the 1901–1920 mean (Haywood et al., 2013). Weakened meridional temperature gradients are shown by all GCM simulations, and have significant implications for the stability of polar ice sheets and sea level (see Box 5.1 and Section 5.6). SST gradients and the Pacific Ocean thermocline gradient along the equator were greatly reduced compared to present (Fedorov et al., 2013) (Section 5.4). Vegetation reconstructions (Salzmann et al., 2008) imply that the global extent of arid deserts decreased and boreal forests replaced tundra, and GCMs predict an enhanced hydrological cycle, but with large inter-model spread (Haywood et al., 2013). The East Asian Summer Monsoon, as well as other monsoon systems, may have been enhanced at this time (e.g., Wan et al., 2010).

Climate reconstructions for the warm periods of the Cenozoic also provide an opportunity to assess Earth-system and equilibrium climate sensitivities. Uncertainties on both global temperature and CO_2 reconstructions preclude deriving robust quantitative estimates from the available PETM data. The limited number of models for MPWP, which take into account slow feedbacks such as ice sheets and the carbon cycle, imply with *medium confidence* that Earth-system sensitivity may be up to two times the model equilibrium climate sensitivity (ECS) (Lunt et al., 2010; Pagani et al., 2010; Haywood et al., 2013). However, if the slow amplifying feedbacks associated with ice sheets and CO_2 are considered as forcings rather than feedbacks, climate records of the past 65 Myr yield an estimate of 1.1°C to 7°C (95% confidence interval) for ECS (PALAEOSENS Project Members, 2012) (see also Section 5.3.3.2).

5.3.2 Glacial–Interglacial Dynamics

5.3.2.1 Role of Carbon Dioxide in Glacial Cycles

Recent modelling work provides strong support for the important role of variations in the Earth's orbital parameters in generating long-term climate variability. In particular, new simulations with GCMs (Carlson et al., 2012; Herrington and Poulsen, 2012) support the fundamental premise of the Milankovitch theory that a reduction in NH summer insolation generates sufficient cooling to initiate ice sheet growth. Climate–ice sheet models with varying degrees of complexity and forced by variations in orbital parameters and reconstructed atmospheric CO_2 concentrations simulate ice volume variations and other climate characteristics during the last and several previous glacial cycles consistent with paleoclimate records (Abe-Ouchi et al., 2007; Bonelli et al., 2009; Ganopolski et al., 2010) (see Figure 5.3).

There is *high confidence* that orbital forcing is the primary external driver of glacial cycles (Kawamura et al,. 2007; Cheng et al., 2009; Lisiecki, 2010; Huybers, 2011). However, atmospheric CO_2 content plays an important internal feedback role. Orbital-scale variability in CO_2 concentrations over the last several hundred thousand years covaries (Figure 5.3) with variability in proxy records including reconstructions of global ice volume (Lisiecki and Raymo, 2005), climatic conditions in central Asia (Prokopenko et al., 2006), tropical (Herbert et al., 2010) and Southern Ocean SST (Pahnke et al., 2003; Lang and Wolff, 2011), Antarctic temperature (Parrenin et al., 2013), deep-ocean temperature (Elderfield et al., 2010), biogeochemical conditions in the North Pacific (Jaccard et al., 2010) and deep-ocean ventilation (Lisiecki et al., 2008). Such close linkages between CO_2 concentration and climate variability are consistent with modelling results suggesting with *high confidence* that glacial–interglacial variations of CO_2 and other GHGs explain a considerable fraction of glacial–interglacial climate variability in regions not directly affected by the NH continental ice sheets (Timmermann et al., 2009; Shakun et al., 2012).

5.3.2.2 Last Glacial Termination

It is *very likely* that global mean surface temperature increased by 3°C to 8°C over the last deglaciation (see Table 5.2), which gives a *very likely* average rate of change of 0.3 to 0.8°C kyr^{-1}. Deglacial global warming occurred in two main steps from 17.5 to 14.5 ka and 13.0 to 10.0 ka that *likely* reached maximum rates of change between 1°C kyr^{-1} and 1.5°C kyr^{-1} at the millennial time scale (cf. Shakun et al., 2012; Figure 5.3i), although regionally and on shorter time scales higher rates may have occurred, in particular during a sequence of abrupt climate change events (see Section 5.7).

For the last glacial termination, a large-scale temperature reconstruction (Shakun et al., 2012) documents that temperature change in the SH lead NH temperature change. This lead can be explained by the bipolar thermal seesaw concept (Stocker and Johnsen, 2003) (see also Section 5.7) and the related changes in the inter-hemispheric ocean heat transport, caused by weakening of the Atlantic Ocean meridional overturning circulation (AMOC) during the last glacial termination (Ganopolski and Roche, 2009). SH warming prior to NH warming can also be explained by the fast sea ice response to changes in austral spring insolation (Stott et al., 2007; Timmermann et al., 2009). According to these mechanisms, SH temperature lead over the NH is fully consistent with the NH orbital forcing of deglacial ice volume changes (*high confidence*) and the importance of the climate–carbon cycle feedbacks in glacial–interglacial transitions. The tight coupling is further highlighted by the near-zero lag between the deglacial rise in CO_2 and averaged deglacial Antarctic temperature recently reported from improved estimates of gas-ice age differences (Pedro et al., 2012; Parrenin et al., 2013). Previous studies (Monnin et al., 2001; Table 5.A.4)

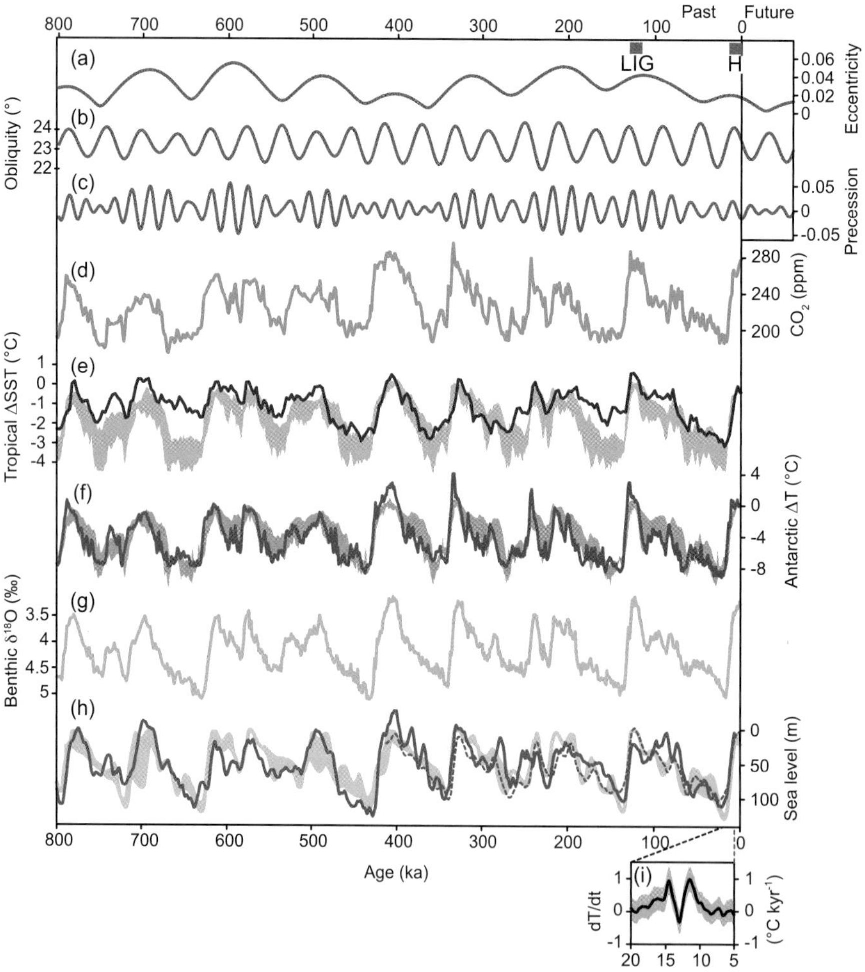

Figure 5.3 | Orbital parameters and proxy records over the past 800 kyr. (a) Eccentricity. (b) Obliquity. (c) Precessional parameter (Berger and Loutre, 1991). (d) Atmospheric concentration of CO_2 from Antarctic ice cores (Petit et al., 1999; Siegenthaler et al., 2005; Ahn and Brook, 2008; Lüthi et al., 2008). (e) Tropical sea surface temperature stack (Herbert et al., 2010). (f) Antarctic temperature stack based on up to seven different ice cores (Petit et al., 1999; Blunier and Brook, 2001; Watanabe et al., 2003; European Project for Ice Coring in Antarctica (EPICA) Community Members, 2006; Jouzel et al., 2007; Stenni et al., 2011). (g) Stack of benthic $\delta^{18}O$, a proxy for global ice volume and deep-ocean temperature (Lisiecki and Raymo, 2005). (h) Reconstructed sea level (dashed line: Rohling et al., 2010; solid line: Elderfield et al., 2012). Lines represent orbital forcing and proxy records, shaded areas represent the range of simulations with climate models (Grid Enabled Integrated Earth System Model-1, GENIE-1, Holden et al., 2010a; Bern3D, Ritz et al., 2011), climate–ice sheet models of intermediate complexity (CLIMate and BiosphERe model, CLIMBER-2, Ganopolski and Calov, 2011) and an ice sheet model (ICe sheet model for Integrated Earth system studies, IcIES, Abe-Ouchi et al., 2007) forced by variations of the orbital parameters and the atmospheric concentrations of the major greenhouse gases. (i) Rate of changes of global mean temperature during Termination I based on Shakun et al. (2012).

suggesting a temperature lead of 800 ± 600 years over the deglacial CO_2 rise probably overestimated gas-ice age differences.

5.3.2.3 Monsoon Systems

Since AR4, new high-resolution hydroclimate reconstructions using speleothems (Sinha et al., 2007; Hu et al., 2008; Wang et al., 2008; Cruz et al., 2009; Asmerom et al., 2010; Berkelhammer et al., 2010; Stríkis et al., 2011; Kanner et al., 2012), lake sediments (Shanahan et al., 2009; Stager et al., 2009; Wolff et al., 2011), marine sediments (Weldeab et al., 2007b; Mulitza et al., 2008; Tjallingii et al., 2008; Ponton et al., 2012) and tree-ring chronologies (Buckley et al., 2010; Cook et al., 2010a) have provided a more comprehensive view on the dynamics of monsoon systems on a variety of time scales. Water isotope-enabled modelling experiments (LeGrande and Schmidt, 2009; Lewis et al., 2010; Pausata et al., 2011) and evaluation of marine and terrestrial data (Clemens et al., 2010) document that speleothem $\delta^{18}O$ variations in some monsoon regions can be explained as a combination of changes in local precipitation and large-scale moisture transport.

This subsection focuses on the response of monsoon systems to orbital forcing on glacial–interglacial time scales. Proxy data including speleothem $\delta^{18}O$ from southeastern China (Wang et al., 2008), northern Borneo (Meckler et al., 2013), eastern Brazil (Cruz et al., 2005) and the Arabian Peninsula (Bar-Matthews et al., 2003), along with marine-based records off northwestern Africa (Weldeab et al., 2007a) and from the Arabian Sea (Schulz et al., 1998) document hydrological changes that are dominated by eccentricity-modulated precessional cycles. Increasing boreal summer insolation can generate a strong inter-hemispheric surface temperature gradient that leads to large-scale decreases in precipitation in the SH summer monsoon systems and increased hydrological cycle in the NH tropics (Figure 5.4a, d, g). Qualitatively

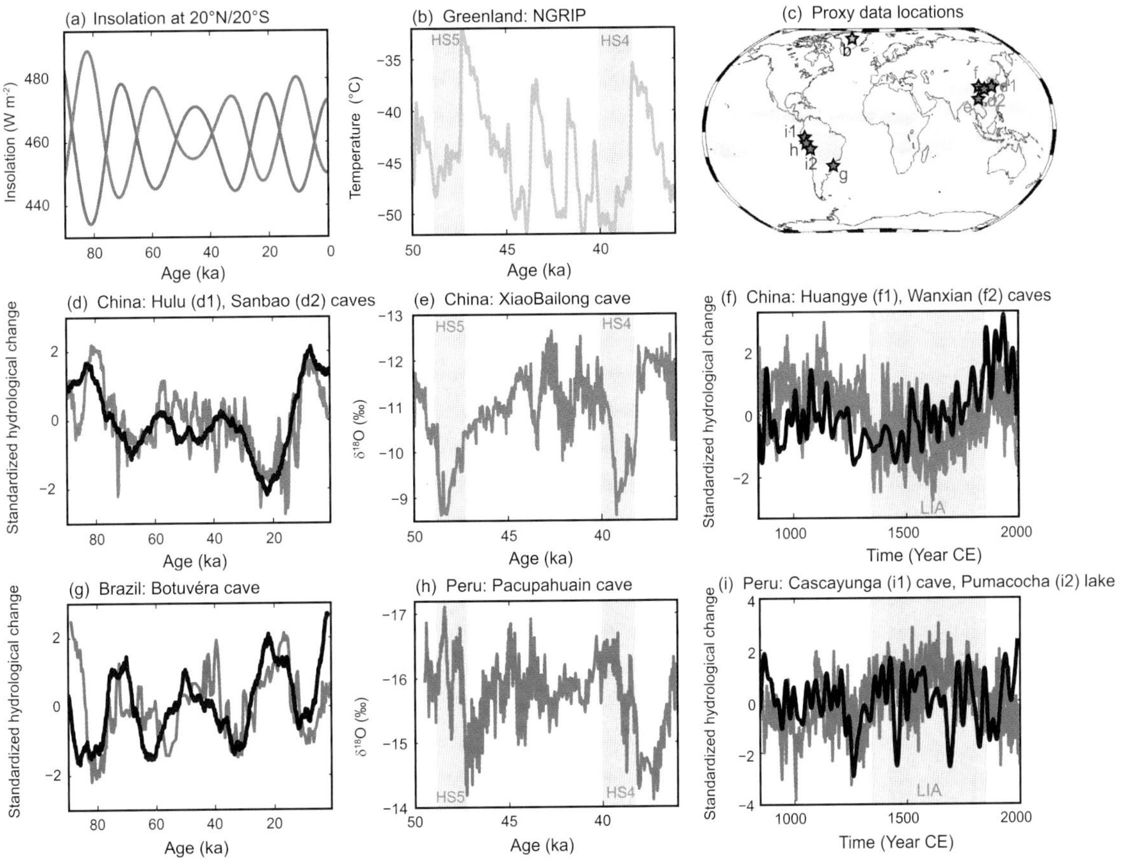

Figure 5.4 | Inter-hemispheric response of monsoon systems at orbital, millennial and centennial scales. (a) Boreal summer insolation changes at 20°N (red) (W m $^{-2}$) and austral summer insolation changes at 20°S (blue). (b) Temperature changes in Greenland (degrees Celsius) reconstructed from North Greenland Ice Core Project (NGRIP) ice core on SS09 time scale (Huber et al., 2006), location indicated by orange star in c. (c) Location of proxy records displayed in panels a, b, d–i in relation to the global monsoon regions (cyan shading) (Wang and Ding, 2008). (d) Reconstructed (red) standardized negative $\delta^{18}O$ anomaly in East Asian Summer Monsoon region derived from Hulu (Wang et al., 2001) and Sanbao (Wang et al., 2008) cave speleothem records, China and simulated standardized multi-model average (black) of annual mean rainfall anomalies averaged over region 108°E to 123°E and 25°N to 40°N using the transient runs conducted with LOch–Vecode-Ecbilt-CLio-aglsm Model (LOVECLIM, Timm et al., 2008), FAst Met Office/UK Universities Simulator (FAMOUS, Smith and Gregory, 2012), and the Hadley Centre Coupled Model (HadCM3) snapshot simulations (Singarayer and Valdes, 2010). (e) $\delta^{18}O$ from Xiaobailong cave, China (Cai et al., 2010). (f) Standardized negative $\delta^{18}O$ anomalies (red) in Huangye (Tan et al., 2011) and Wanxian (Zhang et al., 2008) caves, China and simulated standardized annual mean and 30-year low-pass filtered rainfall anomalies (black) in region 100°E to 110°E, 20°N to 35°N, ensemble averaged over externally forced Atmosphere-Ocean General Circulation Model (AOGCM) experiments conducted with Community Climate System Model-4 (CCSM4), ECHAM4+HOPE-G (ECHO-G), Max Planck Institute Earth System Model (MPI-ESM), Commonwealth Scientific and Industrial Research Organisation model (CSIRO-Mk3L-1-2), Model for Interdisciplinary Research on Climate (MIROC), HadCM3 (Table 5.A.1). (g) Standardized negative $\delta^{18}O$ anomaly (blue) from Botuvéra speleothem, Brazil (Cruz et al., 2005) and simulated standardized multi-model average (black) of annual mean rainfall anomalies averaged over region 45°W to 60°W and 35°S to 15°S using same experiments as in panel d. (h) Standardized $\delta^{18}O$ anomaly (blue) from Pacupahuain cave, Peru (Kanner et al., 2012). (i) Standardized negative $\delta^{18}O$ anomalies (blue) from Cascayunga Cave, Peru (Reuter et al., 2009) and Pumacocha Lake, Peru (Bird et al., 2011) and simulated standardized annual mean and 30-year low-pass filtered rainfall anomalies (black) in region 76°W to 70°W, 16°S to 8°S, ensemble averaged over the same model simulations as in f. HS4/5 denote Heinrich stadials 4 and 5, and LIA denotes Little Ice Age (Table 5.1).

similar, out-of-phase inter-hemispheric responses to insolation forcing have also been documented in coupled time-slice and transient GCM simulations (Braconnot et al., 2008; Kutzbach et al., 2008) (see also Figure 5.4d, g). The similarity in response in both proxy records and models provide *high confidence* that orbital forcing induces inter-hemispheric rainfall variability. Across longitudes, the response of precipitation may, however, be different for the same orbital forcing (Shin et al., 2006; Marzin and Braconnot, 2009). For example, in the mid-Holocene drier conditions occurred in central North America (Diffenbaugh et al., 2006) and wetter conditions in northern Africa (Liu et al., 2007b; Hély et al., 2009; Tierney et al., 2011). There is further evidence for east–west shifts of precipitation in response to orbital forcing in South America (Cruz et al., 2009).

Box 5.2 | Climate-Ice Sheet Interactions

Ice sheets have played an essential role in the Earth's climate history (see Sections 5.3, 5.6 and 5.7). They interact with the atmosphere, the ocean–sea ice system, the lithosphere and the surrounding vegetation (see Box 5.2, Figure 1). They serve as nonlinear filters and integrators of climate effects caused by orbital and GHG forcings (Ganopolski and Calov, 2011), while at the same time affecting the global climate system on a variety of time scales (see Section 5.7).

Ice sheets form when annual snow accumulation exceeds melting. Growing ice sheets expand on previously vegetated areas, thus leading to an increase of surface albedo, further cooling and an increase in net surface mass balance. As ice sheets grow in height and area, surface temperatures drop further as a result of the lapse-rate effect, but also snow accumulation decreases because colder air holds less moisture (inlay in Box 5.2, Figure 1). This so-called elevation-desert effect (Oerlemans, 1980) is an important negative feedback for ice sheets which limits their growth. Higher elevation ice sheets can be associated with enhanced calving at their margins, because the ice flow will be accelerated directly by increased surface slopes and indirectly by lubrication at the base of the ice sheet. Calving, grounding line processes, basal lubrication and other forms of thermo-mechanical coupling may have played important roles in accelerating glacial terminations following phases of relatively slow ice sheet growth, hence contributing to the temporal saw-tooth structure of the recent glacial–interglacial cycles (Figure 5.3).

Large glacial ice sheets also deflect the path of the extratropical NH westerly winds (Cook and Held, 1988), generating anticyclonic circulation anomalies (Box 5.2, Figure 1), which tend to warm the western side of the ice sheet and cool the remainder (e.g., Roe and Lindzen, 2001). Furthermore, the orographic effects of ice sheets lead to reorganizations of the global atmosphere circulation by changing the major stationary wave patterns (e.g., Abe-Ouchi et al., 2007; Yin et al., 2008) and trade wind systems (Timmermann et al., 2004). This allows for a fast transmission of ice sheet signals to remote regions.

The enormous weight of ice sheets depresses the underlying bedrocks causing a drop in ice sheet height and a surface warming as a result of the lapse-rate effect. The lithospheric adjustment has been shown to play an important role in modulating the ice sheet response to orbital forcing (Birchfield et al., 1981; van den Berg et al., 2008). The presence of terrestrial sedimentary materials (regolith) on top of the unweathered bedrock affects the friction at the base of an ice sheet, and may further alter the response of continental ice sheets to external forcings, with impacts on the dominant periodicities of glacial cycles (Clark and Pollard, 1998).

An area of very active research is the interaction between ice sheets, ice shelves and the ocean (see Sections 4.4, 13.4.3 and 13.4.4). The mass balance of marine ice sheets is strongly determined by ocean temperatures (Joughin and Alley, 2011). Advection of warmer waters below ice shelves can cause ice shelf instabilities, reduced buttressing, accelerated ice stream flow (De Angelis and Skvarca, 2003) and grounding line retreat in regions with retrograde bedrock slopes (Schoof, 2012), such as West Antarctica. On orbital and millennial time scales such processes may have played an essential role in driving ice volume changes of the West Antarctic ice sheet (Pollard and DeConto, 2009) and the Laurentide ice sheet (Alvarez-Solas et al., 2010). Massive freshwater release from retreating ice sheets, can feed back to the climate system by altering sea level, oceanic deep convection, ocean circulation, heat transport, sea ice and the global atmospheric circulation (Sections 5.6.3 and 5.7).

Whereas the initial response of ice sheets to external forcings can be quite fast, involving for instance ice shelf processes and outlet glaciers (10 to 10^3 years), their long-term adjustment can take much longer (10^4 to 10^5 years) (see Section 12.5.5.3). As a result, the climate–cryosphere system is not even in full equilibrium with the orbital forcing. This also implies that future anthropogenic radiative perturbations over the next century can determine the evolution of the Greenland (Charbit et al., 2008) and Antarctic ice sheets for centuries and millennia to come with a potential commitment to significant global sea level rise (Section 5.8). *(continued on next page)*

Box 5.2 (continued)

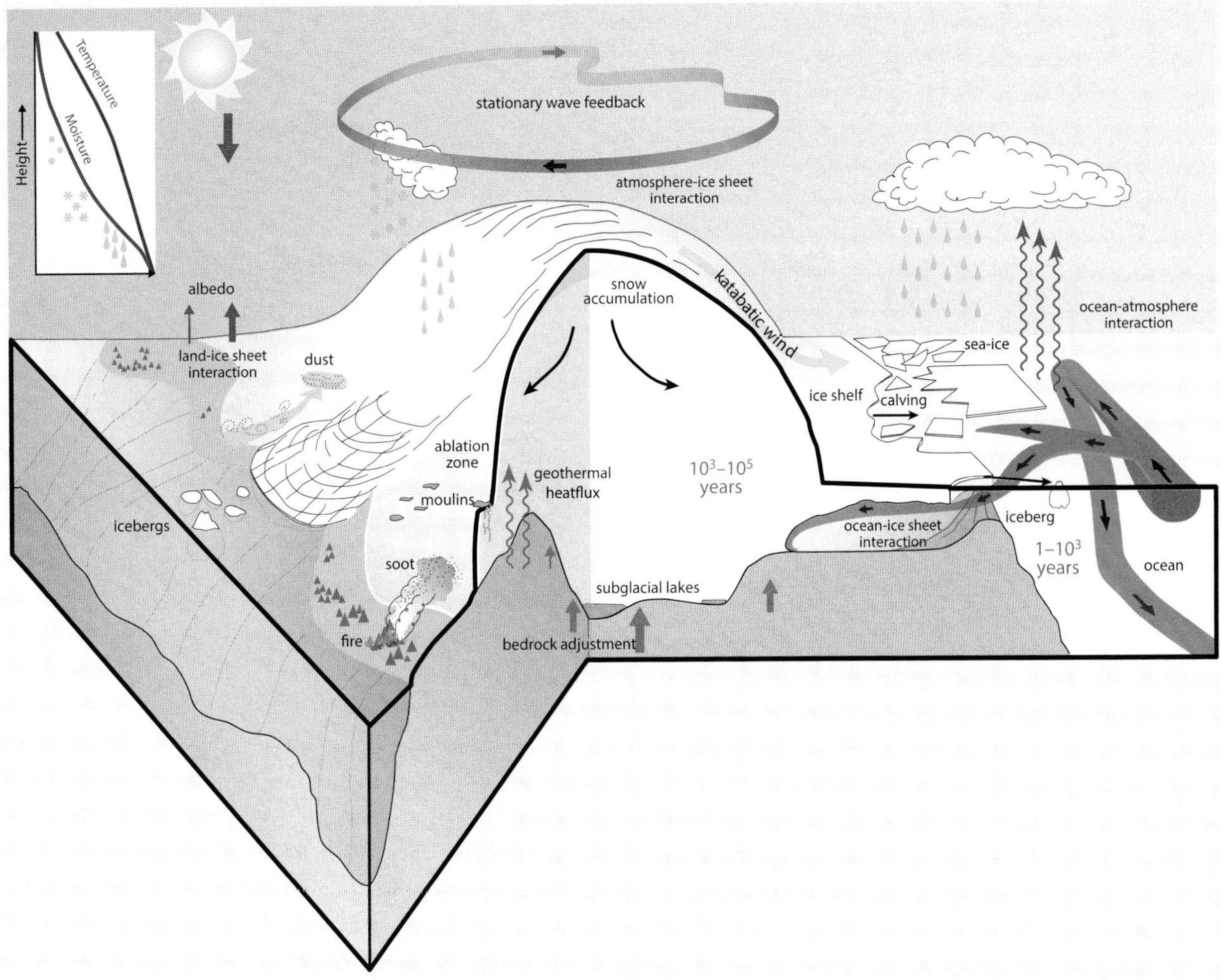

Box 5.2, Figure 1 | Schematic illustration of multiple interactions between ice sheets, solid earth and the climate system which can drive internal variability and affect the coupled ice sheet–climate response to external forcings on time scales of months to millions of years. The inlay figure represents a typical height profile of atmospheric temperature and moisture in the troposphere.

5.3.3 Last Glacial Maximum and Equilibrium Climate Sensitivity

The LGM is characterized by a large temperature response (Section 5.3.3.1) to relatively well-defined radiative perturbations (Section 5.2), linked to atmospheric CO_2 concentration around 200 ppm (Section 5.2.2) and large ice sheets covering northern Europe and North America. This can be used to evaluate climate models (Braconnot et al. (2012b); see Sections 9.7 and 10.8) and to estimate ECS from the combined use of proxy information and simulations (Section 5.3.3.2).

5.3.3.1 Last Glacial Maximum Climate

Since AR4, synthesis of proxy LGM temperature estimates was completed for SST (MARGO Project Members, 2009), and for land SAT (Bartlein et al., 2011) (Box 5.1, Figure 1). The Multiproxy Approach for the Reconstruction of the Glacial Ocean Surface (MARGO) SST synthesis expanded earlier work (CLIMAP Project Members, 1976, 1981; Sarnthein et al., 2003a; Sarnthein et al., 2003b) by using multiple proxies (Table 5.2). The land SAT synthesis is based on pollen data, following the Cooperative Holocene Mapping Project (COHMAP Members, 1988).

Climate models and proxy data consistently show that mean annual SST change (relative to pre-industrial) is largest in the mid-latitude North Atlantic (up to –10°C), and the Mediterranean (about –6°C) (MARGO Project Members, 2009, Box 5.1, Table 5.2). Warming and seasonally ice-free conditions are reconstructed, however, in the north-eastern North Atlantic, in the eastern Nordic Seas and north Pacific, albeit with large uncertainty because of the different interpretation of proxy data (de Vernal et al., 2006). SAT reconstructions generally shows year-round cooling, with regional exceptions such as Alaska (Bartlein et al., 2011). Modelling studies show how atmospheric dynamics influenced by ice sheets affect regional temperature patterns in the North

Table 5.2 | Summary of Last Glacial Maximum (LGM) sea surface temperature (SST) and surface air temperature (SAT) reconstructions (anomalies with respect to pre-industrial climate) using proxy data and model ensemble constrained by proxy data. Cooling ranges indicate 90% confidence intervals (C.I., where available).

Region	Cooling (°C) 90% C.I.	Methods	Reference and remarks
Sea Surface Temperature (SST)			
Global	0.7–2.7	Multi-proxy	MARGO Project Members (2009)
Mid-latitude North Atlantic	up to 10	Multi-proxy	MARGO Project Members (2009)
Southern Ocean	2–6	Multi-proxy	MARGO Project Members (2009)
Low-latitude (30°S to 30°N)	0.3–2.7	Multi-proxy	MARGO Project Members (2009) 1.7°C ± 1°C: 15°S to 15°N 2.9°C ± 1.3°C: Atlantic 15°S to 15°N 1.2°C ± 1.1°C : Pacific 15°S to 15°N (1.2°C ± 1°C based on microfossil assemblages; 2.5°C ± 1°C based on Mg/Ca ratios and alkenones)
Low-latitude (30°S to 30°N)	2.2–3.2	Multi-proxy	Ballantyne et al. (2005)
Low-latitude (western and eastern tropical Pacific)	2–3	Multi-proxy	Lea et al. (2000); de Garidel-Thoron et al. (2007); Leduc et al. (2007); Pahnke et al. (2007); Stott et al. (2007); Koutavas and Sachs (2008); Steinke et al. (2008); Linsley et al. (2010)
Surface Air Temperature (SAT)			
Eastern Antarctica	7–10	Water stable isotopes from ice core	Stenni et al. (2010); Uemura et al. (2012)
Central Greenland	21–25	Borehole paleothermometry	Cuffey et al. (1995); Johnsen et al. (1995); Dahl-Jensen et al. (1998)
Global	4.4–7.2	Single-EMIC ensemble with microfossil-assemblage derived tropical Atlantic SST	Schneider von Deimling et al. (2006)
Global	4.6–8.3	Single-EMIC ensemble with multi-proxy derived tropical SST	Holden et al. (2010a)
Global	1.7–3.7	Single-EMIC ensemble with global multi-proxy data	Schmittner et al. (2011)
Global	3.9–4.6	Multi-proxy	Shakun et al. (2012); for the interval 17.5–9.5 ka
Global	3.4–4.6	Multi-AOGCM ensemble with global multi-proxy data	Annan and Hargreaves (2013)
Global	3.1–5.9	Multi-AOGCM ensemble	PMIP2 and PMIP3/CMIP5

Notes:

AOGCM = Atmosphere-Ocean General Circulation Model; CMIP5 = Coupled Model Intercomparison Project Phase 5; EMIC = Earth System Model of Intermediate Complexity; MARGO = Multiproxy Approach for the Reconstruction of the Glacial Ocean surface; PMIP2 and PMIP3 = Paleoclimate Modelling Intercomparison Project Phase II and III, respectively.

5

Atlantic region (Lainé et al., 2009; Pausata et al., 2011; Unterman et al., 2011; Hofer et al., 2013), and in the north Pacific region (Yanase and Abe-Ouchi, 2010). Larger cooling over land compared to ocean is a robust feature of observations and multiple atmosphere–ocean general circulation models (AOGCM) (Izumi et al., 2013). As in AR4, central Greenland temperature change during the LGM is underestimated by PMIP3/CMIP5 simulations, which show 2°C to 18°C cooling, compared to 21°C to 25°C cooling reconstructed from ice core data (Table 5.2). A mismatch between reconstructions and model results may arise because of missing Earth system feedbacks (dust, vegetation) (see Section 5.2.2.3) or insufficient integration time to reach an equilibrium for LGM boundary conditions.

Uncertainties remain on the magnitude of tropical SST cooling during the LGM. Previous estimates of tropical cooling (2.7°C ± 0.5°C, Ballantyne et al., 2005) are greater than more recent estimates (1.5°C ± 1.2°C, MARGO Project Members, 2009). Such discrepancies may arise from seasonal productivity (Leduc et al., 2010) and habitat depth (Telford et al., 2013) biases. In the western and eastern tropical Pacific, many proxy records show 2°C to 3°C cooling relative to the pre-industrial period (Lea et al., 2000, 2006; de Garidel-Thoron et al., 2007; Leduc et al., 2007; Pahnke et al., 2007; Stott et al., 2007; Koutavas and Sachs, 2008; Steinke et al., 2008; Linsley et al., 2010). AOGCMs tend to underestimate longitudinal patterns of tropical SST (Otto-Bliesner et al., 2009) and atmospheric circulation (DiNezio et al., 2011).

Larger sea ice seasonality is reconstructed for the LGM compared to the pre-industrial period around Antarctica (Gersonde et al., 2005). Climate models underestimate this feature, as well as the magnitude of Southern Ocean cooling (Roche et al., 2012) (see Box 5.1, Figure 1). In Antarctica 7°C to 10°C cooling relative to the pre-industrial period is reconstructed from ice cores (Stenni et al., 2010; Uemura et al., 2012) and captured in most PMIP3/CMIP5 simulations (Figure 5.5d).

The combined use of proxy reconstructions, with incomplete spatial coverage, and model simulations is used to estimate LGM global mean temperature change (Table 5.2). One recent such study, combining multi-proxy data with multiple AOGCMs, estimates LGM global cooling at 4.0°C ± 0.8°C (95% confidence interval) (Annan and Hargreaves, 2013). This result contrasts with the wider range of global cooling

(1.7°C to 8.3°C) obtained using Earth-system models of intermediate complexity (EMIC) (Schneider von Deimling et al., 2006; Holden et al., 2010b; Schmittner et al., 2011). The source of these differences in the estimate of LGM cooling may be the result of (i) the proxy data used to constrain the simulations, such as data sets associated with mild cooling in the tropics and the North Atlantic; (ii) model resolution and structure, which affects their ability to resolve the land–sea contrast (Annan and Hargreaves, 2013) and polar amplification (Fyke and Eby, 2012); (iii) the experimental design of the simulations, where the lack of dust and vegetation feedbacks (Section 5.2.2.3) and insufficient integration time. Based on the results and the caveats in the studies assessed here (Table 5.2), it is *very likely* that global mean surface temperature during the LGM was cooler than pre-industrial by 3°C to 8°C.

Some recent AOGCM simulations produce a stronger AMOC under LGM conditions, leading to mild cooling over the North Atlantic and GIS (Otto-Bliesner et al., 2007; Weber et al., 2007). This finding contrasts with proxy-based information (Lynch-Stieglitz et al., 2007; Hesse et al., 2011). Changes in deep-ocean temperature and salinity during the LGM have been constrained by pore-water chemistry in deep-sea sediments. For example, pore-water data indicate that deep water in the Atlantic Ocean cooled by between –1.7°C ± 0.9°C and –4.5°C ± 0.2°C, and became saltier by up to 2.4 ± 0.2 psu in the South Atlantic and at least 0.95 ± 0.07 psu in the North Atlantic (cf. Adkins et al., 2002). The magnitude of deep-water cooling is supported by other marine proxy data (e.g., Dwyer et al., 2000; Martin et al., 2002; Elderfield et al., 2010), while the increase in salinity is consistent with independent estimates based on $\delta^{18}O$ (Duplessy et al., 2002; Waelbroeck et al., 2002). Average salinity increased due to storage of freshwater in ice sheets. The much larger than average salinity increase in deep Southern Ocean compared to the North Atlantic is probably due to increased salt rejection through sea ice freezing processes around Antarctica (Miller et al., 2012b) (see also Section 9.4.2.3.2).

As a result of prevailing LGM modelling uncertainties (Chavaillaz et al., 2013; Rojas, 2013) and ambiguities in proxy interpretations (Kohfeld et al., 2013), it cannot be determined robustly whether LGM SH westerlies changed in amplitude and position relative to today.

5.3.3.2 Last Glacial Maximum Constraints on Equilibrium Climate Sensitivity

Temperature change recorded in proxies results from various feedback processes, and external forcings vary before equilibrium of the whole Earth system is reached. Nevertheless, the equilibrium climate sensitivity (ECS) can be estimated from past temperatures by explicitly counting the slow components of the processes (e.g., ice sheets) as forcings, rather than as feedbacks (PALAEOSENS Project Members, 2012). This is achieved in three fundamentally different ways (Edwards et al., 2007); see also Sections 9.7.3.2 and 10.8.2.4.

In the first approach, ECS is estimated by scaling the reconstructed global mean temperature change in the past with the RF difference of the past and 2 × CO_2 (Hansen et al., 2008; Köhler et al., 2010) (Table 5.3). The results are subject to uncertainties in the estimate of global mean surface temperature based on proxy records of incomplete spatial coverage (see Section 5.3.3.1). Additional uncertainty is introduced when the sensitivity to LGM forcing is scaled to the sensitivity to 2 × CO_2 forcing, as some but not all (Brady et al., 2013) models show that these sensitivities differ due to the difference in cloud feedbacks (Crucifix, 2006; Hargreaves et al., 2007; Yoshimori et al., 2011) (Figure 5.5a, b).

In the second approach, an ensemble of LGM simulations is carried out using a single climate model in which each ensemble member differs in model parameters and the ensemble covers a range of ECS (Annan et al., 2005; Schneider von Deimling et al., 2006; Holden et al., 2010a; Schmittner et al., 2011) (Table 5.3). Model parameters are then constrained by comparison with LGM temperature proxy information, generating a probability distribution of ECS. Although EMICSs are often used in order to attain sufficiently large ensemble size, the uncertainty arising from asymmetric cloud feedbacks cannot be addressed because they are parameterized in the EMICs.

In the third approach, multiple GCM simulations are compared to proxy data, and performance of the models and indirectly their ECSs are assessed (Otto-Bliesner et al., 2009; Braconnot et al., 2012b). Because the cross-model correlation between simulated LGM global cooling

5

Table 5.3 | Summary of equilibrium climate sensitivity (ECS) estimates based on Last Glacial Maximum (LGM) climate. Uncertainty ranges are 5 to 95% confidence intervals, with the exception of Multiproxy Approach for the Reconstruction of the Glacial Ocean surface (MARGO) Project Members (2009), where the published interval is reported here.

Method (number follows the "approach" in the text)	ECS Estimate (°C)	Reference and Model Name
1. Proxy data	1.0–3.6	MARGO Project Members (2009)
	1.4–5.2	Köhler et al. (2010)
2. Single-model ensemble constrained by proxy data	<6	Annan et al. (2005), MIROC3.2 AOGCM
	1.2–4.3	Schneider von Deimling et al. (2006), CLIMBER-2
	2.0–5.0	Holden et al. (2010a), GENIE-1
	1.4–2.8	Schmittner et al. (2011), UVic
	~3.6	Fyke and Eby (2012), UVic
3. Multi-GCM ensemble constrained by proxy data	1.2–4.2	Hargreaves et al. (2012), updated with addition of PMIP3/CMIP5 AOGCMs
	1.6–4.5	Hargreaves et al. (2012), updated with addition of PMIP3/CMIP5 AOGCMs[a]

Notes:

[a] Temperature constraints in the tropics were lowered by 0.4°C according to Annan and Hargreaves (2013).

AOGCM = Atmosphere-Ocean General Circulation Model; CMIP5 = Coupled Model Intercomparison Project Phase 5; CLIMBER-2 = CLIMate and BiosphERe model-2; GENIE-1 = Grid ENabled Integrated Earth system model, version 1; MIROC3.2 = Model for Interdisciplinary Research on Climate 3.2; UVic = University of Victoria Earth system model; PMIP3 = Paleoclimate Modelling Intercomparison Project Phase III.

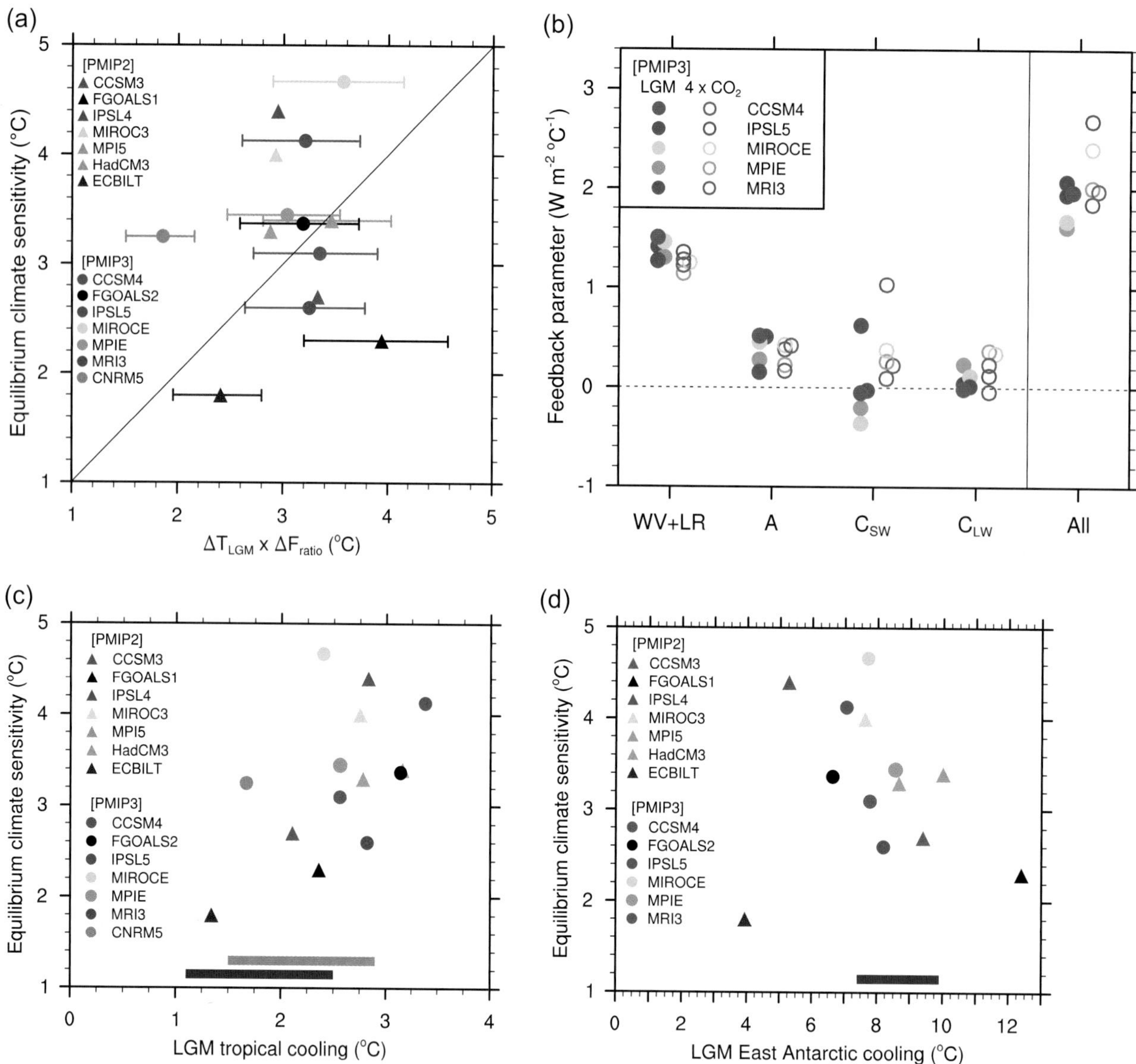

Figure 5.5 | (a) Relation between equilibrium climate sensitivity (ECS) estimated from Last Glacial Maximum (LGM) simulations and that estimated from equilibrium 2 × CO_2 or abrupt 4 × CO_2 experiments. All experiments are referenced to pre-industrial simulations. Flexible Global Ocean Atmosphere Land System model version 1 (FGOALS1), Institut Pierre Simon Laplace version 4 (IPSL4), Max Planck Institute version 5 (MPI5), and ECBILT of Paleoclimate Modelling Intercomparison Project Phase II (PMIP2) models stand for FGOALS-1.0g, IPSL-CM4-V1-MR, European Centre Hamburg Model 5 – and Max Planck Institute Ocean Model – Lund-Potsdam-Jena Dynamic Global Model (LPJ), and Coupled Atmosphere Ocean Model from de Bilt (ECBilt) with Coupled Large-scale Ice-Ocean model (CLIO), respectively. Flexible Global Ocean Atmosphere Land System model-2 (FGOALS2), Institut Pierre Simon Laplace 5 (IPSL5), Model for Interdisciplinary Research on Climate (MIROC3), Max Planck Institute für Meteorologie Earth system model (MPIE), Meteorological Research Institute of Japan Meteorological Agency version 3 (MRI3) and Centre National de Recherches Météorologiques version 5 (CNRM5) of Paleoclimate Modelling Intercomparison Project Phase III (PMIP3) models stand for FGOALS-g2, IPSL-CM5A-LR, MIROC-ESM, MPI-ESM-P, MRI-CGCM3, and CNRM-CM5, respectively. ECS based on LGM simulations (abscissa) was derived by multiplying the LGM global mean temperature anomaly (ΔT_{LGM}) and the ratio of radiative forcing between 2 × CO_2 and LGM (ΔF_{ratio}). ΔF_{ratio} for three PMIP2 models (Community Climate System Model-3 (CCSM3), Met Office Hadley Centre climate prediction models-3 (HadCM3) and IPSL4) were taken from Crucifix (2006), and it was taken from Yoshimori et al. (2009) for MIROC3. Its range, –0.80 to –0.56, with a mean of –0.69 was used for other PMIP2 and all PMIP3 models. ECS of PMIP2 models (ordinate) was taken from Hargreaves et al. (2012). ECS of PMIP3 models was taken from Andrews et al. (2012) and Brady et al. (2013), or computed using the method of Andrews et al. (2012) for FGOALS2. Also plotted is a one-to-one line. (b) Strength of individual feedbacks for the PMIP3/CMIP5 abrupt 4 × CO_2 (131 to 150 years) and LGM (stable states) experiments following the method in Yoshimori et al. (2011). WV+LR, A, C_{SW} and C_{LW} denote water vapour plus lapse rate, surface albedo, shortwave cloud, and longwave cloud feedbacks, respectively. 'All' denotes the sum of all feedbacks except for the Planck response. Feedback parameter here is defined as the change in net radiation at the top of the atmosphere due to the change in individual fields, such as water vapour, with respect to the pre-industrial simulations. It was normalized by the global mean surface air temperature change. Positive (negative) value indicates that the feedback amplifies (damp) the initial temperature response. Only models with all necessary data available for the analysis are displayed. (c) Relation between LGM tropical (20°S to 30°N) surface air temperature anomaly from pre-industrial simulations and ECS across models. A dark blue bar represents a 90% confidence interval for the estimate of reconstructed temperature anomaly of Annan and Hargreaves (2013). A light blue bar represents the same with additional 0.4°C anomaly increase according to the result of the sensitivity experiment conducted in Annan and Hargreaves (2013), in which additional 1°C lowering of tropical SST proxy data was assumed. (d) Same as in (c) but for the average of East Antarctica ice core sites of Dome F, Vostok, European Project for Ice Coring in Antarctica Dome C and Droning Maud Land ice cores. A dark blue bar represents a range of reconstructed temperature anomaly based on stable isotopes at these core sites (Stenni et al., 2010; Uemura et al., 2012). A value below zero is not displayed (the result of CNRM5). Note that the Antarctic ice sheet used for PMIP3 simulations, based on several different methods (Tarasov and Peltier, 2007; Argus and Peltier, 2010; Lambeck et al., 2010) differs in elevation from that used in PMIP2 (d).

and ECS is poor (Figure 5.5a), ECS cannot be constrained by the LGM global mean cooling. Models that show weaker sensitivity to LGM forcing than 4 × CO_2 forcing in Figure 5.5a tend to show weaker shortwave cloud feedback and hence weaker total feedback under LGM forcing than 4 × CO_2 forcing (Figure 5.5b), consistent with previous studies (Crucifix, 2006). The relation found between simulated LGM tropical cooling and ECS across multi-GCM (Hargreaves et al., 2012; Figure 5.5c) has been used to constrain the ECS with the reconstructed LGM tropical cooling (Table 5.3). Again, the main caveats of this approach are the uncertainties in proxy reconstructions and missing dust and vegetation effects which may lead to underestimated LGM cooling in the simulations.

Based on these different approaches, estimates of ECS yield low probability for values outside the range 1°C to 5°C (Table 5.3). Even though there is some uncertainty in these studies owing to problems in both the paleoclimate data and paleoclimate modelling as discussed in this section, it is *very likely* that ECS is greater than 1°C, and *very unlikely* that ECS exceeds 6°C.

5.3.4 Past Interglacials

Past interglacials are characterized by different combinations of orbital forcing (Section 5.2.1.1), atmospheric composition (Section 5.2.2.1) and climate responses (Tzedakis et al., 2009; Lang and Wolff, 2011). Documenting natural interglacial climate variability in the past provides a deeper understanding of the physical climate responses to orbital forcing. This section reports on interglacials of the past 800 kyr, with emphasis on the Last Interglacial (LIG, Table 5.1) which has more data and modelling studies for assessing regional and global temperature changes than earlier interglacials. The LIG sea level responses are assessed in Section 5.6.2. Section 5.5 is devoted to the current interglacial, the Holocene.

The phasing and strengths of the precessional parameter and obliquity varied over past interglacials (Figure 5.3b, c), influencing their timing, duration, and intensity (Tzedakis et al., 2012b; Yin and Berger, 2012) (Figure 5.3e, f, h). Since 800 ka, atmospheric CO_2 concentrations during interglacials were systematically higher than during glacial periods. Prior to ~430 ka, ice cores from Antarctica record lower interglacial CO_2 concentrations than for the subsequent interglacial periods (Section 5.2.2.1; Figure 5.3d). While LIG WMGHG concentrations were similar to the pre-industrial Holocene values, orbital conditions were very different with larger latitudinal and seasonal insolation variations. Large eccentricity and the phasing of precession and obliquity (Figure 5.3a–c) during the LIG resulted in July 65°N insolation peaking at ~126 ka and staying above the Holocene maximum values from ~129 to 123 ka. The high obliquity (Figure 5.3b) contributed to small, but positive annual insolation anomalies at high latitudes in both hemispheres and negative anomalies at low latitudes.

New data and syntheses from marine and terrestrial archives, with updated age models, have provided an expanded view of temperature patterns during interglacials since 800 ka (Masson-Delmotte et al., 2010a; Lang and Wolff, 2011; Rohling et al., 2012). There is currently no consensus on whether interglacials changed intensity after ~430 ka. EPICA Dome C Antarctic ice cores record warmer temperatures after this transition (Jouzel et al., 2007) and marine records of deep-water temperatures are characterized by generally higher values during later interglacials than earlier interglacials (Lang and Wolff, 2011). In contrast, similar interglacial magnitudes are observed across the ~430 ka boundary in some terrestrial archives from Eurasia (Prokopenko et al., 2002; Tzedakis et al., 2006; Candy et al., 2010). Simulations with an EMIC relate global and southern high latitude mean annual surface temperature variations to changes in CO_2 variations, while orbital forcing and associated feedbacks of vegetation and sea ice have a major impact on the simulated northern high-latitude mean annual surface temperature (Yin and Berger, 2012). The highest and lowest interglacial temperatures occur in models when WMGHG concentrations and local insolation reinforce each other (Yin and Berger, 2010, 2012; Herold et al., 2012).

At the time of the AR4, a compilation of Arctic records and two AOGCM simulations allowed an assessment of LIG summer temperature changes. New quantitative data syntheses (Figure 5.6a) now allow estimation of maximum annual surface temperatures around the globe for the LIG (Turney and Jones, 2010; McKay et al., 2011). A caveat is that these data syntheses assume that the warmest phases were globally synchronous (see Figure 5.6 legend for details). However, there is *high confidence* that warming in the Southern Ocean (Cortese et al., 2007; Schneider Mor et al., 2012) and over Antarctica (Masson-Delmotte et al., 2010b) occurred prior to peak warmth in the North Atlantic, Nordic Seas, and Greenland (Bauch et al., 2011; Govin et al., 2012; North Greenland Eemian Ice Drilling (NEEM) community members, 2013). Overall, higher annual temperatures than pre-industrial are reconstructed for high latitudes of both hemispheres. At ~128 ka, East Antarctic ice cores record early peak temperatures ~5°C above the present (Jouzel et al., 2007; Sime et al., 2009; Stenni et al., 2010). Higher temperatures are derived for northern Eurasia and Alaska, with sites near the Arctic coast in Northeast Siberia indicating warming of more than 10°C as compared to late Holocene (Velichko et al., 2008). Greenland warming of 8°C ± 4°C at 126 ka is estimated from the new Greenland NEEM ice core, after accounting for ice sheet elevation changes (NEEM community members, 2013). Seasonally open waters off northern Greenland and in the central Arctic are recorded during the LIG (Nørgaard-Pedersen et al., 2007; Adler et al., 2009). Changes in Arctic sea ice cover (Sime et al., 2013) may have affected the Greenland water stable isotope – temperature relationship, adding some uncertainty to LIG Greenland temperature reconstructions. Marine proxies from the Atlantic indicate warmer than late Holocene year-round SSTs north of 30°N, whereas SST changes were more variable south of this latitude (McKay et al., 2011).

Transient LIG simulations with EMICs and low-resolution AOGCMs display peak NH summer warmth between 128 ka and 125 ka in response to orbital and WMGHG forcings. This warming is delayed when NH ice sheets are allowed to evolve (Bakker et al., 2013). Time-slice climate simulations run by 13 modelling groups with a hierarchy of climate models forced with orbital and WMGHG changes for 128 to 125 ka (Figure 5.6b) simulate the reconstructed pattern of NH annual warming (Figure 5.6a). Positive feedbacks with the cryosphere (sea ice and snow cover) provide the memory that allows simulated NH high-latitude warming, annually as well as seasonally, in response to the seasonal orbital forcing (Schurgers et al., 2007; Yin and Berger, 2012). The

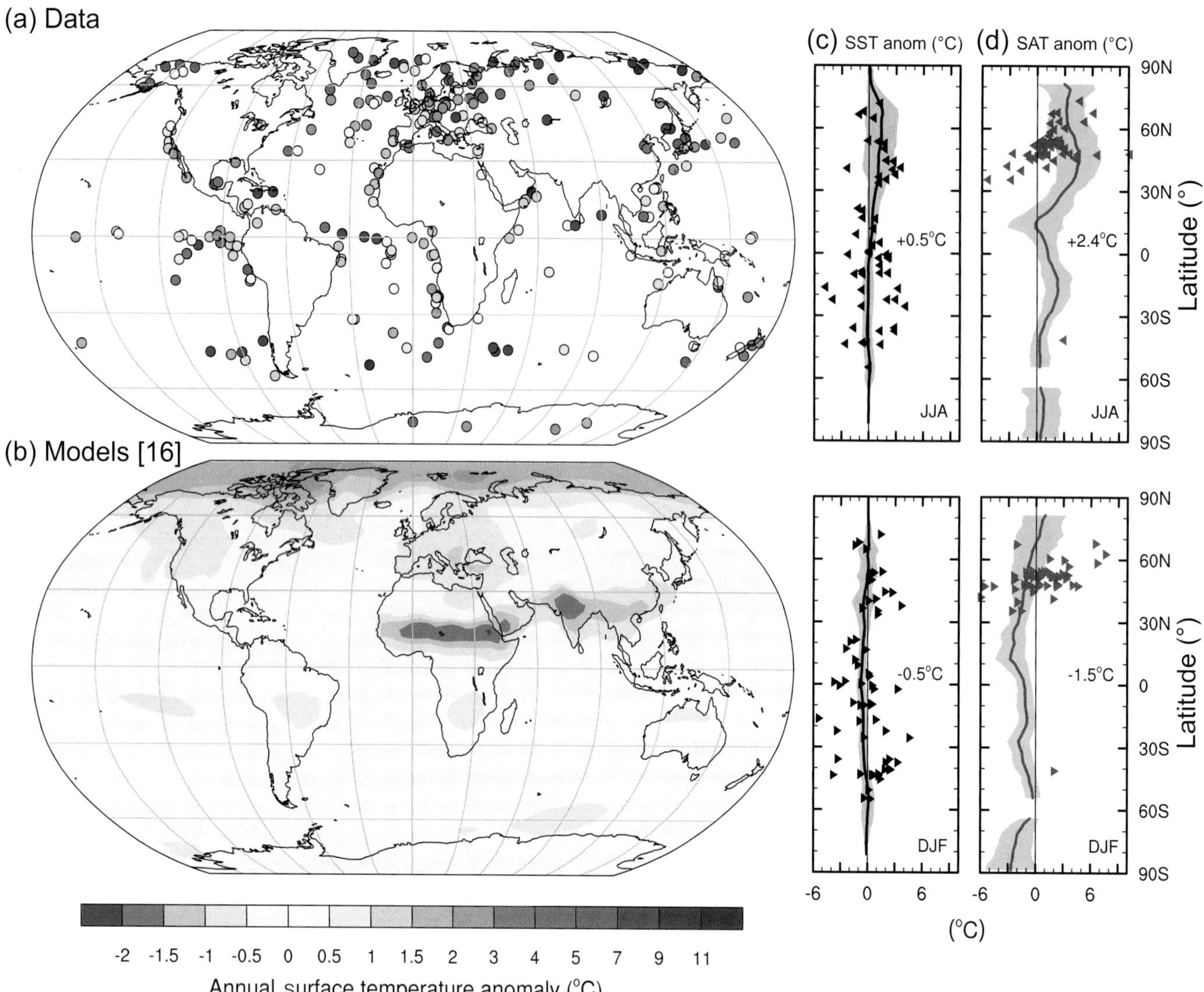

Figure 5.6 | Changes in surface temperature for the Last Interglacial (LIG) as reconstructed from data and simulated by an ensemble of climate model experiments in response to orbital and well-mixed greenhouse gas (WMGHG) forcings. (a) Proxy data syntheses of annual surface temperature anomalies as published by Turney and Jones (2010) and McKay et al. (2011). McKay et al., (2011) calculated an annual anomaly for each record as the average sea surface temperature (SST) of the 5-kyr period centred on the warmest temperature between 135 ka and 118 ka and then subtracting the average SST of the late Holocene (last 5 kyr). Turney and Jones (2010) calculated the annual temperature anomalies relative to 1961–1990 by averaging the LIG temperature estimates across the isotopic plateau in the marine and ice records and the period of maximum warmth in the terrestrial records (assuming globally synchronous terrestrial warmth). (b) Multi-model average of annual surface air temperature anomalies simulated for the LIG computed with respect to pre-industrial. The results for the LIG are obtained from 16 simulations for 128 to 125 ka conducted by 13 modelling groups (Lunt et al., 2013). (c) Seasonal SST anomalies. Multi-model zonal averages are shown as solid line with shaded bands indicating 2 standard deviations. Plotted values are the respective seasonal multi-mean global average. Symbols are individual proxy records of seasonal SST anomalies from McKay et al. (2011). (d) Seasonal terrestrial surface temperature anomalies (SAT). As in (c) but with symbols representing terrestrial proxy records as compiled from published literature (Table 5.A.5). Observed seasonal terrestrial anomalies larger than 10°C or less than –6°C are not shown. In (c) and (d) JJA denotes June – July – August and DJF December – January – February, respectively.

magnitude of observed NH annual warming though is only reached in summer in the simulations (Lunt et al., 2013) (Figure 5.6c, d). The reasons for this discrepancy are not yet fully determined. Error bars on temperature reconstructions vary significantly between methods and regions, due to the effects of seasonality and resolution (Kienast et al., 2011; McKay et al., 2011; Tarasov et al., 2011). Differences may also be related to model representations of cloud and sea ice processes (Born et al., 2010; Fischer and Jungclaus, 2010; Kim et al., 2010; Otto-Bliesner et al., 2013), and that most LIG simulations set the vegetation and ice sheets to their pre-industrial states (Schurgers et al., 2007; Holden et al., 2010b; Bradley et al., 2013). Simulations accounting for the bipolar seesaw response to persistent iceberg melting at high northern latitudes (Govin et al., 2012) and disintegration of the WAIS (Overpeck et al., 2006) (see Section 5.6.2.3) are better able to reproduce the early LIG Antarctic warming (Holden et al., 2010b).

From data synthesis, the LIG global mean annual surface temperature is estimated to be ~1°C to 2°C warmer than pre-industrial *(medium confidence)* (Turney and Jones, 2010; Otto-Bliesner et al., 2013), albeit proxy reconstructions may overestimate the global temperature change. High latitude surface temperature, averaged over several thousand years, was at least 2°C warmer than present (*high confidence*). In response to orbital forcing and WMGHG concentration changes, time slice simulations for 128 to 125 ka exhibit global mean annual surface temperature changes of 0.0°C ± 0.5°C as compared to pre-industrial. Data and models suggest a land–ocean contrast in the responses to the LIG forcing (Figure 5.6c, d). Peak global annual SST warming is estimated from data to be 0.7°C ± 0.6°C (*medium confidence*) (McKay et al., 2011). Models give more confidence to the lower bound. The ensemble of climate model simulations gives a large range of global annual land temperature change relative to pre-industrial, –0.4°C to 1.7°C, when sampled at the data locations and cooler than when averaged for all model land areas, pointing to difficulties in estimating global mean annual surface temperature with current spatial data coverage (Otto-Bliesner et al., 2013).

5.3.5 Temperature Variations During the Last 2000 Years

The last two millennia allow comparison of instrumental records with multi-decadal-to-centennial variability arising from external forcings and internal climate variability. The assessment benefits from high-resolution proxy records and reconstructions of natural and anthropogenic forcings back to at least 850 (Section 5.2), used as boundary conditions for transient GCM simulations. Since AR4, expanded proxy data networks and better understanding of reconstruction methods have supported new reconstructions of surface temperature changes during the last 2000 years (Section 5.3.5.1) and their associated uncertainties (Section 5.3.5.2), and supported more extensive comparisons with GCM simulations (Section 5.3.5.3).

5.3.5.1 Recent Warming in the Context of New Reconstructions

New paleoclimate reconstruction efforts since AR4 (Figure 5.7; Table 5.4; Appendix 5.A.1) have provided further insights into the characteristics of the Medieval Climate Anomaly (MCA; Table 5.1) and the Little Ice Age (LIA; Table 5.1). The timing and spatial structure of the MCA and LIA are complex (see Box 6.4 in AR4 and Diaz et al., 2011; and Section 5.5), with different reconstructions exhibiting warm and cold conditions at different times for different regions and seasons. The median of the NH temperature reconstructions (Figure 5.7) indicates mostly warm conditions from about 950 to about 1250 and colder conditions from about 1450 to about 1850; these time intervals are chosen here to represent the MCA and the LIA, respectively.

Based on multiple lines of evidence (using different statistical methods or different compilations of proxy records; see Appendix 5.A.1 for a description of reconstructions and selection criteria), published reconstructions and their uncertainty estimates indicate, with *high confidence*, that the mean NH temperature of the last 30 or 50 years *very likely* exceeded any previous 30- or 50-year mean during the past 800

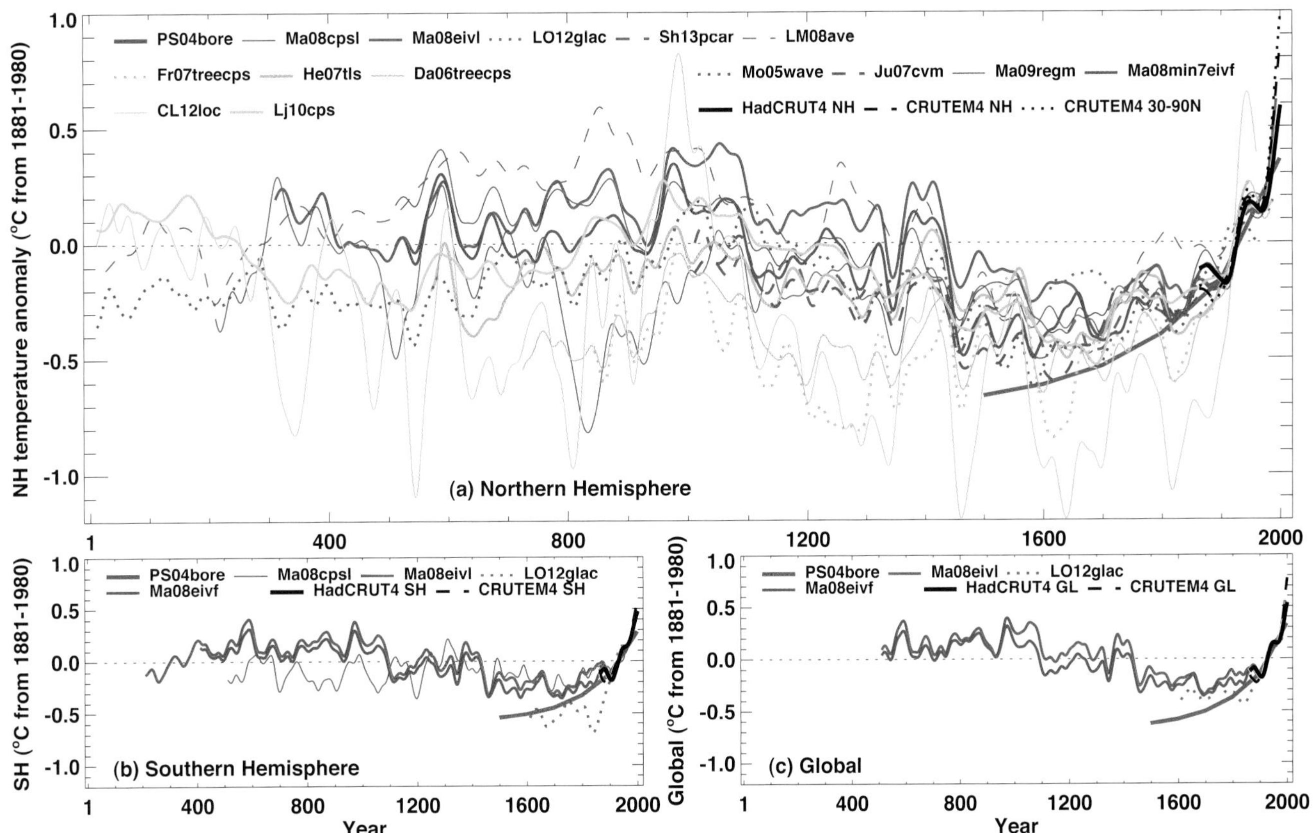

Figure 5.7 | Reconstructed (a) Northern Hemisphere and (b) Southern Hemisphere, and (c) global annual temperatures during the last 2000 years. Individual reconstructions (see Appendix 5.A.1 for further information about each one) are shown as indicated in the legends, grouped by colour according to their spatial representation (red: land-only all latitudes; orange: land-only extratropical latitudes; light blue: land and sea extra-tropical latitudes; dark blue: land and sea all latitudes) and instrumental temperatures shown in black (Hadley Centre/ Climatic Research Unit (CRU) gridded surface temperature-4 data set (HadCRUT4) land and sea, and CRU Gridded Dataset of Global Historical Near-Surface Air TEMperature Anomalies Over Land version 4 (CRUTEM4) land-only; Morice et al., 2012). All series represent anomalies (°C) from the 1881–1980 mean (horizontal dashed line) and have been smoothed with a filter that reduces variations on time scales less than about 50 years.

years (Table 5.4). The timing of warm and cold periods is mostly consistent across reconstructions (in some cases this is because they use similar proxy compilations) but the magnitude of the changes is clearly sensitive to the statistical method and to the target domain (land or land and sea; the full hemisphere or only the extra-tropics; Figure 5.7a). Even accounting for these uncertainties, almost all reconstructions agree that each 30-year (50-year) period from 1200 to 1899 was *very likely* colder in the NH than the 1983–2012 (1963–2012) instrumental temperature. NH reconstructions covering part or all of the first millennium suggest that some earlier 50-year periods might have been as warm as the 1963–2012 mean instrumental temperature, but the higher temperature of the last 30 years appear to be at least *likely* the warmest 30-year period in all reconstructions (Table 5.4). However, the confidence in this finding is lower prior to 1200, because the evidence is less reliable and there are fewer independent lines of evidence. There are fewer proxy records, thus yielding less independence among the

Table 5.4 | Comparison of recent hemispheric and global temperature estimates with earlier reconstructed values, using published uncertainty ranges to assess likelihood of unusual warmth. Each reconstructed N-year mean temperature within the indicated period is compared with both the warmest N-year mean reconstructed after 1900 and with the most recent N-year mean instrumental temperature, for $N = 30$ and $N = 50$ years. Blue symbols indicate the periods and reconstructions where the reconstructed temperatures are *very likely* cooler than the post-1900 reconstruction (■), or otherwise *very likely* (✱) or *likely* (☒) cooler than the most recent instrumental temperatures; ☐ indicates that some reconstructed temperatures were as likely warmer or colder than recent temperatures.

Region	NH														SH			Global		
Domain	Land & Sea					Land				Extratropics					Land			Land		
Study	1	2	3	4	5	6	7	8	9	10	11	12	13	14	7	6	9	7	9	15
50-year means																				
1600–1899	■	✱	✱	■	✱	■	✱	■	■	■	■	■	■	■	✱	■	■	✱	■	■
1400–1899	■	✱	✱	■	✱	■	✱	■		■	■	■	■	■	✱	■		✱		
1200–1899	■	✱	✱	■	✱	■	✱	✱		■	■	✱	■	■	✱	✱		✱		
1000–1899	✱	✱	✱	✱	☒	■	✱	✱		✱	■	✱	■	✱	✱	✱		✱		
800–1899	✱	☒	☒		☐	■	✱			✱	☐	✱	■		☒	✱		✱		
600–1899	✱	☒	☒		☐	■	✱			✱	☐	✱			☒	✱		✱		
400–1899	✱	☒	☐		☐	✱	✱			✱	☐				☐					
200–1899	✱				☐	✱				✱	☐				☐					
1–1899	✱									✱	☐									
30-year means																				
1600–1899	■	✱	✱	■	✱	■	✱	■	■	■		■	■	■	✱	■	■	✱	■	
1400–1899	■	✱	✱	■	✱	■	✱	■		■		■	■	■	✱	■		✱		
1200–1899	■	✱	✱	■	✱	■	✱	■		■		■	■	■	✱	✱		✱		
1000–1899	✱	✱	✱	✱	✱	■	✱	✱		✱		■	■	✱	✱	✱		✱		
800–1899	✱	✱	✱		☒	■	✱			✱		■	■		✱	✱		✱		
600–1899	✱	✱	✱		☒	■	✱			✱		■			✱	✱		✱		
400–1899	✱	✱			☒	✱	✱			✱					✱					
200–1899	✱				☒	✱				✱					✱					
1–1899	✱									✱										

Notes:

Symbols indicate the likelihood (based on the published multi-decadal uncertainty ranges) that each N-year mean of the reconstructed temperature during the indicated period was colder than the warmest N-year mean after 1900. A reconstructed mean temperature X is considered to be *likely* (*very likely*) colder than a modern temperature Y if $X+aE < Y$, where E is the reconstruction standard error and $a = 0.42$ (1.29) corresponding to a 66% (90%) one-tailed confidence interval assuming the reconstruction error is normally distributed. Symbols indicate that the reconstructed temperatures were either:

- ☒ *likely* colder than the 1983–2012 or 1963–2012 mean instrumental temperature;
- ✱ *very likely* colder than the 1983–2012 or 1963–2012 mean instrumental temperature;
- ■ *very likely* colder than the 1983–2012 or 1963–2012 mean instrumental temperature and additionally *very likely* colder than the warmest 30- or 50-year mean of the post-1900 reconstruction (which is typically not as warm as the end of the instrumental record);
- ☐ indicates that at least one N-year reconstructed mean is about as likely colder or warmer than the 1983–2012 or 1963–2012 mean instrumental temperature.

No symbol is given where the reconstruction does not fully cover the indicated period.

Identification and further information for each study is given in Table 5.A.6 of Appendix 5.A.1:

1 = Mo05wave; 2 = Ma08eivf; 3 = Ma09regm; 4 = Ju07cvm; 5 = LM08ave; 6 = Ma08cpsl; 7 = Ma08eivl; 8 = Sh13pcar; 9 = LO12gla; 10 = Lj10cps; 11 = CL12loc; 12 = He07tls; 13 = Da06treecps; 14 = Fr07treecps; 15 = PS04bore.

reconstructions while making them more susceptible to errors in individual proxy records. The published uncertainty ranges do not include all sources of error (Section 5.3.5.2), and some proxy records and uncertainty estimates do not fully represent variations on time scales as short as the 30 years considered in Table 5.4. Considering these caveats, there is *medium confidence* that the last 30 years were *likely* the warmest 30-year period of the last 1400 years.

Increasing numbers of proxy records and regional reconstructions are being developed for the SH (see Section 5.5), but few reconstructions of SH or global mean temperatures have been published (Figure 5.7b, c). The SH and global reconstructions with published uncertainty estimates indicate that each 30- or 50-year interval during the last four centuries was *very likely* colder than the warmest 30- or 50-year interval after 1900 (Table 5.4). However, there is only limited proxy evidence and therefore *low confidence* that the recent warming has exceeded the range of reconstructed temperatures for the SH and global scales.

5.3.5.2 Reconstruction Methods, Limitations and Uncertainties

Reconstructing NH, SH or global-mean temperature variations over the last 2000 years remains a challenge due to limitations of spatial sampling, uncertainties in individual proxy records and challenges associated with the statistical methods used to calibrate and integrate multi-proxy information (Hughes and Ammann, 2009; Jones et al., 2009; Frank et al., 2010a). Since AR4, new assessments of the statistical methods used to reconstruct either global/hemispheric temperature averages or spatial fields of past temperature anomalies have been published. The former include approaches for simple compositing and scaling of local or regional proxy records into global and hemispheric averages using uniform or proxy-dependent weighting (Hegerl et al., 2007; Juckes et al., 2007; Mann et al., 2008; Christiansen and Ljungqvist, 2012). The latter correspond to improvements in climate field reconstruction methods (Mann et al., 2009; Smerdon et al., 2011) that apply temporal and spatial relationships between instrumental and proxy records to the pre-instrumental period. New developments for both reconstruction approaches include implementations of Bayesian inference (Li et al., 2010a; Tingley and Huybers, 2010, 2012; McShane and Wyner, 2011; Werner et al., 2013). In particular, Bayesian hierarchical models enable a more explicit representation of the underlying processes that relate proxy (and instrumental) records to climate, allowing a more systematic treatment of the multiple uncertainties that affect the climate reconstruction process. This is done by specifying simple parametric forms for the proxy-temperature relationships that are then used to estimate a probability distribution of the reconstructed temperature evolution that is compatible with the available data (Tingley et al., 2012).

An improved understanding of potential uncertainties and biases associated with reconstruction methods has been achieved, particularly by using millennial GCM simulations as a surrogate reality in which pseudo-proxy records are created and reconstruction methods are replicated and tested (Smerdon, 2012). A key finding is that the methods used for many published reconstructions can underestimate the amplitude of the low-frequency variability (Lee et al., 2008; Christiansen et al., 2009; Smerdon et al., 2010). The magnitude of this amplitude attenuation in real-world reconstructions is uncertain, but for affected methods the problem will be larger: (i) for cases with weaker correlation between instrumental temperatures and proxies (Lee et al., 2008; Christiansen et al., 2009; Smerdon et al., 2011); (ii) if errors in the proxy data are not incorporated correctly (Hegerl et al., 2007; Ammann et al., 2010); or (iii) if the data are detrended in the calibration phase (Lee et al., 2008; Christiansen et al., 2009). The 20th-century trends in proxies may contain relevant temperature information (Ammann and Wahl, 2007) but calibration with detrended or undetrended data has been an issue of debate (von Storch et al., 2006; Wahl et al., 2006; Mann et al., 2007) because trends in proxy records can be induced by other (non-temperature) climate and non-climatic influences (Jones et al., 2009; Gagen et al., 2011). Recent developments mitigate the loss of low-frequency variance in global and hemispheric reconstructions by increasing the correlation between proxies and temperature through temporal smoothing (Lee et al., 2008) or by correctly attributing part or all of the temperature-proxy differences to imperfect proxy data (Hegerl et al., 2007; Juckes et al., 2007; Mann et al., 2008). Pseudoproxy experiments have shown that the latter approach used with a site-by-site calibration (Christiansen, 2011; Christiansen and Ljungqvist, 2012) can also avoid attenuation of low-frequency variability, though it is debated whether it might instead inflate the variability and thus constitute an upper bound for low-frequency variability (Moberg, 2013). Even those field reconstruction methods that do not attenuate the low-frequency variability of global or hemispheric means may still suffer from attenuation and other errors at regional scales (Smerdon et al., 2011; Annan and Hargreaves, 2012; Smerdon, 2012; Werner et al., 2013).

The fundamental limitations for deriving past temperature variability at global/hemispheric scales are the relatively short instrumental period and the number, temporal and geographical distribution, reliability and climate signal of proxy records (Jones et al., 2009). The database of high-resolution proxies has been expanded since AR4 (Mann et al., 2008; Wahl et al., 2010; Neukom and Gergis, 2011; PAGES 2k Consortium, 2013), but data are still sparse in the tropics, SH and over the oceans (see new developments in Section 5.5). Integration of low-resolution records (e.g., marine or some lake sediment cores and some speleothem records) with high-resolution tree-ring, ice core and coral records in global/hemispheric reconstructions is still challenging. Dating uncertainty, limited replication and the possibility of temporal lags in low-resolution records (Jones et al., 2009) make regression-based calibration particularly difficult (Christiansen et al., 2009) and can be potentially best addressed in the future with Bayesian hierarchical models (Tingley et al., 2012). The short instrumental period and the paucity of proxy data in specific regions may preclude obtaining accurate estimates of the covariance of temperature and proxy records (Juckes et al., 2007), impacting the selection and weighting of proxy records in global/hemispheric reconstructions (Bürger, 2007; Osborn and Briffa, 2007; Emile-Geay et al., 2013b) and resulting in regional errors in climate field reconstructions (Smerdon et al., 2011).

Two further sources of uncertainty have been only partially considered in the published literature. First, some studies have used multiple statistical models (Mann et al., 2008) or generated ensembles of reconstructions by sampling parameter space (Frank et al., 2010b), but this type of structural and parameter uncertainty needs further examination (Christiansen et al., 2009; Smerdon et al., 2011). Second, proxy-temperature relationships may change over time due to the effect of other climate and non-climate influences on a proxy, a prominent example being the divergence between some tree-ring width and density chronologies and instrumental temperature trends during the last decades of the 20th century (Briffa et al., 1998). In cases that do show divergence, a number of factors may be responsible, such as direct temperature or drought stress on trees, delayed snowmelt, changes in seasonality and reductions in solar radiation (Lloyd and Bunn, 2007; D'Arrigo et al., 2008; Porter and Pisaric, 2011). However, this phenomenon does not affect all tree-ring records (Wilson et al., 2007; Esper and Frank, 2009) and in some cases where divergence is apparent it may arise from the use of inappropriate statistical standardization of the data (Melvin and Briffa, 2008; Briffa and Melvin, 2011) and not from a genuine change in the proxy–temperature relationship. For the European Alps and Siberia, Büntgen et al., (2008) and Esper et al. (2010) demonstrate that divergence can be avoided by careful selection of sites and standardization methods together with large sample replication.

Limitations in proxy data and reconstruction methods suggest that published uncertainties will underestimate the full range of uncertainties of large-scale temperature reconstructions (see Section 5.3.5.1). While this has fostered debate about the extent to which proxy-based reconstructions provide useful climate information (e.g., McShane and Wyner, 2011 and associated comments and rejoinder), it is well established that temperature and external forcing signals are detectable in proxy reconstructions (Sections 5.3.5.3 and 10.7.2). Recently, model experiments assuming a nonlinear sensitivity of tree-rings to climate (Mann et al., 2012) have been used to suggest that the tree-ring response to volcanic cooling may be attenuated and lagged. Tree-ring data and additional tree-growth model assessments (Anchukaitis et al., 2012; Esper et al., 2013) have challenged this interpretation and analyses of instrumental data suggest hemispheric temperature reconstructions agree well with the degree of volcanic cooling during early 19th-century volcanic events (Brohan et al., 2012; see Section 5.3.5.3). These lines of evidence leave the representation of volcanic events in tree-ring records and associated hemispheric scale temperature reconstructions as an emerging area of investigation.

5.3.5.3 Comparing Reconstructions and Simulations

The number of GCM simulations of the last millennium has increased since AR4 (Fernández-Donado et al., 2013). The simulations have used different estimates of natural and anthropogenic forcings (Table 5.A.1). In particular, the PMIP3/CMIP5 simulations are driven by smaller long-term changes in TSI (Section 5.2.1; Figure 5.1b): TSI increases by ≤0.10% from the Late Maunder Minimum (LMM; 1675–1715) to the late 20th century (Schmidt et al., 2011), while most previous simulations use increases between 0.23% and 0.29% (Fernández-Donado et al., 2013). Simulated NH temperatures during the last millennium lie mostly within the uncertainties of the available reconstructions (Figure 5.8a). This agreement between GCM simulations and reconstructions provides neither strong constraints on forcings nor on model sensitivities because internal variability and uncertainties in the forcings and reconstructions are considerable factors.

Data have also been assimilated into climate models (see Sections 5.5 and 10.7, Figure 10.19) by either nudging simulations to follow local or regional proxy-based reconstructions (Widmann et al., 2010) or by selecting simulations from decade-by-decade ensembles to obtain the closest match to reconstructed climate patterns (Annan and Hargreaves, 2012; Goosse et al., 2012a). The resulting simulations provide insight into the relative roles of internal variability and external forcing (Goosse et al., 2012b), and processes that may account for the spatial distribution of past climate anomalies (Crespin et al., 2009; Palastanga et al., 2011).

Figure 5.8b–d provides additional tests of model-data agreement by compositing the temperature response to a number of distinct forcing events. The models simulate a significant NH cooling in response to volcanic events (Figure 5.8b; peaks between 0.1°C and 0.5°C depending on model) that lasts 3 to 5 years, overlapping with the signal inferred from reconstructions with annual resolution (0.05°C to 0.3°C). CMIP5 simulations tend to overestimate cooling following the major 1809 and 1815 eruptions relative to early instrumental data (Brohan et al., 2012). Such differences could arise from uncertainties in volcanic forcing (Section 5.2.1.3) and its implementation in climate models (Joshi and Jones, 2009) or from errors in the reconstructions (Section 5.3.5.2). Since many reconstructions do not have annual resolution, similar composites (Figure 5.8c) are formed to show the response to changes in multi-decadal volcanic forcings (representing clusters of eruptions). Both the simulated and reconstructed responses are significant and comparable in magnitude, although simulations show a faster recovery (<5 years) than reconstructions. Solar forcing estimated over the last millennium shows weaker variations than volcanic forcing (Figure 5.8d), even at multi-decadal time scales. Compositing the response to multi-decadal fluctuations in solar irradiance shows cooling in simulations and reconstructions of NH temperature between 0.0°C and 0.15°C. In both cases, the cooling may be partly a response to concurrent variations in volcanic forcing (green line in Figure 5.8d).

Temperature differences between the warmest and coolest centennial or multi-centennial periods provide an additional comparison of the amplitude of NH temperature variations in the reconstructions and simulations: between 950 and 1250 (nominally the MCA) and 1450–1850 (nominally the LIA; Figures 5.8e; 5.9a–c) and between the LIA and the 20th century (Figures 5.8f; 5.9g–i). Despite similar multi-model and multi-reconstruction means for the warming from the LIA to the present, the range of individual results is very wide (see Sections 9.5.3.1 and 10.7.1 for a comparison of reconstructed and simulated variability across various frequency ranges) and there is no clear difference between runs with weaker or stronger solar forcing (Figure 5.8f; Section 10.7.2). The difference between the MCA and LIA temperatures, however, has a smaller range for the model simulations than the reconstructions, and the simulations (especially those with weaker solar forcing) lie within the lower half of the reconstructed range of temperature changes (Figure 5.8e). Recent studies have assessed the consistency of model simulations and temperature reconstructions at the hemispheric scale. Hind and Moberg (2012) found closer data-model agreement for simulations with 0.1% TSI increase than 0.24% TSI increase, but the result is sensitive to the reconstruction uncertainty and the climate sensitivity of the model. Simulations with an EMIC using a much stronger solar forcing (0.44% TSI increase from LMM to present, Shapiro et al., 2011) appear to be incompatible with most temperature reconstructions (Feulner, 2011).

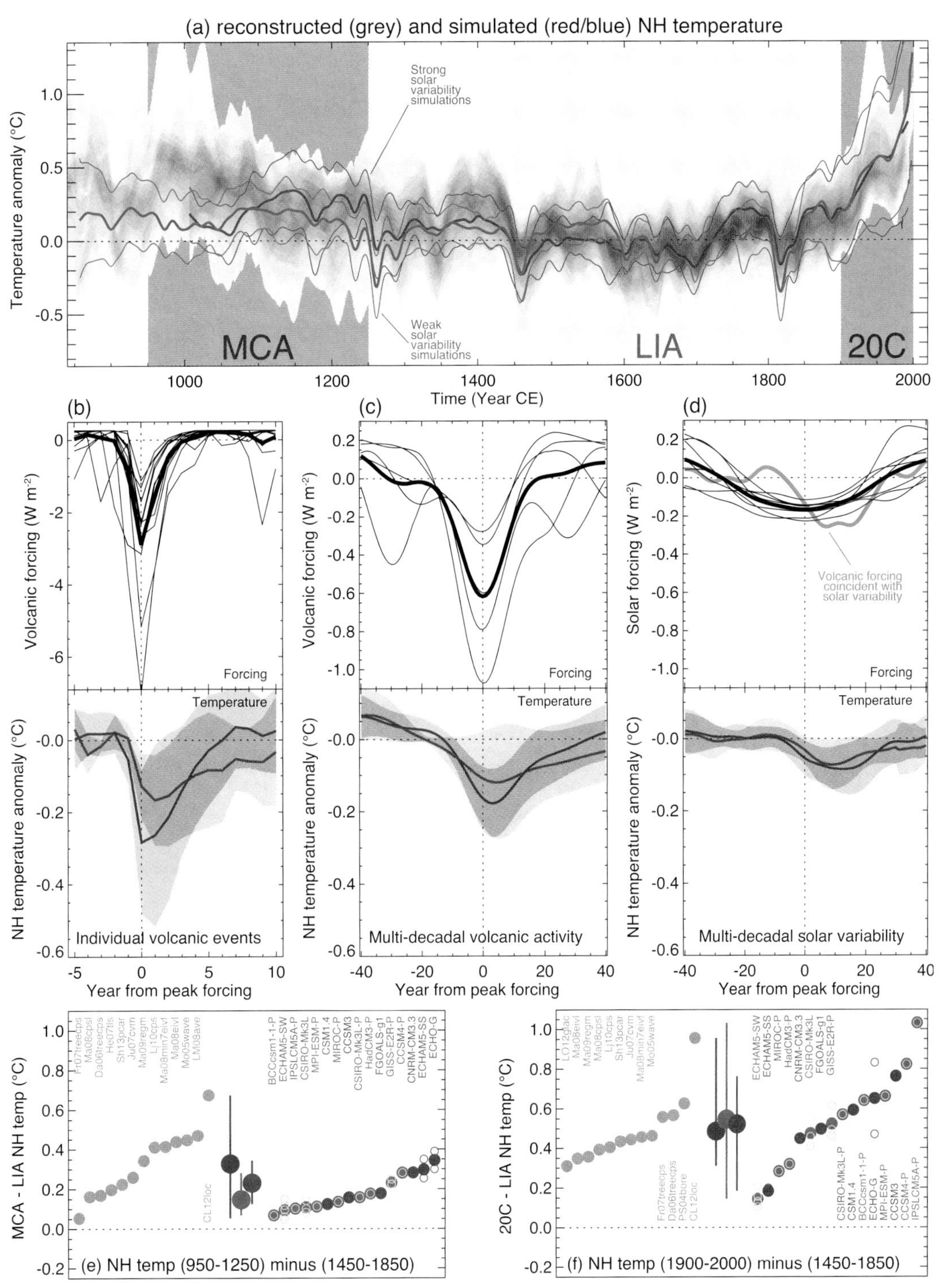

Figure 5.8 | Comparisons of simulated and reconstructed NH temperature changes. (a) Changes over the last millennium (Medieval Climate Anomaly, MCA; Little Ice Age, LIA; 20th century, 20C) (b) Response to individual volcanic events. (c) Response to multi-decadal periods of volcanic activity. (d) Response to multi-decadal variations in solar activity. (e) Mean change from the MCA to the LIA. (f) Mean change from 20th century to LIA. Note that some reconstructions represent a smaller spatial domain than the full Northern Hemisphere (NH) or a specific season, while annual temperatures for the full NH mean are shown for the simulations. (a) Simulations shown by coloured lines (thick lines: multi-model-mean; thin lines: multi-model 90% range; red/blue lines: models forced by stronger/weaker solar variability, though other forcings and model sensitivities also differ between the red and blue groups); overlap of reconstructed temperatures shown by grey shading; all data are expressed as anomalies from their 1500–1850 mean and smoothed with a 30-year filter. Superposed composites (time segments from selected periods positioned so that the years with peak negative forcing are aligned) of the forcing and temperature response to: (b) 12 of the strongest individual volcanic forcing events after 1400 (the data shown are not smoothed); (c) multi-decadal changes in volcanic activity; (d) multi-decadal changes in solar irradiance. Upper panels show volcanic or solar forcing for the individual selected periods together with the composite mean (thick line); in (d), the composite mean of volcanic forcing (green) during the solar composite is also shown. Lower panels show the NH temperature composite means and 90% range of spread between simulations (red line, pink shading) or reconstructions (grey line and shading), with overlap indicated by darker shading. Mean NH temperature difference between (e) MCA (950–1250) and LIA (1450–1850) and (f) 20th century (1900–2000) and LIA, from reconstructions (grey), multi-reconstruction mean and range (dark grey), multi-model mean and range and individual simulations (red/blue for models forced by stronger/weaker solar variability). Where an ensemble of simulations is available from one model, the ensemble mean is shown in solid and the individual ensemble members by open circles. Results are sorted into ascending order and labelled. Reconstructions, models and further details are given in Appendix 5.A.1 and Tables 5.A.1 and 5.A.6.

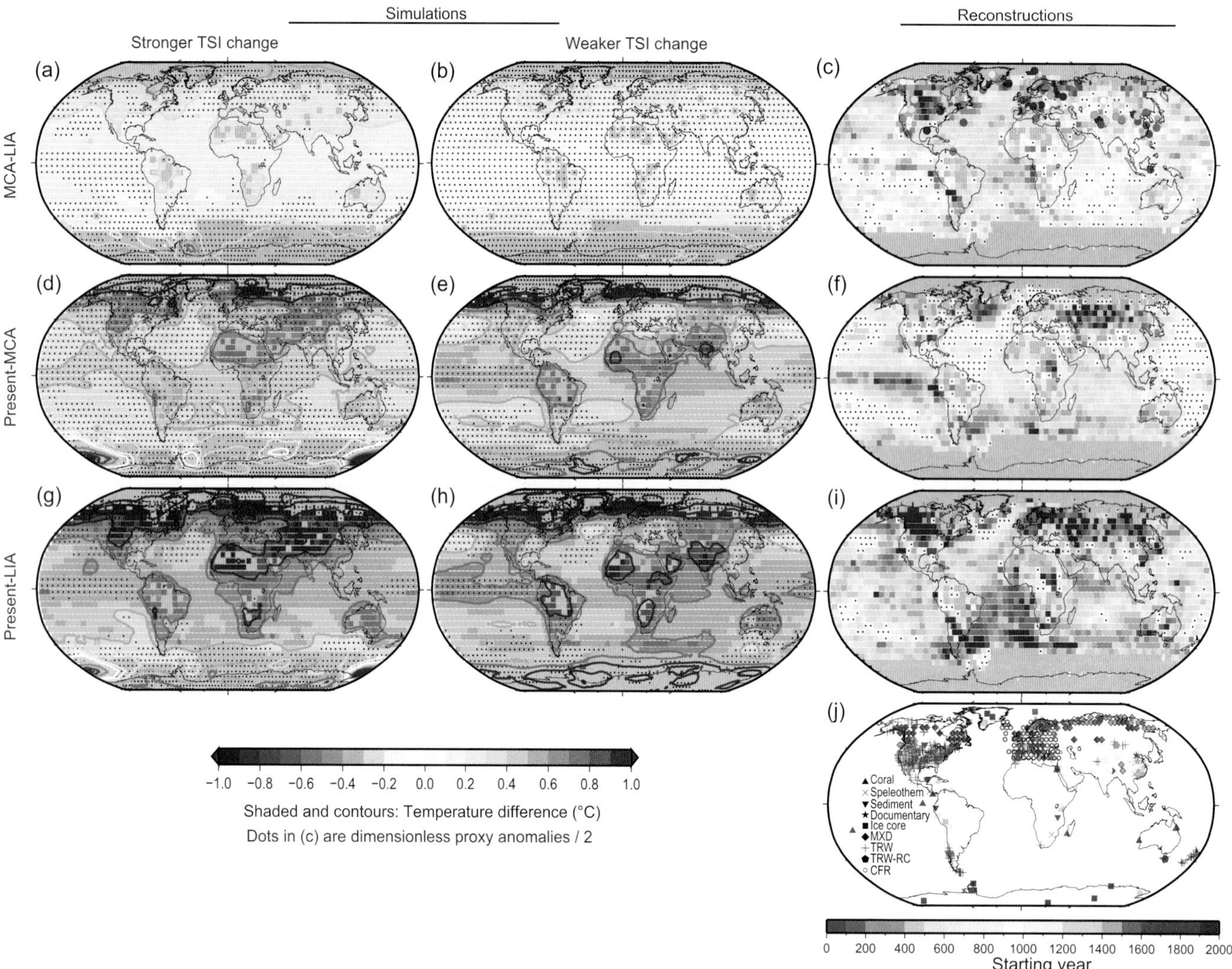

Figure 5.9 | Simulated and reconstructed temperature changes for key periods in the last millennium. Annual temperature differences for: (a) to (c) Medieval Climate Anomaly (MCA, 950–1250) minus Little Ice Age (LIA, 1450–1850); (d) to (f) present (1950–2000) minus MCA; (g) to (i) present minus LIA. Model temperature differences (left and middle columns) are average temperature changes in the ensemble of available model simulations of the last millennium, grouped into those using stronger (total solar irradiance (TSI) change from the Late Maunder Minimum (LMM) to present >0.23%; left column; SS in Table 5.A.1) or weaker solar forcing changes (TSI change from the LMM to present <0.1%; middle column; SW in Table 5.A.1). Right column panels (c, f, i) show differences (shading) for the Mann et al. (2009) field reconstruction. In (c), dots represent additionally proxy differences from Ljungqvist et al. (2012), scaled by 0.5 for display purposes. The distribution, type and temporal span of the input data used in the field reconstruction of Mann et al. (2009) are shown in (j); proxy types are included in the legend, acronyms stand for: tree-ring maximum latewood density (MXD); tree-ring width (TRW); regional TRW composite (TRW-RC); and multi-proxy climate field reconstruction (CFR). Dotted grid-cells indicate non-significant differences (<0.05 level) in reconstructed fields (right) or that <80% of the simulations showed significant changes of the same sign (left and middle). For simulations starting after 950, the period 1000–1250 was used to estimate MCA values. Grid cells outside the domain of the Mann et al. (2009) reconstruction are shaded grey in the model panels to enable easier comparison, though contours (interval 0.2 K) illustrate model output over the complete global domain. Only simulations spanning the whole millennium and including at least solar, volcanic and greenhouse gas forcing have been used (Table 5.A.1): BCC-csm1-1 (1), CCSM3 (1), CCSM4 (1), CNRM-CM3.3, CSIRO-mk3L-1-2 (4), CSM1.4 (1), ECHAM5-MPIOM (8), ECHO-G (1), MPI-ESM-P (1), FGOALS-gl (1), GISS-E2-R (3), HadCM3 (1), IPSL-CM5A-LR (1). Averages for each model are calculated first, to avoid models with multiple simulations having greater influence on the ensemble means shown here.

The spatial distributions of simulated and reconstructed (Mann et al., 2009; Ljungqvist et al., 2012) temperature changes between the MCA, LIA and 20th century are shown in Figure 5.9. Simulated changes tend to be larger, particularly with stronger TSI forcing, over the continents and ice/snow-covered regions, showing polar amplification (see Box 5.1). The largest simulated and reconstructed changes are between the LIA and present, with reconstructions (Figure 5.9i) indicating widespread warming except for the cooling south of Greenland. Models also simulate overall warming between the MCA and present (Figure 5.9d, e), whereas the reconstructions indicate significant regional cooling (in the North Atlantic, southeastern North America, and the mid-latitudes of the Pacific Ocean). This is not surprising because greater regional variability is expected in the reconstructions compared with the mean of multiple model simulations, though reconstructed changes for such areas with few or no proxy data (Figure 5.9i) should also be interpreted with caution (Smerdon et al., 2011). The reconstructed temperature differences between MCA and LIA (Figure 5.9c) indicate higher medieval temperatures over the NH continents in agreement with simulations (Figure 5.9a, b). The reconstructed MCA warming is higher than in the simulations, even for stronger TSI changes and individual simulations (Fernández-Donado et al., 2013). Simulations with proxy assimilation show that this pattern of change is compatible with a direct response to a relatively weak solar forcing and internal variability patterns similar to a positive Northern Annular Mode (NAM) phase and northward shifts of the Kuroshio and Gulf Stream currents (Goosse et al., 2012b). For the tropical regions, an enhanced zonal SST gradient produced by either a warmer Indian Ocean (Graham et al., 2011) or a cooler eastern Pacific (La Niña-like state) (Seager et al., 2007; Mann et al., 2009) could explain the reconstructed MCA patterns (Figure 5.9c). However, the enhanced gradients are not reproduced by model simulations (Figure 5.9a, b) and are not robust when considering the reconstruction uncertainties and the limited proxy records in these tropical ocean regions (Emile-Geay et al., 2013b) (Sections 5.4.1 and 5.5.1). This precludes an assessment of the role of external forcing and/or internal variability in these reconstructed patterns.

5.4 Modes of Climate Variability

Since AR4, new proxy reconstructions and model simulations have provided additional insights into the forced and unforced behaviour of modes of climate variability. This section focuses only on the interannual ENSO, the NAM and NAO, the Southern Annular Mode (SAM) and longer term variability associated with the Atlantic Multidecadal Oscillation (AMO) (see Glossary and Chapter 14 for definitions and illustrations and Box 2.5). It is organized from low to high latitudes and from interannual to decadal-scale modes of variability.

5.4.1 Tropical Modes

During the MPWP, climate conditions in the equatorial Pacific were characterized by weaker zonal (Wara et al., 2005) and cross-equatorial (Steph et al., 2010) SST gradients, consistent with the absence of an eastern equatorial cold tongue. This state still supported interannual variability, according to proxy records (Scroxton et al., 2011; Watanabe et al., 2011). These results together with recent GCM experiments (Haywood et al., 2007) indicate (*medium confidence*) that interannual ENSO variability existed, at least sporadically, during the warm background state of the Pliocene (Section 5.3.1).

LGM GCM simulations display wide ranges in the behaviour of ENSO and the eastern equatorial Pacific annual cycle of SST with little consistency (Liu et al., 2007a; Zheng et al., 2008) (Figure 5.10). Currently ENSO variance reconstructions for the LGM are too uncertain to help constrain the simulated responses of the annual cycle and ENSO to LGM boundary conditions. GCMs show that a reduced AMOC *very likely* induces intensification of ENSO amplitude and for the majority of climate models also a reduction of the amplitude of the SST annual cycle in the eastern equatorial Pacific (Timmermann et al., 2007; Merkel et al., 2010; Braconnot et al., 2012a) (Figure 5.10). About 75% of the PMIP2 and PMIP3/CMIP5 mid-Holocene simulations exhibit a weakening of interannual SST amplitude in the eastern equatorial Pacific relative to pre-industrial conditions. More than 87% of these simulations also show a concomitant substantial weakening in the amplitude of the annual cycle of eastern equatorial Pacific SST. Model results are consistent with a reduction of total variance of $\delta^{18}O$ variations of individual foraminifera in the eastern equatorial Pacific, indicative of an orbital effect on eastern equatorial Pacific SST variance (Koutavas and Joanides, 2012). In contrast to these findings, a recent proxy study using sub-annually resolved $\delta^{18}O$ from central equatorial Pacific coral segments (Cobb et al., 2013) reveals no evidence for orbitally-induced changes in interannual ENSO amplitude throughout the last 7 ka (*high confidence*), which is consistent with the weak reduction in mid-Holocene ENSO amplitude of only ~10% simulated by the majority of climate models (Fig. 5.10), but contrasts with reconstructions reported in AR4 that showed a reduction in ENSO variance during the first half of the Holocene. The same study revealed an ENSO system that experienced very large internal variance changes on decadal and centennial time scales. This latter finding is also confirmed by the analysis of about 2000 years of annually varved lake sediments (Wolff et al., 2011) in the ENSO-teleconnected region of equatorial East Africa. Furthermore, Cobb et al. (2013) identify the late 20th century as a period of anomalously high, although not unprecedented, ENSO variability relative to the average reconstructed variance over the last 7000 years.

Reconstructions of ENSO for the last millennium also document multi-decadal-to-centennial variations in the amplitude of reconstructed interannual eastern equatorial Pacific SST anomalies (McGregor et al., 2010; Wilson et al., 2010; Li et al., 2011; Emile-Geay et al., 2013a). Statistical efforts to determine ENSO variance changes in different annually resolved ENSO proxies (D'Arrigo et al., 2005; Braganza et al., 2009; McGregor et al., 2010; Fowler et al., 2012; Hereid et al., 2013) and from documentary sources (Garcia-Herrera et al., 2008; Gergis and Fowler, 2009) reveal (*medium confidence*) extended periods of low ENSO activity during parts of the LIA compared to the 20th century. Direct TSI effects on reconstructed multi-decadal ENSO variance changes cannot be identified (McGregor et al., 2010). According to reconstructions of volcanic events (Section 5.2.1.3) and some ENSO proxies, a slightly increased probability exists (*medium confidence*) for the occurrence of El Niño events 1 to 2 years after major volcanic eruptions (Adams et al., 2003; McGregor et al., 2010; Wilson et al., 2010). This response is not captured robustly by GCMs (McGregor and Timmermann, 2010; Ohba et al., 2013).

5.4.2 Extratropical Modes

Robust evidence from LGM simulations indicates a weakening of the NAM variability, connected with stronger planetary wave activity (Lü et al., 2010). A significant but model-dependent distortion of the simulated LGM NAO pattern may result from the strong topographic ice sheet forcing (Justino and Peltier, 2005; Handorf et al., 2009; Pausata et al., 2009; Riviere et al., 2010). A multimodel analysis of NAO behaviour in mid-Holocene GCM simulations (Gladstone et al., 2005) reveals an NAO structure, similar to its pre-industrial state, but a tendency for more positive NAO values during the early Holocene (Rimbu et al., 2003), with no consistent change in its interannual variability. Robust proxy evidence to test these model-based results has not yet been established. A new 5200-year-long lake sediment record from southwestern Greenland (Olsen et al., 2012) suggests that around 4500 and 650 years ago variability associated with the NAO changed from generally positive to variable, intermittently negative conditions. Since AR4, a few cold-season NAO reconstructions for the last centuries have been published. They are based on long instrumental pressure series (Cornes et al., 2012), a combination of instrumental and ship log-book data (Küttel et al., 2010) and two proxy records (Trouet et al., 2009). Whereas these and earlier NAO reconstructions (Cook et al., 2002;

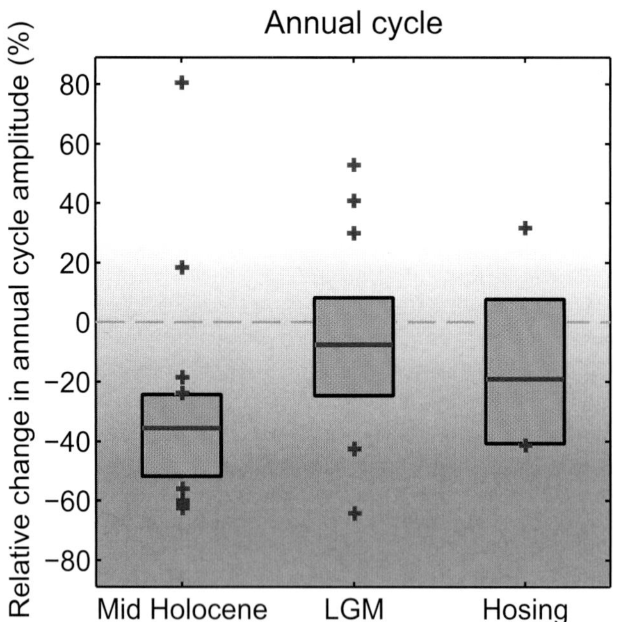

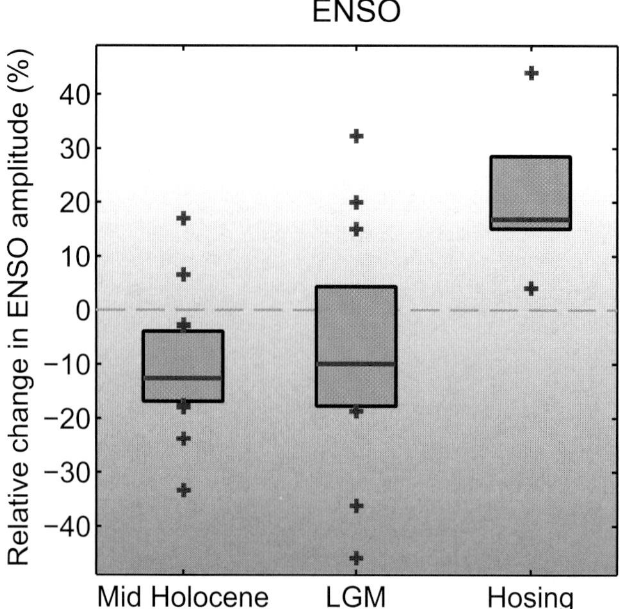

Figure 5.10 | Relative changes in amplitude of the annual cycle of sea surface temperature (SST) in Niño 3 region (average over 5°S to 5°N and 150°W to 90°W) (left) and in amplitude of interannual SST anomalies in the Niño 3.4 region (average over 5°S to 5°N and 170°W to 120°W) (right) simulated by an ensemble of climate model experiments in response to external forcing. Left: Multi-model average of relative changes (%) in amplitude of the mean seasonal cycle of Niño 3 SST for mid Holocene (MH) and Last Glacial Maximum (LGM) time-slice experiments and for freshwater perturbation experiments (Hosing) that lead to a weakening of the Atlantic Ocean meridional overturning circulation (AMOC) by more than 50%. Bars encompass the 25 and 75 percentiles, with the red horizontal lines indicating the median in the respective multi-model ensemble, red crosses are values in the upper and lower quartile of the distribution; Right: same as left, but for the SST anomalies in the Niño 3.4 region, representing El Niño-Southern Oscillation (ENSO) variability. The MH ensemble includes 4 experiments performed by models participating in Paleoclimate Modelling Intercomparison Project Phase II (PMIP2) (FGOALS1.0g, IPSL-CM4, MIROC3.2 medres, CCSM3.0) and 7 experiments (mid-Holocene) performed by models participating in PMIP3/CMIP5 (CCSM4.0, CSIRO-Mk3-6-0, HadGEM2-CC, HadGEM2-ES, MIROC-ESM, MPI-ESM-P, MRI-CGCM3). The LGM ensemble includes 5 experiments performed by models participating in PMIP2 (FGOALS1.0g, IPSL-CM4, MIROC3.2 medres, CCSM3.0, HadCM3) and 5 experiments (LGM) performed by models participating in PMIP3/CMIP5 (CCSM4, GISS-E2-R, IPSL-CM5A-LR, MIROC-ESM, and MPI-ESM-P). The changes in response to MH and LGM forcing are computed with respect to the pre-industrial control simulations coordinated by PMIP2 and PMIP3/CMIP5. The results for Hosing are obtained from freshwater perturbation experiments conducted with CCSM2.0, CCSM3.0, HadCM3, ECHAM5-MPIOM, GFDL-CM2.1 (Timmermann et al., 2007), CSM1.4 (Bozbiyik et al., 2011) for pre-industrial or present-day conditions and with CCSM3 for glacial conditions (Merkel et al., 2010). The changes in response to fresh water forcing are computed with respect the portion of simulations when the AMOC is high.

Luterbacher et al., 2002; Timm et al., 2004; Pinto and Raible, 2012) differ in several aspects, and taking into consideration associated reconstruction uncertainties, they demonstrate with *high confidence* that the strong positive NAO phases of the early 20th century and the mid-1990s are not unusual in the context of the past half millennium. Trouet et al. (2009) presented a winter NAO reconstruction that yielded a persistent positive phase during the MCA in contrast to higher frequency variability during the LIA. This is not consistent with the strong NAO imprint in Greenland ice core data (Vinther et al., 2010) and recent results from transient model simulations that neither support such a persistent positive NAO during the MCA, nor a strong NAO phase shift during the LIA (Lehner et al., 2012; Yiou et al., 2012). A recent pseudo-proxy-based assessment of low-frequency NAO behaviour (Lehner et al., 2012) infers weaknesses in the reconstruction method used by Trouet et al. (2009). Last millennium GCM simulations reveal no significant response of the NAO to solar forcing (Yiou et al., 2012), except for the GISS-ER coupled model which includes ozone photochemistry, extends into the middle atmosphere and exhibits changes in NAO that are weak during the MCA compared to the LIA (Mann et al., 2009).

Changes in the SAM modulate the strength and position of the mean SH westerlies, and leave an important signature on SH present-day surface climate (Gillett et al., 2006) past tree-ring growth (e.g., Urrutia et al., 2011), wildfires (Holz and Veblen, 2011) as well as on LGM climate (Justino and Peltier, 2008). A first hemispheric-wide, tree-ring-based reconstruction of the austral summer SAM (Villalba et al., 2012) indicates that the late 20th century positive trend may have been anomalous in the context of the last 600 years, thus supporting earlier South American proxy evidence for the last 400 years (e.g., Lara et al., 2008) and GCM (Wilmes et al., 2012). Hence, there is *medium confidence* that the positive trend in SAM since 1950 may be anomalous compared to the last 400 years.

The AMO (Delworth and Mann, 2000; Knight et al., 2005) (see also Sections 9.5.3.3.2 and 14.7.6) has been reconstructed using marine (Black et al., 2007; Kilbourne et al., 2008; Sicre et al., 2008; Chiessi et al., 2009; Saenger et al., 2009) and terrestrial proxy records (Gray et al., 2004; Shanahan et al., 2009) from different locations. Correlations among different AMO reconstructions decrease rapidly prior to 1900 (Winter et al., 2011). An 8000-year long AMO reconstruction (Knudsen et al., 2011) shows no correlation with TSI changes, and is interpreted as internally generated ocean-atmosphere variability. However, GCM experiments (Waple et al., 2002; Ottera et al., 2010) using solar and/or volcanic forcing reconstructions indicate that external forcings may have played a role in driving or at least acting as pacemaker for AMO variations.

5.5 Regional Changes During the Holocene

Reconstructions and simulations of regional changes that have emerged since AR4 are assessed. Most emphasis is on the last 2000 years, which has the best data coverage.

5.5.1 Temperature

5.5.1.1 Northern Hemisphere Mid to High Latitudes

New studies confirm the spatial patterns of SAT and SST distribution as summarised in AR4 (Jansen et al., 2007). According to a recent compilation of proxy data, the global mean annual temperatures around 8 to 6 ka were about 0.7°C higher, and extratropical NH temperatures were about 1°C higher than for pre-industrial conditions (Marcott et al., 2013). Spatial variability in the temperature anomalies and the timing of the thermal maximum implicate atmospheric or oceanic dynamical feedbacks including effects from remaining ice sheets (e.g., Wanner et al., 2008; Leduc et al., 2010; Bartlein et al., 2011; Renssen et al., 2012). The peak early-to-mid-Holocene North Atlantic and sub-Arctic SST anomalies are reconstructed and simulated to primarily occur in summer and in the stratified uppermost surface-ocean layer (Hald et al., 2007; Andersson et al., 2010). Terrestrial MH (~6 ka, Table 5.1) summer-season temperatures were higher than modern in the mid-to-high latitudes of the NH, consistent with minimum glacier extents (Section 5.5.3) and PMIP2 and PMIP3/CMIP5 simulated responses to orbital forcing (Figure 5.11) (Braconnot et al., 2007; Bartlein et al., 2011; Izumi et al., 2013). There is also robust evidence for warmer MH winters compared to the late 20th century (e.g., Wanner et al., 2008; Sundqvist et al., 2010; Bartlein et al., 2011) (Figure 5.11), but the simulated high latitude winter warming is model dependent and is sensitive to ocean and sea-ice changes (Otto et al., 2009; Zhang et al., 2010). Overall, models underestimate the reduction in the latitudinal gradient of European winter temperatures during the MH (Brewer et al., 2007). There is a general, gradual NH cooling after ~5 ka, linked to orbital forcing, and increased amplitude of millennial-scale variability (Wanner et al., 2008; Vinther et al., 2009; Kobashi et al., 2011; Marcott et al., 2013).

Since AR4, regional temperature reconstructions have been produced for the last 2 kyr (Figure 5.12; PAGES 2k Consortium, 2013). A recent multi-proxy 2000-year Arctic temperature reconstruction shows that temperatures during the first centuries were comparable or even higher than during the 20th century (Hanhijärvi et al., 2013; PAGES 2k Consortium, 2013). During the MCA, portions of the Arctic and sub-Arctic experienced periods warmer than any subsequent period, except for the most recent 50 years (Figure 5.12) (Kaufman et al., 2009; Kobashi et al., 2010, 2011; Vinther et al., 2010; Spielhagen et al., 2011). Tingley and Huybers (2013) provided a statistical analysis of northern high-latitude temperature reconstructions back to 1400 and found that recent extreme hot summers are unprecedented over this time span. Marine proxy records indicate anomalously high SSTs north of Iceland and the Norwegian Sea from 900 to 1300, followed by a generally colder period that ended in the early 20th century. Modern SSTs in this region may still be lower than the warmest intervals of the 900–1300 period (Cunningham et al., 2013). Further north, in Fram Strait, modern SSTs from Atlantic Water appear warmer than those reconstructed from foraminifera for any prior period of the last 2000 years (Spielhagen et al., 2011). However, different results are obtained using dinocysts from the same sediment core (Bonnet et al. (2010) showing a cooling trend over the last 2000 years without a 20th century rise, and warmest intervals centered at years 100 and 600.

Tree-ring data and lake sediment information from the North American treeline (McKay et al., 2008; Bird et al., 2009; D'Arrigo et al., 2009; Anchukaitis et al., 2012) suggest common variability between different records during the last centuries. An annually resolved, tree-ring based 800-year temperature reconstruction over temperate North America (PAGES 2k Consortium, 2013) and a 1500-year long pollen-based temperature estimate (Viau et al., 2012) show cool periods 500–700 and 1200–1900 as well as a warm period between 750 and 1100. The generally colder conditions until 1900 are in broad agreement with other pollen, tree-ring and lake-sediment evidence from northwest Canada, the Canadian Rockies and Colorado (Luckman and Wilson, 2005; Salzer and Kipfmueller, 2005; Loso, 2009; MacDonald et al., 2009; Thomas and Briner, 2009). It is *very likely* that the most recent decades have been, on average, the warmest across mid-latitude western and temperate North America over at least 500 years (Wahl and Smerdon, 2012; PAGES 2k Consortium, 2013; Figure 5.12).

New warm-season temperature reconstructions (PAGES 2k Consortium, 2013; Figure 5.12) covering the past 2 millennia show that warm European summer conditions were prevalent during 1st century, followed by cooler conditions from the 4th to the 7th century. Persistent warm conditions also occurred during the 8th–11th centuries, peaking throughout Europe during the 10th century. Prominent periods with cold summers occurred in the mid-15th and early 19th centuries. There is *high confidence* that northern Fennoscandia from 900 to 1100 was as warm as the mid-to-late 20th century (Helama et al., 2010; Linderholm et al., 2010; Büntgen et al., 2011a; Esper et al., 2012a; 2012b; McCarroll et al., 2013; Melvin et al., 2013). The evidence also suggests warm conditions during the 1st century, but comparison with recent temperatures is restricted because long-term temperature trends from tree-ring data are uncertain (Esper et al., 2012a). In the European Alps region, tree-ring based summer temperature reconstructions (Büntgen et al., 2005; Nicolussi et al., 2009; Corona et al., 2010, 2011; Büntgen et al., 2011b) show higher temperatures in the last decades than during any time in the MCA, while reconstructions based on lake sediments (Larocque-Tobler et al., 2010; Trachsel et al., 2012) show as high, or slightly higher temperatures during parts of the MCA compared to most recent decades. The longest summer temperature reconstructions from parts of the Alps show several intervals during Roman and earlier times as warm (or warmer) than most of the 20th century (Büntgen et al., 2011b; Stewart et al., 2011).

Since AR4, new temperature reconstructions have also been generated for Asia. A tree-ring based summer temperature reconstruction for temperate East Asia back to 800 indicates warm conditions during the period 850–1050, followed by cooler conditions during 1350–1880 and a subsequent 20th century warming (Cook et al., 2012; PAGES 2k Consortium, 2013; Figure 5.12). Tree-ring reconstructions from the western Himalayas, Tibetan Plateau, Tianshan Mountains and western High Asia depict warm conditions from the 10th to the 15th centuries, lower temperature afterwards and a 20th century warming (Esper et al., 2007a; Zhu et al., 2008; Zhang et al., 2009; Yadav et al., 2011).

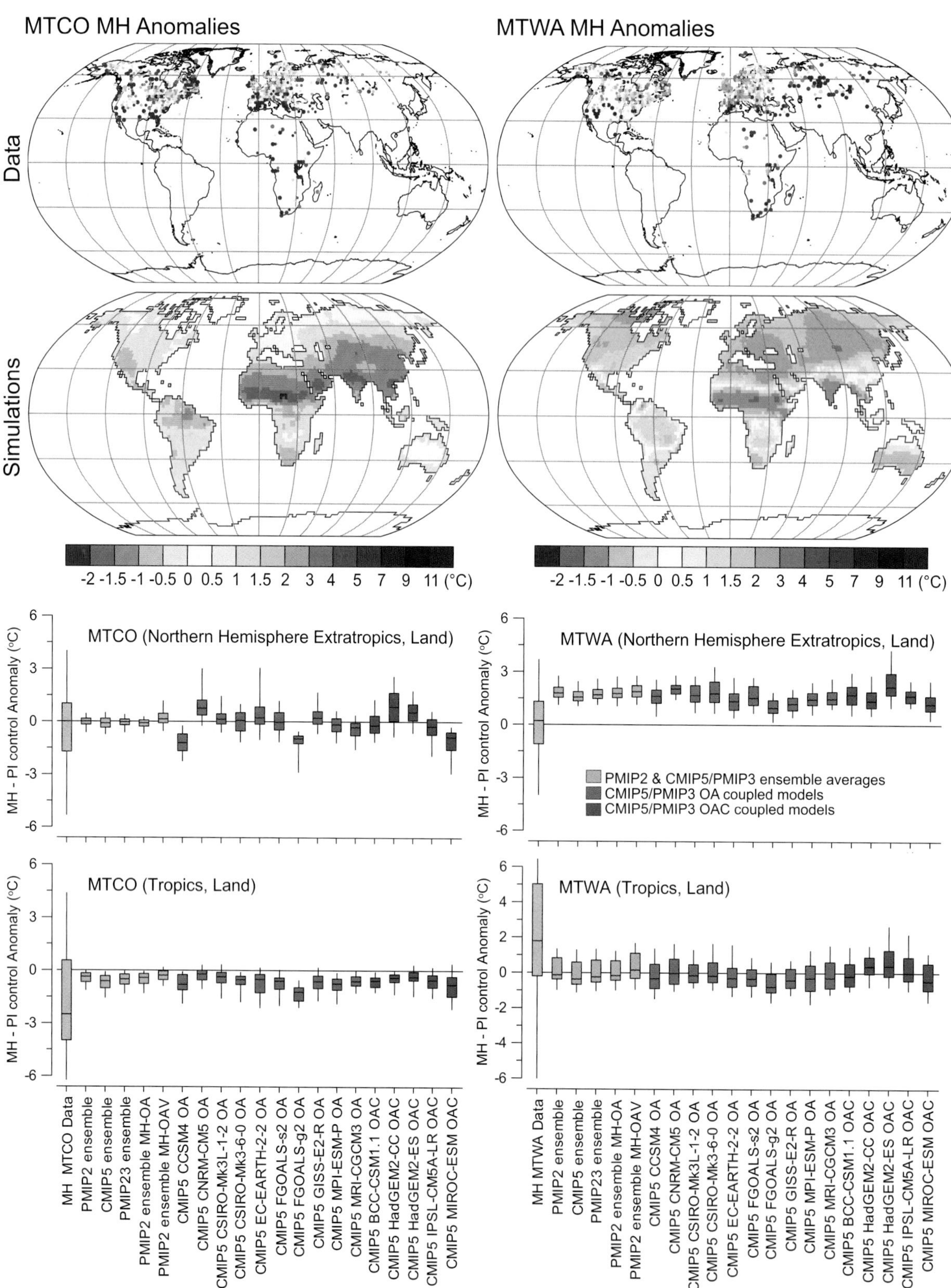

Figure 5.11 | Model-data comparison of surface temperature anomalies for the mid-Holocene (6 ka). MTCO is the mean temperature of the coldest month; MTWA is the mean temperature of the warmest month. Top panels are pollen-based reconstructions of Bartlein et al., (2011) with anomalies defined as compared to modern, which varies among the records. The bulk of the records fall within the range of 5.5 to 6.5 ka, with only 3.5% falling outside this range. Middle panels are corresponding surface temperature anomalies simulated by the Paleoclimate Modelling Intercomparison Project Phase II (PMIP2) and Paleoclimate Modelling Intercomparison Project Phase III (PMIP3)/ Coupled Model Intercomparison Project Phase 5 (CMIP5) models for 6 ka as compared to pre-industrial. Bottom panels contain boxplots for reconstructions (grey), for model ensembles and for the individual CMIP5 models interpolated to the locations of the reconstructions. Included are OA (ocean–atmosphere), OAV (ocean–atmosphere–vegetation), and OAC (ocean–atmosphere–carbon cycle) models. The boxes are drawn using the 25th, 50th and 75th percentiles (bottom, middle and top of the box, respectively), and whiskers extend to the 5th and 95th percentiles of data or model results within each area. The northern extratropics are defined as 30°N to 90°N and the tropics as 30°S to 30°N. For additional model–data comparisons for the mid-Holocene, see Section 9.4.1.4 and Figures 9.11 and 9.12.

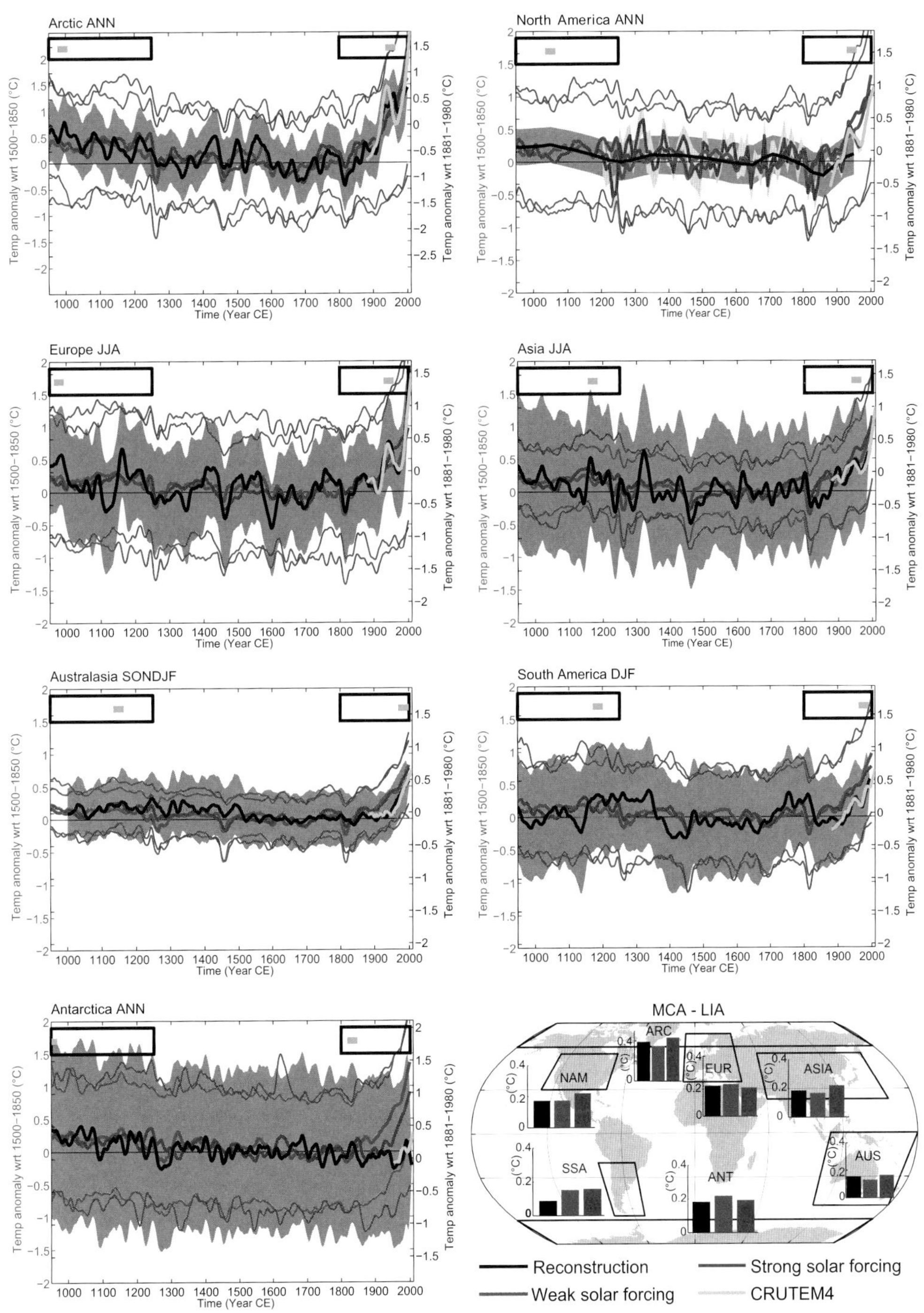

Figure 5.12 | Regional temperature reconstructions, comparison with model simulations over the past millennium (950–2010). Temperature anomalies (thick black line), and uncertainty estimated provided by each individual reconstruction (grey envelope). Uncertainties: Arctic: 90% confidence bands. Antarctica, Australasia, North American pollen and South America: ±2 standard deviation. Asia: ±2 root mean square error. Europe: 95% confidence bands. North American trees: upper/lower 5% bootstrap bounds. Simulations are separated into 2 groups: High solar forcing (red thick line), and weak solar forcing (blue thick line). For each model sub-group, uncertainty is shown as 1.645 times sigma level (light red and blue lines). For comparison with instrumental record, the Climatic Research Unit (CRU) Gridded Dataset of Global Historical Near-Surface Air TEMperature Anomalies Over Land version 4 (CRUTEM4) data set is shown (yellow line). These instrumental data are not necessarily those used in calibration of the reconstructions, and thus may show greater or lesser correspondence with the reconstructions than the instrumental data actually used for calibration; cut-off timing may also lead to end effects for the smoothed data shown. Cf. PAGES 2k Consortium (2013, SOM) in this regard for the North America reconstruction. Green bars in rectangles on top of each panel indicate the 30 warmest years in the 950–1250 period (left rectangle) and 1800–2010 period (right rectangle). All lines are smoothed by applying a 30 year moving average. Map at bottom right shows the individual regions for each reconstruction, and in bars the Medieval Climate Anomaly (MCA, 950-1250) – Little Ice Age (LIA, 1450-1850) differences over those regions. Reconstructions: from PAGES 2k Consortium (2013). Models used: simulations with strong solar forcing (mostly pre-Paleoclimate Modelling Intercomparison Project Phase III (pre-PMIP3) simulations): CCSM3 (1), CNRM-CM3.3 (1), CSM1.4 (1), CSIRO-MK3L-1-2 (3), ECHAM5/MPIOM (3), ECHO-G (1) IPSLCM4 (1), FGOALS-gl (1). Simulations with weak solar forcing (mostly PMIP3/CMIP5 simulations): BCC-csm1-1 (1), CCSM4 (1), CSIRO-MK3L-1-2 (1), GISS-E2-R (3, ensemble members 121, 124, 127), HadCM3 (1), MPI-ESM, ECHAM5/MPIOM (5), IPSL-CM5A-LR (1). In parenthesis are the number of simulations used for each model. All simulations are treated individually, in the time series as well as in the MCA–LIA bars. More information about forcings used in simulations and corresponding references are given in Table 5.A.1. Time periods for averaging are JJA for June – July – August, SONDJF for the months from September to February, and DJF for December – January – February, respectively, while ANN denotes annual mean.

Taking associated uncertainties into consideration, 20th century temperatures in those regions were *likely* not higher than during the first part of the last millennium. In different regions of China, temperatures appear higher during recent decades than during earlier centuries, although with large uncertainties (e.g., Ge et al., 2006, 2010; Wang et al., 2007; Holmes et al., 2009; Yang et al., 2009; Zhang et al., 2009; Cook et al., 2012). In northeast China, an alkenone-based reconstruction indicates that the growing season temperature during the periods 480–860, 1260–1300, 1510–1570 and 1800–1900 was about 1°C lower compared with the 20th century (Chu et al., 2011). These reconstructed NH regional temperature evolutions appear consistent with last millennium GCM simulations using a range of solar forcing estimates (Figure 5.12).

There is *high confidence* that in the extratropical NH, both regionally and on a hemispheric basis, the surface warming of the 20th century reversed the long term cooling trend due to orbital forcing.

5.5.1.2 Tropics

Marcott et al. (2013) provide a compilation of tropical SST reconstructions, showing a gradual warming of about 0.5°C until 5 ka and little change thereafter. Holocene tropical SST trends are regionally heterogeneous and variable in magnitude. Alkenone records from the eastern tropical Pacific, western tropical Atlantic, and the Indonesian archipelago document a warming trend of ~0.5°C to 2°C from the early Holocene to present (Leduc et al., 2010), consistent with local insolation. In contrast, regional trends of planktonic foraminiferal Mg/Ca records are heterogeneous, and imply smaller magnitude SST changes (Leduc et al., 2010; Schneider et al., 2010). Foraminiferal Mg/Ca records in the Indo-Pacific warm-pool region show cooling trends with varying magnitudes (Stott et al., 2004; Linsley et al., 2010). When comparing SST records from different paleoclimate proxies it is important to note that they can have different, and also regionally varying, seasonal biases (Schneider et al., 2010).

Terrestrial temperature reconstruction efforts have mostly focussed on Africa and to some extent on southeast Asia (Figure 5.11), with a lack of syntheses from South America and Australia (Bartlein et al., 2011). The PMIP2 and PMIP3/CMIP5 MH simulations show summer cooling compared to pre-industrial conditions and a shorter growing season in the tropical monsoon regions of Africa and southeast Asia (Figure 5.11), attributed to increased cloudiness and local evaporation (Braconnot et al., 2007). In contrast, MH simulations and reconstructions for the entire tropics (30°S to 30°N) show generally higher mean temperature of the warmest month and lower mean temperature of the coldest month than for the mid-20th century.

5.5.1.3 Southern Hemisphere Mid to High Latitudes

In the high latitude Southern Ocean, Holocene SST trends follow the decrease in austral summer duration, with a cooling trend from the early Holocene into the late Holocene (Kaiser et al., 2008; Shevenell et al., 2011). Similar cooling trends are found in the Australian-New Zealand region (Bostock et al., 2013). Increased amplitude of millennial-to-centennial scale SST variability between 5 ka and 4 ka is recorded in several locations, possibly due to variations in the position and strength of the westerlies (Moros et al., 2009; Euler and Ninnemann, 2010; Shevenell et al., 2011). The Holocene land-surface temperature history in the SH is difficult to assess. Individual reconstructions generally track the trends registered by Antarctic ice core records with peak values at around 12 to 10 ka (Masson-Delmotte et al., 2011b; Marcott et al., 2013; Mathiot et al., 2013). Pollen-based records indicate positive MH temperature anomalies in southern South Africa that are not reproduced in the PMIP3/CMIP5 simulations (Figure 5.11).

Indices for the position of Southern Ocean fronts and the strength and position of the westerlies diverge (Moros et al., 2009; e.g., Shevenell et al., 2011). For the mid-to-late-Holocene, climate models of different complexity consistently show a poleward shift and intensification of the SH westerlies in response to orbital forcing (Varma et al., 2012). However, the magnitude, spatial pattern and seasonal response vary significantly among the models.

New high-resolution, climate reconstructions for the last millennium are based on tree-ring records from the subtropical and central Andes, northern and southern Patagonia, Tierra del Fuego, New Zealand and Tasmania (Cook et al., 2006; Boninsegna et al., 2009; Villalba et al., 2009), ice cores, lake and marine sediments and documentary evidence from southern South America (Prieto and García Herrera, 2009; Vimeux et al., 2009; von Gunten et al., 2009; Tierney et al., 2010; Neukom et al., 2011), terrestrial and shallow marine geological records from eastern Antarctica (Verleyen et al., 2011), ice cores from Antarctica (Goosse et al., 2012c; Abram et al., 2013; Steig et al., 2013), boreholes from western Antarctica (Orsi et al., 2012) and coral records from the Indian and Pacific Oceans (Linsley et al., 2008; Zinke et al., 2009; Lough, 2011; DeLong et al., 2012). There is *medium confidence* that southern South America (Neukom et al., 2011) austral summer temperatures during 950–1350 were warmer than the 20th century. A 1000-year temperature reconstruction for land and ocean representing Australasia indicates a warm period during 1160–1370 though this reconstruction is based on only three records before 1430 (PAGES 2k Consortium, 2013). In Australasia, 1971–2000 temperatures were *very likely* higher than any other 30-year period over the last 580 years (PAGES 2k Consortium, 2013).

Antarctica was *likely* warmer than 1971–2000 during the late 17th century, and during the period from approximately the mid-2nd century to 1250 (PAGES 2k Consortium, 2013).

In conclusion, continental scale surface temperature reconstructions from 950 to 1250 show multi-decadal periods that were in some regions as warm as in the mid-20th century and in others as warm as in the late 20th century (*high confidence*). These regional warm periods were not as synchronous across regions as the warming since the mid-20th century (*high confidence*).

5.5.2 Sea Ice

Since AR4 several new Holocene sea ice reconstructions for the Arctic and sub-Arctic have been made available that resolve multi-decadal to century-scale variability. Proxies of sea-ice extent have been further developed from biomarkers in deep sea sediments (e.g., IP25, Belt et al., 2007; Müller et al., 2011) and from sea-ice biota preserved in sediments (e.g., Justwan and Koç, 2008). Indirect information on sea-ice conditions based on drift wood and beach erosion has also been compiled (Funder et al., 2011). In general, these sea-ice reconstructions parallel regional SST, yet they display spatial heterogeneity, and differences between the methods, making it difficult to provide quantitative estimates of past sea-ice extent. Summer sea-ice cover was reduced compared to late 20th century levels both in the Arctic Ocean and along East Greenland between 8 ka and 6.5 ka (e.g., Moros et al., 2006; Polyak et al., 2010; Funder et al., 2011), a feature which is captured by some MH simulations (Berger et al., 2013). The response of this sea ice cover to summer insolation warming was shown to be central for explaining the reconstructed warmer winter temperatures over the adjacent land (Otto et al., 2009; Zhang et al., 2010). During the last 6 kyr available records show a long-term trend of a more extensive Arctic sea ice cover driven by the orbital forcing (e.g., Polyak et al., 2010), but punctuated by strong century-to-millennial scale variability. Consistent with Arctic temperature changes (see Section 5.5.1), sea ice proxies indicate relatively reduced sea-ice cover from 800 to 1200 followed by a subsequent increase during the LIA (Polyak et al., 2010). Proxy reconstructions document the 20th-century ice loss trend, which is also observed in historical sea ice data sets with a decline since the late 19th century (Divine and Dick, 2006). There is *medium confidence* that the current ice loss was unprecedented and that current SSTs in the Arctic were anomalously high at least in the context of the last 1450 years (England et al., 2008; Kinnard et al., 2008; Kaufman et al., 2009; Macias Fauria et al., 2010; Polyakov et al., 2010; Kinnard et al., 2011; Spielhagen et al., 2011). Fewer high-resolution records exist from the Southern Ocean. Data from the Indian Ocean sector document an increasing sea-ice trend during the Holocene, with a rather abrupt increase between 5 ka and 4 ka, consistent with regional temperatures (see Section 5.5.1.3) (Denis et al., 2010).

5.5.3 Glaciers

Due to the response time of glacier fronts, glacier length variations resolve only decadal- to centennial-scale climate variability. Since AR4 new and improved chronologies of glacier size variations were published (Anderson et al., 2008; Joerin et al., 2008; Yang et al., 2008; Jomelli et al., 2009; Licciardi et al., 2009; Menounos et al., 2009; Schaefer et al., 2009; Wiles et al., 2011; Hughes et al., 2012). Studies of sediments from glacier-fed lakes and marine deposits have allowed new continuous reconstructions of glacier fluctuations (Matthews and Dresser, 2008; Russell et al., 2009; Briner et al., 2010; Bowerman and Clark, 2011; Larsen et al., 2011; Bertrand et al., 2012; Vasskog et al., 2012). Reconstructions of the history of ice shelves and ice sheets/caps have also emerged (Antoniades et al., 2011; Hodgson, 2011; Simms et al., 2011; Smith et al., 2011; Kirshner et al., 2012). New data confirm a general increase of glacier extent in the NH and decrease in the SH during the Holocene (Davis et al., 2009; Menounos et al., 2009), consistent with the local trends in summer insolation and temperatures. Some exceptions exist (e.g., in the eastern Himalayas), where glaciers were most extensive in the early Holocene (Gayer et al., 2006; Seong et al., 2009), potentially due to monsoon changes (Rupper et al., 2009). Due to dating uncertainties, incompleteness and heterogeneity of most existing glacial chronologies, it is difficult to compare glacier variations between regions at centennial and shorter time scales (Heyman et al., 2011; Kirkbride and Winkler, 2012). There are no definitive conclusions regarding potential inter-hemispheric synchronicity of sub-millennial scale glacier fluctuations (Wanner et al., 2008; Jomelli et al., 2009; Licciardi et al., 2009; Schaefer et al., 2009; Winkler and Matthews, 2010; Wanner et al., 2011).

Glacial chronologies for the last 2 kyr are better constrained (Yang et al., 2008; Clague et al., 2010; Wiles et al., 2011; Johnson and Smith, 2012). Multi-centennial glacier variability has been linked with variations in solar activity (Holzhauser et al., 2005; Wiles et al., 2008), volcanic forcing (Anderson et al., 2008) and changes in North Atlantic circulation (Linderholm and Jansson, 2007; Nesje, 2009; Marzeion and Nesje, 2012). Glacier response is more heterogeneous and complex during the MCA than the uniform global glacier recession observed at present (see Section 4.3). Glaciers were smaller during the MCA than in the early 21st Century in the western Antarctic Peninsula (Hall et al., 2010) and Southern Greenland (Larsen et al., 2011). However, prominent advances occurred within the MCA in the Alps (Holzhauser et al., 2005), Patagonia (Luckman and Villalba, 2001), New Zealand (Schaefer et al., 2009), East Greenland (Lowell et al., 2013) and SE Tibet (Yang et al., 2008). Glaciers in northwestern North America were similar in size during the MCA compared to the peak during the LIA, probably driven by increased winter precipitation (Koch and Clague, 2011).

There is *high confidence* that glaciers at times have been smaller than at the end of the 20th century in the Alps (Joerin et al., 2008; Ivy-Ochs et al., 2009; Goehring et al., 2011), Scandinavia (Nesje et al., 2011), Altai in Central Asia (Agatova et al., 2012), Baffin Island (Miller et al., 2005), Greenland (Larsen et al., 2011; Young et al., 2011), Spitsbergen (Humlum et al., 2005), but the precise glacier extent in the previous warm periods of the Holocene is often difficult to assess. While early-to-mid-Holocene glacier minima can be attributed with *high confidence* to high summer insolation (see Section 5.5.1.1), the current glacier retreat, however, occurs within a context of orbital forcing that would be favourable for NH glacier growth. If retreats continue at current rates, most extratropical NH glaciers will shrink to their minimum extent, that existed between 8 ka and 6 ka (*medium confidence*) (e.g., Anderson et al., 2008); and ice shelves on the Antarctic peninsula will retreat to an extent unprecedented through Holocene (Hodgson, 2011; Mulvaney et al., 2012).

5.5.4 Monsoon Systems and Convergence Zones

This subsection focuses on internally and externally driven variability of monsoon systems during the last millennium. Abrupt monsoon changes associated with Dansgaard–Oeschger and Heinrich events (Figure 5.4b, e, h) are further assessed in Section 5.7.1. Orbital-scale monsoon (Figure 5.4a, d, g) changes are evaluated in Section 5.3.2.3.

Hydrological proxy data characterizing the intensity of the East Asian monsoon (South American monsoon) show decreased (increased) hydrological activity during the LIA as compared to the MCA (*medium confidence*) (Figure 5.4f, i) (Zhang et al., 2008; Bird et al., 2011; Vuille et al., 2012). These shifts were accompanied by changes in the occurrence of megadroughts (*high confidence*) in parts of the Asian monsoon region (Buckley et al., 2010; Cook et al., 2010a) (Figure 5.13). Lake sediment data from coastal eastern Africa document dry conditions in the late MCA, a wet LIA, and return toward dry conditions in the 18th or early 19th century (Verschuren et al., 2000; Stager et al., 2005; Verschuren et al., 2009; Tierney et al., 2011; Wolff et al., 2011), qualitatively similar to the South American monsoon proxies in Figure 5.4i, whereas some inland and southern African lakes suggest dry spells during the LIA (Garcin et al., 2007; Anchukaitis and Tierney, 2013). Rainfall patterns associated with the Pacific ITCZ also shifted southward during the MCA/LIA transition in the central equatorial Pacific (Sachs et al., 2009). Extended intervals of monsoon failures and dry spells have been reconstructed for the last few millennia for west Africa (Shanahan et al., 2009), east Africa (Wolff et al., 2011), northern Africa (Esper et al., 2007b; Touchan et al., 2008; Touchan et al., 2011), India and southeastern Asia (Zhang et al., 2008; Berkelhammer et al., 2010; Buckley et al., 2010; Cook et al., 2010a) and Australia (Mohtadi et al., 2011).

On multi-decadal-to-centennial time scales, influences of North Atlantic SST variations have been demonstrated for the North and South African monsoon, and the Indian and East Asian summer monsoons (see Figure 5.4c for monsoon regions), both using proxy reconstructions (Feng and Hu, 2008; Shanahan et al., 2009) and GCM simulations (Lu et al., 2006; Zhang and Delworth, 2006; Wang et al., 2009; Luo et al., 2011). These simulations suggest that solar and volcanic forcing (Fan et al., 2009; Liu et al., 2009a; Man et al., 2012) may exert only weak regional influences on monsoon systems (Figure 5.4f, i). A five-member multi-model ensemble mean of PMIP3/CMIP5 simulations (Table 5.A.1) exhibits decreased standardized monsoon rainfall accompanying periods of reduced solar forcing during the LIA in the East Asian monsoon regions (Figure 5.4f). There is, however, a considerable inter-model spread in the simulated annual mean precipitation response to solar forcing with the multi-model mean, explaining on average only ~25 ± 15% (1 standard deviation) of the variance of the individual model simulations. An assessment of the pre-instrumental response of monsoon systems to volcanic forcing using paleo-proxy data has revealed wetter conditions over southeast Asia in the year of a major volcanic eruption and drier conditions in central Asia (Anchukaitis et al., 2010), in contrast to GCM simulations (Oman et al., 2005; Brovkin et al., 2008; Fan et al., 2009; Schneider et al., 2009).

5.5.5 Megadroughts and Floods

Multiple lines of proxy evidence from tree rings, lake sediments, and speleothems indicate with *high confidence* that decadal or multi-decadal episodes of drought have been a prominent feature of North American Holocene hydroclimate (e.g., Axelson et al., 2009; St. George et al., 2009; Cook et al., 2010a, 2010b; Shuman et al., 2010; Woodhouse et al., 2010; Newby et al., 2011; Oswald and Foster, 2011; Routson et al., 2011; Stahle et al., 2011; Stambaugh et al., 2011; Laird et al., 2012; Ault et al., 2013). During the last millennium, western North America drought reconstructions based on tree ring information (Figure 5.13) show longer and more severe droughts than today, particularly during the MCA in the southwestern and central United States (Meko et al., 2007; Cook et al., 2010b). The mid-14th century cooling coincides in southwestern North America with a shift towards overall wetter conditions (Cook et al., 2010a). In the Pacific Northwest, contrasting results emerge from lake sediment records, indicating wetter conditions during the MCA (Steinman et al., 2013), and tree-ring data showing no substantial change (Zhang and Hebda, 2005; Cook et al., 2010a). In Scandinavia, new tree-ring based reconstructions show a multi-centennial summer drought phase during Medieval times (900–1350) (Helama et al., 2009), while lake sediment proxies from the same region suggest wetter winters (Luoto et al., 2013). New tree-ring reconstructions from the southern-central (Wilson et al., 2013) and southeastern British Isles (Cooper et al., 2013) do not reveal multi-centennial drought during medieval times, but rather alternating multidecades of dry and wet periods. Wilson et al. (2013) reconstructed drier conditions between ~1300 and the early 16th century. Büntgen et al. (2011b) identified exceptionally dry conditions in central Europe from 200 to 350 and between 400 and 600. Numerous tree-ring records from the eastern Mediterranean testify to the regular occurrence of droughts in the past few millennia (e.g., Akkemik et al., 2008; Nicault et al., 2008; Luterbacher et al., 2012). In northern Africa, Esper et al. (2007b) and Touchan et al. (2008; 2011) show severe drought events through the last millennium, particularly prior to 1300, in the 1400s, between 1700 and 1900, and in the most recent instrumental data. Using multiple proxies from Chile, Boucher et al. (2011) inferred wetter conditions during 1000–1250, followed by much drier period until 1400 and wetter conditions similar to present afterwards, while Ledru et al. (2013) reconstructed a dry MCA-LIA transition until 1550. For the South American Altiplano Morales et al. (2012) found periods of drier conditions in the 14th, 16th, and 18th centuries, as well as a modern drying trend.

Reconstruction of past flooding from sedimentary, botanical and historical records (Brázdil et al., 2006; Baker, 2008; Brázdil et al., 2012) provides a means to compare recent large, rare floods, and to analyse links between flooding and climate variability. During the last few millennia, flood records reveal strong decadal to secular variability and non-stationarity in flood frequency and clustering of paleofloods, which varied among regions. In Europe, modern flood magnitudes are not unusual within the context of the last 1000 years (e.g., Brázdil et al., 2012). In Central Europe, the Elbe and the Oder/Odra Rivers show a decrease in the frequency of winter floods during the last 80 to 150 years compared to earlier centuries, while summer floods indicate no significant trend (Mudelsee et al., 2003) (Figure 5.14f–i). In the Alps, paleoflood records derived from lake sediments have shown a higher flood frequency during cool and/or wet phases (Stewart et al., 2011; Giguet-Covex et al., 2012; Wilhelm et al., 2012), a feature also found in Central Europe (Starkel et al., 2006) and the British Isles (Macklin et al., 2012). In the western Mediterranean, winter floods were more frequent during relatively cool and wet climate conditions of the LIA (Benito et al., 2003b; Piccarreta et al., 2011; Luterbacher et al., 2012; Figure 5.14a), whereas autumn floods reflect multi-decadal variations (Benito et al., 2010; Machado et al., 2011; Figure 5.14b, c). In China, extraordinary paleoflood events in the Yellow, Weihe and Qishuihe rivers, occurred synchronously with severe droughts and dust accumulations coinciding with a monsoonal shift, the most severe floods dated at 3.1 ka (Zha et al., 2009; Huang et al., 2012). In India, flood frequencies since 1950 are the largest for the last several hundred years for eight rivers, interpreted as a strengthening of the monsoon conditions after the LIA (Kale, 2008). In southwestern United States, increased frequency of high-magnitude paleofloods coincide with periods of cool, wet climate, whereas warm intervals including the MCA, corresponded with significant decreases in the number of large floods (Ely et al., 1993). In the Great Plains of North America, the frequency of large floods increased significantly around 850 with magnitudes

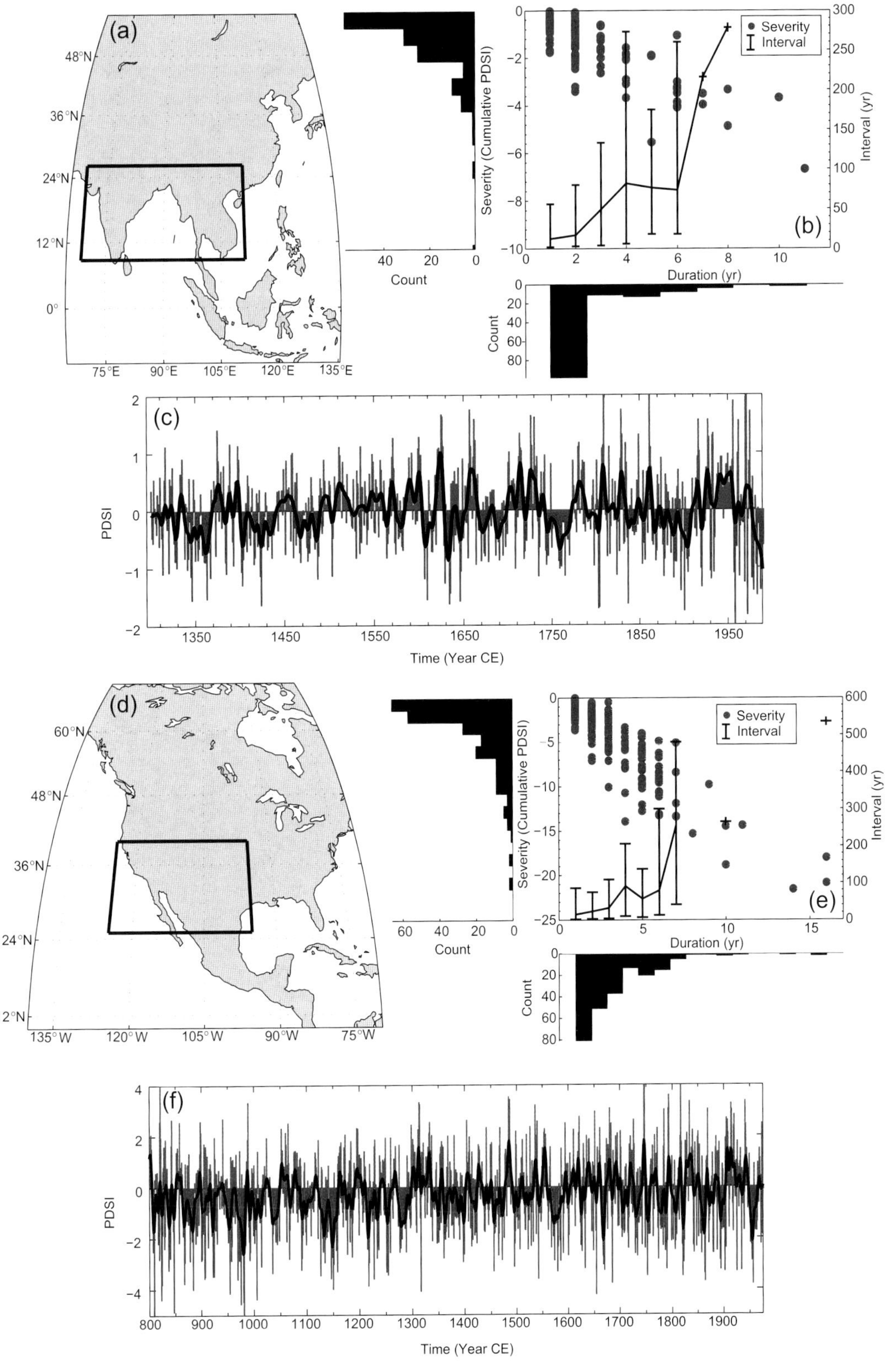

Figure 5.13 | Severity, duration, and frequency of droughts in the Monsoon Asia (Cook et al., 2010b) and North American (Cook et al., 2004) Drought Atlases. The box in (a) and (d) indicates the region over which the tree-ring reconstructed Palmer Drought Severity Index (PDSI) values have been averaged to form the regional mean time series shown in (c) and (f), respectively. Solid black lines in (c) and (f) are a 9-year Gaussian smooth on the annual data shown by the red and blue bars. The covariance of drought (PDSI <0) duration and cumulative severity for each region is shown in panels (b) and (e) by the red circles (corresponding to the left *y*-axes), along with the respective marginal frequency histograms for each quantity. Not shown in b) is an outlier with an apparent duration of 24 years, corresponding to the 'Strange Parallels' drought identified in Cook et al. (2010b). Intervals between droughts of given durations are shown in the same panels and are estimated as the mean interval between their occurrence, with the minimum and maximum reconstructed intervals indicated (corresponding to the right *y*-axes, shown as connected lines and their corresponding range). No error bars are present if there are fewer than three observations of a drought of that duration. The period of analysis is restricted by the availability of tree-ring data to the period 1300–1989 for Monsoon Asia, following Cook et al. (2010a), and from 800 to 1978 for southwestern North America, following Cook et al. (2004).

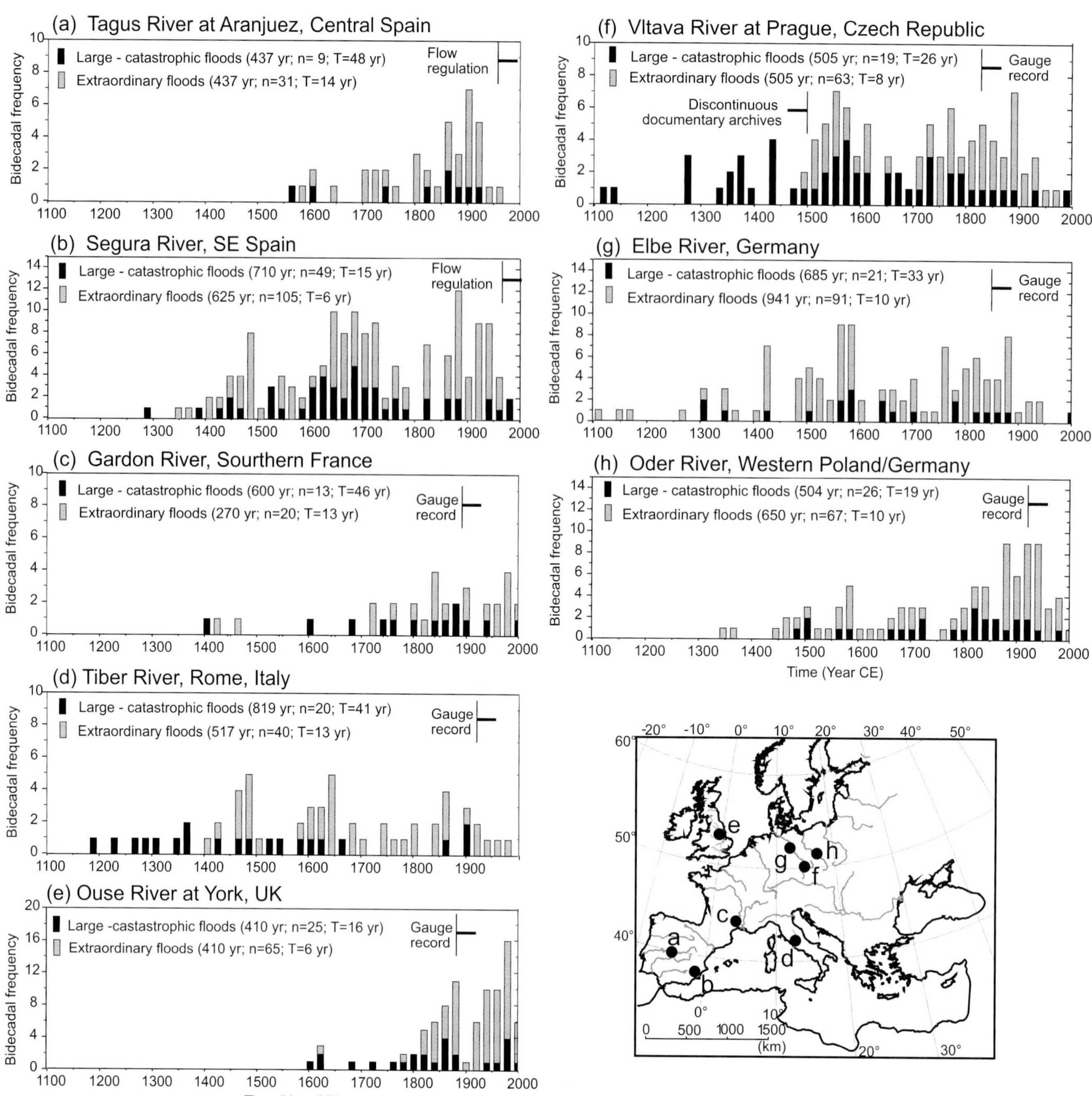

Figure 5.14 | Flood frequency from paleofloods, historical and instrumental records in selected European rivers. Depicted is the number of floods exceeding a particular discharge threshold or flood height over periods of 20 years (bidecadal). Flood categories include rare or catastrophic floods (CAT) associated with high flood discharge or severe damages, and extraordinary floods (EXT) causing inundation of the floodplain with moderate-to-minor damages. Legend at each panel indicates for each category the period of record in years, number of floods (n) over the period, and the average occurrence interval (T in years). (a) Tagus River combined paleoflood, historical and instrumental flood records from Aranjuez and Toledo with thresholds of 100–400 $m^3 s^{-1}$ (EXT) and >400 $m^3 s^{-1}$ (CAT) (data from Benito et al., 2003a; 2003b). (b) Segura River Basin (SE Spain) documentary and instrumental records at Murcia (Barriendos and Rodrigo, 2006; Machado et al., 2011). (c) Gardon River combined discharges from paleofloods at La Baume (Sheffer et al., 2008), documented floods (since the 15th century) and historical and daily water stage readings at Anduze (1741–2005; Neppel et al., 2010). Discharge thresholds referred to Anduze are 1000 to 3000 $m^3 s^{-1}$ (EXT), >3000 $m^3 s^{-1}$ (CAT). At least five floods larger than the 2002 flood (the largest in the gauged record) occurred in the period 1400–1800 (Sheffer et al., 2008). (d) Tiber River floods in Rome from observed historical stages (since 1100; Camuffo and Enzi, 1996; Calenda et al., 2005) and continuous stage readings (1870 to present) at the Ripetta landing (Calenda et al., 2005). Discharge thresholds set at 2300 to 2900 $m^3 s^{-1}$ (EXT) and >2900 $m^3 s^{-1}$ (CAT; >17 m stage at Ripetta). Recent flooding is difficult to evaluate in context due to river regulation structures. (e) River Ouse at York combined documentary and instrumental flood record (Macdonald and Black, 2010). Discharge thresholds for large floods were set at 500 $m^3 s^{-1}$ (CAT) and 350 to 500 $m^3 s^{-1}$ (EXT). (f) Vltava River combined documentary and instrumental flood record at Prague (Brázdil et al., 2005) discharge thresholds: CAT, flood index 2 and 3 or discharge >2900 $m^3 s^{-1}$; EXT flood index 1 or discharge 2000 to 2900 $m^3 s^{-1}$. (g) Elbe River combined documentary and instrumental flood record (Mudelsee et al., 2003). Classes refer to Mudelsee et al. (2003) strong (EXT) and exceptionally strong (CAT) flooding. (h) Oder River combined documentary and instrumental flood record (Mudelsee et al., 2003). The map shows the location of rivers used in the flood frequency plots. Note that flood frequencies obtained from historical sources may be down biased in the early part of the reported periods owing to document loss.

roughly two to three times larger than those of the 1972 flood (Harden et al., 2011). South America large flooding in the Atacama and Peruvian desert streams originated in the highland Altiplano and were particularly intense during El Niño events (Magilligan et al., 2008). In the winter rainfall zone of southern Africa, the frequency of large floods decreased during warmer conditions (e.g., from 1425 to 1600 and after 1925) and increased during wetter, colder conditions (Benito et al., 2011).

In summary, there is *high confidence* that past floods larger than recorded since the 20th century have occurred during the past 500 years in northern and central Europe, western Mediterranean region, and eastern Asia. There is, however, *medium confidence* that in the Near East, India, central North America, modern large floods are comparable to or surpass historical floods in magnitude and/or frequency.

5.6 Past Changes in Sea Level

This section discusses evidence for global mean sea level (GMSL) change from key periods. The MPWP (Table 5.1) has been selected as a period of higher than present sea level (Section 5.6.1), warmer temperature (Section 5.3.1) and 350-450 ppm atmospheric CO_2 concentration (Section 5.2.2.2). Of the recent interglacial periods with evidence for higher than present sea level, the LIG has the best-preserved record (Section 5.6.2). For testing glacio-isostatic adjustment (GIA) models, the principal characteristics of Termination I, including Meltwater Pulse-1A (Section 5.6.3) is assessed. For the Holocene, the emphasis is on the last 6000 years when ice volumes stabilized near present-day values, providing the baseline for discussion of anthropogenic contributions.

5.6.1 Mid-Pliocene Warm Period

Estimates of peak sea levels during the MPWP (Table 5.1, Section 5.3.1) based on a variety of geological records are consistent in suggesting higher-than-present sea levels, but they range widely (10 to 30 m; Miller et al., 2012a), and are each subject to large uncertainties. For example, coastal records (shorelines, continental margin sequences) are influenced by GIA, with magnitudes of the order of 5 m to 30 m for sites in the far and near fields of ice sheets, respectively (Raymo et al., 2011), and global mantle dynamic processes (Moucha et al., 2008; Müller et al., 2008) may contribute up to an additional ±10 m. Consequently, both signals can be as large as the sea level estimate itself and current estimates of their amplitudes are uncertain.

Benthic $\delta^{18}O$ records are better dated than many coastal records and provide a continuous time series, but the $\delta^{18}O$ signal reflects ice volume, temperature and regional hydrographic variability. During the mid-Pliocene warm interval, the 0.1 to 0.25‰ anomalies recorded in the LR04 benthic $\delta^{18}O$ stack (Lisiecki and Raymo, 2005) would translate into ~12 to 31 m higher than present GMSL, if they reflected only ice volume. Conversely, these anomalies could be explained entirely by warmer deep-water temperatures (Dowsett et al., 2009). Attempts to constrain the temperature component in benthic $\delta^{18}O$ records conclude higher than present GMSL during the MPWP with large uncertainties (±10 m) (Dwyer and Chandler, 2009; Naish and Wilson, 2009; Sosdian and Rosenthal, 2009; Miller et al., 2012a).

The first appearance of ice-rafted debris across the entire North Atlantic indicates that continental-scale ice sheets in North America and Eurasia did not develop until about 2.7 Ma (Kleiven et al., 2002). This suggests that MPWP high sea levels were due to mass loss from the GIS, the WAIS and possibly the East Antarctic Ice sheet (EAIS). Sedimentary record from the Ross Sea indicates that the WAIS and the marine margin of EAIS retreated periodically during obliquity-paced interglacial periods of MPWP (Naish et al., 2009a). Reconstructed SSTs for the ice free seasons in the Ross Sea range from 2°C to 8°C (McKay et al., 2012b), with mean values >5°C being, according to one ice sheet model, above the stability threshold for ice shelves and marine portions of the WAIS and EAIS (Pollard and DeConto, 2009; see also Section 5.8.1). A synthesis of the geological evidence from the coastal regions of the Transantarctic Mountains (Barrett, 2013) and an iceberg-rafted debris record offshore of Prydz Bay (Passchier, 2011) also supports coastal thinning and retreat of the EAIS between about 5 to 2.7 Ma. In response to Pliocene climate, ice sheet models consistently produce near-complete deglaciation of GIS (+7 m) and WAIS (+4 m) and retreat of the marine margins of EAIS (+3 m) (Lunt et al., 2008; Pollard and DeConto, 2009; Hill et al., 2010), altogether corresponding to a GMSL rise of up to 14 m.

In summary, there is *high confidence* that GMSL was above present, due to deglaciation of GIS, WAIS and areas of EAIS, and that sea level was not higher than 20 m above present during the interglacials of the MPWP.

5.6.2 The Last Interglacial

Proxy indicators of sea level, including emergent shoreline deposits (Blanchon et al., 2009; Thompson et al., 2011; Dutton and Lambeck, 2012) and foraminiferal $\delta^{18}O$ records (Siddall et al., 2003; Rohling et al., 2008a; Grant et al., 2012) are used to reconstruct LIG sea levels. Implicit in these reconstructions is that geophysical processes affecting the elevation of the sea level indicators (uplift, subsidence, GIA) have been properly modelled and/or that the sea level component of the stable isotope signal has been properly isolated. Particularly important issues with regard to the LIG are (1) the ongoing debate on its initiation and duration (cf. 130 to 116 ka, Stirling et al., 1998; 124 to 119 ka, Thompson and Goldstein, 2005), due to coral geochronology issues; (2) the magnitude of its maximum rise; and (3) sea level variability within the interval. Foraminiferal $\delta^{18}O$ records can constrain (2) and possibly (3). Emergent shorelines on tectonically active coasts can constrain (1) and (3) and possibly (2) if vertical tectonic rates are independently known. Shorelines on tectonically stable coasts can constrain all three issues. Here evidence from emergent shorelines that can be dated directly is emphasized.

5.6.2.1 Magnitude of the Last Interglacial Sea Level Rise

AR4 assessed that global sea level was *likely* between 4 and 6 m higher during the LIG than in the 20th century. Since AR4, two studies (Kopp et al., 2009; Dutton and Lambeck, 2012) have addressed GIA effects from observations of coastal sites.

Kopp et al. (2009) obtained a probabilistic estimate of GMSL based on a large and geographically broadly distributed database of LIG sea level indicators (Figure 5.15a). Their analysis accounted for GIA effects as well as uncertainties in geochronology, the interpretation of sea level indicators, and regional tectonic uplift and subsidence. They concluded that GMSL was *very likely* +6.6 m and *likely* +8.0 m relative to present, and that it is *unlikely* to have exceeded +9.4 m, although some of the most rapid and sustained rates of change occur in the early period when GMSL was still below present (Figure 5.15a).

Dutton and Lambeck (2012) used data from two tectonically stable far-field areas (areas far from the former centres of glaciation), Australia and the Seychelles islands. At these sites, in contrast to sites near the former ice margins, the isostatic signals are less sensitive to the choice of parameters defining the Earth rheology and the glacial ice sheets (Lambeck et al., 2012). On the west coast of Australia, the highest LIG reef elevations are at +3.5 m and the inferred paleo-sea level, allowing for possible reef erosion, is about +5.5 m relative to present. In the Seychelles, LIG coral reefs occur from 0 m to 6 m, but also possibly as high as ~9 m (Israelson and Wohlfarth, 1999, and references therein). Ten of the eleven LIG coral samples from the Seychelles used in Dutton and Lambeck (2012) have reef elevation estimates ranging from +2.1 to +4 m relative to present; whereas a single LIG coral sample has a reef elevation estimate of +6 m. Additional results are needed to support an estimate of a maximum LIG sea level at the Seychelles of + 9 m relative to present.

In conclusion, there is *very high confidence* that the maximum GMSL during the LIG was, for several thousand years, at least 5 m higher than present but that GMSL at this time did not exceed 10 m (*high confidence*). The best estimate from the two available studies is 6 m higher than present.

5.6.2.2 Evidence for Last Interglacial Sea Level Variability

Since AR4, there is evidence for meter-scale variability in local LIG sea level between 126 ka and 120 ka (Thompson and Goldstein, 2005; Hearty et al., 2007; Rohling et al., 2008a; Kopp et al., 2009; Thompson et al., 2011). However, there are considerable differences in the timing and amplitude of the reported fluctuations due to regional sea level variability and uncertainties in sea level proxies and their ages.

Two episodes of reef building during the LIG have been reported on the Yucatan coast (Blanchon et al., 2009) and in the Bahamas (Chen et al., 1991; Thompson et al., 2011). Blanchon et al. (2009) provide evidence of Yucatan reef growth early in the LIG at a relative sea level of +3 m, followed by a later episode at +6 m. Thompson et al. (2011) inferred a +4 m relative sea level at ~123 ka, followed by a fall to near present, and finally a rise to +6 m at ~119 ka. This yields a rate of sea level change in the Bahamas of ~2.6 m kyr^{-1}, although the higher estimate at the end of the interval may reflect GIA effects that result in a rise in relative sea level at these locations (Dutton and Lambeck, 2012).

LIG sea level rise rates of between 1.1 and 2.6 m per century have been estimated based on a foraminiferal $\delta^{18}O$ record from the Red Sea (Rohling et al., 2008a). However, the original Red-Sea chronology was based on a short LIG duration of 124 to 119 ka, after Thompson and Goldstein (2005). The longer LIG duration of 130 to 116 ka indicated by the coral data (Stirling et al., 1998) reduces these rates to 0.4 to 0.9 m per century, and a revised chronology of the Red Sea sea level record adjusted to ages from Soreq Cave yields estimates of sea level rise rates of up to 0.7 m per century when sea level was above present level during the LIG (Grant et al., 2012).

In their probabilistic assessment of LIG sea level, Kopp et al. (2013) concluded that it was *extremely likely* that there were at least two peaks in sea level during the LIG. They further concluded that during the interval following the initial peak at ~126 ka (Figure 5.15a) it is likely that there was a period in which GMSL rose at an average rate exceeding 3 m kyr^{-1}, but unlikely that this rate exceeded 7 m kyr^{-1}.

In summary, there is evidence for two intra-LIG sea level peaks (*high confidence*) during which sea level varied by up to 4 m (*medium confidence*). The millennial-scale rate of sea level rise during these periods exceeded 2 m kyr^{-1} (*high confidence*).

5.6.2.3 Implications for Ice Sheet Loss During the Last Interglacial

The principal sources for the additional LIG meltwater are the GIS, WAIS and the low elevation, marine-based margins of the EAIS. An upper limit for the contributions from mountain glaciers is ~0.42 ± 0.11 m if all present-day mountain glaciers melted (cf. Section 4.3). The estimated LIG ocean thermal expansion contribution is 0.4 ± 0.3 m (McKay et al., 2011).

Sedimentological evidence indicates that southern Greenland was not ice-free during the LIG (Colville et al., 2011). Since AR4, the evidence for LIG ice layers in Greenland ice cores, which was ambiguous from Dye 3 and unequivocal from Summit and NGRIP ice cores (summarised in Masson-Delmotte et al., 2011a), has been strengthened. Data from the new NEEM ice core (NEEM community members, 2013; see Figure 5.16 for locations) point to an unequivocal existence of ice throughout the LIG with elevations differing a few hundred meter from present, possibly decreasing in elevation by ~ 400 m ± 350 m between 128 and 122 ka BP. GIS simulations give an average contribution of ~2.3 m to LIG GMSL (1.5 m, 1.9 m, 1.4 m and 4.3 m respectively for four models illustrated in Figure 5.16). Each model result has been selected from a series of runs within a range of parameter uncertainties that yield predictions consistent with the occurrence of ice at NEEM and the elevation of that ice reconstructed from the ice core record. In summary, the GIS simulations that are consistent with elevation changes from the ice core analysis show limited ice retreat during this period such that this ice sheet *very likely* contributed between 1.4 and 4.3 m sea level equivalent, implying with *medium confidence* a contribution from the Antarctic ice sheet.

One model of WAIS glacial–interglacial variability shows very little difference in ice volumes between the LIG and present (Pollard and DeConto, 2009) (Figure 5.15g), when the surface climate and ocean melt term were parameterised using the global benthic $\delta^{18}O$ record for the last 5 Ma. Direct geological evidence of fluctuations in the extent of WAIS margin during the LIG is equivocal due to inadequate age control on two sediment cores which imply that open-water conditions existed in the southeastern sector of the Ross Ice Shelf at some time in the last 1 Ma (Scherer et al., 1998; McKay et al., 2011; Vaughan et al.,

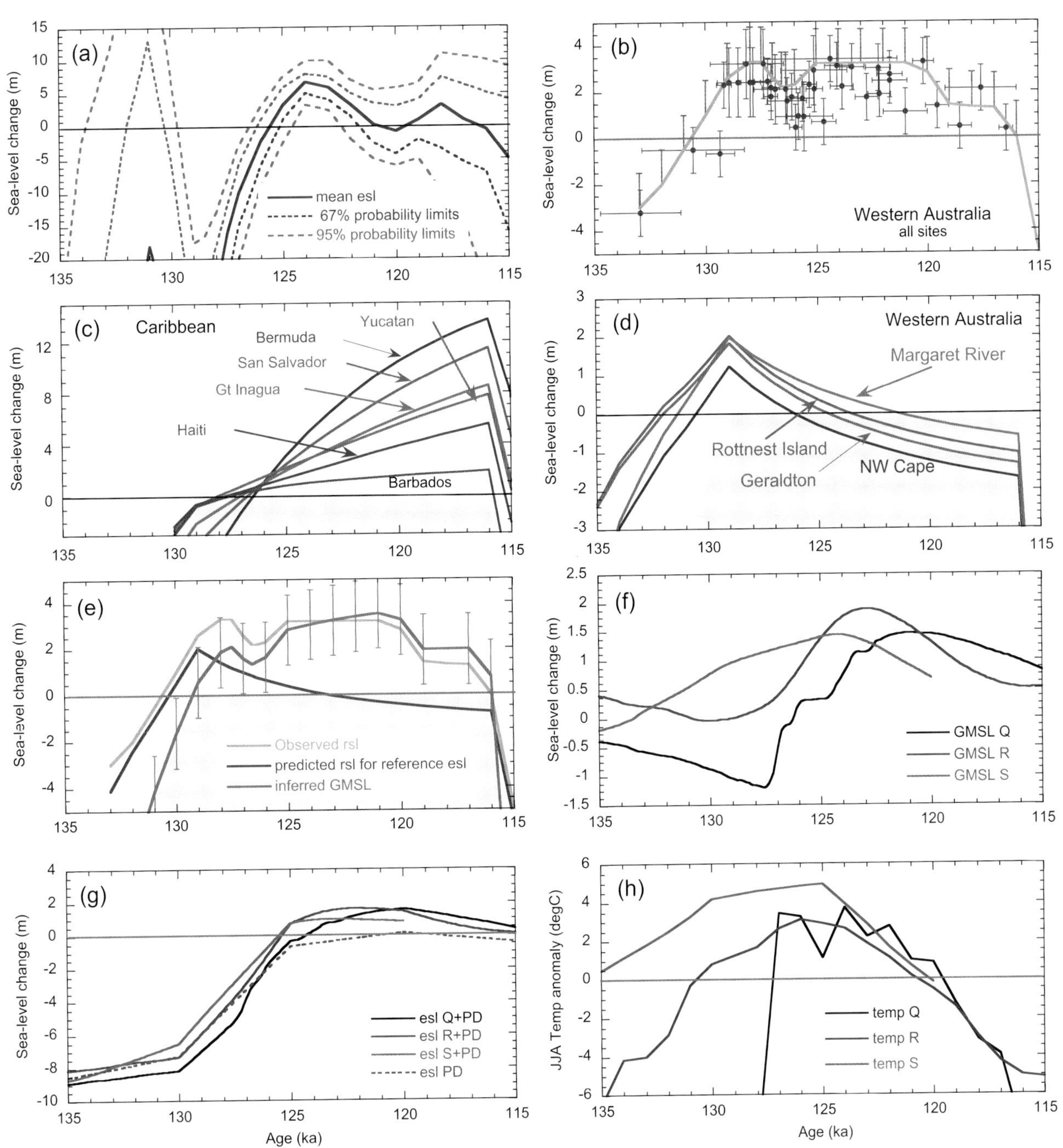

Figure 5.15 | Sea level during the Last Interglacial (LIG) period. (a) Proxy-derived estimate of global mean sea level (GMSL) anomaly from Kopp et al. (2013). Mean GMSL (red line), 67% confidence limits (blue dashed lines) and 95% confidence limits (green dashed lines). The chronology is based on open-system U/Th dates. (b) Local LIG relative sea level reconstructions from Western Australia based on *in situ* coral elevations (red) that pass diagenetic screening (Dutton and Lambeck, 2012). Age error bars correspond to 2 standard deviation uncertainties. All elevations have been normalised to the upper growth limit of corals corresponding to mean low water spring or mean low sea level. The blue line indicates the simplest interpretation of local sea level consistent with reef stratigraphy and should be considered as lower limits by an amount indicated by the blue upper limit error bars. The chronology is based on closed system U/Th dates. (c) Predicted sea levels for selected sites in the Caribbean and North Atlantic in the absence of tectonics with the assumption that ice volumes during the interval from 129 to 116 ka are equal to those of today (Lambeck et al., 2012) illustrating the spatial variability expected across the region due to glacio-isostatic effects of primarily the MIS-6 and MIS-2 deglaciations (see also Raymo and Mitrovica, 2012). The reference ice-volume model for the LIG interval (blue shaded), earth rheology and ice sheet parameters are based on rebound analyses from different regions spanning the interval from Marine Isotope Stage 6 to the present (c.f. Lambeck et al., 2006). LIG sea level observations from these sites contain information on these ice histories and on GMSL. (d) Same as (c) but for different sites along the Western Australia coast contributing to the data set in (b). The dependence on details of the ice sheet and on Earth-model parameters is less important at these sites than for those in (c). Thus data from these locations, assuming tectonic stability, is more appropriate for estimating GMSL. (e) The Western Australian reconstructed evidence (blue) from (b), compared with the model-predicted result (red) for a reference site midway between the northern and southern most localities. The difference between the reconstructed and predicted functions provides an estimate of GMSL (green). Uncertainties in this estimate (67% confidence limits) include the observational uncertainties from (b) and model uncertainties (see e.g., Lambeck et al., 2004a, for a treatment of model errors). (f) Simulated contribution of GIS to GMSL change (black, Q, Quiquet et al. (2013); red, R, Robinson et al. (2011); blue, S, Stone et al. (2013), The Q, R, S correspond to the labels in Figure 5.16). (g) Simulated total GMSL contribution from the GIS (Q, R, S as in panel (f)) and the Antarctic ice sheet contribution (PD) according to Pollard and DeConto (2009). (h) Central Greenland surface-air temperature anomalies for summer (June–August, JJA) used for ice sheet simulations displayed in panel (f) and in Figure 5.16. Anomalies in all panels are calculated relative to present.

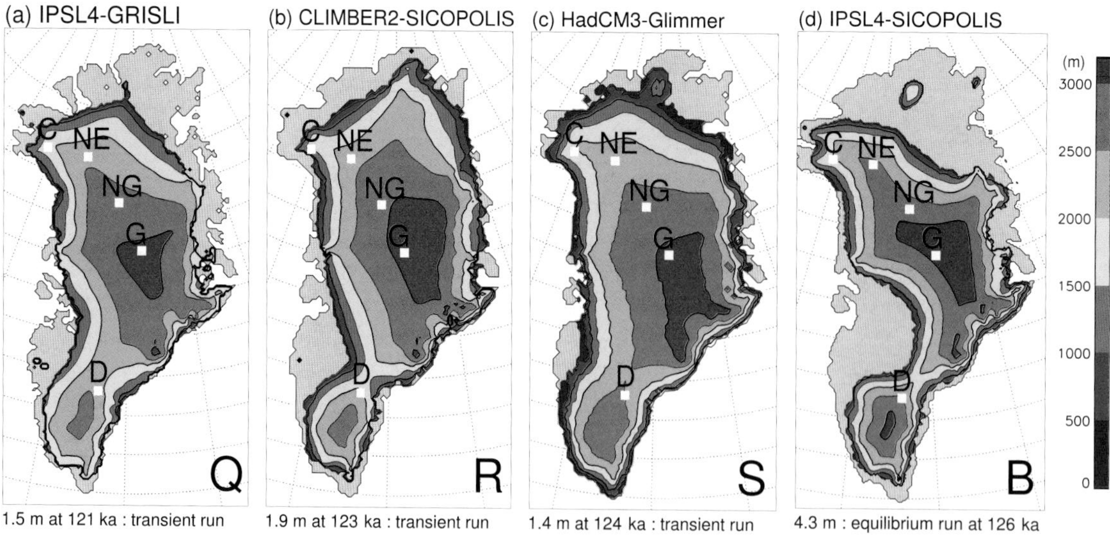

Figure 5.16 | Simulated GIS elevation at the Last Interglacial (LIG) in transient (Q, R, S) and constant-forcing experiments (B). (Q) GRISLI ice sheet model with transient climate forcing derived from IPSL simulations and paleoclimate reconstructions (Quiquet et al., 2013). (R) Simulation, most consistent with independent evidence from ice cores, from ensemble runs SICOPOLIS ice sheet model driven by transient LIG climate simulations downscaled from CLIMBER2 with the regional model REMBO (Robinson et al., 2011). (S) As R but from ensemble simulations with the Glimmer ice sheet model forced with transient climate forcing from 135 to 120 ka with HadCM3 (Stone et al., 2013). (B) SICOPOLIS ice model forced with a constant Eemian climate simulation of IPSL (at 126 ka), running for 6000 years starting from fully glaciated present-day GIS (Born and Nisancioglu, 2012). White squares in each panel show the locations of ice core sites: Greenland Ice Core Project/Greenland Ice Sheet Project (GRIP/GISP) from the summit (G), North Greenland Ice Core Project (NGRIP) (NG), North Greenland Eemian Ice Drilling (NEEM) (NE), Camp Century (C), and Dye3 (D). For ice sheet simulations using transient climate forcing, the minimum in ice volume is illustrated. All panels use original model resolution and grids. Below each panel, maximum contribution to global mean sea level rise and time of minimum ice volume are denoted together with information on experimental design (either "transient" run of the Interglacial period starting from the former glacial or "equilibrium" run of a time slice at the peak interglacial). The differences in model outputs regarding timing and ice elevations result from different methodologies (e.g., transient climate change or equilibrium climate, with the latter assumption leading to the highest estimate of the four models), melt schemes (van de Berg et al., 2011), and the reference climate input (Quiquet et al., 2013).

2011) and in the vicinity of the northwestern Ross Ice Shelf within the last 250,000 years during MIS 7 or LIG (McKay et al., 2012a). Ackert et al. (2011) dated glacial erratics and moraines across the Ohio Mountain Range of the Transantarctic Mountains and concluded that the ice elevations were similar during the LIG and today, but such results cannot be extrapolated beyond this region. East Antarctic ice core LIG data may reflect the impact of a reduced WAIS due to climatic effects (Holden et al., 2010b) but not through isostatic effects (Bradley et al., 2012). Modelling and ice core data suggest EAIS may have retreated in the Wilkes Basin (Bradley et al., 2012).

5

In summary, no reliable quantification of the contribution of the Antarctic ice volume to LIG sea level is currently possible. The only available transient ice sheet model simulation (Pollard and DeConto, 2009) does not have realistic boundary conditions, not enough is known about the subsurface temperatures and there are few direct observational constraints on the location of ice margins during this period. If the above inference of the contribution to GMSL from the GIS (5.6.2.1) is correct, the full GMSL change (section 5.6.2.2) implies significantly less LIG ice in Antarctica than today, but as yet this cannot be supported by the observational and model evidence.

5.6.3 Last Glacial Termination and Holocene

The onset of melting of the LGM ice sheets occurred at approximately 20 ka and was followed by a GMSL rise of ~130 m in ~13 kyr (Lambeck et al., 2002b). Coeval with the onset of the Bølling warming in the NH, a particularly rapid rise of ~ 20 m occurred within ~ 340 years (Meltwater Pulse 1A, MWP-1A), as most recently documented by Deschamps et al. (2012) from a new Tahiti coral record. At this location sea level rose between 14 and 18 m at a rate approaching 5 m per century. The source of MWP-1A continues to be widely debated with most attention being on scenarios in which the Antarctic ice sheet contributed either significantly (Clark et al., 2002, 2009; Bassett et al., 2005) or very little (Bentley et al., 2010; Mackintosh et al., 2011). Evidence of rapid WAIS retreat at around the time of MWP-1A is also indicated by analysis of marine sediment cores (e.g., Kilfeather et al., 2011; Smith et al., 2011).

If the Antarctic ice sheet was the major contributor to MWP-1A then it must have contained at least $7 \cdot 10^6$ km^3 more ice than at present (equivalent to ~17 m GMSL), which is about twice the difference in Antarctic ice volume between the LGM and present found by Whitehouse et al. (2012). Because of the Earth-ocean (including gravitational, deformational and rotational) response to rapid changes in ice volume, the amplitude of the associated sea level change is spatially variable (Clark et al., 2002) and can provide insight into the source region. Based on the comparison of the new Tahiti record with records from Barbados (Fairbanks, 1989) and the Sunda Shelf (Hanebuth et al., 2011), Deschamps et al. (2012) conclude that a significant meltwater contribution to GMSL, of at least 7 m, originated from Antarctica. From ice sheet modelling, Gregoire et al. (2012) argued that the separation of the North American Laurentide and Cordilleran ice sheets may in part be the cause of MWP-1A, contributing ~9 m in 500 years. Another ice sheet modelling study Carlson et al. (2012) suggests a contribution of 6 to 8 m in 500 years from the Laurentide at the onset of the Bølling warming over North America. These studies indicate that there are no glaciological impediments to a major North American contribution to MWP-1A. In contrast, there are as yet no modelling results that show a rapid retreat or partial collapse of the Antarctic ice sheet at that time.

Since AR4, high-resolution sea level records from different localities suggest further periods of rapid ice-mass loss. For example, records from Singapore indicate a rise of ~14 m from ~9.5 to 8.0 ka followed by a short interval of a smaller rise centred on about 7.2 ka (Bird et al., 2010, for Singapore) and records from the US Atlantic (Cronin et al., 2007) and North Sea coasts (Hijma and Cohen, 2010) suggest a rise at around ~9.0–7.5 ka that is possibly punctuated by one or two short intervals of higher rates. These and similar rapid events have to be interpreted against a background of rapid rise that is spatially variable because of the residual isostatic response to the last deglaciation (Milne and Mitrovica, 2008). Different explanations of these short-duration events have been proposed: a multi-stage draining of glacial Lake Agassiz (Hijma and Cohen, 2010), although estimates of the amount of water stored in this lake are less than the required amount; a rapid melting of the Labrador and Baffin ice domes (Carlson et al., 2007; Gregoire et al., 2012); or to Antarctic ice sheet decay (Bird et al., 2007; Cronin et al., 2007).

Ocean volume between about 7 ka and 3 ka is *likely* to have increased by an equivalent sea level rise of 2 to 3 m (Lambeck et al., 2004b, 2010) (Figure 5.17). About 10% of this increase can be attributed to a mid-to-late-Holocene ice reduction over Marie Byrd Land, West Antarctica

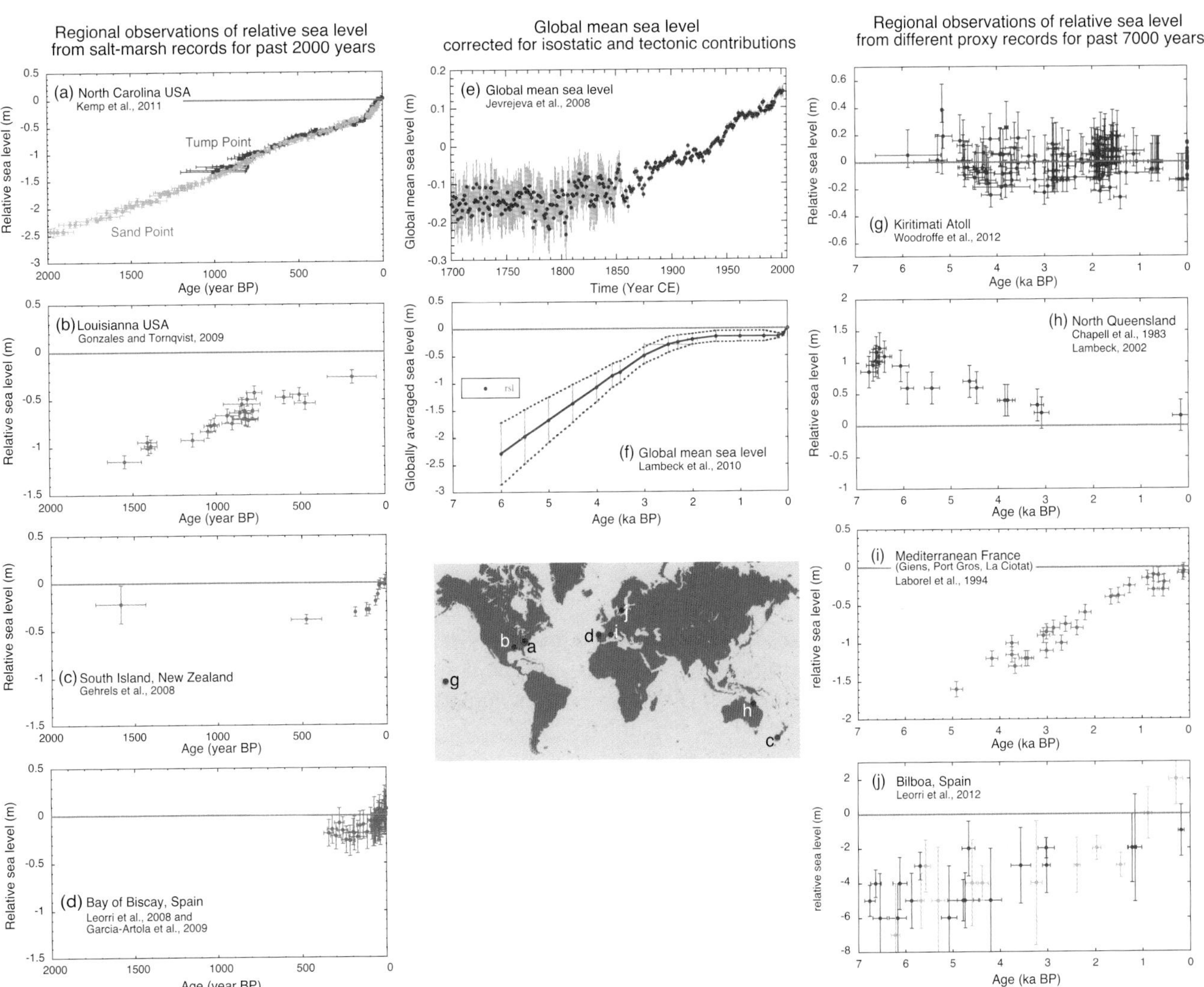

Figure 5.17 | Observational evidence for sea level change in the recent and late Holocene. Left panels (a–d): High-resolution, relative sea level results from salt-marsh data at representative sites, without corrections for glacial isostatic movement of land and sea surfaces. Locations are given on the map. The North Carolina (a) result is based on two nearby locations, Tump Point (dark blue) and Sand Point (light blue). They are representative of other North American Atlantic coast locations (e.g., b; Kemp et al., 2011). The rate of change occurring late in the 19th century are seen in all high resolution salt-marsh records—e.g., in New Zealand (c) (Gehrels et al., 2008; Gehrels and Woodworth, 2013) and in Spain (d) (Leorri et al., 2008; García-Artola et al., 2009)—that extend into modern time and is consistent with Roman archaeological evidence (Lambeck et al., 2004b). The oscillation in sea level seen in the North Carolina record at about 1000 years ago occurs in some (González and Törnqvist, 2009) but not all records (cf. Gehrels et al., 2011; Kemp et al., 2011). Right hand side panels (g–j): Observational evidence for sea level change from lower resolution but longer period records. All records are uncorrected for isostatic effects resulting in spatially variable near-linear trend in sea level over the 7000-year period. The Kiritimati record (Christmas Island) (g) consists of coral microatoll elevations whose fossil elevations are with respect to the growth position of living microatolls (Woodroffe et al., 2012). The North Queensland record (h) is also based on microatoll evidence from several sites on Orpheus Island (Chappell, 1983; Lambeck et al., 2002a). The data from Mediterranean France (i) is based on biological indicators (Laborel et al., 1994) restricted to three nearby locations between which differential isostatic effects are less than the observational errors (Lambeck and Bard, 2000). The Spanish record (j) from estuarine sedimentary deposits is for two nearby localities; Bilboa (dark blue) and Urdaibai (light blue) (Leorri et al., 2012). The two global records (central panels) are estimates of change in global mean sea level from (i) the instrumental record (Jevrejeva et al., 2008) that overlaps the salt-marsh records, and (j) from a range of geological and archaeological indicators from different localities around the world (Lambeck et al., 2010), with the contributing records corrected individually for the isostatic effects at each location.

Frequently Asked Questions

FAQ 5.2 | How Unusual is the Current Sea Level Rate of Change?

The rate of mean global sea level change—averaging 1.7 ± 0.2 mm yr^{-1} for the entire 20th century and between 2.8 and 3.6 mm yr^{-1} since 1993 (Chapter 13)—is unusual in the context of centennial-scale variations of the last two millennia. However, much more rapid rates of sea level change occurred during past periods of rapid ice sheet disintegration, such as transitions between glacial and interglacial periods. Exceptional tectonic effects can also drive very rapid local sea level changes, with local rates exceeding the current global rates of change.

'Sea level' is commonly thought of as the point where the ocean meets the land. Earth scientists define sea level as a measure of the position of the sea surface relative to the land, both of which may be moving relative to the center of the Earth. A measure of sea level therefore reflects a combination of geophysical and climate factors. Geophysical factors affecting sea level include land subsidence or uplift and glacial isostatic adjustments—the earth–ocean system's response to changes in mass distribution on the Earth, specifically ocean water and land ice.

Climate influences include variations in ocean temperatures, which cause sea water to expand or contract, changes in the volume of glaciers and ice sheets, and shifts in ocean currents. Local and regional changes in these climate and geophysical factors produce significant deviations from the global estimate of the mean rate of sea level change. For example, *local* sea level is falling at a rate approaching 10 mm yr^{-1} along the northern Swedish coast (Gulf of Bothnia), due to ongoing uplift caused by continental ice that melted after the last glacial period. In contrast, *local* sea level rose at a rate of ~20 mm yr^{-1} from 1960 to 2005 south of Bangkok, mainly in response to subsidence due to ground water extraction.

For the past ~150 years, sea level change has been recorded at tide gauge stations, and for the past ~20 years, with satellite altimeters. Results of these two data sets are consistent for the overlapping period. The globally averaged rate of sea level rise of ~1.7 ± 0.2 mm yr^{-1} over the 20th century—and about twice that over the past two decades—may seem small compared with observations of wave and tidal oscillations around the globe that can be orders of magnitude larger. However, if these rates persist over long time intervals, the magnitude carries important consequences for heavily populated, low-lying coastal regions, where even a small increase in sea level can inundate large land areas.

Prior to the instrumental period, local rates of sea level change are estimated from indirect measures recorded in sedimentary, fossil and archaeological archives. These proxy records are spatially limited and reflect both local and global conditions. Reconstruction of a global signal is strengthened, though, when individual proxy records from widely different environmental settings converge on a common signal. It is important to note that geologic archives—particularly those before about 20,000 years ago—most commonly only capture millennial-scale changes in sea level. Estimates of century-scale rates of sea level change are therefore based on millennial-scale information, but it must be recognised that such data do not necessarily preclude more rapid rates of century-scale changes in sea level.

Sea level reconstructions for the last two millennia offer an opportunity to use proxy records to overlap with, and extend beyond, the instrumental period. A recent example comes from salt-marsh deposits on the Atlantic Coast of the United States, combined with sea level reconstructions based on tide-gauge data and model predictions, to document an average rate of sea level change since the late 19th century of 2.1 ± 0.2 mm yr^{-1}. This century-long rise exceeds any other century-scale change rate in the entire 2000-year record for this same section of coast.

On longer time scales, much larger rates and amplitudes of sea level changes have sometimes been encountered. Glacial–interglacial climate cycles over the past 500,000 years resulted in global sea level changes of up to about 120 to 140 m. Much of this sea level change occurred in 10,000 to 15,000 years, during the transition from a full glacial period to an interglacial period, at average rates of 10 to 15 mm yr^{-1}. These high rates are only sustainable when the Earth is emerging from periods of extreme glaciation, when large ice sheets contact the oceans. For example, during the transition from the last glacial maximum (about 21,000 years ago) to the present interglacial (Holocene, last 11,650 years), fossil coral reef deposits indicate that global sea level rose abruptly by 14 to 18 m in less than 500 years. This event is known as Meltwater Pulse 1A, in which the rate of sea level rise reached more than 40 mm yr^{-1}.

These examples from longer time scales indicate rates of sea level change greater than observed today, but it should be remembered that they all occurred in special circumstances: at times of transition from full glacial to interglacial condition; at locations where the long-term after-effects of these transitions are still occurring; at locations of

(continued on next page)

FAQ 5.2 (continued)

major tectonic upheavals or in major deltas, where subsidence due to sediment compaction—sometimes amplified by ground-fluid extraction—dominates.

The instrumental and geologic record support the conclusion that the current rate of mean global sea level change is unusual relative to that observed and/or estimated over the last two millennia. Higher rates have been observed in the geological record, especially during times of transition between glacial and interglacial periods.

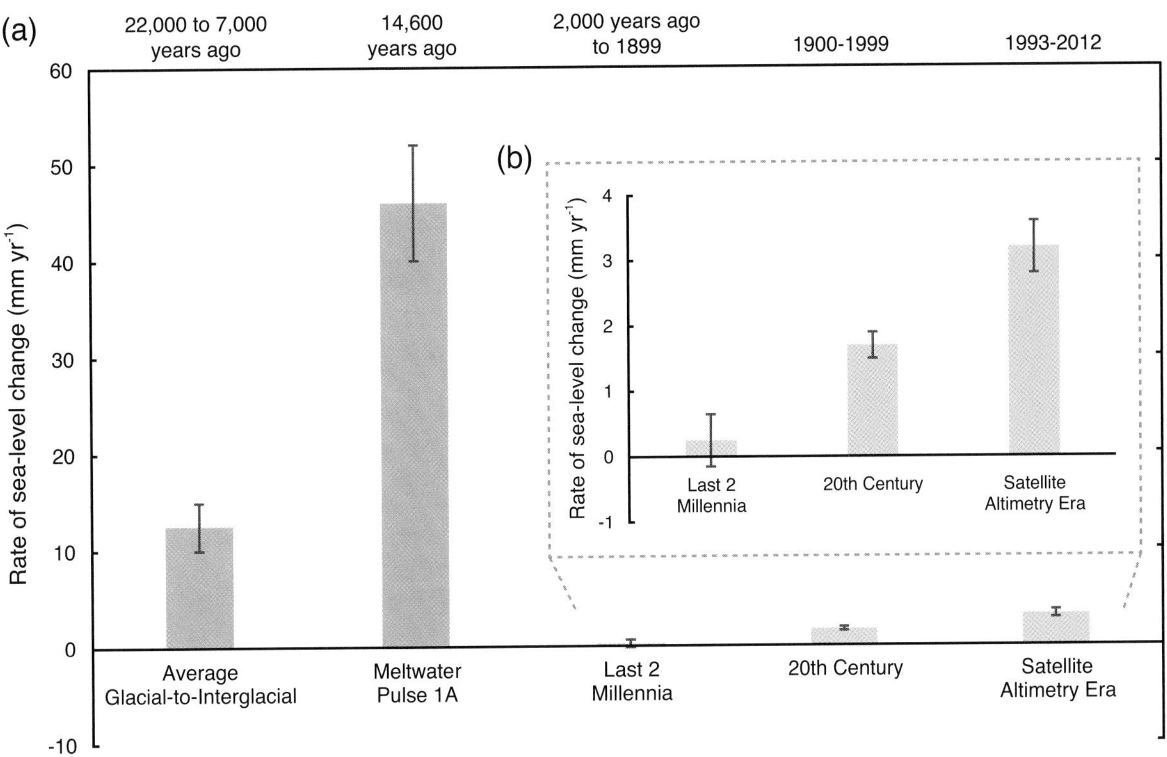

FAQ 5.2, Figure 1 | (a) Estimates of the average rate of global mean sea level change (in mm yr^{-1}) for five selected time intervals: last glacial-to-interglacial transition; Meltwater Pulse 1A; last 2 millennia; 20th century; satellite altimetry era (1993–2012). Blue columns denote time intervals of transition from a glacial to an interglacial period, whereas orange columns denote the current interglacial period. Black bars indicate the range of likely values of the average rate of global mean sea level change. Note the overall higher rates of global mean sea level change characteristic of times of transition between glacial and interglacial periods. (b) Expanded view of the rate of global mean sea level change during three time intervals of the present interglacial.

(Stone et al., 2003). Elevation histories derived from central Greenland ice core data (Vinther et al., 2009; Lecavalier et al., 2013) have presented evidence for thinning from 8 ka to 6 ka but no integrated observation-based estimate for the total ice sheet is available. Contributions from mountain glaciers for this interval are unknown.

Resolving decimeter-scale sea level fluctuations is critical for understanding the causes of sea level change during the last few millennia. Three types of proxies have this capability: salt-marsh plants and microfauna (foraminifera and diatoms) that form distinctive elevation zones reflecting variations in tolerances to the frequency and duration of tidal inundation (Donnelly et al., 2004; Horton and Edwards, 2006; Gehrels et al., 2008; Kemp et al., 2009; Long et al., 2012); coral microatolls found in intertidal environments close to lowest spring tides (Woodroffe and McLean, 1990; Smithers and Woodroffe, 2001; Goodwin and Harvey, 2008); and coastal archaeological features constructed with direct (e.g., fish ponds and certain harbour structures) or indirect (e.g., changes in water-table level in ancient wells) relationships to sea level (Lambeck et al., 2004b; Sivan et al., 2004; Auriemma and Solinas, 2009; Anzidei et al., 2011). Of these, the salt-marsh records are particularly important because they have been validated against regional tide-gauge records and because they can provide near-continuous records. The most robust signal captured in the salt-marsh proxy sea level records from both the NH and SH is an increase in rate, late in the 19th or in the early 20th century (Figure 5.17), that marks a transition from relatively low rates of change during the late Holocene (order tenths of mm yr^{-1}) to modern rates (order mm yr^{-1}) (see also FAQ 5.2). Variability in both the magnitude and the timing (1840–1920) of this acceleration has been reported (Gehrels et al., 2006, 2008, 2011; Kemp et al., 2009, 2011), but Gehrels and Woodworth (2013) have concluded that these mismatches can be reconciled within the observational uncertainties. Combined with the instrumental evidence (see Section 3.7) and with inferences drawn from archaeological evidence from 2000 years ago (Lambeck et al., 2004b), rates of sea level rise exceeded the late Holocene background rate after about 1900 (*high confidence*) (Figure 5.17).

Regionally, as along the US Atlantic coast and Gulf of Mexico coast, the salt-marsh records reveal some consistency in multi-decadal and centennial time scales deviations from the linear trends expected from the GIA signal (see e.g., panels (a) and (b) in Figure 5.17) (van de Plassche et al., 1998; González and Törnqvist, 2009; Kemp et al., 2011) but they have not yet been identified as truly global phenomena. For the past 5 millennia the most complete sea level record from a single location consists of microatoll evidence from Kiritimati (Christmas Island; Pacific Ocean) (Woodroffe et al., 2012) that reveals with *medium confidence* that amplitudes of any fluctuations in GMSL during this interval did not exceed approximately ±25 cm on time scales of a few hundred years. Proxy data from other localities with quasi-continuous records for parts of this pre-industrial period, likewise, do not identify significant global oscillations on centennial time scales (Figure 5.17).

5.7 Evidence and Processes of Abrupt Climate Change

Many paleoclimate archives document climate changes that happened at rates considerably exceeding the average rate of change for longer-term averaging periods prior and after this change (see Glossary for other definition of Abrupt Climate Change). A variety of mechanisms have been suggested to explain the emergence of such abrupt climate changes (see Section 12.5.5). Most of them invoke the existence of nonlinearities or, more specifically, thresholds in the underlying dynamics of one or more Earth-system components. Both internal dynamics and external forcings can generate abrupt changes in the climate state. Documentation of abrupt climate changes in the past using multiple sources of proxy evidence can provide important benchmarks to test instability mechanisms in climate models. This assessment of abrupt climate change on time scales of 10 to 100 years focuses on Dansgaard-Oeschger (DO) events and iceberg/meltwater discharges during Heinrich events, especially the advances since AR4 in reconstructing and understanding their global impacts and in extending the record of millennial-scale variability to about 800 ka.

Twenty-five abrupt DO events (North Greenland Ice Core Project members, 2004) and several centennial-scale events (Capron et al., 2010b) occurred during the last glacial cycle (see Section 5.3.2). DO events in Greenland were marked by an abrupt transition (within a few decades) from a cold phase, referred to as Greenland Stadial (GS) into a warm phase, known as Greenland Interstadial (GI). Subsequently but within a GI, a gradual cooling preceded a rapid jump to GS that lasted for centuries up to millennia. Thermal gas-fractionation methods (Landais et al., 2004; Huber et al., 2006) suggest that for certain DO events Greenland temperatures increased by up to 16°C ± 2.5°C (1 standard deviation) within several decades. Such transitions were also accompanied by abrupt shifts in dust and deuterium excess, indicative of reorganizations in atmospheric circulation (Steffensen et al., 2008; Thomas et al., 2009). Reconstructions from the subtropical Atlantic and Mediterranean reveal concomitant SST changes attaining values up to 5°C (e.g., Martrat et al., 2004; Martrat et al., 2007).

In spite of the visible presence of DO events in many paleoclimate records from both hemispheres, the underlying mechanisms still remain unresolved and range from internally generated atmosphere–ocean–ice sheet events (Timmermann et al., 2003; Ditlevsen and Ditlevsen, 2009), to solar-forced variability (Braun et al., 2008; Braun and Kurths, 2010). However, given the lack of observational evidence for a direct linear modulation of solar irradiance on DO time scales, (Muscheler and Beer, 2006), solar forcing is an improbable candidate to generate DO events. There is robust evidence from multiple lines of paleoceanographic information and modelling that DO variability is often associated with AMOC changes, as suggested by climate models of varying complexity (Ganopolski and Rahmstorf, 2001; Arzel et al., 2009) and marine proxy records (Piotrowski et al., 2005; Kissel et al., 2008; Barker et al., 2010; Roberts et al., 2010); but also potential influences of sea-ice cover (Li et al., 2010b), atmosphere circulation and ice sheet topography (Wunsch, 2006) have been proposed.

The widespread presence of massive layers of ice-rafted detritus in North Atlantic marine sediments provide robust evidence that some DO GS, known as Heinrich stadials, were associated with iceberg discharges originating from the Northern Hemispheric ice sheets. During these periods global sea level rose by up to several tens of meters (Chappell, 2002; Rohling et al., 2008b; Siddall et al., 2008; González and Dupont, 2009; Yokoyama and Esat, 2011), with remaining uncertainties in timing and amplitude of sea level rise, stadial cooling and ocean circulation changes relative to the iceberg discharge (Hall et al., 2006; Arz et al., 2007; Siddall et al., 2008; González and Dupont, 2009; Sierro et al., 2009; Hodell et al., 2010). Internal instabilities of the Laurentide ice sheet can cause massive calving and meltwater events similar to those reconstructed from proxy records (Calov et al., 2002, 2010; Marshall and Koutnik, 2006). Alternatively, an initial weakening of the AMOC can lead to subsurface warming in parts of the North Atlantic (Shaffer et al., 2004) and subsequent basal melting of the Labrador ice shelves, and a resulting acceleration of ice streams and iceberg discharge (Alvarez-Solas et al., 2010; Marcott et al., 2011). At present, unresolved dynamics in ice sheet models and limited proxy information do not allow us to distinguish the two mechanisms with confidence.

Since AR4, climate model simulations (Liu et al., 2009b; Otto-Bliesner and Brady, 2010; Menviel et al., 2011; Kageyama et al., 2013) have further confirmed the finding (*high confidence*) that changes in AMOC strength induce abrupt climate changes with magnitude and patterns resembling reconstructed paleoclimate-proxy data of DO and Heinrich events.

Recent studies have presented a better understanding of the global imprints of DO events and Heinrich events, for various regions. Widespread North Atlantic cooling and sea-ice anomalies during GS induced atmospheric circulation changes (*high confidence*) (Krebs and Timmermann, 2007; Clement and Peterson, 2008; Kageyama et al., 2010; Merkel et al., 2010; Otto-Bliesner and Brady, 2010; Timmermann et al., 2010) which in turn affected inter-hemispheric tropical rainfall patterns, leading to drying in Northern South America (Peterson and Haug, 2006), the Mediterranean (Fletcher and Sánchez Goñi, 2008; Fleitmann et al., 2009), equatorial western Africa and Arabia (Higginson et al., 2004; Ivanochko et al., 2005; Weldeab et al., 2007a; Mulitza et al., 2008; Tjallingii et al., 2008; Itambi et al., 2009; Weldeab, 2012), wide parts of Asia (Wang et al., 2008; Cai et al., 2010) (see Figure 5.4e) as well as in the Australian-Indonesian monsoon region (Mohtadi et al., 2011). Concomitant wetter conditions have been reconstructed for southwestern North America (Asmerom et al., 2010; Wagner et al., 2010) and southern South America (Kanner et al., 2012) (Figure 5.4h). Moreover, atmospheric circulation changes have been invoked (Zhang and Delworth, 2005; Xie et al., 2008; Okumura et al., 2009) to explain temperature variations in the North Pacific that varied in unison with abrupt climate change in the North Atlantic region (Harada et al., 2008, 2012; Pak et al., 2012). Other factors that may have contributed to North Pacific climate anomalies include large-scale Pacific Ocean circulation changes (Saenko et al., 2004; Schmittner et al., 2007; Harada et al., 2009; Okazaki et al., 2010) during phases of a weak AMOC. Recent high-resolution ice core studies (EPICA Community Members, 2006; Capron et al., 2010a, 2010b, 2012; Stenni et al., 2011) show that Antarctica warmed gradually for most GS, reaching maximum values at the time of GS/GI transitions, which is in agreement with the bipolar seesaw concept (Stocker and Johnsen, 2003; Stenni et al., 2011). A recent global temperature compilation (Shakun et al., 2012), Southern Ocean temperature records (Lamy et al., 2007; Barker et al., 2009; De Deckker et al., 2012), evidence from SH terrestrial records (Kaplan et al., 2010; Putnam et al., 2010) and transient climate model experiments (Menviel et al., 2011) provide multiple lines of evidence for the inter-hemispheric character of millennial-scale variability during the last glacial termination and for DO events (*high confidence*).

Newly available marine records (Martrat et al., 2007; Grützner and Higgins, 2010; Margari et al., 2010; Kleiven et al., 2011), Antarctic WMGHG records (Loulergue et al., 2008; Schilt et al., 2010) and statistical analyses of Antarctic ice core data (Siddall et al., 2010; Lambert et al., 2012) combined with bipolar seesaw modelling (Siddall et al., 2006; Barker et al., 2011) document with *high confidence* that abrupt climate change events, similar to the DO events and Heinrich stadials of the last glacial cycle, occurred during previous glacial periods extending back about 800 ka and, with *medium confidence*, to 1100 ka.

5.8 Paleoclimate Perspective on Irreversibility in the Climate System

For an introduction of the concept of irreversibility see Glossary.

5.8.1 Ice Sheets

Modelling studies suggest the existence of multiple equilibrium states for ice sheets with respect to temperature, CO_2 concentration and orbital forcing phase spaces (DeConto and Pollard, 2003; Calov and Ganopolski, 2005; Ridley et al., 2010). This implies a possibility of irreversible changes in the climate-cryosphere system in the past and future.

The existence of threshold behaviour in the EAIS is consistent with an abrupt increase in Antarctic ice volume at the Eocene/Oligocene boundary, 33 Ma, attributed to gradual atmospheric CO_2 concentration decline on geological time scale (Pagani et al., 2005b; Pearson et al., 2009) (Figure 5.2, Section 5.2.2). Ice sheet models produce a hysteresis behaviour of the EAIS with respect to CO_2 concentrations, leading to EAIS glaciation when CO_2 concentration declined to 600–900 ppm (DeConto and Pollard, 2003; Langebroek et al., 2009) and deglaciation for CO_2 above 1200 ppm (Pollard and DeConto, 2009). Proxy records suggest that the WAIS might have collapsed during last interglacials (Naish et al., 2009b; Vaughan et al., 2011) and was absent during warm periods of the Pliocene when CO_2 concentration was 350 to 450 ppm (see Section 5.2.2.2) and global sea level was higher than present (see Section 5.6.1). These reconstructions and one ice sheet model simulation (Pollard and DeConto, 2009) suggest that WAIS is very sensitive to the subsurface ocean temperature. This implies, with *medium confidence*, that a large part of the WAIS will be eventually lost if the atmospheric CO_2 concentration stays within, or above, the range of 350 to 450 ppm for several millennia.

Observational evidence suggest that the GIS was also much smaller than today during the MPWP (see Sections 5.6.1 and 5.2.2), consistent with the results of simulations with ice sheet models (Dolan et al., 2011; Koenig et al., 2011). Ice sheet model simulations and proxy records show that the volume of the GIS was also reduced during the past interglacial period (Section 5.6.2). This supports modelling results that indicate temperature or CO_2 thresholds for melting and re-growth of the GIS may lie in close proximity to the present and future levels (Gregory and Huybrechts, 2006; Lunt et al., 2008) (Section 5.6.1) and that the GIS may have multiple equilibrium states under present-day climate state (Ridley et al., 2010).

Therefore, proxy records and results of model simulations indicate with *medium confidence* that the GIS and WAIS could be destabilized by projected climate changes, although the time scales of the ice sheets response to climate change are very long (several centuries to millennia).

5.8.2 Ocean Circulation

Numerous modelling studies demonstrate that increased freshwater flux into the North Atlantic leads to weakening of the AMOC. Results of EMICs (Rahmstorf et al., 2005) and coupled GCMs also suggest that AMOC may have multiple equilibrium states under present or glacial climate conditions (Hawkins et al., 2011; Hu et al., 2012). Experiments with climate models provide evidence that the sensitivity of the AMOC to freshwater perturbation is larger for glacial boundary conditions than for interglacial conditions (Swingedouw et al., 2009) and that the recovery time scale of the AMOC is longer for LGM conditions than for the Holocene (Bitz et al., 2007).

The abrupt climate-change event at 8.2 ka permits the study of the recovery time of the AMOC to freshwater perturbation under near-modern boundary conditions (Rohling and Pälike, 2005). Since AR4, new proxy records and simulations confirm that the pattern of surface-ocean and atmospheric climate anomalies is consistent with a reduction in the strength of the AMOC (Figure 5.18a, b, d). Available proxy records from the North Atlantic support the hypothesis that freshwater input into the North Atlantic reduced the amount of deep and central water-mass formation, Nordic Seas overflows, intermediate water temperatures and the ventilation state of North Atlantic Deep Water (Figure 5.18c, d) (McManus et al., 2004; Ellison et al., 2006; Kleiven et al., 2008; Bamberg et al., 2010). A concomitant decrease of SST and atmospheric temperatures in the North Atlantic and in Greenland has been observed (Figure 5.18a, b) with the climate anomaly associated with the event lasting 100–160 years (Daley et al., 2011). The additional freshwater that entered the North Atlantic during the 8.2 ka event is estimated between $1.6 \cdot 10^{14}$ m^3 and $8 \cdot 10^{14}$ m^3 (von Grafenstein et al., 1998; Barber et al., 1999; Clarke et al., 2004). The duration of the

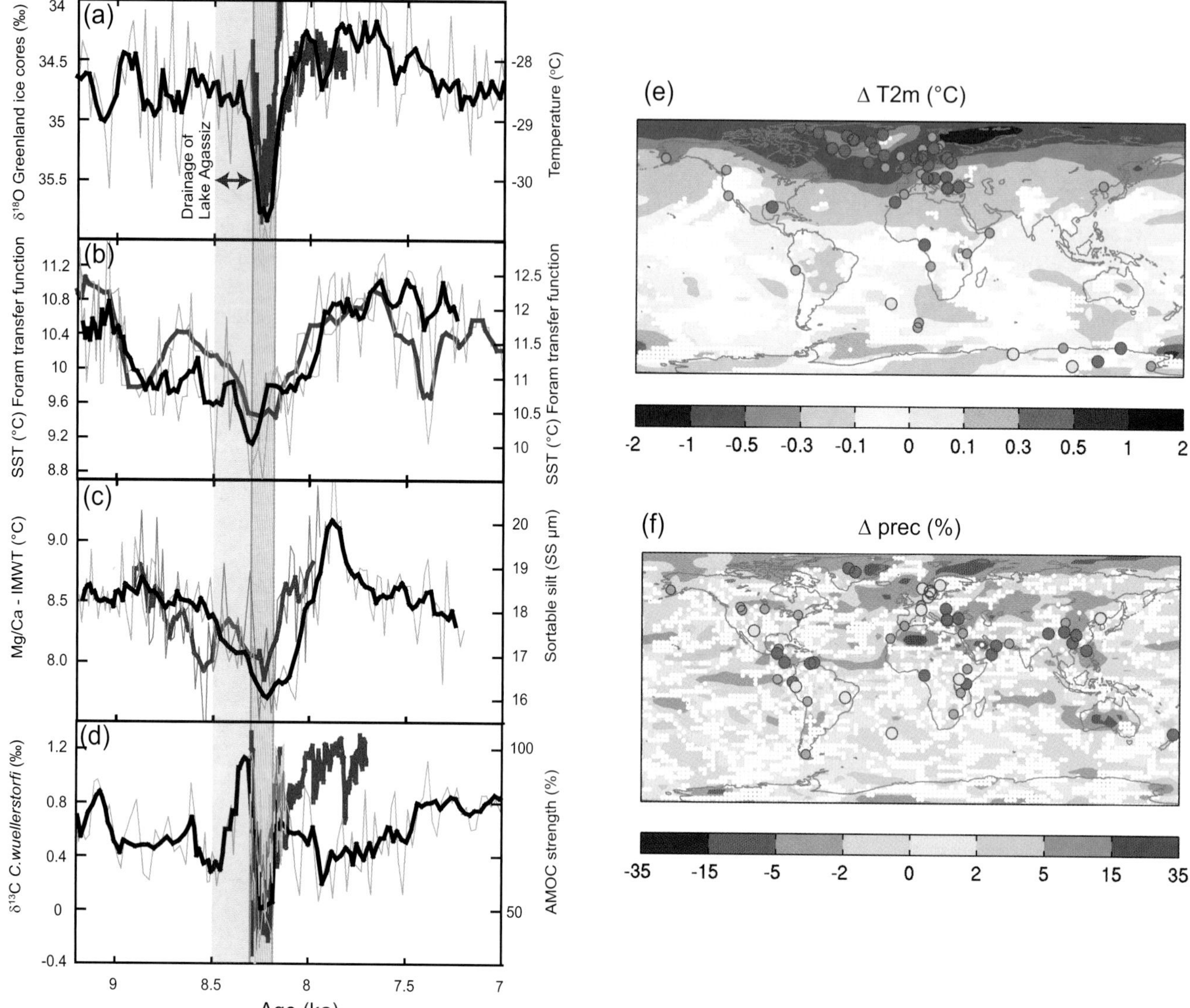

Figure 5.18 | Compilation of selected paleoenvironmental and climate model data for the abrupt Holocene cold event at 8.2 ka, documenting temperature and ocean-circulation changes around the event and the spatial extent of climate anomalies following the event. Published age constraints for the period of release of freshwater from glacier lakes Agassiz and Ojibway are bracketed inside the vertical blue bar. Vertical grey bar denotes the time of the main cold event as found in Greenland ice core records (Thomas et al., 2007). Thick lines in (a–d) denote 5-point running mean of underlying data in thin lines. (a) Black curve: North Greenland Ice Core Project (NGRIP) $\delta^{18}O$ (temperature proxy) from Greenland Summit (North Greenland Ice Core Project members, 2004). Red curve: Simulated Greenland temperature in an 8.2 ka event simulation with the ECBilt-CLIO-VECODE model (Wiersma et al., 2011). Blue curve: Simulated Greenland temperature in an 8.2 ka event simulation with the CCSM3 model (Morrill et al., 2011). (b) North Atlantic/Nordic Seas sea surface temperature (SST) reconstructions, age models are aligned on the peak of the cold-event (less than 100-year adjustment). Blue curve: Nordic Seas (Risebrobakken et al., 2011). Black curve: Gardar Drift south of Iceland (Ellison et al., 2006). (c) Deep- and intermediate-water records. Black curve: Sortable silt (SS) record (overflow strength proxy) from Gardar Drift south of Iceland (Ellison et al., 2006), Atlantic intermediate water temperature reconstruction (Bamberg et al., 2010). (d) Black curve: $\delta^{13}C$ (deep water ventilation proxy) at 3.4 km water depth south of Greenland (Kleiven et al., 2008). Age model is aligned on the minimum overflow strength in (c) (less than 100-year adjustment). Modelled change in the strength of the Atlantic Ocean meridional overturning circulation (AMOC)—Green curve: an 8.2 ka event simulation with the GISS model (LeGrande et al., 2006). Red curve: an 8.2 ka event simulation with the ECBilt-CLIO-VECODE (v. 3) model (Wiersma et al., 2011). Blue curve: an 8.2 ka event simulation with the CCSM3 model (Morrill et al., 2011). (e) Spatial distribution of the 4-member ensemble mean annual mean surface temperature anomaly (°C) compared with the control experiment from model simulations of the effects of a freshwater release at 8.2 ka (based on Morrill et al., 2013a). White dots indicate regions where less than 3 models agree on the sign of change. Coloured circles show paleoclimate data from records resolving the 8.2 ka event: purple = cold anomaly, yellow = warm anomaly, grey = no significant anomaly. Data source and significance thresholds are as summarized by Morrill et al. (2013b). (f) Same as (e) but for annual mean precipitation anomalies in %. Coloured circles show paleoclimate data from records resolving the 8.2 ka event: purple = dry anomaly, yellow = wet anomaly, grey = no significant anomaly.

meltwater release may have been as short as 0.5 years (Clarke et al., 2004), but new drainage estimates indicate an up to 200 year-duration in two separate stages (Gregoire et al., 2012). A four-model ensemble with a one-year freshwater perturbation of 2.5 Sv only gives temperature anomalies half of what has been reconstructed and with a shorter duration than observed, resulting from unresolved processes in models, imprecise representation of the initial climate state or a too short duration of the freshwater forcing (Morrill et al., 2013a). These marine-based reconstructions consistently show that the recovery time scale of the shallow and deep overturning circulation is on the order of 200 years (Ellison et al., 2006; Bamberg et al., 2010) (Figure 5.18c, d), with one record pointing to a partial recovery on a decadal time scale (Kleiven et al., 2008). Both recovery time scale and sensitivity of the AMOC to the freshwater perturbation are generally consistent with model experiments for the 8.2 ka event using coarse-resolution models, GCMs and eddy permitting models (LeGrande and Schmidt, 2008; Spence et al., 2008; Li et al., 2009). The recovery of temperatures out of the cold anomaly appears overprinted with natural variability in the proxy data, and is more gradual in data than in the AOGCM experiments (Figure 5.18c, d). In summary, multiple lines of evidence indicate, with *high confidence*, that the interglacial mode of the AMOC can recover from a short-term freshwater input into the subpolar North Atlantic.

The characteristic teleconnection patterns associated with a colder North Atlantic Ocean as described in Section 5.7 are evident for the 8.2 ka event in both models and proxy data (Figure 5.18e, f).

5.8.3 Next Glacial Inception

Since orbital forcing can be accurately calculated for the future (see Section 5.2.1), efforts can be made to predict the onset of the next glacial period. However, the glaciation threshold depends not only on insolation but also on the atmospheric CO_2 concentration (Archer and Ganopolski, 2005). Models of different complexity have been used to investigate the response to orbital forcing in the future for a range of atmospheric CO_2 levels. These results consistently show that a glacial inception is not expected to happen within the next approximate 50 kyr if either atmospheric CO_2 concentration remains above 300 ppm or cumulative carbon emissions exceed 1000 PgC (Loutre and Berger, 2000; Archer and Ganopolski, 2005; Cochelin et al., 2006). Only if atmospheric CO_2 content was below the pre-industrial level would a glaciation be possible under present orbital configuration (Loutre and Berger, 2000; Cochelin et al., 2006; Kutzbach et al., 2011; Vettoretti and Peltier, 2011; Tzedakis et al., 2012a). Simulations with climate–carbon cycle models show multi-millennial lifetime of the anthropogenic CO_2 in the atmosphere (see Box 6.1). Even for the lowest RCP 2.6 scenario, atmospheric CO_2 concentrations will exceed 300 ppm until the year 3000. It is therefore *virtually certain* that orbital forcing will not trigger a glacial inception before the end of the next millennium.

5.9 Concluding Remarks

The assessments in this chapter are based on a rapidly growing body of new evidence from the peer-review literature. Since AR4, there exists a wide range of new information on past changes in atmospheric composition, sea level, regional climates including droughts and floods, as well as new results from internationally coordinated model experiments on past climates (PMIP3/CMIP5). At the regional scale proxy-based temperature estimates are still scarce for key regions such as Africa, India and parts of the Americas. Syntheses of past precipitation changes were too limited to support regional assessments.

Precise knowledge of past changes in atmospheric concentrations of well-mixed GHGs prior to the period for which ice core records are available remains a strong limitation on assessing longer-term climate change. Key limitations to our knowledge of past climate continues to be associated with uncertainties of the quantitative information derived from climate proxies, in particular due to seasonality effects, the lack of proxy records sensitive to winter temperature, or the precise water depth at which ocean proxies signals form. Moreover, methodological uncertainties associated with regional, hemispheric or global syntheses need to be further investigated and quantified.

Despite progress on developing proxy records of past changes in sea ice it is not yet possible to provide quantitative and spatially coherent assessments of past sea ice cover in both polar oceans.

While this assessment could build on improved reconstructions of abrupt climate changes during glacial periods, key questions remain open regarding the underlying cause of these changes. Large uncertainties remain on the variations experienced by the West and East Antarctic ice sheets over various time scales of the past. Regarding past sea level change, major difficulties are associated with deconvolving changes in ocean geodynamic effects, as well as for inferring global signals from regional reconstructions.

The PMIP3/CMIP5 model framework offers the opportunity to directly incorporate information from paleoclimate data and simulations into assessments of future projections. This is an emerging field for which only preliminary information was available for AR5.

5

Acknowledgements

The compilation of this chapter has benefited greatly from the technical support by the chapter's scientific assistants Vera Bender (Germany), Hiroshi Kawamura (Germany/Japan), and Anna Peregon (France/Russian Federation). We are indebted to Hiroshi and Vera for compiling the various drafts, managing the ever-growing reference list and their skilful stylistic overhaul of figures. Anna is thanked for help with output from PMIP3 simulations, for tracking acronyms, and for identifying entries for the glossary.

References

Abbot, D. S., and E. Tziperman, 2008: A high-latitude convective cloud feedback and equable climates. *Q. J. R. Meteorol. Soc.*, **134**, 165–185.

Abbot, D. S., and E. Tziperman, 2009: Controls on the activation and strength of a high-latitude convective cloud feedback. *J. Atmos. Sci.*, **66**, 519–529.

Abe-Ouchi, A., T. Segawa, and F. Saito, 2007: Climatic Conditions for modelling the Northern Hemisphere ice sheets throughout the ice age cycle. *Clim. Past*, **3**, 423–438.

Abram, N. J., et al., 2013: Acceleration of snow melt in an Antarctic Peninsula ice core during the twentieth century. *Nature Geosci.*, **6**, 404–411.

Ackert Jr, R. P., S. Mukhopadhyay, D. Pollard, R. M. DeConto, A. E. Putnam, and H. W. Borns Jr, 2011: West Antarctic Ice Sheet elevations in the Ohio Range: Geologic constraints and ice sheet modeling prior to the last highstand. *Earth Planet. Sci. Lett.*, **307**, 83–93.

Adams, J. B., M. E. Mann, and C. M. Ammann, 2003: Proxy evidence for an El Niño-like response to volcanic forcing. *Nature*, **426**, 274–278.

Adkins, J. F., K. McIntyre, and D. P. Schrag, 2002: The salinity, temperature, and $\delta^{18}O$ of the glacial deep ocean. *Science*, **298**, 1769–1773.

Adler, R. E., et al., 2009: Sediment record from the western Arctic Ocean with an improved Late Quaternary age resolution: HOTRAX core HLY0503-8JPC, Mendeleev Ridge. *Global Planet. Change*, **68**, 18–29.

Agatova, A. R., A. N. Nazarov, R. K. Nepop, and H. Rodnight, 2012: Holocene glacier fluctuations and climate changes in the southeastern part of the Russian Altai (South Siberia) based on a radiocarbon chronology. *Quat. Sci. Rev.*, **43**, 74–93.

Ahn, J., and E. J. Brook, 2008: Atmospheric CO_2 and climate on millennial time scales during the last glacial period. *Science*, **322**, 83–85.

Ahn, J., E. J. Brook, A. Schmittner, and K. Kreutz, 2012: Abrupt change in atmospheric CO_2 during the last ice age. *Geophys. Res. Lett.*, **39**, L18711.

Akkemik, Ü., R. D'Arrigo, P. Cherubini, N. Köse, and G. C. Jacoby, 2008: Tree-ring reconstructions of precipitation and streamflow for north-western Turkey. *Int. J. Climatol.*, **28**, 173–183.

Alvarez-Solas, J., S. Charbit, C. Ritz, D. Paillard, G. Ramstein, and C. Dumas, 2010: Links between ocean temperature and iceberg discharge during Heinrich events. *Nature Geosci.*, **3**, 122–126.

Ammann, C. M., and E. R. Wahl, 2007: The importance of the geophysical context in statistical evaluations of climate reconstruction procedures. *Clim. Change*, **85**, 71–88.

Ammann, C. M., M. G. Genton, and B. Li, 2010: Technical Note: Correcting for signal attenuation from noisy proxy data in climate reconstructions. *Clim. Past*, **6**, 273–279.

Ammann, C. M., G. A. Meehl, W. M. Washington, and C. S. Zender, 2003: A monthly and latitudinally varying volcanic forcing dataset in simulations of 20th century climate. *Geophys. Res. Lett.*, **30**, 1657.

Ammann, C. M., F. Joos, D. S. Schimel, B. L. Otto-Bliesner, and R. A. Tomas, 2007: Solar influence on climate during the past millennium: Results from transient simulations with the NCAR Climate System Model. *Proc. Natl. Acad. Sci. U.S.A.*, **104**, 3713–3718.

Anchukaitis, K. J., and J. E. Tierney, 2013: Identifying coherent spatiotemporal modes in time-uncertain proxy paleoclimate records. *Clim. Dyn.*, **41**, 1291 - 1306.

Anchukaitis, K. J., B. M. Buckley, E. R. Cook, B. I. Cook, R. D. D'Arrigo, and C. M. Ammann, 2010: Influence of volcanic eruptions on the climate of the Asian monsoon region. *Geophys. Res. Lett.*, **37**, L22703.

Anchukaitis, K. J., et al., 2012: Tree rings and volcanic cooling. *Nature Geosci.*, **5**, 836–837.

Anderson, R. K., G. H. Miller, J. P. Briner, N. A. Lifton, and S. B. DeVogel, 2008: A millennial perspective on Arctic warming from ^{14}C in quartz and plants emerging from beneath ice caps. *Geophys. Res. Lett.*, **35**, L01502.

Andersson, C., F. S. R. Pausata, E. Jansen, B. Risebrobakken, and R. J. Telford, 2010: Holocene trends in the foraminifer record from the Norwegian Sea and the North Atlantic Ocean. *Clim. Past*, **6**, 179–193.

Andreev, A. A., et al., 2004: Late Saalian and Eemian palaeoenvironmental history of the Bol'shoy Lyakhovsky Island (Laptev Sea region, Arctic Siberia). *Boreas*, **33**, 319–348.

Andrews, T., J. M. Gregory, M. J. Webb, and K. E. Taylor, 2012: Forcing, feedbacks and climate sensitivity in CMIP5 coupled atmosphere-ocean climate models. *Geophys. Res. Lett.*, **39**, L09712.

Annan, J. D., and J. C. Hargreaves, 2012: Identification of climatic state with limited proxy data. *Clim. Past*, **8**, 1141–1151.

——, 2013: A new global reconstruction of temperature changes at the Last Glacial Maximum. *Clim. Past*, **9**, 367–376.

Annan, J. D., J. C. Hargreaves, R. Ohgaito, A. Abe-Ouchi, and S. Emori, 2005: Efficiently constraining climate sensitivity with ensembles of paleoclimate simulations. *Sci. Online Lett. Atmos.*, **1**, 181–184.

Antoine, P., et al., 2009: Rapid and cyclic aeolian deposition during the Last Glacial in European loess: a high-resolution record from Nussloch, Germany. *Quat. Sci. Rev.*, **28**, 2955–2973.

Antoniades, D., P. Francus, R. Pienitz, G. St-Onge, and W. F. Vincent, 2011: Holocene dynamics of the Arctic's largest ice shelf. *Proc. Natl. Acad. Sci. U.S.A.*, **108**, 18899–18904.

Anzidei, M., F. Antonioli, A. Benini, K. Lambeck, D. Sivan, E. Serpelloni, and P. Stocchi, 2011: Sea level change and vertical land movements since the last two millennia along the coasts of southwestern Turkey and Israel. *Quat. Int.*, **232**, 13–20.

Archer, D., and A. Ganopolski, 2005: A movable trigger: Fossil fuel CO_2 and the onset of the next glaciation. *Geochem. Geophys., Geosyst.*, **6**, Q05003.

Argus, D. F., and W. R. Peltier, 2010: Constraining models of postglacial rebound using space geodesy: A detailed assessment of model ICE-5G (VM2) and its relatives. *Geophys. J. Int.*, **181**, 697–723.

Arz, H. W., F. Lamy, A. Ganopolski, N. Nowaczyk, and J. Pätzold, 2007: Dominant Northern Hemisphere climate control over millennial-scale glacial sea level variability. *Quat. Sci. Rev.*, **26**, 312–321.

Arzel, O., A. Colin de Verdière, and M. H. England, 2009: The role of oceanic heat transport and wind stress forcing in abrupt millennial-scale climate transitions. *J. Clim.*, **23**, 2233–2256.

Asmerom, Y., V. J. Polyak, and S. J. Burns, 2010: Variable winter moisture in the southwestern United States linked to rapid glacial climate shifts. *Nature Geosci.*, **3**, 114–117.

Ault, T. R., et al., 2013: The continuum of hydroclimate variability in western North America during the last millennium. *J. Clim.*, **26**, 5863-5878.

Auriemma, R., and E. Solinas, 2009: Archaeological remains as sea level change markers: A review. *Quat. Int.*, **206**, 134–146.

Axelson, J. N., D. J. Sauchyn, and J. Barichivich, 2009: New reconstructions of streamflow variability in the South Saskatchewan River Basin from a network of tree ring chronologies, Alberta, Canada. *Water Resourc. Res.*, **45**, W09422.

Baker, V. R., 2008: Paleoflood hydrology: Origin, progress, prospects. *Geomorphology*, **101**, 1–13.

Bakker, P., et al., 2013: Last interglacial temperature evolution – a model inter-comparison. *Clim. Past*, **9**, 605–619.

Ballantyne, A. P., M. Lavine, T. J. Crowley, J. Liu, and P. B. Baker, 2005: Meta-analysis of tropical surface temperatures during the Last Glacial Maximum. *Geophys. Res. Lett.*, **32**, L05712.

Balmaceda, L., N. A. Krivova, and S. K. Solanki, 2007: Reconstruction of solar irradiance using the Group sunspot number. *Adv. Space Res.*, **40**, 986–989.

Bamberg, A., Y. Rosenthal, A. Paul, D. Heslop, S. Mulitza, C. Rühlemann, and M. Schulz, 2010: Reduced north Atlantic central water formation in response to early Holocene ice-sheet melting. *Geophys. Res. Lett.*, **37**, L17705.

Bar-Matthews, M., A. Ayalon, M. Gilmour, A. Matthews, and C. J. Hawkesworth, 2003: Sea–land oxygen isotopic relationships from planktonic foraminifera and speleothems in the Eastern Mediterranean region and their implication for paleorainfall during interglacial intervals. *Geochim Cosmochim. Acta*, **67**, 3181–3199.

Barber, D. C., et al., 1999: Forcing of the cold event of 8,200 years ago by catastrophic drainage of Laurentide lakes. *Nature*, **400**, 344–348.

Bard, E., G. Raisbeck, F. Yiou, and J. Jouzel, 2000: Solar irradiance during the last 1200 years based on cosmogenic nuclides. *Tellus B*, **52**, 985–992.

Barker, S., G. Knorr, M. J. Vautravers, P. Diz, and L. C. Skinner, 2010: Extreme deepening of the Atlantic overturning circulation during deglaciation. *Nature Geosci.*, **3**, 567–571.

Barker, S., P. Diz, M. J. Vautravers, J. Pike, G. Knorr, I. R. Hall, and W. S. Broecker, 2009: Interhemispheric Atlantic seesaw response during the last deglaciation. *Nature*, **457**, 1097–1102.

Barker, S., et al., 2011: 800,000 years of abrupt climate variability. *Science*, **334**, 347–351.

5

Baroni, M., M. H. Thiemens, R. J. Delmas, and J. Savarino, 2007: Mass-independent sulfur isotopic compositions in stratospheric volcanic eruptions. *Science*, **315**, 84–87.

Baroni, M., J. Savarino, J. H. Cole-Dai, V. K. Rai, and M. H. Thiemens, 2008: Anomalous sulfur isotope compositions of volcanic sulfate over the last millennium in Antarctic ice cores. *J. Geophys. Res.*, **113**, D20112.

Barrett, P. J., 2013: Resolving views on Antarctic Neogene glacial history – the Sirius debate. *Trans. R. Soc. Edinburgh, published online 7 May 2013*, CJ02013, doi:10.1017/S175569101300008X.

Barriendos, M., and F. S. Rodrigo, 2006: Study of historical flood events on Spanish rivers using documentary data. *Hydrol. Sci. J.*, **51**, 765–783.

Bartlein, P. J., et al., 2011: Pollen-based continental climate reconstructions at 6 and 21 ka: A global synthesis. *Clim. Dyn.*, **37**, 775–802.

Bartoli, G., B. Hönisch, and R. E. Zeebe, 2011: Atmospheric CO_2 decline during the Pliocene intensification of Northern Hemisphere glaciations. *Paleoceanography*, **26**, PA4213.

Bassett, S. E., G. A. Milne, J. X. Mitrovica, and P. U. Clark, 2005: Ice sheet and solid Earth influences on far-field sea level histories. *Science*, **309**, 925–928.

Battle, M., et al., 1996: Atmospheric gas concentrations over the past century measured in air from firn at the South Pole. *Nature*, **383**, 231–235.

Bauch, H. A., E. S. Kandiano, J. Helmke, N. Andersen, A. Rosell-Melé, and H. Erlenkeuser, 2011: Climatic bisection of the last interglacial warm period in the Polar North Atlantic. *Quat. Sci. Rev.*, **30**, 1813–1818.

Beerling, D. J., and D. L. Royer, 2002: Fossil plants as indicator of the Phanerozoic global carbon cycle. *Annu. Rev. Earth Planet. Sci.*, **30**, 527–556.

Beerling, D. J., and D. L. Royer, 2011: Convergent Cenozoic CO_2 history. *Nature Geosci.*, **4**, 418–420.

Beerling, D. J., A. Fox, and C. W. Anderson, 2009: Quantitative uncertainty analyses of ancient atmospheric CO_2 estimates from fossil leaves. *Am. J. Sci.*, **309**, 775–787.

Beerling, D. J., B. H. Lomax, D. L. Royer, G. R. Upchurch, and L. R. Kump, 2002: An atmospheric pCO_2 reconstruction across the Cretaceous-Tertiary boundary from leaf megafossils. *Proc. Natl. Acad. Sci. U.S.A.*, **99**, 7836–7840.

Beets, D. J., C. J. Beets, and P. Cleveringa, 2006: Age and climate of the late Saalian and early Eemian in the type-area, Amsterdam basin, The Netherlands. *Quat. Sci. Rev.*, **25**, 876–885.

Bekryaev, R. V., I. V. Polyakov, and V. A. Alexeev, 2010: Role of polar amplification in long-term surface air temperature variations and modern Arctic warming. *J. Clim.*, **23**, 3888–3906.

Belt, S. T., G. Massé, S. J. Rowland, M. Poulin, C. Michel, and B. LeBlanc, 2007: A novel chemical fossil of palaeo sea ice: IP_{25}. *Org. Geochem.*, **38**, 16–27.

Benito, G., A. Díez-Herrero, and M. Fernández De Villalta, 2003a: Magnitude and frequency of flooding in the Tagus basin (Central Spain) over the last millennium. *Clim. Change*, **58**, 171–192.

Benito, G., A. Sopeña, Y. Sánchez-Moya, M. J. Machado, and A. Pérez-González, 2003b: Palaeoflood record of the Tagus River (Central Spain) during the Late Pleistocene and Holocene. *Quat. Sci. Rev.*, **22**, 1737–1756.

Benito, G., M. Rico, Y. Sánchez-Moya, A. Sopeña, V. R. Thorndycraft, and M. Barriendos, 2010: The impact of late Holocene climatic variability and land use change on the flood hydrology of the Guadalentín River, southeast Spain. *Global Planet. Change*, **70**, 53–63.

Benito, G., et al., 2011: Hydrological response of a dryland ephemeral river to southern African climatic variability during the last millennium. *Quat. Res.*, **75**, 471–482.

Bentley, M. J., C. J. Fogwill, A. M. Le Brocq, A. L. Hubbard, D. E. Sugden, T. J. Dunai, and S. P. H. T. Freeman, 2010: Deglacial history of the West Antarctic Ice Sheet in the Weddell Sea embayment: Constraints on past ice volume change. *Geology*, **38**, 411–414.

Bereiter, B., D. Lüthi, M. Siegrist, S. Schüpbach, T. F. Stocker, and H. Fischer, 2012: Mode change of millennial CO_2 variability during the last glacial cycle associated with a bipolar marine carbon seesaw. *Proc. Natl. Acad. Sci. U.S.A.*, **109**, 9755–9760.

Berger, A., and M. F. Loutre, 1991: Insolation values for the climate of the last 10 million years. *Quat. Sci. Rev.*, **10**, 297–317.

Berger, A. L., 1978: Long-term variations of daily insolation and Quaternary climatic changes. *J. Atmos. Sci.*, **35**, 2362–2367.

Berger, G. W., and P. M. Anderson, 2000: Extending the geochronometry of Arctic lake cores beyond the radiocarbon limit by using thermoluminescence. *J. Geophys. Res.*, **105**, 15439–15455.

Berger, M., J. Brandefelt, and J. Nilsson, 2013: The sensitivity of the Arctic sea ice to orbitally induced insolation changes: a study of the mid-Holocene Paleoclimate Modelling Intercomparison Project 2 and 3 simulations. *Clim. Past*, **9**, 969–982.

Berkelhammer, M., A. Sinha, M. Mudelsee, H. Cheng, R. L. Edwards, and K. Cannariato, 2010: Persistent multidecadal power of the Indian Summer Monsoon. *Earth Planet. Sci. Lett.*, **290**, 166–172.

Bertrand, S., K. A. Hughen, F. Lamy, J.-B. W. Stuut, F. Torrejón, and C. B. Lange, 2012: Precipitation as the main driver of Neoglacial fluctuations of Gualas glacier, Northern Patagonian Icefield. *Clim. Past*, **8**, 519–534.

Bhatt, U. S., et al., 2010: Circumpolar arctic tundra vegetation change is linked to sea ice decline. *Earth Interact.*, **14**, 1–20.

Bintanja, R., R. G. Graversen, and W. Hazeleger, 2011: Arctic winter warming amplified by the thermal inversion and consequent low infrared cooling to space. *Nature Geosci.*, **4**, 758–761.

Birchfield, G. E., J. Weertman, and A. T. Lunde, 1981: A paleoclimate model of northern hemisphere ice sheets. *Quat. Res.*, **15**, 126–142.

Bird, B. W., M. B. Abbott, B. P. Finney, and B. Kutchko, 2009: A 2000 year varve-based climate record from the central Brooks Range, Alaska. *J. Paleolimnol.*, **41**, 25–41.

Bird, B. W., M. B. Abbott, D. T. Rodbell, and M. Vuille, 2011: Holocene tropical South American hydroclimate revealed from a decadally resolved lake sediment $\delta^{18}O$ record. *Earth Planet. Sci. Lett.*, **310**, 192–202.

Bird, M. I., L. K. Fifield, T. S. Teh, C. H. Chang, N. Shirlaw, and K. Lambeck, 2007: An inflection in the rate of early mid-Holocene eustatic sea level rise: A new sea level curve from Singapore. *Estuar. Coast. Shelf Sci.*, **71**, 523–536.

Bird, M. I., W. E. N. Austin, C. M. Wurster, L. K. Fifield, M. Mojtahid, and C. Sargeant, 2010: Punctuated eustatic sea level rise in the early mid-Holocene. *Geology*, **38**, 803–806.

Bitz, C. M., J. C. H. Chiang, W. Cheng, and J. J. Barsugli, 2007: Rates of thermohaline recovery from freshwater pluses in modern, Last Glacial Maximum, and greenhouse warming climates. *Geophys. Res. Lett.*, **34**, L07708.

Black, D. E., M. A. Abahazi, R. C. Thunell, A. Kaplan, E. J. Tappa, and L. C. Peterson, 2007: An 8–century tropical Atlantic SST record from the Cariaco Basin: Baseline variability, twentieth-century warming, and Atlantic hurricane frequency. *Paleoceanography*, **22**, PA4204.

Blanchon, P., A. Eisenhauer, J. Fietzke, and V. Liebetrau, 2009: Rapid sea level rise and reef back-stepping at the close of the last interglacial highstand. *Nature*, **458**, 881–884.

Blunier, T., and E. Brook, 2001: Timing of millennial-scale climate change in Antarctica and Greenland during the last glacial period. *Science*, **291**, 109–112.

Blunier, T., J. Chappellaz, J. Schwander, B. Stauffer, and D. Raynaud, 1995: Variations in atmospheric methane concentration during the Holocene epoch. *Nature*, **374**, 46–49.

Blunier, T., R. Spahni, J. M. Barnola, J. Chappellaz, L. Loulergue, and J. Schwander, 2007: Synchronization of ice core records via atmospheric gases. *Clim. Past*, **3**, 325–330.

Blunier, T., et al., 1997: Timing of the Antarctic cold reversal and the atmospheric CO_2 increase with respect to the Younger Dryas event. *Geophys. Res. Lett.*, **24**, 2683–2686.

Bonelli, S., S. Charbit, M. Kageyama, M. N. Woillez, G. Ramstein, C. Dumas, and A. Quiquet, 2009: Investigating the evolution of major Northern Hemisphere ice sheets during the last glacial-interglacial cycle. *Clim. Past*, **5**, 329–345.

Böning, C. W., A. Dispert, M. Visbeck, S. R. Rintoul, and F. U. Schwarzkopf, 2008: The response of the Antarctic Circumpolar Current to recent climate change. *Nature Geosci.*, **1**, 864–869.

Boninsegna, J. A., et al., 2009: Dendroclimatological reconstructions in South America: A review. *Palaeogeography, Palaeoclimatol. Palaeoecol.*, **281**, 210–228.

Bonnet, S., A. de Vernal, C. Hillaire-Marcel, T. Radi, and K. Husum, 2010: Variability of sea-surface temperature and sea-ice cover in the Fram Strait over the last two millennia. *Mar. Micropaleontol.*, **74**, 59–74.

Born, A., and K. H. Nisancioglu, 2012: Melting of Northern Greenland during the last interglaciation. *Cryosphere*, **6**, 1239–1250.

Born, A., K. Nisancioglu, and P. Braconnot, 2010: Sea ice induced changes in ocean circulation during the Eemian. *Clim. Dyn.*, **35**, 1361–1371.

Bostock, H. C., et al., 2013: A review of the Australian–New Zealand sector of the Southern Ocean over the last 30 ka (Aus-INTIMATE project). *Quat. Sci. Rev.*, **74**, 35-57.

Boucher, É., J. Guiot, and E. Chapron, 2011: A millennial multi-proxy reconstruction of summer PDSI for Southern South America. *Clim. Past*, **7**, 957–974.

5

Boucher, O., and M. Pham, 2002: History of sulfate aerosol radiative forcings. *Geophys. Res. Lett.*, **29**, 22–1–22–4.

Bowerman, N. D., and D. H. Clark, 2011: Holocene glaciation of the central Sierra Nevada, California. *Quat. Sci. Rev.*, **30**, 1067–1085.

Bozbiyik, A., M. Steinacher, F. Joos, T. F. Stocker, and L. Menviel, 2011: Fingerprints of changes in the terrestrial carbon cycle in response to large reorganizations in ocean circulation. *Clim. Past*, **7**, 319–338.

Braconnot, P., Y. Luan, S. Brewer, and W. Zheng, 2012a: Impact of Earth's orbit and freshwater fluxes on Holocene climate mean seasonal cycle and ENSO characteristics. *Clim. Dyn.*, **38**, 1081–1092.

Braconnot, P., C. Marzin, L. Grégoire, E. Mosquet, and O. Marti, 2008: Monsoon response to changes in Earth's orbital parameters: comparisons between simulations of the Eemian and of the Holocene. *Clim. Past*, **4**, 281–294.

Braconnot, P., et al., 2012b: Evaluation of climate models using palaeoclimatic data. *Nature Clim. Change*, **2**, 417–424.

Braconnot, P., et al., 2007: Results of PMIP2 coupled simulations of the mid-Holocene and Last Glacial Maximum - Part 1: experiments and large-scale features. *Clim. Past*, **3**, 261–277.

Bradley, S. L., M. Siddall, G. A. Milne, V. Masson-Delmotte, and E. Wolff, 2012: Where might we find evidence of a Last Interglacial West Antarctic Ice Sheet collapse in Antarctic ice core records? *Global Planet. Change*, **88–89**, 64–75.

Bradley, S. L., M. Siddall, G. A. Milne, V. Masson-Delmotte, and E. Wolff, 2013: Combining ice core records and ice sheet models to explore the evolution of the East Antarctic Ice sheet during the Last Interglacial period. *Global Planet. Change*, **100**, 278–290.

Brady, E. C., B. L. Otto-Bliesner, J. E. Kay, and N. Rosenbloom, 2013: Sensitivity to Glacial Forcing in the CCSM4. *J. Clim.*, **26**, 1901–1925.

Braganza, K., J. L. Gergis, S. B. Power, J. S. Risbey, and A. M. Fowler, 2009: A multiproxy index of the El Niño–Southern Oscillation, A.D. 1525–1982. *J. Geophys. Res.*, **114**, D05106.

Braun, H., and J. Kurths, 2010: Were Dansgaard-Oeschger events forced by the Sun? *Eur. Phys. J. Special Top.*, **191**, 117–129.

Braun, H., P. Ditlevsen, and D. R. Chialvo, 2008: Solar forced Dansgaard-Oeschger events and their phase relation with solar proxies. *Geophys. Res. Lett.*, **35**, L06703.

Brázdil, R., Z. W. Kundzewicz, and G. Benito, 2006: Historical hydrology for studying flood risk in Europe. *Hydrol. Sci. J.*, **51**, 739–764.

Brázdil, R., C. Pfister, H. Wanner, H. von Storch, and J. Luterbacher, 2005: Historical climatology in Europe —The state of the art. *Clim. Change*, **70**, 363–430.

Brázdil, R., Z. W. Kundzewicz, G. Benito, G. Demaree, N. MacDonald, and L. A. Roald, 2012: Historical floods in Europe in the past millennium. In: *Changes of Flood Risk in Europe* [Z. W. Kundzewicz (ed.)]. CRC Press, Boca Raton, Fl, USA, pp. 121–166.

Breecker, D. O., Z. D. Sharp, and L. D. McFadden, 2010: Atmospheric CO_2 concentrations during ancient greenhouse climates were similar to those predicted for A.D. 2100. *Proc. Natl. Acad. Sci. U.S.A.*, **107**, 576–580.

Bretagnon, P., and G. Francou, 1988: Planetary theories in rectangular and spherical variables - VSOP 87 solutions. *Astronomy & Astrophysics*, **202**, 309–315.

Brewer, S., J. Guiot, and F. Torre, 2007: Mid-Holocene climate change in Europe: A data-model comparison. *Clim. Past*, **3**, 499–512.

Briffa, K. R., and T. M. Melvin, 2011: A closer look at Regional Curve Standardization of tree-ring records: Justification of the need, a warning of some pitfalls, and suggested improvements in its application. In: *Dendroclimatology: Progress and Prospects* [M. K. Hughes, H. F. Diaz, and T. W. Swetnam (eds]. Springer Science+Business Media, Dordrecht, the Netherlands, pp. 113–145.

Briffa, K. R., F. H. Schweingruber, P. D. Jones, T. J. Osborn, S. G. Shiyatov, and E. A. Vaganov, 1998: Reduced sensitivity of recent tree-growth to temperature at high northern latitudes. *Nature*, **391**, 678–682.

Briffa, K. R., T. J. Osborn, F. H. Schweingruber, I. C. Harris, P. D. Jones, S. G. Shiyatov, and E. A. Vaganov, 2001: Low-frequency temperature variations from a northern tree ring density network. *J. Geophys. Res.*, **106**, 2929–2941.

Briner, J. P., H. A. M. Stewart, N. E. Young, W. Philipps, and S. Losee, 2010: Using proglacial-threshold lakes to constrain fluctuations of the Jakobshavn Isbræ ice margin, western Greenland, during the Holocene. *Quat. Sci. Rev.*, **29**, 3861–3874.

Brohan, P., R. Allan, E. Freeman, D. Wheeler, C. Wilkinson, and F. Williamson, 2012: Constraining the temperature history of the past millennium using early instrumental observations. *Clim. Past*, **8**, 1551–1563.

Bromwich, D. H., J. P. Nicolas, A. J. Monaghan, M. A. Lazzara, L. M. Keller, G. A. Weidner, and A. B. Wilson, 2013: Central West Antarctica among the most rapidly warming regions on Earth. *Nature Geosci.*, **6**, 139–145.

Brovkin, V., J.-H. Kim, M. Hofmann, and R. Schneider, 2008: A lowering effect of reconstructed Holocene changes in sea surface temperatures on the atmospheric CO_2 concentration. *Global Biogeochem. Cycles*, **22**, GB1016.

Buckley, B. M., et al., 2010: Climate as a contributing factor in the demise of Angkor, Cambodia. *Proc. Natl. Acad. Sci. U.S.A.*, **107**, 6748–6752.

Büntgen, U., D. C. Frank, D. Nievergelt, and J. Esper, 2006: Summer temperature variations in the European Alps, AD 755–2004. *J. Clim.*, **19**, 5606–5623.

Büntgen, U., J. Esper, D. Frank, K. Nicolussi, and M. Schmidhalter, 2005: A 1052-year tree-ring proxy for Alpine summer temperatures. *Clim. Dyn.*, **25**, 141–153.

Büntgen, U., D. Frank, R. Wilson, M. Carrer, C. Urbinati, and J. Esper, 2008: Testing for tree-ring divergence in the European Alps. *Global Change Biol.*, **14**, 2443–2453.

Büntgen, U., et al., 2011a: Causes and consequences of past and projected Scandinavian summer temperatures, 500–2100 AD. *PLoS ONE*, **6**, e25133.

Büntgen, U., et al., 2011b: 2500 years of european climate variability and human susceptibility. *Science*, **331**, 578–582.

Bürger, G., 2007: Comment on "The spatial extent of 20th-century warmth in the context of the past 1200 years". *Science*, **316**, 1844a.

Cahalan, R. F., G. Wen, J. W. Harder, and P. Pilewskie, 2010: Temperature responses to spectral solar variability on decadal time scales. *Geophys. Res. Lett.*, **37**, L07705.

Cai, Y. J., et al., 2010: The variation of summer monsoon precipitation in central China since the last deglaciation. *Earth Planet. Sci. Lett.*, **291**, 21–31.

Caillon, N., J. P. Severinghaus, J. Jouzel, J.-M. Barnola, J. Kang, and V. Y. Lipenkov, 2003: Timing of atmospheric CO_2 and Antarctic temperature changes across Termination III. *Science*, **299**, 1728–1731.

Calenda, G., C. P. Mancini, and E. Volpi, 2005: Distribution of the extreme peak floods of the Tiber River from the XV century. *Adv. Water Resourc.*, **28**, 615–625.

Calov, R., and A. Ganopolski, 2005: Multistability and hysteresis in the climate-cryosphere system under orbital forcing. *Geophys. Res. Lett.*, **32**, L21717.

Calov, R., A. Ganopolski, V. Petoukhov, M. Claussen, and R. Greve, 2002: Large-scale instabilities of the Laurentide ice sheet simulated in a fully coupled climate-system model. *Geophys. Res. Lett.*, **29**, 2216.

Calov, R., et al., 2010: Results from the Ice-Sheet Model Intercomparison Project-Heinrich Event INtercOmparison (ISMIP HEINO). *J. Glaciol.*, **56**, 371–383.

Camuffo, D., and S. Enzi, 1996: The analysis of two bi-millenary series: Tiber and Po river floods. In: *Climatic Variations and Forcing Mechanisms of the Last 2000 Years* [P. D. Jones, R. S. Bradley, and J. Jouzel (eds.)]. Springer-Verlag, Heidelberg, Germany, and New York, NY, USA, pp. 433–450.

Candy, I., G. R. Coope, J. R. Lee, S. A. Parfitt, R. C. Preece, J. Rose, and D. C. Schreve, 2010: Pronounced warmth during early Middle Pleistocene interglacials: Investigating the Mid-Brunhes Event in the British terrestrial sequence. *Earth Sci. Rev.*, **103**, 183–196.

Capron, E., et al., 2012: A global picture of the first abrupt climatic event occurring during the last glacial inception. *Geophys. Res. Lett.*, **39**, L15703.

Capron, E., et al., 2010a: Synchronising EDML and NorthGRIP ice cores using $\delta^{18}O$ of atmospheric oxygen ($\delta^{18}O_{atm}$) and CH_4 measurements over MIS5 (80–123 kyr). *Quat. Sci. Rev.*, **29**, 222–234.

Capron, E., et al., 2010b: Millennial and sub-millennial scale climatic variations recorded in polar ice cores over the last glacial period. *Clim. Past*, **6**, 345–365.

Carlson, A. E., P. U. Clark, G. M. Raisbeck, and E. J. Brook, 2007: Rapid Holocene deglaciation of the Labrador sector of the Laurentide Ice Sheet. *J. Clim.*, **20**, 5126–5133.

Carlson, A. E., D. J. Ullman, F. S. Anslow, F. He, P. U. Clark, Z. Liu, and B. L. Otto-Bliesner, 2012: Modeling the surface mass-balance response of the Laurentide Ice Sheet to Bølling warming and its contribution to Meltwater Pulse 1A. *Earth Planet. Sci. Lett.*, **315–316**, 24–29.

Cerling, T. E., 1992: Use of carbon isotopes in paleosols as an indicator of the pCO_2 of the paleoatmosphere. *Global Biogeochem. Cycles*, **6**, 307–314.

Chappell, J., 1983: Evidence for smoothly falling sea level relative to north Queensland, Australia, during the past 6,000 yr. *Nature*, **302**, 406–408.

Chappell, J., 2002: Sea level changes forced ice breakouts in the Last Glacial cycle: new results from coral terraces. *Quat. Sci. Rev.*, **21**, 1229–1240.

Charbit, S., D. Paillard, and G. Ramstein, 2008: Amount of CO_2 emissions irreversibly leading to the total melting of Greenland. *Geophys. Res. Lett.*, **35**, L12503.

Chavaillaz, Y., F. Codron, and M. Kageyama, 2013: Southern Westerlies in LGM and future (RCP4.5) climates. *Clim. Past*, **9**, 517–524.

Chen, J. H., H. A. Curran, B. White, and G. J. Wasserburg, 1991: Precise chronology of the last interglacial period: ^{234}U-^{230}Th data from fossil coral reefs in the Bahamas. *Geol. Soc. Am. Bull.*, **103**, 82–97.

Cheng, H., et al., 2009: Ice Age Terminations. *Science*, **326**, 248–252.

Chiessi, C. M., S. Mulitza, J. Pätzold, G. Wefer, and J. A. Marengo, 2009: Possible impact of the Atlantic Multidecadal Oscillation on the South American summer monsoon. *Geophys. Res. Lett.*, **36**, L21707.

Christiansen, B., 2011: Reconstructing the NH mean temperature: Can underestimation of trends and variability be avoided? *J. Clim.*, **24**, 674–692.

Christiansen, B., and F. C. Ljungqvist, 2012: The extra-tropical Northern Hemisphere temperature in the last two millennia: Reconstructions of low-frequency variability. *Clim. Past*, **8**, 765–786.

Christiansen, B., T. Schmith, and P. Thejll, 2009: A surrogate ensemble study of climate reconstruction methods: stochasticity and robustness. *J. Clim.*, **22**, 951–976.

Chu, G., et al., 2011: Seasonal temperature variability during the past 1600 years recorded in historical documents and varved lake sediment profiles from northeastern China. *Holocene*, **22**, 785–792.

Chylek, P., C. K. Folland, G. Lesins, M. K. Dubey, and M. Wang, 2009: Arctic air temperature change amplification and the Atlantic Multidecadal Oscillation. *Geophys. Res. Lett.*, **36**, L14801.

Clague, J. J., J. Koch, and M. Geertsema, 2010: Expansion of outlet glaciers of the Juneau Icefield in northwest British Columbia during the past two millennia. *Holocene*, **20**, 447–461.

Claquin, T., et al., 2003: Radiative forcing of climate by ice-age atmospheric dust. *Clim. Dyn.*, **20**, 193–202.

Clark, P. U., and D. Pollard, 1998: Origin of the middle Pleistocene transition by ice sheet erosion of regolith. *Paleoceanography*, **13**, 1–9.

Clark, P. U., J. X. Mitrovica, G. A. Milne, and M. E. Tamisiea, 2002: Sea level fingerprinting as a direct test for the source of global meltwater pulse IA. *Science*, **295**, 2438–2441.

Clark, P. U., et al., 2009: The Last Glacial Maximum. *Science*, **325**, 710–714.

Clarke, G. K. C., D. W. Leverington, J. T. Teller, and A. S. Dyke, 2004: Paleohydraulics of the last outburst flood from glacial Lake Agassiz and the 8200 BP cold event. *Quat. Sci. Rev.*, **23**, 389–407.

Clemens, S. C., W. L. Prell, and Y. Sun, 2010: Orbital-scale timing and mechanisms driving late Pleistocene Indo-Asian summer monsoons: reinterpreting cave speleothem δ^{18}O. *Paleoceanography*, **25**, PA4207.

Clement, A. C., and L. C. Peterson, 2008: Mechanisms of abrupt climate change of the last glacial period. *Rev. Geophys.*, **46**, RG4002.

CLIMAP Project Members, 1976: The surface of the Ice-Age Earth. *Science*, **191**, 1131–1137.

CLIMAP Project Members, 1981: Seasonal reconstructions of the earth's surface at the last glacial maximum. *Geol. Soc. Am., MC-36*.

Cobb, K. M., et al., 2013: Highly variable El Niño-Southern Oscillation throughout the Holocene. *Science*, **339**, 67–70.

Cochelin, A.-S. B., L. A. Mysak, and Z. Wang, 2006: Simulation of long-term future climate changes with the green McGill paleoclimate model: The next glacial inception. *Clim. Change*, **79**, 381–401.

COHMAP Members, 1988: Climatic changes of the last 18,000 years: observations and model simulations. *Science*, **241**, 1043–1052.

Cole-Dai, J., D. Ferris, A. Lanciki, J. Savarino, M. Baroni, and M. Thiemens, 2009: Cold decade (AD 1810–1819) caused by Tambora (1815) and another (1809) stratospheric volcanic eruption. *Geophys. Res. Lett.*, **36**, L22703.

Colville, E. J., A. E. Carlson, B. L. Beard, R. G. Hatfield, J. S. Stoner, A. V. Reyes, and D. J. Ullman, 2011: Sr-Nd-Pb isotope evidence for ice-sheet presence on southern Greenland during the Last Interglacial. *Science*, **333**, 620–623.

Cook, E. R., R. D. D'Arrigo, and M. E. Mann, 2002: A well-verified, multiproxy reconstruction of the winter North Atlantic Oscillation index since AD 1400. *J. Clim.*, **15**, 1754–1764.

Cook, E. R., C. A. Woodhouse, C. M. Eakin, D. M. Meko, and D. W. Stahle, 2004: Long-term aridity changes in the western United States. *Science*, **306**, 1015–1018.

Cook, E. R., K. J. Anchukaitis, B. M. Buckley, R. D. D'Arrigo, G. C. Jacoby, and W. E. Wright, 2010a: Asian Monsoon Failure and Megadrought During the Last Millennium. *Science*, **328**, 486–489.

Cook, E. R., R. Seager, R. R. Heim Jr, R. S. Vose, C. Herweijer, and C. Woodhouse, 2010b: Megadroughts in North America: placing IPCC projections of hydroclimatic change in a long-term palaeoclimate context. *J. Quat. Sci.*, **25**, 48–61.

Cook, E. R., P. J. Krusic, K. J. Anchukaitis, B. M. Buckley, T. Nakatsuka, and M. Sano, 2012: Tree-ring reconstructed summer temperature anomalies for temperate East Asia since 800 C.E. *Clim. Dyn.*, doi:10.1007/s00382-012-1611-x, 1-16, *published online 5 December 2012*.

Cook, E. R., B. M. Buckley, J. G. Palmer, P. Fenwick, M. J. Peterson, G. Boswijk, and A. Fowler, 2006: Millennia-long tree-ring records from Tasmania and New Zealand: A basis for modelling climate variability and forcing, past, present and future. *J. Quat. Sci.*, **21**, 689–699.

Cook, K. H., and I. M. Held, 1988: Stationary Waves of the Ice Age Climate. *J. Clim.*, **1**, 807–819.

Cooper, R. J., T. M. Melvin, I. Tyers, R. J. S. Wilson, and K. R. Briffa, 2013: A tree-ring reconstruction of East Anglian (UK) hydroclimate variability over the last millennium. *Clim. Dyn.*, **40**, 1019–1039.

Cornes, R. C., P. D. Jones, K. R. Briffa, and T. J. Osborn, 2012: Estimates of the North Atlantic Oscillation back to 1692 using a Paris-London westerly index. *Int. J. Climatol.*, **32**, 1135–1150.

Corona, C., J. Guiot, J.-L. Edouard, F. Chalie, U. Büntgen, P. Nola, and C. Urbinati, 2010: Millennium-long summer temperature variations in the European Alps as reconstructed from tree rings. *Clim. Past*, **6**, 379–400.

Corona, C., J.-L. Edouard, F. Guibal, J. Guiot, S. Bernard, A. Thomas, and N. Denelle, 2011: Long-term summer (AD 751–2008) temperature fluctuation in the French Alps based on tree-ring data. *Boreas*, **40**, 351–366.

Cortese, G., A. Abelmann, and R. Gersonde, 2007: The last five glacial-interglacial transitions: A high-resolution 450,000–year record from the subantarctic Atlantic. *Paleoceanography*, **22**, PA4203.

Cramer, B. S., K. G. Miller, P. J. Barrett, and J. D. Wright, 2011: Late Cretaceous-Neogene trends in deep ocean temperature and continental ice volume: Reconciling records of benthic foraminiferal geochemistry (δ^{18}O and Mg/Ca) with sea level history. *J. Geophys. Res.*, **116**, C12023.

Crespin, E., H. Goosse, T. Fichefet, and M. E. Mann, 2009: The 15th century Arctic warming in coupled model simulations with data assimilation. *Clim. Past*, **5**, 389–401.

Cronin, T. M., P. R. Vogt, D. A. Willard, R. Thunell, J. Halka, M. Berke, and J. Pohlman, 2007: Rapid sea level rise and ice sheet response to 8,200–year climate event. *Geophys. Res. Lett.*, **34**, L20603.

Crouch, A. D., P. Charbonneau, G. Beaubien, and D. Paquin-Ricard, 2008: A model for the total solar irradiance based on active region decay. *Astrophys. J.*, **677**, 723–741.

Crowley, T. J., 2000: Causes of Climate Change Over the Past 1000 Years. *Science*, **289**, 270–277.

Crowley, T. J., and M. B. Unterman, 2013: Technical details concerning development of a 1200–year proxy index for global volcanism. *Earth Syst. Sci. Data*, **5**, 187–197.

Crowley, T. J., S. K. Baum, K.-Y. Kim, G. C. Hegerl, and W. T. Hyde, 2003: Modeling ocean heat content changes during the last millennium. *Geophys. Res. Lett.*, **30**, 1932.

Crucifix, M., 2006: Does the Last Glacial Maximum constrain climate sensitivity? *Geophys. Res. Lett.*, **33**, L18701.

Cruz, F. W., et al., 2005: Insolation-driven changes in atmospheric circulation over the past 116,000 years in subtropical Brazil. *Nature*, **434**, 63–66.

Cruz, F. W., et al., 2009: Orbitally driven east-west antiphasing of South American precipitation. *Nature Geosci.*, **2**, 210–214.

Cuffey, K. M., G. D. Clow, R. B. Alley, M. Stuiver, E. D. Waddington, and R. W. Saltus, 1995: Large Arctic temperature change at the Wisconsin-Holocene glacial transition. *Science*, **270**, 455–458.

Cunningham, L. K., et al., 2013: Reconstructions of surface ocean conditions from the northeast Atlantic and Nordic seas during the last millennium. *Holocene*, **23**, 921-935.

Curry, J. A., J. L. Schramm, and E. E. Ebert, 1995: Sea ice-albedo climate feedback mechanism. *J. Clim.*, **8**, 240–247.

D'Arrigo, R., R. Wilson, and G. Jacoby, 2006: On the long-term context for late twentieth century warming. *J. Geophys. Res.*, **111**, D03103.

D'Arrigo, R., R. Wilson, B. Liepert, and P. Cherubini, 2008: On the 'Divergence Problem' in Northern Forests: A review of the tree-ring evidence and possible causes. *Global Planet. Change*, **60**, 289–305.

D'Arrigo, R., E. R. Cook, R. J. Wilson, R. Allan, and M. E. Mann, 2005: On the variability of ENSO over the past six centuries. *Geophys. Res. Lett.*, **32**, L03711.

D'Arrigo, R., et al., 2009: Tree growth and inferred temperature variability at the North American Arctic treeline. *Global Planet. Change*, **65**, 71–82.

Dahl-Jensen, D., K. Mosegaard, N. Gundestrup, G. D. Clow, S. J. Johnsen, A. W. Hansen, and N. Balling, 1998: Past temperatures directly from the Greenland Ice Sheet. *Science*, **282**, 268–271.

Daley, T. J., et al., 2011: The 8200 yr BP cold event in stable isotope records from the North Atlantic region. *Global Planet. Change*, **79**, 288–302.

Davis, P. T., B. Menounos, and G. Osborn, 2009: Holocene and latest Pleistocene alpine glacier fluctuations: A global perspective. *Quat. Sci. Rev.*, **28**, 2021–2238.

De Angelis, H., and P. Skvarca, 2003: Glacier surge after ice shelf collapse. *Science*, **299**, 1560–1562.

De Deckker, P., M. Moros, K. Perner, and E. Jansen, 2012: Influence of the tropics and southern westerlies on glacial interhemispheric asymmetry. *Nature Geosci.*, **5**, 266–269.

De Deckker, P., M. Norman, I. D. Goodwin, A. Wain, and F. X. Gingele, 2010: Lead isotopic evidence for an Australian source of aeolian dust to Antarctica at times over the last 170,000 years. *Palaeogeogr. Palaeoclimatol. Palaeoecol.*, **285**, 205–223.

de Garidel-Thoron, T., Y. Rosenthal, L. Beaufort, E. Bard, C. Sonzogni, and A. C. Mix, 2007: A multiproxy assessment of the western equatorial Pacific hydrography during the last 30 kyr. *Paleoceanography*, **22**, PA3204.

de Vernal, A., et al., 2006: Comparing proxies for the reconstruction of LGM sea-surface conditions in the northern North Atlantic. *Quat. Sci. Rev.*, **25**, 2820–2834.

DeConto, R. M., and D. Pollard, 2003: Rapid Cenozoic glaciation of Antarctica induced by declining atmospheric CO_2. *Nature*, **421**, 245–249.

DeConto, R. M., et al., 2012: Past extreme warming events linked to massive carbon release from thawing permafrost. *Nature*, **484**, 87–91.

Delaygue, G., and E. Bard, 2011: An Antarctic view of Beryllium-10 and solar activity for the past millennium. *Clim. Dyn.*, **36**, 2201–2218.

DeLong, K. L., T. M. Quinn, F. W. Taylor, K. Lin, and C.-C. Shen, 2012: Sea surface temperature variability in the southwest tropical Pacific since AD 1649. *Nature Clim. Change*, **2**, 799–804.

Delworth, T. L., and M. E. Mann, 2000: Observed and simulated multidecadal variability in the Northern Hemisphere. *Clim. Dyn.*, **16**, 661–676.

Denis, D., X. Crosta, L. Barbara, G. Massé, H. Renssen, O. Ther, and J. Giraudeau, 2010: Sea ice and wind variability during the Holocene in East Antarctica: insight on middle–high latitude coupling. *Quat. Sci. Rev.*, **29**, 3709–3719.

Derbyshire, E., 2003: Loess, and the dust indicators and records of terrestrial and marine palaeoenvironments (DIRTMAP) database. *Quat. Sci. Rev.*, **22**, 1813–1819.

Deschamps, P., et al., 2012: Ice-sheet collapse and sea level rise at the Bølling warming 14,600 years ago. *Nature*, **483**, 559–564.

Diaz, H. F., R. M. Trigo, M. K. Hughes, M. E. Mann, E. Xoplaki, and D. Barriopedro, 2011: Spatial and temporal characteristics of Climate in medieval times revisited. *Bull. Am. Meteorol. Soc.*, **92**, 1487–1500.

Diffenbaugh, N. S., M. Ashfaq, B. Shuman, J. W. Williams, and P. J. Bartlein, 2006: Summer aridity in the United States: Response to mid-Holocene changes in insolation and sea surface temperature. *Geophys. Res. Lett.*, **33**, L22712.

DiNezio, P. N., A. Clement, G. A. Vecchi, B. Soden, A. J. Broccoli, B. L. Otto-Bliesner, and P. Braconnot, 2011: The response of the Walker circulation to Last Glacial Maximum forcing: Implications for detection in proxies. *Paleoceanography*, **26**, PA3217.

Ditlevsen, P. D., and O. D. Ditlevsen, 2009: On the stochastic nature of the rapid climate shifts during the Last Ice Age. *J. Clim.*, **22**, 446–457.

Divine, D. V., and C. Dick, 2006: Historical variability of sea ice edge position in the Nordic Seas. *J. Geophys. Res.*, **111**, C01001.

Dolan, A. M., A. M. Haywood, D. J. Hill, H. J. Dowsett, S. J. Hunter, D. J. Lunt, and S. J. Pickering, 2011: Sensitivity of Pliocene ice sheets to orbital forcing. *Palaeogeogr. Palaeoclimatol. Palaeoecol.* **309**, 98–110.

Donnelly, J. P., P. Cleary, P. Newby, and R. Ettinger, 2004: Coupling instrumental and geological records of sea level change: Evidence from southern New England of an increase in the rate of sea level rise in the late 19th century. *Geophys. Res. Lett.*, **31**, L05203.

Doria, G., D. L. Royer, A. P. Wolfe, A. Fox, J. A. Westgate, and D. J. Beerling, 2011: Declining atmospheric CO_2 during the late Middle Eocene climate transition. *Am. J. Sci.*, **311**, 63–75.

Dowsett, H. J., M. M. Robinson, and K. M. Foley, 2009: Pliocene three-dimensional global ocean temperature reconstruction. *Clim. Past*, **5**, 769–783.

Dowsett, H. J., et al., 2012: Assessing confidence in Pliocene sea surface temperatures to evaluate predictive models. *Nature Clim. Change*, **2**, 365–371.

Dreimanis, A., 1992: Transition from the Sangamon interglaciation to the Wisconsin glaciation along the southeastern margin of the Laurentide Ice Sheet, North America. In: *Start of a Glacial, NATO ASI Series, 13* [G. T. Kukla, and E. Went (eds.)]. Springer-Verlag, Heidelberg, Germany, and New York, NY, USA, pp. 225–251.

Duplessy, J.-C., L. Labeyrie, and C. Waelbroeck, 2002: Constraints on the ocean oxygen isotopic enrichment between the Last Glacial Maximum and the Holocene: Paleoceanographic implications. *Quat. Sci. Rev.*, **21**, 315–330.

Dutton, A., and K. Lambeck, 2012: Ice volume and sea level during the Last Interglacial. *Science*, **337**, 216–219.

Dwyer, G. S., and M. A. Chandler, 2009: Mid-Pliocene sea level and continental ice volume based on coupled benthic Mg/Ca palaeotemperatures and oxygen isotopes. *Philos. Trans. R. Soc. London A*, **367**, 157–168.

Dwyer, G. S., T. M. Cronin, P. A. Baker, and J. Rodriguez-Lazaro, 2000: Changes in North Atlantic deep-sea temperature during climatic fluctuations of the last 25,000 years based on ostracode Mg/Ca ratios. *Geochem. Geophys. Geosyst.* **1**, 1028.

Edwards, T. L., M. Crucifix, and S. P. Harrison, 2007: Using the past to constrain the future: How the palaeorecord can improve estimates of global warming. *Prog. Phys. Geogr.*, **31**, 481–500.

Ekart, D. D., T. E. Cerling, I. P. Montanez, and N. J. Tabor, 1999: A 400 million year carbon isotope record of pedogenic carbonate; implications for paleoatmospheric carbon dioxide. *Am. J. Sci.*, **299**, 805–827.

Elderfield, H., P. Ferretti, M. Greaves, S. Crowhurst, I. N. McCave, D. Hodell, and A. M. Piotrowski, 2012: Evolution of ocean temperature and ice volume through the mid-Pleistocene climate transition. *Science*, **337**, 704–709.

Elderfield, H., et al., 2010: A record of bottom water temperature and seawater $\delta^{18}O$ for the Southern Ocean over the past 440 kyr based on Mg/Ca of benthic foraminiferal *Uvigerina* spp. *Quat. Sci. Rev.*, **29**, 160–169.

Ellison, C. R. W., M. R. Chapman, and I. R. Hall, 2006: Surface and deep ocean interactions during the cold climate event 8200 years ago. *Science*, **312**, 1929–1932.

Ely, L. L., Y. Enzel, V. R. Baker, and D. R. Cayan, 1993: A 5000-year record of extreme floods and climate change in the southwestern United States. *Science*, **262**, 410–412.

Emile-Geay, J., K. M. Cobb, M. E. Mann, and A. T. Wittenberg, 2013a: Estimating central equatorial Pacific SST variability over the past millennium. Part I: Methodology and validation. *J. Clim.*, **26**, 2302–2328.

Emile-Geay, J., K. M. Cobb, M. E. Mann, and A. T. Wittenberg, 2013b: Estimating central equatorial Pacific SST variability over the past millennium. Part II: Reconstructions and implications. *J. Clim.*, **26**, 2329–2352.

England, J. H., T. R. Lakeman, D. S. Lemmen, J. M. Bednarski, T. G. Stewart, and D. J. A. Evans, 2008: A millennial-scale record of Arctic Ocean sea ice variability and the demise of the Ellesmere Island ice shelves. *Geophys. Res. Lett.*, **35**, L19502.

EPICA Community Members, 2006: One-to-one coupling of glacial climate variability in Greenland and Antarctica. *Nature*, **444**, 195–198.

Esper, J., and D. Frank, 2009: Divergence pitfalls in tree-ring research. *Clim. Change*, **94**, 261–266.

Esper, J., U. Büntgen, M. Timonen, and D. C. Frank, 2012a: Variability and extremes of northern Scandinavian summer temperatures over the past two millennia. *Global Planet. Change*, **88–89**, 1–9.

Esper, J., U. Büntgen, J. Luterbacher, and P. J. Krusic, 2013: Testing the hypothesis of globally missing rings in temperature sensitive dendrochronological data. *Dendrochronologia*, **31**, 216-222.

Esper, J., D. Frank, R. Wilson, U. Büntgen, and K. Treydte, 2007a: Uniform growth trends among central Asian low- and high-elevation juniper tree sites. *Trees*, **21**, 141–150.

Esper, J., D. Frank, U. Büntgen, A. Verstege, J. Luterbacher, and E. Xoplaki, 2007b: Long-term drought severity variations in Morocco. *Geophys. Res. Lett.*, **34**, L17702.

Esper, J., D. Frank, U. Büntgen, A. Verstege, R. M. Hantemirov, and A. V. Kirdyanov, 2010: Trends and uncertainties in Siberian indicators of 20th century warming. *Global Change Biol.*, **16**, 386–398.

Esper, J., et al., 2012b: Orbital forcing of tree-ring data. *Nature Clim. Change*, **2**, 862–866.

Etheridge, D. M., L. P. Steele, R. J. Francey, and R. L. Langenfelds, 1998: Atmospheric methane between 1000 A.D. and present: Evidence of anthropogenic emissions and climatic variability. *J. Geophys. Res.*, **103**, 15979–15993.

Etheridge, D. M., L. P. Steele, R. L. Langenfelds, R. J. Francey, J. M. Barnola, and V. I. Morgan, 1996: Natural and anthropogenic changes in atmospheric CO_2 over the last 1000 years from air in Antarctic ice and firn. *J. Geophys. Res.*, **101**, 4115–4128.

Euler, C., and U. S. Ninnemann, 2010: Climate and Antarctic Intermediate Water coupling during the late Holocene. *Geology*, **38**, 647–650.

Fairbanks, R. G., 1989: A 17,000 year glacio-eustatic sea level record: Influence of glacial melting rates on the Younger Dryas event and deep ocean circulation. *Nature*, **342**, 637–642.

Fan, F. X., M. E. Mann, and C. M. Ammann, 2009: Understanding changes in the Asian summer monsoon over the past millennium: Insights from a long-term coupled model simulation. *J. Clim.*, **22**, 1736–1748.

Fedorov, A. V., C. M. Brierley, K. T. Lawrence, Z. Liu, P. S. Dekens, and A. C. Ravelo, 2013: Patterns and mechanisms of early Pliocene warmth. *Nature*, **496**, 43–49.

Feng, S., and Q. Hu, 2008: How the North Atlantic Multidecadal Oscillation may have influenced the Indian summer monsoon during the past two millennia. *Geophys. Res. Lett.*, **35**, L01707.

Fernández-Donado, L., et al., 2013: Large-scale temperature response to external forcing in simulations and reconstructions of the last millennium. *Clim. Past*, **9**, 393–421.

Feulner, G., 2011: Are the most recent estimates for Maunder Minimum solar irradiance in agreement with temperature reconstructions? *Geophys. Res. Lett.*, **38**, L16706.

Fischer, H., M. Wahlen, J. Smith, D. Mastroiani, and B. Deck, 1999: Ice core records of atmospheric CO_2 around the last three glacial terminations. *Science*, **283**, 1712–1714.

Fischer, H., M. L. Siggaard-Andersen, U. Ruth, R. Röthlisberger, and E. Wolff, 2007: Glacial/interglacial changes in mineral dust and sea-salt records in polar ice cores: Sources, transport, and deposition. *Rev. Geophys.*, **45**, RG1002.

Fischer, N., and J. H. Jungclaus, 2010: Effects of orbital forcing on atmosphere and ocean heat transports in Holocene and Eemian climate simulations with a comprehensive Earth system model. *Clim. Past*, **6**, 155–168.

Fleitmann, D., et al., 2009: Timing and climatic impact of Greenland interstadials recorded in stalagmites from northern Turkey. *Geophys. Res. Lett.*, **36**, L19707.

Fletcher, B. J., S. J. Brentnall, C. W. Anderson, R. A. Berner, and D. J. Beerling, 2008: Atmospheric carbon dioxide linked with Mesozoic and early Cenozoic climate change. *Nature Geosci.*, **1**, 43–48.

Fletcher, W. J., and M. F. Sánchez Goñi, 2008: Orbital- and sub-orbital-scale climate impacts on vegetation of the western Mediterranean basin over the last 48,000 yr. *Quat. Res.*, **70**, 451–464.

Flückiger, J., A. Dällenbach, T. Blunier, B. Stauffer, T. F. Stocker, D. Raynaud, and J.-M. Barnola, 1999: Variations in atmospheric N_2O concentration during abrupt climatic changes. *Science*, **285**, 227–230.

Flückiger, J., et al., 2002: High-resolution Holocene N_2O ice core record and its relationship with CH_4 and CO_2. *Global Biogeochem. Cycles*, **16**, 1010.

Foster, G. L., 2008: Seawater pH, pCO_2 and $[CO_3^{2-}]$ variations in the Caribbean Sea over the last 130 kyr: a boron isotope and B/Ca study of planktic forminifera. *Earth Planet. Sci. Lett.*, **271**, 254–266.

Foster, G. L., C. H. Lear, and J. W. B. Rae, 2012: The evolution of pCO_2, ice volume and climate during the middle Miocene. *Earth Planet. Sci. Lett.*, **341–344**, 243–254.

Fowler, A. M., et al., 2012: Multi-centennial tree-ring record of ENSO-related activity in New Zealand. *Nature Clim. Change*, **2**, 172–176.

Frank, D., J. Esper, and E. R. Cook, 2007: Adjustment for proxy number and coherence in a large-scale temperature reconstruction. *Geophys. Res. Lett.*, **34**, L16709.

Frank, D., J. Esper, E. Zorita, and R. Wilson, 2010a: A noodle, hockey stick, and spaghetti plate: a perspective on high-resolution paleoclimatology. *Clim. Change*, **1**, 507–516.

Frank, D. C., J. Esper, C. C. Raible, U. Büntgen, V. Trouet, B. Stocker, and F. Joos, 2010b: Ensemble reconstruction constraints on the global carbon cycle sensitivity to climate. *Nature*, **463**, 527–530.

Fréchette, B., A. P. Wolfe, G. H. Miller, P. J. H. Richard, and A. de Vernal, 2006: Vegetation and climate of the last interglacial on Baffin Island, Arctic Canada. *Palaeogeogr. Palaeoclimatol. Palaeoecol.* **236**, 91–106.

Freeman, K. H., and J. M. Hayes, 1992: Fractionation of carbon isotopes by phytoplankton and estimates of ancient CO_2 levels. *Global Biogeochem. Cycles*, **6**, 185–198.

Funder, S., et al., 2011: A 10,000-year record of Arctic Ocean sea-ice variability—view from the beach. *Science*, **333**, 747–750.

Fyke, J., and M. Eby, 2012: Comment on "Climate sensitivity estimated from temperature reconstructions of the Last Glacial Maximum". *Science*, **337**, 1294.

Gabrielli, P., et al., 2010: A major glacial-interglacial change in aeolian dust composition inferred from Rare Earth Elements in Antarctic ice. *Quat. Sci. Rev.*, **29**, 265–273.

Gagen, M., et al., 2011: Cloud response to summer temperatures in Fennoscandia over the last thousand years. *Geophys. Res. Lett.*, **38**, L05701.

Gaiero, D. M., 2007: Dust provenance in Antarctic ice during glacial periods: From where in southern South America? *Geophys. Res. Lett.*, **34**, L17707.

Ganopolski, A., and S. Rahmstorf, 2001: Rapid changes of glacial climate simulated in a coupled climate model. *Nature*, **409**, 153–158.

Ganopolski, A., and D. M. Roche, 2009: On the nature of lead-lag relationships during glacial-interglacial climate transitions. *Quat. Sci. Rev.*, **28**, 3361–3378.

Ganopolski, A., and R. Calov, 2011: The role of orbital forcing, carbon dioxide and regolith in 100 kyr glacial cycles. *Clim. Past*, **7**, 1415–1425.

Ganopolski, A., R. Calov, and M. Claussen, 2010: Simulation of the last glacial cycle with a coupled climate ice-sheet model of intermediate complexity. *Clim. Past*, **6**, 229–244.

Gao, C., A. Robock, and C. Ammann, 2008: Volcanic forcing of climate over the past 1500 years: An improved ice core-based index for climate models. *J. Geophys. Res.*, **113**, D23111.

——, 2012: Correction to "Volcanic forcing of climate over the past 1500 years: An improved ice core-based index for climate models". *J. Geophys. Res.*, **117**, D16112.

García-Artola, A., A. Cearreta, E. Leorri, M. Irabien, and W. Blake, 2009: Las marismas costeras como archivos geológicos de las variaciones recientes en el nivel marino/Coastal salt-marshes as geological archives of recent sea level changes. *Geogaceta*, **47**, 109–112.

Garcia-Herrera, R., D. Barriopedro, E. Hernández, H. F. Diaz, R. R. Garcia, M. R. Prieto, and R. Moyano, 2008: A chronology of El Niño events from primary documentary sources in northern Peru. *J. Clim.*, **21**, 1948–1962.

Garcin, Y., et al., 2007: Solar and anthropogenic imprints on Lake Masoko (southern Tanzania) during the last 500 years. *J. Paleolimnol.*, **37**, 475–490.

Gayer, E., J. Lavé, R. Pik, and C. France-Lanord, 2006: Monsoonal forcing of Holocene glacier fluctuations in Ganesh Himal (Central Nepal) constrained by cosmogenic ^{3}He exposure ages of garnets. *Earth Planet. Sci. Lett.*, **252**, 275–288.

Ge, Q.-S., J.-Y. Zheng, Z.-X. Hao, X.-M. Shao, W.-C. Wang, and J. Luterbacher, 2010: Temperature variation through 2000 years in China: an uncertainty analysis of reconstruction and regional difference. *Geophys. Res. Lett.*, **37**, L03703.

Ge, Q. S., S. B. Wang, and J. Y. Zheng, 2006: Reconstruction of temperature series in China for the last 5000 years. *Prog. Nat. Sci.*, **16**, 838–845.

Gehrels, W. R., and P. L. Woodworth, 2013: When did modern rates of sea level rise start? *Global Planet. Change*, **100**, 263–277.

Gehrels, W. R., B. W. Hayward, R. M. Newnham, and K. E. Southall, 2008: A 20th century acceleration of sea level rise in New Zealand. *Geophys. Res. Lett.*, **35**, L02717.

Gehrels, W. R., B. P. Horton, A. C. Kemp, and D. Sivan, 2011: Two millennia of sea level data: The key to predicting change. *Eos Trans. AGU*, **92**, 289–290.

Gehrels, W. R., et al., 2006: Rapid sea level rise in the North Atlantic Ocean since the first half of the nineteenth century. *The Holocene*, **16**, 949–965.

Gergis, J. L., and A. M. Fowler, 2009: A history of ENSO events since A.D. 1525: Implications for future climate change. *Clim. Change*, **92**, 343–387.

Gersonde, R., X. Crosta, A. Abelmann, and L. Armand, 2005: Sea-surface temperature and sea ice distribution of the Southern Ocean at the EPILOG Last Glacial Maximum—a circum-Antarctic view based on siliceous microfossil records. *Quat. Sci. Rev.*, **24**, 869–896.

Ghatak, D., A. Frei, G. Gong, J. Stroeve, and D. Robinson, 2010: On the emergence of an Arctic amplification signal in terrestrial Arctic snow extent. *J. Geophys. Res.*, **115**, D24105.

Giguet-Covex, C., et al., 2012: Frequency and intensity of high-altitude floods over the last 3.5 ka in northwestern French Alps (Lake Anterne). *Quat. Res.*, **77**, 12–22.

Gille, S. T., 2008: Decadal-scale temperature trends in the southern hemisphere ocean. *J. Clim.*, **21**, 4749–4765.

Gillett, N., T. Kell, and P. Jones, 2006: Regional climate impacts of the Southern Annular Mode. *Geophys. Res. Lett.*, **33**, L23704.

Gillett, N., et al., 2008: Attribution of polar warming to human influence. *Nature Geosci.*, **1**, 750–754.

Gladstone, R. M., et al., 2005: Mid-Holocene NAO: A PMIP2 model intercomparison. *Geophys. Res. Lett.*, **32**, L16707.

Goehring, B. M., et al., 2011: The Rhone Glacier was smaller than today for most of the Holocene. *Geology*, **39**, 679–682.

Goldewijk, K. K., 2001: Estimating global land use change over the past 300 years: The HYDE Database. *Global Biogeochem. Cycles*, **15**, 417–433.

González-Rouco, F., H. von Storch, and E. Zorita, 2003: Deep soil temperature as proxy for surface air-temperature in a coupled model simulation of the last thousand years. *Geophys. Res. Lett.*, **30**, 2116.

González-Rouco, J. F., H. Beltrami, E. Zorita, and H. von Storch, 2006: Simulation and inversion of borehole temperature profiles in surrogate climates: Spatial distribution and surface coupling. *Geophys. Res. Lett.*, **33**, L01703.

González, C., and L. Dupont, 2009: Tropical salt marsh succession as sea level indicator during Heinrich events. *Quat. Sci. Rev.*, **28**, 939–946.

González, J. L., and T. E. Törnqvist, 2009: A new Late Holocene sea level record from the Mississippi Delta: evidence for a climate/sea level connection? *Quat. Sci. Rev.*, **28**, 1737–1749.

Goodwin, I. D., and N. Harvey, 2008: Subtropical sea level history from coral microatolls in the Southern Cook Islands, since 300 AD. *Mar. Geol.*, **253**, 14–25.

Goosse, H., J. Guiot, M. E. Mann, S. Dubinkina, and Y. Sallaz-Damaz, 2012a: The medieval climate anomaly in Europe: Comparison of the summer and annual mean signals in two reconstructions and in simulations with data assimilation. *Global Planet. Change*, **84–85**, 35–47.

Goosse, H., et al., 2012b: The role of forcing and internal dynamics in explaining the "Medieval Climate Anomaly". *Clim. Dyn.*, **39**, 2847–2866.

Goosse, H., et al., 2012c: Antarctic temperature changes during the last millennium: evaluation of simulations and reconstructions. *Quat. Sci. Rev.*, **55**, 75–90.

Govin, A., et al., 2012: Persistent influence of ice sheet melting on high northern latitude climate during the early Last Interglacial. *Clim. Past*, **8**, 483–507.

Grachev, A. M., E. J. Brook, J. P. Severinghaus, and N. G. Pisias, 2009: Relative timing and variability of atmospheric methane and GISP2 oxygen isotopes between 68 and 86 ka. *Global Biogeochem. Cycles*, **23**, GB2009.

Graham, N., C. Ammann, D. Fleitmann, K. Cobb, and J. Luterbacher, 2011: Support for global climate reorganization during the "Medieval Climate Anomaly". *Clim. Dyn.*, **37**, 1217–1245.

Grant, K. M., et al., 2012: Rapid coupling between ice volume and polar temperature over the past 150,000 years. *Nature*, **491**, 744–747.

Graversen, R. G., and M. H. Wang, 2009: Polar amplification in a coupled climate model with locked albedo. *Clim. Dyn.*, **33**, 629–643.

Gray, S. T., L. J. Graumlich, J. L. Betancourt, and G. T. Pederson, 2004: A tree-ring based reconstruction of the Atlantic Multidecadal Oscillation since 1567 A.D. *Geophys. Res. Lett.*, **31**, L12205.

Greenwood, D. R., M. J. Scarr, and D. C. Christophel, 2003: Leaf stomatal frequency in the Australian tropical rainforest tree *Neolitsea dealbata* (Lauraceae) as a proxy measure of atmospheric pCO_2. *Palaeogeogr. Palaeoclimatol. Palaeoecol.* **196**, 375–393.

Gregoire, L. J., A. J. Payne, and P. J. Valdes, 2012: Deglacial rapid sea level rises caused by ice-sheet saddle collapses. *Nature*, **487**, 219–222.

Gregory, J. M., and P. Huybrechts, 2006: Ice-sheet contributions to future sea level change. *Philos. Trans. R. Soc. A*, **364**, 1709–1732.

Grichuk, V. P., 1985: Reconstructed climatic indexes by means of floristic data and an estimation of their accuracy. In: *Metody reconstruktsii paleoklimatov* [A. A. Velichko and Y. Y. Gurtovaya (eds.)]. Nauka-press, St. Petersburg, Russian Federation, pp. 20–28 (in Russian).

Grützner, J., and S. M. Higgins, 2010: Threshold behavior of millennial scale variability in deep water hydrography inferred from a 1.1 Ma long record of sediment provenance at the southern Gardar Drift. *Paleoceanography*, **25**, PA4204.

Haigh, J. D., 1996: The impact of solar variability on climate. *Science*, **272**, 981–984.

Haigh, J. D., A. R. Winning, R. Toumi, and J. W. Harder, 2010: An influence of solar spectral variations on radiative forcing of climate. *Nature*, **467**, 696–699.

Hald, M., et al., 2007: Variations in temperature and extent of Atlantic Water in the northern North Atlantic during the Holocene. *Quat. Sci. Rev.*, **26**, 3423–3440.

Hall, B. L., T. Koffman, and G. H. Denton, 2010: Reduced ice extent on the western Antarctic Peninsula at 700–970 cal. yr B.P. *Geology*, **38**, 635–638.

Hall, I. R., S. B. Moran, R. Zahn, P. C. Knutz, C. C. Shen, and R. L. Edwards, 2006: Accelerated drawdown of meridional overturning in the late-glacial Atlantic triggered by transient pre-H event freshwater perturbation. *Geophys. Res. Lett.*, **33**, L16616.

Handorf, D., K. Dethloff, A. G. Marshall, and A. Lynch, 2009: Climate regime variability for past and present time slices simulated by the Fast Ocean Atmosphere Model. *J. Clim.*, **22**, 58–70.

Hanebuth, T. J. J., H. K. Voris, Y. Yokoyama, Y. Saito, and J. i. Okuno, 2011: Formation and fate of sedimentary depocentres on Southeast Asia's Sunda Shelf over the past sea level cycle and biogeographic implications. *Earth Sci. Rev.*, **104**, 92–110.

Hanhijärvi, S., M. P. Tingley, and A. Korhola, 2013: Pairwise comparisons to reconstruct mean temperature in the Arctic Atlantic Region over the last 2,000 years. *Clim. Dyn.*, **41**, 2039-2060.

Hansen, J., and M. Sato, 2004: Greenhouse gas growth rates. *Proc. Natl. Acad. Sci. U.S.A.*, **101**, 16109–16114.

Hansen, J., et al., 2008: Target atmospheric CO_2: Where should humanity aim? *Open Atmos. Sci. J.*, **2**, 217–231.

Harada, N., M. Sato, and T. Sakamoto, 2008: Freshwater impacts recorded in tetraunsaturated alkenones and alkenone sea surface temperatures from the Okhotsk Sea across millennial-scale cycles. *Paleoceanography*, **23**, PA3201.

Harada, N., K. Kimoto, Y. Okazaki, K. Nagashima, A. Timmermann, and A. Abe-Ouchi, 2009: Millennial time scale changes in surface to intermediate-deep layer circulation recorded in sediment cores from the northwestern North Pacific. *Quat. Res. (Daiyonki-Kenkyu)*, **48**, 179–194.

Harada, N., et al., 2012: Sea surface temperature changes in the Okhotsk Sea and adjacent North Pacific during the last glacial maximum and deglaciation. *Deep-Sea Res. Pt. II*, **61–64**, 93–105.

Harden, T. M., J. E. O'Connor, D. G. Driscoll, and J. F. Stamm, 2011: Flood-frequency analyses from paleoflood investigations for Spring, Rapid, Boxelder, and Elk Creeks, Black Hills, western South Dakota. *U.S. Geological Survey Scientific Investigations Report 2011–5131*, 136 pp.

Harder, J. W., J. M. Fontenla, P. Pilewskie, E. C. Richard, and T. N. Woods, 2009: Trends in solar spectral irradiance variability in the visible and infrared. *Geophys. Res. Lett.*, **36**, L07801.

Hargreaves, J., A. Abe-Ouchi, and J. Annan, 2007: Linking glacial and future climates through an ensemble of GCM simulations. *Clim. Past*, **3**, 77–87.

Hargreaves, J. C., J. D. Annan, M. Yoshimori, and A. Abe-Ouchi, 2012: Can the Last Glacial Maximum constrain climate sensitivity? *Geophys. Res. Lett.*, **39**, L24702.

Hawkins, E., R. S. Smith, L. C. Allison, J. M. Gregory, T. J. Woollings, H. Pohlmann, and B. de Cuevas, 2011: Bistability of the Atlantic overturning circulation in a global climate model and links to ocean freshwater transport. *Geophys. Res. Lett.*, **38**, L10605.

Haywood, A. M., P. J. Valdes, and V. L. Peck, 2007: A permanent El Niño-like state during the Pliocene? *Paleoceanography*, **22**, PA1213.

Haywood, A. M., et al., 2013: Large-scale features of Pliocene climate: results from the Pliocene Model Intercomparison Project. *Clim. Past*, **9**, 191–209.

Hearty, P. J., J. T. Hollin, A. C. Neumann, M. J. O'Leary, and M. McCulloch, 2007: Global sea level fluctuations during the Last Interglaciation (MIS 5e). *Quat. Sci. Rev.*, **26**, 2090–2112.

Hegerl, G., T. Crowley, W. Hyde, and D. Frame, 2006: Climate sensitivity constrained by temperature reconstructions over the past seven centuries. *Nature*, **440**, 1029–1032.

Hegerl, G. C., T. J. Crowley, M. Allen, W. T. Hyde, H. N. Pollack, J. Smerdon, and E. Zorita, 2007: Detection of human influence on a new, validated 1500–year temperature reconstruction. *J. Clim.*, **20**, 650–666.

Helama, S., J. Meriläinen, and H. Tuomenvirta, 2009: Multicentennial megadrought in northern Europe coincided with a global El Niño–Southern Oscillation drought pattern during the Medieval Climate Anomaly. *Geology*, **37**, 175–178.

Helama, S., M. M. Fauria, K. Mielikäinen, M. Timonen, and M. Eronen, 2010: Sub-Milankovitch solar forcing of past climates: mid and late Holocene perspectives. *Geol. Soc. Am. Bull.*, **122**, 1981–1988.

Hély, C., P. Braconnot, J. Watrin, and W. Zheng, 2009: Climate and vegetation: Simulating the African humid period. *C. R. Geosci.*, **341**, 671–688.

Hemming, S. R., 2004: Heinrich events: Massive late Pleistocene detritus layers of the North Atlantic and their global climate imprint. *Rev. Geophys.*, **42**, RG1005.

Henderiks, J., and M. Pagani, 2007: Refining ancient carbon dioxide estimates: Significance of coccolithophore cell size for alkenone-based pCO_2 records. *Paleoceanography*, **22**, PA3202.

Herbert, T. D., L. C. Peterson, K. T. Lawrence, and Z. Liu, 2010: Tropical ocean temperatures over the past 3.5 million years. *Science*, **328**, 1530–1534.

Hereid, K. A., T. M. Quinn, F. W. Taylor, C.-C. Shen, R. L. Edwards, and H. Cheng, 2013: Coral record of reduced El Niño activity in the early 15th to middle 17th century. *Geology*, **41**, 51–54.

Herold, N., Q. Z. Yin, M. P. Karami, and A. Berger, 2012: Modeling the diversity of the warm interglacials. *Clim. Dyn.*, **56**, 126–141.

5

Herrington, A., and C. Poulsen, 2012: Terminating the Last Interglacial: the role of ice sheet-climate feedbacks in a GCM asynchronously coupled to an Ice Sheet Model. *J. Clim.*, **25**, 1871–1882.

Hesse, T., M. Butzin, T. Bickert, and G. Lohmann, 2011: A model-data comparison of $\delta^{13}C$ in the glacial Atlantic Ocean. *Paleoceanography*, **26**, PA3220.

Heusser, C. J., and L. E. Heusser, 1990: Long continental pollen sequence from Washington State (U.S.A.): Correlation of upper levels with marine pollen-oxygen isotope stratigraphy through substage 5e. *Palaeogeogr. Palaeoclimatol. Palaeoecol.*, **79**, 63–71.

Heyman, J., A. P. Stroeven, J. M. Harbor, and M. W. Caffee, 2011: Too young or too old: evaluating cosmogenic exposure dating based on an analysis of compiled boulder exposure ages. *Earth Planet. Sci. Lett.*, **302**, 71–80.

Higginson, M. J., M. A. Altabet, D. W. Murray, R. W. Murray, and T. D. Herbert, 2004: Geochemical evidence for abrupt changes in relative strength of the Arabian monsoons during a stadial/interstadial climate transition. *Geochim Cosmochim. Acta*, **68**, 3807–3826.

Hijma, M. P., and K. M. Cohen, 2010: Timing and magnitude of the sea level jump preluding the 8200 yr event. *Geology*, **38**, 275–278.

Hill, D. J., A. M. Dolan, A. M. Haywood, S. J. Hunter, and D. K. Stoll, 2010: Sensitivity of the Greenland Ice Sheet to Pliocene sea surface temperatures. *Stratigraphy*, **7**, 111 – 122.

Hind, A., and A. Moberg, 2012: Past millennial solar forcing magnitude: A statistical hemispheric-scale climate model versus proxy data comparison. *Clim. Dyn.*, doi:10.1007/s00382–012–1526–6, *published online 22 September 2012.*

Hodell, D. A., H. F. Evans, J. E. T. Channell, and J. H. Curtis, 2010: Phase relationships of North Atlantic ice-rafted debris and surface-deep climate proxies during the last glacial period. *Quat. Sci. Rev.*, **29**, 3875–3886.

Hodgson, D. A., 2011: First synchronous retreat of ice shelves marks a new phase of polar deglaciation. *Proc. Natl. Acad. Sci. U.S.A.*, **108**, 18859–18860.

Hofer, D., C. Raible, and T. Stocker, 2011: Variations of the Atlantic Meridional circulation in control and transient simulations of the last millennium. *Clim. Past*, **7**, 133–150.

Hofer, D., C. C. Raible, N. Merz, A. Dehnert, and J. Kuhlemann, 2013: Simulated winter circulation types in the North Atlantic and European region for preindustrial and glacial conditions. *Geophys. Res. Lett.*, **39**, L15805.

Holden, P., N. Edwards, K. Oliver, T. Lenton, and R. Wilkinson, 2010a: A probabilistic calibration of climate sensitivity and terrestrial carbon change in GENIE-1. *Clim. Dyn.*, **35**, 785–806.

Holden, P. B., N. R. Edwards, E. W. Wolff, N. J. Lang, J. S. Singarayer, P. J. Valdes, and T. F. Stocker, 2010b: Interhemispheric coupling, the West Antarctic Ice Sheet and warm Antarctic interglacials. *Clim. Past*, **6**, 431–443.

Hollis, C. J., et al., 2012: Early Paleogene temperature history of the Southwestern Pacific Ocean: reconciling proxies and models. *Earth Planet. Sci. Lett.*, **349–350**, 53–66.

Holmes, J. A., E. R. Cook, and B. Yang, 2009: Climate change over the past 2000 years in Western China. *Quaternary International*, **194**, 91–107.

Holz, A., and T. T. Veblen, 2011: Variability in the Southern Annular Mode determines wildfire activity in Patagonia. *Geophys. Res. Lett.*, **38**, L14710.

Holzhauser, H., M. Magny, and H. J. Zumbühl, 2005: Glacier and lake-level variations in west-central Europe over the last 3500 years. *Holocene*, **15**, 789–801.

Hönisch, B., and N. G. Hemming, 2005: Surface ocean pH response to variations in pCO_2 through two full glacial cycles. *Earth Planet. Sci. Lett.*, **236**, 305–314.

Hönisch, B., N. G. Hemming, D. Archer, M. Siddall, and J. F. McManus, 2009: Atmospheric carbon dioxide concentration across the Mid-Pleistocene transition. *Science*, **324**, 1551–1554.

Horton, B., and R. Edwards, 2006: Quantifying Holocene Sea Level Change Using Intertidal Foraminifera: Lessons from the British Isles. *Journal of Foraminiferal Research, Special publication* **40**, 1–97.

Hu, A. X., et al., 2012: Role of the Bering Strait on the hysteresis of the ocean conveyor belt circulation and glacial climate stability. *Proc. Natl. Acad. Sci. U.S.A.*, **109**, 6417–6422.

Hu, C., G. M. Henderson, J. Huang, S. Xie, Y. Sun, and K. R. Johnson, 2008: Quantification of Holocene Asian monsoon rainfall from spatially separated cave records. *Earth Planet. Sci. Lett.*, **266**, 221–232.

Huang, C. C., J. Pang, X. Zha, Y. Zhou, H. Su, H. Wan, and B. Ge, 2012: Sedimentary records of extraordinary floods at the ending of the mid-Holocene climatic optimum along the Upper Weihe River, China. *Holocene*, **22**, 675–686.

Huber, C., et al., 2006: Isotope calibrated Greenland temperature record over Marine Isotope Stage 3 and its relation to CH_4. *Earth Planet. Sci. Lett.*, **243**, 504–519.

Huber, M., and R. Caballero, 2011: The early Eocene equable climate problem revisited. *Clim. Past*, **7**, 603–633.

Hughes, A. L. C., E. Rainsley, T. Murray, C. J. Fogwill, C. Schnabel, and S. Xu, 2012: Rapid response of Helheim Glacier, southeast Greenland, to early Holocene climate warming. *Geology*, **40**, 427–430.

Hughes, M. K., and C. M. Ammann, 2009: The future of the past—an Earth system framework for high resolution paleoclimatology: editorial essay. *Clim. Change*, **94**, 247–259.

Humlum, O., B. Elberling, A. Hormes, K. Fjordheim, O. H. Hansen, and J. Heinemeier, 2005: Late-Holocene glacier growth in Svalbard, documented by subglacial relict vegetation and living soil microbes. *Holocene*, **15**, 396–407.

Hurtt, G. C., et al., 2006: The underpinnings of land-use history: three centuries of global gridded land-use transitions, wood-harvest activity, and resulting secondary lands. *Global Change Biol.*, **12**, 1208–1229.

Huybers, P., 2011: Combined obliquity and precession pacing of Late Pleistocene deglaciations. *Nature*, **480**, 229–232.

Israelson, C., and B. Wohlfarth, 1999: Timing of the last-interglacial high sea level on the Seychelles Islands, Indian Ocean. *Quat. Res.*, **51**, 306–316.

Itambi, A. C., T. von Dobeneck, S. Mulitza, T. Bickert, and D. Heslop, 2009: Millennial-scale northwest African droughts related to Heinrich events and Dansgaard-Oeschger cycles: Evidence in marine sediments from offshore Senegal. *Paleoceanography*, **24**, PA1205.

Ivanochko, T. S., R. S. Ganeshram, G.-J. A. Brummer, G. Ganssen, S. J. A. Jung, S. G. Moreton, and D. Kroon, 2005: Variations in tropical convection as an amplifier of global climate change at the millennial scale. *Earth Planet. Sci. Lett.*, **235**, 302–314.

Ivy-Ochs, S., H. Kerschner, M. Maisch, M. Christl, P. W. Kubik, and C. Schlüchter, 2009: Latest Pleistocene and Holocene glacier variations in the European Alps. *Quat. Sci. Rev.*, **28**, 2137–2149.

Izumi, K., P. J. Bartlein, and S. P. Harrison, 2013: Consistent large-scale temperature responses in warm and cold climates. *Geophys. Res. Lett.*, **40**, 1817-1823.

Jaccard, S. L., E. D. Galbraith, D. M. Sigman, and G. H. Haug, 2010: A pervasive link between Antarctic ice core and subarctic Pacific sediment records over the past 800 kyrs. *Quat. Sci. Rev.*, **29**, 206–212.

Jansen, E., et al., 2007: Palaeoclimate. In: *Climate Change 2007: The Physical Science Basis. Contribution of Working Group I to the Fourth Assessment Report of the Intergovernmental Panel on Climate Change* [Solomon, S., D. Qin, M. Manning, Z. Chen, M. Marquis, K. B. Averyt, M. Tignor and H. L. Miller (eds.)] Cambridge University Press, Cambridge, United Kingdom and New York, NY, USA, pp. 433–497.

Jevrejeva, S., J. C. Moore, A. Grinsted, and P. L. Woodworth, 2008: Recent global sea level acceleration started over 200 years ago? *Geophys. Res. Lett.*, **35**, L08715.

Joerin, U. E., K. Nicolussi, A. Fischer, T. F. Stocker, and C. Schlüchter, 2008: Holocene optimum events inferred from subglacial sediments at Tschierva Glacier, Eastern Swiss Alps. *Quat. Sci. Rev.*, **27**, 337–350.

Johns, T. C., et al., 2003: Anthropogenic climate change for 1860 to 2100 simulated with the HadCM3 model under updated emissions scenarios. *Clim. Dyn.*, **20**, 583–612.

Johnsen, S. J., D. Dahl-Jensen, W. Dansgaard, and N. Gundestrup, 1995: Greenland palaeotemperatures derived from GRIP bore hole temperature and ice core isotope profiles. *Tellus B*, **47**, 624–629.

Johnson, K., and D. J. Smith, 2012: Dendroglaciological reconstruction of late-Holocene glacier activity at White and South Flat glaciers, Boundary Range, northern British Columbia Coast Mountains, Canada. *Holocene*, **22**, 987–995.

Jomelli, V., V. Favier, A. Rabatel, D. Brunstein, G. Hoffmann, and B. Francou, 2009: Fluctuations of glaciers in the tropical Andes over the last millennium and palaeoclimatic implications: A review. *Palaeogeogr. Palaeoclimatol. Palaeoecol.* **281**, 269–282.

Jones, P. D., D. H. Lister, T. J. Osborn, C. Harpham, M. Salmon, and C. P. Morice, 2012: Hemispheric and large-scale land-surface air temperature variations: an extensive revision and an update to 2010. *J. Geophys. Res.*, **117**, D05127.

Jones, P. D., et al., 2009: High-resolution palaeoclimatology of the last millennium: A review of current status and future prospects. *Holocene*, **19**, 3–49.

Joos, F., and R. Spahni, 2008: Rates of change in natural and anthropogenic radiative forcing over the past 20,000 years. *Proc. Natl. Acad. Sci. U.S.A.*, **105**, 1425–1430.

Joos, F., et al., 2001: Global warming feedbacks on terrestrial carbon uptake under the Intergovernmental Panel on Climate Change (IPCC) Emission Scenarios. *Global Biogeochem. Cycles*, **15**, 891–907.

Joshi, M. M., and G. S. Jones, 2009: The climatic effects of the direct injection of water vapour into the stratosphere by large volcanic eruptions. *Atmos. Chem. Phys.*, **9**, 6109–6118.

Joughin, I., and R. B. Alley, 2011: Stability of the West Antarctic ice sheet in a warming world. *Nature Geosci.*, **4**, 506–513.

Jouzel, J., et al., 2007: Orbital and millennial Antarctic climate variability over the past 800,000 years. *Science*, **317**, 793–796.

Juckes, M. N., et al., 2007: Millennial temperature reconstruction intercomparison and evaluation. *Clim. Past*, **3**, 591–609.

Jungclaus, J. H., et al., 2010: Climate and carbon-cycle variability over the last millennium. *Clim. Past*, **6**, 723–737.

Justino, F., and W. R. Peltier, 2005: The glacial North Atlantic Oscillation. *Geophys. Res. Lett.*, **32**, L21803.

Justino, F., and W. R. Peltier, 2008: Climate anomalies induced by the arctic and antarctic oscillations: glacial maximum and present-day perspectives. *J. Clim.*, **21**, 459–475.

Justwan, A., and N. Koç, 2008: A diatom based transfer function for reconstructing sea ice concentrations in the North Atlantic. *Mar. Micropaleontol.*, **66**, 264–278.

Kageyama, M., A. Paul, D. M. Roche, and C. J. Van Meerbeeck, 2010: Modelling glacial climatic millennial-scale variability related to changes in the Atlantic meridional overturning circulation: a review. *Quat. Sci. Rev.*, **29**, 2931–2956.

Kageyama, M., et al., 2013: Climatic impacts of fresh water hosing under Last Glacial Maximum conditions: a multi-model study. *Clim. Past*, **9**, 935–953.

Kaiser, J., E. Schefuß, F. Lamy, M. Mohtadi, and D. Hebbeln, 2008: Glacial to Holocene changes in sea surface temperature and coastal vegetation in north central Chile: high versus low latitude forcing. *Quat. Sci. Rev.*, **27**, 2064–2075.

Kale, V. S., 2008: Palaeoflood hydrology in the Indian context. *J. Geol. Soc. India*, **71**, 56–66.

Kanner, L. C., S. J. Burns, H. Cheng, and R. L. Edwards, 2012: High-latitude forcing of the South American Summer Monsoon during the Last Glacial. *Science*, **335**, 570–573.

Kaplan, J. O., K. M. Krumhardt, E. C. Ellis, W. F. Ruddiman, C. Lemmen, and K. K. Goldewijk, 2011: Holocene carbon emissions as a result of anthropogenic land cover change. *Holocene*, **21**, 775–791.

Kaplan, M. R., et al., 2010: Glacier retreat in New Zealand during the Younger Dryas stadial. *Nature*, **467**, 194–197.

Kaufman, D. S., et al., 2009: Recent warming reverses long-term Arctic cooling. *Science*, **325**, 1236–1239.

Kawamura, K., et al., 2007: Northern Hemisphere forcing of climatic cycles in Antarctica over the past 360,000 years. *Nature*, **448**, 912–916.

Kemp, A. C., B. P. Horton, J. P. Donnelly, M. E. Mann, M. Vermeer, and S. Rahmstorf, 2011: Climate related sea level variations over the past two millennia. *Proc. Natl. Acad. Sci. U.S.A.*, **108**, 11017–11022.

Kemp, A. C., et al., 2009: Timing and magnitude of recent accelerated sea level rise (North Carolina, United States). *Geology*, **37**, 1035–1038.

Kienast, F., et al., 2011: Paleontological records indicate the occurrence of open woodlands in a dry inland climate at the present-day Arctic coast in western Beringia during the Last Interglacial. *Quat. Sci. Rev.*, **30**, 2134–2159.

Kilbourne, K. H., T. M. Quinn, R. Webb, T. Guilderson, J. Nyberg, and A. Winter, 2008: Paleoclimate proxy perspective on Caribbean climate since the year 1751: Evidence of cooler temperatures and multidecadal variability. *Paleoceanography*, **23**, PA3220.

Kilfeather, A. A., C. Ó Cofaigh, J. M. Lloyd, J. A. Dowdeswell, S. Xu, and S. G. Moreton, 2011: Ice-stream retreat and ice-shelf history in Marguerite Trough, Antarctic Peninsula: Sedimentological and foraminiferal signatures. *Geol. Soc. Am. Bull.*, **123**, 997–1015.

Kim, S. J., et al., 2010: Climate response over Asia/Arctic to change in orbital parameters for the last interglacial maximum. *Geosci. J.*, **14**, 173–190.

Kinnard, C., C. M. Zdanowicz, R. M. Koerner, and D. A. Fisher, 2008: A changing Arctic seasonal ice zone: Observations from 1870–2003 and possible oceanographic consequences. *Geophys. Res. Lett.*, **35**, L02507.

Kinnard, C., C. M. Zdanowicz, D. A. Fisher, E. Isaksson, A. de Vernal, and L. G. Thompson, 2011: Reconstructed changes in Arctic sea ice over the past 1,450 years. *Nature*, **479**, 509–512.

Kirkbride, M. P., and S. Winkler, 2012: Correlation of Late Quaternary moraines: Impact of climate variability, glacier response, and chronological resolution. *Quat. Sci. Rev.*, **46**, 1–29.

Kirshner, A. E., J. B. Anderson, M. Jakobsson, M. O'Regan, W. Majewski, and F. O. Nitsche, 2012: Post-LGM deglaciation in Pine Island Bay, West Antarctica. *Quat. Sci. Rev.*, **38**, 11–26.

Kissel, C., C. Laj, A. M. Piotrowski, S. L. Goldstein, and S. R. Hemming, 2008: Millennial-scale propagation of Atlantic deep waters to the glacial Southern Ocean. *Paleoceanography*, **23**, PA2102.

Kleiven, H. F., E. Jansen, T. Fronval, and T. M. Smith, 2002: Intensification of Northern Hemisphere glaciations in the circum Atlantic region (3.5–2.4 Ma)—ice-rafted detritus evidence. *Palaeogeogr. Palaeoclimatol. Palaeoecol.*, **184**, 213–223.

Kleiven, H. F., I. R. Hall, I. N. McCave, G. Knorr, and E. Jansen, 2011: Coupled deep-water flow and climate variability in the middle Pleistocene North Atlantic. *Geology*, **39**, 343–346.

Kleiven, H. F., C. Kissel, C. Laj, U. S. Ninnemann, T. O. Richter, and E. Cortijo, 2008: Reduced North Atlantic Deep Water coeval with the glacial Lake Agassiz freshwater outburst. *Science*, **319**, 60–64.

Klotz, S., J. Guiot, and V. Mosbrugger, 2003: Continental European Eemian and early Würmian climate evolution: comparing signals using different quantitative reconstruction approaches based on pollen. *Global Planet. Change*, **36**, 277–294.

Knight, J. R., R. J. Allan, C. K. Folland, M. Vellinga, and M. E. Mann, 2005: A signature of persistent natural thermohaline circulation cycles in observed climate. *Geophys. Res. Lett.*, **32**, L20708.

Knudsen, M. F., M.-S. Seidenkrantz, B. H. Jacobsen, and A. Kuijpers, 2011: Tracking the Atlantic Multidecadal Oscillation through the last 8,000 years. *Nature Commun.*, **2**, 178.

Kobashi, T., J. P. Severinghaus, J. M. Barnola, K. Kawamura, T. Carter, and T. Nakaegawa, 2010: Persistent multi-decadal Greenland temperature fluctuation through the last millennium. *Clim. Change*, **100**, 733–756.

Kobashi, T., et al., 2011: High variability of Greenland surface temperature over the past 4000 years estimated from trapped air in an ice core. *Geophys. Res. Lett.*, **38**, L21501.

Koch, J., and J. Clague, 2011: Extensive glaciers in northwest North America during medieval time. *Clim. Change*, **107**, 593–613.

Koch, P. L., J. C. Zachos, and P. D. Gingerich, 1992: Correlation between isotope records in marine and continental carbon reservoirs near the Palaeocene/Eocene boundary. *Nature*, **358**, 319–322.

Koenig, S. J., R. M. DeConto, and D. Pollard, 2011: Late Pliocene to Pleistocene sensitivity of the Greenland Ice Sheet in response to external forcing and internal feedbacks. *Clim. Dyn.*, **37**, 1247–1268.

Kohfeld, K. E., R. M. Graham, A. M. de Boer, L. C. Sime, E. W. Wolff, C. Le Quéré, and L. Bopp, 2013: Southern hemisphere westerly wind changes during the Last Glacial Maximum: paleo-data synthesis. *Quat. Sci. Rev.*, **68**, 76–95.

Köhler, P., G. Knorr, D. Buiron, A. Lourantou, and J. Chappellaz, 2011: Abrupt rise in atmospheric CO_2 at the onset of the Bølling/Allerød: in-situ ice core data versus true atmospheric signals. *Clim. Past*, **7**, 473–486.

Köhler, P., R. Bintanja, H. Fischer, F. Joos, R. Knutti, G. Lohmann, and V. Masson-Delmotte, 2010: What caused Earth's temperature variations during the last 800,000 years? Data-based evidence on radiative forcing and constraints on climate sensitivity. *Quat. Sci. Rev.*, **29**, 129–145.

Kopp, R. E., F. J. Simons, J. X. Mitrovica, A. C. Maloof, and M. Oppenheimer, 2009: Probabilistic assessment of sea level during the last interglacial stage. *Nature*, **462**, 863–867.

Kopp, R. E., F. J. Simons, J. X. Mitrovica, A. C. Maloof, and M. Oppenheimer, 2013: A probabilistic assessment of sea level variations within the last interglacial stage. *Geophys. J. Int.*, **193**, 711–716.

Koutavas, A., and J. P. Sachs, 2008: Northern timing of deglaciation in the eastern equatorial Pacific from alkenone paleothermometry. *Paleoceanography*, **23**, PA4205.

Koutavas, A., and S. Joanides, 2012: El Niño-Southern Oscillation extrema in the Holocene and Last Glacial Maximum. *Paleoceanography*, **27**, PA4208.

Kravitz, B., and A. Robock, 2011: Climate effects of high-latitude volcanic eruptions: Role of the time of year. *J. Geophys. Res.*, **116**, D01105.

Krebs, U., and A. Timmermann, 2007: Tropical air-sea interactions accelerate the recovery of the Atlantic Meridional Overturning Circulation after a major shutdown. *J. Clim.*, **20**, 4940–4956.

Krivova, N., and S. Solanki, 2008: Models of solar irradiance variations: Current status. *J. Astrophys. Astron.*, **29**, 151–158.

Krivova, N., L. Balmaceda, and S. Solanki, 2007: Reconstruction of solar total irradiance since 1700 from the surface magnetic flux. *Astron. Astrophys.*, **467**, 335–346.

Krivova, N., S. Solanki, and Y. Unruh, 2011: Towards a long-term record of solar total and spectral irradiance. *J. Atmos. Solar-Terres. Phys.*, **73**, 223–234.

Kühl, N., 2003: *Die Bestimmung botanisch-klimatologischer Transferfunktionen und die Rekonstruktion des bodennahen Klimazustandes in Europa während der Eem-Warmzeit.* Vol. 375, Dissertationes Botanicae, Cramer, Berlin, 149 pp.

Kürschner, W. M., 1996: Leaf stomata as biosensors of paleoatmospheric CO_2 levels. *LPP Contributions Series*, **5**, 1–153.

Kürschner, W. M., Z. Kvaček, and D. L. Dilcher, 2008: The impact of Miocene atmospheric carbon dioxide fluctuations on climate and the evolution of terrestrial ecosystems. *Proc. Natl. Acad. Sci. U.S.A.*, **105**, 449–453.

Kürschner, W. M., F. Wagner, D. L. Dilcher, and H. Visscher, 2001: Using fossil leaves for the reconstruction of Cenozoic paleoatmospheric CO_2 concentrations. In: *Geological Perspectives of Global Climate Change: APPG Studies in Geology 47*, Tulsa, [L. C. Gerhard, W. E. Harrison, and B. M. Hanson (eds.)]. The American Association of Petroleum Geologists, pp. 169–189.

Küttel, M., et al., 2010: The importance of ship log data: Reconstructing North Atlantic, European and Mediterranean sea level pressure fields back to 1750. *Clim. Dyn.*, **34**, 1115–1128.

Kutzbach, J. E., X. D. Liu, Z. Y. Liu, and G. S. Chen, 2008: Simulation of the evolutionary response of global summer monsoons to orbital forcing over the past 280,000 years. *Clim. Dyn.*, **30**, 567–579.

Kutzbach, J. E., S. J. Vavrus, W. F. Ruddiman, and G. Philippon-Berthier, 2011: Comparisons of atmosphere–ocean simulations of greenhouse gas-induced climate change for pre-industrial and hypothetical 'no-anthropogenic' radiative forcing, relative to present day. *Holocene*, **21**, 793–801.

Laborel, J., C. Morhange, R. Lafont, J. Le Campion, F. Laborel-Deguen, and S. Sartoretto, 1994: Biological evidence of sea level rise during the last 4500 years on the rocky coasts of continental southwestern France and Corsica. *Mar. Geol.*, **120**, 203–223.

Lainé, A., et al., 2009: Northern hemisphere storm tracks during the last glacial maximum in the PMIP2 ocean-atmosphere coupled models: Energetic study, seasonal cycle, precipitation. *Clim. Dyn.*, **32**, 593–614.

Laird, K. R., et al., 2012: Expanded spatial extent of the Medieval Climate Anomaly revealed in lake-sediment records across the boreal region in northwest Ontario. *Global Change Biol.*, **18**, 2869–2881.

Lamarque, J. F., et al., 2010: Historical (1850–2000) gridded anthropogenic and biomass burning emissions of reactive gases and aerosols: methodology and application. *Atmos. Chem. Phys.*, **10**, 7017–7039.

Lambeck, K., and E. Bard, 2000: Sea level change along the French Mediterranean coast for the past 30 000 years. *Earth Planet. Sci. Lett.*, **175**, 203–222.

Lambeck, K., Y. Yokoyama, and T. Purcell, 2002a: Into and out of the Last Glacial Maximum: sea level change during oxygen isotope stages 3 and 2. *Quat. Sci. Rev.*, **21**, 343–360.

Lambeck, K., T. Esat, and E. Potter, 2002b: Links between climate and sea levels for the past three million years. *Nature*, **419**, 199–206.

Lambeck, K., A. Purcell, and A. Dutton, 2012: The anatomy of interglacial sea levels: The relationship between sea levels and ice volumes during the Last Interglacial. *Earth Planet. Sci. Lett.*, **315–316**, 4–11.

Lambeck, K., F. Antonioli, A. Purcell, and S. Silenzi, 2004a: Sea level change along the Italian coast for the past 10,000 yr. *Quat. Sci. Rev.*, **23**, 1567–1598.

Lambeck, K., M. Anzidei, F. Antonioli, A. Benini, and A. Esposito, 2004b: Sea level in Roman time in the Central Mediterranean and implications for recent change. *Earth Planet. Sci. Lett.*, **224**, 563–575.

Lambeck, K., A. Purcell, S. Funder, K. H. Kjær, E. Larsen, and P. E. R. Moller, 2006: Constraints on the late Saalian to early middle Weichselian ice sheet of Eurasia from field data and rebound modelling. *Boreas*, **35**, 539–575.

Lambeck, K., C. D. Woodroffe, F. Antonioli, M. Anzidei, W. R. Gehrels, J. Laborel, and A. J. Wright, 2010: Paleoenvironmental records, geophysical modelling, and reconstruction of sea level trends and variability on centennial and longer timescales. In: *Understanding Sea Level Rise and Variability* [J. A. Church, P. L. Woodworth, T. Aarup, and W. S. Wilson (eds.)]. Wiley-Blackwell, Hoboken, NJ, USA, pp. 61–121.

Lambert, F., M. Bigler, J. P. Steffensen, M. A. Hutterli, and H. Fischer, 2012: Centennial mineral dust variability in high-resolution ice core data from Done C, Antarctica. *Clim. Past*, **8**, 609–623.

Lambert, F., et al., 2013: The role of mineral dust aerosols in polar amplification. *Nature Clim. Change*, **3**, 487–491.

Lambert, F., et al., 2008: Dust-climate couplings over the past 800,000 years from the EPICA Dome C ice core. *Nature*, **452**, 616–619.

Lamy, F., et al., 2007: Modulation of the bipolar seesaw in the southeast Pacific during Termination 1. *Earth Planet. Sci. Lett.*, **259**, 400–413.

Lanciki, A., J. Cole-Dai, M. H. Thiemens, and J. Savarino, 2012: Sulfur isotope evidence of little or no stratospheric impact by the 1783 Laki volcanic eruption. *Geophys. Res. Lett.*, **39**, L01806.

Landais, A., et al., 2004: A continuous record of temperature evolution over a sequence of Dansgaard-Oeschger events during marine isotopic stage 4 (76 to 62 kyr BP). *Geophys. Res. Lett.*, **31**, L22211.

Landrum, L., B. L. Otto-Bliesner, E. R. Wahl, A. Conley, P. J. Lawrence, and H. Teng, 2013: Last millennium climate and its variability in CCSM4. *J. Clim.*, **26**, 1085–1111.

Lang, N., and E. W. Wolff, 2011: Interglacial and glacial variability from the last 800 ka in marine, ice and terrestrial archives. *Clim. Past*, **7**, 361–380.

Langebroek, P. M., A. Paul, and M. Schulz, 2009: Antarctic ice-sheet response to atmospheric CO_2 and insolation in the Middle Miocene. *Clim. Past*, **5**, 633–646.

Lara, A., R. Villalba, and R. Urrutia, 2008: A 400–year tree-ring record of the Puelo river summer–fall streamflow in the valdivian rainforest eco-region, Chile. *Clim. Change*, **86**, 331–356.

Larocque-Tobler, I., M. Grosjean, O. Heiri, M. Trachsel, and C. Kamenik, 2010: Thousand years of climate change reconstructed from chironomid subfossils preserved in varved lake Silvaplana, Engadine, Switzerland. *Quat. Sci. Rev.*, **29**, 1940–1949.

Larocque-Tobler, I., M. M. Stewart, R. Quinlan, M. Trachsel, C. Kamenik, and M. Grosjean, 2012: A last millennium temperature reconstruction using chironomids preserved in sediments of anoxic Seebergsee (Switzerland): Consensus at local, regional and central European scales. *Quat. Sci. Rev.*, **41**, 49–56.

Larsen, N. K., K. H. Kjær, J. Olsen, S. Funder, K. K. Kjeldsen, and N. Nørgaard-Pedersen, 2011: Restricted impact of Holocene climate variations on the southern Greenland Ice Sheet. *Quat. Sci. Rev.*, **30**, 3171–3180.

Laskar, J., P. Robutel, F. Joutel, M. Gastineau, A. C. M. Correia, and B. Levrard, 2004: A long-term numerical solution for the insolation quantities of the earth. *Astron. Astrophys.*, **428**, 261–285.

Lea, D. W., D. K. Pak, and H. J. Spero, 2000: Climate impact of late Quaternary equatorial Pacific sea surface temperature variations. *Science*, **289**, 1719–1724.

Lea, D. W., D. K. Pak, C. L. Belanger, H. J. Spero, M. A. Hall, and N. J. Shackleton, 2006: Paleoclimate history of Galápagos surface waters over the last 135,000 yr. *Quat. Sci. Rev.*, **25**, 1152–1167.

Lean, J., J. Beer, and R. Bradley, 1995a: Reconstruction of solar irradiance since 1610: implications for climate change. *Geophys. Res. Lett.*, **22**, 3195–3198.

Lean, J. L., O. R. White, and A. Skumanich, 1995b: On the solar ultraviolet spectral irradiance during the Maunder Minimum. *Global Biogeochem. Cycles*, **9**, 171–182.

Lean, J. L., T. N. Woods, F. G. Eparvier, R. R. Meier, D. J. Strickland, J. T. Correira, and J. S. Evans, 2011: Solar extreme ultraviolet irradiance: Present, past, and future. *J. Geophys. Res.*, **116**, A01102.

Lecavalier, B. S., G. A. Milne , B. M. Vinther, D. A. Fisher, A. S. Dyke, and M. J. R. Simpson, 2013: Revised estimates of Greenland ice sheet thinning histories based on ice-core records. *Quat. Sci. Rev.*, **63**, 73–82.

Leclercq, P. W., and J. Oerlemans, 2012: Global and Hemispheric temperature reconstruction from glacier length fluctuations. *Clim. Dyn.*, **38**, 1065–1079.

Ledru, M. P., V. Jomelli, P. Samaniego, M. Vuille, S. Hidalgo, M. Herrera, and C. Ceron, 2013: The Medieval Climate Anomaly and the Little Ice Age in the eastern Ecuadorian Andes. *Clim. Past*, **9**, 307–321.

Leduc, G., R. Schneider, J. H. Kim, and G. Lohmann, 2010: Holocene and Eemian sea surface temperature trends as revealed by alkenone and Mg/Ca paleothermometry. *Quat. Sci. Rev.*, **29**, 989–1004.

Leduc, G., L. Vidal, K. Tachikawa, F. Rostek, C. Sonzogni, L. Beaufort, and E. Bard, 2007: Moisture transport across Central America as a positive feedback on abrupt climatic changes. *Nature*, **445**, 908–911.

Lee, T. C. K., F. W. Zwiers, and M. Tsao, 2008: Evaluation of proxy-based millennial reconstruction methods. *Clim. Dyn.*, **31**, 263–281.

Lefohn, A. S., J. D. Husar, and R. B. Husar, 1999: Estimating historical anthropogenic global sulfur emission patterns for the period 1850–1990. *Atmos. Environ.*, **33**, 3435–3444.

LeGrande, A. N., and G. A. Schmidt, 2008: Ensemble, water isotope-enabled, coupled general circulation modeling insights into the 8.2 ka event. *Paleoceanography*, **23**, PA3207.

LeGrande, A. N., and G. A. Schmidt, 2009: Sources of Holocene variability of oxygen isotopes in paleoclimate archives. *Clim. Past*, **5**, 441–455.

LeGrande, A. N., et al., 2006: Consistent simulation of multiple proxy responses to an abrupt climate change event. *Proc. Natl. Acad. Sci. U.S.A.*, **103**, 10527–10527.

Lehner, F., C. C. Raible, and T. F. Stocker, 2012: Testing the robustness of a precipitation proxy-based North Atlantic Oscillation reconstruction. *Quat. Sci. Rev.*, **45**, 85–94.

Lemieux-Dudon, B., et al., 2010: Consistent dating for Antarctic and Greenland ice cores. *Quat. Sci. Rev.*, **29**, 8–20.

Leorri, E., B. P. Horton, and A. Cearreta, 2008: Development of a foraminifera-based transfer function in the Basque marshes, N. Spain: implications for sea level studies in the Bay of Biscay. *Mar. Geol.*, **251**, 60–74.

Leorri, E., A. Cearreta, and G. Milne, 2012: Field observations and modelling of Holocene sea level changes in the southern Bay of Biscay: implication for understanding current rates of relative sea level change and vertical land motion along the Atlantic coast of SW Europe. *Quat. Sci. Rev.*, **42**, 59–73.

Lewis, S. C., A. N. LeGrande, M. Kelley, and G. A. Schmidt, 2010: Water vapour source impacts on oxygen isotope variability in tropical precipitation during Heinrich events. *Clim. Past*, **6**, 325–343.

Li, B., D. W. Nychka, and C. M. Ammann, 2010a: The value of multiproxy reconstruction of past climate. *J. Am. Stat. Assoc.*, **105**, 883–895.

Li, C., D. S. Battisti, and C. M. Bitz, 2010b: Can North Atlantic sea ice anomalies account for Dansgaard-Oeschger climate signals? *J. Clim.*, **23**, 5457–5475.

Li, J., et al., 2011: Interdecadal modulation of El Niño amplitude during the past millennium. *Nature Clim. Change*, **1**, 114–118.

Li, Y. X., H. Renssen, A. P. Wiersma, and T. E. Törnqvist, 2009: Investigating the impact of Lake Agassiz drainage routes on the 8.2 ka cold event with a climate model. *Clim. Past*, **5**, 471–480.

Licciardi, J. M., J. M. Schaefer, J. R. Taggart, and D. C. Lund, 2009: Holocene glacier fluctuations in the Peruvian Andes indicate northern climate linkages. *Science*, **325**, 1677–1679.

Linderholm, H. W., and P. Jansson, 2007: Proxy data reconstructions of the Storglaciaren (Sweden) mass-balance record back to AD 1500 on annual to decadal timescales. *Ann. Glaciol.*, **46**, 261–267.

Linderholm, H. W., et al., 2010: Dendroclimatology in Fennoscandia—from past accomplishments to future potential. *Clim. Past*, **5**, 1415–1462.

Linsley, B. K., Y. Rosenthal, and D. W. Oppo, 2010: Holocene evolution of the Indonesian throughflow and the western Pacific warm pool. *Nature Geosci.*, **3**, 578–583.

Linsley, B. K., P. P. Zhang, A. Kaplan, S. S. Howe, and G. M. Wellington, 2008: Interdecadal-decadal climate variability from multicoral oxygen isotope records in the South Pacific convergence zone region since 1650 AD. *Paleoceanography*, **23**, PA2219.

Lisiecki, L. E., and M. E. Raymo, 2005: A Pliocene-Pleistocene stack of 57 globally distributed benthic $\delta^{18}O$ records. *Paleoceanography*, **20**, PA1003.

Lisiecki, L. E., M. E. Raymo, and W. B. Curry, 2008: Atlantic overturning responses to late Pleistocene climate forcings. *Nature*, **456**, 85–88.

Lisiecki, L.E., 2010: Links between eccentricity forcing and the 100,000-year glacial cycle. *Nature Geoscience*, **3**, 349–352.

Liu, J., B. Wang, Q. Ding, X. Kuang, W. Soon, and E. Zorita, 2009a: Centennial variations of the global monsoon precipitation in the last millennium: results from ECHO-G model. *J. Clim.*, **22**, 2356–2371.

Liu, X. D., Z. Y. Liu, S. Clemens, W. Prell, and J. Kutzbach, 2007a: A coupled model study of glacial Asian monsoon variability and Indian ocean dipole. *J. Meteorol. Soc. Jpn.*, **85**, 1–10.

Liu, Z., et al., 2007b: Simulating the transient evolution and abrupt change of Northern Africa atmosphere-ocean-terrestrial ecosystem in the Holocene. *Quat. Sci. Rev.*, **26**, 1818–1837.

Liu, Z., et al., 2009b: Transient simulation of last deglaciation with a new mechanism for Bølling-Allerød warming. *Science*, **325**, 310–314.

Ljungqvist, F. C., 2010: A new reconstruction of temperature variability in the extra-tropical northern hemisphere during the last two millennia. *Geograf. Annal. A*, **92**, 339–351.

Ljungqvist, F. C., P. J. Krusic, G. Brattström, and H. S. Sundqvist, 2012: Northern hemisphere temperature patterns in the last 12 centuries. *Clim. Past*, **8**, 227–249.

Lloyd, A. H., and A. G. Bunn, 2007: Responses of the circumpolar boreal forest to 20th century climate variability. *Environ. Res. Lett.*, **2**, 045013.

Lockwood, M., and M. J. Owens, 2011: Centennial changes in the heliospheric magnetic field and open solar flux: the consensus view from geomagnetic data and cosmogenic isotopes and its implications. *J. Geophys. Res.*, **116**, A04109.

Loehle, C., and J. H. McCulloch, 2008: Correction to: A 2000-year global temperature reconstruction based on non-tree ring proxies. *Energy Environ.*, **19**, 93–100.

Long, A. J., S. A. Woodroffe, G. A. Milne, C. L. Bryant, M. J. R. Simpson, and L. M. Wake, 2012: Relative sea level change in Greenland during the last 700-yrs and ice sheet response to the Little Ice Age. *Earth Planet. Sci. Lett.*, **315–316**, 76–85.

Loso, M. G., 2009: Summer temperatures during the Medieval Warm Period and Little Ice Age inferred from varved proglacial lake sediments in southern Alaska. *J. Paleolimnol.*, **41**, 117–128.

Lough, J. M., 2011: Great Barrier Reef coral luminescence reveals rainfall variability over northeastern Australia since the 17th century. *Paleoceanography*, **26**, PA2201.

Loulergue, L., et al., 2008: Orbital and millennial-scale features of atmospheric CH_4 over the past 800,000 years. *Nature*, **453**, 383–386.

Loutre, M. F., and A. Berger, 2000: Future climatic changes: are we entering an exceptionally long interglacial? *Clim. Change*, **46**, 61–90.

Lowell, T. V., et al., 2013: Late Holocene expansion of Istorvet ice cap, Liverpool Land, east Greenland. *Quat. Sci. Rev.*, **63**, 128–140.

Lowenstein, T. K., and R. V. Demicco, 2006: Elevated Eocene atmospheric CO_2 and its subsequent decline. *Science*, **313**, 1928–1928.

Lozhkin, A. V., and P. A. Anderson, 2006: A reconstruction of the climate and vegetation of northeastern Siberia based on lake sediments. *Paleontol. J.*, **40**, 622–628.

Lu, J., and M. Cai, 2009: Seasonality of polar surface warming amplification in climate simulations. *Geophys. Res. Lett.*, **36**, L16704.

Lü, J. M., S. J. Kim, A. Abe-Ouchi, Y. Q. Yu, and R. Ohgaito, 2010: Arctic oscillation during the mid-Holocene and Last Glacial Maximum from PMIP2 coupled model simulations. *J. Clim.*, **23**, 3792–3813.

Lu, R., B. Dong, and H. Ding, 2006: Impact of the Atlantic multidecadal oscillation on the Asian summer monsoon. *Geophys. Res. Lett.*, **33**, L24701.

Luckman, B. H., and R. Villalba, 2001: Assessing the synchroneity of glacier fluctuations in the western Cordillera of the Americas during the last millenium. In: *Interhemispheric Climate Linkages* [V. Markgraf (ed.)]. Academic Press, San Diego, CA, USA, pp. 119–140.

Luckman, B. H., and R. J. S. Wilson, 2005: Summer temperatures in the Canadian Rockies during the last millennium: a revised record. *Clim. Dyn.*, **24**, 131–144.

Lunt, D. J., G. L. Foster, A. M. Haywood, and E. J. Stone, 2008: Late Pliocene Greenland glaciation controlled by a decline in atmospheric CO_2 levels. *Nature*, **454**, 1102–1105.

Lunt, D. J., A. M. Haywood, G. A. Schmidt, U. Salzmann, P. J. Valdes, and H. J. Dowsett, 2010: Earth system sensitivity inferred from Pliocene modelling and data. *Nature Geosci.*, **3**, 60–64.

Lunt, D. J., T. Dunkleay Jones, M. Heinemann, M. Huber, A. Legrande, A. Winguth, C. Lopston, J. Marotzke, C.D. Roberts, J. Tindall, P. Valdes, C. Winguth, 2012: A model-data comparison for a multi-model ensemble of early Eocene atmosphere-ocean simulations: EoMIP. *Climate of the Past*, **8**, 1717–1736.

Lunt, D. J., et al., 2013: A multi-model assessment of last interglacial temperatures. *Clim. Past*, **9**, 699–717.

Luo, F. F., S. L. Li, and T. Furevik, 2011: The connection between the Atlantic multidecadal oscillation and the Indian summer monsoon in Bergen climate model version 2.0. *J. Geophys. Res.*, **116**, D19117.

Luoto, T. P., S. Helama, and L. Nevalainen, 2013: Stream flow intensity of the Saavanjoki River, eastern Finland, during the past 1500 years reflected by mayfly and caddisfly mandibles in adjacent lake sediments. *J. Hydrol.*, **476**, 147–153.

Luterbacher, J., et al., 2002: Reconstruction of sea level pressure fields over the Eastern North Atlantic and Europe back to 1500. *Clim. Dyn.*, **18**, 545–561.

Luterbacher, J., et al., 2012: A review of 2000 years of paleoclimatic evidence in the Mediterranean. In: *The Climate of the Mediterranean Region: From the Past to the Future* [P. Lionello (ed.)]. Elsevier, Philadelphia, PA, USA, pp. 87–185.

Lüthi, D., et al., 2008: High-resolution carbon dioxide concentration record 650,000–800,000 years before present. *Nature*, **453**, 379–382.

Lynch-Stieglitz, J., et al., 2007: Atlantic meridional overturning circulation during the Last Glacial Maximum. *Science*, **316**, 66–69.

MacDonald, G. M., D. F. Porinchu, N. Rolland, K. V. Kremenetsky, and D. S. Kaufman, 2009: Paleolimnological evidence of the response of the central Canadian treeline zone to radiative forcing and hemispheric patterns of temperature change over the past 2000 years. *J. Paleolimnol.*, **41**, 129–141.

Macdonald, N., and A. R. Black, 2010: Reassessment of flood frequency using historical information for the River Ouse at York, UK (1200–2000). *Hydrol. Sci. J.*, **55**, 1152–1162.

5

MacFarling Meure, C. M., et al., 2006: Law Dome CO_2, CH_4 and N_2O ice core records extended to 2000 years BP. *Geophys. Res. Lett.*, **10**, L14810.

Machado, M. J., G. Benito, M. Barriendos, and F. S. Rodrigo, 2011: 500 years of rainfall variability and extreme hydrological events in southeastern Spain drylands. *J. Arid Environ.*, **75**, 1244–1253.

Machida, T., T. Nakazawa, Y. Fujii, S. Aoki, and O. Watanabe, 1995: Increase in the atmospheric nitrous oxide concentration during the last 250 years. *Geophys. Res. Lett.*, **22**, 2921–2924.

Macias Fauria, M., et al., 2010: Unprecedented low twentieth century winter sea ice extent in the western Nordic Seas since AD 1200. *Clim. Dyn.*, **34**, 781–795.

Mackintosh, A., et al., 2011: Retreat of the East Antarctic ice sheet during the last glacial termination. *Nature Geosci.*, **4**, 195–202.

Macklin, M. G., J. Lewin, and J. C. Woodward, 2012: The fluvial record of climate change. *Philos. Trans. R. Soc. London A*, **370**, 2143–2172.

Magilligan, F. J., P. S. Goldstein, G. B. Fisher, B. C. Bostick, and R. B. Manners, 2008: Late Quaternary hydroclimatology of a hyper-arid Andean watershed: climate change, floods, and hydrologic responses to the El Niño-Southern Oscillation in the Atacama Desert. *Geomorphology*, **101**, 14–32.

Maher, B. A., J. M. Prospero, D. Mackie, D. Gaiero, P. P. Hesse, and Y. Balkanski, 2010: Global connections between aeolian dust, climate and ocean biogeochemistry at the present day and at the last glacial maximum. *Earth Sci. Rev.*, **99**, 61–97.

Mahowald, N., S. Albani, S. Engelstaedter, G. Winckler, and M. Goman, 2011: Model insight into glacial–interglacial paleodust records. *Quat. Sci. Rev.*, **30**, 832–854.

Mahowald, N. M., M. Yoshioka, W. D. Collins, A. J. Conley, D. W. Fillmore, and D. B. Coleman, 2006: Climate response and radiative forcing from mineral aerosols during the last glacial maximum, pre-industrial, current and doubled-carbon dioxide climates. *Geophys. Res. Lett.*, **33**, L20705.

Man, W. M., T. J. Zhou, and J. H. Jungclaus, 2012: Simulation of the East Asian Summer Monsoon during the last millennium with the MPI Earth System Model. *J. Clim.*, **25**, 7852–7866.

Mann, M. E., J. D. Fuentes, and S. Rutherford, 2012: Underestimation of volcanic cooling in tree-ring-based reconstructions of hemispheric temperatures. *Nature Geosci.*, **5**, 202–205.

Mann, M. E., S. Rutherford, E. R. Wahl, and C. Ammann, 2007: Robustness of proxy-based climate field reconstruction methods. *J. Geophys. Res.*, **112**, D12109.

Mann, M. E., Z. H. Zhang, M. K. Hughes, R. S. Bradley, S. K. Miller, S. Rutherford, and F. B. Ni, 2008: Proxy-based reconstructions of hemispheric and global surface temperature variations over the past two millennia. *Proc. Natl. Acad. Sci. U.S.A.*, **105**, 13252–13257.

Mann, M. E., et al., 2009: Global signatures and dynamical origins of the Little Ice Age and Medieval Climate Anomaly. *Science*, **326**, 1256–1260.

Marcott, S. A., J. D. Shakun, P. U. Clark, and A. C. Mix, 2013: A reconstruction of regional and global temperature for the past 11,300 years. *Science*, **339**, 1198–1201.

Marcott, S. A., et al., 2011: Ice-shelf collapse from subsurface warming as a trigger for Heinrich events. *Proc. Natl. Acad. Sci. U.S.A.*, **108**, 13415–13419.

Margari, V., L. C. Skinner, P. C. Tzedakis, A. Ganopolski, M. Vautravers, and N. J. Shackleton, 2010: The nature of millennial-scale climate variability during the past two glacial periods. *Nature Geosci.*, **3**, 127–131.

MARGO Project Members, 2009: Constraints on the magnitude and patterns of ocean cooling at the Last Glacial Maximum. *Nature Geosci.*, **2**, 127–132.

Marra, M. J., 2003: Last interglacial beetle fauna from New Zealand. *Quat. Res.*, **59**, 122–131.

Marshall, S. J., and M. R. Koutnik, 2006: Ice sheet action versus reaction: distinguishing between Heinrich events and Dansgaard-Oeschger cycles in the North Atlantic. *Paleoceanography*, **21**, PA2021.

Martin, P. A., D. W. Lea, Y. Rosenthal, N. J. Shackleton, M. Sarnthein, and T. Papenfuss, 2002: Quaternary deep sea temperature histories derived from benthic foraminiferal Mg/Ca. *Earth Planet. Sci. Lett.*, **198**, 193–209.

Martínez-Garcia, A., A. Rosell-Melé, S. L. Jaccard, W. Geibert, D. M. Sigman, and G. H. Haug, 2011: Southern Ocean dust-climate coupling over the past four million years. *Nature*, **476**, 312–315.

Martrat, B., J. O. Grimalt, N. J. Shackleton, L. de Abreu, M. A. Hutterli, and T. F. Stocker, 2007: Four climate cycles of recurring deep and surface water destabilizations on the Iberian margin. *Science*, **317**, 502–507.

Martrat, B., et al., 2004: Abrupt temperature changes in the western Mediterranean over the past 250,000 years. *Science*, **306**, 1762–1765.

Marzeion, B., and A. Nesje, 2012: Spatial patterns of North Atlantic Oscillation influence on mass balance variability of European glaciers. *Cryosphere*, **6**, 661–673.

Marzin, C., and P. Braconnot, 2009: Variations of Indian and African monsoons induced by insolation changes at 6 and 9.5 kyr BP. *Clim. Dyn.*, **33**, 215–231.

Masson-Delmotte, V., et al., 2011a: Sensitivity of interglacial Greenland temperature and $\delta^{18}O$: ice core data, orbital and increased CO_2 climate simulations. *Clim. Past*, **7**, 1041–1059.

Masson-Delmotte, V., et al., 2010a: EPICA Dome C record of glacial and interglacial intensities. *Quat. Sci. Rev.*, **29**, 113–128.

Masson-Delmotte, V., et al., 2011b: A comparison of the present and last interglacial periods in six Antarctic ice cores. *Clim. Past*, **7**, 397–423.

Masson-Delmotte, V., et al., 2010b: Abrupt change of Antarctic moisture origin at the end of Termination II. *Proc. Natl. Acad. Sci. U.S.A.*, **107**, 12091–12094.

Mathiot, P., et al., 2013: Using data assimilation to investigate the causes of Southern Hemisphere high latitude cooling from 10 to 8 ka BP. *Clim. Past*, **9**, 887–901.

Matthews, J. A., and P. Q. Dresser, 2008: Holocene glacier variation chronology of the Smørstabbtindan massif, Jotunheimen, southern Norway, and the recognition of century- to millennial-scale European Neoglacial Events. *Holocene*, **18**, 181–201.

McCarroll, D., et al., 2013: A 1200–year multiproxy record of tree growth and summer temperature at the northern pine forest limit of Europe. *Holocene*, **23**, 471–484.

McElwain, J. C., 1998: Do fossil plants signal palaeoatmospheric CO_2 concentration in the geological past? *Philos. Trans. R. Soc. London B*, **353**, 83–96.

McGee, D., W. S. Broecker, and G. Winckler, 2010: Gustiness: the driver of glacial dustiness? *Quat. Sci. Rev.*, **29**, 2340–2350.

McGregor, S., and A. Timmermann, 2010: The effect of explosive tropical volcanism on ENSO. *J. Clim.*, **24**, 2178–2191.

McGregor, S., A. Timmermann, and O. Timm, 2010: A unified proxy for ENSO and PDO variability since 1650. *Clim. Past*, **6**, 1–17.

McInerney, F. A., and S. L. Wing, 2011: The Paleocene-Eocene Thermal Maximum: A perturbation of carbon cycle, climate, and biosphere with implications for the future. *Annu. Rev. Earth Planet. Sci.*, **39**, 489–516.

McKay, N. P., D. S. Kaufman, and N. Michelutti, 2008: Biogenic silica concentration as a high-resolution, quantitative temperature proxy at Hallet Lake, south-central Alaska. *Geophys. Res. Lett.*, **35**, L05709.

McKay, N. P., J. T. Overpeck, and B. L. Otto-Bliesner, 2011: The role of ocean thermal expansion in Last Interglacial sea level rise. *Geophys. Res. Lett.*, **38**, L14605.

McKay, R., et al., 2012a: Pleistocene variability of Antarctic ice sheet extent in the Ross embayment. *Quat. Sci. Rev.*, **34**, 93–112.

McKay, R., et al., 2012b: Antarctic and Southern Ocean influences on Late Pliocene global cooling. *Proc. Natl. Acad. Sci. U.S.A.*, **109**, 6423-6428.

McManus, J., R. Francois, J. Gherardi, L. Keigwin, and S. Brown-Leger, 2004: Collapse and rapid resumption of Atlantic meridional circulation linked to deglacial climate changes. *Nature*, **428**, 834–837.

McShane, B. B., and A. J. Wyner, 2011: A statistical analysis of multiple temperature proxies: Are reconstructions of surface temperatures over the last 1000 years reliable? *Ann. Appl. Stat.*, **5**, 5–44.

Meckler, A. N., M. O. Clarkson, K. M. Cobb, H. Sodemann, and J. F. Adkins, 2013: Interglacial hydroclimate in the tropical West Pacific through the Late Pleistocene. *Science*, **336**, 1301–1304.

Meko, D. M., C. A. Woodhouse, C. A. Baisan, T. Knight, J. J. Lukas, M. K. Hughes, and M. W. Salzer, 2007: Medieval drought in the upper Colorado River Basin. *Geophys. Res. Lett.*, **34**, L10705.

Melvin, T. M., and K. R. Briffa, 2008: A "signal-free" approach to dendroclimatic standardisation. *Dendrochronologia*, **26**, 71–86.

Melvin, T. M., H. Grudd, and K. R. Briffa, 2013: Potential bias in 'updating' tree-ring chronologies using regional curve standardisation: Re-processing 1500 years of Torneträsk density and ring-width data. *Holocene*, **23**, 364–373.

Menounos, B., G. Osborn, J. Clague, and B. Luckman, 2009: Latest Pleistocene and Holocene glacier fluctuations in western Canada. *Quat. Sci. Rev.*, **28**, 2049–2074.

Menviel, L., A. Timmermann, O. E. Timm, and A. Mouchet, 2011: Deconstructing the last glacial termination: the role of millennial and orbital-scale forcings. *Quat. Sci. Rev.*, **30**, 1155–1172.

Merkel, U., M. Prange, and M. Schulz, 2010: ENSO variability and teleconnections during glacial climates. *Quat. Sci. Rev.*, **29**, 86–100.

Miller, G. H., A. P. Wolfe, J. P. Briner, P. E. Sauer, and A. Nesje, 2005: Holocene glaciation and climate evolution of Baffin Island, Arctic Canada. *Quat. Sci. Rev.*, **24**, 1703–1721.

Miller, G. H., et al., 1999: Stratified interglacial lacustrine sediments from Baffin Island, Arctic Canada: Chronology and paleoenvironmental implications. *Quat. Sci. Rev.*, **18**, 789–810.

Miller, K. G., et al., 2012a: High tide of the warm Pliocene: implications of global sea level for Antarctic deglaciation. *Geology*, **40**, 407–410.

Miller, M. D., J. F. Adkins, D. Menemenlis, and M. P. Schodlok, 2012b: The role of ocean cooling in setting glacial southern source bottom water salinity. *Paleoceanography*, **27**, PA3207.

Milne, G., and J. Mitrovica, 2008: Searching for eustasy in deglacial sea level histories. *Quat. Sci. Rev.*, **27**, 2292–2302.

Mischler, J. A., et al., 2009: Carbon and hydrogen isotopic composition of methane over the last 1000 years. *Global Biogeochem. Cycles*, **23**, GB4024.

Moberg, A., 2013: Comments on "Reconstruction of the extra-tropical NH mean temperature over the last millennium with a method that preserves low-frequency variability". *J. Clim.*, **25**, 7991–7997.

Moberg, A., D. M. Sonechkin, K. Holmgren, N. M. Datsenko, and W. Karlén, 2005: Highly variable Northern Hemisphere temperatures reconstructed from low- and high-resolution proxy data. *Nature*, **433**, 613–617.

Mohtadi, M., D. W. Oppo, S. Steinke, J.-B. W. Stuut, R. De Pol-Holz, D. Hebbeln, and A. Lückge, 2011: Glacial to Holocene swings of the Australian-Indonesian monsoon. *Nature Geosci.*, **4**, 540–544.

Monnin, E., et al., 2001: Atmospheric CO_2 concentrations over the last glacial termination. *Science*, **291**, 112–114.

Moore, J. C., E. Beaudon, S. Kang, D. Divine, E. Isaksson, V. A. Pohjola, and R. S. W. van de Wal, 2012: Statistical extraction of volcanic sulphate from nonpolar ice cores. *J. Geophys. Res.*, **117**, D03306.

Morales, M. S., et al., 2012: Precipitation changes in the South American Altiplano since 1300 AD reconstructed by tree-rings. *Clim. Past*, **8**, 653–666.

Morice, C. P., J. J. Kennedy, N. A. Rayner, and P. D. Jones, 2012: Quantifying uncertainties in global and regional temperature change using an ensemble of observational estimates: The HadCRUT4 data set. *J. Geophys. Res.*, **117**, D08101.

Moros, M., J. T. Andrews, D. D. Eberl, and E. Jansen, 2006: Holocene history of drift ice in the northern North Atlantic: Evidence for different spatial and temporal modes. *Paleoceanography*, **21**, PA2017.

Moros, M., P. De Deckker, E. Jansen, K. Perner, and R. J. Telford, 2009: Holocene climate variability in the Southern Ocean recorded in a deep-sea sediment core off South Australia. *Quat. Sci. Rev.*, **28**, 1932–1940.

Morrill, C., A. J. Wagner, B. L. Otto-Bliesner, and N. Rosenbloom, 2011: Evidence for significant climate impacts in monsoonal Asia at 8.2 ka from multiple proxies and model simulations. *J. Earth Environ.*, **2**, 426–441.

Morrill, C., A. N. LeGrande, H. Renssen, P. Bakker, and B. L. Otto-Bliesner, 2013a: Model sensitivity to North Atlantic freshwater forcing at 8.2 ka. *Clim. Past*, **9**, 955–968.

Morrill, C., et al., 2013b: Proxy benchmarks for intercomparison of 8.2 ka simulations. *Clim. Past*, **9**, 423–432.

Moucha, R., A. M. Forte, J. X. Mitrovica, D. B. Rowley, S. Quéré, N. A. Simmons, and S. P. Grand, 2008: Dynamic topography and long-term sea level variations: There is no such thing as a stable continental platform. *Earth Planet. Sci. Lett.*, **271**, 101–108.

Mudelsee, M., 2001: The phase relations among atmospheric CO_2 content, temperature and global ice volume over the past 420 ka. *Quat. Sci. Rev.*, **20**, 583–589.

Mudelsee, M., and M. E. Raymo, 2005: Slow dynamics of the Northern Hemisphere glaciation. *Paleoceanography*, **20**, PA4022.

Mudelsee, M., J. Fohlmeister, and D. Scholz, 2012: Effects of dating errors on nonparametric trend analyses of speleothem time series. *Clim. Past*, **8**, 1637–1648.

Mudelsee, M., M. Börngen, G. Tetzlaff, and U. Grünewald, 2003: No upward trends in the occurrence of extreme floods in central Europe. *Nature*, **425**, 166–169.

Mulitza, S., et al., 2008: Sahel megadroughts triggered by glacial slowdowns of Atlantic meridional overturning. *Paleoceanography*, **23**, PA4206.

Müller, J., A. Wagner, K. Fahl, R. Stein, M. Prange, and G. Lohmann, 2011: Towards quantitative sea ice reconstructions in the northern North Atlantic: A combined biomarker and numerical modelling approach. *Earth Planet. Sci. Lett.*, **306**, 137–148.

Müller, R. D., M. Sdrolias, C. Gaina, B. Steinberger, and C. Heine, 2008: Long-term sea level fluctuations driven by ocean basin dynamics. *Science*, **319**, 1357–1362.

Müller, U., 2001: *Die Vegetations-und Klimaentwicklung im jüngeren Quartär anhand ausgewählter Profile aus dem südwestdeutschen Alpenvorland.* Tübinger Geowissenschaftliche Arbeiten D7, Geographisches Institut der Universität Tübingen, 118 pp.

Mulvaney, R., et al., 2012: Recent Antarctic Peninsula warming relative to Holocene climate and ice-shelf history. *Nature*, **489**, 141–144.

Muscheler, R., and J. Beer, 2006: Solar forced Dansgaard/Oeschger events? *Geophys. Res. Lett.*, **33**, L20706.

Muscheler, R., F. Joos, J. Beer, S. A. Müller, M. Vonmoos, and I. Snowball, 2007: Solar activity during the last 1000 yr inferred from radionuclide records. *Quat. Sci. Rev.*, **26**, 82–97.

Naish, T., et al., 2009a: Obliquity-paced Pliocene West Antarctic ice sheet oscillations. *Nature*, **458**, 322–328.

Naish, T. R., and G. S. Wilson, 2009: Constraints on the amplitude of mid-Pliocene (3.6–2.4Ma) eustatic sea level fluctuations from the New Zealand shallow-marine sediment record. *Philos. Trans. R. Soc. London A*, **367**, 169–187.

Naish, T. R., L. Carter, E. Wolff, D. Pollard, and R. D. Powell, 2009b: Late Pliocene–Pleistocene Antarctic climate variability at orbital and suborbital scale: Ice sheet, ocean and atmospheric interactions. In: *Developments in Earth & Environmental Sciences* [F. Florindo and S. M. (eds.)]. Elsevier, Philadelphia, PA, USA, pp. 465–529.

Nakagawa, T., et al., 2008: Regulation of the monsoon climate by two different orbital rhythms and forcing mechanisms. *Geology*, **36**, 491–494.

NEEM community members, 2013: Eemian interglacial reconstructed from Greenland folded ice core. *Nature*, **493**, 489–494.

Neppel, L., et al., 2010: Flood frequency analysis using historical data: Accounting for random and systematic errors. *Hydrol. Sci. J. J. Sci. Hydrol.*, **55**, 192–208.

Nesje, A., 2009: Latest Pleistocene and Holocene alpine glacier fluctuations in Scandinavia. *Quat. Sci. Rev.*, **28**, 2119–2136.

Nesje, A., et al., 2011: The climatic significance of artefacts related to prehistoric reindeer hunting exposed at melting ice patches in southern Norway. *Holocene*, **22**, 485–496.

Neukom, R., and J. Gergis, 2011: Southern Hemisphere high-resolution palaeoclimate records of the last 2000 years. *Holocene*, **22**, 501–524.

Neukom, R., et al., 2011: Multiproxy summer and winter surface air temperature field reconstructions for southern South America covering the past centuries. *Clim. Dyn.*, **37**, 35–51.

Newby, P. E., B. N. Shuman, J. P. Donnelly, and D. MacDonald, 2011: Repeated century-scale droughts over the past 13,000 yr near the Hudson River watershed, USA. *Quat. Res.*, **75**, 523–530.

Nicault, A., S. Alleaume, S. Brewer, M. Carrer, P. Nola, and J. Guiot, 2008: Mediterranean drought fluctuation during the last 500 years based on tree-ring data. *Clim. Dyn.*, **31**, 227–245.

Nicolussi, K., M. Kaufmann, T. M. Melvin, J. van der Plicht, P. Schießling, and A. Thurner, 2009: A 9111 year long conifer tree-ring chronology for the European Alps: a base for environmental and climatic investigations. *Holocene*, **19**, 909–920.

Nordt, L., S. Atchley, and S. I. Dworkin, 2002: Paleosol barometer indicates extreme fluctuations in atmospheric CO_2 across the Cretaceous-Tertiary boundary. *Geology*, **30**, 703–706.

Nørgaard-Pedersen, N., N. Mikkelsen, S. J. Lassen, Y. Kristoffersen, and E. Sheldon, 2007: Reduced sea ice concentrations in the Arctic Ocean during the last interglacial period revealed by sediment cores off northern Greenland. *Paleoceanography*, **22**, PA1218.

North Greenland Ice Core Project members, 2004: High-resolution record of Northern Hemisphere climate extending into the last interglacial period. *Nature*, **431**, 147–151.

Novenko, E. Y., M. Seifert-Eulen, T. Boettger, and F. W. Junge, 2008: Eemian and early Weichselian vegetation and climate history in Central Europe: a case study from the Klinge section (Lusatia, eastern Germany). *Rev. Palaeobot. Palynol.*, **151**, 72–78.

O'Donnell, R., N. Lewis, S. McIntyre, and J. Condon, 2010: Improved methods for PCA-based reconstructions: case study using the Steig et al. (2009) Antarctic temperature reconstruction. *J. Clim.*, **24**, 2099–2115.

Oerlemans, J., 1980: Model experiments on the 100,000–yr glacial cycle. *Nature*, **287**, 430–432.

Ohba, M., H. Shiogama, T. Yokohata, and M. Watanabe, 2013: Impact of strong tropical volcanic eruptions on ENSO simulated in a coupled GCM. *J. Clim.*, **26**, 5169-5182.

Okazaki, Y., et al., 2010: Deepwater formation in the north Pacific during the Last Glacial Termination. *Science*, **329**, 200–204.

Okumura, Y. M., C. Deser, A. Hu, A. Timmermann, and S. P. Xie, 2009: North Pacific climate response to freshwater forcing in the subarctic North Atlantic: Oceanic and atmospheric pathways. *J. Clim.*, **22**, 1424–1445.

Olsen, J., N. J. Anderson, and M. F. Knudsen, 2012: Variability of the North Atlantic Oscillation over the past 5,200 years. *Nature Geosci.*, **5**, 808–812.

Oman, L., A. Robock, G. Stenchikov, G. A. Schmidt, and R. Ruedy, 2005: Climatic response to high-latitude volcanic eruptions. *J. Geophys. Res.*, **110**, D13103.

Orsi, A. J., B. D. Cornuelle, and J. P. Severinghaus, 2012: Little Ice Age cold interval in West Antarctica: Evidence from borehole temperature at the West Antarctic Ice Sheet (WAIS) Divide. *Geophys. Res. Lett.*, **39**, L09710.

Osborn, T., and K. Briffa, 2007: Response to comment on "The spatial extent of 20th-century warmth in the context of the past 1200 years". *Science*, **316**, 1844.

Osborn, T., S. Raper, and K. Briffa, 2006: Simulated climate change during the last 1,000 years: Comparing the ECHO-G general circulation model with the MAGICC simple climate model. *Clim. Dyn.*, **27**, 185–197.

Oswald, W. W., and D. R. Foster, 2011: A record of late-Holocene environmental change from southern New England, USA. *Quat. Res.*, **76**, 314–318.

Ottera, O. H., M. Bentsen, H. Drange, and L. L. Suo, 2010: External forcing as a metronome for Atlantic multidecadal variability. *Nature Geosci.*, **3**, 688–694.

Otto-Bliesner, B., et al., 2009: A comparison of PMIP2 model simulations and the MARGO proxy reconstruction for tropical sea surface temperatures at Last Glacial Maximum. *Clim. Dyn.*, **32**, 799–815.

Otto-Bliesner, B. L., and E. C. Brady, 2010: The sensitivity of the climate response to the magnitude and location of freshwater forcing: Last glacial maximum experiments. *Quat. Sci. Rev.*, **29**, 56–73.

Otto-Bliesner, B. L., N. Rosenbloom, E. J. Stone, N. McKay, D. Lunt, E. C. Brady, and J. T. Overpeck, 2013: How warm was the last interglacial? New model-data comparisons. *Philos. Trans. R. Soc. London A*, **371**, 20130097, *published online 16 September 2013*.

Otto-Bliesner, B. L., et al., 2007: Last Glacial Maximum ocean thermohaline circulation: PMIP2 model intercomparisons and data constraints. *Geophys. Res. Lett.*, **34**, L12706.

Otto, J., T. Raddatz, M. Claussen, V. Brovkin, and V. Gayler, 2009: Separation of atmosphere-ocean-vegetation feedbacks and synergies for mid-Holocene climate. *Geophys. Res. Lett.*, **36**, L09701.

Overpeck, J., B. Otto-Bliesner, G. Miller, D. Muhs, R. Alley, and J. Kiehl, 2006: Paleoclimatic evidence for future ice-sheet instability and rapid sea level rise. *Science*, **311**, 1747–1750.

Pagani, M., K. H. Freeman, and M. A. Arthur, 1999a: Late Miocene Atmospheric CO_2 concentrations and the expansion of C4 grasses. *Science*, **285**, 876–879.

Pagani, M., M. A. Arthur, and K. H. Freeman, 1999b: Miocene evolution of atmospheric carbon dioxide. *Paleoceanography*, **14**, 273–292.

Pagani, M., D. Lemarchand, A. Spivack, and J. Gaillardet, 2005a: A critical evaluation of the boron isotope-pH proxy: The accuracy of ancient ocean pH estimates. *Geochim. Cosmochim. Acta*, **69**, 953–961.

Pagani, M., Z. H. Liu, J. LaRiviere, and A. C. Ravelo, 2010: High Earth-system climate sensitivity determined from Pliocene carbon dioxide concentrations. *Nature Geosci.*, **3**, 27–30.

Pagani, M., J. C. Zachos, K. H. Freeman, B. Tipple, and S. Bohaty, 2005b: Marked decline in atmospheric carbon dioxide concentrations during the Paleogene. *Science*, **309**, 600–603.

Pagani, M., et al., 2011: The role of carbon dioxide during the onset of Antarctic glaciation. *Science*, **334**, 1261–1264.

PAGES 2k Consortium, 2013: Continental-scale temperature variability during the last two millennia. *Nature Geosci.*, **6**, 339–346.

Pahnke, K., R. Zahn, H. Elderfield, and M. Schulz, 2003: 340,000-year centennial-scale marine record of Southern Hemisphere climatic oscillation. *Science*, **301**, 948–952.

Pahnke, K., J. P. Sachs, L. Keigwin, A. Timmermann, and S. P. Xie, 2007: Eastern tropical Pacific hydrologic changes during the past 27,000 years from D/H ratios in alkenones. *Paleoceanography*, **22**, PA4214.

Pak, D. K., D. W. Lea, and J. P. Kennett, 2012: Millennial scale changes in sea surface temperature and ocean circulation in the northeast Pacific, 10–60 kyr BP. *Paleoceanography*, **27**, PA1212.

PALAEOSENS Project Members, 2012: Making sense of palaeoclimate sensitivity. *Nature*, **491**, 683–691.

Palastanga, V., G. van der Schrier, S. Weber, T. Kleinen, K. Briffa, and T. Osborn, 2011: Atmosphere and ocean dynamics: contributors to the European Little Ice Age? *Clim. Dyn.*, **36**, 973–987.

Panchuk, K., A. Ridgwell, and L. R. Kump, 2008: Sedimentary response to Paleocene-Eocene Thermal Maximum carbon release: a model-data comparison. *Geology*, **36**, 315–318.

Parrenin, F., et al., 2013: Synchronous change of atmospheric CO_2 and Antarctic temperature during the last deglacial warming. *Science*, **339**, 1060–1063.

Passchier, S., 2011: Linkages between East Antarctic ice sheet extent and Southern Ocean temperatures based on a Pliocene high-resolution record of ice-rafted debris off Prydz Bay, East Antarctica. *Paleoceanography*, **26**, PA4204.

Patadia, F., E.-S. Yang, and S. A. Christopher, 2009: Does dust change the clear sky top of atmosphere shortwave flux over high surface reflectance regions? *Geophys. Res. Lett.*, **36**, L15825.

Pausata, F. S. R., C. Li, J. J. Wettstein, K. H. Nisancioglu, and D. S. Battisti, 2009: Changes in atmospheric variability in a glacial climate and the impacts on proxy data: A model intercomparison. *Clim. Past*, **5**, 489–502.

Pausata, F. S. R., C. Li, J. J. Wettstein, M. Kageyama, and K. H. Nisancioglu, 2011: The key role of topography in altering North Atlantic atmospheric circulation during the last glacial period. *Clim. Past*, **7**, 1089–1101.

Pearson, P. N., G. L. Foster, and B. S. Wade, 2009: Atmospheric carbon dioxide through the Eocene-Oligocene climate transition. *Nature*, **461**, 1110–1113.

Pedro, J., et al., 2011: The last deglaciation: timing the bipolar seesaw. *Clim. Past*, **7**, 671–683.

Pedro, J. B., S. O. Rasmussen, and T. D. van Ommen, 2012: Tightened constraints on the time-lag between Antarctic temperature and CO_2 during the last deglaciation. *Clim. Past*, **8**, 1213–1221.

Pépin, L., D. Raynaud, J.-M. Barnola, and M. F. Loutre, 2001: Hemispheric roles of climate forcings during glacial-interglacial transitions as deduced from the Vostok record and LLN-2D model experiments. *J. Geophys. Res.*, **106**, 31885–31892.

Peschke, P., C. Hannss, and S. Klotz, 2000: Neuere Ergebnisse aus der Banquette von Barraux (Grésivaudan, französische Nordalpen) zur spätpleistozänen Vegetationsentwicklung mit Beiträgen zur Reliefgenese und Klimarekonstruktion. *Eiszeitalter Gegenwart*, **50**, 1–24.

Peterson, L. C., and G. H. Haug, 2006: Variability in the mean latitude of the Atlantic Intertropical Convergence Zone as recorded by riverine input of sediments to the Cariaco Basin (Venezuela). *Palaeogeogr. Palaeoclimatol. Palaeoecol.*, **234**, 97–113.

Petit, J. R., and B. Delmonte, 2009: A model for large glacial–interglacial climate-induced changes in dust and sea salt concentrations in deep ice cores (central Antarctica): palaeoclimatic implications and prospects for refining ice core chronologies. *Tellus B*, **61B**, 768–790.

Petit, J. R., et al., 1999: Climate and atmospheric history of the past 420,000 years from the Vostok ice core, Antarctica. *Nature*, **399**, 429–436.

Phipps, S., et al., 2013: Palaeoclimate data-model comparison and the role of climate forcings over the past 1500 years. *J. Clim.*, **26**, 6915-6936.

Piccarreta, M., M. Caldara, D. Capolongo, and F. Boenzi, 2011: Holocene geomorphic activity related to climatic change and human impact in Basilicata, Southern Italy. *Geomorphology*, **128**, 137–147.

Pinto, J. G., and C. C. Raible, 2012: Past and recent changes in the North Atlantic Oscillation. *WIREs Clim. Change*, **3**, 79–90.

Piotrowski, A. M., S. L. Goldstein, S. R. Hemming, and R. G. Fairbanks, 2005: Temporal relationships of carbon cycling and ocean circulation at glacial boundaries. *Science*, **307**, 1933–1938.

Plummer, C. T., et al., 2012: An independently dated 2000–yr volcanic record from Law Dome, East Antarctica, including a new perspective on the dating of the 1450s CE eruption of Kuwae, Vanuatu. *Clim. Past*, **8**, 1929–1940.

Pollack, H. N., and J. E. Smerdon, 2004: Borehole climate reconstructions: Spatial structure and hemispheric averages. *J. Geophys. Res.*, **109**, D11106.

Pollard, D., and R. M. DeConto, 2009: Modelling West Antarctic ice sheet growth and collapse through the past five million years. *Nature*, **458**, 329–332.

Polyak, L., et al., 2010: History of sea ice in the Arctic. *Quat. Sci. Rev.*, **29**, 1757–1778.

Polyakov, I. V., et al., 2010: Arctic Ocean warming contributes to reduced polar ice cap. *J. Phys. Oceanogr.*, **40**, 2743–2756.

Pongratz, J., C. Reick, T. Raddatz, and M. Claussen, 2008: A reconstruction of global agricultural areas and land cover for the last millennium. *Global Biogeochem. Cycles*, **22**, GB3018.

Pongratz, J., T. Raddatz, C. H. Reick, M. Esch, and M. Claussen, 2009: Radiative forcing from anthropogenic land cover change since A.D. 800. *Geophys. Res. Lett.*, **36**, L02709.

Ponton, C., L. Giosan, T. I. Eglinton, D. Q. Fuller, J. E. Johnson, P. Kumar, and T. S. Collett, 2012: Holocene aridification of India. *Geophys. Res. Lett.*, **39**, L03704.

Porter, T. J., and M. F. J. Pisaric, 2011: Temperature-growth divergence in white spruce forests of Old Crow Flats, Yukon Territory, and adjacent regions of northwestern North America. *Global Change Biol.*, **17**, 3418–3430.

5

Prieto, M. d. R., and R. García Herrera, 2009: Documentary sources from South America: Potential for climate reconstruction. *Palaeogeogr. Palaeoclimatol. Palaeoecol.*, **281**, 196–209.

Prokopenko, A., L. Hinnov, D. Williams, and M. Kuzmin, 2006: Orbital forcing of continental climate during the Pleistocene: A complete astronomically tuned climatic record from Lake Baikal, SE Siberia. *Quat. Sci. Rev.*, **25**, 3431–3457.

Prokopenko, A. A., D. F. Williams, M. I. Kuzmin, E. B. Karabanov, G. K. Khursevich, and J. A. Peck, 2002: Muted climate variations in continental Siberia during the mid-Pleistocene epoch. *Nature*, **418**, 65–68.

Putnam, A. E., et al., 2010: Glacier advance in southern middle-latitudes during the Antarctic Cold Reversal. *Nature Geosci.*, **3**, 700–704.

Quiquet, A., C. Ritz, H. J. Punge, and D. Salas y Mélia, 2013: Greenland ice sheet contribution to sea level rise during the last interglacial period: A modelling study driven and constrained by ice core data. *Clim. Past*, **8**, 353–366.

Rahmstorf, S., et al., 2005: Thermohaline circulation hysteresis: A model intercomparison. *Geophys. Res. Lett.*, **32**, L23605.

Ramankutty, N., and J. A. Foley, 1999: Estimating historical changes in global land cover: Croplands from 1700 to 1992. *Global Biogeochem. Cycles*, **13**, 997–1027.

Rasmussen, S. O., et al., 2006: A new Greenland ice core chronology for the last glacial termination. *J. Geophys. Res.*, **111**, D06102.

Raymo, M. E., and J. X. Mitrovica, 2012: Collapse of polar ice sheets during the stage 11 interglacial. *Nature*, **483**, 453–456.

Raymo, M. E., J. X. Mitrovica, M. J. O'Leary, R. M. DeConto, and P. J. Hearty, 2011: Departures from eustasy in Pliocene sea level records. *Nature Geosci.*, **4**, 328–332.

Renssen, H., H. Seppä, X. Crosta, H. Goosse, and D. M. Roche, 2012: Global characterization of the Holocene Thermal Maximum. *Quat. Sci. Rev.*, **48**, 7–19.

Retallack, G. J., 2009a: Refining a pedogenic-carbonate CO_2 paleobarometer to quantify a middle Miocene greenhouse spike. *Palaeogeogr. Palaeoclimatol. Palaeoecol.* **281**, 57–65.

Retallack, G. J, 2009b: Greenhouse crises of the past 300 million years. *Geol. Soc. Am. Bull.*, **121**, 1441–1455.

Reuter, J., L. Stott, D. Khider, A. Sinha, H. Cheng, and R. L. Edwards, 2009: A new perspective on the hydroclimate variability in northern South America during the Little Ice Age. *Geophys. Res. Lett.*, **36**, L21706.

Ridgwell, A., and D. N. Schmidt, 2010: Past constraints on the vulnerability of marine calcifiers to massive carbon dioxide release. *Nature Geosci.*, **3**, 196–200.

Ridley, J., J. Gregory, P. Huybrechts, and J. Lowe, 2010: Thresholds for irreversible decline of the Greenland ice sheet. *Clim. Dyn.*, **35**, 1049–1057.

Rimbu, N., G. Lohmann, J. H. Kim, H. W. Arz, and R. Schneider, 2003: Arctic/North Atlantic Oscillation signature in Holocene sea surface temperature trends as obtained from alkenone data. *Geophys. Res. Lett.*, **30**, 4.

Risebrobakken, B., T. Dokken, L. H. Smedsrud, C. Andersson, E. Jansen, M. Moros, and E. V. Ivanova, 2011: Early Holocene temperature variability in the Nordic Seas: The role of oceanic heat advection versus changes in orbital forcing. *Paleoceanography*, **26**, PA4206.

Ritz, S., T. Stocker, and F. Joos, 2011: A coupled dynamical ocean-energy balance atmosphere model for paleoclimate studies. *J. Clim.*, **24**, 349–375.

Riviere, G., A. Laîné, G. Lapeyre, D. Salas-Melia, and M. Kageyama, 2010: Links between Rossby wave breaking and the North Atlantic Oscillation-Arctic Oscillation in present-day and Last Glacial Maximum climate simulations. *J. Clim.*, **23**, 2987–3008.

Roberts, A. P., E. J. Rohling, K. M. Grant, J. C. Larrasoaña, and Q. Liu, 2011: Atmospheric dust variability from Arabia and China over the last 500,000 years. *Quat. Sci. Rev.*, **30**, 3537–3541.

Roberts, N. L., A. M. Piotrowski, J. F. McManus, and L. D. Keigwin, 2010: Synchronous deglacial overturning and water mass source changes. *Science*, **327**, 75–78.

Robertson, A., et al., 2001: Hypothesized climate forcing time series for the last 500 years. *J. Geophys. Res.*, **106**, 14783–14803.

Robinson, A., R. Calov, and A. Ganopolski, 2011: Greenland ice sheet model parameters constrained using simulations of the Eemian Interglacial. *Clim. Past*, **7**, 381–396.

Roche, D. M., X. Crosta, and H. Renssen, 2012: Evaluating Southern Ocean sea-ice for the Last Glacial Maximum and pre-industrial climates: PMIP-2 models and data evidence. *Quat. Sci. Rev.*, **56**, 99–106.

Roe, G. H., and R. S. Lindzen, 2001: The mutual interaction between continental-scale ice sheets and atmospheric stationary waves. *J. Clim.*, **14**, 1450–1465.

Roeckner, E., L. Bengtsson, J. Feichter, J. Lelieveld, and H. Rodhe, 1999: Transient climate change simulations with a coupled atmosphere–ocean GCM including the tropospheric sulfur cycle. *J. Clim.*, **12**, 3004–3032.

Rohling, E. J., and H. Pälike, 2005: Centennial-scale climate cooling with a sudden cold event around 8,200 years ago. *Nature*, **434**, 975–979.

Rohling, E. J., M. Medina-Elizalde, J. G. Shepherd, M. Siddall, and J. D. Stanford, 2012: Sea surface and high-latitude temperature sensitivity to radiative forcing of climate over several glacial cycles. *J. Clim.*, **25**, 1635–1656.

Rohling, E. J., K. Grant, C. Hemleben, M. Siddall, B. A. A. Hoogakker, M. Bolshaw, and M. Kucera, 2008a: High rates of sea level rise during the last interglacial period. *Nature Geosci.*, **1**, 38–42.

Rohling, E. J., K. Grant, M. Bolshaw, A. P. Roberts, M. Siddall, C. Hemleben, and M. Kucera, 2009: Antarctic temperature and global sea level closely coupled over the past five glacial cycles. *Nature Geosci.*, **2**, 500–504.

Rohling, E. J., K. Braun, K. Grant, M. Kucera, A. P. Roberts, M. Siddall, and G. Trommer, 2010: Comparison between Holocene and Marine Isotope Stage-11 sea level histories. *Earth Planet. Sci. Lett.*, **291**, 97–105.

Rohling, E. J., et al., 2008b: New constraints on the timing of sea level fluctuations during early to middle marine isotope stage 3. *Paleoceanography*, **23**, PA3219.

Rojas, M., 2013: Sensitivity of Southern Hemisphere circulation to LGM and 4 × CO2 climates. *Geophys. Res. Lett.*, **40**, 965–970.

Rousseau, D.-D., C. Hatté, D. Duzer, P. Schevin, G. Kukla, and J. Guiot, 2007: Estimates of temperature and precipitation variations during the Eemian interglacial: New data from the grande pile record (GP XXI). In: *Developments in Quaternary Sciences* [F. Sirocko, M. Claussen, M. F. Sánchez Goñi, and T. Litt (eds.)]. Elsevier, Philadelphia, PA, USA, pp. 231–238.

Routson, C. C., C. A. Woodhouse, and J. T. Overpeck, 2011: Second century megadrought in the Rio Grande headwaters, Colorado: How unusual was medieval drought? *Geophys. Res. Lett.*, **38**, L22703.

Royer, D. L., 2003: Estimating latest Cretaceous and Tertiary atmospheric CO_2 from stomatal indices. In: *Causes and Consequences of Globally Warm Climates in the Early Paleogen* [S. L. Wing, P. D. Gingerich, B. Schmitz and E. Thomas (eds.)]. Geological Society of America Special Paper 369, pp.79–93.

Royer, D. L., R. A. Berner, and D. J. Beerling, 2001a: Phanerozoic atmospheric CO_2 change: evaluating geochemical and paleobiological approaches. *Earth Sci. Rev.*, **54**, 349–392.

Royer, D. L., S. L. Wing, D. J. Beerling, D. W. Jolley, P. L. Koch, L. J. Hickey, and R. A. Berner, 2001b: Paleobotanical evidence for near present-day levels of atmospheric CO_2 during part of the Tertiary. *Science*, **292**, 2310–2313.

Rupper, S., G. Roe, and A. Gillespie, 2009: Spatial patterns of Holocene glacier advance and retreat in Central Asia. *Quat. Res.*, **72**, 337–346.

Russell, J., H. Eggermont, R. Taylor, and D. Verschuren, 2009: Paleolimnological records of recent glacier recession in the Rwenzori Mountains, Uganda-D. R. Congo. *J. Paleolimnol.*, **41**, 253–271.

Ruth, U., et al., 2007: Ice core evidence for a very tight link between North Atlantic and east Asian glacial climate. *Geophys. Res. Lett.*, **34**, L03706.

Sachs, J. P., D. Sachse, R. H. Smittenberg, Z. H. Zhang, D. S. Battisti, and S. Golubic, 2009: Southward movement of the Pacific intertropical convergence zone AD 1400–1850. *Nature Geosci.*, **2**, 519–525.

Saenger, C., A. Cohen, D. Oppo, R. Halley, and J. Carilli, 2009: Surface-temperature trends and variability in the low-latitude North Atlantic since 1552. *Nature Geosci.*, **2**, 492–495.

Saenko, O. A., A. Schmittner, and A. J. Weaver, 2004: The Atlantic-Pacific Seesaw. *J. Clim.*, **17**, 2033–2038.

Salisbury, E. J., 1928: On the causes and ecological significance of stomatal frequency, with special reference to the woodland flora. *Philos. Trans. R. Soc. B*, **216**, 1–65.

Salzer, M., and K. Kipfmueller, 2005: Reconstructed temperature and precipitation on a millennial timescale from tree-rings in the Southern Colorado Plateau, USA. *Clim. Change*, **70**, 465–487.

Salzmann, U., A. M. Haywood, D. J. Lunt, P. J. Valdes, and D. J. Hill, 2008: A new global biome reconstruction and data-model comparison for the Middle Pliocene. *Global Ecol. Biogeogr.*, **17**, 432–447.

Sarnthein, M., U. Pflaumann, and M. Weinelt, 2003a: Past extent of sea ice in the northern North Atlantic inferred from foraminiferal paleotemperature estimates. *Paleoceanography*, **18**, 1047.

5

Sarnthein, M., S. Van Kreveld, H. Erlenkeuser, P. Grootes, M. Kucera, U. Pflaumann, and M. Schulz, 2003b: Centennial-to-millennial-scale periodicities of Holocene climate and sediment injections off the western Barents shelf, 75°N. *Boreas*, **32**, 447–461.

Schaefer, J. M., et al., 2009: High-frequency Holocene glacier fluctuations in New Zealand differ from the northern signature. *Science*, **324**, 622–625.

Scherer, D., M. Gude, M. Gempeler, and E. Parlow, 1998: Atmospheric and hydrological boundary conditions for slushflow initiation due to snowmelt. *Ann. Glaciol.*, **26**, 377–380.

Schilt, A., M. Baumgartner, T. Blunier, J. Schwander, R. Spahni, H. Fischer, and T. F. Stocker, 2010: Glacial–interglacial and millennial-scale variations in the atmospheric nitrous oxide concentration during the last 800,000 years. *Quat. Sci. Rev.*, **29**, 182–192.

Schmidt, A., T. Thordarson, L. D. Oman, A. Robock, and S. Sell, 2012a: Climatic impact of the long-lasting 1783 Laki eruption: Inapplicability of mass-independent sulfur isotopic composition measurements. *J. Geophys. Res.*, **117**, D23116.

Schmidt, G. A., et al., 2011: Climate forcing reconstructions for use in PMIP simulations of the last millennium (v1.0). *Geoscientif. Model Dev.*, **4**, 33–45.

Schmidt, G. A., et al., 2012b: Climate forcing reconstructions for use in PMIP simulations of the Last Millennium (v1.1). *Geoscientif. Model Dev.*, **5**, 185–191.

Schmittner, A., E. D. Galbraith, S. W. Hostetler, T. F. Pedersen, and R. Zhang, 2007: Large fluctuations of dissolved oxygen in the Indian and Pacific oceans during Dansgaard-Oeschger oscillations caused by variations of North Atlantic Deep Water subduction. *Paleoceanography*, **22**, PA3207.

Schmittner, A., et al., 2011: Climate sensitivity estimated from temperature reconstructions of the Last Glacial Maximum. *Science*, **334**, 1385–1388.

Schneider, B., G. Leduc, and W. Park, 2010: Disentangling seasonal signals in Holocene climate trends by satellite-model-proxy integration. *Paleoceanography*, **25**, PA4217.

Schneider, D. P., C. M. Ammann, B. L. Otto-Bliesner, and D. S. Kaufman, 2009: Climate response to large, high-latitude and low-latitude volcanic eruptions in the Community Climate System Model. *J. Geophys. Res.*, **114**, D15101.

Schneider Mor, A., R. Yam, C. Bianchi, M. Kunz-Pirrung, R. Gersonde, and A. Shemesh, 2012: Variable sequence of events during the past seven terminations in two deep-sea cores from the Southern Ocean. *Quat. Res.*, **77**, 317–325.

Schneider von Deimling, T., H. Held, A. Ganopolski, and S. Rahmstorf, 2006: Climate sensitivity estimated from ensemble simulations of glacial climate. *Clim. Dyn.*, **27**, 149–163.

Schoof, C., 2012: Marine ice sheet stability. *J. Fluid Mech.*, **698**, 62–72.

Schrijver, C. J., W. C. Livingston, T. N. Woods, and R. A. Mewaldt, 2011: The minimal solar activity in 2008–2009 and its implications for long-term climate modeling. *Geophys. Res. Lett.*, **38**, L06701.

Schulz, H., U. von Rad, and H. Erlenkeuser, 1998: Correlation between Arabian Sea and Greenland climate oscillations of the past 110,000 years. *Nature*, **393**, 54–57.

Schurer, A., G. C. Hegerl, M. E. Mann, S. F. B. Tett, and S. J. Phipps, 2013: Separating forced from chaotic climate variability over the past millennium. *J. Clim.*, **26**, 6954-6973.

Schurgers, G., U. Mikolajewicz, M. Gröger, E. Maier-Reimer, M. Vizcaino, and A. Winguth, 2007: The effect of land surface changes on Eemian climate. *Clim. Dyn.*, **29**, 357–373.

Screen, J. A., and I. Simmonds, 2010: The central role of diminishing sea ice in recent Arctic temperature amplification. *Nature*, **464**, 1334–1337.

Scroxton, N., S. G. Bonham, R. E. M. Rickaby, S. H. F. Lawrence, M. Hermoso, and A. M. Haywood, 2011: Persistent El Niño-Southern Oscillation variation during the Pliocene Epoch. *Paleoceanography*, **26**, PA2215.

Seager, R., N. Graham, C. Herweijer, A. Gordon, Y. Kushnir, and E. Cook, 2007: Blueprints for Medieval hydroclimate. *Quat. Sci. Rev.*, **26**, 2322–2336.

Seki, O., G. L. Foster, D. N. Schmidt, A. Mackensen, K. Kawamura, and R. D. Pancost, 2010: Alkenone and boron-based Pliocene pCO_2 records. *Earth Planet. Sci. Lett.*, **292**, 201–211.

Semenov, V. A., M. Latif, D. Dommenget, N. S. Keenlyside, A. Strehz, T. Martin, and W. Park, 2010: The Impact of North Atlantic-Arctic multidecadal variability on northern hemisphere surface air temperature. *J. Clim.*, **23**, 5668–5677.

Seong, Y., L. Owen, C. Yi, and R. Finkel, 2009: Quaternary glaciation of Muztag Ata and Kongur Shan: Evidence for glacier response to rapid climate changes throughout the Late Glacial and Holocene in westernmost Tibet. *Geol. Soc. Am. Bull.*, **129**, 348–365.

Serreze, M. C., and R. G. Barry, 2011: Processes and impacts of Arctic amplification: A research synthesis. *Global Planet. Change*, **77**, 85–96.

Serreze, M. C., A. P. Barrett, J. C. Stroeve, D. N. Kindig, and M. M. Holland, 2009: The emergence of surface-based Arctic amplification. *Cryosphere*, **3**, 11–19.

Servonnat, J., P. Yiou, M. Khodri, D. Swingedouw, and S. Denvil, 2010: Influence of solar variability, CO_2 and orbital forcing between 1000 and 1850 AD in the IPSLCM4 model. *Clim. Past*, **6**, 445–460.

Shackleton, N. J., 2000: The 100,000-year ice-age cycle identified and found to lag temperature, carbon dioxide, and orbital eccentricity. *Science*, **289**, 1897–1902.

Shackleton, N. J., M. F. Sánchez-Goñi, D. Pailler, and Y. Lancelot, 2003: Marine Isotope Substage 5e and the Eemian interglacial. *Global Planet. Change*, **36**, 151–155.

Shaffer, G., S. M. Olsen, and C. J. Bjerrum, 2004: Ocean subsurface warming as a mechanism for coupling Dansgaard-Oeschger climate cycles and ice-rafting events. *Geophys. Res. Lett.*, **31**, L24202.

Shakun, J. D., et al., 2012: Global warming preceded by increasing carbon dioxide concentrations during the last deglaciation. *Nature*, **484**, 49–54.

Shanahan, T. M., et al., 2009: Atlantic forcing of persistent drought in west Africa. *Science*, **324**, 377–380.

Shapiro, A. I., W. Schmutz, E. Rozanov, M. Schoell, M. Haberreiter, A. V. Shapiro, and S. Nyeki, 2011: A new approach to the long-term reconstruction of the solar irradiance leads to large historical solar forcing. *Astron. Astrophys.*, **529**, 1–8.

Sheffer, N. A., M. Rico, Y. Enzel, G. Benito, and T. Grodek, 2008: The Palaeoflood record of the Gardon River, France: A comparison with the extreme 2002 flood event. *Geomorphology*, **98**, 71–83.

Shevenell, A. E., A. E. Ingalls, E. W. Domack, and C. Kelly, 2011: Holocene Southern Ocean surface temperature variability west of the Antarctic Peninsula. *Nature*, **470**, 250–254.

Shi, F., et al., 2013: Northern hemisphere temperature reconstruction during the last millennium using multiple annual proxies. *Clim. Res.*, **56**, 231–244.

Shin, S. I., P. D. Sardeshmukh, R. S. Webb, R. J. Oglesby, and J. J. Barsugli, 2006: Understanding the mid-Holocene climate. *J. Clim.*, **19**, 2801–2817.

Shuman, B., P. Pribyl, T. A. Minckley, and J. J. Shinker, 2010: Rapid hydrologic shifts and prolonged droughts in Rocky Mountain headwaters during the Holocene. *Geophys. Res. Lett.*, **37**, L06701.

Sicre, M. A., et al., 2008: A 4500-year reconstruction of sea surface temperature variability at decadal time-scales off North Iceland. *Quat. Sci. Rev.*, **27**, 2041–2047.

Siddall, M., E. J. Rohling, W. G. Thompson, and C. Waelbroeck, 2008: Marine isotope stage 3 sea level fluctuations: Data synthesis and new outlook. *Rev. Geophys.*, **46**, RG4003.

Siddall, M., E. J. Rohling, T. Blunier, and R. Spahni, 2010: Patterns of millennial variability over the last 500 ka. *Clim. Past*, **6**, 295–303.

Siddall, M., T. F. Stocker, T. Blunier, R. Spahni, J. F. McManus, and E. Bard, 2006: Using a maximum simplicity paleoclimate model to simulate millennial variability during the last four glacial periods. *Quat. Sci. Rev.*, **25**, 3185–3197.

Siddall, M., E. J. Rohling, A. Almogi-Labin, C. Hemleben, D. Meischner, I. Schmelzer, and D. A. Smeed, 2003: Sea level fluctuations during the last glacial cycle. *Nature*, **423**, 853–858.

Siegenthaler, U., et al., 2005: Stable carbon cycle-climate relationship during the late Pleistocene. *Science*, **310**, 1313–1317.

Sierro, F. J., et al., 2009: Phase relationship between sea level and abrupt climate change. *Quat. Sci. Rev.*, **28**, 2867–2881.

Sigl, M., et al., 2013: A new bipolar ice core record of volcanism from WAIS Divide and NEEM and implications for climate forcing of the last 2000 years. *J. Geophys. Res.*, **118**, 1151-1169.

Sime, L. C., E. W. Wolff, K. I. C. Oliver, and J. C. Tindall, 2009: Evidence for warmer interglacials in East Antarctic ice cores. *Nature*, **462**, 342–345.

Sime, L. C., C. Risib, J. C. Tindall, J. Sjolted, E. W. Wolff, V. Masson-Delmotte, and E. Caprona, 2013: Warm climate isotopic simulations: What do we learn about interglacial signals in Greenland ice cores? *Quat. Sci. Rev.*, **67**, 59–80.

Simms, A. R., K. T. Milliken, J. B. Anderson, and J. S. Wellner, 2011: The marine record of deglaciation of the South Shetland Islands, Antarctica since the Last Glacial Maximum. *Quat. Sci. Rev.*, **30**, 1583–1601.

Singarayer, J. S., and P. J. Valdes, 2010: High-latitude climate sensitivity to ice-sheet forcing over the last 120 kyr. *Quat. Sci. Rev.*, **29**, 43–55.

Sinha, A., and L. D. Stott, 1994: New atmospheric pCO_2 estimates from palesols during the late Paleocene/early Eocene global warming interval. *Global Planet. Change*, **9**, 297–307.

Sinha, A., et al., 2007: A 900-year (600 to 1500 A.D.) record of the Indian summer monsoon precipitation from the core monsoon zone of India. *Geophys. Res. Lett.*, **34**, L16707.
Sivan, D., K. Lambeck, R. Toueg, A. Raban, Y. Porath, and B. Shirman, 2004: Ancient coastal wells of Caesarea Maritima, Israel, an indicator for relative sea level changes during the last 2000 years. *Earth Planet. Sci. Lett.*, **222**, 315–330.
Sluijs, A., et al., 2007: Environmental precursors to rapid light carbon injection at the Palaeocene/Eocene boundary. *Nature*, **450**, 1218–1221.
Smerdon, J. E., 2012: Climate models as a test bed for climate reconstruction methods: pseudoproxy experiments. *Rev. Clim. Change*, **3**, 63–77.
Smerdon, J. E., A. Kaplan, D. Chang, and M. N. Evans, 2010: A pseudoproxy evaluation of the CCA and RegEM methods for reconstructing climate fields of the last millennium. *J. Clim.*, **23**, 4856–4880.
Smerdon, J. E., A. Kaplan, E. Zorita, J. F. González-Rouco, and M. N. Evans, 2011: Spatial performance of four climate field reconstruction methods targeting the Common Era. *Geophys. Res. Lett.*, **38**, L11705.
Smith, J. A., et al., 2011: Deglacial history of the West Antarctic Ice Sheet in the western Amundsen Sea Embayment. *Quat. Sci. Rev.*, **30**, 488–505.
Smith, R., and J. Gregory, 2012: The last glacial cycle: Transient simulations with an AOGCM. *Clim. Dyn.*, **38**, 1545–1559.
Smith, R. Y., D. R. Greenwood, and J. F. Basinger, 2010: Estimating paleoatmospheric pCO_2 during the Early Eocene Climatic Optimum from stomatal frequency of Ginkgo, Okanagan Highlands, British Columbia, Canada. *Palaeogeogr. Palaeoclimatol. Palaeoecol.* **293**, 120–131.
Smithers, S. G., and C. D. Woodroffe, 2001: Coral microatolls and 20th century sea level in the eastern Indian Ocean. *Earth Planet. Sci. Lett.*, **191**, 173–184.
Soden, B. J., I. M. Held, R. Colman, K. M. Shell, J. T. Kiehl, and C. A. Shields, 2008: Quantifying climate feedbacks using radiative kernels. *J. Clim.*, **21**, 3504–3520.
Sokolov, S., and S. R. Rintoul, 2009: Circumpolar structure and distribution of the Antarctic Circumpolar Current fronts: 1. Mean circumpolar paths. *J. Geophys. Res.*, **114**, C11018.
Solanki, S. K., I. G. Usoskin, B. Kromer, M. Schussler, and J. Beer, 2004: Unusual activity of the Sun during recent decades compared to the previous 11,000 years. *Nature*, **431**, 1084–1087.
Sosdian, S., and Y. Rosenthal, 2009: Deep-sea temperature and ice volume changes across the Pliocene-Pleistocene climate transitions. *Science*, **325**, 306–310.
Sowers, T., and M. Bender, 1995: Climate records covering the Last Deglaciation. *Science*, **269**, 210–214.
Spence, J. P., M. Eby, and A. J. Weaver, 2008: The sensitivity of the Atlantic Meridional Overturning Circulation to freshwater forcing at eddy-permitting resolutions. *J. Clim.*, **21**, 2697–2710.
Spielhagen, R. F., et al., 2011: Enhanced modern heat transfer to the Arctic by warm Atlantic water. *Science*, **331**, 450–453.
St. George, S., et al., 2009: The tree-ring record of drought on the Canadian prairies. *J. Clim.*, **22**, 689–710.
Stager, J. C., D. Ryves, B. F. Cumming, L. D. Meeker, and J. Beer, 2005: Solar variability and the levels of Lake Victoria, East Africa, during the last millenium. *J. Paleolimnol.*, **33**, 243–251.
Stager, J. C., C. Cocquyt, R. Bonnefille, C. Weyhenmeyer, and N. Bowerman, 2009: A late Holocene paleoclimatic history of Lake Tanganyika, East Africa. *Quat. Res.*, **72**, 47–56.
Stahle, D. W., et al., 2011: Major Mesoamerican droughts of the past millennium. *Geophys. Res. Lett.*, **38**, L05703.
Stambaugh, M. C., R. P. Guyette, E. R. McMurry, E. R. Cook, D. M. Meko, and A. R. Lupo, 2011: Drought duration and frequency in the U.S. Corn Belt during the last millennium (AD 992–2004). *Agr. For. Meteorol.*, **151**, 154–162.
Stanford, J. D., E. J. Rohling, S. Bacon, A. P. Roberts, F. E. Grousset, and M. Bolshaw, 2011: A new concept for the paleoceanographic evolution of Heinrich event 1 in the North Atlantic. *Quat. Sci. Rev.*, **30**, 1047–1066.
Starkel, L., R. Soja, and D. J. Michczyńska, 2006: Past hydrological events reflected in Holocene history of Polish rivers. *CATENA*, **66**, 24–33.
Steffensen, J. P., et al., 2008: High-resolution Greenland ice core data show abrupt climate change happens in few years. *Science*, **321**, 680–684.
Steig, E. J., D. P. Schneider, S. D. Rutherford, M. E. Mann, J. C. Comiso, and D. T. Shindell, 2009: Warming of the Antarctic ice-sheet surface since the 1957 International Geophysical Year. *Nature*, **457**, 459–462.
Steig, E. J., et al., 2013: Recent climate and ice-sheet changes in West Antarctica compared with the past 2,000 years. *Nature Geosci.*, **6**, 372-375.
Steinhilber, F., J. Beer, and C. Fröhlich, 2009: Total solar irradiance during the Holocene. *Geophys. Res. Lett.*, **36**, L19704.
Steinhilber, F., et al., 2012: 9,400 years of cosmic radiation and solar activity from ice cores and tree rings. *Proc. Natl. Acad. Sci. U.S.A.*, **109**, 5967-5971.
Steinke, S., M. Kienast, J. Groeneveld, L.-C. Lin, M.-T. Chen, and R. Rendle-Bühring, 2008: Proxy dependence of the temporal pattern of deglacial warming in the tropical South China Sea: toward resolving seasonality. *Quat. Sci. Rev.*, **27**, 688–700.
Steinman, B. A., M. B. Abbott, M. E. Mann, N. D. Stansell, and B. P. Finney, 2013: 1,500 year quantitative reconstruction of winter precipitation in the Pacific Northwest. *Proc. Natl. Acad. Sci. U.S.A.*, **109**, 11619-11623.
Stendel, M., I. Mogensen, and J. Christensen, 2006: Influence of various forcings on global climate in historical times using a coupled atmosphere–ocean general circulation model. *Clim. Dyn.*, **26**, 1–15.
Stenni, B., et al., 2010: The deuterium excess records of EPICA Dome C and Dronning Maud Land ice cores (East Antarctica). *Quat. Sci. Rev.*, **29**, 146–159.
Stenni, B., et al., 2011: Expression of the bipolar see-saw in Antarctic climate records during the last deglaciation. *Nature Geosci.*, **4**, 46–49.
Steph, S., et al., 2010: Early Pliocene increase in thermohaline overturning: A precondition for the development of the modern equatorial Pacific cold tongue. *Paleoceanography*, **25**, PA2202.
Stewart, M. M., I. Larocque-Tobler, and M. Grosjean, 2011: Quantitative inter-annual and decadal June–July–August temperature variability ca. 570 BC to AD 120 (Iron Age–Roman Period) reconstructed from the varved sediments of Lake Silvaplana, Switzerland. *J. Quat. Sci.*, **26**, 491–501.
Stirling, C., T. Esat, K. Lambeck, and M. McCulloch, 1998: Timing and duration of the Last Interglacial: Evidence for a restricted interval of widespread coral reef growth. *Earth Planet. Sci. Lett.*, **160**, 745–762.
Stocker, T., and S. Johnsen, 2003: A minimum thermodynamic model for the bipolar seesaw. *Paleoceanography*, **18**, 1087.
Stone, E. J., D. J. Lunt, J. D. Annan, and J. C. Hargreaves, 2013: Quantification of the Greenland ice sheet contribution to Last Interglacial sea level rise. *Clim. Past*, **9**, 621–639.
Stone, J. O., G. A. Balco, D. E. Sugden, M. W. Caffee, L. C. Sass, S. G. Cowdery, and C. Siddoway, 2003: Holocene Deglaciation of Marie Byrd Land, West Antarctica. *Science*, **299**, 99–102.
Stott, L., A. Timmermann, and R. Thunell, 2007: Southern hemisphere and deep-sea warming led deglacial atmospheric CO_2 rise and tropical warming. *Science*, **318**, 435–438.
Stott, L., K. Cannariato, R. Thunell, G. H. Haug, A. Koutavas, and S. Lund, 2004: Decline of surface temperature and salinity in the western tropical Pacific Ocean in the Holocene epoch. *Nature*, **431**, 56–59.
Stott, L. D., 1992: Higher temperatures and lower oceanic pCO_2: A climate enigma at the end of the Paleocene epoch. *Paleoceanography*, **7**, 395–404.
Stríkis, N. M., et al., 2011: Abrupt variations in South American monsoon rainfall during the Holocene based on a speleothem record from central-eastern Brazil. *Geology*, **39**, 1075–1078.
Stuiver, M., and T. F. Braziunas, 1993: Sun, ocean, climate and atmospheric $^{14}CO_2$: An evaluation of causal and spectral relationships. *Holocene*, **3**, 289–305.
Sundqvist, H. S., Q. Zhang, A. Moberg, K. Holmgren, H. Körnich, J. Nilsson, and G. Brattström, 2010: Climate change between the mid and late Holocene in northern high latitudes - Part 1: survey of temperature and precipitation proxy data. *Clim. Past*, **6**, 591–608.
Svalgaard, L., and E. W. Cliver, 2010: Heliospheric magnetic field 1835–2009. *J. Geophys. Res.*, **115**, A09111.
Svensson, A., et al., 2008: A 60 000 year Greenland stratigraphic ice core chronology. *Clim. Past*, **4**, 47–57.
Swingedouw, D., J. Mignot, P. Braconnot, E. Mosquet, M. Kageyama, and R. Alkama, 2009: Impact of freshwater release in the North Atlantic under different climate conditions in an OAGCM. *J. Clim.*, **22**, 6377–6403.
Swingedouw, D., L. Terray, C. Cassou, A. Voldoire, D. Salas-Mélia, and J. Servonnat, 2011: Natural forcing of climate during the last millennium: Fingerprint of solar variability. *Clim. Dyn.*, **36**, 1349–1364.
Takemura, T., M. Egashira, K. Matsuzawa, H. Ichijo, R. O'ishi, and A. Abe-Ouchi, 2009: A simulation of the global distribution and radiative forcing of soil dust aerosols at the Last Glacial Maximum. *Atmos. Chem. Phys.*, **9**, 3061–3073.

Tan, L., Y. Cai, R. Edwards, H. Cheng, C. Shen, and H. Zhang, 2011: Centennial- to decadal-scale monsoon precipitation variability in the semi-humid region, northern China during the last 1860 years: Records from stalagmites in Huangye Cave. *Holocene*, **21**, 287–296.

Tarasov, L., and W. R. Peltier, 2007: Coevolution of continental ice cover and permafrost extent over the last glacial-interglacial cycle in North America. *J. Geophys. Res.*, **112**, F02S08.

Tarasov, P., W. Granoszewski, E. Bezrukova, S. Brewer, M. Nita, A. Abzaeva, and H. Oberhänsli, 2005: Quantitative reconstruction of the last interglacial vegetation and climate based on the pollen record from Lake Baikal, Russia. *Clim. Dyn.*, **25**, 625–637.

Tarasov, P. E., et al., 2011: Progress in the reconstruction of Quaternary climate dynamics in the Northwest Pacific: A new modern analogue reference dataset and its application to the 430-kyr pollen record from Lake Biwa. *Earth Sci. Rev.*, **108**, 64–79.

Taylor, K. E., R. J. Stouffer, and G. A. Meehl, 2012: An overview of CMIP5 and the experiment design. *Bull. Am. Meteorol. Soc.*, **93**, 485–498.

Telford, R. J., C. Li, and M. Kucera, 2013: Mismatch between the depth habitat of planktonic foraminifera and the calibration depth of SST transfer functions may bias reconstructions. *Clim. Past*, **9**, 859–870.

Tett, S., et al., 2007: The impact of natural and anthropogenic forcings on climate and hydrology since 1550. *Clim. Dyn.*, **28**, 3–34.

Thomas, E., and J. Briner, 2009: Climate of the past millennium inferred from varved proglacial lake sediments on northeast Baffin Island, Arctic Canada. *J. Paleolimnol.*, **41**, 209–224.

Thomas, E. R., E. W. Wolff, R. Mulvaney, S. J. Johnsen, J. P. Steffensen, and C. Arrowsmith, 2009: Anatomy of a Dansgaard-Oeschger warming transition: high-resolution analysis of the North Greenland Ice Core Project ice core. *J. Geophys. Res.*, **114**, D08102.

Thomas, E. R., et al., 2007: The 8.2 ka event from Greenland ice cores. *Quat. Sci. Rev.*, **26**, 70–81.

Thompson, D. W. J., and S. Solomon, 2002: Interpretation of recent southern hemisphere climate change. *Science*, **296**, 895–899.

Thompson, D. W. J., and S. Solomon, 2009: Understanding recent stratospheric climate change. *J. Clim.*, **22**, 1934–1943.

Thompson, W. G., and S. L. Goldstein, 2005: Open-system coral ages reveal persistent suborbital sea level cycles. *Science*, **308**, 401–404.

Thompson, W. G., H. Allen Curran, M. A. Wilson, and B. White, 2011: Sea level oscillations during the last interglacial highstand recorded by Bahamas corals. *Nature Geosci.*, **4**, 684–687.

Thordarson, T., and S. Self, 2003: Atmospheric and environmental effects of the 1783–1784 Laki eruption: A review and reassessment. *J. Geophys. Res.*, **108**, D14011.

Tierney, J., M. Mayes, N. Meyer, C. Johnson, P. Swarzenski, A. Cohen, and J. Russell, 2010: Late-twentieth-century warming in Lake Tanganyika unprecedented since AD 500. *Nature Geosci.*, **3**, 422–425.

Tierney, J. E., S. C. Lewis, B. I. Cook, A. N. LeGrande, and G. A. Schmidt, 2011: Model, proxy and isotopic perspectives on the east African humid period. *Earth Planet. Sci. Lett.*, **307**, 103–112.

Tiljander, M. I. A., M. Saarnisto, A. E. K. Ojala, and T. Saarinen, 2003: A 3000–year palaeoenvironmental record from annually laminated sediment of Lake Korttajarvi, central Finland. *Boreas*, **32**, 566–577.

Timm, O., E. Ruprecht, and S. Kleppek, 2004: Scale-dependent reconstruction of the NAO index. *J. Clim.*, **17**, 2157–2169.

Timm, O., A. Timmermann, A. Abe-Ouchi, F. Saito, and T. Segawa, 2008: On the definition of seasons in paleoclimate simulations with orbital forcing. *Paleoceanography*, **23**, PA2221.

Timmermann, A., H. Gildor, M. Schulz, and E. Tziperman, 2003: Coherent resonant millennial-scale climate oscillations triggered by massive meltwater pulses. *J. Clim.*, **16**, 2569–2585.

Timmermann, A., O. Timm, L. Stott, and L. Menviel, 2009: The roles of CO_2 and orbital forcing in driving southern hemispheric temperature variations during the last 21 000 yr. *J. Clim.*, **22**, 1626–1640.

Timmermann, A., F. Justino, F. F. Jin, U. Krebs, and H. Goosse, 2004: Surface temperature control in the North and tropical Pacific during the last glacial maximum. *Clim. Dyn.*, **23**, 353–370.

Timmermann, A., et al., 2010: Towards a quantitative understanding of millennial-scale Antarctic warming events. *Quat. Sci. Rev.*, **29**, 74–85.

Timmermann, A., et al., 2007: The influence of a weakening of the Atlantic meridional overturning circulation on ENSO. *J. Clim.*, **20**, 4899–4919.

Timmreck, C., S. J. Lorenz, T. J. Crowley, S. Kinne, T. J. Raddatz, M. A. Thomas, and J. H. Jungclaus, 2009: Limited temperature response to the very large AD 1258 volcanic eruption. *Geophys. Res. Lett.*, **36**, L21708.

Tingley, M. P., and P. Huybers, 2010: A Bayesian algorithm for reconstructing climate anomalies in space and time. Part I: development and applications to paleoclimate reconstruction problems. *J. Clim.*, **23**, 2759–2781.

Tingley, M. P., and P. Huybers, 2013: Recent temperature extremes at high northern latitudes unprecedented in the past 600 years. *Nature*, **496**, 201–205.

Tingley, M. P., P. F. Craigmile, M. Haran, B. Li, E. Mannshardt, and B. Rajaratnam, 2012: Piecing together the past: statistical insights into paleoclimatic reconstructions. *Quat. Sci. Rev.*, **35**, 1–22.

Tjallingii, R., et al., 2008: Coherent high- and low-latitude control of the northwest African hydrological balance. *Nature Geosci.*, **1**, 670–675.

Torrence, C., and G. P. Compo, 1998: A practical guide to wavelet analysis. *Bull. Am. Meteorol. Soc.*, **79**, 61–78.

Touchan, R., K. J. Anchukaitis, D. M. Meko, S. Attalah, C. Baisan, and A. Aloui, 2008: Long term context for recent drought in northwestern Africa. *Geophys. Res. Lett.*, **35**, L13705.

Touchan, R., K. Anchukaitis, D. Meko, M. Sabir, S. Attalah, and A. Aloui, 2011: Spatiotemporal drought variability in northwestern Africa over the last nine centuries. *Clim. Dyn.*, **37**, 237–252.

Trachsel, M., et al., 2012: Multi-archive summer temperature reconstruction for the European Alps, AD 1053–1996. *Quat. Sci. Rev.*, **46**, 66–79.

Tripati, A. K., C. D. Roberts, and R. A. Eagle, 2009: Coupling of CO_2 and ice sheet stability over major climate transitions of the last 20 million years. *Science*, **326**, 1394–1397.

Trouet, V., J. Esper, N. E. Graham, A. Baker, J. D. Scourse, and D. C. Frank, 2009: Persistent positive north Atlantic oscillation mode dominated the Medieval Climate Anomaly. *Science*, **324**, 78–80.

Turney, C. S. M., and R. T. Jones, 2010: Does the Agulhas Current amplify global temperatures during super-interglacials? *J. Quat. Sci.*, **25**, 839–843.

Tzedakis, P. C., H. Hooghiemstra, and H. Pälike, 2006: The last 1.35 million years at Tenaghi Philippon: revised chronostratigraphy and long-term vegetation trends. *Quat. Sci. Rev.*, **25**, 3416–3430.

Tzedakis, P. C., J. E. T. Channell, D. A. Hodell, H. F. Kleiven, and L. C. Skinner, 2012a: Determining the natural length of the current interglacial. *Nature Geosci.*, **5**, 138–141.

Tzedakis, P. C., D. Raynaud, J. F. McManus, A. Berger, V. Brovkin, and T. Kiefer, 2009: Interglacial diversity. *Nature Geosci.*, **2**, 751–755.

Tzedakis, P. C., E. W. Wolff, L. C. Skinner, V. Brovkin, D. A. Hodell, J. F. McManus, and D. Raynaud, 2012b: Can we predict the duration of an interglacial? *Clim. Dyn.*, **8**, 1473–1485.

Uemura, R., V. Masson-Delmotte, J. Jouzel, A. Landais, H. Motoyama, and B. Stenni, 2012: Ranges of moisture-source temperature estimated from Antarctic ice cores stable isotope records over glacial–interglacial cycles. *Clim. Past*, **8**, 1109–1125.

Unterman, M. B., T. J. Crowley, K. I. Hodges, S. J. Kim, and D. J. Erickson, 2011: Paleometeorology: High resolution Northern Hemisphere wintertime mid-latitude dynamics during the Last Glacial Maximum. *Geophys. Res. Lett.*, **38**, L23702.

Urrutia, R., A. Lara, R. Villalba, D. Christie, C. Le Quesne, and A. Cuq, 2011: Multicentury tree ring reconstruction of annual streamflow for the Maule River watershed in south central Chile. *Water Resourc. Res.*, **47**, W06527.

van de Berg, W. J., M. van den Broeke, J. Ettema, E. van Meijgaard, and F. Kaspar, 2011: Significant contribution of insolation to Eemian melting of the Greenland ice sheet. *Nature Geosci.*, **4**, 679–683.

van de Plassche, O., K. van der Borg, and A. F. M. de Jong, 1998: Sea level–climate correlation during the past 1400 yr. *Geology*, **26**, 319–322.

van den Berg, J., R. S. W. van de Wal, G. A. Milne, and J. Oerlemans, 2008: Effect of isostasy on dynamical ice sheet modeling: A case study for Eurasia. *J. Geophys. Res.*, **113**, B05412.

van der Burgh, J., H. Visscher, D. L. Dilcher, and W. M. Kürschner, 1993: Paleoatmospheric signatures in Neogene fossil leaves. *Science*, **260**, 1788–1790.

van Leeuwen, R. J., et al., 2000: Stratigraphy and integrated facies analysis of the Saalian and Eemian sediments in the Amsterdam-Terminal borehole, the Netherlands. *Geolog.Mijnbouw / Netherlands J. Geosci.*, **79**, 161–196.

Varma, V., et al., 2012: Holocene evolution of the Southern Hemisphere westerly winds in transient simulations with global climate models. *Clim. Past*, **8**, 391–402.

Vasskog, K., Ø. Paasche, A. Nesje, J. F. Boyle, and H. J. B. Birks, 2012: A new approach for reconstructing glacier variability based on lake sediments recording input from more than one glacier. *Quat. Res.*, **77**, 192–204.

Vaughan, D. G., D. K. A. Barnes, P. T. Fretwell, and R. G. Bingham, 2011: Potential seaways across West Antarctica. *Geochem., Geophys., Geosyst.*, **12**, Q10004.

Vavrus, S., 2004: The impact of cloud feedbacks on Arctic climate under greenhouse forcing. *J. Clim.*, **17**, 603–615.

Velichko, A. A., O. K. Borisova, and E. M. Zelikson, 2008: Paradoxes of the Last Interglacial climate: Reconstruction of the northern Eurasia climate based on palaeofloristic data. *Boreas*, **37**, 1–19.

Verleyen, E., et al., 2011: Post-glacial regional climate variability along the East Antarctic coastal margin—Evidence from shallow marine and coastal terrestrial records. *Earth Sci. Rev.*, **104**, 199–212.

Verschuren, D., K. Laird, and B. Cumming, 2000: Rainfall and drought in equatorial East Africa during the past 1000 years. *Nature*, **403**, 410–414.

Verschuren, D., J. S. Sinninghe Damste, J. Moernaut, I. Kristen, M. Blaauw, M. Fagot, and G. H. Haug, 2009: Half-precessional dynamics of monsoon rainfall near the East African Equator. *Nature*, **462**, 637–641.

Vettoretti, G., and W. R. Peltier, 2011: The impact of insolation, greenhouse gas forcing and ocean circulation changes on glacial inception. *Holocene*, **21**, 803–817.

Viau, A. E., M. Ladd, and K. Gajewski, 2012: The climate of North America during the past 2000–years reconstructed from pollen data. *Global Planet. Change*, **84–85**, 75–83.

Vieira, L. E., S. K. Solanki , A. V. Krivov, and I. G. Usoskin 2011: Evolution of the solar irradiance during the Holocene. *Astron. Astrophys.*, **531**, A6.

Vieira, L. E. A., and S. K. Solanki, 2010: Evolution of the solar magnetic flux on time scales of years to millenia. *Astron. Astrophys.*, **509**, A100.

Villalba, R., M. Grosjean, and T. Kiefer, 2009: Long-term multi-proxy climate reconstructions and dynamics in South America (LOTRED-SA): State of the art and perspectives. *Palaeogeogr. Palaeoclimatol. Palaeoecol.*, **281**, 175–179.

Villalba, R., et al., 2012: Unusual Southern Hemisphere tree growth patterns induced by changes in the Southern Annular Mode. *Nature Geosci.*, **5**, 793–798.

Vimeux, F., P. Ginot, M. Schwikowski, M. Vuille, G. Hoffmann, L. G. Thompson, and U. Schotterer, 2009: Climate variability during the last 1000 years inferred from Andean ice cores: A review of methodology and recent results. *Palaeogeogr. Palaeoclimatol. Palaeoecol.*, **281**, 229–241.

Vinther, B., P. Jones, K. Briffa, H. Clausen, K. Andersen, D. Dahl-Jensen, and S. Johnsen, 2010: Climatic signals in multiple highly resolved stable isotope records from Greenland. *Quat. Sci. Rev.*, **29**, 522–538.

Vinther, B. M., et al., 2009: Holocene thinning of the Greenland ice sheet. *Nature*, **461**, 385–388.

von Grafenstein, U., E. Erlenkeuser, J. Müller, J. Jouzel, and S. Johnsen, 1998: The cold event 8,200 years ago documented in oxygen isotope records of precipitation in Europe and Greenland. *Clim. Dyn.*, **14**, 73–81.

von Gunten, L., M. Grosjean, B. Rein, R. Urrutia, and P. Appleby, 2009: A quantitative high-resolution summer temperature reconstruction based on sedimentary pigments from Laguna Aculeo, central Chile, back to AD 850. *Holocene*, **19**, 873–881.

von Königswald, W., 2007: Mammalian faunas from the interglacial periods in Central Europe and their stratigraphic correlation. In: *Developments in Quaternary Science* [F. Sirocko, M. Claussen, M. F. Sánchez Goñi and T. Litt (eds.)]. Elsevier, Philadelphia, PA, USA, pp. 445–454.

von Storch, H., E. Zorita, J. Jones, F. González-Rouco, and S. Tett, 2006: Response to comment on "Reconstructing past climate from noisy data". *Science*, **312**, 1872–1873.

Vuille, M., et al., 2012: A review of the South American Monsoon history as recorded in stable isotopic proxies over the past two millennia. *Clim. Past*, **8**, 1309–1321.

Waelbroeck, C., et al., 2002: Sea level and deep water temperature changes derived from benthic foraminifera isotopic records. *Quat. Sci. Rev.*, **21**, 295–305.

Wagner, J. D. M., J. E. Cole, J. W. Beck, P. J. Patchett, G. M. Henderson, and H. R. Barnett, 2010: Moisture variability in the southwestern United States linked to abrupt glacial climate change. *Nature Geosci.*, **3**, 110–113.

Wagner, S., et al., 2007: Transient simulations, empirical reconstructions and forcing mechanisms for the Mid-holocene hydrological climate in southern Patagonia. *Clim. Dyn.*, **29**, 333–355.

Wahl, E., et al., 2010: An archive of high-resolution temperature reconstructions over the past 2+ millennia. *Geochem. Geophys. Geosyst.*, **11**, Q01001.

Wahl, E. R., and J. E. Smerdon, 2012: Comparative performance of paleoclimate field and index reconstructions derived from climate proxies and noise-only predictors. *Geophys. Res. Lett.*, **39**, L06703.

Wahl, E. R., D. M. Ritson, and C. M. Ammann, 2006: Comment on "Reconstructing past climate from noisy data". *Science*, **312**, 529.

Walter, K. M., S. A. Zimov, J. P. Chanton, D. Verbyla, and F. S. Chapin, 2006: Methane bubbling from Siberian thaw lakes as a positive feedback to climate warming. *Nature*, **443**, 71–75.

Wan, S., J. Tian, S. Steinke, A. Li, and T. Li, 2010: Evolution and variability of the East Asian summer monsoon during the Pliocene: Evidence from clay mineral records of the South China Sea. *Palaeogeogr. Palaeoclimatol. Palaeoecol.* **293**, 237–247.

Wang, B., and Q. Ding, 2008: Global monsoon: Dominant mode of annual variation in the tropics. *Dyn. Atmos. Oceans*, **44**, 165–183.

Wang, S., X. Wen, Y. Luo, W. Dong, Z. Zhao, and B. Yang, 2007: Reconstruction of temperature series of China for the last 1000 years. *Chin. Sci. Bull.*, **52**, 3272–3280.

Wang, Y. J., H. Cheng, R. L. Edwards, Z. S. An, J. Y. Wu, C. C. Shen, and J. A. Dorale, 2001: A high-resolution absolute-dated Late Pleistocene monsoon record from Hulu Cave, China. *Science*, **294**, 2345–2348.

Wang, Y. J., et al., 2008: Millennial- and orbital-scale changes in the East Asian monsoon over the past 224,000 years. *Nature*, **451**, 1090–1093.

Wang, Y. M., J. Lean, and N. Sheeley, 2005: Modeling the Sun's magnetic field and irradiance since 1713. *Astrophys. J.*, **625**, 522–538.

Wang, Y. M., S. L. Li, and D. H. Luo, 2009: Seasonal response of Asian monsoonal climate to the Atlantic Multidecadal Oscillation. *J. Geophys. Res.*, **114**, D02112.

Wanner, H., O. Solomina, M. Grosjean, S. P. Ritz, and M. Jetel, 2011: Structure and origin of Holocene cold events. *Quat. Sci. Rev.*, **30**, 3109–3123.

Wanner, H., et al., 2008: Mid- to Late Holocene climate change: an overview. *Quat. Sci. Rev.*, **27**, 1791–1828.

Waple, A. M., M. E. Mann, and R. S. Bradley, 2002: Long-term patterns of solar irradiance forcing in model experiments and proxy based surface temperature reconstructions. *Clim. Dyn.*, **18**, 563–578.

Wara, M. W., A. C. Ravelo, and M. L. Delaney, 2005: Permanent El Niño-like conditions during the Pliocene Warm Period. *Science*, **309**, 758–761.

Watanabe, O., J. Jouzel, S. Johnsen, F. Parrenin, H. Shoji, and N. Yoshida, 2003: Homogeneous climate variability across East Antarctica over the past three glacial cycles. *Nature*, **422**, 509–512.

Watanabe, T., et al., 2011: Permanent El Niño during the Pliocene warm period not supported by coral evidence. *Nature*, **471**, 209–211.

Weber, S. L., et al., 2007: The modern and glacial overturning circulation in the Atlantic ocean in PMIP coupled model simulations. *Clim. Past*, **3**, 51–64.

Wegmüller, S., 1992: *Vegetationsgeschichtliche und stratigraphische Untersuchungen an Schieferkohlen des nördlichen Alpenvorlandes.* Denkschriften der Schweizerischen Akademie der Naturwissenschaften, 102, Birkhauser, Basel, 445–454 pp.

Wegner, A., et al., 2012: Change in dust variability in the Atlantic sector of Antarctica at the end of the last deglaciation. *Clim. Past*, **8**, 135–147.

Wei, L. J., E. Mosley-Thompson, P. Gabrielli, L. G. Thompson, and C. Barbante, 2008: Synchronous deposition of volcanic ash and sulfate aerosols over Greenland in 1783 from the Laki eruption (Iceland). *Geophys. Res. Lett.*, **35**, L16501.

Weldeab, S., 2012: Bipolar modulation of millennial-scale West African monsoon variability during the last glacial (75,000–25,000 years ago). *Quat. Sci. Rev.*, **40**, 21–29.

Weldeab, S., D. W. Lea, R. R. Schneider, and N. Andersen, 2007a: 155,000 years of West African monsoon and ocean thermal evolution. *Science*, **316**, 1303–1307.

Weldeab, S., D. W. Lea, R. R. Schneider, and N. Andersen, 2007b: Centennial scale climate instabilities in a wet early Holocene West African monsoon. *Geophys. Res. Lett.*, **34**, L24702.

Welten, M., 1988: *Neue pollenanalytische Ergebnisse über das jüngere Quartär des nördlichen Alpenvorlandes der Schweiz (Mittel-und Jungpleistozän).* Beiträge zur Geologischen Karte der Schweiz, 162, Stämpfli, 40 pp.

Wenzler, T., S. Solanki, and N. Krivova, 2005: Can surface magnetic fields reproduce solar irradiance variations in cycles 22 and 23? *Astron. Astrophys.*, **432**, 1057–1061.

Wenzler, T., S. K. Solanki, N. A. Krivova, and C. Fröhlich, 2006: Reconstruction of solar irradiance variations in cycles 21–23 based on surface magnetic fields. *Astron. Astrophys.*, **460**, 583–595.

Werner, J. P., J. Luterbacher, and J. E. Smerdon, 2013: A pseudoproxy evaluation of Bayesian hierarchical modelling and canonical correlation analysis for climate field reconstructions over Europe. *J. Clim.*, **26**, 851–867.

Westerhold, T., U. Röhl, J. Laskar, I. Raffi, J. Bowles, L. J. Lourens, and J. C. Zachos, 2007: On the duration of magnetochrons C24r and C25n and the timing of early Eocene global warming events: Implications from the Ocean Drilling Program Leg 208 Walvis Ridge depth transect. *Paleoceanography*, **22**, PA2201.

Whitehouse, P. L., M. J. Bentley, G. A. Milne, M. A. King, and I. D. Thomas, 2012: A new glacial isostatic adjustment model for Antarctica: Calibrated and tested using observations of relative sea level change and present-day uplift rates. *Geophys. J. Int.*, **190**, 1464–1482.

Widmann, M., H. Goosse, G. van der Schrier, R. Schnur, and J. Barkmeijer, 2010: Using data assimilation to study extratropical Northern Hemisphere climate over the last millennium. *Clim. Past*, **6**, 627–644.

Wiersma, A., D. Roche, and H. Renssen, 2011: Fingerprinting the 8.2 ka event climate response in a coupled climate model. *J. Quat. Sci.*, **26**, 118–127.

Wiles, G. C., D. J. Barclay, P. E. Calkin, and T. V. Lowell, 2008: Century to millennial-scale temperature variations for the last two thousand years indicated from glacial geologic records of Southern Alaska. *Global Planet. Change*, **60**, 115–125.

Wiles, G. C., D. E. Lawson, E. Lyon, N. Wiesenberg, and R. D. D'Arrigo, 2011: Tree-ring dates on two pre-Little Ice Age advances in Glacier Bay National Park and Preserve, Alaska, USA. *Quat. Res.*, **76**, 190–195.

Wilhelm, B., et al., 2012: 1400 years of extreme precipitation patterns over the Mediterranean French Alps and possible forcing mechanisms. *Quat. Res.*, **78**, 1–12.

Wilmes, S. B., C. C. Raible, and T. F. Stocker, 2012: Climate variability of the mid- and high-latitudes of the Southern Hemisphere in ensemble simulations from 1500 to 2000 AD. *Clim. Past*, **8**, 373–390.

Wilson, M. F., and A. Henderson-Sellers, 1985: A global archive of land cover and soils data for use in general circulation climate models. *J. Climatol.*, **5**, 119–143.

Wilson, R., E. Cook, R. D'Arrigo, N. Riedwyl, M. N. Evans, A. Tudhope, and R. Allan, 2010: Reconstructing ENSO: the influence of method, proxy data, climate forcing and teleconnections. *J. Quat. Sci.*, **25**, 62–78.

Wilson, R., D. Miles, N. Loader, T. Melvin, L. Cunningham, R. Cooper, and K. Briffa, 2013: A millennial long march–july precipitation reconstruction for southern-central England. *Clim. Dyn.*, **40**, 997–1017.

Wilson, R., et al., 2007: A matter of divergence: Tracking recent warming at hemispheric scales using tree ring data. *J. Geophys. Res.*, **112**, D17103.

Winckler, G., R. F. Anderson, M. Q. Fleisher, D. McGee, and N. Mahowald, 2008: Covariant Glacial-Interglacial Dust Fluxes in the Equatorial Pacific and Antarctica. *Science*, **320**, 93–96.

Winkler, S., and J. Matthews, 2010: Holocene glacier chronologies: Are 'high-resolution' global and inter-hemispheric comparisons possible? *Holocene*, **20**, 1137–1147.

Winter, A., et al., 2011: Evidence for 800 years of North Atlantic multi-decadal variability from a Puerto Rican speleothem. *Earth Planet. Sci. Lett.*, **308**, 23–28.

Wolff, C., et al., 2011: Reduced interannual rainfall variability in east Africa during the Last Ice Age. *Science*, **333**, 743–747.

Wolff, E. W., et al., 2010: Changes in environment over the last 800,000 years from chemical analysis of the EPICA Dome C ice core. *Quat. Sci. Rev.*, **29**, 285–295.

Woodhouse, C. A., D. M. Meko, G. M. MacDonald, D. W. Stahle, and E. R. Cook, 2010: A 1,200–year perspective of 21st century drought in southwestern North America. *Proc. Natl. Acad. Sci. U.S.A.*, **107**, 21283–21288.

Woodroffe, C., and R. McLean, 1990: Microatolls and recent sea level change on coral atolls. *Nature*, **344**, 531–534.

Woodroffe, C. D., H. V. McGregor, K. Lambeck, S. G. Smithers, and D. Fink, 2012: Mid-Pacific microatolls record sea level stability over the past 5000 yr. *Geology*, **40**, 951–954.

Wunsch, C., 2006: Abrupt climate change: An alternative view. *Quat. Res.*, **65**, 191–203.

Xie, S. P., Y. Okumura, T. Miyama, and A. Timmermann, 2008: Influences of Atlantic climate change on the tropical Pacific via the Central American Isthmus. *J. Clim.*, **21**, 3914–3928.

Yadav, R., A. Braeuning, and J. Singh, 2011: Tree ring inferred summer temperature variations over the last millennium in western Himalaya, India. *Clim. Dyn.*, **36**, 1545–1554.

Yanase, W., and A. Abe-Ouchi, 2010: A numerical study on the atmospheric circulation over the midlatitude North Pacific during the Last Glacial Maximum. *J. Clim.*, **23**, 135–151.

Yang, B., A. Bräuning, Z. Dong, Z. Zhang, and J. Keqing, 2008: Late Holocene monsoonal temperate glacier fluctuations on the Tibetan Plateau. *Global Planet. Change*, **60**, 126–140.

Yang, B., J. Wang, A. Bräuning, Z. Dong, and J. Esper, 2009: Late Holocene climatic and environmental changes in and central Asia. *Quat. Int.*, **194**, 68–78.

Yin, Q., and A. Berger, 2012: Individual contribution of insolation and CO_2 to the interglacial climates of the past 800,000 years. *Clim. Dyn.*, **38**, 709–724.

Yin, Q. Z., and A. Berger, 2010: Insolation and CO_2 contribution to the interglacial climate before and after the Mid-Brunhes Event. *Nature Geosci.*, **3**, 243–246.

Yin, Q. Z., A. Berger, E. Driesschaert, H. Goosse, M. F. Loutre, and M. Crucifix, 2008: The Eurasian ice sheet reinforces the East Asian summer monsoon during the interglacial 500 000 years ago. *Clim. Past*, **4**, 79–90.

Yiou, P., J. Servonnat, M. Yoshimori, D. Swingedouw, M. Khodri, and A. Abe-Ouchi, 2012: Stability of weather regimes during the last millennium from climate simulations. *Geophys. Res. Lett.*, **39**, L08703.

Yokoyama, Y., and T. M. Esat, 2011: Global climate and sea level: Enduring variability and rapid fluctuations over the past 150,000 years. *Oceanography*, **24**, 54–69.

Yoshimori, M., T. Yokohata, and A. Abe-Ouchi, 2009: A comparison of climate feedback strength between CO_2 doubling and LGM experiments. *J. Clim.*, **22**, 3374–3395.

Yoshimori, M., J. C. Hargreaves, J. D. Annan, T. Yokohata, and A. Abe-Ouchi, 2011: Dependency of feedbacks on forcing and climate state in physics parameter ensembles. *J. Clim.*, **24**, 6440–6455.

Young, N. E., J. P. Briner, H. A. M. Stewart, Y. Axford, B. Csatho, D. H. Rood, and R. C. Finkel, 2011: Response of Jakobshavn Isbræ, Greenland, to Holocene climate change. *Geology*, **39**, 131–134.

Yue, X., H. Wang, H. Liao, and D. Jiang, 2010: Simulation of the direct radiative effect of mineral dust aerosol on the climate at the Last Glacial Maximum. *J. Clim.*, **24**, 843–858.

Zachos, J. C., G. R. Dickens, and R. E. Zeebe, 2008: An early Cenozoic perspective on greenhouse warming and carbon-cycle dynamics. *Nature*, **451**, 279–283.

Zachos, J. C., et al., 2005: Rapid acidification of the ocean during the Paleocene-Eocene Thermal Maximum. *Science*, **308**, 1611–1615.

Zagwijn, W. H., 1996: An analysis of Eemian climate in western and central Europe. *Quat. Sci. Rev.*, **15**, 451–469.

Zeebe, R. E., J. C. Zachos, and G. R. Dickens, 2009: Carbon dioxide forcing alone insufficient to explain Palaeocene-Eocene Thermal Maximum warming. *Nature Geosci.*, **2**, 576–580.

Zha, X., C. Huang, and J. Pang, 2009: Palaeofloods recorded by slackwater deposits on the Qishuihe river in the middle Yellow river. *J. Geograph. Sci.*, **19**, 681–690.

Zhang, P. Z., et al., 2008: A test of climate, sun, and culture relationships from an 1810–year Chinese cave record. *Science*, **322**, 940–942.

Zhang, Q.-B., and R. J. Hebda, 2005: Abrupt climate change and variability in the past four millennia of the southern Vancouver Island, Canada. *Geophys. Res. Lett.*, **32**, L16708.

Zhang, Q., H. S. Sundqvist, A. Moberg, H. Kornich, J. Nilsson, and K. Holmgren, 2010: Climate change between the mid and late Holocene in northern high latitudes—Part 2: Model-data comparisons. *Clim. Past*, **6**, 609–626.

Zhang, R., and T. L. Delworth, 2005: Simulated tropical response to a substantial weakening of the Atlantic thermohaline circulation. *J. Clim.*, **18**, 1853–1860.

Zhang, R., and T. L. Delworth, 2006: Impact of Atlantic multidecadal oscillations on India/Sahel rainfall and Atlantic hurricanes. *Geophys. Res. Lett.*, **33**, L17712.

Zhang, Y., Z. Kong, S. Yan, Z. Yang, and J. Ni, 2009: "Medieval Warm Period" on the northern slope of central Tianshan Mountains, Xinjiang, NW China. *Geophys. Res. Lett.*, **36**, L11702.

Zheng, W., P. Braconnot, E. Guilyardi, U. Merkel, and Y. Yu, 2008: ENSO at 6ka and 21ka from ocean-atmosphere coupled model simulations. *Clim. Dyn.*, **30**, 745–762.

Zhou, T., B. Li, W. Man, L. Zhang, and J. Zhang, 2011: A comparison of the Medieval Warm Period, Little Ice Age and 20th century warming simulated by the FGOALS climate system model. *Chin. Sci. Bull.*, **56**, 3028–3041.

Zhu, H., F. Zheng, X. Shao, X. Liu, X. Yan, and E. Liang, 2008: Millennial temperature reconstruction based on tree-ring widths of Qilian juniper from Wulan, Qinghai province, China. *Chin. Sci. Bull.*, **53**, 3914–3920.

Zinke, J., M. Pfeiffer, O. Timm, W. C. Dullo, and G. Brummer, 2009: Western Indian Ocean marine and terrestrial records of climate variability: A review and new concepts on land–ocean interactions since AD 1660. *Int. J. Earth Sci.*, **98**, 115–133.

Appendix 5.A: Additional Information on Paleoclimate Archives and Models

Table 5.A.1 | Summary of the Atmosphere-Ocean General Circulation Model (AOGCM) simulations available and assessed for Sections 5.3.5 and 5.5.1. Acronyms describing forcings are: SS (solar forcing, stronger variability), SW (solar forcing, weaker variability), V (volcanic activity), G (greenhouse gases concentration), A (aerosols), L (land use–land cover), and O (orbital). The table is divided into Paleoclimate Modelling Intercomparison Project Phase III (PMIP3) and Coupled Model Intercomparison Project Phase 5 (CMIP5) and non-PMIP3/CMIP5 experiments (Braconnot et al., 2012b; Taylor et al., 2012). Superscript indices in forcing acronyms identify the forcing reconstructions used and are listed in the table footnotes. PMIP3 experiments follow forcing guidelines provided in Schmidt et al. (2011, 2012b). See Fernández-Donado et al. (2013) for more information on pre-PMIP3/CMIP5 forcing configurations. See Chapter 8 and Table 9.1 for the forcing and model specifications of the CMIP5 historical runs. The simulations highlighted in red were excluded from Figures 5.8, 5.9 and 5.12 because they did not include at least solar, volcanic and greenhouse gas forcings, they did not span the whole of the last millennium, or for a reason given in the table notes.

Model	(No. runs) Period	Forcings[a]	Reference
Pre PMIP3/CMIP5 Experiments			
CCSM3	(1×) 1000–2000 (4×) 1500–2000	SS[11]·V[22]·G[30,31,35]	Hofer et al. (2011)
CNRM-CM3.3	(1×) 1001–1999	SS[11]·V[21]·G[30,34,35]·A[44]·L[54]	Swingedouw et al. (2011)
CSM1.4	(1×) 850–1999	SS[10]·V[21]·G[30,31,35]·A[41]	Ammann et al. (2007)
CSIRO-MK3L-1-2	(3×) 1–2001 (3×) 1–2001 (3×) 501–2001	SW[14] SW[14]·G[34]·O[60] SW[14]·V[24]·G[34]·O[60]	Phipps et al. (2013)
ECHAM4/OPYC	(1×) 1500–2000	SS[11]·V[21,26]·G[38]·A[42]·L[55]	Stendel et al. (2006)
ECHAM5/MPIOM	(5×) 800–2005 (3×) 800–2005	SW[13]·V[25]·G[34,39]·A[40]·L[53]·O[61] SS[10]·V[25]·G[34,39]·A[40]·L[53]·O[61]	Jungclaus et al. (2010)
ECHO-G	(1×) 1000–1990 (1×) 1000–1990 (2×) –7000–1998	SS[11]·V[20]·G[31,36,37] SS[11]·V[20]·G[31,36,37] SS[12]·G[30]·O[62]	González-Rouco et al. (2003)[b] González-Rouco et al. (2006) Wagner et al. (2007)
HadCM3	(1×) 1492–1999	SS[11]·V[23]·G[32]·A[43]·L[50,54,55]·O[60]	Tett et al. (2007)
IPSLCM4	(1×) 1001–2000	SS[11]·G[30,34,35]·A[44]·O[63]	Servonnat et al. (2010)
FGOALS-gl	(1×) 1000–1999	SS[11]·V[20]·G[30,31,35]	Zhou et al. (2011)[c]
PMIP3/CMIP5 Experiments			
BCC-csm1-1	(1×) 850–2005	SW[15]·V[24]·G[30,33,34]· A[45]·O[60]	
CCSM4	(1×) 850–2004	SW[15]·V[24]·G[30,33,34]· A[45] ·L[51]·O[60]	Landrum et al. (2013)
CSIRO-MK3L-1-2	(1×) 851–2000	SW[14]·V[25]·G[30,33,34]·O[60]	
GISS-E2-R	(8×) 850–2004	SW[14]·V[25]·G[30,33,34]·A[45]·L[51]·O[60] SW[14]·V[24]·G[30,33,34]·A[45]·L[51]·O[60] SW[14]·G[30,33,34]·A[4]·L[51]·O[60] SW[15]·V[25]·G[30,33,34]·A[45]·L[51]·O[60] SW[15]·V[24]·G[30,33,34]·A[45]·L[52]·O[60] SW[15]·G[30,33,34]·A[4]·L[51]·O[60] SW[15]·V[25]·G[30,33,34]·A[45]·L[52]·O[60] SW[15]·V[24]·G[30,33,34]·A[45]·L[51]·O[60]	[d]
HadCM3	(1×) 800–2000	SW[14]·V[25]·G[30,32,34]·A[43]·L[51]·O[60]	Schurer et al. (2013)
IPSL-CM5A-LR	(1×) 850–2005	SW[15]·V[27]·G[30,33,34]·O[60]	
MIROC-ESM	(1×) 850–2005	SW[16]·V[25]·G[30,34,39]·O[60]	[e]
MPI-ESM-P	(1×) 850–2005	SW[15]·V[25]·G[30,33,34]·A[45]·L[51]·O[60]	

Notes:

[a] Key for superscript indices in forcing acronyms:

[1] Solar:
- [10] Bard et al. (2000)
- [11] Bard et al. (2000) spliced to Lean et al. (1995a)
- [12] Solanki et al. (2004)
- [13] Krivova et al. (2007)
- [14] Steinhilber et al. (2009) spliced to Wang et al. (2005)
- [15] Vieira and Solanki (2010) spliced to Wang et al. (2005)
- [16] Delaygue and Bard (2011) spliced to Wang et al. (2005)

[2] Volcanic:
- [20] Crowley (2000)
- [21] Ammann et al. (2003)
- [22] Total solar irradiances from Crowley (2000) converted to aerosol masses using Ammann et al. (2003) regression coefficients.
- [23] Crowley et al. (2003)
- [24] Gao et al. (2008). In the GISS-E2-R simulations this forcing was implemented twice as large as in Gao et al. (2008).
- [25] Crowley and Unterman (2013)
- [26] Robertson et al. (2001)
- [27] Ammann et al. (2007)

[3] WMGHGs:
- [30] Flückiger et al., (1999; 2002); Machida et al. (1995)
- [31] Etheridge et al. (1996)
- [32] Johns et al. (2003)
- [33] Hansen and Sato (2004)
- [34] MacFarling Meure et al. (2006)
- [35] Blunier et al. (1995)
- [36] Etheridge et al. (1998)
- [37] Battle et al. (1996)
- [38] Robertson et al. (2001)
- [39] CO_2 diagnosed by the model.

[4] Aerosols:
- [40] Lefohn et al. (1999)
- [41] Joos et al. (2001)
- [42] Roeckner et al. (1999)
- [43] Johns et al. (2003)
- [44] Boucher and Pham (2002)
- [45] Lamarque et al. (2010). See Sections 8.2 and 8.3

[5] Land use, land cover:
- [50] Wilson and Henderson-Sellers (1985)
- [51] Pongratz et al. (2009) spliced to Hurtt et al. (2006)
- [52] Kaplan et al. (2011)
- [53] Pongratz et al. (2008)

(continued on next page)

Table 5.A.1 Notes (continued)

[54] Ramankutty and Foley (1999)
[55] Goldewijk (2001)
[6] Orbital:
[60] Berger (1978)
[61] Bretagnon and Francou (1988)
[62] Berger and Loutre (1991)
[63] Laskar et al. (2004)

b This simulation was only used in Figure 5.8, using NH temperature adjusted by Osborn et al. (2006).
c The FGOALS-gl experiment is available in the PMIP3 repository, but the forcing configuration is different from Schmidt et al (2011; 2012b) recommendations so it is included here within the pre-PMIP3 ensemble.
d The GISS-E2-R experiments with Gao et al. (2008) volcanic forcing were not used in Figures 5.8, 5.9 or 5.12. See [24].
e This simulation was only used in Figure 5.8, using drift-corrected NH temperature.

Table 5.A.2 | Summary of atmospheric carbon dioxide (CO_2) proxy methods and confidence assessment of their main assumptions.

Method	Scientific Rationale	Estimated Applicability	Limitations	Main Assumptions (relative confidence)
Alkenone (phytoplankton biomarker) carbon isotopes	Measurements of carbon isotope ratios of marine sedimentary alkenones (or other organic compounds) allows determination of the isotopic fractionation factor during carbon fixation (ε_p) from which pCO_2 can be calculated.	100 to ~4000 ppm; 0 to 100 Ma	Alkenones are often rare in oligotrophic areas and sometimes absent. Method relies on empirical calibration and $\delta^{13}C$ is sensitive to other environmental factors, especially nutrient-related variables. Method has been used successfully to reconstruct glacial–interglacial changes.	Measured alkenone carbon isotope ratio is accurate and precise (*high*). Ambient aqueous partial pressure of carbon dioxide (pCO_2) has a quantifiable relationship with ε_p that can be distinguished from the nutrient-related physiological factors such as algal growth rate, cell size, cell geometry and light-limited growth (*medium*). Aqueous pCO_2 is in equilibrium with atmospheric pCO_2 (*medium*). Carbon isotope fractionation in modern alkenone-producing species is the same in ancient species and constant through time (*medium*). Levels of biological productivity (e.g., dissolved phosphate concentrations) can be calculated (*high*). Carbon isotope ratio of aqueous CO_2 in the mixed layer can be determined (*medium*). Sea surface temperature can be determined (*high*). Atmospheric partial pressure of oxygen (pO_2) is known or assumed (*medium*). Diagenetic effects are minimal, or can be quantified (*medium*).
Boron isotopes in foraminifera	Boron isotope ratios ($\delta^{11}B$) in foraminifera (or other calcifying organisms) give paleo-pH from which pCO_2 can be calculated if a value for a second carbonate system parameter (e.g., alkalinity) is assumed.	100 to ~4000 ppm; 0 to 100 Ma	Calculated pCO_2 is very sensitive to the boron isotope ratio of seawater which is relatively poorly known, especially for the earlier Cenozoic. Effects of foraminiferal preservation are not well understood. Method has been used successfully to reconstruct glacial–interglacial changes.	Measured boron isotope ratio is accurate and precise (*high*). The equilibrium constant for dissociation of boric acid and boron isotopic fractionation between $B(OH)_3$ and $B(OH)_4^-$ are well known (*high*). Boron incorporation into carbonate is predominantly from borate ion (*high*). Boron isotope ratio of foraminifer calcification reflects ambient surface seawater pH (*high*). Aqueous pCO_2 is in equilibrium with atmospheric pCO_2 (*medium*). Habitats of extinct species can be determined (*high*). There is no vital effect fractionation in extinct species, or it can be determined (*medium*). The boron isotope ratio of seawater ($\delta^{11}B_{sw}$) can be determined (*medium*). Ocean alkalinity or concentration of Total Dissolved Inorganic Carbon can be determined (*high*). Sea surface temperature (SST) and salinity (SSS) can be determined (*high*). Diagenetic effects are minimal or can be quantified (*high*).
Carbon isotopes in soil carbonate and organic matter	Atmospheric pCO_2 affects the relationship between the $\delta^{13}C$ of soil CO_2 and the $\delta^{13}C$ of soil organic matter at depth in certain soil types, hence measurement of these parameters in paleosols can be used to calculate past pCO_2.	1000 to ~4000 ppm; 0 to 400 Ma	Method works better for some soil types than others. CO_2 loss is difficult to quantify and method and effects of late diagenesis may be difficult to determine.	Isotopic composition of soil CO_2 is reflected in soil carbonates below a depth of 50 cm. (*medium*). The concentration of respired CO_2 in the soil is known or assumed (*medium*). Isotopic composition of atmospheric CO_2 is known or can be inferred (*low*). Soil carbonates were precipitated in the vadose zone in exchange with atmospheric CO_2 (*high*). The original depth profile of a paleosoil can be determined (*low*). Burial (late) diagenetic effects are minimal or can be quantified (*high*).
Stomata in plant leaves	The relative frequency of stomata on fossil leaves (Stomatal Index; (Salisbury, 1928) can be used to calculate past atmospheric CO_2 levels.	100 to ~1000 ppm; 0 to 400 Ma	Closely related species have very different responses to pCO_2. The assumption that short-term response is the same as the evolutionary response is difficult to test. This and the shape of the calibration curves mean that much greater certainty applies to low pCO_2 and short time scales.	Measured stomatal index is accurate and precise (*high*). Measured stomatal index is representative of the plant (*high*). The target plants adjust their stomatal index of leaves to optimize CO_2 uptake (*medium*). Atmospheric pCO_2 close to the plant is representative of the atmosphere as a whole (*medium*). The quantitative relationship between stomatal index and CO_2 observed on short time scales (ecophenotypic or 'plastic response') applies over evolutionary time (*low*). Environmental factors such as irradiance, atmospheric moisture, water availability, temperature, and nutrient availability do not affect the relationship between stomatal index and CO_2 (*medium*). Stomatal index response to CO_2 of extinct species can be determined or assumed (*low*). Taphonomic processes do not affect stomatal index counts (*high*). Diagenetic processes do not affect stomatal index counts (*high*).

5

Table 5.A.3 | Summary of sea surface temperature (SST) proxy methods and confidence assessment of their main assumptions.

Method	Scientific Rationale	Estimated Applicability	Limitations	Main Assumptions (relative confidence)
$\delta^{18}O$ of mixed-layer planktonic foraminifera	Partitioning of $^{18}O/^{16}O$ from seawater into calcite shells of all foraminifera is temperature dependent. Verified by theoretical, field and laboratory studies. Utilizes extant and extinct species that resided in the photic zone.	0°C to 50°C; 0 to 150 Ma	The $^{18}O/^{16}O$ ratios of recrystallized planktonic foraminifer shells in carbonate-rich sediments are biased toward colder seafloor temperatures, and at most, can only constrain the lower limit of SST. The transition in preservation is progressive with age. Well-preserved forams from clay-rich sequences on continental margins are preferred. Diagenetic calcite is detectable by visual and microscopic techniques.	Analytical errors are negligible (*high*). Sensitivity to T is high and similar to modern descendants (*high*). Seawater $\delta^{18}O$ is known. The uncertainty varies with time depending on presence of continental ice-sheets, though error is negligible in the Pleistocene and during minimal ice periods such as the Eocene (<±0.25°C). Error doubles during periods of Oligocene and early Neogene glaciation because of weak constraints on ice-volume (*medium to high*). Species lives in the mixed-layer and thus records SST (*high*). Local salinity/seawater $\delta^{18}O$ is known (*low to medium*). Carbonate ion/pH is similar to modern (*medium, high*). Foraminifera from clay-rich sequences are well preserved and $^{18}O/^{16}O$ ratios unaffected by diagenesis (*high*). Foraminifera from carbonate-rich pelagic sequences are well preserved and ratios unaffected by diagenesis (*medium to low;* decreasing confidence with age). Biased towards summer SST in polar oceans (*medium*).
Mg/Ca in mixed-layer planktonic foraminifera	Partitioning of Mg/Ca from seawater into calcite shells is temperature dependent. Calibration to T is based on empirical field and laboratory culturing studies, as Mg concentrations of inorganically precipitated calcite are an order of magnitude higher than in biogenic calcite. There is no ice-volume influence on seawater Mg/Ca, though sensitivity does change with seawater Mg concentration.	5°C to 35°C; 0 to 65 Ma	Diagenetic recrystallization of foram shells can bias ratios, though the direction of bias is unknown and comparisons with other proxies suggest it is minor. The Mg/Ca is also slightly sensitive to seawater pH. Long-term changes in seawater Mg/Ca, on the order of a 2–5%/10 Myr, must be constrained via models.	Analytical errors are negligible (*high*). Mg containing oxide and organic contaminants have been removed by oxidative/reductive cleaning (*high*). Sensitivity to T in extinct species is similar to modern species (*medium*). Species lives in the mixed-layer and thus records SST (*high*). Seawater Mg/Ca is known (*high to low*: decreasing confidence with time). Surface water carbonate ion/pH is similar to modern (*medium*). Foraminfera from clay-rich sequences are well preserved and ratios unaffected by diagenesis (*high*). Foraminifera from carbonate-rich pelagic sequences are well preserved and ratios unaffected by diagenesis (*high to low;* decreasing confidence with age). Biased towards summer SST in polar oceans (*medium*).
TEX_{86} index in Archea	The ratio of cyclopentane rings in archaeal tetraether lipids (TEX), i.e., isoprenoid glycerol dibiphytanyl glycerol tetraethers (GDGTs), is sensitive to the temperature of growth environment. The relationship and calibration with temperature is empirical (based on core tops), as the underlying mechanism(s) for this relationship has yet to be identified. Verification of field calibrations with laboratory cultures is still in progress. The compounds are extracted from bulk sediments.	1°C to 40°C; 0 to 150 Ma	The depth from which the bulk of sedimentary GDGT's are produced is assumed to be the mixed-layer though this cannot be verified, for the modern or past. At least two species with differing ecologies appear to be producing the tetraethers. The GDGT signal is ultimately an integrated community signal allowing the potential for evolutionary changes to influence regional signals over time. Tetraethers are found in measurable abundances on continental shelves and/or organic rich sediments.	Analytical errors are small (*high*). Sensitivity to T similar to modern (*medium*). Species that produced tropical sedimentary GDGT's resided mainly in the mixed-layer and thus records SST (*high to medium*). Species that produced the sedimentary GDGT's in the sub-polar to polar regions mainly resided in the mixed-layer and thus records SST (*low*). No alteration of GDGT ratios during degradation of compounds (*medium to low:* decreasing confidence with age). No contamination by GDGT's derived from terrestrial sources (*high to medium* if BIT index <0.3). Biased towards summer SST in polar oceans (*medium*).
UK_{37} Index in Algae	Based on the relative concentration of C_{37} methyl ketones derived from the cells of haptophyte phytoplankton. Calibrations are empirically derived through field and culture studies.	5°C to 28°C; 0 to 50 Ma	The distribution of haptophyte algae ranges from sub-polar to tropical.	Analytical errors are negligible (*high*). Sensitivity to SST similar to modern (*high to medium;* decreasing confidence with time). Species that produced the sedimentary alkenones lived in the mixed-layer and thus record SST (*high*). No alteration of alkenone saturation index during degradation of compounds (*medium;* decreasing confidence with age). Biased towards summer SST in polar oceans (*medium*).
Microfossil census modern analogue techniques	Utilises a statistical correlation between extant planktonic microfossil assemblage data (most commonly foraminifera, but also diatoms and radiolarians) and climate parameters. Most commonly used statistical methods are modern analogue technique (MAT) and artificial neural network (ANN).	0°C to 40°C; 0 to 5 Ma	Dependent on quality, coverage, size and representativeness of the core top modern analogue data base. Extant species reduce with increasing age. This and paleogeographic and ocean circulation differences with age-limit applicability to less than 1 Ma.	The composition of modern assemblages can be correlated to SST (*high*). Sensitivity of paleo-assemblages to SST is similar to modern (*high*, but decreases with increasing age). Eurythermal assemblages responding to non-temperature (e.g., nutrient availability) influences can be identified (*medium*). That the extant species used to reconstruct SST mainly reside in the mixed layer (*medium to high*). Depositional and post-depositional processes have not biased the assemblage (*medium to high*).

Table 5.A.4 | Assessment of leads and lags between Antarctic, hemispheric temperatures and atmospheric CO_2 concentration during terminations. Chronological synthesis of publications, main findings, incorporation in IPCC assessments and key uncertainties.

Reference	Investigated Period	Source CO_2 Data	Source Temperature Data	Lag Quantification Method	Lag Between Temperature and CO_2 (positive, temperature lead)	Key Limitations
TAR: From a detailed study of the last three glacial terminations in the Vostok ice core, Fischer et al. (1999) conclude that CO_2 increases started 600 ± 400 years after the Antarctic warming. However, considering the large uncertainty in the ages of the CO_2 and ice (1000 years or more if we consider the ice accumulation rate uncertainty), Petit et al. (1999) felt it premature to ascertain the sign of the phase relationship between CO_2 and Antarctic temperature at the initiation of the terminations. In any event, CO_2 changes parallel Antarctic temperature changes during deglaciations (Sowers and Bender, 1995; Blunier et al., 1997; Petit et al., 1999).						
Fischer et al. (1999)	Termination I Terminations I, II, III	Taylor Dome, Byrd[a] (CH_4 synchonized age scales) Vostok[a] (gas age scales based on firn modelling)	Byrd $\delta^{18}O$, Vostok δD (CH_4 synchonized age) Vostok[a] δD (GT4 ice age scale)	Maximum at onset of interglacial periods	Antarctica: 600 ± 400 years	Ice core synchronization for Termination I (~300 years). Gas age-ice age difference simulated by firn models for interglacial conditions could be overestimated by ~400 years. Signal-to-noise ratio. Resolution of CO_2 measurements and firnification smoothing (~300 years).
Petit et al. (1999) Pépin et al. (2001)	Terminations I, II, III, IV	Vostok[a] (GT4 gas age scale based on firn modelling)	Vostok δD (GT4 ice age scale)	Onset of transitions	Antarctica: in phase within uncertainties Positive	Gas age-ice age difference simulated by firn models for glacial conditions could be overestimated by up to 1500 years. Resolution of CO_2 measurements and firnification smoothing (~300 years).
Mudelsee (2001)	0–420 ka	Vostok[a] (GT4 gas age scale)	Vostok δD (GT4 ice age scale)	Lagged generalised least square regression with parametric bootstrap resampling, entire record	Antarctica : 1300 ± 1000 years	Signal to noise ratio (1 ice core).
AR4: High-resolution ice core records of temperature proxies and CO_2 during deglaciation indicates that Antarctic temperature starts to rise several hundred years before CO_2 (Monnin et al., 2001; Caillon et al., 2003). During the last deglaciation, and possibly also the three previous ones, the onset of warming at both high southern and northern latitudes preceded by several thousand years the first signals of significant sea level increase resulting from the melting of the northern ice sheets linked with the rapid warming at high northern latitudes (Petit et al., 1999; Shackleton, 2000; Pépin et al., 2001). Current data are not accurate enough to identify whether warming started earlier in the SH or NH, but a major deglacial feature is the difference between North and South in terms of the magnitude and timing of strong reversals in the warming trend, which are not in phase between the hemispheres and are more pronounced in the NH (Blunier and Brook, 2001).						
Monnin et al. (2001)	Termination I	High resolution data from EDC on EDC1 gas age scale (based on firn modelling)	EDC (EPICA (European Project for Ice Coring in Antarctica Dome C) on EDC1 ice age scale	Crossing points of linear fit	Antarctica: 800 ± 600 years	Gas age–ice age difference (±1000 years). Signal to noise ratio (1 ice core).
Caillon et al. (2003)	Termination III	Vostok on GT4 gas age scale	Vostok $\delta^{40}Ar$ on GT4 gas age scale	Maximum lagged correlation	Antarctica: 800 ± 200 years	Relationship between $\delta^{40}Ar$ and temperature assumed to be instantaneous. The 800 years is a minimum CO_2-temperature lag which does not account for a possible delayed response of firn gravitational fractionation to surface temperature change.
AR5: For the last glacial termination, a large-scale temperature reconstruction (Shakun et al., 2012) documents that temperature change in the SH lead NH temperature change. This lead can be explained by the bipolar thermal seesaw concept (Stocker and Johnsen, 2003) (see also Section 5.7) and the related changes in the inter-hemispheric ocean heat transport, caused by weakening of the Atlantic Ocean meridional overturning circulation (AMOC) during the last glacial termination (Ganopolski and Roche, 2009). SH warming prior to NH warming can also be explained by the fast sea ice response to changes in austral spring insolation (Stott et al., 2007; Timmermann et al., 2009). According to these mechanisms, SH temperature lead over the NH is fully consistent with the NH orbital forcing of deglacial ice volume changes (high confidence) and the importance of the climate–carbon cycle feedbacks in glacial–interglacial transitions. The tight coupling is further highlighted by the near-zero lag between the deglacial rise in CO_2 and averaged deglacial Antarctic temperature recently reported from improved estimates of gas-ice age differences (Pedro et al., 2012; Parrenin et al., 2013). Previous studies (Monnin et al., 2001) suggesting a temperature lead of 800 ± 600 years over the deglacial CO_2 rise probably overestimated gas-ice age differences.						
Shakun et al. (2012)	Termination I	EDC age scale synchronized to GICC05[b](Lemieux-Dudon et al., 2010)	NH: stack of 50 records including 2 Greenland ice cores SH: stack of 30 records incl. 4 ice cores (Vostok, EDML, EDC, Dome F)[a] on their original age scale	Lag correlation (20–10 kyr) using Monte-Carlo statistics	SH: 620 ± 660 years NH: –720 ± 660 years Global: –460 ± 340 years	Uncertainties in the original age scales of each record: e.g., reservoir ages of marine sediments, radiocarbon calibration (intCal04), Antarctic gas /ice chronology. Assumption that time scale errors (e.g., from reservoir ages or ice core chronologies) are independent from each other. This could lead to higher-than-reported lag estimation uncertainties. Similar limitations as in earlier studies for Antarctic temperature lead on CO_2. Non stability of the phase lags: global temperature leads CO_2 at the onset of deglacial warming.

(continued on next page)

5

Table 5.A.4 (continued)

Reference	Investigated Period	Source CO_2 Data	Source Temperature Data	Lag Quantification Method	Lag Between Temperature and CO_2 (positive, temperature lead)	Key Limitations
Pedro et al. (2012)		Siple Dome and Byrd, synchronized to GICC05[b] age scale	$\delta^{18}O$ composite (Law Dome, Siple Dome, Byrd, EDML and TALDICE[a] ice cores) synchronized to GICC05[b] using firn modelling (Pedro et al., 2011)	Lag correlation (9–21 kyr) and derivative lag correlation	Antarctica: –60 to 380 years	Uncertainty on gas – ice age difference in high accumulation sites (<300 years) and on synchronization methods to GICC05. Data resolution (145 year for Byrd CO_2, 266 year for Siple CO_2). The CO_2 data were resampled at 20 year resolution prior to the lag analysis, which may lead to an underestimation of the statistical error in the lag determination. Temperature versus other (e.g., elevation, moisture origin) signals in coastal ice core $\delta^{18}O$. Correlation method sensitive to minima, maxima and inflexion points.
(Parrenin et al., 2013)		EDC, new gas age scale produced from the modified EDC3 ice age scale using lock-in depth derived from $\delta^{15}N$ of N_2 and adjusted to be consistent with GICC05[b] gas age scale. Processes affecting the gas lock-in depth such as impurities are implicitly taken into account when using $\delta^{15}N$ (no use of firn models).	Stack temperature profile derived from water isotopes from EDC[a], Vostok[a], Dome Fuji[a], TALDICE[a] and EDML[a] synchronized to a modified EDC3 ice age scale	Monte-Carlo algorithm at linear break points	Antarctica: Warming onset: –10 ± 160 years Bølling onset: 260 ± 130 years Younger Dryas onset: –60 ± 120 years Holocene onset: 500 ± 90 years	Accuracy, resolution and interpolation of $\delta^{15}N$ of N_2; assumption of no firn convective zone at EDC under glacial conditions. Data resolution and noise (e.g., precipitation intermittency biases in stable isotope records).

Notes:

[a] Names of different Antarctic ice cores (Byrd, Taylor Dome, Vostok, Siple Dome, Law Dome, TALDICE, Dome Fuji, EDML, EDC), with different locations, surface climate and firnification conditions. For the most inland sites (Vostok, EDC, Dome Fuji), at a given ice core depth, gas ages are lower than ice ages by 1500 to 2000 years (interglacial conditions) and 5000–5500 years (glacial conditions) while this gas age–ice age difference is lower (400 to 800. years) for coastal, higher accumulation sites (Byrd, Law Dome, Siple Dome).

[b] GICC05: Greenland Ice Core Chronology 2005, based on annual layer counting in Greenland (NGRIP, GRIP and DYE3 ice cores) (Rasmussen et al., 2006), back to 60 ka (Svensson et al., 2008). The synchronism between rapid shifts in Greenland climate and in atmospheric CH_4 variations allows to transfer GICC05 to Greenland and then to Antarctic CH_4 variations (Blunier et al., 2007).

Additional point: CO_2-Antarctic temperature phase during AIM events.

Studies on CO_2 phasing relative to CH_4 during Dansgaard Oeschger (DO) event onsets (Ahn and Brook, 2008; Ahn et al., 2012; Bereiter et al., 2012) suggest a lag of maximum CO_2 concentration relative to the Antarctic Isotope Maxima (AIM) 19, 20, 21, 23 and 24 by 260 ± 220 years during MIS5 and 670 to 870 years ± 360 years relative to AIM 12, 14, 17 during MIS3 (Bereiter et al., 2012). Accordingly, the lag is dependent on the climate state. A lag is not discernible for shorter AIM. This study avoids the ice age–gas age difference problem, but relies on the bipolar seesaw concept, i.e., it assumes that maximum Antarctic temperatures are coincident to the onset of DO events and the concurrent CH_4 increase.

Table 5.A.5 | Summary of seasonal estimates of terrestrial surface temperature anomalies (°C) for the Last Interglacial (LIG) plotted in Figure 5.6. *pdf*-method stands for probability-density function method. Dating methods: AMS=Accelerator mass spectrometry; IRSL=Infrared stimulated luminescence; OSL=Optically stimulated luminescence; TL=thermoluminescence.

Site	Latitude (°N)	Longitude (°E)	Elevation (m asl)	Dating	Proxy	Temperature Anomaly (°C)		References
						July	January	
Netherlands, Amsterdam Terminal	52.38	4.91	1	Eemian, U/Th	Pollen, diatoms, molluscs, foraminifera, dinoflagellates, ostracods, heavy minerals, paleomagnetism, grain-size, trace elements	2	3	(Zagwijn, 1996; van Leeuwen et al., 2000; Beets et al., 2006)
E Canada, Addington Forks, Nova Scotia	45.65	–62.1	50	Uranium-series	Pollen	4		(Dreimanis, 1992)
NW America, Humptulips	47.28	–123.55	100	interpolation with ^{14}C dates (peat) of the same core	Pollen	1		(Heusser and Heusser, 1990)
NE Siberia, Lake El'gygytgyn	67.5	172	492	TL	Pollen	6	14	(Lozhkin and Anderson, 2006)
NW Alaska, Squirrel Lake	67.1	–160.38	91	TL	Pollen, plant macrofossils	1.5	–2	(Berger and Anderson, 2000)
SE Baffin Island, Robinson Lake	63.38	–64.25	170	TL, IRSL	Pollen, diatoms, macrofossils	5		(Miller et al., 1999; Fréchette et al., 2006)
Sweden, Leveäniemi	67.63	21.02	380	125 ka	Pollen, *pdf* method	2.1	6.6	(Kühl, 2003, and ref. therein)
Finland, Evijärvi	63.43	23.33	67	125 ka	Pollen, *pdf* method	2.3	10.3	(Kühl, 2003, and ref. therein)
Finland, Norinkylä	62.58	22.02	110	125 ka	Pollen, *pdf* method	1.3	7.7	(Kühl, 2003, and ref. therein)
Estland, Prangli	59.65	25.08	5	125 ka	Pollen, *pdf* method	1.7	3.2	(Kühl, 2003, and ref. therein)
Estland, Waewa-Ringen	58.33	26.73	50	125 ka	Pollen, *pdf* method	1.3	6.8	(Kühl, 2003, and ref. therein)
Norway, Fjøsanger	60.35	5.33	5	125 ka	Pollen, *pdf* method	2.9	1.6	(Kühl, 2003, and ref. therein)
Denmark, Hollerup	56.7	9.83	40	125 ka	Pollen, *pdf* method	1.1	3.7	(Kühl, 2003, and ref. therein)
Germany, Husum	54.52	9.17	2	125 ka	Pollen, *pdf* method	2.3	–0.3	(Kühl, 2003, and ref. therein)
Germany, Rederstall	54.28	9.25	0	125 ka	Pollen, *pdf* method	0.3	1	(Kühl, 2003, and ref. therein)
Germany, Odderade	54.23	9.28	7	125 ka	Pollen, *pdf* method	1.8	1.4	(Kühl, 2003, and ref. therein)
Germany, Helgoland	53.95	8.85	–1	125 ka	Pollen & macrofossils, *pdf* method	2	0.6	(Kühl, 2003, and ref. therein)
Germany, Oerel	53.48	9.07	12.5	125 ka	Pollen, *pdf* method	1.1	0.6	(Kühl, 2003, and ref. therein)
Germany, Quakenbrück	52.4	7.57	26	125 ka	Pollen, *pdf* method	1.4	0.3	(Kühl, 2003, and ref. therein)
Netherlands, Amersfoort	52.15	5.38	3	125 ka	Pollen, *pdf* method	–0.3	0.5	(Kühl, 2003, and ref. therein)
Germany, Wallensen	52	9.4	160	125 ka	Pollen & macrofossils, *pdf* method	1.9	–0.7	(Kühl, 2003, and ref. therein)
Germany, Neumark-Nord	51.33	11.88	90	125 ka	Pollen & macrofossils, *pdf* method	1.4	0.5	(Kühl, 2003, and ref. therein)
Germany, Grabschütz	51.48	12.28	100	125 ka	Pollen & macrofossils, *pdf* method	1.3	–0.2	(Kühl, 2003, and ref. therein)
Germany, Schönfeld	51.8	13.89	65	125 ka	Pollen, *pdf* method	–0.5	2.6	(Kühl, 2003, and ref. therein)
Germany, Kittlitz	51.43	14.78	150	125 ka	Pollen, *pdf* method	1.4	2.4	(Kühl, 2003, and ref. therein)
Poland, Imbramovice	50.88	16.57	175	125 ka	Pollen & macrofossils, *pdf* method	2.5	3.4	(Kühl, 2003, and ref. therein)
Poland, Zgierz-Rudunki	51.87	19.42	200	125 ka	Pollen & macrofossils, *pdf* method	0.2	2.6	(Kühl, 2003, and ref. therein)
Poland, Wladyslawow	52.13	18.47	100	125 ka	Pollen & macrofossils, *pdf* method	2.4	–0.9	(Kühl, 2003, and ref. therein)
Poland, Glowczyn	52.48	20.21	124	125 ka	Pollen, *pdf* method	1.4	3.6	(Kühl, 2003, and ref. therein)
Poland, Gora Kalwaria	51.98	21.18	100	125 ka	Pollen & macrofossils, *pdf* method	0.3	1.9	(Kühl, 2003, and ref. therein)
Poland, Naklo	53.15	17.6	62	125 ka	Pollen & macrofossils, *pdf* method	0.6	3	(Kühl, 2003, and ref. therein)
Poland, Grudziadz	53.48	18.75	10	125 ka	Pollen, *pdf* method	0.5	1.6	(Kühl, 2003, and ref. therein)
England, Wing	52.62	–0.78	119	125 ka	Pollen, *pdf* method	2.4	–0.5	(Kühl, 2003, and ref. therein)

(continued on next page)

Table 5.A.5 (continued)

Site	Latitude (°N)	Longitude (°E)	Elevation (m asl)	Dating	Proxy	Temperature Anomaly (°C)		References
						July	January	
England, Bobbitshole	52.05	1.15	3	125 ka	Pollen & macrofossils, *pdf* method	2.5	–2.3	(Kühl, 2003, and ref. therein)
England, Selsey	50.42	0.48	0	125 ka	Pollen & macrofossils, *pdf* method	0.7	–2.2	(Kühl, 2003, and ref. therein)
England, Stone	50.42	–1.02	0	125 ka	Pollen & macrofossils, *pdf* method	2.9	–2.3	(Kühl, 2003, and ref. therein)
France, La Grande Pile	47.73	6.5	330	125 ka	Pollen, *pdf* method	0.5	–0.7	(Kühl, 2003, and ref. therein)
Germany, Krumbach	48.04	9.5	606	125 ka	Pollen, *pdf* method	0.5	2.3	(Kühl, 2003, and ref. therein)
Germany, Jammertal	48.1	9.72	578	125 ka	Pollen, *pdf* method	0	0.5	(Kühl, 2003, and ref. therein)
Germany, Samerberg	47.75	12.2	600	125 ka	Pollen & macrofossils, *pdf* method	2.7	4.1	(Kühl, 2003, and ref. therein)
Germany, Zeifen	47.93	12.83	427	125 ka	Pollen & macrofossils, *pdf* method	3.4	2.5	(Kühl, 2003, and ref. therein)
Austria, Mondsee	47.51	13.21	534	125 ka	Pollen & macrofossils, *v*method	4.3	1.3	(Kühl, 2003, and ref. therein)
Germany, Eurach	47.29	11.13	610	125 ka	Pollen, *pdf* method	6.4	4.7	(Kühl, 2003, and ref. therein)
Germany, Füramoos	47.91	9.95	662	125 ka	Pollen, modern analogue vegetation (MAV) and probability mutual climatic spheres (PCS)	–2.8	–1.2	(Müller, 2001)
Swiss, Gondiswil-Seilern	47.12	7.88	639	125 ka	Pollen, *pdf* method	0.1	0.4	(Kühl, 2003, and ref. therein)
Swiss, Meikirch	47	7.37	620	125 ka	Pollen, *pdf* method	0.3	–0.3	(Kühl, 2003, and ref. therein)
Swiss, Meikirch II	47.01	7.33	620	125 ka	Pollen, modern analogue vegetation (MAV) and probability mutual climatic spheres (PCS)	–1.2	–4.5	(Welten, 1988)
Swiss, Beerenmösli	47.06	7.51	649	125 ka	Pollen, modern analogue vegetation (MAV) and probability mutual climatic spheres (PCS)	–1.1	–5.5	(Wegmüller, 1992)
France, Lac Du Bouchet	44.55	3.47	1200	125 ka	Pollen, *pdf* method	1.7	–0.2	(Kühl, 2003, and ref. therein)
Italia, Valle di Castiglione	41.85	12.73	110	125 ka	Pollen, *pdf* method	–3.4	–5.9	(Kühl, 2003, and ref. therein)
Romania, Turbuta	47.25	23.3	275	U/Th, 125 ka	Pollen, *pdf* method	–1.2	2.4	(Kühl, 2003, and ref. therein)
Greece, Tenaghi Phillipon	41.17	24.33	40	125 ka	Pollen, *pdf* method	0.9	2	(Kühl, 2003, and ref. therein)
Greece, Ioannina	39.67	20.85	472	125 ka	Pollen, *pdf* method	–1.9	–1.9	(Kühl, 2003, and ref. therein)
Germany, Bispingen	53.08	9.98	100	TL,125ka	Pollen, *pdf* method	1.2	0.9	(Kühl, 2003, and ref. therein)
Germany, Gröbern	52.02	12.08	95	TL, 125 ka	Pollen & macrofossils, *pdf* method	0.4	1.8	(Kühl, 2003, and ref. therein)
Germany, Klinge	51.75	14.52	10	pollen correlation	Pollen (Grichuk, 1985)	0	2	(Novenko et al., 2008)
Germany, Ober-Rheinebene near Darmstadt	49.82	8.4	90	Eem	Vegetation, mammals			(von Königswald, 2007)
France, La Flachere	45.23	5.58	333	125 ka	Pollen, modern analogue vegetation (MAV) and probability mutual climatic spheres (PCS)	–0.9	–14.4	(Peschke et al., 2000)
France, Lathuile	45.75	6.14	452	125 ka	Pollen, modern analogue vegetation (MAV) and probability mutual climatic spheres (PCS)	–0.5	–2.2	(Klotz et al., 2003)
France, La Grande Pile	47.73	6.5	330	TL, 125 ka	Pollen, carbon isotopes	10	–15	(Rousseau et al., 2007)
Japan, Lake Biwa	35.33	136.17	86	tephrochronological and magnetostratigraphic information	Pollen	–3	–2.5	(Nakagawa et al., 2008)
					Pollen	–5.5	–1.5	(Tarasov et al., 2011)
Siberia, Lake Baikal, Continent Ridge CON01-603-2	53.95	108.9	–386	AMS, 125 ka	Pollen	2	–1	(Tarasov et al., 2005)
Bol'shoi Lyakhovsky Island	73.33	141.5	40	MIS 5, ca. 130–110 ka (IRSL)	Pollen, beetles, chironomids, rhizopods, palaeomagnetic, BMA	4.5		(Andreev et al., 2004)
Wairarapa Valley; New Zealand	–41.37	175.07	10	OSL, MIS 5e	Beetles	2.8 winter	2.1 summer	(Marra, 2003)

5.A.1 Additional Information to Section 5.3.5

Section 5.3.5 assesses knowledge of changes in hemispheric and global temperature over the last 2 ka from a range of studies, reconstructions and simulations. Tables 5.A.1 and 5.A.6 provide further information about the datasets used in Figures 5.7–5.9 and 5.12, and the construction of Figure 5.8 is described in more detail. All reconstructions assessed in, or published since, AR4 were considered, but those that have been superseded by a related study using an expanded proxy dataset and/or updated statistical methods were excluded.

Figure 5.8 compares simulated and reconstructed NH temperature changes (see caption). Some reconstructions represent a smaller spatial domain than the full NH or a specific season, while annual temperatures for the full NH mean are shown for the simulations. Multi-model means and estimated 90% multi-model ranges are shown by the thick and thin lines, respectively, for two groups of simulations (Table 5.A.1): those forced by stronger (weaker) solar variability in red (blue). Note that the strength of the solar variability is not the only difference between these groups: the GCMs and the other forcings are also different between the groups. In Figure 5.7, the reconstructions are shown as deviations from their 1881–1980 means, which allows them to be compared with the instrumental record. In Figure 5.8a, all timeseries are expressed as anomalies from their 1500–1850 mean (prior to smoothing with a 30-year Gaussian-weighted filter, truncated 7 years from the end of each series to reduce end-effects of the filter) because the comparison of simulations and reconstructions is less sensitive to errors in anthropogenic aerosol forcing applied to the models when a pre-industrial reference period is used, and less sensitive to different realisations of internal variability with a multi-century reference period. The grey shading represents a measure of the overlapping reconstruction confidence intervals, with scores of 1 and 2 assigned to temperatures within ±1.645 standard deviation (90% confidence range) or ±1 standard deviation, respectively, then summed over all reconstructions and scaled so that the maximum score is dark grey, and minimum score is pale grey. This allows the multi-model ensembles to be compared with the ensemble of reconstructed NH temperatures, taking into account the published confidence intervals.

The superposed composites (time segments from selected periods positioned so that the years with peak negative forcing are aligned; top panels of Figure 5.8b–d) compare the simulated and reconstructed temperatures (bottom panels) associated with (b) individual volcanic forcing events; (c) multi-decadal changes in volcanic activity; (d) multi-decadal changes in solar irradiance. Only reconstructions capable of resolving (b) interannual or (c, d) interdecadal variations are used. The thick green line in Figure 5.8d shows the composite mean of the volcanic forcing, also band-pass filtered, but constructed using the solar composite periods to demonstrate the changes in volcanic forcing that are coincident with solar variability. The composite of individual volcanic events shown in (b) is formed by aligned time segments centred on the 12 years (1442, 1456, 1600, 1641, 1674, 1696, 1816, 1835, 1884, 1903, 1983 and 1992) during 1400–1999 that the Crowley and Unterman (2013) volcanic forcing history exceeds 1.0 W m^{-2} below the 1500–1899 mean volcanic forcing, excluding events within 7 years (before or after) of a stronger event. The composite of multi-decadal changes in volcanic forcing shown in (c) is formed from 80-year periods centred on the five years (1259, 1456, 1599, 1695 and 1814) during 850–1999 when the Crowley and Unterman (2013) volcanic forcing, smoothed with a 40-year Gaussian-weighted filter, exceeds 0.2 W m^{-2} below the 1500–1899 mean volcanic forcing, except that a year is not selected if it is within 39 years of another year that has a larger negative 40-year smoothed volcanic forcing. The composite of the strongest multidecadal changes in the solar forcing shown in (d) is formed from 80-year periods centred on the seven years (1044, 1177, 1451, 1539, 1673, 1801 and 1905) during 850–1999 when the Ammann et al. (2007) solar forcing, band-pass filtered to retain variations on time scales between 20 and 160 years, is reduced by at least 0.1 W m^{-2} over a 40-year period. Reconstructed and simulated temperature timeseries were smoothed with a 40-year Gaussian-weighted filter in (c) or 20-to-160-year band-pass filtered in (d), and each composite was shifted to have zero mean during the (b) 5 or (c, d) 40 years preceding the peak negative forcing.

5

Table 5.A.6 | Hemispheric and global temperature reconstructions assessed in Table 5.4 and used in Figures 5.7 to 5.9.

Reference [Identifier]	Period (CE)	Resolution	Region[a]	Proxy Coverage[b]				Method & Data
				H	M	L	O	
Briffa et al. (2001) [only used in Figure 5.8b–d due to divergence issue]	1402–1960	Annual (summer)	L 20°N to 90°N	✱	☒	☐	☐	Principal component forward regression of regional composite averages. Tree-ring density network, age effect removed via age-band decomposition.
Christiansen and Ljungqvist (2012) [CL12loc]	1–1973	Annual	L+S 30°N to 90°N	✱	✱	☐	☐	Composite average of local records calibrated by local inverse regression. Multi-proxy network.
D'Arrigo et al. (2006) [Da06treecps]	713–1995	Annual	L 20°N to 90°N	✱	☒	☐	☐	Forward linear regression of composite average. Network of long tree-ring width chronologies, age effect removed by Regional Curve Standardisation.
Frank et al. (2007) [Fr07treecps]	831–1992	Annual	L 20°N to 90°N	☒	☒	☐	☐	Variance matching of composite average, adjusted for artificial changes in variance. Network of long tree-ring width chronologies, age effect removed by Regional Curve Standardisation.
Hegerl et al. (2007) [He07tls]	558–1960	Decadal	L 30°N to 90°N	☒	☒	☐	☐	Total Least Squares regression. Multi-proxy network.
Juckes et al. (2007) [Ju07cvm]	1000–1980	Annual	L+S 0° to 90°N	☒	☒	☐	☐	Variance matching of composite average. Multi-proxy network.
Leclercq and Oerlemans (2012) [LO12glac]	1600–2000	Multidecadal	L 0° to 90°N L 90S to 0° L 90°S to 90°N	☒	✱	☒	☐	Inversion of glacier length response model. 308 glacier records.
Ljungqvist (2010) [Lj10cps]	1–1999	Decadal	L+S 30°N to 90°N	✱	☒	☐	☒	Variance matching of composite average. Multi-proxy network.
Loehle and McCulloch (2008) [LM08ave]	16–1935	Multidecadal	L+S mostly 0° to 90°N	☒	☒	☐	☒	Average of calibrated local records. Multi-proxy network (almost no tree-rings).
Mann et al. (2008) [Ma08cpsl] [Ma08eivl] [Ma08eivf] [Ma08min7eivf]	200–1980	Decadal	L [cpsl/eivl] and L+S [eivf] versions, 0° to 90°N, 0° to 90°S, and 90°S to 90°N	✱	✱	☒	☒	(i) Variance matching of composite average. (ii) Total Least Squares regression. Multi-proxy network.[c]
Mann et al. (2009) [Ma09regm]	500–1849	Decadal	L+S 0 to 90°N	✱	✱	☒	☒	Regularized Expectation Maximization with Truncated Total Least Squares. Multi-proxy network.[c]
Moberg et al. (2005) [Mo05wave]	1–1979	Annual	L+S 0° to 90°N	☒	✱	☒	☐	Variance matching of composites of wavelet decomposed records. Tree-ring width network for short time scales; non-tree-ring network for long time scales.
Pollack and Smerdon (2004) [PS04bore]	1500–2000	Centennial	L 0° to 90°N L 0° to 90°S L 90°S to 90°N	☒	✱	☒	☐	Borehole temperature profiles inversion
Shi et al. (2013) [Sh13pcar]	1000–1998	Annual	L 0 to 90°N	✱	☒	☐	☐	Principal component regression with autoregressive timeseries model. Multi-proxy network (tree-ring and non-tree-ring versions).

Notes:

[a] Region: L = land only, L+S = land and sea, latitude range indicated.

[b] Proxy location and coverage: H = high latitude, M = mid latitude, L = low latitude, O = oceans, ☐ = none or very few, ☒ = limited, ✱ = moderate

[c] These studies also present versions without tree-rings or without seven inhomogeneous proxies (including the Lake Korttajärvi sediment records; Tiljander et al., 2003). The latter version is used in Figure 5.7a (Ma08min7eivf) in preference to the reconstruction from the full network. The impact of these seven proxies on the other NH reconstructions is negligible (MA08cpsl) or results in a slightly warmer pre-900 reconstruction compared to the version without them (Ma09regm).

5

6

Carbon and Other Biogeochemical Cycles

Coordinating Lead Authors:
Philippe Ciais (France), Christopher Sabine (USA)

Lead Authors:
Govindasamy Bala (India), Laurent Bopp (France), Victor Brovkin (Germany/Russian Federation), Josep Canadell (Australia), Abha Chhabra (India), Ruth DeFries (USA), James Galloway (USA), Martin Heimann (Germany), Christopher Jones (UK), Corinne Le Quéré (UK), Ranga B. Myneni (USA), Shilong Piao (China), Peter Thornton (USA)

Contributing Authors:
Anders Ahlström (Sweden), Alessandro Anav (UK/Italy), Oliver Andrews (UK), David Archer (USA), Vivek Arora (Canada), Gordon Bonan (USA), Alberto Vieira Borges (Belgium/Portugal), Philippe Bousquet (France), Lex Bouwman (Netherlands), Lori M. Bruhwiler (USA), Kenneth Caldeira (USA), Long Cao (China), Jérôme Chappellaz (France), Frédéric Chevallier (France), Cory Cleveland (USA), Peter Cox (UK), Frank J. Dentener (EU/Netherlands), Scott C. Doney (USA), Jan Willem Erisman (Netherlands), Eugenie S. Euskirchen (USA), Pierre Friedlingstein (UK/Belgium), Nicolas Gruber (Switzerland), Kevin Gurney (USA), Elisabeth A. Holland (Fiji/USA), Brett Hopwood (USA), Richard A. Houghton (USA), Joanna I. House (UK), Sander Houweling (Netherlands), Stephen Hunter (UK), George Hurtt (USA), Andrew D. Jacobson (USA), Atul Jain (USA), Fortunat Joos (Switzerland), Johann Jungclaus (Germany), Jed O. Kaplan (Switzerland/Belgium/USA), Etsushi Kato (Japan), Ralph Keeling (USA), Samar Khatiwala (USA), Stefanie Kirschke (France/Germany), Kees Klein Goldewijk (Netherlands), Silvia Kloster (Germany), Charles Koven (USA), Carolien Kroeze (Netherlands), Jean-François Lamarque (USA/Belgium), Keith Lassey (New Zealand), Rachel M. Law (Australia), Andrew Lenton (Australia), Mark R. Lomas (UK), Yiqi Luo (USA), Takashi Maki (Japan), Gregg Marland (USA), H. Damon Matthews (Canada), Emilio Mayorga (USA), Joe R. Melton (Canada), Nicolas Metzl (France), Guy Munhoven (Belgium/Luxembourg), Yosuke Niwa (Japan), Richard J. Norby (USA), Fiona O'Connor (UK/Ireland), James Orr (France), Geun-Ha Park (USA), Prabir Patra (Japan/India), Anna Peregon (France/Russian Federation), Wouter Peters (Netherlands), Philippe Peylin (France), Stephen Piper (USA), Julia Pongratz (Germany), Ben Poulter (France/USA), Peter A. Raymond (USA), Peter Rayner (Australia), Andy Ridgwell (UK), Bruno Ringeval (Netherlands/France), Christian Rödenbeck (Germany), Marielle Saunois (France), Andreas Schmittner (USA/Germany), Edward Schuur (USA), Stephen Sitch (UK), Renato Spahni (Switzerland), Benjamin Stocker (Switzerland), Taro Takahashi (USA), Rona L. Thompson (Norway/New Zealand), Jerry Tjiputra (Norway/Indonesia), Guido van der Werf (Netherlands), Detlef van Vuuren (Netherlands), Apostolos Voulgarakis (UK/Greece), Rita Wania (Austria), Sönke Zaehle (Germany), Ning Zeng (USA)

Review Editors:
Christoph Heinze (Norway), Pieter Tans (USA), Timo Vesala (Finland)

This chapter should be cited as:
Ciais, P., C. Sabine, G. Bala, L. Bopp, V. Brovkin, J. Canadell, A. Chhabra, R. DeFries, J. Galloway, M. Heimann, C. Jones, C. Le Quéré, R.B. Myneni, S. Piao and P. Thornton, 2013: Carbon and Other Biogeochemical Cycles. In: *Climate Change 2013: The Physical Science Basis. Contribution of Working Group I to the Fifth Assessment Report of the Intergovernmental Panel on Climate Change* [Stocker, T.F., D. Qin, G.-K. Plattner, M. Tignor, S.K. Allen, J. Boschung, A. Nauels, Y. Xia, V. Bex and P.M. Midgley (eds.)]. Cambridge University Press, Cambridge, United Kingdom and New York, NY, USA.

Table of Contents

Supplementary Material is available in online versions of the report.

Executive Summary

This chapter addresses the biogeochemical cycles of carbon dioxide (CO_2), methane (CH_4) and nitrous oxide (N_2O). The three greenhouse gases (GHGs) have increased in the atmosphere since pre-industrial times, and this increase is the main driving cause of climate change (Chapter 10). CO_2, CH_4 and N_2O altogether amount to 80% of the total radiative forcing from well-mixed GHGs (Chapter 8). The increase of CO_2, CH_4 and N_2O is caused by anthropogenic emissions from the use of fossil fuel as a source of energy and from land use and land use changes, in particular agriculture. The observed change in the atmospheric concentration of CO_2, CH_4 and N_2O results from the dynamic balance between anthropogenic emissions, and the perturbation of natural processes that leads to a partial removal of these gases from the atmosphere. Natural processes are linked to physical conditions, chemical reactions and biological transformations and they respond themselves to perturbed atmospheric composition and climate change. Therefore, the physical climate system and the biogeochemical cycles of CO_2, CH_4 and N_2O are coupled. This chapter addresses the present human-caused perturbation of the biogeochemical cycles of CO_2, CH_4 and N_2O, their variations in the past coupled to climate variations and their projected evolution during this century under future scenarios.

The Human-Caused Perturbation in the Industrial Era

CO_2 increased by 40% from 278 ppm about 1750 to 390.5 ppm in 2011. During the same time interval, CH_4 increased by 150% from 722 ppb to 1803 ppb, and N_2O by 20% from 271 ppb to 324.2 ppb in 2011. It is unequivocal that the current concentrations of atmospheric CO_2, CH_4 and N_2O exceed any level measured for at least the past 800,000 years, the period covered by ice cores. Furthermore, the average rate of increase of these three gases observed over the past century exceeds any observed rate of change over the previous 20,000 years. {2.2, 5.2, 6.1, 6.2}

Anthropogenic CO_2 emissions to the atmosphere were 555 ± 85 PgC (1 PgC = 10^{15} gC) between 1750 and 2011. Of this amount, fossil fuel combustion and cement production contributed 375 ± 30 PgC and land use change (including deforestation, afforestation and reforestation) contributed 180 ± 80 PgC. {6.3.1, Table 6.1}

With a *very high level of confidence*[1], the increase in CO_2 emissions from fossil fuel burning and those arising from land use change are the dominant cause of the observed increase in atmospheric CO_2 concentration. About half of the emissions remained in the atmosphere (240 ± 10 PgC) since 1750. The rest was removed from the atmosphere by sinks and stored in the natural carbon cycle reservoirs. The ocean reservoir stored 155 ± 30 PgC. Vegetation biomass and soils not affected by land use change stored 160 ± 90 PgC. {6.1, 6.3, 6.3.2.3, Table 6.1, Figure 6.8}

Carbon emissions from fossil fuel combustion and cement production increased faster during the 2000–2011 period than during the 1990–1999 period. These emissions were 9.5 ± 0.8 PgC yr^{-1} in 2011, 54% above their 1990 level. Anthropogenic net CO_2 emissions from land use change were 0.9 ± 0.8 PgC yr^{-1} throughout the past decade, and represent about 10% of the total anthropogenic CO_2 emissions. It is *more likely than not*[2] that net CO_2 emissions from land use change decreased during 2000–2011 compared to 1990–1999. {6.3, Table 6.1, Table 6.2, Figure 6.8}

Atmospheric CO_2 concentration increased at an average rate of 2.0 ± 0.1 ppm yr^{-1} during 2002–2011. This decadal rate of increase is higher than during any previous decade since direct atmospheric concentration measurements began in 1958. Globally, the size of the combined natural land and ocean sinks of CO_2 approximately followed the atmospheric rate of increase, removing 55% of the total anthropogenic emissions every year on average during 1958–2011. {6.3, Table 6.1}

After almost one decade of stable CH_4 concentrations since the late 1990s, atmospheric measurements have shown renewed CH_4 concentrations growth since 2007. The drivers of this renewed growth are still debated. The methane budget for the decade of 2000–2009 (bottom-up estimates) is 177 to 284 Tg(CH_4) yr^{-1} for natural wetlands emissions, 187 to 224 Tg(CH_4) yr^{-1} for agriculture and waste (rice, animals and waste), 85 to 105 Tg(CH_4) yr^{-1} for fossil fuel related emissions, 61 to 200 Tg(CH_4) yr^{-1} for other natural emissions including, among other fluxes, geological, termites and fresh water emissions, and 32 to 39 Tg(CH_4) yr^{-1} for biomass and biofuel burning (the range indicates the expanse of literature values). Anthropogenic emissions account for 50 to 65% of total emissions. By including natural geological CH_4 emissions that were not accounted for in previous budgets, the fossil component of the total CH_4 emissions (i.e., anthropogenic emissions related to leaks in the fossil fuel industry and natural geological leaks) is now estimated to amount to about 30% of the total CH_4 emissions (*medium confidence*). Climate driven fluctuations of CH_4 emissions from natural wetlands are the main drivers of the global interannual variability of CH_4 emissions (*high confidence*), with a smaller contribution from the variability in emissions from biomass burning during high fire years. {6.3.3, Figure 6.2, Table 6.8}

The concentration of N_2O increased at a rate of 0.73 ± 0.03 ppb yr^{-1} over the last three decades. Emissions of N_2O to the atmosphere are mostly caused by nitrification and de-nitrification reactions

6

[1] In this Report, the following summary terms are used to describe the available evidence: limited, medium, or robust; and for the degree of agreement: low, medium, or high. A level of confidence is expressed using five qualifiers: very low, low, medium, high, and very high, and typeset in italics, e.g., *medium confidence*. For a given evidence and agreement statement, different confidence levels can be assigned, but increasing levels of evidence and degrees of agreement are correlated with increasing confidence (see Section 1.4 and Box TS.1 for more details).

[2] In this Report, the following terms have been used to indicate the assessed likelihood of an outcome or a result: Virtually certain 99–100% probability, Very likely 90–100%, Likely 66–100%, About as likely as not 33–66%, Unlikely 0–33%, Very unlikely 0–10%, Exceptionally unlikely 0–1%. Additional terms (Extremely likely: 95–100%, More likely than not >50–100%, and Extremely unlikely 0–5%) may also be used when appropriate. Assessed likelihood is typeset in italics, e.g., *very likely* (see Section 1.4 and Box TS.1 for more details).

of reactive nitrogen in soils and in the ocean. Anthropogenic N_2O emissions increased steadily over the last two decades and were 6.9 (2.7 to 11.1) TgN (N_2O) yr^{-1} in 2006. Anthropogenic N_2O emissions are 1.7 to 4.8 TgN (N_2O) yr^{-1} from the application of nitrogenous fertilisers in agriculture, 0.2 to 1.8 TgN (N_2O) yr^{-1} from fossil fuel use and industrial processes, 0.2 to 1.0 TgN (N_2O) yr^{-1} from biomass burning (including biofuels) and 0.4 to 1.3 TgN (N_2O) yr^{-1} from land emissions due to atmospheric nitrogen deposition (the range indicates expand of literature values). Natural N_2O emissions derived from soils, oceans and a small atmospheric source are together 5.4 to 19.6 TgN (N_2O) yr^{-1}. {6.3, 6.3.4, Figure 6.4c, Figure 6.19, Table 6.9}

The human-caused creation of reactive nitrogen in 2010 was at least two times larger than the rate of natural terrestrial creation. The human-caused creation of reactive nitrogen is dominated by the production of ammonia for fertiliser and industry, with important contributions from legume cultivation and combustion of fossil fuels. Once formed, reactive nitrogen can be transferred to waters and the atmosphere. In addition to N_2O, two important nitrogen compounds emitted to the atmosphere are NH_3 and NO_x both of which influence tropospheric O_3 and aerosols through atmospheric chemistry. All of these effects contribute to radiative forcing. It is also *likely* that reactive nitrogen deposition over land currently increases natural CO_2 sinks, in particular forests, but the magnitude of this effect varies between regions. {6.1.3, 6.3, 6.3.2.6.5, 6.3.4, 6.4.6, Figures 6.4a and 6.4b, Table 6.9, Chapter 7}

Before the Human-Caused Perturbation

During the last 7000 years prior to 1750, atmospheric CO_2 from ice cores shows only very slow changes (increase) from 260 ppm to 280 ppm, in contrast to the human-caused increase of CO_2 since pre-industrial times. The contribution of CO_2 emissions from early anthropogenic land use is *unlikely* sufficient to explain the CO_2 increase prior to 1750. Atmospheric CH_4 from ice cores increased by about 100 ppb between 5000 years ago and around 1750. *About as likely as not*, this increase can be attributed to early human activities involving livestock, human-caused fires and rice cultivation. {6.2, Figures 6.6 and 6.7}

Further back in time, during the past 800,000 years prior to 1750, atmospheric CO_2 varied from 180 ppm during glacial (cold) up to 300 ppm during interglacial (warm) periods. This is well established from multiple ice core measurements. Variations in atmospheric CO_2 from glacial to interglacial periods were caused by decreased ocean carbon storage (500 to 1200 PgC), partly compensated by increased land carbon storage (300 to 1000 PgC). {6.2.1, Figure 6.5}

6

Future Projections

With *very high confidence*, ocean carbon uptake of anthropogenic CO_2 emissions will continue under all four Representative Concentration Pathways (RCPs) through to 2100, with higher uptake corresponding to higher concentration pathways. The future evolution of the land carbon uptake is much more uncertain, with a majority of models projecting a continued net carbon uptake under all RCPs, but with some models simulating a net loss of carbon by the land due to the combined effect of climate change and land use change. In view of the large spread of model results and incomplete process representation, there is *low confidence* on the magnitude of modelled future land carbon changes. {6.4.3, Figure 6.24}

There is *high confidence* that climate change will partially offset increases in global land and ocean carbon sinks caused by rising atmospheric CO_2. Yet, there are regional differences among Climate Modelling Intercomparison Project Phase 5 (CMIP5) Earth System Models, in the response of ocean and land CO_2 fluxes to climate. There is a high agreement between models that tropical ecosystems will store less carbon in a warmer climate. There is medium agreement between models that at high latitudes warming will increase land carbon storage, although none of the models account for decomposition of carbon in permafrost, which may offset increased land carbon storage. There is high agreement between CMIP5 Earth System models that ocean warming and circulation changes will reduce the rate of carbon uptake in the Southern Ocean and North Atlantic, but that carbon uptake will nevertheless persist in those regions. {6.4.2, Figures 6.21 and 6.22}

It is *very likely*, based on new experimental results {6.4.6.3} and modelling, that nutrient shortage will limit the effect of rising atmospheric CO_2 on future land carbon sinks, for the four RCP scenarios. There is *high confidence* that low nitrogen availability will limit carbon storage on land, even when considering anthropogenic nitrogen deposition. The role of phosphorus limitation is more uncertain. Models that combine nitrogen limitations with rising CO_2 and changes in temperature and precipitation thus produce a systematically larger increase in projected future atmospheric CO_2, for a given fossil fuel emissions trajectory. {6.4.6, 6.4.6.3, 6.4.8.2, Figure 6.35}

Taking climate and carbon cycle feedbacks into account, we can quantify the fossil fuel emissions compatible with the RCPs. Between 2012 and 2100, the RCP2.6, RCP4.5, RCP6.0, and RCP8.5 scenarios imply cumulative compatible fossil fuel emissions of 270 (140 to 410) PgC, 780 (595 to 1005) PgC, 1060 (840 to 1250) PgC and 1685 (1415 to 1910) PgC respectively (values quoted to nearest 5 PgC, range derived from CMIP5 model results). For RCP2.6, an average 50% (range 14 to 96%) emission reduction is required by 2050 relative to 1990 levels. By the end of the 21st century, about half of the models infer emissions slightly above zero, while the other half infer a net removal of CO_2 from the atmosphere. {6.4.3, Table 6.12, Figure 6.25}

There is *high confidence* that reductions in permafrost extent due to warming will cause thawing of some currently frozen carbon. However, there is *low confidence* on the magnitude of carbon losses through CO_2 and CH_4 emissions to the atmosphere, with a range from 50 to 250 PgC between 2000 and 2100 under the RCP8.5 scenario. The CMIP5 Earth System Models did not include frozen carbon feedbacks. {6.4.3.4, Chapter 12}

There is *medium confidence* that emissions of CH_4 from wetlands are *likely* to increase under elevated CO_2 and a warmer climate. But there is *low confidence* in quantitative projections of these changes. The likelihood of the future release of CH_4 from marine gas hydrates in response to seafloor warming is poorly understood. In the event of a significant release of CH_4 from hydrates in the sea floor by the end of the 21st century, it is *likely* that subsequent emissions to the atmosphere would be in the form of CO_2, due to CH_4 oxidation in the water column. {6.4.7, Figure 6.37}

It is *likely* that N_2O emissions from soils will increase due to the increased demand for feed/food and the reliance of agriculture on nitrogen fertilisers. Climate warming will *likely* amplify agricultural and natural terrestrial N_2O sources, but there is *low confidence* in quantitative projections of these changes. {6.4.6, Figure 6.32}

It is *virtually certain* that the increased storage of carbon by the ocean will increase acidification in the future, continuing the observed trends of the past decades. Ocean acidification in the surface ocean will follow atmospheric CO_2 while it will also increase in the deep ocean as CO_2 continues to penetrate the abyss. The CMIP5 models consistently project worldwide increased ocean acidification to 2100 under all RCPs. The corresponding decrease in surface ocean pH by the end of the 21st century is 0.065 (0.06 to 0.07) for RCP2.6, 0.145 (0.14 to 0.15) for RCP4.5, 0.203 (0.20 to 0.21) for RCP6.0, and 0.31 (0.30 to 0.32) for RCP8.5 (range from CMIP5 models spread). Surface waters become seasonally corrosive to aragonite in parts of the Arctic and in some coastal upwelling systems within a decade, and in parts of the Southern Ocean within 1 to 3 decades in most scenarios. Aragonite undersaturation becomes widespread in these regions at atmospheric CO_2 levels of 500 to 600 ppm. {6.4.4, Figures 6.28 and 6.29}

It is *very likely* that the dissolved oxygen content of the ocean will decrease by a few percent during the 21st century. CMIP5 models suggest that this decrease in dissolved oxygen will predominantly occur in the subsurface mid-latitude oceans, caused by enhanced stratification, reduced ventilation and warming. However, there is no consensus on the future development of the volume of hypoxic and suboxic waters in the open-ocean because of large uncertainties in potential biogeochemical effects and in the evolution of tropical ocean dynamics. {6.4.5, Figure 6.30}

Irreversible Long-Term Impacts of Human-Caused Emissions

With *very high confidence*, the physical, biogeochemical carbon cycle in the ocean and on land will continue to respond to climate change and rising atmospheric CO_2 concentrations created during the 21st century. Ocean acidification will *very likely* continue in the future as long as the oceans take up atmospheric CO_2. Committed land ecosystem carbon cycle changes will manifest themselves further beyond the end of the 21st century. In addition, it is *virtually certain* that large areas of permafrost will experience thawing over multiple centuries. There is, however, *low confidence* in the magnitude of frozen carbon losses to the atmosphere, and the relative contributions of CO_2 and CH_4 emissions. {6.4.4, 6.4.9, Chapter 12}

The magnitude and sign of the response of the natural carbon reservoirs to changes in climate and rising CO_2 vary substantially over different time scales. The response to rising CO_2 is to increase cumulative land and ocean uptake, regardless of the time scale. The response to climate change is variable, depending of the region considered because of different responses of the underlying physical and biological mechanisms at different time scales. {6.4, Table 6.10, Figures 6.14 and 6.17}

The removal of human-emitted CO_2 from the atmosphere by natural processes will take a few hundred thousand years (*high confidence*). Depending on the RCP scenario considered, about 15 to 40% of emitted CO_2 will remain in the atmosphere longer than 1,000 years. This very long time required by sinks to remove anthropogenic CO_2 makes climate change caused by elevated CO_2 irreversible on human time scale. {Box 6.1}

Geoengineering Methods and the Carbon Cycle

Unconventional ways to remove CO_2 from the atmosphere on a large scale are termed Carbon Dioxide Removal (CDR) methods. CDR could in theory be used to reduce CO_2 atmospheric concentrations but these methods have biogeochemical and technological limitations to their potential. Uncertainties make it difficult to quantify how much CO_2 emissions could be offset by CDR on a human time scale, although it is *likely* that CDR would have to be deployed at large-scale for at least one century to be able to significantly reduce atmospheric CO_2. In addition, it is *virtually certain* that the removal of CO_2 by CDR will be partially offset by outgassing of CO_2 from the ocean and land ecosystems. {6.5, Figures 6.39 and 6.40, Table 6.15, Box 6.1, FAQ 7.3}

The level of confidence on the side effects of CDR methods on carbon and other biogeochemical cycles is low. Some of the climatic and environmental effects of CDR methods are associated with altered surface albedo (for afforestation), de-oxygenation and enhanced N_2O emissions (for artificial ocean fertilisation). Solar Radiation Management (SRM) methods (Chapter 7) will not directly interfere with the effects of elevated CO_2 on the carbon cycle, such as ocean acidification, but will impact carbon and other biogeochemical cycles through their climate effects. {6.5.3, 6.5.4, 7.7, Tables 6.14 and 6.15}

6.1 Introduction

The radiative properties of the atmosphere are strongly influenced by the abundance of well-mixed GHGs (see Glossary), mainly carbon dioxide (CO_2), methane (CH_4) and nitrous oxide (N_2O), which have substantially increased since the beginning of the Industrial Era (defined as beginning in the year 1750), due primarily to anthropogenic emissions (see Chapter 2). Well-mixed GHGs represent the gaseous phase of global biogeochemical cycles, which control the complex flows and transformations of the elements between the different components of the Earth System (atmosphere, ocean, land, lithosphere) by biotic and abiotic processes. Since most of these processes are themselves also dependent on the prevailing environment, changes in climate and human impacts on ecosystems (e.g., land use and land use change) also modify the atmospheric concentrations of CO_2, CH_4 and N_2O. During the glacial-interglacial cycles (see Glossary), in absence of significant direct human impacts, long variations in climate also affected CO_2, CH_4 and N_2O and vice versa (see Chapter 5, Section 5.2.2). In the coming century, the situation would be quite different, because of the dominance of anthropogenic emissions that affect global biogeochemical cycles, and in turn, climate change (see Chapter 12). Biogeochemical cycles thus constitute feedbacks in the Earth System.

This chapter summarizes the scientific understanding of atmospheric budgets, variability and trends of the three major biogeochemical greenhouse gases, CO_2, CH_4 and N_2O, their underlying source and sink processes and their perturbations caused by direct human impacts, past and present climate changes as well as future projections of climate change. After the introduction (Section 6.1), Section 6.2 assesses the present understanding of the mechanisms responsible for the variations of CO_2, CH_4 and N_2O in the past emphasizing glacial-interglacial changes, and the smaller variations during the Holocene (see Glossary) since the last glaciation and over the last millennium. Section 6.3 focuses on the Industrial Era addressing the major source and sink processes, and their variability in space and time. This information is then used to evaluate critically the models of the biogeochemical cycles, including their sensitivity to changes in atmospheric composition and climate. Section 6.4 assesses future projections of carbon and other biogeochemical cycles computed, in particular, with CMIP5 Earth System Models. This includes a quantitative assessment of the direction and magnitude of the various feedback mechanisms as represented in current models, as well as additional processes that might become important in the future but which are not yet fully understood. Finally, Section 6.5 addresses the potential effects and uncertainties of deliberate carbon dioxide removal methods (see Glossary) and solar radiation management (see Glossary) on the carbon cycle.

6.1.1 Global Carbon Cycle Overview

6.1.1.1 Carbon Dioxide and the Global Carbon Cycle

Atmospheric CO_2 represents the main atmospheric phase of the global carbon cycle. The global carbon cycle can be viewed as a series of reservoirs of carbon in the Earth System, which are connected by exchange fluxes of carbon. Conceptually, one can distinguish two domains in the global carbon cycle. The first is a fast domain with large exchange fluxes and relatively 'rapid' reservoir turnovers, which consists of carbon in the atmosphere, the ocean, surface ocean sediments and on land in vegetation, soils and freshwaters. Reservoir turnover times, defined as reservoir mass of carbon divided by the exchange flux, range from a few years for the atmosphere to decades to millennia for the major carbon reservoirs of the land vegetation and soil and the various domains in the ocean. A second, slow domain consists of the huge carbon stores in rocks and sediments which exchange carbon with the fast domain through volcanic emissions of CO_2, chemical weathering (see Glossary), erosion and sediment formation on the sea floor (Sundquist, 1986). Turnover times of the (mainly geological) reservoirs of the slow domain are 10,000 years or longer. Natural exchange fluxes between the slow and the fast domain of the carbon cycle are relatively small (<0.3 PgC yr^{-1}, 1 PgC = 10^{15} gC) and can be assumed as approximately constant in time (volcanism, sedimentation) over the last few centuries, although erosion and river fluxes may have been modified by human-induced changes in land use (Raymond and Cole, 2003).

During the Holocene (beginning 11,700 years ago) prior to the Industrial Era the fast domain was close to a steady state, as evidenced by the relatively small variations of atmospheric CO_2 recorded in ice cores (see Section 6.2), despite small emissions from human-caused changes in land use over the last millennia (Pongratz et al., 2009). By contrast, since the beginning of the Industrial Era, fossil fuel extraction from geological reservoirs, and their combustion, has resulted in the transfer of significant amount of fossil carbon from the slow domain into the fast domain, thus causing an unprecedented, major human-induced perturbation in the carbon cycle. A schematic of the global carbon cycle with focus on the fast domain is shown in Figure 6.1. The numbers represent the estimated current pool sizes in PgC and the magnitude of the different exchange fluxes in PgC yr^{-1} averaged over the time period 2000–2009 (see Section 6.3).

In the atmosphere, CO_2 is the dominant carbon bearing trace gas with a current (2011) concentration of approximately 390.5 ppm (Dlugokencky and Tans, 2013a), which corresponds to a mass of 828 PgC (Prather et al., 2012; Joos et al., 2013). Additional trace gases include methane (CH_4, current content mass ~3.7 PgC) and carbon monoxide (CO, current content mass ~0.2 PgC), and still smaller amounts of hydrocarbons, black carbon aerosols and organic compounds.

The terrestrial biosphere reservoir contains carbon in organic compounds in vegetation living biomass (450 to 650 PgC; Prentice et al., 2001) and in dead organic matter in litter and soils (1500 to 2400 PgC; Batjes, 1996). There is an additional amount of old soil carbon in wetland soils (300 to 700 PgC; Bridgham et al., 2006) and in permafrost soils (see Glossary) (~1700 PgC; Tarnocai et al., 2009); albeit some overlap with these two quantities. CO_2 is removed from the atmosphere by plant photosynthesis (Gross Primary Production (GPP), 123±8 PgC yr^{-1}, (Beer et al., 2010)) and carbon fixed into plants is then cycled through plant tissues, litter and soil carbon and can be released back into the atmosphere by autotrophic (plant) and heterotrophic (soil microbial and animal) respiration and additional disturbance processes (e.g., sporadic fires) on a very wide range of time scales (seconds to millennia). Because CO_2 uptake by photosynthesis occurs only during the growing season, whereas CO_2 release by respiration occurs nearly year-round, the greater land mass in the Northern Hemisphere (NH) imparts

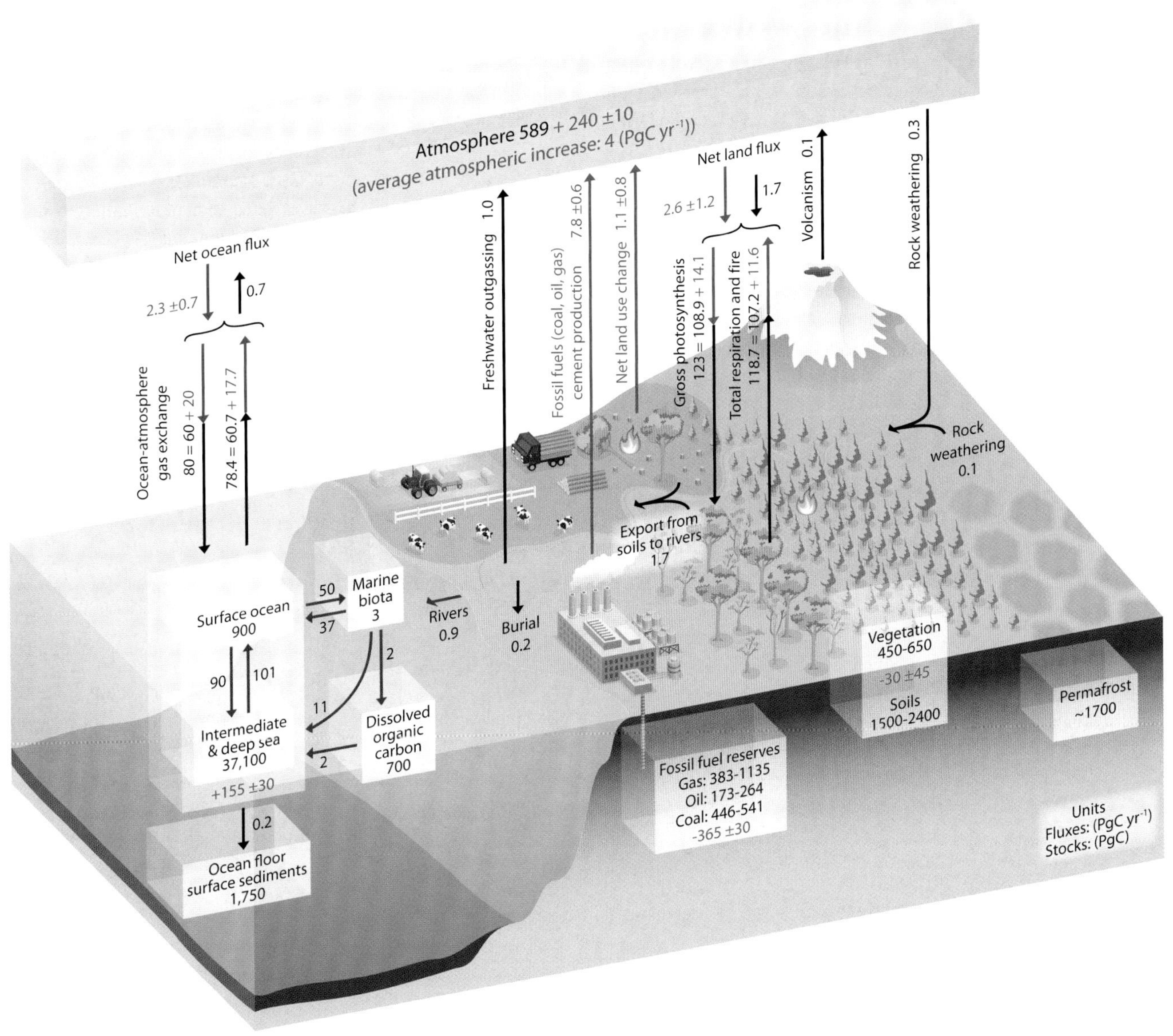

Figure 6.1 | Simplified schematic of the global carbon cycle. Numbers represent reservoir mass, also called 'carbon stocks' in PgC (1 PgC = 10^{15} gC) and annual carbon exchange fluxes (in PgC yr^{-1}). Black numbers and arrows indicate reservoir mass and exchange fluxes estimated for the time prior to the Industrial Era, about 1750 (see Section 6.1.1.1 for references). Fossil fuel reserves are from GEA (2006) and are consistent with numbers used by IPCC WGIII for future scenarios. The sediment storage is a sum of 150 PgC of the organic carbon in the mixed layer (Emerson and Hedges, 1988) and 1600 PgC of the deep-sea $CaCO_3$ sediments available to neutralize fossil fuel CO_2 (Archer et al., 1998). Red arrows and numbers indicate annual 'anthropogenic' fluxes averaged over the 2000–2009 time period. These fluxes are a perturbation of the carbon cycle during Industrial Era post 1750. These fluxes (red arrows) are: *Fossil fuel and cement emissions of CO_2* (Section 6.3.1), *Net land use change* (Section 6.3.2), and the *Average atmospheric increase* of CO_2 in the atmosphere, also called 'CO_2 growth rate' (Section 6.3). The uptake of anthropogenic CO_2 by the ocean and by terrestrial ecosystems, often called 'carbon sinks' are the red arrows part of *Net land flux* and *Net ocean flux*. Red numbers in the reservoirs denote cumulative changes of anthropogenic carbon over the Industrial Period 1750–2011 (column 2 in Table 6.1). By convention, a positive cumulative change means that a reservoir has gained carbon since 1750. The cumulative change of anthropogenic carbon in the terrestrial reservoir is the sum of carbon cumulatively lost through land use change and carbon accumulated since 1750 in other ecosystems (Table 6.1). Note that the mass balance of the two ocean carbon stocks *Surface ocean* and *Intermediate and deep ocean* includes a yearly accumulation of anthropogenic carbon (not shown). Uncertainties are reported as 90% confidence intervals. Emission estimates and land and ocean sinks (in red) are from Table 6.1 in Section 6.3. The change of gross terrestrial fluxes (red arrows of *Gross photosynthesis* and *Total respiration and fires*) has been estimated from CMIP5 model results (Section 6.4). The change in air–sea exchange fluxes (red arrows of ocean atmosphere gas exchange) have been estimated from the difference in atmospheric partial pressure of CO_2 since 1750 (Sarmiento and Gruber, 2006). Individual gross fluxes and their changes since the beginning of the Industrial Era have typical uncertainties of more than 20%, while their differences (*Net land flux* and *Net ocean flux* in the figure) are determined from independent measurements with a much higher accuracy (see Section 6.3). Therefore, to achieve an overall balance, the values of the more uncertain gross fluxes have been adjusted so that their difference matches the *Net land flux* and *Net ocean flux* estimates. Fluxes from volcanic eruptions, rock weathering (silicates and carbonates weathering reactions resulting into a small uptake of atmospheric CO_2), export of carbon from soils to rivers, burial of carbon in freshwater lakes and reservoirs and transport of carbon by rivers to the ocean are all assumed to be pre-industrial fluxes, that is, unchanged during 1750–2011. Some recent studies (Section 6.3) indicate that this assumption is likely not verified, but global estimates of the Industrial Era perturbation of all these fluxes was not available from peer-reviewed literature. The atmospheric inventories have been calculated using a conversion factor of 2.12 PgC per ppm (Prather et al., 2012).

a characteristic 'sawtooth' seasonal cycle in atmospheric CO_2 (Keeling, 1960) (see Figure 6.3). A significant amount of terrestrial carbon (1.7 PgC yr^{-1}; Figure 6.1) is transported from soils to rivers headstreams. A fraction of this carbon is outgassed as CO_2 by rivers and lakes to the atmosphere, a fraction is buried in freshwater organic sediments and the remaining amount (~0.9 PgC yr^{-1}; Figure 6.1) is delivered by rivers to the coastal ocean as dissolved inorganic carbon, dissolved organic carbon and particulate organic carbon (Tranvik et al., 2009).

Atmospheric CO_2 is exchanged with the surface ocean through gas exchange. This exchange flux is driven by the partial CO_2 pressure difference between the air and the sea. In the ocean, carbon is available predominantly as Dissolved Inorganic Carbon (DIC, ~38,000 PgC; Figure 6.1), that is carbonic acid (dissolved CO_2 in water), bicarbonate and carbonate ions, which are tightly coupled via ocean chemistry. In addition, the ocean contains a pool of Dissolved Organic Carbon (DOC, ~700 PgC), of which a substantial fraction has a turnover time of 1000 years or longer (Hansell et al., 2009). The marine biota, predominantly phytoplankton and other microorganisms, represent a small organic carbon pool (~3 PgC), which is turned over very rapidly in days to a few weeks.

Carbon is transported within the ocean by three mechanisms (Figure 6.1): (1) the 'solubility pump' (see Glossary), (2) the 'biological pump' (see Glossary), and (3) the 'marine carbonate pump' that is generated by the formation of calcareous shells of certain oceanic microorganisms in the surface ocean, which, after sinking to depth, are re-mineralized back into DIC and calcium ions. The marine carbonate pump operates counter to the marine biological soft-tissue pump with respect to its effect on CO_2: in the formation of calcareous shells, two bicarbonate ions are split into one carbonate and one dissolved CO_2 molecules, which increases the partial CO_2 pressure in surface waters (driving a release of CO_2 to the atmosphere). Only a small fraction (~0.2 PgC yr^{-1}) of the carbon exported by biological processes (both soft-tissue and carbonate pumps) from the surface reaches the sea floor where it can be stored in sediments for millennia and longer (Denman et al., 2007).

Box 6.1 | Multiple Residence Times for an Excess of Carbon Dioxide Emitted in the Atmosphere

On an average, CO_2 molecules are exchanged between the atmosphere and the Earth surface every few years. This fast CO_2 cycling through the atmosphere is coupled to a slower cycling of carbon through land vegetation, litter and soils and the upper ocean (decades to centuries); deeper soils and the deep sea (centuries to millennia); and geological reservoirs, such as deep-sea carbonate sediments and the upper mantle (up to millions of years) as explained in Section 6.1.1.1. Atmospheric CO_2 represents only a tiny fraction of the carbon in the Earth System, the rest of which is tied up in these other reservoirs. Emission of carbon from fossil fuel reserves, and additionally from land use change (see Section 6.3) is now rapidly increasing atmospheric CO_2 content. The removal of all the human-emitted CO_2 from the atmosphere by natural processes will take a few hundred thousand years (*high confidence*) as shown by the timescales of the removal process shown in the table below (Archer and Brovkin, 2008). For instance, an extremely long atmospheric CO_2 recovery time scale from a large emission pulse of CO_2 has been inferred from geological evidence when during the Paleocene–Eocene thermal maximum event about 55 million years ago a large amount of CO_2 was released to the atmosphere (McInerney and Wing, 2011). Based on the amount of CO_2 remaining in the atmosphere after a pulse of emissions (data from Joos et al. 2013) and on the magnitude of the historical and future emissions for each RCP scenario, we assessed that about 15 to 40% of CO_2 emitted until 2100 will remain in the atmosphere longer than 1000 years.

Box 6.1, Table 1 | The main natural processes that remove CO_2 consecutive to a large emission pulse to the atmosphere, their atmospheric CO_2 adjustment time scales, and main (bio)chemical reactions involved.

Processes	Time scale (years)	Reactions
Land uptake: Photosynthesis–respiration	$1–10^2$	$6CO_2 + 6H_2O + \text{photons} \rightarrow C_6H_{12}O_6 + 6O_2$ $C_6H_{12}O_6 + 6O_2 \rightarrow 6CO_2 + 6H_2O + \text{heat}$
Ocean invasion: Seawater buffer	$10–10^3$	$CO_2 + CO_3^{2-} + H_2O \rightleftharpoons 2HCO_3^-$
Reaction with calcium carbonate	$10^3–10^4$	$CO_2 + CaCO_3 + H_2O \rightarrow Ca^{2+} + 2HCO_3^-$
Silicate weathering	$10^4–10^6$	$CO_2 + CaSiO_3 \rightarrow CaCO_3 + SiO_2$

6

These processes are active on all time scales, but the relative importance of their role in the CO_2 removal is changing with time and depends on the level of emissions. Accordingly, the times of atmospheric CO_2 adjustment to anthropogenic carbon emissions can be divided into three phases associated with increasingly longer time scales.

Phase 1. Within several decades of CO_2 emissions, about a third to half of an initial pulse of anthropogenic CO_2 goes into the land and ocean, while the rest stays in the atmosphere (Box 6.1, Figure 1a). Within a few centuries, most of the anthropogenic CO_2 will be in the form of additional dissolved inorganic carbon in the ocean, thereby decreasing ocean pH (Box 6.1, Figure 1b). Within a thousand years, the remaining atmospheric fraction of the CO_2 emissions (see Section 6.3.2.4) is between 15 and 40%, depending on the amount of carbon released (Archer et al., 2009b). The carbonate buffer capacity of the ocean decreases with higher CO_2, so the larger the cumulative emissions, the higher the remaining atmospheric fraction (Eby et al., 2009; Joos et al., 2013). (continued on next page)

Box 6.1 (continued)

Phase 2. In the second stage, within a few thousands of years, the pH of the ocean that has decreased in Phase 1 will be restored by reaction of ocean dissolved CO_2 and calcium carbonate ($CaCO_3$) of sea floor sediments, partly replenishing the buffer capacity of the ocean and further drawing down atmospheric CO_2 as a new balance is re-established between $CaCO_3$ sedimentation in the ocean and terrestrial weathering (Box 6.1, Figure 1c right). This second phase will pull the remaining atmospheric CO_2 fraction down to 10 to 25% of the original CO_2 pulse after about 10 kyr (Lenton and Britton, 2006; Montenegro et al., 2007; Ridgwell and Hargreaves, 2007; Tyrrell et al., 2007; Archer and Brovkin, 2008).

Phase 3. In the third stage, within several hundred thousand years, the rest of the CO_2 emitted during the initial pulse will be removed from the atmosphere by silicate weathering, a very slow process of CO_2 reaction with calcium silicate ($CaSiO_3$) and other minerals of igneous rocks (e.g., Sundquist, 1990; Walker and Kasting, 1992).

Involvement of extremely long time scale processes into the removal of a pulse of CO_2 emissions into the atmosphere complicates comparison with the cycling of the other GHGs. This is why the concept of a single, characteristic atmospheric lifetime is not applicable to CO_2 (Chapter 8).

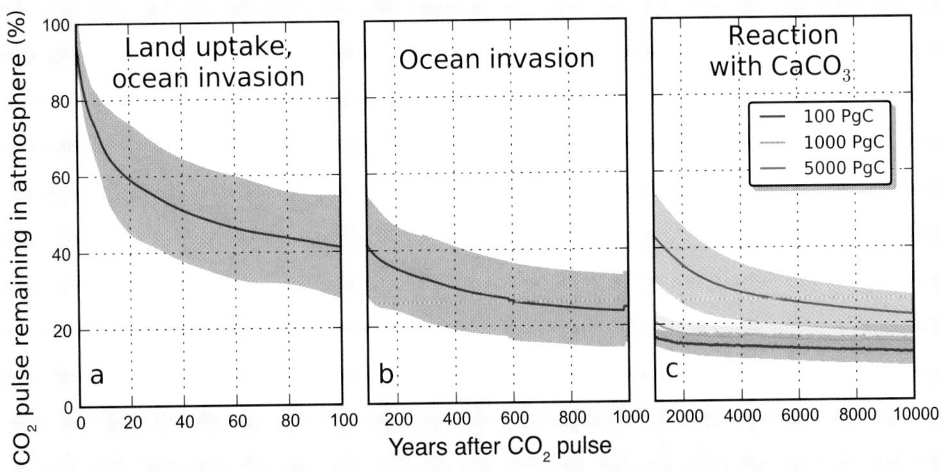

Box 6.1, Figure 1 | A percentage of emitted CO_2 remaining in the atmosphere in response to an idealised instantaneous CO_2 pulse emitted to the atmosphere in year 0 as calculated by a range of coupled climate–carbon cycle models. (Left and middle panels, a and b) Multi-model mean (blue line) and the uncertainty interval (±2 standard deviations, shading) simulated during 1000 years following the instantaneous pulse of 100 PgC (Joos et al., 2013). (Right panel, c) A mean of models with oceanic and terrestrial carbon components and a maximum range of these models (shading) for instantaneous CO_2 pulse in year 0 of 100 PgC (blue), 1000 PgC (orange) and 5000 PgC (red line) on a time interval up to 10 kyr (Archer et al., 2009b). Text at the top of the panels indicates the dominant processes that remove the excess of CO_2 emitted in the atmosphere on the successive time scales. Note that higher pulse of CO_2 emissions leads to higher remaining CO_2 fraction (Section 6.3.2.4) due to reduced carbonate buffer capacity of the ocean and positive climate–carbon cycle feedback (Section 6.3.2.6.6).

6.1.1.2 Methane Cycle

CH_4 absorbs infrared radiation relatively stronger per molecule compared to CO_2 (Chapter 8), and it interacts with photochemistry. On the other hand, the methane turnover time (see Glossary) is less than 10 years in the troposphere (Prather et al., 2012; see Chapter 7). The sources of CH_4 at the surface of the Earth (see Section 6.3.3.2) can be thermogenic including (1) natural emissions of fossil CH_4 from geological sources (marine and terrestrial seepages, geothermal vents and mud volcanoes) and (2) emissions caused by leakages from fossil fuel extraction and use (natural gas, coal and oil industry; Figure 6.2). There are also pyrogenic sources resulting from incomplete burning of fossil fuels and plant biomass (both natural and anthropogenic fires). The biogenic sources include natural biogenic emissions predominantly from wetlands, from termites and very small emissions from the ocean (see Section 6.3.3). Anthropogenic biogenic emissions occur from rice paddy agriculture, ruminant livestock, landfills, man-made lakes and wetlands and waste treatment. In general, biogenic CH_4 is produced from organic matter under low oxygen conditions by fermentation processes of methanogenic microbes (Conrad, 1996). Atmospheric CH_4 is removed primarily by photochemistry, through atmospheric chemistry reactions with the OH radicals. Other smaller removal processes of atmospheric CH_4 take place in the stratosphere through reaction with chlorine and oxygen radicals, by oxidation in well aerated soils, and possibly by reaction with chlorine in the marine boundary layer (Allan et al., 2007; see Section 6.3.3.3).

A very large geological stock (globally 1500 to 7000 PgC, that is 2 x 10^6 to 9.3 x 10^6 Tg(CH_4) in Figure 6.2; Archer (2007); with *low confidence* in estimates) of CH_4 exists in the form of frozen hydrate deposits ('clathrates') in shallow ocean sediments and on the slopes of continental shelves, and permafrost soils. These CH_4 hydrates are stable under conditions of low temperature and high pressure. Warming or changes in pressure could render some of these hydrates unstable with a potential release of CH_4 to the overlying soil/ocean and/or atmosphere. Possible future CH_4 emissions from CH_4 released by gas hydrates are discussed in Section 6.4.7.3.

6.1.2 Industrial Era

6.1.2.1 Carbon Dioxide and the Global Carbon Cycle

Since the beginning of the Industrial Era, humans have been producing energy by burning of fossil fuels (coal, oil and gas), a process that is releasing large amounts of CO_2 into the atmosphere (Rotty, 1983; Boden et al., 2011; see Section 6.3.2.1). The amount of fossil fuel CO_2 emitted to the atmosphere can be estimated with an accuracy of about 5 to 10% for recent decades from statistics of fossil fuel use (Andres et al., 2012). Total cumulative emissions between 1750 and 2011 amount to 375 ± 30 PgC (see Section 6.3.2.1 and Table 6.1), including a contribution of 8 PgC from the production of cement.

The second major source of anthropogenic CO_2 emissions to the atmosphere is caused by changes in land use (mainly deforestation), which causes globally a net reduction in land carbon storage, although recovery from past land use change can cause a net gain in in land

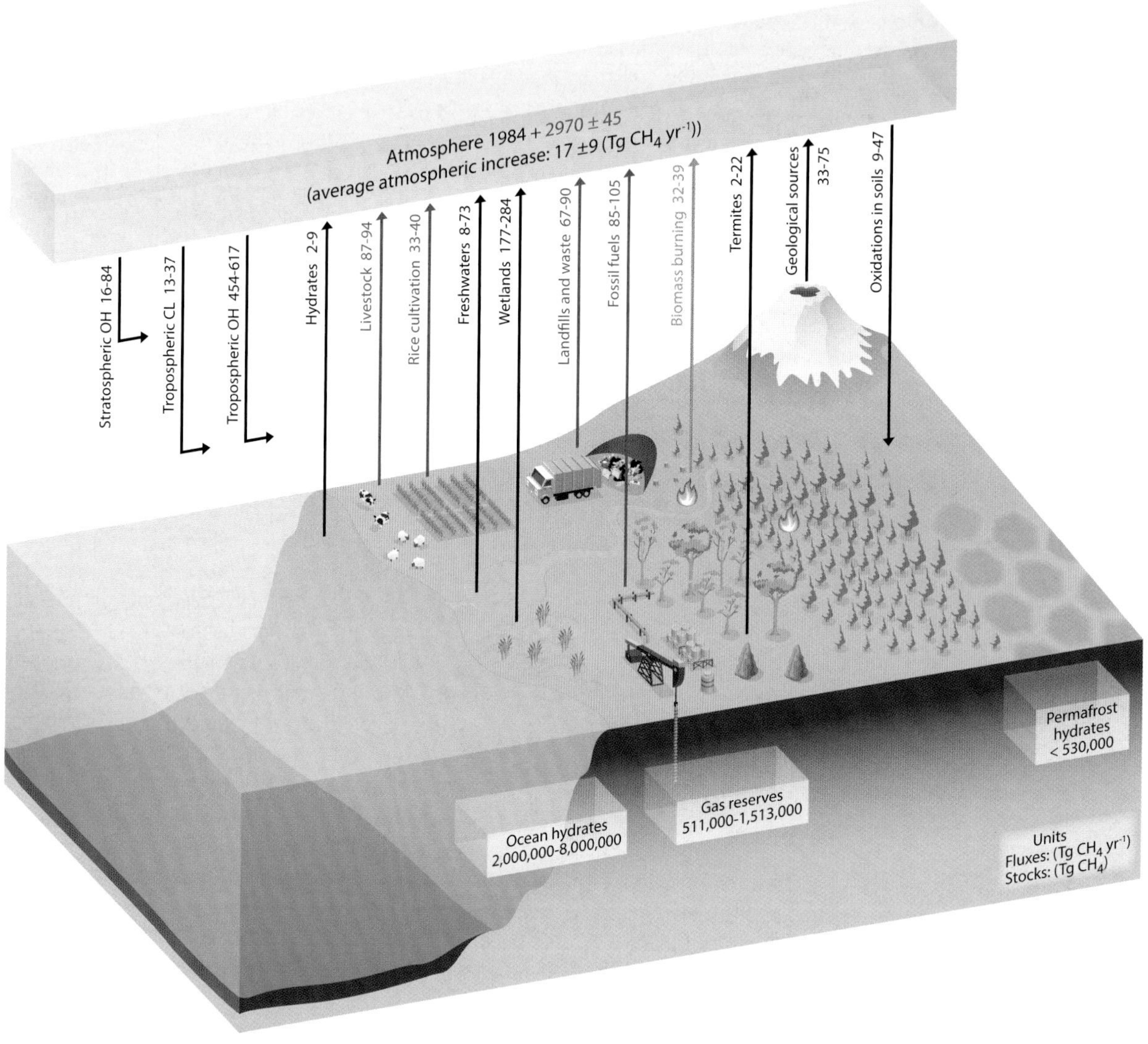

Figure 6.2 | Schematic of the global cycle of CH_4. Numbers represent annual fluxes in Tg(CH_4) yr^{-1} estimated for the time period 2000–2009 and CH_4 reservoirs in Tg (CH_4): the atmosphere and three geological reservoirs (hydrates on land and in the ocean floor and gas reserves) (see Section 6.3.3). Black arrows denote 'natural' fluxes, that is, fluxes that are not directly caused by human activities since 1750, red arrows anthropogenic fluxes, and the light brown arrow denotes a combined natural + anthropogenic flux. Note that human activities (e.g., land use) may have modified indirectly the global magnitude of the natural fluxes (Section 6.3.3). Ranges represent minimum and maximum values from cited references as given in Table 6.8 in Section 6.3.3. Gas reserves are from GEA (2006) and are consistent with numbers used by IPCC WG III for future scenarios. Hydrate reservoir sizes are from Archer et al. (2007). The atmospheric inventories have been calculated using a conversion factor of 2.7476 $TgCH_4$ per ppb (Prather et al., 2012). The assumed preindustrial annual mean globally averaged CH_4 concentration was 722 ± 25 ppb taking the average of the Antarctic Law Dome ice core observations (MacFarling-Meure et al., 2006) and the measurements from the GRIP ice core in Greenland (Blunier et al., 1995; see also Table 2.1). The atmospheric inventory in the year 2011 is based on an annual globally averaged CH_4 concentration of 1803 ± 4 ppb in the year 2011 (see Table 2.1). It is the sum of the atmospheric increase between 1750 and 2011 (in red) and of the pre-industrial inventory (in black). The *average atmospheric increase* each year, also called growth rate, is based on a measured concentration increase of 2.2 ppb yr^{-1} during the time period 2000–2009 (Dlugokencky et al., 2011).

carbon storage in some regions. Estimation of this CO_2 source to the atmosphere requires knowledge of changes in land area as well as estimates of the carbon stored per area before and after the land use change. In addition, longer term effects, such as the decomposition of soil organic matter after land use change, have to be taken into account (see Section 6.3.2.2). Since 1750, anthropogenic land use changes have resulted into about 50 million km^2 being used for cropland and pasture, corresponding to about 38% of the total ice-free land area (Foley et al., 2007, 2011), in contrast to an estimated cropland and pasture area of 7.5 to 9 million km^2 about 1750 (Ramankutty and Foley, 1999; Goldewijk, 2001). The cumulative net CO_2 emissions from land use changes between 1750 and 2011 are estimated at approximately 180 ± 80 PgC (see Section 6.3 and Table 6.1).

Multiple lines of evidence indicate that the observed atmospheric increase in the global CO_2 concentration since 1750 (Figure 6.3) is caused by the anthropogenic CO_2 emissions (see Section 6.3.2.3). The rising atmospheric CO_2 content induces a disequilibrium in the exchange fluxes between the atmosphere and the land and oceans respectively. The rising CO_2 concentration implies a rising atmospheric CO_2 partial pressure (pCO_2) that induces a globally averaged net-air-to-sea flux and thus an ocean sink for CO_2 (see Section 6.3.2.5). On land, the rising atmospheric CO_2 concentration fosters photosynthesis via the CO_2 fertilisation effect (see Section 6.3.2.6). However, the efficacy of these oceanic and terrestrial sinks does also depend on how the excess carbon is transformed and redistributed within these sink reservoirs. The magnitude of the current sinks is shown in Figure 6.1 (averaged over the years 2000–2009, red arrows), together with the cumulative reservoir content changes over the industrial era (1750–2011, red numbers) (see Table 6.1, Section 6.3).

6.1.2.2 Methane Cycle

After 1750, atmospheric CH_4 levels rose almost exponentially with time, reaching 1650 ppb by the mid-1980s and 1803 ppb by 2011. Between the mid-1980s and the mid-2000s the atmospheric growth of CH_4 declined to nearly zero (see Section 6.3.3.1, see also Chapter 2). More recently since 2006, atmospheric CH_4 is observed to increase again (Rigby et al., 2008); however, it is unclear if this is a short-term fluctuation or a new regime for the CH_4 cycle (Dlugokencky et al., 2009).

There is *very high* level of *confidence* that the atmospheric CH_4 increase during the Industrial Era is caused by anthropogenic activities. The massive increase in the number of ruminants (Barnosky, 2008), the emissions from fossil fuel extraction and use, the expansion of rice paddy agriculture and the emissions from landfills and waste are the dominant anthropogenic CH_4 sources. Total anthropogenic sources contribute at present between 50 and 65% of the total CH_4 sources (see Section 6.3.3). The dominance of CH_4 emissions located mostly in the NH (wetlands and anthropogenic emissions) is evidenced by the observed positive north–south gradient in CH_4 concentrations (Figure 6.3). Satellite-based CH_4 concentration measurements averaged over the entire atmospheric column also indicate higher concentrations of CH_4 above and downwind of densely populated and intensive agriculture areas where anthropogenic emissions occur (Frankenberg et al., 2011).

6.1.3 Connections Between Carbon and the Nitrogen and Oxygen Cycles

6.1.3.1 Global Nitrogen Cycle Including Nitrous Oxide

The biogeochemical cycles of nitrogen and carbon are tightly coupled with each other owing to the metabolic needs of organisms for these two elements. Changes in the availability of one element will influence not only biological productivity but also availability and requirements for the other element (Gruber and Galloway, 2008) and in the longer term, the structure and functioning of ecosystems as well.

Before the Industrial Era, the creation of reactive nitrogen Nr (all nitrogen species other than N_2) from non-reactive atmospheric N_2 occurred primarily through two natural processes: lightning and biological nitrogen fixation (BNF). BNF is a set of reactions that convert N_2 to ammonia in a microbially mediated process. This input of Nr to the land and ocean biosphere was in balance with the loss of Nr though denitrification, a process that returns N_2 back to the atmosphere (Ayres et al., 1994). This equilibrium has been broken since the beginning of the Industrial Era. Nr is produced by human activities and delivered to ecosystems. During the last decades, the production of Nr by humans has been much greater than the natural production (Figure 6.4a; Section 6.3.4.3). There are three main anthropogenic sources of Nr: (1) the Haber-Bosch industrial process, used to make NH_3 from N_2, for nitrogen fertilisers and as a feedstock for some industries; (2) the cultivation of legumes and other crops, which increases BNF; and (3) the combustion of fossil fuels, which converts atmospheric N_2 and fossil fuel nitrogen into nitrogen oxides (NO_x) emitted to the atmosphere and re-deposited at the surface (Figure 6.4a). In addition, there is a small flux from the mobilization of sequestered Nr from nitrogen-rich sedimentary rocks (Morford et al., 2011) (not shown in Figure 6.4a).

The amount of anthropogenic Nr converted back to non-reactive N_2 by denitrification is much smaller than the amount of Nr produced each year, that is, about 30 to 60% of the total Nr production, with a large uncertainty (Galloway et al., 2004; Canfield et al., 2010; Bouwman et al., 2013). What is more certain is the amount of N_2O emitted to the atmosphere. Anthropogenic sources of N_2O are about the same size as natural terrestrial sources (see Section 6.3.4 and Table 6.9 for the global N_2O budget). In addition, emissions of Nr to the atmosphere, as NH_3 and NO_x, are caused by agriculture and fossil fuel combustion. A portion of the emitted NH_3 and NO_x is deposited over the continents, while the rest gets transported by winds and deposited over the oceans. This atmospheric deposition flux of Nr over the oceans is comparable to the flux going from soils to rivers and delivered to the coastal ocean (Galloway et al., 2004; Suntharalingam et al., 2012). The increase of Nr creation during the Industrial Era, the connections among its impacts, including on climate and the connections with the carbon cycle are presented in Box 6.2.

For the global ocean, the best BNF estimate is 160 TgN yr^{-1}, which is roughly the midpoint of the minimum estimate of 140 TgN yr^{-1} of Deutsch et al. (2007) and the maximum estimate of 177 TgN yr^{-1} (Großkopf et al., 2012). The probability that this estimate will need an upward revision in the near future is high because several additional processes are not yet considered (Voss et al., 2013).

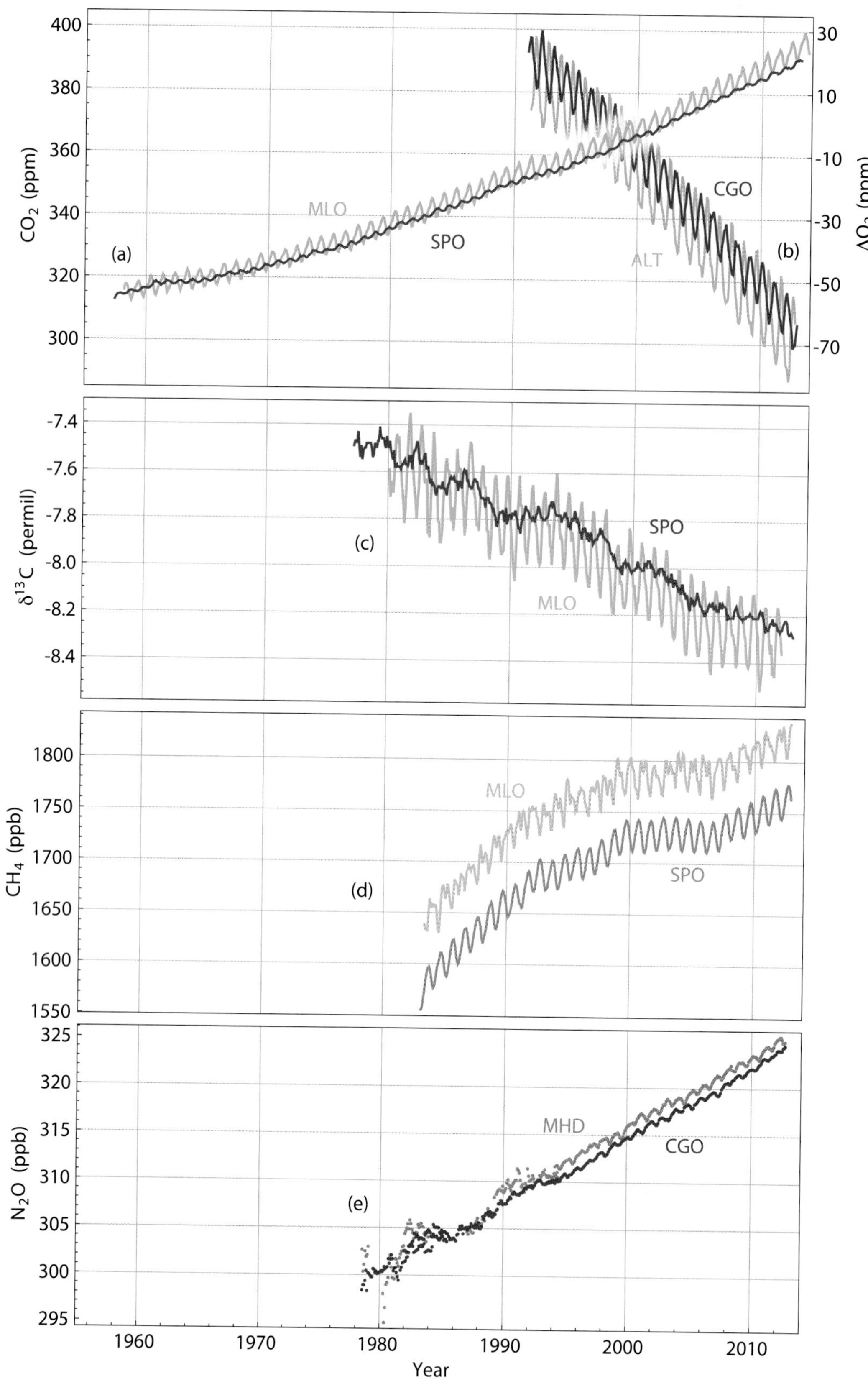

Figure 6.3 | Atmospheric concentration of CO_2, oxygen, $^{13}C/^{12}C$ stable isotope ratio in CO_2, CH_4 and N_2O recorded over the last decades at representative stations (a) CO_2 from Mauna Loa (MLO) Northern Hemisphere and South Pole Southern Hemisphere (SPO) atmospheric stations (Keeling et al., 2005). (b) O_2 from Alert Northern Hemisphere (ALT) and Cape Grim Southern Hemisphere (CGO) stations (http://scrippso2.ucsd.edu/ right axes, expressed relative to a reference standard value). (c) $^{13}C/^{12}C$: Mauna Loa, South Pole (Keeling et al., 2005). (d) CH_4 from Mauna Loa and South Pole stations (Dlugokencky et al., 2012). (e) N_2O from Mace-Head Northern Hemisphere (MHD) and Cape Grim stations (Prinn et al., 2000).

Box 6.2 | Nitrogen Cycle and Climate-Carbon Cycle Feedbacks

Human creation of reactive nitrogen by the Haber–Bosch process (see Sections 6.1.3 and 6.3.4), fossil fuel combustion and agricultural biological nitrogen fixation (BNF) dominate Nr creation relative to biological nitrogen fixation in natural terrestrial ecosystems. This dominance impacts on the radiation balance of the Earth (covered by the IPCC; see, e.g., Chapters 7 and 8), and affects human health and ecosystem health as well (EPA, 2011b; Sutton et al., 2011).

The Nr creation from 1850 to 2005 is shown in Box 6.2 (Figure 1). After mid-1970s, human production of Nr exceeded natural production. During the 2000s food production (mineral fertilisers, legumes) accounts for three-quarters of Nr created by humans, with fossil fuel combustion and industrial uses accounting equally for the remainder (Galloway et al., 2008; Canfield et al., 2010; Sutton et al., 2011).

The three most relevant questions regarding the anthropogenic perturbation of the nitrogen cycle with respect to global change are: (1) What are the interactions with the carbon cycle, and the effects on carbon sinks (see Sections 6.3.2.6.5 and 6.4.2.1), (2) What are the effects of increased Nr on the radiative forcing of nitrate aerosols (Chapter 7, 7.3.2) and tropospheric ozone (Chapters 8), (3) What are the impacts of the excess of Nr on humans and ecosystems (health, biodiversity, eutrophication, not treated in this report, but see, for example, EPA, 2011b; Sutton et al., 2011).

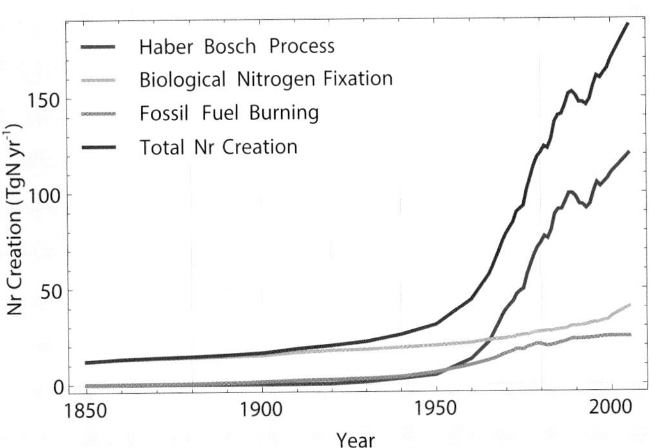

Box 6.2, Figure 1 | Anthropogenic reactive nitrogen (Nr) creation rates (in TgN yr^{-1}) from fossil fuel burning (orange line), cultivation-induced biological nitrogen fixation (blue line), Haber–Bosch process (green line) and total creation (red line). Source: Galloway et al. (2003), Galloway et al. (2008). Note that updates are given in Table 6.9. The only one with significant changes in the more recent literature is cultivation-induced BNF) which Herridge et al. (2008) estimated to be 60 TgN yr^{-1}. The data are only reported since 1850, as no published estimate is available since 1750.

Essentially all of the Nr formed by human activity is spread into the environment, either at the point of creation (i.e., fossil fuel combustion) or after it is used in food production and in industry. Once in the environment, Nr has a number of negative impacts if not converted back into N_2. In addition to its contributions to climate change and stratospheric ozone depletion, Nr contributes to the formation of smog; increases the haziness of the troposphere; contributes to the acidification of soils and freshwaters; and increases the productivity in forests, grasslands, open and coastal waters and open ocean, which can lead to eutrophication and reduction in biodiversity in terrestrial and aquatic ecosystems. In addition, Nr-induced increases in nitrogen oxides, aerosols, tropospheric ozone, and nitrates in drinking water have negative impacts on human health (Galloway et al., 2008; Davidson et al., 2012). Once the nitrogen atoms become reactive (e.g., NH_3, NO_x), any given Nr atom can contribute to all of the impacts noted above in sequence. This is called the nitrogen cascade (Galloway et al., 2003; Box 6.2, Figure 2). The nitrogen cascade is the sequential transfer of the same Nr atom through the atmosphere, terrestrial ecosystems, freshwater ecosystems and marine ecosystems that results in multiple effects in each reservoir. Because of the nitrogen cascade, the creation of any molecule of Nr from N_2, at any location, has the potential to affect climate, either directly or indirectly, as explained in this box This potential exists until the Nr gets converted back to N_2.

The most important processes causing direct links between anthropogenic Nr and climate change include (Erisman et al., 2011): (1) N_2O formation during industrial processes (e.g., fertiliser production), combustion, or microbial conversion of substrate containing nitrogen—notably after fertiliser and manure application to soils. N_2O is a strong greenhouse gas (GHG), (2) emission of anthropogenic NO_x leading to (a) formation of tropospheric O_3, (which is the third most important GHG), (b) a decrease of CH_4 and (c) the formation of nitrate aerosols. Aerosol formation affects radiative forcing, as nitrogen-containing aerosols have a direct cooling effect in addition to an indirect cooling effect through cloud formation and (3) NH_3 emission to the atmosphere which contributes to aerosol formation. The first process has a warming effect. The second has both a warming (as a GHG) and a cooling (through the formation of the OH radical in the troposphere which reacts with CH_4, and through aerosol formation) effect. The net effect of all three NO_x-related contributions is cooling. The third process has a cooling effect.

The most important processes causing an indirect link between anthropogenic Nr and climate change include: (1) nitrogen-dependent changes in soil organic matter decomposition and hence CO_2 emissions, affecting heterotrophic respiration; (2) alteration of the biospheric CO_2 sink due to increased supply of Nr. About half of the carbon that is emitted to the atmosphere is

(continued on next page)

6

Box 6.2 (continued)

taken up by the biosphere; Nr affects net CO_2 uptake from the atmosphere in terrestrial systems, rivers, estuaries and the open ocean in a positive direction (by increasing productivity or reducing the rate of organic matter breakdown) and negative direction (in situations where it accelerates organic matter breakdown). CO_2 uptake in the ocean causes ocean acidification, which reduces CO_2 uptake; (3) changes in marine primary productivity, generally an increase, in response to Nr deposition; and (4) O_3 formed in the troposphere as a result of NO_x and volatile organic compound emissions reduces plant productivity, and therefore reduces CO_2 uptake from the atmosphere. On the global scale the net influence of the direct and indirect contributions of Nr on the radiative balance was estimated to be –0.24 W m^{-2} (with an uncertainty range of +0.2 to –0.5 W m^{-2}) (Erisman et al., 2011).

Nr is required for both plants and soil microorganisms to grow, and plant and microbial processes play important roles in the global carbon cycle. The increasing concentration of atmospheric CO_2 is observed to increase plant photosynthesis (see Box 6.3) and plant growth, which drives an increase of carbon storage in terrestrial ecosystems. Plant growth is, however, constrained by the availability of Nr in soils (see Section 6.3.2.6.5). This means that in some nitrogen-poor ecosystems, insufficient Nr availability will limit carbon sinks, while the deposition of Nr may instead alleviate this limitation and enable larger carbon sinks (see Section 6.3.2.6.5). Therefore, human production of Nr has the potential to mitigate CO_2 emissions by providing additional nutrients for plant growth in some regions. Microbial growth can also be limited by the availability of Nr, particularly in cold, wet environments, so that human production of Nr also has the potential to accelerate the decomposition of organic matter, increasing release of CO_2. The availability of Nr also changes in response to climate change, generally increasing with warmer temperatures and increased precipitation (see Section 6.4.2.1), but with complex interactions in the case of seasonally inundated environments. This complex network of feedbacks is amenable to study through observation and experimentation (Section 6.3) and Earth System modelling (Section 6.4). Even though we do not yet have a thorough understanding of how nitrogen and carbon cycling will interact with climate change, elevated CO_2 and human Nr production in the future, given scenarios of human activity, current observations and model results all indicate that low nitrogen availability will limit carbon storage on land in the 21st century (see Section 6.4.2.1).

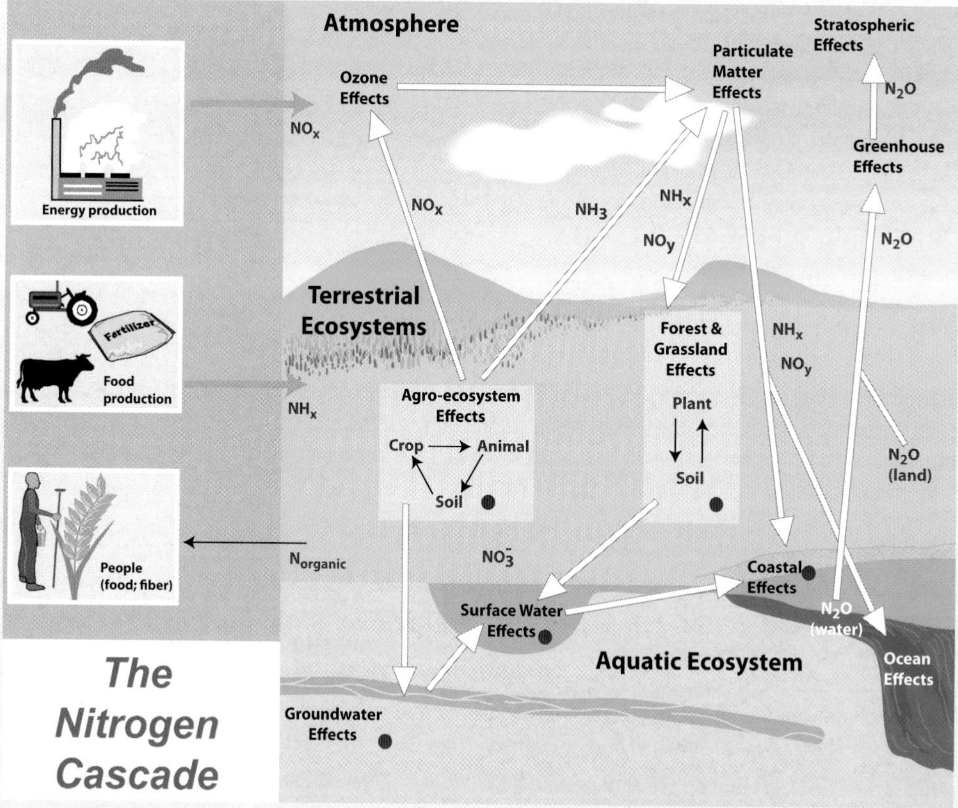

Box 6.2, Figure 2 | Illustration of the nitrogen cascade showing the sequential effects that a single atom of nitrogen in its various molecular forms can have in various reservoirs after it has been converted from nonreactive N_2 to a reactive form by energy and food production (orange arrows). Once created the reactive nitrogen has the potential to continue to contribute to impacts until it is converted back to N_2. The small black circle indicates that there is the potential for denitrification to occur within that reservoir. NH_3 = ammonia; NH_x = ammonia plus ammonium; NO_3^- = nitrate; NO_x = nitrogen oxides; NO_y = NO_x and other combinations of nitrogen and oxygen (except N_2O); N_2O = nitrous oxide. (Adapted with permission from the GEO Yearbook 2003, United Nations Environmental Programme (UNEP), 2004 which was based on Galloway et al., 2003.)

6

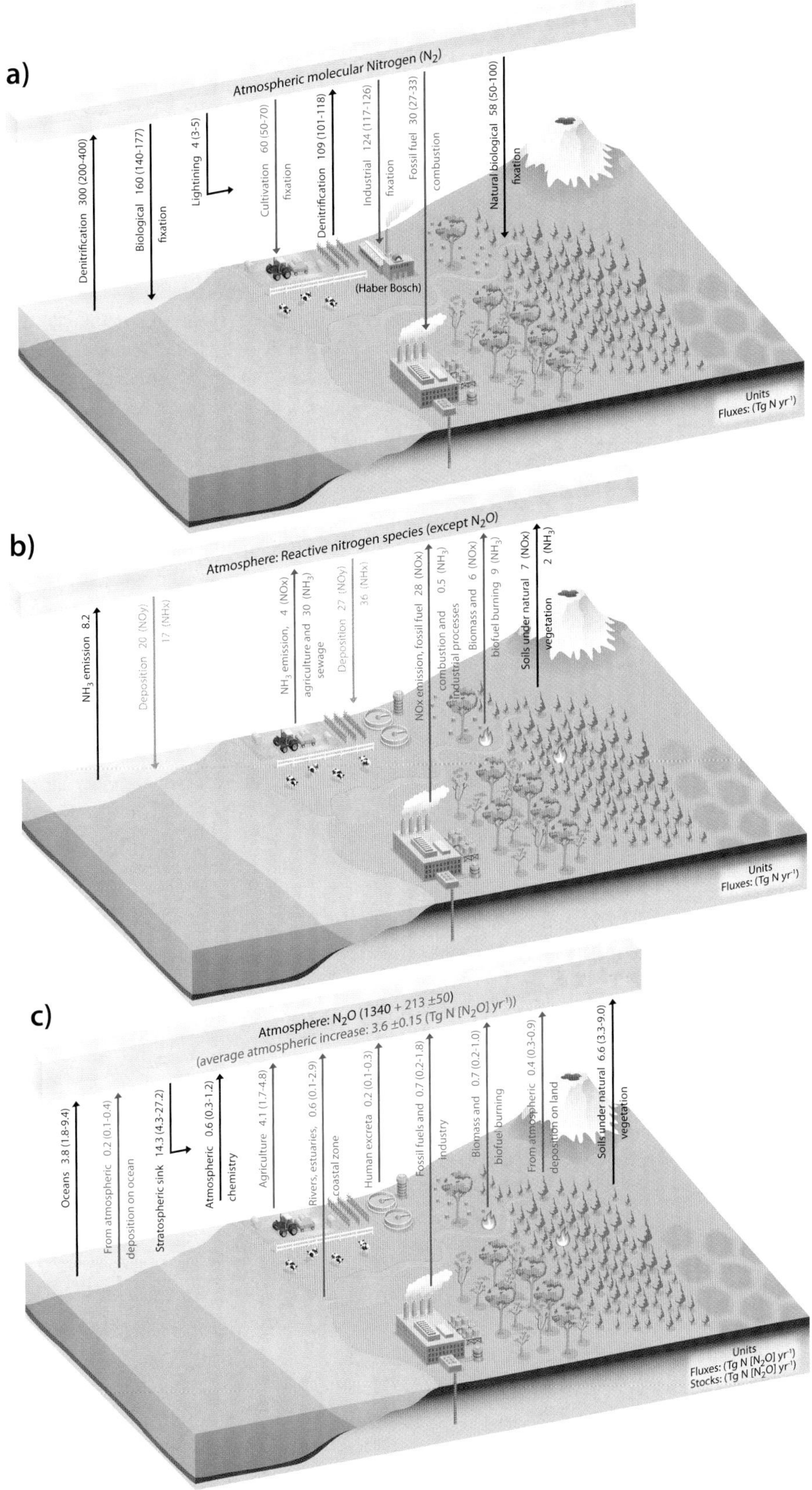

Figure 6.4 | Schematic of the global nitrogen cycle. (a) The natural and anthropogenic processes that create reactive nitrogen and the corresponding rates of denitrification that convert reactive nitrogen back to N_2. (b) The flows of the reactive nitrogen species NO_y and NH_x. (c) The stratospheric sink of N_2O is the sum of losses via photolysis and reaction with O(1D) (oxygen radical in the 1D excited state; Table 6.9). The global magnitude of this sink is adjusted here in order to be equal to the difference between the total sources and the observed growth rate. This value falls within literature estimates (Volk et al., 1997). The atmospheric inventories have been calculated using a conversion factor of 4.79 TgN (N_2O) per ppb (Prather et al., 2012).

A global denitrification rate is much more difficult to constrain than the BNF considering the changing paradigms of nitrogen cycling in the oxygen minimum zones or the unconstrained losses in permeable sediments on the continental shelves (Gao et al., 2012). The coastal ocean may have losses in the range of 100 to 250 (Voss et al., 2011). For the open and distal ocean Codispoti (2007) estimated an upper limit of denitrification of 400 TgN yr^{-1}. Voss et al. (2013) used a conservative estimate of 100 TgN yr^{-1} for the coastal ocean, and 200 to 300 TgN yr^{-1} for the open ocean. Because the upper limit in the global ocean is 400 TgN yr^{-1}, 300 ± 100 TgN yr^{-1} is the best estimate for global ocean losses of reactive nitrogen (Table 6.9).

This chapter does not describe the phosphorus and sulphur biogeochemical cycles, but phosphorus limitations on carbon sinks are briefly addressed in Section 6.4.8.2 and future sulphur deposition in Section 6.4.6.2.

6.1.3.2 Oxygen

Atmospheric oxygen is tightly coupled with the global carbon cycle (sometimes called a mirror of the carbon cycle). The burning of fossil fuels removes oxygen from the atmosphere in a tightly defined stoichiometric ratio depending on fuel carbon content. As a consequence of the burning of fossil fuels, atmospheric O_2 levels have been observed to decrease steadily over the last 20 years (Keeling and Shertz, 1992; Manning and Keeling, 2006) (Figure 6.3b). Compared to the atmospheric oxygen content of about 21% this decrease is very small; however, it provides independent evidence that the rise in CO_2 must be due to an oxidation process, that is, fossil fuel combustion and/or organic carbon oxidation, and is not caused by, for example, volcanic emissions or by outgassing of dissolved CO_2 from a warming ocean. The atmospheric oxygen measurements furthermore also show the north–south concentration O_2 difference (higher in the south and mirroring the CO_2 north–south concentration difference) as expected from the stronger fossil fuel consumption in the NH (Keeling et al., 1996).

On land, during photosynthesis and respiration, O_2 and CO_2 are exchanged in nearly a 1:1 ratio. However, with respect to exchanges with the ocean, O_2 behaves quite differently from CO_2, because compared to the atmosphere only a small amount of O_2 is dissolved in the ocean whereas by contrast the oceanic CO_2 content is much larger due to the carbonate chemistry. This different behaviour of the two gases with respect to ocean exchange provides a powerful method to assess independently the partitioning of the uptake of anthropogenic CO_2 by land and ocean (Manning and Keeling, 2006), Section 6.3.2.3.

6

6.2 Variations in Carbon and Other Biogeochemical Cycles Before the Fossil Fuel Era

The Earth System mechanisms that were responsible for past variations in atmospheric CO_2, CH_4, and N_2O will probably operate in the future as well. Past archives of GHGs and climate therefore provide useful knowledge, including constraints for biogeochemical models applied to the future projections described in Section 6.4. In addition, past archives of GHGs also show with *very high confidence* that the average rates of increase of CO_2, CH_4 and N_2O are larger during the Industrial Era (see Section 6.3) than during any comparable period of at least the past 22,000 years (Joos and Spahni, 2008).

6.2.1 Glacial–Interglacial Greenhouse Gas Changes

6.2.1.1 Processes Controlling Glacial Carbon Dioxide

Ice cores recovered from the Antarctic ice sheet reveal that the concentration of atmospheric CO_2 at the Last Glacial Maximum (LGM; see Glossary) at 21 ka was about one third lower than during the subsequent interglacial (Holocene) period started at 11.7 ka (Delmas et al., 1980; Neftel et al., 1982; Monnin et al., 2001). Longer (to 800 ka) records exhibit similar features, with CO_2 values of ~180 to 200 ppm during glacial intervals (Petit et al., 1999). Prior to 420 ka, interglacial CO_2 values were 240 to 260 ppm rather than 270 to 290 ppm after that date (Lüthi et al., 2008).

A variety of proxy reconstructions as well as models of different complexity from conceptual to complex Earth System Models (ESM; see Glossary) have been used to test hypotheses for the cause of lower LGM atmospheric CO_2 concentrations (e.g., Köhler et al., 2005; Sigman et al., 2010). The mechanisms of the carbon cycle during the LGM which lead to low atmospheric CO_2 can be broken down by individual drivers (Figure 6.5). It should be recognized, however, that this separation is potentially misleading, as many of the component drivers shown in Figure 6.5 may combine nonlinearly (Bouttes et al., 2011). Only well-established individual drivers are quantified (Figure 6.5), and discussed here.

6.2.1.1.1 Reduced land carbon

Despite local evidence of larger carbon storage in permafrost regions during glacial periods (Zimov et al., 2009; Zech et al., 2011), the $\delta^{13}C$ record of ocean waters as preserved in benthic foraminiferal shells has been used to infer that global terrestrial carbon storage was reduced in glacial times, thus opposite to recorded changes in atmospheric CO_2. Data-based estimates of the deficit between LGM and pre-industrial land carbon storage range from a few hundreds to 1000 PgC (e.g., Bird et al., 1996; Ciais et al., 2012). Dynamic vegetation models tend to simulate values at the higher end (~800 PgC) (Kaplan et al., 2002; Otto et al., 2002) and indicate a role for the physiological effects of low CO_2 on photosynthesis at the LGM at least as large as that of colder and dryer climate conditions in determining the past extent of forests (Prentice and Harrison, 2009).

6.2.1.1.2 Lower sea surface temperatures

Reconstructions of sea surface temperatures (SSTs) during the LGM suggest that the global surface ocean was on average 3°C to 5°C cooler compared to the Holocene. Because the solubility of CO_2 increases at colder temperature (Zeebe and Wolf-Gladrow, 2001), a colder glacial ocean will hold more carbon. However, uncertainty in reconstructing the LGM pattern of ocean temperature, particularly in the tropics (Archer et al., 2000; Waelbroeck et al., 2009), together with problems in transforming this pattern to the resolution of models in light of the nonlinear nature of the CO_2–temperature relationship

(Ridgwell, 2001), creates a large spread in modelled estimates, Most ocean general circulation models (OGCM) projections, however, cluster more tightly and suggest that lower ocean temperatures contribute to lower CO_2 values by 25 ppm during the LGM (Figure 6.5).

6.2.1.1.3 Lower sea level and increased salinity

During the LGM, sea level was about ~120 m lower than today, and this change in ocean volume had several well-understood effects on atmospheric CO_2 concentrations. Lower sea level impacts the LGM ocean carbon cycle in two main ways. First, the resulting higher LGM ocean surface salinity causes atmospheric CO_2 to be higher than during the Holocene. Second, the total dissolved inorganic carbon and alkalinity (a measure of the capacity of an aqueous solution to neutralize acid) become more concentrated in equal proportions, and this process also causes atmospheric CO_2 to be higher during the LGM. In total, lower sea level is estimated to contribute to higher CO_2 values by 15 ppm during the LGM (Figure 6.5), implying that other processes must explain the lower CO_2 values measured in ice cores.

6.2.1.1.4 Ocean circulation and sea ice

Reorganization in ocean circulation during glacial periods that promoted the retention of dissolved inorganic carbon in the deep ocean during the LGM has become the focus of most research on the glacial–interglacial CO_2 problem. That ocean circulation plays a key role in low glacial period atmospheric CO_2 concentration is exemplified by the tight coupling observed between reconstructed deep ocean temperatures and atmospheric CO_2 (Shackleton, 2000). Evidence from marine bore hole sites (Adkins et al., 2002) and from marine sediment cores (Jaccard et al., 2005; Skinner et al., 2010) show that the glacial ocean was highly stratified compared to interglacial conditions and may thus have held a larger store of carbon during glacial times. $\delta^{13}CO_2$ ice core records (Lourantou et al., 2010a, 2010b; Schmitt et al., 2012), as well as radiocarbon records from deep-sea corals demonstrate the role of a deep and stratified Southern Ocean in the higher LGM ocean carbon storage. However, conflicting hypotheses exist on the drivers of this increase in the Southern Ocean stratification, for example, northward shift and weakening of Southern Hemisphere (SH) westerly winds (Toggweiler et al., 2006), reduced air–sea buoyancy fluxes (Watson and Garabato, 2006) or massive brine rejections during sea ice formation (Bouttes et al., 2011, 2012). Ocean carbon cycle models have simulated a circulation-induced effect on LGM CO_2 that can explain lower values than during interglacial by 3 ppm (Bopp et al., 2003) to 57 ppm (Toggweiler, 1999).

A long-standing hypothesis is that increased LGM sea ice cover acted as a barrier to air–sea gas exchange and hence reduced the 'leakage' of CO_2 during winter months from the ocean to the atmosphere during glacial periods (Broecker and Peng, 1986). However, concurrent changes in ocean circulation and biological productivity complicate the estimation of the impact of increased sea ice extent on LGM atmospheric CO_2 (Kurahashi-Nakamura et al., 2007). With the exception of the results of an idealised box model (Stephens and Keeling, 2000), ocean carbon models are relatively consistent in projecting a small effect of higher sea ice extent on maintaining atmospheric CO_2 lower during LGM (Archer et al., 2003).

6.2.1.1.5 Iron fertilisation

Both marine and terrestrial sediment records indicate higher rates of deposition of dust and hence iron (Fe) supply at the LGM (Mahowald et al., 2006), implying a potential link between Fe fertilisation of marine productivity and lower glacial CO_2 (Martin, 1990). However, despite the fact that ocean carbon cycle models generally employ similar reconstructions of glacial dust fluxes (i.e., Mahowald et al., 1999; Mahowald et al., 2006), there is considerable disagreement among them in the associated CO_2 change. OGCM that include a description of the Fe cycle tend to cluster at the lower end of simulated CO_2 changes between glacial and interglacial (e.g., Archer at al., 2000; Bopp et al., 2003), whereas box models (e.g., Watson et al., 2000) or Earth System Models of Intermediate Complexity (EMICs, e.g., Brovkin et al., 2007) tend to produce CO_2 changes which are at the higher end (Parekh et al., 2008). An alternative view comes from inferences drawn from the timing and magnitude of changes in dust and CO_2 in ice cores (Röthlisberger et al., 2004), assigning a 20 ppm limit for the lowering of CO_2 during the LGM in response to an Southern Ocean Fe fertilisation effect, and a 8 ppm limit for the same effect in the North Pacific.

6.2.1.1.6 Other glacial carbon dioxide drivers

A number of further aspects of altered climate and biogeochemistry at the LGM are also likely to have affected atmospheric CO_2. Reduced bacterial metabolic rates and remineralization (see Glossary) of organic matter (Matsumoto, 2007; Menviel et al., 2012), increased glacial supply of dissolved silica (required by diatoms to form frustules) (Harrison, 2000), 'silica leakage' (Brzezinski et al., 2002; Matsumoto et al., 2002), changes in net global weathering rates (Berner, 1992; Munhoven, 2002), reduction in coral reef growth and other forms of shallow water $CaCO_3$ accumulation (Berger, 1982), carbonate compensation (Ridgwell and Zeebe, 2005) and changes in the $CaCO_3$ to organic matter 'rain ratio' to the sediments (Archer and Maier-Reimer, 1994) will act to amplify or diminish the effect of many of the aforementioned drivers on glacial CO_2.

6.2.1.1.7 Summary

All of the major drivers of the glacial-to-interglacial atmospheric CO_2 changes (Figure 6.5) are likely to have already been identified. However, Earth System Models have been unable to reproduce the full magnitude of the glacial-to-interglacial CO_2 changes. Significant uncertainties exist in glacial boundary conditions and on some of the primary controls on carbon storage in the ocean and in the land. These uncertainties prevent an unambiguous attribution of individual mechanisms as controllers of the low glacial CO_2 concentrations. Further assessments of the interplay of different mechanisms prior to deglacial transitions or in glacial inceptions will provide additional insights into the drivers and processes that caused the glacial decrease of CO_2. Because several of these identified drivers (e.g., organic matter remineralization, ocean stratification) are sensitive to climate change in general, improved understanding drawn from the glacial–interglacial cycles will help constrain the magnitude of future ocean feedbacks on atmospheric CO_2. Other drivers (e.g., iron fertilisation) are involved in geoengineering methods (see Glossary), such that improved under-

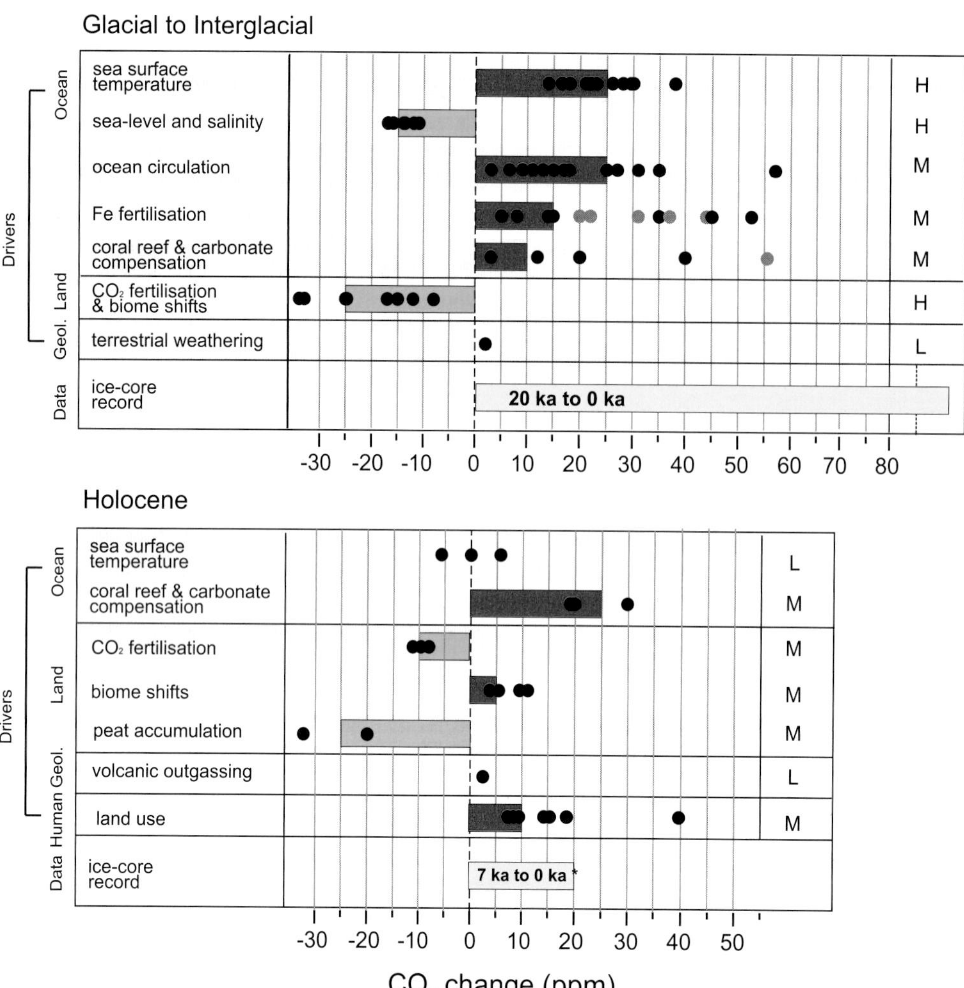

Figure 6.5 | Mechanisms contributing to carbon dioxide concentrations changes from Last Glacial Maximum (LGM) to late Holocene (top) and from early/mid Holocene (7 ka) to late Holocene (bottom). Filled black circles represent individual model-based estimates for individual ocean, land, geological or human mechanisms. Solid colour bars represent expert judgment (to the nearest 5 ppm) rather than a formal statistical average. References for the different model results used for explaining CO_2 changes from LGM to late Holocene are as per (Kohfeld and Ridgwell, 2009) with excluded model projections in grey. References for the different model results used for explaining CO_2 changes during the Holocene are: Joos et al. (2004), Brovkin et al. (2002, 2008), Kleinen et al. (2010, 2012), Broecker et al. (1999), Ridgwell et al. (2003), Schurgers et al. (2006), Yu (2011), Ruddiman (2003, 2007), Strassmann et al. (2008), Olofsson and Hickler (2008), Pongratz et al. (2009), Kaplan et al. (2011), Lemmen (2009), Stocker et al. (2011), Roth and Joos (2012). Confidence levels for each mechanism are indicated in the left column — H for *high confidence*, M for *medium confidence* and L for *low confidence*.

standing could also help constrain the potential and applicability of these methods (see Section 6.5.2).

6.2.1.2 Processes Controlling Glacial Methane and Nitrous Oxide

Ice core measurements show that atmospheric CH_4 and N_2O were much lower under glacial conditions compared to interglacial ones. Their reconstructed history encompasses the last 800 ka (Loulergue et al., 2008; Schilt et al., 2010a). Glacial CH_4 mixing ratios are in the 350 to 400 ppb range during the eight glacial maxima covered by the ice core record. This is about half the levels observed during interglacial conditions. The N_2O concentration amounts to 202 ± 8 ppb at the LGM, compared to the early Holocene levels of about 270 ppb (Flückiger et al., 1999).

CH_4 and N_2O isotopic ratio measurements in ice cores provide important constraints on the mechanisms responsible for their temporal changes. N_2O isotopes suggest a similar increase in marine and terrestrial N_2O emissions during the last deglaciation (Sowers et al., 2003). Marine sediment proxies of ocean oxygenation suggest that most of the observed N_2O deglacial rise was of marine origin (Jaccard and Galbraith, 2012). δD and ^{14}C isotopic composition measurements of CH_4 have shown that catastrophic methane hydrate degassing events are *unlikely* to have caused the last deglaciation CH_4 increase (Sowers, 2006; Petrenko et al., 2009; Bock et al., 2010). $\delta^{13}C$ and δD measurements of CH_4 combined with interpolar atmospheric CH_4 gradient changes (Greenland minus Antarctica ice cores) suggest that most of the deglacial CH_4 increase was caused by increased emissions from boreal and tropical wetlands and an increase in CH_4 atmospheric residence time due to a reduced oxidative capacity of the atmosphere (Fischer et al., 2008). The biomass burning source apparently changed little on the same time scale, whereas this CH_4 source experienced large fluctuations over the last millennium (Mischler et al., 2009; Wang et al., 2010b). Recent modelling studies, however, suggest that changes in the atmospheric oxidising capacity of the atmosphere at the LGM are probably negligible compared to changes in sources (Levine et al., 2011) and that tropical temperature influencing tropical wetlands and global vegetation were the dominant controls for CH_4 atmospheric changes on glacial–interglacial time scales (Konijnendijk et al., 2011).

6.2.1.3 Processes Controlling Changes in Carbon Dioxide, Methane, and Nitrous Oxide During Abrupt Glacial Events

Ice core measurements of CO_2, CH_4 and N_2O show sharp (millennial-scale) changes in the course of glaciations, associated with the so-called Dansgaard/Oeschger (DO) climatic events (see Section 5.7), but their amplitude, shape and timing differ. During these millennial scale climate events, atmospheric CO_2 concentrations varied by about 20 ppm, in phase with Antarctic, but not with Greenland temperatures. CO_2 increased during cold (stadial) periods in Greenland, several thousands years before the time of the rapid warming event in Greenland (Ahn and Brook, 2008). CH_4 and N_2O showed rapid transitions in phase with Greenland temperatures with little or no lag. CH_4 changes are in the 50 to 200 ppb range (Flückiger et al., 2004), in phase with Greenland temperature warming at a decadal time scale (Huber et al., 2006). N_2O changes are large, of same magnitude than glacial–interglacial changes, and for the warmest and longest DO events N_2O starts to increase several centuries before Greenland temperature and CH_4 (Schilt et al., 2010b).

Conflicting hypotheses exist on the drivers of these millennial-scale changes. Some model simulations suggest that both CO_2 and N_2O fluctuations can be explained by changes in the Atlantic meridional overturning ocean circulation (Schmittner and Galbraith, 2008), CO_2 variations being explained mainly by changes in the efficiency of the biological pump which affects deep ocean carbon storage (Bouttes et al., 2011), whereas N_2O variations could be due to changes in productivity and oxygen concentrations in the subsurface ocean (Schmittner and Galbraith, 2008). Other studies, however, suggest that the millennial-scale CO_2 fluctuations can be explained by changes in the land carbon storage (Menviel et al., 2008; Bozbiyik et al., 2011). For CH_4, models have difficulties in reproducing changes in wetland emissions compatible with DO atmospheric variations (Hopcroft et al., 2011), and the changes in the atmospheric oxidizing capacity of the atmosphere during DO events seem to be too weak to explain the CH_4 changes (Levine et al., 2012).

6.2.2 Greenhouse Gas Changes over the Holocene

6.2.2.1 Understanding Processes Underlying Holocene Carbon Dioxide Changes

The evolution of the atmospheric CO_2, CH_4, and N_2O concentrations during the Holocene, the interglacial period which began 11.7 ka, is known with high certainty from ice core measurements (Figure 6.6). A decrease in atmospheric CO_2 of about 7 ppm is measured in ice cores

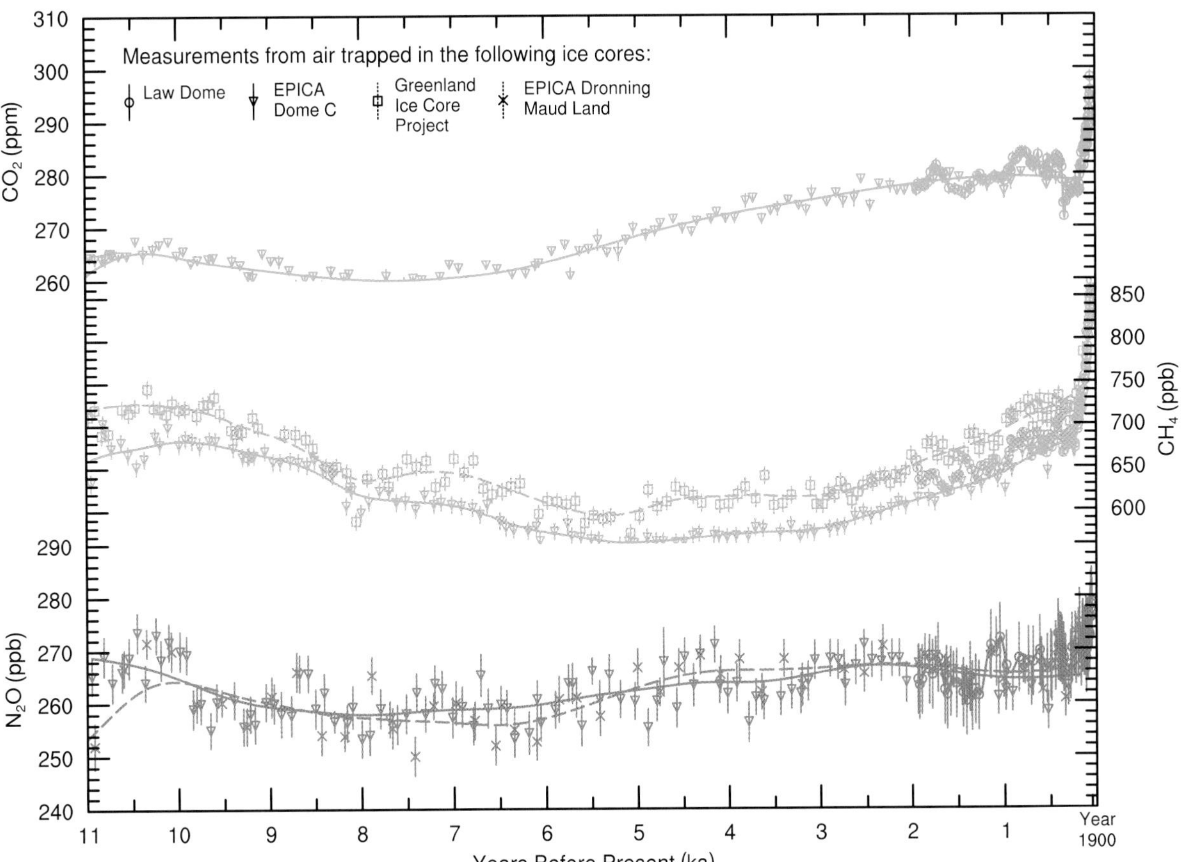

Figure 6.6 | Variations of CO_2, CH_4, and N_2O concentrations during the Holocene. The data are for Antarctic ice cores: European Programme for Ice Coring in Antarctica EPICA Dome C (Flückiger et al., 2002; Monnin et al., 2004), triangles; EPICA Dronning Maud Land (Schilt et al., 2010b), crosses; Law Dome (MacFarling-Meure et al., 2006), circles; and for Greenland Ice Core Project (GRIP) (Blunier et al., 1995), squares. Lines correspond to spline fits.

between 11 and 7 ka, followed by a 20 ppm CO_2 increase until the onset of the Industrial Era in 1750 (Indermühle et al., 1999; Monnin et al., 2004; Elsig et al., 2009). These variations in atmospheric CO_2 over the past 11 kyr preceding industrialisation are more than five times smaller than the CO_2 increase observed during the Industrial Era (see Section 6.3.2.3). Despite the small magnitude of CO_2 variations prior to the Industrial Era, these changes are nevertheless useful for understanding the role of natural forcing in carbon and other biogeochemical cycles during interglacial climate conditions.

Since the IPCC AR4, the mechanisms underlying the observed 20 ppm CO_2 increase between 7 ka and the Industrial Era have been a matter of intensive debate. During three interglacial periods prior to the Holocene, CO_2 did not increase, and this led to a hypothesis that pre-industrial anthropogenic CO_2 emissions could be associated with early land use change and forest clearing (Ruddiman, 2003, 2007). However, ice core CO_2 data (Siegenthaler et al., 2005b) indicate that during Marine Isotope Stage 11 (see Section 5.2.2), an interglacial period that lasted from 400 to 420 ka, CO_2 increased similarly to the Holocene period. Drivers of atmospheric CO_2 changes during the Holocene can be divided into oceanic and terrestrial processes (Figure 6.5) and their roles are examined below.

6.2.2.1.1 Oceanic processes

The change in oceanic carbonate chemistry could explain the slow atmospheric CO_2 increase during the Holocene since 7 ka. Proposed mechanisms include: (1) a shift of oceanic carbonate sedimentation from deep sea to the shallow waters due to sea level rise onto continental shelves causing accumulation of $CaCO_3$ on shelves including coral reef growth, a process that releases CO_2 to the atmosphere (Ridgwell et al., 2003; Kleinen et al., 2010), (2) a 'carbonate compensation' in response to the release of carbon from the deep ocean during deglaciation and to the buildup of terrestrial biosphere in the early Holocene (Broecker et al., 1999; Joos et al., 2004; Elsig et al., 2009; Menviel and Joos, 2012). Proxies for carbonate ion concentration in the deep sea (Yu et al., 2010) and a decrease in modern $CaCO_3$ preservation in equatorial Pacific sediments (Anderson et al., 2008) support the hypothesis that the ocean was a source of CO_2 to the atmosphere during the Holocene. Changes in SSTs over the last 7 kyr (Kim et al., 2004) could have contributed to slightly lower (Brovkin et al., 2008) or higher (Menviel and Joos, 2012) atmospheric CO_2 concentration but, *very likely*, SST-driven CO_2 change represents only a minor contribution to the observed CO_2 increase during the Holocene after 7 ka (Figure 6.5).

6.2.2.1.2 Terrestrial processes

The $\delta^{13}C$ of atmospheric CO_2 trapped in ice cores can be used to infer changes in terrestrial biospheric carbon pools. Calculations based on inferred $\delta^{13}C$ of atmospheric CO_2 during the Holocene suggest an increase in terrestrial carbon storage of about 300 PgC between 11 and 5 ka and small overall terrestrial changes thereafter (Elsig et al., 2009). Modelling studies suggest that CO_2 fertilisation (Box 6.3) in response to increasing atmospheric CO_2 concentration after 7 ka contributed to a substantially increased terrestrial carbon storage (>100 PgC) on Holocene time scales (Kaplan et al., 2002; Joos et al., 2004; Kleinen et al., 2010). Orbitally forced climate variability, including the intensification and decline of the Afro-Asian monsoon and the mid-Holocene warming of the high latitudes of the NH are estimated in models to have caused changes in vegetation distribution and hence of terrestrial carbon storage. These climate-induced carbon storage changes are estimated using models to have been smaller than the increase due to CO_2 fertilisation (Brovkin et al., 2002; Schurgers et al., 2006). The Holocene accumulation of carbon in peatlands has been reconstructed globally, suggesting a land carbon additional storage of several hundred petagrams of carbon between the early Holocene and the Industrial Era, although uncertainties remain on this estimate (Tarnocai et al., 2009; Yu, 2011; Kleinen et al., 2012). Volcanic CO_2 emissions to the atmosphere between 12 and 7 ka were estimated to be two to six times higher than during the last millennium, of about 0.1 PgC yr^{-1} (Huybers and Langmuir, 2009; Roth and Joos, 2012). However, a peak in the inferred volcanic emissions coincides with the period of decreasing atmospheric CO_2 and the *confidence* in changes of volcanic CO_2 emissions is *low*.

Global syntheses of the observational, paleoecological and archaeological records for Holocene land use change are not currently available (Gaillard et al., 2010). Available reconstructions of anthropogenic land use and land cover change (LULCC) prior to the last millennium currently extrapolate using models and assumptions from single regions to changes in all regions of the world (Goldewijk et al., 2011; Kaplan et al., 2011). Because of regional differences in land use systems and uncertainty in historical population estimates, the *confidence* in spatially explicit LULCC reconstructions is *low*.

Some recent studies focused on reconstructing LULCC and making very simple assumptions regarding the effect of land use on carbon (Olofsson and Hickler, 2008; Lemmen, 2009). Other studies relied on more sophisticated terrestrial biosphere models to simulate carbon storage and loss in response to pre-industrial LULCC during the late Holocene (Strassmann et al., 2008; Pongratz et al., 2009; Stocker et al., 2011). The conclusion of the aforementioned studies was that cumulative Holocene carbon emissions as a result of pre-industrial LULCC were not large enough (~50 to 150 PgC during the Holocene before 1850) to have had an influence larger than an increase of ~10 ppm on late Holocene observed CO_2 concentration increase (Figure 6.5). However, a modelling study by Kaplan et al. (2011) suggested that more than 350 PgC could have been released as a result of LULCC between 8 ka and 1850 as a result of a much stronger loss of soil carbon in response to land use change, than in other studies.

In addition to clearing of forests, large-scale biomass burning activity, inferred from synthesized charcoal records and bog sediments has been hypothesized to correlate with the observed Late Holocene atmospheric CO_2 (Carcaillet et al., 2002). A global extensive synthesis of charcoal records for the last 21 kyr (Power et al., 2008) and updates of those shows that fire activity followed climate variability on global (Marlon et al., 2008; Daniau et al., 2012) and regional scale (Archibald et al., 2009; Mooney et al., 2011; Marlon et al., 2012; Power et al., 2013). There is no evidence, however, for a distinct change in fire activity linked to human activity alone as hypothesized from a regional charcoal record synthesis for the tropical Americas (Nevle and Bird, 2008; Nevle et al., 2011). Fire being a newly studied component, no estimate for its role is given in Figure 6.5.

6.2.2.2 Holocene Methane and Nitrous Oxide Drivers

The atmospheric CH_4 levels decreased from the early Holocene to about 6 ka, were lowest at around 5 ka, and increased between 5 ka and year 1750 by about 100 ppb (Figure 6.6). Major Holocene agricultural developments, in particular rice paddy cultivation and widespread domestication of ruminants, have been proposed as an explanation for the Late Holocene CH_4 rise (Ruddiman, 2007). The most recent syntheses of archaeological data point to an increasing anthropogenic CH_4 source from domesticated ruminants after 5 ka and from rice cultivation after 4 ka (Ruddiman, 2007; Fuller et al., 2011). The modelling support for either natural or anthropogenic explanations of the Late Holocene increase in the atmospheric CH_4 concentration is equivocal. A study by Kaplan et al. (2006) suggested that a part of the Late Holocene CH_4 rise could be explained by anthropogenic sources. Natural wetland CH_4 models driven by simulated climate changes are able (Singarayer et al., 2011) or unable (Konijnendijk et al., 2011) to simulate Late Holocene increase in the CH_4 concentration, reflecting a large spread in present-day CH_4 emissions simulated by this type of models (Melton et al., 2013; see Section 6.3.3.2). Consequently, *about as likely as not,* the atmospheric CH_4 increase after 5000 years ago can be attributed to early human activities. The mechanisms causing the N_2O concentration changes during the Holocene are not firmly identified (Flückiger et al., 2002).

6.2.3 Greenhouse Gas Changes over the Last Millennium

6.2.3.1 A Decrease of Carbon Dioxide around Year 1600 and Possible Explanations for this Event

High resolution ice cores records reveal that atmospheric CO_2 during the last millennium varied with a drop in atmospheric CO_2 concentration by 7 to 10 ppm around year 1600, followed by a CO_2 increase during the 17th century (Trudinger et al., 2002; Siegenthaler et al., 2005a; MacFarling-Meure et al., 2006; Ahn et al., 2012). This is shown in Figure 6.7. The CO_2 decrease during the 17th century was used to evaluate the response of atmospheric CO_2 concentration to a century-scale shift in global temperature (Scheffer et al., 2006; Cox and Jones, 2008; Frank et al., 2010) which was found to be dependent on the choice of global temperature reconstructions used in the model.

One of the possible explanations for the drop in atmospheric CO_2 around year 1600 is enhanced land and/or ocean carbon uptake in response to the cooling caused by reduced solar irradiance during the Maunder Minimum (Section 5.3.5.3). However, simulations using Earth System Models of Intermediate Complexity (EMICs)(Gerber et al., 2003; Brovkin et al., 2004) and by complex Earth System Models (ESMs) (Jungclaus et al., 2010) suggest that solar irradiance forcing alone is not sufficient

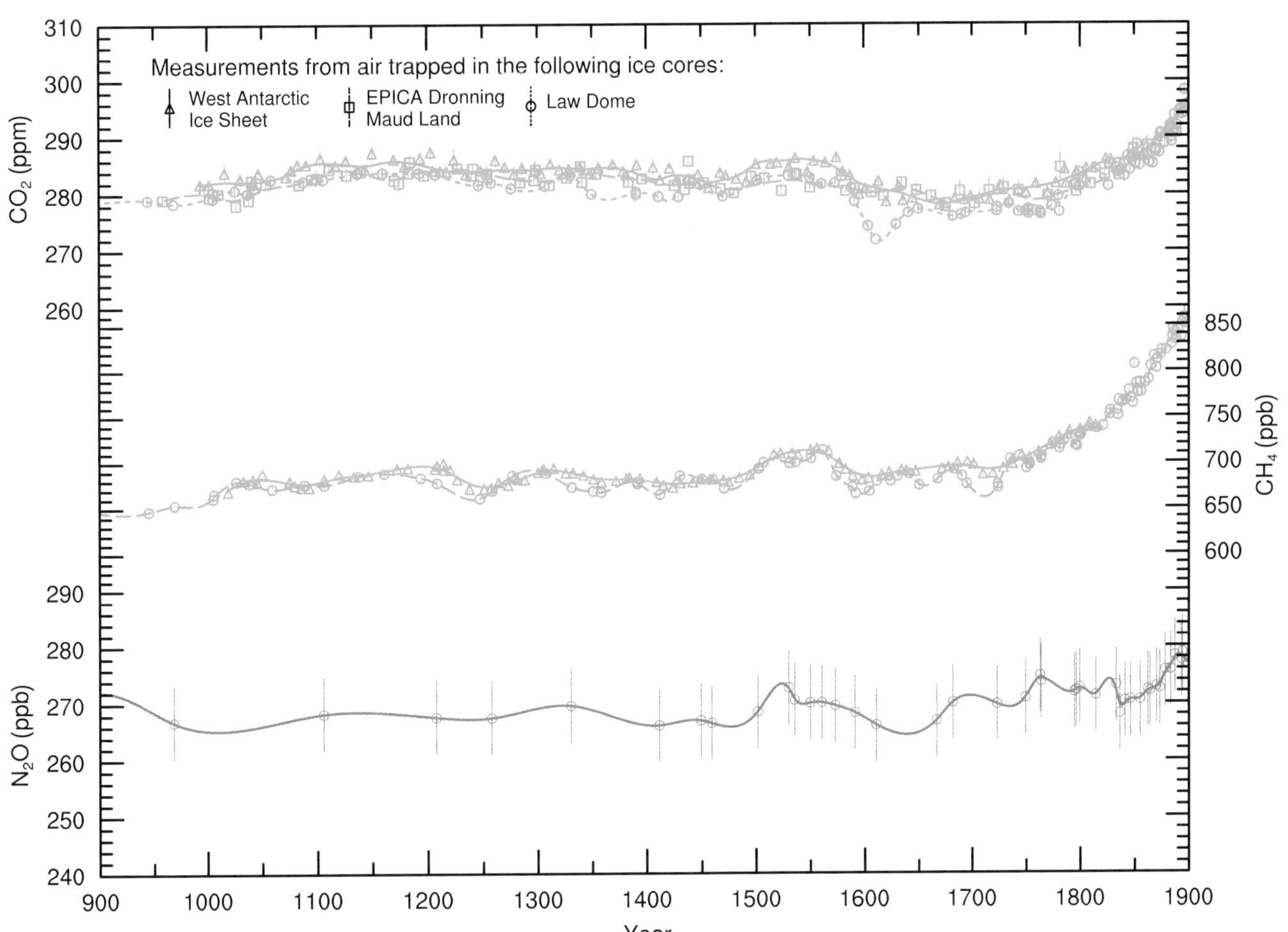

Figure 6.7 | Variations of CO_2, CH_4, and N_2O during 900–1900 from ice cores. The data are for Antarctic ice cores: Law Dome (Etheridge et al., 1996; MacFarling-Meure et al., 2006), circles; West Antarctic Ice Sheet (Mitchell et al., 2011; Ahn et al., 2012), triangles; Dronning Maud Land (Siegenthaler et al., 2005a), squares. Lines are spline fits to individual measurements.

to explain the magnitude of the CO_2 decrease. The drop in atmospheric CO_2 around year 1600 could also be caused by a cooling from increased volcanic eruptions (Jones and Cox, 2001; Brovkin et al., 2010; Frölicher et al., 2011). A third hypothesis calls for a link between CO_2 and epidemics and wars associated with forest regrowth over abandoned lands and increased carbon storage, especially in Central America. Here, results are model and scenario dependent. Simulations by Pongratz et al. (2011a) do not reproduce a decrease in CO_2, while simulations by Kaplan et al. (2011) suggest a considerable increase in land carbon storage around year 1600. The temporal resolution of Central American charcoal and pollen records is insufficient to support or falsify these model results (e.g., Nevle and Bird, 2008; Marlon et al., 2008).

Ensemble simulations over the last 1200 years have been conducted using an ESM (Jungclaus et al., 2010) and EMICs (Eby et al., 2013) including a fully interactive carbon cycle. The sensitivity of atmospheric CO_2 concentration to NH temperature changes in ESM was modeled to be of 2.7 to 4.4 ppm °C^{-1}, while EMICs show on average a higher sensitivity of atmospheric CO_2 to global temperature changes of 8.6 ppm °C^{-1}. These sensitivities fall within the range of 1.7 to 21.4 ppm °C^{-1} of a recent reconstruction based on tree-ring NH temperature reconstructions (Frank et al., 2010).

6.2.3.2 Mechanisms Controlling Methane and Nitrous Oxide during the Last Millennium

Recent high-resolution ice core records confirm a CH_4 decrease in the late 16th century by about 40 ppb (MacFarling-Meure et al., 2006; Mitchell et al., 2011), as shown in Figure 6.7. Correlations between this drop in atmospheric CH_4 and the lower temperatures reconstructed during the 15th and 16th centuries suggest that climate change may have reduced CH_4 emissions by wetlands during this period. In addition to changes in the wetland CH_4 source, changes in biomass burning have been invoked to explain the last millennium CH_4 record (Ferretti et al., 2005; Mischler et al., 2009), ice core CO and CO isotopes (Wang et al., 2010b) and global charcoal depositions (Marlon et al., 2008). Changes in anthropogenic CH_4 emissions during times of war and plague hypothetically contributed to variability in atmospheric CH_4 concentration (Mitchell et al., 2011). Ice core $\delta^{13}CH_4$ measurements suggested pronounced variability in both natural and anthropogenic CH_4 sources over the 1000–1800 period (Sapart et al., 2012). No studies are known about mechanisms of N_2O changes for the last millennium.

6.3 Evolution of Biogeochemical Cycles Since the Industrial Revolution

6.3.1 Carbon Dioxide Emissions and Their Fate Since 1750

Prior to the Industrial Era, that began in 1750, the concentration of atmospheric CO_2 fluctuated roughly between 180 ppm and 290 ppm for at least 2.1 Myr (see Section 5.2.2 and Hönisch et al., 2009; Lüthi et al., 2008; Petit et al., 1999). Between 1750 and 2011, the combustion of fossil fuels (coal, gas, oil and gas flaring) and the production of cement have released 375 ± 30 PgC (1 PgC = 10^{15} gC) to the atmosphere (Table 6. 1; Boden et al., 2011). Land use change activities, mainly deforestation, has released an additional 180 ± 80 PgC (Table 6.1). This carbon released by human activities is called anthropogenic carbon.

Of the 555 ± 85 PgC of anthropogenic carbon emitted to the atmosphere from fossil fuel and cement and land use change, less than half have accumulated in the atmosphere (240 ± 10 PgC) (Table 6.1). The remaining anthropogenic carbon has been absorbed by the ocean and

Table 6.1 | Global anthropogenic CO_2 budget, accumulated since the Industrial Revolution (onset in 1750) and averaged over the 1980s, 1990s, 2000s, as well as the last 10 years until 2011. By convention, a negative ocean or land to atmosphere CO_2 flux is equivalent to a gain of carbon by these reservoirs. The table does not include natural exchanges (e.g., rivers, weathering) between reservoirs. The uncertainty range of 90% confidence interval presented here differs from how uncertainties were reported in AR4 (68%).

	1750–2011 Cumulative PgC	1980–1989 PgC yr^{-1}	1990–1999 PgC yr^{-1}	2000–2009 PgC yr^{-1}	2002–2011 PgC yr^{-1}
Atmospheric increase[a]	240 ± 10[f]	3.4 ± 0.2	3.1 ± 0.2	4.0 ± 0.2	4.3 ± 0.2
Fossil fuel combustion and cement production[b]	375 ± 30[f]	5.5 ± 0.4	6.4 ± 0.5	7.8 ± 0.6	8.3 ± 0.7
Ocean-to-atmosphere flux[c]	−155 ± 30[f]	−2.0 ± 0.7	−2.2 ± 0.7	−2.3 ± 0.7	−2.4 ± 0.7
Land-to-atmosphere flux *Partitioned as follows*	30 ± 45[f]	−0.1 ± 0.8	−1.1 ± 0.9	−1.5 ± 0.9	−1.6 ± 1.0
Net land use change[d]	180 ± 80[f,g]	1.4 ± 0.8	1.5 ± 0.8	1.1 ± 0.8	0.9 ± 0.8
Residual land sink[e]	−160 ± 90[f]	−1.5 ± 1.1	−2.6 ± 1.2	−2.6 ± 1.2	−2.5 ± 1.3

Notes:

[a] Data from Charles D. Keeling, (http://scrippsco2.ucsd.edu/data/data.html), Thomas Conway and Pieter Tans, National Oceanic and Atmospheric Administration–Earth System Research Laboratory (NOAA–ESRL, www.esrl.noaa.gov/gmd/ccgg/trends/) using a conversion factor of 2.120 PgC per ppm (Prather et al., 2012). Prior to the atmospheric record in 1960, ice core data is used (Neftel et al., 1982; Friedli et al., 1986; Etheridge et al., 1996).

[b] Estimated by the Carbon Dioxide Information Analysis Center (CDIAC) based on UN energy statistics for fossil fuel combustion (up to 2009) and US Geological Survey for cement production (Boden et al., 2011), and updated to 2011 using BP energy statistics.

[c] Based on observations for 1990–1999, with the trends based on existing global estimates (see Section 6.3.2.5 and Table 6.4).

[d] Based on the "bookkeeping" land use change flux accounting model of Houghton et al. (2012) until 2010, and assuming constant LUC emissions for 2011, consistent with satellite-based fire emissions (Le Quéré et al., 2013; see Section 6.3.2.2 and Table 6.2).

[e] Calculated as the sum of the *Land-to-atmosphere flux* minus *Net land use change flux*, assuming the errors on each term are independent and added quadratically.

[f] The 1750–2011 estimate and its uncertainty is rounded to the nearest 5 PgC.

[g] Estimated from the cumulative net land use change emissions of Houghton et al. (2012) during 1850–2011 and the average of four publications (Pongratz at al., 2009; van Minnen et al., 2009; Shevliakova et al., 2009; Zaehle et al., 2011) during 1750–1850.

in terrestrial ecosystems: the carbon 'sinks' (Figure 6.8). The ocean stored 155 ± 30 PgC of anthropogenic carbon since 1750 (see Section 6.3.2.5.3 and Box 6.1). Terrestrial ecosystems that have not been affected by land use change since 1750, have accumulated 160 ± 90 PgC of anthropogenic carbon since 1750 (Table 6.1), thus not fully compensating the net CO_2 losses from terrestrial ecosystems to the atmosphere from land use change during the same period estimated of 180 ± 80 PgC (Table 6.1). The net balance of all terrestrial ecosystems, those affected by land use change and the others, is thus close to neutral since 1750, with an average loss of 30 ± 45 (see Figure 6.1). This increased storage in terrestrial ecosystems not affected by land use change is *likely* to be caused by enhanced photosynthesis at higher CO_2 levels and nitrogen deposition, and changes in climate favouring carbon sinks such as longer growing seasons in mid-to-high latitudes. Forest area expansion and increased biomass density of forests that result from changes in land use change are also carbon sinks, and they

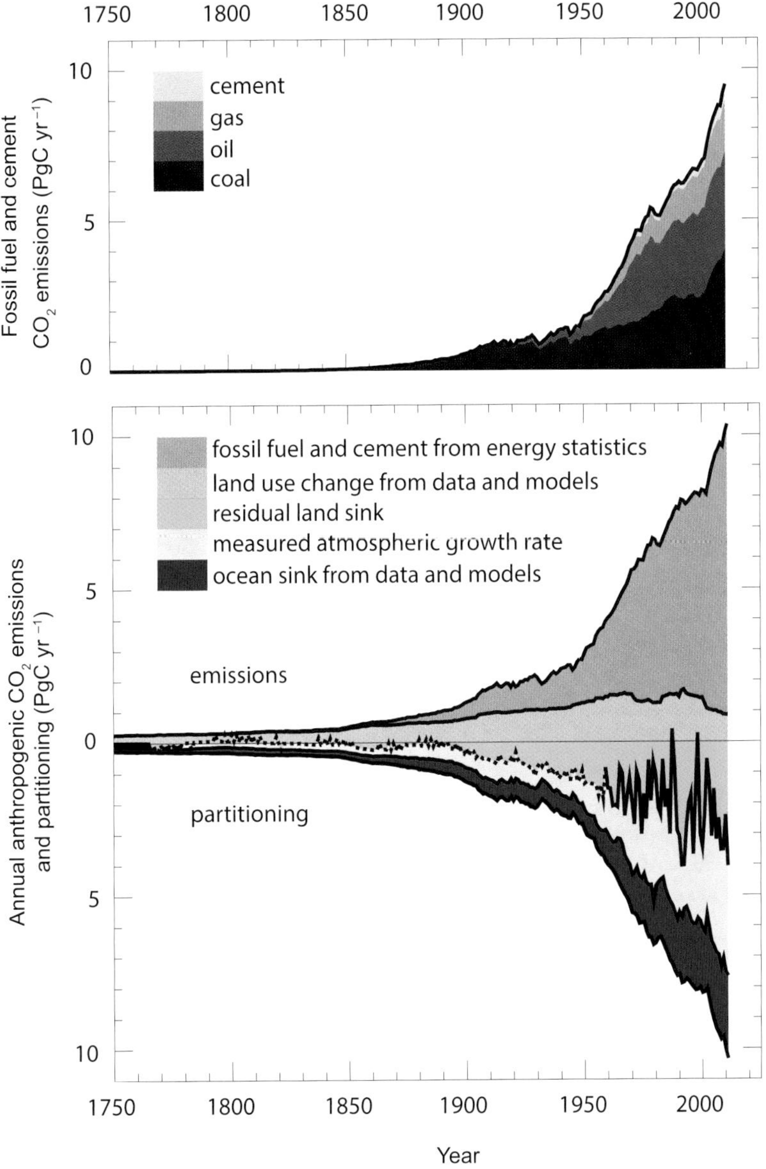

Figure 6.8 | Annual anthropogenic CO_2 emissions and their partitioning among the atmosphere, land and ocean (PgC yr^{-1}) from 1750 to 2011. (Top) Fossil fuel and cement CO_2 emissions by category, estimated by the Carbon Dioxide Information Analysis Center (CDIAC) based on UN energy statistics for fossil fuel combustion and US Geological Survey for cement production (Boden et al., 2011). (Bottom) Fossil fuel and cement CO_2 emissions as above. CO_2 emissions from net land use change, mainly deforestation, are based on land cover change data and estimated for 1750–1850 from the average of four models (Pongratz et al., 2009; Shevliakova et al., 2009; van Minnen et al., 2009; Zaehle et al., 2011) before 1850 and from Houghton et al. (2012) after 1850 (see Table 6.2). The atmospheric CO_2 growth rate (term in light blue 'atmosphere from measurements' in the figure) prior to 1959 is based on a spline fit to ice core observations (Neftel et al., 1982; Friedli et al., 1986; Etheridge et al., 1996) and a synthesis of atmospheric measurements from 1959 (Ballantyne et al., 2012). The fit to ice core observations does not capture the large interannual variability in atmospheric CO_2 and is represented with a dashed line. The ocean CO_2 sink prior to 1959 (term in dark blue 'ocean from indirect observations and models' in the figure) is from Khatiwala et al. (2009) and from a combination of models and observations from 1959 from (Le Quéré et al., 2013). The residual land sink (term in green in the figure) is computed from the residual of the other terms, and represents the sink of anthropogenic CO_2 in natural land ecosystems. The emissions and their partitioning only include the fluxes that have changed since 1750, and not the natural CO_2 fluxes (e.g., atmospheric CO_2 uptake from weathering, outgassing of CO_2 from lakes and rivers, and outgassing of CO_2 by the ocean from carbon delivered by rivers; see Figure 6.1) between the atmosphere, land and ocean reservoirs that existed before that time and still exist today. The uncertainties in the various terms are discussed in the text and reported in Table 6.1 for decadal mean values.

are accounted in Table 6.1 as part of the net flux from land use change. The increased terrestrial carbon storage in ecosystems not affected by land use change is called the *Residual land sink* in Table 6.1 because it is inferred from mass balance as the difference between fossil and net land use change emissions and measured atmospheric and oceanic storage increase.

6.3.2 Global Carbon Dioxide Budget

Since the IPCC AR4 (Denman et al., 2007), a number of new advancements in data availability and data-model synthesis have allowed the establishment of a more constrained anthropogenic CO_2 budget and better attribution of its flux components. The advancements are: (1) revised data on the rates of land use change conversion from country statistics (FAO, 2010) now providing an arguably more robust estimate of the land use change flux (Houghton et al., 2012; Section 6.3.2.2); (2) a new global compilation of forest inventory data that provides an independent estimate of the amount of carbon that has been gained by forests over the past two decades, albeit with very scarce measurements for tropical forest (Pan et al., 2011); (3) over 2 million new observations of the partial pressure of CO_2 (pCO_2) at the ocean surface have been taken and added to the global databases (Takahashi et al., 2009; Pfeil et al., 2013) and used to quantify ocean CO_2 sink variability and trends (Section 6.3.2.5) and to evaluate and constrain models (Schuster et al., 2013; Wanninkhof et al., 2013); and (4) the use of multiple constraints with atmospheric inversions and combined atmosphere–ocean

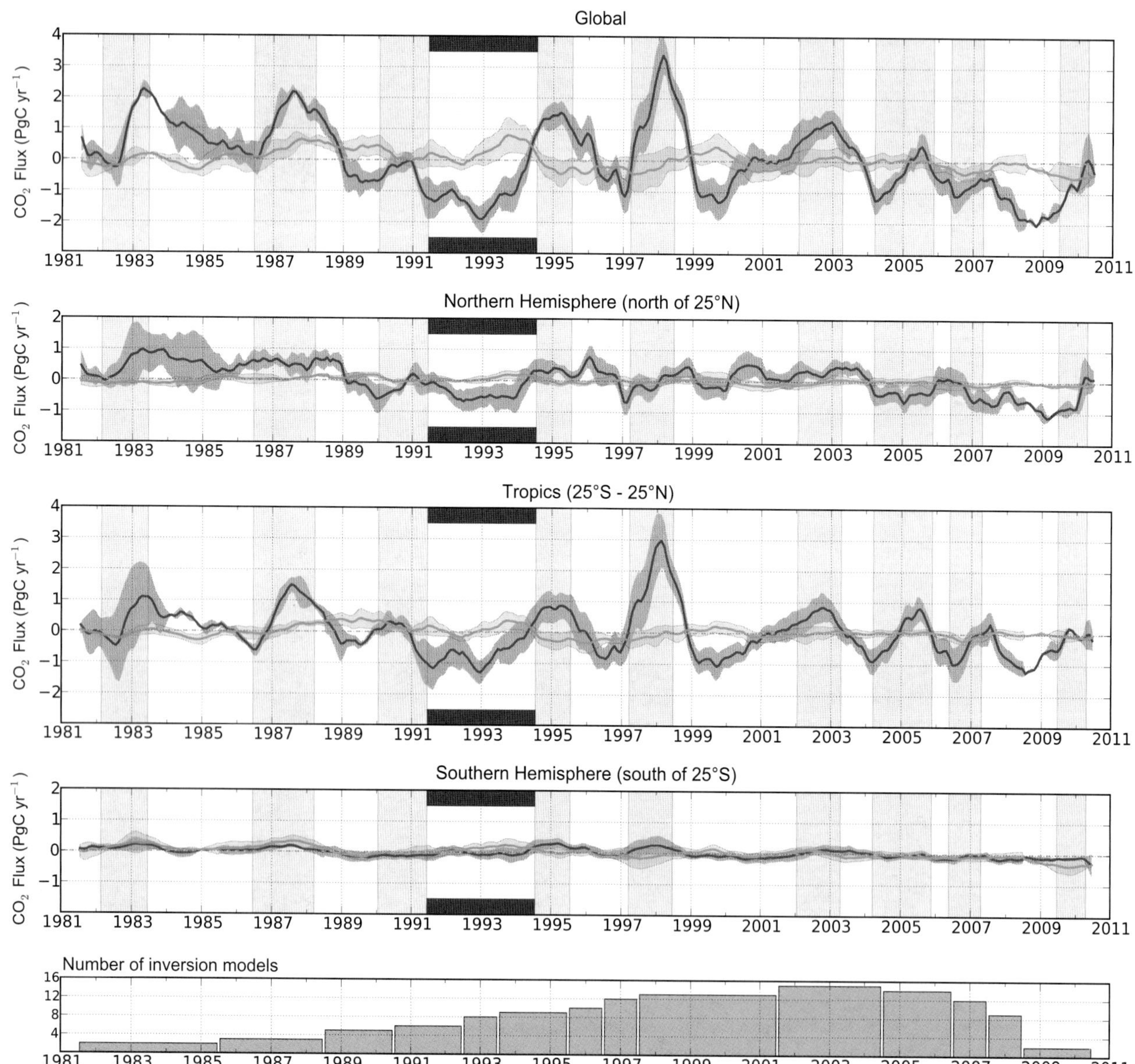

Figure 6.9 | Interannual surface CO_2 flux anomalies from inversions of the TRANSCOM project for the period 1981–2010 (Peylin et al., 2013). The ensemble of inversion results contains up to 17 atmospheric inversion models. The orange bars in the bottom panel indicate the number of available inversion models for each time period. The ensemble mean is bounded by the 1-σ inter-model spread in ocean–atmosphere (blue) and land–atmosphere (green) CO_2 fluxes (PgC yr^{-1}) grouped into large latitude bands, and the global. For each flux and each region, the CO_2 flux anomalies were obtained by subtracting the long-term mean flux from each inversion and removing the seasonal signal. Grey shaded regions indicate El Niño episodes, and the black bars indicate the cooling period following the Mt. Pinatubo eruption, during which the growth rate of CO_2 remained low. A positive flux means a larger than normal source of CO_2 to the atmosphere (or a smaller CO_2 sink).

6

inversions (so called top down approaches; Jacobson et al., 2007) and the up-scaling of reservoir-based observations using models (so called bottom up approaches) provides new coarse scale consistency checks on CO_2 flux estimates for land and ocean regions (McGuire et al., 2009; Piao et al., 2009b; Schulze et al., 2009; Ciais et al., 2010; Schuster et al., 2013). The causes of the year-to-year variability observed in the annual atmospheric CO_2 accumulation shown in Figure 6.8 are estimated with a *medium* to *high confidence* to be mainly driven by terrestrial processes occurring in tropical latitudes as inferred from atmospheric CO_2 inversions and supported by ocean data and models (Bousquet et al., 2000; Raupach et al., 2008; Sarmiento et al., 2010) (Figures 6.9 and 6.13; Section 6.3.2.5) and land models (Figure 6.16; Section 6.3.2.6).

6.3.2.1 Carbon Dioxide Emissions from Fossil Fuel Combustion and Cement Production

Global CO_2 emissions from the combustion of fossil fuels used for this chapter are determined from national energy consumption statistics and converted to emissions by fuel type (Marland and Rotty, 1984). Estimated uncertainty for the annual global emissions are on the order of ±8% (converted from ±10% uncertainty for 95% confidence intervals in Andres et al. (2012) to the 90% confidence intervals used here). The uncertainty has been increasing in recent decades because a larger fraction of the global emissions originate from emerging economies where energy statistics and emission factors per fuel type are more uncertain (Gregg et al., 2008). CO_2 emissions from cement production were 4% of the total emissions during 2000–2009, compared to 3% in the 1990s (Boden et al., 2011). Additional emissions from gas flaring represent <1% of the global emissions.

Global CO_2 emissions from fossil fuel combustion and cement production were 7.8 ± 0.6 PgC yr^{-1} on average during 2000–2009, 6.4 ± 0.5 PgC yr^{-1} during 1990–1999 and 5.5 ± 0.4 PgC yr^{-1} during 1980–1989 (Table 6.1; Figure 6.8). Global fossil fuel CO_2 emissions increased by 3.2% yr^{-1} on average during the decade 2000–2009 compared to 1.0% yr^{-1} in the 1990s and 1.9% yr^{-1} in the 1980s. Francey et al. (2013) recently suggested a cumulative underestimation of 8.8 PgC emissions during the period 1993–2004, which would reduce the contrast in emissions growth rates between the two decades. The global financial crisis in 2008–2009 induced only a short-lived drop in global emissions in 2009 (–0.3%), with the return to high annual growth rates of 5.1% and 3.0% in 2010 and 2011, respectively, and fossil fuel and cement CO_2 emissions of 9.2 ± 0.8 PgC in 2010 and 9.5 ± 0.8 PgC in 2011(Peters et al., 2013).

6.3.2.2 Net Land Use Change Carbon Dioxide Flux

CO_2 is emitted to the atmosphere by land use and land use change activities, in particular deforestation, and taken up from the atmosphere by other land uses such as afforestation (the deliberate creation of new forests) and vegetation regrowth on abandoned lands. A critical distinction in estimating land use change is the existence of gross and net fluxes. Gross fluxes are the individual fluxes from multiple processes involved in land use change that can be either emissions to or removals from the atmosphere occurring at different time scales. For example, gross emissions include instantaneous emissions from deforestation fires and long-term emissions from the decomposition of organic carbon; and they also include the long-term CO_2 uptake by forest regrowth and soil carbon storage on abandoned agricultural lands, afforestation and storage changes of wood products (Houghton et al., 2012; Mason Earles et al., 2012). The net flux of land use change is the balance among all source and sink processes involved in a given timeframe. The net flux of land use change is globally a net source to the atmosphere (Table 6.1; Figure 6.8).

Approaches to estimate global net CO_2 fluxes from land use fall into three categories: (1) the 'bookkeeping' method that tracks carbon in living vegetation, dead plant material, wood products and soils with cultivation, harvesting and reforestation using country-level reports on changes in forest area and biome-averaged biomass values (Houghton, 2003); (2) process-based terrestrial ecosystem models that simulate on a grid-basis the carbon stocks (biomass, soils) and exchange fluxes between vegetation, soil and atmosphere (see references in Table 6.2) and (3) detailed regional (primarily tropical forests) analyses based on satellite data that estimate changes in forest area or biomass (DeFries et al., 2002; Achard et al., 2004; Baccini et al., 2012; Harris et al., 2012). Satellite-derived estimates of CO_2 emissions to the atmosphere from so-called deforestation fires (van der Werf et al., 2010) provide additional constraints on the spatial attribution and variability of land use change gross emissions. Most global estimates do not include emissions from peat burning or decomposition after a land use change, which are estimated to be 0.12 PgC yr^{-1} over 1997–2006 for peat fires (van der Werf et al., 2008) and between 0.10 and 0.23 PgC yr^{-1} from the decomposition of drained peat (Hooijer et al., 2010). The processes and time scales captured by these methods to estimate net land use change CO_2 emissions are diverse, creating difficulties with comparison of different estimates (Houghton et al., 2012; Table 6.2). The bookkeeping method of Houghton et al. (2012) was used for Table 6.1 because it is closest to observations and includes the most extensive set of management practices (Table 6.2). Methods that do not include long-term 'legacy' fluxes from soils caused by deforestation (Table 6.2) underestimate net land use change CO_2 emissions by 13 to 62% depending on the starting year and decade (Ramankutty et al., 2006), and methods that do not include the fate of carbon wood harvest and shifting cultivation underestimate CO_2 emissions by 25 to 35% (Houghton et al., 2012).

Global net CO_2 emissions from land use change are estimated at 1.4, 1.5 and 1.1 PgC yr^{-1} for the 1980s, 1990s and 2000s, respectively, by the bookkeeping method of Houghton et al. (2012) (Table 6.2; Figure 6.10). This estimate is consistent with global emissions simulated by process-based terrestrial ecosystem models using mainly three land cover change data products as input for time-varying maps of land use change (Table 6.2). The bookkeeping method estimate is also generally consistent although higher than the satellite-based methods (tropics only). Part of the discrepancy can be accounted for by emissions from extratropical regions (~0.1 PgC yr^{-1}; Table 6.3) and by legacy fluxes for land cover change prior to 1980s (~0.2 PgC yr^{-1}) that are not covered by satellite-based methods used in Table 6.2, and by the fact that the bookkeeping method accounts for degradation and shifting agriculture CO_2 losses not detected in the satellite-based method reported in Table 6.2. We adopt an uncertainty of ±0.8 PgC yr^{-1} as representative of 90% uncertainty intervals. This is identical to the uncertainty of ±0.5 PgC yr^{-1} representing ±1-σ interval (68% if Gaussian distributed error) from Houghton et al. (2012). This uncertainty of ±0.8 PgC yr^{-1} on net land use change CO_2 fluxes is smaller than the one that was reported in AR4 of 0.5 to 2.7 PgC yr^{-1} for the 1990s (68% confidence interval). In this chapter, uncertainty is estimated based on expert judgment of the available evidence, including improved accuracy of land cover change incorporating satellite data, the larger number of independent methods to quantify emissions and the consistency of the reported results (Table 6.2; Figure 6.10).

Different estimates of net land use change CO_2 emissions are shown in Figure 6.10. The lower net land use change CO_2 emissions reported in the 2000s compared to the 1990s, by 0.5 PgC yr^{-1} in the bookkeeping method based on FAO (2010), and by 0.3 to 0.5 PgC yr^{-1} from five process-based ecosystem models based on the HistorY Database of the global Environment (HYDE) land cover change data updated to 2009 (Goldewijk et al., 2011), are within the error bar of the data and methods. The bookkeeping method suggests that most of the LUC emissions

Table 6.2 | Estimates of net land to atmosphere CO_2 flux from land use change covering recent decades (PgC yr^{-1}). Positive values indicate CO_2 losses to the atmosphere. Various forms of land management are also included in the different estimates, including wood harvest (W), shifting cultivation (C) and harvesting (H) of crops and peat burning and peat drainage (P). All methods include the vegetation degradation after land clearance. Additional processes included are initial biomass loss during the year of deforestation (I), decomposition of slash and soil carbon during the year of initial loss (D), regrowth (R), change in storage in wood products pools (S), the effect of increasing CO_2, (C), the effect of observed climate variability between decades (M) and 'legacy' long-term decomposition flux carried over from land use change transitions prior to start of time period used for reporting in the table (L). In the absence of data on L in the assessed estimates, the studies have either assumed instantaneous loss of all biomass and soil carbon (I, a committed future flux) or did not consider the legacy flux L. Satellite-based methods have examined Land Use Change (LUC) emissions in the tropical regions only. Numbers in parentheses are ranges in uncertainty provided in some studies.

	Data for Land Use Change Area[a]	Biomass Data	Land Management Included	Processes Included	1980–1989 PgC yr^{-1}	1990–1999 PgC yr^{-1}	2000–2009 PgC yr^{-1}
Bookkeeping Method (global)							
Houghton et al. (2012)	FAO-2010	Observed[b]	W, C, H	I, D, R, S, L	1.4	1.5	1.1
Baccini et al. (2012)	FAO-2010	Satellite data	W, C, H	I, D, R, S, L			1.0
Satellite-based Methods (tropics only)							
Achard et al. (2004)	Landsat	Observed[b]		I, D, R, S, C, M		0.9 (0.5–1.4)[c]	
DeFries et al. (2002)	AVHRR	Observed[b]		I, D, R, S[d], C, M	0.6 (0.3–0.8)	0.9 (0.5–1.4)	
Van der Werf et al. (2010)	GFED	CASA[e]	P	I, D, C, M		1.2 (0.6–1.8)[f]	
Process Models (global)							
Shevliakova et al. (2009)	HYDE	LM3V	W, C	I, D, R, S, L, C	1.1	1.1	
Shevliakova et al. (2009)	SAGE	LM3V	W, C	I, D, R, S, L, C	1.4	1.3	
van Minnen et al. (2009)[g]	HYDE	IMAGE 2[e]	W	I, D, R, S, L, C	1.8	1.4	1.2
Strassmann et al. (2008)	HYDE	BernCC[e]		I, D, R, S, L, C	1.3	1.3	
Stocker et al. (2011)[g]	HYDE	BernCC[e]	H	I, D, R, S, L, C	1.4	0.9	0.6
Yang et al. (2010)	SAGE	ISAM[e]	W	I, D, R, S, L, C	1.7	1.7	
Yang et al. (2010)	FAO-2005	ISAM[e]	W	I, D, R, S, L, C	1.7	1.8	
Yang et al. (2010)[g]	HYDE	ISAM[e]	W	I, D, R, S, L, C	2.2	1.5	1.2
Arora and Boer (2010)	SAGE	CTEM[e]	H	I, D, R, S, L, C	1.1[h]	1.1[h]	
Arora and Boer (2010)[g]	HYDE	CTEM[e]	H	I, D, R, S, L, C	0.4[h]	0.4[h]	
Poulter et al. (2010)[g]	HYDE	LPJmL[e]		I, D, R, S, L, C	1.0	0.9	0.5
Kato et al. (2013)[g]	HYDE	VISIT[e]	C	I, D, R, S, L, C	1.2	1.0	0.5
Zaehle et al. (2011)	HYDE	O-CN		I, D, R, S, L, C	1.2	1.0	
Average of process models[i]					1.3 ± 0.7	1.2 ± 0.6	0.8 ± 0.6
Range of process models					[0.4–2.2]	[0.4–1.8]	[0.5–1.2]

Notes:

[a] References for the databases used: FAO (2010) as applied in Houghton et al. (2012); FAO (2005) as applied in Houghton (2003), updated; GFED (van der Werf et al., 2009); HYDE (Goldewijk et al., 2011), SAGE (Ramankutty and Foley, 1999). Landsat and AVHRR are satellite-based data and GFED is derived from satellite products as described in the references.

[b] Based on average estimates by biomes compiled from literature data (see details in corresponding references).

[c] 1990–1997 only.

[d] Legacy fluxes for land cover change prior to 1980 are not included and are estimated to add about 0.2 PgC yr^{-1} to the 1980s and 0.1 PgC yr^{-1} to the 1990s estimates, based on Ramankutty et al. (2006).

[e] The vegetation and soil biomass is computed using a vegetation model described in the reference.

[f] 1997–2006 average based on estimates of carbon emissions from deforestation and degradation fires, including peat fires and oxidation. Estimates were doubled to account for emissions other than fire including respiration of leftover plant materials and soil carbon following deforestation following (Olivier et al., 2005).

[g] Method as described in the reference but updated to 2010 using the land cover change data listed in column 2.

[h] The large variability produced by the calculation method is removed for comparison with other studies by averaging the flux over the two decades.

[i] Average of estimates from all process models and 90% confidence uncertainty interval; note that the spread of the different estimates does not follow a Gaussian distribution. AVHRR = Advanced Very High Resolution Radiometer; FAO = Food and Agriculture Organization (UN); GFED = Global Fire Emissions Database; HYDE = HistorY Database of the global Environment; SAGE = Center for Sustainability and the Global Environment.

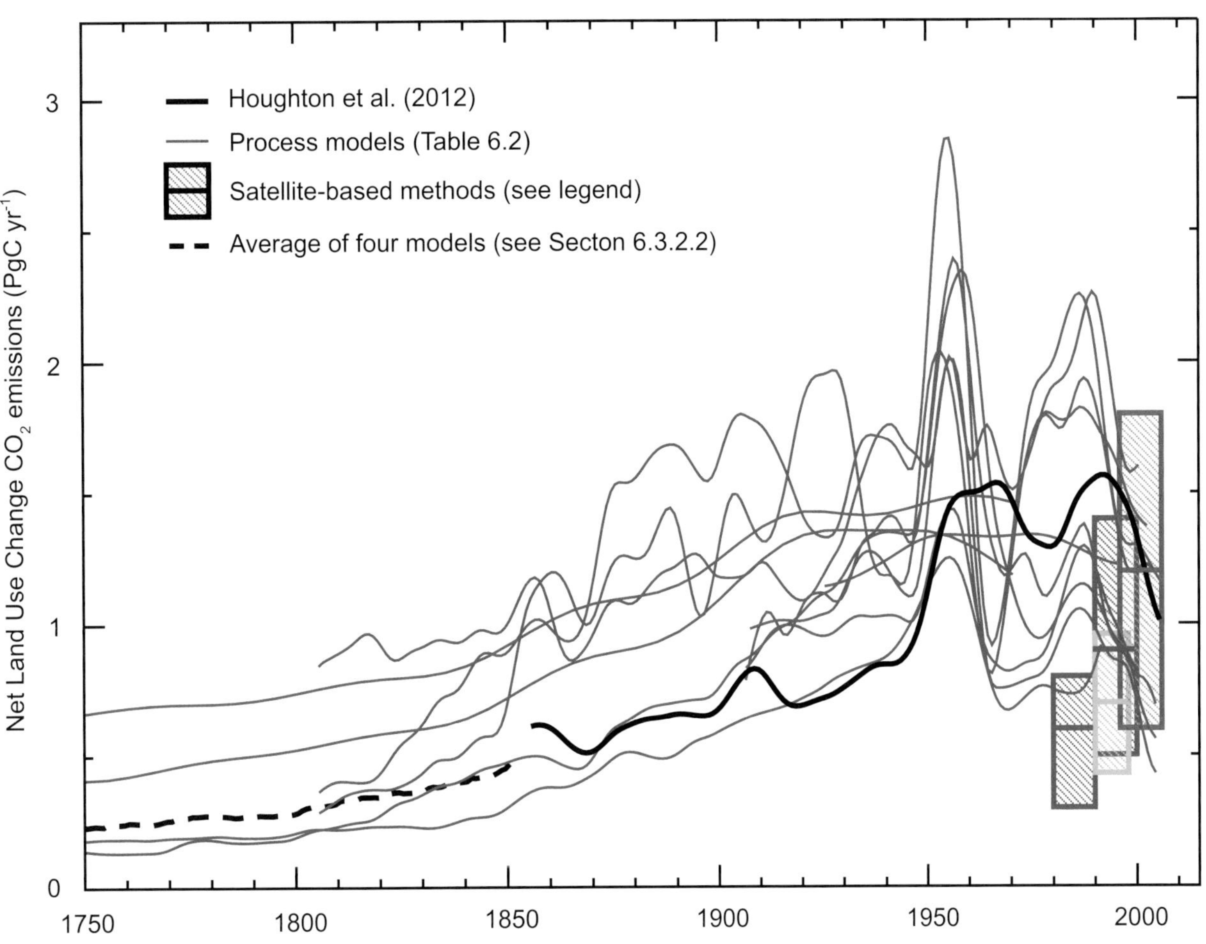

Figure 6.10 | Net land use change CO_2 emissions (PgC yr^{-1}). All methods are based on land cover change data (see Table 6.2) and are smoothed with a 10-year filter to remove interannual variability. The bookkeeping estimate of Houghton et al. (2012) (thick black over 1850–2011) and the average of four process models (dash black) over 1750–1850 (see 6.3.2.2) are used in Table 6.1. The process model results for net land use change CO_2 emissions from Table 6.2 are shown in blue. Satellite-based methods are available for the tropics only, from (red) van der Werf et al. (2010), (blue) DeFries et al. (2002), and (green) Achard et al. (2004). Note that the definitions of land use change fluxes vary between models (Table 6.2). The grey shading shows a constant uncertainty of ±0.8 PgC yr^{-1} around the mean estimate used in Table 6.3.

originate from Central and South America, Africa and Tropical Asia since the 1980s (Table 6.3). The process models based on the HYDE database allocate about 30% of the global land use change emissions to East Asia, but this is difficult to reconcile with the large afforestation programmes reported in this region. Inconsistencies in the available land cover change reconstructions and in the modelling results prevent a firm assessment of recent trends and their partitioning among regions (see regional data in Table 6.3).

In this chapter, we do not assess individual gross fluxes that sum up to make the net land use change CO_2 emission, because there are too few independent studies. Gross emissions from tropical deforestation and degradation were 3.0 ± 0.5 PgC yr^{-1} for the 1990s and 2.8 ± 0.5 PgC yr^{-1} for the 2000s using forest inventory data, FAO (2010) and the bookeeping method (Pan et al., 2011). These gross emissions are about double the net emissions because of the presence of a large regrowth that compensates for about half of the gross emissions. A recent analysis estimated a lower gross deforestation of 0.6 to 1.2 PgC yr^{-1} (Harris et al., 2012). That study primarily estimated permanent deforestation and excluded additional gross emissions from degraded forests, shifting agriculture and some carbon pools. In fact, gross emissions from permanent deforestation are in agreement between the bookkeeping method of Houghton et al. (2012) and the satellite data analysis of Harris et al. (2012).

Over the 1750–2011 period, cumulative net CO_2 emissions from land use change of 180 ± 80 PgC are estimated (Table 6.1). The uncertainty is based on the spread of the available estimates (Figure 6.10). The cumulative net CO_2 emissions from land use change have been dominated by deforestation and other land use change in the mid-northern latitudes prior to 1980s, and in the tropics since the 1980s, largely from deforestation in tropical America and Asia with smaller contributions from tropical Africa. Deforestation from 800 to 1750 has been estimated at 27 PgC using a process-based ecosystem model (Pongratz et al., 2009).

6.3.2.3 Atmospheric Carbon Dioxide Concentration Growth Rate

Since the beginning of the Industrial Era (1750), the concentration of CO_2 in the atmosphere has increased by 40%, from 278 ± 5 ppm to 390.5 ± 0.1 ppm in 2011 (Figure 6.11; updated from Ballantyne et al. (2012), corresponding to an increase in CO_2 of 240 ± 10 PgC in the atmosphere. Atmospheric CO_2 grew at a rate of 3.4 ± 0.2 PgC yr^{-1} in the 1980s, 3.1 ± 0.2 PgC yr^{-1} in the 1990s and 4.0 ± 0.2 PgC yr^{-1} in the 2000s (Conway and Tans, 2011) (Table 6.1). The increase of atmospheric CO_2 between 1750 and 1957, prior to direct measurements in the atmosphere, is established from measurements of CO_2 trapped in air bubbles in ice cores (e.g., Etheridge et al., 1996). After 1957, the increase of atmospheric CO_2 is established from highly precise continuous atmospheric CO_2 concentration measurements at background stations (e.g., Keeling et al., 1976).

The ice core record of atmospheric CO_2 during the past century exhibits interesting variations, which can be related to climate induced-changes in the carbon cycle. Most conspicuous is the interval from about 1940 to 1955, during which atmospheric CO_2 concentration stabilised

Table 6.3 | Estimates of net land to atmosphere flux from land use change (PgC yr^{-1}; except where noted) for decadal periods from 1980s to 2000s by region. Positive values indicate net CO_2 losses from land ecosystems affected by land use change to the atmosphere. Uncertainties are reported as 90% confidence interval (unlike 68% in AR4). Numbers in parentheses are ranges in uncertainty provided in some studies. Tropical Asia includes the Middle East, India and surrounding countries, Indonesia and Papua New Guinea. East Asia includes China, Japan, Mongolia and Korea.

	Land Cover Data	Central and South Americas	Africa	Tropical Asia	North America	Eurasia	East Asia	Oceania
2000s								
van der Werf et al. (2010)[a,b]	GFED	0.33	0.15	0.35				
DeFries and Rosenzweig (2010)[c]	MODIS	0.46	0.08	0.36				
Houghton et al. (2012)	FAO-2010	0.48	0.31[e]	0.25	0.01	–0.07[d]	0.01[e]	
van Minnen et al. (2009)[a]	HYDE	0.45	0.21	0.20	0.09	0.08	0.10	0.03
Stocker et al. (2011)[a]	HYDE	0.19	0.18	0.21	0.019	–0.067	0.12	0.011
Yang et al. (2010)[a]	HYDE	0.14	0.03	0.25	0.25	0.39	0.12	0.02
Poulter et al. (2010)[a]	HYDE	0.09	0.13	0.14	0.01	0.03	0.05	0.00
Kato et al. (2013)[a]	HYDE	0.36	–0.09	0.23	–0.05	–0.04	0.10	0.00
Average		0.31 ± 0.25	0.13 ± 0.20	0.25 ± 0.12	0.05 ± 0.17	0.12 ± 0.31	0.08 ± 0.07	0.01 ± 0.02
1990s								
DeFries et al. (2002)	AVHRR	0.5 (0.2–0.7)	0.1 (0.1–0.2)	0.4 (0.2–0.6)				
Achard et al. (2004)	Landsat	0.3 (0.3–0.4)	0.2 (0.1–0.2)	0.4 (0.3–0.5)				
Houghton et al. (2012)	FAO-2010	0.67	0.32[e]	0.45	0.05	–0.04[d]	0.05[e]	
van Minnen et al. (2009)[a]	HYDE	0.48	0.22	0.34	0.07	0.08	0.20	0.07
Stocker et al. (2011)[a]	HYDE	0.30	0.14	0.19	–0.072	0.11	0.27	0.002
Yang et al. (2010)[a]	HYDE	0.20	0.04	0.31	0.27	0.47	0.19	0.00
Poulter et al. (2010)[a]	HYDE	0.26	0.13	0.12	0.07	0.16	0.11	0.01
Kato et al. (2013)[a]	HYDE	0.53	0.07	0.25	–0.04	–0.01	0.16	0.02
Average		0.41 ± 0.27	0.15 ± 0.15	0.31 ± 0.19	0.08 ± 0.19	0.16 ± 0.30	0.16 ± 0.13	0.02 ± 0.05
1980s								
DeFries et al. (2002)	AVHRR	0.4 (0.2–0.5)	0.1 (0.08–0.14)	0.2 (01–0.3)				
Houghton et al. (2012)	FAO-2010	0.79	0.22[e]	0.32	0.04	0.00[d]	0.07[e]	
van Minnen et al. (2009)[a]	HYDE	0.70	0.18	0.43	0.07	0.06	0.37	0.04
Stocker et al. (2011)[a]	HYDE	0.44	0.16	0.25	0.085	0.11	0.40	0.009
Yang et al. (2010)[a]	HYDE	0.26	0.01	0.34	0.30	0.71	0.59	0.00
Poulter et al. (2010)[a]	HYDE	0.37	0.11	0.19	0.02	0.03	0.29	0.01
Kato et al. (2013)[a]	HYDE	0.61	0.07	0.25	–0.04	–0.02	0.35	0.01
Average		0.51 ± 0.32	0.12 ± 0.12	0.28 ± 0.14	0.08 ± 0.19	0.15 ± 0.46	0.35 ± 0.28	0.01 ± 0.03

Notes:

[a] Method as described in the reference but updated to 2010 using the HYDE land cover change data.

[b] 1997–2006 average based on estimates of CO_2 emissions from deforestation and degradation fires, including peat carbon emissions. Estimates were doubled to account for emissions other than fire including respiration of leftover plant materials and soil carbon following deforestation following (Olivier et al., 2005). Estimates include peat fires and peat soil oxidation. If peat fires are excluded, estimate in tropical Asia is 0.23 and Pan-tropical total is 0.71.

[c] CO_2 estimates were summed for dry and humid tropical forests, converted to C and normalized to annual values. Estimates are based on satellite-derived deforestation area (Hansen et al., 2010), and assume 0.6 fraction of biomass emitted with deforestation. Estimates do not include carbon uptake by regrowth or legacy fluxes from historical deforestation. Estimates cover emissions from 2000 to 2005.

[d] Includes China only.

[e] East Asia and Oceania are averaged in one region. The flux is split in two equally for computing the average; North Africa and the Middle East are combined with Eurasia. AVHRR = Advanced Very High Resolution Radiometer; FAO = Food and Agriculture Organization (UN); GFED = Global Fire Emissions Database; HYDE = HistorY Database of the global Environment; MODIS = Moderate Resolution Imaging Spectrometer.

(Trudinger et al., 2002), and the CH_4 and N_2O growth slowed down (MacFarling-Meure et al., 2006), possibly caused by slightly decreasing temperatures over land in the NH (Rafelski et al., 2009).

There is substantial evidence, for example, from ^{13}C carbon isotopes in atmospheric CO_2 (Keeling et al., 2005) that source/sink processes on land generate most of the interannual variability in the atmospheric CO_2 growth rate (Figure 6.12). The strong positive anomalies of the CO_2 growth rate in El Niño years (e.g., 1986–1987 and 1997–1998) originated in tropical latitudes (see Sections 6.3.6.3 and 6.3.2.5.4), while the anomalies in 2003 and 2005 originated in northern mid-latitudes, perhaps reflecting the European heat wave in 2003 (Ciais et al., 2005). Volcanic forcing also contributes to multi-annual variability in carbon storage on land and in the ocean (Jones and Cox, 2001; Gerber et al., 2003; Brovkin et al., 2010; Frölicher et al., 2011).

With a *very high confidence*, the increase in CO_2 emissions from fossil fuel burning and those arising from land use change are the dominant cause of the observed increase in atmospheric CO_2 concentration. Several lines of evidence support this conclusion:

- The observed decrease in atmospheric O_2 content over past two decades and the lower O_2 content in the northern compared to the SH are consistent with the burning of fossil fuels (see Figure 6.3 and Section 6.1.3.2; Keeling et al., 1996; Manning and Keeling, 2006).

- CO_2 from fossil fuels and from the land biosphere has a lower $^{13}C/^{12}C$ stable isotope ratio than the CO_2 in the atmosphere. This induces a decreasing temporal trend in the atmospheric $^{13}C/^{12}C$ ratio of atmospheric CO_2 concentration as well as, on annual average, slightly lower $^{13}C/^{12}C$ values in the NH (Figure 6.3). These signals are measured in the atmosphere.

- Because fossil fuel CO_2 is devoid of radiocarbon (^{14}C), reconstructions of the $^{14}C/C$ isotopic ratio of atmospheric CO_2 from tree rings

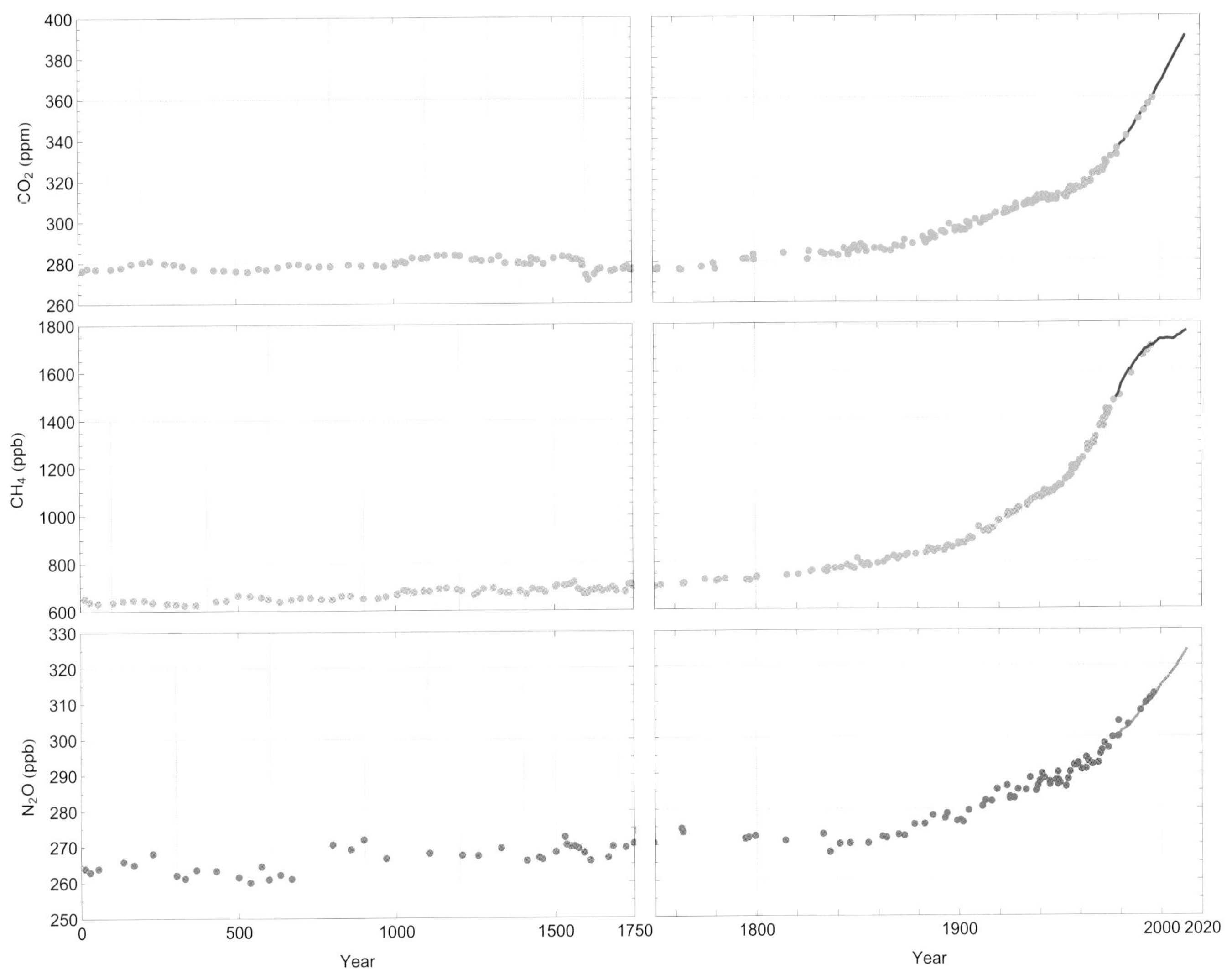

Figure 6.11 | Atmospheric CO_2, CH_4, and N_2O concentrations history over the industrial era (right) and from year 0 to the year 1750 (left), determined from air enclosed in ice cores and firn air (colour symbols) and from direct atmospheric measurements (blue lines, measurements from the Cape Grim observatory) (MacFarling-Meure et al., 2006).

6

show a declining trend, as expected from the addition of fossil CO_2 (Stuiver and Quay, 1981; Levin et al., 2010). Yet nuclear weapon tests in the 1950s and 1960s have been offsetting that declining trend signal by adding ^{14}C to the atmosphere. Since this nuclear weapon induced ^{14}C pulse in the atmosphere has been fading, the $^{14}C/C$ isotopic ratio of atmospheric CO_2 is observed to resume its declining trend (Naegler and Levin, 2009; Graven et al., 2012).

- Most of the fossil fuel CO_2 emissions take place in the industrialised countries north of the equator. Consistent with this, on annual average, atmospheric CO_2 measurement stations in the NH record increasingly higher CO_2 concentrations than stations in the SH, as witnessed by the observations from Mauna Loa, Hawaii, and the South Pole (see Figure 6.3). The annually averaged concentration difference between the two stations has increased in proportion of the estimated increasing difference in fossil fuel combustion emissions between the hemispheres (Figure 6.13; Keeling et al., 1989; Tans et al., 1989; Fan et al., 1999).

- The rate of CO_2 emissions from fossil fuel burning and land use change was almost exponential, and the rate of CO_2 increase in the atmosphere was also almost exponential and about half that of the emissions, consistent with a large body of evidence about changes of carbon inventory in each reservoir of the carbon cycle presented in this chapter.

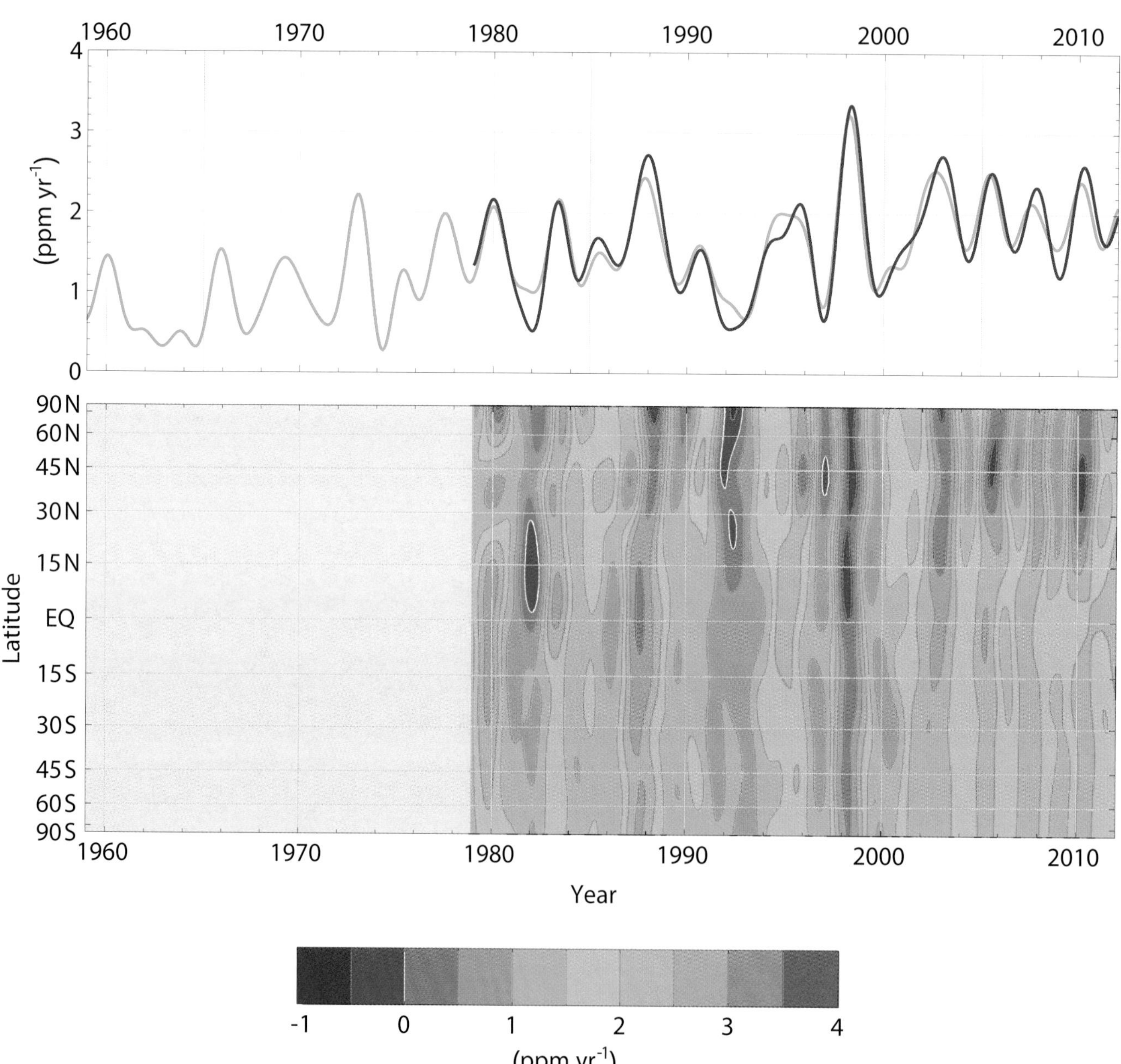

Figure 6.12 | (Top) Global average atmospheric CO_2 growth rate, computed from the observations of the Scripps Institution of Oceanography (SIO) network (light green line: Keeling et al. 2005, updated) and from the marine boundary layer air reference measurements of the National Oceanic and Atmospheric Administration –Global Monitoring Division (NOAA–GMD) network (dark green line: Conway et al., 1994; Dlugokencky and Tans, 2013b). (Bottom) Atmospheric growth rate of CO_2 as a function of latitude determined from the National Oceanic and Atmospheric Administration–Earth System Research Laboratory (NOAA–ESRL) network, representative of stations located in the marine boundary layer at each given latitude (Masarie and Tans, 1995; Dlugokencky and Tans, 2013b). Sufficient observations are available only since 1979.

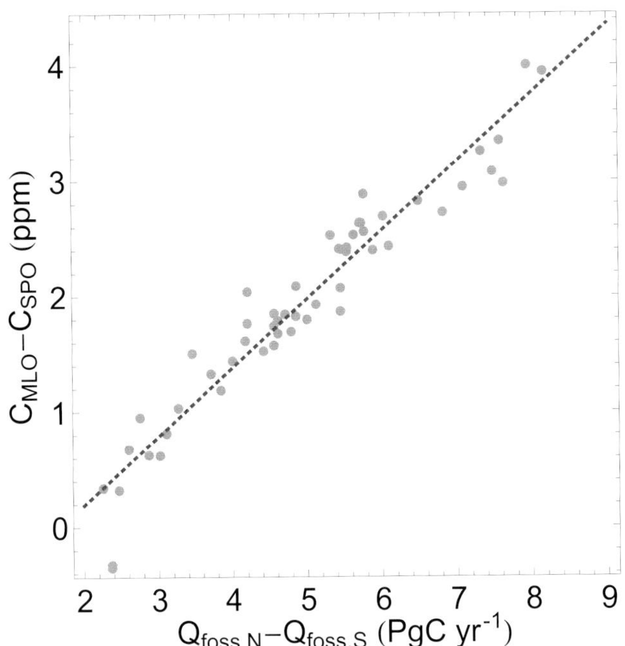

Figure 6.13 | Blue points: Annually averaged CO_2 concentration difference between the station Mauna Loa in the Northern Hemisphere and the station South Pole in the Southern Hemisphere (vertical axis; Keeling et al., 2005, updated) versus the difference in fossil fuel combustion CO_2 emissions between the hemispheres (Boden et al., 2011). Dark red dashed line: regression line fitted to the data points.

6.3.2.4 Carbon Dioxide Airborne Fraction

Until recently, the uncertainty in CO_2 emissions from land use change emissions was large and poorly quantified which led to the use of an airborne fraction (see Glossary) based on CO_2 emissions from fossil fuel only (e.g., Figure 7.4 in AR4 and Figure 6.26 of this chapter). However, reduced uncertainty of emissions from land use change and larger agreement in its trends over time (Section 6.3.2.2) allow making use of an airborne fraction that includes all anthropogenic emissions. The airborne fraction will increase if emissions are too fast for the uptake of CO_2 by the carbon sinks (Bacastow and Keeling, 1979; Gloor et al., 2010; Raupach, 2013). It is thus controlled by changes in emissions rates, and by changes in carbon sinks driven by rising CO_2, changes in climate and all other biogeochemical changes.

A positive trend in airborne fraction of ~0.3% yr^{-1} relative to the mean of 0.44 ±0.06 (or about 0.05 increase over 50 years) was found by all recent studies (Raupach et al., 2008, and related papers; Knorr, 2009; Gloor et al., 2010) using the airborne fraction of total anthropogenic CO_2 emissions over the approximately 1960–2010 period (for which the most accurate atmospheric CO_2 data are available). However, there is no consensus on the significance of the trend because of differences in the treatment of uncertainty and noise (Raupach et al., 2008; Knorr, 2009). There is also no consensus on the cause of the trend (Canadell et al., 2007b; Raupach et al., 2008; Gloor et al., 2010). Land and ocean carbon cycle model results attributing the trends of fluxes to underlying processes suggest that the effect of climate change and variability on ocean and land sinks have had a significant influence (Le Quéré et al., 2009), including the decadal influence of volcanic eruptions (Frölicher et al., 2013).

6.3.2.5 Ocean Carbon Dioxide Sink

6.3.2.5.1 Global ocean sink and decadal change

The estimated mean anthropogenic ocean CO_2 sink assessed in AR4 was 2.2 ± 0.7 PgC yr^{-1} for the 1990s based on observations (McNeil et al., 2003; Manning and Keeling, 2006; Mikaloff-Fletcher et al., 2006), and is supported by several contemporary estimates (see Chapter 3). Note that the uncertainty of ±0.7 PgC yr^{-1} reported here (90% confidence interval) is the same as the ±0.4 PgC yr^{-1} uncertainty reported in AR4 (68% confidence intervals). The uptake of anthropogenic CO_2 by the ocean is primarily a response to increasing CO_2 in the atmos-

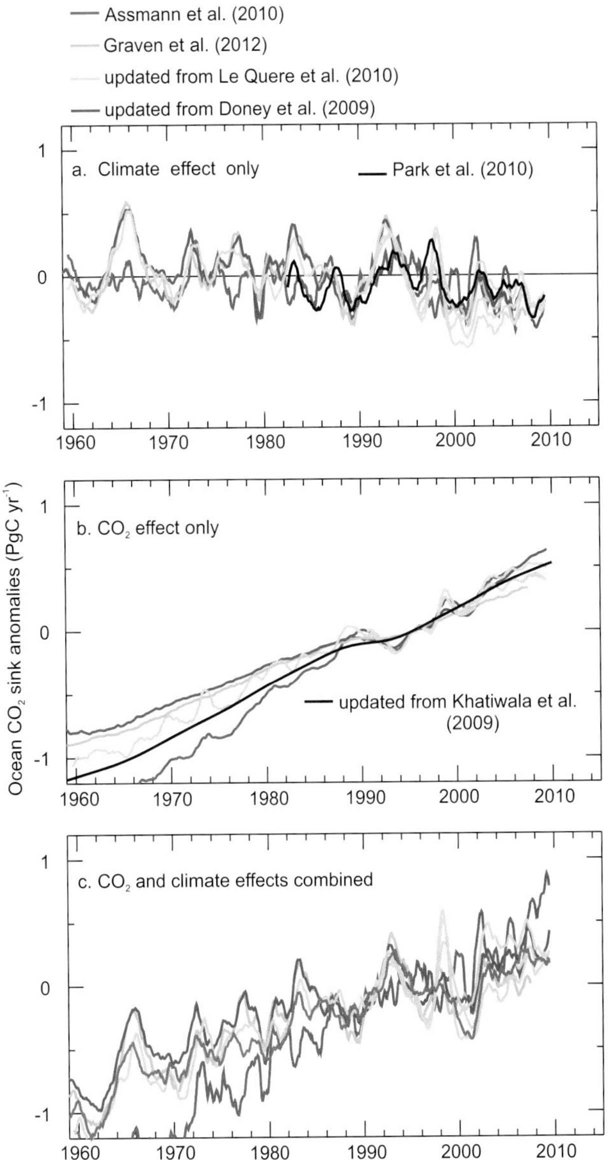

Figure 6.14 | Anomalies in the ocean CO_2 ocean-to-atmosphere flux in response to (a) changes in climate, (b) increasing atmospheric CO_2 and (c) the combined effects of increasing CO_2 and changes in climate (PgC yr^{-1}). All estimates are shown as anomalies with respect to the 1990–2000 averages. Estimates are updates from ocean models (in colours) and from indirect methods based on observations (Khatiwala et al., 2009; Park et al., 2010). A negative ocean-to-atmosphere flux represents a sink of CO_2, as in Table 6.1.

6

Table 6.4 | Decadal changes in the ocean CO_2 sink from models and from data-based methods (a positive change between two decades means an increasing sink with time). It is reminded that the total CO_2 sink for the 1990s is estimated at 2.2 ± 0.7 PgC yr^{-1} based on observations.

	Method	1990s Minus 1980s PgC yr^{-1}	2000s Minus 1990s PgC yr^{-1}
CO_2 effects only			
Khatiwala et al. (2009)	Data-based[c]	0.24	0.20
Mikaloff-Fletcher et al. (2006)[a]	Data-based[d]	0.40	0.44
Assmann et al. (2010) (to 2007 only)	Model	0.28	0.35
Graven et al. (2012)	Model	0.15	0.25
Doney et al. (2009)	Model	0.15	0.39
Le Quéré et al. (2010) NCEP	Model	0.16	0.32
Le Quéré et al. (2010) ECMWF	Model	—	0.39
Le Quéré et al. (2010) JPL	Model	—	0.32
Average[b]		0.23 ± 0.15	0.33 ± 0.13
Climate effects only			
Park et al. (2010)	Data-based[e]	—	–0.15
Assmann et al. (2010) (to 2007 only)	Model	0.07	0.00
Graven et al. (2012)	Model	0.02	–0.27
Doney et al. (2009)	Model	–0.02	–0.21
Le Quéré et al. (2010) NCEP	Model	0.02	–0.27
Le Quéré et al. (2010) ECMWF	Model	—	–0.14
Le Quéré et al. (2010) JPL	Model	—	–0.36
Average[b]		0.02 ± 0.05	–0.19 ± 0.18
CO_2 and climate effects combined		0.25 ± 0.16	0.14 ± 0.22

Notes:

[a] As published by Sarmiento et al. (2010).

[b] Average of all estimates ±90% confidence interval. The average includes results by Le Quéré et al. (2010)–NCEP only because the other Le Quéré et al. model versions do not differ sufficiently to be considered separately.

[c] Based on observed patterns of atmospheric minus oceanic pCO_2, assuming the difference increases with time following the increasing atmospheric CO_2.

[d] Ocean inversion, assuming constant oceanic transport through time.

[e] Based on observed fit between the variability in temperature and pCO_2, and observed variability in temperature.
ECMWF = European Centre for Medium-Range Weather Forecasts; JPL = Jet Propulsion Laboratory; NCEP = National Centers for Environmental Prediction.

phere and is limited mainly by the rate at which anthropogenic CO_2 is transported from the surface waters into the deep ocean (Sarmiento et al., 1992; Graven et al., 2012). This anthropogenic ocean CO_2 sink occurs on top of a very active natural oceanic carbon cycle. Recent climate trends, such as ocean warming, changes in ocean circulation and changes in marine ecosystems and biogeochemical cycles, can have affected both the anthropogenic ocean CO_2 sink as well as the natural air–sea CO_2 fluxes. We report a decadal mean uptake of 2.0 ± 0.7 PgC yr^{-1} for the 1980s and of 2.3 ± 0.7 PgC yr^{-1} for the 2000s (Table 6.4). The methods used are: (1) an empirical Green's function approach fitted to observations of transient ocean tracers (Khatiwala et al., 2009), (2) a model-based Green's function approach fitted to anthropogenic carbon reconstructions (Mikaloff-Fletcher et al., 2006), (3) estimates based on empirical relationships between observed ocean surface pCO_2 and temperature and salinity (Park et al., 2010) and (4) process-based global ocean biogeochemical models forced by observed meteorological fields (Doney et al., 2009; Assmann et al., 2010; Le Quéré et al., 2010; Graven et al., 2012). All these different methods suggest that in the absence of recent climate change and climate variability, the ocean anthropogenic CO_2 sink should have increased by 0.23 ± 0.15 PgC yr^{-1} between the 1980s and the 1990s, and by 0.33 ± 0.13 PgC yr^{-1} between the 1990s and the 2000s (Figure 6.14). The decadal estimates in the ocean CO_2 sink reported in Table 6.4 as 'CO_2 effects only' are entirely explained by the faster rate of increase of atmospheric CO_2 in the later decade. On the other hand, 'climate effects only' in Table 6.4 are assessed to have no noticeable effect on the sink difference between the 1980s and the 1990s (0.02 ± 0.05 PgC yr^{-1}), but are estimated to have reduced the ocean anthropogenic CO_2 sink by 0.19 ± 0.18 PgC yr^{-1} between the 1990s and the 2000s (Table 6.4).

6

6.3.2.5.2 Regional changes in ocean dissolved inorganic carbon

Observational-based estimates for the global ocean inventory of anthropogenic carbon are obtained from shipboard repeated hydrographic cross sections (Sabine et al., 2004; Waugh et al., 2006; Khatiwala et al., 2009). These estimates agree well among each other, with an average value of 155 ± 30 PgC of increased dissolved inorganic carbon for the period 1750–2011 (see Chapter 3). The uptake of anthropogenic carbon into the ocean is observed to be larger in the high latitudes than in the tropics and subtropics over the entire Industrial Era, because of the more vigorous ocean convection in the high latitudes (Khatiwala et al., 2009). A number of ocean cross sections have been repeated over the last decade, and the observed changes

Table 6.5 | Regional rates of change in inorganic carbon storage from shipboard repeated hydrographic cross sections.

Section	Time	Storage Rate (mol C m^{-2} yr^{-1})	Data Source
Global average (used in Table 6.1)	2007–2008	0.5 ± 0.2	Khatiwala et al. (2009)
Pacific Ocean			
Section along 30°S	1992–2003	1.0 ± 0.4	Murata et al. (2007)
N of 50°S, 120°W to 180°W	1974–1996	0.9 ± 0.3	Peng et al. (2003)
154°W, 20°N to 50°S	1991–2006	0.6 ± 0.1	Sabine et al. (2008)
140°E to 170°W, 45°S to 65°S	1968–1991/1996	0.4 ± 0.2	Matear and McNeil (2003)
149° W, 4°S to 10°N	1993–2005	0.3 ± 0.1	Murata et al. (2009)
149° W, 24°N to 30°N	1993–2005	0.6 ± 0.2	Murata et al. (2009)
Northeast Pacific	1973–1991	1.3 ± 0.5	Peng et al. (2003)
~160°E, ~45°N	1997–2008	0.4 ± 0.1	Wakita et al. (2010)
North of 20°N	1994–2004/2005	0.4 ± 0.2	Sabine et al. (2008)
150°W, 20°S to 20°N	1991/1992–2006	0.3 ± 0.1	Sabine et al. (2008)
Indian Ocean			
20°S to 10°S	1978–1995	0.1	Peng et al. (1998)
10°S to 5°N	1978–1995	0.7	Peng et al. (1998)
Section along 20°S	1995–2003/2004	1.0 ± 0.1	Murata et al. (2010)
Atlantic Ocean			
Section along 30°S	1992/1993–2003	0.6 ± 0.1	Murata et al. (2010)
~30°W, 56°S to 15°S	1989–2005	0.8	Wanninkhof et al. (2010)
20°W, 64°N to 15°N	1993–2003	0.6	Wanninkhof et al. (2010)
~25°W, 15°N to 15°S	1993–2003	0.2	Wanninkhof et al. (2010)
40°N to 65°N	1981–1997/1999	2.2 ± 0.7	Friis et al. (2005)
20°N to 40°N	1981–2004	1.2 ± 0.3	Tanhua et al. (2007)
Nordic Seas	1981–2002/2003	0.9 ± 0.2	Olsen et al. (2006)
Sub-decadal variations			
Irminger Sea	1981–1991	0.6 ± 0.4	Pérez et al. (2008)
Irminger Sea	1991–1996	2.3 ± 0.6	Pérez et al. (2008)
Irminger Sea	1997–2006	0.8 ± 0.2	Pérez et al. (2008)

in carbon storage (Table 6.5) suggest that some locations have rates of carbon accumulation that are higher and others that are lower than the global average estimated by Khatiwala et al. (2009). Model results suggest that there may be an effect of climate change and variability in the storage of total inorganic carbon in the ocean (Table 6.4), but that this effect is small (~2 PgC over the past 50 years; Figure 6.14) compared to the cumulative uptake of anthropogenic carbon during the same period.

6.3.2.5.3 Interannual variability in air-sea CO_2 fluxes

The interannual variability in the global ocean CO_2 sink is estimated to be of about ±0.2 PgC yr^{-1} (Wanninkhof et al., 2013) which is small compared to the interannual variability of the terrestrial CO_2 sink (see Sections 6.3.2.3 and 6.3.2.6.3; Figure 6.12). In general, the ocean takes up more CO_2 during El Niño episodes (Park et al., 2010) because of the temporary suppression of the source of CO_2 to the atmosphere over the eastern Pacific upwelling. Interannual variability of ~0.3 PgC yr^{-1} has been reported for the North Atlantic ocean region alone (Watson et al., 2009) but there is no agreement among estimates regarding the exact magnitude of driving factors of air–sea CO_2 flux variability in this region (Schuster et al., 2013). Interannual variability of 0.1 to 0.2 PgC yr^{-1} was also estimated by models and one atmospheric inversion in the Southern Ocean (Le Quéré et al., 2007), possibly driven by the Southern Annular Mode of climate variability (Lenton and Matear, 2007; Lovenduski et al., 2007; Lourantou and Metzl, 2011).

6.3.2.5.4 Regional ocean carbon dioxide partial pressure trends

Observations of the partial pressure of CO_2 at the ocean surface (pCO_2) show that ocean pCO_2 has been increasing generally at about the same rate as CO_2 in the atmosphere when averaged over large ocean regions during the past two to three decades (Yoshikawa-Inoue and Ishii, 2005; Takahashi et al., 2009; McKinley et al., 2011). However, analyses of regional observations highlight substantial regional and temporal variations around the mean trend.

In the North Atlantic, repeated observations show ocean pCO_2 increasing regionally either at the same rate or faster than atmospheric CO_2 between about 1990 and 2006 (Schuster et al., 2009), thus indicating a constant or decreasing sink for CO_2 in that region, in contrast to the increasing sink expected from the response of the ocean to increasing atmospheric CO_2 alone. The anomalous North Atlantic trends appear to be related to sea surface warming and its effect on solubility (Corbière et al., 2007) and/or changes in ocean circulation (Schuster and Watson, 2007; Schuster et al., 2009) and deep convection (Metzl et al., 2010). Recent changes have been associated with decadal variability in the North Atlantic Oscillation (NAO) and the Atlantic Multidecadal Variability (AMV) (Thomas et al., 2007; Ullman et al., 2009; McKinley et al., 2011; Tjiputra et al., 2012). A systematic analysis of trends estimated in this region show no agreement regarding the drivers of change (Schuster et al., 2013).

In the Southern Ocean, an approximately constant sink was inferred from atmospheric (Le Quéré et al., 2007) and oceanic (Metzl, 2009; Takahashi et al., 2009) CO_2 observations but the uncertainties are large (Law et al., 2008). Most ocean biogeochemistry models reproduce the constant sink and attribute it as a response to an increase in Southern Ocean winds driving increased upwards transport of carbon-rich deep waters (Lenton and Matear, 2007; Verdy et al., 2007; Lovenduski et al., 2008; Le Quéré et al., 2010). The increase in winds has been attributed to the depletion of stratospheric ozone (Thompson and Solomon, 2002) with a contribution from GHGs (Fyfe and Saenko, 2006).

Large decadal variability has been observed in the Equatorial Pacific (Ishii et al., 2009) associated with changes in the phasing of the Pacific Decadal Oscillation (see Glossary) and its impact on gas transfer velocity (Feely et al., 2006; Valsala et al., 2012). By contrast, ocean pCO_2 appears to have increased at a slower rate than atmospheric CO_2 (thus a growing ocean CO_2 sink in that region) in the northern North Pacific Ocean (Takahashi et al., 2006). There is less evidence available to attribute the observed changes in other regions to changes in underlying processes or climate change and variability.

6.3.2.5.5 Processes driving variability and trends in air–sea carbon dioxide fluxes

Three type of processes are estimated to have an important effect on the air–sea CO_2 fluxes on century time scales: (1) the dissolution of CO_2 at the ocean surface and its chemical equilibrium with other forms of carbon in the ocean (mainly carbonate and bicarbonate), (2) the transport of carbon between the surface and the intermediate and deep ocean and (3) changes in the cycling of carbon through marine ecosystem processes (the ocean biological pump; see Section 6.1.1.1). The surface dissolution and equilibration of CO_2 with the atmosphere is well understood and quantified. It varies with the surface ocean conditions, in particular with temperature (solubility effect) and alkalinity. The capacity of the ocean to take up additional CO_2 for a given alkalinity decreases at higher temperature (4.23% per degree warming; Takahashi et al., 1993) and at elevated CO_2 concentrations (about 15% per 100 ppm, computed from the so called Revelle factor; Revelle and Suess, 1957).

Recent changes in nutrient supply in the ocean are also thought to have changed the export of organic carbon from biological processes below the surface layer, and thus the ocean CO_2 sink (Duce et al., 2008). Anthropogenic reactive nitrogen Nr (see Box 6.2) entering the ocean via atmospheric deposition or rivers acts as a fertiliser and may enhance carbon export to depth and hence the CO_2 sink. This Nr contribution has been estimated to be between 0.1 and 0.4 PgC yr^{-1} around the year 2000 using models (Duce et al., 2008; Reay et al., 2008; Krishnamurthy et al., 2009; Suntharalingam et al., 2012). Similarly, increases in iron deposition over the ocean from dust generated by human activity is estimated to have enhanced the ocean cumulative CO_2 uptake by 8 PgC during the 20th century (or about 0.05 PgC yr^{-1} in the past decades) (Mahowald et al., 2010). Although changes in ocean circulation and in global biogeochemical drivers have the potential to alter the ocean carbon fluxes through changes in marine ecosystems, modelling studies show only small variability in ocean biological pump, which has not significantly impacted the response of the ocean carbon cycle over the recent period (Bennington et al., 2009).

Model studies suggest that the response of the air–sea CO_2 fluxes to climate change and variability in recent decades has decreased the rate at which anthropogenic CO_2 is absorbed by the ocean (Sarmiento et al. (2010); Figure 6.14 and Table 6.4). This result is robust to the model or climate forcing used (Figure 6.13), but no formal attribution to anthropogenic climate change has been made. There is insufficient data coverage to separate the impact of climate change on the global ocean CO_2 sink directly from observations, though the regional trends described in Section 6.3.2.5.4 suggest that surface ocean pCO_2 responds to changes in ocean properties in a significant and measurable way.

6.3.2.5.6 Model evaluation of global and regional ocean carbon balance

Ocean process-based carbon cycle models are capable of reproducing the mean air–sea fluxes of CO_2 derived from pCO_2 observations (Takahashi et al., 2009), including their general patterns and amplitude (Sarmiento et al., 2000), the anthropogenic uptake of CO_2 (Orr et al., 2001; Wanninkhof et al., 2013) and the regional distribution of air–sea fluxes (Gruber et al., 2009). The spread between different model results for air–sea CO_2 fluxes is the largest in the Southern Ocean (Matsumoto et al., 2004), where intense convection occurs. Tracer observations (Schmittner et al., 2009) and water mass analysis (Iudicone et al., 2011) have been used to reduce the model uncertainty associated with this process and improve the simulation of carbon fluxes. The models reproduce the observed seasonal cycle of pCO_2 in the sub-tropics but generally do poorly in sub-polar regions where the balance of processes is more difficult to simulate well (McKinley et al., 2006; Schuster et al., 2013). Less information is available to evaluate specifically the representation of biological fluxes in the models, outside of their realistic representation of surface ocean chlorophyll distributions. Ocean process-based carbon cycle models used in AR5 reproduce the relatively small interannual variability inferred from observations (Figure 6.12; Wanninkhof et al., 2013). See also Section 9.4.5.

Sensitivity of modelled air–sea fluxes to CO_2. Data-based studies estimated a cumulative carbon uptake of ~155 ± 30 PgC across studies for the 1750–2011 time period (Sabine et al., 2004; Waugh et al., 2006; Khatiwala et al., 2009), a mean anthropogenic CO_2 sink of 2.2 ± 0.7 PgC yr^{-1} for the 1990s, and decadal trends of 0.13 PgC yr^{-1} per decade during the two decades 1990–2009 (Wanninkhof et al., 2013; from atmospheric inversions), respectively. Models that have estimated these quantities give a total ocean uptake of 170 ± 25 PgC for

1750–2011 (from the model ensemble of Orr et al., (2005) until 1994, plus an additional 40 PgC from estimates in Table 6.4 for 1995–2011), a mean anthropogenic CO_2 sink of 2.1 ± 0.6 PgC yr^{-1} for 1990–1999 (Le Quéré et al., 2013) and a decadal trend of 0.14 PgC yr^{-1} per decade for 1990–2009 (Wanninkhof et al., 2013). Therefore, although the ocean models do not reproduce all the details of the regional structure and changes in air–sea CO_2 fluxes, their globally integrated ocean CO_2 sink and decadal rate of change of this sink is in good agreement with the available observations.

Sensitivity of modelled air–sea fluxes to climate. The relationship between air–sea CO_2 flux and climate is strongly dependent on the oceanic region and on the time scale. Ocean carbon cycle models of the type used in AR5 estimate a reduction in cumulative ocean CO_2 uptake of 1.6 to 5.4 PgC over the period 1959–2008 (1.5 to 5.4%) in response to climate change and variability compared to simulations with no changes in climate (Figure 6.14), partly due to changes in the equatorial Pacific and to changes in the Southern Ocean. The only observation-based estimate available to evaluate the climate response of the global air–sea CO_2 flux is from Park et al. (2010), which is at the low end of the model estimate for the past two decades (Table 6.4). However, this estimate does not include the nonlinear effects of changes in ocean circulation and warming on the global air–sea CO_2 flux, which could amplify the response of the ocean CO_2 sink to climate by 20 to 30% (Le Quéré et al., 2010; Zickfeld et al., 2011).

Processes missing in ocean models. The most important processes missing in ocean carbon cycle models used in the AR5 are those representing explicitly small-scale physical circulation (e.g., eddies, brine formation), which are parameterised in models. These processes have an important influence on the vertical transport of water, heat, salt and carbon (Loose and Schlosser, 2011; Sallée et al., 2012). In particular, changes in vertical transport in the Southern Ocean are thought to explain part of the changes in atmospheric CO_2 between glacial and interglacial conditions, a signal that is not entirely reproduced by models (Section 6.2) suggesting that the sensitivity of ocean models could be underestimated.

Processes related to marine ecosystems in global ocean models are also limited to the simulation of lower trophic levels, with crude parameterizations for sinking processes, bacterial and other loss processes at the surface and in the ocean interior and their temperature dependence (Kwon et al., 2009). Projected changes in carbon fluxes from the response of marine ecosystems to changes in temperature (Beaugrand et al., 2010), ocean acidification (Riebesell et al., 2009) (see Glossary) and pressure from fisheries (Pershing et al., 2010) are all considered potentially important, though not yet quantified. Several processes have been specifically identified that could lead to changes in the ocean CO_2 sink, in particular the temperature effects on marine ecosystem processes (Riebesell et al., 2009; Taucher and Oschlies, 2011) and the variable nutrient ratios induced by ocean acidification or ecosystem changes (Tagliabue et al., 2011). Coastal ocean processes are also poorly represented in global and may influence the ocean CO_2 sink. Nevertheless, the fit of ocean model results to the integrated CO_2 sink and decadal trends discussed above suggest that, up to now, the missing processes have not had a dominant effect on ocean CO_2 beyond the limits of the uncertainty of the data.

6.3.2.6 Land Carbon Dioxide Sink

6.3.2.6.1 Global residual land sink and atmosphere-to-land carbon dioxide flux

The residual land CO_2 sink, that is, the uptake of CO_2 in ecosystems excluding the effects of land use change, is 1.5 ± 1.1, 2.6 ± 1.2 and 2.6 ± 1.2 PgC yr^{-1} for the 1980s, 1990s and 2000s, respectively (Table 6.1). After including the net land use change emissions, the atmosphere-to-land flux of CO_2 (Table 6.1) corresponds to a net sink of CO_2 by all terrestrial ecosystems. This sink has intensified globally from a neutral CO_2 flux of 0.1 ± 0.8 PgC yr^{-1} in the 1980s to a net CO_2 sink of 1.1 ± 0.9 PgC yr^{-1} and 1.5 ± 0.9 PgC yr^{-1} during the 1990s and 2000s, respectively (Table 6.1; Sarmiento et al., 2010). This growing land sink is also supported by an atmospheric inversion (Gurney and Eckels, 2011) and by process-based models (Le Quéré et al., 2009).

6.3.2.6.2 Regional atmosphere-to-land carbon dioxide fluxes

The results from atmospheric CO_2 inversions, terrestrial ecosystem models and forest inventories consistently show that there is a large net CO_2 sink in the northern extratropics, albeit the very limited availability of observations in the tropics (Jacobson et al., 2007; Gurney and Eckels, 2011; Pan et al., 2011). Inversion estimates of atmosphere–land CO_2 fluxes show net atmosphere-to-land CO_2 flux estimates ranging from neutral to a net source of 0.5 to 1.0 PgC yr^{-1} (Jacobson et al., 2007; Gurney and Eckels, 2011) (Figure 6.15). However, Stephens et al. (2007) selected from an ensemble of inversion models those that were consistent with independent aircraft cross-validation data, and constrained an atmosphere-to-land CO_2 flux of 0.1 ± 0.8 PgC yr^{-1} during the period 1992–1996, and a NH net CO_2 sink of 1.5 ± 0.6 PgC yr^{-1}. These results shows that after subtracting emissions from land use change, tropical land ecosystems might also be large CO_2 sinks.

Based on repeated forest biomass inventory data, estimated soil carbon changes, and CO_2 emissions from land use change from the bookkeeping method of Houghton et al. (2012), Pan et al. (2011) estimated a global forest carbon accumulation of 0.5 ± 0.1 $PgCyr^{-1}$ in boreal forests, and of 0.8 ± 0.1 PgC yr^{-1} in temperate forests for the period 2000–2007. Tropical forests were found to be near neutral with net emissions from land use change being compensated by sinks in established tropical forests (forests not affected by land use change), therefore consistent with the Stephens et al. (2007) inversion estimate of tropical atmosphere–land CO_2 fluxes.

Since AR4, a number of studies have compared and attempted to reconcile regional atmosphere-to-land CO_2 flux estimates from multiple approaches and so providing further spatial resolution of the regional contributions of carbon sources and sinks (Table 6.6). A synthesis of regional contributions estimated a 1.7 PgC yr^{-1} sink in the NH regions above 20°N with consistent estimates from terrestrial models and inventories (uncertainty: ±0.3 PgC yr^{-1}) and atmospheric CO_2 inversions (uncertainty: ±0.7 PgC yr^{-1}) (Ciais et al., 2010).

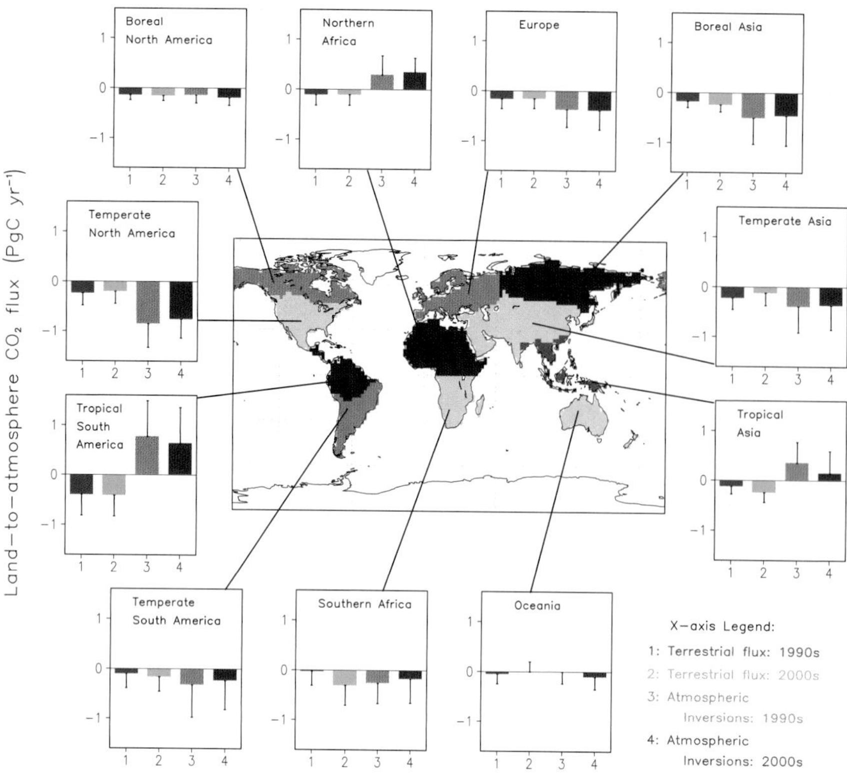

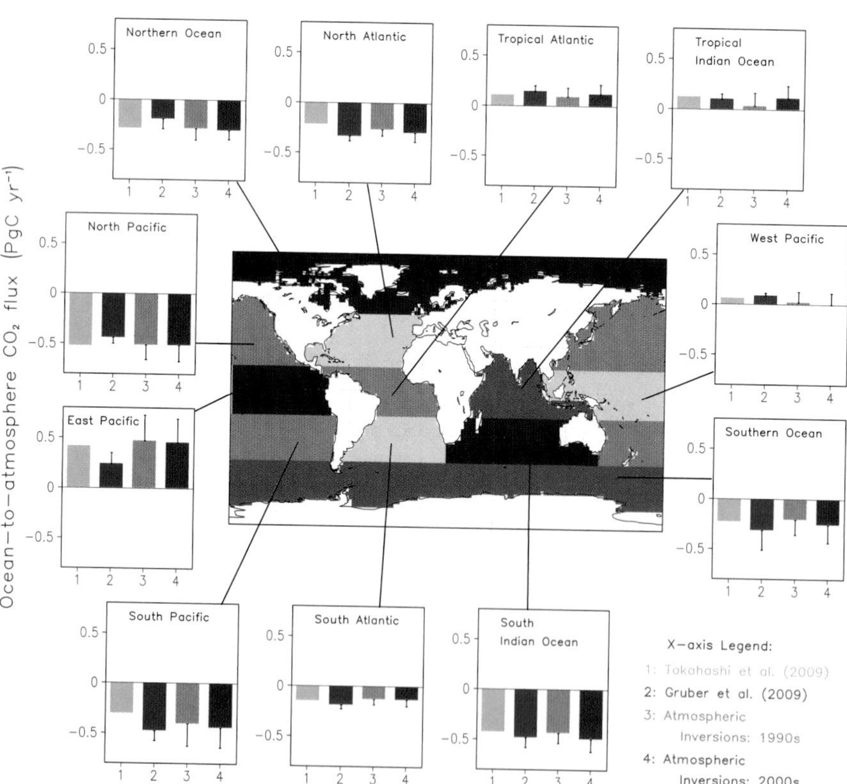

Figure 6.15 | (Top) Bar plots showing decadal average CO_2 fluxes for 11 land regions (1) as estimated by 10 different atmospheric CO_2 inversions for the 1990s (yellow) and 2000s (red) (Peylin et al., 2013; data source: http://transcom.lsce.ipsl.fr/), and (2) as simulated by 10 dynamic vegetation models (DGVMs) for the 1990s (green) and 2000s (light green) (Piao et al., 2013; data source: http://www-lscedods.cea.fr/invsat/RECCAP/). The divisions of land regions are shown in the map. (Bottom) Bar plots showing decadal average CO_2 fluxes for 11 ocean regions (1) as estimated by 10 different atmospheric CO_2 inversions for the 1990s (yellow) and 2000s (red) (data source: http://transcom.lsce.ipsl.fr/), (2) inversion of contemporary interior ocean carbon measurements using 10 ocean transport models (dark blue) (Gruber et al., 2009) and (3) surface ocean pCO_2 measurements based air-sea exchange climatology (Takahashi et al., 2009). The divisions of 11 ocean regions are shown in the map.

Table 6.6 | Regional CO_2 budgets using top-down estimates (atmospheric inversions) and bottom-up estimates (inventory data, biogeochemical modelling, eddy-covariance), excluding fossil fuel emissions. A positive sign indicates a flux from the atmosphere to the land (i.e., a land sink).

Region	CO_2 Sink (PgC yr^{-1})	Uncertainty[a]	Period	Reference
Artic Tundra	0.1	±0.3[b]	2000–2006	McGuire et al. (2012)
Australia	0.04	±0.03	1990–2009	Haverd et al. (2013)
East Asia	0.25	±0.1	1990–2009	Piao et al. (2012)
Europe	0.9	±0.2	2001–2005	Luyssaert et al. (2012)
North America	0.6	±0.02	2000–2005	King et al. (2012)
Russian Federation	0.6	–0.3 to –1.3	1990–2007	Dolman et al. (2012)
South Asia	0.15	±0.24	2000–2009	Patra et al. (2013)
South America	–0.3	±0.3	2000–2005	Gloor et al. (2012)

Notes:

[a] One standard deviation from mean unless indicated otherwise.

[b] Based on range provided.

6.3.2.6.3 Interannual variability in atmosphere-to-land carbon dioxide fluxes

The interannual variability of the residual land sink shown in Figures 6.12 and 6.16 accounts for most of the interannual variability of the atmospheric CO_2 growth rate (see Section 6.3.2.3). Atmospheric CO_2 inversion results suggest that tropical land ecosystems dominate the global CO_2 variability, with positive anomalies during El Niño episodes (Bousquet et al., 2000; Rödenbeck et al., 2003; Baker et al., 2006), which is consistent with the results of one inversion of atmospheric ^{13}C and CO_2 measurements (Rayner et al., 2008). A combined El Niño-Southern Oscillation (ENSO)–Volcanic index time series explains 75% of the observed variability (Raupach et al., 2008). A positive phase of ENSO (El Niño, see Glossary) is generally associated with enhanced land CO_2 source, and a negative phase (La Niña) with enhanced land CO_2 sink (Jones and Cox, 2001; Peylin et al., 2005). Observations from eddy covariance networks suggest that interannual carbon flux variability in the tropics and temperate regions is dominated by precipitation, while boreal ecosystem fluxes are more sensitive to temperature and shortwave radiation variation (Jung et al., 2011), in agreement with the results from process-based terrestrial ecosystem models (Piao et al., 2009a). Terrestrial biogeochemical models suggest that interannual net biome productivity (NBP) variability is dominated by GPP (see Glossary) rather than terrestrial ecosystem respiration (Piao et al., 2009b; Jung et al., 2011).

6.3.2.6.4 Carbon fluxes from inland water

Global analyses estimate that inland waters receive about 1.7 to 2.7 PgC yr^{-1} emitted by soils to rivers headstreams, of which, 0.2 to 0.6 PgC yr^{-1} is buried in aquatic sediments, 0.8 to 1.2 PgC yr^{-1} returns to the atmosphere as CO_2, and 0.9 PgC yr^{-1} is delivered to the ocean (Cole et al., 2007; Battin et al., 2009; Aufdenkampe et al., 2011). Estimates of the transport of carbon from land ecosystems to the coastal ocean by rivers are ~0.2 PgC yr^{-1} for Dissolved Organic Carbon (DOC), 0.3 PgC yr^{-1} for Dissolved Inorganic Carbon (DIC), and 0.1 to 0.4 PgC yr^{-1} for Particulate Organic Carbon (POC) (Seitzinger et al., 2005; Syvitski et al., 2005; Mayorga et al., 2010). For the DIC fluxes, only about two-thirds of it originates from atmospheric CO_2 and the rest of the carbon is supplied by weathered carbonate rocks (Suchet and Probst, 1995; Gaillardet et al., 1999; Oh and Raymond, 2006; Hartmann et al., 2009). Regional DIC concentrations in rivers has increased during the Industrial Era (Oh and Raymond, 2006; Hamilton et al., 2007; Perrin et al., 2008). Agricultural practices coupled with climate change can lead to large increases in regional scale DIC export in watersheds with a large agricultural footprint (Raymond et al., 2008). Furthermore, regional urbanization also elevates DIC fluxes in rivers (Baker et al., 2008; Barnes and Raymond, 2009), which suggests that anthropogenic activities have contributed a significant portion of the annual global river DIC flux to the ocean.

Land clearing and management are thought to produce an acceleration of POC transport, much of which is trapped in alluvial and colluvial deposition zones, lakes, reservoirs and wetlands (Stallard, 1998; Smith et al., 2001b; Syvitski et al., 2005). Numerous studies have demonstrated an increase in the concentration of DOC in rivers in the northeastern United States and northern/central Europe over the past two to four decades (Worrall et al., 2003; Evans et al., 2005; Findlay, 2005; Monteith et al., 2007; Lepistö et al., 2008). Owing to the important role of wetlands in DOC production, the mobilization of DOC due to human-induced changes in wetlands probably represents an important cause of changes in global river DOC fluxes to date (Seitzinger et al., 2005), although a global estimate of this alteration is not available. A robust partitioning between natural and anthropogenic carbon fluxes in freshwater systems is not yet possible, nor a quantification of the ultimate fate of carbon delivered by rivers to the coastal and open oceans.

6.3.2.6.5 Processes driving terrestrial atmosphere-to-land carbon dioxide fluxes

Assessment of experimental data, observations and model results suggests that the main processes responsible for the residual land sink include the CO_2 fertilisation effect on photosynthesis (see Box 6.3), nitrogen fertilisation by increased deposition (Norby, 1998; Thornton et al., 2007; Bonan and Levis, 2010; Zaehle and Dalmonech, 2011) and climate effects (Nemani et al., 2003; Gloor et al., 2009). It is *likely* that reactive nitrogen deposition over land currently increases natural CO_2 in particular in forests, but the magnitude of this effect varies between regions (Norby, 1998; Thornton et al., 2007; Bonan and Levis, 2010; Zaehle and Dalmonech, 2011). Processes responsible for the net atmosphere-to-land CO_2 sink on terrestrial ecosystems include, in addition, forest regrowth and afforestation (Myneni et al., 2001;

Box 6.3 | The Carbon Dioxide Fertilisation Effect

Elevated atmospheric CO_2 concentrations lead to higher leaf photosynthesis and reduced canopy transpiration, which in turn lead to increased plant water use efficiency and reduced fluxes of surface latent heat. The increase in leaf photosynthesis with rising CO_2, the so-called CO_2 fertilisation effect, plays a dominant role in terrestrial biogeochemical models to explain the global land carbon sink (Sitch et al., 2008), yet it is one of most unconstrained process in those models.

Field experiments provide a direct evidence of increased photosynthesis rates and water use efficiency (plant carbon gains per unit of water loss from transpiration) in plants growing under elevated CO_2. These physiological changes translate into a broad range of higher plant carbon accumulation in more than two-thirds of the experiments and with increased net primary productivity (NPP) of about 20 to 25% at double CO_2 from pre-industrial concentrations (Ainsworth and Long, 2004; Luo et al., 2004, 2006; Nowak et al., 2004; Norby et al., 2005;Canadell et al., 2007a; Denman et al., 2007; Ainsworth et al., 2012; Wang et al., 2012a). Since the AR4, new evidence is available from long-term Free-air CO_2 Enrichment (FACE) experiments in temperate ecosystems showing the capacity of ecosystems exposed to elevated CO_2 to sustain higher rates of carbon accumulation over multiple years (Liberloo et al., 2009; McCarthy et al., 2010; Aranjuelo et al., 2011; Dawes et al., 2011; Lee et al., 2011; Zak et al., 2011). However, FACE experiments also show the diminishing or lack of CO_2 fertilisation effect in some ecosystems and for some plant species (Dukes et al., 2005; Adair et al., 2009; Bader et al., 2009; Norby et al., 2010; Newingham et al., 2013). This lack of response occurs despite increased water use efficiency, also confirmed with tree ring evidence (Gedalof and Berg, 2010; Peñuelas et al., 2011).

Nutrient limitation is hypothesized as primary cause for reduced or lack of CO_2 fertilisation effect observed on NPP in some experiments (Luo et al., 2004; Dukes et al., 2005; Finzi et al., 2007; Norby et al., 2010). Nitrogen and phosphorus are *very likely* to play the most important role in this limitation of the CO_2 fertilisation effect on NPP, with nitrogen limitation prevalent in temperate and boreal ecosystems, and phosphorus limitation in the tropics (Luo et al., 2004; Vitousek et al., 2010; Wang et al., 2010a; Goll et al., 2012). Micronutrients interact in diverse ways with other nutrients in constraining NPP such as molybdenum and phosphorus in the tropics (Wurzburger et al., 2012). Thus, with *high confidence*, the CO_2 fertilisation effect will lead to enhanced NPP, but significant uncertainties remain on the magnitude of this effect, given the lack of experiments outside of temperate climates.

Pacala et al., 2001; Houghton, 2010; Bellassen et al., 2011; Williams et al., 2012a), changes in forest management and reduced harvest rates (Nabuurs et al., 2008).

Process attribution of the global land CO_2 sink is difficult due to limited availability of global data sets and biogeochemical models that include all major processes. However, regional studies shed light on key drivers and their interactions. The European and North American carbon sinks are explained by the combination of forest regrowth in abandoned lands and decreased forest harvest along with the fertilisation effects of rising CO_2 and nitrogen deposition (Pacala et al., 2001; Ciais et al., 2008; Sutton et al., 2008; Schulze et al., 2010; Bellassen et al., 2011; Williams et al., 2012a). In the tropics, there is evidence from forest inventories that increasing forest growth rates are not explained by the natural recovery from disturbances, suggesting that increasing atmospheric CO_2 and climate change play a role in the observed sink in established forests (Lewis et al., 2009; Pan et al., 2011). There is also recent evidence of tropical nitrogen deposition becoming more notable although its effects on the net carbon balance have not been assessed (Hietz et al., 2011).

The land carbon cycle is very sensitive to climate changes (e.g., precipitation, temperature, diffuse vs. direct radiation), and thus the changes in the physical climate from increasing GHGs as well as in the diffuse fraction of sunlight are *likely* to be causing significant changes in the carbon cycle (Jones et al., 2001; Friedlingstein et al., 2006; Mercado et al., 2009). Changes in the climate are also associated with disturbances such as fires, insect damage, storms, droughts and heat waves which are already significant processes of interannual variability and possibly trends of regional land carbon fluxes (Page et al., 2002; Ciais et al., 2005; Chambers et al., 2007; Kurz et al., 2008b; Clark et al., 2010; van der Werf et al., 2010; Lewis et al., 2011) (see Section 6.3.2.2).

Warming (and possibly the CO_2 fertilisation effect) has also been correlated with global trends in satellite greenness observations, which resulted in an estimated 6% increase of global NPP, or the accumulation of 3.4 PgC on land over the period 1982–1999 (Nemani et al., 2003). This enhanced NPP was attributed to the relaxation of climatic constraints to plant growth, particularly in high latitudes. Concomitant to the increased of NPP with warming, global soil respiration also increased between 1989 and 2008 (Bond-Lamberty and Thomson, 2010), reducing the magnitude of the net land sink. A recent study suggests a declining NPP trend over 2000–2009 (Zhao and Running, 2010) although the model used to reconstruct NPP trends from satellite observation has not been widely accepted (Medlyn, 2011; Samanta et al., 2011).

6.3.2.6.6 Model evaluation of global and regional terrestrial carbon balance

Evaluation of global process-based land carbon models was performed against ground and satellite observations including (1) measured CO_2

fluxes and carbon storage change at particular sites around the world, in particular sites from the Fluxnet global network (Baldocchi et al., 2001; Jung et al., 2007; Stöckli et al., 2008; Schwalm et al., 2010; Tan et al., 2010), (2) observed spatio-temporal change in leaf area index (LAI) (Lucht et al., 2002; Piao et al., 2006) and (3) interannual and seasonal change in atmospheric CO_2 (Randerson et al., 2009; Cadule et al., 2010).

Figure 6.16 compares the global land CO_2 sink driven by climate change and rising CO_2 as simulated by different process based carbon cycle models (without land use change), with the residual land sink computed as the sum of fossil fuel and cement emissions and land use change emissions minus the sum of CO_2 growth rate and ocean sink (Le Quéré et al., 2009; Friedlingstein et al., 2010). Although these two quantities are not the same, the multi-model mean reproduces well the trend and interannual variability of the residual land sink which is dominated by climate variability and climate trends and CO_2, respectively, both represented in models (Table 6.7). Limited availability of *in situ* measurements, particularly in the tropics, limits the progress towards reducing uncertainty on model parameterizations.

Regional and local measurements can be used to evaluate and improve global models. Regionally, forest inventory data show that the forest carbon sink density over Europe is of –89 ± 19 gC m^{-2} yr^{-1}, which is compatible with model estimates with afforestation (–63 gC m^{-2} yr^{-1}; Luyssaert et al., 2010), while modelled NPP was 43% larger than the inventory estimate. In North America, the ability of 22 terrestrial carbon cycle models to simulate the seasonal cycle of land–atmosphere CO_2 exchange from 44 eddy covariance flux towers was poor with a difference between observations and simulations of 10 times the observational uncertainty (Schwalm et al., 2010). Model shortcomings included spring phenology, soil thaw, snow pack melting and lag responses to extreme climate events (Keenan et al., 2012). In China, the magnitude of the carbon sink estimated by five terrestrial ecosystem models (–0.22 to –0.13 PgC yr^{-1}) was comparable to the observation-based estimate (–0.18 ± 0.73 PgC yr^{-1}; Piao et al., 2009a), but modelled interannual variation was weakly correlated to observed regional land–atmosphere CO_2 fluxes (Piao et al., 2011).

Sensitivity of the terrestrial carbon cycle to rising atmospheric carbon dioxide. An inter-comparison of 10 process-based models showed increased NPP by 3% to 10% over the last three decades, during which CO_2 increased by ~50 ppm (Piao et al., 2013). These results are consistent within the broad range of responses from experimental studies (see Box 6.3). However, Hickler et al. (2008) suggested that currently available FACE results (largely from temperate regions) are not applicable to vegetation globally because there may be large spatial heterogeneity in vegetation responses to CO_2 fertilisation.

Table 6.7 | Estimates of the land CO_2 sink from process-based terrestrial ecosystem models driven by rising CO_2 and by changes in climate. The land sink simulated by these models is close to but not identical to the terrestrial CO_2 sink from Table 6.1 because the models calculate the effect of CO_2 and climate over managed land, and many do not include nitrogen limitation and disturbances.

Model Name	Nitrogen Limitation	Natural Fire CO_2 Emissions	1980–1989	1990–1999	2000–2009
	(Yes/No)	(Yes/No)	PgC yr^{-1}	PgC yr^{-1}	PgC yr^{-1}
CLM4C[b,c]	No	Yes	1.98	2.11	2.64
CLM4CN[b,c]	Yes	Yes	1.27	1.25	1.67
Hyland[d]	No	No	2.21	2.92	3.99
LPJ[e]	No	Yes	1.14	1.90	2.60
LPJ_GUESS[f]	No	Yes	1.15	1.54	2.07
OCN[g]	Yes	No	1.75	2.18	2.36
ORC[h]	No	No	2.08	3.05	3.74
SDGVM[i]	Yes	Yes	1.25	1.95	2.30
TRIFFID[j]	No	No	1.85	2.52	3.00
VEGAS[k]	No	No	1.40	1.68	1.89
Average[a]			1.61 ± 0.65	2.11 ± 0.93	2.63 ± 1.22

Notes:

[a] Average of all models ±90% confidence interval.
[b] Oleson et al. (2010).
[c] Lawrence et al. (2011).
[d] Levy et al. (2004).
[e] Sitch et al. (2003).
[f] Smith et al. (2001a).
[g] Zaehle and Friend (2010).
[h] Krinner et al. (2005).
[i] Woodward and Lomas (2004).
[j] Cox (2001).
[k] Zeng (2003).

All of these models run are forced by rising CO_2 concentration and time-varying historical reconstructed weather and climate fields using the same protocol from the TRENDY project (Piao et al., 2013). (http://www.globalcarbonproject.org/global/pdf/DynamicVegetationModels.pdf).

CLM4C = Community Land Model for Carbon; CLM4CN = Community Land Model for Carbon–Nitrogen; GUESS = General Ecosystem Simulator; LPJ = Lund-Potsdam-Jena Dynamic Global Vegetation Model; OCN = Cycling of Carbon and Nitrogen on land, derived from ORCHIDEE model; ORC = ORCHIDEE, ORganizing Carbon and Hydrology in Dynamic EcosystEms model; SDGVM = Sheffield Dynamic Global Vegetation Model; TRIFFID = Top-down Representation of Interactive Foliage and Flora Including Dynamics; VEGAS = VEgetation-Global-Atmosphere-Soil terrestrial carbon cycle model.

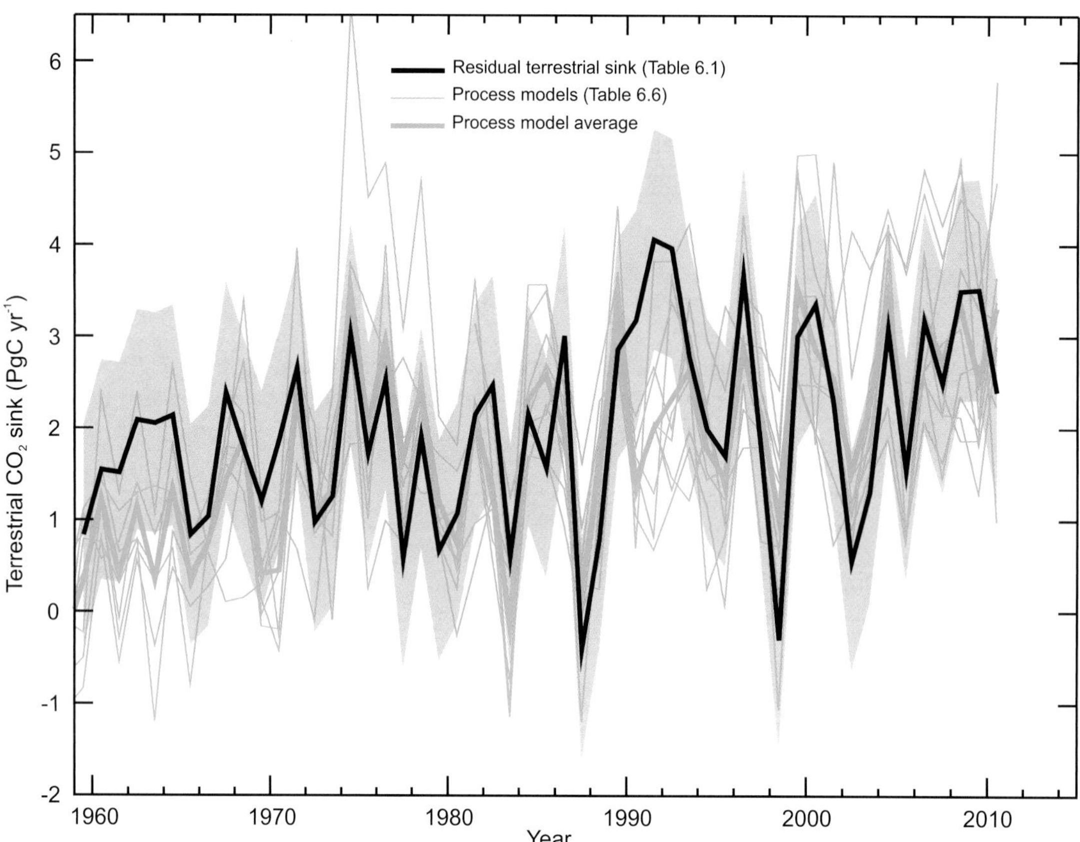

Figure 6.16 | The black line and gray shading represent the estimated value of the residual land sink (PgC yr^{-1}) and its uncertainty from Table 6.1, which is calculated from the difference between emissions from fossil fuel and land use change plus emissions from net land use change, minus the atmospheric growth rate and the ocean sink. The atmosphere-to-land flux simulated by process land ecosystem models from Table 6.7 are shown in thin green, and their average in thick green. A positive atmosphere-to-land flux represents a sink of CO_2. The definition of the atmosphere-to-land flux simulated by these models is close to but not identical to the residual land sink from Table 6.1 (see Table 6.7).

Sensitivity of terrestrial carbon cycle to climate trends and variability. Warming exerts a direct control on the net land–atmosphere CO_2 exchange because both photosynthesis and respiration are sensitive to changes in temperature. From estimates of interannual variations in the residual land sink, 1°C of positive global temperature anomaly leads to a decrease of 4 PgC yr^{-1} of the global land CO_2 sink (Figure 6.17). This observed interannual response is close to the response of the models listed in Table 6.7 (–3.5 ± 1.5 PgC yr^{-1}°C^{-1} in Piao et al., 2013), albeit individual models show a range going from –0.5 to –6.2 PgC yr^{-1} °C^{-1}. The sensitivity of atmospheric CO_2 concentration to century scale temperature change was estimated at about 3.6 to 45.6 PgC °C^{-1} (or 1.7 to 21.4 ppm CO_2 °C^{-1}) using the ice core observed CO_2 drop during the Little Ice Age (see Section 6.2; Frank et al., 2010).

6

Terrestrial carbon cycle models used in AR5 generally underestimate GPP in the water limited regions, implying that these models do not correctly simulate soil moisture conditions, or that they are too sensitive to changes in soil moisture (Jung et al., 2007). Most models (Table 6.7) estimated that the interannual precipitation sensitivity of the global land CO_2 sink to be higher than that of the observed residual land sink (–0.01 PgC yr^{-1} mm^{-1}; Figure 6.17).

Processes missing in terrestrial carbon cycle models. First, many models do not explicitly take into account the various forms of disturbances or ecosystem dynamics: migration, fire, logging, harvesting, insect outbreaks and the resulting variation in forest age structure which is known to affect the net carbon exchange (Kurz et al., 2008c; Bellassen et al., 2010; Higgins and Harte, 2012). Second, many key processes relevant to decomposition of carbon are missing in models (Todd-Brown et al., 2012), and particularly for permafrost carbon and for carbon in boreal and tropical wetlands and peatlands, despite the large amount of carbon stored in these ecosystems and their vulnerability to warming and land use change (Tarnocai et al., 2009; Hooijer et al., 2010; Page et al., 2011). However, progress has been made (Wania et al., 2009; Koven et al., 2011; Schaefer et al., 2011). Third, nutrient dynamics are taken into account only by few models despite the fact it is well established that nutrient constrains NPP and nitrogen deposition enhances NPP (Elser et al., 2007; Magnani et al., 2007; LeBauer and Treseder, 2008); see Section 6.3.2.6.5. Very few models have phosphorus dynamics (Zhang et al., 2011; Goll et al., 2012). Fourth, the negative effects of elevated tropospheric ozone on NPP have not been taken into account by most current carbon cycle models (Sitch et al., 2007). Fifth, transfer of radiation, water and heat in the vegetation–soil–atmosphere continuum are treated very simply in the global ecosystem models. Finally, processes that transport carbon at the surface (e.g., water and tillage erosion; Quinton et al., 2010) and human managements including fertilisation and irrigation (Gervois et al., 2008) are poorly or not represented at all. Broadly, models are still at their early stages in dealing with land use, land use change and forestry.

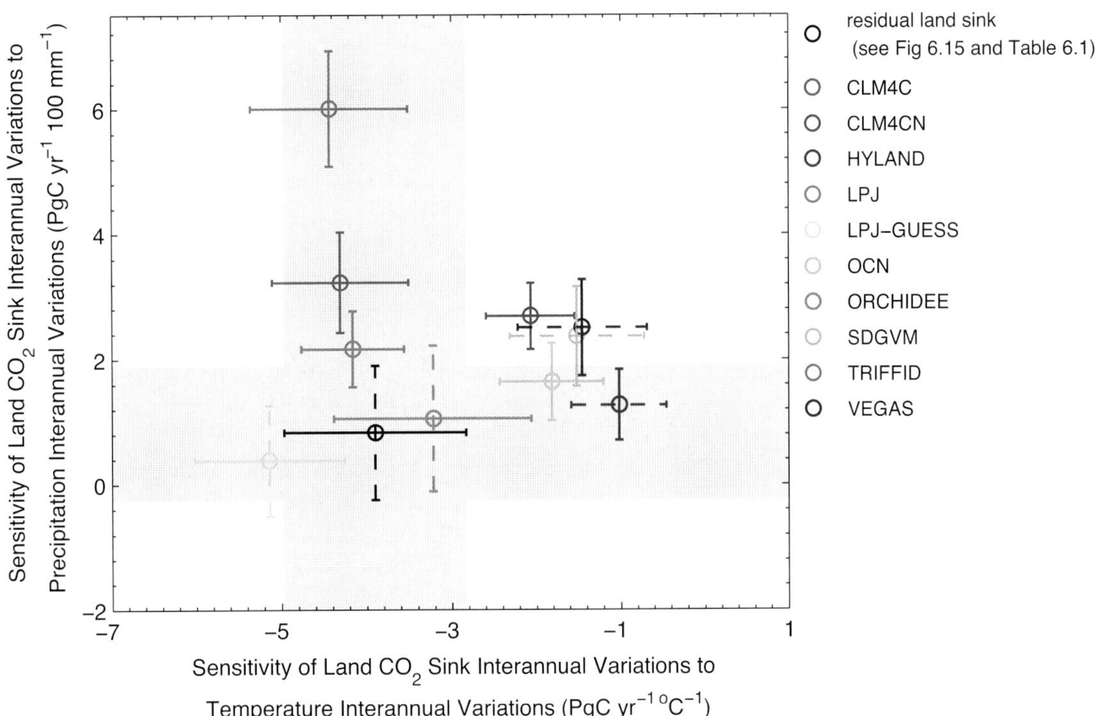

Figure 6.17 | The response of interannual land CO_2 flux anomaly to per 1°C interannual temperature anomaly and per 100 mm interannual precipitation anomaly during 1980–2009. Black circles show climate sensitivity of land CO_2 sink estimated from the residual land sink (see Figure 6.15 and Table 6.1), which is the sum of fossil fuel and cement emissions and land use change emissions minus the sum of observed atmospheric CO_2 growth rate and modeled ocean sink sink (Le Quéré et al., 2009; Friedlingstein and Prentice, 2010). Coloured circles show land CO_2 sink estimated by 10 process-based terrestrial carbon cycle models (CLM4C (Community Land Model for Carbon), CLM4CN (Community Land Model for Carbon–Nitrogen), HYLAND (HYbrid LAND terrestrial ecosystem model), LPJ (Lund-Potsdam-Jena Dynamic Global Vegetation Model), LPJ–GUESS (LPJ–General Ecosystem Simulator, OCN (Cycling of Carbon and Nitrogen on land, derived from ORCHIDEE model), ORCHIDEE (ORganizing Carbon and Hydrology in Dynamic EcosystEms model), SDGVM (Sheffield Dynamic Global Vegetation Model), TRIFFID (Top-down Representation of Interactive Foliage and Flora Including Dynamics) and VEGAS (terrestrial vegetation and carbon model)). Error bars show standard error of the sensitivity estimates. Dashed error bars indicate the estimated sensitivity by the regression approach is statistically insignificant (P > 0.05). Grey area denoted the area bounded by the estimated climate sensitivity of the residual land sink ± the standard error of the estimated climate sensitivity of the residual land sink. The sensitivity of land CO_2 sink interannual variations to interannual variations of temperature (or precipitation) is estimated as the regression coefficient of temperature (or precipitation) in a multiple regression of detrended anomaly of land CO_2 sink against detrended anomaly of annual mean temperature and annual precipitation.

6.3.3 Global Methane Budget

AR5 is the first IPCC assessment report providing a consistent synthesis of the CH_4 budget per decade using multiple atmospheric CH_4 inversion models (top-down) and process-based models and inventories (bottom-up). Table 6.8 shows the budgets for the decades of 1980s, 1990s and 2000s. Uncertainties on emissions and sinks are listed using minimum and maximum of each published estimate for each decade. Bottom-up approaches are used to attribute decadal budgets to individual processes emitting CH_4 (see Section 6.1.1.2 for a general overview). Top-down inversions provide an atmospheric-based constraint mostly for the total CH_4 source per region, and the use of additional observations (e.g., isotopes) allows inferring emissions per source type. Estimates of CH_4 sinks in the troposphere by reaction with tropospheric OH, in soils and in the stratosphere are also presented. Despite significant progress since the AR4, large uncertainties remain in the present knowledge of the budget and its evolution over time.

6.3.3.1 Atmospheric Changes

Since the beginning of the Industrial Era, the atmospheric CH_4 concentration increased by a factor of 2.5 (from 722 ppb to 1803 ppb in 2011). CH_4 is currently measured by a network of more than 100 surface sites (Blake et al., 1982; Cunnold et al., 2002; Langenfelds et al., 2002; Dlugokencky et al., 2011), aircraft profiles (Brenninkmeijer et al., 2007), satellite (Wecht et al., 2012; Worden et al., 2012) and before 1979 from analyses of firn air and ice cores (see Sections 5.2.2 and Section 6.2, and Figure 6.11). The growth of CH_4 in the atmosphere is largely in response to increasing anthropogenic emissions. The vertically averaged atmospheric CH_4 concentration field can be mapped by remote sensing from the surface using Fourier Transform Infrared Spectroscopy (FTIR) instruments (Total Carbon Column Observing Network, TCCON, http://www.tccon.caltech.edu/) and from space by several satellite instruments: Atmospheric Infrared Sounder (AIRS, since 2002; http://airs.jpl.nasa.gov), Tropospheric Emission Spectrometer (TES, since 2004; http://tes.jpl.nasa.gov), Infrared Atmospheric Sounder Interferometer (IASI, since 2006; Crévoisier et al., 2009), Scanning Imaging Spectrometer for Atmospheric Cartography (SCIAMACHY, 2003–2012; Frankenberg et al., 2008), and Greenhouse Gases Observing Satellite-Thermal And Near infrared Sensor for carbon Observation Fourier-Transform Spectrometer (GOSAT-TANSO-FTS, since 2009; Morino et al., 2011). As an example, SCIAMACHY shows the column CH_4 gradient between the two hemispheres as well as increased concentrations over Southeast Asia, due to emissions from agriculture, wetlands, waste and energy production (Frankenberg et al., 2008). *In situ* observations provide very precise measurements (~0.2%) but unevenly located at the surface of the globe. Satellite data offer a global coverage at the cost of a lower precision on individual measurements (~2%) and possible biases (Bergamaschi et al., 2009).

The growth rate of CH_4 has declined since the mid-1980s, and a near zero growth rate (quasi-stable concentrations) was observed during 1999–2006, suggesting an approach to steady state where the sum of emissions are in balance with the sum of sinks (Dlugokencky et al., 2003; Khalil et al., 2007; Patra et al., 2011; Figure 6.18). The reasons for this growth rate decline after the mid-1980s are still debated, and results from various studies provide possible scenarios: (1) a reduction of anthropogenic emitting activities such as coal mining, gas industry and/or animal husbandry, especially in the countries of the former Soviet Union (Dlugokencky et al., 2003; Chen and Prinn, 2006; Savolainen et al., 2009; Simpson et al., 2012); (2) a compensation between increasing anthropogenic emissions and decreasing wetland emissions (Bousquet et al., 2006; Chen and Prinn, 2006); (3) significant (Rigby et al., 2008) to small (Montzka et al., 2011) changes in OH concentrations and/or based on two different $^{13}CH_4$ data sets; (4) reduced emissions from rice paddies attributed to changes in agricultural practices (Kai et al., 2011); or (5) stable microbial and fossil fuel emissions from 1990 to 2005 (Levin et al., 2012).

Since 2007, atmospheric CH_4 has been observed to increase again (Rigby et al., 2008; Dlugokencky et al., 2009) with positive anomalies of emissions of 21 Tg(CH_4) yr^{-1} and 18 Tg(CH_4) yr^{-1} estimated by inversions during 2007 and 2008, respectively (Bousquet et al., 2011) as compared to the 1999–2006 period. The increase of emissions in 2007–2008 was dominated by tropical regions (Bousquet et al., 2011), with a major contribution from tropical wetlands and some contribution from high-latitude wetlands during the 2007 anomaly (Dlugokencky et al., 2009; Bousquet et al., 2011). This increase is suggested by the growth rate over latitude in Figure 6.18 (Dlugokencky et al., 2009). The recent increase of CH_4 concentration since 2007 is also consistent

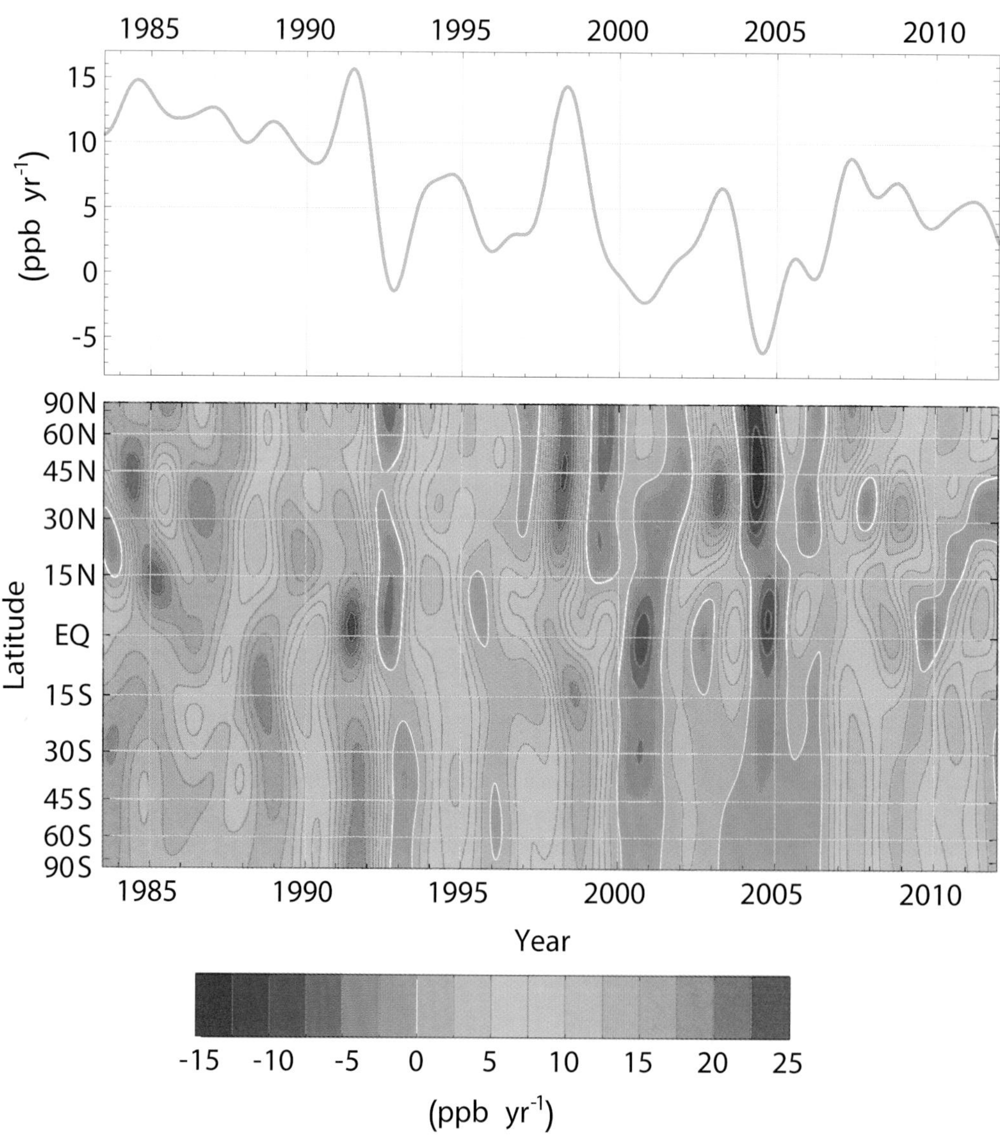

Figure 6.18 | (Top) Globally averaged growth rate of atmospheric CH_4 in ppb yr^{-1} determined from the National Oceanic and Atmospheric Administration–Earth System Research Laboratory (NOAA–ESRL) network, representative for the marine boundary layer. (Bottom) Atmospheric growth rate of CH_4 as a function of latitude (Masarie and Tans, 1995; Dlugokencky and Tans, 2013b).

Table 6.8 | Global CH_4 budget for the past three decades (in Tg(CH_4) yr^{-1}) and present day (2011)[38]. The bottom-up estimates for the decade of 2000–2009 are used in the Executive Summary and in Figure 6.2. T-D stands for Top-Down inversions and B-U for Bottom-Up approaches. Only studies covering at least 5 years of each decade have been used. Reported values correspond to the mean of the cited references and therefore not always equal (max-min)/2; likewise, ranges [in brackets] represent minimum and maximum values of the cited references. The sum of sources and sinks from B-U approaches does not automatically balance the atmospheric changes. For B-U studies, individual source types are also presented. For T-D inversions, the 1980s decade starts in 1984. As some atmospheric inversions did not reference their global sink, balance with the atmosphere and the sum of the sources has been assumed. One biomass burning estimate (Schultz et al., 2007) excludes biofuels (a). Stratospheric loss for B-U is the sum of the loss by OH radicals, a 10 Tg yr^{-1} loss due to O1D radicals (Neef et al., 2010) and a 20 to 35% contribution due to Cl radicals[24] (Allan et al., 2007). Present day budgets[39] adopt a global mean lifetime of 9.14 yr (±10%).

Tg(CH_4) yr^{-1}	1980–1989		1990–1999		2000–2009	
	Top-Down	**Bottom-Up**	**Top-Down**	**Bottom-Up**	**Top-Down**	**Bottom-Up**
Natural Sources	**193 [150–267]**	**355 [244–466]**	**182 [167–197]**	**336 [230–465]**	**218 [179–273]**	**347 [238–484]**
Natural wetlands	**157 [115–231][1,2,3]**	**225 [183–266][4,5]**	**150 [144–160][1,28,29]**	**206 [169–265][4,5,27]**	**175 [142–208][1,29,33,34,35,36]**	**217 [177–284][4,5,27]**
Other sources	**36 [35–36][1,2]**	**130 [61–200]**	**32 [23–37][1,28,29]**	**130 [61–200]**	**43 [37–65][1,29,33,34,35,36]**	**130 [61–200]**
Freshwater (lakes and rivers)		40 [8–73][6,7,8]		40 [8–73][6,7,8]		40 [8–73][6,7,8]
Wild animals		15 [15–15] [9]		15 [15–15][9]		15 [15–15][9]
Wildfires		3 [1–5][9,10,11,12,13]		3 [1–5][9,10,11,12,13]		3 [1–5][9,10,11,12,13]
Termites		11 [2–22][9,10,14,15,x]		11 [2–22][9,10,14,15,x]		11 [2–22][9,10,14,15,x]
Geological (incl. oceans)		54 [33–75][10,16,17]		54 [33–75][10,16,17]		54 [33–75][10,16,17]
Hydrates		6 [2–9][9,18,19]		6 [2–9][9,18,19]		6 [2–9][9,18,19]
Permafrost (excl. lakes and wetlands)		1 [0–1][10]		1 [0–1][10]		1 [0–1][10]
Anthropogenic Sources	**348 [305–383]**	**308 [292–323]**	**372 [290–453]**	**313 [281–347]**	**335 [273–409]**	**331 [304–368]**
Agriculture and waste	**208 [187–220][1,2,3]**	**185 [172–197][20]**	**239 [180–301][1,28,29]**	**187 [177–196][20,30,31]**	**209 [180–241][1,29,33,34,35,36]**	**200 [187–224][20,30,31]**
Rice		45 [41–47][20]		35 [32–37][20,27,30,31]		36 [33–40][20,27,30,31]
Ruminants		85 [81–90][20]		87 [82–91][20,30,31]		89 [87–94][20,30,31]
Landfills and waste		55 [50–60][20]		65 [63–68][20,30,31]		75 [67–90][20,30,31]
Biomass burning (incl. biofuels)	**46 [43–55][1,2,3]**	**34 [31–37][20,21,22a,38]**	**38 [26–45][1,28,29]**	**42 [38–45][13,20,21,22a,32,38]**	**30 [24–45][1,29,33,34,35,36]**	**35 [32–39][13,20,21,32,37,38]**
Fossil fuels	**94 [75–108][1,2,3]**	**89 [89–89][20]**	**95 [84–107][1,28,29]**	**84 [66–96][20,30,31]**	**96 [77–123][1,29,33,34,35,36]**	**96 [85–105][20,30,31]**
Sinks						
Total chemical loss	**490 [450–533][1,2,3]**	**539 [411–671][23,24,25,26]**	**515 [491–554][1,28,29]**	**571 [521–621][23,24,25,26]**	**518 [510–538][1,29,33,34,36]**	**604 [483–738][23,24,25,26]**
Tropospheric OH		468 [382–567][26]		479 [457–501][26]		528 [454–617][25,26]
Stratospheric OH		46 [16–67][23,25,26]		67 [51–83][23,25,26]		51 [16–84][23,25,26]
Tropospheric Cl		25 [13–37][24]		25 [13–37][24]		25 [13–37][24]
Soils	**21 [10–27][1,2,3]**	**28 [9–47][27,34,36]**	**27 [27–27][1]**	**28 [9–47][27,34,36]**	**32 [26–42][1,33,34,35,36]**	**28 [9–47][27,34,36]**
Global						
Sum of sources	541 [500–592]	663 [536–789]	554 [529–596]	649 [511–812]	553 [526–569]	678 [542–852]
Sum of sinks	511 [460–559]	567 [420–718]	542 [518–579]	599 [530–668]	550 [514–560]	632 [592–785]
Imbalance (sources minus sinks)	30 [16–40]		12 [7–17]		3 [-4–19]	
Atmospheric growth rate	34		17		6	
Global top-down (year 2011)	**2011 (AR5)[38]**					
Burden (Tg CH_4)	**4954±10**					
Atmospheric loss (Tg CH_4 yr^{-1})	**542±56**					
Atmos. increase (Tg CH_4 yr^{-1})	**14±3**					
Total source (Tg CH_4 yr^{-1})	**556±56**					
Anthropogenic source (Tg CH_4 yr^{-1})	**354±45**					
Natural source (Tg CH_4 yr^{-1})	**202±35**					

References:
1 Bousquet et al. (2011)
2 Fung et al. (1991)
3 Hein et al. (1997)
4 Hodson et al. (2011)
5 Ringeval et al. (2011)
6 Bastviken et al. (2004)
7 Bastviken et al. (2011)
8 Walter et al. (2007)
9 Denman et al. (2007)
10 EPA (2010)

6

Table 6.8 References (continued)

[11] Hoelzemann et al. (2004)
[12] Ito and Penner (2004)
[13] van der Werf et al. (2010)
[14] Sanderson (1996)
[15] Sugimoto et al. (1998)
[16] Etiope et al. (2008)
[17] Rhee et al. (2009)
[18] Dickens (2003)
[19] Shakhova et al. (2010)
[20] EDGAR4-database (2009)
[21] Mieville et al. (2010)
[22] Schultz et al. (2007) (excluding biofuels)
[23] Neef et al. (2010)
[24] Allan et al. (2007)
[25] Williams et al. (2012b)
[26] Voulgarakis et al. (2013)
[27] Spahni et al. (2011)
[28] Chen and Prinn (2006)
[29] Pison et al. (2009)
[30] Dentener et al. (2005)
[31] EPA (2011a)
[32] van der Werf (2004)
[33] Bergamaschi et al. (2009)
[34] Curry (2007)
[35] Spahni et al. (2011)
[36] Ito and Inatomi (2012)
[37] Wiedinmyer et al. (2011)
[38] Andreae and Merlet (2001)
[39] Prather et al. (2012), updated to 2011 (Table 2.1) and used in Chapter 11 projections; uncertainties evaluated as 68% confidence intervals, see also Annex II.2.2 and II.4.2.

with anthropogenic emission inventories, which show more (EDGAR v4.2) or less (EPA, 2011a) rapidly increasing anthropogenic CH_4 emissions in the period 2000–2008. This is related to increased energy production in growing Asian economies (EDGAR, edgar.jrc.ec.europa.eu; EPA, http://www.epa.gov/nonco2/econ-inv/international.html). The atmospheric increase has continued after 2009, at a rate of 4 to 5 ppb yr^{-1} (Sussmann et al., 2012).

6.3.3.2 Methane Emissions

\The CH_4 growth rate results from the balance between emissions and sinks. Methane emissions around the globe are biogenic, thermogenic or pyrogenic in origin (Neef et al., 2010), and they can be the direct result of either human activities and/or natural processes (see Section 6.1.1.2 and Table 6.8). Biogenic sources are due to degradation of organic matter in anaerobic conditions (natural wetlands, ruminants, waste, landfills, rice paddies, fresh waters, termites). Thermogenic sources come from the slow transformation of organic matter into fossil fuels on geological time scales (natural gas, coal, oil). Pyrogenic sources are due to incomplete combustion of organic matter (biomass and biofuel burning). Some sources can eventually combine a biogenic and a thermogenic origin (e.g., natural geological sources such as oceanic seeps, mud volcanoes or hydrates). Each of these three types of emissions is characterized by ranges in its isotopic composition in^{13}C-CH_4: typically –55 to –70‰ for biogenic, –25 to –45‰ for thermogenic, and –13 to –25‰ for pyrogenic. These isotopic distinctions provide a basis for attempting to separate the relative contribution of different methane sources using the top-down approach (Bousquet et al., 2006; Neef et al., 2010; Monteil et al., 2011).

During the decade of the 2000s, natural sources of CH_4 account for 35 to 50% of the decadal mean global emissions (Table 6.8). The single most dominant CH_4 source of the global flux and interannual variability is CH_4 emissions from wetlands (177 to 284 Tg(CH_4) yr^{-1}). With *high confidence*, climate driven changes of emissions from wetlands are the main drivers of the global inter-annual variability of CH_4 emissions. The term 'wetlands' denotes here a variety of ecosystems emitting CH_4 in the tropics and the high latitudes: wet soils, swamps, bogs and peatlands. These emissions are highly sensitive to climate change and variability, as shown, for instance, from the high CH_4 growth rate in 2007–2008 that coincides with positive precipitation and temperature anomalies (Dlugokencky et al., 2009). Several process-based models of methane emissions from wetlands have been developed and improved since AR4 (Hodson et al., 2011; Ringeval et al., 2011; Spahni et al., 2011; Melton et al., 2013), yet the *confidence* in modeled wetland CH_4 emissions remains *low*, particularly because of limited observational data sets available for model calibration and evaluation. Spatial distribution and temporal variability of wetlands also remains highly unconstrained in spite the existence of some remote sensing products (Papa et al., 2010). It has been observed that wetland CH_4 emissions increase in response to elevated atmospheric CO_2 concentrations (van Groenigen et al., 2011). van Groenigen et al. attribute such an increase in CH_4 emissions from natural wetlands to increasing soil moisture due to the reduced plant demand for water under higher CO_2. However, the sign and magnitude of the CH_4 emission response to changes in temperature and precipitation vary among models but show, on average, a decrease of wetland area and CH_4 flux with increasing temperature, especially in the tropics, and a modest (~4%) increase of wetland area and CH_4 flux with increasing precipitation (Melton et al., 2013).

In AR4, natural geological sources were estimated between 4 and 19 Tg(CH_4) yr^{-1}. Since then, Etiope et al. (2008) provided improved emission estimates from terrestrial (13 to 29 Tg(CH_4) yr^{-1}) and marine (~20 Tg(CH_4) yr^{-1}) seepages, mud volcanoes (6 to 9 Tg(CH_4) yr^{-1}), hydrates (5 to 10 Tg(CH_4) yr^{-1}) and geothermal and volcanic areas (3 to 6 Tg(CH_4) yr^{-1}), which represent altogether between 42 and 64 Tg(CH_4) yr^{-1} (see Table 6.8 for full range of estimates). This contribution from natural, geological and partly fossil CH_4 is larger than in AR4 and consistent with a $^{14}CH_4$ reanalysis showing natural and anthropogenic fossil contributions to the global CH_4 budget to be around 30% (*medium confidence*) (Lassey et al., 2007) and not around 20% as previously estimated (e.g., AR4). However, such a large percentage was not confirmed by an analysis of the global atmospheric record of ethane (Simpson et al., 2012) which is co-emitted with geological CH_4.

Of the natural sources of CH_4, emissions from thawing permafrost and CH_4 hydrates in the northern circumpolar region will become potentially important in the 21st century because they could increase dramatically due to the rapid climate warming of the Arctic and the large carbon pools stored there (Tarnocai et al., 2009; Walter Anthony et al., 2012) (see Section 6.4.3.4). Hydrates are, however, estimated to represent only a very small emission, between 2 and 9 Tg(CH_4) yr^{-1} under the current time period (Table 6.8). Supersaturation of dissolved CH_4 at the bottom and surface waters in the East Siberian Arctic Shelf indicate some CH_4 activity across the region, with a net sea–air flux of 10.5 Tg(CH_4) yr^{-1} which is similar in magnitude to the flux for the entire ocean (Shakhova et al., 2010) but it is not possible to say whether this source has always been present or is a consequence of recent Arctic changes. The ebullition of CH_4 from decomposing, thawing lake sediments in north Siberia with an estimated flux of ~4 Tg(CH_4) yr^{-1} is another demonstration of the activity of this region and of its potential importance in the future (Walter et al., 2006; van Huissteden et al., 2011). The sum of all natural emission estimates other than wetlands is still very uncertain based on bottom-up studies [see Table 6.8, range of 238 to 484 Tg(CH_4) yr^{-1} for 2000–2009].

Pyrogenic sources of CH_4 (biomass burning in Table 6.8) are assessed to have a small contribution in the global flux for the 2000s (32 to 39 Tg(CH_4) yr^{-1}). Biomass burning of tropical and boreal forests (17 to 21 Tg(CH_4) yr^{-1}) play a much smaller role than wetlands in interannual variability of emissions, except during intensive fire periods (Langenfelds et al., 2002; Simpson et al., 2006). Only during the 1997–1998 record strong El Niño, burning of forests and peatland that took place in Indonesia and Malaysia, released ~12 Tg(CH_4) and contributed to the observed growth rate anomaly (Langenfelds et al., 2002; van der Werf et al., 2004). Other smaller fire CH_4 emissions positive anomalies were suggested over the northern mid-latitudes in 2002–2003, in particular over Eastern Siberia in 2003 (van der Werf et al., 2010) and Russia in 2010. Traditional biofuel burning is estimated to be a source of 14 to 17 Tg(CH_4) yr^{-1}(Andreae and Merlet, 2001; Yevich and Logan, 2003).

Keppler at al. (2006) reported that plants under aerobic conditions were able to emit CH_4, and thus potentially could constitute a large additional emission, which had not been previously considered in the global CH_4 budget. Later studies do not support plant emissions as a widespread mechanism (Dueck et al., 2007; Wang et al., 2008; Nisbet et al., 2009) or show small to negligible emissions in the context of the global CH_4 budget (Vigano et al., 2008; Nisbet et al., 2009; Bloom et al., 2010). Alternative mechanisms have been suggested to explain an apparent aerobic CH_4 production, which involve (1) adsorption and desorption (Kirschbaum and Walcroft, 2008; Nisbet et al., 2009), (2) degradation of organic matter under strong ultraviolet (UV) light (Dueck et al., 2007; Nisbet et al., 2009) and (3) methane in the groundwater emitted through internal air spaces in tree bodies (Terazawa et al., 2007). Overall, a significant emission of CH_4 by plants under aerobic conditions is *very unlikely*, and this source is not reported in Table 6.8.

Anthropogenic CH_4 sources are estimated to range between 50% and 65% of the global emissions for the 2000s (Table 6.8). They include rice paddies agriculture, ruminant animals, sewage and waste, landfills, and fossil fuel extraction, storage, transformation, transportation and use (coal mining, gas and oil industries). Anthropogenic sources are dominant over natural sources in top-down inversions (~65%) but they are of the same magnitude in bottom-up models and inventories (Table 6.8). Rice paddies emit between 33 to 40 Tg(CH_4) yr^{-1} and 90% of these emissions come from tropical Asia, with more than 50% from China and India (Yan et al., 2009). Ruminant livestock, such as cattle, sheep, goats, etc. produce CH_4 by food fermentation in their anoxic rumens with a total estimate of between 87 and 94 Tg(CH_4) yr^{-1}. Major regional contributions of this flux come from India, China, Brazil and the USA (EPA, 2006; Olivier and Janssens-Maenhout, 2012), EDGAR v4.2. India, with the world's largest livestock population emitted 11.8 Tg(CH_4) yr^{-1} in 2003, including emission from enteric fermentation (10.7 Tg(CH_4) yr^{-1}) and manure management (1.1 Tg(CH_4) yr^{-1}; Chhabra et al., 2013). Methanogenesis in landfills, livestock manure and waste waters produces between 67 and 90 Tg(CH_4) yr^{-1} due to anoxic conditions and a high availability of acetate, CO_2 and H_2. Loss of natural gas (~90% CH_4) is the largest contributor to fossil fuel related fugitive emissions, estimated between 85 and 105 Tg(CH_4) yr^{-1} in the USA (EPA, 2006; Olivier and Janssens-Maenhout, 2012), EDGAR v4.2.

6.3.3.3 Sinks of Atmospheric Methane

The main sink of atmospheric CH_4 is its oxidation by OH radicals, a chemical reaction that takes place mostly in the troposphere and stratosphere (Table 6.8). OH removes each year an amount of CH_4 equivalent to 90% of all surface emissions (Table 6.8), that is, 9% of the total burden of CH_4 in the atmosphere, which defines a partial atmospheric lifetime with respect to OH of 7 to 11 years for an atmospheric burden of 4800 Tg(CH_4) (4700 to 4900 TgCH_4 as computed by Atmospheric Chemistry and Climate Model Intercomparison Project (ACCMIP) atmospheric chemistry models in Voulgarakis et al. (2013), thus slightly different from Figure 6.2; see Section 8.2.3.3 for ACCMIP models). A recent estimate of the CH_4 lifetime is 9.1 ± 0.9 years (Prather et al., 2012). A small sink of atmospheric CH_4 is suspected, but still debated, in the marine boundary layer due to a chemical reaction with chlorine (Allan et al., 2007). Another small sink is the reaction of CH_4 with Cl radicals and O(1D) in the stratosphere (Shallcross et al., 2007; Neef et al., 2010). Finally, oxidation in upland soils (with oxygen) by methanotrophic bacterias removes about 9 to 47 Tg(CH_4) yr^{-1} (Curry, 2007; Dutaur and Verchot, 2007; Spahni et al., 2011; Ito and Inatomi, 2012).

There have been a number of published estimates of global OH concentrations and variations over the past decade (Prinn et al., 2001; Dentener et al., 2003; Bousquet et al., 2005; Prinn et al., 2005; Rigby et al., 2008; Montzka et al., 2011). The very short lifetime of OH makes it almost impossible to measure directly global OH concentrations in the atmosphere. Chemistry transport models (CTMs), chemistry climate models (CCMs) or proxy methods have to be used to obtain a global mean value and time variations. For the 2000s, CTMs and CCMs (Young et al., 2013) estimate a global chemical loss of methane due to OH of 604 Tg(CH_4) yr^{-1} (509 to 764 Tg(CH_4) yr^{-1}). This loss is larger, albeit compatible considering the large uncertainties, with a recent extensive analysis by Prather et al. (2012) inferring a global chemical loss of 554 ± 56 Tg(CH_4) yr^{-1}. Top-down inversions using methyl-chloroform (MCF) measurements to infer OH provide a smaller chemical loss of 518 Tg(CH_4) yr^{-1} with a more narrow range of 510 to 538 Tg(CH_4) yr^{-1} in the 2000s. However, inversion estimates probably do not account for all sources of uncertainties (Prather et al., 2012).

CCMs and CTMs simulate small interannual variations of OH radicals, typically of 1 to 3% (standard deviation over a decade) due to a high buffering of this radical by atmospheric photochemical reactions (Voulgarakis et al., 2013; Young et al., 2013). Atmospheric inversions show much larger variations for the 1980s and the 1990s (5 to 10%), because of their oversensitivity to uncertainties on MCF emissions, when measurements of this tracer are used to reconstruct OH (Montzka et al., 2011), although reduced variations are inferred after 1998 by Prinn et al. (2005). For the 2000s, the reduction of MCF in the atmosphere, due to the Montreal protocol (1987) and its further amendments, allows a consistent estimate of small OH variations between atmospheric inversions (<±5%) and CCMs/CTMs (<±3%). However, the very low atmospheric values reached by MCF (few ppt in 2010) impose the need to find another tracer to reconstruct global OH in the upcoming years. Finally, evidence for the role of changes in OH concentrations in explaining the increase in atmospheric methane since 2007 is variable, ranging from a significant contribution (Rigby et al., 2008) to only a small role (Bousquet et al., 2011).

6.3.3.4 Global Methane Budget for the 2000s

Based on the inversion of atmospheric measurements of CH_4 from surface stations, global CH_4 emissions for the 2000s are of 553 Tg(CH_4) yr^{-1}, with a range of 526 to 569 Tg(CH_4) yr^{-1} (Table 6.8). The total loss of atmospheric methane is of 550 Tg(CH_4) yr^{-1} with a range of 514 to 560 Tg(CH_4) yr^{-1}, determining a small imbalance of about 3 Tg(CH_4) yr^{-1}, in line with the small growth rate of 6 Tg(CH_4) yr^{-1} observed for the 2000s.

Based on bottom-up models and inventories, a larger global CH_4 emissions of 678 Tg(CH_4) yr^{-1} are found, mostly because of the still debated upward re-evaluation of geological (Etiope et al., 2008) and freshwater (Walter et al., 2007; Bastviken et al., 2011) emission sources. An averaged total loss of 632 Tg(CH_4) yr^{-1} is found, by an ensemble of Atmospheric Chemistry models (Lamarque et al., 2013) leading to an imbalance of about 45 Tg(CH_4) yr^{-1} during the 2000s, as compared to the observed mean growth rate of 6 Tg(CH_4) yr^{-1}(Table 6.8; Dlugokencky et al., 2011). There is no constraint that applies to the sum of emissions in the bottom-up approach, unlike for top-down inversions when these have constrained OH fields (e.g., from MCF). Therefore, top-down inversions can help constrain global CH_4 emissions in the global budget, although they do not resolve the same level of detail in the mix of sources than the bottom-up approaches, and thus provide more limited information about processes (Table 6.8).

6.3.4 Global Nitrogen Budgets and Global Nitrous Oxide Budget in the 1990s

The atmospheric abundance of N_2O has been increasing mainly as a result of agricultural intensification to meet the food demand for a growing human population. Use of synthetic fertiliser (primarily from the Haber–Bosch process) and manure applications increase the production of N_2O in soils and sediments, via nitrification and denitrification pathways, leading to increased N_2O emissions to the atmosphere. Increased emissions occur not only in agricultural fields, but also in aquatic systems after nitrogen leaching and runoff, and in natural soils and ocean surface waters as a result of atmospheric deposition of nitrogen originating from agriculture, fossil fuel combustion and industrial activities. Food production is *likely* responsible for 80% of the increase in atmospheric N_2O (Kroeze et al., 1999; Davidson, 2009; Williams and Crutzen, 2010; Syakila and Kroeze, 2011; Zaehle et al., 2011; Park et al., 2012), via the addition of nitrogen fertilisers. Global emissions of N_2O are difficult to estimate owing to heterogeneity in space and time. Table 6.9 presents global emissions based on upscaling of local flux measurements at the surface. Modelling of the atmospheric lifetime of N_2O and atmospheric inversions constrain global and regional N_2O budgets (Hirsch et al., 2006; Huang et al., 2008; Rhee et al., 2009; Prather et al., 2012), although there is uncertainty in these estimates because of uncertainty in the dominant loss term of N_2O, that is, the destruction of N_2O by photolysis and reaction with O(1D) in the stratosphere. The long atmospheric lifetime of N_2O (118 to 131 years, Volk et al., 1997; Hsu and Prather, 2010; Fleming et al., 2011; see Chapter 8) implies that it will take more than a century before atmospheric abundances stabilise after the stabilization of global emissions. This is of concern not only because of its contribution to the radiative forcing (see Glossary), but also because of the relative importance of N_2O and other GHGs in affecting the ozone layer (Ravishankara et al., 2009; Fleming et al., 2011).

Since AR4 (Table 6.9 for the 1990s), a number of studies allow us to update some of the N_2O emission estimates. First and most importantly, the IPCC Guidelines were revised in 2006 (De Klein et al., 2007) and in particular emission factors for estimating agricultural N_2O emissions. Applying these 2006 emission factors to global agricultural statistics results in higher direct emissions from agriculture (from fertilised soils and animal production) than in AR4, but into indirect emissions (associated with leaching and runoff of Nr resulting in N_2O emissions from groundwater, riparian zones and surface waters) that are considerably lower than reported in AR4 (Table 6.9). It should be noted that emissions of N_2O show large uncertainties when default emission factors are applied at the global scale (Crutzen et al., 2008; Davidson, 2009; Smith et al., 2012). Second, estimates of the anthropogenic source of N_2O from the open ocean have been made for the first time. These emissions result from atmospheric deposition of anthropogenic Nr (nitrogen oxides and ammonia/ammonium) (Duce et al., 2008; Suntharalingam et al., 2012). This anthropogenic ocean N_2O source was implicitly included as part of the natural ocean N_2O source in AR4, but is now given as a separate anthropogenic source of 0.2 (0.1 to 0.4) TgN yr^{-1} in Table 6.9. Finally, a first estimate of global N_2O uptake at the surface is now available (Syakila et al., 2010; Syakila and Kroeze, 2011), based on reviews of measurements of N_2O uptake in soils and sediments (Chapuis-Lardy et al., 2007; Kroeze et al., 2007). The uncertainty in this sink of N_2O is large. On the global scale, this surface sink is negligible, but at the local scale it may not be irrelevant.

6.3.4.1 Atmosphere Nitrous Oxide Burden and Growth Rate

The concentration of N_2O is currently 20% higher than pre-industrial levels (Figure 6.11; MacFarling-Meure et al., 2006). Figure 6.19 shows the annual growth rate of atmospheric N_2O estimated from direct measurements (National Oceanic and Atmospheric Administration – Global Monitoring Division (NOAA–GMD) network of surface stations). On decadal time scales, the concentration of N_2O has been increasing at a rate of 0.73 ± 0.03 ppb yr^{-1}. The interannual variability in mid- to high-latitude N_2O abundance in both the NH and SH was found to correlate with the strength of the stratospheric Brewer–Dobson circulation (Nevison et al., 2011). Variability in stratosphere to troposphere air mass exchange, coupled with the stratospheric N_2O sink is *likely* to be responsible for a fraction of the interannual variability in tropospheric N_2O, but the understanding of this process is poor (Huang et al., 2008). This removal process signal is obscured in the SH by the timing of oceanic thermal and biological ventilation signals (Nevison et al., 2011) and terrestrial sources (Ishijima et al., 2009). These two factors may thus also be important determinants of seasonal and interannual variability of N_2O in the atmosphere. Quantitative understanding of terrestrial N_2O emissions variability is poor, although emissions are known to be sensitive to soil water content (Ishijima et al., 2009). A first process model-based estimate suggests that the mainly climate-driven variability in the terrestrial source may account for only 0.07 ppb yr^{-1} variability in atmospheric N_2O growth rate, which would be difficult to detect in the observed growth rate (Zaehle et al., 2011).

Most N_2O is produced by biological (microbial) processes such as nitrification and denitrification in terrestrial and aquatic systems, including rivers, estuaries, coastal seas and the open ocean (Table 6.9; Freing et al., 2012). In general, more N_2O is formed when more reactive nitrogen is available. The production of N_2O shows large spatial and temporal variability. Emission estimates for tropical regions and for aquatic systems are relatively uncertain. Inverse modelling studies show that the errors in emissions are large, especially in (sub)-tropical regions (e.g., Hirsch et al., 2006; Huang et al., 2008). Emissions from rivers, estuaries and continental shelves have been the subject of debate for many years (Seitzinger and Kroeze, 1998; De Klein et al., 2007). Recent studies confirm that rivers can be important sources of N_2O, which could be a reason to reconsider recent estimates of aquatic N_2O emissions (Beaulieu et al., 2011; Rosamond et al., 2012).

Table 6.9 does not include the formation of atmospheric N_2O from abiotic decomposition of ammonium nitrate in the presence of light, appropriate relative humidity and a surface. This process recently has been proposed as a potentially important source of N_2O (Rubasinghege et al., 2011); however, a global estimate does not yet exist. Table 6.9 indicates that the global N_2O emissions in the mid-1990s amount to 17.5 (8.1 to 30.7) TgN (N_2O) yr^{-1}. The uncertainty range is consistent with that of atmospheric inversions studies (14.1 to 17.8) by Huang et al. (2008). The estimates of anthropogenic N_2O emissions of Table 6.9 are in line with the top-down estimates by Prather et al. (2012) of 6.5 ± 1.3 TgN (N_2O) yr^{-1}, and somewhat higher than their estimates for

Table 6.9 | Section 1 gives the global nitrogen budget (TgN yr^{-1}): (a) creation of reactive nitrogen, (b) emissions of NO_x, NH_3 in 2000s to atmosphere, (c) deposition of nitrogen to continents and oceans, (d) discharge of total nitrogen to coastal ocean and (e) conversion of Nr to N_2 by denitrification. Section 2 gives the N_2O budget for the year 2006, and for the 1990s compared to AR4. Unit: Tg(N_2O-N) yr^{-1}.

SECTION 1 (NO_y and NH_x)			
a. Conversion of N_2 to Nr	**2005**	**2005**	**References**
Anthropogenic sources			
Fossil fuel combustion	30 (27–33)		Fowler et al. (2013)
Haber–Bosch process			
Fertiliser	100 (95–100)		Galloway et al. (2008), Fowler et al. (2013)
Industrial feedstock	24 (22–26)		Galloway et al. (2008), Fowler et al. (2013)
Biological nitrogen fixation (BNF)	60 (50–70)		Herridge et al. (2008)
Anthropogenic total	**210**		
Natural sources			
BNF, terrestrial	58 (50–100)		Vitousek et al. (2013)
BNF, marine	160 (140–177)		Voss et al. (2013), Codispoti (2007)
Lightning	4 (3–5)		AR4
Natural total	**220**		
Total conversion of N_2 to reactive N	**440**		
b. Emissions to Atmosphere			
	NO_x	**NH_3**	
Fossil fuel combustion industrial processes	28.3	0.5	Dentener et al. (2006)
Agriculture	3.7	30.4	Dentener et al. (2006)
Biomass and biofuel burning	5.5	9.2	Dentener et al. (2006)
Anthropogenic total	**37.5**	**40.1**	
Natural sources			
Soils under natural vegetation	7.3 (5–8)	2.4 (1–10)	AR4
Oceans	—	8.2 (3.6)	AR4
Lightning	4 (3–5)	—	AR4
Natural total	**11.3**	**10.6**	AR4
Total sources	**48.8**	**50.7**	
c. Deposition from the Atmosphere			
	NO_y	**NH_x**	
Continents	27.1	36.1	Lamarque et al. (2010)
Oceans	19.8	17.0	Lamarque et al. (2010)
Total	**46.9**	**53.1**	
d. Discharge to Coastal Ocean			
Surface water N flux	45		Mayorga et al. (2010), Seitzinger et al. (2010)
e. Conversion of Nr to N_2 by Denitrification			
Continents	109 (101–118)		Bouwman et al. (2013)

(continued on next page)

Table 6.9 (continued)

SECTION 2 (N_2O)			
	AR5 (2006/2011)	**AR5 (mid-1990s)**	**AR4 (1990s)**
Anthropogenic sources			
Fossil fuel combustion and industrial processes	0.7 (0.2–1.8)[a]	0.7 (0.2–1.8)[a]	0.7 (0.2–1.8)
Agriculture	4.1 (1.7–4.8)[b]	3.7 (1.7–4.8)[b]	2.8(1.7–4.8)
Biomass and biofuel burning	0.7(0.2–1.0)[a]	0.7(0.2–1.0)[a]	0.7(0.2–1.0)
Human excreta	0.2 (0.1–0.3)[a]	0.2 (0.1–0.3)[a]	0.2 (0.1–0.3)
Rivers, estuaries, coastal zones	0.6 (0.1–2.9)[c]	0.6 (0.1–2.9)[c]	1.7(0.5–2.9)
Atmospheric deposition on land	0.4 (0.3–0.9)[d]	0.4 (0.3–0.9)[d]	0.6 (0.3–0.9)
Atmospheric deposition on ocean	0.2 (0.1–0.4)[e]	0.2 (0.1–0.4)[e]	—
Surface sink	–0.01 (0– -1)[f]	–0.01 (0– -1)[f]	—
Total anthropogenic sources	**6.9 (2.7–11.1)**	**6.5 (2.7–11.1)**	**6.7 (2.7–11.1)**
Natural sources[a]			
Soils under natural vegetation	6.6 (3.3–9.0)	6.6 (3.3–9.0)	6.6 (3.3–9.0)
Oceans	3.8(1.8–9.4)	3.8(1.8–9.4)	3.8(1.8–5.8)
Lightning	—	—	—
Atmospheric chemistry	0.6 (0.3–1.2)	0.6 (0.3–1.2)	0.6 (0.3–1.2)
Total natural sources	**11.0 (5.4–19.6)**	**11.0 (5.4–19.6)**	**11.0 (5.4–19.6)**
Total natural + anthropogenic sources	**17.9 (8.1–30.7)**	**17.5 (8.1–30.7)**	**17.7 (8.5–27.7)**
Stratospheric sink	**14.3 (4.3–27.2)**[g]		
Observed growth rate	**3.61 (3.5–3.8)**[h]		
Global top-down (year 2011)[i]			
Burden (Tg N)	**1553**		
Atmospheric Loss	**11.9±0.9**		
Atmospheric Increase	**4.0±0.5**		
Total Source	**15.8±1.0**		
Natural Source	**9.1±1.0**		
Anthropogenic Source	**6.7±1.3**		

Notes:

[a] All units for N_2O fluxes are in TgN (N_2O) yr^{-1} as in AR4 (not based on 2006 IPCC Guidelines). Lower end of range in the natural ocean from Rhee et al. (2009); higher end of the range from Bianchi et al. (2012) and Olivier and Janssens-Maenhout (2012); natural soils in line with Stocker et al. (2013).

[b] Direct soil emissions and emissions from animal production; calculated following 2006 IPCC Guidelines (Syakila and Kroeze, 2011); range from AR4 (Olivier and Janssens-Maenhout, 2012).

[c] Following 2006 IPCC Guidelines (Kroeze et al., 2010; Syakila and Kroeze, 2011). Higher end of range from AR4; lower end of range from 1996 IPCC Guidelines (Mosier et al., 1998). Note that a recent study indicates that emissions from rivers may be underestimated in the IPCC assessments (Beaulieu et al., 2011).

[d] Following 2006 IPCC Guidelines (Syakila and Kroeze, 2011).

[e] Suntharalingam et al. (2012).

[f] Syakila et al. (2010).

[g] The stratospheric sink regroups losses via photolysis and reaction with O(1D) that account for 90% and 10% of the sink, respectively (Minschwaner et al., 1993). The global magnitude of the stratospheric sink was adjusted in order to be equal to the difference between the total sources and the observed growth rate. This value falls within literature estimates (Volk et al., 1997).

[h] Data from Sections 6.1 and 6.3 (see Figure 6.4c). The range on the observed growth rate in this table is given by the 90% confidence interval of Figure 6.4c.

[i] Based on Prather et al. (2012), updated to 2011 (Table 2.1) and used in Chapter 11 projections; uncertainties evaluated as 68% confidence intervals, N_2O budget reduced based on recently published longer lifetimes of 131±10 yrs, see Annex II.2.3 and II.4.3.

natural (9.1 ± 1.3 TgN (N_2O) yr^{-1}) and total (15.7 ± 1.1 TgN (N_2O) yr^{-1}) emissions. Anthropogenic emissions have steadily increased over the last two decades and were 6.9 (2.7 to 11.1) TgN (N_2O) yr^{-1} in 2006, or 6% higher than the value in mid-1990s (Davidson, 2009; Zaehle et al., 2011) (see also Figure 6.4c). Overall, anthropogenic N_2O emissions are now a factor of 8 greater than their estimated level in 1900. These trends are consistent with observed increases in atmospheric N_2O (Syakila et al., 2010). Human activities strongly influence the source of N_2O, as nitrogen fertiliser used in agriculture is now the main source of nitrogen for nitrification and denitrification (Opdyke et al., 2009). Nitrogen stable isotope ratios confirm that fertilised soils are primarily responsible for the historic increase in N_2O (Röckmann and Levin, 2005; Sutka et al., 2006; Park et al., 2012).

6.3.4.2 Sensitivity of Nitrous Oxide Fluxes to Climate and Elevated Carbon Dioxide

Previous studies suggested a considerable positive feedback between N_2O and climate (Khalil and Rasmussen, 1989) supported by observed glacial–interglacial increases of ~70 ppb in atmospheric N_2O (Flückiger et al., 1999). Climate change influences marine and terrestrial N_2O sources, but their individual contribution and even the sign of their response to long-term climate variations are difficult to estimate (see Section 6.2). Simulations by terrestrial biosphere models suggest a moderate increase of global N_2O emissions with recent climate changes, related mainly to changes in land temperature (Zaehle and Dalmonech, 2011; Xu-Ri et al., 2012), thus suggesting a possible positive feedback to the climate system. Nonetheless, the recent change in atmospheric N_2O is largely dominated to anthropogenic reactive nitrogen (Nr) and industrial emissions (Holland et al., 2005; Davidson, 2009; Zaehle and Dalmonech, 2011). Stocker et al. (2013) have found, using a global coupled model of climate and biogeochemical cycles, that future climate change will amplify terrestrial N_2O emissions resulting from anthropogenic Nr additions, consistent with empirical understanding (Butterbach-Bahl and Dannenmann, 2011). This result suggests that the use of constant emission factors might underestimate future N_2O emission trajectories. Significant uncertainty remains in the N_2O–climate feedback from land ecosystems, given the poorly known response of emission processes to the changes in seasonal and frequency distribution of precipitation, and also because agricultural emissions themselves may also be sensitive to climate.

N_2O production will be affected by climate change through the effects on the microbial nitrification and denitrification processes (Barnard et al., 2005; Singh et al., 2010; Butterbach-Bahl and Dannenmann, 2011). Warming experiments tend to show enhanced N_2O emission (Lohila et al., 2010; Brown et al., 2011; Chantarel et al., 2011; Larsen et al., 2011). Elevated CO_2 predominantly increases N_2O emissions(van Groenigen et al., 2011); however, reductions have also been observed (Billings et al., 2002; Mosier et al., 2002), induced by changes in soil moisture, plant productivity and nitrogen uptake, as well as activity and composition of soil microbial and fungal communities (Barnard et al., 2005; Singh et al., 2010). The effect of interacting climate and atmospheric CO_2 change modulates and potentially dampens the individual responses to each driver (Brown et al., 2011). A terrestrial biosphere model that integrates the interacting effects of temperature, moisture and CO_2

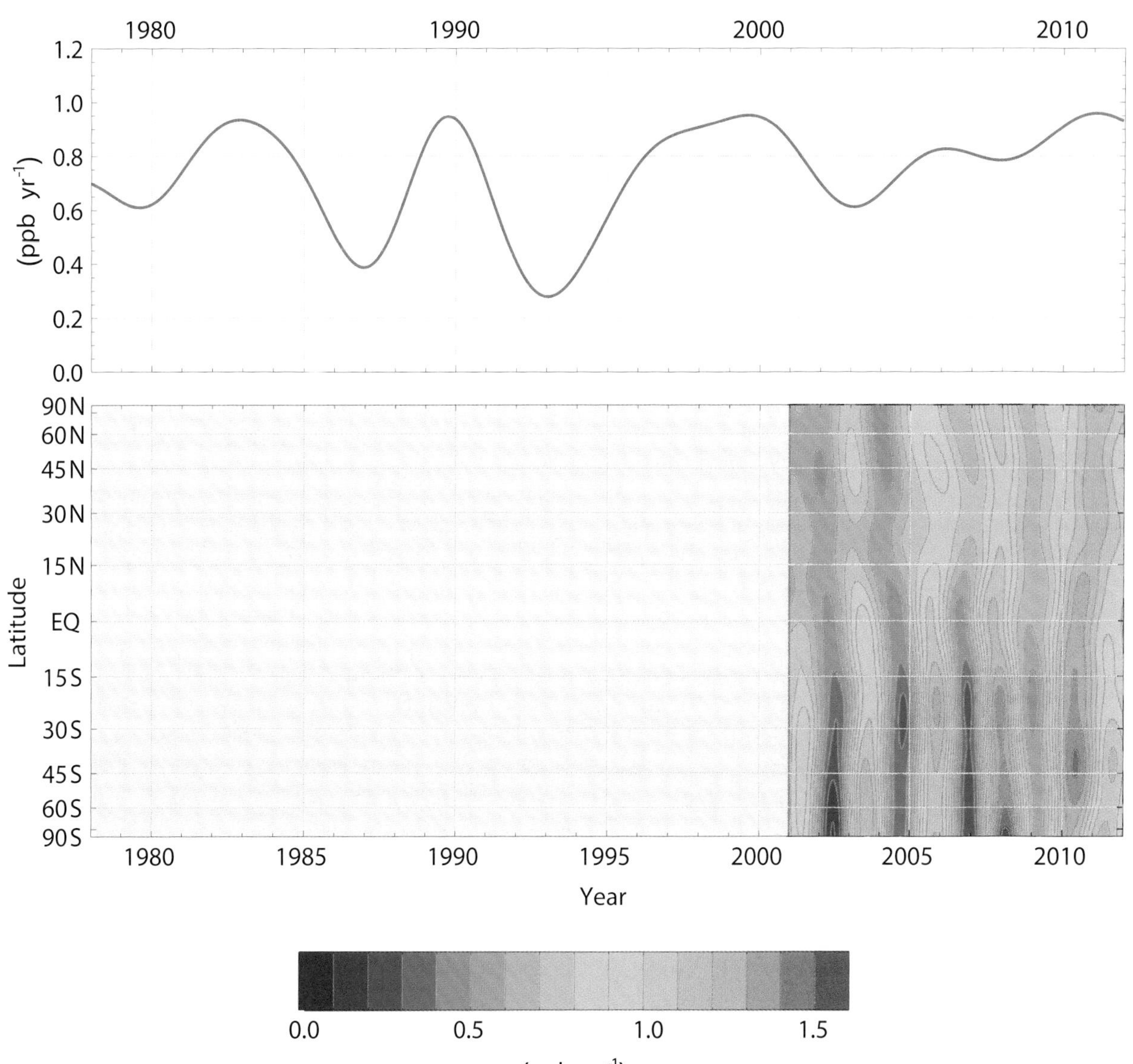

Figure 6.19 | (Top) Globally averaged growth rate of atmospheric N_2O in ppb yr^{-1} representative for the marine boundary layer. (Bottom) Atmospheric growth rate of N_2O as a function of latitude. Sufficient observations are available only since the year 2002. Observations from the National Oceanic and Atmospheric Administration–Earth System Research Laboratory (NOAA–ESRL) network (Masarie and Tans, 1995; Dlugokencky and Tans, 2013b).

changes is capable of qualitatively reproducing the observed sensitivities to these factors and their combinations (Xu-Ri et al., 2012). Thawing permafrost soils under particular hydrological settings may liberate reactive nitrogen and turn into significant sources of N_2O; however, the global significance of this source is not established (Elberling et al., 2010).

6.3.4.3 Global Nitrogen Budget

For base year 2010, anthropogenic activities created ~210 (190 to 230) TgN of reactive nitrogen Nr from N_2. This human-caused creation of reactive nitrogen in 2010 is at least 2 times larger than the rate of natural terrestrial creation of ~58 TgN (50 to 100 TgN yr^{-1}) (Table 6.9, Section 1a). Note that the estimate of natural terrestrial biological fixation (58 TgN yr^{-1}) is lower than former estimates (100 TgN yr^{-1}, Galloway et al., 2004), but the ranges overlap, 50 to 100 TgN yr^{-1}, vs. 90 to 120 TgN yr^{-1}, respectively). Of this created reactive nitrogen, NO_x and NH_3 emissions from anthropogenic sources are about fourfold greater than natural emissions (Table 6.9, Section 1b). A greater portion of the NH_3 emissions is deposited to the continents rather than to the oceans, relative to the deposition of NO_y, due to the longer atmospheric residence time of the latter. These deposition estimates are lower limits, as they do not include organic nitrogen species. New model and measurement information (Kanakidou et al., 2012) suggests that incomplete inclusion of emissions and atmospheric chemistry of reduced and oxidized organic nitrogen components in current models may lead to systematic underestimates of total global reactive nitrogen deposition by up to 35% (Table 6.9, Section 1c). Discharge of reactive nitrogen to the coastal oceans is ~45 TgN yr^{-1} (Table 6.9, Section 1d). Denitrification converts Nr back to atmospheric N_2. The current estimate for the production of atmospheric N_2 is 110 TgN yr^{-1} (Bouwman et al., 2013). Thus of the ~280 TgN yr^{-1} of Nr from anthropogenic and natural sources, ~40% gets converted to N_2 each year. The remaining 60% is stored in terrestrial ecosystems, transported by rivers and by atmospheric transport and deposition to the ocean, or emitted as N_2O (a small fraction of total Nr only despite the important forcing of increasing N_2O emissions for climate change). For the oceans, denitrification producing atmospheric N_2 is 200 to 400 TgN yr^{-1}, which is larger than the current uptake of atmospheric N_2 by ocean biological fixation of 140 to 177 TgN yr^{-1} (Table 6.9 Section 1e; Figure 6.4a).

6.4 Projections of Future Carbon and Other Biogeochemical Cycles

6.4.1 Introduction

In this section, we assess coupled model projections of changes in the evolution of CO_2, CH_4 and N_2O fluxes, and hence the role of carbon and other biogeochemical cycles in future climate under socioeconomic emission scenarios (see Box 6.4). AR4 reported how climate change can affect the natural carbon cycle in a way that could feed back onto climate itself. A comparison of 11 coupled climate–carbon cycle models of different complexity (Coupled Carbon Cycle Climate–Model Intercomparison Project (C4MIP); Friedlingstein et al., 2006) showed that all 11 models simulated a positive feedback. There is substantial quantitative uncertainty in future CO_2 and temperature, both across coupled carbon–climate models (Friedlingstein et al., 2006; Plattner et al., 2008) and within each model parametrizations (Falloon et al., 2011; Booth et al., 2012; Higgins and Harte, 2012). This uncertainty on the coupling between carbon cycle and climate is of comparable magnitude to the uncertainty caused by physical climate processes discussed in Chapter 12 of this Report (Denman et al., 2007; Gregory et al., 2009; Huntingford et al., 2009).

Other biogeochemical cycles and feedbacks play an important role in the future of the climate system, although the carbon cycle represents the strongest of these. Natural CH_4 emissions from wetland and fires are sensitive to climate change (Sections 6.2, 6.4.7 and 6.3.3.2). The fertilising effects of nitrogen deposition and rising CO_2 also affect CH_4 emissions by wetlands through increased plant productivity (Stocker et al., 2013). Changes in the nitrogen cycle, in addition to interactions with CO_2 sources and sinks, are *very likely* to affect the emissions of N_2O both on land and from the ocean (Sections 6.3.4.2 and 6.4.6) and potentially on the rate of CH_4 oxidation (Gärdenäs et al., 2011). A recent review highlighted the complexity of terrestrial biogeochemical feedbacks on climate change (Arneth et al., 2010) and used the methodology of Gregory et al. (2009) to express their magnitude in common units of W m^{-2} $°C^{-1}$ (Figure 6.20). A similar degree of complexity exists in the ocean and in interactions between land, atmosphere and ocean cycles. Many of these processes are not yet represented in coupled climate–biogeochemistry models. Leuzinger et al. (2011) observed a trend from manipulation experiments for higher-order interactions between feedbacks to reduce the magnitude of response. *Confidence* in the magnitude, and sometimes even the sign, of many of these feedbacks between climate and carbon and other biogeochemical cycles is *low*.

The response of land and ocean carbon storage to changes in climate, atmospheric CO_2 and other anthropogenic activities (e.g., land use change; Table 6.2) varies strongly on different time scales. This chapter has assessed carbon cycle changes across many time scales from millennial (see Section 6.2) to interannual and seasonal (see Section 6.3), and these are summarized in Table 6.10. A common result is that an increase in atmospheric CO_2 will *always* lead to an *increase* in land and ocean carbon storage, all other things being held constant. Cox et al. (2013) find an empirical relationship between short-term interannual variability and long-term land tropical carbon cycle sensitivity that may offer an observational constraint on the climate–carbon cycle response over the next century. Generally, however, changes in climate on different time scales do not lead to a consistent sign and magnitude of the response in carbon storage change owing to the many different mechanisms that operate. Thus, changes in carbon cycling on one time scale cannot be extrapolated to make projections on different time scales, but can provide valuable information on the processes at work and can be used to evaluate and improve models.

6.4.2 Carbon Cycle Feedbacks in Climate Modelling Intercomparison Project Phase 5 Models

6.4.2.1 Global Analysis

The carbon cycle response to future climate and CO_2 changes can be viewed as two strong and opposing feedbacks (Gregory et al., 2009).

The climate–carbon response (γ) determines changes in carbon storage due to changes in climate, and the concentration–carbon response (β) determines changes in storage due to elevated CO_2. Climate–carbon cycle feedback responses have been analyzed for eight CMIP5 ESMs that performed idealised simulations involving atmospheric CO_2 increasing at a prescribed rate of 1% yr^{-1} (Arora et al., 2013; Box 6.4). There is *high confidence* that increased atmospheric CO_2 will lead to increased land and ocean carbon uptake but by an uncertain amount. Models agree on the sign of land and ocean response to rising CO_2 but show only medium and low agreement for the magnitude of ocean and land carbon uptake respectively (Figure 6.21). Future climate change will decrease land and ocean carbon uptake compared to the case with constant climate (*medium confidence*). Models agree on the sign, globally, of land and ocean response to climate change but show low agreement on the magnitude of this response, especially for the land. Land and ocean carbon uptake may differ in sign between different regions and between models (Section 6.4.2.3). Inclusion of nitrogen cycle processes in two of the land carbon cycle model components out of these eight reduces the magnitude of the sensitivity to both CO_2 and climate (Section 6.4.6.3) and increases the spread across the CMIP5 ensemble. The CMIP5 spread in ocean sensitivity to CO_2 and climate appears reduced compared with C4MIP.

The role of the idealised experiment presented here is to study model processes and understand what causes the differences between models. Arora et al. (2013) assessed the global carbon budget from these idealised simulations and found that the CO_2 contribution to changes in land and ocean carbon storage sensitivity is typically four to five times larger than the sensitivity to climate across the CMIP5 ESMs. The land carbon-climate response (γ) is larger than the ocean carbon–climate

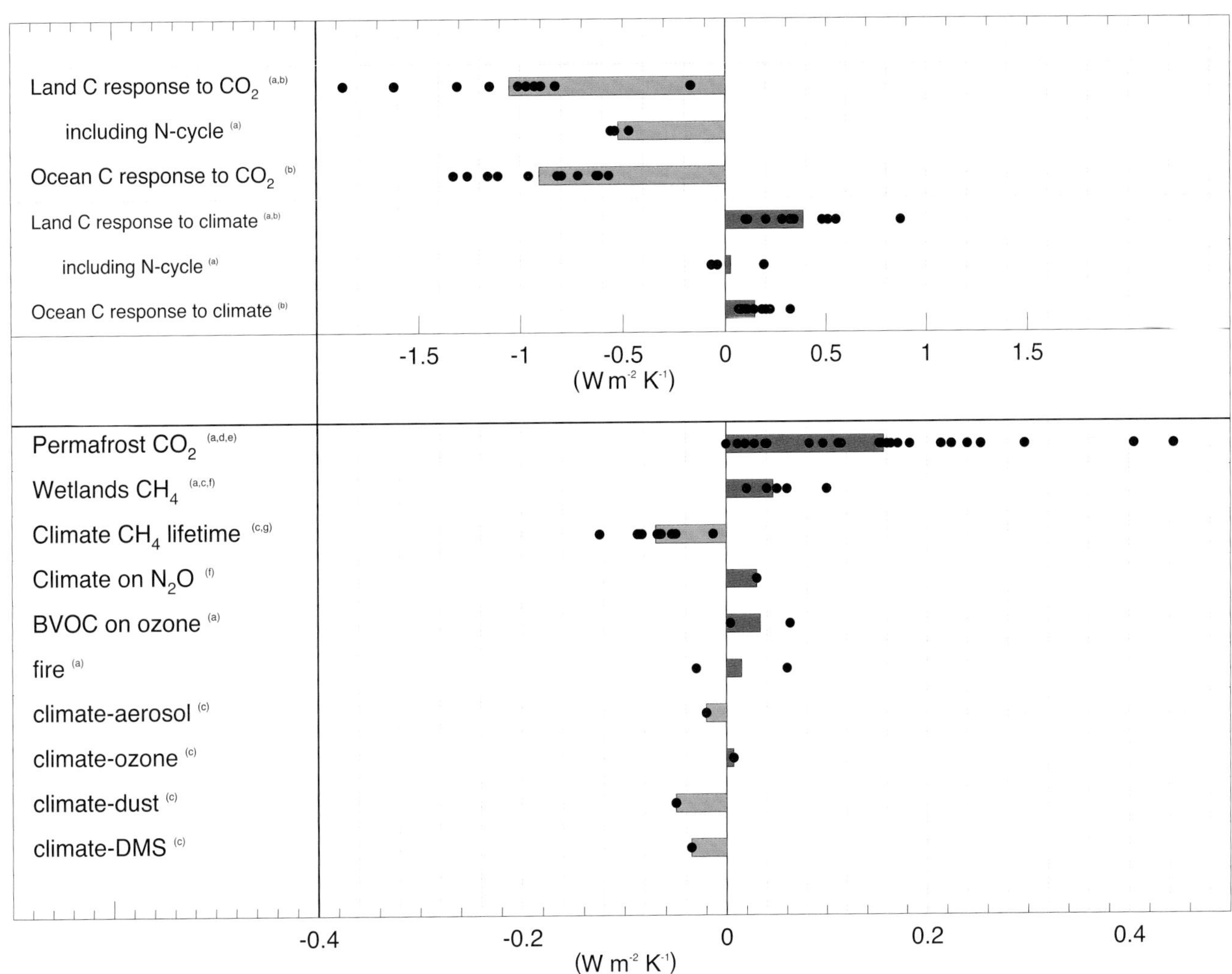

Figure 6.20 | A synthesis of the magnitude of biogeochemical feedbacks on climate. Gregory et al. (2009) proposed a framework for expressing non-climate feedbacks in common units (W m^{-2} °C^{-1}) with physical feedbacks, and Arneth et al. (2010) extended this beyond carbon cycle feedbacks to other terrestrial biogeochemical feedbacks. The figure shows the results compiled by Arneth et al. (2010), with ocean carbon feedbacks from the C4MIP coupled climate–carbon models used for AR4 also added. Some further biogeochemical feedbacks are also shown but this list is not exhaustive. Black dots represent single estimates, and coloured bars denote the simple mean of the dots with no weighting or assessment being made to likelihood of any single estimate. There is *low confidence* in the magnitude of the feedbacks in the lower portion of the figure, especially for those with few, or only one, dot. The role of nitrogen limitation on terrestrial carbon sinks is also shown—this is not a separate feedback, but rather a modulation to the climate–carbon and concentration–carbon feedbacks. These feedback metrics are also to be state or scenario dependent and so cannot always be compared like-for-like (see Section 6.4.2.2). Results have been compiled from (a) Arneth et al. (2010), (b) Friedlingstein et al. (2006), (c) Hadley Centre Global Environmental Model 2-Earth System (HadGEM2-ES, Collins et al., 2011) simulations, (d) Burke et al. (2013), (e) von Deimling et al. (2012), (f) Stocker et al. (2013), (g) Stevenson et al. (2006). Note the different *x*-axis scale for the lower portion of the figure.

Box 6.4 | Climate–Carbon Cycle Models and Experimental Design

What are coupled climate–carbon cycle models and why do we need them?
Atmosphere–Ocean General Circulation Models (AOGCMs; see Glossary) have long been used for making climate projections, and have formed the core of previous IPCC climate projection chapters (e.g., Meehl et al. (2007); see also Chapters 1, 9 and 12). For the 5th Coupled Model Intercomparison Project (CMIP5), many models now have an interactive carbon cycle. What exactly does this mean, how do they work and how does their use differ from previous climate models? AOGCMs typically represent the physical behaviour of the atmosphere and oceans but atmospheric composition, such as the amount of CO_2 in the atmosphere, is prescribed as an input to the model. This approach neglects the fact that changes in climate might affect the natural biogeochemical cycles, which control atmospheric composition, and so there is a need to represent these processes in climate projections.

At the core of coupled climate–carbon cycle models is the physical climate model, but additional components of land and ocean biogeochemistry respond to the changes in the climate conditions to influence in return the atmospheric CO_2 concentration. Input to themodels comes in the form of anthropogenic CO_2 emissions, which can increase the CO_2 and then the natural carbon cycle exchanges CO_2 between the atmosphere and land and ocean components. These 'climate–carbon cycle models' ('Earth System Models', ESMs; see Glossary) provide a predictive link between fossil fuel CO_2 emissions and future CO_2 concentrations and climate and are an important part of the CMIP5 experimental design (Hibbard et al., 2007; Taylor et al., 2012).

Apart from Earth System GCMs, so-called Earth System Models of Intermediate Complexity (EMICs) are often used to perform similar experiments (Claussen et al., 2002; Plattner et al., 2008). EMICs have reduced resolution or complexity but run much more quickly and can be used for longer experiments or large ensembles.

How are these models used?
The capability of ESMs to simulate carbon cycle processes and feedbacks, and in some models other biogeochemical cycles, allows for a greater range of quantities to be simulated such as changes in natural carbon stores, fluxes or ecosystem functioning. There may also be applications where it is desirable for a user to predefine the pathway of atmospheric CO_2 and prescribe it as a forcing to the ESMs. Thus, numerical simulations with ESM models can be either 'concentration driven' or 'emissions driven'.

Concentration-driven simulations follow the 'traditional' approach of prescribing the time-evolution of atmospheric CO_2 as an input to the model. This is shown schematically in Box 6.4 Figure 1 (left-hand side). Atmospheric CO_2 concentration is prescribed as input to the model from a given scenario and follows a predefined pathway regardless of changes in the climate or natural carbon cycle processes. The processes between the horizontal dashed lines in the figure represent the model components which are calculated during the concentration-driven simulation. Externally prescribed changes in atmospheric CO_2 concentration, which drive climate change, affect land and ocean carbon storage. By construction, changes in land and ocean storage, however, do not feed back on the atmospheric CO_2 concentration or on climate. The changes in natural carbon fluxes and stores are output by the model.

So-called 'compatible fossil fuel emissions', *E*, can be diagnosed afterwards from mass conservation by calculating the residual between the prescribed CO_2 pathway and the natural fluxes:

$$E = \frac{dCO_2}{dt}_{prescribed} + (\text{land\_carbon\_uptake} + \text{ocean\_carbon\_uptake}) \tag{6.1}$$

Land use change emissions cannot be diagnosed separately from a single simulation (see Section 6.4.3.2).

Emissions-driven simulations allow the full range of interactions in the models to operate and determine the evolution of atmospheric CO_2 and climate as an internal part of the simulation itself (Box 6.4, Figure 1, right-hand side). In this case emissions of CO_2 are the externally prescribed input to the model and the subsequent changes in atmospheric CO_2 concentration are simulated by it.

In *emissions-driven* experiments, the global atmospheric CO_2 growth rate is calculated within the model as a result of the net balance between the anthropogenic emissions, *E*, and natural fluxes:

$$\frac{dCO_2}{dt}_{simulated} = E - (\text{land\_carbon\_uptake} + \text{ocean\_carbon\_uptake}) \tag{6.2}$$

The effect of climate change on the natural carbon cycle will manifest itself either through changes in atmospheric CO_2 in the *emissions-driven* experiments or in the compatible emissions in the *concentration-driven* experiments.

(continued on next page)

Box 6.4 (continued)

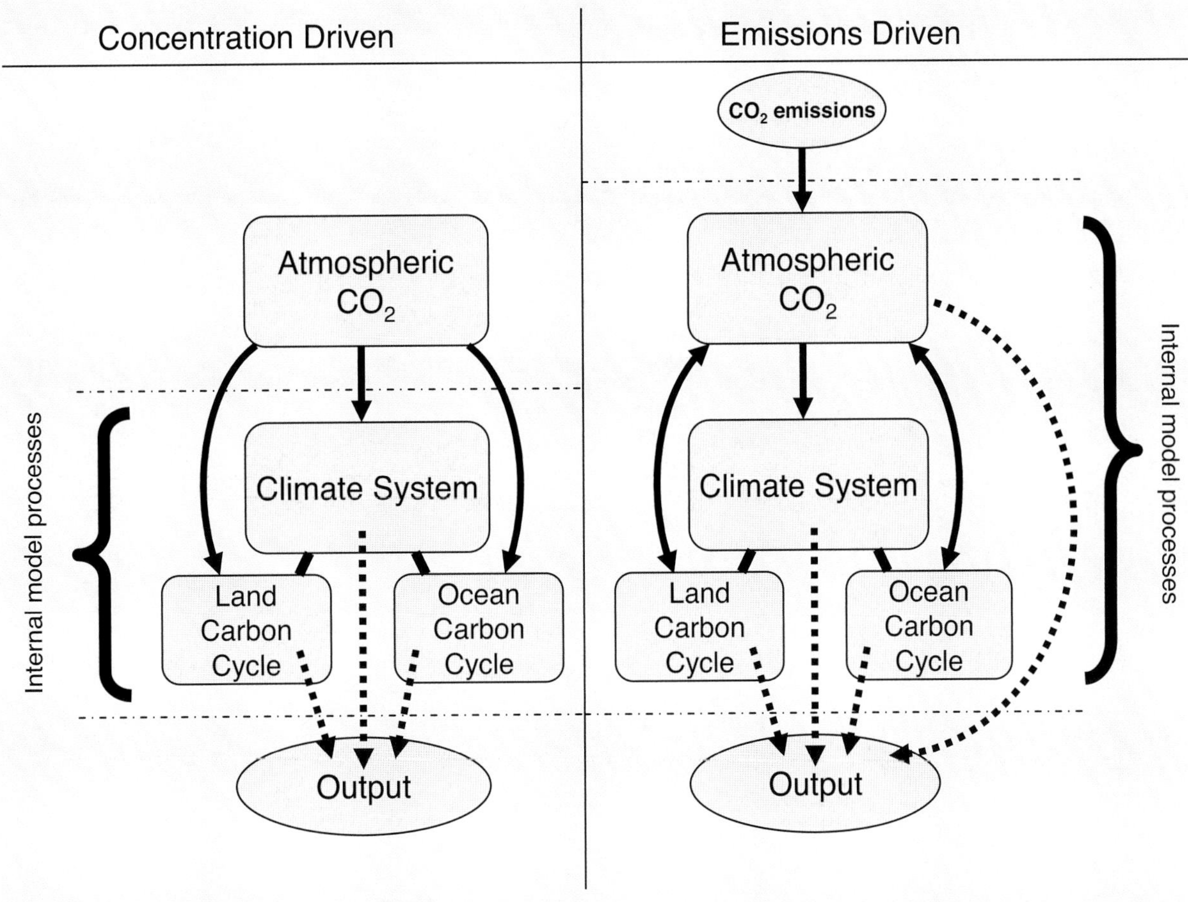

Box 6.4, Figure 1 | Schematic representation of carbon cycle numerical experimental design. Concentration-driven (left) and emissions-driven (right) simulation experiments make use of the same Earth System Models (ESMs), but configured differently. Concentration-driven simulations prescribe atmospheric CO_2 as a pre-defined input to the climate and carbon cycle model components, but their output does not affect the CO_2. Compatible emissions can be calculated from the output of the concentration-driven simulations. Emissions-driven simulations prescribe CO_2 emissions as the input and atmospheric CO_2 is an internally calculated element of the ESM.

Concentration-driven simulation experiments have the advantage that they can also be performed by GCMs without an interactive-carbon cycle and have been used extensively in previous assessments (e.g., Prentice et al., 2001). For this reason, most of the Representative Concentration Pathway (RCP) simulations (see Chapter 1) presented later in this chapter with carbon cycle models and in Chapter 12 with models that do not all have an interactive carbon cycle are performed this way. Emissions-driven simulations have the advantage of representing the full range of interactions in the coupled climate–carbon cycle models. The RCP8.5 pathway was repeated by many ESM models as an emissions-driven simulation (Chapter 12).

Feedback Analysis

The ESMs are made up of many 'components', corresponding to different processes or aspects of the system. To understand their behaviour, techniques have been applied to assess different aspects of the models' sensitivities (Friedlingstein et al., 2003, 2006; Arora et al., 2013). The two dominant emerging interactions are the sensitivity of the carbon cycle to changes in CO_2 and its sensitivity to changes in climate. These can be measured using two metrics: 'beta' (β) measures the strength of changes in carbon fluxes by land or ocean in response to changes in atmospheric CO_2; 'gamma' (γ) measures the strength of changes in carbon fluxes by land or ocean in response to changes in climate. These metrics can be calculated as cumulative changes in carbon storage (as in Friedlingstein et al., 2006) or instantaneous rates of change (Arora et al., 2013).

It is not possible to calculate these sensitivities in a single simulation, so it is necessary to perform 'decoupled' simulations in which some processes in the models are artificially disabled in order to be able to evaluate the changes in other processes. See Table 1 in Box 6.4.

(continued on next page)

6

Box 6.4 (continued)

A large positive value of β denotes that a model responds to increasing CO_2 by simulating large increases in natural carbon sinks. Negative values of γ denote that a model response to climate warming is to reduce CO_2 uptake from the atmosphere, while a positive value means warming acts to increase CO_2 uptake. β and γ values are not specified in a model, but are properties that emerge from the suite of complex processes represented in the model. The values of the β and γ metrics diagnosed from simulations can vary from place to place within the same model (see Section 6.4.2.3), although it is the average over the whole globe that determines the global extent of the climate–carbon cycle feedback.

Such an idealised analysis framework should be seen as a technique for assessing relative sensitivities of models and understanding their differences, rather than as absolute measures of invariant system properties. By design, these experiments exclude land use change.

The complex ESMs have new components and new processes beyond conventional AO GCMs and thus require additional evaluation to assess their ability to make climate projections. Evaluation of the carbon cycle model components of ESMs is presented in Section 6.3.2.5.6 for ocean carbon models and Section 6.3.2.6.6 for land carbon models. Evaluation of the fully coupled ESMs is presented in Chapter 9.

Box 6.4, Table 1 | Configurations of simulations designed for feedback analysis by allowing some carbon–climate interactions to operate but holding others constant. The curves denote whether increasing or constant CO_2 values are input to the radiation and carbon cycle model components. In a fully coupled simulation, the carbon cycle components of the models experience both changes in atmospheric CO_2 (see Box 6.3 on fertilisation) and changes in climate. In 'biogeochemically' coupled experiments, the atmospheric radiation experiences constant CO_2 (i.e., the radiative forcing of increased CO_2 is not activated in the simulation) whereas the carbon cycle model components experience increasing CO_2. This experiment quantifies the strength of the effect of rising CO_2 concentration alone on the carbon cycle (β). In a radiatively coupled experiment, the climate model's radiation scheme experiences an increase in the radiative forcing of CO_2 (and hence produces a change in climate) but CO_2 concentration is kept fixed to pre-industrial value as input to the carbon cycle model components. This simulation quantifies the effect of climate change alone on the carbon cycle (γ).

	CO_2 input to radiation scheme	CO_2 input to carbon-cycle scheme	Reason
Fully coupled			Simulates the fully coupled system
'Biogeochemically' coupled 'esmFixClim'			Isolates the carbon-cycle response to CO_2 (β) for land and oceans
Radiatively coupled 'esmFdbk'			Isolates carbon-cycle response to climate change (γ) for land and for oceans

response in all models. Although land and ocean contribute equally to the total carbon–concentration response (β), the model spread in the land response is greater than for the ocean.

6.4.2.2 Scenario Dependence of Feedbacks

The values of carbon-cycle feedback metrics can vary markedly for different scenarios and as such cannot be used to compare model simulations over different time periods, nor to inter-compare model simulations with different scenarios (Arora et al., 2013). Gregory et al. (2009) demonstrated how sensitive the feedback metrics are to the rate of change of CO_2 for two models: faster rates of CO_2 increase lead to reduced β values as the carbon uptake (especially in the ocean) lags further behind the forcing. γ is much less sensitive to the scenario, as both global temperature and carbon uptake lag the forcing.

6.4.2.3 Regional Feedback Analysis

The linear feedback analysis with the β and γ metrics of Friedlingstein et al. (2006) has been applied at the regional scale to future carbon uptake by Roy et al. (2011) and Yoshikawa et al. (2008). Figure 6.22 shows this analysis extended to land and ocean points for the CMIP5 models under the 1% yr^{-1} CO_2 simulations.

6.4.2.3.1 Regional ocean response

Increased CO_2 is projected by the CMIP5 models to increase oceanic CO_2 sinks almost everywhere (positive β) (*high confidence*) with the exception of some very limited areas (Figure 6.22). The spatial distribution of the CO_2 ocean response, β_o, is consistent between the models and with the Roy et al. (2011) analysis. On average, the regions with

Table 6.10 | Comparison of the sign and magnitude of changes in carbon storage (PgC) by land and ocean over different time scales. These changes are shown as approximate numbers to allow a comparison across time scales. For more details see the indicated chapter section. An indication, where known, of what causes these changes (climate, CO_2, land use change) is also given with an indication of the sign: '+' means that an increase in CO_2 or global-mean temperature is associated with an increase in carbon storage (positive β or γ; see Section 6.4.2), and a '–' means an increase in CO_2 or global-mean temperature is associated with a decrease in carbon storage (negative β or γ). The processes that operate to drive these changes can vary markedly, for example, from seasonal phenology of vegetation to long-term changes in ice sheet cover or ocean circulation impacting carbon reservoirs. Some of these processes are 'reversible' in the context that they can increase and decrease cyclically, whereas some are 'irreversible' in the context that changes in one sense might be much longer than in the opposite direction.

<table>
<tr><th>Time Period</th><th>Duration</th><th colspan="3">Land</th><th colspan="2">Ocean</th><th>Section</th></tr>
<tr><th></th><th></th><th>Climate</th><th>CO₂</th><th>Land Use</th><th>Climate</th><th>CO₂</th><th></th></tr>
<tr><td>Seasonal cycle</td><td>Weeks to months</td><td>3–8[a]
+</td><td></td><td></td><td>2</td><td>1
+</td><td>6.3.2.5.1</td></tr>
<tr><td>Interannual variability</td><td>Months to years</td><td>2–4[b]
–</td><td></td><td></td><td>1
+</td><td>0.2
+</td><td>6.3.2.5.4</td></tr>
<tr><td>Historical (1750–Present)</td><td>Decades to centuries</td><td colspan="2">150[c]
– | +</td><td>–180</td><td>2
?</td><td>155
+</td><td>6.3.2.5.3, Table 6.1</td></tr>
<tr><td>21st Century</td><td>Decades to centuries</td><td colspan="2">100–400[d]
– | +</td><td>–100 to +100[e]</td><td colspan="2">100–600[d]
– | +</td><td>6.4.3</td></tr>
<tr><td>Little Ice Age (LIA) [f]</td><td>Century</td><td colspan="2">+5
– | +</td><td>+2 to +30</td><td colspan="2"></td><td>6.2.3</td></tr>
<tr><td>Holocene</td><td>10 kyr</td><td colspan="2">+300
+ | +</td><td>–50 to –150</td><td colspan="2">+270 to –220[g]</td><td>6.2.2</td></tr>
<tr><td>Last Glacial Maximum/ glacial cycles</td><td>>10 kyr</td><td colspan="2">+300 to +1000[h]
+ | +</td><td></td><td colspan="2">–500 to –1200[h]
– | +</td><td>6.2.1</td></tr>
<tr><td>Pulse [i], 100 PgC</td><td>1 kyr</td><td colspan="2">+0 to +35
| +</td><td>n/a</td><td colspan="2">+48 to +75
– | +</td><td>6.2.2</td></tr>
</table>

Notes:

[a] Dominated by northern mid to high latitudes.

[b] Dominated by the tropics.

[c] 'Residual land sink', Table 6.1.

[d] Varies widely according to scenario. Climate effect estimated separately for RCP4.5 as –157 PgC (combined land and ocean), but not for other scenarios.

[e] Future scenarios may increase or decrease area of anthropogenic land use.

[f] Little Ice Age, 1500–1750.

[g] Shown here are two competing drivers of Holocene ocean carbon changes: carbonate accumulation on shelves (coral growth) and carbonate compensation to pre-Holocene changes. These are discussed in Section 6.2.2.

[h] Defined as positive if increasing from LGM to present, negative if decreasing.

[i] Idealised simulations with models to assess the response of the global carbon cycle to a sudden release of 100 PgC.

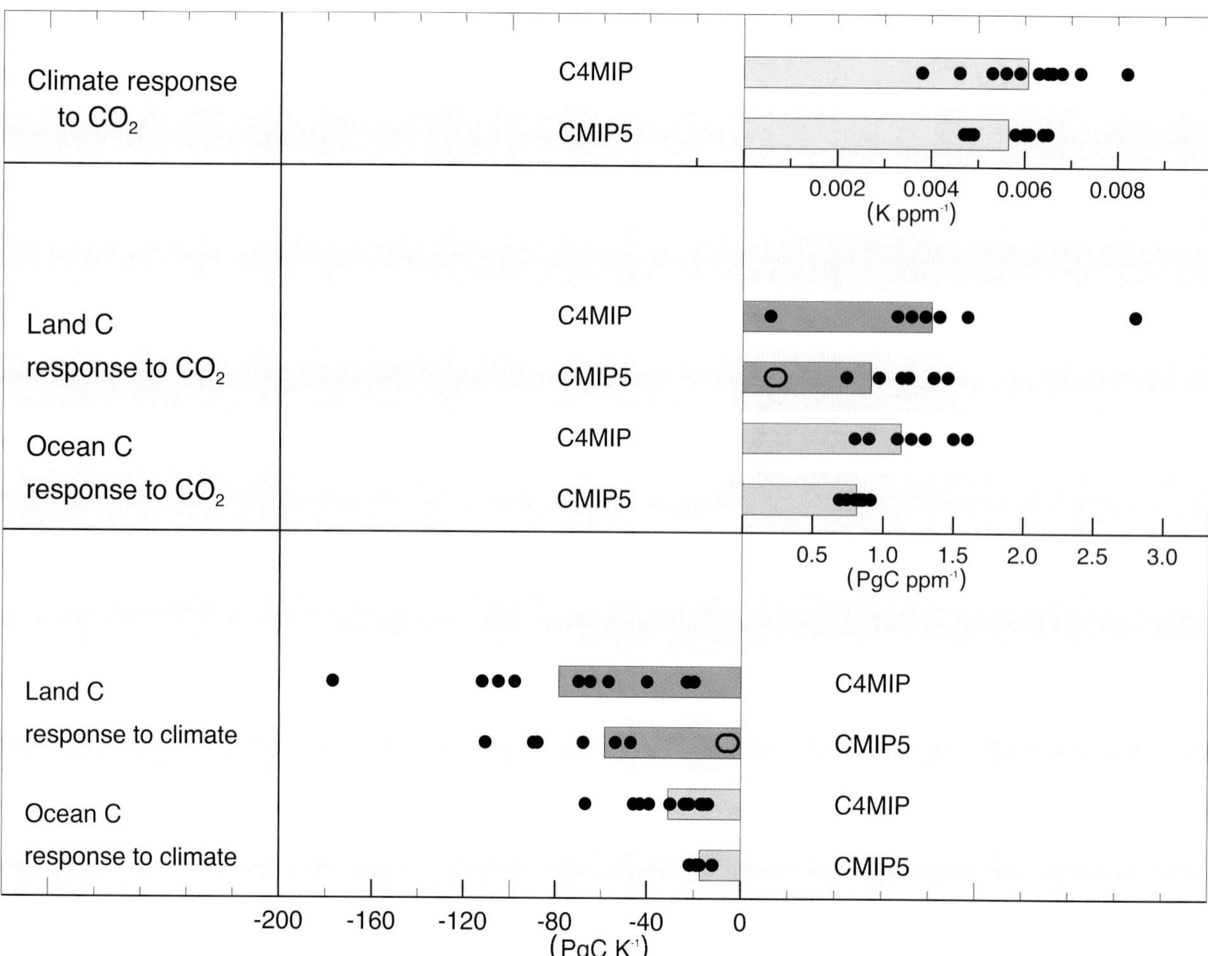

Figure 6.21 | Comparison of carbon cycle feedback metrics between the C4MIP ensemble of seven GCMs and four EMICs under the Special Report on Emission Scenario-A2 (SRES-A2) (Friedlingstein et al., 2006) and the eight CMIP5 models (Arora et al., 2013) under the 140-year 1% CO_2 increase per year scenario. Black dots represent a single model simulation and coloured bars the mean of the multi-model results; grey dots are used for models with a coupled terrestrial nitrogen cycle. The comparison with C4MIP is for context, but these metrics are known to be variable across different scenarios and rates of change (see Section 6.4.2.2). Some of the CMIP5 models are derived from models that contributed to C4MIP and some are new to this analysis. Table 6.11 lists the main attributes of each CMIP5 model used in this analysis. The SRES A2 scenario is closer in rate of change to a 0.5% yr^{-1} scenario and as such it should be expected that the CMIP5 γ terms are comparable, but the β terms are *likely* to be around 20% smaller for CMIP5 than for C4MIP due to lags in the ability of the land and ocean to respond to higher rates of CO_2 increase (Gregory et al., 2009). This dependence on scenario (Section 6.4.2.2) reduces confidence in any quantitative statements of how CMIP5 carbon cycle feedbacks differ from C4MIP. CMIP5 models used: Max Planck Institute–Earth System Model–Low Resolution (MPI–ESM–LR), Beijing Climate Center–Climate System Model 1 (BCC–CSM1), Hadley Centre Global Environmental Model 2–Earth System (HadGEM2–ES), Institute Pierre Simon Laplace–Coupled Model 5A–Low Resolution (IPSL–CM5A–LR), Canadian Earth System Model 2 (CanESM2), Norwegian Earth System Model– intermediate resolution with carbon cycle (NorESM–ME), Community Earth System Model 1–Biogeochemical (CESM1–BGC), Model for Interdisciplinary Research On Climate–Earth System Model (MIROC–ESM).

the strongest increase of oceanic CO_2 sinks in response to higher atmospheric CO_2 are the North Atlantic and the Southern Oceans. The magnitude and distribution of β_o in the ocean closely resemble the distribution of historical anthropogenic CO_2 flux from inversion studies and forward modelling studies (Gruber et al., 2009), with the dominant anthropogenic CO_2 uptake in the Southern Ocean (Section 6.3.2.5).

Climate warming is projected by the CMIP5 models to reduce oceanic carbon uptake in most oceanic regions (negative γ) (*medium confidence*) consistent with the Roy et al. (2011) analysis (Figure 6.22). This sensitivity of ocean CO_2 sinks to climate, γ_o, is mostly negative (i.e., a reduced regional ocean CO_2 sink in response to climate change) but with regions of positive values in the Arctic, the Antarctic and in the equatorial Pacific (i.e., climate change increases ocean CO_2 sink in these regions). The North Atlantic Ocean and the mid-latitude Southern Ocean have the largest negative γ_o values. Reduced CO_2 uptake in response to climate change in the sub-polar Southern Ocean and the tropical regions has been attributed to warming induced decreased CO_2 solubility, reduced CO_2 uptake in the mid latitudes to decreased CO_2 solubility and decreased water mass formation which reduces the absorption of anthropogenic CO_2 in intermediate and deep waters (Roy et al., 2011). Increased uptake in the Arctic Ocean and the polar Southern Ocean is partly associated with a reduction in the fractional sea ice coverage (Roy et al., 2011).

6.4.2.3.2 Regional land response

Increased CO_2 is projected by the CMIP5 models to increase land CO_2 sinks everywhere (positive β) (*medium confidence*). This response, β_L, has the largest values over tropical land, in humid rather than arid

Table 6.11 | CMIP5 model descriptions in terms of carbon cycle attributes and processes.

Model	Modelling Centre	Atmos Resolution	Ocean Resolution	Land-Carbon						Ocean Carbon			Reference
				Model Name	Dynamic Vegetation Cover?	No. of PFTs	Incl. LUC?	Nitrogen-Cycle	Fire	Model Name	No. of Plankton Types	Micronutrients?	
BCC-CSM1.1	BCC	≈2.8°, L26	0.3–1°, L40	BCC_AVIM1.0	N	15		N	N	OCMIP2	n/a	n/a	Wu et al. (2013)
CanESM2	CCCma	T63, L35	1.41° × 0.94°, L40	CTEM	N	9	Y	N	N	CMOC	1	N	Arora et al. (2011)
CESM1-BGC	NSF-DOE-NCAR	FV 0.9 × 1.25	1°	CLM4	N	15	Y	Y	Y	BEC	4	Y	Long et al. (2013)
GFDL-ESM2G	NOAA GFDL	2 × 2.5°, L24	1°, tri-polar, 1/3° at equator, L63.	LM3	Y	5	Y	N	Y	TOPAZ2	6	y	Dunne et al. (2012); Dunne et al. (2013)
GFDL-ESM2M	NOAA GFDL	2 × 2.5°, L24	1°, tri-polar, 1/3° at equator, L50.	LM3	Y	5	Y	N	Y	TOPAZ2	6	y	Dunne et al. (2012); Dunne et al. (2013)
HadGEM2-ES	MOHC	N96 (~ 1.6°), L38	1°, 1/3 ° at equator, L40	JULES	Y	5	Y	N	N	Diat-HadOCC	3	Y	Collins et al. (2011); Jones et al. (2011)
INMCM4	INM												
IPSL-CM5A-LR	IPSL	3.75 × 1.9 , L39	Zonal 2°, Meridional 2°–0.5° L31	ORCHIDEE	N	13	Y	N	Y	PISCES	2	Y	Dufresne et al. (2013)
MIROC-ESM	MIROC	T42, L80	Zonal: 1.4°, Meridional: 0.5–1.7°, Vertical: L43+BBL1	SEIB-DGVM	Y	13	Y	N	N	NPZD (Oschlies, 2001)	2 (Phyto-plankton and Zoolo-plankton)	N	Watanabe et al. (2011)
MPI-ESM-LR	MPI-M	T63 (~ 1.9°), L47	ca.1.5°, L47	JSBACH	Y	12 (8 natural)	Y	N	Y	HAMOCC	2	Y	Raddatz et al. (2007), Brovkin et al. (2009), Maier-Reimer et al. (2005)
NorESM-ME	NCC	1.9 × 2.5°, L26	1°, L53	CLM4	N	16	Y	Y	Y	HAMOCC	2	N	Iversen et al. (2013)

6

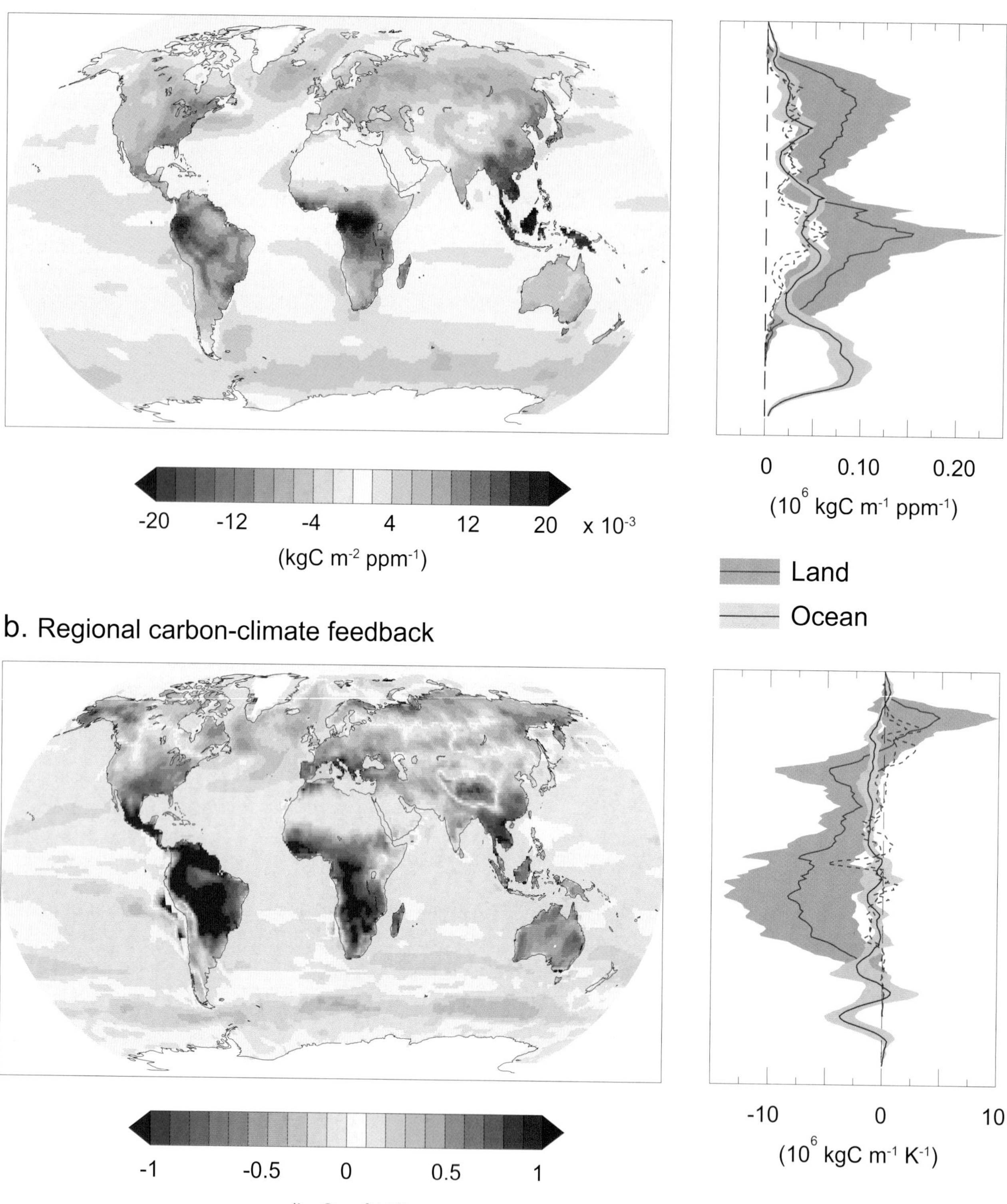

6

Figure 6.22 | The spatial distributions of multi-model-mean land and ocean β and γ for seven CMIP5 models using the *concentration-driven* idealised 1% yr^{-1} CO_2 simulations. For land and ocean, β and γ are defined from changes in terrestrial carbon storage and changes in air–sea integrated fluxes respectively, from 1 × CO_2 to 4 × CO_2, relative to global (not local) CO_2 and temperature change. In the zonal mean plots, the solid lines show the multi-model mean and shaded areas denote ±1 standard deviation. Models used: Beijing Climate Center–Climate System Model 1 (BCC–CSM1), Canadian Earth System Model 2 (CanESM2), Community Earth System Model 1–Biogeochemical (CESM1–BGC), Hadley Centre Global Environmental Model 2–Earth System (HadGEM2–ES), Institute Pierre Simon Laplace– Coupled Model 5A–Low Resolution (IPSL–CM5A-LR), Max Planck Institute–Earth System Model–Low Resolution (MPI–ESM–LR), Norwegian Earth System Model 1 (Emissions capable) (NorESM1–ME). The dashed lines show the models that include a land carbon component with an explicit representation of nitrogen cycle processes (CESM1-BGC, NorESM1-ME).

regions, associated with enhanced carbon uptake in forested areas of already high biomass. In the zonal totals, there is a secondary peak of high β_L values over NH temperate and boreal ecosystems, partly due to a greater land area there but also coincident with large areas of forest. Models agree on the sign of response but have low agreement on the magnitude.

The climate effect alone is projected by the CMIP5 models to reduce land CO_2 sinks in tropics and mid latitudes (negative γ) (*medium confidence*). CMIP5 models show medium agreement that warming may increase land carbon uptake in high latitudes but none of these models include representation of permafrost carbon pools which are projected to decrease in warmer conditions (Section 6.4.3.3); therefore *confidence* is *low* regarding the sign and magnitude of future high-latitude land carbon response to climate change. Matthews et al. (2005) showed that vegetation productivity is the major cause of C4MIP model spread, but this manifests itself as changes in soil organic matter (Jones and Falloon, 2009).

6.4.3 Implications of the Future Projections for the Carbon Cycle and Compatible Emissions

6.4.3.1 The RCP Future Carbon Dioxide Concentration and Emissions Scenarios

The CMIP5 simulations include four future scenarios referred to as Representative Concentration Pathways (RCPs; see Glossary) (Moss et al., 2010): RCP2.6, RCP4.5, RCP6.0, RCP8.5 (see Chapter 1). These future scenarios include CO_2 concentration and emissions, and have been generated by four Integrated Assessment Models (IAMs) and are labelled according to the approximate global radiative forcing level at 2100. These scenarios are described in more detail in Chapter 1 (Box 1.1) and Section 12.3 and also documented in Annex II.

van Vuuren et al. (2011) showed that the basic climate and carbon cycle responses of IAMs is generally consistent with the spread of climate and carbon cycle responses from ESMs. For the physical and biogeochemical components of the RCP scenarios 4.5, 6.0 and 8.5, the underlying IAMs are closely related. Only the Integrated Model to Assess the Global Environment (IMAGE) IAM, which created RCP2.6, differs markedly by using a more sophisticated carbon cycle sub-model for land and ocean. The Model for the Assessment of Greenhouse-gas Induced Climate Change 6 (MAGICC6) simple climate model was subsequently used to generate the CO_2 pathway for all four RCP scenarios using the CO_2 emissions output by the four IAMs (Meinshausen et al., 2011).

6.4.3.2 Land Use Changes in Future Scenarios

ESMs and IAMs use a diversity of approaches for representing land use changes, including different land use classifications, parameter settings, and geographical scales. To implement land use change in a consistent manner across ESMs, a 'harmonized' set of annual gridded land use change during the period 1500–2100 was developed for input to the CMIP5 ESMs (Hurtt et al., 2011).

Not all the CMIP5 ESMs used the full range of information available from the land use change scenarios, such as wood harvest projections or sub-grid scale shifting cultivation. Sensitivity studies indicated that these processes, along with the start date of the simulation, all strongly affect estimated carbon fluxes (Hurtt et al., 2011; Sentman et al., 2011).

Land use has been in the past and will be in the future a significant driver of forest land cover change and terrestrial carbon storage. Land use trajectories in the RCPs show very distinct trends and cover a wide range of projections. These land use trajectories are very sensitive to assumptions made by each individual IAM regarding the amount of land needed for food production (Figure 6.23). The area of cropland and pasture increases in RCP8.5 with the Model for Energy Supply Strategy Alternatives and their General Environmental Impact (MESSAGE) IAM model, mostly driven by an increasing global population, but cropland area also increases in the RCP2.6 with the IMAGE IAM model, as a result of bio-energy production and increased food demand as well. RCP6 with the AIM model shows an expansion of cropland but a decline in pasture land. RCP4.5 with the Global Change Assessment Model (GCAM) IAM is the only scenario to show a decrease in global cropland. Several studies (Wise et al., 2009; Thomson et al., 2010; Tilman et al., 2011) highlight the large sensitivity of future land use requirements to assumptions such as increases in crop yield, changes in diet, or how agricultural technology and intensification is applied.

Within the IAMs, land use change is translated into land use CO_2 emissions as shown in Figure 6.23(b). Cumulative emissions for the 21st century (Figure 6.23c) vary markedly across RCPs, with increasing cropland and pastureland areas in RCP2.6 and RCP8.5 giving rise to the highest emissions from land use change, RCP4.5 to intermediate emissions and RCP6.0 to close to zero net emissions. All scenarios suggest that 21st century land use emissions will be less than half of those from 1850 to the present day as rate of change of land conversion stabilises in future.

The adoption of widely differing approaches among ESMs for the treatment and diagnosis of land use and land cover change (LULCC) processes in terrestrial carbon cycle models leads to substantial between-model variation in the simulated impact on land carbon stocks. It is not yet possible to fully quantify LULCC fluxes from the CMIP5 model simulations. The harmonization process applied to LULCC data sets for CMIP5 has been an important step toward consistency among IAMs; however, among ESMs, and between IAMs and ESMs, assignment of meaningful uncertainty ranges to present-day and future LULCC fluxes and states remains a critical knowledge gap with implications for compatible emissions to achieve CO_2 pathways (Section 6.4.3.3; Jones et al., 2013).

6.4.3.3 Projections of Future Carbon Cycle Response by Earth System Models Under the Representative Concentration Pathway Scenarios

Simulated changes in land and ocean carbon uptake and storage under the four RCP scenarios are presented here using results from CMIP5 ESMs concentration-driven simulations (see Box 6.4). The implications of these changes on atmospheric CO_2 and climate as simulated by CMIP5 emissions-driven simulations are presented in Chapter 12.

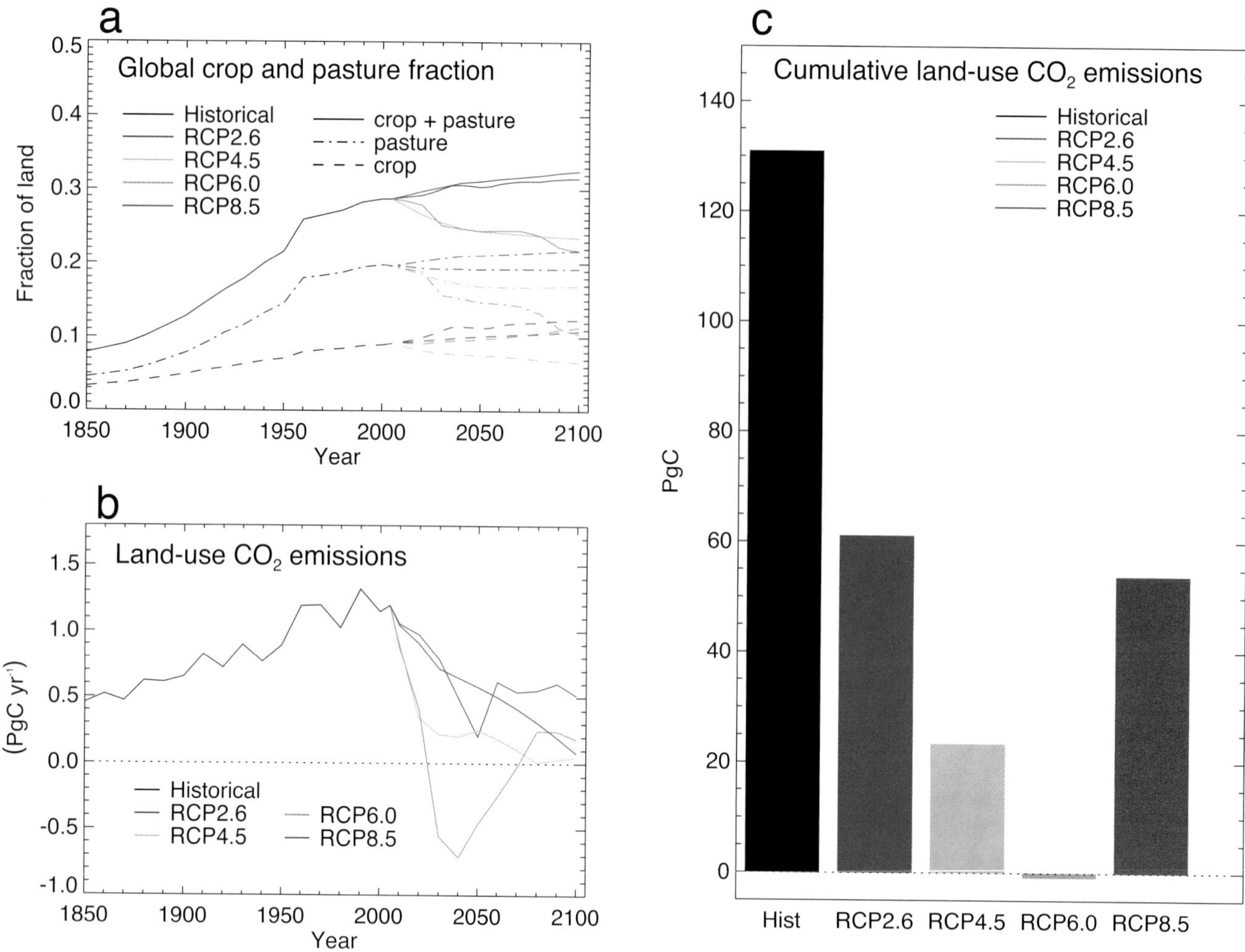

Figure 6.23 | Land use trends and CO_2 emissions according to the four different integrated assessment models (IAMs) used to define the RCP scenarios. Global changes in croplands and pasture from the historical record and the RCP scenarios (top left), and associated annual land use emissions of CO_2 (bottom left). Bars (right panel) show cumulative land use emissions for the historical period (defined here as 1850–2005) and the four RCP scenarios from 2006 to 2100.

The results of the *concentration-driven* CMIP5 ESMs simulations show medium agreement on the magnitude of cumulative ocean carbon uptake from 1850 to 2005 (Figure 6.24a): average 127 ± 28 PgC (1 standard deviation). The models show low agreement on the sign and magnitude of changes in land carbon storage (Figure 6.24a): average 2 ± 74 PgC (1 standard deviation). These central estimates are very close to observational estimates of 125 ± 25 PgC for the ocean and –5 ± 40 PgC for the net cumulative land–atmosphere flux respectively (see Table 6.12), but show a large spread across models. With *very high confidence*, for all four RCP scenarios, all models project continued ocean uptake throughout the 21st century, with higher uptake corresponding to higher concentration pathways. For RCP4.5, all the models also project an increase in land carbon uptake, but for RCP2.6, RCP6.0 and RCP8.5 a minority of models (4 out of 11 for RCP2.6, 1 out of 8 for RCP6.0 and 4 out of 15 for RCP8.5; Jones et al., 2013) project a decrease in land carbon storage at 2100 relative to 2005. Model spread in land carbon projections is much greater than model spread in ocean carbon projections, at least in part due to different treatment of land use change. Decade mean land and ocean fluxes are documented in Annex II, Table AII.3.1a, b. Important processes missing from many or all CMIP5 land carbon cycles include the role of nutrient cycles, permafrost, fire and ecosystem acclimation to changing climate. For this reason we assign *low confidence* to quantitative projections of future land uptake.

The *concentration-driven* ESM simulations can be used to quantify the compatible fossil fuel emissions required to follow the four RCP CO_2 pathways (Jones et al., 2013; see Box 6.4, Figure 6.25, Table 6.12, Annex II, Table AII.2.1a). There is significant spread between ESMs, but general consistency between ESMs and compatible emissions estimated by IAMs to define each RCP scenario. However, for RCP8.5 on average, the CMIP5 models project lower compatible emissions than the MESSAGE IAM. The IMAGE IAM predicts that global negative emissions are required to achieve the RCP2.6 decline in radiative forcing from 3 W m^{-2} to 2.6 W m^{-2} by 2100. All models agree that strong emissions reductions are required to achieve this after about 2020 (Jones et al., 2013). An average emission reduction of 50% (range 14 to 96%) is required by 2050 relative to 1990 levels. There is disagreement between those ESMs that performed this simulation over the necessity for global emissions in the RCP2.6 to become negative by

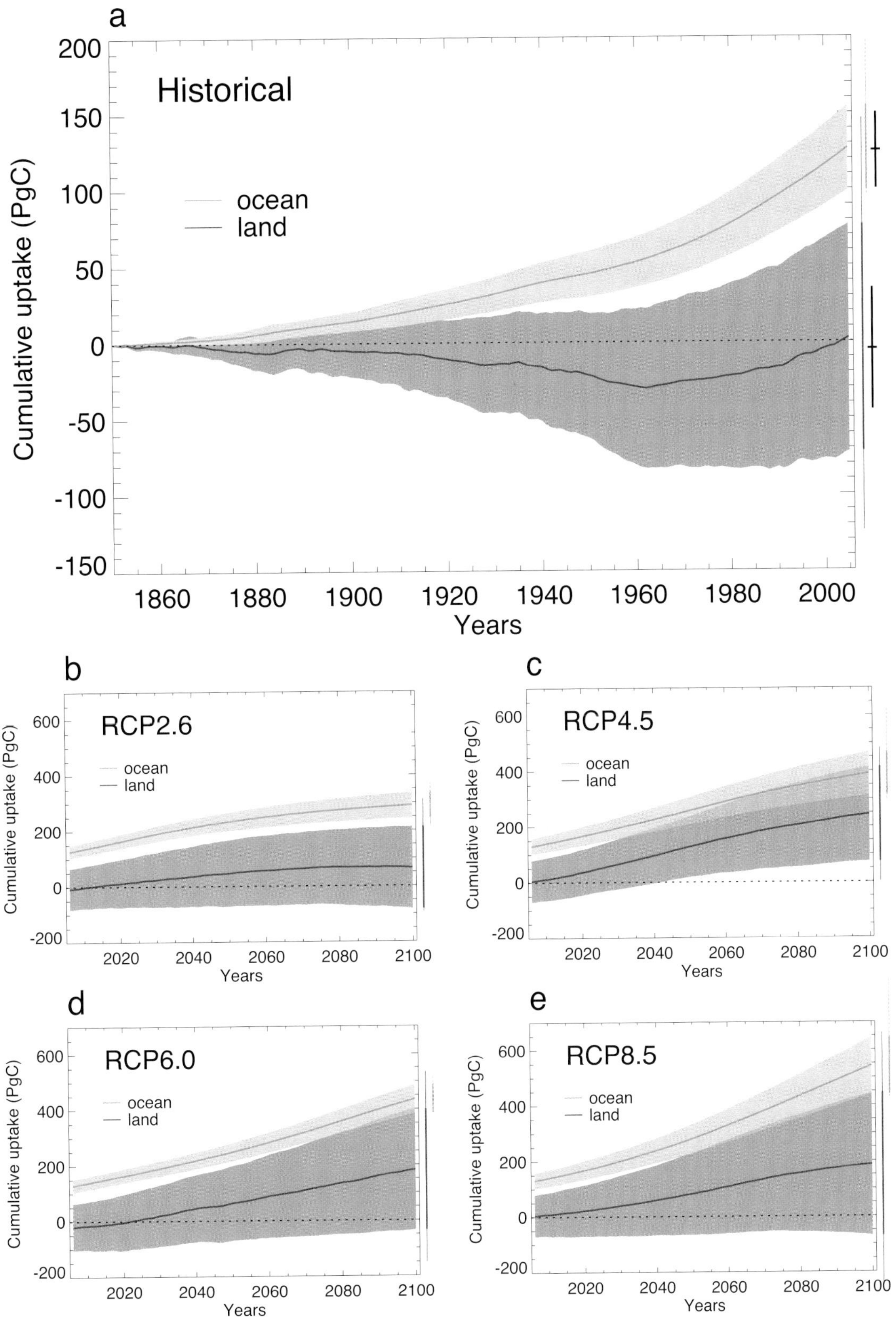

Figure 6.24 | Cumulative land and ocean carbon uptake simulated for the historical period 1850–2005 (top) and for the four RCP scenarios up to 2100 (b–e). Mean (thick line) and 1 standard deviation (shaded). Vertical bars on the right show the full model range as well as standard deviation. Black bars show observationally derived estimates for 2005. Models used: Canadian Earth System Model 2 (CanESM2), Geophysical Fluid Dynamics Laboratory–Earth System Model 2G (GFDL–ESM2G), Geophysical Fluid Dynamics Laboratory–Earth System Model 2M (GFDL–ESM2M), Hadley Centre Global Environmental Model 2–Carbon Cycle (HadGEM2-CC), Hadley Centre Global Environmental Model 2–Earth System (HadGEM2-ES), Institute Pierre Simon Laplace–Coupled Model 5A–Low Resolution (IPSL–CM5A–LR), Institute Pierre Simon Laplace–Coupled Model 5A–Medium Resolution (IPSL–CM5A–MR), Institute Pierre Simon Laplace–Coupled Model 5B–Low Resolution (IPSL–CM5B–LR), Model for Interdisciplinary Research On Climate–Earth System Model (MIROC–ESM–CHEM), Model for Interdisciplinary Research On Climate–Earth System Model (MIROC–ESM), Max Planck Institute–Earth System Model–Low Resolution (MPI–ESM–LR), Norwegian Earth System Model 1 (Emissions capable) (NorESM1–ME), Institute for Numerical Mathematics Coupled Model 4 (INMCM4), Community Earth System Model 1–Biogeochemical (CESM1–BGC), Beijing Climate Center–Climate System Model 1.1 (BCC–CSM1.1). Not every model performed every scenario simulation.

Table 6.12 | The range of compatible fossil fuel emissions (PgC) simulated by the CMIP5 models for the historical period and the four RCP scenarios, expressed as cumulative fossil fuel emission. To be consistent with Table 6.1 budgets are calculated up to 2011 for historical and 2012–2100 for future scenarios, and values are rounded to the nearest 5 PgC.

	Compatible Fossil Fuel Emissions Diagnosed from *Concentration-Driven* CMIP5 Simulations			**Land Carbon Changes**			**Ocean Carbon Changes**		
	Historical / RCP Scenario	CMIP5 ESM Mean	CMIP5 ESM Range	Historical / RCP Scenario	CMIP5 ESM Mean	CMIP5 ESM Range	Historical / RCP Scenario	CMIP5 ESM Mean	CMIP5 ESM Range
1850–2011	375[a]	350	235–455	5 ± 40[b]	10	–125 to 160	140 ± 25[b]	140	110–220
RCP2.6	275	270	140–410	[c]	65	–50 to 195	[c]	150	105–185
RCP4.5	735	780	595–1005		230	55 to 450		250	185–400
RCP6.0	1165	1060	840–1250		200	–80 to 370		295	265–335
RCP8.5	1855	1685	1415–1910		180	–165 to 500		400	320–635

Notes:

[a] Historical estimates of fossil fuel are as prescribed to all CMIP5 ESMs in the emissions-driven simulations (Andres et al., 2011).

[b] Estimate of historical net land and ocean carbon uptake from Table 6.1 but over the shorter 1850–2011 time period.

[c] IAM breakdown of future carbon changes by land and ocean are not available.

the end of the 21st century to achieve this, with six ESMs simulating negative compatible emissions and four ESM models simulating positive emissions from 2080 to 2100. The RCP2.6 scenario achieves this negative emission rate through use of large-scale bio-energy with carbon-capture and storage (BECCS). It is *as likely as not* that sustained globally negative emissions will be required to achieve the reductions in atmospheric CO_2 in the RCP2.6 scenario. This would be classed as a carbon dioxide removal (CDR) form of geoengineering under the definition used in this IPCC report, and is discussed further in Section 6.5.2. The ESMs themselves make no assumptions about how the compatible emissions could or would be achieved, but merely compute the global total emission that is required to follow the CO_2 concentration pathway, accounting for the carbon cycle response to climate and CO_2, and for land use change CO_2 emissions.

The dominant cause of future changes in the airborne fraction of fossil fuel emissions (see Section 6.3.2.4) is the emissions scenario and not carbon cycle feedbacks (Jones et al., 2013; Figure 6.26). Models show high agreement that 21st century cumulative airborne fraction will increase under rapidly increasing CO_2 in RCP8.5 and decreases under the peak-and-decline RCP2.6 scenarios. The airborne fraction declines slightly under RCP4.5 and remains of similar magnitude in the RCP6.0 scenario. Between-model spread in changes in the land-fraction is greater than between-scenario spread. Models show high agreement that the ocean fraction will increase under RCP2.6 and remain of similar magnitude in the other RCP scenarios.

Several studies (Jones et al., 2006; Matthews, 2006; Plattner et al., 2008; Miyama and Kawamiya, 2009) have shown that climate–carbon cycle feedbacks affect the compatible fossil fuel CO_2 emissions that are consistent with a given CO_2 concentration pathway. Using decoupled RCP4.5 simulations (see Box 6.4) five CMIP5 ESMs agree that the climate impact on carbon uptake by both land and oceans will reduce the compatible fossil fuel CO_2 emissions for that scenario by between 6% and 29% between 2006 and 2100 respectively (Figure 6.27), equating to an average of 157 ± 76 PgC (1 standard deviation) less carbon that can be emitted from fossil fuel use if climate feedback (see Glossary) is included. Compatible emissions would be reduced by a greater degree under higher CO_2 scenarios that exhibit a greater degree of climate change (Jones et al., 2006).

6.4.3.4 Permafrost Carbon

Current estimates of permafrost soil carbon stocks are ~1700 PgC (Tarnocai et al., 2009), the single largest component of the terrestrial carbon pool. Terrestrial carbon models project a land CO_2 sink with warming at high northern latitudes; however none of the models participating in C4MIP or CMIP5 included explicit representation of permafrost soil carbon decomposition in response to future warming. Including permafrost carbon processes into an ESM may change the sign of the high northern latitude carbon cycle response to warming from a sink to a source (Koven et al., 2011). Overall, there is *high confidence* that reductions in permafrost extent due to warming will cause thawing of some currently frozen carbon. However, there is *low confidence* on the magnitude of carbon losses through CO_2 and CH_4 emissions to the atmosphere. The magnitude of CO_2 and CH_4 emissions to the atmosphere is assessed to range from 50 to 250 PgC between 2000 and 2100 for RCP8.5. The magnitude of the source of CO_2 to the atmosphere from decomposition of permafrost carbon in response to warming varies widely according to different techniques and scenarios. Process models provide different estimates of the cumulative loss of permafrost carbon: 7 to 17 PgC (Zhuang et al., 2006) (not considered in the range given above because it corresponds only to contemporary tundra soil carbon), 55 to 69 Pg (Koven et al., 2011), 126 to 254 PgC (Schaefer et al., 2011) and 68 to 508 PgC (MacDougall et al., 2012) (not considered in the range given above because this estimate is not obtained from a concentration driven, but for emission driven RCP scenario and it is the only study of that type so far). Combining observed vertical soil carbon profiles with modeled thaw rates provides an estimate that the total quantity of newly thawed soil carbon by 2100 will be 246 PgC for RCP4.5 and 436 PgC for RCP8.5 (Harden et al., 2012), although not all of this amount will be released to the atmosphere on that time scale. Uncertainty estimates suggest the cumulative amount of thawed permafrost carbon could range from 33 to 114 PgC (68% range) under RCP8.5 warming (Schneider von Deimling et al., 2012), or 50 to 270 PgC (5th to 95th percentile range) (Burke et al., 2013).

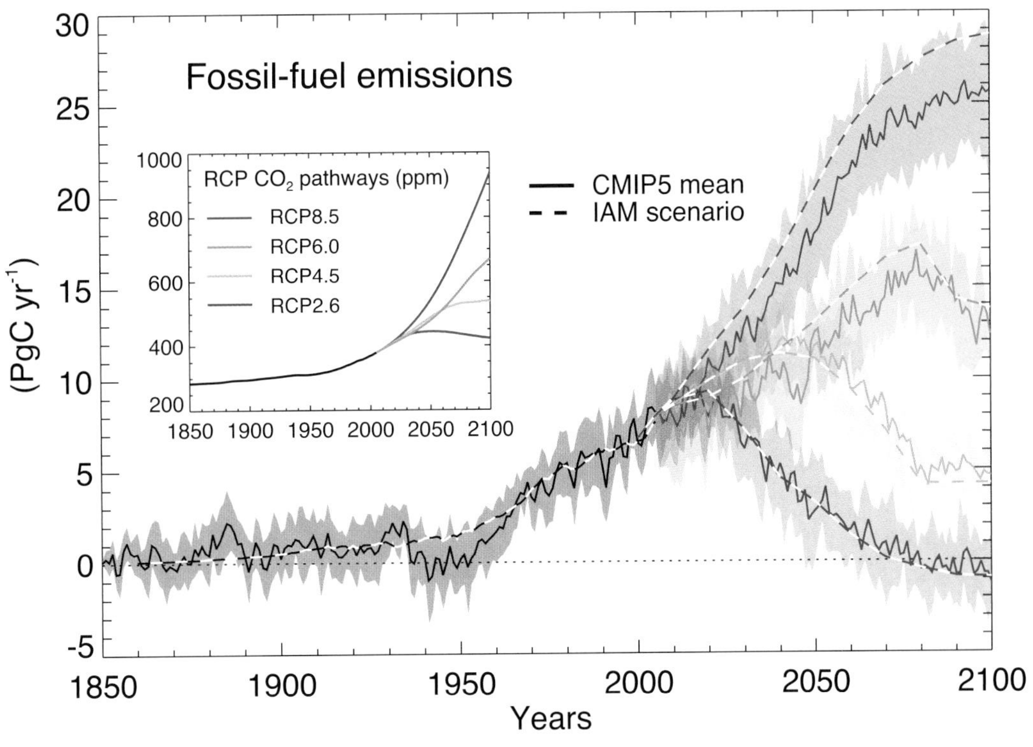

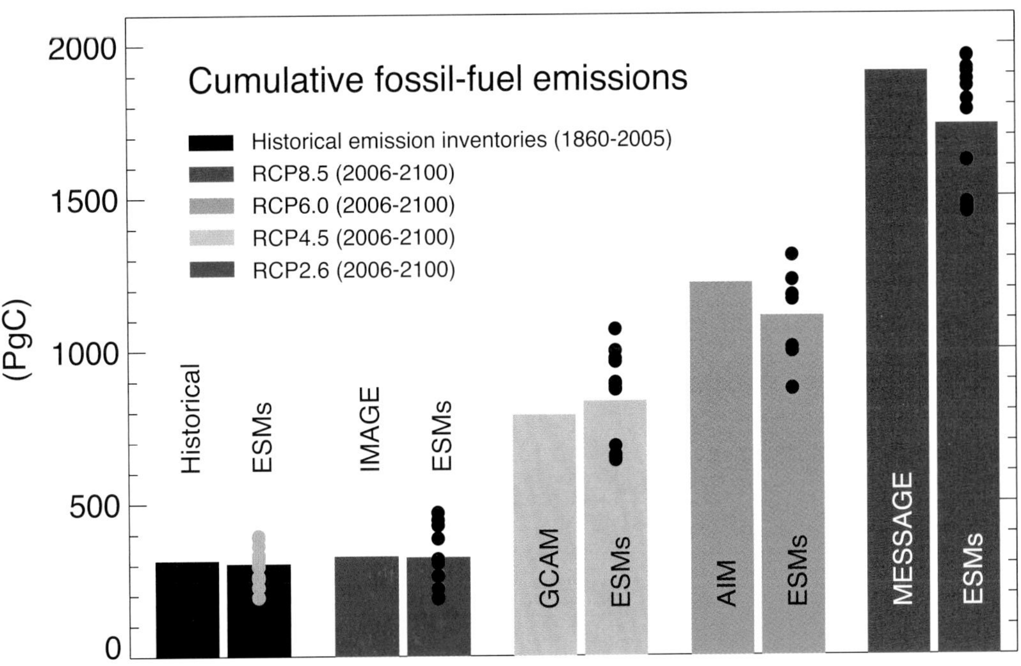

Figure 6.25 | Compatible fossil fuel emissions simulated by the CMIP5 ESMs for the four RCP scenarios. Top: time series of compatible emission rate (PgC yr^{-1}). Dashed lines represent the historical estimates and emissions calculated by the Integrated Assessment Models (IAMs) used to define the RCP scenarios, solid lines and plumes show results from CMIP5 ESMs (model mean, with 1 standard deviation shaded). Bottom: cumulative emissions for the historical period (1860–2005) and 21st century (defined in CMIP5 as 2006–2100) for historical estimates and RCP scenarios. Dots denote individual ESM results, bars show the multi-model mean. In the CMIP5 model results, total carbon in the land–atmosphere–ocean system can be tracked and changes in this total must equal fossil fuel emissions to the system (see Box 6.4). Models used: Canadian Earth System Model 2 (CanESM2), Geophysical Fluid Dynamics Laboratory–Earth System Model 2G (GFDL–ESM2G), Geophysical Fluid Dynamics Laboratory–Earth System Model 2M (GFDL–ESM2M), Hadley Centre Global Environmental Model 2–Carbon Cycle(HadGEM2-CC), Hadley Centre Global Environmental Model 2–Earth System (HadGEM2-ES), Institute Pierre Simon Laplace–Coupled Model 5A–Low Resolution (IPSL–CM5A–LR), Institute Pierre Simon Laplace–Coupled Model 5A–Medium Resolution (IPSL–CM5A–MR), Institute Pierre Simon Laplace–Coupled Model 5B–Low Resolution (IPSL–CM5B–LR), Model for Interdisciplinary Research On Climate–Earth System Model (MIROC–ESM–CHEM), Model for Interdisciplinary Research On Climate–Earth System Model (MIROC–ESM), Max Planck Institute–Earth System Model–Low Resolution (MPI–ESM–LR), Norwegian Earth System Model 1 (Emissions capable) (NorESM1–ME), Institute for Numerical Mathematics Coupled Model 4 (INMCM4), Community Earth System Model 1–Biogeochemical (CESM1–BGC), Beijing Climate Center–Climate System Model 1.1 (BCC–CSM1.1). Not every model performed every scenario simulation.

Sources of uncertainty for the permafrost carbon feedback include the physical thawing rates, the fraction of carbon that is released after being thawed and the time scales of release, possible mitigating nutrient feedbacks and the role of fine-scale processes such as spatial variability in permafrost degradation. It is also uncertain how much thawed carbon will decompose to CO_2 or to CH_4 (see Sections 6.4.7, 12.5.5.4 and 12.4.8.1).

6.4.4 Future Ocean Acidification

A fraction of CO_2 emitted to the atmosphere dissolves in the ocean, reducing surface ocean pH and carbonate ion concentrations. The associated chemistry response to a given change in CO_2 concentration is known with *very high confidence*. Overall, given evidence from Chapter 3 and model results from this chapter, it is *virtually certain* that the increased storage of carbon by the ocean will increase acidification in the future, continuing the observed trends of the past decades. Expected future changes are in line with what is measured at ocean time series stations (see Chapter 3). Multi-model projections using ocean process-based carbon cycle models discussed in AR4 demonstrate large decreases in pH and carbonate ion concentration $[CO_3^{2-}]$ during the 21st century throughout the world oceans (Orr et al., 2005). The largest decrease in surface $[CO_3^{2-}]$ occur in the warmer low and mid-latitudes, which are naturally rich in this ion (Feely et al., 2009). However, it is the low Ω_A waters in the high latitudes and in the upwelling regions that first become undersaturated with respect to aragonite (i.e., $\Omega_A < 1$,

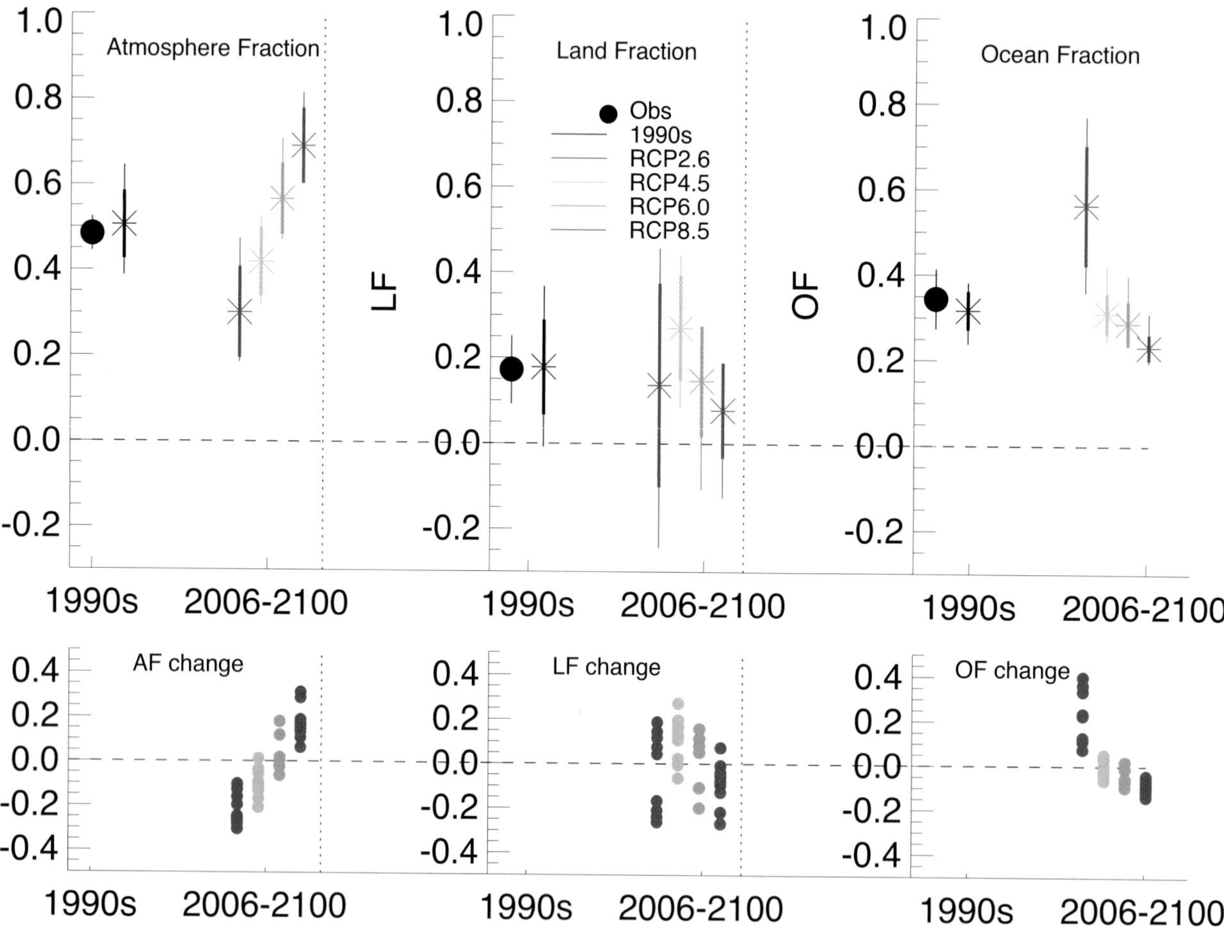

Figure 6.26 | Changes in atmospheric, land and ocean fraction of fossil fuel carbon emissions. The fractions are defined as the changes in storage in each component (atmosphere, land, ocean) divided by the compatible fossil fuel emissions derived from each CMIP5 simulation for the four RCP scenarios. Solid circles show the observed estimate based on Table 6.1 for the 1990s. The coloured bars denote the cumulative uptake fractions for the 21st century under the different RCP scenarios for each model. Multi-model mean values are shown as star symbols and the multi-model range (min-to-max) and standard deviation are shown by thin and thick vertical lines respectively. Owing to the difficulty of estimating land use emissions from the ESMs this figure uses a fossil fuel definition of airborne fraction, rather than the preferred definition of fossil and land use emissions discussed in Section 6.3.2.4. 21st century cumulative atmosphere, land and ocean fractions are shown here in preference to the more commonly shown instantaneous fractions because for RCP2.6 emissions reach and cross zero for some models and so an instantaneous definition of AF becomes singular at that point. Models used: Canadian Earth System Model 2 (CanESM2), Geophysical Fluid Dynamics Laboratory–Earth System Model 2G (GFDL–ESM2G), Geophysical Fluid Dynamics Laboratory–Earth System Model 2M (GFDL–ESM2M), Hadley Centre Global Environmental Model 2–Carbon Cycle (HadGEM2-CC), Hadley Centre Global Environmental Model 2–Earth System (HadGEM2-ES), Institute Pierre Simon Laplace–Coupled Model 5A–Low Resolution (IPSL–CM5A–LR), Institute Pierre Simon Laplace–Coupled Model 5A–Medium Resolution (IPSL–CM5A–MR), Institute Pierre Simon Laplace–Coupled Model 5B–Low Resolution (IPSL–CM5B–LR), Model for Interdisciplinary Research On Climate–Earth System Model (MIROC–ESM), Model for Interdisciplinary Research On Climate–Earth System Model (MIROC–ESM–CHEM), Max Planck Institute–Earth System Model–Low Resolution (MPI–ESM–LR), Norwegian Earth System Model 1 (Emissions capable) (NorESM1–ME), Institute for Numerical Mathematics Coupled Model 4 (INMCM4), Community Earth System Model 1–Biogeochemical (CESM1–BGC). Not every model performed every scenario simulation.

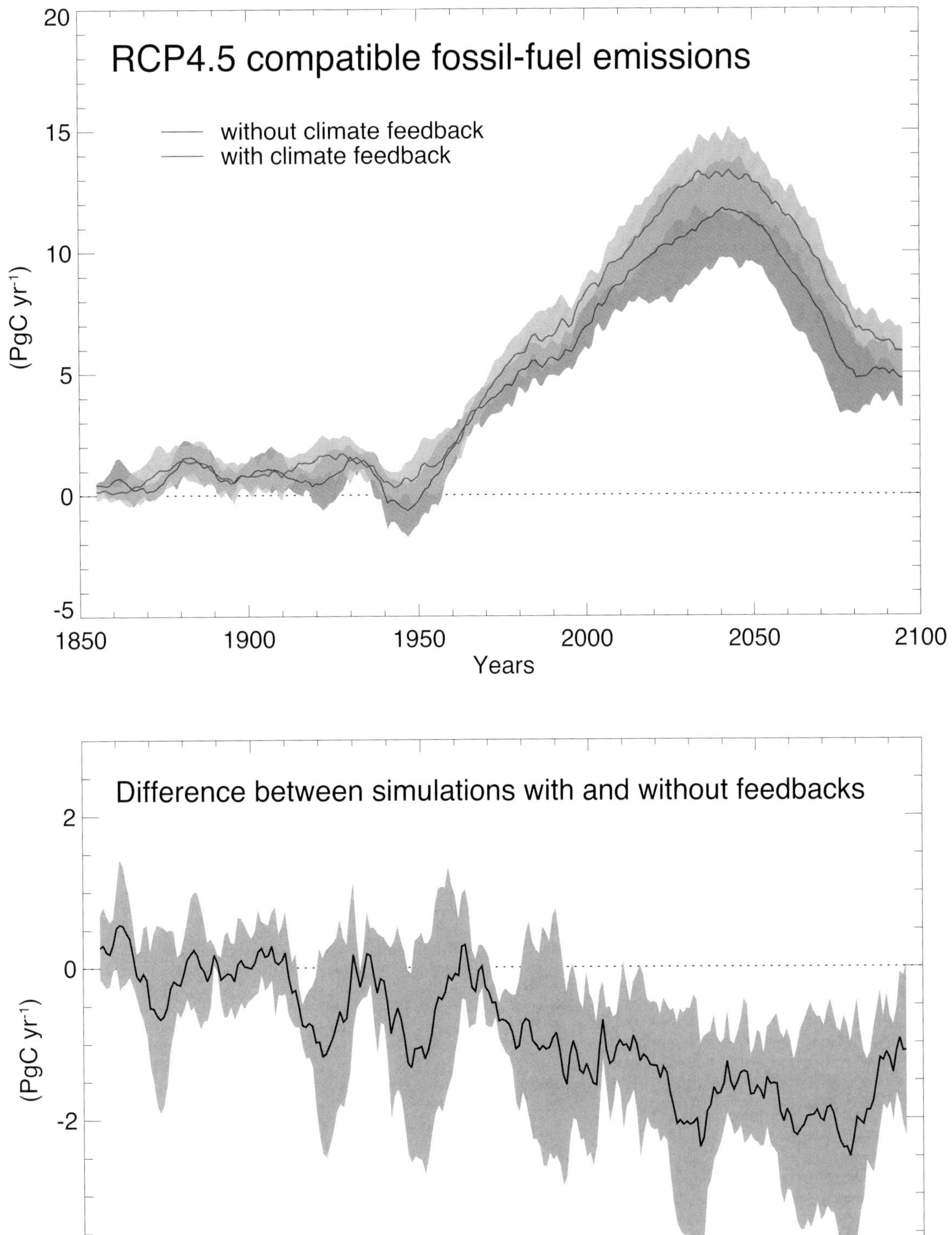

Figure 6.27 | Compatible fossil fuel emissions for the RCP4.5 scenario (top) in the presence (red) and absence (blue) of the climate feedback on the carbon cycle, and the difference between them (bottom). Multi-model mean, 10-year smoothed values are shown, with 1 standard deviation shaded. This shows the impact of climate change on the compatible fossil fuel CO_2 emissions to achieve the RCP4.5 CO_2 concentration pathway. Models used: Canadian Earth System Model 2 (CanESM2), Geophysical Fluid Dynamics Laboratory–Earth System Model 2M (GFDL-ESM2M), Hadley Centre Global Environmental Model 2–Earth System (HadGEM2-ES), Institute Pierre Simon Laplace–Coupled Model 5A–Low Resolution (IPSL-CM5A-LR) and Model for Interdisciplinary Research On Climate–Earth System Model (MIROC–ESM).

Frequently Asked Questions

FAQ 6.1 | Could Rapid Release of Methane and Carbon Dioxide from Thawing Permafrost or Ocean Warming Substantially Increase Warming?

Permafrost is permanently frozen ground, mainly found in the high latitudes of the Arctic. Permafrost, including the sub-sea permafrost on the shallow shelves of the Arctic Ocean, contains old organic carbon deposits. Some are relicts from the last glaciation, and hold at least twice the amount of carbon currently present in the atmosphere as carbon dioxide (CO_2). Should a sizeable fraction of this carbon be released as methane and CO_2, it would increase atmospheric concentrations, which would lead to higher atmospheric temperatures. That in turn would cause yet more methane and CO_2 to be released, creating a positive feedback, which would further amplify global warming.

The Arctic domain presently represents a net sink of CO_2—sequestering around 0.4 ± 0.4 PgC yr^{-1} in growing vegetation representing about 10% of the current global land sink. It is also a modest source of methane (CH_4): between 15 and 50 Tg(CH_4) yr^{-1} are emitted mostly from seasonally unfrozen wetlands corresponding to about 10% of the global wetland methane source. There is no clear evidence yet that thawing contributes significantly to the current global budgets of these two greenhouse gases. However, under sustained Arctic warming, modelling studies and expert judgments indicate with medium agreement that a potential combined release totalling up to 350 PgC as CO_2 equivalent could occur by the year 2100.

Permafrost soils on land, and in ocean shelves, contain large pools of organic carbon, which must be thawed and decomposed by microbes before it can be released—mostly as CO_2. Where oxygen is limited, as in waterlogged soils, some microbes also produce methane.

On land, permafrost is overlain by a surface 'active layer', which thaws during summer and forms part of the tundra ecosystem. If spring and summer temperatures become warmer on average, the active layer will thicken, making more organic carbon available for microbial decomposition. However, warmer summers would also result in greater uptake of carbon dioxide by Arctic vegetation through photosynthesis. That means the net Arctic carbon balance is a delicate one between enhanced uptake and enhanced release of carbon.

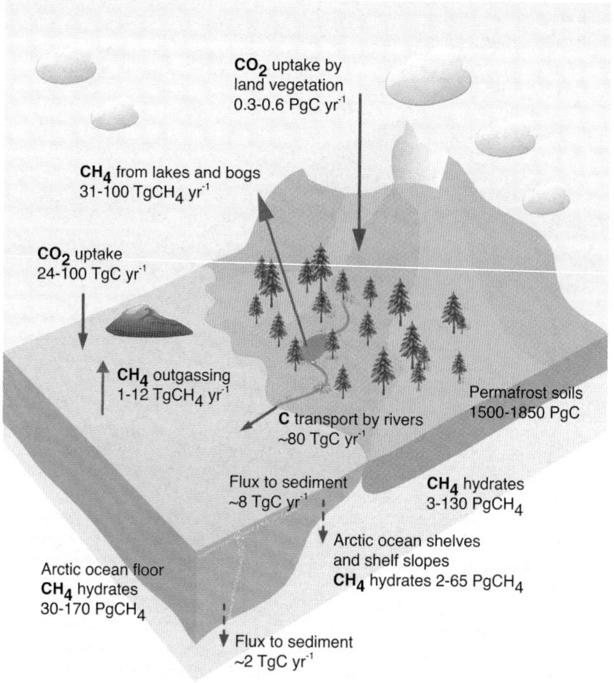

FAQ 6.1, Figure 1 | A simplified graph of current major carbon pools and flows in the Arctic domain, including permafrost on land, continental shelves and ocean. (Adapted from McGuire et al., 2009; and Tarnocai et al., 2009.) TgC = 10^{12} gC, and PgC = 10^{15} gC.

Hydrological conditions during the summer thaw are also important. The melting of bodies of excess ground ice may create standing water conditions in pools and lakes, where lack of oxygen will induce methane production. The complexity of Arctic landscapes under climate warming means we have *low confidence* in which of these different processes might dominate on a regional scale. Heat diffusion and permafrost melting takes time—in fact, the deeper Arctic permafrost can be seen as a relict of the last glaciation, which is still slowly eroding—so any significant loss of permafrost soil carbon will happen over long time scales.

Given enough oxygen, decomposition of organic matter in soil is accompanied by the release of heat by microbes (similar to compost), which, during summer, might stimulate further permafrost thaw. Depending on carbon and ice content of the permafrost, and the hydrological regime, this mechanism could, under warming, trigger relatively fast local permafrost degradation. *(continued on next page)*

FAQ 6.1 (continued)

Modelling studies of permafrost dynamics and greenhouse gas emissions indicate a relatively slow positive feedback, on time scales of hundreds of years. Until the year 2100, up to 250 PgC could be released as CO_2, and up to 5 Pg as CH_4. Given methane's stronger greenhouse warming potential, that corresponds to a further 100 PgC of equivalent CO_2 released until the year 2100. These amounts are similar in magnitude to other biogeochemical feedbacks, for example, the additional CO_2 released by the global warming of terrestrial soils. However, current models do not include the full complexity of Arctic processes that occur when permafrost thaws, such as the formation of lakes and ponds.

Methane hydrates are another form of frozen carbon, occurring in deep permafrost soils, ocean shelves, shelf slopes and deeper ocean bottom sediments. They consist of methane and water molecule clusters, which are only stable in a specific window of low temperatures and high pressures. On land and in the ocean, most of these hydrates originate from marine or terrestrial biogenic carbon, decomposed in the absence of oxygen and trapped in an aquatic environment under suitable temperature–pressure conditions.

Any warming of permafrost soils, ocean waters and sediments and/or changes in pressure could destabilise those hydrates, releasing their CH_4 to the ocean. During larger, more sporadic releases, a fraction of that CH_4 might also be outgassed to the atmosphere. There is a large pool of these hydrates: in the Arctic alone, the amount of CH_4 stored as hydrates could be more than 10 times greater than the CH_4 presently in the global atmosphere.

Like permafrost thawing, liberating hydrates on land is a slow process, taking decades to centuries. The deeper ocean regions and bottom sediments will take still longer—between centuries and millennia to warm enough to destabilise the hydrates within them. Furthermore, methane released in deeper waters has to reach the surface and atmosphere before it can become climatically active, but most is expected to be consumed by microorganisms before it gets there. Only the CH_4 from hydrates in shallow shelves, such as in the Arctic Ocean north of Eastern Siberia, may actually reach the atmosphere to have a climate impact.

Several recent studies have documented locally significant CH_4 emissions over the Arctic Siberian shelf and from Siberian lakes. How much of this CH_4 originates from decomposing organic carbon or from destabilizing hydrates is not known. There is also no evidence available to determine whether these sources have been stimulated by recent regional warming, or whether they have always existed—it may be possible that these CH_4 seepages have been present since the last deglaciation. In any event, these sources make a very small contribution to the global CH_4 budget—less than 5%. This is also confirmed by atmospheric methane concentration observations, which do not show any substantial increases over the Arctic.

However modelling studies and expert judgment indicate that CH_4 and CO_2 emissions will increase under Arctic warming, and that they will provide a positive climate feedback. Over centuries, this feedback will be moderate: of a magnitude similar to other climate–terrestrial ecosystem feedbacks. Over millennia and longer, however, CO_2 and CH_4 releases from permafrost and shelves/shelf slopes are much more important, because of the large carbon and methane hydrate pools involved.

where $\Omega_A = [Ca^{+2}][CO_3^{2-}]/K_{sp}$, where K_{sp} is the solubility product for the metastable form of $CaCO_3$ known as aragonite; a value of $\Omega_A < 1$ thus indicates aragonite undersaturation). This aragonite undersaturation in surface waters is reached before the end of the 21st century in the Southern Ocean as highlighted in AR4, but occurs sooner and is more intense in the Arctic (Steinacher et al., 2009). Ten percent of Arctic surface waters are projected to become undersaturated when atmospheric CO_2 reaches 428 ppm (by 2025 under all IPCC SRES scenarios). That proportion increases to 50% when atmospheric CO_2 reaches 534 ppm (Steinacher et al., 2009). By 2100 under the A2 scenario, much of the Arctic surface is projected to become undersaturated with respect to calcite (Feely et al., 2009). Surface waters would then be corrosive to all $CaCO_3$ minerals. These general trends are confirmed by the latest projections from the CMIP5 Earth System models (Figure 6.28 and 6.29). Between 1986–2005 and 2081–2100, decrease in global-mean surface pH is 0.065 (0.06 to 0.07) for RCP2.6, 0.145 (0.14 to 0.15) for RCP4.5, 0.203 (0.20 to 0.21) for RCP6.0 and 0.31 (0.30 to 0.32) for RCP8.5 (range from CMIP5 models spread).

Surface $CaCO_3$ saturation also varies seasonally, particularly in the high latitudes, where observed saturation is higher in summer and lower in winter (Feely et al., 1988; Merico et al., 2006; Findlay et al.,

2008). Future projections using ocean carbon cycle models indicate that undersaturated conditions will be reached first in winter (Orr et al., 2005). In the Southern Ocean, it is projected that wintertime undersaturation with respect to aragonite will begin when atmospheric CO_2 will reach 450 ppm, within 1-3 decades, which is about 100 ppm sooner (~30 years under the IS92a scenario) than for the annual mean undersaturation (McNeil and Matear, 2008). As well, aragonite undersaturation will be first reached during wintertime in parts (10%) of the Arctic when atmospheric CO_2 will reach 410 ppm, within a decade (Steinacher et al., 2009). Then, aragonite undersaturation will become widespread in these regions at atmospheric CO_2 levels of 500–600 ppm (Figure 6.28).

Although projected changes in pH are generally largest at the surface, the greatest pH changes in the subtropics occur between 200 and 300 m where subsurface increased loads of anthropogenic CO_2 are similar to surface changes but the carbonate buffering capacity is lower (Orr, 2011). This more intense projected subsurface pH reduction is consistent with the observed subsurface changes in pH in the subtropical North Pacific (Dore et al., 2009; Byrne et al., 2010; Ishii et al., 2011). As subsurface saturation states decline, the horizon separating undersaturated waters below from supersaturated waters above is projected to move upward (shoal). By 2100 under the RCP8.5 scenario, the median projection from 11 CMIP5 models is that this interface (aragonite saturation horizon) will shoal from 200 m up to 40 m in the subarctic Pacific, from 1000 m up to the surface in the Southern Ocean, and from 2850 m to 150 m in the North Atlantic (Figure 6.29), consistent with results from previous model comparison (Orr et al., 2005; Orr, 2011). Under the SRES A2 scenario, the volume of ocean with supersaturated waters is projected to decline from 42% in the preindustrial Era to 25% in 2100 (Steinacher et al., 2009). Yet even if atmospheric CO_2 does not go above 450 ppm, most of the deep ocean volume is projected to become undersaturated with respect to both aragonite and calcite after several centuries (Caldeira and Wickett, 2005). Nonetheless, the most recent projections under all RCPs scenarios but RCP8.5 illustrate that limiting atmospheric CO_2 will greatly reduce the level of ocean acidification that will be experienced (Joos et al., 2011).

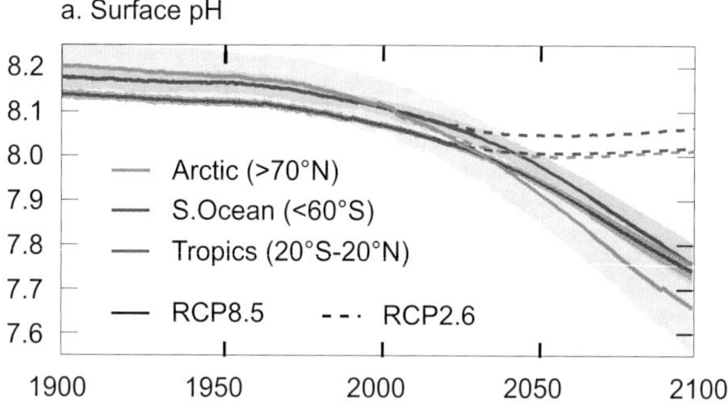

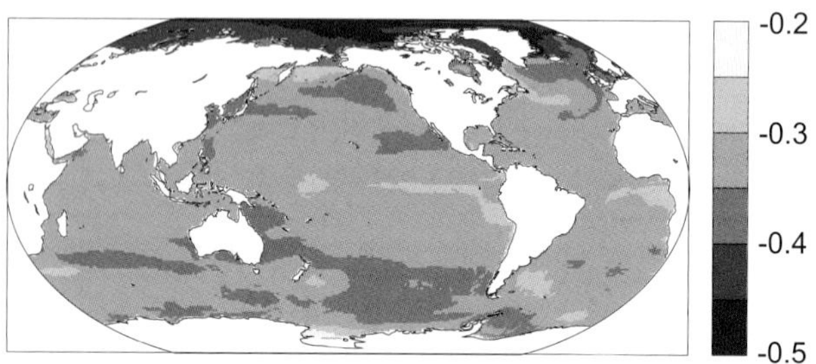

Figure 6.28 | Projected ocean acidification from 11 CMIP5 Earth System Models under RCP8.5 (other RCP scenarios have also been run with the CMIP5 models): (a) Time series of surface pH shown as the mean (solid line) and range of models (filled), given as area-weighted averages over the Arctic Ocean (green), the tropical oceans (red) and the Southern Ocean (blue). (b) Maps of the median model's change in surface pH from 1850 to 2100. Panel (a) also includes mean model results from RCP2.6 (dashed lines). Over most of the ocean, gridded data products of carbonate system variables (Key et al., 2004) are used to correct each model for its present-day bias by subtracting the model-data difference at each grid cell following (Orr et al., 2005). Where gridded data products are unavailable (Arctic Ocean, all marginal seas, and the ocean near Indonesia), the results are shown without bias correction. The bias correction reduces the range of model projections by up to a factor of 4, e.g., in panel (a) compare the large range of model projections for the Arctic (without bias correction) to the smaller range in the Southern Ocean (with bias correction).

In the open ocean, future reductions in surface ocean pH and $CaCO_3$ (calcite and aragonite) saturation states are controlled mostly by the invasion of anthropogenic carbon. Other effects due to future climate change counteract less than 10% of the reductions in $CaCO_3$ saturation induced by the invasion of anthropogenic carbon (Orr et al., 2005; McNeil and Matear, 2006; Cao et al., 2007). Warming dominates other effects from climate-change by reducing CO_2 solubility and thus by enhancing [CO_3^{2-}]. An exception is the Arctic Ocean where reductions in pH and $CaCO_3$ saturation states are projected to be exacerbated by effects from increased freshwater input due to sea ice melt, more precipitation, and greater air–sea CO_2 fluxes due to less sea ice cover (Steinacher et al., 2009; Yamamoto et al., 2012). The projected effect of freshening is consistent with current observations of lower saturation states and lower pH values near river mouths and in areas under substantial fresh-water influence (Salisbury et al., 2008; Chierici and Fransson, 2009; Yamamoto-Kawai et al., 2009).

Regional ocean carbon cycle models project that some nearshore systems are also highly vulnerable to future pH decrease. In the California Current System, an eastern boundary upwelling system, observations and model results show that strong seasonal upwelling of carbon-rich waters (Feely et al., 2008) renders surface waters as vulnerable to future ocean acidification as those in the Southern Ocean (Gruber et al., 2012). In the Northwestern European Shelf Seas, large spatio-temporal variability is enhanced by local effects from river input and organic matter degradation, exacerbating acidification from anthropogenic CO_2 invasion (Artioli et al., 2012). In the Gulf of Mexico and East China Sea, coastal eutrophication, another anthropogenic perturbation, has been shown to enhance subsurface acidification as additional respired carbon accumulates at depth (Cai et al., 2011).

6.4.5 Future Ocean Oxygen Depletion

It is *very likely* that global warming will lead to declines in dissolved O_2 in the ocean interior through warming-induced reduction in O_2 solubility and increased ocean stratification. This will have implications for nutrient and carbon cycling, ocean productivity and marine habitats (Keeling et al., 2010).

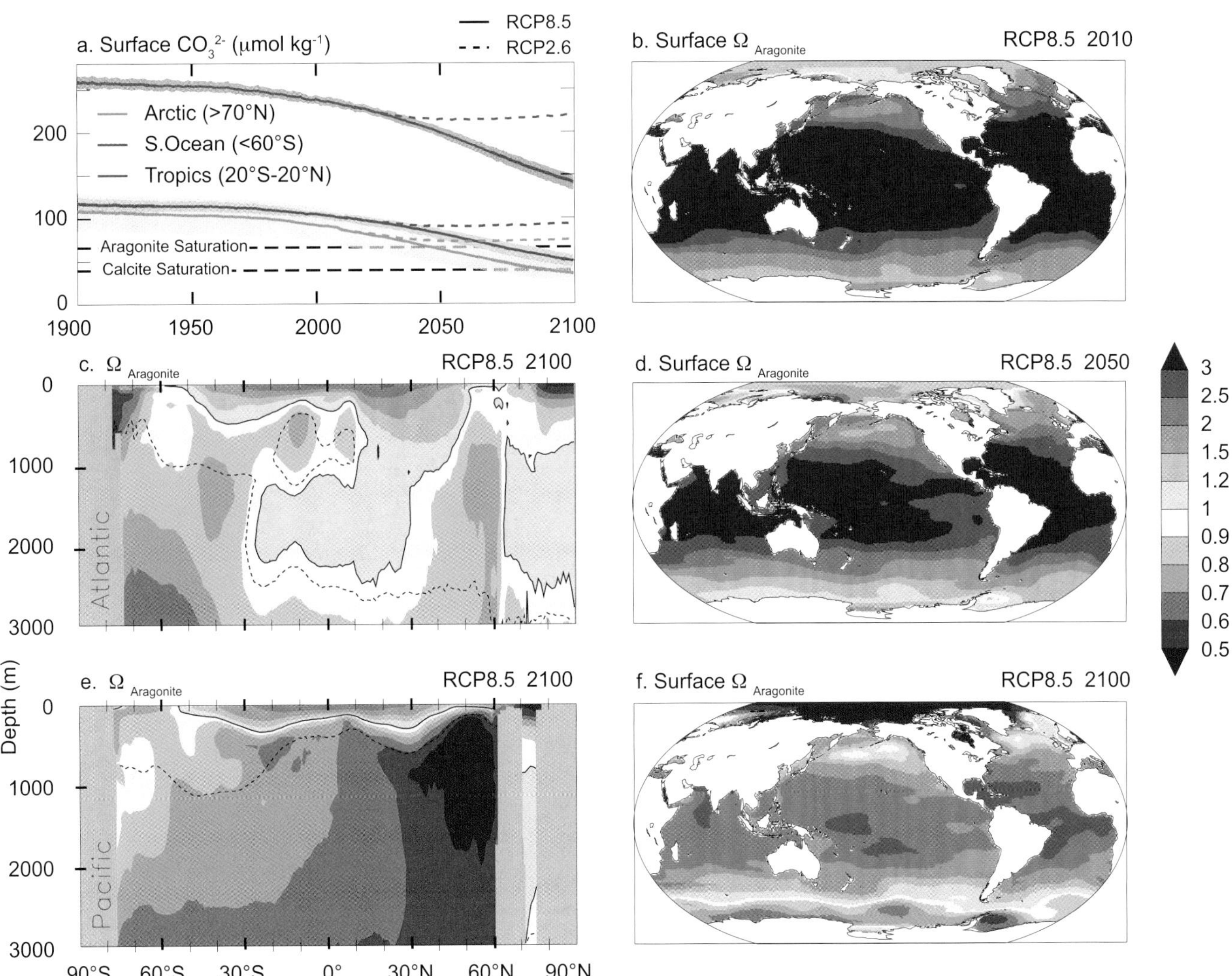

Figure 6.29 | Projected aragonite saturation state from 11 CMIP5 Earth System Models under RCP8.5 scenario: (a) time series of surface carbonate ion concentration shown as the mean (solid line) and range of models (filled), given as area-weighted averages over the Arctic Ocean (green), the tropical oceans (red), and the Southern Ocean (blue); maps of the median model's surface Ω_A in (b) 2010, (d) 2050 and (f) 2100; and zonal mean sections (latitude vs. depth) of Ω_A in 2100 over the (c) Atlantic and (e) Pacific, while the ASH is shown in 2010 (dotted line) as well as 2100 (solid line). Panel (a) also includes mean model results from RCP2.6 (dashed lines). As for Figure 6.28, gridded data products of carbonate system variables (Key et al., 2004) are used to correct each model for its present-day bias by subtracting the model-data difference at each grid cell following (Orr et al., 2005). Where gridded data products are unavailable (Arctic Ocean, all marginal seas, and the ocean near Indonesia), results are shown without bias correction.

Future changes in dissolved O_2 have been investigated using models of various complexity (see references in Table 6.13). The global ocean dissolved oxygen will decline significantly under future scenarios (Cocco et al., 2013). Simulated declines in mean dissolved O_2 concentration for the global ocean range from 6 to 12 µmol kg^{-1} by the year 2100 (Table 6.13), with a projection of 3 to 4 µmol kg^{-1} in one model with low climate sensitivity (Frölicher et al., 2009). This general trend is confirmed by the latest projections from the CMIP5 Earth System models, with reductions in mean dissolved O_2 concentrations from 1.5 to 4% (2.5 to 6.5 µmol kg^{-1}) in 2090s relative to 1990s for all RCPs (Figure 6.30a).

Most modelling studies (Table 6.13) explain the global decline in dissolved oxygen by enhanced surface ocean stratification leading to reductions in convective mixing and deep water formation and by a contribution of 18 to 50% from ocean warming-induced reduction in solubility. These two effects are in part compensated by a small increase in O_2 concentration from projected reductions in biological export production production (Bopp et al., 2001; Steinacher et al., 2010) or changes in ventilation age of the tropical thermocline (Gnanadesikan et al., 2007). The largest regional decreases in oxygen concentration (~20 to 100 µmol kg^{-1}) are projected for the intermediate (200 to 400 m) to deep waters of the North Atlantic, North Pacific and Southern Ocean for 2100 (Plattner et al., 2002; Matear and Hirst, 2003; Frölicher et al., 2009; Matear et al., 2010; Cocco et al., 2013), which is confirmed by the latest CMIP5 projections (Figure 6.30c and 6.30d).

It is *as likely as not* that the extent of open-ocean hypoxic (dissolved oxygen <60 to 80 µmol kg^{-1}) and suboxic (dissolved oxygen <5 µmol kg^{-1}) waters will increase in the coming decades. Most models show even some increase in oxygen in most O_2-poor waters and thus a slight decrease in the extent of suboxic waters under the SRES-A2 scenario (Cocco et al., 2013), as well as under RCP8.5 scenario (see the model-

Table 6.13 | Model configuration and projections for global marine O_2 depletion by 2100 (adapted from Keeling et al. (2010).

Study	Ocean Carbon Cycle Model	Forcing	Mean [O_2] Decrease ($\mu mol\ kg^{-1}$)[a,b]	Solubility Contribution (%)
Sarmiento et al. (1998)	GFDL		7[c]	
Matear et al. (2000)	CSIRO	IS92a		18
Plattner et al. (2002)	Bern 2D	SRES A1	12	35
Bopp et al. (2002)	IPSL	SRES A2[d]	4	25
Matear and Hirst (2003)	CSIRO	IS92a	9	26
Schmittner et al. (2008)	UVic	SRES A2	9	
Oschlies et al. (2008)	UVic	SRES A2	9	
	UVic-variable C:N	SRES A2	12	
Frölicher et al. (2009)	NCAR CSM1.4-CCCM	SRES A2	4	50
		SRES B1	3	
Shaffer et al. (2009)	DCESS	SRES A2	10[e]	

Notes:

[a] Assuming a total ocean mass of 1.48×10^{21} kg.

[b] Relative to pre-industrial baseline in 1750.

[c] Model simulation ends at 2065.

[d] Radiative forcing of non-CO_2 GHGs is excluded from this simulation.

[e] For simulations with reduced ocean exchange.

CCCM = Coupled-Climate-Carbon Model; CSIRO = Commonwealth Scientific and Industrial Research Organisation; DCESS = Danish Center for Earth System Science; GFDL = Geophysical Fluid Dynamics Laboratory; IPSL = Institute Pierre Simon Laplace; NCAR = National Center for Atmospheric Research; IS92 = IPCC scenarios for 1992; SRES = Special Report on Emission Scenarios; UVic = University of Victoria.

mean increase of sub-surface O_2 in large parts of the tropical Indian and Atlantic Oceans, Figure 6.30d). This rise in oxygen in most suboxic waters has been shown to be caused in one model study by an increased supply of oxygen due to lateral diffusion (Gnanadesikan et al., 2012). Given limitations of global ocean models in simulating today's O_2 distribution (Cocco et al., 2013), as well as reproducing the measured changes in O_2 concentrations over the past 50 years (see Chapter 3, and Stramma et al., 2012), the model projections are uncertain, especially concerning the evolution of O_2 in and around oxygen minimum zones.

A number of biogeochemical ocean carbon cycle feedbacks, not yet included in most marine biogeochemical models (including CMIP5 models, see Section 6.3.2.5.6), could also impact future trends of ocean deoxygenation. For example, model experiments which include a pCO_2-sensitive C:N drawdown in primary production, as suggested by some mesocosm experiments (Riebesell et al., 2007), project future increases of up to 50% in the volume of the suboxic waters by 2100 (Oschlies et al., 2008; Tagliabue et al., 2011). In addition, future marine hypoxia could be amplified by changes in the $CaCO_3$ to organic matter 'rain ratio' in response to rising pCO_2 (Hofmann and Schellnhuber, 2009). Reduction in biogenic calcification due to ocean acidification would weaken the strength of $CaCO_3$ mineral ballasting effect, which could lead organic material to be remineralized at a shallower depth exacerbating the future expansion of shallow hypoxic waters.

The modeled estimates do not take into account processes that are specific to the coastal ocean and may amplify deoxygenation. Recent observations for the period 1976–2000 have shown that dissolved O_2 concentrations have declined at a faster rate in the coastal ocean (–0.28 $\mu mol\ kg^{-1}\ yr^{-1}$) than the open ocean (–0.02 $\mu mol\ kg^{-1}\ y^{-1}$, and a faster rate than in the period 1951–1975, indicating a worsening of hypoxia (Gilbert et al., 2010). Hypoxia in the shallow coastal ocean (apart from continental shelves in Eastern Boundary Upwelling Systems) is largely eutrophication driven and is controlled by the anthropogenic flux of nutrients (N and P) and organic matter from rivers. If continued industrialisation and intensification of agriculture yield larger nutrient loads in the future, eutrophication should intensify (Rabalais et al., 2010), and further increase the coastal ocean deoxygenation.

On longer time scales beyond 2100, ocean deoxygenation is projected to increase with some models simulating a tripling in the volume of suboxic waters by 2500 (Schmittner et al., 2008). Ocean deoxygenation and further expansion of suboxic waters could persist on millennial time scales, with average dissolved O_2 concentrations projected to reach minima of up to 56 $\mu mol\ kg^{-1}$ below pre-industrial levels in experiments with high CO_2 emissions and high climate sensitivity (Shaffer et al., 2009).

A potential expansion of hypoxic or suboxic water over large parts of the ocean is *likely* to impact the marine cycling of important nutrients, particularly nitrogen. The intensification of low oxygen waters has been suggested to lead to increases in water column denitrification and N_2O emissions (e.g., Codispoti, 2010; Naqvi et al., 2010). Recent works, however, suggest that oceanic N_2O production is dominated by nitrification with a contribution of 7% by denitrification (Freing et al., 2012), Figure 6.4c) and that ocean deoxygenation in response to anthropogenic climate change could leave N_2O production relatively unchanged (Bianchi et al., 2012).

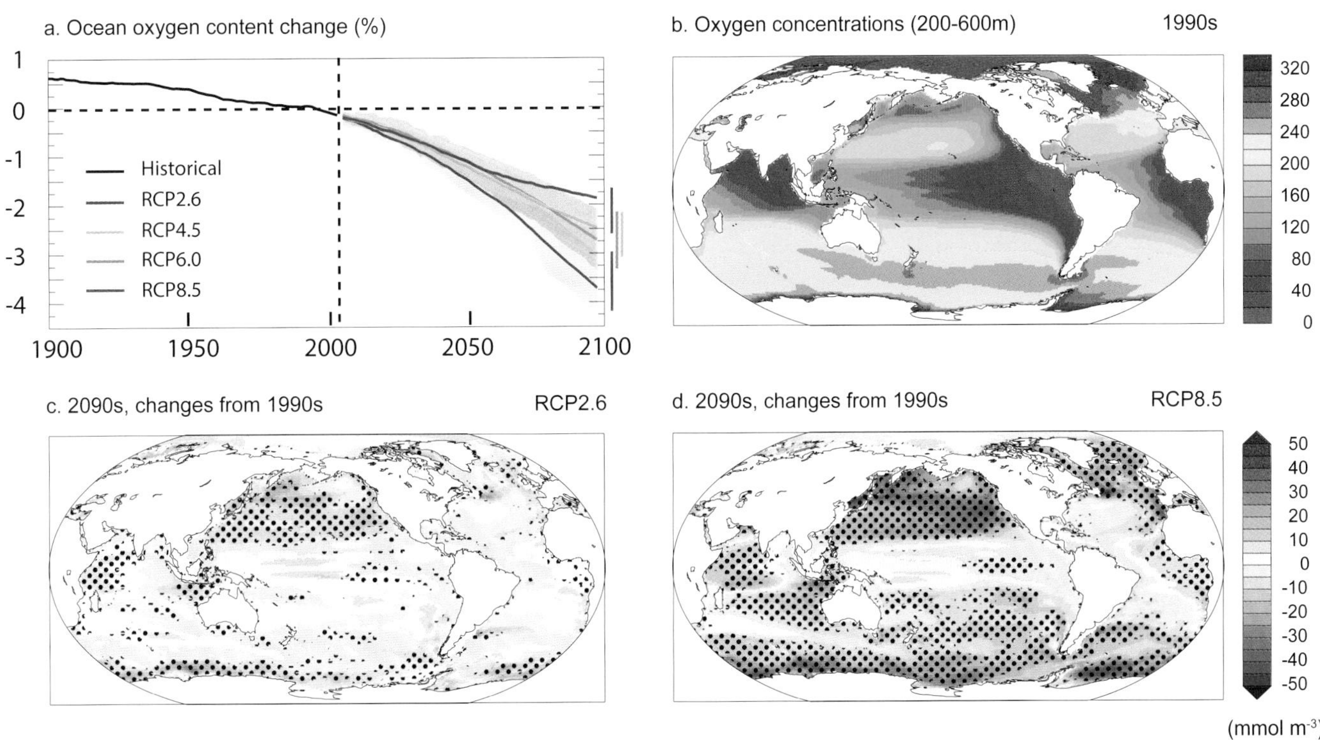

Figure 6.30 | (a) Simulated changes in dissolved O_2 (mean and model range as shading) relative to 1990s for RCP2.6, RCP4.5, RCP6.0 and RCP8.5. (b) Multi-model mean dissolved O_2 (µmol m^{-3}) in the main thermocline (200 to 600 m depth average) for the 1990s, and changes in 2090s relative to 1990s for RCP2.6 (c) and RCP8.5 (d). To indicate consistency in the sign of change, regions are stippled where at least 80% of models agree on the sign of the mean change. These diagnostics are detailed in Cocco et al. (2013) in a previous model intercomparison using the SRES-A2 scenario and have been applied to CMIP5 models here. Models used: Community Earth System Model 1–Biogeochemical (CESM1-BGC), Geophysical Fluid Dynamics Laboratory–Earth System Model 2G (GFDL-ESM2G), Geophysical Fluid Dynamics Laboratory–Earth System Model 2M (GFDL-ESM2M), Hadley Centre Global Environmental Model 2–Earth System (HadGEM2-ES), Institute Pierre Simon Laplace–Coupled Model 5A–Low Resolution (IPSL-CM5A-LR), Institute Pierre Simon Laplace–Coupled Model 5A–Medium Resolution (IPSL-CM5A-MR), Max Planck Institute–Earth System Model–Low Resolution (MPI-ESM-LR), Max Planck Institute–Earth System Model–Medium Resolution (MPI-ESM-MR), Norwegian Earth System Model 1 (Emissions capable) (NorESM1).

6.4.6 Future Trends in the Nitrogen Cycle and Impact on Carbon Fluxes

6.4.6.1 Projections for Formation of Reactive Nitrogen by Human Activity

Since the 1970s, food production, industrial activity and fossil fuel combustion have resulted in the creation of more reactive nitrogen (Nr) than natural terrestrial processes (Section 6.1; Box 6.2, Figure 1). Building on the general description of the set of AR4 Special Report on Emission (SRES) scenarios, Erisman et al. (2008) estimated anthropogenic nitrogen fertiliser consumption throughout the 21st century. Five driving parameters (population growth, consumption of animal protein, agricultural efficiency improvement and additional biofuel production) are used to project future nitrogen demands for four scenarios (A1, B1, A2 and B2) (Figure 6.31). Assigning these drivers to these four SRES scenarios, they estimated a production of Nr for agricultural use of 90 to 190 TgN yr^{-1} by 2100, a range that spans from slightly less to almost twice as much current fertiliser consumption rates (Section 6.1, Figure 6.4a, Figure 1 in Box 6.2).

Despite the uncertainties and the non-inclusion of many important drivers, three of the scenarios generated by the Erisman et al. (2008) model point towards an increase in future production of reactive nitrogen. In particular, the A1 scenario which assumes a world with rapid economic growth, a global population that peaks mid-century and rapid introduction of new and more efficient technologies ends up as the potentially largest contributor to nitrogen use, as a result of large amounts of biofuels required and the fertiliser used to produce it. This increase in nitrogen use is assumed to be largely in line with the RCP2.6 scenario, where it appears to have rather limited adverse effects like increasing N_2O emissions (van Vuuren et al., 2011).

N_2O emissions are projected to increase from increased anthropogenic Nr production. It is thus *likely* that N_2O emissions from soils will increase due to the increased demand for feed/food and the reliance of agriculture on nitrogen fertilisers. This is illustrated by the comparison of emissions from 1900 to those in 2000 and 2050, using the IAM IMAGE model that served to define the RCP2.6 pathway (Figure 6.32). The anthropogenic N_2O emission map IN 2050 shown in Figure 6.32 is established from the RCP4.5 scenario; the RCP8.5 and RCP6 scenarios have much higher emissions, and RCP2.6 much lower (van Vuuren et al., 2011). A spatially explicit inventory of soil nitrogen budgets in livestock and crop production systems using the IMAGE model (Bouwman et al., 2011) shows that between 1900 and 1950, the global soil Nr budget surplus almost doubled to 36 TgN yr^{-1}, and further increased to 138 TgN yr^{-1} between 1950 and 2000. The IMAGE model scenario from Bouwman et al. (2011) shown in Figure 6.32 portrays a world with a

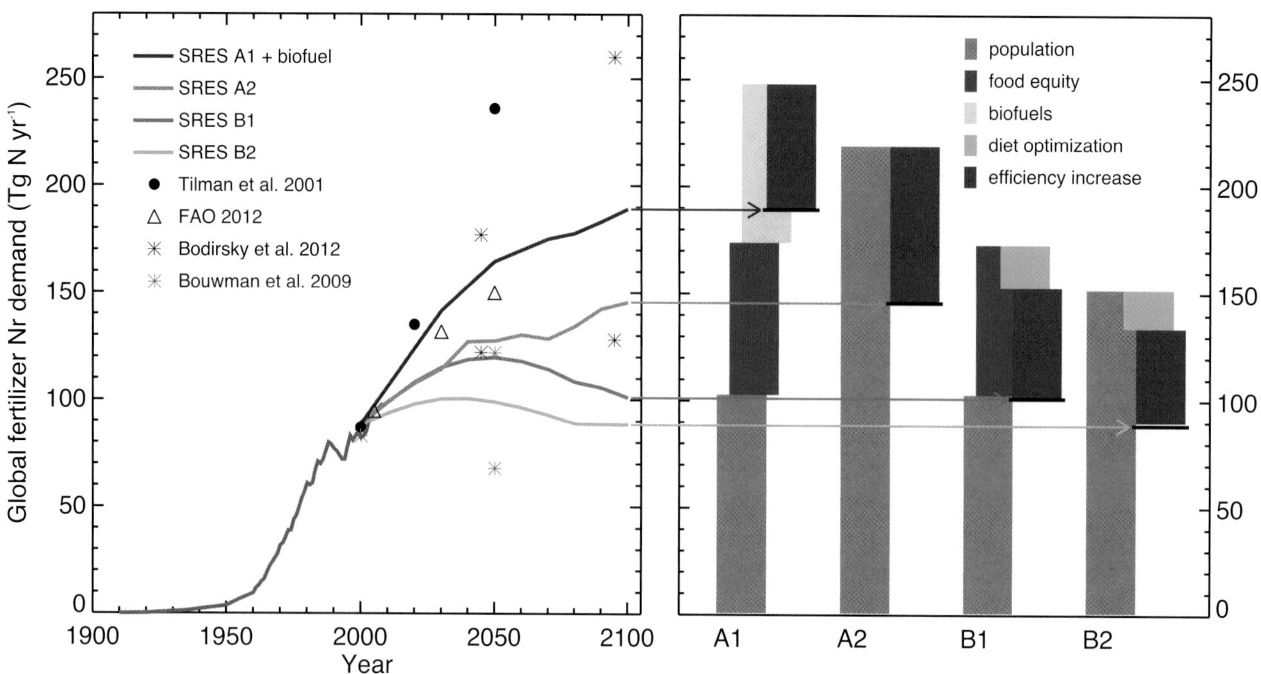

Figure 6.31 | Global nitrogen fertiliser consumption scenarios (left) and the impact of individual drivers on 2100 consumption (right). This resulting consumption is always the sum (denoted at the end points of the respective arrows) of elements increasing as well as decreasing nitrogen consumption. Other relevant estimates are presented for comparison. The A1, B1, A2 and B2 scenarios draw from the assumptions of the IPCC Special Report on Emission Scenarios (SRES) emission scenario storylines as explained in Erisman et al. (2008).

further increasing global crop production (+82% for 2000–2050) and livestock production (+115%). Despite the assumed rapid increase in nitrogen use efficiency in crop (+35%) and livestock (+35%) production, global agricultural Nr surpluses are projected to continue to increase (+23%), and associated emissions of N_2O to triple compared to 1900 levels.

Regional to global scale model simulations suggest a strong effect of climate variability on interannual variability of land N_2O emissions (Tian et al., 2010; Zaehle et al., 2011; Xu-Ri et al., 2012). Kesik et al. (2006) found for European forests that higher temperatures and lower soil moisture will decrease future N_2O emissions under scenarios of climate change, despite local increases of emission rates by up to 20%. Xu-Ri et al. (2012) show that local climate trends result in a spatially diverse pattern of increases and decreases of N_2O emissions, which globally integrated result in a net climate response of N_2O emissions of 1 TgN yr^{-1} per 1°C of land temperature warming. Using a further development of this model, Stocker et al. (2013) estimate increases in terrestrial N_2O from a pre-industrial terrestrial source of 6.9 TgN (N_2O) yr^{-1} to 9.8 to 11.1 TgN (N_2O) yr^{-1} (RCP 2.6) and 14.2 to 17.0 TgN (N_2O) yr^{-1} (RCP 8.5) by 2100. Of these increases, 1.1 to 2.4 TgN (N_2O) yr^{-1} (RCP 2.6) or 4.7 to 7.7 TgN (N_2O) yr^{-1} (RCP 8.5) are due to the interacting effects of climate and CO_2 on N_2O emissions from natural and agricultural ecosystems. An independent modelling study suggested a climate change related increase of N_2O emissions between 1860 and 2100 by 3.1 TgN (N_2O) yr^{-1} for the A2 SRES scenario (Zaehle, 2013) implying a slightly lower sensitivity of soil N_2O emissions to climate of 0.5 TgN (N_2O) yr^{-1} per 1°C warming. While the present-day contribution of these climate-mediated effects on the radiative forcing from N_2O is *likely* to be small (0.016 W m^{-2} °C^{-1}; Zaehle and Dalmonech, 2011). Modelling results (Stocker et al., 2013) suggest that the climate and CO_2-related amplification of terrestrial N_2O emissions imply a larger feedback of 0.03 to 0.05 W m^{-2} °C^{-1} by 2100.

With the continuing increases in the formation of Nr from anthropogenic activities will come increased Nr emissions and distribution of Nr by waters and the atmosphere. For the atmosphere, the main driver of future global nitrogen deposition is the emission trajectories of NO_y and NH_3. For all RCP scenarios except RCP2.6, nitrogen deposition is projected to remain relatively constant globally although there is a projected increase in NH_x deposition and decrease in NO_y deposition. On a regional basis, future decreases of NH_x and NOx are projected in North America and northern Europe, and increases in Asia (Figure 6.33). Spatially, projected changes in total nitrogen deposition driven primarily by increases in NH_x emissions occur over large regions of the world for all RCPs, with generally the largest in RCP8.5 and the smallest in RCP2.6 (Figure 6.33) (Supplementary Material has RCP4.5 and RCP6.0). Previous IPCC scenarios (SRES A2 or IS92a) project a near doubling of atmospheric nitrogen deposition over some world biodiversity hotspots with half of these hotspots subjected to deposition rates greater than 15 kgN ha^{-1} yr^{-1} (critical load threshold value) over at least 10% of their total area (Dentener et al., 2005; Phoenix et al., 2006; Bleeker et al., 2011).

Large uncertainties remain in our understanding and modelling of changes in Nr emissions, atmospheric transport and deposition processes, lead to *low confidence* in the projection of future Nr deposition fluxes, particularly in regions remote from anthropogenic emissions (Dentener et al., 2006). The large spread between atmospheric GCM models associated with precipitation projections confounds extraction of a climate signal in deposition projections (Langner et al., 2005; Hedegaard et al., 2008).

6.4.6.2. Projected Changes in Sulphur Deposition

Given the tight coupling between the atmospheric nitrogen and sulphur cycles, and the impact on climate (Section 7.3) this Chapter also presents scenarios for sulphur deposition. Deposition of SO_x is projected to decrease in all RCP pathways (Figures 6.33 and 6.34). By contrast, scenarios established prior to RCPs indicated decreases of sulphur deposition in North America and Europe, but increases in South America, Africa, South and East Asia (Dentener et al., 2006; Tagaris et al., 2008). In all RCPs, sulphur deposition is lower by 2100 than in 2000 in all regions, with the largest decreases in North America, Europe and Asia (RCP2.6 and RCP 8.5 are seen in Figure 6.34; RCP4.5 and RCP6.0 are in the Supplementary Material) (Lamarque et al., 2011). Future hot spots of deposition are still evident in East and South East Asia, especially for RCP6.0.

Projected future increase of Nr input into terrestrial ecosystems also yields increased flux of Nr from rivers into coastal systems. As illustrated by the Global NEWS 2 model for 2050, by the base year 2000, the discharge of dissolved inorganic nitrogen (DIN) to marine coastal waters was >500 kg N km^{-2} of watershed area for most watershed systems downstream of either high population or extensive agricultural activity (Mayorga et al., 2010; Seitzinger et al., 2010). Additional information and the supporting figure are found in the Supplementary Material.

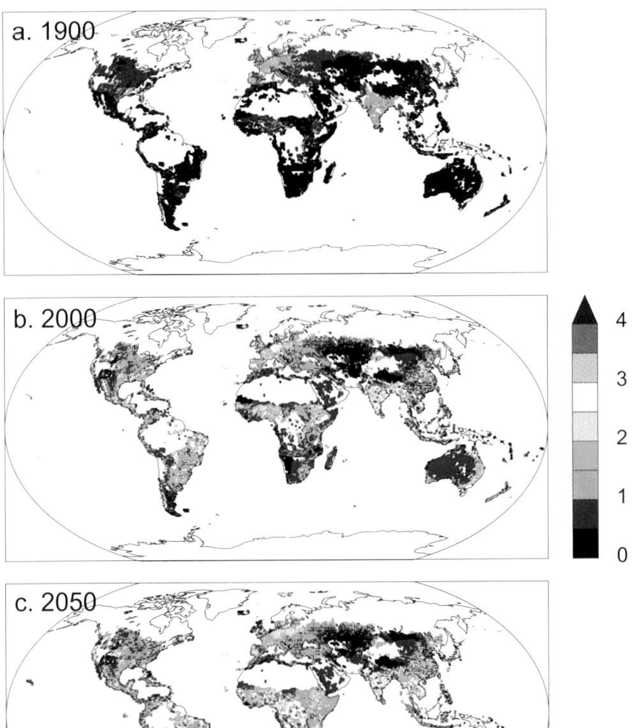

Figure 6.32 | N_2O emissions in 1900, 2000 and projected to 2050 (Bouwman et al., 2011). This spatially explicit soil nutrient budget and nitrogen gas emission scenario was elaborated with the Integrated Model to Assess the Global Environment (IMAGE) model on the basis of the International Assessment of Agricultural Knowledge, Science and Technology for Development (IAASTD) baseline scenario (McIntyre et al., 2009).

6.4.6.3 Impact of Future Changes in Reactive Nitrogen on Carbon Uptake and Storage

Anthropogenic Nr addition and natural nitrogen-cycle responses to global changes will have an important impact on the global carbon cycle. As a principal nutrient for plant growth, nitrogen can both limit future carbon uptake and stimulate it depending on changes in Nr availability. A range of global terrestrial carbon cycle models have been developed since AR4 that integrate nitrogen dynamics into the simulation of land carbon cycling (Thornton et al., 2007; Wang et al., 2007, 2010a; Sokolov et al., 2008; Xu-Ri and Prentice, 2008; Churkina et al., 2009; Jain et al., 2009; Fisher et al., 2010; Gerber et al., 2010; Zaehle and Friend, 2010; Esser et al., 2011). However, only two ESMs in CMIP5 (CESM1-BGC and NorESM1-ME) include a description of nitrogen–carbon interactions.

In response to climate warming, increased decomposition of soil organic matter increases nitrogen mineralisation, (*high confidence*) which can enhance Nr uptake and carbon storage by vegetation. Generally, higher C:N ratio in woody vegetation compared to C:N ratio of soil organic matter causes increased ecosystem carbon storage as increased Nr uptake shifts nitrogen from soil to vegetation (Melillo et al., 2011). In two studies (Sokolov et al., 2008; Thornton et al., 2009), this effect was strong enough to turn the carbon–climate interaction into a small negative feedback, that is, an increased land CO_2 uptake in response to climate warming (positive γ_L values in Figure 6.20), whereas in another study that described carbon–nitrogen interactions (Zaehle et al., 2010b) the carbon–climate interaction was reduced but remained positive, that is, decreased land CO_2 uptake in response to climate change (negative γ_L values in Figures 6.20, 6.21 and 6.22). The two CMIP5 ESMs which include terrestrial carbon–nitrogen interactions (Table 6.11) also simulate a small but positive climate–carbon feedback.

Consistent with the observational evidence (Finzi et al., 2006; Palmroth et al., 2006; Norby et al., 2010), modelling studies have shown a strong effect of Nr availability in limiting the response of plant growth and land carbon storage to elevated atmospheric CO_2 (e.g., Sokolov et al., 2008; Thornton et al., 2009; Zaehle and Friend, 2010). These analyses are affected by the projected future trajectories of anthropogenic Nr deposition. The effects of Nr deposition counteract the nitrogen limitation of CO_2 fertilisation (Churkina et al., 2009; Zaehle et al., 2010a). Estimates of the total net carbon storage on land due to Nr deposition between 1860 and 2100 range between 27 and 66 PgC (Thornton et al., 2009; Zaehle et al., 2010a).

It is *very likely* that, at the global scale, nutrient limitation will reduce the global land carbon storage projected by CMIP5 carbon-cycle only models. Only two of the current CMIP5 ESM models explicitly consider carbon–nitrogen interactions (CESM1-BGC and NorESM1-ME).

The effect of the nitrogen limitations on terrestrial carbon sequestration in the results of the other CMIP5 models may be approximated by comparing the implicit Nr requirement given plausible ranges of terrestrial C:N stoichiometry (Wang and Houlton, 2009) to plausible increases in terrestrial Nr supply due to increased biological nitrogen fixation (Wang and Houlton, 2009) and anthropogenic Nr deposition (Figure 6.35). For the ensemble of CMIP5 projections under the RCP 8.5 scenario, this implies a lack of available nitrogen of 1.3 to 13.1 PgN which would reduce terrestrial C sequestration by an average of 137 PgC over the period 1860–2100, with a range of 41 to 273 PgC among models. This represents an ensemble mean reduction in land carbon sequestration of 55%, with a large spread across models (14 to 196%). Inferred reductions in ensemble-mean land carbon sink over the same period for RCPs 6.0, 4.5 and 2.6 are 109, 117 and 85 PgC, respectively. Between-model variation in these inferred reduced land carbon sinks is similar for all RCPs, with ranges of 57 to 162 PgC, 38 to 208 PgC, and 32 to 171 PgC for RCPs 6.0, 4.5 and 2.6, respectively. The influence of nutrient addition for agriculture and pasture management is not addressed in this analysis. Results from the two CMIP5 models with explicit carbon–nitrogen interactions show even lower land carbon sequestration than obtained by this approximation method (Figure 6.35). More models with explicit carbon–nitrogen interactions are needed to understand between-model variation and construct an ensemble response.

The positive effect on land carbon storage due to climate-increased Nr mineralization is of comparable magnitude to the land carbon storage increase associated with increased anthropogenic Nr deposition. Models disagree, however, which of the two factors is more important, with both effects dependent on the choice of scenario. Crucially, the effect of nitrogen limitation on vegetation growth and carbon storage under elevated CO_2 is the strongest effect of the natural and disturbed nitrogen cycle on terrestrial carbon dynamics (Bonan and Levis, 2010; Zaehle et al., 2010a). In consequence, the projected atmospheric CO_2

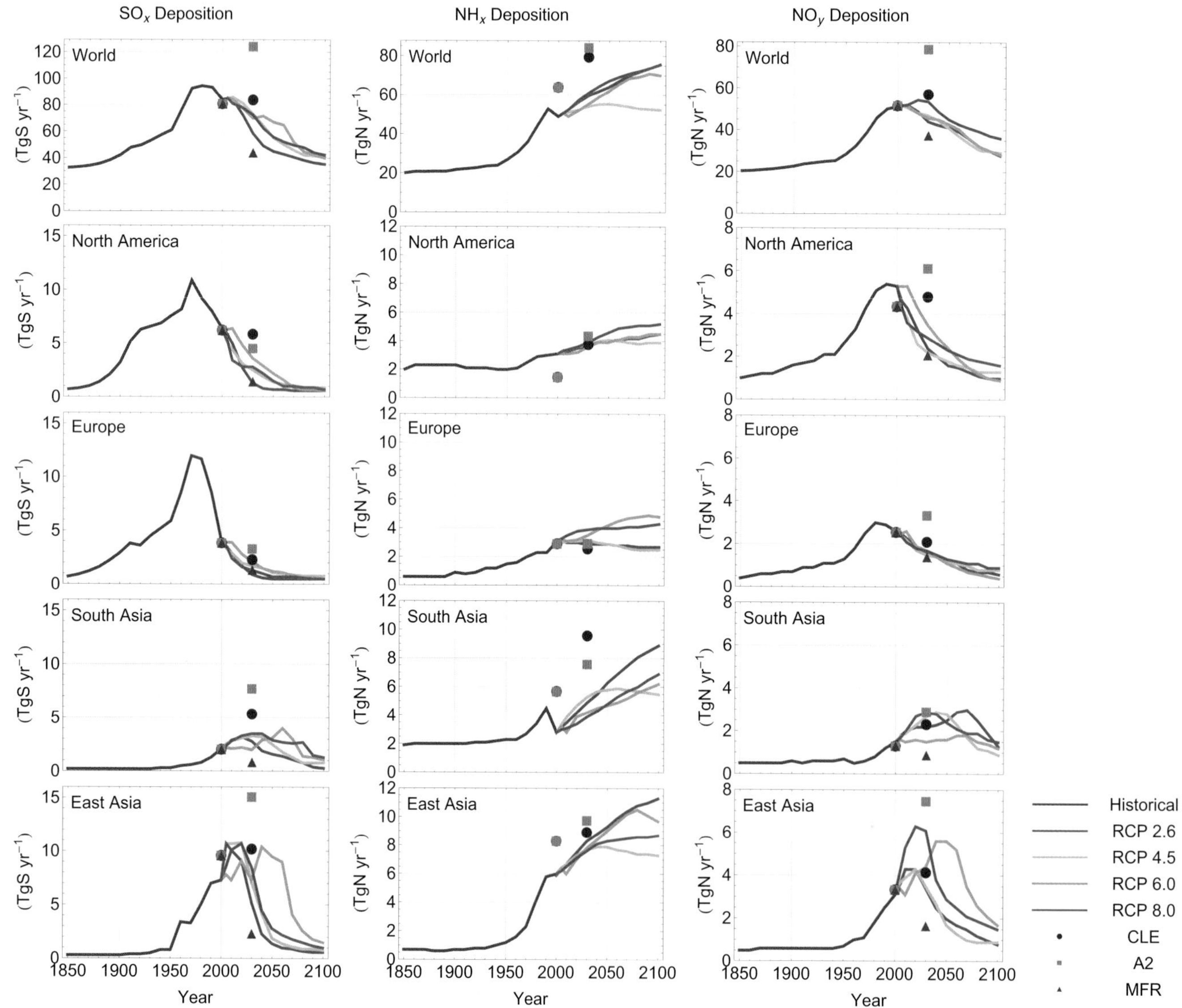

Figure 6.33 | Deposition of SO_x (left, TgS yr^{-1}), NH_x (middle, TgN yr^{-1}) and NO_y (right, TgN yr^{-1}) from 1850 to 2000 and projections of deposition to 2100 under the four RCP emission scenarios (Lamarque et al., 2011; van Vuuren et al., 2011). Also shown are the 2030 scenarios using the SRES B1/A2 energy scenario with assumed current legislation and maximum technically feasible air pollutant reduction controls (Dentener et al., 2006).

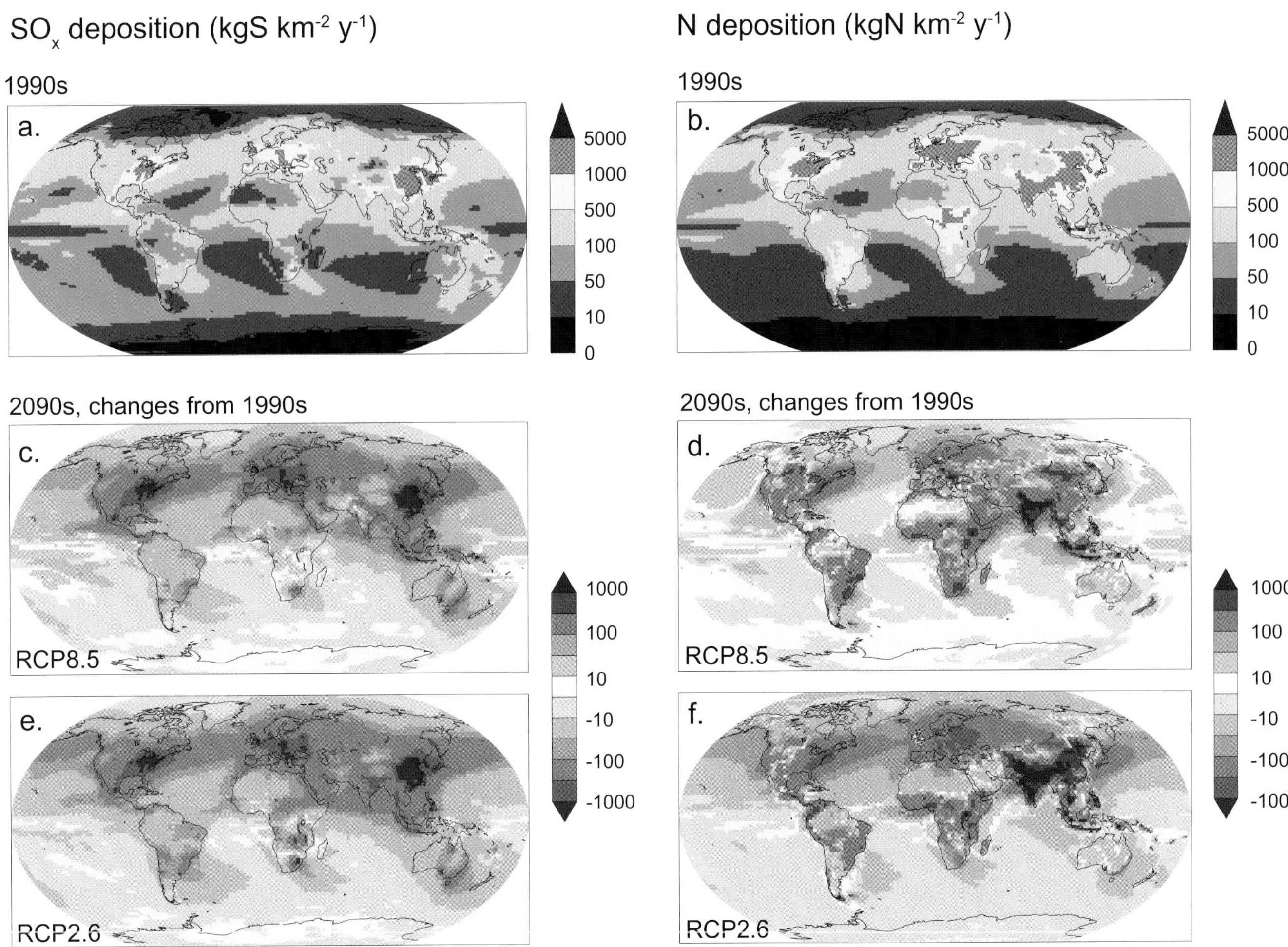

Figure 6.34 | Spatial variability of nitrogen and SO_x deposition in 1990s with projections to the 2090s (shown as difference relative to the 1990s), using the RCP2.6 and RCP8.5 scenarios, kgN km^{-2} yr^{-1}, adapted from Lamarque et al. (2011). Note that no information on the statistical significance of the shown differences is available. This is of particular relevance for areas with small changes. The plots for all four of the RCP scenarios are in the Supplementary Material.

concentrations (and thus degree of climate change) in 2100 are higher in projections with models describing nitrogen limitations than in those same models without these interactions. The influence of current and future nitrogen deposition on the ocean sink for anthropogenic carbon is estimated to be rather small, with less than 5% of the ocean carbon sink in 2100 attributable to fertilisation from anthropogenic Nr deposition over the oceans (Reay et al., 2008).

None of the CMIP5 models include phosphorus as a limiting nutrient for land ecosystems, although this limitation and interactions with Nr availability are observed in many systems (Elser et al., 2007). Limitation by Nr availability alone may act as a partial surrogate for combined nitrogen–phosphorus limitation (Thornton et al., 2009; Section 6.4.8.2), but are *likely* to underestimate the overall nutrient limitation, especially in lowland tropical forest.

6.4.7 Future Changes in Methane Emissions

Future atmospheric CH_4 concentrations are sensitive to changes in both emissions and OH oxidation. Atmospheric chemistry is not covered in this chapter and we assess here future changes in natural CH_4 emissions in response to climate change (e.g., O'Connor et al., 2010; Figure 6.36). Projected increases in future fire occurrence (Section 6.4.8.1) suggest that CH_4 from fires may increase (*low confidence*). Future changes in anthropogenic emissions due to anthropogenic alteration of wetlands (e.g., peatland drainage) may also be important but are not assessed here.

6.4.7.1 Future Methane Emissions from Wetlands

Overall, there is *medium confidence* that emissions of CH_4 from wetlands are *likely* to increase in the future under elevated CO_2 and warmer climate. Wetland extent is determined by geomorphology and soil moisture, which depends on precipitation, evapotranspiration, drainage and runoff. All of these may change in the future. Increasing temperature can lead to higher rates of evapotranspiration, reducing soil moisture and therefore affect wetland extent, and temporary increasing aeration of existing wetlands with further consequences to methane emissions. Regional projections of precipitation changes are especially uncertain (see Chapter 12).

Direct effects on wetland CH_4 emissions include: higher NPP under higher temperature and higher atmospheric CO_2 concentrations leading to more substrate for methanogenesis (White et al., 2008); higher CH_4 production rates under higher temperature; and changes in CH_4 oxidation through changed precipitation that alters water table position (Melton et al., 2013). Wetland CH_4 emissions are also affected by changes in wetland area which may either increase (due to thawing permafrost or reduced evapotranspiration) or decrease (due to reduced precipitation or increased evaporation) regionally. In most models, elevated CO_2 has a stronger enhancement effect on CH_4 emissions than climate change. However, large uncertainties exist concerning the lack of wetland specific plant functional types in most models and the lack of understanding how wetland plants will react to CO_2 fertilisation (e.g., Berendse et al., 2001; Boardman et al., 2011; Heijmans et al., 2001, 2002a, 2002b).

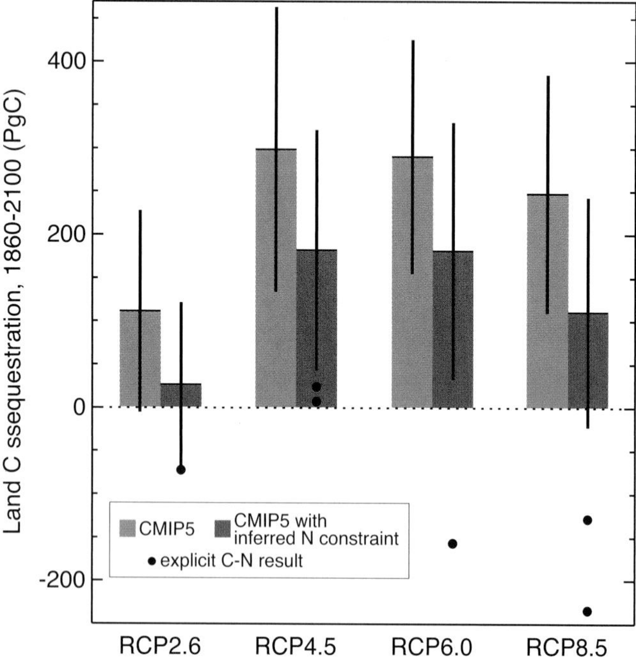

Figure 6.35 | Estimated influence of nitrogen availability on total land carbon sequestration over the period 1860–2100 (based on analysis method of Wang and Houlton (2009). Blue bars show, for each RCP scenario, the multi-model ensemble mean of land carbon sequestration, based on the carbon-only subset of CMIP5 models (Canadian Earth System Model 2 (CanESM2), Geophysical Fluid Dynamics Laboratory–Earth System Model 2G (GFDL-ESM2G), Geophysical Fluid Dynamics Laboratory–Earth System Model 2M (GFDL-ESM2M), Hadley Centre Global Environmental Model 2–Carbon Cycle(HadGEM2-CC), Hadley Centre Global Environmental Model 2–Earth System (HadGEM2-ES), Institute Pierre Simon Laplace–Coupled Model 5A–Low Resolution (IPSL-CM5A-LR), Institute Pierre Simon Laplace– Coupled Model 5A–Medium Resolution (IPSL-CM5A-MR), Institute Pierre Simon Laplace–Coupled Model 5B–Low Resolution (IPSL-CM5B-LR), Max Planck Institute–Earth System Model–Low Resolution (MPI-ESM-LR): not all models produced results for all scenarios). Red bars show, for each scenario, the mean land carbon sequestration from the same ensemble of carbon-only models after correcting for inferred constraints on carbon uptake due to limited availability of nitrogen. Black bars show ± one standard deviation around the means. Black symbols show individual model results from the two CMIP5 models with explicit carbon–nitrogen interactions (Community Earth System Model 1–Biogeochemical (CESM1-BGC) and Norwegian Earth System Model 1 (Emissions capable) (NorESM1-ME)). These two models have nearly identical representations of land carbon–nitrogen dynamics, and differences between them here (for RCP4.5 and RCP8.5, where both models contributed results) are due to differences in coupled system climate. All simulations shown here used prescribed atmospheric CO_2 concentrations.

Since AR4, several modelling studies have attempted to quantify the sensitivity of global wetland CH_4 emissions to environmental changes (see Figure 6.37). The studies cover a wide range of simulation results but there is high agreement between model results that the combined effect of CO_2 increase and climate change by the end of the 21st century will increase wetland CH_4 emissions. Using a common experimental protocol with spatially uniform changes in precipitation, temperature and CO_2 ("WETCHIMP"; Melton et al., 2013) seven models predict that the effect of increased temperature alone (red bars in Figure 6.37) may cause an increase or decrease of wetland CH_4 emissions, while the effect of increased precipitation alone (green bars in Figure 6.37) is always an increase, although generally small. The effect of increased atmospheric CO_2 concentration (fertilisation of NPP; Box 6.3; blue bars in Figure 6.37) always resulted in an increase of emissions (22 to 162%). Other studies assessed the effects of temperature and precipitation together (orange bars in Figure 6.37) and often found an increase in wetland CH_4 emissions (Eliseev et al., 2008; Gedney et al., 2004; Shindell et al., 2004; Volodin, 2008) although Ringeval et al. (2011) found a net decrease. The combined effect of climate and CO_2 resulted in an increase of wetland CH_4 emissions from 40% (Volodin (2008); fixed wetland area) to 68% (Ringeval et al., 2011); variable wetland area).

The models assessed here do not consider changes in soil hydrological properties caused by changes in organic matter content. Positive feedbacks from increased drainage due to organic carbon loss may

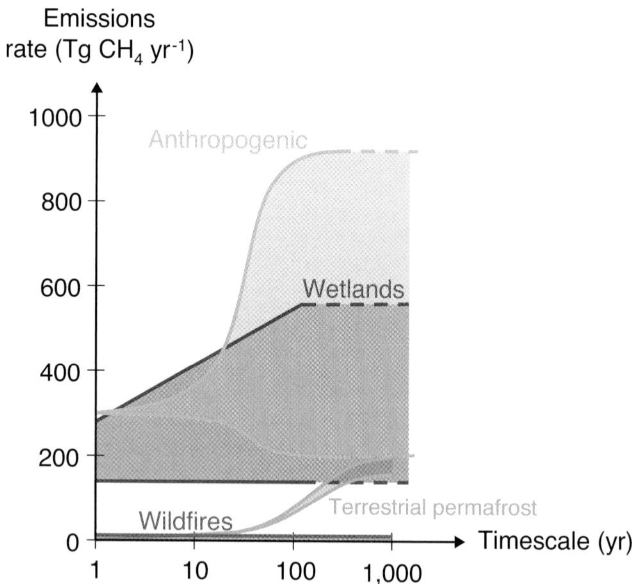

Figure 6.36 | Schematic synthesis of the magnitude and time scales associated with possible future CH_4 emissions (adapted from O'Connor et al., 2010). Uncertainty in these future changes is large, and so this figure demonstrates the relative magnitude of possible future changes. Anthropogenic emissions starting at a present-day level of 300 Tg(CH_4) yr^{-1} (consistent with Table 6.8) and increasing or decreasing according to RCP8.5 and RCP2.6 are shown for reference. Wetland emissions are taken as 140 to 280 Tg(CH_4) yr^{-1} present day values (Table 6.8) and increasing by between 0 and 100% (Section 6.4.7.1; Figure 6.37). Permafrost emissions may become important during the 21st century. CH_4 release from marine hydrates and subsea permafrost may also occur but uncertainty is sufficient to prevent plotting emission rates here. Large CH_4 hydrate release to the atmosphere is not expected during the 21st century. No quantitative estimates of future changes in CH_4 emissions from wildfires exist, so plotted here are continued present-day emissions of 1 to 5 Tg(CH_4) yr^{-1} (Table 6.8).

accelerate peat decomposition rates (Ise et al., 2008). However, carbon accumulation due to elevated NPP in wetland and permafrost regions may to some extent offset CH_4 emissions (Frolking and Roulet, 2007; Turetsky et al., 2007). None of the studies or models assessed here considers CH_4 emissions from mangroves.

The models also do not agree in their simulations of present day wetland extent or CH_4 emissions, and there are not adequate data sets to evaluate them thoroughly at the grid scale (typically 0.5°) (Melton et al., 2013). Hence despite high agreement between models of a strong positive response of wetland CH_4 emission rates to increasing atmospheric CO_2 we assign *low confidence* to quantitative projections of future wetland CH_4 emissions.

Soil CH_4 oxidation of about 30 Tg(CH_4) yr^{-1} (Table 6.8) represents the smallest of the three sinks for atmospheric methane (see Table 6.8) but is also sensitive to future environmental changes. Soil CH_4 oxidation is projected to increase by up to 23% under the SRES A1B due to rising atmospheric CH_4 concentrations, higher soil temperature and lower soil moisture (Curry, 2007, 2009).

6.4.7.2 Future Methane Emissions from Permafrost Areas

Permafrost thaw may lead to increased drainage and a net reduction in lakes and wetlands, a process that has already begun to be seen in lakes in the discontinuous permafrost zone (Smith et al., 2005; Jones et al., 2011) and has been projected to continue under future scenarios (Avis et al., 2011). Alternatively, small lakes or ponds and wetland growth may occur in continuous permafrost areas underlain by ice-rich material subject to thermokarst (Christensen et al., 2004; Jorgenson et al., 2006; Plug and West, 2009; Jones et al., 2011).

There is high agreement between land surface models that permafrost extent is expected to reduce during the 21st century, accompanying particularly rapid warming at high latitudes (Chapter 12). However, estimates vary widely as to the pace of degradation (Lawrence and Slater, 2005; Burn and Nelson, 2006; Lawrence et al., 2008). The LPJ-WHyMe model projected permafrost area loss of 30% (SRES B1) and 47% (SRES A2) by 2100 (Wania, 2007). Marchenko et al. (2008) calculate that by 2100, 57% of Alaska will lose permafrost within the top 2 m. For the RCP scenarios, the CMIP5 multi-model ensemble shows a wide range of projections for permafrost loss: 15 to 87% under RCP4.5 and 30 to 99% under RCP8.5 (Koven et al., 2013).

Hydrological changes may lead to tradeoffs between the CO_2 and CH_4 balance of ecosystems underlain by permafrost, with methane production rates being roughly an order of magnitude less than rates of oxic decomposition to CO_2 but CH_4 having a larger greenhouse warming potential (Frolking and Roulet, 2007). The extent of permafrost thaw simulated by climate models has been used to estimate possible subsequent carbon release (Burke et al., 2013; Harden et al., 2012; Section 6.4.3.4) but few studies explicitly partition this into CO_2 or CH_4 release

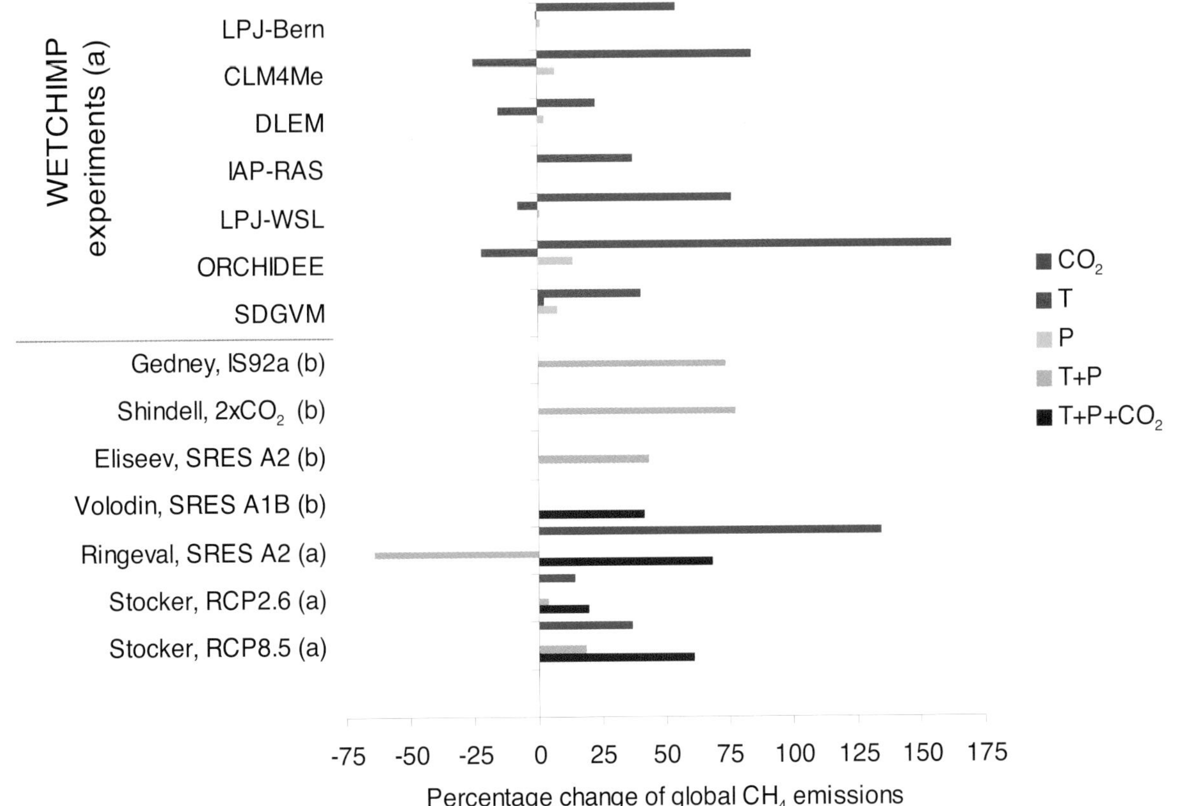

Figure 6.37 | Relative changes of global CH_4 emissions from either pre-industrial (a) or present-day (b) conditions and environmental changes that reflect potential conditions in 2100. The first seven models took part in the WETCHIMP intercomparison project and were run under a common protocol (Melton et al., 2013). Bars represent CH_4 emission changes associated with temperature-only changes (T), precipitation only (P), CO_2 only (CO_2) or combinations of multiple factors. Other studies as listed in the figure used different future scenarios: Eliseev et al. (2008), Gedney et al. (2004), Ringeval et al. (2011), Shindell et al. (2004), Volodin (2008), Stocker et al. (2013).

6

to the atmosphere. Schneider von Deimling et al. (2012) estimate cumulative CH_4 emissions by 2100 between 131 and 533 Tg(CH_4) across the 4 RCPs. CMIP5 projections of permafrost thaw do not consider changes in pond or lake formation. Thawing of unsaturated Yedoma carbon deposits (which contain large, but uncertain amounts of organic carbon in permafrost in northeast Siberia; Schirrmeister et al., 2011) was postulated to produce significant CH_4 emissions (Khvorostyanov et al., 2008), however more recent estimates with Yedoma carbon lability constrained by incubation observations (Dutta et al., 2006) argue for smaller emissions at 2100 (Koven et al., 2011).

6.4.7.3 Future Methane Hydrate Emissions

Substantial quantities of methane are believed to be stored within submarine hydrate deposits at continental margins (see also Section 6.1, FAQ 6.1). There is concern that warming of overlying waters may melt these deposits, releasing CH_4 into the ocean and atmosphere systems. Overall, it is *likely* that subsequent emissions to the atmosphere caused by hydrate destabilisation would be in the form of CO_2, due to CH_4 oxidation in the water column.

Considering a potential warming of bottom waters by 1°C, 3°C and 5°C during the next 100 years, Reagan and Moridis (2007) found that hydrates residing in a typical deep ocean setting (4°C and 1000 m depth) would be stable and in shallow low-latitude settings (6°C and 560 m) any sea floor CH_4 fluxes would be oxidized within the sediments. Only in cold-shallow Arctic settings (0.4°C and 320 m) would CH_4 fluxes exceed rates of benthic sediment oxidation. Simulations of heat penetration through the sediment by Fyke and Weaver (2006) suggest that changes in the gas hydrate stability zone will be small on century time scales except in high-latitude regions of shallow ocean shelves. In the longer term, Archer et al. (2009a) estimated that between 35 and 940 PgC could be released over several thousand years in the future following a 3°C seafloor warming.

Using multiple climate models (Lamarque, 2008), predicted an upper estimate of the global sea floor flux of between 560 and 2140 Tg(CH_4) yr^{-1}, mostly in the high latitudes. Hunter et al. (2013) also found 21st century hydrate dissociation in shallow Arctic waters and comparable in magnitude to Biastoch et al. (2011), although maximum CH_4 sea floor fluxes were smaller than Lamarque (2008), with emissions from 330 to 450 Tg(CH_4) yr^{-1} for RCP 4.5 to RCP8.5. Most of the sea floor flux of CH_4 is expected to be oxidised in the water column into dissolved CO_2. Mau et al. (2007) suggest only 1% might be released to the atmosphere but this fraction depends on the depth of water and ocean conditions. Elliott et al. (2011) demonstrated significant impacts of such sea floor release on marine hypoxia and acidity, although atmospheric CH_4 release was small.

Observations of CH_4 release along the Svalbard margin seafloor (Westbrook et al., 2009) suggest observed regional warming of 1°C during the last 30 years is driving hydrate disassociation, an idea supported by modelling (Reagan and Moridis, 2009). However, these studies do not consider subsea-permafrost hydrates suggested recently to be regionally significant sources of atmospheric CH_4 (Shakhova et al., 2010). There was no positive excursion in the methane concentration recorded in ice cores from the largest known submarine landslide, the Storegga slide of Norway 8200 years ago. Large methane hydrate release due to marine landslides is *unlikely* as any given landslide could release only a tiny fraction of the global inventory (Archer, 2007).

There is *low confidence* in modelling abilities to simulate transient changes in hydrate inventories, but large CH_4 release to the atmosphere during this century is *unlikely*.

6.4.8 Other Drivers of Future Carbon Cycle Changes

6.4.8.1 Changes in Fire under Climate Change/Scenarios of Anthropogenic Fire Changes

Regional studies for boreal regions suggest an increase in future fire risk (e.g., Amiro et al., 2009; Balshi et al., 2009; Flannigan et al., 2009a; Spracklen et al., 2009; Tymstra et al., 2007; Westerling et al., 2011; Wotton et al., 2010) with implications for carbon and nutrient storage (Certini, 2005). Kurz et al. (2008b) and Metsaranta et al. (2010) indicated that increased fire activity has the potential to turn the Canadian forest from a sink to a source of atmospheric CO_2. Models predict spatially variable responses in fire activity, including strong increases and decreases, due to regional variations in the climate–fire relationship, and anthropogenic interference (Scholze et al., 2006; Flannigan et al., 2009b; Krawchuk et al., 2009; Pechony and Shindell, 2010; Kloster et al., 2012). Wetter conditions can reduce fire activity, but increased biomass availability can increase fire emissions (Scholze et al., 2006; Terrier et al., 2013). Using a land surface model and future climate projections from two GCMs, Kloster et al. (2012) projected fire carbon emissions in 2075–2099 that exceed present-day emissions by 17 to 62% (0.3 to 1.0 PgC yr^{-1}) depending on scenario.

Future fire activity will also depend on anthropogenic factors especially related to land use change. For the Amazon it is estimated that at present 58% of the area is too humid to allow deforestation fires but climate change might reduce this area to 37% by 2050 (Le Page et al., 2010). Golding and Betts (2008) estimated that future Amazon forest vulnerability to fire may depend nonlinearly on combined climate change and deforestation.

6.4.8.2 Other Biogeochemical Cycles and Processes Impacting Future Carbon Fluxes

6.4.8.2.1 Phosphorus

On centennial time scales, the phosphoros (P) limitation of terrestrial carbon uptake could become more severe than the nitrogen limitation because of limited phosphorus sources. Model simulations have shown a shift after 2100 from nitrogen to phosphorus limitation at high latitudes (Goll et al., 2012).

6.4.8.2.2 Elevated surface ozone

Plants are known to suffer damage due to exposure to levels of ozone (O_3) above about 40 ppb (Ashmore, 2005). Model simulations of plant O_3 damage on the carbon cycle have found a reduction in terrestrial carbon storage between 2005 and 2100 ranging from 4 to 140 PgC (Felzer et al., 2005) and up to 260 PgC (Sitch et al., 2007).

6.4.8.2.3 Iron deposition to oceans

Changes in iron deposition may have affected ocean carbon uptake in the past (Section 6.2.1.1), but future projections of iron deposition from desert dust over the ocean are uncertain, even about the sign of changes (Tegen et al., 2004; Mahowald et al., 2009). Tagliabue et al. (2008) found relatively little impact of varying aeolian iron input on ocean CO_2 fluxes, but Mahowald et al. (2011) show projected changes in ocean productivity as large as those due to CO_2 increases and climate change.

6.4.8.2.4 Changes in the diffuse fraction of solar radiation at the surface

Mercado et al. (2009) estimated that variations in the diffuse fraction, associated largely with the 'global dimming' period (Stanhill and Cohen, 2001), enhanced the land carbon sink by approximately 25% between 1960 and 1999. Under heavily polluted or dark cloudy skies, plant productivity may decline as the diffuse effect is insufficient to offset decreased surface irradiance (UNEP, 2011). Under future scenarios involving reductions in aerosol emissions (Figures 6.33 and 6.34), the diffuse-radiation enhancement of carbon uptake will decline.

6.4.9 The Long-term Carbon Cycle and Commitments

With *very high confidence*, the physical, biogeochemical carbon cycle in the ocean and on land will continue to respond to climate change and rising atmospheric CO_2 concentrations created during the 21st century. Long-term changes in vegetation structure and induced carbon storage potentially show larger changes beyond 2100 than during the 21st century as the long time scale response of tree growth and ecosystem migrations means that by 2100 only a part of the eventual committed change will be realized (Jones et al., 2009). Holocene changes in tree-line lagged changes in climate by centuries (MacDonald et al., 2008). Long-term 'commitments' to ecosystems migration also carry long-term committed effects to changes in terrestrial carbon storage (Jones et al., 2010; Liddicoat et al., 2013) and permafrost (O'Connor et al., 2010; Sections 6.4.3.3 and 6.4.7).

Warming of high latitudes is common to most climate models (Chapter 12) and this may enable increased productivity and northward expansion of boreal forest ecosystems into present tundra regions depending on nutrient availability (Kellomäki et al., 2008; Kurz et al., 2008a; MacDonald et al., 2008). CMIP5 simulations by two ESMs with dynamic vegetation for extended RCP scenarios to 2300 (Meinshausen et al., 2011) allow analysis of this longer term response of the carbon cycle. Increases in tree cover and terrestrial carbon storage north of 60°N are shown in Figure 6.38.

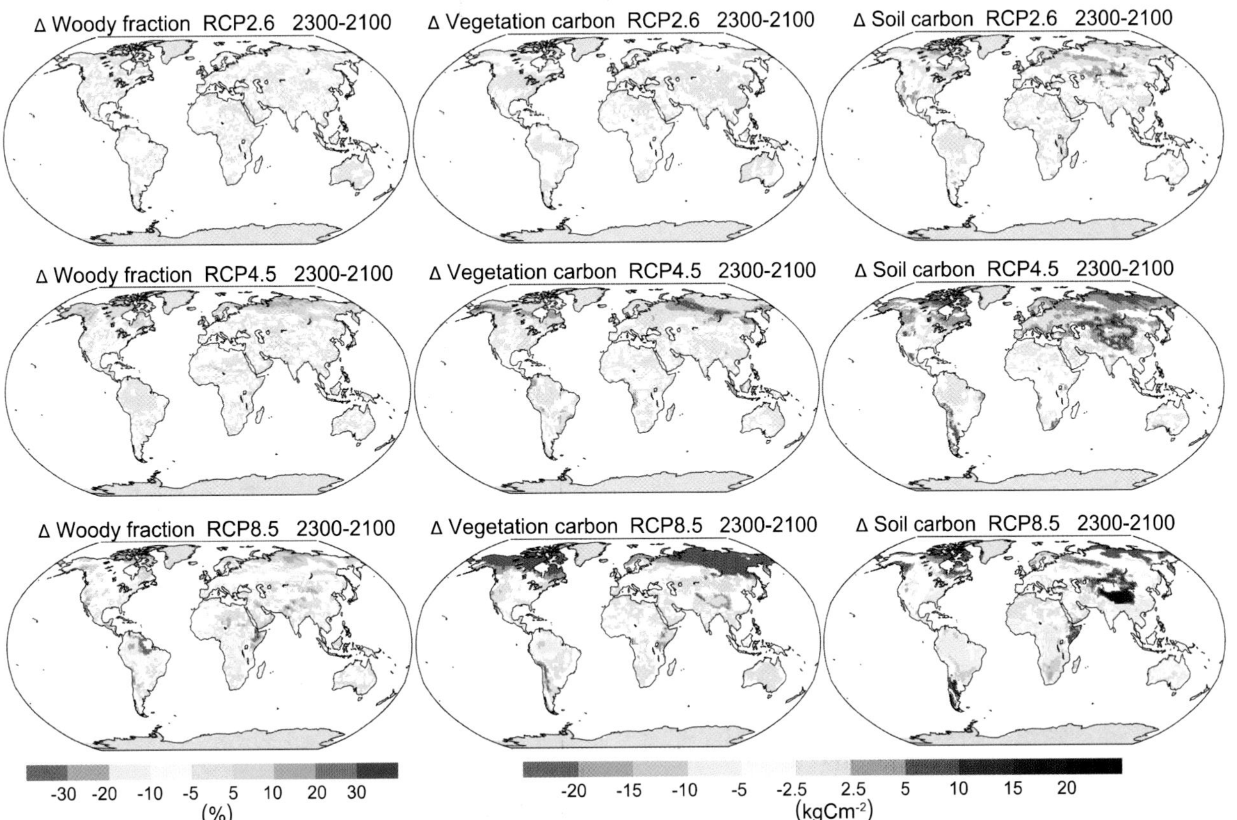

Figure 6.38 | Maps of changes in woody cover fraction, %, (left) and terrestrial carbon storage, kg C m^{-2} (vegetation carbon, middle; soil carbon, right) between years 2100 and 2300 averaged for two models, Hadley Centre Global Environmental Model 2-Earth System (HadGEM2-ES) and Max Planck Institute–Earth System Model (MPI-ESM), which simulate vegetation dynamics for three RCP extension scenarios 2.6 (top), 4.5 (middle), and 8.5 (bottom). Note the RCP6.0 extension was not a CMIP5 required simulation. Model results were interpolated on 1° × 1° grid; white colour indicates areas where models disagree in sign of changes. Anthropogenic land use in these extension scenarios is kept constant at 2100 levels, so these results show the response of natural ecosystems to the climate change.

Frequently Asked Questions

FAQ 6.2 | What Happens to Carbon Dioxide After It Is Emitted into the Atmosphere?

Carbon dioxide (CO_2), after it is emitted into the atmosphere, is firstly rapidly distributed between atmosphere, the upper ocean and vegetation. Subsequently, the carbon continues to be moved between the different reservoirs of the global carbon cycle, such as soils, the deeper ocean and rocks. Some of these exchanges occur very slowly. Depending on the amount of CO_2 released, between 15% and 40% will remain in the atmosphere for up to 2000 years, after which a new balance is established between the atmosphere, the land biosphere and the ocean. Geological processes will take anywhere from tens to hundreds of thousands of years—perhaps longer—to redistribute the carbon further among the geological reservoirs. Higher atmospheric CO_2 concentrations, and associated climate impacts of present emissions, will, therefore, persist for a very long time into the future.

CO_2 is a largely non-reactive gas, which is rapidly mixed throughout the entire troposphere in less than a year. Unlike reactive chemical compounds in the atmosphere that are removed and broken down by sink processes, such as methane, carbon is instead redistributed among the different reservoirs of the global carbon cycle and ultimately recycled back to the atmosphere on a multitude of time scales. FAQ 6.2, Figure 1 shows a simplified diagram of the global carbon cycle. The open arrows indicate typical timeframes for carbon atoms to be transferred through the different reservoirs.

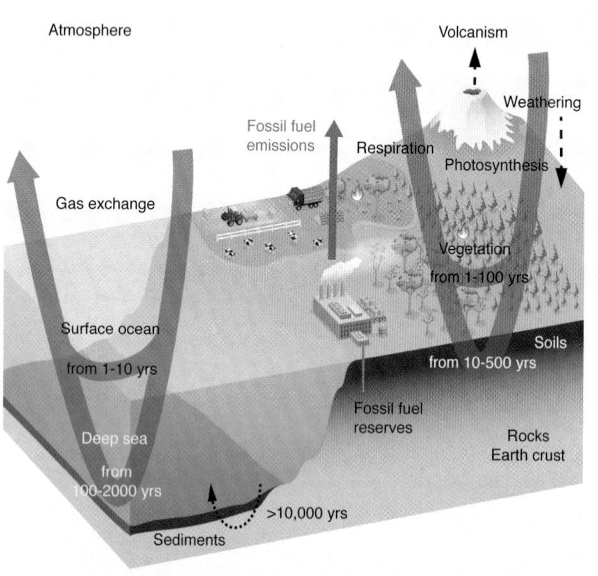

FAQ 6.2, Figure 1 | Simplified schematic of the global carbon cycle showing the typical turnover time scales for carbon transfers through the major reservoirs.

Before the Industrial Era, the global carbon cycle was roughly balanced. This can be inferred from ice core measurements, which show a near constant atmospheric concentration of CO_2 over the last several thousand years prior to the Industrial Era. Anthropogenic emissions of carbon dioxide into the atmosphere, however, have disturbed that equilibrium. As global CO_2 concentrations rise, the exchange processes between CO_2 and the surface ocean and vegetation are altered, as are subsequent exchanges within and among the carbon reservoirs on land, in the ocean and eventually, the Earth crust. In this way, the added carbon is redistributed by the global carbon cycle, until the exchanges of carbon between the different carbon reservoirs have reached a new, approximate balance.

Over the ocean, CO_2 molecules pass through the air-sea interface by gas exchange. In seawater, CO_2 interacts with water molecules to form carbonic acid, which reacts very quickly with the large reservoir of dissolved inorganic carbon—bicarbonate and carbonate ions—in the ocean. Currents and the formation of sinking dense waters transport the carbon between the surface and deeper layers of the ocean. The marine biota also redistribute carbon: marine organisms grow organic tissue and calcareous shells in surface waters, which, after their death, sink to deeper waters, where they are returned to the dissolved inorganic carbon reservoir by dissolution and microbial decomposition. A small fraction reaches the sea floor, and is incorporated into the sediments.

The extra carbon from anthropogenic emissions has the effect of increasing the atmospheric partial pressure of CO_2, which in turn increases the air-to-sea exchange of CO_2 molecules. In the surface ocean, the carbonate chemistry quickly accommodates that extra CO_2. As a result, shallow surface ocean waters reach balance with the atmosphere within 1 or 2 years. Movement of the carbon from the surface into the middle depths and deeper waters takes longer—between decades and many centuries. On still longer time scales, acidification by the invading CO_2 dissolves carbonate sediments on the sea floor, which further enhances ocean uptake. However, current understanding suggests that, unless substantial ocean circulation changes occur, plankton growth remains roughly unchanged because it is limited mostly by environmental factors, such as nutrients and light, and not by the availability of inorganic carbon it does not contribute significantly to the ocean uptake of anthropogenic CO_2. *(continued on next page)*

FAQ 6.2 (continued)

On land, vegetation absorbs CO_2 by photosynthesis and converts it into organic matter. A fraction of this carbon is immediately returned to the atmosphere as CO_2 by plant respiration. Plants use the remainder for growth. Dead plant material is incorporated into soils, eventually to be decomposed by microorganisms and then respired back into the atmosphere as CO_2. In addition, carbon in vegetation and soils is also converted back into CO_2 by fires, insects, herbivores, as well as by harvest of plants and subsequent consumption by livestock or humans. Some organic carbon is furthermore carried into the ocean by streams and rivers.

An increase in atmospheric CO_2 stimulates photosynthesis, and thus carbon uptake. In addition, elevated CO_2 concentrations help plants in dry areas to use ground water more efficiently. This in turn increases the biomass in vegetation and soils and so fosters a carbon sink on land. The magnitude of this sink, however, also depends critically on other factors, such as water and nutrient availability.

Coupled carbon-cycle climate models indicate that less carbon is taken up by the ocean and land as the climate warms constituting a positive climate feedback. Many different factors contribute to this effect: warmer seawater, for instance, has a lower CO_2 solubility, so altered chemical carbon reactions result in less oceanic uptake of excess atmospheric CO_2. On land, higher temperatures foster longer seasonal growth periods in temperate and higher latitudes, but also faster respiration of soil carbon.

The time it takes to reach a new carbon distribution balance depends on the transfer times of carbon through the different reservoirs, and takes place over a multitude of time scales. Carbon is first exchanged among the 'fast' carbon reservoirs, such as the atmosphere, surface ocean, land vegetation and soils, over time scales up to a few thousand years. Over longer time scales, very slow secondary geological processes—dissolution of carbonate sediments and sediment burial into the Earth's crust—become important.

FAQ 6.2, Figure 2 illustrates the decay of a large excess amount of CO_2 (5000 PgC, or about 10 times the cumulative CO_2 emitted so far since the beginning of the industrial Era) emitted into the atmosphere, and how it is redistributed among land and the ocean over time. During the first 200 years, the ocean and land take up similar amounts of carbon. On longer time scales, the ocean uptake dominates mainly because of its larger reservoir size (~38,000 PgC) as compared to land (~4000 PgC) and atmosphere (589 PgC prior to the Industrial Era). Because of ocean chemistry the size of the initial input is important: higher emissions imply that a larger fraction of CO_2 will remain in the atmosphere. After 2000 years, the atmosphere will still contain between 15% and 40% of those initial CO_2 emissions. A further reduction by carbonate sediment dissolution, and reactions with igneous rocks, such as silicate weathering and sediment burial, will take anything from tens to hundreds of thousands of years, or even longer.

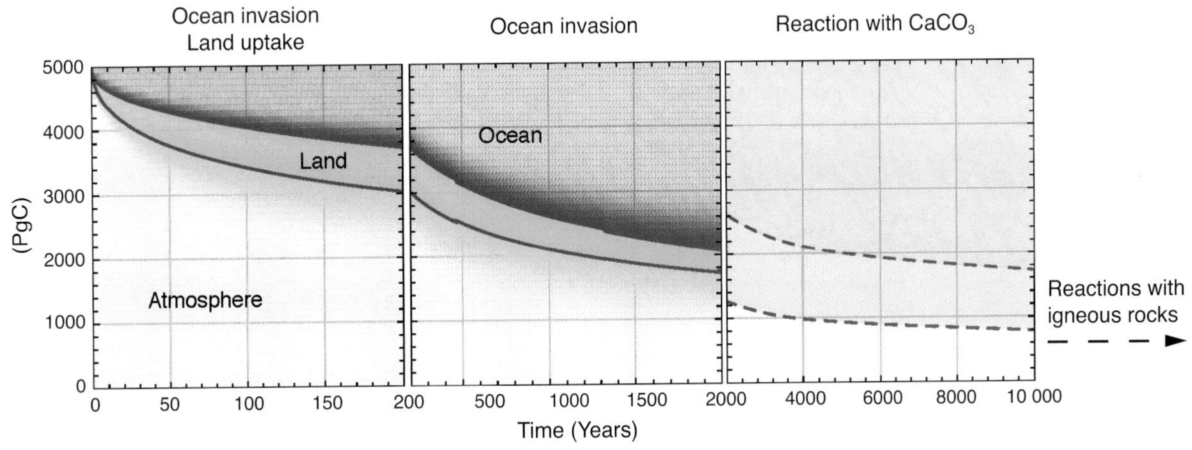

FAQ 6.2, Figure 2 | Decay of a CO_2 excess amount of 5000 PgC emitted at time zero into the atmosphere, and its subsequent redistribution into land and ocean as a function of time, computed by coupled carbon-cycle climate models. The sizes of the colour bands indicate the carbon uptake by the respective reservoir. The first two panels show the multi-model mean from a model intercomparison project (Joos et al., 2013). The last panel shows the longer term redistribution including ocean dissolution of carbonaceous sediments as computed with an Earth System Model of Intermediate Complexity (after Archer et al., 2009b).

Increases in fire disturbance or insect damage may drive loss of forest in temperate regions (Kurz et al., 2008c), but this process is poorly represented or not accounted at all in models. Recent evidence from models (Huntingford et al., 2013) and studies on climate variability (Cox et al., 2013) suggests that large scale loss of tropical forest as previously projected in some models (Cox et al., 2004; Scholze et al., 2006) is *unlikely*, but depends strongly on the predicted future changes in regional temperature (Galbraith et al., 2010) and precipitation (Good et al., 2011, 2013), although both models here simulate reduced tree cover and carbon storage for the RCP8.5 scenario. ESMs also poorly simulate resilience of ecosystems to climate changes and usually do not account for possible existence of alternative ecosystem states such as tropical forest or savannah (Hirota et al., 2011).

Regional specific changes in ecosystem composition and carbon storage are uncertain but it is *very likely* that ecosystems will continue to change for decades to centuries following stabilisation of GHGs and climate change.

6.5 Potential Effects of Carbon Dioxide Removal Methods and Solar Radiation Management on the Carbon Cycle

6.5.1 Introduction to Carbon Dioxide Removal Methods

To slow or perhaps reverse projected increases in atmospheric CO_2 (Section 6.4), several methods have been proposed to increase the removal of atmospheric CO_2 and enhance the storage of carbon in land, ocean and geological reservoirs. These methods are categorized as 'Carbon Dioxide Removal (CDR)' methods (see Glossary). Another class of methods involves the intentional manipulation of planetary solar absorption to counter climate change, and is called the 'Solar Radiation Management (SRM)' (discussed in Chapter 7, Section 7.7; see Glossary). In this section, CDR methods are discussed from the aspect of the carbon cycle processes (Section 6.5.2) and their impacts and side effects on carbon cycle and climate (Section 6.5.3). A brief discussion on the indirect carbon cycle effects of SRM methods is given in Section 6.5.4. Most of the currently proposed CDR methods are summarized in Table 6.14 and some are illustrated schematically in Chapter 7 (Section 7.7; FAQ 7.3 Figure 1). Since some CDR methods might operate on large spatial scales they are also called 'Geoengineering' proposals (Keith, 2001). Removal of CH_4 and N_2O has also been proposed to reduce climate change (Stolaroff et al., 2012). While the science of geoengineering methods is assessed in this section (CDR) and Chapter 7 (SRM), the benefits and risks of SRM are planned to be assessed in Chapter 19 of AR5 WGII report. Further, Chapter 6 of AR5 WGIII report plans to assess the cost and socioeconomic implications of some CDR and SRM methods for climate stabilization pathways.

6

Large-scale industrial methods such as carbon capture and storage (CCS), biofuel energy production (without CCS) and reducing emissions from deforestation and degradation (REDD) cannot be called CDR methods since they *reduce fossil fuel use or land use change CO_2 emissions* to the atmosphere but they do not involve a net removal of CO_2 that is already in the atmosphere. However, direct air capture of CO_2 using industrial methods (Table 6.14; and FAQ 7.3 Figure 1) will remove CO_2 from the atmosphere and is thus considered as a CDR method. The distinction between CDR and mitigation (see Glossary) is not clear and there could be some overlap between the two.

Insofar as the CDR-removed CO_2 is sequestered in a permanent reservoir, CDR methods could potentially reduce direct consequences of high CO_2 levels, including ocean acidification (see Section 6.4.4) (Matthews et al., 2009). However, the effects of CDR methods that propose to manipulate carbon cycle processes are slow (see Box 6.1) and hence the consequent climate effects would be slow. The climate system has a less than 5-years relaxation (e-folding) time scale for an assumed instantaneous reduction in radiative forcing to preindustrial levels (Held et al., 2010). While the climate effect of SRM could be rapid (Shepherd et al., 2009) given this time scale, at present, there is no known CDR method, including industrial direct air capture that can feasibly reduce atmospheric CO_2 to pre-industrial levels within a similar time scale. Therefore, CDR methods do not present an option for rapidly preventing climate change when compared to SRM. It is *likely* that CDR would have to be deployed at large-scale for at least one century to be able to significantly reduce atmospheric CO_2.

Important carbon cycle science considerations for evaluating CDR methods include the associated carbon storage capacity, the permanence of carbon storage and potential adverse side effects (Shepherd et al., 2009). Geological reservoirs could store several thousand PgC and the ocean may be able to store a few thousand PgC of anthropogenic carbon in the long-term (Metz et al., 2005; House et al., 2006; Orr, 2009) (see Box 6.1 and Archer et al., 2009b). The terrestrial biosphere may have a typical potential to store carbon equivalent to the cumulative historical land use loss of 180 ± 80 PgC (Table 6.1; Section 6.5.2.1).

In this assessment, we use "permanence" to refer to time scales larger than tens of thousands of years. CDR methods associated with either permanent or non-permanent carbon sequestration (see Table 6.14) have very different climate implications (Kirschbaum, 2003). Permanent sequestration methods have the potential to reduce the radiative forcing of CO_2 over time. By contrast, non-permanent sequestration methods will release back the temporarily sequestered carbon as CO_2 to the atmosphere, after some delayed time interval (Herzog et al., 2003). As a consequence, elevated levels of atmospheric CO_2 and climate warming will only be delayed and not avoided by the implementation of non-permanent CDR methods (Figure 6.39). Nevertheless, CDR methods that could create a temporary CO_2 removal (Table 6.14) may still have value (Dornburg and Marland, 2008) by reducing the cumulative impact of higher temperature.

Another important carbon cycle consequence of CDR methods is the 'rebound effect' (see Glossary). In the Industrial Era (since 1750) about half of the CO_2 emitted into the atmosphere from fossil fuel emissions has been taken up by land and ocean carbon reservoirs (see Section 6.3 and Table 6.1). As for current CO_2 emissions and the consequent CO_2 rise, which are currently *opposed* by uptake of CO_2 by natural reservoirs, any removal of CO_2 from the atmosphere by CDR will be *opposed* by release of CO_2 from natural reservoirs (Figure 6.40). It is thus *virtually certain* that the removal of CO_2 by CDR will be partially offset by outgassing of CO_2 from the ocean and land ecosystems. Therefore, return-

Table 6.14 | Examples of CDR methods and their implications for carbon cycle and climate. The list is non-exhaustive. A 'rebound' effect and a thermal inertia of climate system are associated with all CDR methods.

Carbon Cycle Process to be Modified Intentionally	CDR Method Name	Nature of CDR Removal Process	Storage Location	Storage Form	Some Carbon Cycle and Climate Implications
Enhanced biological production and storage on land	Afforestation / reforestation[a] Improved forest management[b] Sequestration of wood in buildings[c] Biomass burial[d] No till agriculture[e] Biochar[f] Conservation agriculture[g] Fertilisation of land plants[h] Creation of wetlands[i] Biomass Energy with Carbon Capture and Storage (BECCS)[j]	Biological	[a,b,h] Land (biomass, soils) [d] Land/ocean floor [e, f,g] Land (soils) [i] Land (wetland soils) [j] Ocean / geological formations	[a,b,c,d,e,f,g,h,i] Organic [j] Inorganic	[a,b,c,d,e,f,g,h,i,j] Alters surface albedo and evapotranspiration [a,b,c,e,f,g,h,i] Lack of permanence [d] Potentially permanent if buried on the ocean floor [j] Permanent if stored in geological reservoir
Enhanced biological production and storage in ocean	Ocean iron fertilisation[k] Algae farming and burial[l] Blue carbon (mangrove, kelp farming)[m] Modifying ocean upwelling to bring nutrients from deep ocean to surface ocean[n]	Biological	Ocean	[k,n] Inorganic [l,m] Organic	[k] May lead to expanded regions with low oxygen concentration, increased N_2O production, deep ocean acidification and disruptions to marine ecosystems and regional carbon cycle [n] Disruptions to regional carbon cycle
Accelerated weathering	Enhanced weathering over land[o] Enhanced weathering over ocean[p]	Chemical	[o] Soils and oceans [p] Ocean	[o,p] Inorganic	[o] Permanent removal; *likely* to change pH of soils, rivers, and ocean [p] Permanent removal; *likely* to change pH of ocean
Others	Direct-air capture with storage	Chemical	Ocean/geological formations	Inorganic	Permanent removal if stored in geological reservoirs

Notes

Superscripts in column 2 refer to the corresponding superscripts in columns 4, 5 and 6 of the same row.

ing to pre-industrial CO_2 levels would require permanently sequestering an amount of carbon equal to total anthropogenic CO_2 emissions that have been released before the time of CDR, roughly twice as much the excess of atmospheric CO_2 above pre-industrial level (Lenton and Vaughan, 2009; Cao and Caldeira, 2010b; Matthews, 2010).

6.5.2 Carbon Cycle Processes Involved in Carbon Dioxide Removal Methods

The CDR methods listed in Table 6.14 rely primarily on human management of carbon cycle processes to remove CO_2: (1) enhanced net biological uptake and subsequent sequestration by land ecosystems, (2) enhanced biological production in ocean and subsequent sequestration in the ocean and (3) accelerated chemical weathering reactions over land and ocean. The exceptional CDR method is industrial direct air capture of CO_2, for example, relying on chemistry methods. CO_2 removed by CDR is expected to be stored in organic form on land and in inorganic form in ocean and geological reservoirs (Table 6.14). This management of the carbon cycle however has other implications on ecosystems and biogeochemical cycles. The principle of different CDR methods listed in Table 6.14 is described below and the characteristics of some CDR methods are summarized in Table 6.15.

Some of the RCP scenarios used as a basis for future projections in this Assessment Report already include some CDR methods. To achieve the RCP2.6 CO_2 peak and decline the IMAGE integrated assessment model simulates widespread implementation of BECCS technology to achieve globally negative emissions after around 2080 (see Section 6.4.3). RCP4.5 also assumes some use of BECCS to stabilise CO_2 concentration by 2100. Therefore it should be noted that potentials for CDR assessed in this section cannot be seen as additional potential for CO_2 removal from the low RCPs as this is already included in those scenarios.

6.5.2.1 Enhanced Carbon Sequestration by Land Ecosystems

The key driver of these CDR methods is net primary productivity on land that currently produces biomass at a rate of approximately 50 to 60 PgC yr^{-1} (Nemani et al., 2003). The principle of these CDR methods is to increase net primary productivity and/or store a larger fraction of the biomass produced into ecosystem carbon pools with long turnover times, for example, under the form of wood or refractory organic matter in soils (Table 6.14). One variant is to harvest biomass for energy production and sequester the emitted CO_2 (BECCS). BECCS technology has not been tested at industrial scale, but is commonly included in Integrated Assessment Models and future scenarios that aim to achieve low CO_2 concentrations.

Estimates of the global potential for enhanced primary productivity over land are uncertain because the potential of any specific method will be severely constrained by competing land needs (e.g., agriculture, biofuels, urbanization and conservation) and sociocultural considerations. An order of magnitude of the upper potential of afforestation/reforestation would be the restoration of all the carbon released by

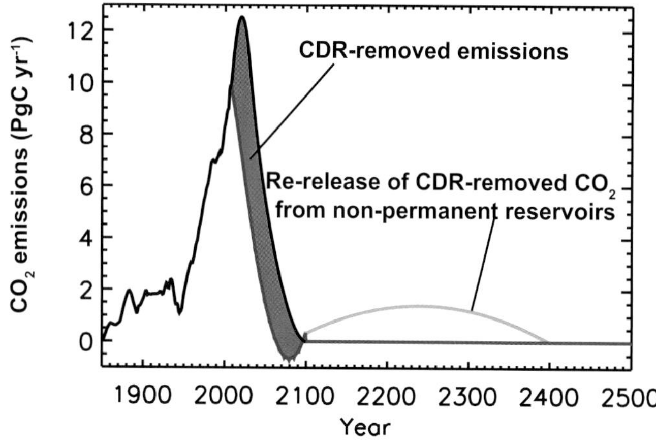

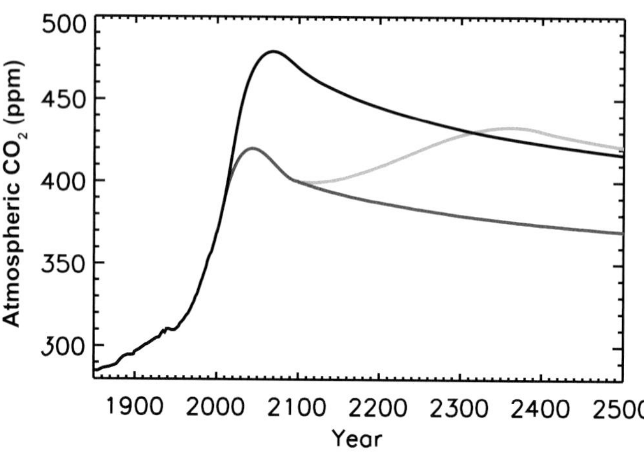

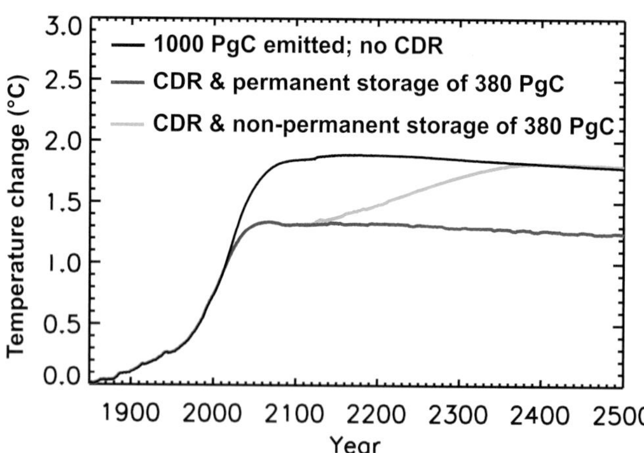

Figure 6.39 | Idealised model simulations (Matthews, 2010) to illustrate the effects of CDR methods associated with either permanent or non-permanent carbon sequestration. There is an emission of 1000 PgC in the reference case (black line) between 1800 and 2100, corresponding approximately to RCP4.5 scenario (Section 6.4). Permanent sequestration of 380 PgC, assuming no leakage of sequestered carbon would reduce climate change (blue line, compared to black line). By contrast, a non-permanent sequestration CDR method where carbon will be sequestered and later on returned to the atmosphere in three centuries would not. In this idealised non-permanent sequestration example scenario, climate change would only be delayed but the eventual magnitude of climate change will be equivalent to the no-sequestration case (green line, compared to black). Figure adapted from Figure 5 of Matthews (2010).

6

historical land use (180 ± 80 PgC; Table 6.1; Section 6.3.2.2). House et al. (2002) estimated that the atmospheric CO_2 concentration by 2100 would be lowered by only about 40 to 70 ppm in that scenario (accounting for the 'rebound' effect).

The capacity for enhancing the soil carbon content on agricultural and degraded lands was estimated by one study at 50 to 60% of the historical soil carbon released, that is 42 to 78 PgC (Lal, 2004a). The proposed agricultural practices are the adoption of conservation tillage with cover crops and crop residue mulch, conversion of marginal lands into restorative land uses and nutrient cycling including the use of

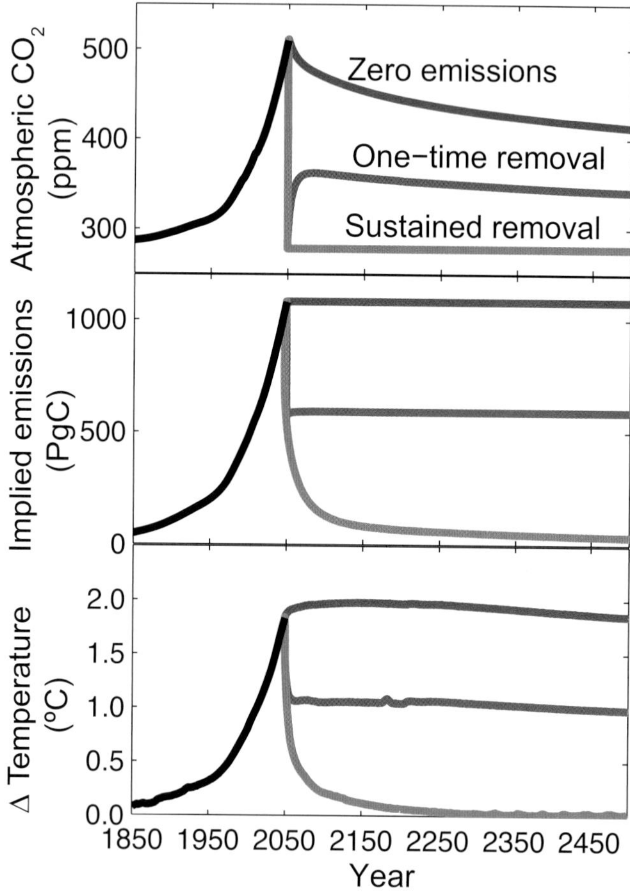

Figure 6.40 | Idealised simulations with a simple global carbon cycle model (Cao and Caldeira, 2010b) to illustrate the 'rebound effect'. Effects of an instantaneous cessation of CO_2 emissions in the year 2050 (amber line), one-time removal of the excess of atmospheric CO_2 over pre-industrial levels (blue line) and removal of this excess of atmospheric CO_2 followed by continued removal of all the CO_2 that degasses from the ocean (green line) are shown. For the years 1850–2010 observed atmospheric CO_2 concentrations are prescribed and CO_2 emissions are calculated from CO_2 concentrations and modeled carbon uptake. For the years 2011–2049, CO_2 emissions are prescribed following the SRES A2 scenario. Starting from year 2050, CO_2 emission is either set to zero or calculated from modeled CO_2 concentrations and CO_2 uptake. To a first approximation, a cessation of emissions would prevent further warming but would not lead to significant cooling on the century time scale. A one-time removal of excess atmospheric CO_2 would eliminate approximately only half of the warming experienced at the time of the removal because of CO_2 that outgases from the ocean (the rebound effect). To bring atmospheric CO_2 back to pre-industrial levels permanently, would require the removal of all previously emitted CO_2, that is, an amount equivalent to approximately twice the excess atmospheric CO_2 above pre-industrial level. (Figure adapted from Cao and Caldeira, 2010b.)

Table 6.15 | Characteristics of some CDR methods from peer-reviewed literature. Note that a variety of economic, environmental, and other constraints could also limit their implementation and net potential.

Carbon Dioxide Removal Method	Means of Removing CO_2 from Atmosphere	Carbon Storage / Form	Time Scale of Carbon Storage	Physical Potential of CO_2 Removed in a Century[a]	Reference	Unintended Side Effects
Afforestation and reforestation	Biological	Land /organic	Decades to centuries	40–70 PgC	House et al. (2002) Canadell and Raupach (2008)	Alters surface energy budget, depending on location; surface warming will be locally increased or decreased; hydrological cycle will be changed
Bio-energy with carbon-capture and storage (BECCS); biomass energy with carbon capture and storage	Biological	Geological or ocean /inorganic	Effectively permanent for geologic, centuries for ocean	125 PgC	See the footnote[b]	Same as above
Biochar creation and storage in soils	Biological	Land /organic	Decades to centuries	130 PgC	Woolf et al. (2010)	Same as above
Ocean fertilisation by adding nutrients to surface waters	Biological	Ocean / inorganic	Centuries to millennia	15–60 PgC 280 PgC	Aumont and Bopp (2006), Jin and Gruber (2003) Zeebe and Archer (2005) Cao and Caldeira (2010a)	Expanded regions with low oxygen concentration; enhanced N_2O emissions; altered production of dimethyl sulphide and non-CO_2 greenhouse gases; possible disruptions to marine ecosystems and regional carbon cycles
Ocean-enhanced upwelling bringing more nutrients to surface waters	Biological	Ocean / inorganic	Centuries to millennia	90 PgC 1–2 PgC	Oschlies et al. (2010a); Lenton and Vaughan (2009), Zhou and Flynn (2005)	*Likely* to cause changes to regional ocean carbon cycle opposing CO_2 removal, e.g., compensatory downwelling in other regions
Land-based increased weathering	Geochemical	Ocean (and some soils) / inorganic	Centuries to millennia for carbonates, permanent for silicate weathering	No determined limit 100 PgC	Kelemen and Matter (2008), Schuiling and Krijgsman (2006) Köhler et al. (2010)	pH of soils and rivers will increase locally, effects on terrestrial/ freshwater ecosystems
Ocean-based increased weathering	Geochemical	Ocean / inorganic	Centuries to millennia for carbonates, permanent for silicate weathering	No determined limit	Rau (2008), Kheshgi (1995)	Increased alkalinity effects on marine ecosystems
Direct air capture	Chemical	Geological or ocean /inorganic	Effectively permanent for geologic, centuries for ocean	No determined limit	Keith et al. (2006), Shaffer (2010)	Not known

Notes:

[a] Physical potential does not account for economic or environmental constraints of CDR methods; for example, the value of the physical potential for afforestation and reforestation does not consider the conflicts with land needed for agricultural production. Potentials for BECCS and biochar are highly speculative.

[b] If 2.5 tC yr^{-1} per hectare can be harvested on a sustainable basis (Kraxner et al., 2003) on about 4% (~500 million hectares, about one tenth of global agricultural land area) of global land (13.4 billion hectares) for BECCS, approximately 1.25 PgC yr^{-1} could be removed or about 125 PgC in this century. Future CO_2 concentration pathways, especially RCP2.6 and RCP4.5 include some CO_2 removal by BECCS (Chapter 6 of AR5 WGIII) and hence the potentials estimated here cannot add on to existing model results (Section 6.4).

compost and manure. Recent estimates suggest a cumulative potential of 30 to 60 PgC of additional storage over 25 to 50 years (Lal, 2004b).

Finally, biochar and biomass burial methods aim to store organic carbon into very long turnover time ecosystem carbon pools. The maximum sustainable technical potential of biochar cumulative sequestration is estimated at 130 PgC over a century by one study (Woolf et al., 2010). The residence time of carbon converted to biochar and the additional effect of biochar on soil productivity are uncertain, and further research is required to assess the potential of this method (Shepherd et al., 2009).

6.5.2.2 Enhanced Carbon Sequestration in the Ocean

The principle here is to enhance the primary productivity of phytoplankton (biological pump; Section 6.1.1) so that a fraction of the extra organic carbon produced gets transported to the deep ocean. Some of the inorganic carbon in the surface ocean that is removed by the export of net primary productivity below the surface layer will be subsequently replaced by CO_2 pumped from the atmosphere, thus removing atmospheric CO_2. Ocean primary productivity is limited by nutrients (e.g., iron, nitrogen and phosphorus). Enhanced biological production in ocean CDR methods (Table 6.14) is obtained by adding nutrients that would otherwise be limiting (Martin, 1990). The expected increase in the downward flux of carbon can be partly sequestered as Dissolved Inorganic Carbon (DIC) after mineralization in the intermediate and deep waters. In other ocean-based CDR methods, algae and kelp farming and burial, carbon would be stored in organic form.

The effectiveness of ocean CDR through iron addition depends on the resulting increase of productivity and the fraction of this extra carbon exported to deep and intermediate waters, and its fate. Small-scale

(~10 km^2) experiments (Boyd et al., 2007) have shown only limited transient effects of iron addition in removing atmospheric CO_2. An increased productivity was indeed observed, but this effect was moderated either by other limiting elements, or by compensatory respiration from increased zooplankton grazing. Most of the carbon produced by primary productivity is oxidized (remineralized into DIC) in the surface layer, so that only a small fraction is exported to the intermediate and deep ocean (Lampitt et al., 2008) although some studies indicate little remineralization in the surface layer (Jacquet et al., 2008). A recent study (Smetacek et al., 2012) finds that at least half the extra carbon in plankton biomass generated by artificial iron addition sank far below a depth of 1000 m, and that a substantial portion is likely to have reached the sea floor. There are some indications that sustained natural iron fertilisation may have a higher efficiency in exporting carbon from surface to intermediate and deep ocean than short term blooms induced by artificial addition of iron (Buesseler et al., 2004; Blain et al., 2007; Pollard et al., 2009). Thus, there is no consensus on the efficiency of iron fertilisation from available field experiments.

Using ocean carbon models (see Section 6.3.2.5.6), the maximum drawdown of atmospheric CO_2 have been estimated from 15 ppm (Zeebe and Archer, 2005) to 33 ppm (Aumont and Bopp, 2006) for an idealised continuous (over 100 years) global ocean iron fertilisation, which is technically unrealistic. In other idealised simulations of ocean fertilisation in the global ocean or only in the Southern Ocean (Joos et al., 1991; Peng and Broecker, 1991; Watson et al., 1994; Cao and Caldeira, 2010a), atmospheric CO_2 was reduced by less than 100 ppm for ideal conditions. Jin and Gruber (2003) obtained an atmospheric drawdown of more than 60 ppm over 100 years from an idealised iron fertilisation scenario over the entire Southern Ocean. The radiative benefit from lower CO_2 could be offset by a few percent to more than 100% from an increase in N_2O emissions (Jin and Gruber, 2003). All the above estimates of maximum potential CO_2 removal account for the rebound effect from oceans but not from the land (thus overestimate the atmospheric CO_2 reduction).

One ocean CDR variant is to artificially supply more nutrients to the surface ocean in upwelling areas (Lovelock and Rapley, 2007; Karl and Letelier). The amount of carbon sequestered by these enhanced upwelling methods critically depends on their location (Yool et al., 2009). Idealised simulations suggest an atmospheric CO_2 removal at a net rate of about 0.9 PgC yr^{-1} (Oschlies et al., 2010b). This ocean-based CDR method has not been tested in the field, unlike iron addition experiments.

6.5.2.3 Accelerated Weathering

The removal of CO_2 by the weathering of silicate and carbonate minerals (Berner et al., 1983; Archer et al., 2009b) occurs on time scales from thousands to tens of thousands of years (see Box 6.1) and at a rate of ~ 0.3 PgC yr^{-1} (Figure 6.1; Gaillardet et al., 1999; Hartmann et al., 2009). This rate is currently much too small to offset the rate at which fossil fuel CO_2 is being emitted (Section 6.3).

The principle of accelerated weathering CDR on land is to dissolve *artificially* silicate minerals so drawdown of atmospheric CO_2 and geochemical equilibrium restoration could proceed on a much faster (century) time scale. For instance, large amounts of silicate minerals such as olivine ($(Mg,Fe)_2SiO_4$) could be mined, crushed, transported to and distributed on agricultural lands, to remove atmospheric CO_2 and form carbonate minerals in soils and/or bicarbonate ions that would be transported to the ocean by rivers (Schuiling and Krijgsman, 2006)l. Alternatively, CO_2 removal by weathering reactions might be enhanced by exposing minerals such as basalt or olivine to elevated CO_2, with potential CO_2 removal rates exceeding 0.25 PgC yr^{-1} (Kelemen and Matter, 2008). In the idealised case where olivine could be spread as a fine powder over all the humid tropics, potential removal rates of up to 1 PgC yr^{-1} have been estimated, despite limitations by the saturation concentration of silicic acid (Köhler et al., 2010). For the United Kingdom, the potential from silicate resources was estimated to be more than 100 PgC (Renforth, 2012).

Fossil fuel CO_2 released to the atmosphere leads to the addition of anthropogenic CO_2 in the ocean (Section 6.3.2.5). This anthropogenic CO_2 will eventually dissolve ocean floor carbonate sediments to reach geochemical equilibrium on a 10 kyr time scale (Archer et al., 1997). The principle of ocean based weathering CDR methods is to accelerate this process. For instance, carbonate rocks could be crushed, reacted with CO_2 (e.g., captured at power plants) to produce bicarbonate ions that would be released to the ocean (Rau and Caldeira, 1999; Caldeira and Rau, 2000; Rau, 2008). Alternatively, carbonate minerals could be directly released into the ocean (Kheshgi, 1995; Harvey, 2008). Strong bases, derived from silicate rocks, could also be released to ocean (House et al., 2007) to increase alkalinity and drawdown of atmospheric CO_2. Carbonate minerals such as limestone could be heated to produce lime ($Ca(OH)_2$); this lime could be added to the ocean to increase alkalinity as well (Kheshgi, 1995). While the level of *confidence* is *very high* for the scientific understanding of weathering chemical reactions, it is *low* for its effects and risks at planetary scale (Section 6.5.3.3).

6.5.2.4 Carbon Dioxide Removal by Direct Industrial Capture of Atmospheric Carbon Dioxide

Direct Air Capture refers to the chemical process by which a pure CO_2 stream is produced by capturing CO_2 from ambient air. The captured CO_2 could be sequestered in geological reservoirs or the deep ocean. At least three methods have been proposed to capture CO_2 from the atmosphere: (1) adsorption on solids (Gray et al., 2008; Lackner, 2009, 2010; Lackner et al., 2012); (2) absorption into highly alkaline solutions (Stolaroff et al., 2008; Mahmoudkhani and Keith, 2009) and (3) absorption into moderate alkaline solution with a catalyst (Bao and Trachtenberg, 2006). The main limitation to direct air capture is the thermodynamic barrier due to the low concentration of CO_2 in ambient air.

6.5.3 Impacts of Carbon Dioxide Removal Methods on Carbon Cycle and Climate

One impact common to all CDR methods is related to the thermal inertia of the climate system. Climate warming will indeed continue for at least decades after CDR is applied. Therefore, temperature (and climate change) will lag a CDR-induced decrease in atmospheric CO_2 (Boucher et al., 2012). Modelling the impacts of CDR on climate change is still in its infancy. Some of the first studies (Wu et al., 2010; Cao et al., 2011) showed that the global hydrological cycle could intensify in response to a reduction in atmospheric CO_2 concentrations.

6.5.3.1 Impacts of Enhanced Land Carbon Sequestration

In the case of land-based CDR, biomass in forests is a non-permanent ecosystem carbon pool and hence there is a risk that this carbon may return to the atmosphere, for example, by disturbances such as fire, or by future land use change. When considering afforestation/reforestation, it is also important to account for biophysical effects on climate that come together with carbon sequestration because afforestation/reforestation changes the albedo (see Glossary), evapotranspiration and the roughness of the surface (Bonan, 2008; Bernier et al., 2011). Modelling studies show that afforestation in seasonally snow covered boreal and temperate regions will decrease the land surface albedo and have a net (biophysical plus biogeochemical) warming effect, whereas afforestation in low latitudes (Tropics) is *likely* to enhance latent heat flux from evapotranspiration and have a net cooling effect (Bonan et al., 1992; Betts, 2000; Bala et al., 2007; Montenegro et al., 2009; Bathiany et al., 2010). Consequently, the location of land ecosystem based CDR methods needs to be considered carefully when evaluating their effects on climate (Bala et al., 2007; Arora and Montenegro, 2011; Lee et al., 2011; Pongratz et al., 2011b). In addition CDR in land ecosystems is *likely* to increase N_2O emissions (Li et al., 2005). Enhanced biomass production may also require more nutrients (fertilisers) which are associated with fossil fuel CO_2 emission from industrial fertiliser production and Nr impacts. Biochar-based CDR could reduce N_2O emissions but may increase CO_2 and CH_4 emissions from agricultural soils (Wang et al., 2012b). Addition of biochar could also promote a rapid loss of forest humus and soil carbon in some ecosystems during the first decades (Wardle et al., 2008).

6.5.3.2 Impacts of Enhanced Carbon Sequestration in the Ocean

In the case of ocean-based CDR using fertilisation, adding macronutrients such as nitrogen and phosphate in the fertilised region could lead to a decrease in production 'downstream' of the fertilised region (Gnanadesikan et al., 2003; Gnanadesikan and Marinov, 2008; Watson et al., 2008). Gnanadesikan et al. (2003) simulated a decline in export production of 30 tC for every ton removed from the atmosphere. A sustained global-ocean iron fertilisation for SRES A2 CO_2 emission scenario was also found to acidify the deep ocean (pH decrease of about 0.1 to 0.2) while mitigating surface pH change by only 0.06 (Cao and Caldeira, 2010a). Other environmental risks associated with ocean fertilisation include expanded regions with low oxygen concentration (Oschlies et al., 2010a), increased N_2O emission (Jin and Gruber, 2003), increased production of dimethylsulphide (DMS), isoprene, CO, N_2O, CH_4 and other non-CO_2 GHGs (Oschlies et al., 2010a) and possible disruptions to marine ecosystems (Denman, 2008).

In the case of enhanced ocean upwelling CDR methods there could be disturbance to the regional carbon balances, since the extra-upwelling will be balanced by extra-downwelling at another location. Along with growth-supporting nutrients, enhanced concentrations of DIC will also be brought to surface waters and partially offset the removal of CO_2 by increased biological pump. Further, in case artificially enhanced upwelling would be stopped, atmospheric CO_2 concentrations could rise rapidly because carbon removed from the atmosphere and stored in soils in the cooler climate caused by artificial upwelling could be rapidly released back (Oschlies et al., 2010b). The level of *confidence* on the impacts of the enhanced upwelling is *low*.

6.5.3.3 Impact of Enhanced Weathering

In the case of weathering-based CDR, the pH and carbonate mineral saturation of soils, rivers and ocean surface waters will increase where CDR is implemented. Köhler et al. (2010) simulated that the pH of the Amazon river would rise by 2.5 units if the dissolution of olivine in the entire Amazon basin was used to remove 0.5 PgC yr^{-1} from the atmosphere. In the marine environment, elevated pH and increased alkalinity could potentially counteract the effects of ocean acidification, which is beneficial. Changes in alkalinity could also modify existing ecosystems. There is uncertainty in our understanding of the net effect on ocean CO_2 uptake but there will be a partial offset of the abiotic effect by calcifying species. As for other CDR methods, the confidence level on the carbon cycle impacts of enhanced weathering is low.

6.5.4 Impacts of Solar Radiation Management on the Carbon Cycle

Solar radiation management (SRM) methods aim to reduce incoming solar radiation at the surface (discussed in Section 7.7 and in AR5, WG2, Chapter 19). Balancing reduced outgoing radiation by reduced incoming radiation may be able to cool global mean temperature but may lead to a less intense global hydrological cycle (Bala et al., 2008) with regionally different climate impacts (Govindasamy et al., 2003; Matthews and Caldeira, 2007; Robock et al., 2008; Irvine et al., 2010; Ricke et al., 2010). Therefore, SRM will not prevent the effects of climate change on the carbon and other biogeochemical cycles.

SRM could reduce climate warming but will not interfere with the direct biogeochemical effects of elevated CO_2 on the carbon cycle. For example, ocean acidification caused by elevated CO_2 (Section 6.4.4) and the CO_2 fertilisation of productivity (Box 6.3) will not be altered by SRM (Govindasamy et al., 2002; Naik et al., 2003; Matthews and Caldeira, 2007). Similarly, SRM will not interfere with the stomatal response of plants to elevated CO_2 (the CO_2-physiological effect) that leads to a decline in evapotranspiration, causing land temperatures to warm and runoff to increase (Gedney et al., 2006; Betts et al., 2007; Matthews and Caldeira, 2007; Piao et al., 2007; Cao et al., 2010; Fyfe et al., 2013).

However, due to carbon–climate feedbacks (Section 6.4), the implementation of SRM could affect the carbon cycle. For instance, carbon uptake by land and ocean could increase in response to SRM by reducing the negative effects of climate change on carbon sinks (Matthews and Caldeira, 2007). For instance, for the SRES A2 scenario with SRM, a lower CO_2 concentration of 110 ppm by year 2100 relative to a baseline case without SRM has been simulated by Matthews and Caldeira (2007). Land carbon sinks may be enhanced by increasing the amount of diffuse relative to direct radiation (Mercado et al., 2009) if SRM causes the fraction of diffuse light to increase (e.g., injection of aerosols into the stratosphere). However, reduction of total incoming solar radiation could decrease terrestrial CO_2 sinks as well.

6.5.5 Synthesis

CDR methods are intentional large scale methods to remove atmospheric CO_2 either by managing the carbon cycle or by direct industrial processes (Table 6.14). In contrast to SRM methods, CDR methods that manage the carbon cycle are *unlikely* to present an option for rapidly preventing climate change. The maximum (idealised) potential for atmospheric CO_2 removal by individual CDR methods is compiled in Table 6.15. In this compilation, note that unrealistic assumptions about the scale of deployment, such as fertilising the entire global ocean, are used, and hence large potentials are simulated. The 'rebound effect' in the natural carbon cycle is *likely* to diminish the effectiveness of all the CDR methods (Figure 6.40). The level of *confidence* on the effects of both CDR and SRM methods on carbon and other biogeochemical cycles is *very low*.

Acknowledgements

We wish to acknowledge Anna Peregon (LSCE, France) for investing countless hours compiling and coordinating input from all of the Chapter 6 Lead Authors. She was involved in the production of every aspect of the chapter and we could not have completed our task on time without her help. We also thank Brett Hopwood (ORNL, USA) for skilful and artistic edits of several graphical figures representing the global biogeochemical cycles in the Chapter 6 Introduction. We also thank Silvana Schott (Max Planck Institute for Biogeochemistry, Germany) for graphics artwork for several of the figures in Chapter 6.

References

Achard, F., H. D. Eva, P. Mayaux, H.-J. Stibig, and A. Belward, 2004: Improved estimates of net carbon emissions from land cover change in the tropics for the 1990s. *Global Biogeochem. Cycles*, **18**, GB2008.

Adair, E. C., P. B. Reich, S. E. Hobbie, and J. M. H. Knops, 2009: Interactive effects of time, CO_2, N, and diversity on total belowground carbon allocation and ecosystem carbon storage in a grassland community. *Ecosystems*, **12**, 1037–1052.

Adkins, J. F., K. McIntyre, and D. P. Schrag, 2002: The salinity, temperature and ?^{18}O of the glacial deep ocean. *Science*, **298**, 1769–1773.

Ahn, J. and E. J. Brook, 2008: Atmospheric CO_2 and climate on millennial time scales during the last glacial period. *Science*, **322**, 83–85.

Ahn, J., et al., 2012: Atmospheric CO_2 over the last 1000 years: A high resolution record from the West Antarctic Ice Sheet (WAIS) Divide ice core. *Global Biogeochem. Cycles*, **26**, GB2027.

Ainsworth, E. A. and S. P. Long, 2004: What have we learned from 15 years of free-air CO_2 enrichment (FACE)? A meta-analytic review of the responses of photosynthesis, canopy properties and plant production to rising CO_2. *New Phytologist*, **165**, 351–372.

Ainsworth, E. A., C. R. Yendrek, S. Sitch, W. J. Collins, and L. D. Emberson, 2012: The effects of tropospheric ozone on net primary productivity and implications for climate change. *Annu. Rev. Plant Biol.*, **63**, 637–661.

Allan, W., H. Struthers, and D. C. Lowe, 2007: Methane carbon isotope effects caused by atomic chlorine in the marine boundary layer: Global model results compared with Southern Hemisphere measurements. *J. Geophys. Res. Atmos.*, **112**, D04306.

Amiro, B. D., A. Cantin, M. D. Flannigan, and W. J. de Groot, 2009: Future emissions from Canadian boreal forest fires. *Can. J. Forest Res.*, **39**, 383–395.

Anderson, R. F., M. Q. Fleisher, Y. Lao, and G. Winckler, 2008: Modern $CaCO_3$ preservation in equatorial Pacific sediments in the context of late-Pleistocene glacial cycles. *Mar. Chem.*, **111**, 30–46.

Andreae, M. O. and P. Merlet, 2001: Emission of trace gases and aerosols from biomass burning. *Global Biogeochem. Cycles*, **15**, 955–966.

Andres, R. J., J. S. Gregg, L. Losey, G. Marland, and T. A. Boden, 2011: Monthly, global emissions of carbon dioxide from fossil fuel consumption. *Tellus B*, **63**, 309–327.

Andres, R. J., et al., 2012: A synthesis of carbon dioxide emissions from fossil-fuel combustion. *Biogeosciences*, **9**, 1845–1871.

Aranjuelo, I., et al., 2011: Maintenance of C sinks sustains enhanced C assimilation during long-term exposure to elevated [CO_2] in Mojave Desert shrubs. *Oecologia*, **167**, 339–354.

Archer, D., 2007: Methane hydrate stability and anthropogenic climate change. *Biogeosciences*, **4**, 521–544.

Archer, D. and E. Maier-Reimer, 1994: Effect of deep-sea sedimentary calcite preservation on atmospheric CO_2 concentration. *Nature*, **367**, 260–263.

Archer, D. and V. Brovkin, 2008: The millennial atmospheric lifetime of anthropogenic CO_2. *Clim. Change*, **90**, 283–297.

Archer, D., H. Kheshgi, and E. Maier-Reimer, 1997: Multiple timescales for neutralization of fossil fuel CO_2. *Geophys. Res. Lett.*, **24**, 405–408.

Archer, D., H. Kheshgi, and E. Maier-Reimer, 1998: Dynamics of fossil fuel CO_2 neutralization by marine $CaCO_3$. *Global Biogeochem. Cycles*, **12**, 259–276.

Archer, D., B. Buffett, and V. Brovkin, 2009a: Ocean methane hydrates as a slow tipping point in the global carbon cycle. *Proc. Natl. Acad. Sci. U.S.A.*, **106**, 20596–20601.

Archer, D., A. Winguth, D. Lea, and N. Mahowald, 2000: What caused the glacial/interglacial atmospheric pCO_2 cycles? *Rev. Geophys.*, **38**, 159–189.

Archer, D., et al., 2009b: Atmospheric lifetime of fossil fuel carbon dioxide. *Annu. Rev. Earth Planet. Sci.*, **37**, 117–134.

Archer, D. E., P. A. Martin, J. Milovich, V. Brovkin, G.-K. Plattner, and C. Ashendel, 2003: Model sensitivity in the effect of Antarctic sea ice and stratification on atmospheric pCO_2. *Paleoceanography*, **18**, 1012.

Archibald, S., D. P. Roy, B. W. van Wilgen, and R. J. Scholes, 2009: What limits fire? An examination of drivers of burnt area in Southern Africa. *Global Change Biol.*, **15**, 613–630.

Arneth, A., et al., 2010: Terrestrial biogeochemical feedbacks in the climate system. *Nature Geosci.*, **3**, 525–532.

Arora, V. K., and G. J. Boer, 2010: Uncertainties in the 20th century carbon budget associated with land use change. *Global Change Biol.*, **16**, 3327–3348.

Arora, V. K., and A. Montenegro, 2011: Small temperature benefits provided by realistic afforestation efforts. *Nature Geosci.*, **4**, 514–518.

Arora, V. K., et al., 2011: Carbon emission limits required to satisfy future representative concentration pathways of greenhouse gases. *Geophys. Res. Lett.*, **38**, L05805.

Arora, V. K., et al., 2013: Carbon-concentration and carbon-climate feedbacks in CMIP5 Earth system models. *J. Clim.*, **26**, 5289-5314.

Artioli, Y., et al., 2012: The carbonate system in the North Sea: Sensitivity and model validation. *J. Mar. Syst.*, **102–104**, 1–13.

Ashmore, M. R., 2005: Assessing the future global impacts of ozone on vegetation. *Plant Cell Environ.*, **28**, 949–964.

Assmann, K. M., M. Bentsen, J. Segschneider, and C. Heinze, 2010: An isopycnic ocean carbon cycle model. *Geosci. Model Dev.*, **3**, 143–167.

Aufdenkampe, A. K., et al., 2011: Rivering coupling of biogeochemical cycles between land, oceans and atmosphere. *Front. Ecol. Environ.*, **9**, 23–60.

Aumont, O., and L. Bopp, 2006: Globalizing results from ocean in situ iron fertilization studies. *Global Biogeochem. Cycles*, **20**, GB2017.

Avis, C. A., A. J. Weaver, and K. J. Meissner, 2011: Reduction in areal extent of high-latitude wetlands in response to permafrost thaw. *Nature Geosci.*, **4**, 444–448.

Ayres, R. U., W. H. Schlesinger, and R. H. Socolow, 1994: Human impacts on the carbon and nitrogen cycles.In: *Industrial Ecology and Global Change* [R. H. Socolow, C. Andrews, F. Berkhout and V. Thomas (eds.)]. Cambridge University Press, Cambridge, United Kingdom, and New York, NY, USA, pp. 121–155.

Bacastow, R. B., and C. D. Keeling, 1979: Models to predict future atmospheric CO_2 concentrations. In: *Workshop on the Global Effects of Carbon Dioxide from Fossil Fuels.* United States Department of Energy, Washington, DC, pp. 72–90.

Baccini, A., et al., 2012: Estimated carbon dioxide emissions from tropical deforestation improved by carbon-density maps. *Nature Clim. Change*, **2**, 182–185.

Bader, M., E. Hiltbrunner, and C. Körner, 2009: Fine root responses of mature deciduous forest trees to free air carbon dioxide enrichment (FACE). *Funct. Ecol.*, **23**, 913–921.

Baker, A., S. Cumberland, and N. Hudson, 2008: Dissolved and total organic and inorganic carbon in some British rivers. *Area*, **40**, 117–127.

Baker, D. F., et al., 2006: TransCom 3 inversion intercomparison: Impact of transport model errors on the interannual variability of regional CO_2 fluxes, 1988–2003. *Global Biogeochem. Cycles*, **20**, GB1002.

Bala, G., P. B. Duffy, and K. E. Taylor, 2008: Impact of geoengineering schemes on the global hydrological cycle. *Proc. Natl. Acad. Sci. U.S.A.*, **105**, 7664–7669.

Bala, G., K. Caldeira, M. Wickett, T. J. Phillips, D. B. Lobell, C. Delire, and A. Mirin, 2007: Combined climate and carbon-cycle effects of large-scale deforestation. *Proc. Natl. Acad. Sci. U.S.A.*, **104**, 6550–6555.

Baldocchi, D. D., et al., 2001: FLUXNET: A new tool to study the temporal and spatial variability of ecosystem-scale carbon dioxide, water vapor and energy flux densities. *Bull. Am. Meteorol. Soc.*, **82**, 2415–2435.

Ballantyne, A. P., C. B. Alden, J. B. Miller, P. P. Tans, and J. W. C. White, 2012: Increase in observed net carbon dioxide uptake by land and oceans during the last 50 years. *Nature*, **488**, 70–72.

Balshi, M. S., A. D. McGuire, P. Duffy, M. D. Flannigan, D. W. Kicklighter, and J. Melillo, 2009: Vulnerability of carbon storage in North American boreal forests to wildfires during the 21st century. *Global Change Biol.*, **15**, 1491–1510.

Bao, L. H., and M. C. Trachtenberg, 2006: Facilitated transport of CO_2 across a liquid membrane: Comparing enzyme, amine, and alkaline. *J. Membr. Sci.*, **280**, 330–334.

Barnard, R., P. W. Leadley, and B. A. Hungate, 2005: Global change, nitrification, and denitrification: A review. *Global Biogeochem. Cycles*, **19**, GB1007.

Barnes, R. T., and P. A. Raymond, 2009: The contribution of agricultural and urban activities to inorganic carbon fluxes within temperate watersheds. *Chem. Geol.*, **266**, 318–327.

Barnosky, A. D., 2008: Colloquium Paper: Megafauna biomass tradeoff as a driver of Quaternary and future extinctions. *Proc. Natl. Acad. Sci. U.S.A.*, **105**, 11543–11548.

Bastviken, D., J. Cole, M. Pace, and L. Tranvik, 2004: Methane emissions from lakes: Dependence of lake characteristics, two regional assessments, and a global estimate. *Global Biogeochem. Cycles*, **18**, GB4009.

Bastviken, D., L. J. Tranvik, J. A. Downing, P. M. Crill, and A. Enrich-Prast, 2011: Freshwater methane emissions offset the continental carbon sink. *Science*, **331**, 50.

Bathiany, S., M. Claussen, V. Brovkin, T. Raddatz, and V. Gayler, 2010: Combined biogeophysical and biogeochemical effects of large-scale forest cover changes in the MPI earth system model. *Biogeosciences*, **7**, 1383–1399.

Batjes, N. H., 1996: Total carbon and nitrogen in the soils of the world. *Eur. J. Soil Sci.*, **47**, 151–163.

Battin, T. J., S. Luyssaert, L. A. Kaplan, A. K. Aufdenkampe, A. Richter, and L. J. Tranvik, 2009: The boundless carbon cycle *Nature Geosci.*, **2**, 598–600.

Beaugrand, G., M. Edwards, and L. Legendre, 2010: Marine biodiversity, ecosystem functioning, and carbon cycles. *Proc. Natl. Acad. Sci. U.S.A.*, **107**, 10120–10124.

Beaulieu, J. J., et al., 2011: Nitrous oxide emission from denitrification in stream and river networks. *Proc. Natl. Acad. Sci. U.S.A.*, **108**, 214–219.

Beer, C., et al., 2010: Terrestrial gross carbon dioxide uptake: Global distribution and covariation with climate. *Science*, **329**, 834-838.

Bellassen, V., G. Le Maire, J. F. Dhote, P. Ciais, and N. Viovy, 2010: Modelling forest management within a global vegetation model. Part 1: Model structure and general behaviour. *Ecol. Model.*, **221**, 2458–2474.

Bellassen, V., N. Viovy, S. Luyssaert, G. Le Maire, M.-J. Schelhaas, and P. Ciais, 2011: Reconstruction and attribution of the carbon sink of European forests between 1950 and 2000. *Global Change Biol.*, **17**, 3274–3292.

Bennington, V., G. A. McKinley, S. Dutkiewicz, and D. Ullman, 2009: What does chlorophyll variability tell us about export and air-sea CO_2 flux variability in the North Atlantic? *Global Biogeochem. Cycles*, **23**, GB3002.

Berendse, F., et al., 2001: Raised atmospheric CO_2 levels and increased N deposition cause shifts in plant species composition and production in Sphagnum bogs. *Global Change Biol.*, **7**, 591–598.

Bergamaschi, P., et al., 2009: Inverse modeling of global and regional CH_4 emissions using SCIAMACHY satellite retrievals. *J. Geophys. Res.*, **114**, D22301.

Berger, W. H., 1982: Increase of carbon dioxide in the atmosphere during deglaciation: The coral-reef hypothesis. *Naturwissenschaften*, **69**, 87–88.

Berner, R. A., 1992: Weathering, plants, and the long-term carbon-cycle. *Geochim. Cosmochim. Acta*, **56**, 3225–3231.

Berner, R. A., A. C. Lasaga, and R. M. Garrels, 1983: The carbonate-silicate geochemical cycle and its effect on atmospheric carbon dioxide over the past 100 million years. *Am. J. Sci.*, **283**, 641–683.

Bernier, P. Y., R. L. Desjardins, Y. Karimi-Zindashty, D. E. Worth, A. Beaudoin, Y. Luo, and S. Wang, 2011: Boreal lichen woodlands: A possible negative feedback to climate change in Eastern North America. *Agr. Forest Meteorol.*, **151**, 521–528.

Betts, R. A., 2000: Offset of the potential carbon sink from boreal forestation by decreases in surface albedo. *Nature*, **408**, 187–190.

Betts, R. A., et al., 2007: Projected increase in continental runoff due to plant responses to increasing carbon dioxide. *Nature*, **448**, 1037–1041.

Bianchi, D., J. P. Dunne, J. L. Sarmiento, and E. D. Galbraith, 2012: Data-based estimates of suboxia, denitrification and N_2O production in the ocean and their sensitivity to dissolved O_2. *Global Biogeochem. Cycles*, **26**, GB2009.

Biastoch, A., et al., 2011: Rising Arctic Ocean temperatures cause gas hydrate destabilization and ocean acidification. *Geophys. Res. Lett.*, **38**, L08602.

Billings, S. A., S. M. Schaeffer, and R. D. Evans, 2002: Trace N gas losses and N mineralization in Mojave desert soils exposed to elevated CO_2. *Soil Biol. Biochem.*, **34**, 1777–1784.

Bird, M. I., J. Lloyd, and G. D. Farquhar, 1996: Terrestrial carbon storage from the last glacial maximum to the present. *Chemosphere*, **33**, 1675–1685.

Blain, S., et al., 2007: Effect of natural iron fertilization on carbon sequestration in the Southern Ocean. *Nature*, **446**, 1070–1074.

Blake, D. R., E. W. Mayer, S. C. Tyler, Y. Makide, D. C. Montague, and F. S. Rowland, 1982: Global increase in atmospheric methane concentrations between 1978 and 1980. *Geophys. Res. Lett.*, **9**, 477–480.

Bleeker, A., W. K. Hicks, F. Dentener, and J. Galloway, 2011: Nitrogen deposition as a threat to the World's protected areas under the Convention on Biological Diversity. *Environ. Pollut.*, **159**, 2280–2288.

Bloom, A. A., J. Lee-Taylor, S. Madronich, D. J. Messenger, P. I. Palmer, D. S. Reay, and A. R. McLeod, 2010: Global methane emission estimates from ultraviolet irradiation of terrestrial plant foliage. *New Phytologist*, **187**, 417–425.

Blunier, T., J. Chappellaz, J. Schwander, B. Stauffer, and D. Raynaud, 1995: Variations in atmospheric methane concentration during the Holocene epoch. *Nature*, **374**, 46–49.

Boardman, C. P., V. Gauci, J. S. Watson, S. Blake, and D. J. Beerling, 2011: Contrasting wetland CH_4 emission responses to simulated glacial atmospheric CO_2 in temperate bogs and fens. *New Phytologist*, **192**, 898–911.

Bock, M., J. Schmitt, L. Moller, R. Spahni, T. Blunier, and H. Fischer, 2010: Hydrogen isotopes preclude marine hydrate CH_4 emissions at the onset of Dansgaard-Oeschger events. *Science*, **328**, 1686–1689.

Boden, T., G. Marland, and R. Andres, 2011: Global CO_2 emissions from fossil-fuel burning, cement manufacture, and gas flaring: 1751–2008 (accessed at 2011.11.10). Oak Ridge National Laboratory, U. S. Department of Energy, Carbon Dioxide Information Analysis Center, Oak Ridge, TN, U.S.A., doi:10.3334/CDIAC/00001_V2011, http://cdiac.ornl.gov/trends/emis/overview_2008.html.

Bonan, G. B., 2008: *Ecological Climatology: Concepts and Applications.* Cambridge University Press, New York, NY, USA.

Bonan, G. B., and S. Levis, 2010: Quantifying carbon-nitrogen feedbacks in the Community Land Model (CLM4). *Geophys. Res. Lett.*, **37**, L07401.

Bonan, G. B., D. Pollard, and S. L. Thompson, 1992: Effects of boreal forest vegetation on global climate. *Nature*, **359**, 716–718.

Bond-Lamberty, B., and A. Thomson, 2010: Temperature-associated increases in the global soil respiration record. *Nature*, **464**, 579–U132.

Booth, B. B. B., et al., 2012: High sensitivity of future global warming to land carbon cycle processes. *Environ. Res. Lett.*, **7**, 024002.

Bopp, L., K. E. Kohfeld, C. Le Quéré, and O. Aumont, 2003: Dust impact on marine biota and atmospheric CO_2 during glacial periods. *Paleoceanography*, **18**, 1046.

Bopp, L., C. Le Quéré, M. Heimann, A. C. Manning, and P. Monfray, 2002: Climate-induced oceanic oxygen fluxes: Implications for the contemporary carbon budget. *Global Biogeochem. Cycles*, **16**, 1022.

Bopp, L., et al., 2001: Potential impact of climate change on marine export production. *Global Biogeochem. Cycles*, **15**, 81–99.

Boucher, O., et al., 2012: Reversibility in an Earth System model in response to CO_2 concentration changes. *Environ. Res. Lett.*, **7**, 024013.

Bousquet, P., D. A. Hauglustaine, P. Peylin, C. Carouge, and P. Ciais, 2005: Two decades of OH variability as inferred by an inversion of atmospheric transport and chemistry of methyl chloroform. *Atmos. Chem. Phys.*, **5**, 2635–2656.

Bousquet, P., P. Peylin, P. Ciais, C. Le Quéré, P. Friedlingstein, and P. P. Tans, 2000: Regional changes in carbon dioxide fluxes of land and oceans since 1980. *Science*, **290**, 1342–1346.

Bousquet, P., et al., 2006: Contribution of anthropogenic and natural sources to atmospheric methane variability. *Nature*, **443**, 439–443.

Bousquet, P., et al., 2011: Source attribution of the changes in atmospheric methane for 2006–2008. *Atmos. Chem. Phys.*, **11**, 3689–3700.

Bouttes, N., D. Paillard, D. M. Roche, V. Brovkin, and L. Bopp, 2011: Last Glacial Maximum CO_2 and d[13]C successfully reconciled. *Geophys. Res. Lett.*, **38**, L02705.

Bouttes, N., et al., 2012: Impact of oceanic processes on the carbon cycle during the last termination. *Clim. Past*, **8**, 149–170.

Bouwman, A. F., et al., 2013: Global trends and uncertainties in terrestrial denitrification and N_2O emissions. *Philos. Trans. R. Soc. London Ser. B, 368*, 20130112.

Bouwman, L., et al., 2011: Exploring global changes in nitrogen and phosphorus cycles in agriculture induced by livestock production over the 1900–2050 period. *Proc. Natl. Acad. Sci. U.S.A.*, doi:10.1073/pnas.1012878108.

Boyd, P. W., et al., 2007: Mesoscale iron enrichment experiments 1993–2005: Synthesis and future directions. *Science*, **315**, 612–617.

Bozbiyik, A., M. Steinacher, F. Joos, T. F. Stocker, and L. Menviel, 2011: Fingerprints of changes in the terrestrial carbon cycle in response to large reorganizations in ocean circulation. *Clim. Past*, **7**, 319–338.

Brenninkmeijer, C. A. M., et al., 2007: Civil Aircraft for the regular investigation of the atmosphere based on an instrumented container: The new CARIBIC system. *Atmos. Chem. Phys.*, **7**, 4953–4976.

Bridgham, S. D., J. P. Megonigal, J. K. Keller, N. B. Bliss, and C. Trettin, 2006: The carbon balance of North American wetlands. *Wetlands*, **26**, 889–916.

Broecker, W. S., and T.-H. Peng, 1986: Carbon cycle: 1985 glacial to interglacial changes in the operation of the global carbon cycle. *Radiocarbon*, **28**, 309–327.

Broecker, W. S., E. Clark, D. C. McCorkle, T.-H. Peng, I. Hajdas, and G. Bonani, 1999: Evidence for a reduction in the carbonate ion content of the deep sea during the course of the Holocene. *Paleoceanography*, **14**, 744–752.

Brovkin, V., A. Ganopolski, D. Archer, and S. Rahmstorf, 2007: Lowering of glacial atmospheric CO_2 in response to changes in oceanic circulation and marine biogeochemistry. *Paleoceanography*, **22**, PA4202.

Brovkin, V., J. H. Kim, M. Hofmann, and R. Schneider, 2008: A lowering effect of reconstructed Holocene changes in sea surface temperatures on the atmospheric CO_2 concentration. *Global Biogeochem. Cycles*, **22**, GB1016.

Brovkin, V., T. Raddatz, C. H. Reick, M. Claussen, and V. Gayler, 2009: Global biogeophysical interactions between forest and climate. *Geophys. Res. Lett.*, **36**, L07405.

Brovkin, V., S. Sitch, W. von Bloh, M. Claussen, E. Bauer, and W. Cramer, 2004: Role of land cover changes for atmospheric CO_2 increase and climate change during the last 150 years. *Global Change Biol.*, **10**, 1253–1266.

Brovkin, V., J. Bendtsen, M. Claussen, A. Ganopolski, C. Kubatzki, V. Petoukhov, and A. Andreev, 2002: Carbon cycle, vegetation, and climate dynamics in the Holocene: Experiments with the CLIMBER-2 model. *Global Biogeochem. Cycles*, **16**, 1139.

Brovkin, V., et al., 2010: Sensitivity of a coupled climate-carbon cycle model to large volcanic eruptions during the last millennium. *Tellus B*, **62**, 674–681.

Brown, J. R., et al., 2011: Effects of multiple global change treatments on soil N_2O fluxes. *Biogeochemistry*, **109**, 85–100.

Brzezinski, M. A., et al., 2002: A switch from $Si(OH)_4$ to NO_3 depletion in the glacial Southern Ocean. *Geophys. Res. Lett.*, **29**, 1564.

Buesseler, K. O., J. E. Andrews, S. M. Pike, and M. A. Charette, 2004: The effects of iron fertilization on carbon sequestration in the Southern Ocean. *Science*, **304**, 414–417.

Burke, E. J., C. D. Jones, and C. D. Koven, 2013: Estimating the permafrost-carbon-climate response in the CMIP5 climate models using a simplified approach. *J. Clim.*, **26**, 4897-4909.

Burn, C. R., and F. E. Nelson, 2006: Comment on "A projection of severe near-surface permafrost degradation during the 21st century" by David M. Lawrence and Andrew G. Slater. *Geophys. Res. Lett.*, **33**, L21503.

Butterbach-Bahl, K., and M. Dannenmann, 2011: Denitrification and associated soil N_2O emissions due to agricultural activities in a changing climate. *Curr. Opin. Environ. Sustain.*, **3**, 389–395.

Byrne, R. H., S. Mecking, R. A. Feely, and X. W. Liu, 2010: Direct observations of basin-wide acidification of the North Pacific Ocean. *Geophys. Res. Lett.*, **37**, L02601.

Cadule, P., et al., 2010: Benchmarking coupled climate-carbon models against long-term atmospheric CO_2 measurements. *Global Biogeochem. Cycles*, **24**, GB2016.

Cai, W.-J., et al., 2011: Acidification of subsurface coastal waters enhanced by eutrophication. *Nature Geosci*, **4**, 766–770.

Caldeira, K., and G. H. Rau, 2000: Accelerating carbonate dissolution to sequester carbon dioxide in the ocean: Geochemical implications. *Geophys. Res. Lett.*, **27**, 225–228.

Caldeira, K., and M. E. Wickett, 2005: Ocean model predictions of chemistry changes from carbon dioxide emissions to the atmosphere and ocean. *J. Geophys. Res. Oceans*, **110**, C09S04.

Canadell, J. G., and M. R. Raupach, 2008: Managing forests for climate change mitigation. *Science*, **320**, 1456–1457.

Canadell, J. G., et al., 2007a: Saturation of the terrestrial carbon sink. In: *Terrestrial Ecosystems in a Changing World.* [J. G. Canadell, D. Pataki and L. Pitelka (eds.)]. The IGBP Series. Springer-Verlag, Berlin and Heidelberg, Germany, pp. 59–78,

Canadell, J. G., et al., 2007b: Contributions to accelerating atmospheric CO_2 growth from economic activity, carbon intensity, and efficiency of natural sinks. *Proc. Natl. Acad. Sci. U.S.A.*, **104**, 18,866–18,870.

Canfield, D. E., A. N. Glazer, and P. G. Falkowski, 2010: The evolution and future of Earth's nitrogen cycle. *Science*, **330**, 192–196.

Cao, L., and K. Caldeira, 2010a: Can ocean iron fertilization mitigate ocean acidification? *Clim. Change*, **99**, 303–311.

Cao, L., and K. Caldeira, 2010b: Atmospheric carbon dioxide removal: Long-term consequences and commitment. *Environ. Res. Lett.*, **5**, 024011.

Cao, L., K. Caldeira, and A. K. Jain, 2007: Effects of carbon dioxide and climate change on ocean acidification and carbonate mineral saturation. *Geophys. Res. Lett.*, **34**, L05607.

Cao, L., G. Bala, and K. Caldeira, 2011: Why is there a short-term increase in global precipitation in response to diminished CO_2 forcing? *Geophys. Res. Lett.*, **38**, L06703.

Cao, L., G. Bala, K. Caldeira, R. Nemani, and G. Ban-Weiss, 2010: Importance of carbon dioxide physiological forcing to future climate change. *Proc. Natl. Acad. Sci. U.S.A.*, **107**, 9513–9518.

Carcaillet, C., et al., 2002: Holocene biomass burning and global dynamics of the carbon cycle. *Chemosphere*, **49**, 845–863.

Certini, G., 2005: Effects of fire on properties of forest soils: A review. *Oecologia*, **143**, 1–10.

Chambers, J. Q., J. I. Fisher, H. Zeng, E. L. Chapman, D. B. Baker, and G. C. Hurtt, 2007: Hurricane Katrina's carbon footprint on U.S. Gulf Coast forests. *Science*, **318**, 1107.

Chantarel, A. M., J. M. G. Bloor, N. Deltroy, and J.-F. Soussana, 2011: Effects of climate change drivers on nitrous oxide fluxes in an upland temperate grassland. *Ecosystems*, **14**, 223–233.

Chapuis-Lardy, L., N. Wrage, A. Metay, J. L. Chotte, and M. Bernoux, 2007: Soils, a sink for N_2O? A review. *Global Change Biol.*, **13**, 1–17.

Chen, Y. H., and R. G. Prinn, 2006: Estimation of atmospheric methane emissions between 1996 and 2001 using a three-dimensional global chemical transport model. *J. Geophys. Res. Atmos.*, **111**, D10307.

Chhabra, A., K. R. Manjunath, S. Panigrahy, and J. S. Parihar, 2013: Greenhouse gas emissions from Indian livestock. *Clim. Change*, **117**, 329–344.

Chierici, M., and A. Fransson, 2009: Calcium carbonate saturation in the surface water of the Arctic Ocean: Undersaturation in freshwater influenced shelves. *Biogeosciences*, **6**, 2421–2431.

Christensen, T. R., et al., 2004: Thawing sub-arctic permafrost: Effects on vegetation and methane emissions. *Geophys. Res. Lett.*, **31**, L04501.

Churkina, G., V. Brovkin, W. von Bloh, K. Trusilova, M. Jung, and F. Dentener, 2009: Synergy of rising nitrogen depositions and atmospheric CO_2 on land carbon uptake moderately offsets global warming. *Global Biogeochem. Cycles*, **23**, GB4027.

Ciais, P., P. Rayner, F. Chevallier, P. Bousquet, M. Logan, P. Peylin, and M. Ramonet, 2010: Atmospheric inversions for estimating CO_2 fluxes: methods and perspectives. *Clim. Change*, **103**, 69–92(24).

Ciais, P., et al., 2012: Large inert carbon pool in the terrestrial biosphere during the Last Glacial Maximum. *Nature Geosci.*, **5**, 74–79.

Ciais, P., et al., 2008: Carbon accumulation in European forests. *Nature Geosci.*, **1**, 425–429.

Ciais, P., et al., 2005: Europe-wide reduction in primary productivity caused by the heat and drought in 2003. *Nature*, **437**, 529–533.

Clark, D. B., D. A. Clark, and S. F. Oberbauer, 2010: Annual wood production in a tropical rain forest in NE Costa Rica linked to climatic variation but not to increasing CO_2. *Global Change Biol.*, **16**, 747–759.

Claussen, M., et al., 2002: Earth system models of intermediate complexity: Closing the gap in the spectrum of climate system models. *Clim. Dyn.*, **18**, 579–586.

Cocco, V., et al., 2013: Oxygen and indicators of stress for marine life in multi-model global warming projections. *Biogeosciences*, **10**, 1849–1868.

Codispoti, L. A., 2007: An oceanic fixed nitrogen sink exceeding 400 Tg N a^{-1} vs the concept of homeostasis in the fixed-nitrogen inventory. *Biogeosciences*, **4**, 233–253.

Codispoti, L. A., 2010: Interesting Times for Marine N_2O. *Science*, **327**, 1339–1340.

Cole, J. J., et al., 2007: Plumbing the global carbon cycle: Integrating inland waters into the terrestrial carbon budget. *Ecosystems*, **10**, 171–184.

Collins, W. J., et al., 2011: Development and evaluation of an Earth-System model—HadGEM2. *Geosci. Model Dev.*, **4**, 1051–1075.

Conrad, R., 1996: Soil microorganisms as controllers of atmospheric trace gases (H_2, CO, CH_4, OCS, N_2O, and NO). *Microbiol. Rev.*, **60**, 609–640.

Conway, T., and P. Tans, 2011: Global CO_2. National Oceanic and Atmospheric Administration, Earth System Research Library, Silver Spring, MD, USA.

Conway, T. J., P. P. Tans, L. S. Waterman, K. W. Thoning, D. R. Kitzis, K. A. Masarie, and N. Zhang, 1994: Evidence for interannual variability of the carbon cycle from the National Oceanic and Atmospheric Administration Climate Monitoring and Diagnostics Laboratory Global Air Sampling Network. *J. Geophys. Res. Atmos.*, **99**, 22831–22855.

Corbière, A., N. Metzl, G. Reverdin, C. Brunet, and A. Takahashi, 2007: Interannual and decadal variability of the oceanic carbon sink in the North Atlantic subpolar gyre. *Tellus B*, **59**, 168–178.

Cox, P., and C. Jones, 2008: Climate change. Illuminating the modern dance of climate and CO_2. *Science*, **321**, 1642–1644.

Cox, P. M., 2001: Description of the TRIFFID dynamic global vegetation model. Technical Note 24. Hadley Centre, Met Office, Exeter, Devon, UK.

Cox, P. M., R. A. Betts, M. Collins, P. P. Harris, C. Huntingford, and C. D. Jones, 2004: Amazonian forest dieback under climate-carbon cycle projections for the 21st century. *Theor. Appl. Climatol.*, **78**, 137–156.

Cox, P. M., D. Pearson, B. B. Booth, P. Friedlingstein, C. Huntingford, C. D. Jones, and C. M. Luke, 2013: Sensitivity of tropical carbon to climate change constrained by carbon dioxide variability. *Nature*, **494**, 341–344.

Crévoisier, C., D. Nobileau, A. M. Fiore, R. Armante, A. Chédin, and N. A. Scott, 2009: Tropospheric methane in the tropics – first year from IASI hyperspectral infrared observations. *Atmos. Chem. Phys.*, **9**, 6337–6350.

Crutzen, P. J., A. R. Mosier, K. A. Smith, and W. Winiwarter, 2008: N_2O release from agro-biofuel production negates global warming reduction by replacing fossil fuels. *Atmos. Chem. Phys.*, **8**, 389–395.

Cunnold, D. M., et al., 2002: In situ measurements of atmospheric methane at GAGE/AGAGE sites during 1985–2000 and resulting source inferences. *J. Geophys. Res. Atmos.*, **107**, ACH20–1, CiteID 4225.

Curry, C. L., 2007: Modeling the soil consumption of methane at the global scale. *Global Biogeochem. Cycles*, **21**, GB4012.

Curry, C. L., 2009: The consumption of atmospheric methane by soil in a simulated future climate. *Biogeosciences*, **6**, 2355–2367.

Daniau, A. L., et al., 2012: Predictability of biomass burning in response to climate changes. *Global Biogeochem. Cycles*, **26**, Gb4007.

Davidson, E. A., 2009: The contribution of manure and fertilizer nitrogen to atmospheric nitrous oxide since 1860. *Nature Geosci.*, **2**, 659–662.

Davidson, E. A., et al., 2012: Excess nitrogen in the U.S. environnement: Trends, risks, and solutions. Issues of Ecology, Report number 15. Ecological Society of America, Washington, DC.

Dawes, M. A., S. Hättenschwiler, P. Bebi, F. Hagedorn, I. T. Handa, C. Körner, and C. Rixen, 2011: Species-specific tree growth responses to 9 years of CO_2 enrichment at the alpine treeline. *J. Ecol.*, **99**, 383–394.

De Klein, C., et al., 2007: N_2O emissions from managed soils, and CO_2 emissions from lime and urea application. In: *2006 IPCC Guidelines for National Greenhouse Gas Inventories*, Vol. 4 [M. Gytarsky, T. Higarashi, W. Irving, T. Krug and J. Penman (eds.)]. Intergovernmental Panel on Climate Change, Geneva, Switzerland, pp. 11.1–11.54.

DeFries, R., and C. Rosenzweig, 2010: Toward a whole-landscape approach for sustainable land use in the tropics. *Proc. Natl. Acad. Sci. U.S.A.*, **107**, 19627–19632.

DeFries, R. S., R. A. Houghton, M. C. Hansen, C. B. Field, D. L. Skole, and J. Townshend, 2002: Carbon emissions from tropical deforestation and regrowth based on satellite observations for the 1980s and 1990s. *Proc. Natl. Acad. Sci. U.S.A.*, **99**, 14256–14261.

Delmas, R. J., J.-M. Ascencio, and M. Legrand, 1980: Polar ice evidence that atmospheric CO_2 20,000–yr BP was 50% of present. *Nature*, **284**, 155–157.

Denman, K. L., 2008: Climate change, ocean processes and ocean iron fertilization. *Mar. Ecol. Prog. Ser.*, **364**, 219–225.

Denman, K. L., et al., 2007: Couplings between changes in the climate system and biogeochemistry. In: *Climate Change 2007: The Physical Science Basis. Contribution of Working Group I to the Fourth Assessment Report of the Intergovernmental Panel on Climate Change* [Solomon, S., D. Qin, M. Manning, Z. Chen, M. Marquis, K. B. Averyt, M. Tignor and H. L. Miller (eds.)] Cambridge University Press, Cambridge, United Kingdom and New York, NY, USA, 499-587.

Dentener, F., W. Peters, M. Krol, M. van Weele, P. Bergamaschi, and J. Lelieveld, 2003: Interannual variability and trend of CH_4 lifetime as a measure for OH changes in the 1979–1993 time period. *J. Geophys. Res. Atmos.*, **108**, 4442.

Dentener, F., et al., 2005: The impact of air pollutant and methane emission controls on tropospheric ozone and radiative forcing: CTM calculations for the period 1990–2030. *Atmos. Chem. Phys.*, **5**, 1731–1755.

Dentener, F., et al., 2006: The global atmospheric environment for the next generation. *Environ. Sci. Technol.*, **40**, 3586–3594.

Deutsch, C., J. L. Sarmiento, D. M. Sigman, N. Gruber, and J. P. Dunne, 2007: Spatial coupling of nitrogen inputs and losses in the ocean. *Nature*, **445**, 163–167.

Dickens, G. R., 2003: A methane trigger for rapid warming? *Science*, **299**, 1017.

Dlugokencky, E., and P. P. Tans, 2013a: Recent CO_2, NOAA, ESRS. Retrieved from www.esrl.noaa.gov/gmd/ccgg/trends/global.html, accessed 01-02-2013.

Dlugokencky, E., and P. P. Tans, 2013b: Globally averaged marine surface annual mean data, NOAA/ESRL. Retrieved from www.esrl.noaa.gov/gmd/ccgg/trends/, accessed 01-02-2013.

Dlugokencky, E. J., E. G. Nisbet, R. Fisher, and D. Lowry, 2011: Global atmospheric methane: Budget, changes and dangers. *Philos. Trans. R. Soc. London Ser. A*, **369**, 2058–2072.

Dlugokencky, E. J., P. M. Lang, A. M. Crotwell, and K. A. Masarie, 2012: Atmospheric methane dry air mole fractions from the NOAA ESRL Carbon Cycle Cooperative Global Air Sampling Network, 1983–2011, version 2012-09-24. Version 2010-08-12 ed.

Dlugokencky, E. J., S. Houweling, L. Bruhwiler, K. A. Masarie, P. M. Lang, J. B. Miller, and P. P. Tans, 2003: Atmospheric methane levels off: Temporary pause or a new steady-state? *Geophys. Res. Lett.*, **30**, 1992.

Dlugokencky, E. J., et al., 2009: Observational constraints on recent increases in the atmospheric CH_4 burden. *Geophys. Res. Lett.*, **36**, L18803.

Dolman, A. J., et al., 2012: An estimate of the terrestrial carbon budget of Russia using inventory based, eddy covariance and inversion methods. *Biogeosciences*, **9**, 5323–5340.

Doney, S. C., et al., 2009: Mechanisms governing interannual variability in upper-ocean inorganic carbon system and air-sea CO_2 fluxes: Physical climate and atmospheric dust. *Deep-Sea Res. Pt. II*, **56**, 640–655.

Dore, J. E., R. Lukas, D. W. Sadler, M. J. Church, and D. M. Karl, 2009: Physical and biogeochemical modulation of ocean acidification in the central North Pacific. *Proc. Natl. Acad. Sci. U.S.A.*, **106**, 12235–12240.

Dornburg, V., and G. Marland, 2008: Temporary storage of carbon in the biosphere does have value for climate change mitigation: A response to the paper by Miko Kirschbaum. *Mitigat. Adapt. Strat. Global Change*, **13**, 211–217.

Duce, R. A., et al., 2008: Impacts of atmospheric anthropogenic nitrogen on the open ocean. *Science*, **320**, 893–897.

Dueck, T. A., et al., 2007: No evidence for substantial aerobic methane emission by terrestrial plants: A ^{13}C-labelling approach. *New Phytologist*, **175**, 29–35.

Dufresne, J.-L., et al., 2013: Climate change projections using the IPSL-CM5 Earth System Model: From CMIP3 to CMIP5. *Clim. Dyn.*, **40**, 2123-2165.

Dukes, J. S., et al., 2005: Responses of grassland production to single and multiple global environmental changes. *PloS Biol.*, **3**, 1829–1837.

Dunne, J. P., et al., 2012: GFDL's ESM2 global coupled climate-carbon Earth System Models Part I: Physical formulation and baseline simulation characteristics. *J. Clim.*, **25**, 6646–6665.

Dunne, J. P., et al., 2013: GFDL's ESM2 global coupled climate-carbon Earth System Models. Part II: Carbon system formation and baseline simulation characteristics. *J. Clim.*, **26**, 2247-2267.

Dutaur, L., and L. V. Verchot, 2007: A global inventory of the soil CH_4 sink. *Global Biogeochem. Cycles*, **21**, GB4013.

Dutta, K., E. A. G. Schuur, J. C. Neff, and S. A. Zimov, 2006: Potential carbon release from permafrost soils of Northeastern Siberia. *Global Change Biol.*, **12**, 2336–2351.

Eby, M., K. Zickfeld, A. Montenegro, D. Archer, K. J. Meissner, and A. J. Weaver, 2009: Lifetime of anthropogenic climate change: Millennial time scales of potential CO_2 and surface temperature perturbations. *J. Clim.*, **22**, 2501–2511.

Eby, M., et al., 2013: Historical and idealized climate model experiments: An EMIC intercomparison. *Clim. Past*, **9**, 1–30.

EDGAR4–database, 2009: Emission Database for Global Atmospheric Research (EDGAR), release version 4.0. http://edgar.jrc.ec.europa.eu, 2009. European Commission. Joint Research Centre (JRC) / Netherlands Environmental Assessment Agency (PBL).

Elberling, B., H. H. Christiansen, and B. U. Hansen, 2010: High nitrous oxide production from thawing permafrost. *Nature Geosci.*, **3**, 332–335.

Eliseev, A. V., I. I. Mokhov, M. M. Arzhanov, P. F. Demchenko, and S. N. Denisov, 2008: Interaction of the methane cycle and processes in wetland ecosystems in a climate model of intermediate complexity. *Izvestiya Atmos. Ocean. Phys.*, **44**, 139–152.

Elliott, S., M. Maltrud, M. Reagan, G. Moridis, and P. Cameron-Smith, 2011: Marine methane cycle simulations for the period of early global warming. *J. Geophys. Res. Biogeosci.*, **116**, G01010.

Elser, J. J., et al., 2007: Global analysis of nitrogen and phosphorus limitation of primary producers in freshwater, marine and terrestrial ecosystems. *Ecol. Lett.*, **10**, 1135–1142.

Elsig, J., et al., 2009: Stable isotope constraints on Holocene carbon cycle changes from an Antarctic ice core. *Nature*, **461**, 507–510.

Emerson, S., and J. I. Hedges, 1988: Processes controlling the organic carbon content of open ocean sediments. *Paleoceanography*, **3**, 621–634.

EPA, 2006: Global anthropogenic non-CO_2 greenhouse gas emissions. United States Environmental Protection Agency (US EPA, Washington, DC) Report EPA-430-R-06-003. Retrieved from http://nepis.epa.gov/EPA/html/DLwait.htm?url=/Adobe/PDF/2000ZL5G.PDF.

EPA, 2010: Methane and nitrous oxide emissions from natural sources. United States Environmental Protection Agency (EPA) Report. Washington, DC. http://www.epa.gov/outreach/pdfs/Methane-and-Nitrous-Oxide-Emissions-From-Natural-Sources.pdf

EPA, 2011a: Global anthropogenic non-CO_2 greenhouse gas emissions: 1990–2030, United States Environmental Protection Agency (US EPA) Report. Washington, DC. http://www.epa.gov/climatechange/Downloads/EPAactivities/EPA_Global_NonCO2_Projections_Dec2012.pdf

EPA, 2011b: Reactive nitrogen in the United States: An analysis of inputs, flows, consequences, and management options. Report EPA-SAB-11-013, Washington, DC, 140 pp. http://yosemite.epa.gov/sab/sabproduct.nsf/WebBOARD/INCFullReport/$File/Final%20INC%20Report_8_19_11%28without%20signatures%29.pdf

Erisman, J. W., M. S. Sutton, J. N. Galloway, Z. Klimont, and W. Winiwarter, 2008: A century of ammonia synthesis. *Nature Geosci.*, **1**, 1–4.

Erisman, J. W., J. Galloway, S. Seitzinger, A. Bleeker, and K. Butterbach-Bahl, 2011: Reactive nitrogen in the environment and its effect on climate change. *Curr. Opin. Environ. Sustain.*, **3**, 281–290.

Esser, G., J. Kattge, and A. Sakalli, 2011: Feedback of carbon and nitrogen cycles enhances carbon sequestration in the terrestrial biosphere. *Global Change Biol.*, **17**, 819–842.

Etheridge, D. M., L. P. Steele, R. L. Langenfelds, R. J. Francey, J.-M. Barnola, and V. I. Morgan, 1996: Natural and anthropogenic changes in atmospheric CO_2 over the last 1000 years from air in Antarctic ice and firn. *J. Geophys. Res.*, **101**, 4115–4128.

Etiope, G., K. R. Lassey, R. W. Klusman, and E. Boschi, 2008: Reappraisal of the fossil methane budget and related emission from geologic sources. *Geophys. Res. Lett.*, **35**, L09307.

Evans, C. D., D. T. Monteith, and D. M. Cooper, 2005: Long-term increses in surface water dissolved organic carbon: Observations, possible causes and environmental impacts. *Environ. Pollut.*, **137**, 55–71.

Falloon, P., C. Jones, M. Ades, and K. Paul, 2011: Direct soil moisture controls of future global soil carbon changes: An important source of uncertainty. *Global Biogeochem. Cycles*, **25**, GB3010.

Fan, S.-M., T. L. Blaine, and J. L. Sarmiento, 1999: Terrestrial carbon sink in the Northern Hemisphere estimated from the atmospheric CO_2 difference between Manna Loa and the South Pole since 1959. *Tellus B*, **51**, 863–870.

FAO, 2005: Global Forest Resource Assessment 2005. Progress toward sustainable forest management. FAO Forestry Paper 147. Food and Agriculture Organization of the United Nations, Rome, Italy, pp. 129–147.

FAO, 2010: Global Forest Resources Assessment 2010. Main report. FAO Forestry Paper 163, Food and Agriculture Organization of the United Nations, Rome, Italy, 340 pp.

Feely, R. A., S. C. Doney, and S. R. Cooley, 2009: Ocean acidification: Present conditions and future changes in a high-CO_2 world. *Oceanography*, **22**, 36–47.

Feely, R. A., C. L. Sabine, J. M. Hernandez-Ayon, D. Ianson, and B. Hales, 2008: Evidence for upwelling of corrosive "acidified" water onto the continental shelf. *Science*, **320**, 1490–1492.

Feely, R. A., R. H. Byrne, J. G. Acker, P. R. Betzer, C.-T. A. Chen, J. F. Gendron, and M. F. Lamb, 1988: Winter-summer variations of calcite and aragonite saturation in the northeast Pacific. *Mar. Chem.*, **25**, 227–241.

Feely, R. A., T. Takahashi, R. Wanninkhof, M. J. McPhaden, C. E. Cosca, S. C. Sutherland, and M. E. Carr, 2006: Decadal variability of the air-sea CO_2 fluxes in the equatorial Pacific Ocean. *J. Geophys. Res. Oceans*, **111**, C08S90.

Felzer, B., et al., 2005: Past and future effects of ozone on carbon sequestration and climate change policy using a global biogeochemical model. *Clim. Change*, **73**, 345–373.

Ferretti, D. F., et al., 2005: Unexpected changes to the global methane budget over the past 2000 years. *Science*, **309**, 1714–1717.

Findlay, H. S., T. Tyrrell, R. G. J. Bellerby, A. Merico, and I. Skjelvan, 2008: Carbon and nutrient mixed layer dynamics in the Norwegian Sea. *Biogeosciences*, **5**, 1395–1410.

Findlay, S. E. G., 2005: Increased carbon transport in the Hudson River: Unexpected consequence of nitrogen deposition? *Front. Ecol. Environ.*, **3**, 133–137.

Finzi, A. C., et al., 2006: Progressive nitrogen limitation of ecosystem processes under elevated CO_2 in a warm-temperate forest. *Ecology*, **87**, 15–25.

Finzi, A. C., et al., 2007: Increases in nitrogen uptake rather than nitrogen-use efficiency support higher rates of temperate forest productivity under elevated CO_2. *Proc. Natl. Acad. Sci. U.S.A.*, **104**, 14014–14019.

Fischer, H., et al., 2008: Changing boreal methane sources and constant biomass burning during the last termination. *Nature*, **452**, 864–867.

Fisher, J. B., S. Sitch, Y. Malhi, R. A. Fisher, C. Huntingford, and S. Y. Tan, 2010: Carbon cost of plant nitrogen acquisition: A mechanistic, globally applicable model of plant nitrogen uptake, retranslocation, and fixation. *Global Biogeochem. Cycles*, **24**, GB1014.

Flannigan, M. D., B. Stocks, M. Turetsky, and M. Wotton, 2009a: Impacts of climate change on fire activity and fire management in the circumboreal forest. *Global Change Biol.*, **15**, 549–560.

Flannigan, M. D., M. A. Krawchuk, W. J. de Groot, B. M. Wotton, and L. M. Gowman, 2009b: Implications of changing climate for global wildland fire. *Int. J. Wildland Fire*, **18**, 483–507.

Fleming, E. L., C. H. Jackman, R. S. Stolarski, and A. R. Douglass, 2011: A model study of the impact of source gas changes on the stratosphere for 1850–2100. *Atmos. Chem. Phys.*, **11**, 8515–8541.

Flückiger, J., A. Dällenbach, T. Blunier, B. Stauffer, T. F. Stocker, D. Raynaud, and J.-M. Barnola, 1999: Variations in atmospheric N_2O concentration during abrupt climate changes. *Science*, **285**, 227–230.

Flückiger, J., et al., 2002: High-resolution Holocene N_2O ice core record and its relationship with CH_4 and CO_2. *Global Biogeochem. Cycles*, **16**, 1–10.

Flückiger, J., et al., 2004: N_2O and CH_4 variations during the last glacial epoch: Insight into global processes. *Global Biogeochem. Cycles*, **18**, GB1020.

Foley, J. A., C. Monfreda, N. Ramankutty, and D. Zaks, 2007: Our share of the planetary pie. *Proc. Natl. Acad. Sci. U.S.A.*, **104**, 12585–12586.

Foley, J. A., et al., 2011: Solutions for a cultivated planet. *Nature*, **478**, 337–342.

Fowler, D., et al., 2013: The global nitrogen cycle in the 21th century. *Philos. Trans. R. Soc. London Ser. B*, **368**, 20130165.

Francey, R. J., et al., 2013: Atmospheric verification of anthropogenic CO_2 emission trends. *Nature Clim. Change*, **3**, 520-524.

Frank, D. C., J. Esper, C. C. Raible, U. Buntgen, V. Trouet, B. Stocker, and F. Joos, 2010: Ensemble reconstruction constraints on the global carbon cycle sensitivity to climate. *Nature*, **463**, 527–U143.

Frankenberg, C., et al., 2011: Global column-averaged methane mixing ratios from 2003 to 2009 as derived from SCIAMACHY: Trends and variability. *J. Geophys. Res.*, **116**, D04302.

Frankenberg, C., et al., 2008: Tropical methane emissions: A revised view from SCIAMACHY onboard ENVISAT. *Geophys. Res. Lett.*, **35**, L15811.

Freing, A., D. W. R. Wallace, and H. W. Bange, 2012: Global oceanic production of nitrous oxide. *Philos. Trans. R. Soc. London Ser. B*, **367**, 1245–1255.

Friedli, H., H. Lötscher, H. Oeschger, U. Siegenthaler, and B. Stauffer, 1986: Ice core record of the $^{13}C/^{12}C$ ratio of atmospheric CO_2 in the past two centuries. *Nature*, **324**, 237–238.

Friedlingstein, P., and I. C. Prentice, 2010: Carbon-climate feedbacks: A review of model and observation based estimates. *Curr. Opin. Environ. Sustain.*, **2**, 251–257.

Friedlingstein, P., J. L. Dufresne, P. M. Cox, and P. Rayner, 2003: How positive is the feedback between climate change and the carbon cycle? *Tellus B*, **55**, 692–700.

Friedlingstein, P., et al., 2010: Update on CO_2 emissions. *Nature Geosci.*, **3**, 811–812.

Friedlingstein, P., et al., 2006: Climate-carbon cycle feedback analysis: Results from the C^{4}MIP model intercomparison. *J. Clim.*, **19**, 3337–3353.

Friis, K., A. Körtzinger, J. Patsch, and D. W. R. Wallace, 2005: On the temporal increase of anthropogenic CO_2 in the subpolar North Atlantic. *Deep-Sea Res. Pt. I*, **52**, 681–698.

Frölicher, T. L., F. Joos, and C. C. Raible, 2011: Sensitivity of atmospheric CO_2 and climate to explosive volcanic eruptions. *Biogeosciences*, **8**, 2317–2339.

Frölicher, T. L., F. Joos, C. C. Raible, and J. L. Sarmiento, 2013: Atmospheric CO_2 response to volcanic eruptions: The role of ENSO, season, and variability. *Global Biogeochem. Cycles*, **27**, 239-251.

Frölicher, T. L., F. Joos, G. K. Plattner, M. Steinacher, and S. C. Doney, 2009: Natural variability and anthropogenic trends in oceanic oxygen in a coupled carbon cycle–climate model ensemble. *Global Biogeochem. Cycles*, **23**, GB1003.

Frolking, S., and N. T. Roulet, 2007: Holocene radiative forcing impact of northern peatland carbon accumulation and methane emissions. *Global Change Biol.*, **13**, 1079–1088.

Fuller, D. Q., et al., 2011: The contribution of rice agriculture and livestock pastoralism to prehistoric methane levels: An archaeological assessment. *The Holocene*, **21**, 743–759.

Fung, I., M. Prather, J. John, J. Lerner, and E. Matthews, 1991: Three-dimensional model synthesis of the global methane cycle. *J. Geophys. Res.*, **96**, 13033–13065.

Fyfe, J. C., and O. A. Saenko, 2006: Simulated changes in the extratropical Southern Hemisphere winds and currents. *Geophys. Res. Lett.*, **33**, L06701.

Fyfe, J. C., J. N. S. Cole, V. K. Arora, and J. F. Scinocca, 2013: Biogeochemical carbon coupling influences global precipitation in geoengineering experiments. *Geophys. Res. Lett.*, **40**, 651–655.

Fyke, J. G., and A. J. Weaver, 2006: The effect of potential future climate change on the marine methane hydrate stability zone. *J. Clim.*, **19**, 5903–5917.

Gaillard, M. J., et al., 2010: Holocene land-cover reconstructions for studies on land cover-climate feedbacks. *Clim. Past*, **6**, 483–499.

Gaillardet, J., B. Dupre, P. Louvat, and C. J. Allegre, 1999: Global silicate weathering and CO_2 consumption rates deduced from the chemistry of large rivers. *Chem. Geol.*, **159**, 3–30.

Galbraith, D., P. E. Levy, S. Sitch, C. Huntingford, P. Cox, M. Williams, and P. Meir, 2010: Multiple mechanisms of Amazonian forest biomass losses in three dynamic global vegetation models under climate change. *New Phytologist*, **187**, 647–665.

Galloway, J. N., J. D. Aber, J. W. Erisman, S. P. Seitzinger, R. W. Howarth, E. B. Cowling, and B. J. Cosby, 2003: The nitrogen cascade. *BioScience*, **53**, 341–356.

Galloway, J. N., et al., 2008: Transformation of the nitrogen cycle: Recent trends, questions, and potential solutions. *Science*, **320**, 889.

Galloway, J. N., et al., 2004: Nitrogen cycles: Past, present and future. *Biogeochemistry*, **70**, 153–226.

Gao, H., et al., 2012: Intensive and extensive nitrogen loss from intertidal permeable sediments of the Wadden Sea *Limnol. Oceanogr.*, **57**, 185–198.

Gärdenäs, A. I., et al., 2011: Knowledge gaps in soil carbon and nitrogen interactions—From molecular to global scale. *Soil Biol. Biochem.*, **43**, 702–717.

GEA, 2006: Energy resources and potentials. In: Global Energy Assessment—Toward a Sustainable Future. Cambridge University Press, Cambridge, United Kingdom, and New York, NY, USA, 425-512.

Gedalof, Z., and A. A. Berg, 2010: Tree ring evidence for limited direct CO_2 fertilization of forests over the 20th century. *Global Biogeochem. Cycles*, **24**, Gb3027.

Gedney, N., P. M. Cox, and C. Huntingford, 2004: Climate feedback from wetland methane emissions. *Geophys. Res. Lett.*, **31**, L20503.

Gedney, N., P. M. Cox, R. A. Betts, O. Boucher, C. Huntingford, and P. A. Stott, 2006: Detection of a direct carbon dioxide effect in continental river runoff records. *Nature*, **439**, 835–838.

Gerber, S., L. O. Hedin, M. Oppenheimer, S. W. Pacala, and E. Shevliakova, 2010: Nitrogen cycling and feedbacks in a global dynamic land model. *Global Biogeochem. Cycles*, **24**, GB1001.

Gerber, S., F. Joos, P. Brugger, T. F. Stocker, M. E. Mann, S. Sitch, and M. Scholze, 2003: Constraining temperature variations over the last millennium by comparing simulated and observed atmospheric CO_2. *Clim. Dyn.*, **20**, 281–299.

Gervois, S., P. Ciais, N. de Noblet-Ducoudre, N. Brisson, N. Vuichard, and N. Viovy, 2008: Carbon and water balance of European croplands throughout the 20th century. *Global Biogeochem. Cycles*, **22**, GB003018.

Gilbert, D., N. N. Rabalais, R. J. Diaz, and J. Zhang, 2010: Evidence for greater oxygen decline rates in the coastal ocean than in the open ocean. *Biogeosciences*, **7**, 2283–2296.

Gloor, M., J. L. Sarmiento, and N. Gruber, 2010: What can be learned about carbon cycle climate feedbacks from the CO_2 airborne fraction? *Atmos. Chem. Phys.*, **10**, 7739–7751.

Gloor, M., et al., 2012: The carbon balance of South America: A review of the status, decadal trends and main determinants. *Biogeosciences*, **9**, 5407–5430.

Gloor, M., et al., 2009: Does the disturbance hypothesis explain the biomass increase in basin-wide Amazon forest plot data? *Global Change Biol.*, **15**, 2418–2430.

Gnanadesikan, A., and I. Marinov, 2008: Export is not enough: Nutrient cycling and carbon sequestration. *Mar. Ecol. Prog. Ser.*, **364**, 289–294.

Gnanadesikan, A., J. L. Sarmiento, and R. D. Slater, 2003: Effects of patchy ocean fertilization on atmospheric carbon dioxide and biological production. *Global Biogeochem. Cycles*, **17**, 1050.

Gnanadesikan, A., J. L. Russell, and F. Zeng, 2007: How does ocean ventilation change under global warming? *Ocean Sci.*, **3**, 43–53.

Gnanadesikan, A., J. P. Dunne, and J. John, 2012: Understanding why the volume of suboxic waters does not increase over centuries of global warming in an Earth System Model. *Biogeosciences*, **9**, 1159–1172.

Goldewijk, K. K., 2001: Estimating global land use change over the past 300 years: The HYDE Database. *Global Biogeochem. Cycles*, **15**, 417–433.

Goldewijk, K. K., A. Beusen, G. van Drecht, and M. de Vos, 2011: The HYDE 3.1 spatially explicit database of human-induced global land-use change over the past 12,000 years. *Global Ecol. Biogeogr.*, **20**, 73–86.

Golding, N., and R. Betts, 2008: Fire risk in Amazonia due to climate change in the HadCM3 climate model: Potential interactions with deforestation. *Global Biogeochem. Cycles*, **22**, GB4007.

Goll, D. S., et al., 2012: Nutrient limitation reduces land carbon uptake in simulations with a model of combined carbon, nitrogen and phosphorus cycling. *Biogeosciences*, **9**, 3547–3569.

Good, P., C. Jones, J. Lowe, R. Betts, and N. Gedney, 2013: Comparing tropical forest projections from two generations of Hadley Centre Earth System Models, HadGEM2-ES and HadCM3LC. *J. Clim.*, **26**, 495–511.

Good, P., C. Jones, J. Lowe, R. Betts, B. Booth, and C. Huntingford, 2011: Quantifying environmental drivers of future tropical forest extent. *J. Clim.*, **24**, 1337–1349.

Govindasamy, B., K. Caldeira, and P. B. Duffy, 2003: Geoengineering Earth's radiation balance to mitigate climate change from a quadrupling of CO_2. *Global Planet. Change*, **37**, 157–168.

Govindasamy, B., S. Thompson, P. B. Duffy, K. Caldeira, and C. Delire, 2002: Impact of geoengineering schemes on the terrestrial biosphere. *Geophys. Res. Lett.*, **29**, 2061.

Graven, H. D., N. Gruber, R. Key, S. Khatiwala, and X. Giraud, 2012: Changing controls on oceanic radiocarbon: New insights on shallow-to-deep ocean exchange and anthropogenic CO_2 uptake. *J. Geophys. Res. Oceans*, **117**, C10005.

Gray, M. L., K. J. Champagne, D. Fauth, J. P. Baltrus, and H. Pennline, 2008: Performance of immobilized tertiary amine solid sorbents for the capture of carbon dioxide. *Int. J. Greenh. Gas Control*, **2**, 3–8.

Gregg, J. S., R. J. Andres, and G. Marland, 2008: China: Emissions pattern of the world leader in CO_2 emissions from fossil fuel consumption and cement production. *Geophys. Res. Lett.*, **35**, L08806.

Gregory, J. M., C. D. Jones, P. Cadule, and P. Friedlingstein, 2009: Quantifying carbon cycle feedbacks. *J. Clim.*, **22**, 5232–5250.

Groszkopf, T., et al., 2012: Doubling of marine dinitrogen-fixation rates based on direct measurements. *Nature*, **488**, 361–364.

Gruber, N., and J. N. Galloway, 2008: An Earth-system perspective of the global nitrogen cycle. *Nature*, **451**, 293–296.

Gruber, N., C. Hauri, Z. Lachkar, D. Loher, T. L. Frölicher, and G. K. Plattner, 2012: Rapid progression of ocean acidification in the California Current System. *Science*, **337**, 220–223.

Gruber, N., et al., 2009: Oceanic sources, sinks, and transport of atmospheric CO_2. *Global Biogeochem. Cycles*, **23**, GB1005.

Gurney, K. R., and W. J. Eckels, 2011: Regional trends in terrestrial carbon exchange and their seasonal signatures. *Tellus B*, **63**, 328–339.

Hamilton, S. K., A. L. Kurzman, C. Arango, L. Jin, and G. P. Robertson, 2007: Evidence for carbon sequestration by agricultural liming. *Global Biogeochem. Cycles*, **21**, GB2021.

Hansell, D. A., C. A. Carlson, D. J. Repeta, and R. Schlitzer, 2009: Dissolved organic matter in the ocean: A controversy stimulates new insights. *Oceanography*, **22**, 202–211.

Hansen, M. C., S. V. Stehman, and P. V. Potapov, 2010: Quantification of global gross forest cover loss. *Proc. Natl. Acad. Sci. U.S.A.*, **107**, 8650–8655.

Harden, J. W., et al., 2012: Field information links permafrost carbon to physical vulnerabilities of thawing. *Geophys. Res. Lett.*, **39**, L15704.

Harris, N. L., et al., 2012: Baseline map of carbon emissions from deforestation in tropical regions. *Science*, **336**, 1573–1576.

Harrison, K. G., 2000: Role of increased marine silica input on paleo-pCO_2 levels. *Paleoceanography*, **15**, 292–298.

Hartmann, J., N. Jansen, H. H. Dürr, S. Kempe, and P. Köhler, 2009: Global CO_{2-} consumption by chemical weathering: What is the contribution of highly active weathering regions? *Global Planet. Change*, **69**, 185–194.

Harvey, L. D. D., 2008: Mitigating the atmospheric CO_2 increase and ocean acidification by adding limestone powder to upwelling regions. *J. Geophys. Res.*, **113**, C04028.

Haverd, V., et al., 2013: The Australian terrestrial carbon budget. *Biogeosciences*, **10**, 851–869.

Hedegaard, G. B., J. Brandt, J. H. Christensen, L. M. Frohn, C. Geels, K. M. Hansen, and M. Stendel, 2008: Impacts of climate change on air pollution levels in the Northern Hemisphere with special focus on Europe and the Arctic. *Atmos. Chem. Phys.*, **8**, 3337–3367.

Heijmans, M. M. P. D., W. J. Arp, and F. Berendse, 2001: Effects of elevated CO_2 and vascular plants on evapotranspiration in bog vegetation. *Global Change Biol.*, **7**, 817–827.

Heijmans, M. M. P. D., H. Klees, and F. Berendse, 2002a: Competition between *Sphagnum magellanicum* and *Eriophorum angustifolium* as affected by raised CO_2 and increased N deposition. *Oikos*, **97**, 415–425.

Heijmans, M. M. P. D., H. Klees, W. de Visser, and F. Berendse, 2002b: Response of a *Sphagnum* bog plant community to elevated CO_2 and N supply. *Plant Ecol.*, **162**, 123–134.

Hein, R., P. J. Crutzen, and M. Heimann, 1997: An inverse modeling approach to investigate the global atmospheric methane cycle. *Global Biogeochem. Cycles*, **11**, 43–76.

Held, I. M., M. Winton, K. Takahashi, T. Delworth, F. Zeng, and G. K. Vallis, 2010: Probing the fast and slow components of global warming by returning abruptly to preindustrial forcing. *J. Clim.*, **23**, 2418–2427.

Herridge, D. F., M. B. Peoples, and R. M. Boddey, 2008: Global inputs of biological nitrogen fixation in agricultural systems. *Plant Soil*, **311**, 1–18.

Herzog, H., K. Caldeira, and J. Reilly, 2003: An issue of permanence: Assessing the effectiveness of temporary carbon storage. *Clim. Change*, **59**, 293–310.

Hibbard, K. A., G. A. Meehl, P. M. Cox, and P. Friedlingstein, 2007: A strategy for climate change stabilization experiments. *EOS Trans. Am. Geophys. Union*, **88**, 217-221.

Hickler, T., B. Smith, I. C. Prentice, K. Mjöfors, P. Miller, A. Arneth, and M. T. Sykes, 2008: CO_2 fertilization in temperate FACE experiments not representative of boreal and trophical forests. *Global Change Biol.*, **14**, 1531–1542.

Hietz, P., B. L. Turner, W. Wanek, A. Richter, C. A. Nock, and S. J. Wright, 2011: Long-term change in the nitrogen cycle of tropical forests. *Science*, **334**, 664-666.

Higgins, P. A. T., and J. Harte, 2012: Carbon cycle uncertainty increases climate change risks and mitigation challenges. *J. Clim.*, **25**, 7660–7668.

Hirota, M., M. Holmgren, E. H. Van Nes, and M. Scheffer, 2011: Global resilience of tropical forest and savanna to critical transitions. *Science*, **334**, 232–235.

Hirsch, A. I., A. M. Michalak, L. M. Bruhwiler, W. Peters, E. J. Dlugokencky, and P. P. Tans, 2006: Inverse modeling estimates of the global nitrous oxide surface flux from 1998 to 2001. *Global Biogeochem. Cycles*, **20**, GB1008.

Hodson, E. L., B. Poulter, N. E. Zimmermann, C. Prigent, and J. O. Kaplan, 2011: The El Nino-Southern Oscillation and wetland methane interannual variability. *Geophys. Res. Lett.*, **38**, L08810.

Hoelzemann, J. J., M. G. Schultz, G. P. Brasseur, C. Granier, and M. Simon, 2004: Global Wildland Fire Emission Model (GWEM): Evaluating the use of global area burnt satellite data. *J. Geophys. Res. Atmos.*, **109**, D14S04.

Hofmann, M., and H.-J. Schellnhuber, 2009: Oceanic acidification affects marine carbon pump and triggers extended marine oxygen holes. *Proc. Natl. Acad. Sci. U.S.A.*, **106**, 3017–3022.

Holland, E. A., J. Lee-Taylor, C. D. Nevison, and J. Sulzman, 2005: Global N cycle: Fluxes and N_2O mixing ratios originating from human activity. Data set. Oak Ridge National Laboratory Distributed Active Archive Center, Oak Ridge National Laboratory, Oak Ridge, TN. Retrieved from http://www.daac.ornl.gov

Hönisch, B., N. G. Hemming, D. Archer, M. Siddall, and J. F. McManus, 2009: Atmospheric carbon dioxide concentration across the mid-Pleistocene transition. *Science*, **324**, 1551–1554.

Hooijer, A., S. Page, J. G. Canadell, M. Silvius, J. Kwadijk, H. Wösten, and J. Jauhiainen, 2010: Current and future CO_2 emissions from drained peatlands in Southeast Asia. *Biogeosciences*, **7**, 1505–1514.

Hopcroft, P. O., P. J. Valdes, and D. J. Beerling, 2011: Simulating idealized Dansgaard-Oeschger events and their potential impacts on the global methane cycle. *Quat. Sci. Rev.*, **30**, 3258–3268.

Houghton, R. A., 2003: Revised estimates of the annual net flux of carbon to the atmosphere from changes in land use and land management 1850–2000. *Tellus B*, **55**, 378–390.

Houghton, R. A., 2010: How well do we know the flux of CO_2 from land-use change? *Tellus B*, **62**, 337–351.

Houghton, R. A., et al., 2012: Carbon emissions from land use and land-cover change. *Biogeosciences*, **9**, 5125–5142.

House, J. I., I. C. Prentice, and C. Le Quéré, 2002: Maximum impacts of future reforestation or deforestation on atmospheric CO_2. *Global Change Biol.*, **8**, 1047–1052.

House, K. Z., D. P. Schrag, C. F. Harvey, and K. S. Lackner, 2006: Permanent carbon dioxide storage in deep-sea sediments. *Proc. Natl. Acad. Sci. U.S.A.*, **103**, 14255.

House, K. Z., C. H. House, D. P. Schrag, and M. J. Aziz, 2007: Electrochemical acceleration of chemical weathering as an energetically feasible approach to mitigating anthropogenic climate change. *Environ. Sci. Technol.*, **41**, 8464–8470.

Hsu, J., and M. J. Prather, 2010: Global long-lived chemical modes excited in a 3-D chemistry transport model: Stratospheric N_2O, NOy, O_3 and CH_4 chemistry. *Geophys. Res. Lett.*, **37**, L07805.

Huang, J., et al., 2008: Estimation of regional emissions of nitrous oxide from 1997 to 2005 using multinetwork measurements: A chemical transport model, and an inverse method. *J. Geophys. Res.*, **113**, D17313.

Huber, C., et al., 2006: Isotope calibrated Greenland temperature record over Marine Isotope Stage 3 and its relation to CH_4. *Earth Planet. Sci. Lett.*, **243**, 504–519.

Hunter, S. J., A. M. Haywood, D. S. Goldobin, A. Ridgwell, and J. G. Rees, 2013: Sensitivity of the global submarine hydrate inventory to scenarious of future climate change. *Earth Planet. Sci. Lett.*, **367**, 105–115.

Huntingford, C., J. A. Lowe, B. B. B. Booth, C. D. Jones, G. R. Harris, L. K. Gohar, and P. Meir, 2009: Contributions of carbon cycle uncertainty to future climate projection spread. *Tellus B*, **61**, 355–360.

Huntingford, C., et al., 2013: Simulated resilience of tropical rainforests to CO_2-induced climate change. *Nature Geosci.*, **6**, 268–273.

Hurtt, G. C., et al., 2011: Harmonization of land-use scenarios for the period 1500–2100: 600 years of global gridded annual land-use transitions, wood harvest, and resulting secondary lands. *Clim. Change*, **109**, 117–161.

Huybers, P., and C. Langmuir, 2009: Feedback between deglaciation, volcanism, and atmospheric CO_2. *Earth Planet. Sci. Lett.*, **286**, 479–491.

Indermühle, A., et al., 1999: Holocene carbon-cycle dynamics based on CO_2 trapped in ice at Taylor Dome, Antarctica. *Nature*, **398**, 121–126.

Irvine, P. J., A. Ridgwell, and D. J. Lunt, 2010: Assessing the regional disparities in geoengineering impacts. *Geophys. Res. Lett.*, **37**, L18702.

Ise, T., A. L. Dunn, S. C. Wofsy, and P. R. Moorcroft, 2008: High sensitivity of peat decomposition to climate change through water-table feedback. *Nature Geosci.*, **1**, 763–766.

Ishii, M., N. Kosugi, D. Sasano, S. Saito, T. Midorikawa, and H. Y. Inoue, 2011: Ocean acidification off the south coast of Japan: A result from time series observations of CO_2 parameters from 1994 to 2008. *J. Geophys. Res. Oceans*, **116**, C06022.

Ishii, M., et al., 2009: Spatial variability and decadal trend of the oceanic CO_2 in the western equatorial Pacific warm/fresh water. *Deep-Sea Res. Pt. II*, **56**, 591–606.

Ishijima, K., T. Nakazawa, and S. Aoki, 2009: Variations of atmospheric nitrous oxide concentration in the northern and western Pacific. *Tellus B*, **61**, 408–415.

Ito, A., and J. E. Penner, 2004: Global estimates of biomass burning emissions based on satellite imagery for the year 2000. *J. Geophys. Res.*, **109**, D14S05.

Ito, A., and M. Inatomi, 2012: Use of a process-based model for assessing the methane budgets of global terrestrial ecosystems and evaluation of uncertainty. *Biogeosciences*, **9**, 759–773.

Iudicone, D., et al., 2011: Water masses as a unifying framework for understanding the Southern Ocean Carbon Cycle. *Biogeosciences*, **8**, 1031–1052.

Iversen, T., et al., 2013: The Norwegian Earth System Model, NorESM1–M. Part 2: Climate response and scenario projections. *Geosci. Model Dev.*, **6**, 389–415.

Jaccard, S. L., and E. D. Galbraith, 2012: Large climate-driven changes of oceanic oxygen concentrations during the last deglaciation. *Nature Geosci.*, **5**, 151-156.

Jaccard, S. L., G. H. Haug, D. M. Sigman, T. F. Pedersen, H. R. Thierstein, and U. R?hl, 2005: Glacial/interglacial changes in subarctic North Pacific stratification. *Science*, **308**, 1003–1006.

Jacobson, A. R., S. E. Mikaloff Fletcher, N. Gruber, J. L. Sarmiento, and M. Gloor, 2007: A joint atmosphere-ocean inversion for surface fluxes of carbon dioxide: 2. Regional results. *Global Biogeochem. Cycles*, **21**, GB1020.

Jacquet, S. H. M., N. Savoye, F. Dehairs, V. H. Strass, and D. Cardinal, 2008: Mesopelagic carbon remineralization during the European Iron Fertilization Experiment. *Global Biogeochem. Cycles*, **22**, 1–9.

Jain, A., X. J. Yang, H. Kheshgi, A. D. McGuire, W. Post, and D. Kicklighter, 2009: Nitrogen attenuation of terrestrial carbon cycle response to global environmental factors. *Global Biogeochem. Cycles*, **23**, GB4028.

Jin, X., and N. Gruber, 2003: Offsetting the radiative benefit of ocean iron fertilization by enhancing N_2O emissions. *Geophys. Res. Lett.*, **30**, 24, 2249.

Jones, C., S. Liddicoat, and J. Lowe, 2010: Role of terrestrial ecosystems in determining CO_2 stabilization and recovery behaviour. *Tellus B*, **62**, 682–699.

Jones, C., M. Collins, P. M. Cox, and S. A. Spall, 2001: The carbon cycle response to ENSO: A coupled climate-carbon cycle model study. *J. Clim.*, **14**, 4113–4129.

Jones, C., J. Lowe, S. Liddicoat, and R. Betts, 2009: Committed terrestrial ecosystem changes due to climate change. *Nature Geosci.*, **2**, 484–487.

Jones, C., et al., 2013: 21th Century compatible CO_2 emissions and airborne fraction simulated by CMIP5 Earth System models under 4 Representative Concentration Pathways. *J. Clim.*, **26**, 4398-4413.

Jones, C. D., and P. M. Cox, 2001: Modeling the volcanic signal in the atmospheric CO_2 record. *Global Biogeochem. Cycles*, **15**, 453–465.

Jones, C. D., and P. Falloon, 2009: Sources of uncertainty in global modelling of future soil organic carbon storage. In: *Uncertainties in Environmental Modelling and Consequences for Policy Making* [P. Bavaye, J. Mysiak and M. Laba (eds.)]. Springer Science+Business Media, New York, NY, USA and Heidelberg, Germany, pp. 283–315.

Jones, C. D., P. M. Cox, and C. Huntingford, 2006: Impact of climate carbon cycle feedbacks on emission scenarios to achieve stabilization. In: *Avoiding Dangerous Climate Change* [H. J. Schellnhuber, W. Cramer, N. Nakicenovic, T. Wigley and G. Yohe (eds.)]. Cambridge University Press, Cambridge, United Kingdom, and New York, NY, USA, pp. 323–332.

Jones, C. D., et al., 2011: The HadGEM2–ES implementation of CMIP5 centennial simulations. *Geosci. Model Dev.*, **4**, 543–570.

Joos, F., and R. Spahni, 2008: Rates of change in natural and anthropogenic radiative forcing over the past 20,000 years. *Proc. Natl. Acad. Sci. U.S.A.*, **105**, 1425–1430.

Joos, F., J. L. Sarmiento, and U. Siegenthaler, 1991: Estimates of the effect of Southern-Ocean iron fertilization on atmospheric CO_2 concentrations. *Nature*, **349**, 772–775.

Joos, F., T. L. Frölicher, M. Steinacher, and G.-K. Plattner, 2011: Impact of climate change mitigation on ocean acidification projections. In: *Ocean Acidification* [J. P. Gattuso and L. Hansson (eds.)]. Oxford University Press, Oxford, United Kingdom, and New York, NY, USA, pp. 273-289.

Joos, F., S. Gerber, I. C. Prentice, B. L. Otto-Bliesner, and P. J. Valdes, 2004: Transient simulations of Holocene atmospheric carbon dioxide and terrestrial carbon since the Last Glacial Maximum. *Global Biogeochem. Cycles*, **18**, Gb2002.

Joos, F., et al., 2013: Carbon dioxide and climate impulse response functions for the computation of greenhouse gas metrics: A multi-model analysis. *Atmos. Chem. Phys.*, **13**, 2793–2825.

Jorgenson, M. T., Y. L. Shur, and E. R. Pullman, 2006: Abrupt increase in permafrost degradation in Arctic Alaska. *Geophys. Res. Lett.*, **33**, L02503.

Jung, M., et al., 2007: Assessing the ability of three land ecosystem models to simulate gross carbon uptake of forests from boreal to Mediterranean climate in Europe. *Biogeosciences*, **4**, 647–656.

Jung, M., et al., 2011: Global patterns of land-atmosphere fluxes of carbon dioxide, latent heat, and sensible heat derived from eddy covariance, satellite, and meteorological observations. *J. Geophys. Res. Biogeosci.*, **116**, G00J07.

Jungclaus, J. H., et al., 2010: Climate and carbon-cycle variability over the last millennium. *Clim. Past*, **6**, 723–737.

Kai, F. M., S. C. Tyler, J. T. Randerson, and D. R. Blake, 2011: Reduced methane growth rate explained by decreased Northern Hemisphere microbial sources. *Nature*, **476**, 194–197.

Kanakidou, M., et al., 2012: Atmospheric fluxes of organic N and P to the global ocean. *Global Biogeochem. Cycles*, **26**, GB3026.

Kaplan, J. O., G. Folberth, and D. A. Hauglustaine, 2006: Role of methane and biogenic volatile organic compound sources in late glacial and Holocene fluctuations of atmospheric methane concentrations. *Global Biogeochem. Cycles*, **20**, Gb2016.

Kaplan, J. O., I. C. Prentice, W. Knorr, and P. J. Valdes, 2002: Modeling the dynamics of terrestrial carbon storage since the Last Glacial Maximum. *Geophys. Res. Lett.*, **29**, 31-1-31-4.

Kaplan, J. O., K. M. Krumhardt, E. C. Ellis, W. F. Ruddiman, C. Lemmen, and K. Klein Goldewijk, 2011: Holocene carbon emissions as a result of anthropogenic land cover change. *Holocene*, **21**, 775–791.

Karl, D. M., and R. M. Letelier, 2008: Nitrogen fixation-enhanced carbon sequestration in low nitrate, low chlorophyll seascapes. *Mar. Ecol. Prog. Ser.*, **364**, 257–268.

Kato, E., T. Kinoshita, A. Ito, M. Kawamiya, and Y. Yamagata, 2013: Evaluation of spatially explicit emission scenario of land-use change and biomass burning using a process-based biogeochemical model. *J. Land Use Sci.*, **8**, 104–122.

Keeling, C. D., 1960: The concentration and isotopic abundances of carbon dioxide in the atmosphere. *Tellus B*, **12**, 200–203.

Keeling, C. D., S. C. Piper, and M. Heimann, 1989: A three dimensional model of atmospheric CO_2 transport based on observed winds: 4. Mean annual gradients and interannual variations. In: *Aspects of Climate Variability in the Pacific and the Western Americas* [D. H. Peterson (ed.)]. Geophysical Monograph Series, Vol. 55. American Geophysical Union, Washington, DC, pp. 305–363.

Keeling, C. D., R. B. Bacastow, A. E. Bainbridge, C. A. Ekdahl, P. R. Guenther, L. S. Waterman, and J. F. S. Chin, 1976: Atmospheric carbon-dioxide variations at Mauna-Loa Observatory, Hawaii. *Tellus*, **28**, 538–551.

Keeling, C. D., S. C. Piper, R. B. Bacastow, M. Wahlen, T. P. Whorf, M. Heimann, and H. A. Meijer, 2005: Atmospheric CO_2 and $^{13}CO_2$ exchange with the terrestrial biosphere and oceans from 1978 to 2000: Observations and carbon cycle implications. In: *A History of Atmospheric CO_2 and Its Effects on Plants, Animals, and Ecosystems* [J. R. Ehleringer, T. E. Cerling and M. D. Dearing (eds.)]. Springer Science+Business Media, New York, NY, USA, and Heidelberg, Germany, pp. 83–113.

Keeling, R. F., and S. R. Shertz, 1992: Seasonal and interannual variations in atmospheric oxygen and implications for the global carbon cycle. *Nature*, **358**, 723–727.

Keeling, R. F., S. C. Piper, and M. Heimann, 1996: Global and hemispheric CO_2 sinks deduced from changes in atmospheric O_2 concentration. *Nature*, **381**, 218–221.

Keeling, R. F., A. Körtzinger, and N. Gruber, 2010: Ocean deoxygenation in a warming world. *Annu. Rev. Mar. Sci.*, **2**, 199–229.

Keenan, T. F., et al., 2012: Terrestrial biosphere model performance for inter-annual variability of land-atmosphere CO_2 exchange. *Global Change Biol.*, **18**, 1971–1987.

Keith, D. W., 2001: Geoengineering. *Nature*, **409**, 420.

Keith, D. W., M. Ha-Duong, and J. K. Stolaroff, 2006: Climate strategy with CO_2 capture from the air. *Clim. Change*, **74**, 17–45.

Kelemen, P. B., and J. Matter, 2008: In situ carbonation of peridotite for CO_2 storage. *Proc. Natl. Acad. Sci. U.S.A.*, **105**, 17295–17300.

Kellomäki, S., H. Peltola, T. Nuutinen, K. T. Korhonen, and H. Strandman, 2008: Sensitivity of managed boreal forests in Finland to climate change, with implications for adaptive management. *Philos. Trans. R. Soc. London Ser. B*, **363**, 2341–2351.

Keppler, F., J. T. G. Hamilton, M. Bra[illegible], and T. Röckmann, 2006: Methane emissions from terrestrial plants under aerobic conditions. *Nature*, **439**, 187–191.

Kesik, M., et al., 2006: Future scenarios of N_2O and NO emissions from European forest soils. *J. Geophys. Res. Biogeosci.*, **111**, G02018.

Key, R. M., et al., 2004: A global ocean carbon climatology: Results from Global Data Analysis Project (GLODAP). *Global Biogeochem. Cycles*, **18**, GB4031.

Khalil, M. A. K., and R. A. Rasmussen, 1989: Climate-induced feedbacks for the global cycles of methane and nitrous oxide. *Tellus B*, **41**, 554–559.

Khalil, M. A. K., C. L. Butenhoff, and R. A. Rasmussen, 2007: Atmospheric methane: Trends and cycles of sources and sinks. *Environ. Sci. Technol.*, **41**, 2131–2137.

Khatiwala, S., F. Primeau, and T. Hall, 2009: Reconstruction of the history of anthropogenic CO_2 concentrations in the ocean. *Nature*, **462**, 346–349.

Kheshgi, H. S., 1995: Sequestering atmospheric carbon-dioxide by increasing ocean alkalinity. *Energy*, **20**, 915–922.

Khvorostyanov, D., P. Ciais, G. Krinner, and S. Zimov, 2008: Vulnerability of east Siberia's frozen carbon stores to future warming. *Geophys. Res. Lett.*, **35**, L10703.

Kim, J. H., et al., 2004: North Pacific and North Atlantic sea-surface temperature variability during the Holocene. *Quat. Sci. Rev.*, **23**, 2141–2154.

King, A. W., D. J. Hayes, D. N. Huntzinger, T. O. West, and W. M. Post, 2012: North America carbon dioxide sources and sinks: Magnitude, attribution, and uncertainty. *Front. Ecol. Environ.*, **10**, 512–519.

Kirschbaum, M. U. F., 2003: Can trees buy time? An assessment of the role of vegetation sinks as part of the global carbon cycle. *Clim. Change*, **58**, 47–71.

Kirschbaum, M. U. F., and A. Walcroft, 2008: No detectable aerobic methane efflux from plant material, nor from adsorption/desorption processes. *Biogeosciences*, **5**, 1551–1558.

Kleinen, T., V. Brovkin, and R. J. Schuldt, 2012: A dynamic model of wetland extent and peat accumulation: Results for the Holocene. *Biogeosciences*, **9**, 235–248.

Kleinen, T., V. Brovkin, W. von Bloh, D. Archer, and G. Munhoven, 2010: Holocene carbon cycle dynamics. *Geophys. Res. Lett.*, **37**, L02705.

Kloster, S., N. M. Mahowald, J. T. Randerson, and P. J. Lawrence, 2012: The impacts of climate, land use, and demography on fires during the 21st century simulated by CLM-CN. *Biogeosciences*, **9**, 509–525.

Knorr, W., 2009: Is the airborne fraction of anthropogenic emissions increasing? *Geophys. Res. Lett.*, **36**, L21710.

Kohfeld, K. E., and A. Ridgwell, 2009: Glacial-interglacial variability in atmospheric CO_2. In: *Surface Ocean–Lower Atmospheres Processes* [C. Le Quéré and E. S. Saltzman (eds.)]. American Geophysical Union, Washington, DC, pp. 251-286.

Köhler, P., J. Hartmann, and D. A. Wolf-Gladrow, 2010: Geoengineering potential of artificially enhanced silicate weathering of olivine. *Proc. Natl. Acad. Sci. U.S.A.*, **107**, 20228–20233.

6

Köhler, P., H. Fischer, G. Munhoven, and R. E. Zeebe, 2005: Quantitative interpretation of atmospheric carbon records over the last glacial termination. *Global Biogeochem. Cycles*, **19**, GB4020.

Konijnendijk, T. Y. M., S. L. Weber, E. Tuenter, and M. van Weele, 2011: Methane variations on orbital timescales: A transient modeling experiment. *Clim. Past*, **7**, 635–648.

Koven, C. D., W. J. Riley, and A. Stern, 2013: Analysis of permafrost thermal dynamics and response to climate change in the CMIP5 Earth System Models. *J. Clim.*, **26**, 1877-1900.

Koven, C. D., et al., 2011: Permafrost carbon-climate feedbacks accelerate global warming. *Proc. Natl. Acad. Sci. U.S.A.*, **108**, 14769–14774.

Krawchuk, M. A., M. A. Moritz, M.-A. Parisien, J. Van Dorn, and K. Hayhoe, 2009: Global pyrogeography: The current and future distribution of wildfire. *PLoS ONE*, **4**, e5102.

Kraxner, F., S. Nilsson, and M. Obersteiner, 2003: Negative emissions from BioEnergy use, carbon capture and sequestration (BECS)—the case of biomass production by sustainable forest management from semi-natural temperate forests. *Biomass Bioenerg.*, **24**, 285–296.

Krinner, G., et al., 2005: A dynamic global vegetation model for studies of the coupled atmosphere-biosphere system. *Global Biogeochem. Cycles*, **19**, GB1015.

Krishnamurthy, A., J. K. Moore, N. Mahowald, C. Luo, S. C. Doney, K. Lindsay, and C. S. Zender, 2009: Impacts of increasing anthropogenic soluble iron and nitrogen deposition on ocean biogeochemistry. *Global Biogeochem. Cycles*, **23**, GB3016.

Kroeze, C., A. Mosier, and L. Bouwman, 1999: Closing the global N_2O budget: A retrospective analysis 1500–1994. *Global Biogeochem. Cycles*, **13**, 1–8.

Kroeze, C., L. Bouwman, and C. P. Slomp, 2007: Sinks for N_2O at the Earth's surface. In: *Greenhouse Gas Sinks* [D. S. Raey , M. Hewitt, J. Grace and K. A. Smith (eds.)]. CAB International, pp. 227–243.

Kroeze, C., E. Dumont, and S. P. Seitzinger, 2010: Future trends in emissions of N_2O from rivers and estuaries. *J. Integrat. Environ. Sci.*, **7**, 71–78.

Kurahashi-Nakamura, T., A. Abe-Ouchi, Y. Yamanaka, and K. Misumi, 2007: Compound effects of Antarctic sea ice on atmospheric pCO_2 change during glacial-interglacial cycle. *Geophys. Res. Lett.*, **34**, L20708.

Kurz, W. A., G. Stinson, and G. Rampley, 2008a: Could increased boreal forest ecosystem productivity offset carbon losses from increased disturbances? *Philos. Trans. R. Soc. London Ser. B*, **363**, 2261–2269.

Kurz, W. A., G. Stinson, G. J. Rampley, C. C. Dymond, and E. T. Neilson, 2008b: Risk of natural disturbances makes future contribution of Canada's forests to the global carbon cycle highly uncertain. *Proc. Natl. Acad. Sci. U.S.A.*, **105**, 1551–1555.

Kurz, W. A., et al., 2008c: Mountain pine beetle and forest carbon feedback to climate change. *Nature*, **452**, 987–990.

Kwon, E. Y., F. Primeau, and J. L. Sarmiento, 2009: The impact of remineralization depth on the air-sea carbon balance. *Nature Geosci.*, **2**, 630–635.

Lackner, K. S., 2009: Capture of carbon dioxide from ambient air. *Eur. Phys. J. Spec. Topics*, **176**, 93–106.

Lackner, K. S., 2010: Washing carbon out of the air. *Sci. Am.*, **302**, 66–71.

Lackner, K. S., S. Brennan, J. M. Matter, A.-H. A. Park, A. Wright, and B. van der Zwaan, 2012: The urgency of the development of CO_2 capture from ambient air. *Proc. Natl. Acad. Sci. U.S.A.*, **109**, 13156–13162.

Lal, R., 2004a: Soil carbon sequestration impacts on global climate change and food security. *Science*, **304**, 1623–1627.

Lal, R., 2004b: Soil carbon sequestration to mitigate climate change. *Geoderma*, **123**, 1–22.

Lamarque, J.-F., 2008: Estimating the potential for methane clathrate instability in the 1%-CO_2 IPCC AR-4 simulations. *Geophys. Res. Lett.*, **35**, L19806.

Lamarque, J.-F., et al., 2010: Historical (1850–2000) gridded anthropogenic and biomass burning emissions of reactive gases and aerosols: Methodology and application. *Atmos. Chem. Phys.*, **10**, 7017–7039.

Lamarque, J.-F., et al., 2013: The Atmospheric Chemistry and Climate Model Intercomparison Project (ACCMIP): Overview and description of models, simulations and climate diagnostics. *Geosci. Model Dev.*, **6**, 179–206.

Lamarque, J. F., et al., 2011: Global and regional evolution of short-lived radiatively-active gases and aerosols in the Representative Concentration Pathways. *Clim. Change*, **109**, 191–212.

Lampitt, R. S., et al., 2008: Ocean fertilization: A potential means of geoengineering? *Philos. Trans. R. Soc. London Ser. A*, **366**, 3919–3945.

Langenfelds, R. L., R. J. Francey, B. C. Pak, L. P. Steele, J. Lloyd, C. M. Trudinger, and C. E. Allison, 2002: Interannual growth rate variations of atmospheric CO_2 and its $d^{13}C$, H_2, CH_4, and CO between 1992 and 1999 linked to biomass burning. *Global Biogeochem. Cycles*, **16**, 1048.

Langner, J., R. Bergstrom, and V. Foltescu, 2005: Impact of climate change on surface ozone and deposition of sulphur and nitrogen in Europe. *Atmos. Environ.*, **39**, 1129–1141.

Larsen, K. S., et al., 2011: Reduced N cycling in response to elevated CO_2, warming, and drought in a Danish heathland: Synthesizing results of the CLIMAITE project after two years of treatments. *Global Change Biol.*, **17**, 1884–1899.

Lassey, K. R., D. C. Lowe, and A. M. Smith, 2007: The atmospheric cycling of radiomethane and the "fossil fraction" of the methane source. *Atmos. Chem. Phys.*, **7**, 2141–2149.

Law, R. M., R. J. Matear, and R. J. Francey, 2008: Comment on "Saturation of the Southern Ocean CO_2 sink due to recent climate change". *Science*, **319**, 570a.

Lawrence, D., et al., 2011: Parameterization improvements and functional and structural advances in version 4 of the Community Land Model. *J. Adv. Model. Earth Syst.*, **3**, M03001, 27 pp.

Lawrence, D. M., and A. G. Slater, 2005: A projection of severe near-surface permafrost degradation during the 21st century. *Geophys. Res. Lett.*, **32**, L24401.

Lawrence, D. M., A. G. Slater, V. E. Romanovsky, and D. J. Nicolsky, 2008: Sensitivity of a model projection of near-surface permafrost degradation to soil column depth and representation of soil organic matter. *J. Geophys. Res. Earth Surf.*, **113**, F02011.

Le Page, Y., G. R. van der Werf, D. C. Morton, and J. M. C. Pereira, 2010: Modeling fire-driven deforestation potential in Amazonia under current and projected climate conditions. *J. Geophys. Res. Biogeosci.*, **115**, G03012.

Le Quéré, C., T. Takahashi, E. T. Buitenhuis, C. Rodenbeck, and S. C. Sutherland, 2010: Impact of climate change and variability on the global oceanic sink of CO_2. *Global Biogeochem. Cycles*, **24**, GB4007.

Le Quéré, C., et al., 2007: Saturation of the southern ocean CO_2 sink due to recent climate change. *Science*, **316**, 1735–1738.

Le Quéré, C., et al., 2009: Trends in the sources and sinks of carbon dioxide. *Nature Geosci.*, **2**, 831–836.

Le Quéré, C., et al., 2013: The global carbon budget 1959–2011. *Earth Syst. Sci. Data*, **5**, 165–186.

LeBauer, D. S., and K. K. Treseder, 2008: Nitrogen limitation of net primary productivity in terrestrial ecosystems is globally distributed. *Ecology*, **89**, 371–379.

Lee, X., et al., 2011: Observed increase in local cooling effect of deforestation at higher latitudes. *Nature*, **479**, 384–387.

Lemmen, C., 2009: World distribution of land cover changes during Pre- and Protohistoric Times and estimation of induced carbon releases. *Geomorphol. Relief Proc. Environ.*, **4**, 303–312.

Lenton, A., and R. J. Matear, 2007: Role of the Southern Annular Mode (SAM) in Southern Ocean CO_2 uptake. *Global Biogeochem. Cycles*, **21**, Gb2016.

Lenton, T. M., and C. Britton, 2006: Enhanced carbonate and silicate weathering accelerates recovery from fossil fuel CO_2 perturbations. *Global Biogeochem. Cycles*, **20**, Gb3009.

Lenton, T. M., and N. E. Vaughan, 2009: The radiative forcing potential of different climate geoengineering options. *Atmos. Chem. Phys.*, **9**, 5539–5561.

Lepistö, A., P. Kortelainen, and T. Mattsson, 2008: Increased organic C and N leaching in a northern boreal river basin in Finland. *Global Biogeochem. Cycles*, **22**, GB3029.

LeQuere, C., T. Takahashi, E. T. Buitenhuis, C. Rodenbeck, and S. C. Sutherland, 2010: Impact of climate change and variability on the global oceanic sink of CO2. *Global Biogeochem. Cycles*, **24**.

Leuzinger, S., Y. Q. Luo, C. Beier, W. Dieleman, S. Vicca, and C. Körner, 2011: Do global change experiments overestimate impacts on terrestrial ecosystems? *Trends Ecol. Evol.*, **26**, 236–241.

Levin, I., et al., 2010: Observations and modelling of the global distribution and long-term trend of atmospheric $^{14}CO_2$. *Tellus B*, **62**, 26–46.

Levin, I., et al., 2012: No inter-hemispheric ⍰$^{13}CH_4$ trend observed. *Nature*, **486**, E3–E4.

Levine, J. G., E. W. Wolff, P. O. Hopcroft, and P. J. Valdes, 2012: Controls on the tropospheric oxidizing capacity during an idealized Dansgaard-Oeschger event, and their implications for the rapid rises in atmospheric methane during the last glacial period. *Geophys. Res. Lett.*, **39**, L12805.

Levine, J. G., et al., 2011: Reconciling the changes in atmospheric methane sources and sinks between the Last Glacial Maximum and the pre-industrial era. *Geophys. Res. Lett.*, **38**, L23804.

Levy, P. E., M. G. R. Cannell, and A. D. Friend, 2004: Modelling the impact of future changes in climate, CO_2 concentration and land use on natural ecosystems and the terrestrial carbon sink. *Global Environ. Change*, **14**, 21–30.

Lewis, S. L., P. M. Brando, O. L. Phillips, G. M. van der Heijden, and D. Nepstad, 2011: The 2010 Amazon drought. *Science*, **331**, 554.

Lewis, S. L., et al., 2009: Increasing carbon storage in intact African tropical forests. *Nature*, **457**, 1003–1006.

Li, C., S. Frolking, and K. Butterbach-Bahl, 2005: Carbon sequestration can increase nitrous oxide emissions. *Clim. Change*, **72**, 321–338.

Liberloo, M., et al., 2009: Coppicing shifts CO_2 stimulation of poplar productivity to above-ground pools: A synthesis of leaf to stand level results from the POP/EUROFACE experiment. *New Phytologist*, **182**, 331–346.

Liddicoat, S., C. Jones, and E. Robertson, 2013: CO_2 emissions determined by HadGEM2–ES to be compatible with the Representative Concentration Pathway scenarious and their extension. *J. Clim.*, **26**, 4381-4397.

Lohila, A., M. Aurela, J. Hatakka, M. Pihlatie, K. Minkkinen, T. Penttilä, and T. Laurila, 2010: Responses of N_2O fluxes to temperature, water table and N deposition in a northern boreal fen. *Eur. J. Soil Sci.*, **61**, 651–661.

Long, M. C., K. Lindsay, S. Peacock, J. K. Moore, and S. C. Doney, 2013: Twentieth-century oceanic carbon uptake and storage in CESM1(BGC). *J. Clim.*, **26**, 6775-6800.

Loose, B., and P. Schlosser, 2011: Sea ice and its effect on CO_2 flux between the atmosphere and the Southern Ocean interior. *J. Geophys. Res. Oceans*, **116**, C11.

Loulergue, L., et al., 2008: Orbital and millennial-scale features of atmospheric CH_4 over the past 800,000 years. *Nature*, **453**, 383–386.

Lourantou, A., and N. Metzl, 2011: Decadal evolution of carbon sink within a strong bloom area in the subantarctic zone. *Geophys. Res. Lett.*, **38**, L23608.

Lourantou, A., J. Chappellaz, J.-M. Barnola, V. Masson-Delmotte, and D. Raynaud, 2010a: Changes in atmospheric CO_2 and its carbon isotopic ratio during the penultimate deglaciation. *Quat. Sci. Rev.*, **29**, 1983–1992.

Lourantou, A., et al., 2010b: Constraint of the CO_2 rise by new atmospheric carbon isotopic measurements during the last deglaciation. *Global Biogeochem. Cycles*, **24**, GB2015.

Lovelock, J. E., and C. G. Rapley, 2007: Ocean pipes could help the Earth to cure itself. *Nature*, **449**, 403–403.

Lovenduski, N. S., N. Gruber, and S. C. Doney, 2008: Towards a mechanistic understanding of the decadal trends in the Southern Ocean carbon sink. *Global Biogeochem. Cycles*, **22**, GB3016.

Lovenduski, N. S., N. Gruber, S. C. Doney, and I. D. Lima, 2007: Enhanced CO_2 outgassing in the Southern Ocean from a positive phase of the Southern Annular Mode. *Global Biogeochem. Cycles*, **21**, Gb2026.

Lucht, W., et al., 2002: Climatic control of the high-latitude vegetation greening trend and Pinatubo effect. *Science*, **296**, 1687–1689.

Luo, Y., D. Hui, and D. Zhang, 2006: Elevated carbon dioxide stimulates net accumulations of carbon and nitrogen in terrestrial ecosystems: A meta-analysis. *Ecology*, **87**, 53–63.

Luo, Y., et al., 2004: Progressive nitrogen limitation of ecosystem responses to rising atmospheric carbon dioxide. *BioScience*, **54**, 731–739.

Lüthi, D., et al., 2008: High-resolution carbon dioxide concentration record 650,000–800,000 years before present. *Nature*, **453**, 379–382.

Luyssaert, S., et al., 2010: The European carbon balance. Part 3: Forests. *Global Change Biol.*, **16**, 1429–1450.

Luyssaert, S., et al., 2012: The European land and inland water CO_2, CO, CH_4 and N_2O balance between 2001 and 2005. *Biogeosciences*, **9**, 3357–3380.

MacDonald, G. M., K. V. Kremenetski, and D. W. Beilman, 2008: Climate change and the northern Russian treeline zone. *Philos. Trans. R. Soc. London Ser. B*, **363**, 2285–2299.

MacDougall, A. H., C. A. Avis, and A. J. Weaver, 2012: Significant contribution to climate warming from the permafrost carbon feedback. *Nature Geosci.*, **5**, 719–721.

MacFarling-Meure, C., et al., 2006: Law Dome CO_2, CH_4 and N_2O ice core records extended to 2000 years BP. *Geophys. Res. Lett.*, **33**, L14810.

Magnani, F., et al., 2007: The human footprint in the carbon cycle of temperate and boreal forests. *Nature*, **447**, 848–850.

Mahmoudkhani, M., and D. W. Keith, 2009: Low-energy sodium hydroxide recovery for CO_2 capture from atmospheric air - Thermodynamic analysis. *Int. J. Greenh. Gas Cont.*, **3**, 376–384.

Mahowald, N., et al., 1999: Dust sources and deposition during the last glacial maximum and current climate: A comparison of model results with paleodata from ice cores and marine sediments. *J. Geophys. Res. Atmos*, **104**, 15895–15916.

Mahowald, N., et al., 2011: Desert dust and anthropogenic aerosol interactions in the Community Climate System Model coupled-carbon-climate model. *Biogeosciences*, **8**, 387–414.

Mahowald, N. M., D. R. Muhs, S. Levis, P. J. Rasch, M. Yoshioka, C. S. Zender, and C. Luo, 2006: Change in atmospheric mineral aerosols in response to climate: Last glacial period, preindustrial, modern, and doubled carbon dioxide climates. *J. Geophys. Res. Atmos.*, **111**, D10202.

Mahowald, N. M., et al., 2009: Atmospheric iron deposition: Global ddistribution, variability, and human perturbations. *Annu. Rev. Mar. Sci.*, **1**, 245–278.

Mahowald, N. M., et al., 2010: Observed 20th century desert dust variability: Impact on climate and biogeochemistry. *Atmos. Chem. Phys.*, **10**, 10875–10893.

Maier-Reimer, E., I. Kriest, J. Segschneider, and P. Wetzel, 2005: The HAMburg Ocean Carbon Cycle model HAMOCC 5.1 – Technical description, Release 1.1. Max-Planck Institute for Meteorology, Hamburg, Germany.

Manning, A. C., and R. F. Keeling, 2006: Global oceanic and land biotic carbon sinks from the Scripps atmospheric oxygen flask sampling network. *Tellus B*, **58**, 95–116.

Marchenko, S. S., V. Romanovsky, and G. S. Tipenko, 2008: Numerical modeling of spatial permafrost dynamics in Alaska, Proceedings of the Ninth International Conference on Permafrost, University of Alaska Fairbanks, June 29–July 3, 2008, 1125–1130.

Marland, G., and R. M. Rotty, 1984: Carbon dioxide emissions from fossil fuels: A procedure for estimation and results for 1950–1982. *Tellus B*, **36**, 232–261.

Marlon, J. R., et al., 2008: Climate and human influences on global biomass burning over the past two millennia. *Nature Geosci.*, **1**, 697–702.

Marlon, J. R., et al., 2012: Long-term perspective on wildfires in the western USA. *Proc. Natl. Acad. Sci. U.S.A.*, **109**, E535–E543.

Martin, J. H., 1990: Glacial-interglacial CO_2 change: The iron hypothesis. *Paleoceanography*, **5**, 1–13.

Masarie, K. A., and P. P. Tans, 1995: Extension and integration of atmospheric carbon-dioxide data into a globally consistent measurement record. *J. Geophys. Res. Atmos.*, **100**, 11593–11610.

Mason Earles, J., S. Yeh, and K. E. Skog, 2012: Timing of carbon emissions from global forest clearance. *Nature Clim. Change*, **2**, 682–685.

Matear, R. J., and B. I. McNeil, 2003: Decadal accumulation of anthropogenic CO_2 in the Southern Ocean: A comparison of CFC-age derived estimates to multiple-linear regression estimates. *Global Biogeochem. Cycles*, **17**, 1113.

Matear, R. J., and A. C. Hirst, 2003: Long-term changes in dissolved oxygen concentrations in the ocean caused by protracted global warming. *Global Biogeochem. Cycles*, **17**, 1125.

Matear, R. J., A. C. Hirst, and B. I. McNeil, 2000: Changes in dissolved oxygen in the Southern Ocean with climate change. *Geochem. Geophys. Geosyst.*, **1**, 1050.

Matear, R. J., Y.-P. Wang, and A. Lenton, 2010: Land and ocean nutrient and carbon cycle interactions. *Curr. Opin. Environ. Sustain.*, **2**, 258–263.

Matsumoto, K., 2007: Biology-mediated temperature control on atmospheric pCO_2 and ocean biogeochemistry. *Geophys. Res. Lett.*, **34**, L20605.

Matsumoto, K., J. L. Sarmiento, and M. A. Brzezinski, 2002: Silicic acid leakage from the Southern Ocean: A possible explanation for glacial atmospheric pCO_2. *Global Biogeochem. Cycles*, **16**, 1031.

Matsumoto, K., et al., 2004: Evaluation of ocean carbon cycle models with data-based metrics. *Geophys. Res. Lett.*, **31**, L007303.

Matthews, H. D., 2006: Emissions targets for CO_2 stabilization as modified by carbon cycle feedbacks. *Tellus B*, **58**, 591–602.

Matthews, H. D., 2010: Can carbon cycle geoengineering be a useful complement to ambitious climate mitigation? *Carbon Management*, **1**, 135–144.

Matthews, H. D., and K. Caldeira, 2007: Transient climate-carbon simulations of planetary geoengineering. *Proc. Natl. Acad. Sci. U.S.A.*, **104**, 9949–9954.

Matthews, H. D., A. J. Weaver, and K. J. Meissner, 2005: Terrestrial carbon cycle dynamics under recent and future climate change. *J. Clim.*, **18**, 1609–1628.

Matthews, H. D., L. Cao, and K. Caldeira, 2009: Sensitivity of ocean acidification to geoengineered climate stabilization. *Geophys. Res. Lett.*, **36**, L10706.

Mau, S., D. Valentine, J. Clark, J. Reed, R. Camilli, and L. Washburn, 2007: Dissolved methane distributions and air-sea flux in the plume of a massive seep field, Coal Oil Point, California. *Geophys. Res. Lett.*, **34**, L22603.

Mayorga, E., et al., 2010: Global nutrient export from WaterSheds 2 (NEWS 2): Model development and implementation. *Environ. Model. Software*, **25**, 837–853.

McCarthy, H. R., et al., 2010: Re-assessment of plant carbon dynamics at the Duke free-air CO_2 enrichment site: Interactions of atmospheric CO_2 with nitrogen and water availability over stand development. *New Phytologist*, **185**, 514–528.

McGuire, A. D., et al., 2009: Sensitivity of the carbon cycle in the Arctic to climate change. *Ecol. Monogr.*, **79**, 523–555.

McGuire, A. D., et al., 2012: An assessment of the carbon balance of Arctic tundra: Comparisons among observations, process models, and atmospheric inversions. *Biogeosciences*, **9**, 3185–3204.

McInerney, F. A., and S. L. Wing, 2011: The Paleocene-Eocene thermal maximum: A perturbation of carbon cycle, climate, and biosphere with implications for the future. *Annu. Rev. Earth Planet. Sci.*, **39**, 489–516.

McIntyre, B. D., H. R. Herren, J. Wakhungu, and R. T. Watson, 2009: International assessment of agricultural knowledge, science and technology for development (IAASTD): Global report. International Assessment of Agricultural Knowledge, Science and Technology for Development, 590 pp.

McKinley, G. A., A. R. Fay, T. Takahashi, and N. Metzl, 2011: Convergence of atmospheric and North Atlantic carbon dioxide trends on multidecadal timescales. *Nature Geosci.*, **4**, 606–610.

McKinley, G. A., et al., 2006: North Pacific carbon cycle response to climate variability on seasonal to decadal timescales. *J. Geophys. Res. Oceans*, **111**, C07s06.

McNeil, B. I., and R. J. Matear, 2006: Projected climate change impact on oceanic acidification. *Carbon Bal. Manag.*, **1**.

McNeil, B. I., and R. J. Matear, 2008: Southern Ocean acidification: A tipping point at 450-ppm atmospheric CO_2. *Proc. Natl. Acad. Sci. U.S.A.*, **105**, 18860–18864.

McNeil, B. I., R. J. Matear, R. M. Key, J. L. Bullister, and J. L. Sarmiento, 2003: Anthropogenic CO_2 uptake by the ocean based on the global chlorofl uorocarbon data set. *Science*, **299**, 235–239.

Medlyn, B. E., 2011: Comment on "Drought-induced reductions in global terrestrial net primary production from 2000 through 2009". *Science*, **333**, 1093.

Meehl, G. H., et al., 2007: Global Climate Projections. In: *Climate Change 2007: The Physical Science Basis. Contribution of Working Group I to the Fourth Assessment Report of the Intergovernmental Panel on Climate Change* [Solomon, S., D. Qin, M. Manning, Z. Chen, M. Marquis, K. B. Averyt, M. Tignor and H. L. Miller (eds.)] Cambridge University Press, Cambridge, United Kingdom and New York, NY, USA, pp. 747–846.

Meinshausen, M., et al., 2011: The RCP greenhouse gas concentrations and their extensions from 1765 to 2300. *Clim. Change*, **109**, 213–241.

Melillo, J. M., et al., 2011: Soil warming, carbon-nitrogen interactions, and forest carbon budgets. *Proc. Natl. Acad. Sci. U.S.A.*, **108**, 9508–9512.

Melton, J. R., et al., 2013: Present state of global wetland extent and wetland methane modelling: Conclusions from a model intercomparison project (WETCHIMP). *Biogeosciences*, **10**, 753-788.

Menviel, L., and F. Joos, 2012: Toward explaining the Holocene carbon dioxide and carbon isotope records: Results from transient ocean carbon cycle-climate simulations. *Paleoceanography*, **27**, PA1207.

Menviel, L., F. Joos, and S. P. Ritz, 2012: Simulating atmospheric CO_2, C^{13} and the marine carbonate cycle during the last Glacial-Interglacial cycle: Possible role for a deepening of the mean remineralization depth and an increase in the oceanic nutrient inventory. *Quat. Sci. Rev.*, **56**, 46–68.

Menviel, L., A. Timmermann, A. Mouchet, and O. Timm, 2008: Meridional reorganizations of marine and terrestrial productivity during Heinrich events. *Paleoceanography*, **23**, PA1203.

Mercado, L. M., N. Bellouin, S. Sitch, O. Boucher, C. Huntingford, M. Wild, and P. M. Cox, 2009: Impact of changes in diffuse radiation on the global land carbon sink. *Nature*, **458**, 1014–1017.

Merico, A., T. Tyrrell, and T. Cokacar, 2006: Is there any relationship between phytoplankton seasonal dynamics and the carbonate system? *J. Mar. Syst.*, **59**, 120–142.

Metsaranta, J. M., W. A. Kurz, E. T. Neilson, and G. Stinson, 2010: Implications of future disturbance regimes on the carbon balance of Canada's managed forest (2010–2100). *Tellus B*, **62**, 719–728.

Metz, B., O. Davidson, H. C. De Coninck, M. Loss, and L. A. Meyer, 2005: *IPCC Special Report on Carbon Dioxide Capture and Storage* Cambridge University Press, Cambridge, United Kingdom, and New York, NY, USA, 442 pp.,

Metzl, N., 2009: Decadal increase of oceanic carbon dioxide in Southern Indian Ocean surface waters (1991–2007). *Deep-Sea Res. Pt. II*, **56**, 607–619.

Metzl, N., et al., 2010: Recent acceleration of the sea surface fCO_2 growth rate in the North Atlantic subpolar gyre (1993–2008) revealed by winter observations. *Global Biogeochem. Cycles*, **24**, GB4004.

Mieville, A., et al., 2010: Emissions of gases and particles from biomass burning during the 20th century using satellite data and an historical reconstruction. *Atmos. Environ.*, **44**, 1469–1477.

Mikaloff-Fletcher, S. E., et al., 2006: Inverse estimates of anthropogenic CO_2 uptake, transport, and storage by the ocean. *Global Biogeochem. Cycles*, **20**, GB2002.

Minschwaner, K., R. J. Salawitch, and M. B. McElroy, 1993: Absorption of solar radiation by O_2: Implications for O_3 and lifetimes of N_2O, $CFCl_3$, and CF_2Cl_2. *J. Geophys. Res. Atmos.*, **98**, 10543–10561.

Mischler, J. A., et al., 2009: Carbon and hydrogen isotopic composition of methane over the last 1000 years. *Global Biogeochem. Cycles*, **23**, GB4024.

Mitchell, L. E., E. J. Brook, T. Sowers, J. R. McConnell, and K. Taylor, 2011: Multidecadal variability of atmospheric methane, 1000–1800 C.E. *J. Geophys. Res. Biogeosci.*, **116**, G02007.

Miyama, T., and M. Kawamiya, 2009: Estimating allowable carbon emission for CO_2 concentration stabilization using a GCM-based Earth system model. *Geophys. Res. Lett.*, **36**, L19709.

Monnin, E., et al., 2001: Atmospheric CO_2 concentrations over the last glacial termination. *Science*, **291**, 112–114.

Monnin, E., et al., 2004: Evidence for substantial accumulation rate variability in Antarctica during the Holocene through synchronization of CO_2 in the Taylor Dome, Dome C and DML ice cores. *Earth Planet. Sci. Lett.*, **224**, 45–54.

Monteil, G., S. Houweling, E. J. Dlugockenky, G. Maenhout, B. H. Vaughn, J. W. C. White, and T. Rockmann, 2011: Interpreting methane variations in the past two decades using measurements of CH_4 mixing ratio and isotopic composition. *Atmos. Chem. Phys.*, **11**, 9141–9153.

Monteith, D. T., et al., 2007: Dissolved organic carbon trends resulting from changes in atmospheric deposition chemistry. *Nature*, **450**, 537–U539.

Montenegro, A., V. Brovkin, M. Eby, D. Archer, and A. J. Weaver, 2007: Long term fate of anthropogenic carbon. *Geophys. Res. Lett.*, **34**, L19707.

Montenegro, A., M. Eby, Q. Z. Mu, M. Mulligan, A. J. Weaver, E. C. Wiebe, and M. S. Zhao, 2009: The net carbon drawdown of small scale afforestation from satellite observations. *Global Planet. Change*, **69**, 195–204.

Montzka, S. A., M. Krol, E. Dlugokencky, B. Hall, P. Joeckel, and J. Lelieveld, 2011: Small interannual variability of global atmospheric hydroxyl. *Science*, **331**, 67–69.

Mooney, S. D., et al., 2011: Late Quaternary fire regimes of Australasia. *Quat. Sci. Rev.*, **30**, 28–46.

Morford, S. L., B. Z. Houlton, and R. A. Dahlgren, 2011: Increased forest ecosystem carbon and nitrogen storage from nitrogen rich bedrock. *Nature*, **477**, 78–81.

Morino, I., et al., 2011: Preliminary validation of column-averaged volume mixing ratios of carbon dioxide and methane retrieved from GOSAT short-wavelength infrared spectra. *Atmos. Measure. Techn.*, **3**, 5613–5643.

Mosier, A., C. Kroeze, C. Nevison, O. Oenema, S. Seitzinger, and O. van Cleemput, 1998: Closing the global N_2O budget: Nitrous oxide emissions through the agricultural nitrogen cycle - OECD/IPCC/IEA phase II development of IPCC guidelines for national greenhouse gas inventory methodology. *Nutr. Cycl. Agroecosyst.*, **52**, 225–248.

Mosier, A. R., J. A. Morgan, J. Y. King, D. R. LeCain, and D. G. Milchunas, 2002: Soil-atmosphere exchange of CH_4, CO_2, NOx, and N_2O in the Colorado shortgrass steppe under elevated CO_2. *Plant Soil*, **240**, 201–211.

Moss, R. H., et al., 2010: The next generation of scenarios for climate change research and assessment. *Nature*, **463**, 747–756.

Munhoven, G., 2002: Glacial-interglacial changes of continental weathering: Estimates of the related CO_2 and HCO^{3-} flux variations and their uncertainties. *Global Planet. Change*, **33**, 155–176.

Murata, A., Y. Kumamoto, S. Watanabe, and M. Fukasawa, 2007: Decadal increases of anthropogenic CO_2 in the South Pacific subtropical ocean along 32 degrees S. *J. Geophys. Res. Oceans*, **112**, C05033.

Murata, A., Y. Kumamoto, K.-i. Sasaki, S. Watanabe, and M. Fukasawa, 2009: Decadal increases of anthropogenic CO_2 along 149 degrees E in the western North Pacific. *J. Geophys. Res. Oceans*, **114**, C04018.

Murata, A., Y. Kumamoto, K. Sasaki, S. Watanabe, and M. Fukasawa, 2010: Decadal increases in anthropogenic CO_2 along 20 degrees S in the South Indian Ocean. *J. Geophys. Res. Oceans*, **115**, C12055.

Myneni, R. B., et al., 2001: A large carbon sink in the woody biomass of Northern forests. *Proc. Natl. Acad. Sci. U.S.A.*, **98**, 14784–14789.

Nabuurs, G. J., et al., 2008: Hotspots of the European forests carbon cycle. *Forest Ecol. Manage.*, **256**, 194–200.

Naegler, T., and I. Levin, 2009: Observation-based global biospheric excess radiocarbon inventory 1963–2005. *J. Geophys. Res.*, **114**, D17302.

Naik, V., D. J. Wuebbles, E. H. De Lucia, and J. A. Foley, 2003: Influence of geoengineered climate on the terrestrial biosphere. *Environ. Manage.*, **32**, 373–381.

Naqvi, S. W. A., H. W. Bange, L. Farias, P. M. S. Monteiro, M. I. Scranton, and J. Zhang, 2010: Coastal hypoxia/anoxia as a source of CH_4 and N_2O. *Biogeosciences*, **7**, 2159–2190.

Neef, L., M. van Weele, and P. van Velthoven, 2010: Optimal estimation of the present-day global methane budget. *Global Biogeochem. Cycles*, **24**, GB4024.

Neftel, A., H. Oeschger, J. Schwander, B. Stauffer, and R. Zumbrunn, 1982: Ice core sample measurements give atmospheric CO_2 content during the past 40,000 yr. *Nature*, **295**, 220–223.

Nemani, R. R., et al., 2003: Climate-driven increases in global terrestrial net primary production from 1982 to 1999. *Science*, **300**, 1560–1563.

Nevison, C. D., et al., 2011: Exploring causes of interannual variability in the seasonal cycles of tropospheric nitrous oxide. *Atmos. Chem. Phys.*, **11**, 3713–3730.

Nevle, R. J., and D. K. Bird, 2008: Effects of syn-pandemic fire reduction and reforestation in the tropical Americas on atmospheric CO_2 during European conquest. *Palaeogeogr. Palaeoclimatol. Palaeoecol.* **264**, 25–38.

Nevle, R. J., D. K. Bird, W. F. Ruddiman, and R. A. and Dull, 2011: Neotropical human landscape interactions, fire, and atmospheric CO_2 during European conquest. *Holocene*, **21**, 853–864.

Newingham, B. A., C. H. Vanier, T. N. Charlet, K. Ogle, S. D. Smith, and R. S. Nowak, 2013: No cumulative effect of ten years of elevated CO_2 on perennial plant biomass components in the Mojave Desert. *Global Change Biol.*, **19**, 2168-2181..

Nisbet, R. E. R., et al., 2009: Emission of methane from plants. *Proc. R. Soc. Ser. B*, **276**, 1347–1354.

Norby, R. J., 1998: Nitrogen deposition: A component of global change analysis. *New Phytologist*, **139**, 189–200.

Norby, R. J., J. M. Warren, C. M. Iversen, B. E. Medlyn, and R. E. McMurtrie, 2010: CO_2 enhancement of forest productivity constrained by limited nitrogen availability. *Proc. Natl. Acad. Sci. U.S.A.*, **107**, 19368–19373.

Norby, R. J., et al., 2005: Forest response to elevated CO_2 is conserved across a broad range of productivity. *Proc. Natl. Acad. Sci. U.S.A.*, **102**, 18052–18056.

Nowak, R. S., D. S. Ellsworth, and S. D. Smith, 2004: Functional responses of plants to elevated atmospheric CO_2—do photosynthetic and productivity data from FACE experiments support early predictions? *New Phytologist*, **162**, 253–280.

O'Connor, F. M., et al., 2010: Possible role of wetlands, permafrost, and methane hydrates in the methane cycle under future climate change: A review. *Rev. Geophys.*, **48**, RG4005.

Oh, N.-H., and P. A. Raymond, 2006: Contribution of agricultural liming to riverine bicarbonate export and CO_2 sequestration in the Ohio River basin. *Global Biogeochem. Cycles*, **20**, GB3012.

Oleson, K. W., et al., 2010: Technical description of version 4.0 of the Community Land Model (CLM), NCAR Technical Note NCAR/TN-478+STR, National Center for Atmospheric Research, Boulder, CO, USA, 257 pp.

Olivier, J., J. Aardenne, F. Dentener, L. Ganzeveld, and J. Peters, 2005: Recent trends in global greenhouse emissions: Regional trends 1970–2000 and spatial distribution of key sources in 2000. *Environ. Sci.*, **2**, 81–99.

Olivier, J. G. J., and G. Janssens-Maenhout, 2012: Part III: Greenhouse gas emissions: 1. Shares and trends in greenhouse gas emissions; 2. Sources and Methods; Total greenhouse gas emissions. In: *CO_2 Emissions from Fuel Combustion, 2012 Edition*. International Energy Agency (IEA), Paris, France, III.1–III.51.

Olofsson, J., and T. Hickler, 2008: Effects of human land-use on the global carbon cycle during the last 6,000 years. *Veget. Hist. Archaeobot.*, **17**, 605–615.

Olsen, A., et al., 2006: Magnitude and origin of the anthropogenic CO_2 increase and ^{13}C Suess effect in the Nordic seas since 1981. *Global Biogeochem. Cycles*, **20**, GB3027.

Opdyke, M. R., N. E. Ostrom, and P. H. Ostrom, 2009: Evidence for the predominance of denitrification as a source of N_2O in temperate agricultural soils based on isotopologue measurements. *Global Biogeochem. Cycles*, **23**, Gb4018.

Orr, F. M. J., 2009: Onshore geologic storage of CO_2. *Science*, **325**, 1656–1658.

Orr, J. C., 2011: Recent and future changes in ocean carbonate chemistry. In: *Ocean Acidification* [J.-P. Gattuso and L. Hansson (eds.)]. Oxford University Press, Oxford, United Kingdom, and New York, NY, USA, pp. 41-66.

Orr, J. C., et al., 2001: Estimates of anthropogenic carbon uptake from four three-dimensional global ocean models. *Global Biogeochem. Cycles*, **15**, 43–60.

Orr, J. C., et al., 2005: Anthropogenic ocean acidification over the twenty-first century and its impact on calcifying organisms. *Nature*, **437**, 681–686.

Oschlies, A., 2001: Model-derived estimates of new production: New results point towards lower values. *Deep-Sea Res. Pt. II*, **48**, 2173–2197.

Oschlies, A., K. G. Schulz, U. Riebesell, and A. Schmittner, 2008: Simulated 21st century's increase in oceanic suboxia by CO_2-enhanced biotic carbon export. *Global Biogeochem. Cycles*, **22**, GB4008.

Oschlies, A., W. Koeve, W. Rickels, and K. Rehdanz, 2010a: Side effects and accounting aspects of hypothetical large-scale Southern Ocean iron fertilization. *Biogeosciences*, **7**, 4017–4035.

Oschlies, A., M. Pahlow, A. Yool, and R. J. Matear, 2010b: Climate engineering by artificial ocean upwelling: Channelling the sorcerer's apprentice. *Geophys. Res. Lett.*, **37**, L04701.

Otto, D., D. Rasse, J. Kaplan, P. Warnant, and L. Francois, 2002: Biospheric carbon stocks reconstructed at the Last Glacial Maximum: Comparison between general circulation models using prescribed and computed sea surface temperatures. *Global Planet. Change*, **33**, 117–138.

Pacala, S. W., et al., 2001: Consistent land- and atmosphere-based U.S. carbon sink estimates. *Science*, **292**, 2316–2320.

Page, S. E., J. O. Rieley, and C. J. Banks, 2011: Global and regional importance of the tropical peatland carbon pool. *Global Change Biol.*, **17**, 798–818.

Page, S. E., F. Siegert, J. O. Rieley, H.-D. V. Boehm, A. Jaya, and S. Limin, 2002: The amount of carbon released from peat and forest fires in Indonesia during 1997. *Nature*, **420**, 61–65.

Palmroth, S., et al., 2006: Aboveground sink strength in forests controls the allocation of carbon below ground and its $[CO_2]$-induced enhancement. *Proc. Natl. Acad. Sci. U.S.A.*, **103**, 19362–19367.

Pan, Y. D., et al., 2011: A large and persistent carbon sink in the world's forests. *Science*, **333**, 988–993.

Papa, F., C. Prigent, F. Aires, C. Jimenez, W. B. Rossow, and E. Matthews, 2010: Interannual variability of surface water extent at the global scale, 1993–2004. *J. Geophys. Res.Atmos.*, **115**, D12111.

Parekh, P., F. Joos, and S. A. Müller, 2008: A modeling assessment of the interplay between aeolian iron fluxes and iron-binding ligands in controlling carbon dioxide fluctuations during Antarctic warm events. *Paleoceanography*, **23**, PA4202.

Park, G.-H., et al., 2010: Variability of global net air-sea CO_2 fluxes over the last three decades using empirical relationships. *Tellus B*, **62**, 352–368.

Park, S., et al., 2012: Trends and seasonal cycles in the isotopic composition of nitrous oxide since 1940. *Nature Geosci.*, **5**, 261–265.

Patra, P. K., et al., 2013: The carbon budget of South Asia. *Biogeosciences*, **10**, 513–527.

Patra, P. K., et al., 2011: TransCom model simulations of CH_4 and related species: Linking transport, surface flux and chemical loss with CH_4 variability in the troposphere and lower stratosphere. *Atmos. Chem. Phys.*, **11**, 12813–12837.

Pechony, O., and D. T. Shindell, 2010: Driving forces of global wildfires over the past millennium and the forthcoming century. *Proc. Natl. Acad. Sci. U.S.A.*, **107**, 19167–19170.

Peng, T.-H., and W. S. Broecker, 1991: Dynamic limitations on the Antarctic iron fertilization strategy. *Nature*, **349**, 227–229.

Peng, T.-H., R. Wanninkhof, J. L. Bullister, R. A. Feely, and T. Takahashi, 1998: Quantification of decadal anthropogenic CO_2 uptake in the ocean based on dissolved inorganic carbon measurements. *Nature*, **396**, 560–563.

Peng, T. H., R. Wanninkhof, and R. A. Feely, 2003: Increase of anthropogenic CO_2 in the Pacific Ocean over the last two decades. *Deep-Sea Res. Pt. II*, **50**, 3065–3082.

Peñuelas, J., J. G. Canadell, and R. Ogaya, 2011: Increased water-use-efficiency during the 20th century did not translate into enhanced tree growth. *Global Ecol. Biogeogr.*, **20**, 597–608.

Pérez, F. F., M. Vázquez-Rodríguez, E. Louarn, X. A. Padin, H. Mercier, and A. F. Rios, 2008: Temporal variability of the anthropogenic CO_2 storage in the Irminger Sea. *Biogeosciences*, **5**, 1669–1679.

Perrin, A.-S., A. Probst, and J.-L. Probst, 2008: Impact of nitrogenous fertilizers on carbonate dissolution in small agricultural catchments: Implications for weathering CO_2 uptake at regional and global scales. *Geochim. Cosmochim. Acta*, **72**, 3105–3123.

Pershing, A. J., L. B. Christensen, N. R. Record, G. D. Sherwood, and P. B. Stetson, 2010: The impact of whaling on the ocean carbon cycle: Why bigger was better. *PLoS ONE*, **5**, e12444.

6

Peters, G. P., et al., 2013: The challenge to keep global warming below 2°C. *Nature Clim. Change*, **3**, 4–6.

Petit, J. R., et al., 1999: Climate and atmospheric history of the past 420,000 years from the Vostok ice core, Antarctica. *Nature*, **399**, 429–436.

Petrenko, V. V., et al., 2009: $^{14}CH_4$ Measurements in Greenland ice: Investigating Last Glacial Termination CH_4 sources. *Science*, **324**, 506–508.

Peylin, P., et al., 2005: Multiple constraints on regional CO_2 flux variations over land and oceans. *Global Biogeochem. Cycles*, **19**, GB1011.

Peylin, P., et al., 2013: Global atmospheric carbon budget: Results from an ensemble of atmospheric CO_2 inversions. *Biogeosci. Discuss.*, **10**, 5301–5360.

Pfeil, G. B., et al., 2013: A uniform, quality controlled Surface Ocean CO_2 Atlas (SOCAT). *Earth Syst. Sci. Data*, **5**, 125–143.

Phoenix, G. K., et al., 2006: Atmospheric nitrogen deposition in world biodiversity hotspots: The need for a greater global perspective in assessing N deposition impacts. *Global Change Biol.*, **12**, 470–476.

Piao, S., et al., 2011: Contribution of climate change and rising CO_2 to terrestrial carbon balance in East Asia: A multi-model analysis. *Global Planet. Change*, **75**, 133–142.

Piao, S., et al., 2013: Evaluation of terrestrial carbon cycle models for their response to climate variability and CO_2 trends. *Global Change Biol.*, **19**, 2117-2132..

Piao, S. L., P. Friedlingstein, P. Ciais, L. M. Zhou, and A. P. Chen, 2006: Effect of climate and CO_2 changes on the greening of the Northern Hemisphere over the past two decades. *Geophys. Res. Lett.*, **33**, L23402.

Piao, S. L., P. Friedlingstein, P. Ciais, N. de Noblet-Ducoudré, D. Labat, and S. Zaehle, 2007: Changes in climate and land use have a larger direct impact than rising CO_2 on global river runoff trends. *Proc. Natl. Acad. Sci. U.S.A.*, **104**, 15242–15247.

Piao, S. L., P. Ciais, P. Friedlingstein, N. de Noblet-Ducoudré, P. Cadule, N. Viovy, and T. Wang, 2009a: Spatiotemporal patterns of terrestrial carbon cycle during the 20th century. *Global Biogeochem. Cycles*, **23**, Gb4026.

Piao, S. L., J. Y. Fang, P. Ciais, P. Peylin, Y. Huang, S. Sitch, and T. Wang, 2009b: The carbon balance of terrestrial ecosystems in China. *Nature*, **458**, 1009–U82.

Piao, S. L., et al., 2012: The carbon budget of terrestrial ecosystems in East Asia over the last two decades. *Biogeosciences*, **9**, 3571–3586.

Pison, I., P. Bousquet, F. Chevallier, S. Szopa, and D. Hauglustaine, 2009: Multi-species inversion of CH_4, CO and H_2 emissions from surface measurements. *Atmos. Chem. Phys.*, **9**, 5281–5297.

Plattner, G.-K., et al., 2008: Long-term climate commitments projected with climate-carbon cycle models. *J. Clim.*, **21**, 2721-2751.

Plattner, G. K., F. Joos, and T. Stocker, 2002: Revision of the global carbon budget due to changing air-sea oxygen fluxes. *Global Biogeochem. Cycles*, **16**, 1096.

Plug, L. J., and J. J. West, 2009: Thaw lake expansion in a two-dimensional coupled model of heat transfer, thaw subsidence, and mass movement. *J. Geophys. Res.*, **114**, F01002.

Pollard, R. T., et al., 2009: Southern Ocean deep-water carbon export enhanced by natural iron fertilization. *Nature*, **457**, 577–580.

Pongratz, J., C. H. Reick, T. Raddatz, and M. Claussen, 2009: Effects of anthropogenic land cover change on the carbon cycle of the last millennium. *Global Biogeochem. Cycles*, **23**, Gb4001.

Pongratz, J., K. Caldeira, C. H. Reick, and M. Claussen, 2011a: Coupled climate-carbon simulations indicate minor global effects of wars and epidemics on atmospheric CO_2 between AD 800 and 1850. *Holocene*, **21**, 843–851.

Pongratz, J., C. H. Reick, T. Raddatz, K. Caldeira, and M. Claussen, 2011b: Past land use decisions have increased mitigation potential of reforestation. *Geophys. Res. Lett.*, **38**, L15701.

Poulter, B., et al., 2010: Net biome production of the Amazon Basin in the 21st century. *Global Change Biol.*, **16**, 2062–2075.

Power, M. J., et al., 2013: Climatic control of the biomass-burning decline in the Americas after AD 1500. *Holocene*, **23**, 3–13.

Power, M. J., et al., 2008: Changes in fire regimes since the Last Glacial Maximum: An assessment based on a global synthesis and analysis of charcoal data. *Clim. Dyn.*, **30**, 887–907.

Prather, M. J., C. D. Holmes, and J. Hsu, 2012: Reactive greenhouse gas scenarios: Systematic exploration of uncertainties and the role of atmospheric chemistry. *Geophys. Res. Lett.*, **39**, L09803.

Prentice, I. C., and S. P. Harrison, 2009: Ecosystem effects of CO_2 concentration: Evidence from past climates. *Clim. Past*, **5**, 297–307.

Prentice, I. C., et al., 2001: The carbon cycle and atmospheric carbon dioxide. In: *Climate Change 2001: The Scientific Basis. Contribution of Working Group I to the Third Assessment Report of the Intergovernmental Panel on Climate Change* [J. T. Houghton, Y. Ding, D. J. Griggs, M. Noquer, P. J. van der Linden, X. Dai, K. Maskell and C. A. Johnson (eds.)]. Cambridge University Press, Cambridge, United Kingdom and New York, NY, USA, pp. 183–237.

Prinn, R. G., et al., 2001: Evidence for substantial variations of atmospheric hydroxyl radicals in the past two decades. *Science*, **292**, 1882–1888.

Prinn, R. G., et al., 2005: Evidence for variability of atmospheric hydroxyl radicals over the past quarter century. *Geophys. Res. Lett.*, **32**, L07809.

Prinn, R. G., et al., 2000: A history of chemically and radiatively important gases in air deduced from ALE/GAGE/AGAGE. *J. Geophys. Res. Atmos.*, **105**, 17751–17792.

Quinton, J. N., G. Govers, K. Van Oost, and R. D. Bardgett, 2010: The impact of agricultural soil erosion on biogeochemical cycling. *Nature Geosci.*, **3**, 311–314.

Rabalais, N. N., R. J. Diaz, L. A. Levin, R. E. Turner, D. Gilbert, and J. Zhang, 2010: Dynamics and distribution of natural and human-caused hypoxia. *Biogeosciences*, **7**, 585–619.

Raddatz, T. J., et al., 2007: Will the tropical land biosphere dominate the climate-carbon cycle feedback during the twenty-first century? *Clim. Dyn.*, **29**, 565–574.

Rafelski, L. E., S. C. Piper, and R. F. Keeling, 2009: Climate effects on atmospheric carbon dioxide over the last century. *Tellus B*, **61**, 718–731.

Ramankutty, N., and J. A. Foley, 1999: Estimating historical changes in global land cover: Croplands from 1700 to 1992. *Global Biogeochem. Cycles*, **13**, 997–1027.

Ramankutty, N., C. Delire, and P. Snyder, 2006: Feedbacks between agriculture and climate: An illustration of the potential unintended consequences of human land use activities. *Global Planet. Change*, **54**, 79–93.

Randerson, J. T., et al., 2009: Systematic assessment of terrestrial biogeochemistry in coupled climate-carbon models. *Global Change Biol.*, **15**, 2462–2484.

Rau, G. H., 2008: Electrochemical splitting of calcium carbonate to increase solution alkalinity: Implications for mitigation of carbon dioxide and ocean acidity. *Environ. Sci. Technol.*, **42**, 8935–8940.

Rau, G. H., and K. Caldeira, 1999: Enhanced carbonate dissolution: A means of sequestering waste CO_2 as ocean bicarbonate. *Energ. Conv. Manage.*, **40**, 1803–1813.

Raupach, M. R., 2013: The exponential eigenmodes of the carbon-climate system, and their implications for ratios of responses to forcings *Earth Syst. Dyn.*, **4**, 31–49.

Raupach, M. R., J. G. Canadell, and C. Le Quéré, 2008: Anthropogenic and biophysical contributions to increasing atmospheric CO_2 growth rate and airborne fraction. *Biogeosciences*, **5**, 1601–1613.

Ravishankara, A. R., J. S. Daniel, and R. W. Portmann, 2009: Nitrous oxide (N_2O): The dominant ozone-depleting substance emitted in the 21st century. *Science*, **326**, 123–125.

Raymond, P. A., and J. J. Cole, 2003: Increase in the export of alkalinity from North America's largest river. *Science*, **301**, 88–91.

Raymond, P. A., N.-H. Oh, R. E. Turner, and W. Broussard, 2008: Anthropogenically enhanced fluxes of water and carbon from the Mississippi River. *Nature*, **451**, 449–452.

Rayner, P. J., R. M. Law, C. E. Allison, R. J. Francey, C. M. Trudinger, and C. Pickett-Heaps, 2008: Interannual variability of the global carbon cycle (1992–2005) inferred by inversion of atmospheric CO_2 and ▯^{13} CO_2 measurements. *Global Biogeochem. Cycles*, **22**, GB3008.

Reagan, M. T., and G. J. Moridis, 2007: Oceanic gas hydrate instability and dissociation under climate change scenarios. *Geophys. Res. Lett.*, **34**, L22709.

Reagan, M. T., and G. J. Moridis, 2009: Large-scale simulation of methane hydrate dissociation along the West Spitsbergen Margin. *Geophys. Res. Lett.*, **36**, L23612.

Reay, D. S., F. Dentener, P. Smith, J. Grace, and R. A. Feely, 2008: Global nitrogen deposition and carbon sinks. *Nature Geosci.*, **1**, 430–437.

Renforth, P., 2012: The potential of enhanced weathering in the UK. *Int. J. Greenh. Gas Cont.*, **10**, 229–243.

Revelle, R., and H. E. Suess, 1957: Carbon dioxide exchange between atmosphere and ocean and the question of an increase of atmospheric CO_2 during the past decades. *Tellus*, **9**, 18–27.

Rhee, T. S., A. J. Kettle, and M. O. Andreae, 2009: Methane and nitrous oxide emissions from the ocean: A reassessment using basin-wide observations in the Atlantic. *J. Geophys. Res.*, **114**, D12304.

Ricke, K. L., M. G. Morgan, and M. R. Allen, 2010: Regional climate response to solar-radiation management. *Nature Geosci.*, **3**, 537–541.

Ridgwell, A., and R. E. Zeebe, 2005: The role of the global carbonate cycle in the regulation and evolution of the Earth system. *Earth Planet. Sci. Lett.*, **234**, 299–315.

Ridgwell, A., and J. C. Hargreaves, 2007: Regulation of atmospheric CO_2 by deep-sea sediments in an Earth system model. *Global Biogeochem. Cycles*, **21**, Gb2008.

Ridgwell, A. J., 2001: Glacial-interglacial perturbations in the global carbon cycle. PhD Thesis, University of East Anglia, Norwich, United Kingdom, 134 pp.

Ridgwell, A. J., A. J. Watson, M. A. Maslin, and J. O. Kaplan, 2003: Implications of coral reef buildup for the controls on atmospheric CO_2 since the Last Glacial Maximum. *Paleoceanography*, **18**, 1083.

Riebesell, U., A. Körtzinger, and A. Oschlies, 2009: Sensitivities of marine carbon fluxes to ocean change. *Proc. Natl. Acad. Sci. U.S.A.*, **106**, 20602–20609.

Riebesell, U., et al., 2007: Enhanced biological carbon consumption in a high CO_2 ocean. *Nature*, **450**, 545–548.

Rigby, M., et al., 2008: Renewed growth of atmospheric methane. *Geophys. Res. Lett.*, **35**, L22805.

Ringeval, B., P. Friedlingstein, C. Koven, P. Ciais, N. de Noblet-Ducoudre, B. Decharme, and P. Cadule, 2011: Climate-CH_4 feedback from wetlands and its interaction with the climate-CO_2 feedback. *Biogeosciences*, **8**, 2137–2157.

Robock, A., L. Oman, and G. L. Stenchikov, 2008: Regional climate responses to geoengineering with tropical and Arctic SO_2 injections. *J. Geophys. Res.*, **113**, D16101.

Röckmann, T., and I. Levin, 2005: High-precision determination of the changing isotopic composition of atmospheric N_2O from 1990 to 2002. *J. Geophys. Res. Atmos.*, **110**, D21304.

Rödenbeck, C., S. Houweling, M. Gloor, and M. Heimann, 2003: CO_2 flux history 1982–2001 inferred from atmospheric data using a global inversion of atmospheric transport. *Atmos. Chem. Phys.*, **3**, 1919–1964.

Rosamond, M. S., S. J. Thuss, and S. L. Schiff, 2012: Dependence of riverine nitrous oxide emissions on dissolved oxygen levels. *Nature Geosci.*, **5**, 715–718.

Roth, R., and F. Joos, 2012: Model limits on the role of volcanic carbon emissions in regulating glacial-interglacial CO_2 variations. *Earth Planet. Sci. Lett.*, **329–330**, 141–149.

Röthlisberger, R., M. Bigler, E. W. Wolff, F. Joos, E. Monnin, and M. A. Hutterli, 2004: Ice core evidence for the extent of past atmospheric CO_2 change due to iron fertilisation. *Geophys. Res. Lett.*, **31**, L16207.

Rotty, R. M., 1983: Distribution of and changes in industrial carbon-cycle production. *J. Geophys. Res. Oceans*, **88**, 1301–1308.

Roy, T., et al., 2011: Regional impacts of climate change and atmospheric CO_2 on future ocean carbon uptake: A multimodel linear feedback analysis. *J. Clim.*, **24**, 2300–2318.

Rubasinghege, G., S. N. Spak, C. O. Stanier, G. R. Carmichael, and V. H. Grassian, 2011: Abiotic mechanism for the formation of atmospheric nitrous oxide from ammonium nitrate. *Environ. Sci. Technol.*, **45**, 2691–2697.

Ruddiman, W. F., 2003: The anthropogenic greenhouse era began thousands of years ago. *Clim. Change*, **61**, 261–293.

Ruddiman, W. F., 2007: The early anthropogenic hypothesis: Challenges and responses. *Rev. Geophys.*, **45**, RG4001.

Sabine, C. L., R. A. Feely, F. J. Millero, A. G. Dickson, C. Langdon, S. Mecking, and D. Greeley, 2008: Decadal changes in Pacific carbon. *J. Geophys. Res. Oceans*, **113**, C07021.

Sabine, C. L., et al., 2004: The oceanic sink for anthropogenic CO_2. *Science*, **305**, 367–371.

Salisbury, J., M. Green, C. Hunt, and J. Campbell, 2008: Coastal acidification by rivers: A threat to shellfish? *EOS Trans. AGU*, **89**, 513.

Sallée, J.-B., R. J. Matear, S. R. Rintoul, and A. Lenton, 2012: Localized subduction of anthropogenic carbon dioxide in the Southern Hemisphere oceans. *Nature Geosci.*, **5**, 579–584.

Samanta, A., M. H. Costa, E. L. Nunes, S. A. Viera, L. Xu, and R. B. Myneni, 2011: Comment on "Drought-induced reduction in global terrestrial net primary production from 2000 through 2009". *Science*, **333**, 1093.

Sanderson, M. G., 1996: Biomass of termites and their emissions of methane and carbon dioxide: A global database. *Global Biogeochem. Cycles*, **10**, 543–557.

Sapart, C. J., et al., 2012: Natural and anthropogenic variations in methane sources during the past two millennia. *Nature*, **490**, 85–88.

Sarmiento, J. L., and N. Gruber, 2006: *Ocean Biogeochemical Dynamics.* Princeton University Press, Princeton, NJ, USA.

Sarmiento, J. L., J. C. Orr, and U. Siegenthaler, 1992: A perturbation simulation of CO_2 uptake in an Ocean General Circulation Model. *J. Geophys. Res.*, **97**, 3621–3645.

Sarmiento, J. L., T. M. C. Hughes, R. J. Stouffer, and S. Manabe, 1998: Simulated response of the ocean carbon cycle to anthropogenic climate warming. *Nature*, **393**, 245–249.

Sarmiento, J. L., P. Monfray, E. Maier-Reimer, O. Aumont, R. Murnane, and J. C. Orr, 2000: Sea-air CO_2 fluxes and carbon transport: A comparison of three ocean general circulation models. *Global Biogeochem. Cycles*, **14**, 1267–1281.

Sarmiento, J. L., et al., 2010: Trends and regional distributions of land and ocean carbon sinks. *Biogeosciences*, **7**, 2351–2367.

Savolainen, I., S. Monni, and S. Syri, 2009: The mitigation of methane emissions from the industrialised countries can explain the atmospheric concentration level-off. *Int. J. Energ. Clean Environ.*, **10**, 193–201.

Schaefer, K., T. Zhang, L. Bruhwiler, and A. P. Barrett, 2011: Amount and timing of permafrost carbon release in response to climate warming. *Tellus B*, **63**, 165–180.

Scheffer, M., V. Brovkin, and P. M. Cox, 2006: Positive feedback between global warming and atmospheric CO_2 concentration inferred from past climate change. *Geophys. Res. Lett.*, **33**, L10702.

Schilt, A., M. Baumgartner, T. Blunier, J. Schwander, R. Spahni, H. Fischer, and T. F. Stocker, 2010a: Glacial-interglacial and millennial-scale variations in the atmospheric nitrous oxide concentration during the last 800,000 years. *Quat. Sci. Rev.*, **29**, 182–192.

Schilt, A., et al., 2010b: Atmospheric nitrous oxide during the last 140,000 years. *Earth Planet. Sci. Lett.*, **300**, 33–43.

Schirrmeister, L., G. Grosse, S. Wetterich, P. P. Overduin, J. Strauss, E. A. G. Schuur, and H.-W. Hubberten, 2011: Fossil organic matter characteristics in permafrost deposits of the northeast Siberian Arctic. *J. Geophys. Res.*, **116**, G00M02.

Schmitt, J., et al., 2012: Carbon isotope constraints on the deglacial CO_2 rise from ice cores. *Science*, **336**, 711–714.

Schmittner, A., and E. D. Galbraith, 2008: Glacial greenhouse-gas fluctuations controlled by ocean circulation changes. *Nature*, **456**, 373–376.

Schmittner, A., A. Oschlies, H. D. Matthews, and E. D. Galbraith, 2008: Future changes in climate, ocean circulation, ecosystems, and biogeochemical cycling simulated for a business-as-usual CO_2 emission scenario until year 4000 AD. *Global Biogeochem. Cycles*, **22**, GB1013.

Schmittner, A., N. M. Urban, K. Keller, and D. Matthews, 2009: Using tracer observations to reduce the uncertainty of ocean diapycnal mixing and climate-carbon cycle projections. *Global Biogeochem. Cycles*, **23**, GB4009.

Schneider von Deimling, T. S., M. Meinshausen, A. Levermann, V. Huber, K. Frieler, D. M. Lawrence, and V. Brovkin, 2012: Estimating the near-surface permafrost-carbon feedback on global warming. *Biogeosciences*, **9**, 649–665.

Scholze, M., W. Knorr, N. W. Arnell, and I. C. Prentice, 2006: A climate-change risk analysis for world ecosystems. *Proc. Natl. Acad. Sci. U.S.A.*, **103**, 13116–13120.

Schuiling, R. D., and P. Krijgsman, 2006: Enhanced weathering: An effective and cheap tool to sequester CO_2. *Clim. Change*, **74**, 349–354.

Schultz, M. G., et al., 2007: Emission data sets and methodologies for estimating emissions. *REanalysis of the TROpospheric chemical composition over the past 40 years.* A long-term global modeling study of tropospheric chemistry funded under the 5th EU framework programme EU-Contract EVK2-CT-2002–00170.

Schulze, E. D., S. Luyssaert, P. Ciais, A. Freibauer, and I. A. Janssens, 2009: Importance of methane and nitrous oxide for Europe's terrestrial greenhouse-gas balance. *Nature Geosci.*, **2**, 842–850.

Schulze, E. D., et al., 2010: The European carbon balance. Part 4: Integration of carbon and other trace-gas fluxes. *Global Change Biol.*, **16**, 1451–1469.

Schurgers, G., U. Mikolajewicz, M. Gröger, E. Maier-Reimer, M. Vizcaino, and A. Winguth, 2006: Dynamics of the terrestrial biosphere, climate and atmospheric CO_2 concentration during interglacials: A comparison between Eemian and Holocene. *Clim. Past*, **2**, 205–220.

Schuster, U., and A. J. Watson, 2007: A variable and decreasing sink for atmospheric CO_2 in the North Atlantic. *J. Geophys. Res. Oceans*, **112**, C11006.

Schuster, U., et al., 2009: Trends in North Atlantic sea-surface fCO_2 from 1990 to 2006. *Deep-Sea Res. Pt. II*, **56**, 620–629.

Schuster, U., et al., 2013: An assessment of the Atlantic and Arctic sea-air CO_2 fluxes, 1990–2009. *Biogeosciences*, **10**, 607–627.

Schwalm, C. R., et al., 2010: A model-data intercomparison of CO_2 exchange across North America: Results from the North American Carbon Program site synthesis. *J. Geophys. Res.*, **115**, G00H05.

Seitzinger, S. P., and C. Kroeze, 1998: Global distribution of nitrous oxide production and N inputs in freshwater and coastal marine ecosystems. *Global Biogeochem. Cycles*, **12**, 93–113.

Seitzinger, S. P., J. A. Harrison, E. Dumont, A. H. W. Beusen, and A. F. Bouwman, 2005: Sources and delivery of carbon, nitrogen, and phosphorus to the coastal zone: An overview of Global Nutrient Export from Watersheds (NEWS) models and their application. *Global Biogeochem. Cycles*, **19**, Gb4s01.

Seitzinger, S. P., et al., 2010: Global river nutrient export: A scenario analysis of past and future trends. *Global Biogeochem. Cycles*, **24**, GB0A08.

Sentman, L. T., E. Shevliakova, R. J. Stouffer, and S. Malyshev, 2011: Time scales of terrestrial carbon response related to land-use application: Implications for initializing an Earth System Model. *Earth Interactions*, **15**, 1–16.

Shackleton, N. J., 2000: The 100,000–year ice-age cycle identified and found to lag temperature, carbon dioxide, and orbital eccentricity. *Science*, **289**, 1897–1902.

Shaffer, G., 2010: Long-term effectiveness and consequences of carbon dioxide sequestration. *Nature Geosci.*, **3**, 464–467.

Shaffer, G., S. M. Olsen, and J. O. P. Pedersen, 2009: Long-term ocean oxygen depletion in response to carbon dioxide emission from fossil fuels. *Nature Geosci.*, **2**, 105–109.

Shakhova, N., I. Semiletov, A. Salyuk, V. Yusupov, D. Kosmach, and O. Gustafsson, 2010: Extensive methane venting to the atmosphere from sediments of the East Siberian Arctic shelf. *Science*, **327**, 1246–1250.

Shallcross, D. E., M. A. K. Khalil, and C. L. Butenhoff, 2007: The atmospheric methane sink. In: *Greenhouse Gas Sinks* [D. Reay (ed.)] CAB International, pp. 171-183.

Shepherd, J., et al., 2009: Geoengineering the climate: Science, governance and uncertainty. Report of the Royal Society, London, 98 pp.

Shevliakova, E., et al., 2009: Carbon cycling under 300 years of land use change: Importance of the secondary vegetation sink. *Global Biogeochem. Cycles*, **23**, GB2022.

Shindell, D. T., B. P. Walter, and G. Faluvegi, 2004: Impacts of climate change on methane emissions from wetlands. *Geophys. Res. Lett.*, **31**, L21202.

Siegenthaler, U., et al., 2005a: Supporting evidence from the EPICA Dronning Maud Land ice core for atmospheric CO_2 changes during the past millennium. *Tellus B*, **57**, 51–57.

Siegenthaler, U., et al., 2005b: Stable carbon cycle-climate relationship during the late Pleistocene. *Science*, **310**, 1313–1317.

Sigman, D. M., M. P. Hain, and G. H. Haug, 2010: The polar ocean and glacial cycles in atmospheric CO_2 concentration. *Nature*, **466**, 47–55.

Simpson, I. J., F. S. Rowland, S. Meinardi, and D. R. Blake, 2006: Influence of biomass burning during recent fluctuations in the slow growth of global tropospheric methane. *Geophys. Res. Lett.*, **33**, L22808.

Simpson, I. J., et al., 2012: Long-term decline of global atmospheric ethane concentrations and implications for methane. *Nature*, **488**, 490–494.

Singarayer, J. S., P. J. Valdes, P. Friedlingstein, S. Nelson, and D. J. Beerling, 2011: Late Holocene methane rise caused by orbitally controlled increase in tropical sources. *Nature*, **470**, 82–85.

Singh, B. K., R. D. Bardgett, P. Smith, and D. S. Reay, 2010: Microorganisms and climate change: Terrestrial feedbacks and mitigation options. *Nature Rev. Microbiol.*, **8**, 779–790.

Sitch, S., P. M. Cox, W. J. Collins, and C. Huntingford, 2007: Indirect radiative forcing of climate change through ozone effects on the land-carbon sink. *Nature*, **448**, 791–794.

Sitch, S., et al., 2003: Evaluation of ecosystem dynamics, plant geography and terrestrial carbon cycling in the LPJ Dynamic Global Vegetation Model. *Global Change Biol.*, **9**, 161–185.

Sitch, S., et al., 2008: Evaluation of the terrestrial carbon cycle, future plant geography and climate-carbon cycle feedbacks using five Dynamic Global Vegetation Models (DGVMs). *Global Change Biol.*, **14**, 2015–2039.

Skinner, L. C., S. Fallon, C. Waelbroeck, E. Michel, and S. Barker, 2010: Ventilation of the deep Southern ocean and deglacial CO_2 rise. *Science*, **328**, 1147–1151.

Smetacek, V., et al., 2012: Deep carbon export from a Southern Ocean iron-fertilized diatom bloom. *Nature*, **487**, 313–319.

Smith, B., I. C. Prentice, and M. T. Sykes, 2001a: Representation of vegetation dynamics in the modelling of terrestrial ecosystems: Comparing two contrasting approaches within European climate space. *Global Ecol. Biogeogr.*, **10**, 621–637.

Smith, K. A., A. R. Mosier, P. J. Crutzen, and W. Winiwarter, 2012: The role of N_2O derived from crop-based biofuels, and from agriculture in general, in Earth's climate. *Philos. Trans. R. Soc. London B*, **367**, 1169–1174.

Smith, L. C., Y. Sheng, G. M. MacDonald, and L. D. Hinzman, 2005: Disappearing Arctic lakes. *Science*, **308**, 1429.

Smith, S. V., W. H. Renwick, R. W. Buddemeier, and C. J. Crossland, 2001b: Budgets of soil erosion and deposition for sediments and sedimentary organic carbon across the conterminous United States. *Global Biogeochem. Cycles*, **15**, 697–707.

Sokolov, A. P., D. W. Kicklighter, J. M. Melillo, B. S. Felzer, C. A. Schlosser, and T. W. Cronin, 2008: Consequences of considering carbon-nitrogen interactions on the feedbacks between climate and the terrestrial carbon cycle. *J. Clim.*, **21**, 3776–3796.

Sowers, T., 2006: Late quaternary atmospheric CH_4 isotope record suggests marine clathrates are stable. *Science*, **311**, 838–840.

Sowers, T., R. B. Alley, and J. Jubenville, 2003: Ice core records of atmospheric N_2O covering the last 106,000 years. *Science*, **301**, 945–948.

Spahni, R., et al., 2011: Constraining global methane emissions and uptake by ecosystems. *Biogeosciences*, **8**, 1643–1665.

Spracklen, D. V., L. J. Mickley, J. A. Logan, R. C. Hudman, R. Yevich, M. D. Flannigan, and A. L. Westerling, 2009: Impacts of climate change from 2000 to 2050 on wildfire activity and carbonaceous aerosol concentrations in the western United States. *J. Geophys. Res. Atmos.*, **114**, D20301.

Stallard, R. F., 1998: Terrestrial sedimentation and the carbon cycle: Coupling weathering and erosion to carbon burial. *Global Biogeochem. Cycles*, **12**, 231–257.

Stanhill, G., and S. Cohen, 2001: Global dimming: A review of the evidence for a widespread and significant reduction in global radiation with discussion of its probable causes and possible agricultural consequences. *Agr. Forest Meteorol.*, **107**, 255–278.

Steinacher, M., F. Joos, T. L. Frölicher, G.-K. Plattner, and S. C. Doney, 2009: Imminent ocean acidification in the Arctic projected with the NCAR global coupled carbon cycle-climate model. *Biogeosciences*, **6**, 515–533.

Steinacher, M., et al., 2010: Projected 21st century decrease in marine productivity: A multi-model analysis. *Biogeosciences*, **7**, 979–1005.

Stephens, B. B., and R. F. Keeling, 2000: The influence of Antarctic sea ice on glacial-interglacial CO_2 variations. *Nature*, **404**, 171–174.

Stephens, B. B., et al., 2007: Weak northern and strong tropical land carbon uptake from vertical profiles of atmospheric CO_2. *Science*, **316**, 1732–1735.

Stevenson, D. S., et al., 2006: Multimodel ensemble simulations of present-day and near-future tropospheric ozone. *J. Geophys. Res.*, **111**, D08301.

Stocker, B. D., K. Strassmann, and F. Joos, 2011: Sensitivity of Holocene atmospheric CO_2 and the modern carbon budget to early human land use: Analyses with a process-based model. *Biogeosciences*, **8**, 69–88.

Stocker, B. D., et al., 2013: Multiple greenhouse gas feedbacks from the land biosphere under future climate change scenarious. *Nature Clim. Change*, **3**, 666-672.

Stöckli, R., et al., 2008: Use of FLUXNET in the Community Land Model development. *J. Geophys. Res.Biogeosci.*, **113**, G01025.

Stolaroff, J. K., D. W. Keith, and G. V. Lowry, 2008: Carbon dioxide capture from atmospheric air using sodium hydroxide spray. *Environ. Sci. Technol.*, **42**, 2728–2735.

Stolaroff, J. K., S. Bhattacharyya, C. A. Smith, W. L. Bourcier, P. J. Cameron-Smith, and R. D. Aines, 2012: Review of methane mitigation technologies with application to rapid release of methane from the Arctic. *Environ. Sci. Technol.*, **46**, 6455–6469.

Stramma, L., A. Oschlies, and S. Schmidtko, 2012: Mismatch between observed and modeled trends in dissolved upper-ocean oxygen over the last 50 years. *Biogeosciences*, **9**, 4045–4057.

Strassmann, K. M., F. Joos, and G. Fischer, 2008: Simulating effects of land use changes on carbon fluxes: Past contributions to atmospheric CO_2 increases and future commitments due to losses of terrestrial sink capacity. *Tellus B*, **60**, 583–603.

Stuiver, M., and P. D. Quay, 1981: Atmospheric ^{14}C changes resulting from fossil-fuel CO_2 release and cosmic-ray flux variability. *Earth Planet. Sci. Lett.*, **53**, 349–362.

Suchet, P. A., and J. L. Probst, 1995: A Global model for present-day atmospheric soil CO_2 consumption by chemical erosion of continental rocks (GEM-CO_2). *Tellus B*, **47**, 273–280.

Sugimoto, A., T. Inoue, N. Kirtibutr, and T. Abe, 1998: Methane oxidation by termite mounds estimated by the carbon isotopic composition of methane. *Global Biogeochem. Cycles*, **12**, 595–605.

Sundquist, E. T., 1986: Geologic analogs: Their value and limitations in carbon dioxide research. In: *The Changing Carbon Cycle* [J. R. Trabalka and D. E. Reichle (eds.)], Springer-Verlag, New York, pp. 371–402.

Sundquist, E. T, 1990: Influence of deep-sea benthic processes on atmospheric CO_2. *Philos. Trans. R. Soc. London Series A*, **331**, 155–165.

Suntharalingam, P., et al., 2012: Quantifying the impact of anthropogenic nitrogen deposition on oceanic nitrous oxide. *Geophys. Res. Lett.*, **39**, L07605.

Sussmann, R., F. Forster, M. Rettinger, and P. Bousquet, 2012: Renewed methane increase for five years (2007–2011) observed by solar FTIR spectrometry. *Atmos. Chem. Phys.*, **112**, 4885–4891.

Sutka, R. L., N. E. Ostrom, P. H. Ostrom, J. A. Breznak, H. Gandhi, A. J. Pitt, and F. Li, 2006: Distinguishing nitrous oxide production from nitrification and denitrification on the basis of isotopomer abundances. *Appl. Environ. Microbiol.*, **72**, 638–644.

Sutton, M. A., D. D. Simpson, P. E. Levy, R. I. Smith, S. Reis, M. Van Oijen, and W. De Vries, 2008: Uncertainties in the relationship between atmospheric nitrogen deposition and forest carbon sequestration. *Global Change Biol.*, **14**, 2057–2063.

Sutton, M. A., et al., 2011: *The European Nitrogen Assessment —Sources, Effects and Policy Perspectives.* Cambridge University Press, Cambridge, United Kingdom, and New York, NY, USA, 664 pp.

Syakila, A., and C. Kroeze, 2011: The global N_2O budget revisited. *Greenh. Gas Measure. Manage.*, **1**, 17–26.

Syakila, A., C. Kroeze, and C. P. Slomp, 2010: Neglecting sinks for N_2O at the earth's surface: Does it matter? *J. Integrat. Environ. Sci.*, **7**, 79–87.

Syvitski, J. P. M., C. J. Vörösmarty, A. J. Kettner, and P. Green, 2005: Impact of humans on the flux of terrestrial sediment to the global coastal ocean. *Science*, **308**, 376–380.

Tagaris, E., K.-J. Liao, K. Manomaiphiboon, J.-H. Woo, S. He, P. Amar, and A. G. Russell, 2008: Impacts of future climate change and emissions reductions on nitrogen and sulfur deposition over the United States. *Geophys. Res. Lett.*, **35**, L08811.

Tagliabue, A., L. Bopp, and O. Aumont, 2008: Ocean biogeochemistry exhibits contrasting responses to a large scale reduction in dust deposition. *Biogeosciences*, **5**, 11–24.

Tagliabue, A., L. Bopp, and M. Gehlen, 2011: The response of marine carbon and nutrient cycles to ocean acidification: Large uncertainties related to phytoplankton physiological assumptions. *Global Biogeochem. Cycles*, **25**, GB3017.

Takahashi, T., S. C. Sutherland, R. A. Feely, and R. Wanninkhof, 2006: Decadal change of the surface water pCO_2 in the North Pacific: A synthesis of 35 years of observations. *J. Geophys. Res. Oceans*, **111**, C07s05.

Takahashi, T., J. Olafsson, J. G. Goddard, D. W. Chipman, and S. C. Sutherland, 1993: Seasonal variation of CO_2 and nutrients in the high-latitude surface oceans—A comparative study. *Global Biogeochem. Cycles*, **7**, 843–878.

Takahashi, T., et al., 2009: Climatological mean and decadal change in surface ocean pCO_2, and net sea-air CO_2 flux over the global oceans. *Deep-Sea Res. Pt. II*, **56**, 554–577.

Tan, K., et al., 2010: Application of the ORCHIDEE global vegetation model to evaluate biomass and soil carbon stocks of Qinghai-Tibetan grasslands. *Global Biogeochem. Cycles*, **24**, GB1013.

Tanhua, T., A. Körtzinger, K. Friis, D. W. Waugh, and D. W. R. Wallace, 2007: An estimate of anthropogenic CO_2 inventory from decadal changes in oceanic carbon content. *Proc. Natl. Acad. Sci. U.S.A.*, **104**, 3037–3042.

Tans, P. P., T. J. Conway, and T. Nakazawa, 1989: Latitudinal distribution of the sources and sinks of atmospheric carbon dioxide derived from surface observations and an atmospheric transport model. *J. Geophys. Res. Atmos.*, **94**, 5151–5172.

Tarnocai, C., J. G. Canadell, E. A. G. Schuur, P. Kuhry, G. Mazhitova, and S. Zimov, 2009: Soil organic carbon pools in the northern circumpolar permafrost region. *Global Biogeochem. Cycles*, **23**, Gb2023.

Taucher, J., and A. Oschlies, 2011: Can we predict the direction of marine primary production change under global warming? *Geophys. Res. Lett.*, **38**, L02603.

Taylor, K. E., R. J. Stouffer, and G. A. Meehl, 2012: An overview of CMIP5 and the experiment design. *Bull. Am. Meteorol. Soc.*, **93**, 485–498.

Tegen, I., M. Werner, S. P. Harrison, and K. E. Kohfeld, 2004: Relative importance of climate and land use in determining present and future global soil dust emission. *Geophys. Res. Lett.*, **31**, L05105.

Terazawa, K., S. Ishizuka, T. Sakata, K. Yamada, and M. Takahashi, 2007: Methane emissions from stems of *Fraxinus mandshurica* var. *japonica* trees in a floodplain forest. *Soil Biology and Biochemistry*, **39**, 2689–2692.

Terrier, A., M. P. Girardin, C. Périé, P. Legendre, and Y. Bergeron, 2013: Potential changes in forest composition could reduce impacts of climate change on boreal wildfires. *Ecol. Appl.*, **23**, 21–35.

Thomas, H., et al., 2007: Rapid decline of the CO_2 buffering capacity in the North Sea and implications for the North Atlantic Ocean. *Global Biogeochem. Cycles*, **21**, GB4001.

Thompson, D. W. J., and S. Solomon, 2002: Interpretation of recent Southern Hemisphere climate change. *Science*, **296**, 895–899.

Thomson, A. M., et al., 2010: Climate mitigation and food production in tropical landscapes. Special feature: Climate mitigation and the future of tropical landscapes. *Proc. Natl. Acad. Sci. U.S.A.*, **107**, 19633–19638.

Thornton, P. E., J.-F. Lamarque, N. A. Rosenbloom, and N. M. Mahowald, 2007: Influence of carbon-nitrogen cycle coupling on land model response to CO_2 fertilization and climate variability. *Global Biogeochem. Cycles*, **21**, Gb4018.

Thornton, P. E., et al., 2009: Carbon-nitrogen interactions regulate climate-carbon cycle feedbacks: Results from an atmosphere-ocean general circulation model. *Biogeosciences*, **6**, 2099–2120.

Tian, H., X. Xu, M. Liu, W. Ren, C. Zhang, G. Chen, and C. Lu, 2010: Spatial and temporal patterns of CH_4 and N_2O fluxes in terrestrial ecosystems of North America during 1979–2008: Application of a global biogeochemistry model. *Biogeosciences*, **7**, 2673–2694.

Tilman, D., C. Balzer, J. Hill, and B. L. Befort, 2011: Global food demand and the sustainable intensification of agriculture. *Proc. Natl. Acad. Sci. U.S.A.*, **108**, 20260–20264.

Tjiputra, J. F., A. Olsen, K. Assmann, B. Pfeil, and C. Heinze, 2012: A model study of the seasonal and long-term North Atlantic surface pCO_2 variability. *Biogeosciences*, **9**, 907–923.

Todd-Brown, K., F. M. Hopkins, S. N. Kivlin, J. M. Talbot, and S. D. Allison, 2012: A framework for representing microbial decomposition in coupled climate models. *Biogeoschemistry*, **109**, 19–33.

Toggweiler, J. R., 1999: Variation of atmospheric CO_2 by ventilation of the ocean's deepest water. *Paleoceanography*, **14**, 571–588.

Toggweiler, J. R., J. L. Russell, and S. R. Carson, 2006: Midlatitude westerlies, atmospheric CO_2, and climate change during the ice ages. *Paleoceanography*, **21**, PA2005.

Tranvik, L. J., et al., 2009: Lakes and reservoirs as regulators of carbon cycling and climate. *Limnol. Oceanogr.*, **54**, 2298–2314.

Trudinger, C. M., I. G. Enting, P. J. Rayner, and R. J. Francey, 2002: Kalman filter analysis of ice core data – 2. Double deconvolution of CO_2 and □13C measurements. *J. Geophys. Res.*, **107**, D20.

Turetsky, M. R., R. K. Wieder, D. H. Vitt, R. J. Evans, and K. D. Scott, 2007: The disappearance of relict permafrost in boreal north America: Effects on peatland carbon storage and fluxes. *Global Change Biol.*, **13**, 1922–1934.

Tymstra, C., M. D. Flannigan, O. B. Armitage, and K. Logan, 2007: Impact of climate change on area burned in Alberta's boreal forest. *Int. J. Wildland Fire*, **16**, 153–160.

Tyrrell, T., J. G. Shepherd, and S. Castle, 2007: The long-term legacy of fossil fuels. *Tellus B*, **59**, 664–672.

Ullman, D. J., G. A. McKinley, V. Bennington, and S. Dutkiewicz, 2009: Trends in the North Atlantic carbon sink: 1992–2006. *Global Biogeochem. Cycles*, **23**, Gb4011.

UNEP, 2011: Integrated assessment of black carbon and tropospheric ozone: Summary for decision makers. United Nations Environment Programme and World Meterological Association, 38 pp.

Valsala, V., S. Maksyutov, M. Telszewski, S. Nakaoka, Y. Nojiri, M. Ikeda, and R. Murtugudde, 2012: Climate impacts on the structures of the North Pacific air-sea CO_2 flux variability. *Biogeosciences*, **9**, 477–492.

van der Werf, G. R., et al., 2004: Continental-scale partitioning of fire emissions during the 1997 to 2001 El Niño/La Niña period. *Science*, **303**, 73–76.

van der Werf, G. R., et al., 2009: CO_2 emissions from forest loss. *Nature Geosci.*, **2**, 737–738.

van der Werf, G. R., et al., 2010: Global fire emissions and the contribution of deforestation, savanna, forest, agricultural, and peat fires (1997–2009). *Atmos. Chem. Phys.*, **10**, 11707–11735.

van der Werf, G. R., et al., 2008: Climate regulation of fire emissions and deforestation in equatorial Asia *Proc. Natl. Acad. Sci. U.S.A.*, **105**, 20350–20355.

van Groenigen, K. J., C. W. Osenberg, and B. A. Hungate, 2011: Increased soil emissions of potent greenhouse gases under increased atmospheric CO_2. *Nature*, **475**, 214–216.

van Huissteden, J., C. Berrittella, F. J. W. Parmentier, Y. Mi, T. C. Maximov, and A. J. Dolman, 2011: Methane emissions from permafrost thaw lakes limited by lake drainage. *Nature Clim. Change*, **1**, 119–123.

van Minnen, J. G., K. Klein Goldewijk, E. Stehfest, B. Eickhout, G. van Drecht, and R. Leemans, 2009: The importance of three centuries of land-use change for the global and regional terrestrial carbon cycle. *Clim. Change*, **97**, 123–144.

van Vuuren, D. P., L. F. Bouwman, S. J. Smith, and F. Dentener, 2011: Global projections for anthropogenic reactive nitrogen emissions to the atmosphere: An assessment of scenarios in the scientific literature. *Curr. Opin. Environ. Sustain.*, **3**, 359–369.

Verdy, A., S. Dutkiewicz, M. J. Follows, J. Marshall, and A. Czaja, 2007: Carbon dioxide and oxygen fluxes in the Southern Ocean: Mechanisms of interannual variability. *Global Biogeochem. Cycles*, **21**, Gb2020.

Vigano, I., H. van Weelden, R. Holzinger, F. Keppler, A. McLeod, and T. Röckmann, 2008: Effect of UV radiation and temperature on the emission of methane from plant biomass and structural components. *Biogeosciences*, **5**, 937–947.

Vitousek, P. M., S. Porder, B. Z. Houlton, and O. A. Chadwick, 2010: Terrestrial phosphorus limitation: Mechanisms, implications, and nitrogen-phosphorus interactions. *Ecol. Appl.*, **20**, 5–15.

Vitousek, P. M., D. N. L. Menge, S. C. Reed, and C. C. Cleveland, 2013: Biological nitrogen fixation: Rates, patterns, and ecological controls in terrestrial ecosystems. *Philos. Trans. R. Soc. London B*,***368***, 20130119.

Volk, C. M., et al., 1997: Evaluation of source gas lifetimes from stratospheric observations. *J. Geophys. Res.: Atmos.*, **102**, 25543–25564.

Volodin, E. M., 2008: Methane cycle in the INM RAS climate model. *Izvestiya Atmos. Ocean. Phys.*, **44**, 153–159.

Voss, M., H. W. Bange, J. W. Dippner, J. J. Middelburg, J. P. Montoya, and B. Ward, 2013: The marine nitrogen cycle: Recent discoveries, uncertainties and the potential relevance of climate change. *Philos. Trans. R. Soc. London B*, **368**, 20130121.

Voss, M., et al., 2011: Nitrogen processes in coastal and marine ecosystems. In: *The European Nitrogen Assessment: Sources, Effects, and Policy Perspectives.* [M. A. Sutton, C. M. Howard, J. W. Erisman, G. Billen, A. Bleeker, P. Grennfelt, H. van Grinsven and B. Grizetti (eds.)]. Cambridge University Press, Cambridge, United Kingdom, and New York, NY, USA, pp. 147–176.

Voulgarakis, A., et al., 2013: Analysis of present day and future OH and methane lifetime in the ACCMIP simulations. *Atmos. Chem. Phys.*, **13**, 2563–2587.

Waelbroeck, C., et al., 2009: Constraints on the magnitude and patterns of ocean cooling at the Last Glacial Maximum. *Nature Geosci.*, **2**, 127–132.

Wakita, M., S. Watanabe, A. Murata, N. Tsurushima, and M. Honda, 2010: Decadal change of dissolved inorganic carbon in the subarctic western North Pacific Ocean. *Tellus B*, **62**, 608–620.

Walker, J. C. G., and J. F. Kasting, 1992: Effects of fuel and forest conservation on future levels of atmospheric carbon dioxide. *Palaeogeogr. Palaeoclimat. Palaeoecol.* (*Global Planet. Change Sect.*), **97**, 151–189.

Walter Anthony, K. M., P. Anthony, G. Grosse, and J. Chanton, 2012: Geologic methane seeps along boundaries of Arctic permafrost thaw and melting glaciers. *Nature Geosci.*, **5**, 419-426.

Walter, K. M., L. C. Smith, and F. Stuart Chapin, 2007: Methane bubbling from northern lakes: Present and future contributions to the global methane budget. *Philos. Trans. R. Soc. A*, **365**, 1657–1676.

Walter, K. M., S. A. Zimov, J. P. Chanton, D. Verbyla, and F. S. I. Chapin, 2006: Methane bubbling from Siberian thaw lakes as a positive feedback to climate warming. *Nature*, **443**, 71–75.

Wang, D., S. A. Heckathorn, X. Wang, and S. M. Philpott, 2012a: A meta-analysis of plant physiological and growth responses to temperature and elevated CO_2. *Oecologia*, **169**, 1–13.

Wang, J., X. Pan, Y. Liu, X. Zhang, and Z. Xiong, 2012b: Effects of biochar amendment in two soils on greenhouse gas emissions and crop production. *Plant Soil*, **360**, 287–298.

Wang, Y.-P., and B. Z. Houlton, 2009: Nitrogen constraints on terrestrial carbon uptake: Implications for the global carbon-climate feedback. *Geophys. Res. Lett.*, **36**, L24403.

Wang, Y. P., B. Z. Houlton, and C. B. Field, 2007: A model of biogeochemical cycles of carbon, nitrogen, and phosphorus including symbiotic nitrogen fixation and phosphatase production. *Global Biogeochem. Cycles*, **21**, GB1018.

Wang, Y. P., R. M. Law, and B. Pak, 2010a: A global model of carbon, nitrogen and phosphorus cycles for the terrestrial biosphere. *Biogeosciences*, **7**, 2261–2282.

Wang, Z., J. Chappellaz, K. Park, and J. E. Mak, 2010b: Large variations in Southern Hemisphere biomass burning during the last 650 years. *Science*, **330**, 1663–1666.

Wang, Z. P., X. G. Han, G. G. Wang, Y. Song, and J. Gulledge, 2008: Aerobic methane emission from plants in the Inner Mongolia steppe. *Environ. Sci. Technol.*, **42**, 62–68.

Wania, R., 2007: *Modelling Northern Peatland Land Surface Processes, Vegetation Dynamics and Methane Emissions*. Ph.D. Thesis, Bristol, UK.

Wania, R., I. Ross, and I. C. Prentice, 2009: Integrating peatlands and permafrost into a dynamic global vegetation model: 1. Evaluation and sensitivity of physical land surface processes. *Global Biogeochem. Cycles*, **23**, GB3014.

Wanninkhof, R., S. C. Doney, J. L. Bullister, N. M. Levine, M. Warner, and N. Gruber, 2010: Detecting anthropogenic CO_2 changes in the interior Atlantic Ocean between 1989 and 2005. *J. Geophys. Res. Oceans*, **115**, C11028.

Wanninkhof, R., et al., 2013: Global ocean carbon uptake: Magnitude, variability and trends. *Biogeosciences*, **10**, 1983–2000.

Wardle, D. A., M.-C. Nilsson, and O. Zackrisson, 2008: Fire-derived charcoal causes loss of forest humus. *Science*, **320**, 629.

Watanabe, S., et al., 2011: MIROC-ESM 2010: Model description and basic results of CMIP5–20c3m experiments. *Geosci. Model Dev.*, **4**, 845–872.

Watson, A. J., and A. C. N. Garabato, 2006: The role of Southern Ocean mixing and upwelling in glacial-interglacial atmospheric CO_2 change. *Tellus B*, **58**, 73–87.

Watson, A. J., D. C. E. Bakker, A. J. Ridgwell, P. W. Boyd, and C. S. Law, 2000: Effect of iron supply on Southern Ocean CO_2 uptake and implications for glacial atmospheric CO_2. *Nature*, **407**, 730–733.

Watson, A. J., P. W. Boyd, S. M. Turner, T. D. Jickells, and P. S. Liss, 2008: Designing the next generation of ocean iron fertilization experiments. *Mar. Ecol. Prog. Ser.*, **364**, 303–309.

Watson, A. J., et al., 1994: Minimal effect of iron fertilization on sea-surface carbon-dioxide concentartions. *Nature*, **371**, 143–145.

Watson, A. J., et al., 2009: Tracking the variable North Atlantic sink for atmospheric CO_2. *Science*, **326**, 1391–1393.

Waugh, D. W., T. M. Hall, B. I. McNeil, R. Key, and R. J. Matear, 2006: Anthropogenic CO_2 in the oceans estimated using transit time distributions. *Tellus B*, **58**, 376–389.

Wecht, K. J., et al., 2012: Validation of TES methane with HIPPO aircraft observations: Implications for inverse modeling of methane sources. *Atmos. Chem. Phys.*, **12**, 1823–1832.

Westbrook, G. K., et al., 2009: Escape of methane gas from the seabed along the West Spitsbergen continental margin. *Geophys. Res. Lett.*, **36**, L15608.

Westerling, A. L., M. G. Turner, E. A. H. Smithwick, W. H. Romme, and M. G. Ryan, 2011: Continued warming could transform Greater Yellowstone fire regimes by mid-21st century. *Proc. Natl. Acad. Sci. U.S.A.*, **108**, 13165–13170.

White, J. R., R. D. Shannon, J. F. Weltzin, J. Pastor, and S. D. Bridgham, 2008: Effects of soil warming and drying on methane cycling in a northern peatland mesocosm study. *J. Geophys. Res. Biogeosci.*, **113**, G00A06.

Wiedinmyer, C., S. K. Akagi, R. J. Yokelson, L. K. Emmons, J. A. Al-Saadi, J. J. Orlando, and A. J. Soja, 2011: The Fire INventory from NCAR (FINN): A high resolution global model to estimate the emissions from open burning. *Geosci. Model Dev.*, **4**, 625–641.

Williams, C. A., G. J. Collatz, J. Masek, and S. N. Goward, 2012a: Carbon consequences of forest disturbance and recovery across the conterminous United States. *Global Biogeochem. Cycles*, **26**, GB1005.

Williams, J., and P. J. Crutzen, 2010: Nitrous oxide from aquaculture. *Nature Geosci.*, **3**, 143.

Williams, J. E., A. Strunk, V. Huijnen, and M. van Weele, 2012b: The application of the Modified Band Approach for the calculation of on-line photodissociation rate constants in TM5: Implications for oxidative capacity. *Geosci. Model Dev.*, **5**, 15–35.

Wise, M., et al., 2009: Implications of limiting CO_2 concentrations for land Uue and energy. *Science*, **324**, 1183–1186.

Woodward, F. I., and M. R. Lomas, 2004: Simulating vegetation processes along the Kalahari transect. *Global Change Biol.*, **10**, 383–392.

Woolf, D., J. E. Amonette, F. A. Street-Perrott, J. Lehmann, and S. Joseph, 2010: Sustainable biochar to mitigate global climate change. *Nature Commun.*, **1**, 1–9.

Worden, J., et al., 2012: Profiles of CH_4, HDO, H_2O, and N_2O with improved lower tropospheric vertical resolution from Aura TES radiances. *Atmos. Measure. Techn.*, **5**, 397–411

Worrall, F., T. Burt, and R. Shedden, 2003: Long term records of riverine dissolved organic matter. *Biogeochemistry*, **64**, 165–178.

Wotton, B. M., C. A. Nock, and M. D. Flannigan, 2010: Forest fire occurrence and climate change in Canada. *Int. J. Wildland Fire*, **19**, 253–271.

Wu, P. L., R. Wood, J. Ridley, and J. Lowe, 2010: Temporary acceleration of the hydrological cycle in response to a CO_2 rampdown. *Geophys. Res. Lett.*, **37**, L12705.

Wu, T., et al., 2013: Global Carbon budgets simulated by the Beijing Climate Center Climate System Model for the last Century. *J. Geophys. Res. Atmos.*, doi:10.1002/jgrd.50320, in press.

Wurzburger, N., J. P. Bellenger, A. M. L. Kraepiel, and L. O. Hedin, 2012: Molybdenum and phosphorus interact to constrain asymbiotic nitrogen fixation in tropical forests. *PLoS ONE*, **7**, e33710.

Xu-Ri, and I. C. Prentice, 2008: Terrestrial nitrogen cycle simulation with a dynamic global vegetation model. *Global Change Biol.*, **14**, 1745–1764.

Xu-Ri, I. C. Prentice, R. Spahni, and H. S. Niu, 2012: Modelling terrestrial nitrous oxide emissions and implications for climate feedback. *New Phytologist*, **196**, 472–488.

Yamamoto-Kawai, M., F. A. McLaughlin, E. C. Carmack, S. Nishino, and K. Shimada, 2009: Aragonite undersaturation in the Arctic Ocean: Effects of ocean acidification and sea ice melt. *Science*, **326**, 1098–1100.

Yamamoto, A., M. Kawamiya, A. Ishida, Y. Yamanaka, and S. Watanabe, 2012: Impact of rapid sea-ice reduction in the Arctic Ocean on the rate of ocean acidification. *Biogeosciences*, **9**, 2365–2375.

Yan, X., H. Akiyama, K. Yagi, and H. Akimoto, 2009: Global estimations of the inventory and mitigation potential of methane emissions from rice cultivation conducted using the 2006 Intergovernmental Panel on Climate Change Guidelines. *Global Biogeochem. Cycles*, **23**, GB2002.

Yang, X., T. K. Richardson, and A. K. Jain, 2010: Contributions of secondary forest and nitrogen dynamics to terrestrial carbon uptake. *Biogeosciences*, **7**, 3041-3050.

Yevich, R., and J. A. Logan, 2003: An assessment of biofuel use and burning of agricultural waste in the developing world. *Global Biogeochem. Cycles*, **17**, 1095.

Yool, A., J. G. Shepherd, H. L. Bryden, and A. Oschlies, 2009: Low efficiency of nutrient translocation for enhancing oceanic uptake of carbon dioxide. *J. Geophys. Res. Oceans*, **114**, C08009.

Yoshikawa-Inoue, H. Y., and M. Ishii, 2005: Variations and trends of CO_2 in the surface seawater in the Southern Ocean south of Australia between 1969 and 2002. *Tellus B*, *57*, 58–69.

Yoshikawa, C., M. Kawamiya, T. Kato, Y. Yamanaka, and T. Matsuno, 2008: Geographical distribution of the feedback between future climate change and the carbon cycle. *J. Geophys. Res. Biogeosci.*, **113**, G03002.

Young, P., et al., 2013: Pre-industrial to end 21st century projections of tropospheric ozone from the Atmospheric Chemistry and Climate Model Intercomparison Project (ACCMIP). *Atmos. Chem. Phys.*, **4**, 2063–2090.

Yu, J., W. S. Broecker, H. Elderfield, Z. Jin, J. McManus, and F. Zhang, 2010: Loss of carbon from the deep sea since the Last Glacial Maximum. *Science*, **330**, 1084–1087.

Yu, Z., 2011: Holocene carbon flux histories of the world's peatlands: Global carbon-cycle implications. *Holocene*, **21**, 761–774.

Zaehle, S., 2013: Terrestrial nitrogen-carbon cycle interactions at the global scale, *Philos. Trans. R. Soc. London B*, **368**, 20130125.

Zaehle, S., and A. D. Friend, 2010: Carbon and nitrogen cycle dynamics in the O-CN land surface model: 1. Model description, site-scale evaluation, and sensitivity to parameter estimates. *Global Biogeochem. Cycles*, **24**, GB1005.

Zaehle, S., and D. Dalmonech, 2011: Carbon-nitrogen interactions on land at global scales: Current understanding in modelling climate biosphere feedbacks. *Curr. Opin. Environ. Sustain.*, **3**, 311–320.

Zaehle, S., P. Friedlingstein, and A. D. Friend, 2010a: Terrestrial nitrogen feedbacks may accelerate future climate change. *Geophys. Res. Lett.*, **37**, L01401.

Zaehle, S., P. Ciais, A. D. Friend, and V. Prieur, 2011: Carbon benefits of anthropogenic reactive nitrogen offset by nitrous oxide emissions. *Nature Geosci.*, **4**, 601–605.

Zaehle, S., A. D. Friend, P. Friedlingstein, F. Dentener, P. Peylin, and M. Schulz, 2010b: Carbon and nitrogen cycle dynamics in the O-CN land surface model: 2. Role of the nitrogen cycle in the historical terrestrial carbon balance. *Global Biogeochem. Cycles*, **24**, GB1006.

Zak, D. R., K. S. Pregitzer, M. E. Kubiske, and A. J. Burton, 2011: Forest productivity under elevated CO_2 and O_3: Positive feedbacks to soil N cycling sustain decade-long net primary productivity enhancement by CO_2. *Ecol. Lett.*, **14**, 1220–1226.

Zech, R., Y. Huang, M. Zech, R. Tarozo, and W. Zech, 2011: High carbon sequestration in Siberian permafrost loess-paleosoils during glacials. *Clim. Past*, **7**, 501–509.

Zeebe, R. E., and D. Wolf-Gladrow, 2001: *CO_2 in Seawater: Equilibrium, Kinetics, Isotopes.* Elsevier Science, Amsterdam, Netherlands, and Philadelphia, PA, USA.

Zeebe, R. E., and D. Archer, 2005: Feasibility of ocean fertilization and its impact on future atmospheric CO_2 levels. *Geophys. Res. Lett.*, **32**, L09703.

Zeng, N., 2003: Glacial-interglacial atmospheric CO_2 change —The glacial burial hypothesis. *Adv. Atmos. Sci.*, **20**, 677–693.

Zhang, Q., Y. P. Wang, A. J. Pitman, and Y. J. Dai, 2011: Limitations of nitrogen and phosphorous on the terrestrial carbon uptake in the 20th century. *Geophys. Res. Lett.*, **38**, L22701.

Zhao, M., and S. W. Running, 2010: Drought-induced reduction in global terrestrial net primary production from 2000 through 2009. *Science*, **329**, 940–943.

Zhou, S., and P. C. Flynn, 2005: Geoengineering downwelling ocean currents: A cost assessment. *Clim. Change*, **71**, 203–220.

Zhuang, Q. L., et al., 2006: CO_2 and CH_4 exchanges between land ecosystems and the atmosphere in northern high latitudes over the 21st century. *Geophys. Res. Lett.*, **33**, L17403.

Zickfeld, K., M. Eby, H. D. Matthews, A. Schmittner, and A. J. Weaver, 2011: Nonlinearity of carbon cycle feedbacks. *J. Clim.*, **24**, 4255–4275.

Zimov, N. S., S. A. Zimov, A. E. Zimova, G. M. Zimova, V. I. Chuprynin, and F. S. Chapin, 2009: Carbon storage in permafrost and soils of the mammoth tundra-steppe biome: Role in the global carbon budget. *Geophys. Res. Lett.*, **36**, L02502.

6

7 Clouds and Aerosols

Coordinating Lead Authors:
Olivier Boucher (France), David Randall (USA)

Lead Authors:
Paulo Artaxo (Brazil), Christopher Bretherton (USA), Graham Feingold (USA), Piers Forster (UK), Veli-Matti Kerminen (Finland), Yutaka Kondo (Japan), Hong Liao (China), Ulrike Lohmann (Switzerland), Philip Rasch (USA), S.K. Satheesh (India), Steven Sherwood (Australia), Bjorn Stevens (Germany), Xiao-Ye Zhang (China)

Contributing Authors:
Govindasamy Bala (India), Nicolas Bellouin (UK), Angela Benedetti (UK), Sandrine Bony (France), Ken Caldeira (USA), Anthony Del Genio (USA), Maria Cristina Facchini (Italy), Mark Flanner (USA), Steven Ghan (USA), Claire Granier (France), Corinna Hoose (Germany), Andy Jones (UK), Makoto Koike (Japan), Ben Kravitz (USA), Benjamin Laken (Spain), Matthew Lebsock (USA), Natalie Mahowald (USA), Gunnar Myhre (Norway), Colin O'Dowd (Ireland), Alan Robock (USA), Bjørn Samset (Norway), Hauke Schmidt (Germany), Michael Schulz (Norway), Graeme Stephens (USA), Philip Stier (UK), Trude Storelvmo (USA), Dave Winker (USA), Matthew Wyant (USA)

Review Editors:
Sandro Fuzzi (Italy), Joyce Penner (USA), Venkatachalam Ramaswamy (USA), Claudia Stubenrauch (France)

This chapter should be cited as:
Boucher, O., D. Randall, P. Artaxo, C. Bretherton, G. Feingold, P. Forster, V.-M. Kerminen, Y. Kondo, H. Liao, U. Lohmann, P. Rasch, S.K. Satheesh, S. Sherwood, B. Stevens and X.Y. Zhang, 2013: Clouds and Aerosols. In: *Climate Change 2013: The Physical Science Basis. Contribution of Working Group I to the Fifth Assessment Report of the Intergovernmental Panel on Climate Change* [Stocker, T.F., D. Qin, G.-K. Plattner, M. Tignor, S.K. Allen, J. Boschung, A. Nauels, Y. Xia, V. Bex and P.M. Midgley (eds.)]. Cambridge University Press, Cambridge, United Kingdom and New York, NY, USA.

Table of Contents

Supplementary Material

Supplementary Material is available in online versions of the report.

Executive Summary

Clouds and aerosols continue to contribute the largest uncertainty to estimates and interpretations of the Earth's changing energy budget. This chapter focuses on process understanding and considers observations, theory and models to assess how clouds and aerosols contribute and respond to climate change. The following conclusions are drawn.

Progress in Understanding

Many of the cloudiness and humidity changes simulated by climate models in warmer climates are now understood as responses to large-scale circulation changes that do not appear to depend strongly on sub-grid scale model processes, increasing confidence in these changes. For example, multiple lines of evidence now indicate positive feedback contributions from circulation-driven changes in both the height of high clouds and the latitudinal distribution of clouds (*medium to high confidence*[1]). However, some aspects of the overall cloud response vary substantially among models, and these appear to depend strongly on sub-grid scale processes in which there is less confidence. {7.2.4, 7.2.5, 7.2.6, Figure 7.11}

Climate-relevant aerosol processes are better understood, and climate-relevant aerosol properties better observed, than at the time of AR4. However, the representation of relevant processes varies greatly in global aerosol and climate models and it remains unclear what level of sophistication is required to model their effect on climate. Globally, between 20 and 40% of aerosol optical depth (*medium confidence*) and between one quarter and two thirds of cloud condensation nucleus concentrations (*low confidence*) are of anthropogenic origin. {7.3, Figures 7.12 to 7.15}

Cosmic rays enhance new particle formation in the free troposphere, but the effect on the concentration of cloud condensation nuclei is too weak to have any detectable climatic influence during a solar cycle or over the last century (*medium evidence, high agreement*). No robust association between changes in cosmic rays and cloudiness has been identified. In the event that such an association existed, a mechanism other than cosmic ray-induced nucleation of new aerosol particles would be needed to explain it. {7.4.6}

Recent research has clarified the importance of distinguishing forcing (instantaneous change in the radiative budget) and rapid adjustments (which modify the radiative budget indirectly through fast atmospheric and surface changes) from feedbacks (which operate through changes in climate variables that are mediated by a change in surface temperature). Furthermore, one can distinguish between the traditional concept of radiative forcing (RF) and the relatively new concept of effective radiative forcing (ERF) that also includes rapid adjustments. For aerosols one can further distinguish forcing processes arising from aerosol–radiation interactions (ari) and aerosol–cloud interactions (aci). {7.1, Figures 7.1 to 7.3}

The quantification of cloud and convective effects in models, and of aerosol–cloud interactions, continues to be a challenge. Climate models are incorporating more of the relevant processes than at the time of AR4, but confidence in the representation of these processes remains weak. Cloud and aerosol properties vary at scales significantly smaller than those resolved in climate models, and cloud-scale processes respond to aerosol in nuanced ways at these scales. Until sub-grid scale parameterizations of clouds and aerosol–cloud interactions are able to address these issues, model estimates of aerosol–cloud interactions and their radiative effects will carry large uncertainties. Satellite-based estimates of aerosol–cloud interactions remain sensitive to the treatment of meteorological influences on clouds and assumptions on what constitutes pre-industrial conditions. {7.3, 7.4, 7.5.3, 7.5.4, 7.6.4, Figures 7.8, 7.12, 7.16}

Precipitation and evaporation are expected to increase on average in a warmer climate, but also undergo global and regional adjustments to carbon dioxide (CO_2) and other forcings that differ from their warming responses. Moreover, there is *high confidence* that, as climate warms, extreme precipitation rates on for example, daily time scales will increase faster than the time average. Changes in average precipitation must remain consistent with changes in the net rate of cooling of the troposphere, which is affected by its temperature but also by greenhouse gases (GHGs) and aerosols. Consequently, while the increase in global mean precipitation would be 1.5 to 3.5% $°C^{-1}$ due to surface temperature change alone, warming caused by CO_2 or absorbing aerosols results in a smaller sensitivity, even more so if it is partially offset by albedo increases. The complexity of land surface and atmospheric processes limits confidence in regional projections of precipitation change, especially over land, although there is a component of a 'wet-get-wetter' and 'dry-get-drier' response over oceans at the large scale. Changes in local extremes on daily and sub-daily time scales are strongly influenced by lower-tropospheric water vapour concentrations, and on average will increase by roughly 5 to 10% per degree Celsius of warming (*medium confidence*). Aerosol–cloud interactions can influence the character of individual storms, but evidence for a systematic aerosol effect on storm or precipitation intensity is more limited and ambiguous. {7.2.4, 7.4, 7.6, Figures 7.20, 7.21}

---

1 In this Report, the following summary terms are used to describe the available evidence: limited, medium, or robust; and for the degree of agreement: low, medium, or high. A level of confidence is expressed using five qualifiers: very low, low, medium, high, and very high, and typeset in italics, e.g., *medium confidence*. For a given evidence and agreement statement, different confidence levels can be assigned, but increasing levels of evidence and degrees of agreement are correlated with increasing confidence (see Section 1.4 and Box TS.1 for more details).

Water Vapour, Cloud and Aerosol Feedbacks

The net feedback from water vapour and lapse rate changes combined, as traditionally defined, is *extremely likely*[2] positive (amplifying global climate changes). The sign of the net radiative feedback due to all cloud types is less certain but *likely* positive. Uncertainty in the sign and magnitude of the cloud feedback is due primarily to continuing uncertainty in the impact of warming on low clouds. We estimate the water vapour plus lapse rate feedback[3] to be +1.1 (+0.9 to +1.3) W m^{-2} °C^{-1} and the cloud feedback from all cloud types to be +0.6 (–0.2 to +2.0) W m^{-2} °C^{-1}. These ranges are broader than those of climate models to account for additional uncertainty associated with processes that may not have been accounted for in those models. The mean values and ranges in climate models are essentially unchanged since AR4, but are now supported by stronger indirect observational evidence and better process understanding, especially for water vapour. Low clouds contribute positive feedback in most models, but that behaviour is not well understood, nor effectively constrained by observations, so we are not confident that it is realistic. {7.2.4, 7.2.5, 7.2.6, Figures 7.9 to 7.11}.

Aerosol–climate feedbacks occur mainly through changes in the source strength of natural aerosols or changes in the sink efficiency of natural and anthropogenic aerosols; a limited number of modelling studies have bracketed the feedback parameter within ±0.2 W m^{-2} °C^{-1} with *low confidence*. There is *medium confidence* for a weak dimethylsulphide–cloud condensation nuclei–cloud albedo feedback due to a weak sensitivity of cloud condensation nuclei population to changes in dimethylsulphide emissions. {7.3.5}

Quantification of climate forcings[4] due to aerosols and clouds

The ERF due to aerosol–radiation interactions that takes rapid adjustments into account (ERFari) is assessed to be –0.45 (–0.95 to +0.05) W m^{-2}. The RF from absorbing aerosol on snow and ice is assessed separately to be +0.04 (+0.02 to +0.09) W m^{-2}. Prior to adjustments taking place, the RF due to aerosol–radiation interactions (RFari) is assessed to be –0.35 (–0.85 to +0.15) W m^{-2}. The assessment for RFari is less negative than reported in AR4 because of a re-evaluation of aerosol absorption. The uncertainty estimate is wider but more robust, based on multiple lines of evidence from models, remotely sensed data, and ground-based measurements. Fossil fuel and biofuel emissions[4] contribute to RFari via sulphate aerosol: –0.4 (–0.6 to –0.2) W m^{-2}, black carbon (BC) aerosol: +0.4 (+0.05 to +0.8) W m^{-2}, and primary and secondary organic aerosol: –0.12 (–0.4 to +0.1) W m^{-2}. Additional RFari contributions occur via biomass burning emissions[5]: +0.0 (–0.2 to +0.2) W m^{-2}, nitrate aerosol: –0.11 (–0.3 to –0.03) W m^{-2}, and mineral dust: –0.1 (–0.3 to +0.1) W m^{-2} although the latter may not be entirely of anthropogenic origin. While there is robust evidence for the existence of rapid adjustment of clouds in response to aerosol absorption, these effects are multiple and not well represented in climate models, leading to large uncertainty. Unlike in the last IPCC assessment, the RF from BC on snow and ice includes the effects on sea ice, accounts for more physical processes and incorporates evidence from both models and observations. This RF has a 2 to 4 times larger global mean surface temperature change per unit forcing than a change in CO_2. {7.3.4, 7.5.2, Figures 7.17, 7.18}

The total ERF due to aerosols (ERFari+aci, excluding the effect of absorbing aerosol on snow and ice) is assessed to be –0.9 (–1.9 to –0.1) W m^{-2} with *medium confidence*. The ERFari+aci estimate includes rapid adjustments, such as changes to the cloud lifetime and aerosol microphysical effects on mixed-phase, ice and convective clouds. This range was obtained from expert judgement guided by climate models that include aerosol effects on mixed-phase and convective clouds in addition to liquid clouds, satellite studies and models that allow cloud-scale responses. This forcing can be much larger regionally but the global mean value is consistent with several new lines of evidence suggesting less negative estimates for the ERF due to aerosol–cloud interactions than in AR4. {7.4, 7.5.3, 7.5.4, Figure 7.19}

Persistent contrails from aviation contribute a RF of +0.01 (+0.005 to +0.03) W m^{-2} for year 2011, and the combined contrail and contrail-cirrus ERF from aviation is assessed to be +0.05 (+0.02 to +0.15) W m^{-2}. This forcing can be much larger regionally but there is now *medium confidence* that it does not produce observable regional effects on either the mean or diurnal range of surface temperature. {7.2.7}

Geoengineering Using Solar Radiation Management Methods

Theory, model studies and observations suggest that some Solar Radiation Management (SRM) methods, if practicable, could substantially offset a global temperature rise and partially offset some other impacts of global warming, but the compensation for the climate change caused by GHGs would be imprecise (*high confidence*). SRM methods are unimplemented and untested. Research on SRM is in its infancy, though it leverages understanding of how the climate responds to forcing more generally. The efficacy of a number of SRM strategies was assessed, and there is *medium confidence* that stratospheric aerosol SRM is scalable to counter the RF from increasing GHGs at least up to approximately 4 W m^{-2}; however,

---

[2] In this Report, the following terms have been used to indicate the assessed likelihood of an outcome or a result: Virtually certain 99–100% probability, Very likely 90–100%, Likely 66–100%, About as likely as not 33–66%, Unlikely 0–33%, Very unlikely 0–10%, Exceptionally unlikely 0–1%. Additional terms (Extremely likely: 95–100%, More likely than not >50–100%, and Extremely unlikely 0–5%) may also be used when appropriate. Assessed likelihood is typeset in italics, e.g., *very likely* (see Section 1.4 and Box TS.1 for more details).

[3] This and all subsequent ranges given with this format are 90% uncertainty ranges unless otherwise specified.

[4] All climate forcings (RFs and ERFs) are anthropogenic and relate to the period 1750–2010 unless otherwise specified.

[5] This species breakdown is less certain than the total RFari and does not sum to the total exactly.

the required injection rate of aerosol precursors remains very uncertain. There is no consensus on whether a similarly large RF could be achieved from cloud brightening SRM owing to uncertainties in understanding and representation of aerosol–cloud interactions. It does not appear that land albedo change SRM can produce a large RF. Limited literature on other SRM methods precludes their assessment. Models consistently suggest that SRM would generally reduce climate differences compared to a world with elevated GHG concentrations and no SRM; however, there would also be residual regional differences in climate (e.g., temperature and rainfall) when compared to a climate without elevated GHGs. {7.4.3, 7.7}

Numerous side effects, risks and shortcomings from SRM have been identified. Several lines of evidence indicate that SRM would produce a small but significant decrease in global precipitation (with larger differences on regional scales) if the global surface temperature were maintained. A number of side effects have been identified. One that is relatively well characterized is the likelihood of modest polar stratospheric ozone depletion associated with stratospheric aerosol SRM. There could also be other as yet unanticipated consequences. As long as GHG concentrations continued to increase, the SRM would require commensurate increase, exacerbating side effects. In addition, scaling SRM to substantial levels would carry the risk that if the SRM were terminated for any reason, there is *high confidence* that surface temperatures would increase rapidly (within a decade or two) to values consistent with the GHG forcing, which would stress systems sensitive to the rate of climate change. Finally, SRM would not compensate for ocean acidification from increasing CO_2. {7.6.3, 7.7, Figures 7.22 to 7.24}

7.1 Introduction

7.1.1 Clouds and Aerosols in the Atmosphere

The atmosphere is composed mostly of gases, but also contains liquid and solid matter in the form of particles. It is usual to distinguish these particles according to their size, chemical composition, water content and fall velocity into atmospheric aerosol particles, cloud particles and falling hydrometeors. Despite their small mass or volume fraction, particles in the atmosphere strongly influence the transfer of radiant energy and the spatial distribution of latent heating through the atmosphere, thereby influencing the weather and climate.

Cloud formation usually takes place in rising air, which expands and cools, thus permitting the activation of aerosol particles into cloud droplets and ice crystals in supersaturated air. Cloud particles are generally larger than aerosol particles and composed mostly of liquid water or ice. The evolution of a cloud is governed by the balance between a number of dynamical, radiative and microphysical processes. Cloud particles of sufficient size become falling hydrometeors, which are categorized as drizzle drops, raindrops, snow crystals, graupel and hailstones. Precipitation is an important and complex climate variable that is influenced by the distribution of moisture and cloudiness, and to a lesser extent by the concentrations and properties of aerosol particles.

Aerosol particles interact with solar radiation through absorption and scattering and, to a lesser extent with terrestrial radiation through absorption, scattering and emission. Aerosols[6] can serve as cloud condensation nuclei (CCN) and ice nuclei (IN) upon which cloud droplets and ice crystals form. They also play a wider role in atmospheric chemistry and biogeochemical cycles in the Earth system, for instance, by carrying nutrients to ocean ecosystems. They can be of natural or anthropogenic origin.

Cloud and aerosol amounts[7] and properties are extremely variable in space and time. The short lifetime of cloud particles in subsaturated air creates relatively sharp cloud edges and fine-scale variations in cloud properties, which is less typical of aerosol layers. While the distinction between aerosols and clouds is generally appropriate and useful, it is not always unambiguous, which can cause interpretational difficulties (e.g., Charlson et al., 2007; Koren et al., 2007).

7.1.2 Rationale for Assessing Clouds, Aerosols and Their Interactions

The representation of cloud processes in climate models has been recognized for decades as a dominant source of uncertainty in our understanding of changes in the climate system (e.g., Arakawa, 1975, 2004; Charney et al., 1979; Cess et al., 1989; Randall et al., 2003; Bony et al., 2006), but has never been systematically assessed by the IPCC before. Clouds respond to climate forcing mechanisms in multiple ways, and inter-model differences in cloud feedbacks constitute by far the primary source of spread of both equilibrium and transient climate responses simulated by climate models (Dufresne and Bony, 2008) despite the fact that most models agree that the feedback is positive (Randall et al., 2007; Section 7.2). Thus confidence in climate projections requires a thorough assessment of how cloud processes have been accounted for.

Aerosols of anthropogenic origin are responsible for a radiative forcing (RF) of climate change through their interaction with radiation, and also as a result of their interaction with clouds. Quantification of this forcing is fraught with uncertainties (Haywood and Boucher, 2000; Lohmann and Feichter, 2005) and aerosols dominate the uncertainty in the total anthropogenic RF (Forster et al., 2007; Haywood and Schulz, 2007; Chapter 8). Furthermore, our inability to better quantify non-greenhouse gas RFs, and primarily those that result from aerosol–cloud interactions, underlie difficulties in constraining climate sensitivity from observations even if we had a perfect knowledge of the temperature record (Andreae et al., 2005). Thus a complete understanding of past and future climate change requires a thorough assessment of aerosol–cloud–radiation interactions.

7.1.3 Forcing, Rapid Adjustments and Feedbacks

Figure 7.1 illustrates key aspects of how clouds and aerosols contribute to climate change, and provides an overview of important terminological distinctions. *Forcings* associated with agents such as greenhouse gases (GHGs) and aerosols act on global mean surface temperature through the global radiative (energy) budget. *Rapid adjustments* (sometimes called rapid responses) arise when forcing agents, by altering flows of energy internal to the system, affect cloud cover or other components of the climate system and thereby alter the global budget indirectly. Because these adjustments do not operate through changes in the global mean surface temperature (ΔT), which are slowed by the massive heat capacity of the oceans, they are generally rapid and most are thought to occur within a few weeks. *Feedbacks* are associated with changes in climate variables that are mediated by a change in global mean surface temperature; they contribute to amplify or damp global temperature changes via their impact on the radiative budget.

In this report, following an emerging consensus in the literature, the traditional concept of radiative forcing (RF, defined as the instantaneous radiative forcing with stratospheric adjustment only) is de-emphasized in favour of an absolute measure of the radiative effects of all responses triggered by the forcing agent that are independent of surface temperature change (see also Section 8.1). This new measure of the forcing includes rapid adjustments and the net forcing with these adjustments included is termed the *effective radiative forcing* (ERF). The climate sensitivity to ERF will differ somewhat from traditional equilibrium climate sensitivity, as the latter include adjustment effects. As shown in Figure 7.1, adjustments can occur through geographic temperature variations, lapse rate changes, cloud changes

6 For convenience the term 'aerosol', which includes both the particles and the suspending gas, is often used in its plural form to mean 'aerosol particles' both in this chapter and the rest of this Report.

7 In this chapter, we use 'cloud amount' as an inexact term to refer to the quantity of clouds, both in the horizontal and vertical directions. The term 'cloud cover' is used in its usual sense and refers to the horizontal cloud cover.

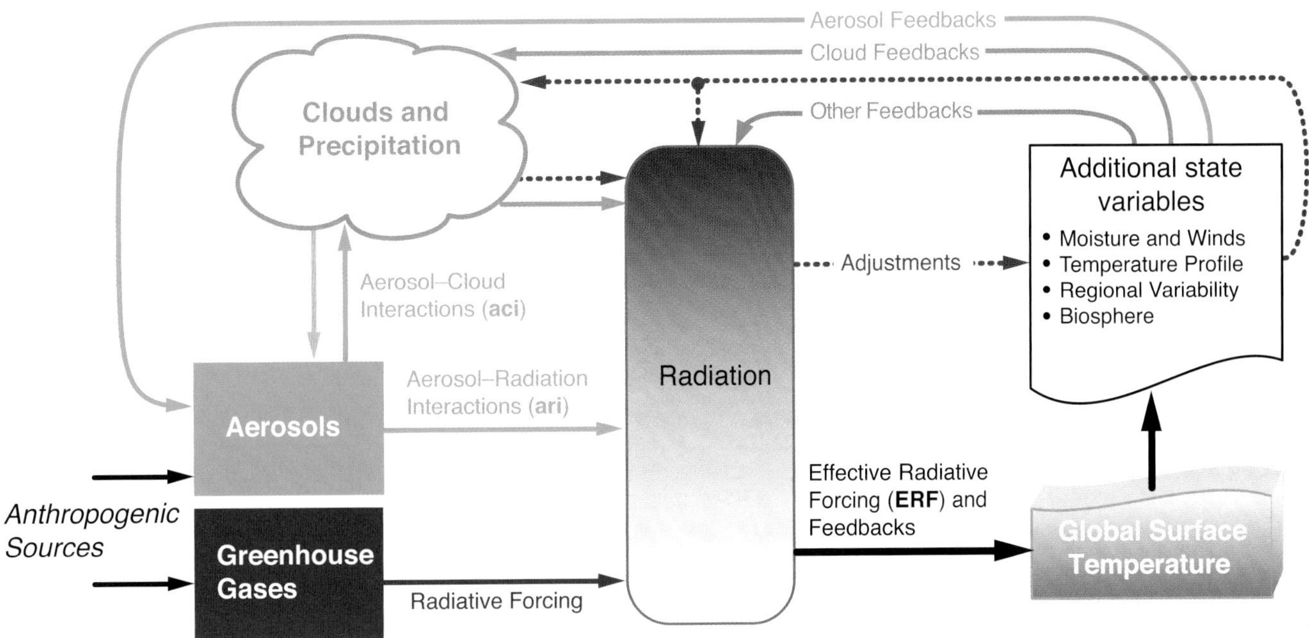

Figure 7.1 | Overview of forcing and feedback pathways involving greenhouse gases, aerosols and clouds. Forcing agents are in the green and dark blue boxes, with forcing mechanisms indicated by the straight green and dark blue arrows. The forcing is modified by rapid adjustments whose pathways are independent of changes in the globally averaged surface temperature and are denoted by brown dashed arrows. Feedback loops, which are ultimately rooted in changes ensuing from changes in the surface temperature, are represented by curving arrows (blue denotes cloud feedbacks; green denotes aerosol feedbacks; and orange denotes other feedback loops such as those involving the lapse rate, water vapour and surface albedo). The final temperature response depends on the effective radiative forcing (ERF) that is felt by the system, that is, after accounting for rapid adjustments, and the feedbacks.

and vegetation effects. Measures of ERF and rapid adjustments have existed in the literature for more than a decade, with a number of different terminologies and calculation methods adopted. These were principally aimed to help quantify the effects of aerosols on clouds (Rotstayn and Penner, 2001; Lohmann et al., 2010) and understand different forcing agent responses (Hansen et al., 2005), but it is now realized that there are rapid adjustments in response to the CO_2 forcing itself (Section 7.2.5.6).

In principle rapid adjustments are independent of ΔT, while feedbacks operate purely through ΔT. Thus, within this framework adjustments are not another type of 'feedback' but rather a non-feedback phenomenon, required in the analysis by the fact that a single scalar ΔT cannot fully characterize the system. This framework brings most efficacies close to unity although they are not necessarily exactly 1 (Hansen et al., 2005; Bond et al., 2013). There is also no clean separation in time scale between rapid adjustments and warming. Although the former occur mostly within a few days of applying a forcing (Dong et al., 2009), some adjustments such as those that occur within the stratosphere and snowpack can take several months or longer. Meanwhile the land surface warms quickly so that a small part of ΔT occurs within days to weeks of an applied forcing. This makes the two phenomena difficult to isolate in model runs. Other drawbacks are that adjustments are difficult to observe, and typically more model-dependent than RF. However, recent work is beginning to meet the challenges of quantifying the adjustments, and has noted advantages of the new framework (e.g., Vial et al., 2013; Zelinka et al., 2013).

There is no perfect method to determine ERF. Two common methods are to regress the net energy imbalance onto ΔT in a transient

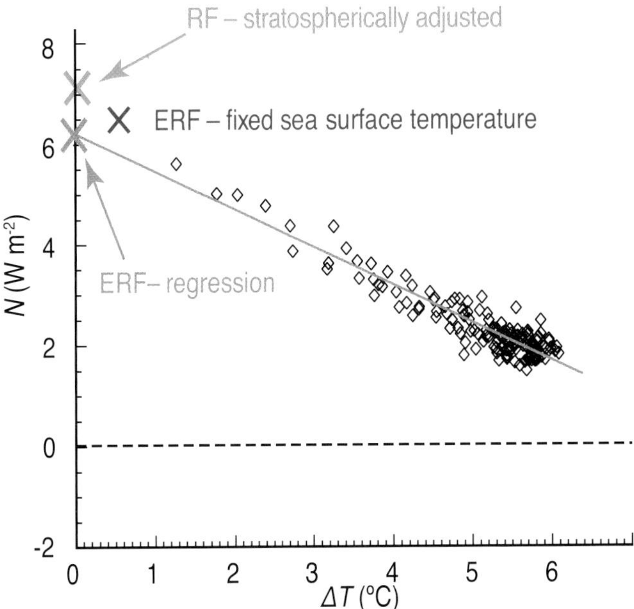

Figure 7.2 | Radiative forcing (RF) and effective radiative forcing (ERF) estimates derived by two methods, for the example of $4 \times CO_2$ experiments in one climate model. N is the net energy imbalance at the top of the atmosphere and ΔT the global mean surface temperature change. The fixed sea surface temperature ERF estimate is from an atmosphere–land model averaged over 30 years. The regression estimate is from 150 years of a coupled model simulation after an instantaneous quadrupling of CO_2, with the N from individual years in this regression shown as black diamonds. The stratospherically adjusted RF is the tropopause energy imbalance from otherwise identical radiation calculations at $1 \times$ and $4 \times CO_2$ concentrations. (Figure follows Andrews et al., 2012.) See also Figure 8.1.

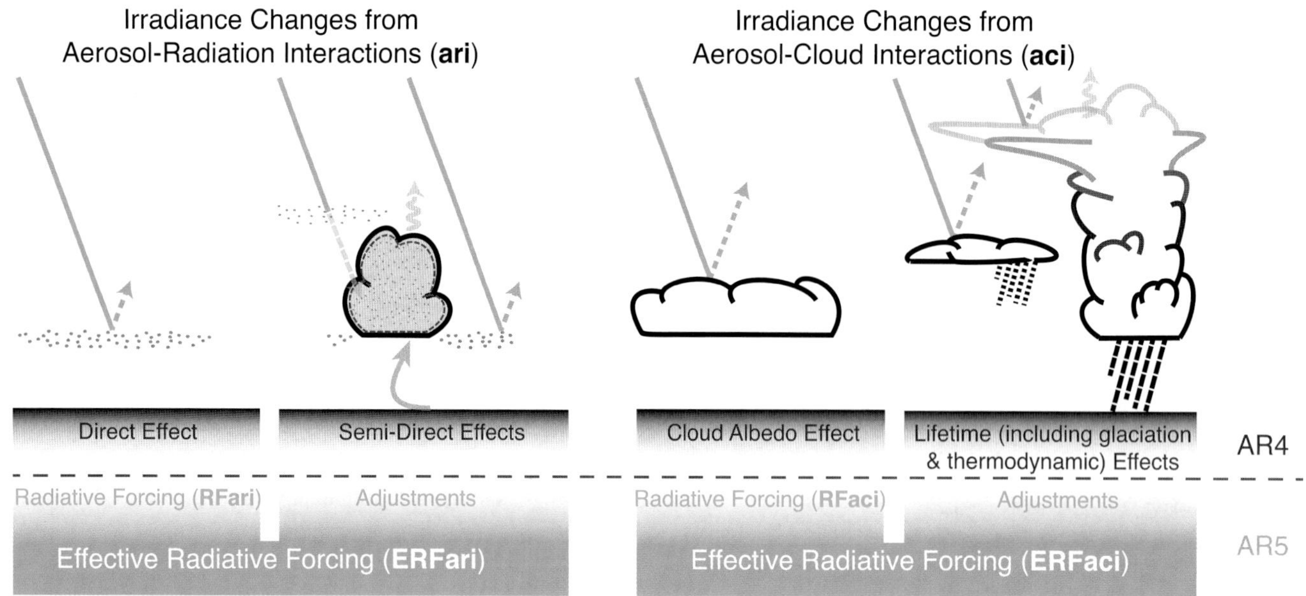

Figure 7.3 | Schematic of the new terminology used in this Assessment Report (AR5) for aerosol–radiation and aerosol–cloud interactions and how they relate to the terminology used in AR4. The blue arrows depict solar radiation, the grey arrows terrestrial radiation and the brown arrow symbolizes the importance of couplings between the surface and the cloud layer for rapid adjustments. See text for further details.

warming simulation (Gregory et al., 2004; Figure 7.2), or to simulate the climate response with sea surface temperatures (SSTs) held fixed (Hansen et al., 2005). The former can be complicated by natural variability or time-varying feedbacks, while the non-zero ΔT from land warming complicates the latter. Both methods are used in this chapter.

Figure 7.3 links the former terminology of aerosol direct, semi-direct and indirect effects with the new terminology used in this chapter and in Chapter 8. The RF from aerosol–radiation interactions (abbreviated RFari) encompasses radiative effects from anthropogenic aerosols before any adjustment takes place and corresponds to what is usually referred to as the aerosol direct effect. Rapid adjustments induced by aerosol radiative effects on the surface energy budget, the atmospheric profile and cloudiness contribute to the ERF from aerosol–radiation interactions (abbreviated ERFari). They include what has earlier been referred to as the semi-direct effect. The RF from aerosol–cloud interactions (abbreviated RFaci) refers to the instantaneous effect on cloud albedo due to changing concentrations of cloud condensation and ice nuclei, also known as the Twomey effect. All subsequent changes to the cloud lifetime and thermodynamics are rapid adjustments, which contribute to the ERF from aerosol–cloud interactions (abbreviated ERFaci). RFaci is a theoretical construct that is not easy to separate from other aerosol–cloud interactions and is therefore not quantified in this chapter.

7.1.4 Chapter Roadmap

For the first time in the IPCC WGI assessment reports, clouds and aerosols are discussed together in a single chapter. Doing so allows us to assess, and place in context, recent developments in a large and growing area of climate change research. In addition to assessing cloud feedbacks and aerosol forcings, which were covered in previous assessment reports in a less unified manner, it becomes possible to assess understanding of the multiple interactions among aerosols, clouds and precipitation and their relevance for climate and climate change. This chapter assesses the climatic roles and feedbacks of water vapour, lapse rate and clouds (Section 7.2), discusses aerosol–radiation (Section 7.3) and aerosol–cloud (Section 7.4) interactions and quantifies the resulting aerosol RF on climate (Section 7.5). It also introduces the physical basis for the precipitation responses to aerosols and climate changes (Section 7.6) noted later in the Report, and assesses geoengineering methods based on solar radiation management (Section 7.7).

7.2 Clouds

This section summarizes our understanding of clouds in the current climate from observations and process models; advances in the representation of cloud processes in climate models since AR4; assessment of cloud, water vapour and lapse rate feedbacks and adjustments; and the RF due to clouds induced by moisture released by two anthropogenic processes (air traffic and irrigation). Aerosol–cloud interactions are assessed in Section 7.4. The fidelity of climate model simulations of clouds in the current climate is assessed in Chapter 9.

7.2.1 Clouds in the Present-Day Climate System

7.2.1.1 Cloud Formation, Cloud Types and Cloud Climatology

To form a cloud, air must cool or moisten until it is sufficiently supersaturated to activate some of the available condensation or freezing nuclei. Clouds may be composed of liquid water (possibly supercooled), ice or both (mixed phase). The nucleated cloud particles are initially very small, but grow by vapour deposition. Other microphysical mechanisms dependent on the cloud phase (e.g., droplet collision and coalescence for liquid clouds, riming and Wegener–Bergeron–Findeisen processes for mixed-phase clouds and crystal aggregation in ice clouds) can produce a broader spectrum of particle sizes and types; turbulent mixing produces further variations in cloud properties on scales from kilometres to less than a centimetre (Davis et al., 1999; Bodenschatz et al., 2010). If and when some of the droplets or ice particles become large enough, these will fall out of the cloud as precipitation.

Atmospheric flows often organize convection and associated clouds into coherent systems having scales from tens to thousands of kilometres, such as cyclones or frontal systems. These represent a significant modelling and theoretical challenge, as they are usually too large to represent within the limited domains of cloud-resolving models (Section 7.2.2.1), but are also not well resolved nor parameterized by most climate models; this gap, however, is beginning to close (Section 7.2.2.2). Finally, clouds and cloud systems are organized by larger-scale circulations into different regimes such as deep convection near the equator, subtropical marine stratocumulus, or mid-latitude storm tracks guided by the tropospheric westerly jets. Figure 7.4 shows a selection of widely occurring cloud regimes schematically and as they might appear in a typical geostationary satellite image.

New satellite sensors and new analysis of previous data sets have given us a clearer picture of the Earth's clouds since AR4. A notable example is the launch in 2006 of two coordinated, active sensors, the Cloud Profiling Radar (CPR) on the CloudSat satellite (Stephens et al., 2002) and the Cloud–Aerosol Lidar with Orthogonal Polarization (CALIOP) on board the Cloud–Aerosol Lidar and Infrared Pathfinder Satellite Observations (CALIPSO) satellite (Winker et al., 2009). These sensors have significantly improved our ability to quantify vertical profiles of cloud occurrence and water content (see Figures 7.5 and 7.6), and complement the detection capabilities of passive multispectral sensors (e.g., Stubenrauch et al., 2010; Chan and Comiso, 2011). Satellite cloud-observing capacities are reviewed by Stubenrauch et al. (2013).

Clouds cover roughly two thirds of the globe (Figure 7.5a, c), with a more precise value depending on both the optical depth threshold used to define cloud and the spatial scale of measurement (Wielicki and Parker, 1992; Stubenrauch et al., 2013). The mid-latitude oceanic storm tracks and tropical precipitation belts are particularly cloudy, while continental desert regions and the central subtropical oceans are relatively cloud-free. Clouds are composed of liquid at temperatures above 0°C, ice below about –38°C (e.g., Koop et al., 2000), and either or both phases at intermediate temperatures (Figure 7.5b). Throughout most of the troposphere, temperatures at any given altitude are usually warmer in the tropics, but clouds also extend higher there such that ice

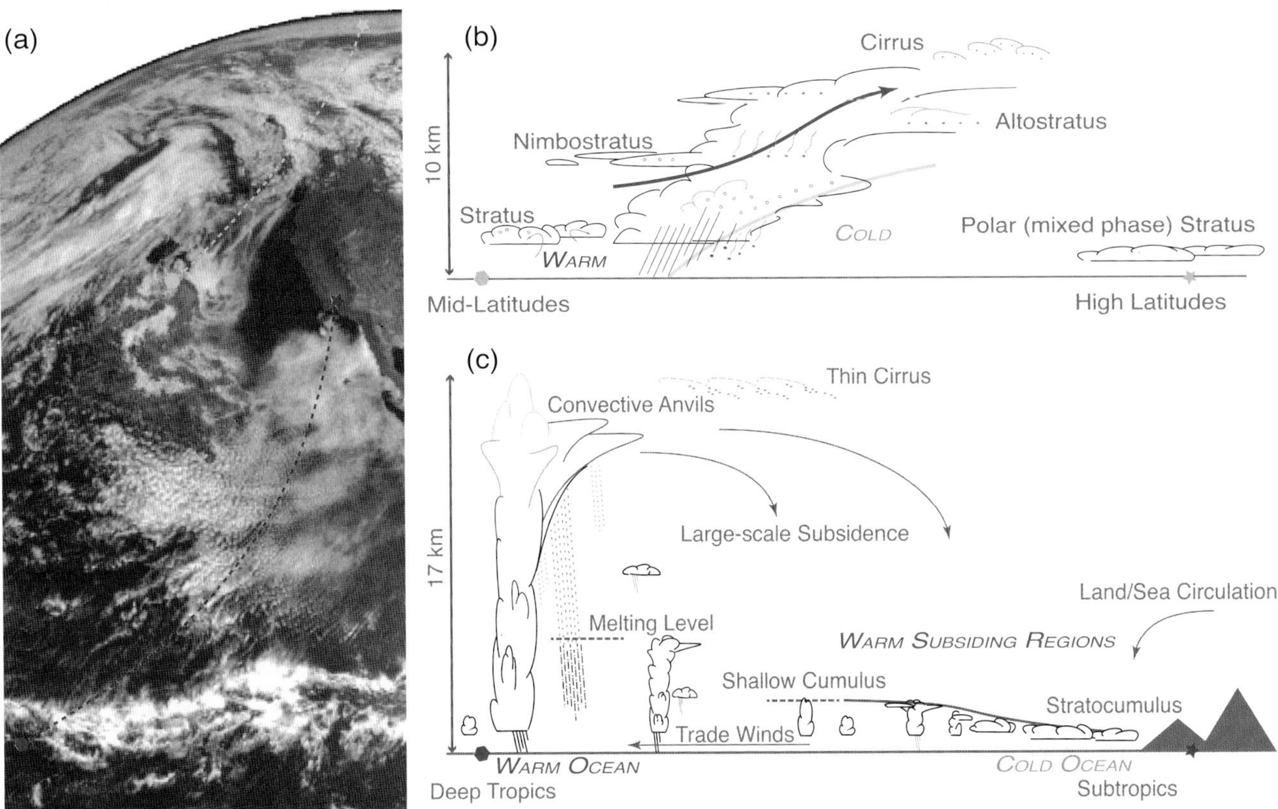

Figure 7.4 | Diverse cloud regimes reflect diverse meteorology. (a) A visible-wavelength geostationary satellite image shows (from top to bottom) expanses and long arcs of cloud associated with extratropical cyclones, subtropical coastal stratocumulus near Baja California breaking up into shallow cumulus clouds in the central Pacific and mesoscale convective systems outlining the Pacific Intertropical Convergence Zone (ITCZ). (b) A schematic section along the dashed line from the orange star to the orange circle in (a), through a typical warm front of an extratropical cyclone. It shows (from right to left) multiple layers of upper-tropospheric ice (cirrus) and mid-tropospheric water (altostratus) cloud in the upper-tropospheric outflow from the frontal zone, an extensive region of nimbostratus associated with frontal uplift and turbulence-driven boundary layer cloud in the warm sector. (c) A schematic section along the dashed line from the red star to the red circle in (a), along the low-level trade wind flow from a subtropical west coast of a continent to the ITCZ. It shows (from right to left) typical low-latitude cloud mixtures, shallow stratocumulus trapped under a strong subsidence inversion above the cool waters of the oceanic upwelling zone near the coast and shallow cumulus over warmer waters further offshore transitioning to precipitating cumulonimbus cloud systems with extensive cirrus anvils associated with rising air motions in the ITCZ.

cloud amounts are no less than those at high latitudes. At any given time, most clouds are not precipitating (Figure 7.5d).

In this chapter cloud above the 440 hPa pressure level is considered 'high', that below the 680 hPa level 'low', and that in-between is considered 'mid-level'. Most high cloud (mainly cirrus and deep cumulus outflows) occurs near the equator and over tropical continents, but can also be seen in the mid-latitude storm track regions and over mid-latitude continents in summer (Figure 7.6a, e); it is produced by the storms generating most of the global rainfall in regions where tropospheric air motion is upward, such that dynamical, rainfall and high-cloud fields closely resemble one another (Figure 7.6d, h). Mid-level cloud (Figure 7.6b, f), comprising a variety of types, is prominent in the storm tracks and some occurs in the Intertropical Convergence Zone (ITCZ). Low cloud (Figure 7.6c, g), including shallow cumulus and stratiform cloud, occurs over essentially all oceans but is most prevalent over cooler subtropical oceans and in polar regions. It is less common over land, except at night and in winter.

Overlap between cloud layers has long been an issue both for satellite (or ground-based) detection and for calculating cloud radiative effects. Active sensors show more clearly that low clouds are prevalent in nearly all types of convective systems, and are often underestimated by models (Chepfer et al., 2008; Naud et al., 2010; Haynes et al., 2011). Cloud layers at different levels overlap less often than typically assumed in General Circulation Models (GCMs), especially over high-latitude continents and subtropical oceans (Naud et al., 2008; Mace et al., 2009), and the common assumption that the radiative effects of precipitating ice can be neglected is not necessarily warranted (Waliser et al., 2011). New observations have led to revised treatments of overlap in some models, which significantly affects cloud radiative effects (Pincus et al., 2006; Shonk et al., 2012). Active sensors have also been useful in detecting low-lying Arctic clouds over sea ice (Kay et al., 2008), improving our ability to test climate model simulations of the interaction between sea ice loss and cloud cover (Kay et al., 2011).

7.2.1.2 Effects of Clouds on the Earth's Radiation Budget

The effect of clouds on the Earth's present-day top of the atmosphere (TOA) radiation budget, or cloud radiative effect (CRE), can be inferred from satellite data by comparing upwelling radiation in cloudy and non-cloudy conditions (Ramanathan et al., 1989). By enhancing the planetary albedo, cloudy conditions exert a global and annual shortwave cloud radiative effect (SWCRE) of approximately –50 W m^{-2} and, by contributing to the greenhouse effect, exert a mean longwave effect (LWCRE) of approximately +30 W m^{-2}, with a range of 10% or less between published satellite estimates (Loeb et al., 2009). Some of the apparent LWCRE comes from the enhanced water vapour coinciding with the natural cloud fluctuations used to measure the effect, so the true cloud LWCRE is about 10% smaller (Sohn et al., 2010). The net global mean CRE of approximately –20 W m^{-2} implies a net cooling

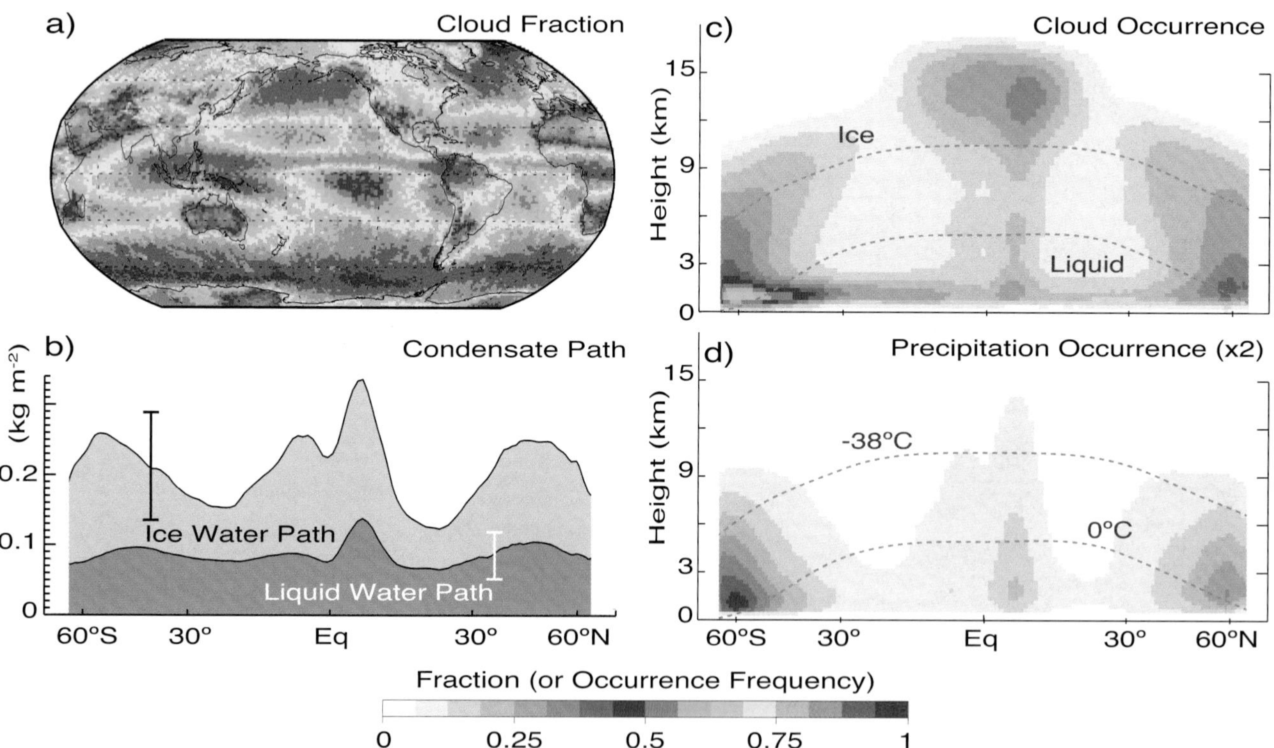

Figure 7.5 | (a) Annual mean cloud fractional occurrence (CloudSat/CALIPSO 2B-GEOPROF-LIDAR data set for 2006–2011; Mace et al., 2009). (b) Annual zonal mean liquid water path (blue shading, microwave radiometer data set for 1988–2005 from O'Dell et al. (2008)) and total water path (ice path shown with grey shading, from CloudSat 2C-ICE data set for 2006–2011 from Deng et al. (2010) over oceans). The 90% uncertainty ranges, assessed to be approximately 60 to 140% of the mean for the liquid and total water paths, are schematically indicated by the error bars. (c–d) latitude-height sections of annual zonal mean cloud (including precipitation falling from cloud) occurrence and precipitation (attenuation-corrected radar reflectivity >0 dBZ) occurrence; the latter has been doubled to make use of a common colour scale (2B-GEOPROF-LIDAR data set). The dashed curves show the annual mean 0°C and –38°C isotherms.

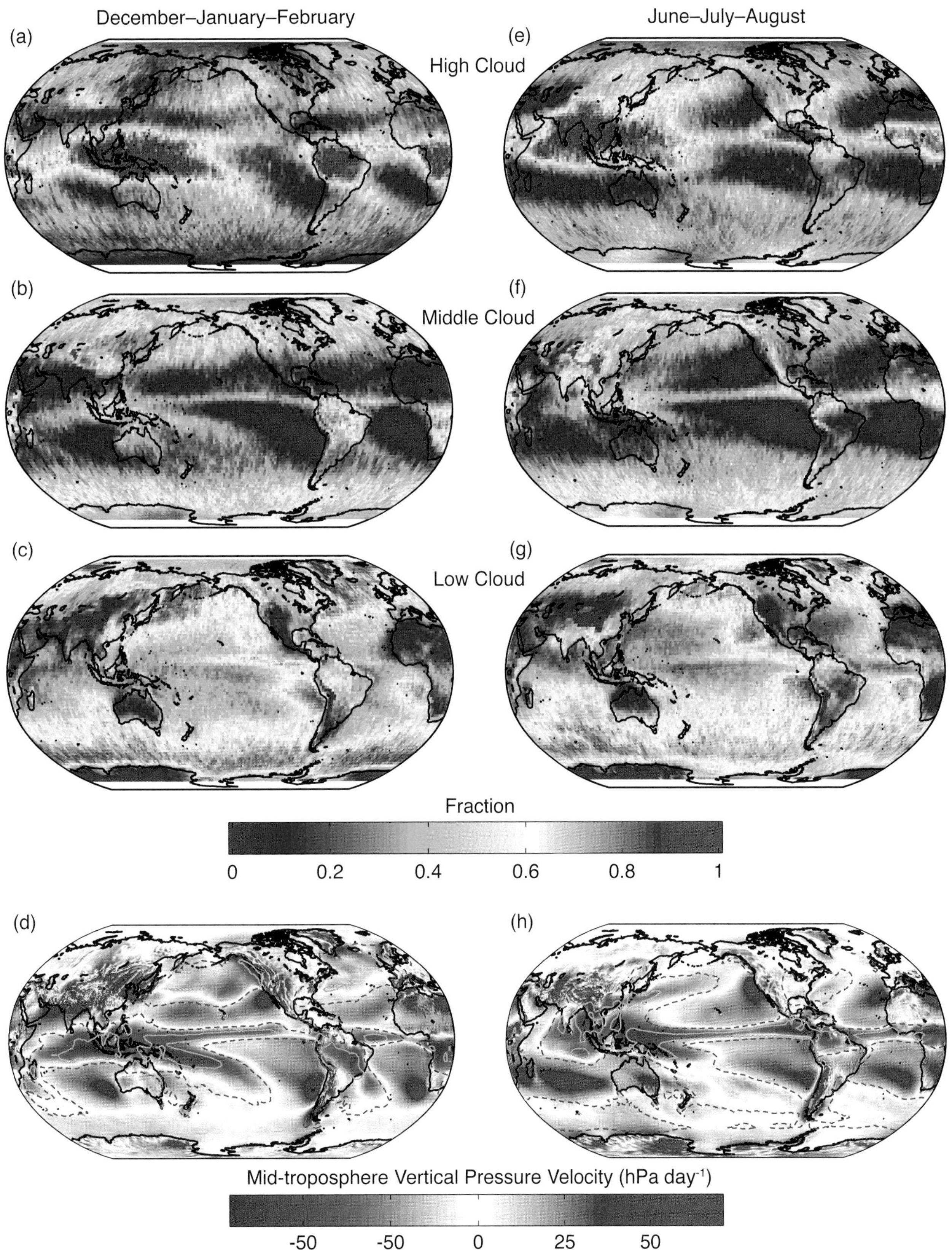

Figure 7.6 | (a–d) December–January–February mean high, middle and low cloud cover from CloudSat/CALIPSO 2B-GEOPROF R04 and 2B-GEOPROF-LIDAR P1.R04 data sets for 2006–2011 (Mace et al., 2009), 500 hPa vertical pressure velocity (colours, from ERA-Interim for 1979–2010; Dee et al., 2011), and Global Precipitation Climatology Project (GPCP) version 2.2 precipitation rate (1981–2010, grey contours at 3 mm day^{-1} in dash and 7 mm day^{-1} in solid); (e–h) same as (a–d), except for June–July–August. For low clouds, the GCM-Oriented CALIPSO Cloud Product (GOCCP) data set for 2007–2010 (Chepfer et al., 2010) is used at locations where it indicates a larger fractional cloud cover, because the GEOPROF data set removes some clouds with tops at altitudes below 750 m. Low cloud amounts are probably underrepresented in regions of high cloud (Chepfer et al., 2008), although not as severely as with earlier satellite instruments.

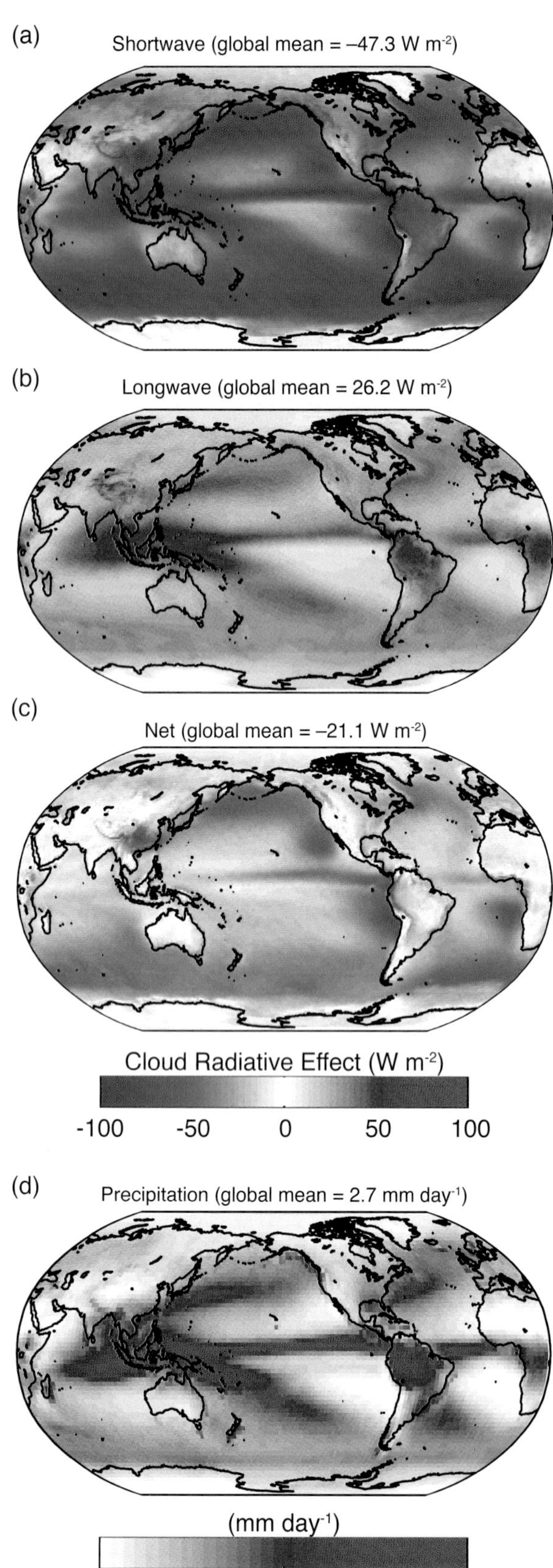

Figure 7.7 | Distribution of annual-mean top of the atmosphere (a) shortwave, (b) longwave, (c) net cloud radiative effects averaged over the period 2001–2011 from the Clouds and the Earth's Radiant Energy System (CERES) Energy Balanced and Filled (EBAF) Ed2.6r data set (Loeb et al., 2009) and (d) precipitation rate (1981–2000 average from the GPCP version 2.2 data set; Adler et al., 2003).

effect of clouds on the current climate. Owing to the large magnitudes of the SWCRE and LWCRE, clouds have the potential to cause significant climate feedback (Section 7.2.5). The sign of this feedback on climate change cannot be determined from the sign of CRE in the current climate, but depends instead on how climate-sensitive the properties are that govern the LWCRE and SWCRE.

The regional patterns of annual-mean TOA CRE (Figure 7.7a, b) reflect those of the altitude-dependent cloud distributions. High clouds, which are cold compared to the clear-sky radiating temperature, dominate patterns of LWCRE, while the SWCRE is sensitive to optically thick clouds at all altitudes. SWCRE also depends on the available sunlight, so for example is sensitive to the diurnal and seasonal cycles of cloudiness. Regions of deep, thick cloud with large positive LWCRE and large negative SWCRE tend to accompany precipitation (Figure 7.7d), showing their intimate connection with the hydrological cycle. The net CRE is negative over most of the globe and most negative in regions of very extensive low-lying reflective stratus and stratocumulus cloud such as the mid-latitude and eastern subtropical oceans, where SWCRE is strong but LWCRE is weak (Figure 7.7c). In these regions, the spatial distribution of net CRE on seasonal time scales correlates strongly with measures of low-level stability or inversion strength (Klein and Hartmann, 1993; Williams et al., 2006; Wood and Bretherton, 2006; Zhang et al., 2010).

Clouds also exert a CRE at the surface and within the troposphere, thus affecting the hydrological cycle and circulation (Section 7.6), though this aspect of CRE has received less attention. The net downward flux of radiation at the surface is sensitive to the vertical and horizontal distribution of clouds. It has been estimated more accurately through radiation budget measurements and cloud profiling (Kato et al., 2011). Based on these observations, the global mean surface downward longwave flux is about 10 W m^{-2} larger than the average in climate models, probably due to insufficient model-simulated cloud cover or lower tropospheric moisture (Stephens et al., 2012). This is consistent with a global mean precipitation rate in the real world somewhat larger than current observational estimates.

7.2.2 Cloud Process Modelling

Cloud formation processes span scales from the sub-micrometre scale of CCN, to cloud-system scales of up to thousands of kilometres. This range of scales is impossible to resolve with numerical simulations on computers, and this is not expected to change in the foreseeable future. Nonetheless progress has been made through a variety of modelling strategies, which are outlined briefly in this section, followed by a discussion in Section 7.2.3 of developments in representing clouds in global models. The implications of these discussions are synthesized in Section 7.2.3.5.

7.2.2.1 Explicit Simulations in Small Domains

High-resolution models in small domains have been widely used to simulate interactions of turbulence with various types of clouds. The grid spacing is chosen to be small enough to resolve explicitly the dominant turbulent eddies that drive cloud heterogeneity, with the effects of smaller-scale phenomena parameterized. Such models can be run in idealized settings, or with boundary conditions for specific observed cases. This strategy is typically called large-eddy simulation (LES) when boundary-layer eddies are resolved, and cloud-resolving model (CRM) when only deep cumulus motions are well resolved. It is useful not only in simulating cloud and precipitation characteristics, but also in understanding how turbulent circulations within clouds transport and process aerosols and chemical constituents. It can be applied to any type of cloud system, on any part of the Earth. Direct numerical simulation (DNS) can be used to study turbulence and cloud microphysics on scales of a few metres or less (e.g., Andrejczuk et al., 2006) but cannot span crucial meteorological scales and is not further considered here.

Cloud microphysics, precipitation and aerosol interactions are treated with varying levels of sophistication, and remain a weak point in all models regardless of resolution. For example, recent comparisons to satellite data show that liquid water clouds in CRMs generally begin to rain too early in the day (Suzuki et al., 2011). Especially for ice clouds, and for interactions between aerosols and clouds, our understanding of the basic micro-scale physics is not yet adequate, although it is improving. Moreover, microphysical effects are quite sensitive to co-variations of velocity and composition down to very small scales. High-resolution models, such as those used for LES, explicitly calculate most of these variations, and so provide much more of the information needed for microphysical calculations, whereas in a GCM they are not explicitly available. For these reasons, low-resolution (e.g., climate) models will have even more trouble representing local aerosol–cloud interactions than will high-resolution models. Parameterizations are under development that could account for the small-scale variations statistically (e.g., Larson and Golaz, 2005) but have not been used in the Coupled Model Intercomparison Project Phase 5 (CMIP5) simulations.

High-resolution models have enhanced our understanding of cloud processes in several ways. First, they can help interpret *in situ* and high-resolution remote sensing observations (e.g., Stevens et al., 2005b; Blossey et al., 2007; Fridlind et al., 2007). Second, they have revealed important influences of small-scale interactions, turbulence, entrainment and precipitation on cloud dynamics that must eventually be accounted for in parameterizations (e.g., Krueger et al., 1995; Derbyshire et al., 2004; Kuang and Bretherton, 2006; Ackerman et al., 2009). Third, they can be used to predict how cloud system properties (such as cloud cover, depth, or radiative effect) may respond to climate changes (e.g., Tompkins and Craig, 1998; Bretherton et al., 2013). Fourth, they have become an important tool in testing and improving parameterizations of cloud-controlling processes such as cumulus convection, turbulent mixing, small-scale horizontal cloud variability and aerosol–cloud interactions (Randall et al., 2003; Rio and Hourdin, 2008; Stevens and Seifert, 2008; Lock, 2009; Del Genio and Wu, 2010; Fletcher and Bretherton, 2010), as well as the interplay between convection and large-scale circulations (Kuang, 2008).

Different aspects of clouds, and cloud types, require different grid resolutions. CRMs of deep convective cloud systems with horizontal resolutions of 2 km or finer (Bryan et al., 2003) can represent some statistical properties of the cloud system, including fractional area coverage of cloud (Xu et al., 2002), vertical thermodynamic structure (Blossey et al., 2007), the distribution of updraughts and downdraughts (Khairoutdinov et al., 2009) and organization into mesoscale convective systems (Grabowski et al., 1998). Modern high-order turbulence closure schemes may allow some statistics of boundary-layer cloud distributions, including cloud fractions and fluxes of moisture and energy, to be reasonably simulated even at horizontal resolution of 1 km or larger (Cheng and Xu, 2006, 2008). Finer grids (down to hundreds of metres) better resolve individual storm characteristics such as vertical velocity or tracer transport. Some cloud ensemble properties remain sensitive to CRM microphysical parameterization assumptions regardless of resolution, particularly the vertical distribution and optical depth of clouds containing ice.

Because of these requirements, it is computationally demanding to run a CRM in a domain large enough to capture convective organisation or perform regional forecasts. Some studies have created smaller regions of CRM-like resolution within realistically forced regional-scale models (e.g., Zhu et al., 2010; Boutle and Abel, 2012; Zhu et al., 2012), a special case of the common 'nesting' approach for regional downscaling (see Section 9.6). One application has been to orographic precipitation, associated both with extratropical cyclones (e.g., Garvert et al., 2005) and with explicitly simulated cumulus convection (e.g., Hohenegger et al., 2008); better resolution of the orography improves the simulation of precipitation initiation and wind drift of falling rain and snow between watersheds.

LES of shallow cumulus cloud fields with horizontal grid spacing of about 100 m and vertical grid spacing of about 40 m produces vertical profiles of cloud fraction, temperature, moisture and turbulent fluxes that agree well with available observations (Siebesma et al., 2003), though the simulated precipitation efficiency still shows some sensitivity to microphysical parameterizations (vanZanten et al., 2011). LES of stratocumulus-topped boundary layers reproduces the turbulence statistics and vertical thermodynamic structure well (e.g., Stevens et al., 2005b; Ackerman et al., 2009), and has been used to study the sensitivity of stratocumulus properties to aerosols (e.g., Savic-Jovcic and Stevens, 2008; Xue et al., 2008) and meteorological conditions. However, the simulated entrainment rate and cloud liquid water path are sensitive to the underlying numerical algorithms, even with vertical grid spacings as small as 5 m, due to poor resolution of the sharp capping inversion (Stevens et al., 2005a).

These grid requirements mean that low-cloud processes dominating the known uncertainty in cloud feedback cannot be explicitly simulated except in very small domains. Thus, notwithstanding all of the above benefits of explicit cloud modeling, these models cannot on their own quantify global cloud feedbacks or aerosol–cloud interactions definitively. They are important, however, in suggesting and testing feedback and adjustment mechanisms (see Sections 7.2.5 and 7.4).

7.2.2.2 Global Models with Explicit Clouds

Since AR4, increasing computer power has led to three types of developments in global atmospheric models. First, models have been run with resolution that is higher than in the past, but not sufficiently high that cumulus clouds can be resolved explicitly. Second, models have been run with resolution high enough to resolve (or 'permit') large individual cumulus clouds over the entire globe. In a third approach, the parameterizations of global models have been replaced by embedded CRMs. The first approach is assessed in Chapter 9. The other two approaches are discussed below.

Global Cloud-Resolving Models (GCRMs) have been run with grid spacings as small as 3.5 km (Tomita et al., 2005; Putman and Suarez, 2011). At present GCRMs can be used only for relatively short simulations of a few simulated months to a year or two on the fastest supercomputers, but in the not-too distant future they may provide climate projections. GCRMs provide a consistent way to couple convective circulations to large-scale dynamics, but must still parameterize the effects of individual clouds, microphysics and boundary-layer circulations.

Because they avoid the use of uncertain cumulus parameterizations, GCRMs better simulate many properties of convective circulations that are very challenging for many current conventional GCMs, including the diurnal cycles of precipitation (Sato et al., 2009) and the Asian summer monsoon (Oouchi et al., 2009). Inoue et al. (2010) showed that the cloudiness simulated by a GCRM is in good agreement with observations from CloudSat and CALIPSO, but the results are sensitive to the parameterizations of turbulence and cloud microphysics (Satoh et al., 2010; Iga et al., 2011; Kodama et al., 2012).

Heterogeneous multiscale methods, in which CRMs are embedded in each grid cell of a larger scale model (Grabowski and Smolarkiewicz, 1999), have also been further developed as a way to realize some of the advantages of GCRMs but at less cost. This approach has come to be known as super-parameterization, because the CRM effectively replaces some of the existing GCM parameterizations (e.g., Khairoutdinov and Randall, 2001; Tao et al., 2009). Super-parameterized models, which are sometimes called multiscale modeling frameworks, occupy a middle ground between high-resolution 'process models' and 'climate models' (see Figure 7.8), in terms of both advantages and cost.

Like GCRMs, super-parameterized models give more realistic simulations of the diurnal cycle of precipitation (Khairoutdinov et al., 2005; Pritchard and Somerville, 2010) and the Madden-Julian Oscillation (Benedict and Randall, 2009) than most conventional GCMs; they can also improve aspects of the Asian monsoon and the El Niño–Southern Oscillation (ENSO; Stan et al., 2010; DeMott et al., 2011). Moreover, because they also begin to resolve cloud-scale circulations, both strategies provide a framework for studying aerosol–cloud interactions that conventional GCMs lack (Wang et al., 2011b). Thus both types of global model provide important insights, but because neither of them fully resolves cloud processes, especially for low clouds (see Section 7.2.2.1), their results must be treated with caution just as with conventional GCMs.

7.2.3 Parameterization of Clouds in Climate Models

7.2.3.1 Challenges of Parameterization

The representation of cloud microphysical processes in climate models is particularly challenging, in part because some of the fundamental details of these microphysical processes are poorly understood (particularly for ice- and mixed-phase clouds), and because spatial heterogeneity of key atmospheric properties occurs at scales significantly smaller than a GCM grid box. Such representation, however,

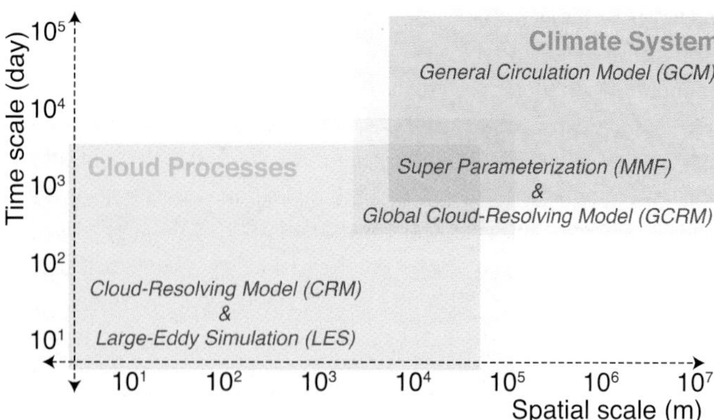

Figure 7.8 | Model and simulation strategy for representing the climate system and climate processes at different space and time scales. Also shown are the ranges of space and time scales usually associated with cloud processes (orange, lower left) and the climate system (blue, upper right). Classes of models are usually defined based on the range of spatial scales they represent, which in the figure is roughly spanned by the text for each model class. The temporal scales simulated by a particular type of model vary more widely. For instance, climate models are often run for a few time steps for diagnostic studies, or can simulate millennia. Hence the figure indicates the typical time scales for which a given model is used. Computational power prevents one model from covering all time and space scales. Since the AR4, the development of Global Cloud Resolving Models (GCRMs), and hybrid approaches such as General Circulation Models (GCMs) using the 'super-parameterization' approach (sometimes called the Multiscale Modelling Framework (MMF)), have helped fill the gap between climate system and cloud process models.

affects many aspects of a model's overall simulated climate including the Hadley circulation, precipitation patterns, and tropical variability. Therefore continuing weakness in these parameterizations affects not only modeled climate sensitivity, but also the fidelity with which these other variables can be simulated or projected.

Most CMIP5 climate model simulations use horizontal resolutions of 100 to 200 km in the atmosphere, with vertical layers varying between 100 m near the surface to more than 1000 m aloft. Within regions of this size in the real world, there is usually enormous small-scale variability in cloud properties, associated with variability in humidity, temperature and vertical motion (Figure 7.16). This variability must be accounted for to accurately simulate cloud–radiation interaction, condensation, evaporation and precipitation and other cloud processes that crucially depend on how cloud condensate is distributed across each grid box (Cahalan et al., 1994; Pincus and Klein, 2000; Larson et al., 2001; Barker et al., 2003).

The simulation of clouds in modern climate models involves several parameterizations that must work in unison. These include parameterization of turbulence, cumulus convection, microphysical processes, radiative transfer and the resulting cloud amount (including the vertical overlap between different grid levels), as well as sub-grid scale transport of aerosol and chemical species. The system of parameterizations must balance simplicity, realism, computational stability and efficiency. Many cloud processes are unrealistic in current GCMs, and as such their cloud response to climate change remains uncertain.

Cloud processes and/or turbulence parameterization are important not only for the GCMs used in climate projections but also for specialized chemistry–aerosol–climate models (see review by Zhang, 2008), for regional climate models, and indeed for the cloud process models described in Section 7.2.2 which must still parameterize small-scale and microphysical effects. The nature of the parameterization problem, however, shifts as model scale decreases. Section 7.2.3.2 briefly assesses recent developments relevant to GCMs.

7.2.3.2 Recent Advances in Representing Cloud Microphysical Processes

7.2.3.2.1 Liquid clouds

Recent development efforts have been focused on the introduction of more complex representations of microphysical processes, with the dual goals of coupling them better to atmospheric aerosols and linking them more consistently to the sub-grid variability assumed by the model for other calculations. For example, most CMIP3 climate models predicted the average cloud and rain water mass in each grid cell only at a given time, diagnosing the droplet concentration using empirical relationships based on aerosol mass (e.g., Boucher and Lohmann, 1995; Menon et al., 2002), or altitude and proximity to land. Many were forced to employ an arbitrary lower bound on droplet concentration to reduce the aerosol RF (Hoose et al., 2009). Such formulations oversimplify microphysically mediated cloud variations.

By contrast, more models participating in CMIP5 predict both mass and number mixing ratios for liquid stratiform cloud. Some determine rain and snow number concentrations and mixing ratios (e.g., Morrison and Gettelman, 2008; Salzmann et al., 2010), allowing treatment of aerosol scavenging and the radiative effect of snow. Some models explicitly treat sub-grid cloud water variability for calculating microphysical process rates (e.g., Morrison and Gettelman, 2008). Cloud droplet activation schemes now account more realistically for particle composition, mixing and size (Abdul-Razzak and Ghan, 2000; Ghan et al., 2011; Liu et al., 2012). Despite such advances in internal consistency, a continuing weakness in GCMs (and to a much lesser extent GCRMs and super-parameterized models) is their inability to fully represent turbulent motions to which microphysical processes are highly sensitive.

7.2.3.2.2 Mixed-phase and ice clouds

Ice treatments are following a path similar to those for liquid water, and face similar but greater challenges because of the greater complexity of ice processes. Many CMIP3 models predicted the condensed water amount in just two categories—cloud and precipitation—with a temperature-dependent partitioning between liquid and ice within either category. Although supersaturation with respect to ice is commonly observed at low temperatures, only one CMIP3 GCM (ECHAM) allowed ice supersaturation (Lohmann and Kärcher, 2002).

Many climate models now include separate, physically based equations for cloud liquid versus cloud ice, and for rain versus snow, allowing a more realistic treatment of mixed-phase processes and ice supersaturation (Liu et al., 2007; Tompkins et al., 2007; Gettelman et al., 2010; Salzmann et al., 2010; see also Section 7.4.4). These new schemes are tested in a single-column model against cases observed in field campaigns (e.g., Klein et al., 2009) or against satellite observations (e.g., Kay et al., 2012), and provide superior simulations of cloud structure than typical CMIP3 parameterizations (Kay et al., 2012). However new observations reveal complexities not correctly captured by even relatively advanced schemes (Ma et al., 2012a). New representations of the Wegener–Bergeron–Findeisen process in mixed-phase clouds (Storelvmo et al., 2008b; Lohmann and Hoose, 2009) compare the rate at which the pre-existing ice crystals deplete the water vapour with the condensation rate for liquid water driven by vertical updraught speed (Korolev, 2007); these are not yet included in CMIP5 models. Climate models are increasingly representing detailed microphysics, including mixed-phase processes, inside convective clouds (Fowler and Randall, 2002; Lohmann, 2008; Song and Zhang, 2011). Such processes can influence storm characteristics like strength and electrification, and are crucial for fully representing aerosol–cloud interactions, but are still not included in most climate models; their representation is moreover subject to all the caveats noted in Section 7.2.3.1.

7.2.3.3 Recent Advances in Parameterizing Moist Turbulence and Convection

Both the mean state and variability in climate models are sensitive to the parameterization of cumulus convection. Since AR4, the development of convective parameterization has been driven largely by rapidly growing use of process models, in particular LES and CRMs, to inform parameterization development (e.g., Hourdin et al., 2013).

Accounting for greater or more state-dependent entrainment of air into deep cumulus updraughts has improved simulations of the Madden–Julian Oscillation, tropical convectively coupled waves and mean rainfall patterns in some models (Bechtold et al., 2008; Song and Zhang, 2009; Chikira and Sugiyama, 2010; Hohenegger and Bretherton, 2011; Mapes and Neale, 2011; Del Genio et al., 2012; Kim et al., 2012) but usually at the expense of a degraded simulation of the mean state. In another model, revised criteria for convective initiation and parameterizations of cumulus momentum fluxes improved ENSO and tropical vertical temperature profiles (Neale et al., 2008; Richter and Rasch, 2008). Since AR4, more climate models have adopted cumulus parameterizations that diagnose the expected vertical velocity in cumulus updraughts (e.g., Del Genio et al., 2007; Park and Bretherton, 2009; Chikira and Sugiyama, 2010; Donner et al., 2011), in principle allowing more complete representations of aerosol activation, cloud microphysical evolution and gravity wave generation by the convection.

Several new parameterizations couple shallow cumulus convection more closely to moist boundary layer turbulence (Siebesma et al., 2007; Neggers, 2009; Neggers et al., 2009; Couvreux et al., 2010) including cold pools generated by nearby deep convection (Grandpeix and Lafore, 2010). Many of these efforts have led to more accurate simulations of boundary-layer cloud radiative properties and vertical structure (e.g., Park and Bretherton, 2009; Köhler et al., 2011), and have ameliorated the common problem of premature deep convective initiation over land in one CMIP5 GCM (Rio et al., 2009).

7.2.3.4 Recent Advances in Parameterizing Cloud Radiative Effects

Some models have improved representation of sub-grid scale cloud variability, which has important effects on grid-mean radiative fluxes and precipitation fluxes, for example, based on the use of probability density functions of thermodynamic variables (Sommeria and Deardorff, 1977; Watanabe et al., 2009). Stochastic approaches for radiative transfer can account for this variability in a computationally efficient way (Barker et al., 2008). New treatments of cloud overlap have been motivated by new observations (Section 7.2.1.1). Despite these advances, the CMIP5 models continue to exhibit the 'too few, too bright' low-cloud problem (Nam et al., 2012), with a systematic overestimation of cloud optical depth and underestimation of cloud cover.

7.2.3.5 Cloud Modelling Synthesis

Global climate models used in CMIP5 have improved their representation of cloud processes relative to CMIP3, but still face challenges and uncertainties, especially regarding details of small-scale variability that are crucial for aerosol–cloud interactions (see Section 7.4). Finer-scale LES and CRM models are much better able to represent this variability and are an important research tool, but still suffer from imperfect representations of aerosol and cloud microphysics and known biases. Most CRM and LES studies do not span the large space and time scales needed to fully determine the interactions among different cloud regimes and the resulting net planetary radiative effects. Thus our assessments in this chapter do not regard any model type on its own as definitive, but weigh the implications of process model studies in assessing the quantitative results of the global models.

7.2.4 Water Vapour and Lapse Rate Feedbacks

Climate feedbacks determine the sensitivity of global surface temperature to external forcing agents. Water vapour, lapse rate and cloud feedbacks each involve moist atmospheric processes closely linked to clouds, and in combination, produce most of the simulated climate feedback and most of its inter-model spread (Section 9.7). The radiative feedback from a given constituent can be quantified as its impact (other constituents remaining equal) on the TOA net downward radiative flux per degree of global surface (or near-surface) temperature increase, and may be compared with the basic 'black-body' response of -3.4 W m^{-2} °C^{-1} (Hansen et al., 1984). This definition assigns positive values to positive feedbacks, in keeping with the literature on this topic but contradictory to the conventions sometimes adopted in other climate research.

7.2.4.1 Water Vapour Response and Feedback

As pointed out in previous reports (Section 8.6.3.1 in Randall et al., 2007), physical arguments and models of all types suggest global water vapour amounts increase in a warmer climate, leading to a positive feedback via its enhanced greenhouse effect. The saturated water vapour mixing ratio (WVMR) increases nearly exponentially and very rapidly with temperature, at 6 to 10% °C^{-1} near the surface, and even more steeply aloft (up to 17% °C^{-1}) where air is colder. Mounting evidence indicates that any changes in relative humidity in warmer climates would have much less impact on specific humidity than the above increases, at least in a global and statistical sense. Hence the overall WVMR is expected to increase at a rate similar to the saturated WVMR.

Because global temperatures have been rising, the above arguments imply WVMR should be rising accordingly, and multiple observing systems indeed show this (Sections 2.5.4 and 2.5.5). A study challenging the water vapour increase (Paltridge et al., 2009) used an old reanalysis product, whose trends are contradicted by newer ones (Dessler and Davis, 2010) and by actual observations (Chapter 2). The study also reported decreasing relative humidity in data from Australian radiosondes, but more complete studies show Australia to be exceptional in this respect (Dai et al., 2011). Thus data remain consistent with the expected global feedback.

Some studies have proposed that the response of upper-level humidity to natural fluctuations in the global mean surface temperature is informative about the feedback. However, small changes to the global mean (primarily from ENSO) involve geographically heterogeneous temperature change patterns, the responses to which may be a poor analogue for global warming (Hurley and Galewsky, 2010a). Most climate models reproduce these natural responses reasonably well (Gettelman and Fu, 2008; Dessler and Wong, 2009), providing additional evidence that they at least represent the key processes.

The 'last-saturation' concept approximates the WVMR of air by its saturation value when it was last in a cloud (see Sherwood et al., 2010a for a review), which can be inferred from trajectory analysis. Studies since the AR4 using a variety of models and observations (including concentrations of water vapour isotopes) support this concept (Sherwood and Meyer, 2006; Galewsky and Hurley, 2010). The concept has clarified what determines relative humidity in the subtropical upper troposphere and placed the water vapour feedback on firmer theoretical footing by directly linking actual and saturation WVMR values (Hurley and Galewsky, 2010b). CRMs show that convection can adopt varying degrees of self-aggregation (e.g., Muller and Held, 2012), which could modify the water vapour or other feedbacks if this were climate sensitive, although observations do not suggest aggregation changes have a large net radiative effect (Tobin et al., 2012).

In a warmer climate, an upward shift of the tropopause and poleward shift of the jets and associated climate zones are expected (Sections 2.7.4 and 2.7.5) and simulated by most GCMs (Section 10.3.3). These changes account, at least qualitatively, for robust regional changes in the relative humidity simulated in warmer climate by GCMs, including decreases in the subtropical troposphere and tropical uppermost troposphere, and increases near the extratropical tropopause and high latitudes (Sherwood et al., 2010b). This pattern may be amplified, however, by non-uniform atmospheric temperature or wind changes (Hurley and Galewsky, 2010b). It is also the apparent cause of most model-predicted changes in mid- and upper-level cloudiness patterns (Wetherald and Manabe, 1980; Sherwood et al., 2010b; see also Section 7.2.5.2). Idealized CRM simulations of warming climates also show upward shifts of the humidity patterns with little change in the mean (e.g., Kuang and Hartmann, 2007; Romps, 2011).

It remains unclear whether stratospheric water vapour contributes significantly to climate feedback. Observations have shown decadal variations in stratospheric water vapour, which may have affected the planetary radiation budget somewhat (Solomon et al., 2010) but are not clearly linked to global temperature (Section 3.4.2.4 in Trenberth et al., 2007). A strong positive feedback from stratospheric water vapour was reported in one GCM, but with parameter settings that produced an unrealistic present climate (Joshi et al., 2010).

7.2.4.2 Relationship Between Water Vapour and Lapse Rate Feedbacks

The lapse rate (decrease of temperature with altitude) should, in the tropics, change roughly as predicted by a moist adiabat, due to the strong restoring influence of convective heating. This restoring influence has now been directly inferred from satellite data (Lebsock et al., 2010), and the near-constancy of tropical atmospheric stability and deep-convective thresholds over recent decades is also now observable in SST and deep convective data (Johnson and Xie, 2010). The stronger warming of the atmosphere relative to the surface produces a negative feedback on global temperature because the warmed system radiates more thermal emission to space for a given increase in surface temperature than in the reference case where the lapse rate is fixed. This feedback varies somewhat among models because lapse rates in middle and high latitudes, which decrease less than in the tropics, do so differently among models (Dessler and Wong, 2009).

As shown by Cess (1975) and discussed in the AR4 (Randall et al., 2007), models with a more negative lapse rate feedback tend to have a more positive water vapour feedback. Cancellation between these is close enough that their sum has a 90% range in CMIP3 models of only +0.96 to +1.22 W m^{-2} °C^{-1} (based on a Gaussian fit to the data of Held and Shell (2012), see Figure 7.9) with essentially the same range in CMIP5 (Section 9.7). The physical reason for this cancellation is that as long as water vapour infrared absorption bands are nearly saturated, outgoing longwave radiation is determined by relative humidity (Ingram, 2010) which exhibits little global systematic change in any model (Section 7.2.4.1). In fact, Held and Shell (2012) and Ingram (2013a) argue that it makes more sense physically to redefine feedbacks in a different analysis framework in which relative humidity, rather than specific humidity, is the feedback variable. Analysed in that framework the inherent stabilization by the Planck response is weaker, but the water vapour and lapse rate feedbacks are also very small; thus the traditional view of large and partially compensating feedbacks has, arguably, arisen from arbitrary choices made when the analysis framework was originally set out, rather than being an intrinsic feature of climate or climate models.

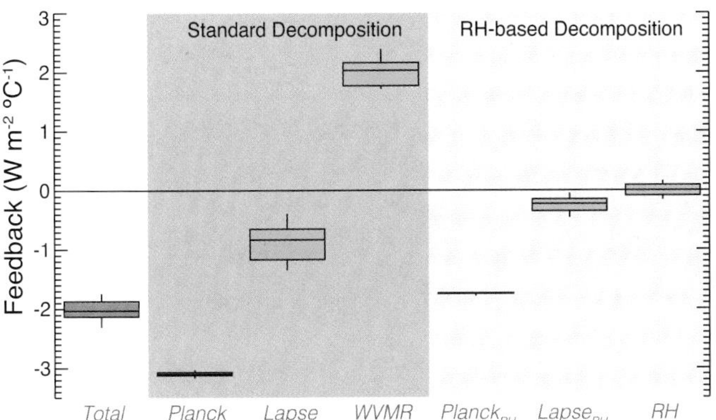

Figure 7.9 | Feedback parameters associated with water vapour or the lapse rate predicted by CMIP3 GCMs, with boxes showing interquartile range and whiskers showing extreme values. At left is shown the total radiative response including the Planck response. In the darker shaded region is shown the traditional breakdown of this into a Planck response and individual feedbacks from water vapour (labelled 'WVMR') and lapse rate (labelled 'Lapse'). In the lighter-shaded region at right are the equivalent three parameters calculated in an alternative, relative humidity-based framework. In this framework all three components are both weaker and more consistent among the models. (Data are from Held and Shell, 2012.)

There is some observational evidence (Section 2.4.4) suggesting tropical lapse rates might have increased in recent decades in a way not simulated by models (Section 9.4.1.4.2). Because the combined lapse rate and water vapour feedback depends on relative humidity change, however, the imputed lapse rate variations would have little influence on the total feedback or climate sensitivity even if they were a real warming response (Ingram, 2013b). In summary, there is increased evidence for a strong, positive feedback (measured in the traditional framework) from the combination of water vapour and lapse rate changes since AR4, with no reliable contradictory evidence.

7.2.5 Cloud Feedbacks and Rapid Adjustments to Carbon Dioxide

The dominant source of spread among GCM climate sensitivities in AR4 was due to diverging cloud feedbacks, particularly due to low clouds, and this continues to be true (Section 9.7). All global models continue to produce a near-zero to moderately strong positive net cloud feedback. Progress has been made since the AR4 in understanding the reasons for positive feedbacks in models and providing a stronger theoretical and observational basis for some mechanisms contributing to them. There has also been progress in quantifying feedbacks—including separating the effects of different cloud types, using radiative-kernel residual methods (Soden et al., 2008) and by computing cloud effects directly (e.g., Zelinka et al., 2012a)—and in distinguishing between feedback and adjustment responses (Section 7.2.5.6).

Until very recently cloud feedbacks have been diagnosed in models by differencing cloud radiative effects in doubled CO_2 and control climates, normalized by the change in global mean surface temperature. Different diagnosis methods do not always agree, and some simple methods can make positive cloud feedbacks look negative by failing to account for the nonlinear interaction between cloud and water vapour (Soden and Held, 2006). Moreover, it is now recognized that some of the cloud changes are induced directly by the atmospheric radiative effects of CO_2 independently of surface warming, and are therefore rapid adjustments rather than feedbacks (Section 7.2.5.6). Most of the published studies available for this assessment did not separate these effects, and only the total response is assessed here unless otherwise noted. It appears that the adjustments are sufficiently small in most models that general conclusions regarding feedbacks are not significantly affected.

Cloud changes cause both longwave (greenhouse warming) and shortwave (reflective cooling) effects, which combine to give the overall cloud feedback or forcing adjustment. Cloud feedback studies point to five aspects of the cloud response to climate change which are distinguished here: changes in high-level cloud altitude, effects of hydrological cycle and storm track changes on cloud systems, changes in low-level cloud amount, microphysically induced opacity (optical depth) changes and changes in high-latitude clouds. Finally, recent research on the rapid cloud adjustments to CO_2 is assessed. Feedbacks involving aerosols (Section 7.3.5) are not considered here, and the discussion focuses only on mechanisms affecting the TOA radiation budget.

7.2.5.1 Feedback Mechanisms Involving the Altitude of High-Level Cloud

A dominant contributor of positive cloud feedback in models is the increase in the height of deep convective outflows tentatively attributed in AR4 to the so-called 'fixed anvil-temperature' mechanism (Hartmann and Larson, 2002). According to this mechanism, the average outflow level from tropical deep convective systems is determined in steady state by the highest point at which water vapour cools the atmosphere significantly through infrared emission; this occurs at a particular water vapour partial pressure, therefore at a similar temperature (higher altitude) as climate warms. A positive feedback results because, since the cloud top temperature does not keep pace with that of the troposphere, its emission to space does not increase at the rate expected for the no-feedback system. This occurs at all latitudes and has long been noted in model simulations (Hansen et al., 1984; Cess et al., 1990). This mechanism, with a small modification to account for lapse rate changes, predicts roughly +0.5 W m^{-2} °C^{-1} of positive longwave feedback in GCMs (Zelinka and Hartmann, 2010), compared to an overall cloud-height feedback of +0.35 (+0.09 to +0.58) W m^{-2} °C^{-1} (Figure 7.10). Importantly, CRMs also reproduce this increase in cloud height (Tompkins and Craig, 1998; Kuang and Hartmann, 2007; Romps, 2011; Harrop and Hartmann, 2012).

On average, natural fluctuations in tropical high cloud amount exert little net TOA radiative effect in the current climate due to near-compensation between their longwave and shortwave cloud radiative effects (Harrison et al., 1990; Figure 7.7). Similar compensation can be seen in the opposing variations of these two components of the high-cloud feedback across GCMs (Figure 7.10). This might suggest that the altitude feedback could be similarly compensated. However, GCMs can reproduce the observed compensation in the present climate (Sherwood et al., 1994) without producing one under global warming. In the above-noted cloud-resolving simulations, the entire cloud field (including the typical base) moved upward, in accord with a general upward shift of tropospheric fields (Singh and O'Gorman, 2012) and with drying at levels near cloud base (Minschwaner et al., 2006; Sherwood et al., 2010b). This supports the prediction of GCMs that the altitude feedback is not compensated by an increase in high-cloud thickness or albedo.

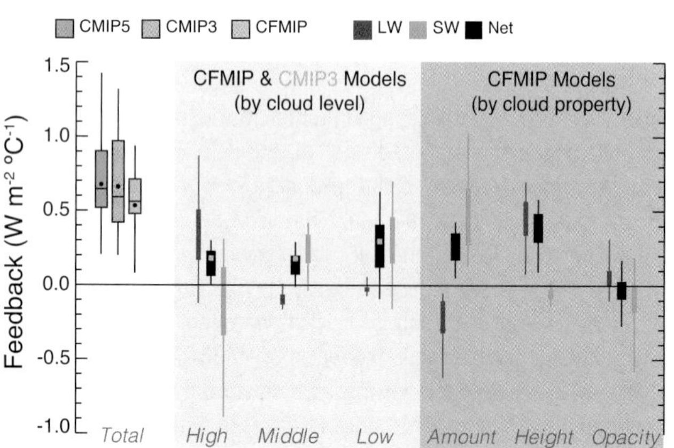

Figure 7.10 | Cloud feedback parameters as predicted by GCMs for responses to CO_2 increase including rapid adjustments. Total feedback shown at left, with centre light-shaded section showing components attributable to clouds in specific height ranges (see Section 7.2.1.1), and right dark-shaded panel those attributable to specific cloud property changes where available. The net feedback parameters are broken down in their longwave (LW) and shortwave (SW) components. Type attribution reported for CMIP3 does not conform exactly to the definition used in the Cloud Feedback Model Intercomparison Project (CFMIP) but is shown for comparison, with their 'mixed' category assigned to middle cloud. CFMIP data (original and CFMIP2) are from Zelinka et al. (2012a, 2012b; 2013); CMIP3 from Soden and Vecchi (2011); and CMIP5 from Tomassini et al. (2013).

The observational record offers limited further support for the altitude increase. The global tropopause is rising as expected (Section 2.7.4). Observed cloud heights change roughly as predicted with regional, seasonal and interannual changes in near-tropopause temperature structure (Xu et al., 2007; Eitzen et al., 2009; Chae and Sherwood, 2010; Zelinka and Hartmann, 2011), although these tests may not be good analogues for global warming. Davies and Molloy (2012) report an apparent recent downward mean cloud height trend but this is probably an artefact (Evan and Norris, 2012); observed cloud height trends do not appear sufficiently reliable to test this cloud-height feedback mechanism (Section 2.5.6).

In summary, the consistency of GCM responses, basic understanding, strong support from process models, and weak further support from observations give us *high confidence* in a positive feedback contribution from increases in high-cloud altitude.

7.2.5.2 Feedback Mechanisms Involving the Amount of Middle and High Cloud

As noted in Section 7.2.5.1, models simulate a range of nearly compensating differences in shortwave and longwave high-cloud feedbacks, consistent with different changes in high-cloud amount, but also show a net positive offset consistent with higher cloud altitude (Figure 7.10). However, there is a tendency in most GCMs toward reduced middle and high cloud amount in warmer climates in low- and mid-latitudes, especially in the subtropics (Trenberth and Fasullo, 2009; Zelinka and Hartmann, 2010). This loss of cloud amount adds a positive shortwave and negative longwave feedback to the model average, which causes the average net positive feedback to appear to come from the shortwave part of the spectrum. The net effect of changes in amount of all cloud types averaged over models is a positive feedback of about +0.2 W m^{-2} °C^{-1}, but this roughly matches the contribution from low clouds (see the following section), implying a near-cancellation of longwave and shortwave effects for the mid- and high-level amount changes.

Changes in predicted cloud cover geographically correlate with simulated subtropical drying (Meehl et al., 2007), suggesting that they are partly tied to large-scale circulation changes including the poleward shifts found in most models (Wetherald and Manabe, 1980; Sherwood et al., 2010b; Section 2.7). Bender et al. (2012) and Eastman and Warren (2013) report poleward shifts in cloud since the 1970s consistent with those reported in other observables (Section 2.5.6) and simulated by most GCMs, albeit with weaker amplitude (Yin, 2005). This shift of clouds to latitudes of weaker sunlight decreases the planetary albedo and would imply a strong positive feedback if it were due to global warming (Bender et al., 2012), although it is probably partly driven by other factors (Section 10.3). The true amount of positive feedback coming from poleward shifts therefore remains highly uncertain, but is underestimated by GCMs if, as suggested by observational comparisons, the shifts are underestimated (Johanson and Fu, 2009; Allen et al., 2012).

The upward mass flux in deep clouds should decrease in a warmer climate (Section 7.6.2), which might contribute to cloudiness decreases in storm tracks or the ITCZ (Chou and Neelin, 2004; Held and Soden, 2006). Tselioudis and Rossow (2006) predict this within the storm tracks based on observed present-day relationships with meteorological variables combined with model-simulated changes to those driving variables but do not infer a large feedback. Most CMIP3 GCMs produce too little storm-track cloud cover in the southern hemisphere compared to nearly overcast conditions in reality, but clouds are also too bright. Arguments have been advanced that such biases could imply either model overestimation or underestimation of feedbacks (Trenberth and Fasullo, 2010; Brient and Bony, 2012).

The role of thin cirrus clouds for cloud feedback is not known and remains a source of possible systematic bias. Unlike high-cloud systems overall, these particular clouds exert a clear net warming effect (Jensen et al., 1994; Chen et al., 2000), making a significant cloud-cover feedback possible in principle (e.g., Rondanelli and Lindzen, 2010). While this does not seem to be important in recent GCMs (Zelinka et al., 2012b), and no specific mechanism has been suggested, the representation of cirrus in GCMs appears to be poor (Eliasson et al., 2011) and such clouds are microphysically complex (Section 7.4.4). This implies significant feedback uncertainty in addition to that already evident from model spread.

Model simulations, physical understanding and observations thus provide *medium confidence* that poleward shifts of cloud distributions will contribute to positive feedback, but by an uncertain amount. Feedbacks from thin cirrus amount cannot be ruled out and are an important source of uncertainty.

7.2.5.3 Feedback Mechanisms Involving Low Cloud

Differences in the response of low clouds to a warming are responsible for most of the spread in model-based estimates of equilibrium climate sensitivity (Randall et al., 2007). Since the AR4 this finding has withstood further scrutiny (e.g., Soden and Vecchi, 2011; Webb et al., 2013), holds in CMIP5 models (Vial et al., 2013) and has been shown to apply also to the transient climate response (e.g., Dufresne and Bony, 2008). This discrepancy in responses occurs over most oceans and cannot be clearly confined to any single region (Trenberth and Fasullo, 2010; Webb et al., 2013), but is usually associated with the representation of shallow cumulus or stratocumulus clouds (Williams and Tselioudis, 2007; Williams and Webb, 2009; Xu et al., 2010). Because the spread of responses emerges in a variety of idealized model formulations (Medeiros et al., 2008; Zhang and Bretherton, 2008; Brient and Bony, 2013), or conditioned on a particular dynamical state (Bony et al., 2004), and is similar in equilibrium or transient simulations (Yokohata et al., 2008), it appears to be attributable to how cloud, convective and boundary layer processes are parameterized in GCMs.

The modelled response of low clouds does not appear to be dominated by a single feedback mechanism, but rather the net effect of several potentially competing mechanisms as elucidated in LES and GCM sensitivity studies (e.g., Zhang and Bretherton, 2008; Blossey et al., 2013; Bretherton et al., 2013). Starting with some proposed negative feedback mechanisms, it has been argued that in a warmer climate, low clouds will be: (1) horizontally more extensive, because changes in the lapse rate of temperature also modify the lower-tropospheric stability (Miller, 1997); (2) optically thicker, because adiabatic ascent is accompanied by a larger condensation rate (Somerville and Remer, 1984); and (3) vertically more extensive, in response to a weakening of the tropical overturning circulation (Caldwell and Bretherton, 2009). While these mechanisms may play some role in subtropical low cloud feedbacks, none of them appears dominant. Regarding (1), dry static stability alone is a misleading predictor with respect to climate changes, as models with comparably good simulations of the current regional distribution and/or relationship to stability of low cloud can produce a broad range of cloud responses to climate perturbations (Wyant et al., 2006). Mechanism (2), discussed briefly in the next section, appears to have a small effect. Mechanism (3) cannot yet be ruled out but does not appear to be the dominant factor in determining subtropical cloud changes in GCMs (Bony and Dufresne, 2005; Zhang and Bretherton, 2008).

Since the AR4, several new positive feedback mechanisms have been proposed, most associated with the marine boundary layer clouds thought to be at the core of the spread in responses. These include the ideas that: warming-induced changes in the absolute humidity lapse rate change the energetics of mixing in ways that demand a reduction in cloud amount or thickness (Webb and Lock, 2013; Bretherton et al., 2013; Brient and Bony, 2013); energetic constraints prevent the surface evaporation from increasing with warming at a rate sufficient to balance expected changes in dry air entrainment, thereby reducing the supply of moisture to form clouds (Rieck et al., 2012; Webb and Lock, 2013); and that increased concentrations of GHGs reduce the radiative cooling that drives stratiform cloud layers and thereby the cloud amount (Caldwell and Bretherton, 2009; Stevens and Brenguier, 2009; Bretherton et al., 2013). These mechanisms, crudely operating through parameterized representations of cloud processes, could explain why climate models consistently produce positive low-cloud feedbacks. Among CFMIP GCMs, the low-cloud feedback ranges from –0.09 to +0.63 W m^{-2} $°C^{-1}$ (Figure 7.10), and is largely associated with a reduction in low-cloud amount, albeit with considerable spatial variability (e.g., Webb et al., 2013). One 'super-parameterized' GCM (Section 7.2.2.2) simulates a negative low-cloud feedback (Wyant et al., 2006, 2009), but that model's representation of low clouds was worse than some conventional GCMs.

The tendency of both GCMs and process models to produce these positive feedback effects suggests that the feedback contribution from changes in low clouds is positive. However, deficient representation of low clouds in GCMs, diverse model results, a lack of reliable observational constraints, and the tentative nature of the suggested mechanisms leave us with *low confidence* in the sign of the low-cloud feedback contribution.

7.2.5.4 Feedbacks Involving Changes in Cloud Opacity

It has long been suggested that cloud water content could increase in a warmer climate simply due to the availability of more vapour for condensation in a warmer atmosphere, yielding a negative feedback (Paltridge, 1980; Somerville and Remer, 1984), but this argument ignores the physics of crucial cloud-regulating processes such as precipitation formation and turbulence. Observational evidence discounting a large effect of this kind was reported in AR4 (Randall et al., 2007).

The global mean net feedback from cloud opacity changes in CFMIP models (Figure 7.10) is approximately zero. Optical depths tend to reduce slightly at low and middle latitudes, but increase poleward of 50°, yielding a positive longwave feedback that roughly offsets the negative shortwave feedback. These latitude-dependent opacity changes may be attributed to phase changes at high latitudes and greater poleward moisture transport (Vavrus et al., 2009), and possibly to poleward shifts of the circulation.

Studies have reported warming-related changes in cloud opacity tied to cloud phase (e.g., Senior and Mitchell, 1993; Tsushima et al., 2006). This might be expected to cause negative feedback, because at mixed-phase temperatures of –38 to 0°C, cloud ice particles have typical diameters of 10 to 100 μm (e.g., Figure 8 in Donovan, 2003), several-fold larger than cloud water drops, so a given mass of cloud water would have less surface area and reflect less sunlight in ice form than in liquid form. As climate warms, a shift from ice to liquid at a given location could increase cloud opacity. An offsetting factor that may explain the absence of this in CFMIP, however, is that mixed-phase clouds may form at higher altitudes, and similar local temperatures, in warmer climates (Section 7.2.5.1). The key physics is in any case not adequately represented in climate models. Thus this particular feedback mechanism is highly uncertain.

7.2.5.5 Feedback from Arctic Cloud Interactions with Sea Ice

Arctic clouds, despite their low altitude, have a net heating effect at the surface in the present climate because their downward emission of infrared radiation over the year outweighs their reflection of sunlight during the short summer season. Their net effect on the atmosphere is cooling, however, so their effect on the energy balance of the whole system is ambiguous and depends on the details of the vertical cloud distribution and the impact of cloud radiative interactions on ice cover (Palm et al., 2010).

Low-cloud amount over the Arctic oceans varies inversely with sea ice amount (open water producing more cloud) as now confirmed since AR4 by visual cloud reports (Eastman and Warren, 2010) and lidar and radar observations (Kay and Gettelman, 2009; Palm et al., 2010). The observed effect is weak in boreal summer, when the melting sea ice is at a similar temperature to open water and stable boundary layers with extensive low cloud are common over both surfaces, and strongest in boreal autumn when cold air flowing over regions of open water stimulates cloud formation by boundary-layer convection (Kay and Gettelman, 2009; Vavrus et al., 2011). Kay et al. (2011) show that a GCM can represent this seasonal sensitivity of low cloud to open water, but doing so depends on the details of how boundary-layer clouds are parameterized. Vavrus et al. (2009) show that in a global warming scenario, GCMs simulate more Arctic low cloud in all seasons, but especially during autumn and the onset of winter when open water and very thin sea ice increase considerably, increasing upward moisture transport to the clouds.

A few studies in the literature suggest negative feedbacks from Arctic clouds, based on spatial correlations of observed warming and cloudiness (Liu et al., 2008) or tree-ring proxies of cloud shortwave effects over the last millenium (Gagen et al., 2011). However, the spatial correlations are not reliable indicators of feedback (Section 7.2.5.7), and the tree-ring evidence (assuming it is a good proxy) applies only to the shortwave effect of summertime cloud cover. The GCM studies would be consistent with warmer climates being cloudier, but have opposite radiative effects and positive feedback during the rest of the year. Even though a small positive feedback is suggested by models, there is overall little evidence for significant feedbacks from Arctic cloud.

7.2.5.6 Rapid Adjustments of Clouds to a Carbon Dioxide Change

It is possible to partition the response of TOA radiation in GCMs to an instantaneous doubling of CO_2 into a 'rapid adjustment' in which the land surface, atmospheric circulations and clouds respond to the radiative effect of the CO_2 increase, and an 'SST-mediated' response that develops more slowly as the oceans warm (see Section 7.1.3). This distinction is important not only to help understand model processes, but because the presence of rapid adjustments would cause clouds to respond slightly differently to a transient climate change (in which SST changes have not caught up to CO_2 changes) or to a climate change caused by other forcings, than they would to the same warming at equilibrium driven by CO_2. There is also a rapid adjustment of the hydrological cycle and precipitation field, discussed in Section 7.6.3.

Gregory and Webb (2008) reported that in some climate models, rapid adjustment of clouds can have TOA radiative effects comparable to those of the ensuing SST-mediated cloud changes, though Andrews and Forster (2008) found a smaller effect. Subsequent studies using more accurate kernel-based techniques find a cloud adjustment of roughly +0.4 to +0.5 W m^{-2} per doubling of CO_2, with standard deviation of about 0.3 W m^{-2} across models (Vial et al., 2013; Zelinka et al., 2013). This would account for about 20% of the overall cloud response in a model with average sensitivity; and because it is not strongly correlated with model sensitivity, it contributes perhaps 20% of the inter-model response spread (Andrews et al., 2012; Webb et al., 2013), which therefore remains dominated by feedbacks. The response occurs due to a general decrease in cloud cover caused by the slight stratification driven by CO_2 warming of the troposphere, which especially for middle and low clouds has a net warming effect (Colman and McAvaney, 2011; Zelinka et al., 2013). As explained at the beginning of Section 7.2.5, feedback numbers given in this report already account for these rapid adjustments.

7.2.5.7 Observational Constraints on Global Cloud Feedback

A number of studies since AR4 have attempted to constrain overall cloud feedback (or climate sensitivity) from observations of natural climate variability; here we discuss those using modern cloud, radiation or other measurements (see a complementary discussion in Section 12.5 based on past temperature data and forcing proxies).

One approach is to seek observable aspects of present-day cloud behaviour that reveal cloud feedback or some component thereof. Varying parameters in a GCM sometimes produces changes in cloud feedback that correlate with the properties of cloud simulated for the present day, but this depends on the GCM (Yokohata et al., 2010; Gettelman et al., 2013), and the resulting relationships do not hold across multiple models such as those from CMIP3 (Collins et al., 2011). Among the AR4 models, net cloud feedback correlates strongly with mid-latitude relative humidity (Volodin, 2008), with TOA radiation at high southern latitudes (Trenberth and Fasullo, 2010), and with humidity at certain latitudes during boreal summer (Fasullo and Trenberth, 2012); if valid each of these regression relations would imply a relatively strong positive cloud feedback in reality, but no mechanism has been proposed to explain or validate them and such apparent skill can arise fortuitously (Klocke et al., 2011). Likewise, Clement et al. (2009) found realistic decadal variations of cloud cover over the North Pacific in only one model (HadGEM1) and argued that the relatively strong cloud feedback in this model should therefore be regarded as more likely, but this finding lacks a mechanistic explanation and may depend on how model output is used (Broccoli and Klein, 2010). Chang and Coakley (2007) examined mid-latitude maritime clouds and found cloud thinning with increasing temperature, consistent with a positive feedback, whereas Gordon and Norris (2010) found the opposite result following a methodology that tried to isolate thermal and advective effects. In summary, there is no evidence of a robust link between any of the noted observables and the global feedback, though some apparent connections are tantalizing and are being studied further.

Several studies have attempted to derive global climate sensitivity from interannual relationships between global mean observations of TOA radiation and surface temperature (see also Section 10.8.2.2). One problem with this is the different spatial character of interannual and long-term warming; another is that the methodology can be confounded by cloud variations not caused by those of surface temperature (Spencer and Braswell, 2008). A range of climate sensitivities has been inferred based on such analyses (Forster and Gregory, 2006; Lindzen and Choi, 2011). Crucially, however, among different GCMs there is no correlation between the interannual and long-term cloud–temperature relationships (Dessler, 2010; Colman and Hanson, 2012), contradicting the basic assumption of these methods. Many but not all atmosphere–ocean GCMs predict relationships that are consistent with observations (Dessler, 2010, 2013). More recently there is interest in relating the time-lagged correlations of cloud and temperature to feedback processes (Spencer and Braswell, 2010) but again these relationships appear to reveal only a model's ability to simulate ENSO or other modes of interannual variability properly, which are not obviously informative about the cloud feedback on long-term global warming (Dessler, 2011).

For a putative observational constraint on climate sensitivity to be accepted, it should have a sound physical basis and its assumptions should be tested appropriately in climate models. No method yet proposed meets both conditions. Moreover, cloud responses to warming can be sensitive to relatively subtle details in the geographic warming pattern, such as the slight hemispheric asymmetry due to the lag of southern ocean warming relative to northern latitudes (Senior and Mitchell, 2000; Yokohata et al., 2008). Cloud responses to specified uniform ocean warming without CO_2 increases are not the same as those to CO_2-induced global warming simulated with more realistic oceans (Ringer et al., 2006), partly because of rapid adjustments (Section 7.2.5.6) and because low clouds also feed back tightly to the underlying surface (Caldwell and Bretherton, 2009). Simulated cloud feedbacks also differ significantly between colder and warmer climates in some models (Crucifix, 2006; Yoshimori et al., 2009) and between volcanic and other forcings (Yokohata et al., 2005). These sensitivities highlight the challenges facing any attempt to infer long-term cloud feedbacks from simple data analyses.

7.2.6 Feedback Synthesis

Together, the water vapour, lapse rate and cloud feedbacks are the principal determinants of equilibrium climate sensitivity. The water vapour and lapse rate feedbacks, as traditionally defined, should be thought of as a single phenomenon rather than in isolation (see Section 7.2.4.2). To estimate a 90% probability range for that feedback, we double the variance of GCM results about the mean to account for the possibility of errors common to all models, to arrive at +1.1 (+0.9 to +1.3) W m^{-2} °C^{-1}. Values in this range are supported by a steadily growing body of observational evidence, model tests, and physical reasoning. As a corollary, the net feedback from water vapour and lapse rate changes combined is *extremely likely* positive, allowing for the possibility of deep uncertainties or a fat-tailed error distribution. Key aspects of the responses of water vapour and clouds to climate warming now appear to be constrained by large-scale dynamical mechanisms that are not sensitive to poorly represented small-scale processes, and as such, are more credible. This feedback is thus known to be positive with *high confidence*, and contributes only a small part of the spread in GCM climate sensitivity (Section 9.7). An alternative framework has recently been proposed in which these feedbacks, and stabilization via thermal emission, are all significantly smaller and more consistent among models; thus the range given above may overstate the true uncertainty.

Several cloud feedback mechanisms now appear consistently in GCMs, summarized in Figure 7.11, most supported by other lines of evidence. Nearly all act in a positive direction. First, high clouds are expected to rise in altitude and thereby exert a stronger greenhouse effect in warmer climates. This altitude feedback mechanism is well understood, has theoretical and observational support, occurs consistently in GCMs and CRMs and explains about half of the mean positive cloud feedback in GCMs. Second, middle and high-level cloud cover tends to decrease in warmer climates even within the ITCZ, although the feedback effect of this is ambiguous and it cannot yet be tested observationally. Third, observations and most models suggest storm tracks shift poleward in a warmer climate, drying the subtropics and moistening the high latitudes, which causes further positive feedback via a net shift of cloud cover to latitudes that receive less sunshine. Finally, most GCMs also

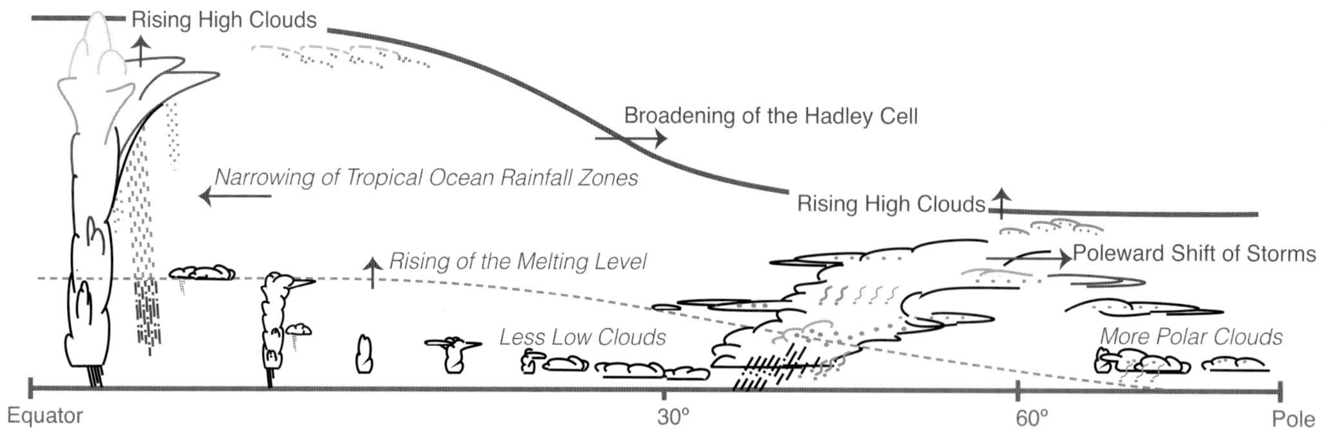

Figure 7.11 | Robust cloud responses to greenhouse warming (those simulated by most models and possessing some kind of independent support or understanding). The tropopause and melting level are shown by the thick solid and thin grey dashed lines, respectively. Changes anticipated in a warmer climate are shown by arrows, with red colour indicating those making a robust positive feedback contribution and grey indicating those where the feedback contribution is small and/or highly uncertain. No robust mechanisms contribute negative feedback. Changes include rising high cloud tops and melting level, and increased polar cloud cover and/or optical thickness (*high confidence*); broadening of the Hadley Cell and/or poleward migration of storm tracks, and narrowing of rainfall zones such as the Intertropical Convergence Zone (*medium confidence*); and reduced low-cloud amount and/or optical thickness (*low confidence*). Confidence assessments are based on degree of GCM consensus, strength of independent lines of evidence from observations or process models and degree of basic understanding.

predict that low cloud amount decreases, especially in the subtropics, another source of positive feedback though one that differs significantly among models and lacks a well-accepted theoretical basis. Over middle and high latitudes, GCMs suggest warming-induced transitions from ice to water clouds may cause clouds to become more opaque, but this appears to have a small systematic net radiative effect in models, possibly because it is offset by cloud altitude changes.

Currently, neither cloud process models (CRMs and LES) nor observations provide clear evidence to contradict or confirm feedback mechanisms involving low clouds. In some cases these models show stronger low-cloud feedbacks than GCMs, but each model type has limitations, and some studies suggest stronger positive feedbacks are more realistic (Section 7.2.5.7). Cloud process models suggest a variety of potentially opposing response mechanisms that may account for the current spread of GCM feedbacks. In summary we find no evidence to contradict either the cloud or water vapour–lapse rate feedback ranges shown by current GCMs, although the many uncertainties mean that true feedback could still lie outside these ranges. In particular, microphysical mechanisms affecting cloud opacity or cirrus amount may well be missing from GCMs. Missing feedbacks, if any, could act in either direction.

Based on the preceding synthesis of cloud behaviour, the net radiative feedback due to all cloud types is judged *likely* to be positive. This is reasoned probabilistically as follows. First, because evidence from observations and process models is mixed as to whether GCM cloud feedback is too strong or too weak overall, and because the positive feedback found in GCMs comes mostly from mechanisms now supported by other lines of evidence, the central (most likely) estimate of the total cloud feedback is taken as the mean from GCMs (+0.6 W m^{-2} $°C^{-1}$). Second, because there is no accepted basis to discredit individual GCMs *a priori*, the probability distribution of the true feedback cannot be any narrower than the distribution of GCM results. Third, since feedback mechanisms are probably missing from GCMs and some CRMs suggest feedbacks outside the range in GCMs, the probable range of the feedback must be broader than its spread in GCMs. We estimate a probability distribution for this feedback by doubling the spread about the mean of all model values in Figure 7.10 (in effect assuming an additional uncertainty about 1.7 times as large as that encapsulated in the GCM range, added to it in quadrature). This yields a 90% (*very likely*) range of –0.2 to +2.0 W m^{-2} $°C^{-1}$, with a 17% probability of a negative feedback.

Note that the assessment of feedbacks in this chapter is independent of constraints on climate sensitivity from observed trends or palaeoclimate information discussed in Box 12.2.

7.2.7 Anthropogenic Sources of Moisture and Cloudiness

Human activity can be a source of additional cloudiness through specific processes involving a source of water vapour in the atmosphere. We discuss here the impact of aviation and irrigation on water vapour and cloudiness. The impact of water vapour sources from combustion at the Earth's surface is thought to be negligible. Changes to the hydrological cycle because of land use change are briefly discussed in Section 12.4.8.

7.2.7.1 Contrails and Contrail-Induced Cirrus

Aviation jet engines emit hot moist air, which can form line shaped persistent condensation trails (contrails) in environments that are supersaturated with respect to ice and colder than about –40°C. The contrails are composed of ice crystals that are typically smaller than those of background cirrus (Heymsfield et al., 2010; Frömming et al., 2011). Their effect on longwave radiation dominates over their shortwave effect (Stuber and Forster, 2007; Rap et al., 2010b; Burkhardt and Kärcher, 2011) but models disagree on the relative importance of the two effects. Contrails have been observed to spread into large cirrus sheets that may persist for several hours, and observational studies confirm their overall positive net RF impact (Haywood et al., 2009).

Frequently Asked Questions

FAQ 7.1 | How Do Clouds Affect Climate and Climate Change?

Clouds strongly affect the current climate, but observations alone cannot yet tell us how they will affect a future, warmer climate. Comprehensive prediction of changes in cloudiness requires a global climate model. Such models simulate cloud fields that roughly resemble those observed, but important errors and uncertainties remain. Different climate models produce different projections of how clouds will change in a warmer climate. Based on all available evidence, it seems likely that the net cloud–climate feedback amplifies global warming. If so, the strength of this amplification remains uncertain.

Since the 1970s, scientists have recognized the critical importance of clouds for the climate system, and for climate change. Clouds affect the climate system in a variety of ways. They produce precipitation (rain and snow) that is necessary for most life on land. They warm the atmosphere as water vapour condenses. Although some of the condensed water re-evaporates, the precipitation that reaches the surface represents a net warming of the air. Clouds strongly affect the flows of both sunlight (warming the planet) and infrared light (cooling the planet as it is radiated to space) through the atmosphere. Finally, clouds contain powerful updraughts that can rapidly carry air from near the surface to great heights. The updraughts carry energy, moisture, momentum, trace gases, and aerosol particles. For decades, climate scientists have been using both observations and models to study how clouds change with the daily weather, with the seasonal cycle, and with year-to-year changes such as those associated with El Niño.

All cloud processes have the potential to change as the climate state changes. Cloud feedbacks are of intense interest in the context of climate change. Any change in a cloud process that is caused by climate change—and in turn influences climate—represents a cloud–climate feedback. Because clouds interact so strongly with both sunlight and infrared light, small changes in cloudiness can have a potent effect on the climate system.

Many possible types of cloud–climate feedbacks have been suggested, involving changes in cloud amount, cloud-top height and/or cloud reflectivity (see FAQ7.1, Figure 1). The literature shows consistently that high clouds amplify global warming as they interact with infrared light emitted by the atmosphere and surface. There is more uncertainty, however, about the feedbacks associated with low-altitude clouds, and about cloud feedbacks associated with amount and reflectivity in general.

Thick high clouds efficiently reflect sunlight, and both thick and thin high clouds strongly reduce the amount of infrared light that the atmosphere and surface emit to space. The compensation between these two effects makes

(continued on next page)

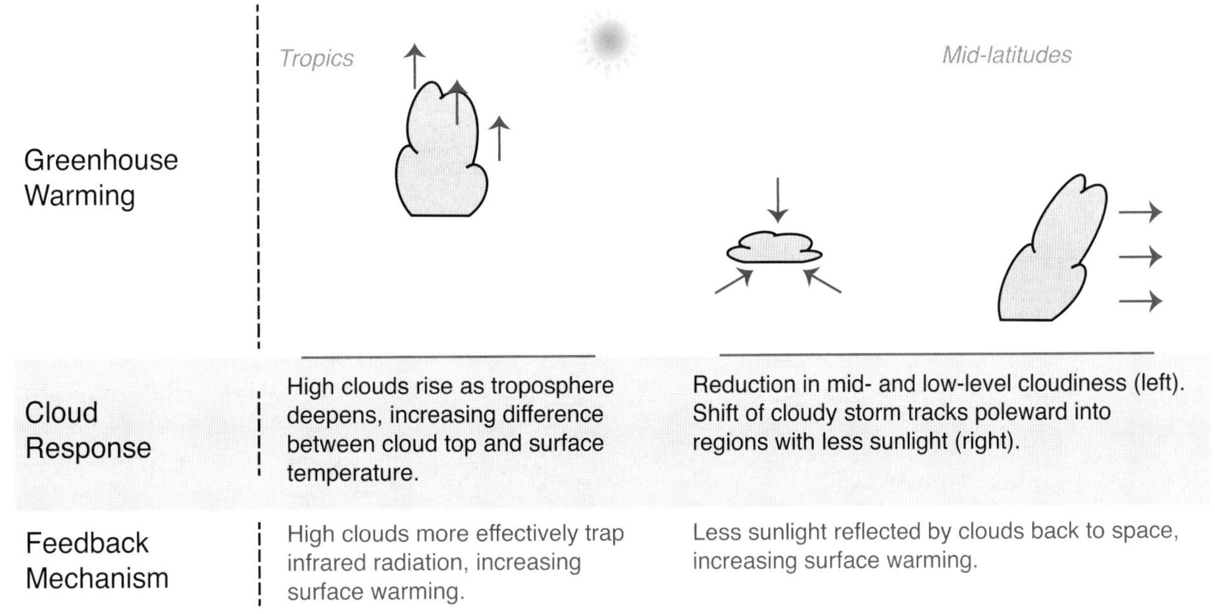

FAQ 7.1, Figure 1 | Schematic of important cloud feedback mechanisms.

FAQ 7.1 (continued)

the surface temperature somewhat less sensitive to changes in high cloud amount than to changes in low cloud amount. This compensation could be disturbed if there were a systematic shift from thick high cloud to thin cirrus cloud or vice versa; while this possibility cannot be ruled out, it is not currently supported by any evidence. On the other hand, changes in the altitude of high clouds (for a given high-cloud amount) can strongly affect surface temperature. An upward shift in high clouds reduces the infrared light that the surface and atmosphere emit to space, but has little effect on the reflected sunlight. There is strong evidence of such a shift in a warmer climate. This amplifies global warming by preventing some of the additional infrared light emitted by the atmosphere and surface from leaving the climate system.

Low clouds reflect a lot of sunlight back to space but, for a given state of the atmosphere and surface, they have only a weak effect on the infrared light that is emitted to space by the Earth. As a result, they have a net cooling effect on the present climate; to a lesser extent, the same holds for mid-level clouds. In a future climate warmed by increasing greenhouse gases, most IPCC-assessed climate models simulate a decrease in low and mid-level cloud amount, which would increase the absorption of sunlight and so tend to increase the warming. The extent of this decrease is quite model-dependent, however.

There are also other ways that clouds may change in a warmer climate. Changes in wind patterns and storm tracks could affect the regional and seasonal patterns of cloudiness and precipitation. Some studies suggest that the signal of one such trend seen in climate models—a poleward migration of the clouds associated with mid-latitude storm tracks—is already detectable in the observational record. By shifting clouds into regions receiving less sunlight, this could also amplify global warming. More clouds may be made of liquid drops, which are small but numerous and reflect more sunlight back to space than a cloud composed of the same mass of larger ice crystals. Thin cirrus cloud, which exerts a net warming effect and is very hard for climate models to simulate, could change in ways not simulated by models although there is no evidence for this. Other processes may be regionally important, for example, interactions between clouds and the surface can change over the ocean where sea ice melts, and over land where plant transpiration is reduced.

There is as yet no broadly accepted way to infer global cloud feedbacks from observations of long-term cloud trends or shorter-time scale variability. Nevertheless, all the models used for the current assessment (and the preceding two IPCC assessments) produce net cloud feedbacks that either enhance anthropogenic greenhouse warming or have little overall effect. Feedbacks are not 'put into' the models, but emerge from the functioning of the clouds in the simulated atmosphere and their effects on the flows and transformations of energy in the climate system. The differences in the strengths of the cloud feedbacks produced by the various models largely account for the different sensitivities of the models to changes in greenhouse gas concentrations.

Aerosol emitted within the aircraft exhaust may also affect high-level cloudiness. This last effect is classified as an aerosol–cloud interaction and is deemed too uncertain to be further assessed here (see also Section 7.4.4). Climate model experiments (Rap et al., 2010a) confirm earlier results (Kalkstein and Balling Jr, 2004; Ponater et al., 2005) that aviation contrails do not have, at current levels of coverage, an observable effect on the mean or diurnal range of surface temperature (*medium confidence*).

Estimates of the RF from persistent (linear) contrails often correspond to different years and need to be corrected for the continuous increase in air traffic. More recent estimates tend to indicate somewhat smaller RF than assessed in the AR4 (see Table 7.SM.1 and text in Supplementary Material). We adopt an RF estimate of +0.01 (+0.005 to +0.03) W m^{-2} for persistent (linear) contrails for 2011, with a *medium confidence* attached to this estimate. An additional RF of +0.003 W m^{-2} is due to emissions of water vapour in the stratosphere by aviation as estimated by Lee et al. (2009).

Forster et al. (2007) quoted Sausen et al. (2005) to update the 2000 forcing for aviation-induced cirrus (including linear contrails) to +0.03 (+0.01 to +0.08) W m^{-2} but did not consider this to be a best estimate because of large uncertainties. Schumann and Graf (2013) constrained their model with observations of the diurnal cycle of contrails and cirrus in a region with high air traffic relative to a region with little air traffic, and estimated a RF of +0.05 (+0.04 to +0.08) W m^{-2} for contrails and contrail-induced cirrus in 2006, but their model has a large shortwave contribution, and larger estimates are possible. An alternative approach was taken by Burkhardt and Kärcher (2011), who estimated a global RF for 2002 of +0.03 W m^{-2} from contrails and contrail cirrus within a climate model (Burkhardt and Kärcher, 2009), after compensating for reduced background cirrus cloudiness in the main traffic areas. Based on these two studies we assess the combined contrail and contrail-induced cirrus ERF for the year 2011 to be +0.05 (+0.02 to +0.15) W m^{-2} to take into uncertainties on spreading rate, optical depth, ice particle shape and radiative transfer and the ongoing increase in air traffic (see also Supplementary Material). A *low confidence* is attached to this estimate.

7.2.7.2 Irrigation-Induced Cloudiness

Boucher et al. (2004) estimated a global ERF due to water vapour from irrigation in the range of +0.03 to +0.10 W m^{-2} but the net climate effect was dominated by the evaporative cooling at the surface and by atmospheric thermal responses to low-level humidification. Regional surface cooling was confirmed by a number of more recent regional and global studies (Kueppers et al., 2007; Lobell et al., 2009). The resulting increase in water vapour may induce a small enhancement in precipitation downwind of the major irrigation areas (Puma and Cook, 2010), as well as some regional circulation patterns (Kueppers et al., 2007). Sacks et al. (2009) reported a 0.001 increase in cloud fraction over land (0.002 over irrigated land). This suggests an ERF no more negative than –0.1 W m^{-2} with *very low confidence*.

7.3 Aerosols

The section assesses the role of aerosols in the climate system, focusing on aerosol processes and properties, as well as other factors, that influence aerosol–radiation and aerosol–cloud interactions. Processes directly relevant to aerosol–cloud interactions are discussed in Section 7.4, and estimates of aerosol RFs and ERFs are assessed in Section 7.5. The time evolution of aerosols and their forcings are discussed in Chapters 2 and 8, with Chapter 8 also covering changes in natural volcanic aerosols.

7.3.1 Aerosols in the Present-Day Climate System

7.3.1.1 Aerosol Formation and Aerosol Types

Atmospheric aerosols, whether natural or anthropogenic, originate from two different pathways: emissions of primary particulate matter and formation of secondary particulate matter from gaseous precursors (Figure 7.12). The main constituents of the atmospheric aerosol are inorganic species (such as sulphate, nitrate, ammonium, sea salt), organic species (also termed organic aerosol or OA), black carbon (BC, a distinct type of carbonaceous material formed from the incomplete combustion of fossil and biomass based fuels under certain conditions), mineral species (mostly desert dust) and primary biological aerosol particles (PBAPs). Mineral dust, sea salt, BC and PBAPs are introduced into the atmosphere as primary particles, whereas non-sea-salt sulphate, nitrate and ammonium are predominantly from secondary aerosol formation processes. The OA has both significant primary and secondary sources. In the present-day atmosphere, the majority of BC, sulphate, nitrate and ammonium come from anthropogenic sources, whereas sea salt, most mineral dust and PBAPs are predominantly of natural origin. Primary and secondary organic aerosols (POA and SOA) are influenced by both natural and anthropogenic sources. Emission rates of aerosols and aerosol precursors are summarized in Table 7.1. The characteristics and role of the main aerosol species are listed in Table 7.2.

7.3.1.2 Aerosol Observations and Climatology

New and improved observational aerosol data sets have emerged since AR4. A number of field experiments have taken place such as the Intercontinental Chemical Transport Experiment (INTEX, Bergstrom et al., 2010; Logan et al., 2010), African Monsoon Multidisciplinary Analysis (AMMA; Jeong et al., 2008; Hansell et al., 2010), Integrated Campaign for Aerosols, gases and Radiation Budget (ICARB; Moorthy et al., 2008 and references therein), Megacity Impact on Regional and Global Environments field experiment (MILAGRO; Paredes-Miranda et al., 2009), Geostationary Earth Radiation Budget Inter-comparisons of Longwave and Shortwave (GERBILS, Christopher et al., 2009), Arctic Research of the Composition of the Troposphere from Aircraft and Satellites

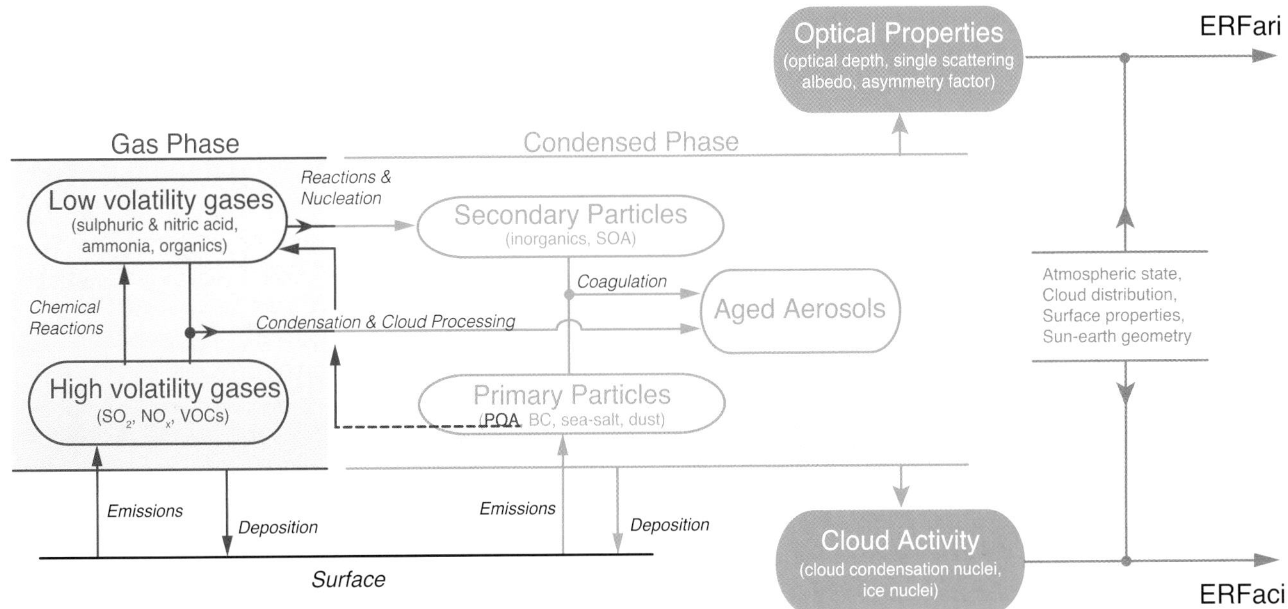

Figure 7.12 | Overview of atmospheric aerosol and environmental variables and processes influencing aerosol–radiation and aerosol–cloud interactions. Gas-phase variables and processes are highlighted in red while particulate-phase variables and processes appear in green. Although this figure shows a linear chain of processes from aerosols to forcings (ERFari and ERFaci), it is increasingly recognized that aerosols and clouds form a coupled system with two-way interactions (see Figure 7.16).

Table 7.1 | (a) Global and regional anthropogenic emissions of aerosols and aerosol precursors. The average, minimum and maximum values are from a range of available inventories (Cao et al., 2006; European Commission et al., 2009; Sofiev et al., 2009; Lu et al., 2010, 2011; Granier et al., 2011 and references therein; Knorr et al., 2012). It should be noted that the minimum to maximum range is not a measure of uncertainties which are often difficult to quantify. Units are Tg yr^{-1} and TgS yr^{-1} for sulphur dioxide (SO_2). NMVOCs stand for non-methane volatile organic compounds. (b) Global natural emissions of aerosols and aerosol precursors. Dust and sea-spray estimates span the range in the historical CMIP5 simulations. The ranges for monoterpenes and isoprene are from Arneth et al. (2008). There are other biogenic volatile organic compounds (BVOCs) such as sesquiterpenes, alcohols and aldehydes which are not listed here. Marine primary organic aerosol (POA) and terrestrial primary biological aerosol particle (PBAP) emission ranges are from Gantt et al. (2011) and Burrows et al. (2009), respectively. Note that emission fluxes from mineral dust, sea spray and terrestrial PBAPs are highly sensitive to the cut-off radius. The conversion rate of BVOCs to secondary organic aerosol (SOA) is also indicated using the range from Spracklen et al. (2011) and a lower bound from Kanakidou et al. (2005). Units are Tg yr^{-1} except for BVOCs (monoterpenes and isoprene), in TgC yr^{-1}, and dimethysulphide (DMS), in TgS yr^{-1}.

(a)

Year 2000 Emissions Tg yr^{-1} or TgS yr^{-1}	Anthropogenic NMVOCs			Anthropogenic Black Carbon			Anthropogenic POA			Anthropogenic SO_2			Anthropogenic NH_3			Biomass Burning Aerosols		
	Avg	Min	Max	Avg	Min	Max	Avg	Min	Max	Avg	Min	Max	Avg	Min	Max	Avg	Min	Max
Total	126.9	98.2	157.9	4.8	3.6	6.0	10.5	6.3	15.3	55.2	43.3	77.9	41.6	34.5	49.6	49.1	29.0	85.3
Western Europe	11.0	9.2	14.3	0.4	0.3	0.4	0.4	0.3	0.4	4.0	3.0	7.0	4.2	3.4	4.5	0.4	0.1	0.8
Central Europe	2.9	2.3	3.5	0.1	0.1	0.2	0.3	0.2	0.4	3.0	2.3	5.0	1.2	1.1	1.2	0.3	0.1	0.4
Former Soviet Union	9.8	6.5	15.2	0.3	0.2	0.4	0.7	0.5	0.9	5.2	3.0	7.0	1.7	1.5	2.0	5.4	3.0	7.9
Middle East	13.0	9.9	15.0	0.1	0.1	0.2	0.2	0.2	0.3	3.6	3.2	4.1	1.4	1.4	1.4	0.3	0.0	1.3
North America	17.8	14.5	20.9	0.4	0.3	0.4	0.5	0.4	0.6	8.7	7.8	10.4	4.6	3.8	5.5	2.0	0.8	4.4
Central America	3.8	2.9	4.4	0.1	0.1	0.1	0.3	0.2	0.3	2.1	1.9	2.8	1.1	1.1	1.2	1.44	0.3	2.7
South America	8.6	8.2	9.2	0.3	0.2	0.3	0.6	0.3	0.8	2.5	1.9	3.6	3.4	3.4	3.5	5.9	2.6	10.9
Africa	13.2	9.9	15.0	0.5	0.4	0.6	1.4	1.0	1.9	3.1	2.6	4.4	2.4	2.3	2.4	23.9	18.5	35.3
China	16.4	11.5	24.5	1.2	0.7	1.5	2.4	1.1	3.1	11.7	9.6	17.0	10.9	8.9	12.7	1.1	0.3	2.3
India	8.9	7.3	10.8	0.7	0.5	1.0	1.9	1.0	3.3	2.9	2.6	3.9	5.8	3.7	8.5	0.5	0.1	0.9
Rest of Asia	18.1	14.1	23.9	0.6	0.5	0.7	1.7	0.8	3.0	3.9	2.2	5.7	4.1	3.2	5.9	2.0	0.4	3.4
Oceania	1.2	1.0	1.5	0.03	0.03	0.04	0.05	0.04	0.08	1.2	0.9	1.4	0.7	0.7	0.7	5.8	2.7	16.8
International Shipping	2.1	1.3	3.0	0.1	0.1	0.1	0.1	0.1	0.1	3.3	2.1	5.5						

(b)

Source	Natural Global	
	Min	**Max**
Sea spray	1400	6800
including marine POA	2	20
Mineral dust	1000	4000
Terrestrial PBAPs	50	1000
including spores		28
Dimethylsulphide (DMS)	10	40
Monoterpenes	30	120
Isoprene	410	600
SOA production from all BVOCs	20	380

(ARCTAS, Lyapustin et al., 2010), the Amazonian Aerosol Characterization Experiment 2008 (AMAZE-08, Martin et al., 2010b), the Integrated project on Aerosol Cloud Climate and Air Quality interactions (EUCAARI, Kulmala et al., 2011) and Atmospheric Brown Clouds (ABC, Nakajima et al., 2007), which have improved our understanding of regional aerosol properties.

Long-term aerosol mass concentrations are also measured more systematically at the surface by global and regional networks (see Section 2.2.3), and there are institutional efforts to improve the coordination and quality assurance of the measurements (e.g., GAW, 2011). A survey of the main aerosol types can be constructed from such measurements (e.g., Jimenez et al., 2009; Zhang et al., 2012b; Figure 7.13). Such analyses show a wide spatial variability in aerosol mass concentration, dominant aerosol type, and aerosol composition. Mineral dust dominates the aerosol mass over some continental regions with relatively higher concentrations especially in urban South Asia and China, accounting for about 35% of the total aerosol mass with diameter smaller than 10 μm. In the urban North America and South America, organic carbon (OC) contributes the largest mass fraction to the atmospheric aerosol (i.e., 20% or more), while in other areas of the world the OC fraction ranks second or third with a mean of about 16%. Sulphate normally accounts for about 10 to 30% by mass, except for the areas in rural Africa, urban Oceania and South America, where it is less than about 10%. The mass fractions of nitrate and ammonium are only around 6% and 4% on average, respectively. In most areas, elemental carbon (EC, which refers to a particular way of measuring BC) represents less than 5% of the aerosol mass, although this percentage may be larger (about 12%) in South America, urban Africa, urban Europe, South, Southeast and East Asia and urban Oceania due to the larger impact of combustion sources. Sea salt can be dominant at oceanic remote sites with 50 to 70% of aerosol mass.

Aerosol optical depth (AOD), which is related to the column-integrated aerosol amount, is measured by the Aerosol Robotic Network (AERONET, Holben et al., 1998), other ground-based networks (e.g.,

Table 7.2 | Key aerosol properties of the main aerosol species in the troposphere. Terrestrial primary biological aerosol particles (PBAPs), brown carbon and marine primary organic aerosols (POA) are particular types of organic aerosols (OA) but are treated here as separate components because of their specific properties. The estimated lifetimes in the troposphere are based on the AeroCom models, except for terrestrial PBAPs which are treated by analogy to other coarse mode aerosol types.

Aerosol Species	Size Distribution	Main Sources	Main Sinks	Tropospheric Lifetime	Key Climate Relevant Properties
Sulphate	Primary: Aitken, accumulation and coarse modes Secondary: Nucleation, Aitken, and accumulation modes	Primary: marine and volcanic emissions. Secondary: oxidation of SO_2 and other S gases from natural and anthropogenic sources	Wet deposition Dry deposition	~ 1 week	Light scattering. Very hygroscopic. Enhances absorption when deposited as a coating on black carbon. Cloud condensation nuclei (CCN) active.
Nitrate	Accumulation and coarse modes	Oxidation of NO_x	Wet deposition Dry deposition	~ 1 week	Light scattering. Hygroscopic. CCN active.
Black carbon	Freshly emitted: <100 nm Aged: accumulation mode	Combustion of fossil fuels, biofuels and biomass	Wet deposition Dry deposition	1 week to 10 days	Large mass absorption efficiency in the shortwave. CCN active when coated. May be ice nuclei (IN) active.
Organic aerosol	POA: Aitken and accumulation modes. SOA: nucleation, Aitken and mostly accumulation modes. Aged OA: accumulation mode	Combustion of fossil fuel, biofuel and biomass. Continental and marine ecosystems. Some anthropogenic and biogenic non-combustion sources	Wet deposition Dry deposition	~ 1 week	Light scattering. Enhances absorption when deposited as a coating on black carbon. CCN active (depending on aging time and size).
... of which brown carbon	Freshly emitted: 100–400 nm Aged: accumulation mode	Combustion of biofuels and biomass. Natural humic-like substances from the biosphere	Wet deposition Dry deposition	~ 1 week	Medium mass absorption efficiency in the UV and visible. Light scattering.
... of which terrestrial PBAP	Mostly coarse mode	Terrestrial ecosystems	Sedimentation Wet deposition Dry deposition	1 day to 1 week depending on size	May be IN active. May form giant CCN
Mineral dust	Coarse and super-coarse modes, with a small accumulation mode	Wind erosion, soil resuspension. Some agricultural practices and industrial activities (cement)	Sedimentation Dry deposition Wet deposition	1 day to 1 week depending on size	IN active. Light scattering and absorption. Greenhouse effect.
Sea spray	Coarse and accumulation modes	Breaking of air bubbles induced e.g., by wave breaking. Wind erosion.	Sedimentation Wet deposition Dry deposition	1 day to 1 week depending on size	Light scattering. Very hygroscopic. CCN active. Can include primary organic compounds in smaller size range
... of which marine POA	Preferentially Aitken and accumulation modes	Emitted with sea spray in biologically active oceanic regions	Sedimentation Wet deposition Dry deposition	~ 1 week	CCN active.

Bokoye et al., 2001; Che et al., 2009) and a number of satellite-based sensors. Instruments designed for aerosol retrievals such as Moderate Resolution Imaging Spectrometer (MODIS; Remer et al., 2005; Levy et al., 2010; Kleidman et al., 2012), Multi-angle Imaging Spectro-Radiometer (MISR; Kahn et al., 2005; Kahn et al., 2007) and Polarization and Directionality of the Earth's Reflectances (POLDER)/Polarization and Anisotropy of Reflectances for Atmospheric Sciences Coupled with Observations from Lidar (PARASOL) (Tanré et al., 2011) are used preferentially to less specialized instruments such as Advanced Very High Resolution Radiometer (AVHRR; e.g., Zhao et al., 2008a; Mishchenko et al., 2012), Total Ozone Mapping Spectrometer (TOMS; Torres et al., 2002) and Along Track Scanning Radiometer (ATSR)/Advanced Along Track Scanning Radiometer (AATSR) (Thomas et al., 2010) although the latter are useful for building aerosol climatologies because of their long measurement records (see Section 2.2.3). Although each AOD retrieval by satellite sensors shows some skill against more accurate sunphotometer measurements such as those of AERONET, there are still large differences among satellite products in regional and seasonal patterns because of differences and uncertainties in calibration, sampling, cloud screening, treatment of the surface reflectivity and aerosol microphysical properties (e.g., Li et al., 2009; Kokhanovsky et al., 2010). The global but incomplete sampling of satellite measurements can be combined with information from global aerosol models through data assimilation techniques (e.g., Benedetti et al., 2009; Figure 7.14a). Owing to the heterogeneity in their sources, their short lifetime and the dependence of sinks on the meteorology, aerosol distributions show large variations on daily, seasonal and interannual scales.

The CALIPSO spaceborne lidar (Winker et al., 2009) complements existing ground-based lidars. It now provides a climatology of the aerosol extinction coefficient (Figure 7.14b–e), highlighting that over most regions the majority of the optically active aerosol resides in the lowest 1 to 2 km. Yu et al. (2010) and Koffi et al. (2012) found that global aerosol models tend to have a positive bias in the aerosol extinction scale height in some (but not all) regions, due to an overestimate of aerosol concentrations above 6 km. There is less information available on the vertical profile of aerosol number and mass concentrations, although a number of field experiments involving research and commercial aircraft have measured aerosol concentrations (e.g., Heintzenberg et al., 2011). In particular vertical profiles of BC mixing ratios have been measured during the Aerosol Radiative Forcing over India (ARFI) aircraft/high altitude balloon campaigns (Satheesh et al., 2008), Arctic Research of the Composition of the Troposphere from Aircraft and Satellites (ARCTAS; Jacob et al., 2010), Aerosol, Radiation,

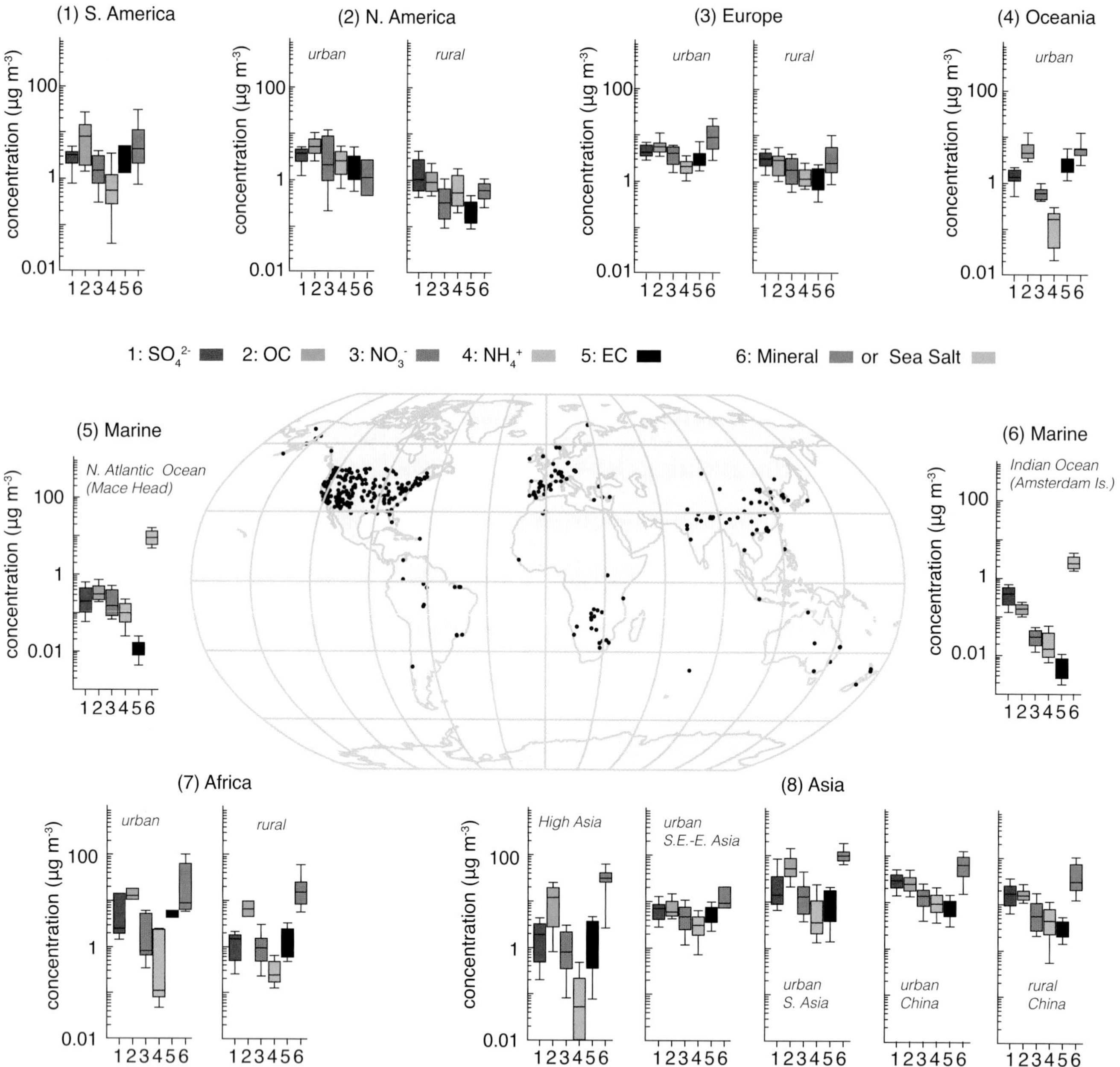

Figure 7.13 | Bar chart plots summarizing the mass concentration (μg m^{-3}) of seven major aerosol components for particles with diameter smaller than 10 μm, from various rural and urban sites (dots on the central world map) in six continental areas of the world with at least an entire year of data and two marine sites. The density of the sites is a qualitative measure of the spatial representativeness of the values for each area. The North Atlantic and Indian Oceans panels correspond to measurements from single sites (Mace Head and Amsterdam Island, respectively) that are not necessarily representative. The relative abundances of different aerosol compounds are considered to reflect the relative importance of emissions of these compounds or their precursors, either anthropogenic or natural, in the different areas. For consistency the mass of organic aerosol (OA) has been converted to that of organic carbon (OC), according to a conversion factor (typically 1.4 to 1.6), as provided in each study. For each area, the panels represent the median, the 25th to 75th percentiles (box), and the 10th to 90th percentiles (whiskers) for each aerosol component. These include: **(1) South America** (Artaxo et al., 1998; Morales et al., 1998; Artaxo et al., 2002; Celis et al., 2004; Bourotte et al., 2007; Fuzzi et al., 2007; Mariani and Mello, 2007; de Souza et al., 2010; Martin et al., 2010a; Gioda et al., 2011); **(2) North America** with **urban United States** (Chow et al., 1993; Kim et al., 2000; Ito et al., 2004; Malm and Schichtel, 2004; Sawant et al., 2004; Liu et al., 2005); and **rural United States** (Chow et al., 1993; Malm et al., 1994; Malm and Schichtel, 2004; Liu et al., 2005); **(3) Europe** with **urban Europe** (Lenschow et al., 2001; Querol et al., 2001, 2004, 2006, 2008; Roosli et al., 2001; Rodriguez et al., 2002, 2004; Putaud et al., 2004; Hueglin et al., 2005; Lonati et al., 2005; Viana et al., 2006, 2007; Perez et al., 2008; Yin and Harrison, 2008; Lodhi et al., 2009); and **rural Europe** (Gullu et al., 2000; Querol et al., 2001, 2004, 2009; Rodriguez et al., 2002; Putaud et al., 2004; Puxbaum et al., 2004;Rodrıguez et al., 2004; Hueglin et al., 2005; Kocak et al., 2007; Salvador et al., 2007; Yttri, 2007; Viana et al., 2008; Yin and Harrison, 2008;Theodosi et al., 2010); **(4) urban Oceania** (Chan et al., 1997; Maenhaut et al., 2000; Wang and Shooter, 2001; Wang et al., 2005a; Radhi et al., 2010); **(5) marine northern Atlantic Ocean** (Rinaldi et al., 2009; Ovadnevaite et al., 2011); **(6) marine Indian Ocea**n (Sciare et al., 2009; Rinaldi et al., 2011); **(7) Africa** with **urban Africa** (Favez et al., 2008; Mkoma, 2008; Mkoma et al., 2009a); and **rural Africa** (Maenhaut et al., 1996; Nyanganyura et al., 2007; Mkoma, 2008, 2009a, 2009b; Weinstein et al., 2010); **(8) Asia** with **high Asia**, with altitude larger than 1680 m (Shresth et al., 2000; Zhang et al., 2001, 2008, 2012b; Carrico et al., 2003; Rastogi and Sarin, 2005; Ming et al., 2007a; Rengarajan et al., 2007; Qu et al., 2008; Decesari et al., 2010; Ram et al., 2010); **urban Southeast and East Asia** (Lee and Kang, 2001; Oanh et al., 2006; Kim et al., 2007; Han et al., 2008; Khan et al., 2010); **urban South Asia** (Rastogi and Sarin, 2005; Kumar et al., 2007; Lodhi et al., 2009; Chakraborty and Gupta, 2010; Khare and Baruah, 2010; Raman et al., 2010; Safai et al., 2010; Stone et al., 2010; Sahu et al., 2011); **urban China** (Cheng et al., 2000; Yao et al., 2002; Zhang et al., 2002; Wang et al., 2003, 2005b, 2006; Ye et al., 2003; Xiao and Liu, 2004; Hagler et al., 2006; Oanh et al., 2006; Zhang et al., 2011, 2012b); and **rural China** (Hu et al., 2002; Zhang et al., 2002; Hagler et al., 2006; Zhang et al., 2012b).

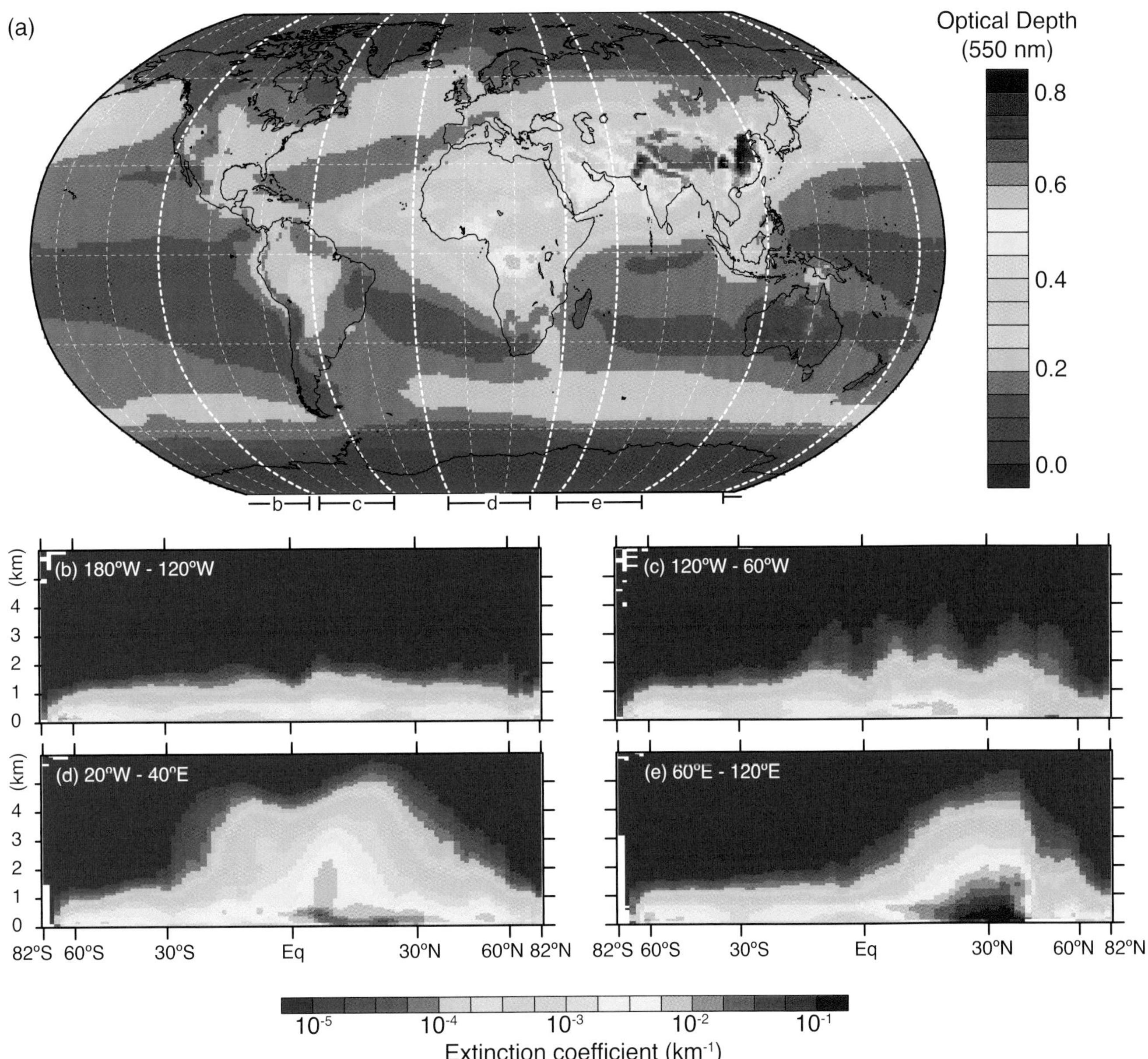

Figure 7.14 | (a) Spatial distribution of the 550 nm aerosol optical depth (AOD, unitless) from the European Centre for Medium Range Weather Forecasts (ECMWF) Integrated Forecast System model with assimilation of Moderate Resolution Imaging Spectrometer (MODIS) aerosol optical depth (Benedetti et al., 2009; Morcrette et al., 2009) averaged over the period 2003–2010; (b–e) latitudinal vertical cross sections of the 532 nm aerosol extinction coefficient (km^{-1}) for four longitudinal bands (180°W to 120°W, 120°W to 60°W, 20°W to 40°E, and 60°E to 120°E, respectively) from the Cloud–Aerosol Lidar with Orthogonal Polarization (CALIOP) instrument for the year 2010 (nighttime all-sky data, version 3; Winker et al., 2013).

and Cloud Processes affecting Arctic Climate (ARCPAC; Warneke et al., 2010), Aerosol Radiative Forcing in East Asia (A-FORCE; Oshima et al., 2012) and HIAPER Pole-to-Pole Observations (HIPPO1; Schwarz et al., 2010) campaigns. Comparison between models and observations have shown that aerosol models tend to underestimate BC mass concentrations in some outflow regions, especially in Asia, but overestimate concentrations in remote regions, especially at altitudes (Koch et al., 2009b; Figure 7.15), which make estimates of their RFari uncertain (see Section 7.5.2) given the large dependence of RFari on the vertical distribution of BC (Ban-Weiss et al., 2012). Absorption AOD can be retrieved from sun photometer measurements (Dubovik et al., 2002) or a combination of ground-based transmittance and satellite reflectance measurements (Lee et al., 2007) in situations where AOD is larger than about 0.4. Koch et al. (2009b) and Bond et al. (2013) used AERONET-based retrievals of absorption AOD to show that most AeroCom models underestimate absorption in many regions, but there remain representativeness issues when comparing point observations to a model climatology.

7.3.2 Aerosol Sources and Processes

7.3.2.1 Aerosol Sources

Sea spray is produced at the sea surface by bubble bursting induced mostly, but not exclusively, by breaking waves. The effective emission flux of sea spray particles to the atmosphere depends on the surface

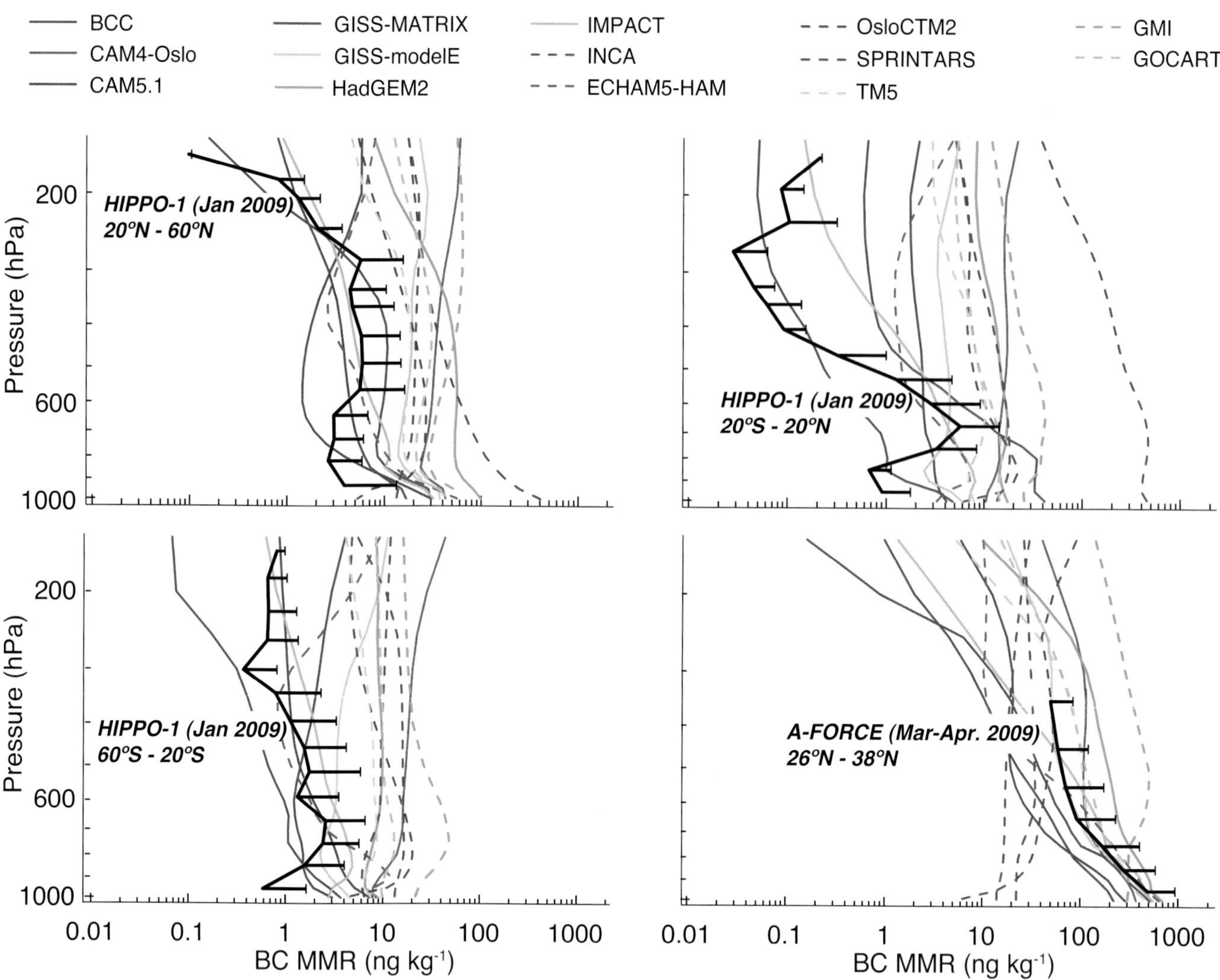

Figure 7.15 | Comparison of vertical profiles of black carbon (BC) mass mixing ratios (MMR, in ng kg^{-1}) as measured by airborne single particle soot photometer (SP2) instruments during the HIAPER Pole-to-Pole Observations (HIPPO1; Schwarz et al., 2010) and Aerosol Radiative Forcing in East Asia (A-FORCE; Oshima et al., 2012) aircraft campaigns and simulated by a range of AeroCom II models (Schulz et al., 2009). The black solid lines are averages of a number of vertical profiles in each latitude zone with the horizontal lines representing the standard deviation of the measurements at particular height ranges. Each HIPPO1 profile is the average of about 20 vertical profiles over the mid-Pacific in a two-week period in January 2009. The A-FORCE profile is the average of 120 vertical profiles measured over the East China Sea and Yellow Sea downstream of the Asian continent in March to April 2009. The model values (colour lines) are monthly averages corresponding to the measurement location and month, using meteorology and emissions corresponding to the year 2006.

wind speed, sea state and atmospheric stability, and to a lesser extent on the temperature and composition of the sea water. Our understanding of sea spray emissions has increased substantially since AR4; however, process-based estimates of the total mass and size distribution of emitted sea spray particles continue to have large uncertainties (de Leeuw et al., 2011; Table 7.1). Sea spray particles are composed of sea salt and marine primary organic matter, the latter being found preferentially in particles smaller than 200 nm in diameter (Leck and Bigg, 2008; Russell et al., 2010). The emission rate of marine POA depends on biological activity in ocean waters (Facchini et al., 2008) and its global emission rate has been estimated to be in the range 2 to 20 Tg yr^{-1} (Gantt et al., 2011). Uncertainty in the source and composition of sea spray translates into a significant uncertainty in the aerosol number concentration in the marine atmosphere that, unlike aerosol optical depth and mass concentrations, can only be constrained by *in situ* observations (Heintzenberg et al., 2000; Jaeglé et al., 2011).

Mineral dust particles are produced mainly by disintegration of aggregates following creeping and saltation of larger soil particles over desert and other arid surfaces (e.g., Zhao et al., 2006; Kok, 2011). The magnitude of dust emissions to the atmosphere depends on the surface wind speed and many soil-related factors such as its texture, moisture and vegetation cover. The range of estimates for the global dust emissions spans a factor of about five (Huneeus et al., 2011; Table 7.1). Anthropogenic sources, including road dust and mineral dust due to human land use change, remain ill quantified although some recent satellite observations suggest the fraction of mineral dust due to the latter source could be 20 to 25% of the total (Ginoux et al., 2012a, 2012b).

The sources of biomass burning aerosols at the global scale are usually inferred from satellite retrieval of burned areas and/or active fires, but inventories continue to suffer from the lack of sensitivity of satellite data to small fires (Randerson et al., 2012) and uncertainties in emission factors. Terrestrial sources of PBAPs include bacteria, pollen, fungal spores, lichen, viruses and fragments of animals and plants (Després et al., 2012). Most of these particles are emitted in the coarse mode (Pöschl et al., 2010) and the contribution to the accumulation mode is thought to be small. There are only a few estimates of the global flux of PBAPs and these are poorly constrained (Burrows et al., 2009; Heald and Spracklen, 2009; see Table 7.1).

The main natural aerosol precursors are dimethylsulphide (DMS) emitted by the oceans and biogenic volatile organic compounds (BVOC) emitted mainly by the terrestrial biosphere. BVOC emissions depend on the amount and type of vegetation, temperature, radiation, the ambient CO_2 concentration and soil humidity (Grote and Niinemets, 2008; Pacifico et al., 2009; Peñuelas and Staudt, 2010). While speciated BVOC emission inventories have been derived for some continental regions, global emission inventories or schemes are available only for isoprene, monoterpenes and a few other compounds (Müller et al., 2008; Guenther et al., 2012). The total global BVOC emissions have large uncertainties, despite the apparent convergence in different model-based estimates (Arneth et al., 2008).

The ratio of secondary to primary organic aerosol is larger than previously thought, but has remained somewhat ambiguous due to atmospheric transformation processes affecting both these components (Robinson et al., 2007; Jimenez et al., 2009; Pye and Seinfeld, 2010). Globally, most of the atmospheric SOA is expected to originate from biogenic sources, even though anthropogenic sources could be equally important at northern mid-latitudes (de Gouw and Jimenez, 2009; Lin et al., 2012). Recent studies suggest that the SOA formation from BVOCs may be enhanced substantially by anthropogenic pollution due to (1) high concentrations of nitrogen oxides (NO_x) enhancing BVOC oxidation, and (2) high anthropogenic POA concentrations that facilitate transformation of oxidized volatile organic compounds (VOCs) to the particle phase (Carlton et al., 2010; Heald et al., 2011; Hoyle et al., 2011). The uncertainty range of atmospheric SOA formation is still large and estimated to be approximately 20 to 380 Tg yr^{-1} (Hallquist et al., 2009; Farina et al., 2010; Heald et al., 2010; Spracklen et al., 2011; Table 7.1).

Anthropogenic sources of aerosol particles (BC, POA) and aerosol precursors (sulphur dioxide, ammonia, NO_x and NMVOCs, hereafter also referred to as VOCs for simplicity) can be inferred from *a priori* emission inventories (Table 7.1). They are generally better constrained than natural sources, exceptions being anthropogenic sources of BC, which could be underestimated (Bond et al., 2013), and anthropogenic emissions of some VOCs, fly-ash and dust which are still poorly known. Since AR4, remote sensing by satellites has been increasingly used to constrain natural and anthropogenic aerosol and aerosol precursor emissions (e.g., Dubovik et al., 2008; Jaeglé et al., 2011; Huneeus et al., 2012).

7.3.2.2 Aerosol Processes

New particle formation is the process by which low-volatility vapours nucleate into stable molecular clusters, which under certain condensable vapour regimes can grow rapidly to produce nanometre-sized aerosol particles. Since AR4, substantial progress in our understanding of atmospheric nucleation and new particle formation has been made (e.g., Zhang et al., 2012a). Multiple lines of evidence indicate that while sulphuric acid is the main driver of nucleation (Kerminen et al., 2010; Sipilä et al., 2010), the nucleation rate is affected by ammonia and amines (Kurten et al., 2008; Smith et al., 2010; Kirkby et al., 2011; Yu et al., 2012) as well as low-volatility organic vapours (Metzger et al., 2010; Paasonen et al., 2010; Wang et al., 2010a). Nucleation pathways involving only uncharged molecules are expected to dominate over nucleation induced by ionization of atmospheric molecules in continental boundary layers, but the situation might be different in the free atmosphere (Kazil et al., 2010; Hirsikko et al., 2011).

Condensation is the main process transferring low-volatility vapours to aerosol particles, and also usually the dominant process for growth to larger sizes. The growth of the smallest particles depends crucially on the condensation of organic vapours (Donahue et al., 2011b; Riipinen et al., 2011; Yu, 2011) and is therefore tied strongly with atmospheric SOA formation discussed in Section 7.3.3.1. The treatment of condensation of semi-volatile compounds, such as ammonia, nitric acid and most organic vapours, remains a challenge in climate modeling. In addition, small aerosol particles collide with one another and stick (coagulate), one of the processes contributing to aerosol internal mixing. Coagulation is an important sink for sub-micrometre size particles, typically under high concentrations near sources and at lower concentrations in locations where the aerosol lifetime is long and amount of condensable vapours is low. It is the main sink for the smallest aerosol particles (Pierce and Adams, 2007).

Since AR4, observations of atmospheric nucleation and subsequent growth of nucleated particles to larger sizes have been increasingly reported in different atmospheric environments (Kulmala and Kerminen, 2008; Manninen et al., 2010; O'Dowd et al., 2010). Nucleation and growth enhance atmospheric CCN concentrations (Spracklen et al., 2008; Merikanto et al., 2009; Pierce and Adams, 2009a; Yu and Luo, 2009) and potentially affect aerosol–cloud interactions (Wang and Penner, 2009; Kazil et al., 2010; Makkonen et al., 2012a). However, CCN concentrations may be fairly insensitive to changes in nucleation rate because the growth of nucleated particles to larger sizes is limited by coagulation (see Sections 7.3.3.3 and 7.4.6.2).

Aerosols also evolve due to cloud processing, followed by the aerosol release upon evaporation of cloud particles, affecting the number concentration, composition, size and mixing state of atmospheric aerosol particles. This occurs via aqueous-phase chemistry taking place inside clouds, via altering aerosol precursor chemistry around and below clouds, and via different aerosol–hydrometeor interactions. These processes are discussed further in Section 7.4.1.2.

The understanding and modelling of aerosol sinks has seen less progress since AR4 in comparison to other aerosol processes. Improved dry deposition models, which depend on the particle size as well as the characteristics of the Earth's surface, have been developed and are increasingly being used in global aerosol models (Kerkweg et al., 2006; Feng, 2008; Petroff and Zhang, 2010). Sedimentation throughout the atmosphere and its role in dry deposition at the surface are important for the largest particles in the coarse mode. The uncertainty in the estimate of wet deposition by nucleation and impaction scavenging is controlled by the uncertainties in the prediction of the amount, frequency and areal extent of precipitation, as well as the size and chemical composition of aerosol particles. For insoluble primary particles like BC and dust, nucleation scavenging also depends strongly on their degree of mixing with soluble compounds. Parameterization of aerosol wet deposition remains a key source of uncertainty in aerosol models, which affects the vertical and horizontal distributions of aerosols (Prospero et al., 2010; Vignati et al., 2010; Lee et al., 2011), with further impact on model estimates of aerosol forcings.

7.3.3 Progress and Gaps in Understanding Climate Relevant Aerosol Properties

The climate effects of atmospheric aerosol particles depend on their atmospheric distribution, along with their hygroscopicity, optical properties and ability to act as CCN and IN. Key quantities for aerosol optical and cloud forming properties are the particle number size distribution, chemical composition, mixing state and morphology. These properties are determined by a complex interplay between their sources, atmospheric transformation processes and their removal from the atmosphere (Section 7.3.2, Figures 7.12 and 7.16). Since AR4, measurement of some of the key aerosol properties has been greatly improved in laboratory and field experiments using advanced instrumentation, which allows for instance the analysis of individual particles. These experimental studies have in turn stimulated improvement in the model representations of the aerosol physical, chemical and optical properties (Ghan and Schwartz, 2007). We focus our assessment on some of the key issues where there has been progress since AR4.

7.3.3.1 Chemical Composition and Mixing State

Research on the climate impacts of aerosols has moved beyond the simple cases of externally mixed sulphate, BC emitted from fossil fuel combustion and biomass burning aerosols. Although the role of inorganic aerosols as an important anthropogenic contributor to aerosol–radiation and aerosol–cloud interactions has not been questioned, BC has received increasing attention because of its high absorption as has SOA because of its ubiquitous nature and ability to mix with other aerosol types.

The physical properties of BC (strongly light-absorbing, refractory with a vaporization temperature near 4000 K, aggregate in morphology, insoluble in most organic solvents) allow a strict definition in principle (Bond et al., 2013). Direct measurement of individual BC-containing particles is possible with laser-induced incandescence (single particle soot photometer, also called SP2; Gao et al., 2007; Schwarz et al., 2008a; Moteki and Kondo, 2010), which has enabled accurate measurements of the size of BC cores, as well as total BC mass concentrations. Condensation of gas-phase compounds on BC and coagulation with other particles alter the mixing state of BC (e.g., Li et al., 2003; Pósfai et al., 2003; Adachi et al., 2010), which can produce internally mixed BC in polluted urban air on a time scale of about 12 hours (Moteki et al., 2007; McMeeking et al., 2010). The resulting BC-containing particles can become hygroscopic, which can lead to reduced lifetime and atmospheric loading (Stier et al., 2006).

Formation processes of OA still remain highly uncertain, which is a major weakness in the present understanding and model representation of atmospheric aerosols (Kanakidou et al., 2005; Hallquist et al., 2009; Ziemann and Atkinson, 2012). Measurements by aerosol mass spectrometers have provided some insights into sources and atmospheric processing of OA (Zhang et al., 2005b; Lanz et al., 2007; Ulbrich et al., 2009). Observations at continental mid-latitudes including urban and rural/remote air suggest that the majority of SOA is probably oxygenated OA (Zhang et al., 2005a, 2007a). Experiments within and downstream of urban air indicate that under most circumstances SOA substantially contributes to the total OA mass (de Gouw et al., 2005; Volkamer et al., 2006; Zhang et al., 2007a).

There is a large range in the complexity with which OA is represented in global aerosol models. Some complex, yet still parameterized, chemical schemes have been developed recently that account for multigenerational oxidation (Robinson et al., 2007; Jimenez et al., 2009; Donahue et al., 2011a). Since AR4, some regional and global model have used a new scheme based on lumping VOCs into volatility bins (Robinson et al., 2007), which is an improved representation of the two-product absorptive partitioning scheme (Kroll and Seinfeld, 2008) for the formation and aging of SOA. This new framework includes organic compounds of different volatility, produced from parent VOCs by multi-generation oxidation processes and partitioned between the aerosol and gas phases (Farina et al., 2010; Tsimpidi et al., 2010; Yu, 2011), which improves the agreement between observed and modeled SOA in urban areas (Hodzic et al., 2010; Shrivastava et al., 2011). Field observations and laboratory studies suggest that OA is also formed efficiently in aerosol and cloud and liquid water contributing a substantial fraction of the organic aerosol mass (Sorooshian et al., 2007; Miyazaki et al., 2009; Lim et al., 2010; Ervens et al., 2011a). Chemical reactions in the aerosol phase (e.g., oligomerization) also make OA less volatile and more hygroscopic, influencing aerosol–radiation and aerosol–cloud interactions (Jimenez et al., 2009). As a consequence, OA concentrations are probably underestimated in many global aerosol models that do not include these chemical processes (Hallquist et al., 2009).

Some of the OA is light absorbing and can be referred to as brown carbon (BrC; Kirchstetter et al., 2004; Andreae and Gelencser, 2006). A fraction of the SOA formed in cloud and aerosol water is light-absorbing in the visible (e.g., Shapiro et al., 2009), while SOA produced from gas-phase oxidation of VOCs absorbs ultraviolet radiation (e.g., Nakayama et al., 2010).

Multiple observations show co-existence of external and internal mixtures relatively soon after emission (e.g., Hara et al., 2003; Schwarz et al., 2008b; Twohy and Anderson, 2008). In biomass burning aerosol, organic compounds and BC are frequently internally mixed with ammonium, nitrate, and sulphate (Deboudt et al., 2010; Pratt and Prather, 2010). Over urban locations, as much as 90% of the particles are internally mixed with secondary inorganic species (Bi et al., 2011). Likewise mineral dust and biomass burning aerosols can become internally mixed when these aerosol types age together (Hand et al., 2010). The aerosol mixing state can alter particle size distribution and hygroscopicity and hence the aerosol optical properties and ability to act as CCN (Wex et al., 2010). Global aerosol models can now approximate the aerosol mixing state using size-resolving bin or modal schemes (e.g., Stier et al., 2005; Kim et al., 2008; Mann et al., 2010).

7.3.3.2 Size Distribution and Optical Properties

Aerosol size distribution is a key parameter determining both the aerosol optical and CCN properties. Since the AR4, much effort has been put into measuring and simulating the aerosol number rather than volume size distribution. For instance, number size distributions in the submicron range (30 to 500 nm) were measured at 24 sites in Europe for two years (Asmi et al., 2011), although systematic measurements are still limited in other regions. Although validation studies show agreement between column-averaged volume size distribution from sunphotometer measurements and direct *in situ* (surface as well as aircraft-based) measurements at some locations (Gerasopoulos et al., 2007; Radhi et al., 2010; Haywood et al., 2011), these inversion products have not been systematically validated. Satellite measurements produce valuable but more limited information on aerosol size.

The aerosol scattering, absorption and extinction coefficients depend on the aerosol size distribution, aerosol refractive index and mixing state. The humidification of internally mixed aerosols further influences their light scattering and absorption properties, through changes in particle shape, size and refractive index (Freney et al., 2010). Aerosol absorption is a key climate-relevant aerosol property and earlier *in situ* methods to measure it suffered from significant uncertainties (Moosmüller et al., 2009), partly due to the lack of proper reference material for instrument calibration and development (Baumgardner et al., 2012). Recent measurements using photo-acoustic methods and laser-induced incandescence methods are more accurate but remain sparse. The mass absorption cross sections for freshly generated BC were measured to be 7.5 ± 1.2 $m^2 g^{-1}$ at 550 nm (Bond et al., 2013). Laboratory measurements conducted under well controlled conditions show that thick coating of soluble material over BC cores enhance the mass absorption cross section by a factor of 1.8 to 2 (Cross et al., 2010; Shiraiwa et al., 2010). It is more difficult to measure the enhancement factor in mass absorption cross sections for ambient BC, partly owing to the necessity of removing coatings of BC. Knox et al. (2009) observed enhancement by a factor 1.2 to 1.6 near source regions. A much lower enhancement factor was observed by Cappa et al. (2012) by a new measurement technique, in contradiction to the laboratory experiments and theoretical calculations. These results may not be representative and would require confirmation by independent measurement methods.

As discussed in Section 7.3.1.2, the global mean AOD is not well constrained from satellite-based measurements and remains a significant source of uncertainty when estimating aerosol–radiation interactions (Su et al., 2013). This is also true of the anthropogenic fraction of AOD which is more difficult to constrain from observations. AeroCom phase II models simulate an anthropogenic AOD at 550 nm of 0.03 ± 0.01 (with the range corresponding to one standard deviation) relative to 1850, which represents 24 ± 6% of the total AOD (Myhre et al., 2013). This is less than suggested by some satellite-based studies, i.e., 0.03 over the ocean only in Kaufman et al. (2005), and about 0.06 as a global average in Loeb and Su (2010) and Bellouin et al. (2013), but more than in the CMIP5 models (see Figure 9.29). Overall there is *medium confidence* that between 20 and 40% of the global mean AOD is of anthropogenic origin. There is agreement that the anthropogenic aerosol is smaller in size and more absorbing than the natural aerosol (Myhre, 2009; Loeb and Su, 2010), but there is disagreement on the anthropogenic absorption AOD and its contribution to the total absorption AOD, that is, 0.0015 ± 0.0007 (one standard deviation) relative to 1850 in Myhre et al. (2013) but about 0.004 and half of the total absorption AOD in Bellouin et al. (2013).

7.3.3.3 Cloud Condensation Nuclei

A subset of aerosol particles acts as CCN (see Table 7.2). The ability of an aerosol particle to take up water and subsequently activate, thereby acting as a CCN at a given supersaturation, is determined by its size and composition. Common CCN in the atmosphere are composed of sea salt, sulphates and sulphuric acid, nitrate and nitric acid and some organics. The uptake of water vapour by hygroscopic aerosols strongly affects their RFari.

CCN activity of inorganic aerosols is relatively well understood, and lately most attention has been paid to the CCN activity of mixed organic/inorganic aerosols (e.g., King et al., 2010; Prisle et al., 2010). Uncertainties in our current understanding of CCN properties are due primarily to SOA (Good et al., 2010), mainly because OA is still poorly characterized (Jimenez et al., 2009). The important effect of the formation of SOA is that internally mixed SOA contributes to the mass of aerosol particles, and therefore to their sizes. The size of the CCN has been found to be more important than their chemical composition at two continental locations as larger particles are more readily activated than smaller particles because they require a lower critical supersaturation (Dusek et al., 2006; Ervens et al., 2007). However, the chemical composition may be important in other locations such as the marine environment, where primary organic particles (hydrogels) have been shown to be exceptionally good CCN (Orellana et al., 2011; Ovadnevaite et al., 2011). For SOA it is not clear how important surface tension effects and bulk-to-surface partitioning of surfactants are, and if the water activity coefficient changes significantly as a function of the solute concentration (Prisle et al., 2008; Good et al., 2010).

The bulk hygroscopicity parameter κ has been introduced as a concise measure of how effectively an aerosol particle acts as a CCN (Rissler et al., 2004, 2010; Petters and Kreidenweis, 2007). It can be measured experimentally and is increasingly being used as a way to characterize aerosol properties. Pringle et al. (2010) used surface and aircraft measurements to evaluate the κ distributions simulated by a global aerosol model, and found generally good agreement. When the aerosol is dominated by organics, discrepancies between values of κ obtained directly from both CCN activity measurements and sub-saturated particle water uptake measurements have been observed in some instances (e.g., Prenni et al., 2007; Irwin et al., 2010; Roberts et al., 2010), whereas in other studies closure has been obtained (e.g., Duplissy et al., 2008; Kammermann et al., 2010; Rose et al., 2011). Adsorption theory (Kumar et al., 2011) replaces κ-theory for CCN activation for insoluble particles (e.g., mineral dust) while alternative theories are still required for explanation of marine POA that seem to have peculiar gel-like properties (Ovadnevaite et al., 2011).

Available modelling studies (Pierce and Adams, 2009a; Wang and Penner, 2009; Schmidt et al., 2012a) disagree on the anthropogenic fraction of CCN (taken here at 0.2% supersaturation). Based on these studies we assess this fraction to be between one fourth and two thirds in the global mean with *low confidence*, and highlight large interhemispheric and regional variations. Models that neglect or underestimate volcanic and natural organic aerosols would overestimate this fraction.

7.3.3.4 Ice Nuclei

Aerosols that act as IN are solid substances at atmospheric temperatures and supersaturations. Mineral dust, volcanic ash and primary bioaerosols such as bacteria, fungal spores and pollen, are typically known as good IN (Vali, 1985; Hoose and Möhler, 2012). Conflicting evidence has been presented for the ability of BC, organic, organic semi-solid/glassy organic and biomass burning particles to act as IN (Hoose and Möhler, 2012; Murray et al., 2012). The importance of biological particles acting as IN is unclear. A new study finds evidence of a large fraction of submicron particles in the middle-to-upper troposphere to be composed of biological particles (DeLeon-Rodriguez et al., 2013); however global modelling studies suggest that their concentrations are not sufficient to play an important role for ice formation (Hoose et al., 2010a; Sesartic et al., 2012). Because BC has anthropogenic sources, its increase since pre-industrial times may have caused changes to the lifetime of mixed-phase clouds (Section 7.4.4) and thus to RF (Lohmann, 2002b; Section 7.5).

Four heterogeneous ice-nucleation modes are distinguished in the literature: immersion freezing (initiated from within a cloud droplet), condensation freezing (freezing during droplet formation), contact freezing (due to collision with an IN) and deposition nucleation (that refers to the direct deposition of vapour onto IN). Lidar observations reveal that liquid cloud droplets are present before ice crystals form via heterogeneous freezing mechanisms (Ansmann et al., 2008; de Boer et al., 2011), indicating that deposition nucleation does not seem to be important for mixed-phase clouds. IN can either be bare or mixed with other substances. As bare particles age in the atmosphere, they acquire liquid surface coatings by condensing soluble species and water vapour or by coagulating with soluble particles, which may transform IN from deposition or contact nuclei into possible immersion nuclei. A change from contact to immersion freezing implies activation at colder temperatures, with consequences for the lifetime and radiative effect of mixed-phase clouds (Sections 7.4.4 and 7.5.3).

The atmospheric concentrations of IN are very uncertain because of the aforementioned uncertainties in freezing mechanisms and the difficulty of measuring IN in the upper troposphere. The anthropogenic fraction cannot be estimated at this point because of a lack of knowledge about the anthropogenic fractions of BC and mineral dust acting as IN and the contributions of PBAPs, other organic aerosols and other aerosols acting as IN.

7.3.4 Aerosol–Radiation Interactions

7.3.4.1 Radiative Effects due to Aerosol–Radiation Interactions

The radiative effect due to aerosol–radiation interactions (REari), formerly known as direct radiative effect, is the change in radiative flux caused by the combined scattering and absorption of radiation by anthropogenic and natural aerosols. The REari results from well-understood physics and is close to being an observable quantity, yet our knowledge of aerosol and environmental characteristics needed to quantify the REari at the global scale remains incomplete (Anderson et al., 2005; Satheesh and Moorthy, 2005; Jaeglé et al., 2011). The REari requires knowledge of the spectrally varying aerosol extinction coefficient, single scattering albedo, and phase function, which can in principle be estimated from the aerosol size distribution, shape, chemical composition and mixing state (Figure 7.12). Radiative properties of the surface, atmospheric trace gases and clouds also influence the REari. In the solar spectrum under cloud-free conditions the REari is typically negative at the TOA, but it weakens and can become positive with increasing aerosol absorption, decreasing upscatter fraction or increasing albedo of the underlying surface. REari is weaker in cloudy conditions, except when the cloud layer is thin or when absorbing aerosols are located above or between clouds (e.g., Chand et al., 2009). The REari at the surface is negative and can be much stronger than the REari at the TOA over regions where aerosols are absorbing (Li et al., 2010). In the longwave part of the spectrum, TOA REari is generally positive and mainly exerted by coarse-mode aerosols, such as sea spray and desert dust (Reddy et al., 2005), and by stratospheric aerosols (McCormick et al., 1995).

There have been many measurement-based estimates of shortwave REari (e.g., Yu et al., 2006; Bergamo et al., 2008; Di Biagio et al., 2010; Bauer et al., 2011) although some studies involve some degree of modelling. In contrast, estimates of longwave REari remain limited (e.g., Bharmal et al., 2009). Observed and calculated shortwave radiative fluxes agree within measurement uncertainty when aerosol properties are known (e.g., Osborne et al., 2011). Global observational estimates of the REari rely on satellite remote sensing of aerosol properties and/or measurements of the Earth's radiative budget (Chen et al., 2011; Kahn, 2012). Estimates of shortwave TOA REari annually averaged over cloud-free oceans range from –4 to –6 W m^{-2}, mainly contributed by sea spray (Bellouin et al., 2005; Loeb and Manalo-Smith, 2005; Yu et al., 2006; Myhre et al., 2007). However, REari can reach tens of W m^{-2} locally. Estimates over land are more difficult as the surface is less well characterized (Chen et al., 2009; Jethva et al., 2009) despite recent progress in aerosol inversion algorithms (e.g., Dubovik et al., 2011). Attempts to estimate the REari in cloudy sky remain elusive (e.g., Peters et al., 2011b), although passive and active remote sensing of aerosols over clouds is now possible (Torres et al., 2007; Omar et al., 2009; Waquet et al., 2009; de Graaf et al., 2012). Notable areas of positive TOA REari exerted by absorbing aerosols include the Arctic over ice surfaces (Stone et al., 2008) and seasonally over southeastern Atlantic stratocumulus clouds (Chand et al., 2009; de Graaf et al., 2012). While AOD and aerosol size are relatively well constrained, uncertainties in the aerosol single-scattering albedo (McComiskey et al., 2008; Loeb and Su, 2010) and vertical profile (e.g., Zarzycki and Bond, 2010) contribute significantly to the overall uncertainties in REari. Consequently, diversity in large-scale numerical model estimates of REari increases with aerosol absorption and between cloud-free and cloudy conditions (Stier et al., 2013).

7.3.4.2 Rapid Adjustments to Aerosol–Radiation Interactions

Aerosol–radiation interactions give rise to rapid adjustments (see Section 7.1), which are particularly pronounced for absorbing aerosols such as BC. Associated cloud changes are often referred to as the semi-direct aerosol effect (see Figure 7.3). The ERF from aerosol–radiation interactions is quantified in Section 7.5.2; only the corresponding processes governing rapid adjustments are discussed here. Impacts on precipitation are discussed in Section 7.6.3.

Since AR4, additional observational studies have found correlations between cloud cover and absorbing aerosols (e.g., Brioude et al., 2009; Wilcox, 2010), and eddy-resolving, regional and global scale modelling studies have helped confirm a causal link. Relationships between cloud and aerosol reveal a more complicated picture than initially anticipated (e.g., Hill and Dobbie, 2008; Koch and Del Genio, 2010; Zhuang et al., 2010; Sakaeda et al., 2011; Ghan et al., 2012).

Absorbing aerosols modify atmospheric stability in the boundary layer and free troposphere (e.g., Wendisch et al., 2008; Babu et al., 2011). The effect of this on cloud cover depends on the height of the aerosol relative to the cloud and the type of cloud (e.g., Yoshimori and Broccoli, 2008; Allen and Sherwood, 2010; Koch and Del Genio, 2010; Persad et al., 2012). Aerosol also reduces the downwelling solar radiation at the surface. Together the changes in atmospheric stability and reduction in surface fluxes provide a means for aerosols to significantly modify the fraction of surface-forced clouds (Feingold et al., 2005; Sakaeda et al., 2011). These changes may also affect precipitation as discussed in Section 7.6.3.

Cloud cover is expected to decrease if absorbing aerosol is embedded in the cloud layer. This has been observed (Koren et al., 2004) and simulated (e.g., Feingold et al., 2005) for clouds over the Amazon forest in the presence of smoke aerosols. In the stratocumulus regime, absorbing aerosol above cloud top strengthens the temperature inversion, reduces entrainment and tends to enhance cloudiness. Satellite observations (Wilcox, 2010) and modelling (Johnson et al., 2004) of marine stratocumulus show a thickening of the cloud layer beneath layers of absorbing smoke aerosol, which induces a local negative forcing. The responses of other cloud types, such as those associated with deep convection, are not well determined.

Absorbing aerosols embedded in cloud drops enhance their absorption, which can affect the dissipation of cloud. The contribution to RFari is small (Stier et al., 2007; Ghan et al., 2012), and there is contradictory evidence regarding the magnitude of the cloud dissipation effect influencing ERFari (Feingold et al., 2005; Ghan et al., 2012; Jacobson, 2012; Bond et al., 2013). Global forcing estimates are necessarily based on global models (see Section 7.5.2), although the accuracy of GCMs in this regard is limited by their ability to represent low cloud processes accurately. This is an area of concern as discussed in Section 7.2 and limits confidence in these estimates.

7.3.5 Aerosol Responses to Climate Change and Feedback

The climate drivers of changes in aerosols can be split into physical changes (temperature, humidity, precipitation, soil moisture, solar radiation, wind speed, sea ice extent, etc.), chemical changes (availability of oxidants) and biological changes (vegetation cover and properties, plankton abundance and speciation, etc). The response of aerosols to climate change may constitute a feedback loop whereby climate processes amplify or dampen the initial perturbation (Carslaw et al., 2010; Raes et al., 2010). We assess here the relevance and strength of aerosol–climate feedbacks in the context of future climate change scenarios.

7.3.5.1 Changes in Sea Spray and Mineral Dust

Concentrations of sea spray will respond to changes in surface wind speed, atmospheric stability, precipitation and sea ice cover (Struthers et al., 2011). Climate models disagree about the balance of effects, with estimates ranging from an overall 19% reduction in global sea salt burden from the present-day to year 2100 (Liao et al., 2006), to little sensitivity (Mahowald et al., 2006a), to a sizeable increase (Jones et al., 2007; Bellouin et al., 2011). In particular there is little understanding of how surface wind speed may change over the ocean in a warmer climate, and observed recent changes (e.g., Young et al., 2011; Section 2.7.2) may not be indicative of future changes. Given that sea spray particles comprise a significant fraction of CCN concentrations over the oceans, such large changes will feed back on climate through changes in cloud droplet number (Korhonen et al., 2010b).

Studies of the effects of climate change on dust loadings also give a wide range of results. Woodward et al. (2005) found a tripling of the dust burden in 2100 relative to present-day because of a large increase in bare soil fraction. A few studies projected moderate (10 to 20%) increases, or decreases (e.g., Tegen et al., 2004; Jacobson and Streets, 2009; Liao et al., 2009). Mahowald et al. (2006b) found a 60% decrease under a doubled CO_2 concentration due to the effect of CO_2 fertilization on vegetation. The large range reflects different responses of the atmosphere and vegetation cover to climate change forcings, and results in *low confidence* in these predictions.

7.3.5.2 Changes in Sulphate, Ammonium and Nitrate Aerosols

The DMS–sulphate–cloud–climate feedback loop could operate in numerous ways through changes in temperature, absorbed solar radiation, ocean mixed layer depth and nutrient recycling, sea ice extent, wind speed, shift in marine ecosystems due to ocean acidification and climate change, and atmospheric processing of DMS into CCN. Although no study has included all the relevant effects, two decades of research have questioned the original formulation of the feedback loop (Leck and Bigg, 2007) and have provided important insights into this complex, coupled system (Ayers and Cainey, 2007; Kloster et al., 2007; Carslaw et al., 2010). There is now *medium confidence* for a weak feedback due to a weak sensitivity of the CCN population to changes in DMS emissions, based on converging evidence from observations and Earth System model simulations (Carslaw et al., 2010; Woodhouse et al., 2010; Quinn and Bates, 2011). Parameterizations of oceanic DMS production nevertheless lack robust mechanistic justification (Halloran et al., 2010) and as a result the sensitivity to ocean acidification and climate change remains uncertain (Bopp et al., 2004; Kim et al., 2010; Cameron-Smith et al., 2011).

Chemical production of sulphate increases with atmospheric temperature (Aw and Kleeman, 2003; Dawson et al., 2007; Kleeman, 2008), but future changes in sulphate are found to be more sensitive to simulated future changes in precipitation removal. Under fixed anthropogenic emissions, most studies to date predict a small (0 to 9%) reduction in global sulphate burden mainly because of future increases in precipitation (Liao et al., 2006; Racherla and Adams, 2006; Unger et al., 2006; Pye et al., 2009). However, Rae et al. (2007) found a small increase in global sulphate burden from 2000 to 2100 because the simulated future precipitation was reduced in regions of high sulphate abundance.

Changes in temperature have a large impact on nitrate aerosol formation through shifting gas–particle equilibria. There is some agreement among global aerosol models that climate change alone will contribute to a decrease in the nitrate concentrations (Liao et al., 2006; Racherla and Adams, 2006; Pye et al., 2009; Bellouin et al., 2011) with the exception of Bauer et al. (2007) who found little change in nitrate for year 2030. It should be noted however that these modeling studies have reported that changes in precursor emissions are expected to increase nitrate concentrations in the future (Section 11.3.5). Besides the changes in meteorological parameters, climate change can also influence ammonium formation by changing concentrations of sulphate and nitrate, but the effect of climate change alone was found to be small (Pye et al., 2009).

7.3.5.3 Changes in Carbonaceous Aerosols

There is evidence that future climate change could lead to increases in the occurrence of wildfires because of changes in fuel availability, readiness of the fuel to burn and ignition sources (Mouillot et al., 2006; Marlon et al., 2008; Spracklen et al., 2009; Kloster et al., 2010; Pechony and Shindell, 2010). However, vegetation dynamics may also play a role that is not well understood. Increased fire occurrence would increase aerosol emissions, but decrease BVOC emissions. This could lead to a small positive or negative net radiative effect and feedback (Carslaw et al., 2010).

A large fraction of SOA forms from the oxidation of isoprene, monoterpenes and sesquiterpenes from biogenic sources (Section 7.3.3.1). Emissions from vegetation can increase in a warmer atmosphere, everything else being constant (Guenther et al., 2006). Global aerosol models simulate an increase in isoprene emissions of 22 to 55% by 2100 in response to temperature change (Sanderson et al., 2003; Liao et al., 2006; Heald et al., 2008) and a change in global SOA burden of –6% to +100% through climate-induced changes in aerosol processes and removal rates (Liao et al., 2006; Tsigaridis and Kanakidou, 2007; Heald et al., 2008). An observationally based study suggest a small global feedback parameter of –0.01 W m^{-2} °C^{-1} despite larger regional effects (Paasonen et al., 2013). Increasing CO_2 concentrations, drought and surface ozone also affect BVOC emissions (Arneth et al., 2007; Peñuelas and Staudt, 2010), which adds significant uncertainty to future global emissions (Makkonen et al., 2012b). Future changes in vegetation cover, whether they are natural or anthropogenic, also introduce large uncertainty in emissions (Lathière et al., 2010; Wu et al., 2012). There is little understanding on how the marine source of organic aerosol may change with climate, notwithstanding the large range of emission estimates for the present day (Carslaw et al., 2010).

7.3.5.4 Synthesis

The emissions, properties and concentrations of aerosols or aerosol precursors could respond significantly to climate change, but there is little consistency across studies in the magnitude or sign of this response. The lack of consistency arises mostly from our limited understanding of processes governing the source of natural aerosols and the complex interplay of aerosols with the hydrological cycle. The feedback parameter as a result of the future changes in emissions of natural aerosols is mostly bracketed within ±0.1 W m^{-2} °C^{-1} (Carslaw et al., 2010). With respect to anthropogenic aerosols, Liao et al. (2009) showed a significant positive feedback (feedback parameter of +0.04 to +0.15 W m^{-2} °C^{-1} on a global mean basis) while Bellouin et al. (2011) simulated a smaller negative feedback of –0.08 to –0.02 W m^{-2} °C^{-1}. Overall we assess that models simulate relatively small feedback parameters (i.e., within ±0.2 W m^{-2} °C^{-1}) with *low confidence*, however regional effects on the aerosol may be important.

7.4 Aerosol–Cloud Interactions

7.4.1 Introduction and Overview of Progress Since AR4

This section assesses our understanding of aerosol–cloud interactions, emphasizing the ways in which anthropogenic aerosol may be affecting the distribution and radiative properties of non- and weakly precipitating clouds. The idea that anthropogenic aerosol is changing cloud properties, thus contributing a substantial forcing to the climate system, has been addressed to varying degrees in all of the previous IPCC assessment reports.

Since AR4, research has continued to articulate new pathways through which the aerosol may affect the radiative properties of clouds, as well as the intensity and spatial patterns of precipitation (e.g., Rosenfeld et al., 2008). Progress can be identified on four fronts: (1) global-scale modelling now represents a greater diversity of aerosol–cloud interactions, and with greater internal consistency; (2) observational studies continue to document strong local correlations between aerosol and cloud properties or precipitation, but have become more quantitative and are increasingly identifying and addressing the methodological challenges associated with such correlations; (3) regional-scale modelling is increasingly being used to assess regional influences of aerosol on cloud field properties and precipitation; (4) fine-scale process models have begun to be used more widely, and among other things have shown how turbulent mixing, cloud and meso-scale circulations may buffer the effects of aerosol perturbations.

This section focuses on the microphysics of aerosol–cloud interactions in liquid, mixed-phase and pure ice clouds. Their radiative implications are quantified in Section 7.5. This section also includes a discussion of aerosol influences on light precipitation in shallow clouds but defers discussion of aerosol effects on more substantial precipitation from mixed-phase clouds to Section 7.6.4.

7.4.1.1 Classification of Hypothesized Aerosol–Cloud Interactions

Denman et al. (2007) catalogued several possible pathways via which the aerosol might affect clouds. Given the number of possible aerosol–cloud interactions, and the difficulty of isolating them individually, there is little value in attempting to assess each effect in isolation, especially because modelling studies suggest that the effects may interact and compensate (Stevens and Feingold, 2009; Morrison and Grabowski, 2011). Instead, all radiative consequences of aerosol–cloud interactions are grouped into an 'effective radiative forcing due to aerosol–cloud interactions', or ERFaci (Figure 7.3). ERFaci accounts for aerosol-related microphysical modifications to the cloud albedo (Twomey, 1977), as well as any secondary effects that result from clouds adjusting rapidly to changes in their environment (i.e., 'lifetime effects'; Albrecht, 1989; Liou and Ou, 1989; Pincus and Baker, 1994). We do assess the physical underpinnings of the cloud albedo effect, but in contrast to previous assessments, no longer distinguish the resultant forcing. Note that ERFaci includes potential radiative adjustments to the cloud system associated with aerosol–cloud interactions but does not include adjustments originating from aerosol–radiation interactions (ERFari). Possible contributions to ERFaci from warm (liquid) clouds are discussed in Section 7.4.3, separately from those associated with adjustments by cold (ice or mixed-phase) clouds (Section 7.4.4). Figure 7.16 shows a schematic of many of the processes to be discussed in Sections 7.4, 7.5 and 7.6.

7.4.1.2 Advances and Challenges in Observing Aerosol–Cloud Interactions

Since AR4, numerous field studies (e.g., Rauber et al., 2007; Wood et al., 2011b; Vogelmann et al., 2012) and laboratory investigations (e.g., Stratmann et al., 2009) of aerosol–cloud interactions have highlighted the numerous ways that the aerosol impacts cloud processes, and how clouds in turn modify the aerosol. The latter occurs along a number of pathways including aqueous chemistry, which adds aerosol mass to droplets (e.g., Schwartz and Freiberg, 1981; Ervens et al., 2011a); coalescence scavenging, whereby drop collision–coalescence diminishes the droplet (and aerosol) number concentration (Hudson, 1993) and changes the mixing state of the aerosol; new particle formation in the vicinity of clouds (Clarke et al., 1999); and aerosol removal by precipitation (see also Section 7.3.2.2).

Satellite-based remote sensing continues to be the primary source of global data for aerosol–cloud interactions but concerns persist regarding how measurement artefacts affect retrievals of both aerosol (Tanré et al., 1996; Tanré et al., 1997; Kahn et al., 2005; Jeong and Li, 2010) and cloud properties (Platnick et al., 2003; Yuekui and Di Girolamo, 2008) in broken cloud fields. Two key issues are that measurements classified as 'cloud-free' may not be, and that aerosol measured in the vicinity of clouds is significantly different than it would be were the cloud field, and its proximate cause (high humidity), not present (e.g., Loeb and Schuster, 2008). The latter results from humidification effects on aerosol optical properties (Charlson et al., 2007; Su et al., 2008; Tackett and Di Girolamo, 2009; Twohy et al., 2009; Chand et al., 2012), contamination by undetectable cloud fragments (Koren et al., 2007) and the remote effects of radiation scattered by cloud edges on aerosol retrieval (Wen et al., 2007; Várnai and Marshak, 2009).

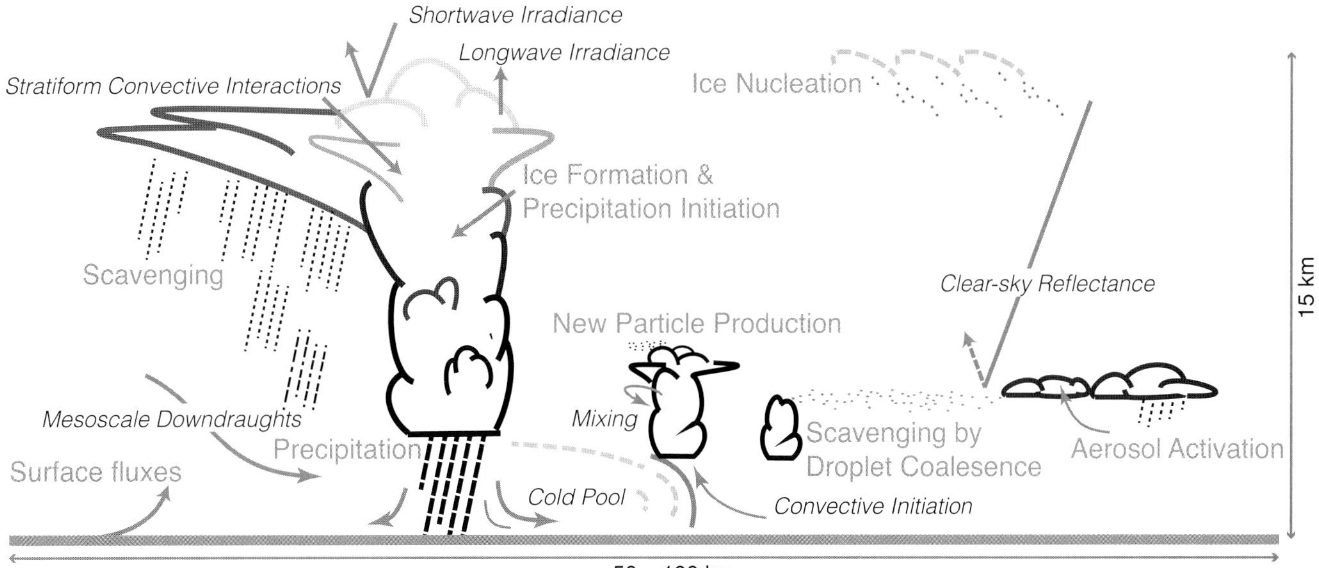

Figure 7.16 | Schematic depicting the myriad aerosol–cloud–precipitation related processes occurring within a typical GCM grid box. The schematic conveys the importance of considering aerosol–cloud–precipitation processes as part of an interactive system encompassing a large range of spatiotemporal scales. Cloud types include low-level stratocumulus and cumulus where research focuses on aerosol activation, mixing between cloudy and environmental air, droplet coalescence and scavenging which results in cloud processing of aerosol particles, and new particle production near clouds; cirrus clouds where a key issue is ice nucleation through homogeneous and heterogeneous freezing; and deep convective clouds where some of the key questions relate to aerosol influences on liquid, ice, and liquid–ice pathways for precipitation formation, cold pool formation and scavenging. These processes influence the shortwave and longwave cloud radiative effect and hence climate. Primary processes that affect aerosol–cloud interactions are labelled in blue while secondary processes that result from and influence aerosol–cloud interactions are in grey.

While most passive satellite retrievals are unable to distinguish aerosol layers above or below clouds from those intermingling with the cloud field, active space-based remote sensing (L'Ecuyer and Jiang, 2010) has begun to address this problem (Stephens et al., 2002; Anderson et al., 2005; Huffman et al., 2007; Chand et al., 2008; Winker et al., 2010). Spectral polarization and multi-angular measurements can discriminate between cloud droplets and aerosol particles and thus improve estimates of aerosol loading and absorption (Deuzé et al., 2001; Chowdhary et al., 2005; Mishchenko et al., 2007; Hasekamp, 2010).

Use of active remote sensing, both from monitoring ground stations (e.g., McComiskey et al., 2009; Li et al., 2011) and satellites (Costantino and Bréon, 2010), as well as aerosol proxies not influenced by cloud contamination of retrievals (Jiang et al., 2008; Berg et al., 2011) have emerged as a particular effective way of identifying whether aerosol and cloud perturbations are intermingled.

Because the aerosol is a strong function of air-mass history and origin, and is strongly influenced by cloud and precipitation processes (Clarke et al., 1999; Petters et al., 2006; Anderson et al., 2009), and both are affected by meteorology (Engström and Ekman, 2010; Boucher and Quaas, 2013), correlations between the aerosol and cloud, or precipitation, should not be taken as generally indicating a cloud response to the aerosol (e.g., Painemal and Zuidema, 2010). Furthermore, attempts to control for other important factors (air-mass history or cloud dynamical processes) are limited by a lack of understanding of large-scale cloud controlling factors in the first place (Anderson et al., 2009; Siebesma et al., 2009; Stevens and Brenguier, 2009). These problems are increasingly being considered in observationally based inferences of aerosol effects on clouds and precipitation, but ascribing changes in cloud properties to changes in the aerosol remains a fundamental challenge.

7.4.1.3 Advances and Challenges in Modelling Aerosol–Cloud Interactions

Modelling of aerosol–cloud interactions must contend with the fact that the key physical processes are fundamentally occurring at the fine scale and cannot be represented adequately based on large-scale fields. There exist two distinct challenges: fundamental understanding of processes and their representation in large-scale models.

Fine-scale LES and CRM models (Section 7.2.2.1) have greatly advanced as a tool for testing the physical mechanisms proposed to govern aerosol–cloud–precipitation interactions (e.g., Ackerman et al., 2009; vanZanten et al., 2011). Their main limitation is that they are idealized, for example, they do not resolve synoptic scale circulations or allow for representation of orography. A general finding from explicit numerical simulations of warm (liquid) clouds is that various aerosol impact mechanisms tend to be mediated (and often buffered) by interactions across scales not included in the idealized albedo and lifetime effects (Stevens and Feingold, 2009). Specific examples involve the interplay between the drop-size distribution and mixing processes that determine cloud macrostructure (Stevens et al., 1998; Ackerman et al., 2004; Bretherton et al., 2007; Wood, 2007; Small et al., 2009), or the dependence of precipitation development in stratiform clouds on details of the vertical structure of the cloud (Wood, 2007). Thus warm clouds may typically be less sensitive to aerosol perturbations in nature than in large-scale models, which do not represent all of these compensating processes. Hints of similar behaviour in mixed-phase (liquid and ice) stratus are beginning to be documented (Section 7.4.4.3) but process-level understanding and representation in models are less advanced.

Regional models include realism in the form of non-idealized meteorology, synoptic scale forcing, variability in land surface, and diurnal/monthly cycles (e.g., Iguchi et al., 2008; Bangert et al., 2011; Seifert et al., 2012; Tao et al., 2012), however, at the expense of resolving fine-scale cloud processes. Regional models have brought to light the possibility of aerosol spatial inhomogeneity causing changes in circulation patterns via numerous mechanisms including changes in the radiative properties of cloud anvils (van den Heever et al., 2011), changes in the spatial distribution of precipitation (Lee, 2012; Section 7.6.2) or gradients in heating rates associated with aerosol–radiation interactions (Lau et al., 2006; Section 7.3.4.2).

GCMs, our primary tool for quantifying global mean forcings, now represent an increasing number of hypothesized aerosol–cloud interactions, but at poor resolution. GCMs are being more closely scrutinized through comparisons to observations and to other models (Quaas et al., 2009; Penner et al., 2012). Historically, aerosol–cloud interactions in GCMs have been based on simple constructs (e.g., Twomey, 1977; Albrecht, 1989; Pincus and Baker, 1994). There has been significant progress on parameterizing aerosol activation (e.g., Ghan et al., 2011) and ice nucleation (Liu and Penner, 2005; Barahona and Nenes, 2008; DeMott et al., 2010; Hoose et al., 2010b); however, these still depend heavily on unresolved quantities such as updraught velocity. Similarly, parameterizations of aerosol influences on cloud usually do not account for known non-monotonic responses of cloud amount and properties to aerosols (Section 7.4.3.2). Global models are now beginning to represent aerosol effects in convective, ice and mixed-phase clouds (e.g., Lohmann, 2008; Song and Zhang, 2011; Section 7.6.4). Nevertheless, for both liquid-only and mixed-phase clouds, these parameterizations are severely limited by the need to parameterize cloud-scale motions over a huge range of spatio-temporal scales (Section 7.2.3).

Although advances have been considerable, the challenges remain daunting. The response of cloud systems to aerosol is nuanced (e.g., vanZanten et al., 2011) and the representation of both clouds and aerosol–cloud interactions in large-scale models remains primitive (Section 7.2.3). Thus it is not surprising that large-scale models exhibit a range of manifestations of aerosol–cloud interactions, which limits quantitative inference (Quaas et al., 2009). This highlights the need to incorporate into GCMs the lessons learned from cloud-scale models, in a physically-consistent way. New 'super-parameterization' and probability distribution function approaches (Golaz et al., 2002; Rio and Hourdin, 2008; Section 7.2.2.2) hold promise, with recent results supporting the notion that aerosol forcing is smaller than simulated by standard climate models (Wang et al., 2011b; see Section 7.5.3).

7.4.1.4 Combined Modelling and Observational Approaches

Combined approaches, which attempt to maximize the respective advantage of models and observations, are beginning to add to understanding of aerosol–cloud interactions. These include inversions of the observed historical record using simplified climate models (e.g., Forest et al., 2006; Aldrin et al., 2012) but also the use of reanalysis and chemical transport models to help interpret satellite records (Chameides et al., 2002; Koren et al., 2010a; Mauger and Norris, 2010), field study data to help constrain fine-scale modelling studies (e.g., Ackerman et al., 2009; vanZanten et al., 2011), or satellite/surface-based climatologies to constrain large-scale modelling (Wang et al., 2012).

7.4.2 Microphysical Underpinnings of Aerosol–Cloud Interactions

7.4.2.1 The Physical Basis

The cloud albedo effect (Twomey, 1977) is the mechanism by which an increase in aerosol number concentration leads to an increase in the albedo of liquid clouds (reflectance of incoming solar radiation) by increasing the cloud droplet number concentration, decreasing the droplet size, and hence increasing total droplet surface area, with the liquid water content and cloud geometrical thickness hypothetically held fixed. Although only the change in the droplet concentration is considered in the original concept of the cloud albedo effect, a change in the shape of the droplet size distribution that is directly induced by the aerosols may also play a role (e.g., Feingold et al., 1997; Liu and Daum, 2002). In the Arctic, anthropogenic aerosol may influence the longwave emissivity of optically thin liquid clouds and generate a positive forcing at the surface (Garrett and Zhao, 2006; Lubin and Vogelmann, 2006; Mauritsen et al., 2011), but TOA forcing is thought to be negligible.

7.4.2.2 Observational Evidence for Aerosol–Cloud Interactions

The physical basis of the albedo effect is fairly well understood, with research since AR4 generally reinforcing earlier work. Detailed *in situ* aircraft observations show that droplet concentrations observed just above the cloud base generally agree with those predicted based on the aerosol properties and updraught velocity observed below the cloud (e.g., Fountoukis et al., 2007). Vertical profiles of cloud droplet effective radius also agree with those predicted by models that take into account the effect of entrainment (Lu et al., 2008), although uncertainties still remain in estimating the shape of the droplet size distribution (Brenguier et al., 2011), and the degree of entrainment mixing within clouds.

At relatively low aerosol loading (AOD less than about 0.3) there is ample observational evidence for increases in aerosol resulting in an increase in droplet concentration and decrease in droplet size (for constant liquid water) but uncertainties remain regarding the magnitude of this effect, and its sensitivity to spatial averaging. Based on simple metrics, there is a large range of physically plausible responses, with aircraft measurements (e.g., Twohy et al., 2005; Lu et al., 2007, 2008;; Hegg et al., 2012) tending to show stronger responses than satellite-derived responses (McComiskey and Feingold, 2008; Nakajima and Schulz, 2009; Grandey and Stier, 2010). At high AOD and high aerosol concentration, droplet concentration tends to saturate (e.g., Verheggen et al., 2007) and, if the aerosol is absorbing, there may be reductions in droplet concentration and cloudiness (Koren et al., 2008). This absorbing effect originates from aerosol–radiation interactions and is therefore part of ERFari (Section 7.3.4.2). Negative correlation between AOD and ice particle size has also been documented in deep convective clouds (e.g., Sherwood, 2002; Jiang et al., 2008).

7.4.2.3 Advances in Process Level Understanding

At the heart of the albedo effect lie two fundamental issues. The first is aerosol activation and its sensitivity to aerosol and dynamical parameters. The primary controls on droplet concentration are the aerosol number concentration (particularly at diameters greater than about 60 nm) and cooling rate (proportional to updraught velocity). Aerosol size distribution can play an important role under high aerosol loadings, whereas aerosol composition tends to be much less important, except perhaps under very polluted conditions and low updraught velocities (e.g., Ervens et al., 2005; McFiggans et al., 2006). This is partially because aging tends to make particles more hygroscopic regardless of their initial composition, but also because more hygroscopic particles lead to faster water vapour uptake, which then lowers supersaturation, limiting the initial increase in activation.

The second issue is that the amount of energy reflected by a cloud system is a strong function of the amount of condensate. Simple arguments show that in a relative sense the amount of reflected energy is approximately two-and-a-half times more sensitive to changes in the liquid water path than to changes in droplet concentration (Boers and Mitchell, 1994). Because both of these parameters experience similar ranges of relative variability, the magnitude of aerosol–cloud related forcing rests mostly on dynamical factors such as turbulent strength and entrainment that control cloud condensate, and a few key aerosol parameters such as aerosol number concentration and size distribution, and to a much lesser extent, composition.

7.4.3 Forcing Associated with Adjustments in Liquid Clouds

7.4.3.1 The Physical Basis for Adjustments in Liquid Clouds

The adjustments giving rise to ERFaci are multi-faceted and are associated with both albedo and so-called 'lifetime' effects (Figure 7.3). However, this old nomenclature is misleading because it assumes a relationship between cloud lifetime and cloud amount or water content. Moreover, the effect of the aerosol on cloud amount may have nothing to do with cloud lifetime *per se* (e.g., Pincus and Baker, 1994).

The traditional view (Albrecht, 1989; Liou and Ou, 1989) has been that adjustment effects associated with aerosol–cloud–precipitation interactions will add to the initial albedo increase by increasing cloud amount. The chain of reasoning involves three steps: that droplet concentrations depend on the number of available CCN; that precipitation development is regulated by the droplet concentration; and that the development of precipitation reduces cloud amount (Stevens and Feingold, 2009). Of the three steps, the first has ample support in both observations and theory (Section 7.4.2.2). More problematic are the last two links in the chain of reasoning. Although increased droplet concentrations inhibit the initial development of precipitation (see Section 7.4.3.2.1), it is not clear that such an effect is sustained in an evolving cloud field. In the trade-cumulus regime, some modelling studies suggest the opposite, with increased aerosol concentrations actually promoting the development of deeper clouds and invigorating precipitation (Stevens and Seifert, 2008; see discussion of similar responses in deep convective clouds in Section 7.6.4). Others have shown alternating cycles of larger *and* smaller cloud water in both aerosol-perturbed stratocumulus (Sandu et al., 2008) and trade cumulus (Lee et al., 2012), pointing to the important role of environmental adjustment. There exists limited unambiguous observational evidence (exceptions to be given below) to support the original hypothesised cloud-amount effects, which are often assumed to hold universally and have dominated GCM parameterization of aerosol–cloud interactions. GCMs lack the more nuanced responses suggested by recent work, which influences their ERFaci estimates.

7.4.3.2 Observational Evidence of Adjustments in Liquid Clouds

Since observed effects generally include both the albedo effect and the adjustments, with few if any means of observing only one or the other in isolation, in this section we discuss and interpret observational findings that reflect both effects. Stratocumulus and trade cumulus regimes are discussed separately.

7.4.3.2.1 Stratocumulus

The cloud albedo effect is best manifested in so-called ship tracks, which are bright lines of clouds behind ships. Many ship tracks are characterized by an increase in the droplet concentration resulting from the increase in aerosol number concentration and an absence of drizzle size drops, which leads to a decrease in the droplet radius and an increase in the cloud albedo (Durkee et al., 2000), all else equal. However, liquid water changes are the primary determinant of albedo changes (Section 7.4.2.3; Chen et al., 2012), therefore adjustments are key to understanding radiative response. Coakley and Walsh (2002) showed that cloud water responses can be either positive or negative. This is supported by more recent shiptrack analyses based on new satellite sensors (Christensen and Stephens, 2011): aerosol intrusions result in weak decreases in liquid water (–6%) in overcast clouds, but significant increases in liquid water (+39%) and increases in cloud fraction in precipitating, broken stratocumulus clouds. The global ERFaci of visible ship tracks has been estimated from satellite and found to be insignificant at about –0.5 mW m^{-2} (Schreier et al., 2007), although this analysis may not have identified all shiptracks. Some observational studies downwind of ship tracks have been unable to distinguish aerosol influences from meteorological influences on cloud microphysical or macrophysical properties (Peters et al., 2011a), although it is not clear whether their methodology had sufficient sensitivity to detect the aerosol effects. Notwithstanding evidence of shiptracks locally increasing the cloud fraction and albedo of broken cloud scenes quite significantly (e.g., Christensen and Stephens, 2011; Goren and Rosenfeld, 2012), their contribution to global ERFaci is thought to be small. These ship track results are consistent with satellite studies of the influence of long-term degassing of low-lying volcanic aerosol on stratocumulus, which point to smaller droplet sizes but ambiguous changes in cloud fraction and cloud water (Gasso, 2008). The lack of clear evidence for a global increase in cloud albedo from shiptracks and volcanic plumes should be borne in mind when considering geoengineering methods that rely on cloud modification (Section 7.7.2.2).

The development of precipitation in stratocumulus, whether due to aerosol or meteorological influence can, in some instances, change a highly reflective closed-cellular cloud field to a weakly reflective broken open-cellular field (Comstock et al., 2005; Stevens et al., 2005a; vanZanten et al., 2005; Sharon et al., 2006; Savic-Jovcic and Stevens, 2008; Wang and Feingold, 2009a). In some cases, compact regions (pockets) of open-cellular convection become surrounded by regions of closed-cellular convection. It is, however, noteworthy that observed precipitation rates can be similar in both open and closed-cell environments (Wood et al., 2011a). The lack of any apparent difference in the large-scale environment of the open cells, versus the surrounding closed cellular convection, suggests the potential for multiple equilibria (Baker and Charlson, 1990; Feingold et al., 2010). Therefore in the stratocumulus regime, the onset of precipitation due to a dearth of aerosol may lead to a chain of events that leads to a large-scale reduction of cloudiness in agreement with Liou and Ou (1989) and Albrecht (1989). The transition may be bidirectional: ship tracks passing through open-cell regions also appear to revert the cloud field to a closed-cell regime inducing a potentially strong ERFaci locally (Christensen and Stephens, 2011; Wang et al., 2011a; Goren and Rosenfeld, 2012).

7.4.3.2.2 Trade-cumulus

Precipitation from trade cumuli proves difficult to observe, as the clouds are small, and not easily observed by space-based remote sensing techniques (Stephens et al., 2008). Satellite remote sensing of trade cumuli influenced by aerosol associated with slow volcanic degassing points to smaller droplet size, decreased precipitation efficiency, increased cloud amount and higher cloud tops (Yuan et al., 2011). Other studies show that in the trade cumulus regime cloud amount tends to increase with precipitation amount: for example, processes that favour precipitation development also favour cloud development (Nuijens et al., 2009); precipitation-driven colliding outflows tend to regenerate clouds; and trade cumuli that support precipitation reach heights where wind shear increases cloud fraction (Zuidema et al., 2012).

While observationally based study of the microphysical aspects of aerosol–cloud interactions has a long history, more recent assessment of the ability of detailed models to reproduce the associated radiative effect in cumulus cloud fields is beginning to provide the important link between aerosol–cloud interactions and total RF (Schmidt et al., 2009).

7.4.3.3 Advances in Process Level Understanding

Central to ERFaci is the question of how susceptible is precipitation to droplet concentration, and by inference, to the available aerosol. Some studies point to the droplet effective radius as a threshold indicator of the onset of drizzle (Rosenfeld and Gutman, 1994; Gerber, 1996; Rosenfeld et al., 2012). Others focus on the sensitivity of the conversion of cloud water to rain water (i.e., autoconversion) to droplet concentration, which is usually in the form of (droplet concentration) to the power $-\alpha$. Both approaches indicate that from the microphysical standpoint, an increase in the aerosol suppresses rainfall. Models and theory show α ranging from ½ (Kostinski, 2008; Seifert and Stevens, 2010) to 2 (Khairoutdinov and Kogan, 2000), while observational studies suggest $\alpha = 1$ (approximately the inverse of drop concentration;

Pawlowska and Brenguier, 2003; Comstock et al., 2005; vanZanten et al., 2005). Note that thicker liquid clouds amplify rain via accretion of cloud droplets by raindrops, a process that is relatively insensitive to droplet concentration, and therefore to aerosol perturbations (e.g., Khairoutdinov and Kogan, 2000). The balance of evidence suggests that $\alpha = \frac{1}{2}$ is more likely and that liquid water path (or cloud depth) has significantly more leverage over precipitation than does droplet concentration. Many GCMs assume a much stronger relationship between precipitation and cloud droplet number concentration (i.e., $\alpha = 2$) (Quaas et al., 2009).

Small-scale studies (Ackerman et al., 2004; Xue et al., 2008; Small et al., 2009) and satellite observations (Lebsock et al., 2008; Christensen and Stephens, 2011) tend to confirm two responses of the cloud liquid water to increasing aerosol. Under clean conditions when clouds are prone to precipitation, an increase in the aerosol tends to increase cloud amount as a result of aerosol suppression of precipitation. Under non-precipitating conditions, clouds tend to thin in response to increasing aerosol through a combination of droplet sedimentation (Bretherton et al., 2007) and evaporation–entrainment adjustments (e.g., Hill et al., 2009). Treatment of the subtlety of these responses and associated detail in small-scale cloud processes is not currently feasible in GCMs, although probability distribution function approaches are promising (Guo et al., 2010).

Since AR4, cloud resolving model simulation has begun to stress the importance of scale interactions when addressing aerosol–cloud interactions. Model domains on the order of 100 km allow mesoscale circulations to develop in response to changes in the aerosol. These dynamical responses may have a significant impact on cloud morphology and RF. Examples include the significant changes in cloud albedo associated with transitions between closed and open cellular states discussed above, and the cloud-free, downdraught 'shadows' that appear alongside ship tracks (Wang and Feingold, 2009b). Similar examples of large-scale changes in circulation associated with aerosol and associated influence on precipitation are discussed in Section 7.6.4. These underscore the large gap between our process level understanding of aerosol–cloud–precipitation interactions and the ability of GCMs to represent them.

7.4.3.4 Advances in and Insights Gained from Large-Scale Modelling Studies

Regional models are increasingly including representation of aerosol–cloud interactions using sophisticated microphysical models (Bangert et al., 2011; Yang et al., 2011; Seifert et al., 2012). Some of these regional models are operational weather forecast models that undergo routine evaluation. Yang et al. (2011) show improved simulations of stratocumulus fields when aerosol–cloud interactions are introduced. Regional models are increasingly being used to provide the meteorological context for satellite observations of aerosol–cloud interactions (see Section 7.4.1.4), with some (e.g., Painemal and Zuidema, 2010) suggesting that droplet concentration differences are driven primarily by synoptic scale influences rather than aerosol.

GCM studies that have explored sensitivity to autoconversion parameterization (Golaz et al., 2011) show that ERFaci can vary by 1 W m^{-2} depending on the parameterization. Elimination of the sensitivity of rain formation to the autoconversion process has begun to be considered in GCMs (Posselt and Lohmann, 2009). Wang et al. (2012) have used satellite observations to constrain autoconversion and find a reduction in ERFaci of about 33% relative to a standard GCM autoconversion parameterization. It is worth reiterating that these uncertainties are not necessarily associated with uncertainties in the physical process itself, but more so by the ability of a GCM to resolve the processes (see Section 7.4.1.3).

7.4.4 Adjustments in Cold Clouds

7.4.4.1 The Physical Basis for Adjustments in Cold Clouds

Mixed-phase clouds, containing both liquid water and ice particles, exist at temperatures between 0°C and –38°C. At warmer temperatures ice melts rapidly, whereas at colder temperatures liquid water freezes homogeneously. The formation of ice in mixed-phase clouds depends on heterogeneous freezing, initiated by IN (Section 7.3.3.4), which are typically solid or crystalline aerosol particles. In spite of their very low concentrations (on the order of 1 per litre), IN have an important influence on mixed-phase clouds. Mineral dust particles have been identified as good IN but far less is known about the IN ability of other aerosol types, and their preferred modes of nucleation. For example, the ice nucleating ability of BC particles remains controversial (Hoose and Möhler, 2012). Soluble matter can hinder glaciation by depressing the freezing temperature of supercooled drops to the point where homogeneous freezing occurs (e.g., Girard et al., 2004; Baker and Peter, 2008). Hence anthropogenic perturbations to the aerosol have the potential to affect glaciation, water and ice optical properties, and their radiative effect.

Because the equilibrium vapour pressure with respect to ice is lower than that with respect to liquid, the initiation of ice in a supercooled liquid cloud will cause vapour to diffuse rapidly toward ice particles at the expense of the liquid water (Wegener–Bergeron–Findeisen process; e.g., Schwarzenbock et al., 2001; Verheggen et al., 2007; Hudson et al., 2010). This favours the depositional growth of ice crystals, the largest of which may sediment away from the water-saturated region of the atmosphere, influencing the subsequent evolution of the cloud. Hence anthropogenic perturbations to the IN can influence the rate at which ice forms, which in turn may regulate cloud amount (Lohmann, 2002b; Storelvmo et al., 2011; see also Section 7.2.3.2.2), cloud optical properties and humidity near the tropopause.

Finally, formation of the ice phase releases latent heat to the environment (influencing cloud dynamics), and provides alternate, complex pathways for precipitation to develop (e.g., Zubler et al., 2011, and Section 7.6.4).

7.4.4.2 Observations of ERFaci in Deep Convective Clouds

As noted in Section 7.4.2.2, observations have demonstrated correlations between aerosol loading and ice crystal size but influence on cloud optical depth is unclear (e.g., Koren et al., 2005). Satellite remote sensing suggests that aerosol-related invigoration of deep convective clouds may generate more extensive anvils that radiate at cooler temperatures, are optically thinner, and generate a positive contribution to ERFaci (Koren et al., 2010b). The global influence on ERFaci is unclear.

7.4.4.3 Observations of Aerosol Effects on Arctic Ice and Mixed-Phase Stratiform Clouds

Arctic mixed-phase clouds have received a great deal of attention since AR4, with major field programs conducted in 2004 (Verlinde et al., 2007) and 2008 (Jacob et al., 2010; Brock et al., 2011; McFarquhar et al., 2011), in addition to long-term monitoring at high northern latitude stations (e.g., Shupe et al., 2008) and analysis of earlier field experiments (Uttal et al., 2002). Mixed-phase Arctic clouds persist for extended periods of time (days and even weeks; Zuidema et al., 2005), in spite of the inherent instability of the ice–water mix (see also Section 7.2.3.2.2). In spite of their low concentrations, IN have an important influence on cloud persistence, with clouds tending to glaciate and disappear rapidly when IN concentrations are relatively high and/or updraught velocities too small to sustain a liquid water layer (e.g., Ovchinnikov et al., 2011). The details of the heterogeneous ice-nucleation mechanism remain controversial but there is increasing evidence that ice forms in Arctic stratus via the liquid phase (immersion freezing) so that the CCN population also plays an important role (de Boer et al., 2011; Lance et al., 2011). If ice indeed forms via the liquid phase this represents a self-regulating feedback that helps sustain the mixed-phase clouds: as ice forms, liquid water (the source of the ice) is depleted, which restricts further ice formation and competition for water vapour via the Wegener–Bergeron–Findeisen process (Morrison et al., 2012).

7.4.4.4 Advances in Process Level Understanding

Since AR4, research on ice-microphysical processes has been very active as evidenced by the abovementioned field experiments (Section 7.4.4.3). The persistence of some mixed-phase stratiform clouds has prompted efforts to explain this phenomenon in a theoretical framework (Korolev and Field, 2008). Predicting cloud persistence may require a high level of understanding of very detailed processes. For example, ice particle growth by vapour diffusion depends strongly on crystal shape (Harrington et al., 2009), the details of which may have similar relative influence on glaciation times to the representation of ice nucleation mechanism (Ervens et al., 2011b). A recent review (Morrison et al., 2012) discusses the myriad processes that create a resilient mixed-phase cloud system, invoking the ideas of 'buffering' seen in liquid clouds (Stevens and Feingold, 2009). Importantly, the Wegener–Bergeron–Findeisen process does not necessarily destabilize the cloud system, unless sufficient ice exists (Korolev, 2007). Bistability has also been observed in the mixed-phase Arctic cloud system; the resilient cloud state is sometimes interrupted by a cloud-free state (Stramler et al., 2011), but there is much uncertainty regarding the meteorological and microphysical conditions determining which of these states is preferred.

Significant effort has been expended on heterogeneous freezing parameterizations employed in cloud or larger-scale models. Some parameterizations are empirical (e.g., Lohmann and Diehl, 2006; Hoose et al., 2008; Phillips et al., 2008; Storelvmo et al., 2008a; DeMott et al., 2010; Gettelman et al., 2010; Salzmann et al., 2010), whereas others attempt to represent the processes explicitly (Jacobson, 2003) or ground the development of parameterizations in concepts derived from classical nucleation theory (Chen et al., 2008; Hoose et al., 2010b). The details of how these processes are treated have important implications for tropical anvils (Ekman et al., 2007; Fan et al., 2010).

Homogeneous ice nucleation in cirrus clouds (at temperatures lower than about –38°C) depends crucially on the cloud updraught velocity and hence the supersaturation with respect to ice. The onset relative humidities for nucleation have been parameterized using results of parcel model simulations (e.g., Sassen and Dodd, 1988; Barahona and Nenes, 2009), airborne measurements in cirrus or wave clouds (Heymsfield and Miloshevich, 1995; Heymsfield et al., 1998), extensions of classical homogeneous ice nucleation theory (Khvorostyanov and Sassen, 1998; Khvorostyanov and Curry, 2009) and data from laboratory measurements (e.g., Bertram et al., 2000; Koop et al., 2000; Mohler et al., 2003; Magee et al., 2006; Friedman et al., 2011). There is new evidence that although ice nucleation in cirrus has traditionally been regarded as homogeneous, the preferred freezing pathway may be heterogeneous because it occurs at lower onset relative humidities (or higher onset temperatures) than homogeneous nucleation (Jensen et al., 2010). The onset relative humidities (or temperatures) for heterogeneous nucleation depend on the type and size of the IN (Section 7.3.3.4).

Cloud resolving modeling of deep convective clouds points to the potential for aerosol-related changes in cirrus anvils (e.g., Morrison and Grabowski, 2011; van den Heever et al., 2011; Storer and van den Heever, 2013), but the physical mechanisms involved and their influence on ERFaci are poorly understood, and their global impact is unclear.

7.4.4.5 Advances in and Insights Gained from Large-Scale Modelling Studies

Since the AR4, mixed-phase and ice clouds have received significant attention, with effort on representation of both heterogeneous (mixed-phase clouds) and homogeneous (cirrus) freezing processes in GCMs (e.g., Lohmann and Kärcher, 2002; Storelvmo et al., 2008a). In GCMs the physics of cirrus clouds usually involves only ice-phase microphysical processes and is somewhat simpler than that of mixed-phase clouds. Nevertheless, representation of aerosol–cloud interactions in mixed-phase and ice clouds is considerably less advanced than that involving liquid-only clouds.

Our poor understanding of the climatology and lifecycle of aerosol particles that can serve as IN complicates attempts to assess what constitutes an anthropogenic perturbation to the IN population, let alone the effect of such a perturbation. BC can impact background (i.e., non contrail) cirrus by affecting ice nucleation properties but the effect remains uncertain (Kärcher et al., 2007). The numerous GCM studies that have evaluated ERFaci for ice clouds are summarised in Section 7.5.4.

7.4.5 Synthesis on Aerosol–Cloud Interactions

Earlier assessments considered the radiative implications of aerosol–cloud interactions as two complementary processes—albedo and lifetime effects—that together amplify forcing. Since then the complexity of cloud system responses to aerosol perturbations has become more fully appreciated. Recent work at the process scale has identified compensating adjustments that make the system less susceptible to perturbation than might have been expected based on the earlier albedo and lifetime effects. Increases in the aerosol can therefore result in either an increase or a decrease in aerosol–cloud related forcing depending on the particular environmental conditions. Because many current GCMs do not include the possibility of compensating effects that are not mediated by the large-scale state, there are grounds for expecting these models to overestimate the magnitude of ERFaci. Nevertheless it is also possible that poorly understood and unrepresented interactions could cause real ERFaci to differ in either direction from that predicted by current models. Forcing estimates are discussed in Section 7.5.3.

7.4.6 Impact of Cosmic Rays on Aerosols and Clouds

Many studies have reported observations that link solar activity to particular aspects of the climate system (e.g., Bond et al., 2001; Dengel et al., 2009; Eichler et al., 2009). Various mechanisms have been proposed that could amplify relatively small variations in total solar irradiance, such as changes in stratospheric and tropospheric circulation induced by changes in the spectral solar irradiance or an effect of the flux of cosmic rays on clouds. We focus in this subsection on the latter hypothesis while Box 10.2 discusses solar influences on the climate system more generally.

Solar activity variations influence the strength and three-dimensional structure of the heliosphere. High solar activity increases the deflection of low energy cosmic rays, which reduces the flux of cosmic rays impinging upon the Earth's atmosphere. It has been suggested that the ionization caused by cosmic rays in the troposphere has an impact on aerosols and clouds (e.g., Dickinson, 1975; Kirkby, 2007). This subsection assesses studies that either seek to establish a causal relationship between cosmic rays and aerosols or clouds by examining empirical correlations, or test one of the physical mechanisms that have been put forward to account for such a relationship.

7.4.6.1 Observed Correlations Between Cosmic Rays and Properties of Aerosols and Clouds

Correlation between the cosmic ray flux and cloud properties has been examined for decadal variations induced by the 11-year solar cycle, shorter variations associated with the quasi-periodic oscillation in solar activity centred on 1.68 years, or sudden and large variations known as Forbush decrease events. It should be noted that long-term variations in cloud properties are difficult to detect (Section 2.5.6) while short-term variations may be difficult to attribute to a particular cause. Moreover, the cosmic ray flux co-varies with other solar parameters such as solar and UV irradiance. This makes any attribution of cloud changes to the cosmic ray flux problematic (Laken et al., 2011).

Some studies have shown co-variation between the cosmic ray flux and low-level cloud cover using global satellite data over periods of typically 5 to 10 years (Marsh and Svensmark, 2000; Pallé Bagó and Butler, 2000). Such correlations have not proved to be robust when extending the time period under consideration (Agee et al., 2012), and restricting the analysis to particular cloud types (Kernthaler et al., 1999) or locations (Udelhofen and Cess, 2001; Usoskin and Kovaltsov, 2008). The purported correlations have also been attributed to ENSO variability (Farrar, 2000; Laken et al., 2012) and artefacts of the satellite data cannot be ruled out (Pallé, 2005). Statistically significant (but weak) correlations between the diffuse fraction of surface solar radiation and the cosmic ray flux have been found at some locations in the UK over the 1951–2000 period (Harrison and Stephenson, 2006). Harrison (2008) also found a unique 1.68-year periodicity in surface radiation for two different UK sites between 1978 and 1990, potentially indicative of a cosmic ray effect of the same periodicity. Svensmark et al. (2009) found large global reductions in the aerosol Ångström exponent, liquid water path, and cloud cover after large Forbush decreases, but these results were not corroborated by other studies that found no statistically significant links between the cosmic ray flux and clouds at the global scale (Čalogović et al., 2010; Laken and Čalogović, 2011). Although some studies found statistically significant correlations between the cosmic ray flux and cloudiness at the regional scale (Laken et al., 2010; Rohs et al., 2010), these correlations were generally weak, cloud changes were small, and the results were sensitive to how the Forbush events were selected and composited (Kristjánsson et al., 2008; Laken et al., 2009).

7.4.6.2 Physical Mechanisms Linking Cosmic Rays to Cloudiness

The most widely studied mechanism proposed to explain the possible influence of the cosmic ray flux on cloudiness is the 'ion-aerosol clear air' mechanism, in which atmospheric ions produced by cosmic rays facilitate aerosol nucleation and growth ultimately impacting CCN concentrations and cloud properties (Carslaw et al., 2002; Usoskin and Kovaltsov, 2008). The variability in atmospheric ionization rates due to changes in cosmic ray flux can be considered relatively well quantified (Bazilevskaya et al., 2008), whereas resulting changes in aerosol nucleation rates are very poorly known (Enghoff and Svensmark, 2008; Kazil et al., 2008). Laboratory experiments indicate that ionization induced by cosmic rays enhances nucleation rates under middle and upper tropospheric conditions, but not necessarily so in the continental boundary layer (Kirkby et al., 2011). Field measurements qualitatively support this view but cannot provide any firm conclusion due to the scarcity and other limitations of free-troposphere measurements (Arnold, 2006; Mirme et al., 2010), and due to difficulties in separating nucleation induced by cosmic rays from other nucleation pathways in the continental boundary layer (Hirsikko et al., 2011). Based on surface aerosol measurements at one site, Kulmala et al. (2010) found no connection between the cosmic ray flux and new particle formation or any other aerosol property over a solar cycle (1996–2008), although particles nucleated in the free troposphere are known to contribute to particle number and CCN concentrations in the boundary layer (Merikanto et al., 2009). Our understanding of the 'ion-aerosol clear air' mechanism as a whole relies on a few model investigations that simulate changes in cosmic ray flux over a solar cycle (Pierce and Adams, 2009b; Snow-Kropla et al., 2011; Kazil et al., 2012) or during strong Forbush decreases (Bondo et al., 2010; Snow-Kropla et al., 2011; Dunne et al., 2012). Changes in CCN concentrations due to variations in the cosmic ray flux appear too weak to cause a significant radiative effect because the aerosol system is insensitive to a small change in the nucleation rate in the presence of pre-existing aerosols (see also Section 7.3.2.2).

A second pathway linking the cosmic ray flux to cloudiness has been proposed through the global electric circuit. A small direct current is able to flow vertically between the ionosphere and the Earth's surface over fair-weather regions because of cosmic-ray-induced atmospheric ionization. Charge can accumulate at the upper and lower cloud boundaries as a result of the effective scavenging of ions by cloud droplets (Tinsley, 2008). This creates conductivity gradients at the cloud edges (Nicoll and Harrison, 2010), and may influence droplet–droplet collisions (Khain et al., 2004), cloud droplet–particle collisions (Tinsley, 2008) and cloud droplet formation processes (Harrison and Ambaum, 2008). These microphysical effects may potentially influence cloud properties both directly and indirectly. Although Harrison and Ambaum (2010) observed a small reduction in downward longwave radiation that they associated with variations in surface current density, supporting observations are extremely limited. Our current understanding of the relationship between cloud properties and the global electric circuit remains very low, and there is no evidence yet that associated cloud processes could be of climatic significance.

7.4.6.3 Synthesis

Correlations between cosmic ray flux and observed aerosol or cloud properties are weak and local at best, and do not prove to be robust on the regional or global scale. Although there is some evidence that ionization from cosmic rays may enhance aerosol nucleation in the free troposphere, there is *medium evidence* and *high agreement* that the cosmic ray-ionization mechanism is too weak to influence global concentrations of CCN or droplets or their change over the last century or during a solar cycle in any climatically significant way.

7.5 Radiative Forcing and Effective Radiative Forcing by Anthropogenic Aerosols

7.5.1 Introduction and Summary of AR4

In this section, aerosol forcing estimates are synthesized and updated from AR4. As depicted in Figure 7.3, RF refers to the radiative forcing due to either aerosol–radiation interactions (ari), formerly known as the direct aerosol forcing, or aerosol–cloud interactions (aci), formerly known as the first indirect aerosol forcing or cloud albedo effect in AR4. ERF refers to the effective radiative forcing and is typically estimated from experiments with fixed SSTs (see Sections 7.1.3 and 8.1). It includes rapid adjustments, such as changes to the cloud lifetime, cloud altitude, changes in lapse rate due to absorbing aerosols and aerosol microphysical effects on mixed-phase, ice and convective clouds.

Chapter 2 of AR4 (Forster et al., 2007) assessed RFari to be -0.5 ± 0.4 W m^{-2} and broke this down into components associated with several species. Land albedo changes associated with BC on snow were assessed to be $+0.1 \pm 0.1$ W m^{-2}. The RFaci was assessed to be -0.70 W m^{-2} with a -1.8 to -0.3 W m^{-2} uncertainty range. These uncertainty estimates were based on a combination of model results and observations from remote sensing. The semi-direct effect and other aerosol indirect effects were assessed in Chapter 7 of AR4 (Denman et al., 2007) to contribute additional uncertainty. The combined total aerosol forcing was given as two distinct ranges: -2.3 to -0.2 W m^{-2} from models and a -1.7 to -0.1 W m^{-2} range from inverse estimates.

As discussed in Section 7.4, it is inherently difficult to separate RFaci from subsequent rapid cloud adjustments either in observations or model calculations (e.g., George and Wood, 2010; Lohmann et al., 2010; Mauger and Norris, 2010; Painemal and Zuidema, 2010). For this reason estimates of RFaci are of limited interest and are not assessed in this report. This chapter estimates RFari, ERFari, and ERFari+aci based purely on *a priori* approaches, and calculates ERFaci as the residual between ERFari+aci and ERFari assuming the two effects are additive. Inverse studies that estimate ERFari+aci from the observed rate of planetary energy uptake and estimates of climate feedbacks and other RFs are discussed in Section 10.8.

For consistency with AR4 and Chapter 8 of this Report, all quoted ranges represent a 5 to 95% uncertainty range unless otherwise stated, and we evaluate the forcings between 1750 and approximately 2010. The reference year of 1750 is chosen to represent pre-industrial times, so changes since then broadly represent the anthropogenic effect on climate, although for several aerosol species (such as biomass burning) this does not quite equate to the anthropogenic effect as emissions started to be influenced by humans before the Industrial Revolution. Many studies estimate aerosol forcings between 1850 and the present day and any conversion to a forcing between 1750 and the present day increases the uncertainty (Bellouin et al., 2008). This section principally discusses global forcing estimates and attributes them to aerosol species. Chapter 8 discusses regional forcings and additionally attributes aerosol forcing to emission sources.

7.5.2 Estimates of Radiative Forcing and Effective Radiative Forcing from Aerosol–Radiation Interactions

Building on our understanding of aerosol processes and their radiative effects (Section 7.3), this subsection assesses RFari and ERFari, but also the forcings from absorbing aerosol (BC and dust) on snow and ice.

7.5.2.1 Radiative Forcing and Effective Radiative Forcing from All Aerosols

Observations can give useful constraints to aspects of the global RFari but cannot measure it directly (Section 7.3.4; Anderson et al., 2005; Kahn, 2012). Remote sensing observations, *in situ* measurements of fine-mode aerosol properties and a better knowledge of bulk aerosol optical properties make the estimate of total RFari more robust than the RF for individual species (see Forster et al., 2007). Estimates of RFari are either taken from global aerosol models directly (Schulz et al., 2009; Myhre et al., 2013) or based mostly on observations, but using supplemental information from models (e.g., Myhre, 2009; Loeb and Su, 2010; Su et al., 2013). A number of studies (Bellouin et al., 2008; Zhao et al., 2008b, 2011; Myhre, 2009) have improved aspects of the satellite-based RFari estimate over those quoted in AR4. Of these, only Myhre (2009) make the necessary adjustments to the observations to account for forcing in cloudy regions and pre-industrial concentrations to estimate a RFari of -0.3 ± 0.2 W m^{-2}.

A second phase of AeroCom model results gives an RFari estimate of –0.35 W m^{-2}, with a model range of about –0.60 to –0.13 W m^{-2}, after their forcings for 1850–2000 have been scaled by emissions to represent 1750–2010 changes (Myhre et al., 2013). Figure 7.17 shows the zonal mean total RFari for AeroCom phase II models for 1850–2000. Robust features are the maximum negative RF around 10°N to 50°N, at latitudes of highest aerosol concentrations, and a positive RF at higher latitudes due to the higher surface albedo there.

For observationally based estimates, a variety of factors are important in constraining the radiative effect of aerosols (McComiskey et al., 2008; Loeb and Su, 2010; Kahn, 2012). Particularly important are the single scattering albedo (especially over land or above clouds) and the AOD (see Section 7.3.4.1). Errors in remotely sensed, retrieved AOD can be 0.05 or larger over land (Remer et al., 2005; Kahn et al., 2010; Levy et al., 2010; Kahn, 2012). Loeb and Su (2010) found that the total uncertainty in forcing was dominated by the uncertainty in single scattering albedo, using single scattering albedo errors of ± 0.06 over ocean and ± 0.03 over land from Dubovik et al. (2000), and assuming errors can be added in quadrature. These retrieval uncertainties could lead to a 0.5 to 1.0 W m^{-2} uncertainty in RFari (Loeb and Su, 2010). However, model sensitivity studies and reanalyses can provide additional constraints leading to a reduced error estimate. Ma et al. (2012b) performed a sensitivity study in one model, finding a best estimate of RFari of –0.41 W m^{-2} with an asymmetrical uncertainty range of –0.61 to –0.08 W m^{-2}, with BC particle size and mixing state having the largest effect of the parameters investigated. In models, assumptions about surface albedo, background cloud distribution and radiative transfer contribute a relative standard deviation of 39% (Stier et al., 2013). Bellouin et al. (2013) quantified uncertainties in RFari using reanalysis data that combined MODIS satellite data over oceans with the global coverage of their model. This approach broke down the uncertainty in aerosol properties into a local and a regional error to find a RFari standard deviation of 0.3 W m^{-2}, not accounting for uncertainty in the pre-industrial reference. When cloudy-sky and pre-industrial corrections were applied an RFari best estimate of –0.4 W m^{-2} was suggested.

The overall forcing uncertainty in RFari consists of the uncertainty in the distribution of aerosol amount, composition and radiative properties (Loeb and Su, 2010; Myhre et al., 2013), the uncertainty in radiative transfer (Randles et al., 2013) and the uncertainty owing to the dependence of the forcing calculation on other uncertain parameters, such as clouds or surface albedos (Stier et al., 2013). To derive a best estimate and range for RFari we combine modelling and observationally based studies. The best estimate is taken as –0.35 W m^{-2}. This is the same as the AeroCom II model estimate, and also the average of the Myhre (2009) observationally based estimate (–0.3 W m^{-2}) and the Bellouin et al. (2013) reanalysis estimate (–0.4 W m^{-2}). Models probably underestimate the positive RFari from BC and the negative forcing from OA aerosol (see Section 7.5.2.2), and currently there is no evidence that one of these opposing biases dominates over the other. The 5 to 95% range of RFari adopted in this assessment employs the Bellouin et al. (2013) uncertainty to account for retrieval error in observational quantities when constrained by global models, giving an uncertainty estimate of ±0.49 W m^{-2}. This is at the low end of the uncertainty analysis of Loeb and Su (2010). However, our uncertainty is partly based on models, and to account for this aspect, it is combined in quadrature with a ±0.1 W m^{-2} uncertainty from non-aerosol related parameters following Stier et al. (2013). This gives an assessed RFari of –0.35 ± 0.5 W m^{-2}. This is a larger range than that exhibited by the AeroCom II models. It is also a smaller magnitude but slightly larger range than in AR4, with a more positive upper bound. This more positive upper bound can be justified by the sensitivity to BC aerosol (Ma et al., 2012b and Section 7.5.2.3). Despite the larger range, there is increased confidence in this assessment due to dedicated modelling sensitivity studies, more robust observationally based estimates and their better agreement with models.

ERFari adds the radiative effects from rapid adjustments onto RFari. Studies have evaluated the rapid adjustments separately as a semi-direct effect (see Section 7.3.4.2) and/or the ERFari has been directly evaluated. Rapid adjustments are caused principally by cloud changes. There is *high confidence* that the local heating caused by absorbing aerosols can alter clouds. However, there is *low confidence* in determining the sign and magnitude of the rapid adjustments at the global scale as current models differ in their responses and are known to inadequately represent some of the important relevant cloud processes (see Section 7.3.4). Existing estimates of ERFari nevertheless rely on such global models. Five GCMs were analysed for RFari and ERFari in Lohmann et al. (2010). Their rapid adjustments ranged from –0.3 to +0.1 W m^{-2}. In a further study, Takemura and Uchida (2011) found a rapid adjustment of +0.06 W m^{-2}. The sensitivity analysis of Ghan et al. (2012) found a –0.1 to +0.1 W m^{-2} range over model variants, where an improved aging of the mixing state led to small negative rapid adjustment of around –0.1 W m^{-2}. Bond et al. (2013) assessed scaled RF and efficacy estimates from seven earlier studies focusing on

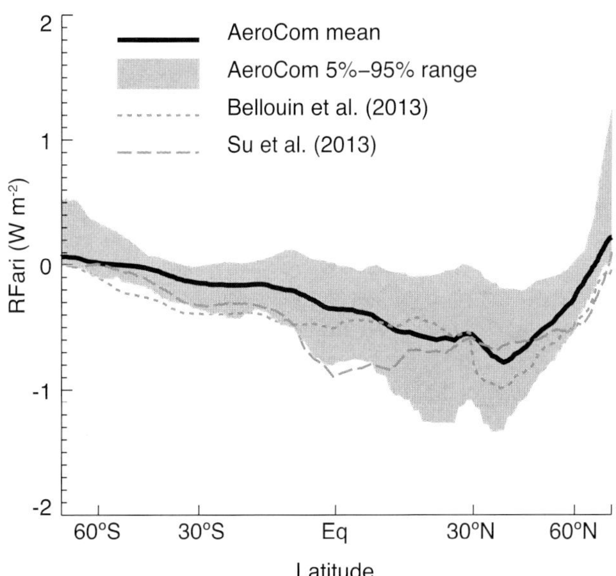

Figure 7.17 | Annual zonal mean top of the atmosphere radiative forcing due to aerosol–radiation interactions (RFari, in W m^{-2}) due to all anthropogenic aerosols from the different AeroCom II models. No adjustment for missing species in certain models has been applied. The multi-model mean and 5th to 95th percentile range from AeroCom II models (Myhre et al., 2013) are shown with a black solid line and grey envelope. The estimates from Bellouin et al. (2013) and Su et al. (2013) are shown with dotted and dashed lines, respectively. The forcings are for the 1850 to 2000 period. See Supplementary Material for a figure with labelled individual AeroCom II model estimates.

BC and found a range of rapid adjustments between –0.2 and –0.01 W m^{-2}. There is a potential additional rapid adjustment term from the effect of cloud drop inclusions (see Section 7.3.4.2). Based on Ghan et al. (2012) and Jacobson (2012), Bond et al. (2013) estimate an additional ERFari term of +0.2 W m^{-2}, with an uncertainty range of –0.1 to +0.9 W m^{-2}; however there is *very low confidence* in the sign or magnitude of this effect and we do not include it in our assessment. Overall a best estimate for the rapid adjustment is taken to be –0.1 W m^{-2}, with a 5 to 95% uncertainty range of –0.3 to +0.1 W m^{-2}. The best estimate is based on Ghan et al. (2012) and the range on Lohmann et al. (2010). The uncertainties are added in quadrature to the estimate of RFari and rounded to give an overall assessment for ERFari of –0.45 ± 0.5 W m^{-2}.

7.5.2.2 Radiative Forcing by Species

AeroCom II studies have calculated aerosol distributions using 1850 and 2000 simulations with the same meteorology to isolate RFari for individual aerosol types (sulphate, BC fossil-fuel plus biofuel, OA fossil-fuel and biofuel, biomass burning or BB, SOA, nitrate). Many of these models account for internal mixing, so that partitioning RFari by species is not straightforward, and different modelling groups adopt different techniques (Myhre et al., 2013). Note also that due to internal mixing of aerosol types the total RFari is not necessarily the sum of the RFari from different types (Ocko et al., 2012). Unless otherwise noted in the text below, the best estimate and 5 to 95% ranges for individual types quoted in Figure 7.18 are solely based on the AeroCom II range (Myhre et al., 2013) and the estimates have been scaled by emissions to derive 1750–2010 RFari values. Note that although global numbers are presented here, these RF estimates all exhibit large regional variations, and individual aerosol species can contribute significantly to regional climate change despite rather small RF estimates (e.g., Wang et al., 2010b).

For sulphate, AeroCom II models give a RF median and 5 to 95% uncertainty range of –0.31 (–0.58 to –0.11) W m^{-2} for the 1850–2000 period, and –0.34 (–0.61 to –0.13) W m^{-2} for the 1750–2010 period. This estimate and uncertainty range are consistent with the AR4 estimate of –0.4 ± 0.2 W m^{-2}, which is retained as the best estimate for AR5.

RF from BC is evaluated in different ways in the literature. The BC RF in this report is from fossil fuel and biofuel sources, while open burning sources are attributed separately to the biomass-burning aerosol, which also includes other organic species (see Section 7.3.2). BC can also affect clouds and surface albedo (see Sections 7.5.2.3 and Chapter 8). Here we only isolate the fossil fuel and biofuel RFari attributable to BC over 1750–2010. Two comprehensive studies have quantified the BC RFari and derive different central estimates and uncertainty ranges. Myhre et al. (2013) quantify RF over 1850–2000 in the AeroCom II generation of models and scale these up using emissions to derive an RF estimate over 1750–2010 of +0.23 (+0.06 to +0.48) W m^{-2} for fossil fuel and biofuel emissions. Bond et al. (2013) employ an observationally weighted scaling of an earlier generation of AeroCom models, regionally scaling BC absorption to match absorption AOD as retrieved at available AERONET sites. They derive a RF of +0.51 (+0.06 to +0.91) W m^{-2} for fossil fuel and biofuel sources. There are known biases in BC RF estimates from aerosol models. BC concentrations are underestimated near source regions, especially in Asia, but overestimated in remote regions and at altitude (Figure 7.15). Models also probably underestimate the mass absorption cross-section probably because enhanced absorption due to internal mixing is insufficiently accounted for (see Section 7.3.3.2). Together these biases are expected to cause the modelled BC RF to be underestimated. The Bond et al. estimate accounted for these biases by scaling model results. However, there are a number of methodological difficulties associated with the absorption AOD retrieval from sunphotometer retrievals (see Section 7.3.3.2), the attribution of absorption AOD to BC, and the distribution and representativeness of AERONET stations for constraining global and relatively coarse-resolution models. Absorption by OA (see Section 7.3.3), which may amount to 20% of fine-mode aerosol absorption (Chung et al., 2012), is included into the BC RF estimate in Bond et al. but is now treated separately in most AeroCom II models, some of which have a global absorption AOD close to the Bond et al. estimate. We use our expert judgement here to adopt a BC RF estimate that is halfway between the two estimates and has a wider uncertainty range from combining distributions. This gives a BC RF estimate from fossil fuel and biofuel of +0.4 (+0.05 to +0.8) W m^{-2}.

The AeroCom II estimate of the SOA RFari is –0.03 (–0.27 to –0.02) W m^{-2} and the primary OA from fossil fuel and biofuel estimate is –0.05 (–0.09 to –0.02) W m^{-2}. An intercomparison of current chemistry–climate models found two models outside of this range for SOA RFari, with one model exhibiting a significant positive forcing from land use and cover changes influencing biogenic emissions (Shindell et al., 2013). We therefore adjust the upper end of the range to account for this, giving an SOA RFari estimate of –0.03 (–0.27 to +0.20) W m^{-2}. Our assessment also scales the AeroComII estimate of the primary OA from fossil fuel and biofuel by 1.74 to –0.09 (–0.16 to –0.03) W m^{-2} to allow for the underestimate of emissions identified in Bond et al. (2013). For OA from natural burning, and for SOA, the natural radiative effects can be an order of magnitude larger than the RF (see Sections 7.3.2 and 7.3.4, and O'Donnell et al., 2011) and they could thus contribute to climate feedback (see Section 7.3.5).

The RFari from biomass burning includes both BC and OA species that contribute RFari of opposite sign, giving a net RFari close to zero (Bond et al., 2013; Myhre et al., 2013). The AeroCom II models give a 1750–2010 RFari of 0.00 (–0.08 to +0.07) W m^{-2}, and an estimate of +0.0 (–0.2 to +0.2) W m^{-2} is adopted in this assessment, doubling the model uncertainty range to account for a probable underestimate of their emissions (Bond et al., 2013). Combining information in Samset et al. (2013) and Myhre et al. (2013) would give a BC RFari contribution from biomass burning of slightly less than +0.1 W m^{-2} over 1850–2000 from the models. However, this ignores a significant contribution expected before 1850 and the probable underestimate in emissions. Our assessment therefore solely relies on Bond et al. (2013), giving an estimate of +0.2 (+0.03 to +0.4) W m^{-2} for the 1750–2010 BC contribution to the biomass burning RFari. This is a 50% larger forcing than the earlier generation of AeroCom models found (Schulz et al., 2006). Note that we also expect an OA RFari of the same magnitude with opposite sign.

The AeroCom II RF estimate for nitrate aerosol gives an RFari of –0.11 (–0.17 to –0.03) W m^{-2}, but comprises a relatively large 1850 to 1750 correction term. In these models ammonium aerosol is included within the sulphate and nitrate estimates. An intercomparison of current chemistry–climate models found an RF range of –0.41 to –0.03 W m^{-2} over 1850–2000. Some of the models with strong RF did not exhibit obvious biases, whereas others did (Shindell et al., 2013). These sets of estimates are in good agreement with earlier estimates (e.g., Adams et al., 2001; Bauer et al., 2007; Myhre et al., 2009). Our assessment of the RFari from nitrate aerosol is –0.11 (–0.3 to –0.03) W m^{-2}. This is based on AeroCom II with an increased lower bound.

Anthropogenic sources of mineral aerosols can result from changes in land use and water use or climate change. Estimates of the RF from anthropogenic mineral aerosols are highly uncertain, because natural and anthropogenic sources of mineral aerosols are often located close to each other (Mahowald et al., 2009; Ginoux et al., 2012b). Using a compilation of observations of dust records over the 20th century with model simulations, Mahowald et al. (2010) deduced a 1750–2000 change in mineral aerosol RFari including both natural and anthropogenic changes of –0.14 ± 0.11 W m^{-2}. This is consistent within the AR4 estimate of –0.1 ± 0.2 W m^{-2} (Forster et al., 2007) which is retained here. Note that part of the dust RF could be due to feedback processes (see Section 7.3.5).

Overall the species breakdown of RFari is less certain than the total RFari. Fossil fuel and biofuel emissions contribute to RFari via sulphate aerosol –0.4 (–0.6 to –0.2) W m^{-2}; black carbon aerosol +0.4 (+0.05 to +0.8) W m^{-2}; and primary and secondary organic aerosol –0.12 (–0.4 to +0.1) W m^{-2} (adding uncertainties in quadrature). Additional RFari contributions are via biomass burning emissions, where black carbon and organic aerosol changes offset each other to give an estimate of +0.0 (–0.2 to +0.2) W m^{-2}; nitrate aerosol –0.11 (–0.3 to –0.03) W m^{-2}; and a contribution from mineral dust of –0.1 (–0.3 to +0.1) W m^{-2} that may not be entirely anthropogenic in origin. The sum of the RFari

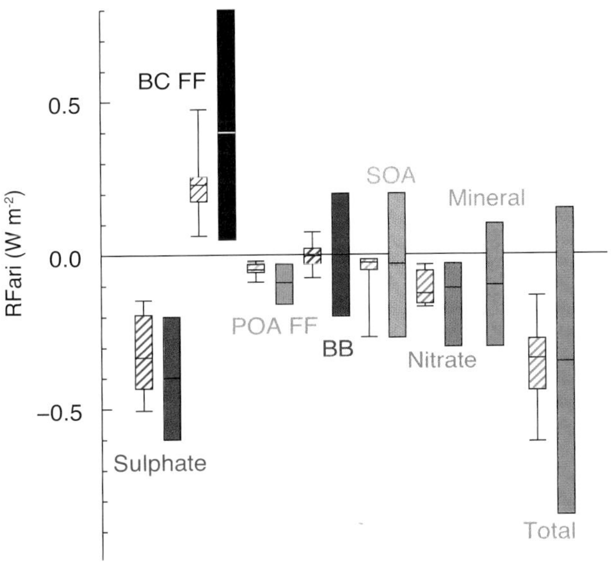

Figure 7.18 | Annual mean top of the atmosphere radiative forcing due to aerosol–radiation interactions (RFari, in W m^{-2}) due to different anthropogenic aerosol types, for the 1750–2010 period. Hatched whisker boxes show median (line), 5th to 95th percentile ranges (box) and min/max values (whiskers) from AeroCom II models (Myhre et al., 2013) corrected for the 1750–2010 period. Solid coloured boxes show the AR5 best estimates and 90% uncertainty ranges. BC FF is for black carbon from fossil fuel and biofuel, POA FF is for primary organic aerosol from fossil fuel and biofuel, BB is for biomass burning aerosols and SOA is for secondary organic aerosols.

from these species agrees with, but is slightly weaker than, the best estimate of the better-constrained total RFari.

7.5.2.3 Absorbing Aerosol on Snow and Sea Ice

Forster et al. (2007) estimated the RF for surface albedo changes from BC deposited on snow to be +0.10 ± 0.10 W m^{-2}, with a low level of understanding, based largely on studies from Hansen and Nazarenko (2004) and Jacobson (2004). Since AR4, observations of BC in snow have been conducted using several different measurement techniques (e.g., McConnell et al., 2007; Forsström et al., 2009; Ming et al., 2009; Xu et al., 2009; Doherty et al., 2010; Huang et al., 2011; Kaspari et al., 2011), providing data with which to constrain models. Laboratory measurements have confirmed the albedo reduction due to BC in snow (Hadley and Kirchstetter, 2012). The albedo effects of non-BC constituents have also been investigated but not rigorously quantified. Remote sensing can inform on snow impurity content in some highly polluted regions. However, it cannot be used to infer global anthropogenic RF because of numerous detection challenges (Warren, 2013).

Global modelling studies since AR4 have quantified present-day radiative effects from BC on snow of +0.01 to +0.08 W m^{-2} (Flanner et al., 2007, 2009; Hansen et al., 2007; Koch et al., 2009a; Rypdal et al., 2009; Skeie et al., 2011; Wang et al., 2011c; Lee et al., 2013). These studies apply different BC emission inventories and atmospheric aerosol representations, include forcing from different combinations of terrestrial snow, sea ice, and snow on sea ice, and some include different rapid adjustment effects such as snow grain size evolution and melt-induced accumulation of impurities at the snow surface, observed on Tibetan glaciers (Xu et al., 2012) and in Arctic snow (Doherty et al., 2013). The forcing operates mostly on terrestrial snow and is largest during March to May, when boreal snow and ice are exposed to strong insolation (Flanner et al., 2007).

All climate modelling studies find that the Arctic warms in response to snow and sea ice forcing. In addition, estimates of the change in global mean surface temperature per unit forcing are 1.7 to 4.5 times greater for snow and sea ice forcing than for CO_2 forcing (Hansen and Nazarenko, 2004; Hansen et al., 2005; Flanner et al., 2007; Flanner et al., 2009; Bellouin and Boucher, 2010). The Koch et al. (2009a) estimate is not included in this range owing to the lack of a clear signal in their study. The greater response of global mean temperature occurs primarily because all of the forcing energy is deposited directly into the cryosphere, whose evolution drives a positive albedo feedback on climate. Key sources of forcing uncertainty include BC concentrations in snow and ice, BC mixing state and optical properties, snow and ice coverage and patchiness, co-presence of other light-absorbing particles in the snow pack, snow effective grain size and its influence on albedo perturbation, the masking of snow surfaces by clouds and vegetation and the accumulation of BC at the top of snowpack caused by melting and sublimation. Bond et al. (2013) derive a 1750–2010 snow and sea ice RF estimate of +0.046 (+0.015 to +0.094) W m^{-2} for BC by (1) considering forcing ranges from all relevant global studies, (2) accounting for biases caused by (a) modelled Arctic BC-in-snow concentrations using measurements from Doherty et al. (2010), and (b) excluding mineral dust, which reduces BC forcing by approximately 20%, (3) combining in quadrature individual uncertainty terms from Flanner et al. (2007)

plus that originating from the co-presence of dust, and (4) scaling the present-day radiative contributions from BB, biofuel and fossil fuel BC emissions according to their 1750–2010 changes. Note that this RF estimate allows for some rapid adjustments in the snowpack but is not a full ERF as it does not account for adjustments in the atmosphere. For this RF, we adopt an estimate of +0.04 (+0.02 to +0.09) W m^{-2} and note that the surface temperature change is roughly three (two to four) times more responsive to this RF relative to CO_2.

7.5.3 Estimate of Effective Radiative Forcing from Combined Aerosol–Radiation and Aerosol–Cloud Interactions

In addition to ERFari, there are changes due to aerosol–cloud interactions (ERFaci). Because of nonlinearities in forcings and rapid adjustments, the total effective forcing ERFari+aci does not necessarily equal the sum of the ERFari and ERFaci calculated separately. Moreover a strict separation is often difficult in either state of the art models or observations. Therefore we first assess ERFari+aci and postpone our assessment of ERFaci to Section 7.5.4. For similar reasons, we focus primarily on ERF rather than RF.

ERFari+aci is defined as the change in the net radiation at the TOA from pre-industrial to present day. Climate model estimates of ERFari+aci in the literature differ for a number of reasons. (1) The reference years for pre-industrial and present-day conditions vary between estimates. Studies can use 1750, 1850, or 1890 for pre-industrial; early estimates of ERFari+aci used present-day emissions for 1985, whereas most newer estimates use emissions for the year 2000. (2) The processes they include also differ: aerosol–cloud interactions in large-scale liquid stratiform clouds are typically included, but studies can also include aerosol–cloud interactions for mixed-phase, convective clouds and/or cirrus clouds. (3) The way in which ERFari+aci is calculated can also differ between models, with some earlier studies only reporting the change in shortwave radiation. Changes in longwave radiation arise from rapid adjustments, or from aerosol–cloud interactions involving mixed-phase or ice clouds (e.g., Storelvmo et al., 2008b, 2010; Ghan et al., 2012), and tend to partially offset changes in shortwave radiation. In the estimates discussed below and those shown in Figure 7.19, we refer to estimates of the change in net (shortwave plus longwave) TOA radiation whenever possible, but report changes in shortwave radiation when changes in net radiation are not available. While this mostly affects earlier studies, the subset of models that we concentrate on all include both shortwave and longwave radiative effects. However, for the sake of comparison, the satellite studies must be adjusted to account for missing longwave contributions as explained below.

Early GCM estimates of ERFari+aci only included aerosol–cloud interactions in liquid phase stratiform clouds; some of these were already considered in AR4. Grouping these early estimates with similar (liquid phase only) estimates from publications since the AR4 yields a median value of ERFari+aci of –1.5 W m^{-2} with a 5 to 95% range between –2.4 and –0.6 W m^{-2} (Figure 7.19). In those studies that attempt a more complete representation of aerosol–cloud interactions, by including aerosol–cloud interactions in mixed-phase and/or convective cloud, the magnitude of the ERF tends to be somewhat smaller (see Figure 7.19). The physical explanation for the mixed-phase reduction in the magnitude of the ERF is that some aerosols also act as IN causing supercooled clouds to glaciate and precipitate more readily. This reduction in cloud cover leads to less reflected shortwave radiation and a less negative ERFari+aci. This effect can however be offset if the IN become coated with soluble material, making them less effective at nucleating ice, leading to less efficient precipitation production and more reflected shortwave radiation (Hoose et al., 2008; Storelvmo et al., 2008a). Models that have begun to incorporate aerosol–cloud interactions in convective clouds also have a tendency to reduce the magnitude of the ERF, but this effect is less systematic (Jacobson, 2003; Lohmann, 2008; Suzuki et al., 2008) and reasons for differences among the models in this category are less well understood.

For our expert judgment of ERFari+aci a subset of GCM studies, which strived for a more complete and consistent treatment of aerosol–cloud interactions (by incorporating either convective or mixed-phase processes) was identified and scrutinized. The ERFari+aci derived from these models is somewhat less negative than in the full suite of models, and ranges from –1.68 and –0.81 W m^{-2} with a median value of –1.38 W m^{-2}. Because in some cases a number of studies have been performed with the same GCM, in what might be described as an evolving effort, our assessment is further restricted to the best (usually most recent) estimate by each modelling group (see black symbols in Figure 7.19 and Table 7.4). This ensures that no single GCM is given a disproportionate weight. Further, we consider only simulations not constrained by the historical temperature rise, motivated by the desire to emphasize a process-based estimate. Although it may be argued that greater uncertainty is introduced by giving special weight to models that only incorporate more comprehensive treatments of aerosol–cloud interactions, and for processes that (as Section 7.4 emphasizes) are on the frontier of understanding, it should be remembered that aerosol–cloud interactions for liquid-phase clouds remain very uncertain. Although the understanding and treatment of aerosol–cloud interactions in convective or mixed-phase clouds are also very uncertain, as discussed in Section 7.4.4, we exercise our best judgment of their influence.

A less negative ERFari+aci (–0.93 to –0.45 W m^{-2} with a median of –0.85 W m^{-2}, Figure 7.19 and Table 7.4) is found in studies that use variability in the present day satellite record to infer aerosol–cloud interactions, or that constrain GCM parameterizations to optimize agreement with satellite observations. Because some groups have published multiple estimates as better information became available, only their latest study was incorporated into this assessment. Moreover, if a study did not report ERFari+aci but only evaluated changes in ERFaci, their individual estimate was combined with the average ERFari of –0.45 W m^{-2} from Section 7.5.2. Likewise, those (all but one) studies that only accounted for changes in shortwave radiation when computing ERFaci were corrected by adding a constant factor of +0.2 W m^{-2}, taken from the lower range of the modeled longwave effects which varied from +0.2 to +0.6 W m^{-2} in the assessed models. These procedures result in the final estimates of ERFari+aci shown as black symbols in Figure 7.19 and in Table 7.4. This resulted in a median ERFari+aci of –0.85 W m^{-2} for satellite-based ERFari+aci estimates. Results of pure satellite-based studies are sensitive to the spatial scale of measurements (Grandey and Stier, 2010; McComiskey and Feingold, 2012; Section 7.4.2.2), as well as to how pre-industrial conditions and variations between pre-industrial and present-day conditions are inferred from the observed

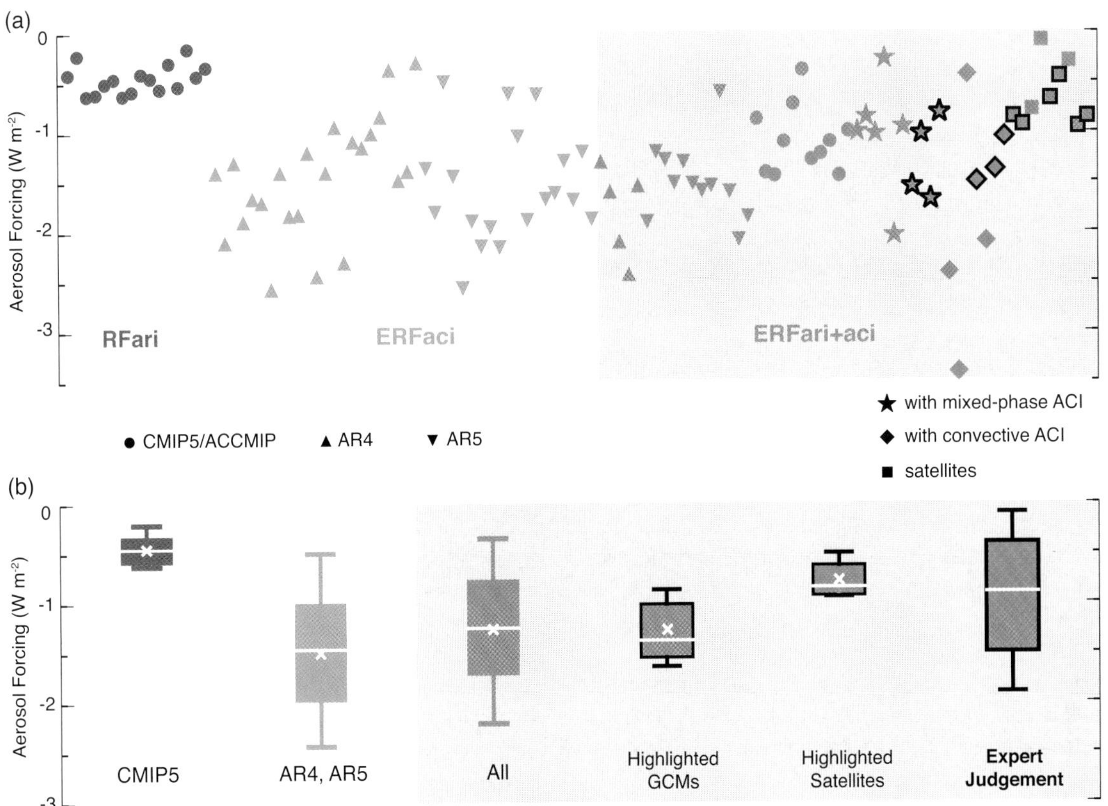

Figure 7.19 | (a) GCM studies and studies involving satellite estimates of RFari (red), ERFaci (green) and ERFari+aci (blue in grey-shaded box). Each symbol represents the best estimate per model and paper (see Table 7.3 for references). The values for RFari are obtained from the CMIP5 models. ERFaci and ERFari+aci studies from GCMs on liquid phase stratiform clouds are divided into those published prior to and included in AR4 (labelled AR4, triangles up), studies published after AR4 (labelled AR5, triangles down) and from the CMIP5/ACCMIP models (filled circles). GCM estimates that include adjustments beyond aerosol–cloud interactions in liquid phase stratiform clouds are divided into those including aerosol–cloud interactions in mixed-phase clouds (stars) and those including aerosol–cloud interactions in convective clouds (diamonds). Studies that take satellite data into account are labelled as 'satellites'. Studies highlighted in black are considered for our expert judgement of ERFari+aci. (b) Whisker boxes from GCM studies and studies involving satellite data of RFari, ERFaci and ERFari+aci. They are grouped into RFari from CMIP5/ACCMIP GCMs (labelled CMIP5 in red), ERFaci from GCMs (labelled AR4, AR5 in green), all estimates of ERFari+aci shown in the upper panel (labelled 'All' in blue), ERFari+aci from GCMs highlighted in the upper panel (labelled 'Highlighted GCMs' in blue), ERFari+aci from satellites highlighted in the upper panel (labelled 'Highlighted Satellites' in blue), and our expert judgement based on estimates of ERFari+aci from these GCM and satellite studies (labelled 'Expert Judgement' in blue). Displayed are the averages (cross sign), median values (middle line), 17th and 83th percentiles (*likely* range shown as box boundaries) and 5th and 95th percentiles (whiskers). References for the individual estimates are provided in Table 7.3. Table 7.4 includes the values of the GCM and satellite studies considered for the expert judgement of ERFari+aci that are highlighted in black.

Table 7.3 | List of references for each category of estimates displayed in Figure 7.19.

Estimate	Acronym	References
Effective radiative forcing due to aerosol–cloud interactions (ERFaci) published prior to and considered in AR4	AR4	Lohmann and Feichter (1997); Rotstayn (1999); Lohmann et al., (2000); Ghan et al. (2001); Jones et al. (2001); Rotstayn and Penner (2001); Williams et al. (2001); Kristjánsson (2002); Lohmann (2002a); Menon et al. (2002); Peng and Lohmann (2003); Penner et al. (2003); Easter et al. (2004); Kristjánsson et al. (2005); Ming et al. (2005); Rotstayn and Liu (2005); Takemura et al. (2005); Johns et al. (2006); Penner et al. (2006); Quaas et al. (2006); Storelvmo et al. (2006)
ERFaci published since AR4	AR5	Menon and Del Genio (2007); Ming et al. (2007b); Kirkevåg et al. (2008); Seland et al. (2008); Storelvmo et al. (2008a); Hoose et al. (2009); Quaas et al. (2009); Rotstayn and Liu (2009); Chen et al. (2010); Ghan et al. (2011); Penner et al. (2011); Makkonen et al. (2012a); Takemura (2012); Kirkevåg et al. (2013)
Effective radiative forcing due to aerosol–radiation and aerosol–cloud interactions (ERFari+aci) in liquid phase stratiform clouds published prior to AR4	AR4	Lohmann and Feichter (2001); Quaas et al. (2004); Menon and Rotstayn (2006); Quaas et al. (2006)
ERFari+aci in liquid phase stratiform clouds published since AR4	AR5	Lohmann et al. (2007); Rotstayn et al. (2007); Posselt and Lohmann (2008); Posselt and Lohmann (2009); Quaas et al. (2009); Salzmann et al. (2010); Bauer and Menon (2012); Gettelman et al. (2012); Ghan et al. (2012); Makkonen et al. (2012a); Takemura (2012); Kirkevåg et al. (2013)
ERFari+aci in liquid and mixed-phase stratiform clouds	with mixed-phase clouds	Lohmann (2004); Jacobson (2006); Lohmann and Diehl (2006); Hoose et al. (2008); Storelvmo et al. (2008a); Lohmann and Hoose (2009); Hoose et al. (2010b); Lohmann and Ferrachat (2010); Salzmann et al. (2010); Storelvmo et al. (2010)
ERFari+aci in stratiform and convective clouds	with convective clouds	Menon and Rotstayn (2006); Menon and Del Genio (2007); Lohmann (2008); Koch et al. (2009a); Unger et al. (2009); Wang et al. (2011b)
ERFari+aci including satellite observations	Satellites	Lohmann and Lesins (2002); Sekiguchi et al. (2003); Quaas et al. (2006); Lebsock et al. (2008); Quaas et al. (2008); Quaas et al. (2009); Bellouin et al. (2013)

7

Table 7.4 | List of ERFari+aci values (W m^{-2}) considered for the expert judgement of ERFari+aci (black symbols in Figure 7.19). For the GCM studies only the best estimate per modelling group is used. For satellite studies the estimates are corrected for the ERFari and for the longwave component of ERFari+aci when these are not included (see text).

Category	Best Estimate	Climate Model and/or Satellite Instrument	Reference
with mixed-phase clouds	–1.55	CAM Oslo	Hoose et al. (2010b)
with mixed-phase clouds	–1.02	ECHAM	Lohmann and Ferrachat (2010)
with mixed-phase clouds	–1.68	GFDL	Salzmann et al. (2010)
with mixed-phase clouds	0.81	CAM Oslo	Storelvmo et al. (2008b; 2010)
with convective clouds	–1.50	ECHAM	Lohmann (2008)
with convective clouds	–1.38	GISS	Koch et al. (2009a)
with convective clouds	–1.05	PNNL-MMF	Wang et al. (2011b)
Satellite-based	–0.85	ECHAM + POLDER	Lohmann and Lesins (2002)
Satellite-based	–0.93	AVHRR	Sekiguchi et al. (2003)
Satellite-based	–0.67	CERES / MODIS	Lebsock et al. (2008)
Satellite-based	–0.45	CERES / MODIS	Quaas et al. (2008)
Satellite-based	–0.95	Model mean + MODIS	Quaas et al. (2009)
Satellite-based	–0.85	MACC + MODIS	Bellouin et al. (2013)

AVHRR = Advanced Very High Resolution Radiometer. MACC = Monitoring Atmospheric Composition and Climate. POLDER = Polarization and Directionality of the Earth's Reflectances. CERES = Clouds and the Earth's Radiant Energy System. MODIS = Moderate Resolution Imaging Spectrometer.

variability in present-day aerosol and cloud properties (Quaas et al., 2011; Penner et al., 2012). In addition, all (model- and satellite-based) estimates of ERFari+aci are very sensitive to the assumed pre-industrial or natural cloud droplet concentration (Hoose et al., 2009). The large spatial scales of satellite measurements relative to *in situ* measurements generally suggest smaller responses in cloud droplet number increases for a given aerosol increase (Section 7.4.2.2). Satellite studies, however, show a strong effect of aerosol on cloud amount, which could be a methodological artefact as GCMs associate clouds with humidity and aerosol swelling (Quaas et al., 2010). There are thus possible biases in both directions, so the sign and magnitude of any net bias is not clear.

In large-scale models for which cloud-scale circulations are not explicitly represented, it is difficult to capture all relevant cloud controlling processes (Section 7.2.2). Because the response of clouds to aerosol perturbations depends critically on the interplay of poorly understood physical processes, global model-based estimates of aerosol–cloud interactions remain uncertain (Section 7.4). Moreover, the connection between the aerosol amount and cloud properties is too direct in the large-scale modelling studies (as it relies heavily on the autoconversion rate). Because of this, GCMs tend to overestimate the magnitude of the aerosol effect on cloud properties (Section 7.4.5; see also discussion in Section 7.5.4). This view has some support from studies that begin to incorporate some cloud, or cloud-system scale responses to aerosol–cloud interactions. For instance, in an attempt to circumvent some of difficulties of parameterizing clouds, some groups (e.g., Wang et al., 2011b) have begun developing modelling frameworks that can explicitly represent cloud-scale circulations, and hence the spatio-temporal covariances of cloud-controlling processes. Another group (Khairoutdinov and Yang, 2013) has used the same cloud-resolving model in a radiative convective equilibrium approach, and compared the relative contribution of aerosol–cloud interactions to warming from the doubling of atmospheric CO_2. In both studies a smaller (–1.1 and –0.8 W m^{-2}, respectively) ERFari+aci than for the average GCM was found. Furthermore, the study best resolving the cloud-scale circulations (Wang et al., 2011b) found little change in cloud amount in response to changes in aerosol, consistent with other fine-scale modelling studies discussed in Section 7.4.

Based on the above considerations, we assess ERFari+aci using expert judgement to be –0.9 W m^{-2} with a 5 to 95% uncertainty range of –1.9 to –0.1 W m^{-2} (*medium confidence*), and a *likely* range of –1.5 to –0.4 W m^{-2}. These ranges account for the GCM results by allowing for an ERFari+aci somewhat stronger than what is estimated by the satellite studies with a longer tail in the direction of stronger effects, but (for reasons given above) give less weight to the early GCM estimates shown in Figure 7.19. The ERFari+aci can be much larger regionally but the global value is consistent with several new lines of evidence suggesting less negative estimates of aerosol–cloud interactions than the corresponding estimate in Chapter 7 of AR4 of –1.2 W m^{-2}. The AR4 estimate was based mainly on GCM studies that did not take secondary processes (such as aerosol effects on mixed-phase and/or convective clouds) into account, did not benefit as much from the use of the recent satellite record, and did not account for the effect of rapid adjustments on the longwave radiative budget. This uncertainty range is slightly smaller than the –2.3 to –0.2 W m^{-2} in AR4, with a less negative upper bound due to the reasons outlined above. The best estimate of ERFari+aci is not only consistent with the studies allowing cloud-scale responses (Wang et al., 2011b; Khairoutdinov and Yang, 2013) but also is in line with the average ERFari+aci from the CMIP5/ACCMIP models (about –1 W m^{-2}, see Table 7.5), which as a whole reproduce the observed warming of the 20th century (see Chapter 10). Studies that infer ERFari+aci from the historical temperature rise are discussed in Section 10.8.

7.5.4 Estimate of Effective Radiative Forcing from Aerosol–Cloud Interactions Alone

ERFaci refers to changes in TOA radiation since pre-industrial times due only to aerosol–cloud interactions, i.e., albedo effects augmented by possible changes in cloud amount and lifetime. As stated in Section 7.5.1, we do not discuss RFaci by itself because it is an academic

Table 7.5 | Estimates of aerosol 1850–2000 effective radiative forcing (ERF, in W m^{-2}) in some of the CMIP5 and ACCMIP models. The ERFs are estimated from fixed-sea-surface temperature (SST) experiments using atmosphere-only version of the models listed. Different models include different aerosol effects. The CMIP5 and ACCMIP protocols differ, hence differences in forcing estimates for one model.

Modelling Group	Model Name	ERFari+aci from All Anthropogenic Aerosols	ERFari+aci from Sulphate Aerosols Only
CCCma	CanESM2	–0.87	–0.90
CSIRO-QCCCE	CSIRO-Mk3-6-0[b]	–1.41	–1.10
GFDL	GFDL-AM3	–1.60 (–1.44[a])	–1.62
GISS	GISS-E2-R[b]	–1.10[a]	–0.61
GISS	GISS-E2-R-TOMAS[b]	–0.76[a]	
IPSL	IPSL-CM5A-LR	–0.72	–0.71
LASG-IAP	FGOALS-s2[c]	–0.38	–0.34
MIROC	MIROC-CHEM[b]	–1.24[a]	
MIROC	MIROC5	–1.28	–1.05
MOHC	HadGEM2-A	–1.22	–1.16
MRI	MRI-CGM3	–1.10	–0.48
NCAR	NCAR-CAM5.1[b]	–1.44[a]	
NCC	NorESM1-M	–0.99	
Ensemble mean		–1.08	
Standard deviation		+0.32	

Notes:

[a] From ACCMIP (Shindell et al., 2013).

[b] These models include the black carbon on snow effect.

[c] This model does not include the ERF from aerosol–cloud interactions.

ACCMIP = Atmospheric Chemistry and Climate Model Intercomparison Project.

CMIP5 = Coupled Model Intercomparison Project Phase 5.

construct. However, processes in GCMs that tend to affect RFaci such as changes to the droplet size distribution breadth (e.g., Rotstayn and Liu, 2005) will also affect ERFaci. Early studies evaluated just the change in shortwave radiation or cloud radiative effect for ERFaci, but lately the emphasis has changed to report changes in net TOA radiation for ERFaci. As discussed in Section 7.5.3, evaluating ERFaci from changes in net TOA radiation is the only correct method, and therefore this is used whenever possible also in this section. However some earlier estimates of ERFaci only reported changes in cloud radiative effect, which we show in Figure 7.19 as the last resort. However, estimates of changes in cloud radiative effect can differ quite substantially from those in net radiation if rapid adjustments to aerosol–cloud interactions induce changes in clear-sky radiation.

Cloud amount and lifetime effects manifest themselves in GCMs via their representation of autoconversion of cloud droplets to rain, a process that is inversely dependent on droplet concentration. Thus, ERFaci and ERFari+aci have been found to be very sensitive to the autoconversion parameterization (Rotstayn, 2000; Golaz et al., 2011; Wang et al., 2012). GCMs probably underestimate the extent to which precipitation is formed via raindrop accretion of cloud droplets (Wood, 2005), a process that is insensitive to aerosol and droplet concentration. Indeed, models that remedy this imbalance in precipitation formation between autoconversion and accretion (Posselt and Lohmann, 2009; Wang et al., 2012) exhibit weaker ERFaci in agreement with small-scale studies that typically do not show a systematic increase in cloud lifetime because of entrainment and because smaller droplets also evaporate more readily (Jiang et al., 2006; Bretherton et al., 2007). Bottom-up estimates of ERFaci are shown in Figure 7.19. Their median estimate of –1.4 W m^{-2} is more negative than our expert judgement of ERFari+aci because of the limitations of these studies discussed above.

There is conflicting evidence for the importance of ERFaci associated with cirrus, ranging from a statistically significant impact on cirrus coverage (Hendricks et al., 2005) to a very small effect (Liu et al., 2009). Penner et al. (2009) obtained a rather large negative RFaci of anthropogenic ice-forming aerosol on upper tropospheric clouds of –0.67 to –0.53 W m^{-2}; however, they ignored potential compensating effects on lower lying clouds. A new study based on two GCMs and different ways to deduce ERFaci on cirrus clouds estimates ERFaci to be +0.27 ± 0.1 W m^{-2} (Gettelman et al., 2012), thus rendering aerosol effects on cirrus clouds smaller than previously estimated and of opposite sign.

One reason for having switched to providing an expert judgment estimate of ERFari+aci rather than of ERFaci is that the individual contributions are very difficult to disentangle. The individual components can be isolated only if linearity of ERFari and ERFaci is assumed but there is no *a priori* reason why the ERFs should be additive because by definition they occur in a system that is constantly readjusting to multiple nonlinear forcings. Nevertheless assuming additivity, ERFaci could be obtained as the difference between ERFari+aci and ERFari. This yields an ERFaci estimate of –0.45 W m^{-2}, that is, much smaller than the median ERFaci value of –1.4 W m^{-2} (see above and Figure 7.19). This discrepancy arises because the GCM estimates of ERFaci do not consider secondary processes and because these studies are not necessarily conducted with the same GCMs that estimate ERFari+aci. This difference could also be a measure of the non-linearity of the ERFs. A 90% uncertainty range of –1.2 to 0 W m^{-2} is adopted for ERFaci, which accounts for the error covariance between ERFari and ERFaci and the larger uncertainty on the lower bound. In summary, there is much less confidence associated with the estimate of ERFaci than with the estimate of ERFari+aci.

Frequently Asked Questions

FAQ 7.2 | How Do Aerosols Affect Climate and Climate Change?

Atmospheric aerosols are composed of small liquid or solid particles suspended in the atmosphere, other than larger cloud and precipitation particles. They come from natural and anthropogenic sources, and can affect the climate in multiple and complex ways through their interactions with radiation and clouds. Overall, models and observations indicate that anthropogenic aerosols have exerted a cooling influence on the Earth since pre-industrial times, which has masked some of the global mean warming from greenhouse gases that would have occurred in their absence. The projected decrease in emissions of anthropogenic aerosols in the future, in response to air quality policies, would eventually unmask this warming.

Atmospheric aerosols have a typical lifetime of one day to two weeks in the troposphere, and about one year in the stratosphere. They vary greatly in size, chemical composition and shape. Some aerosols, such as dust and sea spray, are mostly or entirely of natural origin, while other aerosols, such as sulphates and smoke, come from both natural and anthropogenic sources.

Aerosols affect climate in many ways. First, they scatter and absorb sunlight, which modifies the Earth's radiative balance (see FAQ.7.2, Figure 1). Aerosol scattering generally makes the planet more reflective, and tends to cool the climate, while aerosol absorption has the opposite effect, and tends to warm the climate system. The balance between cooling and warming depends on aerosol properties and environmental conditions. Many observational studies have quantified local radiative effects from anthropogenic and natural aerosols, but determining their

(continued on next page)

Aerosol-radiation interactions

Scattering aerosols

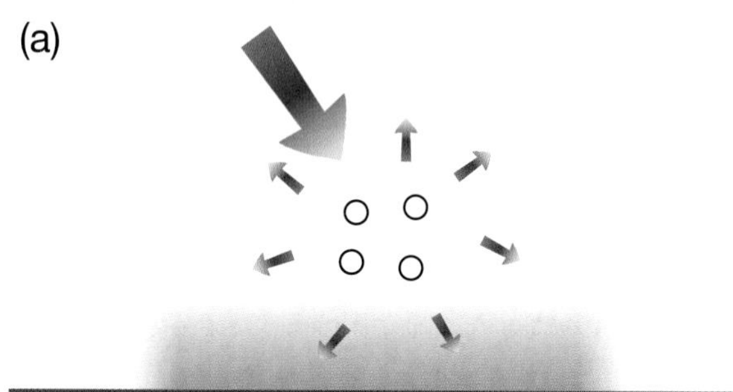

Aerosols scatter solar radiation. Less solar radiation reaches the surface, which leads to a localised cooling.

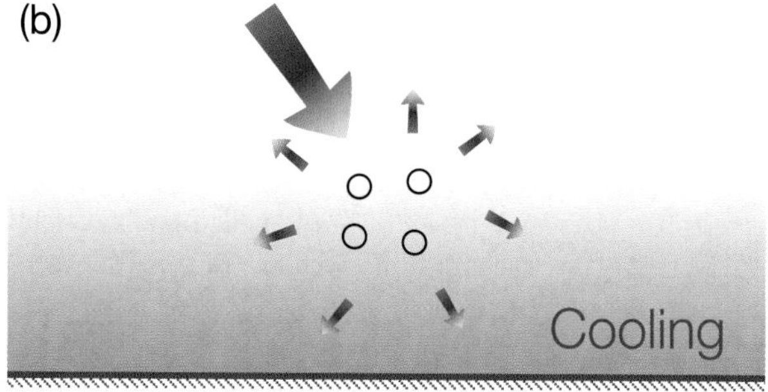

The atmospheric circulation and mixing processes spread the cooling regionally and in the vertical.

Absorbing aerosols

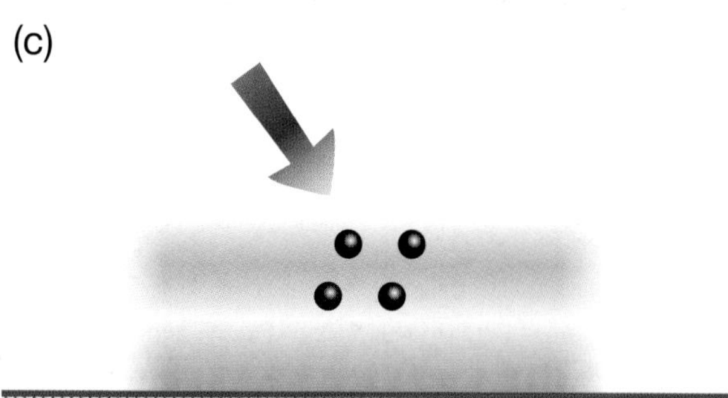

Aerosols absorb solar radiation. This heats the aerosol layer but the surface, which receives less solar radiation, can cool locally.

At the larger scale there is a net warming of the surface and atmosphere because the atmospheric circulation and mixing processes redistribute the thermal energy.

FAQ 7.2, Figure 1 | Overview of interactions between aerosols and solar radiation and their impact on climate. The left panels show the instantaneous radiative effects of aerosols, while the right panels show their overall impact after the climate system has responded to their radiative effects.

FAQ 7.2 (continued)

global impact requires satellite data and models. One of the remaining uncertainties comes from black carbon, an absorbing aerosol that not only is more difficult to measure than scattering aerosols, but also induces a complicated cloud response. Most studies agree, however, that the overall radiative effect from anthropogenic aerosols is to cool the planet.

Aerosols also serve as condensation and ice nucleation sites, on which cloud droplets and ice particles can form (see FAQ.7.2, Figure 2). When influenced by more aerosol particles, clouds of liquid water droplets tend to have more, but smaller droplets, which causes these clouds to reflect more solar radiation. There are however many other pathways for aerosol–cloud interactions, particularly in ice—or mixed liquid and ice—clouds, where phase changes between liquid and ice water are sensitive to aerosol concentrations and properties. The initial view that an increase in aerosol concentration will also increase the amount of low clouds has been challenged because a number of counteracting processes come into play. Quantifying the overall impact of aerosols on cloud amounts and properties is understandably difficult. Available studies, based on climate models and satellite observations, generally indicate that the net effect of anthropogenic aerosols on clouds is to cool the climate system.

Because aerosols are distributed unevenly in the atmosphere, they can heat and cool the climate system in patterns that can drive changes in the weather. These effects are complex, and hard to simulate with current models, but several studies suggest significant effects on precipitation in certain regions.

Aerosol-cloud interactions

(a)

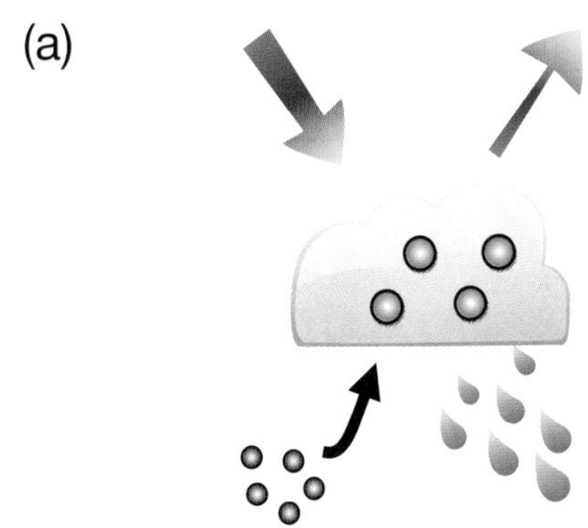

Aerosols serve as cloud condensation nuclei upon which liquid droplets can form.

(b)

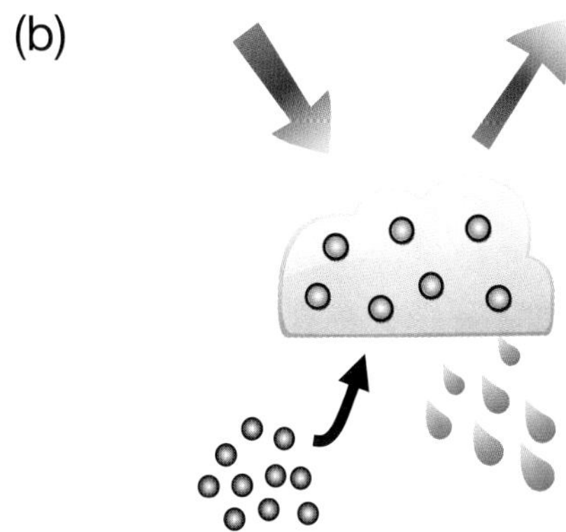

More aerosols result in a larger concentration of smaller droplets, leading to a brighter cloud. However there are many other possible aerosol–cloud–precipitation processes which may amplify or dampen this effect.

FAQ 7.2, Figure 2 | Overview of aerosol–cloud interactions and their impact on climate. Panels (a) and (b) represent a clean and a polluted low-level cloud, respectively.

Because of their short lifetime, the abundance of aerosols—and their climate effects—have varied over time, in rough concert with anthropogenic emissions of aerosols and their precursors in the gas phase such as sulphur dioxide (SO_2) and some volatile organic compounds. Because anthropogenic aerosol emissions have increased substantially over the industrial period, this has counteracted some of the warming that would otherwise have occurred from increased concentrations of well mixed greenhouse gases. Aerosols from large volcanic eruptions that enter the stratosphere, such as those of El Chichón and Pinatubo, have also caused cooling periods that typically last a year or two.

Over the last two decades, anthropogenic aerosol emissions have decreased in some developed countries, but increased in many developing countries. The impact of aerosols on the global mean surface temperature over this particular period is therefore thought to be small. It is projected, however, that emissions of anthropogenic aerosols will ultimately decrease in response to air quality policies, which would suppress their cooling influence on the Earth's surface, thus leading to increased warming.

7.6 Processes Underlying Precipitation Changes

7.6.1 Introduction

In this section we outline some of the main processes thought to control the climatological distribution of precipitation and precipitation extremes. Emphasis is placed on large-scale constraints that relate to processes, such as changes in the water vapour mixing ratio that accompany warming, or changes in atmospheric heating rates that accompany changing GHG and aerosol concentrations, which are discussed earlier in this chapter. The fidelity with which large-scale models represent different aspects of precipitation, ranging from the diurnal cycle to extremes, is discussed in Section 9.4.1. Building on, and adding to, concepts developed here, Section 11.3.2 presents near term projections of changes in regional precipitation features. Projections of changes on longer time scales, again with more emphasis on regionally specific features and the coupling to the land surface, are presented in Section 12.4.5. The effect of processes discussed in this section on specific precipitation systems, such as the monsoon, the intertropical convergence zones, or tropical cyclones are presented in Chapter 14.

Precipitation is sustained by the availability of moisture and energy. In a globally averaged sense the oceans provide an unlimited supply of moisture, so that precipitation formation is energetically limited (Mitchell et al., 1987). Locally precipitation can be greatly modified by limitations in the availability of moisture (for instance over land) and the effect of circulation systems, although these too are subject to local energetic constraints (Neelin and Held, 1987; Raymond et al., 2009). There are many ways to satisfy these constraints, and climate models still exhibit substantial biases in their representation of the spatio-temporal distribution of precipitation (Stephens et al., 2010; Liepert and Previdi, 2012; Section 9.5). Nonetheless, through careful analysis, it is possible to identify robust features in the simulated response of precipitation to changes in precipitation drivers. In almost every case these can be related to well understood processes, as described below.

7.6.2 The Effects of Global Warming on Large-Scale Precipitation Trends

The atmospheric water vapour mixing ratio is expected to increase with temperature roughly following the saturation value (e.g., with increases in surface values ranging from 6 to 10% $°C^{-1}$ and larger increases aloft, see Section 7.2.4.1). Increases in global mean precipitation are, however, constrained by changes in the net radiative cooling rate of the troposphere. GCMs, whose detailed treatment of radiative transfer provides a basis for calculating these energetic limitations, suggest that for the CO_2 forcing, globally-averaged precipitation increases with global mean surface temperature at about 1 to 3% $°C^{-1}$ (Mitchell et al., 1987; Held and Soden, 2006; Richter and Xie, 2008). Precipitation changes evince considerable regional variability about the globally averaged value; generally speaking precipitation is expected to increase in the wettest latitudes, whereas dry latitudes may even see a precipitation decrease (Mitchell et al., 1987; Allen and Ingram, 2002; Held and Soden, 2006). On smaller scales, or near precipitation margins, the response is less clear due to model-specific, and less well understood, regional circulation shifts (Neelin et al., 2006), but there is some evidence that the sub-tropical dry zones are expanding (Section 7.2.5.2 and Section 2.7.5), both as a result of the tropical convergence zones narrowing (Neelin et al., 2006; Chou et al., 2009), and the storm tracks moving poleward (Allen et al., 2012) and strengthening (O'Gorman and Schneider, 2008).

The 'wet-get-wetter' and 'dry-get-drier' response that is evident at large scales over oceans can be understood as a simple consequence of a change in the water vapour content carried by circulations, which otherwise are little changed (Mitchell et al., 1987; Held and Soden, 2006). Wet regions are wet because they import moisture from dry regions, increasingly so with warmer temperatures. These ideas have withstood additional analysis and scrutiny since AR4 (Chou et al., 2009; Seager et al., 2010; Muller and O'Gorman, 2011), are evident in 20th century precipitation trends (Allan and Soden, 2007; Zhang et al., 2007b; see Section 2.5.1) and are assessed on different time scales in Chapters 11 and 12. Because the wet-get-wetter argument implies that precipitation changes associated with warming correlate with the present-day pattern of precipitation, biases in the simulation of present-day precipitation will lead to biases in the projections of future precipitation change (Bony et al., 2013).

The wet-get-wetter and dry-get-drier response pattern is mitigated, particularly in the dry regions, by the anticipated slowdown of the atmospheric circulation (as also discussed in Section 7.2.5.3), as well as by gains from local surface evaporation. The slowdown within the descent regions can be partly understood as a consequence of the change in the dry static stability of the atmosphere with warming. And although this line of argument is most effective for explaining changes over the ocean (Chou et al., 2009; Bony et al., 2013), it can also be used to understand the GCM land responses to some extent (Muller and O'Gorman, 2011).

The non-uniform nature of surface warming induces regional circulation shifts that affect precipitation trends. In the tropics SSTs warm more where winds are weak and thus are less effective in damping surface temperature anomalies, and precipitation systematically shifts to regions that warm more (Xie et al., 2010). The greater warming over land, and its regional variations, also affect the regional distributions of precipitation (Joshi et al., 2008). However, low understanding of soil moisture–precipitation feedbacks complicates interpretations of local responses to warming over land (Hohenegger et al., 2009), so that the effect of warming on precipitation at the scale of individual catchments is not well understood. Some broad-scale responses, particularly over the ocean, are more robust and relatively well understood.

7.6.3 Radiative Forcing of the Hydrological Cycle

In the absence of a compensating temperature change, an increase in well-mixed GHG concentrations tends to reduce the net radiative cooling of the troposphere. This reduces the rainfall rate and the strength of the overturning circulation (Andrews et al., 2009; Bony et al., 2013), such that the increase in global mean precipitation would be 1.5 to 3.5% $°C^{-1}$ due to temperature alone but is reduced by about 0.5% $°C^{-1}$ due to the effect of CO_2 (Lambert and Webb, 2008). The dynamic effects are similar to those that result from the effect of atmospheric warming on the lapse rate, which also reduces the strength of the atmospheric

7

overturning circulation (e.g., Section 7.6.2), and are robustly evident over a wide range of models and model configurations (Bony et al., 2013; see also Figure 7.20). These circulation changes influence the regional response, and are more pronounced over the ocean, because asymmetries in the land-sea response to changing concentrations of GHGs (Joshi et al., 2008) amplify the maritime and dampen or even reverse the terrestrial signal (Wyant et al., 2012; Bony et al., 2013).

The dependence of the intensity of the hydrological cycle on the tropospheric cooling rate helps to explain why perturbations having the same RF do not produce the same precipitation responses. Apart from the relatively small increase in absorption by atmospheric water vapour, increased solar forcing does not directly affect the net tropospheric cooling rate. As a result the hydrological cycle mostly feels the subsequent warming through its influence on the rate of tropospheric cooling (Takahashi, 2009). This is why modeling studies suggest that solar radiation management (geoengineering) methods that maintain a constant surface temperature will lead to a reduction in globally averaged precipitation as well as different regional distributions of precipitation (Schmidt et al., 2012b; Section 7.7.3).

Changes in cloud radiative effects, and aerosol RF can also be effective in changing the net radiative heating rate within the troposphere (Lambert and Webb, 2008; Pendergrass and Hartmann, 2012). Most

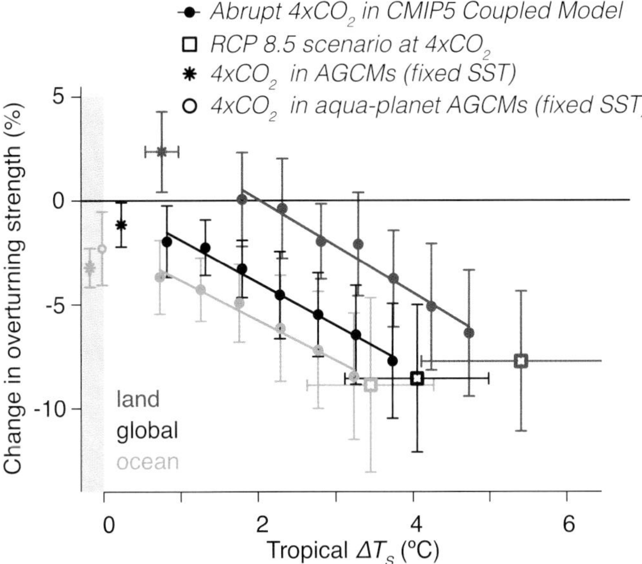

Figure 7.20 | Illustration of the response of the large-scale atmospheric overturning to warming (adapted from Bony et al., 2013). The overturning intensity is shown on the *y*-axis and is measured by the difference between the mean motion in upward moving air and the mean motion in downward moving air. The warming is shown on the *x*-axis and is measured by the change in surface temperature averaged over the Tropics, ΔT_s, after an abrupt quadrupling of atmospheric CO_2. The grey region delineates responses for which ΔT_s is zero by definition. Nearly one half of the final reduction in the intensity of the overturning is evident before any warming is felt, and can be associated with a rapid adjustment of the hydrological cycle to changes in the atmospheric cooling rate accompanying a change in CO_2. With warming the circulation intensity is further reduced. The rapid adjustment, as measured by the change in circulation intensity for zero warming, is different over land and ocean. Over land the increase in CO_2 initially causes an intensification of the circulation. The result is robust in the sense that it is apparent in all of the 15 CMIP5 models analysed, irrespective of the details of their configuration.

prominently, absorption of solar radiation by atmospheric aerosols is understood to reduce the globally averaged precipitation. But this effect may be offset by the tendency of absorbing aerosols to reduce the planetary albedo, thereby raising surface temperature, leading to more precipitation (Andrews et al., 2009). Heterogeneously distributed precipitation drivers such as clouds, aerosols and tropospheric ozone will also induce circulations that may amplify or dampen their local impact on the hydrological cycle (Ming et al., 2010; Allen et al., 2012; Shindell et al., 2012). Such regional effects are discussed further in Chapter 14 for the case of aerosols.

7.6.4 Effects of Aerosol–Cloud Interactions on Precipitation

Aerosol–cloud interactions directly influence the cloud microphysical structure, and only indirectly (if at all) the net atmospheric heating rate, and for this reason have mostly been explored in terms of their effect on the character and spatio-temporal distribution of precipitation, rather than on the globally-averaged amount of precipitation.

The sensitivity of simulated clouds to their microphysical development (e.g., Fovell et al., 2009; Parodi and Emanuel, 2009) suggests that they may be susceptible to the availability of CCN and IN. For instance, an increase in CCN favours smaller cloud droplets, which delays the onset of precipitation and the formation of ice particles in convective clouds (Rosenfeld and Woodley, 2001; Khain et al., 2005). It has been hypothesized that such changes may affect the vertical distribution and total amount of latent heating in ways that would intensify or invigorate convective storms, as measured by the strength and vertical extent of the convective updraughts (Andreae et al., 2004; Rosenfeld et al., 2008; Rosenfeld and Bell, 2011; Tao et al., 2012). Support for the idea that the availability of CCN influences the vigour of convective systems can be found in some modelling studies, but the strength, and even sign, of such an effect has been shown to be contingent on a variety of environmental factors (Seifert and Beheng, 2006; Fan et al., 2009; Khain, 2009; Seifert et al., 2012) as well as on modelling assumptions (Ekman et al., 2011).

Observational studies, based on large data sets that sample many convective systems, report systematic correlations between aerosol amount and cloud-top temperatures (Devasthale et al., 2005; Koren et al., 2010a; Li et al., 2011). Weekly cycles in cloud properties and precipitation, wherein convective intensity, cloud cover or precipitation increases during that part of the week when aerosol concentrations are largest, have also been reported (Bäumer and Vogel, 2007; Bell et al., 2008; Rosenfeld and Bell, 2011). Both types of studies have been interpreted in terms of an aerosol influence on convective cloud systems. However, whether or not these findings demonstrate that a greater availability of CCN systematically invigorates, or otherwise affects, convection remains controversial. Many of the weekly cycle studies are disputed on statistical or other methodological grounds (Barmet et al., 2009; Stjern, 2011; Tuttle and Carbone, 2011; Sanchez-Lorenzo et al., 2012; Yuter et al., 2013). Even in cases where relationships between aerosol amount and some measure of convective intensity appear to be unambiguous, the interpretation that this reflects an aerosol effect on the convection is less clear, as both aerosol properties and convection are strongly influenced by meteorological factors that are not well controlled for (e.g., Boucher and Quaas, 2013). Studies that have used CRMs to consider the net effect of aerosol–cloud interactions integrated over many storms, or in more of a climate context wherein convective heating must balance radiative cooling within the atmosphere, also do not support a strong and systematic invigoration effect resulting from very large (many fold) changes in the ambient aerosol (Morrison and Grabowski, 2011; van den Heever et al., 2011; Seifert et al., 2012; Khairoutdinov and Yang, 2013). Locally in space or time, however, radiative processes are less constraining, leaving open the possibility of stronger effects from localized or transient aerosol perturbations.

Because precipitation development in clouds is a time-dependent process, which proceeds at rates that are partly determined by the cloud microphysical structure (Seifert and Zängl, 2010), aerosol–cloud interactions may lead to shifts in topographic precipitation to the leeward side of mountains when precipitation is suppressed, or to the windward side in cases when it is more readily initiated. Orographic clouds show a reduction in the annual precipitation over topographical barriers downwind of major urban areas in some studies (Givati and Rosenfeld, 2004; Jirak and Cotton, 2006) but not in others (Halfon et al., 2009). Even in cases where effects are reported, the results have proven sensitive to how the data are analysed (Alpert et al., 2008; Levin and Cotton, 2009).

In summary, it is unclear whether changes in aerosol–cloud interactions arising from changes in the availability of CCN or IN can affect, and possibly intensify, the evolution of individual precipitating cloud systems. Some observational and modelling studies suggest such an effect, but are undermined by alternative interpretations of the observational evidence, and a lack of robustness in the modelling studies. The evidence for systematic effects over larger areas and long time periods is, if anything, more limited and ambiguous.

7.6.5 The Physical Basis for Changes in Precipitation Extremes

The physical basis for aerosol microphysical effects on convective intensity was discussed in the previous section. Here we briefly discuss process understanding of the effect of warming on precipitation extremes; observed trends supporting these conclusions are presented in Section 2.6.2.

Precipitation within individual storms is expected to increase with the available moisture content in the atmosphere or near the surface rather than with the global precipitation (Allen and Ingram, 2002; Held and Soden, 2006), which leads to a 6 to 10% °C^{-1} increase, but with longer intervals between storms (O'Gorman and Schneider, 2009). Because GCMs are generally poor at simulating precipitation extremes (Stephens et al., 2010) and predicted changes in a warmer climate vary (Kharin et al., 2007; Sugiyama et al., 2010), they are not usually thought of as a source of reliable information regarding extremes. However, a recent study (O'Gorman, 2012) shows that GCM predictions of extremes can be constrained by observable relationships in the present day climate, and upon doing so become broadly consistent with the idea that extreme precipitation increases by 6 to 10% per °C of warming. Central estimates of sensitivity of extreme (99.9th percentile)

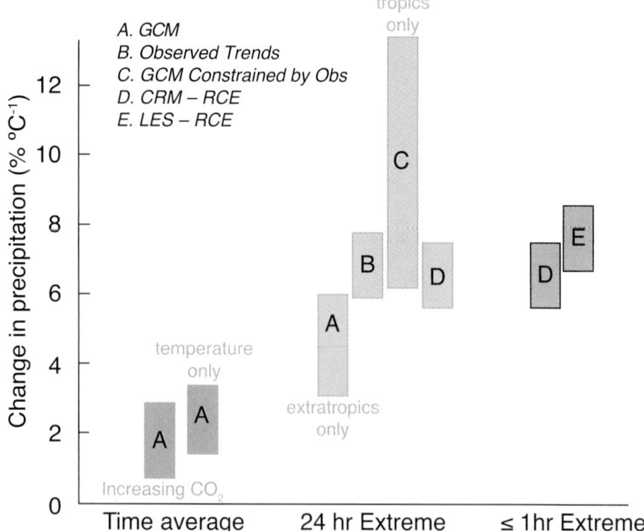

Figure 7.21 | Estimate (5 to 95% range) of the increase in precipitation amount per degree Celsius of global mean surface temperature change. At left (blue) are climate model predictions of changes in time-averaged global precipitation; at centre and right (orange) are predictions or estimates of the typical or average increase in local 99.9th percentile extremes, over 24 hours (centre) and over one hour or less (right). Data are adapted from (A) GCM studies (Allen and Ingram, 2002; and Lambert and Webb, 2008, for time average; O'Gorman and Schneider, 2009 for extremes), (B) long-term trends at many sites globally (Westra et al., 2013), (C) GCMs constrained by present-day observations of extremes (O'Gorman, 2012), (D, E) cloud-resolving model (CRM) and large-eddy simulation (LES) studies of radiative convective equilibrium (Muller et al., 2011; Romps, 2011).

daily rainfall to global temperature from this study were 10% °C^{-1} in the tropics (O'Gorman, 2012), compared to 5% °C^{-1} predicted by the models in the extratropics, where they may be more reliable (O'Gorman and Schneider, 2009). How precipitation extremes depend on temperature has also been explored using cloud resolving simulations (but only for tropical conditions) which produce similar increases in extreme instantaneous rain rate (Romps, 2011) and daily or hourly rain totals (Muller et al., 2011) within storms (Figure 7.21). Because these latter studies are confined to small domains, they may exclude important synoptic or larger-scale dynamical changes such as increases in flow convergence (Chou et al., 2009; Sugiyama et al., 2010).

By taking advantage of natural variability in the present day climate, a number of studies have correlated observed rainfall extremes with local temperature variations. In the extratropics, these studies document sensitivities of extreme precipitation to temperature much higher than those reported above (Lenderink and Van Meijgaard, 2008), but sensitivities vary with temperature (Lenderink et al., 2011), are often negative in the tropics (Hardwick Jones et al., 2010) and usually strengthen at the shortest (e.g., hourly or less) time scales (e.g., Haerter et al., 2010; Hardwick Jones et al., 2010). However, local temperature changes may not be a good proxy for global warming because they tend to co-vary with other meteorological factors (such as humidity, atmospheric stability, or wind direction) in ways that are uncharacteristic of changes in the mean temperature (see Section 7.2.5.7), and these other meteorological factors may dominate the observed signal (e.g., Haerter and Berg, 2009). Thus, the idea that precipitation extremes depend much more strongly on temperature than the 5 to

10% increase per degree Celsius attributable to water vapour changes, remains controversial.

Following the AR4, studies have also continued to show that extremes in precipitation are associated with the coincidence of particular weather patterns (e.g., Lavers et al., 2011). We currently lack an adequate understanding of what controls the return time and persistence of such rare events.

From the aforementioned model and observational evidence, there is *high confidence* that the intensity of extreme precipitation events will increase with warming, at a rate well exceeding that of the mean precipitation. There is *medium confidence* that the increase is roughly 5 to 10% °C^{-1} warming but may vary with time scale, location and season.

7.7 Solar Radiation Management and Related Methods

7.7.1 Introduction

Geoengineering—also called climate engineering—is defined as a broad set of methods and technologies that aim to deliberately alter the climate system in order to alleviate impacts of climate change (Keith, 2000; Izrael et al., 2009; Royal Society, 2009; IPCC, 2011). Two main classes of geoengineering are often considered. Solar Radiation Management (SRM) proposes to counter the warming associated with increasing GHG concentrations by reducing the amount of sunlight absorbed at the surface. A related method seeks to alter high-altitude cirrus clouds to reduce their greenhouse effect. Another class of geoengineering called Carbon Dioxide Removal (CDR) is discussed in Section 6.5. This section assesses how the climate system might respond to some proposed SRM methods and related methods thought to have the potential to influence the global energy budget by at least a few tenths of a W m^{-2} but it does not assess technological or economical feasibility, or consider methods targeting specific climate impacts (MacCracken, 2009). Geoengineering is quite a new field of research, and there are relatively few studies focussed on it. Assessment of SRM is limited by (1) gaps in understanding of some important processes; (2) a relative scarcity of studies; and (3) a scarcity of studies using similar experimental design. This section discusses some aspects of SRM potential to mitigate global warming, outlines robust conclusions where they are apparent, and evaluates uncertainties and potential side effects. Additional impacts of SRM are assessed in Section 19.5.4 of the WGII report, while some of the socio-economic issues are assessed in Chapters 3, 6 and 13 of the WGIII report.

7.7.2 Assessment of Proposed Solar Radiation Management Methods

A number of studies have suggested reducing the amount of sunlight reaching the Earth by placing solid or refractive disks, or dust particles, in outer space (Early, 1989; Mautner, 1991; Angel, 2006; Bewick et al., 2012). Although we do not assess the feasibility of these methods, they provide an easily described mechanism for reducing sunlight reaching the planet, and motivate the idealized studies discussed in Section 7.7.3.

7.7.2.1 Stratospheric Aerosols

Some SRM methods propose increasing the amount of stratospheric aerosol to produce a cooling effect like that observed after strong explosive volcanic eruptions (Budyko, 1974; Crutzen, 2006). Recent studies have used numerical simulations and/or natural analogues to explore the possibility of forming sulphuric acid aerosols by injecting sulphur-containing gases into the stratosphere (Rasch et al., 2008b). Because aerosols eventually sediment out of the stratosphere (within roughly a year or less), these methods require replenishment to maintain a given level of RF. Research has also begun to explore the efficacy of other types of aerosol particles (Crutzen, 2006; Keith, 2010; Ferraro et al., 2011; Kravitz et al., 2012) but the literature is much more limited and not assessed here.

The RF depends on the choice of chemical species (gaseous sulphur dioxide (SO_2), sulphuric acid (H_2SO_4) or sprayed aerosols), location(s), rate and frequency of injection. The injection strategy affects particle size (Rasch et al., 2008a; Heckendorn et al., 2009; Pierce et al., 2010; English et al., 2012), with larger particles producing less RF (per unit mass) and more rapid sedimentation than smaller particles, affecting the efficacy of the method. The aerosol size distribution is controlled by an evolving balance between new particle formation, condensation of vapour on pre-existing particles, evaporation of particles, coagulation and sedimentation. Models that more fully account for aerosol processes (Heckendorn et al., 2009; Pierce et al., 2010; English et al., 2012) found smaller aerosol burdens, larger particles and weaker RF than earlier studies that prescribed the particle size over the particle lifetime. Current modeling studies indicate that injection of sulphate aerosol precursors of at least 10 Mt S (approximately the amount of sulphur injected by the Mount Pinatubo eruption) would be needed annually to maintain a RF of –4 W m^{-2}, roughly equal but opposite to that associated with a doubling of atmospheric CO_2 (Heckendorn et al., 2009; Pierce et al., 2010; Niemeier et al., 2011). Stratospheric aerosols may affect high clouds in the tropopause region, and one study (Kuebbeler et al., 2012) suggests significant negative forcing would result, but this is uncertain given limited understanding of ice nucleation in high clouds (Section 7.4.4.4).

Along with its potential to mitigate some aspects of global warming, the potential side effects of SRM must also be considered. Tilmes et al. (2008; 2009) estimated that stratospheric aerosols SRM might increase chemical ozone loss at high latitudes and delay recovery of the Antarctic ozone hole (expected at the end of this century) by 30 to 70 years, with changes in column ozone of –3 to –10% in polar latitudes and +3 to +5% in the tropics. A high latitude ozone loss is expected to increase UV radiation reaching the surface there, although the effect would be partially compensated by the increase in attenuation by the aerosol itself (Vogelmann et al., 1992; Tilmes et al., 2012). A decrease in direct radiation and increase in diffuse radiation reaching the Earth's surface would occur and would be expected to increase photosynthesis in terrestrial ecosystems (Mercado et al., 2009; see Section 6.5.4) and decrease the efficiency of some solar energy technologies (see WGII AR5, Section 19.5.4). Models indicate that stratospheric aerosol SRM would not pose a surface acidification threat with maximum acid deposition rates estimated to be at least 500 times less than the threshold of concern for the most sensitive land ecosystems (Kuylenstierna et al.,

2001; Kravitz et al., 2009); contributions to ocean acidification are also estimated to represent a very small fraction of that induced by anthropogenic CO_2 emissions (Kravitz et al., 2009). There are other known side effects that remain unquantified, and limited understanding (and limited study) make additional impacts difficult to anticipate.

7.7.2.2 Cloud Brightening

Boundary layer clouds act to cool the planet, and relatively small changes in albedo or areal extent of low cloud can have profound effects on the Earth's radiation budget (Section 7.2.1). Theoretical, modelling and observational studies show that the albedo of these types of cloud systems are susceptible to changes in their droplet concentrations, but the detection and quantification of RF attributable to such effects is difficult to separate from meteorological variability (Section 7.4.3.2). Nonetheless, by systematically introducing CCN into the marine boundary layer, it should be possible to locally increase boundary layer cloud albedo as discussed in Section 7.4.2. These ideas underpin the method of cloud brightening, for instance through the direct injection (seeding) of sea-spray particles into cloud-forming air masses (Latham, 1990). An indirect cloud brightening mechanism through enhanced DMS production has also been proposed (Wingenter et al., 2007) but the efficacy of the DMS mechanism is disputed (Vogt et al., 2008; Woodhouse et al., 2008).

The seeding of cloud layers with a propensity to precipitate may change cloud structure (e.g., from open to closed cells) and/or increase liquid water content (Section 7.4.3.2.1), in either case changing albedo and producing strong negative forcing. A variety of methods have been used to identify which cloud regions are most susceptible to an aerosol change (Oreopoulos and Platnick, 2008; Salter et al., 2008; Alterskjær et al., 2012). Marine stratocumulus clouds with relatively weak precipitation are thought to be an optimal cloud type for brightening because of their relatively low droplet concentrations, their expected increase in cloud water in response to seeding (Section 7.4.3.2.1), and the longer lifetime of sea salt particles in non- or weakly precipitating environments. Relatively strong local ERFaci (–30 to –100 W m^{-2}) would be required to produce a global forcing of –1 to –5 W m^{-2} if only the more susceptible clouds were seeded.

Simple modelling studies suggest that increasing droplet concentrations in marine boundary layer clouds by a factor of five or so (to concentrations of 375 to 1000 cm^{-3}) could produce an ERFaci of about –1 W m^{-2} if 5% of the ocean surface area were seeded, and an ERFaci as strong as –4 W m^{-2} if that fraction were increased to 75% (Latham et al., 2008; Jones et al., 2009; Rasch et al., 2009). Subsequent studies with more complete treatments of aerosol–cloud interactions have produced both stronger (Alterskjær et al., 2012; Partanen et al., 2012) and weaker (Korhonen et al., 2010a) changes. Because the initial response to cloud seeding is local, high-resolution, limited-domain simulations are especially useful to explore the efficacy of seeding. One recent study of this type (Wang et al., 2011a) found that cloud brightening is sensitive to cloud dynamical adjustments that are difficult to treat in current GCMs (Sections 7.4.3 and 7.5.3) and concluded that the seeding rates initially proposed for cloud seeding may be insufficient to produce the desired cloud brightening. Recent studies accounting for clear-sky brightening from increased aerosol concentrations (i.e., ERFari) also found an increase in the amplitude of the ERF by 30 to 50% (Hill and Ming, 2012; Partanen et al., 2012), thereby making the aerosol seeding more effective than previous estimates that neglected that effect.

In summary, evidence that cloud brightening methods are effective and feasible in changing cloud reflectivity is ambiguous and subject to many of the uncertainties associated with aerosol–cloud interactions more broadly. If cloud brightening were to produce large local changes in ERF, those changes would affect the local energy budget, with further impacts on larger-scale oceanic and atmospheric circulations. Possible side effects accompanying such large and spatially heterogeneous changes in ERF have not been systematically studied.

7.7.2.3 Surface Albedo Changes

A few studies have explored how planetary albedo might be increased by engineering local changes to the albedo of urban areas, croplands, grasslands, deserts and the ocean surface. Effects from whitening of urban areas have been estimated to yield a potential RF of –0.17 W m^{-2} (Hamwey, 2007) although subsequent studies (Lenton and Vaughan, 2009; Oleson et al., 2010; Jacobson and Ten Hoeve, 2012) suggest that this estimate may be at the upper end of what is achievable. Larger effects might be achievable by replacing native grassland or cropland with natural or bioengineered species with a larger albedo. A hypothetical 25% increase in grassland albedo could yield a RF as large as –0.5 W m^{-2} (Lenton and Vaughan, 2009), with the maximum effect in the mid-latitudes during summer (Ridgwell et al., 2009; Doughty et al., 2011). The feasibility of increasing crop and grassland albedo remains unknown, and there could be side effects on photosynthetic activity, carbon uptake and biodiversity. The low albedo and large extent of oceanic surfaces mean that only a small increase in albedo, for example, by increasing the concentration of microbubbles in the surface layer of the ocean (Evans et al., 2010; Seitz, 2011), could be sufficient to offset several W m^{-2} of RF by GHGs. Neither the extent of microbubble generation and persistence required for a significant climate impact, nor the potential side effects on the ocean circulation, air-sea fluxes and marine ecosystems have been assessed.

7.7.2.4 Cirrus Thinning

Although not strictly a form of SRM, proposals have been made to cool the planet by reducing the coverage or longwave opacity of high thin cirrus clouds, which act to warm the surface through their greenhouse effect (see Section 7.2.1.2). A proposal for doing so involves adding efficient IN in regions prone to forming thin cirrus cloud (Mitchell and Finnegan, 2009). To the extent such a proposal is feasible, one modelling study suggests that an ERFaci of as strong as –2 W m^{-2} could be achieved (Storelvmo et al., 2013), with further negative forcing caused by a reduction in humidity of the upper troposphere associated with the cloud changes. However, lack of understanding of cirrus cloud formation processes, as well as ice microphysical processes (Section 7.4.4), makes it difficult to judge the feasibility of such a method, particularly in light of the fact that increasing ice nucleation can also increase cirrus opacity, under some circumstances producing an opposite, positive forcing (Storelvmo et al., 2013). Side effects specific to the 'cirrus thinning' method have not been investigated.

7.7.3 Climate Response to Solar Radiation Management Methods

As discussed elsewhere in this and other chapters of this assessment, significant gaps remain in our understanding of the climate response to forcing mechanisms. Geoengineering is also a relatively new field of research. The gaps in understanding, scarcity of studies and diversity in the model experimental design make quantitative model evaluation and intercomparison difficult, hindering an assessment of the efficacy and side effects of SRM. This motivates dividing the discussion into two sections, one that assesses idealized studies that focus on conceptual issues and searches for robust responses to simple changes in the balance between solar irradiance and CO_2 forcing, and another discussing studies that more closely emulate specific SRM methods.

7.7.3.1 Climate Response in Idealized Studies

Perhaps the simplest SRM experiment that can be performed in a climate model consists of a specified reduction of the total solar irradiance, which could approximate the radiative impact of space reflectors. Reductions in solar irradiance in particular regions (over land or ocean, or in polar or tropical regions) could also provide useful information. Idealized simulations often focus on the effects of a complete cancellation of the warming from GHGs, but the rate of warming is occasionally explored by producing a negative RF that partially cancels the anthropogenic forcing (e.g., Eliseev et al., 2009). They can also provide insight into the climate response to other SRM methods, and can provide a simple baseline for examining other SRM techniques.

The most comprehensive and systematic evaluation of idealized SRM to date is the Geoengineering Model Intercomparison Project (Kravitz et al., 2011). Together with earlier model studies, this project found robust surface temperature reductions when the total solar irradiance is reduced: when this reduction compensates for CO_2 RF, residual effects appear regionally, but they are much smaller than the warming due to the CO_2 RF alone (Kravitz et al., 2011; Schmidt et al., 2012b; Figures 7.22a, b and 7.23a–d) The substantial warming from $4 \times CO_2$ at high latitudes (4°C to 18°C) is reduced to a warming of 0°C to 3°C near

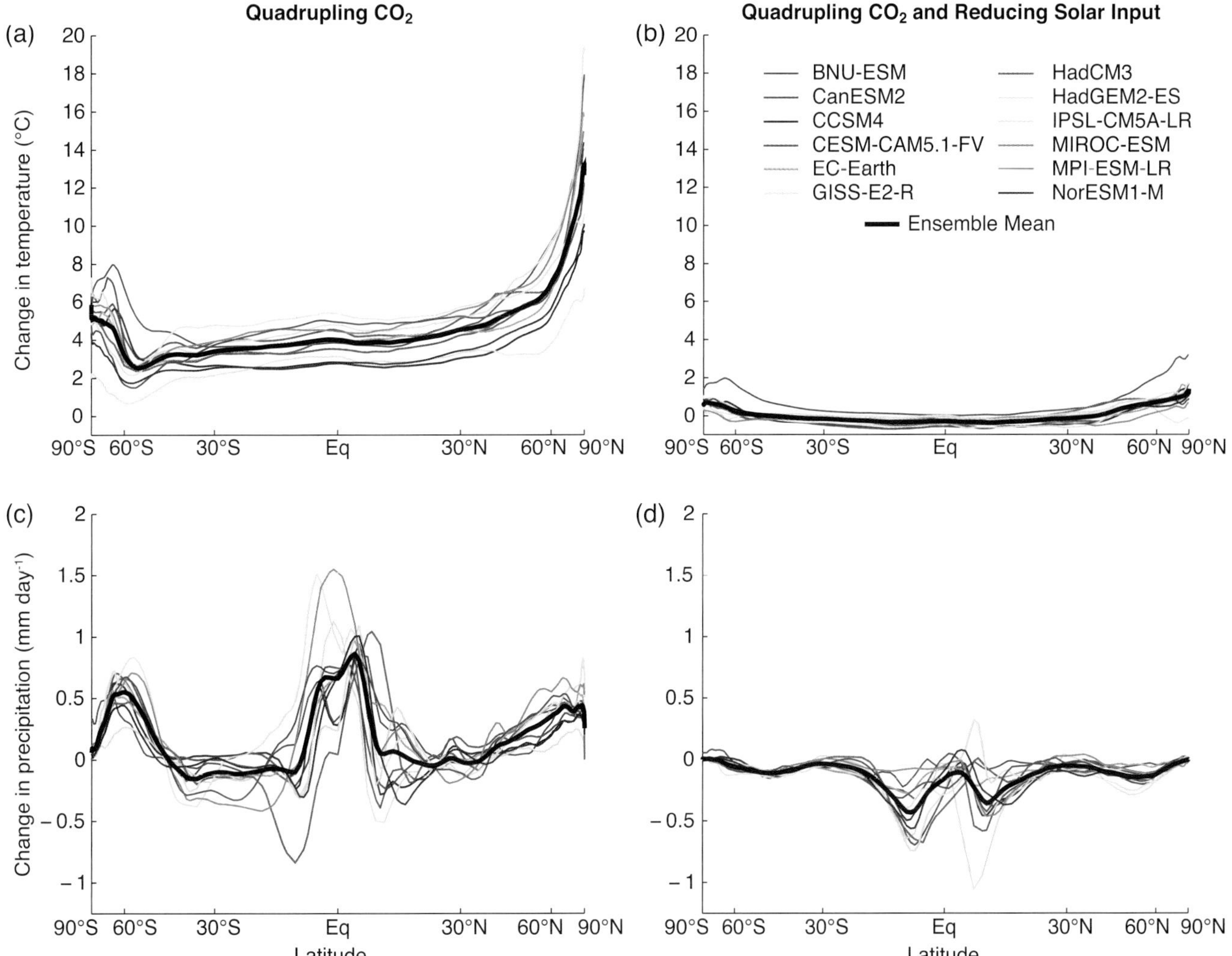

Figure 7.22 | Zonally and annually averaged change in surface air temperature (°C) for (a) an abrupt $4 \times CO_2$ experiment and (b) an experiment where the $4 \times CO_2$ forcing is balanced by a reduction in the total solar irradiance to produce a global top of the atmosphere flux imbalance of less than ±0.1 W m^{-2} during the first 10 years of the simulation (Geoengineering Model Intercomparison Project (GeoMIP) G1 experiment; Kravitz et al., 2011). (c, d) Same as (a) and (b) but for the change in precipitation (mm day^{-1}). The multi-model ensemble mean is shown with a thick black solid line. All changes are relative to the pre-industrial control experiment and averaged over years 11 to 50. The figure extends the results from Schmidt et al. (2012b) and shows the results from an ensemble of 12 coupled ocean–atmosphere general circulation models.

the winter pole. The residual surface temperature changes are generally positive at mid- and high-latitudes, especially over continents, and generally negative in the tropics (Bala et al., 2008; Lunt et al., 2008; Schmidt et al., 2012b). These anomalies can be understood in terms of the difference between the more uniform longwave forcing associated with changes in long-lived GHGs and the less uniform shortwave forcing from SRM that has a stronger variation in latitude and season. The compensation between SRM and CO_2 forcing is inexact in other ways. For example, SRM will change heating rates only during daytime, but increasing greenhouse effect changes temperatures during both day and night, influencing the diurnal cycle of surface temperature even if compensation for the diurnally averaged surface temperature is correct.

Although increasing CO_2 concentrations lead to a positive RF that warms the entire troposphere, SRM produces a negative RF that tends to cool the surface. The combination of RFs produces an increase in

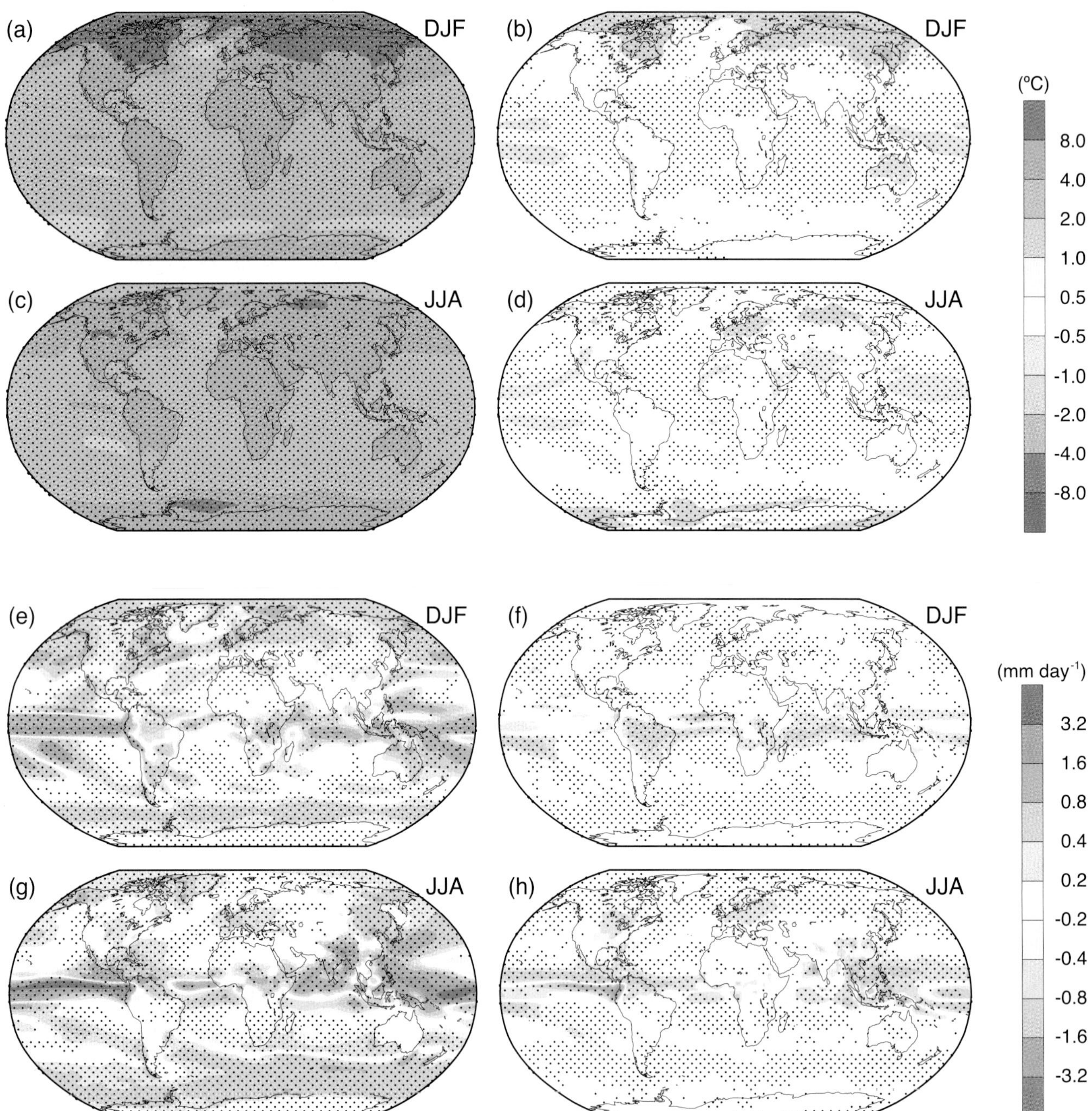

Figure 7.23 | Multi-model mean of the change in surface air temperature (°C) averaged over December, January and February (DJF) for (a) an abrupt 4 × CO_2 simulation and (b) an experiment where the 4 × CO_2 forcing is balanced by a reduction in the total solar irradiance to produce a global top of the atmosphere flux imbalance of less than ±0.1 W m^{-2} during the first 10 years of the simulation (Geoengineering Model Intercomparison Project (GeoMIP) G1 experiment; Kravitz et al., 2011). (c, d) Same as (a-b) but for June, July and August (JJA). (e–h) same as (a–d) but for the change in precipitation (mm day^{-1}). All changes are relative to the pre-industrial control experiment and averaged over years 11 to 50. The figure extends the results from Schmidt et al. (2012b) and shows the results from an ensemble of 12 coupled ocean–atmosphere general circulation models. Stippling denotes agreement on the sign of the anomaly in at least 9 out of the 12 models.

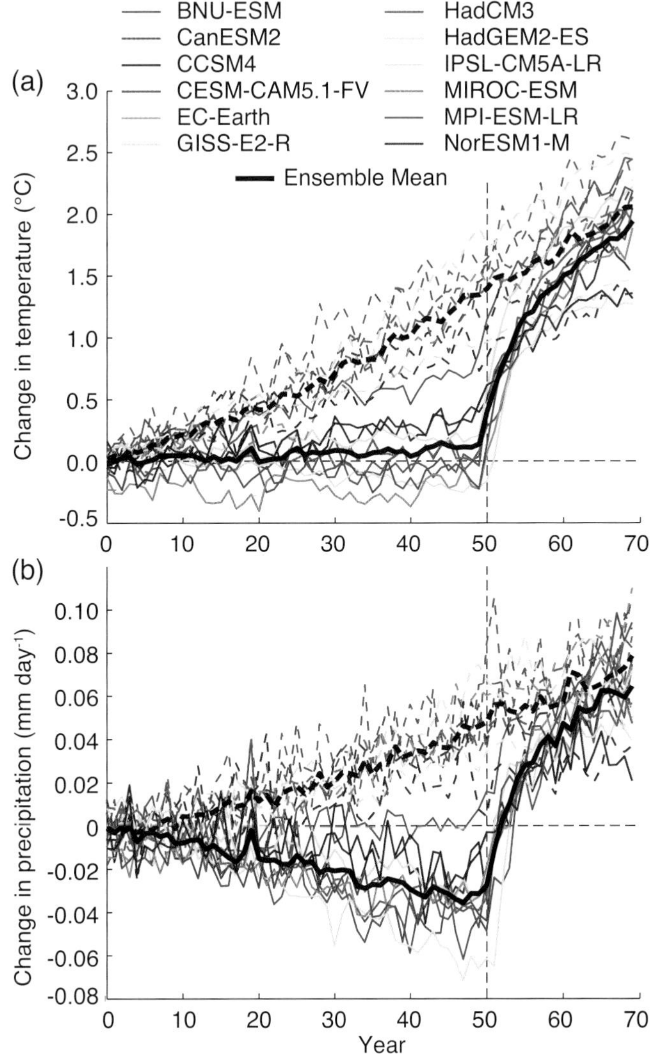

Figure 7.24 | Time series of globally averaged (a) surface temperature (°C) and (b) precipitation (mm day^{-1}) changes relative to each model's 1 × CO_2 reference simulation. Solid lines are for simulations using solar radiation management (SRM) through an increasing reduction of the total solar irradiance to balance a 1% yr^{-1} increase in CO_2 concentration until year 50, after which SRM is stopped for the next 20 years (Geoengineering Model Intercomparison Project (GeoMIP) G2 experiment; Kravitz et al., 2011). Dashed lines are for 1% CO_2 increase simulations with no SRM. The multi-model ensemble mean is shown with thick black lines.

stability that leads to less global precipitation as seen in Figures 7.22c, d and 7.23e–g (Bala et al., 2008; Andrews et al., 2010; Schmidt et al., 2012b) and discussed in Section 7.6.3. The reduction in precipitation shows similarities to the climate response induced by the Pinatubo eruption (Trenberth and Dai, 2007). Although the impact of changes in the total solar irradiance on global mean precipitation is well understood and robust in models, there is less understanding and agreement among models in the spatial pattern of the precipitation changes. Modelling studies suggest that some residual patterns may be robust (e.g., approximately 5% reduction in precipitation over Southeast Asia and the Pacific Warm Pool in June, July and August), but a physical explanation for these changes is lacking. Some model results indicate that an asymmetric hemispheric SRM forcing would induce changes in some regional precipitation patterns (Haywood et al., 2013).

High CO_2 concentrations from anthropogenic emissions will persist in the atmosphere for more than a thousand years in the absence of active efforts to remove atmospheric CO_2 (see Chapter 6). If SRM were used to counter positive forcing, it would be needed as long as the CO_2 concentrations remained high (Boucher et al., 2009). If GHG concentrations continued to increase, then the scale of SRM to offset the resulting warming would need to increase proportionally, amplifying residual effects from increasingly imperfect compensation. Figure 7.24 shows projections of the globally averaged surface temperature and precipitation changes associated with a 1% yr^{-1} CO_2 increase, with and without SRM. The scenario includes a hypothetical, abrupt termination of SRM at year 50, which could happen due to any number of unforeseeable circumstances. After this event, all the simulations predict a return to temperature levels consistent with the CO_2 forcing within one to two decades (*high confidence*), and with a large rate of temperature change (see also Irvine et al., 2012). Precipitation, which drops by 1% over the SRM period, rapidly returns to levels consistent with the CO_2 forcing upon SRM termination. The very rapid warming would probably affect ecosystem and human adaptation, and would also weaken carbon sinks, accelerating atmospheric CO_2 accumulation and contributing to further warming (Matthews and Caldeira, 2007).

Research suggests that this 'termination effect' might be avoided if SRM were used at a modest level and for a relatively short period of time (less than a century) when combined with aggressive CO_2 removal efforts to minimize the probability that the global mean temperature might exceed some threshold (Matthews, 2010; Smith and Rasch, 2012).

7.7.3.2 Climate Response to Specific Solar Radiation Management Methods

Several studies examined the model response to more realistic stratospheric aerosol SRM (Rasch et al., 2008b; Robock et al., 2008; Jones et al., 2010; Fyfe et al., 2013). These studies produced varying aerosol burdens, and RF and model responses also varied more strongly than in idealized experiments. Although these studies differ in details, their climate responses were generally consistent with the idealized experiments described in Section 7.7.3.1.

Studies treating the interaction between the carbon cycle, the hydrologic cycle, and SRM indicate that SRM could affect the temperature-driven suppression of some carbon sinks, and that the increased stomatal resistance with increased CO_2 concentrations combined with less warming, may further affect the hydrological cycle over land (Matthews and Caldeira, 2007; Fyfe et al., 2013), with larger impacts on precipitation for stratospheric aerosol SRM than for a uniform reduction in incoming sunlight.

Coupled ocean–atmosphere–sea ice models have also been used to assess the climate impacts of cloud brightening due to droplet concentration changes (Jones et al., 2009; Rasch et al., 2009; Baughman et al., 2012; Hill and Ming, 2012). The patterns of temperature and precipitation change differ substantially between models. These studies showed larger residual temperature changes than the idealized SRM studies, with more pronounced cooling over the regions of enhanced albedo. The cooling over the seeded regions (the marine stratocumulus regions)

Frequently Asked Questions

FAQ 7.3 | Could Geoengineering Counteract Climate Change and What Side Effects Might Occur?

Geoengineering—also called climate engineering—is defined as a broad set of methods and technologies that aim to deliberately alter the climate system in order to alleviate impacts of climate change. Two distinct categories of geoengineering methods are usually considered: Solar Radiation Management (SRM, assessed in Section 7.7) aims to offset the warming from anthropogenic greenhouse gases by making the planet more reflective while Carbon Dioxide Removal (CDR, assessed in Section 6.5) aims at reducing the atmospheric CO_2 concentration. The two categories operate on different physical principles and on different time scales. Models suggest that if SRM methods were realizable they would be effective in countering increasing temperatures, and would be less, but still, effective in countering some other climate changes. SRM would not counter all effects of climate change, and all proposed geoengineering methods also carry risks and side effects. Additional consequences cannot yet be anticipated as the level of scientific understanding about both SRM and CDR is low. There are also many (political, ethical, and practical) issues involving geoengineering that are beyond the scope of this report.

Carbon Dioxide Removal Methods

CDR methods aim at removing CO_2 from the atmosphere by deliberately modifying carbon cycle processes, or by industrial (e.g., chemical) approaches. The carbon withdrawn from the atmosphere would then be stored in land, ocean or in geological reservoirs. Some CDR methods rely on biological processes, such as large-scale afforestation/reforestation, carbon sequestration in soils through biochar, bioenergy with carbon capture and storage (BECCS) and ocean fertilization. Others would rely on geological processes, such as accelerated weathering of silicate and carbonate rocks—on land or in the ocean (see FAQ.7.3, Figure 1). The CO_2 removed from the atmosphere would

(continued on next page)

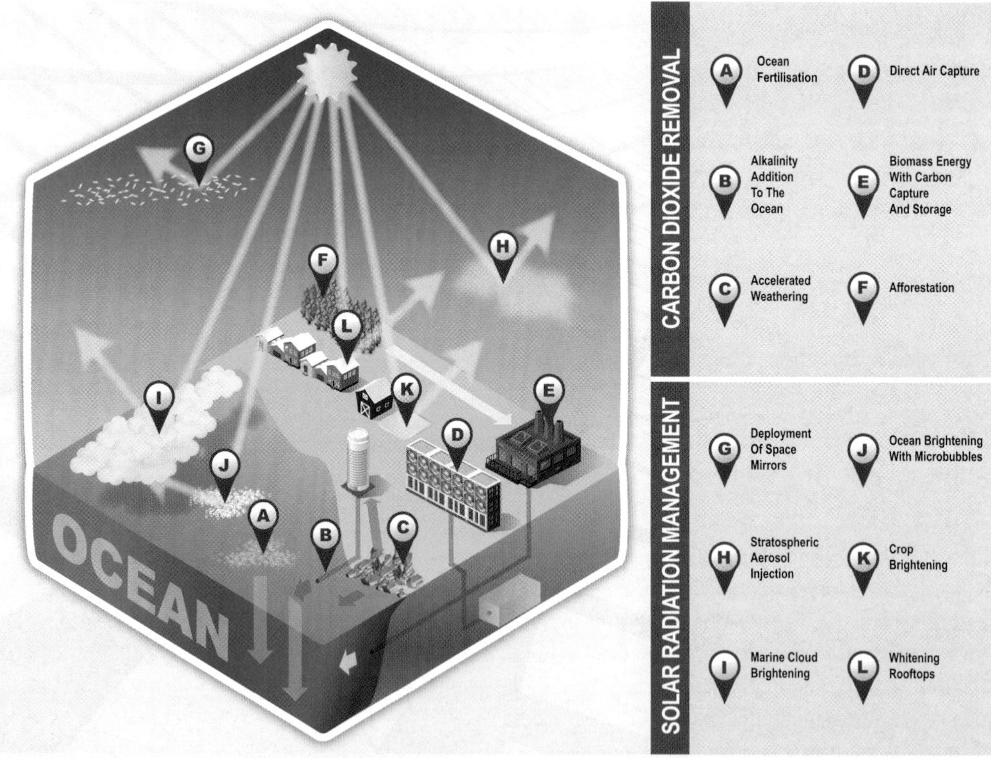

FAQ 7.3, Figure 1 | Overview of some proposed geoengineering methods as they have been suggested. Carbon Dioxide Removal methods (see Section 6.5 for details): (A) nutrients are added to the ocean (ocean fertilization), which increases oceanic productivity in the surface ocean and transports a fraction of the resulting biogenic carbon downward; (B) alkalinity from solid minerals is added to the ocean, which causes more atmospheric CO_2 to dissolve in the ocean; (C) the weathering rate of silicate rocks is increased, and the dissolved carbonate minerals are transported to the ocean; (D) atmospheric CO_2 is captured chemically, and stored either underground or in the ocean; (E) biomass is burned at an electric power plant with carbon capture, and the captured CO_2 is stored either underground or in the ocean; and (F) CO_2 is captured through afforestation and reforestation to be stored in land ecosystems. Solar Radiation Management methods (see Section 7.7 for details): (G) reflectors are placed in space to reflect solar radiation; (H) aerosols are injected in the stratosphere; (I) marine clouds are seeded in order to be made more reflective; (J) microbubbles are produced at the ocean surface to make it more reflective; (K) more reflective crops are grown; and (L) roofs and other built structures are whitened.

FAQ 7.3 (continued)

then be stored in organic form in land reservoirs, or in inorganic form in oceanic and geological reservoirs, where it would have to be stored for at least hundreds of years for CDR to be effective.

CDR methods would reduce the radiative forcing of CO_2 inasmuch as they are effective at removing CO_2 from the atmosphere and keeping the removed carbon away from the atmosphere. Some methods would also reduce ocean acidification (see FAQ 3.2), but other methods involving oceanic storage might instead increase ocean acidification if the carbon is sequestered as dissolved CO_2. A major uncertainty related to the effectiveness of CDR methods is the storage capacity and the permanence of stored carbon. Permanent carbon removal and storage by CDR would decrease climate warming in the long term. However, non-permanent storage strategies would allow CO_2 to return back to the atmosphere where it would once again contribute to warming. An intentional removal of CO_2 by CDR methods will be partially offset by the response of the oceanic and terrestrial carbon reservoirs if the CO_2 atmospheric concentration is reduced. This is because some oceanic and terrestrial carbon reservoirs will outgas to the atmosphere the anthropogenic CO_2 that had previously been stored. To completely offset past anthropogenic CO_2 emissions, CDR techniques would therefore need to remove not just the CO_2 that has accumulated in the atmosphere since pre-industrial times, but also the anthropogenic carbon previously taken up by the terrestrial biosphere and the ocean.

Biological and most chemical weathering CDR methods cannot be scaled up indefinitely and are necessarily limited by various physical or environmental constraints such as competing demands for land. Assuming a maximum CDR sequestration rate of 200 PgC per century from a combination of CDR methods, it would take about one and half centuries to remove the CO_2 emitted in the last 50 years, making it difficult—even for a suite of additive CDR methods—to mitigate climate change rapidly. Direct air capture methods could in principle operate much more rapidly, but may be limited by large-scale implementation, including energy use and environmental constraints.

CDR could also have climatic and environmental side effects. For instance, enhanced vegetation productivity may increase emissions of N_2O, which is a more potent greenhouse gas than CO_2. A large-scale increase in vegetation coverage, for instance through afforestation or energy crops, could alter surface characteristics, such as surface reflectivity and turbulent fluxes. Some modelling studies have shown that afforestation in seasonally snow-covered boreal regions could in fact accelerate global warming, whereas afforestation in the tropics may be more effective at slowing global warming. Ocean-based CDR methods that rely on biological production (i.e., ocean fertilization) would have numerous side effects on ocean ecosystems, ocean acidity and may produce emissions of non-CO_2 greenhouse gases.

Solar Radiation Management Methods

The globally averaged surface temperature of the planet is strongly influenced by the amount of sunlight absorbed by the Earth's atmosphere and surface, which warms the planet, and by the existence of the greenhouse effect, the process by which greenhouse gases and clouds affect the way energy is eventually radiated back to space. An increase in the greenhouse effect leads to a surface temperature rise until a new equilibrium is found. If less incoming sunlight is absorbed because the planet has been made more reflective, or if energy can be emitted to space more effectively because the greenhouse effect is reduced, the average global surface temperature will be reduced.

Suggested geoengineering methods that aim at managing the Earth's incoming and outgoing energy flows are based on this fundamental physical principle. Most of these methods propose to either reduce sunlight reaching the Earth or increase the reflectivity of the planet by making the atmosphere, clouds or the surface brighter (see FAQ 7.3, Figure 1). Another technique proposes to suppress high-level clouds called cirrus, as these clouds have a strong greenhouse effect. Basic physics tells us that if any of these methods change energy flows as expected, then the planet will cool. The picture is complicated, however, because of the many and complex physical processes which govern the interactions between the flow of energy, the atmospheric circulation, weather and the resulting climate.

While the globally averaged surface temperature of the planet will respond to a change in the amount of sunlight reaching the surface or a change in the greenhouse effect, the temperature at any given location and time is influenced by many other factors and the amount of cooling from SRM will not in general equal the amount of warming caused by greenhouse gases. For example, SRM will change heating rates only during daytime, but increasing greenhouse gases can change temperatures during both day and night. This inexact compensation can influence

(continued on next page)

FAQ 7.3 (continued)

the diurnal cycle of surface temperature, even if the average surface temperature is unchanged. As another example, model calculations suggest that a uniform decrease in sunlight reaching the surface might offset global mean CO_2-induced warming, but some regions will cool less than others. Models suggest that if anthropogenic greenhouse warming were completely compensated by stratospheric aerosols, then polar regions would be left with a small residual warming, while tropical regions would become a little cooler than in pre-industrial times.

SRM could theoretically counteract anthropogenic climate change rapidly, cooling the Earth to pre-industrial levels within one or two decades. This is known from climate models but also from the climate records of large volcanic eruptions. The well-observed eruption of Mt Pinatubo in 1991 caused a temporary increase in stratospheric aerosols and a rapid decrease in surface temperature of about 0.5°C.

Climate consists of many factors besides surface temperature. Consequences for other climate features, such as rainfall, soil moisture, river flow, snowpack and sea ice, and ecosystems may also be important. Both models and theory show that compensating an increased greenhouse effect with SRM to stabilize surface temperature would somewhat lower the globally averaged rainfall (see FAQ 7.3, Figure 2 for an idealized model result), and there also could be regional changes. Such imprecise compensation in regional and global climate patterns makes it improbable that SRM will produce a future climate that is 'just like' the one we experience today, or have experienced in the past. However, available climate models indicate that a geoengineered climate with SRM and high atmospheric CO_2 levels would be generally closer to 20th century climate than a future climate with elevated CO_2 concentrations and no SRM.

SRM techniques would probably have other side effects. For example, theory, observation and models suggest that stratospheric sulphate aerosols from volcanic eruptions and natural emissions deplete stratospheric ozone, especially while chlorine from chlorofluorocarbon emissions resides in the atmosphere. Stratospheric aerosols introduced for SRM are expected to have the same effect. Ozone depletion would increase the amount of ultraviolet light reaching the surface damaging terrestrial and marine ecosystems. Stratospheric aerosols would also increase the ratio of direct to diffuse sunlight reaching the surface, which generally increases plant productivity. There has also been some concern that sulphate aerosol SRM would increase acid rain, but model studies suggest that acid rain is probably not a major concern since the rate of acid rain production from stratospheric aerosol SRM would be much smaller than values currently produced by pollution sources. SRM will also not address the ocean acidification associated with increasing atmospheric CO_2 concentrations and its impacts on marine ecosystems.

Without conventional mitigation efforts or potential CDR methods, high CO_2 concentrations from anthropogenic emissions will persist in the atmosphere for as long as a thousand years, and SRM would have to be maintained as long as CO_2 concentrations were high. Stopping SRM while CO_2 concentrations are still high would lead to a very rapid warming over one or two decades (see FAQ7.3, Figure 2), severely stressing ecosystem and human adaptation.

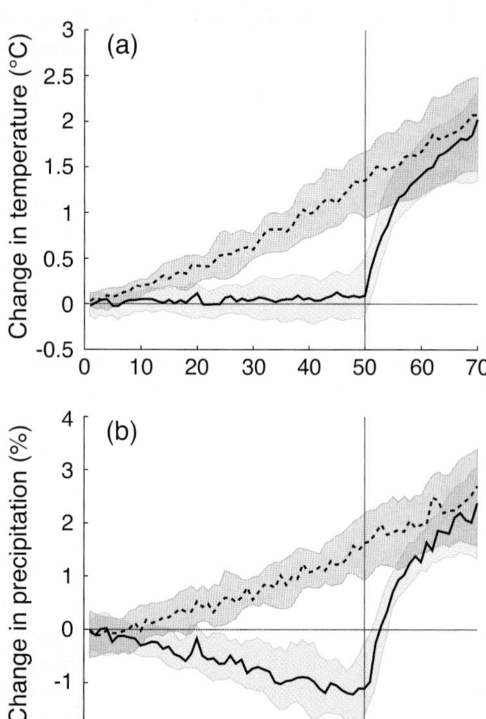

FAQ 7.3, Figure 2 | Change in globally averaged (a) surface temperature (°C) and (b) precipitation (%) in two idealized experiments. Solid lines are for simulations using Solar Radiation Management (SRM) to balance a 1% yr^{-1} increase in CO_2 concentration until year 50, after which SRM is stopped. Dashed lines are for simulations with a 1% yr^{-1} increase in CO_2 concentration and no SRM. The yellow and grey envelopes show the 25th to 75th percentiles from eight different models.

If SRM were used to avoid some consequences of increasing CO_2 concentrations, the risks, side effects and shortcomings would clearly increase as the scale of SRM increase. Approaches have been proposed to use a time-limited amount of SRM along with aggressive strategies for reducing CO_2 concentrations to help avoid transitions across climate thresholds or tipping points that would be unavoidable otherwise; assessment of such approaches would require a very careful risk benefit analysis that goes much beyond this Report.

and a warmer North Pacific adjacent to a cooler northwestern Canada, produced a SST response with a La Niña-like pattern. One study has noted regional shifts in the potential hurricane intensity and hurricane genesis potential index in the Atlantic Ocean and South China Sea in response to cloud brightening (Baughman et al., 2012), due primarily to decreases in vertical wind shear, but overall the investigation and identification of robust side effects has not been extensively explored.

Irvine et al. (2011) tested the impact of increasing desert albedo up to 0.80 in a climate model. This cooled surface temperature by –1.1°C (versus –0.22°C and –0.11°C for their largest crop and urban albedo change) and produced very significant changes in regional precipitation patterns.

7.7.4 Synthesis on Solar Radiation Management Methods

Theory, model studies and observations suggest that some SRM methods may be able to counteract a portion of global warming effects (on temperature, sea ice and precipitation) due to high concentrations of anthropogenic GHGs (*high confidence*). But the level of understanding about SRM is low, and it is difficult to assess feasibility and efficacy because of remaining uncertainties in important climate processes and the interactions among those processes. Although SRM research is still in its infancy, enough is known to identify some potential benefits, which must be weighed against known side effects (there could also be side effects that have not yet been identified). All studies suggest there would be a small but measurable decrease in global precipitation from SRM. Other side effects are specific to specific methods, and a number of research areas remain largely unexplored. There are also features that develop as a consequence of the combination of high CO_2 and SRM (e.g., effects on evapotranspiration and precipitation). SRM counters only some consequences of elevated CO_2 concentrations; it does not in particular address ocean acidification.

Many model studies indicate that stratospheric aerosol SRM could counteract some changes resulting from GHG increases that produce a RF as strong as 4 W m^{-2} (*medium confidence*), but they disagree on details. Marine cloud brightening SRM has received less attention, and there is no consensus on its efficacy, in large part due to the high level of uncertainty about cloud radiative responses to aerosol changes. There have been fewer studies and much less attention focused on all other SRM methods, and it is not currently possible to provide a general assessment of their specific efficacy, scalability, side effects and risks.

There is robust agreement among models and *high confidence* that the compensation between GHG warming and SRM cooling is imprecise. SRM would not produce a future climate identical to the present (or pre-industrial) climate. Nonetheless, although models disagree on details, they consistently suggest that a climate with SRM and high atmospheric CO_2 levels would be closer to that of the last century than a world with elevated CO_2 concentrations and no SRM (Lunt et al., 2008; Ricke et al., 2010; Moreno-Cruz et al., 2011), as long as the SRM could be continuously sustained and calibrated to offset the forcing by GHGs. Aerosol-based methods would, however, require a continuous program of replenishment to achieve this. If CO_2 concentrations and SRM were increased in concert, the risks and residual climate change produced by the imprecise compensation between SRM and CO_2 forcing would also increase. If SRM were terminated for any reason, a rapid increase in surface temperatures (within a decade or two) to values consistent with the high GHG forcing would result (*high confidence*). This rate of climate change would far exceed what would have occurred without geoengineering, causing any impacts related to the rate of change to be correspondingly greater than they would have been without geoengineering. In contrast, SRM in concert with aggressive CO_2 mitigation might conceivably help avoid transitions across climate thresholds or tipping points that would be unavoidable otherwise.

Acknowledgements

Thanks go to Anne-Lise Barbanes (IPSL/CNRS, Paris), Bénédicte Fisset (IPSL/CNRS, Paris) and Edwina Berry for their help in assembling the list of references. Sylvaine Ferrachat (ETH Zürich) is acknowledged for her contribution to drafting Figure 7.19.

References

Abdul-Razzak, H., and S. Ghan, 2000: A parameterization of aerosol activation 2. Multiple aerosol types. *J. Geophys. Res.*, **105**, 6837–6844.

Ackerman, A. S., M. P. Kirkpatrick, D. E. Stevens, and O. B. Toon, 2004: The impact of humidity above stratiform clouds on indirect aerosol climate forcing. *Nature*, **432**, 1014–1017.

Ackerman, A. S., et al., 2009: Large-eddy simulations of a drizzling, stratocumulus-topped marine boundary layer. *Mon. Weather Rev.*, **137**, 1083–1110.

Adachi, K., S. H. Chung, and P. R. Buseck, 2010: Shapes of soot aerosol particles and implications for their effects on climate. *J. Geophys. Res.*, **115**, D15206.

Adams, P. J., J. H. Seinfeld, D. Koch, L. Mickley, and D. Jacob, 2001: General circulation model assessment of direct radiative forcing by the sulfate-nitrate-ammonium-water inorganic aerosol system. *J. Geophys. Res.*, **106**, 1097–1111.

Adler, R. F., et al., 2003: The Version 2 Global Precipitation Climatology Project (GPCP) Monthly Precipitation Analysis (1979–Present). *J. Hydrometeor.*, **4**, 1147–1167.

Agee, E. M., K. Kiefer, and E. Cornett, 2012: Relationship of lower troposphere cloud cover and cosmic rays: An updated perspective. *J. Clim.*, **25**, 1057–1060.

Albrecht, B. A., 1989: Aerosols, cloud microphysics, and fractional cloudiness. *Science*, **245**, 1227–1230.

Aldrin, M., M. Holden, P. Guttorp, R. B. Skeie, G. Myhre, and T. K. Berntsen, 2012: Bayesian estimation of climate sensitivity based on a simple climate model fitted to observations of hemispheric temperatures and global ocean heat content. *Environmetrics*, **23**, 253–271.

Allan, R. P., and B. J. Soden, 2007: Large discrepancy between observed and simulated precipitation trends in the ascending and descending branches of the tropical circulation. *Geophys. Res. Lett.*, **34**, L18705.

Allen, M. R., and W. J. Ingram, 2002: Constraints on future changes in climate and the hydrologic cycle. *Nature*, **419**, 224–232.

Allen, R. J., and S. C. Sherwood, 2010: Aerosol-cloud semi-direct effect and land-sea temperature contrast in a GCM. *Geophys. Res. Lett.*, **37**, L07702.

Allen, R. J., S. C. Sherwood, J. R. Norris, and C. S. Zender, 2012: Recent Northern Hemisphere tropical expansion primarily driven by black carbon and tropospheric ozone. *Nature*, **485**, 350–354.

Alpert, P., N. Halfon, and Z. Levin, 2008: Does air pollution really suppress precipitation in Israel? *J. Appl. Meteor. Climatol.*, **47**, 933–943.

Alterskjær, K., J. E. Kristjánsson, and Ø. Seland, 2012: Sensitivity to deliberate sea salt seeding of marine clouds - observations and model simulations. *Atmos. Chem. Phys.*, **12**, 2795–2807.

Anderson, T. L., et al., 2005: An "A-Train" strategy for quantifying direct climate forcing by anthropogenic aerosols. *Bull. Am. Meteor. Soc.*, **86**, 1795–1809.

Anderson, T. L., et al., 2009: Temporal and spatial variability of clouds and related aerosol. In: *Clouds in the Perturbed Climate System: Their Relationship to Energy Balance, Atmospheric Dynamics, and Precipitation* [R. J. Charlson, and J. Heintzenberg (eds.)]. MIT Press, Cambridge, MA, USA, pp. 127–148.

Andreae, M. O., and A. Gelencser, 2006: Black carbon or brown carbon? The nature of light-absorbing carbonaceous aerosols. *Atmos. Chem. Phys.*, **6**, 3131–3148.

Andreae, M. O., C. D. Jones, and P. M. Cox, 2005: Strong present-day aerosol cooling implies a hot future. *Nature*, **435**, 1187–1190.

Andreae, M. O., D. Rosenfeld, P. Artaxo, A. A. Costa, G. P. Frank, K. M. Longo, and M. A. F. Silva-Dias, 2004: Smoking rain clouds over the Amazon. *Science*, **303**, 1337–1342.

Andrejczuk, M., W. W. Grabowski, S. P. Malinowski, and P. K. Smolarkiewicz, 2006: Numerical simulation of cloud-clear air interfacial mixing: Effects on cloud microphysics. *J. Atmos. Sci.*, **63**, 3204–3225.

Andrews, T., and P. M. Forster, 2008: CO_2 forcing induces semi-direct effects with consequences for climate feedback interpretations. *Geophys. Res. Lett.*, **35**, L04802.

Andrews, T., P. M. Forster, and J. M. Gregory, 2009: A surface energy perspective on climate change. *J. Clim.*, **22**, 2557–2570.

Andrews, T., J. M. Gregory, M. J. Webb, and K. E. Taylor, 2012: Forcing, feedbacks and climate sensitivity in CMIP5 coupled atmosphere-ocean climate models. *Geophys. Res. Lett.*, **39**, L09712.

Andrews, T., P. M. Forster, O. Boucher, N. Bellouin, and A. Jones, 2010: Precipitation, radiative forcing and global temperature change. *Geophys. Res. Lett.*, **37**, L14701.

Angel, R., 2006: Feasibility of cooling the Earth with a cloud of small spacecraft near the inner Lagrange Point (L1). *Proc. Natl. Acad. Sci. U.S.A.*, **103**, 17184–17189.

Ansmann, A., et al., 2008: Influence of Saharan dust on cloud glaciation in southern Morocco during the Saharan Mineral Dust Experiment. *J. Geophys. Res.*, **113**, D04210.

Arakawa, A., 1975: Modeling clouds and cloud processes for use in climate models. In: *The Physical Basis of Climate and Climate Modelling*. ICSU/WMO, GARP Publications Series N° 16, Geneva, Switzerland, pp. 181–197.

Arakawa, A., 2004: The cumulus parameterization problem: Past, present, and future. *J. Clim.*, **17**, 2493–2525.

Arneth, A., R. K. Monson, G. Schurgers, U. Niinemets, and P. I. Palmer, 2008: Why are estimates of global terrestrial isoprene emissions so similar (and why is this not so for monoterpenes)? *Atmos. Chem. Phys.*, **8**, 4605–4620.

Arneth, A., P. A. Miller, M. Scholze, T. Hickler, G. Schurgers, B. Smith, and I. C. Prentice, 2007: CO_2 inhibition of global terrestrial isoprene emissions: Potential implications for atmospheric chemistry. *Geophys. Res. Lett.*, **34**, L18813.

Arnold, F., 2006: Atmospheric aerosol and cloud condensation nuclei formation: A possible influence of cosmic rays? *Space Sci. Rev.*, **125**, 169–186.

Artaxo, P., et al., 1998: Large-scale aerosol source apportionment in Amazonia. *J. Geophys. Res.*, **103**, 31837–31847.

Artaxo, P., et al., 2002: Physical and chemical properties of aerosols in the wet and dry seasons in Rondônia, Amazonia. *J. Geophys. Res.*, **107**, 8081.

Asmi, A., et al., 2011: Number size distributions and seasonality of submicron particles in Europe 2008–2009. *Atmos. Chem. Phys.*, **11**, 5505–5538.

Aw, J., and M. J. Kleeman, 2003: Evaluating the first-order effect of intraannual temperature variability on urban air pollution. *J. Geophys. Res.*, **108**, 4365.

Ayers, G. P., and J. M. Cainey, 2007: The CLAW hypothesis: A review of the major developments. *Environ. Chem.*, **4**, 366–374.

Babu, S. S., et al., 2011: Free tropospheric black carbon aerosol measurements using high altitude balloon: Do BC layers build "their own homes" up in the atmosphere? *Geophys. Res. Lett.*, **38**, L08803.

Baker, M. B., and R. J. Charlson, 1990: Bistability of CCN concentrations and thermodynamics in the cloud-topped boundary-layer. *Nature*, **345**, 142–145.

Baker, M. B., and T. Peter, 2008: Small-scale cloud processes and climate. *Nature*, **451**, 299–300.

Bala, G., P. B. Duffy, and K. E. Taylor, 2008: Impact of geoengineering schemes on the global hydrological cycle. *Proc. Natl. Acad. Sci. U.S.A.*, **105**, 7664–7669.

Ban-Weiss, G., L. Cao, G. Bala, and K. Caldeira, 2012: Dependence of climate forcing and response on the altitude of black carbon aerosols. *Clim. Dyn.*, **38**, 897–911.

Bangert, M., C. Kottmeier, B. Vogel, and H. Vogel, 2011: Regional scale effects of the aerosol cloud interaction simulated with an online coupled comprehensive chemistry model. *Atmos. Chem. Phys.*, **11**, 4411–4423.

Barahona, D., and A. Nenes, 2008: Parameterization of cirrus cloud formation in large-scale models: Homogeneous nucleation. *J. Geophys. Res.*, **113**, D11211.

Barahona, D., and A. Nenes, 2009: Parameterizing the competition between homogeneous and heterogeneous freezing in ice cloud formation—polydisperse ice nuclei. *Atmos. Chem. Phys.*, **9**, 5933–5948.

Barker, H. W., J. N. S. Cole, J.-J. Morcrette, R. Pincus, P. Raisaenen, K. von Salzen, and P. A. Vaillancourt, 2008: The Monte Carlo Independent Column Approximation: An assessment using several global atmospheric models. *Q. J. R. Meteorol. Soc.*, **134**, 1463–1478.

Barker, H. W., et al., 2003: Assessing 1D atmospheric solar radiative transfer models: Interpretation and handling of unresolved clouds. *J. Clim.*, **16**, 2676–2699.

Barmet, P., T. Kuster, A. Muhlbauer, and U. Lohmann, 2009: Weekly cycle in particulate matter versus weekly cycle in precipitation over Switzerland. *J. Geophys. Res.*, **114**, D05206.

Bauer, S., E. Bierwirth, M. Esselborn, A. Petzold, A. Macke, T. Trautmann, and M. Wendisch, 2011: Airborne spectral radiation measurements to derive solar radiative forcing of Saharan dust mixed with biomass burning smoke particles. *Tellus B*, **63**, 742–750.

Bauer, S. E., and S. Menon, 2012: Aerosol direct, indirect, semidirect, and surface albedo effects from sector contributions based on the IPCC AR5 emissions for preindustrial and present-day conditions. *J. Geophys. Res.*, **117**, D01206.

Bauer, S. E., D. Koch, N. Unger, S. M. Metzger, D. T. Shindell, and D. G. Streets, 2007: Nitrate aerosols today and in 2030: |A global simulation including aerosols and tropospheric ozone. *Atmos. Chem. Phys.*, **7**, 5043–5059.

Baughman, E., A. Gnanadesikan, A. Degaetano, and A. Adcroft, 2012: Investigation of the surface and circulation impacts of cloud-brightening geoengineering. *J. Clim.*, **25**, 7527–7543.

Bäumer, D., and B. Vogel, 2007: An unexpected pattern of distinct weekly periodicities in climatological variables in Germany. *Geophys. Res. Lett.*, **34**, L03819.

Baumgardner, D., et al., 2012: Soot reference materials for instrument calibration and intercomparisons: A workshop summary with recommendations. *Atmos. Meas. Tech.*, **5**, 1869–1887.

Bazilevskaya, G. A., et al., 2008: Cosmic ray induced ion production in the atmosphere. *Space Sci. Rev.*, **137**, 149–173.

Bechtold, P., et al., 2008: Advances in simulating atmospheric variability with the ECMWF model: From synoptic to decadal time-scales. *Q. J. R. Meteorol. Soc.*, **134**, 1337–1351.

Bell, T. L., D. Rosenfeld, K.-M. Kim, J.-M. Yoo, M.-I. Lee, and M. Hahnenberger, 2008: Midweek increase in US summer rain and storm heights suggests air pollution invigorates rainstorms. *J. Geophys. Res.*, **113**, D02209.

Bellouin, N., and O. Boucher, 2010: Climate response and efficacy of snow forcing in the HadGEM2–AML climate model. Hadley Centre Technical Note N°82. Met Office, Exeter, Devon, UK.

Bellouin, N., O. Boucher, J. Haywood, and M. S. Reddy, 2005: Global estimate of aerosol direct radiative forcing from satellite measurements. *Nature*, **438**, 1138–1141.

Bellouin, N., A. Jones, J. Haywood, and S. A. Christopher, 2008: Updated estimate of aerosol direct radiative forcing from satellite observations and comparison against the Hadley Centre climate model. *J. Geophys. Res.*, **113**, D10205.

Bellouin, N., J. Quaas, J.-J. Morcrette, and O. Boucher, 2013: Estimates of aerosol radiative forcing from the MACC re-analysis. *Atmos. Chem. Phys.*, **13**, 2045–2062.

Bellouin, N., J. Rae, C. Johnson, J. Haywood, A. Jones, and O. Boucher, 2011: Aerosol forcing in the Climate Model Intercomparison Project (CMIP5) simulations by HadGEM2–ES and the role of ammonium nitrate. *J. Geophys. Res.*, **116**, D20206.

Bender, F. A.-M., V. Ramanathan, and G. Tselioudis, 2012: Changes in extratropical storm track cloudiness 1983–2008: Observational support for a poleward shift. *Clim. Dyn.*, **38**, 2037–2053.

Benedetti, A., et al., 2009: Aerosol analysis and forecast in the ECMWF Integrated Forecast System. Part II : Data assimilation. *J. Geophys. Res.*, **114**, D13205.

Benedict, J. J., and D. A. Randall, 2009: Structure of the Madden-Julian Oscillation in the Superparameterized CAM. *J. Atmos. Sci.*, **66**, 3277–3296.

Berg, L. K., C. M. Berkowitz, J. C. Barnard, G. Senum, and S. R. Springston, 2011: Observations of the first aerosol indirect effect in shallow cumuli. *Geophys. Res. Lett.*, **38**, L03809.

Bergamo, A., A. M. Tafuro, S. Kinne, F. De Tomasi, and M. R. Perrone, 2008: Monthly-averaged anthropogenic aerosol direct radiative forcing over the Mediterranian based on AERONET aerosol properties. *Atmos. Chem. Phys.*, **8**, 6995–7014.

Bergstrom, R. W., et al., 2010: Aerosol spectral absorption in the Mexico City area: Results from airborne measurements during MILAGRO/INTEX B. *Atmos. Chem. Phys.*, **10**, 6333–6343.

Bertram, A. K., T. Koop, L. T. Molina, and M. J. Molina, 2000: Ice formation in $(NH_4)_2SO_4$-H_2O particles. *J. Phys. Chem. A*, **104**, 584–588.

Bewick, R., J. P. Sanchez, and C. R. McInnes, 2012: Gravitationally bound geoengineering dust shade at the inner Lagrange point. *Adv. Space Res.*, **50**, 1405–1410.

Bharmal, N. A., A. Slingo, G. J. Robinson, and J. J. Settle, 2009: Simulation of surface and top of atmosphere thermal fluxes and radiances from the radiative atmospheric divergence using the ARM Mobile Facility, GERB data, and AMMA Stations experiment. *J. Geophys. Res.*, **114**, D00E07.

Bi, X., et al., 2011: Mixing state of biomass burning particles by single particle aerosol mass spectrometer in the urban area of PRD, China. *Atmos. Environ.*, **45**, 3447–3453.

Blossey, P. N., C. S. Bretherton, J. Cetrone, and M. Kharoutdinov, 2007: Cloud-resolving model simulations of KWAJEX: Model sensitivities and comparisons with satellite and radar observations. *J. Atmos. Sci.*, **64**, 1488–1508.

Blossey, P. N., et al., 2013: Marine low cloud sensitivity to an idealized climate change: The CGILS LES intercomparison. *J. Adv. Model. Earth Syst.*,**5**, 234–258.

Bodenschatz, E., S. P. Malinowski, R. A. Shaw, and F. Stratmann, 2010: Can we understand clouds without turbulence? *Science*, **327**, 970–971.

Boers, R., and R. M. Mitchell, 1994: Absorption feedback in stratocumulus clouds—Influence on cloud-top albedo. *Tellus A*, **46**, 229–241.

Bokoye, A. I., A. Royer, N. T. O'Neil, P. Cliche, G. Fedosejevs, P. M. Teillet, and L. J. B. McArthur, 2001: Characterization of atmospheric aerosols across Canada from a ground-based sunphotometer network: AEROCAN. *Atmos. Ocean*, **39**, 429–456.

Bond, G., et al., 2001: Persistent solar influence on North Atlantic climate during the Holocene. *Science*, **294**, 2130–2136.

Bond, T. C., et al., 2013: Bounding the role of black carbon in the climate system: A scientific assessment. *J. Geophys. Res. Atmos.*, **118**, 5380–5552.

Bondo, T., M. B. Enghoff, and H. Svensmark, 2010: Model of optical response of marine aerosols to Forbush decreases. *Atmos. Chem. Phys.*, **10**, 2765–2776.

Bony, S., and J.-L. Dufresne, 2005: Marine boundary layer clouds at the heart of tropical cloud feedback uncertainties in climate models. *Geophys. Res. Lett.*, **32**, L20806.

Bony, S., J.-L. Dufresne, H. Le Treut, J.-J. Morcrette, and C. Senior, 2004: On dynamic and thermodynamic components of cloud changes. *Clim. Dyn.*, **22**, 71–86.

Bony, S., G. Bellon, D. Klocke, S. Sherwood, S. Fermapin, and S. Denvil, 2013: Robust direct effect of carbon dioxide on tropical circulation and regional precipitation. *Nature Geosci.*, **6**, 447–451.

Bony, S., et al., 2006: How well do we understand and evaluate climate change feedback processes? *J. Clim.*, **19**, 3445–3482.

Bopp, L., O. Boucher, O. Aumont, S. Belviso, J.-L. Dufresne, M. Pham, and P. Monfray, 2004: Will marine dimethylsulfide emissions amplify or alleviate global warming? A model study. *Can. J. Fish. Aquat. Sci.*, **61**, 826–835.

Boucher, O., and U. Lohmann, 1995: The sulfate-CCN-cloud albedo effect—A sensitivity study with two general circulation models. *Tellus B*, **47**, 281–300.

Boucher, O., and J. Quaas, 2013: Water vapour affects both rain and aerosol optical depth. *Nature Geosci.*, **6**, 4–5.

Boucher, O., G. Myhre, and A. Myhre, 2004: Direct influence of irrigation on atmospheric water vapour and climate. *Clim. Dyn.*, **22**, 597–603.

Boucher, O., J. Lowe, and C. D. Jones, 2009: Constraints of the carbon cycle on timescales of climate-engineering options. *Clim. Change*, **92**, 261–273.

Bourotte, C., A.-P. Curi-Amarante, M.-C. Forti, L. A. A. Pereira, A. L. Braga, and P. A. Lotufo, 2007: Association between ionic composition of fine and coarse aerosol soluble fraction and peak expiratory flow of asthmatic patients in Sao Paulo city (Brazil). *Atmos. Environ.*, **41**, 2036–2048.

Boutle, I. A., and S. J. Abel, 2012: Microphysical controls on the stratocumulus topped boundary-layer structure during VOCALS-REx. *Atmos. Chem. Phys.*, **12**, 2849–2863.

Brenguier, J.-L., F. Burnet, and O. Geoffroy, 2011: Cloud optical thickness and liquid water path—does the *k* coefficient vary with droplet concentration? *Atmos. Chem. Phys.*, **11**, 9771–9786.

Bretherton, C. S., P. N. Blossey, and J. Uchida, 2007: Cloud droplet sedimentation, entrainment efficiency, and subtropical stratocumulus albedo. *Geophys. Res. Lett.*, **34**, L03813.

Bretherton, C. S., P. N. Blossey, and C. R. Jones, 2013: A large-eddy simulation of mechanisms of boundary layer cloud response to climate change. *J. Adv. Model. Earth Syst.*, **5**, 316–337.

Brient, F., and S. Bony, 2012: How may low-cloud radiative properties simulated in the current climate influence low-cloud feedbacks under global warming? *Geophys. Res. Lett.*, **39**, L20807.

Brient, F., and S. Bony, 2013: Interpretation of the positive low-cloud feedback predicted by a climate model under global warming. *Clim. Dyn.*, **40**, 2415–2431.

Brioude, J., et al., 2009: Effect of biomass burning on marine stratocumulus clouds off the California coast. *Atmos. Chem. Phys.*, **9**, 8841–8856.

Broccoli, A. J., and S. A. Klein, 2010: Comment on "Observational and model evidence for positive low-level cloud feedback". *Science*, **329**, 277–a.

Brock, C. A., et al., 2011: Characteristics, sources, and transport of aerosols measured in spring 2008 during the aerosol, radiation, and cloud processes affecting Arctic Climate (ARCPAC) Project. *Atmos. Chem. Phys.*, **11**, 2423–2453.

Bryan, G. H., J. C. Wyngaard, and J. M. Fritsch, 2003: Resolution requirements for the simulation of deep moist convection. *Mon. Weather Rev.*, **131**, 2394–2416.

Budyko, M. I., 1974: *Izmeniya Klimata*. Gidrometeoroizdat, Leningrad.

Burkhardt, U., and B. Kärcher, 2009: Process-based simulation of contrail cirrus in a global climate model. *J. Geophys. Res.*, **114**, D16201.

Burkhardt, U., and B. Kärcher, 2011: Global radiative forcing from contrail cirrus. *Nature Clim. Change*, **1**, 54–58.

Burrows, S. M., T. Butler, P. Jöckel, H. Tost, A. Kerkweg, U. Pöschl, and M. G. Lawrence, 2009: Bacteria in the global atmosphere—Part 2: Modeling of emissions and transport between different ecosystems. *Atmos. Chem. Phys.*, **9**, 9281–9297.

Cahalan, R. F., W. Ridgway, W. J. Wiscombe, T. L. Bell, and J. B. Snider, 1994: The albedo of fractal stratocumulus clouds. *J. Atmos. Sci.*, **51**, 2434–2455.

Caldwell, P., and C. S. Bretherton, 2009: Response of a subtropical stratocumulus-capped mixed layer to climate and aerosol changes. *J. Clim.*, **22**, 20–38

Čalogović, J., C. Albert, F. Arnold, J. Beer, L. Desorgher, and E. O. Flueckiger, 2010: Sudden cosmic ray decreases: No change of global cloud cover. *Geophys. Res. Lett.*, **37**, L03802.

Cameron-Smith, P., S. Elliott, M. Maltrud, D. Erickson, and O. Wingenter, 2011: Changes in dimethyl sulfide oceanic distribution due to climate change. *Geophys. Res. Lett.*, **38**, L07704.

Cao, G., X. Zhang, and F. Zheng, 2006: Inventory of black carbon and organic carbon emissions from China. *Atmos. Environ.*, **40**, 6516–6527.

Cappa, C. D., et al., 2012: Radiative absorption enhancements due to the mixing state of atmospheric black carbon. *Science*, **337**, 1078–1081.

Carlton, A. G., R. W. Pinder, P. V. Bhave, and G. A. Pouliot, 2010: To what extent can biogenic SOA be controlled? *Environ. Sci. Technol.*, **44**, 3376–3380.

Carrico, C. M., M. H. Bergin, A. B. Shrestha, J. E. Dibb, L. Gomes, and J. M. Harris, 2003: The importance of carbon and mineral dust to seasonal aerosol properties in the Nepal Himalaya. *Atmos. Environ.*, **37**, 2811–2824.

Carslaw, K. S., R. G. Harrison, and J. Kirkby, 2002: Cosmic rays, clouds, and climate. *Science*, **298**, 1732–1737.

Carslaw, K. S., O. Boucher, D. V. Spracklen, G. W. Mann, J. G. L. Rae, S. Woodward, and M. Kulmala, 2010: A review of natural aerosol interactions and feedbacks within the Earth system. *Atmos. Chem. Phys.*, **10**, 1701–1737.

Celis, J. E., J. R. Morales, C. A. Zarorc, and J. C. Inzunza, 2004: A study of the particulate matter PM_{10} composition in the atmosphere of Chillán, Chile. *Chemosphere*, **54**, 541–550.

Cess, R. D., 1975: Global climate change—Investigation of atmospheric feedback mechanisms. *Tellus*, **27**, 193–198.

Cess, R. D., et al., 1989: Interpretation of cloud-climate feedbacks as produced by 14 atmospheric general circulation models. *Science*, **245**, 513–516.

Cess, R. D., et al., 1990: Intercomparison and interpretation of climate feedback processes in 19 atmospheric general circulation models. *J. Geophys. Res.*, **95**, 16601–16615.

Chae, J. H., and S. C. Sherwood, 2010: Insights into cloud-top height and dynamics from the seasonal cycle of cloud-top heights observed by MISR in the West Pacific region. *J. Atmos. Sci.*, **67**, 248–261.

Chakraborty, A., and T. Gupta, 2010: Chemical characterization and source apportionment of submicron (PM1) aerosol in Kanpur region, India. *Aeros. Air Qual. Res.*, **10**, 433–445.

Chameides, W. L., C. Luo, R. Saylor, D. Streets, Y. Huang, M. Bergin, and F. Giorgi, 2002: Correlation between model-calculated anthropogenic aerosols and satellite-derived cloud optical depths: Indication of indirect effect? *J. Geophys. Res.*, **107**, 4085.

Chan, M. A., and J. C. Comiso, 2011: Cloud features detected by MODIS but not by CloudSat and CALIOP. *Geophys. Res. Lett.*, **38**, L24813.

Chan, Y. C., R. W. Simpson, G. H. Mctainsh, P. D. Vowles, D. D. Cohen, and G. M. Bailey, 1997: Characterisation of chemical species in $PM_{2.5}$ and PM_{10} aerosols in Brisbane, Australia. *Atmos. Environ.*, **31**, 3773–3785.

Chand, D., R. Wood, T. L. Anderson, S. K. Satheesh, and R. J. Charlson, 2009: Satellite-derived direct radiative effect of aerosols dependent on cloud cover. *Nature Geosci.*, **2**, 181–184.

Chand, D., T. L. Anderson, R. Wood, R. J. Charlson, Y. Hu, Z. Liu, and M. Vaughan, 2008: Quantifying above-cloud aerosol using spaceborne lidar for improved understanding of cloudy-sky direct climate forcing. *J. Geophys. Res.*, **113**, D13206.

Chand, D., et al., 2012: Aerosol optical depth increase in partly cloudy conditions. *J. Geophys. Res.*, **117**, D17207.

Chang, F. L., and J. A. Coakley, 2007: Relationships between marine stratus cloud optical depth and temperature: Inferences from AVHRR observations. *J. Clim.*, **20**, 2022–2036.

Charlson, R. J., A. S. Ackerman, F. A. M. Bender, T. L. Anderson, and Z. Liu, 2007: On the climate forcing consequences of the albedo continuum between cloudy and clear air. *Tellus B*, **59**, 715–727.

Charney, J. G., et al., 1979: Carbon dioxide and climate: A scientific assessment. Report of an Ad-Hoc Group on Carbon Dioxide and Climate, National Academy of Sciences, Washington D.C., USA, 33 pp.

Che, H., et al., 2009: Instrument calibration and aerosol optical depth validation of the China Aerosol Remote Sensing Network. *J. Geophys. Res.*, **114**, D03206.

Chen, J. P., A. Hazra, and Z. Levin, 2008: Parameterizing ice nucleation rates using contact angle and activation energy derived from laboratory data. *Atmos. Chem. Phys.*, **8**, 7431–7449.

Chen, L., G. Shi, S. Qin, S. Yang, and P. Zhang, 2011: Direct radiative forcing of anthropogenic aerosols over oceans from satellite observations. *Adv. Atmos. Sci.*, **28**, 973–984.

Chen, T., W. B. Rossow, and Y. Zhang, 2000: Radiative effects of cloud-type variations. *J. Clim.*, **13**, 264–286.

Chen, W. T., Y. H. Lee, P. J. Adams, A. Nenes, and J. H. Seinfeld, 2010: Will black carbon mitigation dampen aerosol indirect forcing? *Geophys. Res. Lett.*, **37**, L09801.

Chen, Y., Q. Li, R. A. Kahn, J. T. Randerson, and D. J. Diner, 2009: Quantifying aerosol direct radiative effect with Multiangle Imaging Spectroradiometer observations: Top-of-atmosphere albedo change by aerosols based on land surface types. *J. Geophys. Res.*, **114**, D02109.

Chen, Y. C., M. W. Christensen, L. Xue, A. Sorooshian, G. L. Stephens, R. M. Rasmussen, and J. H. Seinfeld, 2012: Occurrence of lower cloud albedo in ship tracks. *Atmos. Chem. Phys.*, **12**, 8223–8235.

Cheng, A., and K. M. Xu, 2006: Simulation of shallow cumuli and their transition to deep convective clouds by cloud-resolving models with different third-order turbulence closures. *Q. J. R. Meteorol. Soc.*, **132**, 359–382.

Cheng, A., and K.-M. Xu, 2008: Simulation of boundary-layer cumulus and stratocumulus clouds using a cloud-resolving model with low- and third-order turbulence closures. *J. Meteorol. Soc. Jpn*, **86**, 67–86.

Cheng, Z. L., K. S. Lam, L. Y. Chan, and K. K. Cheng, 2000: Chemical characteristics of aerosols at coastal station in Hong Kong. I. Seasonal variation of major ions, halogens and mineral dusts between 1995 and 1996. *Atmos. Environ.*, **34**, 2771–2783.

Chepfer, H., S. Bony, D. Winker, M. Chiriaco, J.-L. Dufresne, and G. Sèze, 2008: Use of CALIPSO lidar observations to evaluate the cloudiness simulated by a climate model. *Geophys. Res. Lett.*, **35**, L15704.

Chepfer, H., et al., 2010: The GCM-Oriented CALIPSO Cloud Product (CALIPSO-GOCCP). *J. Geophys. Res.*, **115**, D00H16.

Chikira, M., and M. Sugiyama, 2010: A cumulus parameterization with state-dependent entrainment rate. Part I: Description and sensitivity to temperature and humidity profiles. *J. Atmos. Sci.*, **67**, 2171–2193.

Chou, C., and J. D. Neelin, 2004: Mechanisms of global warming impacts on regional tropical precipitation. *J. Clim.*, **17**, 2688–2701.

Chou, C., J. D. Neelin, C. A. Chen, and J. Y. Tu, 2009: Evaluating the "Rich-Get-Richer" mechanism in tropical precipitation change under global warming. *J. Clim.*, **22**, 1982–2005.

Chow, J. C., J. G. Waston, D. H. Lowenthal, P. A. Solomon, K. L. Magliano, S. D. Ziman, and L. W. Richards, 1993: PM_{10} and $PM_{2.5}$ compositions in California's San Joaquin Valley. *Aer. Sci. Technol.*, **18**, 105–128.

Chowdhary, J., et al., 2005: Retrieval of aerosol scattering and absorption properties from photopolarimetric observations over the ocean during the CLAMS experiment. *J. Atmos. Sci.*, **62**, 1093–1117.

Christensen, M. W., and G. L. Stephens, 2011: Microphysical and macrophysical responses of marine stratocumulus polluted by underlying ships: Evidence of cloud deepening. *J. Geophys. Res.*, **116**, D03201.

Christopher, S. A., B. Johnson, T. A. Jones, and J. Haywood, 2009: Vertical and spatial distribution of dust from aircraft and satellite measurements during the GERBILS field campaign. *Geophys. Res. Lett.*, **36**, L06806.

Chung, C. E., V. Ramanathan, and D. Decremer, 2012: Observationally constrained estimates of carbonaceous aerosol radiative forcing. *Proc. Natl. Acad. Sci. U.S.A.*, **109**, 11624–11629.

Clarke, A. D., V. N. Kapustin, F. L. Eisele, R. J. Weber, and P. H. McMurry, 1999: Particle production near marine clouds: Sulfuric acid and predictions from classical binary nucleation. *Geophys. Res. Lett.*, **26**, 2425–2428.

Clement, A. C., R. Burgman, and J. R. Norris, 2009: Observational and model evidence for positive low-level cloud feedback. *Science*, **325**, 460–464.

Coakley, J. A., and C. D. Walsh, 2002: Limits to the aerosol indirect radiative effect derived from observations of ship tracks. *J. Atmos. Sci.*, **59**, 668–680.

Collins, M., B. B. Booth, B. Bhaskaran, G. R. Harris, J. M. Murphy, D. M. H. Sexton, and M. J. Webb, 2011: Climate model errors, feedbacks and forcings: A comparison of perturbed physics and multi-model ensembles. *Clim. Dyn.*, **36**, 1737–1766.

Colman, R. A., and B. J. McAvaney, 2011: On tropospheric adjustment to forcing and climate feedbacks. *Clim. Dyn.*, **36**, 1649–1658.

Colman, R. A., and L. I. Hanson, 2012: On atmospheric radiative feedbacks associated with climate variability and change. *Clim. Dyn.*, **40**, 475–492.

Comstock, K. K., C. S. Bretherton, and S. E. Yuter, 2005: Mesoscale variability and drizzle in Southeast Pacific stratocumulus. *J. Atmos. Sci.*, **62**, 3792–3807.

Costantino, L., and F.-M. Bréon, 2010: Analysis of aerosol-cloud interaction from multi-sensor satellite observations. *Geophys. Res. Lett.*, **37**, L11801.

Couvreux, F., F. Hourdin, and C. Rio, 2010: Resolved versus parametrized boundary-layer plumes. Part I: A parametrization-oriented conditional sampling in large-eddy simulations. *Bound. Layer Meteor.*, **134**, 441–458.

Cross, E. S., et al., 2010: Soot Particle Studies—Instrument Inter-Comparison—Project Overview. *Aer. Sci. Technol.*, **44**, 592–611.

Crucifix, M., 2006: Does the Last Glacial Maximum constrain climate sensitivity? *Geophys. Res. Lett.*, **33**, L18701.

Crutzen, P. J., 2006: Albedo enhancement by stratospheric sulfur injections: A contribution to resolve a policy dilemma? *Clim. Change*, **77**, 211–220.

Dai, A. G., J. H. Wang, P. W. Thorne, D. E. Parker, L. Haimberger, and X. L. L. Wang, 2011: A new approach to homogenize daily radiosonde humidity data. *J. Clim.*, **24**, 965–991.

Davies, R., and M. Molloy, 2012: Global cloud height fluctuations measured by MISR on Terra from 2000 to 2010. *Geophys. Res. Lett.*, **39**, L03701.

Davis, A. B., A. Marshak, H. Gerber, and W. J. Wiscombe, 1999: Horizontal structure of marine boundary layer clouds from centimeter to kilometer scales. *J. Geophys. Res.*, **104**, 6123–6144.

Dawson, J. P., P. J. Adams, and S. N. Pandis, 2007: Sensitivity of PM2.5 to climate in the Eastern US: A modeling case study. *Atmos. Chem. Phys.*, **7**, 4295–4309.

de Boer, G., H. Morrison, M. D. Shupe, and R. Hildner, 2011: Evidence of liquid dependent ice nucleation in high-latitude stratiform clouds from surface remote sensors. *Geophys. Res. Lett.*, **38**, L01803.

de Gouw, J., and J. L. Jimenez, 2009: Organic aerosols in the Earth's atmosphere. *Environ. Sci. Technol.*, **43**, 7614–7618.

de Gouw, J. A., et al., 2005: Budget of organic carbon in a polluted atmosphere: Results from the New England Air Quality Study in 2002. *J. Geophys. Res.*, **110**, D16305.

de Graaf, M., L. G. Tilstra, P. Wang, and P. Stammes, 2012: Retrieval of the aerosol direct radiative effect over clouds from spaceborne spectrometry. *J. Geophys. Res.*, **117**, D07207.

de Leeuw, G., et al., 2011: Production flux of sea spray aerosol. *Rev. Geophys.*, **49**, RG2001.

de Souza, P. A., W. Z. de Mello, L. M. Rauda, and S. M. Sella, 2010: Caracterização do material particulado fino e grosso e composição da fração inorgânica solúvel em água em São José Dos Campos (SP). *Química Nova*, **33**, 1247–1253.

Deboudt, K., P. Flament, M. Choel, A. Gloter, S. Sobanska, and C. Colliex, 2010: Mixing state of aerosols and direct observation of carbonaceous and marine coatings on African dust by individual particle analysis. *J. Geophys. Res.*, **115**, D24207.

Decesari, S., et al., 2010: Chemical composition of PM_{10} and PM_1 at the high-altitude Himalayan station Nepal Climate Observatory-Pyramid (NCO-P) (5079 m.a.s.l.). *Atmos. Chem. Phys.*, **10**, 4583–4596.

Dee, D. P., et al., 2011: The ERA-Interim reanalysis: Configuration and performance of the data assimilation system. *Q. J. R. Meteorol. Soc.*, **137**, 553–597.

Del Genio, A. D., and J. B. Wu, 2010: The role of entrainment in the diurnal cycle of continental convection. *J. Clim.*, **23**, 2722–2738.

Del Genio, A. D., M.-S. Yao, and J. Jonas, 2007: Will moist convection be stronger in a warmer climate? *Geophys. Res. Lett.*, **34**, L16703.

Del Genio, A. D., Y.-H. Chen, D. Kim, and M.-S. Yao, 2012: The MJO transition from shallow to deep convection in CloudSat/CALIPSO data and GISS GCM simulations. *J. Clim.*, **25**, 3755–3770.

DeLeon-Rodriguez, N., et al., 2013: Microbiome of the upper troposphere: Species composition and prevalence, effects of tropical storms, and atmospheric implications. *Proc. Natl. Acad. Sci. U.S.A.*, **110**, 2575–2580.

DeMott, C. A., C. Stan, D. A. Randall, J. L. Kinter III, and M. Khairoutdinov, 2011: The Asian Monsoon in the super-parameterized CCSM and its relation to tropical wave activity. *J. Clim.*, **24**, 5134–5156.

DeMott, P. J., et al., 2010: Predicting global atmospheric ice nuclei distributions and their impacts on climate. *Proc. Natl. Acad. Sci. U.S.A.*, **107**, 11217–11222.

Deng, M., G. G. Mace, Z. E. Wang, and H. Okamoto, 2010: Tropical Composition, Cloud and Climate Coupling Experiment validation for cirrus cloud profiling retrieval using CloudSat radar and CALIPSO lidar. *J. Geophys. Res.*, **115**, D00J15.

Dengel, S., D. Aeby, and J. Grace, 2009: A relationship between galactic cosmic radiation and tree rings. *New Phytologist*, **184**, 545–551.

Denman, K. L., et al., 2007: Couplings between changes in the climate system and biogeochemistry. In: *Climate Change 2007: The Physical Science Basis. Contribution of Working Group I to the Fourth Assessment Report of the Intergovernmental Panel on Climate Change* [Solomon, S., D. Qin, M. Manning, Z. Chen, M. Marquis, K. B. Averyt, M. Tignor and H. L. Miller (eds.)] Cambridge University Press, Cambridge, United Kingdom and New York, NY, USA, pp. 499-587.

Derbyshire, S. H., I. Beau, P. Bechtold, J.-Y. Grandpeix, J.-M. Piriou, J.-L. Redelsperger, and P. M. M. Soares, 2004: Sensitivity of moist convection to environmental humidity. *Q. J. R. Meteorol. Soc.*, **130**, 3055–3079.

Després, V. R., et al., 2012: Primary biological aerosol particles in the atmosphere: A review. *Tellus B*, **64**, 15598.

Dessler, A. E., 2010: A determination of the cloud feedback from climate variations over the past decade. *Science*, **330**, 1523–1527.

Dessler, A. E., 2011: Cloud variations and the Earth's energy budget. *Geophys. Res. Lett.*, **38**, L19701.

Dessler, A. E., 2013: Observations of climate feedbacks over 2000–10 and comparisons to climate models. *J. Clim.*, **26**, 333–342.

Dessler, A. E., and S. Wong, 2009: Estimates of the water vapor climate feedback during El Niño-Southern Oscillation. *J. Clim.*, **22**, 6404–6412.

Dessler, A. E., and S. M. Davis, 2010: Trends in tropospheric humidity from reanalysis systems. *J. Geophys. Res.*, **115**, D19127.

Deuzé, J.-L., et al., 2001: Remote sensing of aerosols over land surfaces from POLDER-ADEOS-1 polarized measurements. *J. Geophys. Res.*, **106**, 4913–4926.

Devasthale, A., O. Kruger, and H. Graßl, 2005: Change in cloud-top temperatures over Europe. *IEEE Geosci. Remote Sens. Lett.*, **2**, 333–336.

Di Biagio, C., A. di Sarra, and D. Meloni, 2010: Large atmospheric shortwave radiative forcing by Mediterranean aerosols derived from simultaneous ground-based and spaceborne observations and dependence on the aerosol type and single scattering albedo. *J. Geophys. Res.*, **115**, D10209.

Dickinson, R., 1975: Solar variability and the lower atmosphere. *Bull. Am. Meteor. Soc.*, **56**, 1240–1248.

Doherty, S. J., S. G. Warren, T. C. Grenfell, A. D. Clarke, and R. E. Brandt, 2010: Light-absorbing impurities in Arctic snow. *Atmos. Chem. Phys.*, **10**, 11647–11680.

Doherty, S. J., T. C. Grenfell, S. Forsström, D. L. Hegg, R. E. Brandt, and S. G. Warren, 2013: Observed vertical redistribution of black carbon and other insoluble light-absorbing particles in melting snow. *J. Geophys. Res. Atmos.*, **118**, 5553–5569.

Donahue, N. M., S. A. Epstein, S. N. Pandis, and A. L. Robinson, 2011a: A two-dimensional volatility basis set: 1. Organic-aerosol mixing thermodynamics. *Atmos. Chem. Phys.*, **11**, 3303–3318.

Donahue, N. M., E. R. Trump, J. R. Pierce, and I. Riipinen, 2011b: Theoretical constraints on pure vapor-pressure driven condensation of organics to ultrafine particles. *Geophys. Res. Lett.*, **38**, L16801.

Dong, B.-W., J. M. Gregory, and R. T. Sutton, 2009: Understanding land-sea warming contrast in response to increasing greenhouse gases. Part I: Transient adjustment. *J. Clim.*, **22**, 3079–3097.

Donner, L. J., et al., 2011: The dynamical core, physical parameterizations, and basic simulation characteristics of the atmospheric component AM3 of the GFDL global coupled model CM3. *J. Clim.*, **24**, 3484–3519

Donovan, D. P., 2003: Ice-cloud effective particle size parameterization based on combined lidar, radar reflectivity, and mean Doppler velocity measurements. *J. Geophys. Res.*, **108**, 4573.

Doughty, C. E., C. B. Field, and A. M. S. McMillan, 2011: Can crop albedo be increased through the modification of leaf trichomes, and could this cool regional climate? *Clim. Change*, **104**, 379–387.

Dubovik, O., A. Smirnov, B. N. Holben, M. D. King, Y. J. Kaufman, T. F. Eck, and I. Slutsker, 2000: Accuracy assessments of aerosol optical properties retrieved from Aerosol Robotic Network (AERONET) Sun and sky radiance measurements. *J. Geophys. Res.*, **105**, 9791–9806.

Dubovik, O., T. Lapyonok, Y. J. Kaufman, M. Chin, P. Ginoux, R. A. Kahn, and A. Sinyuk, 2008: Retrieving global aerosol sources from satellites using inverse modeling. *Atmos. Chem. Phys.*, **8**, 209–250.

Dubovik, O., et al., 2002: Variability of absorption and optical properties of key aerosol types observed in worldwide locations. *J. Atmos. Sci.*, **59**, 590–608.

Dubovik, O., et al., 2011: Statistically optimized inversion algorithm for enhanced retrieval of aerosol properties from spectral multi-angle polarimetric satellite observations. *Atmos. Meas. Tech.*, **4**, 975–1018.

Dufresne, J.-L., and S. Bony, 2008: An assessment of the primary sources of spread of global warming estimates from coupled atmosphere-ocean models. *J. Clim.*, **21**, 5135–5144.

Dunne, E. M., L. A. Lee, C. L. Reddington, and K. S. Carslaw, 2012: No statistically significant effect of a short-term decrease in the nucleation rate on atmospheric aerosols. *Atmos. Chem. Phys.*, **12**, 11573–11587.

Duplissy, J., et al., 2008: Cloud forming potential of secondary organic aerosol under near atmospheric conditions. *Geophys. Res. Lett.*, **35**, L03818.

Durkee, P. A., K. J. Noone, and R. T. Bluth, 2000: The Monterey Area Ship Track experiment. *J. Atmos. Sci.*, **57**, 2523–2541.

Dusek, U., et al., 2006: Size matters more than chemistry for cloud-nucleating ability of aerosol particles. *Science*, **312**, 1375–1378.

Early, J. T., 1989: Space-based solar shield to offset greenhouse effect. *J. Br. Interplanet. Soc.*, **42**, 567–569.

Easter, R. C., et al., 2004: MIRAGE: Model description and evaluation of aerosols and trace gases. *J. Geophys. Res.*, **109**, D20210.

Eastman, R., and S. G. Warren, 2010: Interannual variations of Arctic cloud types in relation to sea ice. *J. Clim.*, **23**, 4216–4232.

Eastman, R., and S. G. Warren, 2013: A 39–yr survey of cloud changes from land stations worldwide 1971–2009: Long-term trends, relation to aerosols, and expansion of the tropical belt. *J. Clim.*, **26**, 1286–1303.

Eichler, A., et al., 2009: Temperature response in the Altai region lags solar forcing. *Geophys. Res. Lett.*, **36**, L01808.

Eitzen, Z. A., K. M. Xu, and T. Wong, 2009: Cloud and radiative characteristics of tropical deep convective systems in extended cloud objects from CERES observations. *J. Clim.*, **22**, 5983–6000.

Ekman, A. M. L., A. Engström, and C. Wang, 2007: The effect of aerosol composition and concentration on the development and anvil properties of a continental deep convective cloud. *Q. J. R. Meteorol. Soc.*, **133**, 1439–1452.

Ekman, A. M. L., A. Engström, and A. Söderberg, 2011: Impact of two-way aerosol–cloud interaction and changes in aerosol size distribution on simulated aerosol-induced deep convective cloud sensitivity. *J. Atmos. Sci.*, **68**, 685–698.

Eliasson, S., S. A. Buehler, M. Milz, P. Eriksson, and V. O. John, 2011: Assessing observed and modelled spatial distributions of ice water path using satellite data. *Atmos. Chem. Phys.*, **11**, 375–391.

Eliseev, A. V., A. V. Chernokulsky, A. A. Karpenko, and I. I. Mokhov, 2009: Global warming mitigation by sulphur loading in the stratosphere: Dependence of required emissions on allowable residual warming rate. *Theor. Appl. Climatol.*, **101**, 67–81.

Enghoff, M. B., and H. Svensmark, 2008: The role of atmospheric ions in aerosol nucleation—a review. *Atmos. Chem. Phys.*, **8**, 4911–4923.

English, J. T., O. B. Toon, and M. J. Mills, 2012: Microphysical simulations of sulfur burdens from stratospheric sulfur geoengineering. *Atmos. Phys. Chem.*, **12**, 4775–4793.

Engström, A., and A. M. L. Ekman, 2010: Impact of meteorological factors on the correlation between aerosol optical depth and cloud fraction. *Geophys. Res. Lett.*, **37**, L18814.

Ervens, B., G. Feingold, and S. M. Kreidenweis, 2005: Influence of water-soluble organic carbon on cloud drop number concentration. *J. Geophys. Res.*, **110**, D18211.

Ervens, B., B. J. Turpin, and R. J. Weber, 2011a: Secondary organic aerosol formation in cloud droplets and aqueous particles (aqSOA): A review of laboratory, field and model studies. *Atmos. Chem. Phys.*, **11**, 11069–11102.

Ervens, B., G. Feingold, K. Sulia, and J. Harrington, 2011b: The impact of microphysical parameters, ice nucleation mode, and habit growth on the ice/liquid partitioning in mixed-phase Arctic clouds. *J. Geophys. Res.*, **116**, D17205.

Ervens, B., et al., 2007: Prediction of cloud condensation nucleus number concentration using measurements of aerosol size distributions and composition and light scattering enhancement due to humidity. *J. Geophys. Res.*, **112**, D10S32.

European Commission, Joint Research Centre, and Netherlands Environmental Assessment Agency (PBL), 2009: Emission Database for Global Atmospheric Research (EDGAR), release version 4.0. http://edgar.jrc.ec.europa.eu, last accessed 7 June 2013.

Evan, A. T., and J. R. Norris, 2012: On global changes in effective cloud height. *Geophys. Res. Lett.*, **39**, L19710.

Evans, J. R. G., E. P. J. Stride, M. J. Edirisinghe, D. J. Andrews, and R. R. Simons, 2010: Can oceanic foams limit global warming? *Clim. Res.*, **42**, 155–160.

Facchini, M. C., et al., 2008: Primary submicron marine aerosol dominated by insoluble organic colloids and aggregates. *Geophys. Res. Lett.*, **35**, L17814.

Fan, J., et al., 2009: Dominant role by vertical wind shear in regulating aerosol effects on deep convective clouds. *J. Geophys. Res.*, **114**, D22206.

Fan, J. W., J. M. Comstock, and M. Ovchinnikov, 2010: The cloud condensation nuclei and ice nuclei effects on tropical anvil characteristics and water vapor of the tropical tropopause layer. *Environ. Res. Lett.*, **5**, 6.

Farina, S. C., P. J. Adams, and S. N. Pandis, 2010: Modeling global secondary organic aerosol formation and processing with the volatility basis set: Implications for anthropogenic secondary organic aerosol. *J. Geophys. Res.*, **115**, D09202.

Farrar, P. D., 2000: Are cosmic rays influencing oceanic cloud coverage – or is it only El Niño? *Clim. Change*, **47**, 7–15.

Fasullo, J. T., and K. E. Trenberth, 2012: A less cloudy future: The role of subtropical subsidence in climate sensitivity. *Science*, **338**, 792–794.

Favez, O., H. Cachier, J. Sciarea, S. C. Alfaro, T. M. El-Araby, M. A. Harhash, and Magdy M. Abdelwahab, 2008: Seasonality of major aerosol species and their transformations in Cairo megacity. *Atmos. Environ.*, **42**, 1503–1516.

Feingold, G., H. L. Jiang, and J. Y. Harrington, 2005: On smoke suppression of clouds in Amazonia. *Geophys. Res. Lett.*, **32**, L02804.

Feingold, G., R. Boers, B. Stevens, and W. R. Cotton, 1997: A modeling study of the effect of drizzle on cloud optical depth and susceptibility. *J. Geophys. Res.*, **102**, 13527–13534.

Feingold, G., I. Koren, H. Wang, H. Xue, and W. A. Brewer, 2010: Precipitation-generated oscillations in open cellular cloud fields. *Nature*, **466**, 849–852.

Feng, J., 2008: A size-resolved model and a four-mode parameterization of dry deposition of atmospheric aerosols. *J. Geophys. Res.*, **113**, D12201.

Ferraro, A. J., E. J. Highwood, and A. J. Charlton-Perez, 2011: Stratospheric heating by potential geoengineering aerosols. *Geophys. Res. Lett.*, **38**, L24706.

Flanner, M. G., C. S. Zender, J. T. Randerson, and P. J. Rasch, 2007: Present-day climate forcing and response from black carbon in snow. *J. Geophys. Res.*, **112**, D11202.

Flanner, M. G., C. S. Zender, P. G. Hess, N. M. Mahowald, T. H. Painter, V. Ramanathan, and P. J. Rasch, 2009: Springtime warming and reduced snow cover from carbonaceous particles. *Atmos. Chem. Phys.*, **9**, 2481–2497.

Fletcher, J. K., and C. S. Bretherton, 2010: Evaluating boundary layer-based mass flux closures using cloud-resolving model simulations of deep convection. *J. Atmos. Sci.*, **67**, 2212–2225.

Forest, C. E., P. H. Stone, and A. P. Sokolov, 2006: Estimated PDFs of climate system properties including natural and anthropogenic forcings. *Geophys. Res. Lett.*, **33**, L01705.

Forsström, S., J. Ström, C. A. Pedersen, E. Isaksson, and S. Gerland, 2009: Elemental carbon distribution in Svalbard snow. *J. Geophys. Res.*, **114**, D19112.

Forster, P., et al., 2007: Changes in Atmospheric Constituents and in Radiative Forcing. In: *Climate Change 2007: The Physical Science Basis. Contribution of Working Group I to the Fourth Assessment Report of the Intergovernmental Panel on Climate Change* [Solomon, S., D. Qin, M. Manning, Z. Chen, M. Marquis, K. B. Averyt, M. Tignor and H. L. Miller (eds.)] Cambridge University Press, Cambridge, United Kingdom and New York, NY, USA, pp. 129–234.

Forster, P. M. D., and J. M. Gregory, 2006: The climate sensitivity and its components diagnosed from Earth Radiation Budget data. *J. Clim.*, **19**, 39–52.

Fountoukis, C., et al., 2007: Aerosol-cloud drop concentration closure for clouds sampled during the International Consortium for Atmospheric Research on Transport and Transformation 2004 campaign. *J. Geophys. Res.*, **112**, D10S30.

Fovell, R. G., K. L. Corbosiero, and H.-C. Kuo, 2009: Cloud microphysics impact on hurricane track as revealed in idealized experiments. *J. Atmos. Sci.*, **66**, 1764–1778.

Fowler, L. D., and D. A. Randall, 2002: Interactions between cloud microphysics and cumulus convection in a general circulation model. *J. Atmos. Sci.*, **59**, 3074–3098.

Freney, E. J., K. Adachi, and P. R. Buseck, 2010: Internally mixed atmospheric aerosol particles: Hygroscopic growth and light scattering. *J. Geophys. Res.*, **115**, D19210.

Fridlind, A. M., et al., 2007: Ice properties of single-layer stratocumulus during the Mixed-Phase Arctic Cloud Experiment: 2. Model results. *J. Geophys. Res.*, **112**, D24202.

Friedman, B., G. Kulkarni, J. Beranek, A. Zelenyuk, J. A. Thornton, and D. J. Cziczo, 2011: Ice nucleation and droplet formation by bare and coated soot particles. *J. Geophys. Res.*, **116**, D17203.

Frömming, C., M. Ponater, U. Burkhardt, A. Stenke, S. Pechtl, and R. Sausen, 2011: Sensitivity of contrail coverage and contrail radiative forcing to selected key parameters. *Atmos. Environ.*, **45**, 1483–1490.

Fuzzi, S., et al., 2007: Overview of the inorganic and organic composition of size-segregated aerosol in Rondonia, Brazil, from the biomass-burning period to the onset of the wet season. *J. Geophys. Res.*, **112**, D01201.

Fyfe, J. C., J. N. S. Cole, V. K. Arora, and J. F. Scinocca, 2013: Biogeochemical carbon coupling influences global precipitation in geoengineering experiments. *Geophys. Res. Lett.*, **40**, 651–655.

Gagen, M., et al., 2011: Cloud response to summer temperatures in Fennoscandia over the last thousand years. *Geophys. Res. Lett.*, **38**, L05701.

Galewsky, J., and J. V. Hurley, 2010: An advection-condensation model for subtropical water vapor isotopic ratios. *J. Geophys. Res.*, **115**, D16116.

Gantt, B., N. Meskhidze, M. C. Facchini, M. Rinaldi, D. Ceburnis, and C. D. O'Dowd, 2011: Wind speed dependent size-resolved parameterization for the organic mass fraction of sea spray aerosol. *Atmos. Chem. Phys.*, **11**, 8777–8790.

Gao, R. S., et al., 2007: A novel method for estimating light-scattering properties of soot aerosols using a modified single-particle soot photometer. *Aer. Sci. Technol.*, **41**, 125–135.

Garrett, T. J., and C. F. Zhao, 2006: Increased Arctic cloud longwave emissivity associated with pollution from mid-latitudes. *Nature*, **440**, 787–789.

Garvert, M. F., C. P. Woods, B. A. Colle, C. F. Mass, P. V. Hobbs, M. T. Stoelinga, and J. B. Wolfe, 2005: The 13–14 December 2001 IMPROVE-2 event. Part II: Comparisons of MM5 model simulations of clouds and precipitation with observations. *J. Atmos. Sci.*, **62**, 3520–3534.

Gasso, S., 2008: Satellite observations of the impact of weak volcanic activity on marine clouds. *J. Geophys. Res.*, **113**, D14S19.

GAW, 2011: WMO/GAW Standard Operating Procedures for In-situ Measurements of Aerosol Mass Concentration, Light Scattering and Light Absorption. GAW Report No. 200, World Meteorological Organization, Geneva, Switzerland, 130 pp.

George, R. C., and R. Wood, 2010: Subseasonal variability of low cloud radiative properties over the southeast Pacific Ocean. *Atmos. Chem. Phys.*, **10**, 4047–4063.

Gerasopoulos, E., et al., 2007: Size-segregated mass distributions of aerosols over Eastern Mediterranean: Seasonal variability and comparison with AERONET columnar size-distributions. *Atmos. Chem. Phys.*, **7**, 2551–2561.

Gerber, H., 1996: Microphysics of marine stratocumulus clouds with two drizzle modes. *J. Atmos. Sci.*, **53**, 1649–1662.

Gettelman, A., and Q. Fu, 2008: Observed and simulated upper-tropospheric water vapor feedback. *J. Clim.*, **21**, 3282–3289.

Gettelman, A., J. E. Kay, and J. T. Fasullo, 2013: Spatial decomposition of climate feedbacks in the Community Earth System Model. *J. Clim.*, **26**, 3544–3561.

Gettelman, A., X. Liu, D. Barahona, U. Lohmann, and C.-C. Chen, 2012: Climate impacts of ice nucleation. *J. Geophys. Res.*, **117**, D20201.

Gettelman, A., et al., 2010: Global simulations of ice nucleation and ice supersaturation with an improved cloud scheme in the Community Atmosphere Model. *J. Geophys. Res.*, **115**, D18216.

Ghan, S., R. Easter, J. Hudson, and F.-M. Bréon, 2001: Evaluation of aerosol indirect radiative forcing in MIRAGE. *J. Geophys. Res.*, **106**, 5317–5334.

Ghan, S. J., and S. E. Schwartz, 2007: Aerosol properties and processes - A path from field and laboratory measurements to global climate models. *Bull. Am. Meteor. Soc.*, **88**, 1059–1083.

Ghan, S. J., H. Abdul-Razzak, A. Nenes, Y. Ming, X. Liu, and M. Ovchinnikov, 2011: Droplet nucleation: Physically-based parameterizations and comparative evaluation. *J. Adv. Model. Earth Syst.*, **3**, M10001.

Ghan, S. J., X. Liu, R. C. Easter, R. Zaveri, P. J. Rasch, J.-H. Yoon, and B. Eaton, 2012: Toward a minimal representation of aerosols in climate models: Comparative decomposition of aerosol direct, semi-direct, and indirect radiative forcing. *J. Clim.*, **25**, 6461–6476.

Ginoux, P., J. M. Prospero, T. E. Gill, N. C. Hsu, and M. Zhao, 2012a: Global-scale attribution of anthropogenic and natural dust sources and their emission rates based on MODIS Deep Blue aerosol products. *Rev. Geophys.*, **50**, RG3005.

Ginoux, P., L. Clarisse, C. Clerbaux, P.-F. Coheur, O. Dubovik, N. C. Hsu, and M. Van Damme, 2012b: Mixing of dust and NH_3 observed globally over anthropogenic dust sources *Atmos. Chem. Phys.*, **12**, 7351–7363.

Gioda, A., B. S. Amaral, I. L. G. Monteiro, and T. D. Saint'Pierre, 2011: Chemical composition, sources, solubility, and transport of aerosol trace elements in a tropical region. *J. Environ. Monit.*, **13**, 2134–2142.

Girard, E., J.-P. Blanchet, and Y. Dubois, 2004: Effects of arctic sulphuric acid aerosols on wintertime low-level atmospheric ice crystals, humidity and temperature at Alert, Nunavut. *Atmos. Res.*, **73**, 131–148.

Givati, A., and D. Rosenfeld, 2004: Quantifying precipitation suppression due to air pollution. *J. Appl. Meteorol.*, **43**, 1038–1056.

Golaz, J. C., V. E. Larson, and W. R. Cotton, 2002: A PDF-based model for boundary layer clouds. Part I: Method and model description. *J. Atmos. Sci.*, **59**, 3540–3551.

Golaz, J. C., M. Salzmann, L. J. Donner, L. W. Horowitz, Y. Ming, and M. Zhao, 2011: Sensitivity of the aerosol indirect effect to subgrid variability in the cloud parameterization of the GFDL atmosphere general circulation model AM3. *J. Clim.*, **24**, 3145–3160.

Good, N., et al., 2010: Consistency between parameterisations of aerosol hygroscopicity and CCN activity during the RHaMBLe discovery cruise. *Atmos. Chem. Phys.*, **10**, 3189–3203.

Gordon, N. D., and J. R. Norris, 2010: Cluster analysis of midlatitude oceanic cloud regimes: Mean properties and temperature sensitivity. *Atmos. Chem. Phys.*, **10**, 6435–6459.

Goren, T., and D. Rosenfeld, 2012: Satellite observations of ship emission induced transitions from broken to closed cell marine stratocumulus over large areas. *J. Geophys. Res.*, **117**, D17206.

Grabowski, W. W., and P. K. Smolarkiewicz, 1999: CRCP: A Cloud Resolving Convection Parameterization for modeling the tropical convecting atmosphere. *Physica D*, **133**, 171–178.

Grabowski, W. W., X. Wu, M. W. Moncrieff, and W. D. Hall, 1998: Cloud-resolving modeling of cloud systems during Phase III of GATE. Part II: Effects of resolution and the third spatial dimension. *J. Atmos. Sci.*, **55**, 3264–3282.

Grandey, B. S., and P. Stier, 2010: A critical look at spatial scale choices in satellite-based aerosol indirect effect studies. *Atmos. Chem. Phys.*, **10**, 11459–11470.

Grandpeix, J.-Y., and J.-P. Lafore, 2010: A density current parameterization coupled with Emanuel's convection scheme. Part I: The models. *J. Atmos. Sci.*, **67**, 881–897.

Granier, C., et al., 2011: Evolution of anthropogenic and biomass burning emissions of air pollutants at global and regional scales during the 1980–2010 period. *Clim. Change*, **109**, 163–190.

Gregory, J., and M. Webb, 2008: Tropospheric adjustment induces a cloud component in CO_2 forcing. *J. Clim.*, **21**, 58–71.

Gregory, J. M., et al., 2004: A new method for diagnosing radiative forcing and climate sensitivity. *Geophys. Res. Lett.*, **31**, L03205.

Grote, R., and U. Niinemets, 2008: Modeling volatile isoprenoid emissions - a story with split ends. *Plant Biology*, **10**, 8–28.

Guenther, A., T. Karl, P. Harley, C. Wiedinmyer, P. I. Palmer, and C. Geron, 2006: Estimates of global terrestrial isoprene emissions using MEGAN (Model of Emissions of Gases and Aerosols from Nature). *Atmos. Chem. Phys.*, **6**, 3181–3210.

Guenther, A. B., X. Jiang, C. L. Heald, T. Sakulyanontvittaya, T. Duhl, L. K. Emmons, and X. Wang, 2012: The Model of Emissions of Gases and Aerosols from Nature version 2.1 (MEGAN2.1): An extended and updated framework for modeling biogenic emissions. *Geosci. Model Dev.*, **5**, 1471–1492.

Gullu, H. G., I. Ölmez, and G. Tuncel, 2000: Temporal variability of atmospheric trace element concentrations over the eastern Mediterranean Sea. *Spectrochim. Acta*, **B55**, 1135–1150.

Guo, H., J. C. Golaz, L. J. Donner, V. E. Larson, D. P. Schanen, and B. M. Griffin, 2010: Multi-variate probability density functions with dynamics for cloud droplet activation in large-scale models: Single column tests. *Geosci. Model Dev.*, **3**, 475–486.

Hadley, O. L., and T. W. Kirchstetter, 2012: Black-carbon reduction of snow albedo. *Nature Clim. Change*, **2**, 437–440.

Haerter, J. O., and P. Berg, 2009: Unexpected rise in extreme precipitation caused by a shift in rain type? *Nature Geosci.*, **2**, 372–373.

Haerter, J. O., P. Berg, and S. Hagemann, 2010: Heavy rain intensity distributions on varying time scales and at different temperatures. *J. Geophys. Res.*, **115**, D17102.

Hagler, G. S. W., et al., 2006: Source areas and chemical composition of fine particulate matter in the Pearl River Delta region of China. *Atmos. Environ.*, **40**, 3802–3815..

Halfon, N., Z. Levin, and P. Alpert, 2009: Temporal rainfall fluctuations in Israel and their possible link to urban and air pollution effects. *Environ. Res. Lett.*, **4**, 025001.

Halloran, P. R., T. G. Bell, and I. J. Totterdell, 2010: Can we trust empirical marine DMS parameterisations within projections of future climate? *Biogeosciences*, **7**, 1645–1656.

Hallquist, M., et al., 2009: The formation, properties and impact of secondary organic aerosol: Current and emerging issues. *Atmos. Chem. Phys.*, **9**, 5155–5236.

Hamwey, R. M., 2007: Active amplification of the terrestrial albedo to mitigate climate change: An exploratory study. *Mitigat. Adapt. Strat. Global Change*, **12**, 419–439.

Han, Y.-J., T.-S. Kim, and H. Kim, 2008: Ionic constituents and source analysis of $PM_{2.5}$ in three Korean cities. *Atmos. Environ.*, **42**, 4735–4746.

Hand, V. L., G. Capes, D. J. Vaughan, P. Formenti, J. M. Haywood, and H. Coe, 2010: Evidence of internal mixing of African dust and biomass burning particles by individual particle analysis using electron beam techniques. *J. Geophys. Res.*, **115**, D13301.

Hansell, R. A., et al., 2010: An assessment of the surface longwave direct radiative effect of airborne Saharan dust during the NAMMA field campaign. *J. Atmos. Sci.*, **67**, 1048–1065.

Hansen, J., and L. Nazarenko, 2004: Soot climate forcing via snow and ice albedos. *Proc. Natl. Acad. Sci. U.S.A.*, **101**, 423–428.

Hansen, J., M. Sato, P. Kharecha, G. Russell, D. W. Lea, and M. Siddall, 2007: Climate change and trace gases. *Philos. Trans. R. Soc. London A*, **365**, 1925–1954.

Hansen, J., et al., 1984: Climate sensitivity: Analysis of feedback mechanisms. In: *Climate Processes and Climate Sensitivity*, Geophysical Monograph Series, Vol. 29 [J. E. Hansen and T. Takahashi (eds.)]. American Geophysical Union, Washington, DC, USA, pp. 130–163.

Hansen, J., et al., 2005: Efficacy of climate forcings. *J. Geophys. Res.*, **110**, D18104.

Hara, K., et al., 2003: Mixing states of individual aerosol particles in spring Arctic troposphere during ASTAR 2000 campaign. *J. Geophys. Res.*, **108**, 4209.

Hardwick Jones, R., S. Westra, and A. Sharma, 2010: Observed relationships between extreme sub-daily precipitation, surface temperature, and relative humidity. *Geophys. Res. Lett.*, **37**, L22805.

Harrington, J. Y., D. Lamb, and R. Carver, 2009: Parameterization of surface kinetic effects for bulk microphysical models: Influences on simulated cirrus dynamics and structure. *J. Geophys. Res.*, **114**, D06212.

Harrison, E. F., P. Minnis, B. R. Barkstrom, V. Ramanathan, R. D. Cess, and G. G. Gibson, 1990: Seasonal variation of cloud radiative forcing derrived from the Earth Radiation Budget Experiment. *J. Geophys. Res.*, **95**, 18687–18703.

Harrison, R., and M. Ambaum, 2008: Enhancement of cloud formation by droplet charging. *Proc. R. Soc. London A*, **464**, 2561–2573.

Harrison, R. G., 2008: Discrimination between cosmic ray and solar irradiance effects on clouds, and evidence for geophysical modulation of cloud thickness. *Proc. R. Soc. London A*, **464**, 2575–2590.

Harrison, R. G., and D. B. Stephenson, 2006: Empirical evidence for a nonlinear effect of galactic cosmic rays on clouds. *Proc. R. Soc. London A*, **462**, 1221–1233.

Harrison, R. G., and M. H. P. Ambaum, 2010: Observing Forbush decreases in cloud at Shetland. *J. Atmos. Sol. Terres. Phys.*, **72**, 1408–1414.

Harrop, B. E., and D. L. Hartmann, 2012: Testing the role of radiation in determining tropical cloud-top temperature. *J. Clim.*, **25**, 5731–5747.

Hartmann, D. L., and K. Larson, 2002: An important constraint on tropical cloud-climate feedback. *Geophys. Res. Lett.*, **29**, 1951.

Hasekamp, O. P., 2010: Capability of multi-viewing-angle photo-polarimetric measurements for the simultaneous retrieval of aerosol and cloud properties. *Atmos. Meas. Tech.*, **3**, 839–851.

Haynes, J. M., C. Jakob, W. B. Rossow, G. Tselioudis, and J. Brown, 2011: Major characteristics of Southern Ocean cloud regimes and their effects on the energy budget. *J. Clim.*, **24**, 5061–5080.

Haywood, J., and O. Boucher, 2000: Estimates of the direct and indirect radiative forcing due to tropospheric aerosols: A review. *Rev. Geophys.*, **38**, 513–543.

Haywood, J., and M. Schulz, 2007: Causes of the reduction in uncertainty in the anthropogenic radiative forcing of climate between IPCC (2001) and IPCC (2007). *Geophys. Res. Lett.*, **34**, L20701.

Haywood, J. M., A. Jones, N. Bellouin, and D. Stephenson, 2013: Asymmetric forcing from stratospheric aerosols impacts Sahelian rainfall. *Nature Clim. Change*, **3**, 660–665.

Haywood, J. M., et al., 2009: A case study of the radiative forcing of persistent contrails evolving into contrail-induced cirrus. *J. Geophys. Res.*, **114**, D24201.

Haywood, J. M., et al., 2011: Motivation, rationale and key results from the GERBILS Saharan dust measurement campaign. *Q. J. R. Meteorol. Soc.*, **137**, 1106–1116.

Heald, C. L., and D. V. Spracklen, 2009: Atmospheric budget of primary biological aerosol particles from fungal spores. *Geophys. Res. Lett.*, **36**, L09806.

Heald, C. L., D. A. Ridley, S. M. Kreidenweis, and E. E. Drury, 2010: Satellite observations cap the atmospheric organic aerosol budget. *Geophys. Res. Lett.*, **37**, L24808.

Heald, C. L., et al., 2008: Predicted change in global secondary organic aerosol concentrations in response to future climate, emissions, and land use change. *J. Geophys. Res.*, **113**, D05211.

Heald, C. L., et al., 2011: Exploring the vertical profile of atmospheric organic aerosol: Comparing 17 aircraft field campaigns with a global model. *Atmos. Chem. Phys.*, **11**, 12673–12696.

Heckendorn, P., et al., 2009: The impact of geoengineering aerosols on stratospheric temperature and ozone. *Environ. Res. Lett.*, **4**, 045108.

Hegg, D. A., D. S. Covert, H. H. Jonsson, and R. K. Woods, 2012: A simple relationship between cloud drop number concentration and precursor aerosol concentration for the regions of Earth's large marine stratocumulus decks. *Atmos. Chem. Phys.*, **12**, 1229–1238.

Heintzenberg, J., D. C. Covert, and R. Van Dingenen, 2000: Size distribution and chemical composition of marine aerosols: A compilation and review. *Tellus*, **52**, 1104–1122.

Heintzenberg, J., et al., 2011: Near-global aerosol mapping in the upper troposphere and lowermost stratosphere with data from the CARIBIC project. *Tellus B*, **63**, 875–890.

Held, I. M., and B. J. Soden, 2006: Robust responses of the hydrological cycle to global warming. *J. Clim.*, **19**, 5686–5699.

Held, I. M., and K. M. Shell, 2012: Using relative humidity as a state variable in climate feedback analysis. *J. Clim.*, **25**, 2578–2582.

Hendricks, J., B. Karcher, U. Lohmann, and M. Ponater, 2005: Do aircraft black carbon emissions affect cirrus clouds on the global scale? *Geophys. Res. Lett.*, **32**, L12814.

Heymsfield, A., D. Baumgardner, P. DeMott, P. Forster, K. Gierens, and B. Kärcher, 2010: Contrail microphysics. *Bull. Am. Meteor. Soc.*, **91**, 465–472.

Heymsfield, A. J., and L. M. Miloshevich, 1995: Relative humidity and temperature influences on cirrus formation and evolution: Observations from wave clouds and FIRE II. *J. Atmos. Sci.*, **52**, 4302–4326.

Heymsfield, A. J., et al., 1998: Cloud properties leading to highly reflective tropical cirrus: Interpretations from CEPEX, TOGA COARE, and Kwajalein, Marshall Islands. *J. Geophys. Res.*, **103**, 8805–8812.

Hill, A. A., and S. Dobbie, 2008: The impact of aerosols on non-precipitating marine stratocumulus. II: The semi-direct effect. *Q. J. R. Meteorol. Soc.*, **134**, 1155–1165.

Hill, A. A., G. Feingold, and H. Jiang, 2009: The influence of entrainment and mixing assumption on aerosol-cloud interactions in marine stratocumulus. *J. Atmos. Sci.*, **66**, 1450–1464.

Hill, S., and Y. Ming, 2012: Nonlinear climate response to regional brightening of tropical marine stratocumulus. *Geophys. Res. Lett.*, **39**, L15707.

Hirsikko, A., et al., 2011: Atmospheric ions and nucleation: A review of observations. *Atmos. Chem. Phys.*, **11**, 767–798.

Hodzic, A., J. L. Jimenez, S. Madronich, M. R. Canagaratna, P. F. DeCarlo, L. Kleinman, and J. Fast, 2010: Modeling organic aerosols in a megacity: Potential contribution of semi-volatile and intermediate volatility primary organic compounds to secondary organic aerosol formation. *Atmos. Chem. Phys.*, **10**, 5491–5514.

Hohenegger, C., and C. S. Bretherton, 2011: Simulating deep convection with a shallow convection scheme. *Atmos. Chem. Phys.*, **11**, 10389–10406.

Hohenegger, C., P. Brockhaus, and C. Schar, 2008: Towards climate simulations at cloud-resolving scales. *Meteorol. Z.*, **17**, 383–394.

Hohenegger, C., P. Brockhaus, C. S. Bretherton, and C. Schär, 2009: The soil moisture–precipitation feedback in simulations with explicit and parameterized convection. *J. Clim.*, **22**, 5003–5020.

Holben, B. N., et al., 1998: AERONET - A federated instrument network and data archive for aerosol characterization. *Remote Sens. Environ.*, **66**, 1–16.

Hoose, C., and O. Möhler, 2012: Heterogeneous ice nucleation on atmospheric aerosols: A review of results from laboratory experiments. *Atmos. Chem. Phys.*, **12**, 9817–9854.

Hoose, C., J. E. Kristjánsson, and S. M. Burrows, 2010a: How important is biological ice nucleation in clouds on a global scale? *Environ. Res. Lett.*, **5**, 024009.

Hoose, C., U. Lohmann, R. Erdin, and I. Tegen, 2008: The global influence of dust mineralogical composition on heterogeneous ice nucleation in mixed-phase clouds. *Environ. Res. Lett.*, **3**, 025003.

Hoose, C., J. E. Kristjánsson, J. P. Chen, and A. Hazra, 2010b: A classical-theory-based parameterization of heterogeneous ice nucleation by mineral dust, soot, and biological particles in a global climate model. *J. Atmos. Sci.*, **67**, 2483–2503.

Hoose, C., J. E. Kristjánsson, T. Iversen, A. Kirkevåg, Ø. Seland, and A. Gettelman, 2009: Constraining cloud droplet number concentration in GCMs suppresses the aerosol indirect effect. *Geophys. Res. Lett.*, **36**, L12807.

Hourdin, F., et al., 2013: LMDZ5B: The atmospheric component of the IPSL climate model with revisited parameterizations for clouds and convection. *Clim. Dyn.*, **40**, 2193–2222.

Hoyle, C. R., et al., 2011: A review of the anthropogenic influence on biogenic secondary organic aerosol. *Atmos. Chem. Phys.*, **11**, 321–343.

Hu, M., L. Y. He, Y. H. Zhang, M. Wang, Y. Pyo Kim, and K. C. Moon, 2002: Seasonal variation of ionic species in fine particles at Qingdao. *Atmos. Environ.*, **36**, 5853–5859.

Huang, J., Q. Fu, W. Zhang, X. Wang, R. Zhang, H. Ye, and S. G. Warren, 2011: Dust and black carbon in seasonal snow across Northern China. *Bull. Am. Meteor. Soc.*, **92**, 175–181.

Hudson, J. G., 1993: Cloud condensation nuclei near marine cumulus. *J. Geophys. Res.*, **98**, 2693–2702.

Hudson, J. G., S. Noble, and V. Jha, 2010: Comparisons of CCN with supercooled clouds. *J. Atmos. Sci.*, **67**, 3006–3018.

Hueglin, C., R. Gehrig, U. Baltensperger, M. Gysel, C. Monn, and H. Vonmont, 2005: Chemical characterisation of $PM_{2.5}$, PM_{10} and coarse particles at urban, near-city and rural sites in Switzerland. *Atmos. Environ.*, **39**, 637–651.

Huffman, G. J., et al., 2007: The TRMM multisatellite precipitation analysis (TMPA): Quasi-global, multiyear, combined-sensor precipitation estimates at fine scales. *J. Hydrometeor.*, **8**, 38–55.

Huneeus, N., F. Chevallier, and O. Boucher, 2012: Estimating aerosol emissions by assimilating observed aerosol optical depth in a global aerosol model. *Atmos. Chem. Phys.*, **12**, 4585–4606.

Huneeus, N., et al., 2011: Global dust model intercomparison in AeroCom phase I. *Atmos. Chem. Phys.*, **11**, 7781–7816.

Hurley, J. V., and J. Galewsky, 2010a: A last saturation analysis of ENSO humidity variability in the Subtropical Pacific. *J. Clim.*, **23**, 918–931.

Hurley, J. V., and J. Galewsky, 2010b: A last-saturation diagnosis of subtropical water vapor response to global warming. *Geophys. Res. Lett.*, **37**, L06702.

Iga, S., H. Tomita, Y. Tsushima, and M. Satoh, 2011: Sensitivity of Hadley Circulation to physical parameters and resolution through changing upper-tropospheric ice clouds using a global cloud system resolving model. *J. Clim.*, **24**, 2666–2679.

Iguchi, T., T. Nakajima, A. P. Khain, K. Saito, T. Takemura, and K. Suzuki, 2008: Modeling the influence of aerosols on cloud microphysical properties in the east Asia region using a mesoscale model coupled with a bin-based cloud microphysics scheme. *J. Geophys. Res.*, **113**, D14215.

Ingram, W., 2010: A very simple model for the water vapour feedback on climate change. *Q. J. R. Meteorol. Soc.*, **136**, 30–40.

Ingram, W., 2013a: A new way of quantifying GCM water vapour feedback. *Clim. Dyn.*, **40**, 913–924.

Ingram, W., 2013b: Some implications of a new approach to the water vapour feedback. *Clim. Dyn.*, **40**, 925–933.

Inoue, T., M. Satoh, Y. Hagihara, H. Miura, and J. Schmetz, 2010: Comparison of high-level clouds represented in a global cloud system-resolving model with CALIPSO/CloudSat and geostationary satellite observations. *J. Geophys. Res.*, **115**, D00H22.

IPCC, 2011: IPCC Expert Meeting Report on Geoengineering, [Edenhofer O, Field C, Pichs-Madruga R, Sokona Y, Stocker T, Barros V, Dahe Q, Minx J, Mach K, Plattner GK, Schlomer S, Hansen G, Mastrandrea M (eds.)]. IPCC Working Group III Technical Support Unit, Potsdam Institute for Climate Impact Research.

Irvine, P. J., A. Ridgwell, and D. J. Lunt, 2011: Climatic effects of surface albedo geoengineering. *J. Geophys. Res.*, **116**, D24112.

Irvine, P. J., R. L. Sriver, and K. Keller, 2012: Tension between reducing sea-level rise and global warming through solar-radiation management. *Nature Clim. Change*, **2**, 97–100.

Irwin, M., N. Good, J. Crosier, T. W. Choularton, and G. McFiggans, 2010: Reconciliation of measurements of hygroscopic growth and critical supersaturation of aerosol particles in central Germany. *Atmos. Chem. Phys.*, **10**, 11737–11752.

Ito, K., N. Xue, and G. Thurston, 2004: Spatial variation of $PM_{2.5}$ chemical species and source-apportioned mass concentrations in New York City. *Atmos. Environ.*, **38**, 5269–5282.

Izrael, Y. A., A. G. Ryaboshapko, and N. N. Petrov, 2009: Comparative analysis of geo-engineering approaches to climate stabilization. *Russ. Meteorol. Hydrol.*, **34**, 335–347.

Jacob, D. J., et al., 2010: The Arctic Research of the Composition of the Troposphere from Aircraft and Satellites (ARCTAS) mission: Design, execution, and first results. *Atmos. Chem. Phys.*, **10**, 5191–5212.

Jacobson, M. Z., 2003: Development of mixed-phase clouds from multiple aerosol size distributions and the effect of the clouds on aerosol removal. *J. Geophys. Res.*, **108**, 4245.

Jacobson, M. Z., 2004: Climate response of fossil fuel and biofuel soot, accounting for soot's feedback to snow and sea ice albedo and emissivity. *J. Geophys. Res.*, **109**, D21201.

Jacobson, M. Z., 2006: Effects of externally-through-internally-mixed soot inclusions within clouds and precipitation on global climate. *J. Phys. Chem. A*, **110**, 6860–6873.

Jacobson, M. Z., 2012: Investigating cloud absorption effects: Global absorption properties of black carbon, tar balls, and soil dust in clouds and aerosols. *J. Geophys. Res.*, **117**, D06205.

Jacobson, M. Z., and D. G. Streets, 2009: Influence of future anthropogenic emissions on climate, natural emissions, and air quality. *J. Geophys. Res.*, **114**, D08118.

Jacobson, M. Z., and J. E. Ten Hoeve, 2012: Effects of urban surfaces and white roofs on global and regional climate. *J. Clim.*, **25**, 1028–1044.

Jaeglé, L., P. K. Quinn, T. S. Bates, B. Alexander, and J. T. Lin, 2011: Global distribution of sea salt aerosols: New constraints from in situ and remote sensing observations. *Atmos. Chem. Phys.*, **11**, 3137–3157.

Jensen, E. J., S. Kinne, and O. B. Toon, 1994: Tropical cirrus cloud radiative forcing: Sensitivity studies. *Geophys. Res. Lett.*, **21**, 2023–2026.

Jensen, E. J., L. Pfister, T. P. Bui, P. Lawson, and D. Baumgardner, 2010: Ice nucleation and cloud microphysical properties in tropical tropopause layer cirrus. *Atmos. Chem. Phys.*, **10**, 1369–1384.

Jeong, M.-J., and Z. Li, 2010: Separating real and apparent effects of cloud, humidity, and dynamics on aerosol optical thickness near cloud edges. *J. Geophys. Res.*, **115**, D00K32.

Jeong, M. J., S. C. Tsay, Q. Ji, N. C. Hsu, R. A. Hansell, and J. Lee, 2008: Ground-based measurements of airborne Saharan dust in marine environment during the NAMMA field experiment. *Geophys. Res. Lett.*, **35**, L20805.

Jethva, H., S. K. Satheesh, J. Srinivasan, and K. K. Moorthy, 2009: How good is the assumption about visible surface reflectance in MODIS aerosol retrieval over land? A comparison with aircraft measurements over an urban site in India. *IEEE Trans. Geosci. Remote Sens.*, **47**, 1990–1998.

Jiang, H. L., H. W. Xue, A. Teller, G. Feingold, and Z. Levin, 2006: Aerosol effects on the lifetime of shallow cumulus. *Geophys. Res. Lett.*, **33**, L14806.

Jiang, J. H., H. Su, M. Schoeberl, S. T. Massie, P. Colarco, S. Platnick, and N. J. Livesey, 2008: Clean and polluted clouds: Relationships among pollution, ice cloud and precipitation in South America. *Geophys. Res. Lett.*, **35**, L14804.

Jimenez, J. L., et al., 2009: Evolution of organic aerosols in the atmosphere. *Science*, **326**, 1525–1529.

Jirak, I. L., and W. R. Cotton, 2006: Effect of air pollution on precipitation along the front range of the Rocky Mountains. *J. Appl. Meteor. Climatol.*, **45**, 236–245.

Johanson, C. M., and Q. Fu, 2009: Hadley cell widening: Model simulations versus observations. *J. Clim.*, **22**, 2713–2725.

Johns, T. C., et al., 2006: The new Hadley Centre Climate Model (HadGEM1): Evaluation of coupled simulations. *J. Clim.*, **19**, 1327–1353.

Johnson, B. T., K. P. Shine, and P. M. Forster, 2004: The semi-direct aerosol effect: Impact of absorbing aerosols on marine stratocumulus. *Q. J. R. Meteorol. Soc.*, **130**, 1407–1422.

Johnson, N. C., and S. P. Xie, 2010: Changes in the sea surface temperature threshold for tropical convection. *Nature Geosci.*, **3**, 842–845.

Jones, A., J. M. Haywood, and O. Boucher, 2007: Aerosol forcing, climate response and climate sensitivity in the Hadley Centre climate model. *J. Geophys. Res.*, **112**, D20211.

Jones, A., J. Haywood, and O. Boucher, 2009: Climate impacts of geoengineering marine stratocumulus clouds. *J. Geophys. Res.*, **114**, D10106.

Jones, A., D. L. Roberts, M. J. Woodage, and C. E. Johnson, 2001: Indirect sulphate aerosol forcing in a climate model with an interactive sulphur cycle. *J. Geophys. Res.*, **106**, 20293–20310.

Jones, A., J. Haywood, O. Boucher, B. Kravitz, and A. Robock, 2010: Geoengineering by stratospheric SO_2 injection: Results from the Met Office HadGEM2 climate model and comparison with the Goddard Institute for Space Studies ModelE. *Atmos. Chem. Phys.*, **10**, 5999–6006.

Joshi, M. M., M. J. Webb, A. C. Maycock, and M. Collins, 2010: Stratospheric water vapour and high climate sensitivity in a version of the HadSM3 climate model. *Atmos. Chem. Phys.*, **10**, 7161–7167.

Joshi, M. M., J. M. Gregory, M. J. Webb, D. M. H. Sexton, and T. C. Johns, 2008: Mechanisms for the land/sea warming contrast exhibited by simulations of climate change. *Clim. Dyn.*, **30**, 455–465.

Kahn, R., 2012: Reducing the uncertainties in direct aerosol radiative forcing. *Surv. Geophys.*, **33**, 701–721.

Kahn, R. A., B. J. Gaitley, J. V. Martonchik, D. J. Diner, K. A. Crean, and B. Holben, 2005: Multiangle Imaging Spectroradiometer (MISR) global aerosol optical depth validation based on 2 years of coincident Aerosol Robotic Network (AERONET) observations. *J. Geophys. Res.*, **110**, D10S04.

Kahn, R. A., B. J. Gaitley, M. J. Garay, D. J. Diner, T. F. Eck, A. Smirnov, and B. N. Holben, 2010: Multiangle Imaging SpectroRadiometer global aerosol product assessment by comparison with the Aerosol Robotic Network. *J. Geophys. Res.*, **115**, D23209.

Kahn, R. A., et al., 2007: Satellite-derived aerosol optical depth over dark water from MISR and MODIS: Comparisons with AERONET and implications for climatological studies. *J. Geophys. Res.*, **112**, D18205.

Kalkstein, A. J., and R. C. Balling Jr, 2004: Impact of unusually clear weather on United States daily temperature range following 9/11/2001 *Clim. Res.*, **26**, 1–4.

Kammermann, L., et al., 2010: Subarctic atmospheric aerosol composition: 3. Measured and modeled properties of cloud condensation nuclei. *J. Geophys. Res.*, **115**, D04202.

Kanakidou, M., et al., 2005: Organic aerosol and global climate modelling: A review. *Atmos. Chem. Phys.*, **5**, 1053–1123.

Kärcher, B., O. Mohler, P. J. DeMott, S. Pechtl, and F. Yu, 2007: Insights into the role of soot aerosols in cirrus cloud formation. *Atmos. Chem. Phys.*, **7**, 4203–4227.

Kaspari, S. D., M. Schwikowski, M. Gysel, M. G. Flanner, S. Kang, S. Hou, and P. A. Mayewski, 2011: Recent increase in black carbon concentrations from a Mt. Everest ice core spanning 1860–2000 AD. *Geophys. Res. Lett.*, **38**, L04703.

Kato, S., et al., 2011: Improvements of top-of-atmosphere and surface irradiance computations with CALIPSO-, CloudSat-, and MODIS-derived cloud and aerosol properties. *J. Geophys. Res.*, **116**, D19209.

Kaufman, Y. J., O. Boucher, D. Tanré, M. Chin, L. A. Remer, and T. Takemura, 2005: Aerosol anthropogenic component estimated from satellite data. *Geophys. Res. Lett.*, **32**, L17804.

Kay, J. E., and A. Gettelman, 2009: Cloud influence on and response to seasonal Arctic sea ice loss. *J. Geophys. Res.*, **114**, D18204.

Kay, J. E., K. Raeder, A. Gettelman, and J. Anderson, 2011: The boundary layer response to recent Arctic sea ice loss and implications for high-latitude climate feedbacks. *J. Clim.*, **24**, 428–447.

Kay, J. E., T. L'Ecuyer, A. Gettelman, G. Stephens, and C. O'Dell, 2008: The contribution of cloud and radiation anomalies to the 2007 Arctic sea ice extent minimum. *Geophys. Res. Lett.*, **35**, L08503.

Kay, J. E., et al., 2012: Exposing global cloud biases in the Community Atmosphere Model (CAM) using satellite observations and their corresponding instrument simulators. *J. Clim.*, **25**, 5190–5207.

Kazil, J., R. G. Harrison, and E. R. Lovejoy, 2008: Tropospheric new particle formation and the role of ions. *Space Sci. Rev.*, **137**, 241–255.

Kazil, J., K. Zhang, P. Stier, J. Feichter, U. Lohmann, and K. O'Brien, 2012: The present-day decadal solar cycle modulation of Earth's radiative forcing via charged H_2SO_4/H_2O aerosol nucleation. *Geophys. Res. Lett.*, **39**, L02805.

Kazil, J., et al., 2010: Aerosol nucleation and its role for clouds and Earth's radiative forcing in the aerosol-climate model ECHAM5–HAM. *Atmos. Chem. Phys.*, **10**, 10733–10752.

Keith, D. W., 2000: Geoengineering the climate: History and prospect. *Annu. Rev. Energ. Environ.*, **25**, 245–284.

Keith, D. W., 2010: Photophoretic levitation of engineered aerosols for geoengineering. *Proc. Natl. Acad. Sci. U.S.A.*, **107**, 16428–16431.

Kerkweg, A., J. Buchholz, L. Ganzeveld, A. Pozzer, H. Tost, and P. Jöckel, 2006: Technical Note: An implementation of the dry removal processes DRY DEPositionand SEDImentation in the Modular Earth Submodel System (MESSy). *Atmos. Chem. Phys.*, **6**, 4617–4632.

Kerminen, V. M., et al., 2010: Atmospheric nucleation: Highlights of the EUCAARI project and future directions. *Atmos. Chem. Phys.*, **10**, 10829–10848.

Kernthaler, S. C., R. Toumi, and J. D. Haigh, 1999: Some doubts concerning a link between cosmic ray fluxes and global cloudiness. *Geophys. Res. Lett.*, **26**, 863–865.

Khain, A., D. Rosenfeld, and A. Pokrovsky, 2005: Aerosol impact on the dynamics and microphysics of deep convective clouds. *Q. J. R. Meteorol. Soc.*, **131**, 2639–2663.

Khain, A., M. Arkhipov, M. Pinsky, Y. Feldman, and Y. Ryabov, 2004: Rain enhancement and fog elimination by seeding with charged droplets. Part 1: Theory and numerical simulations. *J. Appl. Meteorol.*, **43**, 1513–1529.

Khain, A. P., 2009: Notes on state-of-the-art investigations of aerosol effects on precipitation: A critical review. *Environ. Res. Lett.*, **4**, 015004.

Khairoutdinov, M., and Y. Kogan, 2000: A new cloud physics parameterization in a large-eddy simulation model of marine stratocumulus. *Mon. Weather Rev.*, **128**, 229–243.

Khairoutdinov, M., D. Randall, and C. DeMott, 2005: Simulations of the atmospheric general circulation using a cloud-resolving model as a superparameterization of physical processes. *J. Atmos. Sci.*, **62**, 2136–2154.

Khairoutdinov, M. F., and D. A. Randall, 2001: A cloud resolving model as a cloud parameterization in the NCAR Community Climate System Model: Preliminary results. *Geophys. Res. Lett.*, **28**, 3617–3620.

Khairoutdinov, M. F., and C.-E. Yang, 2013: Cloud-resolving modelling of aerosol indirect effects in idealised radiative-convective equilibrium with interactive and fixed sea surface temperature. *Atmos. Chem. Phys.*, **13**, 4133–4144.

Khairoutdinov, M. F., S. K. Krueger, C.-H. Moeng, P. A. Bogenschutz, and D. A. Randall, 2009: Large-eddy simulation of maritime deep tropical convection. *J. Adv. Model. Earth Syst.*, **1**, 15.

Khan, M. F., Y. Shirasuna, K. Hirano, and S. Masunaga, 2010: Characterization of PM2.5, PM2.5–10 and PMN10 in ambient air, Yokohama, Japan. *Atmos. Res.*, **96**, 159–172.

Khare, P., and B. P. Baruah, 2010: Elemental characterization and source identification of $PM_{2.5}$ using multivariate analysis at the suburban site of North-East India. *Atmos. Res.*, **98**, 148–162.

Kharin, V. V., F. W. Zwiers, X. B. Zhang, and G. C. Hegerl, 2007: Changes in temperature and precipitation extremes in the IPCC ensemble of global coupled model simulations. *J. Clim.*, **20**, 1419–1444.

Khvorostyanov, V., and K. Sassen, 1998: Toward the theory of homogeneous nucleation and its parameterization for cloud models. *Geophys. Res. Lett.*, **25**, 3155–3158.

Khvorostyanov, V. I., and J. A. Curry, 2009: Critical humidities of homogeneous and heterogeneous ice nucleation: Inferences from extended classical nucleation theory. *J. Geophys. Res.*, **114**, D04207.

Kim, B. M., S. Teffera, and M. D. Zeldin, 2000: Characterization of $PM_{2.5}$ and PM_{10} in the South Coast air basin of Southern California: Part 1– Spatial variations. *J. Air Waste Manag. Assoc.*, **50**, 2034–2044.

Kim, D., C. Wang, A. M. L. Ekman, M. C. Barth, and P. J. Rasch, 2008: Distribution and direct radiative forcing of carbonaceous and sulfate aerosols in an interactive size-resolving aerosol–climate model. *J. Geophys. Res.*, **113**, D16309.

Kim, D., et al., 2012: The tropical subseasonal variability simulated in the NASA GISS General Circulation Model. *J. Clim.*, **25**, 4641–4659.

Kim, H.-S., J.-B. Huh, P. K. Hopke, T. M. Holsen, and S.-M. Yi, 2007: Characteristics of the major chemical constituents of $PM_{2.5}$ and smog events in Seoul, Korea in 2003 and 2004. *Atmos. Environ.*, **41**, 6762–6770.

Kim, J. M., et al., 2010: Enhanced production of oceanic dimethylsulfide resulting from CO_2-induced grazing activity in a high CO_2 world. *Environ. Sci. Technol.*, **44**, 8140–8143.

King, S. M., et al., 2010: Cloud droplet activation of mixed organic-sulfate particles produced by the photooxidation of isoprene. *Atmos. Chem. Phys.*, **10**, 3953–3964.

Kirchstetter, T. W., T. Novakov, and P. V. Hobbs, 2004: Evidence that the spectral dependence of light absorption by aerosols is affected by organic carbon. *J. Geophys. Res.*, **109**, D21208.

Kirkby, J., 2007: Cosmic rays and climate. *Surv. Geophys.* **28**, 333–375.

Kirkby, J., et al., 2011: Role of sulphuric acid, ammonia and galactic cosmic rays in atmospheric aerosol nucleation. *Nature*, **476**, 429–433.

Kirkevåg, A., T. Iversen, Ø. Seland, J. B. Debernard, T. Storelvmo, and J. E. Kristjánsson, 2008: Aerosol-cloud-climate interactions in the climate model CAM-Oslo. *Tellus A*, **60**, 492–512.

Kirkevåg, A., et al., 2013: Aerosol–climate interactions in the Norwegian Earth System Model – NorESM1–M. *Geophys. Model Dev.*, **6**, 207–244.

Kleeman, M. J., 2008: A preliminary assessment of the sensitivity of air quality in California to global change. *Clim. Change*, **87**, S273–S292.

Kleidman, R. G., A. Smirnov, R. C. Levy, S. Mattoo, and D. Tanré, 2012: Evaluation and wind speed dependence of MODIS aerosol retrievals over open ocean. *IEEE Trans. Geosci. Remote Sens.*, **50**, 429–435.

Klein, S. A., and D. L. Hartmann, 1993: The seasonal cycle of low stratiform clouds. *J. Clim.*, **6**, 1587–1606.

Klein, S. A., et al., 2009: Intercomparison of model simulations of mixed-phase clouds observed during the ARM Mixed-Phase Arctic Cloud Experiment. I: Single-layer cloud. *Q. J. R. Meteorol. Soc.*, **135**, 979–1002.

Klocke, D., R. Pincus, and J. Quaas, 2011: On constraining estimates of climate sensitivity with present-day observations through model weighting. *J. Clim.*, **24**, 6092–6099.

Kloster, S., et al., 2007: Response of dimethylsulfide (DMS) in the ocean and atmosphere to global warming. *J. Geophys. Res.*, **112**, G03005.

Kloster, S., et al., 2010: Fire dynamics during the 20th century simulated by the Community Land Model. *Biogeosciences*, **7**, 1877–1902.

Knorr, W., V. Lehsten, and A. Arneth, 2012: Determinants and predictability of global wildfire emissions. *Atmos. Chem. Phys.*, **12**, 6845–6861.

Knox, A., et al., 2009: Mass absorption cross-section of ambient black carbon aerosol in relation to chemical age. *Aer. Sci. Technol.*, **43**, 522–532.

Kocak, M., N. Mihalopoulos, and N. Kubilay, 2007: Chemical composition of the fine and coarse fraction of aerosols in the northeastern Mediterranean. *Atmos. Environ.*, **41**, 7351–7368.

Koch, D., and A. D. Del Genio, 2010: Black carbon semi-direct effects on cloud cover: Review and synthesis. *Atmos. Chem. Phys.*, **10**, 7685–7696.

Koch, D., S. Menon, A. Del Genio, R. Ruedy, I. Alienov, and G. A. Schmidt, 2009a: Distinguishing aerosol impacts on climate over the past century. *J. Clim.*, **22**, 2659–2677.

Koch, D., et al., 2009b: Evaluation of black carbon estimations in global aerosol models. *Atmos. Chem. Phys.*, **9**, 9001–9026.

Kodama, C., A. T. Noda, and M. Satoh, 2012: An assessment of the cloud signals simulated by NICAM using ISCCP, CALIPSO, and CloudSat satellite simulators. *J. Geophys. Res.*, **117**, D12210.

Koffi, B., et al., 2012: Application of the CALIOP layer product to evaluate the vertical distribution of aerosols estimated by global models: Part 1. AeroCom phase I results. *J. Geophys. Res.*, **117**, D10201.

Köhler, M., M. Ahlgrimm, and A. Beljaars, 2011: Unified treatment of dry convective and stratocumulus-topped boundary layers in the ECMWF model. *Q. J. R. Meteorol. Soc.*, **137**, 43–57.

Kok, J. F., 2011: A scaling theory for the size distribution of emitted dust aerosols suggests climate models underestimate the size of the global dust cycle. *Proc. Natl. Acad. Sci. U.S.A.*, **108**, 1016–1021.

Kokhanovsky, A. A., et al., 2010: The inter-comparison of major satellite aerosol retrieval algorithms using simulated intensity and polarization characteristics of reflected light. *Atmos. Meas. Tech.*, **3**, 909–932.

Koop, T., B. P. Luo, A. Tsias, and T. Peter, 2000: Water activity as the determinant for homogeneous ice nucleation in aqueous solutions. *Nature*, **406**, 611–614.

Koren, I., G. Feingold, and L. A. Remer, 2010a: The invigoration of deep convective clouds over the Atlantic: Aerosol effect, meteorology or retrieval artifact? *Atmos. Chem. Phys.*, **10**, 8855–8872.

Koren, I., Y. Kaufman, L. Remer, and J. Martins, 2004: Measurement of the effect of Amazon smoke on inhibition of cloud formation. *Science*, **303**, 1342–1345.

Koren, I., J. V. Martins, L. A. Remer, and H. Afargan, 2008: Smoke invigoration versus inhibition of clouds over the Amazon. *Science*, **321**, 946–949.

Koren, I., Y. J. Kaufman, D. Rosenfeld, L. A. Remer, and Y. Rudich, 2005: Aerosol invigoration and restructuring of Atlantic convective clouds. *Geophys. Res. Lett.*, **32**, L14828.

Koren, I., L. A. Remer, Y. J. Kaufman, Y. Rudich, and J. V. Martins, 2007: On the twilight zone between clouds and aerosols. *Geophys. Res. Lett.*, **34**, L08805.

Koren, I., L. A. Remer, O. Altaratz, J. V. Martins, and A. Davidi, 2010b: Aerosol-induced changes of convective cloud anvils produce strong climate warming. *Atmos. Chem. Phys.*, **10**, 5001–5010.

Korhonen, H., K. S. Carslaw, and S. Romakkaniemi, 2010a: Enhancement of marine cloud albedo via controlled sea spray injections: A global model study of the influence of emission rates, microphysics and transport. *Atmos. Chem. Phys.*, **10**, 4133–4143.

Korhonen, H., K. S. Carslaw, P. M. Forster, S. Mikkonen, N. D. Gordon, and H. Kokkola, 2010b: Aerosol climate feedback due to decadal increases in Southern Hemisphere wind speeds. *Geophys. Res. Lett.*, **37**, L02805.

Korolev, A., 2007: Limitations of the Wegener-Bergeron-Findeisen mechanism in the evolution of mixed-phase clouds. *J. Atmos. Sci.*, **64**, 3372–3375.

Korolev, A., and P. R. Field, 2008: The effect of dynamics on mixed-phase clouds: Theoretical considerations. *J. Atmos. Sci.*, **65**, 66–86.

Kostinski, A. B., 2008: Drizzle rates versus cloud depths for marine stratocumuli. *Environ. Res. Lett.*, **3**, 045019.

Kravitz, B., A. Robock, D. Shindell, and M. Miller, 2012: Sensitivity of stratospheric geoengineering with black carbon to aerosol size and altitude of injection. *J. Geophys. Res.*, **117**, D09203.

Kravitz, B., A. Robock, L. Oman, G. Stenchikov, and A. B. Marquardt, 2009: Sulfuric acid deposition from stratospheric geoengineering with sulfate aerosols. *J. Geophys. Res.*, **114**, D14109.

Kravitz, B., A. Robock, O. Boucher, H. Schmidt, K. Taylor, G. Stenchikov, and M. Schulz, 2011: The Geoengineering Model Intercomparison Project (GeoMIP). *Atmos. Sci. Lett.*, **12**, 162–167.

Kristjánsson, J. E., 2002: Studies of the aerosol indirect effect from sulfate and black carbon aerosols. *J. Geophys. Res.*, **107**, 4246.

Kristjánsson, J. E., T. Iversen, A. Kirkevåg, Ø. Seland, and J. Debernard, 2005: Response of the climate system to aerosol direct and indirect forcing: Role of cloud feedbacks. *J. Geophys. Res.*, **110**, D24206.

Kristjánsson, J. E., C. W. Stjern, F. Stordal, A. M. Fjæraa, G. Myhre, and K. Jónasson, 2008: Cosmic rays, cloud condensation nuclei and clouds – a reassessment using MODIS data. *Atmos. Chem. Phys.*, **8**, 7373–7387.

Kroll, J. H., and J. H. Seinfeld, 2008: Chemistry of secondary organic aerosol: Formation and evolution of low-volatility organics in the atmosphere. *Atmos. Environ.*, **42**, 3593–3624.

Krueger, S. K., G. T. McLean, and Q. Fu, 1995: Numerical simulation of the stratus-to-cumulus transition in the subtropical marine boundary layer. 1. Boundary-layer structure. *J. Atmos. Sci.*, **52**, 2839–2850.

Kuang, Z. M., 2008: Modeling the interaction between cumulus convection and linear gravity waves using a limited-domain cloud system-resolving model. *J. Atmos. Sci.*, **65**, 576–591.

Kuang, Z. M., and C. S. Bretherton, 2006: A mass-flux scheme view of a high-resolution simulation of a transition from shallow to deep cumulus convection. *J. Atmos. Sci.*, **63**, 1895–1909.

Kuang, Z. M., and D. L. Hartmann, 2007: Testing the fixed anvil temperature hypothesis in a cloud-resolving model. *J. Clim.*, **20**, 2051–2057.

Kuebbeler, M., U. Lohmann, and J. Feichter, 2012: Effects of stratospheric sulfate aerosol geo-engineering on cirrus clouds. *Geophys. Res. Lett.*, **39**, L23803.

Kueppers, L. M., M. A. Snyder, and L. C. Sloan, 2007: Irrigation cooling effect: Regional climate forcing by land-use change. *Geophys. Res. Lett.*, **34**, L03703.

Kulmala, M., and V. M. Kerminen, 2008: On the formation and growth of atmospheric nanoparticles. *Atmos. Res.*, **90**, 132–150.

Kulmala, M., et al., 2010: Atmospheric data over a solar cycle: No connection between galactic cosmic rays and new particle formation. *Atmos. Chem. Phys.*, **10**, 1885–1898.

Kulmala, M., et al., 2011: General overview: European Integrated project on Aerosol Cloud Climate and Air Quality interactions (EUCAARI) – integrating aerosol research from nano to global scales. *Atmos. Chem. Phys.*, **11**, 13061–13143.

Kumar, P., I. N. Sokolik, and A. Nenes, 2011: Measurements of cloud condensation nuclei activity and droplet activation kinetics of fresh unprocessed regional dust samples and minerals. *Atmos. Chem. Phys.*, **11**, 3527–3541.

Kumar, R., S. S. Srivastava, and K. M. Kumari, 2007: Characteristics of aerosols over suburban and urban site of semiarid region in India: Seasonal and spatial variations. *Aer. Air Qual. Res.*, **7**, 531–549.

Kurten, T., V. Loukonen, H. Vehkamaki, and M. Kulmala, 2008: Amines are likely to enhance neutral and ion-induced sulfuric acid-water nucleation in the atmosphere more effectively than ammonia. *Atmos. Chem. Phys.*, **8**, 4095–4103.

Kuylenstierna, J. C. I., H. Rodhe, S. Cinderby, and K. Hicks, 2001: Acidification in developing countries: Ecosystem sensitivity and the critical load approach on a global scale. *Ambio*, **30**, 20–28.

L'Ecuyer, T. S., and J. H. Jiang, 2010: Touring the atmosphere aboard the A-Train. *Physics Today*, **63**, 36–41.

Laken, B., A. Wolfendale, and D. Kniveton, 2009: Cosmic ray decreases and changes in the liquid water cloud fraction over the oceans. *Geophys. Res. Lett.*, **36**, L23803.

Laken, B., D. Kniveton, and A. Wolfendale, 2011. Forbush decreases, solar irradiance variations, and anomalous cloud changes. *J. Geophys. Res.*, **116**, D09201.

Laken, B., E. Pallé, and H. Miyahara, 2012: A decade of the Moderate Resolution Imaging Spectroradiometer: is a solar–cloud link detectable? *J. Clim.*, **25**, 4430–4440.

Laken, B. A., and J. Čalogović, 2011: Solar irradiance, cosmic rays and cloudiness over daily timescales. *Geophys. Res. Lett.*, **38**, L24811.

Laken, B. A., D. R. Kniveton, and M. R. Frogley, 2010: Cosmic rays linked to rapid mid-latitude cloud changes. *Atmos. Chem. Phys.*, **10**, 10941–10948.

Lambert, F. H., and M. J. Webb, 2008: Dependency of global mean precipitation on surface temperature. *Geophys. Res. Lett.*, **35**, L16706.

Lance, S., et al., 2011: Cloud condensation nuclei as a modulator of ice processes in Arctic mixed-phase clouds. *Atmos. Chem. Phys.*, **11**, 8003–8015.

Lanz, V. A., M. R. Alfarra, U. Baltensperger, B. Buchmann, C. Hueglin, and A. S. H. Prévôt, 2007: Source apportionment of submicron organic aerosols at an urban site by factor analytical modelling of aerosol mass spectra. *Atmos. Chem. Phys.*, **7**, 1503–1522.

Larson, V. E., and J. C. Golaz, 2005: Using probability density functions to derive consistent closure relationships among higher-order moments. *Mon. Weather Rev.*, **133**, 1023–1042.

Larson, V. E., R. Wood, P. R. Field, J.-C. Golaz, T. H. Vonder Haar, and W. R. Cotton, 2001: Systematic biases in the microphysics and thermodynamics of numerical models that ignore subgrid-scale variability. *J. Atmos. Sci.*, **58**, 1117–1128.

Latham, J., 1990: Control of global warming? *Nature*, **347**, 339–340.

Latham, J., et al., 2008: Global temperature stabilization via controlled albedo enhancement of low-level maritime clouds. *Philos. Trans. R. Soc. London A*, **366**, 3969–3987.

Lathière, J., C. N. Hewitt, and D. J. Beerling, 2010: Sensitivity of isoprene emissions from the terrestrial biosphere to 20th century changes in atmospheric CO_2 concentration, climate, and land use. *Global Biogeochem. Cycles*, **24**, GB1004.

Lau, K. M., M. K. Kim, and K. M. Kim, 2006: Asian summer monsoon anomalies induced by aerosol direct forcing: Tthe role of the Tibetan Plateau. *Clim. Dyn.*, **26**, 855–864.

Lavers, D. A., R. P. Allan, E. F. Wood, G. Villarini, D. J. Brayshaw, and A. J. Wade, 2011: Winter floods in Britain are connected to atmospheric rivers. *Geophys. Res. Lett.*, **38**, L23803.

Lebsock, M. D., G. L. Stephens, and C. Kummerow, 2008: Multisensor satellite observations of aerosol effects on warm clouds. *J. Geophys. Res.*, **113**, D15205.

Lebsock, M. D., C. Kummerow, and G. L. Stephens, 2010: An observed tropical oceanic radiative-convective cloud feedback. *J. Clim.*, **23**, 2065–2078.

Leck, C., and E. K. Bigg, 2007: A modified aerosol–cloud–climate feedback hypothesis. *Environ. Chem.*, **4**, 400–403.

Leck, C., and E. K. Bigg, 2008: Comparison of sources and nature of the tropical aerosol with the summer high Arctic aerosol. *Tellus B*, **60**, 118–126.

Lee, D. S., et al., 2009: Aviation and global climate change in the 21st century. *Atmos. Environ.*, **43**, 3520–3537.

Lee, H. S., and B. W. Kang, 2001: Chemical characteristics of principal $PM_{2.5}$ species in Chongju, South Korea. *Atmos. Environ.*, **35**, 739–749.

Lee, K. H., Z. Li, M. S. Wong, J. Xin, Y. Wang, W.-M. Hao, and F. Zhao, 2007: Aerosol single scattering albedo estimated across China from a combination of ground and satellite measurements. *J. Geophys. Res.*, **112**, D22S15.

Lee, L. A., K. S. Carslaw, K. Pringle, G. W. Mann, and D. V. Spracklen, 2011: Emulation of a complex global aerosol model to quantify sensitivity to uncertain parameters. *Atmos. Chem. Phys.*, **11**, 12253–12273.

Lee, S.-S., G. Feingold, and P. Y. Chuang, 2012: Effect of aerosol on cloud-environment interactions in trade cumulus. *J. Atmos. Sci.*, **69**, 3607–3632.

Lee, S. S., 2012: Effect of aerosol on circulations and precipitation in deep convective clouds. *J. Atmos. Sci.*, **69**, 1957–1974.

Lee, Y. H., et al., 2013: Evaluation of preindustrial to present-day black carbon and its albedo forcing from Atmospheric Chemistry and Climate Model Intercomparison Project (ACCMIP). *Atmos. Chem. Phys.*, **13**, 2607–2634.

Lenderink, G., and E. Van Meijgaard, 2008: Increase in hourly precipitation extremes beyond expectations from temperature changes. *Nature Geosci.*, **1**, 511–514.

Lenderink, G., H. Y. Mok, T. C. Lee, and G. J. van Oldenborgh, 2011: Scaling and trends of hourly precipitation extremes in two different climate zones - Hong Kong and the Netherlands. *Hydrol. Earth Syst. Sci.*, **15**, 3033–3041.

Lenschow, P., H. J. Abraham, K. Kutzner, M. Lutz, J. D. Preu, and W. Reichenbacher, 2001: Some ideas about the sources of PM_{10}. *Atmos. Environ.*, **35**, 23–33.

Lenton, T. M., and N. E. Vaughan, 2009: The radiative forcing potential of different climate geoengineering options. *Atmos. Chem. Phys.*, **9**, 5539–5561.

Levin, Z., and W. R. Cotton, 2009: *Aerosol Pollution Impact on Precipitation: A Scientific Review.* Springer Science+Business Media, New York, NY, USA, and Heidelberg, Germany, 386 pp.

Levy, R. C., L. A. Remer, R. G. Kleidman, S. Mattoo, C. Ichoku, R. Kahn, and T. F. Eck, 2010: Global evaluation of the Collection 5 MODIS dark-target aerosol products over land. *Atmos. Chem. Phys.*, **10**, 10399–10420.

Li, J., M. Pósfai, P. V. Hobbs, and P. R. Buseck, 2003: Individual aerosol particles from biomass burning in southern Africa: 2, Compositions and aging of inorganic particles. *J. Geophys. Res.*, **108**, 8484.

Li, Z., K.-H. Lee, Y. Wang, J. Xin, and W.-M. Hao, 2010: First observation-based estimates of cloud-free aerosol radiative forcing across China. *J. Geophys. Res.*, **115**, D00K18.

Li, Z., F. Niu, J. Fan, Y. Liu, D. Rosenfeld, and Y. Ding, 2011: Long-term impacts of aerosols on the vertical development of clouds and precipitation *Nature Geosci.*, **4**, 888–894.

Li, Z., et al., 2009: Uncertainties in satellite remote sensing of aerosols and impact on monitoring its long-term trend: A review and perspective. *Annal. Geophys.*, **27**, 2755–2770.

Liao, H., W. T. Chen, and J. H. Seinfeld, 2006: Role of climate change in global predictions of future tropospheric ozone and aerosols. *J. Geophys. Res.*, **111**, D12304.

Liao, H., Y. Zhang, W. T. Chen, F. Raes, and J. H. Seinfeld, 2009: Effect of chemistry-aerosol-climate coupling on predictions of future climate and future levels of tropospheric ozone and aerosols. *J. Geophys. Res.*, **114**, D10306.

Liepert, B. G., and M. Previdi, 2012: Inter-model variability and biases of the global water cycle in CMIP3 coupled climate models. *Environ. Res. Lett.*, **7**, 014006.

Lim, Y. B., Y. Tan, M. J. Perri, S. P. Seitzinger, and B. J. Turpin, 2010: Aqueous chemistry and its role in secondary organic aerosol (SOA) formation. *Atmos. Chem. Phys.*, **10**, 10521–10539.

Lin, G., J. E. Penner, S. Sillman, D. Taraborrelli, and J. Lelieveld, 2012: Global modeling of SOA formation from dicarbonyls, epoxides, organic nitrates and peroxides. *Atmos. Chem. Phys.*, **12**, 4743–4774.

Lindzen, R. S., and Y.-S. Choi, 2011: On the observational determination of climate sensitivity and its implications. *Asia Pac. J. Atmos. Sci.*, **47**, 377–390.

Liou, K. N., and S. C. Ou, 1989: The role of cloud microphysical processes in climate—An assessment from a one-dimensional perspective. *J. Geophys. Res.*, **94**, 8599–8607.

Liu, W., Y. Wang, A. Russell, and E. S. Edgerton, 2005: Atmospheric aerosol over two urban-rural pairs in the southeastern United States: Chemical composition and possible sources. *Atmos. Environ.*, **39**, 4453–4470.

Liu, X., J. Penner, S. Ghan, and M. Wang, 2007: Inclusion of ice microphysics in the NCAR community atmospheric model version 3 (CAM3). *J. Clim.*, **20**, 4526–4547.

Liu, X., et al., 2012: Toward a minimal representation of aerosols in climate models: Description and evaluation in the Community Atmosphere Model CAM5. *Geosci. Model Dev.*, **5**, 709–739.

Liu, X. H., and J. E. Penner, 2005: Ice nucleation parameterization for global models. *Meteorol. Z.*, **14**, 499–514.

Liu, X. H., J. E. Penner, and M. H. Wang, 2009: Influence of anthropogenic sulfate and black carbon on upper tropospheric clouds in the NCAR CAM3 model coupled to the IMPACT global aerosol model. *J. Geophys. Res.*, **114**, D03204.

Liu, Y., and P. H. Daum, 2002: Anthropogenic aerosols: Indirect warming effect from dispersion forcing. *Nature*, **419**, 580–581.

Liu, Y., J. R. Key, and X. Wang, 2008: The influence of changes in cloud cover on recent surface temperature trends in the Arctic. *J. Clim.*, **21**, 705–715.

Lobell, D., G. Bala, A. Mirin, T. Phillips, R. Maxwell, and D. Rotman, 2009: Regional differences in the influence of irrigation on climate. *J. Clim.*, **22**, 2248–2255.

Lock, A. P., 2009: Factors influencing cloud area at the capping inversion for shallow cumulus clouds. *Q. J. R. Meteorol. Soc.*, **135**, 941–952.

Lodhi, A., B. Ghauri, M. R. Khan, S. Rahmana, and S. Shafiquea, 2009: Particulate matter ($PM_{2.5}$) concentration and source apportionment in Lahore. *J. Brazil. Chem. Soc.*, **20**, 1811–1820.

Loeb, N. G., and N. Manalo-Smith, 2005: Top-of-atmosphere direct radiative effect of aerosols over global oceans from merged CERES and MODIS observations. *J. Clim.*, **18**, 3506–3526.

Loeb, N. G., and G. L. Schuster, 2008: An observational study of the relationship between cloud, aerosol and meteorology in broken low-level cloud conditions. *J. Geophys. Res.*, **113**, D14214.

Loeb, N. G., and W. Y. Su, 2010: Direct aerosol radiative forcing uncertainty based on a radiative perturbation analysis. *J. Clim.*, **23**, 5288–5293.

Loeb, N. G., et al., 2009: Toward optimal closure of the Earth's top-of-atmosphere radiation budget. *J. Clim.*, **22**, 748–766.

Logan, T., B. K. Xi, X. Q. Dong, R. Obrecht, Z. Q. Li, and M. Cribb, 2010: A study of Asian dust plumes using satellite, surface, and aircraft measurements during the INTEX-B field experiment. *J. Geophys. Res.*, **115**, D00K25.

Lohmann, U., 2002a: Possible aerosol effects on ice clouds via contact nucleation. *J. Atmos. Sci.*, **59**, 647–656.
Lohmann, U., 2002b: A glaciation indirect aerosol effect caused by soot aerosols. *Geophys. Res. Lett.*, **29**, 1052.
Lohmann, U., 2004: Can anthropogenic aerosols decrease the snowfall rate? *J. Atmos. Sci.*, **61**, 2457–2468.
Lohmann, U., 2008: Global anthropogenic aerosol effects on convective clouds in ECHAM5–HAM. *Atmos. Chem. Phys.*, **8**, 2115–2131.
Lohmann, U., and J. Feichter, 1997: Impact of sulfate aerosols on albedo and lifetime of clouds: A sensitivity study with the ECHAM4 GCM. *J. Geophys. Res.*, **102**, 13685–13700.
Lohmann, U., and J. Feichter, 2001: Can the direct and semi-direct aerosol effect compete with the indirect effect on a global scale? *Geophys. Res. Lett.*, **28**, 159–161.
Lohmann, U., and B. Kärcher, 2002: First interactive simulations of cirrus clouds formed by homogeneous freezing in the ECHAM general circulation model. *J. Geophys. Res.*, **107**, 4105.
Lohmann, U., and G. Lesins, 2002: Stronger constraints on the anthropogenic indirect aerosol effect. *Science*, **298**, 1012–1015.
Lohmann, U., and J. Feichter, 2005: Global indirect aerosol effects: A review. *Atmos. Chem. Phys.*, **5**, 715–737.
Lohmann, U., and K. Diehl, 2006: Sensitivity studies of the importance of dust ice nuclei for the indirect aerosol effect on stratiform mixed-phase clouds. *J. Atmos. Sci.*, **63**, 968–982.
Lohmann, U., and C. Hoose, 2009: Sensitivity studies of different aerosol indirect effects in mixed-phase clouds. *Atmos. Chem. Phys.*, **9**, 8917–8934.
Lohmann, U., and S. Ferrachat, 2010: Impact of parametric uncertainties on the present-day climate and on the anthropogenic aerosol effect. *Atmos. Chem. Phys.*, **10**, 11373–11383.
Lohmann, U., J. Feichter, J. Penner, and R. Leaitch, 2000: Indirect effect of sulfate and carbonaceous aerosols: A mechanistic treatment. *J. Geophys. Res.*, **105**, 12193–12206.
Lohmann, U., P. Stier, C. Hoose, S. Ferrachat, S. Kloster, E. Roeckner, and J. Zhang, 2007: Cloud microphysics and aerosol indirect effects in the global climate model ECHAM5–HAM. *Atmos. Chem. Phys.*, **7**, 3425–3446.
Lohmann, U., et al., 2010: Total aerosol effect: Radiative forcing or radiative flux perturbation? *Atmos. Chem. Phys.*, **10**, 3235–3246.
Lonati, G., M. Giugliano, P. Butelli, L. Romele, and R. Tardivo, 2005: Major chemical components of $PM_{2.5}$ in Milan (Italy). *Atmos. Environ.*, **39**, 1925–1934.
Lu, M.-L., W. C. Conant, H. H. Jonsson, V. Varutbangkul, R. C. Flagan, and J. H. Seinfeld, 2007: The marine stratus/stratocumulus experiment (MASE): Aerosol-cloud relationships in marine stratocumulus. *J. Geophys. Res.*, **112**, D10209.
Lu, M.-L., G. Feingold, H. H. Jonsson, P. Y. Chuang, H. Gates, R. C. Flagan, and J. H. Seinfeld, 2008: Aerosol-cloud relationships in continental shallow cumulus. *J. Geophys. Res.*, **113**, D15201.
Lu, Z., Q. Zhang, and D. G. Streets, 2011: Sulfur dioxide and primary carbonaceous aerosol emissions in China and India, 1996–2010. *Atmos. Chem. Phys.*, **11**, 9839–9864.
Lu, Z., et al., 2010: Sulfur dioxide emissions in China and sulfur trends in East Asia since 2000. *Atmos. Chem. Phys.*, **10**, 6311–6331.
Lubin, D., and A. M. Vogelmann, 2006: A climatologically significant aerosol longwave indirect effect in the Arctic. *Nature*, **439**, 453–456.
Lunt, D. J., A. Ridgwell, P. J. Valdes, and A. Seale, 2008: "Sunshade World": A fully coupled GCM evaluation of the climatic impacts of geoengineering. *Geophys. Res. Lett.*, **35**, L12710.
Lyapustin, A., et al., 2010: Analysis of snow bidirectional reflectance from ARCTAS Spring-2008 Campaign. *Atmos. Chem. Phys.*, **10**, 4359–4375.
Ma, H. Y., M. Kohler, J. L. F. Li, J. D. Farrara, C. R. Mechoso, R. M. Forbes, and D. E. Waliser, 2012a: Evaluation of an ice cloud parameterization based on a dynamical-microphysical lifetime concept using CloudSat observations and the ERA-Interim reanalysis. *J. Geophys. Res.*, **117**, D05210.
Ma, X., F. Yu, and G. Luo, 2012b: Aerosol direct radiative forcing based on GEOS-Chem-APM and uncertainties. *Atmos. Chem. Phys.*, **12**, 5563–5581.
MacCracken, M. C., 2009: On the possible use of geoengineering to moderate specific climate change impacts. *Environ. Res. Lett.*, **4**, 045107.
Mace, G. G., Q. Q. Zhang, M. Vaughan, R. Marchand, G. Stephens, C. Trepte, and D. Winker, 2009: A description of hydrometeor layer occurrence statistics derived from the first year of merged Cloudsat and CALIPSO data. *J. Geophys. Res.*, **114**, D00A26.
Maenhaut, W., I. Salma, and J. Cafrneyer, 1996: Regional atmospheric aerosol composition and sources in the eastern Transvaal, South Africa, and impact of biomass burning. *J. Geophys. Res.*, **101**, 23613–23650.
Maenhaut, W., M.-T. Fernandez-Jimenez, J. L. Vanderzalm, B. Hooper, M. A. Hooper, and N. J. Tapper, 2000: Aerosol composition at Jabiru, Australia, and impact of biomass burning. *J. Aer. Sci.*, **31**, 745–746.
Magee, N., A. M. Moyle, and D. Lamb, 2006: Experimental determination of the deposition coefficient of small cirrus-like ice crystals near –50°C. *Geophys. Res. Lett.*, **33**, L17813.
Mahowald, N. M., J. F. Lamarque, X. X. Tie, and E. Wolff, 2006a: Sea-salt aerosol response to climate change: Last Glacial Maximum, preindustrial, and doubled carbon dioxide climates. *J. Geophys. Res.*, **111**, D05303.
Mahowald, N. M., D. R. Muhs, S. Levis, P. J. Rasch, M. Yoshioka, C. S. Zender, and C. Luo, 2006b: Change in atmospheric mineral aerosols in response to climate: Last glacial period, preindustrial, modern, and doubled carbon dioxide climates. *J. Geophys. Res.*, **111**, D10202.
Mahowald, N. M., et al., 2009: Atmospheric iron deposition: Global distribution, variability, and human perturbations. *Annu. Rev. Mar. Sci.*, **1**, 245–278.
Mahowald, N. M., et al., 2010: Observed 20th century desert dust variability: impact on climate and biogeochemistry. *Atmos. Chem. Phys.*, **10**, 10875–10893.
Makkonen, R., A. Asmi, V. Kerminen, M. Boy, A. Arneth, P. Hari, and M. Kulmala, 2012a: Air pollution control and decreasing new particle formation lead to strong climate warming. *Atmos. Chem. Phys.*, **12**, 1515–1524.
Makkonen, R., A. Asmi, V. M. Kerminen, M. Boy, A. Arneth, A. Guenther, and M. Kulmala, 2012b: BVOC-aerosol-climate interactions in the global aerosol-climate model ECHAM5.5–HAM2. *Atmos. Chem. Phys.*, **12**, 10077–10096.
Malm, W. C., and B. A. Schichtel, 2004: Spatial and monthly trends in speciated fine particle concentration in the United States. *J. Geophys. Res.*, **109**, D03306.
Malm, W. C., J. F. Sisler, D. Huffman, R. A. Eldred, and T. A. Cahill, 1994: Spatial and seasonal trends in particle concentration and optical extinction in the United States. *J. Geophys. Res.*, **99**, 1347–1370.
Mann, G. W., et al., 2010: Description and evaluation of GLOMAP-mode: A modal global aerosol microphysics model for the UKCA composition-climate model. *Geosci. Model Dev.*, **3**, 519–551.
Manninen, H. E., et al., 2010: EUCAARI ion spectrometer measurements at 12 European sites—analysis of new particle formation events. *Atmos. Chem. Phys.*, **10**, 7907–7927.
Mapes, B., and R. Neale, 2011: Parameterizing convective organization to escape the entrainment dilemma. *J. Adv. Model. Earth Syst.*, **3**, M06004.
Mariani, R. L., and W. Z. d. Mello, 2007: PM2.5–10, PM2.5 and associated water-soluble inorganic species at a coastal urban site in the metropolitan region of Rio de Janeiro. *Atmos. Environ.*, **41**, 2887–2892.
Marlon, J. R., et al., 2008: Climate and human influences on global biomass burning over the past two millennia. *Nature Geosci.*, **1**, 697–702.
Marsh, N. D., and H. Svensmark, 2000: Low cloud properties influenced by cosmic rays. *Phys. Rev. Lett.*, **85**, 5004–5007.
Martin, S. T., et al., 2010a: Sources and properties of Amazonian aerosol particles. *Rev. Geophys.*, **48**, RG2002.
Martin, S. T., et al., 2010b: An overview of the Amazonian Aerosol Characterization Experiment 2008 (AMAZE-08). *Atmos. Chem. Phys.*, **10**, 11415–11438.
Matthews, H. D., 2010: Can carbon cycle geoengineering be a useful complement to ambitious climate mitigation? *Carbon Manag.*, **1**, 135–144.
Matthews, H. D., and K. Caldeira, 2007: Transient climate–carbon simulations of planetary geoengineering. *Proc. Natl. Acad. Sci. U.S.A.*, **104**, 9949–9954.
Mauger, G. S., and J. R. Norris, 2010: Assessing the impact of meteorological history on subtropical cloud fraction. *J. Clim.*, **23**, 2926–2940.
Mauritsen, T., et al., 2011: An Arctic CCN-limited cloud-aerosol regime. *Atmos. Chem. Phys.*, **11**, 165–173.
Mautner, M., 1991: A space-based solar screen against climate warming. *J. Br. Interplanet. Soc.*, **44**, 135–138.
McComiskey, A., and G. Feingold, 2008: Quantifying error in the radiative forcing of the first aerosol indirect effect. *Geophys. Res. Lett.*, **35**, L02810.
McComiskey, A., and G. Feingold, 2012: The scale problem in quantifying aerosol indirect effects. *Atmos. Chem. Phys.*, **12**, 1031–1049.
McComiskey, A., S. Schwartz, B. Schmid, H. Guan, E. Lewis, P. Ricchiazzi, and J. Ogren, 2008: Direct aerosol forcing: Calculation from observables and sensitivities to inputs. *J. Geophys. Res.*, **113**, D09202.
McComiskey, A., et al., 2009: An assessment of aerosol-cloud interactions in marine stratus clouds based on surface remote sensing. *J. Geophys. Res.*, **114**, D09203.

McConnell, J. R., et al., 2007: 20th-century industrial black carbon emissions altered Arctic climate forcing. *Science*, **317**, 1381–1384.

McCormick, M. P., L. W. Thomason, and C. R. Trepte, 1995: Atmospheric effects of the Mt Pinatubo eruption. *Nature*, **373**, 399–404.

McFarquhar, G. M., et al., 2011: Indirect and semi-direct aerosol campaign: The impact of Arctic aerosols on clouds. *Bull. Am. Meteor. Soc.*, **92**, 183–201.

McFiggans, G., et al., 2006: The effect of physical and chemical aerosol properties on warm cloud droplet activation. *Atmos. Chem. Phys.*, **6**, 2593–2649.

McMeeking, G. R., et al., 2010: Black carbon measurements in the boundary layer over western and northern Europe. *Atmos. Chem. Phys.*, **10**, 9393–9414.

Medeiros, B., B. Stevens, I. M. Held, M. Zhao, D. L. Williamson, J. G. Olson, and C. S. Bretherton, 2008: Aquaplanets, climate sensitivity, and low clouds. *J. Clim.*, **21**, 4974–4991.

Meehl, G. A., et al., 2007: Global climate projections. In: *Climate Change 2007: The Physical Science Basis. Contribution of Working Group I to the Fourth Assessment Report of the Intergovernmental Panel on Climate Change* [Solomon, S., D. Qin, M. Manning, Z. Chen, M. Marquis, K. B. Averyt, M. Tignor and H. L. Miller (eds.)] Cambridge University Press, Cambridge, United Kingdom and New York, NY, USA, pp. 747–843.

Menon, S., and L. Rotstayn, 2006: The radiative influence of aerosol effects on liquid-phase cumulus and stratiform clouds based on sensitivity studies with two climate models. *Clim. Dyn.*, **27**, 345–356.

Menon, S., and A. Del Genio, 2007: Evaluating the impacts of carbonaceous aerosols on clouds and climate. In: *Human-Induced Climate Change: An Interdisciplinary Assessment* [M. E. Schlesinger, H. S. Kheshgi, J. Smith, F. C. de la Chesnaye, J. M. Reilly, T. Wilson and C. Kolstad (eds.)]. Cambridge University Press, Cambridge, United Kingdom, and New York, NY, USA, pp. 34–48.

Menon, S., A. D. Del Genio, D. Koch, and G. Tselioudis, 2002: GCM Simulations of the aerosol indirect effect: Sensitivity to cloud parameterization and aerosol burden. *J. Atmos. Sci.*, **59**, 692–713.

Mercado, L. M., N. Bellouin, S. Sitch, O. Boucher, C. Huntingford, M. Wild, and P. M. Cox, 2009: Impact of changes in diffuse radiation on the global land carbon sink. *Nature*, **458**, 1014–1017.

Merikanto, J., D. V. Spracklen, G. W. Mann, S. J. Pickering, and K. S. Carslaw, 2009: Impact of nucleation on global CCN. *Atmos. Chem. Phys.*, **9**, 8601–8616.

Metzger, A., et al., 2010: Evidence for the role of organics in aerosol particle formation under atmospheric conditions. *Proc. Natl. Acad. Sci. U.S.A.*, **107**, 6646–6651.

Miller, R. L., 1997: Tropical thermostats and low cloud cover. *J. Clim.*, **10**, 409–440.

Ming, J., D. Zhang, S. Kang, and W. Tian, 2007a: Aerosol and fresh snow chemistry in the East Rongbuk Glacier on the northern slope of Mt. Qomolangma (Everest). *J. Geophys. Res.*, **112**, D15307.

Ming, J., C. D. Xiao, H. Cachier, D. H. Qin, X. Qin, Z. Q. Li, and J. C. Pu, 2009: Black Carbon (BC) in the snow of glaciers in west China and its potential effects on albedos. *Atmos. Res.*, **92**, 114–123.

Ming, Y., V. Ramaswamy, and G. Persad, 2010: Two opposing effects of absorbing aerosols on global-mean precipitation. *Geophys. Res. Lett.*, **37**, L13701.

Ming, Y., V. Ramaswamy, P. A. Ginoux, L. W. Horowitz, and L. M. Russell, 2005: Geophysical Fluid Dynamics Laboratory general circulation model investigation of the indirect radiative effects of anthropogenic sulfate aerosol. *J. Geophys. Res.*, **110**, D22206.

Ming, Y., V. Ramaswamy, L. J. Donner, V. T. J. Phillips, S. A. Klein, P. A. Ginoux, and L. W. Horowitz, 2007b: Modeling the interactions between aerosols and liquid water clouds with a self-consistent cloud scheme in a general circulation model. *J. Atmos. Sci.*, **64**, 1189–1209.

Minschwaner, K., A. E. Dessler, and P. Sawaengphokhai, 2006: Multimodel analysis of the water vapor feedback in the tropical upper troposphere. *J. Clim.*, **19**, 5455–5464.

Mirme, S., A. Mirme, A. Minikin, A. Petzold, U. Horrak, V. M. Kerminen, and M. Kulmala, 2010: Atmospheric sub-3 nm particles at high altitudes. *Atmos. Chem. Phys.*, **10**, 437–451.

Mishchenko, M. I., et al., 2007: Accurate monitoring of terrestrial aerosols and total solar irradiance—Introducing the Glory mission. *Bull. Am. Meteor. Soc.*, **88**, 677–691.

Mishchenko, M. I., et al., 2012: Aerosol retrievals from channel-1 and -2 AVHRR radiances: Long-term trends updated and revisited. *J. Quant. Spectrosc. Radiat. Transfer*, **113**, 1974–1980.

Mitchell, D. L., and W. Finnegan, 2009: Modification of cirrus clouds to reduce global warming. *Environ. Res. Lett.*, **4**, 045102.

Mitchell, J. F., C. A. Wilson, and W. M. Cunnington, 1987: On CO_2 climate sensitivity and model dependence of results. *Q. J. R. Meteorol. Soc.*, **113**, 293–322.

Miyazaki, Y., S. G. Aggarwal, K. Singh, P. K. Gupta, and K. Kawamura, 2009: Dicarboxylic acids and water-soluble organic carbon in aerosols in New Delhi, India, in winter: Characteristics and formation processes. *J. Geophys. Res.*, **114**, D19206.

Mkoma, S. L., 2008: Physico-chemical characterisation of atmospheric aerosol in Tanzania, with emphasis on the carbonaceous aerosol components and on chemical mass closure. Ph.D. Ghent University, Ghent, Belgium.

Mkoma, S. L., W. Maenhaut, X. G. Chi, W. Wang, and N. Raes, 2009a: Characterisation of PM_{10} atmospheric aerosols for the wet season 2005 at two sites in East Africa. *Atmos. Environ.*, **43**, 631–639.

Mkoma, S. L., W. Maenhaut, X. Chi, W. Wang, and N. Raes, 2009b: Chemical composition and mass closure for PM10 aerosols during the 2005 dry season at a rural site in Morogoro, Tanzania. *X-Ray Spectrom.*, **38**, 293–300.

Mohler, O., et al., 2003: Experimental investigation of homogeneous freezing of sulphuric acid particles in the aerosol chamber AIDA. *Atmos. Chem. Phys.*, **3**, 211–223.

Moorthy, K. K., S. K. Satheesh, S. S. Babu, and C. B. S. Dutt, 2008: Integrated Campaign for Aerosols, gases and Radiation Budget (ICARB): An overview. *J. Earth Syst. Sci.*, **117**, 243–262.

Moosmüller, H., R. Chakrabarty, and W. Arnott, 2009: Aerosol light absorption and its measurement: A review. *J. Quant. Spectrosc. Radiat. Transfer*, **110**, 844–878.

Morales, J. A., D. Pirela, M. G. d. Nava, B. S. d. Borrego, H. Velasquez, and J. Duran, 1998: Inorganic water soluble ions in atmospheric particles over Maracaibo Lake Basin in the western region of Venezuela. *Atmos. Res.*, **46**, 307–320.

Morcrette, J.-J., et al., 2009: Aerosol analysis and forecast in the ECMWF Integrated Forecast System. Part I: Forward modelling. *J. Geophys. Res.*, **114**, D06206.

Moreno-Cruz, J. B., K. W. Ricke, and D. W. Keith, 2011: A simple model to account for regional inequalities in the effectiveness of solar radiation management. *Clim. Change*, **110**, 649–668.

Morrison, H., and A. Gettelman, 2008: A new two-moment bulk stratiform cloud microphysics scheme in the community atmosphere model, version 3 (CAM3). Part I: Description and numerical tests. *J. Clim.*, **21**, 3642–3659.

Morrison, H., and W. W. Grabowski, 2011: Cloud-system resolving model simulations of aerosol indirect effects on tropical deep convection and its thermodynamic environment. *Atmos. Chem. Phys.*, **11**, 10503–10523.

Morrison, H., G. de Boer, G. Feingold, J. Harrington, M. D. Shupe, and K. Sulia, 2012: Resilience of persistent Arctic mixed-phase clouds. *Nature Geosci.*, **5**, 11–17.

Moteki, N., and Y. Kondo, 2010: Dependence of laser-induced incandescence on physical properties of black carbon aerosols: Measurements and theoretical interpretation. *Aer. Sci. Technol.*, **44**, 663–675.

Moteki, N., et al., 2007: Evolution of mixing state of black carbon particles: Aircraft measurements over the western Pacific in March 2004. *Geophys. Res. Lett.*, **34**, L11803.

Mouillot, F., A. Narasimha, Y. Balkanski, J. F. Lamarque, and C. B. Field, 2006: Global carbon emissions from biomass burning in the 20th century. *Geophys. Res. Lett.*, **33**, L01801.

Muller, C. J., and P. A. O'Gorman, 2011: An energetic perspective on the regional response of precipitation to climate change. *Nature Clim. Change*, **1**, 266–271.

Muller, C. J., and I. M. Held, 2012: Detailed investigation of the self-aggregation of convection in cloud-resolving simulations. *J. Atmos. Sci.*, **69**, 2551–2565.

Muller, C. J., P. A. O'Gorman, and L. E. Back, 2011: Intensification of precipitation extremes with warming in a cloud-resolving model. *J. Clim.*, **24**, 2784–2800.

Müller, J.-F., et al., 2008: Global isoprene emissions estimated using MEGAN, ECMWF analyses and a detailed canopy environment model. *Atmos. Chem. Phys.*, **8**, 1329–1341.

Murray, B. J., D. O'Sullivan, J. D. Atkinson, and M. E. Webb, 2012: Ice nucleation by particles immersed in supercooled cloud droplets. *Chem. Soc. Rev.*, **41**, 6519–6554.

Myhre, G., 2009: Consistency between satellite-derived and modeled estimates of the direct aerosol effect. *Science*, **325**, 187–190.

Myhre, G., et al., 2007: Comparison of the radiative properties and direct radiative effect of aerosols from a global aerosol model and remote sensing data over ocean. *Tellus B*, **59**, 115–129.

Myhre, G., et al., 2009: Modelled radiative forcing of the direct aerosol effect with multi-observation evaluation. *Atmos. Chem. Phys.*, **9**, 1365–1392.

Myhre, G., et al., 2013: Radiative forcing of the direct aerosol effect from AeroCom Phase II simulations. *Atmos. Chem. Phys.*, **13**, 1853–1877.

Nakajima, T., and M. Schulz, 2009: What do we know about large scale changes of aerosols, clouds, and the radiative budget? In: *Clouds in the Perturbed Climate System: Their Relationship to Energy Balance, Atmospheric Dynamics, and Precipitation* [R. J. Charlson and J. Heintzenberg (eds.)], MIT Press, Cambridge, MA, USA, pp. 401–430.

Nakajima, T., et al., 2007: Overview of the Atmospheric Brown Cloud East Asian Regional Experiment 2005 and a study of the aerosol direct radiative forcing in East Asia. *J. Geophys. Res.*, **112**, D24S91.

Nakayama, T., Y. Matsumi, K. Sato, T. Imamura, A. Yamazaki, and A. Uchiyama, 2010: Laboratory studies on optical properties of secondary organic aerosols generated during the photooxidation of toluene and the ozonolysis of ⍰-pinene. *J. Geophys. Res.*, **115**, D24204.

Nam, C., S. Bony, J.-L. Dufresne, and H. Chepfer, 2012: The 'too few, too bright' tropical low-cloud problem in CMIP5 models. *Geophys. Res. Lett.*, **39**, L21801.

Naud, C. M., A. D. Del Genio, M. Bauer, and W. Kovari, 2010: Cloud vertical distribution across warm and cold fronts in CloudSat-CALIPSO data and a General Circulation Model. *J. Clim.*, **23**, 3397–3415.

Naud, C. M., A. Del Genio, G. G. Mace, S. Benson, E. E. Clothiaux, and P. Kollias, 2008: Impact of dynamics and atmospheric state on cloud vertical overlap. *J. Clim.*, **21**, 1758–1770.

Neale, R. B., J. H. Richter, and M. Jochum, 2008: The impact of convection on ENSO: From a delayed oscillator to a series of events. *J. Clim.*, **21**, 5904–5924.

Neelin, J. D., and I. M. Held, 1987: Modeling tropical convergence based on the moist static energy budget. *Mon. Weather Rev.*, **115**, 3–12.

Neelin, J. D., M. Münnich, H. Su, J. E. Meyerson, and C. E. Holloway, 2006: Tropical drying trends in global warming models and observations. *Proc. Natl. Acad. Sci. U.S.A.*, **103**, 6110–6115.

Neggers, R. A. J., 2009: A dual mass flux framework for boundary layer convection. Part II: Clouds. *J. Atmos. Sci.*, **66**, 1489–1506.

Neggers, R. A. J., M. Kohler, and A. C. M. Beljaars, 2009: A dual mass flux framework for boundary layer convection. Part I: Transport. *J. Atmos. Sci.*, **66**, 1465–1487.

Nicoll, K., and R. G. Harrison, 2010: Experimental determination of layer cloud edge charging from cosmic ray ionization. *Geophys. Res. Lett.*, **37**, L13802.

Niemeier, U., H. Schmidt, and C. Timmreck, 2011: The dependency of geoengineered sulfate aerosol on the emission strategy. *Atmos. Sci. Lett.*, **12**, 189–194.

Nuijens, L., B. Stevens, and A. P. Siebesma, 2009: The environment of precipitating shallow cumulus convection. *J. Atmos. Sci.*, **66**, 1962–1979.

Nyanganyura, D., W. Maenhaut, M. Mathuthu, A. Makarau, and F. X. Meixner, 2007: The chemical composition of tropospheric aerosols and their contributing sources to a continental background site in northern Zimbabwe from 1994 to 2000. *Atmos. Environ.*, **41**, 2644–2659.

O'Dell, C. W., F. J. Wentz, and R. Bennartz, 2008: Cloud liquid water path from satellite-based passive microwave observations: A new climatology over the global oceans. *J. Clim.*, **21**, 1721–1739.

O'Donnell, D., K. Tsigaridis, and J. Feichter, 2011: Estimating the direct and indirect effects of secondary organic aerosols using ECHAM5–HAM. *Atmos. Chem. Phys.*, **11**, 8635–8659.

O'Dowd, C., C. Monahan, and M. Dall'Osto, 2010: On the occurrence of open ocean particle production and growth events. *Geophys. Res. Lett.*, **37**, L19805.

O'Gorman, P. A., 2012: Sensitivity of tropical precipitation extremes to climate change. *Nature Geosci.*, **5**, 697–700.

O'Gorman, P. A., and T. Schneider, 2008: The hydrological cycle over a wide range of climates simulated with an idealized GCM. *J. Clim.*, **21**, 3815–3832.

O'Gorman, P. A., and T. Schneider, 2009: The physical basis for increases in precipitation extremes in simulations of 21st-century climate change. *Proc. Natl. Acad. Sci. U.S.A.*, **106**, 14773–14777.

Oanh, N. T. K., et al., 2006: Particulate air pollution in six Asian cities: Spatial and temporal distributions, and associated sources. *Atmos. Environ.*, **40**, 3367–3380.

Ocko, I. B., V. Ramaswamy, P. Ginoux, Y. Ming, and L. W. Horowitz, 2012: Sensitivity of scattering and absorbing aerosol direct radiative forcing to physical climate factors. *J. Geophys. Res.*, **117**, D20203.

Oleson, K. W., G. B. Bonan, and J. Feddema, 2010: Effects of white roofs on urban temperature in a global climate model. *Geophys. Res. Lett.*, **37**, L03701.

Omar, A. H., et al., 2009: The CALIPSO automated aerosol classification and lidar ratio selection algorithm. *J. Atmos. Ocean. Technol.*, **26**, 1994–2014.

Oouchi, K., A. T. Noda, M. Satoh, B. Wang, S. P. Xie, H. G. Takahashi, and T. Yasunari, 2009: Asian summer monsoon simulated by a global cloud-system-resolving model: Diurnal to intra-seasonal variability. *Geophys. Res. Lett.*, **36**, L11815.

Orellana, M. V., P. A. Matrai, C. Leck, C. D. Rauschenberg, A. M. Lee, and E. Coz, 2011: Marine microgels as a source of cloud condensation nuclei in the high Arctic. *Proc. Natl. Acad. Sci. U.S.A.*, **108**, 13612–13617.

Oreopoulos, L., and S. Platnick, 2008: Radiative susceptibility of cloudy atmospheres to droplet number perturbations: 2. Global analysis from MODIS. *J. Geophys. Res.*, **113**, D14S21.

Osborne, S. R., A. J. Barana, B. T. Johnson, J. M. Haywood, E. Hesse, and S. Newman, 2011: Short-wave and long-wave radiative properties of Saharan dust aerosol. *Q. J. R. Meteorol. Soc.*, **137**, 1149–1167.

Oshima, N., et al., 2012: Wet removal of black carbon in Asian outflow: Aerosol Radiative Forcing in East Asia (A-FORCE) aircraft campaign. *J. Geophys. Res.*, **117**, D03204.

Ovadnevaite, J., et al., 2011: Primary marine organic aerosol: A dichotomy of low hygroscopicity and high CCN activity. *Geophys. Res. Lett.*, **38**, L21806.

Ovchinnikov, M., A. Korolev, and J. Fan, 2011: Effects of ice number concentration on dynamics of a shallow mixed-phase stratiform cloud. *J. Geophys. Res.*, **116**, D00T06.

Paasonen, P., et al., 2010: On the roles of sulphuric acid and low-volatility organic vapours in the initial steps of atmospheric new particle formation. *Atmos. Chem. Phys.*, **10**, 11223–11242.

Paasonen, P., et al., 2013: Warming-induced increase in aerosol number concentration likely to moderate climate change. *Nature Geosci.*, **6**, 438–442.

Pacifico, F., S. P. Harrison, C. D. Jones, and S. Sitch, 2009: Isoprene emissions and climate. *Atmos. Environ.*, **43**, 6121–6135.

Painemal, D., and P. Zuidema, 2010: Microphysical variability in southeast Pacific Stratocumulus clouds: Synoptic conditions and radiative response. *Atmos. Chem. Phys.*, **10**, 6255–6269.

Pallé Bagó, E., and C. J. Butler, 2000: The influence of cosmic rays on terrestrial clouds and global warming. *Astron. Geophys.*, **41**, 4.18–4.22.

Pallé, E., 2005: Possible satellite perspective effects on the reported correlations between solar activity and clouds. *Geophys. Res. Lett.*, **32**, L03802.

Palm, S. P., S. T. Strey, J. Spinhirne, and T. Markus, 2010: Influence of Arctic sea ice extent on polar cloud fraction and vertical structure and implications for regional climate. *J. Geophys. Res.*, **115**, D21209.

Paltridge, G., A. Arking, and M. Pook, 2009: Trends in middle- and upper-level tropospheric humidity from NCEP reanalysis data. *Theor. Appl. Climatol.*, **98**, 351–359.

Paltridge, G. W., 1980: Cloud-radiation feedback to climate. *Q. J. R. Meteorol. Soc.*, **106**, 895–899.

Paredes-Miranda, G., W. P. Arnott, J. L. Jimenez, A. C. Aiken, J. S. Gaffney, and N. A. Marley, 2009: Primary and secondary contributions to aerosol light scattering and absorption in Mexico City during the MILAGRO 2006 campaign. *Atmos. Chem. Phys.*, **9**, 3721–3730.

Park, S., and C. S. Bretherton, 2009: The University of Washington shallow convection and moist turbulence schemes and their impact on climate simulations with the Community Atmosphere Model. *J. Clim.*, **22**, 3449–3469.

Parodi, A., and K. Emanuel, 2009: A theory for buoyancy and velocity scales in deep moist convection. *J. Atmos. Sci.*, **66**, 3449–3463.

Partanen, A.-I., et al., 2012: Direct and indirect effects of sea spray geoengineering and the role of injected particle size. *J. Geophys. Res.*, **117**, D02203.

Pawlowska, H., and J. L. Brenguier, 2003: An observational study of drizzle formation in stratocumulus clouds for general circulation model (GCM) parameterizations. *J. Geophys. Res.*, **108**, 8630.

Pechony, O., and D. T. Shindell, 2010: Driving forces of global wildfires over the past millennium and the forthcoming century. *Proc. Natl. Acad. Sci. U.S.A.*, **107**, 19167–19170.

Pendergrass, A. G., and D. L. Hartmann, 2012: Global-mean precipitation and black carbon in AR4 simulations. *Geophys. Res. Lett.*, **39**, L01703.

Peng, Y. R., and U. Lohmann, 2003: Sensitivity study of the spectral dispersion of the cloud droplet size distribution on the indirect aerosol effect. *Geophys. Res. Lett.*, **30**, 1507.

Penner, J. E., S. Y. Zhang, and C. C. Chuang, 2003: Soot and smoke aerosol may not warm climate. *J. Geophys. Res.*, **108**, 4657.

Penner, J. E., L. Xu, and M. H. Wang, 2011: Satellite methods underestimate indirect climate forcing by aerosols. *Proc. Natl. Acad. Sci. U.S.A.*, **108**, 13404–13408.

Penner, J. E., C. Zhou, and L. Xu, 2012: Consistent estimates from satellites and models for the first aerosol indirect forcing. *Geophys. Res. Lett.*, **39**, L13810.

Penner, J. E., Y. Chen, M. Wang, and X. Liu, 2009: Possible influence of anthropogenic aerosols on cirrus clouds and anthropogenic forcing. *Atmos. Chem. Phys.*, **9**, 879–896.

Penner, J. E., et al., 2006: Model intercomparison of indirect aerosol effects. *Atmos. Chem. Phys.*, **6**, 3391–3405.

Peñuelas, J., and M. Staudt, 2010: BVOCs and global change. *Trends Plant Sci.*, **15**, 133–144.

Perez, N., J. Pey, X. Querol, A. Alastuey, J. M. Lopez, and M. Viana, 2008: Partitioning of major and trace components in PM10–PM2.5–PM1 at an urban site in Southern Europe. *Atmos. Environ.*, **42**, 1677–1691.

Persad, G. G., Y. Ming, and V. Ramaswamy, 2012: Tropical tropospheric-only responses to absorbing aerosols. *J. Clim.*, **25**, 2471–2480.

Peters, K., J. Quaas, and H. Grassl, 2011a: A search for large-scale effects of ship emissions on clouds and radiation in satellite data. *J. Geophys. Res.*, **116**, D24205.

Peters, K., J. Quaas, and N. Bellouin, 2011b: Effects of absorbing aerosols in cloudy skies: A satellite study over the Atlantic Ocean. *Atmos. Chem. Phys.*, **11**, 1393–1404.

Petroff, A., and L. Zhang, 2010: Development and validation of a size-resolved particle dry deposition scheme for application in aerosol transport models. *Geosci. Model Dev.*, **3**, 753–769.

Petters, M. D., and S. M. Kreidenweis, 2007: A single parameter representation of hygroscopic growth and cloud condensation nucleus activity. *Atmos. Chem. Phys.*, **7**, 1961–1971.

Petters, M. D., J. R. Snider, B. Stevens, G. Vali, I. Faloona, and L. M. Russell, 2006: Accumulation mode aerosol, pockets of open cells, and particle nucleation in the remote subtropical Pacific marine boundary layer. *J. Geophys. Res.*, **111**, D02206.

Phillips, V. T. J., P. J. DeMott, and C. Andronache, 2008: An empirical parameterization of heterogeneous ice nucleation for multiple chemical species of aerosol. *J. Atmos. Sci.*, **65**, 2757–2783.

Pierce, J. R., and P. J. Adams, 2007: Efficiency of cloud condensation nuclei formation from ultrafine particles. *Atmos. Chem. Phys.*, **7**, 1367–1379.

Pierce, J. R., and P. J. Adams, 2009a: Uncertainty in global CCN concentrations from uncertain aerosol nucleation and primary emission rates. *Atmos. Chem. Phys.*, **9**, 1339–1356.

Pierce, J. R., and P. J. Adams, 2009b: Can cosmic rays affect cloud condensation nuclei by altering new particle formation rates? *Geophys. Res. Lett.*, **36**, L09820.

Pierce, J. R., D. K. Weisenstein, P. Heckendorn, T. Peter, and D. W. Keith, 2010: Efficient formation of stratospheric aerosol for climate engineering by emission of condensible vapor from aircraft. *Geophys. Res. Lett.*, **37**, L18805.

Pincus, R., and M. B. Baker, 1994: Effect of precipitation on the albedo susceptibility of clouds in the marine boundary-layer. *Nature*, **372**, 250–252.

Pincus, R., and S. A. Klein, 2000: Unresolved spatial variability and microphysical process rates in large-scale models. *J. Geophys. Res.*, **105**, 27059–27065.

Pincus, R., R. Hemler, and S. A. Klein, 2006: Using stochastically generated subcolumns to represent cloud structure in a large-scale model. *Mon. Weather Rev.*, **134**, 3644–3656.

Platnick, S., M. D. King, S. A. Ackerman, W. P. Menzel, B. A. Baum, J. C. Riedi, and R. A. Frey, 2003: The MODIS cloud products: Algorithms and examples from Terra. *IEEE Trans. Geosci. Remote Sens.*, **41**, 459–473.

Ponater, M., S. Marquart, R. Sausen, and U. Schumann, 2005: On contrail climate sensitivity. *Geophys. Res. Lett.*, **32**, L10706.

Pöschl, U., et al., 2010: Rainforest aerosols as biogenic nuclei of clouds and precipitation in the Amazon. *Science*, **329**, 1513–1516.

Pósfai, M., R. Simonics, J. Li, P. V. Hobbs, and P. R. Buseck, 2003: Individual aerosol particles from biomass burning in southern Africa: 1. Compositions and size distributions of carbonaceous particles. *J. Geophys. Res.*, **108**, 8483.

Posselt, R., and U. Lohmann, 2008: Influence of giant CCN on warm rain processes in the ECHAM5 GCM. *Atmos. Chem. Phys.*, **8**, 3769–3788.

Posselt, R., and U. Lohmann, 2009: Sensitivity of the total anthropogenic aerosol effect to the treatment of rain in a global climate model. *Geophys. Res. Lett.*, **36**, L02805.

Pratt, K. A., and K. A. Prather, 2010: Aircraft measurements of vertical profiles of aerosol mixing states. *J. Geophys. Res.*, **115**, D11305.

Prenni, A. J., et al., 2007: Can ice-nucleating aerosols affect arctic seasonal climate? *Bull. Am. Meteor. Soc.*, **88**, 541–550.

Pringle, K. J., H. Tost, A. Pozzer, U. Pöschl, and J. Lelieveld, 2010: Global distribution of the effective aerosol hygroscopicity parameter for CCN activation. *Atmos. Chem. Phys.*, **10**, 5241–5255.

Prisle, N. L., T. Raatikainen, A. Laaksonen, and M. Bilde, 2010: Surfactants in cloud droplet activation: Mixed organic-inorganic particles. *Atmos. Chem. Phys.*, **10**, 5663–5683.

Prisle, N. L., T. Raatikainen, R. Sorjamaa, B. Svenningsson, A. Laaksonen, and M. Bilde, 2008: Surfactant partitioning in cloud droplet activation: A study of C8, C10, C12 and C14 normal fatty acid sodium salts. *Tellus B*, **60**, 416–431.

Pritchard, M. S., and R. C. J. Somerville, 2010: Assessing the diurnal cycle of precipitation in a multi-scale climate model. *J. Adv. Model. Earth Syst.*, **1**, 12.

Prospero, J. M., W. M. Landing, and M. Schulz, 2010: African dust deposition to Florida: Temporal and spatial variability and comparisons to models. *J. Geophys. Res.*, **115**, D13304.

Puma, M. J., and B. I. Cook, 2010: Effects of irrigation on global climate during the 20th century. *J. Geophys. Res.*, **115**, D16120.

Putaud, J.-P., et al., 2004: European aerosol phenomenology-2: Chemical characteristics of particulate matter at kerbside, urban, rural and background sites in Europe. *Atmos. Environ.*, **38**, 2579–2595.

Putman, W. M., and M. Suarez, 2011: Cloud-system resolving simulations with the NASA Goddard Earth Observing System global atmospheric model (GEOS-5). *Geophys. Res. Lett.*, **38**, L16809.

Puxbaum, H., et al., 2004: A dual site study of PM2.5 and PM10 aerosol chemistry in the larger region of Vienna, Austria. *Atmos. Environ.*, **38**, 3949–3958.

Pye, H. O. T., and J. H. Seinfeld, 2010: A global perspective on aerosol from low-volatility organic compounds. *Atmos. Chem. Phys.*, **10**, 4377–4401.

Pye, H. O. T., H. Liao, S. Wu, L. J. Mickley, D. J. Jacob, D. K. Henze, and J. H. Seinfeld, 2009: Effect of changes in climate and emissions on future sulfate-nitrate-ammonium aerosol levels in the United States. *J. Geophys. Res.*, **114**, D01205.

Qu, W. J., X. Y. Zhang, R. Arimoto, D. Wang, Y. Q. Wang, L. W. Yan, and Y. Li, 2008: Chemical composition of the background aerosol at two sites in southwestern and northwestern China: Potential influences of regional transport. *Tellus B*, **60**, 657–673.

Quaas, J., O. Boucher, and F. M. Bréon, 2004: Aerosol indirect effects in POLDER satellite data and the Laboratoire de Météorologie Dynamique-Zoom (LMDZ) general circulation model. *J. Geophys. Res.*, **109**, D08205.

Quaas, J., O. Boucher, and U. Lohmann, 2006: Constraining the total aerosol indirect effect in the LMDZ and ECHAM4 GCMs using MODIS satellite data. *Atmos. Chem. Phys.*, **6**, 947–955.

Quaas, J., N. Bellouin, and O. Boucher, 2011: Which of satellite-based or model-based estimates are closer to reality for aerosol indirect forcing? Reply to Penner et al. *Proc. Natl. Acad. Sci. U.S.A.*, **108**, E1099.

Quaas, J., O. Boucher, N. Bellouin, and S. Kinne, 2008: Satellite-based estimate of the direct and indirect aerosol climate forcing. *J. Geophys. Res.*, **113**, D05204.

Quaas, J., B. Stevens, P. Stier, and U. Lohmann, 2010: Interpreting the cloud cover – aerosol optical depth relationship found in satellite data using a general circulation model. *Atmos. Chem. Phys.*, **10**, 6129–6135.

Quaas, J., et al., 2009: Aerosol indirect effects—general circulation model intercomparison and evaluation with satellite data. *Atmos. Chem. Phys.*, **9**, 8697–8717.

Querol, X., et al., 2001: PM10 and PM2.5 source apportionment in the Barcelona Metropolitan Area, Catalonia, Spain. *Atmos. Environ.*, **35/36**, 6407–6419.

Querol, X., et al., 2009: Variability in regional background aerosols within the Mediterranean. *Atmos. Chem. Phys.*, **9**, 4575–4591.

Querol, X., et al., 2004: Speciation and origin of PM10 and PM2.5 in selected European cities. *Atmos. Environ.*, **38**, 6547–6555.

Querol, X., et al., 2006: Atmospheric particulate matter in Spain: Levels, composition and source origin. CSIC and Ministerio de Medioambiente, Madrid, Spain, 39 pp.

Querol, X., et al., 2008: Spatial and temporal variations in airborne particulate matter (PM10 and PM2.5) across Spain 1999–2005. *Atmos. Environ.*, **42**, 3964–3979.

Quinn, P. K., and T. S. Bates, 2011: The case against climate regulation via oceanic phytoplankton sulphur emissions. *Nature*, **480**, 51–56.

Racherla, P. N., and P. J. Adams, 2006: Sensitivity of global tropospheric ozone and fine particulate matter concentrations to climate change. *J. Geophys. Res.*, **111**, D24103.

Radhi, M., M. A. Box, G. P. Box, R. M. Mitchell, D. D. Cohen, E. Stelcer, and M. D. Keywood, 2010: Optical, physical and chemical characteristics of Australian continental aerosols: Results from a field experiment. *Atmos. Chem. Phys.*, **10**, 5925–5942.

Rae, J. G. L., C. E. Johnson, N. Bellouin, O. Boucher, J. M. Haywood, and A. Jones, 2007: Sensitivity of global sulphate aerosol production to changes in oxidant concentrations and climate. *J. Geophys. Res.*, **112**, D10312.

Raes, F., H. Liao, W.-T. Chen, and J. H. Seinfeld, 2010: Atmospheric chemistry-climate feedbacks. *J. Geophys. Res.*, **115**, D12121.

Ram, K., M. M. Sarin, and P. Hegde, 2010: Long-term record of aerosol optical properties and chemical composition from a high-altitude site (Manora Peak) in Central Himalaya. *Atmos. Chem. Phys.*, **10**, 11791–11803.

Raman, R. S., S. Ramachandran, and N. Rastogi, 2010: Source identification of ambient aerosols over an urban region in western India. *J. Environ. Monit.*, **12**, 1330–1340.

Ramanathan, V., R. D. Cess, E. F. Harrison, P. Minnis, B. R. Barkstrom, E. Ahmad, and D. Hartmann, 1989: Cloud-radiative forcing and climate: Results from the Earth Radiation Budget Experiment. *Science*, **243**, 57–63.

Randall, D., M. Khairoutdinov, A. Arakawa, and W. Grabowski, 2003: Breaking the cloud parameterization deadlock. *Bull. Am. Meteor. Soc.*, **84**, 1547–1564.

Randall, D. A., et al., 2007: Climate models and their evaluation. In: *Climate Change 2007: The Physical Science Basis. Contribution of Working Group I to the Fourth Assessment Report of the Intergovernmental Panel on Climate Change* [Solomon, S., D. Qin, M. Manning, Z. Chen, M. Marquis, K. B. Averyt, M. Tignor and H. L. Miller (eds.)] Cambridge University Press, Cambridge, United Kingdom and New York, NY, USA, pp. 589–662.

Randerson, J. T., Y. Chen, G. R. van der Werf, B. M. Rogers, and D. C. Morton, 2012: Global burned area and biomass burning emissions from small fires. *J. Geophys. Res.*, **117**, G04012.

Randles, C. A., et al., 2013: Intercomparison of shortwave radiative transfer schemes in global aerosol modeling: Results from the AeroCom Radiative Transfer Experiment. *Atmos. Chem. Phys.*, **13**, 2347–2379.

Rap, A., P. M. Forster, J. M. Haywood, A. Jones, and O. Boucher, 2010a: Estimating the climate impact of linear contrails using the UK Met Office climate model. *Geophys. Res. Lett.*, **37**, L20703.

Rap, A., P. Forster, A. Jones, O. Boucher, J. Haywood, N. Bellouin, and R. De Leon, 2010b: Parameterization of contrails in the UK Met Office Climate Model. *J. Geophys. Res.*, **115**, D10205.

Rasch, P. J., P. J. Crutzen, and D. B. Coleman, 2008a: Exploring the geoengineering of climate using stratospheric sulfate aerosols: The role of particle size. *Geophys. Res. Lett.*, **35**, L02809.

Rasch, P. J., C. C. Chen, and J. L. Latham, 2009: Geo-engineering by cloud seeding: influence on sea-ice and the climate system. *Environ. Res. Lett.*, **4**, 045112.

Rasch, P. J., et al., 2008b: An overview of geoengineering of climate using stratospheric sulphate aerosols. *Philos. Trans. R. Soc. London A*, **366**, 4007–4037.

Rastogi, N., and M. M. Sarin, 2005: Long-term characterization of ionic species in aerosols from urban and high-altitude sites in western India: Role of mineral dust and anthropogenic sources. *Atmos. Environ.*, **39**, 5541–5554.

Rauber, R. M., et al., 2007: Rain in shallow cumulus over the ocean - The RICO campaign. *Bull. Am. Meteor. Soc.*, **88**, 1912–1928.

Raymond, D. J., S. L. Sessions, A. H. Sobel, and Z. Fuchs, 2009: The mechanics of gross moist stability. *J. Adv. Model. Earth Syst.*, **1**, 9.

Reddy, M. S., O. Boucher, Y. Balkanski, and M. Schulz, 2005: Aerosol optical depths and direct radiative perturbations by species and source type. *Geophys. Res. Lett.*, **32**, L12803.

Remer, L. A., et al., 2005: The MODIS aerosol algorithm, products, and validation. *J. Atmos. Sci.*, **62**, 947–973.

Rengarajan, R., M. M. Sarin, and A. K. Sudheer, 2007: Carbonaceous and inorganic species in atmospheric aerosols during wintertime over urban and high-altitude sites in North India. *J. Geophys. Res.*, **112**, D21307.

Richter, I., and S. P. Xie, 2008: Muted precipitation increase in global warming simulations: A surface evaporation perspective. *J. Geophys. Res.*, **113**, D24118.

Richter, J. H., and P. J. Rasch, 2008: Effects of convective momentum transport on the atmospheric circulation in the community atmosphere model, version 3. *J. Clim.*, **21**, 1487–1499.

Ricke, K. L., G. Morgan, and M. R. Allen, 2010: Regional climate response to solar-radiation management. *Nature Geosci.*, **3**, 537–541.

Ridgwell, A., J. S. Singarayer, A. M. Hetherington, and P. J. Valdes, 2009: Tackling regional climate change by leaf albedo bio-geoengineering. *Curr. Biol.*, **19**, 146–150.

Rieck, M., L. Nuijens, and B. Stevens, 2012: Marine boundary-layer cloud feedbacks in a constant relative humidity atmosphere. *J. Atmos. Sci.*, **69**, 2538–2550.

Riipinen, I., et al., 2011: Organic condensation: A vital link connecting aerosol formation to cloud condensation nuclei (CCN) concentrations. *Atmos. Chem. Phys.*, **11**, 3865–3878.

Rinaldi, M., et al., 2009: On the representativeness of coastal aerosol studies to open ocean studies: Mace Head – a case study. *Atmos. Chem. Phys.*, **9**, 9635–9646.

Rinaldi, M., et al., 2011: Evidence of a natural marine source of oxalic acid and a possible link to glyoxal. *J. Geophys. Res.*, **116**, D16204.

Ringer, M. A., et al., 2006: Global mean cloud feedbacks in idealized climate change experiments. *Geophys. Res. Lett.*, **33**, L07718.

Rio, C., and F. Hourdin, 2008: A thermal plume model for the convective boundary layer: Representation of cumulus clouds. *J. Atmos. Sci.*, **65**, 407–425.

Rio, C., F. Hourdin, J.-Y. Grandpeix, and J.-P. Lafore, 2009: Shifting the diurnal cycle of parameterized deep convection over land. *Geophys. Res. Lett.*, **36**, L07809.

Rissler, J., B. Svenningsson, E. O. Fors, M. Bilde, and E. Swietlicki, 2010: An evaluation and comparison of cloud condensation nucleus activity models: Predicting particle critical saturation from growth at subsaturation. *J. Geophys. Res.*, **115**, D22208.

Rissler, J., E. Swietlicki, J. Zhou, G. Roberts, M. O. Andreae, L. V. Gatti, and P. Artaxo, 2004: Physical properties of the sub-micrometer aerosol over the Amazon rain forest during the wet-to-dry season transition—comparison of modeled and measured CCN concentrations. *Atmos. Chem. Phys.*, **4**, 2119–2143.

Roberts, G. C., et al., 2010: Characterization of particle cloud droplet activity and composition in the free troposphere and the boundary layer during INTEX-B. *Atmos. Chem. Phys.*, **10**, 6627–6644.

Robinson, A. L., et al., 2007: Rethinking organic aerosols: Semivolatile emissions and photochemical aging. *Science*, **315**, 1259–1262.

Robock, A., L. Oman, and G. Stenchikov, 2008: Regional climate responses to geoengineering with tropical and Arctic SO_2 injections. *J. Geophys. Res.*, **113**, D16101.

Rodriguez, S., X. Querol, A. Alastuey, and F. Plana, 2002: Sources and processes affecting levels and composition of atmospheric aerosol in the western Mediterranean. *J. Geophys. Res.*, **107**, 4777.

Rodrıguez, S., X. Querol, A. Alastuey, M. M. Viana, M. Alarcon, E. Mantilla, and C. R. Ruiz, 2004: Comparative PM10–PM2.5 source contribution study at rural, urban and industrial sites during PM episodes in Eastern Spain. *Sci. Tot. Environ.*, **328**, 95–113.

Rohs, S., R. Spang, F. Rohrer, C. Schiller, and H. Vos, 2010: A correlation study of high-altitude and midaltitude clouds and galactic cosmic rays by MIPAS-Envisat. *J. Geophys. Res.*, **115**, D14212.

Romps, D. M., 2011: Response of tropical precipitation to global warming. *J. Atmos. Sci.*, **68**, 123–138.

Rondanelli, R., and R. S. Lindzen, 2010: Can thin cirrus clouds in the tropics provide a solution to the faint young Sun paradox? *J. Geophys. Res.*, **115**, D02108.

Roosli, M., et al., 2001: Temporal and spatial variation of the chemical composition of PM10 at urban and rural sites in the Basel area, Switzerland. *Atmos. Environ.*, **35**, 3701–3713.

Rose, D., et al., 2011: Cloud condensation nuclei in polluted air and biomass burning smoke near the mega-city Guangzhou, China -Part 2: Size-resolved aerosol chemical composition, diurnal cycles, and externally mixed weakly CCN-active soot particles. *Atmos. Chem. Phys.*, **11**, 2817–2836.

Rosenfeld, D., and G. Gutman, 1994: Retrieving microphysical properties near the tops of potential rain clouds by multi spectral analysis of AVHRR data. *Atmos. Res.*, **34**, 259–283.

Rosenfeld, D., and W. L. Woodley, 2001: Pollution and clouds. *Physics World*, **14**, 33–37.

Rosenfeld, D., and T. L. Bell, 2011: Why do tornados and hailstorms rest on weekends? *J. Geophys. Res.*, **116**, D20211.

Rosenfeld, D., H. Wang, and P. J. Rasch, 2012: The roles of cloud drop effective radius and LWP in determining rain properties in marine stratocumulus. *Geophys. Res. Lett.*, **39**, L13801.

Rosenfeld, D., et al., 2008: Flood or drought: How do aerosols affect precipitation? *Science*, **321**, 1309–1313.

Rotstayn, L. D., 1999: Indirect forcing by anthropogenic aerosols: A global climate model calculation of the effective-radius and cloud-lifetime effects. *J. Geophys. Res.*, **104**, 9369–9380.

Rotstayn, L. D., 2000: On the "tuning" of autoconversion parameterizations in climate models. *J. Geophys. Res.*, **105**, 15495–15508.

Rotstayn, L. D., and J. E. Penner, 2001: Indirect aerosol forcing, quasi forcing, and climate response. *J. Clim.*, **14**, 2960–2975.

Rotstayn, L. D., and Y. G. Liu, 2005: A smaller global estimate of the second indirect aerosol effect. *Geophys. Res. Lett.*, **32**, L05708.

Rotstayn, L. D., and Y. G. Liu, 2009: Cloud droplet spectral dispersion and the indirect aerosol effect: Comparison of two treatments in a GCM. *Geophys. Res. Lett.*, **36**, L10801.

Rotstayn, L. D., et al., 2007: Have Australian rainfall and cloudiness increased due to the remote effects of Asian anthropogenic aerosols? *J. Geophys. Res.*, **112**, D09202.

Royal Society, 2009: Geoengineering the climate, Science, governance and uncertainty. Report 10/09, Royal Society, London, United Kingdom, 82 pp.

Russell, L. M., L. N. Hawkins, A. A. Frossard, P. K. Quinn, and T. S. Bates, 2010: Carbohydrate-like composition of submicron atmospheric particles and their production from ocean bubble bursting. *Proc. Natl. Acad. Sci. U.S.A.*, **107**, 6652–6657.

Rypdal, K., N. Rive, T. K. Berntsen, Z. Klimont, T. K. Mideksa, G. Myhre, and R. B. Skeie, 2009: Costs and global impacts of black carbon abatement strategies. *Tellus B*, **61**, 625–641.

Sacks, W. J., B. I. Cook, N. Buenning, S. Levis, and J. H. Helkowski, 2009: Effects of global irrigation on the near-surface climate. *Clim. Dyn.*, **33**, 159–175.

Safai, P. D., K. B. Budhavant, P. S. P. Rao, K. Ali, and A. Sinha, 2010: Source characterization for aerosol constituents and changing roles of calcium and ammonium aerosols in the neutralization of aerosol acidity at a semi-urban site in SW India. *Atmos. Res.*, **98**, 78–88.

Sahu, L. K., Y. Kondo, Y. Miyazaki, P. Pongkiatkul, and N. T. Kim Oanh, 2011: Seasonal and diurnal variations of black carbon and organic carbon aerosols in Bangkok. *J. Geophys. Res.*, **116**, D15302.

Sakaeda, N., R. Wood, and P. J. Rasch, 2011: Direct and semidirect aerosol effects of southern African biomass burning aerosol. *J. Geophys. Res.*, **116**, D12205.

Salter, S., G. Sortino, and J. Latham, 2008: Sea-going hardware for the cloud albedo method of reversing global warming. *Philos. Trans. R. Soc. London A*, **366**, 3989–4006.

Salvador, P., B. Artınano, X. Querol, A. Alastuey, and M. Costoya, 2007: Characterisation of local and external contributions of atmospheric particulate matter at a background coastal site. *Atmos. Environ.*, **41**, 1–17.

Salzmann, M., et al., 2010: Two-moment bulk stratiform cloud microphysics in the GFDL AM3 GCM: Description, evaluation, and sensitivity tests. *Atmos. Chem. Phys.*, **10**, 8037–8064.

Samset, B. H., et al., 2013: Black carbon vertical profiles strongly affect its radiative forcing uncertainty. *Atmos. Chem. Phys.*, **13**, 2423–2434.

Sanchez-Lorenzo, A., P. Laux, H. J. Hendricks Franssen, J. Calbó, S. Vogl, A. K. Georgoulias, and J. Quaas, 2012: Assessing large-scale weekly cycles in meteorological variables: A review. *Atmos. Chem. Phys.*, **12**, 5755–5771.

Sanderson, M. G., C. D. Jones, W. J. Collins, C. E. Johnson, and R. G. Derwent, 2003: Effect of climate change on isoprene emissions and surface ozone levels. *Geophys. Res. Lett.*, **30**, 1936.

Sandu, I., J.-L. Brenguier, O. Geoffroy, O. Thouron, and V. Masson, 2008: Aerosol impacts on the diurnal cycle of marine stratocumulus. *J. Atmos. Sci.*, **65**, 2705–2718.

Sassen, K., and G. C. Dodd, 1988: Homogeneous nucleation rate for highly supercooled cirrus cloud droplets. *J. Atmos. Sci.*, **45**, 1357–1369.

Satheesh, S. K., and K. K. Moorthy, 2005: Radiative effects of natural aerosols: A review. *Atmos. Environ.*, **39**, 2089–2110.

Satheesh, S. K., K. K. Moorthy, S. S. Babu, V. Vinoj, and C. B. S. Dutt, 2008: Climate implications of large warming by elevated aerosol over India. *Geophys. Res. Lett.*, **35**, L19809.

Sato, T., H. Miura, M. Satoh, Y. N. Takayabu, and Y. Q. Wang, 2009: Diurnal cycle of precipitation in the Tropics simulated in a global cloud-resolving model. *J. Clim.*, **22**, 4809–4826.

Satoh, M., T. Inoue, and H. Miura, 2010: Evaluations of cloud properties of global and local cloud system resolving models using CALIPSO and CloudSat simulators. *J. Geophys. Res.*, **115**, D00H14.

Sausen, R., et al., 2005: Aviation radiative forcing in 2000: An update on IPCC (1999). *Meteorol. Z.*, **14**, 555–561.

Savic-Jovcic, V., and B. Stevens, 2008: The structure and mesoscale organization of precipitating stratocumulus. *J. Atmos. Sci.*, **65**, 1587–1605.

Sawant, A. A., K. Na, X. Zhu, and D. R. Cocker III, 2004: Chemical characterization of outdoor PM2.5 and gas-phase compounds in Mira Loma, California. *Atmos. Environ.*, **38**, 5517–5528.

Schmidt, A., et al., 2012a: Importance of tropospheric volcanic aerosol for indirect radiative forcing of climate. *Atmos. Chem. Phys.*, **12**, 7321–7339.

Schmidt, H., et al., 2012b: Solar irradiance reduction to counteract radiative forcing from a quadrupling of CO_2: Climate responses simulated by four earth system models. *Earth Syst. Dyn.*, **3**, 63–78.

Schmidt, K. S., G. Feingold, P. Pilewskie, H. Jiang, O. Coddington, and M. Wendisch, 2009: Irradiance in polluted cumulus fields: Measured and modeled cloud-aerosol effects. *Geophys. Res. Lett.*, **36**, L07804.

Schreier, M., H. Mannstein, V. Eyring, and H. Bovensmann, 2007: Global ship track distribution and radiative forcing from 1 year of AATSR data. *Geophys. Res. Lett.*, **34**, L17814.

Schulz, M., M. Chin, and S. Kinne, 2009: The Aerosol Model Comparison Project, AeroCom, phase II: Clearing up diversity. *IGAC Newsletter N° 41*, 2–11.

Schulz, M., et al., 2006: Radiative forcing by aerosols as derived from the AeroCom present-day and pre-industrial simulations. *Atmos. Chem. Phys.*, **6**, 5225–5246.

Schumann, U., and K. Graf, 2013: Aviation-induced cirrus and radiation changes at diurnal timescales. *J. Geophys. Res.*, **118**, 2404–2421.

Schwartz, S. E., and J. E. Freiberg, 1981: Mass-transport limitation to the rate of reaction of gases in liquid droplets - Application to oxidation of SO_2 in aqueous solutions. *Atmos. Environ.*, **15**, 1129–1144.

Schwarz, J. P., et al., 2008a: Coatings and their enhancement of black carbon light absorption in the tropical atmosphere. *J. Geophys. Res.*, **113**, D03203.

Schwarz, J. P., et al., 2008b: Measurement of the mixing state, mass, and optical size of individual black carbon particles in urban and biomass burning emissions. *Geophys. Res. Lett.*, **35**, L13810.

Schwarz, J. P., et al., 2010: Global-scale black carbon profiles observed in the remote atmosphere and compared to models. *Geophys. Res. Lett.*, **37**, L18812.

Schwarzenbock, A., S. Mertes, J. Heintzenberg, W. Wobrock, and P. Laj, 2001: Impact of the Bergeron-Findeisen process on the release of aerosol particles during the evolution of cloud ice. *Atmos. Res.*, **58**, 295–313.

Sciare, J., O. Favez, K. Oikonomou, R. Sarda-Estève, H. Cachier, and V. Kazan, 2009: Long-term observation of carbonaceous aerosols in the Austral Ocean: Evidence of a marine biogenic origin. *J. Geophys. Res.*, **114**, D15302.

Seager, R., N. Naik, and G. A. Vecchi, 2010: Thermodynamic and dynamic mechanisms for large-scale changes in the hydrological cycle in response to global warming. *J. Clim.*, **23**, 4651–4668.

Seifert, A., and K. Beheng, 2006: A two-moment cloud microphysics parameterization for mixed-phase clouds, Part II: Maritime versus continental deep convective storms. *Meteorol. Atmos. Phys.*, **92**, 67–88.

Seifert, A., and B. Stevens, 2010: Microphysical scaling relations in a kinematic model of isolated shallow cumulus clouds. *J. Atmos. Sci.*, **67**, 1575–1590.

Seifert, A., and G. Zängl, 2010: Scaling relations in warm-rain orographic precipitation. *Meteorol. Z.*, **19**, 417–426.

Seifert, A., C. Köhler, and K. D. Beheng, 2012: Aerosol-cloud-precipitation effects over Germany as simulated by a convective-scale numerical weather prediction model. *Atmos. Chem. Phys.*, **12**, 709–725.

Seitz, R., 2011: Bright water: Hydrosols, water conservation and climate change. *Clim. Change*, **105**, 365–381.

Sekiguchi, M., et al., 2003: A study of the direct and indirect effects of aerosols using global satellite data sets of aerosol and cloud parameters. *J. Geophys. Res.*, **108**, 4699.

Seland, Ø., T. Iversen, A. Kirkevåg, and T. Storelvmo, 2008: Aerosol-climate interactions in the CAM-Oslo atmospheric GCM and investigation of associated basic shortcomings. *Tellus A*, **60**, 459–491.

Senior, C. A., and J. F. B. Mitchell, 1993: Carbon dioxide and climate: The impact of cloud parameterization. *J. Clim.*, **6**, 393–418.

Senior, C. A., and J. F. B. Mitchell, 2000: The time-dependence of climate sensitivity. *Geophys. Res. Lett.*, **27**, 2685–2688.

Sesartic, A., U. Lohmann, and T. Storelvmo, 2012: Bacteria in the ECHAM5–HAM global climate model. *Atmos. Chem. Phys.*, **12**, 8645–8661.

Shapiro, E. L., J. Szprengiel, N. Sareen, C. N. Jen, M. R. Giordano, and V. F. McNeill, 2009: Light-absorbing secondary organic material formed by glyoxal in aqueous aerosol mimics. *Atmos. Chem. Phys.*, **9**, 2289–2300.

Sharon, T. M., et al., 2006: Aerosol and cloud microphysical characteristics of rifts and gradients in maritime stratocumulus clouds. *J. Atmos. Sci.*, **63**, 983–997.

Sherwood, S. C., 2002: Aerosols and ice particle size in tropical cumulonimbus. *J. Clim.*, **15**, 1051–1063.

Sherwood, S. C., and C. L. Meyer, 2006: The general circulation and robust relative humidity. *J. Clim.*, **19**, 6278–6290.

Sherwood, S. C., R. Roca, T. M. Weckwerth, and N. G. Andronova, 2010a: Tropospheric water vapor, convection, and climate. *Rev. Geophys.*, **48**, RG2001.

Sherwood, S. C., V. Ramanathan, T. P. Barnett, M. K. Tyree, and E. Roeckner, 1994: Response of an atmospheric general circulation model to radiative forcing of tropical clouds. *J. Geophys. Res.*, **99**, 20829–20845.

Sherwood, S. C., W. Ingram, Y. Tsushima, M. Satoh, M. Roberts, P. L. Vidale, and P. A. O'Gorman, 2010b: Relative humidity changes in a warmer climate. *J. Geophys. Res.*, **115**, D09104.

Shindell, D. T., A. Voulgarakis, G. Faluvegi, and G. Milly, 2012: Precipitation response to regional radiative forcing. *Atmos. Chem. Phys.*, **12**, 6969–6982.

Shindell, D. T., et al., 2013: Radiative forcing in the ACCMIP historical and future climate simulations. *Atmos. Chem. Phys.*, **13**, 2939–2974.

Shiraiwa, M., Y. Kondo, T. Iwamoto, and K. Kita, 2010: Amplification of light absorption of black carbon by organic coating. *Aer. Sci. Technol.*, **44**, 46–54.

Shonk, J. K. P., R. J. Hogan, and J. Manners, 2012: Impact of improved representation of horizontal and vertical cloud structure in a climate model. *Clim. Dyn.*, **38**, 2365–2376.

Shresth, A. B., C. P. Wake, J. E. Dibb, P. A. Mayewski, S. I. Whitlow, G. R. Carmichael, and M. Ferm, 2000: Seasonal variations in aerosol concentrations and compositions in the Nepal Himalaya. *Atmos. Environ.*, **34**, 3349–3363.

Shrivastava, M., et al., 2011: Modeling organic aerosols in a megacity: Comparison of simple and complex representations of the volatility basis set approach. *Atmos. Chem. Phys.*, **11**, 6639–6662.

Shupe, M. D., et al., 2008: A focus on mixed-phase clouds. The status of ground-based observational methods. *Bull. Am. Meteor. Soc.*, **89**, 1549–1562.

Siebesma, A. P., P. M. M. Soares, and J. Teixeira, 2007: A combined eddy-diffusivity mass-flux approach for the convective boundary layer. *J. Atmos. Sci.*, **64**, 1230–1248.

Siebesma, A. P., et al., 2003: A large eddy simulation intercomparison study of shallow cumulus convection. *J. Atmos. Sci.*, **60**, 1201–1219.

Siebesma, A. P., et al., 2009: Cloud-controlling factors. In: *Clouds in the Perturbed Climate System: Their Relationship to Energy Balance, Atmospheric Dynamics, and Precipitation* [J. Heintzenberg and R. J. Charlson (eds.)]. MIT Press, Cambridge, MA, USA, pp. 269–290.

Singh, M. S., and P. A. O'Gorman, 2012: Upward shift of the atmospheric general circulation under global warming: Theory and simulations. *J. Clim.*, **25**, 8259–8276.

Sipilä, M., et al., 2010: The role of sulfuric acid in atmospheric nucleation. *Science*, **327**, 1243–1246.

Skeie, R. B., T. Berntsen, G. Myhre, C. A. Pedersen, J. Ström, S. Gerland, and J. A. Ogren, 2011: Black carbon in the atmosphere and snow, from pre-industrial times until present. *Atmos. Chem. Phys.*, **11**, 6809–6836.

Small, J. D., P. Y. Chuang, G. Feingold, and H. Jiang, 2009: Can aerosol decrease cloud lifetime? *Geophys. Res. Lett.*, **36**, L16806.

Smith, J. N., et al., 2010: Observations of aminium salts in atmospheric nanoparticles and possible climatic implications. *Proc. Natl. Acad. Sci. U.S.A.*, **107**, 6634–6639.

Smith, S. J., and P. J. Rasch, 2012: The long-term policy context for solar radiation management. *Clim. Change*, doi: 10.1007/s10584-012-0577-3.

Snow-Kropla, E. J., J. R. Pierce, D. M. Westervelt, and W. Trivitayanurak, 2011: Cosmic rays, aerosol formation and cloud-condensation nuclei: Sensitivities to model uncertainties. *Atmos. Chem. Phys.*, **11**, 4001–4013.

Soden, B. J., and I. M. Held, 2006: An assessment of climate feedbacks in coupled ocean-atmosphere models. *J. Clim.*, **19**, 3354–3360.

Soden, B. J., and G. A. Vecchi, 2011: The vertical distribution of cloud feedback in coupled ocean-atmosphere models. *Geophys. Res. Lett.*, **38**, L12704.

Soden, B. J., I. M. Held, R. Colman, K. M. Shell, J. T. Kiehl, and C. A. Shields, 2008: Quantifying climate feedbacks using radiative kernels. *J. Clim.*, **21**, 3504–3520.

Sofiev, M., et al., 2009: An operational system for the assimilation of the satellite information on wild-land fires for the needs of air quality modelling and forecasting. *Atmos. Chem. Phys.*, **9**, 6833–6847.

Sohn, B. J., T. Nakajima, M. Satoh, and H. S. Jang, 2010: Impact of different definitions of clear-sky flux on the determination of longwave cloud radiative forcing: NICAM simulation results. *Atmos. Chem. Phys.*, **10**, 11641–11646.

Solomon, S., K. H. Rosenlof, R. W. Portmann, J. S. Daniel, S. M. Davis, T. J. Sanford, and G.-K. Plattner, 2010: Contributions of stratospheric water vapor to decadal changes in the rate of global warming. *Science*, **327**, 1219–1223.

Somerville, R. C. J., and L. A. Remer, 1984: Cloud optical-thickness feedbacks in the CO_2 climate problem. *J. Geophys. Res.*, **89**, 9668–9672.

Sommeria, G., and J. W. Deardorff, 1977: Subgrid-scale condensation in models of nonprecipitating clouds. *J. Atmos. Sci.*, **34**, 344–355.

Song, X., and G. J. Zhang, 2011: Microphysics parameterization for convective clouds in a global climate model: Description and single-column model tests. *J. Geophys. Res.*, **116**, D02201.

Song, X. L., and G. J. Zhang, 2009: Convection parameterization, tropical Pacific double ITCZ, and upper-ocean biases in the NCAR CCSM3. Part I: Climatology and atmospheric feedback. *J. Clim.*, **22**, 4299–4315.

Sorooshian, A., N. L. Ng, A. W. H. Chan, G. Feingold, R. C. Flagan, and J. H. Seinfeld, 2007: Particulate organic acids and overall water-soluble aerosol composition measurements from the 2006 Gulf of Mexico Atmospheric Composition and Climate Study (GoMACCS). *J. Geophys. Res.*, **112**, D13201.

Spencer, R. W., and W. D. Braswell, 2008: Potential biases in feedback diagnosis from observational data: A simple model demonstration. *J. Clim.*, **21**, 5624–5628.

Spencer, R. W., and W. D. Braswell, 2010: On the diagnosis of radiative feedback in the presence of unknown radiative forcing. *J. Geophys. Res.*, **115**, D16109.

Spracklen, D. V., L. J. Mickley, J. A. Logan, R. C. Hudman, R. Yevich, M. D. Flannigan, and A. L. Westerling, 2009: Impacts of climate change from 2000 to 2050 on wildfire activity and carbonaceous aerosol concentrations in the western United States. *J. Geophys. Res.*, **114**, D20301.

Spracklen, D. V., et al., 2008: Contribution of particle formation to global cloud condensation nuclei concentrations. *Geophys. Res. Lett.*, **35**, L06808.

Spracklen, D. V., et al., 2011: Aerosol mass spectrometer constraint on the global secondary organic aerosol budget. *Atmos. Chem. Phys.*, **11**, 12109–12136.

Stan, C., et al., 2010: An ocean-atmosphere climate simulation with an embedded cloud resolving model. *Geophys. Res. Lett.*, **37**, L01702.

Stephens, G. L., M. Wild, P. W. Stackhouse, T. L'Ecuyer, S. Kato, and D. S. Henderson, 2012: The global character of the flux of downward longwave radiation. *J. Clim.*, **25**, 2329–2340.

Stephens, G. L., et al., 2002: The Cloudsat mission and the A-train—A new dimension of space-based observations of clouds and precipitation. *Bull. Am. Meteor. Soc.*, **83**, 1771–1790.

Stephens, G. L., et al., 2008: CloudSat mission: Performance and early science after the first year of operation. *J. Geophys. Res.*, **113**, D00A18.

Stephens, G. L., et al., 2010: Dreary state of precipitation in global models. *J. Geophys. Res.*, **115**, D24211.

Stevens, B., and A. Seifert, 2008: Understanding macrophysical outcomes of microphysical choices in simulations of shallow cumulus convection. *J. Meteorol. Soc. Jpn.*, **86**, 143–162.

Stevens, B., and G. Feingold, 2009: Untangling aerosol effects on clouds and precipitation in a buffered system. *Nature*, **461**, 607–613.

Stevens, B., and J.-L. Brenguier, 2009: Cloud-controlling factors: Low clouds. In: *Clouds in the Perturbed Climate System: Their Relationship to Energy Balance, Atmospheric Dynamics, and Precipitation* [J. Heintzenberg and R. J. Charlson (eds.)]. MIT Press, Cambridge, MA, USA, pp. 173–196.

Stevens, B., W. R. Cotton, G. Feingold, and C.-H. Moeng, 1998: Large-eddy simulations of strongly precipitating, shallow, stratocumulus-topped boundary layers. *J. Atmos. Sci.*, **55**, 3616–3638.

Stevens, B., et al., 2005a: Pockets of open cells and drizzle in marine stratocumulus. *Bull. Am. Meteor. Soc.*, **86**, 51–57.

Stevens, B., et al., 2005b: Evaluation of large-eddy simulations via observations of nocturnal marine stratocumulus. *Mon. Weather Rev.*, **133**, 1443–1462.

Stier, P., J. H. Seinfeld, S. Kinne, and O. Boucher, 2007: Aerosol absorption and radiative forcing. *Atmos. Chem. Phys.*, **7**, 5237–5261.

Stier, P., J. H. Seinfeld, S. Kinne, J. Feichter, and O. Boucher, 2006: Impact of nonabsorbing anthropogenic aerosols on clear-sky atmospheric absorption. *J. Geophys. Res.*, **111**, D18201.

Stier, P., et al., 2005: The Aerosol-Climate Model ECHAM5–HAM. *Atmos. Chem. Phys.*, **5**, 1125–1156.

Stier, P., et al., 2013: Host model uncertainties in aerosol forcing estimates: Results from the AeroCom Prescribed intercomparison study. *Atmos. Chem. Phys.*, **13**, 3245–3270.

Stjern, C. W., 2011: Weekly cycles in precipitation and other meteorological variables in a polluted region of Europe. *Atmos. Chem. Phys.*, **11**, 4095–4104.

Stone, E., J. Schauer, T. A. Quraishi, and A. Mahmood, 2010: Chemical characterization and source apportionment of fine and coarse particulate matter in Lahore, Pakistan. *Atmos. Environ.*, **44**, 1062–1070.

Stone, R. S., et al., 2008: Radiative impact of boreal smoke in the Arctic: Observed and modeled. *J. Geophys. Res.*, **113**, D14S16.

Storelvmo, T., J. E. Kristjánsson, and U. Lohmann, 2008a: Aerosol influence on mixed-phase clouds in CAM-Oslo. *J. Atmos. Sci.*, **65**, 3214–3230.

Storelvmo, T., C. Hoose, and P. Eriksson, 2011: Global modeling of mixed-phase clouds: The albedo and lifetime effects of aerosols. *J. Geophys. Res.*, **116**, D05207.

Storelvmo, T., J. E. Kristjánsson, S. J. Ghan, A. Kirkevåg, Ø. Seland, and T. Iversen, 2006: Predicting cloud droplet number concentration in Community Atmosphere Model (CAM)-Oslo. *J. Geophys. Res.*, **111**, D24208.

Storelvmo, T., J. E. Kristjánsson, U. Lohmann, T. Iversen, A. Kirkevåg, and Ø. Seland, 2008b: Modeling of the Wegener–Bergeron–Findeisen process–implications for aerosol indirect effects. *Environ. Res. Lett.*, **3**, 045001.

Storelvmo, T., J. E. Kristjánsson, U. Lohmann, T. Iversen, A. Kirkevåg, and Ø. Seland, 2010: Corrigendum: Modeling of the Wegener–Bergeron–Findeisen process—implications for aerosol indirect effects. *Environ. Res. Lett.*, **5**, 019801

Storelvmo, T., J. E. Kristjánsson, H. Muri, M. Pfeffer, D. Barahona, and A. Nenes, 2013: Cirrus cloud seeding has potential to cool climate *Geophys. Res. Lett.*, **40**, 178–182.

Storer, R. L., and S. C. van den Heever, 2013: Microphysical processes evident in aerosol forcing of tropical deep convective clouds. *J. Atmos. Sci.*, **70**, 430–446.

Stramler, K., A. D. Del Genio, and W. B. Rossow, 2011: Synoptically driven Arctic winter states. *J. Clim.*, **24**, 1747–1762.

Stratmann, F., O. Moehler, R. Shaw, and W. Heike, 2009: Laboratory cloud simulation: Capabilities and future directions. In: *Clouds in the Perturbed Climate System: Their Relationship to Energy Balance, Atmospheric Dynamics, and Precipitation* [J. Heintzenberg and R. J. Charlson (eds.)]. MIT Press, Cambridge, MA, USA, pp. 149–172.

Struthers, H., et al., 2011: The effect of sea ice loss on sea salt aerosol concentrations and the radiative balance in the Arctic. *Atmos. Chem. Phys.*, **11**, 3459–3477.

Stubenrauch, C. J., S. Cros, A. Guignard, and N. Lamquin, 2010: A 6–year global cloud climatology from the Atmospheric InfraRed Sounder AIRS and a statistical analysis in synergy with CALIPSO and CloudSat. *Atmos. Chem. Phys.*, **10**, 7197–7214.

Stubenrauch, C. J., et al., 2013: Assessment of global cloud datasets from satellites: Project and database initiated by the GEWEX Radiation Panel. *Bull. Am. Meteor. Soc.*, **94**, 1031–1049.

Stuber, N., and P. Forster, 2007: The impact of diurnal variations of air traffic on contrail radiative forcing. *Atmos. Chem. Phys.*, **7**, 3153–3162.

Su, W., N. G. Loeb, G. L. Schuster, M. Chin, and F. G. Rose, 2013: Global all-sky shortwave direct radiative forcing of anthropogenic aerosols from combined satellite observations and GOCART simulations. *J. Geophys. Res.*, **118**, 655–669.

Su, W. Y., et al., 2008: Aerosol and cloud interaction observed from high spectral resolution lidar data. *J. Geophys. Res.*, **113**, D24202.

Sugiyama, M., H. Shiogama, and S. Emori, 2010: Precipitation extreme changes exceeding moisture content increases in MIROC and IPCC climate models. *Proc. Natl. Acad. Sci. U.S.A.*, **107**, 571–575.

Suzuki, K., G. L. Stephens, S. C. van den Heever, and T. Y. Nakajima, 2011: Diagnosis of the warm rain process in cloud-resolving models using joint CloudSat and MODIS observations. *J. Atmos. Sci.*, **68**, 2655–2670.

Suzuki, K., T. Nakajima, M. Satoh, H. Tomita, T. Takemura, T. Y. Nakajima, and G. L. Stephens, 2008: Global cloud-system-resolving simulation of aerosol effect on warm clouds. *Geophys. Res. Lett.*, **35**, L19817.

Svensmark, H., T. Bondo, and J. Svensmark, 2009: Cosmic ray decreases affect atmospheric aerosols and clouds. *Geophys. Res. Lett.*, **36**, L15101.

Tackett, J. L., and L. Di Girolamo, 2009: Enhanced aerosol backscatter adjacent to tropical trade wind clouds revealed by satellite-based lidar. *Geophys. Res. Lett.*, **36**, L14804.

Takahashi, K., 2009: The global hydrological cycle and atmospheric shortwave absorption in climate models under CO_2 forcing. *J. Clim.*, **22**, 5667–5675.

Takemura, T., 2012: Distributions and climate effects of atmospheric aerosols from the preindustrial era to 2100 along Representative Concentration Pathways (RCPs) simulated using the global aerosol model SPRINTARS. *Atmos. Chem. Phys.*, **12**, 11555–11572.

Takemura, T., and T. Uchida, 2011: Global climate modeling of regional changes in cloud, precipitation, and radiation budget due to the aerosol semi-direct effect of black carbon. *Sola*, **7**, 181–184.

Takemura, T., T. Nozawa, S. Emori, T. Y. Nakajima, and T. Nakajima, 2005: Simulation of climate response to aerosol direct and indirect effects with aerosol transport-radiation model. *J. Geophys. Res.*, **110**, D02202.

Tanré, D., M. Herman, and Y. J. Kaufman, 1996: Information on aerosol size distribution contained in solar reflected spectral radiances. *J. Geophys. Res.*, **101**, 19043–19060.

Tanré, D., Y. J. Kaufman, M. Herman, and S. Mattoo, 1997: Remote sensing of aerosol properties over oceans using the MODIS/EOS spectral radiances. *J. Geophys. Res.*, **102**, 16971–16988.

Tanré, D., et al., 2011: Remote sensing of aerosols by using polarized, directional and spectral measurements within the A-Train: The PARASOL mission. *Atmos. Meas. Tech.*, **4**, 1383–1395.

Tao, W.-K., J.-P. Chen, Z. Li, C. Wang, and C. Zhang, 2012: Impact of aerosols on convective clouds and precipitation. *Rev. Geophys.*, **50**, RG2001.

Tao, W.-K., et al., 2009: A Multiscale Modeling System: Developments, applications, and critical issues. *Bull. Am. Meteor. Soc.*, **90**, 515–534.

Tegen, I., M. Werner, S. P. Harrison, and K. E. Kohfeld, 2004: Relative importance of climate and land use in determining present and future global soil dust emission. *Geophys. Res. Lett.*, **31**, L05105.

Theodosi, C., U. Im, A. Bougiatioti, P. Zarmpas, O. Yenigun, and N. Mihalopoulos, 2010: Aerosol chemical composition over Istanbul. *Sci. Tot. Environ.*, **408**, 2482–2491.

Thomas, G. E., et al., 2010: Validation of the GRAPE single view aerosol retrieval for ATSR-2 and insights into the long term global AOD trend over the ocean. *Atmos. Chem. Phys.*, **10**, 4849–4866.

Tilmes, S., R. Müller, and R. Salawitch, 2008: The sensitivity of polar ozone depletion to proposed geoengineering schemes. *Science*, **320**, 1201–1204.

Tilmes, S., R. R. Garcia, E. D. Kinnison, A. Gettelman, and P. J. Rasch, 2009: Impact of geo-engineered aerosols on troposphere and stratosphere. *J. Geophys. Res.*, **114**, D12305.

Tilmes, S., et al., 2012: Impact of very short-lived halogens on stratospheric ozone abundance and UV radiation in a geo-engineered atmosphere. *Atmos. Chem. Phys.*, **12**, 10945–10955.

Tinsley, B. A., 2008: The global atmospheric electric circuit and its effects on cloud microphysics. *Rep. Prog. Phys.*, **71**, 066801.

Tobin, I., S. Bony, and R. Roca, 2012: Observational evidence for relationships between the degree of aggregation of deep convection, water vapor, surface fluxes, and radiation. *J. Clim.*, **25**, 6885–6904.

Tomassini, L., et al., 2013: The respective roles of surface temperature driven feedbacks and tropospheric adjustment to CO_2 in CMIP5 transient climate simulations. *Clim. Dyn.*, doi:10.1007/s00382-013-1682-3.

Tomita, H., H. Miura, S. Iga, T. Nasuno, and M. Satoh, 2005: A global cloud-resolving simulation: Preliminary results from an aqua planet experiment. *Geophys. Res. Lett.*, **32**, L08805.

Tompkins, A. M., and G. C. Craig, 1998: Radiative–convective equilibrium in a three-dimensional cloud-ensemble model. *Q. J. R. Meteorol. Soc.*, **124**, 2073–2097.

Tompkins, A. M., K. Gierens, and G. Radel, 2007: Ice supersaturation in the ECMWF integrated forecast system. *Q. J. R. Meteorol. Soc.*, **133**, 53–63.

Torres, O., P. K. Bhartia, J. R. Herman, A. Sinyuk, P. Ginoux, and B. Holben, 2002: A long-term record of aerosol optical depth from TOMS observations and comparison to AERONET measurements. *J. Atmos. Sci.*, **59**, 398–413.

Torres, O., et al., 2007: Aerosols and surface UV products from Ozone Monitoring Instrument observations: An overview. *J. Geophys. Res.*, **112**, D24S47.

Trenberth, K. E., and A. Dai, 2007: Effects of Mount Pinatubo volcanic eruption on the hydrological cycle as an analog of geoengineering. *Geophys. Res. Lett.*, **34**, L15702.

Trenberth, K. E., and J. T. Fasullo, 2009: Global warming due to increasing absorbed solar radiation. *Geophys. Res. Lett.*, **36**, L07706.

Trenberth, K. E., and J. T. Fasullo, 2010: Simulation of present-day and twenty-first-century energy budgets of the Southern Oceans. *J. Clim.*, **23**, 440–454.

Trenberth, K. E., et al., 2007: Observations: Surface and atmospheric climate change. In: *Climate Change 2007: The Physical Science Basis. Contribution of Working Group I to the Fourth Assessment Report of the Intergovernmental Panel on Climate Change* [Solomon, S., D. Qin, M. Manning, Z. Chen, M. Marquis, K. B. Averyt, M. Tignor and H. L. Miller (eds.)] Cambridge University Press, Cambridge, United Kingdom and New York, NY, USA, pp. 235–336.

Tselioudis, G., and W. B. Rossow, 2006: Climate feedback implied by observed radiation and precipitation changes with midlatitude storm strength and frequency. *Geophys. Res. Lett.*, **33**, L02704.

Tsigaridis, K., and M. Kanakidou, 2007: Secondary organic aerosol importance in the future atmosphere. *Atmos. Environ.*, **41**, 4682–4692.

Tsushima, Y., et al., 2006: Importance of the mixed-phase cloud distribution in the control climatefor assessing the response of clouds to carbon dioxide increase:A multi-model study. *Clim. Dyn.*, **27**, 113–126.

Tsimpidi, A. P., et al., 2010: Evaluation of the volatility basis-set approach for the simulation of organic aerosol formation in the Mexico City metropolitan area. *Atmos. Chem. Phys.*, **10**, 525–546.

Tuttle, J. D., and R. E. Carbone, 2011: Inferences of weekly cycles in summertime rainfall. *J. Geophys. Res.*, **116**, D20213.

Twohy, C. H., and J. R. Anderson, 2008: Droplet nuclei in non-precipitating clouds: Composition and size matter. *Environ. Res. Lett.*, **3**, 045002.

Twohy, C. H., J. A. Coakley, Jr., and W. R. Tahnk, 2009: Effect of changes in relative humidity on aerosol scattering near clouds. *J. Geophys. Res.*, **114**, D05205.

Twohy, C. H., et al., 2005: Evaluation of the aerosol indirect effect in marine stratocumulus clouds: Droplet number, size, liquid water path, and radiative impact. *J. Geophys. Res.*, **110**, D08203.

Twomey, S., 1977: Influence of pollution on shortwave albedo of clouds. *J. Atmos. Sci.*, **34**, 1149–1152.

Udelhofen, P. M., and R. D. Cess, 2001: Cloud cover variations over the United States: An influence of cosmic rays or solar variability? *Geophys. Res. Lett.*, **28**, 2617–2620.

Ulbrich, I. M., M. R. Canagaratna, Q. Zhang, D. R. Worsnop, and J. L. Jimenez, 2009: Interpretation of organic components from Positive Matrix Factorization of aerosol mass spectrometric data. *Atmos. Chem. Phys.*, **9**, 2891–2918.

Unger, N., S. Menon, D. M. Koch, and D. T. Shindell, 2009: Impacts of aerosol-cloud interactions on past and future changes in tropospheric composition. *Atmos. Chem. Phys.*, **9**, 4115–4129.

Unger, N., D. T. Shindell, D. M. Koch, M. Amann, J. Cofala, and D. G. Streets, 2006: Influences of man-made emissions and climate changes on tropospheric ozone, methane, and sulfate at 2030 from a broad range of possible futures. *J. Geophys. Res.*, **111**, D12313.

Usoskin, I. G., and G. A. Kovaltsov, 2008: Cosmic rays and climate of the Earth: Possible connection. *C. R. Geosci.*, **340**, 441–450.

Uttal, T., et al., 2002: Surface heat budget of the Arctic Ocean. *Bull. Am. Meteor. Soc.*, **83**, 255–275.

Vali, G., 1985: Atmospheric ice nucleation—A review. *J. Rech. Atmos.*, **19**, 105–115.

van den Heever, S. C., G. L. Stephens, and N. B. Wood, 2011: Aerosol indirect effects on tropical convection characteristics under conditions of radiative-convective equilibrium. *J. Atmos. Sci.*, **68**, 699–718.

vanZanten, M. C., B. Stevens, G. Vali, and D. H. Lenschow, 2005: Observations of drizzle in nocturnal marine stratocumulus. *J. Atmos. Sci.*, **62**, 88–106.

vanZanten, M. C., et al., 2011: Controls on precipitation and cloudiness in simulations of trade-wind cumulus as observed during RICO. *J. Adv. Model. Earth Syst.*, **3**, M06001.

Várnai, T., and A. Marshak, 2009: MODIS observations of enhanced clear sky reflectance near clouds. *Geophys. Res. Lett.*, **36**, L06807.

Vavrus, S., M. M. Holland, and D. A. Bailey, 2011: Changes in Arctic clouds during intervals of rapid sea ice loss. *Clim. Dyn.*, **36**, 1475–1489.

Vavrus, S., D. Waliser, A. Schweiger, and J. Francis, 2009: Simulations of 20th and 21st century Arctic cloud amount in the global climate models assessed in the IPCC AR4. *Clim. Dyn.*, **33**, 1099–1115.

Verheggen, B., et al., 2007: Aerosol partitioning between the interstitial and the condensed phase in mixed-phase clouds. *J. Geophys. Res.*, **112**, D23202.

Verlinde, J., et al., 2007: The mixed-phase Arctic cloud experiment. *Bull. Am. Meteor. Soc.*, **88**, 205–221.

Vial, J., J.-L. Dufresne, and S. Bony, 2013: On the interpretation of inter-model spread in CMIP5 climate sensitivity estimates. *Climate Dynamics*, doi:10.1007/s00382-013-1725-9.

Viana, M., W. Maenhaut, X. Chi, X. Querol, and A. Iastuey, 2007: Comparative chemical mass closure of fine and coarse aerosols at two sites in South and West Europe: Implications for EU air pollution policies. *Atmos. Environ.*, **41**, 315–326.

Viana, M., X. Chi, W. Maenhaut, X. Querol, A. Alastuey, P. Mikuska, and Z. Vecera, 2006: Organic and elemental carbon concentrations during summer and winter sampling campaigns in Barcelona, Spain. *Atmos. Environ.*, **40**, 2180–2193.

Viana, M., et al., 2008: Characterising exposure to PM aerosols for an epidemiological study. *Atmos. Environ.*, **42**, 1552–1568.

Vignati, E., M. Karl, M. Krol, J. Wilson, P. Stier, and F. Cavalli, 2010: Sources of uncertainties in modelling black carbon at the global scale. *Atmos. Chem. Phys.*, **10**, 2595–2611.

Vogelmann, A. M., T. P. Ackerman, and R. P. Turco, 1992: Enhancements in biologically effective ultraviolet radiation following volcanic eruptions. *Nature*, **359**, 47–49.

Vogelmann, A. M., et al., 2012: RACORO extended-term aircraft observations of boundary layer clouds. *Bull. Am. Meteor. Soc.*, **93**, 861–878.

Vogt, M., S. Vallina, and R. von Glasow, 2008: Correspondence on "Enhancing the natural cycle to slow global warming". *Atmos. Environ.*, **42**, 4803–4805.

Volkamer, R., et al., 2006: Secondary organic aerosol formation from anthropogenic air pollution: Rapid and higher than expected. *Geophys. Res. Lett.*, **33**, L17811.

Volodin, E. M., 2008: Relation between temperature sensitivity to doubled carbon dioxide and the distribution of clouds in current climate models. *Izvestiya Atmos. Ocean. Phys.*, **44**, 288–299.

Waliser, D. E., J. L. F. Li, T. S. L'Ecuyer, and W. T. Chen, 2011: The impact of precipitating ice and snow on the radiation balance in global climate models. *Geophys. Res. Lett.*, **38**, L06802.

Wang, G., H. Wang, Y. Yu, S. Gao, J. Feng, S. Gao, and L. Wang, 2003: Chemical characterization of water-soluble components of PM10 and PM2.5 atmospheric aerosols in five locations of Nanjing, China. *Atmos. Environ.*, **37**, 2893–2902.

Wang, H., and D. Shooter, 2001: Water soluble ions of atmospheric aerosols in three New Zealand cities: Seasonal changes and sources. *Atmos. Environ.*, **35**, 6031–6040.

Wang, H., K. Kawamuraa, and D. Shooter, 2005a: Carbonaceous and ionic components in wintertime atmospheric aerosols from two New Zealand cities: Implications for solid fuel combustion. *Atmos. Environ.*, **39**, 5865–5875.

Wang, H., P. J. Rasch, and G. Feingold, 2011a: Manipulating marine stratocumulus cloud amount and albedo: A process-modelling study of aerosol-cloud-precipitation interactions in response to injection of cloud condensation nuclei. *Atmos. Chem. Phys.*, **11**, 4237–4249.

Wang, H. L., and G. Feingold, 2009a: Modeling mesoscale cellular structures and drizzle in marine stratocumulus. Part I: Impact of drizzle on the formation and evolution of open cells. *J. Atmos. Sci.*, **66**, 3237–3256.

Wang, H. L., and G. Feingold, 2009b: Modeling mesoscale cellular structures and drizzle in marine stratocumulus. Part II: The microphysics and dynamics of the boundary region between open and closed cells. *J. Atmos. Sci.*, **66**, 3257–3275.

Wang, L., A. F. Khalizov, J. Zheng, W. Xu, Y. Ma, V. Lal, and R. Y. Zhang, 2010a: Atmospheric nanoparticles formed from heterogeneous reactions of organics. *Nature Geosci.*, **3**, 238–242.

Wang, M., and J. Penner, 2009: Aerosol indirect forcing in a global model with particle nucleation. *Atmos. Chem. Phys.*, **9**, 239–260.

Wang, M., et al., 2011b: Aerosol indirect effects in a multi-scale aerosol-climate model PNNL-MMF. *Atmos. Chem. Phys.*, **11**, 5431–5455.

Wang, M., et al., 2012: Constraining cloud lifetime effects of aerosols using A-Train satellite observations. *Geophys. Res. Lett.*, **39**, L15709.

Wang, T., S. Li, Y. Shen, J. Deng, and M. Xie, 2010b: Investigations on direct and indirect effect of nitrate on temperature and precipitation in China using a regional climate chemistry modeling system. *J. Geophys. Res.*, **115**, D00K26.

Wang, Y., G. Zhuang, A. Tang, H. Yuan, Y. Sun, S. Chen, and A. Zheng, 2005b: The ion chemistry and the source of PM 2.5 aerosol in Beijing. *Atmos. Environ.* **39**, 3771–3784.

Wang, Y., et al., 2006: The ion chemistry, seasonal cycle, and sources of PM2.5 and TSP aerosol in Shanghai. *Atmos. Environ.*, **40**, 2935–2952.

Wang, Z. L., H. Zhang, and X. S. Shen, 2011c: Radiative forcing and climate response due to black carbon in snow and ice. *Adv. Atmos. Sci.*, **28**, 1336–1344.

Waquet, F., J. Riedi, L. C. Labonnote, P. Goloub, B. Cairns, J.-L. Deuzé, and D. Tanré, 2009: Aerosol remote sensing over clouds using A-Train observations. *J. Atmos. Sci.*, **66**, 2468–2480.

Warneke, C., et al., 2010: An important contribution to springtime Arctic aerosol from biomass burning in Russia. *Geophys. Res. Lett.*, **37**, L01801.

Warren, S. G., 2013: Can black carbon in snow be detected by remote sensing? *J. Geophys. Res.*, **118**, 779–786.

Watanabe, M., S. Emori, M. Satoh, and H. Miura, 2009: A PDF-based hybrid prognostic cloud scheme for general circulation models. *Clim. Dyn.*, **33**, 795–816.

Webb, M. J., and A. Lock, 2013: Coupling between subtropical cloud feedback and the local hydrological cycle in a climate model. *Clim. Dyn.*, **41**, 1923–1939.

Webb, M. J., F. H. Lambert, and J. M. Gregory, 2013: Origins of differences in climate sensitivity, forcing and feedback in climate models. *Clim. Dyn.*, **40**, 677–707.

Weinstein, J. P., S. R. Hedges, and S. Kimbrough, 2010: Characterization and aerosol mass balance of PM2.5 and PM10 collected in Conakry, Guinea during the 2004 Harmattan period. *Chemosphere*, **78**, 980–988.

Wen, G., A. Marshak, R. F. Cahalan, L. A. Remer, and R. G. Kleidman, 2007: 3-D aerosol-cloud radiative interaction observed in collocated MODIS and ASTER images of cumulus cloud fields. *J. Geophys. Res.*, **112**, D13204.

Wendisch, M., et al., 2008: Radiative and dynamic effects of absorbing aerosol particles over the Pearl River Delta, China. *Atmos. Environ.*, **42**, 6405–6416.

Westra, S., L. V. Alexander, and F. W. Zwiers, 2013: Global increasing trends in annual maximum daily precipitation. *J. Clim.*, **26**, 3904–3918.

Wetherald, R. T., and S. Manabe, 1980: Cloud cover and climate sensitivity. *J. Atmos. Sci.*, **37**, 1485–1510.

Wex, H., G. McFiggans, S. Henning, and F. Stratmann, 2010: Influence of the external mixing state of atmospheric aerosol on derived CCN number concentrations. *Geophys. Res. Lett.*, **37**, L10805.

Wielicki, B. A., and L. Parker, 1992: On the determination of cloud cover from satellite sensors: The effect of sensor spatial resolution. *J. Geophys. Res.*, **97**, 12799–12823.

Wilcox, E. M., 2010: Stratocumulus cloud thickening beneath layers of absorbing smoke aerosol. *Atmos. Chem. Phys.*, **10**, 11769–11777.

Williams, K. D., and G. Tselioudis, 2007: GCM intercomparison of global cloud regimes: Present-day evaluation and climate change response. *Clim. Dyn.*, **29**, 231–250.

Williams, K. D., and M. J. Webb, 2009: A quantitative performance assessment of cloud regimes in climate models. *Clim. Dyn.*, **33**, 141–157.

Williams, K. D., A. Jones, D. L. Roberts, C. A. Senior, and M. J. Woodage, 2001: The response of the climate system to the indirect effects of anthropogenic sulfate aerosol. *Clim. Dyn.*, **17**, 845–856.

Williams, K. D., et al., 2006: Evaluation of a component of the cloud response to climate change in an intercomparison of climate models. *Clim. Dyn.*, **26**, 145–165.

Wingenter, O. W., S. M. Elliot, and D. R. Blake, 2007: Enhancing the natural sulfur cycle to slow global warming *Atmos. Environ.*, **41**, 7373–7375.

Winker, D. M., J. L. Tackett, B. J. Getzewich, Z. Liu, M. A. Vaughan, and R. R. Rogers, 2013: The global 3–D distribution of tropospheric aerosols as characterized by CALIOP. *Atmos. Chem. Phys.*, **13**, 3345–3361.

Winker, D. M., et al., 2009: Overview of the CALIPSO mission and CALIOP data processing algorithms. *J. Atmos. Ocean. Technol.*, **26**, 2310–2323.

Winker, D. M., et al., 2010: The CALIPSO mission: A global 3D view of aerosols and clouds. *Bull. Am. Meteor. Soc.*, **91**, 1211–1229.

Wood, R., 2005: Drizzle in stratiform boundary layer clouds. Part II: Microphysical aspects. *J. Atmos. Sci.*, **62**, 3034–3050.

Wood, R., 2007: Cancellation of aerosol indirect effects in marine stratocumulus through cloud thinning. *J. Atmos. Sci.*, **64**, 2657–2669.

Wood, R., and C. S. Bretherton, 2006: On the relationship between stratiform low cloud cover and lower-tropospheric stability. *J. Clim.*, **19**, 6425–6432.

Wood, R., C. S. Bretherton, D. Leon, A. D. Clarke, P. Zuidema, G. Allen, and H. Coe, 2011a: An aircraft case study of the spatial transition from closed to open mesoscale cellular convection over the Southeast Pacific. *Atmos. Chem. Phys.*, **11**, 2341–2370.

Wood, R., et al., 2011b: The VAMOS Ocean-Cloud-Atmosphere-Land Study Regional Experiment (VOCALS-REx): Goals, platforms, and field operations. *Atmos. Chem. Phys.*, **11**, 627–654.

Woodhouse, M. T., G. W. Mann, K. S. Carslaw, and O. Boucher, 2008: The impact of oceanic iron fertilisation on cloud condensation nuclei. *Atmos. Environ.*, **42**, 5728–5730.

Woodhouse, M. T., K. S. Carslaw, G. W. Mann, S. M. Vallina, M. Vogt, P. R. Halloran, and O. Boucher, 2010: Low sensitivity of cloud condensation nuclei to changes in the sea-air flux of dimethyl-sulphide. *Atmos. Chem. Phys.*, **10**, 7545–7559.

Woodward, S., D. L. Roberts, and R. A. Betts, 2005: A simulation of the effect of climate change-induced desertification on mineral dust aerosol. *Geophys. Res. Lett.*, **32**, L18810.

Wu, S., L. J. Mickley, J. O. Kaplan, and D. J. Jacob, 2012: Impacts of changes in land use and land cover on atmospheric chemistry and air quality over the 21st century. *Atmos. Chem. Phys.*, **12**, 1597–1609.

Wyant, M. C., C. S. Bretherton, and P. N. Blossey, 2009: Subtropical low cloud response to a warmer climate in a superparameterized climate model. Part I: Regime sorting and physical mechanisms. *J. Adv. Model. Earth Syst.*, **1**, 7.

Wyant, M. C., C. S. Bretherton, P. N. Blossey, and M. Khairoutdinov, 2012: Fast cloud adjustment to increasing CO_2 in a superparameterized climate model. *J. Adv. Model. Earth Syst.*, **4**, M05001.

Wyant, M. C., et al., 2006: A comparison of low-latitude cloud properties and their response to climate change in three AGCMs sorted into regimes using mid-tropospheric vertical velocity. *Clim. Dyn.*, **27**, 261–279.

Xiao, H.-Y., and C.-Q. Liu, 2004: Chemical characteristics of water-soluble components in TSP over Guiyang, SW China, 2003. *Atmos. Environ.*, **38**, 6297–6306.

Xie, S.-P., C. Deser, G. A. Vecchi, J. Ma, H. Teng, and A. T. Wittenberg, 2010: Global warming pattern formation: Sea surface temperature and rainfall. *J. Clim.*, **23**, 966–986.

Xu, B., J. Cao, D. R. Joswiak, X. Liu, H. Zhao, and J. He, 2012: Post-depositional enrichment of black soot in snow-pack and accelerated melting of Tibetan glaciers. *Environ. Res. Lett.*, **7**, 014022.

Xu, B. Q., et al., 2009: Black soot and the survival of Tibetan glaciers. *Proc. Natl. Acad. Sci. U.S.A.*, **106**, 22114–22118.

Xu, K. M., A. N. Cheng, and M. H. Zhang, 2010: Cloud-resolving simulation of low-cloud feedback to an increase in sea surface temperature. *J. Atmos. Sci.*, **67**, 730–748.

Xu, K. M., T. Wong, B. A. Wielicki, L. Parker, B. Lin, Z. A. Eitzen, and M. Branson, 2007: Statistical analyses of satellite cloud object data from CERES. Part II: Tropical convective cloud objects during 1998 El Niño and evidence for supporting the fixed anvil temperature hypothesis. *J. Clim.*, **20**, 819–842.

Xu, K. M., et al., 2002: An intercomparison of cloud-resolving models with the atmospheric radiation measurement summer 1997 intensive observation period data. *Q. J. R. Meteorol. Soc.*, **128**, 593–624.

Xue, H., G. Feingold, and B. Stevens, 2008: Aerosol effects on clouds, precipitation, and the organization of shallow cumulus convection. *J. Atmos. Sci.*, **65**, 392–406.

Yang, Q., et al., 2011: Assessing regional scale predictions of aerosols, marine stratocumulus, and their interactions during VOCALS-REx using WRF-Chem. *Atmos. Chem. Phys.*, **11**, 11951–11975.

Yao, X., et al., 2002: The water-soluble ionic composition of PM2.5 in Shanghai and Beijing, China. *Atmos. Environ.*, **36**, 4223–4234.

Ye, B., et al., 2003: Concentration and chemical composition of PM2.5 in Shanghai for a 1–year period. *Atmos. Environ.*, **37**, 499–510.

Yin, J., and R. M. Harrison, 2008: Pragmatic mass closure study for PM1.0, PM2.5 and PM10 at roadside, urban background and rural sites. *Atmos. Environ.*, **42**, 980–988.

Yin, J. H., 2005: A consistent poleward shift of the storm tracks in simulations of 21st century climate. *Geophys. Res. Lett.*, **32**, L18701.

Yokohata, T., S. Emori, T. Nozawa, Y. Tsushima, T. Ogura, and M. Kimoto, 2005: Climate response to volcanic forcing: Validation of climate sensitivity of a coupled atmosphere-ocean general circulation model. *Geophys. Res. Lett.*, **32**, L21710.

Yokohata, T., M. J. Webb, M. Collins, K. D. Williams, M. Yoshimori, J. C. Hargreaves, and J. D. Annan, 2010: Structural similarities and differences in climate responses to CO_2 increase between two perturbed physics ensembles. *J. Clim.*, **23**, 1392–1410.

Yokohata, T., et al., 2008: Comparison of equilibrium and transient responses to CO_2 increase in eight state-of-the-art climate models. *Tellus A*, **60**, 946–961.

Yoshimori, M., and A. J. Broccoli, 2008: Equilibrium response of an atmosphere-mixed layer ocean model to different radiative forcing agents: Global and zonal mean response. *J. Clim.*, **21**, 4399–4423.

Yoshimori, M., T. Yokohata, and A. Abe-Ouchi, 2009: A comparison of climate feedback strength between CO_2 doubling and LGM experiments. *J. Clim.*, **22**, 3374–3395.

Young, I. R., S. Zieger, and A. V. Babanin, 2011: Global trends in wind speed and wave height. *Science*, **332**, 451–455.

Yttri, K. E., 2007: Concentrations of particulate matter (PM10, PM2.5) in Norway. Annual and seasonal trends and spatial variability. In: *EMEP Particulate Matter Assessment Report, Part B, report EMEP/CCC-Report 8/2007*, Norwegian Institute for Air Research, Oslo, Norway, pp. 292–307.

Yu, F., 2011: A secondary organic aerosol formation model considering successive oxidation aging and kinetic condensation of organic compounds: Global scale implications. *Atmos. Chem. Phys.*, **11**, 1083–1099.

Yu, F., and G. Luo, 2009: Simulation of particle size distribution with a global aerosol model: Contribution of nucleation to aerosol and CCN number concentrations. *Atmos. Chem. Phys.*, **9**, 7691–7710.

Yu, H., R. McGraw, and S. Lee, 2012: Effects of amines on formation of sub-3 nm particles and their subsequent growth. *Geophys. Res. Lett.*, **39**, L02807.

Yu, H., et al., 2006: A review of measurement-based assessments of the aerosol direct radiative effect and forcing. *Atmos. Chem. Phys.*, **6**, 613–666.

Yu, H. B., M. Chin, D. M. Winker, A. H. Omar, Z. Y. Liu, C. Kittaka, and T. Diehl, 2010: Global view of aerosol vertical distributions from CALIPSO lidar measurements and GOCART simulations: Regional and seasonal variations. *J. Geophys. Res.*, **115**, D00H30.

Yuan, T., L. A. Remer, and H. Yu, 2011: Microphysical, macrophysical and radiative signatures of volcanic aerosols in trade wind cumulus observed by the A-Train. *Atmos. Chem. Phys.*, **11**, 7119–7132.

Yuekui, Y., and L. Di Girolamo, 2008: Impacts of 3–D radiative effects on satellite cloud detection and their consequences on cloud fraction and aerosol optical depth retrievals. *J. Geophys. Res.*, **113**, D04213.

Yuter, S. E., M. A. Miller, M. D. Parker, P. M. Markowski, Y. Richardson, H. Brooks, and J. M. Straka, 2013: Comment on "Why do tornados and hailstorms rest on weekends?" by D. Rosenfeld and T. Bell. *J. Geophys. Res. Atmos.*, **118**, 7332–7338.

Zarzycki, C. M., and T. C. Bond, 2010: How much can the vertical distribution of black carbon affect its global direct radiative forcing? *Geophys. Res. Lett.*, **37**, L20807.

Zelinka, M. D., and D. L. Hartmann, 2010: Why is longwave cloud feedback positive? *J. Geophys. Res.*, **115**, D16117.

Zelinka, M. D., and D. L. Hartmann, 2011: The observed sensitivity of high clouds to mean surface temperature anomalies in the tropics. *J. Geophys. Res.*, **116**, D23103.

Zelinka, M. D., S. A. Klein, and D. L. Hartmann, 2012a: Computing and partitioning cloud feedbacks using cloud property histograms. Part I: Cloud radiative kernels. *J. Clim.*, **25**, 3715–3735.

Zelinka, M. D., S. A. Klein, and D. L. Hartmann, 2012b: Computing and partitioning cloud feedbacks using cloud property histograms. Part II: Attribution to changes in cloud amount, altitude, and optical depth. *J. Clim.*, **25**, 3736–3754.

Zelinka, M. D., S. A. Klein, K. E. Taylor, T. Andrews, M. J. Webb, J. M. Gregory, and P. M. Forster, 2013: Contributions of different cloud types to feedbacks and rapid adjustments in CMIP5. *J. Clim.*, **26**, 5007–5027.

Zhang, G. J., A. M. Vogelmann, M. P. Jensen, W. D. Collins, and E. P. Luke, 2010: Relating satellite-observed cloud properties from MODIS to meteorological conditions for marine boundary layer clouds. *J. Clim.*, **23**, 1374–1391.

Zhang, M. H., and C. Bretherton, 2008: Mechanisms of low cloud-climate feedback in idealized single-column simulations with the Community Atmospheric Model, version 3 (CAM3). *J. Clim.*, **21**, 4859–4878.

Zhang, Q., D. R. Worsnop, M. R. Canagaratna, and J. L. Jimenez, 2005a: Hydrocarbon-like and oxygenated organic aerosols in Pittsburgh: Insights into sources and processes of organic aerosols. *Atmos. Chem. Phys.*, **5**, 3289–3311.

Zhang, Q., M. R. Alfarra, D. R. Worsnop, J. D. Allan, H. Coe, M. R. Canagaratna, and J. L. Jimenez, 2005b: Deconvolution and quantification of hydrocarbon-like and oxygenated organic aerosols based on aerosol mass spectrometry. *Environ. Sci. Technol.*, **39**, 4938–4952.

Zhang, Q., et al., 2007a: Ubiquity and dominance of oxygenated species in organic aerosols in anthropogenically-influenced Northern Hemisphere midlatitudes. *Geophys. Res. Lett.*, **34**, L13801.

Zhang, R. Y., A. Khalizov, L. Wang, M. Hu, and W. Xu, 2012a: Nucleation and growth of nanoparticles in the atmosphere. *Chem. Rev.*, **112**, 1957–2011.

Zhang, X. B., et al., 2007b: Detection of human influence on twentieth-century precipitation trends. *Nature*, **448**, 461–465.

Zhang, X. Y., R. Arimoto, Z. S. An, J. J. Cao, and D. Wang, 2001: Atmospheric dust aerosol over the Tibetian Plateau. *J. Geophys. Res.*, **106**, 18471–18476.

Zhang, X. Y., Y. Q. Wang, X. C. Zhang, W. Guo, and S. L. Gong, 2008: Carbonaceous aerosol composition over various regions of China during 2006. *J. Geophys. Res.*, **113**, D14111.

Zhang, X. Y., J. J. Cao, L. M. Li, R. Arimoto, Y. Cheng, B. Huebert, and D. Wang, 2002: Characterization of atmospheric aerosol over Xian in the south margin of the loess plateau, China. *Atmos. Environ.*, **36**, 4189–4199.

Zhang, X. Y., Y. Q. Wang, T. Niu, X. C. Zhang, S. L. Gong, Y. M. Zhang, and J. Y. Sun, 2012b: Atmospheric aerosol compositions in China: Spatial/temporal variability, chemical signature, regional haze distribution and comparisons with global aerosols. *Atmos. Chem. Phys.*, **12**, 779–799.

Zhang, Y., 2008: Online-coupled meteorology and chemistry models: History, current status, and outlook. *Atmos. Chem. Phys.*, **8**, 2895–2932.

Zhang, Y. M., X. Y. Zhang, J. Y. Sun, W. L. Lin, S. L. Gong, X. J. Shen, and S. Yang, 2011: Characterization of new particle and secondary aerosol formation during summertime in Beijing, China. *Tellus B*, **63**, 382–394.

Zhao, T. L., S. L. Gong, X. Y. Zhang, A. A. Mawgoud, and Y. P. Shao, 2006: An assessment of dust emission schemes in modeling east Asian dust storms. *J. Geophys. Res.*, **111**, D05S90.

Zhao, T. X.-P., et al., 2008a: Study of long-term trend in aerosol optical thickness observed from operational AVHRR satellite instrument. *J. Geophys. Res.*, **113**, D07201.

Zhao, T. X. P., N. G. Loeb, I. Laszlo, and M. Zhou, 2011: Global component aerosol direct radiative effect at the top of atmosphere. *Int. J. Remote Sens.*, **32**, 633–655.

Zhao, T. X. P., H. B. Yu, I. Laszlo, M. Chin, and W. C. Conant, 2008b: Derivation of component aerosol direct radiative forcing at the top of atmosphere for clear-sky oceans. *J. Quant. Spectrosc. Radiat. Transfer*, **109**, 1162–1186.

Zhu, P., B. A. Albrecht, V. P. Ghate, and Z. D. Zhu, 2010: Multiple-scale simulations of stratocumulus clouds. *J. Geophys. Res.*, **115**, D23201.

Zhu, P., et al., 2012: A limited area model (LAM) intercomparison study of a TWP-ICE active monsoon mesoscale convective event. *J. Geophys. Res.*, **117**, D11208.

Zhuang, B. L., L. Liu, F. H. Shen, T. J. Wang, and Y. Han, 2010: Semidirect radiative forcing of internal mixed black carbon cloud droplet and its regional climatic effect over China. *J. Geophys. Res.*, **115**, D00K19.

Ziemann, P. J., and R. Atkinson, 2012: Kinetics, products, and mechanisms of secondary organic aerosol formation. *Chem. Soc. Rev.*, **41**, 6582–6605.

Zubler, E. M., U. Lohmann, D. Lüthi, C. Schär, and A. Muhlbauer, 2011: Statistical analysis of aerosol effects on simulated mixed-phase clouds and precipitation in the Alps. *J. Atmos. Sci.*, **68**, 1474–1492.

Zuidema, P., et al., 2005: An arctic springtime mixed-phase cloudy boundary layer observed during SHEBA. *J. Atmos. Sci.*, **62**, 160–176.

Zuidema, P., et al., 2012: On trade wind cumulus cold pools. *J. Atmos. Sci.*, **69**, 258–280.

Anthropogenic and Natural Radiative Forcing

Coordinating Lead Authors:
Gunnar Myhre (Norway), Drew Shindell (USA)

Lead Authors:
François-Marie Bréon (France), William Collins (UK), Jan Fuglestvedt (Norway), Jianping Huang (China), Dorothy Koch (USA), Jean-François Lamarque (USA), David Lee (UK), Blanca Mendoza (Mexico), Teruyuki Nakajima (Japan), Alan Robock (USA), Graeme Stephens (USA), Toshihiko Takemura (Japan), Hua Zhang (China)

Contributing Authors:
Borgar Aamaas (Norway), Olivier Boucher (France), Stig B. Dalsøren (Norway), John S. Daniel (USA), Piers Forster (UK), Claire Granier (France), Joanna Haigh (UK), Øivind Hodnebrog (Norway), Jed O. Kaplan (Switzerland/Belgium/USA), George Marston (UK), Claus J. Nielsen (Norway), Brian C. O'Neill (USA), Glen P. Peters (Norway), Julia Pongratz (Germany), Michael Prather (USA), Venkatachalam Ramaswamy (USA), Raphael Roth (Switzerland), Leon Rotstayn (Australia), Steven J. Smith (USA), David Stevenson (UK), Jean-Paul Vernier (USA), Oliver Wild (UK), Paul Young (UK)

Review Editors:
Daniel Jacob (USA), A.R. Ravishankara (USA), Keith Shine (UK)

This chapter should be cited as:
Myhre, G., D. Shindell, F.-M. Bréon, W. Collins, J. Fuglestvedt, J. Huang, D. Koch, J.-F. Lamarque, D. Lee, B. Mendoza, T. Nakajima, A. Robock, G. Stephens, T. Takemura and H. Zhang, 2013: Anthropogenic and Natural Radiative Forcing. In: *Climate Change 2013: The Physical Science Basis. Contribution of Working Group I to the Fifth Assessment Report of the Intergovernmental Panel on Climate Change* [Stocker, T.F., D. Qin, G.-K. Plattner, M. Tignor, S.K. Allen, J. Boschung, A. Nauels, Y. Xia, V. Bex and P.M. Midgley (eds.)]. Cambridge University Press, Cambridge, United Kingdom and New York, NY, USA.

Table of Contents

Supplementary Material is available in online versions of the report.

Executive Summary

It is unequivocal that anthropogenic increases in the well-mixed greenhouse gases (WMGHGs) have substantially enhanced the greenhouse effect, and the resulting forcing continues to increase. Aerosols partially offset the forcing of the WMGHGs and dominate the uncertainty associated with the total anthropogenic driving of climate change.

As in previous IPCC assessments, AR5 uses the radiative forcing[1] (RF) concept, but it also introduces effective radiative forcing[2] (ERF). The RF concept has been used for many years and in previous IPCC assessments for evaluating and comparing the strength of the various mechanisms affecting the Earth's radiation balance and thus causing climate change. Whereas in the RF concept all surface and tropospheric conditions are kept fixed, the ERF calculations presented here allow all physical variables to respond to perturbations except for those concerning the ocean and sea ice. The inclusion of these adjustments makes ERF a better indicator of the eventual temperature response. ERF and RF values are significantly different for anthropogenic aerosols owing to their influence on clouds and on snow cover. These changes to clouds are rapid adjustments and occur on a time scale much faster than responses of the ocean (even the upper layer) to forcing. RF and ERF are estimated over the Industrial Era from 1750 to 2011 if other periods are not explicitly stated. {8.1, Box 8.1, Figure 8.1}

Industrial-Era Anthropogenic Forcing

The total anthropogenic ERF over the Industrial Era is 2.3 (1.1 to 3.3) W m^{-2}.[3] It is certain that the total anthropogenic ERF is positive. Total anthropogenic ERF has increased more rapidly since 1970 than during prior decades. The total anthropogenic ERF estimate for 2011 is 43% higher compared to the AR4 RF estimate for the year 2005 owing to reductions in estimated forcing due to aerosols but also to continued growth in greenhouse gas RF. {8.5.1, Figures 8.15, 8.16}

Due to increased concentrations, RF from WMGHGs has increased by 0.20 (0.18 to 0.22) W m^{-2} (8%) since the AR4 estimate for the year 2005. The RF of WMGHG is 2.83 (2.54 to 3.12) W m^{-2}. The majority of this change since AR4 is due to increases in the carbon dioxide (CO_2) RF of nearly 10%. The Industrial Era RF for CO_2 alone is 1.82 (1.63 to 2.01) W m^{-2}, and CO_2 is the component with the largest global mean RF. Over the last decade RF of CO_2 has an average growth rate of 0.27 (0.24 to 0.30) W m^{-2} per decade. Emissions of CO_2 have made the largest contribution to the increased anthropogenic forcing in every decade since the 1960s. The best estimate for ERF of WMGHG is the same as the RF but with a larger uncertainty (±20%). {8.3.2, 8.5.2, Figures 8.6, 8.18}

The net forcing by WMGHGs other than CO_2 shows a small increase since the AR4 estimate for the year 2005. A small growth in the CH_4 concentration has increased its RF by 2% to an AR5 value of 0.48 (0.43 to 0.53) W m^{-2}. RF of nitrous oxide (N_2O) has increased by 6% since AR4 and is now 0.17 (0.14 to 0.20) W m^{-2}. N_2O concentrations continue to rise while those of dichlorodifluoromethane (CFC-12), the third largest WMGHG contributor to RF for several decades, is falling due to its phase-out under the Montreal Protocol and amendments. Since 2011 N_2O has become the third largest WMGHG contributor to RF. The RF from all halocarbons (0.36 W m^{-2}) is very similar to the value in AR4, with a reduced RF from chlorofluorocarbons (CFCs) but increases from many of their substitutes. Four of the halocarbons (trichlorofluoromethane (CFC-11), CFC-12, trichlorotrifluoroethane (CFC-113) and chlorodifluoromethane (HCFC-22)) account for around 85% of the total halocarbon RF. The first three of these compounds have declining RF over the last 5 years but their combined decrease is compensated for by the increased RF from HCFC-22. Since AR4, the RF from all HFCs has nearly doubled but still only amounts to 0.02 W m^{-2}. There is *high confidence*[4] that the overall growth rate in RF from all WMGHG is smaller over the last decade than in the 1970s and 1980s owing to a reduced rate of increase in the combined non-CO_2 RF. {8.3.2; Figure 8.6}

Ozone and stratospheric water vapour contribute substantially to RF. The total RF estimated from modelled ozone changes is 0.35 (0.15 to 0.55) W m^{-2}, with RF due to tropospheric ozone changes of 0.40 (0.20 to 0.60) W m^{-2} and due to stratospheric ozone changes of –0.05 (–0.15 to +0.05) W m^{-2}. Ozone is not emitted directly into the atmosphere but is formed by photochemical reactions. Tropospheric ozone RF is largely attributed to anthropogenic emissions of methane (CH_4), nitrogen oxides (NO_x), carbon monoxide (CO) and non-methane volatile organic compounds (NMVOCs), while stratospheric ozone RF results primarily from ozone depletion by halocarbons. Estimates are also provided attributing RF to emitted compounds. Ozone-depleting substances (ODS) cause ozone RF of –0.15 (–0.30 to 0.0) W m^{-2}, some of which is in the troposphere. Tropospheric ozone precursors cause ozone RF of 0.50 (0.30 to 0.70) W m^{-2}, some of which is in the stratosphere; this value is larger than that in AR4. There is *robust evidence* that tropospheric ozone also has a detrimental impact on vegetation physiology, and therefore on its CO_2 uptake, but there is a *low confidence* on quantitative estimates of the RF owing to this indirect effect. RF for stratospheric water vapour produced by CH_4 oxidation is 0.07 (0.02 to 0.12) W m^{-2}. The RF best estimates for ozone and stratospheric

---

1 Change in net downward radiative flux at the tropopause after allowing for stratospheric temperatures to readjust to radiative equilibrium, while holding surface and tropospheric temperatures and state variables fixed at the unperturbed values.

2 Change in net downward radiative flux at the top of the atmosphere (TOA) after allowing for atmospheric temperatures, water vapour, clouds and land albedo to adjust, but with global mean surface temperature or ocean and sea ice conditions unchanged (calculations presented in this chapter use the fixed ocean conditions method).

3 Uncertainties are given associated with best estimates of forcing. The uncertainty values represent the 5–95% (90%) confidence range.

4 In this Report, the following summary terms are used to describe the available evidence: limited, medium, or robust; and for the degree of agreement: low, medium, or high. A level of confidence is expressed using five qualifiers: very low, low, medium, high, and very high, and typeset in italics, e.g., *medium confidence*. For a given evidence and agreement statement, different confidence levels can be assigned, but increasing levels of evidence and degrees of agreement are correlated with increasing confidence (see Section 1.4 and Box TS.1 for more details).

8

water vapour are either identical or consistent with the range in AR4. {8.2, 8.3.3, Figure 8.7}

The magnitude of the aerosol forcing is reduced relative to AR4. The RF due to aerosol–radiation interactions, sometimes referred to as *direct aerosol effect*, is given a best estimate of –0.35 (–0.85 to +0.15) W m^{-2}, and black carbon (BC) on snow and ice is 0.04 (0.02 to 0.09) W m^{-2}. The ERF due to aerosol–radiation interactions is –0.45 (–0.95 to +0.05) W m^{-2}. A total aerosol–cloud interaction[5] is quantified in terms of the ERF concept with an estimate of –0.45 (–1.2 to 0.0) W m^{-2}. The total aerosol effect (excluding BC on snow and ice) is estimated as ERF of –0.9 (–1.9 to –0.1) W m^{-2}. The large uncertainty in aerosol ERF is the dominant contributor to overall net Industrial Era forcing uncertainty. Since AR4, more aerosol processes have been included in models, and differences between models and observations persist, resulting in similar uncertainty in the aerosol forcing as in AR4. Despite the large uncertainty range, there is a *high confidence* that aerosols have offset a substantial portion of WMGHG global mean forcing. {8.3.4, 8.5.1, Figures 8.15, 8.16}

There is *robust evidence* that anthropogenic land use change has increased the land surface albedo, which leads to an RF of –0.15 ± 0.10 W m^{-2}. There is still a large spread of estimates owing to different assumptions for the albedo of natural and managed surfaces and the fraction of land use changes before 1750. Land use change causes additional modifications that are not radiative, but impact the surface temperature, in particular through the hydrologic cycle. These are more uncertain and they are difficult to quantify, but tend to offset the impact of albedo changes. As a consequence, there is *low agreement* on the sign of the net change in global mean temperature as a result of land use change. {8.3.5}

Attributing forcing to emissions provides a more direct link from human activities to forcing. The RF attributed to methane emissions is *very likely*[6] to be much larger (~1.0 W m^{-2}) than that attributed to methane *concentration* increases (~0.5 W m^{-2}) as concentration changes result from the partially offsetting impact of emissions of multiple species and subsequent chemical reactions. In addition, emissions of CO are *virtually certain* to have had a positive RF, while emissions of NO_X are *likely* to have had a net negative RF at the global scale. Emissions of ozone-depleting halocarbons are *very likely* to have caused a net positive RF as their own positive RF has outweighed the negative RF from the stratospheric ozone depletion that they have induced. {8.3.3, 8.5.1, Figure 8.17, FAQ 8.2}

Forcing agents such as aerosols, ozone and land albedo changes are highly heterogeneous spatially and temporally. These patterns generally track economic development; strong negative aerosol forcing appeared in eastern North America and Europe during the early 20th century, extending to Asia, South America and central Africa by 1980. Emission controls have since reduced aerosol pollution in North America and Europe, but not in much of Asia. Ozone forcing increased throughout the 20th century, with peak positive amplitudes around 15°N to 30°N due to tropospheric pollution but negative values over Antarctica due to stratospheric loss late in the century. The pattern and spatial gradients of forcing affect global and regional temperature responses as well as other aspects of climate response such as the hydrologic cycle. {8.6.2, Figure 8.25}

Natural Forcing

Satellite observations of total solar irradiance (TSI) changes from 1978 to 2011 show that the most recent solar cycle minimum was lower than the prior two. This *very likely* led to a small negative RF of –0.04 (–0.08 to 0.00) W m^{-2} between 1986 and 2008. The best estimate of RF due to TSI changes representative for the 1750 to 2011 period is 0.05 (to 0.10) W m^{-2}. This is substantially smaller than the AR4 estimate due to the addition of the latest solar cycle and inconsistencies in how solar RF has been estimated in earlier IPCC assessments. There is *very low confidence* concerning future solar forcing estimates, but there is *high confidence* that the TSI RF variations will be much smaller than the projected increased forcing due to GHG during the forthcoming decades. {8.4.1, Figures 8.10, 8.11}

The RF of volcanic aerosols is well understood and is greatest for a short period (~2 years) following volcanic eruptions. There have been no major volcanic eruptions since Mt Pinatubo in 1991, but several smaller eruptions have caused a RF for the years 2008–2011 of –0.11 (–0.15 to –0.08) W m^{-2} as compared to 1750 and –0.06 (–0.08 to –0.04) W m^{-2} as compared to 1999–2002. Emissions of CO_2 from volcanic eruptions since 1750 have been at least 100 times smaller than anthropogenic emissions. {8.4.2, 8.5.2, Figures 8.12, 8.13, 8.18}

There is *very high confidence* that industrial-era natural forcing is a small fraction of the anthropogenic forcing except for brief periods following large volcanic eruptions. In particular, *robust evidence* from satellite observations of the solar irradiance and volcanic aerosols demonstrates a near-zero (–0.1 to +0.1 W m^{-2}) change in the natural forcing compared to the anthropogenic ERF increase of 1.0 (0.7 to 1.3) W m^{-2} from 1980 to 2011. The natural forcing over the last 15 years has *likely* offset a substantial fraction (at least 30%) of the anthropogenic forcing. {8.5.2; Figures 8.18, 8.19, 8.20}

Future Anthropogenic Forcing and Emission Metrics

Differences in RF between the emission scenarios considered here[7] are relatively small for year 2030 but become very large by 2100 and are dominated by CO_2. The scenarios show a substantial

[5] The aerosol–cloud interaction represents the portion of rapid adjustments to aerosols initiated by aerosol-cloud interactions, and is defined here as the total aerosol ERF minus the ERF due to aerosol-radiation-interactions (the latter includes cloud responses to the aerosol–radiation interaction RF)

[6] In this Report, the following terms have been used to indicate the assessed likelihood of an outcome or a result: Virtually certain 99–100% probability, Very likely 90–100%, Likely 66–100%, About as likely as not 33–66%, Unlikely 0–33%, Very unlikely 0–10%, Exceptionally unlikely 0–1%. Additional terms (Extremely likely: 95–100%, More likely than not >50–100%, and Extremely unlikely 0–5%) may also be used when appropriate. Assessed likelihood is typeset in italics, e.g., *very likely* (see Section 1.4 and Box TS.1 for more details).

[7] Chapter 1 describes the Representative Concentration Pathways (RCPs) that are the primary scenarios discussed in this report.

weakening of the negative total aerosol ERF. Nitrate aerosols are an exception to this reduction, with a substantial increase, which is a robust feature among the few available models for these scenarios. The scenarios emphasized in this assessment do not span the range of future emissions in the literature, however, particularly for near-term climate forcers. {8.2.2, 8.5.3, Figures 8.2, 8.21, 8.22}

Emission metrics such as Global Warming Potential (GWP) and Global Temperature change Potential (GTP) can be used to quantify and communicate the relative and absolute contributions to climate change of emissions of different substances, and of emissions from regions/countries or sources/sectors. The metric that has been used in policies is the GWP, which integrates the RF of a substance over a chosen time horizon, relative to that of CO_2. The GTP is the ratio of change in global mean surface temperature at a chosen point in time from the substance of interest relative to that from CO_2. There are significant uncertainties related to both GWP and GTP, and the relative uncertainties are larger for GTP. There are also limitations and inconsistencies related to their treatment of indirect effects and feedbacks. The values are very dependent on metric type and time horizon. The choice of metric and time horizon depends on the particular application and which aspects of climate change are considered relevant in a given context. Metrics do not define policies or goals but facilitate evaluation and implementation of multi-component policies to meet particular goals. All choices of metric contain implicit value-related judgements such as type of effect considered and weighting of effects over time. This assessment provides updated values of both GWP and GTP for many compounds. {8.7.1, 8.7.2, Table 8.7, Table 8.A.1, Supplementary Material Table 8.SM.16}

Forcing and temperature response can also be attributed to sectors. From this perspective and with the GTP metric, a single year's worth of current global emissions from the energy and industrial sectors have the largest contributions to global mean warming over the next approximately 50 to 100 years. Household fossil fuel and biofuel, biomass burning and on-road transportation are also relatively large contributors to warming over these time scales, while current emissions from sectors that emit large amounts of CH_4 (animal husbandry, waste/landfills and agriculture) are also important over shorter time horizons (up to 20 years). {8.7.2, Figure 8.34}

8

8.1 Radiative Forcing

There are a variety of ways to examine how various drivers contribute to climate change. In principle, observations of the climate response to a single factor could directly show the impact of that factor, or climate models could be used to study the impact of any single factor. In practice, however, it is usually difficult to find measurements that are influenced by only a single cause, and it is computationally prohibitive to simulate the response to every individual factor of interest. Hence various metrics intermediate between cause and effect are used to provide estimates of the climate impact of individual factors, with applications both in science and policy. Radiative forcing (RF) is one of the most widely used metrics, with most other metrics based on RF. In this chapter, we discuss RF from natural and anthropogenic components during the industrial period, presenting values for 2011 relative to 1750 unless otherwise stated, and projected values through 2100 (see also Annex II). In this section, we present the various definitions of RF used in this chapter, and discuss the utility and limitations of RF. These definitions are used in the subsequent sections quantifying the RF due to specific anthropogenic (Section 8.3) and natural (Section 8.4) causes and integrating RF due to all causes (Sections 8.5 and 8.6). Atmospheric chemistry relevant for RF is discussed in Section 8.2 and used throughout the chapter. Emission metrics using RF that are designed to facilitate rapid evaluation and comparison of the climate effects of emissions are discussed in Section 8.7.

8.1.1 The Radiative Forcing Concept

RF is the net change in the energy balance of the Earth system due to some imposed perturbation. It is usually expressed in watts per square meter averaged over a particular period of time and quantifies the energy imbalance that occurs when the imposed change takes place. Though usually difficult to observe, calculated RF provides a simple quantitative basis for comparing some aspects of the potential climate response to different imposed agents, especially global mean temperature, and hence is widely used in the scientific community. Forcing is often presented as the value due to changes between two particular times, such as pre-industrial to present-day, while its time evolution provides a more complete picture.

8.1.1.1 Defining Radiative Forcing

Alternative definitions of RF have been developed, each with its own advantages and limitations. The instantaneous RF refers to an instantaneous change in net (down minus up) radiative flux (shortwave plus longwave; in W m^{-2}) due to an imposed change. This forcing is usually defined in terms of flux changes at the top of the atmosphere (TOA) or at the climatological tropopause, with the latter being a better indicator of the global mean surface temperature response in cases when they differ.

Climate change takes place when the system responds in order to counteract the flux changes, and all such responses are explicitly excluded from this definition of forcing. The assumed relation between a sustained RF and the equilibrium global mean surface temperature response (ΔT) is $\Delta T = \lambda \text{RF}$ where λ is the climate sensitivity parameter. The relationship between RF and ΔT is an expression of the energy balance of the climate system and a simple reminder that the steady-state global mean climate response to a given forcing is determined both by the forcing and the responses inherent in λ.

Implicit in the concept of RF is the proposition that the change in net irradiance in response to the imposed forcing alone can be separated from all subsequent responses to the forcing. These are not in fact always clearly separable and thus some ambiguity exists in what may be considered a forcing versus what is part of the climate response.

In both the Third Assessment Report (TAR) and AR4, the term radiative forcing (RF, also called stratospherically adjusted RF, as distinct from instantaneous RF) was defined as the change in net irradiance at the tropopause after allowing for stratospheric temperatures to readjust to radiative equilibrium, while holding surface and tropospheric temperatures and state variables such as water vapour and cloud cover fixed at the unperturbed values[8]. RF is generally more indicative of the surface and tropospheric temperature responses than instantaneous RF, especially for agents such as carbon dioxide (CO_2) or ozone (O_3) change that substantially alter stratospheric temperatures. To be consistent with TAR and AR4, RF is hereafter taken to mean the stratospherically adjusted RF.

8.1.1.2 Defining Effective Radiative Forcing

For many forcing agents the RF gives a very useful and appropriate way to compare the relative importance of their potential climate effect. Instantaneous RF or RF is not an accurate indicator of the temperature response for all forcing agents, however. Rapid adjustments in the troposphere can either enhance or reduce the flux perturbations, leading to substantial differences in the forcing driving long-term climate change. In much the same way that allowing for the relatively rapid adjustment of stratospheric temperatures provides a more useful characterization of the forcing due to stratospheric constituent changes, inclusion of rapid tropospheric adjustments has the potential to provide more useful characterization for drivers in the troposphere (see also Section 7.1.3).

Many of the rapid adjustments affect clouds and are not readily included into the RF concept. For example, for aerosols, especially absorbing ones, changes in the temperature distribution above the surface occur due to a variety of effects, including cloud response to changing atmospheric stability (Hansen et al., 2005; see Section 7.3.4.2) and cloud absorption effects (Jacobson, 2012), which affect fluxes but are not strictly part of RF. Similar adjustments take place for many forcings, including CO_2 (see Section 7.2.5.6).

Aerosols also alter cloud properties via microphysical interactions leading to indirect forcings (referred to as aerosol–cloud interactions;

---

[8] Tropospheric variables were fixed except for the impact of aerosols on cloud albedo due to changes in droplet size with constant cloud liquid water which was considered an RF in AR4 but is part of ERF in AR5.

see Section 7.4). Although these adjustments are complex and not fully quantified, they occur both on the microphysical scale of the cloud particles as well as on a more macroscopic scale involving whole cloud systems (e.g., Shine et al., 2003; Penner et al., 2006; Quaas et al., 2009). A portion of these adjustments occurs over a short period, on cloud life cycle time scales, and is not part of a feedback arising from the surface temperature changes. Previously these type of adjustments were sometimes termed 'fast feedbacks' (e.g., Gregory et al., 2004; Hansen et al., 2005), whereas in AR5 they are denoted 'rapid adjustments' to emphasize their distinction from feedbacks involving surface temperature changes. Atmospheric chemistry responses have typically been included under the RF framework, and hence could also be included in a forcing encompassing rapid adjustments, which is important when evaluating forcing attributable to emissions changes (Section 8.1.2) and in the calculation of emission metrics (Section 8.7).

Studies have demonstrated the utility of including rapid adjustment in comparison of forcing agents, especially in allowing quantification of forcing due to aerosol-induced changes in clouds (e.g., effects previously denoted as cloud lifetime or semi-direct effects; see Figure 7.3) that are not amenable to characterization by RF (e.g., Rotstayn and Penner, 2001; Shine et al., 2003; Hansen et al., 2005; Lohmann et al., 2010; Ban-Weiss et al., 2012). Several measures of forcing have been introduced that include rapid adjustments. We term a forcing that accounts for rapid adjustments the effective radiative forcing (ERF). Conceptually, ERF represents the change in net TOA downward radiative flux after allowing for atmospheric temperatures, water vapour and clouds to adjust, but with global mean surface temperature or a portion of surface conditions unchanged. The primary methods in use for such calculations are (1) fixing sea surface temperatures (SSTs) and sea ice cover at climatological values while allowing all other parts of the system to respond until reaching steady state (e.g., Hansen et al., 2005) or (2) analyzing the transient global mean surface temperature response to an instantaneous perturbation and using the regression of the response extrapolated back to the start of the simulation to derive the initial ERF (Gregory et al., 2004; Gregory and Webb, 2008). The ERF calculated using the regression technique has an uncertainty of about 10% (for the 5 to 95% confidence interval) for a single $4 \times CO_2$ simulation (ERF ~7 W m^{-2}) due to internal variability in the transient climate (Andrews et al., 2012a), while given a similar length simulation the uncertainty due to internal variability in ERF calculated using the fixed-SST technique is much smaller and hence the latter may be more suitable for very small forcings. Analysis of both techniques shows that the fixed-SST method yields a smaller spread across models, even in calculations neglecting the uncertainty in the regression fitting procedure (Andrews et al., 2012a). As a portion of land area responses are included in the fixed-SST technique, however, that ERF is slightly less than it would be with surface temperature held fixed everywhere. It is possible to adjust for this in the global mean forcing, though we do not include such a correction here as we examine regional as well as global ERF, but the land response will also introduce artificial gradients in land–sea temperatures that could cause small local climate responses. In contrast, there is no global mean temperature response included in the regression method. Despite the low bias in fixed-SST ERF due to land responses, results from a multi-model analysis of the forcing due to CO_2 are 7% greater using this method than using the regression technique (Andrews et al., 2012a) though this is within the uncertainty range of the calculations. Although each technique has advantages, forcing diagnosed using the fixed-SST method is available for many more forcing agents in the current generation of climate models than forcing diagnosed using the regression method. Hence for practical purposes, ERF is hereafter used for results from the fixed-SST technique unless otherwise stated (see also Box 8.1).

The conceptual relation between instantaneous RF, RF and ERF is illustrated in Figure 8.1. It implies the adjustments to the instantaneous RF involve effects of processes that occur more rapidly than the time scale of the response of the global mean surface temperature to the forcing. However, there is no *a priori* time scale defined for adjustments to be rapid with the fixed-SST method. The majority take place on time scales

Box 8.1 | Definition of Radiative Forcing and Effective Radiative Forcing

The two most commonly used measures of radiative forcing in this chapter are the radiative forcing (RF) and the effective radiative forcing (ERF). RF is defined, as it was in AR4, as the change in net downward radiative flux at the tropopause after allowing for stratospheric temperatures to readjust to radiative equilibrium, while holding surface and tropospheric temperatures and state variables such as water vapor and cloud cover fixed at the unperturbed values.

ERF is the change in net TOA downward radiative flux after allowing for atmospheric temperatures, water vapour and clouds to adjust, but with surface temperature or a portion of surface conditions unchanged. Although there are multiple methods to calculate ERF, we take ERF to mean the method in which sea surface temperatures and sea ice cover are fixed at climatological values unless otherwise specified. Land surface properties (temperature, snow and ice cover and vegetation) are allowed to adjust in this method. Hence ERF includes both the effects of the forcing agent itself and the rapid adjustments to that agent (as does RF, though stratospheric temperature is the only adjustment for the latter). In the case of aerosols, the rapid adjustments of clouds encompass effects that have been referred to as indirect or semi-direct forcings (see Figure 7.3 and Section 7.5), with some of these same cloud responses also taking place for other forcing agents (see Section 7.2). Calculation of ERF requires longer simulations with more complex models than calculation of RF, but the inclusion of the additional rapid adjustments makes ERF a better indicator of the eventual global mean temperature response, especially for aerosols. When forcing is attributed to emissions or used for calculation of emission metrics, additional responses including atmospheric chemistry and the carbon cycle are also included in both RF and ERF (see Section 8.1.2). The general term *forcing* is used to refer to both RF and ERF.

Frequently Asked Questions

FAQ 8.1 | How Important Is Water Vapour to Climate Change?

As the largest contributor to the natural greenhouse effect, water vapour plays an essential role in the Earth's climate. However, the amount of water vapour in the atmosphere is controlled mostly by air temperature, rather than by emissions. For that reason, scientists consider it a feedback agent, rather than a forcing to climate change. Anthropogenic emissions of water vapour through irrigation or power plant cooling have a negligible impact on the global climate.

Water vapour is the primary greenhouse gas in the Earth's atmosphere. The contribution of water vapour to the natural greenhouse effect relative to that of carbon dioxide (CO_2) depends on the accounting method, but can be considered to be approximately two to three times greater. Additional water vapour is injected into the atmosphere from anthropogenic activities, mostly through increased evaporation from irrigated crops, but also through power plant cooling, and marginally through the combustion of fossil fuel. One may therefore question why there is so much focus on CO_2, and not on water vapour, as a forcing to climate change.

Water vapour behaves differently from CO_2 in one fundamental way: it can condense and precipitate. When air with high humidity cools, some of the vapour condenses into water droplets or ice particles and precipitates. The typical residence time of water vapour in the atmosphere is ten days. The flux of water vapour into the atmosphere from anthropogenic sources is considerably less than from 'natural' evaporation. Therefore, it has a negligible impact on overall concentrations, and does not contribute significantly to the long-term greenhouse effect. This is the main reason why tropospheric water vapour (typically below 10 km altitude) is not considered to be an anthropogenic gas contributing to radiative forcing.

Anthropogenic emissions do have a significant impact on water vapour in the stratosphere, which is the part of the atmosphere above about 10 km. Increased concentrations of methane (CH_4) due to human activities lead to an additional source of water, through oxidation, which partly explains the observed changes in that atmospheric layer. That stratospheric water change has a radiative impact, is considered a forcing, and can be evaluated. Stratospheric concentrations of water have varied significantly in past decades. The full extent of these variations is not well understood and is probably less a forcing than a feedback process added to natural variability. The contribution of stratospheric water vapour to warming, both forcing and feedback, is much smaller than from CH_4 or CO_2.

The maximum amount of water vapour in the air is controlled by temperature. A typical column of air extending from the surface to the stratosphere in polar regions may contain only a few kilograms of water vapour per square metre, while a similar column of air in the tropics may contain up to 70 kg. With every extra degree of air temperature, the atmosphere can retain around 7% more water vapour (see upper-left insert in the FAQ 8.1, Figure 1). This increase in concentration amplifies the greenhouse effect, and therefore leads to more warming. This process, referred to as the water vapour feedback, is well understood and quantified. It occurs in all models used to estimate climate change, where its strength is consistent with observations. Although an increase in atmospheric water vapour has been observed, this change is recognized as a climate feedback (from increased atmospheric temperature) and should not be interpreted as a radiative forcing from anthropogenic emissions. *(continued on next page)*

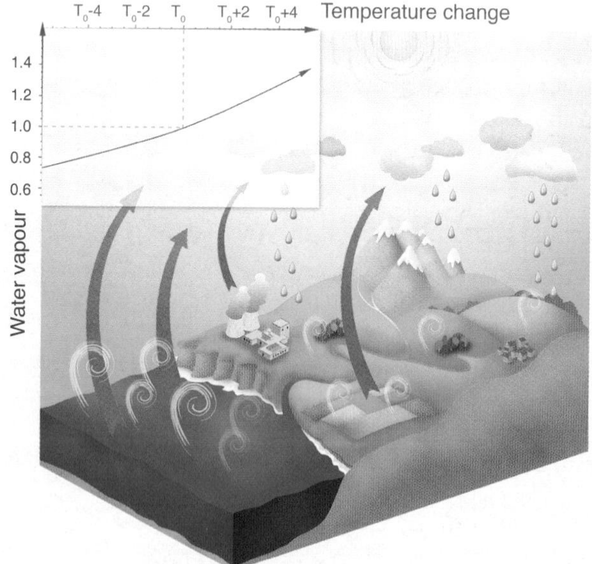

FAQ 8.1, Figure 1 | Illustration of the water cycle and its interaction with the greenhouse effect. The upper-left insert indicates the relative increase of potential water vapour content in the air with an increase of temperature (roughly 7% per degree). The white curls illustrate evaporation, which is compensated by precipitation to close the water budget. The red arrows illustrate the outgoing infrared radiation that is partly absorbed by water vapour and other gases, a process that is one component of the greenhouse effect. The stratospheric processes are not included in this figure.

FAQ 8.1 (continued)

Currently, water vapour has the largest greenhouse effect in the Earth's atmosphere. However, other greenhouse gases, primarily CO_2, are necessary to sustain the presence of water vapour in the atmosphere. Indeed, if these other gases were removed from the atmosphere, its temperature would drop sufficiently to induce a decrease of water vapour, leading to a runaway drop of the greenhouse effect that would plunge the Earth into a frozen state. So greenhouse gases other than water vapour provide the temperature structure that sustains current levels of atmospheric water vapour. Therefore, although CO_2 is the main anthropogenic control knob on climate, water vapour is a strong and fast feedback that amplifies any initial forcing by a typical factor between two and three. Water vapour is not a significant initial forcing, but is nevertheless a fundamental agent of climate change.

of seasons or less, but there is a spectrum of adjustment times. Changes in land ice and snow cover, for instance, may take place over many years. The ERF thus represents that part of the instantaneous RF that is maintained over long time scales and more directly contributes to the steady-state climate response. The RF can be considered a more limited version of ERF. Because the atmospheric temperature has been allowed to adjust, ERF would be nearly identical if calculated at the tropopause instead of the TOA for tropospheric forcing agents, as would RF. Recent work has noted likely advantages of the ERF framework for understanding model responses to CO_2 as well as to more complex forcing agents (see Section 7.2.5.6).

The climate sensitivity parameter λ derived with respect to RF can vary substantially across different forcing agents (Forster et al., 2007). The response to RF from a particular agent relative to the response to RF from CO_2 has been termed the *efficacy* (Hansen et al., 2005). By including many of the rapid adjustments that differ across forcing agents, the ERF concept includes much of their relative efficacy and therefore leads to more uniform climate sensitivity across agents. For example, the influence of clouds on the interaction of aerosols with sunlight and the effect of aerosol heating on cloud formation can lead to very large differences in the response per unit RF from black carbon (BC) located at different altitudes, but the response per unit ERF is nearly uniform with altitude (Hansen et al., 2005; Ming et al., 2010; Ban-Weiss et al., 2012). Hence as we use ERF in this chapter when it differs significantly from RF, efficacy is not used hereinafter. For inhomogeneous forcings, we note that the climate sensitivity parameter may also depend on the horizontal forcing distribution, especially with latitude (Shindell and Faluvegi, 2009; Section 8.6.2).

A combination of RF and ERF will be used in this chapter with RF provided to keep consistency with TAR and AR4, and ERF used to allow quantification of more complex forcing agents and, in some cases, provide a more useful metric than RF.

8.1.1.3 Limitations of Radiative Forcing

Both the RF and ERF concepts have strengths and weaknesses in addition to those discussed previously. Dedicated climate model simulations that are required to diagnose the ERF can be more computationally demanding than those for instantaneous RF or RF because many years are required to reduce the influence of climate variability. The presence of meteorological variability can also make it difficult to isolate the ERF of small forcings that are easily isolated in the pair of radiative transfer calculations performed for RF (Figure 8.1). For RF, on the other hand, a definition of the tropopause is required, which can be ambiguous.

In many cases, however, ERF and RF are nearly equal. Analysis of 11 models from the current Coupled Model Intercomparison Project Phase 5 (CMIP5) generation finds that the rapid adjustments to CO_2 cause fixed-SST-based ERF to be 2% less than RF, with an intermodel standard deviation of 7% (Vial et al., 2013). This is consistent with an earlier study of six GCMs that found a substantial inter-model variation in the rapid tropospheric adjustment to CO_2 using regression analysis in slab ocean models, though the ensemble mean adjustment was less than 5% (Andrews and Forster, 2008). Part of the large uncertainty range arises from the greater noise inherent in regression analyses of single runs in comparison with fixed-SST experiments. Using fixed-SST simulations, Hansen et al. (2005) found that ERF is virtually identical to RF for increased CO_2, tropospheric ozone and solar irradiance, and within 6% for methane (CH_4), nitrous oxide (N_2O), stratospheric aerosols and for the aerosol–radiation interaction of reflective aerosols. Shindell et al. (2013b) also found that RF and ERF are statistically equal for tropospheric ozone. Lohmann et al. (2010) report a small increase in the forcing from CO_2 using ERF instead of RF based on the fixed-SST technique, while finding no substantial difference for CH_4, RF due to aerosol–radiation interactions or aerosol effects on cloud albedo. In the fixed-SST simulations of Hansen et al. (2005), ERF was about 20% less than RF for the atmospheric effects of BC aerosols (not including microphysical aerosol–cloud interactions), and nearly 300% greater for the forcing due to BC snow albedo forcing (Hansen et al., 2007). ERF was slightly greater than RF for stratospheric ozone in Hansen et al. (2005), but the opposite is true for more recent analyses (Shindell et al., 2013b), and hence it seems most appropriate at present to use RF for this small forcing. The various studies demonstrate that RF provides a good estimate of ERF in most cases, as the differences are very small, with the notable exceptions of BC-related forcings (Bond et al., 2013). ERF provides better characterization of those effects, as well as allowing quantification of a broader range of effects including all aerosol–cloud interactions. Hence while RF and ERF are generally quite similar for WMGHGs, ERF typically provides a more useful indication of climate response for near-term climate forcers (see Box 8.2). As the rapid adjustments included in ERF differ in strength across climate models, the uncertainty range for ERF estimates tends to be larger than the range for RF estimates.

Box 8.2 | Grouping Forcing Compounds by Common Properties

As many compounds cause RF when their atmospheric concentration is changed, it can be useful to refer to groups of compounds with similar properties. Here we discuss two primary groupings: well-mixed greenhouse gases (WMGHGs) and near-term climate forcers (NTCFs).

We define as 'well-mixed' those greenhouse gases that are sufficiently mixed throughout the troposphere that concentration measurements from a few remote surface sites can characterize the climate-relevant atmospheric burden; although these gases may still have local variation near sources and sinks and even small hemispheric gradients. Global forcing per unit emission and emission metrics for these gases thus do not depend on the geographic location of the emission, and forcing calculations can assume even horizontal distributions. These gases, or a subset of them, have sometimes been referred to as 'long-lived greenhouse gases' as they are well mixed because their atmospheric lifetimes are much greater than the time scale of a few years for atmospheric mixing, but the physical property that causes the aforementioned common characteristics is more directly associated with their mixing within the atmosphere. WMGHGs include CO_2, N_2O, CH_4, SF_6, and many halogenated species. Conversely, ozone is not a WMGHG.

We define 'near-term climate forcers' (NTCFs) as those compounds whose impact on climate occurs primarily within the first decade after their emission. This set of compounds is composed primarily of those with short lifetimes in the atmosphere compared to WMGHGs, and has been sometimes referred to as short-lived climate forcers or short-lived climate pollutants. However, the common property that is of greatest interest to a climate assessment is the time scale over which their impact on climate is felt. This set of compounds includes methane, which is also a WMGHG, as well as ozone and aerosols, or their precursors, and some halogenated species that are not WMGHGs. These compounds do not accumulate in the atmosphere at decadal to centennial time scales, and so their effect on climate is predominantly in the near term following their emission.

Whereas the global mean ERF provides a useful indication of the eventual change in global mean surface temperature, it does not reflect regional climate changes. This is true for all forcing agents, but is especially the case for the inhomogeneously distributed forcings because they activate climate feedbacks based on their regional distribution. For example, forcings over Northern Hemisphere (NH) middle and high latitudes induce snow and ice albedo feedbacks more than forcings at lower latitudes or in the Southern Hemisphere (SH) (e.g., Shindell and Faluvegi, 2009).

In the case of agents that strongly absorb incoming solar radiation (such as BC, and to a lesser extent organic carbon (OC) and ozone) the TOA forcing provides little indication of the change in solar radiation reaching the surface which can force local changes in evaporation and alter regional and general circulation patterns (e.g., Ramanathan and Carmichael, 2008; Wang et al., 2009). Hence the forcing at the surface, or the atmospheric heating, defined as the difference between surface and tropopause/TOA forcing, might also be useful metrics. Global mean precipitation changes can be related separately to ERF within the atmosphere and to a slower response to global mean temperature changes (Andrews et al., 2010; Ming et al., 2010; Ban-Weiss et al., 2012). Relationships between surface forcing and localized aspects of climate response have not yet been clearly quantified, however.

In general, most widely used definitions of forcing and most forcing-based metrics are intended to be proportional to the eventual temperature response, and most analyses to date have explored the global mean temperature response only. These metrics do not explicitly include impacts such as changes in precipitation, surface sunlight available for photosynthesis, extreme events, and so forth, or regional temperatures, which can differ greatly from the global mean. Hence although they are quite useful for understanding the factors driving global mean temperature change, they provide only an imperfect and limited perspective on the factors driving broader climate change. In addition, a metric based solely on radiative perturbations does not allow comparison of non-RFs, such as effects of land cover change on evapotranspiration or physiological impacts of CO_2 and O_3 except where these cause further impacts on radiation such as through cloud cover changes (e.g., Andrews et al., 2012b).

8.1.2 Calculation of Radiative Forcing due to Concentration or Emission Changes

Analysis of forcing due to observed or modelled concentration changes between pre-industrial, defined here as 1750, and a chosen later year provides an indication of the importance of different forcing agents to climate change during that period. Such analyses have been a mainstay of climate assessments. This perspective has the advantage that observational data are available to accurately quantify the concentration changes for several of the largest forcing components. Atmospheric concentration changes, however, are the net result of variations in emissions of multiple compounds and any climate changes that have influenced processes such as wet removal, atmospheric chemistry or the carbon cycle. Characterizing forcing according to *concentration* changes thus mixes multiple root causes along with climate feedbacks. Policy decisions are better informed by analysis of forcing attributable to *emissions*, which the IPCC first presented in AR4. These analyses can be applied to historical emissions changes in a 'backward-looking' perspective, as done for example, for major WMGHGs (den Elzen et al., 2005; Hohne et al., 2011) and NTCFs (Shindell et al., 2009), or to current

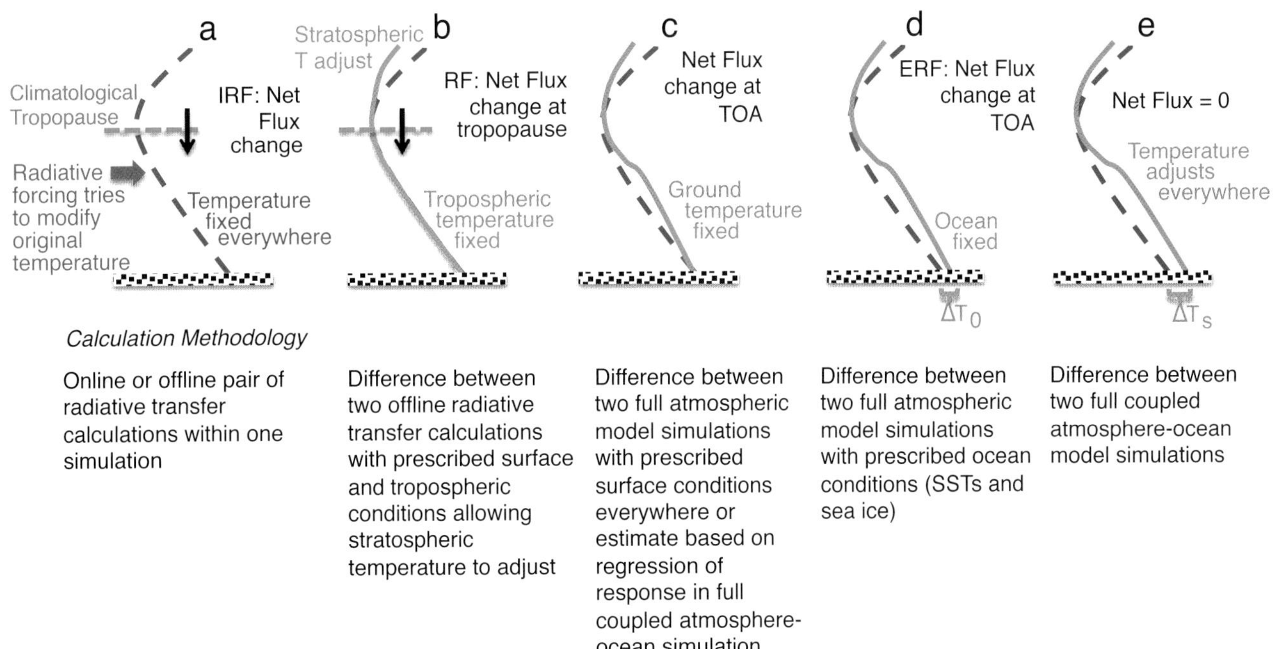

Figure 8.1 | Cartoon comparing (a) instantaneous RF, (b) RF, which allows stratospheric temperature to adjust, (c) flux change when the surface temperature is fixed over the whole Earth (a method of calculating ERF), (d) the ERF calculated allowing atmospheric and land temperature to adjust while ocean conditions are fixed and (e) the equilibrium response to the climate forcing agent. The methodology for calculation of each type of forcing is also outlined. ΔT_o represents the land temperature response, while ΔT_s is the full surface temperature response. (Updated from Hansen et al., 2005.)

or projected future emissions in a 'forward-looking' view (see Section 8.7). Emissions estimates through time typically come from the scientific community, often making use of national reporting for recent decades.

With the greater use of emission-driven models, for example, in CMIP5, it is becoming more natural to estimate ERF resulting from emissions of a particular species rather than concentration-based forcing. Such calculations typically necessitate model simulations with chemical transport models or chemistry–climate models, however, and require careful consideration of which processes are included, especially when comparing results to concentration-based forcings. In particular, simulation of concentration responses to emissions changes requires incorporating models of the carbon cycle and atmospheric chemistry (gas and aerosol phases). The requisite expansion of the modelling realm for emissions-based forcing or emission metrics should in principle be consistent for all drivers. For example, as the response to aerosol or ozone precursor emissions includes atmospheric chemistry, the response to CO_2 emissions should as well. In addition, if the CO_2 concentration responses to CO_2 emissions include the impact of CO_2-induced climate changes on carbon uptake, then the effect of climate changes caused by any other emission on carbon uptake should also be included. Similarly, if the effects of atmospheric CO_2 concentration change on carbon uptake are included, the effects of other atmospheric composition or deposition changes on carbon uptake should be included as well (see also Section 6.4.1). Comparable issues are present for other forcing agents. In practice, the modelling realm used in studies of forcing attributable to emissions has not always been consistent. Furthermore, climate feedbacks have sometimes been included in the calculation of forcing due to ozone or aerosol changes, as when concentrations from a historical transient climate simulation are imposed for an ERF calculation. In this chapter, we endeavour to clarify which processes have been included in the various estimates of forcing attributed to emissions (Sections 8.3 and 8.7).

RF or ERF estimates based on either historical emissions or concentrations provide valuable insight into the relative and absolute contribution of various drivers to historical climate change. Scenarios of changing future emissions and land use are also developed based on various assumptions about socioeconomic trends and societal choices. The forcing resulting from such scenarios is used to understand the drivers of potential future climate changes (Sections 8.5.3 and 8.6). As with historical forcings, the actual impact on climate depends on both the temporal and spatial structure of the forcings and the rate of response of various portions of the climate system.

8.2 Atmospheric Chemistry

8.2.1 Introduction

Most radiatively active compounds in the Earth's atmosphere are chemically active, meaning that atmospheric chemistry plays a large role in determining their burden and residence time. In the atmosphere, a gaseous chemically active compound can be affected by (1) interaction with other species (including aerosols and water) in its immediate vicinity and (2) interaction with solar radiation (photolysis). Physical processes (wet removal and dry deposition) act on some chemical compounds (gas or aerosols) to further define their residence time in the atmosphere. Atmospheric chemistry is characterized by many interactions and patterns of temporal or spatial variability, leading to significant nonlinearities (Kleinman et al., 2001) and a wide range of time scales of importance (Isaksen et al., 2009).

This section assesses updates in understanding of processes, modelling and observations since AR4 (see Section 2.3) on key reactive species contributing to RF. Note that aerosols, including processes responsible for the formation of aerosols, are extensively described in Section 7.3.

8.2.2 Global Chemistry Modelling in Coupled Model Intercomparison Project Phase 5

Because the distribution of NTCFs cannot be estimated from observations alone, coupled chemistry-climate simulations are required to define their evolution and associated RF. While several CMIP5 modeling groups performed simulations with interactive chemistry (i.e., computed simultaneously within the climate model), many models used as input pre-computed distributions of radiatively active gases and/or aerosols. To assess the distributions of chemical species and their respective RF, many research groups participated in the Atmospheric Chemistry and Climate Model Intercomparison Project (ACCMIP).

The ACCMIP simulations (Lamarque et al., 2013) were defined to provide information on the long-term changes in atmospheric composition with a few, well-defined atmospheric simulations. Because of the nature of the simulations (pre-industrial, present-day and future climates), only a limited number of chemistry-transport models (models which require a full definition of the meteorological fields needed to simulate physical processes and transport) participated in the ACCMIP project, which instead drew primarily from the same General Circulation Models (GCMs) as CMIP5 (see Lamarque et al., 2013 for a list of the participating models and their configurations), with extensive model evaluation against observations (Bowman et al., 2013; Lee et al., 2013; Shindell et al., 2013c; Voulgarakis et al., 2013; Young et al., 2013).

In all CMIP5/ACCMIP chemistry simulations, anthropogenic and biomass burning emissions are specified. More specifically, a single set of historical anthropogenic and biomass burning emissions (Lamarque et al., 2010) and one set of emissions for each of the RCPs (van Vuuren et al., 2011) was defined (Figure 8.2). This was designed to increase the comparability of simulations. However, these uniform emission specifications mask the existing uncertainty (e.g., Bond et al., 2007; Lu et al., 2011), so that there is in fact a considerable range in the estimates and time evolution of recent anthropogenic emissions (Granier et al., 2011). Historical reconstructions of biomass burning (wildfires and deforestation) also exhibit quite large uncertainties (Kasischke and Penner, 2004; Ito and Penner, 2005; Schultz et al., 2008; van der Werf et al., 2010). In addition, the RCP biomass burning projections do not include the feedback between climate change and fires discussed in Bowman et al. (2009), Pechony and Shindell (2010) and Thonicke et al. (2010). Finally, the RCP anthropogenic precursor emissions of NTCFs tend to span a smaller range than available from existing scenarios (van Vuuren et al., 2011). The ACCMIP simulations therefore provide an estimate of the uncertainty due to range of representation of physical and chemical processes in models, but do not incorporate uncertainty in emissions.

8.2.3 Chemical Processes and Trace Gas Budgets

8.2.3.1 Tropospheric Ozone

The RF from tropospheric ozone is strongly height- and latitude-dependent through coupling of ozone change with temperature, water vapour and clouds (Lacis et al., 1990; Berntsen et al., 1997; Worden et al., 2008, 2011; Bowman et al., 2013). Consequently, it is necessary to accurately estimate the change in the ozone spatio-temporal structure using global models and observations. It is also well established that surface ozone detrimentally affects plant productivity (Ashmore, 2005; Fishman et al., 2010), albeit estimating this impact on climate, although possibly significant, is still limited to a few studies (Sitch et al., 2007; UNEP, 2011).

Tropospheric ozone is a by-product of the oxidation of carbon monoxide (CO), CH_4, and non-CH_4 hydrocarbons in the presence of nitrogen oxides (NO_x). As emissions of these precursors have increased (Figure 8.2), tropospheric ozone has increased since pre-industrial times (Volz and Kley, 1988; Marenco et al., 1994) and over the last decades (Parrish et al., 2009; Cooper et al., 2010; Logan et al., 2012), but with important regional variations (Section 2.2). Ozone production is usually limited by the supply of HO_x (OH + HO_2) and NO_X (NO + NO_2) (Levy, 1971; Logan et al., 1981). Ozone's major chemical loss pathways in the troposphere are through (1) photolysis (to $O(^1D)$, followed by reaction with water vapour) and (2) reaction with HO_2 (Seinfeld and Pandis, 2006). The former pathway leads to couplings between stratospheric ozone (photolysis rate being a function of the overhead ozone column) and climate change (through water vapour). Observed surface ozone abundances typically range from less than 10 ppb over the tropical Pacific Ocean to more than 100 ppb downwind of highly emitting regions. The lifetime of ozone in the troposphere varies strongly with season and location: it may be as little as a few days in the tropical boundary layer, or as much as 1 year in the upper troposphere. Two recent studies give similar global mean lifetime of ozone: 22.3 ± 2 days (Stevenson et al., 2006) and 23.4 ± 2.2 days (Young et al., 2013).

For present (about 2000) conditions, the various components of the budget of global mean tropospheric ozone are estimated from the ACCMIP simulations and other model simulations since AR4 (Table 8.1). In particular, most recent models define a globally and annually averaged tropospheric ozone burden of (337 ± 23 Tg, 1-σ). Differences in the definition of the tropopause lead to inter-model variations of approximately 10% (Wild, 2007). This multi-model mean estimate of global annual tropospheric ozone burden has not significantly changed since the Stevenson et al. (2006) estimates (344 ± 39 Tg, 1-σ), and is consistent with the most recent satellite-based Ozone Monitoring Instrument–Microwave Limb Sounder (OMI-MLS; Ziemke et al., 2011) and Tropospheric Emission Spectrometer (TES; Osterman et al., 2008) climatologies.

Estimates of the ozone chemical sources and sinks (uncertainty estimates are quoted here as 1-σ) are less robust, with a net chemical production (production *minus* loss) of 618 ± 275 Tg yr^{-1} (Table 8.1), larger than the Atmospheric Composition Change: a European Network (ACCENT) results (442 ± 309 Tg yr^{-1}; Stevenson et al., 2006). Estimates of ozone deposition (1094 ± 264 Tg yr^{-1}) are slightly increased

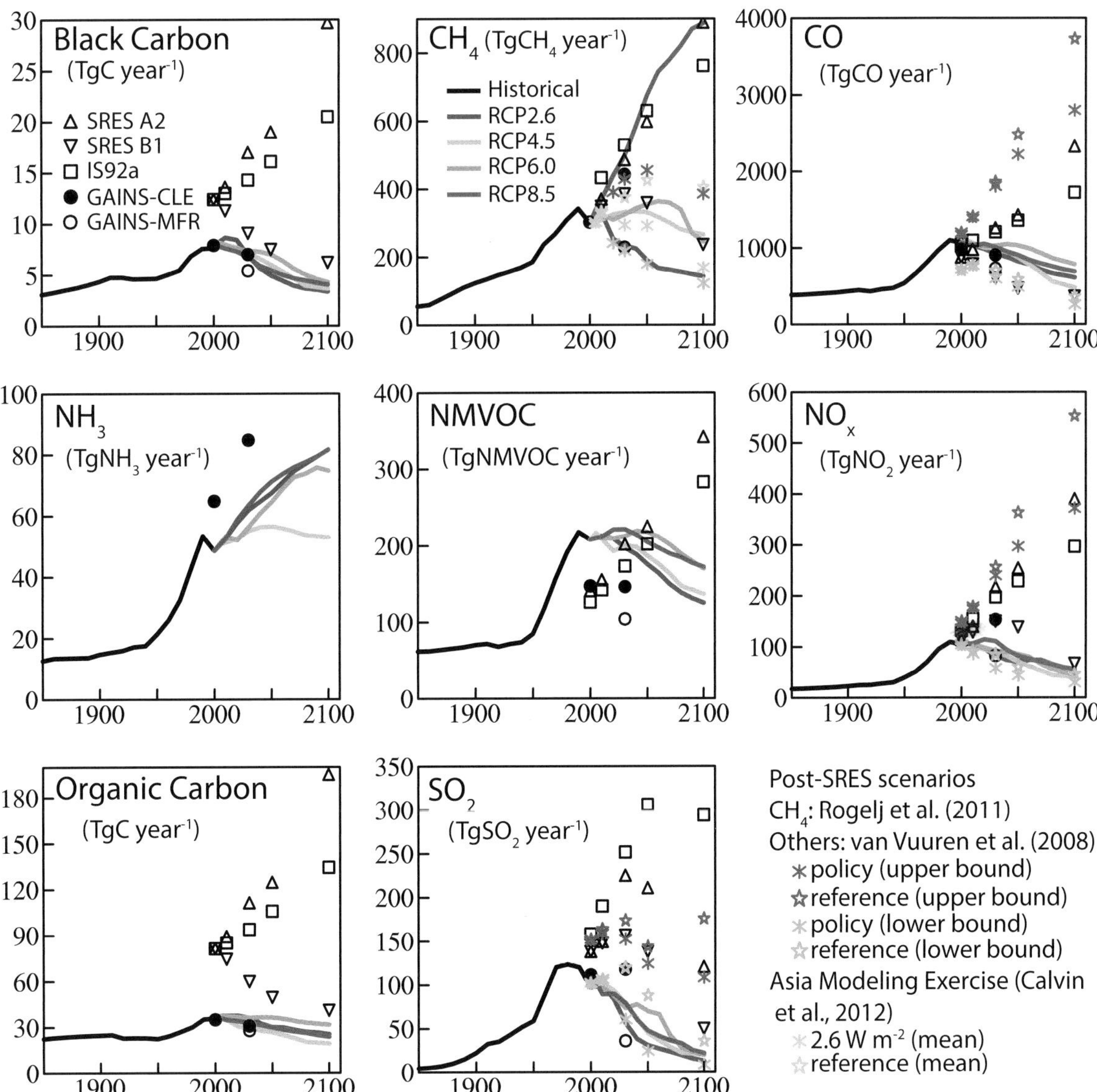

Figure 8.2 | Time evolution of global anthropogenic and biomass burning emissions 1850–2100 used in CMIP5/ACCMIP following each RCP. Historical (1850–2000) values are from Lamarque et al. (2010). RCP values are from van Vuuren et al. (2011). Emissions estimates from Special Report on Emission Scenarios (SRES) are discussed in Annex II; note that black carbon and organic carbon estimates were not part of the SRES and are shown here only for completeness. The Maximum Feasible Reduction (MFR) and Current Legislation (CLE) are discussed in Cofala et al. (2007); as biomass burning emissions are not included in that publication, a fixed amount, equivalent to the value in 2000 from the RCP estimates, is added (see Annex II for more details; Dentener et al., 2006). The post-SRES scenarios are discussed in Van Vuuren et al. (2008) and Rogelj et al. (2011). For those, only the range (minimum to maximum) is shown. Global emissions from the Asian Modelling Exercise are discussed in Calvin et al. (2012). Regional estimates are shown in Supplementary Material Figure 8.SM.1 and Figure 8.SM.2 for the historical and RCPs.

since ACCENT (1003 ± 200 Tg yr^{-1}) while estimates of the net influx of ozone from the stratosphere to the troposphere (477 ± 96 Tg yr^{-1}) have slightly decreased since ACCENT (552 ± 168 Tg yr^{-1}). Additional model estimates of this influx (Hegglin and Shepherd, 2009; Hsu and Prather, 2009) fall within both ranges, as do estimates based on observations (Murphy and Fahey, 1994; Gettelman et al., 1997; Olsen et al., 2002), all estimates being sensitive to their choice of tropopause definition and interannual variability.

Model simulations for present-day conditions or the recent past are evaluated (Figure 8.3) against frequent ozonesonde measurements (Logan, 1999; Tilmes et al., 2012) and additional surface, aircraft and satellite measurements. The ACCMIP model simulations (Figure 8.3) indicate 10 to 20% negative bias at 250 hPa in the SH tropical region, and a slight underestimate in NH tropical region. Comparison with satellite-based estimates of tropospheric ozone column (Ziemke et al., 2011) indicates an annual mean bias of –4.3 ± 29 Tg (with a spatial correlation of 0.87 ± 0.07, 1-σ) for the ACCMIP simulations (Young et al., 2013). Overall, our ability to simulate tropospheric ozone burden for present (about 2000) has not substantially changed since AR4. Evaluation (using a subset of two ACCMIP models) of simulated trends (1960s to present or shorter) in surface ozone against observations at remote surface sites (see Section 2.2) indicates an underestimation, especially in the NH (Lamarque et al., 2010). Although this limits the ability to represent recent ozone changes, it is unclear how this translates into an uncertainty on changes since pre-industrial times.

8

Table 8.1 | Summary of tropospheric ozone global budget model and observation estimates for present (about 2000) conditions. Focus is on modelling studies published since AR4. STE stands for stratosphere–troposphere exchange. All uncertainties quoted as 1 standard deviation (68% confidence interval).

Burden	Production	Loss	Deposition	STE	Reference
Tg	Tg yr^{-1}	Tg yr^{-1}	Tg yr^{-1}	Tg yr^{-1}	
Modelling Studies					
337 ± 23	4877 ± 853	4260 ± 645	1094 ± 264	477 ± 96	Young et al. (2013); ACCMIP
323	N/A	N/A	N/A	N/A	Archibald et al. (2011)
330	4876	4520	916	560	Kawase et al. (2011)
312	4289	3881	829	421	Huijnen et al. (2010)
334	3826	3373	1286	662	Zeng et al. (2010)
324	4870	4570	801	502	Wild and Palmer (2008)
314	N/A	N/A	1035	452	Zeng et al. (2008)
319	4487	3999	N/A	500	Wu et al. (2007)
372	5042	4507	884	345	Horowitz (2006)
349	4384	3972	808	401	Liao et al. (2006)
344 ± 39	5110 ± 606	4668 ± 727	1003 ± 200	552 ± 168	Stevenson et al. (2006); ACCENT
314 ± 33	4465 ± 514	4114 ± 409	949 ± 222	529 ± 105	Wild (2007) (post-2000 studies)
N/A	N/A	N/A	N/A	515	Hsu and Prather (2009)
N/A	N/A	N/A	N/A	655	Hegglin and Shepherd (2009)
N/A	N/A	N/A	N/A	383–451	Clark et al. (2007)
Observational Studies					
333	N/A	N/A	N/A	N/A	Fortuin and Kelder (1998)
327	N/A	N/A	N/A	N/A	Logan (1999)
325	N/A	N/A	N/A	N/A	Ziemke et al. (2011); 60S–60N
319–351	N/A	N/A	N/A	N/A	Osterman et al. (2008); 60S–60N
N/A	N/A	N/A	N/A	449 (192–872)	Murphy and Fahey (1994)
N/A	N/A	N/A	N/A	510 (450–590)	Gettelman et al. (1997)
N/A	N/A	N/A	N/A	500 ± 140	Olsen et al. (2001)

In most studies 'pre-industrial' does not identify a specific year but is usually assumed to correspond to 1850s levels; no observational information on ozone is available for that time period. Using the Lamarque et al. (2010) emissions, the ACCMIP models (Young et al., 2013) are unable to reproduce the low levels of ozone observed at Montsouris 1876–1886 (Volz and Kley, 1988). The other early ozone measurements using the Schönbein paper are controversial (Marenco et al., 1994) and assessed to be of qualitative use only. The main uncertainty in estimating the pre-industrial to present-day change in ozone therefore remains the lack of constraint on emission trends because of the very incomplete knowledge of pre-industrial ozone concentrations, of which no new information is available. The uncertainty on pre-industrial conditions is not confined to ozone but applies to aerosols as well (e.g., Schmidt et al., 2012), although ice and lake core records provide some constraint on pre-industrial aerosol concentrations.

The ACCMIP results provide an estimated tropospheric ozone increase (Figure 8.4) from 1850 to 2000 of 98 ± 17 Tg (model range), similar to AR4 estimates. Skeie et al. (2011a) found an additional 5% increase in the anthropogenic contribution to the ozone burden between 2000 and 2010, which translates into an approximately 1.5% increase in tropospheric ozone burden. A best estimate of the change in ozone since 1850 is assessed at 100 ± 25 Tg (1-σ). Attribution simulations (Stevenson et al., 2013) indicate unequivocally that anthropogenic changes in ozone precursor emissions are responsible for the increase between 1850 and present or into the future.

8.2.3.2 Stratospheric Ozone and Water Vapour

Stratospheric ozone has experienced significant depletion since the 1960s due to bromine and chlorine-containing compounds (Solomon, 1999), leading to an estimated global decrease of stratospheric ozone of 5% between the 1970s and the mid-1990s, the decrease being largest over Antarctica (Fioletov et al., 2002). Most of the ozone loss is associated with the long-lived bromine and chlorine-containing compounds (chlorofluorocarbons and substitutes) released by human activities, in addition to N_2O. This is in addition to a background level of natural emissions of short-lived halogens from oceanic and volcanic sources.

With the advent of the Montreal Protocol and its amendments, emissions of chlorofluorocarbons (CFCs) and replacements have strongly declined (Montzka et al., 2011), and signs of ozone stabilization and even possibly recovery have already occurred (Mader et al., 2010; Salby et al., 2012). A further consequence is that N_2O emissions (Section 8.2.3.4) *likely* dominate all other emissions in terms of ozone-depleting

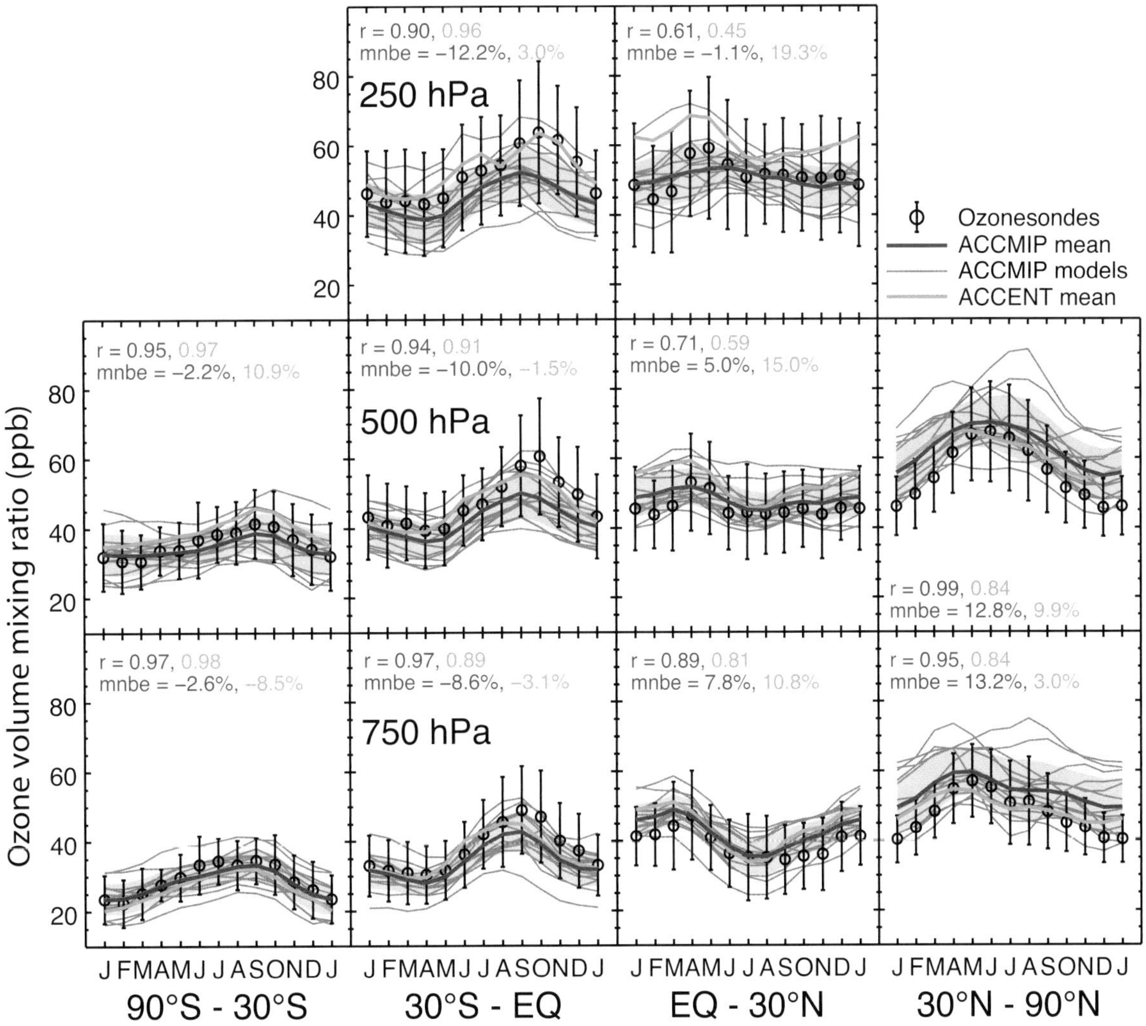

Figure 8.3 | Comparisons between observations and simulations for the monthly mean ozone for ACCMIP results (Young et al., 2013). ACCENT refers to the model results in Stevenson et al. (2006). For each box, the correlation of the seasonal cycle is indicated by the *r* value, while the mean normalized bias estimated is indicated by *mnbe* value.

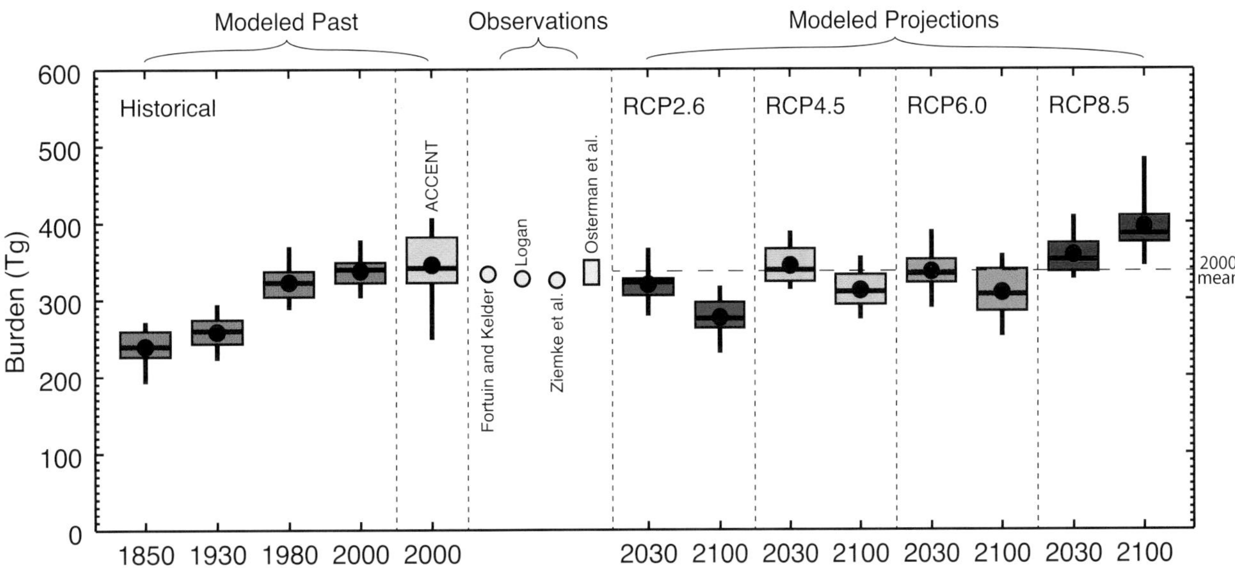

Figure 8.4 | Time evolution of global tropospheric ozone burden (in Tg(O_3)) from 1850 to 2100 from ACCMIP results, ACCENT results (2000 only), and observational estimates (see Table 8.1). The box, whiskers, line and dot show the interquartile range, full range, median and mean burdens and differences, respectively. The dashed line indicates the 2000 ACCMIP mean. (Adapted from Young et al., 2013.)

potential (Ravishankara et al., 2009). Chemistry-climate models with resolved stratospheric chemistry and dynamics recently predicted an estimated global mean total ozone column recovery to 1980 levels to occur in 2032 (multi-model mean value, with a range of 2024 to 2042) under the A1B scenario (Eyring et al., 2010a). Increases in the stratospheric burden and acceleration of the stratospheric circulation leads to an increase in the stratosphere–troposphere flux of ozone (Shindell et al., 2006c; Grewe, 2007; Hegglin and Shepherd, 2009; Zeng et al., 2010). This is also seen in recent RCP8.5 simulations, with the impact of increasing tropospheric burden (Kawase et al., 2011; Lamarque et al., 2011). However, observationally based estimates of recent trends in age of air (Engel et al., 2009; Stiller et al., 2012) do not appear to be consistent with the acceleration of the stratospheric circulation found in model simulations, possibly owing to inherent difficulties with extracting trends from SF_6 observations (Garcia et al., 2011).

Oxidation of CH_4 in the stratosphere (see Section 8.2.3.3) is a significant source of water vapour and hence the long-term increase in CH_4 leads to an anthropogenic forcing (see Section 8.3) in the stratosphere. Stratospheric water vapour abundance increased by an average of 1.0 ± 0.2 (1-σ) ppm during 1980–2010, with CH_4 oxidation explaining approximately 25% of this increase (Hurst et al., 2011). Other factors contributing to the long-term change in water vapour include changes in tropical tropopause temperatures (see Section 2.2.2.1).

8.2.3.3 Methane

The surface mixing ratio of CH_4 has increased by 150% since pre-industrial times (Sections 2.2.1.1.2 and 8.3.2.2), with some projections indicating a further doubling by 2100 (Figure 8.5). Bottom-up estimates of present CH_4 emissions range from 542 to 852 $TgCH_4 yr^{-1}$ (see Table 6.8), while a recent top-down estimate with uncertainty analysis is 554 ± 56 $TgCH_4 yr^{-1}$ (Prather et al., 2012). All quoted uncertainties in Section 8.2.3.3 are defined as 1-σ.

The main sink of CH_4 is through its reaction with the hydroxyl radical (OH) in the troposphere (Ehhalt and Heidt, 1973). A primary source of tropospheric OH is initiated by the photodissociation of ozone, followed by reaction with water vapour (creating sensitivity to humidity, cloud cover and solar radiation) (Levy, 1971; Crutzen, 1973). The other main source of OH is through secondary reactions (Lelieveld et al., 2008), although some of those reactions are still poorly understood (Paulot et al., 2009; Peeters et al., 2009; Taraborrelli et al., 2012). A recent estimate of the CH_4 tropospheric chemical lifetime with respect to OH constrained by methyl chloroform surface observations is 11.2 ± 1.3 years (Prather et al., 2012). In addition, bacterial uptake in soils provides an additional small, less constrained loss (Fung et al., 1991); estimated lifetime = 120 ± 24 years (Prather et al., 2012), with another small loss in the stratosphere (Ehhalt and Heidt, 1973); estimated lifetime = 150 ± 50 years (Prather et al., 2012). Halogen chemistry in the troposphere also contributes to some tropospheric CH_4 loss (Allan et al., 2007), estimated lifetime = 200 ± 100 years (Prather et al., 2012).

The ACCMIP estimate for present CH_4 lifetime with respect to tropospheric OH varies quite widely (9.8 ± 1.6 years (Voulgarakis et al., 2013)), slightly shorter than the 10.2 ± 1.7 years in (Fiore et al. (2009), but much shorter than the methyl chloroform-based estimate of 11.2 ± 1.3 years (Prather et al., 2012). A partial explanation for the range in CH_4 lifetime changes can be found in the degree of representation of chemistry in chemistry–climate models. Indeed, Archibald et al. (2010) showed that the response of OH to increasing nitrogen oxides strongly depends on the treatment of hydrocarbon chemistry in a model. The impact on CH_4 distribution in the ACCMIP simulations is, however, rather limited because most models prescribed CH_4 as a time-varying lower-boundary mixing ratio (Lamarque et al., 2013).

The chemical coupling between OH and CH_4 leads to a significant amplification of an emission impact; that is, increasing CH_4 emissions decreases tropospheric OH which in turn increases the CH_4 lifetime and therefore its burden. The OH-lifetime sensitivity for CH_4, s_OH = –δln(OH)/δln(CH4), was estimated in Chapter 4 of TAR to be 0.32, implying a 0.32% decrease in tropospheric mean OH (as weighted by CH_4 loss) for a 1% increase in CH_4. The Fiore et al. (2009) multi-model (12 models) study provides a slightly smaller value (0.28 ± 0.03). Holmes et al. (2013) gives a range 0.31 ± 0.04 by combining Fiore et al. (2009), Holmes et al. (2011) and three new model results (0.36, 0.31, 0.27). Only two ACCMIP models reported values (0.19 and 0.26; Voulgarakis et al., 2013). The projections of future CH_4 in Chapter 11 use the Holmes et al. (2013) range and uncertainty, which at the 2-σ level covers all but one model result. The feedback factor *f*, the ratio of the

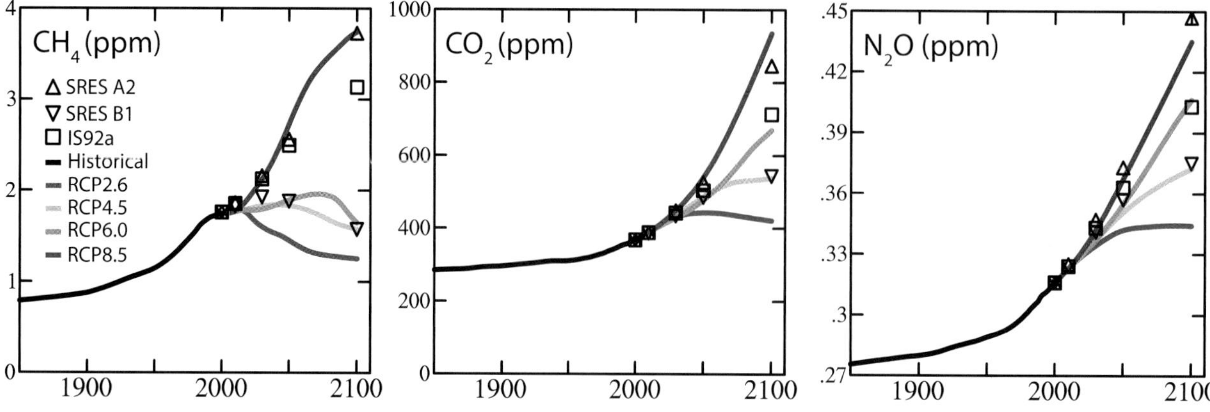

Figure 8.5 | Time evolution of global-averaged mixing ratio of long-lived species1850–2100 following each RCP; blue (RCP2.6), light blue (RCP4.5), orange (RCP6.0) and red (RCP8.5). (Based on Meinshausen et al., 2011b.)

lifetime of a CH_4 perturbation to the lifetime of the total CH_4 burden, is calculated as f = 1/(1-s). Other CH_4 losses, which are relatively insensitive to CH_4 burden, must be included so that $f = 1.34 \pm 0.06$, (slightly larger but within the range of the Stevenson et al. (2006) estimate of 1.29 ± 0.04, based on six models), leading to an overall perturbation lifetime of 12.4 ± 1.4 years, which is used in calculations of metrics in Section 8.7. Additional details are provided in the Supplementary Material Section 8.SM.2.

8.2.3.4 Nitrous Oxide

Nitrous oxide (N_2O) in 2011 has a surface concentration 19% above its 1750 level (Sections 2.2.1.1.3 and 8.3.2.3). Increases in N_2O lead to depletion of mid- to upper-stratospheric ozone and increase in mid-latitude lower stratospheric ozone (as a result of increased photolysis rate from decreased ozone above). This impacts tropospheric chemistry through increase in stratosphere–troposphere exchange of ozone and odd nitrogen species and increase in tropospheric photolysis rates and OH formation (Prather and Hsu, 2010). Anthropogenic emissions represent around 30 to 45% of the present-day global total, and are mostly from agricultural and soil sources (Fowler et al., 2009) and fossil-fuel activities. Natural emissions come mostly from microbial activity in the soil. The main sink for N_2O is through photolysis and oxidation reactions in the stratosphere, leading to an estimated lifetime of 131 ± 10 years (Prather et al., 2012), slightly larger than previous estimates (Prather and Hsu, 2010; Montzka et al., 2011). The addition of N_2O to the atmosphere changes its own lifetime through feedbacks that couple N_2O to stratospheric NO_y and ozone depletion (Prather, 1998; Ravishankara et al., 2009; Prather and Hsu, 2010), so that the lifetime of a perturbation is less than that of the total burden, 121 ± 10 years (1-σ; Prather et al., 2012) and is used in calculations of metrics (Section 8.7).

8.2.3.5 Halogenated Species

Halogenated species can be powerful greenhouse gases (GHGs). Those containing chlorine and bromine also deplete stratospheric ozone and are referred to as ozone-depleting substances (ODSs). Most of those compounds do not have natural emissions and, because of the implementation of the Montreal Protocol and its amendments, total emissions of ODSs have sharply decreased since the 1990s (Montzka et al., 2011). For CFCs, perfluorocarbons (PFCs) and SF_6 the main loss is through photolysis in the stratosphere. The CFC substitutes (hydrochlorofluorocarbons (HCFCs) and hydrofluorocarbons (HFCs)) are destroyed by OH oxidation in the troposphere. Their global concentration has steadily risen over the recent past (see Section 2.2.1.1.4).

8.2.3.6 Aerosols

Aerosol particles are present in the atmosphere with size ranges from a few nanometres to tens of micrometres. They are the results of direct emission (primary aerosols: BC, OC, sea salt, dust) into the atmosphere or as products of chemical reactions (secondary inorganic aerosols: sulphate, nitrate, ammonium; and secondary organic aerosols (SOAs)) occurring in the atmosphere. Secondary inorganic aerosols are the products of reactions involving sulphur dioxide, ammonia and nitric oxide emissions. SOAs are the result of chemical reactions of non-methane hydrocarbons (and their products) with the hydroxyl radical (OH), ozone, nitrate (NO_3) or photolysis (Hallquist et al., 2009). Thus although many hydrocarbons in the atmosphere are of biogenic origin, anthropogenic pollutants can have impacts on their conversion to SOAs. There is tremendous complexity and still much uncertainty in the processes involved in the formation of SOAs (Hallquist et al., 2009; Carslaw et al., 2010). Additional information can be found in Section 7.3.2.

Once generated, the size and composition of aerosol particles can be modified by additional chemical reactions, condensation or evaporation of gaseous species and coagulation (Seinfeld and Pandis, 2006). It is this set of processes that defines their physical, chemical and optical properties, and hence their impact on radiation and clouds, with large regional and global differences (see Section 7.3.3). Furthermore, their distribution is affected by transport and deposition, defining a residence time in the troposphere of usually a few days (Textor et al., 2006).

8.3 Present-Day Anthropogenic Radiative Forcing

Human activity has caused a variety of changes in different forcing agents in the atmosphere or land surface. A large number of GHGs have had a substantial increase over the Industrial Era and some of these gases are entirely of anthropogenic origin. Atmospheric aerosols have diverse and complex influences on the climate. Human activity has modified the land cover and changed the surface albedo. Some of the gases and aerosols are directly emitted to the atmosphere whereas others are secondary products from chemical reactions of emitted species. The lifetimes of these different forcing agents vary substantially. This section discusses all known anthropogenic forcing agents of non-negligible importance and their quantification in terms of RF or ERF based on changes in abundance over the 1750–2011 period.

In this section we determine the RFs for WMGHGs and heterogeneously distributed species in fundamentally different ways. As described in Box 8.2, the concentrations of WMGHGs can be determined from observations at a few surface sites. For the pre-industrial concentrations these are typically from trapped air in polar ice or firn (see Section 2.2.1). Thus the RFs from WMGHGs are determined entirely from observations (Section 8.3.2). In contrast, we do not have sufficient pre-industrial or present-day observations of heterogeneously distributed forcing agents (e.g., ozone and aerosols) to be able to characterize their RF; therefore we instead have to rely on chemistry–climate models (Sections 8.3.3 and 8.3.4).

8.3.1 Updated Understanding of the Spectral Properties of Greenhouse Gases and Radiative Transfer Codes

RF estimates are performed with a combination of radiative transfer codes typical for GCMs as well as more detailed radiative transfer codes. Physical properties are needed in the radiative transfer codes such as spectral properties for gases. The HITRAN (HIgh Resolution TRANsmission molecular absorption) database (Rothman, 2010) is widely used in radiative transfer models. Some researchers studied the difference among different editions of HITRAN databases for diverse uses (Feng et al., 2007; Kratz, 2008; Feng and Zhao, 2009; Fomin and Falaleeva, 2009; Lu et al., 2012). Model calculations have shown that modifications of the spectroscopic characteristics tend to have a modest effect on the determination of RF estimates of order 2 to 3% of the calculated RF attributed to the combined doubling of CO_2, N_2O and CH_4. These results showed that even the largest overall RF induced by differences among the HITRAN databases is considerably smaller than the range reported for the modelled RF estimates; thus the line parameter updates to the HITRAN database are not a significant source for discrepancies in the RF calculations appearing in the IPCC reports. However, the more recent HITRAN data set is still recommended, as the HITRAN process offers internal verification and tends to progress closer to the best laboratory measurements. It is found that the differences among the water vapour continuum absorption formulations tend to be comparable to the differences among the various HITRAN databases (Paynter and Ramaswamy, 2011); but use of the older Robert continuum formula produces significantly larger flux differences, thus, replacement of the older continuum is warranted (Kratz, 2008) and there are still numerous unresolved issues left in the continuum expression, especially related to shortwave radiative transfer (Shine et al., 2012). Differences in absorption data from various HITRAN versions are *very likely* a small contributor to the uncertainty in RF of GHGs.

Line-by-line (LBL) models using the HITRAN data set as an input are the benchmark of radiative transfer models for GHGs. Some researchers compared different LBL models (Zhang et al., 2005; Collins et al., 2006) and line-wing cutoff, line-shape function and gas continuum absorption treatment effects on LBL calculations (Zhang et al., 2008; Fomin and Falaleeva, 2009). The agreement between LBL codes has been investigated in many studies and found to generally be within a few percent (e.g., Collins et al., 2006; Iacono et al., 2008; Forster et al., 2011a) and to compare well to observed radiative fluxes under controlled situations (Oreopoulos et al., 2012). Forster et al. (2011a) evaluated global mean radiatively important properties of chemistry climate models (CCMs) and found that the combined WMGHG global annual mean instantaneous RF at the tropopause is within 30% of LBL models for all CCM radiation codes tested. The accuracies of the LW RF due to CO_2 and tropospheric ozone increase are generally very good and within 10% for most of the participation models, but problems remained in simulating RF for stratospheric water vapour and ozone changes with errors between 3% and 200% compared to LBL models. Whereas the differences in the results from CCM radiation codes were large, the agreement among the LW LBL codes was within 5%, except for stratospheric water vapour changes.

Most intercomparison studies of the RF of GHGs are for clear-sky and aerosol-free conditions; the introduction of clouds would greatly complicate the targets of research and are usually omitted in the intercomparison exercises of GCM radiation codes and LBL codes (e.g., Collins et al., 2006; Iacono et al., 2008). It is shown that clouds can reduce the magnitude of RF due to GHGs by about 25% (Forster et al., 2005; Worden et al., 2011; Zhang et al., 2011), but the influence of clouds on the diversity in RF is found to be within 5% in four detailed radiative transfer schemes with realistic cloud distributions (Forster et al., 2005). Estimates of GHG RF are based on the LBL codes or the radiative transfer codes compared and validated against LBL models, and the uncertainty range from AR4 in the RF of GHG of 10% is retained. We underscore that uncertainty in RF calculations in many GCMs is substantially higher owing both to radiative transfer codes and meteorological data such as clouds adopted in the simulations.

8.3.2 Well-mixed Greenhouse Gases

AR4 assessed the RF from 1750 to 2005 of the WMGHGs to be 2.63 W m^{-2}. The four most important gases were CO_2, CH_4, dichlorodifluoromethane (CFC-12) and N_2O in that order. Halocarbons, comprising CFCs, HCFCs, HFCs, PFCs and SF_6, contributed 0.337 W m^{-2} to the total. Uncertainties (90% confidence ranges) were assessed to be approximately 10% for the WMGHGs. The major changes to the science since AR4 are the updating of the atmospheric concentrations, the inclusion of new species (NF_3 and SO_2F_2) and discussion of ERF for CO_2. Since AR4 N_2O has overtaken CFC-12 as the third largest contributor to RF. The total WMGHG RF is now 2.83 (2.54 to 3.12) W m^{-2}.

The RFs in this section are derived from the observed differences in concentrations of the WMGHGs between 1750 and 2011. The concentrations of CO_2, CH_4 and N_2O vary throughout the pre-industrial era, mostly due to varying climate, with a possible small contribution from anthropogenic emissions (MacFarling Meure et al., 2006). These variations do not contribute to uncertainty in the RF as strictly defined here, but do affect the RF attribution to anthropogenic emissions. On centennial time scales, variations in late Holocene concentrations of CO_2 are around 10 ppm (see note to Table 2.1), much larger than the uncertainty in the 1750 concentration. This would equate to a variation in the RF of 10%. For CH_4 and N_2O the centennial variations are comparable to the uncertainties in the 1750 concentrations and so do not significantly affect the estimate of the 1750 value used in calculating RF.

8.3.2.1 Carbon Dioxide

The tropospheric mixing ratio of CO_2 has increased globally from 278 (276–280) ppm in 1750 to 390.5 (390.3 to 390.7) ppm in 2011 (see Section 2.2.1.1.1). Here we assess the RF due to changes in atmospheric concentration rather than attributing it to anthropogenic emissions. Section 6.3.2.6 describes how only a fraction of the historical CO_2 emissions have remained in the atmosphere. The impact of land use change on CO_2 from 1850 to 2000 was assessed in AR4 to be 12 to 35 ppm (0.17 to 0.51 W m^{-2}).

Using the formula from Table 3 of Myhre et al. (1998), and see Supplementary Material Table 8.SM.1, the CO_2 RF (as defined in Section 8.1) from 1750 to 2011 is 1.82 (1.63 to 2.01) W m^{-2}. The uncertainty is dominated by the radiative transfer modelling which is assessed to be 10% (Section 8.3.1). The uncertainty in the knowledge of 1750 concentrations contributes only 2% (see Supplementary Material Table 8.SM.2)

Table 8.2 shows the concentrations and RF in AR4 (2005) and 2011 for the most important WMGHGs. Figure 8.6 shows the time evolution of RF and its rate of change. Since AR4, the RF of CO_2 has increased by 0.16 W m^{-2} and continues the rate noted in AR4 of almost 0.3 W m^{-2} per decade. As shown in Figure 8.6(d) the rate of increase in the RF from the WMGHGs over the last 15 years has been dominated by CO_2. Since AR4, CO_2 has accounted for more than 80% of the WMGHG RF increase. The interannual variability in the rate of increase in the CO_2 RF is due largely to variation in the natural land uptake whereas the trend is driven by increasing anthropogenic emissions (see Figure 6.8 in Section 6.3.1).

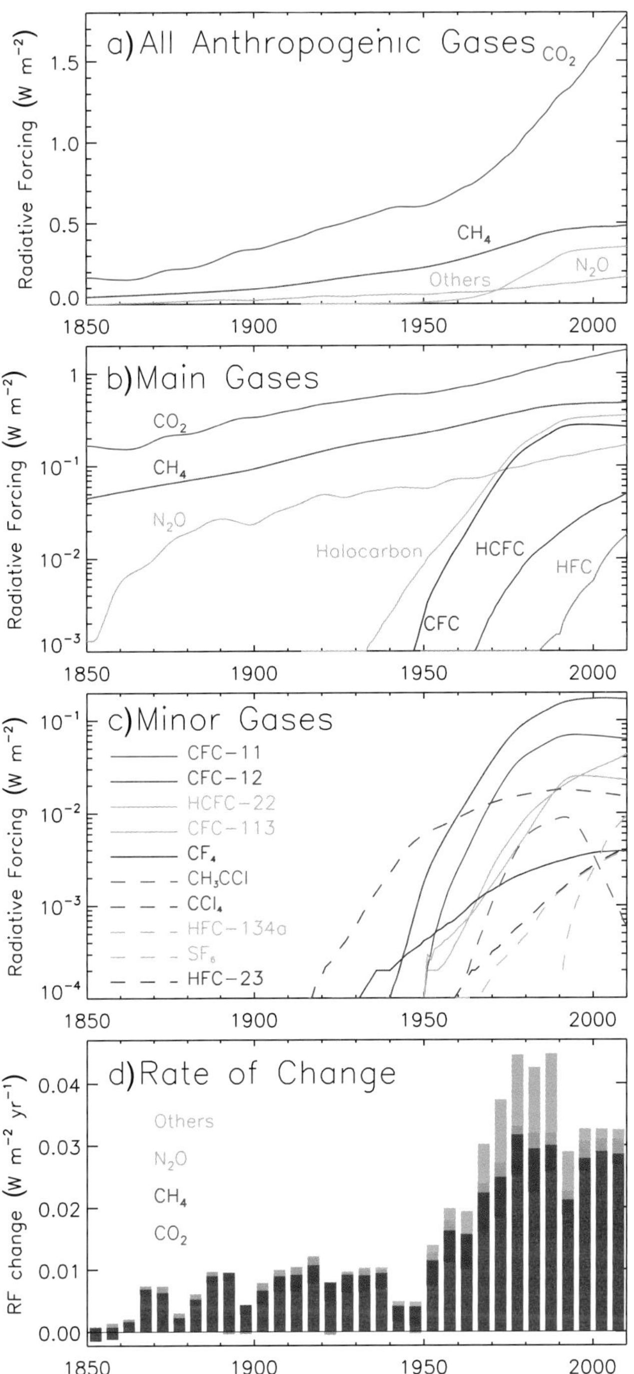

Figure 8.6 | (a) Radiative forcing (RF) from the major well-mixed greenhouse gases (WMGHGs) and groups of halocarbons from 1850 to 2011 (data from Tables A.II.1.1 and A.II.4.16), (b) as (a) but with a logarithmic scale, (c) RF from the minor WMGHGs from 1850 to 2011 (logarithmic scale). (d) Rate of change in forcing from the major WMGHGs and groups of halocarbons from 1850 to 2011.

As described in Section 8.1.1.3, CO_2 can also affect climate through physical effects on lapse rates and clouds, leading to an ERF that will be different from the RF. Analysis of CMIP5 models (Vial et al., 2013) found a large negative contribution to the ERF (20%) from the increase in land surface temperatures which was compensated for by positive contributions from the combined effects on water vapour, lapse rate, albedo and clouds. It is therefore not possible to conclude with the current information whether the ERF for CO_2 is higher or lower than the RF. Therefore we assess the ratio ERF/RF to be 1.0 and assess our uncertainty in the CO_2 ERF to be (–20% to 20%). We have *medium confidence* in this based on our understanding that the physical processes responsible for the differences between ERF and RF are small enough to be covered within the 20% uncertainty.

There are additional effects mediated through plant physiology, reducing the conductance of the plant stomata and hence the transpiration of water. Andrews et al. (2012b) find a physiological enhancement of the adjusted forcing by 3.5% due mainly to reductions in low cloud. This is smaller than a study with an earlier model by Doutriaux-Boucher et al. (2009) which found a 10% effect. Longer-term impacts of CO_2 on vegetation distributions also affect climate (O'ishi et al., 2009; Andrews et al., 2012b) but because of the longer time scale we choose to class these as feedbacks rather than rapid adjustments.

8.3.2.2 Methane

Globally averaged surface CH_4 concentrations have risen from 722 ± 25 ppb in 1750 to 1803 ± 2 ppb by 2011 (see Section 2.2.1.1.2). Over that time scale the rise has been due predominantly to changes in anthropogenic-related CH_4. Anthropogenic emissions of other compounds have also affected CH_4 concentrations by changing its removal rate (Section 8.2.3.3). Using the formula from Myhre et al. (1998) (see Supplementary Material Table 8.SM.1) the RF for CH_4 from 1750 to 2011 is 0.48 ± 0.05 W m^{-2}, with an uncertainty dominated by the radiative transfer calculation. This increase of 0.01 W m^{-2} since AR4 is due to the 29 ppb increase in the CH_4 mixing ratio. This is much larger than the 11 ppb increase between TAR and AR4, and has been driven by increases in net natural and anthropogenic emissions, but the relative contributions are not well quantified. Recent trends in CH_4 and their causes are discussed in Sections 2.2.1.1.2 and 6.3.3.1. CH_4 concentrations do vary with latitude and decrease above the tropopause; however, this variation contributes only 2% to the uncertainty in RF (Freckleton et al., 1998).

In this section only the direct forcing from changing CH_4 concentrations is addressed. CH_4 emissions can also have indirect effects on climate through impacts on CO_2, stratospheric water vapour, ozone, sulphate aerosol and lifetimes of HFCs and HCFCs (Boucher et al., 2009; Shindell et al., 2009; Collins et al., 2010). Some of these are discussed further in Sections 8.3.3, 8.5.1 and 8.7.2.

8.3.2.3 Nitrous Oxide

Concentrations of nitrous oxide have risen from 270 ± 7 ppb in 1750 to 324.2 ± 0.1 ppb in 2011, an increase of 5 ppb since 2005 (see Section 2.2.1.1.3). N_2O now has the third largest forcing of the anthropogenic gases, at 0.17 ± 0.03 W m^{-2} an increase of 6% since 2005 (see Table

8

8.2) where the uncertainty is due approximately equally to the pre-industrial concentration and radiative transfer. Only the direct RF from changing nitrous oxide concentrations is included. Indirect effects of N_2O emissions on stratospheric ozone are not taken into account here but are discussed briefly in Section 8.7.2.

8.3.2.4 Other Well-mixed Greenhouse Gases

RFs of the other WMGHG are shown in Figure 8.6 (b and c) and Table 8.2. The contribution of groups of halocarbons to the rate of change of WMGHG RF is shown in Figure 8.6 (d). Between 1970 and 1990 halocarbons made a significant contribution to the rate of change of RF. The rate of change in the total WMGHG RF was higher in 1970 to 1990 with *high confidence* compared to the present owing to higher contribution from non-CO_2 gases especially the halocarbons. Since the Montreal Protocol and its amendments, the rate of change of RF from halocarbons and related compounds has been much less, but still just positive (total RF of 0.360 W m^{-2} in 2011 compared to 0.351 W m^{-2} in 2005) as the growth of HCFCs, HFCs, PFCs and other halogens (SF_6, SO_2F_2, NF_3) RFs (total 0.022 W m^{-2} since 2005) more than compensates

Table 8.2 | Present-day mole fractions (in ppt(pmol mol^{-1}) except where specified) and RF (in W m^{-2}) for the WMGHGs. Concentration data are averages of National Oceanic and Atmospheric Administration (NOAA) and Advanced Global Atmospheric Gases Experiment (AGAGE) observations where available. CO_2 concentrations are the average of NOAA and SIO. See Table 2.1 for more details of the data sources. The data for 2005 (the time of the AR4 estimates) are also shown. Some of the concentrations vary slightly from those reported in AR4 owing to averaging different data sources. Radiative efficiencies for the minor gases are given in Table 8.A.1. Uncertainties in the RF for all gases are dominated by the uncertainties in the radiative efficiencies. We assume the uncertainties in the radiative efficiencies to be perfectly correlated between the gases, and the uncertainties in the present day and 1750 concentrations to be uncorrelated.

	Concentrations (ppt)		Radiative forcing[a] (W m^{-2})	
Species	**2011**	**2005**	**2011**	**2005**
CO_2 (ppm)	391 ± 0.2	379	1.82 ± 0.19	1.66
CH_4 (ppb)	1803 ± 2	1774	0.48 ± 0.05	0.47[e]
N_2O (ppb)	324 ± 0.1	319	0.17 ± 0.03	0.16
CFC-11	238 ± 0.8	251	0.062	0.065
CFC-12	528 ± 1	542	0.17	0.17
CFC-13	2.7		0.0007	
CFC-113	74.3 ± 0.1	78.6	0.022	0.024
CFC-115	8.37	8.36	0.0017	0.0017
HCFC-22	213 ± 0.1	169	0.0447	0.0355
HCFC-141b	21.4 ± 0.1	17.7	0.0034	0.0028
HCFC-142b	21.2 ± 0.2	15.5	0.0040	0.0029
HFC-23	24.0 ± 0.3	18.8	0.0043	0.0034
HFC-32	4.92	1.15	0.0005	0.0001
HFC-125	9.58 ± 0.04	3.69	0.0022	0.0008
HFC-134a	62.7 ± 0.3	34.3	0.0100	0.0055
HFC-143a	12.0 ± 0.1	5.6	0.0019	0.0009
HFC-152a	6.4 ± 0.1	3.4	0.0006	0.0003
SF_6	7.28 ± 0.03	5.64	0.0041	0.0032
SO_2F_2	1.71	1.35	0.0003	0.0003
NF_3	0.9	0.4	0.0002	0.0001
CF_4	79.0 ± 0.1	75.0	0.0040	0.0036
C_2F_6	4.16 ± 0.02	3.66	0.0010	0.0009
CH_3CCl_3	6.32 ± 0.07	18.32	0.0004	0.0013
CCl_4	85.8 ± 0.8	93.1	0.0146	0.0158
CFCs			0.263 ± 0.026[b]	0.273[c]
HCFCs			0.052 ± 0.005	0.041
Montreal gases[d]			0.330 ± 0.033	0.331
Total halogens			0.360 ± 0.036	0.351[f]
Total			2.83 ± 0.029	2.64

Notes:

[a] Pre-industrial values are zero except for CO_2 (278 ppm), CH_4 (722 ppb), N_2O (270 ppb) and CF_4 (35 ppt).

[b] Total includes 0.007 W m^{-2} to account for CFC-114, Halon-1211 and Halon-1301.

[c] Total includes 0.009 W m^{-2} forcing (as in AR4) to account for CFC-13, CFC-114, CFC-115, Halon-1211 and Halon-1301.

[d] Defined here as CFCs + HCFCs + CH_3CCl_3 + CCl_4.

[e] The value for the 1750 methane concentrations has been updated from AR4 in this report, thus the 2005 methane RF is slightly lower than reported in AR4.

[f] Estimates for halocarbons given in the table may have changed from estimates reported in AR4 owing to updates in radiative efficiencies and concentrations.

for the decline in the CFCs, CH_3CCl_3 and CCl_4 RFs (–0.013 W m^{-2} since 2005). The total halocarbon RF is dominated by four gases, namely CFC-12, trichlorofluoromethane (CFC-11), chlorodifluoromethane (HCFC-22) and trichlorofluoroethane (CFC-113) in that order, which account for about 85% of the total halocarbon RF (see Table 8.2) . The indirect RF from the impacts of ODSs is discussed in Section 8.3.3.2.

8.3.2.4.1 Chlorofluorocarbons and hydrochlorofluorocarbons

The CFCs and HCFCs contribute approximately 11% of the WMGHG RF. Although emissions have been drastically reduced for CFCs, their long lifetimes mean that reductions take substantial time to affect their concentrations. The RF from CFCs has declined since 2005 (mainly due to a reduction in the concentrations of CFC-11 and CFC-12), whereas the RF from HCFCs is still rising (mainly due to HCFC-22).

8.3.2.4.2 Hydrofluorocarbons

The RF of HFCs is 0.02 W m^{-2} and has close to doubled since AR4 (2005 concentrations). HFC-134a is the dominant contributor to RF of the HFCs, with an RF of 0.01 W m^{-2}.

8.3.2.4.3 Perfluorocarbons and sulphur hexafluoride

These gases have lifetimes of thousands to tens of thousands of years (Table 8.A.1); therefore emissions essentially accumulate in the atmosphere on the time scales considered here. CF_4 has a natural source and a 1750 concentration of 35 ppt (see Section 2.2.1.1.4). These gases currently contribute 0.01 W m^{-2} of the total WMGHG RF.

8.3.2.4.4 New species

Nitrogen trifluoride (NF_3) is used in the electronics industry and sulfuryl fluoride (SO_2F_2) is used as a fumigant. Both have rapidly increasing emissions and high GWPs, but currently contribute only around 0.0002 W m^{-2} and 0.0003 W m^{-2} to anthropogenic RF, respectively (Weiss et al., 2008; Andersen et al., 2009; Muhle et al., 2009; Arnold et al., 2013).

8.3.3 Ozone and Stratospheric Water Vapour

Unlike for the WMGHGs, the estimate of the tropospheric and stratospheric ozone concentration changes are almost entirely model based for the full pre-industrial to present-day interval (though, especially for the stratosphere, more robust observational evidence on changes is available for recent decades; see Section 2.2).

AR4 assessed the RF (for 1750–2005) from tropospheric ozone to be 0.35 W m^{-2} from multi-model studies with a high 95th percentile of 0.65 W m^{-2} to allow for the possibility of model overestimates of the pre-industrial tropospheric ozone levels. The stratospheric ozone RF was assessed from observational trends from 1979 to 1998 to be –0.05 ± 0.1 W m^{-2}, with the 90% confidence range increased to reflect uncertainty in the trend prior to 1979 and since 1998. In AR4 the RF from stratospheric water vapour generated by CH_4 oxidation was assessed to be +0.07 ± 0.05 W m^{-2} based on Hansen et al. (2005).

Since AR4, there have been a few individual studies of tropospheric or stratospheric ozone forcing (Shindell et al., 2006a, 2006c, 2013a; Skeie et al., 2011a; Søvde et al., 2011), a multi-model study of stratospheric ozone RF in the 2010 WMO stratospheric ozone assessment (Forster et al., 2011b), and the ACCMIP multi-model study of tropospheric and tropospheric + stratospheric chemistry models (Conley et al., 2013; Stevenson et al., 2013). There is now greater understanding of how tropospheric ozone precursors can affect stratospheric ozone, and how ODSs can affect tropospheric ozone (Shindell et al., 2013a). We assess the total ozone RF to be +0.35 (0.15 to 0.55) W m^{-2}. This can be split according to altitude or by emitted species (Shindell et al., 2013a). We assess these contributions to be 0.40 (0.20 to 0.60) W m^{-2} for ozone in the troposphere and –0.05 ± 0.10 W m^{-2} for ozone in the stratosphere based on the studies presented in Table 8.3. Alternatively, the contributions to the total ozone forcing can be attributed as 0.50 (0.30 to 0.70) W m^{-2} from ozone precursors and –0.15 (–0.3 to 0.0) W m^{-2} from the effect of ODSs. The value attributed to ODSs is assessed to be slightly smaller in magnitude than in the two studies quoted in Table 8.3 (Søvde et al., 2011; Shindell et al., 2013a) because the models used for these had stratospheric ozone RFs with higher magnitudes than the ACCMIP mean (Conley et al., 2013). Differences between the ERFs and RFs for tropospheric and stratospheric ozone are *likely* to be small compared to the uncertainties in the RFs (Shindell et al., 2013b), so the assessed values for the ERFs are the same as those for the RFs.

The influence of climate change is typically included in ozone RF estimates as those are based on modelled concentration changes, but the available literature provides insufficient evidence for the sign and magnitude of the impact and we therefore refrain from giving an estimate except to assess that it is *very likely* to be smaller than the overall uncertainty in the total RF. Unlike the WMGHGs, there are significant latitudinal variations in the RFs from changes in tropospheric and stratospheric ozone. The implications of inhomogeneous RFs are explored in more detail in Section 8.6.

There has been one study since AR4 (Myhre et al., 2007) on the RF from water vapour formed from the stratospheric oxidation of CH_4 (Section 8.3.3.3). This is consistent with the AR4 value and so has not led to any change in the recommended value of 0.07 (0.02 to 0.12) W m^{-2} since AR4.

8.3.3.1 Tropospheric Ozone

Ozone is formed in the troposphere by photochemical reactions of natural and anthropogenic precursor species (Section 8.2.3.1). Changes in ozone above the tropopause due to emissions of stratospheric ODSs can also affect ozone in the troposphere either by transport across the tropopause or modification of photolysis rates. Changes in climate have also affected tropospheric ozone concentrations (medium evidence, low agreement) through changes in chemistry, natural emissions and transport from the stratosphere (Isaksen et al., 2009).

The most recent estimates of tropospheric ozone RF come from multi-model studies under ACCMIP (Conley et al., 2013; Lamarque et al., 2013; Stevenson et al., 2013). The model ensemble reported only 1850–2000 RFs (0.34 W m^{-2}) so the single-model results from Skeie et

8

al. (2011a) were used to expand the timespan to 1750–2010, adding 0.04 W m^{-2}, and 0.02 W m^{-2} to account for the periods 1750–1850 and 2000–2010 respectively. The best estimate of tropospheric ozone RF taking into account the ACCMIP models and the Søvde et al. (2011) results (the Skeie et al. (2011a) and Shindell et al. (2013a) models are included in ACCMIP) is 0.40 (0.20 to 0.60) W m^{-2}. The quantifiable uncertainties come from the inter-model spread (–0.11 to 0.11 W m^{-2}) and the differences between radiative transfer models (–0.07 to 0.07 W m^{-2}); all 5 to 95% confidence interval. Additional uncertainties arise from the lack of knowledge of pre-industrial emissions and the representation of chemical and physical processes beyond those included in the current models. The tropospheric ozone RF is sensitive to the assumed 'pre-industrial' levels. As described in Section 8.2.3.1, very limited late 19th and early 20th century observations of surface ozone concentrations are lower than the ACCMIP models for the same period; however, we assess that those observations are very uncertain. Skeie et al. (2011a) and Stevenson et al. (2013) increase their uncertainty ranges to 30% for 1 standard deviation which is equivalent to (–50% to +50%) for the 5 to 95% confidence range and we adopt this for AR5. The overall *confidence* in the tropospheric ozone RF is assessed as *high*.

Because we have *low confidence* in the pre-industrial ozone observations, and these were extremely limited in spatial coverage, it is not possible to calculate a purely observationally based ozone RF. However, modern observations can be used to assess the performance of the chemistry models. Bowman et al. (2013) used satellite retrievals from the TES instrument to constrain the RF from the ACCMIP models. This reduced the inter-model uncertainty by 30%; however, we still maintain overall the (–50% to +50%) 5 to 95% confidence range for AR5.

The time evolution of the tropospheric ozone forcing is shown in Figure 8.7. There is a noticeable acceleration in the forcing after 1950 and a deceleration in the 1990s reflecting the time evolution of anthropogenic precursor emissions. Observational evidence for trends in ozone concentrations is discussed in Section 2.2.2.3.

It can be useful to calculate a normalized radiative forcing (NRF) which is an RF per change in ozone column in W m^{-2} DU^{-1} or W mol^{-1}. This is only an approximation as the NRF is sensitive to the vertical profile of the ozone change and to the latitudinal profile to a smaller extent. From Table 8.3 we assess the NRF to be 0.042 (0.037 to 0.047) W m^{-2} DU^{-1} (94 (83 to 105) W mol^{-1}) similar to the value of 0.042 W m^{-2} DU^{-1} (94 W mol^{-1}) in TAR (Ramaswamy et al., 2001).

A small number of studies have looked at attributing the ozone changes among the anthropogenically emitted species. Søvde et al. (2011) report a tropospheric ozone RF of 0.38 W m^{-2}, 0.44 W m^{-2} from ozone precursors and –0.06 W m^{-2} from the impact of stratospheric ozone depletion on the troposphere. Shindell et al. (2013a) also calculate that ODSs are responsible for about –0.06 W m^{-2} of the tropospheric ozone RF, and ozone precursors for about 0.41 W m^{-2}. Six of the models in Stevenson et al. (2013) and Shindell et al. (2009) performed experiments to attribute the ozone RF to the individual precursor emissions. An average of these seven model results leads to attributions of 0.24 ± 0.13 W m^{-2} due to CH_4 emissions, 0.14 ± 0.09 W m^{-2} from NO_X emissions, 0.07 ± 0.03 W m^{-2} from CO, and 0.04 ± 0.03W m^{-2} from non-methane volatile organic compounds (NMVOCs). These results were calculated by reducing the precursor emissions individually from 2000 to pre-industrial levels. The results were scaled by the total ozone RFs attributed to ozone precursors (0.50 W m^{-2}) to give the contributions to the full 1750–2010 RF. Because of the nonlinearity of the chemistry an alternative method of starting from pre-industrial conditions and increasing precursor emissions singly may give a different result. Note that as well as inducing an ozone RF, these ozone precursor species can also strongly affect the concentrations of CH_4 and aerosols, adding extra terms (both positive and negative) to their total indirect forcings. The contributions to the 1750–2010 CH_4 RF are again based on Stevenson et al. (2013) and Shindell et al. (2009). The Stevenson et al. (2013) values are for 1850–2000 rather than 1750 to 2011 so for these we distribute the CH_4 RF for 1750–1850 and 2000–2011 (0.06 W m^{-2}) by scaling the CH_4 and CO contributions (assuming these were the most significant contributors over those time periods). This gives contributions of 0.58 ± 0.08, –0.29 ± 0.18, 0.07 ± 0.02 and 0.02 ± 0.02 W m^{-2} for changes from historical to present day emissions of CH_4 (inferred emissions), NO_x, CO and VOCs respectively (uncertainties are 5 to 95% confidence intervals). The difference between the total CH_4 RF attributed to ozone precursors here (0.38 W m^{-2}) and the value calculated from CH_4 concentration changes in Table 8.2 (0.48 W m^{-2}) is due to nonlinearities in the CH_4 chemistry because large single-step changes were used. To allow an easier comparison between the concentration-based and emission-based approaches in Section 8.5.1 the nonlinear term (+0.1 W m^{-2}) is distributed between the four emitted species according to their absolute magnitude so that they total 0.48 W m^{-2}. The scaled results still lie within the uncertainty bounds of the values quoted above. The impact of climate change over the historical period on CH_4 oxidation is not accounted for in these calculations.

Tropospheric ozone can also affect the natural uptake of CO_2 by decreasing plant productivity (see Sections 6.4.8.2 and 8.2.3.1) and it is found that this indirect effect could have contributed to the total CO_2 RF (Section 8.3.2.1; Sitch et al., 2007), roughly doubling the overall RF attributed to ozone precursors. Although we assess there to be

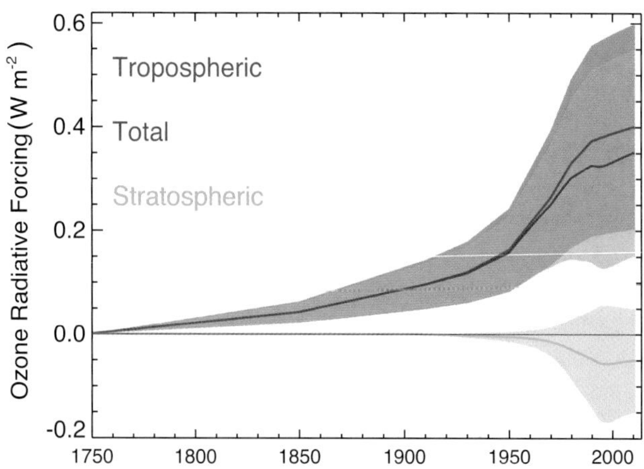

Figure 8.7 | Time evolution of the radiative forcing from tropospheric and stratospheric ozone from 1750 to 2010. Tropospheric ozone data are from Stevenson et al. (2013) scaled to give 0.40 W m^{-2} at 2010. The stratospheric ozone RF follow the functional shape of the Effective Equivalent Stratospheric Chlorine assuming a 3-year age of air (Daniel et al., 2010) scaled to give –0.05 W m^{-2} at 2010.

Table 8.3 | Contributions of tropospheric and stratospheric ozone changes to radiative forcing (W m^{-2}) from 1750 to 2011.

	Troposphere				Stratosphere		
	Longwave	Shortwave	Total	Normalized Radiative Forcing m W m^{-2} DU^{-1}	Longwave	Shortwave	Total
AR4 (Forster et al. (2007)			0.35 (0.25 to 0.65)				−0.05 (−0.15 to 0.05)
Shindell et al. (2013a)[f]			0.33 (0.31 to 0.35)				−0.08 (−0.10 to −0.06)
WMO (Forster et al., 2011b)							−0.03[a] (−0.23 to +0.17) +0.03[b]
Søvde et al. (2011)			0.45[c]	40			−0.12
			0.38[d]	39			−0.12
Skeie et al. (2011a)			0.41 (0.21 to 0.61)	38			
ACCMIP[e]	0.33 (0.24 to 0.42)	0.08 (0.06 to 0.10)	0.41 (0.21 to 0.61)	42 (37 to 47)	−0.13 (−0.26 to 0)	0.11 (0.03 to 0.19)	−0.02 (−0.09 to 0.05)
AR5			0.40 (0.20 to 0.60)	42 (37 to 47)			−0.05 (-0.15 to 0.05)

Notes:

a From multi-model results.

b From Randel and Wu (2007) observation-based data set.

c Using the REF chemistry, see Søvde et al. (2011).

d Using the R2 chemistry.

e The Atmospheric Chemistry and Climate Model Intercomparison Project (ACCMIP) tropospheric ozone RFs are from Stevenson et al. (2013). The stratospheric ozone values are from Conley et al. (2013) calculations for 1850–2005 disregarding the Modèle de Chimie Atmosphérique a Grande Echelle (MOCAGE) model which showed excessive ozone depletion.

f Only the Goddard Institute for Space Studies (GISS)-E2-R results (including bias correction) from the Shindell et al. (2013a) study are shown here rather than the multi-model result presented in that paper.

robust evidence of an effect, we make no assessment of the magnitude because of lack of further corroborating studies.

8.3.3.2 Stratospheric Ozone

The decreases in stratospheric ozone due to anthropogenic emissions of ODSs have a positive RF in the shortwave (increasing the flux into the troposphere) and a negative RF in the longwave. This leaves a residual forcing that is the difference of two larger terms. In the lower stratosphere the longwave effect tends to be larger, whereas in the upper stratosphere the shortwave dominates. Thus whether stratospheric ozone depletion has contributed an overall net positive or negative forcing depends on the vertical profile of the change (Forster and Shine, 1997). WMO (2011) assessed the RF from 1979 to 2005 from observed ozone changes (Randel and Wu, 2007) and results from 16 models for the 1970s average to 2004. The observed and modelled mean ozone changes gave RF values of different signs (see Table 8.3). Negative net RFs arise from models with ozone decline in the lowermost stratosphere, particularly at or near the tropopause.

The ACCMIP study also included some models with stratospheric chemistry (Conley et al., 2013). One model in that study stood out as having excessive ozone depletion. Removing that model leaves a stratospheric ozone RF of −0.02 (−0.09 to 0.05) W m^{-2}. These results are in good agreement with the model studies from WMO (2011). Forster et al. (2007) in AR4 calculated a forcing of −0.05 W m^{-2} from observations over the period 1979–1998 and increased the uncertainty to 0.10 W m^{-2} to encompass changes between the pre-industrial period and 2005. The RF from stratospheric ozone due to changes in emissions of ozone precursors and ODSs is here assessed to be −0.05 (−0.15 to 0.05) taking into account all the studies listed in Table 8.3. This is in agreement with AR4, although derived from different data. The timeline of stratospheric ozone forcing is shown in Figure 8.7, making the assumption that it follows the trajectory of the changes in EESC. It reaches a minimum in the late 1990s and starts to recover after that.

The net global RF from ODSs taking into account the compensating effects on ozone and their direct effects as WMGHGs is 0.18 (0.03 to 0.33) W m^{-2}. The patterns of RF for these two effects are different so the small net global RF comprises areas of positive and negative RF.

8.3.3.3 Stratospheric Water Vapour

Stratospheric water vapour is dependent on the amount entering from the tropical troposphere and from direct injection by volcanic plumes (Joshi and Jones, 2009) and aircraft, and the *in situ* chemical production from the oxidation of CH_4 and hydrogen. This contrasts with tropospheric water vapour which is almost entirely controlled by the balance between evaporation and precipitation (see FAQ 8.1). We consider trends in the transport (for instance, due to the Brewer–Dobson circulation or tropopause temperature changes) to be climate feedback rather than a forcing so the anthropogenic RFs come from oxidation of CH_4 and hydrogen, and emissions from stratospheric aircraft.

Myhre et al. (2007) used observations of the vertical profile of CH_4 to deduce a contribution from oxidation of anthropogenic CH_4 of 0.083

W m^{-2} which compares with the value of 0.07 W m^{-2} from calculations in a 2D model in Hansen et al. (2005). Both of these values are consistent with AR4 which obtained the stratospheric water vapour forcing by scaling the CH_4 direct forcing by 15%. Thus the time evolution of this forcing is also obtained by scaling the CH_4 forcing by 15%. The best estimate and uncertainty range from AR4 of 0.07 (0.02 to 0.12) W m^{-2} remain unchanged and the large uncertainty range is due to large differences found in the intercomparison studies of radiative transfer modelling for changes in stratospheric water vapour (see Section 8.3.1).

RF from the current aircraft fleet through stratospheric water vapour emissions is very small. Wilcox et al. (2012) estimate a contribution from civilian aircraft in 2005 of 0.0009 (0.0003 to 0.0013) W m^{-2} with *high confidence* in the upper limit. Water vapour emissions from aircraft in the troposphere also contribute to contrails which are discussed in Section 8.3.4.5.

8.3.4 Aerosols and Cloud Effects

8.3.4.1 Introduction and Summary of AR4

In AR4 (Forster et al., 2007), RF estimates were provided for three aerosol effects. These were the RF of aerosol–radiation interaction (previously denoted as direct aerosol effect), RF of the aerosol–cloud interaction (previously denoted as the cloud albedo effect), and the impact of BC on snow and ice surface albedo. See Chapter 7 and Figure 7.3 for an explanation of the change in terminology between AR4 and AR5. The RF due to aerosol–radiation interaction is scattering and absorption of shortwave and longwave radiation by atmospheric aerosols. Several different aerosol types from various sources are present in the atmosphere (see Section 8.2). Most of the aerosols primarily scatter solar radiation, but some components absorb solar radiation to various extents with BC as the most absorbing component. RF of aerosols in the troposphere is often calculated at the TOA because it is similar to tropopause values (Forster et al., 2007). A best estimate RF of –0.5 ± 0.4 W m^{-2} was given in AR4 for the change in the net aerosol–radiation interaction between 1750 and 2005 and a medium to low level of scientific understanding (LOSU).

An increase in the hygroscopic aerosol abundance may enhance the concentration of cloud condensation nuclei (CCN). This may increase the cloud albedo and under the assumption of fixed cloud water content this effect was given a best estimate of –0.7 W m^{-2} (range from –1.8 to –0.3) in AR4 and a low LOSU.

BC in the snow or ice can lead to a decrease of the surface albedo. This leads to a positive RF. In AR4 this mechanism was given a best RF estimate of 0.1 ± 0.1 W m^{-2} and a low LOSU.

Impacts on clouds from the ERF of aerosol–cloud interaction (including both effects previously denoted as cloud lifetime and cloud albedo effect) and the ERF of aerosol–radiation interaction (including both effects previously denoted as direct aerosol effect and semi-direct effect) were not strictly in accordance with the RF concept, because they involve tropospheric changes in variables other than the forcing agent at least in the available model estimates, so no best RF estimates were provided in AR4 (see Section 8.1). However, the ERF of aerosol–cloud and aerosol–radiation interactions were included in the discussion of total aerosol effect in Chapter 7 in AR4 (Denman et al., 2007). The mechanisms influenced by anthropogenic aerosol including the aerosol cloud interactions are discussed in detail in this assessment in Section 7.5 and summarized in the subsections that follow.

8.3.4.2 Radiation Forcing of the Aerosol–Radiation Interaction by Component

Based on a combination of global aerosol models and observation-based methods, the best RF estimate of the aerosol–radiation interaction in AR5 is –0.35 (–0.85 to +0.15) W m^{-2} (see Section 7.5). This estimate is thus smaller in magnitude than in AR4, however; with larger uncertainty range. Overall, the estimate compared to AR4 is more robust because the agreement between estimates from models and observation-based methods is much greater (see Section 7.5). The larger range arises primarily from analysis by observation-based methods (see Section 7.5).

The main source of the model estimate is based on updated simulations in AeroCom (Myhre et al., 2013), which is an intercomparison exercise of a large set of global aerosol models that includes extensive evaluation against measurements. The assessment in Chapter 7 relies to a large extent on this study for the separation in the various aerosol components, except for BC where the assessment in Chapter 7 relies in addition on Bond et al. (2013). The RF of aerosol–radiation interaction is separated into seven components in this report; namely sulphate, BC from fossil fuel and biofuel, OA from fossil fuel and biofuel, BC and OA combined from biomass burning (BB), nitrate, SOA and mineral dust. BC and OA from biomass burning are combined due to the joint sources, whereas treated separately for fossil fuel and biofuel because there is larger variability in the ratio of BC to OA in the fossil fuel and biofuel emissions. This approach is consistent with TAR and AR4. Table 8.4 compares the best estimates of RF due to aerosol–radiation interaction for various components in this report with values in SAR, TAR and AR4. In magnitude the sulphate and BC from use of fossil fuel and biofuel dominate. It is important to note that the BB RF is small in magnitude but consists of larger, offsetting terms in magnitude from OA and BC (see Section 7.5.2). Changes in the estimates of RF due to aerosol–radiation interaction of the various components have been rather modest compared to AR4, except for BC from fossil fuel and biofuel (see Section 7.5). SOA is a new component compared to AR4. Anthropogenic SOA precursors contribute only modestly to the anthropogenic change in SOA. The increase in SOA is mostly from biogenic precursors and enhanced partitioning of SOA into existing particles from anthropogenic sources and changes in the atmospheric oxidation (Carlton et al., 2010). This change in SOA is therefore of anthropogenic origin, but natural emission of SOA precursors is important (Hoyle et al., 2011).

Note that the best estimate and the uncertainty for the total is not equal to the sum of the aerosol components because the total is estimated based on a combination of methods (models and observation-based methods), whereas the estimates for the components rely mostly on model estimates.

Table 8.4 | Global and annual mean RF (W m^{-2}) due to aerosol–radiation interaction between 1750 and 2011 of seven aerosol components for AR5. Values and uncertainties from SAR, TAR, AR4 and AR5 are provided when available. Note that for SAR, TAR and AR4 the end year is somewhat different than for AR5 with 1993, 1998 and 2005, respectively.

	Global Mean Radiative Forcing (W m^{-2})			
	SAR	**TAR**	**AR4**	**AR5**
Sulphate aerosol	–0.40 (–0.80 to –0.20)	–0.40 (–0.80 to –0.20)	–0.40 (–0.60 to –0.20)	–0.40 (–0.60 to –0.20)
Black carbon aerosol from fossil fuel and biofuel	+0.10 (+0.03 to +0.30)	+0.20 (+0.10 to +0.40)	+0.20 (+0.05 to +0.35)	+0.40 (+0.05 to +0.80)
Primary organic aerosol from fossil fuel and biofuel	Not estimated	–0.10 (–0.30 to –0.03)	–0.05 (0.00 to –0.10)	–0.09 (–0.16 to –0.03)
Biomass burning	–0.20 (–0.60 to –0.07)	–0.20 (–0.60 to –0.07)	+0.03(–0.09 to +0.15)	–0.0 (–0.20 to +0.20)
Secondary organic aerosol	Not estimated	Not estimated	Not estimated	–0.03 (–0.27 to +0.20)
Nitrate	Not estimated	Not estimated	–0.10 (–0.20 to 0.00)	–0.11 (–0.30 to –0.03)
Dust	Not estimated	–0.60 to +0.40	–0.10 (–0.30 to +0.10)	–0.10 (–0.30 to +0.10)
Total	Not estimated	Not estimated	–0.50 (–0.90 to –0.10)	–0.35 (–0.85 to +0.15)

The RF due to aerosol–radiation interaction during some time periods is more uncertain than the current RF. Improvements in the observations of aerosols have been substantial with availability of remote sensing from the ground-based optical observational network AErosol RObotic NETwork (AERONET) and the launch of the Moderate Resolution Imaging Spectrometer (MODIS) and Multi-angle Imaging Spectro-Radiometer (MISR) instruments (starting in 2000) as well as other satellite data. This has contributed to constraining the current RF using aerosol observations. The aerosol observations are very limited backward in time, although there is growing constraint coming from new ice and lake core records, and uncertainties in the historical emission of aerosols and their precursors used in the global aerosol modeling are larger than for current conditions. Emissions of carbonaceous aerosols are particularly uncertain in the 1800s due to a significant biofuel source in this period, in contrast to the SO_2 emissions which were very small until the end of the 1800s. The uncertainty in the biomass burning emissions also increases backward in time. Note that, for 1850, the biomass burning emissions from Lamarque et al. (2010) are quite different from the previous estimates, but RF due to aerosol–radiation interaction is close to zero for this component. Figure 8.8 shows an example of the time evolution of the RF due to aerosol–radiation interaction as a total and separated into six aerosol components. From 1950 to 1990 there was a strengthening of the total RF due to aerosol–radiation interaction, mainly due to a strong enhancement of the sulphate RF. After 1990 the change has been small with even a weakening of the RF due to aerosol–radiation interaction, mainly due to a stronger BC RF as a result of increased emissions in East and Southeast Asia.

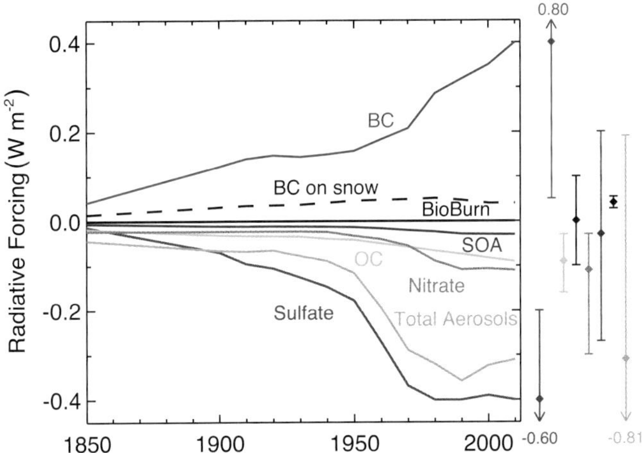

Figure 8.8 | Time evolution of RF due to aerosol–radiation interaction and BC on snow and ice. Multi-model results for 1850, 1930, 1980 and 2000 from ACCMIP for aerosol–radiation interaction (Shindell et al., 2013c) and BC on snow and ice (Lee et al., 2013) are combined with higher temporal-resolution results from the Goddard Institute for Space Studies (GISS)-E2 and Oslo-Chemical Transport Model 2 (OsloCTM2) models (aerosol–radiation interaction) and OsloCTM2 (BC on snow and ice). Uncertainty ranges (5 to 95%) for year 2010 are shown with vertical lines. Values next to the uncertainty lines are for cases where uncertainties go beyond the scale. The total includes the RF due to aerosol–radiation interaction for six aerosol components and RF due to BC on snow and ice. All values have been scaled to the best estimates for 2011 given in Table 8.4. Note that time evolution for mineral dust is not included and the total RF due to aerosol–radiation interaction is estimated based on simulations of the six other aerosol components.

8.3.4.3 Aerosol–Cloud Interactions

The RF by aerosol effects on cloud albedo was previously referred to as the Twomey or cloud albedo effect (see Section 7.1). Although this RF can be calculated, no estimate of this forcing is given because it has heuristic value only and does not simply translate to the ERF due to aerosol–cloud interaction. The total aerosol ERF, namely ERF due to aerosol–radiation and aerosol–cloud interactions (excluding BC on snow and ice) provided in Chapter 7 is estimated with a 5 to 95% uncertainty between –1.9 and –0.1 W m^{-2} with a best estimate value of –0.9 W m^{-2} (*medium confidence*). The *likely* range of this forcing is between –1.5 and –0.4 W m^{-2}. The estimate of ERF due to aerosol–radiation and aerosol-cloud interaction is lower (i.e., less negative) than the corresponding AR4 RF estimate of –1.2 W m^{-2} because the latter was based mainly on GCM studies that did not take secondary processes (such as aerosol effects on mixed-phase and/or convective clouds and effects on longwave radiation) into account. This new best estimate of ERF due to aerosol–radiation and aerosol–cloud interaction is also consistent with the studies allowing cloud-scale processes and related responses and with the lower estimates of this forcing inferred from satellite observations.

Frequently Asked Questions

FAQ 8.2 | Do Improvements in Air Quality Have an Effect on Climate Change?

Yes they do, but depending on which pollutant(s) they limit, they can either cool or warm the climate. For example, whereas a reduction in sulphur dioxide (SO_2) emissions leads to more warming, nitrogen oxide (NO_x) emission control has both a cooling (through reducing of tropospheric ozone) and a warming effect (due to its impact on methane lifetime and aerosol production). Air pollution can also affect precipitation patterns.

Air quality is nominally a measure of airborne surface pollutants, such as ozone, carbon monoxide, NO_x and aerosols (solid or liquid particulate matter). Exposure to such pollutants exacerbates respiratory and cardiovascular diseases, harms plants and damages buildings. For these reasons, most major urban centres try to control discharges of airborne pollutants.

Unlike carbon dioxide (CO_2) and other well-mixed greenhouse gases, tropospheric ozone and aerosols may last in the atmosphere only for a few days to a few weeks, though indirect couplings within the Earth system can prolong their impact. These pollutants are usually most potent near their area of emission or formation, where they can force local or regional perturbations to climate, even if their globally averaged effect is small.

Air pollutants affect climate differently according to their physical and chemical characteristics. Pollution-generated greenhouse gases will impact climate primarily through shortwave and longwave radiation, while aerosols can in addition affect climate through cloud–aerosol interactions.

Controls on anthropogenic emissions of methane (FAQ 8.2, Figure 1) to lower surface ozone have been identified as 'win–win' situations. Consequences of controlling other ozone precursors are not always as clear. NO_x emission controls, for instance, might be expected to have a cooling effect as they reduce tropospheric ozone, but their impact on CH_4 lifetime and aerosol formation is more likely instead to cause overall warming.

Satellite observations have identified increasing atmospheric concentrations of SO_2 (the primary precursor to scattering sulphate aerosols) from coal-burning power plants over eastern Asia during the last few decades. The most recent power plants use scrubbers to reduce such emissions (albeit not the concurrent CO_2 emissions and associated long-term climate warming). This improves air quality, but also reduces the cooling effect of sulphate aerosols and therefore exacerbates warming. Aerosol cooling occurs through aerosol–radiation and aerosol–cloud interactions and is estimated at –0.9 W m^{-2} (all aerosols combined, Section 8.3.4.3) since pre-industrial, having grown especially during the second half of the 20th century when anthropogenic emissions rose sharply. *(continued on next page)*

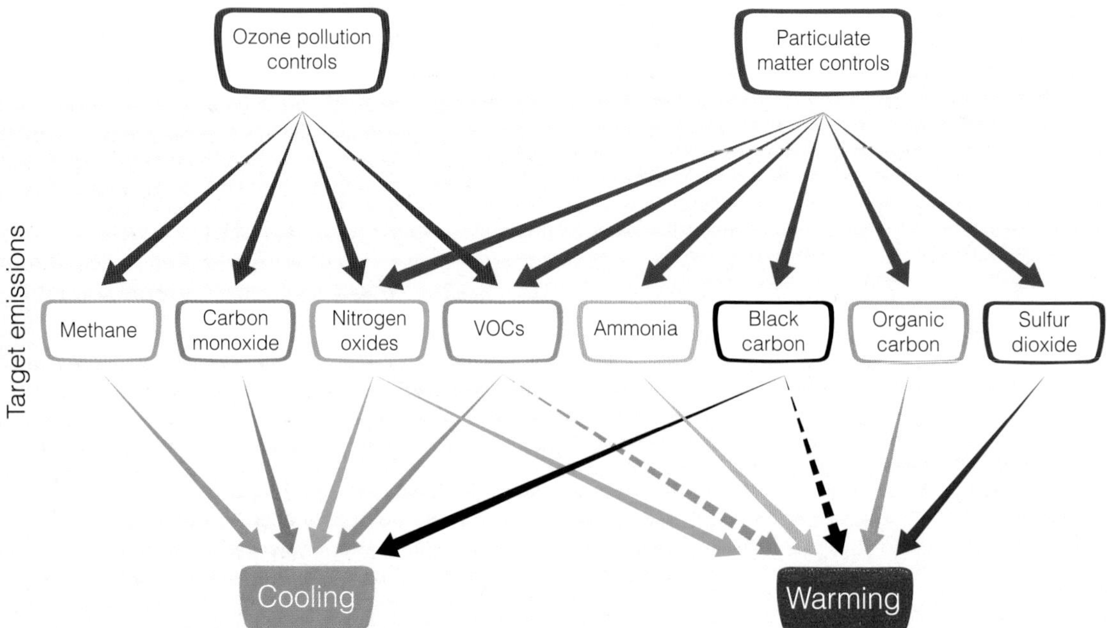

FAQ 8.2, Figure 1 | Schematic diagram of the impact of pollution controls on specific emissions and climate impact. Solid black line indicates known impact; dashed line indicates uncertain impact.

FAQ 8.2 (continued)

Black carbon or soot, on the other hand, absorbs heat in the atmosphere (leading to a 0.4 W m^{-2} radiative forcing from anthropogenic fossil and biofuel emissions) and, when deposited on snow, reduces its albedo, or ability to reflect sunlight. Reductions of black carbon emissions can therefore have a cooling effect, but the additional interaction of black carbon with clouds is uncertain and could lead to some counteracting warming.

Air quality controls might also target a specific anthropogenic activity sector, such as transportation or energy production. In that case, co-emitted species within the targeted sector lead to a complex mix of chemistry and climate perturbations. For example, smoke from biofuel combustion contains a mixture of both absorbing and scattering particles as well as ozone precursors, for which the combined climate impact can be difficult to ascertain.

Thus, surface air quality controls will have some consequences on climate. Some couplings between the targeted emissions and climate are still poorly understood or identified, including the effects of air pollutants on precipitation patterns, making it difficult to fully quantify these consequences. There is an important twist, too, in the potential effect of climate change on air quality. In particular, an observed correlation between surface ozone and temperature in polluted regions indicates that higher temperatures from climate change alone could worsen summertime pollution, suggesting a 'climate penalty'. This penalty implies stricter surface ozone controls will be required to achieve a specific target. In addition, projected changes in the frequency and duration of stagnation events could impact air quality conditions. These features will be regionally variable and difficult to assess, but better understanding, quantification and modelling of these processes will clarify the overall interaction between air pollutants and climate.

One reason an expert judgment estimate of ERF due to aerosol–radiation and aerosol–cloud interaction is provided rather than ERF due to aerosol–cloud interaction specifically is that the individual contributions are very difficult to disentangle. These contributions are the response of processes that are the outputs from a system that is constantly readjusting to multiple nonlinear forcings. Assumptions of independence and linearity are required to deduce ERF due to aerosol–radiation interaction and ERF due to aerosol–cloud interaction (although there is no *a priori* reason why the individual ERFs should be simply additive). Under these assumptions, ERF due to aerosol–cloud interaction is deduced as the difference between ERF due to aerosol–radiation and aerosol–cloud interaction and ERF due to aerosol–radiation interaction alone. This yields an ERF due to aerosol–cloud interaction estimate of –0.45 W m^{-2} which is much smaller in magnitude than the –1.4 W m^{-2} median forcing value of the models summarized in Figure 7.19 and is also smaller in magnitude than the AR4 estimates of –0.7 W m^{-2} for RF due to aerosol–cloud interaction.

8.3.4.4 Black Carbon Deposition in Snow and Ice

Because absorption by ice is very weak at visible and ultraviolet (UV) wavelengths, BC in snow makes the snow darker and increases absorption. This is not enough darkening to be seen by eye, but it is enough to be important for climate (Warren and Wiscombe, 1980; Clarke and Noone, 1985). Several studies since AR4 have re-examined this issue and find that the RF may be weaker than the estimates of Hansen and Nazarenko (2004) in AR4 (Flanner et al., 2007; Koch et al., 2009a; Rypdal et al., 2009; Lee et al., 2013). The anthropogenic BC on snow/ice is assessed to have a positive global and annual mean RF of +0.04 W m^{-2}, with a 0.02–0.09 W m^{-2} 5 to 95% uncertainty range (see further description in Section 7.5.2.3). This RF has a two to four times larger global mean surface temperature change per unit forcing than a change in CO_2.

In Figure 8.8, the time evolution of global mean RF due to BC on snow and ice is shown based on multi-model simulations in ACCMIP (Lee et al., 2013) for 1850, 1930, 1980 and 2000. The results show a maximum in the RF in 1980 with a small increase since 1850 and a 20% lower RF in 2000 compared to 1980. Those results are supported by observations. The BC concentration in the Arctic atmosphere is observed to be declining since 1990, at least in the Western Hemisphere portion (Sharma et al., 2004), which should lead to less deposition of BC on the snow surface. Surveys across Arctic during 1998 and 2005 to 2009 showed that the BC content of Arctic snow appears to be lower than in 1984 (Doherty et al., 2010) and found BC concentrations in Canada, Alaska and the Arctic Ocean (e.g., Hegg et al., 2009), about a factor of 2 lower than measured in the 1980s (e.g., Clarke and Noone, 1985). Large-area field campaigns (Huang et al., 2011; Ye et al., 2012) found that the BC content of snow in northeast China is comparable to values found in Europe. The steep drop off in BC content of snow with latitude in northeast China may indicate that there is not much BC in the Arctic coming from China (Huang et al., 2011; Ye et al., 2012; Wang et al., 2013). The change in the spatial pattern of emission of BC is a main cause for the difference in the temporal development of RF due to BC on snow and ice compared to the BC from RF due to aerosol–radiation interaction over the last decades.

8

8.3.4.5 Contrails and Contrail-Induced Cirrus

AR4 assessed the RF of contrails (persistent linear contrails) as +0.01 (–0.007 to +0.02) W m^{-2} and provided no estimate for contrail induced cirrus. In AR5, Chapter 7 gives a best estimate of RF due to contrails of +0.01 (+0.005 to +0.03) W m^{-2} and an ERF estimate of the combined contrails and contrail-induced cirrus of +0.05 (+0.02 to +0.15) W m^{-2}. Since AR4, the evidence for contrail-induced cirrus has increased because of observational studies (for further details see Section 7.2.7).

8.3.5 Land Surface Changes

8.3.5.1 Introduction

Anthropogenic land cover change has a direct impact on the Earth radiation budget through a change in the surface albedo. It also impacts the climate through modifications in the surface roughness, latent heat flux and river runoff. In addition, human activity may change the water cycle through irrigation and power plant cooling, and also generate direct input of heat to the atmosphere by consuming energy. Land use change, and in particular deforestation, also has significant impacts on WMGHG concentration, which are discussed in Section 6.3.2.2. Potential geo-engineering techniques that aim at increasing the surface albedo are discussed in Section 7.7.2.3.

AR4 referenced a large number of RF estimates resulting from a change in land cover albedo. It discussed the uncertainties due to the reconstruction of historical vegetation, the characterization of present-day vegetation and the surface radiation processes. On this basis, AR4 gave a best estimate of RF relative to 1750 due to land use related surface albedo at –0.2 ± 0.2 W m^{-2} with a level of scientific understanding at medium-low.

8.3.5.2 Land Cover Changes

Hurtt et al. (2006) estimates that 42 to 68% of the global land surface was impacted by land use activities (crop, pasture, wood harvest) during the 1700–2000 period. Until the mid-20th century most land use change took place over the temperate regions of the NH. Since then, reforestation is observed in Western Europe, North America and China as a result of land abandonment and afforestation efforts, while deforestation is concentrated in the tropics. After a rapid increase of the rate of deforestation during the 1980s and 1990s, satellite data indicate a slowdown in the past decade (FAO, 2012).

Since AR4, Pongratz et al. (2008) and Kaplan et al. (2011) extended existing reconstructions on land use back in time to the past millennium, accounting for the progress of agriculture technique and historical events such as the black death or war invasions. As agriculture was already widespread over Europe and South Asia by 1750, the RF, which is defined with respect to this date, is weaker than the radiative flux change from the state of natural vegetation cover (see Figure 8.9). Deforestation in Europe and Asia during the last millennium led to a significant regional negative forcing. Betts et al. (2007) and Goosse et al. (2006) argue that it probably contributed to the 'Little Ice Age', together with natural solar and volcanic activity components, before the increase in GHG concentration led to temperatures similar to those experienced in the early part of the second millennium. There is still significant uncertainty in the anthropogenic land cover change, and in particular its time evolution (Gaillard et al., 2010).

8.3.5.3 Surface Albedo and Radiative Forcing

Surface albedo is the ratio between reflected and incident solar flux at the surface. It varies with the surface cover. Most forests are darker (i.e., lower albedo) than grasses and croplands, which are darker than barren land and desert. As a consequence, deforestation tends to increase the Earth albedo (negative RF) while cultivation of some bright surfaces may have the opposite effect. Deforestation also leads to a large increase in surface albedo in case of snow cover as low vegetation accumulates continuous snow cover more readily in early winter allowing it to persist longer in spring. This causes average winter albedo in deforested areas to be generally much higher than that of a tree-covered landscape (Bernier et al., 2011).

The pre-industrial impact of the Earth albedo increase due to land use change, including the reduced snow masking by tall vegetation, is estimated to be on the order of –0.05 W m^{-2} (Pongratz et al., 2009). Since then, the increase in world population and agriculture development led to additional forcing. Based on reconstruction of land use since the beginning of the Industrial Era, Betts et al. (2007) and Pongratz et al. (2009) computed spatially and temporally distributed estimates of the land use RF. They estimate that the shortwave flux change induced by the albedo variation, from fully natural vegetation state to 1992, is on the order of –0.2 W m^{-2} (range –0.24 to –0.21W m^{-2}). The RF, defined with respect to 1750, is in the range –0.17 to –0.18 W m^{-2}. A slightly stronger value (–0.22 W m^{-2}) was found by Davin et al. (2007) for the period 1860–1992.

In recent years, the availability of global scale MODIS data (Schaaf et al., 2002) has improved surface albedo estimates (Rechid et al., 2009). These data have been used by Myhre et al (2005a) and Kvalevag et al. (2010). They argue that the observed albedo difference between natural vegetation and croplands is less than usually assumed in climate simulations, so that the RF due to land use change is weaker than in estimates that do not use the satellite data. On the other hand, Nair et al. (2007) show observational evidence of an underestimate of the surface albedo change in land use analysis in southwest Australia. Overall, there is still a significant range of RF estimates for the albedo component of land use forcing. This is mostly due to the range of albedo change as a result of land use change, as shown in an inter-comparison of seven atmosphere–land models (de Noblet-Ducoudre et al., 2012).

Deforestation has a direct impact on the atmospheric CO_2 concentration and therefore contributes to the WMGHG RF as quantified in Section 8.3.2. Conversely, afforestation is a climate mitigation strategy to limit the CO_2 concentration increase. Several authors have compared the radiative impact of deforestation/afforestation that results from the albedo change with the greenhouse effect of CO_2 released/sequestered. Pongratz et al. (2010) shows that the historic land use change has had a warming impact (i.e., greenhouse effect dominates) at the global scale and over most regions with the exception of Europe and India. Bala et al. (2007) results show latitudinal contrast where the greenhouse effect dominates for low-latitude deforestation while

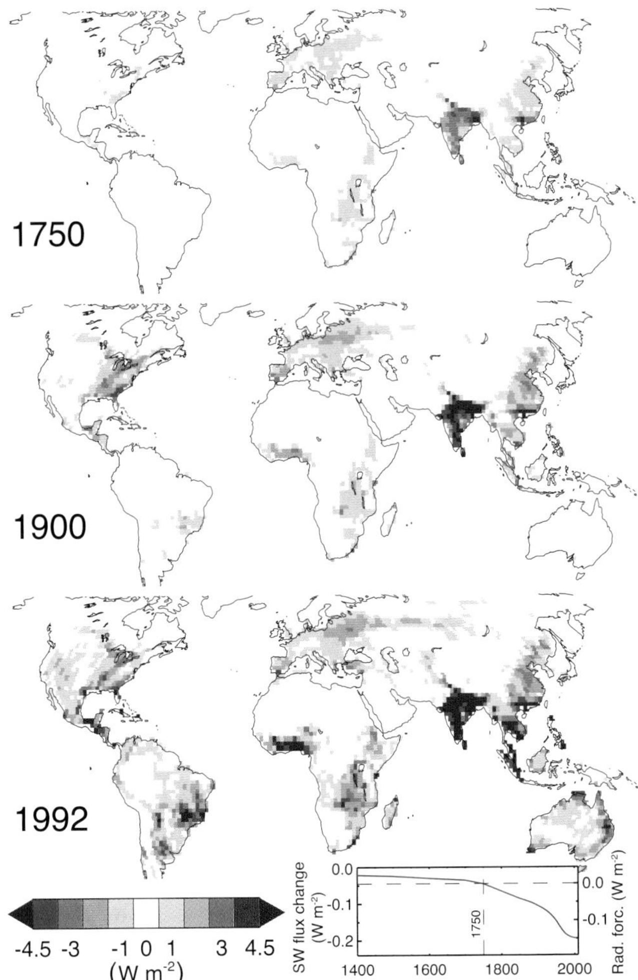

Figure 8.9 | Change in top of the atmosphere (TOA) shortwave (SW) flux (W m^{-2}) following the change in albedo as a result of anthropogenic Land Use Change for three periods (1750, 1900 and 1992 from top to bottom). By definition, the RF is with respect to 1750, but some anthropogenic changes had already occurred in 1750. The lower right inset shows the globally averaged impact of the surface albedo change to the TOA SW flux (left scale) as well as the corresponding RF (right scale) after normalization to the 1750 value. Based on simulations by Pongratz et al. (2009).

the combined effect of albedo and evapotranspiration impact does at high latitude. These results are also supported by Bathiany et al. (2010). Similarly, Lohila et al. (2010) shows that the afforestation of boreal peatlands results in a balanced RF between the albedo and greenhouse effect. Overall, because of the opposite impacts, the potential of afforestation to mitigate climate change is limited (Arora and Montenegro, 2011) while it may have undesired impacts on the atmospheric circulation, shifting precipitation patterns (Swann et al., 2012).

8.3.5.4 Other Impacts of Land Cover Change on the Earth's Albedo

Burn scars resulting from agriculture practices, uncontrolled fires or deforestation (Bowman et al., 2009) have a lower albedo than unperturbed vegetation (Jin and Roy, 2005). On the other hand, at high latitude, burnt areas are more easily covered by snow, which may result in an overall increase of the surface albedo. Surface blackening of natural vegetation due to fire is relatively short lived and typically disappears within one to a few years (Jin et al., 2012). Myhre et al. (2005b) estimates a global albedo-related radiative effect due to African fires of 0.015 W m^{-2}.

Over semi-arid areas, the development of agriculture favours the generation of dust. Mulitza et al. (2010) demonstrates a very large increase of dust emission and deposition in the Sahel concomitant with the development of agriculture in this area. This, together with the analysis of dust sources (Ginoux et al., 2010), suggests that a significant fraction of the dust that is transported over the Atlantic has an anthropogenic origin and impacts the Earth albedo. There is no full estimate of the resulting RF, however. The dust RF estimate in Section 8.3.4.2 includes both land use contributions and change in wind-driven emissions. Both dust and biomass burning aerosol may impact the Earth surface albedo as these particles can be deposed on snow, which has a large impact on its absorption, in particular for soot. This is discussed in Section 8.3.4.4.

Urban areas have an albedo that is 0.01 to 0.02 smaller than adjacent croplands (Jin et al., 2005). There is the potential for a strong increase through white roof coating with the objective of mitigating the heat island effect (Oleson et al., 2010). Although the global scale impact is small, local effects can be very large, as shown by Campra et al. (2008) that reports a regional (260 km^2) 0.09 increase in albedo and –20 W m^{-2} RF as a consequence of greenhouse horticulture development.

8.3.5.5 Impacts of Surface Change on Climate

Davin et al. (2007) argues that the climate sensitivity to land use forcing is lower than that for other forcings, due to its spatial distribution but also the role of non-radiative processes. Indeed, in addition to the impact on the surface albedo, land use change also modifies the evaporation and surface roughness, with counterbalancing consequences on the lower atmosphere temperature. There is increasing evidence that the impact of land use on evapotranspiration—a non-RF on climate—is comparable to, but of opposite sign than, the albedo effect, so that RF is not as useful a metric as it is for gases and aerosols. For instance, Findell et al. (2007) climate simulations show a negligible impact of land use change on the global mean temperature, although there are some significant regional changes.

Numerical climate experiments demonstrate that the impact of land use on climate is much more complex than just the RF. This is due in part to the very heterogeneous nature of land use change (Barnes and Roy, 2008), but mostly due to the impact on the hydrological cycle through evapotranspiration, root depth and cloudiness (van der Molen et al., 2011). As a consequence, the forcing on climate is not purely radiative and the net impact on the surface temperature may be either positive or negative depending on the latitude (Bala et al., 2007). Davin and de Noblet-Ducoudre (2010) analyses the impact on climate of large-scale deforestation; the albedo cooling effect dominates for high latitude whereas reduced evapotranspiration dominates in the tropics. This latitudinal trend is confirmed by observations of the temperature difference between open land and nearby forested land (Lee et al., 2011).

Irrigated areas have continuously increased during the 20th century although a slowdown has been observed in recent decades (Bonfils and Lobell, 2007). There is clear evidence that irrigation leads to local cooling of several degrees (Kueppers et al., 2007). Irrigation also affects cloudiness and precipitation (Puma and Cook, 2010). In the United States, DeAngelis et al. (2010) found that irrigation in the Great Plains in the summer produced enhanced precipitation in the Midwest 1000 km to the northeast.

8.3.5.6 Conclusions

There is still a rather wide range of estimates of the albedo change due to anthropogenic land use change, and its RF. Although most published studies provide an estimate close to –0.2 W m^{-2}, there is convincing evidence that it may be somewhat weaker as the albedo difference between natural and anthropogenic land cover may have been overestimated. In addition, non-radiative impact of land use have a similar magnitude, and may be of opposite sign, as the albedo effect (though these are not part of RF). A comparison of the impact of land use change according to seven climate models showed a wide range of results (Pitman et al., 2009), partly due to difference in the implementation of land cover change, but mostly due to different assumptions on ecosystem albedo, plant phenology and evapotranspiration. There is no agreement on the sign of the temperature change induced by anthropogenic land use change. It is *very likely* that land use change led to an increase of the Earth albedo with a RF of –0.15 ± 0.10 W m^{-2}, but a net cooling of the surface—accounting for processes that are not limited to the albedo—is *about as likely as not*.

8.4 Natural Radiative Forcing Changes: Solar and Volcanic

Several natural drivers of climate change operate on multiple time scales. Solar variability takes place at many time scales that include centennial and millennial scales (Helama et al., 2010), as the radiant energy output of the Sun changes. Also, variations in the astronomical alignment of the Sun and the Earth (Milankovitch cycles) induce cyclical changes in RF, but this is substantial only at millennial and longer time scales (see Section 5.2.1.1). Volcanic forcing is highly episodic, but can have dramatic, rapid impacts on climate. No major asteroid impacts occurred during the reference period (1750–2012) and thus this effect is not considered here. This section discusses solar and volcanic forcings, the two dominant natural contributors of climate change since the pre-industrial time.

8.4.1 Solar Irradiance

In earlier IPCC reports the forcing was estimated as the instantaneous RF at TOA. However, due to wavelength-albedo dependence, solar radiation-wavelength dependence and absorption within the stratosphere and the resulting stratospheric adjustment, the RF is reduced to about 78% of the TOA instantaneous RF (Gray et al., 2009). There is *low confidence* in the exact value of this number, which can be model and time scale dependent (Gregory et al., 2004; Hansen et al., 2005). AR4 gives an 11-year running mean instantaneous TOA RF between 1750 and the present of 0.12 W m^{-2} with a range of estimates of 0.06 to 0.30 W m^{-2}, equivalent to a RF of 0.09 W m^{-2} with a range of 0.05 to 0.23 W m^{-2}. For a consistent treatment of all forcing agents, hereafter we use RF while numbers quoted from AR4 will be provided both as RF and instantaneous RF at TOA.

8.4.1.1 Satellite Measurements of Total Solar Irradiance

Total solar irradiance (TSI) measured by the Total Irradiance Monitor (TIM) on the spaceborne Solar Radiation and Climate Experiment (SORCE) is 1360.8 ± 0.5 W m^{-2} during 2008 (Kopp and Lean, 2011) which is ~4.5 W m^{-2} lower than the Physikalisch-Meteorologisches Observatorium Davos (PMOD) TSI composite during 2008 (Frohlich, 2009).The difference is probably due to instrumental biases in measurements prior to TIM. Measurements with the PREcision MOnitor Sensor (PREMOS) instrument support the TIM absolute values (Kopp and Lean, 2011). The TIM calibration is also better linked to national standards which provides further support that it is the most accurate (see Supplementary Material Section 8.SM.6). Given the lower TIM TSI values relative to currently used standards, most general circulation models are calibrated to incorrectly high values. However, the few tenths of a percent bias in the absolute TSI value has minimal consequences for climate simulations because the larger uncertainties in cloud properties have a greater effect on the radiative balance. As the maximum-to-minimum TSI relative change is well-constrained from observations, and historical variations are calculated as changes relative to modern values, a revision of the absolute value of TSI affects RF by the same fraction as it affects TSI. The downward revision of TIM TSI with respect to PMOD, being 0.3%, thus has a negligible impact on RF, which is given with a relative uncertainty of several tenths of a percent.

Since 1978, several independent space-based instruments have directly measured the TSI. Three main composite series were constructed, referred to as the Active Cavity Radiometer Irradiance Monitor (ACRIM) (Willson and Mordvinov, 2003), the Royal Meteorological Institute of Belgium (RMIB) (Dewitte et al., 2004) and the PMOD (Frohlich, 2006) series. There are two major differences between ACRIM and PMOD. The first is the rapid drift in calibration between PMOD and ACRIM before 1981. This arises because both composites employ the Hickey–Frieden (HF) radiometer data for this interval, while a re-evaluation of the early HF degradation has been implemented by PMOD but not by ACRIM. The second one, involving also RMIB, is the bridging of the gap between the end of ACRIM I (mid-1989) and the beginning of ACRIM II (late 1991) observations, as it is possible that a change in HF data occurred during this gap. This possibility is neglected in ACRIM and thus its TSI increases by more than 0.5 W m^{-2} during solar cycle (SC) 22. These differences lead to different long-term TSI trends in the three composites (see Figure 8.10): ACRIM rises until 1996 and subsequently declines, RMIB has an upward trend through 2008 and PMOD shows a decline since 1986 which unlike the other two composites, follows the solar-cycle-averaged sunspot number (Lockwood, 2010). Moreover, the ACRIM trend implies that the TSI on time scales longer than the SC is positively correlated with the cosmic ray variation indicating a decline in TSI throughout most of the 20th century (the opposite to most TSI reconstructions produced to date; see Section 8.4.1.2). Furthermore, extrapolating the ACRIM TSI long-term drift would imply a brighter Sun in the Maunder minimum (MM) than now, again opposite to most TSI reconstructions (Lockwood and Frohlich, 2008). Finally, analysis of instrument degradation and pointing issues (Lee et al., 1995) and independent modeling based on solar magnetograms (Wenzler et al.,

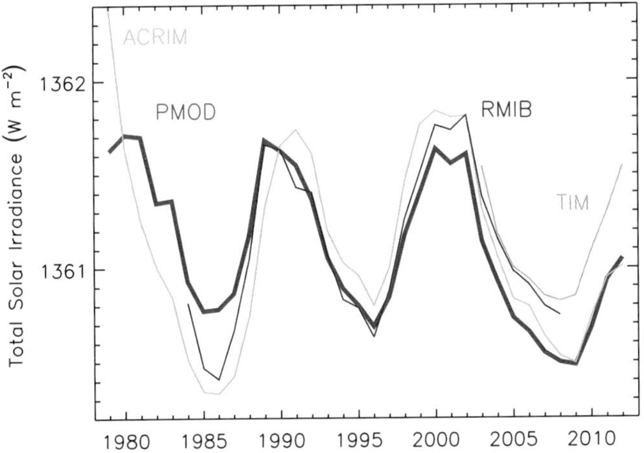

Figure 8.10 | Annual average composites of measured total solar irradiance: The Active Cavity Radiometer Irradiance Monitor (ACRIM) (Willson and Mordvinov, 2003), the Physikalisch-Meteorologisches Observatorium Davos (PMOD) (Frohlich, 2006) and the Royal Meteorological Institute of Belgium (RMIB) (Dewitte et al., 2004).These composites are standardized to the annual average (2003–2012) Total Irradiance Monitor (TIM) (Kopp and Lean, 2011) measurements that are also shown.

2009; Ball et al., 2012), confirm the need for correction of HF data, and we conclude that PMOD is more accurate than the other composites.

TSI variations of approximately 0.1% were observed between the maximum and minimum of the 11-year SC in the three composites mentioned above (Kopp and Lean, 2011). This variation is mainly due to an interplay between relatively dark sunspots, bright faculae and bright network elements (Foukal and Lean, 1988; see Section 5.2.1.2). A declining trend since 1986 in PMOD solar minima is evidenced in Figure 8.10. Considering the PMOD solar minima values of 1986 and 2008, the RF is −0.04 W m^{-2}. Our assessment of the uncertainty range of changes in TSI between 1986 and 2008 is –0.08 to 0.0 W m^{-2} and thus *very likely* negative, and includes the uncertainty in the PMOD data (Frohlich, 2009; see Supplementary Material Section 8.SM.6) but is extended to also take into account the uncertainty of combining the satellite data.

For incorporation of TIM data with the previous and overlapping data, in Figure 8.10 we have standardized the composite time series to the TIM series (over 2003–2012, the procedure is explained in Supplementary Material Section 8.SM.6. Moreover as we consider annual averages, ACRIM and PMOD start at 1979 because for 1978 both composites have only two months of data.

8.4.1.2 Total Solar Irradiance Variations Since Preindustrial Time

The year 1750, which is used as the preindustrial reference for estimating RF, corresponds to a maximum of the 11-year SC. Trend analysis are usually performed over the minima of the solar cycles that are more stable. For such trend estimates, it is then better to use the closest SC minimum, which is in 1745. To avoid trends caused by comparing different portions of the solar cycle, we analyze TSI changes using multi-year running means. For the best estimate we use a recent TSI reconstruction by Krivova et al. (2010) between 1745 and 1973 and from 1974 to 2012 by Ball et al. (2012). The reconstruction is based on physical modeling of the evolution of solar surface magnetic flux, and its relationship with sunspot group number (before 1974) and sunspot umbra and penumbra and faculae afterwards. This provides a more detailed reconstruction than other models (see the time series in Supplementary Material Table 8.SM.3). The best estimate from our assessment of the most reliable TSI reconstruction gives a 7-year running mean RF between the minima of 1745 and 2008 of 0.05 W m^{-2}. Our assessment of the range of RF from TSI changes is 0.0 to 0.10 W m^{-2} which covers several updated reconstructions using the same 7-year running mean past-to-present minima years (Wang et al., 2005; Steinhilber et al., 2009; Delaygue and Bard, 2011), see Supplementary Material Table 8.SM.4. All reconstructions rely on indirect proxies that inherently do not give consistent results. There are relatively large discrepancies among the models (see Figure 8.11).With these considerations, we adopt this value and range for AR5. This RF is almost half of that in AR4, in part because the AR4 estimate was based on the previous solar cycle minimum while the AR5 estimate includes the drop of TSI in 2008 compared to the previous two SC minima (see 8.4.1). Concerning the uncertainty range, in AR4 the upper limit corresponded to the reconstruction of Lean (2000), based on the reduced brightness of non-cycling Sun-like stars assumed typical of a Maunder minimum (MM) state. The use of such stellar analogues was based on the work of Baliunas and Jastrow (1990), but more recent surveys have not reproduced their results and suggest that the selection of the original set was flawed (Hall and Lockwood, 2004; Wright, 2004); the lower limit from 1750 to present in AR4 was due to the assumed increase in the amplitude of the 11-year cycle only. Thus the RF and uncertainty range have been obtained in a different way in AR5 compared to AR4. Maxima to maxima RF give a higher estimate than minima to minima RF, but the latter is more relevant for changes in solar activity. Given the *medium agreement* and *medium evidence*, this RF value has a *medium confidence level* (although confidence is higher for the last three decades). Figure 8.11 shows several TSI reconstructions modelled using sunspot group numbers (Wang et al., 2005; Krivova et al., 2010;

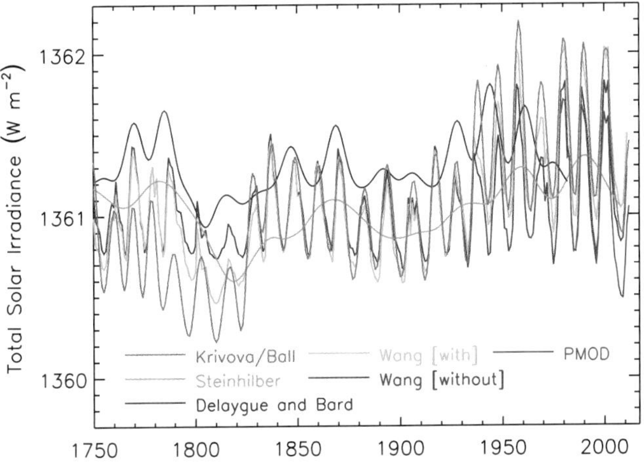

Figure 8.11 | Reconstructions of total solar irradiance since1745; annual resolution series from Wang et al. (2005) with and without an independent change in the background level of irradiance, Krivova et al. (2010) combined with Ball et al. (2012) and 5-year time resolution series from Steinhilber et al. (2009) and Delaygue and Bard (2011). The series are standardized to the Physikalisch-Meteorologisches Observatorium Davos (PMOD) measurements of solar cycle 23 (1996–2008) (PMOD is already standardized to Total Irradiance Monitor).

Ball et al., 2012) and sunspot umbra and penumbra and faculae (Ball et al., 2012), or cosmogenic isotopes (Steinhilber et al., 2009; Delaygue and Bard, 2011). These reconstructions are standardized to PMOD SC 23 (1996–2008) (see also Supplementary Material Section 8.SM.6).

For the MM-to-present AR4 gives a TOA instantaneous RF range of 0.1 to 0.28 W m^{-2}, equivalent to 0.08 to 0.22 W m^{-2} with the RF definition used here. The reconstructions in Schmidt et al. (2011) indicate a MM-to-present RF range of 0.08 to 0.18 W m^{-2}, which is within the AR4 range although narrower. As discussed above, the estimates based on irradiance changes in Sun-like stars are not included in this range because the methodology has been shown to be flawed. A more detailed explanation of this is found in Supplementary Material Section 8.SM.6. For details about TSI reconstructions on millennia time scales see Section 5.2.1.2.

8.4.1.3 Attempts to Estimate Future Centennial Trends of Total Solar Irradiance

Cosmogenic isotope and sunspot data (Rigozo et al., 2001; Solanki and Krivova, 2004; Abreu et al., 2008) reveal that currently the Sun is in a grand activity maximum that began about 1920 (20th century grand maximum). However, SC 23 showed an activity decline not previously seen in the satellite era (McComas et al., 2008; Smith and Balogh, 2008; Russell et al., 2010). Most current estimations suggest that the forthcoming solar cycles will have lower TSI than those for the past 30 years (Abreu et al., 2008; Lockwood et al., 2009; Rigozo et al., 2010; Russell et al., 2010). Also there are indications that the mean magnetic field in sunspots may be diminishing on decadal level. A linear expansion of the current trend may indicate that of the order of half the sunspot activity may disappear by about 2015 (Penn and Livingston, 2006). These studies only suggest that the Sun may have left the 20th century grand maximum and not that it is entering another grand minimum. But other works propose a grand minimum during the 21st century, estimating an RF within a range of -0.16 to 0.12 W m^{-2} between this future minimum and the present-day TSI (Jones et al., 2012). However, much more evidence is needed and at present there is *very low confidence* concerning future solar forcing estimates.

Nevertheless, even if there is such decrease in the solar activity, there is a *high confidence* that the TSI RF variations will be much smaller in magnitude than the projected increased forcing due to GHG (see Section 12.3.1).

8.4.1.4 Variations in Spectral Irradiance

8.4.1.4.1 Impacts of ultraviolet variations on the stratosphere

Ozone is the main gas involved in stratospheric radiative heating. Ozone production rate variations are largely due to solar UV irradiance changes (HAIGH, 1994), with observations showing statistically significant variations in the upper stratosphere of 2 to 4% along the SC (Soukharev and Hood, 2006). UV variations may also produce transport-induced ozone changes due to indirect effects on circulation (Shindell et al., 2006b). In addition, statistically significant evidence for an 11-year variation in stratospheric temperature and zonal winds is attributed to UV radiation (Frame and Gray, 2010). The direct UV heating of the background ozone is dominant and over twice as large as the ozone heating in the upper stratosphere and above, while indirect solar and terrestrial radiation through the SC-induced ozone change is dominant below about 5 hPa (Shibata and Kodera, 2005). The RF due to solar-induced ozone changes is a small fraction of the solar RF discussed in Section 8.4.1.1 (Gray et al., 2009).

8.4.1.4.2 Measurements of spectral irradiance

Solar spectral irradiance (SSI) variations in the far (120 to 200 nm) and middle (200 to 300 nm) ultraviolet (UV) are the primary driver for heating, composition, and dynamic changes of the stratosphere, and although these wavelengths compose a small portion of the incoming radiation they show large relative variations between the maximum and minimum of the SC compared to the corresponding TSI changes. As UV heating of the stratosphere over a SC has the potential to influence the troposphere indirectly, through dynamic coupling, and therefore climate (Haigh, 1996; Gray et al., 2010), the UV may have a more significant impact on climate than changes in TSI alone would suggest. Although this indicates that metrics based only on TSI are not appropriate, UV measurements present several controversial issues and modelling is not yet robust.

Multiple space-based measurements made in the past 30 years indicated that UV variations account for about 30% of the SC TSI variations, while about 70% were produced within the visible and infrared (Rottman, 2006). However, current models and data provide the range of 30 to 90% for the contribution of the UV variability below 400 nm to TSI changes (Ermolli et al., 2013), with a more probable value of ~60% (Morrill et al., 2011; Ermolli et al., 2013). The Spectral Irradiance Monitor (SIM) on board SORCE (Harder et al., 2009) shows, over the SC 23 declining phase, measurements that are rather inconsistent with prior understanding, indicating that additional validation and uncertainty estimates are needed (DeLand and Cebula, 2012; Lean and Deland, 2012). A wider exposition can be found in Supplementary Material Section 8.SM.6.

8.4.1.4.3 Reconstructions of preindustrial ultraviolet variations

The Krivova et al. (2010) reconstruction is based on what is known about spectral contrasts of different surface magnetic features and the relationship between TSI and magnetic fields. The authors interpolated backwards to the year 1610 based on sunspot group numbers and magnetic information. The Lean (2000) model is based on historical sunspot number and area and is scaled in the UV using measurements from the Solar Stellar Irradiance Comparison Experiment (SOLSTICE) on board the Upper Atmosphere Research Satellite (UARS). The results show smoothed 11-year UV SSI changes between 1750 and the present of about 25% at about 120 nm, about 8% at 130 to 175 nm, ~4% at 175 to 200 nm, and about 0.5% at 200 to 350 nm. Thus, the UV SSI appears to have generally increased over the past four centuries, with larger trends at shorter wavelengths. As few reconstructions are available, and recent measurements suggest a poor understanding of UV variations and their relationship with solar activity, there is *very low confidence* in these values.

8.4.1.5 The Effects of Cosmic Rays on Clouds

Changing cloud amount or properties modify the Earth's albedo and therefore affect climate. It has been hypothesized that cosmic ray flux create atmospheric ions which facilitates aerosol nucleation and new particle formation with a further impact on cloud formation (Dickinson, 1975; Kirkby, 2007). High solar activity means a stronger heliospheric magnetic field and thus a more efficient screen against cosmic rays. Under the hypothesis underlined above, the reduced cosmic ray flux would promote fewer clouds amplifying the warming effect expected from high solar activity. There is evidence from laboratory, field and modelling studies that ionization from cosmic ray flux may enhance aerosol nucleation in the free troposphere (Merikanto et al., 2009; Mirme et al., 2010; Kirkby et al., 2011). However, there is *high confidence* (*medium evidence* and *high agreement*) that the cosmic ray–ionization mechanism is too weak to influence global concentrations of cloud condensation nuclei or their change over the last century or during a SC in a climatically significant way (Harrison and Ambaum, 2010; Erlykin and Wolfendale, 2011; Snow-Kropla et al., 2011). A detailed exposition is found in Section 7.4.6.

8.4.2 Volcanic Radiative Forcing

8.4.2.1 Introduction

Volcanic eruptions that inject substantial amounts of SO_2 gas into the stratosphere are the dominant natural cause of externally forced climate change on the annual and multi-decadal time scales, both because of the multi-decadal variability of eruptions and the time scale of the climate system response, and can explain much of the pre-industrial climate change of the last millennium (Schneider et al., 2009; Brovkin et al., 2010; Legras et al., 2010; Miller et al., 2012). Although volcanic eruptions inject both mineral particles (called ash or tephra) and sulphate aerosol precursor gases (predominantly SO_2) into the atmosphere, it is the sulphate aerosols, which because of their small size are effective scatterers of sunlight and have long lifetimes, that are responsible for RF important for climate. Global annually averaged emissions of CO_2 from volcanic eruptions since 1750 have been at least 100 times smaller than anthropogenic emissions and inconsequential for climate on millennial and shorter time scales (Gerlach, 2011). To be important for climate change, sulphur must be injected into the stratosphere, as the lifetime of aerosols in the troposphere is only about one week, whereas sulphate aerosols in the stratosphere from tropical eruptions have a lifetime of about one year, and those from high-latitude eruptions last several months. Most stratospheric aerosols are from explosive eruptions that directly put sulphur into the stratosphere, but Bourassa et al. (2012, 2013) showed that sulphur injected into the upper troposphere can then be lifted into the stratosphere over the next month or two by deep convection and large scale Asian summer monsoon circulation, although Vernier et al. (2013) and Fromm et al. (2013) suggested that direct injection was also important. Robock (2000), AR4 (Forster et al., 2007) and Timmreck (2012) provide summaries of this relatively well understood forcing agent.

There have been no major volcanic eruptions since Mt Pinatubo in 1991 (Figure 8.12), but several smaller eruptions have caused a RF for the years 2008–2011 of –0.11 (–0.15 to –0.08) W m^{-2}, approximately twice the magnitude of the 1999–2002 RF of –0.06 (–0.08 to –0.04) W m^{-2}, consistent with the trends noted in Solomon et al. (2011). However, the CMIP5 simulations discussed elsewhere in this report did not include the recent small volcanic forcing in their calculations. New work has also produced a better understanding of high latitude eruptions, the hydrological response to volcanic eruptions (Trenberth and Dai, 2007; Anchukaitis et al., 2010), better long-term records of past volcanism and better understanding of the effects of very large eruptions.

There are several ways to measure both the SO_2 precursor and sulphate aerosols in the stratosphere, using balloons, airplanes, and both ground- and satellite-based remote sensing. Both the infrared and ultraviolet signals sensed by satellite instruments can measure SO_2, and stratospheric aerosol measurements by space-based sensors have been made on a continuous basis since 1978 by a number of instruments employing solar and stellar occultation, limb scattering, limb emission, and lidar strategies (Thomason and Peter, 2006; Kravitz et al., 2011; Solomon et al., 2011).

Forster et al. (2007) described four mechanisms by which volcanic forcing influences climate: RF due to aerosol–radiation interaction; differential (vertical or horizontal) heating, producing gradients and changes in circulation; interactions with other modes of circulation, such as El Niño-Southern Oscillation (ENSO); and ozone depletion with its effects on stratospheric heating, which depends on anthropogenic chlorine (stratospheric ozone would increase with a volcanic eruption under low-chlorine conditions). In addition, the enhanced diffuse light from volcanic aerosol clouds impacts vegetation and hence the carbon cycle (Mercado et al., 2009) and aerosol–cloud interaction of sulphate aerosols on clouds in the troposphere can also be important (Schmidt et al., 2010), though Frolicher et al. (2011) showed that the impacts of the 1991 Mt Pinatubo eruption on the carbon cycle were small.

8.4.2.2 Recent Eruptions

The background stratospheric aerosol concentration was affected by several small eruptions in the first decade of the 21st century (Nagai et al., 2010; Vernier et al., 2011; Neely et al., 2013; see also Figure 8.13), with a very small contribution from tropospheric pollution (Siddaway and Petelina, 2011; Vernier et al., 2011), and had a small impact on RF (Solomon et al., 2011). Two recent high-latitude eruptions, of Kasatochi Volcano (52.1°N, 175.3°W) on August 8, 2008 and of Sarychev Volcano (48.1°N, 153.2°E) on June 12–16, 2009, each injected ~1.5 Tg(SO_2) into the stratosphere, but did not produce detectable climate response. Their eruptions, however, led to better understanding of the dependence of the amount of material and time of year of high-latitude injections to produce climate impacts (Haywood et al., 2010; Kravitz et al., 2010, 2011). The RF from high-latitude eruptions is a function of seasonal distribution of insolation and the 3- to 4-month lifetime of high-latitude volcanic aerosols. Kravitz and Robock (2011) showed that high-latitude eruptions must inject at least 5 Tg(SO_2) into the lower stratosphere in the spring or summer, and much more in fall or winter, to have a detectible climatic response.

On April 14, 2010 the Eyjafjallajökull volcano in Iceland (63.6°N, 19.6°W) began an explosive eruption phase that shut down air traffic in Europe for 6 days and continued to disrupt it for another month. The climatic impact of Eyjafjallajökull was about 10,000 times less than that of Mt Pinatubo; however, because it emitted less than 50 ktonnes SO_2 and its lifetime in the troposphere was 50 times less than if it had been injected into the stratosphere, and was therefore undetectable amidst the chaotic weather noise in the atmosphere (Robock, 2010). 2011 saw the continuation of a number of small eruptions with significant tropospheric SO_2 and ash injections, including Puyehue-Cordón Caulle in Chile, Nabro in Eritrea, and Grimsvötn in Iceland. None have been shown to have produced an important RF, but the June 13, 2011 Nabro eruption resulted in the largest stratospheric aerosol cloud since the 1991 Mt Pinatubo eruption (Bourassa et al., 2012), more than 1.5 Tg(SO_2).

Figure 8.12 shows reconstructions of volcanic aerosol optical depth since 1750. Figure 8.13 shows details of the vertical distribution of stratospheric aerosols in the tropics since 1985. The numerous small eruptions in the past decade are evident, but some of them were at higher latitudes and their full extent is not captured in this plot.

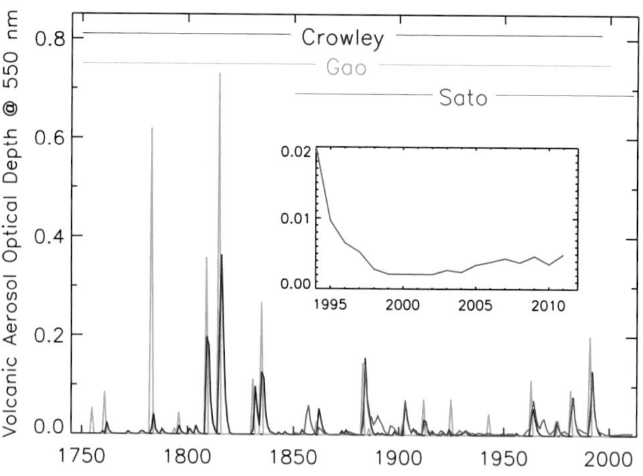

Figure 8.12 | Volcanic reconstructions of global mean aerosol optical depth (at 550 nm). Gao et al. (2008) and Crowley and Unterman (2013) are from ice core data, and end in 2000 for Gao et al. (2008) and 1996 for Crowley and Unterman (2013). Sato et al. (1993) includes data from surface and satellite observations, and has been updated through 2011. (Updated from Schmidt et al., 2011.)

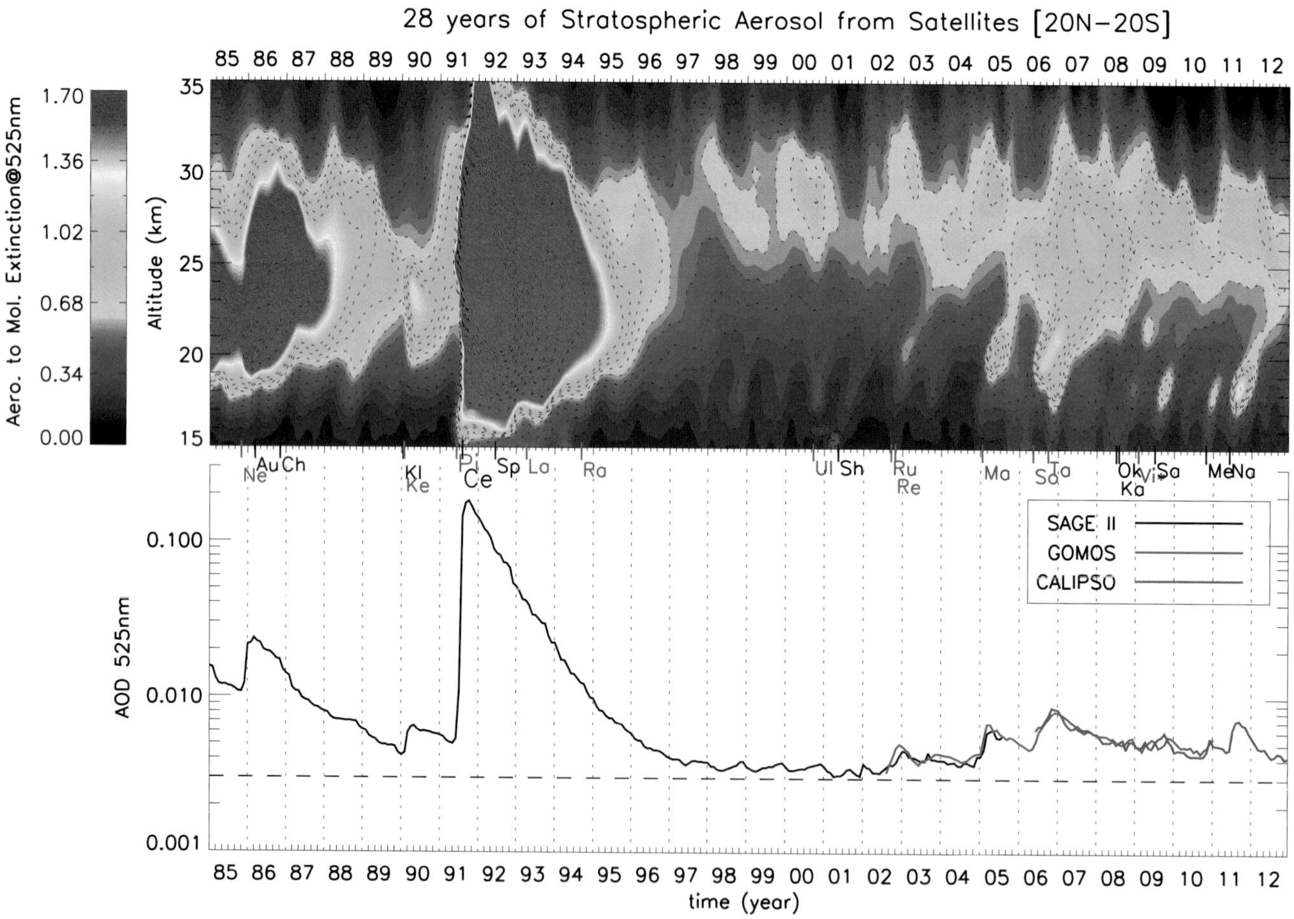

Figure 8.13 | (Top) Monthly mean extinction ratio (525 nm) profile evolution in the tropics [20°N to 20°S] from January 1985 through December 2012 derived from Stratospheric Aerosol and Gas Experiment (SAGE) II extinction in 1985–2005 and Cloud-Aerosol Lidar and Infrared Pathfinder Satellite Observation (CALIPSO) scattering ratio in 2006–2012, after removing clouds below 18 km based on their wavelength dependence (SAGE II) and depolarization properties (CALIPSO) compared to aerosols. Black contours represent the extinction ratio in log-scale from 0.1 to 100. The position of each volcanic eruption occurring during the period is displayed with its first two letters on the horizontal axis, where tropical eruptions are noted in red. The eruptions were Nevado del Ruiz (Ne), Augustine (Au), Chikurachki (Ch), Kliuchevskoi (Kl), Kelut (Ke), Pinatubo (Pi), Cerro Hudson (Ce), Spur (Sp), Lascar (La), Rabaul (Ra), Ulawun (Ul), Shiveluch (Sh), Ruang (Ru), Reventador (Re), Manam (Ma), Soufrière Hills (So), Tavurvur (Ta), Okmok (Ok), Kasatochi (Ka), Victoria (Vi*—forest fires with stratospheric aerosol injection), Sarychev (Sa), Merapi (Me), Nabro (Na). (Updated from Figure 1 from Vernier et al., 2011.) (Bottom) Mean stratospheric aerosol optical depth (AOD) in the tropics [20°N to 20°S] between the tropopause and 40 km since 1985 from the SAGE II (black line), the Global Ozone Monitoring by Occultation of Stars (GOMOS) (red line), and CALIPSO (blue line). (Updated from Figure 5 from Vernier et al., 2011.)

Box 8.3 | Volcanic Eruptions as Analogues

Volcanic eruptions provide a natural experiment of a stratospheric aerosol cloud that can serve to inform us of the impacts of the proposed production of such a cloud as a means to control the climate, which is one method of geoengineering (Rasch et al., 2008); see Section 7.7. For example, Trenberth and Dai (2007) showed that the Asian and African summer monsoon, as well as the global hydrological cycle, was weaker for the year following the 1991 Mt Pinatubo eruption, which is consistent with climate model simulations (Robock et al., 2008). MacMynowski et al. (2011) showed that because the climate system response of the hydrological cycle is rapid, forcing from volcanic eruptions, which typically last about a year, can serve as good analogues for longer-lived forcing. The formation of sulphate aerosols, their transport and removal, their impacts on ozone chemistry, their RF, and the impacts on whitening skies all also serve as good analogues for geoengineering proposals. Volcanic impacts on the carbon cycle because of more diffuse radiation (Mercado et al., 2009) and on remote sensing can also be useful analogues, and the impacts of contrail-generated sub-visual cirrus (Long et al., 2009) can be used to test the long-term impacts of a permanent stratospheric cloud.

Smoke from fires generated by nuclear explosions on cities and industrial areas, which could be lofted into the stratosphere, would cause surface cooling and a reduction of stratospheric ozone (Mills et al., 2008). Volcanic eruptions that produce substantial stratospheric aerosol clouds also serve as an analogue that supports climate model simulations of the transport and removal of stratospheric aerosols, their impacts on ozone chemistry, their RF, and the climate response. The use of the current global nuclear arsenal still has the potential to produce nuclear winter, with continental temperatures below freezing in summer (Robock et al., 2007a; Toon et al., 2008), and the use of only 100 nuclear weapons could produce climate change unprecedented in recorded human history (Robock et al., 2007b), with significant impacts on global agriculture (Özdoğan et al., 2013; Xia and Robock, 2013).

8.4.2.3 Records of Past Volcanism and Effects of Very Large Eruptions

Although the effects of volcanic eruptions on climate are largest in the 2 years following a large stratospheric injection, and the winter warming effect in the NH has been supported by long-term records (Fischer et al., 2007), there is new work indicating extended volcanic impacts via long-term memory in the ocean heat content and sea level (Stenchikov et al., 2009; Gregory, 2010; Otterä et al., 2010). Zanchettin et al. (2012) found changes in the North Atlantic Ocean circulation that imply strengthened northward oceanic heat transport a decade after major eruptions, which contributes to the emergence of extensive winter warming over the continental NH along with persistent cooling over Arctic regions on decadal time scales, in agreement with Zhong et al. (2011) and Miller et al. (2012).

New work on the mechanisms by which a supereruption (Self and Blake, 2008) could force climate has focused on the 74,000 BP eruption of the Toba volcano (2.5°N, 99.0°E). Robock et al. (2009) used simulations of up to 900 times the 1991 Mt Pinatubo sulphate injection to show that the forcing is weaker than that predicted based on a linear relationship with the sulphate aerosol injection. The results agreed with a previous simulation by Jones et al. (2005). They also showed that chemical interactions with ozone had small impacts on the forcing and that the idea of Bekki et al. (1996) that water vapour would limit and prolong the growth of aerosols was not supported. Timmreck et al. (2010) however, incorporating the idea of Pinto et al. (1989) that aerosols would grow and therefore both have less RF per unit mass and fall out of the atmosphere more quickly, found much less of a radiative impact from such a large stratospheric input.

8.4.2.4 Future Effects

We expect large eruptions over the next century but cannot predict when. Ammann and Naveau (2003) and Stothers (2007) suggested an 80-year periodicity in past eruptions, but the data record is quite short and imperfect, and there is no mechanism proposed that would cause this. While the period 1912–1963 was unusual for the past 500 years in having no large volcanic eruptions, and the period 1250–1300 had the most globally climatically significant eruptions in the past 1500 years (Gao et al., 2008), current knowledge only allows us to predict such periods on a statistical basis, assuming that the recent past distributions are stationary. Ammann and Naveau (2003), Gusev (2008), and Deligne et al. (2010) studied these statistical properties and Ammann and Naveau (2010) showed how they could be used to produce a statistical distribution for future simulations. Although the future forcing from volcanic eruptions will depend only on the stratospheric aerosol loading for most forcing mechanisms, the future effects on reducing ozone will diminish as ozone depleting substances diminish in the future (Eyring et al., 2010b).

8.5 Synthesis of Global Mean Radiative Forcing, Past and Future

The RF can be used to quantify the various agents that drive climate change over the Industrial Era or the various contributions to future climate change. There are multiple ways in which RF can be attributed to underlying causes, each providing various perspectives on the importance of the different factors driving climate change. This section evaluates the RF with respect to emitted component and with respect to the ultimate atmospheric concentrations. The uncertainties in the RF agents vary and the confidence levels for these are presented in this section. Finally, this section shows historical and scenarios of future time evolution of RF.

8.5.1 Summary of Radiative Forcing by Species and Uncertainties

Table 8.5 has an overview of the RF agents considered here and each of them is given a confidence level for the change in RF over the Industrial Era to the present day. The confidence level is based on the evidence (robust, medium, and limited) and the agreement (high, medium, and low; see further description in Chapter 1). The confidence level of the forcing agents goes beyond the numerical values available in estimates and is an assessment for a particular forcing agent to have a real value within the estimated range. Some of the RF agents have robust evidence such as WMGHG with well documented increases based on high precision measurements as well as contrails as additional clouds which can be seen by direct observations. However, for some forcing agents the evidence is more limited regarding their existence such as aerosol influence on cloud cover. The consistency in the findings for a particular forcing agent determines the evaluation of the evidence. A combination of different methods, for example, observations and modeling, and thus the understanding of the processes causing the forcing is important for this evaluation. The agreement is a qualitative judgment of the difference between the various estimates for a particular RF agent. Figure 1.11 shows how the combined evidence and agreement results in five levels for the confidence level.

Table 8.5 | Confidence level for the forcing estimate associated with each forcing agent for the 1750–2011 period. The confidence level is based on the evidence and the agreement as given in the table. The basis for the confidence level and change since AR4 is provided. See Figure 1.11 for further description of the evidence, agreement and confidence level. The colours are adopted based on the evidence and agreement shown in Figure 1.11. Dark green is "High agreement and Robust evidence", light green is either "High agreement and Medium evidence" or "Medium agreement and Robust evidence", yellow is either "High agreement and limited evidence" or "Medium agreement and Medium evidence" or "Low agreement and Robust evidence", orange is either "Medium agreement and Limited evidence" or "Low agreement and Medium evidence" and finally red is "Low agreement and Limited evidence". Note, that the confidence levels given in Table 8.5 are for 2011 relative to 1750 and for some of the agents the confidence level may be different for certain portions of the Industrial Era.

	Evidence	Agreement	Confidence Level	Basis for Uncertainty Estimates (more certain / less certain)	Change in Understanding Since AR4
Well-mixed greenhouse gases	Robust	High	*Very high*	Measured trends from different observed data sets and differences between radiative transfer models	No major change
Tropospheric ozone	Robust	Medium	*High*	Observed trends of ozone in the troposphere and model results for the industrial era/Differences between model estimates of RF	No major change
Stratospheric ozone	Robust	Medium	*High*	Observed trends in stratospheric and total ozone and modelling of ozone depletion/Differences between estimates of RF	No major change
Stratospheric water vapour from CH_4	Robust	Low	*Medium*	Similarities in results of independent methods to estimate the RF/Known uncertainty in RF calculations	Elevated owing to more studies
Aerosol–radiation interactions	Robust	Medium	*High*	A large set of observations and converging independent estimates of RF/Differences between model estimates of RF	Elevated owing to more robust estimates from independent methods
Aerosol–cloud interactions	Medium	Low	*Low*	Variety of different observational evidence and modelling activities/Spread in model estimates of ERF and differences between observations and model results	ERF in AR5 has a similar confidence level to RF in AR4
Rapid adjustment aerosol–radiation interactions	Medium	Low	*Low*	Observational evidence combined with results from different types of models/Large spread in model estimates	Elevated owing to increased evidence
Total aerosol effect	Medium	Medium	*Medium*	A large set of observations and model results, independent methods to derive ERF estimates/Aerosol–cloud interaction processes and anthropogenic fraction of CCN still fairly uncertain	Not provided previously
Surface albedo (land use)	Robust	Medium	*High*	Estimates of deforestation for agricultural purposes and well known physical processes/Spread in model estimates of RF	Elevated owing to the availability of high-quality satellite data
Surface albedo (BC aerosol on snow and ice)	Medium	Low	*Low*	Observations of snow samples and the link between BC content in snow and albedo/Large spread in model estimates of RF	No major change
Contrails	Robust	Low	*Medium*	Contrails observations , large number of model estimates/Spread in model estimates of RF and uncertainties in contrail optical properties	Elevated owing to more studies
Contrail- induced cirrus	Medium	Low	*Low*	Observations of a few events of contrail induced cirrus/Extent of events uncertain and large spread in estimates of ERF	Elevated owing to additional studies increasing the evidence
Solar irradiance	Medium	Medium	*Medium*	Satellite information over recent decades and small uncertainty in radiative transfer calculations/Large relative spread in reconstructions based on proxy data	Elevated owing to better agreement of a weak RF
Volcanic aerosol	Robust	Medium	*High*	Observations of recent volcanic eruptions/Reconstructions of past eruptions	Elevated owing to improved understanding

Notes:

The confidence level for aerosol–cloud interactions includes rapid adjustments (which include what was previously denoted as cloud lifetime effect or second indirect aerosol effect). The separate confidence level for the rapid adjustment for aerosol–cloud interactions is very low. For aerosol–radiation interaction the table provides separate confidence levels for RF due to aerosol–radiation interaction and rapid adjustment associated with aerosol–radiation interaction.

Evidence is robust for several of the RF agents because of long term observations of trends over the industrial era and well defined links between atmospheric or land surfaced changes and their radiative effects. Evidence is medium for a few agents where the anthropogenic changes or the link between the forcing agent and its radiative effect are less certain. Medium evidence can be assigned in cases where observations or modelling provide a diversity of information and thus not a consistent picture for a given forcing agent. We assess the evidence to be limited only for rapid adjusment associated with aerosol–cloud interaction where model studies in some cases indicate changes but direct observations of cloud alterations are scarce. High agreement is given only for the WMGHG where the relative uncertainties in the RF estimates are much smaller than for the other RF agents. Low agreement can either be due to large diversity in estimates of the magnitude of the forcing or from the fact that the method to estimate the forcing has a large uncertainty. Stratospheric water vapour is an example of the latter with modest difference in the few available estimates but a known large uncertainty in the radiative transfer calculations (see further description in Section 8.3.1).

Figure 8.14 shows the development of the confidence level over the last four IPCC assessments for the various RF mechanisms. In the previous IPCC reports level of scientific understanding (LOSU) has been used instead of confidence level. For comparison with previous IPCC assessments the LOSU is converted approximately to confidence level. Note that LOSU and confidence level use different terms for their rankings. The figure shows generally increasing confidence levels but also that more RF mechanisms have been included over time. The confidence levels for the RF due to aerosol–radiation interactions, surface albedo due to land use and volcanic aerosols have been raised and are now at the same ranking as those for change in stratospheric and tropospheric ozone. This is due to an increased understanding of key parameteres and their uncertainties for the elevated RF agents. For tropospheric and stratospheric ozone changes, research has shown further complexities with changes primarily influencing the troposphere or the stratosphere being linked to some extent (see Section 8.3.3). The rapid adjustment associated with aerosol–cloud interactions is given the *confidence* level *very low* and had a similar level in AR4. For rapid adjustment associated with aerosol–radiation interactions (previously denoted as semi-direct effect) the *confidence* level is *low* and is raised compared to AR4, as the evidence is improved and is now *medium* (see Section 7.5.2).

Table 8.6 shows the best estimate of the RF and ERF (for AR5 only) for the various RF agents from the various IPCC assessments. The RF due to WMGHG has increased by 16% and 8% since TAR and AR4,

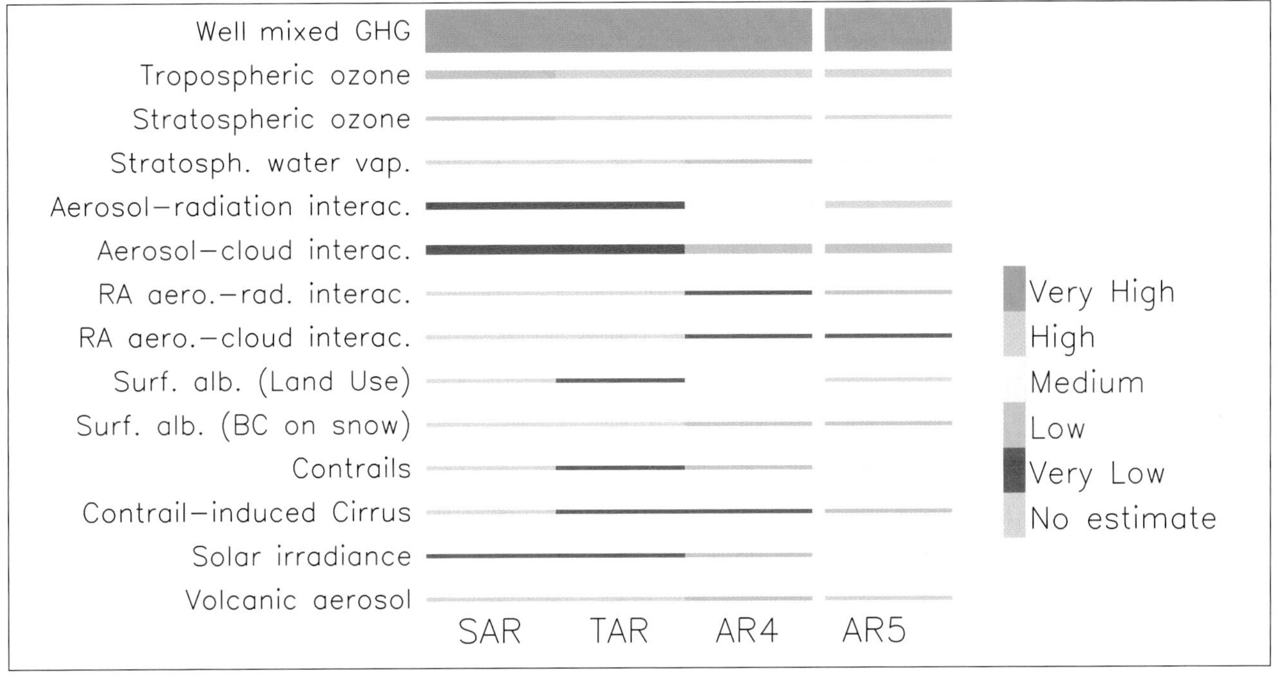

Figure 8.14 | Confidence level of the forcing mechanisms in the 4 last IPCC assessments. In the previous IPCC assessments the level of scientific understanding (LOSU) has been adopted instead of confidence level, but for comparison with previous IPCC assessments the LOSU is converted approximately to confidence level. The thickness of the bars represents the relative magnitude of the current forcing (with a minimum value for clarity of presentation). LOSU for the RF mechanisms was not available in the first IPCC Assessment (Houghton et al., 1990). Rapid adjustments associated with aerosol–cloud interactions (shown as RA aero. –cloud interac.) which include what was previously referred to as the second indirect aerosol effect or cloud lifetime effect whereas rapid adjustments associated with aerosol–radiation interactions (shown as RA aero.-rad. interac.) were previously referred to as the semi-direct effect (see Figure 7.3). In AR4 the confidence level for aerosol–cloud interaction was given both for RF due to aerosol–cloud interaction and rapid adjustment associated with aerosol–cloud interaction. Generally the aerosol–cloud interaction is not separated into various components in AR5, hence the confidence levels for ERF due to aerosol–cloud interaction in AR5 and for RF due to aerosol–cloud interaction from previous IPCC reports are compared. The confidence level for the rapid adjustment associated with aerosol–cloud interraction is comparable for AR4 and AR5. The colours are adopted based on the evidence and agreement shown in Figure 1.11. Dark green is "High agreement and Robust evidence", light green is either "High agreement and Medium evidence" or "Medium agreement and Robust evidence", yellow is either "High agreement and limited evidence" or "Medium agreement and Medium evidence" or "Low agreement and Robust evidence", orange is either "Medium agreement and Limited evidence" or "Low agreement and Medium evidence" and finally red is "Low agreement and Limited evidence".

respectively. This is due mainly to increased concentrations (see Section 8.3.2), whereas the other changes for the anthropogenic RF agents compared to AR4 are due to re-evaluations and in some cases from improved understanding. An increased number of studies, additional observational data and better agreement between models and observations can be the causes for such re-evaluations. The best estimates for RF due to aerosol–radiation interactions, BC on snow and solar irradiance are all substantially decreased in magnitude compared to AR4; otherwise the modifications to the best estimates are rather small. For the RF due to aerosol–radiation interaction and BC on snow the changes in the estimates are based on additional new studies since AR4 (see Section 8.3.4 and Section 7.5). For the change in the estimate of the solar irradiance it is a combination on how the RF is calculated, new evidence showing some larger earlier estimates were incorrect, and a downward trend observed during recent years in the solar activity that has been taken into account (see Section 8.4.1). The estimate for ERF due to to aerosol–cloud interaction includes rapid adjustment but still this ERF is smaller in magnitude than the AR4 RF estimate due to aerosol–cloud interactions without rapid adjustments (a theoretical construct not quantified in AR5). The uncertainties for ERF due to CO_2 increase when compared to RF (see Section 8.3.2). We assume the relative ERF uncertainties for CO_2 apply to all WMGHG. For the short-lived GHG we do not have sufficient information to include separate ERF uncertainty to each of these forcing agents (see Section 8.1.1.3). However, for these forcing mechanisms the RF uncertainties are larger than for the WMGHG and thus it is *unlikely* that rapid adjustments change the uncertainties substantially.

Figure 8.15 shows the RF for agents listed in Table 8.6 over the 1750–2011 period. The methods for calculation of forcing estimates are described in Section 8.3 and 8.4. For some of the components the forcing estimates are based on observed abundance whereas some are estimated from a combination of model simulations and observations and for others are purely model based. Solid bars are given for ERF, whereas RF values are given as (additional) hatched bars. Similarly the uncertainties are given for ERF in solid lines and dotted lines for RF. An important assumption is that different forcing mechanisms can be treated additively to calculate the total forcing (see Boucher and Haywood, 2001; Forster et al., 2007; Haywood and Schulz, 2007). Total ERF over the Industrial Era calculated from Monte Carlo simulations are shown in Figure 8.16, with a best estimate of 2.29 W m^{-2}. For each of the forcing agents a probability density function (PDF) is generated based on uncertainties provided in Table 8.6. The combination of the individual RF agents to derive total forcing follows the same approach as in AR4 (Forster et al., 2007) which is based on the method in Boucher and Haywood (2001). The PDF of the GHGs (sum of WMGHG, ozone and stratospheric water vapour) has a more narrow shape than the PDF for the aerosols owing to the much lower relative uncertainty.

Table 8.6 | Summary table of RF estimates for AR5 and comparison with the three previous IPCC assessment reports. ERF values for AR5 are included. For AR5 the values are given for the period 1750–2011, whereas earlier final years have been adopted in the previous IPCC assessment reports.

	Global Mean Radiative Forcing (W m^{-2})					ERF (W m^{-2})
	SAR (1750–1993)	TAR (1750–1998)	AR4 (1750–2005)	AR5 (1750–2011)	Comment	AR5
Well-mixed greenhouse gases (CO_2, CH_4, N_2O, and halocarbons)	2.45 (2.08 to 2.82)	2.43 (2.19 to 2.67)	2.63 (2.37 to 2.89)	2.83 (2.54 to 3.12)	Change due to increase in concentrations	2.83 (2.26 to 3.40)
Tropospheric ozone	+0.40 (0.20 to 0.60)	+0.35 (0.20 to 0.50)	+0.35 (0.25 to 0.65)	+0.40 (0.20 to 0.60)	Slightly modified estimate	
Stratospheric ozone	–0.1 (–0.2 to –0.05)	–0.15 (–0.25 to –0.05)	–0.05 (–0.15 to +0.05)	–0.05 (–0.15 to +0.05)	Estimate unchanged	
Stratospheric water vapour from CH_4	Not estimated	+0.01 to +0.03	+0.07 (+0.02, +0.12)	+0.07 (+0.02 to +0.12)	Estimate unchanged	
Aerosol–radiation interactions	Not estimated	Not estimated	–0.50 (–0.90 to –0.10)	0.35 (0.85 to +0.15)	Re-evaluated to be smaller in magnitude	–0.45 (–0.95 to +0.05)
Aerosol–cloud interactions	0 to –1.5 (sulphate only)	0 to –2.0 (all aerosols)	–0.70 (–1.80 to –0.30) (all aerosols)	Not estimated	Replaced by ERF and re-evaluated to be smaller in magnitude	–0.45 (–1.2 to 0.0)
Surface albedo (land use)	Not estimated	–0.20 (–0.40 to 0.0)	–0.20 (–0.40 to 0.0)	–0.15 (–0.25 to –0.05)	Re-evaluated to be slightly smaller in magnitude	
Surface albedo (black carbon aerosol on snow and ice)	Not estimated	Not estimated	+0.10 (0.0 to +0.20)	+0.04 (+0.02 to +0.09)	Re-evaluated to be weaker	
Contrails	Not estimated	+0.02 (+0.006 to +0.07)	+0.01 (+0.003 to +0.03)	+0.01 (+0.005 to +0.03)	No major change	
Combined contrails and contrail-induced cirrus	Not estimated	0 to +0.04	Not estimated	Not estimated		0.05 (0.02 to 0.15)
Total anthropogenic	Not estimated	Not estimated	1.6 (0.6 to 2.4)	Not estimated	Stronger positive due to changes in various forcing agents	2.3 (1.1 to 3.3)
Solar irradiance	+0.30 (+0.10 to +0.50)	+0.30 (+0.10 to +0.50)	+0.12 (+0.06 to +0.30)	+0.05 (0.0 to +0.10)	Re-evaluated to be weaker	

Notes:

Volcanic RF is not added to the table due to the periodic nature of volcanic eruptions, which makes it difficult to compare to the other forcing mechanisms.

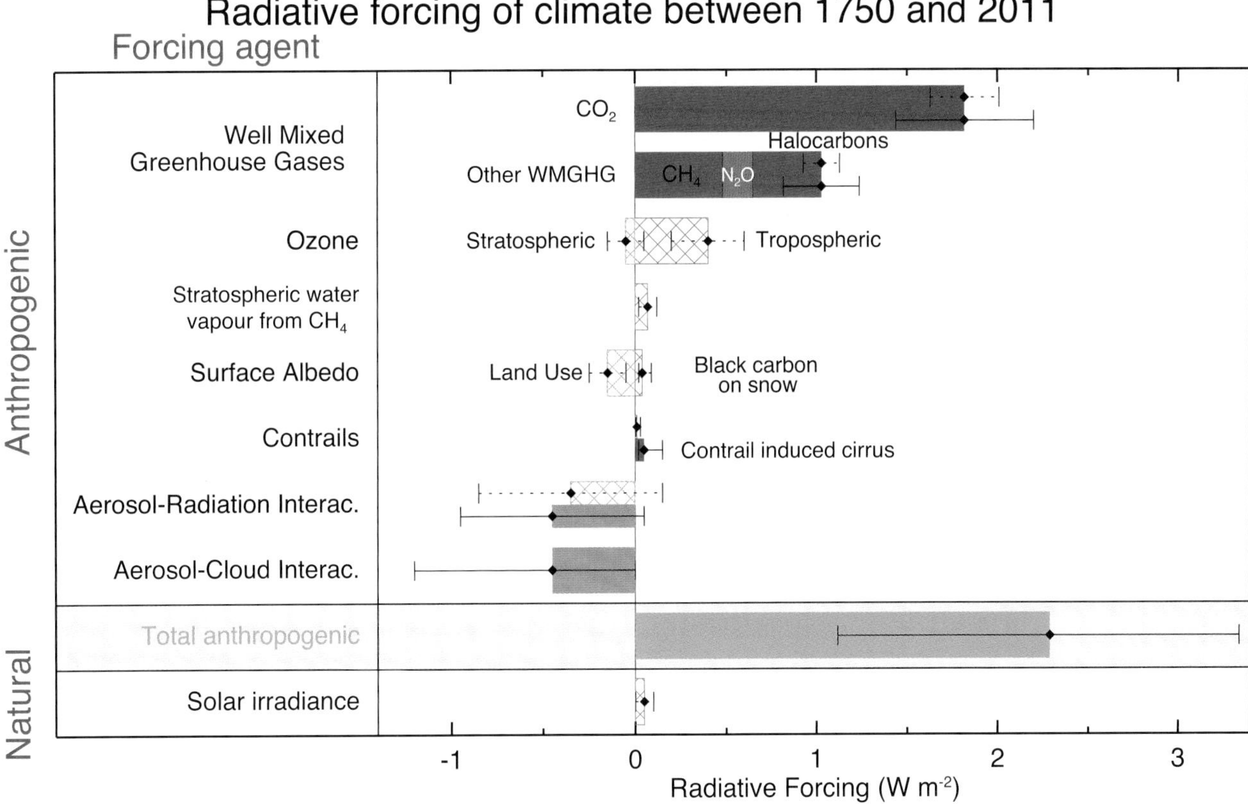

Figure 8.15 | Bar chart for RF (hatched) and ERF (solid) for the period 1750–2011, where the total ERF is derived from Figure 8.16. Uncertainties (5 to 95% confidence range) are given for RF (dotted lines) and ERF (solid lines).

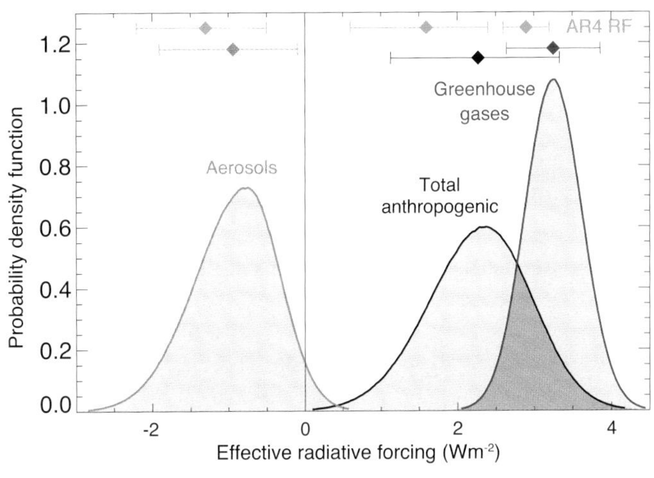

Figure 8.16 | Probability density function (PDF) of ERF due to total GHG, aerosol forcing and total anthropogenic forcing. The GHG consists of WMGHG, ozone and stratospheric water vapour. The PDFs are generated based on uncertainties provided in Table 8.6. The combination of the individual RF agents to derive total forcing over the Industrial Era are done by Monte Carlo simulations and based on the method in Boucher and Haywood (2001). PDF of the ERF from surface albedo changes and combined contrails and contrail-induced cirrus are included in the total anthropogenic forcing, but not shown as a separate PDF. We currently do not have ERF estimates for some forcing mechanisms: ozone, land use, solar, etc. For these forcings we assume that the RF is representative of the ERF and for the ERF uncertainty an additional uncertainty of 17% has been included in quadrature to the RF uncertainty. See Supplementary Material Section 8.SM.7 and Table 8.SM.4 for further description on method and values used in the calculations. Lines at the top of the figure compare the best estimates and uncertainty ranges (5 to 95% confidence range) with RF estimates from AR4.

Therefore, the large uncertainty in the aerosol forcing is the main cause of the large uncertainty in the total anthropogenic ERF. The total anthropogenic forcing is *virtually certain* to be positive with the probability for a negative value less than 0.1%. Compared to AR4 the total anthropogenic ERF is more strongly positive with an increase of 43%. This is caused by a combination of growth in GHG concentration, and thus strengthening in forcing of WMGHG, and weaker ERF estimates of aerosols (aerosol–radiation and aerosol–cloud interactions) as a result of new assessments of these effects.

Figure 8.17 shows the forcing over the Industrial Era by emitted compounds (see Supplementary Material Tables 8.SM.6 and 8.SM.7 for actual numbers and references). It is more complex to view the RF by emitted species than by change in atmospheric abundance (Figure 8.15) since the number of emitted compounds and changes leading to RF is larger than the number of compounds causing RF directly (see Section 8.3.3). The main reason for this is the indirect effect of several compounds and in particular components involved in atmospheric chemistry (see Section 8.2). To estimate the RF by the emitted compounds in some cases the emission over the entire Industrial Era is needed (e.g., for CO_2) whereas for other compounds (such as ozone and CH_4) quite complex simulations are required (see Section 8.3.3). CO_2 is the dominant positive forcing both by abundance and by emitted compound. Emissions of CH_4, CO, and NMVOC all lead to excess CO_2 as one end product if the carbon is of fossil origin and is the reason why the RF of direct CO_2 emissions is slightly lower than the RF of abundance change of CO_2. For CH_4 the contribution from emission is estimated to be almost twice as large as that from the CH_4 concentration change, 0.97 (0.80 to 1.14) W m^{-2} versus 0.48 (0.43 to 0.53) W m^{-2}, respectively. This is because emission of CH_4 leads to ozone production, stratospheric water vapour, CO_2 (as mentioned above), and importantly affects its own lifetime (Section 8.2). Actually, emissions of CH_4 would lead to a stronger RF via the direct CH_4 greenhouse effect (0.64 W m^{-2}) than the RF from abundance change of CH_4 (0.48 W m^{-2}). This is because other compounds have influenced the lifetime of CH_4 and reduced the abundance of CH_4, most notably NO_x. Emissions of CO (0.23 (0.18 to 0.29) W m^{-2}) and NMVOC (0.10 (0.06 to 0.14) W m^{-2}) have only indirect effects on RF through ozone production, CH_4 and CO_2 and thus contribute an overall positive RF. Emissions of NO_X, on the other hand, have indirect effects that lead to positive RF through ozone production and also effects that lead to negative RF through reduction of CH_4 lifetime and thus its concentration, and through contributions to nitrate aerosol formation. The best estimate of the overall effect of anthropogenic emissions of NO_X is a negative RF (-0.15 (-0.34 to +0.02) W m^{-2}). Emissions of ammonia also contribute to nitrate aerosol formation, with a small offset due to compensating changes in sulphate aerosols. Additionally indirect effects from sulphate on atmospheric compounds are not included here as models typically simulate a small effect, but there are large relative differences in the response between models. Impacts of emissions other than CO_2 on the carbon cycle via changes in atmospheric composition (ozone or aerosols) are also not shown owing to the limited amount of available information.

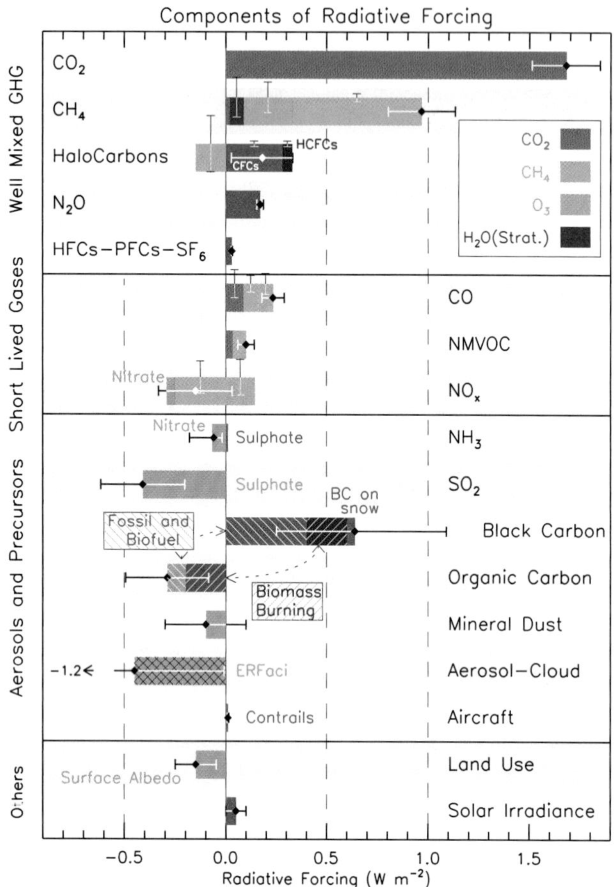

Figure 8.17 | RF bar chart for the period 1750–2011 based on emitted compounds (gases, aerosols or aerosol precursors) or other changes. Numerical values and their uncertainties are shown in Supplementary Material Tables 8.SM.6 and 8.SM.7. Note that a certain part of CH_4 attribution is not straightforward and discussed further in Section 8.3.3. Red (positive RF) and blue (negative forcing) are used for emitted components which affect few forcing agents, whereas for emitted components affecting many compounds several colours are used as indicated in the inset at the upper part the figure. The vertical bars indicate the relative uncertainty of the RF induced by each component. Their length is proportional to the thickness of the bar, that is, the full length is equal to the bar thickness for a ±50% uncertainty. The net impact of the individual contributions is shown by a diamond symbol and its uncertainty (5 to 95% confidence range) is given by the horizontal error bar. ERFaci is ERF due to aerosol–cloud interaction. BC and OC are co-emitted, especially for biomass burning emissions (given as Biomass Burning in the figure) and to a large extent also for fossil and biofuel emissions (given as Fossil and Biofuel in the figure where biofuel refers to solid biomass fuels). SOA have not been included because the formation depends on a variety of factors not currently sufficiently quantified.

For the WMGHG, the ERF best estimate is the same as the RF. The uncertainty range is slightly larger, however. The total emission-based ERF of WMGHG is 3.00 (2.22 to 3.78) W m^{-2}. That of CO_2 is 1.68 (1.33 to 2.03) W m^{-2}; that of CH_4 is 0.97 (0.74 to 1.20) W m^{-2}; that of stratospheric ozone-depleting halocarbons is 0.18 (0.01 to 0.35) W m^{-2}.

Emissions of BC have a positive RF through aerosol–radiation interactions and BC on snow (0.64 W m^{-2}, see Section 8.3.4 and Section 7.5). The emissions from the various compounds are co-emitted; this is in particular the case for BC and OC from biomass burning aerosols. The net RF of biomass burning emissions for aerosol–radiation interactions is close to zero, but with rather strong positive RF from BC and negative RF from OC (see Sections 8.3.4 and 7.5). The ERF due to aerosol–cloud interactions is caused by primary anthropogenic emissions of BC, OC and dust as well as secondary aerosol from anthropogenic emissions of SO_2, NO_X and NH_3. However, quantification of the contribution from the various components to the ERF due to aerosol–cloud interactions has not been attempted in this assessment.

8.5.2 Time Evolution of Historical Forcing

The time evolution of global mean forcing is shown in Figure 8.18 for the Industrial Era. Over all time periods during the Industrial Era CO_2 and other WMGHG have been the dominant term, except for shorter periods with strong volcanic eruptions. The time evolution shows an almost continuous increase in the magnitude of anthropogenic ERF. This is the case both for CO_2 and other WMGHGs as well as several individual aerosol components. The forcing from CO_2 and other WMGHGs has increased somewhat faster since the 1960s. Emissions of CO_2 have made the largest contribution to the increased anthropogenic forcing in every decade since the 1960s. The total aerosol ERF (aerosol–radiation interaction and aerosol–cloud interaction) has the strongest negative forcing (except for brief periods with large volcanic forcing), with a strengthening in the magnitude similar to many of the other anthropogenic forcing mechanisms with time. The global mean forcing of aerosol–radiation interactions was rather weak until 1950 but strengthened in the latter half of the last century and in particular in the period between 1950 and 1980. The RF due to aerosol–radiation interaction by aerosol component is shown in Section 8.3.4 (Figure 8.8).

Although there is *high confidence* for a substantial enhancement in the negative aerosol forcing in the period 1950–1980, there is much more uncertainty in the relative change in global mean aerosol forcing over the last two decades (1990–2010). Over the last two decades there has been a strong geographic shift in aerosol and aerosol precursor emissions (see Section 2.2.3), and there are some uncertainties in these emissions (Granier et al., 2011). In addition to the regional changes in the aerosol forcing there is also likely a competition between various aerosol effects. Emission data indicate a small increase in the BC emissions (Granier et al., 2011) but model studies also indicate a weak enhancement of other aerosol types. Therefore, the net aerosol forcing depends on the balance between absorbing and scattering aerosols for aerosol–radiation interaction as well as balance between the changes in aerosol–radiation and aerosol–cloud interactions. In the ACCMIP models, for example, the RF due to aerosol–radiation interaction becomes less negative during 1980 to 2000, but total aerosol ERF becomes more negative (Shindell et al., 2013c). There is a *very low confidence* for the trend in the total aerosol forcing during the past two to three decades, even the sign; however, there is *high confidence* that the offset from aerosol forcing to WMGHG forcing during this period was much smaller than over the 1950–1980 period.

The volcanic RF has a very irregular temporal pattern and for certain years has a strongly negative RF. There has not been a major volcanic eruption in the past decade, but some weaker eruptions give a current RF that is slightly negative relative to 1750 and slightly stronger in magnitude compared to 1999–2002 (see Section 8.4.2).

Figure 8.19 shows linear trends in forcing (anthropogenic, natural and total) over four different time periods. Three of the periods are the same as chosen in Box 9.2 (1984–1998, 1998–2011 and 1951–2011) and the period 1970–2011 is shown in Box 13.1. Monte Carlo simulations are performed to derive uncertainties in the forcing based on ranges given in Table 8.6 and the derived linear trends. Further, these uncertainties are combined with uncertainties derived from shifting time periods ±2 years and the full 90% confidence range is shown in Figure 8.19 (in Box 9.2 only the total forcing is shown with uncertainties derived from the forcing uncertainty without sensitivity to time period). For the anthropogenic forcing sensitivity to the selection of time periods is very small with a maximum contribution to the uncertainties shown in Figure 8.19 of 2%. However, for the natural forcing the sensitivity to time periods is the dominant contributor to the overall uncertainty (see Supplementary Material Figure 8.SM.3) for the relatively short periods 1998–2011 and 1984–1998, whereas this is not the case for the longer periods. For the 1998–2011 period the natural forcing is *very likely* negative and has offset 2 to 89% of the anthropogenic forcing. It is *likely* that the natural forcing change has offset at least 30% of the anthropogenic forcing increase and *very likely* that it has offset at least 10% of the anthropogenic increase. For the 1998–2011 period both the volcanic and solar forcings contribute to this negative natural forcing, with the latter dominating. For the other periods shown in Figure 8.19 the best estimate of the natural is much smaller in magnitude than the anthropogenic forcing, but note that the natural forcing is very dependent on the selection of time period near the 1984–1998 interval. Over the period 1951–2011 the trend in anthropogenic forcing is almost 0.3 W m^{-2} per decade and thus anthropogenic forcing over this period is more than 1.5 W m^{-2}. The anthropogenic forcing for 1998–2011 is 30% higher and with smaller uncertainty than for the 1951–2011 period. Note that due to large WMGHG forcing (Section 8.3.2) the anthropogenic forcing was similar in the late 1970s and early 1980s to the 1998–2011 period. The reason for the reduced uncertainty in the 1998–2011 anthropogenic forcing is the larger domination of WMGHG forcing and smaller contribution from aerosol forcing compared to previous periods. Similar to the results for 1970–2011 in Figure 8.19, Box 13.1 shows that the global energy budget is dominated by anthropogenic forcing compared to the natural forcing, except for the two major volcanic eruption in this period as can be easily seen in Figure 8.18.

Figure 8.20 shows the forcing between 1980 and 2011. Compared to the whole Industrial Era the dominance of the CO_2 is larger for this recent period both with respect to other WMGHG and the total anthropogenic RF. The forcing due to aerosols is rather weak leading

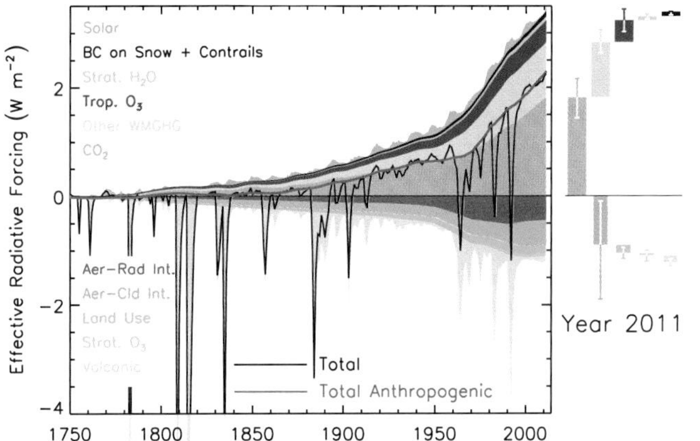

Figure 8.18 | Time evolution of forcing for anthropogenic and natural forcing mechanisms. Bars with the forcing and uncertainty ranges (5 to 95% confidence range) at present are given in the right part of the figure. For aerosol the ERF due to aerosol–radiation interaction and total aerosol ERF are shown. The uncertainty ranges are for present (2011 versus 1750) and are given in Table 8.6. For aerosols, only the uncertainty in the total aerosol ERF is given. For several of the forcing agents the relative uncertainty may be larger for certain time periods compared to present. See Supplementary Material Table 8.SM.8 for further information on the forcing time evolutions. Forcing numbers provided in Annex II. The total antropogenic forcing was 0.57 (0.29 to 0.85) W m^{-2} in 1950, 1.25 (0.64 to 1.86) W m^{-2} in 1980 and 2.29 (1.13 to 3.33) W m^{-2} in 2011.

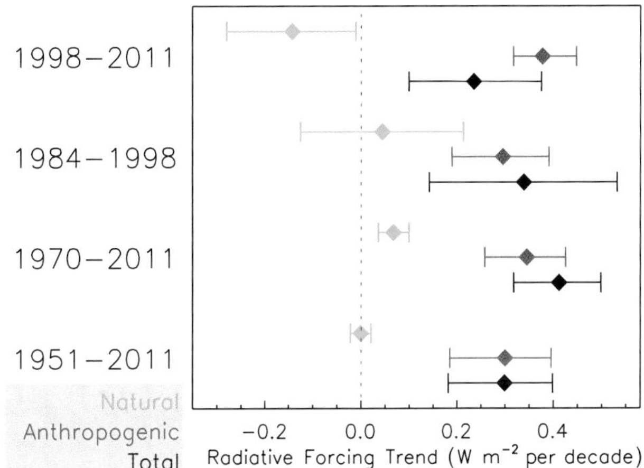

Figure 8.19 | Linear trend in anthropogenic, natural and total forcing for the indicated time periods. The uncertainty ranges (5 to 95% confidence range) are combined from uncertainties in the forcing values (from Table 8.6) and the uncertainties in selection of time period. Monte Carlo simulations were performed to derive uncertainties in the forcing based on ranges given in Table 8.6 and linear trends in forcing. The sensitivity to time periods has been derived from changing the time periods by ±2 years.

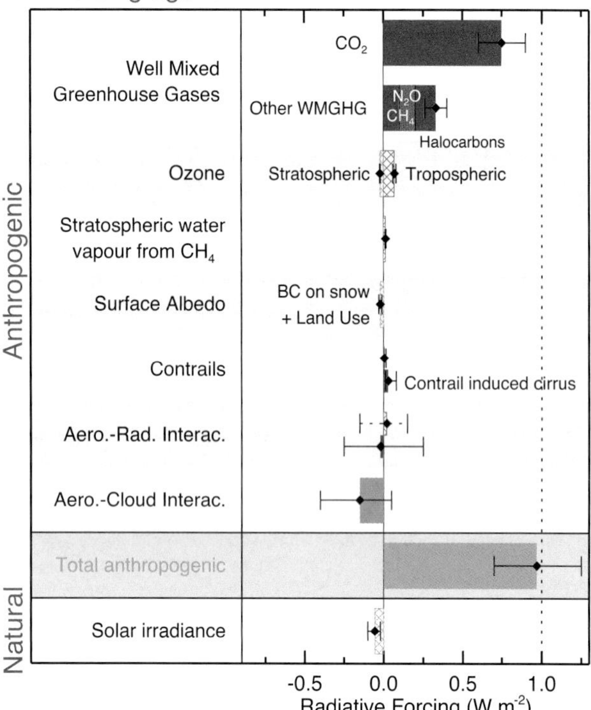

Figure 8.20 | Bar chart for RF (hatched) and ERF (solid) for the period 1980–2011, where the total anthropogenic ERF are derived from Monte-Carlo simulations similar to Figure 8.16. Uncertainties (5 to 95% confidence range) are given for RF (dotted lines) and ERF (solid lines).

to a very strong net positive ERF for the 1980–2011 period. More than 40% of the total anthropogenic ERF has occurred over the 1980–2011 period with a value close to 1.0 (0.7 to 1.3) W m^{-2}. The major contribution to the uncertainties in the time evolution of the anthropogenic forcing is associated with the aerosols (see Section 8.5.1). Despite this, anthropogenic ERF is *very likely* considerably more positive than the natural RF over the decadal time periods since 1950. This is in particular the case after 1980, where satellite data are available that provide important measurements to constrain the natural RF mechanisms (e.g., the volcanic RF change between 2007–2011 and 1978–1982 is 0.06 W m^{-2} and the representative change in solar irradiance over the 1980–2011 period is –0.06 W m^{-2}) with total natural RF of 0.0 (-0.1 to +0.1) W m^{-2}.

8.5.3 Future Radiative Forcing

Projections of global mean RF are assessed based on results from multiple sources examining the RF due to RCP emissions: the ACCMIP initiative (see Section 8.2) provides analysis of the RF or ERF due to aerosols and ozone (Shindell et al., 2013c), while WMGHG, land use and stratospheric water RFs are taken from the results of calculations with the reduced-complexity Model for the Assessment of Greenhouse-gas Induced Climate Change 6 (MAGICC6) driven by the RCP emissions and land use (Meinshausen et al., 2011a). While MAGICC6 also estimated ozone and aerosol RF, those values differ substantially from the ACCMIP values and are considered less realistic. Additional discussion of biases in the MAGICC6 results due to the simplified representations of atmospheric chemistry and the carbon cycle, along with further discussion on the representativeness of the RCP projections in context with the broader set of scenarios in the literature, is presented in Section 11.3.5 and Section 12.3 (also see Section 8.2). As the ACCMIP project provided projected forcings primarily at 2030 and 2100, we hereafter highlight those times. Although understanding the relative contributions of various processes to the overall effect of aerosols on forcing is useful, we emphasize the total aerosol ERF, which includes all aerosol–radiation and aerosol–cloud interactions, as this is the most indicative of the aerosol forcing driving climate change. We also present traditional RF due to aerosol–radiation interaction (previously called direct aerosol effect) but do not examine further the various components of aerosol ERF. Aerosol forcing estimates, both mean and uncertainty ranges, are derived from the 10 ACCMIP models, 8 of which are also CMIP5 models. We analyze forcing during the 21st century (relative to 2000), and hence the WMGHG forcing changes are in addition to persistent forcing from historical WMGHG increases.

Analysis of forcing at 2030 relative to 2000 shows that under RCP2.6, total ozone (tropospheric and stratospheric) forcing is near zero, RF due to aerosol–radiation interaction is positive but small, and hence WMGHG forcing dominates changes over this time period (Figure 8.21). WMGHG forcing is dominated by increasing CO_2, as declining CH_4 and increasing N_2O have nearly offsetting small contributions to forcing. Aerosol ERF was not evaluated for this RCP under ACCMIP, and values cannot be readily inferred from RF due to aerosol–radiation interaction as these are not directly proportional. Under RCP8.5, RF due to aerosol–radiation interaction in 2030 is weakly negative, aerosol ERF is positive with a fairly small value and large uncertainty range, total ozone forcing is positive but small (~0.1 W m^{-2}), and thus WMGHG forcing again dominates with a value exceeding 1 W m^{-2}. As with RCP2.6, WMGHG forcing is dominated by CO_2, but under this scenario the other WMGHGs all contribute additional positive forcing. Going to 2100, ozone forcing diverges in sign between the two scenarios, consistent with changes in the tropospheric ozone burden (Figure 8.4) which are largely attributable to projected CH_4 emissions, but is small in either case. Ozone RF is the net impact of a positive forcing from stratospheric ozone recovery owing to reductions in anthropogenic ozone-depleting halocarbon emissions in both scenarios and a larger impact from changes in tropospheric precursors (Shindell et al., 2013c) which have a negative forcing in RCP2.6 and a positive forcing in RCP8.5.

The two scenarios are fairly consistent in their trends in RF due to aerosol–radiation interaction by component (Figure 8.21). There is positive RF due to aerosol–radiation interaction due to reductions in sulfate aerosol. This is largely offset by negative RF due to aerosol–radiation interaction by primary carbonaceous aerosols and especially by nitrate (though nearly all CMIP5 models did not include nitrate), leaving net aerosol RF due to aerosol–radiation interaction values that are very small, 0.1 W m^{-2} or less in magnitude, in either scenario at 2030 and 2100. Nitrate aerosols continue to increase through 2100 as ammonia emissions rise steadily due to increased use of agricultural fertilizer even as all other aerosol precursor emissions decrease (Figure 8.2), including sulphur dioxide which drives the reduction in sulphate aerosol that also contributes to additional formation of nitrate aerosols in the future (Bauer et al., 2007; Bellouin et al., 2011). Aerosol ERF is *likely* similar at this time in all scenarios given that they all have greatly reduced emissions of all aerosols and aerosol precursors other than ammonia. Aerosol ERF shows a large positive value at 2100 relative to 2000, nearly returning to its 1850 levels (the 2100 versus 1850 ERF represents a decrease in ERF of 91% relative to the 2000 versus 1850 value), as is expected given the RCP emissions. Thus although some models project large increases in nitrate RF in the future, the reduction in overall aerosol loading appears to lead to such a strong reduction in aerosol ERF that the impact of aerosols becomes very small under these RCPs. Of course the projections of drastic reductions in primary aerosol as well as aerosol and ozone precursor emissions may be overly optimistic as they assume virtually all nations in the world become wealthy and that emissions reductions are directly dependent on wealth. The RCPs also contain substantially lower projected growth in HFC emissions than in some studies (e.g., Velders et al., 2009).

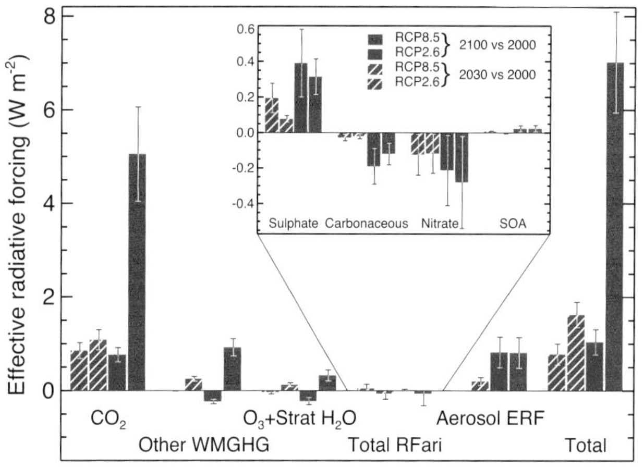

Figure 8.21 | Radiative forcing relative to 2000 due to anthropogenic composition changes based on ACCMIP models for aerosols (with aerosol ERF scaled to match the best estimate of present-day forcing) and total ozone and RCP WMGHG forcings. Ranges are one standard deviation in the ACCMIP models and assessed relative uncertainty for WMGHGs and stratospheric water vapor. Carbonaceous aerosols refer to primary carbonaceous, while SOA are secondary organic aerosols. Note that 2030 ERF for RCP2.6 was not available, and hence the total shown for that scenario is not perfectly comparable to the other total values. RFari is RF due to aerosol–radiation interaction.

Although aerosol ERF becomes less negative by nearly 1 W m^{-2} from 2000 to 2100, this change is still small compared with the increased WMGHG forcing under RCP8.5, which is roughly 6 W m^{-2} during this time (Figure 8.21). Roughly 5 W m^{-2} of this WMGHG forcing comes from CO_2, with substantial additional forcing from increases in both CH_4 and nitrous oxide and only a very small negative forcing from reductions in halocarbons. Under RCP2.6, the WMGHG forcing is only about 0.5 W m^{-2} during this time, as relatively strong decreases in CH_4 and halocarbon forcing offset roughly 40% of the increased CO_2 forcing, which is itself far less than under RCP8.5. Hence under this scenario, the projected future forcing due to aerosol reductions is actually stronger than the WMGHG forcing. Viewing the timeseries of the various forcings, however, indicates that aerosol ERF is returning to its pre-industrial levels, so that net forcing becomes increasingly dominated by WMGHGs regardless of scenario during the 21st century (Figure 8.22). As the forcing is so heavily dominated by WMGHGs at 2100, and the WMGHG concentrations (CO_2) or emissions (others) were chosen to match forcing targets, all the scenarios show net forcing values at that time that are fairly close to the scenarios' target values. The reduced aerosol forcing, with its large uncertainty, leads to a pronounced decrease in the uncertainty of the total net forcing by 2100. Based on the spread across ACCMIP models (using ERF for aerosols and converting to ERF for GHGs), the 90% confidence interval (CI) is about 20% for the 2100 net forcing, versus 26% for 2030 under RCP8.5 and 45–61% for 1980 and 2000 (Shindell et al., 2013c). The total ERF due to all causes has been independently estimated based on the transient response in the CMIP5 models and a linear forcing-response relationship derived through regression of the modelled response to an instantaneous increase in CO_2 (Forster et al., 2013). Uncertainties based on model spread behave similarly, with the 90% CI for net total ERF decreasing from 53% for 2003 to only 24 to 34% for 2100. Forcing relative to 2000 due to land use (via albedo only) and stratospheric water vapor changes are not shown separately as their projected values under the four RCPs are quite small: –0.09 to 0.00 and –0.03 to 0.10 W m^{-2}, respectively.

The CMIP5 forcing estimates (Forster et al., 2013) for the total projected 2030 and 2100 ERF are slightly smaller than the results obtained from the ACCMIP models (or the RCP targets; see Section 12.3.3). Examining the subset of models included in both this regression analysis and in ACCMIP shows that the ACCMIP subset show forcings on the low side of the mean value obtained from the full set of CMIP5 analyzed, indicating that the discrepancy between the methods is not related to analysis of a different set of models. Instead, it may reflect nonlinearities in the response to forcing that are not represented by the regression analysis of the response to abrupt CO_2 increase experiments (Long and Collins, 2013) or differences in the response to other forcing agents relative to the response to CO_2 used in deriving the CMIP5 estimates (see also 12.3.3).

Natural forcings will also change in the future. The magnitudes cannot be reliably projected, but are *likely* to be small at multi-decadal scales (see Section 8.4). Brief episodic volcanic forcing could be large, however.

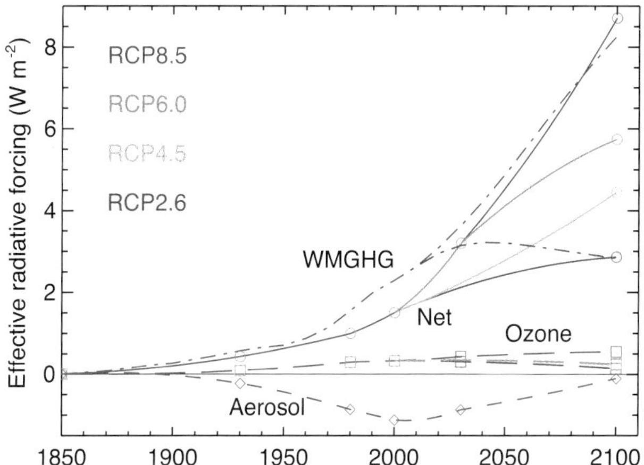

Figure 8.22 | Global mean anthropogenic forcing with symbols indicating the times at which ACCMIP simulations were performed (solid lines with circles are net; long dashes with squares are ozone; short dashes with diamonds are aerosol; dash-dot are WMGHG; colours indicate the RCPs with red for RCP8.5, orange RCP6.0, light blue RCP4.5, and dark blue RCP2.6). RCPs 2.6, 4.5 and 6.0 net forcings at 2100 are approximate values using aerosol ERF projected for RCP8.5 (modified from Shindell et al., 2013c). Some individual components are omitted for some RCPs for visual clarity.

8

8.6 Geographic Distribution of Radiative Forcing

The forcing spatial pattern of the various RF mechanisms varies substantially in space and in time, especially for the NTCFs. The spatial pattern is of interest to the extent that it may influence climate response (Section 8.6.2.2) as is being particularly investigated in the ACCMIP simulations.

8.6.1 Spatial Distribution of Current Radiative Forcing

The WMGHGs such as CO_2 have the largest forcing in the subtropics, decreasing toward the poles, with the largest forcing in warm and dry regions and smaller values in moist regions and in high-altitude regions (Taylor et al., 2011). For the NTCFs (Box 8.2) their concentration spatial pattern and therefore their RF pattern are highly inhomogeneous, and again meteorological factors such as temperature, humidity, clouds, and surface albedo influence how concentration translates to RF.

Figure 8.23 shows the RF spatial distribution of the major NTCFs together with standard deviation among the ACCMIP models (Shindell et al., 2013c) the net anthropogenic composition (WMGHG+ozone+aerosol) forcing is also shown (lower left panel). These models used unified anthropogenic emissions of aerosol and ozone precursors (Supplementary Material Figure 8.SM.2), so that the model diversity in RF is due only to differences in model chemical and climate features and natural emissions, and would be larger if uncertainty in the anthropogenic emissions were also included. In general, the confidence in geographical distribution is lower than for global mean, due to uncertainties in chemistry, transport and removal of species.

The negative RF due to aerosol–radiation interaction (first row; defined in Figure 7.3) is greatest in the NH and near populated and biomass burning regions. The standard deviation for the net RF due to aerosol–radiation interaction is typically largest over regions where vegetation changes are largest (e.g., South Asia and central Africa), due to uncertainties in biomass burning aerosol optical properties and in treatment of secondary organic aerosols. Carbonaceous aerosol forcing (second row) is greatest in South and East Asia and can be negative in biomass burning regions due to large weakly absorbing organic components. Absorbing aerosols also have enhanced positive forcing when they overlie high albedo surfaces such as cryosphere, desert or clouds, with as much as 50% of BC RF resulting from BC above clouds (Zarzycki and Bond, 2010).

Figure 8.24 compares the aerosol RFs for ACCMIP (Shindell et al., 2013c), which are representative of the CMIP5 experiments, with those from the AeroCom model intercomparison (Myhre et al., 2013) which includes sixteen models that used unified meteorology and are more extensively compared to measurements (e.g., Koch et al., 2009b; Koffi et al., 2012). The forcing results are very similar, establishing the representativeness and validity of the ACCMIP aerosol simulations.

The net aerosol ERF (Figure 8.23; third row), includes both aerosol–radiation and aerosol–cloud interactions. The spatial pattern correlates with the RF (first row), except with stronger effect in the outflow regions over oceans. The flux change is larger in the NH than the SH (e.g., by nearly a factor of 3; Ming et al., 2007). Rapid adjustment associated with aerosol–radiation and aerosol–cloud interactions may enhance or reduce cloud cover depending on the region, cloud dynamics and aerosol loading (e.g., Randles and Ramaswamy, 2008; Koch and Del Genio, 2010; Persad et al., 2012). In general, the ocean-land forcing pattern differs from that reported in AR4, where the forcing due to aerosol–cloud interaction were larger over land than ocean (Forster et al., 2007), and this continues to be a source of uncertainty. Since AR4, Quaas et al. (2009) showed using satellite retrievals that the correlation between AOD changes and droplet number changes is stronger over oceans than over land and that models tend to overestimate the strength of the relation over land. Penner et al. (2011) showed that satellite retrievals, due to their dependence on present-day conditions, may underestimate the forcing due to aerosol–cloud interaction, especially over land, although this model analysis may overestimate the cloud condensation nucleus to AOD relation (Quaas et al., 2011). Wang and Penner (2009) also showed that if models include boundary layer nucleation and increase the fraction of sulphur emitted as a primary particle, the effect over land is increased relative to over ocean (see also Section 7.5.3). The aerosol ERF standard deviation is large in biomass burning regions, as for the RF, and in regions where cloud effects differ among models (e.g., northern North America, northeast Asia, Amazonia). The spread in aerosol ERF is much larger than for the RF alone, although the relative standard deviation is no larger (Shindell et al., 2013c).

For components that primarily scatter radiation, the radiative effect at the surface is similar to the RF (according to the definition in Section 8.1.1). However for components that absorb radiation in the atmosphere the radiation reaching the surface is reduced (Forster et al., 2007; Ramanathan and Carmichael, 2008; Andrews et al., 2010). This absorption of incoming solar radiation alters the vertical temperature profile in the atmospheric column and can thus change atmospheric circulation and cloud formation. The aerosol atmospheric absorption (Figure 8.23, bottom right), or the difference between ERF and the analogous radiative flux reaching the surface including rapid adjustments, has a spatial pattern that to lowest order tracks the carbonaceous aerosol forcing, but is also affected by cloud changes, where e.g., cloud loss could enhance atmospheric absorption. Atmospheric aerosol absorption patterns thus mirror the ERF due to aerosol–cloud interaction pattern, with larger forcing over continents.

Ozone RF is calculated using the methodology described in Shindell et al. (2013c), but applied to the larger set of models in ACCMIP (Stevenson et al., 2013). The net ozone RF (Figure 8.23; fourth row) is largest in subtropical latitudes, and is more positive in the NH than the SH. Pollution in the NH accounts for positive tropospheric forcing; stratospheric ozone loss has caused negative SH polar forcing. Model standard deviation is largest in the polar regions where lower stratosphere/upper troposphere changes differ in the models (Young et al., 2013).

Overall, the *confidence* in aerosol and ozone RF spatial patterns is *medium* and lower than that for the global mean due to the large regional standard deviations (Figure 8.23), and is exacerbated in aerosol ERF patterns due to uncertainty in cloud responses.

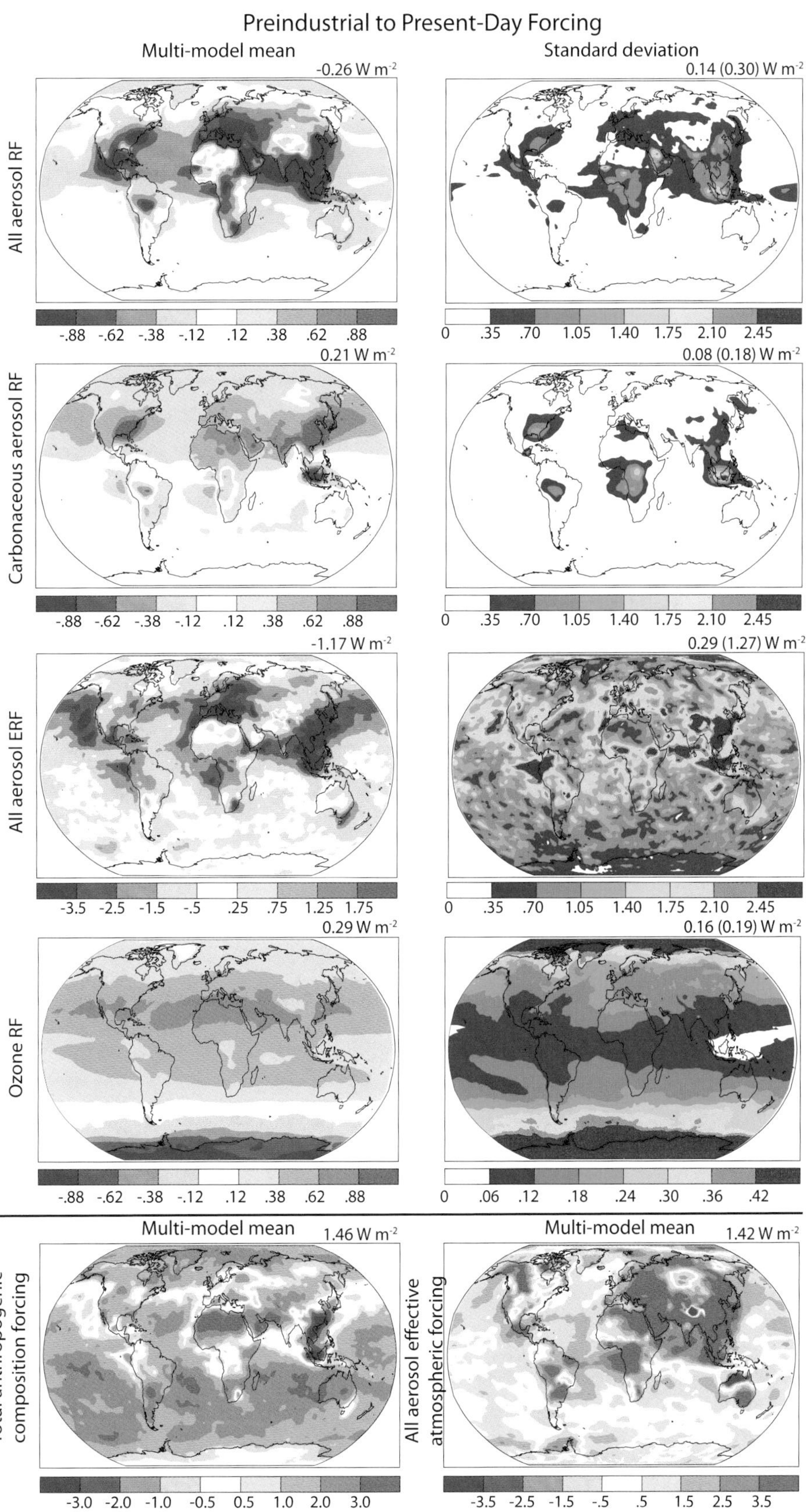

Figure 8.23 | Spatial pattern of ACCMIP models 1850 to 2000 forcings, mean values (left) and standard deviation (right) for aerosols and ozone (top four rows). Values above are the average of the area-weighted global means, with the area weighted mean of the standard deviation of models at each point provided in parenthesis. Shown are net aerosol RF due to aerosol–radiation interaction (top, 10 models), carbonaceous aerosol RF due to aerosol-radiation interaction (2nd row, 7 models), aerosol ERF (3rd row, 8 models), ozone (4th row, 11 models), total anthropogenic composition forcing (WMGHG+ozone+aerosols; bottom left), aerosol atmospheric absorption including rapid adjustment (bottom right, 6 models). Note that RF and ERF means are shown with different colour scales, and standard deviation colour scales vary among rows.

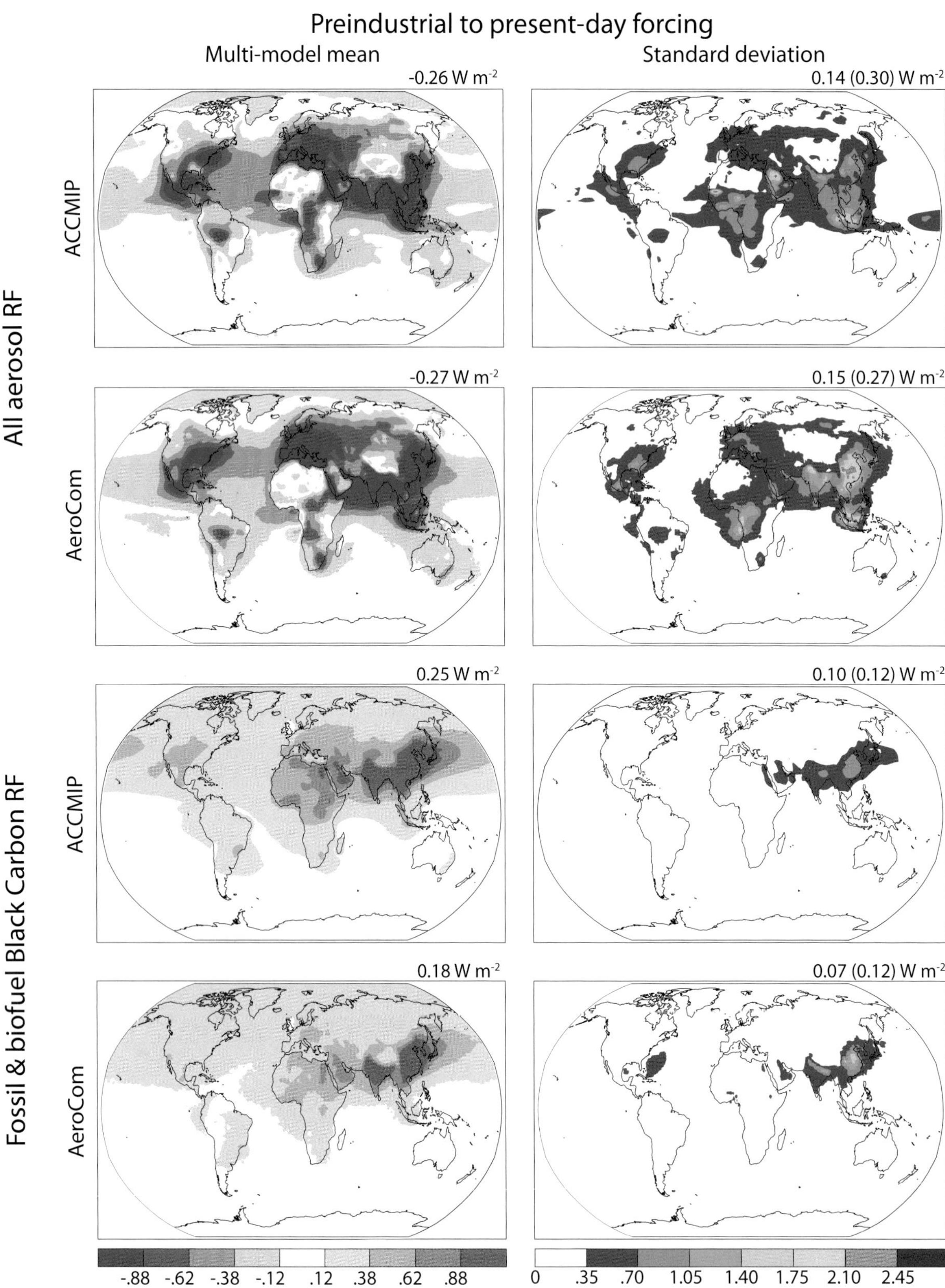

Figure 8.24 | Spatial pattern of ACCMIP and 16 AeroCom models 1850 to 2000 RF due to aerosol–radiation interaction, mean values (left) and standard deviation (right). Note that different carbonaceous aerosol diagnostics are used here compared to Figure 8.23, due to available AeroCom fields. Values above are the average of the area-weighted global means, with the area weighted mean of the standard deviation of models at each point provided in parentheses.

8.6.2 Spatial Evolution of Radiative Forcing and Response over the Industrial Era

8.6.2.1 Regional Forcing Changes During the Industrial Era

The spatial distribution of the WMGHG RF has shifted only slightly over the industrial period; however the RF spatial distributions for NTCFs has shifted with emissions, due to the timing of regional development and implementation of pollution standards (Supplementary Material Figures 8.SM.1 and 8.SM.2 show regional trends and emissions maps; Lamarque et al., 2013). Figure 8.25 shows how the distributions of aerosol and ozone forcings are modelled to have changed up to 1930, 1980 and 2000. Substantial industrial coal-burning in the early part of the 20th century occurred in the northeastern United States and Western Europe, leading to stronger sulphate and BC forcing near those regions (Figure 8.25, left). Between 1950 and 1970, coal burning for power generation increased while coal burning for other purposes was replaced by oil and natural gas and motor vehicle usage grew rapidly in these regions, leading to more sulphate and less BC. Peak aerosol forcing in North America and Europe occurred around 1970–1980 (Figure 8.25, second column), while Asian development led to increased biofuel and fossil fuel sources of aerosols and ozone precursors toward the end of the century. During the final decades of the century, desulphurization controls reduced sulphur emissions from North America and Europe, resulting in reduced negative forcing in these regions and positive Arctic aerosol forcing. The SH ozone hole developed during the final three decades, with negative forcing over high latitudes. Biomass burning generated ozone and carbonaceous aerosols in NH high-latitudes early in the century, with increased tropical burning from mid to late century.

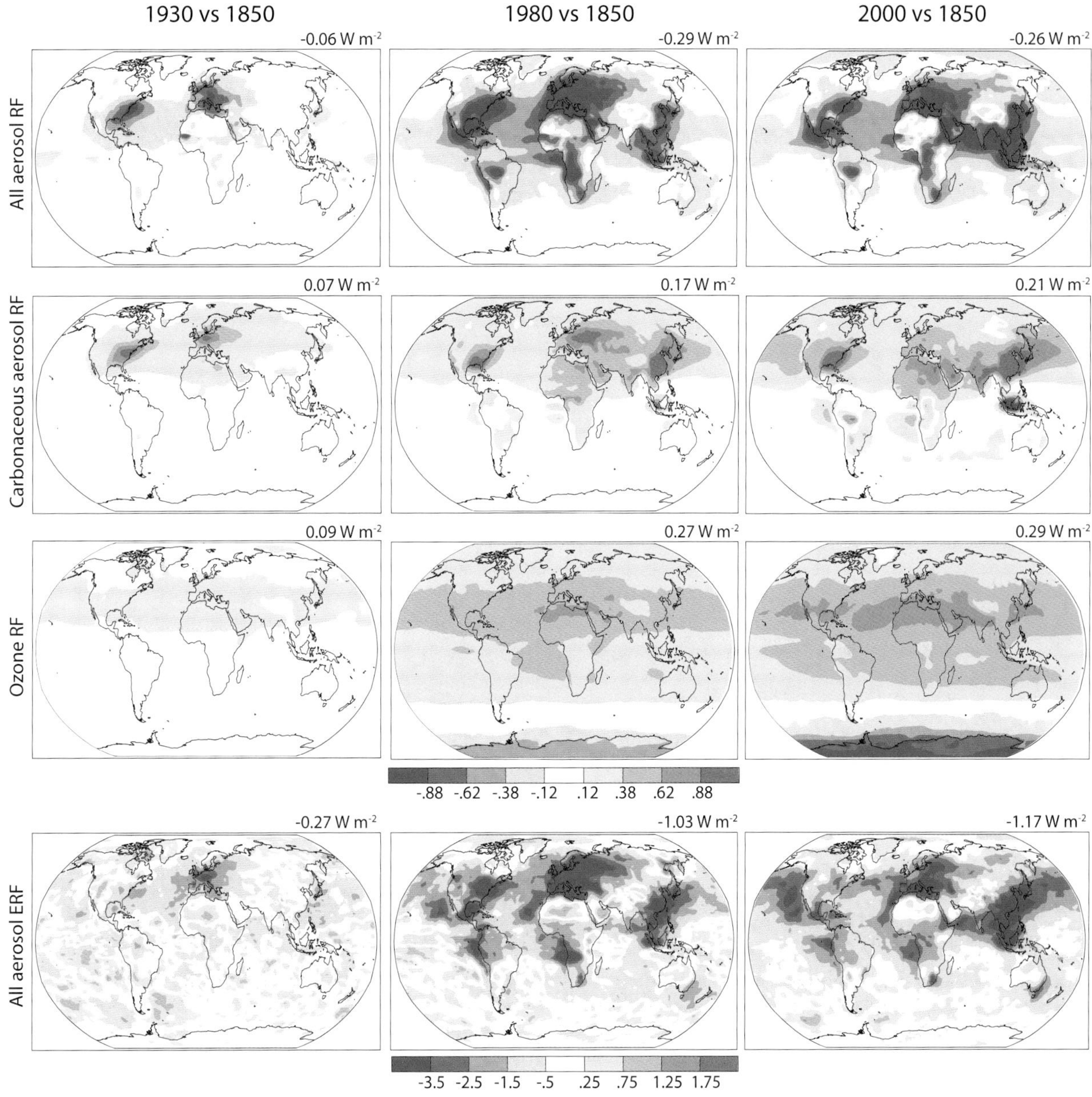

Figure 8.25 | Multi-model mean RF due to aerosol–radiation interaction of all aerosols, carbonaceous aerosols, ozone, and aerosol ERF (W m^{-2}) for the indicated times based on the ACCMIP simulations. Global area-weighted means are given in the upper right.

Aerosol ERF grew rapidly from 1930 to 1980, as did RF due to aerosol–radiation interaction, with a spatial structure reflecting both the influence of aerosol–radiation and aerosol–cloud interactions that are especially strong over pollution outflow regions and over areas with high surface albedo. From 1980 to 2000, aerosol ERF continued to become more negative even as negative RF due to aerosol–radiation interaction grew weaker, with the spatial pattern showing strengthening of aerosol ERF over Asia and weakening of aerosol ERF over North America and Europe.

Soil dust has changed since the pre-industrial due to land disturbance and resulting desertification (a forcing) and to changes in climate (a feedback). Mahowald et al. (2010) showed approximate doubling in dust loading over the 20th century (–0.1 W m^{-2}; consistent with the best estimate in Section 7.5.2; Section 8.3.4.2), primarily from the Saharan and Middle Eastern Deserts, with largest increase from the 1950s to the 1980s (–0.3 W m^{-2}), followed by a leveling. The increased dustiness reduces model precipitation within the Saharan source region, improving agreement with observed precipitation.

Aerosol loading changes during the past century have impacted radiation at the surface (Section 2.3.3), with peak radiation reductions in North America and Europe in the 1980s, and ongoing reduction in South and East Asia (Wild, 2009). The AR4 and CMIP5 models simulated these trends but underestimated their magnitude, the decadal temperature variations and the diurnal temperature range over land (Wild, 2009; see Chapter 9).

Changes in spatial patterns of species and their forcing over the century are difficult to validate due to sparse observations of short-lived species. Some constraint comes from limited historical observations in ice core records and from shorter trends beginning in late century from satellite and surface-based site measurements. The emissions estimates for historical species are very uncertain, especially for carbonaceous aerosols and dust. Therefore, the *confidence* in the historical forcing pattern changes is *low* for RF due to aerosol–radiation interaction and ozone, and *very low* for ERF, carbonaceous aerosols and dust.

8.6.2.2 Relationship Between Regional Forcing Patterns and Climate Response During the Industrial Era

An increasing body of research considers how spatial variations in RF affect climate response. Detection and attribution methods have had limited success in discerning statistically significant regional climate signals from regional forcing, due to large internal climate variability at regional scales, uncertainty in model processes and sparse regional observational records (Chapter 10). Meanwhile, research including model sensitivity studies for NTCFs, which vary strongly in space in time, explores climate response patterns.

In AR4 (Forster et al., 2007; Knutti et al., 2008) it was argued that the spatial pattern of forcing is not indicative of the pattern of climate response. Rather, the response is linked more closely to TOA flux resulting from the climate feedback spatial patterns (Boer and Yu, 2003; Taylor et al., 2011; Ming and Ramaswamy, 2012), with the lapse rate, surface albedo and cloud feedbacks explaining most of the temperature response. Yet Crook and Forster (2011) showed that both the spatial distribution of climate feedbacks and of heterogeneous forcing played important roles in the patterns of 20th century temperature changes. Other studies since AR4 have probed relationships between forcing patterns and climate responses.

Broad links between forcing and climate response have been identified. Shindell et al. (2010) used multiple models to show that surface temperature changes are much more sensitive to latitudinal than longitudinal variations in forcing. Shindell and Faluvegi (2009) used a model inverse approach to infer that NH aerosol reduction was associated with more than 70% of Arctic warming from the 1970s to the 2000s, and that Arctic and much of the SH surface temperature changes are strongly affected by remote forcing changes (alsoSection 10.3.1.1.4). Voulgarakis and Shindell (2010) defined a regional transient temperature sensitivity parameter, or temperature response per unit forcing for each 4-degree latitude band. Using observed surface air temperature changes they showed that the parameter is best constrained from 50°S to 25°N, where the value is 0.35°C (W m^{-2})$^{-1}$, smaller than at northern higher latitudes, and 35% smaller than in AR4 models.

Some aerosol model studies have demonstrated highly localized climate response to regional forcing. Significant regional cooling and hydrological shifts in the eastern USA and in Eastern Asia during the last half of the 20th century were modelled and attributed to local aerosols (Leibensperger et al., 2008, 2012a, 2012b; Chang et al., 2009) and localized warming projected for aerosol reductions (Mickley et al., 2012). Observations have also linked historical trends in aerosols and temperature (Ruckstuhl et al., 2008; Philipona et al., 2009).

Since AR4, there has been new research on aerosol influences on the hydrologic cycle (also Sections 7.4, 7.6.4, 10.3.3.1 and 11.3.2.4.3). Increased aerosol loading, with greater surface energy flux reduction in the NH, has been implicated in the observed southward shift of the Intertropical Convergence Zone (ITCZ) towards the hemisphere with smaller surface energy reduction: southward up to the 1980s with a reversal since (e.g., Denman et al., 2007; Zhang et al., 2007). Several studies have modelled an associated reduction in NH precipitation and associated shifts in the Hadley circulation (e.g., Rotstayn et al., 2000; Williams et al., 2001; Ming et al., 2011). The ITCZ shift may in turn be responsible for broad regional precipitation changes, including drying of the Sahel (e.g., Rotstayn and Lohmann, 2002; Biasutti and Giannini, 2006; Kawase et al., 2010; Ackerley et al., 2011) and northwestern Brazil (Cox et al., 2008), both of which peaked in the 1980s. These hemispheric asymmetric ITCZ effects are overlaid on thermodynamic aerosol effects which moisten subtropical regions, countering GHG-induced drying of these regions (Ming et al., 2011). Studies indicate that aerosols are more effective than an equivalent WMGHG forcing for shifting precipitation, and that historical trends in several areas cannot be explained without including aerosol forcing (Bollasina et al., 2011; Booth et al., 2012; Shindell et al., 2012a; Shindell et al., 2012b). However, *confidence* in attribution of any human influence on zonal shifts in precipitation distribution is only *medium* (Section 10.3.2.2).

There is increasing evidence but limited agreement that absorbing aerosols influence cloud distributions (Section 7.3.4.2). Absorbing aerosols apparently have complex influences on precipitation in monsoon regions. Model studies of Stephens et al. (2004) and Miller et al. (2004) showed that dust absorption over Africa enhances low-level convergence, vertical velocities and therefore local monsoon circulation and precipitation. On the other hand, Kawase et al. (2010) showed that biomass burning BC may cause the decreasing precipitation trend seen in tropical Africa during austral summer, due to reduction in evaporation and enhanced subsidence. The aerosol effects on the Indian monsoon are similarly complex, and have been the subject of numerous studies (e.g., Ramanathan et al., 2005; Chung and Ramanathan, 2006; Lau et al., 2006; Wang et al., 2009; Bollasina et al., (2011), but a clear picture of how the regional aerosol forcing correlates with responses has not yet fully emerged. Attribution of changes in monsoon to human influence generally has *low confidence* (Section 10.3).

Stratospheric ozone loss modelling has demonstrated an effect on the SH stratosphere similar to increased GHGs, cooling stratospheric temperatures, strengthening the polar vortex and shifting the westerly jet poleward; however causing cooler Antarctic surface temperatures, with larger influence on austral summer conditions (Son et al., 2009; McLandress et al., 2011; Thompson et al., 2011; see also Sections 10.3.3 and 11.3.2.4.3.) In the troposphere, models indicate that increased tropospheric ozone has caused warming, proportionally more in the NH and notably to the Arctic during winter, mainly during the second half of the 20th century (Shindell et al., 2006a).

Albedo changes due to land use and land cover changes exert a heterogeneous climate forcing (Figure 8.9). The surface albedo brightened on the one hand due to a shift from forest to brighter croplands, causing local cooling (e.g., Eliseev and Mokhov, 2011; Lee et al., 2011), but also darkened due to the re-expansion of forests to higher latitudes (Esper and Schweingruber, 2004) and increased vegetation height in snowy regions (Bonfils et al., 2012; also Section 8.3.5). Model studies have shown cooling from land use and land cover changes, especially over NH continents, although without demonstrating a detectable signal in observations (Matthews et al., 2004).

In addition to land use and climate-induced vegetation changes, CO_2 affects vegetation forcing indirectly, reducing transpiration from plants as stomata open less with increasing CO_2, resulting in localized atmospheric drying and warming (Section 11.3.2.3.1; Joshi and Gregory, 2008). These are not included in the standard RF (Section 8.1) and may be considered feedbacks (Section 8.3.2). This is modelled to be largest over the Amazon, the central African forest, and to some extent over boreal and temperate forests (Andrews et al., 2011). In the coupled climate modelling study of Lawrence and Chase (2010), the vegetation changes caused significant reduction in evapotranspiration, drying and warming in tropical and subtropical regions, with insignificant cooling at higher latitudes. Overall, vegetation changes may have caused modest cooling at high latitudes and warming at low latitudes, but the uncertainties are large and *confidence* is *very low*.

Deposition of BC on snow and ice, and loss of snow and ice darken the surface, reduces albedo, and enhances climate warming. Substantial snow-cover reduction of North America leads to warmer North American summertime temperature in models having a strong snow albedo feedback. These forcings can also have non-local impacts that result from enhanced land-ocean temperature contrast, increasing surface convergence over land and divergence over oceans. A poleward intensification of the high pressure patterns and subtropical jet may also result (Fletcher et al., 2009). BC contributions to snow darkening reduces snow cover, however the magnitude of the effect is very uncertain (see Sections 7.5.2.3 and 8.3.4.4). A model study calculated BC-albedo reduction to cause about 20% Arctic snow/ice cover reduction and 20% of Arctic warming over the previous century (Koch et al., 2011). However, reductions in Arctic soot during the past two decades (e.g., Hegg et al., 2009) have *likely* reversed that trend (e.g., Koch et al., 2011; Skeie et al., 2011b; Lee et al., 2013). Cryospheric feedbacks and atmospheric dynamical responses in models have an associated poleward shift in the temperature response to aerosol–cloud interactions (Kristjansson et al., 2005; Koch et al., 2009a; Chen et al., 2010).

Solar spectral (UV) irradiance variations along the solar cycle induce ozone responses by modifying the ozone production rate through photolysis of molecular oxygen (Section 8.4.1.4.1), and the resulting differential heating can drive circulation anomalies that lead to regional temperature and precipitation changes (Haigh, 1999; Shindell et al., 2006b; Frame and Gray, 2010; Gray et al., 2010). Such solar forcing may influence natural modes of circulation such as the Northern Annular Mode (e.g., Shindell et al., 2001; de la Torre et al., 2006; Ineson et al., 2011), the South Asian Summer Monsoon (Fan et al., 2009), the Southern Annular Mode (Kuroda and Kodera, 2005; Roscoe and Haigh, 2007) or the ENSO (Mann et al., 2005). The pattern of temperature response is less uniform than the forcing, for example, warming in the NH, but little response in the SH due to temperature moderation by wind speed enhancement effects on ocean circulation (Swingedouw et al., 2011). Regional responses to solar forcing are mediated by the stratosphere, so that reproducing such change requires spectrally varying solar forcing rather than TSI forcing (Lee et al., 2009; Section 8.4.1.4).

Stratospheric aerosol clouds (also Section 8.4.2.2) from tropical eruptions spread poleward and can cover an entire hemisphere or the globe, depending on the initial latitudinal spread. The aerosol eruption cloud from the 1963 Agung was confined mainly to the SH; the 1982 El Chichón mainly to the NH; and the 1991 Pinatubo covered the globe, all with an *e*-folding lifetime of about 1 year (e.g., Antuña et al., 2003). High-latitude eruptions typically stay confined to the high-latitude regions with shorter lifetimes of 2 to 4 months (Kravitz and Robock, 2011). Volcanic aerosols primarily scatter solar radiation back to space, but also absorb longwave radiation with the former larger by an order of magnitude. Stratospheric aerosol absorption heats the layer where they reside and produces distinct vertical and horizontal distributions of the heating rate. The temperature and chemical effects of the aerosols also enhance ozone destruction, which somewhat counteracts the radiative heating (Stenchikov et al., 2002). For tropical eruptions, this may affect atmospheric dynamics, with a stronger polar vortex, a positive mode of the Arctic Oscillation, and winter warming over NH continents (Robock, 2000). Climate responses to solar and volcanic forcings are further discussed in the context of detection and attribution of millennial climate change (see Section 10.7).

The study of how climate responds to regionally varying patterns of forcing is critical for understanding how local activities impact regional climate; however, the studies are exploratory and generally evoke *very low confidence*. However there is *medium* to *high confidence* in some qualitative but robust features, such as the damped warming of the NH and shifting of the ITCZ from aerosols, and positive feedbacks from high-latitude snow and ice albedo changes.

8.6.3 Spatial Evolution of Radiative Forcing and Response for the Future

Most components of aerosols and ozone precursors are estimated to decrease toward the end of this century in the RCPs except CH_4 in RCP8.5 (Figure 8.2) and nitrate aerosols, though some species reach the maximum amounts of emissions around the mid-21st century (Figure 8.2). The RCPs therefore contrast with the emission scenarios for TAR and AR4, which were based on Special Report on Emissions Scenarios (SRES) and have future projections of larger increase in the near-term climate forcers (NTCFs). It has been questioned whether such low emission of NTCFs is possible in the future given the current policies (Pozzer et al., 2012). This section surveys spatial differences in the RF of aerosols and ozone for the future based on the RCPs.

Figure 8.26 shows the global distributions of changes in aerosol and ozone forcings in 2030 and 2100 relative to 2000 for RCP2.6 and 8.5 (Shindell et al., 2013c). Both scenarios indicate reduced aerosol loading, and thus positive forcing over Europe, North America and Asia by 2100 where RF is above +0.5 W m^{-2} because of substantial reduction of scattering aerosols. The global mean RF due to aerosol–radiation interaction is estimated to be +0.12 and +0.08 W m^{-2} for RCP2.6 and 8.5, respectively, in 2100. Though the RF by total anthropogenic aerosols is positive, reduced BC contributes substantial negative forcing especially over the similar regions. The global mean carbonaceous RF including both the effects of BC and OC is estimated to be –0.20 and –0.11 W m^{-2} for RCP2.6 and 8.5, respectively, in 2100. Early in the century, on the other hand, both scenarios indicate increased negative aerosol forcing over South Asia, with reversal between 2030 and 2100. Emissions of BC, OC and SO_2 will reach their maximums early and middle in the century for RCP2.6 and 8.5, respectively in India. In RCP6, high emission levels of SO_2 in China persist until the mid-21st century (Supplementary Material Figure 8.SM.1), and then it is predicted to keep a high negative RF due to aerosol–radiation interaction over East Asia. The RF due to aerosol–radiation interaction for carbonaceous aerosol is positive over East and South Asia in 2030 relative to 2000 for RCP8.5 because BC emission is also larger in 2030. Over central and southern Africa, a change in the future RF due to aerosol–radiation interaction based on RCPs is not clear mainly because of uncertainties in the wildfires emissions (see Section 7.3.5.3). The global mean total RF due to aerosol–radiation interaction in the future is rather small due to offsetting effects, with reductions in BC, increases in nitrate aerosols, and reductions in scattering aerosols each causing substantially more forcing than the net.

Emissions and atmospheric loadings of natural aerosols are affected by climate change. There is, however, no consensus among studies on future trends of their changes for major natural aerosols, mineral dust and sea salt, as indicated in Section 7.3.5.1. The spatial pattern of the aerosol forcing may be influenced by natural aerosols due to reduction in sea ice cover leading to increased emission of high-latitude sea salt (Struthers et al., 2011; Takemura, 2012) and SOA from vegetation changes (Tsigaridis and Kanakidou, 2007).

The simulations applying the RCPs indicate that the latitude of maximum emission of NTCFs, and therefore of maximum RF, is projected to shift somewhat southward for the next few decades (in 2030 of Figure 8.26). The shift of peak aerosol loading southward is expected to cause the ITCZ to continue to shift northward. This, in combination with warming and drying over tropical land, has been modelled to lead to greatly enhanced drought conditions in the Amazon (Cox et al., 2008). On the other hand, if the low-latitude aerosol is sufficiently absorbing, broadening of the ITCZ convergence region and enhanced cloud cover could result, as modelled for dust (Perlwitz and Miller, 2010).

Reductions in high-latitude BC are expected to contribute to reducing Arctic forcing (e.g., Koch et al., 2011), due to reduction in BC deposition on snow as well as in absorption of sunlight over bright surface. On the other hand, reduction in mid-high-latitude scattering aerosols may offset all or part of the impact of the local Arctic forcing change (Shindell et al., 2010; Koch et al., 2011).

Figure 8.26 also shows the ozone RF in 2030 and 2100 relative to 2000, which includes changes both in tropospheric and stratospheric ozone. Recovery of ozone in the stratosphere in the 21st century will result in positive forcing in the SH high latitudes in comparison with the year 2000 for both the pathways. This is because of the reduced emissions of ozone-depleting substances controlled under the Montreal Protocol, with a small additional effect from a feedback of changes in temperature and in the vertical circulation due to changes in stratospheric compositions (Kawase et al., 2011; Lamarque et al., 2011). In the troposphere, on the other hand, a large difference in the CH_4 emissions between RCP8.5 and the other pathways shown in Figure 8.2 leads to a different RF trend outside the SH high latitudes. Ozone recovery in the stratosphere and ozone increase in the troposphere leads to a positive RF all over the globe in RCP8.5 with a mean of +0.26 W m^{-2} in 2100. The cancellation between positive RF due to ozone increase in the stratosphere and negative RF due to ozone decrease in the troposphere results in a global mean RF of –0.12 W m^{-2} in RCP2.6.

Figure 8.26 also shows the global distributions of changes in ERF due to both aerosol–radiation and aerosol–cloud interactions in 2030 and 2100 relative to 2000 for RCP8.5. Although the ERF includes rapid adjustments and therefore its magnitude is much larger than that of RF due to aerosol–radiation interaction, the spatial pattern is generally similar to RF. The ERF in 2100 shows positive values relative to 2000 in North America, Europe and Asia even with RCP8.5, which indicates the aerosol forcing is projected to approach to the pre-industrial level.

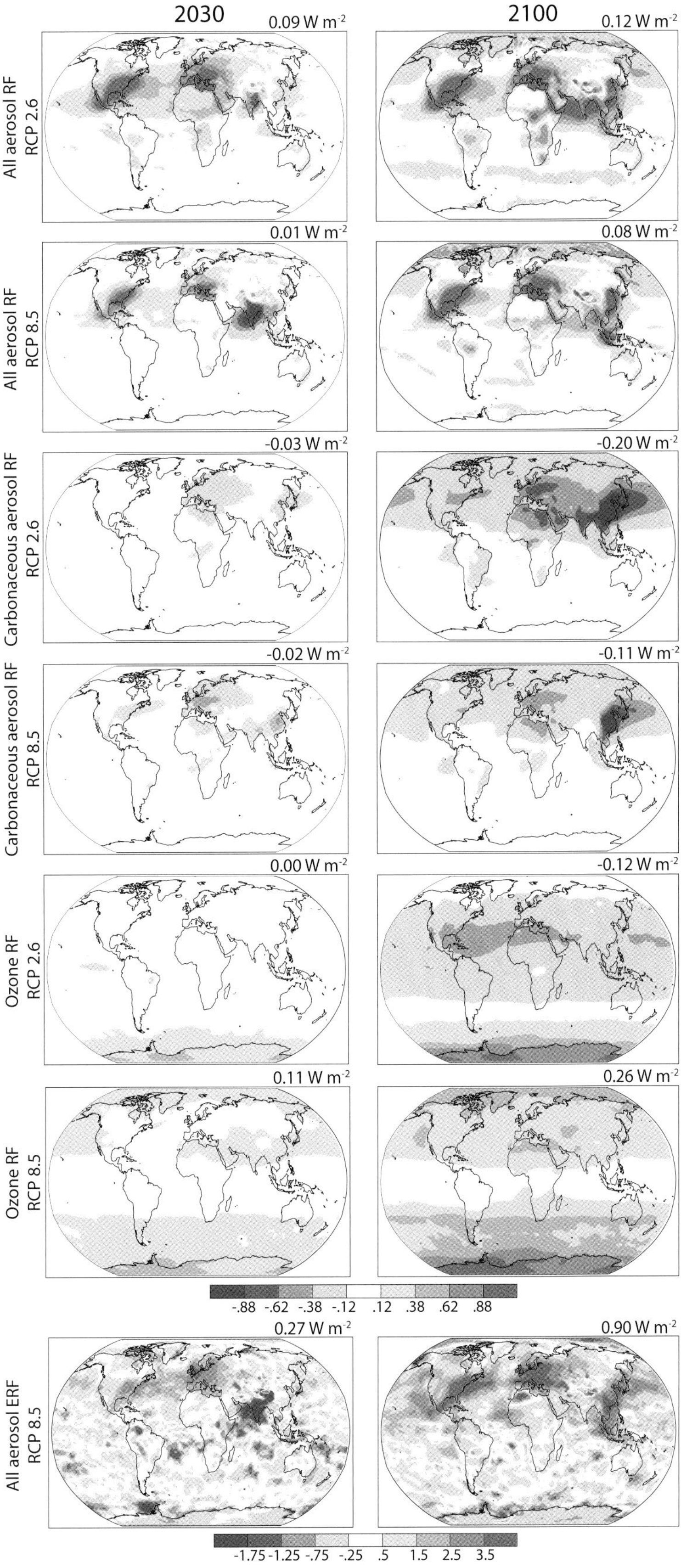

Figure 8.26 | Multi-model mean RF (W m^{-2}) due to aerosol–radiation interaction of all anthropogenic aerosols (first and second rows) and anthropogenic carbonaceous (BC+OC) aerosols (third and fourth rows), and total ozone (fifth and sixth rows) in 2030 (left) and 2100 (right) relative to 2000 for RCP2.6 (top each) and RCP8.5 (bottom each) based on the ACCMIP simulations. The seventh row shows multi-model mean ERF (W m^{-2}) by all anthropogenic aerosols in 2030 (left) and 2100 (right) relative to 2000 for RCP8.5. Global area-weighted means are given in the upper right of each panel.

8

8.7 Emission Metrics

8.7.1 Metric Concepts

8.7.1.1 Introduction

To quantify and compare the climate impacts of various emissions, it is necessary to choose a climate parameter by which to measure the effects; that is, RF, temperature response, and so forth. Thus, various choices are needed for the steps down the cause–effect chain from emissions to climate change and impacts (Figure 8.27 and Box 8.4). Each step in the cause effect chain requires a modelling framework. For assessments and evaluation one may—as an alternative to models that explicitly include physical processes resulting in forcing and responses—apply simpler measures or *metrics* that are based on results from complex models. Metrics are used to quantify the contributions to climate change of emissions of different substances and can thus act as 'exchange rates' in multi-component policies or comparisons of emissions from regions/countries or sources/sectors. Metrics are also used in areas such as Life Cycle Assessments and Integrated Assessment Modelling (e.g., by IPCC WGIII).

Metrics can be given in *absolute* terms (e.g., K kg^{-1}) or in *relative* terms by normalizing to a reference gas — usually CO_2. To transform the effects of different emissions to a common scale — often called 'CO_2 equivalent emissions'—the emission (E_i) of component *i* can be multiplied with the adopted normalized metric (M_i): $M_i \times E_i = CO_2\text{-eq}_i$. Ideally, the climate effects of the calculated CO_2 equivalent emissions should be the same regardless of the mix of components emitted. However, different components have different physical properties, and a metric that establishes equivalence with regard to one effect cannot guarantee equivalence with regard to other effects and over extended time periods, for example, Lauder et al. (2013), O'Neill (2000), Smith and Wigley (2000), Fuglestvedt et al. (2003).

Metrics do not define goals and policy—they are tools that enable evaluation and implementation of multi-component policies (i.e., which emissions to abate). The most appropriate metric will depend on which aspects of climate change are most important to a particular application, and different climate policy goals may lead to different conclusions about what is the most suitable metric with which to implement that policy, for example, Plattner et al. (2009); Tol et al. (2012). Metrics that have been proposed include physical metrics as well as more comprehensive metrics that account for both physical and economic dimensions (see 8.7.1.5 and WGIII, Chapter 3).

This section provides an assessment that focuses on the scientific aspects and utility of emission metrics. Extending such an assessment to include more policy-oriented aspects of their performance and usage such as simplicity, transparency, continuity, economic implications of usage of one metric over another, and so forth, is not given here as this is beyond the scope of WGI. However, consideration of such aspects is vital for user-assessments. In the following, the focus is on the more well-known Global Warming Potential (GWP) and Global Temperature change Potential (GTP), though other concepts are also briefly discussed.

8.7.1.2 The Global Warming Potential Concept

The Global Warming Potential (GWP) is defined as the time-integrated RF due to a pulse emission of a given component, relative to a pulse emission of an equal mass of CO_2 (Figure 8.28a and formula). The GWP was presented in the First IPCC Assessment (Houghton et al., 1990), stating 'It must be stressed that there is no universally accepted methodology for combining all the relevant factors into a single global warming potential for greenhouse gas emissions. A simple approach has been adopted here to illustrate the difficulties inherent in the concept, ...'. Further, the First IPCC Assessment gave no clear physical interpretation of the GWP.

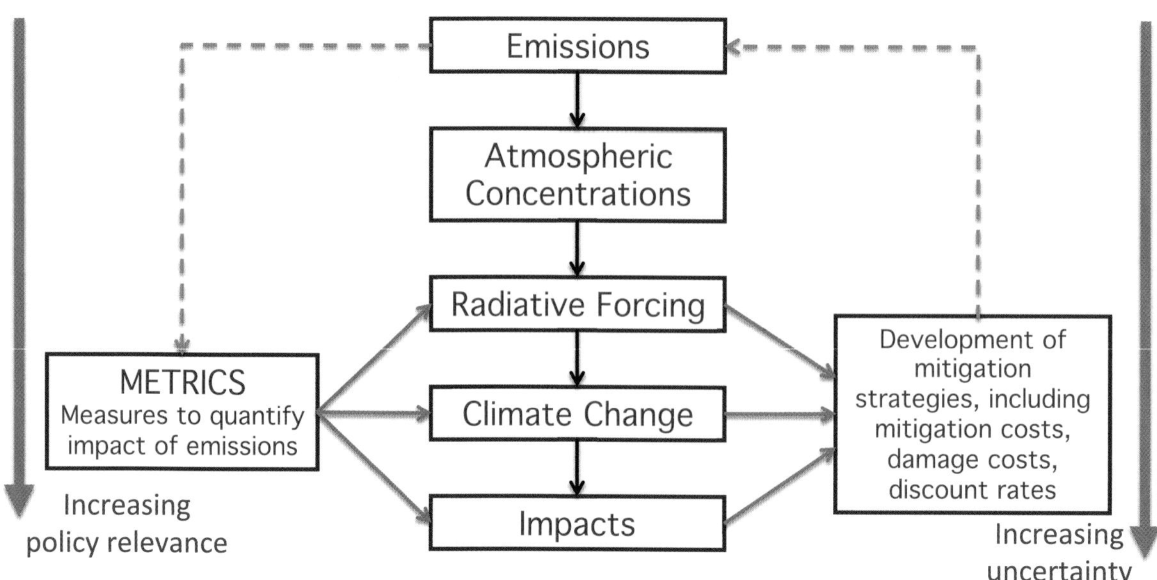

Figure 8.27 | The cause–effect chain from emissions to climate change and impacts showing how metrics can be defined to estimate responses to emissions (left) and for development of multi-component mitigation (right). The relevance of the various effects increases downwards but at the same time the uncertainty also increases. The dotted line on the left indicates that effects and impacts can be estimated directly from emissions, while the arrows on the right side indicate how these estimates can be used in development of strategies for reducing emissions. (Adapted from Fuglestvedt et al., 2003, and Plattner et al., 2009.)

8

Box 8.4 | Choices Required When Using Emission Metrics

Time frames: One can apply a *backward-looking* (i.e., historical) or a *forward-looking* perspective on the responses to emissions. In the forward-looking case one may use pulses of emissions, sustained emissions or emission scenarios. All choices of emission perturbations are somewhat artificial and idealized, and different choices serve different purposes. One may use the *level* (e.g., degrees Celsius) or *rate* of change (e.g., degrees Celsius per decade). Furthermore, the effects of emissions may be estimated at a particular time or be integrated over time up to a chosen time horizon. Alternatively, discounting of future effects may be introduced (i.e., a weighting of effects over time).

Type of effect or end-point: Radiative forcing, temperature change or sea level change, for example, could be examined (Figure 8.27). Metrics may also include eco/biological or socioeconomic damages. The choice of climate impact parameters is related to which aspects of climate change are considered relevant for interpretation of 'dangerous anthropogenic interference with the climate system' (UNFCCC Article 2).

Spatial dimension for emission and response: Equal-mass emissions of NTCFs from different regions can induce varying global mean climate responses, and the climate response also has a regional component irrespective of the regional variation in emissions. Thus, metrics may be given for region of *emission* as well as region of *response*.

Some of the choices involved in metrics are scientific (e.g., type of model, and how processes are included or parameterized in the models). Choices of time frames and climate impact are policy-related and cannot be based on science alone, but scientific studies can be used to analyse different approaches and policy choices.

A direct interpretation is that the GWP is an index of the total energy added to the climate system by a component in question relative to that added by CO_2. However, the GWP does not lead to equivalence with temperature or other climate variables (Fuglestvedt et al., 2000, 2003; O'Neill, 2000; Daniel et al., 2012; Smith and Wigley, 2000; Tanaka et al., 2009). Thus, the name 'Global Warming Potential' may be somewhat misleading, and 'relative cumulative forcing index' would be more appropriate. It can be shown that the GWP is approximately equal to the ratio (normalizing by the similar expression for CO_2) of the *equilibrium temperature response due to a sustained emission* of the species or to the *integrated temperature response for a pulse* emission (assuming efficacies are equal for the gases that are compared; O'Neill, 2000; Prather, 2002; Shine et al., 2005a; Peters et al., 2011a; Azar and Johansson, 2012).

The GWP has become the default metric for transferring emissions of different gases to a common scale; often called 'CO_2 equivalent emissions' (e.g., Shine, 2009). It has usually been integrated over 20, 100 or 500 years consistent with Houghton et al. (1990). Note, however that Houghton et al. presented these time horizons as 'candidates for discussion [that] should not be considered as having any special significance'. The GWP for a time horizon of 100 years was later adopted as a metric to implement the multi-gas approach embedded in the United Nations Framework Convention on Climate Change (UNFCCC) and made operational in the 1997 Kyoto Protocol. The choice of time horizon has a strong effect on the GWP values — and thus also on the calculated contributions of CO_2 equivalent emissions by component, sector or nation. There is no scientific argument for selecting 100 years compared with other choices (Fuglestvedt et al., 2003; Shine, 2009). The choice of time horizon is a value judgement because it depends

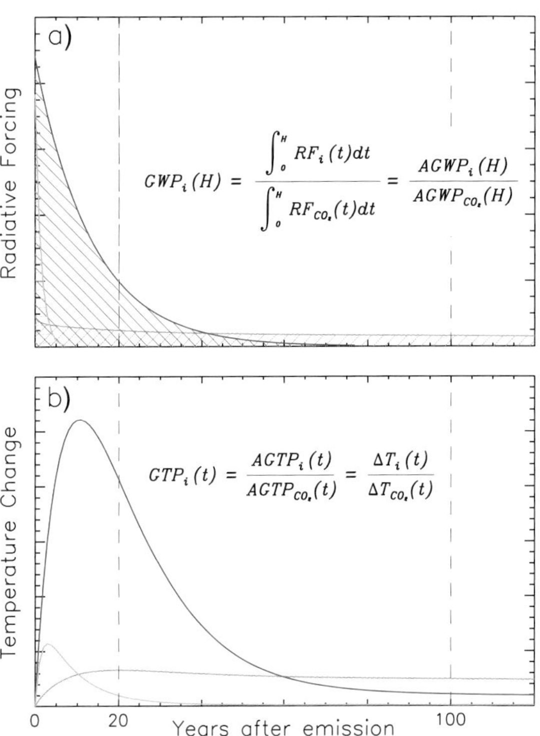

Figure 8.28 | (a) The Absolute Global Warming Potential (AGWP) is calculated by integrating the RF due to emission pulses over a chosen time horizon; for example, 20 and 100 years (vertical lines). The GWP is the ratio of AGWP for component *i* over AGWP for the reference gas CO_2. The blue hatched field represents the integrated RF from a pulse of CO_2, while the green and red fields represent example gases with 1.5 and 13 years lifetimes, respectively. (b) The Global Temperature change Potential (GTP) is based on the temperature response at a selected year after pulse emission of the same gases; e.g., 20 or 100 years (vertical lines). See Supplementary Material Section 8.SM.11 for equations for calculations of GWP and GTP.

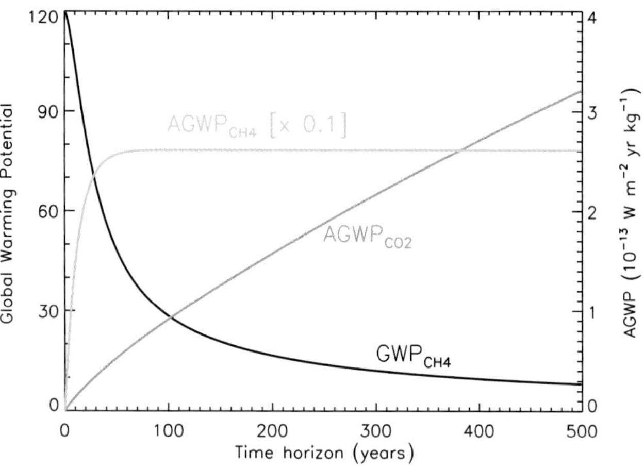

Figure 8.29 | Development of AGWP-CO_2, AGWP-CH_4 and GWP-CH_4 with time horizon. The yellow and blue curves show how the AGWPs changes with increasing time horizon. Because of the integrative nature the AGWP for CH_4 (yellow curve) reaches a constant level after about five decades. The AGWP for CO_2 continues to increase for centuries. Thus the ratio which is the GWP (black curve) falls with increasing time horizon.

on the relative weight assigned to effects at different times. Other important choices include the background atmosphere on which the GWP calculations are superimposed, and the way indirect effects and feedbacks are included (see Section 8.7.1.4).

For some gases the variation in GWP with time horizon mainly reflects properties of the reference gas, not the gas for which the GWP is calculated. The GWP for NTCFs decreases with increasing time horizon, as GWP is defined with the integrated RF of CO_2 in the denominator. As shown in Figure 8.29, after about five decades the development in the GWP for CH_4 is almost entirely determined by CO_2. However, for long-lived gases (e.g., SF_6) the development in GWP is controlled by both the increasing integrals of RF from the long-lived gas and CO_2.

8.7.1.3 The Global Temperature change Potential Concept

Compared to the GWP, the Global Temperature change Potential (GTP; Shine et al., 2005a) goes one step further down the cause–effect chain (Figure 8.27) and is defined as the *change in global mean surface temperature at a chosen point in time* in response to an emission pulse—relative to that of CO_2. Whereas GWP is integrated in time (Figure 8.28a), GTP is an end-point metric that is based on temperature change for a selected year, *t*, (see Figure 8.28b with formula). Like for the GWP, the impact from CO_2 is normally used as reference, hence, for a component *i*, $GTP(t)_i = AGTP(t)_i / AGTP(t)_{CO2} = \Delta T((t)_i / \Delta T(t)_{CO2}$, where AGTP is the absolute GTP giving temperature change per unit emission (see Supplementary Material Section 8.SM.11 for equations and parameter values). Shine et al. (2005a) presented the GTP for both pulse and sustained emission changes based on an energy balance model as well as analytical equations. A modification was later introduced (Shine et al., 2007) in which the time horizon is determined by the proximity to a target year as calculated by using scenarios and climate models (see Section 8.7.1.5).

Like GWP, the GTP values can be used for weighting the emissions to obtain 'CO_2 equivalents' (see Section 8.7.1.1). This gives the temperature effects of emissions relative to that of CO_2 for the chosen time horizon. As for GWP, the choice of time horizon has a strong effect on the metric values and the calculated contributions to warming.

In addition, the AGTP can be used to calculate the global mean temperature change due to any given emission scenario (assuming linearity) using a convolution of the emission scenarios and $AGTP_i$:

$$\Delta T(t) = \sum_i \int_0^t E_i(s) AGTP_i(t-s) ds \qquad (8.1)$$

where *i* is component, *t* is time, and *s* is time of emission (Berntsen and Fuglestvedt, 2008; Peters et al., 2011b; Shindell et al., 2011).

By accounting for the climate sensitivity and the exchange of heat between the atmosphere and the ocean, the GTP includes physical processes that the GWP does not. The GTP accounts for the slow response of the (deep) ocean, thereby prolonging the response to emissions beyond what is controlled by the decay time of the atmospheric concentration. Thus the GTP includes both the atmospheric adjustment time scale of the component considered and the response time scale of the climate system.

The GWP and GTP are fundamentally different by construction and different numerical values can be expected. In particular, the GWPs for NTCFs, over the same time frames, are higher than GTPs due to the integrative nature of the metric. The GTP values can be significantly affected by assumptions about the climate sensitivity and heat uptake by the ocean. Thus, the relative uncertainty ranges are wider for the GTP compared to GWP (see Section 8.7.1.4). The additional uncertainty is a typical trade-off when moving along the cause–effect chain to an effect of greater societal relevance (Figure 8.27). The formulation of the ocean response in the GTP has a substantial effect on the values; thus its characterization also represents a trade-off between simplicity and accuracy. As for GWP, the GTP is also influenced by the background atmosphere, and the way indirect effects and feedbacks are included (see Section 8.7.1.4).

8.7.1.4 Uncertainties and Limitations related to Global Warming Potential and Global Temperature change Potential

The uncertainty in the numerator of GWP; that is, the $AGWP_i$ (see formula in Figure 8.28a) is determined by uncertainties in lifetimes (or perturbation lifetimes) and radiative efficiency. Inclusion of indirect effects increases uncertainties (see below). For the reference gas CO_2, the uncertainty is dominated by uncertainties in the *impulse response function* (IRF) that describes the development in atmospheric concentration that follows from an emission pulse (Joos et al., 2013); see Box 6.2 and Supplementary Material Section 8.SM.12. The IRF is sensitive to model representation of the carbon cycle, pulse size and background CO_2 concentrations and climate.

Based on a multi-model study, Joos et al. (2013) estimate uncertainty ranges for the time-integrated IRF for CO_2 to be ±15% and ±25% (5 to 95% uncertainty range) for 20- and 100-year time horizons, respectively. Assuming quadratic error propagation, and ±10% uncertainty in radiative efficiency, the uncertainty ranges in AGWP for CO_2 were estimated to be ±18% and ±26% for 20 and 100 years. These uncertainties affect all metrics that use CO_2 as reference. Reisinger et al. (2010) and Joos et al. (2013) show that these uncertainties increase with time horizon.

The same factors contribute to uncertainties in the GTP, with an additional contribution from the parameters describing the ocean heat uptake and climate sensitivity. In the first presentation of the GTP, Shine et al. (2005a) used one time constant for the climate response in their analytical expression. Improved approaches were used by Boucher and Reddy (2008), Collins et al. (2010) and Berntsen and Fuglestvedt (2008) that include more explicit representations of the deep ocean that increased the long-term response to a pulse forcing. Over the range of climate sensitivities from AR4, GTP_{50} for BC was found to vary by a factor of 2, the CH_4 GTP_{50} varied by about 50%, while for N_2O essentially no dependence was found (Fuglestvedt et al., 2010). AGTPs for CO_2 were also calculated in the multi-model study by Joos et al. (2013). They found uncertainty ranges in AGTP that are much larger than for AGWP; ±45% and ±90% for 20 and 100 years (5 to 95% uncertainty range). These uncertainty ranges also reflect the signal-to-noise ratio, and not only uncertainty in the physical mechanisms.

There are studies combining uncertainties in various input parameters. Reisinger et al. (2011) estimated the uncertainty in the GWP for CH_4 and found an uncertainty of –30 to +40% for the GWP_{100} and –50 to +75% for GTP_{100} of CH_4 (for 5 to 95% of the range). Boucher (2012) performed a Monte Carlo analysis with uncertainties in perturbation lifetime and radiative efficiency, and for GWP_{100} for CH_4 (assuming a constant background atmosphere) he found ±20%, and –40 to +65 for GTP_{100} (for 5 to 95% uncertainty range).

Here we estimate uncertainties in GWP values based on the uncertainties given for radiative efficiencies (Section 8.3.1), perturbation lifetimes, indirect effects and in the AGWP for the reference gas CO_2 (see Supplementary Material Section 8.SM.12). For CH_4 GWP we estimate an uncertainty of ±30% and ±40% for 20- and 100-year time horizons, respectively (for 5 to 95% uncertainty range). The uncertainty is dominated by AGWP for CO_2 and indirect effects. For gases with lifetimes of a century or more the uncertainties are of the order of ±20% and ±30% for 20- and 100-year horizons. The uncertainty in GWPs for gases with lifetimes of a few decades is estimated to be of the order of ±25% and ±35% for 20 and 100 years. For shorter-lived gases, the uncertainties in GWPs will be larger (see Supplementary Material Section 8.SM.12 for a discussion of contributions to the total uncertainty.) For GTP, few uncertainty estimates are available in the literature. Based on the results from Joos et al. (2013), Reisinger et al. (2010) and Boucher (2012) we assess the uncertainty to be of the order of ±75% for the CH_4 GTP_{100}.

The metric values are also strongly dependent on which processes are included in the definition of a metric. Ideally all indirect effects (Sections 8.2 and 8.3) should be taken into account in the calculation of metrics. The indirect effects of CH_4 on its own lifetime, tropospheric ozone and stratospheric water have been traditionally included in its GWP. Boucher et al. (2009) have quantified an indirect effect on CO_2 when fossil fuel CH_4 is oxidized in the atmosphere. Shindell et al. (2009) estimated the impact of reactive species emissions on both gaseous and aerosol forcing species and found that ozone precursors, including CH_4, had an additional substantial climate effect because they increased or decreased the rate of oxidation of SO_2 to sulphate aerosol. Studies with different sulphur cycle formulations have found lower sensitivity (Collins et al., 2010; Fry et al., 2012). Collins et al. (2010) postulated an additional component to their GWPs and GTPs for ozone precursors due to the decreased productivity of plants under higher levels of surface ozone. This was estimated to have the same magnitude as the ozone and CH_4 effects. This effect, however, has so far only been examined with one model. In a complex and interconnected system, feedbacks can become increasingly complex, and uncertainty of the magnitude and even direction of feedback increases the further one departs from the primary perturbation, resulting in a trade-off between completeness and robustness, and hence utility for decision-making.

Gillett and Matthews (2010) included climate–carbon feedbacks in calculations of GWP for CH_4 and N_2O and found that this increased the values by about 20% for 100 years. For GTP of CH_4 they found an increase of ~80%. They used numerical models for their studies and suggest that climate–carbon feedbacks should be considered and parameterized when used in simple models to derive metrics. Collins et al. (2013) parameterize the climate-carbon feedback based on Friedlingstein et al. (2006) and Arora et al. (2013) and find that this more than doubles the GTP_{100} for CH_4. Enhancement of the GTP for CH_4 due to carbon–climate feedbacks may also explain the higher GTP values found by Reisinger et al. (2010).

The inclusion of indirect effects and feedbacks in metric values has been inconsistent in the IPCC reports. In SAR and TAR, a carbon model without a coupling to a climate model was used for calculation of IRF for CO_2 (Joos et al., 1996), while in AR4 climate-carbon feedbacks were included for the CO_2 IRF (Plattner et al., 2008). For the time horizons 20 and 100 years, the $AGWP_{CO2}$ calculated with the Bern3D-LPJ model is, depending on the pulse size, 4 to 5% and 13 to 15% lower, respectively, when carbon cycle–climate feedbacks are not included (Joos et al., 2013). While the AGWP for the reference gas CO_2 included climate–carbon feedbacks, this is not the case for the non-CO_2 gas in the numerator of GWP, as recognized by Gillett and Matthews (2010), Joos et al. (2013), Collins et al. (2013) and Sarofim (2012). This means that the GWPs presented in AR4 may underestimate the relative impacts of non-CO_2 gases. The different inclusions of feedbacks partially represent the current state of knowledge, but also reflect inconsistent and ambiguous definitions. In calculations of AGWP for CO_2 in AR5 we use the IRF for CO_2 from Joos et al. (2013) which includes climate–carbon feedbacks. Metric values in AR5 are presented both with and without including climate–carbon feedbacks for non-CO_2 gases. This feedback is based on the carbon-cycle response in a similar set of models (Arora et al., 2013) as used for the reference gas (Collins et al., 2013).

The effect of including this feedback for the non-reference gas increases with time horizon due to the long-lived nature of the initiated CO_2 perturbation (Table 8.7). The relative importance also increases with decreasing lifetime of the component, and is larger for GTP than GWP due to the integrative nature of GWP. We calculate an increase in the CH_4 GWP_{100} of 20%. For GTP_{100}, however, the changes are much larger; of the order of 160%. For the shorter time horizons (e.g., 20 years) the effect of including this feedback is small (<5%) for both GWP

Table 8.7 | GWP and GTP with and without inclusion of climate–carbon feedbacks (cc fb) in response to emissions of the indicated non-CO_2 gases (climate-carbon feedbacks in response to the reference gas CO_2 are always included).

	Lifetime (years)		GWP_{20}	GWP_{100}	GTP_{20}	GTP_{100}
CH_4[b]	12.4[a]	No cc fb	84	28	67	4
		With cc fb	86	34	70	11
HFC-134a	13.4	No cc fb	3710	1300	3050	201
		With cc fb	3790	1550	3170	530
CFC-11	45.0	No cc fb	6900	4660	6890	2340
		With cc fb	7020	5350	7080	3490
N_2O	121.0[a]	No cc fb	264	265	277	234
		With cc fb	268	298	284	297
CF_4	50,000.0	No cc fb	4880	6630	5270	8040
		With cc fb	4950	7350	5400	9560

Notes:

Uncertainties related to the climate–carbon feedback are large, comparable in magnitude to the strength of the feedback for a single gas.

[a] Perturbation lifetime is used in the calculation of metrics.

[b] These values do not include CO_2 from methane oxidation. Values for fossil methane are higher by 1 and 2 for the 20 and 100 year metrics, respectively (Table 8.A.1).

and GTP. For the more long-lived gases the GWP_{100} values increase by 10 to 12%, while for GTP_{100} the increase is 20 to 30%. Table 8.A.1 gives metric values including the climate–carbon feedback for CO_2 only, while Supplementary Material Table 8.SM.16 gives values for all halocarbons that include the climate–carbon feedback. Though uncertainties in the carbon cycle are substantial, it is *likely* that including the climate–carbon feedback for non-CO_2 gases as well as for CO_2 provides a better estimate of the metric value than including it only for CO_2.

Emission metrics can be estimated based on a constant or variable background climate and this influences both the adjustment times and the concentration–forcing–temperature relationships. Thus, all metric values will need updating due to changing atmospheric conditions as well as improved input data. In AR5 we define the metric values with respect to a constant present-day condition of concentrations and climate. However, under non-constant background, Joos et al. (2013) found decreasing CO_2 $AGWP_{100}$ for increasing background levels (up to 23% for RCP8.5). This means that GWP for all non-CO_2 gases (except CH_4 and N_2O) would increase by roughly the same magnitude. Reisinger et al. (2011) found a reduction in AGWP for CO_2 of 36% for RCP8.5 from 2000 to 2100 and that the CH_4 radiative efficiency and AGWP also decrease with increasing CH_4 concentration. Accounting for both effects, the GWP_{100} for CH_4 would increase by 10 to 20% under low and mid-range RCPs by 2100, but would decrease by up to 10% by mid-century under the highest RCP. While these studies have focused on the background levels of GHGs, the same issues apply for temperature. Olivié et al. (2012) find different temperature IRFs depending on the background climate (and experimental set up).

User related choices (see Box 8.4) such as the time horizon can greatly affect the numerical values obtained for CO_2 equivalents. For a change in time horizon from 20 to 100 years, the GWP for CH_4 decreases by a factor of approximately 3 and its GTP by more than a factor of 10. Short-lived species are most sensitive to this choice. Some approaches have removed the time horizon from the metrics (e.g., Boucher, 2012), but discounting is usually introduced which means that a discount rate r (for the weighting function e^{-rt}) must be chosen instead. The choice of discount rate is also value based (see WGIII, Chapter 3).

For NTCFs the metric values also depend on the location and timing of emission and whether regional or global metrics are used for these gases is also a choice for the users. Metrics are usually calculated for pulses, but some studies also give metric values that assume constant emissions over the full time horizon (e.g., Shine et al., 2005a; Jacobson, 2010). It is important to be aware of the idealized assumption about constant future emissions (or change in emissions) of the compound being considered if metrics for sustained emissions are used.

8.7.1.5 New Metric Concepts

New metric concepts have been developed both to modify physical metrics to address shortcomings as well as to replace them with metrics that account for economic dimensions of problems to which metrics are applied. Modifications to physical metrics have been proposed to better represent CO_2 emissions from bioenergy, regional patterns of response, and for peak temperature limits.

Emissions of CO_2 from the combustion of biomass for energy in national emission inventories are currently assumed to have no net RF, based on the assumption that these emissions are compensated by biomass regrowth (IPCC, 1996). However, there is a time lag between combustion and regrowth, and while the CO_2 is resident in the atmosphere it leads to an additional RF. Modifications of the GWP and GTP for bioenergy (GWP_{bio}, GTP_{bio}) have been developed (Cherubini et al., 2011; Cherubini et al., 2012). The GWP_{bio} give values generally between zero (current default for bioenergy) and one (current for fossil fuel emissions) (Cherubini et al., 2011), and negative values are possible for GTP_{bio} due to the fast time scale of atmospheric–ocean CO_2 exchange relative to the growth cycle of biomass (Cherubini et al., 2012). GWP_{bio} and GTP_{bio} have been used in only a few applications, and more research is needed to assess their robustness and applicability. Metrics for biogeophysical effects, such as albedo changes, have been proposed (Betts, 2000; Rotenberg and Yakir, 2010) , but as for NTCFs regional variations are important (Claussen et al., 2001) and the RF concept may not be adequate (Davin et al., 2007).

New concepts have also been developed to capture information about regional patterns of responses and cancelling effects that are lost when global mean metrics are used. The use of nonlinear damage functions to capture information on the spatial pattern of responses has been explored (Shine et al., 2005b; Lund et al., 2012). In addition, the Absolute Regional Temperature Potential (ARTP) (Shindell, 2012; Collins et al., 2013) has been developed to provide estimates of impacts at a sub-global scale. ARTP gives the time-dependent temperature response in four latitude bands as a function of the regional forcing imposed in all bands. These metrics, as well as new regional precipitation metrics (Shindell et al., 2012b), require additional studies to determine their robustness.

Alternatives to the single basket approach adopted by the Kyoto Protocol are a component-by-component approach or a multi-basket approach (Rypdal et al., 2005; Daniel et al., 2012; Sarofim, 2012; Jackson, 2009). Smith et al. (2012) show how peak temperature change is constrained by *cumulative emissions* (see 12.5.4) for gases with long lifetimes and *emissions rates* for shorter-lived gases (including CH_4). Thus, they divide gases into two baskets and present two metrics that can be used for estimating peak temperature for various emission scenarios. This division of gases into the two baskets is sensitive to the time of peak temperature in the different scenarios. The approach uses time invariant metrics that do not account for the timing of emissions relative to the target year. The choice of time horizon is implicit in the scenario assumed and this approach works only for a peak scenario.

A number of new metrics have been developed to add economic dimensions to purely physically based metrics such as the GWP and GTP. The use of physical metrics in policy contexts has been criticized by economists (Reilly and Richards, 1993; Schmalensee, 1993; Hammitt et al., 1996; Reilly et al., 1999; Bradford, 2001; De Cara et al., 2008). A prominent use of metrics is to set relative prices of gases when implementing a multi-gas policy. Once a particular policy has been agreed on, economic metrics can address policy goals more directly than physical metrics by accounting not only for physical dimensions but also for economic dimensions such as mitigation costs, damage costs and discount rates (see WGIII, Chapter 3; Deuber et al., 2013).

For example, if mitigation policy is set within a *cost-effectiveness* framework with the aim of making the least cost mix of emissions reductions across components to meet a global temperature target, the 'price ratio' (Manne and Richels, 2001), also called the Global Cost Potential (GCP) (Tol et al., 2012), most directly addresses the goal. The choice of target is a policy decision; metric values can then be calculated based on an agreed upon target. Similarly, if policy is set within a *cost–benefit* framework, the metric that directly addresses the policy goal is the ratio of the marginal damages from the emission of a gas (i.e., the damage costs to society resulting from an incremental increase in emissions) relative to the marginal damages of an emission of CO_2, known as the Global Damage Potential (GDP) (Kandlikar, 1995). Both types of metrics are typically determined within an integrated climate–economy model, since they are affected both by the response of the climate system as well as by economic factors.

If other indexes, such as the GWP, are used instead of an economic cost-minimizing index, costs to society will increase. Cost implications at the project or country level could be substantial under some circumstances (Godal and Fuglestvedt, 2002; Shine, 2009; Reisinger et al., 2013). However, under idealized conditions of full participation in mitigation policy, the increase is relatively small at the global level, particularly when compared to the cost savings resulting from a multi- (as opposed to single-) gas mitigation strategy even when based on an imperfect metric (O'Neill, 2003; Aaheim et al., 2006; Johansson et al., 2006; Johansson, 2012; Reisinger et al., 2013; Smith et al., 2013).

Purely physical metrics continue to be used in many contexts due at least in part to the added uncertainties in mitigation and damage costs, and therefore in the values of economic metrics (Boucher, 2012). Efforts have been made to view purely physical metrics such as GWPs and GTPs as approximations of economic indexes. GTPs, for example, can be interpreted as an approximation of a Global Cost Potential designed for use in a cost-effectiveness setting (Shine et al., 2007; Tol et al., 2012). Quantitative values for time-dependent GTPs reproduce in broad terms several features of the Global Cost Potential such as the rising value of metrics for short-lived gases as a climate policy target is approached (Tanaka et al., 2013). Figure 8.30 shows how contributions of N_2O, CH_4 and BC to warming in the target year changes over time. The contributions are given relative to CO_2 and show the effects of emission occurring at various times. Similarly, GWPs can be interpreted as approximations of the Global Damage Potential designed for a cost–benefit framework (Tol et al., 2012). These interpretations of the GTP and GWP imply that using even a purely physical metric in an economic policy context involves an implicit economic valuation.

In both cases, a number of simplifying assumptions must be made for these approximations to hold (Tol et al., 2012). For example, in the case of the GWP, the influence of emissions on RF, and therefore implicitly on costs to society, beyond the time horizon is not taken into account, and there are substantial numerical differences between GWP and GDP values (Marten and Newbold, 2012). In the case of the GTP, the influence of emissions on temperature change (and costs) is

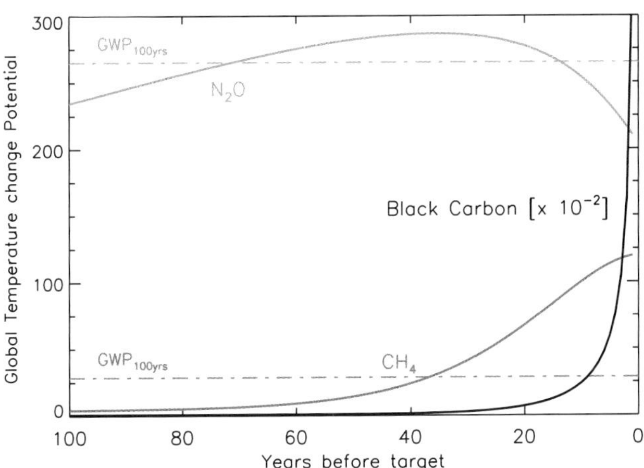

Figure 8.30 | Global Temperature change Potential (GTP(*t*)) for CH_4, nitrous oxide and BC for each year from year of emission to the time at which the temperature change target is reached. The (time-invariant) GWP_{100} is also shown for N_2O and CH_4 for comparison.

8

included only at the time the target is reached, but not before nor after. Other metrics have been developed to more closely approximate GCPs or GDPs. The Cost-Effective Temperature Potential (CETP) reproduces values of the GCP more closely than does the GTP (Johansson, 2012). It is similar to the GTP but accounts for post-target temperature effects based on an assumption about how to value costs beyond the time the target is reached. Metrics have also been proposed that take into account forcing or temperature effects that result from emissions trajectories over broad time spans, and that behave similarly to GCP and GTP (Tanaka et al., 2009; Manning and Reisinger, 2011) or to GWP (e.g., O'Neill, 2000; Peters et al., 2011a; Gillett and Matthews, 2010; Azar and Johansson, 2012).

8.7.1.6 Synthesis

In the application and evaluation of metrics, it is important to distinguish between two main sources of variation in metric values. While scientific choices of input data have to be made, there are also choices involving value judgements. For some metrics such choices are not always explicit and transparent. The choice of metric type and time horizon will for many components have a much larger effect than improved estimates of input parameters and can have strong effects on perceived impacts of emissions and abatement strategies.

In addition to progress in understanding of GWP, new concepts have been introduced or further explored since AR4. Time variant metrics introduce more dynamical views of the temporal contributions that accounts for the proximity to a prescribed target (in contrast to the traditional static GWP). Time variant metrics can be presented in a format that makes changing metric values over time predictable.

As metrics use parameters further down the cause effect chain the metrics become in general more policy relevant, but at the same time the uncertainties increase. Furthermore, metrics that account for regional variations in sensitivity to emissions or regional variation in response could give a very different emphasis to various emissions. Many species, especially NTCFs, produce distinctly regionally heterogeneous RF and climate response patterns. These aspects are not accounted for in the commonly used global scale metrics.

The GWPs and GTPs have had inconsistent treatment of indirect effects and feedbacks. The GWPs reported in AR4 include climate–carbon feedbacks for the reference gas CO_2 but not for the non-CO_2 gases. Such feedbacks may have significant impacts on metrics and should be treated consistently. More studies are needed to assess the importance of consistent treatment of indirect effects/feedbacks in metrics.

The weighting of effects over time—choice of time horizon in the case of GWP and GTP—is value based. Discounting is an alternative, which also includes value judgements and is equally controversial. The weighting used in the GWP is a weight equal to one up to the time horizon and zero thereafter, which is not in line with common approaches for evaluation of future effects in economics (e.g., as in WGIII, Chapter 3). Adoption of a fixed horizon of e.g., 20, 100 or 500 years will inevitably put no weight on the long-term effect of CO_2 beyond the time horizon (Figure 8.28 and Box 6.1). While GWP integrates the effects up to a chosen time horizon the GTP gives the temperature just for one chosen year with no weight on years before or after. The most appropriate metric depends on the particular application and which aspect of climate change is considered relevant in a given context. The GWP is not directly related to a temperature limit such as the 2°C target (Manne and Richels, 2001; Shine et al., 2007; Manning and Reisinger, 2011; Smith et al., 2012; Tol et al., 2012; Tanaka et al., 2013), whereas some economic metrics and physical end-point metrics like the GTP may be more suitable for this purpose.

To provide metrics that can be useful to the users and policymakers a more effective dialog and discussion on three topics is needed: (1) which applications particular metrics are meant to serve; (2) how comprehensive metrics need to be in terms of indirect effects and feedbacks, and economic dimensions; and—related to this (3) how important it is to have simple and transparent metrics (given by analytical formulations) versus more complex model-based and thus model-dependent metrics. These issues are also important to consider in a wider disciplinary context (e.g., across the IPCC Working Groups). Finally, it is important to be aware that all metric choices, even 'traditional' or 'widely used' metrics, contain implicit value judgements as well as large uncertainties.

8.7.2 Application of Metrics

8.7.2.1 Metrics for Carbon Dioxide, Methane, Nitrous Oxide, Halocarbons and Related Compounds

Updated (A)GWP and (A)GTP values for CO_2, CH_4, N_2O, CFCs, HCFCs, bromofluorocarbons, halons, HFCs, PFCs, SF_6, NF_3, and related halogen-containing compounds are given for some illustrative and tentative time horizons in Tables 8.7, 8.A.1 and Supplementary Material Table 8.SM.16. The input data and methods for calculations of GWPs and GTPs are documented in the Supplementary Material Section 8.SM.13. Indirect GWPs that account for the RF caused by depletion of stratospheric ozone (consistent with Section 8.3.3) are given for selected gases in Table 8.A.2.

The *confidence* in the ability to provide useful metrics at time scales of several centuries is *very low* due to nonlinear effects, large uncertainties for multi-century processes and strong assumptions of constant background conditions. Thus, we do not give metric values for longer time scales than 100 years (see discussion in Supplementary Material Section 8.SM.11). However, these time scales are important to consider for gases such as CO_2, SF_6 and PFCs. For CO_2, as much as 20 to 40% of the initial increase in concentration remains after 500 years. For PFC-14, 99% of an emission is still in the atmosphere after 500 years. The effects of emissions on these time scales are discussed in Chapter 12.

The GWP values have changed from previous assessments due to new estimates of lifetimes, impulse response functions and radiative efficiencies. These are updated due to improved knowledge and/or changed background levels. Because CO_2 is used as reference, any changes for this gas will affect all metric values via AGWP changes. Figure 8.31 shows how the values of radiative efficiency (RE), integrated impulse response function (IRF) and consequentially AGWP for CO_2 have changed from earlier assessments relative to AR5 values. The net effect of change in RE and IRF is an increase of approximately 1% and

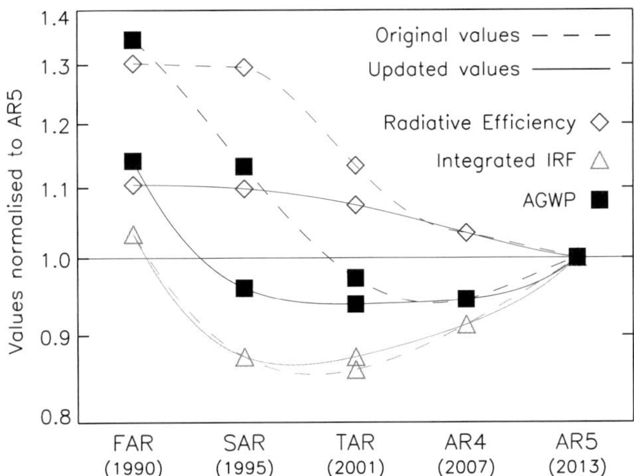

Figure 8.31 | Changes in the radiative efficiency (RE), integrated impulse response function (IRF) and Absolute Global Warming Potential (AGWP) for CO_2 for 100 years from earlier IPCC Assessment Reports normalized relative to the values given in AR5. The 'original' values are calculated based on the methods explained or value reported in each IPCC Assessment Report. The 'updated' values are calculated based on the methods used in AR5, but the input values from each Assessment Report. The difference is primarily in the formula for the RE, which was updated in TAR. The different integrated IRF in TAR relates to a different parameterisation of the same IRF (WMO, 1999). Changes represent both changes in scientific understanding and a changing background atmospheric CO_2 concentration (note that *y*-axis starts from 0.8). The lines connecting individual points are meant as a visual guide and not to represent the values between different Assessment Reports.

6% from AR4 to AR5 in AGWP for CO_2 for 20 and 100 years, respectively (see Supplementary Material Section 8.SM.12). These increases in the AGWP of the reference gas lead to corresponding decreases in the GWPs for all non-CO_2 gases. Continued increases in the atmospheric levels of CO_2 will lead to further changes in GWPs (and GTPs) in the future.

To understand the factors contributing to changes relative to AR4, comparisons are made here using the AR5 values that include climate–carbon feedbacks for CO_2 only. Relative to AR4 the CH_4 AGWP has changed due to changes in perturbation lifetime, a minor change in RE due to an increase in background concentration, and changes in the estimates of indirect effects. The indirect effects on O_3 and stratospheric H_2O are accounted for by increasing the effect of CH_4 by 50% and 15%, respectively (see Supplementary Material Table 8.SM.12). The ozone effect has doubled since AR4 taking into account more recent studies as detailed in Sections 8.3.3 and 8.5.1. Together with the changes in AGWP for CO_2 the net effect is increased GWP values of CH_4.

The GWPs for N_2O are lower here compared to AR4. A longer perturbation lifetime is used in AR5, while the radiative efficiency is lower due to increased abundances of CH_4 and N_2O. In addition, the reduction in CH_4 via stratospheric O_3, UV fluxes and OH levels due to increased N_2O abundance is included in GWPs and GTP. Owing to large uncertainties related to altitude of changes, we do not include the RF from stratospheric ozone changes as an indirect effect of N_2O.

Lifetimes for most of the halocarbons are taken from WMO (2011) and many of these have changed from AR4. The lifetimes of CFC-114, CFC-115 and HCF-161 are reduced by approximately 40%, while HFC-152 is reduced by one third. Among the hydrofluoroethers (HFEs) there are also several large changes in lifetimes. In addition, substantial updates of radiative efficiencies are made for several important gases; CFC-11, CFC-115, HCFC-124, HCFC-225cb, HFC-143a, HFC-245fa, CCl_4, $CHCl_3$, and SF_6. The radiative efficiency for carbon tetrachloride (CCl_4) is higher now and the GWP_{100} has increased by almost 25% from AR4. Uncertainties in metric values are given in Section 8.7.1.4. See also Supplementary Material Section 8.SM.12 and footnote to Table 8.A.1. As can be seen from Table 8.A.2, some ODS have strong indirect effects through stratospheric ozone forcing, which for some of the gases reduce their net GWP_{100} values substantially (and for the halons, to large negative values). Note that, consistent with Section 8.3.3, the uncertainties are large; ±100% for this indirect effect.

When climate-carbon feedbacks are included for both the non-CO_2 and reference gases, all metric values increase relative to the methodology used in AR4, sometimes greatly (Table 8.7, Supplementary Material Table 8.SM.16). Though the uncertainties range for these metric values is greater, as uncertainties in climate-carbon feedbacks are substantial, these calculations provide a more consistent methodology.

8.7.2.2 Metrics for Near-Term Climate Forcers

The GWP concept was initially used for the WMGHGs, but later for NTCFs as well. There are, however, substantial challenges related to calculations of GWP (and GTP) values for these components, which is reflected in the large ranges of values in the literature. Below we present and assess the current status of knowledge and quantification of metrics for various NTCFs.

8.7.2.2.1 Nitrogen oxides

Metric values for NO_X usually include the short-lived ozone effect, CH_4 changes and the CH_4-controlled O_3 response. NO_X also causes RF through nitrate formation, and via CH_4 it affects stratospheric H_2O and through ozone it influences CO_2. In addition, NO_x affects CO_2 through nitrogen deposition (fertilization effect). Due to high reactivity and the many nonlinear chemical interactions operating on different time scales, as well as heterogeneous emission patterns, calculation of *net* climate effects of NO_X is difficult. The net effect is a balance of large opposing effects with very different temporal behaviours. There is also a large spread in values among the regions due to variations in chemical and physical characteristics of the atmosphere.

As shown in Table 8.A.3 the GTP and GWP values are very different. This is due to the fundamentally different nature of these two metrics (see Figure 8.28) and the way they capture the temporal behaviour of responses to NO_x emissions. Time variation of GTP for NO_X is complex, which is not directly seen by the somewhat arbitrary choices of time horizon, and the net GTP is a fine balance between the contributing terms. The general pattern for NO_X is that the short-lived ozone forcing is always positive, while the CH_4-induced ozone forcing and CH_4 forcing are always negative (see Section 8.5.1). Nitrate aerosols from NO_x emission are not included in Table 8.A.3. For the GTP, all estimates for NO_X from surface sources give a negative net effect. As discussed in Section 8.7.1.4 Collins et al. (2010) and Shindell et al. (2009) implemented further indirect effects, but these are not included in Table

8.A.3 due to large uncertainties. The metric estimates for NO_X reflect the level of knowledge, but they also depend on experimental design, treatment of transport processes, and modelling of background levels. The multi-model study by Fry et al. (2012) shows the gaseous chemistry response to NO_X is relatively robust for European emissions, but that the uncertainty is so large that for some regions of emissions it is not possible to conclude whether NO_X causes cooling or warming.

8.7.2.2.2 Carbon monoxide and volatile organic compounds

Emissions of carbon monoxide (CO) and volatile organic compounds (VOCs) lead to production of ozone on short time scales. By affecting OH and thereby the levels of CH_4 they also initiate a positive long-term ozone effect. With its lifetime of 2 to 3 months, the effect of CO emissions is less dependent on location than is the case for NO_X (see Table 8.A.4). There is also less variation across models. However, Collins et al. (2010) found that inclusion of vegetation effects of O_3 increased the GTP values for CO by 20 to 50%. By including aerosol responses Shindell et al. (2009) found an increase in GWP_{100} by a factor of ~2.5. CO of fossil origin will also have a forcing effect by contributing to CO_2 levels. This effect adds 1.4 to 1.6 to the GWP_{100} for CO (Daniel and Solomon, 1998; Derwent et al., 2001). (The vegetation and aerosol effects are not included in the numbers in Table 8.A.4.)

VOC is not a well-defined group of hydrocarbons. This group of gases with different lifetimes is treated differently across models by lumping or using representative key species. However, the spread in metric values in Table 8.A.5 is moderate across regions, with highest values for emissions in South Asia (of the four regions studied). The effects via ozone and CH_4 cause warming, and the additional effects via interactions with aerosols and via the O_3–CO_2 link increase the warming effect further. Thus, the net effects of CO and VOC are less uncertain than for NO_X for which the net is a residual between larger terms of opposite sign. However, the formation of SOAs is usually not included in metric calculations for VOC, which introduces a cooling effect and increased uncertainty.

8.7.2.2.3 Black carbon and organic carbon

Most of the metric values for BC in the literature include the aerosol–radiation interaction and the snow/ice albedo effect of BC, though whether external or internal mixing is used varies between the studies. Bond et al. (2011) calculate GWPs and find that when the albedo effect is included the values increase by 5 to 15%. Studies have shown, however, that the climate response per unit forcing to this mechanism is stronger than for WMGHG (see Section 7.5).

Bond et al. (2013) assessed the current understanding of BC effects and calculated GWP and GTP for BC that includes aerosol–radiation interaction, aerosol–cloud interactions and albedo. As shown in Table 8.A.6 the uncertainties are wide for both metrics (for 90% uncertainty range) reflecting the current challenges related to understanding and quantifying the various effects (see Sections 7.5, 8.3.4 and 8.5.1). Their aerosol–radiation interaction effect is about 65% of the total effect while the albedo effect is approximately 20% of the aerosol–radiation interaction effect. Based on two studies (Rypdal et al., 2009; Bond et al., 2011), the GWP and GTP metrics were found to vary with the region where BC is emitted by about ±30% . For larger regions of emissions, Collins et al. (2013) calculated GWPs and GTPs for the direct effect of BC and found somewhat lower variations among the regions.

Several studies have focused on the effects of emissions of BC and OC from different regions (Bauer et al., 2007; Koch et al., 2007; Naik et al., 2007; Reddy and Boucher, 2007; Rypdal et al., 2009). However, examination of results from these models (Fuglestvedt et al., 2010) reveals that there is not a robust relationship between the region of emission and the metric value — hence, regions that yield the highest metric value in one study, do not, in general, do so in the other studies.

The metric values for OC are quite consistent across studies, but fewer studies are available (see Table 8.A.6). A brief overview of metric values for other components is given in the Supplementary Material Section 8.SM.14.

8.7.2.2.4 Summary of status of metrics for near-term climate forcers

The metrics provide a format for comparing the magnitudes of the various emissions as well as for comparing effects of emissions from different regions. They can also be used for comparing results from different studies. Much of the spread in results is due to differences in experimental design and how the models treat physical and chemical processes. Unlike most of the WMGHGs, many of the NTCFs are tightly coupled to the hydrologic cycle and atmospheric chemistry, leading to a much larger spread in results as these are highly complex processes that are difficult to validate on the requisite small spatial and short temporal scales. The confidence level is lower for many of the NTCF compared to WMGHG and much lower where aerosol–cloud interactions are important (see Section 8.5.1). There are particular difficulties for NO_X, because the net impact is a small residual of opposing effects with quite different spatial distributions and temporal behaviour. Although climate–carbon feedbacks for non-CO_2 emissions have not been included in the NTCF metrics (other than CH_4) presented here, they can greatly increase those values (Collins et al., 2013) and likely provide more realistic results.

8.7.2.3 Impact by Emitted Component

We now use the metrics evaluated here to estimate climate impacts of various components (in a forward looking perspective). Figure 8.32 shows global anthropogenic emissions of some selected components weighted by the GWP and GTP. The time horizons are chosen as examples and illustrate how the perceived impacts of components—relative to the impact of the reference gas—vary strongly as function of impact parameter (integrated RF in GWP or end-point temperature in GTP) and with time horizon.

We may also calculate the temporal development of the temperature responses to pulse or sustained emissions using the AGTP metric. Figure 8.33 shows that for a one-year pulse the impacts of NTCF decay quickly owing to their atmospheric adjustment times even if effects are prolonged due to climate response time (in the case of constant emissions the effects reach approximately constant levels since the emissions are replenished each year, except for CO_2, which has a fraction

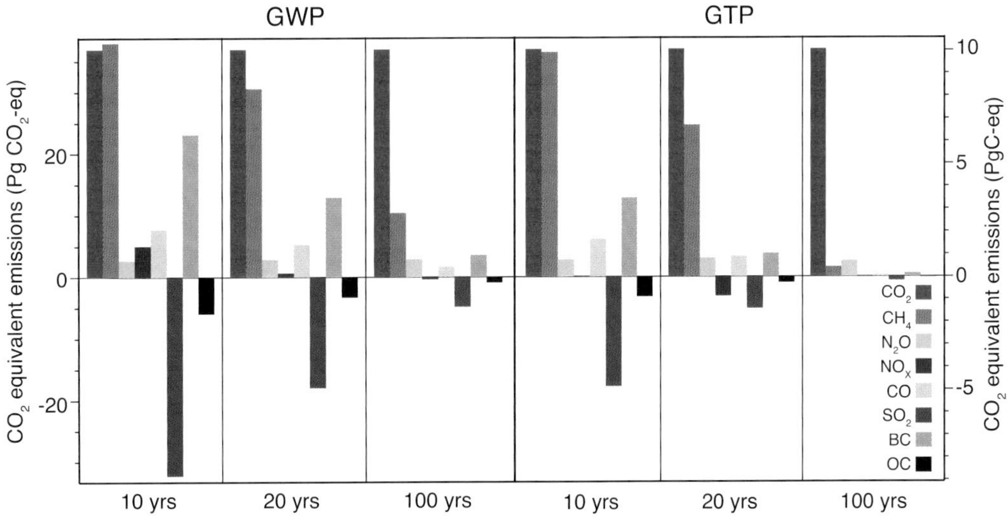

Figure 8.32 | Global anthropogenic emissions weighted by GWP and GTP for chosen time horizons (aerosol–cloud interactions are not included). Emission data for 2008 are taken from the EDGAR database. For BC and OC emissions for 2005 are from Shindell et al. (2012a). The units are 'CO_2 equivalents' which reflects equivalence only in the impact parameter of the chosen metric (integrated RF over the chosen time horizon for GWP; temperature change at the chosen point in time for GTP), given as $Pg(CO_2)_{eq}$ (left axis) and given as PgC_{eq} (right axis). There are large uncertainties related to the metric values and consequentially also to the calculated CO_2 equivalents (see text).

remaining in the atmosphere on time scales of centuries). Figure 8.33 also shows how some components have strong short-lived effects of both signs while CO_2 has a weaker initial effect but one that persists to create a long-lived warming effect. Note that there are large uncertainties related to the metric values (as discussed in Section 8.7.1.4); especially for the NTCFs.

These examples show that the outcome of comparisons of effects of emissions depends strongly on choice of time horizon and metric type. Such end-user choices will have a strong influence on the calculated contributions from NTCFs versus WMGHGs or non-CO_2 versus CO_2 emissions. Thus, each specific analysis should use a design chosen in light of the context and questions being asked.

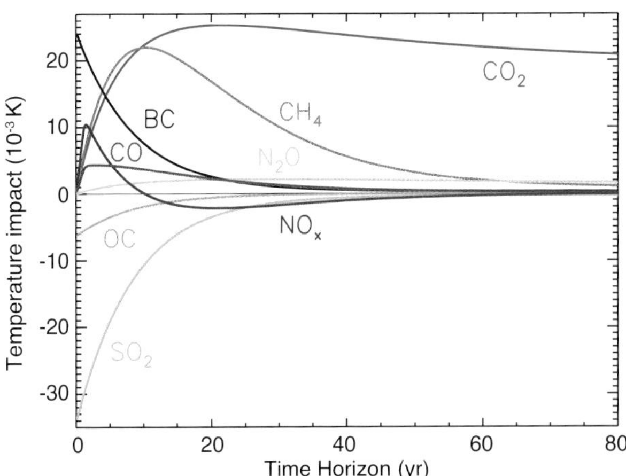

Figure 8.33 | Temperature response by component for total anthropogenic emissions for a 1-year pulse. Emission data for 2008 are taken from the EDGAR database and for BC and OC for 2005 from Shindell et al. (2012a). There are large uncertainties related to the AGTP values and consequentially also to the calculated temperature responses (see text).

8.7.2.4 Metrics and Impacts by Sector

While the emissions of WMGHGs vary strongly between sectors, the climate impacts of these gases are independent of sector. The latter is not the case for chemically active and short-lived components, due to the dependence of their impact on the emission location. Since most sectors have multiple co-emissions, and for NTCFs some of these are warming while others are cooling, the net impact of a given sector requires explicit calculations. Since AR4, there has been significant progress in the understanding and quantification of climate impacts of NTCFs from sectors such as transportation, power production and biomass burning (Berntsen and Fuglestvedt, 2008; Skeie et al., 2009; Stevenson and Derwent, 2009; Lee et al., 2010; Unger et al., 2010; Dahlmann et al., 2011). Supplementary Material Table 8.SM.18 gives an overview of recent published metric values for various components by sector.

The impact from sectors depends on choice of metric, time horizon, pulse versus sustained emissions and forward versus backward looking perspective (see Section 8.7.1 and Box 8.4). Unger et al. (2010) calculated RF for a set of components emitted from each sector. RF at chosen points in time (20 and 100 years) for *sustained* emissions was used by Unger et al. (2010) as the metric for comparison. This is comparable to using integrated RF up to the chosen times for *pulse* emissions (as in GWPs). Such studies are relevant for policymaking that focuses on regulating the *total activity* of a sector or for understanding the contribution from a sector to climate change. On the other hand, the fixed mix of emissions makes it less general and relevant for emission scenarios. Alternatively, one may adopt a component-by-component view which is relevant for policies directed towards specific components (or sets of components, as controlling an individual pollutant in isolation is usually not practical). But this view will not capture interactions and non-linearities within the suite of components emitted by most sectors. The effects of specific emission control technologies or policies or projected societal changes on the mix of emissions is probably the most relevant type of analysis, but there are an enormous number of possible actions and regional details that could be investigated. Henze et al. (2012) demonstrate a method for providing highly spatially resolved

8

estimates of forcing per component, and caution that RF aggregated over regions or sectors may not represent the impacts of emissions changes on finer scales.

Metrics for individual land-based sectors are often similar to the global mean metric values (Shindell et al., 2008). In contrast, metrics for emissions from aviation and shipping usually show large differences from global mean metric values (Table 8.A.3 versus Table 8.SM.18). Though there can sometimes be substantial variation in the impact of land-based sectors across regions, and for a particular region even from one sector to another, variability between different land-based sources is generally smaller than between land, sea and air emissions.

NO_x from aviation is one example where the metric type is especially important. GWP_{20} values are positive due to the strong response of short-lived ozone. Reported GWP_{100} and GTP_{100} values are of either sign, however, due to the differences in balance between the individual effects modelled. Even if the models agree on the net effect of NO_X, the individual contributions can differ significantly, with large uncertainties stemming from the relative magnitudes of the CH_4 and O_3 responses (Myhre et al., 2011) and the background tropospheric concentrations of NO_X (Holmes et al., 2011; Stevenson and Derwent, 2009). Köhler et al. (2013), find strong regional sensitivity of ozone and CH_4 to NO_X particularly at cruise altitude. Generally, they find the strongest effects at low latitudes. For the aviation sector contrails and contrail induced cirrus are also important. Based on detailed studies in the literature, Fuglestvedt et al. (2010) produced GWP and GTP for contrails, water vapor and contrail-induced cirrus.

The GWP and GTPs for NO_X from shipping are strongly negative for all time horizons. The strong positive effect via O_3 due to the low-NO_X environment into which ships generally emit NO_X is outweighed by the stronger effect on CH_4 destruction due to the relatively lower latitudes of these emissions compared to land-based sources.

In addition to having large emissions of NO_X the shipping sector has large emission of SO_2. The direct GWP_{100} for shipping ranges from –11 to –43 (see Supplementary Material Table 8.SM.18). Lauer et al. (2007) reported detailed calculations of the indirect forcing specifically for this sector and found a wide spread of values depending on the emission inventory. Righi et al. (2011) and Peters et al. (2012) calculate indirect effects that are 30 to 50% lower than the indirect forcing reported by Lauer et al. (2007). The values from Shindell and Faluvegi (2010) for SO_2 from power generation are similar to those for shipping.

Although the various land transport sectors often are treated as one aggregate (e.g., road transport) there are important subdivisions. For instance, Bond et al. (2013) points out that among the BC-rich sectors they examined, diesel vehicles have the most clearly positive net impact on forcing. Studies delving even further have shown substantial differences between trucks and cars, gasoline and diesel vehicles, and low-sulphur versus high-sulphur fuels. Similarly, for power production there are important differences depending on fuel type (coal, oil, gas; e.g., Shindell and Faluvegi, 2010).

In the assessment of climate impacts of current emissions by sectors we give examples and apply a forward-looking perspective on effects in terms of temperature change. The AGTP concept can be used to study the effects of the various components for chosen time horizons. A single year's worth of current global emissions from the energy and industrial sectors have the largest contributions to warming after 100 years (see Figure 8.34a). Household fossil fuel and biofuel, biomass burning and on-road transportation are also relatively large contributors to warming over 100-year time scales. Those same sectors, along with sectors that emit large amounts of CH_4 (animal husbandry, waste/landfills and agriculture), are most important over shorter time horizons (about 20 years; see Figure 8.34b).

Analysing climate change impacts by using the net effect of particular activities or sectors may—compared to other perspectives—provide more insight into how societal actions influence climate. Owing to large variations in mix of short- and long-lived components, as well as cooling and warming effects, the results will also in these cases depend strongly on choice of time horizon and climate impact parameter. Improved understanding of aerosol–cloud interactions, and how those are attributed to individual components is clearly necessary to refine estimates of sectoral or emitted component impacts.

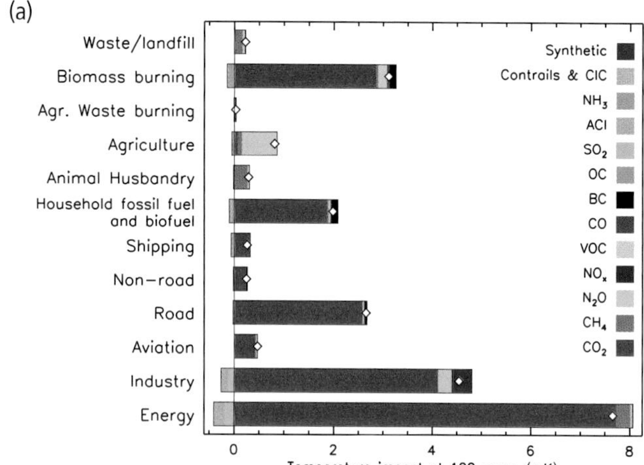

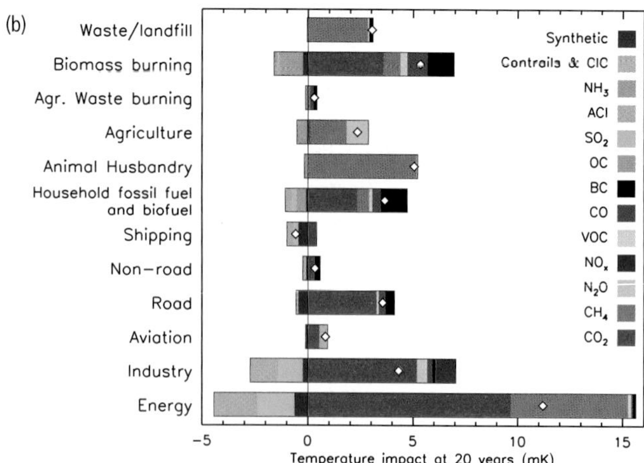

Figure 8.34 | Net global mean temperature change by source sector after (a) 100 and (b) 20 years (for 1-year pulse emissions). Emission data for 2008 are taken from the EDGAR database. For BC and OC anthropogenic emissions are from Shindell et al. (2012a) and biomass burning emissions are from Lamarque et al. (2010), see Supplementary Material Section 8.SM.17. There are large uncertainties related to the AGTP values and consequentially also to the calculated temperature responses (see text).

References

Aaheim, A., J. Fuglestvedt, and O. Godal, 2006: Costs savings of a flexible multi-gas climate policy. *Energy J.* (**Special Issue No. 3**), 485–501.

Abreu, J., J. Beer, F. Steinhilber, S. Tobias, and N. Weiss, 2008: For how long will the current grand maximum of solar activity persist? *Geophys. Res. Lett.*, **35**, L20109.

Ackerley, D., B. B. B. Booth, S. H. E. Knight, E. J. Highwood, D. J. Frame, M. R. Allen, and D. P. Rowell, 2011: Sensitivity of twentieth-century Sahel rainfall to sulfate aerosol and CO_2 forcing. *J. Clim.*, **24**, 4999–5014.

Allan, W., H. Struthers, and D. C. Lowe, 2007: Methane carbon isotope effects caused by atomic chlorine in the marine boundary layer: Global model results compared with Southern Hemisphere measurements. *J. Geophys. Res. Atmos.*, **112**, D04306.

Ammann, C. M., and P. Naveau, 2003: Statistical analysis of tropical explosive volcanism occurrences over the last 6 centuries. *Geophys. Res. Lett.*, **30**, 1210.

Ammann, C. M., and P. Naveau, 2010: A statistical volcanic forcing scenario generator for climate simulations. *J. Geophys. Res. Atmos.*, **115**, D05107.

Anchukaitis, K. J., B. M. Buckley, E. R. Cook, B. I. Cook, R. D. D'Arrigo, and C. M. Ammann, 2010: Influence of volcanic eruptions on the climate of the Asian monsoon region. *Geophys. Res. Lett.*, **37**, L22703.

Andersen, M., D. Blake, F. Rowland, M. Hurley, and T. Wallington, 2009: Atmospheric chemistry of sulfuryl fluoride: Reaction with OH radicals, Cl atoms and O_3, atmospheric lifetime, IR spectrum, and global warming potential. *Environ. Sci. Technol.*, **43**, 1067–1070.

Andersen, M., V. Andersen, O. Nielsen, S. Sander, and T. Wallington, 2010: Atmospheric chemistry of $HCF_2O(CF_2CF_2O)(x)CF_2H$ (x=2–4): Kinetics and mechanisms of the chlorine-atom-initiated oxidation. *Chemphyschem*, **11**, 4035–4041.

Andrews, T., and P. M. Forster, 2008: CO_2 forcing induces semi-direct effects with consequences for climate feedback interpretations. *Geophys. Res. Lett.*, **35**, L04802.

Andrews, T., M. Doutriaux-Boucher, O. Boucher, and P. M. Forster, 2011: A regional and global analysis of carbon dioxide physiological forcing and its impact on climate. *Clim. Dyn.*, **36**, 783–792.

Andrews, T., J. Gregory, M. Webb, and K. Taylor, 2012a: Forcing, feedbacks and climate sensitivity in CMIP5 coupled atmosphere-ocean climate models. *Geophys. Res. Lett.*, **39**, L09712.

Andrews, T., P. Forster, O. Boucher, N. Bellouin, and A. Jones, 2010: Precipitation, radiative forcing and global temperature change. *Geophys. Res. Lett.*, **37**, doi:10.1029/2010GL043991, L14701.

Andrews, T., M. Ringer, M. Doutriaux-Boucher, M. Webb, and W. Collins, 2012b: Sensitivity of an Earth system climate model to idealized radiative forcing. *Geophys. Res. Lett.*, **39**, L10702.

Antuña, J. C., A. Robock, G. Stenchikov, J. Zhou, C. David, J. Barnes, and L. Thomason, 2003: Spatial and temporal variability of the stratospheric aerosol cloud produced by the 1991 Mount Pinatubo eruption. *J. Geophys. Res. Atmos.*, **108**, 4624.

Archibald, A. T., M. E. Jenkin, and D. E. Shallcross, 2010: An isoprene mechanism intercomparison. *Atmos. Environ.*, **44**, 5356–5364.

Archibald, A. T., et al., 2011: Impacts of HO(x) regeneration and recycling in the oxidation of isoprene: Consequences for the composition of past, present and future atmospheres. *Geophys. Res. Lett.*, **38**, L05804.

Arnold, T., et al., 2013: Nitrogen trifluoride global emissions estimated from updated atmospheric measurements, Proc. Natl. Acad. Sci. U.S.A, **110**, 2029-2034.

Arora, V. K., and A. Montenegro, 2011: Small temperature benefits provided by realistic afforestation efforts. *Nature Geosci.*, **4**, 514–518.

Arora, V. K., et al., 2013: Carbon-concentration and carbon-climate feedbacks in CMIP5 Earth system models. *J. Clim.*, **26**, 5289-5314.

Ashmore, M. R., 2005: Assessing the future global impacts of ozone on vegetation. *Plant Cell Environ.*, **28**, 949–964.

Azar, C., and D. J. A. Johansson, 2012: On the relationship between metrics to compare greenhouse gases—the case of IGTP, GWP and SGTP. *Earth Syst. Dynam.*, **3**, 139–147.

Baasandorj, M., A. R. Ravishankara, and J. B. Burkholder, 2011: Atmospheric chemistry of (Z)-CF_3CH=$CHCF_3$: OH radical reaction rate coefficient and global warming potential. *J. Phys. Chem. A*, **115**, 10539–10549.

Baasandorj, M., G. Knight, V. Papadimitriou, R. Talukdar, A. Ravishankara, and J. Burkholder, 2010: Rate coefficients for the gas-phase reaction of the hydroxyl radical with CH_2 = CHF and CH_2 = CF_2. *J. Phys. Chem. A*, **114**, 4619–4633.

Bala, G., K. Caldeira, M. Wickett, T. J. Phillips, D. B. Lobell, C. Delire, and A. Mirin, 2007: Combined climate and carbon-cycle effects of large-scale deforestation. *Proc. Natl. Acad. Sci. U.S.A.*, **104**, 6550–6555.

Baliunas, S., and R. Jastrow, 1990: Evidence for long-term brightness changes of solar-type stars. *Nature*, **348**, 520–523.

Ball, W., Y. Unruh, N. Krivova, S. Solanki, T. Wenzler, D. Mortlock, and A. Jaffe, 2012: Reconstruction of total solar irradiance 1974–2009. *Astron. Astrophys.*, **541**, A27.

Ban-Weiss, G., L. Cao, G. Bala, and K. Caldeira, 2012: Dependence of climate forcing and response on the altitude of black carbon aerosols. *Clim. Dyn.*, **38**, 897–911.

Barnes, C. A., and D. P. Roy, 2008: Radiative forcing over the conterminous United States due to contemporary land cover land use albedo change. *Geophys. Res. Lett.*, **35**, L09706.

Bathiany, S., M. Claussen, V. Brovkin, T. Raddatz, and V. Gayler, 2010: Combined biogeophysical and biogeochemical effects of large-scale forest cover changes in the MPI earth system model. *Biogeosciences*, **7**, 1383–1399.

Bauer, S., D. Koch, N. Unger, S. Metzger, D. Shindell, and D. Streets, 2007: Nitrate aerosols today and in 2030: A global simulation including aerosols and tropospheric ozone. *Atmos. Chem. Phys.*, **7**, 5043–5059.

Bekki, S., J. A. Pyle, W. Zhong, R. Toumi, J. D. Haigh, and D. M. Pyle, 1996: The role of microphysical and chemical processes in prolonging the climate forcing of the Toba eruption. *Geophys. Res. Lett.*, **23**, 2669–2672.

Bellouin, N., J. Rae, A. Jones, C. Johnson, J. Haywood, and O. Boucher, 2011: Aerosol forcing in the Climate Model Intercomparison Project (CMIP5) simulations by HadGEM2–ES and the role of ammonium nitrate. *J. Geophys. Res. Atmos.*, **116**, D20206.

Bernier, P. Y., R. L. Desjardins, Y. Karimi-Zindashty, D. Worth, A. Beaudoin, Y. Luo, and S. Wang, 2011: Boreal lichen woodlands: A possible negative feedback to climate change in eastern North America. *Agr. Forest Meteorol.*, **151**, 521–528.

Berntsen, T., and J. Fuglestvedt, 2008: Global temperature responses to current emissions from the transport sectors. *Proc. Natl. Acad. Sci. U.S.A.*, **105**, 19154–19159.

Berntsen, T. K., et al., 1997: Effects of anthropogenic emissions on tropospheric ozone and its radiative forcing. *J. Geophys. Res. Atmos.*, **102**, 28101–28126.

Betts, R., 2000: Offset of the potential carbon sink from boreal forestation by decreases in surface albedo. *Nature*, **408**, 187–190.

Betts, R. A., P. D. Falloon, K. K. Goldewijk, and N. Ramankutty, 2007: Biogeophysical effects of land use on climate: Model simulations of radiative forcing and large-scale temperature change. *Agr. Forest Meteorol.*, **142**, 216–233.

Biasutti, M., and A. Giannini, 2006: Robust Sahel drying in response to late 20th century forcings. *Geophys. Res. Lett.*, **33**, L11706.

Blowers, P., K. F. Tetrault, and Y. Trujillo-Morehead, 2008: Global warming potential predictions for hydrofluoroethers with two carbon atoms. *Theor. Chem. Acc.*, **119**, 369–381.

Blowers, P., D. Moline, K. Tetrault, R. Wheeler, and S. Tuchawena, 2007: Prediction of radiative forcing values for hydrofluoroethers using density functional theory methods. *J. Geophys. Res. Atmos.*, **112**, D15108.

Boer, G. J., and B. Yu, 2003: Climate sensitivity and response. *Clim. Dyn.*, **20**, 415–429.

Bollasina, M. A., Y. Ming, and V. Ramaswamy, 2011: Anthropogenic aerosols and the weakening of the South Asian summer monsoon. *Science*, **334**, 502–505.

Bond, T., C. Zarzycki, M. Flanner, and D. Koch, 2011: Quantifying immediate radiative forcing by black carbon and organic matter with the Specific Forcing Pulse. *Atmos. Chem. Phys.*, **11**, 1505–1525.

Bond, T. C., et al., 2007: Historical emissions of black and organic carbon aerosol from energy-related combustion, 1850–2000. *Global Biogeochem. Cycles*, **21**, Gb2018.

Bond, T. C., et al., 2013: Bounding the role of black carbon in the climate system: A scientific assessment. *J. Geophys. Res. Atmos.*, **118**, doi:10.1002/jgrd.50171, 5380-5552.

Bonfils, C., and D. Lobell, 2007: Empirical evidence for a recent slowdown in irrigation-induced cooling. *Proc. Natl. Acad. Sci. U.S.A.*, **104**, 13582–13587.

Bonfils, C. J. W., T. J. Phillips, D. M. Lawrence, P. Cameron-Smith, W. J. Riley, and Z. M. Subin, 2012: On the influence of shrub height and expansion on northern high latitude climate. *Environ. Res. Lett.*, **7**, 015503.

Booth, B., N. Dunstone, P. Halloran, T. Andrews, and N. Bellouin, 2012: Aerosols implicated as a prime driver of twentieth-century North Atlantic climate variability. *Nature*, **485**, 534–534.

Boucher, O., 2012: Comparison of physically- and economically-based CO_2–equivalences for methane. *Earth Syst. Dyn.*, **3**, 49–61.

Boucher, O., and J. Haywood, 2001: On summing the components of radiative forcing of climate change. *Clim. Dyn.*, **18**, 297–302.

Boucher, O., and M. Reddy, 2008: Climate trade-off between black carbon and carbon dioxide emissions. *Energy Policy*, **36**, 193–200.

Boucher, O., P. Friedlingstein, B. Collins, and K. P. Shine, 2009: The indirect global warming potential and global temperature change potential due to methane oxidation. *Environ. Res. Lett.*, **4**, 044007.

Bourassa, A. E., et al., 2012: Large volcanic aerosol load in the stratosphere linked to Asian monsoon transport. *Science*, **337**, 78–81.

Bourassa, A. E., et al., 2013: Response to Comments on "Large Volcanic Aerosol Load in the Stratosphere Linked to Asian Monsoon Transport". *Science*, **339**, 6120.

Bowman, D., et al., 2009: Fire in the Earth System. *Science*, **324**, 481–484.

Bowman, K. W., et al., 2013: Evaluation of ACCMIP outgoing longwave radiation from tropospheric ozone using TES satellite observations. *Atmos. Chem. Phys.*, **13**, 4057–4072.

Bradford, D., 2001: Global change – Time, money and tradeoffs. *Nature*, **410**, 649–650.

Bravo, I., et al., 2010: Infrared absorption spectra, radiative efficiencies, and global warming potentials of perfluorocarbons: Comparison between experiment and theory. *J. Geophys. Res. Atmos.*, **115**, D24317.

Brovkin, V., et al., 2010: Sensitivity of a coupled climate-carbon cycle model to large volcanic eruptions during the last millennium. *Tellus B*, **62**, 674–681.

Calvin, K., et al., 2012: The role of Asia in mitigating climate change: Results from the Asia modeling exercise. *Energ. Econ.*, **34**, S251–S260.

Campra, P., M. Garcia, Y. Canton, and A. Palacios-Orueta, 2008: Surface temperature cooling trends and negative radiative forcing due to land use change toward greenhouse farming in southeastern Spain. *J. Geophys. Res. Atmos.*, **113**, D18109.

Carlton, A. G., R. W. Pinder, P. V. Bhave, and G. A. Pouliot, 2010: To what extent can biogenic SOA be controlled? *Environ. Sci. Technol.*, **44**, 3376–3380.

Carslaw, K. S., O. Boucher, D. V. Spracklen, G. W. Mann, J. G. L. Rae, S. Woodward, and M. Kulmala, 2010: A review of natural aerosol interactions and feedbacks within the Earth system. *Atmos. Chem. Phys.*, **10**, 1701–1737.

Chang, W. Y., H. Liao, and H. J. Wang, 2009: Climate responses to direct radiative forcing of anthropogenic aerosols, tropospheric ozone, and long-lived greenhouse gases in Eastern China over 1951–2000. *Adv. Atmos. Sci.*, **26**, 748–762.

Chen, W. T., A. Nenes, H. Liao, P. J. Adams, J. L. F. Li, and J. H. Seinfeld, 2010: Global climate response to anthropogenic aerosol indirect effects: Present day and year 2100. *J. Geophys. Res. Atmos.*, **115**, D12207.

Cherubini, F., G. Guest, and A. Strømman, 2012: Application of probablity distributions to the modelleing of biogenic CO_2 fluxes in life cycle assessment. *Global Change Biol.* , **4**, doi:10.1111/j.1757–1707.2011.01156.x, 784-798.

Cherubini, F., G. Peters, T. Berntsen, A. Stromman, and E. Hertwich, 2011: CO_2 emissions from biomass combustion for bioenergy: Atmospheric decay and contribution to global warming. *Global Change Biol. Bioenerg.*, **3**, 413–426.

Chung, C. E., and V. Ramanathan, 2006: Weakening of North Indian SST gradients and the monsoon rainfall in India and the Sahel. *J. Clim.*, **19**, 2036–2045.

Clark, H. L., M. L. Cathala, H. Teyssedre, J. P. Cammas, and V. H. Peuch, 2007: Cross-tropopause fluxes of ozone using assimilation of MOZAIC observations in a global CTM. *Tellus B*, **59**, 39–49.

Clarke, A. D., and K. J. Noone, 1985: Soot in the Arctic Snowpack—A cause for peturbations in radiative-transfer. *Atmos. Environ.*, **19**, 2045–2053.

Claussen, M., V. Brovkin, and A. Ganopolski, 2001: Biogeophysical versus biogeochemical feedbacks of large-scale land cover change. *Geophys. Res. Lett.*, **28**, 1011–1014.

Cofala, J., M. Amann, Z. Klimont, K. Kupiainen, and L. Hoglund-Isaksson, 2007: Scenarios of global anthropogenic emissions of air pollutants and methane until 2030. *Atmos. Environ.*, **41**, 8486–8499.

Collins, W., R. Derwent, C. Johnson, and D. Stevenson, 2002: The oxidation of organic compounds in the troposphere and their global warming potentials. *Clim. Change*, **52**, 453–479.

Collins, W. D., et al., 2006: Radiative forcing by well-mixed greenhouse gases: Estimates from climate models in the Intergovernmental Panel on Climate Change (IPCC) Fourth Assessment Report (AR4). *J. Geophys. Res. Atmos.*, **111**, D14317.

Collins, W. J., S. Sitch, and O. Boucher, 2010: How vegetation impacts affect climate metrics for ozone precursors. *J. Geophys. Res. Atmos.*, **115**, D23308.

Collins, W. J., M. M. Fry, H. Yu, J. S. Fuglestvedt, D. T. Shindell, and J. J. West, 2013: Global and regional temperature-change potentials for near-term climate forcers. *Atmos. Chem. Phys.*, **13**, 2471–2485.

Conley, A. J., J. F. Lamarque, F. Vitt, W. D. Collins, and J. Kiehl, 2013: PORT, a CESM tool for the diagnosis of radiative forcing. *Geosci. Model Dev.*, **6**, 469–476.

Cooper, O. R., et al., 2010: Increasing springtime ozone mixing ratios in the free troposphere over western North America. *Nature*, **463**, 344–348.

Cox, P. M., et al., 2008: Increasing risk of Amazonian drought due to decreasing aerosol pollution. *Nature*, **453**, 212–215.

Crook, J., and P. Forster, 2011: A balance between radiative forcing and climate feedback in the modeled 20th century temperature response. *J. Geophys. Res. Atmos.*, **116**, D17108.

Crowley, T. J., and M. B. Unterman, 2013: Technical details concerning development of a 1200 yr proxy index for global volcanism. *Earth Syst. Sci. Data*, **5**, 187-197.

Crutzen, P., 1973: Discussion of chemistry of some minor constituents in stratosphere and troposhere. *Pure Appl. Geophys.*, **106**, 1385–1399.

Dahlmann, K., V. Grewe, M. Ponater, and S. Matthes, 2011: Quantifying the contributions of individual NO_x sources to the trend in ozone radiative forcing. *Atmos. Environ.*, **45**, 2860–2868.

Daniel, J., and S. Solomon, 1998: On the climate forcing of carbon monoxide. *J. Geophys. Res. Atmos.*, **103**, 13249–13260.

Daniel, J., S. Solomon, and D. Abritton, 1995: On the evaluation of halocarbon radiative forcing and global warming potentials. *J. Geophys. Res. Atmos.*, **100**, 1271–1285.

Daniel, J., E. Fleming, R. Portmann, G. Velders, C. Jackman, and A. Ravishankara, 2010: Options to accelerate ozone recovery: Ozone and climate benefits. *Atmos. Chem. Phys.*, **10**, 7697–7707.

Daniel, J., S. Solomon, T. Sanford, M. McFarland, J. Fuglestvedt, and P. Friedlingstein, 2012: Limitations of single-basket trading: Lessons from the Montreal Protocol for climate policy. *Clim. Change*, **111**, 241–248.

Davin, E., N. de Noblet-Ducoudre, and P. Friedlingstein, 2007: Impact of land cover change on surface climate: Relevance of the radiative forcing concept. *Geophys. Res. Lett.*, **34**, L13702.

Davin, E. L., and N. de Noblet-Ducoudre, 2010: Climatic impact of global-scale deforestation: Radiative versus nonradiative orocesses. *J. Clim.*, **23**, 97–112.

De Cara, S., E. Galko, and P. Jayet, 2008: The global warming potential paradox: Implications for the design of climate policy. In: *Design of Climate Policy* [R. Guesnerie and H. Tulkens (eds.)]. The MIT Press, Cambridge, MA, USA, pp. 359–384.

de la Torre, L., et al., 2006: Solar influence on Northern Annular Mode spatial structure and QBO modulation. *Part. Accel. Space Plasma Phys. Sol. Radiat. Earth. Atmos. Clim.*, **37**, 1635–1639.

de Noblet-Ducoudre, N., et al., 2012: Determining robust impacts of land-use-induced land cover changes on surface climate over North America and Eurasia: Results from the first set of LUCID experiments. *J. Clim.*, **25**, 3261–3281.

DeAngelis, A., F. Dominguez, Y. Fan, A. Robock, M. D. Kustu, and D. Robinson, 2010: Evidence of enhanced precipitation due to irrigation over the Great Plains of the United States. *J. Geophys. Res. Atmos.*, **115**, D15115.

DeLand, M., and R. Cebula, 2012: Solar UV variations during the decline of Cycle 23. *J. Atmos. Sol. Terres. Phys.*, **77**, 225–234.

Delaygue, G., and E. Bard, 2011: An Antarctic view of Beryllium-10 and solar activity for the past millennium. *Clim. Dyn.*, **36**, 2201–2218.

Deligne, N. I., S. G. Coles, and R. S. J. Sparks, 2010: Recurrence rates of large explosive volcanic eruptions. *J. Geophys. Res. Sol. Earth*, **115**, B06203.

den Elzen, M., et al., 2005: Analysing countries' contribution to climate change: Scientific and policy-related choices. *Environ. Sci. Policy*, **8**, 614–636.

Denman, K. L., et al., 2007: Couplings between changes in the climate system and biogeochemistry. In: *Climate Change 2007: The Physical Science Basis. Contribution of Working Group I to the Fourth Assessment Report of the Intergovernmental Panel on Climate Change* [Solomon, S., D. Qin, M. Manning, Z. Chen, M. Marquis, K. B. Averyt, M. Tignor and H. L. Miller (eds.)] Cambridge University Press, Cambridge, United Kingdom and New York, NY, USA, 499-587.

Dentener, F., et al., 2006: The global atmospheric environment for the next generation. *Environ. Sci. Technol.*, **40**, 3586–3594.

Derwent, R., W. Collins, C. Johnson, and D. Stevenson, 2001: Transient behaviour of tropospheric ozone precursors in a global 3-D CTM and their indirect greenhouse effects. *Clim. Change*, **49**, 463–487.

Deuber, O., G. Luderer, and O. Edenhofer, 2013: Physico-economic evaluation of climate metrics: A conceptual framework. *Environ. Sci. Policy*, **29**, 37–45.

Dewitte, S., D. Crommelynck, S. Mekaoui, and A. Joukoff, 2004: Measurement and uncertainty of the long-term total solar irradiance trend. *Solar Phys.*, **224**, 209–216.

Dickinson, R., 1975: Solar variability and lower atmosphere. *Bull. Am. Meteorol. Soc.*, **56**, 1240–1248.

Doherty, S. J., S. G. Warren, T. C. Grenfell, A. D. Clarke, and R. E. Brandt, 2010: Light-absorbing impurities in Arctic snow. *Atmos. Chem. Phys.*, **10**, 11647–11680.

Doutriaux-Boucher, M., M. Webb, J. Gregory, and O. Boucher, 2009: Carbon dioxide induced stomatal closure increases radiative forcing via a rapid reduction in low cloud. *Geophys. Res. Lett.*, **36**, doi:10.1029/2008GL036273, L02703.

Ehhalt, D. H., and L. E. Heidt, 1973: Vertical profiles of CH_4 in troposphere and stratosphere. *J. Geophys. Res.*, **78**, 5265–5271.

Eliseev, A. V., and I.I. Mokhov, 2011: Effect of including land-use driven radiative forcing of the surface albedo of land on climate response in the 16th-21st centuries. *Izvestiya Atmos. Ocean. Phys.*, **47**, 15–30.

Engel, A., et al., 2009: Age of stratospheric air unchanged within uncertainties over the past 30 years. *Nature Geosci.*, **2**, 28–31.

Erlykin, A., and A. Wolfendale, 2011: Cosmic ray effects on cloud cover and their relevance to climate change. *J. Atmos. Sol. Terres. Phys.*, **73**, 1681–1686.

Ermolli, I., K. Matthes, T. Dudok de Wit, N. A. Krivova, K. Tourpali, M. Weber, Y. C. Unruh, L. Gray, U. Langematz, P. Pilewskie, E. Rozanov, W. Schmutz, A. Shapiro, S. K. Solanki, and T. N. Woods, 2013: Recent variability of the solar spectral irradiance and its impact on climate modelling, *Atmospheric Chemistry and Physics*, **13**, 3945-3977.

Esper, J., and F. H. Schweingruber, 2004: Large-scale treeline changes recorded in Siberia. *Geophys. Res. Lett.*, **31**, L06202.

Eyring, V., et al., 2010a: Sensitivity of 21st century stratospheric ozone to greenhouse gas scenarios. *Geophys. Res. Lett.*, **37**, L16807.

Eyring, V., et al., 2010b: Multi-model assessment of stratospheric ozone return dates and ozone recovery in CCMVal-2 models. *Atmos. Chem. Phys.*, **10**, 9451–9472.

Fan, F. X., M. E. Mann, and C. M. Ammann, 2009: Understanding changes in the Asian summer monsoon over the past millennium: Insights from a long-term coupled model simulation. *J. Clim.*, **22**, 1736–1748.

FAO, 2012: State of the world's forests. Food and Agriculture Organization of the United Nations, Rome, Italy, 60 pp.

Feng, X., and F. Zhao, 2009: Effect of changes of the HITRAN database on transmittance calculations in the near-infrared region. *J. Quant. Spectrosc. Radiat. Transfer*, **110**, 247–255.

Feng, X., F. Zhao, and W. Gao, 2007: Effect of the improvement of the HITIRAN database on the radiative transfer calculation. *J. Quant. Spectrosc. Radiat. Transfer*, **108**, 308–318.

Findell, K. L., E. Shevliakova, P. C. D. Milly, and R. J. Stouffer, 2007: Modeled impact of anthropogenic land cover change on climate. *J. Clim.*, **20**, 3621–3634.

Fioletov, V. E., G. E. Bodeker, A. J. Miller, R. D. McPeters, and R. Stolarski, 2002: Global and zonal total ozone variations estimated from ground-based and satellite measurements: 1964–2000. *J. Geophys. Res. Atmos.*, **107**, 4647.

Fiore, A. M., et al., 2009: Multimodel estimates of intercontinental source-receptor relationships for ozone pollution. *J. Geophys. Res. Atmos.*, **114**, D04301.

Fischer, E. M., J. Luterbacher, E. Zorita, S. F. B. Tett, C. Casty, and H. Wanner, 2007: European climate response to tropical volcanic eruptions over the last half millennium. *Geophys. Res. Lett.*, **34**, L05707.

Fishman, J., et al., 2010: An investigation of widespread ozone damage to the soybean crop in the upper Midwest determined from ground-based and satellite measurements. *Atmos. Environ.*, **44**, 2248–2256.

Flanner, M. G., C. S. Zender, J. T. Randerson, and P. J. Rasch, 2007: Present-day climate forcing and response from black carbon in snow. *J. Geophys. Res. Atmos.*, **112**, D11202.

Fletcher, C. G., P. J. Kushner, A. Hall, and X. Qu, 2009: Circulation responses to snow albedo feedback in climate change. *Geophys. Res. Lett.*, **36**, L09702.

Fomin, B. A., and V. A. Falaleeva, 2009: Recent progress in spectroscopy and its effect on line-by-line calculations for the validation of radiation codes for climate models. *Atmos. Oceanic Opt.*, **22**, 626–629.

Forster, P., and K. Shine, 1997: Radiative forcing and temperature trends from stratospheric ozone changes. *J. Geophys. Res. Atmos.*, **102**, 10841–10855.

Forster, P., et al., 2005: Resolution of the uncertainties in the radiative forcing of HFC-134a. *J. Quant. Spectrosc. Radiat. Transfer*, **93**, 447–460.

Forster, P., et al., 2007: Changes in Atmospheric Constituents and in Radiative Forcing. In: *Climate Change 2007: The Physical Science Basis. Contribution of Working Group I to the Fourth Assessment Report of the Intergovernmental Panel on Climate Change* [Solomon, S., D. Qin, M. Manning, Z. Chen, M. Marquis, K. B. Averyt, M. Tignor and H. L. Miller (eds.)] Cambridge University Press, Cambridge, United Kingdom and New York, NY, USA, 129-234.

Forster, P., et al., 2011a: Evaluation of radiation scheme performance within chemistry climate models. *J. Geophys. Res. Atmos.*, **116**, D10302.

Forster, P. M., T. Andrews, P. Good, J. M. Gregory, L. S. Jackson, and M. Zelinka, 2013: Evaluating adjusted forcing and model spread for historical and future scenarios in the CMIP5 generation of climate models. *J. Geophys. Res. Atmos.*, **118**, 1139–1150.

Forster, P. M., et al., 2011b: Stratospheric changes and climate. In: *Scientific Assessment of Ozone Depletion: 2010*. Global Ozone Research and Monitoring Project–Report No. 52, World Meteorological Organization, Geneva, Switzerland, 516 pp.

Fortuin, J. P. F., and H. Kelder, 1998: An ozone climatology based on ozonesonde and satellite measurements. *J. Geophys. Res. Atmos.*, **103**, 31709–31734.

Foukal, P., and J. Lean, 1988: Magnetic modulation of solar luminosity by phtospheric activity. *Astrophys. J.*, **328**, 347–357.

Fowler, D., et al., 2009: Atmospheric composition change: Ecosystems-atmosphere interactions. *Atmos. Environ.*, **43**, 5193–5267.

Frame, T., and L. Gray, 2010: The 11-yr solar cycle in ERA-40 data: An update to 2008. *J. Clim.*, **23**, 2213–2222.

Freckleton, R., E. Highwood, K. Shine, O. Wild, K. Law, and M. Sanderson, 1998: Greenhouse gas radiative forcing: Effects of averaging and inhomogeneities in trace gas distribution. *Q. J. R. Meteorol. Soc.*, **124**, 2099–2127.

Friedlingstein, P., et al., 2006: Climate-carbon cycle feedback analysis: Results from the C(4)MIP model intercomparison. *J. Clim.*, **19**, 3337–3353.

Frohlich, C., 2006: Solar irradiance variability since 1978—Revision of the PMOD composite during solar cycle 21. *Space Sci. Rev.*, **125**, 53–65.

Frohlich, C., 2009: Evidence of a long-term trend in total solar irradiance. *Astron. Astrophys.*, **501**, L27–L30.

Frolicher, T. L., F. Joos, and C. C. Raible, 2011: Sensitivity of atmospheric CO_2 and climate to explosive volcanic eruptions. *Biogeosciences*, **8**, 2317–2339.

Fromm, M., G. Nedoluha, and Z. Charvat, 2013: Comment on "Large Volcanic Aerosol Load in the Stratosphere Linked to Asian Monsoon Transport". *Science*, **339**, 647–c.

Fry, M., et al., 2012: The influence of ozone precursor emissions from four world regions on tropospheric composition and radiative climate forcing. *J. Geophys. Res. Atmos.*, **117**, D07306.

Fuglestvedt, J., T. Berntsen, O. Godal, and T. Skodvin, 2000: Climate implications of GWP-based reductions in greenhouse gas emissions. *Geophys. Res. Lett.*, **27**, 409–412.

Fuglestvedt, J., T. Berntsen, O. Godal, R. Sausen, K. Shine, and T. Skodvin, 2003: Metrics of climate change: Assessing radiative forcing and emission indices. *Clim. Change*, **58**, 267–331.

Fuglestvedt, J. S., et al., 2010: Transport impacts on atmosphere and climate: Metrics. *Atmos. Environ.*, **44**, 4648–4677.

Fung, I., J. John, J. Lerner, E. Matthews, M. Prather, L. P. Steele, and P. J. Fraser, 1991: 3-Dimensional model synthesis of the global methane cycle. *J. Geophys. Res. Atmos.*, **96**, 13033–13065.

Gaillard, M. J., et al., 2010: Holocene land-cover reconstructions for studies on land cover-climate feedbacks. *Clim. Past*, **6**, 483–499.

Gao, C. C., A. Robock, and C. Ammann, 2008: Volcanic forcing of climate over the past 1500 years: An improved ice core-based index for climate models. *J. Geophys. Res. Atmos.*, **113**, D23111.

Garcia, R. R., W. J. Randel, and D. E. Kinnison, 2011: On the determination of age of air trends from atmospheric trace species. *J. Atmos. Sci.*, **68**, 139–154.

Gerlach, T., 2011: Volcanic versus anthropogenic carbon dioxide. *Eos*, **92**, 201–202.

Gettelman, A., J. Holton, and K. Rosenlof, 1997: Mass fluxes of O_3, CH_4, N_2O and CF_2Cl_2 in the lower stratosphere calculated from observational data. *J. Geophys. Res. Atmos.*, **102**, 19149–19159.

Gillett, N., and H. Matthews, 2010: Accounting for carbon cycle feedbacks in a comparison of the global warming effects of greenhouse gases. *Environ. Res. Lett.*, **5**, 034011.

Ginoux, P., D. Garbuzov, and N. C. Hsu, 2010: Identification of anthropogenic and natural dust sources using Moderate Resolution Imaging Spectroradiometer (MODIS) Deep Blue level 2 data. *J. Geophys. Res. Atmos.*, **115**, D05204.

Godal, O., and J. Fuglestvedt, 2002: Testing 100-year global warming potentials: Impacts on compliance costs and abatement profile. *Clim. Change*, **52**, 93–127.

Goosse, H., et al., 2006: The origin of the European "Medieval Warm Period". *Clim. Past*, **2**, 99–113.

Granier, C., et al., 2011: Evolution of anthropogenic and biomass burning emissions of air pollutants at global and regional scales during the 1980–2010 period. *Clim. Change*, **109**, 163–190.

Gray, L., S. Rumbold, and K. Shine, 2009: Stratospheric temperature and radiative forcing response to 11-year solar cycle changes in irradiance and ozone. *J. Atmos. Sci.*, **66**, 2402–2417.

Gray, L., et al., 2010: Solar influences on climate. *Rev. Geophys.*, **48**, RG4001.

Gregory, J., and M. Webb, 2008: Tropospheric adjustment induces a cloud component in CO_2 forcing. *J. Clim.*, **21**, 58–71.

Gregory, J., et al., 2004: A new method for diagnosing radiative forcing and climate sensitivity. *Geophys. Res. Lett.*, **31**, L03205.

Gregory, J. M., 2010: Long-term effect of volcanic forcing on ocean heat content. *Geophys. Res. Lett.*, **37**, L22701.

Grewe, V., 2007: Impact of climate variability on tropospheric ozone. *Sci. Tot. Environ.*, **374**, 167–181.

Gusev, A. A., 2008: Temporal structure of the global sequence of volcanic eruptions: Order clustering and intermittent discharge rate. *Phys. Earth Planet. Inter.*, **166**, 203–218.

Haigh, J., 1994: The role of stratospheric ozone in modulating the solar radiative forcing of climate. *Nature*, **370**, 544–546.

Haigh, J., 1999: A GCM study of climate change in response to the 11-year solar cycle. *Q. J. R. Meteorol. Soc.*, **125**, 871–892.

Haigh, J. D., 1996: The impact of solar variability on climate. *Science*, **272**, 981–984.

Hall, J., and G. Lockwood, 2004: The chromospheric activity and variability of cycling and flat activity solar-analog stars. *Astrophys. J.*, **614**, 942–946.

Hallquist, M., et al., 2009: The formation, properties and impact of secondary organic aerosol: Current and emerging issues. *Atmos. Chem. Phys.*, **9**, 5155–5236.

Hammitt, J., A. Jain, J. Adams, and D. Wuebbles, 1996: A welfare-based index for assessing environmental effects of greenhouse-gas emissions. *Nature*, **381**, 301–303.

Hansen, J., and L. Nazarenko, 2004: Soot climate forcing via snow and ice albedos. *Proc. Natl. Acad. Sci. U.S.A.*, **101**, 423–428.

Hansen, J., et al., 2005: Efficacy of climate forcings. *J. Geophys. Res. Atmos.*, **110**, D18104.

Hansen, J., et al., 2007: Climate simulations for 1880–2003 with GISS modelE. *Clim. Dyn.*, **29**, 661–696.

Harder, J., J. Fontenla, P. Pilewskie, E. Richard, and T. Woods, 2009: Trends in solar spectral irradiance variability in the visible and infrared. *Geophys. Res. Lett.*, **36**, L07801.

Harrison, R., and M. Ambaum, 2010: Observing Forbush decreases in cloud at Shetland. *J. Atmos. Sol. Terres. Phys.*, **72**, 1408–1414.

Haywood, J., and M. Schulz, 2007: Causes of the reduction in uncertainty in the anthropogenic radiative forcing of climate between IPCC (2001) and IPCC (2007). *Geophys. Res. Lett.*, **34**, L20701.

Haywood, J. M., et al., 2010: Observations of the eruption of the Sarychev volcano and simulations using the HadGEM2 climate model. *J. Geophys. Res. Atmos.*, **115**, D21212.

Heathfield, A., C. Anastasi, A. McCulloch, and F. Nicolaisen, 1998: Integrated infrared absorption coefficients of several partially fluorinated ether compounds: CF_3OCF_2H, CF_2HOCF_2H, $CH_3OCF_2CF_2H$, CH_3OCF_2CFClH, $CH_3CH_2OCF_2CF_2H$, $CF_3CH_2OCF_2CF_2H$ AND $CH_2{=}CHCH_2OCF_2CF_2H$. *Atmos. Environ.*, **32**, 2825–2833.

Hegg, D. A., S. G. Warren, T. C. Grenfell, S. J. Doherty, T. V. Larson, and A. D. Clarke, 2009: Source attribution of black carbon in Arctic snow. *Environ. Sci. Technol.*, **43**, 4016–4021.

Hegglin, M. I., and T. G. Shepherd, 2009: Large climate-induced changes in ultraviolet index and stratosphere-to-troposphere ozone flux. *Nature Geosci.*, **2**, 687–691.

Helama, S., M. Fauria, K. Mielikainen, M. Timonen, and M. Eronen, 2010: Sub-Milankovitch solar forcing of past climates: Mid and late Holocene perspectives. *Geol. Soc. Am. Bull.*, **122**, 1981–1988.

Henze, D. K., et al., 2012: Spatially refined aerosol direct radiative forcing efficiencies. *Environ. Sci. Technol.*, **46**, 9511–9518.

Hohne, N., et al., 2011: Contributions of individual countries' emissions to climate change and their uncertainty. *Clim. Change*, **106**, 359–391.

Holmes, C., Q. Tang, and M. Prather, 2011: Uncertainties in climate assessment for the case of aviation NO. *Proc. Natl. Acad. Sci. U.S.A.*, **108**, 10997–11002.

Holmes, C. D., M. J. Prather, O. A. Sovde, and G. Myhre, 2013: Future methane, hydroxyl, and their uncertainties: Key climate and emission parameters for future predictions. *Atmos. Chem. Phys.*, **13**, 285–302.

Horowitz, L. W., 2006: Past, present, and future concentrations of tropospheric ozone and aerosols: Methodology, ozone evaluation, and sensitivity to aerosol wet removal. *J. Geophys. Res. Atmos.*, **111**, D22211.

Houghton, J. T., G. J. Jenkins, and J. J. Ephraums (eds.), 1990: *Climate Change. The IPCC Scientific Assessment*. Cambridge University Press, Cambridge, United Kingdom and New York, NY, USA, 364 pp.

Hoyle, C., et al., 2011: A review of the anthropogenic influence on biogenic secondary organic aerosol. *Atmos. Chem. Phys.*, **11**, 321–343.

Hsu, J., and M. J. Prather, 2009: Stratospheric variability and tropospheric ozone. *J. Geophys. Res. Atmos.*, **114**, D06102.

Huang, J., Q. Fu, W. Zhang, X. Wang, R. Zhang, H. Ye, and S. Warren, 2011: Dust and black carbon in seasonal snow across northern China. *Bull. Am. Meteorol. Soc.*, **92**, 175–181.

Huijnen, V., et al., 2010: The global chemistry transport model TM5: Description and evaluation of the tropospheric chemistry version 3.0. *Geosci. Model Dev.*, **3**, 445–473.

Hurst, D. F., et al., 2011: Stratospheric water vapor trends over Boulder, Colorado: Analysis of the 30 year Boulder record. *J. Geophys. Res. Atmos.*, **116**, D02306.

Hurtt, G. C., et al., 2006: The underpinnings of land-use history: Three centuries of global gridded land-use transitions, wood-harvest activity, and resulting secondary lands. *Global Change Biol.*, **12**, 1208–1229.

Iacono, M. J., J. S. Delamere, E. J. Mlawer, M. W. Shephard, S. A. Clough, and W. D. Collins, 2008: Radiative forcing by long-lived greenhouse gases: Calculations with the AER radiative transfer models. *J. Geophys. Res. Atmos.*, **113**, D13103.

Ineson, S., A. A. Scaife, J. R. Knight, J. C. Manners, N. J. Dunstone, L. J. Gray, and J. D. Haigh, 2011: Solar forcing of winter climate variability in the Northern Hemisphere. *Nature Geosci.*, **4**, 753–757.

IPCC, 1996: *Revised 1996 IPCC Guidelines for National Greenhouse Gas Inventories*. Intergovernmental Panel of Climate Change.

Isaksen, I., et al., 2009: Atmospheric composition change: Climate-chemistry interactions. *Atmos. Environ.*, **43**, 5138–5192.

Ito, A., and J. E. Penner, 2005: Historical emissions of carbonaceous aerosols from biomass and fossil fuel burning for the period 1870–2000. *Global Biogeochem. Cycles*, **19**, Gb2028.

Jackson, S., 2009: Parallel pursuit of near-term and long-term climate mitigation. *Science*, **326**, 526–527.

Jacobson, M., 2010: Short-term effects of controlling fossil-fuel soot, biofuel soot and gases, and methane on climate, Arctic ice, and air pollution health. *J. Geophys. Res. Atmos.*, **115**, D14209.

Jacobson, M., 2012: Investigating cloud absorption effects: Global absorption properties of black carbon, tar balls, and soil dust in clouds and aerosols. *J. Geophys. Res. Atmos.*, **117**, D06205.

Javadi, M., O. Nielsen, T. Wallington, M. Hurley, and J. Owens, 2007: Atmospheric chemistry of 2-ethoxy-3,3,4,4,5–pentafluorotetra-hydro-2,5-bis[1,2,2,2-tetrafluoro-1-(trifluoromethyl)ethyl]-furan: Kinetics, mechanisms, and products of CL atom and OH radical initiated oxidation. *Environ. Sci. Technol.*, **41**, 7389–7395.

Jin, M. L., R. E. Dickinson, and D. L. Zhang, 2005: The footprint of urban areas on global climate as characterized by MODIS. *J. Clim.*, **18**, 1551–1565.

Jin, Y., and D. P. Roy, 2005: Fire-induced albedo change and its radiative forcing at the surface in northern Australia. *Geophys. Res. Lett.*, **32**, L13401.

Jin, Y. F., J. T. Randerson, M. L. Goulden, and S. J. Goetz, 2012: Post-fire changes in net shortwave radiation along a latitudinal gradient in boreal North America. *Geophys. Res. Lett.*, **39**, L13403.

Johansson, D., 2012: Economics- and physical-based metrics for comparing greenhouse gases. *Clim. Change*, **110**, 123–141.

Johansson, D., U. Persson, and C. Azar, 2006: The cost of using global warming potentials: Analysing the trade off between CO_2, CH_4 and N_2O. *Clim. Change*, **77**, doi:10.1007/s10584-006-9054-1, 291–309.

Jones, G., M. Lockwood, and P. Stott, 2012: What influence will future solar activity changes over the 21st century have on projected global near-surface temperature changes? *J. Geophys. Res. Atmos.*, **117**, D05103.

Jones, G. S., J. M. Gregory, P. A. Stott, S. F. B. Tett, and R. B. Thorpe, 2005: An AOGCM simulation of the climate response to a volcanic super-eruption. *Clim. Dyn.*, **25**, 725–738.

Joos, F., M. Bruno, R. Fink, U. Siegenthaler, T. Stocker, and C. LeQuere, 1996: An efficient and accurate representation of complex oceanic and biospheric models of anthropogenic carbon uptake. *Tellus B*, **48**, 397–417.

Joos, F., et al., 2013: Carbon dioxide and climate impulse response functions for the computation of greenhouse gas metrics: A multi-model analysis. *Atmos. Chem. Phys.*, **13**, 2793–2825.

Joshi, M., and J. Gregory, 2008: Dependence of the land-sea contrast in surface climate response on the nature of the forcing. *Geophys. Res. Lett.*, **35**, L24802.

Joshi, M. M., and G. S. Jones, 2009: The climatic effects of the direct injection of water vapour into the stratosphere by large volcanic eruptions. *Atmos. Chem. Phys.*, **9**, 6109–6118.

Kandlikar, M., 1995: The relative role of trace gas emissions in greenhouse abatement policies. *Energ. Policy*, **23**, 879–883.

Kaplan, J. O., K. M. Krumhardt, E. C. Ellis, W. F. Ruddiman, C. Lemmen, and K. K. Goldewijk, 2011: Holocene carbon emissions as a result of anthropogenic land cover change. *Holocene*, **21**, 775–791.

Kasischke, E. S., and J. E. Penner, 2004: Improving global estimates of atmospheric emissions from biomass burning. *J. Geophys. Res. Atmos.*, **109**, D14S01.

Kawase, H., T. Nagashima, K. Sudo, and T. Nozawa, 2011: Future changes in tropospheric ozone under Representative Concentration Pathways (RCPs). *Geophys. Res. Lett.*, **38**, L05801.

Kawase, H., M. Abe, Y. Yamada, T. Takemura, T. Yokohata, and T. Nozawa, 2010: Physical mechanism of long-term drying trend over tropical North Africa. *Geophys. Res. Lett.*, **37**, L09706.

Kirkby, J., 2007: Cosmic rays and climate. *Surv. Geophys.*, **28**, 333–375.

Kirkby, J., et al., 2011: Role of sulphuric acid, ammonia and galactic cosmic rays in atmospheric aerosol nucleation. *Nature*, **476**, 429–433.

Kleinman, L. I., P. H. Daum, Y. N. Lee, L. J. Nunnermacker, S. R. Springston, J. Weinstein-Lloyd, and J. Rudolph, 2001: Sensitivity of ozone production rate to ozone precursors. *Geophys. Res. Lett.*, **28**, 2903–2906.

Knutti, R., et al., 2008: A review of uncertainties in global temperature projections over the twenty-first century. *J. Clim.*, **21**, 2651–2663.

Koch, D., and A. D. Del Genio, 2010: Black carbon semi-direct effects on cloud cover: Review and synthesis. *Atmos. Chem. Phys.*, **10**, 7685–7696.

Koch, D., T. Bond, D. Streets, N. Unger, and G. van der Werf, 2007: Global impacts of aerosols from particular source regions and sectors. *J. Geophys. Res. Atmos.*, **112**, D02205.

Koch, D., S. Menon, A. Del Genio, R. Ruedy, I. Alienov, and G. A. Schmidt, 2009a: Distinguishing aerosol impacts on climate over the past century. *J. Clim.*, **22**, 2659–2677.

Koch, D., et al., 2011: Coupled aerosol-chemistry-climate twentieth-century transient model investigation: Trends in short-lived species and climate responses. *J. Clim.*, **24**, 2693–2714.

Koch, D., et al., 2009b: Evaluation of black carbon estimations in global aerosol models. *Atmos. Chem. Phys.*, **9**, 9001–9026.

Koehler, M. O., G. Raedel, K. P. Shine, H. L. Rogers, and J. A. Pyle, 2013: Latitudinal variation of the effect of aviation NO_x emissions on atmospheric ozone and methane and related climate metrics. *Atmos. Environ.*, **64**, 1–9.

Koffi, B., et al., 2012: Application of the CALIOP layer product to evaluate the vertical distribution of aerosols estimated by global models: AeroCom phase I results. *J. Geophys. Res. Atmos.*, **117**, D10201.

Kopp, G., and J. Lean, 2011: A new, lower value of total solar irradiance: Evidence and climate significance. *Geophys. Res. Lett.*, **38**, L01706.

Kratz, D., 2008: The sensitivity of radiative transfer calculations to the changes in the HITRAN database from 1982 to 2004. *J. Quant. Spectrosc. Radiat. Transfer*, **109**, 1060–1080.

Kravitz, B., and A. Robock, 2011: Climate effects of high-latitude volcanic eruptions: Role of the time of year. *J. Geophys. Res. Atmos.*, **116**, D01105.

Kravitz, B., A. Robock, and A. Bourassa, 2010: Negligible climatic effects from the 2008 Okmok and Kasatochi volcanic eruptions. *J. Geophys. Res. Atmos.*, **115**, D00L05.

Kravitz, B., et al., 2011: Simulation and observations of stratospheric aerosols from the 2009 Sarychev volcanic eruption. *J. Geophys. Res. Atmos.*, **116**, D18211.

Kristjansson, J. E., T. Iversen, A. Kirkevag, O. Seland, and J. Debernard, 2005: Response of the climate system to aerosol direct and indirect forcing: Role of cloud feedbacks. *J. Geophys. Res. Atmos.*, **110**, D24206.

Krivova, N., L. Vieira, and S. Solanki, 2010: Reconstruction of solar spectral irradiance since the Maunder minimum. *J. Geophys. Res. Space Phys.*, **115**, A12112.

Kueppers, L. M., M. A. Snyder, and L. C. Sloan, 2007: Irrigation cooling effect: Regional climate forcing by land-use change. *Geophys. Res. Lett.*, **34**, L03703.

Kuroda, Y., and K. Kodera, 2005: Solar cycle modulation of the southern annular mode. *Geophys. Res. Lett.*, **32**, L13802.

Kvalevag, M. M., G. Myhre, G. Bonan, and S. Levis, 2010: Anthropogenic land cover changes in a GCM with surface albedo changes based on MODIS data. *Int. J. Climatol.*, **30**, 2105–2117.

Lacis, A. A., D. J. Wuebbles, J. A. Logan, 1990: Radiative forcing of climate by changes in the vertical-distribution of ozone, J. Geophys. Res., **95**, 9971-9981.

Lamarque, J., et al., 2011: Global and regional evolution of short-lived radiatively-active gases and aerosols in the Representative Concentration Pathways. *Clim. Change*, **109**, 191–212.

Lamarque, J., et al., 2010: Historical (1850–2000) gridded anthropogenic and biomass burning emissions of reactive gases and aerosols: Methodology and application. *Atmos. Chem. Phys.*, **10**, 7017–7039.

Lamarque, J. F., et al., 2013: The Atmospheric Chemistry and Climate Model Intercomparison Project (ACCMIP): Overview and description of models, simulations and climate diagnostics. *Geosci. Model Dev.*, **6**, 179–206.

Lau, K. M., M. K. Kim, and K. M. Kim, 2006: Asian summer monsoon anomalies induced by aerosol direct forcing: The role of the Tibetan Plateau. *Clim. Dyn.*, **26**, 855–864.

Lauder, A. R., I. G. Enting, J. O. Carter, N. Clisby, A. L. Cowie, B. K. Henry, and M. R. Raupach, 2013: Offsetting methane emissions—An alternative to emission equivalence metrics. *Int. J. Greenh. Gas Control*, **12**, 419–429.

Lauer, A., V. Eyring, J. Hendricks, P. Jockel, and U. Lohmann, 2007: Global model simulations of the impact of ocean-going ships on aerosols, clouds, and the radiation budget. *Atmos. Chem. Phys.*, **7**, 5061–5079.

Lawrence, P. J., and T. N. Chase, 2010: Investigating the climate impacts of global land cover change in the community climate system model. *Int. J. Climatol.*, **30**, 2066–2087.

Lean, J., 2000: Evolution of the sun's spectral irradiance since the Maunder Minimum. *Geophys. Res. Lett.*, **27**, 2425–2428.

Lean, J., and M. Deland, 2012: How does the sun's spectrum vary? *J. Clim.*, **25**, 2555–2560.

Lee, D. S., et al., 2010: Transport impacts on atmosphere and climate: Aviation. *Atmos. Environ.*, **44**, 4678–4734.

Lee, J., D. Shindell, and S. Hameed, 2009: The influence of solar forcing on tropical circulation. *J. Clim.*, **22**, 5870–5885.

Lee, R., M. Gibson, R. Wilson, and S. Thomas, 1995: Long-term total solar irradiance variability during sunspot cycle-22. *J. Geophys. Res. Space Phys.*, **100**, 1667–1675.

Lee, X., et al., 2011: Observed increase in local cooling effect of deforestation at higher latitudes. *Nature*, **479**, 384–387.

Lee, Y. H., et al., 2013: Evaluation of preindustrial to present-day black carbon and its albedo forcing from Atmospheric Chemistry and Climate Model Intercomparison Project (ACCMIP). *Atmos. Chem. Phys.*, **13**, 2607–2634.

Legras, B., O. Mestre, E. Bard, and P. Yiou, 2010: A critical look at solar-climate relationships from long temperature series. *Clim. Past*, **6**, 745–758.

Leibensperger, E. M., L. J. Mickley, and D. J. Jacob, 2008: Sensitivity of US air quality to mid-latitude cyclone frequency and implications of 1980–2006 climate change. *Atmos. Chem. Phys.*, **8**, 7075–7086.

Leibensperger, E. M., et al., 2012a: Climatic effects of 1950–2050 changes in US anthropogenic aerosols—Part 1: Aerosol trends and radiative forcing. *Atmos. Chem. Phys.*, **12**, 3333–3348.

Leibensperger, E. M., et al., 2012b: Climatic effects of 1950–2050 changes in US anthropogenic aerosols—Part 2: Climate response. *Atmos. Chem. Phys.*, **12**, 3349–3362.

Lelieveld, J., et al., 2008: Atmospheric oxidation capacity sustained by a tropical forest. *Nature*, **452**, 737–740.

Levy, H., 1971: Normal atmosphere—Large radical and formaldehyde concentrations predicted. *Science*, **173**, 141–143.

Liao, H., W. T. Chen, and J. H. Seinfeld, 2006: Role of climate change in global predictions of future tropospheric ozone and aerosols. *J. Geophys. Res. Atmos.*, **111**, D12304.

8

Lockwood, M., 2010: Solar change and climate: An update in the light of the current exceptional solar minimum. *Proc. R. Soc. London A*, **466**, 303–329.

Lockwood, M., and C. Frohlich, 2008: Recent oppositely directed trends in solar climate forcings and the global mean surface air temperature. II. Different reconstructions of the total solar irradiance variation and dependence on response time scale. *Proc. R. Soc. London A*, **464**, 1367–1385.

Lockwood, M., A. Rouillard, and I. Finch, 2009: The rise and fall of open solar flux during the current grand solar maximum. *Astrophys. J.*, **700**, 937–944.

Logan, J. A., 1999: An analysis of ozonesonde data for the troposphere: Recommendations for testing 3-D models and development of a gridded climatology for tropospheric ozone. *J. Geophys. Res. Atmos.*, **104**, 16115–16149.

Logan, J. A., M. J. Prather, S. C. Wofsy, and M. B. McElroy, 1981: Tropospheric chemistry—A global perspective. *J. Geophys. Res. Oceans Atmos.*, **86**, 7210–7254.

Logan, J. A., et al., 2012: Changes in ozone over Europe: Analysis of ozone measurements from sondes, regular aircraft (MOZAIC) and alpine surface sites. *J. Geophys. Res. Atmos.*, **117**, D09301.

Lohila, A., et al., 2010: Forestation of boreal peatlands: Impacts of changing albedo and greenhouse gas fluxes on radiative forcing. *J. Geophys. Res. Biogeosci.*, **115**, G04011.

Lohmann, U., et al., 2010: Total aerosol effect: Radiative forcing or radiative flux perturbation? *Atmos. Chem. Phys.*, **10**, 3235–3246.

Long, C. N., E. G. Dutton, J. A. Augustine, W. Wiscombe, M. Wild, S. A. McFarlane, and C. J. Flynn, 2009: Significant decadal brightening of downwelling shortwave in the continental United States. *J. Geophys. Res. Atmos.*, **114**, D00D06.

Long, D., and M. Collins, 2013: Quantifying global climate feedbacks, responses and forcing under abrupt and gradual CO_2 forcing. *Clim. Dyn.*, **41**, 2471-2479.

Lu, P., H. Zhang, and X. Jing, 2012: The effects of different HITRAN versions on calculated long-wave radiation and uncertainty evaluation. *Acta Meteorol. Sin.*, **26**, 389–398.

Lu, Z., Q. Zhang, and D. G. Streets, 2011: Sulfur dioxide and primary carbonaceous aerosol emissions in China and India, 1996–2010. *Atmos. Chem. Phys.*, **11**, 9839–9864.

Lund, M., T. Berntsen, J. Fuglestvedt, M. Ponater, and K. Shine, 2012: How much information is lost by using global-mean climate metrics? an example using the transport sector. *Clim. Change*, **113**, 949–963.

MacFarling Meure, C., et al., 2006: Law Dome CO_2, CH_4 and N_2O ice core records extended to 2000 years BP. *Geophys. Res. Lett.*, **33**, L14810.

MacMynowski, D., H. Shin, and K. Caldeira, 2011: The frequency response of temperature and precipitation in a climate model. *Geophys. Res. Lett.*, **38**, L16711.

Mader, J. A., J. Staehelin, T. Peter, D. Brunner, H. E. Rieder, and W. A. Stahel, 2010: Evidence for the effectiveness of the Montreal Protocol to protect the ozone layer. *Atmos. Chem. Phys.*, **10**, 12161–12171.

Mahowald, N. M., et al., 2010: Observed 20th century desert dust variability: Impact on climate and biogeochemistry. *Atmos. Chem. Phys.*, **10**, 10875–10893.

Mann, M., M. Cane, S. Zebiak, and A. Clement, 2005: Volcanic and solar forcing of the tropical Pacific over the past 1000 years. *J. Clim.*, **18**, 447–456.

Manne, A., and R. Richels, 2001: An alternative approach to establishing trade-offs among greenhouse gases. *Nature*, **410**, 675–677.

Manning, M., and A. Reisinger, 2011: Broader perspectives for comparing different greenhouse gases. *Philos. Trans. R. Soc. London A*, **369**, 1891–1905.

Marenco, A., H. Gouget, P. Nedelec, J. P. Pages, and F. Karcher, 1994: Evidence of a long-term increase in troposheric ozone from PIC Du Midi Data Series—Consequences—Positive radiative forcing. *J. Geophys. Res. Atmos.*, **99**, 16617–16632.

Marten, A. L., and S. C. Newbold, 2012: Estimating the social cost of non-CO2 GHG emissions: Methane and nitrous oxide. *Energ. Policy*, **51**, 957–972.

Matthews, H. D., A. J. Weaver, K. J. Meissner, N. P. Gillett, and M. Eby, 2004: Natural and anthropogenic climate change: Incorporating historical land cover change, vegetation dynamics and the global carbon cycle. *Clim. Dyn.*, **22**, 461–479.

McComas, D., R. Ebert, H. Elliott, B. Goldstein, J. Gosling, N. Schwadron, and R. Skoug, 2008: Weaker solar wind from the polar coronal holes and the whole Sun. *Geophys. Res. Lett.*, **35**, L18103.

McLandress, C., T. G. Shepherd, J. F. Scinocca, D. A. Plummer, M. Sigmond, A. I. Jonsson, and M. C. Reader, 2011: Separating the dynamical effects of climate change and ozone depletion. Part II. Southern Hemisphere troposphere. *J. Clim.*, **24**, 1850–1868.

Meinshausen, M., T. Wigley, and S. Raper, 2011a: Emulating atmosphere-ocean and carbon cycle models with a simpler model, MAGICC6–Part 2: Applications. *Atmos. Chem. Phys.*, **11**, 1457–1471.

Meinshausen, M., et al., 2011b: The RCP greenhouse gas concentrations and their extensions from 1765 to 2300. *Clim. Change*, **109**, 213–241.

Mercado, L. M., N. Bellouin, S. Sitch, O. Boucher, C. Huntingford, M. Wild, and P. M. Cox, 2009: Impact of changes in diffuse radiation on the global land carbon sink. *Nature*, **458**, 1014–1017.

Merikanto, J., D. Spracklen, G. Mann, S. Pickering, and K. Carslaw, 2009: Impact of nucleation on global CCN. *Atmos. Chem. Phys.*, **9**, 8601–8616.

Mickley, L. J., E. M. Leibensperger, D. J. Jacob, and D. Rind, 2012: Regional warming from aerosol removal over the United States: Results from a transient 2010–2050 climate simulation. *Atmos. Environ.*, **46**, 545–553.

Miller, G. H., et al., 2012: Abrupt onset of the Little Ice Age triggered by volcanism and sustained by sea-ice/ocean feedbacks. *Geophys. Res. Lett.*, **39**, L02708.

Miller, R. L., I. Tegen, and J. Perlwitz, 2004: Surface radiative forcing by soil dust aerosols and the hydrologic cycle. *J. Geophys. Res. Atmos.*, **109**, D04203.

Mills, M. J., O. B. Toon, R. P. Turco, D. E. Kinnison, and R. R. Garcia, 2008: Massive global ozone loss predicted following regional nuclear conflict. *Proc. Natl. Acad. Sci. U.S.A.*, **105**, 5307–5312.

Ming, Y., and V. Ramaswamy, 2012: Nonlocal component of radiative flux perturbation. *Geophys. Res. Lett.*, **39**, L22706.

Ming, Y., V. Ramaswamy, and G. Persad, 2010: Two opposing effects of absorbing aerosols on global-mean precipitation. *Geophys. Res. Lett.*, **37**, L13701.

Ming, Y., V. Ramaswamy, and G. Chen, 2011: A model investigation of aerosol-induced changes in boreal winter extratropical circulation. *J. Clim.*, **24**, 6077–6091.

Ming, Y., V. Ramaswamy, L. J. Donner, V. T. J. Phillips, S. A. Klein, P. A. Ginoux, and L. W. Horowitz, 2007: Modeling the interactions between aerosols and liquid water clouds with a self-consistent cloud scheme in a general circulation model. *J. Atmos. Sci.*, **64**, 1189–1209.

Mirme, S., A. Mirme, A. Minikin, A. Petzold, U. Horrak, V.-M. Kerminen, and M. Kulmala, 2010: Atmospheric sub-3 nm particles at high altitudes. *Atm. Chem. Phys.*, **10**, 437–451.

Montzka, S. A., E. J. Dlugokencky, and J. H. Butler, 2011: Non-CO_2 greenhouse gases and climate change. *Nature*, **476**, 43–50.

Morrill, J., L. Floyd, and D. McMullin, 2011: The solar ultraviolet spectrum estimated using the Mg II Index and Ca II K disk activity. *Solar Physics*, **269**, 253–267.

Muhle, J., et al., 2009: Sulfuryl fluoride in the global atmosphere. *J. Geophys. Res. Atmos.*, **114**, D05306.

Mulitza, S., et al., 2010: Increase in African dust flux at the onset of commercial agriculture in the Sahel region. *Nature*, **466**, 226–228.

Murphy, D., and D. Fahey, 1994: An estimate of the flux of stratospheric reactive nitrogen and ozone into the troposphere. *J. Geophys. Res. Atmos.*, **99**, 5325–5332.

Myhre, G., M. M. Kvalevag, and C. B. Schaaf, 2005a: Radiative forcing due to anthropogenic vegetation change based on MODIS surface albedo data. *Geophys. Res. Lett.*, **32**, L21410.

Myhre, G., E. J. Highwood, K. P. Shine, and F. Stordal, 1998: New estimates of radiative forcing due to well mixed greenhouse gases. *Geophys. Res. Lett.*, **25**, 2715–2718.

Myhre, G., Y. Govaerts, J. M. Haywood, T. K. Berntsen, and A. Lattanzio, 2005b: Radiative effect of surface albedo change from biomass burning. *Geophys. Res. Lett.*, **32**, L20812.

Myhre, G., J. Nilsen, L. Gulstad, K. Shine, B. Rognerud, and I. Isaksen, 2007: Radiative forcing due to stratospheric water vapour from CH_4 oxidation. *Geophys. Res. Lett.*, **34**, L01807.

Myhre, G., et al., 2011: Radiative forcing due to changes in ozone and methane caused by the transport sector. *Atmos. Environ.*, **45**, 387–394.

Myhre, G., et al., 2013: Radiative forcing of the direct aerosol effect from AeroCom Phase II simulations. *Atmos. Chem. Phys.*, **13**, 1853–1877.

Nagai, T., B. Liley, T. Sakai, T. Shibata, and O. Uchino, 2010: Post-Pinatubo evolution and subsequent trend of the stratospheric aerosol layer observed by mid-latitude lidars in both hemispheres. *Sola*, **6**, 69–72.

Naik, V., D. L. Mauzerall, L. W. Horowitz, M. D. Schwarzkopf, V. Ramaswamy, and M. Oppenheimer, 2007: On the sensitivity of radiative forcing from biomass burning aerosols and ozone to emission location. *Geophys. Res. Lett.*, **34**, L03818.

Nair, U. S., D. K. Ray, J. Wang, S. A. Christopher, T. J. Lyons, R. M. Welch, and R. A. Pielke, 2007: Observational estimates of radiative forcing due to land use change in southwest Australia. *J. Geophys. Res. Atmos.*, **112**, D09117.

Neely, R. R., et al., 2013: Recent anthropogenic increases in SO_2 from Asia have minimal impact on stratospheric aerosol. *Geophys. Res. Lett.*, **40**, 999-1004.

O'ishi, R., A. Abe-Ouchi, I. Prentice, and S. Sitch, 2009: Vegetation dynamics and plant CO_2 responses as positive feedbacks in a greenhouse world. *Geophys. Res. Lett.*, **36**, L11706.

O'Neill, B., 2000: The jury is still out on global warming potentials. *Clim. Change*, **44**, 427–443.

O'Neill, B., 2003: Economics, natural science, and the costs of global warming potentials—An editorial comment. *Clim. Change*, **58**, 251–260.

Oleson, K. W., G. B. Bonan, and J. Feddema, 2010: Effects of white roofs on urban temperature in a global climate model. *Geophys. Res. Lett.*, **37**, L03701.

Olivié, D. J. L., G. Peters, and D. Saint-Martin, 2012: Atmosphere response time scales estimated from AOGCM experiments. *J. Climate*, **25**, 7956–7972.

Olsen, S. C., C. A. McLinden, and M. J. Prather, 2001: Stratospheric N_2O–NOy system: Testing uncertainties in a three-dimensional framework, *J. Geophys. Res.*, **106**, 28771.

Oreopoulos, L., et al., 2012: The Continual Intercomparison of Radiation Codes: Results from Phase I. *J. Geophys. Res. Atmos.*, **117**, D06118.

Osterman, G. B., et al., 2008: Validation of Tropospheric Emission Spectrometer (TES) measurements of the total, stratospheric, and tropospheric column abundance of ozone. *J. Geophys. Res. Atmos.*, **113**, D15S16.

Otterä, O. H., M. Bentsen, H. Drange, and L. L. Suo, 2010: External forcing as a metronome for Atlantic multidecadal variability. *Nature Geosci.*, **3**, 688–694.

Özdoğan, M., A. Robock, and C. J. Kucharik, 2013: Impacts of a nuclear war in South Asia on soybean and maize production in the Midwest United States. *Clim. Change*, **116**, 373–387.

Parrish, D. D., D. B. Millet, and A. H. Goldstein, 2009: Increasing ozone in marine boundary layer inflow at the west coasts of North America and Europe. *Atmos. Chem. Phys.*, **9**, 1303–1323.

Paulot, F., J. D. Crounse, H. G. Kjaergaard, A. Kurten, J. M. St Clair, J. H. Seinfeld, and P. O. Wennberg, 2009: Unexpected epoxide formation in the gas-phase photooxidation of isoprene. *Science*, **325**, 730–733.

Paynter, D., and V. Ramaswamy, 2011: An assessment of recent water vapor continuum measurements upon longwave and shortwave radiative transfer. *J. Geophys. Res. Atmos.*, **116**, D20302.

Pechony, O., and D. Shindell, 2010: Driving forces of global wildfires over the past millennium and the forthcoming century. *Proc. Natl. Acad. Sci. U.S.A.*, **107**, 19167–19170.

Peeters, J., T. L. Nguyen, and L. Vereecken, 2009: HO(x) radical regeneration in the oxidation of isoprene. *Phys. Chem. Chem. Phys.*, **11**, 5935–5939.

Penn, M., and W. Livingston, 2006: Temporal changes in sunspot umbral magnetic fields and temperatures. *Astrophys. J.*, **649**, L45–L48.

Penner, J., L. Xu, and M. Wang, 2011: Satellite methods underestimate indirect climate forcing by aerosols. *Proc. Natl. Acad. Sci. U.S.A.*, **108**, 13404–13408.

Penner, J., et al., 2006: Model intercomparison of indirect aerosol effects. *Atmos. Chem. Phys.*, **6**, 3391–3405.

Perlwitz, J., and R. L. Miller, 2010: Cloud cover increase with increasing aerosol absorptivity: A counterexample to the conventional semidirect aerosol effect. *J. Geophys. Res. Atmos.*, **115**, D08203.

Persad, G. G., Y. Ming, and V. Ramaswamy, 2012: Tropical tropospheric-only responses to absorbing aerosols. *J. Clim.*, **25**, 2471–2480.

Peters, G., B. Aamaas, T. Berntsen, and J. Fuglestvedt, 2011a: The integrated global temperature change potential (iGTP) and relationships between emission metrics. *Environ. Res. Lett.*, **6**, 044021.

Peters, G. P., B. Aamaas, M. T. Lund, C. Solli, and J. S. Fuglestvedt, 2011b: Alternative "Global Warming" metrics in life cycle assessment: A case study with existing transportation data. *Environ. Sci. Technol.*, **45**, 8633–8641.

Peters, K., P. Stier, J. Quaas, and H. Grassl, 2012: Aerosol indirect effects from shipping emissions: Sensitivity studies with the global aerosol-climate model ECHAM-HAM. *Atmos. Chem. Phys.*, **12**, 5985–6007.

Philipona, R., K. Behrens, and C. Ruckstuhl, 2009: How declining aerosols and rising greenhouse gases forced rapid warming in Europe since the 1980s. *Geophys. Res. Lett.*, **36**, L02806.

Pinto, J. P., R. P. Turco, and O. B. Toon, 1989: Self-limiting physical and chemical effects in volcanic-eruption clouds. *J. Geophys. Res. Atmos.*, **94**, 11165–11174.

Pitman, A. J., et al., 2009: Uncertainties in climate responses to past land cover change: First results from the LUCID intercomparison study. *Geophys. Res. Lett.*, **36**, L14814.

Plattner, G.-K., T. Stocker, P. Midgley, and M. Tignor, 2009: IPCC Expert Meeting on the Science of Alternative Metrics: Meeting Report. IPCC Working Group I, Technical Support Unit.

Plattner, G. K., et al., 2008: Long-term climate commitments projected with climate-carbon cycle models. *J. Clim.*, **21**, 2721–2751.

Pongratz, J., C. Reick, T. Raddatz, and M. Claussen, 2008: A reconstruction of global agricultural areas and land cover for the last millennium. *Global Biogeochem. Cycles*, **22**, Gb3018.

Pongratz, J., C. H. Reick, T. Raddatz, and M. Claussen, 2010: Biogeophysical versus biogeochemical climate response to historical anthropogenic land cover change. *Geophys. Res. Lett.*, **37**, L08702.

Pongratz, J., T. Raddatz, C. H. Reick, M. Esch, and M. Claussen, 2009: Radiative forcing from anthropogenic land cover change since AD 800. *Geophys. Res. Lett.*, **36**, L02709.

Pozzer, A., et al., 2012: Effects of business-as-usual anthropogenic emissions on air quality. *Atmos. Chem. Phys.*, **12**, 6915–6937.

Prather, M., 2002: Lifetimes of atmospheric species: Integrating environmental impacts. *Geophys. Res. Lett.*, **29**, 2063.

Prather, M., and J. Hsu, 2010: Coupling of nitrous oxide and methane by global atmospheric chemistry. *Science*, **330**, 952–954.

Prather, M. J., 1998: Time scales in atmospheric chemistry: Coupled perturbations to N_2O, NOy, and O_3. *Science*, **279**, 1339–1341.

Prather, M. J., C. D. Holmes, and J. Hsu, 2012: Reactive greenhouse gas scenarios: Systematic exploration of uncertainties and the role of atmospheric chemistry. *Geophys. Res. Lett.*, **39**, L09803.

Puma, M. J., and B. I. Cook, 2010: Effects of irrigation on global climate during the 20th century. *J. Geophys. Res. Atmos.*, **115**, D16120.

Quaas, J., O. Boucher, N. Bellouin, and S. Kinne, 2011: Which of satellite- or model-based estimates is closer to reality for aerosol indirect forcing? *Proc. Natl. Acad. Sci. U.S.A.*, **108**, E1099.

Quaas, J., et al., 2009: Aerosol indirect effects—general circulation model intercomparison and evaluation with satellite data. *Atmos. Chem. Phys.*, **9**, 8697–8717.

Rajakumar, B., R. Portmann, J. Burkholder, and A. Ravishankara, 2006: Rate coefficients for the reactions of OH with $CF_3CH_2CH_3$ (HFC-263fb), CF_3CHFCH_2F (HFC-245eb), and $CHF_2CHFCHF_2$ (HFC-245ea) between 238 and 375 K. *J. Phys. Chem. A*, **110**, 6724–6731.

Ramanathan, V., and G. Carmichael, 2008: Global and regional climate changes due to black carbon. *Nature Geosci.*, **1**, 221–227.

Ramanathan, V., et al., 2005: Atmospheric brown clouds: Impacts on South Asian climate and hydrological cycle. *Proc. Natl. Acad. Sci. U.S.A.*, **102**, 5326–5333.

Ramaswamy, V., et al., 2001: Radiative forcing of climate change. In: *Climate Change 2001: The Scientific Basis. Contribution of Working Group I to the Third Assessment Report of the Intergovernmntal Panel on Climate Change* [J. T. Houghton, Y. Ding, D. J. Griggs, M. Noquer, P. J. van der Linden, X. Dai, K. Maskell and C. A. Johnson (eds.)]. Cambride University Press, Cambridge, United Kingdom and New York, NY, USA, 349-416.

Randel, W., and F. Wu, 2007: A stratospheric ozone profile data set for 1979–2005: Variability, trends, and comparisons with column ozone data. *J. Geophys. Res. Atmos.*, **112**, D06313.

Randles, C. A., and V. Ramaswamy, 2008: Absorbing aerosols over Asia: A Geophysical Fluid Dynamics Laboratory general circulation model sensitivity study of model response to aerosol optical depth and aerosol absorption. *J. Geophys. Res. Atmos.*, **113**, D21203.

Rasch, P. J., et al., 2008: An overview of geoengineering of climate using stratospheric sulphate aerosols. *Philos. Trans. R. Soc. A*, **366**, 4007–4037.

Ravishankara, A. R., J. S. Daniel, and R. W. Portmann, 2009: Nitrous oxide (N_2O): The dominant ozone-depleting substance emitted in the 21st century. *Science*, **326**, 123–125.

Rechid, D., T. Raddatz, and D. Jacob, 2009: Parameterization of snow-free land surface albedo as a function of vegetation phenology based on MODIS data and applied in climate modelling. *Theor. Appl. Climatol.*, **95**, 245–255.

Reddy, M., and O. Boucher, 2007: Climate impact of black carbon emitted from energy consumption in the world's regions. *Geophys. Res. Lett.*, **34**, L11802.

Reilly, J., et al., 1999: Multi-gas assessment of the Kyoto Protocol. *Nature*, **401**, 549–555.

Reilly, J. M., and K. R. Richards, 1993: Climate-change damage and the trace-gas-index issue. *Environ. Resour. Econ.*, **3**, 41–61.

Reisinger, A., M. Meinshausen, and M. Manning, 2011: Future changes in global warming potentials under representative concentration pathways. *Environ. Res. Lett.*, **6**, 024020.

Reisinger, A., M. Meinshausen, M. Manning, and G. Bodeker, 2010: Uncertainties of global warming metrics: CO_2 and CH_4. *Geophys. Res. Lett.*, **37**, L14707.

Reisinger, A., P. Havlik, K. Riahi, O. Vliet, M. Obersteiner, and M. Herrero, 2013: Implications of alternative metrics for global mitigation costs and greenhouse gas emissions from agriculture. *Clim. Change*, **117**, 677-690.

Righi, M., C. Klinger, V. Eyring, J. Hendricks, A. Lauer, and A. Petzold, 2011: Climate impact of biofuels in shipping: global model studies of the aerosol indirect effect. *Environ. Sci. Technol.*, **45**, 3519–3525.

Rigozo, N., E. Echer, L. Vieira, and D. Nordemann, 2001: Reconstruction of Wolf sunspot numbers on the basis of spectral characteristics and estimates of associated radio flux and solar wind parameters for the last millennium. *Sol. Phys.*, **203**, 179–191.

Rigozo, N., D. Nordemann, E. Echer, M. Echer, and H. Silva, 2010: Prediction of solar minimum and maximum epochs on the basis of spectral characteristics for the next millennium. *Planet. Space Sci.*, **58**, 1971–1976.

Robock, A., 2000: Volcanic eruptions and climate. *Rev. Geophys.*, **38**, 191–219.

Robock, A., 2010: New START, Eyjafjallajökull, and Nuclear Winter. *Eos*, **91**, 444–445.

Robock, A., L. Oman, and G. L. Stenchikov, 2007a: Nuclear winter revisited with a modern climate model and current nuclear arsenals: Still catastrophic consequences. *J. Geophys. Res. Atmos.*, **112**, D13107.

Robock, A., L. Oman, and G. L. Stenchikov, 2008: Regional climate responses to geoengineering with tropical and Arctic SO_2 injections. *J. Geophys. Res. Atmos.*, **113**, D16101.

Robock, A., L. Oman, G. L. Stenchikov, O. B. Toon, C. Bardeen, and R. P. Turco, 2007b: Climatic consequences of regional nuclear conflicts. *Atmos. Chem. Phys.*, **7**, 2003–2012.

Robock, A., C. M. Ammann, L. Oman, D. Shindell, S. Levis, and G. Stenchikov, 2009: Did the Toba volcanic eruption of similar to 74 ka BP produce widespread glaciation? *J. Geophys. Res. Atmos.*, **114**, D10107.

Rogelj, J., et al., 2011: Emission pathways consistent with a 2 degrees C global temperature limit. *Nature Clim. Change*, **1**, 413–418.

Roscoe, H. K., and J. D. Haigh, 2007: Influences of ozone depletion, the solar cycle and the QBO on the Southern Annular Mode. *Q. J. R. Meteorol. Soc.*, **133**, 1855–1864.

Rotenberg, E., and D. Yakir, 2010: Contribution of semi-arid forests to the climate system. *Science*, **327**, 451–454.

Rothman, L., 2010: The evolution and impact of the HITRAN molecular spectroscopic database. *J. Quant. Spectrosc. Radiat. Transfer*, **111**, 1565–1567.

Rotstayn, L., and J. Penner, 2001: Indirect aerosol forcing, quasi forcing, and climate response. *J. Clim.*, **14**, 2960–2975.

Rotstayn, L. D., and U. Lohmann, 2002: Tropical rainfall trends and the indirect aerosol effect. *J. Clim.*, **15**, 2103–2116.

Rotstayn, L. D., B. F. Ryan, and J. E. Penner, 2000: Precipitation changes in a GCM resulting from the indirect effects of anthropogenic aerosols. *Geophys. Res. Lett.*, **27**, 3045–3048.

Rottman, G., 2006: Measurement of total and spectral solar irradiance. *Space Sci. Rev.*, **125**, 39–51.

Ruckstuhl, C., et al., 2008: Aerosol and cloud effects on solar brightening and the recent rapid warming. *Geophys. Res. Lett.*, **35**, L12708.

Russell, C., J. Luhmann, and L. Jian, 2010: How unprecedented a solar minimum? *Rev. Geophys.*, **48**, RG2004.

Rypdal, K., N. Rive, T. Berntsen, Z. Klimont, T. Mideksa, G. Myhre, and R. Skeie, 2009: Costs and global impacts of black carbon abatement strategies. *Tellus B.*, **61**, 625–641.

Rypdal, K., et al., 2005: Tropospheric ozone and aerosols in climate agreements: Scientific and political challenges. *Environ. Sci. Policy*, **8**, 29–43.

Salby, M. L., E. A. Titova, and L. Deschamps, 2012: Changes of the Antarctic ozone hole: Controlling mechanisms, seasonal predictability, and evolution. *J. Geophys. Res. Atmos.*, **117**, D10111.

Sarofim, M., 2012: The GTP of methane: Modeling analysis of temperature impacts of methane and carbon dioxide reductions. *Environ. Model. Assess.*, **17**, 231–239.

Sato, M., J. E. Hansen, M. P. McCormick, and J. B. Pollack, 1993: Stratospheric aerosol optical depths, 1850–1990. *J. Geophys. Res. Atmos.*, **98**, 22987–22994.

Schaaf, C., et al., 2002: First operational BRDF, albedo nadir reflectance products from MODIS. *Remote Sens. Environ.*, **83**, 135–148.

Schmalensee, R., 1993: Comparing greenhouse gases for policy purposes. *Energy J.*, **14**, 245–256.

Schmidt, A., K. S. Carslaw, G. W. Mann, M. Wilson, T. J. Breider, S. J. Pickering, and T. Thordarson, 2010: The impact of the 1783–1784 AD Laki eruption on global aerosol formation processes and cloud condensation nuclei. *Atmos. Chem. Phys.*, **10**, 6025–6041.

Schmidt, A., et al., 2012: Importance of tropospheric volcanic aerosol for indirect radiative forcing of climate. *Atmos. Chem. Phys.*, **12**, 7321–7339.

Schmidt, G., et al., 2011: Climate forcing reconstructions for use in PMIP simulations of the last millennium (v1.0). *Geosci. Model Dev.*, **4**, 33–45.

Schneider, D. P., C. M. Ammann, B. L. Otto-Bliesner, and D. S. Kaufman, 2009: Climate response to large, high-latitude and low-latitude volcanic eruptions in the Community Climate System Model. *J. Geophys. Res. Atmos.*, **114**, D15101.

Schultz, M. G., et al., 2008: Global wildland fire emissions from 1960 to 2000. *Global Biogeochem. Cycles*, **22**, Gb2002.

Seinfeld, J. H., and S. N. Pandis, 2006: *Atmospheric Chemistry and Physics: From Air Pollution to Climate Change.* John Wiley & Sons, Hoboken, NJ, USA.

Self, S., and S. Blake, 2008: Consequences of explosive supereruptions. *Elements*, **4**, 41–46.

Sharma, S., D. Lavoue, H. Cachier, L. Barrie, and S. Gong, 2004: Long-term trends of the black carbon concentrations in the Canadian Arctic. *J. Geophys. Res. Atmos.*, **109**, D15203.

Shibata, K., and K. Kodera, 2005: Simulation of radiative and dynamical responses of the middle atmosphere to the 11-year solar cycle. *J. Atmos. Sol. Terres. Phys.*, **67**, 125–143.

Shindell, D., and G. Faluvegi, 2009: Climate response to regional radiative forcing during the twentieth century. *Nature Geosci.*, **2**, 294–300.

Shindell, D., and G. Faluvegi, 2010: The net climate impact of coal-fired power plant emissions. *Atmos. Chem. Phys.*, **10**, 3247–3260.

Shindell, D., G. Schmidt, M. Mann, D. Rind, and A. Waple, 2001: Solar forcing of regional climate change during the maunder minimum. *Science*, **294**, 2149–2152.

Shindell, D., G. Faluvegi, A. Lacis, J. Hansen, R. Ruedy, and E. Aguilar, 2006a: Role of tropospheric ozone increases in 20th-century climate change. *J. Geophys. Res. Atmos.*, **111**, D08302.

Shindell, D., G. Faluvegi, R. Miller, G. Schmidt, J. Hansen, and S. Sun, 2006b: Solar and anthropogenic forcing of tropical hydrology. *Geophys. Res. Lett.*, **33**, L24706.

Shindell, D., M. Schulz, Y. Ming, T. Takemura, G. Faluvegi, and V. Ramaswamy, 2010: Spatial scales of climate response to inhomogeneous radiative forcing. *J. Geophys. Res. Atmos.*, **115**, D19110.

Shindell, D., et al., 2008: Climate forcing and air quality change due to regional emissions reductions by economic sector. *Atmos. Chem. Phys.*, **8**, 7101–7113.

Shindell, D., et al., 2011: Climate, health, agricultural and economic impacts of tighter vehicle-emission standards. *Nature Clim. Change*, **1**, 59–66.

Shindell, D., et al., 2013a: Attribution of historical ozone forcing to anthropogenic emissions. *Nature Clim. Change*, **3**, 567-570.

Shindell, D., et al., 2012a: Simultaneously mitigating near-term climate change and improving human health and food Security. *Science*, **335**, 183–189.

Shindell, D. T., 2012: Evaluation of the absolute regional temperature potential. *Atmos. Chem. Phys.*, **12**, 7955–7960.

Shindell, D. T., A. Voulgarakis, G. Faluvegi, and G. Milly, 2012b: Precipitation response to regional radiative forcing. *Atmos. Chem. Phys.*, **12**, 6969–6982.

Shindell, D. T., G. Faluvegi, D. M. Koch, G. A. Schmidt, N. Unger, and S. E. Bauer, 2009: Improved attribution of climate forcing to emissions. *Science*, **326**, 716–718.

Shindell, D. T., et al., 2006c: Simulations of preindustrial, present-day, and 2100 conditions in the NASA GISS composition and climate model G-PUCCINI. *Atmos. Chem. Phys.*, **6**, 4427–4459.

Shindell, D. T., et al., 2013b: Interactive ozone and methane chemistry in GISS-E2 historical and future climate simulations. *Atmos. Chem. Phys.*, **13**, 2653–2689.

Shindell, D. T., et al., 2013c: Radiative forcing in the ACCMIP historical and future climate simulations. *Atmos. Chem. Phys.*, **13**, 2939–2974.

Shine, K., 2009: The global warming potential-the need for an interdisciplinary retrial. *Clim. Change*, **96**, 467–472.

Shine, K., J. Cook, E. Highwood, and M. Joshi, 2003: An alternative to radiative forcing for estimating the relative importance of climate change mechanisms. *Geophys. Res. Lett.*, **30**, 2047.

Shine, K., J. Fuglestvedt, K. Hailemariam, and N. Stuber, 2005a: Alternatives to the global warming potential for comparing climate impacts of emissions of greenhouse gases. *Clim. Change*, **68**, 281–302.

Shine, K., T. Berntsen, J. Fuglestvedt, and R. Sausen, 2005b: Scientific issues in the design of metrics for inclusion of oxides of nitrogen in global climate agreements. *Proc. Natl. Acad. Sci. U.S.A.*, **102**, 15768–15773.

Shine, K., T. Berntsen, J. Fuglestvedt, R. Skeie, and N. Stuber, 2007: Comparing the climate effect of emissions of short- and long-lived climate agents. *Philos. Trans. R. Soc. A*, **365**, 1903–1914.

Shine, K. P., I. V. Ptashnik, and G. Raedel, 2012: The water vapour continuum: Brief history and recent developments. *Surv. Geophys.*, **33**, 535–555.

Siddaway, J. M., and S. V. Petelina, 2011: Transport and evolution of the 2009 Australian Black Saturday bushfire smoke in the lower stratosphere observed by OSIRIS on Odin. *J. Geophys. Res. Atmos.*, **116**, D06203.

Sitch, S., P. M. Cox, W. J. Collins, and C. Huntingford, 2007: Indirect radiative forcing of climate change through ozone effects on the land-carbon sink. *Nature*, **448**, 791–794.

Skeie, R., T. Berntsen, G. Myhre, K. Tanaka, M. Kvalevag, and C. Hoyle, 2011a: Anthropogenic radiative forcing time series from pre-industrial times until 2010. *Atmos. Chem. Phys.*, **11**, 11827–11857.

Skeie, R., T. Berntsen, G. Myhre, C. Pedersen, J. Strom, S. Gerland, and J. Ogren, 2011b: Black carbon in the atmosphere and snow, from pre-industrial times until present. *Atmos. Chem. Phys.*, **11**, 6809–6836.

Skeie, R. B., J. Fuglestvedt, T. Berntsen, M. T. Lund, G. Myhre, and K. Rypdal, 2009: Global temperature change from the transport sectors: Historical development and future scenarios. *Atmos. Environ.*, **43**, 6260–6270.

Smith, E., and A. Balogh, 2008: Decrease in heliospheric magnetic flux in this solar minimum: Recent Ulysses magnetic field observations. *Geophys. Res. Lett.*, **35**, L22103.

Smith, S., and M. Wigley, 2000: Global warming potentials: 1. Climatic implications of emissions reductions. *Clim. Change*, **44**, 445–457.

Smith, S., J. Karas, J. Edmonds, J. Eom, and A. Mizrahi, 2013: Sensitivity of multi-gas climate policy to emission metrics. *Clim. Change*, **117**, 663–675.

Smith, S. M., J. A. Lowe, N. H. A. Bowerman, L. K. Gohar, C. Huntingford, and M. R. Allen, 2012: Equivalence of greenhouse-gas emissions for peak temperature limits. *Nature Clim. Change*, **2**, 535–538.

Snow-Kropla, E., J. Pierce, D. Westervelt, and W. Trivitayanurak, 2011: Cosmic rays, aerosol formation and cloud-condensation nuclei: Sensitivities to model uncertainties. *Atmos. Chem. Phys.*, **11**, 4001–4013.

Solanki, S., and N. Krivova, 2004: Solar irradiance variations: From current measurements to long-term estimates. *Sol. Phys.*, **224**, 197–208.

Solomon, S., 1999: Stratospheric ozone depletion: A review of concepts and history. *Rev. Geophys.*, **37**, 275–316.

Solomon, S., J. S. Daniel, R. R. Neely, J. P. Vernier, E. G. Dutton, and L. W. Thomason, 2011: The persistently variable "background" stratospheric aerosol layer and global climate change. *Science*, **333**, 866–870.

Son, S. W., N. F. Tandon, L. M. Polvani, and D. W. Waugh, 2009: Ozone hole and Southern Hemisphere climate change. *Geophys. Res. Lett.*, **36**, L15705.

Soukharev, B., and L. Hood, 2006: Solar cycle variation of stratospheric ozone: Multiple regression analysis of long-term satellite data sets and comparisons with models. *J. Geophys. Res. Atmos.*, **111**, D20314.

Søvde, O., C. Hoyle, G. Myhre, and I. Isaksen, 2011: The HNO_3 forming branch of the HO_2 + NO reaction: Pre-industrial-to-present trends in atmospheric species and radiative forcings. *Atmos. Chem. Phys.*, **11**, 8929–8943.

Steinhilber, F., J. Beer, and C. Frohlich, 2009: Total solar irradiance during the Holocene. *Geophys. Res. Lett.*, **36**, L19704.

Stenchikov, G., A. Robock, V. Ramaswamy, M. D. Schwarzkopf, K. Hamilton, and S. Ramachandran, 2002: Arctic Oscillation response to the 1991 Mount Pinatubo eruption: Effects of volcanic aerosols and ozone depletion. *J. Geophys. Res. Atmos.*, **107**, 4803.

Stenchikov, G., T. L. Delworth, V. Ramaswamy, R. J. Stouffer, A. Wittenberg, and F. R. Zeng, 2009: Volcanic signals in oceans. *J. Geophys. Res. Atmos.*, **114**, D16104.

Stephens, G. L., N. B. Wood, and L. A. Pakula, 2004: On the radiative effects of dust on tropical convection. *Geophys. Res. Lett.*, **31**, L23112.

Stevenson, D., and R. Derwent, 2009: Does the location of aircraft nitrogen oxide emissions affect their climate impact? *Geophys. Res. Lett.*, **36**, L17810.

Stevenson, D. S., et al., 2013: Tropospheric ozone changes, radiative forcing and attribution to emissions in the Atmospheric Chemistry and Climate Model Intercomparison Project (ACCMIP). *Atmos. Chem. Phys.*, **13**, 3063–3085.

Stevenson, D. S., et al., 2006: Multimodel ensemble simulations of present-day and near-future tropospheric ozone. *J. Geophys. Res. Atmos.*, **111**, D08301.

Stiller, G. P., et al., 2012: Observed temporal evolution of global mean age of stratospheric air for the 2002 to 2010 period. *Atmos. Chem. Phys.*, **12**, 3311–3331.

Stothers, R. B., 2007: Three centuries of observation of stratospheric transparency. *Clim. Change*, **83**, 515–521.

Struthers, H., et al., 2011: The effect of sea ice loss on sea salt aerosol concentrations and the radiative balance in the Arctic. *Atmos. Chem. Phys.*, **11**, 3459–3477.

Swann, A. L. S., I. Y. Fung, and J. C. H. Chiang, 2012: Mid-latitude afforestation shifts general circulation and tropical precipitation. *Proc. Natl. Acad. Sci. U.S.A.*, **109**, 712–716.

Swingedouw, D., L. Terray, C. Cassou, A. Voldoire, D. Salas-Melia, and J. Servonnat, 2011: Natural forcing of climate during the last millennium: Fingerprint of solar variability. *Clim. Dyn.*, **36**, 1349–1364.

Takemura, T., 2012: Distributions and climate effects of atmospheric aerosols from the preindustrial era to 2100 along Representative Concentration Pathways (RCPs) simulated using the global aerosol model SPRINTARS. *Atmos. Chem. Phys.*, **12**, 11555–11572.

Tanaka, K., D. Johansson, B. O'Neill, and J. Fuglestvedt, 2013: Emission metrics under the 2°C climate stabilization. *Clim. Change Lett.* , **117**, 933-941.

Tanaka, K., B. O'Neill, D. Rokityanskiy, M. Obersteiner, and R. Tol, 2009: Evaluating Global Warming Potentials with historical temperature. *Clim. Change*, **96**, 443–466.

Taraborrelli, D., et al., 2012: Hydroxyl radical buffered by isoprene oxidation over tropical forests. *Nature Geosci.*, **5**, 190–193.

Taylor, P. C., R. G. Ellingson, and M. Cai, 2011: Seasonal variations of climate feedbacks in the NCAR CCSM3. *J. Clim.*, **24**, 3433–3444.

Textor, C., et al., 2006: Analysis and quantification of the diversities of aerosol life cycles within AeroCom. *Atmos. Chem. Phys.*, **6**, 1777–1813.

Thomason, L., and T. Peter, 2006: Assessment of Stratospheric Aerosol Properties (ASAP). *SPARC Reports*WCRP-124, WMO/TD- No. 1295, SPARC Report No. 4.

Thompson, D. W. J., S. Solomon, P. J. Kushner, M. H. England, K. M. Grise, and D. J. Karoly, 2011: Signatures of the Antarctic ozone hole in Southern Hemisphere surface climate change. *Nature Geosci.*, **4**, 741–749.

Thonicke, K., A. Spessa, I. C. Prentice, S. P. Harrison, L. Dong, and C. Carmona-Moreno, 2010: The influence of vegetation, fire spread and fire behaviour on biomass burning and trace gas emissions: Results from a process-based model. *Biogeosciences*, **7**, 1991–2011.

Tilmes, S., et al., 2012: Technical Note: Ozonesonde climatology between 1995 and 2011: Description, evaluation and applications. *Atmos. Chem. Phys.*, **12**, 7475–7497.

Timmreck, C., 2012: Modeling the climatic effects of large explosive volcanic eruptions. *Climate Change*, **3**, 545–564.

Timmreck, C., et al., 2010: Aerosol size confines climate response to volcanic super-eruptions. *Geophys. Res. Lett.*, **37**, L24705.

Tol, R., T. Berntsen, B. O'Neill, J. Fuglestvedt, and K. Shine, 2012: A unifying framework for metrics for aggregating the climate effect of different emissions. *Environ. Res. Lett.*, **7**, 044006.

Toon, O. B., A. Robock, and R. P. Turco, 2008: Environmental consequences of nuclear war. *Physics Today*, **61**, 37–42.

Trenberth, K. E., and A. Dai, 2007: Effects of Mount Pinatubo volcanic eruption on the hydrological cycle as an analog of geoengineering. *Geophys. Res. Lett.*, **34**, L15702.

Tsigaridis, K., and M. Kanakidou, 2007: Secondary organic aerosol importance in the future atmosphere. *Atmos. Environ.*, **41**, 4682–4692.

UNEP, 2011: Near-term Climate Protection and Clean Air Benefits: Actions for Controlling Short-Lived Climate Forcers. United Nations Environment Programme (UNEP), 78 pp.

Unger, N., T. C. Bond, J. S. Wang, D. M. Koch, S. Menon, D. T. Shindell, and S. Bauer, 2010: Attribution of climate forcing to economic sectors. *Proc. Natl. Acad. Sci. U.S.A.*, **107**, 3382–3387.

van der Molen, M. K., B. J. J. M. van den Hurk, and W. Hazeleger, 2011: A dampened land use change climate response towards the tropics. *Clim. Dyn.*, **37**, 2035–2043.

van der Werf, G. R., et al., 2010: Global fire emissions and the contribution of deforestation, savanna, forest, agricultural, and peat fires (1997–2009). *Atmos. Chem. Phys.*, **10**, 11707–11735.

van Vuuren, D., J. Edmonds, M. Kainuma, K. Riahi, and J. Weyant, 2011: A special issue on the RCPs. *Clim. Change*, **109**, 1–4.

Van Vuuren, D. P., et al., 2008: Temperature increase of 21st century mitigation scenarios. *Proc. Natl. Acad. Sci. U.S.A.*, **105**, 15258–15262.

Vasekova, E., E. Drage, K. Smith, and N. Mason, 2006: FTIR spectroscopy and radiative forcing of octafluorocyclobutane and octofluorocyclopentene. *J. Quant. Spectrosc. Radiat. Transfer*, **102**, 418–424.

Velders, G. J. M., D. W. Fahey, J. S. Daniel, M. McFarland, and S. O. Andersen, 2009: The large contribution of projected HFC emissions to future climate forcing. *Proc. Natl. Acad. Sci. U.S.A.*, **106**, 10949–10954.

Vernier, J. P., L. W. Thomason, T. D. Fairlie, P. Minnis, R. Palikonda, and K. M. Bedka, 2013: Comment on "Large Volcanic Aerosol Load in the Stratosphere Linked to Asian Monsoon Transport". *Science*, **339**, 647-d.

Vernier, J. P., et al., 2011: Major influence of tropical volcanic eruptions on the stratospheric aerosol layer during the last decade. *Geophys. Res. Lett.*, **38**, L12807.

Vial, J., J.-L. Dufresne, and S. Bony, 2013: On the interpretation of inter-model spread in CMIP5 climate sensitivity estimates. *Clim. Dyn.*, doi:10.1007/s00382-013-1725-9, in press.

Volz, A., and D. Kley, 1988: Evaluation of the Montsouris Series of ozone measurements made in the 19th century. *Nature*, **332**, 240–242.

Voulgarakis, A., and D. T. Shindell, 2010: Constraining the sensitivity of regional climate with the use of historical observations. *J. Clim.*, **23**, 6068–6073.

Voulgarakis, A., et al., 2013: Analysis of present day and future OH and methane lifetime in the ACCMIP simulations. *Atmos. Chem. Phys.*, **13**, 2563–2587.

Wang, C., D. Kim, A. Ekman, M. Barth, and P. Rasch, 2009: Impact of anthropogenic aerosols on Indian summer monsoon. *Geophys. Res. Lett.*, **36**, L21704.

Wang, M., and J. E. Penner, 2009: Aerosol indirect forcing in a global model with particle nucleation. *Atmos. Chem. Phys.*, **9**, 239–260.

Wang, X., S. J. Doherty, and J. Huang, 2013: Black carbon and other light-absorbing impurities in snow across Northern China. *J. Geophys. Res. Atmos.*, **118**, 1471–1492.

Wang, Y., J. Lean, and N. Sheeley, 2005: Modeling the sun's magnetic field and irradiance since 1713. *Astrophys. J.*, **625**, 522–538.

Warren, S., and W. Wiscombe, 1980: A model for the spectral albedo of snow. 2. Snow containing atmospheric aerosols. *J. Atmos. Sci.*, **37**, 2734–2745.

Weiss, R., J. Muhle, P. Salameh, and C. Harth, 2008: Nitrogen trifluoride in the global atmosphere. *Geophys. Res. Lett.*, **35**, L20821.

Wenzler, T., S. Solanki, and N. Krivova, 2009: Reconstructed and measured total solar irradiance: Is there a secular trend between 1978 and 2003? *Geophys. Res. Lett.*, **36**, L11102.

Wilcox, L., K. Shine, and B. Hoskins, 2012: Radiative forcing due to aviation water vapour emissions. *Atmos. Environ.*, **63**, 1–13.

Wild, M., 2009: How well do IPCC-AR4/CMIP3 climate models simulate global dimming/brightening and twentieth-century daytime and nighttime warming? *J. Geophys. Res. Atmos.*, **114**, D00D11.

Wild, O., 2007: Modelling the global tropospheric ozone budget: Exploring the variability in current models. *Atmos. Chem. Phys.*, **7**, 2643–2660.

Wild, O., and P. I. Palmer, 2008: How sensitive is tropospheric oxidation to anthropogenic emissions? *Geophys. Res. Lett.*, **35**, L22802.

Wild, O., M. Prather, and H. Akimoto, 2001: Indirect long-term global radiative cooling from NOx emissions. *Geophys. Res. Lett.*, **28**, 1719–1722.

Williams, K. D., A. Jones, D. L. Roberts, C. A. Senior, and M. J. Woodage, 2001: The response of the climate system to the indirect effects of anthropogenic sulfate aerosol. *Clim. Dyn.*, **17**, 845–856.

Willson, R., and A. Mordvinov, 2003: Secular total solar irradiance trend during solar cycles 21–23. *Geophys. Res. Lett.*, **30**, 1199.

WMO, 1999: *Scientific Assessment of Ozone Depletion: 1998*. Global Ozone Research and Monitoring Project. Report No. 44.World Meteorological Organization, Geneva, Switzerland.

WMO, 2011: *Scientific Assessment of Ozone Depletion: 2010*. Global Ozone Research and Monitoring Project-Report. World Meteorological Organisation, Geneva, Switzerland, 516 pp.

Worden, H., K. Bowman, S. Kulawik, and A. Aghedo, 2011: Sensitivity of outgoing longwave radiative flux to the global vertical distribution of ozone characterized by instantaneous radiative kernels from Aura-TES. *J. Geophys. Res. Atmos.*, **116**, D14115.

Worden, H., K. Bowman, J. Worden, A. Eldering, and R. Beer, 2008: Satellite measurements of the clear-sky greenhouse effect from tropospheric ozone. *Nature Geosci.*, **1**, 305–308.

Wright, J., 2004: Do we know of any Maunder minimum stars? *Astron. J.*, **128**, 1273–1278.

Wu, S. L., L. J. Mickley, D. J. Jacob, J. A. Logan, R. M. Yantosca, and D. Rind, 2007: Why are there large differences between models in global budgets of tropospheric ozone? *J. Geophys. Res. Atmos.*, **112**, D05302.

Xia, L. L., and A. Robock, 2013: Impacts of a nuclear war in South Asia on rice production in Mainland China. *Clim. Change*, **116**, 357–372.

Ye, H., R. Zhang, J. Shi, J. Huang, S. G. Warren, and Q. Fu, 2012: Black carbon in seasonal snow across northern Xinjiang in northwestern China. *Environ. Res. Lett.*, **7**, 044002.

Young, P. J., et al., 2013: Pre-industrial to end 21st century projections of tropospheric ozone from the Atmospheric Chemistry and Climate Model Intercomparison Project (ACCMIP). *Atmos. Chem. Phys.*, **13**, 2063–2090.

Zanchettin, D., et al., 2012: Bi-decadal variability excited in the coupled ocean-atmosphere system by strong tropical volcanic eruptions. *Clim. Dyn.*, **39**, 419–444.

Zarzycki, C. M., and T. C. Bond, 2010: How much can the vertical distribution of black carbon affect its global direct radiative forcing? *Geophys. Res. Lett.*, **37**, L20807.

Zeng, G., J. A. Pyle, and P. J. Young, 2008: Impact of climate change on tropospheric ozone and its global budgets. *Atmos. Chem. Phys.*, **8**, 369–387.

Zeng, G., O. Morgenstern, P. Braesicke, and J. A. Pyle, 2010: Impact of stratospheric ozone recovery on tropospheric ozone and its budget. *Geophys. Res. Lett.*, **37**, L09805.

Zhang, H., G. Y. Shi, and Y. Liu, 2005: A comparison between the two line-by-line integration algorithms. *Chin. J. Atmos. Sci.*, **29**, 581–593.

Zhang, H., G. Shi, and Y. Liu, 2008: The effects of line-wing cutoff in LBL integration on radiation calculations. *Acta Meteorol. Sin.*, **22**, 248–255.

Zhang, H., J. Wu, and P. Luc, 2011: A study of the radiative forcing and global warming potentials of hydrofluorocarbons. *J. Quant. Spectrosc. Radiat. Transfer*, **112**, 220–229.

Zhang, X. B., et al., 2007: Detection of human influence on twentieth-century precipitation trends. *Nature*, **448**, 461–465.

Zhong, Y., G. H. Miller, B. L. Otto-Bliesner, M. M. Holland, D. A. Bailey, D. P. Schneider, and A. Geirsdottir, 2011: Centennial-scale climate change from decadally-paced explosive volcanism: A coupled sea ice-ocean mechanism. *Clim. Dyn.*, **37**, 2373–2387.

Ziemke, J. R., S. Chandra, G. J. Labow, P. K. Bhartia, L. Froidevaux, and J. C. Witte, 2011: A global climatology of tropospheric and stratospheric ozone derived from Aura OMI and MLS measurements. *Atmos. Chem. Phys.*, **11**, 9237–9251.

Appendix 8.A: Lifetimes, Radiative Efficiencies and Metric Values

Table 8.A.1 | Radiative efficiencies (REs), lifetimes/adjustment times, AGWP and GWP values for 20 and 100 years, and AGTP and GTP values for 20, 50 and 100 years. Climate–carbon feedbacks are included for CO_2 while no climate feedbacks are included for the other components (see discussion in Sections 8.7.1.4 and 8.7.2.1, Supplementary Material and notes below the table; Supplementary Material Table 8.SM.16 gives analogous values including climate–carbon feedbacks for non-CO_2 emissions). For a complete list of chemical names and CAS numbers, and for accurate replications of metric values, see Supplementary Material Section 8.SM.13 and references therein.

Acronym, Common Name or Chemical Name	Chemical Formula	Lifetime (Years)	Radiative Efficiency (W m^{-2} ppb^{-1})	AGWP 20-year (W m^{-2} yr kg^{-1})	**GWP 20-year**	AGWP 100-year (W m^{-2} yr kg^{-1})	**GWP 100-year**	AGTP 20-year (K kg^{-1})	**GTP 20-year**	AGTP 50-year (K kg^{-1})	**GTP 50-year**	AGTP 100-year (K kg^{-1})	**GTP 100-year**
Carbon dioxide	CO_2	see*	1.37e-5	2.49e-14	1	9.17e-14	1	6.84e-16	1	6.17e-16	1	5.47e-16	1
Methane	CH_4	12.4†	3.63e-4	2.09e-12	84	2.61e-12	28	4.62e-14	67	8.69e-15	14	2.34e-15	4
Fossil methane‡	CH_4	12.4†	3.63e-4	2.11e-12	85	2.73e-12	30	4.68e-14	68	9.55e-15	15	3.11e-15	6
Nitrous Oxide	N_2O	121†	3.00e-3	6.58e-12	264	2.43e-11	265	1.89e-13	277	1.74e-13	282	1.28e-13	234
Chlorofluorocarbons													
CFC-11	CCl_3F	45.0	0.26	1.72e-10	6900	4.28e-10	4660	4.71e-12	6890	3.01e-12	4890	1.28e-12	2340
CFC-12	CCl_2F_2	100.0	0.32	2.69e-10	10,800	9.39e-10	10,200	7.71e-12	11,300	6.75e-12	11,000	4.62e-12	8450
CFC-13	$CClF_3$	640.0	0.25	2.71e-10	10,900	1.27e-09	13,900	7.99e-12	11,700	8.77e-12	14,200	8.71e-12	15,900
CFC-113	CCl_2FCClF_2	85.0	0.30	1.62e-10	6490	5.34e-10	5820	4.60e-12	6730	3.85e-12	6250	2.45e-12	4470
CFC-114	$CClF_2CClF_2$	190.0	0.31	1.92e-10	7710	7.88e-10	8590	5.60e-12	8190	5.56e-12	9020	4.68e-12	8550
CFC-115	$CClF_2CF_3$	1,020.0	0.20	1.46e-10	5860	7.03e-10	7670	4.32e-12	6310	4.81e-12	7810	4.91e-12	8980
Hydrochlorofluorocarbons													
HCFC-21	$CHCl_2F$	1.7	0.15	1.35e-11	543	1.35e-11	148	1.31e-13	192	1.59e-14	26	1.12e-14	20
HCFC-22	$CHClF_2$	11.9	0.21	1.32e-10	5280	1.62e-10	1760	2.87e-12	4200	5.13e-13	832	1.43e-13	262
HCFC-122	$CHCl_2CF_2Cl$	1.0	0.17	5.43e-12	218	5.43e-12	59	4.81e-14	70	6.25e-15	10	4.47e-15	8
HCFC-122a	$CHFClCFCl_2$	3.4	0.21	2.36e-11	945	2.37e-11	258	2.91e-13	426	2.99e-14	48	1.96e-14	36
HCFC-123	$CHCl_2CF_3$	1.3	0.15	7.28e-12	292	7.28e-12	79	6.71e-14	98	8.45e-15	14	6.00e-15	11
HCFC-123a	$CHClFCF_2Cl$	4.0	0.23	3.37e-11	1350	3.39e-11	370	4.51e-13	659	4.44e-14	72	2.81e-14	51
HCFC-124	$CHClFCF_3$	5.9	0.20	4.67e-11	1870	4.83e-11	527	7.63e-13	1120	7.46e-14	121	4.03e-14	74
HCFC-132c	CH_2FCFCl_2	4.3	0.17	3.07e-11	1230	3.10e-11	338	4.27e-13	624	4.14e-14	67	2.58e-14	47
HCFC-141b	CH_3CCl_2F	9.2	0.16	6.36e-11	2550	7.17e-11	782	1.27e-12	1850	1.67e-13	271	6.09e-14	111
HCFC-142b	CH_3CClF_2	17.2	0.19	1.25e-10	5020	1.82e-10	1980	3.01e-12	4390	8.46e-13	1370	1.95e-13	356
HCFC-225ca	$CHCl_2CF_2CF_3$	1.9	0.22	1.17e-11	469	1.17e-11	127	1.17e-13	170	1.38e-14	22	9.65e-15	18
HCFC-225cb	$CHClFCF_2CClF_2$	5.9	0.29	4.65e-11	1860	4.81e-11	525	7.61e-13	1110	7.43e-14	120	4.01e-14	73
(E)-1-Chloro-3,3,3-trifluoroprop-1-ene	trans-CF_3CH=CHCl	26.0 days	0.04	1.37e-13	5	1.37e-13	1	1.09e-15	2	1.54e-16	<1	1.12e-16	<1

(continued on next page)

Table 8.A.1 (continued)

Acronym, Common Name or Chemical Name	Chemical Formula	Lifetime (Years)	Radiative Efficiency (W m^{-2} ppb^{-1})	AGWP 20-year (W m^{-2} yr kg^{-1})	**GWP 20-year**	AGWP 100-year (W m^{-2} yr kg^{-1})	**GWP 100-year**	AGTP 20-year (K kg^{-1})	**GTP 20-year**	AGTP 50-year (K kg^{-1})	**GTP 50-year**	AGTP 100-year (K kg^{-1})	**GTP 100-year**
Hydrofluorocarbons													
HFC-23	CHF_3	222.0	0.18	2.70e-10	**10,800**	1.14e-09	**12,400**	7.88e-12	**11,500**	7.99e-12	**13,000**	6.95e-12	**12,700**
HFC-32	CH_2F_2	5.2	0.11	6.07e-11	**2430**	6.21e-11	**677**	9.32e-13	**1360**	8.93e-14	**145**	5.17e-14	**94**
HFC-41	CH_3F	2.8	0.02	1.07e-11	**427**	1.07e-11	**116**	1.21e-13	**177**	1.31e-14	**21**	8.82e-15	**16**
HFC-125	CHF_2CF_3	28.2	0.23	1.52e-10	**6090**	2.91e-10	**3170**	3.97e-12	**5800**	1.84e-12	**2980**	5.29e-13	**967**
HFC-134	CHF_2CHF_2	9.7	0.19	8.93e-11	**3580**	1.02e-10	**1120**	1.82e-12	**2660**	2.54e-13	**412**	8.73e-14	**160**
HFC-134a	CH_2FCF_3	13.4	0.16	9.26e-11	**3710**	1.19e-10	**1300**	2.09e-12	**3050**	4.33e-13	**703**	1.10e-13	**201**
HFC-143	CH_2FCHF_2	3.5	0.13	3.00e-11	**1200**	3.01e-11	**328**	3.76e-13	**549**	3.82e-14	**62**	2.49e-14	**46**
HFC-143a	CH_3CF_3	47.1	0.16	1.73e-10	**6940**	4.41e-10	**4800**	4.76e-12	**6960**	3.12e-12	**5060**	1.37e-12	**2500**
HFC-152	CH_2FCH_2F	0.4	0.04	1.51e-12	**60**	1.51e-12	**16**	1.25e-14	**18**	1.71e-15	**3**	1.24e-15	**2**
HFC-152a	CH_3CHF_2	1.5	0.10	1.26e-11	**506**	1.26e-11	**138**	1.19e-13	**174**	1.47e-14	**24**	1.04e-14	**19**
HFC-161	CH_3CH_2F	66.0 days	0.02	3.33e-13	**13**	3.33e-13	**4**	2.70e-15	**4**	3.76e-16	**<1**	2.74e-16	**<1**
HFC-227ca	$CF_3CF_2CHF_2$	28.2	0.27	1.27e-10	**5080**	2.42e-10	**2640**	3.31e-12	**4830**	1.53e-12	**2480**	4.41e-13	**806**
HFC-227ea	CF_3CHFCF_3	38.9	0.26	1.34e-10	**5360**	3.07e-10	**3350**	3.61e-12	**5280**	2.12e-12	**3440**	7.98e-13	**1460**
HFC-236cb	$CH_2FCF_2CF_3$	13.1	0.23	8.67e-11	**3480**	1.11e-10	**1210**	1.94e-12	**2840**	3.92e-13	**636**	1.01e-13	**185**
HFC-236ea	CHF_2CHFCF_3	11.0	0.30[a]	1.03e-10	**4110**	1.22e-10	**1330**	2.18e-12	**3190**	3.53e-13	**573**	1.06e-13	**195**
HFC-236fa	$CF_3CH_2CF_3$	242.0	0.24	1.73e-10	**6940**	7.39e-10	**8060**	5.06e-12	**7400**	5.18e-12	**8400**	4.58e-12	**8380**
HFC-245ca	$CH_2FCF_2CHF_2$	6.5	0.24[b]	6.26e-11	**2510**	6.56e-11	**716**	1.07e-12	**1570**	1.09e-13	**176**	5.49e-14	**100**
HFC-245cb	$CF_3CF_2CH_3$	47.1	0.24	1.67e-10	**6680**	4.24e-10	**4620**	4.58e-12	**6690**	3.00e-12	**4870**	1.32e-12	**2410**
HFC-245ea	$CHF_2CHFCHF_2$	3.2	0.16[c]	2.15e-11	**863**	2.16e-11	**235**	2.59e-13	**378**	2.70e-14	**44**	1.79e-14	**33**
HFC-245eb	$CH_2FCHFCF_3$	3.1	0.20[c]	2.66e-11	**1070**	2.66e-11	**290**	3.15e-13	**460**	3.31e-14	**54**	2.20e-14	**40**
HFC-245fa	$CHF_2CH_2CF_3$	7.7	0.24	7.29e-11	**2920**	7.87e-11	**858**	1.35e-12	**1970**	1.51e-13	**245**	6.62e-14	**121**
HFC-263fb	$CH_3CH_2CF_3$	1.2	0.10[c]	6.93e-12	**278**	6.93e-12	**76**	6.31e-14	**92**	8.02e-15	**13**	5.70e-15	**10**
HFC-272ca	$CH_3CF_2CH_3$	2.6	0.07	1.32e-11	**530**	1.32e-11	**144**	1.46e-13	**213**	1.61e-14	**26**	1.09e-14	**20**
HFC-329p	$CHF_2CF_2CF_2CF_3$	28.4	0.31	1.13e-10	**4510**	2.16e-10	**2360**	2.94e-12	**4290**	1.37e-12	**2220**	3.96e-13	**725**
HFC-365mfc	$CH_3CF_2CH_2CF_3$	8.7	0.22	6.64e-11	**2660**	7.38e-11	**804**	1.30e-12	**1890**	1.62e-13	**262**	6.24e-14	**114**
HFC-43-10mee	$CF_3CHFCHFCF_2CF_3$	16.1	0.42[b]	1.08e-10	**4310**	1.51e-10	**1650**	2.54e-12	**3720**	6.62e-13	**1070**	1.54e-13	**281**
HFC-1132a	$CH_2{=}CF_2$	4.0 days	0.004[d]	3.87e-15	**<1**	3.87e-15	**<1**	3.08e-17	**<1**	4.35e-18	**<1**	3.18e-18	**<1**
HFC-1141	$CH_2{=}CHF$	2.1 days	0.002[d]	1.54e-15	**<1**	1.54e-15	**<1**	1.23e-17	**<1**	1.73e-18	**<1**	1.27e-18	**<1**
(Z)-HFC-1225ye	$CF_3CF{=}CHF(Z)$	8.5 days	0.02	2.14e-14	**<1**	2.14e-14	**<1**	1.70e-16	**<1**	2.40e-17	**<1**	1.76e-17	**<1**
(E)-HFC-1225ye	$CF_3CF{=}CHF(E)$	4.9 days	0.01	7.25e-15	**<1**	7.25e-15	**<1**	5.77e-17	**<1**	8.14e-18	**<1**	5.95e-18	**<1**
(Z)-HFC-1234ze	$CF_3CH{=}CHF(Z)$	10.0 days	0.02	2.61e-14	**1**	2.61e-14	**<1**	2.08e-16	**<1**	2.93e-17	**<1**	2.14e-17	**<1**
HFC-1234yf	$CF_3CF{=}CH_2$	10.5 days	0.02	3.22e-14	**1**	3.22e-14	**<1**	2.57e-16	**<1**	3.62e-17	**<1**	2.65e-17	**<1**
(E)-HFC-1234ze	trans-$CF_3CH{=}CHF$	16.4 days	0.04	8.74e-14	**4**	8.74e-14	**<1**	6.98e-16	**<1**	9.82e-17	**<1**	7.18e-17	**<1**
(Z)-HFC-1336	$CF_3CH{=}CHCF_3(Z)$	22.0 days	0.07[d]	1.54e-13	**6**	1.54e-13	**2**	1.23e-15	**2**	1.73e-16	**<1**	1.26e-16	**<1**

(continued on next page)

Table 8.A.1 (continued)

Acronym, Common Name or Chemical Name	Chemical Formula	Lifetime (Years)	Radiative Efficiency ($W m^{-2} ppb^{-1}$)	AGWP 20-year ($W m^{-2} yr kg^{-1}$)	**GWP 20-year**	AGWP 100-year ($W m^{-2} yr kg^{-1}$)	**GWP 100-year**	AGTP 20-year ($K kg^{-1}$)	**GTP 20-year**	AGTP 50-year ($K kg^{-1}$)	**GTP 50-year**	AGTP 100-year ($K kg^{-1}$)	**GTP 100-year**
HFC-1243zf	$CF_3CH=CH_2$	7.0 days	0.01	1.37e-14	**1**	1.37e-14	**<1**	1.09e-16	**<1**	1.53e-17	**<1**	1.12e-17	**<1**
HFC-1345zfc	$C_2F_5CH=CH_2$	7.6 days	0.01	1.15e-14	**<1**	1.15e-14	**<1**	9.19e-17	**<1**	1.30e-17	**<1**	9.48e-18	**<1**
3,3,4,4,5,5,6,6,6-Nonafluorohex-1-ene	$C_4F_9CH=CH_2$	7.6 days	0.03	1.25e-14	**<1**	1.25e-14	**<1**	9.92e-17	**<1**	1.40e-17	**<1**	1.02e-17	**<1**
3,3,4,4,5,5,6,6,7,7,8,8,8-Tridecafluorooct-1-ene	$C_6F_{13}CH=CH_2$	7.6 days	0.03	9.89e-15	**<1**	9.89e-15	**<1**	7.87e-17	**<1**	1.11e-17	**<1**	8.12e-18	**<1**
3,3,4,4,5,5,6,6,7,7,8,8,9,9,10,10,10-Heptadecafluorodec-1-ene	$C_8F_{17}CH=CH_2$	7.6 days	0.03	8.52e-15	**<1**	8.52e-15	**<1**	6.79e-17	**<1**	9.57e-18	**<1**	7.00e-18	**<1**
Chlorocarbons and Hydrochlorocarbons													
Methyl chloroform	CH_3CCl_3	5.0	0.07	1.44e-11	**578**	1.47e-11	**160**	2.17e-13	**317**	2.07e-14	**34**	1.22e-14	**22**
Carbon tetrachloride	CCl_4	26.0	0.17	8.69e-11	**3480**	1.59e-10	**1730**	2.24e-12	**3280**	9.68e-13	**1570**	2.62e-13	**479**
Methyl chloride	CH_3Cl	1.0	0.01[a]	1.12e-12	**45**	1.12e-12	**12**	9.93e-15	**15**	1.29e-15	**2**	9.20e-16	**2**
Methylene chloride	CH_2Cl_2	0.4	0.03[b]	8.18e-13	**33**	8.18e-13	**9**	6.78e-15	**10**	9.26e-16	**2**	6.72e-16	**1**
Chloroform	$CHCl_3$	0.4	0.08	1.50e-12	**60**	1.50e-12	**16**	1.25e-14	**18**	1.70e-15	**3**	1.24e-15	**2**
1,2-Dichloroethane	CH_2ClCH_2Cl	65.0 days	0.01	8.24e-14	**3**	8.24e-14	**<1**	6.67e-16	**<1**	9.29e-17	**<1**	6.77e-17	**<1**
Bromocarbons, Hydrobromocarbons and Halons													
Methyl bromide	CH_3Br	0.8	0.004	2.16e-13	**9**	2.16e-13	**2**	1.87e-15	**3**	2.47e-16	**<1**	1.78e-16	**<1**
Methylene bromide	CH_2Br_2	0.3	0.01	9.31e-14	**4**	9.31e-14	**1**	7.66e-16	**1**	1.05e-16	**<1**	7.65e-17	**<1**
Halon-1201	$CHBrF_2$	5.2	0.15	3.37e-11	**1350**	3.45e-11	**376**	5.17e-13	**756**	4.96e-14	**80**	2.87e-14	**52**
Halon-1202	CBr_2F_2	2.9	0.27	2.12e-11	**848**	2.12e-11	**231**	2.43e-13	**356**	2.61e-14	**42**	1.75e-14	**32**
Halon-1211	$CBrClF_2$	16.0	0.29	1.15e-10	**4590**	1.60e-10	**1750**	2.70e-12	**3950**	6.98e-13	**1130**	1.62e-13	**297**
Halon-1301	$CBrF_3$	65.0	0.30	1.95e-10	**7800**	5.77e-10	**6290**	5.46e-12	**7990**	4.16e-12	**6750**	2.28e-12	**4170**
Halon-2301	CH_2BrCF_3	3.4	0.14	1.59e-11	**635**	1.59e-11	**173**	1.96e-13	**286**	2.01e-14	**33**	1.32e-14	**24**
Halon-2311 / Halothane	$CHBrClCF_3$	1.0	0.13	3.77e-12	**151**	3.77e-12	**41**	3.35e-14	**49**	4.34e-15	**7**	3.10e-15	**6**
Halon-2401	$CHFBrCF_3$	2.9	0.19	1.68e-11	**674**	1.68e-11	**184**	1.94e-13	**283**	2.07e-14	**34**	1.39e-14	**25**
Halon-2402	$CBrF_2CBrF_2$	20.0	0.31	8.59e-11	**3440**	1.35e-10	**1470**	2.12e-12	**3100**	7.08e-13	**1150**	1.66e-13	**304**
Fully Fluorinated Species													
Nitrogen trifluoride	NF_3	500.0	0.20	3.19e-10	**12,800**	1.47e-09	**16,100**	9.39e-12	**13,700**	1.02e-11	**16,500**	9.91e-12	**18,100**
Sulphur hexafluoride	SF_6	3,200.0	0.57	4.37e-10	**17,500**	2.16e-09	**23,500**	1.29e-11	**18,900**	1.47e-11	**23,800**	1.54e-11	**28,200**
(Trifluoromethyl) sulphur pentafluoride	SF_5CF_3	800.0	0.59	3.36e-10	**13,500**	1.60e-09	**17,400**	9.93e-12	**14,500**	1.10e-11	**17,800**	1.11e-11	**20,200**
Sulphuryl fluoride	SO_2F_2	36.0	0.20	1.71e-10	**6840**	3.76e-10	**4090**	4.58e-12	**6690**	2.55e-12	**4140**	9.01e-13	**1650**
PFC-14	CF_4	50,000.0	0.09	1.22e-10	**4880**	6.08e-10	**6630**	3.61e-12	**5270**	4.12e-12	**6690**	4.40e-12	**8040**
PFC-116	C_2F_6	10,000.0	0.25	2.05e-10	**8210**	1.02e-09	**11,100**	6.07e-12	**8880**	6.93e-12	**11,200**	7.36e-12	**13,500**
PFC-c216	$c-C_3F_6$	3,000.0	0.23[e]	1.71e-10	**6850**	8.44e-10	**9200**	5.06e-12	**7400**	5.74e-12	**9310**	6.03e-12	**11,000**
PFC-218	C_3F_8	2,600.0	0.28	1.66e-10	**6640**	8.16e-10	**8900**	4.91e-12	**7180**	5.56e-12	**9010**	5.83e-12	**10,700**
PFC-318	$c-C_4F_8$	3,200.0	0.32	1.77e-10	**7110**	8.75e-10	**9540**	5.25e-12	**7680**	5.96e-12	**9660**	6.27e-12	**11,500**

(continued on next page)

Table 8.A.1 (continued)

Acronym, Common Name or Chemical Name	Chemical Formula	Lifetime (Years)	Radiative Efficiency ($W m^{-2} ppb^{-1}$)	AGWP 20-year ($W m^{-2} yr kg^{-1}$)	**GWP 20-year**	AGWP 100-year ($W m^{-2} yr kg^{-1}$)	**GWP 100-year**	AGTP 20-year ($K kg^{-1}$)	**GTP 20-year**	AGTP 50-year ($K kg^{-1}$)	**GTP 50-year**	AGTP 100-year ($K kg^{-1}$)	**GTP 100-year**
PFC-31-10	C_4F_{10}	2,600.0	0.36	1.71e-10	**6870**	8.44e-10	**9200**	5.08e-12	**7420**	5.75e-12	**9320**	6.02e-12	**11,000**
Perfluorocyclopentene	$c-C_5F_8$	31.0 days	0.08[f]	1.71e-13	**7**	1.71e-13	**2**	1.37e-15	**2**	1.92e-16	**<1**	1.40e-16	**<1**
PFC-41-12	$n-C_5F_{12}$	4,100.0	0.41	1.58e-10	**6350**	7.84e-10	**8550**	4.69e-12	**6860**	5.33e-12	**8650**	5.62e-12	**10,300**
PFC-51-14	$n-C_6F_{14}$	3,100.0	0.44	1.47e-10	**5890**	7.26e-10	**7910**	4.35e-12	**6370**	4.94e-12	**8010**	5.19e-12	**9490**
PFC-61-16	$n-C_7F_{16}$	3,000.0	0.50	1.45e-10	**5830**	7.17e-10	**7820**	4.31e-12	**6290**	4.88e-12	**7920**	5.13e-12	**9380**
PFC-71-18	C_8F_{18}	3,000.0	0.55	1.42e-10	**5680**	6.99e-10	**7620**	4.20e-12	**6130**	4.76e-12	**7710**	5.00e-12	**9140**
PFC-91-18	$C_{10}F_{18}$	2,000.0	0.55	1.34e-10	**5390**	6.59e-10	**7190**	3.98e-12	**5820**	4.49e-12	**7290**	4.68e-12	**8570**
Perfluorodecalin (cis)	$Z-C_{10}F_{18}$	2,000.0	0.56	1.35e-10	**5430**	6.64e-10	**7240**	4.01e-12	**5860**	4.52e-12	**7340**	4.72e-12	**8630**
Perfluorodecalin (trans)	$E-C_{10}F_{18}$	2,000.0	0.48	1.18e-10	**4720**	5.77e-10	**6290**	3.48e-12	**5090**	3.93e-12	**6380**	4.10e-12	**7500**
PFC-1114	$CF_2=CF_2$	1.1 days	0.002	2.68e-16	**<1**	2.68e-16	**<1**	2.13e-18	**<1**	3.00e-19	**<1**	2.20e-19	**<1**
PFC-1216	$CF_3CF=CF_2$	4.9 days	0.01	6.42e-15	**<1**	6.42e-15	**<1**	5.11e-17	**<1**	7.21e-18	**<1**	5.27e-18	**<1**
Perfluorobuta-1,3-diene	$CF_2=CFCF=CF_2$	1.1 days	0.003	3.29e-16	**<1**	3.29e-16	**<1**	2.61e-18	**<1**	3.69e-19	**<1**	2.70e-19	**<1**
Perfluorobut-1-ene	$CF_3CF_2CF=CF_2$	6.0 days	0.02	8.38e-15	**<1**	8.38e-15	**<1**	6.67e-17	**<1**	9.41e-18	**<1**	6.88e-18	**<1**
Perfluorobut-2-ene	$CF_3CF=CFCF_3$	31.0 days	0.07	1.62e-13	**6**	1.62e-13	**2**	1.30e-15	**2**	1.82e-16	**<1**	1.33e-16	**<1**
Halogenated Alcohols and Ethers													
HFE-125	CHF_2OCF_3	119.0	0.41	3.10e-10	**12,400**	1.14e-09	**12,400**	8.91e-12	**13,000**	8.14e-12	**13,200**	5.97e-12	**10,900**
HFE-134 (HG-00)	CHF_2OCHF_2	24.4	0.44	2.90e-10	**11,600**	5.10e-10	**5560**	7.42e-12	**10,800**	3.02e-12	**4900**	7.83e-13	**1430**
HFE-143a	CH_3OCF_3	4.8	0.18	4.72e-11	**1890**	4.80e-11	**523**	6.95e-13	**1020**	6.66e-14	**108**	3.99e-14	**73**
HFE-227ea	$CF_3CHFOCF_3$	51.6	0.44	2.22e-10	**8900**	5.92e-10	**6450**	6.15e-12	**8980**	4.22e-12	**6850**	1.98e-12	**3630**
HCFE-235ca2 (enflurane)	CHF_2OCF_2CHFCl	4.3	0.41	5.30e-11	**2120**	5.35e-11	**583**	7.36e-13	**1080**	7.14e-14	**116**	4.44e-14	**81**
HCFE-235da2 (isoflurane)	$CHF_2OCHClCF_3$	3.5	0.42	4.49e-11	**1800**	4.50e-11	**491**	5.62e-13	**822**	5.72e-14	**93**	3.73e-14	**68**
HFE-236ca	$CHF_2OCF_2CHF_2$	20.8	0.56[g]	2.42e-10	**9710**	3.89e-10	**4240**	6.03e-12	**8820**	2.10e-12	**3400**	4.98e-13	**912**
HFE-236ea2 (desflurane)	$CHF_2OCHFCF_3$	10.8	0.45	1.39e-10	**5550**	1.64e-10	**1790**	2.93e-12	**4280**	4.64e-13	**753**	1.42e-13	**260**
HFE-236fa	$CF_3CH_2OCF_3$	7.5	0.36	8.35e-11	**3350**	8.98e-11	**979**	1.53e-12	**2240**	1.68e-13	**273**	7.54e-14	**138**
HFE-245cb2	$CF_3CF_2OCH_3$	4.9	0.33	5.90e-11	**2360**	6.00e-11	**654**	8.77e-13	**1280**	8.40e-14	**136**	4.99e-14	**91**
HFE-245fa1	$CHF_2CH_2OCF_3$	6.6	0.31	7.22e-11	**2900**	7.59e-11	**828**	1.25e-12	**1820**	1.27e-13	**206**	6.35e-14	**116**
HFE-245fa2	$CHF_2OCH_2CF_3$	5.5	0.36	7.25e-11	**2910**	7.45e-11	**812**	1.15e-12	**1670**	1.10e-13	**179**	6.21e-14	**114**
2,2,3,3,3-Pentafluoropropan-1-ol	$CF_3CF_2CH_2OH$	0.3	0.14	1.72e-12	**69**	1.72e-12	**19**	1.42e-14	**21**	1.95e-15	**3**	1.42e-15	**3**
HFE-254cb1	$CH_3OCF_2CHF_2$	2.5	0.26	2.76e-11	**1110**	2.76e-11	**301**	2.99e-13	**438**	3.34e-14	**54**	2.28e-14	**42**
HFE-263fb2	$CF_3CH_2OCH_3$	23.0 days	0.04	1.22e-13	**5**	1.22e-13	**1**	9.72e-16	**1**	1.37e-16	**<1**	9.98e-17	**<1**
HFE-263m1	$CF_3OCH_2CH_3$	0.4	0.13	2.70e-12	**108**	2.70e-12	**29**	2.25e-14	**33**	3.06e-15	**5**	2.22e-15	**4**
3,3,3-Trifluoropropan-1-ol	$CF_3CH_2CH_2OH$	12.0 days	0.02	3.57e-14	**1**	3.57e-14	**<1**	2.85e-16	**<1**	4.01e-17	**<1**	2.93e-17	**<1**
HFE-329mcc2	$CHF_2CF_2OCF_2CF_3$	22.5	0.53	1.68e-10	**6720**	2.81e-10	**3070**	4.23e-12	**6180**	1.59e-12	**2580**	3.93e-13	**718**
HFE-338mmz1	$(CF_3)_2CHOCHF_2$	21.2	0.44	1.48e-10	**5940**	2.40e-10	**2620**	3.70e-12	**5410**	1.31e-12	**2130**	3.14e-13	**575**

(continued on next page)

Table 8.A.1 (continued)

Acronym, Common Name or Chemical Name	Chemical Formula	Lifetime (Years)	Radiative Efficiency ($W m^{-2} ppb^{-1}$)	AGWP 20-year ($W m^{-2} yr kg^{-1}$)	**GWP 20-year**	AGWP 100-year ($W m^{-2} yr kg^{-1}$)	**GWP 100-year**	AGTP 20-year ($K kg^{-1}$)	**GTP 20-year**	AGTP 50-year ($K kg^{-1}$)	**GTP 50-year**	AGTP 100-year ($K kg^{-1}$)	**GTP 100-year**
HFE-338mcf2	$CF_3CH_2OCF_2CF_3$	7.5	0.44	7.93e-11	**3180**	8.52e-11	**929**	1.45e-12	**2120**	1.60e-13	**259**	7.16e-14	**131**
Sevoflurane (HFE-347mmz1)	$(CF_3)_2CHOCH_2F$	2.2	0.32	1.98e-11	**795**	1.98e-11	**216**	2.06e-13	**302**	2.37e-14	**38**	1.64e-14	**30**
HFE-347mcc3 (HFE-7000)	$CH_3OCF_2CF_2CF_3$	5.0	0.35	4.78e-11	**1910**	4.86e-11	**530**	7.18e-13	**1050**	6.87e-14	**111**	4.05e-14	**74**
HFE-347mcf2	$CHF_2CH_2OCF_2CF_3$	6.6	0.42	7.45e-11	**2990**	7.83e-11	**854**	1.29e-12	**1880**	1.31e-13	**212**	6.55e-14	**120**
HFE-347pcf2	$CHF_2CF_2OCH_2CF_3$	6.0	0.48[h]	7.86e-11	**3150**	8.15e-11	**889**	1.30e-12	**1900**	1.27e-13	**206**	6.81e-14	**124**
HFE-347mmy1	$(CF_3)_2CFOCH_3$	3.7	0.32	3.32e-11	**1330**	3.33e-11	**363**	4.27e-13	**624**	4.28e-14	**69**	2.76e-14	**51**
HFE-356mec3	$CH_3OCF_2CHFCF_3$	3.8	0.30	3.53e-11	**1410**	3.55e-11	**387**	4.60e-13	**673**	4.58e-14	**74**	2.94e-14	**54**
HFE-356mff2	$CF_3CH_2OCH_2CF_3$	105.0 days	0.17	1.54e-12	**62**	1.54e-12	**17**	1.26e-14	**18**	1.74e-15	**3**	1.26e-15	**2**
HFE-356pcf2	$CHF_2CH_2OCF_2CHF_2$	5.7	0.37	6.40e-11	**2560**	6.59e-11	**719**	1.03e-12	**1500**	9.97e-14	**162**	5.50e-14	**101**
HFE-356pcf3	$CHF_2OCH_2CF_2CHF_2$	3.5	0.38	4.08e-11	**1640**	4.09e-11	**446**	5.11e-13	**747**	5.20e-14	**84**	3.39e-14	**62**
HFE-356pcc3	$CH_3OCF_2CF_2CHF_2$	3.8	0.32	3.77e-11	**1510**	3.79e-11	**413**	4.91e-13	**718**	4.89e-14	**79**	3.14e-14	**57**
HFE-356mmz1	$(CF_3)_2CHOCH_3$	97.1 days	0.15	1.25e-12	**50**	1.25e-12	**14**	1.02e-14	**15**	1.41e-15	**2**	1.02e-15	**2**
HFE-365mcf3	$CF_3CF_2CH_2OCH_3$	19.3 days	0.05	8.51e-14	**3**	8.51e-14	**<1**	6.80e-16	**<1**	9.56e-17	**<1**	6.99e-17	**<1**
HFE-365mcf2	$CF_3CF_2OCH_2CH_3$	0.6	0.26[i]	5.35e-12	**215**	5.35e-12	**58**	4.53e-14	**66**	6.10e-15	**10**	4.40e-15	**8**
HFE-374pc2	$CHF_2CF_2OCH_2CH_3$	5.0	0.30	5.65e-11	**2260**	5.75e-11	**627**	8.48e-13	**1240**	8.12e-14	**132**	4.79e-14	**88**
4,4,4-Trifluorobutan-1-ol	$CF_3(CH_2)_2CH_2OH$	4.0 days	0.01	1.73e-15	**<1**	1.73e-15	**<1**	1.38e-17	**<1**	1.94e-18	**<1**	1.42e-18	**<1**
2,2,3,3,4,4,5,5-Octafluorocyclopentanol	$-(CF_2)_4CH(OH)-$	0.3	0.16	1.18e-12	**47**	1.18e-12	**13**	9.67e-15	**14**	1.33e-15	**2**	9.69e-16	**2**
HFE-43-10pccc124 (H-Galden 1040x, HG-11)	$CHF_2OCF_2OC_2F_4OCHF_2$	13.5	1.02	2.00e-10	**8010**	2.58e-10	**2820**	4.52e-12	**6600**	9.46e-13	**1530**	2.38e-13	**436**
HFE-449s1 (HFE-7100)	$C_4F_9OCH_3$	4.7	0.36	3.80e-11	**1530**	3.86e-11	**421**	5.54e-13	**809**	5.32e-14	**86**	3.21e-14	**59**
n-HFE-7100	$n\text{-}C_4F_9OCH_3$	4.7	0.42	4.39e-11	**1760**	4.45e-11	**486**	6.39e-13	**934**	6.14e-14	**99**	3.70e-14	**68**
i-HFE-7100	$i\text{-}C_4F_9OCH_3$	4.7	0.35	3.68e-11	**1480**	3.73e-11	**407**	5.35e-13	**783**	5.14e-14	**83**	3.10e-14	**57**
HFE-569sf2 (HFE-7200)	$C_4F_9OC_2H_5$	0.8	0.30	5.21e-12	**209**	5.21e-12	**57**	4.52e-14	**66**	5.97e-15	**10**	4.29e-15	**8**
n-HFE-7200	$n\text{-}C_4F_9OC_2H_5$	0.8	0.35[i]	5.92e-12	**237**	5.92e-12	**65**	5.14e-14	**75**	6.78e-15	**11**	4.87e-15	**9**
i-HFE-7200	$i\text{-}C_4F_9OC_2H_5$	0.8	0.24	4.06e-12	**163**	4.06e-12	**44**	3.52e-14	**52**	4.65e-15	**8**	3.34e-15	**6**
HFE-236ca12 (HG-10)	$CHF_2OCF_2OCHF_2$	25.0	0.65	2.75e-10	**11,000**	4.91e-10	**5350**	7.06e-12	**10,300**	2.94e-12	**4770**	7.75e-13	**1420**
HFE-338pcc13 (HG-01)	$CHF_2OCF_2CF_2OCHF_2$	12.9	0.86	2.10e-10	**8430**	2.67e-10	**2910**	4.69e-12	**6860**	9.28e-13	**1500**	2.42e-13	**442**
1,1,1,3,3,3-Hexafluoropropan-2-ol	$(CF_3)_2CHOH$	1.9	0.26	1.67e-11	**668**	1.67e-11	**182**	1.66e-13	**243**	1.97e-14	**32**	1.38e-14	**25**
HG-02	$HF_2C–(OCF_2CF_2)_2–OCF_2H$	12.9	1.24[i]	1.97e-10	**7900**	2.50e-10	**2730**	4.40e-12	**6430**	8.70e-13	**1410**	2.27e-13	**415**
HG-03	$HF_2C–(OCF_2CF_2)_3–OCF_2H$	12.9	1.76[i]	2.06e-10	**8270**	2.62e-10	**2850**	4.60e-12	**6730**	9.10e-13	**1480**	2.37e-13	**434**
HG-20	$HF_2C–(OCF_2)_2–OCF_2H$	25.0	0.92[i]	2.73e-10	**10,900**	4.86e-10	**5300**	7.00e-12	**10,200**	2.91e-12	**4730**	7.68e-13	**1400**
HG-21	$HF_2C–OCF_2CF_2OC-F_2OCF_2O–CF_2H$	13.5	1.71[i]	2.76e-10	**11,100**	3.57e-10	**3890**	6.23e-12	**9110**	1.31e-12	**2120**	3.29e-13	**602**

(continued on next page)

Table 8.A.1 (continued)

Acronym, Common Name or Chemical Name	Chemical Formula	Lifetime (Years)	Radiative Efficiency ($W\ m^{-2}\ ppb^{-1}$)	AGWP 20-year ($W\ m^{-2}\ yr\ kg^{-1}$)	GWP 20-year	AGWP 100-year ($W\ m^{-2}\ yr\ kg^{-1}$)	GWP 100-year	AGTP 20-year ($K\ kg^{-1}$)	GTP 20-year	AGTP 50-year ($K\ kg^{-1}$)	GTP 50-year	AGTP 100-year ($K\ kg^{-1}$)	GTP 100-year
HG-30	$HF_2C\text{–}(OCF_2)_3\text{–}OCF_2H$	25.0	1.65[i]	3.77e-10	15,100	6.73e-10	7330	9.68e-12	14,100	4.03e-12	6530	1.06e-12	1940
1-Ethoxy-1,1,2,2,3,3,3-heptafluoropropane	$CF_3CF_2CF_2OCH_2CH_3$	0.8	0.28[i]	5.56e-12	223	5.56e-12	61	4.80e-14	70	6.36e-15	10	4.57e-15	8
Fluoroxene	$CF_3CH_2OCH{=}CH_2$	3.6 days	0.01[i]	4.97e-15	<1	4.97e-15	<1	3.95e-17	<1	5.58e-18	<1	4.08e-18	<1
1,1,2,2-Tetrafluoro-1-(fluoromethoxy)ethane	$CH_2FOCF_2CF_2H$	6.2	0.34[i]	7.68e-11	3080	7.99e-11	871	1.29e-12	1880	1.28e-13	207	6.68e-14	122
2-Ethoxy-3,3,4,4,5-pentafluorotetrahydro-2,5-bis[1,2,2,2-tetrafluoro-1-(trifluoromethyl)ethyl]-furan	$C_{12}H_5F_{19}O_2$	1.0	0.49[i]	5.09e-12	204	5.09e-12	56	4.53e-14	66	5.86e-15	10	4.19e-15	8
Fluoro(methoxy)methane	CH_3OCH_2F	73.0 days	0.07[g]	1.15e-12	46	1.15e-12	13	9.34e-15	14	1.30e-15	2	9.46e-16	2
Difluoro(methoxy)methane	CH_3OCHF_2	1.1	0.17[g]	1.32e-11	528	1.32e-11	144	1.18e-13	173	1.52e-14	25	1.08e-14	20
Fluoro(fluoromethoxy)methane	CH_2FOCH_2F	0.9	0.19[g]	1.20e-11	479	1.20e-11	130	1.05e-13	153	1.37e-14	22	9.84e-15	18
Difluoro(fluoromethoxy)methane	CH_2FOCHF_2	3.3	0.30[g]	5.65e-11	2260	5.66e-11	617	6.88e-13	1010	7.11e-14	115	4.69e-14	86
Trifluoro(fluoromethoxy)methane	CH_2FOCF_3	4.4	0.33[g]	6.82e-11	2730	6.89e-11	751	9.59e-13	1400	9.27e-14	150	5.72e-14	105
HG′-01	$CH_3OCF_2CF_2OCH_3$	2.0	0.29	2.03e-11	815	2.03e-11	222	2.06e-13	301	2.42e-14	39	1.68e-14	31
HG′-02	$CH_3O(CF_2CF_2O)_2CH_3$	2.0	0.56	2.16e-11	868	2.16e-11	236	2.19e-13	320	2.57e-14	42	1.79e-14	33
HG′-03	$CH_3O(CF_2CF_2O)_3CH_3$	2.0	0.76	2.03e-11	812	2.03e-11	221	2.05e-13	299	2.41e-14	39	1.67e-14	31
HFE-329me3	$CF_3CFHCF_2OCF_3$	40.0	0.48	1.79e-10	7170	4.17e-10	4550	4.85e-12	7090	2.89e-12	4690	1.12e-12	2040
3,3,4,4,5,5,6,6,7,7,7-Undecafluoroheptan-1-ol	$CF_3(CF_2)_4CH_2CH_2OH$	20.0 days	0.06	3.91e-14	2	3.91e-14	<1	3.12e-16	<1	4.39e-17	<1	3.21e-17	<1
3,3,4,4,5,5,6,6,7,7,8,8,9,9,9-Pentadecafluorononan-1-ol	$CF_3(CF_2)_6CH_2CH_2OH$	20.0 days	0.07	3.00e-14	1	3.00e-14	<1	2.40e-16	<1	3.37e-17	<1	2.46e-17	<1
3,3,4,4,5,5,6,6,7,7,8,8,9,9,10,10,11,11,11-Nonadecafluoroundecan-1-ol	$CF_3(CF_2)_8CH_2CH_2OH$	20.0 days	0.05	1.72e-14	<1	1.72e-14	<1	1.37e-16	<1	1.93e-17	<1	1.41e-17	<1
2-Chloro-1,1,2-trifluoro-1-methoxyethane	CH_3OCF_2CHFCl	1.4	0.21	1.12e-11	449	1.12e-11	122	1.05e-13	153	1.31e-14	21	9.24e-15	17
PFPMIE (perfluoropolymethylisopropyl ether)	$CF_3OCF(CF_3)CF_2OCF_2OCF_3$	800.0	0.65	1.87e-10	7500	8.90e-10	9710	5.52e-12	8070	6.11e-12	9910	6.15e-12	11,300
HFE-216	$CF_3OCF{=}CF_2$	8.4 days	0.02	1.92e-14	<1	1.92e-14	<1	1.53e-16	<1	2.15e-17	<1	1.58e-17	<1
Trifluoromethyl formate	$HCOOCF_3$	3.5	0.31[i]	5.37e-11	2150	5.39e-11	588	6.73e-13	984	6.85e-14	111	4.47e-14	82
Perfluoroethyl formate	$HCOOCF_2CF_3$	3.5	0.44[i]	5.30e-11	2130	5.32e-11	580	6.64e-13	971	6.76e-14	110	4.41e-14	81
Perfluoropropyl formate	$HCOOCF_2CF_2CF_3$	2.6	0.50[i]	3.45e-11	1380	3.45e-11	376	3.80e-13	555	4.19e-14	68	2.85e-14	52
Perfluorobutyl formate	$HCOOCF_2CF_2CF_2CF_3$	3.0	0.56[i]	3.59e-11	1440	3.59e-11	392	4.19e-13	613	4.45e-14	72	2.97e-14	54
2,2,2-Trifluoroethyl formate	$HCOOCH_2CF_3$	0.4	0.16[i]	3.07e-12	123	3.07e-12	33	2.55e-14	37	3.48e-15	6	2.52e-15	5
3,3,3-Trifluoropropyl formate	$HCOOCH_2CH_2CF_3$	0.3	0.13[i]	1.60e-12	64	1.60e-12	17	1.31e-14	19	1.80e-15	3	1.31e-15	2
1,2,2,2-Tetrafluoroethyl formate	$HCOOCHFCF_3$	3.2	0.35[i]	4.30e-11	1720	4.31e-11	470	5.17e-13	755	5.39e-14	87	3.57e-14	65
1,1,1,3,3,3-Hexafluoropropan-2-yl formate	$HCOOCH(CF_3)_2$	3.2	0.33[i]	3.05e-11	1220	3.05e-11	333	3.66e-13	535	3.81e-14	62	2.53e-14	46
Perfluorobutyl acetate	$CH_3COOCF_2CF_2CF_2CF_3$	21.9 days	0.12[i]	1.52e-13	6	1.52e-13	2	1.21e-15	2	1.71e-16	<1	1.25e-16	<1
Perfluoropropyl acetate	$CH_3COOCF_2CF_2CF_3$	21.9 days	0.11[i]	1.59e-13	6	1.59e-13	2	1.27e-15	2	1.78e-16	<1	1.30e-16	<1
Perfluoroethyl acetate	$CH_3COOCF_2CF_3$	21.9 days	0.10[i]	1.89e-13	8	1.89e-13	2	1.51e-15	2	2.12e-16	<1	1.55e-16	<1
Trifluoromethyl acetate	CH_3COOCF_3	21.9 days	0.07[i]	1.90e-13	8	1.90e-13	2	1.52e-15	2	2.14e-16	<1	1.56e-16	<1

(continued on next page)

Table 8.A.1 (continued)

Acronym, Common Name or Chemical Name	Chemical Formula	Lifetime (Years)	Radiative Efficiency ($W m^{-2} ppb^{-1}$)	AGWP 20-year ($W m^{-2} yr kg^{-1}$)	**GWP 20-year**	AGWP 100-year ($W m^{-2} yr kg^{-1}$)	**GWP 100-year**	AGTP 20-year ($K kg^{-1}$)	**GTP 20-year**	AGTP 50-year ($K kg^{-1}$)	**GTP 50-year**	AGTP 100-year ($K kg^{-1}$)	**GTP 100-year**
Methyl carbonofluoridate	$FCOOCH_3$	1.8	0.07^i	8.74e-12	**350**	8.74e-12	**95**	8.60e-14	**126**	1.03e-14	**17**	7.21e-15	**13**
1,1-Difluoroethyl carbonofluoridate	$FCOOCF_2CH_3$	0.3	0.17^i	2.46e-12	**99**	2.46e-12	**27**	2.02e-14	**30**	2.78e-15	**5**	2.02e-15	**4**
1,1-Difluoroethyl 2,2,2-trifluoroacetate	$CF_3COOCF_2CH_3$	0.3	0.27^i	2.83e-12	**113**	2.83e-12	**31**	2.33e-14	**34**	3.20e-15	**5**	2.32e-15	**4**
Ethyl 2,2,2-trifluoroacetate	$CF_3COOCH_2CH_3$	21.9 days	0.05^i	1.26e-13	**5**	1.26e-13	**1**	1.00e-15	**1**	1.41e-16	**<1**	1.03e-16	**<1**
2,2,2-Trifluoroethyl 2,2,2-trifluoroacetate	$CF_3COOCH_2CF_3$	54.8 days	0.15^i	6.27e-13	**25**	6.27e-13	**7**	5.06e-15	**7**	7.07e-16	**1**	5.15e-16	**<1**
Methyl 2,2,2-trifluoroacetate	CF_3COOCH_3	0.6	0.18^i	4.80e-12	**192**	4.80e-12	**52**	4.08e-14	**60**	5.47e-15	**9**	3.95e-15	**7**
Methyl 2,2-difluoroacetate	HCF_2COOCH_3	40.1 days	0.05^i	3.00e-13	**12**	3.00e-13	**3**	2.41e-15	**4**	3.38e-16	**<1**	2.47e-16	**<1**
Difluoromethyl 2,2,2-trifluoroacetate	$CF_3COOCHF_2$	0.3	0.24^i	2.48e-12	**99**	2.48e-12	**27**	2.04e-14	**30**	2.81e-15	**5**	2.04e-15	**4**
2,2,3,3,4,4,4-Heptafluorobutan-1-ol	$C_3F_7CH_2OH$	0.6	0.20	3.10e-12	**124**	3.10e-12	**34**	2.61e-14	**38**	3.52e-15	**6**	2.55e-15	**5**
1,1,2-Trifluoro-2-(trifluoromethoxy)-ethane	$CHF_2CHFOCF_3$	9.8	0.35	9.91e-11	**3970**	1.14e-10	**1240**	2.03e-12	**2960**	2.88e-13	**467**	9.74e-14	**178**
1-Ethoxy-1,1,2,3,3,3-hexafluoropropane	$CF_3CHFCF_2OCH_2CH_3$	0.4	0.19	2.14e-12	**86**	2.14e-12	**23**	1.77e-14	**26**	2.43e-15	**4**	1.76e-15	**3**
1,1,1,2,2,3,3-Heptafluoro-3-(1,2,2,2-tetrafluoroethoxy)-propane	$CF_3CF_2CF_2OCHFCF_3$	67.0	0.58	1.98e-10	**7940**	5.95e-10	**6490**	5.57e-12	**8140**	4.29e-12	**6960**	2.39e-12	**4380**
2,2,3,3-Tetrafluoro-1-propanol	$CHF_2CF_2CH_2OH$	91.3 days	0.11	1.19e-12	**48**	1.19e-12	**13**	9.72e-15	**14**	1.35e-15	**2**	9.79e-16	**2**
2,2,3,4,4,4-Hexafluoro-1-butanol	$CF_3CHFCF_2CH_2OH$	94.9 days	0.19	1.56e-12	**63**	1.56e-12	**17**	1.27e-14	**19**	1.76e-15	**3**	1.28e-15	**2**
2,2,3,3,4,4,4-Heptafluoro-1-butanol	$CF_3CF_2CF_2CH_2OH$	0.3	0.16	1.49e-12	**60**	1.49e-12	**16**	1.23e-14	**18**	1.69e-15	**3**	1.23e-15	**2**
1,1,2,2-Tetrafluoro-3-methoxy-propane	$CHF_2CF_2CH_2OCH_3$	14.2 days	0.03	4.82e-14	**2**	4.82e-14	**<1**	3.84e-16	**<1**	5.41e-17	**<1**	3.96e-17	**<1**
perfluoro-2-methyl-3-pentanone	$CF_3CF_2C(O)CF(CF_3)_2$	7.0 days	0.03	9.14e-15	**<1**	9.14e-15	**<1**	7.27e-17	**<1**	1.03e-17	**<1**	7.51e-18	**<1**
3,3,3-Trifluoro-propanal	CF_3CH_2CHO	2.0 days	0.004	9.86e-16	**<1**	9.86e-16	**<1**	7.84e-18	**<1**	1.11e-18	**<1**	8.10e-19	**<1**
2-Fluoroethanol	CH_2FCH_2OH	20.4 days	0.02	8.07e-14	**3**	8.07e-14	**<1**	6.45e-16	**<1**	9.07e-17	**<1**	6.63e-17	**<1**
2,2-Difluoroethanol	CHF_2CH_2OH	40.0 days	0.04	2.78e-13	**11**	2.78e-13	**3**	2.23e-15	**3**	3.12e-16	**<1**	2.28e-16	**<1**
2,2,2-Trifluoroethanol	CF_3CH_2OH	0.3	0.10	1.83e-12	**73**	1.83e-12	**20**	1.50e-14	**22**	2.07e-15	**3**	1.50e-15	**3**
1,1′-Oxybis[2-(difluoromethoxy)-1,1,2,2-tetrafluoroethane	$HCF_2O(CF_2CF_2O)_2CF_2H$	26.0	1.15^k	2.47e-10	**9910**	4.51e-10	**4920**	6.38e-12	**9320**	2.75e-12	**4460**	7.45e-13	**1360**
1,1,3,3,4,4,6,6,7,7,9,9,10,10,12,12-hexadecafluoro-2,5,8,11-Tetraoxadodecane	$HCF_2O(CF_2CF_2O)_3CF_2H$	26.0	1.43^k	2.26e-10	**9050**	4.12e-10	**4490**	5.83e-12	**8520**	2.51e-12	**4080**	6.81e-13	**1250**
1,1,3,3,4,4,6,6,7,7,9,9,10,10,12,12,13,13,15,15-eicosafluoro-2,5,8,11,14-Pentaoxapentadecane	$HCF_2O(CF_2CF_2O)_4CF_2H$	26.0	1.46^k	1.83e-10	**7320**	3.33e-10	**3630**	4.71e-12	**6880**	2.03e-12	**3300**	5.50e-13	**1010**

Notes:

For CH_4 we estimate an uncertainty of ±30% and ±40% for 20- and 100-year time horizon, respectively (for 90% uncertainty range). The uncertainty is dominated by AGWP for CO_2 and indirect effects. The uncertainty in GWP for N_2O is estimated to ±20% and ±30% for 20- and 100-year time horizon, with the largest contributions from CO_2. The uncertainty in GWP for HFC-134a is estimated to ±25% and ±35% for 20- and 100-year time horizons while for CFC-11 the GWP the corresponding numbers are approximately ±20% and ±35% (not accounting for the indirect effects). For CFC-12 the corresponding numbers are ±20 and ±30. The uncertainties estimated for HFC-134a and CFC-11 are assessed as representative for most other gases with similar or longer lifetimes. For shorter-lived gases, the uncertainties will be larger. For GTP, few estimates are available in the literature. The uncertainty is assessed to be of the order of ±75% for the methane GTP_{100}.

* No single lifetime can be given. The impulse response function for CO_2 from Joos et al. (2013) has been used. See also Supplementary Material Section 8.SM.11.

† Perturbation lifetime is used in calculation of metrics, not the lifetime of the atmospheric burden.

(continued on next page)

Table 8.A.1 Notes (continued)

‡ Metric values for CH_4 of fossil origin include the oxidation to CO_2 (based on Boucher et al., 2009). In applications of these values, inclusion of the CO_2 effect of fossil methane must be done with caution to avoid any double-counting because CO_2 emissions numbers are often based on total carbon content. Methane values without the CO_2 effect from fossil methane are thus appropriate for fossil methane sources for which the carbon has been accounted for elsewhere, or for biospheric methane sources for which there is abalance between CO_2 taken up by the biosphere and CO_2 produced from CH_4 oxidization. The addition effect on GWP and GTP represents lower limits from Boucher et al. (2009) and assume 50% of the carbon is deposited as formaldehyde to the surface and is then lost. The upper limit in Boucher et al. (2009) made the assumption that this deposited formaldehyde was subsequently further oxidized to CO_2 .

[a] RE is unchanged since AR4.

[b] RE is unchanged since AR4 except the absolute forcing is increased by a factor of 1.04 to account for the change in the recommended RE of CFC-11.

[c] Based on Rajakumar et al. (2006) (lifetime correction factor has been applied to account for non-homogeneous horizontal and vertical mixing).

[d] Based on instantaneous RE from Baasandorj et al. (2010); Baasandorj et al. (2011) (correction factors have been applied to account for stratospheric temperature adjustment and non-homogeneous horizontal and vertical mixing).

[e] Based on instantaneous RE from *ab initio* study of Bravo et al. (2010) (a factor 1.10 has been applied to account for stratospheric temperature adjustment).

[f] Based on average instantaneous RE reported in literature (Vasekova et al., 2006; Bravo et al., 2010) (correction factors have been applied to account for stratospheric temperature adjustment and non-homogeneous horizontal and vertical mixing).

[g] Based on instantaneous RE from *ab initio* studies of Blowers et al. (2007, 2008)(correction factors have been applied to account for stratospheric temperature adjustment and non-homogeneous horizontal and vertical mixing).

[h] Based on instantaneous RE from Heathfield et al. (1998) (correction factors have been applied to account for stratospheric temperature adjustment and non-homogeneous horizontal and vertical mixing).

[i] Note that calculation of RE is based on calculated (*ab initio*) absorption cross-section and uncertainties are therefore larger than for calculations using experimental absorption cross section.

[j] Based on instantaneous RE from Javadi et al. (2007) (correction factors have been applied to account for stratospheric temperature adjustment and non-homogeneous horizontal and vertical mixing).

[k] Based on instantaneous RE from Andersen et al. (2010) (correction factors have been applied to account for stratospheric temperature adjustment and non-homogeneous horizontal and vertical mixing).

The GTP values are calculated with a temperature impulse response function taken from Boucher and Reddy (2008). See also Supplementary Material Section 8.SM.11.

Table 8.A.2 | Halocarbon indirect GWPs from ozone depletion using the EESC-based method described in WMO (2011), adapted from Daniel et al. (1995). A radiative forcing in year 2011 of –0.15 (–0.30 to 0.0) W m^{-2} relative to preindustrial times is used (see Section 8.3.3). Uncertainty on the indirect AGWPs due to the ozone forcing uncertainty is ±100%.

Gas	GWP_{100}
CFC-11	–2640
CFC-12	–2100
CFC-113	–2150
CFC-114	–914
CFC-115	–223
HCFC-22	–98
HCFC-123	–37
HCFC-124	–46
HCFC-141b	–261
HCFC-142b	–152
CH_3CCl_3	–319
CCl_4	–2110
CH_3Br	–1250
Halon-1211	–19,000
Halon-1301	–44,500
Halon-2402	–32,000
HCFC-225ca	–40
HCFC-225cb	–60

Table 8.A.3 | GWP and GTP for NO_X from surface sources for time horizons of 20 and 100 years from the literature. All values are on a per kilogram of nitrogen basis. Uncertainty for numbers from Fry et al. (2012) and Collins et al. (2013) refer to 1-σ. For the reference gas CO_2, RE and IRF from AR4 are used in the calculations. The GWP_{100} and GTP_{100} values can be scaled by 0.94 and 0.92, respectively, to account for updated values for the reference gas CO_2. For 20 years the changes are negligible.

	GWP		GTP	
	H = 20	H = 100	H = 20	H = 100
NO_X East Asia[a]	6.4 (±38.1)	–5.3 (±11.5)	–55.6 (±23.8)	–1.3 (±2.1)
NO_X EU + North Africa[a]	–39.4 (±17.5)	–15.6 (±5.8)	–48.0 (±14.9)	–2.5 (±1.3)
NO_X North America[a]	–2.4 (±30.3)	–8.2 (±10.3)	–61.9 (±27.8)	–1.7 (±2.1)
NO_X South Asia[a]	–40.7 (±88.3)	–25.3 (±29.0)	–124.6 (±67.4)	–4.6 (±5.1)
NO_X four above regions[a]	–15.9 (±32.7)	–11.6 (±10.7)	–62.1 (±26.2)	–2.2 (±2.1)
Mid-latitude NOx[c]	–43 to +23	–18 to +1.6	–55 to –37	–2.9 to –0.02
Tropical NO_x[c]	43 to 130	–28 to –10	–260 to –220	–6.6 to –5.4
NO_X global[b]	19	–11	–87	–2.9
NO_X global[d]	–108 ± 35 –335 ± 110 –560 ± 279	–31 ± 10 –95 ± 31 –159 ± 79		

Notes:

[a] Fry et al. (2012) (updated by including stratospheric H_2O) and Collins et al. (2013).

[b] Fuglestvedt et al. (2010); based on Wild et al. (2001).

[c] Fuglestvedt et al. (2010).

[d] Shindell et al. (2009). Three values are given: First, without aerosols, second, direct aerosol effect included (sulfate and nitrate), third, direct and indirect aerosol effects included. Uncertainty ranges from Shindell et al. (2009) are given for 95% confidence levels.

8

Table 8.A.4 | GWP and GTP for CO for time horizons of 20 and 100 years from the literature. Uncertainty for numbers from Fry et al. (2012) and Collins et al. (2013) refer to 1-σ. For the reference gas CO_2, RE and IRF from AR4 are used in the calculations. The GWP_{100} and GTP_{100} values can be scaled by 0.94 and 0.92, respectively, to account for updated values for the reference gas CO_2. For 20 years the changes are negligible.

	GWP		GTP	
	H = 20	H = 100	H = 20	H = 100
CO East Asia[a]	5.4 (±1.7)	1.8 (±0.6)	3.5 (±1.3)	0.26 (±0.12)
CO EU + North Africa[a]	4.9 (±1.5)	1.6 (±0.5)	3.2 (±1.2)	0.24 (±0.11)
CO North America[a]	5.6 (±1.8)	1.8 (±0.6)	3.7 (±1.3)	0.27 (±0.12)
CO South Asia[a]	5.7 (±1.3)	1.8 (±0.4)	3.4 (±1.0)	0.27 (±0.10)
CO four regions above[a]	5.4 (±1.6)	1.8 (±0.5)	3.5 (±1.2)	0.26 (±0.11)
CO global[b]	6 to 9.3	2 to 3.3	3.7 to 6.1	0.29 to 0.55
CO global[c]	7.8 ± 2.0 11.4 ± 2.9 18.6 ± 8.3	2.2 ± 0.6 3.3 ± 0.8 5.3 ± 2.3		

Notes:

[a] Fry et al. (2012) (updated by including stratospheric H_2O) and Collins et al. (2013).

[b] Fuglestvedt et al. (2010).

[c] Shindell et al. (2009). Three values are given: First, without aerosols, second, direct aerosol effect included, third, direct and indirect aerosol effects included. Uncertainty ranges from Shindell et al. (2009) are given for 95% confidence levels.

Table 8.A.5 | GWP and GTP for VOCs for time horizons of 20 and 100 years from the literature. Uncertainty for numbers from Fry et al. (2012) and Collins et al. (2013) refer to 1-σ. For the reference gas CO_2, RE and IRF from AR4 are used in the calculations. The GWP_{100} and GTP_{100} values can be scaled by 0.94 and 0.92, respectively, to account for updated values for the reference gas CO_2. For 20 years the changes are negligible.

	GWP		GTP	
	H = 20	H = 100	H = 20	H = 100
VOC East Asia[a]	16.3 (±6.4)	5.0 (±2.1)	8.4 (±4.6)	0.7 (±0.4)
VOC EU + North Africa[a]	18.0 (±8.5)	5.6 (±2.8)	9.5 (±6.5)	0.8 (±0.5)
VOC North America[a]	16.2 (±9.2)	5.0 (±3.0)	8.6 (±6.4)	0.7 (±0.5)
VOC South Asia[a]	27.8 (±5.6)	8.8 (±1.9)	15.7 (±5.0)	1.3 (±0.5)
VOC four regions above	18.7 (±7.5)	5.8 (±2.5)	10.0 (±5.7)	0.9 (±0.5)
VOC global[b]	14	4.5	7.5	0.66

Notes:

[a] Fry et al. (2012) (updated by including stratospheric H_2O) and Collins et al. (2013).

[b] Fuglestvedt et al. (2010) based on Collins et al. (2002).

The values are given on a per kilogram of C basis.

Table 8.A.6 | GWP and GTP from the literature for BC and OC for time horizons of 20 and 100 years. For the reference gas CO_2, RE and IRF from AR4 are used in the calculations. The GWP_{100} and GTP_{100} values can be scaled by 0.94 and 0.92, respectively, to account for updated values for the reference gas CO_2. For 20 years the changes are negligible.

	GWP		GTP	
	H = 20	H = 100	H = 20	H = 100
BC total, global[c]	3200 (270 to 6200)	900 (100 to 1700)	920 (95 to 2400)	130 (5 to 340)
BC (four regions)[d]	1200 ± 720	345 ± 207	420 ± 190	56 ± 25
BC global[a]	1600	460	470	64
BC aerosol–radiation interaction +albedo, global[b]	2900 ± 1500	830 ± 440		
OC global[a]	–240	–69	–71	–10
OC global[b]	–160 (–60 to –320)	–46 (–18 to –19)		
OC (4 regions)[d]	–160 ± 68	–46 ± 20	–55 ± 16	–7.3±2.1

Notes:

[a] Fuglestvedt et al. (2010).

[b] Bond et al. (2011). Uncertainties for OC are asymmetric and are presented as ranges.

[c] Bond et al. (2013). Metric values are given for total effect.

[d] Collins et al. (2013). The four regions are East Asia, EU + North Africa, North America and South Asia (as also given in Fry et al., 2012). Only aerosol-radiation interaction is included.

Evaluation of Climate Models

Coordinating Lead Authors:
Gregory Flato (Canada), Jochem Marotzke (Germany)

Lead Authors:
Babatunde Abiodun (South Africa), Pascale Braconnot (France), Sin Chan Chou (Brazil), William Collins (USA), Peter Cox (UK), Fatima Driouech (Morocco), Seita Emori (Japan), Veronika Eyring (Germany), Chris Forest (USA), Peter Gleckler (USA), Eric Guilyardi (France), Christian Jakob (Australia), Vladimir Kattsov (Russian Federation), Chris Reason (South Africa), Markku Rummukainen (Sweden)

Contributing Authors:
Krishna AchutaRao (India), Alessandro Anav (UK), Timothy Andrews (UK), Johanna Baehr (Germany), Nathaniel L. Bindoff (Australia), Alejandro Bodas-Salcedo (UK), Jennifer Catto (Australia), Don Chambers (USA), Ping Chang (USA), Aiguo Dai (USA), Clara Deser (USA), Francisco Doblas-Reyes (Spain), Paul J. Durack (USA/Australia), Michael Eby (Canada), Ramon de Elia (Canada), Thierry Fichefet (Belgium), Piers Forster (UK), David Frame (UK/New Zealand), John Fyfe (Canada), Emiola Gbobaniyi (Sweden/Nigeria), Nathan Gillett (Canada), Jesus Fidel González-Rouco (Spain), Clare Goodess (UK), Stephen Griffies (USA), Alex Hall (USA), Sandy Harrison (Australia), Andreas Hense (Germany), Elizabeth Hunke (USA), Tatiana Ilyina (Germany), Detelina Ivanova (USA), Gregory Johnson (USA), Masa Kageyama (France), Viatcheslav Kharin (Canada), Stephen A. Klein (USA), Jeff Knight (UK), Reto Knutti (Switzerland), Felix Landerer (USA), Tong Lee (USA), Hongmei Li (Germany/China), Natalie Mahowald (USA), Carl Mears (USA), Gerald Meehl (USA), Colin Morice (UK), Rym Msadek (USA), Gunnar Myhre (Norway), J. David Neelin (USA), Jeff Painter (USA), Tatiana Pavlova (Russian Federation), Judith Perlwitz (USA), Jean-Yves Peterschmitt (France), Jouni Räisänen (Finland), Florian Rauser (Germany), Jeffrey Reid (USA), Mark Rodwell (UK), Benjamin Santer (USA), Adam A. Scaife (UK), Jörg Schulz (Germany), John Scinocca (Canada), David Sexton (UK), Drew Shindell (USA), Hideo Shiogama (Japan), Jana Sillmann (Canada), Adrian Simmons (UK), Kenneth Sperber (USA), David Stephenson (UK), Bjorn Stevens (Germany), Peter Stott (UK), Rowan Sutton (UK), Peter W. Thorne (USA/Norway/UK), Geert Jan van Oldenborgh (Netherlands), Gabriel Vecchi (USA), Mark Webb (UK), Keith Williams (UK), Tim Woollings (UK), Shang-Ping Xie (USA), Jianglong Zhang (USA)

Review Editors:
Isaac Held (USA), Andy Pitman (Australia), Serge Planton (France), Zong-Ci Zhao (China)

This chapter should be cited as:
Flato, G., J. Marotzke, B. Abiodun, P. Braconnot, S.C. Chou, W. Collins, P. Cox, F. Driouech, S. Emori, V. Eyring, C. Forest, P. Gleckler, E. Guilyardi, C. Jakob, V. Kattsov, C. Reason and M. Rummukainen, 2013: Evaluation of Climate Models. In: *Climate Change 2013: The Physical Science Basis. Contribution of Working Group I to the Fifth Assessment Report of the Intergovernmental Panel on Climate Change* [Stocker, T.F., D. Qin, G.-K. Plattner, M. Tignor, S.K. Allen, J. Boschung, A. Nauels, Y. Xia, V. Bex and P.M. Midgley (eds.)]. Cambridge University Press, Cambridge, United Kingdom and New York, NY, USA.

Table of Contents

Executive Summary

Climate models have continued to be developed and improved since the AR4, and many models have been extended into Earth System models by including the representation of biogeochemical cycles important to climate change. These models allow for policy-relevant calculations such as the carbon dioxide (CO_2) emissions compatible with a specified climate stabilization target. In addition, the range of climate variables and processes that have been evaluated has greatly expanded, and differences between models and observations are increasingly quantified using 'performance metrics'. In this chapter, model evaluation covers simulation of the mean climate, of historical climate change, of variability on multiple time scales and of regional modes of variability. This evaluation is based on recent internationally coordinated model experiments, including simulations of historic and paleo climate, specialized experiments designed to provide insight into key climate processes and feedbacks and regional climate downscaling. Figure 9.44 provides an overview of model capabilities as assessed in this chapter, including improvements, or lack thereof, relative to models assessed in the AR4. The chapter concludes with an assessment of recent work connecting model performance to the detection and attribution of climate change as well as to future projections. {9.1.2, 9.8.1, Table 9.1, Figure 9.44}

The ability of climate models to simulate surface temperature has improved in many, though not all, important aspects relative to the generation of models assessed in the AR4. There continues to be *very high confidence*[1] that models reproduce observed large-scale mean surface temperature patterns (pattern correlation of ~0.99), though systematic errors of several degrees are found in some regions, particularly over high topography, near the ice edge in the North Atlantic, and over regions of ocean upwelling near the equator. On regional scales (sub-continental and smaller), the confidence in model capability to simulate surface temperature is less than for the larger scales; however, regional biases are near zero on average, with intermodel spread of roughly ±3°C. There is *high confidence* that regional-scale surface temperature is better simulated than at the time of the AR4. Current models are also able to reproduce the large-scale patterns of temperature during the Last Glacial Maximum (LGM), indicating an ability to simulate a climate state much different from the present. {9.4.1, 9.6.1, Figures 9.2, 9.6, 9.39, 9.40}

There is *very high confidence* that models reproduce the general features of the global-scale annual mean surface temperature increase over the historical period, including the more rapid warming in the second half of the 20th century, and the cooling immediately following large volcanic eruptions. Most simulations of the historical period do not reproduce the observed reduction in global mean surface warming trend over the last 10 to 15 years. There is *medium confidence* that the trend difference between models and observations during 1998–2012 is to a substantial degree caused by internal variability, with possible contributions from forcing error and some models overestimating the response to increasing greenhouse gas (GHG) forcing. Most, though not all, models overestimate the observed warming trend in the tropical troposphere over the last 30 years, and tend to underestimate the long-term lower stratospheric cooling trend. {9.4.1, Box 9.2, Figure 9.8}

The simulation of large-scale patterns of precipitation has improved somewhat since the AR4, although models continue to perform less well for precipitation than for surface temperature. The spatial pattern correlation between modelled and observed annual mean precipitation has increased from 0.77 for models available at the time of the AR4 to 0.82 for current models. At regional scales, precipitation is not simulated as well, and the assessment remains difficult owing to observational uncertainties. {9.4.1, 9.6.1, Figure 9.6}

9

The simulation of clouds in climate models remains challenging. There is *very high confidence* that uncertainties in cloud processes explain much of the spread in modelled climate sensitivity. However, the simulation of clouds in climate models has shown modest improvement relative to models available at the time of the AR4, and this has been aided by new evaluation techniques and new observations for clouds. Nevertheless, biases in cloud simulation lead to regional errors on cloud radiative effect of several tens of watts per square meter. {9.2.1, 9.4.1, 9.7.2, Figures 9.5, 9.43}

Models are able to capture the general characteristics of storm tracks and extratropical cyclones, and there is some evidence of improvement since the AR4. Storm track biases in the North Atlantic have improved slightly, but models still produce a storm track that is too zonal and underestimate cyclone intensity. {9.4.1}

Many models are able to reproduce the observed changes in upper ocean heat content from 1961 to 2005 with the multi-model mean time series falling within the range of the available observational estimates for most of the period. The ability of models to simulate ocean heat uptake, including variations imposed by large volcanic eruptions, adds confidence to their use in assessing the global energy budget and simulating the thermal component of sea level rise. {9.4.2, Figure 9.17}

The simulation of the tropical Pacific Ocean mean state has improved since the AR4, with a 30% reduction in the spurious westward extension of the cold tongue near the equator, a pervasive bias of coupled models. The simulation of the tropical Atlantic remains deficient with many models unable to reproduce the basic east–west temperature gradient. {9.4.2, Figure 9.14}

---

[1] In this Report, the following summary terms are used to describe the available evidence: limited, medium, or robust; and for the degree of agreement: low, medium, or high. A level of confidence is expressed using five qualifiers: very low, low, medium, high, and very high, and typeset in italics, e.g., *medium confidence*. For a given evidence and agreement statement, different confidence levels can be assigned, but increasing levels of evidence and degrees of agreement are correlated with increasing confidence (see Section 1.4 and Box TS.1 for more details).

Current climate models reproduce the seasonal cycle of Arctic sea ice extent with a multi-model mean error of less than about 10% for any given month. There is *robust evidence* that the downward trend in Arctic summer sea ice extent is better simulated than at the time of the AR4, with about one quarter of the simulations showing a trend as strong as, or stronger, than in observations over the satellite era (since 1979). There is a tendency for models to slightly overestimate sea ice extent in the Arctic (by about 10%) in winter and spring. In the Antarctic, the multi-model mean seasonal cycle agrees well with observations, but inter-model spread is roughly double that for the Arctic. Most models simulate a small decreasing trend in Antarctic sea ice extent, albeit with large inter-model spread, in contrast to the small increasing trend in observations. {9.4.3, Figures 9.22, 9.24}

9

Models are able to reproduce many features of the observed global and Northern Hemispher (NH) mean temperature variance on interannual to centennial time scales (*high confidence*), and most models are now able to reproduce the observed peak in variability associated with the El Niño (2- to 7-year period) in the Tropical Pacific. The ability to assess variability from millennial simulations is new since the AR4 and allows quantitative evaluation of model estimates of low-frequency climate variability. This is important when using climate models to separate signal and noise in detection and attribution studies (Chapter 10). {9.5.3, Figures 9.33, 9.35}

Many important modes of climate variability and intraseasonal to seasonal phenomena are reproduced by models, with some improvements evident since the AR4. The statistics of the global monsoon, the North Atlantic Oscillation, the El Niño-Southern Oscillation (ENSO), the Indian Ocean Dipole and the Quasi-Biennial Oscillation are simulated well by several models, although this assessment is tempered by the limited scope of analysis published so far, or by limited observations. There are also modes of variability that are not simulated well. These include modes of Atlantic Ocean variability of relevance to near term projections in Chapter 11 and ENSO teleconnections outside the tropical Pacific, of relevance to Chapter 14. There is *high confidence* that the multi-model statistics of monsoon and ENSO have improved since the AR4. However, this improvement does not occur in all models, and process-based analysis shows that biases remain in the background state and in the strength of associated feedbacks. {9.5.3, Figures 9.32, 9.35, 9.36}

There has been substantial progress since the AR4 in the assessment of model simulations of extreme events. Based on assessment of a suite of indices, the inter-model range of simulated climate extremes is similar to the spread amongst observationally based estimates in most regions. In addition, changes in the frequency of extreme warm and cold days and nights over the second half of the 20th century are consistent between models and observations, with the ensemble global mean time series generally falling within the range of observational estimates. The majority of models underestimate the sensitivity of extreme precipitation to temperature variability or trends, especially in the tropics, which implies that models may underestimate the projected increase in extreme precipitation in the future. Some high-resolution atmospheric models have been shown to reproduce observed year-to-year variability of Atlantic hurricane counts when forced with observed sea surface temperatures, though so far only a few studies of this kind are available. {9.5.4, Figure 9.37}

An important development since the AR4 is the more widespread use of Earth System models, which include an interactive carbon cycle. In the majority of these models, the simulated global land and ocean carbon sinks over the latter part of the 20th century fall within the range of observational estimates. However, the regional patterns of carbon uptake and release are less well reproduced, especially for NH land where models systematically underestimate the sink implied by atmospheric inversion techniques The ability of models to simulate carbon fluxes is important because these models are used to estimate 'compatible emissions' (carbon dioxide emission pathways compatible with a particular climate change target; see Chapter 6). {9.4.5, Figure 9.27}

The majority of Earth System models now include an interactive representation of aerosols, and make use of a consistent specification of anthropogenic sulphur dioxide emissions. However, uncertainties in sulphur cycle processes and natural sources and sinks remain and so, for example, the simulated aerosol optical depth over oceans ranges from 0.08 to 0.22 with roughly equal numbers of models over- and under-estimating the satellite-estimated value of 0.12. {9.1.2, 9.4.6, Table 9.1, Figure 9.29}

Time-varying ozone is now included in the latest suite of models, either prescribed or calculated interactively. Although in some models there is only *medium agreement* with observed changes in total column ozone, the inclusion of time-varying stratospheric ozone constitutes a substantial improvement since the AR4 where half of the models prescribed a constant climatology. As a result, there is *robust evidence* that the representation of climate forcing by stratospheric ozone has improved since the AR4. {9.4.1, Figure 9.10}

Regional downscaling methods are used to provide climate information at the smaller scales needed for many climate impact studies, and there is *high confidence* that downscaling adds value both in regions with highly variable topography and for various small-scale phenomena. Regional models necessarily inherit biases from the global models used to provide boundary conditions. Furthermore, the ability to systematically evaluate regional climate models, and statistical downscaling schemes, is hampered because coordinated intercomparison studies are still emerging. However, several studies have demonstrated that added value arises from higher resolution of stationary features like topography and coastlines, and from improved representation of small-scale processes like convective precipitation. {9.6.4}

Earth system Models of Intermediate Complexity (EMICs) provide simulations of millennial time-scale climate change, and are used as tools to interpret and expand upon the results of more comprehensive models. Although they are limited in the scope and resolution of information provided, EMIC simulations of global mean surface temperature, ocean heat content and carbon cycle response over the 20th century are consistent with the historical records and with more comprehensive models, suggesting that they can be used to provide calibrated projections of long-term transient climate response and stabilization, as well as large ensembles and alternative, policy-relevant, scenarios. {9.4.1, 9.4.2, 9.4.5, Figures 9.8, 9.17, 9.27}

The Coupled Model Intercomparison Project Phase 5 (CMIP5) model spread in equilibrium climate sensitivity ranges from 2.1°C to 4.7°C and is very similar to the assessment in the AR4. No correlation is found between biases in global mean surface temperature and equilibrium climate sensitivity, and so mean temperature biases do not obviously affect the modelled response to GHG forcing. There is *very high confidence* that the primary factor contributing to the spread in equilibrium climate sensitivity continues to be the cloud feedback. This applies to both the modern climate and the LGM. There is likewise *very high confidence* that, consistent with observations, models show a strong positive correlation between tropospheric temperature and water vapour on regional to global scales, implying a positive water vapour feedback in both models and observations. {9.4.1, 9.7.2, Figures 9.9, 9.42, 9.43}

Climate and Earth System models are based on physical principles, and they reproduce many important aspects of observed climate. Both aspects contribute to our confidence in the models' suitability for their application in detection and attribution studies (Chapter 10) and for quantitative future predictions and projections (Chapters 11 to 14). In general, there is no direct means of translating quantitative measures of past performance into confident statements about fidelity of future climate projections. However, there is increasing evidence that some aspects of observed variability or trends are well correlated with inter-model differences in model projections for quantities such as Arctic summertime sea ice trends, snow albedo feedback, and the carbon loss from tropical land. These relationships provide a way, in principle, to transform an observable quantity into a constraint on future projections, but the application of such constraints remains an area of emerging research. There has been substantial progress since the AR4 in the methodology to assess the reliability of a multi-model ensemble, and various approaches to improve the precision of multi-model projections are being explored. However, there is still no universal strategy for weighting the projections from different models based on their historical performance. {9.8.3, Figure 9.45}

9

9.1 Climate Models and Their Characteristics

9.1.1 Scope and Overview of this Chapter

Climate models are the primary tools available for investigating the response of the climate system to various forcings, for making climate predictions on seasonal to decadal time scales and for making projections of future climate over the coming century and beyond. It is crucial therefore to evaluate the performance of these models, both individually and collectively. The focus of this chapter is primarily on the models whose results will be used in the detection and attribution Chapter 10 and the chapters that present and assess projections (Chapters 11 to 14; Annex I), and so this is necessarily an incomplete evaluation. In particular, this chapter draws heavily on model results collected as part of the Coupled Model Intercomparison Projects (CMIP3 and CMIP5) (Meehl et al., 2007; Taylor et al., 2012b), as these constitute a set of coordinated and thus consistent and increasingly well-documented climate model experiments. Other intercomparison efforts, such as those dealing with Regional Climate Models (RCMs) and those dealing with Earth System Models of Intermediate Complexity (EMICs) are also used. It should be noted that the CMIP3 model archive has been extensively evaluated, and much of that evaluation has taken place subsequent to the AR4. By comparison, the CMIP5 models are only now being evaluated and so there is less published literature available. Where possible we show results from both CMIP3 and CMIP5 models so as to illustrate changes in model performance over time; however, where only CMIP3 results are available, they still constitute a useful evaluation of model performance in that for many quantities, the CMIP3 and CMIP5 model performances are broadly similar.

The direct approach to model evaluation is to compare model output with observations and analyze the resulting difference. This requires knowledge of the errors and uncertainties in the observations, which have been discussed in Chapters 2 through 6. Where possible, averages over the same time period in both models and observations are compared, although for many quantities the observational record is rather short, or only observationally based estimates of the climatological mean are available. In cases where observations are lacking, we resort to intercomparison of model results to provide at least some quantification of model uncertainty via inter-model spread.

After a more thorough discussion of the climate models and methods for evaluation in Sections 9.1 and 9.2, we describe climate model experiments in Section 9.3, evaluate recent and longer-term records as simulated by climate models in Section 9.4, variability and extremes in Section 9.5, and regional-scale climate simulation including downscaling in Section 9.6. We conclude with a discussion of model performance and climate sensitivity in Section 9.7, and the relation between model performance and the credibility of future climate projections in Section 9.8.

9.1.2 Overview of Model Types to Be Evaluated

The models used in climate research range from simple energy balance models to complex Earth System Models (ESMs) requiring state of the art high-performance computing. The choice of model depends directly on the scientific question being addressed (Held, 2005; Collins et al., 2006d). Applications include simulating palaeo or historical climate, sensitivity and process studies for attribution and physical understanding, predicting near-term climate variability and change on seasonal to decadal time scales, making projections of future climate change over the coming century or more and downscaling such projections to provide more detail at the regional and local scale. Computational cost is a factor in all of these, and so simplified models (with reduced complexity or spatial resolution) can be used when larger ensembles or longer integrations are required. Examples include exploration of parameter sensitivity or simulations of climate change on the millennial or longer time scale. Here, we provide a brief overview of the climate models evaluated in this chapter.

9.1.2.1 Atmosphere–Ocean General Circulation Models

Atmosphere–Ocean General Circulation Models (AOGCMs) were the 'standard' climate models assessed in the AR4. Their primary function is to understand the dynamics of the physical components of the climate system (atmosphere, ocean, land and sea ice), and for making projections based on future greenhouse gas (GHG) and aerosol forcing. These models continue to be extensively used, and in particular are run (sometimes at higher resolution) for seasonal to decadal climate prediction applications in which biogeochemical feedbacks are not critical (see Chapter 11). In addition, high-resolution or variable-resolution AOGCMs are often used in process studies or applications with a focus on a particular region. An overview of the AOGCMs assessed in this chapter can be found in Table 9.1 and the details in Table 9.A.1. For some specific applications, an atmospheric component of such a model is used on its own.

9.1.2.2 Earth System Models

ESMs are the current state-of-the-art models, and they expand on AOGCMs to include representation of various biogeochemical cycles such as those involved in the carbon cycle, the sulphur cycle, or ozone (Flato, 2011). These models provide the most comprehensive tools available for simulating past and future response of the climate system to external forcing, in which biogeochemical feedbacks play an important role. An overview of the ESMs assessed in this chapter can be found in Table 9.1 and details in Table 9.A.1.

9.1.2.3 Earth System Models of Intermediate Complexity

EMICs attempt to include relevant components of the Earth system, but often in an idealized manner or at lower resolution than the models described above. These models are applied to certain scientific questions such as understanding climate feedbacks on millennial time scales or exploring sensitivities in which long model integrations or large ensembles are required (Claussen et al., 2002; Petoukhov et al., 2005). This class of models often includes Earth system components not yet included in all ESMs (e.g., ice sheets). As computing power increases, this model class has continued to advance in terms of resolution and complexity. An overview of EMICs assessed in this chapter and in the AR5 WG1 is provided in Table 9.2 with additional details in Table 9.A.2.

Table 9.1 | Main features of the Atmosphere–Ocean General Circulation Models (AOGCMs) and Earth System Models (ESMs) participating in Coupled Model Intercomparison Project Phase 5 (CMIP5), and a comparison with Coupled Model Intercomparison Project Phase 3 (CMIP3), including components and resolution of the atmosphere and the ocean models. Detailed CMIP5 model description can be found in Table 9.A.1 (* refers to Table 9.A.1 for more details). Official CMIP model names are used. HT stands for High-Top atmosphere, which has a fully resolved stratosphere with a model top above the stratopause. AMIP stands for models with atmosphere and land surface only, using observed sea surface temperature and sea ice extent. A component is coloured when it includes at least a physically based prognostic equation and at least a two-way coupling with another component, allowing climate feedbacks. For aerosols, lighter shading means 'semi-interactive' and darker shading means 'fully interactive'. The resolution of the land surface usually follows that of the atmosphere, and the resolution of the sea ice follows that of the ocean. In moving from CMIP3 to CMIP5, note the increased complexity and resolution as well as the absence of artificial flux correction (FC) used in some CMIP3 models.

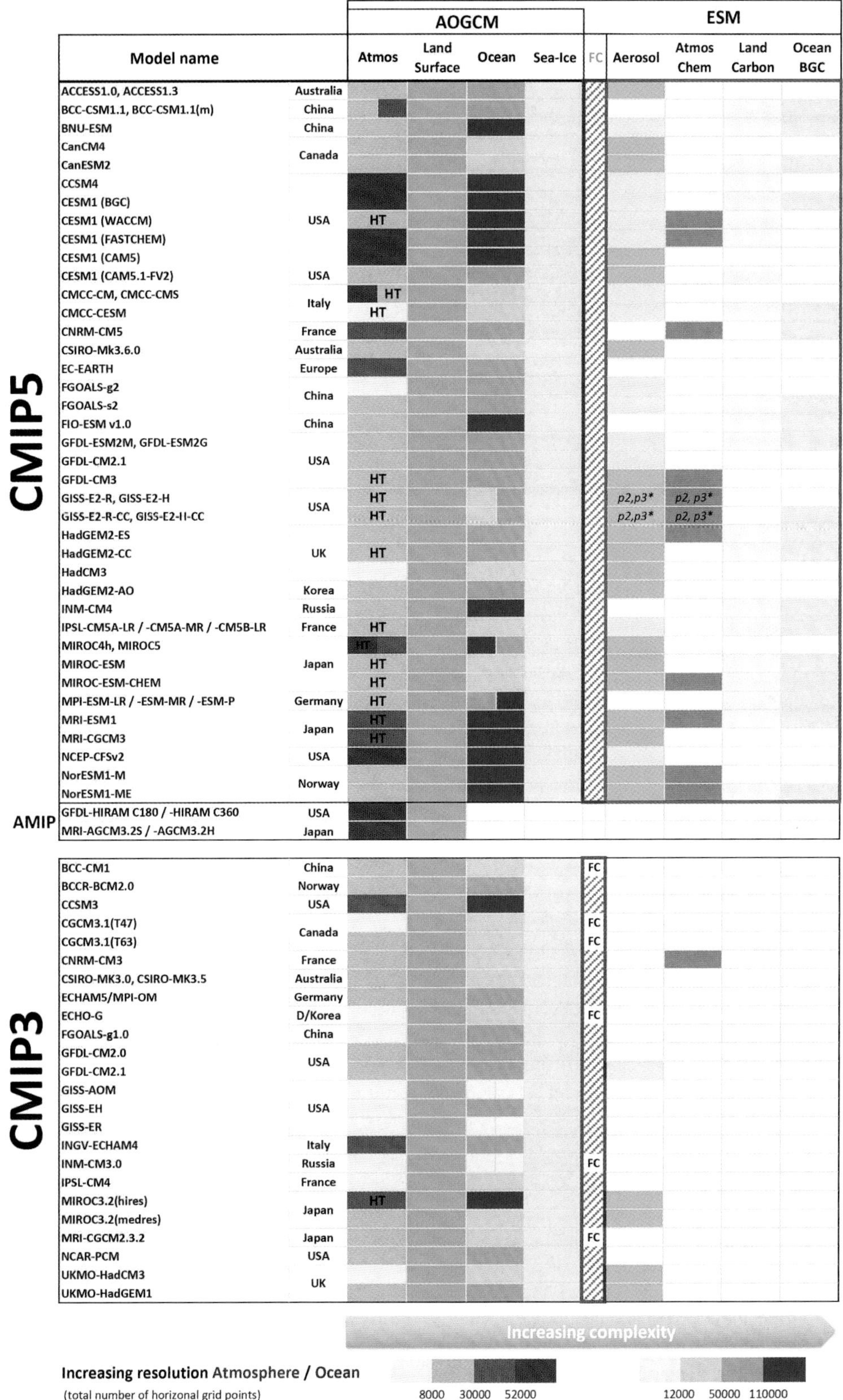

	Model name		AOGCM: Atmos	AOGCM: Land Surface	AOGCM: Ocean	AOGCM: Sea-Ice	FC	ESM: Aerosol	ESM: Atmos Chem	ESM: Land Carbon	ESM: Ocean BGC
CMIP5	ACCESS1.0, ACCESS1.3	Australia									
	BCC-CSM1.1, BCC-CSM1.1(m)	China									
	BNU-ESM	China									
	CanCM4	Canada									
	CanESM2										
	CCSM4	USA									
	CESM1 (BGC)										
	CESM1 (WACCM)		HT								
	CESM1 (FASTCHEM)										
	CESM1 (CAM5)										
	CESM1 (CAM5.1-FV2)	USA									
	CMCC-CM, CMCC-CMS	Italy	HT								
	CMCC-CESM		HT								
	CNRM-CM5	France									
	CSIRO-Mk3.6.0	Australia									
	EC-EARTH	Europe									
	FGOALS-g2	China									
	FGOALS-s2										
	FIO-ESM v1.0	China									
	GFDL-ESM2M, GFDL-ESM2G	USA									
	GFDL-CM2.1										
	GFDL-CM3		HT								
	GISS-E2-R, GISS-E2-H	USA	HT					p2,p3*	p2, p3*		
	GISS-E2-R-CC, GISS-E2-H-CC		HT					p2,p3*	p2, p3*		
	HadGEM2-ES	UK									
	HadGEM2-CC		HT								
	HadCM3										
	HadGEM2-AO	Korea									
	INM-CM4	Russia									
	IPSL-CM5A-LR / -CM5A-MR / -CM5B-LR	France	HT								
	MIROC4h, MIROC5	Japan	HT								
	MIROC-ESM		HT								
	MIROC-ESM-CHEM		HT								
	MPI-ESM-LR / -ESM-MR / -ESM-P	Germany	HT								
	MRI-ESM1	Japan	HT								
	MRI-CGCM3		HT								
	NCEP-CFSv2	USA									
	NorESM1-M	Norway									
	NorESM1-ME										
AMIP	GFDL-HIRAM C180 / -HIRAM C360	USA									
	MRI-AGCM3.2S / -AGCM3.2H	Japan									
CMIP3	BCC-CM1	China					FC				
	BCCR-BCM2.0	Norway									
	CCSM3	USA									
	CGCM3.1(T47)	Canada					FC				
	CGCM3.1(T63)						FC				
	CNRM-CM3	France									
	CSIRO-MK3.0, CSIRO-MK3.5	Australia									
	ECHAM5/MPI-OM	Germany									
	ECHO-G	D/Korea					FC				
	FGOALS-g1.0	China									
	GFDL-CM2.0	USA									
	GFDL-CM2.1										
	GISS-AOM	USA									
	GISS-EH										
	GISS-ER										
	INGV-ECHAM4	Italy									
	INM-CM3.0	Russia					FC				
	IPSL-CM4	France									
	MIROC3.2(hires)	Japan	HT								
	MIROC3.2(medres)										
	MRI-CGCM2.3.2	Japan					FC				
	NCAR-PCM	USA									
	UKMO-HadCM3	UK									
	UKMO-HadGEM1										

Table 9.2 | Main features of the EMICs assessed in the AR5, including components and complexity of the models. Model complexity for four components is indicated by colour shading. Further detailed descriptions of the models are contained in Table 9.A.2.

Model name		Atmos	Ocean	Land Surface	Sea Ice	Coupling	Biosphere	Ice Sheets	Sediment & Weathering
Bern3D	Switzerland								
CLIMBER2	Germany								
CLIMBER3	Germany								
DCESS	Denmark								
FAMOUS	UK								
GENIE	UK								
IAP RAS CM	Russia								
IGSM2	USA								
LOVECLIM1.2	Netherlands								
MESMO	USA								
MIROC-lite	Japan								
MIROC-lite-LCM	Japan								
SPEEDO	Netherlands								
UMD	USA								
Uvic	Canada								

Increasing Complexity (light to dark)					
EMBM	2-Box	NST/NSM			None
SD	Q-flux ML	LST/NSM			BO
QG	FG	LST/BSM			BO,BT
PE	PE	LST/CSM			BO,BT,BV

Significant advances in EMIC capabilities are inclusion of ice sheets (UVic 2.9, Weaver et al., 2001; CLIMBER-2.4, Petoukhov et al., 2000; LOVECLIM, Goosse et al., 2010) and ocean sediment models (DCESS, Shaffer et al., 2008; UVic 2.9, Weaver et al., 2001; Bern3D-LPJ, Ritz et al., 2011). These additional interactive components provide critical feedbacks involved in sea level rise estimates and carbon cycle response on millennial time scales (Zickfeld et al., 2013). Further, the flexibility and efficiency of EMICs allow calibration to specific climate change events to remove potential biases.

9.1.2.4 Regional Climate Models

RCMs are limited-area models with representations of climate processes comparable to those in the atmospheric and land surface components of AOGCMs, though typically run without interactive ocean and sea ice. RCMs are often used to dynamically 'downscale' global model simulations for some particular geographical region to provide more detailed information (Laprise, 2008; Rummukainen, 2010). By contrast, empirical and statistical downscaling methods constitute a range of techniques to provide similar regional or local detail.

9.1.3 Model Improvements

The climate models assessed in this report have seen a number of improvements since the AR4. Model development is a complex and iterative task: improved physical process descriptions are developed, new model components are introduced and the resolution of the models is improved. After assembly of all model components, model parameters are adjusted, or tuned, to provide a stable model climate. The overall approach to model development and tuning is summarized in Box 9.1.

9.1.3.1 Parameterizations

Parameterizations are included in all model components to represent processes that cannot be explicitly resolved; they are evaluated both in isolation and in the context of the full model. The purpose of this section is to highlight recent developments in the parameterizations employed in each model component. Some details for individual models are listed in Table 9.1.

9.1.3.1.1 Atmosphere

Atmospheric models must parameterize a wide range of processes, including those associated with atmospheric convection and clouds, cloud-microphysical and aerosol processes and their interaction, boundary layer processes, as well as radiation and the treatment of unresolved gravity waves. Advances made in the representation of cloud processes, including aerosol–cloud and cloud–radiation interactions, and atmospheric convection are described in Sections 7.2.3 and 7.4.

Improvements in representing the atmospheric boundary layer since the AR4 have focussed on basic boundary layer processes, the representation of the stable boundary layer, and boundary layer clouds (Teixeira et al., 2008). Several global models have successfully adopted new approaches to the parameterization of shallow cumulus convection and moist boundary layer turbulence that acknowledge their

Box 9.1 | Climate Model Development and Tuning

The Atmosphere–Ocean General Circulation Models, Earth System Models and Regional Climate Models evaluated here are based on fundamental laws of nature (e.g., energy, mass and momentum conservation). The development of climate models involves several principal steps:

1. Expressing the system's physical laws in mathematical terms. This requires theoretical and observational work in deriving and simplifying mathematical expressions that best describe the system.

2. Implementing these mathematical expressions on a computer. This requires developing numerical methods that allow the solution of the discretized mathematical expressions, usually implemented on some form of grid such as the latitude–longitude–height grid for atmospheric or oceanic models.

3. Building and implementing conceptual models (usually referred to as parameterizations) for those processes that cannot be represented explicitly, either because of their complexity (e.g., biochemical processes in vegetation) or because the spatial and/or temporal scales on which they occur are not resolved by the discretized model equations (e.g., cloud processes and turbulence). The development of parameterizations has become very complex (e.g., Jakob, 2010) and is often achieved by developing conceptual models of the process of interest in isolation using observations and comprehensive process models. The complexity of each process representation is constrained by observations, computational resources and current knowledge (e.g., Randall et al., 2007).

The application of state-of-the-art climate models requires significant supercomputing resources. Limitations in those resources lead to additional constraints. Even when using the most powerful computers, compromises need to be made in three main areas:

1. Numerical implementations allow for a choice of grid spacing and time step, usually referred to as 'model resolution'. Higher model resolution generally leads to mathematically more accurate models (although not necessarily more reliable simulations) but also to higher computational costs. The finite resolution of climate models implies that the effects of certain processes must be represented through parameterizations (e.g., the carbon cycle or cloud and precipitation processes; see Chapters 6 and 7).

2. The climate system contains many processes, the relative importance of which varies with the time scale of interest (e.g., the carbon cycle). Hence compromises to include or exclude certain processes or components in a model must be made, recognizing that an increase in complexity generally leads to an increase in computational cost (Hurrell et al., 2009).

3. Owing to uncertainties in the model formulation and the initial state, any individual simulation represents only one of the possible pathways the climate system might follow. To allow some evaluation of these uncertainties, it is necessary to carry out a number of simulations either with several models or by using an ensemble of simulations with a single model, both of which increase computational cost.

Trade-offs amongst the various considerations outlined above are guided by the intended model application and lead to the several classes of models introduced in Section 9.1.2.

Individual model components (e.g., the atmosphere, the ocean, etc.) are typically first evaluated in isolation as part of the model development process. For instance, the atmospheric component can be evaluated by prescribing sea surface temperature (SST) (Gates et al., 1999) or the ocean and land components by prescribing atmospheric conditions (Barnier et al., 2006; Griffies et al., 2009). Subsequently, the various components are assembled into a comprehensive model, which then undergoes a systematic evaluation. At this stage, a small subset of model parameters remains to be adjusted so that the model adheres to large-scale observational constraints (often global averages). This final parameter adjustment procedure is usually referred to as 'model tuning'. Model tuning aims to match observed climate system behaviour and so is connected to judgements as to what constitutes a skilful representation of the Earth's climate. For instance, maintaining the global mean top of the atmosphere (TOA) energy balance in a simulation of pre-industrial climate is essential to prevent the climate system from drifting to an unrealistic state. The models used in this report almost universally contain adjustments to parameters in their treatment of clouds to fulfil this important constraint of the climate system (Watanabe et al., 2010; Donner et al., 2011; Gent et al., 2011; Golaz et al., 2011; Martin et al., 2011; Hazeleger et al., 2012; Mauritsen et al., 2012; Hourdin et al., 2013).

With very few exceptions (Mauritsen et al., 2012; Hourdin et al., 2013) modelling centres do not routinely describe in detail how they tune their models. Therefore the complete list of observational constraints toward which a particular model is tuned is generally not

(continued on next page)

Box 9.1 (continued)

available. However, it is clear that tuning involves trade-offs; this keeps the number of constraints that can be used small and usually focuses on global mean measures related to budgets of energy, mass and momentum. It has been shown for at least one model that the tuning process does not necessarily lead to a single, unique set of parameters for a given model, but that different combinations of parameters can yield equally plausible models (Mauritsen et al., 2012). Hence the need for model tuning may increase model uncertainty. There have been recent efforts to develop systematic parameter optimization methods, but owing to model complexity they cannot yet be applied to fully coupled climate models (Neelin et al., 2010).

Model tuning directly influences the evaluation of climate models, as the quantities that are tuned cannot be used in model evaluation. Quantities closely related to those tuned will provide only weak tests of model performance. Nonetheless, by focusing on those quantities not generally involved in model tuning while discounting metrics clearly related to it, it is possible to gain insight into model performance. Model quality is tested most rigorously through the concurrent use of many model quantities, evaluation techniques, and performance metrics that together cover a wide range of emergent (or un-tuned) model behaviour.

The requirement for model tuning raises the question of whether climate models are reliable for future climate projections. Models are not tuned to match a particular future; they are tuned to reproduce a small subset of global mean observationally based constraints. What emerges is that the models that plausibly reproduce the past, universally display significant warming under increasing greenhouse gas concentrations, consistent with our physical understanding.

close mutual coupling. One new development is the Eddy-Diffusivity-Mass-Flux (EDMF) approach (Siebesma et al., 2007; Rio and Hourdin, 2008; Neggers, 2009; Neggers et al., 2009; Rio et al., 2010). The EDMF approach, like the shallow cumulus scheme of Park and Bretherton (2009), determines the cumulus-base mass flux from the statistical distribution of boundary layer updraft properties, a conceptual advance over the *ad hoc* closure assumptions used in the past. Realistic treatment of the stable boundary layer remains difficult (Beare et al., 2006; Cuxart et al., 2006; Svensson and Holtslag, 2009) with implications for modelling of the diurnal cycle of temperature even under clear skies (Svensson et al., 2011).

Parameterizations of unresolved orographic and non-orographic gravity-wave drag (GWD) have seen only a few changes since the AR4 (e.g., Richter et al., 2010; Geller et al., 2011). In addition to new formulations, the estimation of the parameters used in the GWD schemes has recently been advanced through the availability of satellite and ground-based observations of gravity-wave momentum fluxes, high-resolution numerical modelling, and focussed process studies (Alexander et al., 2010). Evidence from the Numerical Weather Prediction community that important terrain-generated features of the atmospheric circulation are better represented at higher model resolution has recently been confirmed (Watanabe et al., 2008; Jung et al., 2012).

9.1.3.1.2 Ocean

Ocean components in contemporary climate models generally have horizontal resolutions that are too coarse to admit mesoscale eddies. Consequently, such models typically employ some version of the Redi (Redi, 1982) neutral diffusion and Gent and McWilliams (Gent and McWilliams, 1990) eddy advection parameterization (see also Gent et al., 1995; McDougall and McIntosh, 2001). Since the AR4, a focus has been on how parameterized mesoscale eddy fluxes in the ocean interior interact with boundary layer turbulence (Gnanadesikan et al., 2007; Danabasoglu et al., 2008; Ferrari et al., 2008, 2010). Another focus concerns eddy diffusivity, with many CMIP5 models employing flow-dependent schemes. Both of these refinements are important for the mean state and the response to changing forcing, especially in the Southern Ocean (Hallberg and Gnanadesikan, 2006; Boning et al., 2008; Farneti et al., 2010; Farneti and Gent, 2011; Gent and Danabasoglu, 2011; Hofmann and Morales Maqueda, 2011).

In addition to mesoscale eddies, there has been a growing awareness of the role that sub-mesoscale eddies and fronts play in restratifying the mixed layer (Boccaletti et al., 2007; Fox-Kemper et al., 2008; Klein and Lapeyre, 2009), and the parameterization of Fox-Kemper et al. (2011) is now used in some CMIP5 models.

There is an active research effort on the representation of dianeutral mixing associated with breaking gravity waves (MacKinnon et al., 2009), with this work adding rigour to the prototype energetically consistent abyssal tidal mixing parameterization of Simmons et al. (2004) now used in several climate models (e.g., Jayne, 2009; Danabasoglu et al., 2012). The transport of dense water down-slope by gravity currents (e.g., Legg et al., 2008, 2009) has also been the subject of focussed efforts, with associated parameterizations making their way into some CMIP5 models (Jackson et al., 2008b; Legg et al., 2009; Danabasoglu et al., 2010).

9.1.3.1.3 Land

Land surface properties such as vegetation, soil type and the amount of water stored on the land as soil moisture, snow and groundwater all strongly influence climate, particularly through their effects on surface albedo and evapotranspiration. These climatic effects can be profound; for example, it has been suggested that changes in the state of the land surface may have played an important part in the severity and length of the 2003 European drought (Fischer et al., 2007), and that more than 60% of the projected increase in interannual summer temperature variability in Europe is due to soil moisture–temperature feedbacks (Seneviratne et al., 2006).

Land surface schemes calculate the fluxes of heat, water and momentum between the land and the atmosphere. At the time of the AR4, even the more advanced land surface schemes suffered from obvious simplifications, such as the need to prescribe rather than simulate the vegetation cover and a tendency to ignore lateral flows of water and sub-gridscale heterogeneity in soil moisture (Pitman, 2003). Since the AR4, a number of climate models have included some representation of vegetation dynamics (see Sections 9.1.3.2.5 and 9.4.4.3), land–atmosphere CO_2 exchange (see Section 9.4.5), sub-gridscale hydrology (Oleson et al., 2008b) and changes in land use (see Section 9.4.4.4).

9.1.3.1.4 Sea ice

Most large-scale sea ice processes, such as basic thermodynamics and dynamics, are well understood and well represented in models (Hunke et al., 2010). However, important details of sea ice dynamics and deformation are not captured, especially at small scales (Coon et al., 2007; Girard et al., 2009; Hutchings et al., 2011). Currently, sea ice model development is focussed mainly on (1) more precise descriptions of physical processes such as microstructure evolution and anisotropy and (2) including biological and chemical species. Many models now include some representation of sub-grid-scale thickness variations, along with a description of mechanical redistribution that converts thinner ice to thicker ice under deformation (Hunke et al., 2010).

Sea ice albedo has long been recognized as a critical aspect of the global heat balance. The average ice surface albedo on the scale of a climate model grid cell is (as on land) the result of a mixture of surface types: bare ice, melting ice, snow-covered ice, open water, etc. Many sea ice models use a relatively simple albedo parameterization that specifies four albedo values: cold snow; warm, melting snow; cold, bare ice; and warm, melting ice, and the specific values may be subject to tuning (e.g., Losch et al., 2010). Some parameterizations take into account the ice and snow thickness, spectral band and surface melt (e.g., Pedersen et al., 2009; Vancoppenolle et al., 2009). Solar radiation may be distributed within the ice column assuming exponential decay or via a more complex multiple-scattering radiative transfer scheme (Briegleb and Light, 2007).

Snow model development for sea ice has lagged behind terrestrial snow models. Lecomte et al. (2011) introduced vertically varying snow temperature, density and conductivity to improve vertical heat conduction and melting in a 1D model intended for climate simulation, but many physical processes affecting the evolution of the snow pack, such as redistribution by wind, moisture transport (including flooding and snow ice formation) and snow grain size evolution, still are not included in most climate models.

Salinity affects the thermodynamic properties of sea ice, and is used in the calculation of fresh water and salt exchanges at the ice–ocean interface (Hunke et al., 2011). Some models allow the salinity to vary in time (Schramm et al., 1997), while others assume a salinity profile that is constant (e.g., Bitz and Lipscomb, 1999). Another new thrust is the inclusion of chemistry and biogeochemistry (Piot and von Glasow, 2008; Zhao et al., 2008; Vancoppenolle et al., 2010; Hunke et al., 2011), with dependencies on the ice microstructure and salinity profile.

Melt ponds can drain through interconnected brine channels when the ice becomes warm and permeable. This flushing can effectively clean the ice of salt, nutrients, and other inclusions, which affect the albedo, conductivity and biogeochemical processes and thereby play a role in climate change. Advanced parameterizations for melt ponds are making their way into sea ice components of global climate models (e.g., Flocco et al., 2012; Hunke et al., 2013).

9.1.3.2 New Components and Couplings: Emergence of Earth System Modelling

9.1.3.2.1 Carbon cycle

The omission of internally consistent feedbacks among the physical, chemical and biogeochemical processes in the Earth's climate system is a limitation of AOGCMs. The conceptual issue is that the physical climate influences natural sources and sinks of CO_2 and methane (CH_4), the two most important long-lived GHGs. ESMs incorporate many of the important biogeochemical processes, making it possible to simulate the evolution of these radiatively active species based on their emissions from natural and anthropogenic sources together with their interactions with the rest of the Earth system. Alternatively, when forced with specified concentrations, a model can be used to diagnose these sources with feedbacks included (Hibbard et al., 2007). Given the large natural sources and sinks of CO_2 relative to anthropogenic emissions, and given the primacy of CO_2 among anthropogenic GHGs, some of the most important enhancements are the addition of terrestrial and oceanic carbon cycles. These cycles have been incorporated into many models (Christian et al., 2010; Tjiputra et al., 2010) used to make projections of climate change (Schurgers et al., 2008; Jungclaus et al., 2010). Several ESMs now include coupled carbon and nitrogen cycles (Thornton et al., 2007; Gerber et al., 2010; Zaehle and Dalmonech, 2011) in order to simulate the interactions of nitrogen compounds with ecosystem productivity, GHGs including nitrous oxide (N_2O) and ozone (O_3), and global carbon sequestration (Zaehle and Dalmonech, 2011).

Oceanic uptake of CO_2 is highly variable in space and time, and is determined by the interplay between the biogeochemical and physical processes in the ocean. About half of CMIP5 models make use of schemes that partition marine ecosystems into nutrients, plankton, zooplankton and detritus (hence called NPZD-type models) while others use a more simplified representation of ocean biogeochemistry (see Table 9.A.1). These NPZD-type models allow simulation of some of the important feedbacks between climate and oceanic CO_2 uptake, but are limited by the lack of marine ecosystem dynamics. Some efforts have been made to include more plankton groups or plankton functional types in the models (Le Quere et al., 2005) with as-yet uncertain implications for Earth system response.

Ocean acidification and the associated decrease in calcification in many marine organisms provides a negative feedback on atmospheric CO_2 increase (Ridgwell et al., 2007a). New-generation models therefore include various parameterizations of calcium carbonate ($CaCO_3$)

9

production as a function of the saturation state of seawater with respect to calcite (Gehlen et al., 2007; Ridgwell et al., 2007a; Ilyina et al., 2009) or partial pressure CO_2 (pCO_2; Heinze, 2004). On centennial to multi-millennial scales, deep-sea carbonate sediments neutralize atmospheric CO_2. Some CMIP5 models include the sediment carbon reservoir, and progress has been made toward refined sediment representation in the models (Heinze et al., 2009).

9

9.1.3.2.2 Aerosol particles

The treatment of aerosol particles has advanced since the AR4. Many AOGCMs and ESMs now include the basic features of the sulphur cycle and so represent both the direct effect of sulphate aerosol, along with some of the more complex indirect effects involving cloud droplet number and size. Further, several AOGCMs and ESMs are currently capable of simulating the mass, number, size distribution and mixing state of interacting multi-component aerosol particles (Bauer et al., 2008b; Liu et al., 2012b). The incorporation of more physically complete representations of aerosol often improves the simulated climate under historical and present-day conditions, including the mean pattern and interannual variability in continental rainfall (Rotstayn et al., 2010, 2011). However, despite the addition of aerosol–cloud interactions to many AOGCMs and ESMs since the AR4, the representation of aerosol particles and their interaction with clouds and radiative transfer remains an important source of uncertainty (see Sections 7.3.5 and 7.4). Additional aerosol-related topics that have received attention include the connection between dust aerosol and ocean biogeochemistry, the production of oceanic dimethylsulphide (DMS, a natural source of sulphate aerosol), and vegetation interactions with organic atmospheric chemistry (Collins et al., 2011).

9.1.3.2.3 Methane cycle and permafrost

In addition to CO_2, an increasing number of ESMs and EMICs are also incorporating components of the CH_4 cycle, for example, atmospheric CH_4 chemistry and wetland emissions, to quantify some of the feedbacks from changes in CH_4 sources and sinks under a warming climate (Stocker et al., 2012). Some models now simulate the evolution of the permafrost carbon stock (Khvorostyanov et al., 2008a, 2008b), and in some cases this is integrated with the representation of terrestrial and oceanic CH_4 cycles (Volodin, 2008b; Volodin et al., 2010).

9.1.3.2.4 Dynamic global vegetation models and wildland fires

One of the potentially more significant effects of climate change is the alteration of the distribution, speciation and life cycle of vegetated ecosystems (Bergengren et al., 2001, 2011). Vegetation has a significant influence on the surface energy balance, exchanges of non-CO_2 GHGs and the terrestrial carbon sink. Systematic shifts in vegetation, for example, northward migration of boreal forests, would therefore impose biogeophysical feedbacks on the physical climate system (Clark et al., 2011). In order to include these effects in projections of climate change, several dynamic global vegetation models (DGVMs) have been developed and deployed in ESMs (Cramer et al., 2001; Sitch et al., 2008; Ostle et al., 2009). Although agriculture and managed forests are not yet generally incorporated, DGVMs can simulate the interactions among natural and anthropogenic drivers of global warming, the state of terrestrial ecosystems and ecological feedbacks on further climate change. The incorporation of DGVMs has required considerable improvement in the physics of coupled models to produce stable and realistic distributions of flora (Oleson et al., 2008b). The improvements include better treatments of surface, subsurface and soil hydrological processes; the exchange of water with the atmosphere; and the discharge of water into rivers and streams. Whereas the first DGVMs have been coupled primarily to the carbon cycle, the current generation of DGVMs are being extended to include ecological sources and sinks of other non-CO_2 trace gases including CH_4, N_2O, biogenic volatile organic compounds (BVOCs) and nitrogen oxides collectively known as NO_x (Arneth et al., 2010). BVOCs and NO_x can alter the lifetime of some GHGs and act as precursors for secondary organic aerosols (SOAs) and ozone. Disturbance of the natural landscape by fire has significant climatic effects through its impact on vegetation and through its emissions of GHGs, aerosols and aerosol precursors. Because the frequency of wildland fires increases rapidly with increases in ambient temperature (Westerling et al., 2006), the effects of fires are projected to grow over the 21st century (Kloster et al., 2012). The interactions of fires with the rest of the climate system are now being introduced into ESMs (Arora and Boer, 2005; Pechony and Shindell, 2009; Shevliakova et al., 2009).

9.1.3.2.5 Land use/land cover change

The impacts of land use and land cover change on the environment and climate are explicitly included as part of the Representative Concentration Pathways (RCPs; cf. Chapters 1 and 12) used for climate projections to be assessed in later chapters (Moss et al., 2010). Several important types of land use and land cover change include effects of agriculture and changing agricultural practices, including the potential for widespread introduction of biofuel crops; the management of forests for preservation, wood harvest and production of woody biofuel stock; and the global trends toward greater urbanization. ESMs include increasingly detailed treatments of crops and their interaction with the landscape (Arora and Boer, 2010; Smith et al., 2010a, 2010b), forest management (Bellassen et al., 2010, 2011) and the interactions between urban areas and the surrounding climate systems (Oleson et al., 2008a).

9.1.3.2.6 Chemistry–climate interactions and stratosphere–troposphere coupling

Important chemistry–climate interactions such as the impact of the ozone hole and recovery on Southern Hemisphere (SH) climate or the radiative effects of stratospheric water vapour changes on surface temperature have been confirmed in multiple studies (SPARC-CCMVal, 2010; WMO, 2011). In the majority of the CMIP5 simulations stratospheric ozone is prescribed. The main advance since the AR4 is that time-varying rather than constant stratospheric ozone is now generally used. In addition, several CMIP5 models treat stratospheric chemistry interactively, thus prognostically calculating stratospheric ozone and other chemical constituents. Important chemistry–climate interactions such as an increased influx of stratospheric ozone in a warmer climate that results in higher ozone burdens in the troposphere have also been identified (Young et al., 2013). Ten of the CMIP5 models simulate tropospheric chemistry interactively whereas it is prescribed in the remaining models (see Table 9.1 and Eyring et al. (2013)).

It is now widely accepted that in addition to the influence of tropospheric circulation and climate change on the stratosphere, stratospheric dynamics can in turn influence the tropospheric circulation and its variability (SPARC-CCMVal, 2010; WMO, 2011). As a result, many climate models now have the ability to include a fully resolved stratosphere with a model top above the stratopause, located at around 50 km. The subset of CMIP5 models with high-top configurations is compared to the set of low-top models with a model top below the stratopause in several studies (Charlton-Perez et al., 2012; Hardiman et al., 2012; Wilcox et al., 2012), although other factors such as differences in tropospheric warming or ozone could affect the two sub-ensembles.

9.1.3.2.7 Land ice sheets

The rate of melt water release from the Greenland and Antarctic ice sheets in response to climate change remains a major source of uncertainty in projections of sea level rise (see Sections 13.4.3 and 13.4.4). Until recently, the long-term response of these ice sheets to alterations in the surrounding atmosphere and ocean has been simulated using offline models. Several ESMs currently have the capability to have ice sheet component models coupled to the rest of the climate system (Driesschaert et al., 2007; Charbit et al., 2008; Vizcaino et al., 2008; Huybrechts et al., 2011; Robinson et al., 2012) although these capabilities are not exercised for CMIP5.

9.1.3.2.8 Additional features in ocean–atmosphere coupling

Several features in the coupling between the ocean and the atmosphere have become more widespread since the AR4. The bulk formulae used to compute the turbulent fluxes of heat, water and momentum at the air–sea interface, have been revised. A number of models now consider the ocean surface current when calculating wind stress (e.g., Luo et al., 2005; Jungclaus et al., 2006). The coupling frequency has been increased in some cases to include the diurnal cycle, which was shown to reduce the SST bias in the tropical Pacific (Schmidt et al., 2006; Bernie et al., 2008; Ham et al., 2010). Several models now represent the coupling between the penetration of the solar radiation into the ocean and light-absorbing chlorophyll, with some implications on the representation of the mean climate and climate variability (Murtugudde et al., 2002; Wetzel et al., 2006). This coupling is achieved either by prescribing the chlorophyll distribution from observations, or by computing the chlorophyll distribution with an ocean biogeochemical model (e.g., Arora et al., 2009).

9.1.3.3 Resolution

The typical horizontal resolution (defined here as horizontal grid spacing) for current AOGCMs and ESMs is roughly 1 to 2 degrees for the atmospheric component and around 1 degree for the ocean (Table 9.1). The typical number of vertical layers is around 30 to 40 for the atmosphere and around 30 to 60 for the ocean (note that some 'high-top' models may have 80 or more vertical levels in the atmosphere). There has been some modest increase in model resolution since the AR4, especially for the near-term simulations (e.g., around 0.5 degree for the atmosphere in some cases), based on increased availability of more powerful computers. For the models used in long-term simulations with interactive biogeochemistry, the resolution has not increased substantially due to the trade-off against higher complexity in such models. Since the AR4, typical regional climate model resolution has increased from around 50 km to around 25 km (see Section 9.6.2.2), and the impact of this has been explored with multi-decadal regional simulations (e.g., Christensen et al., 2010). In some cases, RCMs are being run at 10 km resolution or higher (e.g., Kanada et al., 2008; Kusaka et al., 2010; van Roosmalen et al., 2010; Kendon et al., 2012).

Higher resolution can sometimes lead to a stepwise, rather than incremental, improvement in model performance (e.g., Roberts et al., 2004; Shaffrey et al., 2009). For example, ocean models undergo a transition from laminar to eddy-permitting when the computational grid contains more than one or two grid points per first baroclinic Rossby radius (i.e., finer than 50 km at low latitudes and 10 km at high latitudes) (Smith et al., 2000; McWilliams, 2008). Such mesoscale eddy-permitting ocean models better capture the large amount of energy contained in fronts, boundary currents, and time dependent eddy features (e.g., McClean et al., 2006). Models run at such resolution have been used for some climate simulations, though much work remains before they are as mature as the coarser models currently in use (Bryan et al., 2007; Bryan et al., 2010; Farneti et al., 2010; McClean et al., 2011; Delworth et al., 2012).

Similarly, atmospheric models with grids that allow the explicit representation of convective cloud systems (i.e., finer than a few kilometres) avoid employing a parameterization of their effects—a long-standing source of uncertainty in climate models. For example, Kendon et al. (2012) simulated the climate of the UK region over a 20-year period at 1.5 km resolution, and demonstrated several improvements of errors typical of coarser resolution models. Further discussion of this is provided in Section 7.2.2.

9.2 Techniques for Assessing Model Performance

Systematic evaluation of models through comparisons with observations is a prerequisite to applying them confidently. Several significant developments in model evaluation have occurred since the AR4 and are assessed in this section. This is followed by a description of the overall approach to evaluation taken in this chapter and a discussion of its known limitations.

9.2.1 New Developments in Model Evaluation Approaches

9.2.1.1 Evaluating the Overall Model Results

The most straightforward approach to evaluate models is to compare simulated quantities (e.g., global distributions of temperature, precipitation, radiation etc.) with corresponding observationally based estimates (e.g., Gleckler et al., 2008; Pincus et al., 2008; Reichler and Kim, 2008). A significant development since the AR4 is the increased use of quantitative statistical measures, referred to as performance metrics. The use of such metrics simplifies synthesis and visualization of model performance (Gleckler et al., 2008; Pincus et al., 2008; Waugh and Eyring, 2008; Cadule et al., 2010; Sahany et al., 2012) and enables the quantitative assessment of model improvements over time (Reichler and Kim, 2008). Recent work has addressed redundancy of multiple performance metrics through methods such as cluster analysis (Yokoi et al., 2011; Nishii et al., 2012).

9.2.1.2 Isolating Processes

To understand the cause of model errors it is necessary to evaluate the representation of processes both in the context of the full model and in isolation. A number of evaluation techniques to achieve both process and component isolation have been developed. One involves the so-called 'regime-oriented' approach to process evaluation. Instead of averaging model results in time (e.g., seasonal averages) or space (e.g., global averages), results are averaged within categories that describe physically distinct regimes of the system. Applications of this approach since the AR4 include the use of circulation regimes (Bellucci et al., 2010; Brown et al., 2010b; Brient and Bony, 2012; Ichikawa et al., 2012), cloud regimes (Williams and Brooks, 2008; Chen and Del Genio, 2009; Williams and Webb, 2009; Tsushima et al., 2013) and thermodynamic states (Sahany et al., 2012; Su, 2012). The application of new observations, such as vertically resolved cloud and water vapour information from satellites (Jiang et al., 2012a; Konsta et al., 2012; Quaas, 2012) and water isotopes (Risi et al., 2012a; Risi et al., 2012b), has also enhanced the ability to evaluate processes in climate models.

Another approach involves the isolation of model components or parameterizations in off-line simulations, such as Single Column Models of the atmosphere. Results of such simulations are compared to measurements from field studies or to results of more detailed process models (Randall et al., 2003). Numerous process evaluation data sets have been collected since the AR4 (Redelsperger et al., 2006; Illingworth et al., 2007; Verlinde et al., 2007; May et al., 2008; Wood et al., 2011) and have been applied to the evaluation of climate model processes (Xie et al., 2008; Boone et al., 2009; Boyle and Klein, 2010; Hourdin et al., 2010). These studies are crucial to test the realism of the process formulations that underpin climate models.

9.2.1.3 Instrument Simulators

Satellites provide nearly global coverage, sampling across many meteorological conditions. This makes them powerful tools for model evaluation. The conventional approach has been to convert satellite-observed radiation information to 'model-equivalents' (Stephens and Kummerow, 2007), and these have been used in numerous studies (Allan et al., 2007; Gleckler et al., 2008; Li et al., 2008; Pincus et al., 2008; Waliser et al., 2009b; Li et al., 2011a, 2012a; Jiang et al., 2012a). A challenge is that limitations of the satellite sensors demand various assumptions in order to convert a satellite measurement into a 'model equivalent' climate variable.

An alternative approach is to calculate 'observation-equivalents' from models using radiative transfer calculations to simulate what the satellite would provide if the satellite system were 'observing' the model. This approach is usually referred to as an 'instrument simulator'. Microphysical assumptions (which differ from model to model) can be included in the simulators, avoiding inconsistencies. A simulator for cloud properties from the International Cloud Satellite Climatology Project (ISCCP) (Yu et al., 1996; Klein and Jakob, 1999; Webb et al., 2001) has been widely used for model evaluation since the AR4 (Chen and Del Genio, 2009; Marchand et al., 2009; Wyant et al., 2009; Yokohata et al., 2010), often in conjunction with statistical techniques to separate model clouds into cloud regimes (e.g., Field et al., 2008; Williams and Brooks, 2008; Williams and Webb, 2009). New simulators for other satellite products have also been developed and are increasingly applied for model evaluation (Bodas-Salcedo et al., 2011). Although often focussed on clouds and precipitation, the simulator approach has also been used successfully for other variables such as upper tropospheric humidity (Allan et al., 2003; Iacono et al., 2003; Ringer et al., 2003; Brogniez et al., 2005; Brogniez and Pierrehumbert, 2007; Zhang et al., 2008b; Bodas-Salcedo et al., 2011). Although providing an alternative to the use of model-equivalents from observations, instrument simulators have limitations (Pincus et al., 2012) and are best applied in combination with other model evaluation techniques.

9.2.1.4 Initial Value Techniques

To be able to forecast the weather a few days ahead, knowledge of the present state of the atmosphere is of primary importance. In contrast, climate predictions and projections simulate the statistics of weather seasons to centuries in advance. Despite their differences, both weather predictions and projections of future climate are performed with very similar atmospheric model components. The atmospheric component of climate models can be integrated as a weather prediction model if initialized appropriately (Phillips et al., 2004). This allows testing parameterized sub-grid scale processes without the complication of feedbacks substantially altering the underlying state of the atmosphere.

The application of these techniques since the AR4 has led to some new insights. For example, many of the systematic errors in the modelled climate develop within a few days of simulation, highlighting the important role of fast, parameterized processes (Klein et al., 2006; Boyle et al., 2008; Xie et al., 2012). Errors in cloud properties for example were shown to be present within a few days in a forecast in at least some models (Williams and Brooks, 2008), although this was not the case in another model (Boyle and Klein, 2010; Zhang et al., 2010b). Other studies have highlighted the advantage of such methodologies for the detailed evaluation of model processes using observations that are available only for limited locations and times (Williamson and Olson, 2007; Bodas-Salcedo et al., 2008; Xie et al., 2008; Hannay et al., 2009; Boyle and Klein, 2010), an approach that is difficult to apply to long-term climate simulations.

9.2.2 Ensemble Approaches for Model Evaluation

Ensemble methods are used to explore the uncertainty in climate model simulations that arise from internal variability, boundary conditions, parameter values for a given model structure or structural uncertainty due to different model formulations (Tebaldi and Knutti, 2007; Hawkins and Sutton, 2009; Knutti et al., 2010a). Since the AR4, techniques have been designed to specifically evaluate model performance of individual ensemble members. Although this is typically done to better characterize uncertainties, the methods and insights are applicable to model evaluation in general. The ensembles are generally of two types: Multi-model Ensembles (MMEs) and Perturbed Parameter (or Physics) Ensembles (PPEs).

9.2.2.1 Multi-Model Ensembles

The MME is created from existing model simulations from multiple climate modelling centres. MMEs sample structural uncertainty and internal variability. However, the sample size of MMEs is small, and is confounded because some climate models have been developed by sharing model components leading to shared biases (Masson and Knutti, 2011a). Thus, MME members cannot be treated as purely independent, which implies a reduction in the effective number of independent models (Tebaldi and Knutti, 2007; Jun et al., 2008; Knutti, 2010; Knutti et al., 2010a; Pennell and Reichler, 2011).

9.2.2.2 Perturbed-Parameter Ensembles

In contrast, PPEs are created to assess uncertainty based on a single model and benefit from the explicit control on parameter perturbations. This allows statistical methods to determine which parameters are the main drivers of uncertainty across the ensemble (e.g., Rougier et al., 2009). PPEs have been used frequently in simpler models such as EMICs (Xiao et al., 1998; Forest et al., 2002, 2006, 2008; Stott and Forest, 2007; Knutti and Tomassini, 2008; Sokolov et al., 2009; Loutre et al., 2011) and are now being applied to more complex models (Murphy et al., 2004; Annan et al., 2005; Stainforth et al., 2005; Collins et al., 2006a, 2007; Jackson et al., 2008a; Brierley et al., 2010; Klocke et al., 2011; Lambert et al., 2012). The disadvantage of PPEs is that they do not explore structural uncertainty, and thus the estimated uncertainty will depend on the underlying model that is perturbed (Yokohata et al., 2010) and may be too narrow (Sakaguchi et al., 2012). Several studies (Sexton et al., 2012; Sanderson, 2013) recognize the importance of sampling both parametric and structural uncertainty by combining information from both MMEs and PPEs. However, even these approaches cannot account for the effect on uncertainty of systematic errors.

9.2.2.3 Statistical Methods Applied to Ensembles

The most common approach to characterize MME results is to calculate the arithmetic mean of the individual model results, referred to as an unweighted multi-model mean. This approach of 'one vote per model' gives equal weight to each climate model regardless of (1) how many simulations each model has contributed, (2) how interdependent the models are or (3) how well each model has fared in objective evaluation. The multi-model mean will be used often in this chapter. Some climate models share a common lineage and so share common biases (Frame et al., 2005; Tebaldi and Knutti, 2007; Jun et al., 2008; Knutti, 2010; Knutti et al., 2010a, 2013; Annan and Hargreaves, 2011; Pennell and Reichler, 2011; Knutti and Sedlácek, 2013). As a result, collections such as the CMIP5 MME cannot be considered a random sample of independent models. This complexity creates challenges for how best to make quantitative inferences of future climate as discussed further in Chapter 12 (Knutti et al., 2010a; Collins et al., 2012; Stephenson et al., 2012; Sansom et al., 2013).

Annan and Hargreaves (2010) have proposed a 'rank histogram' approach to evaluate model ensembles as a whole, rather than individual models, by diagnosing whether observations can be considered statistically indistinguishable from a model ensemble. Studies based on this approach have suggested that MMEs (CMIP3/5) are 'reliable' in that they are not too narrow or too dispersive as a sample of possible models, but existing single-model-based ensembles tend to be too narrow (Yokohata et al., 2012, 2013). Although initial work has analysed the current mean climate state, further work is required to study the relationships between simulation errors and uncertainties in ensembles of future projections (Collins et al., 2012).

Bayesian methods offer insights into how to account for model inadequacies and combine information from several metrics in both MME and PPE approaches (Sexton and Murphy, 2012; Sexton et al., 2012), but they are complex. A simpler strategy of screening out some model variants on the basis of some observational comparison has been used with some PPEs (Lambert et al., 2012; Shiogama et al., 2012). Edwards et al. (2011) provided a statistical framework for 'pre-calibrating' out such poor model variants. Screening techniques have also been used with MMEs (Santer et al., 2009).

Additional Bayesian methods are applied to the MMEs so that past model performance is combined with prior distributions to estimate uncertainty from the MME (Furrer et al., 2007; Tebaldi and Knutti, 2007; Milliff et al., 2011). Similar to Bayesian PPE methods, common biases can be assessed within the MME to determine effective independence of the climate models (Knutti et al., 2013) (see Section 12.2.2 for a discussion of the assumptions in the Bayesian approaches).

9.2.3 The Model Evaluation Approach Used in this Chapter and Its Limitations

This chapter applies a variety of evaluation techniques ranging from visual comparison of observations and the multi-model ensemble and its mean, to application of quantitative performance metrics (see Section 9.2.2). No individual evaluation technique or performance measure is considered superior; rather, it is the combined use of many techniques and measures that provides a comprehensive overview of model performance.

Although crucial, the evaluation of climate models based on past climate observations has some important limitations. By necessity, it is limited to those variables and phenomena for which observations exist. Table 9.3 provides an overview of the observations used in this chapter. In many cases, the lack or insufficient quality of long-term observations, be it a specific variable, an important processes, or a particular region (e.g., polar areas, the upper troposphere/lower stratosphere (UTLS), and the deep ocean), remains an impediment. In addition, owing to observational uncertainties and the presence of internal variability, the observational record against which models are assessed is 'imperfect'. These limitations can be reduced, but not entirely eliminated, through the use of multiple independent observations of the same variable as well as the use of model ensembles.

The approach to model evaluation taken in the chapter reflects the need for climate models to represent the observed behaviour of past climate as a necessary condition to be considered a viable tool for future projections. This does not, however, provide an answer to the much more difficult question of determining how well a model must agree with observations before projections made with it can be deemed reliable. Since the AR4, there are a few examples of emergent constraints where observations are used to constrain multi-model ensemble projections. These examples, which are discussed further in Section 9.8.3, remain part of an area of active and as yet inconclusive research.

Table 9.3 | Overview of observations that are used to evaluate climate models in this chapter. The quantity and CMIP5 output variable name are given along with references for the observations. Superscript (1) indicates this observations-based data set is obtained from atmospheric reanalysis. Superscript (D) indicates default reference; superscript (A) alternate reference.

Quantity	CMIP5 Output Variable Name	Observations (Default/Alternates)	Reference	Figure and Section Number(s)
ATMOSPHERE				
Surface (2 m) Air Temperature (°C)	Tas (2 m)	ERA-Interim[1]	Dee et al. (2011)	Figures 9.2, 9.3, 9.6[D], 9.7[D], Section 9.4.1; Figures 9.38, 9.40, Section 9.6.1
		NCEP-NCAR[1]	Kalnay et al. (1996)	Figures 9.6[A], 9.7[A], Section 9.4.1
		ERA40[1]	Uppala et al. (2005)	Figure 9.38, Section 9.6.1
		CRU TS 3.10	Mitchell and Jones (2005)	Figures 9.38, 9.39, Section 9.6.1
		HadCRUT4	Morice et al. (2012)	Figure 9.8, Section 9.4.1
		GISTEMP	Hansen et al. (2010)	Figure 9.8, Section 9.4.1
		MLOST	Vose et al. (2012)	Figure 9.8, Section 9.4.1
Temperature (°C)	Ta (200, 850 hPa)	ERA-Interim[1]	Dee et al. (2011)	Figure 9.9[D] Section 9.4.
		NCEP-NCAR[1]	Kalnay et al. (1996)	Figure 9.9[A] Section 9.4.1
Zonal mean wind (m s^{-1})	Ua (200, 850 hPa)	ERA-Interim[1]	Dee et al. (2011)	Figure 9.7[D], Section 9.4.1
		NCEP-NCAR[1]	Kalnay et al. (1996)	Figure 9.7[A], Section 9.4.1
Zonal wind stress (m s^{-1})	Tauu	QuikSCAT satellite measurements	Risien and Chelton (2008)	Figures 9.19–9.20, Section 9.4.2
		NCEP/NCAR reanalysis	Kalnay et al. (1996)	Figures 9.19–9.20, Section 9.4.2
		ERA-Interim	Dee et al. (2011)	Figures 9.19–9.20, Section 9.4.2
Meridional wind (m s^{-1})	Va (200, 850 hPa)	ERA-Interim[1]	Dee et al. (2011)	Figure 9.7[D], Section 9.4.1
		NCEP-NCAR[1]	Kalnay et al. (1996)	Figure 9.7[A], Section 9.4.1
Geopotential height (m)	Zg (500 hPa)	ERA-Interim[1]	Dee et al. (2011)	Figure 9.7[D], Section 9.4.1
		NCEP-NCAR[1]	Kalnay et al. (1996)	Figure 9.7[A], Section 9.4.1
TOA reflected short-wave radiation (W m^{-2})	Rsut	CERES EBAF 2.6	Loeb et al. (2009)	Figure 9.9[D] Section 9.4.1
		ERBE	Barkstrom (1984)	Figure 9.9[A], Section 9.4.1
TOA longwave radiation (W m^{-2})	Rlut	CERES EBAF 2.6	Loeb et al. (2009)	Figure 9.9[D] Section 9.4.1
		ERBE	Barkstrom (1984)	Figure 9.9[A], Section 9.4.1
Clear sky TOA short-wave cloud radiative effect (W m^{-2})	SW CRE	CERES EBAF 2.6	Loeb et al. (2009)	Figures 9.5[D], 9.6[D], 9.7[D], Section 9.4.1
	Derived from CMIP5 rsut and rsutcs	CERES ES-4 ERBE	Loeb et al. (2009)	Figure 9.5[A], Section 9.4.1
			Barkstrom (1984)	Figure 9.7[A], Section 9.4.1
Clear sky TOA long-wave cloud radiative effect (W m^{-2})	LW CRE	CERES EBAF 2.6	Loeb et al. (2009)	Figure 9.9[D], Section 9.4.1
	Derived from CMIP5 rsut and rsutcs	CERES ES-4 ERBE	Loeb et al. (2009)	Figure 9.5[A], Section 9.4.1
Total precipitation (mm day^{-1})	Pr	GPCP	Adler et al. (2003)	Figures 9.4, 9.6[D], 9.7[D], Section 9.4.1; Figures 9.38, 9.40, Section 9.6.1
		CMAP	Xie and Arkin (1997)	Figures 9.6[A], 9.7[A], Section 9.4.1; Figures 9.38, 9.40, Section 9.6.1
		CRU TS3.10.1	Mitchell and Jones (2005)	Figures 9.38, 9.39, Section 9.6.1

(continued on next page)

Table 9.3 (continued)

Quantity	CMIP5 Output Variable Name	Observations (Default/Alternates)	Reference	Figure and Section Number(s)
ATMOSPHERE (continued)				
Precipitable water	PRW	RSS V7 SSM/I ERA-INTERIM	Wentz et al. (2007) Dee et al. (2011)	Figure 9.9, Section 9.4.1
		MERRA	Rienecker et al. (2011)	
		JRA-25	Onogi et al. (2007)	
Lower-tropospheric temperature	TLT	RSS V3.3 MSU/AMSU	Mears et al. (2011)	Figure 9.9, Section 9.4.1
		UAH V5.4 MSU/AMSU	Christy et al. (2007)	
		ERA-INTERIM	Dee et al. (2011)	
		MERRA	Rienecker et al. (2011)	
		JRA-25	Onogi et al. (2007)	
Snow albedo feedback (%/K)	tas, rsds, rsus	Advanced Very High Resolution Radiometer (AVHRR), Polar Pathfinder-x (APP-x), all-sky albedo and ERA40 surface air temperature	Hall and Qu (2006); Fernandes et al. (2009)	Figure 9.43, Section 9.8.3
Reconstruction of bio-climatic variables for the mid-Holocene and the Last Glacial Maximum	Tas, pr,, tcold, twarm, GDD5, alpha		Bartlein et al. (2010)	Figure 9.11, Section 9.4.1 Figure 9.12, Section 9.4.1
OZONE and AEROSOLS				
Aerosol optical depth	aod	MODIS	Shi et al. (2011)	Figures 9.28, 9.29, Section 9.4.6
		MISR	Zhang and Reid (2010); Stevens and Schwartz (2012)	Figure 9.29, Section 9.4.6
Total column ozone (DU)	tro3	Ground-based measurements	updated from Fioletov et al. (2002)	Figure 9.10, Section 9.4.1
		NASA TOMS/OMI/SBUV(/2) merged satellite data	Stolarski and Frith (2006)	
		NIWA combined total column ozone database	Bodeker et al. (2005)	
		Solar Backscatter Ultraviolet (SBUV, SBUV/2) retrievals	Updated from Miller et al. (2002)	
		DLR GOME/SCIA/GOME-2	Loyola et al. (2009); Loyola and Coldewey-Egbers (2012)	
CARBON CYCLE				
Atmospheric CO_2 (ppmv)	co2		Masarie and Tans (1995); Meinshausen et al. (2011)	Figure 9.45, Section 9.8.3
Global Land Carbon Sink (PgC yr^{-1})	NBP	GCP	Le Quere et al. (2009)	Figure 9.26, 9.27, Section 9.4.5
Global Ocean Carbon Sink (PgC yr^{-1})	fgCO2	GCP	Le Quere et al. (2009)	Figure 9.26, 9.27, Section 9.4.5
Regional Land Sinks (PgC yr^{-1})	NBP	JAM	Gurney et al. (2003)	Figure 9.27, Section 9.4.5
Regional Ocean Sinks (PgC yr^{-1})	fgCO2	JAM	Gurney et al. (2003); Takahashi et al. (2009)	Figure 9.27, Section 9.4.5

(continued on next page)

9

9

Table 9.3 (continued)

Quantity	CMIP5 Output Variable Name	Observations (Default/Alternates)	Reference	Figure and Section Number(s)
OCEAN				
Annual mean temperature	thetao		Levitus et al. (2009)	Figure 9.13, Section 9.4.2
Annual mean salinity	so		Antonov et al. (2010)	Figure 9.13, Section 9.4.2
Sea Surface Temperature	tos	HadISST1.1	Rayner et al. (2003)	Figure 9.14, Section 9.4.2
		HadCRU 4	Jones et al. (2012)	Figure 9.35, Section 9.5.3
		ERA40	Uppala et al. (2005)	Figure 9.36, Section 9.5.3
Global ocean heat content (0 to 700 m)	OHC	Levitus	Levitus et al. (2009)	Figure 9.17, Section 9.4.2
		Ishii Domingues	Ishii and Kimoto, 2009) Domingues et al. (2008)	
Dynamic Sea surface height	SSH	AVISO	AVISO	Figure 9.16, Section 9.4.2
Meridional heat transport	hfnorth	(1) Using surface and TOA heat fluxes:	Trenberth and Fasullo (2008) Large and Yeager (2009)	Figure 9.21, Section 9.4.2
		NCEP/NCAR	Kalnay et al. (1996)	
		ERA40	Uppala et al. (2005)	
		Updated NCEP reanalysis	Kistler et al. (2001)	
		(2) Direct estimates using WOCE and inverse models	Ganachaud and Wunsch (2003)	
Annual mean temperature and salinity		Palaeoclimate reconstruction of temperature and salinity	Adkins et al. (2002)	Figure 9.18, Section 9.4.2
Total area (km²) of grid cells where Sea Ice Area Fraction (%) is >15%. Boundary of sea ice where Sea Ice Area Fraction (%) is >15%		HadISST	Rayner et al. (2003)	Figure 9.22, Section 9.4.3
		NSIDC	Fetterer et al. (2002)	Figure 9.23, Section 9.4.3
		NASA	Comiso and Nishio (2008)	Figure 9.24, Section 9.4.3
MISC				
Total area (km²) of grid cells where Surface Snow Area Fraction (%) is 15% or Surface Snow Amount (kg m⁻²) is >5 kg m⁻²			Robinson and Frei (2000)	Figure 9.25, Section 9.4.4
3-hour precipitation fields		15,000 stations and corrected Ta from COADS (Dai and Deser, 1999; Dai, 2001	Dai (2006)	Figure 9.30, Section 9.5.2
Absolute value of MJJAS minus NDJFM precipitation exceeding 375 mm		GPCP (Adler et al., 2003)	Wang et al. (2011a)	Figure 9.32, Section 9.5.2
EXTREMES				
Daily maximum and minimum surface air temperature fields (°C) Daily precipitation fields (mm day⁻¹) **for calculating extremes indices**	tas, precip	ERA40	Uppala et al. (2005)	Figure 9.37, Section 9.5.4
		ERA-Interim,	Dee et al. (2011)	
		NCEP/NCAR Reanalysis 1,	Kistler et al. (2001)	
		NCEP-DOE, Reanalysis 2	Kanamitsu et al. (2002)	
			Calculation of indices is based on Sillmann et al. (2013)	
Temperature extremes indices based on station observations		HadEX2	Donat et al. (2013)	Figure 9.37, Section 9.5.4

Notes: [1] This observationally constrained data set is obtained from atmospheric reanalysis.

[D] Default reference.

[A] Alternate reference.

9.3 Experimental Strategies in Support of Climate Model Evaluation

9.3.1 The Role of Model Intercomparisons

Systematic model evaluation requires a coordinated and well-documented suite of model simulations. Organized Model Intercomparison Projects (MIPs) provide this via standard or benchmark experiments that represent critical tests of a model's ability to simulate the observed climate. When modelling centres perform a common experiment, it offers the possibility to compare their results not just with observations, but with other models as well. This intercomparison enables researchers to explore the range of model behaviours, to isolate the various strengths and weaknesses of different models in a controlled setting, and to interpret, through idealized experiments, the inter-model differences. Benchmark MIP experiments offer a way to distinguish between errors particular to an individual model and those that might be more universal and should become priority targets for model improvement.

9.3.2 Experimental Strategy for Coupled Model Intercomparison Project Phase 5

9.3.2.1 Experiments Utilized for Model Evaluation

CMIP5 includes a much more comprehensive suite of model experiments than was available in the preceding CMIP3 results assessed in the AR4 (Meehl et al., 2007). In addition to a better constrained specification of historical forcing, the CMIP5 collection also includes initialized decadal-length predictions and long-term experiments using both ESMs and AOGCMs (Taylor et al., 2012b) (Figure 9.1). The CO_2 forcing of these experiments is prescribed as a time series of either global mean concentrations or spatially resolved anthropogenic emissions (Section 9.3.2.2). The analyses of model performance in this chapter are based on the concentration-based experiments with the exception of the evaluation of the carbon cycle (see Section 9.4.5).

Most of the model diagnostics are derived from the historical simulations that span the period 1850– 2005. In some cases, these historical simulations are augmented by results from a scenario run, either RCP4.5 or RCP8.5 (see Section 9.3.2.2), so as to facilitate comparison with more recent observations. CMIP5/Paleoclimate Modelling Intercomparison Project version 3 (PMIP3) simulations for the mid-Holocene and last glacial maximum are used to evaluate model response to palaeoclimatic conditions. Historical emissions-driven simulations are used to evaluate the prognostic carbon cycle. The analysis of global surface temperature variability is based in part on long pre-industrial control runs to facilitate calculation of variability on decadal to centennial time scales. Idealized simulations with 1% per year increases in CO_2 are utilized to derive transient climate response. Equilibrium climate sensitivities are derived using results of specialized experiments, with fourfold CO_2 increase, designed specifically for this purpose.

9.3.2.2 Forcing of the Historical Experiments

Under the protocols adopted for CMIP5 and previous assessments, the transient climate experiments are conducted in three phases. The first phase covers the start of the modern industrial period through to the present day, years 1850–2005 (van Vuuren et al., 2011). The second phase covers the future, 2006–2100, and is described by a collection of RCPs (Moss et al., 2010). As detailed in Chapter 12, the third phase is described by a corresponding collection of Extension Concentration

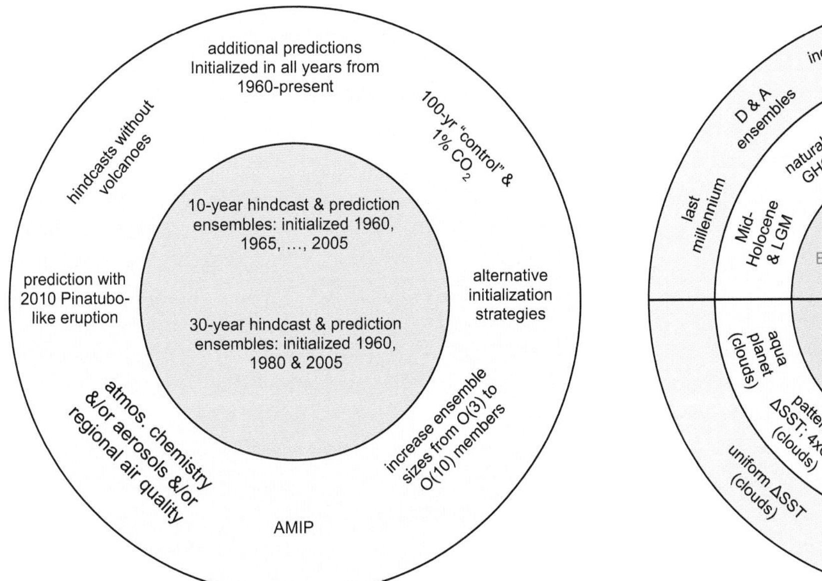

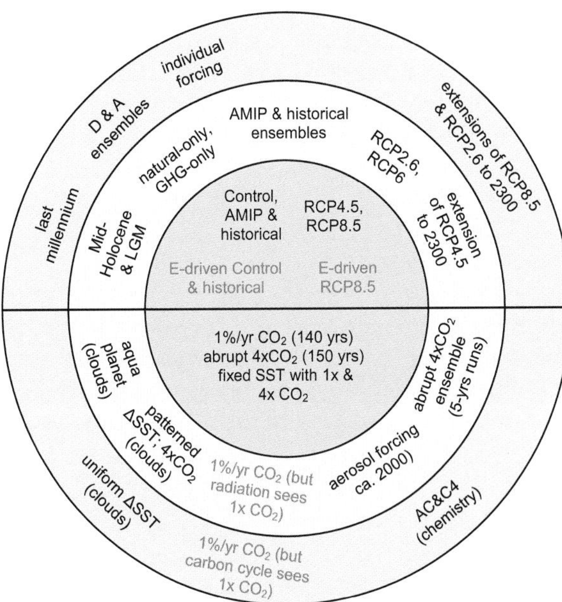

Figure 9.1 | Left: Schematic summary of CMIP5 short-term experiments with tier 1 experiments (yellow background) organized around a central core (pink background). (From Taylor et al., 2012b, their Figure 2). Right: Schematic summary of CMIP5 long-term experiments with tier 1 experiments (yellow background) and tier 2 experiments (green background) organized around a central core (pink background). Green font indicates simulations to be performed only by models with carbon cycle representations, and 'E-driven' means 'emission-driven'. Experiments in the upper semicircle either are suitable for comparison with observations or provide projections, whereas those in the lower semicircle are either idealized or diagnostic in nature, and aim to provide better understanding of the climate system and model behaviour. (From Taylor et al., 2012b, their Figure 3.)

Pathways (Meinshausen et al., 2011). The forcings for the historical simulations evaluated in this section and are described briefly here (with more details in Annex II).

In the CMIP3 20th century experiments, the forcings from radiatively active species other than long-lived GHGs and sulphate aerosols were left to the discretion of the individual modelling groups (IPCC, 2007). By contrast, a comprehensive set of historical anthropogenic emissions and land use and land cover change data have been assembled for the CMIP5 experiments in order to produce a relatively homogeneous ensemble of historical simulations with common time series of forcing agents. Emissions of natural aerosols including soil dust, sea salt and volcanic species are still left to the discretion of the individual modelling groups.

9

For AOGCMs without chemical and biogeochemical cycles, the forcing agents are prescribed as a set of concentrations. The concentrations for GHGs and related compounds include CO_2, CH_4, N_2O, all fluorinated gases controlled under the Kyoto Protocol (hydrofluorocarbons (HFCs), perfluorocarbons (PFCs), and sulphur hexafluoride (SF_6)), and ozone-depleting substances controlled under the Montreal Protocol (chlorofluorocarbons (CFCs), hydrochlorofluorocarbons (HCFCs), Halons, carbon tetrachloride (CCl_4), methyl bromide (CH_3Br), methyl chloride (CH_3Cl)). The concentrations for aerosol species include sulphate (SO_4), ammonium nitrate (NH_4NO_3), hydrophobic and hydrophilic black carbon, hydrophobic and hydrophilic organic carbon, secondary organic aerosols (SOAs) and four size categories of dust and sea salt. For ESMs that include chemical and biogeochemical cycles, the forcing agents are prescribed both as a set of concentrations and as a set of emissions with provisions to separate the forcing by natural and anthropogenic CO_2 (Hibbard et al., 2007). The emissions include time-dependent spatially resolved fluxes of CH_4, NO_X, CO, NH_3, black and organic carbon, and volatile organic compounds (VOCs). For models that treat the chemical processes associated with biomass burning, emissions of additional species such as C_2H_4O (acetaldehyde), C_2H_5OH (ethanol), C_2H_6S (dimethylsulphide) and C_3H_6O (acetone) are also prescribed. Historical land use and land cover change is described in terms of the time-evolving partitioning of land surface area among cropland, pasture, primary land and secondary (recovering) land, including the effects of wood harvest and shifting cultivation, as well as land use changes and transitions from/to urban land (Hurtt et al., 2009). These emissions data are aggregated from empirical reconstructions of grassland and forest fires (Schultz et al., 2008; Mieville et al., 2010); international shipping (Eyring et al., 2010); aviation (Lee et al., 2009), sulphur (Smith et al., 2011b), black and organic carbon (Bond et al., 2007); and NO_X, CO, CH_4 and non methane volatile organic compounds (NMVOCs) (Lamarque et al., 2010) contributed by all other sectors.

For the natural forcings a recommended monthly averaged total solar irradiance time series was given, but there was no recommended treatment of volcanic forcing. Both integrated solar irradiance and its spectrum were available, but not all CMIP5 models used the spectral data. The data employed an 1850-2008 reconstruction of the solar cycle and its secular trend using observations of sunspots and faculae, the 10.7 cm solar irradiance measurements and satellite observations (Frohlich and Lean, 2004).For volcanic forcing CMIP5 models typically employed one of two prescribed volcanic aerosol data sets (Sato et al., 1993) or (Ammann et al., 2003) but at least one ESM employed interactive aerosol injection (Driscoll et al., 2012). The prescribed data sets did not incorporate injection from explosive volcanoes after 2000.

9.3.2.3 Relationship of Decadal and Longer-Term Simulations

The CMIP5 archive also includes a new class of decadal-prediction experiments (Meehl et al., 2009, 2013b) (Figure 9.1). The goal is to understand the relative roles of forced changes and internal variability in historical and near-term climate variables, and to assess the predictability that might be realized on decadal time scales. These experiments comprise two sets of hindcast and prediction ensembles with initial conditions spanning 1960 through 2005. The set of 10-year ensembles are initialized starting at 1960 in 1-year increments through the year 2005 while the 30-year ensembles are initialized at 1960, 1980 and 2005. The same physical models are often used for both the short-term and long-term experiments (Figure 9.1) despite the different initialization of these two sets of simulations. Results from the short-term experiments are described in detail in Chapter 11.

9.4 Simulation of Recent and Longer-Term Records in Global Models

9.4.1 Atmosphere

Many aspects of the atmosphere have been more extensively evaluated than other climate model components. One reason is the availability of near-global observationally based data for energy fluxes at the TOA, cloud cover and cloud condensate, temperature, winds, moisture, ozone and other important properties. As discussed in Box 2.3, atmospheric reanalyses have also enabled integrating independent observations in a physically consistent manner. In this section we use this diversity of data (see Table 9.3) to evaluate the large-scale atmospheric behaviour.

9.4.1.1 Temperature and Precipitation Spatial Patterns of the Mean State

Surface temperature is perhaps the most routinely examined quantity in atmospheric models. Many processes must be adequately represented in order for a model to realistically capture the observed temperature distribution. The dominant external influence is incoming solar radiation, but many aspects of the simulated climate play an important role in modulating regional temperature such as the presence of clouds and the complex interactions between the atmosphere and the underlying land, ocean, snow, ice and biosphere.

The annual mean surface air temperature (at 2 m) is shown in Figure 9.2(a) for the mean of all available CMIP5 models, and the error, relative to an observationally constrained reanalysis (ECMWF reanalysis of the global atmosphere and surface conditions (ERA)-Interim; Dee et al., 2011) is shown in Figure 9.2(b). In most areas the multi-model mean agrees with the reanalysis to within 2°C, but there are several locations where the biases are much larger, particularly at high elevations over the Himalayas and parts of both Greenland and Antarctica, near the ice edge in the North Atlantic, and over ocean upwelling regions off the west coasts of South America and Africa. Averaging the absolute error of the individual CMIP5 models (Figure 9.2c) yields similar magnitude as the multi-model mean bias (Figure 9.2b), implying that compensating errors across models is limited. The inconsistency across the three available global reanalyses (Figure 9.2d) that have assimilated temperature data at two metres (Onogi et al., 2007; Simmons et al., 2010) provides an indication of observational uncertainty. Although the reanalysis inconsistency is smaller than the mean absolute bias in almost all regions, areas where inconsistency is largest (typically where observations are sparse) tend to be the same regions where the CMIP5 models show largest mean absolute error.

Seasonal performance of models can be evaluated by examining the difference between means for December–January–February (DJF) and June–July–August (JJA). Figures 9.3(a) and (b) show the CMIP5 mean model seasonal cycle amplitude in surface air temperature (as measured by the difference between the DJF and JJA and the absolute value of this difference). The seasonal cycle amplitude is much larger over land where the thermal inertia is much smaller than over the oceans, and it is generally larger at higher latitudes as a result of the larger seasonal amplitude in insolation. Figures 9.3(c) and (d) show the mean model bias of the seasonal cycle relative to the ERA-Interim reanalysis (Dee et al., 2011). The largest biases correspond to areas of large seasonal amplitude, notably high latitudes over land, but relatively large biases are also evident in some lower latitude regions such as over northern India. Over most land areas the amplitude of the modelled seasonal cycle is larger than observed, whereas over much of the extratropical oceans the modelled amplitude is too small.

The simulation of precipitation is a more stringent test for models as it depends heavily on processes that must be parameterized. Challenges are compounded by the link to surface fields (topography, coastline, vegetation) that lead to much greater spatial heterogeneity at regional scales. Figure 9.4 shows the mean precipitation rate simulated by the CMIP5 multi-model ensemble, along with measures of error relative to precipitation analyses from the Global Precipitation Climatology Project (Adler et al., 2003). The magnitude of observational uncertainty for precipitation varies with region, which is why many studies make use

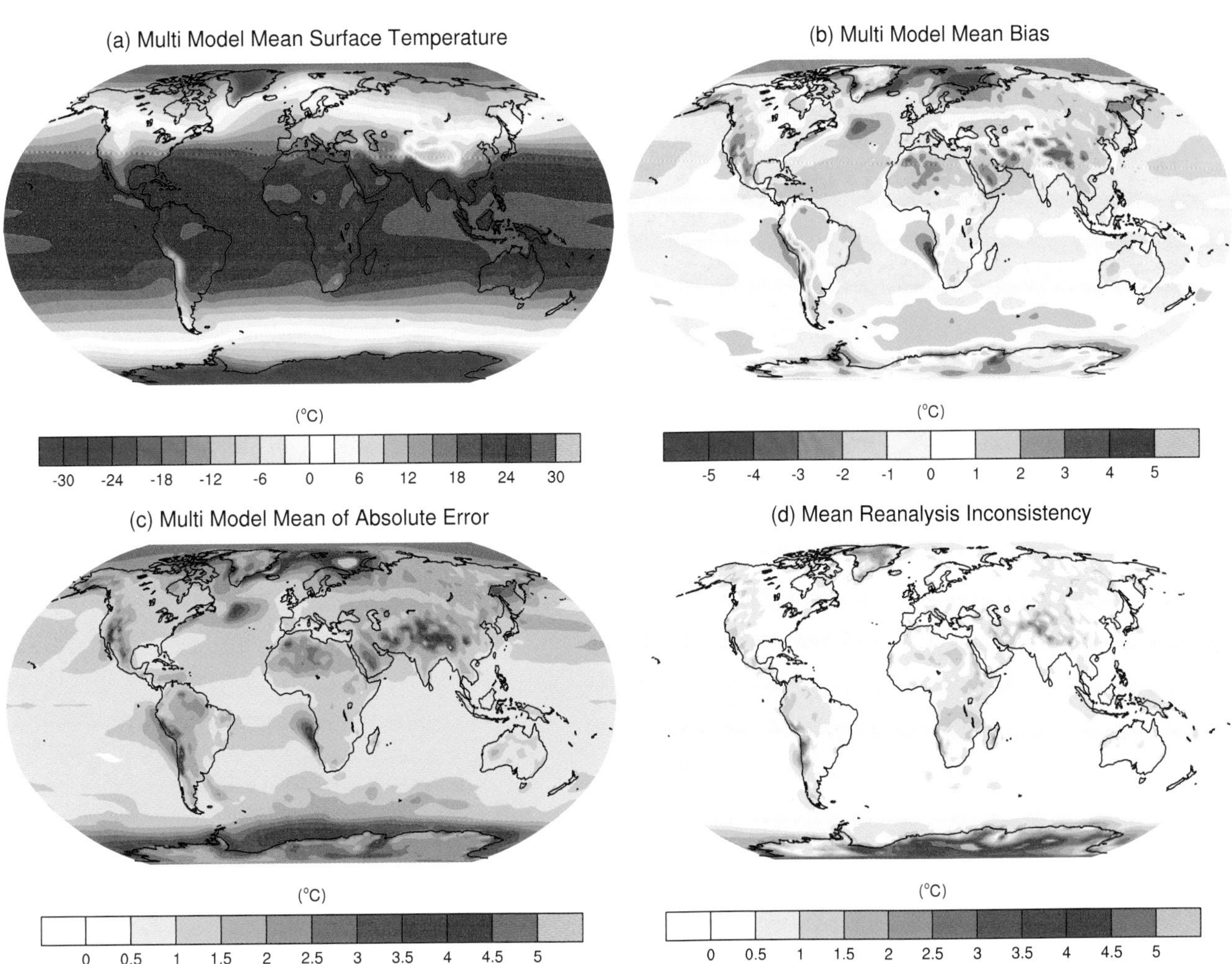

Figure 9.2 | Annual-mean surface (2 m) air temperature (°C) for the period 1980–2005. (a) Multi-model (ensemble) mean constructed with one realization of all available models used in the CMIP5 historical experiment. (b) Multi-model-mean bias as the difference between the CMIP5 multi-model mean and the climatology from ECMWF reanalysis of the global atmosphere and surface conditions (ERA)-Interim (Dee et al., 2011); see Table 9.3. (c) Mean absolute model error with respect to the climatology from ERA-Interim. (d) Mean inconsistency between ERA-Interim, ERA 40-year reanalysis (ERA40) and Japanese 25-year ReAnalysis (JRA-25) products as the mean of the absolute pairwise differences between those fields for their common period (1979–2001).

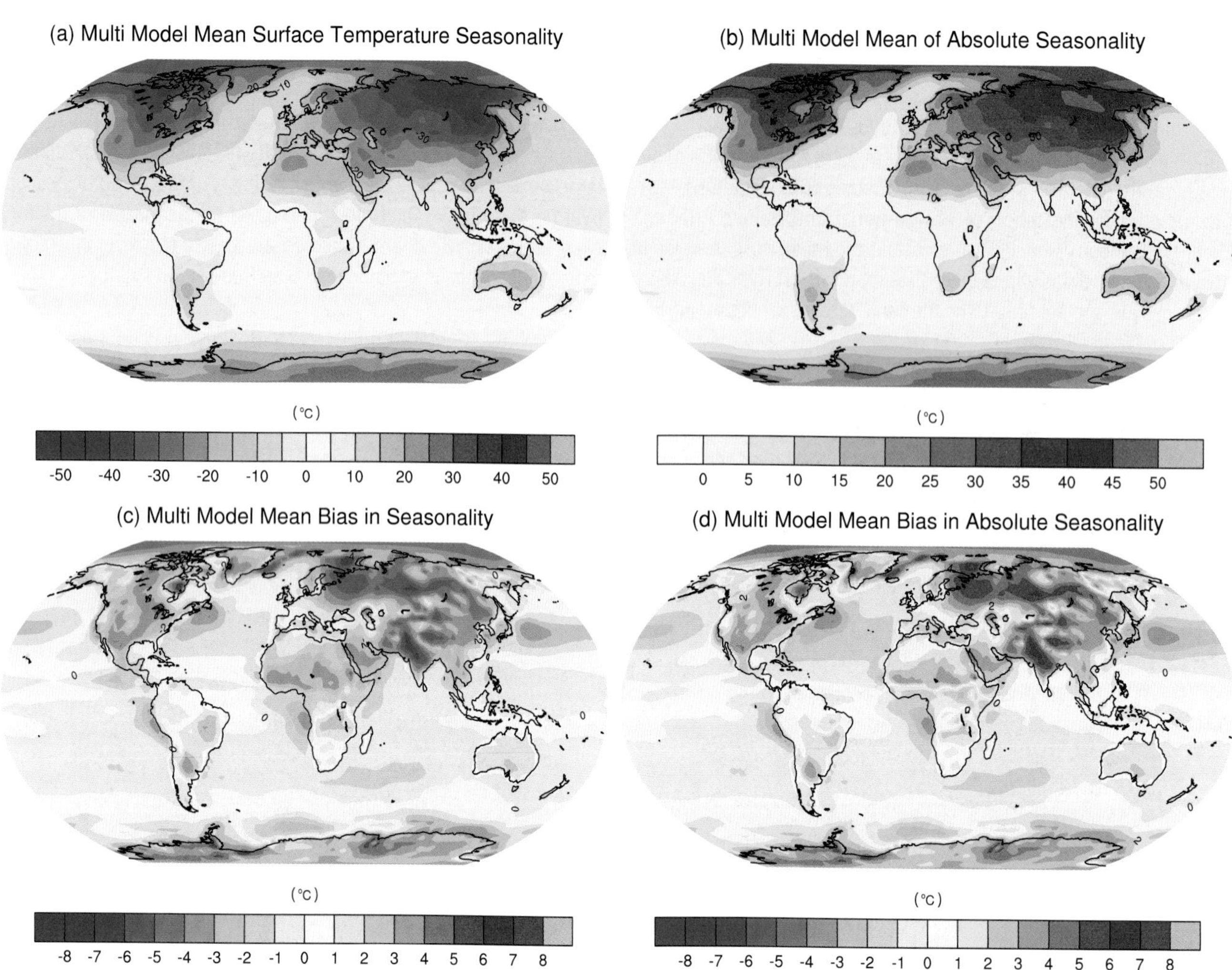

Figure 9.3 | Seasonality (December–January–February minus June–July–August) of surface (2 m) air temperature (°C) for the period 1980–2005. (a) Multi-model mean, calculated from one realization of all available CMIP5 models for the historical experiment. (b) Multi-model mean of absolute seasonality. (c) Difference between the multi-model mean and the ECMWF reanalysis of the global atmosphere and surface conditions (ERA)-Interim seasonality. (d) Difference between the multi-model mean and the ERA-Interim absolute seasonality.

of several estimates of precipitation. Known large-scale features are reproduced by the multi-model mean, such as a maximum precipitation just north of the equator in the central and eastern tropical Pacific, dry areas over the eastern subtropical ocean basins, and the minimum rainfall in Northern Africa (Dai, 2006). While many large-scale features of the tropical circulation are reasonably well simulated, there are persistent biases. These include too low precipitation along the equator in the Western Pacific associated with ocean–atmosphere feedbacks maintaining the equatorial cold tongue (Collins et al., 2010) and excessive precipitation in tropical convergence zones south of the equator in the Atlantic and the Eastern Pacific (Lin, 2007; Pincus et al., 2008). Other errors occurring in several models include an overly zonal orientation of the South-Pacific Convergence Zone (Brown et al., 2013) as well as an overestimate of the frequency of occurrence of light rain events (Stephens et al., 2010). Regional-scale precipitation simulation has strong parameter dependence (Rougier et al., 2009; Chen et al., 2010; Neelin et al., 2010), and in some models substantial improvements have been shown through increases in resolution (Delworth et al., 2012) and improved representations of sub-gridscale processes, particularly convection (Neale et al., 2008). Judged by similarity with the spatial pattern of observations, the overall quality of the simulation of the mean state of precipitation in the CMIP5 ensemble is slightly better than in the CMIP3 ensemble (see FAQ 9.1 and Figure 9.6).

In summary, there is *high confidence* that large-scale patterns of surface temperature are well simulated by the CMIP5 models. In certain regions this agreement with observations is limited, particularly at elevations over the Himalayas and parts of both Greenland and Antarctica. The broad-scale features of precipitation as simulated by the CMIP5 models are in modest agreement with observations, but there are systematic errors in the Tropics.

9.4.1.2 Atmospheric Moisture, Clouds and Radiation

The global annual mean precipitable water is a measure of the total moisture content of the atmosphere. For the CMIP3 ensemble, the values of precipitable water agreed with one another and with multiple estimates from the National Centers for Environmental Prediction/National Center for Atmospheric Research (NCEP/NCAR) and ECMWF

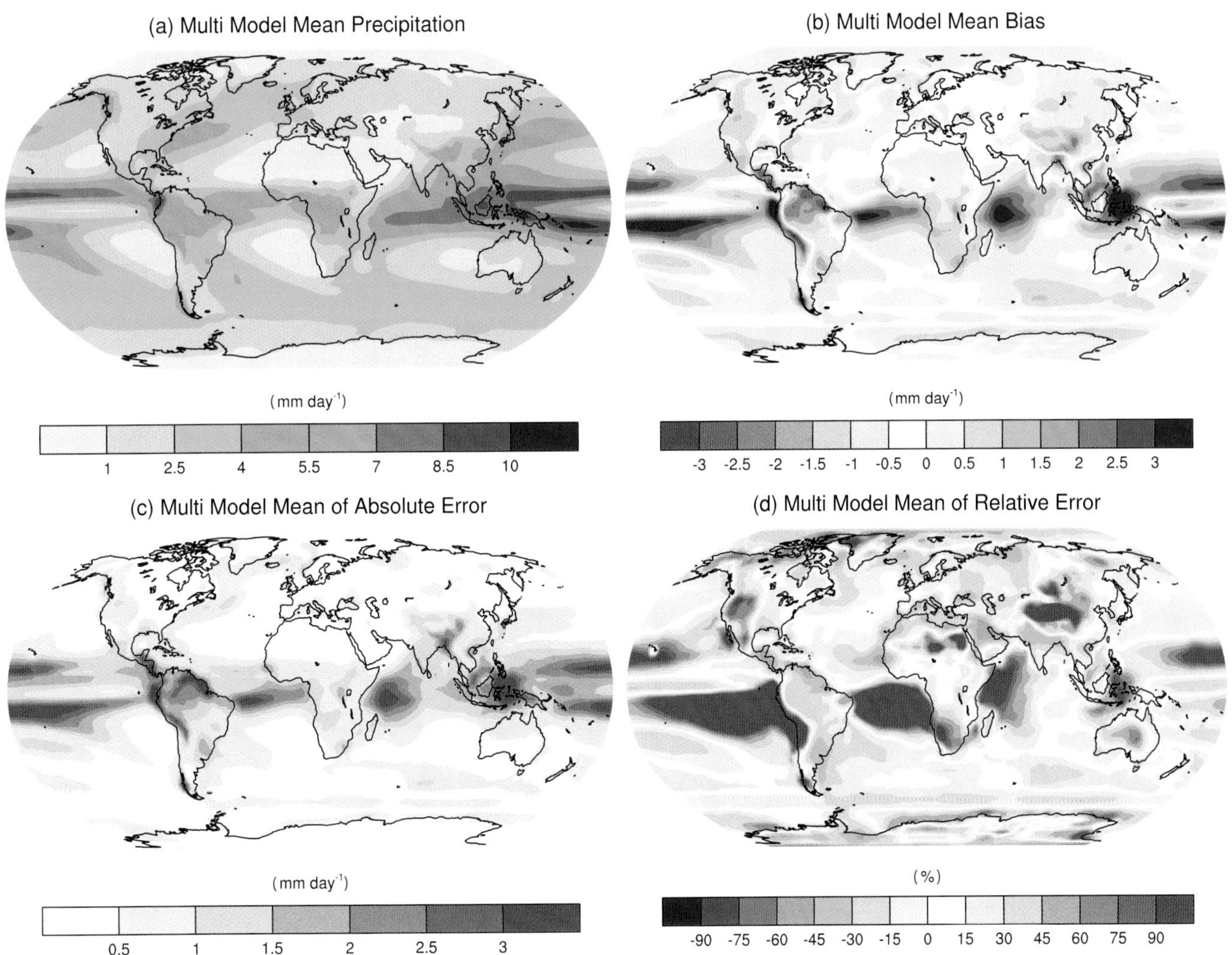

Figure 9.4 | Annual-mean precipitation rate (mm day^{-1}) for the period 1980–2005. (a) Multi-model-mean constructed with one realization of all available AOGCMs used in the CMIP5 historical experiment. (b) Difference between multi-model mean and precipitation analyses from the Global Precipitation Climatology Project (Adler et al., 2003). (c) Multi-model-mean absolute error with respect to observations. (d) Multi-model-mean error relative to the multi-model-mean precipitation itself.

ERA40 meteorological reanalyses to within approximately 10% (Waliser et al., 2007). Initial analysis of the CMIP5 ensemble shows the model results are within the uncertainties of the observations (Jiang et al., 2012a).

Modelling the vertical structure of water vapour is subject to greater uncertainty since the humidity profile is governed by a variety of processes. The CMIP3 models exhibited a significant dry bias of up to 25% in the boundary layer and a significant moist bias in the free troposphere of up to 100% (John and Soden, 2007). Upper tropospheric water vapour varied by a factor of three across the multi-model ensemble (Su et al., 2006). Many models have large biases in lower stratospheric water vapour (Gettelman et al., 2010), which could have implications for surface temperature change (Solomon et al., 2010). The limited number of studies available for the CMIP5 model ensemble broadly confirms the results from the earlier model generation. In tropical regions, the models are too dry in the lower troposphere and too moist in the upper troposphere, whereas in the extratropics they are too moist throughout the troposphere (Tian et al., 2013). However, many of the model values lie within the observational uncertainties. Jiang et al. (2012a) show that the largest biases occur in the upper troposphere, with model values up to twice that observed, while in the middle and lower troposphere models simulate water vapour to within 10% of the observations.

The spatial patterns and seasonal cycle of the radiative fluxes at the TOA are fundamental energy balance quantities. Both the CMIP3 and CMIP5 model ensembles reproduce these patterns with considerable fidelity relative to the National Aeronautics and Space Adminsitration (NASA) Clouds and the Earth's Radiant Energy System (CERES) data sets (Pincus et al., 2008; Wang and Su, 2013). Globally averaged TOA shortwave and longwave components of the radiative fluxes in 12 atmosphere-only versions of the CMIP5 models were within 2.5 W m^{-2} of the observed values (Wang and Su, 2013).

Comparisons against surface components of radiative fluxes show that, on average, the CMIP5 models overestimate the global mean downward all-sky shortwave flux at the surface by 2 ± 6 W m^{-2} (1 ± 3%) and underestimate the global downward longwave flux by 6 ± 9 W m^{-2} (2 ± 2%) (Stephens et al., 2012). Although in tropical regions between 1 and 3 W m^{-2} of the bias may be due to systematic omission of precipitating and/or convective core ice hydrometeors (Waliser et al., 2011), the correlation between the biases in the all-sky and clear-sky downwelling fluxes suggests that systematic errors in clear-sky radiative transfer calculations may be a primary cause for these biases. This is consistent with an analysis of the global annual mean estimates of clear-sky atmospheric absorption from the CMIP3 ensemble and the systematic underestimation of clear-sky solar absorption by radiative transfer codes (Oreopoulos et al., 2012). The underestimation of absorption can be attributed to the omission or underestimation of

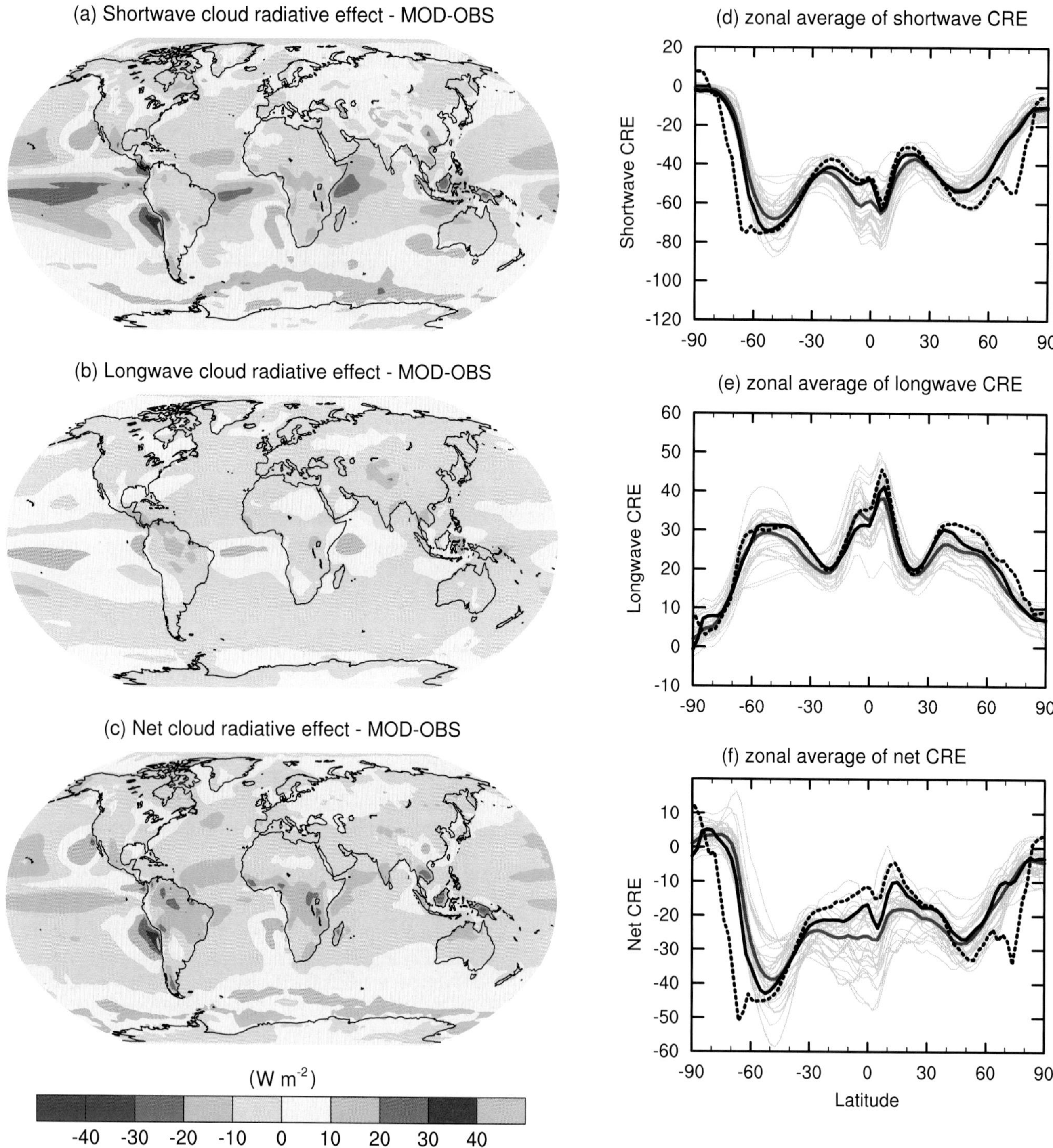

Figure 9.5 | Annual-mean cloud radiative effects of the CMIP5 models compared against the Clouds and the Earth's Radiant Energy System Energy Balanced and Filled 2.6 (CERES EBAF 2.6) data set (in W m^{-2}; top row: shortwave effect; middle row: longwave effect; bottom row: net effect). On the left are the global distributions of the multi-model-mean biases, and on the right are the zonal averages of the cloud radiative effects from observations (solid black: CERES EBAF 2.6; dashed black: CERES ES-4), individual models (thin grey lines), and the multi-model mean (thick red line). Model results are for the period 1985–2005, while the available CERES data are for 2001–2011. For a definition and maps of cloud radiative effect, see Section 7.2.1.2 and Figure 7.7.

absorbing aerosols, in particular carbonaceous species (Kim and Ramanathan, 2008), or to the omission of weak-line (Collins et al., 2006b) or continuum (Ptashnik et al., 2011) absorption by water vapour (Wild et al., 2006).

One of the major influences on radiative fluxes in the atmosphere is the presence of clouds and their radiative properties. To measure the influence of clouds on model deficiencies in the TOA radiation budget, Figure 9.5 shows maps of deviations from observations in annual mean shortwave (top left), longwave (middle left) and net (bottom left) cloud radiative effect (CRE) for the CMIP5 multi-model mean. The figure (right panels) also shows zonal averages of the same quantities from two sets of observations, the individual CMIP5 models, and the multi-model average. The definition of CRE and observed mean fields for these quantities can be found in Chapter 7 (Section 7.2.1.2, Figure 7.7).

Models show large regional biases in CRE in the shortwave component, and these are particularly pronounced in the subtropics with too weak an effect (positive error) of model clouds on shortwave radiation in the stratocumulus regions and too strong an effect (negative error) in the trade cumulus regions. This error has been shown to largely result from an overestimation of cloud reflectance, rather than cloud cover (Nam et al., 2012). A too weak cloud influence on shortwave radiation is evident over the subpolar oceans of both hemispheres and the Northern Hemisphere (NH) land areas. It is evident in the zonal mean graphs that there is a wide range in both longwave and shortwave CRE between individual models. As is also evident, a significant reduction in the difference between models and observations has resulted from changes in the observational estimates of CRE, in particular at polar and subpolar as well as subtropical latitudes (Loeb et al., 2009).

Understanding the biases in CRE in models requires a more in-depth analysis of the biases in cloud properties, including the fractional coverage of clouds, their vertical distribution as well as their liquid water and ice content. Major progress in this area has resulted from both the availability of new observational data sets and improved diagnostic techniques, including the increased use of instrument simulators (e.g., Cesana and Chepfer, 2012; Jiang et al., 2012a). Many models have particular difficulties simulating upper tropospheric clouds (Jiang et al., 2012a), and low and mid-level cloud occurrence are frequently underestimated (Cesana and Chepfer, 2012; Nam et al., 2012; Tsushima et al., 2013). Global mean values of both simulated ice and liquid water path vary by factors of 2 to 10 between models (Jiang et al., 2012a; Li et al., 2012a). The global mean fraction of clouds that can be detected with confidence from satellites (optical thickness >1.3, Pincus et al. (2012)) is underestimated by 5 to 10 % (Klein et al., 2013). Some of the above errors in clouds compensate to provide the global mean balance in radiation required by model tuning (Tsushima et al., 2013; Wang and Su, 2013; Box 9.1).

In-depth analysis of several global and regional models (Karlsson et al., 2008; Teixeira et al., 2011) has shown that the interaction of boundary layer and cloud processes with the larger scale circulation systems that ultimately drive the observed subtropical cloud distribution remains poorly simulated. Large errors in subtropical clouds have been shown to negatively affect SST patterns in coupled model simulations (Hu

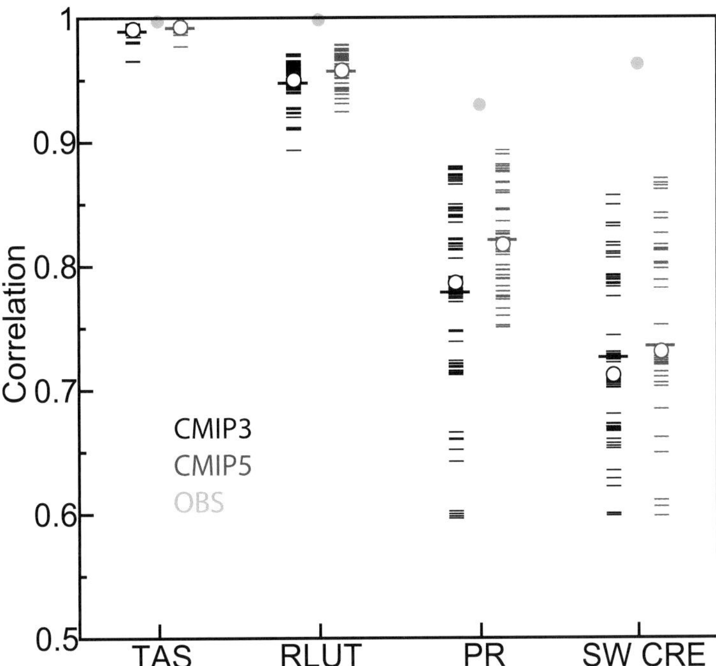

Figure 9.6 | Centred pattern correlations between models and observations for the annual mean climatology over the period 1980–1999. Results are shown for individual CMIP3 (black) and CMIP5 (blue) models as thin dashes, along with the corresponding ensemble average (thick dash) and median (open circle). The four variables shown are surface air temperature (TAS), top of the atmosphere (TOA) outgoing longwave radiation (RLUT), precipitation (PR) and TOA shortwave cloud radiative effect (SW CRE). The observations used for each variable are the default products and climatological periods identified in Table 9.3. The correlations between the default and alternate (Table 9.3) observations are also shown (solid green circles). To ensure a fair comparison across a range of model resolutions, the pattern correlations are computed at a resolution of 4° in longitude and 5° in latitude. Only one realization is used from each model from the CMIP3 20C3M and CMIP5 historical simulations.

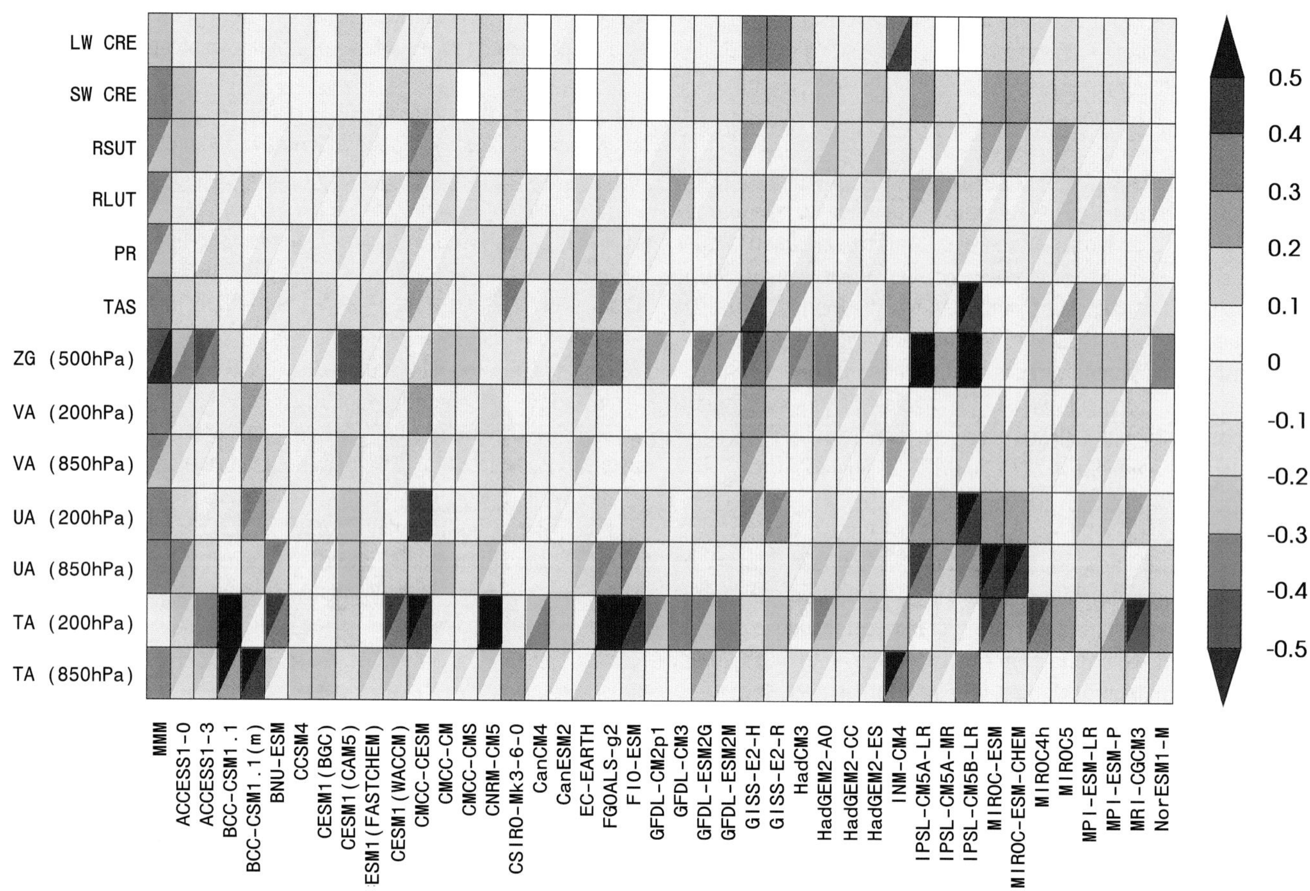

Figure 9.7 | Relative error measures of CMIP5 model performance, based on the global seasonal-cycle climatology (1980–2005) computed from the historical experiments. Rows and columns represent individual variables and models, respectively. The error measure is a space–time root-mean-square error (RMSE), which, treating each variable separately, is portrayed as a relative error by normalizing the result by the median error of all model results (Gleckler et al., 2008). For example, a value of 0.20 indicates that a model's RMSE is 20% larger than the median CMIP5 error for that variable, whereas a value of –0.20 means the error is 20% smaller than the median error. No colour (white) indicates that model results are currently unavailable. A diagonal split of a grid square shows the relative error with respect to both the default reference data set (upper left triangle) and the alternate (lower right triangle). The relative errors are calculated independently for the default and alternate data sets. All reference data used in the diagram are summarized in Table 9.3.

et al., 2011; Wahl et al., 2011). Several studies have highlighted the potential importance and poor simulation of subpolar clouds in the Arctic and Southern Oceans (Karlsson and Svensson, 2010; Trenberth and Fasullo, 2010b; Haynes et al., 2011; Bodas-Salcedo et al., 2012). A particular challenge for models is the simulation of the correct phase of the cloud condensate, although very few observations are available to evaluate models particularly with respect to their representation of cloud ice (Waliser et al., 2009b; Li et al., 2012a). Regime-oriented approaches to the evaluation of model clouds (see Section 9.2.1) have identified that compensating errors in the CRE are largely a result of misrepresentations of the frequency of occurrence of key observed cloud regimes, while the radiative properties of the individual regimes contribute less to the overall model deficiencies (Tsushima et al., 2013).

Several studies have identified progress in the simulation of clouds in the CMIP5 models compared to their CMIP3 counterparts. Particular examples include the improved simulation of vertically integrated ice water path (Jiang et al., 2012a; Li et al., 2012a) as well as a reduction of overabundant optically thick clouds in the mid-latitudes (Klein et al., 2013; Tsushima et al., 2013).

In summary, despite modest improvements there remain significant errors in the model simulation of clouds. There is *very high confidence* that these errors contribute significantly to the uncertainties in estimates of cloud feedbacks (see Section 9.7.2.3; Section 7.2.5, Figure 7.10) and hence the spread in climate change projections reported in Chapter 12.

9.4.1.3 Quantifying Model Performance with Metrics

Performance metrics were used to some extent in the Third Assessment Report (TAR) and the Fourth Assessment Report (AR4), and are expanded upon here because of their increased appearance in the recent literature. As a simple example, Figure 9.6 illustrates how the pattern correlation between the observed and simulated climatological annual mean spatial patterns depends very much on the quantity examined. All CMIP3 and CMIP5 models capture the mean surface temperature distribution quite well, with correlations above 0.95, which are largely determined by the meridional temperature gradient. Correlations for outgoing longwave radiation are somewhat lower. For precipitation and the TOA shortwave cloud radiative effect, the correlations between models and observations are below 0.90, and there is considerable scatter among model results. This example quantifies how some aspects of the simulated large-scale climate agree with observations better than others. Some of these differences are attributable to smoothly varying fields (e.g., temperature, water vapour) often agreeing better with observations than fields that exhibit fine structure (e.g., precipitation) (see also Section 9.6.1.1). Incremental improvement in each field is also evident in Figure 9.6, as gauged by the mean and median results in the CMIP5 ensemble having higher correlations than CMIP3. This multi-variate quantification of model improvement across development cycles is evident in several studies (e.g., Reichler and Kim, 2008; Knutti et al., 2013)

Figure 9.7 (following Gleckler et al., 2008) depicts the space–time root-mean-square error (RMSE) for the 1980–2005 climatological seasonal cycle of the historically forced CMIP5 simulations. For each of the fields examined, this 'portrait plot' depicts relative performance, with blue shading indicating performance being better, and red shading worse, than the median of all model results. In each case, two observations-based estimates are used to demonstrate the impact of the selection of reference data on the results. Some models consistently compare better with observations than others, some exhibit mixed performance and some stand out with relatively poor agreement with observations. For most fields, the choice of the observational data set does not substantially change the result for global error measures (e.g., between a state-of-the-art and an older-generation reanalysis), indicating that inter-model differences are substantially larger than the differences between the two reference data sets or the impact of two different climatological periods (e.g., for radiation fields: Earth Radiation Budget Experiment (ERBE) 1984–1988; CERES EBAF, 2001–2011). Nevertheless, it is important to recognize that different data sets often rely on the same source of measurements, and that the results in this figure can have some sensitivity to a variety of factors such as instrument uncertainty, sampling errors (e.g., limited record length of observations), the spatial scale of comparison, the domain considered and the choice of metric.

Another notable feature of Figure 9.7 is that in most cases the multi-model mean agrees more favourably with observations than any individual model. This has been long recognized to hold for surface temperature and precipitation (e.g., Lambert and Boer, 2001). However, since the AR4, it has become clear that this holds for a broad range of climatological fields (Gleckler et al., 2008; Pincus et al., 2008; Knutti et al., 2010a) and is theoretically better understood (Annan and Hargreaves, 2011). It is worth noting that when most models suffer from a common error, such as the cold bias at high latitudes in the upper troposphere (see TA 200 hPa of Figure 9.7), individual models can agree better with observations than the multi-model mean.

Correlations between the relative errors for different quantities in Figure 9.7 are known to exist, reflecting physical relationships in the model formulations and in the real world. Cluster analysis methods have recently been used in an attempt to reduce this redundancy (e.g., Yokoi et al., 2011; Nishii et al., 2012), thereby providing more succinct summaries of model performance. Some studies have attempted an overall skill score by averaging together the results from multiple metrics (e.g., Reichler and Kim, 2008). Although this averaging process is largely arbitrary, combining the results of multiple metrics can reduce the chance that a poorer performing model will score well for the wrong reasons. Recent work (Nishii et al., 2012) has demonstrated that different methods used to produce a multi-variate skill measure for the CMIP3 models did not substantially alter the conclusions about the better and lesser performing models.

Large scale performance metrics are a typical first-step toward quantifying model agreement with observations, and summarizing broad characteristics of model performance that are not focussed on a particular application. More specialized performance tests target aspects of a simulation believed to be especially important for constraining model projections, although to date the connections between particular performance metrics and reliability of future projections are not well established. This important topic is addressed in Section 9.8.3, which highlights several identified relationships between model performance and projection responses.

9.4.1.4 Long-Term Global-Scale Changes

The comparison of observed and simulated climate change is complicated by the fact that the simulation results depend on both model formulation and the time-varying external forcings imposed on the models (Allen et al., 2000; Santer et al., 2007). De-convolving the importance of model and forcing differences in the historical simulations is an important topic that is addressed in Chapter 10; however, in this section a direct comparison is made to illustrate the ability of models to reproduce past changes.

9.4.1.4.1 Global surface temperature

Figure 9.8 compares the observational record of 20th century changes in global surface temperature to that simulated by each CMIP5 and EMIC model and the respective multi-model means. The inset on the right of the figure shows the climatological mean temperature for each model, averaged over the 1961–1990 reference period. Although biases in mean temperature are apparent, there is less confidence in observational estimates of climatological temperature than in variations about this mean (Jones et al. (1999). For the CMIP5 models, interannual variability in most of the simulations is qualitatively similar to that observed although there are several exceptions. The magnitude of interannual variations in the observations is noticeably larger than the multi-model mean because the averaging of multiple model results acts to filter much of the simulated variability. On the other hand, the episodic volcanic forcing that is applied to most models (see Section 9.3.2.2) is evident in the multi-model agreement with the observed cooling particularly noticeable after the 1991 Pinatubo eruption. The gradual warming evident in the observational record, particularly in the more recent decades, is also evident in the simulations, with the multi-model mean tracking the observed value closely over most of the century, and individual model results departing by less than about 0.5°C. Because the interpretation of differences in model behaviour can be confounded by internal variability and forcing, some studies have attempted to identify and remove dominant factors such as El Niño-Southern Oscillation (ENSO) and the impacts of volcanic eruptions (e.g., Fyfe et al., 2010). Figure 9.8 shows the similar capability for EMICs to simulate the global-scale response to the 20th century forcings (Eby et al. 2013). These results demonstrate a level of consistency between the EMICs with both the observations and the CMIP5 ensemble.

In summary, there is *very high confidence* that models reproduce the general features of the global-scale annual mean surface temperature increase over the historical period, including the more rapid warming in the second half of the 20th century, and the cooling immediately following large volcanic eruptions. The disagreement apparent over the most recent 10 to 15 years is discussed in detail in Box 9.2.

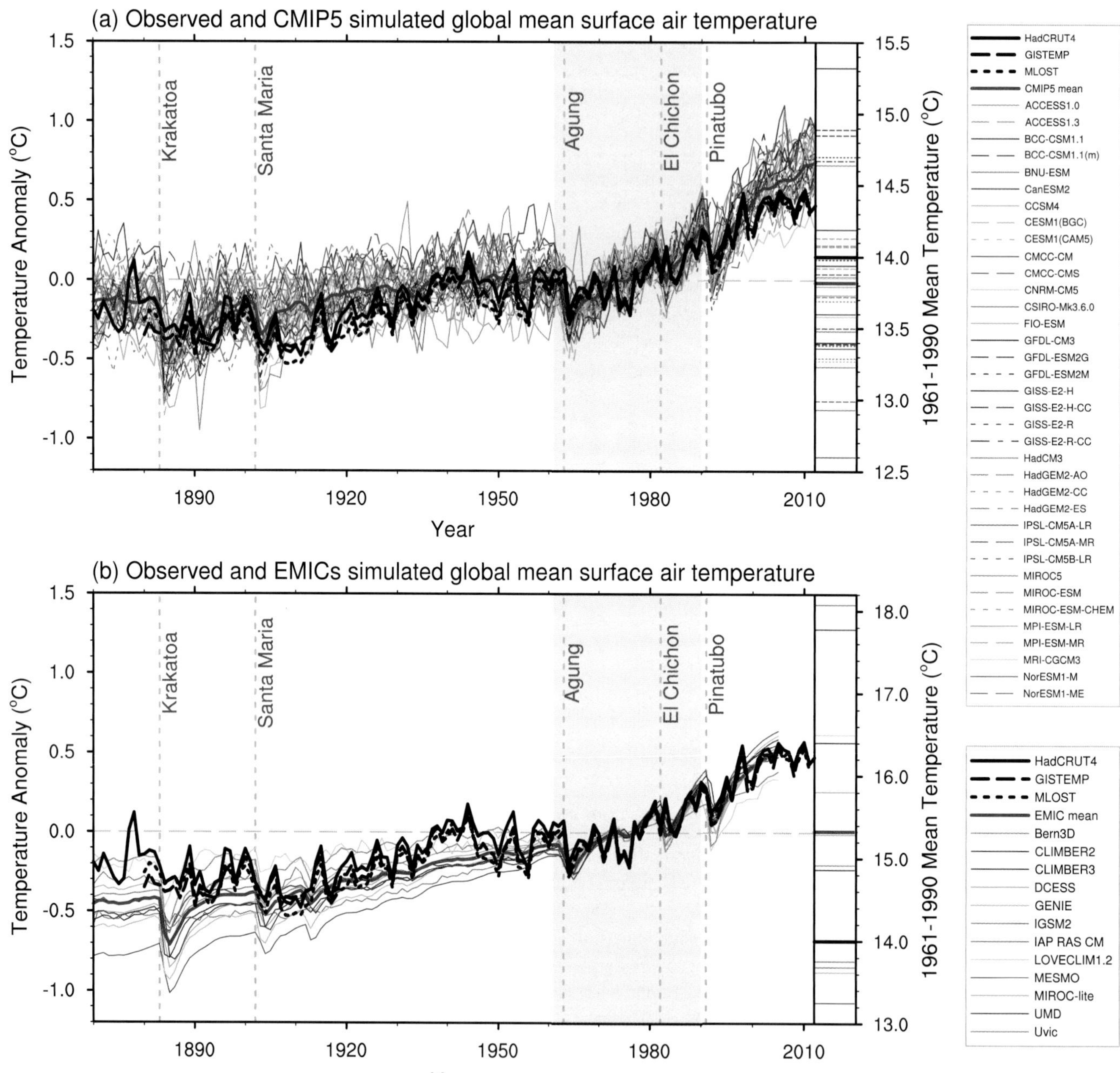

Figure 9.8 | Observed and simulated time series of the anomalies in annual and global mean surface temperature. All anomalies are differences from the 1961–1990 time-mean of each individual time series. The reference period 1961–1990 is indicated by yellow shading; vertical dashed grey lines represent times of major volcanic eruptions. (a) Single simulations for CMIP5 models (thin lines); multi-model mean (thick red line); different observations (thick black lines). Observational data (see Chapter 2) are Hadley Centre/Climatic Research Unit gridded surface temperature data set 4 (HadCRUT4; Morice et al., 2012), Goddard Institute for Space Studies Surface Temperature Analysis (GISTEMP; Hansen et al., 2010) and Merged Land–Ocean Surface Temperature Analysis (MLOST; Vose et al., 2012) and are merged surface temperature (2 m height over land and surface temperature over the ocean). All model results have been sub-sampled using the HadCRUT4 observational data mask (see Chapter 10). Following the CMIP5 protocol (Taylor et al., 2012b), all simulations use specified historical forcings up to and including 2005 and use RCP4.5 after 2005 (see Figure 10.1 and note different reference period used there; results will differ slightly when using alternative RCP scenarios for the post-2005 period). (a) Inset: the global mean surface temperature for the reference period 1961–1990, for each individual model (colours), the CMIP5 multi-model mean (thick red), and the observations (thick black: Jones et al., 1999). (Bottom) Single simulations from available EMIC simulations (thin lines), from Eby et al. (2013). Observational data are the same as in (a). All EMIC simulations ended in 2005 and use the CMIP5 historical forcing scenario. (b) Inset: Same as in (a) but for the EMICs.

Box 9.2 | Climate Models and the Hiatus in Global Mean Surface Warming of the Past 15 Years

The observed global mean surface temperature (GMST) has shown a much smaller increasing linear trend over the past 15 years than over the past 30 to 60 years (Section 2.4.3, Figure 2.20, Table 2.7; Figure 9.8; Box 9.2 Figure 1a, c). Depending on the observational data set, the GMST trend over 1998–2012 is estimated to be around one-third to one-half of the trend over 1951–2012 (Section 2.4.3, Table 2.7; Box 9.2 Figure 1a, c). For example, in HadCRUT4 the trend is 0.04°C per decade over 1998–2012, compared to 0.11°C per decade over 1951–2012. The reduction in observed GMST trend is most marked in Northern Hemisphere winter (Section 2.4.3; Cohen et al., 2012). Even with this "hiatus" in GMST trend, the decade of the 2000s has been the warmest in the instrumental record of GMST (Section 2.4.3, Figure 2.19). Nevertheless, the occurrence of the hiatus in GMST trend during the past 15 years raises the two related questions of what has caused it and whether climate models are able to reproduce it.

Figure 9.8 demonstrates that 15-year-long hiatus periods are common in both the observed and CMIP5 historical GMST time series (see also Section 2.4.3, Figure 2.20; Easterling and Wehner, 2009; Liebmann et al., 2010). However, an analysis of the full suite of CMIP5 historical simulations (augmented for the period 2006–2012 by RCP4.5 simulations, Section 9.3.2) reveals that 111 out of 114 realizations show a GMST trend over 1998–2012 that is higher than the entire HadCRUT4 trend ensemble (Box 9.2 Figure 1a; CMIP5 ensemble mean trend is 0.21°C per decade). This difference between simulated and observed trends could be caused by some combination of (a) internal climate variability, (b) missing or incorrect radiative forcing and (c) model response error. These potential sources of the difference, which are not mutually exclusive, are assessed below, as is the cause of the observed GMST trend hiatus.

Internal Climate Variability

Hiatus periods of 10 to 15 years can arise as a manifestation of internal decadal climate variability, which sometimes enhances and sometimes counteracts the long-term externally forced trend. Internal variability thus diminishes the relevance of trends over periods as short as 10 to 15 years for long-term climate change (Box 2.2, Section 2.4.3). Furthermore, the timing of internal decadal climate variability is not expected to be matched by the CMIP5 historical simulations, owing to the predictability horizon of at most 10 to 20 years (Section 11.2.2; CMIP5 historical simulations are typically started around nominally 1850 from a control run). However, climate models exhibit individual decades of GMST trend hiatus even during a prolonged phase of energy uptake of the climate system (e.g., Figure 9.8; Easterling and Wehner, 2009; Knight et al., 2009), in which case the energy budget would be balanced by increasing subsurface–ocean heat uptake (Meehl et al., 2011, 2013a; Guemas et al., 2013).

Owing to sampling limitations, it is uncertain whether an increase in the rate of subsurface–ocean heat uptake occurred during the past 15 years (Section 3.2.4). However, it is *very likely*[2] that the climate system, including the ocean below 700 m depth, has continued to accumulate energy over the period 1998–2010 (Section 3.2.4, Box 3.1). Consistent with this energy accumulation, global mean sea level has continued to rise during 1998–2012, at a rate only slightly and insignificantly lower than during 1993–2012 (Section 3.7). The consistency between observed heat-content and sea level changes yields *high confidence* in the assessment of continued ocean energy accumulation, which is in turn consistent with the positive radiative imbalance of the climate system (Section 8.5.1; Section 13.3, Box 13.1). By contrast, there is *limited evidence* that the hiatus in GMST trend has been accompanied by a slower rate of increase in ocean heat content over the depth range 0 to 700 m, when comparing the period 2003–2010 against 1971–2010. There is *low agreement* on this slowdown, since three of five analyses show a slowdown in the rate of increase while the other two show the increase continuing unabated (Section 3.2.3, Figure 3.2).

During the 15-year period beginning in 1998, the ensemble of HadCRUT4 GMST trends lies below almost all model-simulated trends (Box 9.2 Figure 1a), whereas during the 15-year period ending in 1998, it lies above 93 out of 114 modelled trends (Box 9.2 Figure 1b; HadCRUT4 ensemble-mean trend 0.26°C per decade, CMIP5 ensemble-mean trend 0.16°C per decade). Over the 62-year period 1951–2012, observed and CMIP5 ensemble-mean trends agree to within 0.02°C per decade (Box 9.2 Figure 1c; CMIP5 ensemble-mean trend 0.13°C per decade). There is hence *very high confidence* that the CMIP5 models show long-term GMST trends consistent with observations, despite the disagreement over the most recent 15-year period. Due to internal climate variability, in any given 15-year period the observed GMST trend sometimes lies near one end of a model ensemble (Box 9.2, Figure 1a, b; Easterling and Wehner, 2009), an effect that is pronounced in Box 9.2, Figure 1a, b because GMST was influenced by a very strong El Niño event in 1998.

(continued on next page)

[2] In this Report, the following terms have been used to indicate the assessed likelihood of an outcome or a result: Virtually certain 99–100% probability, Very likely 90–100%, Likely 66–100%, About as likely as not 33–66%, Unlikely 0–33%, Very unlikely 0–10%, Exceptionally unlikely 0–1%. Additional terms (Extremely likely: 95–100%, More likely than not >50–100%, and Extremely unlikely 0–5%) may also be used when appropriate. Assessed likelihood is typeset in italics, e.g., *very likely* (see Section 1.4 and Box TS.1 for more details).

Box 9.2 (continued)

Unlike the CMIP5 historical simulations referred to above, some CMIP5 predictions were initialized from the observed climate state during the late 1990s and the early 21st century (Section 11.1, Box 11.1; Section 11.2). There is *medium evidence* that these initialized predictions show a GMST lower by about 0.05°C to 0.1°C compared to the historical (uninitialized) simulations and maintain this lower GMST during the first few years of the simulation (Section 11.2.3.4, Figure 11.3 top left; Doblas-Reyes et al., 2013; Guemas et al., 2013). In some initialized models this lower GMST occurs in part because they correctly simulate a shift, around 2000, from a positive to a negative phase of the Interdecadal Pacific Oscillation (IPO, Box 2.5; e.g., Meehl and Teng, 2012; Meehl et al., 2013a). However, the improvement of this phasing of the IPO through initialization is not universal across the CMIP5 predictions (cf. Section 11.2.3.4). Moreover, while part of the GMST reduction through initialization indeed results from initializing at the correct phase of internal variability, another part may result from correcting a model bias that was caused by incorrect past forcing or incorrect model response to past forcing, especially in the ocean. The relative magnitudes of these effects are at present unknown (Meehl and Teng, 2012); moreover, the quality of a forecasting system cannot be evaluated from a single prediction (here, a 10-year prediction within the period 1998–2012; Section 11.2.3). Overall, there is *medium confidence* that initialization leads to simulations of GMST during 1998–2012 that are more consistent with the observed trend hiatus than are the uninitialized CMIP5 historical simulations, and that the hiatus is in part a consequence of internal variability that is predictable on the multi-year time scale.

Radiative Forcing

On decadal to interdecadal time scales and under continually increasing effective radiative forcing (ERF), the forced component of the GMST trend responds to the ERF trend relatively rapidly and almost linearly (*medium confidence*, e.g., Gregory and Forster, 2008; Held et al., 2010; Forster et al., 2013). The expected forced-response GMST trend is related to the ERF trend by a factor that has been estimated for the 1% per year CO_2 increases in the CMIP5 ensemble as 2.0 [1.3 to 2.7] W m^{-2} °C^{-1} (90% uncertainty range; Forster et al., 2013). Hence, an ERF trend can be approximately converted to a forced-response GMST trend, permitting an assessment of how much of the change in the GMST trends shown in Box 9.2 Figure 1 is due to a change in ERF trend.

The AR5 best-estimate ERF trend over 1998–2011 is 0.22 [0.10 to 0.34] W m^{-2} per decade (90% uncertainty range), which is substantially lower than the trend over 1984–1998 (0.32 [0.22 to 0.42] W m^{-2} per decade; note that there was a strong volcanic eruption in 1982) and the trend over 1951–2011 (0.31 [0.19 to 0.40] W m^{-2} per decade; Box 9.2, Figure 1d–f; numbers based on Section 8.5.2, Figure 8.18; the end year 2011 is chosen because data availability is more limited than for GMST). The resulting forced-response GMST trend would approximately be 0.12 [0.05 to 0.29] °C per decade, 0.19 [0.09 to 0.39] °C per decade, and 0.18 [0.08 to 0.37] °C per decade for the periods 1998–2011, 1984–1998 and 1951–2011, respectively (the uncertainty ranges assume that the range of the conversion factor to GMST trend and the range of ERF trend itself are independent). The AR5 best-estimate ERF forcing trend difference between 1998–2011 and 1951–2011 thus might explain about one-half (0.05°C per decade) of the observed GMST trend difference between these periods (0.06 to 0.08°C per decade, depending on observational data set).

The reduction in AR5 best-estimate ERF trend over 1998–2011 compared to both 1984–1998 and 1951–2011 is mostly due to decreasing trends in the natural forcings,–0.16 [–0.27 to –0.06] W m^{-2} per decade over 1998–2011 compared to 0.01 [–0.00 to 0.01] W m^{-2} per decade over 1951–2011 (Section 8.5.2, Figure 8.19). Solar forcing went from a relative maximum in 2000 to a relative minimum in 2009, with a peak-to-peak difference of around 0.15 W m^{-2} and a linear trend over 1998–2011 of around –0.10 W m^{-2} per decade (cf. Section 10.3.1, Box 10.2). Furthermore, a series of small volcanic eruptions has increased the observed stratospheric aerosol loading after 2000, leading to an additional negative ERF linear-trend contribution of around –0.06 W m^{-2} per decade over 1998–2011 (cf. Section 8.4.2.2, Section 8.5.2, Figure 8.19; Box 9.2 Figure 1d, f). By contrast, satellite-derived estimates of tropospheric aerosol optical depth (AOD) suggests little overall trend in global mean AOD over the last 10 years, implying little change in ERF due to aerosol-radiative interaction (*low confidence* because of *low confidence* in AOD trend itself, Section 2.2.3; Section 8.5.1; Murphy, 2013). Moreover, because there is only *low confidence* in estimates of ERF due to aerosol–cloud interaction (Section 8.5.1, Table 8.5), there is likewise *low confidence* in its trend over the last 15 years.

For the periods 1984–1998 and 1951–2011, the CMIP5 ensemble-mean ERF trend deviates from the AR5 best-estimate ERF trend by only 0.01 W m^{-2} per decade (Box 9.2 Figure 1e, f). After 1998, however, some contributions to a decreasing ERF trend are missing in the CMIP5 models, such as the increasing stratospheric aerosol loading after 2000 and the unusually low solar minimum in 2009. Nonetheless, over 1998–2011 the CMIP5 ensemble-mean ERF trend is lower than the AR5 best-estimate ERF trend by 0.03 W m^{-2} per decade (Box 9.2 Figure 1d). Furthermore, global mean AOD in the CMIP5 models shows little trend over 1998–2012, similar to the observations (Figure 9.29). Although the forcing uncertainties are substantial, there are no apparent incorrect or missing global mean forcings in the CMIP5 models over the last 15 years that could explain the model–observations difference during the warming hiatus.

(continued on next page)

Box 9.2 (continued)

Model Response Error

The discrepancy between simulated and observed GMST trends during 1998–2012 could be explained in part by a tendency for some CMIP5 models to simulate stronger warming in response to increases in greenhouse gas (GHG) concentration than is consistent with observations (Section 10.3.1.1.3, Figure 10.4). Averaged over the ensembles of models assessed in Section 10.3.1.1.3, the best-estimate GHG and other anthropogenic (OA) scaling factors are less than one (though not significantly so, Figure 10.4), indicating that the model-mean GHG and OA responses should be scaled down to best match observations. This finding provides evidence that some CMIP5 models show a larger response to GHGs and other anthropogenic factors (dominated by the effects of aerosols) than the real world (*medium confidence*). As a consequence, it is argued in Chapter 11 that near-term model projections of GMST increase should be scaled down by about 10% (Section 11.3.6.3). This downward scaling is, however, not sufficient to explain the model-mean overestimate of GMST trend over the hiatus period.

Another possible source of model error is the poor representation of water vapour in the upper atmosphere (Section 9.4.1.2). It has been suggested that a reduction in stratospheric water vapour after 2000 caused a reduction in downward longwave radiation and hence a surface-cooling contribution (Solomon et al., 2010), possibly missed by the models, However, this effect is assessed here to be small, because there was a recovery in stratospheric water vapour after 2005 (Section 2.2.2.1, Figure 2.5). (continued on next page)

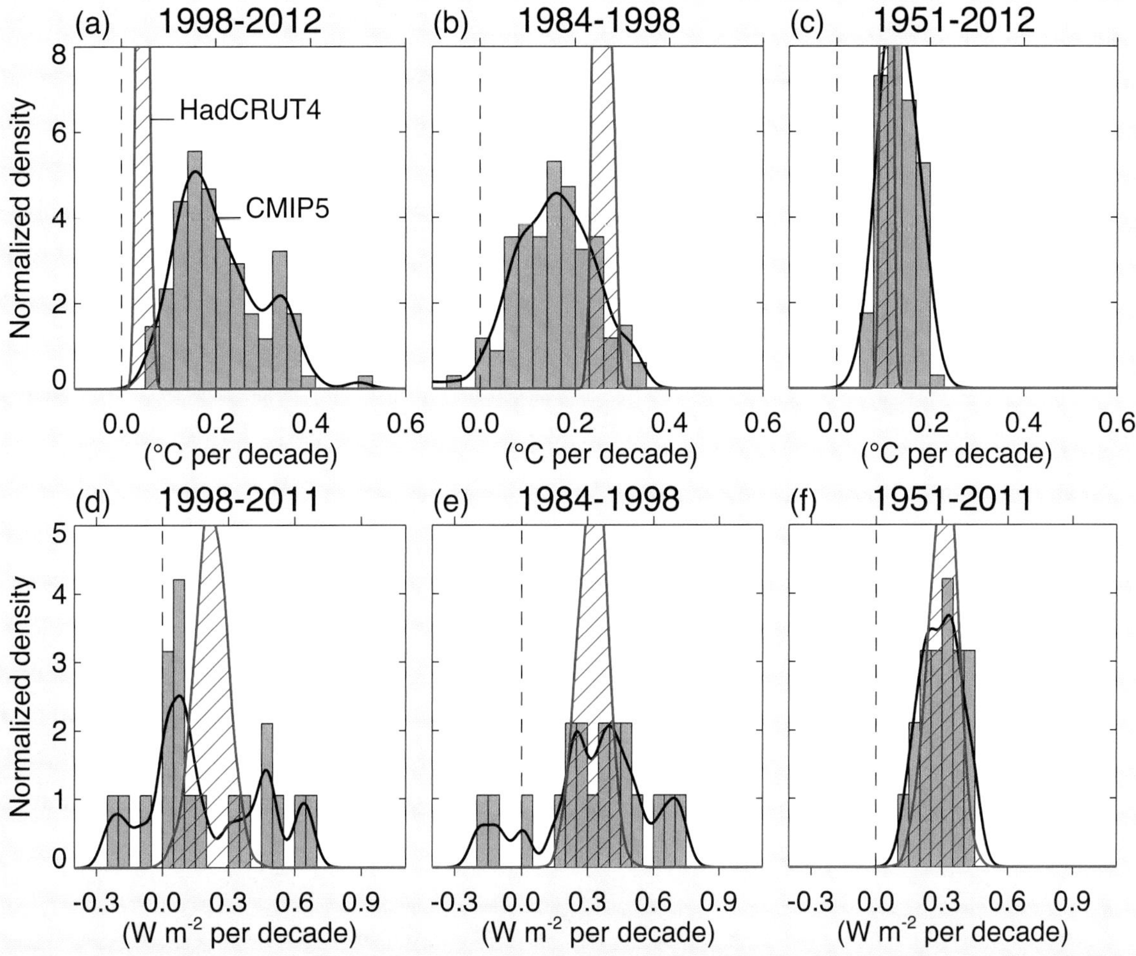

Box 9.2, Figure 1 | (Top) Observed and simulated global mean surface temperature (GMST) trends in degrees Celsius per decade, over the periods 1998–2012 (a), 1984–1998 (b), and 1951–2012 (c). For the observations, 100 realizations of the Hadley Centre/Climatic Research Unit gridded surface temperature data set 4 (HadCRUT4) ensemble are shown (red, hatched: Morice et al., 2012). The uncertainty displayed by the ensemble width is that of the statistical construction of the global average only, in contrast to the trend uncertainties quoted in Section 2.4.3, which include an estimate of internal climate variability. Here, by contrast, internal variability is characterized through the width of the model ensemble. For the models, all 114 available CMIP5 historical realizations are shown, extended after 2005 with the RCP4.5 scenario and through 2012 (grey, shaded: after Fyfe et al., 2010). (Bottom) Trends in effective radiative forcing (ERF, in W m^{-2} per decade) over the periods 1998–2011 (d), 1984–1998 (e), and 1951–2011 (f). The figure shows AR5 best-estimate ERF trends (red, hatched; Section 8.5.2, Figure 8.18) and CMIP5 ERF (grey, shaded: from Forster et al., 2013). Black lines are smoothed versions of the histograms. Each histogram is normalized so that its area sums up to one.

Box 9.2 (continued)

In summary, the observed recent warming hiatus, defined as the reduction in GMST trend during 1998–2012 as compared to the trend during 1951–2012, is attributable in roughly equal measure to a cooling contribution from internal variability and a reduced trend in external forcing (expert judgment, *medium confidence*). The forcing trend reduction is primarily due to a negative forcing trend from both volcanic eruptions and the downward phase of the solar cycle. However, there is *low confidence* in quantifying the role of forcing trend in causing the hiatus, because of uncertainty in the magnitude of the volcanic forcing trend and *low confidence* in the aerosol forcing trend.

Almost all CMIP5 historical simulations do not reproduce the observed recent warming hiatus. There is *medium confidence* that the GMST trend difference between models and observations during 1998–2012 is to a substantial degree caused by internal variability, with possible contributions from forcing error and some CMIP5 models overestimating the response to increasing GHG and other anthropogenic forcing. The CMIP5 model trend in ERF shows no apparent bias against the AR5 best estimate over 1998–2012. However, *confidence* in this assessment of CMIP5 ERF trend is *low*, primarily because of the uncertainties in model aerosol forcing and processes, which through spatial heterogeneity might well cause an undetected global mean ERF trend error even in the absence of a trend in the global mean aerosol loading.

The causes of both the observed GMST trend hiatus and of the model–observation GMST trend difference during 1998–2012 imply that, barring a major volcanic eruption, most 15-year GMST trends in the near-term future will be larger than during 1998–2012 (*high confidence*; see 11.3.6.3. for a full assessment of near-term projections of GMST). The reasons for this implication are fourfold: first, anthropogenic greenhouse-gas concentrations are expected to rise further in all RCP scenarios; second, anthropogenic aerosol concentration is expected to decline in all RCP scenarios, and so is the resulting cooling effect; third, the trend in solar forcing is expected to be larger over most near-term 15-year periods than over 1998–2012 (*medium confidence*), because 1998–2012 contained the full downward phase of the solar cycle; and fourth, it is *more likely than not* that internal climate variability in the near-term will enhance and not counteract the surface warming expected to arise from the increasing anthropogenic forcing.

9.4.1.4.2 Tropical tropospheric temperature trends

Most climate model simulations show a larger warming in the tropical troposphere than is found in observational data sets (e.g., McKitrick et al., 2010; Santer et al., 2013). There has been an extensive and sometimes controversial debate in the published literature as to whether this difference is statistically significant, once observational uncertainties and natural variability are taken into account (e.g., Douglass et al., 2008; Santer et al., 2008, 2013; Christy et al., 2010; McKitrick et al., 2010; Bengtsson and Hodges, 2011; Fu et al., 2011; McKitrick et al., 2011; Thorne et al., 2011). For the period 1979–2012, the various observational data sets find, in the tropical lower troposphere (LT), a linear warming trend ranging from 0.06°C to 0.13°C per decade (Section 2.4.4, Figure 2.27). In the tropical middle troposphere (MT), the linear warming trend ranges from 0.02°C to 0.12°C per decade (Section 2.4.4, Figure 2.27). Uncertainty in these trend values arises from different methodological choices made by the groups deriving satellite products (Mears et al., 2011) and radiosonde compilations (Thorne et al., 2011), and from fitting a linear trend to a time series containing substantial interannual and decadal variability (Box 2.2; Section 2.4.4; (Santer et al., 2008; McKitrick et al., 2010)). Although there have been substantial methodological debates about the calculation of trends and their uncertainty, a 95% confidence interval of around ±0.1°C per decade has been obtained consistently for both LT and MT (e.g., Section 2.4.4; McKitrick et al., 2010). In summary, despite unanimous agreement on the sign of the observed trends, there exists substantial disagreement between available estimates as to the rate of temperature changes in the tropical troposphere, and there is only *low confidence* in the rate of change and its vertical structure (Section 2.4.4).

For the 30-year period 1979–2009 (sometimes updated through 2010 or 2011), the CMIP3 models simulate a tropical warming trend ranging from 0.1°C to somewhat above 0.4°C per decade for both LT and MT (McKitrick et al., 2010), while the CMIP5 models simulate a tropical warming trend ranging from slightly below 0.15°C to somewhat above 0.4°C per decade for both LT and MT (Santer et al., 2013; see also Po-Chedley and Fu, 2012, who considered the period 1979–2005). Both model ensembles show trends that on average are higher than in the observational estimates, although both model ensembles overlap the observational ensemble. Because the differences between the various observational estimates are largely systematic and structural (Section 2.4.4; Mears et al., 2011), the uncertainty in the observed trends cannot be reduced by averaging the observations as if the differences between the data sets were purely random. Likewise, to properly represent internal climate variability, the full model ensemble spread must be used in a comparison against the observations (e.g., Box 9.2; Section 11.2.3.2; Raftery et al., 2005; Wilks, 2006; Jolliffe and Stephenson, 2011). The very high significance levels of model–observation discrepancies in LT and MT trends that were obtained in some studies (e.g., Douglass et al., 2008; McKitrick et al., 2010) thus arose to a substantial degree from using the standard error of the model ensemble mean as a measure of uncertainty, instead of the ensemble standard deviation or some other appropriate measure for uncertainty arising from internal climate variability (e.g., Box 9.2; Section 11.2.3.2; Raftery et al., 2005; Wilks, 2006; Jolliffe and Stephenson, 2011). Nevertheless, almost all model ensemble members show a warming trend in both LT and MT larger than observational estimates (McKitrick et al., 2010; Po-Chedley and Fu, 2012; Santer et al., 2013).

The CMIP3 models show a 1979–2010 tropical SST trend of 0.19°C per decade in the multi-model mean, significantly larger than the various observational trend estimates ranging from 0.10°C to 0.14°C per decade (including the 95% confidence interval; Fu et al., 2011). As a consequence, simulated tropospheric temperature trends are also too large because models attempt to maintain static stability. By contrast, atmospheric models that are forced with the observed SST are in better agreement with observations, as was found in the CMIP3 model ECHAM5 (Bengtsson and Hodges, 2011) and the CMIP5 atmosphere-only runs. In the latter, the LT trend range for the period 1981–2008 is 0.13 to 0.19ºC per decade—less than in the CMIP5 coupled models, but still an overestimate (Po-Chedley and Fu, 2012). The influence of SST trend errors on the analysis can be reduced by considering trends in tropospheric static stability, measured by the amplification of MT trends against LT trends; another approach is to consider the amplification of tropospheric trends against SST trends. The results of such analyses strongly depend on the time scale considered. Month-to-month variations are consistent between observations and models concerning amplification aloft against SST variations (Santer et al., 2005) and concerning amplification of MT against LT variations (Po-Chedley and Fu, 2012). By contrast, the 30-year trend in tropical static stability has been found to be larger than in the satellite observations for almost all ensemble members in both CMIP3 (Fu et al., 2011) and CMIP5 (Po-Chedley and Fu, 2012). However, if the radiosonde compilations are used for the comparison, the trends in static stability in the CMIP3 models agree much better with the observations, and inconsistency cannot be diagnosed unambiguously (Seidel et al., 2012) . What caused the remaining trend overestimate in static stability is not clear but has been argued recently to result from an upward propagation of bias in the model climatology (O'Gorman and Singh, 2013).

In summary, most, though not all, CMIP3 and CMIP5 models overestimate the observed warming trend in the tropical troposphere during the satellite period 1979–2012. Roughly one-half to two-thirds of this difference from the observed trend is due to an overestimate of the SST trend, which is propagated upward because models attempt to maintain static stability. There is *low confidence* in these assessments, however, due to the *low confidence* in observed tropical tropospheric trend rates and vertical structure (Section 2.4.4).

9.4.1.4.3 Extratropical circulation

The AR4 concluded that models, when forced with observed SSTs, are capable of producing the spatial distribution of storm tracks, but generally show deficiencies in the numbers and depth of cyclones and the exact locations of the storm tracks. The ability to represent extratropical cyclones in climate models has been improving, partly due to increases in horizontal resolution.

Storm track biases over the North Atlantic have decreased in CMIP5 models compared to CMIP3 (Zappa et al., 2013) although models still produce too zonal a storm track in this region and most models underestimate cyclone intensity (Colle et al., 2013; Zappa et al., 2013). Chang et al. (2012) also find the storm tracks in the CMIP5 models to be too weak and too equatorwards in their position, similar to the CMIP3 models. The performance of the CMIP5 models in representing North Atlantic cyclones was found to be strongly dependent on model resolution (Colle et al., 2013). Studies based on individual models typically find that models capture the general characteristics of storm tracks and extratropical cyclones (Ulbrich et al., 2008; Catto et al., 2010) and their associated fronts (Catto et al., 2013) and show improvements over earlier model versions (Loptien et al., 2008). However, some models have deficiencies in capturing the location of storm tracks (Greeves et al., 2007; Catto et al., 2011), in part owing to problems related to the location of warm waters such as the Gulf Stream and Kuroshio Current (Greeves et al., 2007; Keeley et al., 2012). This is an important issue because future projections of storm tracks are sensitive to changes in SSTs (Catto et al., 2011; Laine et al., 2011; McDonald, 2011; Woollings et al., 2012). Some studies find that storm track and cyclone biases are strongly related to atmospheric processes and parameterizations (Bauer et al., 2008a; Boer and Lambert, 2008; Zappa et al., 2013). Representation of the Mediterranean storm track has been shown to be particularly dependent on model resolution (Pinto et al., 2006; Raible et al., 2007; Bengtsson et al., 2009; Ulbrich et al., 2009), as is the representation of storm intensity and associated extremes in this area (Champion et al., 2011). Most studies have focussed on NH storm tracks. However, recently two CMIP3 models were found to differ significantly in their simulation of extratropical cyclones affecting Australia (Dowdy et al., 2013) and only about a third of the CMIP3 models were able to capture the observed changes and trends in Southern Hemisphere (SH) baroclinicity responsible for a reduction in the growth rate of the leading winter storm track modes (Frederiksen et al., 2011). There is still a lack of information on SH storm track evaluation for the CMIP5 models.

9.4.1.4.4 Tropical circulation

Earlier assessments of a weakening Walker circulation (Vecchi et al., 2006; Vecchi and Soden, 2007; DiNezio et al., 2009) from models and reanalyses (Yu and Zwiers, 2010) have been tempered by subsequent evidence that tropical Pacific Trade winds may have strengthened since the early 1990s (e.g., Merrifield and Maltrud, 2011). Models suggest that the width of the Hadley cell should increase (Frierson et al., 2007; Lu et al., 2007), and there are indications that this has been observed over the past 25 years (Seidel et al., 2008) but at an apparent rate (2 to 5 degrees of latitude since 1979) that is faster than in the CMIP3 models (Johanson and Fu, 2009).

The tendency in a warming climate for wet areas to receive more precipitation and subtropical dry areas to receive less, often termed the 'rich-get richer' mechanism (Chou et al., 2006; Held and Soden, 2006) is simulated in CMIP3 models (Chou and Tu, 2008), and observational support for this is found from ocean salinity observations (Durack et al., 2012) and precipitation gauge data over land (Zhang et al., 2007). There is *medium confidence* that models are correct in simulating precipitation increases in wet areas and decreases in dry areas on broad spatial scales in a warming climate based on agreement among models and some evidence that this has been detected in observed trends (see Section 2.5.1).

Several recent studies have examined the co-variability of tropical climate variables as a further means of evaluating climate models. Specifically, there are observed relationships between lower tropospheric temperature and total column precipitable water (Mears et al., 2007), and between surface temperature and relative humidity (Willett et al., 2010). Figure 9.9 (updated from Mears et al., 2007) shows the relationship between 25-year (1988–2012) linear trends in tropical precipitable water and lower tropospheric temperature for individual historical simulations (extended by appending RCP8.5 simulations after 2005, see Santer et al., 2013). As described by Mears et al. (2007), the ratio between changes in these two quantities is fairly tightly constrained in the model simulations and similar across a range of time scales, indicating that relative humidity is close to invariant in each model. In the updated figure, the Remote Sensing System (RSS) observations are in fairly good agreement with model expectations, and the University of Alabama in Huntsville (UAH) observations less so. The points associated with two of the reanalyses are also relatively far from the line, consistent with long-term changes in relative humidity. It is not known whether these discrepancies are due to remaining inhomogeneity in the observational data and/or reanalysis results, or due to problems with the climate simulations. All of the observational and reanalysis points lie at the lower end of the model distribution, consistent with the findings of (Santer et al., 2013).

9.4.1.4.5 Ozone and lower stratospheric temperature trends

Stratospheric ozone has been subject to a major perturbation since the late 1970s due to anthropogenic emissions of ozone-depleting substances (see also Section 2.2.2.2 and Figure 2.6). Since the AR4, there is increasing evidence that the ozone hole has led to a poleward shift and strengthening of the SH mid-latitude tropospheric jet during summer (Perlwitz et al., 2008; Son et al., 2008, 2010; SPARC-CCMVal, 2010; McLandress et al., 2011; Polvani et al., 2011; WMO, 2011; Swart and Fyfe, 2012b). These trends are well captured in both chemistry–climate models (CCMs) with interactive stratospheric chemistry and in CMIP3 models with prescribed time-varying ozone (Son et al., 2010; SPARC-CCMVal, 2010). However, around half of the CMIP3 models prescribe ozone as a fixed climatological value, and so these models are not able to simulate trends in surface climate attributable to changing stratospheric ozone amount (Karpechko et al., 2008; Son et al., 2008, 2010; Fogt et al., 2009). For CMIP5, a new time-varying ozone data set (Cionni et al., 2011) was developed and prescribed in the majority of models without interactive chemistry. This zonal mean data set is based on observations by Randel and Wu (2007) and CCM projections in the future (SPARC-CCMVal, 2010). Further, nine of the CMIP5 models include interactive chemistry and so compute their own ozone evolution. As a result, all CMIP5 models consider stratospheric ozone depletion and capture associated effects on SH surface climate, a significant advance over CMIP3. Figure 9.10 shows the global annual mean and Antarctic October mean of total column ozone in the CMIP5 models. The simulated trends in total column ozone are in medium agreement with observations, noting that some models that calculate ozone interactively show significant deviations from observation (Eyring et al., 2013). The multi-model mean agrees well with observations, and there is *robust evidence* that this constitutes a significant improvement over CMIP3, where around half of the models did not include stratospheric

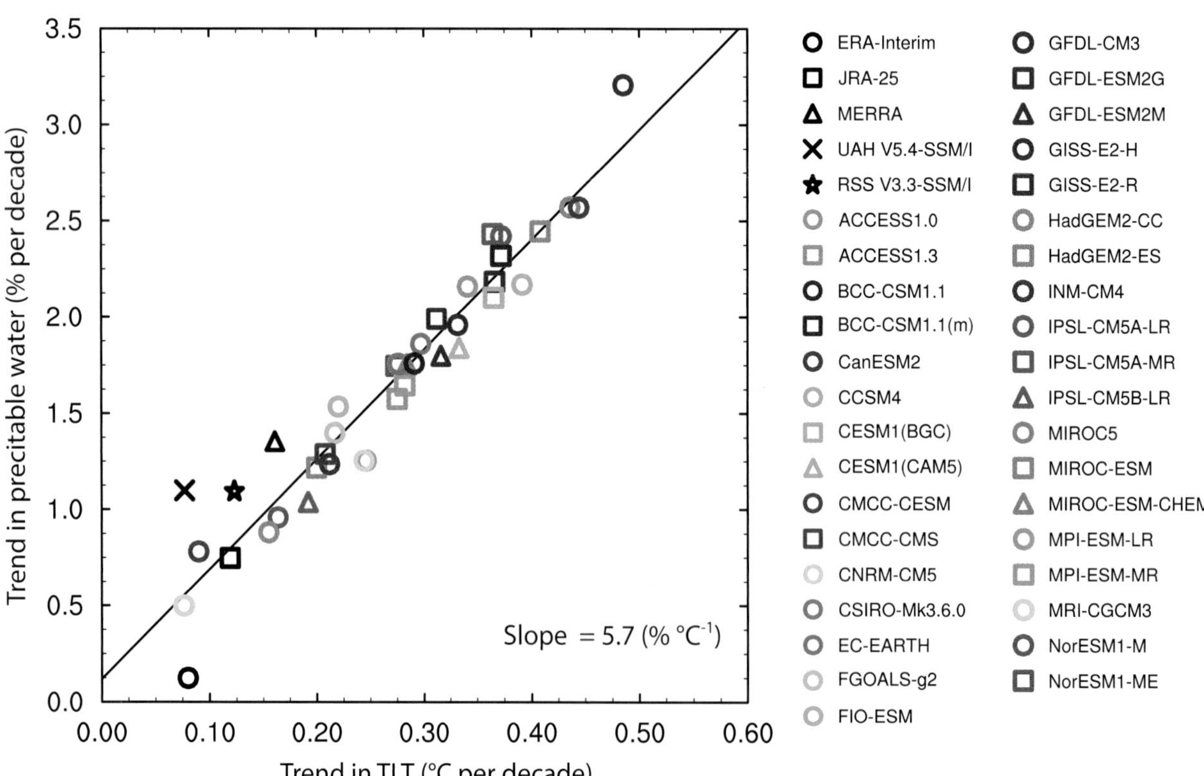

Figure 9.9 | Scatter plot of decadal trends in tropical (20°S to 20°N) precipitable water as a function of trends in lower tropospheric temperature (TLT) over the world's oceans. Coloured symbols are from CMIP5 models; black symbols are from satellite observations or from reanalysis output. Trends are calculated over the 1988–2012 period, so CMIP5 historical runs, which typically end in December 2005, were extended using RCP8.5 simulations initialized using these historical runs. Figure updated from Mears et al. (2007).

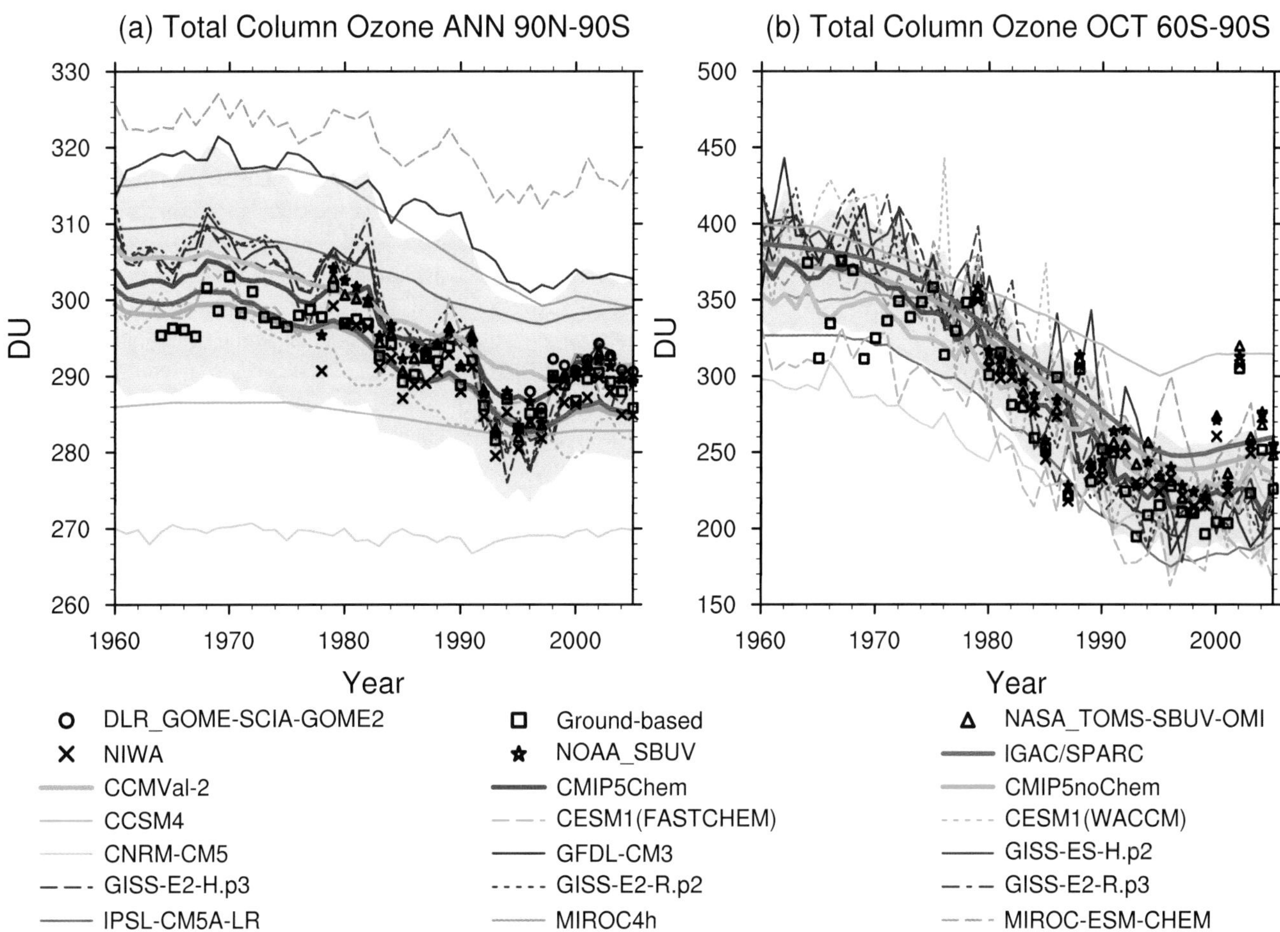

Figure 9.10 | Time series of area-weighted total column ozone from 1960 to 2005 for (a) annual and global mean (90°S to 90°N) and (b) Antarctic October mean (60°S to 90°S). Individual CMIP5 models with interactive or semi-interactive chemistry are shown in thin coloured lines, their multi-model mean (CMIP5Chem) in thick red and their standard deviation as the blue shaded area. Further shown are the multi-model mean of the CMIP5 models that prescribe ozone (CMIP5noChem, thick green), the International Global Atmospheric Chemistry/Stratospheric Processes and their Role in Climate (IGAC/SPARC) ozone database (thick pink), the Chemistry Climate Model Validation-2 (CCMVal-2) multi-model mean (thick orange), and observations from five different sources (black symbols). These sources include ground-based measurements (updated from Fioletov et al., 2002), National Aeronautics and Space Administration (NASA) Total Ozone Mapping Spectrometer/Ozone Monitoring Instrument/Solar Backscatter Ultraviolet(/2) (TOMS/OMI/SBUV(/2)) merged satellite data (Stolarski and Frith, 2006), the National Institute of Water and Atmospheric Research (NIWA) combined total column ozone database (Bodeker et al., 2005), Solar Backscatter Ultraviolet (SBUV, SBUV/2) retrievals (updated from Miller et al. 2002), and Deutsches Zentrum für Luft- und Raumfahrt/ Global Ozone Monitoring Experiment/ SCanning Imaging Absorption spectrometer for atmospheric chartography /GOME-2 (DLR GOME/SCIA/GOME-2; Loyola et al., 2009; Loyola and Coldewey-Egbers, 2012). Note that the IGAC/SPARC database over Antarctica (and thus the majority of the CMIP5noChem models) is based on ozonesonde measurements at the vortex edge (69°S) and as a result underestimates Antarctic ozone depletion compared to the observations shown. Ozone depletion was more pronounced after 1960 as equivalent stratospheric chlorine values steadily increased throughout the stratosphere. (Adapted from Figure 2 of Eyring et al., 2013.)

ozone trends. Correspondingly, there is *high confidence* that the representation of associated effects on high-latitude surface climate and lower stratospheric cooling trends has improved compared to CMIP3.

Lower stratospheric temperature change is affected by ozone, and since 1958 the change is characterized by a long-term global cooling trend interrupted by three 2-year warming episodes following large volcanic eruptions (Figure 2.24). During the satellite era (since 1979) the cooling occurred mainly in two step-like transitions in the aftermath of the El Chichón eruption in 1982 and the Mt Pinatubo eruption in 1991, with each cooling transition followed by a period of relatively steady temperatures (Randel et al., 2009; Seidel et al., 2011). This specific evolution of global lower stratosphere temperatures since 1979 is well captured in the CMIP5 models when forced with both natural and anthropogenic climate forcings, although the models tend to underestimate the long-term cooling trend (Charlton-Perez et al., 2012; Eyring et al., 2013; Santer et al., 2013) (see Chapter 10).

Tropospheric ozone is an important GHG and as such needs to be well represented in climate simulations. In the historical period it has increased due to increases in ozone precursor emissions from anthropogenic activities (see Chapters 2 and 8). Since the AR4, a new emission data set has been developed (Lamarque et al., 2010), which has led to some differences in tropospheric ozone burden compared to previous studies, mainly due to biomass burning emissions (Lamarque et al., 2010; Cionni et al., 2011; Young et al., 2013). Climatological mean tropospheric ozone in the CMIP5 simulations generally agrees well with satellite observations and ozonesondes, although as in the stratosphere, biases exist for individual models (Eyring et al., 2013; Young et al., 2013) (see also Chapter 8).

9.4.1.5 Model Simulations of the Last Glacial Maximum and the Mid-Holocene

Simulations of past climate can be used to test a model's response to forcings larger than those of the 20th century (see Chapter 5), and the CMIP5 protocol includes palaeoclimate simulations referred to as PMIP3 (Paleoclimate Model Intercomparison Project, version 3) (Taylor et al., 2012b). Specifically, the Last Glacial Maximum (LGM, 21000 years BP) allows testing of the modelled climate response to the presence of a large ice sheet in the NH and to lower concentrations of radiatively active trace gases, whereas the mid-Holocene (MH, 6000 years BP) tests the response to changes in seasonality of insolation in the NH (see Chapter 5). For these periods, palaeoclimate reconstructions allow quantitative model assessment (Braconnot et al., 2012). In addition the CMIP5/PMIP3 simulations can compared to previous palaeoclimate intercomparisons (Joussaume and Taylor, 1995; Braconnot et al., 2007c).

Figure 9.11 compares model results to palaeoclimate reconstructions for both LGM (left) and MH (right). For most models the simulated LGM cooling is within the range of the climate reconstructions (Braconnot et al., 2007c; Izumi et al., 2013), however Hargreaves et al. (2011) find a global mean model warm bias over the ocean of about 1°C for this period (Hargreaves et al., 2011). LGM simulations tend to overestimate tropical cooling and underestimate mid-latitude cooling (Kageyama et al., 2006; Otto-Bliesner et al., 2009). They thus underestimate polar amplification which is a feature also found for the mid-Holocene (Masson-Delmotte et al., 2006; Zhang et al., 2010a) and other climatic con-

9

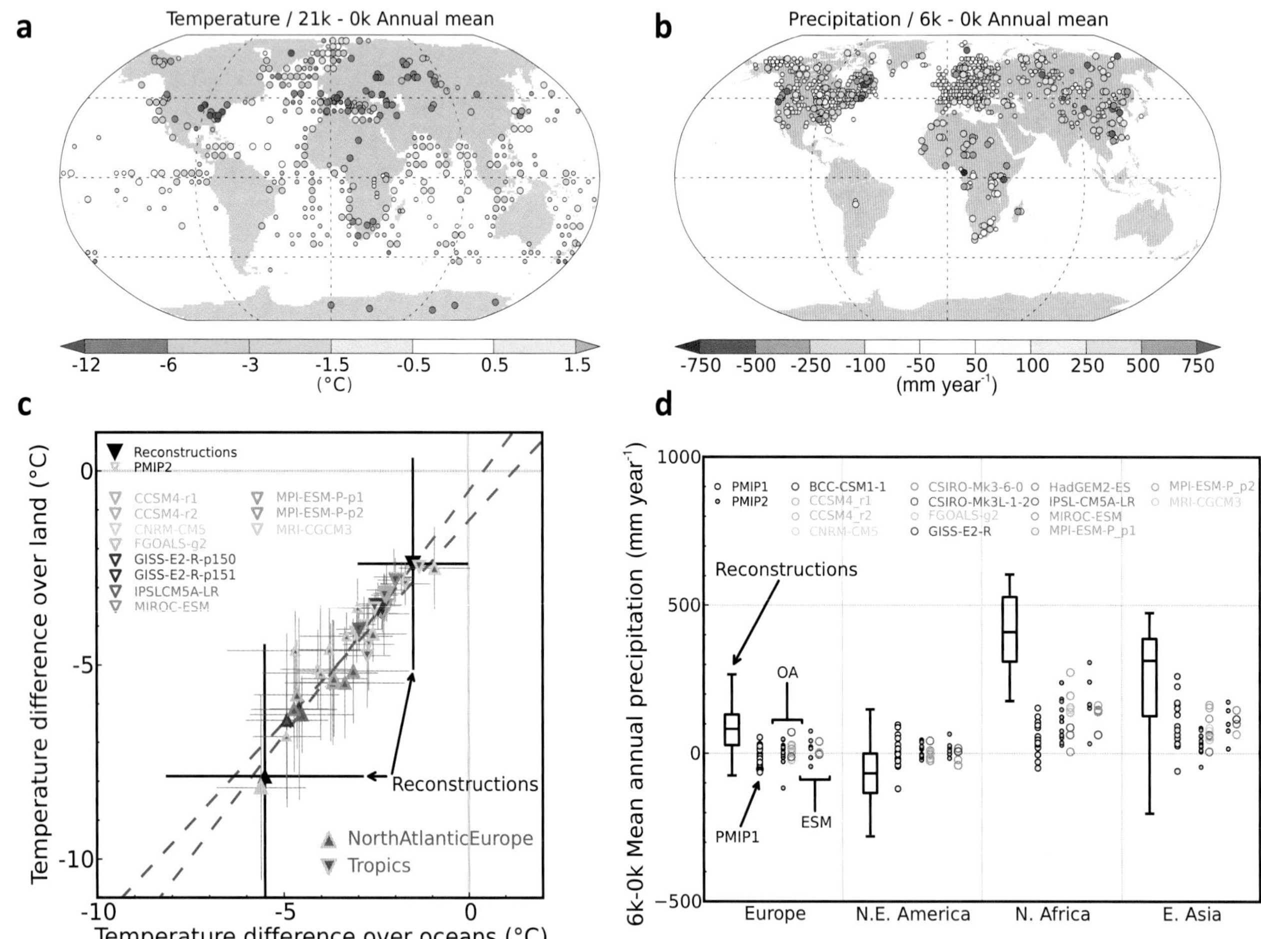

Figure 9.11 | Reconstructed and simulated conditions for the Last Glacial Maximum (LGM, 21,000 years BP, left) and the mid-Holocene (MH, 6000 years BP, right). (a) LGM change in annual mean surface temperature (°C) over land as shown by palaeo-environmental climate reconstructions from pollen, macrofossils, and ice cores (Bartlein et al., 2010; Braconnot et al., 2012), and in annual mean sea surface temperature (°C) over the ocean from different type of marine records (Waelbroeck et al., 2009). (b) MH change in annual mean precipitation (mm yr^{-1}) over land (Bartlein et al., 2010). In (a) and (b), the size of the dots is proportional to the uncertainties at the different sites as provided in the reconstructions. (c) Annual mean temperature changes over land against changes over the ocean, in the tropics (downward triangles) and over the North Atlantic and Europe (upward triangles). The mean and range of the reconstructions are shown in black, the Paleoclimate Modelling Intercomparison Project version 2 (PMIP2) simulations as grey triangles, and the CMIP5/PMIP3 simulations as coloured triangles. The 5 to 95% model ranges are in red for the tropics and in blue for the North Atlantic/Europe. (d) Changes in annual mean precipitation in different data-rich regions. Box plots for reconstructions provide the range of reconstructed values for the region. For models, the individual model average over the region is plotted for PMIP2 (small grey circle) and CMIP5/PMIP3 simulations (coloured circles). Note that in PMIP2, 'ESM' indicates that vegetation is computed using a dynamical vegetation model, whereas in CMIP5/PMIP3 it indicates that models have an interactive carbon cycle with different complexity in dynamical vegetation (see Table 9.A.1). The limits of the boxes are as follows: Western Europe (40°N to 50°N, 10°W to 30°E); northeast America (35°N to 60°N, 95°W to 60°W); North Africa (10°N to 25°N, 20°W to 30°W), and East Asia (25°N to 40°N, 75°E to 105°E). (Adapted from Braconnot et al., 2012.)

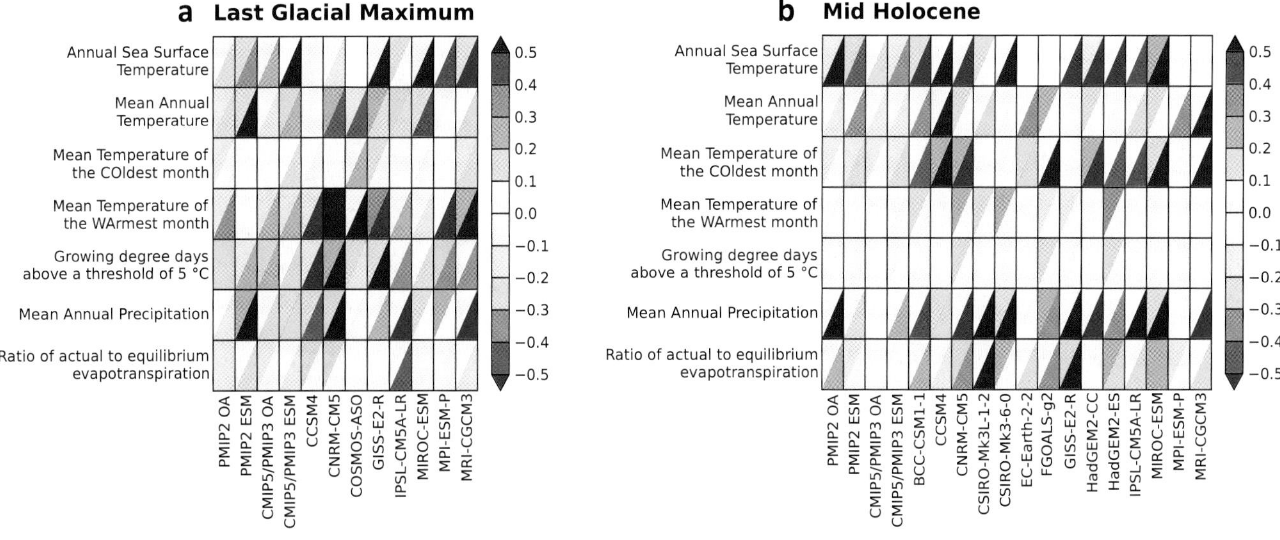

Figure 9.12 | Relative model performance for the Last Glacial Maximum (LGM, about 21,000 yr BP) and the mid-Holocene (MH, about 6000 yr BP) for seven bioclimatic variables: annual mean sea surface temperature, mean annual temperature (over land), mean temperature of the coldest month, mean temperature of the warmest month, growing degree days above a threshold of 5°C, and ratio of actual to equilibrium evapotranspiration. Model output is compared to the Bartlein et al. (2010) data set over land, including ice core data over Greenland and Antarctica (Braconnot et al., 2012) and the Margo data set (Waelbroeck et al., 2009) over the ocean. The CMIP5/Paleoclimate Modelling Intercomparison Project version 3 (PMIP3) ensemble of Ocean–Atmosphere (OA) and Earth System Model (ESM) simulations are compared to the respective PMIP2 ensembles in the first four columns of each panel. A diagonal divides each cell in two parts to show in the upper triangle a measure of the distance between model and data, taking into account the uncertainties in the palaeoclimate reconstructions (Guiot et al., 1999), and in the lower triangle the normalized mean-square error (NMSE) that indicates how well the spatial pattern is represented. In this graph all the values have been normalized following (Gleckler et al., 2008) using the median of the CMIP5/PMIP3 ensemble. The colour scale is such that blue colours mean that the result is better than the median CMIP5 model and red means that it is worse.

texts (Masson-Delmotte et al., 2010). Part of this can be attributed to uncertainties in the representation of sea ice and vegetation feedbacks that have been shown to amplify the response at the LGM and the MH in these latitudes (Braconnot et al., 2007b; Otto et al., 2009; O'ishi and Abe-Ouchi, 2011). Biases in the representation of the coupling between vegetation and soil moisture are also responsible for excessive continental drying at the LGM (Wohlfahrt et al., 2008) and uncertainties in vegetation feedback in monsoon regions (Wang et al., 2008; Dallmeyer et al., 2010). Nevertheless, the ratio between the simulated change in temperature over land and over the ocean (Figure 9.11c) is rather similar in different models, resulting mainly from simulation of the hydrological cycle over land and ocean (Sutton et al., 2007; Laine et al., 2009). At a regional scale, models tend to underestimate the changes in the north-south temperature gradient over Europe both at the LGM (Ramstein et al., 2007) and at the mid-Holocene (Brewer et al., 2007; Davis and Brewer, 2009).

The large-scale pattern of precipitation change during the MH (Figure 9.11d) is reproduced, but models tend to underestimate the magnitude of precipitation change in most regions. In the SH (not shown in the figure), the simulated change in atmospheric circulation is consistent with precipitation records in Patagonia and New Zealand, even though the differences between model results are large and the reconstructions have large uncertainties (Rojas et al., 2009; Rojas and Moreno, 2011).

A wider range of model performance metrics is provided in Figure 9.12 (Guiot et al., 1999; Brewer et al., 2007; Annan and Hargreaves, 2011; Izumi et al., 2013). Results for the MH are less reliable than for the LGM, because the forcing is weaker and involves smaller scale responses over the continent (Hargreaves et al., 2013). As is the case for the simulations of present day climate, there is only modest improvement between the results of the more recent models (CMIP5/PMIP3) and those of earlier model versions (PMIP2) despite higher resolution and sophistication.

9.4.2 Ocean

Accurate simulation of the ocean in climate models is essential for the correct estimation of transient ocean heat uptake and transient climate response, ocean CO_2 uptake, sea level rise, and coupled climate modes such as ENSO. In this section model performance is assessed for the mean state of ocean properties, surface fluxes and their impact on the simulation of ocean heat content and sea level, and aspects of importance for climate variability. Simulations of both the recent and more distant past are evaluated against available data. Following Chapter 3, ocean reanalyses are not used for model evaluation as many of their properties depend on the model used to build the reanalysis.

9.4.2.1 Simulation of Mean Temperature and Salinity Structure

Potential temperature and salinity are the main ocean state variables and their zonal distribution offers an evaluation of climate models in different parts of the ocean (upper ocean, thermocline, deep ocean). Over most latitudes, at depths ranging from 200 m to 2000 m, the CMIP5 multi-model mean zonally averaged ocean temperature is too warm (Figure 9.13a), albeit with a cooler deep ocean. Similar biases were evident in the CMIP3 multi-model mean. Above 200 m, however,

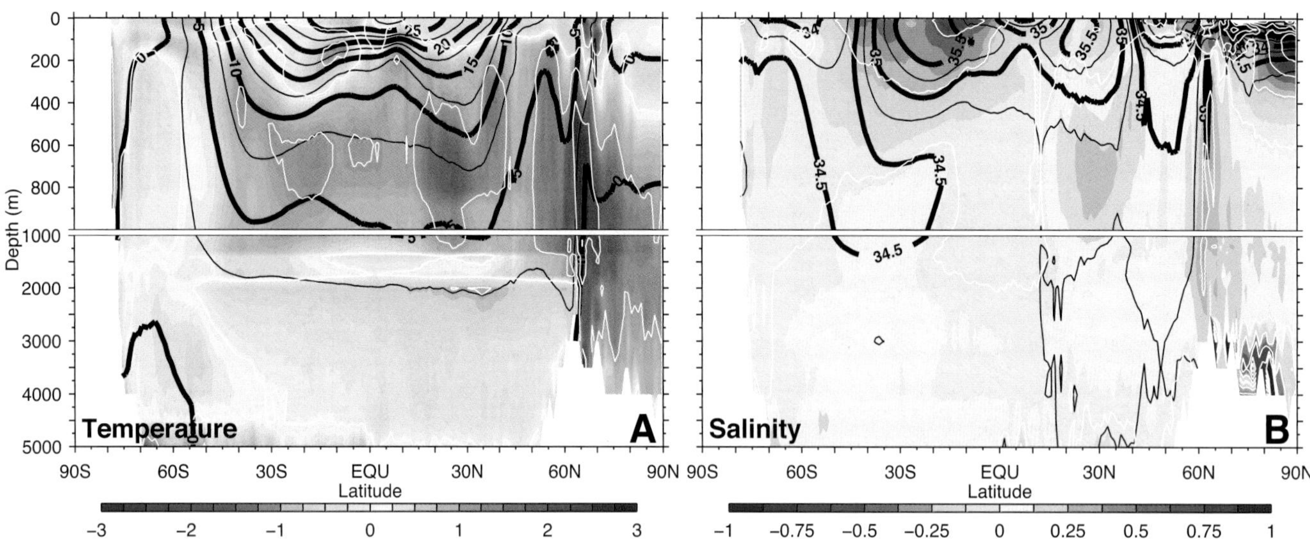

Figure 9.13 | (a) Potential temperature (oC) and (b) salinity (PSS-78); shown in colour are the time-mean differences between the CMIP5 ensemble mean and observations, zonally averaged for the global ocean (excluding marginal and regional seas). The observed climatological values are sourced from the World Ocean Atlas 2009 (WOA09; Prepared by the Ocean Climate Laboratory, National Oceanographic Data Center, Silver Spring, MD, USA), and are shown as labelled black contours. White contours show regions in (a) where potential temperature differences exceed positive or negative 1, 2 or 3°C, and in (b) where salinity differences exceed positive or negative 0.25, 0.5, 0.75 or 1 (PSS-78). The simulated annual mean climatologies are obtained for 1975 to 2005 from available historical simulations, whereas WOA09 synthesizes observed data from 1874 to 2008 in calculations of the annual mean; however, the median time for gridded observations most closely resembles the 1980–2010 period (Durack and Wijffels, 2012). Multiple realizations from individual models are first averaged to form a single-model climatology, before the construction of the multi-model ensemble mean. A total of 43 available CMIP5 models have contributed to the temperature panel (a) and 41 models to the salinity panel (b).

the CMIP5 (and CMIP3) multi-model mean is too cold, with maximum cold bias (more than 1°C) near the surface at mid-latitudes of the NH and near 200 m at 15°S. Zonal salinity errors (Figure 9.13b) exhibit a different pattern from those of the potential temperature indicating that most do not occur via density compensation. Some near surface structures in the tropics and in the northern mid-latitude are indicative of density compensation and are presumably due to surface fluxes errors. At intermediate depths, errors in water mass formation translate into errors in both salinity and potential temperature.

In the AR4 it was noted that the largest errors in SST in CMIP3 were found in mid and high latitudes. While this is still the case in CMIP5, there is marginal improvement with fewer individual models exhibiting serious bias—the inter-model zonal mean SST error standard deviation is significantly reduced at all latitudes north of 40°S—even though the multi-model mean is only slightly improved (Figure 9.14a, c). Near the equator, the cold tongue error in the Pacific (see Section 9.4.2.5.1) is reduced by 30% in CMIP5; the Atlantic still exhibits serious errors and the Indian is still well simulated (Figure 9.14b,d). In the Tropics, Li and Xie (2012) have shown that SST errors could be classified into those exhibiting broad meridional structures that are due to cloud errors, and those associated with Pacific and Atlantic cold tongue errors that are due to thermocline depth errors.

Sea surface salinity (SSS) is more challenging to observe, even though the last decade has seen substantial improvements in the development of global salinity observations, such as those from the Array for Real-time Geostrophic Oceanography (ARGO) network (see Chapter 3). Whereas SST is strongly constrained by air–sea interactions, the sources of SSS variations (surface forcing via evaporation minus precipitation, sea ice formation/melt and river runoff) are only loosely related to the SSS itself, allowing errors to develop unchecked in coupled models. An analysis of CMIP3 models showed that, whereas the historical trend in global mean SSS is well captured by the models, regional SSS biases are as high as ±2.5 psu (Terray et al., 2012). Comparisons of modelled versus observed estimates of evaporation minus precipitation suggest that model biases in surface freshwater flux play a role in some regions (e.g., double Intertropical Convergence Zone (ITCZ) in the East Pacific; Lin, 2007) or over the Indian ocean (Pokhrel et al., 2012).

The performance of coupled climate models in simulating hydrographic structure and variability were assessed in two important regions, the Labrador and Irminger Seas and the Southern Ocean (de Jong et al., 2009) and (Sloyan and Kamenkovich, 2007). Eight CMIP3 models produce simulations of the intermediate and deep layers in the Labrador and Irminger Seas that are generally too warm and saline, with biases up to 0.7 psu and 2.9°C. The biases arise because the convective regime is restricted to the upper 500 m; thus, intermediate water that in reality is formed by convection is, in the models, partly replaced by warmer water from the south. In the Southern Ocean, Subantarctic Mode Water (SAMW) and Antarctic Intermediate Water (AAIW), two water masses indicating very efficient ocean ventilation, are found to be well simulated in some CMIP3 and CMIP5 models but not in others, some having a significant fresh bias (Sloyan and Kamenkovich, 2007; Salle et al., 2013). McClean and Carman (2011) found biases in the properties of the North Atlantic mode waters and their formation rates in the CMIP3 models. Errors in Subtropical Mode Water (STMW) formation rate and volume produce a turnover time of 1 to 2 years, approximately half of that observed. Bottom water properties assessment in CMIP5 shows that about half of the models create dense water on

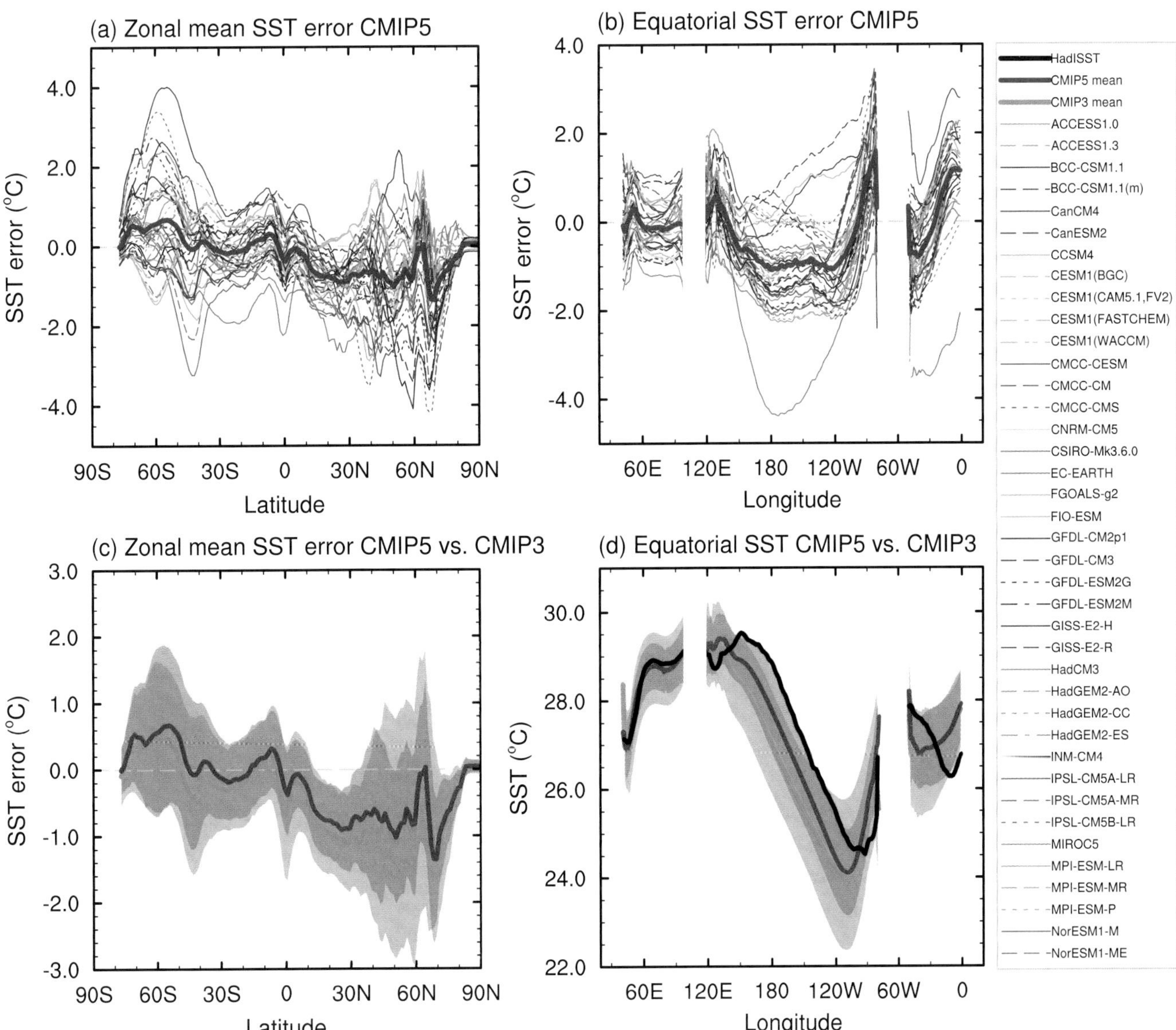

Figure 9.14 | (a) Zonally averaged sea surface temperature (SST) error in CMIP5 models. (b) Equatorial SST error in CMIP5 models. (c) Zonally averaged multi-model mean SST error for CMIP5 (red curve) and CMIP3 (blue curve), together with inter-model standard deviation (shading). (d) Equatorial multi-model mean SST in CMIP5 (red curve), CMIP3 (blue curve) together with inter-model standard deviation (shading) and observations (black). Model climatologies are derived from the 1979–1999 mean of the historical simulations. The Hadley Centre Sea Ice and Sea Surface Temperature (HadISST) (Rayner et al., 2003) observational climatology for 1979–1999 is used as reference for the error calculation (a), (b), and (c); and for observations in (d).

the Antarctic shelf, but it mixes with lighter water and is not exported as bottom water. Instead most models create deep water by open ocean deep convection, a process occurring rarely in reality (Heuzé et al., 2013) which leads to errors in deep water formation and properties in the Southern Ocean as shown in Figure 9.15.

Few studies have assessed the performance of models in simulating Mixed Layer Depth (MLD). In the North East Pacific region, Jang et al. (2011) found that the CMIP3 models exhibit the observed deep MLD in the Kuroshio Extension, though with a deep bias and only one large deep MLD region, rather than the observed two localized maxima. Other studies have noted MLD biases near sea ice edges (Capotondi et al., 2012).

9.4.2.2 Simulation of Sea Level and Ocean Heat Content

Steric and dynamic components of the mean dynamic topography (MDT) and sea surface height (SSH) patterns can be compared to observations (Maximenko et al., 2009). Pattern correlations between simulated and observed MDT are above 0.95 for all of the CMIP5 models (Figure 9.16), an improvement compared to CMIP3. MDT biases over tropical ocean regions are consistent with surface wind stress biases (Lee et al., 2013). Over the Antarctic Circumpolar Current, the parameterization of eddy-induced transports is essential for the models' density structure and thus MDT (Kuhlbrodt et al., 2012). High-resolution eddy resolving ocean models show improved SSH simulations over coarser resolution versions (McClean et al., 2006). Chapter 13 provides a more extensive assessment of sea level changes in CMIP5 simulations, including comparisons with century-scale historical records.

Ocean heat content (OHC) depends only on ocean temperature, whereas absolute changes in sea level are also influenced by processes that are only now being incorporated into global models (e.g., mass loss from large ice sheets discussed in Section 9.1.3.2.7). However, global-scale changes in OHC are highly correlated with the thermosteric contribution to global SSH changes (Domingues et al., 2008). Approximately half of the historical CMIP3 simulations did not include the effects of volcanic eruptions, resulting in substantially greater than observed ocean heat uptake during the late 20th century (Gleckler et

Figure 9.15 | Time-mean bottom potential temperature in the Southern Ocean, observed (a) and the differences between individual CMIP5 models and observations (b–p); left colour bar corresponds to the observations, right colour bar to the differences between model and observations (same unit). Thick dashed black line is the mean August sea ice extent (concentration >15%); thick continuous black line is the mean February sea ice extent (concentration >15%). Numbers indicate the area-weighted root-mean-square (RMS) error for all depths between the model and the climatology (unit °C); mean RMS error = 0.97 °C. (After Heuzé et al., 2013.)

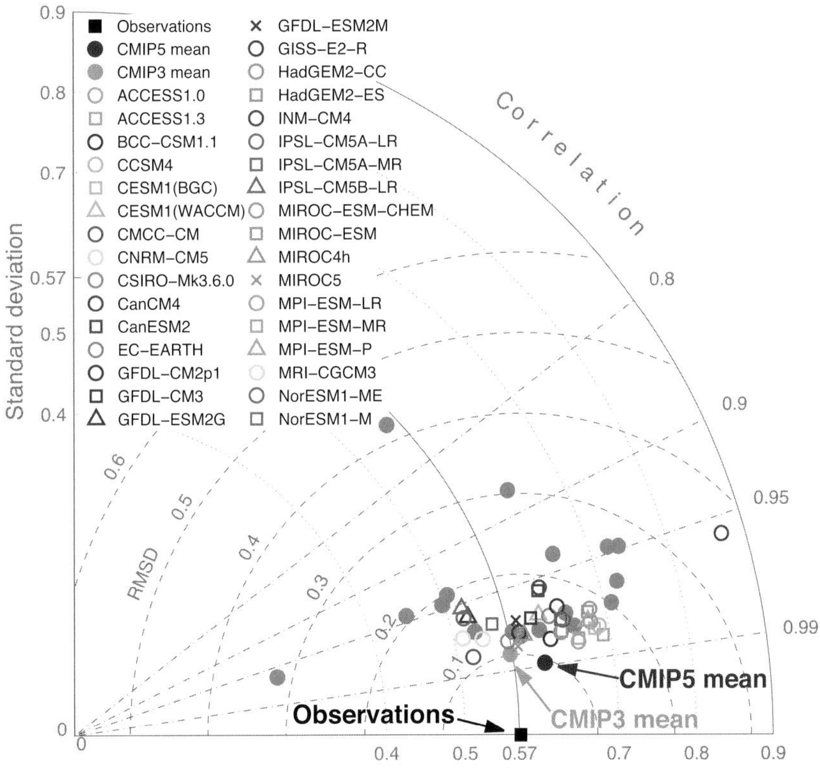

Figure 9.16 | Taylor diagram for the dynamic sea surface height climatology (1987–2000). The radial coordinate shows the standard deviation of the spatial pattern, normalized by the observed standard deviation. The azimuthal variable shows the correlation of the modelled spatial pattern with the observed spatial pattern. The root-mean square error with bias removed is indicated by the dashed grey circles about the observational point. Analysis is for the global ocean, 50°S to 50°N. The reference data set is Archiving, Validation and Interpretation of Satellite Oceanographic data (AVISO), a merged satellite product (Ducet et al., 2000), which is described in Chapter 3. One realization per model is shown for each CMIP5 and CMIP3 model result. Grey filled circles are for individual CMIP3 models; other symbols as in legend.

al., 2006; Domingues et al., 2008). Figure 9.17 shows observed and simulated global 0 to 700 m and total OHC changes during the overlap period of the observational record and the CMIP5 historical experiment (1961–2005). Three upper-ocean observational estimates, assessed in Chapter 3, are also shown to indicate observational uncertainty. The CMIP5 multi-model mean falls within the range of observations for most of the period, and the intermodel spread is reduced relative to CMIP3 (Gleckler et al., 2006; Domingues et al., 2008). This may result from most CMIP5 models including volcanic forcings. When the deep ocean is included, the CMIP5 multi-model mean also agrees well with the observations, although the deeper ocean estimates are much more uncertain (Chapter 3). There is *high confidence* that many CMIP5 models reproduce the observed increase in ocean heat content since 1960.

EMIC results for changes in total OHC are also compared with observations in Figure 9.17. (Note: results in this figure are based on Eby et al. (2013) who show OHC changes for 0 to 2000 m, whereas here the time-integrated net heat flux into the ocean surface is shown to compare with CMIP5 results (Figure 9.17b)). There is a tendency for the EMICs to overestimate total OHC changes and this could alter the temperature related feedbacks on the oceanic carbon cycle, and affect the long-term millennium projections in Chapter 12. However, it should be noted that high OHC changes can compensate for biases in climate sensitivity or RF so as to reproduce surface temperature changes over the 20th century. This will result in biased thermosteric sea level rise for millennial projections. Calibrated EMICs (Meinshausen et al., 2009; Sokolov et al., 2010) would remove such biases.

In idealized CMIP5 experiments (CO_2 increasing 1% yr^{-1}), the heat uptake efficiency of the CMIP5 models varies by a factor of two, explaining about 50% of the model spread (Kuhlbrodt and Gregory, 2012). Despite observational uncertainties, this recent work also provides *limited evidence* that in the upper 2000 m, most CMIP5 models are less stratified (in the global mean) than is observed, which suggests that these models transport heat downwards more efficiently than the real ocean. These results are consistent with earlier studies (Forest et al., 2006, 2008; Boe et al., 2009a; Sokolov et al., 2010) that conclude the CMIP3 models may overestimate oceanic mixing efficiency and therefore underestimate the Transient Climate Response (TCR) and its impact on future surface warming. However, Kuhlbrodt and Gregory (2012) also find that this apparent bias explains very little of the model spread in TCR. Although some progress has been made in understanding mixing deficiencies in ocean models (Griffies and Greatbatch, 2012; Ilicak et al., 2012), this remains a key challenge in improving the representation of physical processes that impact the evolution of ocean heat content and thermal expansion.

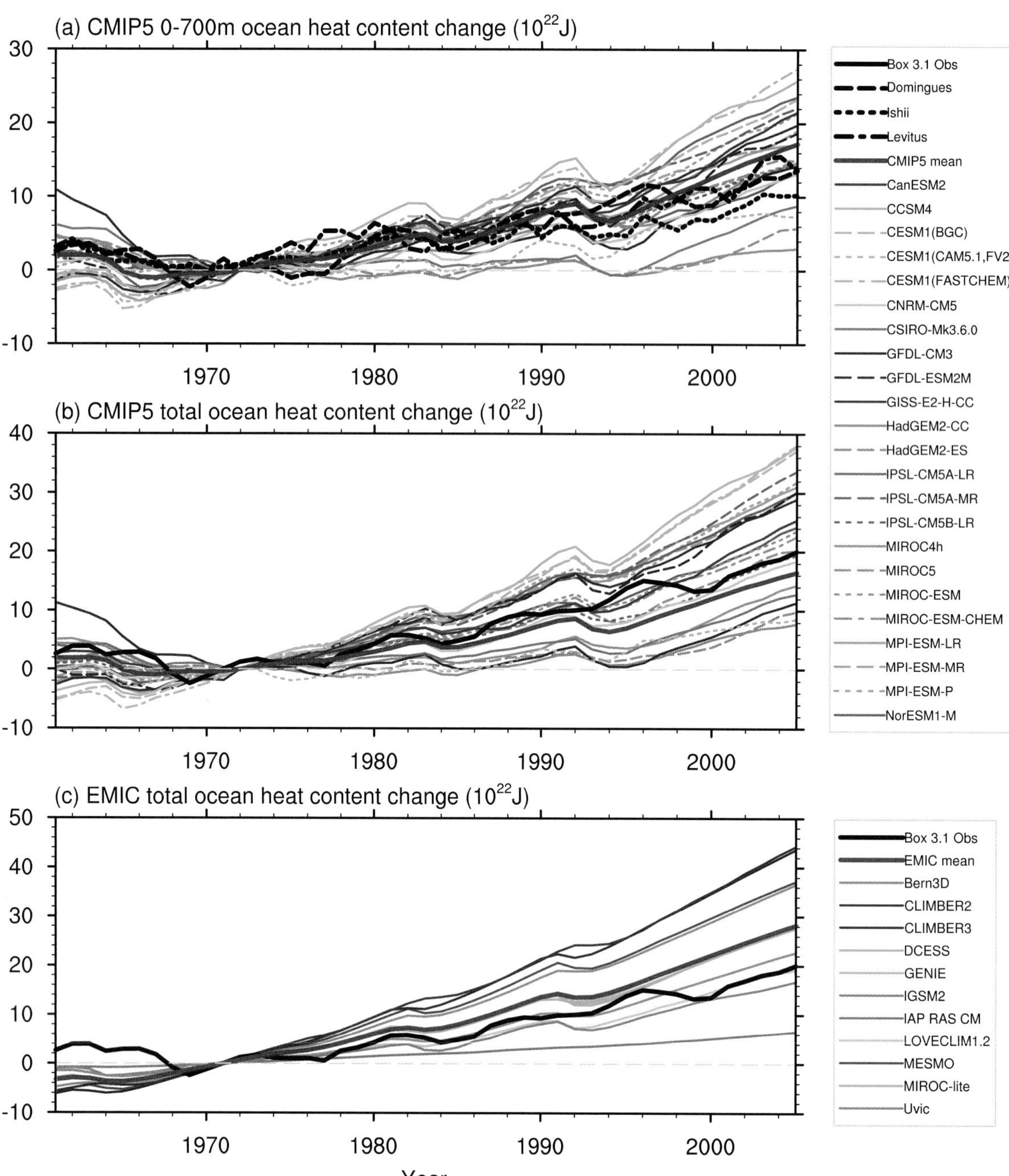

Figure 9.17 | Time series of simulated and observed global ocean heat content anomalies (with respect to 1971). CMIP5 historical simulations and observations for both the upper 700 meters of the ocean (a) as well as for the total ocean heat content (b). Total ocean heat content results are also shown for EMICs and observations (c). EMIC estimates are based on time-integrated surface heat flux into the ocean. The 0 to 700 m and total heat content observational estimates (thick lines) are respectively described in Figure 3.2 and Box 3.1, Figure 1. Simulation drift has been removed from all CMIP5 runs with a contemporaneous portion of a quadratic fit to each corresponding pre-industrial control run (Gleckler et al., 2012). Units are 10^{22} Joules.

9.4.2.3 Simulation of Circulation Features Important for Climate Response

9.4.2.3.1 Simulation of recent ocean circulation

Atlantic Meridional Overturning Circulation

The Atlantic Meridional Overturning Circulation (AMOC) consists of northward transport of shallow warm water overlying a southward transport of deep cold water and is responsible for a considerable part of the northward oceanic heat transport. Long-term AMOC estimates have had to be inferred from hydrographic measurements sporadically available over the last decades (e.g., Bryden et al., 2005; Lumpkin et al., 2008, Chapter 3.6.3). Continuous AMOC monitoring at 26.5°N was started in 2004 (Cunningham et al., 2007) and now provides a 5-year mean value of 18.5 Sv with annual means having a standard deviation of 1 Sv (McCarthy et al., 2012). The ability of models to simulate this important circulation feature is tied to the credibility of simulated AMOC weakening during the 21st century because the magnitude of the weakening is correlated with the initial AMOC strength (Gregory et al., 2005). The mean AMOC strength in CMIP5 models ranges from 15 to 30 Sv for the historical period which is comparable to the CMIP3 models (Weaver et al., 2012; see Figure 12.35). The variability of the AMOC is assessed in Section 9.5.3.3.1.

Southern Ocean circulation

The Southern Ocean is an important driver for the meridional overturning circulation and is closely linked to the zonally continuous Antarctic Circumpolar Current (ACC). Gupta et al. (2009) noted that relatively small deficiencies in the position of the ACC lead to more obvious biases in the SST in the models. The ability of CMIP3 models to adequately represent Southern Ocean circulation and water masses seems to be affected by several factors (Russell et al., 2006). The most important are the strength of the westerlies at the latitude of the Drake Passage, the heat flux gradient over this region, and the change in salinity with depth across the ACC. Kuhlbrodt et al. (2012) found that the strongest influence on ACC transport in the CMIP3 models was the Gent-McWilliams thickness diffusivity. The ACC has a typical transport through the Drake Passage of about 135 Sv (e.g., Cunningham et al., 2003). A comparison of CMIP5 models (Meijers et al., 2012) shows that, firstly, the ACC transport through Drake Passage is improved as compared to the CMIP3 models, and secondly, that the inter-model range in the zonal mean ACC position is smaller than in the CMIP3 ensemble (in CMIP5, the mean transport is 148 Sv and the standard deviation is 50 Sv across an ensemble of 21 models).

Simulation of glacial ocean conditions

Reconstructions of the last glacial maximum from sediment cores discussed in Chapter 5 indicate that the regions of deep water formation in the North Atlantic were shifted southward, that the boundary between North Atlantic Deep Water (NADW) and Antarctic Bottom Water (AABW) was substantially shallower than today, and that NADW formation was less intense (Duplessy et al., 1988; Dokken and Jansen, 1999; McManus et al., 2004; Curry and Oppo, 2005). This signal, although estimated from a limited number of sites, is robust (see Chapter 5). The AR4 reported that model simulations showed a wide range of AMOC response to LGM forcing (Weber et al., 2007), with some models exhibiting reduced strength of the AMOC and its extension at depth and other showing no change or an increase. Figure 9.18 provides an update of the diagnosis proposed by Otto-Bliesner et al. (2007) to compare model results with the deep ocean data from Adkins et al. (2002) using PMIP2 and CMIP5/PMIP3 pre-industrial and LGM simulations (Braconnot et al., 2012). These models reproduce the modern deep ocean temperature–salinity (T–S) structure in the Atlantic basin, but most of them do not capture the cold and salty bottom

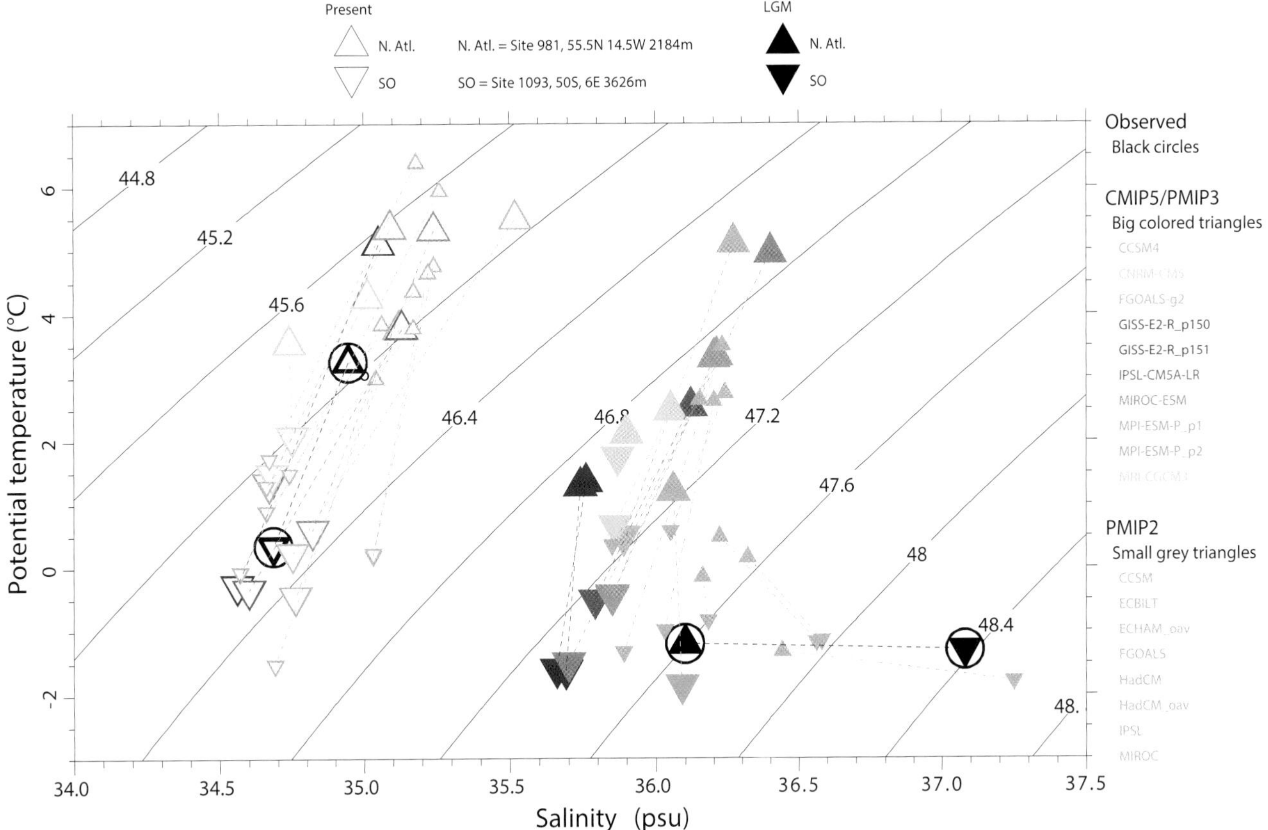

Figure 9.18 | Temperature and salinity for the modern period (open symbols) and the Last Glacial Maximum (LGM, filled symbols) as estimated from proxy data at Ocean Drilling Program (ODP) sites (black symbols, from Adkins et al., 2002) and simulated by the Paleoclimate Modelling Intercomparison Project version 2 (PMIP2, small triangles) and PMIP3/CMIP5 (big triangles) models. The isolines represent lines of equal density. Site 981 (triangles) is located in the North Atlantic (Feni Drift, 55°N, 15°W, 2184 m). Site 1093 (upside-down triangles) is located in the South Atlantic (Shona Rise, 50°S, 6°E, 3626 m). In PMIP2, only Community Climate System Model (CCSM) included a 1 psu adjustment of ocean salinity at initialization to account for freshwater frozen into LGM ice sheets; the other PMIP2 model-simulated salinities have been adjusted to allow a comparison. In PMIP3, all simulations include the 1 psu adjustment as required in the PMIP2/CMIP5 protocol (Braconnot et al., 2012). The dotted lines allow a comparison of the values at the NH and SH sites for a same model. This figure is adapted from Otto-Bliesner et al. (2007).

water suggested by the LGM reconstructions, providing evidence that processes responsible for such palaeoclimate changes may not be well reproduced in contemporary climate models. This is expected to also affect projected changes in deep ocean properties.

9.4.2.4 Simulation of Surface Fluxes and Meridional Transports

Surface fluxes play a large part in determining the fidelity of ocean simulations. As noted in the AR4, large uncertainties in surface heat and fresh water flux observations (usually obtained indirectly) do not allow useful evaluation of models. This is still the case and so the focus here is on an integrated quantity, meridional heat transport, which is less prone to errors. Surface wind stress is better observed and models are evaluated against observed products below.

The zonal component of wind stress is particularly important in driving ocean surface currents; modelled and observed values are shown in Figure 9.19. At middle to high latitudes, the model-simulated wind stress maximum lies 5 to 10 degree equatorward of that in the observationally based estimates, and so mid-latitude westerly winds are

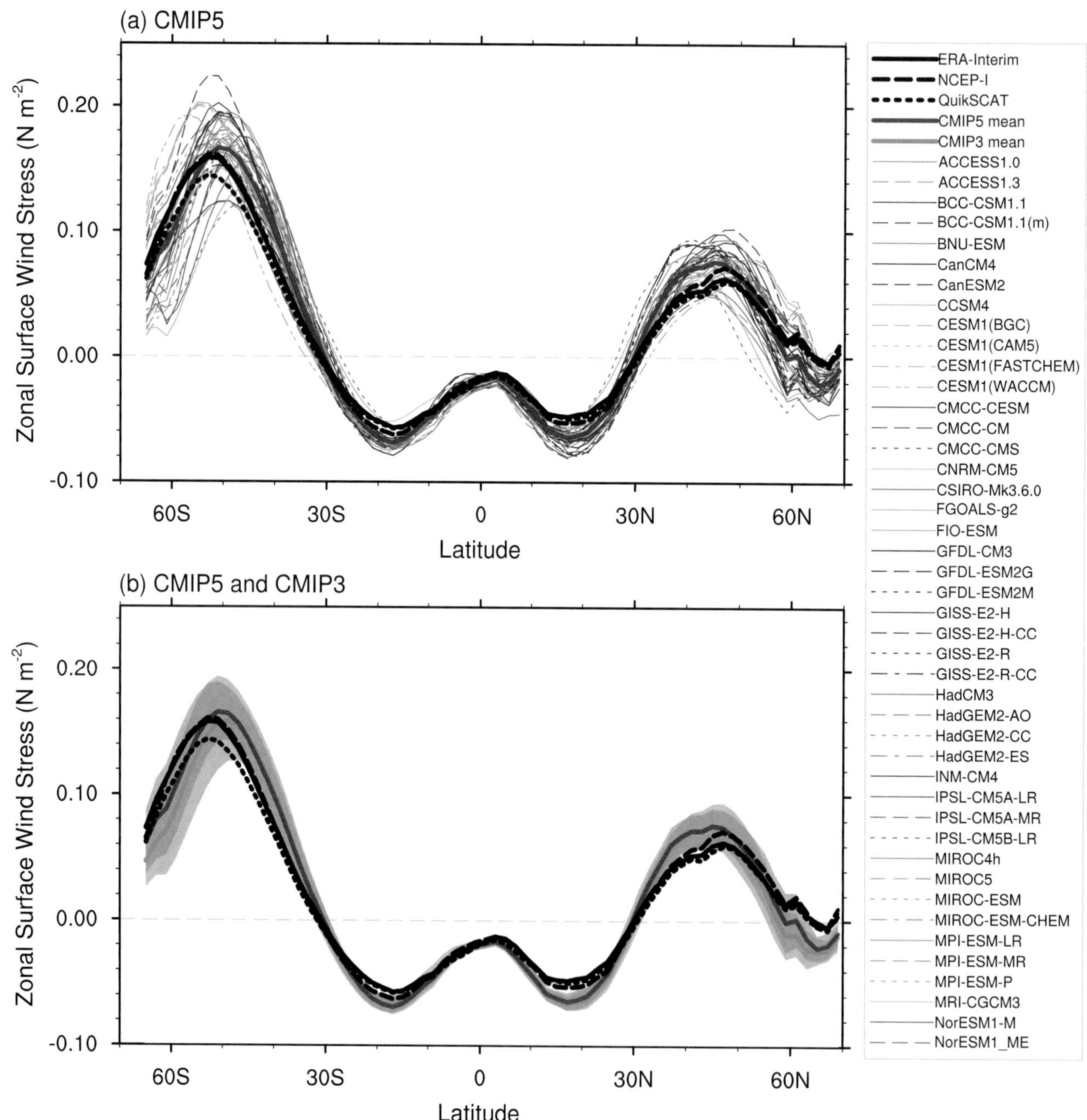

Figure 9.19 | Zonal-mean zonal wind stress over the oceans in (a) CMIP5 models and (b) multi-model mean comparison with CMIP3. Shown is the time-mean of the period 1970–1999 from the historical simulations. The black solid, dashed, and dotted curves represent ECMWF reanalysis of the global atmosphere and surface conditions (ERA)-Interim (Dee et al., 2011), National Centers for Environmental Prediction/National Center for Atmospheric Research (NCEP/NCAR) reanalysis I (Kalnay et al., 1996), and QuikSCAT satellite measurements (Risien and Chelton, 2008), respectively. In (b) the shading indicates the inter-model standard deviation.

too strong in models. This equatorward shift in the southern ocean is slightly reduced in CMIP5 relative to CMIP3. At these latitudes, the largest near surface wind speed biases in CMIP5 are located over the Pacific sector and the smallest are in the Atlantic sector (Bracegirdle et al., 2013). Such wind stress errors may adversely affect oceanic heat and carbon uptake (Swart and Fyfe, 2012a). At middle to low latitudes, the CMIP3 and CMIP5 model spreads are smaller than at high latitudes, although near the equator this can occur through compensating errors (Figure 9.20). The simulated multi-model mean equatorial zonal wind stress is too weak in the Atlantic and Indian Oceans and too strong in the western Pacific, with no major improvement from CMIP3 to CMIP5.

The CMIP5 model simulations qualitatively agree with the various observational estimates on the most important features of ocean heat transport (Figure 9.21) and, in a multi-model sense, no major change from CMIP3 can be seen. All CMIP5 models are able to the represent the strong north-south asymmetry, with the largest values in the NH, consistent with the observational estimates. At most latitudes the majority of CMIP5 model results fall within the range of observational estimates,

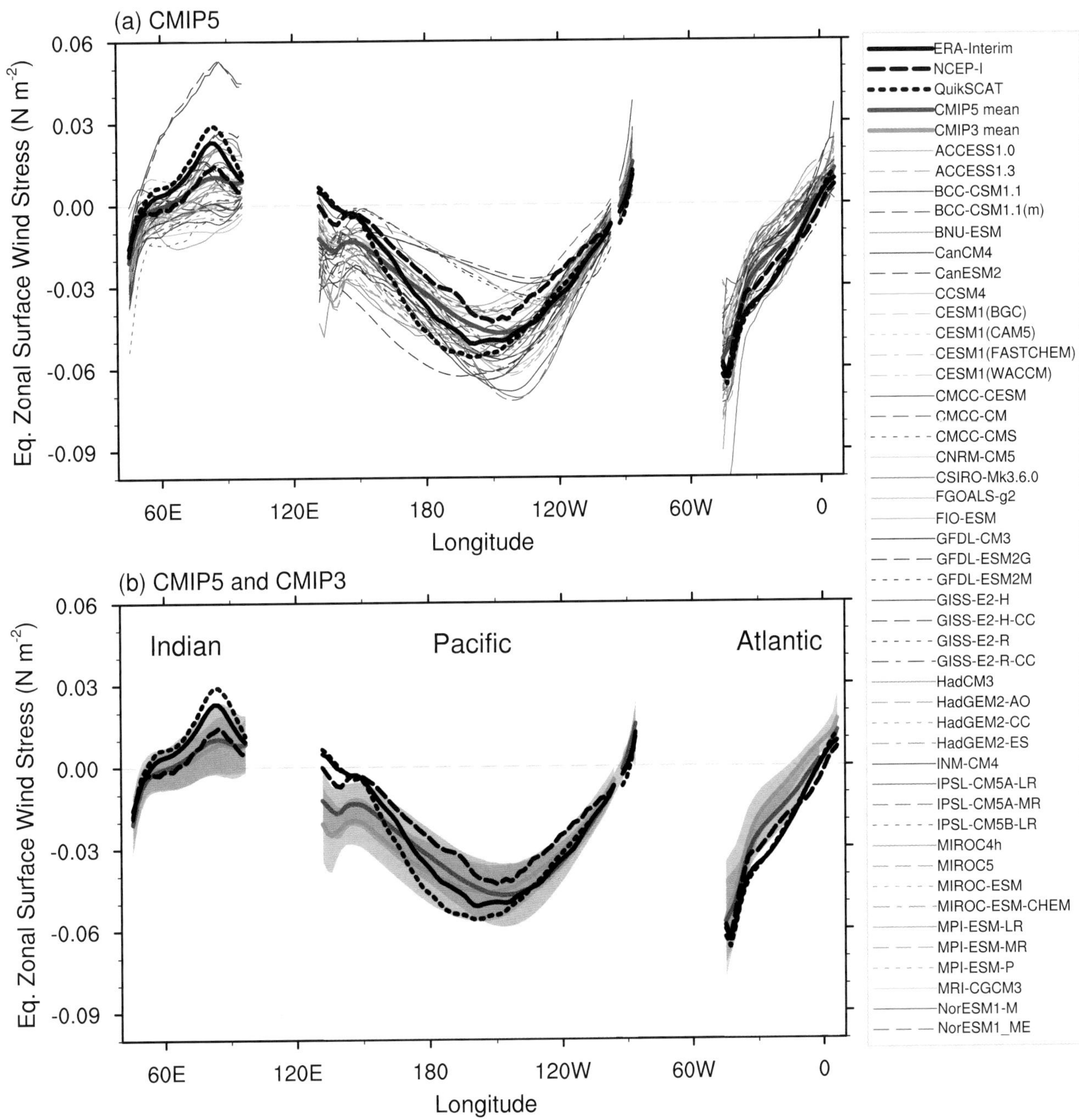

Figure 9.20 | Equatorial (2°S to 2°N averaged) zonal wind stress for the Indian, Pacific, and Atlantic oceans in (a) CMIP5 models and (b) multi-model mean comparison with CMIP3. Shown is the time-mean of the period 1970–1999 from the historical simulations. The black solid, dashed, and dotted curves represent ERA-Interim (Dee et al., 2011), National Centers for Environmental Prediction/National Center for Atmospheric Research (NCEP/NCAR) reanalysis I (Kalnay et al., 1996) and QuikSCAT satellite measurements (Risien and Chelton, 2008), respectively. In (b) the shading indicates the inter-model standard deviation.

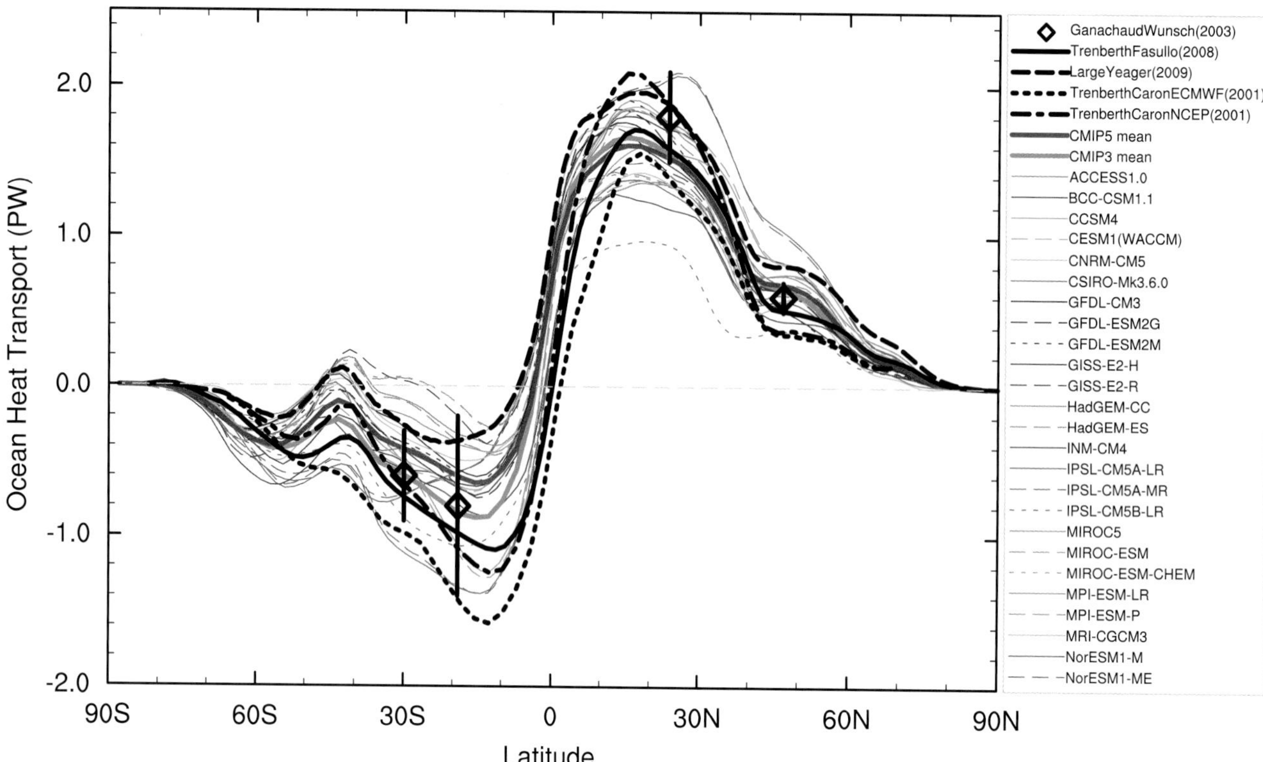

Figure 9.21 | Annual- and zonal-mean oceanic heat transport implied by net heat flux imbalances at the sea surface for CMIP5 simulations, under an assumption of negligible changes in oceanic heat content. Observational estimates include: the data set from Trenberth and Caron (2001) for the period February 1985 to April 1989, derived from reanalysis products from the National Centers for Environmental Prediction/National Center for Atmospheric Research (NCEP/NCAR; Kalnay et al., 1996; dash-dotted black) and European Centre for Medium Range Weather Forecasts 40-year reanalysis (ERA40; Uppala et al., 2005; short-dashed black), an updated version by Trenberth and Fasullo (2008) with improved top of the atmosphere (TOA) radiation data from the Clouds and Earth's Radiant Energy System (CERES) for March 2000 to May 2004, and updated NCEP reanalysis (Kistler et al., 2001) up to 2006 (solid black), the Large and Yeager (2009) analysis based on the range of annual mean transport estimated over the years 1984–2006, computed from air–sea surface fluxes adjusted to agree in the mean with a variety of satellite and *in situ* measurements (long-dashed black), and direct estimates by Ganachaud and Wunsch (2003) obtained from hydrographic sections during the World Ocean Circulation Experiment combined with inverse models (black diamonds). The model climatologies are derived from the years 1986 to 2005 in the historical simulations in CMIP5. The multi-model mean is shown as a thick red line. The CMIP3 multi-model mean is added as a thick blue line.

although there is some suggestion of modest underestimate between 15°N and 25°N and south of about 60°S. Some models show an equatorward transport at Southern-Hemisphere mid-latitudes that is also featured in the observation estimate of Large and Yeager (2009). This highlights the difficulties in representing large-scale energy processes in the Southern ocean as discussed by Trenberth and Fasullo (2010b). Note that climate models should exhibit a vanishing net energy balance when long time averages are considered but unphysical sources and sinks lead to energy biases (Trenberth and Fasullo, 2009, 2010a; Lucarini and Ragone, 2011) that are also found in reanalysis constrained by observations (Trenberth et al., 2009). When correcting for the imperfect closure of the energy cycle, as done here, comparison between models and observational estimates become possible.

9.4.2.5 Simulation of Tropical Mean State

9.4.2.5.1 Tropical Pacific Ocean

Although the basic east–west structure of the tropical Pacific is well captured, models have shown persistent biases in important properties of the mean state (AchutaRao and Sperber, 2002; Randall et al., 2007; Guilyardi et al., 2009b) with severe local impacts (Brown et al., 2012). Among these biases are the mean thermocline depth and slope along the equator, the structure of the equatorial current system, and the excessive equatorial cold tongue (Reichler and Kim, 2008; Brown et al., 2010a; Zheng et al., 2012). Many reasons for these biases have been proposed, such as: too strong trade winds; a too diffusive thermocline; deficient horizontally isotropic mixing coefficients; insufficient penetration of solar radiation; and too weak tropical instability waves (Meehl et al., 2001; Wittenberg et al., 2006; Lin, 2007). It is noteworthy that CMIP5 models exhibit some improvements in the western equatorial Pacific when compared to CMIP3, with reduced SST and trade wind errors (Figures 9.14 and 9.20). Because of strong interactions between the processes involved, it is difficult to identify the ultimate source of these errors, although new approaches using the rapid adjustment of initialized simulations hold promise (Vannière et al., 2011).

A particular problem in simulating the seasonal cycle in the tropical Pacific arises from the 'double ITCZ', defined as the appearance of a spurious ITCZ in the SH associated with excessive tropical precipitation. Further problems are too strong a seasonal cycle in simulated SST and winds in the eastern Pacific and the appearance of a spurious semi-annual cycle. The latter has been attributed to meridional asymmetry in the background state that is too weak, possibly in conjunction with incorrect regional water vapour feedbacks (Li and Philander, 1996; Guilyardi, 2006; Timmermann et al., 2007; De Szoeke and Xie, 2008; Wu et al., 2008a; Hirota et al., 2011).

A further persistent problem is insufficient marine stratocumulus cloud in the eastern tropical Pacific, caused presumably by weak coastal upwelling off South America leading to a warm SST bias (Lin, 2007). Although the problem persists, improvements are being made (AchutaRao and Sperber, 2006).

9.4.2.5.2 Tropical Atlantic Ocean

CMIP3 and CMIP5 models exhibit severe biases in the tropical Atlantic Ocean, so severe that some of the most fundamental features—the east–west SST gradient and the eastward shoaling thermocline along the equator—cannot be reproduced (Figure 9.14; (Chang et al., 2007; Chang et al., 2008; Richter and Xie, 2008; Richter et al., 2013). In many models, the warm SST bias along the Benguela coast is in excess of 5°C and the Atlantic warm pool in the western basin is grossly underestimated (Liu et al., 2013a). As in the Pacific, CMIP3 models suffer the double ITCZ error in the Atlantic. Hypotheses for the complex Atlantic bias problem tend to draw on the fact that the Atlantic Ocean has a far smaller basin, and thus encourages a tighter and more complex land–atmosphere–ocean interaction. A recent study using a high-resolution coupled model suggests that the warm eastern equatorial Atlantic SST bias is more sensitive to the local rather than basin-wide trade wind bias and to a wet Congo basin instead of a dry Amazon—a finding that differs from previous studies (Patricola et al., 2012). Recent ocean model studies show that a warm subsurface temperature bias in the eastern equatorial Atlantic is common to virtually all ocean models forced with 'best estimated' surface momentum and heat fluxes, owing to problems in parameterization of vertical mixing (Hazeleger and Haarsma, 2005). Toniazzo and Woolnough (2013) show that among a variety of causes for the initial bias development, ocean–atmosphere coupling is key for their maintenance.

9.4.2.5.3 Tropical Indian Ocean

CMIP3 and CMIP5 models simulate equatorial Indian Ocean climate reasonably well (e.g., Figure 9.14), though most models produce weak westerly winds and a flat thermocline on the equator. The models show a large spread in the modelled depth of the 20°C isotherm in the eastern equatorial Indian Ocean (Saji et al., 2006). The reasons are unclear but may be related to differences in the various parameterizations of vertical mixing as well as the wind structure (Schott et al., 2009).

CMIP3 models generally simulate the Seychelles Chagos thermocline ridge in the Southwest Indian Ocean, a feature important for the Indian monsoon and tropical cyclone activity in this basin (Xie et al., 2002). The models, however, have significant problems in accurately representing its seasonal cycle because of the difficulty in capturing the asymmetric nature of the monsoonal winds over the basin, resulting in too weak a semi-annual harmonic in the local Ekman pumping over the ridge region compared to observations (Yokoi et al., 2009b). In about half of the models, the thermocline ridge is displaced eastward associated with the easterly wind biases on the equator (Nagura et al., 2013).

9.4.2.6 Summary

There is *high confidence* that the CMIP3 and CMIP5 models simulate the main physical and dynamical processes at play during transient ocean heat uptake, sea level rise, and coupled modes of variability. There is little evidence that CMIP5 models differ significantly from CMIP3, although there is some evidence of modest improvement. Many improvements are seen in individual CMIP5 ocean components (some now including interactive ocean biogeochemistry) and the number of relatively poor-performing models has been reduced (thereby reducing inter-model spread). New since the AR4, process-based model evaluation is now helping identify the cause of some specific biases, helping to overcome the limits set by the short observational records available.

9.4.3 Sea Ice

Evaluation of sea ice performance requires accurate information on ice concentration, thickness, velocity, salinity, snow cover and other factors. The most reliably measured characteristic of sea ice remains sea ice extent (usually understood as the area covered by ice with a concentration above 15%). Caveats, however, exist related to the uneven reliability of different sources of sea ice extent estimates (e.g., satellite vs. pre-satellite observations; see Chapter 4), as well as to limitations of this characteristic as a metric of model performance (Notz et al., 2013).

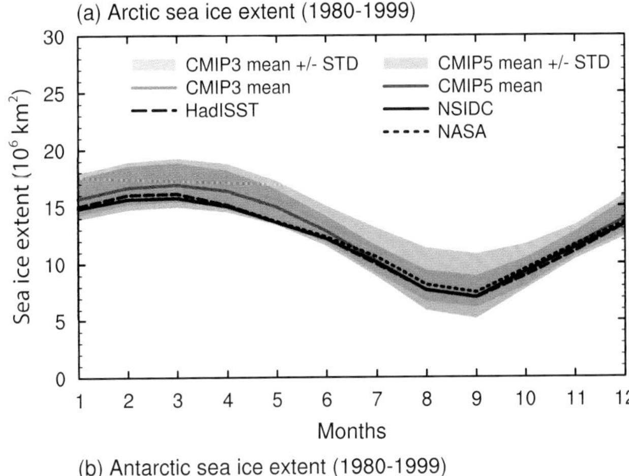

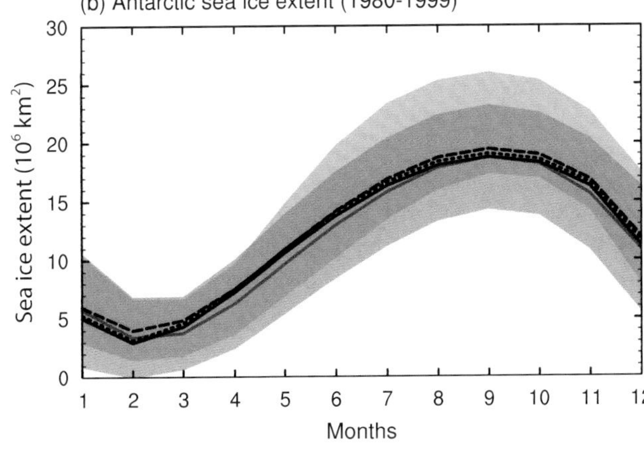

Figure 9.22 | Mean (1980–1999) seasonal cycle of sea ice extent (the ocean area with a sea ice concentration of at least 15%) in the Northern Hemisphere (upper) and the Southern Hemisphere (lower) as simulated by 42 CMIP5 and 17 CMIP3 models. Each model is represented with a single simulation. The observed seasonal cycles (1980–1999) are based on the Hadley Centre Sea Ice and Sea Surface Temperature (HadISST; Rayner et al., 2003), National Aeronautics and Space Administration (NASA; Comiso and Nishio, 2008) and the National Snow and Ice Data Center (NSIDC; Fetterer et al., 2002) data sets. The shaded areas show the inter-model standard deviation for each ensemble. (Adapted from Pavlova et al., 2011.)

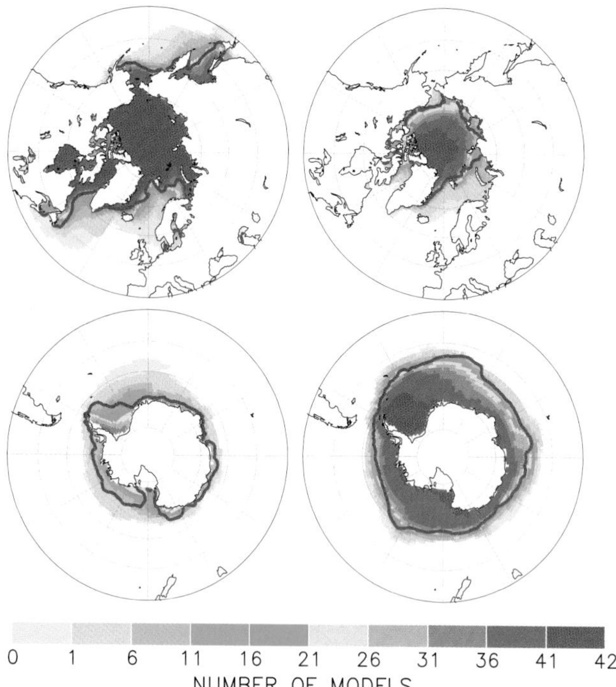

Figure 9.23 | Sea ice distribution (1986–2005) in the Northern Hemisphere (upper panels) and the Southern Hemisphere (lower panels) for February (left) and September (right). AR5 baseline climate (1986–2005) simulated by 42 CMIP5 AOGCMs. Each model is represented with a single simulation. For each 1° × 1° longitude-latitude grid cell, the figure indicates the number of models that simulate at least 15% of the area covered by sea ice. The observed 15% concentration boundaries (red line) are based on the Hadley Centre Sea Ice and Sea Surface Temperature (HadISST) data set (Rayner et al., 2003). (Adapted from Pavlova et al., 2011.)

The CMIP5 multi-model ensemble exhibits improvements over CMIP3 in simulation of sea ice extent in the both hemispheres (Figure 9.22). In the Arctic, the multi-model mean error do not exceed 10% of the observationally based estimates for any given month. In the Antarctic, the corresponding multi-model mean error exceeds 10% (but is less than 20%) near the annual minimum of sea ice extent; around the annual maximum, the CMIP5 multi-model mean shows a clear improvement over CMIP3.

In many models the regional distribution of sea ice concentration is poorly simulated, even if the hemispheric extent is approximately correct. In Figure 9.23, however, one can see that the median ice edge position (indicated by the colour at which half of the models have ice of 15% concentration) agrees reasonably well with observations in both hemispheres (except austral summer in Antarctica), as was the case for the CMIP3 models.

A widely discussed feature of the CMIP3 models as a group is a pronounced underestimation of the trend in the September (annual minimum) sea ice extent in the Arctic over the past several decades (e.g., Stroeve et al., 2007; Zhang, 2010; Rampal et al., 2011; Winton, 2011). Possible reasons for the discrepancy include variability inherent to high latitudes, model shortcomings, and observational uncertainties (e.g., Kattsov et al., 2010; Kay et al., 2011; Day et al., 2012). Compared to CMIP3, the CMIP5 models better simulate the observed trend of September Arctic ice extent (Figure 9.24). It has been suggested (Stroeve et al., 2012) that in some cases model improvements, such as new sea ice albedo parameterization schemes (e.g., Pedersen et al., 2009; Holland et al., 2012), have been responsible. (Holland et al., 2010) show that models with initially thicker ice generally retain more extensive ice throughout the 21st century, and indeed several of the CMIP5 models start the 20th century with rather thin winter ice cover promoting more rapid melt (Stroeve et al., 2012). Notz et al. (2013) caution, however, against direct comparison of modelled trends with observations unless the models' internal variability is carefully taken into account. Their analysis of the MPI-ESM ensemble shows that internal variability in the Arctic can result in individual model realizations exhibiting a range of trends (negative, or even positive) for the 29-year-long period starting in 1979, even if the background climate is warming. According to the distribution of sea ice extent trends over the period 1979–2010 obtained in an ensemble of simulations with individual CMIP5 models (Figure 9.24) about one quarter of the simulations shows a September trend in the Arctic as strong as, or stronger, than in observations.

The majority of CMIP5 (and CMIP3) models exhibit a decreasing trend in SH austral summer sea ice extent over the satellite era, in contrast to the observed weak but significant increase (see Chapter 4). A large spread in the modelled trends is present, and a comparison of multiple ensemble members from the same model suggests large internal variability during the late 20th century and the first decade of the 21st century (e.g., Landrum et al., 2012; Zunz et al., 2013). Compared to observations, CMIP5 models strongly overestimate the variability of sea ice extent, at least in austral winter (Zunz et al., 2013).Therefore, using the models to assess the potential role of the internal variability in the trend of sea ice extent in the Southern Ocean over the last three decades presents a significant challenge.

Sea ice is a product of atmosphere–ocean interaction. There are a number of ways in which sea ice is influenced by and interacts with the atmosphere and ocean, and some of these feedbacks are still poorly quantified. As noted in the AR4, among the primary causes of biases in simulated sea ice extent, especially its geographical distribution, are problems with simulating high-latitude winds, ocean heat advection and mixing. For example, Koldunov et al. (2010) have shown, for a particular CMIP3 model, that significant ice thickness errors originate from biases in the atmospheric component. Similarly, Melsom et al. (2009) note sea ice improvements associated with improved description of heat transport by ocean currents. Biases imparted on modelled sea ice, common to many models, may also be related to representation of high-latitude processes (e.g., polar clouds) or processes not yet commonly included in models (e.g., deposition of carbonaceous aerosols on snow and ice). Some CMIP5 models show improvements in simulation of sea ice that are connected to improvements in simulation of the atmosphere (e.g., Notz et al., 2013).

9.4.3.1 Summary

CMIP5 models reproduce the seasonal cycle of sea ice extent in both hemispheres. There is *robust evidence* that the downward trend in Arctic summer sea ice extent is better simulated than at the time of the AR4, with about one quarter of the simulations showing a trend as strong as, or stronger than, that observed over the satellite era. The performance improvements are not only a result of improvements in

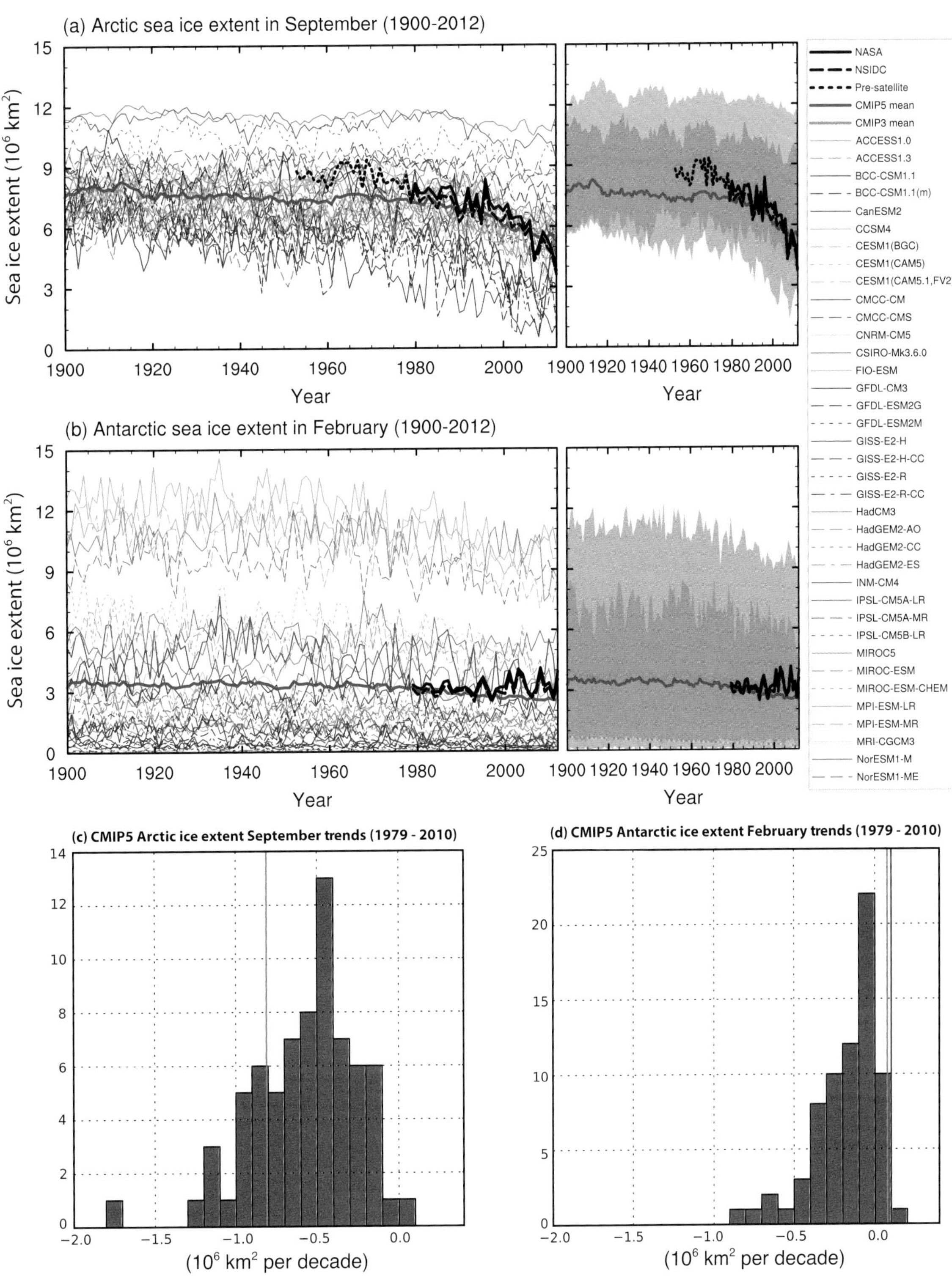

Figure 9.24 | (Top and middle rows) Time series of sea ice extent from 1900 to 2012 for (a) the Arctic in September and (b) the Antarctic in February, as modelled in CMIP5 (coloured lines) and observations-based (NASA; Comiso and Nishio, 2008) and NSIDC; (Fetterer et al., 2002), solid and dashed thick black lines, respectively). The CMIP5 multi-model ensemble mean (thick red line) is based on 37 CMIP5 models (historical simulations extended after 2005 with RCP4.5 projections). Each model is represented with a single simulation. The dotted black line for the Arctic in (a) relates to the pre-satellite period of observation-based time series (Stroeve et al., 2012). In (a) and (b) the panels on the right are based on the corresponding 37-member ensemble means from CMIP5 (thick red lines) and 12-model ensemble means from CMIP3 (thick blue lines). The CMIP3 12-model means are based on CMIP3 historical simulations extended after 1999 with Special Report on Emission Scenarios (SRES) A2 projections. The pink and light blue shadings denote the 5 to 95 percentile range for the corresponding ensembles. Note that these are monthly means, not yearly minima. (Adapted from Pavlova et al., 2011.) (Bottom row) CMIP5 sea ice extent trend distributions over the period 1979–2010 for (c) the Arctic in September and (d) the Antarctic in February. Altogether 66 realizations are shown from 26 different models (historical simulations extended after 2005 with RCP4.5 projections). They are compared against the observations-based estimates of the trends (green vertical lines in (c) and (d) from Comiso and Nishio (2008); blue vertical line in (d) from Parkinson and Cavalieri (2012)). In (c), the observations-based estimates (Cavalieri and Parkinson, 2012; Comiso and Nishio, 2008) coincide.

sea ice components themselves but also in atmospheric circulation. Most CMIP5 models simulate a decrease in Antarctic sea ice extent over the past few decades compared to the small but significant increase observed.

9.4.4 Land Surface, Fluxes and Hydrology

The land surface determines the partitioning of precipitation into evapotranspiration and runoff, and the partitioning of surface net radiation into sensible and latent heat fluxes. Land surface processes therefore impact strongly on both the climate and hydrological resources. This subsection summarizes recent studies on the evaluation of land surface models, wherever possible emphasizing their performance in CMIP3 and CMIP5 climate models.

9.4.4.1 Snow Cover and Near-Surface Permafrost

The modelling of snow and near-surface permafrost (NSP) processes has received increased attention since the AR4, in part because of the recognition that these processes can provide significant feedbacks on climate change (e.g., Koven et al., 2011; Lawrence et al., 2011). The SnowMIP2 project compared results from 33 snowpack models of varying complexity, including some snow models that are used in AOGCMs, using driving data from five NH locations (Rutter et al., 2009). Most snow models were found to be consistent with observations at open sites, but there was much greater discrepancy at forested sites due to the complex interactions between plant canopy and snow cover. Despite these difficulties, the CMIP5 multi-model ensemble reproduces key features of the large-scale snow cover (Figure 9.25). In the NH, models are able to simulate the seasonal cycle of snow cover over the northern parts of continents, with more disagreement in southerly regions where snow cover is sparse, particularly over China and Mongolia (Brutel-Vuilmet et al., 2013). The latter weaknesses are associated with incorrect timing of the snow onset and melt, and possibly with the choice of thresholds for diagnosing snow cover in the model output. In spite of the good performance of the multi-model mean, there is a significant inter-model scatter of spring snow cover extent in some regions. There is a strong linear correlation between Northern-Hemisphere spring snow cover extent and annual mean surface air temperature in the models, consistent with available observations. The recent negative trend in spring snow cover is underestimated by the CMIP5 (and CMIP3) models (Derksen and Brown, 2012), which is associated with an underestimate of the boreal land surface warming (Brutel-Vuilmet et al., 2013).

Some CMIP5 models now represent NSP and frozen soil processes (Koven et al., 2013), but this is not generally the case. Therefore it is difficult to make a direct quantitative evaluation of most CMIP5 models against permafrost observations. A less direct but more inclusive approach is to diagnose NSP extent using snow depths and skin temperatures generated by climate models to drive a stand-alone multi-layer permafrost model (Pavlova et al., 2007). Figure 9.25 shows the result of using this approach on the CMIP5 ensemble. The multi-model mean is able to simulate the approximate location of the NSP boundary (as indicated by the 0°C soil temperature isotherm). However, the range of present-day (1986–2005) NSP area inferred from individual models spans a factor of more than six (~4 to 25 × 10^6 km^2) due to differences in simulated surface climate and to varying abilities of the underlying land surface models. Even though many CMIP5 models include some representation of soil freezing in mineral soils, very few include key processes necessary to accurately model NSP changes, such as the distinct properties of organic soils, the existence of local water tables and the heat released by microbial respiration (Nicolsky et al., 2007; Wania et al., 2009; Koven et al., 2011, 2013).

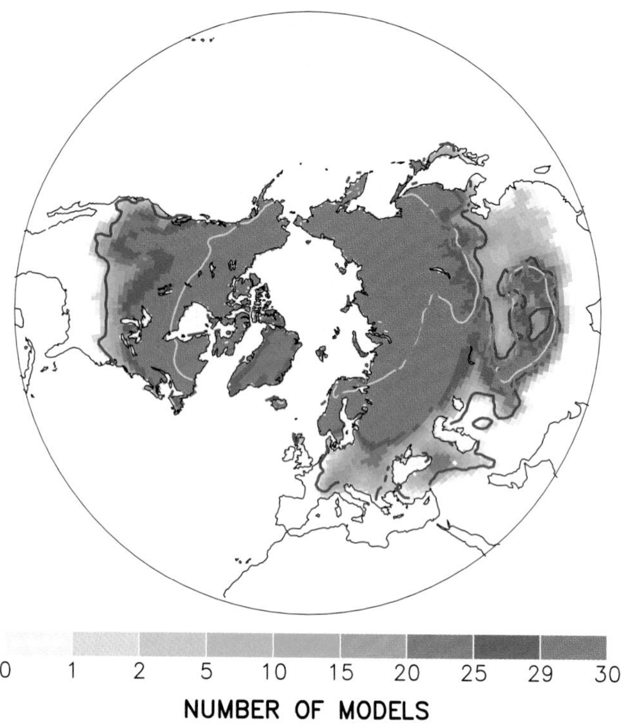

Figure 9.25 | Terrestrial snow cover distribution (1986–2005) in the Northern Hemisphere (NH) as simulated by 30 CMIP5 models for February, updated for CMIP5 from Pavlova et al. (2007). For each 1° × 1° longitude-latitude grid cell, the figure indicates the number of models that simulate at least 5 kg m^{-2} of snow-water equivalent. The observations-based boundaries (red line) mark the territory with at least 20% of the days per month with snow cover (Robinson and Frei, 2000) over the period 1986–2005. The annual mean 0°C isotherm at 3.3 m depth averaged across 24 CMIP5 models (yellow line) is a proxy for the near-surface permafrost boundary. Observed permafrost extent in the NH (magenta line) is based on Brown et al. (1997, 1998).

Despite large differences in the absolute NSP area, the relationship between the decrease in NSP area and the warming air temperature over the present-day NSP region is similar, and approximately linear, in many models (Slater and Lawrence, 2013).

9.4.4.2 Soil Moisture and Surface Hydrology

The partitioning of precipitation into evapotranspiration and runoff is highly dependent on the moisture status of the land surface, especially the amount of soil moisture available for evapotranspiration, which in turn depends on properties of the land cover such as the rooting depth of plants.

There has been a long history of off-line evaluation of land surface schemes, aided more recently by the increasing availability of site-specific data (Friend et al., 2007; Blyth et al., 2010). Throughout this time, representations of the land surface have significantly increased in complexity, allowing the representation of key processes such as links between stomatal conductance and photosynthesis, but at the cost of increasing the number of poorly known internal model parameters. These more sophisticated land surface models are based on physical principles that should make them more appropriate for projections of future climate and increased CO_2. However for specific data-rich sites, current land surface models still struggle to perform as well as statistical models in predicting year-to-year variations in latent and sensible heat fluxes (Abramowitz et al., 2008) and runoff (Materia et al., 2010).

There are few evaluations of the performance of land surface schemes in coupled climate models, but those that have been undertaken find major limitations associated with the atmospheric forcing rather than the land surface schemes themselves. For example, an evaluation of the soil moisture simulations of CMIP3 models found that long-term soil moisture trends could only be reproduced in models that simulated the reduction in solar radiation at the surface associated with 'global dimming' (Li et al., 2007). A comparison of simulated evapotranspiration fluxes from CMIP3 against large-scale observation-based estimates, showed underestimates in India and parts of eastern South America, and overestimates in the western USA, Australia and China (Mueller et al., 2011).

Land–atmosphere coupling determines the ability of climate models to simulate the influence of soil moisture anomalies on rainfall, droughts and high-temperature extremes (Fischer et al., 2007; Lorenz et al., 2012). The coupling strength depends both on the sensitivity of evapotranspiration to soil moisture, which is determined by the land surface scheme, and the sensitivity of precipitation to evapotranspiration, which is determined by the atmospheric model (Koster et al., 2004; Seneviratne et al., 2010). Comparison of climate model simulations to observations suggests that the models correctly represent the soil-moisture impacts on temperature extremes in southeastern Europe, but overestimate them in central Europe (Hirschi et al., 2011). The influence of soil moisture on rainfall varies significantly with region, and with the lead-time between a soil moisture anomaly and a rainfall event (Seneviratne et al., 2010). In some regions, such as the Sahel, enhanced precipitation can even be induced by dry anomalies (Taylor et al., 2011). Recent analyses of CMIP5 models reveals considerable spread in the ability of the models to reproduce observed correlations between precipitation and soil moisture in the tropics (Williams et al., 2012), and a systematic failure to simulate the positive impact of dry soil moisture anomalies on rainfall in the Sahel (Taylor et al., 2012a).

9.4.4.3 Dynamic Global Vegetation and Nitrogen Cycling

At the time of the AR4 very few climate models included dynamic vegetation, with vegetation being prescribed and fixed in all but a handful of coupled climate–carbon cycle models (Friedlingstein et al., 2006). Dynamic Global Vegetation Models (DGVMs) certainly existed at the time of the AR4 (Cramer et al., 2001) but these were not typically incorporated in climate models. Since the AR4 there has been continual development of offline DGVMs, and some climate models incorporate dynamic vegetation in at least a subset of the runs submitted to CMIP5 (also see Section 9.1.3.2.4), with likely consequences for climate model biases and regional climate projection (Martin and Levine, 2012).

DGVMs are designed to simulate the large-scale geographical distribution of plant functional types and how these patterns will change in response to climate change, CO_2 increases, and other forcing factors (Cramer et al., 2001). These models typically include rather detailed representations of plant photosynthesis but less sophisticated treatments of soil carbon, with a varying number of soil carbon pools. In the absence of nitrogen limitations on CO_2 fertilization, offline DGVMs agree qualitatively that CO_2 increase alone will tend to enhance carbon uptake on the land while the associated climate change will tend to reduce it. There is also good agreement on the degree of CO_2 fertilization in the case of no nutrient limitation (Sitch et al., 2008). However, under more extreme emissions scenarios the responses of the DGVMs diverge markedly. Large uncertainties are associated with the responses of tropical and boreal ecosystems to elevated temperatures and changing soil moisture status. Particular areas of uncertainty are the high-temperature response of photosynthesis (Galbraith et al., 2010), and the extent of CO_2 fertilization (Rammig et al., 2010) in the Amazonian rainforest.

Most of the land surface models and DGVMs used in the CMIP5 models continue to neglect nutrient limitations on plant growth (see Section 6.4.6.2), even though these may significantly moderate the response of photosynthesis to CO_2 (Wang and Houlton, 2009). Recent extensions of two land surface models to include nitrogen limitations improve the fit to 'Free-Air CO_2 Enrichment Experiments', and suggest that models without these limitations are expected to overestimate the land carbon sink in the nitrogen-limited mid and high latitudes (Thornton et al., 2007; Zaehle et al., 2010).

9.4.4.4 Land Use Change

A major innovation in the land component of ESMs since the AR4 is the inclusion of the effects of land use change associated with the spread of agriculture, urbanization and deforestation. These affect climate by altering the biophysical properties of the land surface, such as its albedo, aerodynamic roughness and water-holding capacity (Bondeau et al., 2007; Bonan, 2008; Bathiany et al., 2010; Levis, 2010). Land use change has also contributed almost 30% of total anthropogenic CO_2 emissions since 1850 (see Table 6.1), and affects emissions of trace gases, and VOCs such as isoprene. The latest ESMs used in CMIP5 attempt to model the CO_2 emissions implied by prescribed land use change and many also simulate the associated changes in the biophysical properties of the land surface. This represents a major advance on the CMIP3 models which typically neglected land use change, aside from its assumed contribution to anthropogenic CO_2 emissions.

However, the increasing sophistication of the modelling of the impacts of land use change has introduced additional spread in climate model projections. The first systematic model intercomparison demonstrated that large-scale land cover change can significantly affect regional climate (Pitman et al., 2009) and showed a large spread in the response of different models to the same imposed land cover change (de Noblet-Ducoudre et al., 2012). This uncertainty arises from the often counteracting effects of evapotranspiration and albedo changes (Boisier et al., 2012) and has consequences for the simulation of temperature and rainfall extremes (Pitman et al., 2012b).

9.4.5 Carbon Cycle

An important development since the AR4 is the more widespread implementation of ESMs that include an interactive carbon cycle. Coupled climate-carbon cycle models are used extensively for the projections presented in Chapter 12. The evaluation of the carbon cycle within coupled models is discussed here, while the performance of the individual land and ocean carbon models, together with the detailed analysis of climate–carbon cycle feedbacks, is presented in Chapter 6 (Section 6.4 and Box 6.4).

The transition from AOGCMs to ESMs was motivated in part by the results from the first generation coupled climate–carbon cycle models, which suggested that feedbacks between the climate and the carbon cycle were uncertain but potentially very important in the context of 21st century climate change (Cox et al., 2000; Friedlingstein et al., 2001). The first-generation models used in the Coupled Climate Carbon Cycle Model Intercomparison Project (C^4MIP) included both extended AOGCMS and EMICs. The C^4MIP experimental design involved running each model under a common emission scenario (SRES A2) and calculating the evolution of the global atmospheric CO_2 concentration interactively within the model. The impacts of climate–carbon cycle feedbacks were diagnosed by carrying out parallel "uncoupled" simulations in which increases in atmospheric CO_2 did not influence climate. Analysis of the C^4MIP runs highlighted a greater than 200 ppmv range in the CO_2 concentration by 2100 due to uncertainties in climate–carbon cycle feedbacks, and that the largest uncertainties were associated with the response of land ecosystems to climate and CO_2 (Friedlingstein et al., 2006).

For CMIP5 a different experimental design was proposed in which the core simulations use prescribed RCPs of atmospheric CO_2 and other GHGs (Moss et al., 2010). Under a prescribed CO_2 scenario, ESMs calculate land and ocean carbon fluxes interactively, but these fluxes do not affect the evolution of atmospheric CO_2. Instead the modelled land and ocean fluxes, along with the prescribed increase in atmospheric CO_2, can be used to diagnose the 'compatible' emissions of CO_2 consistent with the simulation (see Section 6.3; Miyama and Kawamiya, 2009; Arora et al., 2011). The compatible emissions for each model can then be evaluated against the best estimates of the actual historical CO_2 emissions. Parallel model experiments in which the carbon cycle does not respond to the simulated climate change (which are equivalent to the 'uncoupled' simulations in C^4MIP) provide a means to diagnose climate–carbon cycle feedbacks in terms of their impact on the compatible emissions of CO_2 (Hibbard et al., 2007).

Carbon cycle model evaluation is limited by the availability of direct observations at appropriately large spatial scales. Field studies and eddy-covariance flux measurements provide detailed information on the land carbon cycle over short-time scales and for specific locations, and ocean inventories are able to constrain the long-term uptake of anthropogenic CO_2 by the ocean (Sabine et al., 2004; Takahashi et al., 2009). However the stores of carbon on the land are less well-known, even though these are important determinants of the CO_2 fluxes from land use change. ESM simulations vary by a factor of at least six in global soil carbon (Anav et al., 2013; Todd-Brown et al., 2013) and by a factor of four in global vegetation carbon, although about two thirds of models are within 50% of the uncertain observational estimates (Anav et al., 2013).

Large-scale land–atmosphere and global atmosphere fluxes are not directly measured, but global estimates can be made from the carbon balance, and large-scale regional fluxes can be estimated from the inversion of atmospheric CO_2 measurements (see Section 6.3.2). Figure 9.26 shows modelled annual mean ocean–atmosphere and net land–atmosphere CO_2 fluxes from the historical simulations in the CMIP5 archive (Anav et al., 2013). Also shown are estimates derived from offline ocean carbon cycle models, measurements of atmospheric CO_2, and best estimates of the CO_2 fluxes from fossil fuels and land use change (Le Quere et al., 2009). Uncertainties in these latter annual estimates are approximately ± 0.5 PgC yr^{-1}, arising predominantly from the uncertainty in the model-derived ocean CO_2 uptake. The confidence limits for the ensemble mean are derived by assuming that the CMIP5 models form a *t*-distribution centred on the ensemble mean (Anav et al., 2013).

The evolution of the global ocean carbon sink is shown in the top panel of Figure 9.26. The CMIP5 ensemble mean global ocean uptake ($\pm$ standard deviation of the multi-model ensemble), computed using all the 23 models that reported ocean CO_2 fluxes, increases from 0.47 $\pm$ 0.32 PgC yr^{-1} over the period 1901–1930 to 1.53 $\pm$ 0.36 PgC yr^{-1} for the period 1960–2005. For comparison, the Global Carbon Project (GCP) estimates a stronger ocean carbon sink of 1.92 $\pm$ 0.3 PgC yr^{-1} for 1960–2005 (Anav et al., 2013). The bottom panel of Figure 9.26 shows the variability in global land carbon uptake evident in the GCP estimates, with the global land carbon sink being strongest during La Niña years and after volcanoes, and turning into a source during El Niño years. The CMIP5 models cannot be expected to precisely reproduce this year-to-year variability as these models will naturally simulate chaotic ENSO variability that is out of phase with the historical variability. However, the ensemble mean does successfully simulate a strengthening global land carbon sink during the 1990s, especially after the Mt Pinatubo eruption in 1991. The CMIP5 ensemble mean land–atmosphere flux ($\pm$ standard deviation of the multi-model ensemble) evolves from a small source of -0.34 ± 0.49 PgC yr^{-1} over the period 1901–1930, predominantly due to land use change, to a sink of 0.47 $\pm$ 0.72 PgC yr^{-1} in the period 1960–2005. The GCP estimates give a weaker sink of 0.36 $\pm$ 1 PgC yr^{-1} for the 1960–2005 period.

Figure 9.27 shows the ocean–atmosphere fluxes (top panel) and mean land–atmosphere fluxes (bottom panel) simulated by ESMs and EMICs (Eby et al., 2013) for the period 1986–2005, and compares these to observation-based estimates from GCP and Atmospheric Tracer Transport Model Intercomparison Project (TRANSCOM3) atmospheric inversions (Gurney et al., 2003). Unlike Figure 9.26, only models that reported both land and ocean carbon fluxes are included in this figure. The atmospheric inversions results are taken from the Japanese Meteorological Agency (JMA) as this was the only TRANSCOM3 model that reported results for all years of the 1986–2005 reference period. The error bars on the observational estimates (red triangles) and the ESM simulations (black diamonds) represent the interannual variation in the form of the standard deviation of the annual fluxes. EMICs do not typically simulate interannual variability, so only mean values are shown for these models (green boxes). Here, as in Figure 9.26, the net

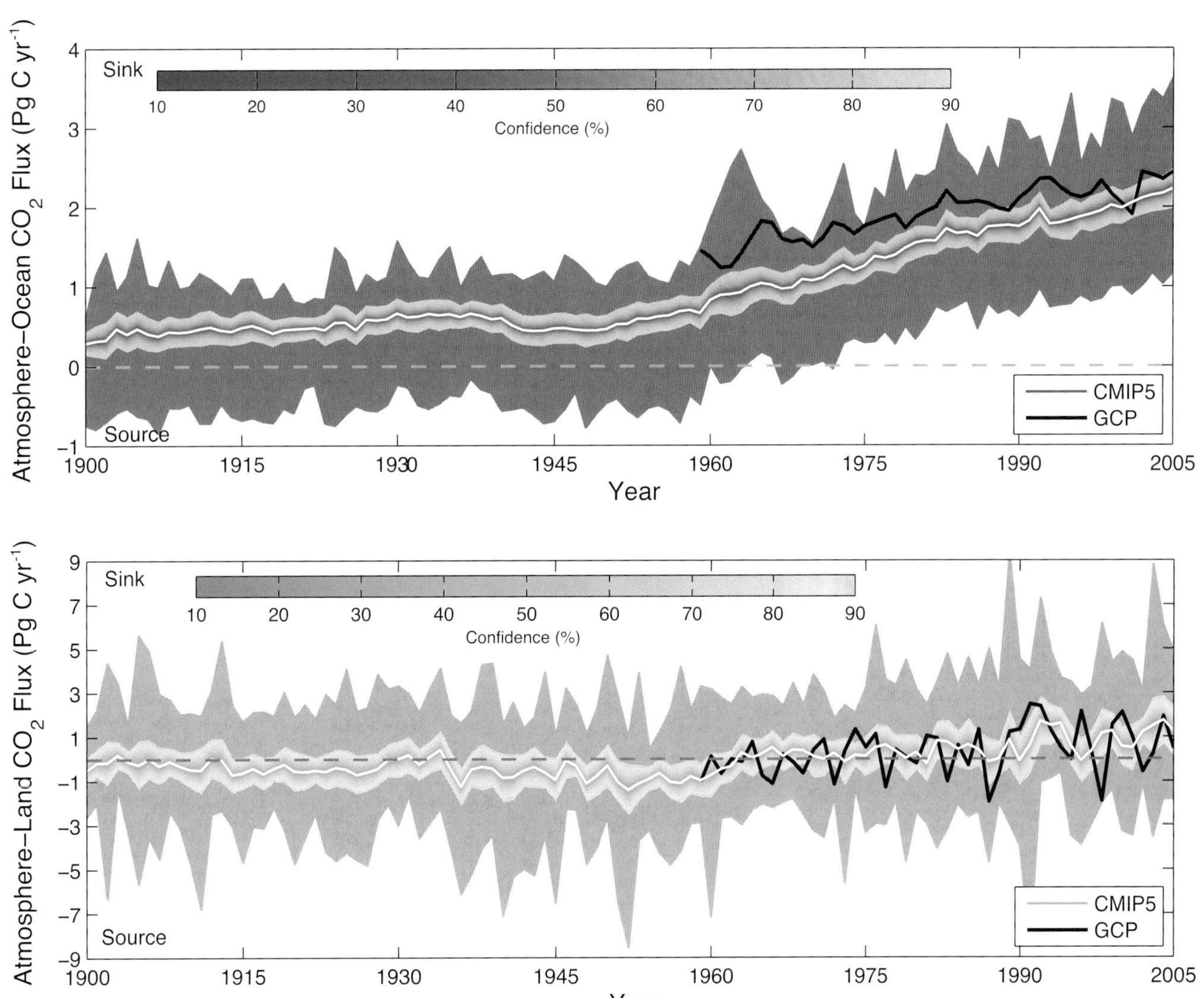

Figure 9.26 | Ensemble-mean global ocean carbon uptake (top) and global land carbon uptake (bottom) in the CMIP5 ESMs for the historical period 1900–2005. For comparison, the observation-based estimates provided by the Global Carbon Project (Le Quere et al., 2009) are also shown (thick black line). The confidence limits on the ensemble mean are derived by assuming that the CMIP5 models come from a *t*-distribution. The grey areas show the range of annual mean fluxes simulated across the model ensemble. This figure includes results from all CMIP5 models that reported land CO_2 fluxes, ocean CO_2 fluxes, or both (Anav et al., 2013).

land–atmosphere flux is 'Net Biome Productivity (NBP)' which includes the net CO_2 emissions from land use change as well as the changing carbon balance of undisturbed ecosystems.

For the period 1986–2005 the observation-based estimates of the global ocean carbon sink are 1.71 PgC yr^{-1} (JMA), 2.19 PgC yr^{-1} (GCP) and 2.33 PgC yr^{-1} (Takahashi et al., 2009). Taking into account the uncertainties in the mean values of these fluxes associated with interannual variability, the observationally constrained range is approximately 1.4 to 2.4 PgC yr^{-1}. All of the ESMs, and all but one of the EMICs, simulate ocean sinks within this range. The observation-based estimates of GCP and JMA agree well on the mean global land carbon sink over the period 1986–2005, and most ESMs fit within the uncertainty bounds of these estimates (i.e., 1.17 ± 1.06 PgC yr^{-1} for JMA). The exceptions are two ESMs sharing common atmosphere and land components (CESM1-BGC and NorESM1-ME) which model a net land carbon source rather than a sink over this period. The EMICs tend to systematically underestimate the contemporary land carbon sink (Eby et al., 2013). Some ESMs (notably GFDL-ESM2M and GFDL-ESM2G) significantly overestimate the interannual variation in the global land–atmosphere CO_2 flux, with a possible consequence being an overestimate of the vulnerability of tropical ecosystems to future climate change (Cox et al., 2013), and see Figure 9.45). All ESMs qualitatively simulate the expected pattern of ocean CO_2 fluxes, with outgassing in the tropics and uptake in the mid and high latitudes (Anav et al., 2013). However, there are systematic differences between the ESMs and the JMA inversion estimates for the zonal land CO_2 fluxes, with the ESMs tending to produce weaker uptake in the NH, and simulating a net land carbon sink rather than a source in the tropics.

In summary, there is *high confidence* that CMIP5 ESMs can simulate the global mean land and ocean carbon sinks within the range of observation-based estimates. Overall, EMICs reproduce the recent global ocean CO_2 fluxes uptake as well as ESMs, but estimate a lower

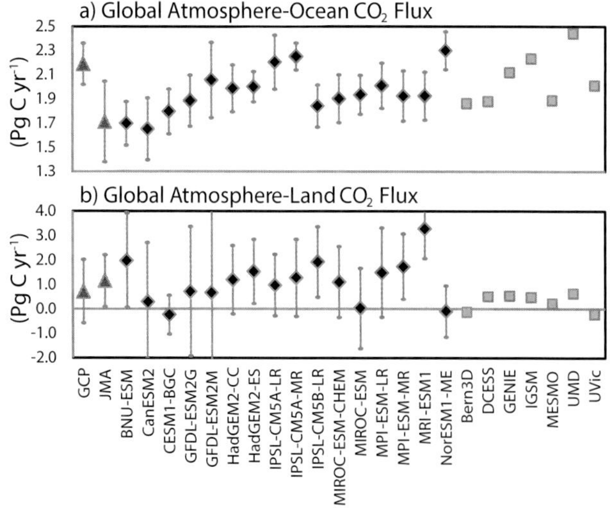

Figure 9.27 | Simulation of global mean (a) atmosphere–ocean CO_2 fluxes ('fgCO2') and (b) net atmosphere–land CO_2 fluxes ('NBP'), by ESMs (black diamonds) and EMICs (green boxes), for the period 1986–2005. For comparison, the observation-based estimates provided by Global Carbon Project (GCP; Le Quere et al., 2009), and the Japanese Meteorological Agency (JMA) atmospheric inversion (Gurney et al., 2003) are also shown as the red triangles. The error bars for the ESMs and observations represent interannual variability in the fluxes, calculated as the standard deviation of the annual means over the period 1986–2005.

land carbon sink compared with most ESMs while remaining consistent with the observations (Eby et al., 2013). With few exceptions, the CMIP5 ESMs also reproduce the large-scale pattern of ocean–atmosphere CO_2 fluxes, with uptake in the Southern Ocean and northern mid-latitudes, and outgassing in the tropics. However, the geographical pattern of simulated land–atmosphere fluxes agrees much less well with inversion estimates, which suggest a larger sink in the northern mid-latitudes, and a net source rather than a sink in the tropics. While there are also inherent uncertainties in atmospheric inversions, discrepancies like this might be expected from known deficiencies in the CMIP5 generation of ESMs—namely the failure to correctly simulate nitrogen fertilization in the mid-latitudes, and a rudimentary treatment of the net CO_2 emissions arising from land use change and forest regrowth.

9.4.6 Aerosol Burdens and Effects on Insolation

9.4.6.1 Recent Trends in Global Aerosol Burdens and Effects on Insolation

The ability of CMIP5 models to simulate the current burden of tropospheric aerosol and the decadal trends in this burden can be assessed using observations of aerosol optical depth (AOD, see Section 7.3.1.2). The historical data used to drive the CMIP5 20th century simulations reflect recent trends in anthropogenic SO_2 emissions, and hence these trends should be manifested in the modelled and observed AOD. During the last three decades, anthropogenic emissions of SO_2 from North America and Europe have declined due to the imposition of emission controls, while the emissions from Asia have increased. The combination of the European, North American, and Asians trends has yielded a global reduction in SO_2 emissions of 20 Gg(SO_2), or 15% between 1970 and 2000 although emissions subsequently increased by 9 Gg(SO_2) between 2000 and 2005 (Smith et al., 2011b). For the period 2001 to 2005, CMIP5 models underestimate the mean AOD at 550 nm relative to satellite-retrieved AOD by at least 20% over virtually all land surfaces (Figure 9.28). The differences between the modelled and measured AODs exceed the errors in the Multi-angle Imaging Spectro-Radiometer (MISR) retrievals over land of ±0.05 or 0.2×AOD (Kahn et al., 2005) and the RMS errors in the corrected Moderate Resolution Imaging Spectrometer (MODIS) retrievals over ocean of 0.061(Shi et al., 2011).

The effects of sulphate and other aerosol species on surface insolation through direct and indirect forcing appear to be one of the principal causes of the 'global dimming' between the 1960s and 1980s and subsequent 'global brightening' in the last two decades (see Section 2.3.3.1). This inference is supported by trends in aerosol optical depth and trends in surface insolation under cloud-free conditions. Thirteen out of 14 CMIP3 models examined by Ruckstuhl and Norris (2009) produce a transition from "dimming" to 'brightening' that is consistent with the timing of the transition from increasing to decreasing global anthropogenic aerosol emissions. The transition from 'dimming' to 'brightening' in both Europe and North America is well simulated with the HadGEM2 model (Haywood et al., 2011).

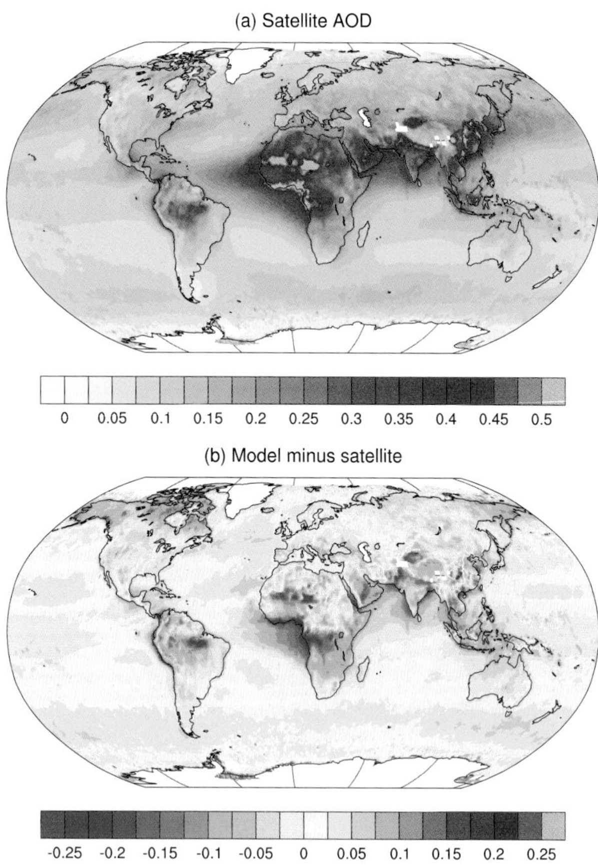

Figure 9.28 | (a): Annual mean visible aerosol optical depth (AOD) for 2001 through 2005 using the Moderate Resolution Imaging Spectrometer (MODIS) version 5 satellite retrievals for ocean regions (Remer et al., 2008) with corrections (Zhang et al., 2008a; Shi et al., 2011) and version 31 of MISR retrievals over land (Zhang and Reid, 2010; Stevens and Schwartz, 2012). (b): The absolute error in visible AOD from the median of a subset of CMIP5 models' historical simulations relative to the satellite retrievals of AOD shown in (a). The model outputs for 2001 through 2005 are from 21 CMIP5 models with interactive or semi-interactive aerosol representation.

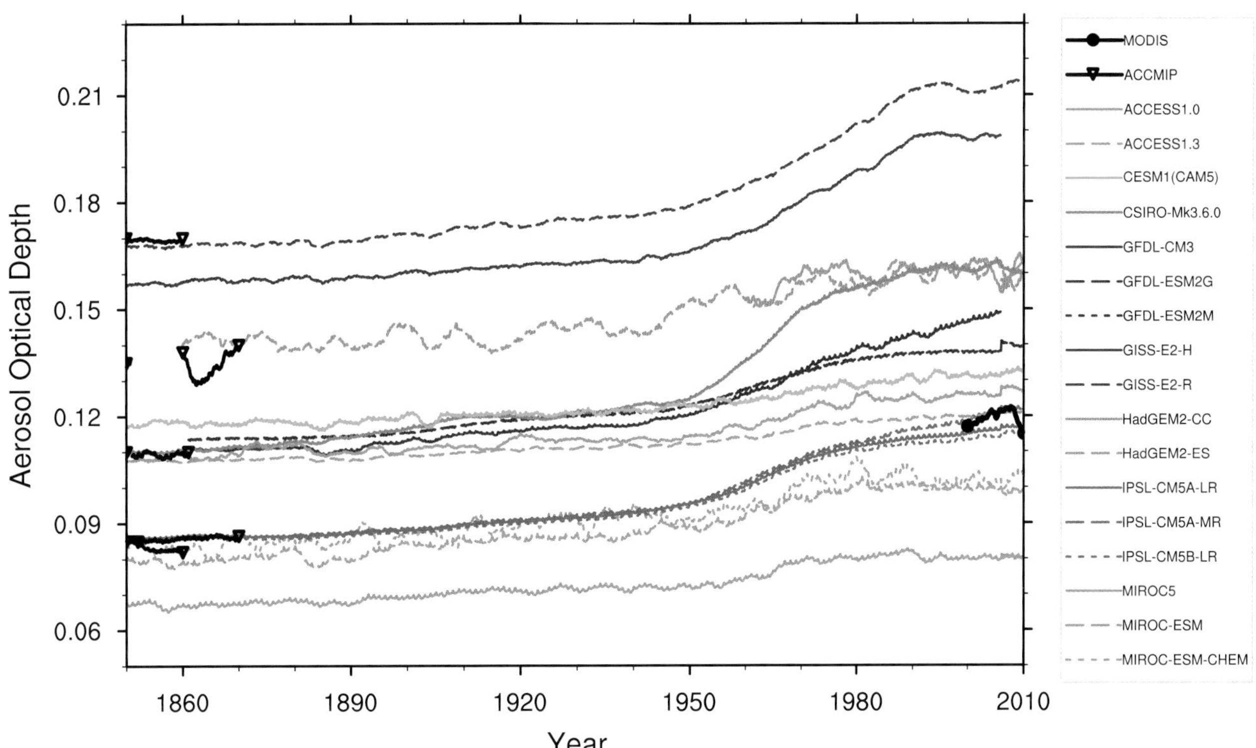

Figure 9.29 | Time series of global oceanic mean aerosol optical depth (AOD) from individual CMIP5 models' historical (1850–2005) and RCP4.5 (2006–2010) simulations, corrected Moderate Resolution Imaging Spectrometer (MODIS) satellite observations by Shi et al. (2011) and Zhang et al. (2008a), and the Atmospheric Chemistry and Climate Model Intercomparison Project (ACCMIP) simulations for the 1850s by Shindell et al. (2013b).

These recent trends are superimposed on a general upward trend in aerosol loading since 1850 reflected by an increase in global mean oceanic AODs from the CMIP5 historical and RCP 4.5 simulations from 1850 to 2010 (Figure 9.29). Despite the use of common anthropogenic aerosol emissions for the historical simulations (Lamarque et al., 2010), the simulated oceanic AODs for 2010 range from 0.08 to 0.215, with nearly equal numbers of models over and underestimating the satellite retrieved AOD of 0.12 (Figure 9.29). This range in AODs results from differing estimates of the trends and of the initial global mean oceanic AOD at 1850 across the CMIP5 ensemble (Figure 9.29).

9.4.6.2 Principal Sources of Uncertainty in Projections of Sulphate Burdens

Natural sources of sulphate from oxidation of dimethylsulphide (DMS) emissions from the ocean surface are not specified under the RCP protocol and therefore represent a source of uncertainty in the sulphur cycle simulated by the CMIP5 ensemble. In simulations of present-day conditions, DMS emissions span a 5 to 95% confidence interval of 10.7 to 28.1 TgS yr^{-1} (Faloona, 2009). After chemical processing, DMS contributes between 18 and 42% of the global atmospheric sulphate burden and up to 80% of the sulphate burden over most the SH (Carslaw et al., 2010). Several CMIP5 models include prognostic calculation of the biogenic DMS source; however, the effects from differences in DMS emissions on modelled sulphate burdens remain to be quantified.

In contrast to CMIP3, the models in the CMIP5 ensemble are provided with a single internally consistent set of future anthropogenic SO_2 emissions. The use of a single set of emissions removes an important, but not dominant, source of uncertainty in the AR5 simulations of the sulphur cycle. In experiments based on a single chemistry–climate model with perturbations to both emissions and sulphur-cycle processes, uncertainties in emissions accounted for 53.3% of the ensemble variance (Ackerley et al., 2009). The next largest source of uncertainty was associated with the wet scavenging of sulphate (see Section 7.3.2), which accounted for 29.5% of the intra-ensemble variance and represents the source/sink term with the largest relative range in the aerosol models evaluated by Faloona (2009). Similarly, simulations run with heterogeneous or harmonized emissions data sets yielded approximately the same intermodel standard deviation in sulphate burden of 25 Tg. These results show that a dominant source of the spread among the sulphate burdens is associated with differences in the treatment of chemical production, transport, and removal from the atmosphere (Liu et al., 2007; Textor et al., 2007). Errors in modelled aerosol burden systematically affect anthropogenic RF (Shindell et al., 2013b).

9.5 Simulation of Variability and Extremes

9.5.1 Importance of Simulating Climate Variability

The ability of a model to simulate the mean climate, and the slow, externally forced change in that mean state, was evaluated in the previous section. However, the ability to simulate climate variability, both unforced internal variability and forced variability (e.g., diurnal and seasonal cycles) is also important. This has implications for the signal-to-noise estimates inherent in climate change detection and attribution studies where low-frequency climate variability must be estimated, at least in part, from long control integrations of climate models (Section 10.2). It also has implications for the ability of models to make quantitative projections of changes in climate variability and the statistics of extreme events under a warming climate. In many cases, the impacts of climate change will be experienced more profoundly in terms of the frequency, intensity or duration of extreme events (e.g., heat waves, droughts, extreme rainfall events; see Section 12.4). The ability to simulate climate variability is also central to achieving skill in climate prediction by initializing models from the observed climate state (Sections 11.1 and 11.2).

9

Evaluating model simulations of climate variability also provides a means to explore the representation of certain processes, such as the coupled processes underlying the ENSO and other important modes of variability. A model's representation of the diurnal or seasonal cycle—both of which represent responses to external (rotational or orbital) forcing – may also provide some insight into a model's 'sensitivity' and by extension, the ability to respond correctly to GHG, aerosol, volcanic and solar forcing.

9.5.2 Diurnal-to-Seasonal Variability

9.5.2.1 Diurnal Cycles of Temperature and Precipitation

The diurnally varying solar radiation received at a given location drives, through complex interactions with the atmosphere, land surface and upper ocean, easily observable diurnal variations in surface and near-surface temperature, precipitation, level stability and winds. The AR4 noted that climate models simulated the global pattern of the diurnal temperature range, zonally and annually averaged over the continent, but tended to underestimate its magnitude in many regions (Randall et al., 2007). New analyses over land indicate that model deficiencies in surface–atmosphere interactions and the planetary boundary layer are also expected to contribute to some of the diurnal cycle errors and that model agreement with observations depends on region, vegetation type and season (Lindvall et al., 2012). Analyses of CMIP3 simulations show that the diurnal amplitude of precipitation is realistic, but most models tend to start moist convection prematurely over land (Dai, 2006; Wang et al., 2011a). Many CMIP5 models also have peak precipitation several hours too early compared to surface observations and TRMM satellite observations (Figure 9.30). This and the so-called 'drizzling bias' (Dai, 2006) can have large adverse impacts on surface evaporation and runoff (Qian et al., 2006). Over the ocean, models often rain too frequently and underestimated the diurnal amplitude (Stephens et al., 2010). It has also been suggested that a weak diurnal cycle of surface air temperature is produced over the ocean because of a lack of diurnal variations in SST (Bernie et al., 2008), and most models have difficulty with this due to coarse vertical resolution and coupling frequency (Dai and Trenberth, 2004; Danabasoglu et al., 2006).

Improved representation of the diurnal cycle has been found with increased atmospheric resolution (Sato et al., 2009; Ploshay and Lau, 2010) or with improved representation of cloud physics (Khairoutdinov et al., 2005), but the reasons for these improvements remain poorly understood. Other changes such as the representation of entrainment in deep convection (Stratton and Stirling, 2012), improved coupling between shallow and deep convection, and inclusion of density currents (Peterson et al., 2009) have been shown to greatly improve the diurnal cycle of convection over tropical land and provide a good representation of the timing of convection over land in coupled ocean–atmosphere simulations (Hourdin et al., 2013). Thanks to improvements like this, the best performing models in Figure 9.30 appear now to be able to capture the land and ocean diurnal phase and amplitude quite well.

9.5.2.2 Blocking

In the mid latitudes, climate is often characterized by weather regimes (see Chapter 2), amongst which blocking regimes play a role in the occurrence of extreme weather events (Buehler et al., 2011; Sillmann et al., 2011; Pfahl and Wernli, 2012). During blocking, the prevailing mid-latitude westerly winds and storm systems are interrupted by a local reversal of the zonal flow. Climate models in the past have universally underestimated the occurrence of blocking, in particular in the Euro-Atlantic sector (Scaife et al., 2010).

There are important differences in methods used to identify blocking (Barriopedro et al., 2010a), and the diagnosed blocking frequency can be very sensitive to details such as the choice of latitude (Barnes et al., 2012). Blocking indices can be sensitive to biases in the representation of mean state (Scaife et al., 2010) or in variability (Barriopedro et al., 2010b; Vial and Osborn, 2012). When blocking is measured via anomaly fields, rather than reversed absolute fields, model skill can be high even in relatively low-resolution models (e.g., Sillmann and Croci-Maspoli, 2009).

Recent work has shown that models with high horizontal (Matsueda, 2009; Matsueda et al., 2009, 2010) or vertical resolution (Anstey et al., 2013) are better able to simulate blocking. These improvement arise from increased representation of orography and atmospheric dynamics (Woollings et al., 2010b; Jung et al., 2012; Berckmans et al., 2013), as well as reduced ocean surface temperature errors in the extra tropics (Scaife et al., 2011). Improved physical parameterizations have also been shown to improve simulations of blocking (Jung et al., 2010). However, as in CMIP3 (Scaife et al., 2010; Barnes et al., 2012), most of the CMIP5 models still significantly underestimate winter Euro-Atlantic blocking (Anstey et al., 2013; Masato et al., 2012; Dunn-Sigouin and Son, 2013). These new results show that the representation of blocking events is improving in models, even though the overall quality of CMIP5 ensemble is medium. There is *high confidence* that model representation of blocking is improved through increases in model resolution.

9.5.2.3 Madden–Julian Oscillation

During the boreal winter the eastward propagating feature known as the Madden–Julian Oscillation (MJO; (Madden and Julian, 1972, 1994) predominantly affects the deep tropics, while during the boreal summer there is also northward propagation over much of southern Asia (Annamalai and Sperber, 2005). The MJO has received much attention given the prominent role it plays in tropical climate variability (e.g., monsoons, ENSO, and mid-latitudes; Lau and Waliser, 2011)

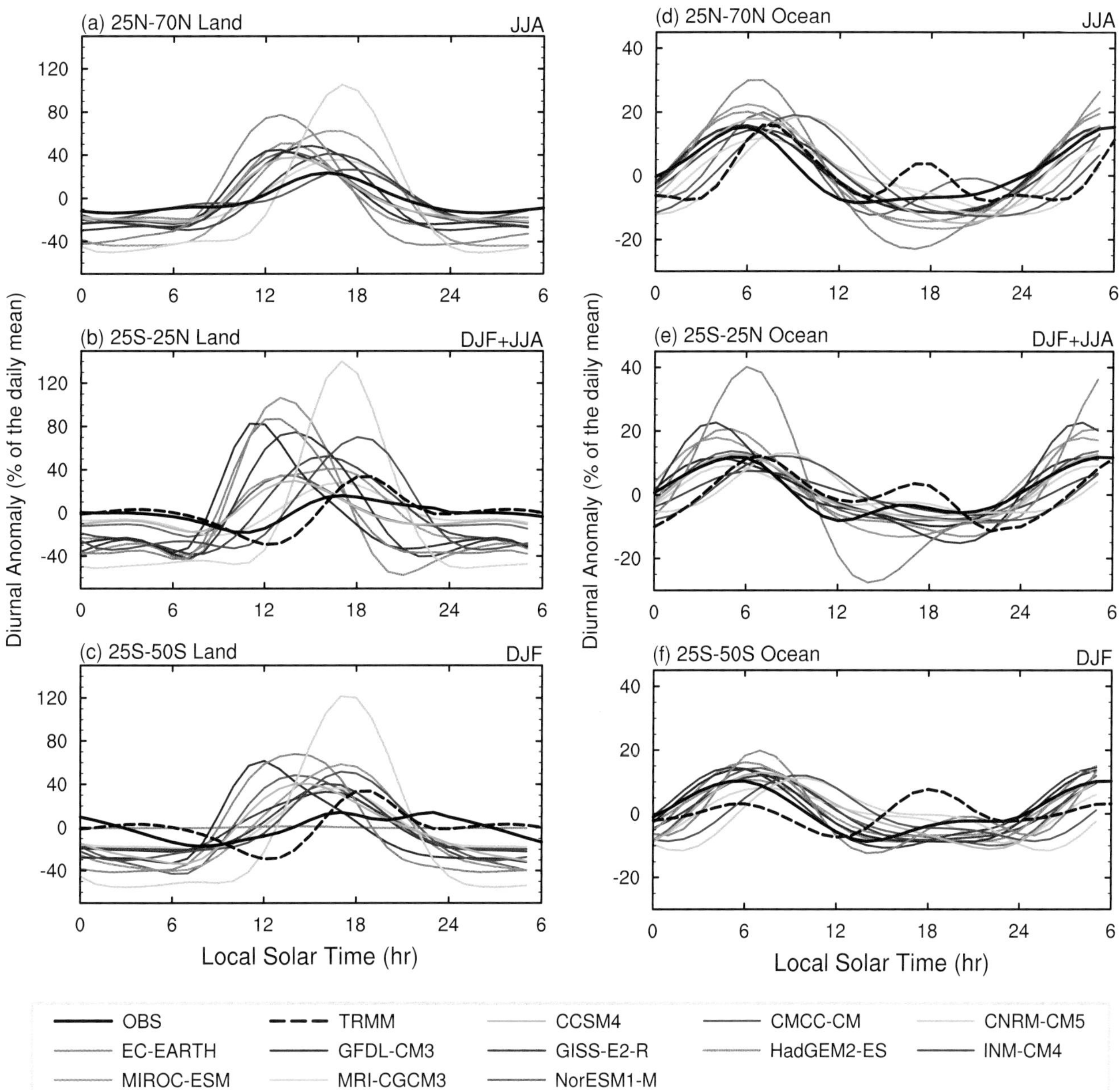

Figure 9.30 | Composite diurnal cycle of precipitation averaged over land (left) and ocean (right) for three different latitude bands at each local time and season (June–July–August (JJA), December–January–February (DJF), or their sum). For most of the CMIP5 models, data from 1980–2005 from the historical runs were averaged to derived the composite cycle; however, a few models had the required 3-hourly data only for 1990–2005 or 1996–2005. For comparison with the model results, a similar diagnosis from observations are shown (black solid line: surface-observed precipitation frequency; black dashed line: TRMM 3B42 data set, 1998–2003 mean). (Update of Figure 17 of Dai, 2006.)

Phenomenological diagnostics (Waliser et al., 2009a) and process-oriented diagnostics (e.g., Xavier, 2012) have been used to evaluate MJO in climate models. An important reason for model errors in representing the MJO is that convection parameterizations do not provide sufficient build-up of moisture in the atmosphere for the large scale organized convection to occur (Kim et al., 2012; Mizuta et al., 2012). Biases in the model mean state also contribute to poor MJO simulation (Inness et al., 2003). High-frequency coupling with the ocean is also an important factor (Bernie et al., 2008). While new parameterizations of convection may improve the MJO (Hourdin et al., 2013), this sometimes occurs at the expense of a good simulation of the mean tropical climate (Kim et al., 2012). In addition, high resolution models with an improved diurnal cycle do not necessarily produce an improved MJO (Mizuta et al., 2012).

Most models underestimate the strength and the coherence of convection and wind variability (Lin et al., 2006; Lin and Li, 2008). The simplified metric shown in Figure 9.31 provides a synthesis of CMIP3 and CMIP5 model results (Sperber and Kim, 2012). It shows that simulation of the MJO is still a challenge for climate models (Lin et al., 2006; Kim et al., 2009; Xavier et al., 2010). Most models have weak coherence in their MJO propagation (smaller maximum positive correlation). Even so, relative to CMIP3 there has been improvement in CMIP5 in simulating the eastward propagation of boreal winter MJO convection from

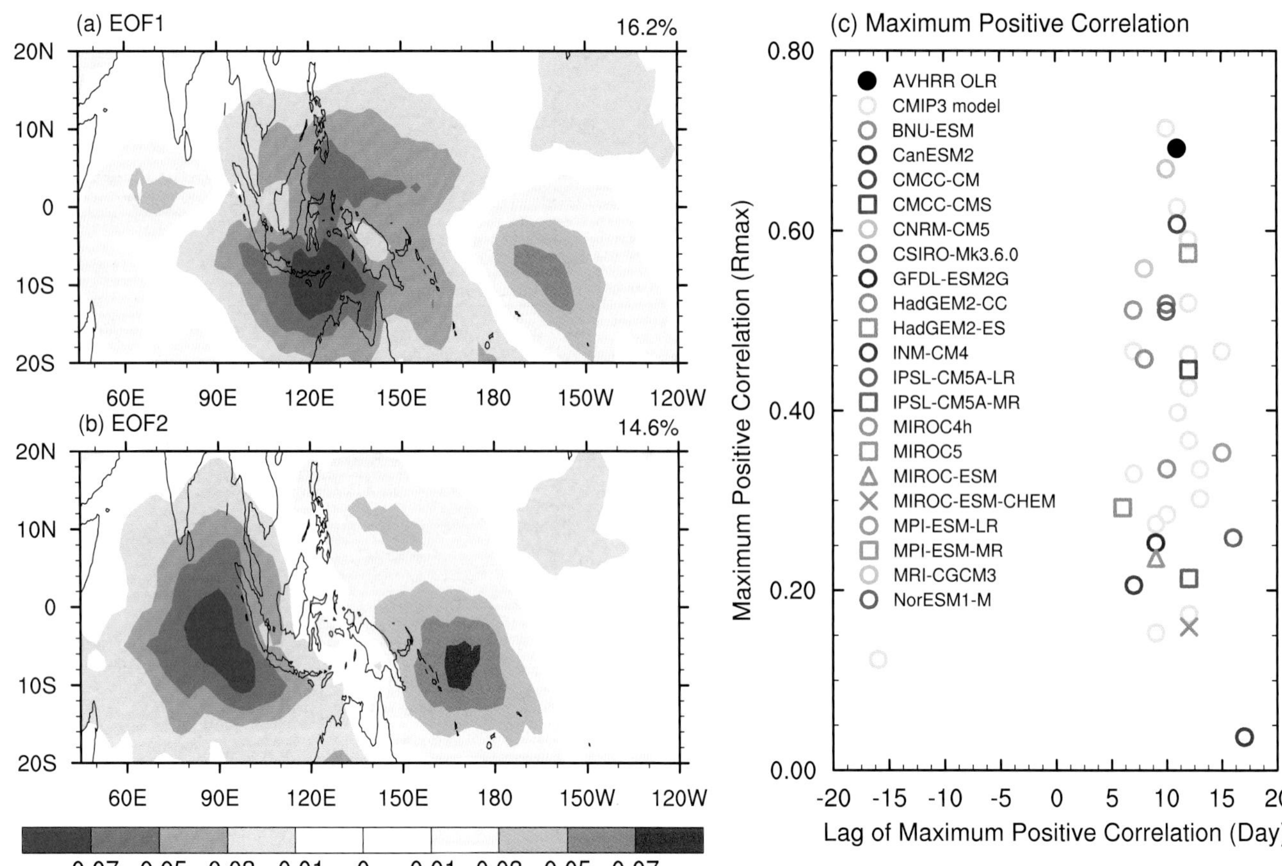

Figure 9.31 | (a, b) The two leading Empirical Orthogonal Functions (EOFs) of outgoing longwave radiation (OLR) from years of strong Madden–Julian Oscillation (MJO) variability computed following Sperber (2003). The 20- to 100-day filtered OLR from observations and each of the CMIP5 historical simulations and the CMIP3 simulations of 20th century climate is projected on these two leading EOFs to obtain MJO Principal Component time series. The scatterplot (c) shows the maximum positive correlation between the resulting MJO Principal Components and the time lag at which it occurred for all winters (November to March). The maximum positive correlation is an indication of the coherence with which the MJO convection propagates from the Indian Ocean to the Maritime Continent/western Pacific, and the time lag is approximately one fourth of the period of the MJO. (Constructed following Sperber and Kim, 2012.)

the Indian Ocean into the western Pacific (Hung et al., 2013) and northward propagation during boreal summer (Sperber et al., 2012). In addition there is evidence that models reproduce MJO characteristics in the east Pacific (Jiang et al., 2012b), and that, overall, there is improvement compared to previous generations of climate models (Waliser et al., 2003; Lin et al., 2006; Sperber and Annamalai, 2008).

9.5.2.4 Large-Scale Monsoon Rainfall and Circulation

Monsoons are the dominant modes of annual variation in the tropics (Trenberth et al., 2000; Wang and Ding, 2008), and affect weather and climate in numerous regions (Chapter 14). High-fidelity simulation of the mean monsoon and its variability is of great importance for simulating future climate impacts (Wang, 2006; Sperber et al., 2010; Colman et al., 2011; Turner and Annamalai, 2012). The monsoon is characterized by an annual reversal of the low level winds and well defined dry and wet seasons (Wang and Ding, 2008), and its variability is primarily connected to the MJO and ENSO (Section 9.5.3). The AR4 reported that most CMIP3 models poorly represent the characteristics of the monsoon and monsoon teleconnections (Randall et al., 2007), though improvement in CMIP5 has been noted for the mean climate, seasonal cycle, intraseasonal and interannual variability (Sperber et al., 2012).

The different monsoon systems are connected through the large-scale tropical circulation, offering the possibility to evaluate a models' representation of monsoon domain and intensity through the global monsoon concept (Wang and Ding, 2008; Wang et al., 2011a). The CMIP5 multi-model ensemble generally reproduces the observed spatial patterns but somewhat underestimates the extent and intensity, especially over Asia and North America (Figure 9.32). The best model has similar performance to the multi-model mean, whereas the poorest models fail to capture the monsoon precipitation domain and intensity over Asia and the western Pacific, Central America, and Australia. Fan et al. (2010) also show that CMIP3 simulations capture the observed trend of weakening of the South Asian summer circulation over the past half century, but are unable to reproduce the magnitude of the observed trend in precipitation. On longer time scales, mid-Holocene simulations show that even though models capture the sign of the monsoon precipitation changes, they tend to underestimate its magnitude (Braconnot et al., 2007b; Zhao and Harrison, 2012)

Poor simulation of the monsoon has been attributed to cold SST biases over the Arabian Sea (Levine and Turner, 2012), a weak meridional temperature gradient (Joseph et al., 2012), unrealistic development of the Indian Ocean dipole (Achuthavarier et al., 2012; Boschat et al.,

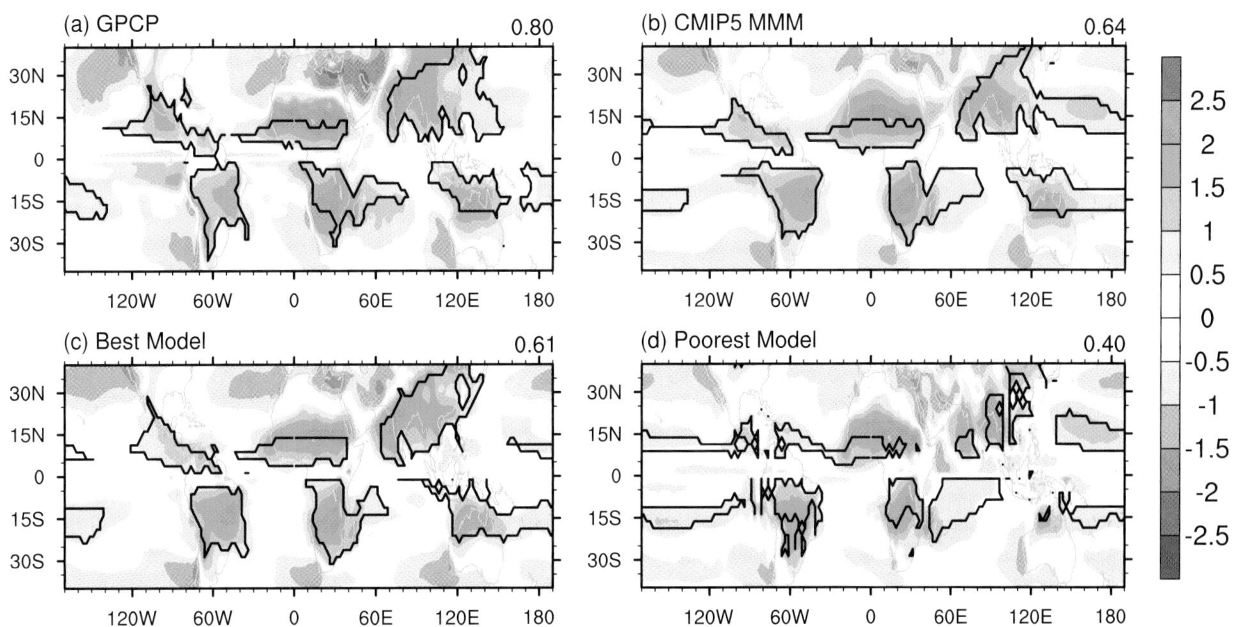

Figure 9.32 | Monsoon precipitation intensity (shading, dimensionless) and monsoon precipitation domain (lines) are shown for (a) observation-based estimates from Global Precipitation Climatology Project (GPCP), (b) the CMIP5 multi-model mean, (c) the best model and (d) the worst model in terms of the threat score for this diagnostic. These measures are based on the seasonal range of precipitation using hemispheric summer (May through September in the Northern Hemisphere (NH)) minus winter (November through March in the NH) values. The monsoon precipitation domain is defined where the annual range is >2.5 mm day^{-1}, and the monsoon precipitation intensity is the seasonal range divided by the annual mean. The threat scores (Wilks, 1995) indicate how well the models represent the monsoon precipitation domain compared to the GPCP data. The threat score in panel (a) is between GPCP and CMAP rainfall to indicate observational uncertainty, whereas in the other panel it is between the simulations and the GPCP observational data set. A threat score of 1.0 would indicate perfect agreement between the two data sets. See Wang and Ding (2008),Wang et al. (2011a), and Kim et al. (2011) for details of the calculations.

2012) and changes to the circulation through excessive precipitation over the southwest equatorial Indian Ocean (Bollasina and Ming, 2013). These biases lead to too weak inland moisture transport and an underestimate of monsoon precipitation over India. Similar SST biases contribute to model-data mismatch in the simulation of the mid-Holocene Asian monsoon (Ohgaito and Abe-Ouchi, 2009), even though the representation of atmospheric processes such as convection seems to dominate the model spread in this region (Ohgaito and Abe-Ouchi, 2009) or over Africa (Zheng and Braconnot, 2013). Factors that have contributed to improved representation of the monsoon in some CMIP5 models include better simulation of topography-related monsoon precipitation due to higher horizontal resolution (Mizuta et al., 2012), a more realistic ENSO–monsoon teleconnection (Meehl et al., 2012) and improved propagation of intraseasonal variations (Sperber and Kim, 2012). The impact of aerosols on monsoon precipitation and its variability is the subject of ongoing investigation (Lau et al., 2008).

These results provide *robust evidence* that CMIP5 models simulate more realistic monsoon climatology and variability than their CMIP3 predecessors, but they still suffer from biases in the representation of the monsoon domain and intensity leading to medium model quality at the global scale and declining quality at the regional scale.

9.5.3 Interannual-to-Centennial Variability

In addition to the annual, intra-seasonal and diurnal cycles described above, a number of other modes of variability arise on multi-annual to multi-decadal time scales (see also Box 2.5). Most of these modes have a particular regional manifestation whose amplitude can be larger than that of human-induced climate change. The observational record is usually too short to fully evaluate the representation of variability in models and this motivates the use of reanalysis or proxies, even though these have their own limitations.

9.5.3.1 Global Surface Temperature Multi-Decadal Variability

The AR4 concluded that modelled global temperature variance at decadal to inter-decadal time scales was consistent with 20th century observations. In addition, results from the last millennium suggest that simulated variability is consistent with indirect estimates (Hegerl et al., 2007).

Figure 9.33a shows simulated internal variability of mean surface temperature from CMIP5 pre-industrial control simulations. Model spread is largest in the tropics and mid to high latitudes (Jones et al., 2012), where variability is also large; however, compared to CMIP3, the spread is smaller in the tropics owing to improved representation of ENSO variability (Jones et al., 2012). The power spectral density of global mean temperature variance in the historical simulations is shown in Figure 9.33b and is generally consistent with the observational estimates.

At longer time scale of the spectra estimated from last millennium simulations, performed with a subset of the CMIP5 models, can be assessed by comparison with different NH temperature proxy records (Figure 9.33c; see Chapter 5 for details). The CMIP5 millennium simulations include natural and anthropogenic forcings (solar, volcanic, GHGs, land use) (Schmidt et al., 2012). Significant differences between unforced and forced simulations are seen for time scale larger than

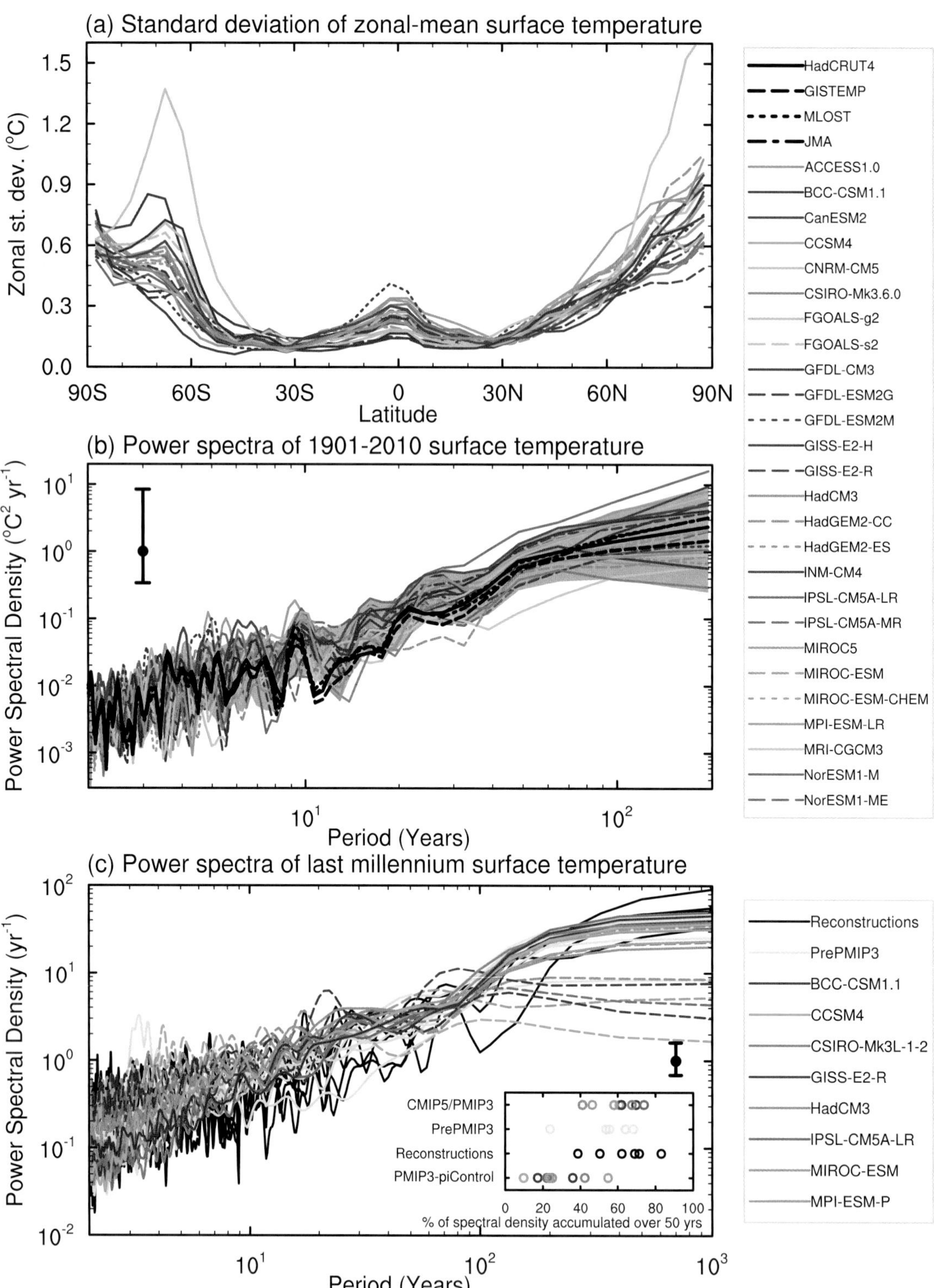

Figure 9.33 | Global climate variability as represented by: (a) Standard deviation of zonal-mean surface temperature of the CMIP5 pre-industrial control simulations (after Jones et al., 2012). (b) Power spectral density for 1901–2010 global mean surface temperature for both historical CMIP5 simulations and the observations (after Jones et al., 2012). The grey shading provides the 5 to 95% range of the simulations. (c) Power spectral density for Northern Hemisphere surface temperature from the CMIP5/ Paleoclimate Modelling Intercomparison Project version 3 (PMIP3) last-millennium simulations (colour, solid) using common external forcing configurations (Schmidt et al., 2012), together with the corresponding pre-industrial simulations (colour, dashed), previous last-millennium AOGCM simulations (black: Fernandez-Donado et al., 2013), and temperature reconstructions from different proxy records (see Section 5.3.5). For comparison between model results and proxy records, the spectra in (c) have been computed from normalized Northern Hemisphere time series. The small panel included in the bottom panel shows for the different models and reconstructions the percentage of spectral density cumulated for periods longer than 50 years, to highlight the differences between unforced (pre-industrial control) and forced (PMIP3 and pre-PMIP3) simulations, compared to temperature reconstruction for the longer time scales. In (b) and (c) the spectra have been computed using a Tukey–Hanning filter of width 97 and 100 years, respectively. The model outputs were not detrended, except for the MIROC-ESM millennium simulation. The 5 to 95% intervals (vertical lines) provide the accuracy of the power spectra estimated given a typical length of 110 years for (b) and 1150 years for (c).

50 years, indicating the importance of forced variability at these time scales (Fernandez-Donado et al., 2013). It should be noted that a few models exhibit slow background climate drift which increases the spread in variance estimates at multi-century time scales. Nevertheless, the lines of evidence above suggest with *high confidence* that models reproduce global and NH temperature variability on a wide range of time scales.

9.5.3.2 Extratropical Circulation, North Atlantic Oscillation and Other Related Dipolar and Annular Modes

Based on CMIP3 models, Gerber et al. (2008) confirmed the AR4 assessment that climate models are able to capture the broad spatial and temporal features of the North Atlantic Oscillation (NAO), but there are substantial differences in the spatial patterns amongst individual models (Casado and Pastor, 2012; Handorf and Dethloff, 2012). Climate models tend to have patterns of variability that are more annular in character than observed (Xin et al., 2008). Models substantially overestimate persistence on sub-seasonal and seasonal time scales, and have difficulty simulating the seasonal cycle of annular mode time scales found in reanalyses (Gerber et al., 2008). The unrealistically long time scale of variability is worse in models with particularly strong equatorward biases in the mean jet location, a result which has been found to hold in the North Atlantic and in the SH (Barnes and Hartmann, 2010; Kidston and Gerber, 2010).

As described in the AR4, climate models have generally been unable to simulate changes as strong as the observed NAO trend over the period 1965–1995, either in coupled mode (Gillett, 2005; Stephenson et al., 2006; Stoner et al., 2009) or forced with observed boundary conditions (Scaife et al. (2009). However, there are a few exceptions to this (e.g., Selten et al., 2004; Semenov et al., 2008), so it is unclear to what extent the underestimation of late 20th century trends reflects model shortcomings versus internal variability. Further evidence has emerged of the coupling of NAO variability between the troposphere and the stratosphere, and even models with improved stratospheric resolution appear to underestimate the vertical coupling (Morgenstern et al., 2010) with consequences for the NAO response to anthropogenic forcing (Sigmond and Scinocca, 2010; Karpechko and Manzini, 2012; Scaife et al., 2012).

The Pacific basin analogue of the NAO, the North Pacific Oscillation (NPO) is a prominent pattern of wintertime atmospheric circulation variability characterized by a north–south dipole in sea level pressure (Linkin and Nigam, 2008). Although climate models simulate the main spatial features of the NPO, many of them are unable to capture the observed linkages with tropical variability and the ocean (Furtado et al., 2011).

Raphael and Holland (2006) showed that coupled models produce a clear Southern Annular Mode (SAM) but that there are relatively large differences between models in terms of the exact shape and orientation of this pattern. Karpechko et al. (2009) found that the CMIP3 models have problems representing linkages between the SAM and SST, surface air temperature, precipitation and particularly sea ice in the Antarctic region.

9.5.3.3 Atlantic Modes

9.5.3.3.1 Atlantic Meridional Overturning Circulation variability

Previous comparisons of the observed and simulated AMOC were restricted to its mean strength, as it had only been sporadically observed (see Chapter 3 and Section 9.4.2.3.1). Continuous AMOC time series now exist for latitudes 41°N (reconstructions since 1993) and 26.5°N (estimate based on direct observations since 2004) (Cunningham et al., 2010; Willis, 2010). At 26.5°N, CMIP3 and CMIP5 model simulations show total AMOC variability that is within the observational uncertainty (Baehr et al., 2009; Marsh et al., 2009; Balan Sarojini et al., 2011; Msadek et al., 2013). However, the total AMOC is the sum of a wind-driven component and a mid-ocean geostrophic component. While both CMIP3 and CMIP5 models tend to overestimate the wind-driven variability, they tend to underestimate the mid-ocean geostrophic variability (Baehr et al., 2009; Balan Sarojini et al., 2011; Msadek et al., 2013). The latter is suggested to result from deficiencies in the simulation of the hydrographic characteristics (Baehr et al., 2009), specifically the Nordic Seas overflows (Yeager and Danabasoglu, 2012; Msadek et al., 2013).

9.5.3.3.2 Atlantic multi-decadal variability/Atlantic Multi-decadal Oscillation

The Atlantic Multi-decadal Variability (AMV), also known as Atlantic Multi-decadal Oscillation (AMO), is a mode of climate variability with an apparent period of about 70 years, and a pattern centred in the North Atlantic Ocean (see Section 14.7.6). In the AR4, it was shown that a number of climate models produced AMO-like multidecadal variability in the North Atlantic linked to variability in the strength of the AMOC. Subsequent analyses has confirmed this, however simulated time scales range from 40 to 60 years (Frankcombe et al., 2010; Park and Latif, 2010; Kavvada et al., 2013), to a century or more (Msadek and Frankignoul, 2009; Menary et al., 2012). In addition, the spatial patterns of variability related to the AMOC differ in many respects from one model to another as shown in Figure 9.34.

The presence of AMO-like variability in unforced simulations, and the fact that forced 20th century simulations in the CMIP3 multi-model ensemble produce AMO variability that is not in phase with that observed, implies the AMO is not predominantly a result of the forcings imposed on the models (Kravtsov and Spannagle, 2008; Knight, 2009; Ting et al., 2009). Results from the CMIP5 models also show a key role for internal variability, alongside a contribution from external forcings in recent decades (Terray, 2012). Historical AMO fluctuations have been better reproduced in a model with a more sophisticated aerosol treatment than was typically used in CMIP3 (Booth et al., 2012a), albeit at the expense of introducing other observational inconsistencies (Zhang et al., 2013). This could suggest that at least part of the AMO may in fact be forced, and that aerosols play a role. In addition to tropospheric aerosols, Otterå et al. (2010) showed the potential for simulated volcanic forcing to have influenced AMO fluctuations over the last 600 years.

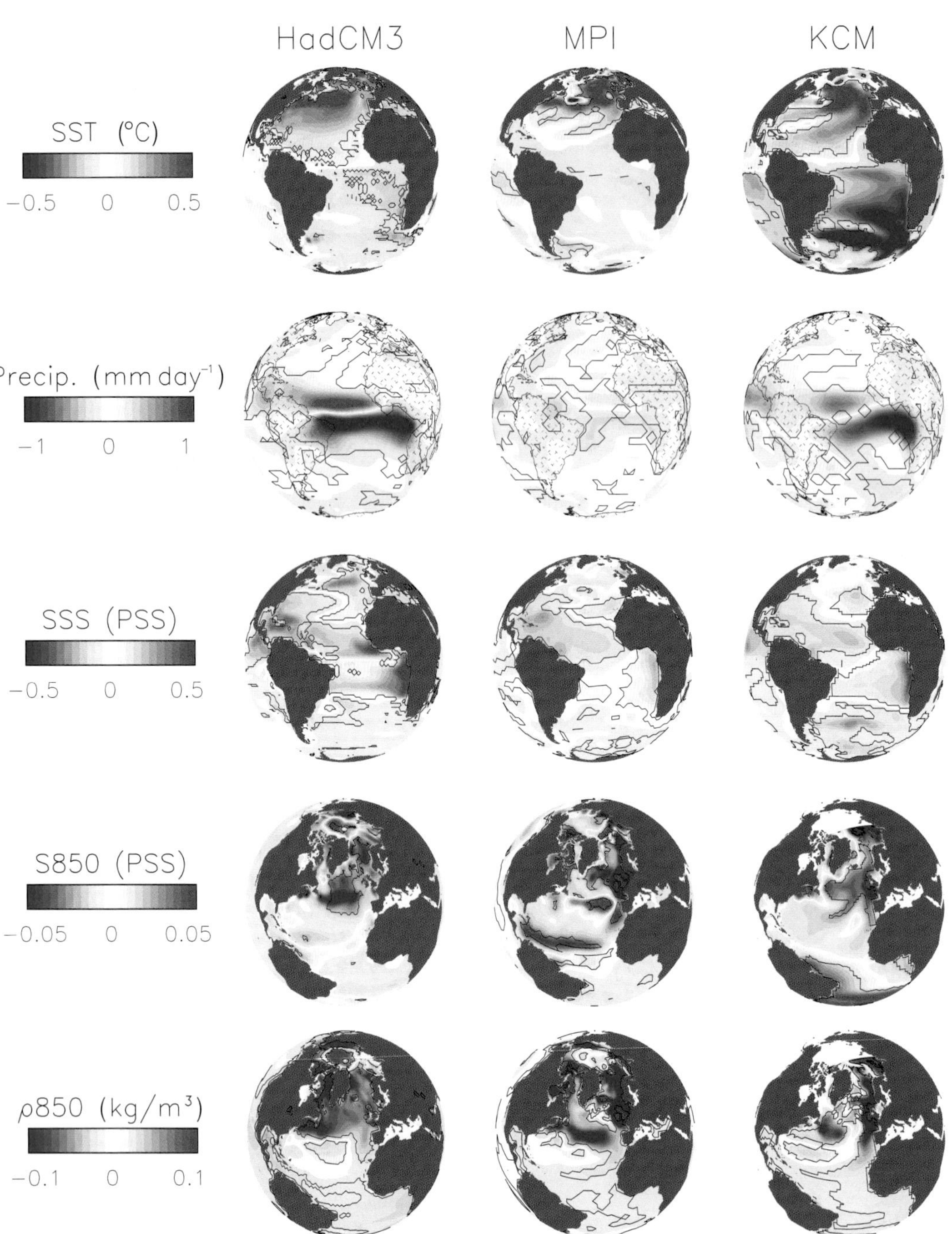

Figure 9.34 | Sequence of physical links postulated to connect Atlantic Meridional Overturning Circulation (AMOC) and Atlantic Multi-decadal Variability (AMV), and how they are represented in three climate models. Shown are regression patters for the following quantities (from top to bottom): sea surface temperature (SST) composites using AMOC time series; precipitation composites using cross-equatorial SST difference time series; equatorial salinity composites using Intertropical Convergence Zone (ITCZ)-strength time series; subpolar-gyre depth-averaged salinity (top 800 to 1000 m) using equatorial salinity time series; subpolar gyre depth averaged density using subpolar gyre depth averaged salinity time series. From left to right: the two CMIP3 models HadCM3 and ECHAM/MPI-OM (MPI), and the non-CMIP model KCM. Black outlining signifies areas statistically significant at the 5% level for a two-tailed t test using the moving-blocks bootstrapping technique (Wilks, 1995). (Figure 3 from Menary et al., 2012.)

9.5.3.3.3 Tropical zonal and meridional modes

The Atlantic Meridional Mode (AMM) is the dominant mode of interannual variability in the tropical Atlantic, is characterized by an anomalous meridional shift in the ITCZ (Chiang and Vimont, 2004), and has impacts on hurricane tracks over the North Atlantic (Xie et al., 2005; Smirnov and Vimont, 2011). Virtually all CMIP models simulate AMM-like SST variability in their 20th century climate simulations. However, most models underestimate the SST variance associated with the AMM, and position the north tropical Atlantic SST anomaly too far equatorward. More problematic is the fact that the development of the AMM in many models is led by a zonal mode during boreal winter—a feature that is not observed in nature (Breugem et al., 2006). This spurious AMM behaviour in the models is expected to be associated with the severe model biases in simulating the ITCZ (see Section 9.4.2.5.2).

Atlantic Niño

CMIP3 models have considerable difficulty simulating Atlantic Niño in their 20th century climate simulations. For many models the so-called 'Atl-3' SST index (20°W to 0°W, 3°S to 3°N) displays the wrong seasonality, with the maximum value in either DJF or SON instead of JJA as is observed (Breugem et al., 2006). Despite large biases in the simulated climatology (Section 9.4.2.5.2), about one third of CMIP5 models capture some aspects of Atlantic Niño variability, including amplitude, spatial pattern and seasonality (Richter et al., 2013). This represents an improvement over CMIP3.

9.5.3.4 Indo-Pacific Modes

9.5.3.4.1 El Niño-Southern Oscillation

The ENSO phenomenon is the dominant mode of climate variability on seasonal to interannual time scales (see Wang and Picaut (2004) and Chapter 14). The representation of ENSO in climate models has steadily improved and now bears considerable similarity to observed ENSO properties (AchutaRao and Sperber, 2002; Randall et al., 2007; Guilyardi et al., 2009b). However, as was the case in the AR4, simulations

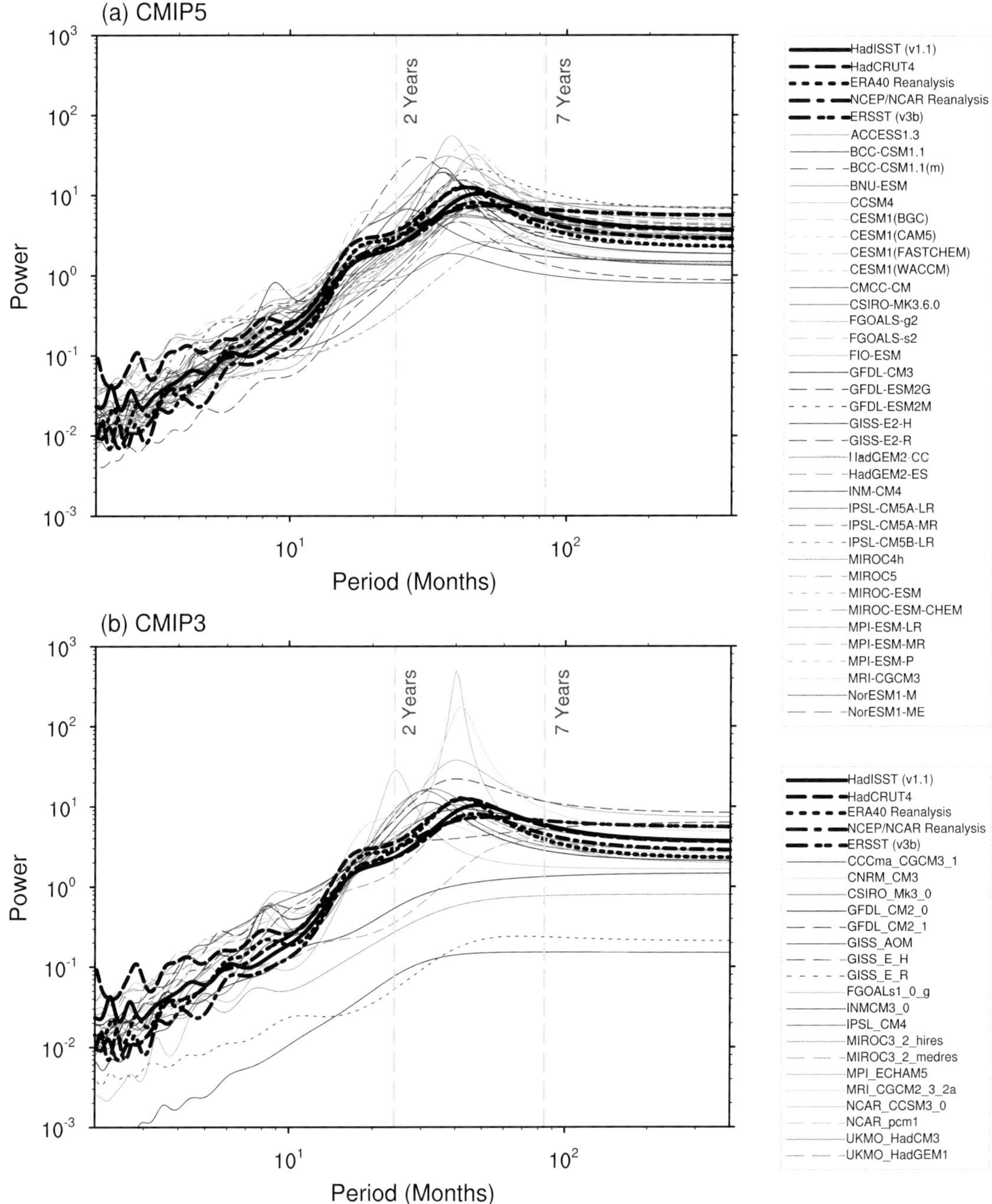

Figure 9.35 | Maximum entropy power spectra of surface air temperature averaged over the NINO3 region (5°N to 5°S, 150°W to 90°W) for (a) the CMIP5 models and (b) the CMIP3 models. ECMWF reanalysis in (a) refers to the European Centre for Medium Range Weather Forecasts (ECMWF) 15-year reanalysis (ERA15). The vertical lines correspond to periods of two and seven years. The power spectra from the reanalyses and for SST from the Hadley Centre Sea Ice and Sea Surface Temperature (HadISST) version 1.1, Hadley Centre/Climatic Research Unit gridded surface temperature data set 4 (HadCRU 4), ECMWF 40-year reanalysis (ERA40) and National Centers for Environmental Prediction/National Center for Atmospheric Research (NCEP/NCAR) data set are given by the series of black curves. (Adapted from AchutaRao and Sperber, 2006.)

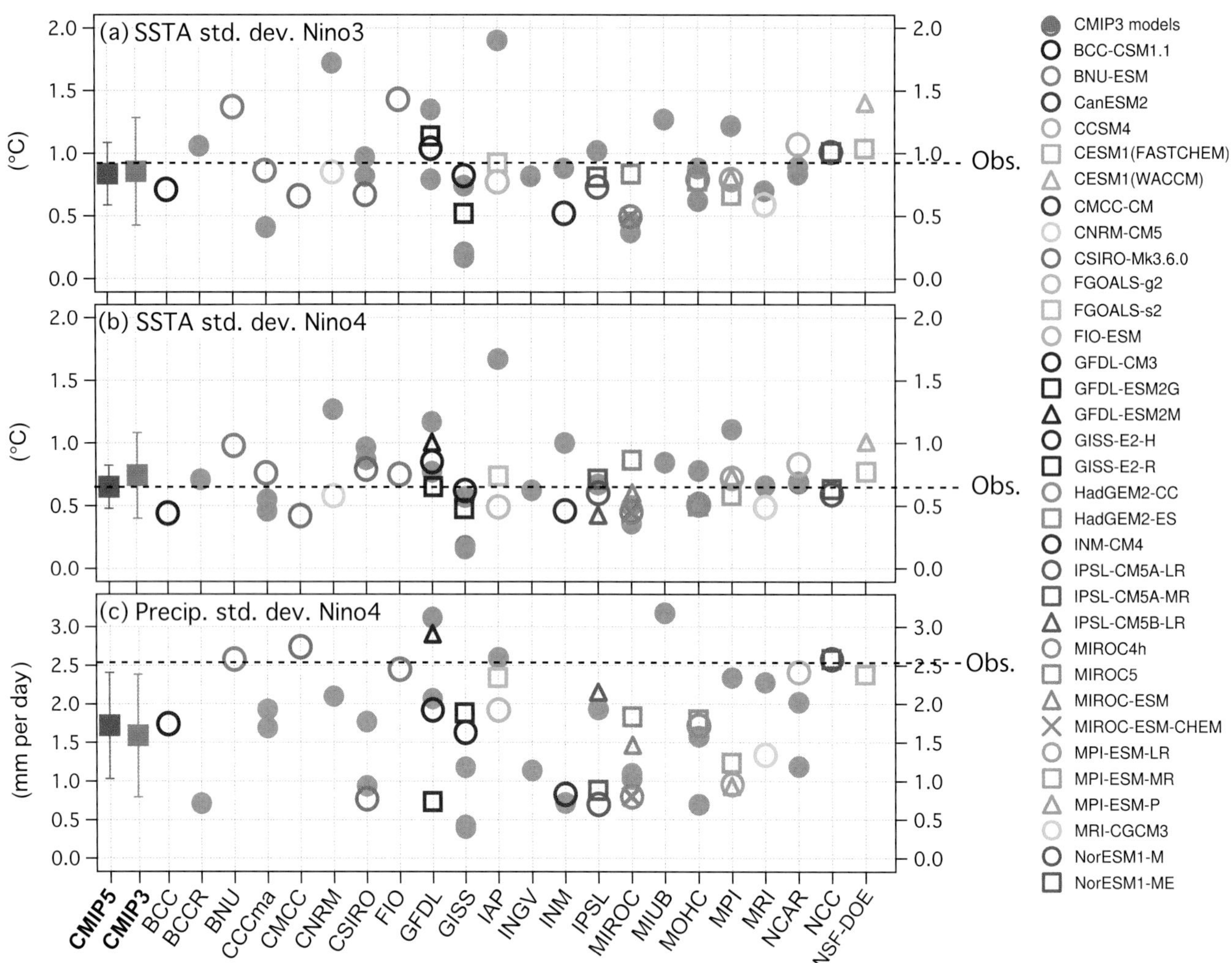

Figure 9.36 | ENSO metrics for pre-industrial control simulations in CMIP3 and CMIP5. (a) and (b): SST anomaly standard deviation (°C) in Niño 3 and Niño 4, respectively, (c) precipitation response (standard deviation, mm/day) in Niño4. Reference data sets, shown as dashed lines: Hadley Centre Sea Ice and Sea Surface Temperature (HadISST) version 1.1 for (a) and (b), CPC Merged Analysis of Precipitation (CMAP) for (c). The CMIP5 and CMIP3 multi-model means are shown as squares on the left of each panel with the whiskers representing the model standard deviation. Individual CMIP3 models shown as filled grey circles, and individual CMIP5 models are identified in the legend.

of both background climate (time mean and seasonal cycle, see Section 9.4.2.5.1) and internal variability exhibit serious systematic errors (van Oldenborgh et al., 2005; Capotondi et al., 2006; Guilyardi, 2006; Wittenberg et al., 2006; Watanabe et al., 2011; Stevenson et al., 2012; Yeh et al., 2012), many of which can be traced to the representation of deep convection, trade wind strength and cloud feedbacks, with little improvement from CMIP3 to CMIP5 (Braconnot et al., 2007a; L'Ecuyer and Stephens, 2007; Guilyardi et al., 2009a; Lloyd et al., 2009, 2010; Sun et al., 2009; Zhang and Jin, 2012).

While a number of CMIP3 models do not exhibit an ENSO variability maximum at the observed 2- to 7- year time scale, most CMIP5 models do have a maximum near the observed range and fewer models have the tendency for biennial oscillations (Figure 9.35; see also Stevenson, 2012). In CMIP3 the amplitude of El Niño ranged from less than half to more than double the observed amplitude (van Oldenborgh et al., 2005; AchutaRao and Sperber, 2006; Guilyardi, 2006; Guilyardi et al., 2009b). By contrast, the CMIP5 models show less inter-model spread (Figure 9.36; Kim and Yu, 2012). The CMIP5 models still exhibit errors in ENSO amplitude, period, irregularity, skewness, spatial patterns (Lin, 2007; Leloup et al., 2008; Guilyardi et al., 2009b; Ohba et al., 2010; Yu and Kim, 2011; Su and Jiang, 2012) or teleconnections (Watanabe et al., 2012; Weller and Cai, 2013a).

Since AR4, new analysis methods have emerged and are now being applied. For example, Jin et al. (2006) and Kim and Jin (2011a) identified five different feedbacks affecting the Bjerknes (or BJ) index, which in turn characterizes ENSO stability. Kim and Jin (2011b) applied this process-based analysis to the CMIP3 multi-model ensemble and demonstrated a significant positive correlation between ENSO amplitude and the BJ index. When respective components of the BJ index obtained from the coupled models were compared with those from observations, it was shown that most coupled models underestimated the negative thermal damping feedback (Lloyd et al., 2012; Chen et al., 2013) and the positive zonal advective and thermocline feedbacks.

Detailed quantitative evaluation of ENSO performance is hampered by the short observational record of key processes (Wittenberg, 2009; Li et al., 2011b; Deser et al., 2012) and the complexity and diversity of the processes involved (Wang and Picaut, 2004). While shortcomings remain (Guilyardi et al., 2009b), the CMIP5 model ensemble shows some improvement compared to CMIP3, but there has been no major breakthrough and the multi-model improvement is mostly due to a reduced number of poor-performing models.

9.5.3.4.2 Indian Ocean basin and dipole modes

Indian Ocean SST displays a basin-wide warming following El Niño (Klein et al., 1999). This Indian Ocean basin (IOB) mode peaks in boreal spring and persists through the following summer. Most CMIP5 models capture this IOB mode, an improvement over CMIP3 (Du et al., 2013). However, only about half the CMIP5 models capture its long temporal persistence, and these models tend to simulate ENSO-forced ocean Rossby waves in the tropical south Indian Ocean (Zheng et al., 2011).

The Indian Ocean zonal dipole mode (IOD) (Saji et al., 1999; Webster et al., 1999) appears to be part of a hemispheric response to tropical atmospheric forcing (Fauchereau et al., 2003; Hermes and Reason, 2005). Most CMIP3 models are able to reproduce the general features of the IOD, including its phase lock onto the July to November season (Saji et al., 2006). The modelled SST anomalies, however, tend to show too strong a westward extension along the equator in the eastern Indian Ocean. CMIP3 models exhibit considerable spread in IOD amplitude, some of which can be explained by differences in the strength of the simulated Bjerknes feedback (Liu et al., 2011; Cai and Cowan, 2013). No substantial change is seen in CMIP5 (Weller and Cai, 2013a).

Many models simulate the observed correlation between IOD and ENSO. The magnitude of this correlation varies substantially between models, but is apparently not tied to the amplitude of ENSO (Saji et al., 2006). A subset of CMIP3 models show a spurious correlation with ENSO following the decay of ENSO events, instead of during the ENSO developing phase, possibly due to erroneous representation of oceanic pathways connecting the equatorial Pacific and Indian Oceans (Cai et al., 2011).

9.5.3.4.3 Tropospheric biennial oscillation

The tropospheric biennial oscillation (TBO, Section 14.7.4) is a biennial tendency of many phenomena in the Indo-Pacific region that affects droughts and floods over large areas of south Asia and Australia (e.g., Chang and Li, 2000; Li et al., 2001; Meehl et al., 2003). The IOD involves regional patterns of SST anomalies in the TBO in the Indian Ocean during the northern fall season following the south Asian monsoon (Loschnigg et al., 2003). The TBO has been simulated in a number of global coupled climate model simulations (e.g., Ogasawara et al., 1999; Loschnigg et al., 2003; Nanjundiah et al., 2005; Turner et al., 2007; Meehl and Arblaster, 2011).

9.5.3.5 Indo-Pacific Teleconnections

Tropical SST variability provides a significant forcing of atmospheric teleconnections and drives a large portion of the climate variability over land (Goddard and Mason, 2002; Shin et al., 2010). Although local forcings and feedbacks can play an important role (Pitman et al., 2012a), the simulation of land surface temperatures and precipitation requires accurate predictions of SST patterns (Compo and Sardeshmukh, 2009; Shin et al., 2010) as well as zonal wind variability patterns (Handorf and Dethloff, 2012). Teleconnections hence play a central role in regional climate change (see Chapter 14).

9.5.3.5.1 Teleconnections affecting North America

The Pacific North American (PNA) pattern is a wavetrain-like pattern in mid-level geopotential heights. The majority of CMIP3 models simulate the spatial structure of the PNA pattern in wintertime (Stoner et al., 2009). The PNA pattern has contributions from both internal atmospheric variability (Johnson and Feldstein, 2010) and ENSO and PDO teleconnections (Deser et al., 2004). The power spectrum of this temporal behaviour is generally captured by the CMIP3 models, although the level of year-to-year autocorrelation varies according to the strength of the simulated ENSO and PDO (Stoner et al., 2009).

9.5.3.5.2 Tropical ENSO teleconnections

These moist teleconnection pathways involve mechanisms related to those at play in the precipitation response to global warming (Chiang and Sobel, 2002; Neelin et al., 2003) and provide challenging test statistics for model precipitation response. Compared to earlier generation climate models, CMIP3 and CMIP5 models tend to do somewhat better (Neelin, 2007; Cai et al., 2009; Coelho and Goddard, 2009; Langenbrunner and Neelin, 2013) at precipitation reductions associated with El Niño over equatorial South America and the Western Pacific, although CMIP5 offers little further improvement over CMIP3 (see for instance the standard deviation of precipitation in the western Pacific in Figure 9.36). CMIP5 models simulate the sign of the precipitation change over broad regions, and do well at predicting the amplitude of the change (for a given SST forcing) (Langenbrunner and Neelin, 2013).

A regression of the West African monsoon precipitation index with global SSTs reveals two major teleconnections (Fontaine and Janicot, 1996). The first highlights the strong influence of ENSO, while the second reveals a relationship between the SST in the Gulf of Guinea and the northward migration of the monsoon rain belt over West Africa. Most CMIP3 models show a single dominant Pacific teleconnection, which is, however, of the wrong sign for half of the models (Joly et al., 2007). Only one model shows a significant second mode, emphasizing the difficulty in simulating the response of the African rain belt to Atlantic SST anomalies that are not synchronous with Pacific anomalies.

Both CMIP3 and CMIP5 models have been evaluated and found to vary in their abilities to represent both the seasonal cycle of correlations between the Niño 3.4 and North Australian SSTs (Catto et al., 2012a, 2012b) with little change in quality from CMIP3 to CMIP5. Generally the models do not capture the strength of the negative correlations during the second half of the year. The models also still struggle to capture the SST evolution in the North Australian region during El Niño and La Niña. Teleconnection patterns from both ENSO and the Indian Ocean Dipole to precipitation over Australia are reasonably well simulated in the key September-November season (Cai et al., 2009; Weller and Cai, 2013b) in the CMIP3 and CMIP5 multi-model mean.

9.5.3.6 Pacific Decadal Oscillation and Interdecadal Pacific Oscillation

The Pacific Decadal Oscillation (PDO) refers to a mode of variability involving SST anomalies over the North Pacific (north of 20°N) (Mantua et al., 1997). Although the PDO time series exhibits considerable decadal variability, it is difficult to ascertain whether there are any robust spectral peaks given the relatively short observational record (Minobe, 1997, 1999; Pierce, 2001; Deser et al., 2004). The ability of climate models to represent the PDO has been assessed by Stoner et al. (2009) and Furtado et al. (2011). Their results indicate that approximately half of the CMIP3 models simulate the observed spatial pattern and temporal behaviour (e.g., enhanced variance at low frequencies); however, spectral peaks are consistently higher in frequency than those suggested by the short observational record. The modelled PDO correlation with SST anomalies in the tropical Indo-Pacific are strongly underestimated by the CMIP3 models (Wang et al., 2010; Deser et al., 2011; Furtado et al., 2011; Lienert et al., 2011). Climate models have been shown to simulate features of the closely related Interdecadal Pacific Oscillation (IPO, based on SSTs over the entire Pacific basin; see Section 14.7.3; Power and Colman, 2006; Power et al., 2006; Meehl et al., 2009), although deficiencies remain in the strength of the tropical–extratropical connections.

9.5.3.7 The Quasi-Biennial Oscillation

Significant progress has been made in recent years to model and understand the impacts of the Quasi-Biennial Oscillation (QBO; Baldwin et al., 2001). Many climate models have now increased their vertical domain and/or improved their physical parameterizations (see Tables 9.1 and 9.A.1), and some of these reproduce a QBO (e.g., HadGEM2, MPI-ESM-LR, MIROC). Many features of the QBO such as its width and phase asymmetry also appear spontaneously in these simulations due to internal dynamics (Dunkerton, 1991; Scaife et al., 2002; Haynes, 2006). Some of the QBO effects on the extratropical climate (Holton and Tan, 1980; Hamilton, 1998; Naoe and Shibata, 2010) as well as ozone (Butchart et al., 2003; Shibata and Deushi, 2005) are also reproduced in models.

9.5.3.8 Summary

In summary, most modes of interannual to interdecadal variability are now present in climate models. As in AR4, their assessment presents a varied picture and CMIP5 models only show a modest improvement over CMIP3, mostly due to fewer poor-performing models. New since the AR4, process-based model evaluation is now helping identify sources of specific biases, although the observational record is often too short or inaccurate to offer strong constraints. The assessment of modes and patterns is summarized in Table 9.4.

9.5.4 Extreme Events

Extreme events are realizations of the tail of the probability distribution of weather and climate variability. They are higher-order statistics and thus generally more difficult to realistically represent in climate models. Shorter time scale extreme events are often associated with smaller scale spatial structure, which may be better represented as model resolution increases. In the AR4, it was concluded that models could simulate the statistics of extreme events better than expected from the generally coarse resolution of the models at that time, especially for temperature extremes (Randall et al., 2007).

Table 9.4 | Summary of assessment of interannual to interdecadal variability in climate models. See also Figure 9.44.

	Short Name	Level of Confidence	Level of Evidence for Evaluation	Degree of Agreement	Model Quality	Difference with AR4 (including CMIP5 vs. CMIP3)	Section
Global sea surface temperature (SST) variability	SST-var	*High*	*Robust*	*Medium*	Medium	Slight improvement in the tropics	9.5.3.1
North Atlantic Oscillation and Northern Annular Mode	NAO	*Medium*	*Medium*	*Medium*	High	No assessment	9.5.3.2
Southern Annular Mode	SAM	*Low*	*Limited*	*Medium*	Medium	No assessment	9.5.3.2
Atlantic Meridional Overturning Circulation Variability	AMOC-var	*Low*	*Limited*	*Medium*	Medium	No improvement	9.5.3.3
Atlantic Multi-decadal Variability	AMO	*Low*	*Limited*	*Medium*	Medium	No improvement	9.5.3.3
Atlantic Meridional Mode	AMM	*High*	*Medium*	*High*	Low	No assessment	9.5.3.3
Atlantic Niño	AN	*Low*	*Limited*	*Medium*	Low	Slight improvement	9.5.3.3
El Niño Southern Oscillation	ENSO	*High*	*Medium*	*High*	Medium	Slight improvement	9.5.3.4
Indian Ocean Basin mode	IOB	*Medium*	*Medium*	*Medium*	High	Slight improvement	9.5.3.4
Indian Ocean Dipole	IOD	*Medium*	*Medium*	*Medium*	Medium	No improvement	9.5.3.4
Pacific North American	PNA	*High*	*Medium*	*High*	Medium	Slight improvement	9.5.3.5
Tropical ENSO teleconnections	ENSOtele	*High*	*Robust*	*Medium*	Medium	No improvement	9.5.3.5
Pacific Decadal Oscillation	PDO	*Low*	*Limited*	*Medium*	Medium	No assessment	9.5.3.6
Interdecadal Pacific Oscillation	IPO	*Low*	*Limited*	*Medium*	High	No assessment	9.5.3.6
Quasi-Biennial Oscillation	QBO	*Medium*	*Medium*	*Medium*	High	No assessment	9.5.3.7

The IPCC has conducted an assessment of extreme events in the context of climate change—the Special Report on Managing the Risks of Extreme Events and Disasters to Advance Climate Change Adaptation (SREX) (IPCC, 2012). Although there is no comprehensive climate model evaluation with respect to extreme events in SREX, there is some consideration of model performance taken into account in assessing uncertainties in projections.

9.5.4.1 Extreme Temperature

Since the AR4, evaluation of CMIP3 and CMIP5 models has been undertaken with respect to temperature extremes. Both model ensembles simulate present-day warm extremes, in terms of 20-year return values, reasonably well, with errors typically within a few degrees Celsius over most of the globe (Kharin et al., 2007; Kharin et al., 2012). The CMIP5 and CMIP3 models perform comparably for various temperature extreme indices, but with smaller inter-model spread in CMIP5. The inter-model range of simulated indices is similar to the spread amongst observationally based estimates in most regions (Sillmann et al., 2013). Figure 9.37 shows relative error estimates of available CMIP5 models for various extreme indices based on Sillmann et al. (2013). Although the relative performance of an individual model may depend on the choice of the reference data set (four different reanalyses are used), the mean and median models tend to outperform individual models. According to the standardized multi-model median errors ($RMSE_{std}$) for CMIP3 and CMIP5 shown on the right side of Figure 9.37, the performance of the two ensembles is similar.

In terms of historical trends, CMIP3 and CMIP5 models generally capture observed trends in temperature extremes in the second half of the 20th century (Sillmann et al., 2013), as illustrated in Figure 9.37. The modelled trends are consistent with both reanalyses and station-based estimates. It is also clear in the figure that model-based indices respond coherently to major volcanic eruptions. Detection and attribution studies based on CMIP3 models suggest that models tend to overestimate the observed warming of warm temperature extremes and underestimate the warming of cold extremes in the second half of 20th century (Christidis et al., 2011; Zwiers et al., 2011) as noted in SREX (Seneviratne et al., 2012). See also Chapter 10. This is not as obvious in the CMIP5 model evaluation (Figure 9.37 and Sillmann et al. (2013)) and needs further investigation.

9.5.4.2 Extreme Precipitation

For extreme precipitation, observational uncertainty is much larger than for temperature, making model evaluation more challenging. Discrepancies between different reanalyses for extreme precipitation are substantial, whereas station-based observations have limited spatial coverage (Kharin et al., 2007, 2012; Sillmann et al., 2013). Moreover, a station-based observational data set, which is interpolated from station measurements, has a potential mismatch of spatial scale when compared to model results or reanalyses (Chen and Knutson, 2008). Uncertainties are especially large in the tropics. In the extratropics, precipitation extremes in terms of 20-year return values simulated by CMIP3 and CMIP5 models compared relatively well with the observational data sets, with typical discrepancies in the 20% range (Kharin et al., 2007, 2012). Figure 9.37 shows relative errors of CMIP5 models for five precipitation-related indices. Darker grey shadings in the RMSE columns for precipitation indicate larger discrepancies between models and reanalyses for precipitation extremes in general. Sillmann et al. (2013) found that the CMIP5 models tend to simulate more intense precipitation and fewer consecutive wet days than the CMIP3, and thus are closer to the observationally based indices.

It is known from sensitivity studies that simulated extreme precipitation is strongly dependent on model resolution. Growing evidence has shown that high-resolution models (50 km or finer in the atmosphere) can reproduce the observed intensity of extreme precipitation (Wehner et al., 2010; Endo et al., 2012; Sakamoto et al., 2012), though some of these results are based on models with observationally constrained surface or lateral boundary conditions (i.e., Atmospheric General Circulation Models (AGCMs) or Regional Climate Models (RCMs)).

In terms of historical trends, a detection and attribution study by Min et al. (2011) found consistency in sign between the observed increase in heavy precipitation over NH land areas in the second half of the 20th century and that simulated by CMIP3 models, but they found that the models tend to underestimate the magnitude of the trend (see also Chapter 10). Related to this, it has been pointed out from comparisons to satellite-based data sets that the majority of models underestimate the sensitivity of extreme precipitation intensity to temperature in the tropics (Allan and Soden, 2008; Allan et al., 2010; O'Gorman, 2012) and globally (Liu et al., 2009; Shiu et al., 2012). O'Gorman (2012) showed that this implies possible underestimation of the projected future increase in extreme precipitation in the tropics.

9.5.4.3 Tropical Cyclones

It was concluded in the AR4 that high-resolution AGCMs generally reproduced the frequency and distribution, but underestimated intensity of tropical cyclones (Randall et al., 2007). Since then, Mizuta et al. (2012) have shown that a newer version of the MRI-AGCM with improved parameterizations (at 20 km horizontal resolution) simulates tropical cyclones as intense as those observed with improved distribution as well. Another remarkable finding since the AR4 is that the observed year-to-year count variability of Atlantic hurricanes can be well simulated by modestly high resolution (100 km or finer) AGCMs forced by observed SST, though with less skill in other basins (Larow et al., 2008; Zhao et al., 2009; Strachan et al., 2013). Vortices that have some characteristics of tropical cyclones can also be detected and tracked in AOGCMs in CMIP3 and 5, but their intensities are generally too weak (Yokoi et al., 2009a; Yokoi et al., 2012; Tory et al., 2013; Walsh et al., 2013).

9.5.4.4 Droughts

Drought is caused by long time scale (months or longer) variability of both precipitation and evaporation. Sheffield and Wood (2008) found that models in the CMIP3 ensemble simulated large-scale droughts in the 20th century reasonably well, in the sense that multi-model spread includes the observational estimate in each of several regions. However, it should be noted that there are various definitions of drought (see Chapter 2 and Seneviratne et al., 2012) and the performance of simulated drought can depend on the definition. Moreover, different models

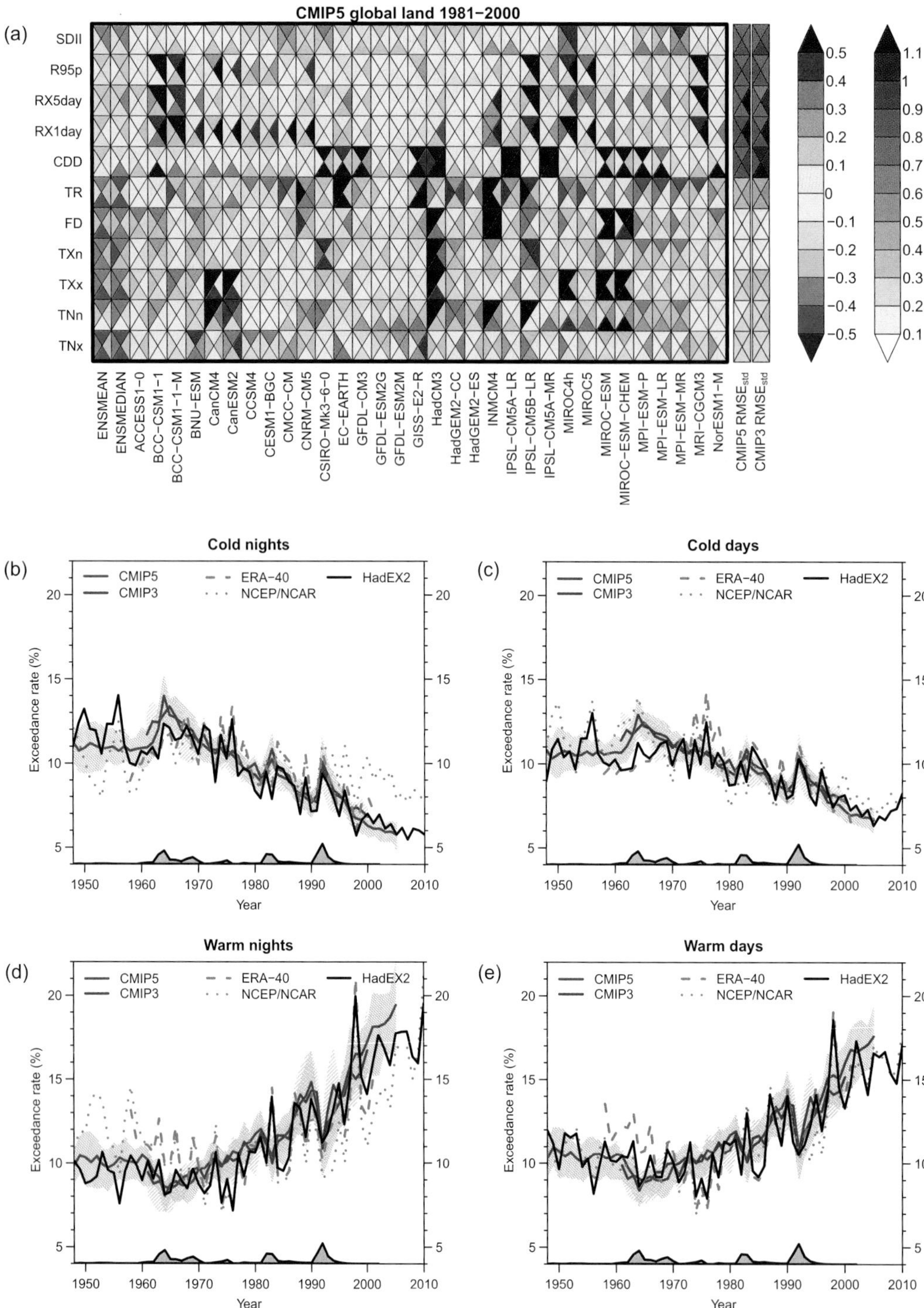

Figure 9.37 | (a) Portrait plot of relative error metrics for the CMIP5 temperature and precipitation extreme indices based on Sillmann et al. (2013). (b)–(e) Time series of global mean temperature extreme indices over land from 1948 to 2010 for CMIP3 (blue) and CMIP5 (red) models, ECMWF 40-year reanalysis (ERA40, green dashed) and National Centers for Environmental Prediction/National Center for Atmospheric Research (NCEP/NCAR, green dotted) reanalyses and HadEX2 station-based observational data set (black) based on Sillmann et al. (2013). In (a), reddish and bluish colours indicate, respectively, larger and smaller root-mean-square (RMS) errors for an individual model relative to the median model. The relative error is calculated for each observational data set separately. The grey-shaded columns on the right side indicate the RMS error for the multi-model median standardized by the spatial standard deviation of the index climatology in the reanalysis, representing absolute errors for CMIP3 and CMIP5 ensembles. Results for four different reference data sets, ERA-interim (top), ERA40 (left), NCEP/NCAR (right) and NCEP- Department of Energy (DOE) (bottom) reanalyses, are shown in each box. The analysis period is 1981–2000, and only land areas are considered. The indices shown are simple daily precipitation intensity index (SDII), very wet days (R95p), annual maximum 5-day/1-day precipitation (RX5day/RX1day), consecutive dry days (CDD), tropical nights (TR), frost days (FD), annual minimum/maximum daily maximum surface air temperature (TXn/TXx), and annual minimum/maximum daily minimum surface air temperature (TNn/TNx). See Box 2.4 for the definitions of indices. Note that only a small selection of the indices analysed in Sillmann et al. (2013) is shown, preferentially those that appear in Chapters 2, 10, 11, 12, 14. Also note that the NCEP/NCAR reanalysis has a known defect for TXx (Sillmann et al., 2013), but its impact on this figure is small. In (b)–(e), shading for model results indicates the 25th to 75th quantile range of inter-model spread. Grey shading along the horizontal axis indicates the evolution of globally averaged volcanic forcing according to Sato et al. (1993). The indices shown are the frequency of daily minimum/maximum surface air temperature below the 10th percentile (b: Cold nights/c: Cold days) and that above 90th percentile (d: Warm nights/e: Warm days) of the 1961–1990 base period. Note that, as these indices essentially represent changes relative to the base period, they are particularly suitable for being shown in time series and not straightforward for being shown in (a).

can simulate drought with different mechanisms (McCrary and Randall, 2010; Taylor et al., 2012a). A comprehensive evaluation of CMIP5 models for drought is currently not available, although Sillmann et al. (2013) found that consecutive dry days simulated by CMIP5 models are comparable to observations in magnitude and distribution.

9.5.4.5 Summary

There is *medium evidence* (i.e., a few multi-model studies) and *high agreement* that the global distribution of temperature extremes are represented well by CMIP3 and CMIP5 models. The observed global warming trend of temperature extremes in the second half of the 20th century is reproduced in models, but there is *medium evidence* (a few CMIP3 studies) and *medium agreement* (not evident in a preliminary look at CMIP5) that models tend to overestimate the warming of warm temperature extremes and underestimate the warming of cold temperature extremes.

There is *medium evidence* (single multi-model study) and *medium agreement* (as inter-model difference is large) that CMIP5 models tend to simulate more intense and thus more realistic precipitation extremes than CMIP3, which could be partly due to generally higher horizontal resolution. There is *medium evidence* and *high agreement* that CMIP3 models tend to underestimate the sensitivity of extreme precipitation intensity to temperature. There is *medium evidence* and *high agreement* that high resolution (50 km or finer) AGCMs tend to simulate the intensity of extreme precipitation comparable to observational estimates.

There is *medium evidence* and *high agreement* that year-to-year count variability of Atlantic hurricanes can be well simulated by modestly high resolution (100 km or finer) AGCMs forced by observed SSTs. There is *medium evidence* and *medium agreement* (as inter-model difference is large) that the intensity of tropical cyclones is too weak in CMIP3 and CMIP5 models. Finally, there is *medium evidence* (a few multi-model studies) and *medium agreement* (as it might depend on definitions of drought) that models can simulate aspects of large-scale drought.

Box 9.3 | Understanding Model Performance

This Box provides a synthesis of findings on understanding model performance based on the model evaluations discussed in this chapter.

Uncertainty in Process Representation

Some model errors can be traced to uncertainty in representation of processes (parameterizations). Some of these are long-standing issues in climate modelling, reflecting our limited, though gradually increasing, understanding of very complex processes and the inherent challenges in mathematically representing them. For the atmosphere, cloud processes, including convection and its interaction with boundary layer and larger-scale circulation, remain major sources of uncertainty (Section 9.4.1). These in turn cause errors or uncertainties in radiation which propagate through the coupled climate system. Distribution of aerosols is also a source of uncertainty arising from modelled microphysical processes and transport (Sections 9.4.1 and 9.4.6). Ocean models are subject to uncertainty in parameterizations of vertical and horizontal mixing and convection (Sections 9.4.2, 9.5.2 and 9.5.3), and ocean errors in turn affect the atmosphere through resulting SST biases. Simulation of sea ice is also affected by errors in both the atmosphere and the ocean as well as the parameterization of sea ice itself (Section 9.4.3). With respect to biogeochemical components in Earth System Models (ESMs), parameterizations of nitrogen limitation and forest fires are thought to be important for simulating the carbon cycle, but very few ESMs incorporate these so far (Sections 9.4.4 and 9.4.5).

Error Propagation

Causes of one model bias can sometimes be associated with another. Although the root cause of those biases is often unclear, knowledge of the causal chain or a set of interrelated biases can provide a key to further understanding and improvement of model performance. For example, biases in storm track position are partly due to a SST biases in the Gulf Stream and Kuroshio Current (Section 9.4.1). Some biases in variability or trend can be partly traced back to biases in mean states. The decreasing trend in September Arctic ice extent tends to be underestimated when sea ice thickness is overestimated (Section 9.4.3). In such cases, improvement of the mean state may improve simulated variability or trend.

Sensitivity to Resolution

Some phenomena or aspects of climate are found to be better simulated with models run at higher horizontal and/or vertical resolution. In particular, increased resolution in the atmosphere has improved, at least in some models, storm track and extratropical cyclones (Section 9.4.1), diurnal variation of precipitation over land (Section 9.5.2), extreme precipitation, and tropical cyclone intensity and structure (Section 9.5.4). Similarly, increased horizontal resolution in the ocean is shown to improve sea surface height variability, western boundary currents, tropical instability waves and coastal upwelling (Section 9.4.2), and variability of Atlantic meridional overturning circulation (Section 9.5.3). High vertical resolution and a high model top, as well as high horizontal resolution, are important for simulating lower stratospheric climate variability (Section 9.4.1), blocking (Section 9.5.2), the Quasi-Biennial Oscillation and the North Atlantic Oscillation (Section 9.5.3). *(continued on next page)*

Box 9.3 (continued)

Uncertainty in Observational Data

In some cases, insufficient length or quality of observational data makes model evaluation challenging, and is a frequent problem in the evaluation of simulated variability or trends. This is evident for evaluation of upper tropical tropospheric temperature, tropical atmospheric circulation (Section 9.4.1), the Atlantic meridional overturning circulation, the North Atlantic Oscillation and the Pacific Decadal Oscillation (Section 9.5.3). Data quality has been pointed out as an issue for arctic cloud properties (Section 9.4.1), ocean heat content, heat and fresh water fluxes over the ocean (Section 9.4.2) and extreme precipitation (Section 9.5.4). Palaeoclimate reconstructions also have large inherent uncertainties (Section 9.5.2). It is clear therefore that updated or newly available data affect model evaluation conclusions.

Other Factors

Model evaluation can be affected by how models are forced. Uncertainties in specified greenhouse gases, aerosols emissions, land use change, etc. will all affect model results and hence evaluation of model quality. Different statistical methods used in model evaluation can also lead to subtle or substantive differences in the assessment of model quality.

9.6 Downscaling and Simulation of Regional-Scale Climate

Regional-scale climate information can be obtained directly from global models; however, their horizontal resolution is often too low to resolve features that are important at regional scales. High-resolution AGCMs, variable-resolution global models, and statistical and dynamical downscaling (i.e., regional climate modelling) are used to complement AOGCMs, and to generate region-specific climate information. These approaches are evaluated in the following.

9.6.1 Global Models

9.6.1.1 Regional-Scale Simulation by Atmosphere–Ocean General Circulation Models

A comparison of CMIP3 and CMIP5 seasonal cycles of temperature and precipitation for different regions (Figure 9.38) shows that temperature is generally better simulated than precipitation in terms of the amplitude and phase of the seasonal cycle. The multi-model mean is closer to observations than most of the individual models. The systematic difference between the CMIP5 and CMIP3 ensembles is small in most regions, although there is evident improvement in South Asia (SAS) and Tropical South America (TSA) in the rainy seasons. In some cases the spread amongst observational estimates can be of comparable magnitude to the model spread, e.g., winter in the Europe and Mediterranean (EUM) region.

There are as yet rather few published studies in which regional behaviour of the CMIP5 models is evaluated in great detail. Cattiaux et al. (2013) obtained results for Europe similar to those discussed above. Joetzjer et al. (2013) considered 13 models that participated in both CMIP3 and in CMIP5 and found that the seasonal cycle of precipitation over the Amazon improved in the latter.

Based on the CMIP archives, regional biases in seasonal and annual mean temperature and precipitation are shown for several land regions in Figure 9.39, and for polar and oceanic regions in Figure 9.40. The CMIP5 median temperature biases range from about –3°C to 1.5°C. Substantial cold biases over NH regions are more prevalent in winter (December to February) than summer (June to August). The median biases appear in most cases slightly less negative for CMIP5 than CMIP3. The spread amongst models, as characterized by the 25 to 75% and 5 to 95% ranges, is slightly reduced from CMIP3 to CMIP5 in a majority of the regions and is roughly ±3°C. The RMS error of individual CMIP5 models is smaller than that for CMIP3 in 24 of the 26 regions in Figure 9.39 in DJF, JJA and the annual mean. The absolute value of the ensemble mean bias has also been reduced in most cases. The inter-model spread remains large, particularly in high-latitude regions in winter and in regions with steep orography (such as CAS, SAS, TIB and WSA). The inter-model temperature spread has decreased from CMIP3 to CMIP5 over most of the oceans and over the Arctic and Antarctic land regions. The cold winter bias over the Arctic has been reduced. There is little systematic inter-ensemble difference in temperature over lower latitude oceans.

Biases in seasonal total precipitation are shown in the right column of Figures 9.39 and 9.40 for the NH winter (October to March) and summer (April to September) half years as well as the annual mean. The largest systematic biases over land regions occur in ALA, WSA and TIB, where the annual precipitation exceeds that observed in all CMIP5 models, with a median bias on the order of 100%. All these regions are characterized by high orography and / or a large fraction of solid precipitation, both of which are expected to introduce a negative bias in gauge-based precipitation (Yang and Ohata, 2001; Adam et al., 2006) that may amplify the model-observation discrepancy. A large negative relative bias in SAH occurs in October to March, but it is of negligible magnitude in absolute terms. In nearly all other seasonal and regional cases over land, the observational estimate falls within the range of the CMIP5 simulations. Compared with CMIP3, the CMIP5 median precipitation is slightly higher in most regions. In contrast with temperature, the seasonal and annual mean ensemble mean and the root-mean square precipitation biases are larger for CMIP5 than for CMIP3 in a slight majority of land regions (Figure 9.39) and in most of the

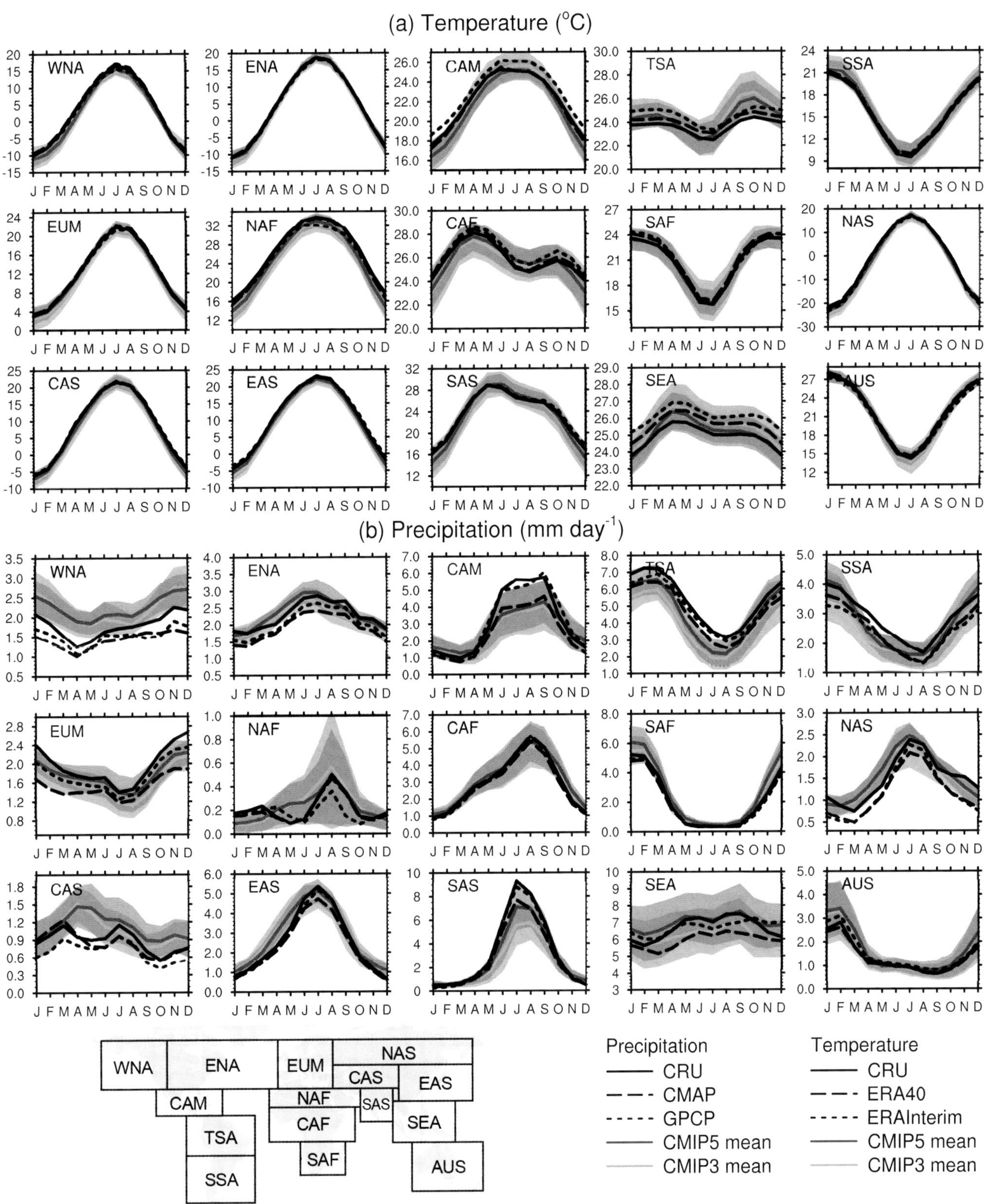

Figure 9.38 | Mean seasonal cycle of (a) temperature (°C) and (b) precipitation (mm day^{-1}). The average is taken over land areas within the indicated regions, and over the period 1980–1999. The red line is the average over 45 CMIP5 models; the blue line is the average over 22 CMIP3 models. The standard deviation of the respective data set is indicated with shading. The different line styles in black refer to observational and reanalysis data: Climatic Research Unit (CRU) TS3.10, ECMWF 40-year reanalysis (ERA40) and ERA-Interim for temperature; CRU TS3.10.1, Global Precipitation Climatology Project (GPCP), and CPC Merged Analysis of Precipitation (CMAP) for precipitation. Note the different axis-ranges for some of the sub-plots. The 15 regions shown are: Western North America (WNA), Eastern North America (ENA), Central America (CAM), Tropical South America (TSA), Southern South America (SSA), Europe and Mediterranean (EUM), North Africa (NAF), Central Africa (CAF), South Africa (SAF), North Asia (NAS), Central Asia (CAS), East Asia (EAS), South Asia (SAS), Southeast Asia (SEA) and Australia (AUS).

14 other regions (Figure 9.40). However, considering the observational uncertainty, the performance of the CMIP3 and CMIP5 ensembles is assessed to be broadly similar. The inter-model spreads are similar and typically largest in arid areas when expressed in relative terms.

Especially over the oceans and polar regions (Figure 9.40), the scarcity of observations and their uncertainty complicates the evaluation of simulated precipitation. Of two commonly used data sets, CMAP indicates systematically more precipitation than GPCP over low-latitude

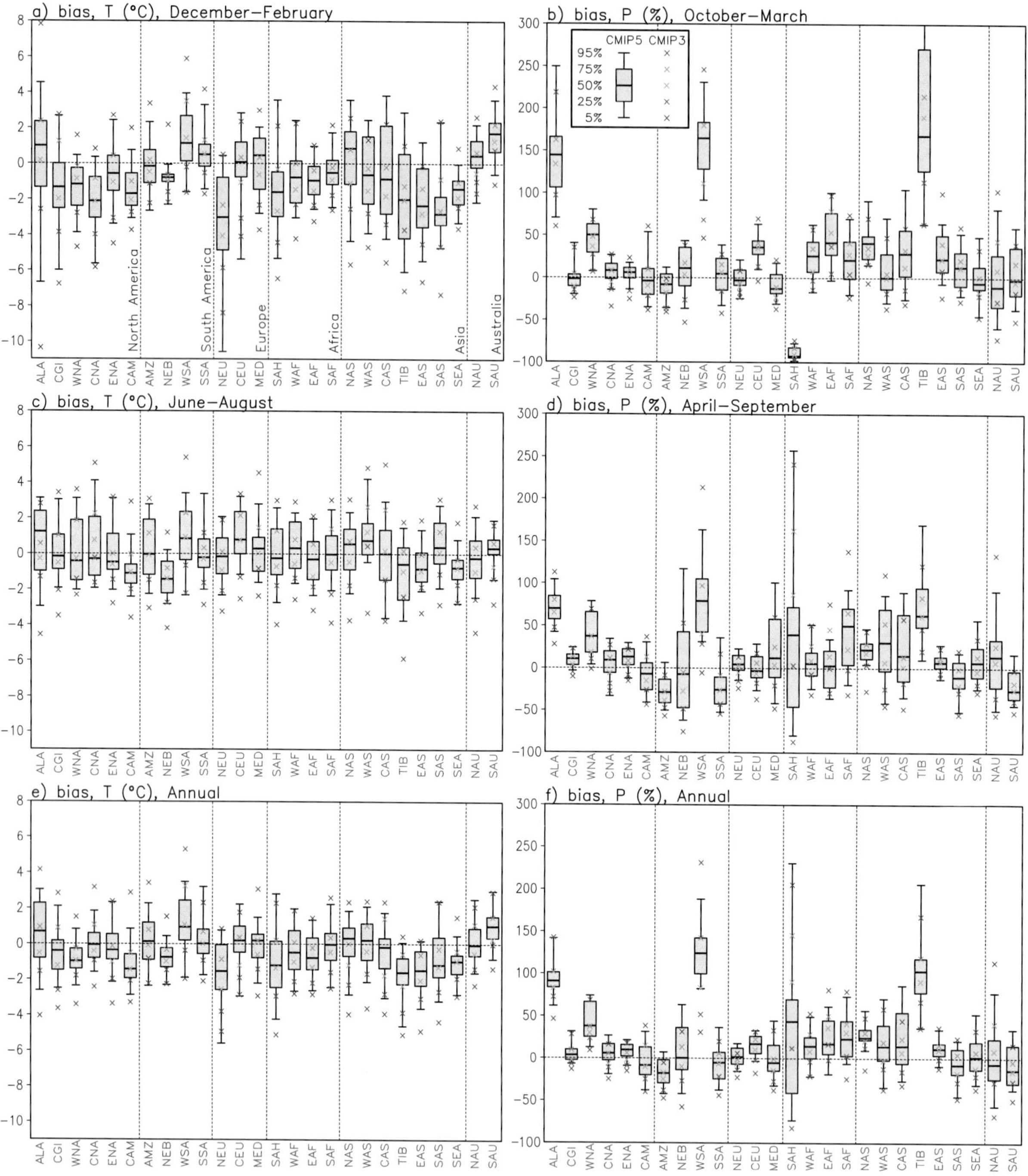

Figure 9.39 | Seasonal- and annual mean biases of (left) temperature (°C) and (right) precipitation (%) in the IPCC Special Report on Managing the Risks of Extreme Events and Disasters to Advance Climate Change Adaptation (SREX) land regions (cf. Seneviratne et al., 2012, p. 12. The region's coordinates can be found from their online Appendix 3.A). The 5th, 25th, 50th, 75th and 95th percentiles of the biases in 42 CMIP5 models are shown in box-and-whisker format, and corresponding values for 23 CMIP3 models with crosses. The CMIP3 20C3M simulations are complemented with the corresponding A1B runs for the 2001–2005 period. The biases are calculated over 1986–2005, using Climatic Research Unit (CRU) T3.10 as the reference for temperature and CRU TS 3.10.01 for precipitation. The regions are labelled with red when the root-mean-square error for the individual CMIP5 models is larger than that for CMIP3 and blue when it is smaller. The regions are: Alaska/NW Canada (ALA), Eastern Canada/Greenland/Iceland (CGI), Western North America (WNA), Central North America (CNA), Eastern North America (ENA), Central America/Mexico (CAM), Amazon (AMZ), NE Brazil (NEB), West Coast South America (WSA), South-Eastern South America (SSA), Northern Europe (NEU), Central Europe (CEU), Southern Europe/the Mediterranean (MED), Sahara (SAH), Western Africa (WAF), Eastern Africa (EAF), Southern Africa (SAF), Northern Asia (NAS), Western Asia (WAS), Central Asia (CAS), Tibetan Plateau (TIB), Eastern Asia (EAS), Southern Asia (SAS), Southeast Asia (SEA), Northern Australia (NAS) and Southern Australia/New Zealand (SAU). Note that the region WSA is poorly resolved in the models.

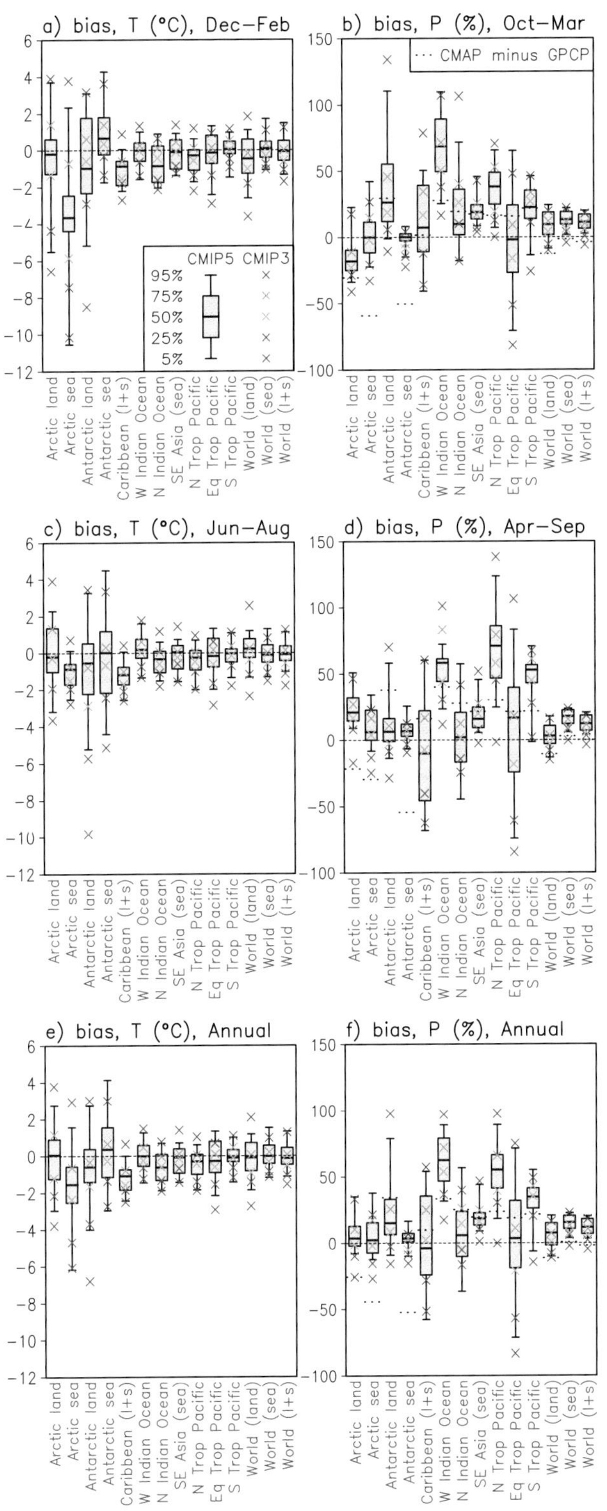

Figure 9.40 | As Figure 9.39, but for polar and ocean regions, with ECMWF reanalysis of the global atmosphere and surface conditions (ERA)-Interim reanalysis as the reference for temperature and Global Precipitation Climatology Project (GPCP) for precipitation. Global land, ocean and overall means are also shown. The regions are: Arctic: 67.5 to 90°N, Caribbean (area defined by the following coordinates): 68.8°W, 11.4°N; 85.8°W, 25°N; 60°W, 25°N, 60°W, 11.44°N; Western Indian Ocean: 25°S to 5°N, 52°E to 75°E; Northern Indian Ocean: 5°N to 30°N, 60°E to 95°E; Northern Tropical Pacific: 5°N to 25°N, 155°E to 150°W; Equatorial Tropical Pacific: 5°S to 5°N, 155°E to 130°W; Southern Tropical Pacific: 5°S to 25°S, 155°E to 130°W; Antarctic: 50°S to 90°S. The normalized difference between CPC Merged Analysis of Precipitation (CMAP) and GPCP precipitation is shown with dotted lines.

oceans and less over many high-latitude regions (Yin et al., 2004; Shin et al., 2011). Over most low-latitude ocean regions, annual precipitation in most CMIP3 and CMIP5 models exceeds GPCP. The difference relative to CMAP is smaller although mostly of the same sign. In Arctic and Antarctic Ocean areas, simulated precipitation is much above CMAP, but more similar to GPCP. Over Antarctic land, precipitation in most models is below CMAP, but close to or above GPCP.

Continental to sub-continental mean values may not be representative for smaller-scale biases, as biases generally increase with decreasing spatial averaging (Masson and Knutti, 2011b; Raisanen and Ylhaisi, 2011). A typical order of magnitude for grid-box-scale annual mean biases in individual CMIP3 models was 2°C for temperature and 1 mm day^{-1} for precipitation (Raisanen, 2007; Masson and Knutti, 2011b), with some geographical variation. This has been noted also in studies on how much spatial averaging would be needed in order to filter out the most unreliable small-scale features (e.g., Räisänen and Ylhäisi, 2011). In order to reduce such errors while still retaining information on small scales, Masson and Knutti (2011b) found, depending on the variable and the region, that smoothing needed to vary from the grid-point scale to around 2000 km.

On the whole, based on analysis of both ensemble means and inter-model spread, there is *high confidence* that the CMIP5 models simulate regional-scale temperature distributions somewhat better than the CMIP3 models did. This improvement is evident for most regions. For precipitation, there is *medium confidence* that there is no systematic change in model performance. In many regions, precipitation biases relative to CRU TS 3.10.01 and GPCP (and CMAP) are larger for CMIP5 than for CMIP3, but given observational uncertainty, the two ensembles are broadly similar.

9.6.1.2 Regional-Scale Simulation by Atmospheric General Circulation Models

Stand-alone global atmospheric models (AGCMs) run at higher resolution than AOGCMs provide complementary regional-scale climate information, sometimes referred to as 'global downscaling'. One important example of this is the simulation of tropical cyclones (e.g., Zhao et al., 2009, 2012; Murakami and Sugi, 2010; Murakami et al., 2012). A number of advantages of high-resolution AGCMs have been identified, including improved regional precipitation (Zhao et al., 2009; Kusunoki et al., 2011) and blocking (Matsueda et al., 2009, 2010). As AGCMs do not simulate interactions with the ocean, their ability to capture some high-resolution phenomena, such as the cold wake in the surface ocean after a tropical cyclone, is limited (e.g., Hasegawa and Emori, 2007). As in lower-resolution models, performance is affected by the quality of physical parameterizations (Lin et al., 2012; Mizuta et al., 2012; Zhao et al., 2012).

9.6.1.3 Regional-Scale Simulation by Variable-Resolution Global Climate Models

An alternative to global high resolution is the use of variable resolution (so-called 'stretched grid') models with higher resolution over the region of interest. Some examples are Abiodun et al. (2011) who showed that such simulations improve the simulation of West African monsoon systems and African easterly jets, and White et al. (2013) who demonstrated improvements in temperature and precipitation related extreme indices. Fox-Rabinovitz et al. (2008) showed that regional biases in the high-resolution portion of a stretched grid model were similar to that of a global model with the same high resolution everywhere. Markovic et al. (2010) and Déqué (2010) reported similar results. Although not widely used, such methods can complement more conventional climate models.

9

9.6.2 Regional Climate Downscaling

Regional Climate Models (RCMs) are applied over a limited-area domain with boundary conditions either from global reanalyses or global climate model output. The use of RCMs for 'dynamical downscaling' has grown since the AR4, their resolution has increased, process-descriptions have developed further, new components have been added, and coordinated experimentation has become more widespread (Laprise, 2008; Rummukainen, 2010). Statistical downscaling (SD) involves deriving empirical relationships linking large-scale atmospheric variables (predictors) and local/regional climate variables (predictands). These relationships may then be applied to equivalent predictors from global models. SD methods have also been applied to RCM output (e.g., Boe et al., 2007; Déqué, 2007; Segui et al., 2010; Paeth, 2011; van Vliet et al., 2011). A significant constraint in a comprehensive evaluation of regional downscaling is that available studies often involve different methods, regions, periods and observational data for evaluation. Thus, evaluation results are difficult to generalize.

9.6.2.1 Recent Developments of Statistical Methods

The development of SD since the AR4 has been quite vigorous (e.g., Fowler et al., 2007; Maraun et al., 2010b), and many state-of-the-art approaches combine different methods (e.g., Vrac and Naveau, 2008; van Vliet et al., 2011). There is an increasing number of studies on extremes (e.g., Vrac and Naveau, 2008; Wang and Zhang, 2008), and on features such as hurricanes (Emanuel et al., 2008), river flow and discharge, sediment, soil erosion and crop yields (e.g., Zhang, 2007; Prudhomme and Davies, 2009; Lewis and Lamoureux, 2010). Techniques have also been developed to consider multiple climatic variables simultaneously in order to preserve some physical consistency (e.g., Zhang and Georgakakos, 2011). The methods used to evaluate SD approaches vary with the downscaled variable and include metrics related to intensities (e.g., Ning et al., 2011; Tryhorn and DeGaetano, 2011), temporal behaviour (e.g., May, 2007; Timbal and Jones, 2008; Maraun et al., 2010a; Brands et al., 2011), and physical processes (Lenderink and Van Meijgaard, 2008; Maraun et al., 2010a). SD capabilities are also examined through secondary variables like runoff, river discharge and stream flow (e.g., Boe et al., 2007; Teutschbein et al., 2011).

9.6.2.2 Recent Developments of Dynamical Methods

Since the AR4, typical RCM resolution has increased from around 50 km to around 25 km (e.g., Christensen et al., 2010). Long RCM runs at very high resolution are still, however, rather few (e.g., Yasutaka et al., 2008; Chan et al., 2012; Kendon et al., 2012). Coupled RCMs, with interactive ocean and, when appropriate, also sea ice have also been developed (Somot et al., 2008; Dorn et al., 2009; Artale et al., 2010; Doscher et al., 2010). Smith et al. (2011a) added vegetation dynamics–ecosystem biogeochemistry in an RCM.

At the time of the AR4, RCMs were typically used for time-slice experiments. Since then, multi-decadal and centennial RCM simulations have emerged in larger numbers (e.g., Diffenbaugh et al., 2011; Kjellstrom et al., 2011; de Elia et al., 2013). Coordinated RCM experiments and ensembles have also become much more common and today, with domains covering Europe (e.g., Christensen et al., 2010; Vautard et al., 2013), North America (e.g., Gutowski et al., 2010; Lucas-Picher et al., 2012a; Mearns et al., 2012), South America (e.g., Menendez et al., 2010; Chou et al., 2012; Krüger et al., 2012), Africa (e.g., Druyan et al., 2010; Ruti et al., 2011; Nikulin et al., 2012; Paeth et al., 2012; Hernández-Díaz et al., 2013), the Arctic (e.g., Inoue et al., 2006) and Asian regions (e.g., Feng and Fu, 2006; Shkolnik et al., 2007; Feng et al., 2011; Ozturk et al., 2012; Suh et al., 2012).

9.6.3 Skill of Downscaling Methods

Downscaling skill varies with location, season, parameter and boundary conditions (see Section 9.6.5) (e.g., Schmidli et al., 2007; Maurer and Hidalgo, 2008). Although there are indications that model skill increases with higher resolution, it does not do so linearly. Rojas (2006) found more improvement when increasing resolution from 135 km to 45 km than from 45 km to 15 km. Walther et al. (2013) found that the diurnal precipitation cycle and light precipitation improved more when going from 12 km to 6 km resolution than when going from 50 km to 25 km or from 25 km to 12 km. Higher resolution does enable better simulation of extremes (Seneviratne et al., 2012). For example, Pryor et al. (2012) noted that an increase in RCM resolution from 50 km to 6 km increased extreme wind speeds more than the mean wind speed. Kawazoe and Gutowski (2013) compared six RCMs and the two GCMs to high resolution observations, concluding that precipitation extremes were more representative in the RCMs than in the GCMs. Vautard et al. (2013) found that warm extremes in Europe were generally better simulated in RCMs with 12 km resolution compared to 50 km. Kendon et al. (2012) and Chan et al. (2012) found mixed results in daily precipitation simulated at 12 km and 1.5 km resolution, although the latter had improved sub-daily features, perhaps as convection could be explicitly resolved.

Coupled RCMs, with an interactive ocean, offer further improvements. Döscher et al. (2010) reproduced empirical relationships between Arctic sea ice extent and sea ice thickness and NAO in a coupled RCM. Zou and Zhou (2013) found that a regional ocean–atmosphere model improved the simulation of precipitation over the western North Pacific compared to an uncoupled model. Samuelsson et al. (2010) showed that coupling a lake model with an RCM captured the effect of lakes on the air temperature over adjacent land. Lenaerts et al. (2012) added drifting snow in an RCM run for the Antarctica, which increased the area of ablation and improved the fit to observations. Smith et al. (2011a) added vegetation dynamics–ecosystem biogeochemistry into an RCM, and found some evidence of local feedback to air temperature.

Applying an RCM developed for a specific region to other regions exposes it to a wider range of conditions and therefore provides an opportunity for more rigorous evaluation. Transferability experiments target this by running RCMs for different regions while holding their process-descriptions constant (cf. Takle et al., 2007; Gbobaniyi et al., 2011; Jacob et al., 2012). Suh et al. (2012) noted that 10 RCMs run for Africa did well overall for average and maximum temperature, but systematically overestimated the daily minimum temperature. Precipitation was generally simulated betted for wet regions than for dry regions. Similarly, Nikulin et al. (2012) reported on 10 RCMs over Africa, run with boundary conditions from ERA-Interim, and evaluated against different observational data sets. Many of the RCMs simulated precipitation better than the ERA-Interim reanalysis itself.

Christensen et al. (2010) examined a range metrics related to simulation of extremes, mesoscale features, trends, aspects of variability and consistency with the driving boundary conditions. Only one of these metrics led to clear differentiation among RCMs (Lenderink, 2010). This may imply a general skilfulness of models, but may also simply indicate that the metrics were not very informative. Nevertheless, using some of these metrics, Coppola et al. (2010) and Kjellström et al. (2010) found that weighted sets of RCMs outperformed sets without weighting for both temperature and precipitation. Sobolowski and Pavelsky (2012) demonstrated a similar impact.

9.6.4 Value Added through RCMs

RCMs are regularly tested to evaluate whether they show improvements over global models (Laprise et al., 2008), that is, whether they do indeed 'add value'. In essence, added value is a measure of the extent to which the downscaled climate is closer to observations than the model from which the boundary conditions were obtained. Differences between RCM and GCM simulations are not always very obvious for time-averaged quantities on larger scales or in fairly homogeneous regions. RCM fields are, however, richer in spatial and temporal detail. Indeed, the added value of RCMs is mainly expected in the simulation of topography-influenced phenomena and extremes with relatively small spatial or short temporal character (e.g., Feser et al., 2011; Feser and Barcikowska, 2012; Shkol'nik et al., 2012). As an example, RCM downscaling led to better large-scale monsoon precipitation patterns (Gao et al., 2012) for East Asia than in the global models used for boundary conditions. In the few instances where RCMs have been interactively coupled to global models (i.e., 'two-way' coupling), the effects of improved small scales propagate to larger scales and this has been found to improve the simulation of larger scale phenomena (Lorenz and Jacob, 2005; Inatsu and Kimoto, 2009; Inatsu et al., 2012).

Other examples include improved simulation of convective precipitation (Rauscher et al., 2010), near-surface temperature (Feser, 2006), near-surface temperature and wind (Kanamaru and Kanamitsu, 2007), temperature and precipitation (Lucas-Picher et al., 2012b), extreme precipitation (Kanada et al., 2008), coastal climate features (Winterfeldt and Weisse, 2009; Winterfeldt et al., 2011; Kawazoe and Gutowski, 2013; Vautard et al., 2013), Atlantic hurricanes (Bender et al., 2010), European storm damage (Donat et al., 2010), strong mesoscale cyclones (Cavicchia and Storch, 2011), cutoff lows (Grose et al., 2012), polar lows (Zahn and von Storch, 2008) and higher statistical moments of the water budget (e.g., Bresson and Laprise, 2011).

In summary, there is *high confidence* that downscaling adds value to the simulation of spatial climate detail in regions with highly variable topography (e.g., distinct orography, coastlines) and for mesoscale phenomena and extremes. Regional downscaling is therefore complementary to results obtained directly from global climate models. These results are from a variety of distinct studies with different RCMs.

9.6.5 Sources of Model Errors and Uncertainties

In addition to issues related to resolution and model complexity (see Section 9.6.3), errors and uncertainties arise from observational uncertainty in evaluation data and parameterizations (see Box 9.3), choice of model domain and application of boundary conditions (driving data).

In the case of SD, sources of model errors and uncertainties depend on the choice of method, including the choice of the predictors, the estimation of empirical relationships between predictors and predictands from limited data sets, and also the data used to estimate the predictors (Frost et al., 2011). There are numerous different SD methods, and the findings are difficult to generalize.

Small domains allow less freedom for RCMs to generate the small-scale features that give rise to added value (e.g., Leduc and Laprise, 2009). Therefore large domains –covering entire continents– have become more common. Køltzow et al. (2008) found improvements with the use of a larger domain, but the RCM solution can become increasingly 'decoupled' from the driving data (e.g., Rockel et al., 2008), which can introduce inconsistencies. Large domains also introduce large internal variability, which can significantly contaminate interannual variability of seasonal means (Kanamitsu et al., 2010). Techniques such as spectral nudging (Misra, 2007; Separovic et al., 2012) can be used to constrain such inconsistencies (Feser et al., 2011). Winterfeldt and Weisse (2009) concluded that nudging improved the simulation of marine wind climate, while Otte et al. (2012) demonstrated improvements in temperature and precipitation. Nudging may, however, also lead to deterioration of features such as precipitation extremes (Alexandru et al., 2009; Kawazoe and Gutowski, 2013). Veljovic et al. (2010) showed that an RCM can in fact improve the large scales with respect to those inherent in the boundary conditions, and argued that nudging may be undesirable.

The quality of RCM results may vary according to the synoptic situation, season, and the geographic location of the lateral boundaries (Alexandru et al., 2007; Xue et al., 2007; Laprise et al., 2008; Separovic et al., 2008; Leduc and Laprise, 2009; Nikiema and Laprise, 2010; Rapaić et al., 2010). In addition to lateral boundary conditions, RCMs also need sea surface information. Few studies have explored the dependency of RCM results on the treatment of the SSTs and sea ice, although Koltzow et al. (2011) found that the specification of SSTs was less influential than was the domain or the lateral boundaries. Woollings et al. (2010a) investigated the effect of specified SST on the simulation of the Atlantic storm track and found that it was better simulated with high-resolution SSTs, whereas increasing temporal resolution gave mixed results.

As is the case in global models, RCM errors are directly related to shortcomings in process parameterizations. Examples include the representation of clouds, convection and land surface–atmosphere interactions, the planetary boundary layer, horizontal diffusion, and microphysics (Tjernstrom et al., 2008; Wyser et al., 2008; Lynn et al., 2009; Pfeiffer and Zängl, 2010; Axelsson et al., 2011; Crétat et al., 2012; Evans et al., 2012; Roy et al., 2012; Solman and Pessacg, 2012). The representation of land surface and atmosphere coupling is also important, particularly for simulating monsoon regions (Cha et al., 2008; Yhang and Hong, 2008; Boone et al., 2010; Druyan et al., 2010; van den Hurk and van Meijgaard, 2010).

9.6.6 Relating Downscaling Performance to Credibility of Regional Climate Information

A fundamental issue is how the performance of a downscaling method relates to its ability to provide credible future projections (Raisanen, 2007). This subject is discussed further in Section 9.8. The credibility of downscaled information of course depends on the quality of the downscaling method itself (e.g., Dawson et al., 2012; Déqué et al., 2012; Eum et al., 2012), and on the quality of the global climate models providing the large-scale boundary conditions (e.g., van Oldenborgh et al., 2009; Diaconescu and Laprise, 2013).

Specific to SD is the statistical stationarity hypothesis, that is, that the relationships inferred from historical data remain valid under a changing climate (Maraun, 2012). Vecchi et al. (2008) note that a statistical method that captures interannual hurricane variability gives very different results for projections compared to RCMs. Such results suggest that good performance of statistical downscaling as assessed against observations does not guarantee credible regional climate information. Some recent studies have proposed ways to evaluate SD approaches using RCM outputs (e.g., Vrac and Naveau, 2008; Driouech et al., 2010) or long series of observations (e.g. Schmith, 2008).

Giorgi and Coppola (2010) argued that regional-scale climate projections over land in the CMIP3 models were not sensitive to their temperature biases. For precipitation, the same was found for about two thirds of the global land area. However, there is some recent evidence that regional biases may be nonlinear for temperature extremes (Christensen et al., 2008; Boberg and Christensen, 2012; Christensen and Boberg, 2013) in both global and regional models. A mechanism at play may be that models tend to dry out the soil too effectively at high temperatures, which can lead to systematic biases in projected warm summertime conditions (Christensen et al., 2008; Kostopoulou et al., 2009). This is illustrated in Figure 9.41 for the Mediterranean region, which suggests a tendency in RCMs, CMIP3 and CMIP5 models towards an enhanced warm bias in the warmer months. The implication is that the typically large warming signal in these regions could be biased (Boberg and Christensen, 2012; Mearns et al., 2012). Findings such as these stress the importance of a thorough assessment of models' biases when they are applied for projections (e.g., de Elia and Cote, 2010; Boberg and Christensen, 2012; Christensen and Boberg, 2013).

Di Luca et al. (2012) analysed downscaled climate change projections from six RCMs run over North America. The climate change signals for seasonal precipitation and temperature were similar to those in the driving AOGCMs, and the spatial detail gained by downscaling was comparable in both present and future climate. Déqué et al. (2012) studied projections with several combinations of AOGCM and RCM for Europe. A larger part of the spread in winter temperature and

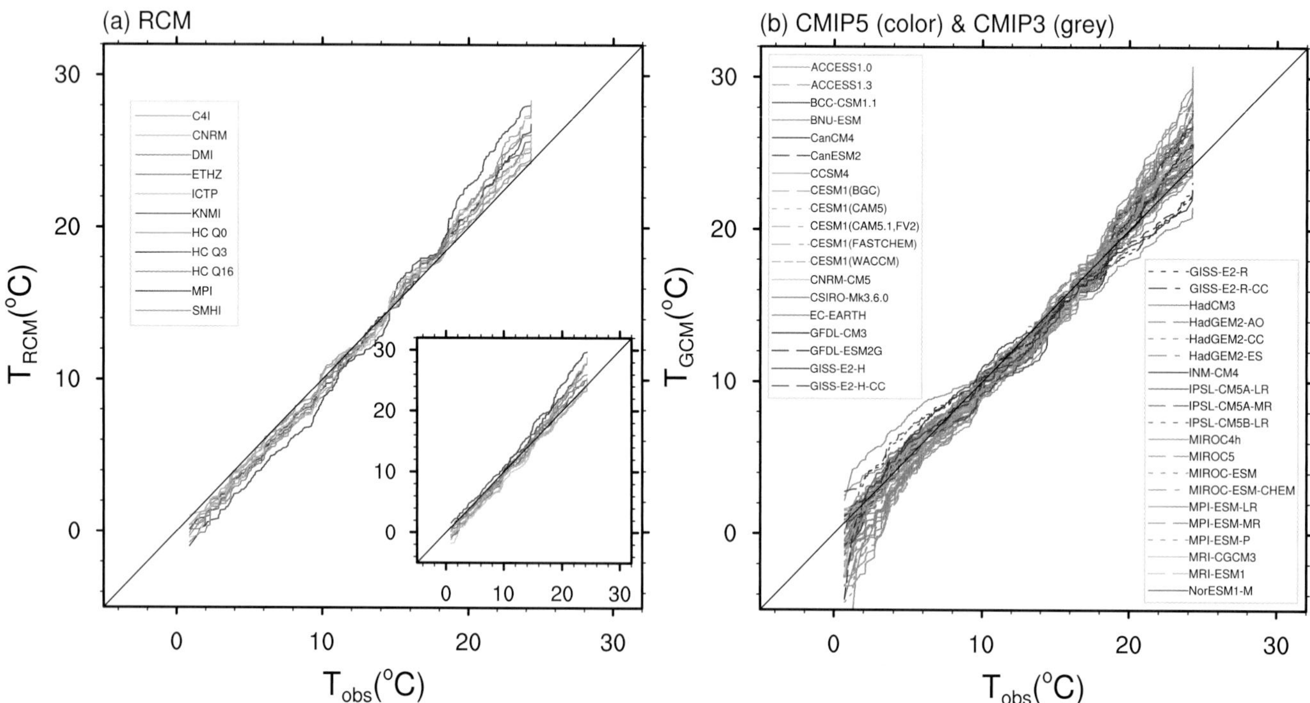

Figure 9.41 | Ranked modelled versus observed monthly mean temperature for the Mediterranean region for the 1961–2000 period. The Regional Climate Model (RCM) data (a) are from Christensen et al. (2008) and are adjusted to get a zero mean in model temperature with respect to the diagonal. The smaller insert shows uncentred data. The General Circulation Model (GCM) data (b) are from CMIP5 and CMIP3 and adjusted in the same way. (After Boberg and Christensen, 2012.)

precipitation projections was explained by the differences in global model boundary conditions, although much of the spread in projected summer precipitation was explained by RCM. This underlines the importance of both the quality of the boundary conditions and the downscaling method.

9.7 Climate Sensitivity and Climate Feedbacks

An overall assessment of climate sensitivity and transient climate response is given in Box 12.2. Observational constraints based on observed warming over the last century are discussed in Section 10.8.2 and shown in Box 12.2, Figure 2.

9.7.1 Equilibrium Climate Sensitivity, Idealized Radiative Forcing, and Transient Climate Response in the Coupled Model Intercomparison Project Phase 5 Ensemble

Equilibrium climate sensitivity (ECS) is the equilibrium change in global and annual mean surface air temperature after doubling the atmospheric concentration of CO_2 relative to pre-industrial levels. In the AR4, the range in equilibrium climate sensitivity of the CMIP3 models was 2.1°C to 4.4°C, and the single largest contributor to this spread was differences among modelled cloud feedbacks. These assessments carry over to the CMIP5 ensemble without any substantial change (Table 9.5).

The method of diagnosing climate sensitivity in CMIP5 differs fundamentally from the method employed in CMIP3 and assessed in the AR4 (Randall et al., 2007). In CMIP3, an AGCM was coupled to a non-dynamic mixed-layer (slab) ocean model with prescribed ocean heat transport convergence. CO_2 concentration was then instantaneously doubled, and the model was integrated to a new equilibrium with unchanged implied ocean heat transport. While computationally efficient, this method had the disadvantage of employing a different model from that used for the historical simulations and climate projections. However, in the few comparisons that were made, the resulting disagreement in ECS was less than about 10% (Boer and Yu, 2003; Williams et al., 2008; Danabasoglu and Gent, 2009; Li et al., 2013a). In CMIP5, climate sensitivity is diagnosed directly from the AOGCMs following the approach of Gregory et al. (2004). In this case the CO_2 concentration is instantaneously quadrupled and kept constant for 150 years of simulation, and both equilibrium climate sensitivity and RF are diagnosed from a linear fit of perturbations in global mean surface temperature to the instantaneous radiative imbalance at the TOA.

The transient climate response (TCR) is the change in global and annual mean surface temperature from an experiment in which the CO_2 concentration is increased by 1% yr^{-1}, and calculated using the difference between the start of the experiment and a 20-year period centred on the time of CO_2 doubling. TCR is smaller than ECS because ocean heat uptake delays surface warming. TCR is linearly correlated with ECS in the CMIP5 ensemble (Figure 9.42), although the relationship may be nonlinear outside the range spanned in Table 9.5 (Knutti et al., 2005).

Based on the methods outlined above and explained in Section 9.7.2 below, Table 9.5 shows effective ERF, ECS, TCR and feedback strengths for the CMIP5 ensemble. The two estimates of ERF agree with each other to within 5% for six models (CanESM2, INM-CM4, IPSL-CM5A-LR, MIROC5, MPI-ESM-LR and MPI-ESM-P), although the deviation exceeds 10% for four models (CCSM4, CSIRO-Mk3-6-0, HadGEM2-ES, and MRI-CGCM3) and is indicative of deviations from the basic assumptions underlying one or both ERF estimation methods. However, the mean difference of 0.3 W m^{-2} between the two methods for diagnosing ERF is only about half of the ensemble standard deviation of 0.5 W m^{-2}, or 15% of the mean value for ERF by CO_2 using fixed SSTs. ECS and TCR vary across the ensemble by a factor of approximately 2. The multi-model ensemble mean in ECS is 3.2°C, a value nearly identical to that for CMIP3, while the CMIP5 ensemble range is 2.1°C to 4.7°C, a spread which is also nearly indistinguishable from that for CMIP3. While every CMIP5 model whose heritage can

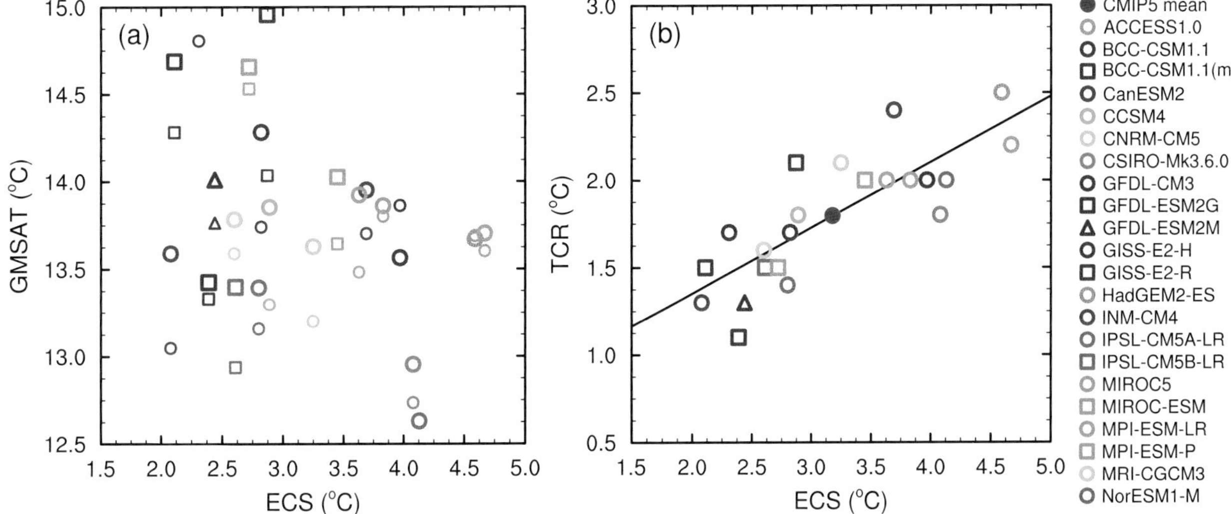

Figure 9.42 | (a) Equilibrium climate sensitivity (ECS) against the global mean surface temperature of CMIP5 models, both for the period 1961–1990 (larger symbols, cf. Figure 9.8, Table 9.5) and for the pre-industrial control runs (smaller symbols). (b) Equilibrium climate sensitivity against transient climate response (TCR). The ECS and TCR information are based on Andrews et al. (2012) and Forster et al. (2013) and updated from the CMIP5 archive.

9

Table 9.5 | Effective radiative forcing (ERF), climate sensitivity and climate feedbacks estimated for the CMIP5 AOGCMs (see Table 9.1 for model details). ERF, equilibrium climate sensitivity (ECS) and transient climate response (TCR) are based on Andrews et al. (2012) and Forster et al. (2013) and updated from the CMIP5 archive. The ERF entries are calculated according to Hansen et al. (2005) using fixed sea surface temperatures (SSTs) and Gregory et al. (2004) using regression. ECS is calculated using regressions following Gregory et al. (2004). TCR is calculated from the CMIP5 simulations with 1% CO_2 increase per year (Taylor et al., 2012b), using the 20-year mean centred on the year of CO_2 doubling. The climate sensitivity parameter and its inverse, the climate feedback parameter, are calculated from the regression-based ERF and the ECS. Strengths of the individual feedbacks are taken from Vial et al. (2013), following Soden et al. (2008) and using radiative kernel methods with two different kernels. The sign convention is such that a positive entry for an individual feedback marks a positive feedback; the sum of individual feedback strengths must hence be multiplied by –1 to make it comparable to the climate feedback parameter. The entries for radiative forcing and equilibrium climate sensitivity were obtained by dividing by two the original results, which were obtained for CO_2 quadrupling. ERF and ECS for BNU-ESM are from Vial et al. (2013).

Model	Effective Radiative Forcing 2 × CO_2 (W m^{-2})		Equilibrium Climate Sensitivity (°C)	Transient Climate Response (°C)	Climate Sensitivity Parameter (°C (W m^{-2})$^{-1}$)	Climate Feed-back Parameter (W m^{-2} °C^{-1})	Planck Feedback (W m^{-2} °C^{-1})	Water Vapour Feedback (W m^{-2} °C^{-1})	Lapse Rate Feedback (W m^{-2} °C^{-1})	Surface Albedo Feedback (W m^{-2} °C^{-1})	Cloud Feedback (W m^{-2} °C^{-1})
	Fixed SST	Regression									
ACCESS1.0	n.a.	3.0	3.8	2.0	1.3	0.8	n.a.	n.a.	n.a.	n.a.	n.a.
ACCESS1.3	n.a.	n.a.	n.a.	1.7	n.a.	n.a.	n.a.	n.a.	n.a.	n.a.	n.a.
BCC–CSM1.1	n.a.	3.2	2.8	1.7	0.9	1.1	n.a.	n.a.	n.a.	n.a.	n.a.
BCC–CSM1.1(m)	n.a.	3.6	2.9	2.1	0.8	1.2	n.a.	n.a.	n.a.	n.a.	n.a.
BNU–ESM	n.a.	3.9	4.1	2.6	1.1	1.0	–3.1	1.4	–0.2	0.4	0.1
CanESM2	3.7	3.8	3.7	2.4	1.0	1.0	–3.2	1.7	–0.6	0.3	0.5
CCSM4	4.4	3.6	2.9	1.8	0.8	1.2	–3.2	1.5	–0.4	0.4	–0.4
CESM1(BGC)	n.a.	n.a.	n.a.	1.7	n.a.	n.a.	n.a.	n.a.	n.a.	n.a.	n.a.
CESM1(CAM5)	n.a.	n.a.	n.a.	2.3	n.a.	n.a.	n.a.	n.a.	n.a.	n.a.	n.a.
CNRM–CM5	n.a.	3.7	3.3	2.1	0.9	1.1	n.a.	n.a.	n.a.	n.a.	n.a.
CSIRO–Mk3.6.0	3.1	2.6	4.1	1.8	1.6	0.6	n.a.	n.a.	n.a.	n.a.	n.a.
FGOALS–g2	n.a.	n.a.	n.a.	1.4	n.a.	n.a.	n.a.	n.a.	n.a.	n.a.	n.a.
GFDL–CM3	n.a.	3.0	4.0	2.0	1.3	0.8	n.a.	n.a.	n.a.	n.a.	n.a.
GFDL–ESM2G	n.a.	3.1	2.4	1.1	0.8	1.3	n.a.	n.a.	n.a.	n.a.	n.a.
GFDL–ESM2M	n.a.	3.4	2.4	1.3	0.7	1.4	n.a.	n.a.	n.a.	n.a.	n.a.
GISS–E2–H	n.a.	3.8	2.3	1.7	0.6	1.7	n.a.	n.a.	n.a.	n.a.	n.a.
GISS–E2–R	n.a.	3.8	2.1	1.5	0.6	1.8	n.a.	n.a.	n.a.	n.a.	n.a.
HadGEM2–ES	3.5	2.9	4.6	2.5	1.6	0.6	–3.2	1.4	–0.5	0.3	0.4
INM–CM4	3.1	3.0	2.1	1.3	0.7	1.4	–3.2	1.7	–0.7	0.3	0
IPSL–CM5A–LR	3.2	3.1	4.1	2.0	1.3	0.8	–3.3	1.9	–1	0.2	1.2
IPSL–CM5A–MR	n.a.	n.a.	n.a.	2.0	n.a.	n.a.	n.a.	n.a.	n.a.	n.a.	n.a.
IPSL–CM5B–LR	n.a.	2.7	2.6	1.5	1.0	1.0	n.a.	n.a.	n.a.	n.a.	n.a.
MIROC5	4.0	4.1	2.7	1.5	0.7	1.5	–3.2	1.7	–0.6	0.3	0.1
MIROC–ESM	n.a.	4.3	4.7	2.2	1.1	0.9	n.a.	n.a.	n.a.	n.a.	n.a.
MPI–ESM–LR	4.3	4.1	3.6	2.0	0.9	1.1	–3.3	1.8	–0.9	0.3	0.5
MPI–ESM–MR	n.a.	n.a.	n.a.	2.0	n.a.	n.a.	n.a.	n.a.	n.a.	n.a.	n.a.
MPI–ESM–P	4.3	4.3	3.5	2.0	0.8	1.2	n.a.	n.a.	n.a.	n.a.	n.a.
MRI–CGCM3	3.6	3.2	2.6	1.6	0.8	1.2	–3.2	1.6	–0.6	0.3	0.2
NorESM1–M	n.a.	3.1	2.8	1.4	0.9	1.1	–3.2	1.6	–0.5	0.3	0.2
NorESM1–ME	n.a.	n.a.	n.a.	1.6	n.a.	n.a.	n.a.	n.a.	n.a.	n.a.	n.a.
Model mean	**3.7**	**3.4**	**3.2**	**1.8**	**1.0**	**1.1**	**–3.2**	**1.6**	**–0.6**	**0.3**	**0.3**
90% uncertainty	**±0.8**	**±0.8**	**±1.3**	**±0.6**	**±0.5**	**±0.5**	**±0.1**	**±0.3**	**±0.4**	**±0.1**	**±0.7**

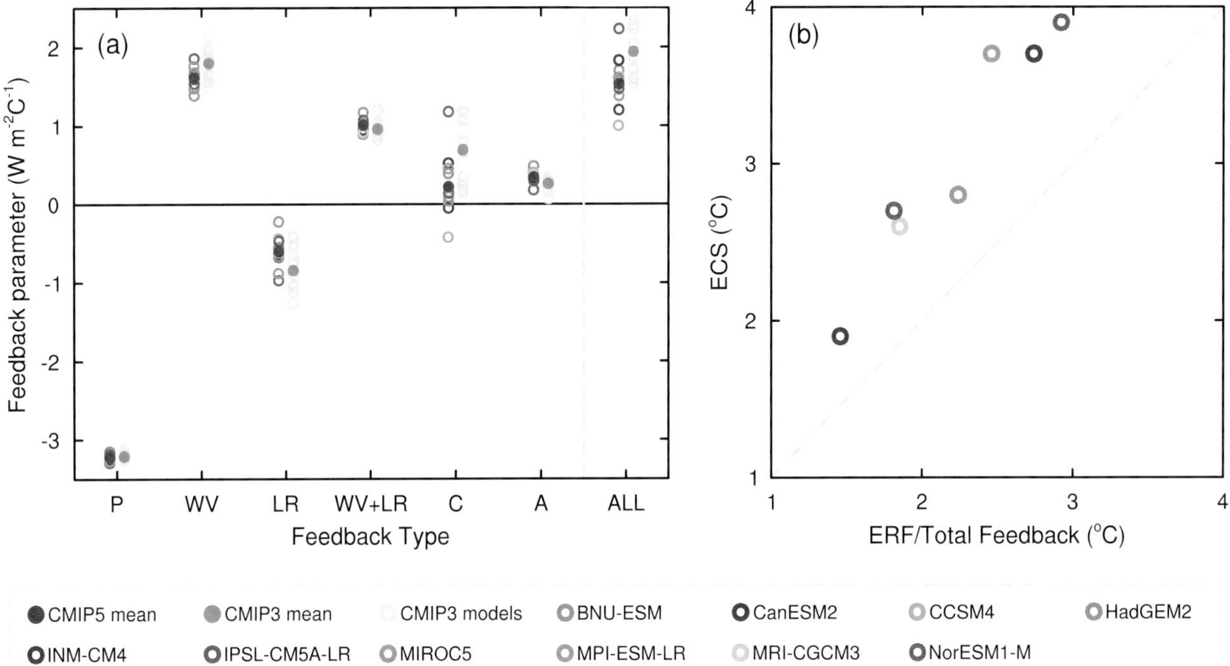

Figure 9.43 | (a) Strengths of individual feedbacks for CMIP3 and CMIP5 models (left and right columns of symbols) for Planck (P), water vapour (WV), clouds (C), albedo (A), lapse rate (LR), combination of water vapour and lapse rate (WV+LR) and sum of all feedbacks except Planck (ALL), from Soden and Held (2006) and Vial et al. (2013), following Soden et al. (2008). CMIP5 feedbacks are derived from CMIP5 simulations for abrupt fourfold increases in CO_2 concentrations (4 × CO_2). (b) ECS obtained using regression techniques by Andrews et al. (2012) against ECS estimated from the ratio of CO_2 ERF to the sum of all feedbacks. The CO_2 ERF is one-half the 4 × CO_2 forcings from Andrews et al. (2012), and the total feedback (ALL + Planck) is from Vial et al. (2013).

be traced to CMIP3 shows some change in ECS, there is no discernible systematic tendency. This broad similarity between CMIP3 and CMIP5 and the good agreement between different methods where they were applied to the same atmospheric GCM indicate that the uncertainty in methodology is minor compared to the overall spread in ECS. The change in TCR from CMIP3 to CMIP5 is generally of the same sign but of smaller magnitude compared to the change in ECS. The relationship between ECS and an estimates derived from total feedbacks are discussed in Section 9.7.2.

Although ECS can vary with global mean surface temperature owing to the temperature dependencies of the various feedbacks (Colman and McAvaney, 2009; cf. Section 9.7.2), Figure 9.42 shows no discernible correlation for the CMIP5 historical temperature ranges, a fact that suggests that ECS is less sensitive to errors in the current climate than to other sources of uncertainty.

9.7.2 Understanding the Range in Model Climate Sensitivity: Climate Feedbacks

The strengths of individual feedbacks for the CMIP3 and CMIP5 models are compared in Figure 9.43. The feedbacks are generally similar between CMIP3 and CMIP5, and the water vapour, lapse rate, and cloud feedbacks are assessed in detail in Chapter 7. The surface albedo feedback is assessed here to be *likely* positive. There is *high confidence* that the sum of all feedbacks (excluding the Planck feedback) is positive. Advances in estimating and understanding each of the feedback parameters in Table 9.5 are described in detail below (see also Chapters 7 and 8).

9.7.2.1 Role of Humidity and Lapse Rate Feedbacks in Climate Sensitivity

The compensation between the water vapour and lapse-rate feedbacks noted in the CMIP3 models is still present in the CMIP5 models, and possible explanations of the compensation have been developed (Ingram, 2010; Ingram, 2013). New formulations of the feedbacks, replacing specific with relative humidity, eliminate most of the cancellation between the water vapour and lapse rate feedbacks and reduce the inter-model scatter in the individual feedback terms (Held and Shell, 2012).

9.7.2.2 Role of Surface Albedo in Climate Sensitivity

Analysis of observed declines in sea ice and snow coverage from 1979 to 2008 suggests that the NH albedo feedback is between 0.3 and 1.1 W m^{-2} °C^{-1} (Flanner et al., 2011). This range is substantially above the global feedback of 0.3 ± 0.1 W m^{-2} °C^{-1} of the CMIP5 models analysed in Table 9.5. One possible explanation is that the CMIP5 models underestimate the strength of the feedback as did the CMIP3 models based upon the systematic errors in simulated sea ice coverage decline relative to observed rates (Boe et al., 2009b).

9.7.2.3 Role of Cloud Feedbacks in Climate Sensitivity

Cloud feedbacks represent the main cause for the range in modelled climate sensitivity (Chapter 7). The spread due to inter-model differences in cloud feedbacks is approximately 3 times larger than the spread contributed by feedbacks due to variations in water vapour and lapse rate combined (Dufresne and Bony, 2008), and is a primary factor governing the range of climate sensitivity across the CMIP3 ensemble (Volodin, 2008a). Differences in equilibrium and effective climate sensitivity are due primarily to differences in the shortwave cloud feedback (Yokohata et al., 2008).

In perturbed ensembles of three different models, the primary cloud-related factor contributing to the spread in equilibrium climate sensitivity is the low-level shortwave cloud feedback (Yokohata et al., 2010; Klocke et al., 2011). Changes in the high-altitude clouds also induce climate feedbacks due to the large areal extent and significant longwave cloud radiative effects of tropical convective cloud systems. In experiments with perturbed physics ensembles of AOGCMs, the parameterization of ice fall speed also emerges as one of the most important determinants of climate sensitivity (Sanderson et al., 2008a, 2010; Sexton et al., 2012). Other non-microphysical feedback mechanisms are detailed in Chapter 7.

Cloud feedbacks in AOGCMs are generally positive or near neutral (Shell et al., 2008; Soden et al., 2008), as evidenced by the net positive or neutral cloud feedbacks in all of the models examined in a multi-thousand member ensemble of AOGCMs constructed by parameter perturbations (Sanderson et al., 2010). The sign of cloud feedbacks in the current climate deduced from observed relationships between SSTs and TOA radiative fluxes are discussed further in Section 7.2.5.7.

9.7.2.4 Relationship of Feedbacks to Modelled Climate Sensitivity

The ECS can be estimated from the ratio of forcing to the total climate feedback parameter. This approach is applicable to simulations in which the net radiative balance is much smaller than the forcing and hence the modelled climate system is essentially in equilibrium. This approach can also serve to check the internal consistency of estimates of the ECS, forcing, and feedback parameters obtained using independent methods. The relationship between ECS from Andrews et al. (2012) and estimates of ECS obtained from the ratio of forcings to feedbacks is shown in Figure 9.43b. The forcings are estimated using both regression and fixed SST techniques (Gregory et al., 2004; Hansen et al., 2005) by Andrews et al. (2012) and the feedbacks are calculated using radiative kernels (Soden et al., 2008). On average, the ECS from forcing to feedback ratios underestimate the ECS from Andrews et al. (2012) by 25% and 35%, or up to 50% for individual models, using fixed-SST and regression forcings, respectively.

9.7.2.5 Relationship of Feedbacks to Uncertainty in Modelled Climate Sensitivity

Objective methods for perturbing uncertain model parameters to optimize performance relative to a set of observational metrics have shown a tendency toward an increase in the mean and a narrowing of the spread of estimated climate sensitivity (Jackson et al., 2008a). This tendency is opposed by the effects of structural biases related to incomplete process representations in GCMs. If common structural biases are replicated across models in a MME (cf. Section 9.2.2.7), the most likely sensitivity for the MME tends to shift towards lower sensitivities while the possibility of larger sensitivities increases at the same time (Lemoine, 2010). Following Schlesinger and Mitchell (1987), Roe and Baker (2007) suggest that symmetrically distributed uncertainties in feedbacks lead to inherently asymmetrical uncertainties in climate sensitivity with increased probability in extreme positive values of the sensitivity. Roe and Baker (2007) conclude that this relationship makes it extremely difficult to reduce uncertainties in climate sensitivity through incremental improvements in the specification of feedback parameters. While subsequent analysis has suggested that this finding could be an artifact of the statistical formulation (Hannart et al., 2009) and linearization (Zaliapin and Ghil, 2010) of the relationship between feedback and sensitivity adopted by (Roe and Baker, 2007), these issues remain unsettled (Roe and Armour, 2011; Roe and Baker, 2011).

9.7.3 Climate Sensitivity and Model Performance

Despite the range in equilibrium sensitivity of 2.1°C to 4.4°C for CMIP3 models, they reproduce the global surface air temperature anomaly of 0.76°C over 1850–2005 to within 25% relative error. The relatively small range of historical climate response suggests that there is another mechanism, for example a compensating non-GHG forcing, present in the historical simulations that counteracts the relatively large range in sensitivity obtained from idealized experiments forced only by increasing CO_2. One possible mechanism is a systematic negative correlation across the multi-model ensemble between ECS and anthropogenic aerosol forcing (Kiehl, 2007; Knutti, 2008; Anderson et al., 2010). A second possible mechanism is a systematic overestimate of the mixing between the oceanic mixed layer and the full depth ocean underneath (Hansen et al., 2011). However, despite the same range of ECS in the CMIP5 models as in the CMIP3 models, there is no significant relationship across the CMIP5 ensemble between ECS and the 20th-century ERF applied to each individual model (Forster et al., 2013). This indicates a lesser role of compensating ERF trends from GHGs and aerosols in CMIP5 historical simulations than in CMIP3. Differences in ocean heat uptake also do not appreciably affect the spread in projected changes in global mean temperature by 2095 (Forster et al., 2013).

9.7.3.1 Constraints on Climate Sensitivity from Earth System Models of Intermediate Complexity

An EMIC intercomparison (Eby et al., 2013; Zickfeld et al., 2013) allows an assessment of model response characteristics, including ECS, TCR, and heat uptake efficiency (Table 9.6). In addition, Bayesian methods applied to PPE experiments using EMICs have estimated uncertainty in model response characteristics (see Box 12.2) based on simulated climate change in 20th century, past millennia, and LGM scenarios. Here, the range of response metrics (Table 9.6) described for default model configurations (Eby et al., 2013) indicates consistency with the CMIP5 ensemble.

9.7.3.2 Climate Sensitivity During the Last Glacial Maximum

Climate sensitivity can also be explored in another climatic context. The AR4 assessed attempts to relate simulated LGM changes in tropical SST to global climate sensitivity (Hegerl et al., 2007; Knutti and Hegerl, 2008). LGM temperature changes in the tropics (Hargreaves et al., 2007), but not in Antarctica (Hargreaves et al., 2012), have been

Table 9.6 | Model response metrics for EMICs in Table 9.2. TCR_{2X}, TCR_{4X} and ECS_{4X} are the changes in global average model surface air temperature from the decades centred at years 70, 140 and 995 respectively, from the idealized 1% increase to 4 × CO_2 experiment. The ocean heat uptake efficiency, κ_{4X}, is calculated from the global average heat flux divided by TCR_{4X} for the decade centred at year 140, from the same idealized experiment. ECS_{2x} was calculated from the decade centred about year 995 from a 2 × CO_2 pulse experiment. (Data from Eby et al., 2013.)

Model	TCR_{2X} (°C)	ECS_{2x}(°C)	TCR_{4X} (°C)	ECS_{4X} (°C)	κ_{4X} (W m^{-2} °C^{-1})
Bern3D	2.0	3.3	4.6	6.8	0.58
CLIMBER2	2.1	3.0	4.7	5.8	0.84
CLIMBER3	1.9	3.2	4.5	5.9	0.93
DCESS	2.1	2.8	3.9	4.8	0.72
FAMOUS	2.3	3.5	5.2	8.0	0.55
GENIE	2.5	4.0	5.4	7.0	0.51
IAP RAS CM	1.6	—	3.7	4.3	—
IGSM2	1.5	1.9	3.7	4.5	—
LOVECLIM1.2	1.2	2.0	2.1	3.5	1.17
MESMO	2.4	3.7	5.3	6.9	0.55
MIROC-lite	1.6	2.4	3.6	4.6	0.66
MIROC-lite-LCM	1.6	2.8	3.7	5.5	1.00
SPEEDO	0.8	3.6	2.9	5.2	0.84
UMD	1.6	2.2	3.2	4.3	—
Uvic	1.9	3.5	4.3	6.6	0.92
EMIC mean	1.8	3.0	4.0	5.6	0.8
EMIC range	0.8–2.5	1.9–4.0	2.1–5.4	3.5–8.0	0.5–1.2

shown to scale well with climate sensitivity because the signal is mostly dominated by CO_2 forcing in these regions (Braconnot et al., 2007b; Jansen et al., 2007). The analogy between the LGM climate sensitivity and future climate sensitivity is, however, not perfect (Crucifix, 2006). In a single-model ensemble of simulations, the magnitudes of the LGM cooling and the warming induced by a doubling of CO_2 are nonlinear in the forcings applied (Hargreaves et al., 2007). Differences in the cloud radiative feedback are at the origin of this asymmetric response to equivalent positive and negative forcings (Yoshimori et al., 2009). There is thus still *low confidence* that the regional LGM model-data comparisons can be used to evaluate model climate sensitivity. However, even if the results do not scale perfectly with equilibrium or transient climate sensitivity, the LGM simulations allow the identification of the different feedback factors that contributed to the LGM global cooling (Yoshimori et al., 2011) and model spread in these feedbacks. The largest spread in LGM model feedbacks is found for the shortwave cloud feedback, just as for the modern climate. This correspondence between LGM and modern climates adds to the *high confidence* that the shortwave cloud feedback is the dominant source of model spread in climate sensitivity (cf. Section 5.3.3).

9.7.3.3 Constraints on Equilibrium Climate Sensitivity from Climate-Model Ensembles and Observations

The large scale climatological information available has so far been insufficient to constrain model behaviour to a range tighter than CMIP3, at least on a global scale. Sanderson and Knutti (2012) suggest that much of the available and commonly used large scale observations have already been used to develop and evaluate models and are therefore of limited value to further constrain climate sensitivity or TCR. The assessed literature suggests that the range of climate sensitivities and transient responses covered by CMIP3/5 cannot be narrowed significantly by constraining the models with observations of the mean climate and variability, consistent with the difficulty of constraining the cloud feedbacks from observations (see Chapter 7). Studies based on PPE and CMIP3 support the conclusion that a credible representation of the mean climate and variability is very difficult to achieve with equilibrium climate sensitivities below 2°C (Piani et al., 2005; Stainforth et al., 2005; Sanderson et al., 2008a, 2008b; Huber et al., 2011; Klocke et al., 2011; Fasullo and Trenberth, 2012). High climate sensitivity values above 5°C (in some cases above 10°C) are found in the PPE based on HadAM/HadCM3. Several recent studies find that such high values cannot be excluded based on climatological constraints, but comparison with observations shows the smallest errors for many fields if ECS is between 3 and 4°C (Piani et al., 2005; Knutti et al., 2006; Rodwell and Palmer, 2007; Sanderson et al., 2008a, 2008b, 2010; Sanderson, 2011, 2013).

9.8 Relating Model Performance to Credibility of Model Applications

9.8.1 Synthesis Assessment of Model Performance

This chapter has assessed the performance of individual climate models as well as the multi-model mean. In addition, changes between models available now and those that were available at the time of the AR4 have been documented. The models display a range of abilities to simulate climate characteristics, underlying processes, and phenomena. No model scores high or low in all performance metrics, but some models perform substantially better than others for specific climate variables or phenomena. For a few climate characteristics, the assessment has shown that some classes of models, for example, those with higher horizontal resolution, higher model top or a more complete representation of the carbon cycle, aerosols or chemistry, agree better with observations, although this is not universally true.

Figure 9.44 provides a synthesis of key model evaluation results for AOGCMs and ESMs. The figure makes use of the calibrated language as defined in Mastrandrea et al. (2011). The *x*-axis refers to the level of confidence which increases towards the right as suggested by the increasing strength of shading. The level of confidence is a combination of the level of evidence and the degree of agreement. The level of evidence includes the number of studies and quality of observational data. Generally, evidence is most robust when there are multiple, independent studies that evaluate multiple models using high-quality observations. The degree of agreement measures whether different studies come to the same conclusions or not. The figure shows that several important aspects of the climate are simulated well by contemporary models, with varying levels of confidence. The colour coding provides an indication of how model quality has changed from CMIP3 to CMIP5. For example, there is *high confidence* that the model performance for global mean surface air temperature (TAS) is high, and it is shown in green because there is *robust evidence* of improvement since CMIP3. By contrast, the diurnal cycle of global mean surface air temperature (TAS-diur) is simulated with medium performance, but there is *low confidence* in this assessment owing to as yet limited analyses. It should be noted that there are no instances in the figure for

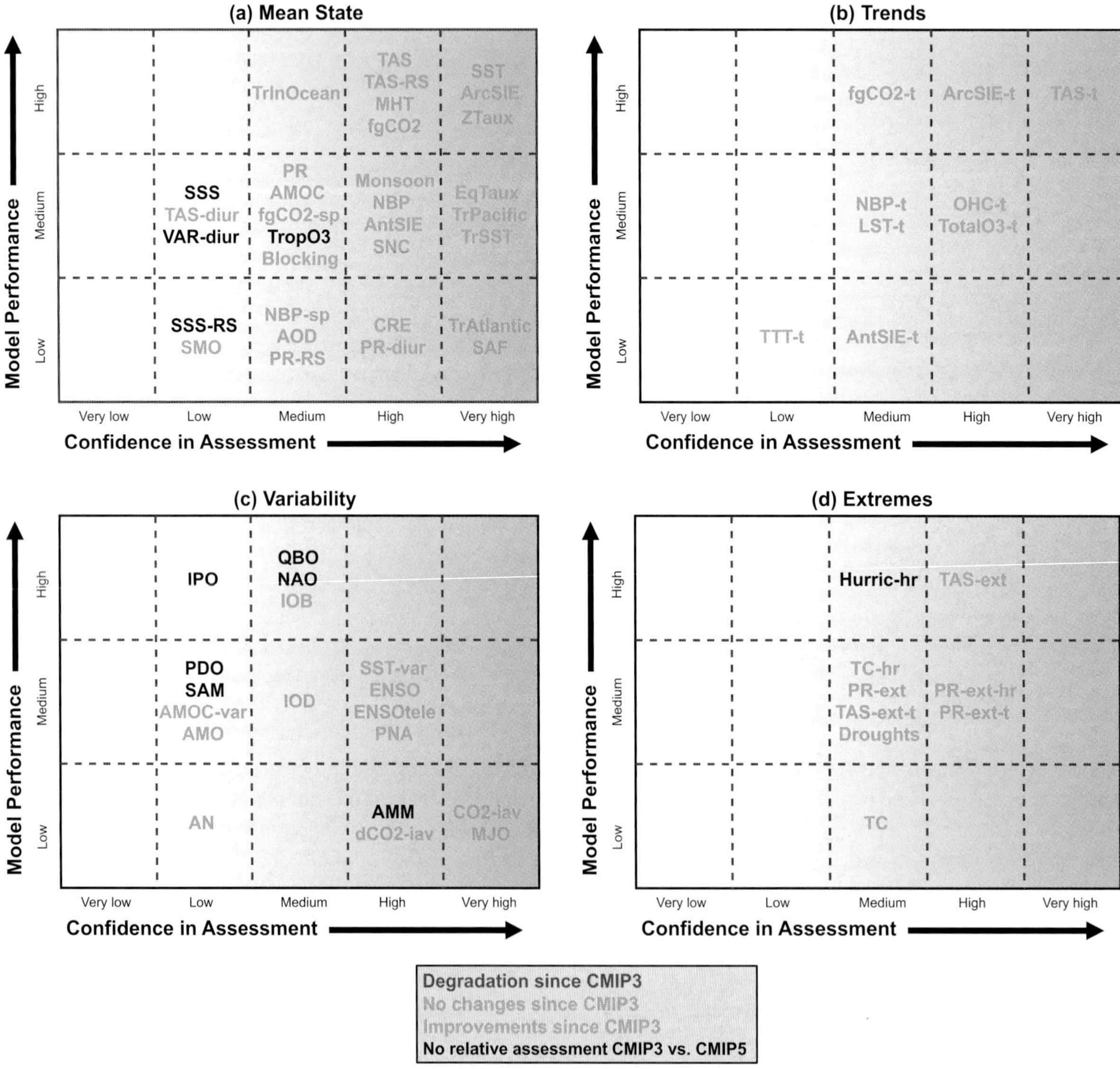

Figure 9.44 | Summary of the findings of Chapter 9 with respect to how well the CMIP5 models simulate important features of the climate of the 20th century. Confidence in the assessment increases towards the right as suggested by the increasing strength of shading. Model performance improves from bottom to top. The colour coding indicates changes since CMIP3 (or models of that generation) to CMIP5. The assessment of model performance is expert judgment based on the agreement with observations of the multi-model mean and distribution of individual models around the mean, taking into account internal climate variability. Note that assessed model performance is simplified for representation in the figure and it is referred to the text for details of each assessment. The figure highlights the following key features, with the sections that back up the assessment added in parentheses:

PANEL a:

AMOC	Atlantic Meridional Overturning Circulation mean (Section 9.4.2.3)
AntSIE	Seasonal cycle Antarctic sea ice extent (Section 9.4.3)
AOD	Aerosol Optical Depth (Section 9.4.6)
ArctSIE	Seasonal cycle Arctic sea ice extent (Section 9.4.3)
Blocking	Blocking events (Section 9.5.2.2)
CRE	Cloud radiative effects (Section 9.4.1.2)
EqTaux	Equatorial zonal wind stress (Section 9.4.2.4)
fgCO2	Global ocean carbon sink (Section 9.4.5)
fgCO2-sp	Spatial pattern of ocean–atmosphere CO_2 fluxes (Section 9.4.5)
MHT	Meridional heat transport (Section 9.4.2.4)
Monsoon	Global monsoon (Section 9.5.2.4)
NBP	Global land carbon sink (Section 9.4.5)
NBP-sp	Spatial pattern of land–atmosphere CO_2 fluxes (Section 9.4.5)
PR	Large scale precipitation (Sections 9.4.1.1, 9.4.1.3)
PR-diur	Diurnal cycle precipitation (Section 9.5.2.1)
PR-RS	Regional scale precipitation (Section 9.6.1.1)
SAF	Snow albedo feedbacks (Section 9.8.3)
SMO	Soil moisture (Section 9.4.4)
SNC	Snow cover (Section 9.4.4)
SSS	Sea surface salinity (Section 9.4.2.1)
SSS-RS	Regional Sea surface salinity (Section 9.4.2.1)
SST	Sea surface temperature (Section 9.4.2.1)
TAS	Large scale surface air temperature (Sections 9.4.1.1, 9.4.1.3)
TAS-diur	Diurnal cycle surface air temperature (Section 9.5.2.1)
TAS-RS	Regional scale surface air temperature (Section 9.6.1.1)
TrSST	Tropical sea surface temperature (Section 9.4.2.1)
TropO3	Tropospheric column ozone climatology (Section 9.4.1.4.5)
TrAtlantic	Tropical Atlantic mean state (Section 9.4.2.5)
TrInOcean	Tropical Indian Ocean mean state (Section 9.4.2.5)
TrPacific	Tropical Pacific mean state (Section 9.4.2.5)
VAR-diur	Diurnal cycle other variables (Section 9.5.2.1)
ZTaux	Zonal mean zonal wind stress (Section 9.4.2.4)

PANEL b (Trends)

AntSIE-t	Trend in Antarctic sea ice extent (Section 9.4.3)
ArctSIE-t	Trend in Arctic sea ice extent (Section 9.4.3)
fgCO2-t	Global ocean carbon sink trends (Section 9.4.5)
LST-t	Lower stratospheric temperature trends (Section 9.4.1.4.5)
NBP-t	Global land carbon sink trends (Section 9.4.5)
OHC-t	Global ocean heat content trends (Section 9.4.2.2)
TotalO3-t	Total column ozone trends (Section 9.4.1.4.5)
TAS-t	Surface air temperature trends (Section 9.4.1.4.1)
TTT-t	Tropical tropospheric temperature trends (Section 9.4.1.4.2)

PANEL c (Variability):

AMM	Atlantic Meridional Mode (Section 9.5.3.3)
AMO	Atlantic Multi-decadal Variability (Section 9.5.3.3)
AMOC-var	Atlantic Meridional Overturning Circulation (Section 9.5.3.3)
AN	Atlantic Niño (Section 9.5.3.3)
CO2-iav	Interannual variability of atmospheric CO_2 (Section 9.8.3)
dCO2-iav	Sensitivity of CO_2 growth rate to tropical temperature (Section 9.8.3)
ENSO	El Niño Southern Oscillation (Section 9.5.3.4)
ENSOtele	Tropical ENSO teleconnections (Section 9.5.3.5)
IOB	Indian Ocean basin mode (Section 9.5.3.4)
IOD	Indian Ocean dipole (Section 9.5.3.4)
IPO	Interdecadal Pacific Oscillation (Section 9.5.3.6)
MJO	Madden-Julian Oscillation (Section 9.5.2.3)
NAO	North Atlantic Oscillation and Northern annular mode (Section 9.5.3.2)
PDO	Pacific Decadal Oscillation (Section 9.5.3.6)
PNA	Pacific North American (Section 9.5.3.5)
QBO	Quasi-Biennial Oscillation (Section 9.5.3.7)
SAM	Southern Annular Mode (Section 9.5.3.2)
SST-var	Global sea surface temperature variability (Section 9.5.3.1)

PANEL d (Extremes):

Hurric-hr	Year-to-year counts of Atlantic hurricanes in high-resolution AGCMs (Section 9.5.4.3)
PR-ext	Global distributions of precipitation extremes (Section 9.5.4.2)
PR-ext-hr	Global distribution of precipitation extremes in high-resolution AGCMs (Section 9.5.4.2)
PR-ext-t	Global trends in precipitation extremes (Section 9.5.4.2)
TAS-ext	Global distributions of surface air temperature extremes (Section 9.5.4.1)
TAS-ext-t	Global trends in surface air temperature extremes (Section 9.5.4.1)
TC	Tropical cyclone tracks and intensity (Section 9.5.4.3)
TC-hr	Tropical cyclone tracks and intensity in high-resolution AGCMs (Section 9.5.4.3)
Droughts	Droughts (Section 9.5.4.4)

which CMIP5 models perform worse than CMIP3 models (something that would have been indicated by the red colour). A description that explains the expert judgment for each of the results presented in Figure 9.44 can be found in the body of this chapter, with a link to the specific sections given in the figure caption.

EMICs have also been evaluated to some extent in this chapter as they are used to provide long-term projections (in Chapter 12) beyond year 2300, and to provide large ensembles emulating the response of more comprehensive ESMs and allowing probabilistic estimates. Results from the EMIC intercomparison project (Eby et al., 2013; Zickfeld et al., 2013) illustrate the ability to reproduce the large-scale climate changes in GMST (Figure 9.8) and OHC (Figure 9.17) during the 20th century. The models also estimate CO_2 fluxes for land and oceans, which are as consistent with observations as are fluxes estimated by ESMs (Figure 9.27). This gives confidence that the EMICs, albeit limited in the scope and resolution of information they can provide, can be used for long-term projections compatible with those of ESMs (Plattner et al., 2008; Eby et al., 2013). Overall, these studies imply that EMICs are well suited for simulations extending beyond the CMIP5 ensemble.

Frequently Asked Questions

FAQ 9.1 | Are Climate Models Getting Better, and How Would We Know?

Climate models are extremely sophisticated computer programs that encapsulate our understanding of the climate system and simulate, with as much fidelity as currently feasible, the complex interactions between the atmosphere, ocean, land surface, snow and ice, the global ecosystem and a variety of chemical and biological processes.

The complexity of climate models—the representation of physical processes like clouds, land surface interactions and the representation of the global carbon and sulphur cycles in many models—has increased substantially since the IPCC First Assessment Report in 1990, so in that sense, current Earth System Models are vastly 'better' than the models of that era. This development has continued since the Fourth Assessment, while other factors have also contributed to model improvement. More powerful supercomputers allow current models to resolve finer spatial detail. Today's models also reflect improved understanding of how climate processes work—understanding that has come from ongoing research and analysis, along with new and improved observations.

Climate models of today are, in principle, better than their predecessors. However, every bit of added complexity, while intended to improve some aspect of simulated climate, also introduces new sources of possible error (e.g., via uncertain parameters) and new interactions between model components that may, if only temporarily, degrade a model's simulation of other aspects of the climate system. Furthermore, despite the progress that has been made, scientific uncertainty regarding the details of many processes remains.

An important consideration is that model performance can be evaluated only relative to past observations, taking into account natural internal variability. To have confidence in the future projections of such models, historical climate—and its variability and change—must be well simulated. The scope of model evaluation, in terms of the kind and quantity of observations available, the availability of better coordinated model experiments, and the expanded use of various performance metrics, has provided much more quantitative information about model performance. But this alone may not be sufficient. Whereas weather and seasonal climate predictions can be regularly verified, climate projections spanning a century or more cannot. This is particularly the case as anthropogenic forcing is driving the climate system toward conditions not previously observed in the instrumental record, and it will always be a limitation.

Quantifying model performance is a topic that has featured in all previous IPCC Working Group I Reports. Reading back over these earlier assessments provides a general sense of the improvements that have been made. Past reports have typically provided a rather broad survey of model performance, showing differences between model-calculated versions of various climate quantities and corresponding observational estimates.

Inevitably, some models perform better than others for certain climate variables, but no individual model clearly emerges as 'the best' overall. Recently, there has been progress in computing various performance metrics, which synthesize model performance relative to a range of different observations according to a simple numerical score. Of course, the definition of such a score, how it is computed, the observations used (which have their

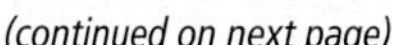
(continued on next page)

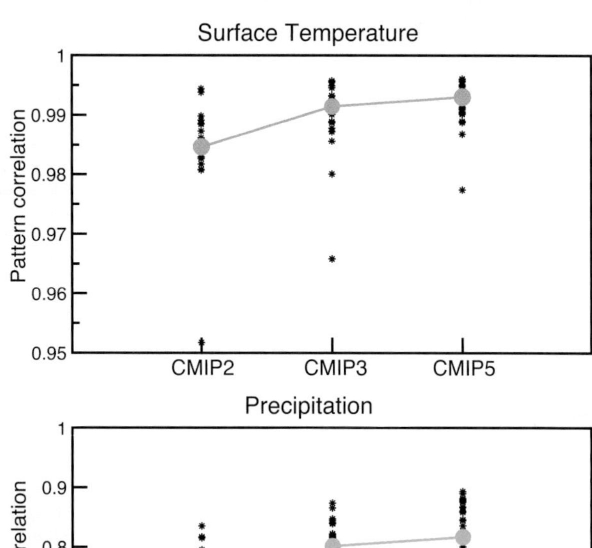

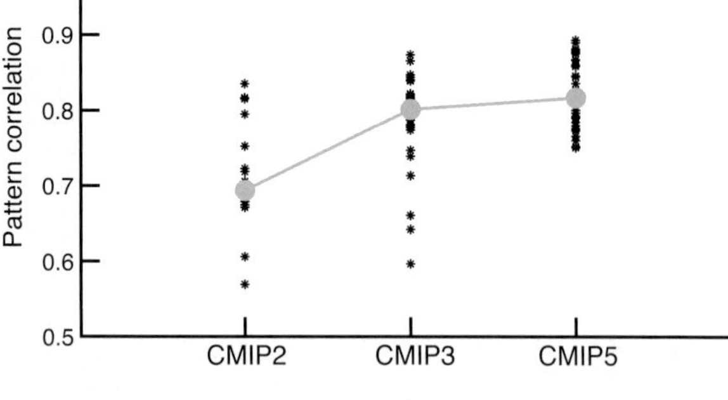

FAQ 9.1, Figure 1 | Model capability in simulating annual mean temperature and precipitation patterns as illustrated by results of three recent phases of the Coupled Model Intercomparison Project (CMIP2, models from about year 2000; CMIP3, models from about 2005; and CMIP5, the current generation of models). The figure shows the correlation (a measure of pattern similarity) between observed and modelled temperature (upper panel) and precipitation (lower panel). Larger values indicate better correspondence between modelled and observed spatial patterns. The black symbols indicate correlation coefficient for individual models, and the large green symbols indicate the median value (i.e., half of the model results lie above and the other half below this value). Improvement in model performance is evident by the increase in correlation for successive model generations.

FAQ 9.1 (continued)

own uncertainties), and the manner in which various scores are combined are all important, and will affect the end result.

Nevertheless, if the metric is computed consistently, one can compare different generations of models. Results of such comparisons generally show that, although each generation exhibits a range in performance, the average model performance index has improved steadily between each generation. An example of changes in model performance over time is shown in FAQ 9.1, Figure 1, and illustrates the ongoing, albeit modest, improvement. It is interesting to note that both the poorest and best performing models demonstrate improvement, and that this improvement comes in parallel with increasing model complexity and an elimination of artificial adjustments to atmosphere and ocean coupling (so-called 'flux adjustment'). Some of the reasons for this improvement include increased understanding of various climate processes and better representation of these processes in climate models. More comprehensive Earth observations are also driving improvements.

So, yes, climate models are getting better, and we can demonstrate this with quantitative performance metrics based on historical observations. Although future climate projections cannot be directly evaluated, climate models are based, to a large extent, on verifiable physical principles and are able to reproduce many important aspects of past response to external forcing. In this way, they provide a scientifically sound preview of the climate response to different scenarios of anthropogenic forcing.

9.8.2 Implications of Model Evaluation for Climate Change Detection and Attribution

The evaluation of model simulations of historical climate is of direct relevance to detection and attribution (D&A) studies (Chapter 10) since these rely on model-derived patterns (or 'fingerprints') of climate response to external forcing, and on the ability of models to simulate decadal and longer-time scale internal variability (Hegerl and Zwiers, 2011). Conversely, D&A research contributes to model evaluation through estimation of the amplitude of modeled response to various forcings (Section 10.3.1.1.3). The estimated fingerprint for some variables such as water vapor is governed by basic physical processes that are well represented in models and are rather insensitive to model uncertainties (Santer et al., 2009). Figure 9.44 illustrates slight improvements in the representation of some of the modes of variability and climate phenomena discussed in Sections 9.5.2 and 9.5.3, suggesting with *medium confidence* that models now better reproduce internal variability. On the other hand, biases that affect D&A studies remain. An example is the warm bias of lower-stratosphere temperature trends during the satellite period (Section 9.4.1.4.5) that can be linked to uncertainties in stratospheric ozone forcing (Solomon et al., 2012; Santer et al., 2013). Recent studies of climate extremes (Section 9.5.4) also provide evidence that models have reasonable skill in these important attributes of a changing climate; however, there is an indication that models have difficulties in reproducing the right balance between historical changes in cold and warm extremes. They also confirm that resolution affects the confidence that can be placed in the analyses of extreme in precipitation. D&A studies focussed on extreme events are therefore constrained by current model limitations. Lastly, some D&A studies have incorporated model quality results by repeating a multi-model analysis with only the models that agree best with observations (Santer et al., 2009). This model discrimination or weighting is less problematic for D&A analysis than it is for model projections of future climate (Section 9.8.3), because D&A research is focussed on historical and control-run simulations which can be directly evaluated against observations.

9.8.3 Implications of Model Evaluation for Model Projections of Future Climate

Confidence in climate model projections is based on physical understanding of the climate system and its representation in climate models, and on a demonstration of how well models represent a wide range of processes and climate characteristics on various spatial and temporal scales (Knutti et al., 2010b). A climate model's credibility is increased if the model is able to simulate past variations in climate, such as trends over the 20th century and palaeoclimatic changes. Projections from previous IPCC assessments can also be directly compared to observations (see Figures 1.4 and 1.5), with the caveat that these projections were not intended to be predictions over the short time scales for which observations are available to date. Unlike shorter lead forecasts, longer-term climate change projections push models into conditions outside the range observed in the historical period used for evaluation.

In some cases, the spread in climate projections can be reduced by weighting of models according to their ability to reproduce past observed climate. Several studies have explored the use of unequally weighted means, with the weights based on the models' performance in simulating past variations in climate, typically using some performance metric or collection of metrics (Connolley and Bracegirdle, 2007; Murphy et al., 2007; Waugh and Eyring, 2008; Pierce et al., 2009; Reifen and Toumi, 2009; Christensen et al., 2010; Knutti et al., 2010b; Raisanen et al., 2010; Abe et al., 2011; Shiogama et al., 2011; Watterson and Whetton, 2011; Tsushima et al., 2013). When applied to projections of Arctic sea ice, averages in which extra weight is given to models with the most realistic historical sea ice do give different results than the unweighted mean (Stroeve et al., 2007, 2012; Scherrer, 2011; Massonnet et al., 2012; Wang and Overland, 2012; Overland and Wang, 2013). Another frequently used approach is the re-calibration of model outputs to a given observed value (Boe et al., 2009b; Mahlstein and Knutti, 2012; Wang and Overland, 2012), see further discussion in Section 12.4.6.1. Some studies explicitly formulate a statistical frameworks that relate future observables to climate model output (reviewed in Knutti et al. (2010b) and Stephenson et al. (2012)). Such frameworks not only provide weights for the mean response but also allow the uncertainty in the predicted response to be quantified (Bracegirdle and Stephenson, 2012).

There are several encouraging examples of 'emergent constraints', which are relationships across an ensemble of models between some aspect of Earth System sensitivity and an observable trend or variation in the contemporary climate (Allen and Ingram, 2002; Hall and Qu, 2006; Eyring et al., 2007; Boe et al., 2009a, 2009b; Mahlstein and Knutti, 2010; Son et al., 2010; Huber et al., 2011; Schaller et al., 2011; Bracegirdle and Stephenson, 2012; Fasullo and Trenberth, 2012; O'Gorman, 2012). For example, analyzing the CMIP3 ensemble, Hall and Qu (2006) showed that inter-model variations of snow albedo feedback in the contemporary seasonal cycle strongly correlate with comparably large inter-model variations in this feedback under future climate change. An update of this analysis with CMIP5 models added is shown in Figure 9.45 (left panel). This relationship presumably arises from the fact that surface albedo values in areas covered by snow vary widely across the models, particularly in the heavily vegetated boreal forest zone. Models with higher surface albedos in these areas have a larger contrast between snow-covered and snow-free areas, and hence a stronger snow albedo feedback whether the context is the seasonal variation in sunshine or anthropogenic forcing. Comparison with an observational estimate of snow albedo feedback reveals a large spread with both high and low biases.

The right panel of Figure 9.45 shows another example of an emergent constraint, where the sensitivity of tropical land carbon to warming (i.e., without CO_2 fertilization effects) is related to the sensitivity of the annual CO_2 growth rate to tropical temperature anomalies (Cox et al., 2013)). The horizontal axis is the regression of the atmospheric CO_2 growth rate on the tropical temperature anomaly for each model. The strong statistical relationship between these two variables is consistent with the fact that interannual variability in the CO_2 growth-rate is known to be dominated by the response of tropical land to climatic anomalies, associated particularly with ENSO. Thus the relationship has a physical as well as a statistical basis. The interannual sensitivity of the CO_2 growth rate to tropical temperature can be estimated from observational data. Like the snow albedo feedback example, this inter-model relationship provides a credible means to reduce model spread in the sensitivity of tropical land carbon to tropical climate change.

On the other hand, many studies have failed to find strong relationships between observables and projections. Whetton et al. (2007) and Knutti et al. (2010a) found that correlations between local to regional climatological values and projected changes are small except for a few regions. Scherrer (2011) finds no robust relationship between the ability of the CMIP3 models to represent interannual variability of near-surface air temperature and the amplitude of future warming.Raisanen et al. (2010) report only small (10–20%) reductions in cross-validation error of simulated 21st century temperature changes when weighting the CMIP3 models based on their simulation of the present-day climatology. The main difficulties are sparse coverage in

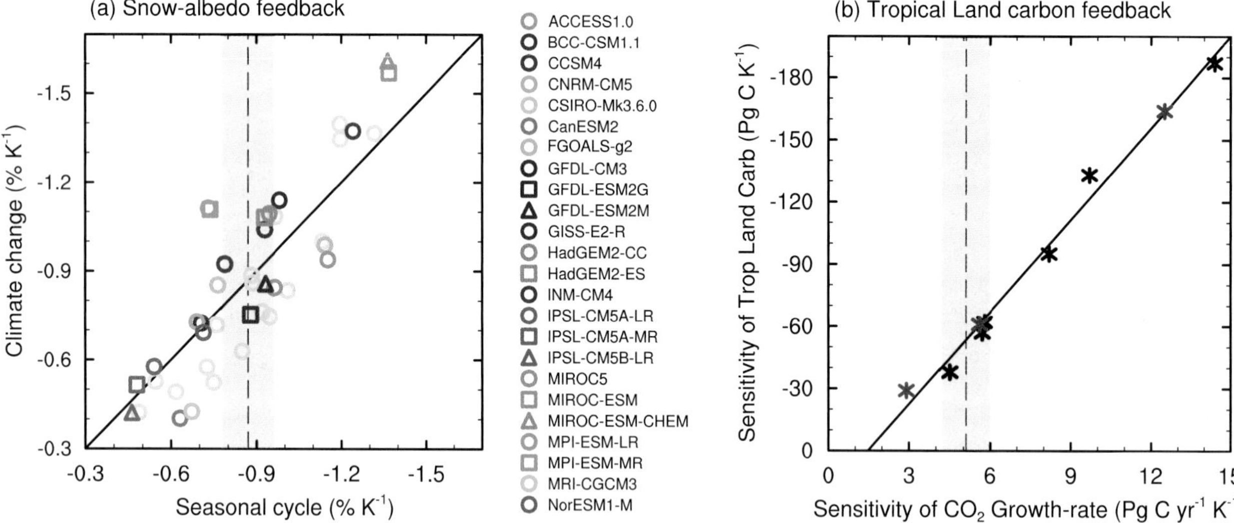

Figure 9.45 | (Left) Scatterplot of simulated springtime snow–albedo feedback ($\Delta\alpha_s/\Delta T_s$) values in climate change (*y*-axis) versus simulated springtime $\Delta\alpha_s/\Delta T_s$ values in the seasonal cycle (*x*-axis) in transient climate change experiments from 17 CMIP3 (blue) and 24 CMIP5 models (α_s and T_s are surface albedo and surface air temperature, respectively). (Adapted from Hall and Qu, 2006.) (Right) Constraint on the climate sensitivity of land carbon in the tropics (30°N to 30°S) from interannual variability in the growth rate of global atmospheric CO_2 (Cox et al., 2013). This is based on results from Earth System Models (ESMs) with free-running CO_2; Coupled Climate Carbon Cycle Model Intercomparison Project General Circulation Models (C4MIP GCMs, black labels; Friedlingstein et al., 2006), and three land carbon 'physics ensembles' with HadCM3 (red labels; Booth et al., 2012b) . The values on the *y*-axis are calculated over the period 1960–2099 inclusive, and those on the *x*-axis over the period 1960–2010 inclusive. In both cases the temperature used is the mean (land+ocean) temperature over 30°N to 30°S. The width of the vertical yellow bands in both (a) and (b) shows the observation-based estimate of the variable on the *x*-axis.

many observed variables, short time series for observed trends, lack of correlation between observed quantities and projected past or future trends, and systematic errors in the models (Tebaldi and Knutti, 2007; Jun et al., 2008; Knutti, 2010; Knutti et al., 2010a), the ambiguity of possible performance metrics and the difficulty of associating them with predictive skill (Parker et al., 2007; Gleckler et al., 2008; Pincus et al., 2008; Reichler and Kim, 2008; Pierce et al., 2009; Knutti et al., 2010a).

Emergent constraints can be difficult to identify if climate models are structurally similar and share common biases, thereby reducing the effective ensemble size. Comparison of emergent constraints in MMEs from different modelling experiments can help reveal which constraints are robust (Massonnet et al., 2012; Bracegirdle and Stephenson, 2013). Another issue is that testing of large numbers of predictors will find statistically significant correlations that do not remain significant in a different ensemble. This is particularly important if many predictors are tested using only small ensembles like CMIP3 (DelSole and Shukla, 2009; Raisanen et al., 2010; Huber et al., 2011; Masson and Knutti, 2013). All of these potential pitfalls underscore the need for analysis of the mechanism underpinning the statistical relationship between current and future climate parameters for any proposed emergent constraint.

References

Abe, M., H. Shiogama, T. Nozawa, and S. Emori, 2011: Estimation of future surface temperature changes constrained using the future-present correlated modes in inter-model variability of CMIP3 multimodel simulations. *J. Geophys. Res. Atmos.*, **116**, D18104.

Abiodun, B., W. Gutowski, A. Abatan, and J. Prusa, 2011: CAM-EULAG: A non-hydrostatic atmospheric climate model with grid stretching. *Acta Geophys.*, **59**, 1158–1167.

Abramowitz, G., R. Leuning, M. Clark, and A. Pitman, 2008: Evaluating the performance of land surface models. *J. Clim.*, **21**, 5468–5481.

AchutaRao, K., and K. Sperber, 2002: Simulation of the El Niño Southern Oscillation: Results from the coupled model intercomparison project. *Clim. Dyn.*, **19**, 191–209.

AchutaRao, K., and K. Sperber, 2006: ENSO simulations in coupled ocean-atmosphere models: Are the current models better? *Clim. Dyn.*, **27**, 1–16.

Achuthavarier, D., V. Krishnamurthy, B. P. Kirtman, and B. H. Huang, 2012: Role of the Indian Ocean in the ENSO-Indian Summer Monsoon Teleconnection in the NCEP Climate Forecast System. *J. Clim.*, **25**, 2490–2508.

Ackerley, D., E. J. Highwood, and D. J. Frame, 2009: Quantifying the effects of perturbing the physics of an interactive sulfur scheme using an ensemble of GCMs on the climateprediction.net platform. *J. Geophys. Res. Atmos.*, **114**, D01203

Adachi, Y., et al., 2013: Basic performance of a new earth system model of the Meteorological Research Institute (MRI-ESM1). *Papers Meteorol. Geophys.*, doi:10.2467/mripapers.64.1.

Adam, J., E. Clark, D. Lettenmaier, and E. Wood, 2006: Correction of global precipitation products for orographic effects. *J. Clim.*, **19**, 15–38.

Adkins, J. F., K. McIntyre, and D. P. Schrag, 2002: The salinity, temperature, and delta O-18 of the glacial deep ocean. *Science*, **298**, 1769–1773.

Adler, R. F., et al., 2003: The Version 2 Global Precipitation Climatology Project (GPCP) Monthly Precipitation Analysis (1979–Present). *J. Hydrometeor.*, **4**, 1147–1167.

Alekseev, V. A., E. M. Volodin, V. Y. Galin, V. P. Dymnikov, and V. N. Lykossov, 1998: Modeling of the present-day climate by the atmospheric model of INM RAS DNM GCM. Description of the model version A5421 and results of AMIP2 simulations. Institute of Numerical Mathematics, Moscow, Russia, 200 pp.

Alessandri, A., P. G. Fogli, M. Vichi, and N. Zeng, 2012: Strengthening of the hydrological cycle in future scenarios: Atmospheric energy and water balance perspective. *Earth Syst. Dyn.*, **3**, 199–212.

Alexander, M. J., et al., 2010: Recent developments in gravity-wave effects in climate models and the global distribution of gravity-wave momentum flux from observations and models. *Q. J. R. Meteorol. Soc.*, **136**, 1103–1124.

Alexandru, A., R. de Elia, and R. Laprise, 2007: Internal variability in regional climate downscaling at the seasonal scale. *Mon. Weather Rev.*, **135** 3221–3238.

Alexandru, A., R. de Elia, R. Laprise, L. Separovic, and S. Biner, 2009: Sensitivity study of regional climate model simulations to large-scale nudging parameters. *Mon. Weather Rev.*, **137**, 1666–1686.

Allan, R. P., and B. J. Soden, 2008: Atmospheric warming and the amplification of precipitation extremes. *Science*, **321**, 1481–1484.

Allan, R. P., M. A. Ringer, and A. Slingo, 2003: Evaluation of moisture in the Hadley Centre climate model using simulations of HIRS water-vapour channel radiances. *Q. J. R. Meteorol. Soc.*, **129**, 3371–3389.

Allan, R. P., A. Slingo, S. F. Milton, and M. E. Brooks, 2007: Evaluation of the Met Office global forecast model using Geostationary Earth Radiation Budget (GERB) data. *Q. J. R. Meteorol. Soc.*, **133**, 1993–2010.

Allan, R. P., B. J. Soden, V. O. John, W. Ingram, and P. Good, 2010: Current changes in tropical precipitation. *Environ. Res. Lett.*, **5**, 025205.

Allen, M., P. Stott, J. Mitchell, R. Schnur, and T. Delworth, 2000: Quantifying the uncertainty in forecasts of anthropogenic climate change. *Nature*, **407**, 617–620.

Allen, M. R., and W. J. Ingram, 2002: Constraints on future changes in climate and the hydrologic cycle. *Nature*, **419**, 224–232.

Ammann, C. M., G. A. Meehl, W. M. Washington, and C. S. Zender, 2003: A monthly and latitudinally varying volcanic forcing dataset in simulations of 20th century climate. *Geophys. Res. Lett.*, **30**, 1657.

Anav, A., et al., 2013: Evaluating the land and ocean components of the global carbon cycle in the CMIP5 Earth System Models. *J. Clim.*, **26**, 6801–6843.

Anderson, B. T., J. R. Knight, M. A. Ringer, C. Deser, A. S. Phillips, J. H. Yoon, and A. Cherchi, 2010: Climate forcings and climate sensitivities diagnosed from atmospheric global circulation models. *Clim. Dyn.*, **35**, 1461–1475.

Andrews, T., J. M. Gregory, M. J. Webb, and K. E. Taylor, 2012: Forcing, feedbacks and climate sensitivity in CMIP5 coupled atmosphere-ocean climate models. *Geophys. Res. Lett.*, **39**, L09712.

Annamalai, H., and K. R. Sperber, 2005: Regional heat sources and the active and break phases of boreal summer intraseasonal (30–50 day) variability. *J. Atmos. Sci.*, **62**, 2726–2748.

Annan, J., and J. Hargreaves, 2011: Understanding the CMIP3 Multimodel Ensemble. *J. Clim.*, **24**, 4529–4538.

Annan, J. D., and J. C. Hargreaves, 2010: Reliability of the CMIP3 ensemble. *Geophys. Res. Lett.*, **37**, L02703.

Annan, J. D., D. J. Lunt, J. C. Hargreaves, and P. J. Valdes, 2005: Parameter estimation in an atmospheric GCM using the Ensemble Kalman Filter. *Nonlin. Proc. Geophys.*, **12**, 363–371.

Anstey, J. A., et al., 2013: Multi-model analysis of Northern Hemisphere winter blocking and its relation to the stratosphere. *J. Geophys. Res. Atmos.*, **118**, 3956–3971.

Antonov, J. I., et al., 2010: *World Ocean Atlas 2009*, Vol. 2: *Salinity*. [S. Levitus (eds.)]. NOAA Atlas NESDIS 69, U.S. Gov. Printing Office, Washington, D.C., 184 pp.

Archer, D. E., G. Eshel, A. Winguth, W. Broecker, R. Pierrehumbert, M. Tobis, and R. Jacob, 2000: Atmospheric pCO(2) sensitivity to the biological pump in the ocean. *Global Biogeochem. Cycles*, **14**, 1219–1230.

Arneth, A., et al., 2010: From biota to chemistry and climate: Towards a comprehensive description of trace gas exchange between the biosphere and atmosphere. *Biogeosciences*, **7**, 121–149.

Arora, V. K., and G. J. Boer, 2005: Fire as an interactive component of dynamic vegetation models. *J. Geophys. Res.-Biogeosciences*, **110**, G02008.

Arora, V. K., and G. J. Boer, 2010: Uncertainties in the 20th century carbon budget associated with land use change. *Global Change Biol.*, **16**, 3327–3348.

Arora, V. K., et al., 2011: Carbon emission limits required to satisfy future representative concentration pathways of greenhouse gases. *Geophys. Res. Lett.*, **38**, L05805.

Arora, V. K., et al., 2009: The effect of terrestrial photosynthesis down regulation on the twentieth-century carbon budget simulated with the CCCma Earth System Model. *J. Clim.*, **22**, 6066–6088.

Artale, V., et al., 2010: An atmosphere–ocean regional climate model for the Mediterranean area: Assessment of a present climate simulation. *Clim. Dyn.*, **35**, 721–740.

Arzhanov, M. M., P. F. Demchenko, A. V. Eliseev, and I. I. Mokhov, 2008: Simulation of characteristics of thermal and hydrologic soil regimes in equilibrium numerical experiments with a Climate Model of Intermediate Complexity. *Izvestiya Atmos. Ocean. Phys.*, **44**, 548–566.

Assmann, K. M., M. Bentsen, J. Segschneider, and C. Heinze, 2010: An isopycnic ocean carbon cycle model. *Geosci. Model Dev.*, **3**, 143–167.

Aumont, O., and L. Bopp, 2006: Globalizing results from ocean in situ iron fertilization studies. *Global Biogeochem. Cycles*, **20**, Gb2017.

Aumont, O., E. Maier-Reimer, S. Blain, and P. Monfray, 2003: An ecosystem model of the global ocean including Fe, Si, P colimitations. *Global Biogeochem. Cycles*, **17**, 1060.

Austin, J., and R. J. Wilson, 2006: Ensemble simulations of the decline and recovery of stratospheric ozone. *J. Geophys. Res. Atmos.*, **111**, D16314.

Axelsson, P., M. Tjernström, S. Söderberg, and G. Svensson, 2011: An ensemble of Arctic simulations of the AOE-2001 field experiment. *Atmosphere*, **2**, 146–170.

Baehr, J., S. Cunnningham, H. Haak, P. Heimbach, T. Kanzow, and J. Marotzke, 2009: Observed and simulated estimates of the meridional overturning circulation at 26.5 N in the Atlantic. *Ocean Sci.*, **5**, 575–589.

Balan Sarojini, B., et al., 2011: High frequency variability of the Atlantic meridional overturning circulation. *Ocean Science*, **7**, 471–486.

Baldwin, M. P., et al., 2001: The quasi-biennial oscillation. *Rev. Geophys.*, **39**, 179–229.

Balsamo, G., P. Viterbo, A. Beljaars, B. van den Hurk, M. Hirschi, A. K. Betts, and K. Scipal, 2009: A revised hydrology for the ECMWF Model: Verification from field site to terrestrial water storage and impact in the Integrated Forecast System. *J. Hydrometeorol.*, **10**, 623–643.

Bao, Q., G. Wu, Y. Liu, J. Yang, Z. Wang, and T. Zhou, 2010: An introduction to the coupled model FGOALS1.1-s and its performance in East Asia. *Adv. Atmos. Sci.*, **27**, 1131–1142.

Bao, Q., et al., 2013: The Flexible Global Ocean-Atmosphere-Land System model Version: FGOALS-s2. *Adv. Atmos. Sci.*, doi:10.1007/s00376-012-2113-9.

Bao, Y., F. L. Qiao, and Z. Y. Song, 2012: Historical simulation and twenty-first century prediction of oceanic CO_2 sink and pH change. *Acta Ocean. Sin.*, **31**, 87–97.

Barkstrom, B. R., 1984: The Earth Radiation Budget Experiment (ERBE). *Bull. Am. Meteorol. Soc.*, **65**, 1170–1185.

Barnes, E. A., and D. L. Hartmann, 2010: Influence of eddy-driven jet latitude on North Atlantic jet persistence and blocking frequency in CMIP3 integrations. *Geophys. Res. Lett.*, **37**, L23802.

Barnes, E. A., J. Slingo, and T. Woollings, 2012: A methodology for the comparison of blocking climatologies across indices, models and climate scenarios. *Clim. Dyn.*, **38**, 2467–2481.

Barnier, B., et al., 2006: Impact of partial steps and momentum advection schemes in a global ocean circulation model at eddy-permitting resolution. *Ocean Dyn.*, **56**, 543–567.

Barriopedro, D., R. Garcia-Herrera, and R. M. Trigo, 2010a: Application of blocking diagnosis methods to General Circulation Models. Part I: A novel detection scheme. *Clim. Dyn.*, **35**, 1373–1391.

Barriopedro, D., R. Garcia-Herrera, J. F. Gonzalez-Rouco, and R. M. Trigo, 2010b: Application of blocking diagnosis methods to General Circulation Models. Part II: Model simulations. *Clim. Dyn.*, **35**, 1393–1409.

Bartlein, P. J., et al., 2010: Pollen-based continental climate reconstructions at 6 and 21 ka: A global synthesis. *Clim. Dyn.*, **37**, 775–802.

Bathiany, S., M. Claussen, V. Brovkin, T. Raddatz, and V. Gayler, 2010: Combined biogeophysical and biogeochemical effects of large-scale forest cover changes in the MPI earth system model. *Biogeosciences*, **7**, 1383–1399.

Bauer, H. S., V. Wulfmeyer, and L. Bengtsson, 2008a: The representation of a synoptic-scale weather system in a thermodynamically adjusted version of the ECHAM4 general circulation model. *Meteorol. Atmos. Phys.*, **99**, 129–153.

Bauer, S. E., D. Koch, N. Unger, S. M. Metzger, D. T. Shindell, and D. G. Streets, 2007: Nitrate aerosols today and in 2030: A global simulation including aerosols and tropospheric ozone. *Atmos. Chem. Phys.*, **7**, 5043–5059.

Bauer, S. E., et al., 2008b: MATRIX (Multiconfiguration Aerosol TRacker of mIXing state): An aerosol microphysical module for global atmospheric models. *Atmos. Chem. Phys.*, **8**, 6003–6035.

Beare, R., et al., 2006: An intercomparison of large-eddy simulations of the Stable Boundary Layer. *Boundary-Layer Meteorol.*, **118**, 247–272.

Bellassen, V., G. Le Maire, J. F. Dhote, P. Ciais, and N. Viovy, 2010: Modelling forest management within a global vegetation model Part 1: Model structure and general behaviour. *Ecol. Model.*, **221**, 2458–2474.

Bellassen, V., G. le Maire, O. Guin, J. F. Dhote, P. Ciais, and N. Viovy, 2011: Modelling forest management within a global vegetation model-Part 2: Model validation from a tree to a continental scale. *Ecol. Model.*, **222**, 57–75.

Bellouin, N., J. Rae, A. Jones, C. Johnson, J. Haywood, and O. Boucher, 2011: Aerosol forcing in the Climate Model Intercomparison Project (CMIP5) simulations by HadGEM2–ES and the role of ammonium nitrate. *J. Geophys. Res.*, **116**, 1–25.

Bellucci, A., S. Gualdi, and A. Navarra, 2010: The Double-ITCZ Syndrome in Coupled General Circulation Models: The role of large-scale vertical circulation regimes. *J. Clim.*, **23**, 1127–1145.

Bender, M. A., T. R. Knutson, R. E. Tuleya, J. J. Sirutis, G. A. Vecchi, S. T. Garner, and I. M. Held, 2010: Modeled impact of anthropogenic warming on the frequency of intense Atlantic hurricanes. *Science*, **327**, 454–458.

Bengtsson, L., and K. Hodges, 2011: On the evaluation of temperature trends in the tropical troposphere. *Clim. Dyn.*, **36**, 419–430.

Bengtsson, L., K. I. Hodges, and N. Keenlyside, 2009: Will extratropical storms intensify in a warmer climate? *J. Clim.*, **22**, 2276–2301.

Berckmans, J., T. Woollings, M.-E. Demory, P.-L. Vidal, and M. Roberts, 2013: Atmospheric blocking in a high resolution climate model: Influences of mean state, orography and eddy forcing. *Atmos. Sci. Lett.*, **14**, 34–40.

Bergengren, J., D. Waliser, and Y. Yung, 2011: Ecological sensitivity: A biospheric view of climate change. *Clim. Change*, **107**, 433–457.

Bergengren, J., S. Thompson, D. Pollard, and R. DeConto, 2001: Modeling global climate-vegetation interactions in a doubled CO_2 world. *Clim. Change*, **50**, 31–75.

Bernie, D. J., E. Guilyardi, G. Madec, J. M. Slingo, S. Woolnough, and J. Cole, 2008: Impact of resolving the diurnal cycle in an ocean-atmosphere GCM. Part 2: A diurnally coupled CGCM. *Clim. Dyn.*, **31**, 909–925.

Bi, D., et al., 2013a: ACCESS-OM: The Ocean and Sea ice Core of the ACCESS Coupled Model. *Aust. Meteorol. Oceanogr. J.*, **63**, 213–232.

Bi, D., et al., 2013b: The ACCESS Coupled Model: Description, control climate and evaluation. *Aust. Meteorol. Oceanogr. J.*, **63**, 41–64.

Bitz, C. M., and W. H. Lipscomb, 1999: An energy-conserving thermodynamic sea ice model for climate study. *J. Geophys. Res.. Oceans*, **104**, 15669–15677.

Blyth, E., J. Gash, A. Lloyd, M. Pryor, G. P. Weedon, and J. Shuttleworth, 2010: Evaluating the JULES Land Surface Model Energy Fluxes Using FLUXNET Data. *J. Hydrometeorol.*, **11**, 509–519.

Boberg, F., and J. H. Christensen, 2012: Overestimation of Mediterranean summer temperature projections due to model deficiencies. *Nature Clim. Change*, **2**, 433–436.

Boccaletti, G., R. Ferrari, and B. Fox-Kemper, 2007: Mixed layer instabilities and restratification. *J. Phys. Oceanogr.*, **37**, 2228–2250.

Bodas-Salcedo, A., K. D. Williams, P. R. Field, and A. P. Lock, 2012: The surface downwelling solar radiation surplus over the Southern Ocean in the Met Office Model: The role of midlatitude cyclone clouds. *J. Clim.*, **25**, 7467–7486.

Bodas-Salcedo, A., M. Webb, M. Brooks, M. Ringer, K. Williams, S. Milton, and D. Wilson, 2008: Evaluating cloud systems in the Met Office global forecast model using simulated CloudSat radar reflectivities. *J. Geophys. Res. Atmos.*, **113**, D00A13.

Bodas-Salcedo, A., et al., 2011: COSP: Satellite simulation software for model assessment. *Bull. Am. Meteorol. Soc.*, **92**, 1023–1043.

Bodeker, G., H. Shiona, and H. Eskes, 2005: Indicators of Antarctic ozone depletion. *Atmos. Chem. Phys.*, **5**, 2603–2615.

Boe, J., A. Hall, and X. Qu, 2009a: Deep ocean heat uptake as a major source of spread in transient climate change simulations. *Geophys. Res. Lett.*, **36**, L22701.

Boe, J., L. Terray, F. Habets, and E. Martin, 2007: Statistical and dynamical downscaling of the Seine basin climate for hydro meteorological studies. *Int. J. Climatol.*, **27**, 1643–1655.

Boe, J. L., A. Hall, and X. Qu, 2009b: September sea-ice cover in the Arctic Ocean projected to vanish by 2100. *Nature Geosci.*, **2**, 341–343.

Boer, G., and S. Lambert, 2008: The energy cycle in atmospheric models. *Clim. Dyn.*, **30**, 371–390.

Boer, G. J., and B. Yu, 2003: Climate sensitivity and climate state. *Clim. Dyn.*, **21**, 167–176.

Boisier, J.-P., et al., 2012: Attributing the biogeophysical impacts of land-use induced Land-Cover Changes on surface climate to specific causes. Results from the first LUCID set of simulations. *J. Geophys. Res.*, **117**, D12116.

Bollasina, M. A., and Y. Ming, 2013: The general circulation model precipitation bias over the southwestern equatorial Indian Ocean and its implications for simulating the South Asian monsoon. *Clim. Dyn.*, **40**, 823–838.

Bonan, G. B., 2008: Forests and climate change: Forcings, feedbacks, and the climate benefits of forests. *Science*, **320**, 1444–1449.

Bond, T. C., et al., 2007: Historical emissions of black and organic carbon aerosol from energy-related combustion, 1850–2000. *Global Biogeochem. Cycles*, **21**, GB2018.

Bondeau, A., P. C. Smith, S. Zaehle, S. Schaphoff, W. Lucht, W. Cramer, and D. Gerten, 2007: Modelling the role of agriculture for the 20th century global terrestrial carbon balance. *Global Change Biol.*, **13**, 679–706.

Boning, C. W., A. Dispert, M. Visbeck, S. R. Rintoul, and F. U. Schwarzkopf, 2008: The response of the Antarctic Circumpolar Current to recent climate change. *Nature Geosci.*, **1**, 864–869.

Boone, A., et al., 2009: THE AMMA Land Surface Model Intercomparison Project (ALMIP). *Bull. Am. Meteorol. Soc.*, **90**, 1865–1880.

Boone, A. A., I. Poccard-Leclercq, Y. K. Xue, J. M. Feng, and P. de Rosnay, 2010: Evaluation of the WAMME model surface fluxes using results from the AMMA land-surface model intercomparison project. *Clim. Dyn.*, **35**, 127–142.

Booth, B. B. B., N. J. Dunstone, P. R. Halloran, T. Andrews, and N. Bellouin, 2012a: Aerosols implicated as a prime driver of twentieth-century North Atlantic climate variability. *Nature*, **484**, 228–232.

Booth, B. B. B., et al., 2012b: High sensitivity of future global warming to land carbon cycle processes. *Environ. Res. Lett.*, **7**, 024002.

Boschat, G., P. Terray, and S. Masson, 2012: Robustness of SST teleconnections and precursory patterns associated with the Indian summer monsoon. *Clim. Dyn.*, **38**, 2143–2165.

Boyle, J., and S. A. Klein, 2010: Impact of horizontal resolution on climate model forecasts of tropical precipitation and diabatic heating for the TWP-ICE period. *J. Geophys. Res.*, **115**, D23113.

Boyle, J., S. Klein, G. Zhang, S. Xie, and X. Wei, 2008: Climate Model Forecast Experiments for TOGA COARE. *Mon. Weather Rev.*, **136**, 808–832.

Bracegirdle, T., et al., 2013: Assessment of surface winds over the Atlantic, Indian and Pacific Ocean sectors of the Southern Hemisphere in CMIP5 models: Historical bias, forcing response, and state dependence. *J. Geophys. Res. Atmos.*, doi:10.1002/jgrd.50153.

Bracegirdle, T. J., and D. B. Stephenson, 2012: Higher precision estimates of regional polar warming by ensemble regression of climate model projections. *Clim. Dyn.*, **39**, 2805–2821.

Bracegirdle, T. J., and D. B. Stephenson, 2013: On the robustness of emergent constraints used in multi-model climate change projections of Arctic warming. *J. Clim.*, **26**, 669–678.

Braconnot, P., F. Hourdin, S. Bony, J. Dufresne, J. Grandpeix, and O. Marti, 2007a: Impact of different convective cloud schemes on the simulation of the tropical seasonal cycle in a coupled ocean-atmosphere model. *Clim. Dyn.*, **29**, 501–520.

Braconnot, P., et al., 2012: Evaluation of climate models using palaeoclimatic data. *Nature Clim. Change*, **2**, 417–424.

Braconnot, P., et al., 2007b: Results of PMIP2 coupled simulations of the Mid-Holocene and Last Glacial Maximum - Part 2: Feedbacks with emphasis on the location of the ITCZ and mid- and high latitudes heat budget. *Clim. Past*, **3**, 279–296.

Braconnot, P., et al., 2007c: Results of PMIP2 coupled simulations of the Mid-Holocene and Last Glacial Maximum - Part 1: Experiments and large-scale features. *Clim. Past*, **3**, 261–277.

Brands, S., J. Taboada, A. Cofino, T. Sauter, and C. Schneider, 2011: Statistical downscaling of daily temperatures in the NW Iberian Peninsula from global climate models: Validation and future scenarios. *Clim. Res.*, **48**, 163–176.

Bresson, R., and R. Laprise, 2011: Scale-decomposed atmospheric water budget over North America as simulated by the Canadian Regional Climate Model for current and future climates. *Clim. Dyn.*, **36**, 365–384.

Breugem, W. P., W. Hazeleger, and R. J. Haarsma, 2006: Multimodel study of tropical Atlantic variability and change. *Geophys. Res. Lett.*, **33**, L23706.

Brewer, S., J. Guiot, and F. Torre, 2007: Mid-Holocene climate change in Europe: A data-model comparison. *Clim. Past*, **3**, 499–512.

Briegleb, B. P., and B. Light, 2007: A Delta-Eddington multiple scattering parameterization for solar radiation in the sea ice component of the Community Climate System Model. *NCAR Technical Note*, National Center for Atmospheric Research, 100 pp.

Briegleb, B. P., C. M. Blitz, E. C. Hunke, W. H. Lipscomb, M. M. Holland, J. L. Schramm, and R. E. Moritz, 2004: Scientific description of the sea ice component in the Community Climate System Model, Version 3. *NCAR Technical Note,* National Center for Atmospheric Research, 70 pp.

Brient, F., and S. Bony, 2012: Interpretation of the positive low-cloud feedback predicted by a climate model under global warming. *Clim. Dyn.*, doi:10.1007/s00382–011–1279–7.

Brierley, C. M., M. Collins, and A. J. Thorpe, 2010: The impact of perturbations to ocean-model parameters on climate and climate change in a coupled model. *Clim. Dyn.*, **34**, 325–343.

Brogniez, H., and R. T. Pierrehumbert, 2007: Intercomparison of tropical tropospheric humidity in GCMs with AMSU-B water vapor data. *Geophys. Res. Lett.*, **34**, L17812

Brogniez, H., R. Roca, and L. Picon, 2005: Evaluation of the distribution of subtropical free tropospheric humidity in AMIP-2 simulations using METEOSAT water vapor channel data. *Geophys. Res. Lett.*, **32**, L19708.

Brovkin, V., J. Bendtsen, M. Claussen, A. Ganopolski, C. Kubatzki, V. Petoukhov, and A. Andreev, 2002: Carbon cycle, vegetation, and climate dynamics in the Holocene: Experiments with the CLIMBER-2 model. *Global Biogeochem. Cycles*, **16**, 1139.

Brown, A., S. Milton, M. Cullen, B. Golding, J. Mitchell, and A. Shelly, 2012: Unified modeling and prediction of weather and climate: A 25-year journey. *Bull. Am. Meteorol. Soc.*, **93**, 1865–1877.

Brown, J., A. Fedorov, and E. Guilyardi, 2010a: How well do coupled models replicate ocean energetics relevant to ENSO? *Clim. Dyn.*, **36**, 2147–2158.

Brown, J., O. J. Ferrians, J. A. Heginbottom, and E. S. E.S. Melnikov, 1997: *International Permafrost Association Circum-Arctic Map of Permafrost and Ground Ice Conditions*. Geological Survey (U.S.), Denver, CO, USA.

Brown, J., O. J. Ferrians, J. A. Heginbottom, and E. S. E.S. Melnikov, 1998: Digital circum-arctic map of permafrost and ground ice conditions. In: *Circumpolar Active-Layer Permafrost System (CAPS)*. CD-ROM. 1.0 ed., University of Colorado at Boulder National Snow and Ice Data Center. Boulder, CO, USA.

Brown, J. R., C. Jakob, and J. M. Haynes, 2010b: An evaluation of rainfall frequency and intensity over the Australian region in a Global Climate Model. *J. Clim.*, **23**, 6504–6525.

Brown, J. R., A. F. Moise, and R. A. Colman, 2013: The South Pacific Convergence Zone in CMIP5 simulations of historical and future climate. *Clim. Dyn.*, doi:10.1007/s00382-012-1591–x.

Brutel-Vuilmet, C., M. Menegoz, and G. Krinner, 2013: An analysis of present and future seasonal Northern Hemisphere land snow cover simulated by CMIP5 coupled climate models. *Cryosphere*, **7**, 67–80.

Bryan, F. O., M. W. Hecht, and R. D. Smith, 2007: Resolution convergence and sensitivity studies with North Atlantic circulation models. Part I: The western boundary current system. *Ocean Model.*, **16**, 141–159.

Bryan, F. O., R. Tomas, J. M. Dennis, D. B. Chelton, N. G. Loeb, and J. L. McClean, 2010: Frontal scale air-sea interaction in high-resolution coupled climate models. *J. Clim.*, **23**, 6277–6291.

Bryan, K., and L. J. Lewis, 1979: Water mass model of the world ocean. *J. Geophys. Res. Oceans Atmos.*, **84**, 2503–2517.

Bryden, H. L., H. R. Longworth, and S. A. Cunningham, 2005: Slowing of the Atlantic meridional overturning circulation at 25° N. *Nature*, **438**, 655–657.

Buehler, T., C. C. Raible, and T. F. Stocker, 2011: The relationship of winter season North Atlantic blocking frequencies to extreme cold or dry spells in the ERA-40. *Tellus A*, **63**, 212–222.

Butchart, N., A. A. Scaife, J. Austin, S. H. E. Hare, and J. R. Knight, 2003: Quasi-biennial oscillation in ozone in a coupled chemistry-climate model. *J. Geophys. Res.*, **108**, 4486.

Cadule, P., et al., 2010: Benchmarking coupled climate-carbon models against long-term atmospheric CO_2 measurements. *Global Biogeochem. Cycles*, **24**, Gb2016.

Cai, W., and T. Cowan, 2013: Why is the amplitude of the Indian Ocean Dipole overly large in CMIP3 and CMIP5 climate models? . *Geophys. Res. Lett.*, doi:10.1002/grl.50208.

Cai, W., A. Sullivan, and T. Cowan, 2011: Interactions of ENSO, the IOD, and the SAM in CMIP3 Models. *J. Clim.*, **24**, 1688–1704.

Cai, W. J., A. Sullivan, and T. Cowan, 2009: Rainfall teleconnections with Indo-Pacific variability in the WCRP CMIP3 models. *J. Clim.*, **22**, 5046–5071.

Calov, R., A. Ganopolski, V. Petoukhov, M. Claussen, and R. Greve, 2002: Large-scale instabilities of the Laurentide ice sheet simulated in a fully coupled climate-system model. *Geophys. Res. Lett.*, **29**, 2216.

Cameron-Smith, P., J. F. Lamarque, P. Connell, C. Chuang, and F. Vitt, 2006: Toward an Earth system model: Atmospheric chemistry, coupling, and petascale computing. *Scidac 2006: Scientific Discovery through Advanced Computing* [W. M. Tang (ed.)]. Journal of Physics: Conference Series, Vol. 46, Denver, Colorado, USA.

Capotondi, A., A. Wittenberg, and S. Masina, 2006: Spatial and temporal structure of Tropical Pacific interannual variability in 20th century coupled simulations. *Ocean Model.*, **15**, 274–298.

Capotondi, A., M. A. Alexander, N. A. Bond, E. N. Curchitser, and J. D. Scott, 2012: Enhanced upper ocean stratification with climate change in the CMIP3 models. *J. Geophys. Res. Oceans* **117**, C04031.

Cariolle, D., and H. Teyssedre, 2007: A revised linear ozone photochemistry parameterization for use in transport and general circulation models: Multi-annual simulations. *Atmos. Chem. Phys.*, **7**, 2183–2196.

Carslaw, K. S., O. Boucher, D. V. Spracklen, G. W. Mann, J. G. L. Rae, S. Woodward, and M. Kulmala, 2010: A review of natural aerosol interactions and feedbacks within the Earth system. *Atmos. Chem. Phys.*, **10**, 1701–1737.

Casado, M. J., and M. A. Pastor, 2012: Use of variability modes to evaluate AR4 climate models over the Euro-Atlantic region. *Clim. Dyn.*, **38**, 225–237.

Cattiaux, J., H. Douville, and Y. Peings, 2013: European temperatures in CMIP5: Origins of present-day biases and future uncertainties. *Clim. Dyn.*, doi:10.1007/s00382-013-1731-y.

Catto, J., N. Nicholls, and C. Jakob, 2012a: North Australian sea surface temperatures and the El Niño Southern Oscillation in observations and models. *J. Clim.*, **25**, 5011–5029.

Catto, J., N. Nicholls, and C. Jakob, 2012b: North Australian sea surface temperatures and the El Niño Southern Oscillation in the CMIP5 models. *J. Clim.*, **25**, 6375–6382.

Catto, J. L., L. C. Shaffrey, and K. I. Hodges, 2010: Can climate models capture the structure of extratropical cyclones? *J. Clim.*, **23**, 1621–1635.

Catto, J. L., L. C. Shaffrey, and K. I. Hodges, 2011: Northern Hemisphere Extratropical cyclones in a warming climate in the HiGEM high-resolution climate Model. *J. Clim.*, **24**, 5336–5352.

Catto, J. L., C. Jakob, and N. Nicholls, 2013: A global evaluation of fronts and precipitation in the ACCESS model. *Aust. Meteorol. Oceanogr. J.*, **63**,191-203.

Cavalieri, D. J., and C. L. Parkinson, 2012: Arctic sea ice variability and trends, 1979–2010. *Cryosphere*, **6**, 881–889.

Cavicchia, L., and H. von Storch, 2011: The simulation of medicanes in a high-resolution regional climate model. *Clim. Dyn.*, **39** 2273–2290.

Cesana, G., and H. Chepfer, 2012: How well do climate models simulate cloud vertical structure? A comparison between CALIPSO-GOCCP satellite observations and CMIP5 models. *Geophys. Res. Lett.*, **39**, L20803.

Cha, D., D. Lee, and S. Hong, 2008: Impact of boundary layer processes on seasonal simulation of the East Asian summer monsoon using a Regional Climate Model. *Meteorol. Atmos. Phys.*, **100**, 53–72.

Champion, A. J., K. I. Hodges, L. O. Bengtsson, N. S. Keenlyside, and M. Esch, 2011: Impact of increasing resolution and a warmer climate on extreme weather from Northern Hemisphere extratropical cyclones. *Tellus A*, **63**, 893–890.

Chan, S. C., E. J. Kendon, H. J. Fowler, S. Blenkinsop, C. A. T. Ferro, and D. B. Stephenson, 2012: Does increasing resolution improve the simulation of United Kingdom daily precipitation in a regional climate model? *Clim. Dyn.*, doi:10.1007/s00382-012-1568-9.

Chang, C. P., and T. Li, 2000: A theory for the tropical tropospheric biennial oscillation. *J. Atmos. Sci.*, **57**, 2209–2224.

Chang, C. Y., S. Nigam, and J. A. Carton, 2008: Origin of the springtime westerly bias in equatorial Atlantic surface winds in the Community Atmosphere Model version 3 (CAM3) simulation. *J. Clim.*, **21**, 4766–4778.

Chang, C. Y., J. A. Carton, S. A. Grodsky, and S. Nigam, 2007: Seasonal climate of the tropical Atlantic sector in the NCAR community climate system model 3: Error structure and probable causes of errors. *J. Clim.*, **20**, 1053–1070.

Chang, E. K. M., Y. Guo, and X. Xia, 2012: CMIP5 multi-model ensemble projection of storm track change under global warming. *J. Geophys. Res.*, **117**, D23118.

Charbit, S., D. Paillard, and G. Ramstein, 2008: Amount of CO_2 emissions irreversibly leading to the total melting of Greenland. *Geophys. Res. Lett.*, **35**, L12503.

Charlton-Perez, A. J., et al., *2012*: Mean climate and variability of the stratosphere in CMIP5 models. *J. Geophys. Res.*, doi:10.1002/jgrd.50125.

Chen, C. T., and T. Knutson, 2008: On the verification and comparison of extreme rainfall indices from climate models. *J. Clim.*, **21**, 1605–1621.

Chen, H. M., T. J. Zhou, R. B. Neale, X. Q. Wu, and G. J. Zhang, 2010: Performance of the New NCAR CAM3.5 in East Asian summer monsoon simulations: Sensitivity to modifications of the Convection Scheme. *J. Clim.*, **23**, 3657–3675.

Chen, L., Y. Yu, and D. Sun, 2013: Cloud and water vapor feedbacks to the El Niño warming: Are they still biased in CMIP5 models? *J. Clim.*, doi:10.1175/JCLI-D-12-00575.1.

Chen, Y. H., and A. D. Del Genio, 2009: Evaluation of tropical cloud regimes in observations and a general circulation model. *Clim. Dyn.*, **32**, 355–369.

Chiang, J. C. H., and A. H. Sobel, 2002: Tropical tropospheric temperature variations caused by ENSO and their influence on the remote tropical climate. *J. Clim.*, **15**, 2616–2631.

Chiang, J. C. H., and D. J. Vimont, 2004: Analogous Pacific and Atlantic meridional modes of tropical atmosphere-ocean variability. *J. Clim.*, **17**, 4143–4158.

Chou, C., and J. Y. Tu, 2008: Hemispherical asymmetry of tropical precipitation in ECHAM5/MPI-OM during El Niño and under global warming. *J. Clim.*, **21**, 1309–1332.

Chou, C., J. D. Neelin, J. Y. Tu, and C. T. Chen, 2006: Regional tropical precipitation change mechanisms in ECHAM4/OPYC3 under global warming. *J. Clim.*, **19**, 4207–4223.

Chou, S., et al., 2012: Downscaling of South America present climate driven by 4-member HadCM3 runs. *Clim. Dyn.*, **38**, 635–653.

Christensen, J., F. Boberg, O. Christensen, and P. Lucas-Picher, 2008: On the need for bias correction of regional climate change projections of temperature and precipitation. *Geophys. Res. Lett.*, **35**, L20709.

Christensen, J., E. Kjellstrom, F. Giorgi, G. Lenderink, and M. Rummukainen, 2010: Weight assignment in regional climate models. *Clim. Res.*, **44**, 179–194.

Christensen, J. H., and F. Boberg, 2013: Temperature dependent climate projection deficiencies in CMIP5 models. *Geophys. Res. Lett.*, **39**, L24705.

Christian, J. R., et al., 2010: The global carbon cycle in the Canadian Earth system model (CanESM1): Preindustrial control simulation. *J. Geophys. Res. Biogeosci.*, **115**, G03014

Christidis, N., P. A. Stott, and S. J. Brown, 2011: The role of human activity in the recent warming of extremely warm daytime temperatures. *J. Clim.*, **24**, 1922–1930.

Christy, J. R., W. B. Norris, R. W. Spencer, and J. J. Hnilo, 2007: Tropospheric temperature change since 1979 from tropical radiosonde and satellite measurements. *J. Geophys. Res. Atmos.*, **112**, D06102.

Christy, J. R., et al., 2010: What do observational datasets say about modeled tropospheric temperature trends since 1979? *Remote Sens.*, **2**, 2148–2169.

Cimatoribus, A. A., S. S. Drijfhout, and H. A. Dijkstra, 2012: A global hybrid coupled model based on atmosphere-SST feedbacks. *Clim. Dyn.*, **38**, 745–760.

Cionni, I., et al., 2011: Ozone database in support of CMIP5 simulations: Results and corresponding radiative forcing. *Atmos. Chem. Phys.*, **11**, 11267–11292.

Clark, D. B., et al., 2011: The Joint UK Land Environment Simulator (JULES), model description - Part 2: Carbon fluxes and vegetation dynamics. *Geosci. Model Dev.*, **4**, 701–722.

Claussen, M., et al., 2002: Earth system models of intermediate complexity: Closing the gap in the spectrum of climate system models. *Clim. Dyn.*, **18**, 579–586.

Coelho, C. A. S., and L. Goddard, 2009: El Niño-induced tropical droughts in climate change projections. *J. Clim.*, **22**, 6456–6476.

Cohen, J. L., J. C. Furtado, M. Barlow, V. A. Alexeev, and J. E. Cherry, 2012: Asymmetric seasonal temperature trends. *Geophys. Res. Lett.*, **39**, L04705.

Collatz, G. J., M. Ribas-Carbo, and J. A. Berry, 1992: Coupled photosynthesis-stomatal conductance model for leaves of C4 Plants. *Aust. J. Plant Physiol.*, **19**, 519–538.

Collatz, G. J., J. T. Ball, C. Grivet, and J. A. Berry, 1991: Physiological and environmental regulation of stomatal conductance, photosynthesis and transpiration: A model that includes a laminar boundary layer. *Agr. Forest Meteorol.*, **54**, 107–136.

Colle, B. A., Z. Zhang, K. A. Lombardo, E. Chang, P. Liu, and M. Zhang, 2013: Historical evaluation and future prediction of eastern North America and western Atlantic extratropical cyclones in the CMIP5 models during the cool season. *J. Clim.*, doi:10.1175/JCLI-D-12–00498.1.

Collins, M., S. Tett, and C. Cooper, 2001: The internal climate variability of HadCM3, a version of the Hadley Centre coupled model without flux adjustments. *Clim. Dyn.*, **17**, 61–81.

Collins, M., C. M. Brierley, M. MacVean, B. B. B. Booth, and G. R. Harris, 2007: The sensitivity of the rate of transient climate change to ocean physics perturbations. *J. Clim.*, **20**, 2315–2320.

Collins, M., B. B. B. Booth, G. R. Harris, J. M. Murphy, D. M. H. Sexton, and M. J. Webb, 2006a: Towards quantifying uncertainty in transient climate change. *Clim. Dyn.*, **27**, 127–147.

Collins, M., R. Chandler, P. Cox, J. Huthnance, J. Rougier, and D. Stephenson, 2012: Quantifying future climate change. *Nature Clim. Change*, **2**, 403–409.

Collins, M., B. Booth, B. Bhaskaran, G. Harris, J. Murphy, D. Sexton, and M. Webb, 2010: Climate model errors, feedbacks and forcings: A comparison of perturbed physics and multi-model ensembles. *Clim. Dyn.*, **36**, 1737–1766.

Collins, W. D., J. M. Lee-Taylor, D. P. Edwards, and G. L. Francis, 2006b: Effects of increased near-infrared absorption by water vapor on the climate system. *J. Geophys. Res. Atmos.*, **111**, D18109.

Collins, W. D., et al., 2006c: The formulation and atmospheric simulation of the Community Atmosphere Model version 3 (CAM3). *J. Clim.*, **19**, 2144–2161.

Collins, W. D., et al., 2006d: The Community Climate System Model version 3 (CCSM3). *J. Clim.*, **19**, 2122–2143.

Collins, W. J., et al., 2011: Development and evaluation of an Earth-System model-HadGEM2. *Geosci. Model Dev.*, **4**, 1051–1075.

Colman, R., and B. McAvaney, 2009: Climate feedbacks under a very broad range of forcing. *Geophys. Res. Lett.*, **36**, L01702.

Colman, R. A., A. F. Moise, and L. I. Hanson, 2011: Tropical Australian climate and the Australian monsoon as simulated by 23 CMIP3 models. *J. Geophys. Res. Atmos.*, **116**, D10116.

Comiso, J. C., and F. Nishio, 2008: Trends in the sea ice cover using enhanced and compatible AMSR-E, SSM/I, and SMMR data. *J. Geophys. Res. Oceans*, **113**, C02s07.

Compo, G. P., and P. D. Sardeshmukh, 2009: Oceanic influences on recent continental warming. *Clim. Dyn.*, **32**, 333–342.

Connolley, W., and T. Bracegirdle, 2007: An Antarctic assessment of IPCC AR4 coupled models. *Geophys. Res. Lett.*, **34** L22505.

Coon, M., R. Kwok, G. Levy, M. Pruis, H. Schreyer, and D. Sulsky, 2007: Arctic Ice Dynamics Joint Experiment (AIDJEX) assumptions revisited and found inadequate. *J. Geophys. Res.*, **112**, C11S90.

Coppola, E., F. Giorgi, S. Rauscher, and C. Piani, 2010: Model weighting based on mesoscale structures in precipitation and temperature in an ensemble of regional climate models. *Clim. Res.*, **44** 121–134.

Cox, P., 2001: Description of the "TRIFFID" Dynamic Global Vegetation Model Hadley Centre, Met Office Hadley Centre, Berks, United Kingdom, 16 pp.

Cox, P. M., R. A. Betts, C. D. Jones, S. A. Spall, and I. J. Totterdell, 2000: Acceleration of global warming due to carbon-cycle feedbacks in a coupled climate model. *Nature*, **408**, 184–187.

Cox, P. M., R. A. Betts, C. B. Bunton, R. L. H. Essery, P. R. Rowntree, and J. Smith, 1999: The impact of new land surface physics on the GCM simulation of climate and climate sensitivity. *Clim. Dyn.*, **15**, 183–203.

Cox, P. M., D. Pearson, B. B. B. Booth, P. Friedlingstein, C. Huntingford, C. D. Jones, and C. M. Luke, 2013: Sensitivity of tropical carbon to climate change constrained by carbon dioxide variability. *Nature*, **494**, 341–344.

Cramer, W., et al., 2001: Global response of terrestrial ecosystem structure and function to CO_2 and climate change: Results from six dynamic global vegetation models. *Global Change Biol.*, **7**, 357–373.

Crétat, J., B. Pohl, Y. Richard, and P. Drobinski, 2012: Uncertainties in simulating regional climate of Southern Africa: Sensitivity to physical parameterizations using WRF. *Clim. Dyn.*, **38**, 613–634.

Croft, B., U. Lohmann, and K. von Salzen, 2005: Black carbon ageing in the Canadian Centre for Climate modelling and analysis atmospheric general circulation model. *Atmos. Chem. Phys.*, **5**, 1931–1949.

Crucifix, M., 2006: Does the Last Glacial Maximum constrain climate sensitivity? *Geophys. Res. Lett.*, **33**, L18701.

Cunningham, S., et al., 2010: The present and future system for measuring the Atlantic meridional overturning circulation and heat transport. In: *Proceedings of OceanObs'09: Sustained Ocean Observations and Information for Society* (Vol. 2), Venice, Italy, 21–25 September 2009, ESA Publication.

Cunningham, S. A., S. G. Alderson, B. A. King, and M. A. Brandon, 2003: Transport and variability of the Antarctic Circumpolar Current in Drake Passage. *J. Geophys. Res.-Oceans*, **108**, 8084.

Cunningham, S. A., et al., 2007: Temporal variability of the Atlantic meridional overturning circulation at 26.5°N. *Science*, **317**, 935–938.

Curry, W. B., and D. W. Oppo, 2005: Glacial water mass geometry and the distribution of delta C-13 of sigma CO_2 in the western Atlantic Ocean. *Paleoceanography*, **20**, Pa1017.

Cuxart, J., et al., 2006: Single-column model intercomparison for a stably stratified atmospheric boundary layer. *Boundary-Layer Meteorol.*, **118**, 273–303.

Dai, A., 2001: Global precipitation and thunderstorm frequencies. Part II: Diurnal variations. *J. Clim.*, **14**, 1112–1128.

Dai, A., 2006: Precipitation characteristics in eighteen coupled climate models. *J. Clim.*, **19**, 4605–4630.

Dai, A., and C. Deser, 1999: Diurnal and semidiurnal variations in global surface wind and divergence fields. *J. Geophys. Res. Atmos.*, **104**, 31109–31125.

Dai, A., and K. E. Trenberth, 2004: The diurnal cycle and its depiction in the Community Climate System Model. *J. Clim.*, **17**, 930–951.

Dai, Y. J., R. E. Dickinson, and Y. P. Wang, 2004: A two-big-leaf model for canopy temperature, photosynthesis, and stomatal conductance. *J. Clim.*, **17**, 2281–2299.

Dai, Y. J., et al., 2003: The Common Land Model. *Bull. Am. Meteorol. Soc.*, **84**, 1013–1023.

Dallmeyer, A., M. Claussen, and J. Otto, 2010: Contribution of oceanic and vegetation feedbacks to Holocene climate change in monsoonal Asia. *Clim. Past*, **6**, 195–218.

Danabasoglu, G., and P. R. Gent, 2009: Equilibrium climate sensitivity: Is it accurate to use a Slab Ocean Model? *J. Clim.*, **22**, 2494–2499.

Danabasoglu, G., R. Ferrari, and J. C. McWilliams, 2008: Sensitivity of an ocean general circulation model to a parameterization of near-surface eddy fluxes. *J. Clim.*, **21**, 1192–1208.

Danabasoglu, G., W. G. Large, and B. P. Briegleb, 2010: Climate impacts of parameterized Nordic Sea overflows. *J. Geophys. Res. Oceans*, **115**, C11005.

Danabasoglu, G., W. G. Large, J. J. Tribbia, P. R. Gent, B. P. Briegleb, and J. C. McWilliams, 2006: Diurnal coupling in the tropical oceans of CCSM3. *J. Clim.*, **19**, 2347–2365.

Danabasoglu, G., et al., 2012: The CCSM4 Ocean Component. *J. Clim.*, **25**, 1361–1389.

Davies, T., M. J. P. Cullen, A. J. Malcolm, M. H. Mawson, A. Staniforth, A. A. White, and N. Wood, 2005: A new dynamical core for the Met Office's global and regional modelling of the atmosphere. *Q. J. R. Meteorol. Soc.*, **131**, 1759–1782.

Davis, B. A. S., and S. Brewer, 2009: Orbital forcing and role of the latitudinal insolation/temperature gradient. *Clim. Dyn.*, **32**, 143–165.

Dawson, A., T. N. Palmer, and S. Corti, 2012: Simulating regime structures in weather and climate prediction models. *Geophys. Res. Lett.*, **39**, L21805.

Day, J. J., J. C. Hargreaves, J. D. Annan, and A. Abe-Ouchi, 2012: Sources of multi-decadal variability in Arctic sea ice extent. *Environ. Res. Lett.*, **7**, 034011.

de Elia, R., and H. Cote, 2010: Climate and climate change sensitivity to model configuration in the Canadian RCM over North America. *Meteorol. Z.*, **19**, 325–339.

de Elia, R., S. Biner, and A. Frigon, 2013: Interannual variability and expected regional climate change over North America. *Clim. Dyn.*, doi:10.1007/s00382-013-1717-9.

de Jong, M. F., S. S. Drijfhout, W. Hazeleger, H. M. van Aken, and C. A. Severijns, 2009: Simulations of hydrographic properties in the Northwestern North Atlantic Ocean in Coupled Climate Models. *J. Clim.*, **22**, 1767–1786.

de Noblet-Ducoudre, N., et al., 2012: Determining robust impacts of land-use induced land-cover changes on surface climate over North America and Eurasia; Results from the first set of LUCID experiments. *J. Clim.*, **25**, 3261–3281.

De Szoeke, S. P., and S. P. Xie, 2008: The tropical eastern Pacific seasonal cycle: Assessment of errors and mechanisms in IPCC AR4 coupled ocean - atmosphere general circulation models. *J. Clim.*, **21**, 2573–2590.

Dee, D. P., et al., 2011: The ERA-Interim reanalysis: Configuration and performance of the data assimilation system. *Q. J. R. Meteorol. Soc.*, **137**, 553–597.

DelSole, T., and J. Shukla, 2009: Artificial skill due to predictor screening. *J. Clim.*, **22**, 331–345.

Delworth, T. L., et al., 2012: Simulated climate and climate change in the GFDL CM2.5 High-Resolution Coupled Climate Model. *J. Clim.*, **25**, 2755–2781.

Delworth, T. L., et al., 2006: GFDL's CM2 global coupled climate models. Part I: Formulation and simulation characteristics. *J. Clim.*, **19**, 643–674.

Déqué, M., 2007: Frequency of precipitation and temperature extremes over France in an anthropogenic scenario: Model results and statistical correction according to observed values. *Global Planet. Change*, **57**, 16–26.

Déqué, M., 2010: Regional climate simulation with a mosaic of RCMs. *Meteorol. Z.*, **19**, 259–266.

Déqué, M., C. Dreveton, A. Braun, and D. Cariolle, 1994: The ARPEGE/IFS atmosphere model: A contribution to the French community climate modelling. *Clim. Dyn.*, **10**, 249–266.

Déqué, M., S. Somot, E. Sanchez-Gomez, C. Goodess, D. Jacob, G. Lenderink, and O. Christensen, 2012: The spread amongst ENSEMBLES regional scenarios: Regional climate models, driving general circulation models and interannual variability. *Clim. Dyn.*, **38**, 951–964.

Derksen, C., and R. Brown, 2012: Spring snow cover extent reductions in the 2008–2012 period exceeding climate model projections. *Geophys. Res. Lett.*, **39**, L19504.

Deser, C., A. S. Phillips, and J. W. Hurrell, 2004: Pacific interdecadal climate variability: Linkages between the tropics and the North Pacific during boreal winter since 1900. *J. Clim.*, **17**, 3109–3124.

Deser, C., A. S. Phillips, V. Bourdette, and H. Teng, 2011: Uncertainty in climate change projections: The role of internal variability. *Clim. Dyn.*, **38**, 527–546.

Deser, C., et al., 2012: ENSO and Pacific decadal variability in Community Climate System Model Version 4. *J. Clim.*, **25**, 2622–2651.

Deushi, M., and K. Shibata, 2011: Development of a Meteorological Research Institute Chemistry-Climate Model version 2 for the Study of Tropospheric and Stratospheric Chemistry. *Papers Meteorol. Geophys.*, **62**, 1–46.

Di Luca, A., R. Elía, and R. Laprise, 2012: Potential for small scale added value of RCM's downscaled climate change signal. *Clim. Dyn.*, **40**, 601–618.

Diaconescu, E. P., and R. Laprise, 2013: Can added value be expected in RCM-simulated large scales? *Clim. Dyn.*, doi:10.1007/s00382-012-1649-9.

Diffenbaugh, N., M. Ashfaq, and M. Scherer, 2011: Transient regional climate change: Analysis of the summer climate response in a high-resolution, century-scale ensemble experiment over the continental United States. *J. Geophys. Res. Atmos.*, **116**, D24111.

DiNezio, P. N., A. C. Clement, G. A. Vecchi, B. J. Soden, and B. P. Kirtman, 2009: Climate response of the equatorial Pacific to global warming. *J. Clim.*, **22**, 4873–4892.

Dix, M., et al., 2013: The ACCESS Coupled Model: Documentation of core CMIP5 simulations and initial results. *Aust. Meteorol. Oceanogr. J.*, **63**, 83–99.

Doblas-Reyes, F. J., et al., 2013: Initialized near-term regional climate change prediction. *Nature Commun.*, **4**, 1715.

Dokken, T. M., and E. Jansen, 1999: Rapid changes in the mechanism of ocean convection during the last glacial period. *Nature*, **401**, 458–461.

Domingues, C., J. Church, N. White, P. Gleckler, S. Wijffels, P. Barker, and J. Dunn, 2008: Improved estimates of upper-ocean warming and multi-decadal sea-level rise. *Nature*, **453**, 1090–1093.

Donat, M., G. Leckebusch, S. Wild, and U. Ulbrich, 2010: Benefits and limitations of regional multi-model ensembles for storm loss estimations. *Clim. Res.*, **44**, 211–225.

Donat, M. G., et al., 2013: Updated analyses of temperature and precipitation extreme indices since the beginning of the twentieth century: The HadEX2 dataset. *J. Geophys. Res.*, doi:10.1002/2012JD018606.

Donner, L. J., et al., 2011: The dynamical core, physical parameterizations, and basic simulation characteristics of the atmospheric component AM3 of the GFDL Global Coupled Model CM3. *J. Clim.*, **24**, 3484–3519.

Dorn, W., K. Dethloff, and A. Rinke, 2009: Improved simulation of feedbacks between atmosphere and sea ice over the Arctic Ocean in a coupled regional climate model. *Ocean Model.*, **29**, 103–114.

Doscher, R., K. Wyser, H. E. M. Meier, M. W. Qian, and R. Redler, 2010: Quantifying Arctic contributions to climate predictability in a regional coupled ocean-ice-atmosphere model. *Clim. Dyn.*, **34**, 1157–1176.

Douglass, D., J. Christy, B. Pearson, and S. Singer, 2008: A comparison of tropical temperature trends with model predictions. *Int. J. Climatol.* , **28**, 1693–1701.

Dowdy, A. J., G. A. Mills, B. Timbal, and Y. Wang, 2013: Changes in the risk of extratropical cyclones in Eastern Australia. *J. Clim.*, **26**, 1403–1417.

Driesschaert, E., et al., 2007: Modeling the influence of Greenland ice sheet melting on the Atlantic meridional overturning circulation during the next millennia. *Geophys. Res. Lett.*, **34**, L10707.

Driouech, F., M. Deque, and E. Sanchez-Gomez, 2010: Weather regimes-Moroccan precipitation link in a regional climate change simulation. *Global Planet. Change*, **72**, 1–10.

Driscoll, S., A. Bozzo, L. J. Gray, A. Robock, and G. Stenchikov, 2012: Coupled Model Intercomparison Project 5 (CMIP5) simulations of climate following volcanic eruptions. *J. Geophys. Res. Atmos.*, **117**, D17105.

Druyan, L. M., et al., 2010: The WAMME regional model intercomparison study. *Clim. Dyn.*, **35**, 175–192.

Du, Y., S.-P. Xie, Y.-L. Yang, X.-T. Zheng, L. Liu, and G. Huang, 2013: Indian Ocean variability in the CMIP5 multi-model ensemble: The basin mode. *J. Clim.*, **26**, 7240–7266.

Ducet, N., P. Y. Le Traon, and G. Reverdin, 2000: Global high-resolution mapping of ocean circulation from TOPEX/Poseidon and ERS-1 and-2. *J. Geophys. Res. Oceans*, **105**, 19477–19498.

Dufresne, J.-L., et al., 2012: Climate change projections using the IPSL-CM5 Earth System Model: From CMIP3 to CMIP5. *Clim. Dyn.*, doi:10.1007/s00382-012-1636-1.

Dufresne, J. L., and S. Bony, 2008: An assessment of the primary sources of spread of global warming estimates from coupled atmosphere-ocean models. *J. Clim.*, **21**, 5135–5144.

Dunkerton, T. J., 1991: Nonlinear propagation of zonal winds in an atmosphere with Newtonian cooling and equatorial wavedriving. *J. Atmos. Sci.*, **48**, 236–263.

Dunn-Sigouin, E., and S.-W. Son, 2013: Northern Hemisphere blocking frequency and duration in the CMIP5 models. *J. Geophys. Res.*, **118**, 1179–1188.

Dunne, J. P., et al., 2013: GFDL's ESM2 global coupled climate-carbon Earth System Models Part II: Carbon system formulation and baseline simulation characteristics. *J. Clim.*, doi:10.1175/JCLI-D-12-00150.1.

Dunne, J. P., et al., 2012: GFDL's ESM2 Global coupled climate-carbon Earth System models. Part I: Physical formulation and baseline simulation characteristics. *J. Clim.*, **25**, 6646–6665.

Duplessy, J. C., N. J. Shackleton, R. Fairbanks, L. Labeyrie, D. Oppo, and N. Kallel, 1988: Deep water source variation during the last climatic cycle and their impact on th global deep water circulation. *Paleoceanography*, **3**, 343–360.

Durack, P. J., and S. E. Wijffels, 2010: Fifty-year trends in global ocean salinities and their relationship to broad-scale warming. *J. Clim.*, **23**, 4342-4362.

Durack, P. J., S. E. Wijffels, and R. J. Matear, 2012: Ocean salinities reveal strong global water cycle intensification during 1950 to 2000. *Science*, **336**, 455–458.

Easterling, D. R., and M. F. Wehner, 2009: Is the climate warming or cooling? *Geophys. Res. Lett.*, **36**, L08706.

Eby, M., K. Zickfeld, A. Montenegro, D. Archer, K. J. Meissner, and A. J. Weaver, 2009: Lifetime of anthropogenic climate change: Millennial time scales of potential CO_2 and surface temperature perturbations. *J. Clim.*, **22**, 2501–2511.

Eby, M., et al., 2013: Historical and idealized climate model experiments: An EMIC intercomparison. *Clim. Past*, **9**, 1111–1140.

Edwards, N., and R. Marsh, 2005: Uncertainties due to transport-parameter sensitivity in an efficient 3-D ocean-climate model. *Clim. Dyn.*, **24**, 415–433.

Edwards, N. R., D. Cameron, and J. Rougier, 2011: Precalibrating an intermediate complexity climate model. *Clim. Dyn.*, **37**, 1469–1482.

Ek, M. B., et al., 2003: Implementation of Noah land surface model advances in the National Centers for Environmental Prediction operational mesoscale Eta model. *J. Geophys. Res. Atmos.*, **108**, 8851.

Eliseev, A. V., and I. I. Mokhov, 2011: Uncertainty of climate response to natural and anthropogenic forcings due to different land use scenarios. *Adv. Atmos. Sci.*, **28**, 1215–1232.

Emanuel, K., R. Sundararajan, and J. Williams, 2008: Hurricanes and global warming—Results from downscaling IPCC AR4 simulations. *Bull. Am. Meteorol. Soc.*, **89**, 347–367.

Endo, H., A. Kitoh, T. Ose, R. Mizuta, and S. Kusunoki, 2012: Future changes and uncertainties in Asian precipitation simulated by multiphysics and multi-sea surface temperature ensemble experiments with high-resolution Meteorological Research Institute atmospheric general circulation models (MRI-AGCMs). *J. Geophys. Res. Atmos.*, **117**, D16118.

Essery, R. L. H., M. J. Best, R. A. Betts, P. M. Cox, and C. M. Taylor, 2003: Explicit representation of subgrid heterogeneity in a GCM land surface scheme. *J. Hydrometeorol.*, **4**, 530–543.

Eum, H., P. Gachon, R. Laprise, and T. Ouarda, 2012: Evaluation of regional climate model simulations versus gridded observed and regional reanalysis products using a combined weighting scheme. *Clim. Dyn.*, **38**, 1433–1457.

Evans, J. P., M. Ekstroem, and F. Ji, 2012: Evaluating the performance of a WRF physics ensemble over South-East Australia. *Clim. Dyn.*, **39**, 1241–1258.

Eyring, V., et al., 2010: Transport impacts on atmosphere and climate: Shipping. *Atmos. Environ.*, **44**, 4735–4771.

Eyring, V., et al., 2013: Long-term ozone changes and associated climate impacts in CMIP5 simulations. *J. Geophys. Res.*, doi:10.1002/jgrd.50316.

Eyring, V., et al., 2007: Multimodel projections of stratospheric ozone in the 21st century. *J. Geophys. Res. Atmos.*, **112**, D16303.

Faloona, I., 2009: Sulfur processing in the marine atmospheric boundary layer: A review and critical assessment of modeling uncertainties. *Atmos. Environ.*, **43**, 2841–2854.

Fan, F. X., M. E. Mann, S. Lee, and J. L. Evans, 2010: Observed and modeled changes in the South Asian summer monsoon over the Historical Period. *J. Clim.*, **23**, 5193–5205.

Fanning, A. F., and A. J. Weaver, 1996: An atmospheric energy-moisture balance model: Climatology, interpentadal climate change, and coupling to an ocean general circulation model. *J. Geophys. Res. Atmos.*, **101**, 15111–15128.

Farneti, R., and P. R. Gent, 2011: The effects of the eddy-induced advection coefficient in a coarse-resolution coupled climate model. *Ocean Model.*, **39**, 135–145.

Farneti, R., T. L. Delworth, A. J. Rosati, S. M. Griffies, and F. R. Zeng, 2010: The role of mesoscale eddies in the rectification of the Southern Ocean response to climate change. *J. Phys. Oceanogr.*, **40**, 1539–1557.

Fasullo, J., and K. E. Trenberth, 2012: A less cloudy future: The role of subtropical subsidence in climate sensitivity. *Science*, **338**, 792–794.

Fauchereau, N., S. Trzaska, Y. Richard, P. Roucou, and P. Camberlin, 2003: Sea-surface temperature co-variability in the southern Atlantic and Indian Oceans and its connections with the atmospheric circulation in the Southern Hemisphere. *Int. J. Climatol.* , **23**, 663–677.

Felzer, B., D. Kicklighter, J. Melillo, C. Wang, Q. Zhuang, and R. Prinn, 2004: Effects of ozone on net primary production and carbon sequestration in the conterminous United States using a biogeochemistry model. *Tellus B*, **56**, 230–248.

Feng, J., and C. Fu, 2006: Inter-comparison of 10–year precipitation simulated by several RCMs for Asia. *Adv. Atmos. Sci.*, **23** 531–542.

Feng, J., et al., 2011: Comparison of four ensemble methods combining regional climate simulations over Asia. *Meteorol. Atmos. Phys.*, **111**, 41–53.

Fernandes, R., H. X. Zhao, X. J. Wang, J. Key, X. Qu, and A. Hall, 2009: Controls on Northern Hemisphere snow albedo feedback quantified using satellite Earth observations. *Geophys. Res. Lett.*, **36**, L21702.

Fernandez-Donado, L., et al., 2013: Large-scale temperature response to external forcing in simulations and reconstructions of the last millennium. *Clim. Past*, **9**, 393–421.

Ferrari, R., J. C. McWilliams, V. M. Canuto, and M. Dubovikov, 2008: Parameterization of eddy fluxes near oceanic boundaries. *J. Clim.*, **21**, 2770–2789.

Ferrari, R., S. M. Griffies, A. J. G. Nurser, and G. K. Vallis, 2010: A boundary-value problem for the parameterized mesoscale eddy transport. *Ocean Model.*, **32**, 143–156.

Feser, F., 2006: Enhanced detectability of added value in limited-area model results separated into different spatial scales. *Mon. Weather Rev.*, **134**, 2180–2190.

Feser, F., and M. Barcikowska, 2012: The influence of spectral nudging on typhoon formation in regional climate models. *Environ. Res. Lett.*, **7**, 014024.

Feser, F., B. Rockel, H. von Storch, J. Winterfeldt, and M. Zahn, 2011: Regional climate models add value to global model data: A review and selected examples. *Bull. Am. Meteorol. Soc.*, **92**, 1181–1192.

Fetterer, F., K. Knowles, W. Meier, and M. Savoie, 2002: Sea Ice Index. National Snow and Ice Data Center. Boulder, CO, USA.

9 Fichefet, T., and M. A. M. Maqueda, 1997: Sensitivity of a global sea ice model to the treatment of ice thermodynamics and dynamics. *J. Geophys. Res.*, **102**, 12609–12646.

Fichefet, T., and M. A. M. Maqueda, 1999: Modelling the influence of snow accumulation and snow-ice formation on the seasonal cycle of the Antarctic sea-ice cover. *Clim. Dyn.*, **15**, 251–268.

Field, P. R., A. Gettelman, R. B. Neale, R. Wood, P. J. Rasch, and H. Morrison, 2008: Midlatitude cyclone compositing to constrain climate model behavior using satellite observations. *J. Clim.*, **21**, 5887–5903.

Fioletov, V., G. Bodeker, A. Miller, R. McPeters, and R. Stolarski, 2002: Global and zonal total ozone variations estimated from ground-based and satellite measurements: 1964–2000. *J. Geophys. Res. Atmos.*, **107**, 4647.

Fischer, E. M., S. I. Seneviratne, D. Lüthi, and C. Schär, 2007: Contribution of land-atmosphere coupling to recent European summer heat waves. *Geophys. Res. Lett.*, **34**, L06707.

Flanner, M. G., K. M. Shell, M. Barlage, D. K. Perovich, and M. A. Tschudi, 2011: Radiative forcing and albedo feedback from the Northern Hemisphere cryosphere between 1979 and 2008. *Nature Geosci.*, **4**, 151–155.

Flato, G., 2011: Earth system models: an overview. *Wiley Interdisciplinary Reviews, Climate Change*, **2**, 783–800.

Flocco, D., D. Schroeder, D. L. Feltham, and E. C. Hunke, 2012: Impact of melt ponds on Arctic sea ice simulations from 1990 to 2007. *J. Geophys. Res.Oceans*, **117**, C09032.

Fogli, P. G., et al., 2009: INGV-CMCC Carbon (ICC): A Carbon Cycle Earth System Model. *CMCC Res. Papers*. Euro-Mediterranean Center on Climate Change, Bologna, Italy, 31 pp.

Fogt, R. L., J. Perlwitz, A. J. Monaghan, D. H. Bromwich, J. M. Jones, and G. J. Marshall, 2009: Historical SAM variability. Part II: Twentieth-century variability and trends from reconstructions, observations, and the IPCC AR4 models. *J. Clim.*, **22**, 5346–5365.

Fontaine, B., and S. Janicot, 1996: Sea surface temperature fields associated with West African rainfall anomaly types. *J. Clim.*, **9**, 2935–2940.

Forest, C. E., P. H. Stone, and A. P. Sokolov, 2006: Estimated PDFs of climate system properties including natural and anthropogenic forcings. *Geophys. Res. Lett.*, **33**, L01705.

Forest, C. E., P. H. Stone, and A. P. Sokolov, 2008: Constraining climate model parameters from observed 20th century changes. *Tellus A*, **60**, 911–920.

Forest, C. E., P. H. Stone, A. P. Sokolov, M. R. Allen, and M. D. Webster, 2002: Quantifying uncertainties in climate system properties with the use of recent climate observations. *Science*, **295**, 113–117.

Forster, P. M., T. Andrews, P. Good, J. M. Gregory, L. S. Jackson, and M. Zelinka, 2013: Evaluating adjusted forcing and model spread for historical and future scenarios in the CMIP5 generation of climate models. *J. Geophys. Res. Atmos.*, **118**, 1139–1150.

Fowler, H., S. Blenkinsop, and C. Tebaldi, 2007: Linking climate change modelling to impacts studies: Recent advances in downscaling techniques for hydrological modelling. *Int. J. Climatol.*, **27**, 1547–1578.

Fox-Kemper, B., R. Ferrari, and R. Hallberg, 2008: Parameterization of mixed layer eddies. Part I: Theory and diagnosis. *J. Phys. Oceanogr.*, **38**, 1145–1165.

Fox-Kemper, B., et al., 2011: Parameterization of mixed layer eddies. III: Implementation and impact in global ocean climate simulations. *Ocean Model.*, **39**, 61–78.

Fox-Rabinovitz, M., J. Cote, B. Dugas, M. Deque, J. McGregor, and A. Belochitski, 2008: Stretched-grid Model Intercomparison Project: Decadal regional climate simulations with enhanced variable and uniform-resolution GCMs. *Meteorol. Atmos. Phys.*, **100**, 159–177.

Frame, D., B. Booth, J. Kettleborough, D. Stainforth, J. Gregory, M. Collins, and M. Allen, 2005: Constraining climate forecasts: The role of prior assumptions. *Geophys. Res. Lett.*, **32**, L09702.

Frankcombe, L. M., A. von der Heydt, and H. A. Dijkstra, 2010: North Atlantic multidecadal climate variability: An investigation of dominant time scales and processes. *J. Clim.*, **23**, 3626–3638.

Frederiksen, C. S., J. S. Frederiksen, J. M. Sisson, and S. L. Osbrough, 2011: Australian winter circulation and rainfall changes and projections. *Int. J. Clim. Change Strat. Manage.*, **3**, 170–188.

Friedlingstein, P., et al., 2001: Positive feedback between future climate change and the carbon cycle. *Geophys. Res. Lett.*, **28**, 1543–1546.

Friedlingstein, P., et al., 2006: Climate-carbon cycle feedback analysis: Results from the (CMIP)-M-4 model intercomparison. *J. Clim.*, **19**, 3337–3353.

Friend, A. D., et al., 2007: FLUXNET and modelling the global carbon cycle. *Global Change Biol.*, **13**, 610–633.

Frierson, D. M. W., J. Lu, and G. Chen, 2007: Width of the Hadley cell in simple and comprehensive general circulation models. *Geophys. Res. Lett.*, **34**, L18804.

Frohlich, C., and J. Lean, 2004: Solar radiative output and its variability: Evidence and mechanisms. *Astron. Astrophys. Rev.*, **12**, 273–320.

Frost, A. J., et al., 2011: A comparison of multi-site daily rainfall downscaling techniques under Australian conditions. *J. Hydrol.*, **408**, 1–18.

Fu, Q., S. Manabe, and C. M. Johanson, 2011: On the warming in the tropical upper troposphere: Models versus observations. *Geophys. Res. Lett.*, **38**, L15704.

Furrer, R., R. Knutti, S. Sain, D. Nychka, and G. Meehl, 2007: Spatial patterns of probabilistic temperature change projections from a multivariate Bayesian analysis. *Geophys. Res. Lett.*, **34**, L06711.

Furtado, J., E. Di Lorenzo, N. Schneider, and N. A. Bond, 2011: North Pacific decadal variability and climate change in the IPCC AR4 models. *J. Clim.*, **24**, 3049–3067

Fyfe, J. C., N. P. Gillett, and D. W. J. Thompson, 2010: Comparing variability and trends in observed and modelled global-mean surface temperature. *Geophys. Res. Lett.*, **37**, L16802.

Fyke, J. G., A. J. Weaver, D. Pollard, M. Eby, L. Carter, and A. Mackintosh, 2011: A new coupled ice sheet/climate model: Description and sensitivity to model physics under Eemian, Last Glacial Maximum, late Holocene and modern climate conditions. *Geosci. Model Dev.*, **4**, 117–136.

Galbraith, D., P. E. Levy, S. Sitch, C. Huntingford, P. Cox, M. Williams, and P. Meir, 2010: Multiple mechanisms of Amazonian forest biomass losses in three dynamic global vegetation models under climate change. *New Phytologist*, **187**, 647–665.

Ganachaud, A., and C. Wunsch, 2003: Large-scale ocean heat and freshwater transports during the World Ocean Circulation Experiment. *J. Clim.*, **16**, 696–705.

Gangsto, R., F. Joos, and M. Gehlen, 2011: Sensitivity of pelagic calcification to ocean acidification. *Biogeosciences*, **8**, 433–458.

Gao, X., Y. Shi, D. Zhang, J. Wu, F. Giorgi, Z. Ji, and Y. Wang, 2012: Uncertainties in monsoon precipitation projections over China: Results from two high-resolution RCM simulations. *Clim. Res.*, **52**, 213–226.

Gates, W. L., et al., 1999: An overview of the results of the Atmospheric Model Intercomparison Project (AMIP I). *Bull. Am. Meteorol. Soc.*, **80**, 29–55.

Gbobaniyi, E. O., B. J. Abiodun, M. A. Tadross, B. C. Hewitson, and W. J. Gutowski, 2011: The coupling of cloud base height and surface fluxes: A transferability intercomparison. *Theor. Appl. Climatol.*, **106**, 189–210.

Gehlen, M., R. Gangsto, B. Schneider, L. Bopp, O. Aumont, and C. Ethe, 2007: The fate of pelagic $CaCO_3$ production in a high CO_2 ocean: A model study. *Biogeosciences*, **4**, 505–519.

Geller, M. A., et al., 2011: New gravity wave treatments for GISS climate models. *J. Clim.*, **24**, 3989–4002.

Gent, P. R., and J. C. McWilliams, 1990: Isopycnal mixing in ocean circulation models. *J. Phys. Oceanogr.*, **20**, 150–155.

Gent, P. R., and G. Danabasoglu, 2011: Response to increasing Southern Hemisphere winds in CCSM4. *J. Clim.*, **24**, 4992–4998.

Gent, P. R., J. Willebrand, T. J. McDougall, and J. C. McWilliams, 1995: Parameterizing eddy-induced tracer transports in ocean circulation models. *J. Phys. Oceanogr.*, **25**, 463–474.

Gent, P. R., et al., 2011: The Community Climate System Model Version 4. *J. Clim.*, **24**, 4973–4991.

Gerber, E. P., L. M. Polvani, and D. Ancukiewicz, 2008: Annular mode time scales in the Intergovernmental Panel on Climate Change Fourth Assessment Report models. *Geophys. Res. Lett.*, **35**, L22707.

Gerber, S., L. O. Hedin, M. Oppenheimer, S. W. Pacala, and E. Shevliakova, 2010: Nitrogen cycling and feedbacks in a global dynamic land model. *Global Biogeochem. Cycles*, **24**, Gb1001.

Gettelman, A., et al., 2010: Multimodel assessment of the upper troposphere and lower stratosphere: Tropics and global trends. *J. Geophys. Res. Atmos.*, **115**, D00m08.

Ghan, S., X. Liu, R. Easter, P. Rasch, J. Yoon, and B. Eaton, 2012: Toward a minimal representation of aerosols in climate models: Comparative decomposition of aerosol direct, semi-direct and indirect radiative forcing. *J. Clim.*, doi:10.1175/JCLI-D-11-00650.1.

Gillett, N. P., 2005: Climate modelling —Northern Hemisphere circulation. *Nature*, **437**, 496–496.

Giorgi, F., and E. Coppola, 2010: Does the model regional bias affect the projected regional climate change? An analysis of global model projections. *Clim. Change*, **100**, 787–795.

Girard, L., J. Weiss, J. M. Molines, B. Barnier, and S. Bouillon, 2009: Evaluation of high-resolution sea ice models on the basis of statistical and scaling properties of Arctic sea ice drift and deformation. *J. Geophys. Res.*, **114**, C08015.

Gleckler, P., K. Taylor, and C. Doutriaux, 2008: Performance metrics for climate models. *J. Geophys. Res. Atmos.*, **113**, D06104.

Gleckler, P., K. AchutaRao, J. Gregory, B. Santer, K. Taylor, and T. Wigley, 2006: Krakatoa lives: The effect of volcanic eruptions on ocean heat content and thermal expansion. *Geophys. Res. Lett.*, **33**, L17702.

Gleckler, P. J., et al., 2012: Human-induced global ocean warming on multidecadal timescales, *Nature Climate Change*, **2**, 524–529.

Gnanadesikan, A., S. M. Griffies, and B. L. Samuels, 2007: Effects in a climate model of slope tapering in neutral physics schemes. *Ocean Model.*, **16**, 1–16.

Goddard, L., and S. J. Mason, 2002: Sensitivity of seasonal climate forecasts to persisted SST anomalies. *Clim. Dyn.*, **19**, 619–631.

Golaz, J.-C., M. Salzmann, L. J. Donner, L. W. Horowitz, Y. Ming, and M. Zhao, 2011: Sensitivity of the aerosol indirect effect to subgrid variability in the cloud parameterization of the GFDL Atmosphere General Circulation Model AM3. *J. Clim.*, **24**, 3145–3160.

Goosse, H., and T. Fichefet, 1999: Importance of ice-ocean interactions for the global ocean circulation: A model study. *J. Geophys. Res. Oceans*, **104**, 23337–23355.

Goosse, H., et al., 2010: Description of the Earth system model of intermediate complexity LOVECLIM version 1.2. *Geosci. Model Dev.*, **3**, 603–633.

Gordon, C., et al., 2000: The simulation of SST, sea ice extents and ocean heat transports in a version of the Hadley Centre coupled model without flux adjustments. *Clim. Dyn.*, **16**, 147–168.

Gordon, H., et al., 2010: The CSIRO Mk3.5 Climate Model. *CAWCR Technical Report*, **21**, 1–74.

Gordon, H. B., et al., 2002: The CSIRO Mk3 Climate System Model. Technical Paper No. 60. CSIRO Atmospheric Research, Aspendale, Vic., Australia.

Greeves, C. Z., V. D. Pope, R. A. Stratton, and G. M. Martin, 2007: Representation of Northern Hemisphere winter storm tracks in climate models. *Clim. Dyn.*, **28**, 683–702.

Gregory, J. M., and P. M. Forster, 2008: Transient climate response estimated from radiative forcing and observed temperature change. *J. Geophys. Res. Atmos.*, **113**, D23105.

Gregory, J. M., et al., 2004: A new method for diagnosing radiative forcing and climate sensitivity. *Geophys. Res. Lett.*, **31**, L03205.

Gregory, J. M., et al., 2005: A model intercomparison of changes in the Atlantic thermohaline circulation in response to increasing atmospheric CO_2 concentration. *Geophys. Res. Lett.*, **32**, L12703.

Griffies, S. M., 2009: *Elements of MOM4p1*. GFDL Ocean Group Technical Report No. 6. NOAA/GFDL. Princeton, USA, 371 pp.

Griffies, S. M., and R. J. Greatbatch, 2012: Physical processes that impact the evolution of global mean sea level in ocean climate models. *Ocean Model.*, **51**, 37–72.

Griffies, S. M., M. J. Harrison, R. C. Pacanowski, and A. Rosati, 2004: *A Technical Guide to MOM4*. GFDL Ocean Group Technical Report No. 5, Princeton, USA, 337 pp.

Griffies, S. M., et al., 2005: Formulation of an ocean model for global climate simulations. *Ocean Sci.*, **1**, 45–79.

Griffies, S. M., et al., 2009: Coordinated Ocean-ice Reference Experiments (COREs). *Ocean Model.*, **26**, 1–46.

Grose, M., M. Pook, P. McIntosh, J. Risbey, and N. Bindoff, 2012: The simulation of cutoff lows in a regional climate model: Reliability and future trends. *Clim. Dyn.*, **39**, 445–459.

Guemas, V., F. J. Doblas-Reyes, I. Andreu-Burillo, and M. Asif, 2013: Retrospective prediction of the global warming slowdown in the last decade. *Nature Clim. Change*, doi:10.1038/nclimate1863.

Guilyardi, E., 2006: El Niño - mean state - seasonal cycle interactions in a multi-model ensemble. *Clim. Dyn.*, **26**, 229–348.

Guilyardi, E., P. Braconnot, F. F. Jin, S. T. Kim, M. Kolasinski, T. Li, and I. Musat, 2009a: Atmosphere feedbacks during ENSO in a coupled GCM with a modified atmospheric convection scheme. *J. Clim.*, **22**, 5698–5718.

Guilyardi, E., et al., 2009b: Understanding El Niño in ocean–atmosphere general circulation models: Progress and challenges. *Bull. Am. Meteorol. Soc.*, **90**, 325–340.

Guiot, J., J. J. Boreux, P. Braconnot, F. Torre, and P. Participants, 1999: Data-model comparison using fuzzy logic in paleoclimatology. *Clim. Dyn.*, **15**, 569–581.

Gupta, A. S., A. Santoso, A. S. Taschetto, C. C. Ummenhofer, J. Trevena, and M. H. England, 2009: Projected changes to the Southern Hemisphere ocean and sea ice in the IPCC AR4 climate models. *J. Clim.*, **22**, 3047–3078.

Gurney, K. R., et al., 2003: TransCom 3 CO_2 inversion intercomparison: 1. Annual mean control results and sensitivity to transport and prior flux information. *Tellus B*, **55**, 555–579.

Gutowski, W., et al., 2010: Regional extreme monthly precipitation simulated by NARCCAP RCMs. *J. Hydrometeorol.*, **11**, 1373–1379.

Hall, A., and X. Qu, 2006: Using the current seasonal cycle to constrain snow albedo feedback in future climate change. *Geophys. Res. Lett.*, **33**, L03502.

Hallberg, R., and A. Gnanadesikan, 2006: The role of eddies in determining the structure and response of the wind-driven southern hemisphere overturning: Results from the Modeling Eddies in the Southern Ocean (MESO) project. *J. Phys. Oceanogr.*, **36**, 2232–2252.

Hallberg, R., and A. Adcroft, 2009: Reconciling estimates of the free surface height in Lagrangian vertical coordinate ocean models with mode-split time stepping. *Ocean Model.*, **29**, 15–26.

Halloran, P. R., 2012: Does atmospheric CO_2 seasonality play an important role in governing the air-sea flux of CO_2? *Biogeosciences*, **9**, 2311–2323.

Ham, Y.-G., J. S. Kug, I. S. Kang, F. F. Jin, and A. Timmermann, 2010: Impact of diurnal atmospher-ocean coupling on tropical climate simulations using a coupled GCM. *Clim. Dyn.*, **34**, 905–917.

Hamilton, K., 1998: Effects of an imposed Quasi-Biennial Oscillation in a comprehensive troposphere-stratosphere-mesosphere general circulation model. *J. Atmos. Sci.*, **55**, 2393– 2418.

Handorf, D., and K. Dethloff, 2012: How well do state-of-the-art atmosphere-ocean general circulation models reproduce atmospheric teleconnection patterns? *Tellus A*, **64**, 19777.

Hannart, A., J. L. Dufresne, and P. Naveau, 2009: Why climate sensitivity may not be so unpredictable. *Geophys. Res. Lett.*, **36**, L16707.

Hannay, C., et al., 2009: Evaluation of Forecasted Southeast Pacific Stratocumulus in the NCAR, GFDL, and ECMWF Models. *J. Clim.*, **22**, 2871–2889.

Hansen, J., R. Ruedy, M. Sato, and K. Lo, 2010: Global surface temperature change. *Rev. Geophys.*, **48**, Rg4004.

Hansen, J., M. Sato, P. Kharecha, and K. von Schuckmann, 2011: Earth's energy imbalance and implications. *Atmos. Chem. Phys.*, **11**, 13421–13449.

Hansen, J., et al., 1983: Efficient Three-Dimensional Global Models for Climate Studies: Models I and II. *Mon. Weath. Rev.*, **111**, 609–662.

Hansen, J., et al., 1984: Climate Sensitivity: Analysis of Feedback Mechanisms. *Clim. Proc. Clim. Sens. Geophys. Monogr.*, **29**, 130–163.

Hansen, J., et al., 2005: Efficacy of climate forcings. *J. Geophys. Res. Atmos.*, **110**, D18104.

Hardiman, S. C., N. Butchart, T. J. Hinton, S. M. Osprey, and L. J. Gray, 2012: The effect of a well-resolved stratosphere on surface climate: Differences between CMIP5 simulations with high and low top versions of the Met Office Climate Model. *J. Clim.*, **25**, 7083–7099.

Hargreaves, J. C., A. Abe-Ouchi, and J. D. Annan, 2007: Linking glacial and future climates through an ensemble of GCM simulations. *Clim. Past*, **3**, 77–87.

Hargreaves, J. C., J. D. Annan, M. Yoshimori, and A. Abe-Ouchi, 2012: Can the Last Glacial Maximum constrain climate sensitivity? *Geophys. Res. Lett.*, **39**, L24702.

Hargreaves, J. C., A. Paul, R. Ohgaito, A. Abe-Ouchi, and J. D. Annan, 2011: Are paleoclimate model ensembles consistent with the MARGO data synthesis? *Clim. Past*, **7**, 917–933.

Hargreaves, J. C., J. D. Annan, R. Ohgaito, A. Paul, and A. Abe-Ouchi, 2013: Skill and reliability of climate model ensembles at the Last Glacial Maximum and mid Holocene. *Clim. Past*, **9**, 811–823.

Hasegawa, A., and S. Emori, 2007: Effect of air-sea coupling in the assessment of CO_2-induced intensification of tropical cyclone activity. *Geophys. Res. Lett.*, **34**, L05701.

Hasumi, H., 2006: *CCSR Ocean Component Model (COCO)* Version 4.0. CCSR Report. Centre for Climate System Research, University of Tokyo, Tokyo, Japan, 68 pp.

Hasumi, H., and S. Emori, 2004: K-1 Coupled GCM (MIROC) Description. Center for Climate System Research, University of Tokyo, Tokyo, Japan, 34 pp.

Hawkins, E., and R. Sutton, 2009: The potential to narrow uncertainty in regional climate predictions. *Bull. Am. Meteorol. Soc.*, **90**, 1095–1107.

Haynes, J. M., C. Jakob, W. B. Rossow, G. Tselioudis, and J. Brown, 2011: Major characteristics of Southern Ocean cloud regimes and their effects on the energy budget. *J. Clim.*, **24**, 5061–5080.

Haynes, P. H., 2006: The latitudinal structure of the QBO. *Q. J. R. Meteorol. Soc.*, **124**, 2645–2670.

Haywood, J. M., N. Bellouin, A. Jones, O. Boucher, M. Wild, and K. P. Shine, 2011: The roles of aerosol, water vapor and cloud in future global dimming/brightening. *J. Geophys. Res. Atmos.*, **116**, D20203.

Hazeleger, W., and R. J. Haarsma, 2005: Sensitivity of tropical Atlantic climate to mixing in a coupled ocean-atmosphere model. *Clim. Dyn.*, **25**, 387–399.

Hazeleger, W., et al., 2012: EC-Earth V2.2: Description and validation of a new seamless earth system prediction model. *Clim. Dyn.*, **39**, 2611–2629.

Hegerl, G., and F. Zwiers, 2011: Use of models in detection and attribution of climate change. *Clim. Change*, **2**, 570–591.

Hegerl, G. C., et al., 2007: Understanding and attributing climate change. In: *Climate Change 2007: The Physical Science Basis. Contribution of Working Group I to the Fourth Assessment Report of the Intergovernmental Panel on Climate Change* [Solomon, S., D. Qin, M. Manning, Z. Chen, M. Marquis, K. B. Averyt, M. Tignor and H. L. Miller (eds.)] Cambridge University Press, Cambridge, United Kingdom and New York, NY, USA, pp. 665–775.

Heinze, C., 2004: Simulating oceanic $CaCO_3$ export production in the greenhouse. *Geophys. Res. Lett.*, **31**, L16308.

Heinze, C., I. Kriest, and E. Maier-Reimer, 2009: Age offsets among different biogenic and lithogenic components of sediment cores revealed by numerical modeling. *Paleoceanography*, **24**, PA4214.

Held, I. M., 2005: The gap between simulation and understanding in climate modeling. *Bull. Am. Meteorol. Soc.*, **86**, 1609–1614.

Held, I. M., and B. J. Soden, 2006: Robust responses of the hydrological cycle to global warming. *J. Clim.*, **19**, 5686–5699.

Held, I. M., and K. M. Shell, 2012: Using Relative Humidity as a State Variable in Climate Feedback Analysis. *J. Clim.*, **25**, 2578–2582.

Held, I. M., M. Winton, K. Takahashi, T. Delworth, F. R. Zeng, and G. K. Vallis, 2010: Probing the fast and slow components of global warming by returning abruptly to preindustrial forcing. *J. Clim.*, **23**, 2418–2427.

Henson, S. A., D. Raitsos, J. P. Dunne, and A. McQuatters-Gollop, 2009: Decadal variability in biogeochemical models: Comparison with a 50-year ocean colour dataset. *Geophys. Res. Lett.*, **36**, L21601.

Hermes, J. C., and C. J. C. Reason, 2005: Ocean model diagnosis of interannual coevolving SST variability in the South Indian and South Atlantic Oceans. *J. Clim.*, **18**, 2864–2882.

Hernández-Díaz, L., R. Laprise, L. Sushama, A. Martynov, K. Winger, and B. Dugas, 2013: Climate simulation over CORDEX Africa domain using the fifth-generation Canadian Regional Climate Model (CRCM5). *Clim. Dyn.*, **40**, 1415–1433.

Heuzé, C., K. J. Heywood, D. P. Stevens, and J. K. Ridley, 2013: Southern Ocean bottom water characteristics in CMIP5 models. *Geophys. Res. Lett.*, doi:10.1002/grl.50287.

Hewitt, H. T., et al., 2011: Design and implementation of the infrastructure of HadGEM3: The next-generation Met Office climate modelling system. *Geosci. Model Dev.*, **4**, 223–253.

Hibbard, K. A., G. A. Meehl, P. M. Cox, and P. Friedlingstein, 2007: A strategy for climate change stabilization experiments. *Eos Trans. Am. Geophys. Union*, **88**, 217–221.

Hirai, M., T. Sakashita, H. Kitagawa, T. Tsuyuki, M. Hosaka, and M. Oh'Izumi, 2007: Development and validation of a new land surface model for JMA's operational global model using the CEOP observation dataset. *J. Meteorol. Soc. Jpn.*, **85A**, 1–24.

Hirota, N., Y. N. Takayabu, M. Watanabe, and M. Kimoto, 2011: Precipitation reproducibility over tropical oceans and its relationship to the double ITCZ problem in CMIP3 and MIROC5 climate models. *J. Clim.*, **24**, 4859–4873.

Hirschi, M., et al., 2011: Observational evidence for soil-moisture impact on hot extremes in southeastern Europe. *Nature Geosci.*, **4**, 17–21.

Hofmann, M., and M. A. Morales Maqueda, 2011: The response of Southern Ocean eddies to increased midlatitude westerlies: A non-eddy resolving model study. *Geophys. Res. Lett.*, **38**, L03605.

Holden, P. B., N. R. Edwards, D. Gerten, and S. Schaphoff, 2013: A model based constraint of CO_2 fertilisation. *Biogeosciences*, **10**, 339–355.

Holian, G. L., A. P. Sokolov, and R. G. Prinn, 2001: Uncertainty in atmospheric CO_2 predictions from a parametric uncertainty analysis of a Global Ocean Carbon Cycle Model. *Joint Program Report Series*. MIT Joint Program on the Science and Policy of Global Change, Cambridge, MA, USA, 25 pp.

Holland, M., D. Bailey, B. Briegleb, B. Light, and E. Hunke, 2012: Improved sea ice shortwave radiation physics in CCSM4: The impact of melt ponds and aerosols on arctic aea ice. *J. Clim.*, **25**, 1413–1430.

Holland, M. M., M. C. Serreze, and J. Stroeve, 2010: The sea ice mass budget of the Arctic and its future change as simulated by coupled climate models. *Clim. Dyn.*, doi:10.1007/s00382-008-0493-4.

Holton, J. R., and H. C. Tan, 1980: The influence of the equatorial Quasi-Biennial Oscillation on the global circulation at 50 mb. *J. Atmos. Sci.*, **37**, 2200–2208.

Horowitz, L. W., et al., 2003: A global simulation of tropospheric ozone and related tracers: Description and evaluation of MOZART, version 2. *J. Geophys. Res. Atmos.*, **108**, 4784.

Hourdin, F., et al., 2012: Impact of the LMDZ atmospheric grid configuration on the climate and sensitivity of the IPSL-CM5A coupled model. *Clim. Dyn.*, doi:10.1007/s00382–012–1411–3.

Hourdin, F., et al., 2013: LMDZ5B: The atmospheric component of the IPSL climate model with revisited parameterizations for clouds and convection. *Clim. Dyn.*, **40**, 2193–2222.

Hourdin, F., et al., 2010: AMMA-Model Intercomparison Project. *Bull. Am. Meteorol. Soc.*, **91**, 95–104.

Hu, Z.-Z., B. Huang, Y.-T. Hou, W. Wang, F. Yang, C. Stan, and E. Schneider, 2011: Sensitivity of tropical climate to low-level clouds in the NCEP climate forecast system. *Clim. Dyn.*, **36**, 1795–1811.

Huang, C. J., F. Qiao, Q. Shu, and Z. Song, 2012: Evaluating austral summer mixed-layer response to surface wave-induced mixing in the Southern Ocean. *J. Geophys. Res.-Oceans*, **117**, C00j18.

Huber, M., I. Mahlstein, M. Wild, J. Fasullo, and R. Knutti, 2011: Constraints on climate sensitivity from radiation patterns in climate models. *J. Clim.*, **24**, 1034–1052.

Hung, M., J. Lin, W. Wang, D. Kim, T. Shinoda, and S. Weaver, 2013: MJO and convectively coupled equatorial waves simulated by CMIP5 climate models. *J. Clim.*, doi:10.1175/JCLI-D-12-00541.1.

Hunke, E. C., and J. K. Dukowicz, 1997: An elastic-viscous-plastic model for sea ice dynamics. *J. Phys. Oceanogr.*, **27**, 1849–1867.

Hunke, E. C., and W. H. Lipscomb, 2008: CICE: The Los Alamos Sea Ice ModelDocumentation and Software User's ManualVersion 4.1. Los Alamos National Laboratory, Los Alamos, NM, USA, 76 pp.

Hunke, E. C., W. H. Lipscomb, and A. K. Turner, 2010: Sea ice models for climate study: Retorspective and new directions. *J. Glaciol.*, **56**, 1162–1172.

Hunke, E. C., D. A. Hebert, and O. Lecomte, 2013: Level-ice melt ponds in the Los Alamos sea ice model, CICE. *Ocean Model.*, doi:10.1016/j.ocemod.2012.11.008.

Hunke, E. C., D. Notz, A. K. Turner, and M. Vancoppenolle, 2011: The multiphase physics of sea ice : A review for model developers. *Cryosphere*, **5**, 989–1009.

Hurrell, J., G. A. Meehl, D. Bader, T. L. Delworth, B. Kirtman, and B. Wielicki, 2009: A unified modeling approach to climate system prediction. *Bull. Am. Meteorol. Soc.*, **90**, 1819–1832.

Hurrell, J., et al., 2013: The Community Earth System Model: A framework for collaborative research. *Bull. Am. Meteorol. Soc.*, doi:10.1175/BAMS-D-12–00121.

Hurtt, G. C., et al., 2009: Harmonization of global land-use scenarios for the period 1500–2100 for IPCC-AR5. *iLEAPS Newsl.*, **7**, 6–8.

Hutchings, J. K., A. Roberts, C. A. Geiger, and J. Richter-Menge, 2011: Spatial and temporal characterization of sea-ice deformation. *Ann. Glaciol.*, **52**, 360–368.

Huybrechts, P., 2002: Sea-level changes at the LGM from ice-dynamic reconstructions of the Greenland and Antarctic ice sheets during the glacial cycles. *Quat. Sci. Rev.*, **21**, 203–231.

Huybrechts, P., H. Goelzer, I. Janssens, E. Driesschaert, T. Fichefet, H. Goosse, and M. F. Loutre, 2011: Response of the Greenland and Antarctic ice sheets to multi-millennial greenhouse warming in the Earth System Model of Intermediate Complexity LOVECLIM. *Surv. Geophys.*, **32**, 397–416.

Iacono, M. J., J. S. Delamere, E. J. Mlawer, and S. A. Clough, 2003: Evaluation of upper tropospheric water vapor in the NCAR Community Climate Model (CCM3) using modeled and observed HIRS radiances. *J. Geophys. Res.*, **108**, 4037.

Ichikawa, H., H. Masunaga, Y. Tsushima, and H. Kanzawa, 2012: Reproducibility by climate models of cloud radiative forcing associated with tropical convection. *J. Clim.*, **25**, 1247–1262.

Ilicak, M., A. J. Adcroft, S. M. Griffies, and R. W. Hallberg, 2012: Spurious dianeutral mixing and the role of momentum closure. *Ocean Model.*, **45–46**, 37–58.

Illingworth, A. J., et al., 2007: Cloudnet. *Bull. Am. Meteor. Soc.*, **88**, 883–898.

Ilyina, T., R. E. Zeebe, E. Maier-Reimer, and C. Heinze, 2009: Early detection of ocean acidification effects on marine calcification. *Global Biogeochem. Cycles*, **23**, Gb1008.

Ilyina, T., K. Six, J. Segschneider, J. Maier-Reimer, H. Li, and I. Nunez-Riboni, 2013: The global ocean biogeochemistry model HAMOCC: Model architecture and performance as component of the MPI-Earth System Model in different CMIP5 experimental realizations. *J. Adv. Model. Earth Syst.*, **5**, 287–315.

Inatsu, M., and M. Kimoto, 2009: A scale interaction study on East Asian cyclogenesis using a General Circulation Model coupled with an Interactively Nested Regional Model. *Mon. Weather Rev.*, **137**, 2851–2868.

Inatsu, M., Y. Satake, M. Kimoto, and N. Yasutomi, 2012: GCM bias of the western Pacific summer monsoon and its correction by two-way nesting system. *J. Meteorol. Soc. Jpn.*, **90B**, 1–10.

Ingram, W., 2010: A very simple model for the water vapour feedback on climate change. *Q. J. R. Meteorol. Soc.*, **136**, 30–40.

Ingram, W., 2013: Some implications of a new approach to the water vapour feedback. *Clim. Dyn.*, **40**, 925–933.

Inness, P. M., J. M. Slingo, E. Guilyardi, and J. Cole, 2003: Simulation of the Madden-Julian oscillation in a coupled general circulation model. Part II: The role of the basic state. *J. Clim.*, **16**, 365–382.

Inoue, J., J. P. Liu, J. O. Pinto, and J. A. Curry, 2006: Intercomparison of Arctic Regional Climate Models: Modeling clouds and radiation for SHEBA in May 1998. *J. Clim.*, **19**, 4167–4178.

IPCC, 2007: *Climate Change 2007: The Physical Science Basis. Contribution of Working Group I to the Fourth Assessment Report of the Intergovernmental Panel on Climate Change* [Solomon, S., D. Qin, M. Manning, Z. Chen, M. Marquis, K. B. Averyt, M. Tignor and H. L. Miller (eds.)] Cambridge University Press, Cambridge, United Kingdom and New York, NY, USA, 996 pp.

IPCC, 2012: *IPCC WGI/WGII Special Report on Managing the Risks of Extreme Events and Disasters to Advance Climate Change Adaptation (SREX).* [Field, C.B., V. Barros, T.F. Stocker, D. Qin, D.J. Dokken, K.L. Ebi, M.D. Mastrandrea, K.J. Mach, G.-K. Plattner, S.K. Allen, M. Tignor, and P.M. Midgley (Eds.)]. Cambridge University Press, The Edinburgh Building, Shaftesbury Road, Cambridge CB2 8RU ENGLAND, 582 pp.

Ishii, M., and M. Kimoto, 2009: Reevaluation of historical ocean heat content variations with time-varying XBT and MBT depth bias corrections. *J. Oceanogr.*, **65**, 287–299.

Ito, A., and T. Oikawa, 2002: A simulation model of the carbon cycle in land ecosystems (Sim-CYCLE): A description based on dry-matter production theory and plot-scale validation. *Ecol. Model.*, **151**, 143–176.

Iversen, T., et al., 2013: The Norwegian Earth System Model, NorESM1–M. Part 2: Climate response and scenario projections. *Geosci. Model Dev.*, **6**, 1–27.

Izumi, K., P. J. Bartlein, and S. P. Harrison, 2013: Consistent large-scale temperature responses in warm and cold climates. *Geophys. Res. Lett.*, doi:2013GL055097.

Jackson, C. S., M. K. Sen, G. Huerta, Y. Deng, and K. P. Bowman, 2008a: Error reduction and convergence in climate prediction. *J. Clim.*, **21**, 6698–6709.

Jackson, L., R. Hallberg, and S. Legg, 2008b: A parameterization of shear-driven turbulence for ocean climate models. *J. Phys. Oceanogr.*, **38**, 1033–1053.

Jacob, D., et al., 2012: Assessing the transferability of the Regional Climate Model REMO to Different COordinated Regional Climate Downscaling EXperiment (CORDEX) regions. *Atmosphere*, **3**, 181–199.

Jakob, C., 2010: Accelerating progress in Global Atmospheric Model development through improved parameterizations: Challenges, opportunities, and strategies. *Bull. Am. Meteorol. Soc.*, **91**, 869–875.

Jang, C. J., J. Park, T. Park, and S. Yoo, 2011: Response of the ocean mixed layer depth to global warming and its impact on primary production: A case for the North Pacific Ocean. *Ices J. Mar. Sci.*, **68**, 996–1007.

Jansen, E., et al., 2007: Paleoclimate. In: *Climate Change 2007: The Physical Science Basis. Contribution of Working Group I to the Fourth Assessment Report of the Intergovernmental Panel on Climate Change* [Solomon, S., D. Qin, M. Manning, Z. Chen, M. Marquis, K. B. Averyt, M. Tignor and H. L. Miller (eds.)] Cambridge University Press, Cambridge, United Kingdom and New York, NY, USA, pp. 433–498.

Jayne, S. R., 2009: The impact of abyssal mixing parameterizations in an ocean General Circulation Model. *J. Phys. Oceanogr.*, **39**, 1756–1775.

Ji, J., M. Huang, and K. Li, 2008: Prediction of carbon exchanges between China terrestrial ecosystem and atmosphere in 21st century. *Sci. China D*, **51**, 885–898.

Ji, J. J., 1995: A climate-vegetation interaction model: Simulating physical and biological processes at the surface. *J. Biogeogr.*, **22**, 445–451.

Jiang, J. H., et al., 2012a: Evaluation of cloud and water vapor simulations in CMIP5 climate models using NASA "A-Train" satellite observations. *J. Geophys. Res.*, **117**, D14105.

Jiang, X., et al., 2012b: Simulation of the intraseasonal variability over the Eastern Pacific ITCZ in climate models. *Clim. Dyn.*, **39**, 617–636.

Jin, F. F., S. T. Kim, and L. Bejarano, 2006: A coupled-stability index for ENSO. *Geophys. Res. Let.*, **33**, L23708.

Joetzjer, E., H. Douville, C. Delire, and P. Ciais, 2013: Present-day and future Amazonian precipitation in global climate models: CMIP5 versus CMIP3. *Clim. Dyn.*, doi: 10.1007/s00382–012-1644-1.

Johanson, C. M., and Q. Fu, 2009: Hadley Cell Widening: Model Simulations versus Observations. *J. Clim.*, **22**, 2713–2725.

John, V., and B. Soden, 2007: Temperature and humidity biases in global climate models and their impact on climate feedbacks. *Geophys. Res. Lett.*, **34**, L18704.

Johns, T. C., et al., 2003: Anthropogenic climate change for 1860 to 2100 simulated with the HadCM3 model under updated emissions scenarios. *Clim. Dyn.*, **20**, 583–612.

Johns, T. C., et al., 2006: The new Hadley Centre Climate Model (HadGEM1): Evaluation of coupled simulations. *J. Clim.*, **19**, 1327–1353

Johnson, N. C., and S. B. Feldstein, 2010: The continuum of north Pacific sea level pressure patterns: Intraseasonal, interannual, and interdecadal variability. *J. Clim.*, **23**, 851–867.

Jolliffe, I. T., and D. B. Stephenson, 2011: *Forecast Verification: A Practitioner's Guide in Atmospheric Science.* 2nd ed. John Wiley & Sons, Hoboken, NJ, 292 pp.

Joly, M., A. Voldoire, H. Douville, P. Terray, and J. F. Royer, 2007: African monsoon teleconnections with tropical SSTs: Validation and evolution in a set of IPCC4 simulations. *Clim. Dyn.*, **29**, 1–20.

Jones, A., D. L. Roberts, M. J. Woodage, and C. E. Johnson, 2001: Indirect sulphate aerosol forcing in a climate model with an interactive sulphur cycle. *J. Geophys. Res. Atmos.*, **106**, 20293–20310.

Jones, G. S., P. A. Stott, and N. Christidis, 2012: Attribution of observed historical near surface temperature variations to anthropogenic and natural causes using CMIP5 simulations. *J. Geophys. Res.*, doi:10.1002/jgrd.50239.

Jones, P. D., M. New, D. E. Parker, S. Martin, and I. G. Rigor, 1999: Surface air temperature and its changes over the past 150 years. *Rev. Geophys.*, **37**, 173–199.

Joseph, S., A. K. Sahai, B. N. Goswami, P. Terray, S. Masson, and J. J. Luo, 2012: Possible role of warm SST bias in the simulation of boreal summer monsoon in SINTEX-F2 coupled model. *Clim. Dyn.*, **38**, 1561–1576.

Joussaume, S., and K. E. Taylor, 1995: Status of the Paleoclimate Modeling Intercomparison Project. In: *Proceedings of the first international AMIP scientific conference, WCRP-92, Monterey, USA*, 425–430.

Jun, M., R. Knutti, and D. Nychka, 2008: Spatial analysis to quantify numerical model bias and dependence: How many climate models are there? *J. Am. Stat. Assoc.*, **103**, 934–947.

Jung, T., et al., 2010: The ECMWF model climate: Recent progress through improved physical parametrizations. *Q. J. R. Meteorol. Soc.*, **136**, 1145–1160.

Jung, T., et al., 2012: High-resolution global climate simulations with the ECMWF Model in Project Athena: Experimental design, model climate, and seasonal forecast skill. *J. Clim.*, **25**, 3155–3172.

Jungclaus, J. H., et al., 2006: Ocean circulation and tropical variability in the coupled model ECHAM5/MPI-OM. *J. Clim.*, **19**, 3952–3972.

Jungclaus, J. H., et al., 2013: Characteristics of the ocean simulations in MPIOM, the ocean componentof the MPI-Earth System Model. *J. Adv. Model. Earth Syst.*, doi:10.1002/jame.20023.

Jungclaus, J. H., et al., 2010: Climate and carbon-cycle variability over the last millennium. *Clim. Past*, **6**, 723–737.

Kageyama, M., et al., 2006: Last Glacial Maximum temperatures over the North Atlantic, Europe and western Siberia: A comparison between PMIP models, MARGO sea-surface temperatures and pollen-based reconstructions. *Quat. Sci. Rev.*, **25**, 2082–2102.

Kahn, R. A., B. J. Gaitley, J. V. Martonchik, D. J. Diner, K. A. Crean, and B. Holben, 2005: Multiangle Imaging Spectroradiometer (MISR) global aerosol optical depth validation based on 2 years of coincident Aerosol Robotic Network (AERONET) observations. *J. Geophys. Res. Atmos.*, **110**, D10s04.

Kalnay, E., et al., 1996: The NCEP/NCAR 40-year reanalysis project. *Bull. Am. Meteorol. Soc.*, **77**, 437–471.

Kanada, S., M. Nakano, S. Hayashi, T. Kato, M. Nakamura, K. Kurihara, and A. Kitoh, 2008: Reproducibility of Maximum Daily Precipitation Amount over Japan by a High-resolution Non-hydrostatic Model. *Sola*, **4**, 105–108.

Kanamaru, H., and M. Kanamitsu, 2007: Fifty-seven-year California reanalysis downscaling at 10 km (CaRD10). Part II: Comparison with North American regional reanalysis. *J. Clim.*, **20**, 5572–5592.

Kanamitsu, M., K. Yoshimura, Y. B. Yhang, and S. Y. Hong, 2010: Errors of interannual variability and trend in dynamical downscaling of reanalysis. *J. Geophys. Res. Atmos.*, **115**, D17115.

Kanamitsu, M., W. Ebisuzaki, J. Woollen, S. K. Yang, J. J. Hnilo, M. Fiorino, and G. L. Potter, 2002: NCEP-DOE AMIP-II reanalysis (R-2). *Bull. Am. Meteorol. Soc.*, **83**, 1631–1643.

Karlsson, J., and G. Svensson, 2010: The simulation of Arctic clouds and their influence on the winter surface temperature in present-day climate in the CMIP3 multi-model dataset. *Clim. Dyn.*, **36**, 623–635.

Karlsson, J., G. Svensson, and H. Rodhe, 2008: Cloud radiative forcing of subtropical low level clouds in global models. *Clim. Dyn.*, **30**, 779–788.

Karpechko, A., N. Gillett, G. Marshall, and A. Scaife, 2008: Stratospheric influence on circulation changes in the Southern Hemisphere troposphere in coupled climate models. *Geophys. Res. Lett.*, **35**, L20806.

Karpechko, A. Y., and E. Manzini, 2012: Stratospheric influence on tropospheric climate change in the Northern Hemisphere. *J. Geophys. Res. Atmos.*, **117**, D05133.

Karpechko, A. Y., N. P. Gillett, G. J. Marshall, and J. A. Screen, 2009: Climate impacts of the southern annular mode simulated by the CMIP3 models. *J. Clim.*, **22**, 6149–6150.

Kattsov, V. M., et al., 2010: Arctic sea-ice change: A grand challenge of climate science. *J. Glaciol.*, **56**, 1115–1121.

Kavvada, A., A. Ruiz-Barradas, and S. Nigam, 2013: AMO's structure and climate footprint in observations and IPCC AR5 climate simulations. *Clim. Dyn.*, doi:10.1007/s00382-013-1712-1.

Kawazoe, S., and W. Gutowski, 2013: Regional, very heavy daily precipitation in NARCCAP simulations. *J. Hydrometeorol.*, doi:10.1175/JHM-D-12-068.1.

Kay, J. E., M. M. Holland, and A. Jahn, 2011: Inter-annual to multi-decadal Arctic sea ice extent trends in a warming world. *Geophys. Res. Lett.*, **38**, L15708.

Keeley, S. P. E., R. T. Sutton, and L. C. Shaffrey, 2012: The impact of North Atlantic sea surface temperature errors on the simulation of North Atlantic European region climate. *Q. J. R. Meteorol. Soc.*, doi:10.1002/qj.1912.

Kendon, E. J., N. M. Roberts, C. A. Senior, and M. J. Roberts, 2012: Realism of rainfall in a very high resolution regional climate model. *J. Clim.*, **25**, 5791–5806.

Khairoutdinov, M. F., D. A. Randall, and C. DeMott, 2005: Simulations of the Atmospheric general circulation using a cloud-resolving model as a superparameterization of physical processes. *J. Atmos. Sci.*, **62**, 2136–2154.

Kharin, V. V., F. W. Zwiers, X. B. Zhang, and G. C. Hegerl, 2007: Changes in temperature and precipitation extremes in the IPCC ensemble of global coupled model simulations. *J. Clim.*, **20**, 1419–1444.

Kharin, V. V., F. W. Zwiers, X. Zhang, and M. Wehner, 2012: Changes in temperature and precipitation extremes in the CMIP5 ensemble. *Clim. Change*, doi:10.1007/s10584-013-0705-8.

Khvorostyanov, D. V., G. Krinner, P. Ciais, M. Heimann, and S. A. Zimov, 2008a: Vulnerability of permafrost carbon to global warming. Part I: Model description and role of heat generated by organic matter decomposition. *Tellus B*, **60**, 250–264.

Khvorostyanov, D. V., P. Ciais, G. Krinner, S. A. Zimov, C. Corradi, and G. Guggenberger, 2008b: Vulnerability of permafrost carbon to global warming. Part II: Sensitivity of permafrost carbon stock to global warming. *Tellus B*, **60**, 265–275.

Kidston, J., and E. P. Gerber, 2010: Intermodel variability of the poleward shift of the austral jet stream in the CMIP3 integrations linked to biases in 20th century climatology. *Geophys. Res. Lett.*, **37**, L09708.

Kiehl, J. T., 2007: Twentieth century climate model response and climate sensitivity. *Geophys. Res. Lett.*, **34**.

Kim, D., and V. Ramanathan, 2008: Solar radiation budget and radiative forcing due to aerosols and clouds. *J. Geophys. Res. Atmos.*, **113**, D02203.

Kim, D., et al., 2012: The tropical subseasonal variability simulated in the NASA GISS general circulation model. *J. Clim.*, **25**, 4641–4659.

Kim, D., et al., 2009: Application of MJO simulation diagnostics to climate models. *J. Clim.*, **22**, 6413–6436.

Kim, H.-J., K. Takata, B. Wang, M. Watanabe, M. Kimoto, T. Yokohata, and T. Yasunari, 2011: Global monsoon, El Niño, and their interannual linkage simulated by MIROC5 and the CMIP3 CGCMs. *J. Clim.*, **24**, 5604–5618.

Kim, S., and F.-F. Jin, 2011a: An ENSO stability analysis. Part I: Results from a hybrid coupled model. *Clim. Dyn.*, **36**, 1593–1607.

Kim, S., and F.-F. Jin, 2011b: An ENSO stability analysis. Part II: Results from the twentieth and twenty-first century simulations of the CMIP3 models. *Clim. Dyn.*, **36**, 1609–1627.

Kim, S. T., and J.-Y. Yu, 2012: The two types of ENSO in CMIP5 models. *Geophys. Res. Lett.*, **39**, L11704.

Kirkevåg, K., et al., 2013: Aerosol-climate interactions in the Norwegian Earth System Model – NorESM1–M. *Geophys. Model Dev.*, **6**, 207–244.

Kistler, R., et al., 2001: The NCEP-NCAR 50–year reanalysis: Monthly means CD-ROM and documentation. *Bull. Am. Meteorol. Soc.*, **82**, 247–267.

Kjellstrom, E., G. Nikulin, U. Hansson, G. Strandberg, and A. Ullerstig, 2011: 21st century changes in the European climate: Uncertainties derived from an ensemble of regional climate model simulations. *Tellus A*, **63**, 24–40.

Kjellstrom, E., F. Boberg, M. Castro, J. Christensen, G. Nikulin, and E. Sanchez, 2010: Daily and monthly temperature and precipitation statistics as performance indicators for regional climate models. *Clim. Res.*, **44** 135–150.

Klein, P., and G. Lapeyre, 2009: The oceanic vertical pump induced by mesoscale and submesoscale turbulence. *Annu. Rev. Mar. Sci.*, **1**, 351–375.

Klein, S. A., and C. Jakob, 1999: Validation and sensitivities of frontal clouds simulated by the ECMWF model. *Mon. Weather Rev.*, **127**, 2514–2531.

Klein, S. A., B. J. Soden, and N. C. Lau, 1999: Remote sea surface temperature variations during ENSO: Evidence for a tropical atmospheric bridge. *J. Clim.*, **12**, 917–932.

Klein, S. A., X. Jiang, J. Boyle, S. Malyshev, and S. Xie, 2006: Diagnosis of the summertime warm and dry bias over the U.S. Southern Great Plains in the GFDL climate model using a weather forecasting approach. *Geophys. Res. Lett.*, **33**, L18805.

Klein, S. A., Y. Zhang, M. D. Zelinka, R. Pincus, J. S. Boyle, and P. J. Glecker, 2013: Are climate model simulations of clouds improving? An evaluation using the ISCCP simulator. *J. Geophys. Res.*, doi:10.1002/jgrd.50141.

Klocke, D., R. Pincus, and J. Quaas, 2011: On constraining estimates of climate sensitivity with present-day observations through model weighting. *J. Clim.*, **24**, 6092–6099.

Kloster, S., N. M. Mahowald, J. T. Randerson, and P. J. Lawrence, 2012: The impacts of climate, land use, and demography on fires during the 21st century simulated by CLM-CN. *Biogeosciences*, **9**, 509–525.

Knight, J., et al., 2009: Do global temperature trends over the last decade falsify climate predictions? [In: State of the Climate in 2008]. *Bull. Am. Meteorol. Soc.*, **90**, S22–S23.

Knight, J. R., 2009: The Atlantic Multidecadal Oscillation inferred from the forced climate response in Coupled General Circulation Models. *J. Clim.*, **22**, 1610–1625.

Knutti, R., 2008: Why are climate models reproducing the observed global surface warming so well? *Geophys. Res. Lett.*, **35**, L18704

Knutti, R., 2010: The end of model democracy? *Clim. Change*, **102**, 395–404.

Knutti, R., and G. C. Hegerl, 2008: The equilibrium sensitivity of the Earth's temperature to radiation changes. *Nature Geosci.*, **1**, 735–743.

Knutti, R., and L. Tomassini, 2008: Constraints on the transient climate response from observed global temperature and ocean heat uptake. *Geophys. Res. Lett.*, **35**, L09701.

Knutti, R., and J. Sedlácek, 2013: Robustness and uncertainties in the new CMIP5 climate model projections. *Nature Clim. Change*, **3**, 369–373.

Knutti, R., D. Masson, and A. Gettelman, 2013: Climate model genealogy: Generation CMIP5 and how we got there. *Geophys. Res. Lett.*, **40**, 1194–1199.

Knutti, R., G. A. Meehl, M. R. Allen, and D. A. Stainforth, 2006: Constraining climate sensitivity from the seasonal cycle in surface temperature. *J. Clim.*, **19**, 4224–4233.

Knutti, R., F. Joos, S. A. Muller, G. K. Plattner, and T. F. Stocker, 2005: Probabilistic climate change projections for CO_2 stabilization profiles. *Geophys. Res. Lett.*, **32**, L20707.

Knutti, R., R. Furrer, C. Tebaldi, J. Cermak, and G. A. Meehl, 2010a: Challenges in combining projections from multiple climate models. *J. Clim.*, **23**, 2739–2758.

Knutti, R., G. Abramowitz, M. Collins, V. Eyring, P. J. Gleckler, B. Hewitson, and L. Mearns, 2010b: Good practice guidance paper on assessing and combining multi model climate projections. In: *Meeting Report of the Intergovernmental Panel on Climate Change Expert Meeting on Assessing and Combining Multi Model Climate Projections* [T. F. Stocker, T.F., D. Qin, G.-K. Plattner, M. Tignor, and P.M. Midgley (eds.)]. IPCC Working Group I Technical Support Unit, University of Bern, Bern, Switzerland.

Koch, D., et al., 2011: Coupled Aerosol-Chemistry-Climate Twentieth-Century Transient Model investigation: Trends in short-lived species and climate responses. *J. Clim.*, **24**, 2693–2714.

Koldunov, N. V., D. Stammer, and J. Marotzke, 2010: Present-day Arctic sea ice variability in the coupled ECHAM5/MPI-OM model. *J. Clim.*, **23**, 2520–2543.

Koltzow, M., T. Iversen, and J. Haugen, 2008: Extended Big-Brother experiments: The role of lateral boundary data quality and size of integration domain in regional climate modelling. *Tellus A*, **60**, 398–410.

Koltzow, M. A. O., T. Iversen, and J. E. Haugen, 2011: The importance of lateral boundaries, surface forcing and choice of domain size for dynamical downscaling of global climate simulations. *Atmosphere*, **2**, 67–95.

Komuro, Y., et al., 2012: Sea-ice in twentieth-century simulations by new MIROC Coupled Models: A comparison between models with high resolution and with ice thickness distribution. *J. Meteorol. Soc. Jpn,*, **90A**, 213–232.

Konsta, D., H. Chepfer, and J.-L. Dufresne, 2012: A process oriented characterization of tropical oceanic clouds for climate model evaluation, based on a statistical analysis of daytime A-train observations. *Clim. Dyn.*, **39**, 2091–2108.

Koster, R., et al., 2004: Regions of strong coupling between soil moisture and precipitation. *Science*, **305**, 1138–1140.

Kostopoulou, E., K. Tolika, I. Tegoulias, C. Giannakopoulos, S. Somot, C. Anagnostopoulou, and P. Maheras, 2009: Evaluation of a regional climate model using in situ temperature observations over the Balkan Peninsula. *Tellus A*, **61**, 357–370.

Koven, C. D., W. J. Riley, and A. Stern, 2013: Analysis of permafrost thermal dynamics and response to climate change in the CMIP5 Earth System Models. *J. Clim.*, **26**, 1877–1900.

Koven, C. J., et al., 2011: Permafrost carbon-climate feedbacks accelerate global warming. *Proc. Natl. Acad. Sci. U.S.A.*, **108**, 14769–14774.

Kowalczyk, E. A., Y. P. Wang, R. M. Law, H. L. Davies, J. L. McGregor, and G. Abramowitz 2006: The CSIRO Atmosphere Biosphere Land Exchange (CABLE) model for use in climate models and as an offline model. CSIRO Marine and Atmospheric Research paper 013, Victoria, Australia, 37 pp.

Kowalczyk, E. A., et al., 2013: The land surface model component of ACCESS: Description and impact on the simulated surface climatology. *Aust. Meteorol. Oceanogr. J.*, **63**, 65–82.

Kravtsov, S., and C. Spannagle, 2008: Multidecadal climate variability in observed and modeled surface temperatures. *J. Clim.*, **21**, 1104–1121.

Krinner, G., et al., 2005: A dynamic global vegetation model for studies of the coupled atmosphere-biosphere system. *Global Biogeochem. Cycles*, **19**, GB1015.

Krüger, L., R. da Rocha, M. Reboita, and T. Ambrizzi, 2012: RegCM3 nested in HadAM3 scenarios A2 and B2: Projected changes in extratropical cyclogenesis, temperature and precipitation over the South Atlantic Ocean. *Clim. Change*, **113**, 599–621.

Kuhlbrodt, T., and J. Gregory, 2012: Ocean heat uptake and its consequences for the magnitude of sea level rise and climate change. *Geophys. Res. Lett.*, doi:10.1029/2012GL052952.

Kuhlbrodt, T., R. S. Smith, Z. Wang, and J. M. Gregory, 2012: The influence of eddy parameterizations on the transport of the Antarctic Circumpolar Current in coupled climate models. *Ocean Model.*, **52–53**, 1–8.

Kusaka, H., T. Takata, and Y. Takane, 2010: Reproducibility of regional climate in central Japan using the 4-km Resolution WRF Model. *Sola*, **6**, 113–116.

Kusunoki, S., R. Mizuta, and M. Matsueda, 2011: Future changes in the East Asian rain band projected by global atmospheric models with 20-km and 60-km grid size. *Clim. Dyn.*, **37**, 2481–2493.

L'Ecuyer, T., and G. Stephens, 2007: The tropical atmospheric energy budget from the TRMM perspective. Part II: Evaluating GCM representations of the sensitivity of regional energy and water cycles to the 1998–99 ENSO Cycle. *J. Clim.*, **20**, 4548–4571.

Laine, A., G. Lapeyre, and G. Riviere, 2011: A quasigeostrophic model for moist storm tracks. *J. Atmos. Sci.*, **68**, 1306–1322.

Laine, A., M. Kageyama, P. Braconnot, and R. Alkama, 2009: Impact of greenhouse gas concentration changes on surface energetics in IPSL-CM4: Regional warming patterns, land-sea warming ratios, and glacial-interglacial differences. *J. Clim.*, **22**, 4621–4635.

Lamarque, J. F., et al., 2012: CAM-chem: Description and evaluation of interactive atmospheric chemistry in the Community Earth System Model. *Geosci. Model Dev.*, **5**, 369–411.

Lamarque, J. F., et al., 2010: Historical (1850–2000) gridded anthropogenic and biomass burning emissions of reactive gases and aerosols: Methodology and application. *Atmos. Chem. Phys.*, **10**, 7017–7039.

Lambert, F. H., G. R. Harris, M. Collins, J. M. Murphy, D. M. H. Sexton, and B. B. B. Booth, 2012: Interactions between perturbations to different Earth system components simulated by a fully-coupled climate model. *Clim. Dyn.*, doi:10.1007/s00382-012-1618-3.

Lambert, S., and G. Boer, 2001: CMIP1 evaluation and intercomparison of coupled climate models. *Clim. Dyn.*, **17**, 83–106.

Landrum, L., M. M. Holland, D. P. Schneider, and E. Hunke, 2012: Antarctic sea ice climatology, variability and late 20th century change in CCSM4. *J. Clim.*, **25**, 4817–4838.

Langenbrunner, B., and J. D. Neelin, 2013: Analyzing ENSO teleconnections in CMIP models as a measure of model fidelity in simulating precipitation. *J. Clim.*, doi:10.1175/JCLI-D-12-00542.1.

Laprise, R., 2008: Regional climate modelling. *J. Comput. Phys.*, **227**, 3641–3666.

Laprise, R., et al., 2008: Challenging some tenets of regional climate modelling. *Meteorol. Atmos. Phys.*, **100**, 3–22.

Large, W., and S. Yeager, 2009: The global climatology of an interannually varying air-sea flux data set. *Clim. Dyn.*, **33**, 341–364.

Larow, T. E., Y. K. Lim, D. W. Shin, E. P. Chassignet, and S. Cocke, 2008: Atlantic basin seasonal hurricane simulations. *J. Clim.*, **21**, 3191–3206.

Lau, K. M., et al., 2008: The Joint Aerosol-Monsoon Experiment —A new challenge for monsoon climate research. *Bull. Am. Meteorol. Soc.*, **89**, 369–383.

Lau, W. K. M., and D. E. Waliser, 2011: *Intraseasonal Variability of the Atmosphere-Ocean Climate System.* Springer Science+Business Media, New York, NY, USA, and Heidelberg, Germany.

Lawrence, D. M., et al., 2012: The CCSM4 Land Simulation, 1850–2005: Assessment of surface climate and new capabilities. *J. Clim.*, **25**, 2240–2260.

Lawrence, D. M., et al., 2011: Parameterization improvements and functional and structural advances in version 4 of the Community Land Model. *J. Adv. Model. Earth Syst.*, **3**, 2011MS000045.

Le Quere, C., et al., 2005: Ecosystem dynamics based on plankton functional types for global ocean biogeochemistry models. *Global Change Biol.*, **11**, 2016–2040.

Le Quere, C., et al., 2009: Trends in the sources and sinks of carbon dioxide. *Nature Geosci.*, **2**, 831–836.

Lecomte, O., T. Fichefet, M. Vancoppenolle, and M. Nicolaus, 2011: A new snow thermodynamic scheme for large-scale sea-ice models. *Ann. Glaciol.*, **52**, 337–346.

Leduc, M., and R. Laprise, 2009: Regional climate model sensitivity to domain size. *Clim. Dyn.*, **32**, 833–854.

Lee, D. S., et al., 2009: Aviation and global climate change in the 21st century. *Atmos. Environ.*, **43**, 3520–3537.

Lee, T., D. E. Waliser, J.-L. F. Li, F. W. Landerer, and M. M. Gierach, 2013: Evaluation of CMIP3 and CMIP5 wind stress climatology using satellite measurements and atmospheric reanalysis products. *J. Clim.*, doi:10.1175/JCLI-D-12-00591.1.

Legg, S., L. Jackson, and R. W. Hallberg, 2008: Eddy-resolving modeling of overflows. In: *Eddy Resolving Ocean Models*, 177 ed. [M. Hecht, and H. Hasumi (eds.)]. American Geophysical Union, Washington, DC, pp. 63–82.

9

Legg, S., et al., 2009: Improving oceanic overflow representation in climate models: The Gravity Current Entrainment Climate Process Team. *Bull. Am. Meteorol. Soc.*, **90**, 657–670.
Leloup, J., M. Lengaigne, and J.-P. Boulanger, 2008: Twentieth century ENSO characteristics in the IPCC database. *Clim. Dyn.*, **30**, 277–291.
Lemoine, D. M., 2010: Climate sensitivity distributions dependence on the possibility that models share biases. *J. Clim.*, **23**, 4395–4415.
Lenaerts, J., M. van den Broeke, S. Dery, E. van Meijgaard, W. van de Berg, S. Palm, and J. Rodrigo, 2012: Modeling drifting snow in Antarctica with a regional climate model: 1. Methods and model evaluation. *J. Geophys. Res. Atmos.*, **117**, D05108.
Lenderink, G., 2010: Exploring metrics of extreme daily precipitation in a large ensemble of regional climate model simulations. *Clim. Res.*, **44** 151–166.
Lenderink, G., and E. Van Meijgaard, 2008: Increase in hourly precipitation extremes beyond expectations from temperature changes. *Nature Geosci.*, **1**, 511–514.
Levine, R. C., and A. G. Turner, 2012: Dependence of Indian monsoon rainfall on moisture fluxes across the Arabian Sea and the impact of coupled model sea surface temperature biases. *Clim. Dyn.*, **38**, 2167–2190.
Levis, S., 2010: Modeling vegetation and land use in models of the Earth System. *Clim. Change*, **1**, 840–856.
Levitus, S., J. I. Antonov, T. P. Boyer, R. A. Locarnini, H. E. Garcia, and A. V. Mishonov, 2009: Global ocean heat content 1955–2008 in light of recently revealed instrumentation problems. *Geophys. Res. Lett.*, **36**, L07608
Levy, H., L. W. Horowitz, M. D. Schwarzkopf, Y. Ming, J.-C. Golaz, V. Naik, and V. Ramaswamy, 2013: The roles of aerosol direct and indirect effects in past and future climate change. *J. Geophys. Res.*, doi:10.1002/jgrd.50192.
Lewis, T., and S. Lamoureux, 2010: Twenty-first century discharge and sediment yield predictions in a small high Arctic watershed. *Global Planet. Change*, **71**, 27–41.
Li, C., J.-S. von Storch, and J. Marotzke, 2013a: Deep-ocean heat uptake and equilibrium climate response. *Clim. Dyn.*, **40**, 1071–1086.
Li, G., and S.-P. Xie, 2012: Origins of tropical-wide SST biases in CMIP multi-model ensembles. *Geophys. Res. Lett.*, **39**, L22703.
Li, H. B., A. Robock, and M. Wild, 2007: Evaluation of Intergovernmental Panel on Climate Change Fourth Assessment soil moisture simulations for the second half of the twentieth century. *J. Geophys. Res. Atmos.*, **112**, D06106
Li, J.-L. F., D. E. Waliser, and J. H. Jiang, 2011a: Correction to "Comparisons of satellites liquid water estimates to ECMWF and GMAO analyses, 20th century IPCC AR4 climate simulations, and GCM simulations". *Geophys. Res. Lett.*, **38**, L24807.
Li, J.-L. F., et al., 2008: Comparisons of satellites liquid water estimates to ECMWF and GMAO analyses, 20th century IPCC AR4 climate simulations, and GCM simulations. *Geophys. Res. Lett.*, **35**, L19710.
Li, J., S.-P. X. and A. Mestas-Nunez, E. R. C. and Gang Huang, R. D'Arrigo, F. Liu, J. Ma, and X. Zheng, 2011b: Interdecadal modulation of ENSO amplitude during the last millennium. *Nature Clim. Change*, **1**, 114–118.
Li, J. L. F., et al., 2012a: An observationally-based evaluation of cloud ice water in CMIP3 and CMIP5 GCMs and contemporary reanalyses using contemporary satellite data. *J. Geophys. Res.*, **117**, D16105.
Li, L., et al., 2013b: Development and Evaluation of Grid-point Atmospheric Model of IAP LASG, Version 2.0 (GAMIL 2.0). *Adv. Atmos. Sci.*, **30**, 855–867.
Li, L., et al., 2012b: The Flexible Global Ocean-Atmosphere-Land System Model: Grid-point Version 2: FGOALS-g2. *Adv. Atmos. Sci.*, doi:10.1007/s00376-012-2140-6.
Li, T., and G. H. Philander, 1996: On the annual cycle in the eastern equatorial Pacific. *J. Clim.*, **9**, 2986–2998.
Li, T., C. W. Tham, and C. P. Chang, 2001: A coupled air-sea-monsoon oscillator for the tropospheric biennial oscillation. *J. Clim.*, **14**, 752–764.
Liebmann, B., R. M. Dole, C. Jones, I. Blade, and D. Allured, 2010: Influence of choice of time period on global surface temperature trend wstimates. *Bull. Am. Meteorol. Soc.*, **91**, 1485–1491.
Lienert, f., J. C. Fyfe, and W. J. Merryfield, 2011: Do climate models capture the tropical influences on North Pacific sea surface temperature variability? *J. Clim.*, **24**, 6203–6209.
Lin, A. L., and T. Li, 2008: Energy spectrum characteristics of Boreal Summer Intraseasonal Oscillations: Climatology and variations during the ENSO developing and decaying phases. *J. Clim.*, **21**, 6304–6320.
Lin, J.-L., 2007: The double-ITCZ problem in IPCC AR4 Coupled GCMs: Ocean-atmosphere feedback analysis. *J. Clim.*, **20**, 4497–4525.
Lin, J. L., et al., 2006: Tropical intraseasonal variability in 14 IPCC AR4 climate models. Part I: Convective signals. *J. Clim.*, **19**, 2665–2690.
Lin, P., Y. Yongqiang, and H. Liu, 2013: Long-term stability and oceanic mean state simulated by the coupled model FGOALS-s2. *Adv. Atmos. Sci.*, **30**, 175–192.
Lin, Y., et al., 2012: TWP-ICE global atmospheric model intercomparison: Convection responsiveness and resolution impact. *J. Geophys. Res.*, **117**, D09111.
Lindvall, J., G. Svensson, and C. Hannay, 2012: Evaluation of near-surface parameters in the two versions of the atmospheric model in CESM1 using flux station observations. *J. Clim.*, **26** 26–44.
Linkin, M., and S. Nigam, 2008: The north pacific oscillation-west Pacific teleconnection pattern: Mature-phase structure and winter impacts. *J. Clim.*, **21**, 1979–1997.
Liu, H., C. Wang, S. K. Lee, and D. Enfield, 2013a: Atlantic Warm Pool Variability in the CMIP5 Simulations. *J. Clim.*, doi:10.1175/JCLI-D-12-00556.1.
Liu, H. L., P. F. Lin, Y. Q. Yu, and X. H. Zhang, 2012a: The baseline evaluation of LASG/IAP Climate system Ocean Model (LICOM) version 2.0. *Acta Meteorol. Sin.*, **26**, 318–329.
Liu, J., 2010: Sensitivity of sea ice and ocean simulations to sea ice salinity in a coupled global climate model. *Science China Earth Sci.*, **53**, 911–918.
Liu, L., W. Yu, and T. Li, 2011: Dynamic and thermodynamic air–sea coupling associated with the Indian Ocean dipole diagnosed from 23 WCRP CMIP3 Models. *J. Clim.*, **24**, 4941–4958.
Liu, S. C., C. B. Fu, C. J. Shiu, J. P. Chen, and F. T. Wu, 2009: Temperature dependence of global precipitation extremes. *Geophys. Res. Lett.*, **36**, L17702.
Liu, X., et al., 2012b: Toward a minimal representation of aerosols in climate models: Description and evaluation in the Community Atmosphere Model CAM5. *Geophys. Model Dev.*, **5**, 709–739.
Liu, X. H., et al., 2007: Uncertainties in global aerosol simulations: Assessment using three meteorological data sets. *J. Geophys. Res. Atmos.*, **112**, D11212
Liu, Y., 1996: Modeling the emissions of nitrous oxide and methane from the terrestrial biosphere to the atmosphere. In: *Joint Program Report Series*. MIT Joint Program on the Science and Policy of Global Change, Cambridge, MA, USA, 219 pp.
Liu, Y., J. Hu, B. He, Q. Bao, A. Duan, and G. X. Wu, 2013b: Seasonal evolution of subtropical anticyclones in the Climate System Model FGOALS-s2. *Adv. Atmos. Sci.*, **30**, 593–606.
Lloyd, J., E. Guilyardi, and H. Weller, 2010: The role of atmosphere feedbacks during ENSO in the CMIP3 models. Part II: Using AMIP runs to understand the heat flux feedback mechanisms. *Clim. Dyn.*, **37**, 1271–1292.
Lloyd, J., E. Guilyardi, and H. Weller, 2012: The role of atmosphere feedbacks during ENSO in the CMIP3 Models. Part III: The Shortwave Flux Feedback. *J. Clim.*, **25**, 4275–4293.
Lloyd, J., E. Guilyardi, H. Weller, and J. Slingo, 2009: The role of atmosphere feedbacks during ENSO in the CMIP3 models. *Atmos. Sci. Lett.*, **10**, 170–176.
Loeb, N. G., et al., 2009: Toward optimal closure of the Earth's top-of-atmosphere radiation budget. *J. Clim.*, **22**, 748–766.
Lohmann, U., K. von Salzen, N. McFarlane, H. G. Leighton, and J. Feichter, 1999: Tropospheric sulfur cycle in the Canadian general circulation model. *J. Geophys. Res. Atmos.*, **104**, 26833–26858.
Long, M. C., K. Lindsay, S. Peacock, J. K. Moore, and S. C. Doney, 2012: Twentieth-century oceanic carbon uptake and storage in CESM1(BGC). *J. Clim.*, doi:10.1175/JCLI-D-12-00184.1.
Loptien, U., O. Zolina, S. Gulev, M. Latif, and V. Soloviov, 2008: Cyclone life cycle characteristics over the Northern Hemisphere in coupled GCMs. *Clim. Dyn.*, **31**, 507–532.
Lorenz, P., and D. Jacob, 2005: Influence of regional scale information on the global circulation: A two-way nesting climate simulation. *Geophys. Res. Lett.*, **32**, L18706.
Lorenz, R., E. L. Davin, and S. I. Seneviratne, 2012: Modeling land-climate coupling in Europe: Impact of land surface representation on climate variability and extremes. *J. Geophys. Res.*, **117**, doi:10.1029/2012JD017755.
Losch, M., D. Menemenlis, J.-M. Campin, P. Heimbach, and C. Hill, 2010: On the formulation of sea-ice models. Part 1: Effects of different solver implementations and parameterizations. *Ocean Model.*, **33**, 129–144.
Loschnigg, J., G. A. Meehl, P. J. Webster, J. M. Arblaster, and G. P. Compo, 2003: The Asian monsoon, the tropospheric biennial oscillation, and the Indian Ocean zonal mode in the NCAR CSM. *J. Clim.*, **16**, 1617–1642.
Loutre, M. F., A. Mouchet, T. Fichefet, H. Goosse, H. Goelzer, and P. Huybrechts, 2011: Evaluating climate model performance with various parameter sets using observations over the recent past. *Clim. Past*, **7**, 511–526.
Loyola, D., and M. Coldewey-Egbers, 2012: Multi-sensor data merging with stacked neural networks for the creation of satellite long-term climate data records. *Eurasip J. Adv. Signal Proc.*, doi:10.1186/1687-6180-2012-91.

Loyola, D., et al., 2009: Global long-term monitoring of the ozone layer—a prerequisite for predictions. *Int. J. Remote Sens.*, **30**, 4295–4318.

Lu, J., G. A. Vecchi, and T. Reichler, 2007: Expansion of the Hadley cell under global warming. *Geophys. Res. Lett.*, **34**, L06805.

Lu, J. H., and J. J. Ji, 2006: A simulation and mechanism analysis of long-term variations at land surface over arid/semi-arid area in north China. *J. Geophys. Res. Atmos.*, **111**, D09306.

Lucarini, V., and F. Ragone, 2011: Energetics of climate models: Net energy balance and meridional enthalpy transport. *Rev. Geophys.*, **49**, RG1001.

Lucas-Picher, P., S. Somot, M. Déqué, B. Decharme, and A. Alias, 2012a: Evaluation of the regional climate model ALADIN to simulate the climate over North America in the CORDEX framework. *Clim. Dyn.*, doi:10.1007/s00382-012-1613-8.

Lucas-Picher, P., M. Wulff-Nielsen, J. Christensen, G. Adalgeirsdottir, R. Mottram, and S. Simonsen, 2012b: Very high resolution regional climate model simulations over Greenland: Identifying added value. *J. Geophys. Res. Atmos.*, **117**, D02108.

Lumpkin, R., K. G. Speer, and K. P. Koltermann, 2008: Transport across 48°N in the Atlantic Ocean. *J. Phys. Oceanogr.*, **38**, 733–752.

Luo, J. J., S. Masson, E. Roeckner, G. Madec, and T. Yamagata, 2005: Reducing climatology bias in an ocean-atmosphere CGCM with improved coupling physics. *J. Clim.*, **18**, 2344–2360.

Lynn, B., R. Healy, and L. Druyan, 2009: Quantifying the sensitivity of simulated climate change to model configuration. *Clim. Change*, **92**, 275–298.

MacKinnon, J., et al., 2009: Using global arrays to investigate internal-waves and mixing. In: *OceanObs09: Sustained Ocean Observations and Information for Society*, Venice, Italy, ESA.

Madden, R. A., and P. R. Julian, 1972: Description of global-scale circulation ells in tropics with a 40–50 day period. *J. Atmos. Sci.*, **29**, 1109–1123.

Madden, R. A., and P. R. Julian, 1994: Observations of the 40–50-Day Tropical Oscillation—a Review. *Mon. Weather Rev.*, **122**, 814–837.

Madec, G., 2008: NEMO ocean engine. Technical Note. Institut Pierre-Simon Laplace (IPSL), France, 300pp.

Madec, G., P. Delecluse, M. Imbard, and C. Levy, 1998: OPA 8.1 ocean general circulation model reference manual. *IPSL Note du Pole de Modelisation*, Institut Pierre-Simon Laplace (IPSL), France, 91 pp.

Mahlstein, I., and R. Knutti, 2010: Regional climate change patterns identified by cluster analysis. *Clim. Dyn.*, **35**, 587–600.

Mahlstein, I., and R. Knutti, 2012: September Arctic sea ice predicted to disappear near 2C global warming above present. *J. Geophys. Res.*, **117**, D06104.

Maier-Reimer, E., I. Kriest, J. Segschneider, and P. Wetze, 2005: The HAMburg Ocean Carbon Cycle Model HAMOCC 5.1-Technical Description Release 1.1. Tech. Rep. 14, *Rep. Earth Syst. Sci.*, Max Planck Institute for Meteorology, Hamburg, Germany, 50 pp.

Mantua, N. J., S. R. Hare, Y. Zhang, J. M. Wallace, and R. C. Francis, 1997: A Pacific interdecadal climate oscillation with impacts on salmon production. *Bull. Am. Meteorol. Soc.*, **78**, 1069–1079.

Manzini, E., C. Cagnazzo, P. G. Fogli, A. Bellucci, and W. A. Muller, 2012: Stratosphere-troposphere coupling at inter-decadal time scales: Implications for the North Atlantic Ocean. *Geophys. Res. Lett.*, **39**, L05801.

Maraun, D., 2012: Nonstationarities of regional climate model biases in European seasonal mean temperature and precipitation sums. *Geophys. Res. Lett.*, **39**, L06706.

Maraun, D., H. Rust, and T. Osborn, 2010a: Synoptic airflow and UK daily precipitation extremes: Development and validation of a vector generalised linear model. *Extremes*, **13**, 133–153.

Maraun, D., et al., 2010b: Precipitation downscaling under climate change: Recent developments to bridge the gap between dynamical models and the end user. *Rev. Geophys.*, **48**, RG3003.

Marchand, R., N. Beagley, and T. P. Ackerman, 2009: Evaluation of hydrometeor occurrence profiles in the Multiscale Modeling Framework Climate Model using atmospheric classification. *J. Clim.*, **22**, 4557–4573.

Markovic, M., H. Lin, and K. Winger, 2010: Simulating global and North American climate using the Global Environmental Multiscale Model with a Variable-Resolution Modeling Approach. *Mon. Weather Rev.*, **138**, 3967–3987.

Marsh, R., S. A. Mueller, A. Yool, and N. R. Edwards, 2011: Incorporation of the C-GOLDSTEIN efficient climate model into the GENIE framework: "eb_go_gs" configurations of GENIE. *Geophys. Model Dev.*, **4**, 957–992.

Marsh, R., et al., 2009: Recent changes in the North Atlantic circulation simulated with eddy-permitting and eddy-resolving ocean models. *Ocean Model.*, **28**, 226–239.

Marsland, S. J., et al., 2013: Evaluation of ACCESS Climate Model ocean diagnostics in CMIP5 simulations. *Aust. Meteorol. and Oceanogr. J.*, **63**,101–119.

Martin, G. M., and R. C. Levine, 2012: The influence of dynamic vegetation on the present-day simulation and future projectons of the South Asian summer monsoon in the HadGEM2 family. *Earth Syst. Dyn.*, **2**, 245–261.

Martin, G. M., et al., 2011: The HadGEM2 family of Met Office Unified Model climate configurations. *Geophys. Model Dev.*, **4**, 723–757.

Masarie, K. A., and P. P. Tans, 1995: Extension and integration of atmospheric carbon dioxide data into a globally consistent measurement record. *J. Geophys. Res. Atmos.*, **100**, 11593–11610.

Masato, G., B. Hoskins, and T. Woollings, 2012: Winter and summer Northern Hemisphere blocking in CMIP5 models. *J. Clim.*, doi:10.1175/JCLI-D-12-00466.1.

Masson-Delmotte, V., et al., 2010: EPICA Dome C record of glacial and interglacial intensities. *Quat. Sci. Rev.*, **29**, 113–128.

Masson-Delmotte, V., et al., 2006: Past and future polar amplification of climate change: Climate model intercomparisons and ice-core constraints. *Clim. Dyn.*, **27**, 437–440.

Masson, D., and R. Knutti, 2011a: Climate model genealogy. *Geophys. Res. Lett.*, **38**, L08703.

Masson, D., and R. Knutti, 2011b: Spatial-scale dependence of climate model performance in the CMIP3 ensemble. *J. Clim.*, **24**, 2680–2692.

Masson, D., and R. Knutti, 2013: Predictor screening, calibration and observational constraints in climate model ensembles: An illustration using climate sensitivity. *J. Clim.*, **26**, 887–898.

Massonnet, F., T. Fichefet, H. Goosse, C. M. Bitz, G. Philippon-Berthier, M. M. Holland, and P.-Y. Barriat, 2012: Constraining projections of summer Arctic sea ice. *Cryosphere*, **6**, 1383–1394.

Mastrandrea, M. D., et al., 2011: Guidance Note for Lead Authors of the IPCC Fifth Assessment Report on Consistent Treatment of Uncertainties. Intergovernmental Panel on Climate Change (IPCC). IPCC guidance note, Jasper Ridge, CA, USA, 7 pp.

Materia, S., P. A. Dirmeyer, Z. C. Guo, A. Alessandri, and A. Navarra, 2010: The sensitivity of simulated river discharge to land surface representation and meteorological forcings. *J. Hydrometeorol.*, **11**, 334–351.

Matsueda, M., 2009: Blocking predictability in operational medium-range ensemble forecasts. *Sola*, **5**, 113–116.

Matsueda, M., R. Mizuta, and S. Kusunoki, 2009: Future change in wintertime atmospheric blocking simulated using a 20-km-mesh atmospheric global circulation model. *J. Geophys. Res. Atmos.*, **114**, D12114.

Matsueda, M., H. Endo, and R. Mizuta, 2010: Future change in Southern Hemisphere summertime and wintertime atmospheric blockings simulated using a 20-km-mesh AGCM. *Geophys. Res. Lett.*, **37**, L02803.

Matsumoto, K., K. S. Tokos, A. R. Price, and S. J. Cox, 2008: First description of the Minnesota Earth System Model for Ocean biogeochemistry (MESMO 1.0). *Geophys. Model Dev.*, **1**, 1–15.

Maurer, E., and H. Hidalgo, 2008: Utility of daily vs. monthly large-scale climate data: An intercomparison of two statistical downscaling methods. *Hydrol. Earth Syst. Sci.*, **12**, 551–563.

Mauritsen, T., et al., 2012: Tuning the climate of a global model. *J. Adv. Model. Earth Syst.*, **4**, M00A01.

Maximenko, N., et al., 2009: Mean dynamic topography of the ocean derived from satellite and drifting buoy data using three different techniques. *J. Atmos. Ocean. Technol.*, **26**, 1910–1919.

May, P. T., J. H. Mather, G. Vaughan, K. N. Bower, C. Jakob, G. M. McFarquhar, and G. G. Mace, 2008: The Tropical Warm Pool International Cloud Experiment. *Bull. Am. Meteorol. Soc.*, **89**, 629–645.

May, W., 2007: The simulation of the variability and extremes of daily precipitation over Europe by the HIRHAM regional climate model. *Global Planet. Change*, **57**, 59–82.

McCarthy, G., et al., 2012: Observed interannual variability of the Atlantic meridional overturning circulation at 26.5 degrees N. *Geophys. Res. Lett.*, **39**, L19609.

McClean, J. L., and J. C. Carman, 2011: Investigation of IPCC AR4 coupled climate model North Atlantic modewater formation. *Ocean Model.*, **40**, 14–34.

McClean, J. L., M. E. Maltrud, and F. O. Bryan, 2006: Measures of the fidelity of eddying ocean models. *Oceanography*, **19**, 104–117.

McClean, J. L., et al., 2011: A prototype two-decade fully-coupled fine-resolution CCSM simulation. *Ocean Model.* **39**, 10–30.

McCormack, J. P., S. D. Eckermann, D. E. Siskind, and T. J. McGee, 2006: CHEM2D-OPP: A new linearized gas-phase ozone photochemistry parameterization for high-altitude NWP and climate models. *Atmos. Chem. Phys.*, **6**, 4943–4972.

McCrary, R. R., and D. A. Randall, 2010: Great plains drought in simulations of the twentieth century. *J. Clim.*, **23**, 2178–2196.

McDonald, R. E., 2011: Understanding the impact of climate change on Northern Hemisphere extra-tropical cyclones. *Clim. Dyn.*, **37**, 1399–1425.

McDougall, T. J., and P. C. McIntosh, 2001: The temporal-residual-mean velocity. Part II: Isopycnal interpretation and the tracer and momentum equations. *J. Phys. Oceanogr.*, **31**, 1222–1246.

McKitrick, R., S. McIntyre, and C. Herman, 2010: Panel and multivariate methods for tests of trend equivalence in climate data series. *Atmos. Sci. Lett.*, **11**, 270–277.

McKitrick, R., S. McIntyre, and C. Herman, 2011: Panel and multivariate methods for tests of trend equivalence in climate data series. *Atmos. Sci. Lett.*, **12**, 386–388.

McLandress, C., T. Shepherd, J. Scinocca, D. Plummer, M. Sigmond, A. Jonsson, and M. Reader, 2011: Separating the dynamical effects of climate change and ozone depletion. Part II Southern Hemisphere troposphere. *J. Clim.*, **24**, 1850–1868.

McLaren, A. J., et al., 2006: Evaluation of the sea ice simulation in a new coupled atmosphere-ocean climate model (HadGEM1). *J. Geophys. Res. Oceans*, **111**, C12014.

McManus, J. F., R. Francois, J. M. Gherardi, L. D. Keigwin, and S. Brown-Leger, 2004: Collapse and rapid resumption of Atlantic meridional circulation linked to deglacial climate changes. *Nature*, **428**, 834–837.

McWilliams, J. C., 2008: The nature and consequences of oceanic eddies. In: *Ocean Modeling in an Eddying Regime* [M. Hecht and H. Hasumi (eds.)]. American Geophysical Union, Washington, DC, pp. 5–15.

Mearns, L. O., et al., 2012: The North American Regional Climate Change Assessment Program: Overview of Phase I Results. *Bull. Am. Meteorol. Soc.*, **93**, 1337–1362.

Mears, C. A., F. J. Wentz, P. Thorne, and D. Bernie, 2011: Assessing uncertainty in estimates of atmospheric temperature changes from MSU and AMSU using a Monte-Carlo estimation technique. *J. Geophys. Res.*, **116**, D08112.

Mears, C. A., B. D. Santer, F. J. Wentz, K. E. Taylor, and M. F. Wehner, 2007: Relationship between temperature and precipitable water changes over tropical oceans. *Geophys. Res. Lett.*, **34**, L24709.

Meehl, G. A., and J. M. Arblaster, 2011: Decadal variability of Asian-Australian Monsoon-ENSO-TBO relationships. *J. Clim.*, **24**, 4925–4940.

Meehl, G. A., and H. Teng, 2012: Case studies for initialized decadal hindcasts and predictions for the Pacific region. *Geophys. Res. Lett.*, **39**, L22705.

Meehl, G. A., J. M. Arblaster, and J. Loschnigg, 2003: Coupled ocean-atmosphere dynamical processes in the tropical Indian and Pacific Oceans and the TBO. *J. Clim.*, **16**, 2138–2158.

Meehl, G. A., J. M. Arblaster, J. T. Fasullo, A. X. Hu, and K. E. Trenberth, 2011: Model-based evidence of deep-ocean heat uptake during surface-temperature hiatus periods. *Nature Clim. Change*, **1**, 360–364.

Meehl, G. A., A. Hu, J. Arblaster, J. Fasullo, and K. E. Trenberth, 2013a: Externally forced and internally generated decadal climate variability associated with the Interdecadal Pacific Oscillation. *J. Clim.*, doi:10.1175/JCLI-D-12-00548.1.

Meehl, G. A., P. R. Gent, J. M. Arblaster, B. L. Otto-Bliesner, E. C. Brady, and A. Craig, 2001: Factors that affect the amplitude of El Niño in global coupled climate models. *Clim. Dyn.*, **17**, 515–526.

Meehl, G. A., J. M. Arblaster, J. M. Caron, H. Annamalai, M. Jochum, A. Chakraborty, and R. Murtugudde, 2012: Monsoon regimes and processes in CCSM4. Part I: The Asian-Australian Monsoon. *J. Clim.*, **25**, 2583–2608.

Meehl, G. A., et al., 2007: The WCRP CMIP3 multimodel dataset —A new era in climate change research. *Bull. Am. Meteorol. Soc.*, **88**, 1383–1394.

Meehl, G. A., et al., 2009: Decadal prediction: Can it be skillful? *Bull. Am. Meteorol. Soc.*, **90**, 1467–1485.

Meehl, G. A., et al., 2013b: Decadal climate prediction: An update from the trenches. *Bull. Am. Meteorol. Soc.*, doi:10.1175/BAMS-D-12-00241.1.

Meijers, A., E. Shuckburgh, N. Bruneau, J.-B. Sallee, T. Bracegirdle, and Z. Wang, 2012: Representation of the Antarctic Circumpolar Current in the CMIP5 climate models and future changes under warming scenarios. *J. Geophys. Res. Oceans*, **117**, C12008.

Meinshausen, M., et al., 2009: Greenhouse-gas emission targets for limiting global warming to 2 degrees C. *Nature*, **458**, 1158–1162.

Meinshausen, M., et al., 2011: The RCP greenhouse gas concentrations and their extensions from 1765 to 2300. *Clim. Change*, **109**, 213–241.

Meissner, K. J., A. J. Weaver, H. D. Matthews, and P. M. Cox, 2003: The role of land surface dynamics in glacial inception: a study with the UVic Earth System Model. *Clim. Dyn.*, **21**, 515–537.

Melillo, J. M., A. D. McGuire, D. W. Kicklighter, B. Moore, C. J. Vorosmarty, and A. L. Schloss, 1993: Global climate-change and terrestrial net primary production. *Nature*, **363**, 234–240.

Melsom, A., V. Lien, and W. P. Budgell, 2009: Using the Regional Ocean Modeling System (ROMS) to improve the ocean circulation from a GCM 20th century simulation. *Ocean Dyn.*, **59**, 969–981.

Menary, M., W. Park, K. Lohmann, M. Vellinga, D. Palmer, M. Latif, and J. H. Jungclaus, 2012: A multimodel comparison of centennial Atlantic meridional overturning circulation variability. *Clim. Dyn.*, **38**, 2377–2388.

Menendez, C., M. de Castro, A. Sorensson, J. Boulanger, and C. M. Grp, 2010: CLARIS Project: Towards climate downscaling in South America. *Meteorol. Z.*, **19**, 357–362.

Menon, S., D. Koch, G. Beig, S. Sahu, J. Fasullo, and D. Orlikowski, 2010: Black carbon aerosols and the third polar ice cap. *Atmos. Chem. Phys.*, **10**, 4559–4571.

Mercado, L. M., C. Huntingford, J. H. C. Gash, P. M. Cox, and V. Jogireddy, 2007: Improving the representation of radiation interception and photosynthesis for climate model applications. *Tellus B*, **59**, 553–565.

Merrifield, M. A., and M. E. Maltrud, 2011: Regional sea level trends due to a Pacific trade wind intensification. *Geophys. Res. Lett.*, **38**, L21605.

Merryfield, W. J., et al., 2013: The Canadian Seasonal to Interannual Prediction System. Part I: Models and Initialization. *Mon. Weather Rev.*, doi:10.1175/MWR-D-12-00216.1.

Mieville, A., et al., 2010: Emissions of gases and particles from biomass burning during the 20th century using satellite data and an historical reconstruction. *Atmos. Environ.*, **44**, 1469–1477.

Miller, A., et al., 2002: A cohesive total ozone data set from the SBUV(/2) satellite system. *J. Geophys. Res. Atmos.*, **107**, 4701.

Milliff, R., A. Bonazzi, C. Wikle, N. Pinardi, and L. Berliner, 2011: Ocean ensemble forecasting. Part I: Ensemble Mediterranean winds from a Bayesian hierarchical model. *Q. J. R. Meteorol. Soc.*, **137**, 858–878.

Milly, P. C. D., and A. B. Shmakin, 2002: Global modeling of land water and energy balances. Part I: the land dynamics (LaD) model. *J. Hydrometeorol.*, **3**, 283–299.

Min, S. K., X. B. Zhang, F. W. Zwiers, and G. C. Hegerl, 2011: Human contribution to more-intense precipitation extremes. *Nature*, **470**, 376–379.

Minobe, S., 1997: A 50–70 year climatic oscillation over the North Pacific and North America. *Geophys. Res. Lett.*, **24**, 683–686.

Minobe, S., 1999: Resonance in bidecadal and pentadecadal climate oscillations over the North Pacific: Role in climatic regime shifts. *Geophys. Res. Lett.*, **26**, 855–858.

Misra, V., 2007: Addressing the issue of systematic errors in a regional climate model. *J. Clim.*, **20**, 801–818.

Mitchell, T. D., and P. D. Jones, 2005: An improved method of constructing a database of monthly climate observations and associated high-resolution grids. *Int. J. Climatol.* , **25**, 693–712.

Miyama, T., and M. Kawamiya, 2009: Estimating allowable carbon emission for CO(2) concentration stabilization using a GCM-based Earth system model. *Geophys. Res. Lett.*, **36**, L19709.

Mizuta, R., et al., 2012: Climate simulations using MRI-AGCM3.2 with 20-km grid. *J. Meteorol. Soc. Jpn.*, **90A**, 233–258.

Molteni, F., 2003: Atmospheric simulations using a GCM with simplified physical parameterizations. I: Model climatology and variability in multi-decadal experiments. *Clim. Dyn.*, **20**, 175–191.

Montoya, M., A. Griesel, A. Levermann, J. Mignot, M. Hofmann, A. Ganopolski, and S. Rahmstorf, 2005: The earth system model of intermediate complexity CLIMBER-3 alpha. Part 1: description and performance for present-day conditions. *Clim. Dyn.*, **25**, 237–263.

Morgenstern, O., et al., 2010: Anthropogenic forcing of the Northern Annular Mode in CCMVal-2 models. *J. Geophys. Res.*, **115**, D00M03.

Morice, C. P., J. J. Kennedy, N. A. Rayner, and P. D. Jones, 2012: Quantifying uncertainties in global and regional temperature change using an ensemble of observational estimates: The HadCRUT4 data set. *J. Geophys. Res. Atmos.*, **117**, D08101.

Moss, R. H., et al., 2010: The next generation of scenarios for climate change research and assessment. *Nature*, **463**, 747–756.

Mouchet, A., and L. M. François, 1996: Sensitivity of a global oceanic carbon cycle model to the circulation and the fate of organic matter: Preliminary results. *Phys. Chem. Earth*, **21**, 511–516.

Msadek, R., and C. Frankignoul, 2009: Atlantic multidecadal oceanic variability and its influence on the atmosphere in a climate model. *Clim. Dyn.*, **33**, 45–62.

Msadek, R., W. E. Johns, S. G. Yeager, G. Danabasoglu, T. Delworth, and T. Rosati, 2013: The Atlantic meridional heat transport at 26.5°N and its relationship with the MOC in the RAPID-array and GFDL and NCAR coupled models. *J. Clim.*, doi:10.1175/JCLI-D-12–00081.1.

Mueller, B., et al., 2011: Evaluation of global observations-based evapotranspiraion datasets and IPCC AR4 simulations. *Geophys. Res. Lett.*, **38**, L06402.

Muller, S. A., F. Joos, N. R. Edwards, and T. F. Stocker, 2006: Water mass distribution and ventilation time scales in a cost-efficient, three-dimensional ocean model. *J. Clim.*, **19**, 5479–5499.

Murakami, H., and M. Sugi, 2010: Effect of model resolution on tropical cyclone climate projections. *Sola*, **6**, 73–76.

Murakami, H., et al., 2012: Future changes in tropical cyclone activity projected by the new high-resolution MRI-AGCM. *J. Clim.*, **25**, 3237–3260.

Murphy, D. M., 2013: Little net clear-sky radiative forcing from recent regional redistribution of aerosols. *Nature Geosci.*, **6**, 258–262.

Murphy, J., B. Booth, M. Collins, G. Harris, D. Sexton, and M. Webb, 2007: A methodology for probabilistic predictions of regional climate change from perturbed physics ensembles. *Philos.Trans. R. Soc. London A*, **365** 1993–2028.

Murphy, J. M., D. M. H. Sexton, D. N. Barnett, G. S. Jones, M. J. Webb, M. Collins, and D. A. Stainforth, 2004: Quantification of modelling uncertainties in a large ensemble of climate change simulations. *Nature*, **430**, 768–772.

Murtugudde, R., J. Beauchamp, C. R. McClain, M. Lewis, and A. J. Busalacchi, 2002: Effects of penetrative radiation on the upper tropical ocean circulation. *J. Clim.*, **15**, 470–486.

Muryshev, K. E., A. V. Eliseev, I. I. Mokhov, and N. A. Diansky, 2009: Validating and assessing the sensitivity of the climate model with an ocean general circulation model developed at the Institute of Atmospheric Physics, Russian Academy of Sciences. *Izvestiya Atmos. Ocean. Phys.*, **45**, 416–433.

Nagura, M., W. Sasaki, T. Tozuka, J.-J. Luo, S. K. Behera, and T. Yamagata, 2013: Longitudinal biases in the Seychelles Dome simulated by 35 ocean-atmosphere coupled general circulation models. *J. Geophys. Res.*, doi:10.1029/2012JC008352.

Nakano, H., H. Tsujino, M. Hirabara, T. Yasuda, T. Motoi, M. Ishii, and G. Yamanaka, 2011: Uptake mechanism of anthropogenic CO_2 in the Kuroshio Extension region in an ocean general circulation model. *J. Oceanogr.*, **67**, 765–783.

Nam, C., S. Bony, J. L. Dufresne, and H. Chepfer, 2012: The 'too few, too bright' tropical low-cloud problem in CMIP5 models. *Geophys. Res. Lett.*, **39**, L21801.

Nanjundiah, R. S., V. Vidyunmala, and J. Srinivasan, 2005: The impact of increase in CO_2 on the simulation of tropical biennial oscillations (TBO) in 12 coupled general circulation models. *Atmos. Sci. Lett.*, **6**, 183–191.

Naoe, H., and K. Shibata, 2010: Equatorial quasi-biennial oscillation influence on northern winter extratropical circulation. *J. Geophys. Res. Atmos.*, **115**, D19102.

Neale, R. B., J. H. Richter, and M. Jochum, 2008: The Impact of Convection on ENSO: From a delayed oscillator to a series of events. *J. Clim.*, **21**, 5904–5924.

Neale, R. B., J. Richter, S. Park, P. H. Lauritzen, S. J. Vavrus, P. J. Rasch, and M. Zhang, 2013: The Mean Climate of the Community Atmosphere Model (CAM4) in forced SST and fully coupled experiments. *J. Clim.*, doi:10.1175/JCLI-D-12-00236.1.

Neale, R. B., et al., 2010: Description of the NCAR Community Atmosphere Model (CAM 4.0). NCAR Technical Note NCAR/TN-486+STR, National Center for Atmopsheric Research, Boulder, CO, 268 pp.

Neelin, J. D., 2007: Moist dynamics of tropical convection zones in monsoons, teleconnections and global warming. In: T*he Global Circulation of the Atmosphere* [T. Schneider and A. Sobel (eds.)]. Princeton University Press, Princeton, NJ. 385 pp.

Neelin, J. D., and N. Zeng, 2000: A quasi-equilibrium tropical circulation model—Formulation. *J. Atmos. Sci.*, **57**, 1741–1766.

Neelin, J. D., C. Chou, and H. Su, 2003: Tropical drought regions in global warming and El Niño teleconnections. *Geophys. Res. Lett.*, **30**, 2275.

Neelin, J. D., A. Bracco, H. Luo, J. C. McWilliams, and J. E. Meyerson, 2010: Considerations for parameter optimization and sensitivity in climate models. *Proc. Nat. Acad. Sci. U.S.A.*, **107**, 21349–21354.

Neggers, R. A. J., 2009: A dual mass flux framework for boundary layer convection. Part II: Clouds. *J. Atmos. Sci.*, **66**, 1489–1506.

Neggers, R. A. J., M. Kohler, and A. C. M. Beljaars, 2009: A dual mass flux framework for boundary layer convection. Part I: Transport. *J. Atmos. Sci.*, **66**, 1465–1487.

Nicolsky, D. J., V. E. Romanovsky, V. A. Alexeev, and D. M. Lawrence, 2007: Improved modeling of permafrost dynamics in a GCM land-surface scheme. *J. Geophys. Res.*, **34**, L08501.

Nikiema, O., and R. Laprise, 2010: Diagnostic budget study of the internal variability in ensemble simulations of the Canadian RCM. *Clim. Dyn.*, **36** 2313–2337.

Nikulin, G., et al., 2012: Precipitation climatology in an ensemble of CORDEX-Africa regional climate simulations. *J. Clim.*, doi:10.1175/jcli-d-11–00375.1.

Ning, L., M. E. Mann, R. Crane, and T. Wagener, 2011: Probabilistic projections of climate change for the Mid-Atlantic region of the United States—Validation of precipitation downscaling during the Historical Era. *J. Clim.*, **25**, 509–526.

Nishii, K., et al., 2012: Relationship of the reproducibility of multiple variables among Global Climate Models. *J. Meteorol. Soc. Jpn.*, **90A**, 87–100.

Notz, D., F. A. Haumann, H. Haak, J. H. Jungclaus, and J. Marotzke, 2013: Sea-ice evolution in the Arctic as modeled by MPI-ESM. *J. Adv. Model. Earth Syst.*, doi:10.1002/jame.20016.

O'Connor, F. M., C. E. Johnson, O. Morgenstern, and W. J. Collins, 2009: Interactions between tropospheric chemistry and climate model temperature and humidity biases. *Geophys. Res. Lett.*, **36**, L16801.

O'Farrell, S. P., 1998: Investigation of the dynamic sea ice component of a coupled atmosphere sea ice general circulation model. *J. Geophys. Res.-Oceans*, **103**, 15751–15782.

O'Gorman, P. A., 2012: Sensitivity of tropical precipitation extremes to climate change. *Nature Geosci.*, **5**, 697–700.

O'Gorman, P. A., and M. S. Singh, 2013: Vertical structure of warming consistent with an upward shift in the middle and upper troposphere. *Geophys. Res. Lett.*, doi:10.1002/grl.50328.

O'ishi, R., and A. Abe-Ouchi, 2011: Polar amplification in the mid-Holocene derived from dynamical vegetation change with a GCM. *Geophys. Res. Lett.*, **38**, L14702.

Ogasawara, N., A. Kitoh, T. Yasunari, and A. Noda, 1999: Tropospheric biennial oscillation of ENSO-monsoon system in the MRI coupled GCM. *J. Meteorol. Soc. Jpn.*, **77**, 1247–1270.

Ohba, M., D. Nohara, and H. Ueda, 2010: Simulation of asymmetric ENSO transition in WCRP CMIP3 Multimodel Experiments. *J. Clim.*, **23**, 6051–6067.

Ohgaito, R., and A. Abe-Ouchi, 2009: The effect of sea surface temperature bias in the PMIP2 AOGCMs on mid-Holocene Asian monsoon enhancement. *Clim. Dyn.*, **33**, 975–983.

Oka, A., E. Tajika, A. Abe-Ouchi, and K. Kubota, 2011: Role of the ocean in controlling atmospheric CO_2 concentration in the course of global glaciations. *Clim. Dyn.*, **37**, 1755–1770.

Oleson, K. W., 2004: Technical description of the Community Land Model (CLM). NCAR Technical Note NCAR/TN-461+STR, National Center for Atmospheric Research, Boulder, CO, 174 pp.

Oleson, K. W., G. B. Bonan, J. Feddema, M. Vertenstein, and C. S. B. Grimmond, 2008a: An urban parameterization for a global climate model. Part I: Formulation and evaluation for two cities. *J. Appl. Meteorol. Climatol.*, **47**, 1038–1060.

Oleson, K. W., et al., 2010: Technical Description of version 4.0 of the Community Land Model (CLM) NCAR Technical Note NCAR/TN-478+STR, National Center for Atmospheric Research, Boulder, CO, 257 pp.

Oleson, K. W., et al., 2008b: Improvements to the Community Land Model and their impact on the hydrological cycle. *J. Geophys. Res. Biogeosci.*, **113**, G01021

Onogi, K., et al., 2007: The JRA-25 reanalysis. *J. Meteorol. Soc. Jpn.*, **85**, 369–432.

Opsteegh, J. D., R. J. Haarsma, F. M. Selten, and A. Kattenberg, 1998: ECBILT: A dynamic alternative to mixed boundary conditions in ocean models. *Tellus A*, **50**, 348–367.

Oreopoulos, L., et al., 2012: The continual intercomparison of radiation codes: Results from Phase I. *J. Geophys. Res. Atmos.*, **117**, D06118.

Ostle, N. J., et al., 2009: Integrating plant-soil interactions into global carbon cycle models. *J. Ecol.*, **97**, 851–863.

Otte, T. L., C. G. Nolte, M. J. Otte, and J. H. Bowden, 2012: Does nudging squelch the extremes in Regional Climate Modeling? *J. Clim.*, **25**, 7046–7066.

Ottera, O. H., M. Bentsen, H. Drange, and L. Suo, 2010: External forcing as a metronome for Atlantic multidecadal variability. *Nature Geosci.*, **3**, 688–694.

Otto-Bliesner, B. L., et al., 2007: Last Glacial Maximum ocean thermohaline circulation: PMIP2 model intercomparisons and data constraints. *Geophys. Res. Lett.*, **34**, L12706.

Otto-Bliesner, B. L., et al., 2009: A comparison of PMIP2 model simulations and the MARGO proxy reconstruction for tropical sea surface temperatures at last glacial maximum. *Clim. Dyn.*, **32**, 799–815.

Otto, J., T. Raddatz, M. Claussen, V. Brovkin, and V. Gayler, 2009: Separation of atmosphere-ocean-vegetation feedbacks and synergies for mid-Holocene climate. *Global Biogeochem. Cycles*, **23**, L09701.

Overland, J. E., and M. Wang, 2013: When will the summer Arctic be nearly sea ice free? *Geophys. Res. Lett.*, doi:10.1002/grl.50316, doi:10.1002/grl.50316.

Ozturk, T., H. Altinsoy, M. Turkes, and M. Kurnaz, 2012: Simulation of temperature and precipitation climatology for the central Asia CORDEX domain using RegCM 4.0. *Clim. Res.*, **52**, 63–76.

Paeth, H., 2011: Postprocessing of simulated precipitation for impact research in West Africa. Part I: Model output statistics for monthly data. *Clim. Dyn.*, **36**, 1321–1336.

Paeth, H., et al., 2012: Progress in regional downscaling of west African precipitation. *Atmos. Sci. Lett.*, **12**, 75–82.

Palmer, J. R., and I. J. Totterdell, 2001: Production and export in a global ocean ecosystem model. *Deep-Sea R. Pt. I*, **48**, 1169–1198.

Parekh, P., F. Joos, and S. A. Muller, 2008: A modeling assessment of the interplay between aeolian iron fluxes and iron-binding ligands in controlling carbon dioxide fluctuations during Antarctic warm events. *Paleoceanography*, **23**, Pa4202.

Park, S., and C. S. Bretherton, 2009: The University of Washington Shallow Convection and Moist Turbulence schemes and their impact on climate simulations with the Community Atmosphere Model. *J. Clim.*, **22**, 3449–3469.

Park, W., and M. Latif, 2010: Pacific and Atlantic multidecadal variability in the Kiel Climate Model. *Geophys. Res. Lett.*, **37**, L24702.

Parker, D., C. Folland, A. Scaife, J. Knight, A. Colman, P. Baines, and B. W. Dong, 2007: Decadal to multidecadal variability and the climate change background. *J. Geophys. Res. Atmos.*, **112**, D18115.

Parkinson, C. L., and D. J. Cavalieri, 2012: Antarctic sea ice variability and trends, 1979–2010. *Cryosphere*, **6**, 871–880.

Patricola, C. M., M. Li, Z. Xu, P. Chang, R. Saravanan, and J.-S. Hsieh, 2012: An investigation of tropical Atlantic bias in a high-resolution Coupled Regional Climate Model. *Clim. Dyn.*, doi:10.1007/s00382-012-1320-5.

Pavlova, T. V., V. M. Kattsov, and V. A. Govorkova, 2011: Sea ice in CMIP5 models: Closer to reality? *Trudy GGO (MGO Proc., in Russian)*, **564**, 7–18.

Pavlova, T. V., V. M. Kattsov, Y. D. Nadyozhina, P. V. Sporyshev, and V. A. Govorkova, 2007: Terrestrial cryosphere evolution through the 20th and 21st centuries as simulated with the new generation of global climate models. *Earth Cryosphere (in Russian)*, **11**, 3–13.

Pechony, O., and D. T. Shindell, 2009: Fire parameterization on a global scale. *J. Geophys. Res. Atmos.*, **114**, D16115

Pedersen, C. A., E. Roeckner, M. Lüthje, and J. Winther, 2009: A new sea ice albedo scheme including melt ponds for ECHAM5 general circulation model. *J. Geophys. Res.*, **114**, D08101.

Pennell, C., and T. Reichler, 2011: On the effective number of climate models. *J. Clim.*, **24** 2358–2367

Perlwitz, J., S. Pawson, R. Fogt, J. Nielsen, and W. Neff, 2008: Impact of stratospheric ozone hole recovery on Antarctic climate. *Geophys. Res. Lett.*, **35**, L08714.

Peterson, T. C., et al., 2009: State of the Climate in 2008. *Bull. Am. Meteorol. Soc.*, **90**, S1–S196.

Petoukhov, V., I. I. Mokhov, A. V. Eliseev, and V. A. Semenov, 1998: The IAP RAS global climate model. *Dialogue-MSU*, Moscow, Russia.

Petoukhov, V., A. Ganopolski, V. Brovkin, M. Claussen, A. Eliseev, C. Kubatzki, and S. Rahmstorf, 2000: CLIMBER-2: a climate system model of intermediate complexity. Part I: Model description and performance for present climate. *Clim. Dyn.*, **16**, 1–17.

Petoukhov, V., et al., 2005: EMIC Intercomparison Project (EMIP-CO2): Comparative analysis of EMIC simulations of climate, and of equilibrium and transient responses to atmospheric CO_2 doubling. *Clim. Dyn.*, **25**, 363–385.

Pfahl, S., and H. Wernli, 2012: Quantifying the relevance of atmospheric blocking for co-located temperature extremes in the Northern Hemisphere on (sub-)daily time scales. *Geophys. Res. Lett.*, **39**, L12807.

Pfeiffer, A., and G. Zängl, 2010: Validation of climate-mode MM5–simulations for the European Alpine Region. *Theor. Appl. Climatol.*, **101**, 93–108.

Phillips, T. J., et al., 2004: Evaluating parameterizations in General Circulation Models: Climate simulation meets weather prediction. *Bull. Am. Meteorol. Soc.*, **85**, 1903–1915.

Piani, C., D. J. Frame, D. A. Stainforth, and M. R. Allen, 2005: Constraints on climate change from a multi-thousand member ensemble of simulations. *Geophys. Res. Lett.*, **32**, L23825.

Pierce, D. W., 2001: Distinguishing coupled ocean-atmosphere interactions from background noise in the North Pacific. *Prog. Oceanogr.*, **49**, 331–352.

Pierce, D. W., T. P. Barnett, B. D. Santer, and P. J. Gleckler, 2009: Selecting global climate models for regional climate change studies. *Proc. Natl. Acad. Sci. U.S.A.*, **106**, 8441–8446.

Pincus, R., C. P. Batstone, R. J. P. Hofmann, K. E. Taylor, and P. J. Glecker, 2008: Evaluating the present-day simulation of clouds, precipitation, and radiation in climate models. *J. Geophys. Res. Atmos.*, **113**, D14209

Pincus, R., S. Platnick, S. A. Ackerman, R. S. Hemler, and R. J. P. Hofmann, 2012: Reconciling simulated and observed views of clouds: MODIS, ISCCP, and the limits of instrument simulators. *J. Clim.*, **25**, 4699–4720.

Pinto, J. G., T. Spangehl, U. Ulbrich, and P. Speth, 2006: Assessment of winter cyclone activity in a transient ECHAM4–OPYC3 GHG experiment. *Meteorol. Z.*, **15**, 279–291.

Piot, M., and R. von Glasow, 2008: The potential importance of frost flowers, recycling on snow, and open leads for ozone depletion events. *Atmos. Chem. Phys.*, **8**, 2437–2467.

Pitman, A., A. Arneth, and L. Ganzeveld, 2012a: Regionalizing global climate models. *Int. J. Climatol.* , **32**, 321–337.

Pitman, A. J., 2003: The evolution of, and revolution in, land surface schemes designed for climate models. *Int. J. Climatol.* , **23**, 479–510.

Pitman, A. J., et al., 2012b: Effects of land cover change on temperature and rainfall extremes in multi-model ensemble simulations. *Earth Syst. Dyn.*, **13**, 213–231.

Pitman, A. J., et al., 2009: Uncertainties in climate responses to past land cover change: First results from the LUCID intercomparison study. *Geophys. Res. Lett.*, **36**, L14814.

Plattner, G. K., et al., 2008: Long-term climate commitments projected with climate-carbon cycle models. *J. Clim.*, **21**, 2721–2751.

Ploshay, J. J., and N.-C. Lau, 2010: Simulation of the diurnal cycle in tropical rainfall and circulation during Boreal Summer with a high-resolution GCM. *Mon. Weather Rev.*, **138**, 3434–3453.

Po-Chedley, S., and Q. Fu, 2012: Discrepancies in tropical upper tropospheric warming between atmospheric circulation models and satellites. *Environ. Res. Lett.*, **7**, 044018.

Pokhrel, S., H. Rahaman, A. Parekh, S. K. Saha, A. Dhakate, H. S. Chaudhari, and R. M. Gairola, 2012: Evaporation-precipitation variability over Indian Ocean and its assessment in NCEP Climate Forecast System (CFSv2). *Clim. Dyn.*, **39**, 2585–2608.

Polvani, L., D. Waugh, G. Correa, and S. Son, 2011: Stratospheric ozone depletion: The main driver of twentieth-century atmospheric circulation changes in the Southern Hemisphere. *J. Clim.*, **24**, 795–812.

Pope, V. D., M. L. Gallani, P. R. Rowntree, and R. A. Stratton, 2000: The impact of new physical parametrizations in the Hadley Centre climate model: HadAM3. *Clim. Dyn.*, **16**, 123–146.

Power, S., and R. Colman, 2006: Multi-year predictability in a coupled general circulation model. *Clim. Dyn.*, **26**, 247–272

Power, S., M. Haylock, R. Colman, and X. D. Wang, 2006: The predictability of interdecadal changes in ENSO activity and ENSO teleconnections. *J. Clim.*, **19**, 4755–4771.

Prudhomme, C., and H. Davies, 2009: Assessing uncertainties in climate change impact analyses on the river flow regimes in the UK. Part 1: Baseline climate. *Clim. Change*, **93**, 177–195.

Pryor, S., G. Nikulin, and C. Jones, 2012: Influence of spatial resolution on regional climate model derived wind climates. *J. Geophys. Res. Atmos.*, **117**, D03117.

Ptashnik, I. V., R. A. McPheat, K. P. Shine, K. M. Smith, and R. G. Williams, 2011: Water vapor self-continuum absorption in near-infrared windows derived from laboratory measurements. *J. Geophys. Res. Atmos.*, **116**, D16305.

Qian, T. T., A. Dai, K. E. Trenberth, and K. W. Oleson, 2006: Simulation of global land surface conditions from 1948 to 2004. Part I: Forcing data and evaluations. *J. Hydrometeorol.*, **7**, 953–975.

Qiao, F., Y. Yuan, Y. Yang, Q. Zheng, C. Xia, and J. Ma, 2004: Wave-induced mixing in the upper ocean: Distribution and application to a global ocean circulation model. *Geophys. Res. Lett.*, **31**, L11303.

Quaas, J., 2012: Evaluating the "critical relative humidity" as a measure of subgrid-scale variability of humidity in general circulation model cloud cover parameterizations using satellite data. *J. Geophys. Res. Atmos.*, **117**, D09208.

Raftery, A. E., T. Gneiting, F. Balabdaoui, and M. Polakowski, 2005: Using Bayesian model averaging to calibrate forecast ensembles. *Mon. Weather Rev.*, **133**, 1155–1174.

Raible, C. C., M. Yoshimori, T. F. Stocker, and C. Casty, 2007: Extreme midlatitude cyclones and their implications for precipitation and wind speed extremes in simulations of the Maunder Minimum versus present day conditions. *Clim. Dyn.*, **28**, 409–423.

Raisanen, J., 2007: How reliable are climate models? *Tellus A*, **59**, 2–29.

Raisanen, J., and J. S. Ylhaisi, 2011: How much should climate model output be smoothed in space? *J. Clim.*, **24**, 867–880.

Raisanen, J., L. Ruokolainen, and J. Ylhaisi, 2010: Weighting of model results for improving best estimates of climate change. *Clim. Dyn.*, **35**, 407–422.

Rammig, A., et al., 2010: Estimating the risk of Amazonian forest dieback. *New Phytologist*, **187**, 694–706.

Rampal, P., J. Weiss, C. Dubois, and J. M. Campin, 2011: IPCC climate models do not capture Arctic sea ice drift acceleration: Consequences in terms of projected sea ice thinning and decline. *J. Geophys. Res. Oceans*, **116**, C00d07.

Ramstein, G., M. Kageyama, J. Guiot, H. Wu, C. Hely, G. Krinner, and S. Brewer, 2007: How cold was Europe at the Last Glacial Maximum? A synthesis of the progress achieved since the first PMIP model-data comparison. *Clim. Past*, **3**, 331–339.

Randall, D. A., M. F. Khairoutdinov, A. Arakawa, and W. W. Grabowski, 2003: Breaking the cloud parameterization deadlock. *Bull. Am. Meteorol. Soc.*, **84**, 1547–1564.

Randall, D. A., et al., 2007: Climate models and their evaluation. In: *Climate Change 2007: The Physical Science Basis. Contribution of Working Group I to the Fourth Assessment Report of the Intergovernmental Panel on Climate Change* [Solomon, S., D. Qin, M. Manning, Z. Chen, M. Marquis, K. B. Averyt, M. Tignor and H. L. Miller (eds.)] Cambridge University Press, Cambridge, United Kingdom and New York, NY, USA, pp. 589–662.

Randel, W., and F. Wu, 2007: A stratospheric ozone profile data set for 1979–2005: Variability, trends, and comparisons with column ozone data. *J. Geophys. Res. Atmos.*, **12**, D06313.

Randel, W. J., et al., 2009: An update of observed stratospheric temperature trends. *J. Geophys. Res. Atmos.*, **114**, D02107.

Rapaić, M., M. Leduc, and R. Laprise, 2010: Evaluation of the internal variability and estimation of the downscaling ability of the Canadian Regional Climate Model for different domain sizes over the north Atlantic region using the Big-Brother experimental approach. *Clim. Dyn.*, **36** 1979–2001.

Raphael, M. N., and M. M. Holland, 2006: Twentieth century simulation of the Southern Hemisphere climate in coupled models. Part 1: Large scale circulation variability. *Clim. Dyn.*, **26**, 217–228.

Rashid, H. A., A. C. Hirst, and M. Dix, 2013: Atmospheric circulation features in the ACCESS model simulations for CMIP5: Historical simulation and future projections *Aust. Meteorol. Oceanogr. J.*, **63**, 145–160.

Rauscher, S. A., E. Coppola, C. Piani, and F. Giorgi, 2010: Resolution effects on regional climate model simulations of seasonal precipitation over Europe. *Clim. Dyn.*, **35**, 685–711.

Rayner, N. A., et al., 2003: Global analysis of sea surface temperature, sea ice, and night marine air temperature since the late ninteeth century. *J. Geophys. Res.*, **108**, 4407.

Redelsperger, J.-L., C. D. Thorncroft, A. Diedhiou, T. Lebel, D. J. Parker, and J. Polcher, 2006: African Monsoon Multidisciplinary Analysis: An international research project and field campaign. *Bull. Am. Meteorol. Soc.*, **87**, 1739–1746.

Redi, M. H., 1982: Oceanic isopycnal mixing by coordinate rotation. *J. Phys. Oceanogr.*, **12**, 1154–1158.

Reichler, T., and J. Kim, 2008: How well do coupled models simulate today's climate? *Bull. Am. Meteorol. Soc.*, **89**, 303–311.

Reick, C. H., T. Raddatz, V. Brovkin, and V. Gayler, 2013: The representation of natural and anthropogenic land cover change in MPI-ESM. *J. Adv. Model. Earth Syst.*, doi:10.1002/jame.20022.

Reifen, C., and R. Toumi, 2009: Climate projections: Past performance no guarantee of future skill? *Geophys. Res. Lett.*, **36**, L13704.

Remer, L. A., et al., 2008: Global aerosol climatology from the MODIS satellite sensors. *J. Geophys. Res. Atmos.*, **113**, D14s07.

Richter, I., and S.-P. Xie, 2008: On the origin of equatorial Atlantic biases in coupled general circulation models. *Clim. Dyn.*, **31**, 587–598.

Richter, I., S.-P. Xie, S. K. Behera, T. Doi, and Y. Masumoto, 2013: Equatorial Atlantic variability and its relation to mean state biases in CMIP5. *Clim. Dyn.*, doi:10.1007/s00382-012-1624-5.

Richter, J. H., F. Sassi, and R. R. Garcia, 2010: Toward a physically based gravity wave source parameterization in a General Circulation Model. *J. Atmos. Sci.*, **67**, 136–156.

Ridgwell, A., and J. C. Hargreaves, 2007: Regulation of atmospheric CO(2) by deep-sea sediments in an Earth system model. *Global Biogeochem. Cycles*, **21**, Gb2008.

Ridgwell, A., I. Zondervan, J. C. Hargreaves, J. Bijma, and T. M. Lenton, 2007a: Assessing the potential long-term increase of oceanic fossil fuel CO_2 uptake due to CO_2–calcification feedback. *Biogeosciences*, **4**, 481–492.

Ridgwell, A., et al., 2007b: Marine geochemical data assimilation in an efficient Earth System Model of global biogeochemical cycling. *Biogeosciences*, **4**, 87–104.

Rienecker, M. M., et al., 2011: MERRA: NASA's modern-era retrospective analysis for research and applications. *J. Clim.*, **24**, 3624–3648.

Ringer, M. A., J. M. Edwards, and A. Slingo, 2003: Simulation of satellite channel radiances in the Met Office Unified Model. *Q. J. R. Meteorol. Soc.*, **129**, 1169–1190.

Rio, C., and F. Hourdin, 2008: A thermal plume model for the convective boundary layer: Representation of cumulus clouds. *J. Atmos. Sci.*, **65**, 407–425.

Rio, C., F. Hourdin, F. Couvreux, and A. Jam, 2010: Resolved versus parametrized boundary-layer plumes. Part II: Continuous formulations of mixing rates for mass-flux schemes. *Boundary-Layer Meteorol.*, **135**, 469–483.

Risi, C., et al., 2012a: Process-evaluation of tropospheric humidity simulated by general circulation models using water vapor isotopic observations: 2. Using isotopic diagnostics to understand the mid and upper tropospheric moist bias in the tropics and subtropics. *J. Geophys. Res. Atmos.*, **117**, D05304.

Risi, C., et al., 2012b: Process-evaluation of tropospheric humidity simulated by general circulation models using water vapor isotopologues: 1. Comparison between models and observations. *J. Geophys. Res. Atmos.*, **117**, D05303.

Risien, C. M., and D. B. Chelton, 2008: A global climatology of surface wind and wind stress fields from eight years of QuikSCAT Scatterometer data. *J. Phys. Oceanogr.*, **38**, 2379–2413.

Ritz, S. P., T. F. Stocker, and F. Joos, 2011: A coupled dynamical ocean-energy balance atmosphere model for paleoclimate studies. *J. Clim.*, **24**, 349–375.

Roberts, M. J., et al., 2004: Impact of an eddy-permitting ocean resolution on control and climate change simulations with a global coupled GCM. *J. Clim.*, **17**, 3–20.

Robinson, A., R. Calov, and A. Ganopolski, 2012: Multistability and critical thresholds of the Greenland ice sheet. *Nature Clim. Change*, **2**, 429–432.

Robinson, D. A., and A. Frei, 2000: Seasonal variability of northern hemisphere snow extent using visible satellite data. *Prof. Geograph.*, **51**, 307–314.

Rockel, B., C. L. Castro, R. A. Pielke, H. von Storch, and G. Leoncini, 2008: Dynamical downscaling: Assessment of model system dependent retained and added variability for two different regional climate models. *J. Geophys. Res. Atmos.*, **113**, D21107.

Rodwell, M., and T. Palmer, 2007: Using numerical weather prediction to assess climate models. *Q. J. R. Meteorol. Soc.*, **133**, 129–146.

Roe, G. H., and M. B. Baker, 2007: Why is climate sensitivity so unpredictable? *Science*, **318**, 629–632.

Roe, G. H., and K. C. Armour, 2011: How sensitive is climate sensitivity? *Geophys. Res. Lett.*, **38**, L14708.

Roe, G. H., and M. B. Baker, 2011: Comment on "Another look at climate sensitivity" by Zaliapin and Ghil (2010). *Nonlin.Proc. Geophys.*, **18**, 125–127.

Roeckner, E., et al., 2006: Sensitivity of simulated climate to horizontal and vertical resolution in the ECHAM5 atmosphere model. *J. Clim.*, **19**, 3771–3791.

Rojas, M., 2006: Multiply nested regional climate simulation for southern South America: Sensitivity to model resolution. *Mon. Weather Rev.*, **134**, 2208–2223.

Rojas, M., and P. I. Moreno, 2011: Atmospheric circulation changes and neoglacial conditions in the Southern Hemisphere mid-latitudes: Insights from PMIP2 simulations at 6 kyr. *Clim. Dyn.*, **37**, 357–375.

Rojas, M., et al., 2009: The Southern Westerlies during the last glacial maximum in PMIP2 simulations. *Clim. Dyn.*, **32**, 525–548.

Romanou, A., et al., 2013: Natural air–sea flux of CO_2 in simulations of the NASA-GISS climate model: Sensitivity to the physical ocean model formulation. *Ocean Model.*, doi:10.1016/j.ocemod.2013.01.008.

Rotstayn, L. D., and U. Lohmann, 2002: Simulation of the tropospheric sulfur cycle in a global model with a physically based cloud scheme. *J. Geophys. Res.*, **107**, 4592.

Rotstayn, L. D., M. A. Collier, R. M. Mitchell, Y. Qin, S. K. Campbell, and S. M. Dravitzki, 2011: Simulated enhancement of ENSO-related rainfall variability due to Australian dust. *Atmos. Chem. Phys.*, **11**, 6575–6592.

Rotstayn, L. D., S. J. Jeffrey, M. A. Collier, S. M. Dravitzki, A. C. Hirst, J. I. Syktus, and K. K. Wong, 2012: Aerosol- and greenhouse gas-induced changes in summer rainfall and circulation in the Australasian region: A study using single-forcing climate simulations. *Atmos. Chem. Phys.*, **12**, 6377–6404.

Rotstayn, L. D., et al., 2010: Improved simulation of Australian climate and ENSO-related rainfall variability in a global climate model with an interactive aerosol treatment. *Int. J. Climatol.* , **30**, 1067–1088.

Rougier, J., D. M. H. Sexton, J. M. Murphy, and D. Stainforth, 2009: Analyzing the climate sensitivity of the HadSM3 climate model using ensembles from different but related experiments. *J. Clim.*, **22**, 3540–3557.

Roy, P., P. Gachon, and R. Laprise, 2012: Assessment of summer extremes and climate variability over the north-east of North America as simulated by the Canadian Regional Climate Model. *Int. J. Climatol.* , **32** 1615–1627.

Ruckstuhl, C., and J. R. Norris, 2009: How do aerosol histories affect solar "dimming" and "brightening" over Europe?: IPCC-AR4 models versus observations. *J. Geophys. Res. Atmos.*, **114**, D00d04.

Rummukainen, M., 2010: State-of-the-art with regional climate models. *Clim. Change*, **1**, 82–96.

Russell, J. L., R. J. Stouffer, and K. W. Dixon, 2006: Intercomparison of the Southern Ocean circulations in IPCC coupled model control simulations. *J. Clim.*, **19**, 4560–4575.

Ruti, P. M., et al., 2011: The West African climate system: A review of the AMMA model inter-comparison initiatives. *Atmos. Sci. Lett.*, **12** 116–122

Rutter, N., et al., 2009: Evaluation of forest snow processes models (SnowMIP2). *J. Geophys. Res. Atmos.*, **114**, D06111.

Sabine, C. L., et al., 2004: The oceanic sink for anthropogenic CO_2. *Science*, **305**, 367–371.

Saha, S., et al., 2010: The NCEP Climate Forecast System Reanalysis. *Bull. Am. Meteorol. Soc.*, **91**, 1015–105.

Sahany, S., J. D. Neelin, K. Hales, and R. B. Neale, 2012: Temperature–moisture dependence of the Deep Convective Transition as a constraint on entrainment in climate models. *J. Atmos. Sci.*, **69**, 1340–1358.

Saji, N. H., S. P. Xie, and T. Yamagata, 2006: Tropical Indian Ocean variability in the IPCC twentieth-century climate simulations. *J. Clim.*, **19**, 4397–4417.

Saji, N. H., B. N. Goswami, P. N. Vinayachandran, and T. Yamagata, 1999: A dipole mode in the tropical Indian Ocean. *Nature*, **401**, 360–363.

Sakaguchi, K., X. B. Zeng, and M. A. Brunke, 2012: The hindcast skill of the CMIP ensembles for the surface air temperature trend. *J. Geophys. Res. Atmos.*, **117**, D16113.

Sakamoto, T. T., et al., 2012: MIROC4h – a new high-resolution atmosphere-ocean coupled general circulation model. *J. Meteorol. Soc. Jpn.*, **90**, 325–359.

Salas-Melia, D., 2002: A global coupled sea ice-ocean model. *Ocean Model.*, **4**, 137–172

Salle, J. B., E. Shuckburgh, N. Bruneau, A. J. S. Meijers, T. J. Bracegirdle, Z. Wang, and T. Roy, 2013: Assessment of Southern Ocean water mass circulation and characteristics in CMIP5 models: Historical bias and forcing response. *J. Geophys. Res. Oceans*, doi:10.1002/jgrc.20135.

Samuelsson, P., E. Kourzeneva, and D. Mironov, 2010: The impact of lakes on the European climate as simulated by a regional climate model. *Boreal Environ. Res.*, **15**, 113–129.

Sander, S. P., 2006: *Chemical Kinetics and Photochemical Data for Use in Atmospheric Studies*. Evaluation 15. JPL Publications, Pasadena, CA, USA, 523 pp.

Sanderson, B. M., 2011: A multimodel study of parametric uncertainty in predictions of climate response to rising greenhouse gas concentrations. *J. Clim.*, **25**, 1362–1377.

Sanderson, B. M., 2013: On the estimation of systematic error in regression-based predictions of climate sensitivity. *Clim. Change*, doi:10.1007/s10584-012-0671-6.

Sanderson, B. M., and R. Knutti, 2012: On the interpretation of constrained climate model ensembles. *Geophys. Res. Lett.*, **39**, L16708.

Sanderson, B. M., K. M. Shell, and W. Ingram, 2010: Climate feedbacks determined using radiative kernels in a multi-thousand member ensemble of AOGCMs. *Clim. Dyn.*, **35**, 1219–1236.

Sanderson, B. M., C. Piani, W. J. Ingram, D. A. Stone, and M. R. Allen, 2008a: Towards constraining climate sensitivity by linear analysis of feedback patterns in thousands of perturbed-physics GCM simulations. *Clim. Dyn.*, **30**, 175–190.

Sanderson, B. M., et al., 2008b: Constraints on model response to greenhouse gas forcing and the role of subgrid-scale processes. *J. Clim.*, **21**, 2384–2400.

Sansom, P. G., D. B. Stephenson, C. A. T. Ferro, G. Zappa, and L. Shaffrey, 2013: Simple uncertainty frameworks for selecting weighting schemes and interpreting multi-model ensemble climate change experiments doi:10.1175/JCLI-D-12-00462.1.

Santer, B., et al., 2009: Incorporating model quality information in climate change detection and attribution studies. *Proc. Natl. Acad. Sci. U.S.A.*, **106**, 14778–14783.

Santer, B., et al., 2007: Identification of human-induced changes in atmospheric moisture content. *Proc. Natl. Acad. Sci. U.S.A.*, **104**, 15248–15253.

Santer, B., et al., 2008: Consistency of modelled and observed temperature trends in the tropical troposphere. *Int. J. Climatol.* , **28**, 1703–1722.

Santer, B., et al., 2005: Amplification of surface temperature trends and variability in the tropical atmosphere. *Science*, **309**, 1551–1556.

Santer, B. D., et al., 2013: Identifying human influences on atmospheric temperature. *Proc. Natl. Acad. Sci. U.S.A*, **110**, 26–33.

Sato, M., J. E. Hansen, M. P. McCormick, and J. B. Pollack, 1993: Stratospheric aerosol optical depth, 1850– 1990. *J. Geophys. Res. Atmos.*, **98**, 22987–22994

Sato, T., H. Miura, M. Satoh, Y. N. Takayabu, and Y. Q. Wang, 2009: Diurnal cycle of precipitation in the Tropics simulated in a global cloud-resolving model. *J. Clim.*, **22**, 4809–4826.

Scaife, A. A., N. Butchart, C. D. Warner, and R. Swinbank, 2002: Impact of a spectral gravity wave parameterization on the stratosphere in the met office unified model. *J. Atmos. Sci.*, **59**, 1473–1489.

Scaife, A. A., T. Woollings, J. Knight, G. Martin, and T. Hinton, 2010: Atmospheric blocking and mean biases in climate models. *J. Clim.*, **23**, 6143–6152.

Scaife, A. A., et al., 2011: Improved Atlantic winter blocking in a climate model. *Geophys. Res. Lett.*, **38**, L23703.

Scaife, A. A., et al., 2012: Climate change and stratosphere-troposphere interaction. *Clim. Dyn.*, **38**, 2089–2097.

Scaife, A. A., et al., 2009: The CLIVAR C20C project: Selected twentieth century climate events. *Clim. Dyn.*, **33**, 603–614.

Schaller, N., I. Mahlstein, J. Cermak, and R. Knutti, 2011: Analyzing precipitation projections: A comparison of different approaches to climate model evaluation. *J. Geophys. Res. Atmos.*, **116**, D10118.

Scherrer, S., 2011: Present-day interannual variability of surface climate in CMIP3 models and its relation to future warming. *Int. J. Climatol.* , **31**, 1518–1529.

Schlesinger, M. E., and J. F. B. Mitchell, 1987: Climate model simulations of the equilibrium climatic response to increased carbon-dioxide. *Rev. Geophys.*, **25**, 760–798.

Schmidli, J., C. Goodess, C. Frei, M. Haylock, Y. Hundecha, J. Ribalaygua, and T. Schmith, 2007: Statistical and dynamical downscaling of precipitation: An evaluation and comparison of scenarios for the European Alps. *J. Geophys. Res. Atmos.*, **112**, D04105.

Schmidt, G. A., et al., 2012: Climate forcing reconstructions for use in PMIP simulations of the Last Millennium (v1.1). *Geophys. Model Dev.*, **5**, 185–191.

Schmidt, G. A., et al., 2006: Present day atmospheric simulations using GISS ModelE: Comparison to in-situ, satellite and reanalysis data. *J. Clim.*, **19**, 153–192.

Schmith, T., 2008: Stationarity of regression relationships: Application to empirical downscaling. *J. Clim.*, **21**, 4529–4537.

Schmittner, A., A. Oschlies, X. Giraud, M. Eby, and H. L. Simmons, 2005: A global model of the marine ecosystem for long-term simulations: Sensitivity to ocean mixing, buoyancy forcing, particle sinking, and dissolved organic matter cycling. *Global Biogeochem. Cycles*, **19**, Gb3004.

Schott, F. A., S.-P. Xie, and J. P. McCreary, Jr., 2009: Indian Ocean circulation and climate variability. *Rev. Geophys.*, **47**, RG1002.

Schramm, J. L., M. M. Holland, J. A. Curry, and E. E. Ebert, 1997: Modeling the thermodynamics of a sea ice thickness 1. Sensitivity to ice thickness resolution. *J. Geophys. Res.*, **102**, 23079–23091.

Schultz, M. G., et al., 2008: Global wildland fire emissions from 1960 to 2000. *Global Biogeochem. Cycles*, **22**, GB2002.

Schurgers, G., U. Mikolajewicz, M. Groger, E. Maier-Reimer, M. Vizcaino, and A. Winguth, 2008: Long-term effects of biogeophysical and biogeochemical interactions between terrestrial biosphere and climate under anthropogenic climate change. *Global Planet. Change*, **64**, 26–37.

Scoccimarro, E., et al., 2011: Effects of tropical cyclones on ocean heat transport in a high resolution Coupled General Circulation Model. *J. Clim.*, **24**, 4368–4384.

Séférian, R., et al., 2013: Skill assessment of three earth system models with common marine biogeochemistry. *Clim. Dyn.*, **40**, 2549–2573.

Segui, P., A. Ribes, E. Martin, F. Habets, and J. Boe, 2010: Comparison of three downscaling methods in simulating the impact of climate change on the hydrology of Mediterranean basins. *J. Hydrol.*, **383**, 111–124.

Seidel, D. J., M. Free, and J. S. Wang, 2012: Reexamining the warming in the tropical upper troposphere: Models versus radiosonde observations. *Geophys. Res. Lett.*, **39**, L22701.

Seidel, D. J., Q. Fu, W. J. Randel, and T. J. Reichler, 2008: Widening of the tropical belt in a changing climate. *Nature Geosci.*, **1**, 21–24.

Seidel, D. J., N. P. Gillett, J. R. Lanzante, K. P. Shine, and P. W. Thorne, 2011: Stratospheric temperature trends: Our evolving understanding. *Clim. Change*, **2**, 592–616.

Selten, F. M., G. W. Branstator, H. A. Dijkstra, and M. Kliphuis, 2004: Tropical origins for recent and future Northern Hemisphere climate change. *Geophys. Res. Lett.*, **31**, L21205.

Semenov, V. A., M. Latif, J. H. Jungclaus, and W. Park, 2008: Is the observed NAO variability during the instrumental record unusual? *Geophys. Res. Lett.*, **35**, L11701.

Seneviratne, S., et al., 2012: Changes in climate extremes and their impacts on the natural physical environment. In: *IPCC WGI/WGII Special Report on Managing the Risks of Extreme Events and Disasters to Advance Climate Change Adaptation (SREX)*, [Field, C.B., V. Barros, T.F. Stocker, D. Qin, D.J. Dokken, K.L. Ebi, M.D. Mastrandrea, K.J. Mach, G.-K. Plattner, S.K. Allen, M. Tignor, and P.M. Midgley (Eds.)]. Cambridge University Press, The Edinburgh Building, Shaftesbury Road, Cambridge CB2 8RU ENGLAND, pp. 109–230.

Seneviratne, S. I., D. Luethi, M. Litschi, and C. Schaer, 2006: Land-atmosphere coupling and climate change in Europe. *Nature*, **443**, 205–209.

Seneviratne, S. I., et al., 2010: Investigating soil moisture-climate interactions in a changing climate: A review. *Earth Sci. Rev.*, **99**, 125–161.

Separovic, L., R. De Elia, and R. Laprise, 2008: Reproducible and irreproducible components in ensemble simulations with a Regional Climate Model. *Mon. Weather Rev.*, **136**, 4942–4961.

Separovic, L., R. Elía, and R. Laprise, 2012: Impact of spectral nudging and domain size in studies of RCM response to parameter modification. *Clim. Dyn.*, **38**, 1325–1343.

Severijns, C. A., and W. Hazeleger, 2010: The efficient global primitive equation climate model SPEEDO V2.0. *Geophys. Model Dev.*, **3**, 105–122.

Sexton, D. M. H., and J. M. Murphy, 2012: Multivariate probabilistic projections using imperfect climate models. Part II: robustness of methodological choices and consequences for climate sensitivity. *Clim. Dyn.*, **38**, 2543–2558.

Sexton, D. M. H., J. M. Murphy, M. Collins, and M. J. Webb, 2012: Multivariate probabilistic projections using imperfect climate models part I: Outline of methodology. *Clim. Dyn.*, **38**, 2513–2542.

Shaffer, G., and J. L. Sarmiento, 1995: Biogeochemical cycling in the global ocean. 1. A new, analytical model with continuous vertical resolution and high-latitude dynamics. *J. Geophys. Res. Oceans*, **100**, 2659–2672.

Shaffer, G., S. M. Olsen, and J. O. P. Pedersen, 2008: Presentation, calibration and validation of the low-order, DCESS Earth System Model (Version 1). *Geophys. Model Dev.*, **1**, 17–51.

Shaffrey, L. C., et al., 2009: UK HiGEM: The new UK High-Resolution Global Environment Model—Model description and basic evaluation. *J. Clim.*, **22**, 1861–1896.

Sheffield, J., and E. F. Wood, 2008: Projected changes in drought occurrence under future global warming from multi-model, multi-scenario, IPCC AR4 simulations. *Clim. Dyn.*, **31**, 79–105.

Shell, K. M., J. T. Kiehl, and C. A. Shields, 2008: Using the radiative kernel technique to calculate climate feedbacks in NCAR's Community Atmospheric Model. *J. Clim.*, **21**, 2269–2282.

Shevliakova, E., et al., 2009: Carbon cycling under 300 years of land use change: Importance of the secondary vegetation sink. *Global Biogeochem. Cycles*, **23**, GB2022

Shi, Y., J. Zhang, J. S. Reid, B. Holben, E. J. Hyer, and C. Curtis, 2011: An analysis of the collection 5 MODIS over-ocean aerosol optical depth product for its implication in aerosol assimilation. *Atmos. Chem. Phys.*, **11**, 557–565.

Shibata, K., and M. Deushi, 2005: Radiative effect of ozone on the quasi-biennial oscillation in the equatorial stratosphere. *Geophys. Res. Lett.*, **32**, L24802.

Shin, D., J. Kim, and H. Park, 2011: Agreement between monthly precipitation estimates from TRMM satellite, NCEP reanalysis, and merged gauge-satellite analysis. *J. Geophys. Res. Atmos.*, **116**, D16105.

Shin, S. I., D. Sardeshmukh, and K. Pegion, 2010: Realism of local and remote feedbacks on tropical sea surface temperatures in climate models. *J. Geophys. Res. Atmos.*, **115**, D21110.

Shindell, D. T., et al., 2013a: Interactive ozone and methane chemistry in GISS-E2 historical and future climate simulations *Atmos. Chem. Phys.*, **13**, 2653–2689.

Shindell, D. T., et al., 2013b: Radiative forcing in the ACCMIP historical and future climate simulations. *Atmos. Chem. Phys.*, **13**, 2939–2974.

Shiogama, H., S. Emori, N. Hanasaki, M. Abe, Y. Masutomi, K. Takahashi, and T. Nozawa, 2011: Observational constraints indicate risk of drying in the Amazon basin. *Nature Commun.*, **2**, 253.

Shiogama, H., et al., 2012: Perturbed physics ensemble using the MIROC5 coupled atmosphere–ocean GCM without flux corrections: Experimental design and results. *Clim. Dyn.*, **39**, 3041–3056.

Shiu, C.-J., S. C. Liu, C. Fu, A. Dai, and Y. Sun, 2012: How much do precipitation extremes change in a warming climate? *Geophys. Res. Lett.*, **39**, L17707.

Shkol'nik, I., V. Meleshko, S. Efimov, and E. Stafeeva, 2012: Changes in climate extremes on the territory of Siberia by the middle of the 21st century: An ensemble forecast based on the MGO regional climate model. *Russ. Meteorol. Hydrol.*, **37**, 71–84.

Shkolnik, I., V. Meleshko, and V. Kattsov, 2007: The MGO climate model for Siberia. *Russ. Meteorol. Hydrol.*, **32**, 351–359.

Siebesma, A. P., P. M. M. Soares, and J. Teixeira, 2007: A combined eddy-diffusivity mass-flux approach for the convective boundary layer. *J. Atmos. Sci.*, **64**, 1230–1248.

Sigmond, M., and J. F. Scinocca, 2010: The influence of the basic state on the Northern Hemisphere circulation response to climate change. *J. Clim.*, **23**, 1434–1446.

Sillmann, J., and M. Croci-Maspoli, 2009: Present and future atmospheric blocking and its impact on European mean and extreme climate. *Geophys. Res. Lett.*, **36**, L10702.

Sillmann, J., M. Croci-Maspoli, M. Kallache, and R. W. Katz, 2011: Extreme cold winter temperatures in Europe under the influence of North Atlantic atmospheric blocking. *J. Clim.*, **24**, 5899–5913.

Sillmann, J., V. V. Kharin, X. Zhang, and F. W. Zwiers, 2013: Climate extreme indices in the CMIP5 multi-model ensemble. Part 1: Model evaluation in the present climate. *J. Geophys. Res.*, doi:10.1029/2012JD018390.

Simmons, A. J., K. M. Willett, P. D. Jones, P. W. Thorne, and D. P. Dee, 2010: Low-frequency variations in surface atmospheric humidity, temperature, and precipitation: Inferences from reanalyses and monthly gridded observational data sets. *J. Geophys. Res. Atmos.*, **115**, D01110.

Simmons, H. L., S. R. Jayne, L. C. St Laurent, and A. J. Weaver, 2004: Tidally driven mixing in a numerical model of the ocean general circulation. *Ocean Model.*, **6**, 245–263.

Sitch, S., et al., 2003: Evaluation of ecosystem dynamics, plant geography and terrestrial carbon cycling in the LPJ dynamic global vegetation model. *Global Change Biol.*, **9**, 161–185.

Sitch, S., et al., 2008: Evaluation of the terrestrial carbon cycle, future plant geography and climate-carbon cycle feedbacks using five Dynamic Global Vegetation Models (DGVMs). *Global Change Biol.*, **14**, 2015–2039.

Six, K. D., and E. Maier-Reimer, 1996: Effects of plankton dynamics on seasonal carbon fluxes in an Ocean General Circulation Model. *Global Biogeochem. Cycles*, **10**, 559–583.

Slater, A. G., and D. M. Lawrence, 2013: Diagnosing present and future permafrost from climate models. *J. Clim.*, doi:10.1175/JCLI-D-12-00341.1.

Sloyan, B. M., and I. V. Kamenkovich, 2007: Simulation of Subantarctic Mode and Antarctic Intermediate Waters in climate models. *J. Clim.*, **20**, 5061–5080.

Smirnov, D., and D. J. Vimont, 2011: Variability of the Atlantic Meridional Mode during the Atlantic Hurricane Season. *J. Clim.*, **24**, 1409–1424.

Smith, B., P. Samuelsson, A. Wramneby, and M. Rummukainen, 2011a: A model of the coupled dynamics of climate, vegetation and terrestrial ecosystem biogeochemistry for regional applications. *Tellus A*, **63**, 87–106.

Smith, P. C., N. De Noblet-Ducoudre, P. Ciais, P. Peylin, N. Viovy, Y. Meurdesoif, and A. Bondeau, 2010a: European-wide simulations of croplands using an improved terrestrial biosphere model: Phenology and productivity. *J. Geophys. Res. Biogeosci.*, **115**, G01014.

Smith, P. C., P. Ciais, P. Peylin, N. De Noblet-Ducoudre, N. Viovy, Y. Meurdesoif, and A. Bondeau, 2010b: European-wide simulations of croplands using an improved terrestrial biosphere model: 2. Interannual yields and anomalous CO_2 fluxes in 2003. *J. Geophys. Res. Biogeosci.*, **115**, G04028

Smith, R. D., M. E. Maltrud, F. O. Bryan, and M. W. Hecht, 2000: Numerical simulation of the North Atlantic Ocean at 1/10 degrees. *J. Phys. Oceanogr.*, **30**, 1532–1561.
Smith, R. S., J. M. Gregory, and A. Osprey, 2008: A description of the FAMOUS (version XDBUA) climate model and control run. *Geophys. Model Dev.*, **1**, 53–68.
Smith, S. J., J. van Aardenne, Z. Klimont, R. J. Andres, A. Volke, and S. D. Arias, 2011b: Anthropogenic sulfur dioxide emissions: 1850–2005. *Atmos. Chem. Phys.*, **11**, 1101–1116.
Sobolowski, S., and T. Pavelsky, 2012: Evaluation of present and future North American Regional Climate Change Assessment Program (NARCCAP) regional climate simulations over the southeast United States. *J. Geophys. Res. Atmos.*, **117**, D01101.
Soden, B. J., and I. M. Held, 2006: An assessment of climate feedbacks in coupled ocean-atmosphere models. *J. Clim.*, **19**, 3354–3360.
Soden, B. J., I. M. Held, R. Colman, K. M. Shell, J. T. Kiehl, and C. A. Shields, 2008: Quantifying climate feedbacks using radiative kernels. *J. Clim.*, **21**, 3504–3520.
Sokolov, A. P., and P. H. Stone, 1998: A flexible climate model for use in integrated assessments. *Clim. Dyn.*, **14**, 291–303.
Sokolov, A. P., C. E. Forest, and P. H. Stone, 2010: Sensitivity of climate change projections to uncertainties in the estimates of observed changes in deep-ocean heat content. *Clim. Dyn.*, **34**, 735–745.
Sokolov, A. P., et al., 2009: Probabilistic forecast for twenty-first-century climate based on uncertainties in emissions (without policy) and climate parameters. *J. Clim.*, **22**, 5175–5204.
Sokolov, A. P., et al., 2005: The MIT Integrated Global System Model (IGSM) Version 2: Model description and baseline evaluation. MIT JP Report 124. MIT, Cambridge, MA.
Solman, S., and N. Pessacg, 2012: Regional climate simulations over South America: Sensitivity to model physics and to the treatment of lateral boundary conditions using the MM5 model. *Clim. Dyn.*, **38**, 281–300.
Solomon, S., P. J. Young, and B. Hassler, 2012: Uncertainties in the evolution of stratospheric ozone and implications for recent temperature changes in the tropical lower stratosphere. *Geophys. Res. Lett.*, **39** L17706.
Solomon, S., K. H. Rosenlof, R. W. Portmann, J. S. Daniel, S. M. Davis, T. J. Sanford, and G. K. Plattner, 2010: Contributions of stratospheric water vapor to decadal changes in the rate of global warming. *Science*, **327**, 1219–1223.
Somot, S., F. Sevault, M. Deque, and M. Crepon, 2008: 21st century climate change scenario for the Mediterranean using a coupled atmosphere-ocean regional climate model. *Global Planet. Change*, **63** 112–126.
Son, S., et al., 2008: The impact of stratospheric ozone recovery on the Southern Hemisphere westerly jet. *Science*, **320**, 1486–1489.
Son, S., et al., 2010: Impact of stratospheric ozone on Southern Hemisphere circulation change: A multimodel assessment. *J. Geophys. Res. Atmos.*, **115**, D00M07.
Song, Z., F. Qiao, and Y. Song, 2012: Response of the equatorial basin-wide SST to non-breakingsurface wave-induced mixing in a climate model: An amendment to tropical bias. *J. Geophys. Res.*, doi:10.1029/2012JC007931.
SPARC-CCMVal, 2010: SPARC Report on the Evaluation of Chemistry-Climate Models [V. Eyring, T.G. Shepherd, D.W. Waugh (eds.)], SPARC Report No. 5, WCRP-132, WMO/TD-No. 1526.
Sperber, K., and D. Kim, 2012: Simplified metrics for the identification of the Madden-Julian oscillation in models. *Atmos. Sci. Let.*, doi:10.1002/asl.378.
Sperber, K., et al., 2012: The Asian summer monsoon: An intercomparison of CMIP-5 vs. CMIP-3 simulations of the late 20th century. *Clim. Dyn.*, doi:10.1007/s00382-012-1607-6.
Sperber, K. R., 2003: Propagation and the vertical structure of the Madden-Julian oscillation. *Mon. Weather Rev.*, **131**, 3018–3037.
Sperber, K. R., and H. Annamalai, 2008: Coupled model simulations of boreal summer intraseasonal (30–50 day) variability, Part 1: Systematic errors and caution on use of metrics. *Clim. Dyn.*, **31**, 345–372.
Sperber, K. R., et al., 2010: Monsoon Fact Sheet: CLIVAR Asian-Australian Monsoon Panel.
Stainforth, D. A., et al., 2005: Uncertainty in predictions of the climate response to rising levels of greenhouse gases. *Nature*, **433**, 403–406.
Stephens, G. L., and C. D. Kummerow, 2007: The remote sensing of clouds and precipitation from space: A review. *J. Atmos. Sci.*, **64**, 3742–3765.
Stephens, G. L., et al., 2010: Dreary state of precipitation in global models. *J. Geophys. Res.*, **115**, D24211.
Stephens, G. L., et al., 2012: An Update on the Earth's energy balance in light of new global observations. *Nature Geosci.*, **5**, 691–696.
Stephenson, D. B., M. Collins, J. C. Rougier, and R. E. Chandler, 2012: Statistical problems in the probabilistic prediction of climate change. *Environmetrics*, **23**, 364–372.
Stephenson, D. B., V. Pavan, M. Collins, M. M. Junge, R. Quadrelli, and C. M. G. Participating, 2006: North Atlantic Oscillation response to transient greenhouse gas forcing and the impact on European winter climate: A CMIP2 multi-model assessment. *Clim. Dyn.*, **27**, 401–420.
Stevens, B., and S. E. Schwartz, 2012: Observing and modeling Earth's energy flows. *Surv. Geophys.*, **33**, 779–816.
Stevens, B., et al., 2012: The atmospheric component of the MPI-M Earth System Model: ECHAM6. *J. Adv. Model. Earth Syst.*, doi:10.1002/jame.20015.
Stevenson, S., 2012: Significant changes to ENSO strength and impacts in the twenty-first century: Results from CMIP5. *Geophys. Res. Lett.*, doi:10.1029/2012GL052759.
Stevenson, S., B. Fox-Kemper, M. Jochum, R. Neale, C. Deser, and G. Meehl, 2012: Will there be a significant change to El Niño in the twenty-first century? *J. Clim.*, **25**, 2129–2145.
Stocker, B. D., K. Strassmann, and F. Joos, 2011: Sensitivity of Holocene atmospheric CO_2 and the modern carbon budget to early human land use: Analyses with a process-based model. *Biogeosciences*, **8**, 69–88.
Stocker, B. D., et al., 2012: Multiple greenhouse gas feedbacks from the land biosphere under future climate change scenarios. *Nature Clim. Change*, doi:10.1038/nclimate1864.
Stolarski, R., and S. Frith, 2006: Search for evidence of trend slow-down in the long-term TOMS/SBUV total ozone data record: The importance of instrument drift uncertainty. *Atmos. Chem. Phys.*, **6**, 4057–4065.
Stoner, A. M. K., K. Hayhoe, and D. J. Wuebbles, 2009: Assessing General Circulation Model simulations of atmospheric teleconnection patterns. *J. Clim.*, **22**, 4348–4372.
Stott, P. A., and C. E. Forest, 2007: Ensemble climate predictions using climate models and observational constraints. *Philos. R. Soc. London A*, **365**, 2029–2052.
Strachan, J., P. L. Vidale, K. Hodges, M. Roberts, and M.-E. Demory, 2013: Investigating global tropical cyclone activity with a hierarchy of AGCMs: The role of model resolution. *J. Clim.*, **26**, 133–152.
Strassmann, K. M., F. Joos, and G. Fischer, 2008: Simulating effects of land use changes on carbon fluxes: Past contributions to atmospheric CO_2 increases and future commitments due to losses of terrestrial sink capacity. *Tellus B*, **60**, 583–603.
Stratton, R. A., and A. J. Stirling, 2012: Improving the diurnal cycle of convection in GCMs. *Q. J. R. Meteorol. Soc.*, **138**, 1121–1134.
Stroeve, J., M. Holland, W. Meier, T. Scambos, and M. Serreze, 2007: Arctic sea ice decline: Faster than forecast. *Geophys. Res. Lett.*, **34**, L09501.
Stroeve, J. C., V. Kattsov, A. Barrett, M. Serreze, T. Pavlova, M. Holland, and W. N. Meier, 2012: Trends in Arctic sea ice extent from CMIP5, CMIP3 and observations. *Geophys. Res. Lett.*, **39**, L16502.
Su, H., and J. H. Jiang, 2012: Tropical clouds and circulation changes during the 2006–07 and 2009–10 El Niños. *J. Clim.*, doi:10.1175/JCLI-D-1200152.1.
Su, H., D. E. Waliser, J. H. Jiang, J. L. Li, W. G. Read, J. W. Waters, and A. M. Tompkins, 2006: Relationships of upper tropospheric water vapor, clouds and SST: MLS observations, ECMWF analyses and GCM simulations. *Geophys. Res. Lett.*, **33**, L22802
Su, H., et al., 2012: Diagnosis of regime-dependent cloud simulation errors in CMIP5 models using "A-Train" satellite observations and reanalysis data. *J. Geophys. Res.*, doi:10.1029/2012JD018575.
Sudo, K., M. Takahashi, J. Kurokawa, and H. Akimoto, 2002: CHASER: A global chemical model of the troposphere - 1. Model description. *J. Geophys. Res. Atmos.*, **107**, 4339.
Suh, M., S. Oh, D. Lee, D. Cha, S. Choi, C. Jin, and S. Hong, 2012: Development of new ensemble methods based on the performance skills of regional climate models over South Korea. *J. Clim.*, **25**, 7067–7082.
Sun, D.-Z., Y. Yu, and T. Zhang, 2009: Tropical water vapor and cloud feedbacks in climate models: A further assessment using coupled simulations. *J. Clim.*, **22**, 1287–1304.
Sutton, R. T., B. W. Dong, and J. M. Gregory, 2007: Land/sea warming ratio in response to climate change: IPCC AR4 model results and comparison with observations. *Geophys. Res. Lett.*, **34**, L02701
Svensson, G., and A. Holtslag, 2009: Analysis of model results for the turning of the wind and related momentum fluxes in the stable boundary layer. *Boundary-Layer Meteorol.*, **132**, 261–277.

Svensson, G., et al., 2011: Evaluation of the diurnal cycle in the atmospheric boundary layer over land as represented by a variety of single-column models: The Second GABLS Experiment. *Boundary-Layer Meteorol.*, **140**, 177–206.
Swart, N. C., and J. C. Fyfe, 2012a: Ocean carbon uptake and storage influenced by wind bias in global climate models. *Nature Clim. Change*, **2**, 47–52.
Swart, N. C., and J. C. Fyfe, 2012b: Observed and simulated changes in the Southern Hemisphere surface westerly wind-stress. *Geophys. Res. Lett.*, doi:10.1029/2012GL052810.
Tachiiri, K., J. C. Hargreaves, J. D. Annan, A. Oka, A. Abe-Ouchi, and M. Kawamiya, 2010: Development of a system emulating the global carbon cycle in Earth system models. *Geophys. Model Dev.*, **3**, 365–376.
Takahashi, T., et al., 2009: Climatological mean and decadal change in surface ocean pCO(2), and net sea-air CO_2 flux over the global oceans. *Deep-Sea Res. Pt.*, **56**, 554–577.
Takata, K., S. Emori, and T. Watanabe, 2003: Development of the minimal advanced treatments of surface interaction and runoff. *Global Planet. Change*, **38**, 209–222.
Takemura, T., T. Nakajima, O. Dubovik, B. N. Holben, and S. Kinne, 2002: Single-scattering albedo and radiative forcing of various aerosol species with a global three-dimensional model. *J. Clim.*, **15**, 333–352.
Takemura, T., T. Nozawa, S. Emori, T. Y. Nakajima, and T. Nakajima, 2005: Simulation of climate response to aerosol direct and indirect effects with aerosol transport-radiation model. *J. Geophys. Res. Atmos.*, **110**, D02202.
Takemura, T., H. Okamoto, Y. Maruyama, A. Numaguti, A. Higurashi, and T. Nakajima, 2000: Global three-dimensional simulation of aerosol optical thickness distribution of various origins. *J. Geophys. Res. Atmos.*, **105**, 17853–17873.
Takemura, T., M. Egashira, K. Matsuzawa, H. Ichijo, R. O'Ishi, and A. Abe-Ouchi, 2009: A simulation of the global distribution and radiative forcing of soil dust aerosols at the Last Glacial Maximum. *Atmos. Chem. Phys.*, **9**, 3061–3073.
Takle, E. S., et al., 2007: Transferability intercomparison—An opportunity for new insight on the global water cycle and energy budget. *Bull. Am. Meteorol. Soc.*, **88**, 375–384.
Taylor, C. M., R. A. M. de Jeu, F. Guichard, P. P. Harris, and W. A. Dorigo, 2012a: Afternoon rain more likely over drier soils. *Nature*, **489**, 423–426.
Taylor, C. M., A. Gounou, F. Guichard, P. P. Harris, R. J. Ellis, F. Couvreux, and M. De Kauwe, 2011: Frequency of Sahelian storm initiation enhanced over mesoscale soil-moisture patterns. *Nature Geosci.*, **4**, 430–433.
Taylor, K. E., R. J. Stouffer, and G. A. Meehl, 2012b: An overview of CMIP5 and the experiment design. *Bull. Am. Meteorol. Soc.*, **93**, 485–498.
Tebaldi, C., and R. Knutti, 2007: The use of the multi-model ensemble in probabilistic climate projections. *Philos. Trans. R. Soc. London A*, **365** 2053–2075.
Teixeira, J., et al., 2008: Parameterization of the atmospheric boundary layer. *Bull. Am. Meteorol. Soc.*, **89**, 453–458.
Teixeira, J., et al., 2011: Tropical and subtropical cloud transitions in weather and climate prediction models: The GCSS/WGNE Pacific Cross-Section Intercomparison (GPCI). *J. Clim.*, **24**, 5223–5256.
Terray, L., 2012: Evidence for multiple drivers of North Atlantic multi-decadal climate variability. *Geophys. Res. Lett.*, **39**, L19712.
Terray, L., L. Corre, S. Cravatte, T. Delcroix, G. Reverdin, and A. Ribes, 2012: Near-surface salinity as nature's rain gauge to detect human influence on the tropical water cycle. *J. Clim.*, **25**, 958–977.
Teutschbein, C., F. Wetterhall, and J. Seibert, 2011: Evaluation of different downscaling techniques for hydrological climate-change impact studies at the catchment scale. *Clim. Dyn.*, **37**, 2087–2105.
Textor, C., et al., 2007: The effect of harmonized emissions on aerosol properties in global models—an AeroCom experiment. *Atmos. Chem. Phys.*, **7**, 4489–4501.
Thorndike, A. S., D. A. Rothrock, G. A. Maykut, and R. Colony, 1975: Thickness distribution of sea ice *J. Geophys. Res. Oceans Atmos.*, **80**, 4501–4513.
Thorne, P. W., et al., 2011: A quantification of uncertainties in historical tropical tropospheric temperature trends from radiosondes. *J. Geophys. Res.*, **116**, D12116.
Thornton, P. E., J. F. Lamarque, N. A. Rosenbloom, and N. M. Mahowald, 2007: Influence of carbon-nitrogen cycle coupling on land model response to CO_2 fertilization and climate variability. *Global Biogeochem. Cycles*, **21**, GB4018.
Tian, B., E. J. Fetzer, B. H. Kahn, J. Teixeira, E. Manning, and T. Hearty, 2013: Evaluating CMIP5 models using AIRS tropospheric air temperature and specific humidity climatology. *J. Geophys. Res. Atmos.*, **118**, 114–134.
Timbal, B., and D. Jones, 2008: Future projections of winter rainfall in southeast Australia using a statistical downscaling technique. *Clim. Change*, **86**, 165–187.
Timmermann, A., S. Lorenz, S.-I. An, A. Clement, and S.-P. Xie, 2007: The effect of orbital forcing on the mean climate and variability of the tropical Pacific. *J. Clim.*, **20**, 4147–4159.
Timmermann, R., H. Goosse, G. Madec, T. Fichefet, C. Ethe, and V. Duliere, 2005: On the representation of high latitude processes in the ORCA-LIM global coupled sea ice-ocean model. *Ocean Model.*, **8**, 175–201.
Ting, M., Y. Kushnir, R. Seager, and C. Li, 2009: Forced and internal twentieth-century SST trends in the north Atlantic. *J. Clim.*, **22**, 1469–1481.
Tjernstrom, M., J. Sedlar, and M. Shupe, 2008: How well do regional climate models reproduce radiation and clouds in the Arctic? An evaluation of ARCMIP simulations. *J. Appl. Meteorol. Climatol.*, **47**, 2405–2422.
Tjiputra, J. F., K. Assmann, M. Bentsen, I. Bethke, O. H. Ottera, C. Sturm, and C. Heinze, 2010: Bergen Earth system model (BCM-C): Model description and regional climate-carbon cycle feedbacks assessment. *Geophys. Model Dev.*, **3**, 123–141.
Tjiputra, J. F., et al., 2013: Evaluation of the carbon cycle components inthe Norwegian Earth System Model (NorESM). *Geophys. Model Dev.*, **6**, 301–325.
Todd-Brown, K. E. O., J. T. Randerson, W. M. Post, F. M. Hoffman, C. Tarnocai, E. A. G. Schuur, and S. D. Allison, 2013: Causes of variation in soil carbon simulations from CMIP5 Earth system models and comparison with observations. *Biogeosciences*, **10**, 1717–1736.
Toniazzo, T., and S. Woolnough, 2013: Development of warm SST errors in the southern tropical Atlantic in CMIP5 decadal hindcasts. *Clim. Dyn.*, doi:10.1007/s00382-013-1691-2.
Tory, K., S. Chand, R. Dare, and J. McBride, 2013: An assessment of a model-, grid- and basin-independent tropical cyclone detection scheme in selected CMIP3 global climate models. *J. Clim.*, doi:10.1175/JCLI-D-12-00511.1.
Trenberth, K. E., and J. M. Caron, 2001: Estimates of meridional atmosphere and ocean heat transports. *J. Clim.*, **14**, 3433–3443.
Trenberth, K. E., and J. T. Fasullo, 2008: An observational estimate of inferred ocean energy divergence. *J. Phys. Oceanogr.*, **38**, 984–999.
Trenberth, K. E., and J. T. Fasullo, 2009: Global warming due to increasing absorbed solar radiation. *Geophys. Res. Lett.*, **36**, L07706.
Trenberth, K. E., and J. T. Fasullo, 2010a: Climate change: Tracking Earth's energy. *Science*, **328**, 316–317.
Trenberth, K. E., and J. T. Fasullo, 2010b: Simulation of present-day and twenty-first-century energy budgets of the Southern Oceans. *J. Clim.*, **23**, 440–454.
Trenberth, K. E., D. P. Stepaniak, and J. M. Caron, 2000: The global monsoon as seen through the divergent atmospheric circulation. *J. Clim.*, **13**, 3969–3993.
Trenberth, K. E., J. T. Fasullo, and J. Kiehl, 2009: Earth's global energy budget. *Bull. Am. Meteorol. Soc.*, **90**, 311–323.
Tryhorn, L., and A. DeGaetano, 2011: A comparison of techniques for downscaling extreme precipitation over the northeastern United States. *Int. J. Climatol.* , **31**, 1975–1989.
Tschumi, T., F. Joos, and P. Parekh, 2008: How important are Southern Hemisphere wind changes for low glacial carbon dioxide? A model study. *Paleoceanography*, **23**, PA4208.
Tschumi, T., F. Joos, M. Gehlen, and C. Heinze, 2011: Deep ocean ventilation, carbon isotopes, marine sedimentation and the deglacial CO(2) rise. *Clim. Past*, **7**, 771–800.
Tsigaridis, K., and M. Kanakidou, 2007: Secondary organic aerosol importance in the future atmosphere. *Atmos. Environ.*, **41**, 4682–4692.
Tsujino, H., M. Hirabara, H. Nakano, T. Yasuda, T. Motoi, and G. Yamanaka, 2011: Simulating present climate of the global ocean–ice system using the Meteorological Research Institute Community Ocean Model (MRI. COM): Simulation characteristics and variability in the Pacific sector. *J. Oceanogr.*, **67**, 449–479.
Tsushima, Y., M. Ringer, M. Webb, and K. Williams, 2013: Quantitative evaluation of the seasonal variations in climate model cloud regimes. *Clim. Dyn.*, doi:10.1007/s00382-012-1609-4.
Turner, A. G., and H. Annamalai, 2012: Climate change and the south Asian summer monsoon. *Nature Clim. Change*, **2**, 1–9.
Turner, A. G., P. M. Inness, and J. M. Slingo, 2007: The effect of doubled CO_2 and model basic state biases on the monsoon-ENSO system. II: Changing ENSO regimes. *Q. J. R. Meteorol. Soc.*, **133**, 1159–1173.
Ulbrich, U., G. C. Leckebusch, and J. G. Pinto, 2009: Extra-tropical cyclones in the present and future climate: a review. *Theor. Appl. Climatol.*, **96**, 117–131.
Ulbrich, U., J. G. Pinto, H. Kupfer, G. C. Leckebusch, T. Spangehl, and M. Reyers, 2008: Changing northern hemisphere storm tracks in an ensemble of IPCC climate change simulations. *J. Clim.*, **21**, 1669–1679.

UNESCO, 1981: Tenth report of the joint panel on oceanographic tables and standards UNESCO.

Uotila, P., S. O'Farrell, S. J. Marsland, and D. Bi, 2012: A sea-ice sensitivity study with a global ocean-ice model. *Ocean Model.*, **51**, 1–18.

Uotila, P., S. O'Farrell, S. J. Marsland, and D. Bi, 2013: The sea-ice performance of the Australian climate models participating in the CMIP5. *Aust. Meteorol. Oceanogr. J.*, **63**, 121–143.

Uppala, S. M., et al., 2005: The ERA-40 re-analysis. *Q. J. R. Meteorol. Soc.*, **131**, 2961–3012.

van den Hurk, B., and E. van Meijgaard, 2010: Diagnosing land-atmosphere interaction from a regional climate model simulation over West Africa. *J. Hydrometeorol.*, **11**, 467–481.

van Oldenborgh, G., et al., 2009: Western Europe is warming much faster than expected. *Clim. Past*, **5**, 1–12.

van Oldenborgh, G. J., S. Y. Philip, and M. Collins, 2005: El Niño in a changing climate: A multi-model study. *Ocean Sci.*, **1**, 81–95.

van Roosmalen, L., J. H. Christensen, M. B. Butts, K. H. Jensen, and J. C. Refsgaard, 2010: An intercomparison of regional climate model data for hydrological impact studies in Denmark. *J. Hydrol.*, **380**, 406–419.

van Vliet, M., S. Blenkinsop, A. Burton, C. Harpham, H. Broers, and H. Fowler, 2011: A multi-model ensemble of downscaled spatial climate change scenarios for the Dommel catchment, Western Europe. *Clim. Change*, **111**, 249–277.

van Vuuren, D., et al., 2011: The representative concentration pathways: An overview. *Clim. Change*, **109**, 5–31.

Vancoppenolle, M., T. Fichefet, H. Goosse, S. Bouillon, G. Madec, and M. A. M. Maqueda, 2009: Simulating the mass balance and salinity of Arctic and Antarctic sea ice. 1. Model description and validation. *Ocean Model.*, **27**, 33–53.

Vancoppenolle, M., H. Goosse, A. de Montety, T. Fichefet, B. Tremblay, and J. L. Tison, 2010: Interactions between brine motion, nutrients and primary production in sea ice. *J. Geophys. Res.*, **115**, C02005.

Vannière, B., E. Guilyardi, G. Madec, F. J. Doblas-Reyes, and S. Woolnough, 2011: Using seasonal hindcasts to understand the origin of the equatorial cold tongue bias in CGCMs and its impact on ENSO. *Clim. Dyn.*, **40**, 963–981.

Vautard, R., et al., 2013: The simulation of European heat waves from an ensemble of regional climate models within the EURO-CORDEX project. *Clim. Dyn.*, doi:10.1007/s00382-013-1714-z.

Vecchi, G. A., and B. J. Soden, 2007: Global warming and the weakening of the tropical circulation. *J. Clim.*, **20**, 4316–4340.

Vecchi, G. A., K. L. Swanson, and B. J. Soden, 2008: Climate Change: Whither hurricane activity? *Science*, **322**, 687–689.

Vecchi, G. A., B. J. Soden, A. T. Wittenberg, I. M. Held, A. Leetmaa, and M. J. Harrison, 2006: Weakening of tropical Pacific atmospheric circulation due to anthropogenic forcing. *Nature*, **327**, 216–-219.

Veljovic, K., B. Rajkovic, M. J. Fennessy, E. L. Altshuler, and F. Mesinger, 2010: Regional climate modeling: Should one attempt improving on the large scales? Lateral boundary condition scheme: Any impact? *Meteorol. Z.*, **19**, 237–246.

Verlinde, J., et al., 2007: The Mixed-Phase Arctic Cloud Experiment. *Bull. Am. Meteorol. Soc.*, **88**, 205–221.

Verseghy, D. L., 2000: The Canadian Land Surface Scheme (CLASS): Its history and future. *Atmos. Ocean*, **38**, 1–13.

Vial, J., and T. J. Osborn, 2012: Assessment of atmosphere-ocean general circulation model simulations of winter northern hemisphere atmospheric blocking. *Clim. Dyn.*, **39**, 95–112.

Vial, J., J.-L. Dufresne, and S. Bony, 2013: On the interpretation of inter-model spread in CMIP5 climate sensitivity estimates. *Clim. Dyn.*, doi:10.1007/s00382-013-1725-9.

Vichi, M., S. Masina, and A. Navarra, 2007: A generalized model of pelagic biogeochemistry for the global ocean ecosystem. Part II: Numerical simulations. *J. Mar. Syst.*, **64**, 110–134.

Vichi, M., et al., 2011: Global and regional ocean carbon uptake and climate change: Sensitivity to a substantial mitigation scenario. *Clim. Dyn.*, **37**, 1929–1947.

Vizcaino, M., U. Mikolajewicz, M. Groger, E. Maier-Reimer, G. Schurgers, and A. M. E. Winguth, 2008: Long-term ice sheet-climate interactions under anthropogenic greenhouse forcing simulated with a complex Earth System Model. *Clim. Dyn.*, **31**, 665–690.

Voldoire, A., et al., 2013: The CNRM-CM5.1 global climate model : Description and basic evaluation. *Clim. Dyn.*, **40**, 2091–2121.

Volodin, E. M., 2007: Atmosphere-ocean general circulation model with the carbon cycle. *Izvestiya Atmos. Ocean. Phys.*, **43**, 298–313.

Volodin, E. M., 2008a: Relation between temperature sensitivity to doubled carbon dioxide and the distribution of clouds in current climate models. *Izvestiya Atmos. Ocean. Phys.*, **44**, 288–299.

Volodin, E. M, 2008b: Methane cycle in the INM RAS climate model. *Izvestiya Atmos. Ocean. Phys.*, **44**, 153–159.

Volodin, E. M., and V. N. Lykosov, 1998: Parametrization of heat and moisture transfer in the soil-vegetation system for use in atmospheric general circulation models: 1. Formulation and simulations based on local observational data. *Izvestiya Akad. Nauk Fizik. Atmosf. Okean.*, **34**, 453–465.

Volodin, E. M., N. A. Dianskii, and A. V. Gusev, 2010: Simulating present-day climate with the INMCM4.0 coupled model of the atmospheric and oceanic general circulations. *Izvestiya Atmos. Ocean. Phys.*, **46**, 414–431.

von Salzen, K., et al., 2013: The Canadian Fourth Generation Atmospheric Global Climate Model (CanAM4). Part I: Representation of physical processes. *Atmos. Ocean*, **51**, 104–125.

Vose, R. S., et al., 2012: NOAA'S merged land-ocean surface temperature analysis. *Bull. Am. Meteorol. Soc.*, **93**, 1677–1685.

Vrac, M., and P. Naveau, 2008: Stochastic downscaling of precipitation: From dry events to heavy rainfall *Water Resour. Res.*, **43**, W07402.

Waelbroeck, C., et al., 2009: Constraints on the magnitude and patterns of ocean cooling at the Last Glacial Maximum. *Nature Geosci.*, **2**, 127–132.

Wahl, S., M. Latif, W. Park, and N. Keenlyside, 2011: On the Tropical Atlantic SST warm bias in the Kiel Climate Model. *Clim. Dyn.*, **36**, 891–906.

Waliser, D., K. W. Seo, S. Schubert, and E. Njoku, 2007: Global water cycle agreement in the climate models assessed in the IPCC AR4. *Geophys. Res. Lett.*, **34**, L16705

Waliser, D., et al., 2009a: MJO simulation diagnostics. *J. Clim.*, **22**, 3006–3030.

Waliser, D. E., J. L. F. Li, T. S. L'Ecuyer, and W. T. Chen, 2011: The impact of precipitating ice and snow on the radiation balance in global climate models. *Geophys. Res. Lett.*, **38**, L06802.

Waliser, D. E., et al., 2003: AGCM simulations of intraseasonal variability associated with the Asian summer monsoon. *Clim. Dyn.*, **21**, 423–446.

Waliser, D. E., et al., 2009b: Cloud ice: A climate model challenge with signs and expectations of progress. *J. Geophys. Res.*, **114**, D00A21.

Walsh, K., S. Lavender, E. Scoccimarro, and H. Murakami, 2013: Resolution dependence of tropical cyclone formation in CMIP3 and finer resolution models. *Clim. Dyn.*, **40**, 585–599.

Walther, A., J.-H. Jeong, G. Nikulin, C. Jones, and D. Chen, 2013: Evaluation of the warm season diurnal cycle of precipitation over Sweden simulated by the Rossby Centre regional climate model RCA3. *Atmos. Res.*, **119**, 131–139.

Wang, B., 2006: *The Asian Monsoon.* Springer Science+Business Media, Praxis, New York, NY, USA, 787 pp.

Wang, B., and Q. H. Ding, 2008: Global monsoon: Dominant mode of annual variation in the tropics. *Dyn. Atmos. Oceans*, **44**, 165–183.

Wang, B., H. J. Kim, K. Kikuchi, and A. Kitoh, 2011a: Diagnostic metrics for evaluation of annual and diurnal cycles. *Clim. Dyn.*, **37**, 941–955.

Wang, B., H. Wan, Z. Z. Ji, X. Zhang, R. C. Yu, Y. Q. Yu, and H. T. Liu, 2004: Design of a new dynamical core for global atmospheric models based on some efficient numerical methods. *Sci. China A*, **47**, 4–21.

Wang, C., and J. Picaut, 2004: Understanding ENSO physics —A review. In: *Earth's Climate: The Ocean-Atmosphere Interaction* [C. Wang, S.-P. Xie and J.A. Carton (eds.)]. American Geophysical Union, Washington, DC, pp. 21–48.

Wang, H., and W. Su, 2013: Evaluating and understanding top of the atmosphere cloud radiative effects. In Intergovernmental Panel on Climate Change (IPCC) Fifth Assessment Report (AR5) Coupled Model Intercomparison Project Phase 5 (CMIP5) models using satellite observations. *J. Geophys. Res. Atmos.*, **118**, 683–699.

Wang, J., Q. Bao, N. Zeng, Y. Liu, G. Wu, and D. Ji, 2013: The Earth System Model FGOALS-s2: Coupling a dynamic global vegetation and terrestrial carbon model with the physical climate. *Adv. Atmos. Sci.*, doi:10.1007/s00376-013-2169-1.

Wang, J. F., and X. B. Zhang, 2008: Downscaling and projection of winter extreme daily precipitation over North America. *J. Clim.*, **21**, 923–937.

Wang, M., and J. E. Overland, 2012: A sea ice free summer Arctic within 30 years: An update from CMIP5 models. *Geophys. Res. Lett.*, **39**, L18501.

Wang, M., J. E. Overland, and N. A. Bond, 2010: Climate projections for selected large marine ecosystems. *J. Mar. Syst.*, **79**, 258–266.

Wang, Y., M. Notaro, Z. Liu, R. Gallimore, S. Levis, and J. E. Kutzbach, 2008: Detecting vegetation-precipitation feedbacks in mid-Holocene North Africa from two climate models. *Clim. Past*, **4**, 59–67.

Wang, Y. P., and B. Z. Houlton, 2009: Nitrogen constraints on terrestrial carbon uptake: Implications for the global carbon-climate feedback. *Geophys. Res. Lett.*, **36**, L24403.

Wang, Y. P., et al., 2011b: Diagnosing errors in a land surface model (CABLE) in the time and frequency domains. *J. Geophys. Res. Biogeosci.*, **116**, G01034.

Wania, R., I. Ross, and I. C. Prentice, 2009: Integrating peatlands and permafrost into a dynamic global vegetation model: 1. Evaluation and sensitivity of physical land surface processes. *Global Biogeochem. Cycles*, **23**, Gb3014.

Watanabe, M., 2008: Two regimes of the equatorial warm pool. Part I: A simple tropical climate model. *J. Clim.*, **21**, 3533–3544.

Watanabe, M., M. Chikira, Y. Imada, and M. Kimoto, 2011: Convective control of ENSO simulated in MIROC. *J. Clim.*, **24**, 543–562.

Watanabe, M., J. S. Kug, F. F. Jin, M. Collins, M. Ohba, and A. T. Wittenberg, 2012: Uncertainty in the ENSO amplitude change from the past to the future. *Geophys. Res. Lett.*, **39**, L20703.

Watanabe, M., et al., 2010: Improved climate simulation by MIROC5: Mean states, variability, and climate sensitivity. *J. Clim.*, **23**, 6312–6335.

Watanabe, S., Y. Kawatani, Y. Tomikawa, K. Miyazaki, M. Takahashi, and K. Sato, 2008: General aspects of a T213L256 middle atmosphere general circulation model. *J. Geophys. Res. Atmos.*, **113**, D12110.

Watterson, I., and P. Whetton, 2011: Distributions of decadal means of temperature and precipitation change under global warming. *J. Geophys. Res. Atmos.*, **116**, D07101.

Waugh, D., and V. Eyring, 2008: Quantitative performance metrics for stratospheric-resolving chemistry-climate models. *Atmos. Chem. Phys.*, **8**, 5699–5713.

Weaver, A. J., et al., 2001: The UVic Earth System Climate Model: Model description, climatology, and applications to past, present and future climates. *Atmos. Ocean*, **39**, 361–428.

Weaver, A. J., et al., 2012: Stability of the Atlantic meridional overturning circulation: A model intercomparison. *Geophys. Res. Lett.*, **39**, L20709.

Webb, M., C. Senior, S. Bony, and J. J. Morcrette, 2001: Combining ERBE and ISCCP data to assess clouds in the Hadley Centre, ECMWF and LMD atmospheric climate models. *Clim. Dyn.*, **17**, 905–922.

Weber, S. L., et al., 2007: The modern and glacial overturning circulation in the Atlantic Ocean in PMIP coupled model simulations. *Clim. Past*, **3**, 51–64.

Webster, P. J., A. M. Moore, J. P. Loschnigg, and R. R. Leben, 1999: Coupled ocean-atmosphere dynamics in the Indian Ocean during 1997–98. *Nature*, **401**, 356–360.

Wehner, M. F., R. L. Smith, G. Bala, and P. Duffy, 2010: The effect of horizontal resolution on simulation of very extreme US precipitation events in a global atmosphere model. *Clim. Dyn.*, **34**, 241–247.

Weller, E., and W. Cai, 2013a: Realism of the Indian Ocean Dipole in CMIP5 models: The implication for 1 climate projections. *J. Clim.*, doi:10.1175/JCLI-D-12-00807.1.

Weller, E., and W. Cai, 2013b: Asymmetry in the IOD and ENSO teleconnection in a CMIP5 model ensemble and its relevance to regional rainfall. *J. Clim.*, doi:10.1175/JCLI-D-12-00789.1.

Wentz, F. J., L. Ricciardulli, K. Hilburn, and C. Mears, 2007: How much more rain will global warming bring? *Science*, **317**, 233–235.

Westerling, A. L., H. G. Hidalgo, D. R. Cayan, and T. W. Swetnam, 2006: Warming and earlier spring increase western US forest wildfire activity. *Science*, **313**, 940–943.

Wetzel, P., E. Maier-Reimer, M. Botzet, J. H. Jungclaus, N. Keenlyside, and M. Latif, 2006: Effects of ocean biology on the penetrative radiation in a Coupled Climate Model. *J. Clim.*, **19**, 3973–3987.

Whetton, P., I. Macadam, J. Bathols, and J. O'Grady, 2007: Assessment of the use of current climate patterns to evaluate regional enhanced greenhouse response patterns of climate models. *Geophys. Res. Lett.*, **34**, L14701.

White, C. J., et al., 2013: On regional dynamical downscaling for the assessment and projection of temperature and precipitation extremes across Tasmania, Australia. *Clim. Dyn.*, doi:10.1007/s00382-013-1718-8.

Wilcox, L. J., A. J. Charlton-Perez, and L. J. Gray, 2012: Trends in Austral jet position in ensembles of high- and low-top CMIP5 models. *J. Geophys. Res.*, ***117***, D13115.

Wild, M., C. N. Long, and A. Ohmura, 2006: Evaluation of clear-sky solar fluxes in GCMs participating in AMIP and IPCC-AR4 from a surface perspective. *J. Geophys. Res. Atmos.*, **111**, D01104

Wilks, D. S., 1995: *Statistical Methods in the Atmospheric Sciences.* Vol. 59, Academic Press, San Diego, CA, USA, 467 pp.

Wilks, D. S., 2006: *Statistical Methods in the Atmospheric Sciences.* Vol. 91, Academic Press, Elsevier, San Diego, CA, USA, 627 pp.

Willett, K., P. Jones, P. Thorne, and N. Gillett, 2010: A comparison of large scale changes in surface humidity over land in observations and CMIP3 general circulation models. *Environ. Res. Lett.*, **5**, 025210.

Williams, C. J. R., R. P. Allan, and D. R. Kniveton, 2012: Diagnosing atmosphere-land feedbacks in CMIP5 climate models. *Environ. Res. Lett.*, **7**, 044003.

Williams, K., and M. Webb, 2009: A quantitative performance assessment of cloud regimes in climate models. *Clim. Dyn.*, **33** 141–157.

Williams, K. D., and M. E. Brooks, 2008: Initial tendencies of cloud regimes in the Met Office unified model. *J. Clim.*, **21**, 833–840.

Williams, K. D., W. J. Ingram, and J. M. Gregory, 2008: Time variation of effective climate sensitivity in GCMs. *J. Clim.*, **21**, 5076–5090.

Williamson, D. L., and J. G. Olson, 2007: A comparison of forecast errors in CAM2 and CAM3 at the ARM Southern Great Plains site. *J. Clim.*, **20**, 4572–4585.

Williamson, M. S., T. M. Lenton, J. G. Shepherd, and N. R. Edwards, 2006: An efficient numerical terrestrial scheme (ENTS) for Earth system modelling. *Ecol. Model.*, **198**, 362–374.

Willis, J. K., 2010: Can in situ floats and satellite altimeters detect long-term changes in Atlantic Ocean overturning? *Geophys. Res. Lett.*, **37**, L06602.

Winterfeldt, J., and R. Weisse, 2009: Assessment of value added for surface marine wind speed obtained from two regional climate models. *Mon. Weather Rev.*, **137**, 2955–2965.

Winterfeldt, J., B. Geyer, and R. Weisse, 2011: Using QuikSCAT in the added value assessment of dynamically downscaled wind speed. *Int. J. Climatol.* , **31**, 1028–1039.

Winton, M., 2000: A reformulated three-layer sea ice model. *J. Atmos. Ocean. Technol.*, **17**, 525–531.

Winton, M., 2011: Do climate models underestimate the sensitivity of Northern Hemisphere sea ice cover? *J. Clim.*, **24**, 3924–3934.

Wittenberg, A. T., 2009: Are historical records sufficient to constrain ENSO simulations? *Geophys. Res. Lett.*, **36**, L12702.

Wittenberg, A. T., A. Rosati, N. C. Lau, and J. J. Ploshay, 2006: GFDL's CM2 Global Coupled Climate Models. Part III: Tropical Pacific climate and ENSO. *J. Clim.*, **19**, 698–722.

WMO, 2011: *Scientific Assessment of Ozone Depletion: 2010*. Global Ozone Research and Monitoring Project–Report. World Meteorological Organisation, Geneva, Switzerland.

Wohlfahrt, J., et al., 2008: Evaluation of coupled ocean-atmosphere simulations of the mid-Holocene using palaeovegetation data from the northern hemisphere extratropics. *Clim. Dyn.*, **31**, 871–890.

Wood, R., et al., 2011: The VAMOS Ocean-Cloud-Atmosphere-Land Study Regional Experiment (VOCALS-REx): Goals, platforms, and field operations. *Atmos. Chem. Phys.*, **11**, 627–654.

Woollings, T., B. Hoskins, M. Blackburn, D. Hassell, and K. Hodges, 2010a: Storm track sensitivity to sea surface temperature resolution in a regional atmosphere model. *Clim. Dyn.*, **35**, 341–353.

Woollings, T., A. Charlton-Perez, S. Ineson, A. G. Marshall, and G. Masato, 2010b: Associations between stratospheric variability and tropospheric blocking. *J. Geophys. Res. Atmos.*, **115**, D06108.

Woollings, T., J. M. Gregory, J. G. Pinto, M. Reyers, and D. J. Brayshaw, 2012: Response of the North Atlantic storm track to climate change shaped by ocean-atmosphere coupling. *Nature Geosci.*, **5**, 313–317.

Wright, D. G., and T. F. Stocker, 1992: Sensitivities of a zonally averaged Global Ocean Circulation Model.. *J. Geophys. Res. Oceans*, **97**, 12707–12730.

Wu, Q. G., D. J. Karoly, and G. R. North, 2008a: Role of water vapor feedback on the amplitude of season cycle in the global mean surface air temperature. *Geophys. Res. Lett.*, **35**, L08711.

Wu, T., 2012: A mass-flux cumulus parameterization scheme for large-scale models: Description and test with observations. *Clim. Dyn.*, **38**, 725–744.

Wu, T., R. Yu, and F. Zhang, 2008b: A modified dynamic framework for the atmospheric spectral model and its application. *J. Atmos. Sci.*, **65**, 2235–2253.

Wu, T., et al., 2010a: Erratum—The Beijing Climate Center atmospheric general circulation model: Description and its performance for the present-day climate. *Clim. Dyn.*, **34**, 149–150.

Wu, T., et al., 2010b: The Beijing Climate Center atmospheric general circulation model: Description and its performance for the present-day climate. *Clim. Dyn.*, **34**, 123–147.

Wyant, M. C., C. S. Bretherton, and P. N. Blossey, 2009: Subtropical low cloud response to a warmer climate in a superparameterized climate model. Part I: Regime sorting and physical mechanisms. *J. Adv. Model. Earth Syst.*, **1**, 7.

Wyser, K., et al., 2008: An evaluation of Arctic cloud and radiation processes during the SHEBA year: Simulation results from eight Arctic regional climate models. *Clim. Dyn.*, **30**, 203–223.

Xavier, P. K., 2012: Intraseasonal convective moistening in CMIP3 models. *J. Clim.*, **25**, 2569–2577.

Xavier, P. K., J. P. Duvel, P. Braconnot, and F. J. Doblas-Reyes, 2010: An evaluation metric for intraseasonal variability and its application to CMIP3 twentieth-century simulations. *J. Clim.*, **23**, 3497–3508.

Xiao, X., et al., 1998: Transient climate change and net ecosystem production of the terrestrial biosphere. *Global Biogeochem. Cycles*, **12**, 345–360.

Xie, L., T. Z. Yan, L. J. Pietrafesa, J. M. Morrison, and T. Karl, 2005: Climatology and interannual variability of North Atlantic hurricane tracks. *J. Clim.*, **18**, 5370–5381.

Xie, P., and P. A. Arkin, 1997: Global Precipitation: A 17-year monthly analysis based on gauge observations, satellite estimates, and numerical model outputs. *Bull. Am. Meteorol. Soc.*, **78**, 2539–2558.

Xie, S., J. Boyle, S. A. Klein, X. Liu, and S. Ghan, 2008: Simulations of Arctic mixed-phase clouds in forecasts with CAM3 and AM2 for M-PACE. *J. Geophys. Res.*, **113**, D04211.

Xie, S., H.-Y. Ma, J. S. Boyle, S. A. Klein, and Y. Zhang, 2012: On the correspondence between short- and long- timescale systematic errors in CAM4/CAM5 for the years of tropical convection. *J. Clim.*, **25**, 7937–7955.

Xie, S. P., H. Annamalai, F. A. Schott, and J. P. McCreary, 2002: Structure and mechanisms of South Indian Ocean climate variability. *J. Clim.*, **15**, 864–878.

Xin, X.-G., T.-J. Zhou, and R.-C. Yu, 2008: The Arctic Oscillation in coupled climate models. *Chin. J. Geophys. Chinese Edition*, **51**, 337–351.

Xin, X., L. Zhang, J. Zhang, T. Wu, and Y. Fang, 2013: Climate change projections over East Asia with BCC_CSM1.1 climate model under RCP scenarios. *J. Meteorol. Soc. Jpn.*, **91**, 413–429.

Xin, X., T. Wu, J. Li, Z. Wang, W. Li, and F. Wu, 2012: How well does BCC_CSM1.1 reproduce the 20th century climate change over China? . *Atmos. Ocean. Sci. Lett.*, **6**, 21–26.

Xu, Y. F., Y. Huang, and Y. C. Li, 2012: Summary of recent climate change studies on the carbon and nitrogen cycles in the terrestrial ecosystem and ocean in China. *Adv. Atmos. Sci.*, **29**, 1027–1047.

Xue, Y. K., R. Vasic, Z. Janjic, F. Mesinger, and K. E. Mitchell, 2007: Assessment of dynamic downscaling of the continental US regional climate using the Eta/SSiB regional climate model. *J. Clim.*, **20**, 4172–4193.

Yakovlev, N. G., 2009: Reproduction of the large-scale state of water and sea ice in the Arctic Ocean in 1948–2002: Part I. Numerical model. *Izvestiya Atmos. Ocean. Phys.*, **45**, 628–641.

Yang, D., and T. Ohata, 2001: A bias-corrected Siberian regional precipitation climatology. *J. Hydrometeorol.*, **2**, 122–139.

Yasutaka, W., N. Masaomi, K. Sachie, and M. Chiashi, 2008: Climatological reproducibility evaluation and future climate projection of extreme precipitation events in the Baiu season using a high-resolution non-hydrostatic RCM in comparison with an AGCM. *J. Meteorol. Soc. Jpn.*, **86**, 951–967.

Yeager, S., and G. Danabasoglu, 2012: Sensitivity of Atlantic Meridional Overturning Circulation variability to parameterized Nordic Sea overflows in CCSM4. *J. Clim.*, **25**, 2077–2103.

Yeh, S. W., Y. G. Ham, and J. Y. Lee, 2012: Changes in the tropical Pacific SST trend from CMIP3 to CMIP5 and its implication of ENSO. *J. Clim.*, **25**, 7764–7771.

Yhang, Y. B., and S. Y. Hong, 2008: Improved physical processes in a regional climate model and their impact on the simulated summer monsoon circulations over East Asia. *J. Clim.*, **21**, 963–979.

Yin, X., A. Gruber, and P. Arkin, 2004: Comparison of the GPCP and CMAP merged gauge-satellite monthly precipitation products for the period 1979–2001. *J. Hydrometeorol.*, **5**, 1207–1222.

Yokohata, T., M. J. Webb, M. Collins, K. D. Williams, M. Yoshimori, J. C. Hargreaves, and J. D. Annan, 2010: Structural similarities and differences in climate responses to CO_2 increase between two perturbed physics ensembles. *J. Clim.*, **23**, 1392–1410.

Yokohata, T., J. Annan, M. Collins, C. Jackson, M. Tobis, M. Webb, and J. Hargreaves, 2012: Reliability of multi-model and structurally different single-model ensembles. *Clim. Dyn.*, **39**, 599–616.

Yokohata, T., et al., 2013: Reliability and importance of structural diversity of climate model ensembles. *Clim. Dyn.*, doi:10.1007/s00382-013-1733-9.

Yokohata, T., et al., 2008: Comparison of equilibrium and transient responses to CO_2 increase in eight state-of-the-art climate models. *Tellus A*, **60**, 946–961.

Yokoi, S., Y. N. Takayabu, and J. C. L. Chan, 2009a: Tropical cyclone genesis frequency over the western North Pacific simulated in medium-resolution coupled general circulation models. *Clim. Dyn.*, **33**, 665–683.

Yokoi, S., C. Takahashi, K. Yasunaga, and R. Shirooka, 2012: Multi-model projection of tropical cyclone genesis frequency over the western North Pacific: CMIP5 results. *Sola*, **8**, 137–140.

Yokoi, S., et al., 2011: Application of cluster analysis to climate model performance metrics. *J. Appl. Meteorol. Climatol.*, **50**, 1666–1675.

Yokoi, T., T. Tozuka, and T. Yamagata, 2009b: Seasonal variations of the Seychelles Dome simulated in the CMIP3 models. *J. Phys. Oceanogr.*, **39**, 449–457.

Yoshimori, M., T. Yokohata, and A. Abe-Ouchi, 2009: A comparison of climate feedback strength between CO_2 doubling and LGM experiments. *J. Clim.*, **22**, 3374–3395.

Yoshimori, M., J. C. Hargreaves, J. D. Annan, T. Yokohata, and A. Abe-Ouchi, 2011: Dependency of feedbacks on forcing and climate state in physics parameter ensembles. *J. Clim.*, **24**, 6440–6455.

Young, P. J., et al., 2013: Pre-industrial to end 21st century projections of tropospheric ozone from the Atmospheric Chemistry and Climate Model Intercomparison Project (ACCMIP). *Atmos. Chem. Phys.*, **13**, 2063–2090.

Yu, B., and F. W. Zwiers, 2010: Changes in equatorial atmospheric zonal circulations in recent decades. *Geophys. Res. Lett.*, **37**, L05701.

Yu, J.-Y., and S. T. Kim, 2011: Reversed spatial asymmetries between El Niño and La Niña and their linkage to decadal ENSO modulation in CMIP3 models. *J. Clim.*, **24**, 5423–5434.

Yu, W., M. Doutriaux, G. Seze, H. LeTreut, and M. Desbois, 1996: A methodology study of the validation of clouds in GCMs using ISCCP satellite observations. *Clim. Dyn.*, **12**, 389–401.

Yukimoto, S., et al., 2011: Meteorological Research Institute-Earth System Model v1 (MRI-ESM1)—Model Description. Technical Report of MRI. Ibaraki, Japan, 88 pp.

Yukimoto, S., et al., 2012: A new global climate model of the Meteorological Research Institute: MRI-CGCM3–Model description and basic performance. *J. Meteorol. Soc. Jpn.*, **90A**, 23–64.

Zaehle, S., and D. Dalmonech, 2011: Carbon-nitrogen interactions on land at global scales: Current understanding in modelling climate biosphere feedbacks. *Curr. Opin. Environ. Sustain.*, **3**, 311–320.

Zaehle, S., P. Friedlingstein, and A. D. Friend, 2010: Terrestrial nitrogen feedbacks may accelerate future climate change. *Geophys. Res. Lett.*, **37**, L01401.

Zahn, M., and H. von Storch, 2008: A long-term climatology of North Atlantic polar lows. *Geophys. Res. Lett.*, **35**, L22702.

Zalesny, V. B., et al., 2010: Numerical simulation of large-scale ocean circulation based on the multicomponent splitting method. *Russ. J. Numer. Anal. Math. Model.*, **25**, 581–609.

Zaliapin, I., and M. Ghil, 2010: Another look at climate sensitivity. *Nonlin. Proc. Geophys.*, **17**, 113–122.

Zappa, G., L. C. Shaffrey, and K. I. Hodges, 2013: The ability of CMIP5 models to simulate North Atlantic extratropical cyclones. *J. Clim.*, doi:10.1175/JCLI-D-12-00501.1.

Zeng, N., 2003: Glacial-interglacial atmospheric CO_2 change—The glacial burial hypothesis. *Adv. Atmos. Sci.*, **20**, 677–693.

Zeng, N., 2006: Quasi-100ky glacial-interglacial cycles triggered by subglacial burial carbon release. *Clim. Past*, **2**, 371–397.

Zeng, N., J. D. Neelin, and C. Chou, 2000: A quasi-equilibrium tropical circulation model—Implementation and simulation. *J. Atmos. Sci.*, **57**, 1767–1796.

Zeng, N., A. Mariotti, and P. Wetzel, 2005: Terrestrial mechanisms of interannual CO_2 variability. *Global Biogeochem. Cycles*, **19**, GB1016.

Zeng, N., H. F. Qian, E. Munoz, and R. Iacono, 2004: How strong is carbon cycle-climate feedback under global warming? *Geophys. Res. Lett.*, **31**, L20203.

Zhang, F., and A. Georgakakos, 2011: Joint variable spatial downscaling. *Clim. Change*, **111**, 945–972

Zhang, J., and J. S. Reid, 2010: A decadal regional and global trend analysis of the aerosol optical depth using a data-assimilation grade over-water MODIS and Level 2 MISR aerosol products. *Atmos. Chem. Phys.*, **10**, 10949–10963.

Zhang, J., J. S. Reid, D. L. Westphal, N. L. Baker, and E. J. Hyer, 2008a: A system for operational aerosol optical depth data assimilation over global oceans. *J. Geophys. Res. Atmos.*, **113**, D10208.

Zhang, Q., H. S. Sundqvist, A. Moberg, H. Kornich, J. Nilsson, and K. Holmgren, 2010a: Climate change between the mid and late Holocene in northern high latitudes—Part 2: Model-data comparisons. *Clim. Past*, **6**, 609–626.

Zhang, R., et al., 2013: Have aerosols caused the observed Atlantic multidecadal variability? *J. Atmos. Sci.*, doi:10.1175/JAS-D-12-0331.1.

Zhang, W., and F.-F. Jin, 2012: Improvements in the CMIP5 simulations of ENSO-SSTA meridional width. *Geophys. Res. Lett.*, **39**, L23704.

Zhang, X., 2007: A comparison of explicit and implicit spatial downscaling of GCM output for soil erosion and crop production assessments. *Clim. Change*, **84**, 337–363.

Zhang, X., 2010: Sensitivity of arctic summer sea ice coverage to global warming forcing: towards reducing uncertainty in arctic climate change projections. *Tellus A*, **62**, 220–227.

Zhang, X., et al., 2007: Detection of human influence on twentieth-century precipitation trends. *Nature*, **448**, 461–465.

Zhang, Y., S. A. Klein, J. Boyle, and G. G. Mace, 2010b: Evaluation of tropical cloud and precipitation statistics of Community Atmosphere Model version 3 using CloudSat and CALIPSO data. *J. Geophys. Res.*, **115**, D12205.

Zhang, Y., et al., 2008b: On the diurnal cycle of deep convection, high-level cloud, and upper troposphere water vapor in the Multiscale Modeling Framework. *J. Geophys. Res.*, **113**, D16105.

Zhao, M., I. M. Held, and S.-J. Lin, 2012: Some counterintuitive dependencies of tropical cyclone frequency on parameters in a GCM. *J. Atmos. Sci.*, **69**, 2272–2283.

Zhao, M., I. M. Held, S. J. Lin, and G. A. Vecchi, 2009: Simulations of global hurricane climatology, interannual variability, and response to global warming using a 50-km resolution GCM. *J. Clim.*, **22**, 6653-6678.

Zhao, T. L., et al., 2008: A three-dimensional model study on the production of BrO and Arctic boundary layer ozone depletion. *J. Geophys. Res.*, **113**, D24304.

Zhao, Y., and S. P. Harrison, 2012: Mid-Holocene monsoons: A multi-model analysis of the inter-hemispheric differences in the responses to orbital forcing and ocean feedbacks. *Clim. Dyn.*, **39**, 1457-1487.

Zheng, W. P., and P. Braconnot, 2013: Characterization of model spread in PMIP2 Mid-Holocene simulations of the African monsoon. *J. Clim.*, **26**, 1192-1210.

Zheng, X.-T., S.-P. Xie, and Q. Liu, 2011: Response of the Indian Ocean basin mode and its capacitor effect to global warming. *J. Clim.*, **24**, 6146-6164.

Zheng, Y., J.-L. Lin, and T. Shinoda, 2012: The equatorial Pacific cold tongue simulated by IPCC AR4 coupled GCMs: Upper ocean heat budget and feedback analysis. *J. Geophys. Res. Oceans*, **117**, C05024.

Zickfeld, K., et al., 2013: Long-term climate change commitment and reversibility: An EMIC intercomparison. *J. Clim.*, doi:10.1175/JCLI-D-12-00584.1.

Zou, L., and T. Zhou, 2013: Can a regional ocean-atmosphere coupled model improve the simulation of the interannual variability of the western North Pacific summer monsoon? *J. Clim.*, **26**, 2353–2367.

Zunz, V., H. Goosse, and F. Massonnet, 2013: How does internal variability influence the ability of CMIP5 models to reproduce the recent trend in Southern Ocean sea ice extent? *Cryosphere*, **7**, 451-468.

Zwiers, F. W., X. Zhang, and Y. Feng, 2011: Anthropogenic influence on long return period daily temperature extremes at regional scales. *J. Clim.*, **24**, 881-892.

Appendix 9.A: Climate Models Assessed in Chapter 9

Table 9.A.1 | Salient features of the Atmosphere–Ocean General Circulation Models (AOGCMs) and Earth System Models (ESMs) participating in CMIP5 (see also Table 9.1). Column 1: Official CMIP5 model name along with the calendar year ('vintage') of the first publication for each model; Column 2: sponsoring institution(s), main reference(s); subsequent columns for each of the model components, with names and main component reference(s). In addition, there are standard entries for the atmosphere component: horizontal grid resolution, number of vertical levels, grid top (low or high top); and for the ocean component: horizontal grid resolution, number of vertical levels, top level, vertical coordinate type, ocean free surface type ('Top BC'). This table information was initially extracted from the CMIP5 online questionnaire (http://q.cmip5.ceda.ac.uk/) as of January 2013. A blank entry indicates that information was not available.

(1) Model Name (2) Vintage	**(1) Institution** (2) Main Reference(s)	**Atmosphere** (1) Component Name (2) Horizontal Grid (3) Number of Vert Levels (4) Grid Top (5) References	**Aerosol** (1) Component Name or type (2) References	**Atmos Chemistry** (1) Component Name (2) References	**Land Surface** (1) Component Name (2) References	**Ocean** (1) Component Name (2) Horizontal Resolution (3) Number of Vertical Levels (4) Top Level (5) z Co-ord (6) Top BC (7) References	**Ocean Biogeochemistry** (1) Component Name (2) References	**Sea Ice** (1) Component Name (2) References
(1) ACCESS1.0 (2) 2011	(1) Commonwealth Scientific and Industrial Research Organization (CSIRO) and Bureau of Meteorology (BOM), Australia (2) (Bi et al., 2013b; Dix et al., 2013)	(1) Included (as in HadGEM2 (r1.1)) (2) 192 × 145 N96 (3) 38 (4) 39,255 m (5) (Martin et al., 2011; Bi et al., 2013b; Rashid et al., 2013)	(1) CLASSIC (2) (Bellouin et al., 2011; Dix et al., 2013)	Not implemented	(1) MOSES2.2 (2) (Cox et al., 1999; Essery et al., 2003; Kowalczyk et al., 2013)	(1) ACCESS-OM (MOM4p1) (2) primarily 1° latitude/longitude tripolar with enhanced resolution near equator and at high latitudes (3) 50 (4) 0–10 m (5) z* (6) nonlinear split-explicit (7) (Bi et al., 2013a; Marsland et al., 2013)	Not implemented	(1) CICE4.1 (2) (Uotila et al., 2012; Bi et al., 2013a; Uotila et al., 2013)
(1) ACCESS1.3 (2) 2011	(1) Commonwealth Scientific and Industrial Research Organization (CSIRO) and Bureau of Meteorology (BOM), Australia (2) (Bi et al., 2013b; Dix et al., 2013)	(1) Included (similar to UK Met Office Global Atmosphere 1.0) (2) 192 × 145 N96 (3) 38 (4) 39,255 m (5) (Hewitt et al., 2011; Bi et al., 2013b; Rashid et al., 2013)	(1) CLASSIC (2) (Bellouin et al., 2011; Dix et al., 2013)	Not implemented	(1) CABLE (2) (Kowalczyk et al., 2006; Wang et al., 2011b; Kowalczyk et al., 2013)	(1) ACCESS-OM (MOM4p1) (2) primarily 1° latitude/longitude tripolar with enhanced resolution near equator and at high latitudes (3) 50 (4) 0–10 m (5) z* (6) nonlinear split-explicit (7) (Bi et al., 2013a; Marsland et al., 2013)	Not implemented	(1) CICE4.1 (2) (Uotila et al., 2012; Bi et al., 2013a; Uotila et al., 2013)
(1) BCC-CSM1.1 (2) 2011	(1) Beijing Climate Center, China Meteorological Administration (2) (Wu, 2012; Xin et al., 2012; Xin et al., 2013)	(1) BCC_AGCM2.1 (2) T42 T42L26 (3) 26 (4) 2.917 hPa (5) (Wu et al., 2008b; Wu et al., 2010b, 2010a; Wu, 2012)	Prescribed	Not implemented	(1) BCC-AVIM1.0 (2) (Ji, 1995; Lu and Ji, 2006; Ji et al., 2008; Wu, 2012)	(1) MOM4-L40 (2) 1° with enhanced resolution in the meridional direction in the tropics (1/3° meridional resolution at the equator) tripolar (3) 40 (4) 25 m (5) z (6) linear split-explicit (7) (Griffies et al., 2005)	(1) Included (2) Based on the protocols from the Ocean Carbon Cycle Model Intercomparison Project–Phase 2 (OCMIP2, http://www.ipsl.jussieu.fr/OCMIP/ phase2/)	(1) GFDL Sea Ice Simulator (SIS) (2) (Winton, 2000)
(1) BCC-CSM1.1(m) (2) 2011	(1) Beijing Climate Center, China Meteorological Administration (2) (Wu, 2012; Xin et al., 2012; Xin et al., 2013)	(1) BCC_AGCM2.1 (2) T106 (3) 26 (4) 2.917 hPa (5) (Wu et al., 2008b; Wu et al., 2010b, 2010a; Wu, 2012)	Prescribed	Not implemented	(1) BCC-AVIM1.0 (2) (Ji, 1995; Lu and Ji, 2006; Ji et al., 2008; Wu, 2012)	(1) MOM4-L40 (2) Tri-polar: 1° with enhanced resolution in the meridional direction in the tropics (1/3° meridional resolution at the equator) (3) 40 (4) 25 m (5) z (6) implicit (7) (Griffies et al., 2005)	(1) Included (2) Based on the protocols from the Ocean Carbon Cycle Model Intercomparison Project–Phase 2 (OCMIP2, http://www.ipsl.jussieu.fr/OCMIP/ phase2/)	(1) GFDL Sea Ice Simulator (SIS) (2) (Winton, 2000)

(continued on next page)

Table 9.A.1 (continued)

(1) Model Name (2) Vintage	**(1) Institution** (2) Main Reference(s)	**Atmosphere** (1) Component Name (2) Horizontal Grid (3) Number of Vert Levels (4) Grid Top (5) References	**Aerosol** (1) Component Name or type (2) References	**Atmos Chemistry** (1) Component Name (2) References	**Land Surface** (1) Component Name (2) References	**Ocean** (1) Component Name (2) Horizontal Resolution (3) Number of Vertical Levels (4) Top Level (5) z Co-ord (6) Top BC (7) References	**Ocean Biogeo-chemistry** (1) Component Name (2) References	**Sea Ice** (1) Component Name (2) References
(1) BNU-ESM (2) 2011	(1) Beijing Normal University (2)	(1) CAM3.5 (2) T42 (3) 26 (4) 2.194 hPa	Semi-interactive	Not implemented	(1)CoLM+B-NUDGVM(C/N) (2) (Dai et al., 2003; Dai et al., 2004)	(1) MOM4p1 (2) 200(lat) × 360(lon) (3) 50	IBGC	CICE4.1
(1) CanCM4 (2) 2010	(1) Canadian Center for Climate Modelling and Analysis (2) (von Salzen et al., 2013)	(1) Included (2) Spectral T63 (3) 35 levels (4) 0.5 hPa (5) (von Salzen et al., 2013)	(1) Interactive (2) (Lohmann et al., 1999; Croft et al., 2005; von Salzen et al., 2013)	(1) Included (2) (von Salzen et al., 2013)	(1) CLASS 2.7 (2) (Verseghy, 2000; von Salzen et al., 2013)	(1) Included (2) 256 × 192 (3) 40 (4) 0 m (5) depth (6) rigid lid (7) (Merryfield et al., 2013)	Not implemented	(1) Included (2) (Merryfield et al., 2013)
(1) CanESM2 (2) 2010	(1) Canadian Center for Climate Modelling and Analysis (2) (Arora et al., 2011; von Salzen et al., 2013)	(1) Included (2) Spectral T63 (3) 35 levels (4) 0.5 hPa (5) (von Salzen et al., 2013)	(1) Interactive (2) (Lohmann et al., 1999; Croft et al., 2005; von Salzen et al., 2013)	(1) Included (2) (von Salzen et al., 2013)	(1) CLASS 2.7; CTEM (2) (Verseghy, 2000) (Arora et al., 2009; von Salzen et al., 2013)	(1) Included (2) 256 × 192 (3) 40 (4) 0 m (5) depth (6) rigid lid (7) (Merryfield et al., 2013)	(1) CMOC (2) (Arora et al., 2009; Christian et al., 2010)	(1) Included (2) (Merryfield et al., 2013)
(1) CCSM4 (2) 2010	(1) US National Centre for Atmospheric Research (2) (Gent et al., 2011)	(1) CAM4 (2) 0.9° ×1.25° (3) 27 (4) 2.194067 hPa (5) (Neale et al., 2010; Neale et al., 2013)	(1) Interactive (2) (Neale et al., 2010; Oleson et al., 2010; Holland et al., 2012)	Not implemented	(1) Community Land Model 4 (CLM4) (2) (Oleson et al., 2010; Lawrence et al., 2011; Lawrence et al., 2012)	(1) POP2 with modifications (2) Nominal 1° (1.125° in longitude, 0.27–0.64° variable in latitude) (3) 60 (4) 10 m thick with sur-face variables at 5 m (5) depth (level) (6) linearized, implicit free surface with constant-volume ocean (7) (Danabasoglu et al., 2012)	Not implemented	(1) CICE4 with modifications (2) (Hunke and Lipscomb, 2008; Holland et al., 2012)
(1) CESM1(BGC) (2) 2010	(1) NSF-DOE-NCAR (2) (Long et al., 2012; Hurrell et al., 2013)	(1) CAM4 (2) 0.9° ×1.25° (3) 27 (4) 2.194067 hPa (5) (Neale et al., 2010; Neale et al., 2013)	(1) Semi-interactive (2) (Neale et al., 2010; Oleson et al., 2010; Holland et al., 2012)	Not implemented	(1) CLM4 (2) (Oleson et al., 2010; Lawrence et al., 2011; Lawrence et al., 2012)	(1) POP2 with modifications (2) Nominal 1° (1.125° in longitude, 0.27–0.64° variable in latitude) (3) 60 (4) 10 m with surface variables at 5 m (5) depth (level) (6) linearized, implicit free surface with constant-volume ocean (7) (Danabasoglu et al., 2012)	(1) Biogeochemical Elemental Cycling (BEC)	(1) CICE4 with modifications (2) (Hunke and Lipscomb, 2008; Holland et al., 2012)

(continued on next page)

9

Table 9.A.1 (continued)

(1) Model Name (2) Vintage	**(1) Institution** (2) Main Reference(s)	**Atmosphere** (1) Component Name (2) Horizontal Grid (3) Number of Vert Levels (4) Grid Top (5) References	**Aerosol** (1) Component Name or type (2) References	**Atmos Chemistry** (1) Component Name (2) References	**Land Surface** (1) Component Name (2) References	**Ocean** (1) Component Name (2) Horizontal Resolution (3) Number of Vertical Levels (4) Top Level (5) z Co-ord (6) Top BC (7) References	**Ocean Biogeo-chemistry** (1) Component Name (2) References	**Sea Ice** (1) Component Name (2) References
(1) CESM1(CAM5) (2) 2010	(1) NSF-DOE-NCAR (2) (Hurrell et al., 2013)	(1) Community Atmosphere Model 5 (CAM5) (2) 0.9° × 1.25° (3) 27 (4) 2.194067 hPa (5) (Neale et al., 2010; Neale et al., 2013)	(1) Semi-interactive (2) (Neale et al., 2010; Oleson et al., 2010; Holland et al., 2012)	Not implemented	(1) CLM4 (2) (Oleson et al., 2010) (Lawrence et al., 2011; Lawrence et al., 2012)	Same as CESM1 (BGC)	Not implemented	(1) CICE4 with modifications (2) (Hunke and Lipscomb, 2008; Holland et al., 2012)
(1) CESM1(CAM5.1.FV2) (2) 2012	(1) NSF-DOE-NCAR (2) (Hurrell et al., 2013)	(1) Community Atmosphere Model (CAM5.1) (2) 1.9° × 2.0° (3) 30 (4) 10 hPa (5) (Neale et al., 2013)	(1) Modal Aerosol Module (MAM3) (2) (Ghan et al., 2012; Liu et al., 2012b)	Not implemented	(1) Community Land Model (CLM4) (2) (Oleson et al., 2010; Lawrence et al., 2011)	Same as CESM1 (BGC)	Not implemented	(1) CICE4 with modifications (2) (Hunke and Lipscomb, 2008; Holland et al., 2012)
(1) CESM1(WACCM) (2) 2010	(1) NSF-DOE-NCAR (2) (Hurrell et al., 2013)	(1) WACCM4 (2) 1.9° × 2.5° (3) 66 (4) 5.1 × 10^{-6} hPa	Semi-interactive	Included	(1) CLM4 (2) (Oleson et al., 2010; Lawrence et al., 2011; Lawrence et al., 2012)	Same as CESM1 (BGC)	Not implemented	(1) CICE4 with modifications (2) (Hunke and Lipscomb, 2008; Holland et al., 2012)
(1) CESM1(FASTCHEM) (2) 2010	(1) NSF-DOE-NCAR (2) (Cameron-Smith et al., 2006; Eyring et al., 2013; Hurrell et al., 2013)	(1) Included, CAM4-CHEM (2) 0.9° × 1.25° (3) 27 (4) 2.194067 hPa (5) (Neale et al., 2010; Lamarque et al., 2012; Neale et al., 2013)	(1) Interactive (2) (Neale et al., 2010; Oleson et al., 2010; Holland et al., 2012; Lamarque et al., 2012)	(1) Included, CAM-CHEM (2) (Lamarque et al., 2012)	(1) Community Land Model 4 (CLM4) (2) (Oleson et al., 2010; Lawrence et al., 2011; Lawrence et al., 2012)	Same as CESM1 (BGC)	Not implemented	(1) CICE4 with modifications (2) (Hunke and Lipscomb, 2008; Holland et al., 2012)
(1) CMCC-CESM (2) 2009	(1) Centro Euro-Mediterraneo per I Cambiamenti Climatici (2) (Fogli et al., 2009; Vichi et al., 2011)	(1) ECHAM5 (2) 3.75° × 3.75° (T31) (3) 39 (4) 0.01 hPa (5) (Roeckner et al., 2006; Manzini et al., 2012)	Semi-interactive	Not implemented	(1) SILVA (2) (Alessandri et al., 2012)	Same as CMCC-CM	(1) PELAGOS (2) (Vichi et al., 2007)	(1) LIM2 (2) (Timmermann et al., 2005)
(1) CMCC-CM (2) 2009	(1) Centro Euro-Mediterraneo per I Cambiamenti Climatici (2) (Fogli et al., 2009; Scoccimarro et al., 2011)	(1) ECHAM5 (2) 0.75° × 0.75° (T159) (3) 31 (4) 10 hPa (5) (Roeckner et al., 2006)	Semi-interactive	Not implemented	Not implemented	(1) OPA8.2 (2) 2° average, 0.5° at the equator (ORCA2) (3) 31 (4) 5 m (5) depth (z-level) (6) linear implicit (7) (Madec et al., 1998)	Not implemented	(1) LIM2 (2) (Timmermann et al., 2005)
(1) CMCC-CMS (2) 2009	(1) Centro Euro-Mediterraneo per I Cambiamenti Climatici (2) (Fogli et al., 2009)	(1) ECHAM5 (2) 1.875° × 1.875° (T63) (3) 95 (4) 0.01 hPa (5) (Roeckner et al., 2006; Manzini et al., 2012)	Semi-interactive	Not implemented	Not implemented	Same as CMCC-CM	Not implemented	(1) LIM2 (2) (Timmermann et al., 2005)

(continued on next page)

Table 9.A.1 (continued)

(1) Model Name (2) Vintage	**(1) Institution** (2) Main Reference(s)	**Atmosphere** (1) Component Name (2) Horizontal Grid (3) Number of Vert Levels (4) Grid Top (5) References	**Aerosol** (1) Component Name or type (2) References	**Atmos Chemistry** (1) Component Name (2) References	**Land Surface** (1) Component Name (2) References	**Ocean** (1) Component Name (2) Horizontal Resolution (3) Number of Vertical Levels (4) Top Level (5) z Co-ord (6) Top BC (7) References	**Ocean Biogeo-chemistry** (1) Component Name (2) References	**Sea Ice** (1) Component Name (2) References
(1) CNRM-CM5[1] (2) 2010	(1) Centre National de Recherches Meteorologiques and Centre Europeen de Recherche et Formation Avancees en Calcul Scientifique. (2) (Voldoire et al., 2013)	(1) ARPEGE-Climat (2) TL127 (3) 31 (4) 10 hPa (5) (Déqué et al., 1994; Voldoire et al., 2013)	Prescribed	(1) (3-D linear ozone chemistry model) (2) (Cariolle and Teyssedre, 2007)	(1) SURFEX (Land and Ocean Surface) (2) (Voldoire et al., 2013)	(1) NEMO (2) 0.7° on average ORCA1 (3) 42 (4) 5 m (5) z-coordinate (6) linear filtered (7) (Madec, 2008)	(1) PISCES (2) (Aumont and Bopp, 2006; Séférian et al., 2013)	(1) Gelato5 (Sea Ice) (2) (Salas-Melia, 2002; Voldoire et al., 2013)
(1) CSIRO-Mk3.6.0 (2) 2009	(1) Queensland Climate Change Centre of Excellence and Commonwealth Scientific and Industrial Research Organisation (2) (Rotstayn et al., 2012)	(1) Included (2) ~1.875° × 1.875° (spectral T63) (3) 18 (4) ~4.5 hPa (5) (Gordon et al., 2002; Gordon et al., 2010; Rotstayn et al., 2012)	(1) Interactive (2) (Rotstayn and Lohmann, 2002; Rotstayn et al., 2011; Rotstayn et al., 2012)	Not implemented	(1) Included (2) (Gordon et al., 2002; Gordon et al., 2010)	(1) Modified MOM2.2 (2) ~0.9 × 1.875 (3) 31 (4) 5 m (5) depth (6) rigid lid (7) (Gordon et al., 2002; Gordon et al., 2010)	Not implemented	(1) Included (2) (O'Farrell, 1998; Gordon et al., 2010)
(1) EC-EARTH (2) 2010	(1) Europe (2) (Hazeleger et al., 2012)	(1) IFS c31r1 (2) 1.125° longitudinal spacing, Gaussian grid T159L62 (3) 62 (4) 1 hPa (5) (Hazeleger et al., 2012)	Prescribed	Not implemented	(1) HTESSEL (2) (Balsamo et al., 2009)	(1) NEMO_ecmwf (2) The grid is a tripolar curvilinear grid with a 1° resolution. ORCA1 (3) 31 (4) 1 m (5) z (6) free surface linear filtered (7) (Hazeleger et al., 2012)	Not implemented	(1) LIM2 (2) (Fichefet and Maqueda, 1999)
(1) FGOALS-g2 (2) 2011	(1) LASG (Institute of Atmospheric Physics)-CESS(Tsinghua University) (2) (Li et al., 2012b)	(1) GAMIL2 (2) 2.8125° × 2.8125° (3) 26 layers (4) 2.194 hPa (5) (Wang et al., 2004; Li et al., 2013b)	Semi-interactive	Not implemented	(1) CLM3 (2) (Oleson et al., 2010)	(1) LICOM2 (2) 1 × 1° with 0.5 meridional degree in the tropical region (3) 30 (4) 10 m (5) eta co-ordinate (6) (7) (Liu et al., 2012a)	Not implemented	(1) CICE4-LASG (2) (Wang and Houlton, 2009; Liu, 2010)
(1) FGOALS-s2 (2) 2011	(1) The State Key Laboratory of Numerical Modeling for Atmospheric Sciences and Geophysical Fluid Dynamics, The Institute of Atmospheric Physics (2) (Bao et al., 2010; Bao et al., 2013)	(1) SAMIL2.4.7 (2)R42 (2.81° × 1.66°) (3) 26 (4) 2.19hPa (5) (Bao et al., 2010; Liu et al., 2013b)	Semi-interactive	Not implemented	(1) CLM3.0 (2) (Oleson, 2004; Zeng et al., 2005; Wang et al., 2013)	(1) LICOM (2) The zonal resolution is 1°. The meridional resolution is 0.5° between 10°S and 10°N and increases from 0.5° to 1° from 10° (3) 30 layers (4) 10 m (for vertical velocity and pressure) and 5 meter (for Temperature and salinity, zonal and meridional velocity) (5) depth (6) linear split-explicit (7) (Lin et al., 2013)	(1) IAP-OBM (2) (Xu et al., 2012)	(1) CSIM5 (2) (Briegleb et al., 2004)

(continued on next page)

[1] A CNRM-CM5-2 version exists that only differs from CNRM-CM5 in the treatment of volcanoes

Table 9.A.1 (continued)

(1) Model Name (2) Vintage	**(1) Institution** (2) Main Reference(s)	**Atmosphere** (1) Component Name (2) Horizontal Grid (3) Number of Vert Levels (4) Grid Top (5) References	**Aerosol** (1) Component Name or type (2) References	**Atmos Chemistry** (1) Component Name (2) References	**Land Surface** (1) Component Name (2) References	**Ocean** (1) Component Name (2) Horizontal Resolution (3) Number of Vertical Levels (4) Top Level (5) z Co-ord (6) Top BC (7) References	**Ocean Biogeo-chemistry** (1) Component Name (2) References	**Sea Ice** (1) Component Name (2) References
(1) FIO-ESM v1.0 (2) 2011	(1) The First Institute of Oceanography, State Oceanic Administration, China	(1) CAM3.0 (2) T42 (3) 26 (4) 3.545 hPa (5) (Collins et al., 2006c)	Prescribed	Not implemented	(1) CLM3.5 (2) (Oleson et al., 2008b)	(1) Modified POP2.0 through incorporating the non-breaking surface wave-induced mixing (2) 1.125° in longitude, 0.27–0.64° variable in latitude (3) 40 (4) 10 m with surface variables at 5 m (5) depth (6) linear implicit (7) (Huang et al., 2012)	(1) Improved OCMIP-2 biogeochemical model (2) (Bao et al., 2012)	(1) CICE4.0 (2) (Hunke and Lipscomb, 2008)
(1) GFDL-CM2.1 (2) 2006	(2) (Qiao et al., 2004; Song et al., 2012)	(1) Included (2) 2.5° longitude, 2° latitude M45L24 (3) 24 (4) midpoint of top box is 3.65 hPa (5) (Delworth et al., 2006)	Semi-interactive	Not implemented	Included	(1) Included (2) 1° tripolar360 × 200L50 (3) 50 (4) 0 m (5) depth (6) nonlinear split-explicit (7)	Not implemented	(1) SIS (2) (Winton, 2000; Delworth et al., 2006)
(1) GFDL-CM3 (2) 2011	(1) NOAA Geophysical Fluid Dynamics Laboratory (2) (Delworth et al., 2006; Donner et al., 2011)	(1) Included (2) ~200 km C48L48 (3) 48 (4) 0.01 hPa (5) (Donner et al., 2011)	(1) Interactive (2) (Levy et al., 2013)	(1) Atmospheric Chemistry (2) (Horowitz et al., 2003; Austin and Wilson, 2006; Sander, 2006)	(1) Included (2) (Milly and Shmakin, 2002; Shevliakova et al., 2009)	(1) MOM4.1 (2) 1° tripolar360 × 200L50 (3) 50 (4) 0 m (5) z* (6) non-linear split-explicit (7) (Griffies and Greatbatch, 2012)	Not implemented	(1) SIS (2) (Griffies and Greatbatch, 2012)
(1) GFDL-ESM2G (2) 2012	(1) NOAA Geophysical Fluid Dynamics Laboratory (2) (Dunne et al., 2012; Dunne et al., 2013)	(1) Included (2) 2.5° longitude, 2° latitude M45L24 (3) 24 (4) midpoint of top box is 3.65 hPa (5) (Delworth et al., 2006)	Semi-interactive	Not implemented	(1) Included (2) (Milly and Shmakin, 2002; Shevliakova et al., 2009; Donner et al., 2011)	(1) GOLD (2) 1° tripolar 360 × 2 10L63 (3) 63 (4) 0 m (5) Isopycnic (6) nonlinear split-explicit (7) (Hallberg and Adcroft, 2009; Dunne et al., 2012)	(1) TOPAZ (2) (Henson et al., 2009; Dunne et al., 2013)	(1) SIS (2) (Winton, 2000; Delworth et al., 2006)
(1) GFDL-ESM2M (2) 2011	(1) NOAA Geophysical Fluid Dynamics Laboratory (2) (Dunne et al., 2012; Dunne et al., 2013)	(1) Included (2) 2.5° longitude, 2° latitude M45L24 (3) 24 (4) midpoint of top box is 3.65 hPa (5) (Delworth et al., 2006)	Semi-interactive	Not implemented	(1) Included (2) (Milly and Shmakin, 2002; Shevliakova et al., 2009; Donner et al., 2011)	(1) MOM4.1 (2) 1° tripolar 360 × 200L50 (3) 50 (4) 0 m (5) z* (6) nonlinear split-explicit (7) (Griffies, 2009; Dunne et al., 2012)	(1) TOPAZ (2) (Henson et al., 2009; Dunne et al., 2013)	(1) SIS (2) (Winton, 2000; Delworth et al., 2006)
(1) GFDL-HIRAM-C180 (2) 2011	(1) NOAA Geophysical Fluid Dynamics Laboratory (2) (Delworth et al., 2006; Donner et al., 2011)	(1) Included (2) Averaged cell size: approximately 50 × 50 km. C180L32 (3) 32 (4) 2.164 hPa (5) (Donner et al., 2011)	Prescribed	Not implemented	(1) Included (2) (Milly and Shmakin, 2002; Shevliakova et al., 2009)	Not implemented	Not implemented	Not implemented

(continued on next page)

Table 9.A.1 (continued)

(1) Model Name (2) Vintage	**(1) Institution** (2) Main Reference(s)	**Atmosphere** (1) Component Name (2) Horizontal Grid (3) Number of Vert Levels (4) Grid Top (5) References	**Aerosol** (1) Component Name or type (2) References	**Atmos Chemistry** (1) Component Name (2) References	**Land Surface** (1) Component Name (2) References	**Ocean** (1) Component Name (2) Horizontal Resolution (3) Number of Vertical Levels (4) Top Level (5) z Co-ord (6) Top BC (7) References	**Ocean Biogeo-chemistry** (1) Component Name (2) References	**Sea Ice** (1) Component Name (2) References
(1) GFDL-HIRAM-C360 (2)	(1) NOAA Geophysical Fluid Dynamics Laboratory (2) (Delworth et al., 2006; Donner et al., 2011)	(1) Included (2) Averaged cell size: approximately 25 × 25 km. C360L32 (3) 32 (4) 2.164 hPa (5) (Donner et al., 2011)	Prescribed	Not implemented	(1) Included (2) (Milly and Shmakin, 2002; Shevliakova et al., 2009)	Not implemented	Not implemented	Not implemented
(1) GISS-E2-H (2) 2004	(1) NASA Goddard Institute for Space Studies USA (2) (Schmidt et al., 2006) *Note: all GISS models come in three flavours: p1 = non-interactive composition, p2= interactive composition, p3 = interactive composition + interactive AIE*	(1) Included (2) 2° latitude × 2.5°longitude F (3) 40 (4) 0.1 hPa	(1) Interactive (2) (Bauer et al., 2007; Tsigaridis and Kanakidou, 2007; Menon et al., 2010; Koch et al., 2011) *Note: Aerosol is "fully interactive" for p2 and p3, "semi interactive" for p1*	(1) G-PUCCINI (2) (Shindell et al., 2013a) *Note: Atmos Chem is "fully interactive" for p2 and p3, "semi interactive" for p1*	Included	(1) HYCOM Ocean (2) 0.2 to 1° latitude × 1° longitude HYCOM (3) 26 (4) 0 m (5) hybrid z isopycnic (6) nonlinear split-explicit (7)	Not implemented	Included
(1) GISS-E2-H-CC (2) 2011	(1) NASA Goddard Institute for Space Studies USA (2) (Schmidt et al., 2006) *Note: p1 only*	(1) Included (2) Nominally 1° (3) 40 (4) 0.1 hPa	(1) Interactive (p1 only) (2) (Bauer et al., 2007; Tsigaridis and Kanakidou, 2007; Menon et al., 2010; Koch et al., 2011)	(1) G-PUCCINI (2) (Shindell et al., 2013a)	Included	(1) HYCOM Ocean (2) 0.2 to 1° latitude × 1° longitude HYCOM (3) 26 (4) 0 m (5) hybrid z isopycnic (6) nonlinear split-explicit (7)	(1) Included (2) (Romanou et al., 2013)	Included
(1) GISS-E2-R (2) 2011	(1) NASA Goddard Institute for Space Studies USA (2) (Schmidt et al., 2006) *See note for GISS-E2-H*	(1) Included (2) 2° latitude × 2.5° longitude F (3) 40 (4) 0.1 hPa	(1) Interactive (2) (Bauer et al., 2007; Tsigaridis and Kanakidou, 2007; Menon et al., 2010; Koch et al., 2011) *Note: Aerosol is "fully interactive" for p2 and p3, "semi interactive" for p1*	(1) G-PUCCINI (2) (Shindell et al., 2013a) *Note: Atmos Chem is "fully interactive" for p2 and p3, "semi interactive" for p1*	Included	(1) Russell Ocean (2) 1° latitude × 1.25° longitude Russell 1 × 1Q (3) 32 (4) 0 m (5) z*-coordinate (6) other (7)	Not implemented	Included
(1) GISS-E2-R-CC (2) 2011	(1) NASA Goddard Institute for Space Studies USA (2) (Schmidt et al., 2006) *Note: p1 only*	(1) Included (2) Nominally 1° (3) 40 (4) 0.1 hPa	(1) Interactive (p1 only) (2) (Bauer et al., 2007; Tsigaridis and Kanakidou, 2007; Menon et al., 2010; Koch et al., 2011)	(1) G-PUCCINI (2) (Shindell et al., 2013a)	Included	(1) Russell Ocean (2) 1° latitude × 1.25° longitude Russell 1×1Q (3) 32 (4) 0 m (5) z*-coordinate (6) other (7)	(1) Included (2) (Romanou et al., 2013)	Included

(continued on next page)

Table 9.A.1 (continued)

(1) Model Name (2) Vintage	**(1) Institution** (2) Main Reference(s)	**Atmosphere** (1) Component Name (2) Horizontal Grid (3) Number of Vert Levels (4) Grid Top (5) References	**Aerosol** (1) Component Name or type (2) References	**Atmos Chemistry** (1) Component Name (2) References	**Land Surface** (1) Component Name (2) References	**Ocean** (1) Component Name (2) Horizontal Resolution (3) Number of Vertical Levels (4) Top Level (5) z Co-ord (6) Top BC (7) References	**Ocean Biogeo-chemistry** (1) Component Name (2) References	**Sea Ice** (1) Component Name (2) References
(1) HadCM3 (2) 1998	(1) UK Met Office Hadley Centre (2) (Gordon et al., 2000; Pope et al., 2000; Collins et al., 2001; Johns et al., 2003)	(1) HadAM3 (2) N48L19 3.75 × 2.5° (3) 19 (4) 0.005 hPa (5) (Pope et al., 2000)	(1) Interactive (2) (Jones et al., 2001)	Not implemented	(1) Included (2) (Collatz et al., 1991; Collatz et al., 1992; Cox et al., 1999; Cox, 2001; Mercado et al., 2007)	(1) HadOM (lat: 1.25 lon: 1.25 L20) (2) 1.25° in longitude by 1.25° in latitude N144 (3) 20 (4) 5.0 m (5) depth (6) linear implicit (7) (UNESCO, 1981)	Not implemented	Included
(1) HadGEM2-AO (2) 2009	(1) National Institute of Meteorological Research/ Korea Meteorological Administration (2) (Collins et al., 2011; Martin et al., 2011)	(1) HadGAM2 (2) 1.875° in longitude by 1.25° in latitude N96 (3) 60 (4) 84132.439 m (5) (Davies et al., 2005)	(1) Interactive (2) (Bellouin et al., 2011)	Not implemented	(1) Included (2) (Cox et al., 1999; Essery et al., 2003)	(1) Included (2) 1.875° in longitude by 1.25° in latitude N96 (3) (4) (5) z (6) linear implicit (7) (Bryan and Lewis, 1979; Johns et al., 2006);	Not implemented	(1) Included (2) (Thorndike et al., 1975; McLaren et al., 2006)
(1) HadGEM2-CC (2) 2010	(1) UK Met Office Hadley Centre (2) (Collins et al., 2011; Martin et al., 2011)	(1) HadGAM2 (2) 1.875° in longitude by 1.25°in latitude N96 (3) 60 (4) 84132.439 m (5) (Davies et al., 2005)	(1) Interactive (2) (Bellouin et al., 2011)	(1) Atmospheric Chemistry (2) (Jones et al., 2001; Martin et al., 2011)	(1) Included (2) (Cox et al., 1999; Essery et al., 2003)	(1) Included (2) 1.875° in longitude by 1.25° in latitude N96 (3) (4) (5) z (6) linear implicit (7) (Bryan and Lewis, 1979; Johns et al., 2006)	(1) Included (2) (Palmer and Totterdell, 2001; Halloran, 2012)	(1) Included (2) (Thorndike et al., 1975; McLaren et al., 2006)
(1) HadGEM2-ES (2) 2009	(1) UK Met Office Hadley Centre (2) (Collins et al., 2011; Martin et al., 2011)	(1) HadGAM2 (2) 1.875° in longitude by 1.25° in latitude N96 (3) 38 (4) 39254.8 m (5) (Davies et al., 2005)	(1) Interactive (2) (Bellouin et al., 2011)	(1) Atmospheric Chemistry (2) (O'Connor et al., 2009)	(1) Included (2) (Cox et al., 1999; Essery et al., 2003)	(1) Included (2) 1° by 1° between 30 N/S and the poles; meridional resolution increases to 1/3° at the equator N180 (3) 40 (4) 5.0 m (5) z (6) linear implicit (7) (Bryan and Lewis, 1979; Johns et al., 2006)	(1) Included (2) (Palmer and Totterdell, 2001; Halloran, 2012)	(1) Included (2) (Thorndike et al., 1975; McLaren et al., 2006)
(1) INM-CM4 (2) 2009	(1) Russian Institute for Numerical Mathematics (2) (Volodin et al., 2010)	(1) Included (2) 2 ×1.5° in longitude and latitude latitude-longitude (3) 21 (4) sigma = 0.01	Prescribed	Not implemented	(1) Included (2) (Alekseev et al., 1998; Volodin and Lykosov, 1998)	(1) Included (2) 1 × 0.5° in longitude and latitude generalized spherical coordinates with poles displaced outside ocean (3) 40 (4) sigma = 0.0010426 (5) sigma (6) linear implicit (7) (Volodin et al., 2010; Zalesny et al., 2010)	(1) Included (2) (Volodin, 2007)	(1) Included (2) (Yakovlev, 2009)

(continued on next page)

Table 9.A.1 (continued)

(1) Model Name (2) Vintage	**(1) Institution** (2) Main Reference(s)	**Atmosphere** (1) Component Name (2) Horizontal Grid (3) Number of Vert Levels (4) Grid Top (5) References	**Aerosol** (1) Component Name or type (2) References	**Atmos Chemistry** (1) Component Name (2) References	**Land Surface** (1) Component Name (2) References	**Ocean** (1) Component Name (2) Horizontal Resolution (3) Number of Vertical Levels (4) Top Level (5) z Co-ord (6) Top BC (7) References	**Ocean Biogeo-chemistry** (1) Component Name (2) References	**Sea Ice** (1) Component Name (2) References
(1) IPSL-CM5A-LR (2) 2010	(1) Institut Pierre Simon Laplace (2) (Dufresne et al., 2012)	(1) LMDZ5 (2) 96 × 95 equivalent to 1.9° × 3.75° LMDZ96 × 95 (3) 39 (4) 0.04 hPa (5)(Hourdin et al., 2012)	Semi-interactive	Not implemented	(1) Included (2) (Krinner et al., 2005)	(1) Included (2) 2 × 2-0.5° ORCA2 (3) 31 (4) 0m (5) depth (6) linear filtered (7) (Madec, 2008)	(1) PISCES (2) (Aumont et al., 2003; Aumont and Bopp, 2006)	(1) LIM2 (2) (Fichefet and Maqueda, 1999)
(1) IPSL-CM5A-MR (2) 2009	(1) Institut Pierre Simon Laplace (2) (Dufresne et al., 2012)	(1) LMDZ5 (2) 144 × 143 equivalent to 1,25° × 2.5° LMDZ144 × 143 (3) 39 (4) 0.04 hPa (5) (Hourdin et al., 2012)	Semi-interactive	Not implemented	(1) Included (2) (Krinner et al., 2005)	(1) Included (2) 2 × 2-0.5° ORCA2 (3) 31 (4) 0 m (5) depth (6) linear filtered (7) (Madec, 2008)	(1) PISCES (2) (Aumont et al., 2003; Aumont and Bopp, 2006)	(1) Included (2) (Fichefet and Maqueda, 1999)
(1) IPSL-CM5B-LR (2) 2010	(1) Institut Pierre Simon Laplace (2) (Dufresne et al., 2012)	(1) LMDZ5 (2) 96 × 95 equivalent to 1.9° × 3.75° LMDZ96 × 95 (3) 39 (4) 0.04 hPa (5)(Hourdin et al., 2013)	Semi-interactive	Not implemented	(1) Included (2) (Krinner et al., 2005)	(1) Included (2) 2 × 2-0.5° ORCA2 (3) 31 (4) 0 m (5) depth (6) linear filtered (7) (Madec, 2008)	(1) PISCES (2) (Aumont et al., 2003; Aumont and Bopp, 2006)	(1) Included (2) (Fichefet and Maqueda, 1999)
(1) MIROC4h (2) 2009	(1) University of Tokyo, National Institute for Environmental Studies, and Japan Agency for Marine-Earth Science and Technology (2) (Sakamoto et al., 2012)	(1) CCSR / NIES / FRCGC AGCM5.7 (2) 0.5625 × 0.5625° T213 (3) 56 (4) about 0.9 hPa	(1) SPRINTARS (2) (Takemura et al., 2000; Takemura et al., 2002)	Not implemented	(1) MATSIRO (2) (Takata et al., 2003)	(1) COCO3.4 (2) 1/4° by 1/6° (average grid spacing is 0.28° and 0.19° for zonal and meridional directions) (3) 48 (4) 1.25 m (5) hybrid z-s (6) nonlinear split-explicit (7) (Hasumi and Emori, 2004)	Not implemented	Included
(1) MIROC5 (2) 2010	(1) University of Tokyo, National Institute for Environmental Studies, and Japan Agency for Marine-Earth Science and Technology (2) (Watanabe et al., 2010)	(1) CCSR/NIES/ FRCGC AGCM6 (2) 1.40625 × 1.40625° T85 (3) 40 (4) about 2.9 hPa	(1) SPRINTARS (2) (Takemura et al., 2005; Takemura et al., 2009)	Not implemented	(1) MATSIRO (2) (Takata et al., 2003)	(1) COCO4.5 (2) 1.4° (zonally) × 0.5–1.4° (meridionally) (3) 50 (4) 1.25 m (5) hybrid z-s (6) linear split-explicit (7) (Hasumi and Emori, 2004)	Not implemented	(1) Included (2) (Komuro et al., 2012)
(1) MIROC-ESM (2) 2010	(1) University of Tokyo, National Institute for Environmental Studies, and Japan Agency for Marine-Earth Science and Technology (2) (Watanabe et al., 2011)	(1) MIROC-AGCM (2) 2.8125 × 2.8125° T42 (3) 80 (4) 0.003 hPa (5) (Watanabe, 2008)	(1) SPRINTARS (2) (Takemura et al., 2005; Takemura et al., 2009)	Not implemented	(1) MATSIRO (2) (Takata et al., 2003)	(1) COCO3.4 (2) 1.4° (zonally) × 0.5–1.4° (meridionally) (3) 44 (4) 1.25 m (5) hybrid z-s (6) linear split-explicit (7) (Hasumi and Emori, 2004)	(1) NPZD-type (2) (Schmittner et al., 2005)	Included

(continued on next page)

Table 9.A.1 (continued)

(1) Model Name (2) Vintage	**(1) Institution** (2) Main Reference(s)	**Atmosphere** (1) Component Name (2) Horizontal Grid (3) Number of Vert Levels (4) Grid Top (5) References	**Aerosol** (1) Component Name or type (2) References	**Atmos Chemistry** (1) Component Name (2) References	**Land Surface** (1) Component Name (2) References	**Ocean** (1) Component Name (2) Horizontal Resolution (3) Number of Vertical Levels (4) Top Level (5) z Co-ord (6) Top BC (7) References	**Ocean Biogeo-chemistry** (1) Component Name (2) References	**Sea Ice** (1) Component Name (2) References
(1) MIROC-ESM-CHEM (2) 2010	(1) University of Tokyo, National Institute for Environmental Studies, and Japan Agency for Marine-Earth Science and Technology (2) (Watanabe et al., 2011)	(1) MIROC-AGCM (2) 2.8125 × 2.8125° T42 (3) 80 (4) 0.003 hPa (5) (Watanabe, 2008)	(1) SPRINTARS (2) (Takemura et al., 2005; Takemura et al., 2009)	(1) CHASER (2) (Sudo et al., 2002)	(1) MATSIRO (2) (Takata et al., 2003)	(1) COCO3.4 (2) 1.4° (zonally) × 0.5–1.4° (meridionally) (3) 44 (4) 1.25 m (5) hybrid z-s (6) linear split-explicit (7) (Hasumi and Emori, 2004)	(1) NPZD-type (2) (Schmittner et al., 2005)	Included
(1) MPI-ESM-LR (2) 2009	(1) Max Planck Institute for Meteorology (2)	(1) ECHAM6 (2) approx. 1.8° T63 (3) 47 (4) 0.01 hPa (5) (Stevens et al., 2012)	Prescribed	Not implemented	(1) JSBACH (2) (Reick et al., 2013)	(1) MPIOM (2) average 1.5° GR15 (3) 40 (4) 6 m (5) depth (6) linear implicit (7) (Jungclaus et al., 2013)	(1) HAMOCC (2) (Maier-Reimer et al., 2005; Ilyina et al., 2013)	(1) Included (2) (Notz et al., 2013)
(1) MPI-ESM-MR (2) 2009	(1) Max Planck Institute for Meteorology (2)	(1) ECHAM6 (2) approx. 1.8° T63 (3) 95 (4) 0.01 hPa (5) (Stevens et al., 2012)	Prescribed	Not implemented	(1) JSBACH (2) (Reick et al., 2013)	(1) MPIOM (2) approx. 0.4° TP04 (3) 40 (4) 6 m (5) depth (6) linear implicit (7) (Jungclaus et al., 2013)	(1) HAMOCC (2) (Maier-Reimer et al., 2005; Ilyina et al., 2013)	(1) Included (2) (Notz et al., 2013)
(1) MPI-ESM-P (2) 2009	(1) Max Planck Institute for Meteorology (2)	(1) ECHAM6 (2) approx. 1.8° T63 (3) 47 (4) 0.01 hPa (5) (Stevens et al., 2012)	Prescribed	Not implemented	(1) JSBACH (2) (Reick et al., 2013)	(1) MPIOM (2) average 1.5° GR15 (3) 40 (4) 6 m (5) depth (6) linear implicit (7) (Jungclaus et al., 2013)	(1) HAMOCC (2) (Maier-Reimer et al., 2005; Ilyina et al., 2013)	(1) Included (2) (Notz et al., 2013)
(1) MRI-AGCM3.2H (2) 2009	(1) Meteorological Research Institute (2) (Mizuta et al., 2012)	(1) Included (2) 640 × 320 TL319 (3) 64 (4) 0.01 hPa	Prescribed	Not implemented	(1) SiB0109 (2) (Hirai et al., 2007; Yukimoto et al., 2011; Yukimoto et al., 2012)	Not implemented	Not implemented	Not implemented
(1) MRI-AGCM3.2S (2) 2009	(1) Meteorological Research Institute (2) (Mizuta et al., 2012)	(1) Included (2) 1920 × 960 TL959 (3) 64 (4) 0.01 hPa (5) (Mizuta et al., 2012)	Prescribed	Not implemented	(1) SiB0109 (2) (Hirai et al., 2007; Yukimoto et al., 2011; Yukimoto et al., 2012)	Not implemented	Not implemented	Not implemented
(1) MRI-CGCM3 (2) 2011	(1) Meteorological Research Institute (2) (Yukimoto et al., 2011; Yukimoto et al., 2012)	(1) MRI-AGCM3.3 (2) 320 × 160 TL159 (3) 48 (4) 0.01 hPa (5) (Yukimoto et al., 2011; Yukimoto et al., 2012)	(1) MASINGAR mk-2 (2) (Yukimoto et al., 2011; Yukimoto et al., 2012; Adachi et al., 2013)	Not implemented	(1) HAL (2) (Yukimoto et al., 2011; Yukimoto et al., 2012)	(1) MRI.COM3 (2) 1 × 0.5 (3) 50 + 1 Bottom Boundary Layer (4) 0 m (5) hybrid sigma-z (6) nonlinear split-explicit (7) (Tsujino et al., 2011; Yukimoto et al., 2011; Yukimoto et al., 2012)	Not implemented	(1) Included (MRI. COM3) (2) (Tsujino et al., 2011; Yukimoto et al., 2011; Yukimoto et al., 2012)

(continued on next page)

Table 9.A.1 (continued)

(1) Model Name (2) Vintage	**(1) Institution** (2) Main Reference(s)	**Atmosphere** (1) Component Name (2) Horizontal Grid (3) Number of Vert Levels (4) Grid Top (5) References	**Aerosol** (1) Component Name or type (2) References	**Atmos Chemistry** (1) Component Name (2) References	**Land Surface** (1) Component Name (2) References	**Ocean** (1) Component Name (2) Horizontal Resolution (3) Number of Vertical Levels (4) Top Level (5) z Co-ord (6) Top BC (7) References	**Ocean Biogeo-chemistry** (1) Component Name (2) References	**Sea Ice** (1) Component Name (2) References
(1) MRI-ESM1 (2) 2011	(1) Meteorological Research Institute (2) (Yukimoto et al., 2011; Yukimoto et al., 2012; Adachi et al., 2013)	(1) MRI-AGCM3.3 (2) TL159(320 × 160) (3) 48 (4) 0.01 hPa (5) (Yukimoto et al., 2011; Yukimoto et al., 2012; Adachi et al., 2013)	(1) MASINGAR mk-2 (2) (Yukimoto et al., 2011; Yukimoto et al., 2012; Adachi et al., 2013)	(1) MRI-CCM2 (2) (Deushi and Shibata, 2011; Yukimoto et al., 2011; Adachi et al., 2013)	(1) HAL (2) (Yukimoto et al., 2011; Yukimoto et al., 2012)	(1) MRI.COM3 (2) 1x0.5 (3) 50 + 1 Bottom Boundary Layer (4) 0m (5) hybrid sigma-z (6) non-linear split-explicit (7)(Tsujino et al., 2011; Yukimoto et al., 2011; Yukimoto et al., 2012)	(1) Included (MRI.COM3) (2) (Nakano et al., 2011; Adachi et al., 2013)	(1) Included (MRI. COM3) (2) (Tsujino et al., 2011; Yukimoto et al., 2011; Yukimoto et al., 2012)
(1) NCEP-CFSv2 (2) 2011	(1) National Centers for Environmental Prediction	(1) Global Forecast Model (2) 0.9375 T126 (3) 64 (4) 0.03 hPa (5) (Saha et al., 2010)	Semi-interactive	(1) Ozone chemistry (2) (McCormack et al., 2006)	(1) Noah Land Surface Model (2) (Ek et al., 2003)	(1) MOM4 (2) 0.5° zonal resolution, meridional resolution varying from 0.25° at the equator to 0.5° north/south of 10N/10S. Tripolar. (3) 40 (4) 5.0 m (5) depth (6) nonlinear split explicit (7) (Griffies et al., 2004)	Not implemented	(1) SIS (2) (Hunke and Dukowicz, 1997; Winton, 2000)
(1) NorESM1-M (2) 2011	(1) Norwegian Climate Centre (2) (Iversen et al., 2013)	(1) CAM4-Oslo (2) Finite Volume 1.9° latitude, 2.5° longitude (3) 26 (4) 2.194067 hPa (5) (Neale et al., 2010; Kirkevåg et al., 2013)	(1) CAM4-Oslo (2) (Kirkevåg et al., 2013)	(1) CAM4-Oslo (2) (Kirkevåg et al., 2013)	(1) CLM4 (2) (Oleson et al., 2010; Lawrence et al., 2011)	(1) NorESM-Ocean (2) 1.125° along the equator (3) 53 (4) 1 m (5) hybrid z isopycnic (6) nonlinear split-explicit (7)	Not implemented	(1)CICE4 (2)(Hunke and Lipscomb, 2008; Holland et al., 2012)
(1) NorESM1-ME (2) 2012	(1) Norwegian Climate Centre (2) (Tjiputra et al., 2013)	(1) CAM4-Oslo (2) Finite Volume 1.9° latitude, 2.5° longitude (3) 26 (4) 2.194067 hPa (5) (Neale et al., 2010; Kirkevåg et al., 2013)	(1) CAM4-Oslo (2) (Kirkevåg et al., 2013)	(1) CAM4-Oslo (2) (Kirkevåg et al., 2013)	(1) CLM4 (2) (Oleson et al., 2010; Lawrence et al., 2011)	(1) NorESM-Ocean (2) 1.125° along the equator (3) 53 (4) 1 m (5) hybrid z isopycnic (6) nonlinear split-explicit (7)	(1) HAMOCC5 (2) (Maier-Reimer et al., 2005; Assmann et al., 2010; Tjiputra et al., 2013)	(1) CICE4 (2) (Hunke and Lipscomb, 2008; Holland et al., 2012)

Table 9.A.2 | Salient features of the Earth system Models of Intermediate Complexity (EMICs) assessed in the AR5 (see also Table 9.2). Column 1: Model name used in WG1 and the official model version along with the first publication for each model; subsequent columns for each of the eight component models with specific information and the related references are provided. This information was initially gathered for the EMIC intercomparison project in Eby et al. (2013).

(1) Model name (2) Model version (3) Main reference	**Atmosphere**[a] (1) Model type (2) Dimensions (3) Resolution (4) Radiation and cloudiness (5) References	**Ocean**[b] (1) Model type (2) Dimensions (3) Resolution (4) Parametrizations (5) References	**Sea Ice**[c] (1) Schemes (2) References	**Coupling**[d] (1) Flux adjustment (2) References	**Land Surface**[e] (1) Soil schemes (2) References	**Biosphere**[f] (1) Ocean and references (2) Land and references (3) Vegetation and references	**Ice Sheets**[g] (1) Model type (2) Dimensions (3) Resolution (4) References	**Sediment and Weathering**[h] (1) Model type (2) References
(1) Bern3D (2) Bern3D-LPJ (3) (Ritz et al., 2011)	(1) EMBM (2) 2-D(ϕ, λ) (3) 10° × (3–19)° (4) NCL (5)	(1) FG with parameterized zonal pressure gradient (2) 3-D (3) 10° × (3–19)°, L32 (4) RL, ISO, MESO (5) (Muller et al., 2006)	(1) 0-LT, DOC, 2-LIT	(1) PM, NH, RW	(1) Bern3D: 1-LST, NSM, RIV LPJ: 8-LST, CSM with uncoupled hydrology (2) (Wania et al., 2009)	(1) BO (Parekh et al., 2008; Tschumi et al., 2008; Gangsto et al., 2011) (2) BT (Sitch et al., 2003; Strassmann et al., 2008; Stocker et al., 2011) (3) BV (Sitch et al., 2003)	N/A	(1) CS, SW (2) (Tschumi et al., 2011)
(1) CLIMBER2 (2) CLIMBER-2.4 (3) (Petoukhov et al., 2000)	(1) SD (2) 3-D (3) 10° × 51°, L10 (4) CRAD, ICL (5)	(1) FG, (2) 2-D(ϕ,z) (3) 2.5°, L21 (4) RL (5) (Wright and Stocker, 1992)	(1) 1-LT, PD, 2-LIT (2) (Petoukhov et al., 2000)	(1) NM, NH, NW (2) (Petoukhov et al., 2000)	(1) 1-LST, CSM, RIV (2) (Petoukhov et al., 2000)	(1,2,3) BO, BT, BV (Brovkin et al., 2002)	(1) TM (2) 3-D (3) 0.75° × 1.5°, L20 (4) (Calov et al., 2002)	N/A
(1) CLIMBER3 (2) CLIMBER-3α (3) (Montoya et al., 2005)	(1) SD (2) 3-D (3) 7.5° × 22.5°, L10 (4) CRAD, ICL (5) (Petoukhov et al., 2000)	(1) PE (2) 3-D (3) 3.75° × 3.75°, L24 (4) FS, ISO, MESO, TCS, DC	(1) 2-LT, R, 2-LIT (2) (Fichefet and Morales Maqueda, 1997)	(1) AM, NH, RW	(1) 1-LST, CSM, RIV (2) (Petoukhov et al., 2000)	(1) BO (Six and Maier-Reimer, 1996) (2,3) BT, BV (Brovkin et al., 2002)	N/A	N/A
(1) DCESS (2) DCESS (3) (Shaffer et al., 2008)	(1) EMBM (2) 2-box in ϕ, (3) (4) LRAD, CHEM (5) (Shaffer et al., 2008)	(1) 2-box in ϕ (2) (3) L55 (4) parameterized circulation and exchange, MESO (5) (Shaffer and Sarmiento, 1995)	(1) Parameterized from surface temperature (2) (Shaffer et al., 2008)	(1) NH, NW (2) (Shaffer et al., 2008)	(1) NST, NSM (2) (Shaffer et al., 2008)	(1,2) BO, BT (Shaffer et al., 2008)	N/A	(1) CS, SW (2) (Shaffer et al., 2008)
(1) FAMOUS (2) FAMOUS XDBUA (3) (Smith et al., 2008)	(1) PE (2) 3-D (3) 5° × 7.5°, L11 (4) CRAD, ICL (5) (Pope et al., 2000)	(1) PE (2) 3-D (3) 2.5° × 3.75°, L20 (4) RL, ISO, MESO (5) (Gordon et al., 2000)	(1) 0-LT, DOC, 2-LIT	(1) NM, NH, NW	(1) 4-LST, CSM, RIV (2) (Cox et al., 1999)	(1) BO (Palmer and Totterdell, 2001)	N/A	N/A
(1) GENIE (2) GENIE (3) (Holden et al., 2013)	(1) EMBM (2) 2-D(ϕ, λ) (3) 10° × (3–19)° (4) NCL (5) (Marsh et al., 2011)	(1) FG (2) 3-D (3) 10° × (3–19) °, L16 (4) RL, ISO, MESO (5) (Marsh et al., 2011)	(1) 1-LT, DOC, 2-LIT (2) (Marsh et al., 2011)	(1) PM, NH, RW (2) (Marsh et al., 2011)	(1) 1-LST, BSM, RIV (2) (Williamson et al., 2006)	(1,2) BO, BT (Williamson et al., 2006; Ridgwell et al., 2007b; Holden et al., 2013)	N/A	(1) CS, SW (2) (Ridgwell and Hargreaves, 2007)
(1) IAP RAS CM (2) IAP RAS CM (3) (Eliseev and Mokhov, 2011)	(1) SD (2) 3-D (3) 4.5° × 6°, L8 (4) CRAD, ICL (5) (Petoukhov et al., 1998)	(1) PE (2) 3-D (3) 3.5° × 3.5°, L32 (4) RL, ISO, TCS (5) (Muryshev et al., 2009)	(1) 0-LT, 2-LIT (2) (Muryshev et al., 2009)	(1) NM, NH, NW (2) (Muryshev et al., 2009)	(1) 240-LST, CSM (2) (Arzhanov et al., 2008)	(2) BT (Eliseev and Mokhov, 2011)	N/A	N/A

(continued on next page)

Table 9.A.2 (continued)

(1) Model name (2) Model version (3) Main reference	**Atmosphere**[a] (1) Model type (2) Dimensions (3) Resolution (4) Radiation and cloudiness (5) References	**Ocean**[b] (1) Model type (2) Dimensions (3) Resolution (4) Parametrizations (5) References	**Sea Ice**[c] (1) Schemes (2) References	**Coupling**[d] (1) Flux adjustment (2) References	**Land Surface**[e] (1) Soil schemes (2) References	**Biosphere**[f] (1) Ocean and references (2) Land and references (3) Vegetation and references	**Ice Sheets**[g] (1) Model type (2) Dimensions (3) Resolution (4) References	**Sediment and Weathering**[h] (1) Model type (2) References
(1) IGSM2 (2) IGSM 2.2 (3) (Sokolov et al., 2005)	(1) SD (2) 2-D(ϕ, Z) (3) 4° × 360° , L11 (4) ICL, CHEM (5) (Sokolov and Stone, 1998)	(1) Q-flux mixed-layer, anomaly diffusing, (2) 3-D (3) 4° × 5°, L11 (4) (5) (Hansen et al., 1984)	(1) 2-LT (2) (Hansen et al., 1984)	(1) Q-flux (2) (Sokolov et al., 2005)	(1) CSM (2) (Oleson et al., 2008b)	(1) BO (Holian et al., 2001) (2) BT (Melillo et al., 1993; Liu, 1996; Felzer et al., 2004)	N/A	N/A
(1) LOVECLIM1.2 (2) LOVECLIM1.2 (3) (Goosse et al., 2010)	(1) QG (2) 3-D (3) 5.6° × 5.6°, L3 (4) LRAD, NCL (5) (Opsteegh et al., 1998)	(1) PE (2) 3-D (3) 3° × 3°, L30 (4) FS, ISO, MESO, TCS, DC (5) (Goosse and Fichefet, 1999)	(1) 3-LT, R, 2-LIT (2) (Fichefet and Morales Maqueda, 1997)	(1) NM, NH, RW (2) (Goosse et al., 2010)	(1) 1-LST, BSM, RIV (2) (Goosse et al., 2010)	(1) BO (Mouchet and François, 1996) (2,3) BT, BV (Brovkin et al., 2002)	(1) TM (2) 3-D (3) 10 km × 10 km, L30 (4) (Huybrechts, 2002)	N/A
(1) MESMO (2) MESMO 1.0 (3) (Matsumoto et al., 2008)	(1) EMBM (2) 2-D(ϕ, λ) (3) 10° × (3–19)° (4) NCL, (5) (Fanning and Weaver, 1996)	(1) FG (2) 3-D (3) 10° × (3–19)°, L16 (4) RL, ISO, MESO (5) (Edwards and Marsh, 2005)	(1) 0-LT, DOC, 2-LIT (2) (Edwards and Marsh, 2005)	(1) PM, NH, RW	(1) NST, NSM, RIV (2) (Edwards and Marsh, 2005)	(1) BO (Matsumoto et al., 2008)	N/A	N/A
(1) MIROC-lite (2) MIROC-lite (3) (Oka et al., 2011)	(1) EMBM (2) 2-D(ϕ, λ) (3) 4° × 4° (4) NCL (5) (Oka et al., 2011)	(1) PE (2) 3-D (3) 4° × 4° (4) FS, ISO, MESO, TCS (5) (Hasumi, 2006)	(1) 0-LT, R, 2-LIT (2) (Hasumi, 2006)	(1) PM, NH, NW (2) (Oka et al., 2011)	(1) 1-LST, BSM (2) (Oka et al., 2011)	N/A	N/A	N/A
(1) MIROC-lite-LCM (2) MIROC-lite-LCM (3) (Tachiiri et al., 2010)	(1) EMBM, tuned for 3 K equilibrium climate sensitivity (2) 2-D(ϕ, λ) (3) 6° × 6° (4) NCL (5) (Oka et al., 2011)	(1) PE (2) 3-D (3) 6° × 6°, L15 (4) FS, ISO, MESO, TCS (5) (Hasumi, 2006)	(1) 0-LT, R, 2-LIT (2) (Hasumi, 2006)	(1) NM, NH RW (2) (Oka et al., 2011) (Tachiiri et al., 2010)	(1) 1-LST, BSM (2) (Oka et al., 2011)	(1) BO (Palmer and Totterdell, 2001) (2) loosely coupled BT (Ito and Oikawa, 2002)	N/A	N/A
(1) SPEEDO (2) SPEEDO V2.0 (3) (Severijns and Hazeleger, 2010)	(1) PE (2) 3-D (3) T30, L8 (4) LRAD, IDL, (5) (Molteni, 2003)	(1) PE (2) 3-D (3) 3° × 3°, L20 (4) FS, ISO, MESO, TCS, DC (5) (Goosse and Fichefet, 1999)	(1) 3-LT, R, 2-LIT (2) (Fichefet and Morales Maqueda, 1997)	(1)NM, NH, NW (2) (Cimatoribus et al., 2012)	(1) 1-LST, BSM, RIV (2) (Opsteegh et al., 1998)	N/A	N/A	N/A
(1) UMD (2) UMD 2.0 (3) (Zeng et al., 2004)	(1) QG (2) 3-D (3) 3.75° × 5.625°, L2 (4) LRAD, ICL (5) (Neelin and Zeng, 2000; Zeng et al., 2000)	(1) Q-flux mixed-layer (2) 2-D surface, deep ocean box model (3) 3.75° × 5.625° (5) (Hansen et al., 1983),	N/A	(1) Energy and water exchange only (2) (Zeng et al., 2004)	(1) 2-LST with 2-layer soil moisture (2) (Zeng et al., 2000)	(1) BO (Archer et al., 2000) (2,3) BT, BV (Zeng, 2003; Zeng et al., 2005; Zeng, 2006)	N/A	N/A

(continued on next page)

Table 9.A.2 (continued)

(1) Model name (2) Model version (3) Main reference	**Atmosphere**[a] (1) Model type (2) Dimensions (3) Resolution (4) Radiation and cloudiness (5) References	**Ocean**[b] (1) Model type (2) Dimensions (3) Resolution (4) Parametrizations (5) References	**Sea Ice**[c] (1) Schemes (2) References	**Coupling**[d] (1) Flux adjustment (2) References	**Land Surface**[e] (1) Soil schemes (2) References	**Biosphere**[f] (1) Ocean and references (2) Land and references (3) Vegetation and references	**Ice Sheets**[g] (1) Model type (2) Dimensions (3) Resolution (4) References	**Sediment and Weathering**[h] (1) Model type (2) References
(1) Uvic (2) UVic 2.9 (3) (Weaver et al., 2001)	(1) DEMBM (2) 2-D(ϕ, λ) (3) 1.8° × 3.6° (4) NCL (5) (Weaver et al., 2001)	(1) PE (2) 3-D (3) 1.8° × 3.6°, L19 (4) RL, ISO, MESO (5) (Weaver et al., 2001)	(1) 0-LT, R, 2-LIT (2) (Weaver et al., 2001)	(1) AM, NH, NW (2) (Weaver et al., 2001)	(1) 1-LST, CSM, RIV (2) (Meissner et al., 2003)	(1) BO (Schmittner et al., 2005) (2,3) BT, BV (Cox, 2001)	(1) TM (2) 3-D (3) 20 km × 20 km, L10 (4) (Fyke et al., 2011)	(1) CS, SW (2) (Eby et al., 2009)

Notes:

(a) EMBM = energy moisture balance model; DEMBM = energy moisture balance model including some dynamics; SD = statistical-dynamical model; QG = quasi-geostrophic model; 2-D(ϕ, λ) = vertically averaged; 3-D = three-dimensional; LRAD = linearized radiation scheme; CRAD = comprehensive radiation scheme; NCL = non-interactive cloudiness; ICL = interactive cloudiness; CHEM = chemistry module; $n° \times m°$ = n degrees latitude by m degrees longitude horizontal resolution; Lp = p vertical levels.

(b) FG = frictional geostrophic model; PE = primitive equation model; 2-D(ϕ, z) = zonally averaged; 3-D = three-dimensional; RL = rigid lid; FS = free surface; ISO = isopycnal diffusion; MESO = parameterization of the effect of mesoscale eddies on tracer distribution; TCS = complex turbulence closure scheme; DC = parameterization of density-driven downward-sloping currents; $n° \times m°$ = n degrees latitude by m degrees longitude horizontal resolution; Lp = p vertical levels.

(c) n-LT = n-layer thermodynamic scheme; PD = prescribed drift; DOC = drift with oceanic currents; R = viscous-plastic or elastic-viscous-plastic rheology; 2-LIT = two-level ice thickness distribution (level ice and leads).

(d) PM = prescribed momentum flux; AM = momentum flux anomalies relative to the control run are computed and added to climatological data; NM = no momentum flux adjustment; NH = no heat flux adjustment; RW = regional freshwater flux adjustment; NW = no freshwater flux adjustment.

(e) NST = no explicit computation of soil temperature; n-LST = n-layer soil temperature scheme; NSM = no moisture storage in soil; BSM = bucket model for soil moisture; CSM = complex model for soil moisture; RIV = river routing scheme.

(f) BO = model of oceanic carbon dynamics; BT = model of terrestrial carbon dynamics; BV = dynamical vegetation model.

(g) TM = thermomechanical model; 3-D = three-dimensional; $n° \times m°$ = n degrees latitude by m degrees longitude horizontal resolution; n km × m km = horizontal resolution in kilometres; Lp = p vertical levels.

(h) CS = complex ocean sediment model; SW = simple, specified or diagnostic weathering model.

Detection and Attribution of Climate Change: from Global to Regional

Coordinating Lead Authors:
Nathaniel L. Bindoff (Australia), Peter A. Stott (UK)

Lead Authors:
Krishna Mirle AchutaRao (India), Myles R. Allen (UK), Nathan Gillett (Canada), David Gutzler (USA), Kabumbwe Hansingo (Zambia), Gabriele Hegerl (UK/Germany), Yongyun Hu (China), Suman Jain (Zambia), Igor I. Mokhov (Russian Federation), James Overland (USA), Judith Perlwitz (USA), Rachid Sebbari (Morocco), Xuebin Zhang (Canada)

Contributing Authors:
Magne Aldrin (Norway), Beena Balan Sarojini (UK/India), Jürg Beer (Switzerland), Olivier Boucher (France), Pascale Braconnot (France), Oliver Browne (UK), Ping Chang (USA), Nikolaos Christidis (UK), Tim DelSole (USA), Catia M. Domingues (Australia/Brazil), Paul J. Durack (USA/Australia), Alexey Eliseev (Russian Federation), Kerry Emanuel (USA), Graham Feingold (USA), Chris Forest (USA), Jesus Fidel González Rouco (Spain), Hugues Goosse (Belgium), Lesley Gray (UK), Jonathan Gregory (UK), Isaac Held (USA), Greg Holland (USA), Jara Imbers Quintana (UK), William Ingram (UK), Johann Jungclaus (Germany), Georg Kaser (Austria), Veli-Matti Kerminen (Finland), Thomas Knutson (USA), Reto Knutti (Switzerland), James Kossin (USA), Mike Lockwood (UK), Ulrike Lohmann (Switzerland), Fraser Lott (UK), Jian Lu (USA/Canada), Irina Mahlstein (Switzerland), Valérie Masson-Delmotte (France), Damon Matthews (Canada), Gerald Meehl (USA), Blanca Mendoza (Mexico), Viviane Vasconcellos de Menezes (Australia/Brazil), Seung-Ki Min (Republic of Korea), Daniel Mitchell (UK), Thomas Mölg (Germany/Austria), Simone Morak (UK), Timothy Osborn (UK), Alexander Otto (UK), Friederike Otto (UK), David Pierce (USA), Debbie Polson (UK), Aurélien Ribes (France), Joeri Rogelj (Switzerland/Belgium), Andrew Schurer (UK), Vladimir Semenov (Russian Federation), Drew Shindell (USA), Dmitry Smirnov (Russian Federation), Peter W. Thorne (USA/Norway/UK), Muyin Wang (USA), Martin Wild (Switzerland), Rong Zhang (USA)

Review Editors:
Judit Bartholy (Hungary), Robert Vautard (France), Tetsuzo Yasunari (Japan)

This chapter should be cited as:
Bindoff, N.L., P.A. Stott, K.M. AchutaRao, M.R. Allen, N. Gillett, D. Gutzler, K. Hansingo, G. Hegerl, Y. Hu, S. Jain, I.I. Mokhov, J. Overland, J. Perlwitz, R. Sebbari and X. Zhang, 2013: Detection and Attribution of Climate Change: from Global to Regional. In: Climate Change 2013: *The Physical Science Basis. Contribution of Working Group I to the Fifth Assessment Report of the Intergovernmental Panel on Climate Change* [Stocker, T.F., D. Qin, G.-K. Plattner, M. Tignor, S.K. Allen, J. Boschung, A. Nauels, Y. Xia, V. Bex and P.M. Midgley (eds.)]. Cambridge University Press, Cambridge, United Kingdom and New York, NY, USA.

Table of Contents

Executive Summary

Atmospheric Temperatures

More than half of the observed increase in global mean surface temperature (GMST) from 1951 to 2010 is *very likely*[1] due to the observed anthropogenic increase in greenhouse gas (GHG) concentrations. The consistency of observed and modeled changes across the climate system, including warming of the atmosphere and ocean, sea level rise, ocean acidification and changes in the water cycle, the cryosphere and climate extremes points to a large-scale warming resulting primarily from anthropogenic increases in GHG concentrations. Solar forcing is the only known natural forcing acting to warm the climate over this period but it has increased much less than GHG forcing, and the observed pattern of long-term tropospheric warming and stratospheric cooling is not consistent with the expected response to solar irradiance variations. The Atlantic Multi-decadal Oscillation (AMO) could be a confounding influence but studies that find a significant role for the AMO show that this does not project strongly onto 1951–2010 temperature trends. {10.3.1, Table 10.1}

It is *extremely likely* that human activities caused more than half of the observed increase in GMST from 1951 to 2010. This assessment is supported by robust evidence from multiple studies using different methods. Observational uncertainty has been explored much more thoroughly than previously and the assessment now considers observations from the first decade of the 21st century and simulations from a new generation of climate models whose ability to simulate historical climate has improved in many respects relative to the previous generation of models considered in AR4. Uncertainties in forcings and in climate models' temperature responses to individual forcings and difficulty in distinguishing the patterns of temperature response due to GHGs and other anthropogenic forcings prevent a more precise quantification of the temperature changes attributable to GHGs. {9.4.1, 9.5.3, 10.3.1, Figure 10.5, Table 10.1}

GHGs contributed a global mean surface warming *likely* to be between 0.5°C and 1.3°C over the period 1951–2010, with the contributions from other anthropogenic forcings *likely* to be between –0.6°C and 0.1°C, from natural forcings *likely* to be between –0.1°C and 0.1°C, and from internal variability *likely* to be between –0.1°C and 0.1°C. Together these assessed contributions are consistent with the observed warming of approximately 0.6°C over this period. {10.3.1, Figure 10.5}

It is *virtually certain* that internal variability alone cannot account for the observed global warming since 1951. The observed global-scale warming since 1951 is large compared to climate model estimates of internal variability on 60-year time scales. The Northern Hemisphere (NH) warming over the same period is far outside the range of any similar length trends in residuals from reconstructions of the past millennium. The spatial pattern of observed warming differs from those associated with internal variability. The model-based simulations of internal variability are assessed to be adequate to make this assessment. {9.5.3, 10.3.1, 10.7.5, Table 10.1}

It is *likely* that anthropogenic forcings, dominated by GHGs, have contributed to the warming of the troposphere since 1961 and *very likely* that anthropogenic forcings, dominated by the depletion of the ozone layer due to ozone-depleting substances, have contributed to the cooling of the lower stratosphere since 1979. Observational uncertainties in estimates of tropospheric temperatures have now been assessed more thoroughly than at the time of AR4. The structure of stratospheric temperature trends and multi-year to decadal variations are well represented by models and physical understanding is consistent with the observed and modelled evolution of stratospheric temperatures. Uncertainties in radiosonde and satellite records make assessment of causes of observed trends in the upper troposphere less confident than an assessment of the overall atmospheric temperature changes. {2.4.4, 9.4.1, 10.3.1, Table 10.1}

10

Further evidence has accumulated of the detection and attribution of anthropogenic influence on temperature change in different parts of the world. Over every continental region, except Antarctica, it is *likely* that anthropogenic influence has made a substantial contribution to surface temperature increases since the mid-20th century. The robust detection of human influence on continental scales is consistent with the global attribution of widespread warming over land to human influence. It is *likely* that there has been an anthropogenic contribution to the very substantial Arctic warming over the past 50 years. For Antarctica large observational uncertainties result in *low confidence*[2] that anthropogenic influence has contributed to the observed warming averaged over available stations. Anthropogenic influence has *likely* contributed to temperature change in many sub-continental regions. {2.4.1, 10.3.1, Table 10.1}

Robustness of detection and attribution of global-scale warming is subject to models correctly simulating internal variability. Although estimates of multi-decadal internal variability of GMST need to be obtained indirectly from the observational record because the observed record contains the effects of external forcings (meaning the combination of natural and anthropogenic forcings), the standard deviation of internal variability would have to be underestimated in climate models by a factor of at least three to account for the observed warming in the absence of anthropogenic influence. Comparison with observations provides no indication of such a large difference between climate models and observations. {9.5.3, Figures 9.33, 10.2, 10.3.1, Table 10.1}

---

1 In this Report, the following terms have been used to indicate the assessed likelihood of an outcome or a result: Virtually certain 99–100% probability, Very likely 90–100%, Likely 66–100%, About as likely as not 33–66%, Unlikely 0–33%, Very unlikely 0-10%, Exceptionally unlikely 0–1%. Additional terms (Extremely likely: 95–100%, More likely than not >50–100%, and Extremely unlikely 0–5%) may also be used when appropriate. Assessed likelihood is typeset in italics, e.g., *very likely* (see Section 1.4 and Box TS.1 for more details).

2 In this Report, the following summary terms are used to describe the available evidence: limited, medium, or robust; and for the degree of agreement: low, medium, or high. A level of confidence is expressed using five qualifiers: very low, low, medium, high, and very high, and typeset in italics, e.g., *medium confidence*. For a given evidence and agreement statement, different confidence levels can be assigned, but increasing levels of evidence and degrees of agreement are correlated with increasing confidence (see Section 1.4 and Box TS.1 for more details).

The observed recent warming hiatus, defined as the reduction in GMST trend during 1998–2012 as compared to the trend during 1951–2012, is attributable in roughly equal measure to a cooling contribution from internal variability and a reduced trend in external forcing (expert judgement, *medium confidence*). The forcing trend reduction is primarily due to a negative forcing trend from both volcanic eruptions and the downward phase of the solar cycle. However, there is *low confidence* in quantifying the role of forcing trend in causing the hiatus because of uncertainty in the magnitude of the volcanic forcing trends and *low confidence* in the aerosol forcing trend. Many factors, in addition to GHGs, including changes in tropospheric and stratospheric aerosols, stratospheric water vapour, and solar output, as well as internal modes of variability, contribute to the year-to-year and decade- to-decade variability of GMST. {Box 9.2, 10.3.1, Figure 10.6}

Ocean Temperatures and Sea Level Rise

10

It is *very likely* that anthropogenic forcings have made a substantial contribution to upper ocean warming (above 700 m) observed since the 1970s. This anthropogenic ocean warming has contributed to global sea level rise over this period through thermal expansion. New understanding since AR4 of measurement errors and their correction in the temperature data sets have increased the agreement in estimates of ocean warming. Observations of ocean warming are consistent with climate model simulations that include anthropogenic and volcanic forcings but are inconsistent with simulations that exclude anthropogenic forcings. Simulations that include both anthropogenic and natural forcings have decadal variability that is consistent with observations. These results are a major advance on AR4. {3.2.3, 10.4.1, Table 10.1}

It is *very likely* that there is a substantial contribution from anthropogenic forcings to the global mean sea level rise since the 1970s. It is *likely* that sea level rise has an anthropogenic contribution from Greenland melt since 1990 and from glacier mass loss since 1960s. Observations since 1971 indicate with *high confidence* that thermal expansion and glaciers (excluding the glaciers in Antarctica) explain 75% of the observed rise. {10.4.1, 10.4.3, 10.5.2, Table 10.1, 13.3.6}

Ocean Acidification and Oxygen Change

It is *very likely* that oceanic uptake of anthropogenic carbon dioxide has resulted in acidification of surface waters which is observed to be between –0.0014 and –0.0024 pH units per year. There is *medium confidence* that the observed global pattern of decrease in oxygen dissolved in the oceans from the 1960s to the 1990s can be attributed in part to human influences. {3.8.2, Box 3.2, 10.4.4, Table 10.1}

The Water Cycle

New evidence is emerging for an anthropogenic influence on global land precipitation changes, on precipitation increases in high northern latitudes, and on increases in atmospheric humidity. There is *medium confidence* that there is an anthropogenic contribution to observed increases in atmospheric specific humidity since 1973 and to global scale changes in precipitation patterns over land since 1950, including increases in NH mid to high latitudes. Remaining observational and modelling uncertainties, and the large internal variability in precipitation, preclude a more confident assessment at this stage. {2.5.1, 2.5.4, 10.3.2, Table 10.1}

It is *very likely* that anthropogenic forcings have made a discernible contribution to surface and subsurface oceanic salinity changes since the 1960s. More than 40 studies of regional and global surface and subsurface salinity show patterns consistent with understanding of anthropogenic changes in the water cycle and ocean circulation. The expected pattern of anthropogenic amplification of climatological salinity patterns derived from climate models is detected in the observations although there remains incomplete understanding of the observed internal variability of the surface and sub-surface salinity fields. {3.3.2, 10.4.2, Table 10.1}

It is *likely* that human influence has affected the global water cycle since 1960. This assessment is based on the combined evidence from the atmosphere and oceans of observed systematic changes that are attributed to human influence in terrestrial precipitation, atmospheric humidity and oceanic surface salinity through its connection to precipitation and evaporation. This is a major advance since AR4. {3.3.2, 10.3.2, 10.4.2, Table 10.1}

Cryosphere

Anthropogenic forcings are *very likely* to have contributed to Arctic sea ice loss since 1979. There is a robust set of results from simulations that show the observed decline in sea ice extent is simulated only when models include anthropogenic forcings. There is *low confidence* in the scientific understanding of the observed increase in Antarctic sea ice extent since 1979 owing to the incomplete and competing scientific explanations for the causes of change and *low confidence* in estimates of internal variability. {10.5.1, Table 10.1}

Ice sheets and glaciers are melting, and anthropogenic influences are *likely* to have contributed to the surface melting of Greenland since 1993 and to the retreat of glaciers since the 1960s. Since 2007, internal variability is *likely* to have further enhanced the melt over Greenland. For glaciers there is a high level of scientific understanding from robust estimates of observed mass loss, internal variability and glacier response to climatic drivers. Owing to a low level of scientific understanding there is *low confidence* in attributing the causes of the observed loss of mass from the Antarctic ice sheet since 1993. {4.3.3, 10.5.2, Table 10.1}

It is *likely* that there has been an anthropogenic component to observed reductions in NH snow cover since 1970. There is high agreement across observations studies and attribution studies find a human influence at both continental and regional scales. {10.5.3, Table 10.1}

Climate Extremes

There has been a strengthening of the evidence for human influence on temperature extremes since the AR4 and IPCC Special Report on Managing the Risks of Extreme Events and Disasters to Advance Climate Change Adaptation (SREX) reports. It is *very likely* that anthropogenic forcing has contributed to the observed changes in the frequency and intensity of daily temperature extremes on the global scale since the mid-20th century. Attribution of changes in temperature extremes to anthropogenic influence is robustly seen in independent analyses using different methods and different data sets. It is *likely* that human influence has substantially increased the probability of occurrence of heatwaves in some locations. {10.6.1, 10.6.2, Table 10.1}

In land regions where observational coverage is sufficient for assessment, there is *medium confidence* that anthropogenic forcing has contributed to a global-scale intensification of heavy precipitation over the second half of the 20th century. There is *low confidence* in attributing changes in drought over global land areas since the mid-20th century to human influence owing to observational uncertainties and difficulties in distinguishing decadal-scale variability in drought from long-term trends. {10.6.1, Table 10.1}

There is *low confidence* in attribution of changes in tropical cyclone activity to human influence owing to insufficient observational evidence, lack of physical understanding of the links between anthropogenic drivers of climate and tropical cyclone activity and the low level of agreement between studies as to the relative importance of internal variability, and anthropogenic and natural forcings. This assessment is consistent with that of SREX. {10.6.1, Table 10.1}

Atmospheric Circulation

It is *likely* that human influence has altered sea level pressure patterns globally. Detectable anthropogenic influence on changes in sea level pressure patterns is found in several studies. Changes in atmospheric circulation are important for local climate change since they could lead to greater or smaller changes in climate in a particular region than elsewhere. There is *medium confidence* that stratospheric ozone depletion has contributed to the observed poleward shift of the southern Hadley Cell border during austral summer. There are large uncertainties in the magnitude of this poleward shift. It is *likely* that stratospheric ozone depletion has contributed to the positive trend in the Southern Annular Mode seen in austral summer since the mid-20th century which corresponds to sea level pressure reductions over the high latitudes and an increase in the subtropics. There is *medium confidence* that GHGs have also played a role in these trends of the southern Hadley Cell border and the Southern Annular Mode in Austral summer. {10.3.3, Table 10.1}

A Millennia to Multi-Century Perspective

Taking a longer term perspective shows the substantial role played by anthropogenic and natural forcings in driving climate variability on hemispheric scales prior to the twentieth century. It is *very unlikely* that NH temperature variations from 1400 to 1850 can be explained by internal variability alone. There is *medium confidence* that external forcing contributed to NH temperature variability from 850 to 1400 and that external forcing contributed to European temperature variations over the last five centuries. {10.7.2, 10.7.5, Table 10.1}

Climate System Properties

The extended record of observed climate change has allowed a better characterization of the basic properties of the climate system that have implications for future warming. New evidence from 21st century observations and stronger evidence from a wider range of studies have strengthened the constraint on the transient climate response (TCR) which is estimated with *high confidence* to be *likely* between 1°C and 2.5°C and *extremely unlikely* to be greater than 3°C. The Transient Climate Response to Cumulative CO_2 Emissions (TCRE) is estimated with *high confidence* to be *likely* between 0.8°C and 2.5°C per 1000 PgC for cumulative CO_2 emissions less than about 2000 PgC until the time at which temperatures peak. Estimates of the Equilibrium Climate Sensitivity (ECS) based on multiple and partly independent lines of evidence from observed climate change indicate that there is *high confidence* that ECS is *extremely unlikely* to be less than 1°C and *medium confidence* that the ECS is *likely* to be between 1.5°C and 4.5°C and *very unlikely* greater than 6°C. These assessments are consistent with the overall assessment in Chapter 12, where the inclusion of additional lines of evidence increases confidence in the assessed *likely* range for ECS. {10.8.1, 10.8.2, 10.8.4, Box 12.2}

10

Combination of Evidence

Human influence has been detected in the major assessed components of the climate system. Taken together, the combined evidence increases the level of confidence in the attribution of observed climate change, and reduces the uncertainties associated with assessment based on a single climate variable. From this combined evidence it is *virtually certain* that human influence has warmed the global climate system. Anthropogenic influence has been identified in changes in temperature near the surface of the Earth, in the atmosphere and in the oceans, as well as changes in the cryosphere, the water cycle and some extremes. There is strong evidence that excludes solar forcing, volcanoes and internal variability as the strongest drivers of warming since 1950. {10.9.2, Table 10.1}

10.1 Introduction

This chapter assesses the causes of observed changes assessed in Chapters 2 to 5 and uses understanding of physical processes, climate models and statistical approaches. The chapter adopts the terminology for detection and attribution proposed by the IPCC good practice guidance paper on detection and attribution (Hegerl et al., 2010) and for uncertainty Mastrandrea et al. (2011). Detection and attribution of impacts of climate changes are assessed by Working Group II, where Chapter 18 assesses the extent to which atmospheric and oceanic changes influence ecosystems, infrastructure, human health and activities in economic sectors.

Evidence of a human influence on climate has grown stronger over the period of the four previous assessment reports of the IPCC. There was little observational evidence for a detectable human influence on climate at the time of the First IPCC Assessment Report. By the time of the second report there was sufficient additional evidence for it to conclude that 'the balance of evidence suggests a discernible human influence on global climate'. The Third Assessment Report found that a distinct greenhouse gas (GHG) signal was robustly detected in the observed temperature record and that 'most of the observed warming over the last fifty years is *likely* to have been due to the increase in greenhouse gas concentrations.'

With the additional evidence available by the time of the Fourth Assessment Report, the conclusions were further strengthened. This evidence included a wider range of observational data, a greater variety of more sophisticated climate models including improved representations of forcings and processes and a wider variety of analysis techniques. This enabled the AR4 report to conclude that 'most of the observed increase in global average temperatures since the mid-20th century is *very likely* due to the observed increase in anthropogenic greenhouse gas concentrations'. The AR4 also concluded that 'discernible human influences now extend to other aspects of climate, including ocean warming, continental-average temperatures, temperature extremes and wind patterns.'

A number of uncertainties remained at the time of AR4. For example, the observed variability of ocean temperatures appeared inconsistent with climate models, thereby reducing the confidence with which observed ocean warming could be attributed to human influence. Also, although observed changes in global rainfall patterns and increases in heavy precipitation were assessed to be qualitatively consistent with expectations of the response to anthropogenic forcings, detection and attribution studies had not been carried out. Since the AR4, improvements have been made to observational data sets, taking more complete account of systematic biases and inhomogeneities in observational systems, further developing uncertainty estimates, and correcting detected data problems (Chapters 2 and 3). A new set of simulations from a greater number of AOGCMs have been performed as part of the Coupled Model Intercomparison Project Phase 5 (CMIP5). These new simulations have several advantages over the CMIP3 simulations assessed in the AR4 (Hegerl et al., 2007b). They incorporate some moderate increases in resolution, improved parameterizations, and better representation of aerosols (Chapter 9). Importantly for attribution, in which it is necessary to partition the response of the climate system to different forcings, most CMIP5 models include simulations of the response to natural forcings only, and the response to increases in well mixed GHGs only (Taylor et al., 2012).

The advances enabled by this greater wealth of observational and model data are assessed in this chapter. In this assessment, there is increased focus on the extent to which the climate system as a whole is responding in a coherent way across a suite of climate variables such as surface mean temperature, temperature extremes, ocean heat content, ocean salinity and precipitation change. There is also a global to regional perspective, assessing the extent to which not just global mean changes but also spatial patterns of change across the globe can be attributed to anthropogenic and natural forcings.

10.2 Evaluation of Detection and Attribution Methodologies

Detection and attribution methods have been discussed in previous assessment reports (Hegerl et al., 2007b) and the IPCC Good Practice Guidance Paper (Hegerl et al., 2010), to which we refer. This section reiterates key points and discusses new developments and challenges.

10.2.1 The Context of Detection and Attribution

In IPCC Assessments, detection and attribution involve quantifying the evidence for a causal link between external drivers of climate change and observed changes in climatic variables. It provides the central, although not the only (see Section 1.2.3) line of evidence that has supported statements such as 'the balance of evidence suggests a discernible human influence on global climate' or 'most of the observed increase in global average temperatures since the mid-20th century is *very likely* due to the observed increase in anthropogenic greenhouse gas concentrations.'

The definition of detection and attribution used here follows the terminology in the IPCC guidance paper (Hegerl et al., 2010). '*Detection* of change is defined as the process of demonstrating that climate or a system affected by climate has changed in some defined statistical sense without providing a reason for that change. An identified change is detected in observations if its likelihood of occurrence by chance due to internal variability alone is determined to be small' (Hegerl et al., 2010). *Attribution* is defined as 'the process of evaluating the relative contributions of multiple causal factors to a change or event with an assignment of statistical confidence'. As this wording implies, attribution is more complex than detection, combining statistical analysis with physical understanding (Allen et al., 2006; Hegerl and Zwiers, 2011). In general, a component of an observed change is attributed to a specific causal factor if the observations can be shown to be consistent with results from a process-based model that includes the causal factor in question, and inconsistent with an alternate, otherwise identical, model that excludes this factor. The evaluation of this consistency in both of these cases takes into account internal chaotic variability and known uncertainties in the observations and responses to external causal factors.

Attribution does not require, and nor does it imply, that every aspect of the response to the causal factor in question is simulated correctly. Suppose, for example, the global cooling following a large volcano matches the cooling simulated by a model, but the model underestimates the magnitude of this cooling: the observed global cooling can still be attributed to that volcano, although the error in magnitude would suggest that details of the model response may be unreliable. Physical understanding is required to assess what constitutes a plausible discrepancy above that expected from internal variability. Even with complete consistency between models and data, attribution statements can never be made with 100% certainty because of the presence of internal variability.

This definition of attribution can be extended to include antecedent conditions and internal variability among the multiple causal factors contributing to an observed change or event. Understanding the relative importance of internal versus external factors is important in the analysis of individual weather events (Section 10.6.2), but the primary focus of this chapter will be on attribution to factors external to the climate system, like rising GHG levels, solar variability and volcanic activity.

There are four core elements to any detection and attribution study:

1. Observations of one or more climate variables, such as surface temperature, that are understood, on physical grounds, to be relevant to the process in question

2. An estimate of how external drivers of climate change have evolved before and during the period under investigation, including both the driver whose influence is being investigated (such as rising GHG levels) and potential confounding influences (such as solar activity)

3. A quantitative physically based understanding, normally encapsulated in a model, of how these external drivers are thought to have affected these observed climate variables

4. An estimate, often but not always derived from a physically based model, of the characteristics of variability expected in these observed climate variables due to random, quasi-periodic and chaotic fluctuations generated in the climate system that are not due to externally driven climate change

A climate model driven with external forcing alone is not expected to replicate the observed evolution of internal variability, because of the chaotic nature of the climate system, but it should be able to capture the statistics of this variability (often referred to as 'noise'). The reliability of forecasts of short-term variability is also a useful test of the representation of relevant processes in the models used for attribution, but forecast skill is not necessary for attribution: attribution focuses on changes in the underlying moments of the 'weather attractor', meaning the expected weather and its variability, while prediction focuses on the actual trajectory of the weather around this attractor.

In proposing that 'the process of attribution requires the detection of a change in the observed variable *or closely associated variables*' (Hegerl et al., 2010), the new guidance recognized that it may be possible, in some instances, to attribute a change in a particular variable to some external factor before that change could actually be detected in the variable itself, provided there is a strong body of knowledge that links a change in that variable to some other variable in which a change can be detected and attributed. For example, it is impossible in principle to detect a trend in the frequency of 1-in-100-year events in a 100-year record, yet if the probability of occurrence of these events is physically related to large-scale temperature changes, and we detect and attribute a large-scale warming, then the new guidance allows attribution of a change in probability of occurrence before such a change can be detected in observations of these events alone. This was introduced to draw on the strength of attribution statements from, for example, time-averaged temperatures, to attribute changes in closely related variables.

Attribution of observed changes is not possible without some kind of model of the relationship between external climate drivers and observable variables. We cannot observe a world in which either anthropogenic or natural forcing is absent, so some kind of model is needed to set up and evaluate quantitative hypotheses: to provide estimates of how we would expect such a world to behave and to respond to anthropogenic and natural forcings (Hegerl and Zwiers, 2011). Models may be very simple, just a set of statistical assumptions, or very complex, complete global climate models: it is not necessary, or possible, for them to be correct in all respects, but they must provide a physically consistent representation of processes and scales relevant to the attribution problem in question.

One of the simplest approaches to detection and attribution is to compare observations with model simulations driven with natural forcings alone, and with simulations driven with all relevant natural and anthropogenic forcings. If observed changes are consistent with simulations that include human influence, and inconsistent with those that do not, this would be sufficient for attribution providing there were no other confounding influences and it is assumed that models are simulating the responses to all external forcings correctly. This is a strong assumption, and most attribution studies avoid relying on it. Instead, they typically assume that models simulate the *shape* of the response to external forcings (meaning the large-scale pattern in space and/or time) correctly, but do not assume that models simulate the *magnitude* of the response correctly. This is justified by our fundamental understanding of the origins of errors in climate modelling. Although there is uncertainty in the size of key forcings and the climate response, the overall shape of the response is better known: it is set in time by the timing of emissions and set in space (in the case of surface temperatures) by the geography of the continents and differential responses of land and ocean (see Section 10.3.1.1.2).

So-called 'fingerprint' detection and attribution studies characterize their results in terms of a best estimate and uncertainty range for 'scaling factors' by which the model-simulated responses to individual forcings can be scaled up or scaled down while still remaining consistent with the observations, accounting for similarities between the patterns of response to different forcings and uncertainty due to internal climate variability. If a scaling factor is significantly larger than zero (at some significance level), then the response to that forcing, as simulated by that model and given that estimate of internal variability and other potentially confounding responses, is detectable in these observations, whereas if the scaling factor is consistent with unity, then that model-simulated response is consistent with observed changes. Studies do not require scaling factors to be consistent with unity for attribution, but any discrepancy from unity should be understandable in terms of known uncertainties in forcing or response: a scaling factor of 10, for example, might suggest the presence of a confounding factor, calling into question any attribution claim. Scaling factors are estimated by fitting model-simulated responses to observations, so results are unaffected, at least to first order, if the model has a transient climate response, or aerosol forcing, that is too low or high. Conversely, if the spatial or temporal *pattern* of forcing or response is wrong, results can be affected: see Box 10.1 and further discussion in Section 10.3.1.1 and Hegerl and Zwiers (2011) and Hegerl et al. (2011b). Sensitivity of results to the pattern of forcing or response can be assessed by comparing results across multiple models or by representing pattern uncertainty explicitly (Huntingford et al., 2006), but errors that are common to all models (through limited vertical resolution, for example) will not be addressed in this way and are accounted for in this assessment by downgrading overall assessed likelihoods to be generally more conservative than the quantitative likelihoods provided by individual studies.

Attribution studies must compromise between estimating responses to different forcings separately, which allows for the possibility of different errors affecting different responses (errors in aerosol forcing that do not affect the response to GHGs, for example), and estimating responses to combined forcings, which typically gives smaller uncertainties because it avoids the issue of 'degeneracy': if two responses have very similar shapes in space and time, then it may be impossible to estimate the magnitude of both from a single set of observations because amplification of one may be almost exactly compensated for by amplification or diminution of the other (Allen et al., 2006). Many studies find it is possible to estimate the magnitude of the responses to GHG and other anthropogenic forcings separately, particularly when spatial information is included. This is important, because it means the estimated response to GHG increase is not dependent on the uncertain magnitude of forcing and response due to aerosols (Hegerl et al., 2011b).

The simplest way of fitting model-simulated responses to observations is to assume that the responses to different forcings add linearly, so the response to any one forcing can be scaled up or down without affecting any of the others and that internal climate variability is independent of the response to external forcing. Under these conditions, attribution can be expressed as a variant of linear regression (see Box 10.1). The additivity assumption has been tested and found to hold for large-scale temperature changes (Meehl et al., 2003; Gillett et al., 2004) but it might not hold for other variables like precipitation (Hegerl et al., 2007b; Hegerl and Zwiers, 2011; Shiogama et al., 2012), nor for regional temperature changes (Terray, 2012). In principle, additivity is not required for detection and attribution, but to date non-additive approaches have not been widely adopted.

The estimated properties of internal climate variability play a central role in this assessment. These are either estimated empirically from the observations (Section 10.2.2) or from paleoclimate reconstructions (Section 10.7.1) (Esper et al., 2012) or derived from control simulations of coupled models (Section 10.2.3). The majority of studies use modelled variability and routinely check that the residual variability from observations is consistent with modelled internal variability used over time scales shorter than the length of the instrumental record (Allen and Tett, 1999). Assessing the accuracy of model-simulated variability on longer time scales using paleoclimate reconstructions is complicated by the fact that some reconstructions may not capture the full spectrum of variability because of limitations of proxies and reconstruction methods, and by the unknown role of external forcing in the pre-instrumental record. In general, however, paleoclimate reconstructions provide no clear evidence either way whether models are over- or underestimating internal variability on time scales relevant for attribution (Esper et al., 2012; Schurer et al., 2013).

10.2.2 Time Series Methods, Causality and Separating Signal from Noise

Some studies attempt to distinguish between externally driven climate change and changes due to internal variability minimizing the use of climate models, for example, by separating signal and noise by time scale (Schneider and Held, 2001), spatial pattern (Thompson et al., 2009) or both. Other studies use model control simulations to identify patterns of maximum predictability and contrast these with the forced component in climate model simulations (DelSole et al., 2011): see Section 10.3.1. Conclusions of most studies are consistent with those based on fingerprint detection and attribution, while using a different set of assumptions (see review in Hegerl and Zwiers, 2011).

A number of studies have applied methods developed in the econometrics literature (Engle and Granger, 1987) to assess the evidence for a causal link between external drivers of climate and observed climate change, using the observations themselves to estimate the expected properties of internal climate variability (e.g., Kaufmann and Stern, 1997). The advantage of these approaches is that they do not depend on the accuracy of any complex global climate model, but they nevertheless have to assume some kind of model, or restricted class of models, of the properties of the variables under investigation. Attribution is impossible without a model: although this model may be implicit in the statistical framework used, it is important to assess its physical consistency (Kaufmann et al., 2013). Many of these time series methods can be cast in the overall framework of co-integration and error correction (Kaufmann et al., 2011), which is an approach to analysing relationships between stationary and non-stationary time series. If there is a consistent causal relationship between two or more possibly non-stationary time series, then it should be possible to find a linear combination such that the residual is stationary (contains no stochastic trend) over time (Kaufmann and Stern, 2002; Kaufmann et al., 2006; Mills, 2009). Co-integration methods are thus similar in overall principle to regression-based approaches (e.g., Douglass et al., 2004; Stone and Allen, 2005; Lean, 2006) to the extent that regression studies take into account the expected time series properties of the data—the example described in Box 10.1 might be characterized as looking for a linear combination of anthropogenic and natural forcings such that the observed residuals were consistent with internal climate variability as simulated by the CMIP5 models. Co-integration and error correction methods, however, generally make more explicit use of time

Box 10.1 | How Attribution Studies Work

This box presents an idealized demonstration of the concepts underlying most current approaches to detection and attribution of climate change and how these relate to conventional linear regression. The coloured dots in Box 10.1a, Figure 1 show observed annual GMST from 1861 to 2012, with warmer years coloured red and colder years coloured blue. Observations alone indicate, unequivocally, that the Earth has warmed, but to quantify how different external factors have contributed to this warming, studies must compare such observations with the expected responses to these external factors. The orange line shows an estimate of the GMST response to anthropogenic (GHG and aerosol) forcing obtained from the mean of the CMIP3 and CMIP5 ensembles, while the blue line shows the CMIP3/CMIP5 ensemble mean response to natural (solar and volcanic) forcing.

In statistical terms, attribution involves finding the combination of these anthropogenic and natural responses that best fits these observations: this is shown by the black line in panel (a). To show how this fit is obtained in non-technical terms, the data are plotted against model-simulated anthropogenic warming, instead of time, in panel (b). There is a strong correlation between observed temperatures and model-simulated anthropogenic warming, but because of the presence of natural factors and internal climate variability, correlation alone is not enough for attribution.

To quantify how much of the observed warming is attributable to human influence, panel (c) shows observed temperatures plotted against the model-simulated response to anthropogenic forcings in one direction and natural forcings in the other. Observed temperatures increase with both natural and anthropogenic model-simulated warming: the warmest years are in the far corner of the box. A flat surface through these points (here obtained by an ordinary least-squares fit), indicated by the coloured mesh, slopes up away from the viewer.

The orientation of this surface indicates how model-simulated responses to natural and anthropogenic forcing need to be scaled to reproduce the observations. The best-fit gradient in the direction of anthropogenic warming (visible on the rear left face of the box) is 0.9, indicating the CMIP3/CMIP5 ensemble average overestimates the magnitude of the observed response to anthropogenic forcing by about 10%. The best-fit gradient in the direction of natural changes (visible on the rear right face) is 0.7, indicating that the observed response to natural forcing is 70% of the average model-simulated response. The black line shows the points on this flat surface that are directly above or below the observations: each 'pin' corresponds to a different year. When re-plotted against time, indicated by the years on the rear left face of the box, this black line gives the black line previously seen in panel (a). The length of the pins indicates 'residual' temperature fluctuations due to internal variability.

The timing of these residual temperature fluctuations is unpredictable, representing an inescapable source of uncertainty. We can quantify this uncertainty by asking how the gradients of the best-fit surface might vary if El Niño events, for example, had occurred in different years in the observed temperature record. To do this, we repeat the analysis in panel (c), replacing observed temperatures with samples of simulated internal climate variability from control runs of coupled climate models. Grey diamonds in panel (d) show the results: these gradients cluster around zero, because control runs have no anthropogenic or natural forcing, but there is still some scatter. Assuming that internal variability in global temperature simply adds to the response to external forcing, this scatter provides an estimate of uncertainty in the gradients, or scaling factors, required to reproduce the observations, shown by the red cross and ellipse.

The red cross and ellipse are clearly separated from the origin, which means that the slope of the best-fit surface through the observations cannot be accounted for by internal variability: some climate change is detected in these observations. Moreover, it is also separated from both the vertical and horizontal axes, which means that the responses to both anthropogenic and natural factors are individually detectable.

The magnitude of observed temperature change is consistent with the CMIP3/CMIP5 ensemble average response to anthropogenic forcing (uncertainty in this scaling factor spans unity) but is significantly lower than the model-average response to natural forcing (this 5 to 95% confidence interval excludes unity). There are, however, reasons why these models may be underestimating the response to volcanic forcing (e.g., Driscoll et al, 2012), so this discrepancy does not preclude detection and attribution of both anthropogenic and natural influence, as simulated by the CMIP3/CMIP5 ensemble average, in the observed GMST record.

The top axis in panel (d) indicates the attributable anthropogenic warming over 1951–2010, estimated from the anthropogenic warming in the CMIP3/CMIP5 ensemble average, or the gradient of the orange line in panel (a) over this period. Because the model-simulated responses have been scaled to fit the observations, the attributable anthropogenic warming in this example is 0.6°C to 0.9°C and does not depend on the magnitude of the raw model-simulated changes. Hence an attribution statement based on such an analysis,

(continued on next page)

Box 10.1 (continued)

such as 'most of the warming over the past 50 years is attributable to anthropogenic drivers', depends only on the shape, or time history, not the size, of the model-simulated warming, and hence does not depend on the models' sensitivity to rising GHG levels.

Formal attribution studies like this example provide objective estimates of how much recent warming is attributable to human influence. Attribution is not, however, a purely statistical exercise. It also requires an assessment that there are no confounding factors that could have caused a large part of the 'attributed' change. Statistical tests can be used to check that observed residual temperature fluctuations (the lengths and clustering of the pins in panel (c)) are consistent with internal variability expected from coupled models, but ultimately these tests must complement physical arguments that the combination of responses to anthropogenic and natural forcing is the only available consistent explanation of recent observed temperature change.

This demonstration assumes, for visualization purposes, that there are only two candidate contributors to the observed warming, anthropogenic and natural, and that only GMST is available. More complex attribution problems can be undertaken using the same principles, such as aiming to separate the response to GHGs from other anthropogenic factors by also including spatial information. These require, in effect, an extension of panel (c), with additional dimensions corresponding to additional causal factors, and additional points corresponding to temperatures in different regions.

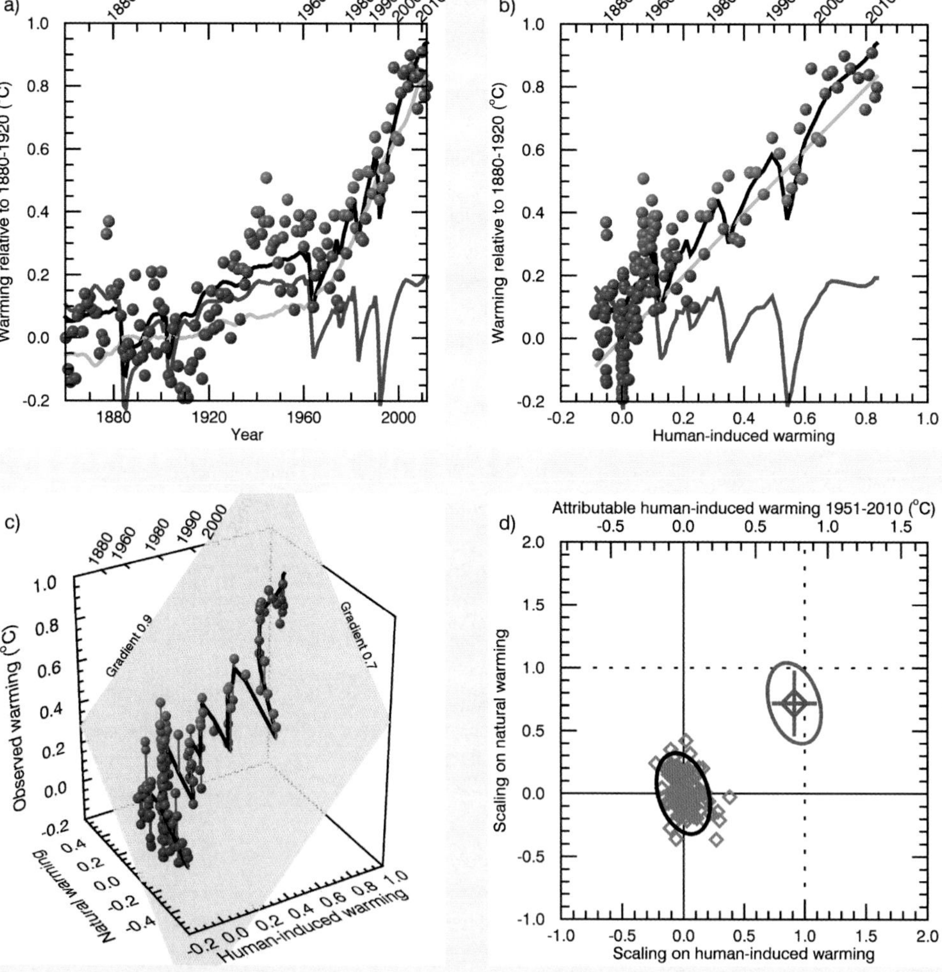

Box 10.1, Figure 1 | Example of a simplified detection and attribution study. (a) Observed global annual mean temperatures relative to 1880–1920 (coloured dots) compared with CMIP3/CMIP5 ensemble-mean response to anthropogenic forcing (orange), natural forcing (blue) and best-fit linear combination (black). (b) As (a) but all data plotted against model-simulated anthropogenic warming in place of time. Selected years (increasing nonlinearly) shown on top axis. (c) Observed temperatures versus model-simulated anthropogenic and natural temperature changes, with best-fit plane shown by coloured mesh. (d) Gradient of best-fit plane in (c), or scaling on model-simulated responses required to fit observations (red diamond) with uncertainty estimate (red ellipse and cross) based on CMIP5 control integrations (grey diamonds). Implied attributable anthropogenic warming over the period 1951–2010 is indicated by the top axis. Anthropogenic and natural responses are noise-reduced with 5-point running means, with no smoothing over years with major volcanoes.

series properties (notice how date information is effectively discarded in panel (b) of Box 10.1, Figure 1) and require fewer assumptions about the stationarity of the input series.

All of these approaches are subject to the issue of confounding factors identified by Hegerl and Zwiers (2011). For example, Beenstock et al. (2012) fail to find a consistent co-integrating relationship between atmospheric carbon dioxide (CO_2) concentrations and GMST using polynomial cointegration tests, but the fact that CO_2 concentrations are derived from different sources in different periods (ice cores prior to the mid-20th-century, atmospheric observations thereafter) makes it difficult to assess the physical significance of their result, particularly in the light of evidence for co-integration between temperature and radiative forcing (RF) reported by Kaufmann et al. (2011) using tests of linear cointegration, and also the results of Gay-Garcia et al. (2009), who find evidence for external forcing of climate using time series properties.

The assumptions of the statistical model employed can also influence results. For example, Schlesinger and Ramankutty (1994) and Zhou and Tung (2013a) show that GMST are consistent with a linear anthropogenic trend, enhanced variability due to an approximately 70-year Atlantic Meridional Oscillation (AMO) and shorter-term variability. If, however, there are physical grounds to expect a nonlinear anthropogenic trend (see Box 10.1 Figure 1a), the assumption of a linear trend can itself enhance the variance assigned to a low-frequency oscillation. The fact that the AMO index is estimated from detrended historical temperature observations further increases the risk that its variance may be overestimated, because regressors and regressands are not independent. Folland et al. (2013), using a physically based estimate of the anthropogenic trend, find a smaller role for the AMO in recent warming.

Time series methods ultimately depend on the structural adequacy of the statistical model employed. Many such studies, for example, use models that assume a single exponential decay time for the response to both external forcing and stochastic fluctuations. This can lead to an overemphasis on short-term fluctuations, and is not consistent with the response of more complex models (Knutti et al., 2008). Smirnov and Mokhov (2009) propose an alternative characterization that allows them to distinguish a 'long-term causality' that focuses on low-frequency changes. Trends that appear significant when tested against an AR(1) model may not be significant when tested against a process that supports this 'long-range dependence' (Franzke, 2010). Although the evidence for long-range dependence in global temperature data remains a topic of debate (Mann, 2011; Rea et al., 2011) , it is generally desirable to explore sensitivity of results to the specification of the statistical model, and also to other methods of estimating the properties of internal variability, such as more complex climate models, discussed next. For example, Imbers et al. (2013) demonstrate that the detection of the influence of increasing GHGs in the global temperature record is robust to the assumption of a Fractional Differencing (FD) model of internal variability, which supports long-range dependence.

10.2.3 Methods Based on General Circulation Models and Optimal Fingerprinting

Fingerprinting methods use climate model simulations to provide more complete information about the expected response to different external drivers, including spatial information, and the properties of internal climate variability. This can help to separate patterns of forced change both from each other and from internal variability. The price, however, is that results depend to some degree on the accuracy of the shape of model-simulated responses to external factors (e.g., North and Stevens, 1998), which is assessed by comparing results obtained with expected responses estimated from different climate models. When the signal-to-noise (S/N) ratio is low, as can be the case for some regional indicators and some variables other than temperature, the accuracy of the specification of variability becomes a central factor in the reliability of any detection and attribution study. Many studies of such variables inflate the variability estimate from models to determine if results are sensitive to, for example, doubling of variance in the control (e.g., Zhang et al., 2007), although Imbers et al. (2013) note that errors in the spectral properties of simulated variability may also be important.

A full description of optimal fingerprinting is provided in Appendix 9.A of Hegerl et al. (2007b) and further discussion is to be found in Hasselmann (1997), Allen and Tett (1999), Allen et al. (2006), and Hegerl and Zwiers (2011). Box 10.1 provides a simple example of 'fingerprinting' based on GMST alone. In a typical fingerprint analysis, model-simulated spatio-temporal patterns of response to different combinations of external forcings, including segments of control integrations with no forcing, are 'observed' in a similar manner to the historical record (masking out times and regions where observations are absent). The magnitudes of the model-simulated responses are then estimated in the observations using a variant of linear regression, possibly allowing for signals being contaminated by internal variability (Allen and Stott, 2003) and structural model uncertainty (Huntingford et al., 2006).

In 'optimal' fingerprinting, model-simulated responses and observations are normalized by internal variability to improve the S/N ratio. This requires an estimate of the inverse noise covariance estimated from the sample covariance matrix of a set of unforced (control) simulations (Hasselmann, 1997), or from variations within an initial-condition ensemble. Because these control runs are generally too short to estimate the full covariance matrix, a truncated version is used, retaining only a small number, typically of order 10 to 20, of high-variance principal components. Sensitivity analyses are essential to ensure results are robust to this, relatively arbitrary, choice of truncation (Allen and Tett, 1999; Ribes and Terray, 2013; Jones et al., 2013). Ribes et al. (2009) use a regularized estimate of the covariance matrix, meaning a linear combination of the sample covariance matrix and a unit matrix that has been shown (Ledoit and Wolf, 2004) to provide a more accurate estimate of the true covariance, thereby avoiding dependence on truncation. Optimization of S/N ratio is not, however, essential for many attribution results (see, e.g., Box 10.1) and uncertainty analysis in conventional optimal fingerprinting does not require the covariance matrix to be inverted, so although regularization may help in some cases, it is not essential. Ribes et al. (2010) also propose a hybrid of the model-based optimal fingerprinting and time series approaches, referred to as 'temporal optimal detection', under which each signal is assumed to consist of a single spatial pattern modulated by a smoothly varying time series estimated from a climate model (see also Santer et al., 1994).

The final statistical step in an attribution study is to check that the residual variability, after the responses to external drivers have been estimated and removed, is consistent with the expected properties of internal climate variability, to ensure that the variability used for uncertainty analysis is realistic, and that there is no evidence that a potentially confounding factor has been omitted. Many studies use a standard *F*-test of residual consistency for this purpose (Allen and Tett, 1999). Ribes et al. (2013) raise some issues with this test, but key results are not found to be sensitive to different formulations. A more important issue is that the *F*-test is relatively weak (Berliner et al., 2000; Allen et al., 2006; Terray, 2012), so 'passing' this test is not a safeguard against unrealistic variability, which is why estimates of internal variability are discussed in detail in this chapter and in Chapter 9.

A further consistency check often used in optimal fingerprinting is whether the estimated magnitude of the externally driven responses are consistent between model and observations (scaling factors consistent with unity in Box 10.1): if they are not, attribution is still possible provided the discrepancy is explicable in terms of known uncertainties in the magnitude of either forcing or response. As is emphasized in Section 10.2.1 and Box 10.1, attribution is not a purely statistical assessment: physical judgment is required to assess whether the combination of responses considered allows for all major potential confounding factors and whether any remaining discrepancies are consistent with a physically based understanding of the responses to external forcing and internal climate variability.

10.2.4 Single-Step and Multi-Step Attribution and the Role of the Null Hypothesis

Attribution studies have traditionally involved explicit simulation of the response to external forcing of an observable variable, such as surface temperature, and comparison with corresponding observations of that variable. This so-called 'single-step attribution' has the advantage of simplicity, but restricts attention to variables for which long and consistent time series of observations are available and that can be simulated explicitly in current models driven solely with external climate forcing.

To address attribution questions for variables for which these conditions are not satisfied, Hegerl et al. (2010) introduced the notation of 'multi-step attribution', formalizing existing practice (e.g., Stott et al., 2004). In a multi-step attribution study, the attributable change in a variable such as large-scale surface temperature is estimated with a single-step procedure, along with its associated uncertainty, and the implications of this change are then explored in a further (physically or statistically based) modelling step. Overall conclusions can only be as robust as the least certain link in the multi-step procedure. As the focus shifts towards more noisy regional changes, it can be difficult to separate the effect of different external forcings. In such cases, it can be useful to detect the response to all external forcings, and then determine the most important factors underlying the attribution results by reference to a closely related variable for which a full attribution analysis is available (e.g., Morak et al., 2011).

Attribution results are typically expressed in terms of conventional 'frequentist' confidence intervals or results of hypothesis tests: when it is reported that the response to anthropogenic GHG increase is *very likely* greater than half the total observed warming, it means that the null hypothesis that the GHG-induced warming is less than half the total can be rejected with the data available at the 10% significance level. Expert judgment is required in frequentist attribution assessments, but its role is limited to the assessment of whether internal variability and potential confounding factors have been adequately accounted for, and to downgrade nominal significance levels to account for remaining uncertainties. Uncertainties may, in some cases, be further reduced if prior expectations regarding attribution results themselves are incorporated, using a Bayesian approach, but this not currently the usual practice.

This traditional emphasis on single-step studies and placing lower bounds on the magnitude of signals under investigation means that, very often, the communication of attribution results tends to be conservative, with attention focussing on whether or not human influence in a particular variable might be zero, rather than the upper end of the confidence interval, which might suggest a possible response much bigger than current model-simulated changes. Consistent with previous Assessments and the majority of the literature, this chapter adopts this conservative emphasis. It should, however, be borne in mind that this means that positive attribution results will tend to be biased towards well-observed, well-modelled variables and regions, which should be taken into account in the compilation of global impact assessments (Allen, 2011; Trenberth, 2011a).

10.3 Atmosphere and Surface

This section assesses causes of change in the atmosphere and at the surface over land and ocean.

10.3.1 Temperature

Temperature is first assessed near the surface of the Earth in Section 10.3.1.1 and then in the free atmosphere in Section 10.3.1.2.

10.3.1.1 Surface (Air Temperature and Sea Surface Temperature)

10.3.1.1.1 Observations of surface temperature change

GMST warmed strongly over the period 1900–1940, followed by a period with little trend, and strong warming since the mid-1970s (Section 2.4.3, Figure 10.1). Almost all observed locations have warmed since 1901 whereas over the satellite period since 1979 most regions have warmed while a few regions have cooled (Section 2.4.3; Figure 10.2). Although this picture is supported by all available global near-surface temperature data sets, there are some differences in detail between them, but these are much smaller than both interannual variability and the long-term trend (Section 2.4.3). Since 1998 the trend in GMST has been small (see Section 2.4.3, Box 9.2). Urbanization is *unlikely* to have caused more than 10% of the measured centennial trend in land mean surface temperature, though it may have contributed substantially more to regional mean surface temperature trends in rapidly developing regions (Section 2.4.1.3).

10.3.1.1.2 Simulations of surface temperature change

As discussed in Section 10.1, the CMIP5 simulations have several advantages compared to the CMIP3 simulations assessed by (Hegerl et al., 2007b) for the detection and attribution of climate change. Figure 10.1a shows that when the effects of anthropogenic and natural external forcings are included in the CMIP5 simulations the spread of simulated GMST anomalies spans the observational estimates of GMST anomaly in almost every year whereas this is not the case for simulations in which only natural forcings are included (Figure 10.1b) (see also Jones et al., 2013; Knutson et al., 2013). Anomalies are shown relative to 1880–1919 rather than absolute temperatures. Showing anomalies is necessary to prevent changes in observational coverage being reflected in the calculated global mean and is reasonable

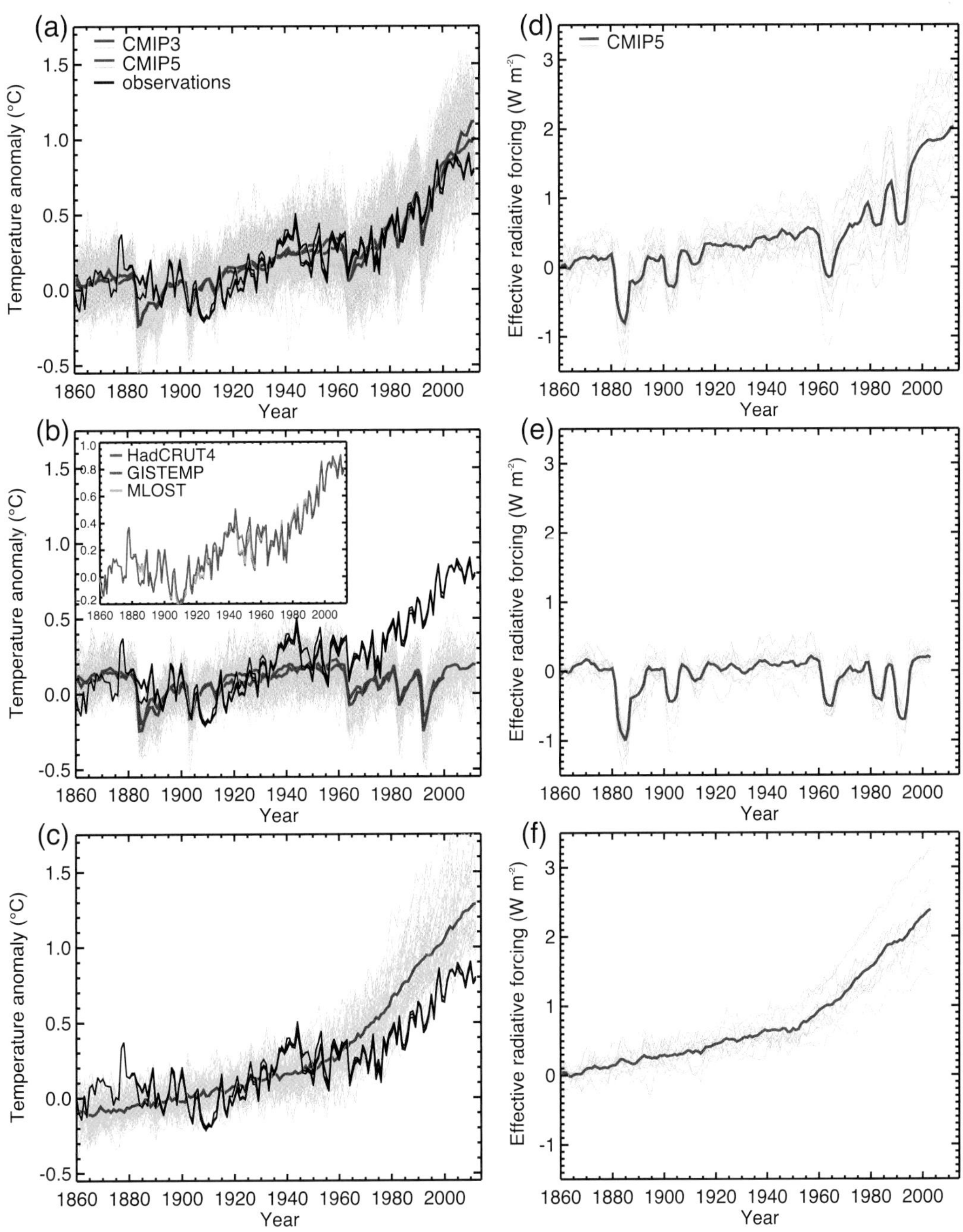

Figure 10.1 | (Left-hand column) Three observational estimates of global mean surface temperature (GMST, black lines) from Hadley Centre/Climatic Research Unit gridded surface temperature data set 4 (HadCRUT4), Goddard Institute of Space Studies Surface Temperature Analysis (GISTEMP), and Merged Land–Ocean Surface Temperature Analysis (MLOST), compared to model simulations [CMIP3 models – thin blue lines and CMIP5 models – thin yellow lines] with anthropogenic and natural forcings (a), natural forcings only (b) and greenhouse gas (GHG) forcing only (c). Thick red and blue lines are averages across all available CMIP5 and CMIP3 simulations respectively. CMIP3 simulations were not available for GHG forcing only (c). All simulated and observed data were masked using the HadCRUT4 coverage (as this data set has the most restricted spatial coverage), and global average anomalies are shown with respect to 1880–1919, where all data are first calculated as anomalies relative to 1961–1990 in each grid box. Inset to (b) shows the three observational data sets distinguished by different colours. (Adapted from Jones et al., 2013.) (Right-hand column) Net adjusted forcing in CMIP5 models due to anthropogenic and natural forcings (d), natural forcings only (e) and GHGs only (f). (From Forster et al., 2013.) Individual ensemble members are shown by thin yellow lines, and CMIP5 multi-model means are shown as thick red lines.

because climate sensitivity is not a strong function of the bias in GMST in the CMIP5 models (Section 9.7.1; Figure 9.42). Simulations with GHG changes only, and no changes in aerosols or other forcings, tend to simulate more warming than observed (Figure 10.1c), as expected. Better agreement between models and observations when the models include anthropogenic forcings is also seen in the CMIP3 simulations (Figure 10.1, thin blue lines). RF in the simulations including anthropogenic and natural forcings differs considerably among models (Figure 10.1d), and forcing differences explain much of the differences in temperature response between models over the historical period (Forster et al., 2013). Differences between observed GMST based on three observational data sets are small compared to forced changes (Figure 10.1).

As discussed in Section 10.2, detection and attribution assessments are more robust if they consider more than simple consistency arguments. Analyses that allow for the possibility that models might be consistently over- or underestimating the magnitude of the response to climate forcings are assessed in Section 10.3.1.1.3, the conclusions from which are not affected by evidence that model spread in GMST in CMIP3, is smaller than implied by the uncertainty in RF (Schwartz et al., 2007). Although there is evidence that CMIP3 models with a higher climate sensitivity tend to have a smaller increase in RF over the historical period (Kiehl, 2007; Knutti, 2008; Huybers, 2010), no such relationship was found in CMIP5 (Forster et al., 2013) which may explain the wider spread of the CMIP5 ensemble compared to the CMIP3 ensemble (Figure 10.1a). Climate model parameters are typically chosen primarily to reproduce features of the mean climate and variability (Box 9.1), and CMIP5 aerosol emissions are standardized across models and based on historical emissions (Lamarque et al., 2010; Section 8.2.2), rather than being chosen by each modelling group independently (Curry and Webster, 2011; Hegerl et al., 2011c).

Figure 10.2a shows the pattern of annual mean surface temperature trends observed over the period 1901–2010, based on Hadley Centre/Climatic Research Unit gridded surface temperature data set 4 (HadCRUT4). Warming has been observed at almost all locations with sufficient observations available since 1901. Rates of warming are generally higher over land areas compared to oceans, as is also apparent over the 1951–2010 period (Figure 10.2c), which simulations indicate is due mainly to differences in local feedbacks and a net anomalous heat transport from oceans to land under GHG forcing, rather than differences in thermal inertia (e.g., Boer, 2011). Figure 10.2e demonstrates that a similar pattern of warming is simulated in the CMIP5 simulations with natural and anthropogenic forcing over the 1901–2010 period. Over most regions, observed trends fall between the 5th and 95th percentiles of simulated trends, and van Oldenborgh et al. (2013) find that over the 1950–2011 period the pattern of observed grid cell trends agrees with CMIP5 simulated trends to within a combination of

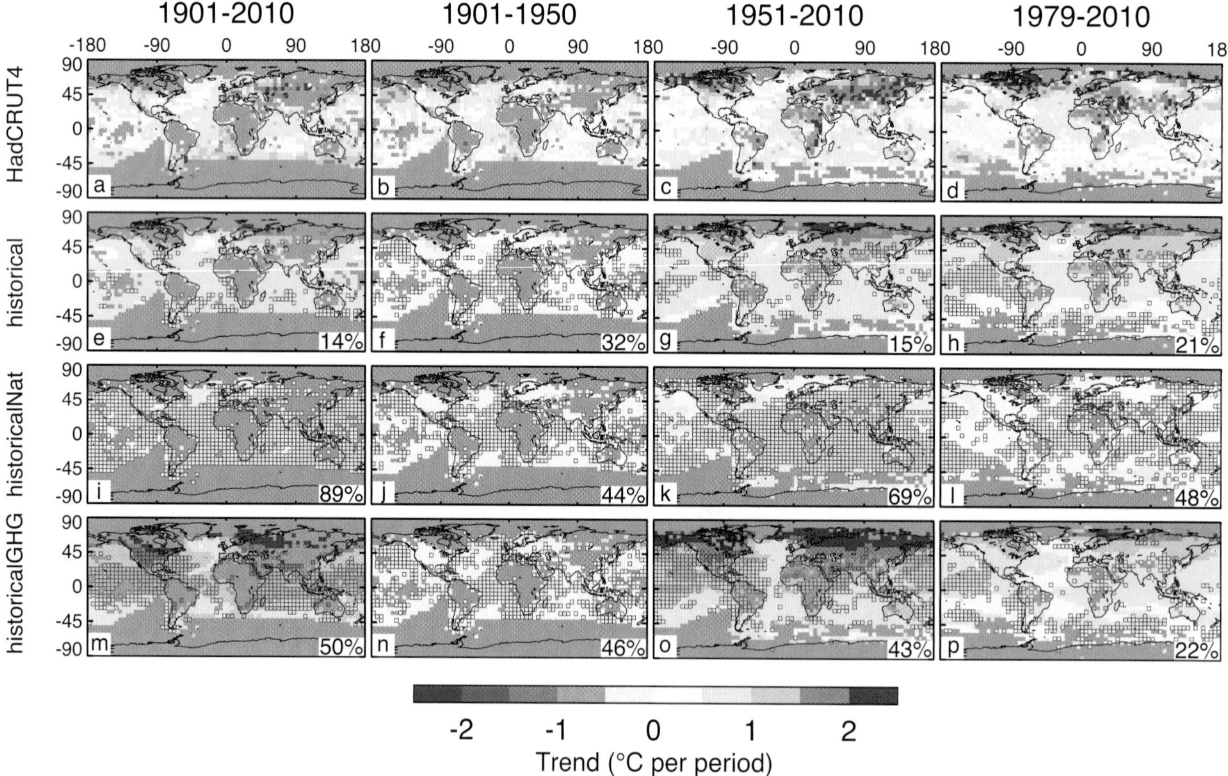

Figure 10.2 | Trends in observed and simulated temperatures (K over the period shown) over the 1901–2010 (a, e, i, m), 1901–1950 (b, f, j, n), 1951–2010 (c, g, k, o) and 1979–2010 (d, h, l, p) periods. Trends in observed temperatures from the Hadley Centre/Climatic Research Unit gridded surface temperature data set 4 (HadCRUT4) (a–d), CMIP3 and CMIP5 model simulations including anthropogenic and natural forcings (e–h), CMIP3 and CMIP5 model simulations including natural forcings only (i–l) and CMIP3 and CMIP5 model simulations including greenhouse gas forcing only (m–p). Trends are shown only where sufficient observational data are available in the HadCRUT4 data set, and grid cells with insufficient observations to derive trends are shown in grey. Boxes in (e–p) show where the observed trend lies outside the 5 to 95th percentile range of simulated trends, and the ratio of the number of such grid cells to the total number of grid cells with sufficient data is shown as a percentage in the lower right of each panel. (Adapted from Jones et al., 2013.)

model spread and internal variability. Areas of disagreement over the 1901–2010 period include parts of Asia and the Southern Hemisphere (SH) mid-latitudes, where the simulations warm less than the observations, and parts of the tropical Pacific, where the simulations warm more than the observations (Jones et al., 2013; Knutson et al., 2013). Stronger warming in observations than models over parts of East Asia could in part be explained by uncorrected urbanization influence in the observations (Section 2.4.1.3), or by an overestimate of the response to aerosol increases. Trends simulated in response to natural forcings only are generally close to zero, and inconsistent with observed trends in most locations (Figure 10.2i) (see also Knutson et al., 2013). Trends simulated in response to GHG changes only over the 1901–2010 period are larger than those observed at most locations, and in many cases significantly so (Figure 10.2m). This is expected because these simulations do not include the cooling effects of aerosols. Differences in patterns of simulated and observed seasonal mean temperature trends and possible causes are considered in more detail in Box 11.2.

Over the period 1979–2010 most observed regions exhibited warming (Figure 10.2d), but much of the eastern Pacific and Southern Oceans cooled. These regions of cooling are not seen in the simulated trends over this period in response to anthropogenic and natural forcing (Figure 10.2h), which show significantly more warming in much of these regions (Jones et al., 2013; Knutson et al., 2013). This cooling and reduced warming in observations over the Southern Hemisphere mid-latitudes over the 1979–2010 period can also be seen in the zonal mean trends (Figure 10.3d), which also shows that the models tend to warm too much in this region over this period. However, there is no discrepancy in zonal mean temperature trends over the longer 1901–2010 period in this region (Figure 10.3a), suggesting that the discrepancy over the 1979–2010 period either may be an unusually strong manifestation of internal variability in the observations or relate to regionally important forcings over the past three decades which are not included in most CMIP5 simulations, such as sea salt aerosol increases due to strengthened high latitude winds (Korhonen et al., 2010), or sea ice extent increases driven by freshwater input from ice shelf melting (Bintanja et al., 2013). Except at high latitudes, zonal mean trends over the 1901–2010 period in all three data sets are inconsistent with naturally forced trends, indicating a detectable anthropogenic signal in most zonal means over this period (Figure 10.3a). McKitrick and Tole (2012) find that few CMIP3 models have significant explanatory power when fitting the spatial pattern of 1979–2002 trends in surface temperature over land, by which they mean that these models add little or no skill to a fit including the spatial pattern of tropospheric temperature trends as well as the major atmospheric oscillations. This is to be expected, as temperatures in the troposphere are well correlated in the vertical, and local temperature trends over so short a period are dominated by internal variability.

CMIP5 models generally exhibit realistic variability in GMST on decadal to multi-decadal time scales (Jones et al., 2013; Knutson et al., 2013; Section 9.5.3.1, Figure 9.33), although it is difficult to evaluate internal variability on multi-decadal time scales in observations given the shortness of the observational record and the presence of external forcing. The observed trend in GMST since the 1950s is very large compared to model estimates of internal variability (Stott et al., 2010; Drost et al., 2012; Drost and Karoly, 2012). Knutson et al. (2013) compare observed

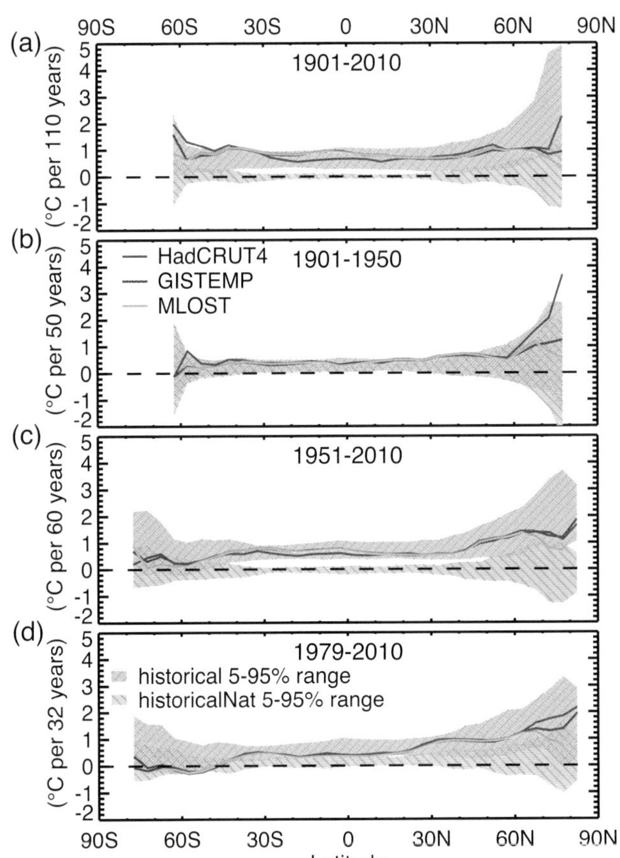

Figure 10.3 | Zonal mean temperature trends over the 1901–2010 (a), 1901–1950 (b), 1951–2010 (c) and 1979–2010 (d) periods. Solid lines show Hadley Centre/Climatic Research Unit gridded surface temperature data set 4 (HadCRUT4, red), Goddard Institute of Space Studies Surface Temperature Analysis (GISTEMP, brown) and Merged Land–Ocean Surface Temperature Analysis (MLOST, green) observational data sets, orange hatching represents the 90% central range of CMIP3 and CMIP5 simulations with anthropogenic and natural forcings, and blue hatching represents the 90% central range of CMIP3 and CMIP5 simulations with natural forcings only. All model and observations data are masked to have the same coverage as HadCRUT4. (Adapted from Jones et al., 2013.)

trends in GMST with a combination of simulated internal variability and the response to natural forcings and find that the observed trend would still be detected for trends over this period even if the magnitude of the simulated natural variability (i.e., the standard deviation of trends) were tripled.

10.3.1.1.3 Attribution of observed global-scale temperature changes

The evolution of temperature since the start of the global instrumental record

Since the AR4, detection and attribution studies have been carried out using new model simulations with more realistic forcings, and new observational data sets with improved representation of uncertainty (Christidis et al., 2010; Jones et al., 2011, 2013; Gillett et al., 2012, 2013; Stott and Jones, 2012; Knutson et al., 2013; Ribes and Terray, 2013). Although some inconsistencies between the simulated and observed responses to forcings in individual models were identified (Gillett et al., 2013; Jones et al., 2013; Ribes and Terray, 2013) overall these results support the AR4 assessment that GHG increases *very likely* caused most (>50%) of the observed GMST increase since the mid-20th century (Hegerl et al., 2007b).

The results of multiple regression analyses of observed temperature changes onto the simulated responses to GHG, other anthropogenic and natural forcings are shown in Figure 10.4 (Gillett et al., 2013; Jones et al., 2013; Ribes and Terray, 2013). The results, based on HadCRUT4 and a multi-model average, show robustly detected responses to GHG in the observational record whether data from 1861–2010 or only from 1951–2010 are analysed (Figure 10.4b). The advantage of analysing the longer period is that more information on observed and modelled changes is included, while a disadvantage is that it is difficult to validate climate models' estimates of internal variability over such a long period. Individual model results exhibit considerable spread among scaling factors, with estimates of warming attributable to each forcing sensitive to the model used for the analsys (Figure 10.4; Gillett et al., 2013; Jones et al., 2013; Ribes and Terray, 2013), the period over which the analysis is applied (Figure 10.4; Gillett et al., 2013; Jones et al., 2013), and the Empirical Orthogonal Function (EOF) truncation or degree of spatial filtering (Jones et al., 2013; Ribes and Terray, 2013). In some cases the GHG response is not detectable in regressions using individual models (Figure 10.4; Gillett et al., 2013; Jones et al., 2013; Ribes and Terray, 2013), or a residual test is failed (Gillett et al., 2013; Jones et al., 2013; Ribes and Terray, 2013), indicating a poor fit between the simulated response and observed changes. Such cases are probably due largely to errors in the spatio-temporal pattern of responses to forcings simulated in individual models (Ribes and Terray, 2013), although observational error and internal variability errors could also play a role. Nonetheless, analyses in which responses are averaged across multiple models generally show much less sensitivity to period and EOF trucation (Gillett et al., 2013; Jones et al., 2013), and more consistent residuals (Gillett et al., 2013), which may be because model response errors are smaller in a multi-model mean.

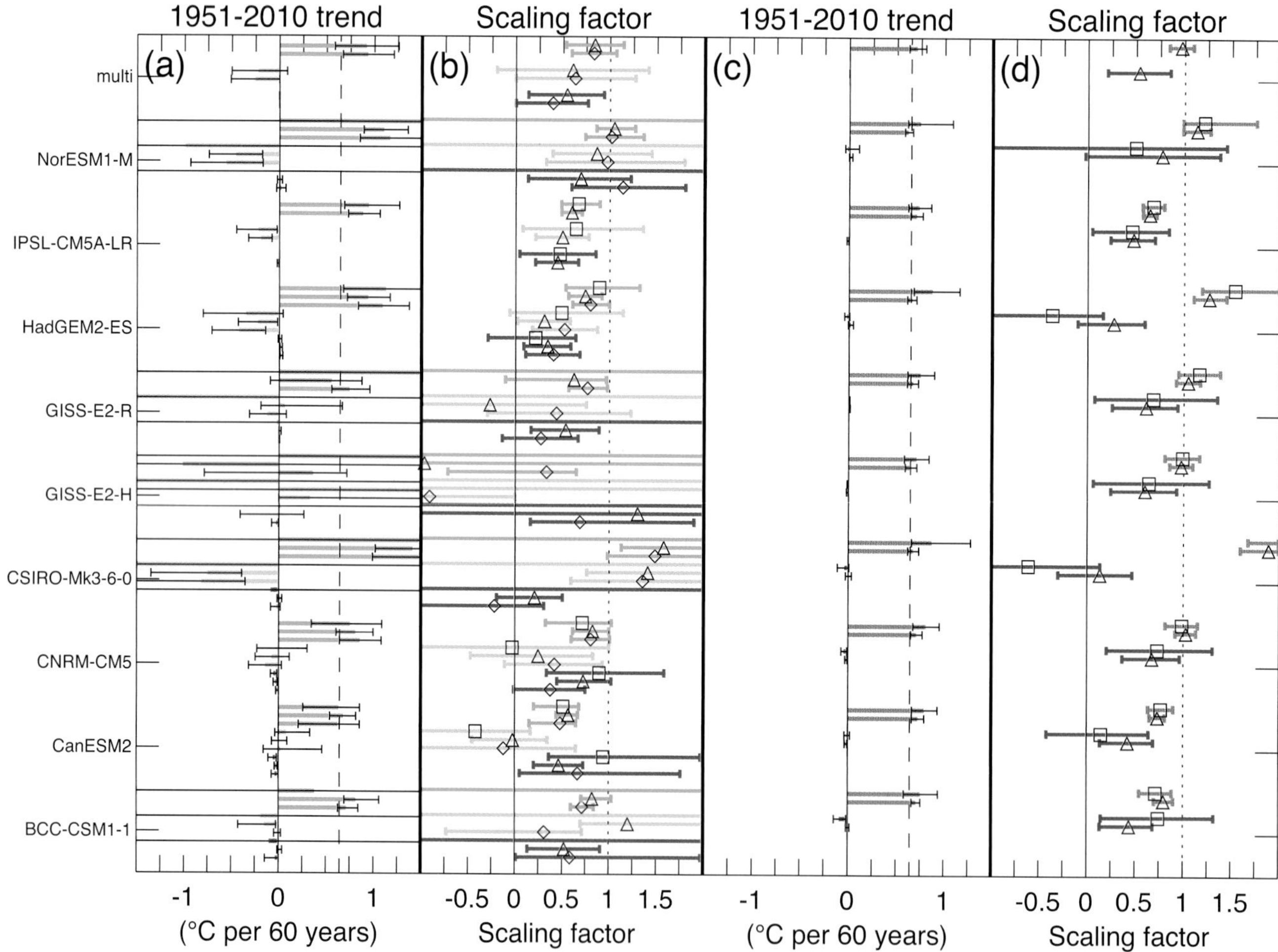

Figure 10.4 | (a) Estimated contributions of greenhouse gas (GHG, green), other anthropogenic (yellow) and natural (blue) forcing components to observed global mean surface temperature (GMST) changes over the 1951–2010 period. (b) Corresponding scaling factors by which simulated responses to GHG (green), other anthropogenic (yellow) and natural forcings (blue) must be multiplied to obtain the best fit to Hadley Centre/Climatic Research Unit gridded surface temperature data set 4 (HadCRUT4; Morice et al., 2012) observations based on multiple regressions using response patterns from nine climate models individually and multi-model averages (multi). Results are shown based on an analysis over the 1901–2010 period (squares, Ribes and Terray, 2013), an analysis over the 1861–2010 period (triangles, Gillett et al., 2013) and an analysis over the 1951–2010 period (diamonds, Jones et al., 2013). (c, d) As for (a) and (b) but based on multiple regressions estimating the contributions of total anthropogenic forcings (brown) and natural forcings (blue) based on an analysis over 1901–2010 period (squares, Ribes and Terray, 2013) and an analysis over the 1861–2010 period (triangles, Gillett et al., 2013). Coloured bars show best estimates of the attributable trends (a and c) and 5 to 95% confidence ranges of scaling factors (b and d). Vertical dashed lines in (a) and (c) show the best estimate HadCRUT4 observed trend over the period concerned. Vertical dotted lines in (b) and d) denote a scaling factor of unity.

We derive assessed ranges for the attributable contributions of GHGs, other anthropogenic forcings and natural forcings by taking the smallest ranges with a precision of one decimal place that span the 5 to 95% ranges of attributable trends over the 1951–2010 period from the Jones et al. (2013) weighted multi-model analysis and the Gillett et al. (2013) multi-model analysis considering observational uncertainty (Figure 10.4a). The assessed range for the attributable contribution of combined anthropogenic forcings was derived in the same way from the Gillett et al. (2013) multi-model attributable trend and shown in Figure 10.4c. We moderate our likelihood assessment and report *likely* ranges rather than the *very likely* ranges directly implied by these studies in order to account for residual sources of uncertainty including sensitivity to EOF truncation and analysis period (e.g., Ribes and Terray, 2013). In this context, GHGs means well-mixed greenhouse gases (WMGHGs), other anthropogenic forcings means aerosol changes, and in most models ozone changes and land use changes, and natural forcings means solar irradiance changes and volcanic aerosols. Over the 1951–2010 period, the observed GMST increased by approximately 0.6°C. GHG increases *likely* contributed 0.5°C to 1.3°C, other anthropogenic forcings *likely* contributed –0.6°C to 0.1°C and natural forcings *likely* contributed –0.1°C to 0.1°C to observed GMST trends over this period. Internal variability *likely* contributed –0.1°C to 0.1°C to observed trends over this period (Knutson et al., 2013). This assessment is shown schematically in Figure 10.5. The assessment is supported additionally by a complementary analysis in which the parameters of an Earth System Model of Intermediate Complexity (EMIC) were constrained using observations of near-surface temperature and ocean heat content, as well as prior information on the magnitudes of forcings, and which concluded that GHGs have caused 0.6°C to 1.1°C (5 to 95% uncertainty) warming since the mid-20th century (Huber and Knutti, 2011); an analysis by Wigley and Santer (2013), who used an energy balance model and RF and climate sensitivity estimates from AR4, and they concluded that there was about a 93% chance that GHGs caused a warming greater than observed over the 1950–2005 period; and earlier detection and attribution studies assessed in the AR4 (Hegerl et al., 2007b).

The inclusion of additional data to 2010 (AR4 analyses stopped at 1999; Hegerl et al. (2007b)) helps to better constrain the magnitude of the GHG-attributable warming (Drost et al., 2012; Gillett et al., 2012; Stott and Jones, 2012; Gillett et al., 2013), as does the inclusion of spatial information (Stott et al., 2006; Gillett et al., 2013), though Ribes and Terray (2013) caution that in some cases there are inconsistencies between observed spatial patterns of response and those simulated in indvidual models. While Hegerl et al. (2007b) assessed that a significant cooling of about 0.2 °C was attributable to natural forcings over the 1950–1999 period, the temperature trend attributable to natural forcings over the 1951–2010 period is very small (<0.1°C). This is because, while Mt Pinatubo cooled global temperatures in the early 1990s, there have been no large volcanic eruptions since, resulting in small simulated trends in response to natural forcings over the 1951–2010 period (Figure 10.1b). Regression coefficients for natural forcings tend to be smaller than one, suggesting that the response to natural forcings may be overestimated by the CMIP5 models on average (Figure 10.4; Gillett et al., 2013; Knutson et al., 2013). Attribution of observed changes is robust to observational uncertainty which is comparably important to internal climate variability as a source of uncertainty in GHG-attributable warming and aerosol-attributable cooling (Jones and Stott, 2011; Gillett et al., 2013; Knutson et al., 2013). The response to GHGs was detected using Hadley Centre new Global Environmental Model 2-Earth System (HadGEM2-ES; Stott and Jones, 2012), Canadian Earth System Model 2 (CanESM2; Gillett et al., 2012) and other CMIP5 models except for Goddard Institute for Space Studies-E2-H (GISS-E2-H; Gillett et al., 2013; Jones et al., 2013) (Figure 10.4). However, the influence of other anthropogenic forcings was detected only in some CMIP5 models (Figure 10.4). This lack of detection of other anthropogenic forcings compared to detection of an aerosol response using four CMIP3 models over the period 1900–1999 (Hegerl et al., 2007b) does not only relate to the use of data to 2010 rather than 2000 (Stott and Jones, 2012), although this could play a role (Gillett et al., 2013; Ribes and Terray, 2013). Whether it is associated with a cancellation of aerosol cooling by ozone and black carbon (BC) warming in the CMIP5 simulations, making the signal harder to detect, or by some aspect of the response to other anthropogenic forcings that is less realistic in these models is not clear.

Although closely constraining the GHG and other anthropogenic contributions to observed warming remains challenging owing to their degeneracy and sensitivity to methodological choices (Jones et al., 2013; Ribes and Terray, 2013), a total anthropogenic contribution to warming can be much more robustly constrained by a regression of observed temperature changes onto the simulated responses to all anthropogenic forcings and natural forcings (Figure 10.4; Gillett et al., 2013; Ribes and Terray, 2013). Robust detection of anthropogenic influence is also found if a new optimal detection methodology, the Regularised Optimal Fingerprint approach (see Section 10.2; Ribes et al., 2013), is applied (Ribes and Terray, 2013). A better constrained estimate of the total anthropogenic contribution to warming since the mid-20th century than the GHG contribution is also found by Wigley and Santer (2013). Knutson et al. (2013) demonstrate that observed trends in GMST are inconsistent with the simulated response to natural forcings alone, but consistent with the simulated response to natural and anthropogenic forcings for all periods beginning between 1880 and 1990 and ending in 2010, which they interpret as evidence that warming is in part attributable to anthropogenic influence over these periods. Based on the well-constrained attributable anthropogenic trends shown in Figure 10.4 we assess that anthropogenic forcings *likely* contributed 0.6°C to 0.8°C to the observed warming over the 1951–2010 period (Figure 10.5).

There are some inconsistencies in the simulated and observed magnitudes of responses to forcing for some CMIP5 models (Figure 10.4); for example, CanESM2 has a GHG regression coefficient significantly less than 1 and a regression coefficient for other anthropogenic forcings also significantly less than 1 (Gillett et al., 2012; Gillett et al., 2013; Jones et al., 2013; Ribes and Terray, 2013), indicating that this model overestimates the magnitude of the response to GHGs and to other anthropogenic forcings. Averaged over the ensembles of models considered by Gillett et al. (2013) and Jones et al. (2013), the best-estimate GHG and OA scaling factors are less than 1 (Figure 10.4), indicating that the model mean GHG and OA responses should be scaled down to best match observations. The best-estimate GHG scaling factors are larger than the best-estimate OA scaling factors, although the discrepancy from 1 is not significant in either case and the ranges of the GHG

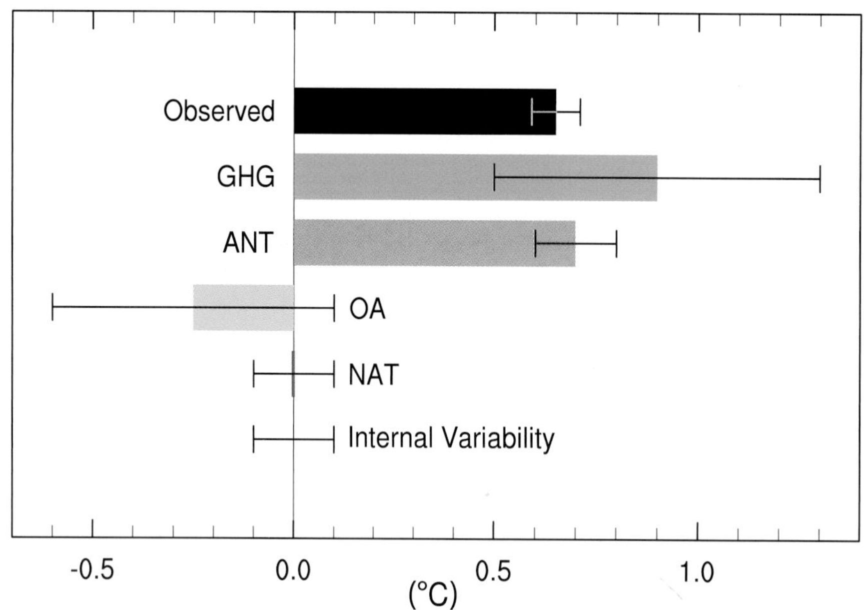

Figure 10.5 | Assessed *likely* ranges (whiskers) and their mid-points (bars) for attributable warming trends over the 1951–2010 period due to well-mixed greenhouse gases, other anthropogenic forcings (OA), natural forcings (NAT), combined anthropogenic forcings (ANT) and internal variability. The Hadley Centre/Climatic Research Unit gridded surface temperature data set 4 (HadCRUT4) observations are shown in black with the 5 to 95% uncertainty range due to observational uncertainty in this record (Morice et al., 2012).

and OA scaling factors are overlapping. Overall there is some evidence that some CMIP5 models have a higher transient response to GHGs and a larger response to other anthropogenic forcings (dominated by the effects of aerosols) than the real world (*medium confidence*). Inconsistencies between simulated and observed trends in GMST were also identified in several CMIP3 models by Fyfe et al. (2010) after removing volcanic, El Niño-Southern Oscillation (ENSO), and Cold Ocean/Warm Land pattern (COWL) signals from GMST, although uncertainties may have been underestimated because residuals were modelled by a first-order autoregressive processes. A longer observational record and a better understanding of the temporal changes in forcing should make it easier to identify discrepancies between the magnitude of the observed response to a forcing, and the magnitude of the response simulated in individual models. To the extent that inconsistencies between simulated and observed changes are independent between models, this issue may be addressed by basing our assessment on attribution analyses using the mean response from multiple models, and by accounting for model uncertainty when making such assessments.

In conclusion, although some inconsistencies in the forced responses of individual models and observations have been identified, the detection of the global temperature response to GHG increases using average responses from multiple models is robust to observational uncertainty and methodological choices. It is supported by basic physical arguments. We conclude, consistent with Hegerl et al. (2007b), that more than half of the observed increase in GMST from 1951 to 2010 is *very likely* due to the observed anthropogenic increase in GHG concentrations.

The influence of BC aerosols (from fossil and biofuel sources) has been detected in the recent global temperature record in one analysis, although the warming attributable to BC by Jones et al. (2011) is small compared to that attributable to GHG increases. This warming is simulated mainly over the Northern Hemisphere (NH) with a sufficiently distinct spatio-temporal pattern that it could be separated from the response to other forcings in this study.

Several recent studies have used techniques other than regression-based detection and attribution analyses to address the causes of recent global temperature changes. Drost and Karoly (2012) demonstrated that observed GMST, land–ocean temperature contrast, meridional temperature gradient and annual cycle amplitude exhibited trends over the period 1956–2005 that were outside the 5 to 95% range of simulated internal variability in eight CMIP5 models, based on three different observational data sets. They also found that observed trends in GMST and land–ocean temperature contrast were larger than those simulated in any of 36 CMIP5 simulations with natural forcing only. Drost et al. (2012) found that 1961–2010 trends in GMST and land–ocean temperature contrast were significantly larger than simulated internal variability in eight CMIP3 models. By comparing observed GMST with simple statistical models, Zorita et al. (2008) concluded that there is a very low probability that observed clustering of very warm years in the last decade occurred by chance. Smirnov and Mokhov (2009), adopting an approach that allowed them to distinguish between conventional Granger causality and a 'long-term causality' that focuses on low-frequency changes (see Section 10.2), found that increasing CO_2 concentrations are the principal determining factor in the rise of GMST over recent decades. Sedlacek and Knutti (2012) found that the spatial patterns of sea surface temperature (SST) trends from simulations forced with increases in GHGs and other anthropogenic forcings agree well with observations but differ from warming patterns associated with internal variability.

Several studies that have aimed to separate forced surface temperature variations from those associated with internal variability have identified the North Atlantic as a dominant centre of multi-decadal internal variability, and in particular modes of variability related to the Atlantic Multi-decadal Oscillation (AMO; Section 14.7.6). The AMO index is defined as an area average of North Atlantic SSTs, and it has an apparent period of around 70 years, which is long compared to the length of observational record making it difficult to deduce robust conclusions about the role of the AMO from only two cycles. Nevertheless, several studies claim a role for internal variability associated with the AMO in driving enhanced warming in the 1980s and 1990s as well as the recent slow down in warming (Box 9.2), while attributing long-term warming to anthropogenically forced variations either by analysing time series of GMST, forcings and indices of the AMO (Rohde et al., 2013; Tung and Zhou, 2013; Zhou and Tung, 2013a) or by analysing both spatial and temporal patterns of temperature (Swanson et al., 2009; DelSole et al., 2011; Wu et al., 2011). Studies based on global mean time series could risk falsely attributing variability to the AMO when variations in external forcings, for example, associated with aerosols, could also cause similar variability. In contrast, studies using space–time patterns seek to distinguish the spatial structure of temperature anomalies associated with the AMO from those associated with forced variability. Unforced climate simulations indicate that internal multi-decadal variability in the Atlantic is characterized by surface anomalies of the same sign from the equator to the high latitudes, with maximum amplitudes in subpolar regions (Delworth and Mann, 2000; Latif et al., 2004; Knight et al., 2005; DelSole et al., 2011) while the net response to anthropogenic and natural forcing over the 20th century, such as observed temperature change, is characterized by warming nearly everywhere on the globe, but with minimum warming or even cooling in the subpolar regions of the North Atlantic (Figure 10.2; Ting et al., 2009; DelSole et al., 2011).

Some studies implicate tropospheric aerosols in driving decadal variations in Atlantic SST (Evan et al., 2011; Booth et al., 2012; Terray, 2012), and temperature variations in eastern North America (Leibensperger et al., 2012). Booth et al. (2012) find that most multi-decadal variability in North Atlantic SSTs is simulated in one model mainly in response to aerosol variations, although its simulated changes in North Atlantic ocean heat content and salinity have been shown to be inconsistent with observations (Zhang et al., 2012). To the extent that climate models simulate realistic internal variability in the AMO (Section 9.5.3.3.2), AMO variability is accounted for in uncertainty estimates from regression-based detection and attribution studies (e.g., Figure 10.4).

To summarize, recent studies using spatial features of observed temperature variations to separate AMO variability from externally forced changes find that detection of external influence on global temperatures is not compromised by accounting for AMO-congruent variability (*high confidence*). There remains some uncertainty about how much decadal variability of GMST that is attributed to AMO in some studies is actually related to forcing, notably from aerosols. There is agreement among studies that the contribution of the AMO to global warming since 1951 is very small (considerably less than 0.1°C; see also Figure 10.6) and given that observed warming since 1951 is very large compared to climate model estimates of internal variability (Section 10.3.1.1.2), which are assessed to be adequate at global scale (Section 9.5.3.1), we conclude that it is *virtually certain* that internal variability alone cannot account for the observed global warming since 1951.

Box 10.2 | The Sun's Influence on the Earth's Climate

A number of studies since AR4 have addressed the possible influences of long-term fluctuations of solar irradiance on past climates, particularly related to the relative warmth of the Medieval Climate Anomaly (MCA) and the relative coolness in the Little Ice Age (LIA). There is *medium confidence* that both external solar and volcanic forcing, and internal variability, contributed substantially to the spatial patterns of surface temperature changes between the MCA and the LIA, but *very low confidence* in quantitative estimates of their relative contributions (Sections 5.3.5.3 and 5.5.1). The combined influence of volcanism, solar forcing and a small drop in greenhouse gases (GHGs) *likely* contributed to Northern Hemisphere cooling during the LIA (Section 10.7.2). Solar radiative forcing (RF) from the Maunder Minimum (1745) to the satellite era (average of 1976–2006) has been estimated to be +0.08 to +0.18 W m^{-2} (*low confidence*, Section 8.4.1.2). This may have contributed to early 20th century warming (*low confidence*, Section 10.3.1).

More recently, it is *extremely unlikely* that the contribution from solar forcing to the observed global warming since 1950 was larger than that from GHGs (Section 10.3.1.1.3). It is *very likely* that there has been a small decrease in solar forcing of –0.04 [–0.08 to 0.00] W m^{-2} over a period with direct satellite measurements of solar output from 1986 to 2008 (Section 8.4.1.1). There is *high confidence* that changes in total solar irradiance have not contributed to global warming during that period.

Since AR4, there has been considerable new research that has connected solar forcing to climate. The effect of solar forcing on GMST trends has been found to be small, with less than 0.1°C warming attributable to combined solar and volcanic forcing over the 1951–2010 period (Section 10.3.1), although the 11-year cycle of solar variability has been found to have some influence on GMST variability over the 20th century. GMST changes between solar maxima and minima are estimated to be of order 0.1°C from some regression studies of GMST and forcing estimates (Figure 10.6), although several studies have suggested these results may be too large owing to issues including degeneracy between forcing and with internal variability, overfitting of forcing indices and underestimated uncertainties in responses (Ingram, 2007; Benestad and Schmidt, 2009; Stott and Jones, 2009). Climate models generally show less than half this variability (Jones et al., 2012). *(continued on next page)*

Box 10.2 (continued)

Variability associated with the 11-year solar cycle has also been shown to produce measurable short-term regional and seasonal climate anomalies (Miyazaki and Yasunari, 2008; Gray et al., 2010; Lockwood, 2012; National Research Council, 2012) particularly in the Indo-Pacific, Northern Asia and North Atlantic regions (*medium evidence*). For example, studies have suggested an 11-year solar response in the Indo-Pacific region in which the equatorial eastern Pacific sea surface temperatures (SSTs) tend to be below normal, the sea level pressure (SLP) in the Gulf of Alaska and the South Pacific above normal, and the tropical convergence zones on both hemispheres strengthened and displaced polewards under solar maximum conditions, although it can be difficult to discriminate the solar-forced signal from the El Niño-Southern Oscillation (ENSO) signal (van Loon et al., 2007; van Loon and Meehl, 2008; White and Liu, 2008; Meehl and Arblaster, 2009; Roy and Haigh, 2010, 2012; Tung and Zhou, 2010; Bal et al., 2011; Haam and Tung, 2012; Hood and Soukharev, 2012; Misios and Schmidt, 2012). For northern summer, there is evidence that for peaks in the 11-year solar cycle, the Indian monsoon is intensified (Kodera, 2004; van Loon and Meehl, 2012), with solar variability affecting interannual connections between the Indian and Pacific sectors due to a shift in the location of the descending branch of the Walker Circulation (Kodera et al., 2007). In addition, model sensitivity experiments (Ineson et al., 2011) suggest that the negative phase of the North Atlantic Oscillation (NAO) is more prevalent during solar minima and there is some evidence of this in observations, including an indication of increased frequency of high-pressure 'blocking' events over Europe in winter (Barriopedro et al., 2008; Lockwood et al., 2010; Woollings et al., 2010).

Two mechanisms have been identified in observations and simulated with climate models that could explain these low amplitude regional responses (Gray et al., 2010; *medium evidence*). These mechanisms are additive and may reinforce one another so that the response to an initial small change in solar irradiance is amplified regionally (Meehl et al., 2009). The first mechanism is a top-down mechanism first noted by Haigh (1996) where greater solar ultraviolet radiation (UV) in peak solar years warms the stratosphere directly via increased radiation and indirectly via increased ozone production. This can result in a chain of processes that influences deep tropical convection (Balachandran et al., 1999; Shindell et al., 1999; Kodera and Kuroda, 2002; Haigh et al., 2005; Kodera, 2006; Matthes et al., 2006). In addition, there is less heating than average in the tropical upper stratosphere under solar minimum conditions which weakens the equator-to-pole temperature gradient. This signal can propagate downward to weaken the tropospheric mid-latitude westerlies, thus favoring a negative phase of the Arctic Oscillation (AO) or NAO. This response has been shown in several models (e.g., Shindell et al., 2001; Ineson et al., 2011) though there is no significant AO or NAO response to solar irradiance variations on average in the CMIP5 models (Gillett and Fyfe, 2013).

The second mechanism is a bottom-up mechanism that involves coupled air–sea radiative processes in the tropical and subtropical Pacific that also influence convection in the deep tropics (Meehl et al., 2003, 2008; Rind et al., 2008; Bal et al., 2011; Cai and Tung, 2012; Zhou and Tung, 2013b). Such mechanisms have also been shown to influence regional temperatures over longer time scales (decades to centuries), and can help explain patterns of regional temperature changes seen in paleoclimate data (e.g., Section 10.7.2; Mann et al., 2009; Goosse et al., 2012b) although they have little effect on global or hemispheric mean temperatures at either short or long time scales.

A possible amplifying mechanism linking solar variability and the Earth's climate system via cosmic rays has been postulated. It is proposed that variations in the cosmic ray flux associated with changes in solar magnetic activity affect ion-induced aerosol nucleation and cloud condensation nuclei (CCN) production in the troposphere (Section 7.4.6). A strong solar magnetic field would deflect cosmic rays and lead to fewer CCN and less cloudiness, thereby allowing for more solar energy into the system. Since AR4, there has been further evidence to disprove the importance of this amplifying mechanism. Correlations between cosmic ray flux and observed aerosol or cloud properties are weak and local at best, and do not prove to be robust on the regional or global scale (Section 7.4.6). Although there is some evidence that ionization from cosmic rays may enhance aerosol nucleation in the free troposphere, there is *medium evidence and high agreement* that the cosmic ray–ionization mechanism is too weak to influence global concentrations of CCN or their change over the last century or during a solar cycle in any climatically significant way (Sections 7.4.6 and 8.4.1.5). The lack of trend in cosmic ray intensity over the 1960–2005 period (McCracken and Beer, 2007) provides another argument against the hypothesis of a major contribution of cosmic ray variations to the observed warming over that period given the existence of short time scales in the climate system response.

Thus, although there is *medium confidence* that solar variability has made contributions to past climate fluctuations, since the mid-20th century there has been little trend in solar forcing. There are at least two amplifying mechanisms that have been proposed and simulated in some models that could explain small observed regional and seasonal climate anomalies associated with the 11-year solar cycle, mostly in the Indo-Pacific region and northern mid to high latitudes.

Regarding possible future influences of the sun on the Earth's climate, there is *very low confidence* in our ability to predict future solar output, but there is *high confidence* that the effects from solar irradiance variations will be much smaller than the projected climate changes from increased RF due to GHGs (Sections 8.4.1.3 and 11.3.6.2.2).

Based on a range of detection and attribution analyses using multiple solar irradiance reconstructions and models, Hegerl et al. (2007b) concluded that it is *very likely* that GHGs caused more global warming than solar irradiance variations over the 1950–1999 period. Detection and attribution analyses applied to the CMIP5 simulations (Figure 10.4) indicate less than 0.1°C temperature change attributable to combined solar and volcanic forcing over the 1951–2010 period. Based on a regression of paleo temperatures onto the response to solar forcing simulated by an energy balance model, Scafetta and West (2007) find that up to 50% of the warming since 1900 may be solar-induced, but Benestad and Schmidt (2009) show this conclusion is not robust, being based on disregarding forcings other than solar in the preindustrial period, and assuming a high and precisely known value for climate sensitivity. Despite claims that more than half the warming since 1970 can be ascribed to solar variability (Loehle and Scaffetta, 2011) , a conclusion based on an incorrect assumption of no anthropogenic influence before 1950 and a 60-year solar cycle influence on global temperature (see also Mazzarella and Scafetta, 2012), several studies show that solar variations cannot explain global mean surface warming over the past 25 years, because solar irradiance has declined over this period (Lockwood and Fröhlich, 2007, 2008; Lockwood, 2008, 2012). Lean and Rind (2008) conclude that solar forcing explains only 10% of the warming over the past 100 years, while contributing a small cooling over the past 25 years. Thus while there is some evidence for solar influences on regional climate variability (Box 10.2) solar forcing has only had a small effect on GMST. Overall, we conclude that it is *extremely unlikely* that the contribution from solar forcing to the warming since 1950 was larger than that from GHGs.

A range of studies have used statistical methods to separate out the influence of known sources of internal variability, including ENSO and, in some cases, the AMO, from the response to external drivers, including volcanoes, solar variability and anthropogenic influence, in the recent GMST record: see, for example, Lockwood (2008), Lean and Rind (2009), Folland et al. (2013), Foster and Rahmstorf (2011) and Kaufmann et al. (2011). Representative results, as summarized in Imbers et al. (2013), are shown in Figure 10.6. These consistently attribute most of the warming over the past 50 years to anthropogenic influence, even allowing for potential confounding factors like the AMO. While results of such statistical approaches are sensitive to assumptions regarding the properties of both responses to external drivers and internal variability (Imbers et al., 2013), they provide a complementary approach to attribution studies based on global climate models.

Overall, given that the anthropogenic increase in GHGs *likely* caused 0.5°C to 1.3°C warming over 1951–2010, with other anthropogenic forcings probably contributing counteracting cooling, that the effects of natural forcings and natural internal variability are estimated to be small, and that well-constrained and robust estimates of net anthropogenic warming are substantially more than half the observed warming (Figure 10.4) we conclude that it is *extremely likely* that human activities caused more than half of the observed increase in GMST from 1951 to 2010.

The early 20th century warming

The instrumental GMST record shows a pronounced warming during the first half of the 20th century (Figure 10.1a). Correction of residual biases in SST observations leads to a higher estimate of 1950s temperatures, but does not substantially change the warming between 1900 and 1940 (Morice et al., 2012). The AR4 concluded that 'the early 20th century warming is *very likely* in part due to external forcing' (Hegerl et al., 2007b), and that it is *likely* that anthropogenic forcing contributed to this warming. This assessment was based on studies including Shiogama et al. (2006) who find a contribution from solar and volcanic forcing to observed warming to 1949, and Min and Hense (2006), who find strong evidence for a forced (either natural or combined natural and anthropogenic) contribution to global warming from 1900 to 1949. Ring et al. (2012) estimate, based on time series analysis, that part of the early 20th century warming was due to GHG increases (see also Figure 10.6), but find a dominant contribution by internal variability. CMIP5 model simulations of the historical period show forced warming over the early 20th century (Figure 10.1a), consistent with earlier detection and attribution analyses highlighted in the AR4 and TAR. The early 20th century contributes to the detection of external forcings over the 20th century estimated by detection and attribution results (Figure 10.4; Gillett et al., 2013; Ribes and Terray, 2013) and to the detected change over the last millennium to 1950 (see Figure 10.19; Schurer et al., 2013).

The pattern of warming and residual differences between models and observations indicate a role for circulation changes as a contributor to early 20th cenury warming (Figure 10.2), and the contribution of internal variability to the early 20th century warming has been analysed in several publications since the AR4. Crook and Forster (2011) find that the observed 1918–1940 warming was significantly greater than that simulated by most of the CMIP3 models. A distinguishing feature of the early 20th century warming is its pattern (Brönnimann, 2009) which shows the most pronounced warming in the Arctic during the cold season, followed by North America during the warm season, the North Atlantic Ocean and the tropics. In contrast, there was no unusual warming in Australia among other regions (see Figure 10.2b). Such a pronounced pattern points to a role for circulation change as a contributing factor to the regional anomalies contributing to this warming. Some studies have suggested that the warming is a response to the AMO (Schlesinger and Ramankutty, 1994; Polyakov et al., 2005; Knight et al., 2006; Tung and Zhou, 2013), or a large but random expression of internal variability (Bengtsson et al., 2006; Wood and Overland, 2010). Knight et al. (2009) diagnose a shift from the negative to the positive phase of the AMO from 1910 to 1940, a mode of circulation that is estimated to contribute approximately 0.1°C, trough to peak, to GMST (Knight et al., 2005). Nonetheless, these studies do not challenge the AR4 assessment that external forcing *very likely* made a contribution to the warming over this period. In conclusion, the early 20th century warming is *very unlikely* to be due to internal variability alone. It remains difficult to quantify the contribution to this warming from internal variability, natural forcing and anthropogenic forcing, due to forcing and response uncertainties and incomplete observational coverage.

Year-to-year and decade-to-decade variability of global mean surface temperature

Time series analyses, such as those shown in Figure 10.6, seek to partition the variability of GMST into components attributable to anthropogenic and natural forcings and modes of internal variability such as ENSO and the AMO. Although such time series analyses support the major role of anthropogenic forcings, particularly due to increasing GHG concentrations, in contributing to the overall warming over the last 60 years, many factors, in addition to GHGs, including changes in tropospheric and stratospheric aerosols, stratospheric water vapour and solar output, as well as internal modes of variability, contribute to the year-to-year and decade-to-decade variability of GMST (Figure 10.6). Detailed discussion of the evolution of GMST of the past 15 years since 1998 is contained in Box 9.2.

10.3.1.1.4 Attribution of regional surface temperature change

Anthropogenic influence on climate has been robustly detected on the global scale, but for many applications an estimate of the anthropogenic contribution to recent temperature trends over a particular region is more useful. However, detection and attribution of climate change at continental and smaller scales is more difficult than on the global scale for several reasons (Hegerl et al., 2007b; Stott et al., 2010).

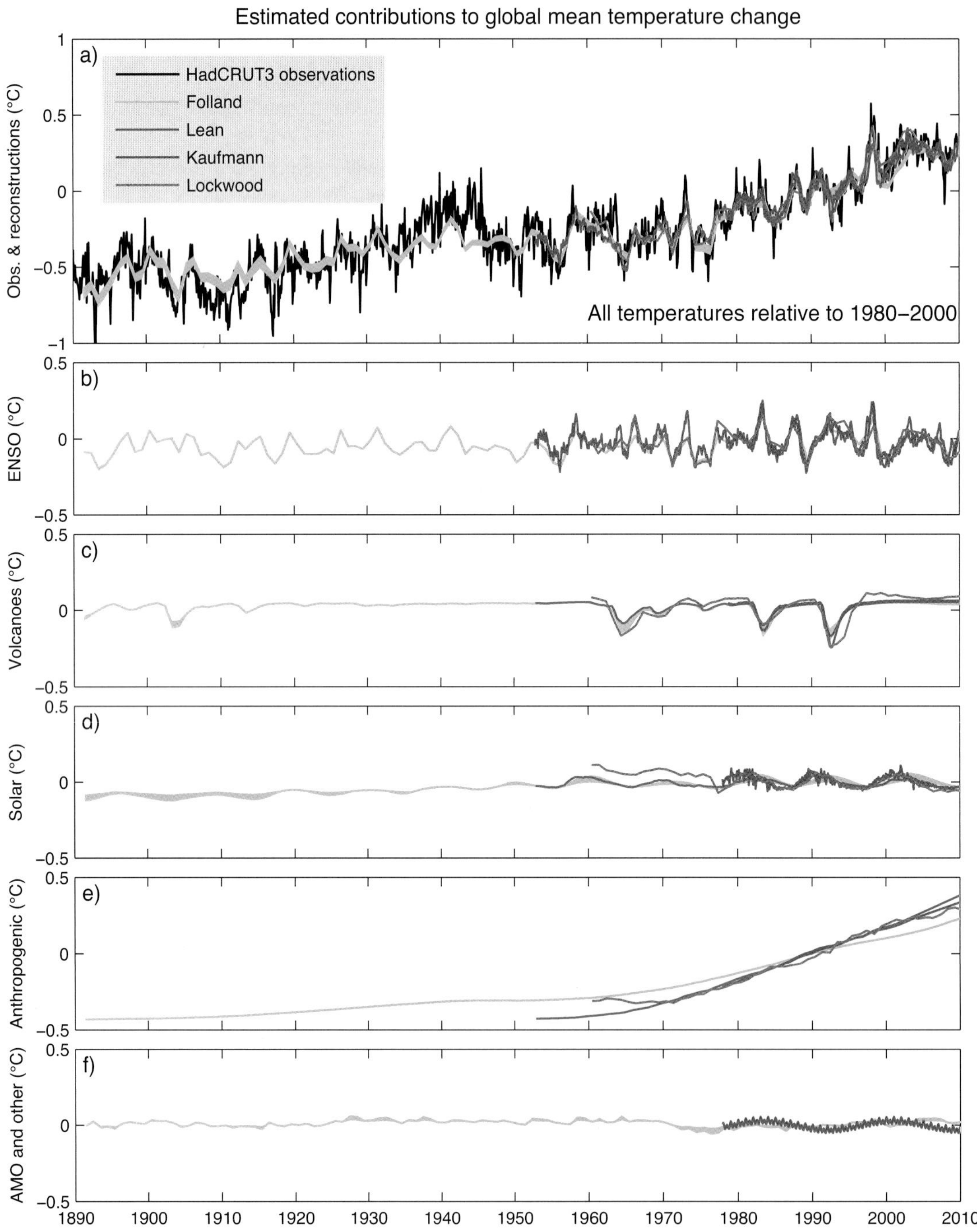

Figure 10.6 | (Top) The variations of the observed global mean surface temperature (GMST) anomaly from Hadley Centre/Climatic Research Unit gridded surface temperature data set version 3 (HadCRUT3, black line) and the best multivariate fits using the method of Lean (red line), Lockwood (pink line), Folland (green line) and Kaufmann (blue line). (Below) The contributions to the fit from (a) El Niño-Southern Oscillation (ENSO), (b) volcanoes, (c) solar forcing, (d) anthropogenic forcing and (e) other factors (Atlantic Multi-decadal Oscillation (AMO) for Folland and a 17.5-year cycle, semi-annual oscillation (SAO), and Arctic Oscillation (AO) from Lean). (From Lockwood (2008), Lean and Rind (2009), Folland et al. (2013) and Kaufmann et al. (2011), as summarized in Imbers et al. (2013).)

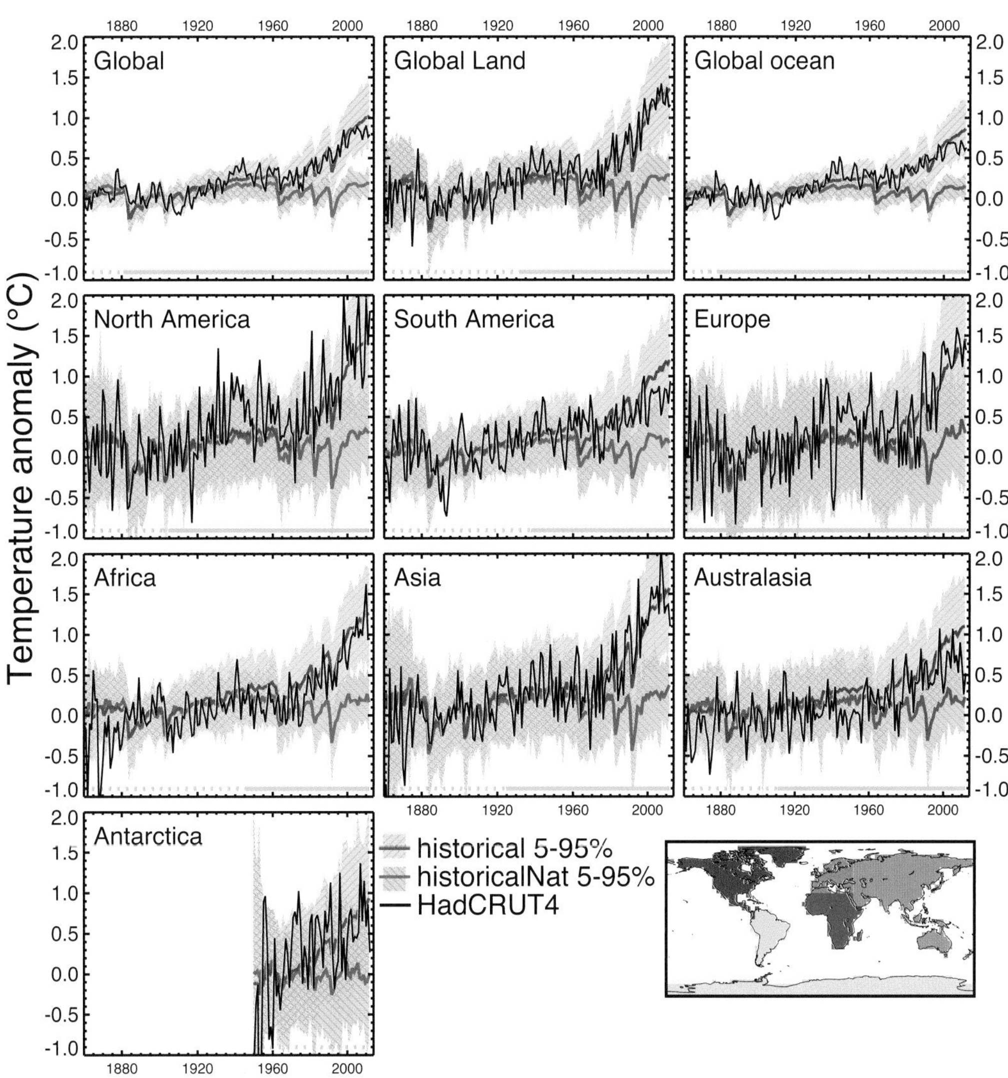

Figure 10.7 | Global, land, ocean and continental annual mean temperatures for CMIP3 and CMIP5 historical (red) and historicalNat (blue) simulations (multi-model means shown as thick lines, and 5 to 95% ranges shown as thin light lines) and for Hadley Centre/Climatic Research Unit gridded surface temperature data set 4 (HadCRUT4, black). Mean temperatures are shown for Antarctica and six continental regions formed by combining the sub-continental scale regions defined by Seneviratne et al. (2012). Temperatures are shown with respect to 1880–1919 for all regions apart from Antarctica where temperatures are shown with respect to 1950–2010. (Adapted from Jones et al., 2013.)

First, the relative contribution of internal variability compared to the forced response to observed changes tends to be larger on smaller scales, as spatial differences in internal variations are averaged out in large-scale means. Second, because the patterns of response to climate forcings tend to be large scale, there is less spatial information to help distinguish between the responses to different forcings when attention is restricted to a sub-global area. Third, forcings omitted in some global climate model simulations may be important on regional scales, such as land use change or BC aerosol. Lastly, simulated internal variability and responses to forcings may be less reliable on smaller scales than on the global scale. Knutson et al. (2013) find a tendency for CMIP5 models to overestimate decadal variability in the NH extratropics in individual grid cells and underestimate it elsewhere, although Karoly and Wu (2005) and Wu and Karoly (2007) find that variability is not generally underestimated in earlier generation models.

Based on several studies, Hegerl et al. (2007b) concluded that 'it is *likely* that there has been a substantial anthropogenic contribution to surface temperature increases in every continent except Antarctica since the middle of the 20th century'. Figure 10.7 shows comparisons of observed continental scale temperatures (Morice et al., 2012) with CMIP5 simulations including both anthropogenic and natural forcings (red lines) and including just natural forcings (blue lines). Observed temperatures are largely within the range of simulations with anthropogenic forcings for all regions and outside the range of simulations with only natural forcings for all regions except Antarctica (Jones et al., 2013). Averaging over all observed locations, Antarctica has warmed over the 1950–2008 period (Section 2.4.1.1; Gillett et al., 2008b; Jones et al., 2013), even though some individual locations have cooled, particularly in summer and autumn, and over the shorter 1960–1999 period (Thompson and Solomon, 2002; Turner et al., 2005). When temperature changes associated with changes in the Southern Annular Mode are removed by regression, both observations and model simulations indicate warming at all observed locations except the South Pole over the 1950–1999 period (Gillett et al., 2008b). An analysis of Antarctic land temperatures over the period 1950–1999 detected separate natural and anthropogenic responses of consistent magnitude in simulations and observations (Gillett et al., 2008b). Thus anthropogenic influence on climate has now been detected on all seven continents. However the evidence for human influence on Antarctic temperature is much weaker than for the other six continental regions. There is only one attribution study for this region, and there is greater observational uncertainty than the other regions, with very few data before 1950, and sparse coverage that is mainly limited to the coast and the Antarctic Peninsula. As a result of the observational uncertainties, there is *low confidence* in Antarctic region land surface air temperatures changes (Section 2.4.1.1) and we conclude for Antarctica there is *low confidence* that anthropogenic influence has contributed to the observed warming averaged over available stations.

Since the publication of the AR4 several other studies have applied attribution analyses to continental and sub-continental scale regions. Min and Hense (2007) applied a Bayesian decision analysis technique to continental-scale temperatures using the CMIP3 multi-model ensemble and concluded that forcing combinations including GHG increases provide the best explanation of 20th century observed changes in temperature on every inhabited continent except Europe, where the observational evidence is not decisive in their analysis. Jones et al. (2008) detected anthropogenic influence on summer temperatures over all NH continents and in many subcontinental NH land regions in an optimal detection analysis that considered the temperature responses to anthropogenic and natural forcings. Christidis et al. (2010) used a multi-model ensemble constrained by global-scale observed temperature changes to estimate the changes in probability of occurrence of warming or cooling trends over the 1950–1997 period over various sub-continental scale regions. They concluded that the probability of occurrence of warming trends had been at least doubled by anthropogenic forcing over all such regions except Central North America. The estimated distribution of warming trends over the Central North America region was approximately centred on the observed trend, so no inconsistency between simulated and observed trends was identified there. Knutson et al. (2013) demonstrated that observed temperature trends from the beginning of the observational record to 2010 averaged over Europe, Africa, Northern Asia, Southern Asia, Australia and South America are all inconsistent with the simulated response to natural forcings alone, and consistent with the simulated response to combined natural and anthropogenic forcings in the CMIP5 models. They reached a similar conclusion for the major ocean basins with the exception of the North Atlantic, where variability is high.

Several recent studies have applied attribution analyses to specific sub-continental regions. Anthropogenic influence has been found in winter minimum temperature over the western USA (Bonfils et al., 2008; Pierce et al., 2009), a conclusion that is found to be robust to weighting models according to various aspects of their climatology (Pierce et al., 2009); anthropogenic influence has been found in temperature trends over New Zealand (Dean and Stott, 2009) after circulation-related variability is removed as in Gillett et al. (2000); and anthropogenic influence has been found in temperature trends over France, using a first-order autoregressive model of internal variability (Ribes et al., 2010). Increases in anthropogenic GHGs were found to be the main driver of the 20th-century SST increases in both Atlantic and Pacific tropical cyclogenesis regions (Santer et al., 2006; Gillett et al., 2008a). Over both regions, the response to anthropogenic forcings is detected when the response to natural forcings is also included in the analysis (Gillett et al., 2008a). Knutson et al. (2013) detect an anthropogenic influence over Canada, but not over the continental USA, Alaska or Mexico.

Gillett et al. (2008b) detect anthropogenic influence on near-surface Arctic temperatures over land, with a consistent magnitude in simulations and observations. Wang et al. (2007) also find that observed Arctic warming is inconsistent with simulated internal variability. Both studies ascribe Arctic warmth in the 1930s and 1940s largely to internal variability. Shindell and Faluvegi (2009) infer a large contribution to both mid-century Arctic cooling and late century warming from aerosol forcing changes, with GHGs the dominant driver of long-term warming, though they infer aerosol forcing changes from temperature changes using an inverse approach which may lead to some changes associated with internal variability being attributed to aerosol forcing. We therefore conclude that despite the uncertainties introduced by limited observational coverage, high internal variability, modelling uncertainties (Crook et al., 2011) and poorly understood local forcings, such as the effect of BC on snow, there is sufficiently strong evidence to conclude that it is *likely* that there has been an anthropogenic contribution to the very substantial warming in Arctic land surface temperatures over the past 50 years.

Some attribution analyses have considered temperature trends at the climate model grid box scale. At these spatial scales robust attribution is difficult to obtain, since climate models often lack the processes needed to simulate regional details realistically, regionally important forcings may be missing in some models and observational uncertainties are very large for some regions of the world at grid box scale (Hegerl et al., 2007b; Stott et al., 2010). Nevertheless an attribution analysis has been carried out on Central England temperature, a record that extends back to 1659 and is sufficiently long to demonstrate that the representation of multi-decadal variability in the single grid box in the model used, Hadley Centre climate prediction model 3 (HadCM3) is adequate for detection (Karoly and Stott, 2006). The observed trend in Central England Temperature is inconsistent with either internal variability or the simulated response to natural forcings, but is consistent with the simulated response when anthropogenic forcings are included (Karoly and Stott, 2006).

Observed 20th century grid cell trends from Hadley Centre/Climatic Research Unit gridded surface temperature data set 2v (HadCRUT2v; Jones et al., 2001) are inconsistent with simulated internal variability at the 10% significance level in around 80% of grid cells even using HadCM2 which was found to overestimate variability in 5-year mean temperatures at most latitudes (Karoly and Wu, 2005). Sixty percent of grid cells were found to exhibit significant warming trends since 1951, a much larger number than expected by chance (Karoly and Wu, 2005; Wu and Karoly, 2007), and similar results apply when circulation-related variability is first regressed out (Wu and Karoly, 2007). However, as discussed in the AR4 (Hegerl et al., 2007b), when a global field significance test is applied, this becomes a global detection study; since not all grid cells exhibit significant warming trends the overall interpretation of the results in terms of attribution at individual locations remains problematic. Mahlstein et al. (2012) find significant changes in summer season temperatures in about 40% of low-latitude and about 20% of extratropical land grid cells with sufficient observations, when testing against the null hypothesis of no change in the distribution of summer temperatures. Observed grid cell trends are compared with CMIP5 simulated trends in Figure 10.2i, which shows that in the great majority (89%) of grid cells with sufficient observational coverage, observed trends over the 1901–2010 period are inconsistent with a combination of simulated internal variability and the response to natural forcings (Jones et al., 2013). Knutson et al. (2013) find some deficiencies in the simulation of multi-decadal variability at the grid cell scale in CMIP5 models, but demonstrate that trends at more than 75% of individual grid cells with sufficient observational coverage in HadCRUT4 are inconsistent with the simulated response to natural forcings alone, and consistent or larger than the simulated response to combined anthropogenic and natural forcings in CMIP5 models.

In summary, it is *likely* that anthropogenic forcing has made a substantial contribution to the warming of each of the inhabited continents since 1950. For Antarctica large observational uncertainties result in *low confidence* that anthropogenic influence has contributed to the observed warming averaged over available stations. Anthropogenic influence has *likely* contributed to temperature change in many sub-continental regions. Detection and attribution of climate change at continental and smaller scales is more difficult than at the global scale due to the greater contribution of internal variability, the greater difficulty of distinguishing between different causal factors, and greater errors in climate models' representation of regional details. Nevertheless, statistically significant warming trends are observed at a majority of grid cells, and the observed warming is inconsistent with estimates of possible warming due to natural causes at the great majority of grid cells with sufficient observational coverage.

10.3.1.2 Atmosphere

This section presents an assessment of the causes of global and regional temperature changes in the free atmosphere. In AR4, Hegerl et al. (2007b) concluded that 'the observed pattern of tropospheric warming and stratospheric cooling is *very likely* due to the influence of anthropogenic forcing, particularly greenhouse gases and stratospheric ozone depletion.' Since AR4, insight has been gained into regional aspects of free tropospheric trends and the causes of observed changes in stratospheric temperature.

Atmospheric temperature trends through the depth of the atmosphere offer the possibility of separating the effects of multiple climate forcings, as climate model simulations indicate that each external forcing produces a different characteristic vertical and zonal pattern of temperature response (Hansen et al., 2005b; Hegerl et al., 2007b; Penner et al., 2007; Yoshimori and Broccoli, 2008). GHG forcing is expected to warm the troposphere and cool the stratosphere. Stratospheric ozone depletion cools the stratosphere, with the cooling being most pronounced in the polar regions. Its effect on tropospheric temperatures is small, which is consistent with a small estimated RF of stratospheric ozone changes (SPARC CCMVal, 2010; McLandress et al., 2012). Tropospheric ozone increase, on the other hand, causes tropospheric warming. Reflective aerosols like sulphate cool the troposphere while absorbing aerosols like BC have a warming effect. Free atmosphere temperatures are also affected by natural forcings: solar irradiance increases cause a general warming of the atmosphere and volcanic aerosol ejected into the stratosphere causes tropospheric cooling and stratospheric warming (Hegerl et al., 2007b).

10.3.1.2.1 Tropospheric temperature change

Chapter 2 concludes that it is *virtually certain* that globally the troposphere has warmed since the mid-twentieth century with only *medium* (NH extratropics) to *low confidence* (tropics and SH extratropics) in the rate and vertical structure of these changes. During the satellite era CMIP3 and CMIP5 models tend to warm faster than observations specifically in the tropics (McKitrick et al., 2010; Fu et al., 2011; Po-Chedley and Fu, 2012; Santer et al., 2013); however, because of the large uncertainties in observed tropical temperature trends (Section 2.4.4; Seidel et al. (2012); Figures 2.26 and Figure 2.27) there is only *low confidence* in this assessment (Section 9.4.1.4.2). Outside the tropics, and over the period of the radiosonde record beginning in 1961, the discrepancy between simulated and observed trends is smaller (Thorne et al., 2011; Lott et al., 2013; Santer et al., 2013). Specifically there is better agreement between observed trends and CMIP5 model trends for the NH extratropics (Lott et al., 2013). Factors other than observational uncertainties that contribute to inconsistencies between observed and simulated free troposphere warming include specific manifestation of natural variability in the observed coupled atmosphere–ocean system, forcing errors incorporated in the historical simulations and model response errors (Santer et al., 2013).

Utilizing a subset of CMIP5 models with single forcing experiments extending until 2010, Lott et al. (2013) detect influences of both human induced GHG increase and other anthropogenic forcings (e.g., ozone and aerosols) in the spatio-temporal changes in tropospheric temperatures from 1961 to 2010 estimated from radiosonde observations. Figure 10.8 illustrates that a subsample of CMIP5 models (see Supplementary Material for model selection) forced with both anthropogenic and natural climate drivers (red profiles) exhibit trends that are consistent with radiosonde records in the troposphere up to about 300 hPa, albeit with a tendency for this subset of models to warm more than the observations. This finding is seen in near-globally averaged data (where there is sufficient observational coverage to make a meaningful comparison: 60°S to 60°N) (right panel), as well as in latitudinal bands of the SH extratropics (Figure 10.8, first panel), tropics (Figure 10.8, second panel) and the NH extratropics (Figure 10.8, third panel). Figure 10.8 also illustrates that it is *very unlikely* that natural forcings alone could have caused the observed warming of tropospheric temperatures (blue profiles). The ensembles with both anthropogenic and natural forcings (red) and with GHG forcings only (green) are not clearly separated. This could be due to cancellation of the effects of increases in reflecting aerosols, which cool the troposphere, and absorbing aerosol (Penner et al., 2007) and tropospheric ozone, which both warm the troposphere. Above 300 hPa the three radiosonde data sets exhibit a larger spread as a result of larger uncertainties in the observational record (Thorne et al., 2011; Section 2.4.4). In this region of the upper troposphere simulated CMIP5 temperature trends tend to be more positive than observed trends (Figure 10.8). Further, an assessment of causes of observed trends in the upper troposphere is less confident than an assessment of overall atmospheric temperature changes because of observational uncertainties and potential remain-

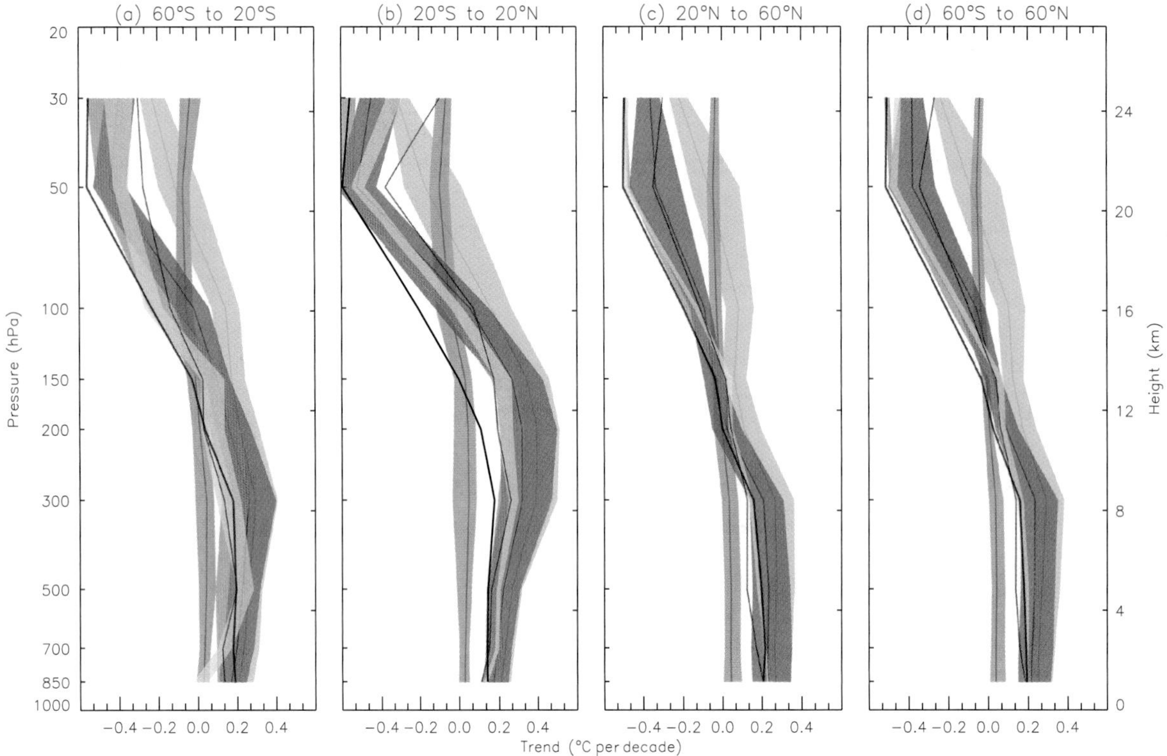

Figure 10.8 | Observed and simulated zonal mean temperatures trends from 1961 to 2010 for CMIP5 simulations containing both anthropogenic and natural forcings (red), natural forcings only (blue) and greenhouse gas forcing only (green) where the 5 to 95th percentile ranges of the ensembles are shown. Three radiosonde observations are shown (thick black line: Hadley Centre Atmospheric Temperature data set 2 (HadAT2), thin black line: RAdiosone OBservation COrrection using REanalyses 1.5 (RAOBCORE 1.5), dark grey band: Radiosonde Innovation Composite Homogenization (RICH)-obs 1.5 ensemble and light grey: RICH- τ 1.5 ensemble. (After Lott et al., 2013.)

ing systematic biases in observational data sets in this region (Thorne et al., 2011; Haimberger et al., 2012). An analysis of contributions of natural and anthropogenic forcings to more recent trends from 1979 to 2010 (Supplementary Material, Figure S.A.1) is less robust because of increased uncertainty in observed trends (consistent with Seidel et al. (2012)) as well as decreased capability to separate between individual forcings ensembles.

One approach to identify a climate change signal in a time series is the analysis of the ratio between the amplitude of the observed signal of change divided by the magnitude of internal variability, in other words the S/N ratio of the data record. The S/N ratio represents the result of a non-optimal fingerprint analysis (in contrast to optimal fingerprint analyses where model-simulated responses and observations are normalized by internal variability to improve the S/N ratio (see Section 10.2.3). For changes in the lower stratospheric temperature between 1979 and 2011, S/N ratios vary from 26 to 36, depending on the choice of observational data set. In the lower troposphere, the fingerprint strength in observations is smaller, but S/N ratios are still significant at the 1% level or better, and range from 3 to 8. There is no evidence that these ratios are spuriously inflated by model variability errors. After all global mean signals are removed, model fingerprints remain identifiable in 70% of the tests involving tropospheric temperature changes (Santer et al., 2013).

Hegerl et al. (2007a) concluded that increasing GHGs are the main cause for warming of the troposphere. This result is supported by a subsample of CMIP5 models that also suggest that the warming effect of well mixed GHGs is partly offset by the combined effects of reflecting aerosols and other forcings. Our understanding has been increased regarding the time scale of detectability of global scale troposphere temperature. Taken together with increased understanding of the uncertainties in observational records of tropospheric temperatures (including residual systematic biases; Section 2.4.4) the assessment remains as it was for AR4 that it is *likely* that anthropogenic forcing has led to a detectable warming of tropospheric temperatures since 1961.

10.3.1.2.2 Stratospheric temperature change

Lower stratospheric temperatures have not evolved uniformly over the period since 1958 when the stratosphere has been observed with sufficient regularity and spatial coverage. A long-term global cooling trend is interrupted by three 2-year warming episodes following large volcanic eruptions (Section 2.4.4). During the satellite period the cooling evolved mainly in two steps occurring in the aftermath of the El Chichón eruption in 1982 and the Mt Pinatubo eruption of 1991, with each cooling transition being followed by a period of relatively steady temperatures (Randel et al., 2009; Seidel et al., 2011). Since the mid-1990s little net change has occurred in lower stratospheric temperatures (Section 2.4.4).

Since AR4, progress has been made in simulating the observed evolution of global mean lower stratospheric temperature. On the one hand, this has been achieved by using models with an improved representation of stratospheric processes (chemistry–climate models and some CMIP5 models). It is found that in these models which have an upper boundary above the stratopause with an altitude of about 50 km (so-called high-top models) and improved stratospheric physics, variability of lower stratosphere climate in general is well simulated (Butchart et al., 2011; Gillett et al., 2011; Charlton-Perez et al., 2013) whereas in so-called low-top models (including models participating in CMIP3) it is generally underestimated (Cordero and Forster, 2006; Charlton-Perez et al., 2013). On the other hand, CMIP5 models all include changes in stratospheric ozone (Eyring et al., 2013) whereas only about half of the models participating in CMIP3 include stratospheric ozone changes (Section 9.4.1.4.5). A comparison of a low-top and high-top version of the HadGEM2 model shows detectable differences in modelled temperature changes, particularly in the lower tropical stratosphere, with the high-top version's simulation of temperature trends in the tropical troposphere in better agreement with radiosondes and reanalyses over 1981–2010 (Mitchell et al., 2013).

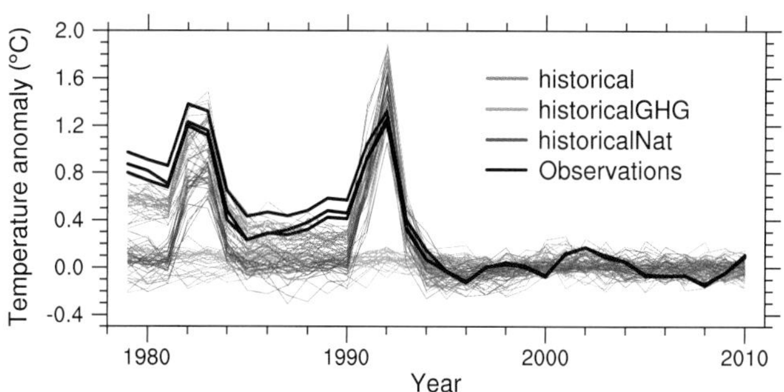

Figure 10.9 | Time series (1979–2010) of observed (black) and simulated global mean (82.5°S to 82.5°N) Microwave Sounding Unit (MSU) lower stratosphere temperature anomalies in a subset of CMIP5 simulations (simulations with both anthropogenic and natural forcings (red), simulations with well-mixed greenhouse gases (green), simulations with natural forcings (blue)). Anomalies are calculated relative to 1996–2010. (Adapted from Ramaswamy et al., 2006.)

CMIP5 models forced with changes in WMGHGs and stratospheric ozone as well as with changes in solar irradiance and volcanic aerosol forcings simulate the evolution of observed global mean lower stratospheric temperatures over the satellite era reasonably well although they tend to underestimate the long-term cooling trend (Charlton-Perez et al., 2013; Santer et al., 2013). Compared with radiosonde data the cooling trend is also underestimated in a subset of CMIP5 simulations over the period 1961–2010 (Figure 10.8) and in CMIP3 models over the 1958–1999 period (Cordero and Forster, 2006). Potential causes for biases in lower stratosphere temperature trends are observational uncertainties (Section 2.4.4) and forcing errors related to prescribed stratospheric aerosol loadings and stratospheric ozone changes affecting the tropical lower stratosphere (Free and Lanzante, 2009; Solomon et al., 2012; Santer et al., 2013).

Since AR4, attribution studies have improved our knowledge of the role of anthropogenic and natural forcings in observed lower stratospheric temperature change. Gillett et al. (2011) use the suite of chemistry climate model simulations carried out as part of the Chemistry Climate Model Validation (CCMVal) activity phase 2 for an attribution study of observed changes in stratospheric zonal mean temperatures. Chemistry–climate models prescribe changes in ozone-depleting substances (ODS) and ozone changes are calculated interactively. Gillett et al. (2011) partition 1979–2005 Microwave Sounding Unit (MSU) lower stratospheric temperature trends into ODS-induced and GHG-induced changes and find that both ODSs and natural forcing contributed to the observed stratospheric cooling in the lower stratosphere with the impact of ODS dominating. The influence of GHGs on stratospheric temperature could not be detected independently of ODSs.

The step-like cooling of the lower stratosphere can only be explained by the combined effects of changes in both anthropogenic and natural factors (Figure 10.9; Eyring et al., 2006; Ramaswamy et al., 2006). Although the anthropogenic factors (ozone depletion and increases in WMGHGs) cause the overall cooling, the natural factors (solar irradiance variations and volcanic aerosols) modulate the evolution of the cooling (Figure 10.9; Ramaswamy et al., 2006; Dall'Amico et al., 2010) with temporal variability of global mean ozone contributing to the step-like temperature evolution (Thompson and Solomon, 2009).

Models disagree with observations for seasonally varying changes in the strength of the Brewer–Dobson circulation in the lower stratosphere (Ray et al., 2010) which has been linked to zonal and seasonal patterns of changes in lower stratospheric temperatures (Thompson and Solomon, 2009; Fu et al., 2010; Lin et al., 2010b; Forster et al., 2011; Free, 2011). One robust feature is the observed cooling in spring over the Antarctic, which is simulated in response to stratospheric ozone depletion in climate models (Young et al., 2012), although this has not been the subject of a formal detection and attribution study.

Since AR4, progress has been made in simulating the response of global mean lower stratosphere temperatures to natural and anthropogenic forcings by improving the representation of climate forcings and utilizing models that include more stratospheric processes. New detection and attribution studies of lower stratospheric temperature changes made since AR4 support an assessment that it is *very likely* that anthropogenic forcing, dominated by stratospheric ozone depletion due to ozone-depleting substances, has led to a detectable cooling of the lower stratosphere since 1979.

10.3.1.2.3 Overall atmospheric temperature change

When temperature trends from the troposphere and stratosphere are analysed together, detection and attribution studies using CMIP5 models show robust detections of the effects of GHGs and other anthropogenic forcings on the distinctive fingerprint of tropospheric warming and stratospheric cooling seen since 1961 in radiosonde data (Lott et al., 2013; Mitchell et al., 2013). Combining the evidence from free atmosphere changes from both troposphere and stratosphere shows an increased confidence in the attribution of free atmosphere temperature changes compared to AR4 owing to improved understanding of stratospheric temperature changes. There is therefore stronger evidence than at the time of AR4 to support the conclusion that it is *very likely* that anthropogenic forcing, particularly GHGs and stratospheric ozone depletion, has led to a detectable observed pattern of tropospheric warming and lower stratospheric cooling since 1961.

Frequently Asked Questions

FAQ 10.1 | Climate Is Always Changing. How Do We Determine the Causes of Observed Changes?

The causes of observed long-term changes in climate (on time scales longer than a decade) are assessed by determining whether the expected 'fingerprints' of different causes of climate change are present in the historical record. These fingerprints are derived from computer model simulations of the different patterns of climate change caused by individual climate forcings. On multi-decade time scales, these forcings include processes such as greenhouse gas increases or changes in solar brightness. By comparing the simulated fingerprint patterns with observed climate changes, we can determine whether observed changes are best explained by those fingerprint patterns, or by natural variability, which occurs without any forcing.

The fingerprint of human-caused greenhouse gas increases is clearly apparent in the pattern of observed 20th century climate change. The observed change cannot be otherwise explained by the fingerprints of natural forcings or natural variability simulated by climate models. Attribution studies therefore support the conclusion that 'it is extremely likely *that human activities have caused more than half of the observed increase in global mean surface temperatures from 1951 to 2010.'*

The Earth's climate is always changing, and that can occur for many reasons. To determine the principal causes of observed changes, we must first ascertain whether an observed change in climate is different from other fluctuations that occur without any forcing at all. Climate variability without forcing—called internal variability—is the consequence of processes within the climate system. Large-scale oceanic variability, such as El Niño-Southern Oscillation (ENSO) fluctuations in the Pacific Ocean, is the dominant source of internal climate variability on decadal to centennial time scales.

Climate change can also result from natural forcings external to the climate system, such as volcanic eruptions, or changes in the brightness of the sun. Forcings such as these are responsible for the huge changes in climate that are clearly documented in the geological record. Human-caused forcings include greenhouse gas emissions or atmospheric particulate pollution. Any of these forcings, natural or human caused, could affect internal variability as well as causing a change in average climate. Attribution studies attempt to determine the causes of a detected change in observed climate. Over the past century we know that global average temperature has increased, so if the observed change is forced then the principal forcing must be one that causes warming, not cooling.

Formal climate change attribution studies are carried out using controlled experiments with climate models. The model-simulated responses to specific climate forcings are often called the fingerprints of those forcings. A climate model must reliably simulate the fingerprint patterns associated with individual forcings, as well as the patterns of unforced internal variability, in order to yield a meaningful climate change attribution assessment. No model can perfectly reproduce all features of climate, but many detailed studies indicate that simulations using current models are indeed sufficiently reliable to carry out attribution assessments.

FAQ 10.1, Figure 1 illustrates part of a fingerprint assessment of global temperature change at the surface during the late 20th century. The observed change in the latter half of the 20th century, shown by the black time series in the left panels, is larger than expected from just internal variability. Simulations driven only by natural forcings (yellow and blue lines in the upper left panel) fail to reproduce late 20th century global warming at the surface with a spatial pattern of change (upper right) completely different from the observed pattern of change (middle right). Simulations including both natural and human-caused forcings provide a much better representation of the time rate of change (lower left) and spatial pattern (lower right) of observed surface temperature change.

Both panels on the left show that computer models reproduce the naturally forced surface cooling observed for a year or two after major volcanic eruptions, such as occurred in 1982 and 1991. Natural forcing simulations capture the short-lived temperature changes following eruptions, but only the natural + human caused forcing simulations simulate the longer-lived warming trend.

A more complete attribution assessment would examine temperature above the surface, and possibly other climate variables, in addition to the surface temperature results shown in FAQ 10.1, Figure 1. The fingerprint patterns associated with individual forcings become easier to distinguish when more variables are considered in the assessment.

(continued on next page)

FAQ 10.1 (continued)

Overall, FAQ 10.1, Figure 1 shows that the pattern of observed temperature change is significantly different than the pattern of response to natural forcings alone. The simulated response to all forcings, including human-caused forcings, provides a good match to the observed changes at the surface. We cannot correctly simulate recent observed climate change without including the response to human-caused forcings, including greenhouse gases, stratospheric ozone, and aerosols. Natural causes of change are still at work in the climate system, but recent trends in temperature are largely attributable to human-caused forcing.

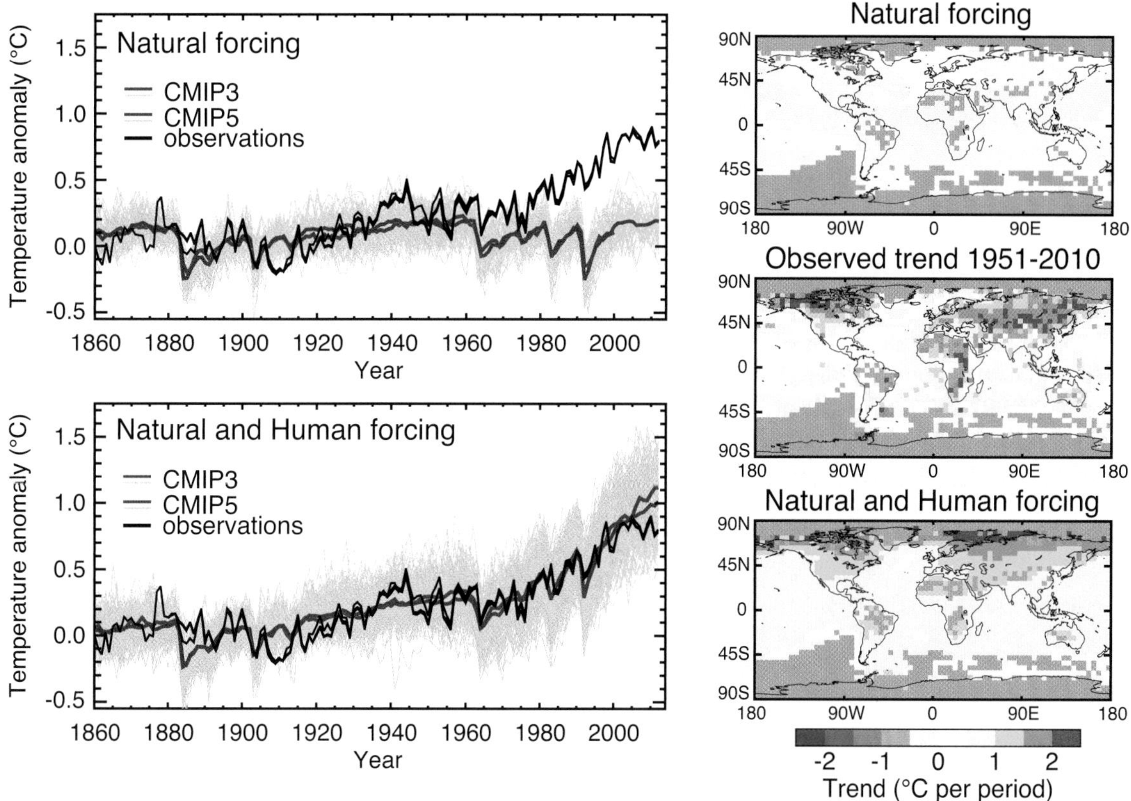

FAQ 10.1, Figure 1 | (Left) Time series of global and annual-averaged surface temperature change from 1860 to 2010. The top left panel shows results from two ensemble of climate models driven with just natural forcings, shown as thin blue and yellow lines; ensemble average temperature changes are thick blue and red lines. Three different observed estimates are shown as black lines. The lower left panel shows simulations by the same models, but driven with both natural forcing and human-induced changes in greenhouse gases and aerosols. (Right) Spatial patterns of local surface temperature trends from 1951 to 2010. The upper panel shows the pattern of trends from a large ensemble of Coupled Model Intercomparison Project Phase 5 (CMIP5) simulations driven with just natural forcings. The bottom panel shows trends from a corresponding ensemble of simulations driven with natural + human forcings. The middle panel shows the pattern of observed trends from the Hadley Centre/Climatic Research Unit gridded surface temperature data set 4 (HadCRUT4) during this period.

10.3.2 Water Cycle

Detection and attribution studies of anthropogenic change in hydrologic variables are challenged by the length and quality of observed data sets, and by the difficulty in simulating hydrologic variables in dynamical models. AR4 cautiously noted that the observed increase in atmospheric water vapour over oceans was consistent with warming of SSTs attributed to anthropogenic influence, and that observed changes in the latitudinal distribution of precipitation, and increased incidence of drought, were suggestive of a possible human influence. Many of the published studies cited in AR4, and some of the studies cited in this section, use less formal detection and attribution criteria than are often used for assessments of temperature change, owing to difficulties defining large-scale fingerprint patterns of hydrologic change in models and isolating those fingerprints in data. For example, correlations between observed hydrologic changes and the patterns of change in models forced by increasing GHGs can provide suggestive evidence towards attribution of change.

Since the publication of AR4, *in situ* hydrologic data sets have been reanalysed with more stringent quality control. Satellite-derived data records of worldwide water vapour and precipitation variations have lengthened. Formal detection and attribution studies have been carried out with newer models that potentially offer better simulations of natural variability. Reviews of detection and attribution of trends in various components of the water cycle have been published by Stott et al. (2010) and Trenberth (2011b).

10.3.2.1 Changes in Atmospheric Water Vapour

In situ surface humidity measurements have been reprocessed since AR4 to create new gridded analyses for climatic research, as discussed in Chapter 2. The HadCRUH Surface Humidity data set (Willett et al., 2008) indicates significant increases in surface specific humidity between 1973 and 2003 averaged over the globe, the tropics, and the NH, with consistently larger trends in the tropics and in the NH during summer, and negative or non significant trends in relative humidity. These results are consistent with the hypothesis that the distribution of relative humidity should remain roughly constant under climate change (see Section 2.5). Simulations of the response to historical anthropogenic and natural forcings robustly generate an increase in atmospheric humidity consistent with observations (Santer et al., 2007; Willett et al., 2007; Figure 9.9). A recent cessation of the upward trend in specific humidity is observed over multiple continental areas in HadCRUH and is also found in the European Centre for Medium range Weather Forecast (ECMWF) interim reanalysis of the global atmosphere and surface conditions (ERA-Interim; Simmons et al. 2010). This change in the specific humidity trend is temporally correlated with a levelling off of global ocean temperatures following the 1997–1998 El Niño event (Simmons et al., 2010).

10

The anthropogenic water vapour fingerprint simulated by an ensemble of 22 climate models has been identified in lower tropospheric moisture content estimates derived from Special Sensor Microwave/Imager (SSM/I) data covering the period 1988–2006 (Santer et al., 2007). Santer et al. (2009) find that detection of an anthropogenic response in column water vapour is insensitive to the set of models used. They rank models based on their ability to simulate the observed mean total column water vapour, and its annual cycle and variability associated with ENSO. They report no appreciable differences between the fingerprints or detection results derived from the best or worst performing models, and so conclude that attribution of water vapour changes to anthropogenic forcing is not sensitive to the choice of models used for the assessment.

In summary, an anthropogenic contribution to increases in specific humidity at and near the Earth's surface is found with *medium confidence*. Evidence of a recent levelling off of the long-term surface atmospheric moistening trend over land needs to be better understood and simulated as a prerequisite to increased confidence in attribution studies of water vapour changes. Length and quality of observational humidity data sets, especially above the surface, continue to limit detection and attribution studies of atmospheric water vapour.

10.3.2.2 Changes in Precipitation

Analysis of CMIP5 model simulations yields clear global and regional scale changes associated with anthropogenic forcing (e.g., Scheff and Frierson, 2012a, 2012b), with patterns broadly similar to those identified from CMIP3 models (e.g., Polson et al., 2013). The AR4 concluded that 'the latitudinal pattern of change in land precipitation and observed increases in heavy precipitation over the 20th century appear to be consistent with the anticipated response to anthropogenic forcing'. Detection and attribution of regional precipitation changes generally focuses on continental areas using *in situ* data because observational coverage over oceans is limited to a few island stations (Arkin et al., 2010; Liu et al., 2012; Noake et al., 2012) , although model-data comparisons over continents also illustrate large observational uncertainties (Tapiador, 2010; Noake et al., 2012; Balan Sarojini et al., 2012; Polson et al., 2013). Available satellite data sets that could supplement oceanic studies are short and their long-term homogeneity is still unclear (Chapter 2); hence they have not yet been used for detection and attribution of changes. Continuing uncertainties in climate model simulations of precipitation make quantitative model/data comparisons difficult (e.g., Stephens et al., 2010), which also limits confidence in detection and attribution. Furthermore, sparse observational coverage of precipitation across much of the planet makes the fingerprint of precipitation change challenging to isolate in observational records (Balan Sarojini et al., 2012; Wan et al., 2013).

Considering just land regions with sufficient observations, the largest signal of differences between models with and without anthropogenic forcings is in the high latitudes of the NH, where increases in precipitation are a robust feature of climate model simulations (Scheff and Frierson, 2012a, 2012b). Such increases have been observed (Figure 10.10) in several different observational data sets (Min et al., 2008a; Noake et al., 2012; Polson et al., 2013), although high-latitude trends vary between data sets and with coverage (e.g., Polson et al., 2013).

Attribution of zonally averaged precipitation trends has been attempted using different observational products and ensembles of forced simulations from both the CMIP3 and CMIP5 archives, for annual-averaged (Zhang et al., 2007; Min et al., 2008a) and season-specific (Noake et al., 2012; Polson et al., 2013) results (Figure 10.11). Zhang et al. (2007) identify the fingerprint of anthropogenic changes in observed annual zonal mean precipitation averaged over the periods 1925–1999 and 1950–1999, and separate the anthropogenic fingerprint from the influence of natural forcing. The fingerprint of external forcing is also detected in seasonal means for boreal spring in all data sets assessed by Noake et al. (2012), and in all but one data set assessed by Polson et al. (2013) (Figure 10.11), and in boreal winter in all but one data set (Noake et al., 2012), over the period 1951–1999 and to 2005. The fingerprint features increasing high-latitude precipitation, and decreasing precipitation trends in parts of the tropics that are reasonably robustly observed in all four data sets considered albeit with large observational uncertainties north of 60°N (Figure 10.11). Detection of seasonal-average precipitation change is less convincing for June, July, August (JJA) and September, October, November (SON) and results vary with observation data set (Noake et al., 2012; Polson et al., 2013). Although Zhang et al. (2007) detect anthropogenic changes even if a separate fingerprint for natural forcings is considered, Polson et al. (2013) find that this result is sensitive to the data set used and that the fingerprints can be separated robustly only for the data set most closely constrained by station data. The analysis also finds that model simulated precipitation variability is smaller than observed variability in the tropics (Zhang et al., 2007;

Polson et al., 2013) which is addressed by increasing the estimate of variance from models (Figure 10.11).

Another detection and attribution study focussed on precipitation in the NH high latitudes and found an attributable human influence (Min et al., 2008a). Both Min et al. (2008a) and Zhang et al. (2007) find that the observed changes are significantly larger than the model simulated changes. However, Noake et al. (2012) and Polson et al. (2013) find that the difference between models and observations decreases if changes are expressed as a percentage of climatological precipitation and that the observed and simulated changes are largely consistent between CMIP5 models and observations given data uncertainty. Use of additional data sets illustrates remaining observational uncertainty in high latitudes of the NH (Figure 10.11). Regional-scale attribution of precipitation change is still problematic although regional climate models have yielded simulations consistent with observed wintertime changes for northern Europe (Bhend and von Storch, 2008; Tapiador, 2010).

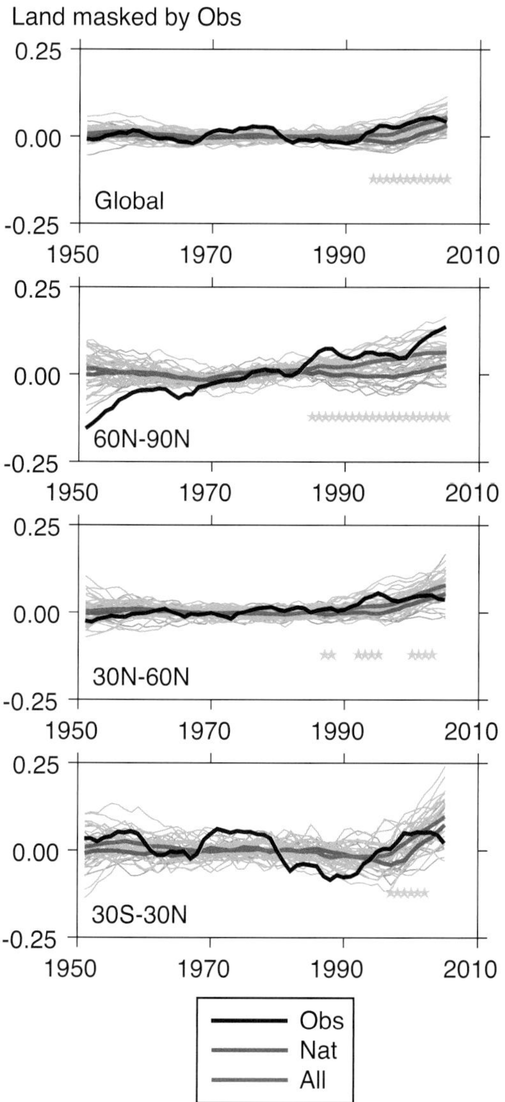

Figure 10.10 | Global and zonal average changes in annual mean precipitation (mm day^{-1}) over areas of land where there are observations, expressed relative to the baseline period of 1961–1990, simulated by CMIP5 models forced with both anthropogenic and natural forcings (red lines) and natural forcings only (blue lines) for the global mean and for three latitude bands. Multi-model means are shown in thick solid lines. Observations (gridded values derived from Global Historical Climatology Network station data, updated from Zhang et al. (2007) are shown as a black solid line. An 11-year smoothing is applied to both simulations and observations. Green stars show statistically significant changes at 5% level (p value <0.05) between the ensemble of runs with both anthropogenic and natural forcings (red lines) and the ensemble of runs with just natural forcings (blue lines) using a two-sample two-tailed t-test for the last 30 years of the time series. (From Balan Sarojini et al., 2012.) Results for the Climate Research Unit (CRU) TS3.1 data set are shown in Figure 10.A.2.

Precipitation change over ocean has been attributed to human influence by Fyfe et al. (2012) for the high-latitude SH in austral summer, where zonally averaged precipitation has declined around 45°S and increased around 60°S since 1957, consistent with CMIP5 historical simulations, with the magnitude of the half-century trend outside the range of simulated natural variability. Confidence in this attribution result, despite limitations in precipitation observations, is enhanced by its consistency with trends in large-scale sea level pressure data (see Section 10.3.3).

In summary, there is *medium confidence* that human influence has contributed to large-scale changes in precipitation patterns over land. The expected anthropogenic fingerprints of change in zonal mean precipitation—reductions in low latitudes and increases in NH mid to high latitudes—have been detected in annual and some seasonal data. Observational uncertainties including limited global coverage and large natural variability, in addition to challenges in precipitation modeling, limit confidence in assessment of climatic changes in precipitation.

10.3.2.3 Changes in Surface Hydrologic Variables

This subsection assesses recent research on detection and attribution of long-term changes in continental surface hydrologic variables, including soil moisture, evapotranspiration and streamflow. Streamflows are often subject to large non-climatic human influence, such as diversions and land use changes, that must be accounted for in order to attribute detected hydrologic changes to climate change. Cryospheric aspects of surface hydrology are discussed in Section 10.5; extremes in surface hydrology (such as drought) and precipitation are covered in Section 10.6.1. The variables discussed here are subject to large modeling uncertainties (Chapter 9) and observational challenges (Chapter 2), which in combination place severe limits on climate change detection and attribution.

Direct observational records of soil moisture and surface fluxes tend to be sparse and/or short, thus limiting recent assessments of change in these variables (Jung et al., 2010). Assimilated land surface data sets and new satellite observations (Chapter 2) are promising tools, but assessment of past and future climate change of these variables (Hoekema and Sridhar, 2011) is still generally carried out on derived quantities such as the Palmer Drought Severity Index, as discussed more fully in Section 10.6.1. Recent observations (Jung et al., 2010) show regional trends towards drier soils. An optimal detection analysis of reconstructed evapotranspiration identifies the effects of anthropogenic forcing on evapotranspiration, with the Centre National de Recherches Météorologiques (CNRM)-CM5 model simulating changes consistent with those estimated to have occurred (Douville et al., 2013).

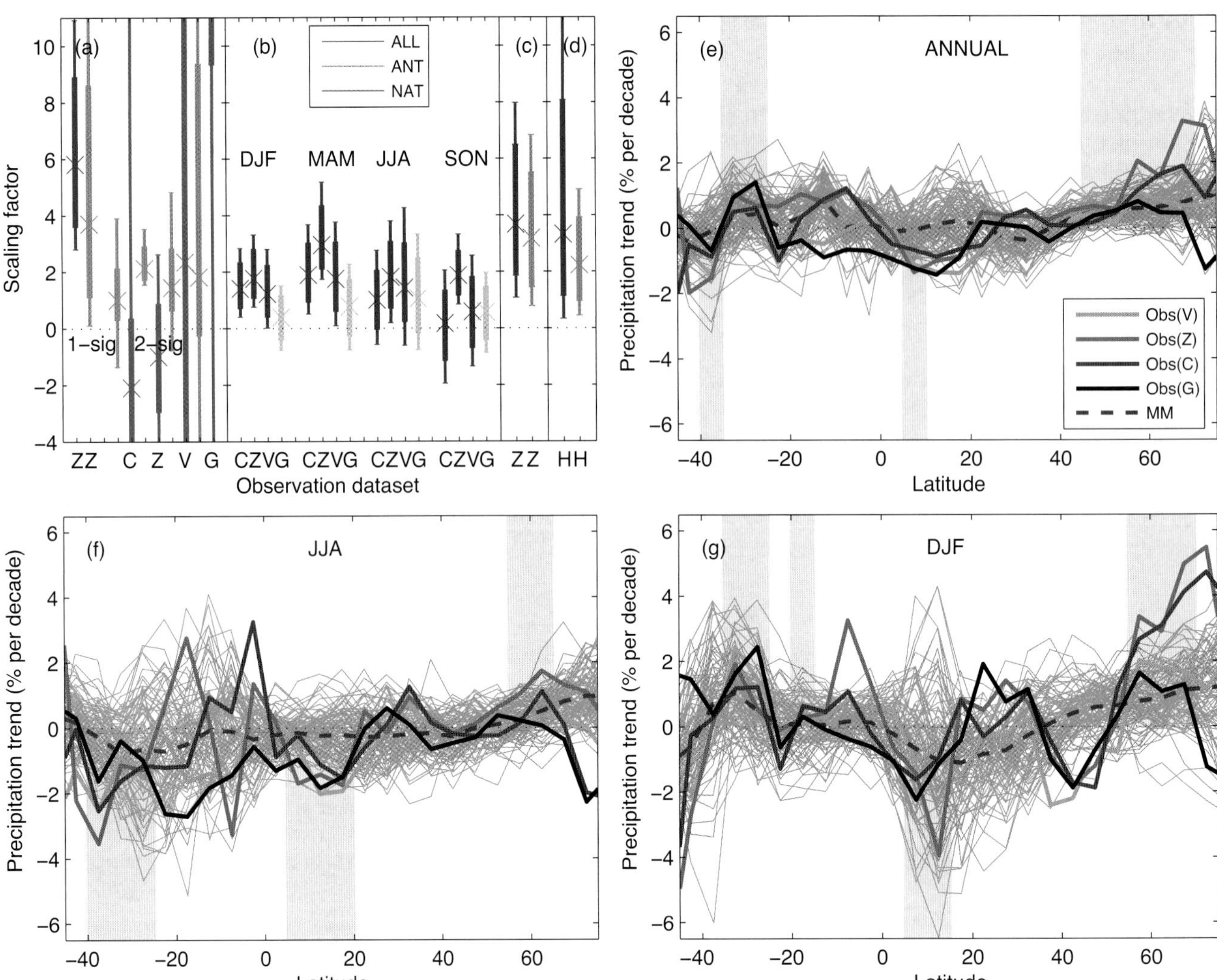

Figure 10.11 | Detection and attribution results for zonal land precipitation trends in the second half of the 20th century. (Top left) Scaling factors for precipitation changes. (Top right and bottom) Zonally averaged precipitation changes over continents from models and observations. (a) Crosses show the best-guess scaling factor derived from multi-model means. Thick bars show the 5 to 95% uncertainty range derived from model-simulated variability, and thin bars show the uncertainty range if doubling the multi-model variance. Red bars indicate scaling factors for the estimated response to all forcings, blue bars for natural-only forcing and brown bars for anthropogenic-only forcing. Labels on the *x*-axis identify results from four different observational data sets (Z is Zhang et al. (2007), C is Climate Research Unit (CRU), V is Variability Analyses of Surface Climate Observations (VasClimO), G is Global Precipitation Climatology Centre (GPCC), H is Hadley Centre gridded data set of temperature and precipitation extremes (HadEX)). (a) Detection and attribution results for annual averages, both single fingerprint ("1-sig"; 1950–1999) and two fingerprint results ("2-sig"; Z, C, G (1951–2005), V (1952–2000)). (b) Scaling factors resulting from single-fingerprint analyses for seasonally averaged precipitation (Z, C, G (1951–2005), V (1952–2000); the latter in pink as not designed for long-term homogeneity) for four different seasons. (c) Scaling factors for spatial pattern of Arctic precipitation trends (1951–1999). (d) Scaling factors for changes in large-scale intense precipitation (1951–1999). (e) Thick solid lines show observed zonally and annually averaged trends (% per decade) for four different observed data sets. Corresponding results from individual simulations from 33 different climate models are shown as thin solid lines, with the multimodel mean shown as a red dashed line. Model results are masked to match the spatial and temporal coverage of the GPCC data set (denoted G in the seasonal scaling factor panel). Grey shading indicates latitude bands within which >75% of simulations yield positive or negative trends. (f, g) Like (e) but showing zonally averaged precipitation changes for (f) June, July, August (JJA) and (g) December, January, February (DJF) seasons. Scaling factors (c) and (d) adapted from Min et al. (2008a) and Min et al. (2011), respectively; other results adapted from Zhang et al. (2007) and Polson et al. (2013).

Trends towards earlier timing of snowmelt-driven streamflows in western North America since 1950 have been demonstrated to be different from natural variability (Hidalgo et al., 2009). Similarly, internal variability associated with natural decade-scale fluctuations could not account for recent observed declines of northern Rocky Mountain streamflow (St Jacques et al., 2010). Statistical analyses of streamflows demonstrate regionally varying changes that are consistent with changes expected from increasing temperature, in Scandinavia (Wilson et al., 2010), Europe (Stahl et al., 2010) and the USA (Krakauer and Fung, 2008; Wang and Hejazi, 2011). Observed increases in Arctic river discharge, which could be a good integrator for monitoring changes in precipitation in high latitudes, are found to be explainable only if model simulations include anthropogenic forcings (Min et al., 2008a).

Barnett et al. (2008) analysed changes in the surface hydrology of the western USA, considering snow pack (measured as snow water equivalent), the seasonal timing of streamflow in major rivers, and average January to March daily minimum temperature over the region, the two hydrological variables they studied being closely related to temperature. Observed changes were compared with the output of a regional hydrologic model forced by the Parallel Climate Model (PCM) and Model for Interdisciplinary Research On Climate (MIROC) climate models. They derived a fingerprint of anthropogenic changes from the two climate models and found that the observations, when projected onto the fingerprint of anthropogenic changes, show a positive signal strength consistent with the model simulations that falls outside the range expected from internal variability as estimated from 1600 years of downscaled climate model data. They conclude that there is a detectable and attributable anthropogenic signature on the hydrology of this region.

In summary, there is *medium confidence* that human influence on climate has affected stream flow and evapotranspiration in limited regions of middle and high latitudes of the NH. Detection and attribution studies have been applied only to limited regions and using a few models. Observational uncertainties are large and in the case of evapotranspiration depend on reconstructions using land surface models.

10.3.3 Atmospheric Circulation and Patterns of Variability

The atmospheric circulation is driven by various processes including the uneven heating of the Earth's surface by solar radiation, land–sea contrast and orography. The circulation transports heat from warm to cold regions and thereby acts to reduce temperature contrasts. Thus, changes in circulation and in patterns of variability are of critical importance for the climate system, influencing regional climate and regional climate variability. Any such changes are important for local climate change because they could act to reinforce or counteract the effects of external forcings on climate in a particular region. Observed changes in atmospheric circulation and patterns of variability are assessed in Section 2.7.5. Although new and improved data sets are now available, changes in patterns of variability remain difficult to detect because of large variability on interannual to decadal time scales (Section 2.7).

Since AR4, progress has been made in understanding the causes of changes in circulation-related climate phenomena and modes of variability such as the width of the tropical circulation, and the Southern Annular Mode (SAM). For other climate phenomena, such as ENSO, Indian Ocean Dipole (IOD), Pacific Decadal Oscillation (PDO), and monsoons, there are large observational and modelling uncertainties (see Section 9.5 and Chapter 14), and there is *low confidence* that changes in these phenomena, if observed, can be attributed to human-induced influence.

10.3.3.1 Tropical Circulation

Various indicators of the width of the tropical belt based on independent data sets suggest that the tropical belt as a whole has widened since 1979; however, the magnitude of this change is very uncertain (Fu et al., 2006; Hudson et al., 2006; Hu and Fu, 2007; Seidel and Randel, 2007; Seidel et al., 2008; Lu et al., 2009; Fu and Lin, 2011; Hu et al., 2011; Davis and Rosenlof, 2012; Lucas et al., 2012; Wilcox et al., 2012; Nguyen et al., 2013) (Section 2.7.5). CMIP3 and CMIP5 simulations suggest that anthropogenic forcings have contributed to the observed widening of the tropical belt since 1979 (Johanson and Fu, 2009; Hu et al., 2013). On average the poleward expansion of the Hadley circulation and other indicators of the width of the tropical belt is greater than determined from CMIP3 and CMIP5 simulations (Seidel et al., 2008; Johanson and Fu, 2009; Hu et al., 2013; Figure 10.12). The causes as to why models underestimate the observed poleward expansion of the tropical belt are not fully understood. Potential factors are lack of understanding of the magnitude of natural variability as well as changes in observing systems that also affect reanalysis products (Thorne and Vose, 2010; Lucas et al., 2012; Box 2.3).

Climate model simulations suggest that Antarctic ozone depletion is a major factor in causing poleward expansion of the southern Hadley cell during austral summer over the last three to five decades with GHGs also playing a role (Son et al., 2008, 2009, 2010; McLandress et al., 2011; Polvani et al., 2011; Hu et al., 2013). In reanalysis data a detectable signal of ozone forcing is separable from other external forcing including GHGs when utilizing both CMIP5 and CMIP3 simulations combined (Min and Son, 2013). An analysis of CMIP3 simulations suggests that BC aerosols and tropospheric ozone were the main drivers of the observed poleward expansion of the northern Hadley cell in boreal summer (Allen et al., 2012). It is found that global greenhouse warming causes increase in static stability, such that the onset of baroclinicity is shifted poleward, leading to poleward expansion of the Hadley circulation (Frierson, 2006; Frierson et al., 2007; Hu and Fu, 2007; Lu et al., 2007, 2008). Tropical SST increase may also contribute to a widening of the Hadley circulation (Hu et al., 2011; Staten et al., 2012). Althoughe some Atmospheric General Circulation Model (AGCM) simulations forced by observed time-varying SSTs yield a widening by about 1° in latitude over 1979–2002 (Hu et al., 2011), other simulations suggest that SST changes have little effect on the tropical expansion when based on the tropopause metric of the tropical width (Lu et al., 2009). However, it is found that the tropopause metric is not

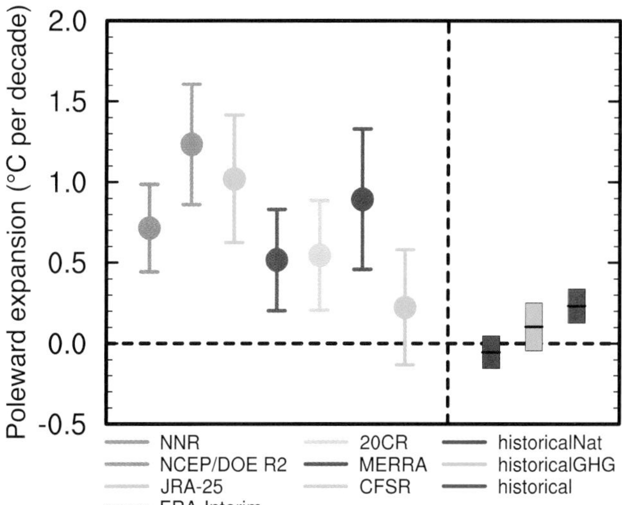

Figure 10.12 | December to February mean change of southern border of the Hadley circulation. Unit is degree in latitude per decade. Reanalysis data sets (see also Box 2.3) are marked with different colours. Trends are all calculated over the period of 1979–2005. The terms historicalNAT, historicalGHG, and historical denote CMIP5 simulations with natural forcing, with greenhouse gas forcing and with both anthropogenic and natural forcings, respectively. For each reanalysis data set, the error bars indicate the 95% confidence level of the standard *t*-test. For CMIP5 simulations, trends are first calculated for each model, and all ensemble members of simulations are used. Then, trends are averaged for multi-model ensembles. Trend uncertainty is estimated from multi-model ensembles, as twice the standard error. (Updated from Hu et al., 2013.)

very reliable because of the use of arbitrary thresholds (Birner, 2010; Davis and Rosenlof, 2012).

In summary, there are multiple lines of evidence that the Hadley cell and the tropical belt as a whole have widened since at least 1979; however, the magnitude of the widening is very uncertain. Based on modelling studies there is *medium confidence* that stratospheric ozone depletion has contributed to the observed poleward shift of the southern Hadley cell border during austral summer, with GHGs also playing a role. The contribution of internal climate variability to the observed poleward expansion of the Hadley circulation remains very uncertain.

10.3.3.2 Northern Annular Mode/North Atlantic Oscillation

The NAO, which exhibited a positive trend from the 1960s to the 1990s, has since exhibited lower values, with exceptionally low anomalies in the winters of 2009/2010 and 2010/2011 (Section 2.7.8). This means that the positive trend in the NAO discussed in the AR4 has considerably weakened when evaluated up to 2011. Similar results apply to the closely related Northern Annular Mode (NAM), with its upward trend over the past 60 years in the 20th Century Reanalysis (Compo et al., 2011) and in Hadley Centre Sea Level Pressure data set 2r (HadSLP2r; Allan and Ansell, 2006) not being significant compared to internal variability (Figure 10.13). An analysis of CMIP5 models shows that they simulate positive trends in NAM in the DJF season over this period, albeit not as large as those observed which are still within the range of natural internal variability (Figure 10.13).

Other work (Woollings, 2008) demonstrates that while the NAM is largely barotropic in structure, the simulated response to anthropogenic forcing has a strong baroclinic component, with an opposite geopotential height trends in the mid-troposphere compared to the surface in many models. Thus while the circulation response to anthropogenic forcing may project onto the NAM, it is not entirely captured by the NAM index.

Consistent with previous findings (Hegerl et al., 2007b), Gillett and Fyfe (2013) find that GHGs tend to drive a positive NAM response in the CMIP5 models. Recent modelling work also indicates that ozone changes drive a small positive NAM response in spring (Morgenstern et al., 2010; Gillett and Fyfe, 2013).

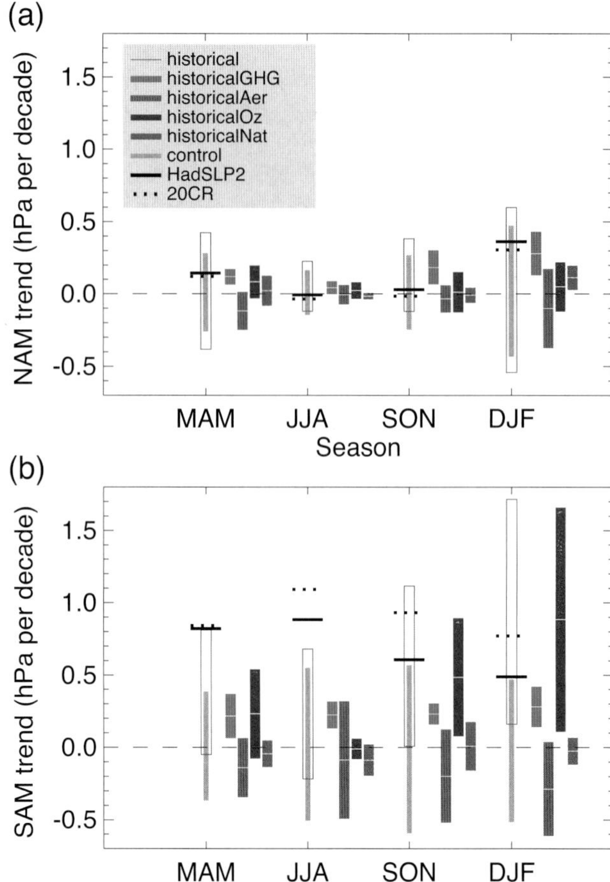

Figure 10.13 | Simulated and observed 1951–2011 trends in the Northern Annular Mode (NAM) index (a) and Southern Annular Mode (SAM) index (b) by season. The NAM is a Li and Wang (2003) index based on the difference between zonal mean seal level pressure (SLP) at 35°N and 65°N. and the, and the SAM index is a difference between zonal mean SLP at 40°S and 65°S (Gong and Wang, 1999). Both indices are defined without normalization, so that the magnitudes of simulated and observed trends can be compared. Black lines show observed trends from the HadSLP2r data set (Allan and Ansell, 2006) (solid), and the 20th Century Reanalysis (Compo et al., 2011) (dotted). Grey bars and red boxes show 5 to 95% ranges of trends in CMIP5 control and historical simulations respectively. Ensemble mean trends and their 5 to 95% uncertainties are shown for the response to greenhouse gases (light green), aerosols (dark green), ozone (magenta) and natural (blue) forcing changes, based on CMIP5 individual forcing simulations. (Adapted from Gillett and Fyfe, 2013.)

10.3.3.3 Southern Annular Mode

The Southern Annular Mode (SAM) index has remained mainly positive since the publication of the AR4, although it has not been as strongly positive as in the late 1990s. Nonetheless, an index of the SAM shows a significant positive trend in most seasons and data sets over the 1951–2011 period (Figure 10.13; Table 2.14). Recent modelling studies confirm earlier findings that the increase in GHG concentrations tends to lead to a strengthening and poleward shift of the SH eddy-driven polar jet (Karpechko et al., 2008; Son et al., 2008, 2010; Sigmond et al., 2011; Staten et al., 2012; Swart and Fyfe, 2012; Eyring et al., 2013; Gillett and Fyfe, 2013) which projects onto the positive phase of the SAM. Stratospheric ozone depletion also induces a strengthening and poleward shift of the polar jet in models, with the largest response in austral summer (Karpechko et al., 2008; Son et al., 2008, 2010; McLandress et al., 2011; Polvani et al., 2011; Sigmond et al., 2011; Gillett and Fyfe, 2013). Sigmond et al. (2011) find approximately equal contributions to simulated annual mean SAM trends from GHGs and stratospheric ozone depletion up to the present. Fogt et al. (2009) demonstrate that observed SAM trends over the period 1957–2005 are positive in all seasons, but only statistically significant in DJF and March, April, May (MAM), based on simulated internal variability. Roscoe and Haigh (2007) apply a regression-based approach and find that stratospheric ozone changes are the primary driver of observed trends in the SAM. Observed trends are also consistent with CMIP3 simulations including stratospheric ozone changes in all seasons, though in MAM observed trends are roughly twice as large as those simulated (Miller et al., 2006). Broadly consistent results are found when comparing observed trends and CMIP5 simulations (Figure 10.13), with a station-based SAM index showing a significant positive trend in MAM, JJA and DJF, compared to simulated internal variability over the 1951–2010 period. Fogt et al. (2009) find that the largest forced response has likely occurred in DJF, the season in which stratospheric ozone depletion has been the dominant contributor to the observed trends.

Taking these findings together, it is *likely* that the positive trend in the SAM seen in austral summer since the mid-20th century is due in part to stratospheric ozone depletion. There is *medium confidence* that GHGs have also played a role.

10.3.3.4 Change in Global Sea Level Pressure Patterns

A number of studies have applied formal detection and attribution studies to global fields of atmospheric SLP finding detection of human influence on global patterns of SLP (Gillett et al., 2003, 2005; Gillett and Stott, 2009). Analysing the contributions of different forcings to observed changes in SLP, Gillett and Stott (2009) find separately detectable influences of anthropogenic and natural forcings in zonal mean seasonal mean SLP, strengthening evidence for a human influence on SLP. Based on the robustness of the evidence from multiple models we conclude that it is *likely* that human influence has altered SLP patterns globally since 1951.

10.4 Changes in Ocean Properties

This section assesses the causes of oceanic changes in the main properties of interest for climate change: ocean heat content, ocean salinity and freshwater fluxes, sea level, oxygen and ocean acidification.

10.4.1 Ocean Temperature and Heat Content

The oceans are a key part of the Earth's energy balance (Boxes 3.1 and 13.1). Observational studies continue to demonstrate that the ocean heat content has increased in the upper layers of the ocean during the second half of the 20th century and early 21st century (Section 3.2; Bindoff et al., 2007), and that this increase is consistent with a net positive radiative imbalance in the climate system. It is of significance that this heat content increase is an order of magnitude larger than the increase in energy content of any other component of the Earth's ocean–atmosphere–cryosphere system and accounts for more than 90% of the Earth's energy increase between 1971 and 2010 (e.g., Boxes 3.1 and 13.1; Bindoff et al., 2007; Church et al., 2011; Hansen et al., 2011).

Despite the evidence for anthropogenic warming of the ocean, the level of confidence in the conclusions of the AR4 report—that the warming of the upper several hundred meters of the ocean during the second half of the 20th century was *likely* to be due to anthropogenic forcing—reflected the level of uncertainties at that time. The major uncertainty was an apparently large decadal variability (warming in the 1970s and cooling in the early 1980s) in the observational estimates that was not simulated by climate models (Hegerl et al., 2007b, see their Table 9.4). The large decadal variability in observations raised concerns about the capacity of climate models to simulate observed variability. There were also lingering concerns about the presence of non-climate–related biases in the observations of ocean heat content change (Gregory et al., 2004; AchutaRao et al., 2006). After the IPCC AR4 report in 2007, time-and depth-dependent systematic errors in bathythermograph temperatures were discovered (Gouretski and Koltermann, 2007; Section 3.2). Bathythermograph data account for a large fraction of the historical temperature observations and are therefore a source of bias in ocean heat content studies. Bias corrections were then developed and applied to observations. With the newer bias-corrected estimates (Domingues et al., 2008; Wijffels et al., 2008; Ishii and Kimoto, 2009; Levitus et al., 2009), it became obvious that the large decadal variability in earlier estimates of global upper-ocean heat content was an observational artefact (Section 3.2).

The interannual to decadal variability of ocean temperature simulated by the CMIP3 models agrees better with observations when the model data is sampled using the observational data mask (AchutaRao et al., 2007). In the upper 700 m, CMIP3 model simulations agreed more closely with observational estimates of global ocean heat content based on bias-corrected ocean temperature data, both in terms of the decadal variability and multi-decadal trend (Figure 10.14a) when forced with the most complete set of natural and anthropogenic forcings (Domingues et al., 2008). For the simulations with the most complete set of forcings, the multi-model ensemble mean trend was only 10% smaller than observed for 1961–1999. Model simulations that included only anthropogenic forcing (i.e., no solar or volcanic forcing) significantly overestimate the multi-decadal trend and underestimate decadal variability. This overestimate of the trend is partially caused by the ocean's response to volcanic eruptions, which results in rapid cooling followed by decadal or longer time variations during the recovery phase. Although it has been suggested (Gregory, 2010) that the cooling trend from successive volcanic events is an artefact because models were not spun up with volcanic forcing, this discrepancy is not expected to be as significant in the upper ocean as in the deeper layers where longer term adjustments take place (Gregory et al., 2012). Thus for the upper ocean, there is *high confidence* that the more frequent eruptions during the second half of the 20th century have caused a multi-decadal cooling that partially offsets the anthropogenic warming and contributes to the apparent decadal variability (Church et al., 2005; Delworth et al., 2005; Fyfe, 2006; Gleckler et al., 2006; Gregory et al., 2006; AchutaRao et al., 2007; Domingues et al., 2008; Palmer et al., 2009; Stenchikov et al., 2009).

Gleckler et al. (2012) examined the detection and attribution of upper-ocean warming in the context of uncertainties in the underlying observational data sets, models and methods. Using three bias-corrected observational estimates of upper-ocean temperature changes (Domingues et al., 2008; Ishii and Kimoto, 2009; Levitus et al., 2009) and models from the CMIP3 multi-model archive, they found that multi-decadal trends in the observations were best understood by including contributions from both natural and anthropogenic forcings. The anthropogenic fingerprint in observed upper-ocean warming, driven by global mean and basin-scale pattern changes, was also detected. The strength of this signal (estimated from successively longer trend periods of ocean heat content starting from 1970) crossed the 5% and 1% significance threshold in 1980 and progressively becomes more strongly detected for longer trend periods (Figure 10.14b), for all ocean heat content time series. This stronger detection for longer periods occurs because the noise (standard deviation of trends in the unforced changes in pattern similarity from model control runs) tends to decrease for longer trend lengths. On decadal time scales, there is *limited evidence* that basin scale space-time variability structure of CMIP3 models is approximately 25% lower than the (poorly constrained) observations, this underestimate is far less than the factor of 2 needed to throw the anthropogenic fingerprint into question. This result is robust to a number of known observational, model, methodological and structural uncertainties.

An analysis of upper-ocean (0 to 700 m) temperature changes for 1955–2004, using bias-corrected observations and 20 global climate models from CMIP5 (Pierce et al., 2012) builds on previous detection and attribution studies of ocean temperature (Barnett et al., 2001, 2005; Pierce et al., 2006). This analysis found that observed temperature changes during the above period are inconsistent with the effects of natural climate variability. That is signal strengths are separated from zero at the 5% significance level, and the probability that the

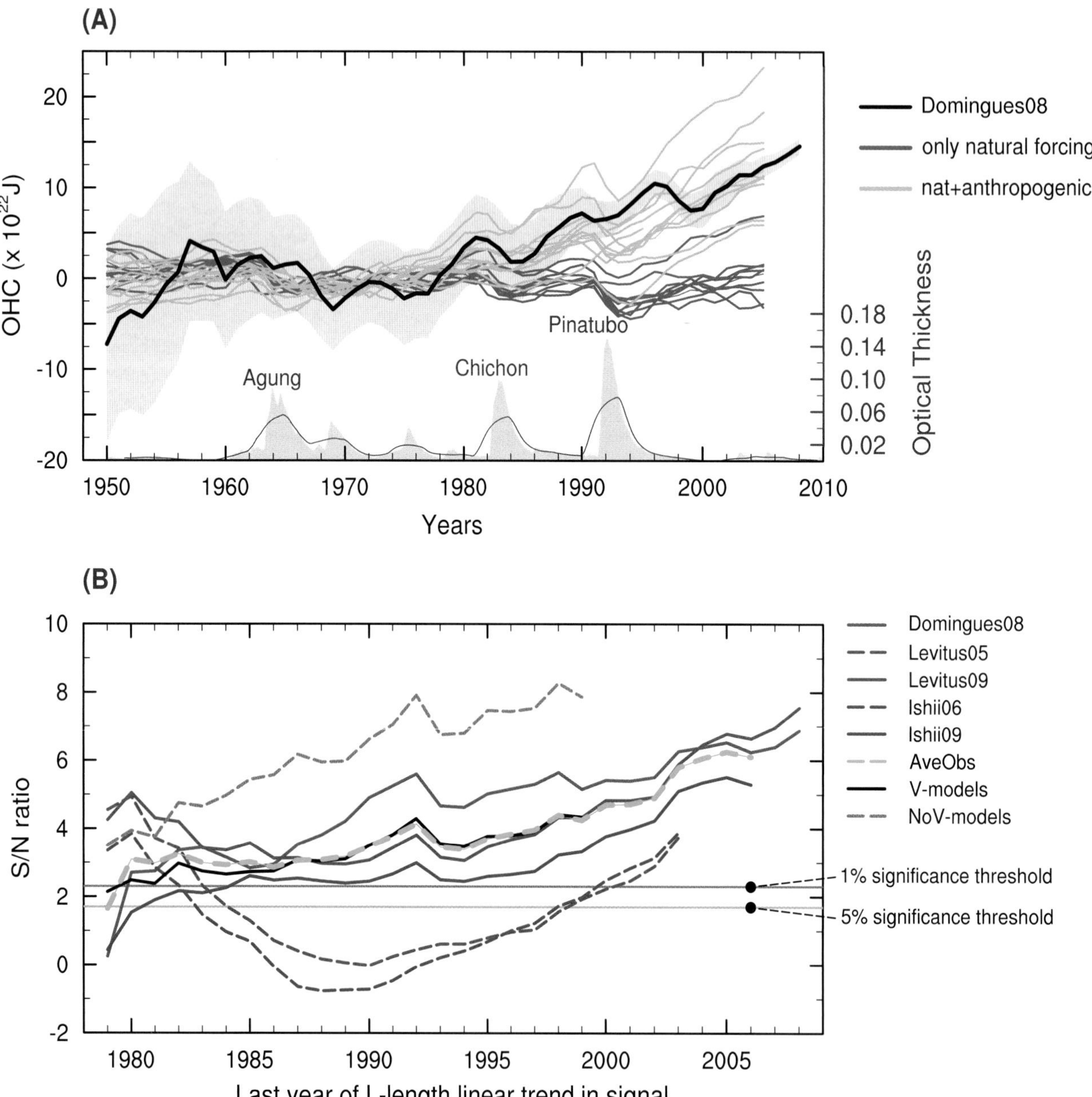

Figure 10.14 | (A) Comparison of observed global ocean heat content for the upper 700 m (updated from Domingues et al. 2008) with simulations from ten CMIP5 models that included only natural forcings ('HistoricalNat' runs shown in blue lines) and simulations that included natural and anthropogenic forcings ('Historical' runs in pink lines). Grey shading shows observational uncertainty. The global mean stratospheric optical depth (Sato et al., 1993) in beige at the bottom indicates the major volcanic eruptions and the brown curve is a 3-year running average of these values. (B) Signal-to-noise (S/N) ratio (plotted as a function of increasing trend length L) of basin-scale changes in volume averaged temperature of newer, expendable bathythermograph (XBT)-corrected data (solid red, purple and blue lines), older, uncorrected data (dashed red and blue lines); the average of the three corrected observational sets (AveObs; dashed cyan line); and simulations that include volcanic (V) or exclude volcanic eruptions (NoV) (black solid and grey dashed lines respectively). The start date for the calculation of signal trends is 1970 and the initial trend length is 10 years. The 1% and 5% significance thresholds are shown (as horizontal grey lines) and assume a Gaussian distribution of noise trends in the V-models control-run pseudo-principal components. The detection time is defined as the year at which S/N exceeds and remains above 1% or 5% significance threshold (Gleckler et al., 2012).

null hypothesis of observed changes being consistent with natural variability is less than 0.05 from variability either internal to the climate system alone, or externally forced by solar fluctuations and volcanic eruptions. However, the observed ocean changes are consistent with those expected from anthropogenically induced atmospheric changes from GHGs and aerosol concentrations.

Attribution to anthropogenic warming from recent detection and attribution studies (Gleckler et al., 2012; Pierce et al., 2012) have made use of new bias-corrected observations and have systematically explored methodological uncertainties, yielding more confidence in the results. With greater consistency and agreement across observational data sets and resolution of structural issues, the major uncertainties at the time of AR4 have now largely been resolved. The high levels of confidence and the increased understanding of the contributions from both natural and anthropogenic sources across the many studies mean that it is *very likely* that the increase in global ocean heat content observed in the upper 700 m since the 1970s has a substantial contribution from anthropogenic forcing.

Although there is *high confidence* in understanding the causes of global heat content increases, attribution of regional heat content changes are less certain. Earlier regional studies used a fixed depth data and only considered basin-scale averages (Barnett et al., 2005). At regional scales, however, changes in advection of ocean heat are important and need to be isolated from changes due to air–sea heat fluxes (Palmer et al., 2009; Grist et al., 2010). Their fixed isotherm (rather than fixed depth) approach to optimal detection analysis, in addition to being largely insensitive to observational biases, is designed to separate the ocean's response to air–sea flux changes from advective changes. Air–sea fluxes are the primary mechanism by which the oceans are expected to respond to externally forced anthropogenic and natural volcanic influences. The finer temporal resolution of the analysis allowed Palmer et al. (2009) to attribute distinct short-lived cooling episodes to major volcanic eruptions while, at multi-decadal time scales, a more spatially uniform near-surface (~ upper 200 m) warming pattern was detected across all ocean basins (except in high latitudes where the isotherm approach has limitations due to outcropping of isotherms at the ocean surface) and attributed to anthropogenic causes at the 5% significance level. Considering that individual ocean basins are affected by different observational and modelling uncertainties and that internal variability is larger at smaller scales, detection of significant anthropogenic forcing through space and time studies (Palmer et al., 2009; Pierce et al., 2012) provides more compelling evidence of human influence at regional scales of near-surface ocean warming observed during the second half of the 20th century.

10.4.2 Ocean Salinity and Freshwater Fluxes

There is increasing recognition of the importance of ocean salinity as an essential climate variable (Doherty et al., 2009), particularly for understanding the hydrological cycle. In the IPCC Fourth Assessment Report observed ocean salinity change indicated that there was a systematic pattern of increased salinity in the shallow subtropics and a tendency to freshening of waters that originate in the polar regions (Bindoff et al., 2007; Hegerl et al., 2007b) (Figure 10.15a, upper and lower panels). New atlases and revisions of the earlier work based on the increasing number of the Array for Real-time Geostrophic Oceanography (ARGO) profile data, and historical data have extended the observational salinity data sets allowing the examination of the long-term changes at the surface and in the interior of the ocean (Section 3.3) and supporting analyses of precipitation changes over land (see Sections 10.3.2.2 and 2.5.1).

Patterns of subsurface salinity changes largely follow the existing mean salinity pattern at the surface and within the ocean. For example, the inter-basin contrast between the Atlantic (salty) and Pacific Oceans (fresh) has intensified over the observed record (Boyer et al., 2005; Hosoda et al., 2009; Roemmich and Gilson, 2009; von Schuckmann et al., 2009; Durack and Wijffels, 2010). In the Southern Ocean, many studies show a coherent freshening of Antarctic Intermediate Water that is subducted at about 50°S (Johnson and Orsi, 1997; Wong et al., 1999; Bindoff and McDougall, 2000; Curry et al., 2003; Boyer et al., 2005; Roemmich and Gilson, 2009; Durack and Wijffels, 2010; Helm et al., 2010; Kobayashi et al., 2012). There is also a clear increase in salinity of the high-salinity subtropical waters (Durack and Wijffels, 2010; Helm et al., 2010).

The 50-year trends in surface salinity show that there is a strong positive correlation between the mean climate of the surface salinity and its temporal changes from 1950 to 2000 (see Figures 3.4 and 10.15b 'ocean obs' point). The correlation between the climate and the trends in surface salinity of 0.7 implies that fresh surface waters get fresher, and salty waters get saltier (Durack et al., 2012). Such patterns of surface salinity change are also found in Atmosphere–Ocean General Circulation Models (AOGCM) simulations both for the 20th century and projected future changes into the 21st century (Figure 10.15b). The pattern of temporal change in observations from CMIP3 simulations is particularly strong for those projections using Special Report on Emission Scenarios (SRES) with larger global warming changes (Figure 10.15b). For the period 1950–2000 the observed amplification of the surface salinity is 16 ± 10% per °C of warming and is twice the simulated surface salinity change in CMIP3 models. This difference between the surface salinity amplification is plausibly caused by the tendency of CMIP3 ocean models mixing surface salinity into deeper layers and consequently surface salinity increases at a slower rate than observed (Durack et al., 2012).

Although there are now many established observed long-term trends of salinity change at the ocean surface and within the interior ocean at regional and global scales (Section 3.3), there are relatively few studies that attribute these changes formally to anthropogenic forcing. Analysis at the regional scale of the observed recent surface salinity increases in the North Atlantic (20°N to 50°N) show a small signal that could be attributed to anthropogenic forcings but for this ocean is not significant compared with internal variability (Stott et al., 2008a; Terray et al., 2012; and Figure 10.15c). On a larger spatial scale, the surface salinity patterns in the band from 30°S to 50°N show anthropogenic contributions that are larger than the 5 to 95% uncertainty range (Terray et al., 2012). The strongest signals that can be attributed to anthropogenic forcing are in the tropics (TRO, 30°S to 30°N) and the western Pacific. These results also show the salinity contrast between the Pacific and Atlantic oceans is also enhanced with significant contributions from anthropogenic forcing.

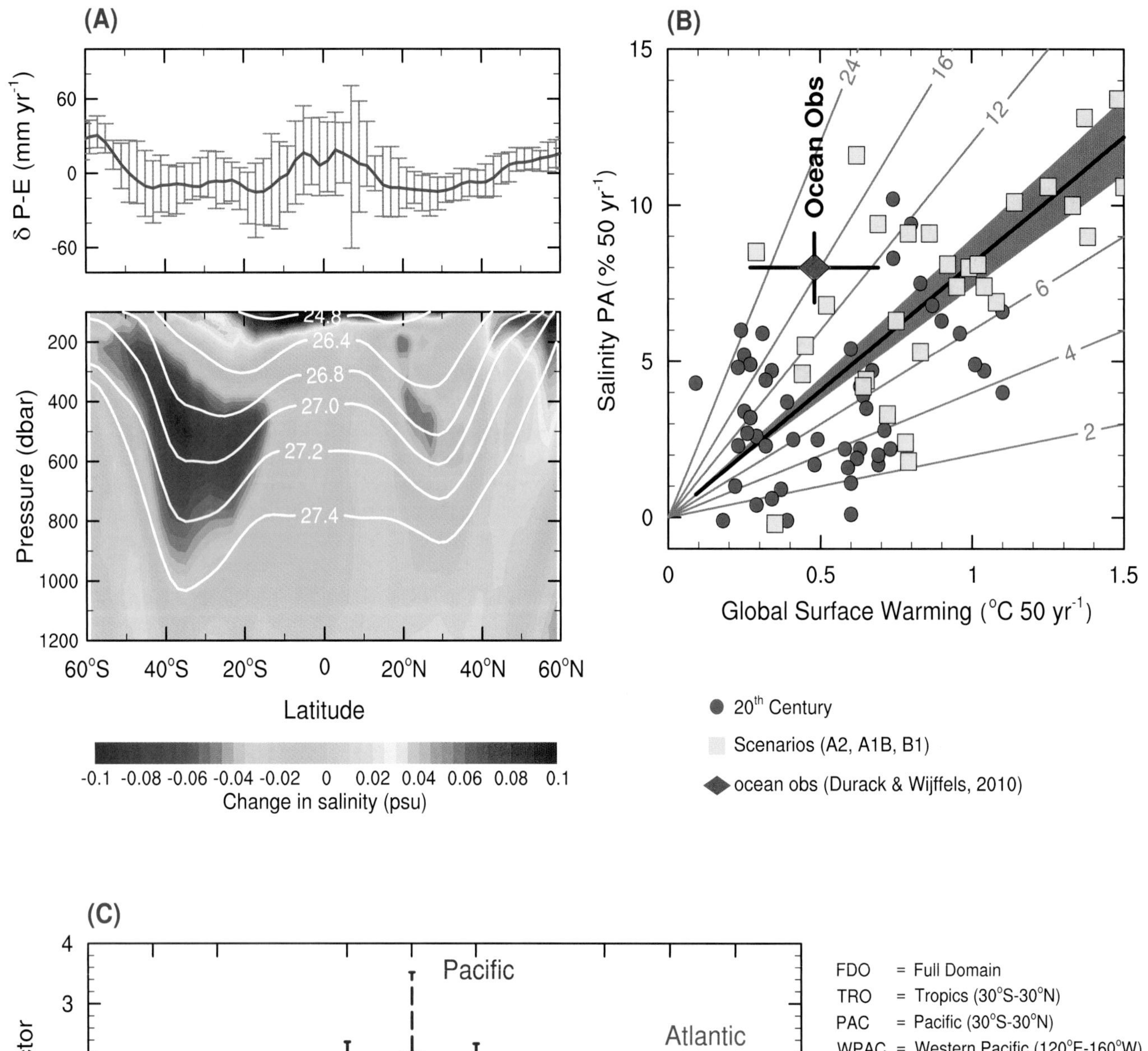

Figure 10.15 | Ocean salinity change and hydrologic cycle. (A) Ocean salinity change observed in the interior of the ocean (A, lower panel in practical salinity units or psu, and white lines are surfaces of constant density) and comparison with ten CMIP3 model projections of precipitation minus evaporation δ (P – E) in mm yr^{-1} for the same period as the observed changes (1970 to 1990s) (A, top panel, red line is the mean of the simulations and error bars are the simulated range). (B) The amplification of the current surface salinity pattern over a 50-year period as a function of global temperature change. Ocean surface salinity pattern amplification has a 16% increase for the 1950–2000 period (red diamond, see text and Section 3.3). Also on this panel CMIP3 simulations from Special Report on Emission Scenarios (SRES) (yellow squares) and from 20th century simulations (blue circles). A total of 93 simulations have been used. (C) Regional detection and attribution in the equatorial Pacific and Atlantic Oceans for 1970 to 2002. Scaling factors for all forcings (anthropogenic) fingerprint are shown (see Box 10.1) with their 5 to 95% uncertainty range, estimated using the total least square approach. Full domain (FDO, 30°S to 50°N), Tropics (TRO, 30°S to 30°N), Pacific (PAC, 30°S to 30°N), west Pacific (WPAC, 120°E to 160°W), east Pacific (EPAC, 160°W to 80°W), Atlantic (ATL, 30°S to 50°N), subtropical north Atlantic (NATL, 20°N to 40°N) and equatorial Atlantic (EATL, 20°S to 20°N) factors are shown. Black filled dots indicate when the residual consistency test passes with a truncation of 16 whereas empty circles indicate a higher truncation was needed to pass the consistency test. Horizontal dashed lines indicate scaling factor of 0 or 1. (A, B and C are adapted from Helm et al. (2010), Durack et al. (2012) and Terray et al. (2012), respectively.)

On a global scale surface and subsurface salinity changes (1955–2004) over the upper 250 m of the water column cannot be explained by natural variability (probability is <0.05) (Pierce et al., 2012). However, the observed salinity changes match the model distribution of forced changes (GHG and tropospheric aerosols), with the observations typically falling between the 25th and 75th percentile of the model distribution at all depth levels for salinity (and temperature). Natural external variability taken from the simulations with just solar and volcanic variations in forcing do not match the observations at all, thus excluding the hypothesis that observed trends can be explained by just solar or volcanic variations.

The results from surface salinity trends and changes are consistent with the results from studies of precipitation over the tropical ocean from the shorter satellite record (Wentz et al., 2007; Allan et al., 2010). These surface salinity results are also consistent with our understanding of the thermodynamic response of the atmosphere to warming (Held and Soden, 2006; Stephens and Hu, 2010) and the amplification of the water cycle. The large number of studies showing patterns of change consistent with amplification of the water cycle, and the detection and attribution studies for the tropical oceans (Terray et al., 2012) and the global pattern of ocean salinity change (Pierce et al., 2012), when combined with our understanding of the physics of the water cycle and estimates of internal climate variability, give *high confidence* in our understanding of the drivers of surface and near surface salinity changes. It is *very likely* that these salinity changes have a discernable contribution from anthropogenic forcing since the 1960s.

10.4.3 Sea Level

At the time of the AR4, the historical sea level rise budget had not been closed (within uncertainties), and there were few studies quantifying the contribution of anthropogenic forcing to the observed sea level rise and glacier melting. Relying on expert assessment, the AR4 had concluded based on modelling and ocean heat content studies that ocean warming and glacier mass loss had *very likely* contributed to sea level rise during the latter half of the 20th century. The AR4 had reported that climate models that included anthropogenic and natural forcings simulated the observed thermal expansion since 1961 reasonably well, and that it is *very unlikely* that the warming during the past half century is due only to known natural causes (Hegerl et al., 2007b).

Since the AR4, corrections applied to instrumental errors in ocean temperature measurements have considerably improved estimates of upper-ocean heat content (see Sections 3.2 and 10.4.1), and therefore ocean thermal expansion. Closure of the global mean sea level rise budget as an evolving time series since the early1970s (Church et al., 2011) indicates that the two major contributions to the rate of global mean sea level rise have been thermal expansion and glacier melting with additional contributions from Greenland and Antarctic ice sheets. Observations since 1971 indicate with *high confidence* that thermal expansion and glaciers (excluding the glaciers in Antarctica) explain 75% of the observed rise (see Section 13.3.6). Ice sheet contributions remain the greatest source of uncertainty over this period and on longer time scales. Over the 20th century, the global mean sea level rise budget (Gregory et al., 2012) has been another important step in understanding the relative contributions of different drivers. The observed contribution from thermal expansion is well captured in climate model simulations with historical forcings as are contributions from glacier melt when simulated by glacier models driven by climate model simulations of historical climate (Church et al., 2013; Table 13.1). The model results indicate that most of the variation in the contributions of thermal expansion and glacier melt to global mean sea level is in response to natural and anthropogenic RFs (Domingues et al., 2008; Palmer et al., 2009; Church et al., 2013).

The strong physical relationship between thermosteric sea level and ocean heat content (through the equation of state for seawater) means that the anthropogenic ocean warming (Section 10.4.1) has contributed to global sea level rise over this period through thermal expansion. As Section 10.5.2 concludes, it is *likely* that the observed substantial mass loss of glaciers is due to human influence and that it is *likely* that anthropogenic forcing and internal variability are both contributors to recent observed changes on the Greenland ice sheet. The causes of recently observed Antarctic ice sheet contribution to sea level are less clear due to the short observational record and incomplete understanding of natural variability. Taking the causes of Greenland ice sheet melt and glacier mass loss together (see Section 10.5.2), it is concluded with *high confidence* that it is *likely* that anthropogenic forcing has contributed to sea level rise from melting glaciers and ice sheets. Combining the evidence from ocean warming and mass loss of glaciers we conclude that it is *very likely* that there is a substantial contribution from anthropogenic forcing to the global mean sea level rise since the 1970s.

On ocean basin scales, detection and attribution studies do show the emergence of detectable signals in the thermosteric component of sea level that can be largely attributed to human influence (Barnett et al., 2005; Pierce et al., 2012). Regional changes in sea level at the sub-ocean basin scales and finer exhibit more complex variations associated with natural (dynamical) modes of climate variability (Section 13.6). In some regions, sea level trends have been observed to differ significantly from global mean trends. These have been related to thermosteric changes in some areas and in others to changing wind fields and resulting changes in the ocean circulation (Han et al., 2010; Timmermann et al., 2010; Merrifield and Maltrud, 2011). The regional variability on decadal and longer time scales can be quite large (and is not well quantified in currently available observations) compared to secular changes in the winds that influence sea level. Detection of human influences on sea level at the regional scale (that is smaller than sub-ocean basin scales) is currently limited by the relatively small anthropogenic contributions compared to natural variability (Meyssignac et al., 2012) and the need for more sophisticated approaches than currently available.

10.4.4 Oxygen and Ocean Acidity

Oxygen is an important physical and biological tracer in the ocean (Section 3.8.3) and is projected to decline by 3 to 6% by 2100 in response to surface warming (see Section 6.4.5). Oxygen decreases are also observed in the atmosphere and linked to burning of fossil fuels (Section 6.1.3.2). Despite the relatively few observational studies of oxygen change in the oceans (Bindoff and McDougall, 2000; Ono et al., 2001; Keeling and Garcia, 2002; Emerson et al., 2004; Aoki et al., 2005;

10

Mecking et al., 2006; Nakanowatari et al., 2007; Brandt et al., 2010) they all show a pattern of change consistent with the known ocean circulation and surface ventilation. A recent global analysis of oxygen data from the 1960s to 1990s for change confirm these earlier results and extends the spatial coverage from local to global scales (Helm et al., 2011). The strongest decreases in oxygen occur in the mid-latitudes of both hemispheres, near regions where there is strong water renewal and exchange between the ocean interior and surface waters. The attribution study of oxygen decreases using two Earth System Models (ESMs) concluded that observed changes for the Atlantic Ocean are 'indistinguishable from natural internal variability'; however, the changes of the global zonal mean to external forcing (all forcings including GHGs) has a detectable influence at the 10% significance level (Andrews et al., 2013). The chief sources of uncertainty are the paucity of oxygen observations, particularly in time, the precise role of the biological pump and changes in ocean productivity in the models (see Sections 3.8.3 and 6.4.5), and model circulation biases particularly near the oxygen minimum zone in tropical waters (Brandt et al., 2010; Keeling et al., 2010; Stramma et al., 2010). These results of observed changes in oxygen and the attribution studies of oxygen changes (Andrews et al., 2013), along with the attribution of human influences on the physical factors that affect oxygen in the oceans such as surface temperatures changes (Section 10.3.2), increased ocean heat content (Section 10.4.1) and observed increased in ocean stratification (Section 3.2.2) provides evidence for human influence on oxygen. When these lines of evidence are taken together it is concluded that with *medium confidence* or *about as likely as not* that the observed oxygen decreases can be attributed in part to human influences.

The observed trends (since the 1980s) for ocean acidification and its cause from rising CO_2 concentrations is discussed in Section 3.8.2 (Box 3.2 and Table 10.1). There is *very high* confidence that anthropogenic CO_2 has resulted in the acidification of surface waters of between –0.0015 and –0.0024 pH units per year.

10.5 Cryosphere

This section considers changes in sea ice, ice sheets and ice shelves, glaciers, snow cover. The assessment of attribution of human influences on temperature over the Arctic and Antarctica is in Section 10.3.1.

10.5.1 Sea Ice

10.5.1.1 Arctic and Antarctic Sea Ice

The Arctic cryosphere shows large observed changes over the last decade as noted in Chapter 4 and many of these shifts are indicators of major regional and global feedback processes (Kattsov et al., 2010). An assessment of sea ice models' capacity to simulate Arctic and Antarctic sea ice extent is given in Section 9.4.3. Of principal importance is 'Arctic Amplification' (see Box 5.1) where surface temperatures in the Arctic are increasing faster than elsewhere in the world.

The rate of decline of Arctic sea ice thickness and September sea ice extent has increased considerably in the first decade of the 21st century (Maslanik et al., 2007; Nghiem et al., 2007; Comiso and Nishio, 2008; Deser and Teng, 2008; Zhang et al., 2008; Alekseev et al., 2009; Comiso, 2012; Polyakov et al., 2012). Based on a sea ice reanalysis and verified by ice thickness estimates from satellite sensors, it is estimated that three quarters of summer Arctic sea ice volume has been lost since the 1980s (Schweiger et al., 2011; Maslowski et al., 2012; Laxon et al., 2013; Overland and Wang, 2013). There was also a rapid reduction in ice extent, to 37% less in September 2007 and to 49% less in September 2012 relative to the 1979–2000 climatology (Figure 4.11, Section 4.2.2). Unlike the loss record set in 2007 that was dominated by a major shift in climatological winds, sea ice loss in 2012 was more due to a general thinning of the sea ice (Lindsay et al., 2009; Wang et al., 2009a; Zhang et al., 2013). All recent years have ice extents that fall at least two standard deviations below the long-term sea ice trend.

The amount of old, thick multi-year sea ice in the Arctic has decreased by 50% from 2005 through 2012 (Giles et al., 2008; Kwok et al., 2009; Kwok and Untersteiner, 2011 and Figures 4.13 and 4.14). Sea ice has also become more mobile (Gascard et al., 2008). We now have seven years of data that show sea ice conditions are substantially different to that observed prior to 2006. The relatively large increase in the percentage of first year sea ice across the Arctic basin can be considered 'a new normal.'

Confidence in detection of change comes in part from the consistency of multiple lines of evidence. Since AR4, evidence has continued to accumulate from a range of observational studies that systematic changes are occurring in the Arctic. Persistent trends in many Arctic variables, including sea ice, the timing of spring snow melt, increased shrubbiness in tundra regions, changes in permafrost, increased area of forest fires, changes in ecosystems, as well as Arctic-wide increases in air temperatures, can no longer be associated solely with the dominant climate variability patterns such as the Arctic Oscillation, Pacific North American pattern or Atlantic Meridional Oscillation (AMO) (Quadrelli and Wallace, 2004; Vorosmarty et al., 2008; Overland, 2009; Brown and Robinson, 2011; Mahajan et al., 2011; Oza et al., 2011a; Wassmann et al., 2011; Nagato and Tanaka, 2012). Duarte et al. (2012) completed a meta-analysis showing evidence from multiple indicators of detectable climate change signals in the Arctic.

The increase in the magnitude of recent Arctic temperature and decrease in sea ice volume and extent are hypothesized to be due to coupled Arctic amplification mechanisms (Serreze and Francis, 2006; Miller et al., 2010). These feedbacks in the Arctic climate system suggest that the Arctic is sensitive to external forcing (Mahlstein and Knutti, 2012). Historically, changes were damped by the rapid formation of sea ice in autumn causing a negative feedback and a rapid seasonal cooling. But recently, the increased mobility and loss of multi-year sea ice, combined with enhanced heat storage in the sea ice-free regions of the Arctic Ocean form a connected set of processes with positive feedbacks causing an increase in Arctic temperatures and a decrease in sea ice extent (Manabe and Wetherald, 1975; Gascard et al., 2008; Serreze et al., 2009; Stroeve et al., 2012a, 2012b) . In addition to the well known *ice albedo* feedback where decreased sea ice cover decreases the amount of *insolation* reflected from the surface, there is a late summer/early autumn positive *ice insulation* feedback due to additional ocean heat storage in areas previously covered by sea ice

(Jackson et al., 2010). Arctic amplification may also have a contribution from poleward heat transport in the atmosphere and ocean (Langen and Alexeev, 2007; Graversen and Wang, 2009; Doscher et al., 2010; Yang et al., 2010).

It appears that recent Arctic changes are in response to a combination of global-scale warming, from warm anomalies from internal climate variability on different time scales, and are amplified from the multiple feedbacks described above. For example, when the 2007 sea ice minimum occurred, Arctic temperatures had been rising and sea ice extent had been decreasing over the previous two decades (Stroeve et al., 2008; Screen and Simmonds, 2010). Nevertheless, it took unusually persistent southerly winds along the dateline over the summer months to initiate the sea ice loss event in 2007 (Zhang et al., 2008; Wang et al., 2009b). Similar southerly wind patterns in previous years did not initiate major reductions in sea ice extent because the sea ice was too thick to respond (Overland et al., 2008). Increased oceanic heat transport through the Barents Sea in the first decade of the 21st century and the AMO on longer time scales may also have played a role in determining sea ice anomalies in the Atlantic Arctic (Dickson et al., 2000; Semenov, 2008; Zhang et al., 2008; Day et al., 2012) . Based on the evidence in the previous paragraphs there is *high confidence* that these Arctic amplification mechanisms are currently affecting regional Arctic climate. But it also suggests that the timing of future major sea ice loss events will be difficult to project. There is evidence therefore that internal variability of climate, long-term warming, and Arctic Amplification feedbacks have all contributed to recent decreases in Arctic sea ice (Kay et al., 2011b; Kinnard et al., 2011; Overland et al., 2011; Notz and Marotzke, 2012).

Turning to model-based attribution studies, Min et al. (2008b) compared the seasonal evolution of Arctic sea ice extent from observations with those simulated by multiple General Circulation Models (GCMs) for 1953–2006. Comparing changes in both the amplitude and shape of the annual cycle of the sea ice extent reduces the chance of spurious detection due to coincidental agreement between the response to anthropogenic forcing and other factors, such as slow internal variability. They found that human influence on the sea ice extent changes has been robustly detected since the early 1990s. The anthropogenic signal is also detectable for individual months from May to December, suggesting that human influence, strongest in late summer, now also extends into colder seasons. Kay et al. (2011b), Jahn et al. (2012) and Schweiger et al. (2011) used the Community Climate System Model 4 (CCSM4) to investigate the influence of anthropogenic forcing on late 20th century and early 21st century Arctic sea ice extent and volume trends. On all time scales examined (2 to 50+ years), the most extreme negative extent trends observed in the late 20th century cannot be explained by modeled internal variability alone. Comparing trends from the CCSM4 ensemble to observed trends suggests that internal variability could account for approximately half of the observed 1979–2005 September Arctic sea ice extent loss. Attribution of anthropogenic forcing is also shown by comparing September sea ice extent as projected by seven models from the set of CMIP5 models' hindcasts to control runs without anthropogenic forcing (Figure 10.16a; Wang and Overland, 2009). The mean of sea ice extents in seven models' ensemble members are below the level of their control runs by about 1995, similar to the result of Min et al. (2008b).

A question as recently as 6 years ago was whether the recent Arctic warming and sea ice loss was unique in the instrumental record and whether the observed trend would continue (Serreze et al., 2007). Arctic temperature anomalies in the 1930s were apparently as large as those in the 1990s and 2000s. There is still considerable discussion of the ultimate causes of the warm temperature anomalies that occurred in the Arctic in the 1920s and 1930s (Ahlmann, 1948; Veryard, 1963; Hegerl et al., 2007a, 2007b). The early 20th century warm period, while reflected in the hemispheric average air temperature record (Brohan et al., 2006), did not appear consistently in the mid-latitudes nor on the Pacific side of the Arctic (Johannessen et al., 2004; Wood and Overland, 2010). Polyakov et al. (2003) argued that the Arctic air temperature records reflected a natural cycle of about 50 to 80 years. However, many authors (Bengtsson et al., 2004; Grant et al., 2009; Wood and Overland, 2010; Brönnimann et al., 2012) instead link the 1930s temperatures to internal variability in the North Atlantic atmospheric and ocean circulation as a single episode that was sustained by ocean and sea ice processes in the Arctic and north Atlantic. The Arctic-wide increases of temperature in the last decade contrast with the episodic regional increases in the early 20th century, suggesting that it is unlikely that recent increases are due to the same primary climate process as the early 20th century.

In the case of the Arctic we have *high confidence* in observations since 1979, from models (see Section 9.4.3 and from simulations comparing with and without anthropogenic forcing), and from physical understanding of the dominant processes; taking these three factors together it is *very likely* that anthropogenic forcing has contributed to the observed decreases in Arctic sea ice since 1979.

Whereas sea ice extent in the Arctic has decreased, sea ice extent in the Antarctic has *very likely* increased (Section 4.2.3). Sea ice extent across the SH over the year as a whole increased by 1.3 to 1.67% per decade from 1979 to 2012, with the largest increase in the Ross Sea during the autumn, while sea ice extent decreased in the Amundsen-Bellingshausen Sea (Comiso and Nishio, 2008; Turner et al., 2009, 2013; Section 4.2.3; Oza et al., 2011b). The observed upward trend in Antarctic sea ice extent is found to be inconsistent with internal variability based on the residuals from a linear trend fitted to the observations, though this approach could underestimate multi-decadal variability (Section 4.2.3; Turner et al., 2013; Section 4.2.3; Zunz et al., 2013). The CMIP5 simulations on average simulate a decrease in Antarctic sea ice extent (Turner et al., 2013; Zunz et al., 2013; Figure 10.16b), though Turner et al. (2013) find that approximately 10% of CMIP5 simulations exhibit an increasing trend in Antarctic sea ice extent larger than observed over the 1979–2005 period. However, Antarctic sea ice extent variability appears on average to be too large in the CMIP5 models (Turner et al., 2013; Zunz et al., 2013). Overall, the shortness of the observed record and differences in simulated and observed variability preclude an assessment of whether or not the observed increase in Antarctic sea ice extent is inconsistent with internal variability. Based on Figure 10.16b and Meehl et al. (2007b), the trend of Antarctic sea ice loss in simulations due to changes in forcing is weak (relative to the Arctic) and the internal variability is high, and thus the time necessary for detection is longer than in the Arctic.

Several recent studies have investigated the possible causes of Antarctic sea ice trends. Early studies suggested that stratospheric ozone depletion may have driven increasing trends in Antarctic ice extent (Goosse et al., 2009; Turner et al., 2009; WMO (World Meteorological Organization), 2011), but recent studies demonstrate that simulated sea ice extent decreases in response to prescribed changes in stratospheric ozone (Sigmond and Fyfe, 2010; Bitz and Polvani, 2012). An alternative explanation for the lack of melting of Antarctic sea ice is that sub-surface ocean warming, and enhanced freshwater input possibly in part from ice shelf melting, have made the high-latitude Southern Ocean fresher (see Section 3.3) and more stratified, decreasing the upward heat flux and driving more sea ice formation (Zhang, 2007; Goosse et al., 2009; Bintanja et al., 2013). An idealized simulation of the response to freshwater input similar to that estimate due to ice shelf melting exhibited an increase in sea ice extent (Bintanja et al., 2013), but this result has yet to be reproduced with other models. Overall we conclude that there is *low confidence* in the scientific understanding of the observed increase in Antarctic sea ice extent since 1979, owing to

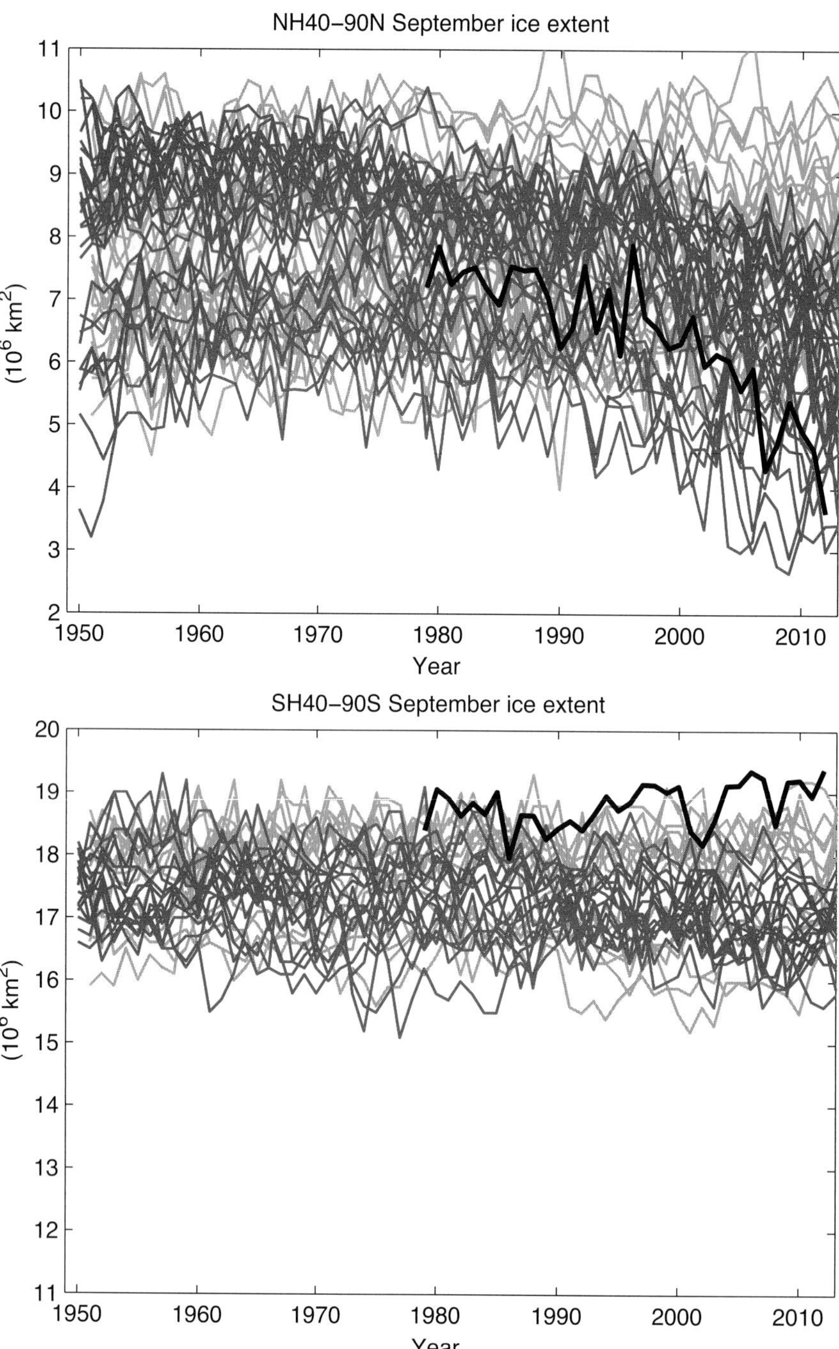

Figure 10.16 | September sea ice extent for Arctic (top) and Antarctic (bottom) adapted from (Wang and Overland, 2012). Only CMIP5 models that simulated seasonal mean and magnitude of seasonal cycle in reasonable agreement with observations are included in the plot. The grey lines are the runs from the pre-industrial control runs, and the red lines are from Historical simulations runs patched with RCP8.5 runs for the period of 2005–2012. The black line is based on data from National Snow and Ice Data Center (NSIDC). There are 24 ensemble members from 11 models for the Arctic and 21 members from 6 models for the Antarctic plot. See Supplementary Material for the precise models used in the top and bottom panel.

the larger differences between sea ice simulations from CMIP5 models and to the incomplete and competing scientific explanations for the causes of change and *low confidence* in estimates of internal variability (Section 9.4.3).

10.5.2 Ice Sheets, Ice Shelves and Glaciers

10.5.2.1 Greenland and Antarctic Ice Sheets

The Greenland and Antarctic ice sheets are important to regional and global climate because (along with other cryospheric elements) they cause a polar amplification of surface temperatures, a source of fresh water to the ocean, and represent a source of potentially irreversible change to the state of the Earth system (Hansen and Lebedeff, 1987). These two ice sheets are important contributors to sea level rise representing two-thirds of the contributions from all ice covered regions (Jacob et al., 2012; Pritchard et al., 2012; see Sections 4.4 and 13.3.3). Observations of surface mass balance (increased ablation versus increased snowfall) are dealt with in Section 4.4.3 and ice sheet models are discussed in Sections 13.3 and 13.5.

Attribution of change is difficult as ice sheet and glacier changes are local and ice sheet processes are not generally well represented in climate models thus precluding formal single-step detection and attribution studies. However, Greenland observational records show large recent changes. Section 13.3 concludes that regional models for Greenland can reproduce trends in the surface mass balance loss quite well if they are forced with the observed meteorological record, but not with forcings from a Global Climate Model. Regional model simulations (Fettweis et al., 2013) show that Greenland surface melt increases nonlinearly with rising temperatures due to the positive feedback between surface albedo and melt.

There have been exceptional changes in Greenland since 2007 marked by record-setting high air temperatures, ice loss by melting and marine-terminating glacier area loss (Hanna et al., 2013; Section 4.4. 4). Along Greenland's west coast temperatures in 2010 and 2011were the warmest since record keeping began in 1873 resulting in the highest observed melt rates in this region since 1958 (Fettweis et al., 2011). The annual rate of area loss in marine-terminating glaciers was 3.4 times that of the previous 8 years, when regular observations became available. In 2012, a new record for summertime ice mass loss was two standard deviations below the 2003–2012 mean, as estimated from the Gravity Recovery and Climate Experiment (GRACE) satellite (Tedesco et al., 2012). The trend of summer mass change during 2003–2012 is rather uniform over this period at -29 ± 11 Gt yr^{-1}.

Record surface melts during 2007–2012 summers are linked to persistent atmospheric circulation that favored warm air advection over Greenland. These persistent events have changed in frequency since the beginning of the 2000s (L'Heureux et al., 2010; Fettweis et al., 2011). Hanna et al. (2013) show a weak relation of Greenland temperatures and ice sheet runoff with the AMO; they more strongly correlate with a Greenland atmospheric blocking index. Overland et al. (2012) and Francis and Vavrus (2012) suggest that the increased frequency of the Greenland blocking pattern is related to broader scale Arctic changes. Since 2007, internal variability is *likely* to have further enhanced the melt over Greenland. Mass loss and melt is also occurring in Greenland through the intrusion of warm water into the major glaciers such as Jacobshaven Glacier (Holland et al., 2008; Walker et al., 2009).

Hanna et al. (2008) attribute increased Greenland runoff and melt since 1990 to global warming; southern Greenland coastal and NH summer temperatures were uncorrelated between the 1960s and early 1990s but correlated significantly positively thereafter. This relationship was modulated by the NAO, whose summer index significantly negatively correlated with southern Greenland summer temperatures until the early 1990s but not thereafter. Regional modelling and observations tell a consistent story of the response of Greenland temperatures and ice sheet runoff to shifts in recent regional atmospheric circulation associated with larger scale flow patterns and global temperature increases. It is *likely* that anthropogenic forcing has contributed to surface melting of the Greenland ice sheet since 1993.

There is clear evidence that the West Antarctic ice sheet is contributing to sea level rise (Bromwich et al., 2013). Estimates of ice mass in Antarctic since 2000 show that the greatest losses are at the edges (see Section 4.4). An analysis of observations underneath a floating ice shelf off West Antarctica shows that ocean warming and more transport of heat by ocean circulation are largely responsible for increasing melt rates (Jacobs et al., 2011; Joughin and Alley, 2011; Mankoff et al., 2012; Pritchard et al., 2012).

Antarctica has regionally dependent decadal variability in surface temperature with variations in these trends depending on the strength of the SAM climate pattern. Recent warming in continental west Antarctica has been linked to SST changes in the tropical Pacific (Ding et al., 2011). As with Antarctic sea ice, changes in Antarctic ice sheets have complex causes (Section 4.4.3). The observational record of Antarctic mass loss is short and the internal variability of the ice sheet is poorly understood. Due to a low level of scientific understanding there is *low confidence* in attributing the causes of the observed loss of mass from the Antarctic ice sheet since 1993. Possible future instabilities in the west Antarctic ice sheet cannot be ruled out, but projection of future climate changes over West Antarctica remains subject to considerable uncertainty (Steig and Orsi, 2013).

10.5.2.2 Glaciers

In the 20th century, there is robust evidence that large-scale internal climate variability governs interannual to decadal variability in glacier mass (Hodge et al., 1998; Nesje et al., 2000; Vuille et al., 2008; Huss et al., 2010; Marzeion and Nesje, 2012) and, along with glacier dynamics, impacts glacier length as well (Chinn et al., 2005). On time periods longer than years and decades, there is now evidence of recent ice loss (see Section 4.3.3) due to increased ambient temperatures and associated regional moisture changes. However, few studies evaluate the direct attribution of the current observed mass loss to anthropogenic forcing, owing to the difficulty associated with contrasting scales between glaciers and the large-scale atmospheric circulation (Mölg et al., 2012). Reichert et al. (2002) show for two sample sites at mid and high latitude that internal climate variability over multiple millennia as represented in a GCM would not result in such short glacier lengths as observed in the 20th century. For a sample site at low latitude using multi-step attribution, Mölg et al. (2009) (and references therein) found a close relation between glacier mass loss and the externally forced atmosphere–ocean circulation in the Indian Ocean since the late 19th century. A second, larger group of studies makes use of century-scale glacier records (mostly glacier length but mass balance as well) to extract evidence for external drivers. These include local and regional changes in precipitation and air temperature, and related parameters (such as melt factors and solid/liquid precipitation ratio) estimated from the observed change in glaciers. In general these studies show that the glacier changes reveal unique departures since the 1970s, and that the inferred climatic drivers in the 20th century and particularly in most recent decades, exceed the variability of the earlier parts of the records (Oerlemans, 2005; Yamaguchi et al., 2008; Huss and Bauder, 2009; Huss et al., 2010; Leclercq and Oerlemans, 2011). These results underline the contrast to former centuries where observed glacier fluctuations can be explained by internal climate variability (Reichert et al., 2002; Roe and O'Neal, 2009; Nussbaumer and Zumbühl, 2012). Anthropogenic land cover change is an unresolved forcing, but a first assessment suggests that it does not confound the impacts of recent temperature and precipitation changes if the land cover changes are of local nature (Mölg et al., 2012). The robustness of the estimates of observed mass loss since the 1960s (Section 4.3, Figure 4.11), the confidence we have in estimates of natural variations and internal variability from long-term glacier records, and our understanding of glacier response to climatic drivers provides robust evidence and, therefore, *high confidence* that a substantial part of the mass loss of glaciers is *likely* due to human influence.

10.5.3 Snow Cover

Both satellite and *in situ* observations show significant reductions in the NH snow cover extent (SCE) over the past 90 years, with most reduction occurring in the 1980s (see Section 4.5). Formal detection and attribution studies have indicated anthropogenic influence on NH SCE (Rupp et al., 2013) and western USA snow water equivalent (SWE, Pierce et al., 2008). Pierce et al. (2008) detected anthropogenic influence in the ratio of 1 April SWE over October to March precipitation over the period 1950–1999. These reductions could not be explained by natural internal climate variability alone, nor by changes in solar and volcanic forcing. In their analysis of NH SCE using 13 CMIP5 simulations over the 1922–2005 period, Rupp et al. (2013) showed that some CMIP5 simulations with natural external and anthropogenic forcings could explain the observed decrease in spring SEC though the CMIP5 simulations with all forcing as a whole could only explain half of the magnitude of decrease, and that volcanic and solar variations (from four CMIP5 simulations) were inconsistent with observations. We conclude with *high confidence* in the observational and modelling evidence that the decrease in NH snow extent since the 1970s is *likely* to be caused by all external forcings and has an anthropogenic contribution (see Table 10.1).

10.6 Extremes

Because many of the impacts of climate changes may manifest themselves through weather and climate extremes, there is increasing interest in quantifying the role of human and other external influences on those extremes. SREX assessed causes of changes in different types of extremes including temperature and precipitation, phenomena that influence the occurrence of extremes (e.g., storms, tropical cyclones), and impacts on the natural physical environment such as drought (Seneviratne et al., 2012). This section assesses current understanding of causes of changes in weather and climate extremes, using AR4 as a starting point. Any changes or modifications to SREX assessment are highlighted.

10.6.1 Attribution of Changes in Frequency/Occurrence and Intensity of Extremes

This sub-section assesses attribution of changes in the characteristics of extremes including frequency and intensity of extremes. Many of the extremes discussed in this sub-section are moderate extreme events that occur more than once in a year (see Box 2.4 for detailed discussion). Attribution of changes in the risk of specific extreme events, which are also very rare in general, is assessed in the next sub-section.

10.6.1.1 Temperature Extremes

AR4 concluded that 'surface temperature extremes have *likely* been affected by anthropogenic forcing'. Many indicators of climate extremes and variability showed changes consistent with warming, including a widespread reduction in number of frost days in mid-latitude regions and evidence that in many regions warm extremes had become warmer and cold extremes had become less cold. We next assess new studies made since AR4.

Relatively warm seasonal mean temperatures (e.g., those that have a recurrence once in 10 years) have seen a rapid increase in frequency for many regions worldwide (Jones et al., 2008; Stott et al., 2011; Hansen et al., 2012) and an increase in the occurrence frequencies of unusually warm seasonal and annual mean temperatures has been attributed in part to human influence (Stott et al., 2011; Christidis et al., 2012a, 2012b).

A large amount of evidence supports changes in daily data based temperature extreme indices consistent with warming, despite different data sets or different methods for data processing having been used (Section 2.6). The effects of human influence on daily temperature extremes is suggested by both qualitative and quantitative comparisons between observed and CMIP3 based modelled values of warm days and warm nights (the number of days exceeding the 90th percentile of daily maximum and daily minimum temperatures referred to as TX90p and TN90p, see also Section 2.7) and cold days and cold nights (the number of days with daily maximum and daily minimum temperatures below the 10th percentile referred to as TX10p and TN10p; see also Section 2.7). Trends in temperature extreme indices computed for Australia (Alexander and Arblaster, 2009) and the USA (Meehl et al., 2007a) using observations and simulations of the 20th century with nine GCMs that include both anthropogenic and natural forcings are found to be consistent. Both observations and model simulations show a decrease in the number of frost days, and an increase in the growing season length, heatwave duration and TN90p in the second half of the 20th century. Two of the models (PCM and CCSM3) with simulations that include only anthropogenic or only natural forcings indicate that the observed changes are simulated with anthropogenic forcings, but not with natural forcings (even though there are some differences in the details of the forcings). Morak et al. (2011) found that over many sub-continental regions, the number of warm nights (TN90p) shows detectable changes over the second half of the 20th century that are consistent with model simulated changes in response to historical external forcings. They also found detectable changes in indices of temperature extremes when the data were analysed over the globe as a whole. As much of the long-term change in TN90p can be predicted based on the interannual correlation of TN90p with mean temperature, Morak et al. (2013) conclude that the detectable changes are attributed in a multi-step approach (see Section 10.2.4) in part to GHG increases. Morak et al. (2013) have extended this analysis to TX10p, TN10p, TX90p as well as TN90p, using fingerprints from HadGEM1 and find detectable changes on global scales and in many regions (Figure 10.17).

Human influence has also been detected in two different measures of the intensity of extreme daily temperatures in a year. Zwiers et al. (2011) compared four extreme temperature variables including warmest daily maximum and minimum temperatures (annual maximum daily maximum and minimum temperatures, referred to as TXx, TNx) and coldest daily maximum and minimum temperatures (annual minimum daily maximum and minimum temperatures, referred to as TXn, TNn) from observations and from simulations with anthropogenic forcing or anthropogenic and natural external forcings from seven GCMs. They consider these extreme daily temperatures to follow generalized extreme value (GEV) distributions with location, shape and scale parameters. They fit GEV distributions to the observed extreme temperatures with location parameters as linear functions of signals obtained from the model simulation. They found that both anthropogenic influence and combined influence of anthropogenic and natural forcing can be detected in all four extreme temperature variables at the global scale over the land, and also regionally over many large land areas (Figure 10.17). In a complementary study, Christidis et al. (2011) used an optimal fingerprint method to compare observed and modelled time-varying location parameters of extreme temperature distributions. They detected the effects of anthropogenic forcing on warmest daily temperatures in a single fingerprint analysis, and were able to separate the effects of natural from anthropogenic forcings in a two fingerprint analysis.

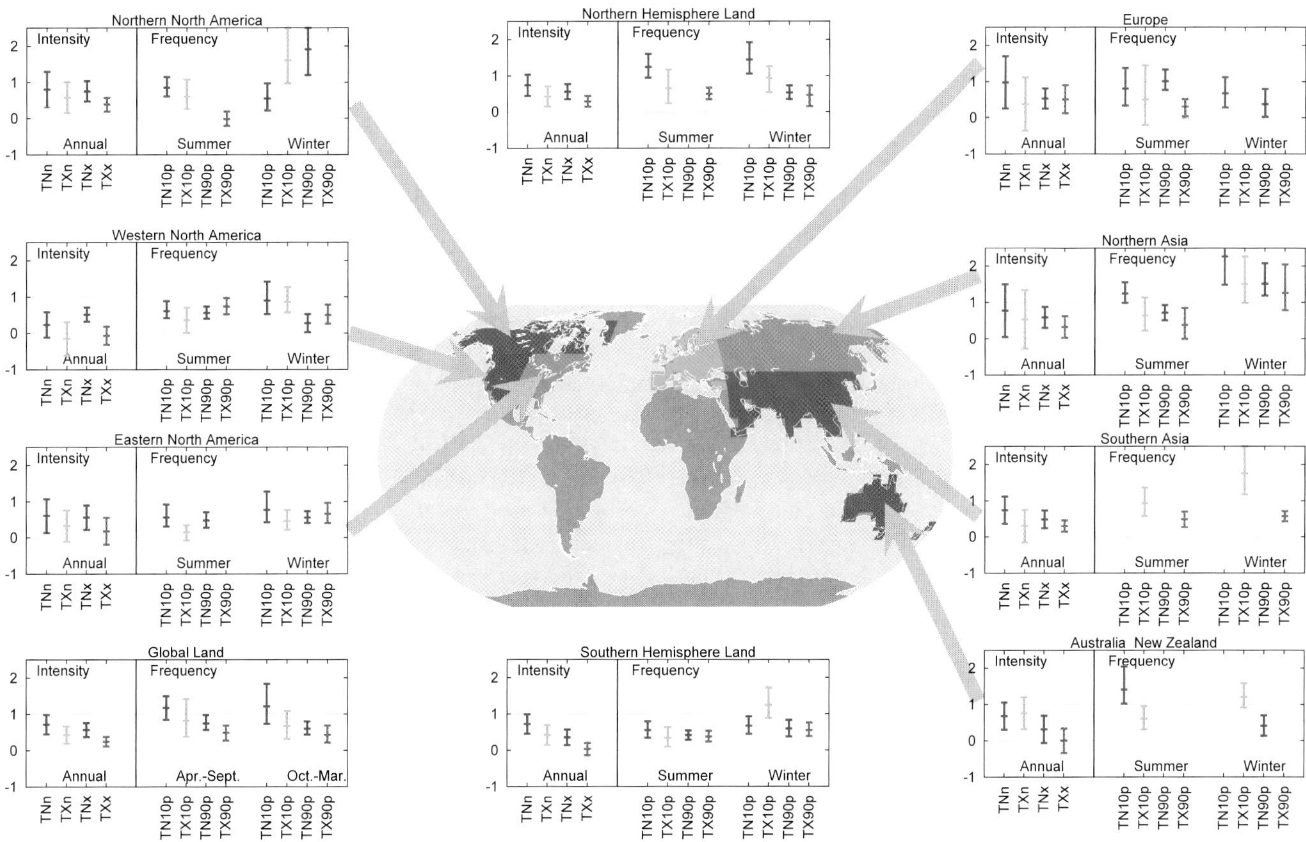

Figure 10.17 | Detection results for changes in intensity and frequency of extreme events. The left side of each panel shows scaling factors and their 90% confidence intervals for intensity of annual extreme temperatures in response to external forcings for the period 1951–2000. TNn and TXn represent coldest daily minimum and maximum temperatures, respectively, while TNx and TXx represent warmest daily minimum and maximum temperatures (updated from Zwiers et al., 2011). Fingerprints are based on simulations of climate models with both anthropogenic and natural forcings. Right-hand sides of each panel show scaling factors and their 90% confidence intervals for changes in the frequency of temperature extremes for winter (October to March for the Northern Hemisphere and April to September for the Southern Hemisphere), and summer half years. TN10p, TX10p are respectively the frequency of cold nights and days (daily minimum and daily maximum temperatures falling below their 10th percentiles for the base period 1961–1990). TN90p and TX90p are the frequency of warm nights and days (daily minimum and daily maximum temperatures above their respective 90th percentiles calculated for the 1961–1990 base period (Morak et al., 2013) with fingerprints based on simulations of Hadley Centre Global Environmental Model 1 (HadGEM1) with both anthropogenic and natural forcings. Detection is claimed at the 5% significance level if the 90% confidence interval of a scaling factor is entirely above the zero line. Grey represents regions with insufficient data.

Human influence on annual extremes of daily temperatures may be detected separately from natural forcing at the global scale (Christidis et al., 2011) and also at continental and sub-continental scales (Min et al., 2013). Over China, Wen et al. (2013) showed that anthropogenic influence may be separately detected from that of natural forcing in daily extreme temperatures (TNn, TNx, TXn and TXx), although the influence of natural forcing is not detected, and they also showed that the influence of GHGs in these indices may be separately detected from other anthropogenic forcings. Christidis et al. (2013) found that on a quasi-global scale, the cooling effect due to the decrease in tree cover and increase in grass cover since pre-industrial times as simulated by one ESM is detectable in the observed change of warm extremes. Urbanization may have also affected extreme temperatures in some regions; for example Zhou and Ren (2011) found that extreme temperature warms more in rural stations than in urban sites in China. The effect of land use change and urban heat Island is found to be small in GMST (Section 2.4.1.3). Consequently, this effect on extreme temperature is also expected to be small in the global average.

10

These new studies show that there is stronger evidence for anthropogenic forcing on changes in extreme temperatures than at the time of the SREX assessment. New evidence since SREX includes the separation of the influence of anthropogenic forcings from that of natural forcings on extreme daily temperatures at the global scale and to some extent at continental and sub-continental scales in some regions. These new results suggest more clearly the role of anthropogenic forcing on temperature extremes compared to results at the time of the SREX assessment. We assess that it is *very likely* that human influence has contributed to the observed changes in the frequency and intensity of daily temperature extremes on the global scale since the mid-20th century.

10.6.1.2 Precipitation Extremes

Observations have showed a general increase in heavy precipitation at the global scale. This appears to be consistent with the expected response to anthropogenic forcing as a result of an enhanced moisture content in the atmosphere but a direct cause-and-effect relationship between changes in external forcing and extreme precipitation had not been established at the time of the AR4. As a result, the AR4 concluded that increases in heavy precipitation were *more likely than not* consistent with anthropogenic influence during the latter half of the 20th century (Hegerl et al., 2007b).

Extreme precipitation is expected to increase with warming. A combination of evidence leads to this conclusion though by how much remains uncertain and may vary with time scale (Section 7.6.5). Observations and model projected future changes both indicate increase in extreme precipitation associated with warming. Analysis of observed annual maximum 1-day precipitation (RX1day) over global land areas with sufficient data smaples indicates a significant increase in extreme percipitation globally, with a median increase about 7% $°C^{-1}$ GMST increase (Westra et al., 2013). CMIP3 and CMIP5 simulations project an increase in the globally averaged 20-year return values of annual maximum 24-hour precipitation amounts of about 6 to 7% with each degree Celsius of global mean warming, with the bulk of models simulating values in the range of 4 to 10% $°C^{-1}$(Kharin et al., 2007; Kharin et al., 2013). Anthropogenic influence has been detected on various aspects of the global hydrological cycle (Stott et al., 2010), which is directly relevant to extreme precipitation changes. An anthropogenic influence on increasing atmospheric moisture content has been detected (see Section 10.3.2). A higher moisture content in the atmosphere would be expected to lead to stronger extreme precipitation as extreme precipitation typically scales with total column moisture if circulation does not change. An observational analysis shows that winter maximum daily precipitation in North America has statistically significant positive correlations with local atmospheric moisture (Wang and Zhang, 2008).

There is only a modest body of direct evidence that natural or anthropogenic forcing has affected global mean precipitation (see Section 10.3.2 and Figure 10.10), despite a robust expectation of increased precipitation (Balan Sarojini et al., 2012) and precipitation extremes (see Section 7.6.5). However, mean precipitation is expected to increase less than extreme precipitation because of energy constraints (e.g., Allen and Ingram, 2002). A perfect model analysis with an ensemble of GCM simulations shows that anthropogenic influence should be detectable in precipitation extremes in the second half of the 20th century at global and hemispheric scales, and at continental scale as well but less robustly (Min et al., 2008c), see also Hegerl et al. (2004). One study has also linked the observed intensification of precipitation extremes (including RX1day and annual maximum 5-day precipitation (RX5day)) over NH land areas to human influence using a limited set of climate models and observations (Min et al., 2011). However, the detection was less robust if using the fingerprint for combined anthropogenic and natural influences compared to that for anthropogenic influences only, possibly due to a number of factors including weak S/N ratio and uncertainties in observation and model simulations. Also, models still have difficulties in simulating extreme daily precipitation directly comparable with those observed at the station level, which has been addressed to some extent by Min et al. (2011) by independently transforming annual precipitation extremes in models and observations onto a dimensionless scale that may be more comparable between the two. Detection of anthropogenic influence on smaller spatial scales is more difficult due to the increased level of noise and uncertainties and confounding factors on local scales. Fowler and Wilby (2010) suggested that there may have only been a 50% likelihood of detecting anthropogenic influence on UK extreme precipitation in winter at that time, and a very small likelihood of detecting it in other seasons.

Given the evidence of anthropogenic influence on various aspects of the global hydrological cycle that implies that extreme precipitation would be expected to have increased and some limited direct evidence of anthropogenic influence on extreme precipitation, but given also the difficulties in simulating extreme precipitation by climate models and limited observational coverage, we assess, consistent with SREX (Seneviratne et al., 2012) that there is *medium confidence* that anthropogenic forcing has contributed to a global scale intensification of heavy precipitation over the second half of the 20th century in land regions where observational coverage is sufficient for assessment.

10.6.1.3 Drought

AR4 concluded that that an increased risk of drought was *more likely than not* due to anthropogenic forcing during the second half of the

20th century. This assessment was based on one detection study that identified an anthropogenic fingerprint in a global Palmer Drought Severity Index (PDSI) data set (Burke et al., 2006) and studies of some regions which indicated that droughts in those regions were linked to SST changes or to a circulation response to anthropogenic forcing. SREX (Seneviratne et al., 2012) assessed that there was *medium confidence* that anthropogenic influence has contributed to some changes in the drought patterns observed in the second half of the 20th century based on attributed impact of anthropogenic forcing on precipitation and temperature changes, and that there was *low confidence* in the assessment of changes in drought at the level of single regions.

Drought is a complex phenomenon that is affected by precipitation predominantly, as well as by other climate variables including temperature, wind speed and solar radiation (e.g., Seneviratne, 2012; Sheffield et al., 2012). It is also affected by non-atmospheric conditions such as antecedent soil moisture and land surface conditions. Trends in two important drought-related climate variables (precipitation and temperature) are consistent with the expected responses to anthropogenic forcing over the globe. However, there is large uncertainty in observed changes in drought (Section 2.6.2.3) and its attribution to causes globally. The evidence for changes in soil moisture indices and drought indices over the period since 1950 globally is conflicting (Hoerling et al., 2012; Sheffield et al., 2012; Dai, 2013), possibly due to the examination of different time periods, different forcing fields used to drive land surface models and uncertainties in land surface models (Pitman et al., 2009; Seneviratne et al., 2010; Sheffield et al., 2012). In a recent study, Sheffield et al. (2012) identify the representation of potential evaporation as solely dependent on temperature (using the Thornthwaite-based formulation) as a possible explanation for their finding that PDSI-based estimates might overestimate historical drought trends. This stands in partial contradiction to previous assessments suggesting that using a more sophisticated formulation (Penman-Monteith) for potential evaporation did not affect the results of respective PDSI trends (Dai, 2011; van der Schrier et al., 2011). Sheffield et al. (2012) argue that issues with the treatment of spurious trends in atmospheric forcing data sets and/or the choice of calibration periods explain these conflicting results. These conflicting results point out the challenges in quantitatively defining and detecting long-term changes in a multivariable phenomenon such as drought.

Recent long-term droughts in western North America cannot definitively be shown to lie outside the very large envelope of natural precipitation variability in this region (Cayan et al., 2010; Seager et al., 2010), particularly given new evidence of the history of high-magnitude natural drought and pluvial episodes suggested by palaeoclimatic reconstructions (see Chapter 5). Low-frequency tropical ocean temperature anomalies in all ocean basins appear to force circulation changes that promote regional drought (Hoerling and Kumar, 2003; Seager et al., 2005; Dai, 2011). Uniform increases in SST are not particularly effective in this regard (Schubert et al., 2009; Hoerling et al., 2012). Therefore, the reliable separation of natural variability and forced climate change will require simulations that accurately reproduce changes in large-scale SST gradients at all time scales.

In summary, assessment of new observational evidence, in conjunction with updated simulations of natural and forced climate variability, indicates that the AR4 conclusions regarding global increasing trends in droughts since the 1970s should be tempered. There is not enough evidence to support *medium* or *high confidence* of attribution of increasing trends to anthropogenic forcings as a result of observational uncertainties and variable results from region to region (Section 2.6.2.3). Combined with difficulties described above in distinguishing decadal scale variability in drought from long-term climate change we conclude consistent with SREX that there is *low confidence* in detection and attribution of changes in drought over global land areas since the mid-20th century.

10.6.1.4 Extratropical Cyclones

AR4 concluded that an anthropogenic influence on extratropical cyclones was not formally detected, owing to large internal variability and problems due to changes in observing systems. Although there is evidence that there has been a poleward shift in the storm tracks (see Section 2.6.4), various causal factors have been cited including oceanic heating (Butler et al., 2010) and changes in large-scale circulation due to effects of external forcings (Section 10.3.3). Increases in mid-latitude SST gradients generally lead to stronger storm tracks that are shifted poleward and increases in subtropical SST gradients may lead to storm tracks shifting towards the equator (Brayshaw et al., 2008; Semmler et al., 2008; Kodama and Iwasaki, 2009; Graff and LaCasce, 2012). However, changes in storm-track intensity are much more complicated, as they are sensitive to the competing effects of changes in temperature gradients and static stability at different levels and are thus not linked to GMST in a simple way (Ulbrich et al., 2009; O'Gorman, 2010). Overall global average cyclone activity is expected to change little under moderate GHG forcing (O'Gorman and Schneider, 2008; Ulbrich et al., 2009; Bengtsson and Hodges, 2011), although in one study, human influence has been detected in geostrophic wind energy and ocean wave heights derived from sea level pressure data (Wang et al., 2009b).

10.6.1.5 Tropical Cyclones

AR4 concluded that 'anthropogenic factors *more likely than not* have contributed to an increase in tropical cyclone intensity' (Hegerl et al., 2007b). Evidence that supports this assessment was the strong correlation between the Power Dissipation Index (PDI, an index of the destructiveness of tropical cyclones) and tropical Atlantic SSTs (Emanuel, 2005; Elsner, 2006) and the association between Atlantic warming and the increase in GMST (Mann and Emanuel, 2006; Trenberth and Shea, 2006). Observations suggest an increase globally in the intensities of the strongest tropical cyclones (Elsner et al., 2008) but it is difficult to attribute such changes to particular causes (Knutson et al., 2010). The US Climate Change Science Program (CCSP; Kunkel et al., 2008) discussed human contributions to recent hurricane activity based on a two-step attribution approach. They concluded merely that it is very likely (Knutson et al., 2010) that human-induced increase in GHGs has contributed to the increase in SSTs in the hurricane formation regions and that over the past 50 years there has been a strong statistical connection between tropical Atlantic SSTs and Atlantic hurricane activity as measured by the PDI. Knutson et al. (2010), assessed that '...it remains uncertain whether past changes in tropical cyclone activity have exceeded the variability expected from natural causes.' Seneviratne et al. (2012) concurred with this finding. Section 14.6.1 gives a detailed account of past and future changes in tropical cyclones. This section assesses causes of observed changes.

Studies that directly attribute tropical cyclone activity changes to anthropogenic GHG emission are lacking. Among many factors that may affect tropical cyclone activity, tropical SSTs have increased and this increase has been attributed at least in part to anthropogenic forcing (Gillett et al., 2008a). However, there are diverse views on the connection between tropical cyclone activity and SST (see Section 14.6.1 for details). Strong correlation between the PDI and tropical Atlantic SSTs (Emanuel, 2005; Elsner, 2006) would suggest an anthropogenic influence on tropical cyclone activity. However, recent studies also suggest that regional potential intensity correlates with the difference between regional SSTs and spatially averaged SSTs in the tropics (Vecchi and Soden, 2007; Xie et al., 2010; Ramsay and Sobel, 2011) and projections are uncertain on whether the relative SST will increase over the 21st century under GHG forcing (Vecchi et al., 2008; Xie et al., 2010; Villarini and Vecchi, 2012, 2013) . Analyses of CMIP5 simulations suggest that while PDI over the North Atlantic is projected to increase towards late 21st century no detectable change in PDI should be present in the 20th century (Villarini and Vecchi, 2013) . On the other hand, Emanuel et al. (2013) point out that while GCM hindcasts indeed predict little change over the 20th century, downscaling driving by reanalysis data that incorporate historical observations are in much better accord with observations and do indicate a late 20th century increase.

10

Some recent studies suggest that the reduction in the aerosol forcing (both anthropogenic and natural) over the Atlantic since the 1970s may have contributed to the increase in tropical cyclone activity in the region (see Section 14.6.1 for details), and similarly that aerosols may have acted to reduce tropical cyclone activity in the Atlantic in earlier years when aerosol forcing was increasing (Villarini and Vecchi, 2013). However, there are different views on the relative contribution of aerosols and decadal natural variability of the climate system to the observed changes in Atlantic tropical cyclone activity among these studies. Some studies indicate that aerosol changes have been the main driver (Mann and Emanuel, 2006; Evan et al., 2009; Booth et al., 2012; Villarini and Vecchi, 2012, 2013). Other studies infer the influence of natural variability to be as large as or larger than that from aerosols (Zhang and Delworth, 2009; Villarini and Vecchi, 2012, 2013).

Globally, there is *low confidence* in any long-term increases in tropical cyclone activity (Section 2.6.3) and we assess that there is *low confidence* in attributing global changes to any particular cause. In the North Atlantic region there is *medium confidence* that a reduction in aerosol forcing over the North Atlantic has contributed at least in part to the observed increase in tropical cyclone activity since the 1970s. There remains substantial disagreement on the relative importance of internal variability, GHG forcing and aerosols for this observed trend. It remains uncertain whether past changes in tropical cyclone activity are outside the range of natural internal variability.

10.6.2 Attribution of Weather and Climate Events

Since many of the impacts of climate change are likely to manifest themselves through extreme weather, there is increasing interest in quantifying the role of human and other external influences on climate in specific weather events. This presents particular challenges for both science and the communication of results. It has so far been attempted for a relatively small number of specific events (e.g., Stott et al., 2004; Pall et al., 2011) although Peterson et al. (2012) attempt, for the first time, a coordinated assessment to place different high-impact weather events of the previous year in a climate perspective. In this assessment, selected studies are used to illustrate the essential principles of event attribution: see Stott et al. (2013) for a more exhaustive review.

Two distinct ways have emerged of framing the question of how an external climate driver like increased GHG levels may have contributed to an observed weather event. First, the 'attributable risk' approach considers the event as a whole, and asks how the external driver may have increased or decreased the probability of occurrence of an event of comparable magnitude. Second, the 'attributable magnitude' approach considers how different external factors contributed to the event or, more specifically, how the external driver may have increased the magnitude of an event of comparable occurrence probability. Hoerling et al. (2013) uses both methods to infer changes in magnitude and likelihood of the 2011 Texas heat wave.

Quantifying the absolute risk or probability of an extreme weather event in the absence of human influence on climate is particularly challenging. Many of the most extreme events occur because a self-reinforcing process that occurs only under extreme conditions amplifies an initial anomaly (e.g., Fischer et al., 2007). Hence the probability of occurrence of such events cannot, in general, be estimated simply by extrapolating from the distribution of less extreme events that are sampled in the historical record. Proxy records of pre-industrial climate generally do not resolve high-frequency weather, so inferring changes in probabilities requires a combination of hard-to-test distributional assumptions and extreme value theory. Quantifying absolute probabilities with climate models is also difficult because of known biases in their simulation of extreme events. Hence, with only a couple of exceptions (e.g., Hansen et al., 2012), studies have focussed on how risks have changed or how different factors have contributed to an observed event, rather than claiming that the absolute probability of occurrence of that event would have been extremely low in the absence of human influence on climate.

Even without considering absolute probabilities, there remain considerable uncertainties in quantifying changes in probabilities. The assessment of such changes will depend on the selected indicator, time period and spatial scale on which the event is analysed, and the way in which the event-attribution question is framed can substantially affect apparent conclusions . If an event occurs in the tail of the distribution, then a small shift in the distribution as a whole can result in a large increase in the probability of an event of a given magnitude: hence it is possible for the same event to be both 'mostly natural' in terms of attributable magnitude (if the shift in the distribution due to human influence is small compared to the anomaly in the natural variability that was the primary cause) and 'mostly anthropogenic' in terms of

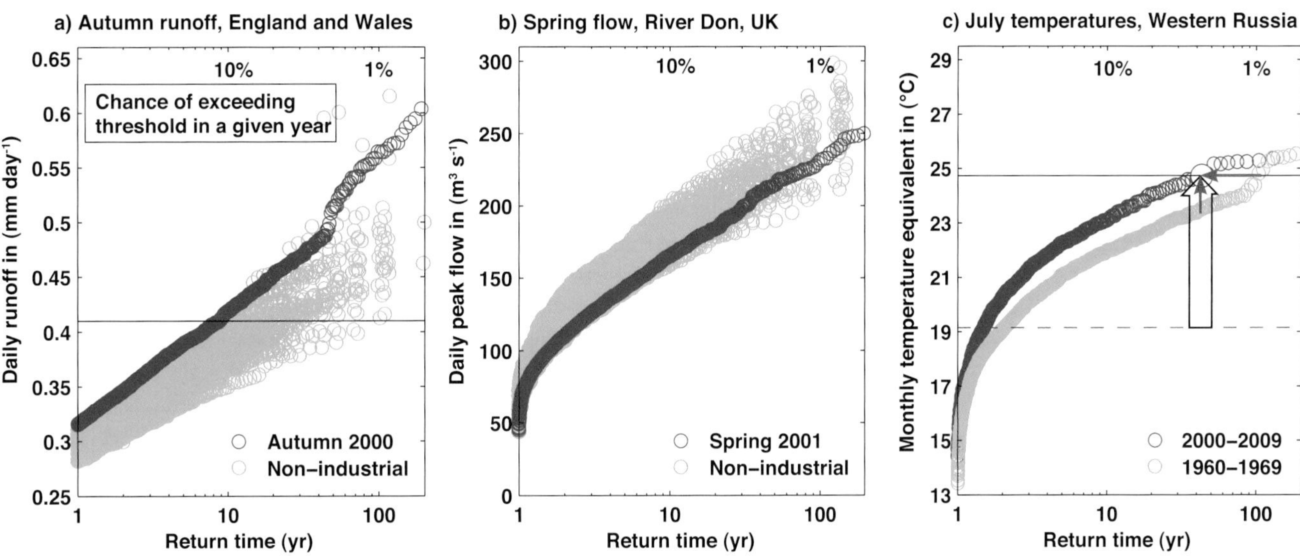

Figure 10.18 | Return times for precipitation-induced floods aggregated over England and Wales for (a) conditions corresponding to September to November 2000 with boundary conditions as observed (blue) and under a range of simulations of the conditions that would have obtained in the absence of anthropogenic greenhouse warming over the 20th century (green) with different AOGCMs used to define the greenhouse signal, black horizontal line corresponds to the threshold exceeded in autumn 2000 (from Pall et al., 2011); (b) corresponding to January to March 2001 with boundary conditions as observed (blue) and under a range of simulations of the condition that would have obtained in the absence of anthropogenic greenhouse warming over the 20th century (green) adapted from Kay et al. (2011a); (c) return periods of temperature-geopotential height conditions in the model simulations for the 1960s (green) and the 2000s (blue). The vertical black arrow shows the anomaly of the 2010 Russian heat wave (black horizontal line) compared to the July mean temperatures of the 1960s (dashed line). The vertical red arrow gives the increase in temperature for the event whereas the horizontal red arrow shows the change in the return period (from Otto et al., 2012).

attributable risk (if human influence has increased its probability of occurrence by more than a factor of 2). These issues are discussed further using the example of the 2010 Russian heat wave below.

The majority of studies have focussed on quantifying attributable risk. Formally, risk is a function of both hazard and vulnerability (IPCC, 2012), although most studies attempting to quantify risk in the context of extreme weather do not explicitly use this definition, which is discussed further in Chapter 19 of WGII, but use the term as a shorthand for the probability of the occurrence of an event of a given magnitude. Any assessment of change in risk depends on an assumption of 'all other things being equal', including natural drivers of climate change and vulnerability. Given this assumption, the change in hazard is proportional to the change in risk, so we will follow the published literature and continue to refer to Fraction Attributable Risk, defined as FAR = 1 – P0/P1, P0 being the probability of an event occurring in the absence of human influence on climate, and P1 the corresponding probability in a world in which human influence is included. FAR is thus the fraction of the risk that is attributable to human influence (or, potentially, any other external driver of climate change) and does not require knowledge of absolute values of P0 and P1, only their ratio.

For individual events with return times greater than the time scale over which the signal of human influence is emerging (30 to 50 years, meaning P0 and P1 less than 2 to 3% in any given year), it is impossible to observe a change in occurrence frequency directly because of the shortness of the observed record, so attribution is necessarily a multi-step procedure. Either a trend in occurrence frequency of more frequent events is attributed to human influence and a statistical model is then used to extrapolate to the implications for P0 and P1; or an attributable trend is identified in some other variable, such as surface temperature, and a physically based weather model is used to assess the implications for extreme weather risk. Neither approach is free of assumptions: no atmospheric model is perfect, but statistical extrapolation may also be misleading for reasons given above.

Pall et al. (2011) provide an example of multi-step assessment of attributable risk using a physically based model, applied to the floods that occurred in the UK in the autumn of 2000, the wettest autumn to have occurred in England and Wales since records began. To assess the contribution of the anthropogenic increase in GHGs to the risk of these floods, a several thousand member ensemble of atmospheric models with realistic atmospheric composition, SST and sea ice boundary conditions imposed was compared with a second ensemble with composition and surface temperatures and sea ice boundary conditions modified to simulate conditions that would have occurred had there been no anthropogenic increase in GHGs since 1900. Simulated daily precipitation from these two ensembles was fed into an empirical rainfall-runoff model and daily England and Wales runoff used as a proxy for flood risk. Results (Figure 10.18a) show that including the influence of anthropogenic greenhouse warming increases flood risk at the threshold relevant to autumn 2000 by around a factor of two in the majority of cases, but with a broad range of uncertainty: in 10% of cases the increase in risk is less than 20%.

Kay et al. (2011a), analysing the same ensembles but using a more sophisticated hydrological model found a reduction in the risk of snow melt–induced flooding in the spring season (Figure 10.18b) which, aggregated over the entire year, largely compensated for the increased risk of precipitation-induced flooding in autumn. This illustrates an important general point: even if a particular flood event may have been made more likely by human influence on climate, there is no certainty that all kinds of flood events in that location, country or region have been made more likely.

Rahmstorf and Coumou (2011) provide an example of an empirical approach to the estimation of attributable risk applied to the 2010 Russian heat wave. They fit a nonlinear trend to central Russian temperatures and show that the warming that has occurred in this region since the 1960s has increased the risk of a heat wave of the magnitude observed in 2010 by around a factor of 5, corresponding to an FAR of 0.8. They do not address what has caused the trend since 1960, although they note that other studies have attributed most of the large-scale warming over this period to the anthropogenic increase in GHG concentrations.

Dole et al. (2011) take a different approach to the 2010 Russian heat wave, focussing on attributable magnitude, analysing contributions from various external factors, and conclude that this event was 'mainly natural in origin'. First, observations show no evidence of a trend in occurrence frequency of hot Julys in western Russia, and despite the warming that has occurred since the 1960s, mean July temperatures in that region actually display a (statistically insignificant) cooling trend over the century as a whole, in contrast to the case for central and southern European summer temperatures (Stott et al., 2004). Members of the CMIP3 multi-model ensemble likewise show no evidence of a trend towards warming summers in central Russia. Second, Dole et al. (2011) note that the 2010 Russian event was associated with a strong blocking atmospheric flow anomaly, and even the complete 2010 boundary conditions are insufficient to increase the probability of a prolonged blocking event in this region, in contrast again to the situation in Europe in 2003. This anomaly in the large-scale atmospheric flow led to low-pressure systems being redirected around the blocking over Russia causing severe flooding in Pakistan which could so far not be attributed to anthropogenic causes (van Oldenborgh et al., 2012), highlighting that a global perspective is necessary to unravel the different factors influencing individual extreme events (Trenberth and Fasullo, 2012).

Otto et al. (2012) argue that it is possible to reconcile the results of Rahmstorf and Coumou (2011) with those of Dole et al. (2011) by relating the attributable risk and attributable magnitude approaches to framing the event attribution question. This is illustrated in Figure 10.18c, which shows return times of July temperatures in western Russia in a large ensemble of atmospheric model simulations for the 1960s (in green) and 2000s (in blue). The threshold exceeded in 2010 is shown by the solid horizontal line which is almost 6°C above 1960s mean July temperatures, shown by the dashed line. The difference between the green and blue lines could be characterized as a 1.5°C increase in the magnitude of a 30-year event (the vertical red arrow, which is substantially smaller than the size of the anomaly itself, supporting the assertion that the event was 'mainly natural' in terms of attributable magnitude. Alternatively, it could be characterized as a threefold increase in the risk of the 2010 threshold being exceeded, supporting the assertion that risk of the event occurring was mainly attributable to the external trend, consistent with Rahmstorf and Coumou (2011). Rupp et al. (2012) and Hoerling et al. (2013) reach similar conclusions about the 2011 Texas heat wave, both noting the importance of La Niña conditions in the Pacific, with anthropogenic warming making a relatively small contribution to the magnitude of the event, but a more substantial contribution to the risk of temperatures exceeding a high threshold. This shows that the quantification of attributable risks and and changes in magnitude are affected by modelling error (e.g., Visser and Petersen, 2012) as they depend on the atmospheric model's ability to simulate the observed anomalies in the general circulation (Chapter 9).

Because much of the magnitude of these two heat waves is attributable to atmospheric flow anomalies, any evidence of a causal link between rising GHGs and the occurrence or persistence of flow anomalies such as blocking would have a very substantial impact on attribution claims. Pall et al. (2011) argue that, although flow anomalies played a substantial role in the autumn 2000 floods in the UK, thermodynamic mechanisms were primarily responsible for the change in risk between their ensembles. Regardless of whether the statistics of flow regimes themselves have changed, observed temperatures in recent years in Europe are distinctly warmer than would be expected for analogous atmospheric flow regimes in the past, affecting both warm and cold extremes (Yiou et al., 2007; Cattiaux et al., 2010).

In summary, increasing numbers of studies are finding that the probability of occurrence of events associated with extremely high temperatures has increased substantially due to the large-scale warming since the mid-20th century. Because most of this large-scale warming is *very likely* due to the increase in atmospheric GHG concentrations, it is possible to attribute, via a multi-step procedure, some of the increase in probability of these regional events to human influence on climate. Such an increase in probability is consistent with the implications of single-step attribution studies looking at the overall implications of increasing mean temperatures for the probabilities of exceeding temperature thresholds in some regions. We conclude that it is *likely* that human influence has substantially increased the probability of occurrence of heat waves in some locations. It is expected that attributable risks for extreme precipitation events are generally smaller and more uncertain, consistent with the findings in Kay et al. (2011a) and Pall et al. (2011). The science of event attribution is still confined to case studies, often using a single model, and typically focussing on high-impact events for which the issue of human influence has already arisen. While the increasing risk of heat waves measured as the occurrence of a previous temperature record being exceeded can simply be explained by natural variability superimposed by globally increasing temperature, conclusions for holistic events including general circulation patterns are specific to the events that have been considered so far and rely on the representation of relevant processes in the model.

Anthropogenic warming remains a relatively small contributor to the overall magnitude of any individual short-term event because its magnitude is small relative to natural random weather variability on short time scales (Dole et al., 2011; Hoerling et al., 2013). Because of this random variability, weather events continue to occur that have been made less likely by human influence on climate, such as extreme winter cold events (Massey et al., 2012), or whose probability of occurrence has not been significantly affected either way. Quantifying how different external factors contribute to current risks, and how risks are changing, is possible with much higher confidence than quantifying absolute risk. Biases in climate models, uncertainty in the probability distribution of the most extreme events and the ambiguity of paleoclimatic records for short-term events mean that it is not yet possible to quantify the absolute probability of occurrence of any observed weather event in a hypothetical pristine climate. At present, therefore, the evidence does not support the claim that we are observing weather events that would, individually, have been *extremely unlikely* in the absence of human-induced climate change, although observed trends in the concurrence of large numbers of events (see Section 10.6.1) may be more easily attributable to external factors. The most important development since AR4 is an emerging consensus that the role of external drivers of climate change in specific extreme weather events, including events that might have occurred in a pre-industrial climate, can be quantified using a probabilistic approach.

10.7 Multi-century to Millennia Perspective

Evaluating the causes of climate change before the 20th century is important to test and improve our understanding of the role of internal and forced natural climate variability for the recent past. This section draws on assessment of temperature reconstructions of climate change over the past millennium and their uncertainty in Chapter 5 (Table 5.A.1; Sections 5.3.5 and 5.5.1 for regional records), and on comparisons of models and data over the pre-instrumental period in Chapters 5 and 9 (Sections 5.3.5, 5.5.1 and 9.5.3), and focuses on the evidence for the contribution by radiatively forced climate change to reconstructions and early instrumental records. In addition, the residual variability that is not explained by forcing from palaeoclimatic records provides a useful comparison to estimates of climate model internal variability. The model dependence of estimates of internal variability is an important uncertainty in detection and attribution results.

The inputs for detection and attribution studies for periods covered by indirect, or proxy, data are affected by more uncertainty than those from the instrumental period (see Chapter 5), owing to the sparse data coverage, particularly further back in time, and uncertainty in the link between proxy data and, for example, temperature. Records of past radiative influences on climate are also uncertain (Section 5.2; see Schmidt et al., 2011; Schmidt et al., 2012). For the preindustrial part of the last millennium changes in solar, volcanic, GHG forcing, and land use change, along with a small orbital forcing are potentially important external drivers of climate change. Estimates of solar forcing (Figure 5.1a; Box 10.2) are uncertain, particularly in their amplitude, as well as in modelling, for example, of the influence of solar forcing on atmospheric circulation involving stratospheric dynamics (see Box 10.2; Gray et al., 2010). Estimates of past volcanism are reasonably well established in their timing, but the magnitude of the RF of individual eruptions is uncertain (Figure 5.1a). It is possible that large eruptions had a more moderated climate effect than simulated by many climate models due to faster fallout associated with larger particle size (Timmreck et al., 2009), or increased amounts of injected water vapour (Joshi and Jones, 2009). Reconstructed changes in land cover and its effect on climate are also uncertain (Kaplan et al., 2009; Pongratz et al., 2009). Forcing of WMGHGs shows only very subtle variations over the last millennium up to 1750. This includes a small drop and partial recovery in the 17th century (Section 6.2.3, Figure 6.7), followed by increases in GHG concentrations with industrialization since the middle of the 18th century (middle of the 19th century for N_2O, Figure 6.11).

When interpreting reconstructions of past climate change with the help of climate models driven with estimates of past forcing, it helps that the uncertainties in reconstructions and forcing are independent from each other. Thus, uncertainties in forcing and reconstructions combined should lead to less, rather than more similarity between fingerprints of forced climate change and reconstructions, making it improbable that the response to external drivers is spuriously detected. However, this is the case only if all relevant forcings and their uncertainties are considered, reducing the risk of misattribution due to spurious correlations between external forcings, and if the data are homogeneous and statistical tests properly applied (e.g., Legras et al., 2010). Hence this section focuses on work that considers all relevant forcings simultaneously.

10.7.1 Causes of Change in Large-Scale Temperature over the Past Millennium

10

Despite the uncertainties in reconstructions of past NH mean temperatures, there are well-defined climatic episodes in the last millennium that can be robustly identified (Chapter 5, see also Figure 10.19). Chapter 5 concludes that in response to solar, volcanic and anthropogenic RFs, climate models simulate temperature changes in the NH which are generally consistent in magnitude and timing with reconstructions, within their broad uncertainty ranges (Section 5.3.5).

10.7.1.1 Role of External Forcing in the Last Millennium

The AR4 concluded that 'A substantial fraction of the reconstructed NH inter-decadal temperature variability of the seven centuries prior to 1950 is *very likely* attributable to natural external forcing'. The literature since the AR4, and the availability of more simulations of the last millennium with more complete forcing (see Schmidt et al., 2012), including solar, volcanic and GHG influences, and generally also land use change and orbital forcing) and more sophisticated models, to a much larger extent coupled climate or coupled ESMs (Chapter 9), some of them with interactive carbon cycle, strengthens these conclusions.

Most reconstructions show correlations with external forcing that are similar to those found between pre-Paleoclimate Modelling Intercomparison Project Phase 3 (PMIP3) simulations of the last millennium and forcing, suggesting an influence by external forcing (Fernández-Donado et al., 2013). From a global scale average of new regional reconstructions, Past Global Changes 2k (PAGES 2k) Consortium (2013) find that periods with strong volcanic and solar forcing combined occurring over the last millennium show significantly cooler conditions than randomly selected periods from the last two millennia. Detection analyses based on PMIP3 and CMIP5 model simulations for the years from 850 to 1950 and also from 850 to 1850 find that the fingerprint of external forcing is detectable in all reconstructions of NH mean temperature considered (Schurer et al., 2013; see Figure 10.19), but only in about half the cases considered does detection also occur prior to 1400. The authors find a smaller response to forcing in reconstructions than simulated, but this discrepancy is consistent with uncertainties in forcing or proxy response to it, particularly associated with volcanism. The discrepancy is reduced when using more strongly smoothed data or omitting major volcanic eruptions from the analysis. The level of agreement between fingerprints from multiple models in response to forcing and reconstructions decreases earlier in time, and the forced signal is detected only in about half the cases considered when analysing the period 851 to 1401. This may be partly due to weaker forcing and larger forcing uncertainty early in the millennium and partly due to increased uncertainty in reconstructions. Detection results indicate a contribution by external drivers to the warm conditions in the 11th to 12th century, but cannot explain the warmth around the 10th century in some of the reconstructions (Figure 10.19). This detection of a role of external forcing extends work reported in AR4 back into to the 9th century CE.

Detection and attribution studies support results from modelling studies that infer a strong role of external forcing in the cooling of NH temperatures during the Little Ice Age (LIA; see Chapter 5 and Glossary). Both model simulations (Jungclaus et al., 2010) and results from detection and attribution studies (Hegerl et al., 2007a; Schurer et al., 2013) suggest that a small drop in GHG concentrations may have contributed to the cool conditions during the 16th and 17th centuries. Note, however, that centennial variations of GHG during the late Holocene are very small relative to their increases since pre-industrial times (Section 6.2.3). The role of solar forcing is less clear except for decreased agreement if using very large solar forcing (e.g., Ammann et al., 2007; Feulner, 2011). Palastanga et al. (2011) demonstrate that neither a slowdown of the thermohaline circulation nor a persistently negative NAO alone can explain the reconstructed temperature pattern over Europe during the periods 1675–1715 and 1790–1820.

10

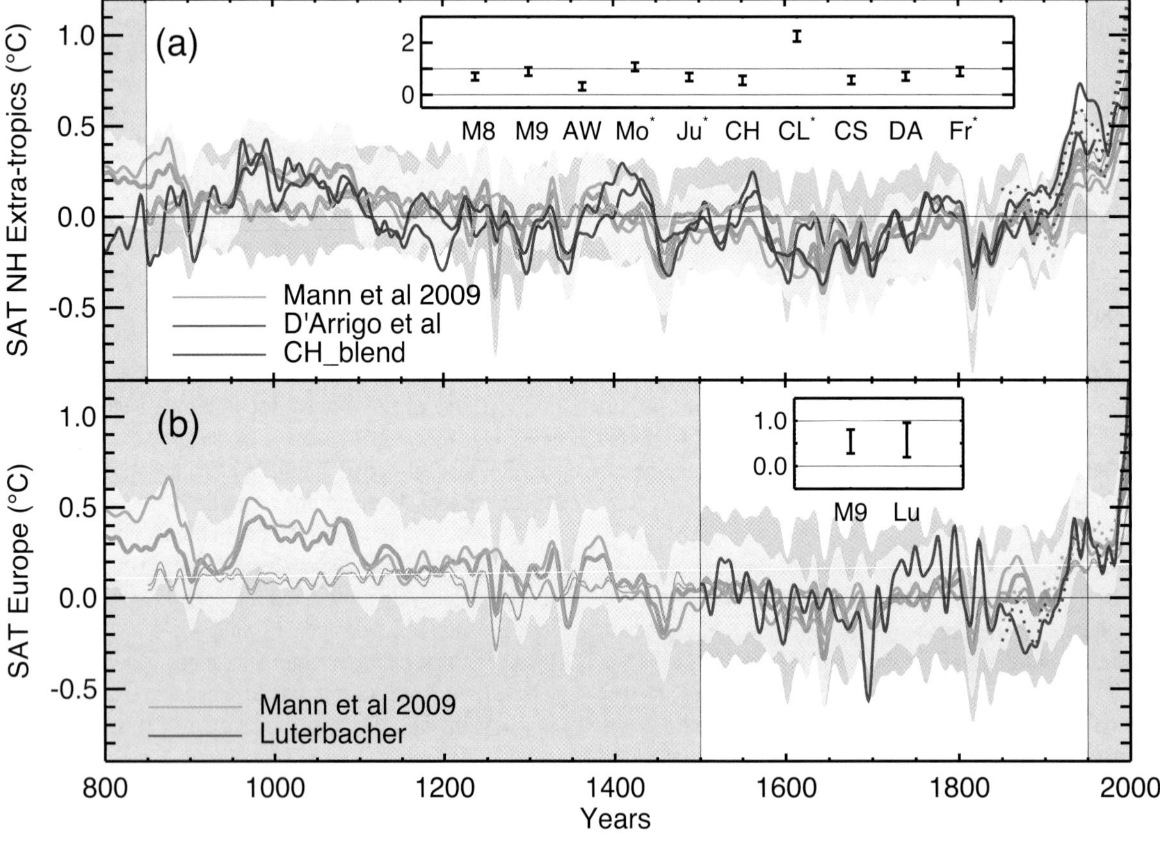

Figure 10.19 | The top panel compares the mean annual Northern Hemisphere (NH) surface air temperature from a multi-model ensemble to several NH temperature reconstructions. These reconstructions are: CH-blend from Hegerl et al. (2007a) in purple, which is a reconstruction of 30°N to 90°N land only (Mann et al., 2009), plotted for the region 30°N to 90°N land and sea (green) and D'Arrigo et al. (2006) in red, which is a reconstruction of 20°N to 90°N land only. The dotted coloured lines show the corresponding instrumental data. The multi-model mean for the reconstructed domain is scaled to fit each reconstruction in turn, using a total least squares (TLS) method. The best estimate of the detected forced signal is shown in orange (as an individual line for each reconstruction; lines overlap closely) with light orange shading indicating the range expected if accounting for internal variability. The best fit scaling values for each reconstruction are given in the insert as well as the detection results for six other reconstructions (M8; M9 (Mann et al., 2008, 2009); AW (Ammann and Wahl, 2007); Mo (Moberg et al., 2005); Ju (Juckes et al., 2007); CH (Hegerl et al., 2007a); CL (Christiansen and Ljungqvist, 2011) and inverse regressed onto the instrumental record CS; DA (D'Arrigo et al., 2006); Fr (Frank et al., 2007). An asterisk next to the reconstruction name indicates that the residuals (over the more robustly reconstructed period 1401–1950) are inconsistent with the internal variability generated by the combined control simulations of all climate models investigated (for details see Schurer et al., 2013). The ensemble average of a data-assimilation simulation (Goosse et al., 2012b) is plotted in blue, for the region 30°N to 90°N land and sea, with the error range shown in light blue shading. The bottom panel is similar to the top panel, but showing the European region, following Hegerl et al. (2011a) but using the simulations and method in Schurer et al. (2013). The detection analysis is performed for the period 1500–1950 for two reconstructions: Luterbacher et al. (2004)(representing the region 35°N to 70°N,25°W to 40°E, "land only, labelled 'Lu' in the insert") shown in red, and Mann et al. (2009) (averaged over the region 25°N to 65°N, 0° to 60°E, land and sea, labelled 'M9' in the insert), shown in green. As in the top panel, best fit estimates are shown in dark orange with uncertainty range due to internal variability shown in light orange. The data assimilation from Goosse et al. (2012a), constrained by the Mann et al. (2009) reconstruction is shown in blue, with error range in light blue. All data are shown with respect to the mean of the period covered by the white part of the figure (850–1950 for the NH, 1500–1950 for European mean data).

Data assimilation studies support the conclusion that external forcing, together with internal climate variability, provides a consistent explanation of climate change over the last millennium. Goosse et al. (2010, 2012a, 2012b) select, from a very large ensemble with an EMIC, the individual simulations that are closest to the spatial reconstructions of temperature between 30°N and 60°N by Mann et al. (2009) accounting for reconstruction uncertainties. The method also varies the external forcing within uncertainties, determining a combined realization of the forced response and internal variability that best matches the data. Results (Figure 10.19) show that simulations reproduce the target reconstruction within the uncertainty range, increasing confidence in the consistency of the reconstruction and the forcing. The results suggest that long-term circulation anomalies may help to explain the hemispheric warmth early in the millennium, although results vary dependent on input parameters of the method.

10.7.1.2 Role of Individual Forcings

Volcanic forcing shows a detectable influence on large-scale temperature (see AR4; Chapter 5), and volcanic forcing plays an important role in explaining past cool episodes, for example, in the late 17th and early 19th centuries (see Chapter 5 and 9; Hegerl et al., 2007b; Jungclaus et al., 2010; Miller et al., 2012) . Schurer et al. (2013) separately detect the response to GHG variations between 1400 and 1900 in most NH reconstructions considered, and that of solar and volcanic forcing combined in all reconstructions considered.

Even the multi-century perspective makes it difficult to distinguish century-scale variations in NH temperature due to solar forcing alone from the response to other forcings, due to the few degrees of freedom constraining this forcing (see Box 10.2). Hegerl et al. (2003, 2007a) found solar forcing detectable in some cases. Simulations with higher than best guess solar forcing may reproduce the warm period around 1000 more closely, but the peak warming occurs about a century earlier in reconstructions than in solar forcing and with it model simulations (Jungclaus et al., 2010; Figure 5.8; Fernández-Donado et al., 2013). Even if solar forcing were on the high end of estimates for the last millennium, it would not be able to explain the recent warming according both to model simulations (Ammann et al., 2007; Tett et al., 2007; Feulner, 2011) and detection and attribution approaches that scale the temporal fingerprint of solar forcing to best match the data (Hegerl et al., 2007a; Schurer et al., 2013; Figure 10.19). Some studies suggest that particularly for millennial and multi-millenial time scales orbital forcing may be important globally (Marcott et al., 2013) and for high-latitude trends (Kaufman et al., 2009) based on a comparison of the correspondence between long-term Arctic cooling in models and data though the last millennium up to about 1750 (see also PAGES 2k Consortium, 2013).

10.7.1.3 Estimates of Internal Climate Variability

The interdecadal and longer-term variability in large-scale temperatures in climate model simulations with and without past external forcing is quite different (Tett et al., 2007; Jungclaus et al., 2010), consistent with the finding that a large fraction of temperature variance in the last millennium has been externally driven. The residual variability in past climate that is not explained by changes in RF provides an estimate of internal variability for NH mean temperature that is not directly derived from climate model simulation. This residual variability is somewhat larger than control simulation variability for some reconstructions if the comparison is extended to the full period since 850 CE (Schurer et al., 2013), However, when extracting 50- and 60-year trends from this residual variability, the distribution of these trends is similar to the multi-model control simulation ensemble used in Schurer et al. (2013). In all cases considered, the most recent 50-and 60-year trend from instrumental data is far outside the range of any 50-year trend in residuals from reconstructions of NH mean temperature of the past millennium.

10.7.2 Changes of Past Regional Temperature

Several reconstructions of European regional temperature variability are available (Section 5.5). While Bengtsson et al. (2006) emphasized the role of internal variability in pre-industrial European climate as reconstructed by Luterbacher et al. (2004), Hegerl et al. (2011a) find a detectable response to external forcing in summer temperatures in the period 1500–1900, for winter temperatures during 1500–1950 and 1500–2000; and throughout the record for spring. The fingerprint of the forced response shows coherent time evolution between models and reconstructed temperatures over the entire analysed period (compare to annual results in Figure 10.19, using a larger multi-model ensemble). This suggests that the cold European winter conditions in the late 17th and early 19th century and the warming in between were at least partly externally driven.

Data assimilation results focussing on the European sector suggests that the explanation of forced response combined with internal variability is self-consistent (Goosse et al., 2012a, Figure 10.19). The assimilated simulations reproduce the warmth of the MCA better than the forced only simulations do. The response to individual forcings is difficult to distinguish from each other in noisier regional reconstructions. An epoch analysis of years immediately following strong, largely tropical, volcanic eruptions shows that European summers show detectable fingerprints of volcanic response , while winters show a noisy response of warming in northern Europe and cooling in southern Europe (Hegerl et al., 2011a). Landrum et al. (2013) suggest similar volcanic responses for North America, with warming in the north of the continent and cooling in the south. There is also evidence for a decrease in SSTs following tropical volcanic forcing in tropical reconstructions over the past 450 years (D'Arrigo et al., 2009). There is also substantial literature suggesting solar influences on regional climate reconstructions, possibly due to circulation changes, for example, changes in Northern Annular Modes (e.g., Kobashi et al., 2013; see Box 10.2).

10.7.3 Summary: Lessons from the Past

Detection and attribution studies strengthen results from AR4 that external forcing contributed to past climate variability and change prior to the 20th century. Ocean–Atmosphere General Circulation Models (OAGCMs) simulate similar changes on hemispheric and annual scales as those by simpler models used earlier, and enable detection of regional and seasonal changes. Results suggest that volcanic forcing and GHG forcing in particular are important for explaining past changes in NH temperatures. Results from data assimilation runs confirm that the combination of internal variability and external forcing provides a consistent explanation of the last millennium and suggest that changes in circulation may have further contributed to climate anomalies. The role of external forcing extends to regional records, for example, European seasonal temperatures. In summary, it is *very unlikely* that NH temperature variations from 1400 to 1850 can be explained by internal variability alone. There is *medium confidence* that external forcing contributed to NH temperature variability from 850 to 1400. There is *medium confidence* that external forcing (anthropogenic and natural forcings together) contributed to European temperatures of the last five centuries.

10.8 Implications for Climate System Properties and Projections

Detection and Attribution results can be used to constrain predictions of future climate change (see Chapters 11 and 12) and key climate system properties. These properties include: the Equilibrium Climate Sensitivity (ECS), which determines the long-term equilibrium warming response to stable atmospheric composition, but not accounting for vegetation or ice sheet changes (Section 12.5.3; see Box 12.2); the transient climate response (TCR), which is a measure of the magnitude of transient warming while the climate system, particularly the deep ocean, is not in equilibrium; and the transient climate response to cumulative CO_2 emissions (TCRE), which is a measure of the transient warming response to a given mass of CO_2 injected into the atmosphere, and combines information on both the carbon cycle and climate response. TCR is more tightly constrained by the observations of transient warming than ECS. The observational constraints on TCR, ECS and TCRE assessed here focus on information provided by recent observed climate change, complementing analysis of feedbacks and climate modelling information, which are assessed in Chapter 9. The assessment in this chapter also incorporates observational constraints based on palaeoclimatic information, building on Chapter 5, and contributes to the overall synthesis assessment in Chapter 12 (Box 12.2).

Because neither ECS nor TCR is directly observed, any inference about them requires some form of climate model, ranging in complexity from a simple zero-dimensional energy balance box model to OAGCMs (Hegerl and Zwiers, 2011). Constraints on estimates of long-term climate change and equilibrium climate change from recent warming hinge on the rate at which the ocean has taken up heat (Section 3.2), and by the extent to which recent warming has been reduced by cooling from aerosol forcing. Therefore, attempts to estimate climate sensitivity (transient or equilibrium) often also estimate the total aerosol forcing and the rate of ocean heat uptake, which are discussed in Section 10.8.3. The AR4 contained a detailed discussion on estimating quantities relevant for projections, and included an appendix with the relevant estimation methods. Here, we build on this assessment, repeating information and discussion only where necessary to provide context.

10.8.1 Transient Climate Response

The AR4 discussed for the first time estimates of the TCR. TCR was originally defined as the warming at the time of CO_2 doubling (i.e., after 70 years) in a 1% yr^{-1} increasing CO_2 experiment (see Hegerl et al., 2007b), but like ECS, it can also be thought of as a generic property of the climate system that determines the global temperature response ΔT to any gradual increase in RF, ΔF, taking place over an approximately 70-year time scale, normalized by the ratio of the forcing change to the forcing due to doubling CO_2, $F_{2\times CO2}$: TCR = $F_{2\times CO2}\ \Delta T/\Delta F$ (Frame et al., 2006; Gregory and Forster, 2008; Held et al., 2010; Otto et al., 2013). This generic definition of the TCR has also been called the 'Transient Climate Sensitivity' (Held et al., 2010). TCR is related to ECS and the global energy budget as follows: ECS = $F_{2\times CO2}/\alpha$, where α is the sensitivity parameter representing the net increase in energy flux to space per degree of warming given all feedbacks operating on these time scales. Hence, by conservation of energy, ECS = $F_{2\times CO2}\ \Delta T/(\Delta F - \Delta Q)$, where ΔQ is the change in the rate of increase of climate system heat content in response to the forcing ΔF. On these time scales, deep ocean heat exchange affects the surface temperature response as if it were an enhanced radiative damping, introducing a slow, or 'recalcitrant', component of the response which would not be reversed for many decades even if it were possible to return RF to pre-industrial values (Held et al., 2010): hence the difficulty of placing an upper bound on ECS from observed surface warming alone (Forest et al., 2002; Frame et al., 2006). Because ΔQ is always positive at the end of a period of increasing forcing, before the climate system has re-equilibrated, TCR is always less than ECS, and since ΔQ is uncertain, TCR is generally better constrained by observations of recent climate change than ECS.

Because TCR focuses on the short- and medium-term response, constraining TCR with observations is a key step in narrowing estimates of future global temperature change in the relatively short term and under scenarios where forcing continues to increase or peaks and declines (Frame et al., 2006). After stabilization, the ECS eventually becomes the relevant climate system property. Based on observational constraints alone, the AR4 concluded that TCR is *very likely* to be larger than 1°C and *very unlikely* to be greater than 3.5°C (Hegerl et al., 2007b). This supported the overall assessment that the transient climate response is *very unlikely* greater than 3°C and *very likely* greater than 1°C (Meehl et al., 2007a). New estimates of the TCR are now available.

Scaling factors derived from detection and attribution studies (see Section 10.2) express how model responses to GHGs and aerosols need to be scaled to match the observations over the historical period. These scaled responses were used in AR4 to provide probabilistic projections of both TCR and future changes in global temperature in response to these forcings under various scenarios (Allen et al., 2000; Stott and Kettleborough, 2002; Stott et al., 2006, 2008b; Kettleborough et al., 2007; Meehl et al., 2007b; Stott and Forest, 2007). Allen et al. (2000), Frame et al. (2006) and Kettleborough et al. (2007) demonstrate a near linear relationship between 20th century warming, TCR and warming by the mid-21st century as parameters are varied in Energy Balance Models, justifying this approach. Forster et al. (2013) show how the ratio $\Delta T/\Delta F$ does depend on the forcing history, with very rapid increases in forcing giving lower values: hence any inference from past attributable warming to future warming or TCR depends on a model (which may be simple or complex, but ideally physically based) to relate these quantities. Such inferences also depend on forcing estimates and projections. Recent revisions to RF (see Chapter 8) suggest higher net anthropogenic forcing over the 20th century, and hence a smaller estimated

TCR. Stott et al. (2008b) demonstrated that optimal detection analysis of 20th century temperature changes (using HadCM3) are able to exclude the very high and low temperature responses to aerosol forcing. Consequently, projected 21st century warming may be more closely constrained than if the full range of aerosol forcings is used (Andreae et al., 2005). Stott and Forest (2007) demonstrate that projections obtained from such an approach are similar to those obtained by constraining EMIC parameters from observations. Stott et al. (2011), using HadGEM2-ES, and Gillett et al. (2012), using CanESM2, both show that the inclusion of observations between 2000 and 2010 in such an analysis reduces the uncertainties in projected warming in the 21st century, and tends to constrain the maximum projected warming to below that projected using data to 2000 only (Stott et al, 2006). Such an improvement is consistent with prior expectations of how additional data will narrow uncertainties (Stott and Kettleborough, 2002).

TCR estimates have been derived using a variety of methods (Figure 10.20a). Knutti and Tomassini (2008) compare EMIC simulations with 20th century surface and ocean temperatures to derive a probability density function for TCR skewed slightly towards lower values with a 5 to 95% range of 1.1°C to 2.3°C. Libardoni and Forest (2011) take a similar approach with a different EMIC and include atmospheric data and, under a variety of assumptions, obtain 5 to 95% ranges for TCR spanning 0.9°C to 2.4°C. Updating this study to include data to 2004 gives results that are essentially unchanged. Using a single model and observations from 1851 to 2010 Gillett et al. (2012) derive a 5 to 95% range of 1.3°C to 1.8°C and using a single model, but using multiple sets of observations and analysis periods ending in 2010 and beginning in 1910 or earlier, Stott et al. (2011) derive 5 to 95% ranges that were generally between 1°C and 3°C. Both Stott et al. (2011) and Gillett et al. (2012) find that the inclusion of data between 2000 and 2010 helps to constrain the upper bound of TCR. Gillett et al. (2012) find that the inclusion of data prior to 1900 also helps to constrain TCR, though Stott et al. (2011) do not. Gillett et al. (2013) account for a broader range of model and observational uncertainties, in particular addressing the efficacy of non-CO_2 gases, and find a range of 0.9°C to 2.3°C. Several of the estimates of TCR that were cited by Hegerl et al. (2007b) may have underestimated non-CO_2 efficacies relative to the more recent estimates in Forster et al. (2007). Because observationally constrained estimates of TCR are based on the ratio between past attributable warming and past forcing, this could account for a high bias in some of the inputs used for the AR4 TCR estimate.

Held et al. (2010) show that a two-box model originally proposed by Gregory (2000), distinguishing the 'fast' and 'recalcitrant' responses, fits both historical simulations and instantaneous doubled CO_2 simulations of the GFDL coupled model CM2.1. The fast response has a relaxation time of 3 to 5 years, and the historical simulation is almost completely described by this fast component of warming. Padilla et al. (2011) use this simple model to derive an observationally constrained estimate of the TCR of 1.3°C to 2.6°C. Schwartz (2012) uses this two-time scale formulation to obtain TCR estimates ranging from 0.9°C to 1.9°C, the lower values arising from higher estimates of forcing over the 20th century. Otto et al. (2013) update the analysis of Gregory et al. (2002) and Gregory and Forster (2008) using forcing estimates from Forster et al. (2013) to obtain a 5 to 95% range for TCR of 0.9°C to 2.0°C comparing the decade 2000–2009 with the period 1860–1879. They note, however, the danger of overinterpreting a single, possibly anomalous, decade, and report a larger TCR range of 0.7°C to 2.5°C replacing the 2000s with the 40 years 1970–2009.

Tung et al. (2008) examine the response to the 11-year solar cycle using discriminant analysis, and find a high range for TCR: >2.5°C to 3.6°C However, this estimate may be affected by different mechanisms by which solar forcing affects climate (see Box 10.2). The authors attempt to minimize possible aliasing with the response to other forcings in the 20th century and with internal climate variability, although some influence by them cannot be ruled out.

Rogelj et al. (2012) take a somewhat different approach, using a simple climate model to match the distribution of TCR to observational constraints and a consensus distribution of ECS (which will itself have been informed by recent climate change), following Meinshausen et al. (2009). Harris et al. (2013) estimate a distribution for TCR based on a large sample of emulated GCM equilibrium responses, constrained by multiannual mean observations of recent climate and adjusted to account for additional uncertainty associated with model structural deficiencies (Sexton et al., 2012). The equilibrium responses are scaled by global temperature changes associated with the sampled model variants, reweighting the projections based on the likelihood that they correctly replicate observed historical changes in surface temperature, to predict the TCR distribution. Both of these studies represent a combination of multiple lines of evidence, although still strongly informed by recent observed climate change, and hence are assessed here for completeness.

Based on this evidence, including the new 21st century observations that were not yet available to AR4, we conclude that, on the basis of constraints provided by recent observed climate change, TCR is *likely* to lie in the range 1°C to 2.5°C and *extremely unlikely* to be greater than 3°C. This range for TCR is smaller than given at the time of AR4, due to the stronger observational constraints and the wider range of studies now available. Our greater confidence in excluding high values of TCR arises primarily from higher and more confident estimates of past forcing: estimates of TCR are not strongly dependent on observations of ocean heat uptake.

10.8.2 Constraints on Long-Term Climate Change and the Equilibrium Climate Sensitivity

The equilibrium climate sensitivity (ECS) is defined as the warming in response to a sustained doubling of carbon dioxide in the atmosphere relative to pre-industrial levels (see AR4). The equilibrium to which the ECS refers to is generally assumed to be an equilibrium involving the ocean–atmosphere system, which does not include Earth system feedbacks such as long-term melting of ice sheets and ice caps, dust forcing or vegetation changes (see Chapter 5 and Section 12.5.3). The ECS cannot be directly deduced from transient warming attributable to GHGs, or from TCR, as the role of ocean heat uptake has to be taken into account (see Forest et al., 2000; Frame et al., 2005; Knutti and Hegerl, 2008). Estimating the ECS generally relies on the paradigm of a comparison of observed change with results from a physically based climate model, sometimes a very simple one, given uncertainty in the model, data, RF and due to internal variability.

For example, estimates can be based on the simple box model introduced in Section 10.8.1, ECS = $F_{2\times CO2}$ $\Delta T/(\Delta F - \Delta Q)$. Simple energy balance calculations rely on a very limited representation of climate response time scales, and cannot account for nonlinearities in the climate system that may lead to changes in feedbacks for different forcings (see Chapter 9). Alternative approaches are estimates that use climate model ensembles with varying parameters that evaluate the ECS of individual models and then infer the probability density function (PDF) for the ECS from the model–data agreement or by using optimization methods (Tanaka et al., 2009).

As discussed in the AR4, the probabilistic estimates available in the literature for climate system parameters, such as ECS and TCR have all been based, implicitly or explicitly, on adopting a Bayesian approach and therefore, even if it is not explicitly stated, involve using some kind of prior information. The shape of the prior has been derived from expert judgement in some studies, observational or experimental evidence in others or from the distribution of the sample of models available. In all cases the constraint by data, for example, from transient warming, or observations related to feedbacks is fairly weak on the upper tail of ECS (e.g., Frame et al., 2005). Therefore, results are sensitive to the prior assumptions (Tomassini et al., 2007; Knutti and Hegerl, 2008; Sanso and Forest, 2009; Aldrin et al., 2012). When the prior distribution fails to taper off for high sensitivities, as is the case for uniform priors (Frame et al., 2005), this leads to long tails (Frame et al., 2005; Annan and Hargreaves, 2011; Lewis, 2013). Uniform priors have been criticized (e.g., Annan and Hargreaves, 2011; Pueyo, 2012; Lewis, 2013) since results assuming a uniform prior in ECS translates instead into a strongly structured prior on climate feedback parameter and vice versa (Frame et al., 2005; Pueyo, 2012). Objective Bayesian analyses attempt to avoid this paradox by using a prior distribution that is invariant to parameter transforms and rescaling, for example, a Jeffreys prior (Lewis, 2013). Estimated probability densities based on priors that are strongly non-uniform in the vicinity of the best fit to the data, as is typically the case for the Jeffreys prior in this instance, can peak at values very different from the location of the best fit, and hence need to be interpreted carefully. To what extent results are sensitive to priors can be evaluated by using different priors, and this has been done more consistently in studies than at the time of AR4 (see Figure 10.20b) and is assessed where available, illustrated in Figure 10.20. Results will also be sensitive to the extent to which uncertainties in forcing (Tanaka et al., 2009), models and observations and internal climate variability are taken into account, and can be acutely sensitive to relatively arbitrary choices of observation period, choice of truncation in estimated covariance matrices and so forth (Lewis, 2013), illustrating the importance of sensitivity studies. Analyses that make a more complete effort to estimate all uncertainties affecting the model–data comparison lead to more trustworthy results, but end up with larger uncertainties (Knutti and Hegerl, 2008).

The detection and attribution chapter in AR4 (Hegerl et al., 2007b) concluded that 'Estimates based on observational constraints indicate that it is *very likely* that the equilibrium climate sensitivity is larger than 1.5°C with a most likely value between 2°C and 3°C'. The following sections discuss evidence since AR4 from several lines of evidence, followed by an overall assessment of ECS based on observed climate changes, and a subset of available new estimates is shown in Figure 10.20b.

10.8.2.1 Estimates from Recent Temperature Change

As estimates of ECS based on recent temperature change can only sample atmospheric feedbacks that occur with presently evolving climate change, they provide information on the 'effective climate sensitivity' (e.g., Forest et al., 2008). As discussed in AR4, analyses based on global scale data find that within data uncertainties, a strong aerosol forcing or a large ocean heat uptake might have masked a strong greenhouse warming (see, e.g., Forest et al., 2002; Frame et al., 2005; Stern, 2006; Roe and Baker, 2007; Hannart et al., 2009; Urban and Keller, 2009; Church et al., 2011). This is consistent with the finding that a set of models with a large range of ECS and aerosol forcing could be consistent with the observed warming (Kiehl, 2007). Consequently, such analyses find that constraints on aerosol forcing are essential to provide tighter constraints on future warming (Tanaka et al., 2009; Schwartz et al., 2010). Aldrin et al. (2012) analyse the observed record from 1850 to 2007 for hemispheric means of surface temperature, and upper 700 m ocean heat content since 1955. The authors use a simple climate model and a Markov Chain Monte Carlo Bayesian technique for analysis. The authors find a quite narrow range of ECS, which narrows further if using a uniform prior in 1/ECS rather than ECS (Figure 10.20). If observations are updated to 2010 and forcing estimates including further indirect aerosol effects are used (following Skeie et al., 2011), this yields a reduced upper tail (see Figure 10.20b, dash dotted). However, this estimate involves a rather simple model for internal variability, hence may underestimate uncertainties. Olson et al. (2012) use similar global scale constraints and surface temperature to 2006, and ocean data to 2003 and arrive at a wide range if using a uniform prior in ECS, and a quite well constrained range if using a prior derived from current mean climate and Last Glacial Maximum (LGM) constraints (see Figure 10.20b). Some of the differences between Olson et al. (2012) and Aldrin et al. (2012) may be due to structural differences in the model used (Aldrin et al. use a simple EBM while Olson use the UVIC EMIC), some due to different statistical methods and some due to use of global rather than hemispheric temperatures in the latter work. An approach based on regressing forcing histories used in 20th century simulations on observed surface temperatures (Schwartz, 2012) estimates ranges of ECS that encompass the AR4 ranges if accounting for data uncertainty (Figure 10.20). Otto et al. (2013) updated the Gregory et al. (2002) global energy balance analysis (see equation above), using temperature and ocean heat content data to 2009 and estimates of RF that are approximately consistent with estimates from Chapters 7 and 8, and ocean heat uptake estimates that are consistent with Chapter 3 and find that inclusion of recent deep ocean heat uptake and temperature data considerably narrow estimates of ECS compared to results using data to the less recent past.

Estimates of ECS and TCR that make use of both spatial and temporal information, or separate the GHG attributable warming using fingerprint methods, can yield tighter estimates (e.g., Frame et al., 2005; Forest et al., 2008; Libardoni and Forest, 2011). The resulting GHG attributable warming tends to be reasonably robust to uncertainties in aerosol forcing (Section 10.3.1.1.3). Forest et al. (2008) have updated their earlier study using a newer version of the MIT model and five different surface temperature data sets (Libardoni and Forest, 2011). Correction of statistical errors in estimation procedure pointed out by Lewis (see Lewis, 2013) changes their result only slightly (Libardoni and Forest, 2013). The overarching 5 to 95% range of effective climate sensitivity widens to 1.2°C to 5.3°C when all five data sets are used, and constraints on effective ocean diffusivity become very weak (Forest et al., 2008). Uncertainties would likely further increase if estimates of forcing uncertainty, for example, due to natural forcings, are also included (Forest et al., 2006). Lewis (2013) reanalysed the data used in Forest et al. (2006) using an objective Bayesian method (see discussion at top of section). The author finds that use of a Jeffreys prior narrows the upper tail considerably, to 3.6°C for the 95th percentile. When revising the method, omitting upper air data, and adding 6 more years of data a much reduced 5 to 95% range of 1.2°C to 2.2°C results (see Figure 10.20), similar to estimates by Ring et al. (2012) using data to 2008. Lewis's upper limit extends to 3.0°C if accounting for forcing and surface temperature uncertainty (Lewis, 2013). Lewis (2013) also reports a range of 1.1°C to 2.9°C using his revised diagnostics and the Forest et al. (2006) statistical method, whereas adding 9 more years to the Libardoni and Forest (2013) corrected diagnostic (after Libardoni and Forest, 2011; Figure 10.20; using an expert prior in both cases), does not change results much (Figure 10.20b). The differences between results reported in Forest et al. (2008); Libardoni and Forest (2011); Lewis (2013); Libardoni and Forest (2013) are still not fully understood, but appear to be due to a combination of sensitivity of results to the choice of analysis period as well as differences in diagnostics and statistical approach.

In summary, analyses that use the most recent decade find a tightening of the range of ECS based on a combination of recent heat uptake and surface temperature data. Results consistently give low probability to ECS values under 1.0°C (Figure 10.20). The mode of the PDFs varies considerably with period considered as expected from the influence of internal variability on the single realization of observed climate change. Estimates including the most recent data tend to have reduced upper tails (Libardoni and Forest, 2011; Aldrin et al., 2012 and update; Ring et al., 2012 and update cf. Figure 10.20; Lewis, 2013; Otto et al., 2013), although further uncertainty in statistical assumptions and structural uncertainties in simple models used, as well as neglected uncertainties, for example, in forcings, increase assessed uncertainty.

10.8.2.2 Estimates Based on Top of the Atmosphere Radiative Balance

With the satellite era, measurements are now long enough to allow direct estimates of variations in the energy budget of the planet, although the measurements are not sufficiently accurate to determine absolute top of the atmosphere (TOA) fluxes or trends (see Section 2.3 and Box 13.1). Using a simple energy balance relationship between net energy flow towards the Earth, net forcing and a climate feedback parameter and the satellite measurements Murphy et al. (2009) made direct estimates of the climate feedback parameter as the regression coefficient of radiative response against GMST. The feedback parameter in turn is inversely proportional to the ECS (see above, also Forster and Gregory, 2006). Such regression based estimates are, however, subject to uncertainties (see Section 7.2.5.7; see also, Gregory and Forster, 2008; Murphy and Forster, 2010). Lindzen and Choi (2009) used data from the radiative budget and simple energy balance models over the tropics to investigate feedbacks in climate models. Their result suggests that climate models overestimate the outgoing shortwave radiation compared to Earth Radiation Budget Experiment (ERBE) data, but this result was found unreliable owing to use of a limited sample of periods and of a domain limited to low latitudes (Murphy and Forster, 2010). Lindzen and Choi (2011) address some of these criticisms (Chung et al., 2010; Trenberth et al., 2010), but the results remains uncertain. For example, the lag-lead relationship between TOA balance and SST (Lindzen and Choi, 2011) is replicated by Atmospheric Model Intercomparison Project (AMIP) simulations where SST cannot respond (Dessler, 2011). Hence, as discussed in Section 7.2.5.7, the influence of internal temperature variations on short time scales seriously affects such estimates of feedbacks. In addition, the energy budget changes that are used to derive feedbacks are also affected by RF, which Lindzen and Choi (2009) do not account for. Murphy and Forster (2010) further question if estimates of the feedback parameter are suitable to estimate the ECS, as multiple time scales are involved in feedbacks that contribute to climate sensitivity (Knutti and Hegerl, 2008; Dessler, 2010). Lin et al. (2010a) use data over the 20th century combined with an estimate of present TOA imbalance based on modelling (Hansen et al., 2005a) to estimate the energy budget of the planet and give a best estimate of ECS of 3.1°C, but do not attempt to estimate a distribution that accounts fully for uncertainties. In conclusion, measurement and methodological uncertainties in estimates of the feedback parameter and the ECS from short-term variations in the satellite period preclude strong constraints on ECS. When accounting for these uncertainties, estimates of ECS based on the TOA radiation budget appear consistent with those from other lines of evidence within large uncertainties (e.g., Forster and Gregory, 2006; Figure 10.20b).

10.8.2.3 Estimates Based on Response to Volcanic Forcing or Internal Variability

Some analyses used in AR4 were based on the well observed forcing and responses to major volcanic eruptions during the 20th century. The constraint is fairly weak because the peak response to short-term volcanic forcing depends nonlinearly on ECS (Wigley et al., 2005; Boer et al., 2007). Recently, Bender et al. (2010) re-evaluated the constraint and found a close relationship in 9 out of 10 AR4 models between the shortwave TOA imbalance, the simulated response to the eruption of Mt Pinatubo and the ECS. Applying the constraint from observations suggests a range of ECS of 1.7°C to 4.1°C. This range for ECS is subject to observational uncertainty and uncertainty due to internal climate variability, and is derived from a limited sample of models. Schwartz (2007) tried to relate the ECS to the strength of natural variability using the fluctuation dissipation theorem but studies suggest that the observations are too short to support a well constrained and reliable estimate and would yield an underestimate of sensitivity (Kirk-Davidoff, 2009); and that assuming single time scales is too simplistic for the climate system (Knutti and Hegerl, 2008) . Thus, credible estimates of ECS from the response to natural and internal variability do not disagree with other estimates, but at present cannot provide more reliable estimates of ECS.

10.8.2.4 Paleoclimatic Evidence

Palaeoclimatic evidence is promising for estimating ECS (Edwards et al., 2007). This section reports on probabilistic estimates of ECS derived from paleoclimatic data by drawing on Chapter 5 information on forcing and temperature changes. For periods of past climate, which were close to radiative balance or when climate was changing slowly, for example, the LGM, radiative imbalance and with it ocean heat uptake is less important than for the present (Sections 5.3.3.1 and 5.3.3.2). Treating the RF due to ice sheets, dust and CO_2 as forcings rather than feedbacks implies that the corresponding RF contributions are associated with considerable uncertainties (see Section 5.2.2.3). Koehler et al. (2010) used an estimate of LGM cooling along with its uncertainties together with estimates of LGM RF and its uncertainty to derive an overall estimate of climate sensitivity. This method accounts for the effect of changes in feedbacks for this very different climatic state using published estimates of changes in feedback factors (see Section 5.3.3.2; Hargreaves et al., 2007; Otto-Bliesner et al., 2009). The authors find a best estimate of 2.4°C and a 5 to 95% range of ECS from 1.4°C to 5.2°C, with sensitivities beyond 6°C difficult to reconcile with the data. In contrast, Chylek and Lohmann (2008b) estimate the ECS to be 1.3°C to 2.3°C based on data for the transition from the LGM to the Holocene. However, the true uncertainties are likely larger due to uncertainties in relating local proxies to large-scale temperature change observed over a limited time (Ganopolski and von Deimling, 2008; Hargreaves and Annan, 2009). The authors also use an aerosol RF estimate that may be high (see response by Chylek and Lohmann, 2008a; Ganopolski and von Deimling, 2008).

At the time of the AR4, several studies were assessed in which parameters in climate models had been perturbed systematically in order to estimate ECS, and further studies have been published since, some making use of expanded data for LGM climate change (see Section 5.3.3.2, Table 5.3). Sometimes substantial differences between estimates based on similar data reflect not only differences in assumptions on forcing and use of data, but also structural model uncertainties, for example, in how feedbacks change between different climatic states (e.g., Schneider von Deimling et al., 2006; Hargreaves et al., 2007; (see also Otto-Bliesner et al., 2009). Holden et al. (2010) analysed which versions of the EMIC Genie are consistent with LGM tropical SSTs and find a 90% range of 2.0°C to 5.0°C. Recently, new data synthesis products have become available for assessment with climate model simulations of the LGM which together with further data cover much more of the LGM ocean and land areas, although there are still substantial gaps and substantial data uncertainty (Section 5.3.3). An analysis of the recent SST and land temperature reconstructions for the LGM compared to simulations with an EMIC suggests a 90% range of 1.4°C to 2.8°C for ECS, with SST data providing a narrower range and lower values than land data only (see Figure 10.20; Schmittner et al., 2011). However, structural model uncertainty as well as data uncertainty may increase this range substantially (Fyke and Eby, 2012; Schmittner et al., 2012). Hargreaves et al. (2012) derived a relationship between ECS and LGM response for seven model simulations from PMIP2 simulations and found a linear relationship between tropical cooling and ECS (see Section 5.3.3.2) which has been used to derive an estimate of ECS (Figure 10.20); and has been updated using PMIP3 simulations (Section 5.3.3.2). However, uncertainties remain as the relationship is dependent on the ensemble of models used.

Estimates of ECS from other, more distant paleoclimate periods (e.g., Royer et al., 2007; Royer, 2008; Pagani et al., 2009; Lunt et al., 2010) are difficult to directly compare, as climatic conditions were very different from today and as climate sensitivity can be state dependent, as discussed above. Also, the response on very long time scales is determined by the Earth System Sensitivity, which includes very slow feedbacks by ice sheets and vegetation (see Section 12.5.3). Paleosens Members (2012) reanalysed the relationship between RF and temperature response from paleoclimatic studies, considering Earth system feedbacks as forcings in order to derive an estimate of ECS that is limited to atmospheric feecbacks (sometimes referred to as Charney sensitivity and directly comparable to ECS), and find that resulting estimates are reasonably consistent over the past 65 million years (see detailed discussion in Section 5.3.1). They estimate a 95% range of 1.1°C to 7.0°C, largely based on the past 800,000 years. However, uncertainties in paleoclimate estimates of ECS are likely to be larger than from the instrumental record, for example, due to changes in feedbacks between different climatic states. In conclusion, estimates of ECS have continued to emerge from palaeoclimatic periods that indicate that ECS is *very likely* less than 6°C and *very likely* greater than 1.0°C (see Section 5.3.3).

10.8.2.5 Combining Evidence and Overall Assessment

Most studies find a lower 5% limit for ECS between 1°C and 2°C (Figure 10.20). The combined evidence thus indicates that the net feedbacks to RF are significantly positive. At present, there is no credible individual line of evidence that yields very high or very low climate sensitivity as best estimate. Some recent studies suggest a low climate sensitivity (Chylek et al., 2007; Schwartz et al., 2007; Lindzen and Choi, 2009). However, these are based on problematic assumptions, for example, about the climate's response time, the cause of climate fluctuations, or neglect uncertainty in forcing, observations and internal variability (as discussed in Foster et al., 2008; Knutti and Hegerl, 2008; Murphy and Forster, 2010). In some cases the estimates of the ECS have been refuted by testing the method of estimation with a climate model of known sensitivity (e.g., Kirk-Davidoff, 2009).

Several authors (Annan and Hargreaves, 2006; Hegerl et al., 2006; Annan and Hargreaves, 2010) had proposed combining estimates of climate sensitivity from different lines of evidence by the time of AR4; these and recent work is shown in the panel 'combined' in Figure 10.20. Aldrin et al. (2012) combined the Hegerl et al. (2006) estimate based on the last millennium with their estimate based on the 20th century; and Olson et al. (2012) combined weak constraints from climatology and the LGM in their prior, updated by data on temperature changes. This approach is robust only if the lines of evidence used are truly independent. The latter is hard to evaluate when using prior distributions based on expert knowledge (e.g., Libardoni and Forest, 2011). If lines of evidence are not independent, overly confident assessments of equilibrium climate sensitivity may result (Henriksson et al., 2010; Annan and Hargreaves, 2011).

In conclusion, estimates of the Equilibrium Climate Sensitivity (ECS) based on multiple and partly independent lines of evidence from observed climate change, including estimates using longer records of surface temperature change and new palaeoclimatic evidence, indicate that there is *high confidence* that ECS is *extremely unlikely* less than 1°C and *medium confidence* that the ECS is *likely* between 1.5°C and 4.5°C and *very unlikely* greater than 6°C. They complement the

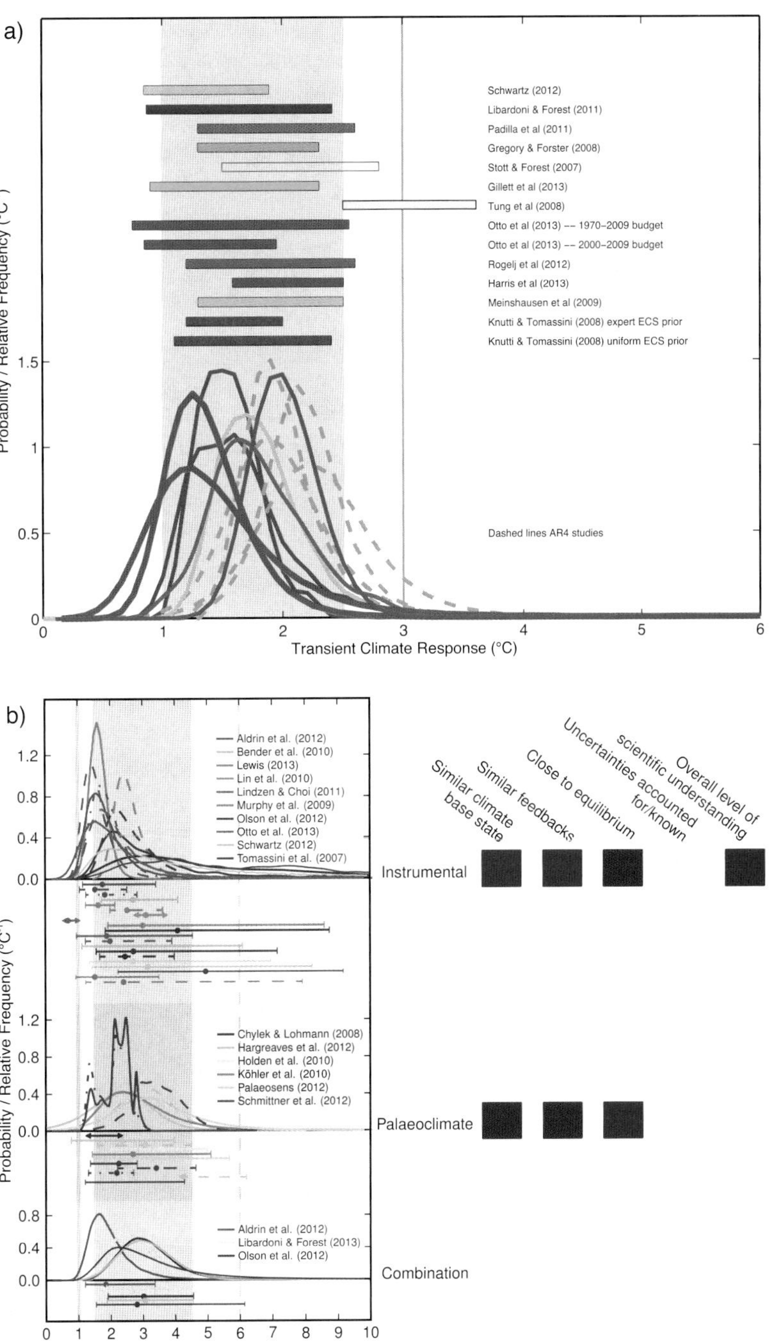

Figure 10.20 | (a) Examples of distributions of the transient climate response (TCR, top) and the equilibrium climate sensitivity (ECS, bottom) estimated from observational constraints. Probability density functions (PDFs), and ranges (5 to 95%) for the TCR estimated by different studies (see text). The grey shaded range marks the *very likely* range of 1°C to 2.5°C for TCR and the grey solid line represents the *extremely unlikely* <3°C upper bound as assessed in this section. Representative distributions from AR4 shown as dashed lines and open bar. (b) Estimates of ECS are compared to overall assessed *likely* range (solid grey), with solid line at 1°C and a dashed line at 6°C. The figure compares some selected old estimates used in AR4 (no labels, thin lines; for references see Supplementary Material) with new estimates available since AR4 (labelled, thicker lines). Distributions are shown where available, together with 5 to 95% ranges and median values (circles). Ranges that are assessed as being incomplete are marked by arrows; note that in contrast to the other estimates Schwartz (2012), shows a sampling range and Chylek and Lohmann a 95% range. Estimates are based on changes over the instrumental period (top row); and changes from palaeoclimatic data (2nd row). Studies that combine multiple lines of evidence are shown in the bottom panel. The boxes on the right-hand side indicate limitations and strengths of each line of evidence, for example, if a period has a similar climatic base state, if feedbacks are similar to those operating under CO_2 doubling, if the observed change is close to equilibrium, if, between all lines of evidence plotted, uncertainty is accounted for relatively completely, and summarizes the level of scientific understanding of this line of evidence overall. A blue box indicates an overall line of evidence that is well understood, has small uncertainty, or many studies and overall *high confidence*. Pale yellow indicates *medium*, and dark red *low*, *confidence* (i.e., poorly understood,very few studies, poor agreement, unknown limitations, after Knutti and Hegerl, 2008). Where available, results are shown using several different prior distributions; for example for Aldrin et al. (2012) solid shows the result using a uniform *prior* in ECS, which is shown as updated to 2010 in dash-dots; dashed: uniform *prior* in 1/ECS; and in bottom panel, result combining with Hegerl et al. (2006) *prior*, For Lewis (2013), dashed shows results using the Forest et al. (2006) diagnostic and an objective Bayesian *prior*, solid a revised diagnostic. For Otto et al. (2013), solid is an estimate using change to 1979–2009, dashed using the change to 2000–2009. Palaeoclimate: Hargreaves et al. (2012) is shown in solid, with dashed showing an update based on PMIP3 simulations (see Chapter 5); For Schmittner et al. (2011), solid is land-and-ocean, dashed land-only, and dash-dotted is ocean-only diagnostic.

evaluation in Chapter 9 and support the overall assessment in Chapter 12 that concludes between all lines of evidence with *high confidence* that ECS is *likely* in the range 1.5°C to 4.5°C. Earth system feedbacks can lead to different, probably larger, warming than indicated by ECS on very long time scales.

10.8.3 Consequences for Aerosol Forcing and Ocean Heat Uptake

Some estimates of ECS also yield estimates of aerosol forcing that are consistent with observational records, which we briefly mention here. Note that the estimate will reflect any forcings with a time or time–space pattern resembling aerosol forcing that is not explicitly included in the overall estimate (see discussion in Olson et al., 2012), for example, BC on snow; and should hence be interpreted as an estimate of aerosol plus neglected forcings. Estimates will also vary with the method applied and diagnostics used (e.g., analyses including spatial information will yield stronger results). Murphy et al. (2009) use correlations between surface temperature and outgoing shortwave and longwave flux over the satellite period to estimate how much of the total recent forcing has been reduced by aerosol total reflection, which they estimate as –1.1 ± 0.4 W m^{-2} from 1970 to 2000 (1 standard deviation), while Libardoni and Forest (2011), see also Forest et al. (2008), based on the 20th century, find somewhat lower estimates, namely a 90% bound of –0.83 to –0.19 W m^{-2} for the 1980s relative to preindustrial. Lewis (2013), using similar diagnostics but an objective Bayesian method, estimates a total aerosol forcing of about –0.6 to –0.1W m^{-2} or –0.6 to 0.0 W m^{-2} dependent on diagnostic used. The range of the aerosol forcing estimates that are based on the observed climate change are in-line with the expert judgement of the effective RF by aerosol radiation and aerosol cloud interactions combined (ERFaci+ari; Chapter 7) of –0.9 W m^{-2} with a range from –1.9 to –0.1 W m^{-2} that has been guided by climate models that include aerosol effects on mixed-phase and convective clouds in addition to liquid clouds, satellite studies and models that allow cloud-scale responses (see Section 7.5.2).

Several estimates of ECS also estimate a parameter that describe the efficiency with which the ocean takes up heat, e.g., effective global vertical ocean diffusivity (e.g., Tomassini et al., 2007; Forest et al., 2008; Olson et al., 2012; Lewis, 2013). Forest and Reynolds (2008) find that the effective global ocean diffusivity K_v in many of the CMIP3 models lies above the median value based on observational constraints, resulting in a positive bias in their ocean heat uptake. Lewis (2013) similarly finds better agreement for small values of effective ocean diffusivity. However, such a finding was very sensitive to data sets used for surface temperature (Libardoni and Forest, 2011) and ocean data (Sokolov et al., 2010), is somewhat sensitive to the diagnostic applied (Lewis, 2013), and limited by difficulties observing heat uptake in the deep ocean (see, e.g., Chapters 3 and 13). Olson et al. (2012) and Tomassini et al. (2007) find that data over the historical period provide only a weak constraint on background ocean effective diffusivity. Comparison of the vertical profiles of temperature and of historical warming in models and observations suggests that the ocean heat uptake efficiency may be typically too large (Kuhlbrodt and Gregory, 2012; Section 13.4.1; see also Sections 9.4.2, 10.4.1, 10.4.3). If effective diffusivity were high in models this might lead to a tendency to bias ocean warming high relative to surface warming; but this uncertainty makes only a small contribution to uncertainty in TCR (Knutti and Tomassini, 2008; Kuhlbrodt and Gregory, 2012; see Section 13.4.1). Nonetheless, ocean thermal expansion and heat content change simulated in CMIP5 models show relatively good agreement with observations, although this might also be due to a compensation between ocean heat uptake efficiency and atmospheric feedbacks (Kuhlbrodt and Gregory, 2012). In summary, constraints on effective ocean diffusivity are presently not conclusive.

10.8.4 Earth System Properties

A number of papers have found the global warming response to CO_2 emissions to be determined primarily by total cumulative emissions of CO_2, irrespective of the timing of those emissions over a broad range of scenarios (Allen et al., 2009; Matthews et al., 2009; Zickfeld et al., 2009; Section 12.5.4.2), although Bowerman et al. (2011) find that, when scenarios with persistent 'emission floors' are included, the strongest predictor of peak warming is cumulative emissions to 2200. Moreover, the ratio of global warming to cumulative carbon emissions, known variously as the Absolute Global Temperature Change Potential (AGTP; defined for an infinitesimal pulse emission) (Shine et al., 2005), the Cumulative Warming Commitment (defined based on peak warming in response to a finite injection; CWC) (Allen et al., 2009) or the Carbon Climate Response (CCR) (Matthews et al., 2009), is approximately scenario-independent and constant in time.

The ratio of CO_2-induced warming realized by a given year to cumulative carbon emissions to that year is known as the Transient Climate Response to cumulative CO_2 Emissions (TCRE, see Chapter 12). TCRE depends on TCR and the Cumulative Airborne Fraction (CAF), which is the ratio of the increased mass of CO_2 in the atmosphere to cumulative CO_2 emissions (not including natural fluxes and those arising from Earth system feedbacks) over a long period, typically since pre-industrial times (Gregory et al., 2009): $TCRE = TCR \times CAF/C_0$, where C_0 is the mass of carbon (in the form of CO_2) in the pre-industrial atmosphere (590 PgC). Given estimates of CAF to the time of CO_2 doubling of 0.4 to 0.7 (Zickfeld et al., 2013), we therefore expect values of TCRE, if expressed in units of °C per 1000 PgC, to be similar to or slightly lower than, and more uncertain than, values of TCR (Gillett et al., 2013).

TCRE may be estimated from observations by dividing an estimate of warming to date attributable to CO_2 by historical cumulative carbon emissions, which gives a 5 to 95% range of 0.7°C to 2.0°C per 1000 PgC (Gillett et al., 2013), 1.0°C to 2.1°C per 1000 PgC (Matthews et al., 2009) or 1.4°C to 2.5°C per 1000 PgC (Allen et al., 2009), the higher range in the latter study reflecting a higher estimate of CO_2-attributable warming to 2000. The peak warming induced by a given total cumulative carbon emission (Peak Response to Cumulative Emissions (PRCE)) is less well constrained, since warming may continue even after a complete cessation of CO_2 emissions, particularly in high-response models or scenarios. Using a combination of observations and models to constrain temperature and carbon cycle parameters in a simple climate-carbon-cycle model, (Allen et al., 2009), obtain a PRCE 5 to 95% confidence interval of 1.3°C to 3.9°C per 1000 PgC. They also report that (Meinshausen et al., 2009) obtain a 5 to 95% range in PRCE of 1.1°C to 2.7°C per 1000 PgC using a Bayesian approach with a different simple model, with climate parameters constrained by observed warming and carbon cycle parameters constrained by the C4MIP simulations (Friedlingstein et al., 2006).

The ratio of warming to cumulative emissions, the TCRE, is assessed to be *likely* between 0.8°C and 2.5°C per 1000 PgC based on observational constraints. This implies that, for warming due to CO_2 emissions alone to be *likely* less than 2°C at the time CO_2 emissions cease, total cumulative emissions from all anthropogenic sources over the entire industrial era would need to be limited to about 1000 PgC, or one trillion tonnes of carbon (see Section 12.5.4).

10.9 Synthesis

The evidence has grown since the Fourth Assessment Report that widespread changes observed in the climate system since the 1950s are attributable to anthropogenic influences. This evidence is documented in the preceding sections of this chapter, including for near surface temperatures (Section 10.3.1.1), free atmosphere temperatures (Section 10.3.1.2), atmospheric moisture content (Section 10.3.2.1), precipitation over land (Section 10.3.2.2), ocean heat content (Section 10.4.1), ocean salinity (Section 10.4.2), sea level (Section 10.4.3), Arctic sea ice (Section 10.5.1), climate extremes (Section 10.6) and evidence from the last millenium (Section 10.7). These results strengthen the conclusion that human influence on climate has played the dominant role in observed warming since the 1950s. However, the approach taken so far in this chapter has been to examine each aspect of the climate system—the atmosphere, oceans, cryosphere, extremes, and from paleoclimate archives—separately in each section and sub-section. In this section we look across the whole climate system to assess the extent that a consistent picture emerges across sub-systems and climate variables.

10.9.1 Multi-variable Approaches

Multi-variable studies provide one approach to gain a more comprehensive view across the climate system, although there have been relatively few applications of multi-variable detection and attribution studies in the literature. A combined analysis of near-surface temperature from weather stations and free atmosphere temperatures from radiosondes detected an anthropogenic influence on the joint changes in temperatures near the surface and aloft (Jones et al., 2003). In a Bayesian application of detection and attribution Schnur and Hasselmann (2005) combined surface temperature, diurnal temperature range and precipitation into a single analysis and showed strong net evidence for detection of anthropogenic forcings despite low likelihood ratios for diurnal temperature range and precipitation on their own. Barnett et al. (2008) applied a multi-variable approach in analysing changes in the hydrology of the Western United States (see also Section 10.3.2.3).

The potential for a multi-variable analysis to have greater power to discriminate between forced changes and internal variability has been demonstrated by Stott and Jones (2009) and Pierce et al. (2012). In the former case, they showed that a multi-variable fingerprint consisting of the responses of GMST and sub-tropical Atlantic salinity has a higher S/N ratio than the fingerprints of each variable separately. They found reduced detection times as a result of low correlations between the two variables in the control simulation although the detection result depends on the ability of the models to represent the co-variability of the variables concerned. Multi-variable attribution studies potentially provide a stronger test of climate models than single variable attribution studies although there can be sensitivity to weighting of different components of the multi-variable fingerprint. In an analysis of ocean variables, Pierce et al. (2012) found that the joint analysis of temperature and salinity changes yielded a stronger signal of climate change than 'either salinity or temperature alone'.

Further insights can be gained by considering a synthesis of evidence across the climate system. This is the subject of the next subsection.

10.9.2 Whole Climate System

To demonstrate how observed changes across the climate system can be understood in terms of natural and anthropogenic causes Figure 10.21 compares observed and modelled changes in the atmosphere, ocean and cryosphere. The instrumental records associated with each element of the climate system are generally independent (see FAQ 2.1), and consequently joint interpretations across observations from the main components of the climate system increases the confidence to higher levels than from any single study or component of the climate system. The ability of climate models to replicate observed changes (to within internal variability) across a wide suite of climate indicators also builds confidence in the capacity of the models to simulate the Earth's climate.

The coherence of observed changes for the variables shown in Figure 10.21 with climate model simulations that include anthropogenic and natural forcing is remarkable. Surface temperatures over land, SSTs and ocean heat content changes show emerging anthropogenic and natural signals with a clear separation between the observed changes and the alternative hypothesis of just natural variations (Figure 10.21, Global panels). These signals appear not just in the global means, but also at continental and ocean basin scales in these variables. Sea ice emerges strongly from the range expected from natural variability for the Arctic and Antarctica remains broadly within the range of natural variability consistent with expectations from model simulations including anthropogenic forcings.

Table 10.1 illustrates a larger suite of detection and attribution results across the climate system than summarized in Figure 10.21. These results include observations from both the instrumental record and paleo-reconstructions on a range of time scales ranging from daily extreme precipitation events to variability over millennium time scales.

From up in the stratosphere, down through the troposphere to the surface of the Earth and into the depths of the oceans there are detectable signals of change such that the assessed likelihood of a detectable, and often quantifiable, human contribution ranges from *likely* to *extremely likely* for many climate variables (Table 10.1). Indeed to successfully describe the observed warming trends in the atmosphere, ocean and at the surface over the past 50 years, contributions from both anthropogenic and natural forcings are required (e.g., results 1, 2, 3, 4, 5, 7, 9 in Table 10.1). This is consistent with anthropogenic forcings warming the surface of the Earth, troposphere and oceans superimposed with cooling events caused by the three large explosive volcanic eruptions since the 1960's. These two effects (anthropogenic warming and vol-

10

Frequently Asked Questions

FAQ 10.2 | When Will Human Influences on Climate Become Obvious on Local Scales?

Human-caused warming is already becoming locally obvious on land in some tropical regions, especially during the warm part of the year. Warming should become obvious in middle latitudes—during summer at first—within the next several decades. The trend is expected to emerge more slowly there, especially during winter, because natural climate variability increases with distance from the equator and during the cold season. Temperature trends already detected in many regions have been attributed to human influence. Temperature-sensitive climate variables, such as Arctic sea ice, also show detected trends attributable to human influence.

Warming trends associated with global change are generally more evident in averages of global temperature than in time series of local temperature ('local' here refers generally to individual locations, or small regional averages). This is because most of the local variability of local climate is averaged away in the global mean. Multi-decadal warming trends detected in many regions are considered to be outside the range of trends one might expect from natural internal variability of the climate system, but such trends will only become obvious when the local mean climate emerges from the 'noise' of year-to-year variability. How quickly this happens depends on both the rate of the warming trend and the amount of local variability. Future warming trends cannot be predicted precisely, especially at local scales, so estimates of the future time of emergence of a warming trend cannot be made with precision.

10

In some tropical regions, the warming trend has already emerged from local variability (FAQ 10.2, Figure 1). This happens more quickly in the tropics because there is less temperature variability there than in other parts of the globe. Projected warming may not emerge in middle latitudes until the mid-21st century—even though warming trends there are larger—because local temperature variability is substantially greater there than in the tropics. On a seasonal basis, local temperature variability tends to be smaller in summer than in winter. Warming therefore tends to emerge first in the warm part of the year, even in regions where the warming trend is larger in winter, such as in central Eurasia in FAQ 10.2, Figure 1.

Variables other than land surface temperature, including some oceanic regions, also show rates of long-term change different from natural variability. For example, Arctic sea ice extent is declining very rapidly, and already shows a human influence. On the other hand, local precipitation trends are very hard to detect because at most locations the variability in precipitation is quite large. The probability of record-setting warm summer temperatures has increased throughout much of the Northern Hemisphere . High temperatures presently considered extreme are projected to become closer to the norm over the coming decades. The probabilities of other extreme events, including some cold spells, have lessened.

In the present climate, individual extreme weather events cannot be unambiguously ascribed to climate change, since such events could have happened in an unchanged climate. However the probability of occurrence of such events could have changed significantly at a particular location. Human-induced increases in greenhouse gases are estimated to have contributed substantially to the probability of some heatwaves. Similarly, climate model studies suggest that increased greenhouse gases have contributed to the observed intensification of heavy precipitation events found over parts of the Northern Hemisphere. However, the probability of many other extreme weather events may not have changed substantially. Therefore, it is incorrect to ascribe every new weather record to climate change.

The date of future emergence of projected warming trends also depends on local climate variability, which can temporarily increase or decrease temperatures. Furthermore, the projected local temperature curves shown in FAQ 10.2, Figure 1 are based on multiple climate model simulations forced by the same assumed future emissions scenario. A different rate of atmospheric greenhouse gas accumulation would cause a different warming trend, so the spread of model warming projections (the coloured shading in FAQ 10.2, Figure 1) would be wider if the figure included a spread of greenhouse gas emissions scenarios. The increase required for summer temperature change to emerge from 20th century local variability (regardless of the rate of change) is depicted on the central map in FAQ 10.2, Figure 1.

A full answer to the question of when human influence on local climate will become obvious depends on the strength of evidence one considers sufficient to render something 'obvious'. The most convincing scientific evidence for the effect of climate change on local scales comes from analysing the global picture, and from the wealth of evidence from across the climate system linking many observed changes to human influence. *(continued on next page)*

FAQ 10.2 (continued)

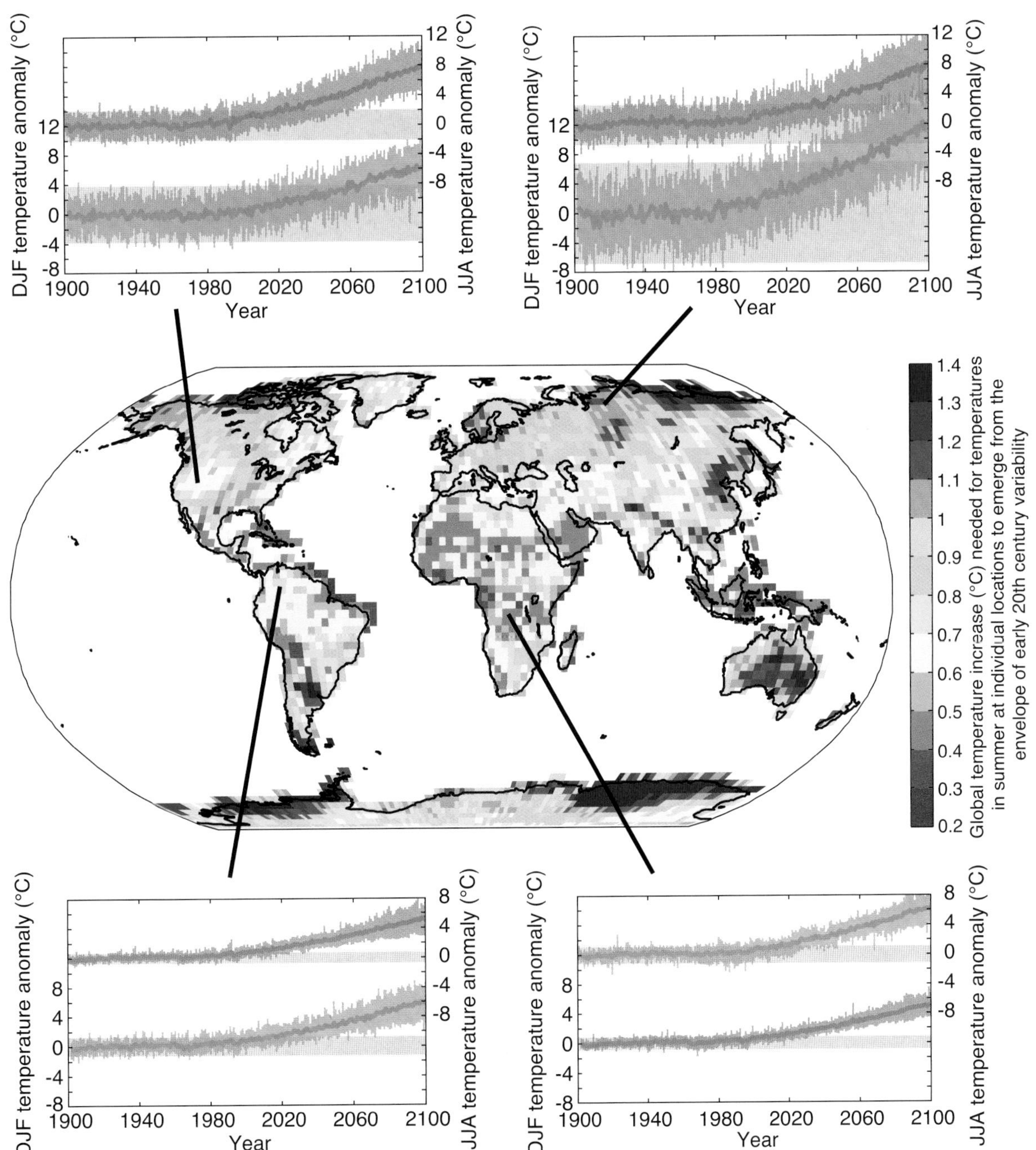

FAQ 10.2, Figure 1 | Time series of projected temperature change shown at four representative locations for summer (red curves, representing June, July and August at sites in the tropics and Northern Hemisphere or December, January and February in the Southern Hemisphere) and winter (blue curves). Each time series is surrounded by an envelope of projected changes (pink for the local warm season, blue for the local cold season) yielded by 24 different model simulations, emerging from a grey envelope of natural local variability simulated by the models using early 20th century conditions. The warming signal emerges first in the tropics during summer. The central map shows the global temperature increase (°C) needed for temperatures in summer at individual locations to emerge from the envelope of early 20th century variability. Note that warm colours denote the smallest needed temperature increase, hence earliest time of emergence. All calculations are based on Coupled Model Intercomparison Project Phase 5 (CMIP5) global climate model simulations forced by the Representative Concentration Pathway 8.5 (RCP8.5) emissions scenario. Envelopes of projected change and natural variability are defined as ±2 standard deviations. (Adapted and updated from Mahlstein et al., 2011.)

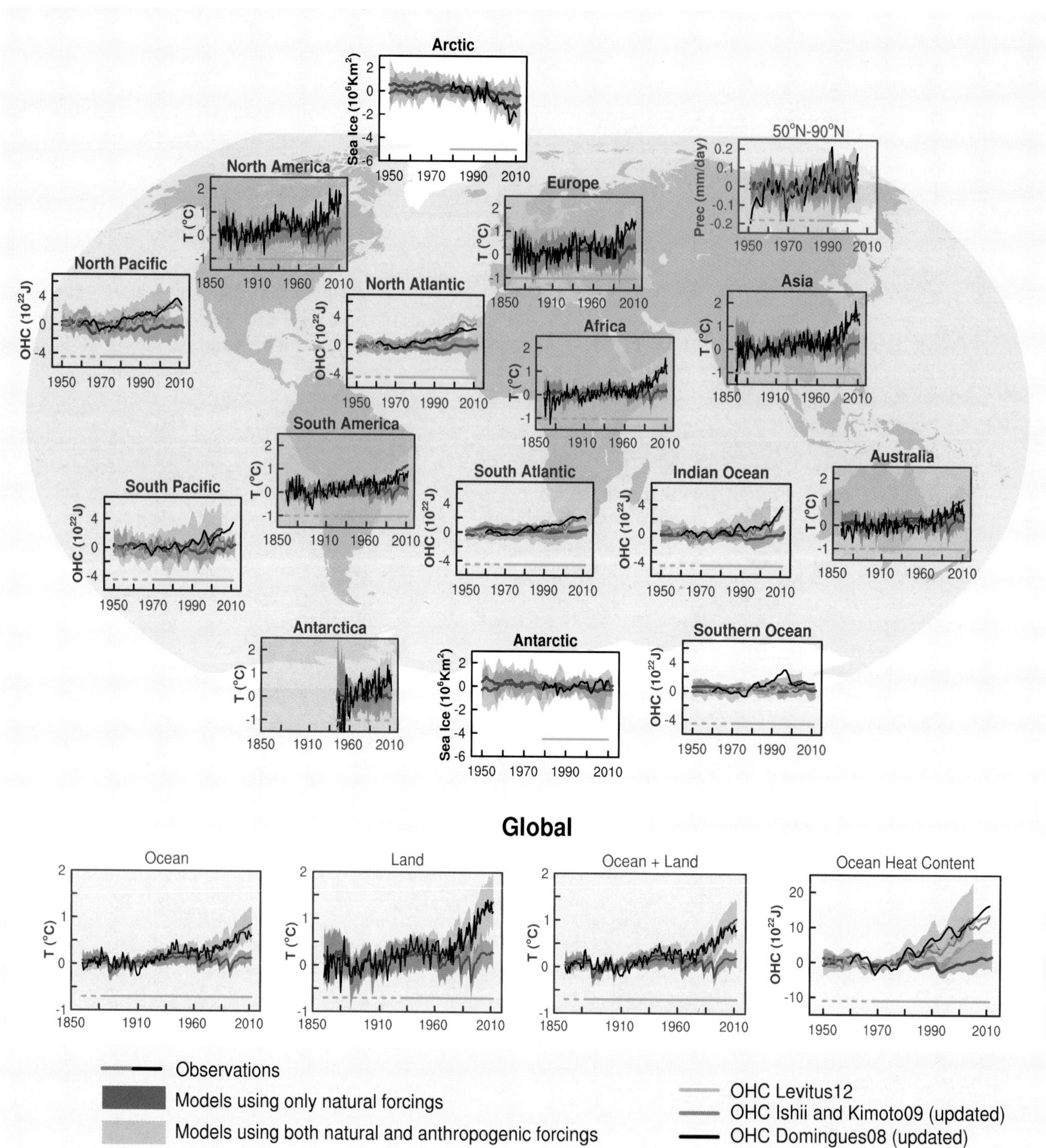

Figure 10.21 | Detection and attribution signals in some elements of the climate system, at regional scales (top panels) and global scales (bottom four panels). Brown panels are land surface temperature time series, green panels are precipitation time series, blue panels are ocean heat content time series and white panels are sea ice time series. Observations are shown on each panel in black or black and shades of grey. Blue shading is the model time series for natural forcing simulations and pink shading is the combined natural and anthropogenic forcings. The dark blue and dark red lines are the ensemble means from the model simulations. All panels show the 5 to 95% intervals of the natural forcing simulations, and the natural and anthropogenic forcing simulations. For surface temperature the results are from Jones et al. (2013) (and Figure 10.1). The observed surface temperature is from Hadley Centre/Climatic Research Unit gridded surface temperature data set 4 (HadCRUT4). Observed precipitation is from Zhang et al. (2007) (black line) and CRU TS 3.0 updated (grey line). Three observed records of ocean heat content (OHC) are shown. Sea ice anomalies (rather than absolute values) are plotted and based on models in Figure 10.16. The green horizontal lines indicate quality of the observations and estimates. For land and ocean surface temperatures panels and precipitation panels, solid green lines at bottom of panels indicate where data spatial coverage being examined is above 50% coverage and dashed green lines where coverage is below 50%. For example, data coverage of Antarctica never goes above 50% of the land area of the continent. For ocean heat content and sea ice panels the solid green line is where the coverage of data is good and higher in quality, and the dashed green line is where the data coverage is only adequate. More details of the sources of model simulations and observations are given in the Supplementary Material (10.SM.1).

canic eruptions) cause much of the observed response (see also Figures 10.5, 10.6, 10.9, 10.14a and 10.21). Both natural and anthropogenic forcings are required to understand fully the variability of the Earth system during the past 50 years.

Water in the free atmosphere is expected to increase, as a consequence of warming of the atmosphere (Section 10.6.1), and atmospheric circulation controls the global distribution of precipitation and evaporation. Simulations show that GHGs increase moisture in the atmosphere and change its transport in such a way as to produce patterns of precipitation and evaporation that are quite distinct from the observed patterns of warming. Our assessment shows that anthropogenic forcings have contributed to observed increases in moisture content in the atmosphere (result 16, *medium confidence*, Table 10.1), to global scale changes in precipitation patterns over land (result 14, *medium confidence*), to a global scale intensification of heavy precipitation in land regions where there observational coverage is sufficient to make an assessment (result 15, *medium confidence*), and to changes in surface and sub-surface ocean salinity (result 11, *very likely*). Combining evidence from both atmosphere and ocean that systematic changes in precipitation over land and ocean salinity can be attributed to human influence supports an assessment that it is *likely* that human influence has affected the global water cycle since 1960.

Warming of the atmosphere and the oceans affects the cryosphere, and in the case of snow and sea ice warming leads to positive feedbacks that amplify the warming response in the atmosphere and oceans. Retreat of mountain glaciers has been observed with an anthropogenic influence detected (result 17, *likely*, Table 10.1), Greenland ice sheet has melted at the edges and accumulating snow at the higher elevations is consistent with GHG warming supporting an assessment for an anthropogenic influence on the negative surface mass balance of Greenland's ice sheet (result 18, *likely*, Table 10.1). Our level of scientific understanding is too low to provide a quantifiable explanation of the observed mass loss of the Antarctic ice sheet (*low confidence*, result 19, Table 10.1). Sea ice in the Arctic is decreasing rapidly and the changes now exceed internal variability and with an anthropogenic contribution detected (result 20, *very likely*, Table 10.1). Antarctic sea ice extent has grown overall over the last 30 years but there is low scientific understanding of the spatial variability and changes in Antarctic sea ice extent (result 21, Table 10.1). There is evidence for an anthropogenic component to observed reductions in NH snow cover since the 1970s (*likely*, result 22, Table 10.1).

Anthropogenic forcing has also affected temperature on continental scales, with human influences having made a substantial contribution to warming in each of the inhabited continents (results 28, *likely*, Table 10.1), and having contributed to the very substantial Arctic warming over the past 50 years (result 29, *likely*, Table 10.1) while because of large observational uncertainties there is *low confidence* in attribution of warming averaged over available stations over Antarctica (result 30, Table 10.1). There is also evidence that anthropogenic forcings have contributed to temperature change in many sub-continental regions (result 32, *likely*, Table 10.1) and that anthropogenic forcings have contributed to the observed changes in the frequency and intensity of daily temperature extremes on the global scale since the mid-20th century (result 8, *very likely*, Table 10.1). Furthermore there is evidence that human influence has substantially increased the probability of occurrence of heat waves in some locations (result 33, *likely*, Table 10.1).

An analysis of these results (from Table 10.1) shows that there is *high confidence* in attributing many aspects of changes in the climate system to human influence including from atmospheric measurements of temperature. Synthesizing the results in Table 10.1 shows that the combined evidence from across the climate system increases the level of confidence in the attribution of observed climate change to human influence and reduces the uncertainties associated with assessments based on a single variable. From this combined evidence, it is *virtually certain* that human influence has warmed the global climate system.

Acknowledgements

We acknowledge the major contributions of the following scientists who took a substantial part in the production of key figures: Beena Balan Sarojini, Oliver Browne, Jara Imbers Quintana, Gareth Jones, Fraser Lott, Irina Mahlstein, Alexander Otto, Debbie Polson, Andrew Schurer, Lijun Tao, and Muyin Wang. We also acknowledge the contributions of Viviane Vasconcellos de Menezes for her work on the production of figures and for her meticulous management of the bibliography database used for this chapter.

Table 10.1 | Synthesis of detection and attribution results across the climate system from this chapter. Note that we follow the guidance note for lead authors of the IPCC AR5 on consistent treatment of uncertainties (Mastrandrea et al., 2011). Where the confidence is *medium* or less there is no assessment given of the quantified likelihood measure, and the table cell is marked not applicable (N/A).

Result	(1) Statement about variable or property: time, season	(2) Confidence (*Very high, High, medium* or *low, very low*)	(3) Quantified measure of uncertainty where the probability of the outcome can be quantified (Likelihood given generally only if high or very high confidence)	(4) Data sources Observational evidence (Chapters 2 to 5); Models (Chapter 9)	(5) Type, amount, quality, consistency of evidence from attribution studies and degree of agreement of studies.	(6) Factors contributing to the assessments including physical understanding, observational and modelling uncertainty, and caveats.
Global Scale Atmospheric Temperature Changes						
1	More than half of the observed increase in global mean surface temperatures from 1951 to 2010 is due to the observed anthropogenic increase in greenhouse gas (GHG) concentrations.	*High*	*Very likely*	Four global surface temperature series (HadCRUT3, HadCRUT4, MLOST, GISTEMP). CMIP3 and CMIP5 models.	· Many formal attribution studies, including optimal fingerprint time-space studies and time series based studies. · Robust evidence. Attribution of more than half of warming since 1950 to GHGs seen in multiple independent analyses using different observational data sets and climate models. · High agreement. Studies agree in robust detection of GHG contribution to observed warming that is larger than any other factor including internal variability.	The observed warming is well understood in terms of contributions of anthropogenic forcings such as greenhouse gases (GHGs) and tropospheric aerosols and natural forcings from volcanic eruptions. Solar forcing is the only other forcing that could explain long-term warming but pattern of warming is not consistent with observed pattern of change in time, vertical change and estimated to be small. AMO could be confounding influence but studies that find significant role for AMO show this does not project strongly onto 60-year trends. (Section 10.3.1.1, Figures 10.4 and 10.5)
2	More than half of the observed increase in global mean surface temperatures from 1951 to 2010 is due to human influence on climate.	*High*	*Extremely likely*	Mutliple CMIP5 models and multiple methodologies.	· Formal attribution studies including different optimal detection methodologies and time series based studies. · Robust evidence of well-constrained estimates of net anthropogenic warming estimated in optimal detection studies. · High agreement. Both optimal detection and time series studies agree in robust detection of anthropogenic influence that is substantially more than half of the observed warming.	The observed warming is well understood in terms of contributions of anthropogenic and natural forcings. Solar forcing and AMO could be confounding influence but are estimated to be smaller than the net effects of human influence. (Section 10.3.1.1, Figures 10.4, 10.5, 10.6)
3	Early 20th century warming is due in part to external forcing.	*High*	*Very likely*	Instrumental global surface temperature series and reconstructions of the last millenium. CMIP3 and CMIP5 models.	· Formal detection and attribution studies looking at early century warming and studies for the last few hundred years. · High agreement across a number of studies in detecting external forcings when including early 20th century period although they vary in contributions from different forcings.	Modelling studies show contribution from external forcings to early century warming. Residual differences between models and observations indicate role for circulation changes as contributor. (Section 10.3.1.1, Figures 10.1, 10.2, 10.6)
4	Warming since 1950 cannot be explained without external forcing.	*High*	*Virtually certain*	Estimates of internal variability from CMIP3 and CMIP5 models, observation based time series and space pattern analyses, and estimating residuals of the non-forced component from paleo data.	· Many, including optimal fingerprint time-space studies, observation based time series and space pattern analyses and paleo data studies. · Robust evidence and high agreement. · Detection of anthropogenic fingerprint robustly seen in independent analyses using different observational data sets, climate models, and methodological approaches.	Based on all evidence above combined. Observed warming since 1950 is very large compared to climate model estimates of internal variability, which are assessed to be adequate at global scale. The Northern Hemisphere (NH) mean warming since 1950 is far outside the range of any similar length trend in residuals from reconstructions of NH mean temperature of the past millennium. The spatial pattern of observed warming differs from those associated with internal variability. (Sections 9.5.3.1, 10.3.1.1, 10.7.1)

(continued on next page)

Table 10.1 *(continued)*

Result	(1) Statement about variable or property: time, season	(2) Confidence (*Very high*, *High*, *medium* or *low*, *very low*)	(3) Quantified measure of uncertainty where the probability of the outcome can be quantified (Likelihood given generally only if high or very high confidence)	(4) Data sources Observational evidence (Chapters 2 to 5); Models (Chapter 9)	(5) Type, amount, quality, consistency of evidence from attribution studies and degree of agreement of studies.	(6) Factors contributing to the assessments including physical understanding, observational and modelling uncertainty, and caveats.
5	Anthropogenic forcing has led to a detectable warming of troposphere temperatures since 1961.	*High*	*Likely*	Multiple radiosonde data sets from 1958 and satellite data sets from 1979 to present. CMIP3 and CMIP5 models.	· Formal attribution studies with CMIP3 models (assessed in AR4) and CMIP5 models. · Robust detection and attribution of anthropogenic influence on tropospheric warming with large signal-to-noise (S/N) ratios estimated. · Studies agree in detecting an anthropogenic influence on tropospheric warming trends.	Observational uncertainties in radiosondes are now much better documented than at time of AR4. It is virtually certain that the troposphere has warmed since the mid-20th century but there is only *medium confidence* in the rate and vertical structure of those changes in the NH extratropical troposphere and *low confidence* elsewhere. Most, though not all, CMIP3 and CMIP5 models overestimate the observed warming trend in the tropical troposphere during the satellite period although observational uncertainties are large and outside the tropics and over the period of the radiosonde record beginning in 1961 there is better agreement between simulated and observed trends. (Sections 2.4.4, 9.4.1.4.2, 10.3.1.2, Figure 10.8)
6	Anthropogenic forcing dominated by the depletion of the ozone layer due to ozone depleting substances, has led to a detectable cooling of lower stratosphere temperatures since 1979.	*High*	*Very Likely*	Radiosonde data from 1958 and satellite data from 1979 to present. CCMVal, CMIP3 and CMIP5 simulations.	· A formal optimal detection attribution study using stratosphere resolving chemistry climate models and a detection study analysing the S/N ratio of the data record together with many separate modelling studies and observational studies. · Physical reasoning and model studies show very consistent understanding of observed evolution of stratospheric temperatures, consistent with formal detection and attribution results. · Studies agree in showing very strong cooling in stratosphere that can be explained only by anthropogenic forcings dominated by ozone depleting substances.	New generation of stratosphere resolving models appear to have adequate representation of lower stratospheric variability. Structure of stratospheric temperature trends and variations is reasonably well represented by models. CMIP5 models all include changes in stratospheric ozone while only about half of the models participating in CMIP3 include stratospheric ozone changes. (Sections 9.4.1.4.5, 10.3.1.2.2, Figures10.8 and 10.9)
7	Anthropogenic forcing, particularly GHGs and stratospheric ozone depletion has led to a detectable observed pattern of tropospheric warming and lower stratospheric cooling since 1961.	*High*	*Very likely*	Radiosonde data from 1958 and satellite data from 1979 to present.	· Attribution studies using CMIP3 and CMIP5 models. · Physical reasoning and modelling supports robust expectation of fingerprint of anthropogenic influence of tropospheric warming and lower stratospheric cooling which is robustly detected in multiple observational records. · Fingerprint of anthropogenic influence is detected in different measures of free atmosphere temperature changes including tropospheric warming, and a very clear identification of stratospheric cooling in models that include anthropogenic forcings.	Fingerprint of changes expected from physical understanding and as simulated by models is detected in observations. Understanding of stratospheric changes has improved since AR4. Understanding of observational uncertainty has improved although uncertainties remain particularly in the tropical upper troposphere. (Sections 2.4.4, 10.3.1.2.3, Figures 10.8, 10.9)

(continued on next page)

Table 10.1 (continued)

Result	(1) Statement about variable or property: time, season	(2) Confidence (*Very high*, *High*, *medium* or *low*, *very low*)	(3) Quantified measure of uncertainty where the probability of the outcome can be quantified (Likelihood given generally only if high or very high confidence)	(4) Data sources Observational evidence (Chapters 2 to 5); Models (Chapter 9)	(5) Type, amount, quality, consistency of evidence from attribution studies and degree of agreement of studies.	(6) Factors contributing to the assessments including physical understanding, observational and modelling uncertainty, and caveats.
8	Anthropogenic forcing has contributed to the observed changes in the frequency and intensity of daily temperatures extremes on the global scale since the mid-20th century.	*High*	*Very Likely*	Indices for frequency and intensity of extreme temperatures including annual maximum and annual minimum daily temperatures, over land areas of the World except parts of Africa, South America and Antarctica. CMIP3 and CMIP5 simulations, 1950–2005.	· Several studies including fingerprint time–space studies. · Detection of anthropogenic influence robustly seen in independent analysis using different statistical methods and different indices.	Expected from physical principles that changes in mean temperature should bring changes in extremes, confirmed by detection and attribution studies. New evidence since AR4 for detection of human influence on extremely warm daytime maximum temperatures and new evidence that influence of anthropogenic forcing can be separately detected from natural forcing. More limited observational data and greater observational uncertainties than for mean temperatures. (Section 10.6.1.1, Figure 10.17)
Oceans						
9	Anthropogenic forcings have made a substantial contribution to upper ocean warming (above 700 m) observed since the 1970s. This anthropogenic ocean warming has contributed to global sea level rise over this period through thermal expansion.	*High*	*Very likely*	Several observational data sets since the 1970s. CMIP3 and CMIP5 models.	· Several new attribution studies detect role of anthropogenic forcing on observed increase in ocean's global heat content with volcanic forcing also contributing to observed variability. · The evidence is very robust, and tested against known structural deficiencies in the observations, and in the models. · High levels of agreement across attribution studies and observation and model comparison studies. The strong physical relationship between thermosteric sea level and ocean heat content means that the anthropogenic ocean warming has contributed to global sea level rise over this period through thermal expansion.	New understanding of the structural errors in the temperature data sets has led to their correction which means that the unexplained multi-decadal scale variability reported in AR4 has largely been resolved as being spurious. The observations and climate simulations have similar trends (including anthropogenic and volcanic forcings) and similar decadal variability. The detection is well above S/N levels required at 1 and 5% significance levels. The new results show the conclusions to be very robust to structural uncertainties in observational data sets and transient climate simulations. (Sections 3.2.5, 10.4.1, 10.4.3, 13.3.6, Figure 10.14)
10	Anthropogenic forcing has contributed to sea level rise through melting glaciers and Greenland ice sheet.	*High*	Likely	Observational evidence of melting glaciers (Section 4.3) and ice sheets (Section 4.4). Global mean sea level budget closure to within uncertainties. (Section13.3.6)	· Several new mass balance studies quantifying glacier and ice sheet melt rates (Section 10.5.2) and their contributions to sea level rise. (Section 13.3)	Strong observational evidence of contribution from melting glaciers and *high confidence* in attribution of glacier melt to human influence. Increasing rates of ice sheet contributions albeit from short observational record (especially of Antarctic mass loss). Current climate models do not represent glacier and ice sheet processes. Natural variability of glaciers and ice sheets not fully understood. (Sections 10.4.3, 10.5.2)

(continued on next page)

Table 10.1 (continued)

Result	(1) Statement about variable or property: time, season	(2) Confidence (*Very high, High, medium* or *low, very low*)	(3) Quantified measure of uncertainty where the probability of the outcome can be quantified (Likelihood given generally only if high or very high confidence)	(4) Data sources Observational evidence (Chapters 2 to 5); Models (Chapter 9)	(5) Type, amount, quality, consistency of evidence from attribution studies and degree of agreement of studies.	(6) Factors contributing to the assessments including physical understanding, observational and modelling uncertainty, and caveats.
11	The observed ocean surface and sub-surface salinity changes since the 1960s are due, in part, to anthropogenic forcing.	*High*	*Very likely*	Oceans chapter (Section 3.3) and attribution studies in Section 10.4.2.	· Robust observational evidence for amplification of climatological patterns of surface salinity. · CMIP3 simulations show patterns of salinity change consistent with observations, but there are only a few formal attributions studies that include a full characterization of internal variability. · Physical understanding of expected patterns of change in salinity due to changes in water cycle support results from detection and attriution studies.	More than 40 studies of regional, global surface and subsurface salinity observations show patterns of change consistent with acceleration of hydrological water cycle. Climate models that include anthropgenic forcings show the same consistent pattern of surface salinity change. (Sections 3.3.5, 10.4.2, Figure 10.15)
12	Observed increase in surface ocean acidification since 1980s is a resulted of rising atmospheric CO_2	*Very high*	*Very likely*	Evidence from Section 3.8.2 and Box 3.2, Figure 3.18.	· Based on ocean chemistry, expert judgement, and many analyses of time series and other indirect measurements · Robust evidence from time series measurements. Measurements have a high degree of certainty (see Table 3.2) and instrumental records show increase in ocean acidity. · High agreement of the observed trends.	*Very high confidence*, based on the number of studies, the updates to earlier results in AR4, and the very well established physical understanding of gas exchange between atmosphere and surface ocean, and the sources of excess carbon dioxide in the atmosphere. Alternative processes and hypotheses can be excluded. (Section 3.8.2, Box 3.2, Section 10.4.4)
13	Observed pattern of decrease in oxygen content is, in part, attributable to anthropogenic forcing." To correctly read: Observed pattern of decrease in oxygen content from the 1960s to the 1990s is, in part, attributable to anthropogenic forcing.	*Medium*	*About as likely as not*	Evidence from Section 3.8.3 and attribution studies in Section 10.4.4.	· Qualitative expert judgement based on comparison of observed and expected changes in response to increasing CO_2. · Medium evidence. One specific global ocean study, many studies of hydrographic sections and repeat station data, high agreement across observational studies. · Medium agreement. One attribution study, and only limited regional and large-scale modelling and observation comparisons.	Physical understanding of ocean circulation and ventilation, and from the global carbon cycle, and from simulations of ocean oxygen concentrations from coupled bio-geochemical models with OAGCMs. Main uncertainty is observed decadal variability which is not well understood in global and regional inventories of dissolved oxygen in the oceans. (Section 10.4.4)
Water Cycle						
14	Global scale precipitation patterns over land have changed due to anthropogenic forcings including increases in NH mid to high latitudes.	*Medium*	N/A	Multiple observational data sets based on rain gauges over land, with coverage dominated by the NH. CMIP3 and CMIP5 models.	· Several land precipitation studies examining annual and seasonal precipitation. · Evidence for consistency between observed and modelled changes in global precipitation patterns over land regions with sufficient observations. · Medium degree of agreement of studies. Expected anthropogenic fingerprints of changes in zonal mean precipitation found in annual and some seasonal data with some sensitivity of attribution results to observational data set used.	Increases of precipitation at high latitudes of the NH are a robust feature of climate model simulations and are expected from process understanding. Global-land average long-term changes small at present time, whereas decadal variability over some land areas is large. Observations are very uncertain and poor coverage of precipitation expected to make fingerprint of changes much more indistinct. (Sections 2.5.1, 10.3.2.2, Figures 10.10 and 10.11)

(continued on next page)

Table 10.1 (continued)

Result	(1) Statement about variable or property: time, season	(2) Confidence (*Very high*, *High*, *medium* or *low*, *very low*)	(3) Quantified measure of uncertainty where the probability of the outcome can be quantified (Likelihood given generally only if high or very high confidence)	(4) Data sources Observational evidence (Chapters 2 to 5); Models (Chapter 9)	(5) Type, amount, quality, consistency of evidence from attribution studies and degree of agreement of studies.	(6) Factors contributing to the assessments including physical understanding, observational and modelling uncertainty, and caveats.
15	In land regions where observational coverage is sufficient for assessment, anthropogenic forcing has contributed to global-scale intensification of heavy precipitation over the second half of the 20th century.	*Medium*	N/A	Wettest 1-day and 5-day precipitation in a year obtained from rain guage observations, CMIP3 simulations.	· Only one detection and attribution study restricted to NH land where observations were available. · Study found stronger detectability for models with natural forcings but not able to differentiate anthropogenic from natural forcings. · Although only one formal detection and attribution study, observations of a general increase in heavy precipitation at the global scale agree with physical expectations.	Evidence for anthropogenic influence on various aspects of the hydrological cycle that implies extreme precipitation would be expected to increase. There are large observational uncertainties and poor global coverage which makes assessment difficult. (Section 10.6.1.2, Figure 10.11)
16	Anthropogenic contribution to atmospheric specific humidity since 1973.	*Medium*	N/A	Observations of atmospheric moisture content over ocean from satellite; observations of surface humidity from weather stations and radiosondes over land.	· Detection and attribution studies of both surface humidity from weather stations over land and atmospheric moisture content over oceans from satellites. · Detection of anthropogenic influence on atmospheric moisture content over oceans robust to choice of models. · Studies looking at different variables agree in detecting specific humidity changes.	Recent reductions in relative humidity over land and levelling off of specific humidity not fully understood. Length and quality of observational data sets limit detection and attribution and assimilated analyses not judged sufficiently reliable for detection and attribution. (Section 10.3.2.1)
Hemispheric Scale Changes; Basin Scale Changes						
Cryosphere						
17	A substantial part of glaciers mass loss since the 1960's is due to human influence.	*High*	*Likely*	Robust agreement from long-term glacier records. (Section 4.3.3)	· Several new recent studies since last assessment. · High agreement across a limited number of studies.	Well established records of glacier length, and better methods of estimating glacier volumes and mass loss. Better characterization of internal variability, and better understanding of the response to natural variability, and local land cover change. (Sections 4.3.3, 10.5.2)
18	Anthropogenic forcing has contributed to surface melting of the Greenland ice sheet since 1993.	*High*	*Likely*	Robust agreement across *in situ* and satellite derived estimates of surface mass balance (Section 4.4). Nested or downscaled model simulations show pattern of change consistent with warming.	· Several new studies since last assessment. · Robust evidence from different sources. · High agreement across a limited number of studies.	Documented evidence of surface mass loss. Uncertainty caused by poor characterization of the internal variability of the surface mass balance (strong dependence on atmospheric variability) that is not well represented in CMIP5 models. (Section 4.4.2, 10.5.2.1)
19	Antarctic ice sheet mass balance loss has a contribution from anthropogenic forcing.	*Low*	N/A	Observational evidence for Antarctic mass sheet loss is well established across a broad range of studies. (Section 4.4)	· No formal studies exist. Processes for mass loss for Antarctica are not well understood. Regional warming and changed wind patterns (increased westerlies, increase in the Southern Annular Mode (SAM)) could contribute to enhanced melt of Antarctica. High agreement in observational studies.	*Low confidence* assessment based on low scientific understanding. (Sections 4.4.2, 13.4, 10.5.2)
20	Anthropogenic forcing has contributed to the Arctic sea ice loss since 1979.	*High*	*Very likely*	Robust agreement across all observations. (Section 4.2)	· Multiple detection and attribution studies, large number of model simulations and data comparisons for instrumental record. · Robust set of studies of simulations of sea ice and observed sea ice extent. · High agreement between studies of sea ice simulations and observed sea ice extent.	*High confidence* based on documented observations of ice extent loss, and also good evidence for a significant reduction in sea ice volume. The physics of Arctic sea ice is well understood and consistent with the observed warming in the region, and from simulations of Arctic sea ice extent with anthropogenic forcing. (Sections 9.4.3, 10.5.1)

(continued on next page)

Table 10.1 *(continued)*

Result	(1) Statement about variable or property: time, season	(2) Confidence (*Very high, High, medium* or *low, very low*)	(3) Quantified measure of uncertainty where the probability of the outcome can be quantified (Likelihood given generally only if high or very high confidence)	(4) Data sources Observational evidence (Chapters 2 to 5); Models (Chapter 9)	(5) Type, amount, quality, consistency of evidence from attribution studies and degree of agreement of studies.	(6) Factors contributing to the assessments including physical understanding, observational and modelling uncertainty, and caveats.
21	Incomplete scientific explanations of the observed increase in Antarctic sea ice extent precludes attribution at this time.	N/A	N/A	The increase in sea ice extent in observations is robust, based on satellite measurements and ship-based measurements. (Section 4.5.2)	· No formal attribution studies. · Estimates of internal variability from CMIP5 simulations exceed observed sea ice variability. · Modelling studies have a low level of agreement for observed increase, and there are competing scientific explanations.	*Low confidence* based on low scientific understanding of the spatial variability and changes in the Antarctic sea ice. (Sections 4.5.2, 10.5.1, 9.4.3)
22	There is an anthropogenic component to observed reductions in NH snow cover since 1970	*High*	*Likely*	Observations show decrease in NH snow cover.	· Two snow cover attribution studies. · Decrease in snow cover in the observations are consistent among many studies. (Section 4.5.2, 4.5.3) · Reductions in observed snow cover inconsistent with internal variability and can be explained only by climate models that include anthropogenic forcings.	Expert judgement and attribution studies support the human influence on reduction in snow cover extent. (Sections 4.5.2, 4.5.3, 10.5.3)
Atmospheric Circulation and Patterns of Variability						
23	Human influence has altered sea level pressure patterns globally since 1951.	*High*	*Likely*	An observational gridded data set and reanalyses. Multiple climate models.	· A number of studies find detectable anthropogenic influence on sea level pressure patterns. · Detection of anthropogenic influence is found to be robust to currently sampled modelling and observational uncertainty.	Detectable anthropogenic influence on changes in sea level pressure patterns is found in several attribution studies that sample observational and modelling uncertainty. Observational uncertainties not fully sampled as results based largely on variants of one gridded data set although analyses based on reanalyses also support the finding of a detectable anthropogenic influence. (Section 10.3.3.4)
24	The positive trend in the SAM seen in austral summer since 1951 is due in part to stratospheric ozone depletion.	*High*	*Likely*	Measurements since 1957. Clear signal of SAM trend in December, January and February (DJF) is robust to observational uncertainty.	· Many studies comparing consistency of observed and modelled trends, and consistency of observed trend with simulated internal variability. · Observed trends are consistent with CMIP3 and CMIP5 simulations that include stratospheric ozone depletion. · Several studies show that the observed increase in the DJF SAM is inconsistent with simulated internal variability. High agreement of modelling studies that ozone depletion drives an increase in the DJF SAM index. There is *medium confidence* that GHGs have also played a role.	Consistent result of modelling studies is that the main aspect of the anthropogenically forced response on the DJF SAM is the impact of ozone depletion. The observational record is relatively short, observational uncertainties remain, and the DJF SAM trend since 1951 is only marginally inconsistent with internal variability in some data sets. (Section 10.3.3.3, Figure 10.13)
25	Stratospheric ozone depletion has contributed to the observed poleward shift of the Southern Hadley cell during austral summer.	*Medium*	N/A	Multiple observational lines of evidence for widening but large spread in the magnitude. Reanalysis suggest a southward shift of southern Hadley cell border during DJF which is also seen in CMIP3 and CMIP5 models.	· Consistent evidence for effects of stratospheric ozone depletion. · Evidence from modelling studies is robust that stratospheric ozone drives a poleward shift of the southern Hadley Cell border during austral summer. The magnitude of the shift is very uncertain and appears to be underestimated by models. There is *medium confidence* that GHGs have also played a role.	The observed magnitude of the tropical belt widening is uncertain. The contribution of internal climate variability to the observed poleward expansion of the Hadley circulation remains very uncertain. (Section 10.3.3.1, Figure 10.12)

(continued on next page)

Table 10.1 *(continued)*

Result	(1) Statement about variable or property: time, season	(2) Confidence (*Very high*, *High*, *medium* or *low*, *very low*)	(3) Quantified measure of uncertainty where the probability of the outcome can be quantified (Likelihood given generally only if high or very high confidence)	(4) Data sources Observational evidence (Chapters 2 to 5); Models (Chapter 9)	(5) Type, amount, quality, consistency of evidence from attribution studies and degree of agreement of studies.	(6) Factors contributing to the assessments including physical understanding, observational and modelling uncertainty, and caveats.
26	Attribution of changes in tropical cyclone activity to human influence.	*Low*	N/A	Incomplete and short observational records in most basins.	· Formal attribution studies on SSTs in tropics. However, mechanisms linking anthropogenically induced SST increases to changes in tropical cyclone activity poorly understood. · Attribution assessments depend on multi-step attribution linking anthropogenic influence to large-scale drivers and thence to tropical cyclone activity. · Low agreement between studies, medium evidence.	Insufficient observational evidence of multi-decadal scale variability. Physical understanding lacking. There remains substantial disagreement on the relative importance of internal variability, GHG forcing, and aerosols. (Sections 10.6.1.5, 14.6.1)
Millennium Time Scale						
27	External forcing contributed to NH temperature variability from 1400 to 1850, and from 850 to 1400.	*High* for period from 1400 to 1850; *medium* for period from 850 to 1400.	*Very likely* for period from 1400 to 1850.	See Chapter 5 for reconstructions; simulations from PMIP3 and CMIP5 models, with more robust detection results for 1400 onwards.	· A small number of detection and attribution studies and further evidence from climate modelling studies; comparison of models with reconstructions and results from data assimilation. · Robust agreement from a number of studies using a range of reconstructions and models (EBMs to ESMs) that models are able to reproduce key features of last seven centuries. · Detection results and simulations indicate a contribution by external drivers to the warm conditions in the 11th to 12th century, but cannot explain the warmth around the 10th century in some reconstructions.	Large uncertainty in reconstructions particularly for the first half of the millennium but good agreement between reconstructed and simulated large scale features from 1400. Detection of forced influence robust for a large range of reconstructions. Difficult to separate role of individual forcings. Results prior to 1400 much more uncertain, partly due to larger data and forcing uncertainty. (Sections 10.7.1, 10.7.2, 10.7.3)
Continental to Regional Scale Changes						
28	Anthropogenic forcing has made a substantial contribution to warming to each of the inhabited continents.	*High*	*Likely*	Robust observational evidence except for Africa due to poor sampling. Detection and attribution studies with CMIP3 and CMIP5 models.	· New studies since AR4 detect anthropogenic warming on continental and sub-continental scales. · Robust detection of human influence on continental scales agrees with global attribution of widespread warming over land to human influence. · Studies agree in detecting human influence on continental scales.	Anthropogenic pattern of warming widespread across all inhabited continents. Lower S/N ratios at continental scales than global scales. Separation of response to forcings more difficult at these scales. Models have greater errors in representation of regional details. (Section 10.3.1.1,4, Box 11.2)
29	Anthropogenic contribution to very substantial Arctic warming over the past 50 years.	*High*	*Likely*	Adequate observational coverage since 1950s. Detection and attribution analysis with CMIP3 models.	· Multiple models show amplification of Arctic temperatures from anthropogenic forcing. · Large positive Arctic-wide temperature anomalies in observations over last decade and models are consistent only when they include external forcing.	Large temperature signal relative to mid-latitudes but also larger internal variability and poorer observational coverage than at lower latitudes. Known multiple processes including albedo shifts and added heat storage contribute to faster warming than at lower latitudes. (Sections 10.3.1.1.4. 10.5.1.1)

(continued on next page)

Table 10.1 (continued)

Result	(1) Statement about variable or property: time, season	(2) Confidence (*Very high, High, medium* or *low, very low*)	(3) Quantified measure of uncertainty where the probability of the outcome can be quantified (Likelihood given generally only if high or very high confidence)	(4) Data sources Observational evidence (Chapters 2 to 5); Models (Chapter 9)	(5) Type, amount, quality, consistency of evidence from attribution studies and degree of agreement of studies.	(6) Factors contributing to the assessments including physical understanding, observational and modelling uncertainty, and caveats.
30	Human contribution to observed warming averaged over available stations over Antarctica.	*Low*	N/A	Poor observational coverage of Antarctica with most observations around the coast. Detection and attribution studies with CMIP3 and CMIP5 models.	· One optimal detection study, and some modelling studies. · Clear detection in one optimal detection study.	Possible contribution to changes from SAM increase. Residual when SAM induced changes are removed shows warming consistent with expectation due to anthropogenic forcing. High observational uncertainty and sparse data coverage (individual stations only mostly around the coast). (Sections 10.3.1.1.4, 2.4.1.1)
31	Contribution by forcing to reconstructed European temperature variability over last five centuries.	*Medium*	N/A	European seasonal temperatures from 1500 onwards.	· One detection and attribution study and several modelling studies. · Clear detection of external forcings in one study; robust volcanic signal seen in several studies (see also Chapter 5).	Robust volcanic response detected in Epoch analyses in several studies. Models reproduce low-frequency evolution when include external forcings. Uncertainty in overall level of variability, uncertainty in reconstruction particularly prior to late 17th century. (Sections 10.7.2, 5.5.1)
32	Anthropogenic forcing has contributed to temperature change in many sub-continental regions of the world.	*High*	*Likely*	Good observational coverage for many regions (e.g., Europe) and poor for others (e.g., Africa, Arctic). Detection and attribution studies with CMIP3 and CMIP5 models.	· A number of detection and attribution studies have analysed temperatures on scales from Giorgi regions to climate model grid box scale. · Many studies agree in showing that an anthropogenic signal is apparent in many sub-continental scale regions. In some sub-continental-scale regions circulation changes may have played a bigger role.	Larger role of internal variability at smaller scales relative to signal of climate change. In some regions observational coverage is poor. Local forcings and feedbacks as well as circulation changes are important in many regions and may not be well simulated in all regions. (Section 10.3.1.1.4, Box 11.2)
33	Human influence has substantially increased the probability of occurrence of heat waves in some locations.	*High*	*Likely*	Good observational coverage for some regions and poor for others (thus biasing studies to regions where observational coverage is good). Coupled modeling studies examining the effects of anthropogonic warming and the probability of occurrence of very warm seasonal temperatures and targeted experiments with models forced with prescribed sea surface temperatures.	· Multi-step attribution studies of some events including the Europe 2003, Western Russia 2010, and Texas 2011 heatwaves have shown an anthropogenic contribution to their occurrence probability, backed up by studies looking at the overall implications of increasing mean temperatures for the probability of exceeding seasonal mean temperature thresholds in some regions. · To infer the probability of a heatwave, extrapolation has to be made from the scales on which most attribution studies have been carried out to the spatial and temporal scales of heatwaves. · Studies agree in finding robust evidence for anthropogenic influence on increase in probability of occurrence of extreme seasonal mean temperatures in many regions.	In some instances, circulation changes could be more important than thermodynamic changes. This could be a possible confounding influence since much of the magnitude (as opposed to the probability of occurrence) of many heat waves is attributable to atmospheric flow anomalies. (Sections 10.6.1, 10.6.2)

10

References

AchutaRao, K. M., B. D. Santer, P. J. Gleckler, K. E. Taylor, D. W. Pierce, T. P. Barnett, and T. M. L. Wigley, 2006: Variability of ocean heat uptake: Reconciling observations and models. *J. Geophys. Res. Oceans*, **111**, C05019.

AchutaRao, K. M., et al., 2007: Simulated and observed variability in ocean temperature and heat content. *Proc. Natl. Acad. Sci. U.S.A.*, **104**, 10768–10773.

Ahlmann, H. W., 1948: The present climatic fluctuation. *Geogr. J.*, **112**, 165–195.

Aldrin, M., M. Holden, P. Guttorp, R. B. Skeie, G. Myhre, and T. K. Berntsen, 2012: Bayesian estimation of climate sensitivity based on a simple climate model fitted to observations of hemispheric temperatures and global ocean heat content. *Environmetrics*, **23**, 253–271.

Alekseev, G. V., A. I. Danilov, V. M. Kattsov, S. I. Kuz'mina, and N. E. Ivanov, 2009: Changes in the climate and sea ice of the Northern Hemisphere in the 20th and 21st centuries from data of observations and modeling. *Izvestiya Atmospheric and Oceanic Physics*, **45**, 675–686.

Alexander, L. V., and J. M. Arblaster, 2009: Assessing trends in observed and modelled climate extremes over Australia in relation to future projections. *Int. J. Climatol.*, **29**, 417–435.

Allan, R., and T. Ansell, 2006: A new globally complete monthly historical gridded mean sea level pressure dataset (HadSLP2): 1850–2004. *J. Clim.*, **19**, 5816–5842.

Allan, R. P., B. J. Soden, V. O. John, W. Ingram, and P. Good, 2010: Current changes in tropical precipitation. *Environ. Res. Lett.*, **5**, 025205.

Allen, M., 2011: In defense of the traditional null hypothesis: Remarks on the Trenberth and Curry WIREs opinion articles. *WIREs Clim. Change*, **2**, 931–934.

Allen, M. R., and S. F. B. Tett, 1999: Checking for model consistency in optimal fingerprinting. *Clim. Dyn.*, **15**, 419–434.

Allen, M. R., and W. J. Ingram, 2002: Constraints on future changes in climate and the hydrologic cycle. *Nature*, **419**, 224–232.

Allen, M. R., and P. A. Stott, 2003: Estimating signal amplitudes in optimal fingerprinting. Part I: Theory. *Clim. Dyn.*, **21**, 477–491.

Allen, M. R., P. A. Stott, J. F. B. Mitchell, R. Schnur, and T. L. Delworth, 2000: Quantifying the uncertainty in forecasts of anthropogenic climate change. *Nature*, **407**, 617–620.

Allen, M. R., D. J. Frame, C. Huntingford, C. D. Jones, J. A. Lowe, M. Meinshausen, and N. Meinshausen, 2009: Warming caused by cumulative carbon emissions towards the trillionth tonne. *Nature*, **458**, 1163–1166.

Allen, M. R., et al., 2006: Quantifying anthropogenic influence on recent near-surface temperature change. *Surv. Geophys.*, **27**, 491–544.

Allen, R. J., S. C. Sherwood, J. R. Norris, and C. S. Zender, 2012: Recent Northern Hemisphere tropical expansion primarily driven by black carbon and tropospheric ozone. *Nature*, **485**, 350–354.

Ammann, C. M., and E. R. Wahl, 2007: The importance of the geophysical context in statistical evaluations of climate reconstruction procedures. *Clim. Change*, **85**, 71–88.

Ammann, C. M., F. Joos, D. S. Schimel, B. L. Otto-Bliesner, and R. A. Tomas, 2007: Solar influence on climate during the past millennium: Results from transient simulations with the NCAR Climate System Model. *Proc. Natl. Acad. Sci. U.S.A.*, **104**, 3713–3718.

Andreae, M. O., C. D. Jones, and P. M. Cox, 2005: Strong present-day aerosol cooling implies a hot future. *Nature*, **435**, 1187–1190.

Andrews, O. D., N. L. Bindoff, P. R. Halloran, T. Ilyina, and C. Le Quéré, 2013: Detecting an external influence on recent changes in oceanic oxygen using an optimal fingerprinting method. *Biogeosciences*, **10**, 1799–1813.

Annan, J. D., and J. C. Hargreaves, 2006: Using multiple observationally-based constraints to estimate climate sensitivity. *Geophys. Res. Lett.*, **33**, L06704.

Annan, J. D., and J. C. Hargreaves, 2010: Reliability of the CMIP3 ensemble. *Geophys. Res. Lett.*, **37**, L02703.

Annan, J. D., and J. C. Hargreaves, 2011: On the generation and interpretation of probabilistic estimates of climate sensitivity. *Clim. Change*, **104**, 423–436.

Aoki, S., N. L. Bindoff, and J. A. Church, 2005: Interdecadal water mass changes in the Southern Ocean between 30 degrees E and 160 degrees E. *Geophys. Res. Lett.*, **32**, L07607.

Arkin, P. A., T. M. Smith, M. R. P. Sapiano, and J. Janowiak, 2010: The observed sensitivity of the global hydrological cycle to changes in surface temperature. *Environ. Res. Lett.*, **5**, 035201.

Bal, S., S. Schimanke, T. Spangehl, and U. Cubasch, 2011: On the robustness of the solar cycle signal in the Pacific region. *Geophys. Res. Lett.*, **38**, L14809.

Balachandran, N. K., D. Rind, P. Lonergan, and D. T. Shindell, 1999: Effects of solar cycle variability on the lower stratosphere and the troposphere. *J. Geophys. Res. Atmos.*, **104**, 27321–27339.

Balan Sarojini, B., P. Stott, E. Black, and D. Polson, 2012 Fingerprints of changes in annual and seasonal precipitation from CMIP5 models over land and ocean. *Geophys. Res. Lett.*, **39**, L23706.

Barnett, T. P., D. W. Pierce, and R. Schnur, 2001: Detection of anthropogenic climate change in the world's oceans. *Science*, **292**, 270–274.

Barnett, T. P., D. W. Pierce, K. Achutarao, P. Gleckler, B. Santer, J. Gregory, and W. Washington, 2005: Penetration of human-induced warming into the world's oceans. *Science*, **309**, 284–287.

Barnett, T. P., et al., 2008: Human-induced changes in the hydrology of the western United States. *Science*, **319**, 1080–1083.

Barriopedro, D., R. Garcia-Herrera, and R. Huth, 2008: Solar modulation of Northern Hemisphere blocking. *J. Geophys. Res. Atmos.*, **113**, D14118.

Beenstock, M., Y. Reingewertz, and N. Paldor, 2012: Polynomial cointegration tests of anthropogenic impact on global warming. *Earth Syst. Dyn. Discuss.*, **3**, 561–596.

Bender, F. A. M., A. M. L. Ekman, and H. Rodhe, 2010: Response to the eruption of Mount Pinatubo in relation to climate sensitivity in the CMIP3 models. *Clim. Dyn.*, **35**, 875–886.

Benestad, R. E., and G. A. Schmidt, 2009: Solar trends and global warming. *J. Geophys. Res. Atmos.*, **114**, D14101.

Bengtsson, L., and K. I. Hodges, 2011: On the evaluation of temperature trends in the tropical troposphere. *Clim. Dyn.*, **36** 419–430.

Bengtsson, L., V. A. Semenov, and O. M. Johannessen, 2004: The early twentieth-century warming in the Arctic—A possible mechanism. *J. Clim.*, **17**, 4045–4057.

Bengtsson, L., K. I. Hodges, E. Roeckner, and R. Brokopf, 2006: On the natural variability of the pre-industrial European climate. *Clim. Dyn.*, **27**, 743–760.

Berliner, L. M., A. L. Richard, and J. S. Dennis, 2000: Bayesian climate change assessment. *J. Clim.*, **13**, 3805–3820.

Bhend, J., and H. von Storch, 2008: Consistency of observed winter precipitation trends in northern Europe with regional climate change projections. *Clim. Dyn.*, **31**, 17–28.

Bindoff, N. L., and T. J. McDougall, 2000: Decadal changes along an Indian ocean section at 32 degrees S and their interpretation. *J. Phys. Oceanogr.*, **30**, 1207–1222.

Bindoff, N. L., et al., 2007: Observations: Oceanic climate change and sea level. In: *Climate Change 2007: The Physical Science Basis. Contribution of Working Group I to the Fourth Assessment Report of the Intergovernmental Panel on Climate Change* [Solomon, S., D. Qin, M. Manning, Z. Chen, M. Marquis, K. B. Averyt, M. Tignor and H. L. Miller (eds.)] Cambridge University Press, Cambridge, United Kingdom and New York, NY, USA, pp. 385–432.

Bintanja, R., G. J. van Oldenborgh, S. S. Drijfhout, B. Wouters, and C. A. Katsman, 2013: Important role for ocean warming and increased ice-shelf melt in Antarctic sea-ice expansion. *Nature Geosci.*, **6**, 376–379.

Birner, T., 2010: Recent widening of the tropical belt from global tropopause statistics: Sensitivities. *J. Geophys. Res. Atmos.*, **115**, D23109.

Bitz, C. M., and L. M. Polvani, 2012: Antarctic climate response to stratospheric ozone depletion in a fine resolution ocean climate model. *Geophys. Res. Lett.*, **39**, L20705.

Boer, G. J., 2011: The ratio of land to ocean temperature change under global warming *Clim. Dyn.*, **37**, 2253–2270.

Boer, G. J., M. Stowasser, and K. Hamilton, 2007: Inferring climate sensitivity from volcanic events. *Clim. Dyn.*, **28**, 481–502.

Bonfils, C., P. B. Duffy, B. D. Santer, T. M. L. Wigley, D. B. Lobell, T. J. Phillips, and C. Doutriaux, 2008: Identification of external influences on temperatures in California. *Clim. Change*, **87**, S43–S55.

Booth, B. B. B., N. J. Dunstone, P. R. Halloran, T. Andrews, and N. Bellouin, 2012: Aerosols implicated as a prime driver of twentieth-century North Atlantic climate variability. *Nature*, **484**, 228–232.

Bowerman, N. H. A., D. J. Frame, C. Huntingford, J. A. Lowe, and M. R. Allen, 2011: Cumulative carbon emissions, emissions floors and short-term rates of warming: Implications for policy. *Philos. Trans. R. Soc. A*, **369**, 45–66.

Boyer, T. P., S. Levitus, J. I. Antonov, R. A. Locarnini, and H. E. Garcia, 2005: Linear trends in salinity for the World Ocean, 1955–1998. *Geophys. Res. Lett.*, **32**, L01604.

Brandt, P., et al., 2010: Changes in the ventilation of the oxygen minimum zone of the tropical North Atlantic. *J. Phys. Oceanogr.*, **40**, 1784–1801.

Brayshaw, D. J., B. Hoskins, and M. Blackburn, 2008: The storm-track response to idealized SST perturbations in an aquaplanet GCM. *J. Atmos. Sci.*, **65**, 2842–2860.

Brohan, P., J. J. Kennedy, I. Harris, S. F. B. Tett, and P. D. Jones, 2006: Uncertainty estimates in regional and global observed temperature changes: A new data set from 1850. *J. Geophys. Res. Atmos.*, **111**, D12106.

Bromwich, D. H., J. P. Nicolas, A. J. Monaghan, M. A. Lazzara, L. M. Keller, G. A. Weidner, and A. B. Wilson, 2013: Central West Antarctica among the most rapidly warming regions on Earth. *Nature Geosci.*, **6**, 139–145.

Brönnimann, S., 2009: Early twentieth-century warming. *Nature Geosci.*, **2**, 735–736.

Brönnimann, S., et al., 2012: A multi-data set comparison of the vertical structure of temperature variability and change over the Arctic during the past 100 year. *Clim. Dyn.*, **39** 1577–1598.

Brown, R. D., and D. A. Robinson, 2011: Northern Hemisphere spring snow cover variability and change over 1922–2010 including an assessment of uncertainty. *Cryosphere*, **5**, 219–229.

Burke, E. J., S. J. Brown, and N. Christidis, 2006: Modeling the recent evolution of global drought and projections for the twenty-first century with the Hadley Centre climate model. *J. Hydrometeorol.*, **7** 1113–1125.

Butchart, N., et al., 2011: Multimodel climate and variability of the stratosphere. *J. Geophys. Res. Atmos.*, **116**, D05102.

Butler, A. H., D. W. Thompson, and R. Heikes, 2010: The steady-state atmospheric circulation response to climate change–like thermal forcings in a simple general circulation model. *J. Clim.*, **23**, 3474–3496.

Cai, M., and K.-K. Tung, 2012: Robustness of dynamical feedbacks from radiative forcing: 2% solar versus 2XCO2 experiments in an idealized GCM. *J. Atmos. Sci.*, **69**, 2256–2271.

Cattiaux, J., R. Vautard, C. Cassou, P. Yiou, V. Masson-Delmotte, and F. Codron, 2010: Winter 2010 in Europe: A cold extreme in a warming climate. *Geophys. Res. Lett.*, **37**, L20704.

Cayan, D. R., T. Das, D. W. Pierce, T. P. Barnett, M. Tyree, and A. Gershunov, 2010: Future dryness in the southwest US and the hydrology of the early 21st century drought. *Proc. Natl. Acad. Sci. U.S.A.*, **107**, 21271–21276.

Charlton-Perez, A. J., et al., 2013: On the lack of stratospheric dynamical variability in low-top versions of the CMIP5 models. *J. Geophys. Res. Atmos.*, **118**, 2494–2505.

Chinn, T., S. Winkler, M. J. Salinger, and N. Haakensen, 2005: Recent glacier advances in Norway and New Zealand: A comparison of their glaciological and meteorological causes. *Geograf. Annal. A*, **87**, 141–157.

Christiansen, B., and F. C. Ljungqvist, 2011: Reconstruction of the extratropical NH mean temperature over the last millennium with a method that preserves low-frequency variability. *J. Clim.*, **24**, 6013–6034.

Christidis, N., P. A. Stott, and S. J. Brown, 2011: The role of human activity in the recent warming of extremely warm daytime temperatures. *J. Clim.*, **24**, 1922–1930.

Christidis, N., P. A. Stott, G. C. Hegerl, and R. A. Betts, 2013: The role of land use change in the recent warming of daily extreme temperatures. *Geophys. Res. Lett.*, **40**, 589–594.

Christidis, N., P. A. Stott, F. W. Zwiers, H. Shiogama, and T. Nozawa, 2010: Probabilistic estimates of recent changes in temperature: A multi-scale attribution analysis. *Clim. Dyn.*, **34**, 1139–1156.

Christidis, N., P. A. Stott, F. W. Zwiers, H. Shiogama, and T. Nozawa, 2012a: The contribution of anthropogenic forcings to regional changes in temperature during the last decade. *Clim. Dyn.*, **39**, 1259–1274.

Christidis, N., P. A. Stott, G. S. Jones, H. Shiogama, T. Nozawa, and J. Luterbacher, 2012b: Human activity and anomalously warm seasons in Europe. *Int. J. Climatol.*, **32**, 225–239.

Chung, E. S., B. J. Soden, and B. J. Sohn, 2010: Revisiting the determination of climate sensitivity from relationships between surface temperature and radiative fluxes. *Geophys. Res. Lett.*, **37**, L10703.

Church, J., N. White, and J. Arblaster, 2005: Significant decadal-scale impact of volcanic eruptions on sea level and ocean heat content. *Nature*, **438**, 74–77.

Church, J. A., D. Monselesan, J. M. Gregory, and B. Marzeion, 2013: Evaluating the ability of process based models to project sealevel change. *Environ. Res. Lett.*, **8**, 014051.

Church, J. A., et al., 2011: Revisiting the Earth's sea-level and energy budgets from 1961 to 2008. *Geophys. Res. Lett.*, **38**, L18601.

Chylek, P., and U. Lohmann, 2008a: Reply to comment by Andrey Ganopolski and Thomas Schneider von Deimling on "Aerosol radiative forcing and climate sensitivity deduced from the Last Glacial Maximum to Holocene transition". *Geophys. Res. Lett.*, **35**, L23704.

Chylek, P., and U. Lohmann, 2008b: Aerosol radiative forcing and climate sensitivity deduced from the last glacial maximum to Holocene transition. *Geophys. Res. Lett.*, **35**, L04804.

Chylek, P., U. Lohmann, M. Dubey, M. Mishchenko, R. Kahn, and A. Ohmura, 2007: Limits on climate sensitivity derived from recent satellite and surface observations. *J. Geophys. Res. Atmos.*, **112**, D24S04.

Comiso, J. C., 2012: Large decadal decline in Arctic multiyear ice cover. *J. Clim.*, **25**, 1176–1193.

Comiso, J. C., and F. Nishio, 2008: Trends in the sea ice cover using enhanced and compatible AMSR-E, SSM/I, and SMMR data. *J. Geophys. Res. Oceans*, **113**, C02S07.

Compo, G. P., et al., 2011: The twentieth century reanalysis project. *Q. J. R. Meteorol. Soc.*, **137**, 1–28.

Cordero, E. C., and P. M. D. Forster, 2006: Stratospheric variability and trends in models used for the IPCC AR4. *Atmos. Chem. Phys.*, **6**, 5369–5380.

Crook, J. A., and P. M. Forster, 2011: A balance between radiative forcing and climate feedback in the modeled 20th century temperature response. *J. Geophys. Res. Atmos.*, **116**, D17108.

Crook, J. A., P. M. Forster, and N. Stuber, 2011: Spatial patterns of modeled climate feedback and contributions to temperature response and polar amplification. *J. Clim.*, **24**, 3575–3592.

Curry, J. A., and P. J. Webster, 2011: Climate science and the uncertainty monster. *Bull. Am. Meteorol. Soc.*, **92**, 1667–1682.

Curry, R., B. Dickson, and I. Yashayaev, 2003: A change in the freshwater balance of the Atlantic Ocean over the past four decades. *Nature*, **426**, 826–829

D'Arrigo, R., R. Wilson, and G. Jacoby, 2006: On the long-term context for late twentieth century warming. *J. Geophys. Res. Atmos.*, **111** D03103.

D'Arrigo, R., R. Wilson, and A. Tudhope, 2009: The impact of volcanic forcing on tropical temperatures during the past four centuries. *Nature Geosci.*, **2**, 51–56.

Dai, A., 2011: Drought under global warming: A review. *WIREs Clim. Change*, **2**, 45–65.

Dai, A., 2013: Increasing drought under global warming in observations and models. *Nature Clim. Change*, **3**, 52–58.

Dall'Amico, M., L. J. Gray, K. H. Rosenlof, A. A. Scaife, K. P. Shine, and P. A. Stott, 2010: Stratospheric temperature trends: Impact of ozone variability and the QBO. *Clim. Dyn.*, **34**, 381–398.

Davis, S. M., and K. H. Rosenlof, 2012: A multidiagnostic intercomparison of tropical-width time series using reanalyses and satellite observations. *J. Clim.*, **25**, 1061–1078.

Day, J. J., J. C. Hargreaves, J. D. Annan, and A. Abe-Ouchi, 2012: Sources of multi-decadal variability in Arctic sea ice extent. *Environ. Res. Lett.*, **7**, 034011.

Dean, S. M., and P. A. Stott, 2009: The effect of local circulation variability on the detection and attribution of New Zealand temperature trends. *J. Clim.*, **22**, 6217–6229.

DelSole, T., M. K. Tippett, and J. Shukla, 2011: A significant component of unforced multidecadal variability in the recent acceleration of global warming. *J. Clim.*, **24**, 909–926.

Delworth, T., V. Ramaswamy, and G. Stenchikov, 2005: The impact of aerosols on simulated ocean temperature and heat content in the 20th century. *Geophys. Res. Lett.*, **32**, L24709.

Delworth, T. L., and M. E. Mann, 2000: Observed and simulated multidecadal variability in the Northern Hemisphere. *Clim. Dyn.*, **16**, 661–676.

Deser, C., and H. Teng, 2008: Evolution of Arctic sea ice concentration trends and the role of atmospheric circulation forcing, 1979–2007. *Geophys. Res. Lett.*, **35**, L02504.

Dessler, A. E., 2010: A determination of the cloud feedback from climate variations over the past decade. *Science*, **330**, 1523–1527.

Dessler, A. E., 2011: Cloud variations and the Earth's energy budget. *Geophys. Res. Lett.*, **38**, L19701.

Dickson, R. R., et al., 2000: The Arctic Ocean response to the North Atlantic oscillation. *J. Clim.*, **13**, 2671–2696.

Ding, Q. H., E. J. Steig, D. S. Battisti, and M. Kuttel, 2011: Winter warming in West Antarctica caused by central tropical Pacific warming. *Nature Geosci.*, **4**, 398–403.

Doherty, S. J., et al., 2009: Lessons learned from IPCC AR4 scientific developments needed to understand, predict, and respond to climate change. *Bull. Am. Meteorol. Soc.*, **90**, 497–513.

Dole, R., et al., 2011: Was there a basis for anticipating the 2010 Russian heat wave? *Geophys. Res. Lett.*, **38**, L06702.

Domingues, C., J. Church, N. White, P. Gleckler, S. Wijffels, P. Barker, and J. Dunn, 2008: Improved estimates of upper-ocean warming and multi-decadal sea-level rise. *Nature*, **453**, 1090–1093.

Doscher, R., K. Wyser, H. E. M. Meier, M. W. Qian, and R. Redler, 2010: Quantifying Arctic contributions to climate predictability in a regional coupled ocean-ice-atmosphere model. *Clim. Dyn.*, **34**, 1157–1176.

Douglass, D. H., E. G. Blackman, and R. S. Knox, 2004: Corrigendum to: Temperature response of Earth to the annual solar irradiance cycle [Phys. Lett. A 323 (2004) 315]. *Phys. Lett. A*, **325**, 175–176.

Douville, H., A. Ribes, B. Decharme, R. Alkama, and J. Sheffield, 2013: Anthropogenic influence on multidecadal changes in reconstructed global evapotranspiration. *Nature Clim. Change*, **3**, 59–62.

Driscoll, S., A. Bozzo, L. J. Gray, A. Robock, and G. Stenchikov, 2012: Coupled Model Intercomparison Project 5 (CMIP5) simulations of climate following volcanic eruptions. *J. Geophys. Res.*, **117**, D17105.

Drost, F., and D. Karoly, 2012 Evaluating global climate responses to different forcings using simple indices. *Geophys. Res. Lett.*, **39**, L16701.

Drost, F., D. Karoly, and K. Braganza, 2012: Communicating global climate change using simple indices: An update *Clim. Dyn.*, **39**, 989–999.

Duarte, C. M., T. M. Lenton, P. Wadhams, and P. Wassmann, 2012: Abrupt climate change in the Arctic. *Nature Clim. Change*, **2**, 60–62.

Durack, P., S. Wijffels, and R. Matear, 2012: Ocean salinities reveal strong global water cycle intensification during 1950 to 2000. *Science*, **336**, 455–458.

Durack, P. J., and S. E. Wijffels, 2010: Fifty-year trends in global ocean salinities and their relationship to broad-scale warming. *J. Clim.*, **23**, 4342–4362.

Edwards, T. L., M. Crucifix, and S. P. Harrison, 2007: Using the past to constrain the future: How the palaeorecord can improve estimates of global warming. *Prog. Phys. Geogr.*, **31**, 481–500.

Elsner, J. B., 2006: Evidence in support of the climate change—Atlantic hurricane hypothesis. *Geophys. Res. Lett.*, **33**, L16705.

Elsner, J. B., J. P. Kossin, and T. H. Jagger, 2008: The increasing intensity of the strongest tropical cyclones. *Nature*, **455**, 92–95.

Emanuel, K., 2005: Increasing destructiveness of tropical cyclones over the past 30 years. *Nature*, **436**, 686–688.

Emanuel, K., S. Solomon, D. Folini, S. Davis, and C. Cagnazzo, 2013: Influence of tropical tropopause layer cooling on Atlantic hurricane activity. *J. Clim.*, **26**, 2288–2301.

Emerson, S., Y. W. Watanabe, T. Ono, and S. Mecking, 2004: Temporal trends in apparent oxygen utilization in the upper pycnocline of the North Pacific: 1980–2000. *J. Oceanogr.*, **60**, 139–147.

Engle, R. F., and C. W. J. Granger, 1987: Co-integration and error correction: Representation, estimation, and testing. *Econometrica*, **55**, 251–276.

Esper, J., et al., 2012: Orbital forcing of tree-ring data. *Nature Clim. Change*, **2**, 862–866.

Evan, A. T., G. R. Foltz, D. X. Zhang, and D. J. Vimont, 2011: Influence of African dust on ocean-atmosphere variability in the tropical Atlantic. *Nature Geosci.*, **4**, 762–765.

Evan, A. T., D. J. Vimont, A. K. Heidinger, J. P. Kossin, and R. Bennartz, 2009: The role of aerosols in the evolution of tropical North Atlantic Ocean temperature anomalies. *Science*, **324**, 778–781.

Eyring, V., et al., 2013: Long-term changes in tropospheric and stratospheric ozone and associated climate impacts in CMIP5 simulations. *J. Geophys. Res. Atmos.*, doi:10.1002/jgrd.50316.

Eyring, V., et al., 2006: Assessment of temperature, trace species, and ozone in chemistry-climate model simulations of the recent past. *J. Geophys. Res. Atmos.*, **111**, D22308.

Fernández-Donado, L., et al., 2013: Large-scale temperature response to external forcing in simulations and reconstructions of the last millennium. *Clim. Past*, **9**, 393–421.

Fettweis, X., G. Mabille, M. Erpicum, S. Nicolay, and M. Van den Broeke, 2011: The 1958–2009 Greenland ice sheet surface melt and the mid-tropospheric atmospheric circulation. *Clim. Dyn.*, **36**, 139–159.

Fettweis, X., B. Franco, M. Tedesco, J. H. van Angelen, J. T. M. Lenaerts, M. R. van den Broeke, and H. Gallée, 2013: Estimating the Greenland ice sheet surface mass balance contribution to future sea level rise using the regional atmospheric climate model MAR. *Cryosphere*, **7**, 469–489.

Feulner, G., 2011: Are the most recent estimates for Maunder Minimum solar irradiance in agreement with temperature reconstructions? *Geophys. Res. Lett.*, **38**, L16706.

Fischer, E. M., S. I. Seneviratne, P. L. Vidale, D. Luthi, and C. Schar, 2007: Soil moisture –atmosphere interactions during the 2003 European summer heat wave. *J. Clim.*, **20**, 5081–5099.

Fogt, R. L., J. Perlwitz, A. J. Monaghan, D. H. Bromwich, J. M. Jones, and G. J. Marshall, 2009: Historical SAM variability. Part II: Twentieth-century variability and trends from reconstructions, observations, and the IPCC AR4 models. *J. Clim.*, **22**, 5346–5365.

Folland, C. K., et al., 2013 High predictive skill of global surface temperature a year ahead. *Geophys. Res. Lett.*, **40**, 761–767.

Forest, C. E., and R. W. Reynolds, 2008: Climate change—Hot questions of temperature bias. *Nature*, **453**, 601–602.

Forest, C. E., P. H. Stone, and A. P. Sokolov, 2006: Estimated PDFs of climate system properties including natural and anthropogenic forcings. *Geophys. Res. Lett.*, **33**, L01705.

Forest, C. E., P. H. Stone, and A. P. Sokolov, 2008: Constraining climate model parameters from observed 20th century changes. *Tellus A*, **60**, 911–920.

Forest, C. E., M. R. Allen, P. H. Stone, and A. P. Sokolov, 2000: Constraining uncertainties in climate models using climate change detection techniques. *Geophys. Res. Lett.*, **27**, 569–572.

Forest, C. E., P. H. Stone, A. P. Sokolov, M. R. Allen, and M. D. Webster, 2002: Quantifying uncertainties in climate system properties with the use of recent climate observations. *Science*, **295**, 113–117.

Forster, P., et al., 2007: Changes in atmospheric constituents and in radiative forcing. In: *Climate Change 2007: The Physical Science Basis. Contribution of Working Group I to the Fourth Assessment Report of the Intergovernmental Panel on Climate Change* [Solomon, S., D. Qin, M. Manning, Z. Chen, M. Marquis, K. B. Averyt, M. Tignor and H. L. Miller (eds.)] Cambridge University Press, Cambridge, United Kingdom and New York, NY, USA, pp. 129–234.

Forster, P. M., T. Andrews, P. Good, J. M. Gregory, L. S. Jackson, and M. Zelinka, 2013 Evaluating adjusted forcing and model spread for historical and future scenarios in the CMIP5 generation of climate models. *J. Geophys. Res. Atmos.*, **118**, 1139–1150.

Forster, P. M., et al., 2011: Stratospheric changes and climate. In: *Scientific Assessment of Ozone Depletion: 2010*. Global Ozone Research and Monitoring Project-Report No. 52 [P. M. Forster and D. W. J. Thompson (eds.)]. World Meteorological Organization, Geneva, Switzerland, 516 pp.

Forster, P. M. D., and J. M. Gregory, 2006: The climate sensitivity and its components diagnosed from Earth Radiation Budget data. *J. Clim.*, **19**, 39–52.

Foster, G., and S. Rahmstorf, 2011: Global temperature evolution 1979–2010. *Environ. Res. Lett.*, **6**, 044022.

Foster, G., J. D. Annan, G. A. Schmidt, and M. E. Mann, 2008: Comment on "Heat capacity, time constant, and sensitivity of Earth's climate system" by S. E. Schwartz. *J. Geophys. Res. Atmos.*, **113**, D15102.

Fowler, H. J., and R. L. Wilby, 2010: Detecting changes in seasonal precipitation extremes using regional climate model projections: Implications for managing fluvial flood risk. *Water Resour. Res.*, **46**, W03525.

Frame, D. J., D. A. Stone, P. A. Stott, and M. R. Allen, 2006: Alternatives to stabilization scenarios. *Geophys. Res. Lett.*, **33**, L14707.

Frame, D. J., B. B. B. Booth, J. A. Kettleborough, D. A. Stainforth, J. M. Gregory, M. Collins, and M. R. Allen, 2005: Constraining climate forecasts: The role of prior assumptions. *Geophys. Res. Lett.*, **32**, L09702.

Francis, J. A., and S. J. Vavrus 2012: Evidence linking Arctic amplification to extreme weather in mid-latitudes. *Geophys. Res. Lett.*, **39**, L06801.

Frank, D., J. Esper, and E. R. Cook, 2007: Adjustment for proxy number and coherence in a large-scale temperature reconstruction. *Geophys. Res. Lett.*, **34**, L16709.

Frank, D. C., J. Esper, C. C. Raible, U. Buntgen, V. Trouet, B. Stocker, and F. Joos, 2010: Ensemble reconstruction constraints on the global carbon cycle sensitivity to climate. *Nature*, **463**, 527–530.

Franzke, C., 2010: Long-range dependence and climate noise characteristics of Antarctic temperature data. *J. Clim.*, **23**, 6074–6081.

Free, M., 2011: The seasonal structure of temperature trends in the tropical lower stratosphere. *J. Clim.*, **24**, 859–866.

Free, M., and J. Lanzante, 2009: Effect of volcanic eruptions on the vertical temperature profile in radiosonde data and climate models. *J. Clim.*, **22**, 2925–2939.

Friedlingstein, P., et al., 2006: Climate–carbon cycle feedback analysis: Results from the C4MIP model intercomparison. *J. Clim.*, **19**, 3337–3353.

Frierson, D. M. W., 2006: Robust increases in midlatitude static stability in simulations of global warming. *Geophys. Res. Lett.*, **33**, L24816.

Frierson, D. M. W., J. Lu, and G. Chen, 2007: Width of the Hadley cell in simple and comprehensive general circulation models. *Geophys. Res. Lett.*, **34**, L18804.

Fu, Q., and P. Lin, 2011: Poleward shift of subtropical jets inferred from satellite-observed lower stratospheric temperatures. *J. Clim.*, **24**, 5597–5603.

Fu, Q., S. Solomon, and P. Lin, 2010: On the seasonal dependence of tropical lower-stratospheric temperature trends. *Atmos. Chem. Phys.*, **10**, 2643–2653.

Fu, Q., S. Manabe, and C. M. Johanson, 2011: On the warming in the tropical upper troposphere: Models versus observations. *Geophys. Res. Lett.*, **38**, L15704.

Fu, Q., C. M. Johanson, J. M. Wallace, and T. Reichler, 2006: Enhanced mid-latitude tropospheric warming in satellite measurements. *Science*, **312**, 1179–1179.

Fyfe, J. C., 2006: Southern Ocean warming due to human influence. *Geophys. Res. Lett.*, **33**, L19701.

Fyfe, J. C., N. P. Gillett, and D. W. J. Thompson, 2010: Comparing variability and trends in observed and modelled global-mean surface temperature. *Geophys. Res. Lett.*, **37**, L16802.

Fyfe, J. C., N. P. Gillett, and G. J. Marshal, 2012: Human influence on extratropical Southern Hemisphere summer precipitation. *Geophys. Res. Lett.*, **39**, L23711.

Fyke, J., and M. Eby, 2012: Comment on "Climate sensitivity estimated from temperature reconstructions of the Last Glacial Maximum". *Science*, **337**, 1294.

Ganopolski, A., and T. S. von Deimling, 2008: Comment on "Aerosol radiative forcing and climate sensitivity deduced from the Last Glacial Maximum to Holocene transition" by Petr Chylek and Ulrike Lohmann. *Geophys. Res. Lett.*, **35**, L23703.

Gascard, J. C., et al., 2008: Exploring Arctic transpolar drift during dramatic sea ice retreat. *Eos Trans. Am. Geophys. Union*, **89**, 21–22.

Gay-Garcia, C., F. Estrada, and A. Sanchez, 2009: Global and hemispheric temperature revisited. *Clim. Change*, **94**, 333–349

Giles, K. A., S. W. Laxon, and A. L. Ridout, 2008: Circumpolar thinning of Arctic sea ice following the 2007 record ice extent minimum. *Geophys. Res. Lett.*, **35**, L22502.

Gillett, N. P., and P. A. Stott, 2009: Attribution of anthropogenic influence on seasonal sea level pressure. *Geophys. Res. Lett.*, **36**, L23709.

Gillett, N. A., and J. C. Fyfe, 2013: Annular Mode change in the CMIP5 simulations. *Geophys. Res. Lett.*, **40**, 1189–1193.

Gillett, N. P., R. J. Allan, and T. J. Ansell, 2005: Detection of external influence on sea level pressure with a multi-model ensemble. *Geophys. Res. Lett.*, **32**, L19714.

Gillett, N. P., P. A. Stott, and B. D. Santer, 2008a: Attribution of cyclogenesis region sea surface temperature change to anthropogenic influence. *Geophys. Res. Lett.*, **35**, L09707.

Gillett, N. P., G. C. Hegerl, M. R. Allen, and P. A. Stott, 2000: Implications of changes in the Northern Hemisphere circulation for the detection of anthropogenic climate change. *Geophys. Res. Lett.*, **27**, 993–996.

Gillett, N. P., F. W. Zwiers, A. J. Weaver, and P. A. Stott, 2003: Detection of human influence on sea-level pressure. *Nature*, **422**, 292–294.

Gillett, N. P., M. F. Wehner, S. F. B. Tett, and A. J. Weaver, 2004: Testing the linearity of the response to combined greenhouse and sulfate aerosol forcing. *Geophys. Res. Lett.*, **31**, L14201.

Gillett, N. P., V. K. Arora, G. M. Flato, J. F. Scinocca, and K. von Salzen, 2012: Improved constraints on 21st-century warming derived using 160 years of temperature observations. *Geophys. Res. Lett.*, **39**, L01704.

Gillett, N. P., V. K. Arora, D. Matthews, P. A. Stott, and M. R. Allen, 2013 Constraining the ratio of global warming to cumulative CO_2 emissions using CMIP5 simulations. *J. Clim.*, doi:10.1175/JCLI-D-12–00476.1.

Gillett, N. P., et al., 2008b: Attribution of polar warming to human influence. *Nature Geosci.*, **1**, 750–754.

Gillett, N. P., et al., 2011: Attribution of observed changes in stratospheric ozone and temperature. *Atmos. Chem. Phys.*, **11** 599–609.

Gleckler, P. J., T. M. L. Wigley, B. D. Santer, J. M. Gregory, K. AchutaRao, and K. E. Taylor, 2006: Volcanoes and climate: Krakatoa's signature persists in the ocean. *Nature*, **439**, 675–675.

Gleckler, P. J., et al., 2012: Human-induced global ocean warming on multidecadal timescales. *Nature Clim. Change*, **2**, 524–529.

Gong, D., and S. Wang, 1999: Definition of Antarctic oscillation index. *Geophys. Res. Lett.*, **26**, 459–462.

Goosse, H., W. Lefebvre, A. de Montety, E. Crespin, and A. H. Orsi, 2009: Consistent past half-century trends in the atmosphere, the sea ice and the ocean at high southern latitudes. *Clim. Dyn.*, **33**, 999–1016.

Goosse, H., J. Guiot, M. E. Mann, S. Dubinkina, and Y. Sallaz-Damaz, 2012a: The medieval climate anomaly in Europe: Comparison of the summer and annual mean signals in two reconstructions and in simulations with data assimilation. *Global Planet. Change*, **84–85**, 35–47.

Goosse, H., E. Crespin, A. de Montety, M. E. Mann, H. Renssen, and A. Timmermann, 2010: Reconstructing surface temperature changes over the past 600 years using climate model simulations with data assimilation. *J. Geophys. Res. Atmos.*, **115**, D09108.

Goosse, H., et al., 2012b: The role of forcing and internal dynamics in explaining the ''Medieval Climate Anomaly'. *Clim. Dyn.*, **39**, 2847–2866.

Gouretski, V., and K. Koltermann, 2007: How much is the ocean really warming? *Geophys. Res. Lett.*, **34**, L01610.

Graff, L. S., and J. H. LaCasce, 2012: Changes in the extratropical storm tracks in response to changes in SST in an GCM. *J. Clim.*, **25**, 1854–1870.

Grant, A. N., S. Bronnimann, T. Ewen, T. Griesser, and A. Stickler, 2009: The early twentieth century warm period in the European Arctic. *Meteorol. Z.*, **18** 425–432.

Graversen, R. G., and M. H. Wang, 2009: Polar amplification in a coupled climate model with locked albedo. *Clim. Dyn.*, **33**, 629–643.

Gray, L. J., et al., 2010: Solar influences on climate. *Rev. Geophys.*, **48** RG4001.

Gregory, J. M., 2000: Vertical heat transports in the ocean and their effect an time-dependent climate change. *Clim. Dyn.*, **16**, 501–515.

Gregory, J. M., 2010: Long-term effect of volcanic forcing on ocean heat content. *Geophys. Res. Lett.*, **37**, L22701.

Gregory, J. M., and P. M. Forster, 2008: Transient climate response estimated from radiative forcing and observed temperature change. *J. Geophys. Res. Atmos.*, **113**, D23105.

Gregory, J. M., J. A. Lowe, and S. F. B. Tett, 2006: Simulated global-mean sea level changes over the last half-millennium. *J. Clim.*, **19**, 4576–4591.

Gregory, J. M., C. D. Jones, P. Cadule, and P. Friedlingstein, 2009: Quantifying carbon cycle feedbacks. *J. Clim.*, **22**, 5232–5250.

Gregory, J. M., R. J. Stouffer, S. C. B. Raper, P. A. Stott, and N. A. Rayner, 2002: An observationally based estimate of the climate sensitivity. *J. Clim.*, **15**, 3117–3121.

Gregory, J. M., H. T. Banks, P. A. Stott, J. A. Lowe, and M. D. Palmer, 2004: Simulated and observed decadal variability in ocean heat content. *Geophys. Res. Lett.*, **31**, L15312.

Gregory, J. M., et al., 2012 Twentieth-century global-mean sea-level rise: Is the whole greater than the sum of the parts? *J. Clim.*, doi:10.1175/JCLI-D-12-00319.1.

Grist, J., et al., 2010: The roles of surface heat flux and ocean heat transport convergence in determining Atlantic Ocean temperature variability. *Ocean Dyn.*, **60**, 771–790.

Haam, E., and K. K. Tung, 2012: Statistics of solar cycle-La Niña connection: Correlation of two autocorrelated time series. *J. Atmos. Sci.*, **69** 2934–2939.

Haigh, J., M. Blackburn, and R. Day, 2005: The response of tropospheric circulation to perturbations in lower-stratospheric temperature. *J. Clim.*, **18**, 3672–3685.

Haigh, J. D., 1996: The impact of solar variability on climate. *Science*, **272** 981–984.

Haimberger, L., C. Tavolato, and S. Sperka, 2012: Homogenization of the global radiosonde temperature dataset through combined comparison with reanalysis background series and neighboring stations. *J. Clim.*, **25**, 8108–8131.

Han, W., et al., 2010: Patterns of Indian Ocean sea-level change in a warming climate. *Nature Geosci.*, **3**, 546–550.

Hanna, E., J. M. Jones, J. Cappelen, S. H. Mernild, L. Wood, K. Steffen, and P. Huybrechts, 2013: The influence of North Atlantic atmospheric and oceanic forcing effects on 1900–2010 Greenland summer climate and ice melt/runoff. *Int. J. Climatol.*, **33**, 862–880.

Hanna, E., et al., 2008: Increased runoff from melt from the Greenland ice sheet: A response to global warming. *J. Clim.*, **21**, 331–341.

Hannart, A., J. L. Dufresne, and P. Naveau, 2009: Why climate sensitivity may not be so unpredictable. *Geophys. Res. Lett.*, **36**, L16707.

Hansen, J., and S. Lebedeff, 1987: Global trends of measured surface air-temperature. *J. Geophys. Res. Atmos.*, **92**, 13345–13372.

Hansen, J., M. Sato, and R. Ruedy, 2012: Perception of climate change. *Proc. Natl. Acad. Sci. U.S.A.*, **109**, 14726–14727.

Hansen, J., M. Sato, P. Kharecha, and K. von Schuckmann, 2011: Earth's energy imbalance and implications. *Atmos. Chem. Phys.*, **11**, 13421–13449.

Hansen, J., et al., 2005a: Earth's energy imbalance: Confirmation and implications. *Science*, **308**, 1431–1435.

Hansen, J., et al., 2005b: Efficacy of climate forcings. *J. Geophys. Res. Atmos.*, **110**, D18104.

Hargreaves, J. C., and J. D. Annan, 2009: Comment on 'Aerosol radiative forcing and climate sensitivity deduced from the Last Glacial Maximum to Holocene transition', by P. Chylek and U. Lohmann, Geophys. Res. Lett., doi:10.1029/2007GL032759., 2008. Clim. Past, 5, 143–145.

Hargreaves, J. C., A. Abe-Ouchi, and J. D. Annan, 2007: Linking glacial and future climates through an ensemble of GCM simulations. *Clim. Past*, **3**, 77–87.

Hargreaves, J. C., J. D. Annan, M. Yoshimori, and A. Abe-Ouchi, 2012: Can the Last Glacial Maximum constrain climate sensitivity? *Geophys. Res. Lett.*, **39**, L24702.

Harris, G. R., D. M. H. Sexton, B. B. B. Booth, M. Collins, and J. M. Murphy, 2013: Probabilistic projections of transient climate change. *Clim. Dyn.*, doi:10.1007/s00382-012-1647-y.

Hasselmann, K., 1997: Multi-pattern fingerprint method for detection and attribution of climate change. *Clim. Dyn.*, **13**, 601–611.

Hegerl, G., and F. Zwiers, 2011: Use of models in detection and attribution of climate change. *WIREs Clim. Change*, **2**, 570–591.

Hegerl, G., J. Luterbacher, F. Gonzalez-Rouco, S. F. B. Tett, T. Crowley, and E. Xoplaki, 2011a: Influence of human and natural forcing on European seasonal temperatures. *Nature Geosci.*, **4**, 99–103.

Hegerl, G. C., F. W. Zwiers, and C. Tebaldi, 2011b: Patterns of change: Whose fingerprint is seen in global warming? *Environ. Res. Lett.*, **6**, 044025.

Hegerl, G. C., F. W. Zwiers, P. A. Stott, and V. V. Kharin, 2004: Detectability of anthropogenic changes in annual temperature and precipitation extremes. *J. Clim.*, **17**, 3683–3700.

Hegerl, G. C., T. J. Crowley, W. T. Hyde, and D. J. Frame, 2006: Climate sensitivity constrained by temperature reconstructions over the past seven centuries. *Nature*, **440**, 1029–1032.

Hegerl, G. C., P. Stott, S. Solomon, and F. W. Zwiers, 2011c: Comment on "Climate science and the uncertainty monster by J.A. Curry and P.J. Webster". *Bull. Am. Meteorol. Soc.*, **92**, 1683–1685.

Hegerl, G. C., T. J. Crowley, S. K. Baum, K. Y. Kim, and W. T. Hyde, 2003: Detection of volcanic, solar and greenhouse gas signals in paleo-reconstructions of Northern Hemispheric temperature. *Geophys. Res. Lett.*, **30**, 1242.

Hegerl, G. C., T. J. Crowley, M. Allen, W. T. Hyde, H. N. Pollack, J. Smerdon, and E. Zorita, 2007a: Detection of human influence on a new, validated 1500-year temperature reconstruction. *J. Clim.*, **20**, 650–666.

Hegerl, G. C., et al., 2010: Good practice guidance paper on detection and attribution related to anthropogenic climate change. In: *Meeting Report of the Intergovernmental Panel on Climate Change Expert Meeting on Detection and Attribution of Anthropogenic Climate Change* [T. F. Stocker, et al. (eds.)]. IPCC Working Group I Technical Support Unit, University of Bern, Bern, Switzerland, 8 pp.

Hegerl, G. C., et al., 2007b: Understanding and attributing climate change. In: *Climate Change 2007: The Physical Science Basis. Contribution of Working Group I to the Fourth Assessment Report of the Intergovernmental Panel on Climate Change* [Solomon, S., D. Qin, M. Manning, Z. Chen, M. Marquis, K. B. Averyt, M. Tignor and H. L. Miller (eds.)] Cambridge University Press, Cambridge, United Kingdom and New York, NY, USA, pp. 663–745.

Held, I. M., and B. J. Soden, 2006: Robust responses of the hydrological cycle to global warming. *J. Clim.*, **19**, 5686–5699.

Held, I. M., M. Winton, K. Takahashi, T. Delworth, F. R. Zeng, and G. K. Vallis, 2010: Probing the fast and slow components of global warming by returning abruptly to preindustrial forcing. *J. Clim.*, **23**, 2418–2427.

Helm, K. P., N. L. Bindoff, and J. A. Church, 2010: Changes in the global hydrological-cycle inferred from ocean salinity. *Geophys. Res. Lett.*, **37**, L18701.

Helm, K. P., N. L. Bindoff, and J. A. Church, 2011: Observed decreases in oxygen content of the global ocean. *Geophys. Res. Lett.*, **38**, L23602.

Henriksson, S. V., E. Arjas, M. Laine, J. Tamminen, and A. Laaksonen, 2010: Comment on 'Using multiple observationally-based constraints to estimate climate sensitivity' by J. D. Annan and J. C. Hargreaves, *Geophys. Res. Lett.*, doi:10.1029/2005GL025259, 2006. *Clim. Past*, **6**, 411–414.

Hidalgo, H. G., et al., 2009: Detection and attribution of streamflow timing changes to climate change in the Western United States. *J. Clim.*, **22**, 3838–3855.

Hodge, S. M., D. C. Trabant, R. M. Krimmel, T. A. Heinrichs, R. S. March, and E. G. Josberger, 1998: Climate variations and changes in mass of three glaciers in western North America. *J. Clim.*, **11**, 2161–2179.

Hoekema, D. J., and V. Sridhar, 2011: Relating climatic attributes and water resources allocation: A study using surface water supply and soil moisture indices in the Snake River basin, Idaho. *Water Resour. Res.*, **47**, W07536.

Hoerling, M., and A. Kumar, 2003: The perfect ocean for drought. *Science*, **299**, 691–694.

Hoerling, M., et al., 2013: Anatomy of an extreme event. *J. Clim.*, **26**, 2811–2832.

Hoerling, M. P., J. K. Eischeid, X.-W. Quan, H. F. Diaz, R. S. Webb, R. M. Dole, and D. R. Easterling, 2012: Is a transition to semipermanent drought conditions imminent in the U.S. great plains? *J. Clim.*, **25**, 8380–8386.

Holden, P. B., N. R. Edwards, K. I. C. Oliver, T. M. Lenton, and R. D. Wilkinson, 2010: A probabilistic calibration of climate sensitivity and terrestrial carbon change in GENIE-1. *Clim. Dyn.*, **35**, 785–806.

Holland, D. M., R. H. Thomas, B. De Young, M. H. Ribergaard, and B. Lyberth, 2008: Acceleration of Jakobshavn Isbrae triggered by warm subsurface ocean waters. *Nature Geosci.*, **1**, 659–664.

Hood, L. L., and R. E. Soukharev, 2012: The lower-stratospheric response to 11-yr solar forcing: Coupling to the troposphere-ocean response. *J. Atmos. Sci.*, **69**, 1841–1864.

Hosoda, S., T. Suga, N. Shikama, and K. Mizuno, 2009: Global surface layer salinity change detected by Argo and its implication for hydrological cycle intensification. *J. Oceanogr.*, **65**, 579–586.

Hu, Y., and Q. Fu, 2007: Observed poleward expansion of the Hadley circulation since 1979. *Atmos. Chem. Phys.*, **7**, 5229–5236.

Hu, Y. Y., C. Zhou, and J. P. Liu, 2011: Observational evidence for poleward expansion of the Hadley circulation. *Adv. Atmos. Sci.*, **28**, 33–44.

Hu, Y. Y., L. J. Tao, and J. P. Liu, 2013: Poleward expansion of the Hadley circulation in CMIP5 simulations. *Adv. Atmos. Sci.*, **30**, 790–795.

Huber, M., and R. Knutti, 2011: Anthropogenic and natural warming inferred from changes in Earth/'s energy balance. *Nature Geosci.*, **5**, 31–36.

Hudson, R. D., M. F. Andrade, M. B. Follette, and A. D. Frolov, 2006: The total ozone field separated into meteorological regimes—Part II: Northern Hemisphere mid-latitude total ozone trends. *Atmos. Chem. Phys.*, **6**, 5183–5191.

Huntingford, C., P. A. Stott, M. R. Allen, and F. H. Lambert, 2006: Incorporating model uncertainty into attribution of observed temperature change. *Geophys. Res. Lett.*, **33**, L05710.

Huss, M., and A. Bauder, 2009: 20th-century climate change inferred from four long-term point observations of seasonal mass balance. *Ann. Glaciol.*, **50**, 207–214.

Huss, M., R. Hock, A. Bauder, and M. Funk, 2010: 100-year mass changes in the Swiss Alps linked to the Atlantic Multidecadal Oscillation. *Geophys. Res. Lett.*, **37**, L10501.

Huybers, P., 2010: Compensation between model feedbacks and curtailment of climate sensitivity. *J. Clim.*, **23**, 3009–3018.

Imbers, J., A. Lopez, C. Huntingford, and M. R. Allen, 2013: Testing the robustness of the anthropogenic climate change detection statements using different empirical models. *J. Geophys. Res. Atmos.*, doi:10.1002/jgrd.50296.

Ineson, S., A. A. Scaife, J. R. Knight, J. C. Manners, N. M. Dunstone, L. J. Gray, and J. D. Haigh, 2011: Solar forcing of winter climate variability in the Northern Hemisphere. *Nature Geosci.*, **4**, 753–757.

Ingram, W. J., 2007: Detection and attribution of climate change, and understanding solar influence on climate. In: *Solar Variability and Planetary Climates* [Y. Calisesi, R.-M. Bonnet , L. Gray , J. Langen, and M. Lockwood (eds.)]. Springer Science+Business Media, New York, NY, USA, and Heidelberg, Germany, pp. 199–211.

IPCC, 2012: Managing the risks of extreme events and disasters to advance climate change adaptation. *A Special Report of Working Groups I and II of the Intergovernmental Panel on Climate Change* [C. B. Field et al. (eds.)]. *Cambridge University Press, Cambridge, UK, and New York, NY, USA,* 582.

Ishii, M., and M. Kimoto, 2009: Reevaluation of historical ocean heat content variations with time-varying XBT and MBT depth bias corrections. *J. Oceanogr.*, **65**, 287–299.

Jackson, J. M., E. C. Carmack, F. A. McLaughlin, S. E. Allen, and R. G. Ingram, 2010: Identification, characterization, and change of the near-surface temperature maximum in the Canada Basin, 1993–2008. *J. Geophys. Res.J. Geophys. Res. Oceans*, **115**, C05021.

Jacob, T., J. Wahr, W. T. Pfeffer, and S. Swenson, 2012: Recent contributions of glaciers and ice caps to sea level rise. *Nature*, **482** 514–518.

Jacobs, S. S., A. Jenkins, C. F. Giulivi, and P. Dutrieux, 2011: Stronger ocean circulation and increased melting under Pine Island Glacier ice shelf. *Nature Geosci.*, **4**, 519–523.

10

Jahn, A., et al., 2012: Late-twentieth-century simulation of Arctic sea-ice and ocean properties in the CCSM4. *J. Clim.*, **25**, 1431–1452.

Johannessen, O. M., et al., 2004: Arctic climate change: Observed and modelled temperature and sea-ice variability *Tellus A*, **56**, 559–560.

Johanson, C. M., and Q. Fu, 2009: Hadley cell widening: Model simulations versus observations. *J. Clim.*, **22**, 2713–2725.

Johnson, G. C., and A. H. Orsi, 1997: Southwest Pacific Ocean water-mass changes between 1968/69 and 1990/91. *J. Clim.*, **10**, 306–316.

Jones, G. S., and P. A. Stott, 2011: Sensitivity of the attribution of near surface temperature warming to the choice of observational dataset. *Geophys. Res. Lett.*, **38**, L21702.

Jones, G. S., S. F. B. Tett, and P. A. Stott, 2003: Causes of atmospheric temperature change 1960–2000: A combined attribution analysis. *Geophys. Res. Lett.*, **30**, 1228.

Jones, G. S., P. A. Stott, and N. Christidis, 2008: Human contribution to rapidly increasing frequency of very warm Northern Hemisphere summers. *J. Geophys. Res. Atmos.*, **113**, D02109.

Jones, G. S., N. Christidis, and P. A. Stott, 2011: Detecting the influence of fossil fuel and bio-fuel black carbon aerosols on near surface temperature changes. *Atmos. Chem. Phys.*, **11** 799–816.

Jones, G. S., M. Lockwood, and P. A. Stott, 2012: What influence will future solar activity changes over the 21st century have on projected global near surface temperature changes ? *J. Geophys. Res. Atmos.*, **117**, D05103.

Jones, G. S., P. A. Stott, and N. Christidis, 2013 Attribution of observed historical near surface temperature variations to anthropogenic and natural causes using CMIP5 simulations. *J. Geophys. Res. Atmos.*, doi:10.1002/jgrd.50239.

Jones, P. D., et al., 2001: Adjusting for sampling density in grid box land and ocean surface temperature time series. *J. Geophys. Res. Atmos.*, **106**, 3371–3380.

Joshi, M. M., and G. S. Jones, 2009: The climatic effects of the direct injection of water vapour into the stratosphere by large volcanic eruptions. *Atmos. Chem. Phys.*, **9**, 6109–6118.

Joughin, I., and R. B. Alley, 2011: Stability of the West Antarctic ice sheet in a warming world. *Nature Geosci.*, **4**, 506–513.

Juckes, M. N., et al., 2007: Millennial temperature reconstruction intercomparison and evaluation. *Clim. Past*, **3**, 591–609.

Jung, M., et al., 2010: Recent decline in the global land evapotranspiration trend due to limited moisture supply. *Nature*, **467**, 951–954.

Jungclaus, J. H., et al., 2010: Climate and carbon-cycle variability over the last millennium. *Clim. Past*, **6**, 723–737.

Kaplan, J. O., K. M. Krumhardt, and N. Zimmermann, 2009: The prehistoric and preindustrial deforestation of Europe. *Quat. Sci. Rev.*, **28**, 3016–3034.

Karoly, D. J., and Q. G. Wu, 2005: Detection of regional surface temperature trends. *J. Clim.*, **18**, 4337–4343.

Karoly, D. J., and P. A. Stott, 2006: Anthropogenic warming of central England temperature. *Atmos. Sci. Lett.*, **7** 81–85.

Karpechko, A. Y., N. P. Gillett, G. J. Marshall, and A. A. Scaife, 2008: Stratospheric influence on circulation changes in the Southern Hemisphere troposphere in coupled climate models. *Geophys. Res. Lett.*, **35**, L20806.

Kattsov, V. M., et al., 2010: Arctic sea-ice change: A grand challenge of climate science. *J. Glaciol.*, **56**, 1115–1121.

Kaufman, D. S., et al., 2009: Recent warming reverses long-term arctic cooling. *Science*, **325**, 1236–1239.

Kaufmann, R. K., and D. I. Stern, 1997: Evidence for human influence on climate from hemispheric temperature relations. *Nature*, **388**, 39–44.

Kaufmann, R. K., and D. I. Stern, 2002: Cointegration analysis of hemispheric temperature relations. *J. Geophys. Res. Atmos.*, **107**, 4012.

Kaufmann, R. K., H. Kauppi, and J. H. Stock, 2006: Emission, concentrations, & temperature: A time series analysis. *Clim. Change*, **77**, 249–278.

Kaufmann, R. K., H. Kauppi, M. L. Mann, and J. H. Stock, 2011: Reconciling anthropogenic climate change with observed temperature 1998–2008. *Proc. Natl. Acad. Sci. U.S.A.*, **108**, 11790–11793.

Kaufmann, R. K., H. Kauppi, M. L. Mann, and J. H. Stock, 2013: Does temperature contain a stochastic trend: Linking statistical results to physical mechanisms. *Clim. Change*, doi:10.1007/s10584-012-0683-2.

Kay, A. L., S. M. Crooks, P. Pall, and D. A. Stone, 2011a: Attribution of Autumn/Winter 2000 flood risk in England to anthropogenic climate change: A catchment-based study. *J. Hydrol.*, **406**, 97–112.

Kay, J. E., M. M. Holland, and A. Jahn, 2011b: Inter-annual to multi-decadal Arctic sea ice extent trends in a warming world. *Geophys. Res. Lett.*, **38**, L15708.

Keeling, R. F., and H. E. Garcia, 2002: The change in oceanic O_2 inventory associated with recent global warming. *Proc. Natl. Acad. Sci. U.S.A.*, **99**, 7848–7853.

Keeling, R. F., A. Kortzinger, and N. Gruber, 2010: Ocean deoxygenation in a warming world. *Annu. Rev. Mar. Sci.*, **2**, 199–229.

Kettleborough, J. A., B. B. B. Booth, P. A. Stott, and M. R. Allen, 2007: Estimates of uncertainty in predictions of global mean surface temperature. *J. Clim.*, **20**, 843–855.

Kharin, V. V., F. W. Zwiers, X. Zhang, and G. C. Hegerl, 2007: Changes in temperature and precipitation extremes in the IPCC ensemble of global coupled model simulations. *J. Clim.*, **20**, 1419–1444.

Kharin, V. V., F. W. Zwiers, X. Zhang, and M. Wehner, 2013: Changes in temperature and precipitation extremes in the CMIP5 ensemble. *Clim. Change*, doi:10.1007/s10584-013-0705-8.

Kiehl, J. T., 2007: Twentieth century climate model response and climate sensitivity. *Geophys. Res. Lett.*, **34**, L22710.

Kinnard, C., C. M. Zdanowicz, D. A. Fisher, E. Isaksson, A. Vernal, and L. G. Thompson, 2011: Reconstructed changes in Arctic sea ice cover over the past 1450 years. *Nature*, **479**, 509–513.

Kirk-Davidoff, D. B., 2009: On the diagnosis of climate sensitivity using observations of fluctuations. *Atmos. Chem. Phys.*, **9**, 813–822.

Knight, J., et al., 2009: Do global temperature trends over the last decade falsify climate predictions? In: State of the Climate in 2008. *Bull. Am. Meteorol. Soc.*, **90**, S22–S23.

Knight, J. R., C. K. Folland, and A. A. Scaife, 2006: Climate impacts of the Atlantic Multidecadal Oscillation. *Geophys. Res. Lett.*, **33**, L17706.

Knight, J. R., R. J. Allan, C. K. Folland, M. Vellinga, and M. E. Mann, 2005: A signature of persistent natural thermohaline circulation cycles in observed climate. *Geophys. Res. Lett.*, **32**, L20708.

Knutson, T. R., F. Zeng, and A. T. Wittenberg, 2013: Multi-model assessment of regional surface temperature trends. *J. Clim.*, doi:10.1175/JCLI-D-12-00567.1.

Knutson, T. R., et al., 2010: Tropical cyclones and climate change. *Nature Geosci.*, **3**, 157–163.

Knutti, R., 2008: Why are climate models reproducing the observed global surface warming so well? *Geophys. Res. Lett.*, **35**, L18704.

Knutti, R., and G. C. Hegerl, 2008: The equilibrium sensitivity of the Earth's temperature to radiation changes. *Nature Geosci.*, **1**, 735–743.

Knutti, R., and L. Tomassini, 2008: Constraints on the transient climate response from observed global temperature and ocean heat uptake. *Geophys. Res. Lett.*, **35**, L09701.

Knutti, R., S. Krähenmann, D. J. Frame, and M. R. Allen, 2008: Comment on "Heat capacity, time constant, and sensitivity of Earth's climate system" by S. E. Schwartz. *J. Geophys. Res. Atmos.*, **113**, D15103.

Kobashi, T., D. T. Shindell, K. Kodera, J. E. Box, T. Nakaegawa, and K. Kawamura, 2013: On the origin of multidecadal to centennial Greenland temperature anomalies over the past 800 yr. *Clim. Past*, **9**, 583–596.

Kobayashi, T., K. Mizuno, and T. Suga, 2012: Long-term variations of surface and intermediate waters in the southern Indian Ocean along 32°S. *J. Oceanogr.*, **68**, 243–265.

Kodama, C., and T. Iwasaki, 2009: Influence of the SST rise on baroclinic instability wave activity under an aquaplanet condition. *J. Atmos. Sci.*, **66**, 2272–2287.

Kodera, K., 2004: Solar influence on the Indian Ocean monsoon through dynamical processes. *Geophys. Res. Lett.*, **31**, L24209.

Kodera, K., 2006: The role of dynamics in solar forcing. *Space Sci. Rev.*, **23** 319–330.

Kodera, K., and Y. Kuroda, 2002: Dynamical response to the solar cycle. *J. Geophys. Res. Atmos.*, **107**, 4749.

Kodera, K., K. Coughlin, and O. Arakawa, 2007: Possible modulation of the connection between the Pacific and Indian Ocean variability by the solar cycle. *Geophys. Res. Lett.*, **34**, L03710.

Koehler, P., R. Bintanja, H. Fischer, F. Joos, R. Knutti, G. Lohmann, and V. Masson-Delmotte, 2010: What caused Earth's temperature variations during the last 800,000 years? Data-based evidence on radiative forcing and constraints on climate sensitivity. *Quat. Sci. Rev.*, **29**, 129–145.

Korhonen, H., K. S. Carslaw, P. M. Forster, S. Mikkonen, N. D. Gordon, and H. Kokkola, 2010: Aerosol climate feedback due to decadal increases in Southern Hemisphere wind speeds. *Geophys. Res. Lett.*, **37**, L02805.

Krakauer, N. Y., and I. Fung, 2008: Mapping and attribution of change in streamflow in the coterminous United States. *Hydrol. Earth Syst. Sci.*, **12**, 1111–1120.

Kuhlbrodt, T., and J. M. Gregory, 2012: Ocean heat uptake and its consequences for the magnitude of sea level rise and climate change. *Geophys. Res. Lett.*, **39**, L18608.

Kunkel, K. E., et al., 2008: Observed changes in weather and climate extremes. In: *Weather and Climate Extremes in a Changing Climate. Regions of Focus: North America, Hawaii, Caribbean, and U.S. Pacific Islands* [G. A. M. T. R. Karl, C. D. Miller, S. J. Hassol, A. M. Waple, and W. L. Murray (eds.)]. A Report by the U.S. Climate Change Science Program and the Subcommittee on Global Change Research, Washington, DC, pp. 35–80.

Kwok, R., and N. Untersteiner, 2011: The thinning of Arctic sea ice. *Physics Today*, **64**, 36–41.

Kwok, R., G. F. Cunningham, M. Wensnahan, I. Rigor, H. J. Zwally, and D. Yi, 2009: Thinning and volume loss of the Arctic Ocean sea ice cover: 2003–2008. *J. Geophys. Res.J. Geophys. Res. Oceans*, **114**, C07005.

L'Heureux, M., A. H. Butler, B. Jha, A. Kumar, and W. Q. Wang, 2010: Unusual extremes in the negative phase of the Arctic Oscillation during 2009. *Geophys. Res. Lett.*, **37**, L10704.

Lamarque, J.-F., et al., 2010: Historical (1850–2000) gridded anthropogenic and biomass burning emissions of reactive gases and aerosols: Methodology and application. *Atmos. Chem. Phys.*, **10**, 7017–7039.

Landrum, L., B. L. Otto-Bliesner, E. R. Wahl, A. Conley, P. J. Lawrence, N. Rosenbloom, and H. Teng, 2013: Last millennium climate and its variability in CCSM4. *J. Clim.*, **26**, 1085–1111.

Langen, P. L., and V. A. Alexeev, 2007: Polar amplification as a preferred response in an idealized aquaplanet GCM. *Clim. Dyn.*, **29**, 305–317.

Latif, M., et al., 2004: Reconstructing, monitoring, and predicting multidecadal-scale changes in the North Atlantic thermohaline circulation with sea surface temperature. *J. Clim.*, **17**, 1605–1614.

Laxon, S. W., et al., 2013: CryoSat-2 estimates of Arctic sea ice thickness and volume. *Geophys. Res. Lett.*, **40**, 732–737.

Lean, J. L., 2006: Comment on "Estimated solar contribution to the global surface warming using the ACRIM TSI satellite composite" by N. Scafetta and B. J. West. *Geophys. Res. Lett.*, **33**, L15701.

Lean, J. L., and D. H. Rind, 2008: How natural and anthropogenic influences alter global and regional surface temperatures: 1889 to 2006. *Geophys. Res. Lett.*, **35**, L18701.

Lean, J. L., and D. H. Rind, 2009: How will Earth's surface temperature change in future decades? *Geophys. Res. Lett.*, **36**, L15708.

Leclercq, P. W., and J. Oerlemans, 2011: Global and hemispheric temperature reconstruction from glacier length fluctuations. *Clim. Dyn.*, **38**, 1065–1079.

Ledoit, O., and M. Wolf, 2004: A well-conditioned estimator for large-dimensional covariance matrices. *J. Multivar. Anal.*, **88**, 365–411.

Legras, B., O. Mestre, E. Bard, and P. Yiou, 2010: A critical look at solar-climate relationships from long temperature series. *Clim. Past*, **6**, 745–758.

Leibensperger, E. M., et al., 2012: Climatic effects of 1950–2050 changes in US anthropogenic aerosols—Part 1: Aerosol trends and radiative forcing. *Atmos. Chem. Phys.*, **12**, 3333–3348.

Levitus, S., J. I. Antonov, T. P. Boyer, R. A. Locarnini, H. E. Garcia, and A. V. Mishonov, 2009: Global ocean heat content 1955–2008 in light of recently revealed instrumentation problems. *Geophys. Res. Lett.*, **36**, L07608.

Lewis, N., 2013: An objective Bayesian, improved approach for applying optimal fingerprint techniques to estimate climate sensitivity. *J. Clim.*, doi:10.1175/JCLI-D-12-00473.1.

Li, J. P., and J. L. X. L. Wang, 2003: A modified zonal index and its physical sense. *Geophys. Res. Lett.*, **30**, 1632.

Libardoni, A. G., and C. E. Forest, 2011: Sensitivity of distributions of climate system properties to the surface temperature dataset. *Geophys. Res. Lett.*, **38**, L22705.

Libardoni, A. G., and C. E. Forest, 2013: Correction to "Sensitivity of distributions of climate system properties to the surface temperature dataset". *Geophys. Res. Lett.*, doi:10.1002/grl.50480.

Lin, B., et al., 2010a: Estimations of climate sensitivity based on top-of-atmosphere radiation imbalance. *Atmos. Chem. Phys.*, **10**, 1923–1930.

Lin, P., Q. A. Fu, S. Solomon, and J. M. Wallace, 2010b: Temperature trend patterns in Southern Hemisphere high latitudes: Novel indicators of stratospheric change *J. Clim.*, **22**, 6325–6341.

Lindsay, R. W., J. Zhang, A. Schweiger, M. Steele, and H. Stern, 2009: Arctic sea ice retreat in 2007 follows thinning trend. *J. Clim.*, **22**, 165–176.

Lindzen, R. S., and Y. S. Choi, 2009: On the determination of climate feedbacks from ERBE data. *Geophys. Res. Lett.*, **36**, L16705.

Lindzen, R. S., and Y. S. Choi, 2011: On the observational determination of climate sensitivity and its implications. *Asia-Pacific J. Atmos. Sci.*, **47**, 377–390.

Liu, C., R. P. Allan, and G. J. Huffman, 2012: Co-variation of temperature and precipitation in CMIP5 models and satellite observations. *Geophys. Res. Lett.*, **39**, L13803.

Lockwood, M., 2008: Recent changes in solar outputs and the global mean surface temperature. III. Analysis of contributions to global mean air surface temperature rise. *Proc. R. Soc. London A*, **464**, 1387–1404.

Lockwood, M., 2012: Solar influence on global and regional climates. *Surv. Geophys.*, **33**, 503–534.

Lockwood, M., and C. Fröhlich, 2007: Recent oppositely directed trends in solar climate forcings and the global mean surface air temperature *Proc. R. Soc. London A*, **463**, 2447–2460.

Lockwood, M., and C. Fröhlich, 2008: Recent oppositely directed trends in solar climate forcings and the global mean surface air temperature: II. Different reconstructions of the total solar irradiance variation and dependence on response time scale. *Proc. R. Soc. London A*, **464**, 1367–1385.

Lockwood, M., R. G. Harrison, T. Woollings, and S. K. Solanki, 2010: Are cold winters in Europe associated with low solar activity? *Environ. Res. Lett.*, **5**, 024001.

Loehle, C., and N. Scaffetta, 2011: Climate change attribution using empirical decomposition of climatic data. *Open Atmos. Sci. J.*, **5**, 74–86.

Lott, F. C., et al., 2013: Models versus radiosondes in the free atmosphere: A new detection and attribution analysis of temperature. *J. Geophys. Res. Atmos.*, **118**, 2609–2619.

Lu, J., G. A. Vecchi, and T. Reichler, 2007: Expansion of the Hadley cell under global warming. *Geophys. Res. Lett.*, **34**, L06805.

Lu, J., G. Chen, and D. M. W. Frierson, 2008: Response of the zonal mean atmospheric circulation to El Nino versus global warming. *J. Clim.*, **21**, 5835–5851.

Lu, J., C. Deser, and T. Reichler, 2009: Cause of the widening of the tropical belt since 1958. *Geophys. Res. Lett.*, **36**, L03803.

Lucas, C., H. Nguyen, and B. Timbal, 2012: An observational analysis of Southern Hemisphere tropical expansion. *J. Geophys. Res. Atmos.*, **117**, D17112.

Lunt, D. J., A. M. Haywood, G. A. Schmidt, U. Salzmann, P. J. Valdes, and H. J. Dowsett, 2010: Earth system sensitivity inferred from Pliocene modelling and data. *Nature Geosci.*, **3**, 60–64.

Luterbacher, J., D. Dietrich, E. Xoplaki, M. Grosjean, and H. Wanner, 2004: European seasonal and annual temperature variability, trend, and extremes since 1500. *Science*, **303**, 1499–1503.

Mahajan, S., R. Zhang, and T. L. Delworth, 2011: Impact of the Atlantic Meridional Overturning Circulation (AMOC) on Arctic surface air temperature and sea ice variability. *J. Clim.*, **24**, 6573–6581.

Mahlstein, I., and R. Knutti, 2012 September Arctic sea ice predicted to disappear for 2oC global warming above present. *J. Geophys. Res. Atmos.*, **117**, D06104.

Mahlstein, I., G. Hegerl, and S. Solomon, 2012: Emerging local warming signals in observational data. *Geophys. Res. Lett.*, **39**, L21711.

Mahlstein, I., R. Knutti, S. Solomon, and R. W. Portmann, 2011: Early onset of significant local warming in low latitude countries. *Environ. Res. Lett.*, **6**, 034009.

Manabe, S., and R. T. Wetherald, 1975: The effects of doubling the CO_2 concentration on the climate of a General Circulation Model. *J. Atmos. Sci.*, **32**, 3–15.

Mankoff, K. D., S. S. Jacobs, S. M. Tulaczyk, and S. E. Stammerjohn, 2012: The role of Pine Island Glacier ice shelf basal channels in deep water upwelling, polynyas and ocean circulation in Pine Island Bay, Antarctica. *Ann. Glaciol.*, **53**, 123–128.

Mann, M. E., 2011: On long range temperature dependence in global surface temperature series. *Clim. Change*, **107**, 267–276.

Mann, M. E., and K. A. Emanuel, 2006: Atlantic hurricane trends linked to climate change. *Eos Trans. Am. Geophys. Union*, **87**, 233–238.

Mann, M. E., Z. H. Zhang, M. K. Hughes, R. S. Bradley, S. K. Miller, S. Rutherford, and F. B. Ni, 2008: Proxy-based reconstructions of hemispheric and global surface temperature variations over the past two millennia. *Proc. Natl. Acad. Sci. U.S.A.*, **105**, 13252–13257.

Mann, M. E., et al., 2009: Global signatures and dynamical origins of the Little Ice age and medieval climate anomaly. *Science*, **326**, 1256–1260.

Marcott, S. A., J. D. Shakun, P. U. Clark, and A. C. Mix, 2013: A reconstruction of regional and global temperature for the past 11,300 years. *Science*, **339**, 1198–1201

Marzeion, B., and A. Nesje, 2012: Spatial patterns of North Atlantic Oscillation influence on mass balance variability of European glaciers. *Cryosphere*, **6**, 661–673.

Maslanik, J. A., C. Fowler, J. Stroeve, S. Drobot, J. Zwally, D. Yi, and W. Emery, 2007: A younger, thinner Arctic ice cover: Increased potential for rapid, extensive sea-ice loss. *Geophys. Res. Lett.*, **34**, L24501.

Maslowski, W., J. C. Kinney, M. Higgins, and A. Roberts, 2012: The future of Arctic sea ice. *Annu. Rev. Earth Planet. Sci.*, **40**, 625–654.

Massey, N., T. Anna, R. C., F. E. L. Otto, S. Wilson, R. G. Jones, and M. R. Allen, 2012: Have the odds of warm November temperatures and of cold December temperatures in central england changed? *Bull. Am. Meteorol. Soc.*, **93**, 1057–1059.

Mastrandrea, M. D., et al., 2011: Guidance note for lead authors of the IPCC Fifth Assessment Report on consistent treatment of uncertainties. Intergovernmental Panel on Climate Change (IPCC), Geneva, Switzerland.

Matthes, K., Y. Kuroda, K. Kodera, and U. Langematz, 2006: Transfer of the solar signal from the stratosphere to the troposphere: Northern winter. *J. Geophys. Res. Atmos.*, **111** D06108.

Matthews, H. D., N. P. Gillett, P. A. Stott, and K. Zickfeld, 2009: The proportionality of global warming to cumulative carbon emissions. *Nature*, **459**, 829–U3.

Mazzarella, A., and N. Scafetta, 2012: Evidences for a quasi 60-year North Atlantic Oscillation since 1700 and its meaning for global climate change. *Theor. Appl. Climatol.*, **107**, 599–609.

McCracken, K. G., and J. Beer, 2007: Long-term changes in the cosmic ray intensity at Earth, 1428–2005. *J. Geophys. Res. Space Physics*, **112**, A10101.

McKitrick, R., and L. Tole, 2012: Evaluating explanatory models of the spatial pattern of surface climate trends using model selection and Bayesian averaging methods. *Clim. Dyn.*, **39**, 2867–2882.

McKitrick, R., S. McIntyre, and C. Herman, 2010: Panel and multivariate methods for tests of trend equivalence in climate data series. *Atmos. Sci. Lett.*, **11**, 270–277.

McLandress, C., J. Perlwitz, and T. G. Shepherd, 2012: Comment on "Tropospheric temperature response to stratospheric ozone recovery in the 21st century" by Hu et al. , 2011. *Atmos. Chem. Phys.*, **12**, 2533–2540.

McLandress, C., T. G. Shepherd, J. F. Scinocca, D. A. Plummer, M. Sigmond, A. I. Jonsson, and M. C. Reader, 2011: Separating the dynamical effects of climate change and ozone depletion. Part II: Southern Hemisphere troposphere. *J. Clim.*, **24**, 1850–1868.

Mecking, S., M. J. Warner, and J. L. Bullister, 2006: Temporal changes in pCFC-12 ages and AOU along two hydrographic sections in the eastern subtropical North Pacific. *Deep-Sea Res. Pt. I*, **53**, 169–187.

Meehl, G. A., and J. M. Arblaster, 2009: A lagged warm event-like response to peaks in solar forcing in the Pacific region. *J. Clim.*, **22**, 3647–3660.

Meehl, G. A., J. M. Arblaster, and C. Tebaldi, 2007a: Contributions of natural and anthropogenic forcing to changes in temperature extremes over the U.S. *Geophys. Res. Lett.*, **34**, L19709.

Meehl, G. A., J. M. Arblaster, G. Branstator, and H. van Loon, 2008: A coupled air-sea response mechanism to solar forcing in the Pacific region. *J. Clim.*, **21** 2883–2897.

Meehl, G. A., W. M. Washington, T. M. L. Wigley, J. M. Arblaster, and A. Dai, 2003: Solar and greenhouse gas forcing and climate response in the 20th century. *J. Clim.*, **16** 426–444.

Meehl, G. A., J. M. Arblaster, K. Matthes, F. Sassi, and H. van Loon, 2009: Amplifying the Pacific climate system response to a small 11-747year solar cycle forcing. *Science*, **325** 1114–1118.

Meehl, G. A., et al., 2007b: Global climate projections. In: *Climate Change 2007: The Physical Science Basis. Contribution of Working Group I to the Fourth Assessment Report of the Intergovernmental Panel on Climate Change* [Solomon, S., D. Qin, M. Manning, Z. Chen, M. Marquis, K. B. Averyt, M. Tignor and H. L. Miller (eds.)] Cambridge University Press, Cambridge, United Kingdom and New York, NY, USA, pp. 747–846.

Meinshausen, M., et al., 2009: Greenhouse-gas emission targets for limiting global warming to 2 °C. *Nature*, **458**, 1158–1162.

Merrifield, M., and M. Maltrud, 2011: Regional sea level trends due to a Pacific trade wind intensification. *Geophys. Res. Lett.*, **38**, L21605.

Meyssignac, B., D. Salas y Melia, M. Becker, W. Llovel, and A. Cazenave, 2012: Tropical Pacific spatial trend patterns in observed sea level: Internal variability and/or anthropogenic signature? *Clim. Past*, **8**, 787–802.

Miller, G. H., R. B. Alley, J. Brigham-Grette, J. J. Fitzpatrick, L. Polyak, M. C. Serreze, and J. W. C. White, 2010: Arctic amplification: Can the past constrain the future? *Quat. Sci. Rev.*, **29**, 1779–1790.

Miller, G. H., et al., 2012: Abrupt onset of the Little Ice Age triggered by volcanism and sustained by sea-ice/ocean feedbacks. *Geophys. Res. Lett.*, **39**, L02708.

Miller, R. L., G. A. Schmidt, and D. T. Shindell, 2006: Forced annular variations in the 20th century intergovernmental panel on climate change fourth assessment report models. *J. Geophys. Res. Atmos.*, **111**, D18101.

Mills, T. C., 2009: How robust is the long-run relationship between temperature and radiative forcing? *Clim. Change*, **94**, 351–361.

Min, S.-K., and A. Hense, 2006: A Bayesian assessment of climate change using multimodel ensembles. Part I: Global mean surface temperature. *J. Clim.*, **19**, 3237–3256.

Min, S.-K., and A. Hense, 2007: A Bayesian assessment of climate change using multimodel ensembles. Part II: Regional and seasonal mean surface temperatures. *J. Clim.*, **20**, 2769–2790.

Min, S.-K., and S.-W. Son, 2013: Multi-model attribution of the Southern Hemisphere Hadley cell widening: Major role of ozone depletion. *J. Geophys. Res. Atmos.*, **118**, 3007–3015.

Min, S.-K., X. B. Zhang, and F. Zwiers, 2008a: Human-induced arctic moistening. *Science*, **320**, 518–520.

Min, S.-K., X. B. Zhang, F. W. Zwiers, and T. Agnew, 2008b: Human influence on Arctic sea ice detectable from early 1990s onwards. *Geophys. Res. Lett.*, **35**, L21701.

Min, S.-K., X. Zhang, F. W. Zwiers, and G. C. Hegerl, 2011: Human contribution to more intense precipitation extremes. *Nature*, **470**, 378–381.

Min, S.-K., X. Zhang, F. W. Zwiers, P. Friederichs, and A. Hense, 2008c: Signal detectability in extreme precipitation changes assessed from twentieth century climate simulations. *Clim. Dyn.*, **32**, 95–111.

Min, S.-K., X. Zhang, F. Zwiers, H. Shiogama, Y.-S. Tung, and M. Wehner, 2013: Multi-model detection and attribution of extreme temperature changes. *J. Clim.*, doi:10.1175/JCLI-D-12-00551.w.

Misios, S., and H. Schmidt, 2012: Mechanisms involved in the amplification of the 11-yr solar cycle signal in the Tropical Pacific ocean. *J. Clim.*, **25**, 5102–5118.

Mitchell, D. M., P. A. Stott, L. J. Gray, F. C. Lott, N. Butchart, S. C. Hardiman, and S. M. Osprey, 2013: The impact of stratospheric resolution on the detectability of climate change signals in the free atmosphere. *Geophys. Res. Lett.*, **40**, 937–942.

Miyazaki, C., and T. Yasunari, 2008: Dominant interannual and decadal variability of winter surface air temperature over Asia and the surrounding oceans. *J. Clim.*, **21**, 1371–1386.

Moberg, A., D. M. Sonechkin, K. Holmgren, N. M. Datsenko, and W. Karlen, 2005: Highly variable Northern Hemisphere temperatures reconstructed from low- and high-resolution proxy data. *Nature*, **433**, 613–617.

Mölg, T., N. J. Cullen, D. R. Hardy, M. Winkler, and G. Kaser, 2009: Quantifying climate change in the tropical midtroposphere over East Africa from glacier shrinkage on Kilimanjaro. *J. Clim.*, **22**, 4162–4181.

Mölg, T., M. Großhauser, A. Hemp, M. Hofer, and B. Marzeion, 2012: Limited forcing of glacier loss through land-cover change on Kilimanjaro. *Nature Clim. Change*, **2**, 254–258.

Morak, S., G. C. Hegerl, and J. Kenyon, 2011: Detectable regional changes in the number of warm nights. *Geophys. Res. Lett.*, **38**, L17703.

Morak, S., G. C. Hegerl, and N. Christidis, 2013: Detectable changes in the frequency of temperature extremes. *J. Clim.*, **26**, 1561–1574.

Morgenstern, O., et al., 2010: Anthropogenic forcing of the Northern Annular Mode in CCMVal-2 models. *J. Geophys. Res. Atmos.*, **115**, D00M03.

Morice, C. P., J. J. Kennedy, N. A. Rayner, and P. D. Jones, 2012: Quantifying uncertainties in global and regional temperature change using an ensemble of observational estimates: The HadCRUT4 data set. *J. Geophys. Res. Atmos.*, **117**, D08101.

Murphy, D. M., and P. M. Forster, 2010: On the accuracy of deriving climate feedback parameters from correlations between surface temperature and outgoing radiation. *J. Clim.*, **23** 4983–4988.

Murphy, D. M., S. Solomon, R. W. Portmann, K. H. Rosenlof, P. M. Forster, and T. Wong, 2009: An observationally based energy balance for the Earth since 1950. *J. Geophys. Res. Atmos.*, **114**, D17107.

Nagato, Y., and H. L. Tanaka, 2012: Global warming trend without the contribution from decadal variability of the Arctic oscillation. *Polar Sci.*, **6**, 15–22.

Nakanowatari, T., K. I. Ohshima, and M. Wakatsuchi, 2007: Warming and oxygen decrease of intermediate water in the northwestern North Pacific, originating from the Sea of Okhotsk, 1955–2004. *Geophys. Res. Lett.*, **34**, L04602.

National Research Council, 2012: *The Effects of Solar Variability on Earth's Climate: A Workshop Report.* The National Academies Press, Washington, DC, 70 pp.

Nesje, A., O. Lie, and S. O. Dahl, 2000: Is the North Atlantic Oscillation reflected in Scandinavian glacier mass balance records? *J. Quat. Sci.*, **15**, 587–601.

Nghiem, S. V., I. G. Rigor, D. K. Perovich, P. Clemente-Colon, J. W. Weatherly, and G. Neumann, 2007: Rapid reduction of Arctic perennial sea ice. *Geophys. Res. Lett.*, **34**, L19504.

Nguyen, H., B. Timbal, A. Evans, C. Lucas, and I. Smith, 2013: The Hadley circulation in reanalyses: Climatology, variability and change. *J. Clim.*, **26**, 3357–3376.

Noake, K., D. Polson, G. Hegerl, and X. Zhang, 2012: Changes in seasonal land precipitation during the latter 20th Century. *Geophys. Res. Lett.*, **39**, L03706.

North, G. R., and M. J. Stevens, 1998: Detecting climate signals in the surface temperature record. *J. Clim.*, **11**, 563–577.

Notz, D., and J. Marotzke, 2012: Observations reveal external driver for Arctic sea-ice retreat. *Geophys. Res. Lett.*, **39**, L08502.

Nussbaumer, S. U., and H. J. Zumbühl, 2012: The Little Ice Age history of the Glacier des Bossons (Mont Blanc massif, France): A new high-resolution glacier length curve based on historical documents. *Clim. Change*, **111**, 301–334.

O'Gorman, P. A., and T. Schneider, 2008: Energy of midlatitude transient eddies in idealized simulations of changed climates. *J. Clim.*, **21**, 5797–5806.

O'Gorman, P. A., 2010: Understanding the varied response of the extratropical storm tracks to climate change. *Proc. Natl. Acad. Sci. U.S.A.*, **107**, 19176–19180.

Oerlemans, J., 2005: Extracting a climate signal from 169 glacier records. *Science*, **308**, 675–677.

Olson, R., R. Sriver, M. Goes, N. M. Urban, H. D. Matthews, M. Haran, and K. Keller, 2012: A climate sensitivity estimate using Bayesian fusion of instrumental observations and an Earth System model. *J. Geophys. Res. Atmos.*, **117**, D04103.

Ono, T., T. Midorikawa, Y. W. Watanabe, K. Tadokoro, and T. Saino, 2001: Temporal increases of phosphate and apparent oxygen utilization in the subsurface waters of western subarctic Pacific from 1968 to 1998. *Geophys. Res. Lett.*, **28**, 3285–3288.

Otto-Bliesner, B. L., et al., 2009: A comparison of PMIP2 model simulations and the MARGO proxy reconstruction for tropical sea surface temperatures at last glacial maximum. *Clim. Dyn.*, **32**, 799–815.

Otto, A., et al., 2013: Energy budget constraints on climate response. *Nature Geosci.*, **6**, 415–416.

Otto, F. E. L., N. Massey, G. J. van Oldenborgh, R. G. Jones, and M. R. Allen, 2012: Reconciling two approaches to attribution of the 2010 Russian heat wave. *Geophys. Res. Lett.*, **39**, L04702.

Overland, J. E., 2009: The case for global warming in the Arctic. In: *Influence of Climate Change on the Changing Arctic and Sub-Arctic Conditions. NATO Science for Peace and Security Series C: Environmental Security* [J. C. J. Nihoul and A. G. Kostianoy (eds.)]. Springer Science+Business Media, Dordrecht, Netherlands, pp. 13–23.

Overland, J. E., and M. Wang, 2013: When will the summer arctic be nearly sea ice free? *Geophys. Res. Lett.*, doi:10.1002/grl.50316.

Overland, J. E., M. Wang, and S. Salo, 2008: The recent Arctic warm period. *Tellus A*, **60**, 589–597.

Overland, J. E., K. R. Wood, and M. Wang, 2011: Warm Arctic-cold continents: Climate impacts of the newley open Arctic sea. *Polar Res.*, **30**, 15787.

Overland, J. E., J. A. Francis, E. Hanna, and W. M., 2012: The recent shift in early summer Arctic atmospheric circulation. *Geophys. Res. Lett.*, **39**, L19804.

Oza, S. R., R. K. K. Singh, N. K. Vyas, and A. Sarkar, 2011a: Spatio-Temporal analysis of melting onset dates of sea–ice in the Arctic. *Indian J. Geo-Mar. Sci.*, **40**, 497–501.

Oza, S. R., R. K. K. Singh, A. Srivastava, M. K. Dash, I. M. L. Das, and N. K. Vyas, 2011b: Inter-annual variations observed in spring and summer antarctic sea ice extent in recent decade. *Mausam*, **62**, 633–640.

Padilla, L. E., G. K. Vallis, and C. W. Rowley, 2011: Probabilistic estimates of transient climate sensitivity subject to uncertainty in forcing and natural variability. *J. Clim.*, **24**, 5521–5537.

Pagani, M., K. Caldeira, R. Berner, and D. J. Beerling, 2009: The role of terrestrial plants in limiting atmospheric CO_2 decline over the past 24 million years. *Nature*, **460**, 85–88.

PAGES 2k Consortium, 2013: Continental-scale temperature variability during the past two millennia. *Nature Geosci.*, **6**, 339–346.

Palastanga, V., G. van der Schrier, S. L. Weber, T. Kleinen, K. R. Briffa, and T. J. Osborn, 2011: Atmosphere and ocean dynamics: Contributors to the Little Ice Age and Medieval Climate Anomaly. *Clim. Dyn.*, **36**, 973–987.

Paleosens Members, 2012: Making sense of palaeoclimate sensitivity. *Nature*, **491**, 683–691.

Pall, P., et al., 2011: Anthropogenic greenhouse gas contribution to UK autumn flood risk. *Nature*, **470**, 382–385.

Palmer, M. D., S. A. Good, K. Haines, N. A. Rayner, and P. A. Stott, 2009: A new perspective on warming of the global oceans. *Geophys. Res. Lett.*, **36**, L20709.

Penner, J. E., M. Wang, A. Kumar, L. Rotstayn, and B. Santer, 2007: Effect of black carbon on mid-troposphere and surface temperature trends. In: *Human-Induced Climate Change: An Interdisciplinary Assessment* [M. Schlesinger, et al. (ed.)], Cambridge University Press, Cambridge, United Kingdom, and New York, NY, USA, pp. 18–33.

Peterson, T. C., P. A. Stott, and S. Herring, 2012: Explaining extreme events of 2011 from a climate perspective. *Bull. Am. Meteorol. Soc.*, **93**, 1041–1067.

Pierce, D. W., T. P. Barnett, B. D. Santer, and P. J. Gleckler, 2009: Selecting global climate models for regional climate change studies. *Proc. Natl. Acad. Sci. U.S.A.*, **106**, 8441–8446.

Pierce, D. W., P. J. Gleckler, T. P. Barnett, B. D. Santer, and P. J. Durack, 2012: The fingerprint of human-induced changes in the ocean's salinity and temperature fields. *Geophys. Res. Lett.*, **39**, L21704.

Pierce, D. W., T. P. Barnett, K. AchutaRao, P. Gleckler, J. Gregory, and W. Washington, 2006: Anthropogenic warming of the oceans: Observations and model results. *J. Clim.*, **19**, 1873–1900.

Pierce, D. W., et al., 2008: Attribution of declining Western U.S. snowpack to human effects. *J. Clim.*, **21**, 6425–6444.

Pitman, A. J., et al., 2009: Uncertainties in climate responses to past land cover change: First results from the LUCID intercomparison study. *Geophys. Res. Lett.*, **36**, L14814.

Po-Chedley, S., and Q. Fu, 2012: Discrepancies in tropical upper tropospheric warming between atmospheric circulation models and satellites. *Environ. Res. Lett.*, **7**, 044018.

Polson, D., G. C. Hegerl, X. Zhang, and T. J. Osborn, 2013: Causes of robust seasonal land precipitation changes. *J. Clim.*, doi:10.1175/JCLI-D-12-00474.1.

Polvani, L. M., D. W. Waugh, G. J. P. Correa, and S. W. Son, 2011: Stratospheric ozone depletion: The main driver of twentieth-century atmospheric circulation changes in the southern hemisphere. *J. Clim.*, **24**, 795–812.

Polyakov, I. V., J. E. Walsh, and R. Kwok, 2012: Recent changes of Arctic multiyear sea ice coverage and the likely causes. *Bull. Am. Meteorol. Soc.*, **93**, 145–151.

Polyakov, I. V., U. S. Bhatt, H. L. Simmons, D. Walsh, J. E. Walsh, and X. Zhang, 2005: Multidecadal variability of North Atlantic temperature and salinity during the twentieth century. *J. Clim.*, **18**, 4562–4581.

Polyakov, I. V., et al., 2003: Variability and trends of air temperature and pressure in the maritime Arctic, 1875–2000. *J. Clim.*, **16**, 2067–2077.

Pongratz, J., C. H. Reick, T. Raddatz, and M. Claussen, 2009: Effects of anthropogenic land cover change on the carbon cycle of the last millennium. *Global Biogeochem. Cycles*, **23**, GB4001.

Pritchard, H. D., S. R. M. Ligtenberg, H. A. Fricker, D. G. Vaughan, M. R. van den Broeke, and L. Padman, 2012: Antarctic ice sheet loss driven by basal melting of ice shelves. *Nature*, **484**, 502–505.

Pueyo, S., 2012: Solution to the paradox of climate sensitivity. *Clim. Change*, **113**, 163–179

Quadrelli, R., and J. M. Wallace, 2004: A simplified linear framework for interpreting patterns of Northern Hemisphere wintertime climate variability. *J. Clim.*, **17**, 3728–3744.

Rahmstorf, S., and D. Coumou, 2011: Increase of extreme events in a warming world. *Proc. Natl. Acad. Sci. U.S.A.*, **108**, 17905–17909.

Ramaswamy, V., M. D. Schwarzkopf, W. J. Randel, B. D. Santer, B. J. Soden, and G. L. Stenchikov, 2006: Anthropogenic and natural influences in the evolution of lower stratospheric cooling. *Science*, **311**, 1138–1141.

Ramsay, H. A., and A. H. Sobel, 2011: The effects of relative and absolute sea surface temperature on tropical cyclone potential intensity using a single column model. *J. Clim.*, **24**, 183–193.

Randel, W. J., et al., 2009: An update of observed stratospheric temperature trends. *J. Geophys. Res. Atmos.*, **114**, D02107.

Ray, E. A., et al., 2010: Evidence for changes in stratospheric transport and mixing over the past three decades based on multiple data sets and tropical leaky pipe analysis. *J. Geophys. Res. Atmos.*, **115**, D21304.

Rea, W., M. Reale, and J. Brown, 2011: Long memory in temperature reconstructions. *Clim. Change*, **107**, 247–265.

Reichert, B. K., L. Bengtsson, and J. Oerlemans, 2002: Recent glacier retreat exceeds internal variability. *J. Clim.*, **15**, 3069–3081.

Ribes, A., and L. Terray, 2013: Application of regularised optimal fingerprint analysis for attribution. Part II: Application to global near-surface temperature *Clim. Dyn.*, doi:10.1007/s00382-013-1736-6.

Ribes, A., J. M. Azais, and S. Planton, 2009: Adaptation of the optimal fingerprint method for climate change detection using a well-conditioned covariance matrix estimate. *Clim. Dyn.*, **33**, 707–722.

Ribes, A., J. M. Azais, and S. Planton, 2010: A method for regional climate change detection using smooth temporal patterns. *Clim. Dyn.*, **35**, 391–406.

Ribes, A., S. Planton, and L. Terray, 2013: Application of regularised optimal fingerprint for attribution. Part I: Method, properties and idealised analysis. *Clim. Dyn.*, doi:10.1007/s00382-013-1735-7.

Rind, D., J. Lean, J. Lerner, P. Lonergan, and A. Leboisitier, 2008: Exploring the stratospheric/troposheric response to solar forcing. *J. Geophys. Res. Atmos.*, **113** D24103.

Ring, M. J., D. Lindner, E. F. Cross, and M. E. Schlesinger, 2012: Causes of the global warming observed since the 19th century. atmospheric and climate sciences *Atmos. Clim. Sci.*, **2**, 401–415.

Roe, G. H., and M. B. Baker, 2007: Why is climate sensitivity so unpredictable? *Science*, **318**, 629–632.

Roe, G. H., and M. A. O'Neal, 2009: The response of glaciers to intrinsic climate variability: Observations and models of late-Holocene variations in the Pacific Northwest. *J. Glaciol.*, **55**, 839–854.

Roemmich, D., and J. Gilson, 2009: The 2004–2008 mean and annual cycle of temperature, salinity, and steric height in the global ocean from the Argo Program. *Prog. Oceanogr.*, **82**, 81–100.

Rogelj, J., M. Meinshausen, and R. Knutti, 2012: Global warming under old and new scenarios using IPCC climate sensitivity range estimates. *Nature Clim. Change*, **2**, 248–253.

Rohde, R., et al., 2013: A new estimate of the average Earth surface land temperature spanning 1753 to 2011. *Geoinf. Geostat. Overview*, **1**, 1.

Roscoe, H. K., and J. D. Haigh, 2007: Influences of ozone depletion, the solar cycle and the QBO on the Southern Annular Mode. *Q. J. R. Meteorol. Soc.*, **133**, 1855–1864.

Roy, I., and J. D. Haigh, 2010: Solar cycle signals in sea level pressure and sea surface temperature. *Atmos. Chem. Phys.*, **10** 3147–3153.

Roy, I., and J. D. Haigh, 2012: Solar cycle signals in the Pacific and the issue of timings. *J. Atmos. Sci.*, **69** 1446–1451.

Royer, D. L., 2008: Linkages between CO_2, climate, and evolution in deep time. *Proc. Natl. Acad. Sci. U.S.A.*, **105**, 407–408.

Royer, D. L., R. A. Berner, and J. Park, 2007: Climate sensitivity constrained by CO_2 concentrations over the past 420 million years. *Nature*, **446**, 530–532.

Rupp, D. E., P. W. Mote, N. L. Bindoff, P. A. Stott, and D. A. Robinson, 2013: Detection and attribution of observed changes in Northern Hemisphere spring snow cover. *J. Clim.*, doi:10.1175/JCLI-D-12-00563.1.

Rupp, D. E., P. W. Mote, N. Massey, J. R. Cameron, R. Jones, and M. R. Allen, 2012: Did human influence on climate make the 2011 Texas drought more probable? *Bull. Am. Meteorol. Soc.*, **93**, 1052–1054.

Sanso, B., and C. Forest, 2009: Statistical calibration of climate system properties. *J. R. Stat. Soc. C*, **58**, 485–503.

Santer, B. D., W. Bruggemann, U. Cubasch, K. Hasselmann, H. Hock, E. Maierreimer, and U. Mikolajewicz, 1994: Signal-to-noise analysis of time-dependent greenhouse warming experiments: 1. Pattern-analysis. *Clim. Dyn.*, **9**, 267–285.

Santer, B. D., et al., 2009: Incorporating model quality information in climate change detection and attribution studies. *Proc. Natl. Acad. Sci. U.S.A.*, **106** 14778–14783.

Santer, B. D., et al., 2007: Identification of human-induced changes in atmospheric moisture content. *Proc. Natl. Acad. Sci. U.S.A.*, **104**, 15248–15253.

Santer, B. D., et al., 2006: Forced and unforced ocean temperature changes in Atlantic and Pacific tropical cyclogenesis regions. *Proc. Natl. Acad. Sci. U.S.A.*, **103**, 13905–13910.

Santer, B. D., et al., 2013: Identifying human influences on atmospheric temperature. *Proc. Natl. Acad. Sci. U.S.A.*, **110**, 26–33.

Sato, M., J. E. Hansen, M. P. McCormick, and J. B. Pollack, 1993: Stratospheric aerosol optical depth, 1850–1990. *J. Geophys. Res. Atmos.*, **98**, 22987–22994.

Scafetta, N., and B. J. West, 2007: Phenomenological reconstructions of the solar signature in the Northern Hemisphere surface temperature records since 1600. *J. Geophys. Res. Atmos.*, **112**, D24S03.

Scheff, J., and D. M. W. Frierson, 2012a: Robust future precipitation declines in CMIP5 largely reflect the poleward explansion of the model subtropical dry zones. *Geophys. Res. Lett.*, **39**, L18704.

Scheff, J., and D. M. W. Frierson, 2012b: Twenty-first-century multimodel subtropical precipitation declines are mostly midlatitude shifts. *J. Clim.*, **25**, 4330–434.

Schlesinger, M. E., and N. Ramankutty, 1994: An oscillation in the global climate system of period 65–70 years. *Nature*, **367**, 723–726.

Schmidt, G., et al., 2012: Climate forcing reconstructions for use in PMIP simulations of the last millennium (v1.1). *Geosci. Model Dev.*, **5**, 185–191.

Schmidt, G. A., et al., 2011: Climate forcing reconstructions for use in PMIP simulations of the last millennium (v1.0). *Geosci. Model Dev.*, **4**, 33–45.

Schmittner, A., et al., 2011: Climate sensitivity estimated from temperature reconstructions of the last glacial maximum. *Science*, **334**, 1385–1388.

Schmittner, A., et al., 2012: Response to comment on "Climate sensitivity estimated from temperature reconstructions of the Last Glacial Maximum". *Science*, **337** 1294

Schneider, T., and I. M. Held, 2001: Discriminants of twentieth-century changes in earth surface temperatures. *J. Clim.*, **14** 249–254.

Schneider von Deimling, T., H. Held, A. Ganopolski, and S. Rahmstorf, 2006: Climate sensitivity estimated from ensemble simulations of glacial climate. *Clim. Dyn.*, **27**, 149–163.

Schnur, R., and K. I. Hasselmann, 2005: Optimal filtering for Bayesian detection and attribution of climate change. *Clim. Dyn.*, **24**, 45–55.

Schubert, S., et al., 2009: A US CLIVAR project to assess and compare the responses of global climate models to drought-related sst forcing patterns: Overview and results. *J. Clim.*, **22**, 5251–5272.

Schurer, A., G. Hegerl, M. E. Mann, S. F. B. Tett, and S. J. Phipps, 2013: Separating forced from chaotic climate variability over the past millennium. *J. Clim.*, doi:10.1175/JCLI-D-12-00826.1.

Schwartz, S. E., 2007: Heat capacity, time constant, and sensitivity of Earth's climate system. *J. Geophys. Res. Atmos.*, **112**, D24S05.

Schwartz, S. E., 2012: Determination of Earth's transient and equilibrium climate sensitivities from observations over the twentieth century: Strong dependence on assumed forcing. *Surv. Geophys.*, **33** 745–777.

Schwartz, S. E., R. J. Charlson, and H. Rodhe, 2007: Quantifying climate change—too rosy a picture? *Nature Rep. Clim. Change*, doi:10.1038/climate.2007.22, 23–24.

Schwartz, S. E., R. J. Charlson, R. A. Kahn, J. A. Ogren, and H. Rodhe, 2010: Why hasn't Earth warmed as much as expected? *J. Clim.*, **23**, 2453–2464.

Schweiger, A., R. Lindsay, J. Zhang, M. Steele, H. Stern, and R. Kwok, 2011: Uncertainty in modeled Arctic sea ice volume. *J. Geophys. Res.J. Geophys. Res. Oceans*, **116**, C00D06.

Screen, J. A., and I. Simmonds, 2010: Increasing fall-winter energy loss from the Arctic Ocean and its role in Arctic temperature amplification. *Geophys. Res. Lett.*, **37**, L16707.

Seager, R., N. Naik, and G. A. Vecchi, 2010: Thermodynamic and dynamic mechanisms for large-scale changes in the hydrological cycle in response to global warming. *J. Clim.*, **23**, 4651–4668.

Seager, R., Y. Kushnir, C. Herweijer, N. Naik, and J. Velez, 2005: Modeling of tropical forcing of persistent droughts and pluvials over western North America: 1856–2000. *J. Clim.*, **18**, 4065–4088.

Sedlacek, K., and R. Knutti, 2012: Evidence for external forcing on 20th-century climate from combined ocean atmosphere warming patterns. *Geophys. Res. Lett.*, **39**, L20708.

Seidel, D. J., and W. J. Randel, 2007: Recent widening of the tropical belt: Evidence from tropopause observations. *J. Geophys. Res. Atmos.*, **112**, D20113.

Seidel, D. J., Q. Fu, W. J. Randel, and T. J. Reichler, 2008: Widening of the tropical belt in a changing climate. *Nature Geosci.*, **1**, 21–24.

Seidel, D. J., N. P. Gillett, J. R. Lanzante, K. P. Shine, and P. W. Thorne, 2011: Stratospheric temperature trends: Our evolving understanding. *WIREs Clim. Change*, **2**, 592–616.

Seidel, D. J., Y. Zhang, A. Beljaars, J.-C. Golaz, A. R. Jacobson, and B. Medeiros, 2012: Climatology of the planetary boundary layer over the continental United States and Europe. *J. Geophys. Res. Atmos.*, **117**, D17106.

Semenov, V. A., 2008: Influence of oceanic inflow to the Barents Sea on climate variability in the Arctic region. *Doklady Earth Sci.*, **418**, 91–94.

Semmler, T., S. Varghese, R. McGrath, P. Nolan, S. L. Wang, P., and C. O'Dowd, 2008: Regional climate model simulations of NorthAtlantic cyclones: Frequency and intensity changes. *Clim. Res*, **36**, 1–16.

Seneviratne, S. I., 2012: Historical drought trends revisited. *Nature*, **491**, 338–339.

Seneviratne, S. I., et al., 2010: Investigating soil moisture-climate interactions in a changing climate: A review. *Earth Sci. Rev.*, **99**, 125–161.

Seneviratne, S. I., et al., 2012: Changes in climate extremes and their impacts on the natural physical environment. In: *Managing the Risks of Extreme Events and Disasters to Advance Climate Change Adaptation.* A Special Report of Working Groups I and II of the Intergovernmental Panel on Climate Change (IPCC) [C. B. Field et al. (eds.)]. Cambridge University Press, Cambridge, United Kingdom, and New York, NY, USA, pp. 109–230.

Serreze, M. C., and J. A. Francis, 2006: The arctic amplification debate. *Clim. Change*, **76**, 241–264.

Serreze, M. C., M. M. Holland, and J. Stroeve, 2007: Perspectives on the Arctic's shrinking sea-ice cover. *Science*, **315**, 1533–1536.

Serreze, M. C., A. P. Barrett, J. C. Stroeve, D. N. Kindig, and M. M. Holland, 2009: The emergence of surface-based Arctic amplification. *Cryosphere*, **3**, 9.

Sexton, D. M. H., J. M. Murphy, M. Collins, and M. J. Webb, 2012: Multivariate probabilistic projections using imperfect climate models part I: Outline of methodology. *Clim. Dyn.*, **38**, 2513–2542.

Sheffield, J., E. F. Wood, and M. Roderick, 2012: Little change in global drought over the past 60 years *Nature*, **491**, 435–438.

Shindell, D., and G. Faluvegi, 2009: Climate response to regional radiative forcing during the twentieth century. *Nature Geosci.*, **2**, 294–300.

Shindell, D., D. Rind, N. Balachandran, J. Lean, and J. Lonergan, 1999: Solar cycle variability, ozone, and climate. *Science*, **284**, 305–308.

Shindell, D. T., G. A. Schmidt, R. L. Miller, and D. Rind, 2001: Northern Hemisphere winter climate response to greenhouse gas, ozone, solar, and volcanic forcing. *J. Geophys. Res. Atmos.*, **106**, 7193–7210.

Shine, K. P., J. S. Fuglestvedt, K. Hailemariam, and N. Stuber, 2005: Alternatives to the global warming potential for comparing climate impacts of emissions of greenhouse gases. *Clim. Change*, **68**, 281–302.

Shiogama, H., T. Nagashima, T. Yokohata, S. A. Crooks, and T. Nozawa, 2006: Influence of volcanic activity and changes in solar irradiance on surface air temperatures in the early twentieth century. *Geophys. Res. Lett.*, **33**, L09702.

Shiogama, H., D. A. Stone, T. Nagashima, T. Nozawa, and S. Emori, 2012: On the linear additivity of climate forcing-response relationships at global and continental scales. *Int. J. Climatol.*, doi:10.1002/joc.3607.

Sigmond, M., and J. C. Fyfe, 2010: Has the ozone hole contributed to increased Antarctic sea ice extent? *Geophys. Rese. Lett.*, **37**, L18502.

Sigmond, M., M. C. Reader, J. C. Fyfe, and N. P. Gillett, 2011: Drivers of past and future Southern Ocean change: Stratospheric ozone versus greenhouse gas impacts. *Geophys. Res. Lett.*, **38**, L12601.

Simmons, A. J., K. M. Willett, P. D. Jones, P. W. Thorne, and D. P. Dee, 2010: Low-frequency variations in surface atmospheric humidity, temperature, and precipitation: Inferences from reanalyses and monthly gridded observational data sets. *J. Geophys. Res. Atmos.*, **115**, D01110

Skeie, R. B., T. K. Berntsen, G. Myhre, K. Tanaka, M. M. Kvalevåg, and C. R. Hoyle, 2011: Anthropogenic radiative forcing time series from pre-industrial times until 2010. *Atmos. Chem. Phys.*, **11**, 11827–11857.

Smirnov, D. A., and I. I. Mokhov, 2009: From Granger causality to long-term causality: Application to climatic data. *Phys. Rev. E*, **80**, 016208.

Sokolov, A., C. Forest, and P. Stone, 2010: Sensitivity of climate change projections to uncertainties in the estimates of observed changes in deep-ocean heat content. *Clim. Dyn.*, **34**, 735–745.

Solomon, S., P. J. Young, and B. Hassler, 2012: Uncertainties in the evolution of stratospheric ozone and implications for recent temperature changes in the tropical lower stratosphere. *Geophys. Res. Lett.*, **39**, L17706.

Solomon, S., et al., 2007: Technical Summary. In: *Climate Change 2007: The Physical Science Basis. Contribution of Working Group I to the Fourth Assessment Report of the Intergovernmental Panel on Climate Change* [Solomon, S., D. Qin, M. Manning, Z. Chen, M. Marquis, K. B. Averyt, M. Tignor and H. L. Miller (eds.)] Cambridge University Press, Cambridge, United Kingdom and New York, NY, USA, pp. 19–92.

Son, S. W., N. F. Tandon, L. M. Polvani, and D. W. Waugh, 2009: Ozone hole and Southern Hemisphere climate change. *Geophys. Res. Lett.*, **36**, L15705.

Son, S. W., et al., 2008: The impact of stratospheric ozone recovery on the Southern Hemisphere westerly jet. *Science*, **320**, 1486–1489.

Son, S. W., et al., 2010: Impact of stratospheric ozone on Southern Hemisphere circulation change: A multimodel assessment. *J. Geophys. Res. Atmos.*, **115**, D00M07.

SPARC CCMVal, 2010: *SPARC Report on the Evaluation of Chemistry-Climate Models.* SPARC Report No. 5, WCRP-132, WMO/TD-No. 1526, [V. Eyring, T. G. Shepherd and D. W. Waugh (eds.)]. Stratospheric Processes And their Role in Climate. Available at: http://www.atmosp.physics.utoronto.ca/SPARC.

St Jacques, J. M., D. J. Sauchyn, and Y. Zhao, 2010: Northern Rocky Mountain streamflow records: Global warming trends, human impacts or natural variability? *Geophys. Res. Lett.*, **37**, L06407.

Stahl, K., et al., 2010: Streamflow trends in Europe: Evidence from a dataset of near-natural catchments. *Hydrol. Earth Syst. Sci.*, **14**, 2367–2382.

Staten, P. W., J. J. Rutz, T. Reichler, and J. Lu, 2012: Breaking down the tropospheric circulation response by forcing. *Clim. Dyn.*, **39** 2361–2375.

Steig, E. J., and A. J. Orsi, 2013: The heat is on in Antarctica. *Nature Geosci.*, **6** 87–88.

Stenchikov, G., T. L. Delworth, V. Ramaswamy, R. J. Stouffer, A. Wittenberg, and F. Zeng, 2009: Volcanic signals in oceans. *J. Geophys. Res. Atmos.*, **114**, D16104.

Stephens, G. L., and Y. X. Hu, 2010: Are climate-related changes to the character of global-mean precipitation predictable? *Environ. Res. Lett.*, **5**, 025209.

Stephens, G. L., et al., 2010: Dreary state of precipitation in global models. *J. Geophys. Res. Atmos.*, **115**, D24211.

Stern, D. I., 2006: An atmosphere-ocean time series model of global climate change. *Comput. Stat. Data Anal.*, **51**, 1330–1346.

Stone, D. A., and M. R. Allen, 2005: Attribution of global surface warming without dynamical models. *Geophys. Res. Lett.*, **32**, L18711.

Stott, P. A., and J. Kettleborough, 2002: Origins and estimates of uncertainty in predictions of twenty-first century temperature rise *Nature*, **416**, 723–726.

Stott, P. A., and C. E. Forest, 2007: Ensemble climate predictions using climate models and observational constraints. *Philos. Trans. R. Soc. A*, **365**, 2029–2052.

Stott, P. A., and G. S. Jones, 2009: Variability of high latitude amplification of anthropogenic warming. *Geophys. Res. Lett.*, **36**, L10701.

Stott, P. A., and G. S. Jones, 2012: Observed 21st century temperatures further constrain decadal predictions of future warming. *Atmos. Sci. Lett.*, **13**, 151–156.

Stott, P. A., D. A. Stone, and M. R. Allen, 2004: Human contribution to the European heatwave of 2003. *Nature*, **432**, 610–614.

Stott, P. A., R. T. Sutton, and D. M. Smith, 2008a: Detection and attribution of Atlantic salinity changes. *Geophys. Res. Lett.*, **35**, L21702.

Stott, P. A., C. Huntingford, C. D. Jones, and J. A. Kettleborough, 2008b: Observed climate change constrains the likelihood of extreme future global warming. *Tellus B*, **60**, 76–81.

Stott, P. A., G. S. Jones, N. Christidis, F. W. Zwiers, G. Hegerl, and H. Shiogama, 2011: Single-step attribution of increasing frequencies of very warm regional temperatures to human influence. *Atmos. Sci. Lett.*, **12**, 220–227.

Stott, P. A., J. F. B. Mitchell, M. R. Allen, T. L. Delworth, J. M. Gregory, G. A. Meehl, and B. D. Santer, 2006: Observational constraints on past attributable warming and predictions of future global warming. *J. Clim.*, **19**, 3055–3069.

Stott, P. A., N. P. Gillett, G. C. Hegerl, D. J. Karoly, D. A. Stone, X. Zhang, and F. Zwiers, 2010: Detection and attribution of climate change: A regional perspective. *WIREs Clim. Change*, **1**, 192–211.

Stott, P. A., et al., 2013: Attribution of weather and climate-related events. In: *Climate Science for Serving Society: Research, Modelling and Prediction Priorities* [G. R. Asrar and J. W. Hurrell (eds.)]. Springer Science+Business Media, Dordrecht, Netherlands, 477 pp.

Stramma, L., S. Schmidtko, L. Levin, and G. Johnson, 2010: Ocean oxygen minima expansions and their biological impacts. *Deep-Sea Res. Pt. I*, **57**, 587–595.

Stroeve, J., et al., 2008: Arctic Sea ice extent plummeents in 2007. *Eos Trans. Am. Geophys. Union*, **89**, 13–14.

Stroeve, J. C., M. C. Serreze, M. M. Holland, J. E. Kay, W. Meier, and A. P. Barrett, 2012a: The Arctic's rapidly shrinking sea ice cover: A research synthesis. *Clim. Change*, **110** 1005–1027.

Stroeve, J. C., V. Kattsov, A. Barrett, M. Serreze, T. Pavlova, M. Holland, and W. N. Meier, 2012b: Trends in Arctic sea ice extent from CMIP5, CMIP3 and observations. *Geophys. Res. Lett.*, **39**, L16502.

Swanson, K. L., G. Sugihara, and A. A. Tsonis, 2009: Long-term natural variability and 20th century climate change. *Proc. Natl. Acad. Sci. U.S.A.*, **106**, 16120–16123.

Swart, N. C., and J. C. Fyfe, 2012: Observed and simulated changes in the Southern Hemisphere surface westerly wind-stress. *Geophys. Res. Lett.*, **39**, L16711.

Tanaka, K., T. Raddatz, B. C. O'Neill, and C. H. Reick, 2009: Insufficient forcing uncertainty underestimates the risk of high climate sensitivity. *Geophys. Res. Lett.*, **36** L16709.

Tapiador, F. J., 2010: A joint estimate of the precipitation climate signal in Europe using eight regional models and five observational datasets. *J. Clim.*, **23**, 1719–1738.

Taylor, K. E., R. J. Stouffer, and G. A. Meehl, 2012: An overview of CMIP5 and the experiment design. *Bull. Am. Meteorol. Soc.*, **93**, 485–498.

Tedesco, M., J. E. Box, J. Cappellen, T. Mote, R. S. W. van der Wal, and J. Wahr, 2012: Greenland ice sheet. In State of the Climate in 2011. *Bull. Am. Meteorol. Soc.*, **93**, S150–S153.

Terray, L., 2012: Evidence for multiple drivers of North Atlantic multi-decadal climate variability. *Geophys. Res. Lett.*, **39**, L19712.

Terray, L., L. Corre, S. Cravatte, T. Delcroix, G. Reverdin, and A. Ribes, 2012: Near-Surface salinity as nature's rain gauge to detect human influence on the tropical water cycle. *J. Clim.*, **25**, 958–977.

Tett, S. F. B., et al., 2007: The impact of natural and anthropogenic forcings on climate and hydrology since 1550. *Clim. Dyn.*, **28**, 3–34.

Thompson, D. W. J., and S. Solomon, 2002: Interpretation of recent Southern Hemisphere climate change. *Science*, **296**, 895–899.

Thompson, D. W. J., and S. Solomon, 2009: Understanding recent stratospheric climate change. *J. Clim.*, **22**, 1934–1943.

Thompson, D. W. J., J. M. Wallace, P. D. Jones, and J. J. Kennedy, 2009: Identifying signatures of natural climate variability in time series of global-mean surface temperature: Methodology and insights. *J. Clim.*, **22**, 6120–6141.

Thorne, P. W., and R. S. Vose, 2010: Reanalyses suitable for characterizing long-term trends *Bull. Am. Meteorol. Soc.*, **91**, 353–361.

Thorne, P. W., et al., 2011: A quantification of uncertainties in historical tropical tropospheric temperature trends from radiosondes. *J. Geophys. Res. Atmos.*, **116**, D12116.

Timmermann, A., S. McGregor, and F. F. Jin, 2010: Wind effects on past and future regional sea level trends in the southern Indo-Pacific. *J. Clim.*, **23**, 4429–4437.

Timmreck, C., S. J. Lorenz, T. J. Crowley, S. Kinne, T. J. Raddatz, M. A. Thomas, and J. H. Jungclaus, 2009: Limited temperature response to the very large AD 1258 volcanic eruption. *Geophys. Res. Lett.*, **36**, L21708.

Ting, M. F., Y. Kushnir, R. Seager, and C. H. Li, 2009: Forced and internal twentieth-century sst trends in the North Atlantic. *J. Clim.*, **22** 1469–1481.

Tomassini, L., P. Reichert, R. Knutti, T. F. Stocker, and M. E. Borsuk, 2007: Robust bayesian uncertainty analysis of climate system properties using Markov chain Monte Carlo methods. *J. Clim.*, **20**, 1239–1254.

Trenberth, K., 2011a: Attribution of climate variations and trends to human influences and natural variability. *WIREs Clim. Change*, **2**, 925–930.

Trenberth, K., 2011b: Changes in precipitation with climate change. *Clim. Research*, **47**, 123–138.

Trenberth, K. E., and D. J. Shea, 2006: Atlantic hurricanes and natural variability in 2005. *Geophys. Res. Lett.*, **33**, L12704.

Trenberth, K. E., and J. T. Fasullo, 2012: Climate extremes and climate change: The Russian heat wave and other climate extremes of 2010. *J. Geophys. Res. Atmos.*, **117**, D17103.

Trenberth, K. E., J. T. Fasullo, C. O'Dell, and T. Wong, 2010: Relationships between tropical sea surface temperature and top-of-atmosphere radiation. *Geophys. Res. Lett.*, **37**, L03702.

Tung, K.-K., and J. Zhou, 2010: The Pacific's response to surface heating in 130 yr of SST: La Niña-like or El Niño-like? *J. Atmos. Sci.*, **67**, 2649–2657.

Tung, K.-K., and J. Zhou, 2013: Using data to attribute episodes of warming and cooling in instrumental records. *Proc. Natl. Acad. Sci. U.S.A.*, **110** 2058–2063.

Tung, K. K., J. S. Zhou, and C. D. Camp, 2008: Constraining model transient climate response using independent observations of solar-cycle forcing and response. *Geophys. Res. Lett.*, **35** L17707.

Turner, J., T. J. Bracegirdle, T. Phillips, G. J. Marshall, and J. S. Hosking, 2013: An initial assessment of Antarctic sea ice extent in the CMIP5 models. *J. Clim.*, **26**, 1473–1484.

Turner, J., et al., 2005: Antarctic change during the last 50 years. *Int. J. Climatol.*, **25**, 1147–1148.

Turner, J., et al., 2009: Non-annular atmospheric circulation change induced by stratospheric ozone depletion and its role in the recent increase of Antarctic sea ice extent. *Geophys. Res. Lett.*, **36**, L08502.

Ulbrich, U., G. C. Leckebusch, and J. G. Pinto, 2009: Extra-tropical cyclones in the present and future climate: A review. *Theor. Appl. Climatol.*, **96**, 117–131.

Urban, N. M., and K. Keller, 2009: Complementary observational constraints on climate sensitivity. *Geophys. Res. Lett.*, **36**, L04708.

van der Schrier, G., P. D. Jones, and K. R. Briff, 2011: The sensitivity of the PDSI to the Thornthwaite and Penman-Monteith parameterizations for potential evapotranspiration. *J. Geophys. Res. Atmos.*, **116**, D03106.

van Loon, H., and G. A. Meehl, 2008: The response in the Pacific to the sun's decadal peaks and contrasts to cold events in the Southern Oscillation. *J. Atmos. Sol. Terres. Phys.*, **70** 1046–1055.

van Loon, H., and G. A. Meehl, 2012: The Indian summer monsoon during peaks in the 11 year sunspot cycle. *Geophys. Res. Lett.*, **39** L13701.

van Loon, H., G. A. Meehl, and D. J. Shea, 2007: Coupled air-sea response to solar forcing in the Pacific region during northern winter. *J. Geophys. Res. Atmos.*, **112**, D02108.

van Oldenborgh, G. J., A. van Urk, and M. Allen, 2012: The absence of a role of climate change in the 2011 Thailand floods. *Bull. Am. Meteorol. Soc.*, **93**, 1047–1049.

van Oldenborgh, G. J., F. J. Doblas Reyes, S. S. Drijfhout, and E. Hawkins, 2013: Reliability of regional climate model trends. *Environ. Res. Lett.*, **8**, 014055.

Vecchi, G. A., and B. J. Soden, 2007: Global warming and the weakening of the tropical circulation. *J. Clim.*, **20**, 4316–4340.

Vecchi, G. A., K. L. Swanson, and B. J. Soden, 2008: Whither hurricane activity. *Science*, **322**, 687–689

Veryard, H. G., 1963: A review of studies on climate fluctuations during the period of the meteorological. *Changes of Climate: Proceedings of the Rome Symposium Organised by UNESCO and WMO*, pp. 3–15.

Villarini, G., and G. A. Vecchi, 2012: Twenty-first-century projections of North Atlantic tropical storms from CMIP5 models. *Nature Clim. Change*, **2**, 604–607.

Villarini, G., and G. A. Vecchi, 2013: Projected increases in North Atlantic tropical cyclone intensity from CMIP5 models. *J. Clim.*, **26**, 3231–3240.

Visser, H., and A. C. Petersen, 2012: Inference on weather extremes and weather related disasters: A review of statistical methods. *Clim. Past*, **8** 265–286.

von Schuckmann, K., F. Gaillard, and P. Y. Le Traon, 2009: Global hydrographic variability patterns during 2003–2008. *J. Geophys. Res.J. Geophys. Res. Oceans*, **114**, C09007.

Vorosmarty, C., L. Hinzman, and J. Pundsack, 2008: Introduction to special section on changes in the arctic freshwater system: Identification, attribution, and impacts at local and global scales. *J. Geophys. Res. Biogeosci.*, **113**, G01S91.

Vuille, M., G. Kaser, and I. Juen, 2008: Glacier mass balance variability in the Cordillera Blanca, Peru and its relationship with climate and the large-scale circulation. *Global Planet. Change*, **62**, 14–28.

Walker, R. T., T. K. Dupont, D. M. Holland, B. R. Parizek, and R. B. Alley, 2009: Initial effects of oceanic warming on a coupled ocean-ice shelf-ice stream system. *Earth Planet. Sci. Lett.*, **287**, 483–487.

Wan, H., X. Zhang, F. W. Zwiers, and H. Shiogama, 2013: Effect of data coverage on the estimation of mean and variability of precipitation at global and regional scales. *J. Geophys. Res. Atmos.*, **118**, 534–546.

Wang, D. B., and M. Hejazi, 2011: Quantifying the relative contribution of the climate and direct human impacts on mean annual streamflow in the contiguous United States. *Water Resour. Res.*, **47**, W00J12.

Wang, J., and X. Zhang, 2008: Downscaling and projection of winter extreme daily precipitation over North America. *J. Clim.*, **21**, 923–937.

Wang, J., et al., 2009a: Is the Dipole Anomaly a major driver to record lows in Arctic summer sea ice extent? *Geophys. Res. Lett.*, **36**, L05706.

Wang, M., and J. E. Overland, 2012: A sea ice free summer Arctic within 30 years: An update from CMIP5 models. *Geophys. Res. Lett.*, **39**, L18501.

Wang, M. Y., and J. E. Overland, 2009: A sea ice free summer Arctic within 30 years? *Geophys. Res. Lett.*, **36**, L07502.

Wang, M. Y., J. E. Overland, V. Kattsov, J. E. Walsh, X. D. Zhang, and T. Pavlova, 2007: Intrinsic versus forced variation in coupled climate model simulations over the Arctic during the twentieth century. *J. Clim.*, **20** 1093–1107.

Wang, X. L., V. R. Swail, F. W. Zwiers, X. Zhang, and Y. Feng, 2009b: Detection of external influence on trends of atmospheric storminess and northern oceans wave heights. *Clim. Dyn.*, **32**, 189–203.

Wassmann, P., C. M. Duarte, S. Agusti, and M. K. Sejr, 2011: Footprints of climate change in the Arctic marine ecosystem. *Global Change Biol.*, **17**, 1235–1249.

Wen, Q. H., X. Zhang, Y. Xu, and B. Wang, 2013: Detecting human influence on extreme temperatures in China. *Geophys. Res. Lett.*, **40**, 1171–1176.

Wentz, F. J., L. Ricciardulli, K. Hilburn, and C. Mears, 2007: How much more rain will global warming bring? *Science*, **317** 233–235.

Westra, S., L. V. Alexander, and F. W. Zwiers, 2013: Global increasing trends in annual maximum daily precipitation. *J. Clim.*, doi:10.1175/JCLI-D-12-00502.1.

White, W. B., and Z. Y. Liu, 2008: Non-linear alignment of El Nino to the 11-yr solar cycle. *Geophys. Res. Lett.*, **35**, L19607.

Wigley, T. M. L., and B. D. Santer, 2013: A probabilistic quantification of the anthropogenic component of twentieth century global warming. *Clim. Dyn.*, **40**, 1087–1102.

Wigley, T. M. L., C. M. Ammann, B. D. Santer, and K. E. Taylor, 2005: Comment on "Climate forcing by the volcanic eruption of Mount Pinatubo" by David H. Douglass and Robert S. Knox. *Geophys. Res. Lett.*, **32**, L20709.

Wijffels, S., et al., 2008: Changing expendable bathythermograph fall rates and their impact on estimates of thermosteric sea level rise. *J. Clim.*, **21**, 5657–5672.

Wilcox, L. J., B. J. Hoskins, and K. P. Shine, 2012: A global blended tropopause based on ERA data. Part II: Trends and tropical broadening. *Q. J. R. Meteorol. Soc.*, **138**, 576–584.

Willett, K. M., N. P. Gillett, P. D. Jones, and P. W. Thorne, 2007: Attribution of observed surface humidity changes to human influence. *Nature*, **449**, 710–712

Willett, K. M., P. D. Jones, N. P. Gillett, and P. W. Thorne, 2008: Recent changes in surface humidity: Development of the HadCRUH dataset. *J. Clim.*, **21**, 5364–5383.

Wilson, D., H. Hisdal, and D. Lawrence, 2010: Has streamflow changed in the Nordic countries? Recent trends and comparisons to hydrological projections. *J. Hydrol.*, **394**, 334–346.

WMO (World Meteorological Organization), 2011: *Scientific Assessment of Ozone Depletion: 2010*. Global Ozone Research and Monitoring Project–Report No. 52, World Meterological Organization, Geneva, Switzerland, 516 pp.

Wong, A. P. S., N. L. Bindoff, and J. A. Church, 1999: Large-scale freshening of intermediate waters in the Pacific and Indian oceans. *Nature*, **400**, 440–443.

Wood, K. R., and J. E. Overland, 2010: Early 20th century Arctic warming in retrospect. *Int. J. Climatol.*, **30**, 1269–1279.

Woollings, T., 2008: Vertical structure of anthropogenic zonal-mean atmospheric circulation change. *Geophys. Res. Lett.*, **35**, L19702.

Woollings, T., M. Lockwood, G. Masato, C. Bell, and L. J. Gray, 2010: Enhanced signatures of solar variability in Eurasian winter climate. *Geophys. Res. Lett.*, **37** L20805.

Wu, Q. G., and D. J. Karoly, 2007: Implications of changes in the atmospheric circulation on the detection of regional surface air temperature trends. *Geophys. Res. Lett.*, **34**, L08703.

Wu, Z. H., N. E. Huang, J. M. Wallace, B. V. Smoliak, and X. Y. Chen, 2011: On the time-varying trend in global-mean surface temperature. *Clim. Dyn.*, **37** 759–773.

Xie, S.-P., C. Deser, G. A. Vecchi, J. Ma, H. Teng, and A. T. Wittenberg, 2010: Global warming pattern formation: Sea surface temperature and rainfall. *J. Clim.*, **23**, 966–986.

Yamaguchi, S., R. Naruse, and T. Shiraiwa, 2008: Climate reconstruction since the Little Ice Age by modelling Koryto glacier, Kamchatka Peninsula, Russia. *J. Glaciol.*, **54**, 125–130.

Yang, X., J. C. Fyfe, and G. M. Flato, 2010: The role of poleward energy transport in Arctic temperature. *Geophys. Res. Lett.*, **37**, L14803.

Yiou, P., R. Vautard, P. Naveau, and C. Cassou, 2007: Inconsistency between atmospheric dynamics and temperatures during the exceptional 2006/2007 fall/winter and recent warming in Europe. *Geophys. Res. Lett.*, **34**, L21808.

Yoshimori, M., and A. J. Broccoli, 2008: Equilibrium response of an atmosphere-mixed layer ocean model to different radiative forcing agents: Global and zonal mean response. *J. Clim.*, **21**, 4399–4423.

Young, P. J., et al., 2012: Agreement in late twentieth century Southern Hemisphere stratospheric temperature trends in observations and CCMVal-2, CMIP3 and CMIP5 models. *J. Geophys. Res. Atmos.*, **118** 605–613.

Zhang, J., R. Lindsay, A. Schweiger, and M. Steele, 2013: The impact of an intense summer cyclone on 2012 Arctic sea ice retreat. *Geophys. Res. Lett.*, **40**, 720–726.

Zhang, J. L., 2007: Increasing Antarctic sea ice under warming atmospheric and oceanic conditions. *J. Clim.*, **20**, 2515–2529.

Zhang, R., and T. L. Delworth, 2009: A new method for attributing climate variations over the Atlantic Hurricane Basin's main development region. *Geophys. Res. Lett.*, **36**, L06701.

Zhang, R., et al., 2012: Have aerosols caused the observed Atlantic Multidecadal Variability? *J. Atmos. Sci.*, **70**, 1135–1144.

Zhang, X. B., et al., 2007: Detection of human influence on twentieth-century precipitation trends. *Nature*, **448**, 461–465.

Zhang, X. D., A. Sorteberg, J. Zhang, R. Gerdes, and J. C. Comiso, 2008: Recent radical shifts of atmospheric circulations and rapid changes in Arctic climate system. *Geophys. Res. Lett.*, **35**, L22701.

Zhou, J., and K.-K. Tung, 2013a: Deducing multidecadal anthropogenic global warming trends using multiple regression analysis. *J. Atmos. Sci.*, **70**, 3–8.

Zhou, J., and K.-K. Tung, 2013b: Observed tropospheric temperature response to 11-yr solar cycle and what it reveals about mechanisms. *J. Atmos. Sci.*, **70**, 9–14.

Zhou, Y., and G. Ren, 2011: Change in extreme temperature event frequency over mainland China, 1961–2008. *Climate Research*, **50**, 125–139.

Zickfeld, K., M. Eby, H. D. Matthews, and A. J. Weaver, 2009: Setting cumulative emissions targets to reduce the risk of dangerous climate change. *Proc. Natl. Acad. Sci. U.S.A.*, **106**, 16129–16134.

Zickfeld, K., et al., 2013: Long-term climate change commitment and reversibility: An EMIC intercomparison. *J. Clim.*, doi:10.1175/JCLI-D-12–00584.1.

Zorita, E., T. F. Stocker, and H. von Storch, 2008: How unusual is the recent series of warm years? *Geophys. Res. Lett.*, **35**, L24706.

Zunz, V., H. Goosse, and F. Massonnet, 2013: How does internal variability influence the ability of CMIP5 models to reproduce the recent trend in Southern Ocean sea ice extent? *Cryosphere*, **7**, 451–468.

Zwiers, F. W., X. Zhang, and Y. Feng, 2011: Anthropogenic influence on long return period daily temperature extremes at regional scales. *J. Clim.*, **24**, 881–892.

Near-term Climate Change: Projections and Predictability

Coordinating Lead Authors:
Ben Kirtman (USA), Scott B. Power (Australia)

Lead Authors:
Akintayo John Adedoyin (Botswana), George J. Boer (Canada), Roxana Bojariu (Romania), Ines Camilloni (Argentina), Francisco Doblas-Reyes (Spain), Arlene M. Fiore (USA), Masahide Kimoto (Japan), Gerald Meehl (USA), Michael Prather (USA), Abdoulaye Sarr (Senegal), Christoph Schär (Switzerland), Rowan Sutton (UK), Geert Jan van Oldenborgh (Netherlands), Gabriel Vecchi (USA), Hui-Jun Wang (China)

Contributing Authors:
Nathaniel L. Bindoff (Australia), Philip Cameron-Smith (USA/New Zealand), Yoshimitsu Chikamoto (USA/Japan), Olivia Clifton (USA), Susanna Corti (Italy), Paul J. Durack (USA/Australia), Thierry Fichefet (Belgium), Javier García-Serrano (Spain), Paul Ginoux (USA), Lesley Gray (UK), Virginie Guemas (Spain/France), Ed Hawkins (UK), Marika Holland (USA), Christopher Holmes (USA), Johnna Infanti (USA), Masayoshi Ishii (Japan), Daniel Jacob (USA), Jasmin John (USA), Zbigniew Klimont (Austria/Poland), Thomas Knutson (USA), Gerhard Krinner (France), David Lawrence (USA), Jian Lu (USA/Canada), Daniel Murphy (USA), Vaishali Naik (USA/India), Alan Robock (USA), Luis Rodrigues (Spain/Brazil), Jan Sedláček (Switzerland), Andrew Slater (USA/Australia), Doug Smith (UK), David S. Stevenson (UK), Bart van den Hurk (Netherlands), Twan van Noije (Netherlands), Steve Vavrus (USA), Apostolos Voulgarakis (UK/Greece), Antje Weisheimer (UK/Germany), Oliver Wild (UK), Tim Woollings (UK), Paul Young (UK)

Review Editors:
Pascale Delecluse (France), Tim Palmer (UK), Theodore Shepherd (Canada), Francis Zwiers (Canada)

This chapter should be cited as:
Kirtman, B., S.B. Power, J.A. Adedoyin, G.J. Boer, R. Bojariu, I. Camilloni, F.J. Doblas-Reyes, A.M. Fiore, M. Kimoto, G.A. Meehl, M. Prather, A. Sarr, C. Schär, R. Sutton, G.J. van Oldenborgh, G. Vecchi and H.J. Wang, 2013: Near-term Climate Change: Projections and Predictability. In: *Climate Change 2013: The Physical Science Basis. Contribution of Working Group I to the Fifth Assessment Report of the Intergovernmental Panel on Climate Change* [Stocker, T.F., D. Qin, G.-K. Plattner, M. Tignor, S.K. Allen, J. Boschung, A. Nauels, Y. Xia, V. Bex and P.M. Midgley (eds.)]. Cambridge University Press, Cambridge, United Kingdom and New York, NY, USA.

Table of Contents

Executive Summary

This chapter assesses the scientific literature describing expectations for near-term climate (present through mid-century). Unless otherwise stated, 'near-term' change and the projected changes below are for the period 2016–2035 relative to the reference period 1986–2005. Atmospheric composition (apart from CO_2; see Chapter 12) and air quality projections through to 2100 are also assessed.

Decadal Prediction

The nonlinear and chaotic nature of the climate system imposes natural limits on the extent to which skilful predictions of climate statistics may be made. Model-based 'predictability' studies, which probe these limits and investigate the physical mechanisms involved, support the potential for the skilful prediction of annual to decadal average temperature and, to a lesser extent precipitation.

Predictions for averages of temperature, over large regions of the planet and for the global mean, exhibit positive skill when verified against observations for forecast periods up to ten years (*high confidence*[1]). Predictions of precipitation over some land areas also exhibit positive skill. Decadal prediction is a new endeavour in climate science. The level of quality for climate predictions of annual to decadal average quantities is assessed from the past performance of initialized predictions and non-initialized simulations. {11.2.3, Figures 11.3 and 11.4}

In current results, observation-based initialization is the dominant contributor to the skill of predictions of annual mean temperature for the first few years and to the skill of predictions of the global mean surface temperature and the temperature over the North Atlantic, regions of the South Pacific and the tropical Indian Ocean for longer periods (*high confidence*). Beyond the first few years the skill for annual and multi-annual averages of temperature and precipitation is due mainly to the specified radiative forcing (*high confidence*). {Section 11.2.3, Figures 11.3 to 11.5}

Projected Changes in Radiative Forcing of Climate

For greenhouse gas (GHG) forcing, the new Representative Concentration Pathway (RCP) scenarios are similar in magnitude and range to the AR4 Special Report on Emission Scenarios (SRES) scenarios in the near term, but for aerosol and ozone precursor emissions the RCPs are much lower than SRES by factors of 1.2 to 3. For these emissions the spread across RCPs by 2030 is much narrower than between scenarios that considered current legislation and maximum technically feasible emission reductions (factors of 2). In the near term, the SRES Coupled Model Intercomparison Project Phase 3 (CMIP3) results, which did not incorporate current legislation on air pollutants, include up to three times more anthropogenic aerosols than RCP CMIP5 results (*high confidence*), and thus the CMIP5 global mean temperatures may be up to 0.2°C warmer than if forced with SRES aerosol scenarios (*medium confidence*). {10.3.1.1.3, Figure 10.4, 11.3.1.1, 11.3.5.1, 11.3.6.1, Figure 11.25, Tables AII.2.16 to AII.2.22 and AII.6.8}

Including uncertainties for the chemically reactive GHG methane gives a range in concentration that is 30% wider than the spread in RCP concentrations used in CMIP5 models (*likely*[2]). By 2100 this range extends 520 ppb above RCP8.5 and 230 ppb below RCP2.6 (*likely*), reflecting uncertainties in emissions from agricultural, forestry and land use sources, in atmospheric lifetimes, and in chemical feedbacks, but not in natural emissions. {11.3.5}

Emission reductions aimed at decreasing local air pollution could have a near-term impact on climate (*high confidence*). Short-lived air pollutants have opposing effects: cooling from sulphate and nitrate; warming from black carbon (BC) aerosol, carbon monoxide (CO) and methane (CH_4). Anthropogenic CH_4 emission reductions (25%) phased in by 2030 would decrease surface ozone and reduce warming averaged over 2036–2045 by about 0.2°C (*medium confidence*). Combined reductions of BC and co-emitted species (78%) on top of methane reductions (24%) would further reduce warming (*low confidence*), but uncertainties increase. {Section 7.6, Chapter 8, 11.3.6.1, Figure 11.24a, 8.7.2.2.2, Table AII.7.5a}

Projected Changes in Near-term Climate

Projections of near-term climate show modest sensitivity to alternative RCP scenarios on global scales, but aerosols are an important source of uncertainty on both global and regional scales. {11.3.1, 11.3.6.1}

11

Projected Changes in Near-term Temperature

The projected change in global mean surface air temperature will *likely* be in the range 0.3 to 0.7°C (*medium confidence*). This projection is valid for the four RCP scenarios and assumes there will be no major volcanic eruptions or secular changes in total solar irradiance before 2035. A future volcanic eruption similar to the 1991 eruption of Mt Pinatubo would cause a rapid drop in global mean surface air temperature of several tenths °C in the following year, with recovery over the next few years. Possible future changes in solar irradiance

---

1 In this Report, the following summary terms are used to describe the available evidence: limited, medium, or robust; and for the degree of agreement: low, medium, or high. A level of confidence is expressed using five qualifiers: very low, low, medium, high, and very high, and typeset in italics, e.g., *medium confidence*. For a given evidence and agreement statement, different confidence levels can be assigned, but increasing levels of evidence and degrees of agreement are correlated with increasing confidence (see Section 1.4 and Box TS.1 for more details).

2 In this Report, the following terms have been used to indicate the assessed likelihood of an outcome or a result: Virtually certain 99–100% probability, Very likely 90–100%, Likely 66–100%, About as likely as not 33–66%, Unlikely 0–33%, Very unlikely 0–10%, Exceptionally unlikely 0–1%. Additional terms (Extremely likely: 95–100%, More likely than not >50–10 0%, and Extremely unlikely 0–5%) may also be used when appropriate. Assessed likelihood is typeset in italics, e.g., *very likely* (see Section 1.4 and Box TS.1 for more details).

could influence the rate at which global mean surface air temperature increases, but there is *high confidence* that this influence will be small in comparison to the influence of increasing concentrations of GHGs in the atmosphere. {11.3.6, Figure 11.25}

It is *more likely than not* that the mean global mean surface air temperature for the period 2016–2035 will be more than 1°C above the mean for 1850–1900, and *very unlikely* that it will be more than 1.5°C above the 1850–1900 mean (*medium confidence*). {11.3.6.3}

In the near term, differences in global mean surface air temperature change across RCP scenarios for a single climate model are typically smaller than differences between climate models under a single RCP scenario. In 2030, the CMIP5 ensemble median values differ by at most 0.2°C between RCP scenarios, whereas the model spread (17 to 83% range) for each RCP is about 0.4°C. The inter-scenario spread increases in time: by 2050 it is 0.8°C, whereas the model spread for each scenario is only 0.6°C. Regionally, the largest differences in surface air temperature between RCP2.6 and RCP8.5 are found in the Arctic. {11.3.2.1.1, 11.3.6.1, 11.3.6.3, Figure 11.24a,b, Table AII.7.5}

It is *very likely* that anthropogenic warming of surface air temperature will proceed more rapidly over land areas than over oceans, and that anthropogenic warming over the Arctic in winter will be greater than the global mean warming over the same period, consistent with the AR4. Relative to natural internal variability, near-term increases in seasonal mean and annual mean temperatures are expected to be larger in the tropics and subtropics than in mid-latitudes (*high confidence*). {11.3.2, Figures 11.10 and 11.11}

11

Projected Changes in the Water Cycle and Atmospheric Circulation

Zonal mean precipitation will *very likely* increase in high and some of the mid latitudes, and will *more likely than not* decrease in the subtropics. At more regional scales precipitation changes may be influenced by anthropogenic aerosol emissions and will be strongly influenced by natural internal variability. {11.3.2, Figures 11.12 and 11.13}

Increases in near-surface specific humidity over land are *very likely*. Increases in evaporation over land are *likely* in many regions. There is *low confidence* in projected changes in soil moisture and surface run off. {11.3.2, Figure 11.14}

It is *likely* that the descending branch of the Hadley Circulation and the Southern Hemisphere (SH) mid-latitude westerlies will shift poleward. It is *likely* that in austral summer the projected recovery of stratospheric ozone and increases in GHG concentrations will have counteracting impacts on the width of the Hadley Circulation and the meridional position of the SH storm track. Therefore, it is *likely* that in the near term the poleward expansion of the descending southern branch of the Hadley Circulation and the SH mid-latitude westerlies in austral summer will be less rapid than in recent decades. {11.3.2}

There is *medium confidence* in near-term projections of a northward shift of Northern Hemisphere storm tracks and westerlies. {11.3.2}

Projected Changes in the Ocean and Cryosphere

It is *very likely* that globally averaged surface and vertically averaged ocean temperatures will increase in the near term. It is *likely* that there will be increases in salinity in the tropical and (especially) subtropical Atlantic, and decreases in the western tropical Pacific over the next few decades. The Atlantic Meridional Overturning Circulation is *likely* to decline by 2050 (*medium confidence*). However, the rate and magnitude of weakening is very uncertain and, due to large internal variability, there may be decades when increases occur. {11.3.3}

It is *very likely* that there will be further shrinking and thinning of Arctic sea ice cover, and decreases of northern high-latitude spring time snow cover and near surface permafrost (see glossary) as global mean surface temperature rises. For high GHG emissions such as those corresponding to RCP8.5, a nearly ice-free Arctic Ocean (sea ice extent less than 1×10^6 km^2 for at least 5 consecutive years) in September is *likely* before mid-century (*medium confidence*). This assessment is based on a subset of models that most closely reproduce the climatological mean state and 1979 to 2012 trend of Arctic sea ice cover. There is *low confidence* in projected near-term decreases in the Antarctic sea ice extent and volume. {11.3.4}

Projected Changes in Extremes

In most land regions the frequency of warm days and warm nights will *likely* increase in the next decades, while that of cold days and cold nights will decrease. Models project near-term increases in the duration, intensity and spatial extent of heat waves and warm spells. These changes may proceed at a different rate than the mean warming. For example, several studies project that European high-percentile summer temperatures warm faster than mean temperatures. {11.3.2.5.1, Figures 11.17 and 11.18}

The frequency and intensity of heavy precipitation events over land will *likely* increase on average in the near term. However, this trend will not be apparent in all regions because of natural variability and possible influences of anthropogenic aerosols. {11.3.2.5.2, Figures 11.17 and 11.18}

There is *low confidence* in basin-scale projections of changes in the intensity and frequency of tropical cyclones (TCs) in all basins to the mid-21st century. This *low confidence* reflects the small number of studies exploring near-term TC activity, the differences across published projections of TC activity, and the large role for natural variability and non-GHG forcing of TC activity up to the mid-21st century. There is *low confidence* in near-term projections for increased TC intensity in the North Atlantic, which is in part due to projected reductions in North Atlantic aerosols loading. {11.3.2.5.3}

Projected Changes in Air Quality

The range in projections of air quality (O_3 and $PM_{2.5}$ in near-surface air) is driven primarily by emissions (including CH_4), rather than by physical climate change (*medium confidence*). The response of air quality to climate-driven changes is more uncertain than the response to emission-driven changes (*high confidence*). Globally, warming decreases background surface O_3 (*high confidence*). High CH_4 levels (RCP8.5, SRES A2) can offset this decrease, raising 2100 background surface O_3 on average by about 8 ppb (25% of current levels) relative to scenarios with small CH_4 changes (RCP4.5, RCP6.0) (*high confidence*). On a continental scale, projected air pollution levels are lower under the new RCP scenarios than under the SRES scenarios because the SRES did not incorporate air quality legislation (*high confidence*). {11.3.5, 11.3.5.2; Figures 11.22 and 11.23ab, AII.4.2, AII.7.1–AII.7.4}

Observational and modelling evidence indicates that, all else being equal, locally higher surface temperatures in polluted regions will trigger regional feedbacks in chemistry and local emissions that will increase peak levels of O_3 and $PM_{2.5}$ (*medium confidence*). Local emissions combined with background levels and with meteorological conditions conducive to the formation and accumulation of pollution are known to produce extreme pollution episodes on local and regional scales. There is *low confidence* in projecting changes in meteorological blocking associated with these extreme episodes. For $PM_{2.5}$, climate change may alter natural aerosol sources (wildfires, wind-lofted dust, biogenic precursors) as well as precipitation scavenging, but no confidence level is attached to the overall impact of climate change on $PM_{2.5}$ distributions. {11.3.5, 11.3.5.2, Box 14.2}

11.1 Introduction

This chapter describes current scientific expectations for 'near-term' climate. Here 'near term' refers to the period from the present to mid-century, during which the climate response to different emissions scenarios is generally similar. Greatest emphasis in this chapter is given to the period 2016–2035, though some information on projected changes before and after this period (up to mid-century) is also assessed. An assessment of the scientific literature relating to atmospheric composition (except carbon dioxide (CO_2), which is addressed in Chapter 12) and air quality for the near-term and beyond to 2100 is also provided.

This emphasis on near-term climate arises from (1) a recognition of its importance to decision makers in government and industry; (2) an increase in the international research effort aimed at improving our understanding of near-term climate; and (3) a recognition that near-term projections are generally less sensitive to differences between future emissions scenarios than are long-term projections. Climate prediction on seasonal to multi-annual time scales require accurate estimates of the initial climate state with less dependence on changes in external forcing[3] over the period. On longer time scales climate projections rely on projections of external forcing with little reliance on the initial state of internal variability. Estimates of near-term climate depend partly on the committed change (caused by the inertia of the oceans as they respond to historical external forcing), the time evolution of internally generated climate variability and the future path of external forcing. Near-term climate is sensitive to rapid changes in some short-lived climate forcing agents (Jacobson and Streets, 2009; Wigley et al., 2009; UNEP and WMO, 2011; Shindell et al., 2012b).

The need for near-term climate information has spawned a new field of climate science: decadal climate prediction (Smith et al., 2007; Meehl et al., 2009b, 2013d). The Coupled Model Intercomparison Project Phase 5 (CMIP5) experimental protocol includes a sequence of near-term predictions (1 to 10 years) where observation-based information is used to initialize the models used to produce the forecasts. The goal is to exploit the predictability of internally generated climate variability as well as that of the externally forced component. The result depends on the ability of current models to reproduce the observed variability as well as on the accurate depiction of the initial state (see Box 11.1). Skilful multi-annual to decadal climate predictions (in the technical sense of 'skilful' as outlined in 11.2.3.2 and FAQ 11.1) are being produced although technical challenges remain that need to be overcome in order to improve skill. These challenges are now being addressed by the scientific community.

Climate change experiments with models that do not depend on initial condition but on the history and projection of climate forcings (often referred to as 'uninitialized' or 'non-initialized' projections or simply as 'projections') are another component of CMIP5. Such projections have been the main focus of assessments of future climate in previous IPCC assessments and are considered in Chapters 12 to 14. The main focus of attention in past assessments has been on the properties of projections for the late 21st century and beyond. Projections also provide valuable information on externally forced changes to near-term climate, however, and are an important source of information that complements information from the predictions. Projections are also assessed in this chapter.

The objectives of this chapter are to assess the state of the science concerning both near-term predictions and near-term projections. CMIP5 results are considered for the near term as are other published near-term predictions and projections. The chapter consists of four major assessments:

1. The scientific basis for near-term prediction as reflected in estimates of predictability (see Box 11.1), and the dynamical and physical mechanisms underpinning predictability, and the processes that limit predictability (see Section 11.2).

2. The current state of knowledge in near-term prediction (see Section 11.2). Here the emphasis is placed on the results from the decadal (10-year) multi-model prediction experiments in the CMIP5 database.

3. The current state of knowledge in near-term projection (see Section 11.3). Here the emphasis is on what the climate in next few decades may look like relative to 1986–2005, based on near-term projections (i.e., the forced climatic response). The focus is on the 'core' near-term period (2016–2035), but some information prior to this period and out to mid-century is also discussed. A key issue is when, where and how the signal of externally forced climate change is expected to emerge from the background of natural climate variability.

4. Projected changes in atmospheric composition and air quality, and their interactions with climate change during the near term and beyond, including new findings from the Atmospheric Chemistry and Climate Model Intercomparison (ACCMIP) initiative.

---

[3] Seasonal-to-interannual predictions typically include the impact of external forcing.

Box 11.1 | Climate Simulation, Projection, Predictability and Prediction

This section outlines some of the ideas and the terminology used in this chapter.

Internally generated and externally forced climate variability

It is useful for purposes of analysis and description to consider the pre-industrial climate system as being in a state of climatic equilibrium with a fixed atmospheric composition and an unchanging Sun. In this idealized state, naturally occurring processes and interactions within the climate system give rise to 'internally generated' climate variability on many time scales (as discussed in Chapter 1). Variations in climate may also result due to features 'external' to this idealized system. Forcing factors, such as volcanic eruptions, solar variations, anthropogenic changes in the composition of the atmosphere, land use change etc., give rise to 'externally forced' climate variations. In this sense climate system variables such as annual mean temperatures (as in Box 11.1, Figure 1 for instance) may be characterized as a combination of externally forced and internally generated components with $T(t) = T_f(t) + T_i(t)$. This separation of T, and other climate variables, into components is useful when analysing climate behaviour but does not, of course, mean that the climate system is linear or that externally forced and internally generated components do not interact.

Climate simulation

A climate simulation is a model-based representation of the temporal behaviour of the climate system under specified external forcing and boundary conditions. The result is the modelled response to the imposed external forcing combined with internally generated variability. The thin yellow lines in Box 11.1, Figure 1 represent an ensemble of climate simulations begun from pre-industrial conditions with imposed historical external forcing. The imposed external conditions are the same for each ensemble member and differences among the simulations reflect differences in the evolutions of the internally generated component. Simulations are not intended to be forecasts of the observed evolution of the system (the black line in Box 11.1, Figure 1) but to be possible evolutions that are consistent with the external forcings.

In practice, and in Box 11.1, Figure 1, the forced component of the temperature variation is estimated by averaging over the different simulations of $T(t)$ with $T_f(t)$ the component that survives ensemble averaging (the red curve) while $T_i(t)$ averages to near zero for a large enough ensemble. The spread among individual ensemble members (from these or pre-industrial simulations) and their behaviour with time provides some information on the statistics of the internally generated variability. *(continued on next page)*

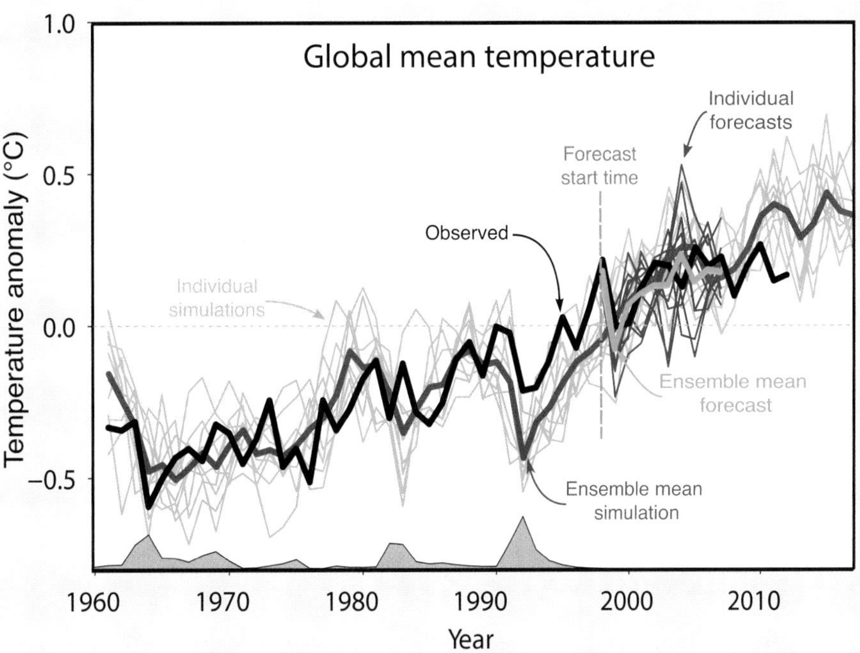

Box 11.1, Figure 1 | The evolution of observation-based global mean temperature T (the black line) as the difference from the 1986–2005 average together with an ensemble of externally forced simulations to 2005 and projections based on the RCP4.5 scenario thereafter (the yellow lines). The model-based estimate of the externally forced component T_f (the red line) is the average over the ensemble of simulations. To the extent that the red line correctly estimates the forced component, the difference between the black and red lines is the internally generated component T_i for global mean temperature. An ensemble of forecasts of global annual mean temperature, initialized in 1998, is plotted as thin purple lines and their average, the ensemble mean forecast, as the thick green line. The grey areas along the axis indicate the presence of external forcing associated with volcanoes.

11

Box 11.1 (continued)

Climate projection

A climate projection is a climate simulation that extends into the future based on a scenario of future external forcing. The simulations in Box 11.1, Figure 1 become climate projections for the period beyond 2005 where the results are based on the RCP4.5 forcing scenario (see Chapters 1 and 8 for a discussion of forcing scenarios).

Climate prediction, climate forecast

A climate prediction or climate forecast is a statement about the future evolution of some aspect of the climate system encompassing both forced and internally generated components. Climate predictions do not attempt to forecast the actual day-to-day progression of the system but instead the evolution of some climate statistic such as seasonal, annual or decadal averages or extremes, which may be for a particular location, or a regional or global average. Climate predictions are often made with models that are the same as, or similar to, those used to produce climate simulations and projections (assessed in Chapter 9). A climate prediction typically proceeds by integrating the governing equations forward in time from observation-based initial conditions. A decadal climate prediction combines aspects of both a forced and an initial condition problem as illustrated in Box 11.1, Figure 2. At short time scales the evolution is largely dominated by the initial state while at longer time scales the influence of the initial conditions decreases and the importance of the forcing increases as illustrated in Box 11.1, Figure 4. Climate predictions may also be made using statistical methods which relate current to future conditions using statistical relationships derived from past system behaviour.

Because of the chaotic and nonlinear nature of the climate system small differences, in initial conditions or in the formulation of the forecast model, result in different evolutions of forecasts with time. This is illustrated in Box 11.1, Figure 1, which displays an ensemble of forecasts of global annual mean temperature (the thin purple lines) initiated in 1998. The individual forecasts are begun from slightly different initial conditions, which are observation-based estimates of the state of the climate system. The thick green line is the average of these forecasts and is an attempt to predict the most probable outcome and to maximize forecast skill. In this schematic example, the 1998 initial conditions for the forecasts are warmer than the average of the simulations. The individual and ensemble mean forecasts exhibit a decline in global temperature before beginning to rise again. In this case, initialization has resulted in more realistic values for the forecasts than for the corresponding simulation, at least for short lead times in the forecast. As the individual forecasts evolve they diverge from one another and begin to resemble the projection results.

A probabilistic view of forecast behaviour is depicted schematically in Box 11.1, Figure 3. The probability distribution associated with the climate simulation of temperature evolves in response to external forcing. By contrast, the probability distribution associated with a climate forecast has a sharply peaked initial distribution representing the comparatively small uncertainty in the observation-based initial state. The forecast probability distribution broadens with time until, ultimately, it becomes indistinguishable from that of an uninitialized climate projection.

Climate predictability

The term 'predictability', as used here, indicates the extent to which even minor imperfections in the knowledge of the current state or of the representation of the system limits knowledge of subsequent states. The rate of separation or divergence of initially close states of the climate system with time (as for the light purple lines in Box 11.1, Figure 1), or the rate of displacement and broadening of its

(continued on next page)

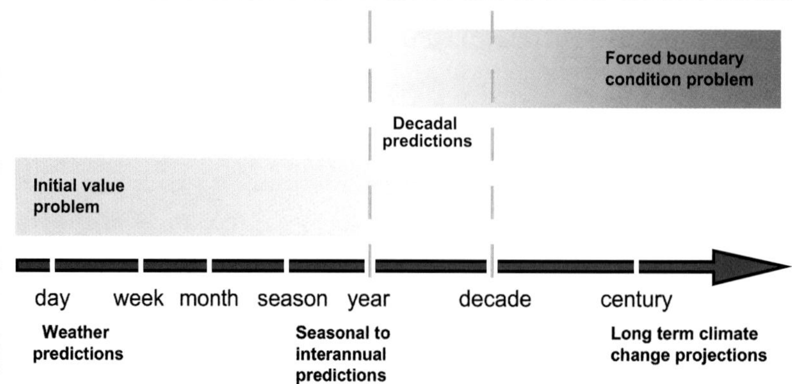

Box 11.1, Figure 2 | A schematic illustrating the progression from an initial-value based prediction at short time scales to the forced boundary-value problem of climate projection at long time scales. Decadal prediction occupies the middle ground between the two. (Based on Meehl et al., 2009b.)

Box 11.1 (continued)

probability distribution (as in Box 11.1, Figure 3) are indications of the system's predictability. If initially close states separate rapidly (or the probability distribution broadens quickly towards the climatological distribution), the predictability of the system is low and vice versa. Formally, predictability in climate science is a feature of the physical system itself, rather than of our 'ability to make skilful predictions in practice'. The latter depends on the accuracy of models and initial conditions and on the correctness with which the external forcing can be treated over the forecast period.

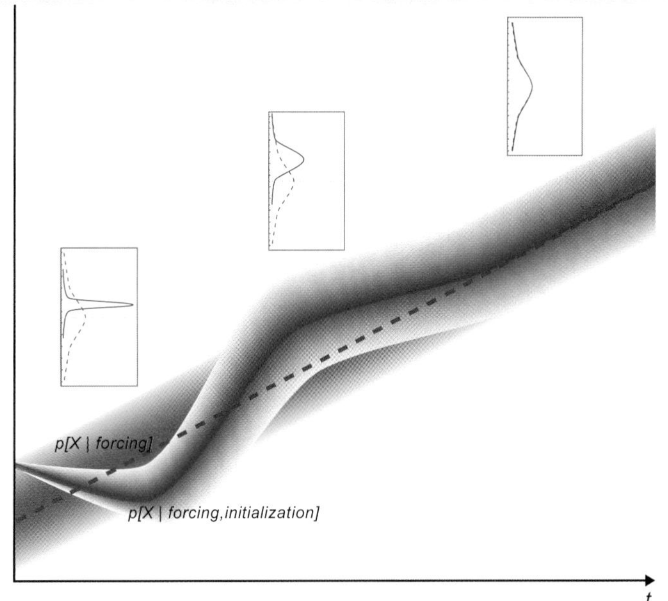

Box 11.1, Figure 3 | A schematic representation of prediction in terms of probability. The probability distribution corresponding to a forced simulation is in red, with the deeper shades indicating higher probability. The probabilistic forecast is in blue. The sharply peaked forecast distribution based on initial conditions broadens with time as the influence of the initial conditions fades until the probability distribution of the initialized prediction approaches that of an uninitialized projection. (Based on Branstator and Teng, 2010.)

Forecast quality, forecast skill

Forecast (or prediction) quality measures the success of a prediction against observation-based information. Forecasts made for past cases, termed retrospective forecasts or hindcasts, may be analysed to give an indication of the quality that may be expected for future forecasts for a particular variable at a particular location.

The relative importance of initial conditions and of external forcing for climate prediction, as depicted schematically in Box 11.1, Figure 2, is further illustrated in the example of Box 11.1, Figure 4 which plots correlation measures of both forecast skill and predictability for temperature averages over the globe ranging from a month to a decade. Initialized forecasts exhibit enhanced values compared to uninitialized simulations for shorter time averages but the advantage declines as averaging time increases and the forced component grows in importance.

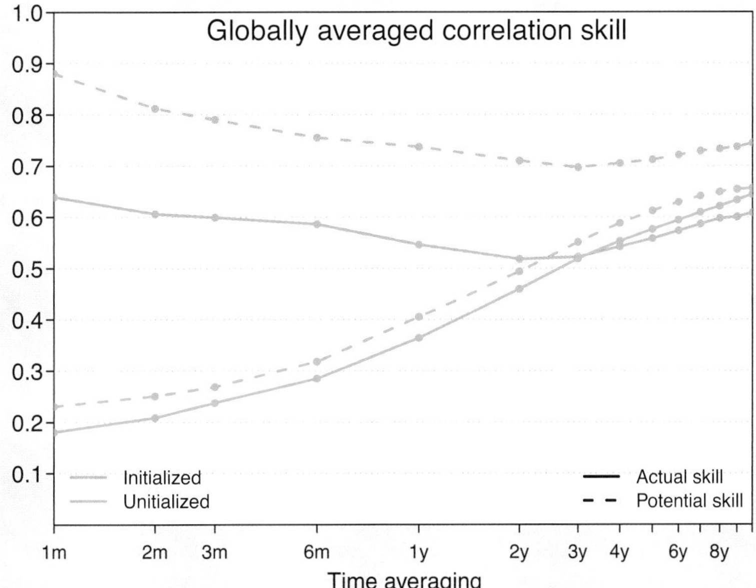

Box 11.1, Figure 4 | An example of the relative importance of initial conditions and external forcing for climate prediction and predictability. The global average of the correlation skill score of ensemble mean initialized forecasts are plotted as solid orange lines and the corresponding model-based predictability measure as dashed orange lines. The green lines are the same quantities but for uninitialized climate simulations. Results are for temperature averaged over periods from a month to a decade. Values plotted for the monthly average correspond to the first month, those for the annual average to the first year and so on up to the decadal average. (Based on Boer et al., 2013.)

11

11.2 Near-term Predictions

11.2.1 Introduction

11.2.1.1 Predictability Studies

The innate behaviour of the climate system imposes limits on the ability to predict its evolution. Small differences in initial conditions, external forcing and/or in the representation of the behaviour of the system produce differences in results that limit useful prediction. Predictability studies estimate predictability limits for different variables and regions.

11.2.1.2 Prognostic Predictability Studies

Prognostic predictability studies analyse the behaviour of models integrated forward in time from perturbed initial conditions. The study of Griffies and Bryan (1997) is one of the earliest studies of the predictability of internally generated decadal variability in a coupled atmosphere–ocean climate model. The study concentrates on the North Atlantic and the subsurface ocean temperature while the subsequent studies of Boer (2000) and Collins (2002) deal mainly with surface temperature. Long time scale temperature variability in the North Atlantic has received considerable attention together with its possible connection to the variability of the Atlantic Meridional Overturning Circulation (AMOC) in predictability studies by Collins and Sinha (2003), Collins et al. (2006), Dunstone and Smith (2010), Dunstone et al. (2011), Grotzner et al. (1999), Hawkins and Sutton (2009), Latif et al. (2006, 2007),)Msadek et al. (2010), Persechino et al. (2012), Pohlmann et al. (2004, 2013), Swingedouw et al. (2013), and Teng et al. (2011). The predictability of the AMOC varies among models and, to some extent, with initial model states, ranging from several to 10 or more years. The predictability values are model-based and the realism of the simulated AMOC in the models cannot be easily judged in the absence of a sufficiently long record of observation-based AMOC values. Many predictability studies are based on perturbations to surface quantities but Sevellec and A. Fedorov (2012) and Zanna (2012) note that small perturbations to deep ocean quantities may also affect upper ocean values. The predictability of the North Atlantic sea surface temperature (SST) is typically weaker than that of the AMOC and the connection between the predictability of the AMOC, and the SST is inconsistent among models.

Prognostic predictability studies of the Pacific are less plentiful although Pacific Decadal Variability (PDV) mechanisms (including the Pacific Decadal Oscillation (PDO) and the Inter-decadal Pacific Oscillation (IPO) have received considerable study (see Chapters 2 and 12). Power and Colman (2006) find predictability on multi-year time scales in SST and on decadal time-scales in the sub-surface ocean temperature in the off-equatorial South Pacific in their model. Power et al. (2006) find no evidence for the predictability of inter-decadal changes in the nature of El Niño-Southern Oscillation (ENSO) impacts on Australian rainfall. Sun and Wang (2006) suggest that some of the temperature variability linked to PDV can be predicted approximately 7 years in advance. Teng et al. (2011) investigate the predictability of the first two Empirical Orthogonal Functions (EOFs) of annual mean SST and upper ocean temperature identified with PDV and find predictability of the order of 6 to 10 years. Meehl et al. (2010) consider the predictability of 19-year filtered Pacific SSTs in terms of low order EOFs and find predictability on these long time scales.

Hermanson and Sutton (2010) report that predictable signals in different regions and for different variables may arise from differing initial conditions and that ocean heat content is more predictable than atmospheric and surface variables. Branstator and Teng (2010) analyse upper ocean temperatures, and some SSTs, for averages over the North Atlantic, North Pacific and the tropical Atlantic and Pacific in the National Center for Atmospheric Research (NCAR) model. Predictability associated with the initial state of the system decreases whereas that due to external forcing increases with time. The 'cross-over' time, when the two contributions are equal, is longer in extratropical (7 to 11 years) compared to tropical (2 years) regions and in the North Atlantic compared to the North Pacific. Boer et al. (2013) estimate surface air (rather than upper ocean) temperature predictability in the Canadian Centre for Climate Modelling and Analysis (CCCma) model and find a cross-over time (using a different measure) on the order of 3 years when averaged over the globe.

11.2.1.3 Diagnostic Predictability Studies

Diagnostic predictability studies are based on analyses of the observed record or the output of climate models. Because long data records are needed, diagnostic multi-annual to decadal predictability studies based on observational data are comparatively few. Newman (2007) and Alexander et al. (2008) develop multivariate empirical Linear Inverse Models (LIMs) from observation-based SSTs and find predictability for ENSO and PDV type patterns that are generally limited to the order of a year although exceeding this in some areas. Zanna (2012) develops a LIM based on Atlantic SSTs and infers the possibility of decadal scale predictability. Hoerling et al. (2011) appeal to forced climate change relative to the 1971–2000 period together with the statistics of natural variability to infer the potential for the prediction of temperature over North America for 2011–2020.

Tziperman et al. (2008) apply LIM-based methods to Geophysical Fluid Dynamics Laboratory (GFDL) model output, as do Hawkins and Sutton (2009) and Hawkins et al. (2011) to Hadley Centre model output and find predictability up to a decade or more for the AMOC and North Atlantic SST. Branstator et al. (2012) use analog and multivariate linear regression methods to quantify the predictability of the internally generated component of upper ocean temperature in results from six coupled models. Results differ considerably across models but offer some areas of commonality. Basin-average estimates indicate predictability for up to a decade in the North Atlantic and somewhat less in the North Pacific. Branstator and Teng (2012) assess the predictability of both the internally generated and forced component of upper ocean temperature in results from 12 coupled models participating in CMIP5. They infer potential predictability from initializing the internally generated component for 5 years in the North Pacific and 9 years in the North Atlantic while the forced component dominates after 6.5 and 8 years in the two basins. Results vary among models, although with some agreement for internal component predictability in subpolar gyre regions.

Studies of 'potential predictability' take a number of forms but broadly assume that overall variability may be separated into a long time scale component of interest and shorter time scale components that are unpredictable on these long time scales, written symbolically as $\sigma^2_X = \sigma^2_v + \sigma^2_\varepsilon$. The fraction $p = \sigma^2_v / \sigma^2_X$ is a measure of potentially predictable variance provided that hypothesis that σ^2_v is zero may be rejected. Small p indicates either a lack of long time scale variability or its smallness as a fraction of the total. Predictability is 'potential' in the sense that the existence of appreciable long time scale variability is not a direct indication that it may be skilfully predicted. There are a number of approaches to estimating potential predictability each with its statistical difficulties (e.g., DelSole and Feng, 2013). At multi-annual time scales the potential predictability of the internally generated component of temperature is studied in Boer (2000), Collins (2002), Pohlmann et al. (2004), Power and Colman (2006) and, in a multi-model context, in Boer (2004) and Boer and Lambert (2008). Power and Colman (2006) report that potential predictability in the ocean tends to increase with latitude and depth. Multi-model results

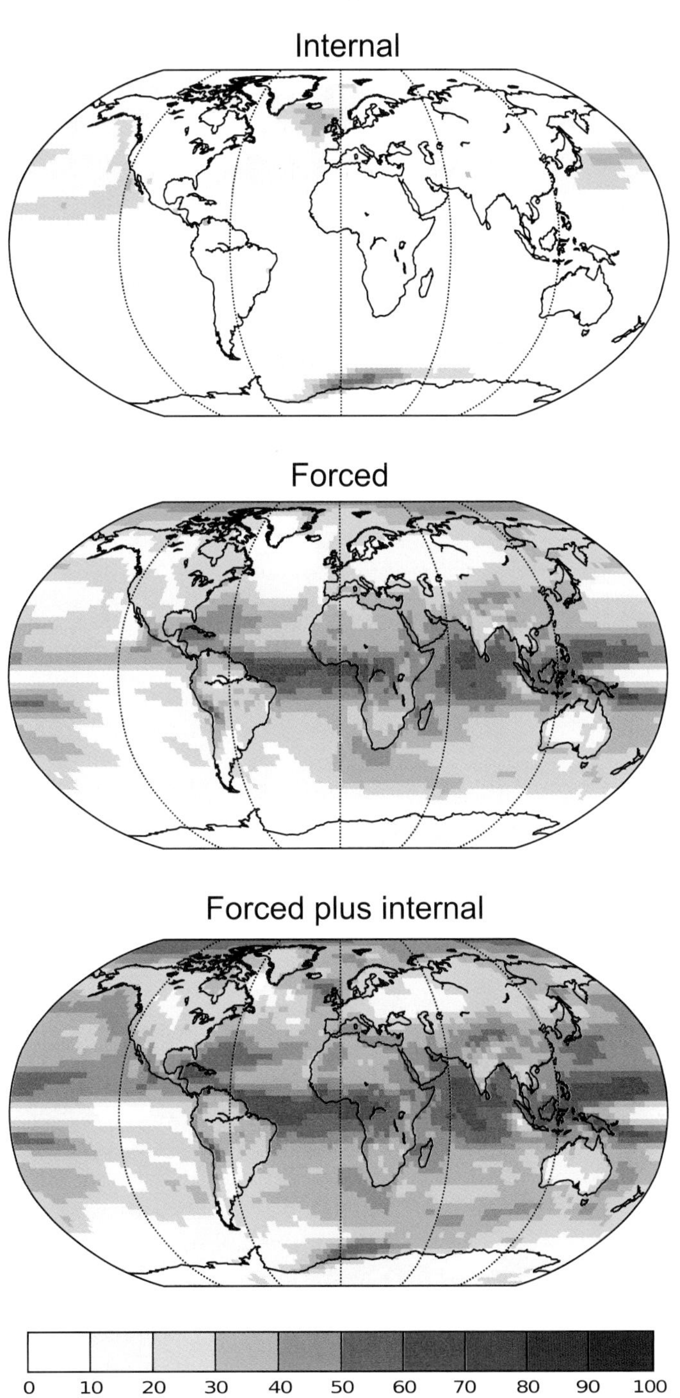

Figure 11.1 | The potential predictability of 5-year means of temperature (lower), the contribution from the forced component (middle) and from the internally generated component (upper). These are multi-model results from CMIP5 RCP4.5 scenario simulations from 17 coupled climate models following the methodology of Boer (2011). The results apply to the early 21st century.

Frequently Asked Questions

FAQ 11.1 | If You Cannot Predict the Weather Next Month, How Can You Predict Climate for the Coming Decade?

Although weather and climate are intertwined, they are in fact different things. Weather is defined as the state of the atmosphere at a given time and place, and can change from hour to hour and day to day. Climate, on the other hand, generally refers to the statistics of weather conditions over a decade or more.

An ability to predict future climate without the need to accurately predict weather is more commonplace that it might first seem. For example, at the end of spring, it can be accurately predicted that the average air temperature over the coming summer in Melbourne (for example) will very likely be higher than the average temperature during the most recent spring—even though the day-to-day weather during the coming summer cannot be predicted with accuracy beyond a week or so. This simple example illustrates that factors exist—in this case the seasonal cycle in solar radiation reaching the Southern Hemisphere—that can underpin skill in predicting changes in climate over a coming period that does not depend on accuracy in predicting weather over the same period.

The statistics of weather conditions used to define climate include long-term averages of air temperature and rainfall, as well as statistics of their variability, such as the standard deviation of year-to-year rainfall variability from the long-term average, or the frequency of days below 5°C. Averages of climate variables over long periods of time are called climatological averages. They can apply to individual months, seasons or the year as a whole. A climate prediction will address questions like: 'How likely will it be that the average temperature during the coming summer will be higher than the long-term average of past summers?' or: 'How likely will it be that the next decade will be warmer than past decades?' More specifically, a climate prediction might provide an answer to the question: 'What is the probability that temperature (in China, for instance) averaged over the next ten years will exceed the temperature in China averaged over the past 30 years?' Climate predictions do not provide forecasts of the detailed day-to-day evolution of future weather. Instead, they provide probabilities of long-term changes to the statistics of future climatic variables.

Weather forecasts, on the other hand, provide predictions of day-to-day weather for specific times in the future. They help to address questions like: 'Will it rain tomorrow?' Sometimes, weather forecasts are given in terms of probabilities. For example, the weather forecast might state that: 'the likelihood of rainfall in Apia tomorrow is 75%'.

To make accurate weather predictions, forecasters need highly detailed information about the current state of the atmosphere. The chaotic nature of the atmosphere means that even the tiniest error in the depiction of 'initial conditions' typically leads to inaccurate forecasts beyond a week or so. This is the so-called 'butterfly effect'.

Climate scientists do not attempt or claim to predict the detailed future evolution of the weather over coming seasons, years or decades. There is, on the other hand, a sound scientific basis for supposing that aspects of climate can be predicted, albeit imprecisely, despite the butterfly effect. For example, increases in long-lived atmospheric greenhouse gas concentrations tend to increase surface temperature in future decades. Thus, information from the past can and does help predict future climate.

Some types of naturally occurring so-called 'internal' variability can—in theory at least—extend the capacity to predict future climate. Internal climatic variability arises from natural instabilities in the climate system. If such variability includes or causes extensive, long-lived, upper ocean temperature anomalies, this will drive changes in the overlying atmosphere, both locally and remotely. The El Niño-Southern Oscillation phenomenon is probably the most famous example of this kind of internal variability. Variability linked to the El Niño-Southern Oscillation unfolds in a partially predictable fashion. The butterfly effect is present, but it takes longer to strongly influence some of the variability linked to the El Nino-Southern Oscillation.

Meteorological services and other agencies have exploited this. They have developed seasonal-to-interannual prediction systems that enable them to routinely predict seasonal climate anomalies with demonstrable predictive skill. The skill varies markedly from place to place and variable to variable. Skill tends to diminish the further the prediction delves into the future and in some locations there is no skill at all. 'Skill' is used here in its technical sense: it is a measure of how much greater the accuracy of a prediction is, compared with the accuracy of some typically simple prediction method like assuming that recent anomalies will persist during the period being predicted.

Weather, seasonal-to-interannual and decadal prediction systems are similar in many ways (e.g., they all incorporate the same mathematical equations for the atmosphere, they all need to specify initial conditions to kick-start

(continued on next page)

FAQ 11.1 (continued)

predictions, and they are all subject to limits on forecast accuracy imposed by the butterfly effect). However, decadal prediction, unlike weather and seasonal-to-interannual prediction, is still in its infancy. Decadal prediction systems nevertheless exhibit a degree of skill in *hindcasting* near-surface temperature over much of the globe out to at least nine years. A 'hindcast' is a prediction of a past event in which only observations prior to the event are fed into the prediction system used to make the prediction. The bulk of this skill is thought to arise from *external forcing*. 'External forcing' is a term used by climate scientists to refer to a forcing agent outside the climate system causing a change in the climate system. This includes increases in the concentration of long-lived greenhouse gases.

Theory indicates that skill in predicting decadal precipitation should be less than the skill in predicting decadal surface temperature, and hindcast performance is consistent with this expectation.

Current research is aimed at improving decadal prediction systems, and increasing the understanding of the reasons for any apparent skill. Ascertaining the degree to which the extra information from internal variability actually translates to increased skill is a key issue. While prediction systems are expected to improve over coming decades, the chaotic nature of the climate system and the resulting butterfly effect will always impose unavoidable limits on predictive skill. Other sources of uncertainty exist. For example, as volcanic eruptions can influence climate but their timing and magnitude cannot be predicted, future eruptions provide one of a number of other sources of uncertainty. Additionally, the shortness of the period with enough oceanic data to initialize and assess decadal predictions presents a major challenge.

Finally, note that decadal prediction systems are designed to exploit both externally forced and internally generated sources of predictability. Climate scientists distinguish between decadal predictions and decadal projections. Projections exploit only the predictive capacity arising from external forcing. While previous IPCC Assessment Reports focussed exclusively on projections, this report also assesses decadal prediction research and its scientific basis.

for both externally forced and internally generated components of the potential predictability of decadal means of surface air temperature in simulations of 21st century climate in CMIP3 model data are analysed in Boer (2011) and results based on CMIP5 model data are shown in Figure 11.2. Potential predictability of 5-year means for internally generated variability is found over extratropical oceans but is generally weak over land while that associated with the decadal change in the forced component is found in tropical areas and over some land areas.

Predictability studies of precipitation on long time scales are comparatively few. Jai and DelSole (2012) identify 'optimally predictable' fractions of internally generated temperature and precipitation variance over land on multi-year time scales in the control simulations of 10 models participating in CMIP5, with results that vary considerably from model to model. Boer and Lambert (2008) find little potential predictability for decadal means of precipitation in the internally generated variability of a collection of CMIP3 model control simulations other than over parts of the North Atlantic. This is also the case for the internally generated component of CMIP3 precipitation in 21st century climate change simulations in Boer (2011) although there is evidence of potential predictability for the forced component of precipitation mainly at higher latitudes and for longer time scales.

11.2.1.4 Summary

Predictability studies suggest that initialized climate forecasts should be able to provide more detailed information on climate evolution, over a few years to a decade, than is available from uninitialized climate simulations alone. Predictability results are, however, based mainly on climate model results and depend on the verisimilitude with which the models reproduce climate system behaviour (Chapter 9). There is evidence of multi-year predictability for both the internally generated and externally forced components of temperature over considerable portions of the globe with the first dominating at shorter and the second at longer time scales. Predictability for precipitation is based on fewer studies, is more modest than for temperature, and appears to be associated mainly with the forced component at longer time scales. Predictability can also vary from location to location.

11.2.2 Climate Prediction on Decadal Time Scales

11.2.2.1 Initial Conditions

A dynamical prediction consists of an ensemble of forecasts produced by integrating a climate model forward in time from a set of observation-based initial conditions. As the forecast range increases, processes in the ocean become increasingly important and the sparseness, non-uniformity and secular change in sub-surface ocean observations is a challenge to analysis and prediction (Meehl et al., 2009b, 2013d; Murphy et al., 2010) and can lead to differences among ocean analyses, that is, quantified descriptions of ocean initial conditions (Stammer, 2006; Keenlyside and Ba, 2010). Approaches to ocean initialization include (as listed in Table 11.1): assimilation only of SSTs to initialize the sub-surface ocean indirectly (Keenlyside et al., 2008;

Dunstone, 2010; Swingedouw et al., 2013); the forcing of the ocean model with atmospheric observations (e.g., Du et al., 2012; Matei et al., 2012b; Yeager et al., 2012) and more sophisticated alternatives based on fully coupled data assimilation schemes (e.g., Zhang et al., 2007a; Sugiura et al., 2009).

Dunstone and Smith (2010) and Zhang et al. (2010a) found an expected improvement in skill when sub-surface information was used as part of the initialization. Assimilation of atmospheric data, on the other hand, is expected to have little impact after the first few months (Balmaseda and Anderson, 2009). The initialization of sea ice, snow cover, frozen soil and soil moisture can potentially contribute to seasonal and sub-seasonal skill (e.g., Koster et al., 2010; Toyoda et al., 2011; Chevallier and Salas-Melia, 2012; Paolino et al., 2012), although an assessment of their benefit at longer time scales has not yet been determined.

11.2.2.2 Ensemble Generation

An ensemble can be generated in many different ways and a wide range of methods have been explored in seasonal prediction (e.g., Stockdale et al., 1998; Stan and Kirtman, 2008) but not yet fully investigated for decadal prediction (Corti et al., 2012). Methods being investigated include adding random perturbations to initial conditions, using atmospheric states displaced in time, using parallel assimilation runs (Doblas-Reyes et al., 2011; Du et al., 2012) and perturbing ocean initial conditions (Zhang et al., 2007a; Mochizuki et al., 2010). Perturbations leading to rapidly growing modes, common in weather forecasting, have also been investigated (Kleeman et al., 2003; Vikhliaev et al., 2007; Hawkins and Sutton, 2009, 2011; Du et al., 2012). The uncertainty associated with the limitations of a model's representation of the climate system may be partially represented by perturbed physics (Stainforth et al., 2005; Murphy et al., 2007) or stochastic physics (Berner et al., 2008), and applied to multi-annual and decadal predictions (Doblas-Reyes et al., 2009; Smith et al., 2010). Weisheimer et al. (2011) compare these three approaches in a seasonal prediction context.

The multi-model approach, which is used widely and most commonly, combines ensembles of predictions from a collection of models, thereby increasing the sampling of both initial conditions and model properties. Multi-model approaches are used across time scales ranging from seasonal–interannual (e.g., DEMETER; Palmer et al. (2004), to seasonal-decadal (e.g., Weisheimer et al., 2011; van Oldenborgh et al., 2012), in climate change simulation (e.g., IPCC, 2007, Chapter 10; Meehl et al., 2007b) and in the ENSEMBLES and CMIP5-based decadal predictions assessed in Section 11.2.3. A problem with the multi-model approach is tha inter-dependence of the climate models used in current forecast systems (Power et al. 2012; Knutti et al. 2013) is expected to lead to co-dependence of forecast error.

11.2.3 Prediction Quality

11.2.3.1 Decadal Prediction Experiments

Decadal predictions for specific variables can be made by exploiting empirical relationships based on past observations and expected physical relationships. Predictions of North Pacific Ocean temperatures have been achieved using prior wind stress observations (Schneider and Miller, 2001). Both global and regional predictions of surface temperature have been made based on projected changes in external forcing and the observed state of the natural variability at the start date (Lean and Rind, 2009; Krueger and von Storch, 2011; Ho et al., 2012a; Newman, 2013). Some of these forecast systems are also used as benchmarks to compare with the dynamical systems under development. Comparisons (Newman (2013) have shown that there is similarity in the temperature skill between a linear inverse method and the CMIP5 hindcasts, pointing at a similarity in their sources of skill. In the future, the combination of information from empirical and dynamical predictions might be explored to provide a unified and more skilful source of information.

Evidence for skilful interannual to decadal temperatures using dynamical models forced only by previous and projected changes in anthropogenic greenhouse gases (GHGs) and aerosols and natural variations in volcanic aerosols and solar irradiance is reported by Lee et al. (2006b), Räisänen and Ruokolainen (2006) and Laepple et al. (2008). Some attempts to predict the 10-year climate over regions have been done using this approach, and include assessments of the role of the internal decadal variability (Hoerling et al., 2011). To be clear, in the context of this report these studies are viewed as projections because no attempt is made to use observational estimates for the initial conditions. Essentially, an uninitialized prediction is synonymous with a projection. These projections or uninitialized predictions are referred to synonymously in the literature as 'NoInit,' or 'NoAssim', referring to the fact that no assimilated observations are used for the specification of the initial conditions.

Additional skill can be realized by initializing the models with observations in order to predict the evolution of the internally generated component and to correct the model's response to previously imposed forcing (Smith et al., 2010; Fyfe et al., 2011; Kharin et al., 2012; Smith et al., 2012). Again, to be clear, the assessment provided here distinguishes between predictions in which attempts are made to initialize the models with observations, and projections. See Box 11.1 and FAQ 11.1 for further details.

The ENSEMBLES project (van Oldenborgh et al., 2012), for example, has conducted a multi-model decadal retrospective prediction study, and the Coupled Model Intercomparison Project phase 5 (CMIP5) proposed a coordinated experiment that focuses on decadal, or near-term, climate prediction (Meehl et al., 2009b; Taylor et al., 2012). Prior to these initiatives, several pioneering attempts at initialized decadal prediction were made (Pierce et al., 2004; Smith et al., 2007; Troccoli and Palmer, 2007; Keenlyside et al., 2008; Pohlmann et al., 2009; Mochizuki et al., 2010). Results from the CMIP5 coordinated experiment (Taylor et al., 2012) are the basis for the assessment reported here.

Because the practice of decadal prediction is in its infancy, details of how to initialize the models included in the CMIP5 near-term experiment were left to the discretion of the modelling groups and are described in Meehl et al. (2013d) and Table 11.1. In CMIP5 experiments, volcanic aerosol and solar cycle variability are prescribed along the integration using observation-based values up to 2005, and assuming a climatological 11-year solar cycle and a background volcanic aerosol load in the future. These forcings are shared with CMIP5 historical runs (i.e., unintialized projections) started from pre-industrial control simulations, enabling an assessment of the impact of initialization. The specification of the volcanic aerosol load and the solar irradiance in the hindcasts gives an optimistic estimate of the forecast quality with respect to an operational prediction system, where no such future information can be used. Table 11.1 summarizes forecast systems contributing to, and the initialization methods used in, the CMIP5 near-term experiment.

The coordinated nature of the ENSEMBLES and CMIP5 experiments also offers a good opportunity to study *multi-model* ensembles (Garcia-Serrano and Doblas-Reyes, 2012; van Oldenborgh et al., 2012) as a means of sampling model uncertainty while some modelling groups have also investigated this using perturbed parameter approaches (Smith et al., 2010). The relative merit of the different approaches for decadal predictions has yet to be assessed.

When initialized with states close to the observations, models 'drift' towards their imperfect climatology (an estimate of the mean climate), leading to biases in the simulations that depend on the forecast time. The time scale of the drift in the atmosphere and upper ocean is, in most cases, a few years (Hazeleger et al., 2013a). Biases can be largely removed using empirical techniques a posteriori (Garcia-Serrano and Doblas-Reyes, 2012; Kharin et al., 2012). The bias correction or adjustment linearly corrects for model drift (e.g., Stockdale, 1997; Garcia-Serrano et al., 2012; Gangstø et al., 2013). The approach assumes that the model bias is stable over the prediction period (from 1960 onward in the CMIP5 experiment). This might not be the case if, for instance, the predicted temperature trend differs from the observed trend (Fyfe et al., 2011; Kharin et al., 2012). Figure 11.2 is an illustration of the time scale of the global SST drift, while at the same time showing the systematic error of several of the forecast systems contributing to CMIP5. It is important to note that the systematic errors illustrated here are

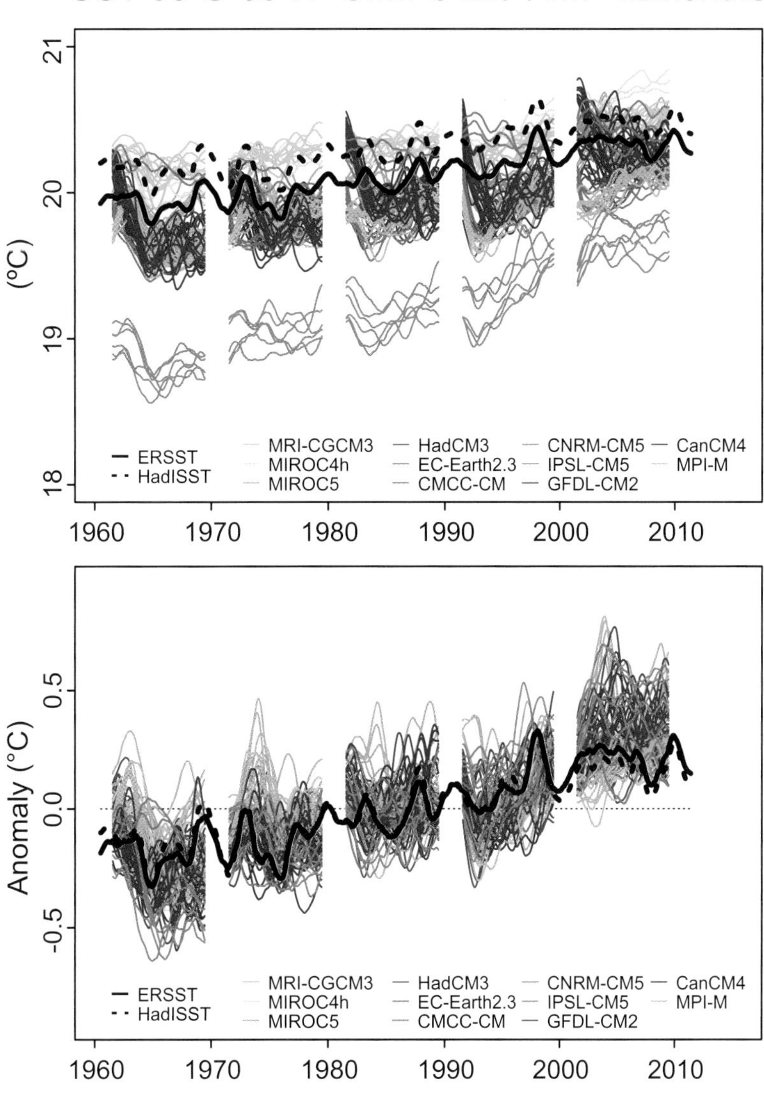

Figure 11.2 | Time series of global mean sea surface temperature from the (a) direct model output and (b) anomalies of the CMIP5 multi-model initialized hindcasts. Results for each forecast system are plotted with a different colour, with each line representing an individual member of the ensemble. Results for the start dates 1961, 1971, 1981, 1991 and 2001 are shown, while the model and observed climatologies to obtain the anomalies in (b) have been estimated using data from start dates every five years. The reference data (ERSST) is drawn in black. All time series have been smoothed with a 24-month centred moving average that filters out the seasonal cycle and removes data for the first and last years of each time series.

Table 11.1 | Initialization methods used in models that entered CMIP5 near-term experiments. (Figures 11.3 to 11.7 have been prepared using those contributions with asterisk on top of the modelling centre's name.).

CMIP5 Near-term Players	CMIP5 official model id	AGCM	OGCM	Initialization				Perturbation		Aerosol		Reference
Name of modeling centre (or group)				Atmosphere/Land	Ocean	Sea Ice	Anomaly Assimilation?	Atmosphere	Ocean	Concentration (C) /Emission (E)	Direct(D)/ Indirect (I1,I2)	
(*) Beijing Climate Center, China Meteorological Administration (BCC) China	BCC-CSM 1.1	2.8°L26	1°L40	No	SST, T&S (SODA)	No	No	Perturbed atmosphere/ ocean/land/sea ice		C	D	Xin et al. (2013)
(*) Canadian Centre for Climate Modelling and Analysis (CCCMA) Canada	CanCM4	2.8°L35	1.4° × 0.9°L40	ERA40/Interim	SST (ERSST&OISST), T&S (SODA & GODAS)	HadISST1.1	No	Ensemble assimilation		E	D, I1	Merryfield et al. (2013)
(*) Centro Euro-Mediterraneo per I Cambiamenti Climatici (CMCC-CM) Italy	CMCC-CM	0.75°L31	0.5 °–2° L31	No	SST, T&S (INGV ocean analysis)	CMCC-CM climatology	No	Ensemble assimilation		C	D, I1	Bellucci et al. (2013)
(*) Centre National de Recherches Metéorologiques, and Centre Européen de Recherche et Formation Avancées en Calcul Scientifique (CNRM-CERFACS) France	CNRM-CM5	1.4°L31	1°L42	No	T&S (NEMOVAR-COMBINE)	No	No	1st day atmospheric conditions	No	C	D, I1	Meehl et al. (2013d)
National Centers for Environmental Prediction and Center for Ocean-Land-Atmosphere Studies (NCEP and COLA) USA	CFSv2-2011	0.9°L64	0.25–0.5°L40	NCEP CFSR reanalysis	NCEP CFSR ocean analysis (NCEP runs) NEMOVAR-S4 ocean analysis (COLA runs)	NCEP CFSR reanalysis	No	No	No	C	D, I1	Saha et al. (2010)
(*) EC-EARTH consortium (EC-EARTH) Europe	EC-EARTH	1.1°L62	1°L42	ERA40/Interim	Ocean assimilation (ORAS4/NEMOVAR S4)	NEMO3.2-LIM2 forced with DFS4.3	No (KNMI & IC3) yes (SMHI)	Start dates and singular vectors	Ensemble ocean assim (NEM-OVAR)	C	D	Du et al. (2012) Hazeleger et al. (2013a)
(*) Institut Pierre-Simon Laplace (IPSL) France	IPSL-CM5A-LR	1.9 × 3.8° L39	2°L31	No	SST anomalies (Reynolds observations)	No	Yes	No	White noise on SST	C	D, I1	Swingedouw et al. (2013)
(*) AORI/NIES/JAMSTEC, Japan	MIROC4h MIROC5	0.6°L56 1.4°L40	0.3°L48 1.4°L50	No	SST, T&S (Ishii and Kimoto, 2009)	No	Yes	Start dates and ensemble assimilation		E	D,I1,I2	Tatebe et al. (2012)
(*) Met Office Hadley Centre (MOHC) UK	HadCM3	3.8°L19	1.3°L20	ERA40/ECMWF operational analysis	SST, T&S (Smith and Murphy, 2007)	HADISST	Yes, also full field	No	SST perturbation	E	D	Smith et al. (2013a)

(continued on next page)

(Table 11.1 continued)

CMIP5 Near-term Players	CMIP5 Official Model ID	AGCM	OGCM	Initialization				Perturbation		Aerosol		Reference
Name of Modeling Centre (or group)				Atmosphere/Land	Ocean	Sea Ice	Anomaly Assimilation?	Atmosphere	Ocean	Concentration (C) /Emission (E)	Direct(D)/ Indirect (I1,I2)	
(*) Max Planck Institute for Meteorology (MPI-M) Germany	MPI-ESM -LR	1.9°L47	1.5°L40	No	T&S from forced OGCM	No	Yes	1-day lagged		C	D	Matei et al. (2012b)
	MPI-ESM -MR	1.9°L95	0.4°L40									
(*) Meteorological Research Institute (MRI) Japan	MRI-CGCM3	1.1°L48	1°L51	No	SST, T&S (Ishii and Kimoto, 2009)	No	Yes	Start dates and ensemble assimilation		E	D,I1,I2	Tatebe et al. (2012)
Global Modeling and Assimilation Office, (NASA) USA	GEOS-5	2.5°×2° L72	1°L50	MERRA	T&S from ocean assimilation (GEOS iODAS)	GEOS iODAS reanalysis	No	Two-sided breeding method		E	D	
(*) National Center for Atmospheric Research (NCAR) USA	CCSM4	1.3°L26	1.0°L60	No	Ocean assimilation (POPDART)	Ice state from forced ocean-ice GCM (strong salinity restoring for POPDART)	No	Single atm from AMIP run	Ensemble assimilation	E	D	
					Ocean state from forced ocean-ice GCM			Staggered atm start dates from uninitialized run	Single member ocean			Yeager et al. (2012)
(*) Geophysical Fluid Dynamics Laboratory (GFDL) USA	GFDL-CM 2.1	2.5°L24	1°L50	NCEP reanalysis	Ocean observations of 3-D T & S & SST	No	No	Coupled EnKF		C	D	Yang et al. (2013)
LASG, Institute of Atmospheric Physics, Chinese Academy of Sciences; and CESS, Tsinghua University China	FGOALS-g2	2.8°L26	1°L30	No	SST, T&S (Ishii et al., 2006)	No	No	A simplified scheme of 3DVar		C	D, I1	Wang et al. (2013)
LASG, Institute of Atmospheric Physics, Chinese Academy of Sciences China, Tsinghua University China	FGOALS-s2	2.8°L26	1°L30	No	T&S (EN3_v2a)	No	Yes	Incremental Analysis Updates (IAU) scheme		C	D	Wu and Zhou (2012)

common to both decadal prediction systems and climate-change projections. The bias adjustment itself is another important source of uncertainty in climate predictions (e.g., Ho et al., 2012b). There may be nonlinear relationships between the mean state and the anomalies, that are neglected in linear bias adjustment techniques. There are also difficulties in estimating the drift in the presence of volcanic eruptions.

It has been recognized that including as many initial states as possible in computing the drift and adjusting the bias is more desirable than a greater number of ensemble members per initial state (Meehl et al., 2013d), although increasing both is desirable to obtain robust forecast quality estimates. A procedure for bias adjustment following the technique outlined above has been recommended for CMIP5 (ICPO, 2011). A suitable adjustment depends also on there being a sufficient number of hindcasts for statistical robustness (Garcia-Serrano et al., 2012; Kharin et al., 2012).

To reduce the impact of the drift many of the early attempts at decadal prediction (Smith et al., 2007; Keenlyside et al., 2008; Pohlmann et al., 2009; Mochizuki et al., 2010) use an approach called anomaly initialization (Schneider et al., 1999; Pierce et al., 2004; Smith et al., 2007). The anomaly initialization approach attempts to circumvent model drift and the need for a time-varying bias correction. The models are initialized by adding observed anomalies to an estimate of the model mean climate. The mean model climate is subsequently subtracted from the predictions to obtain forecast anomalies. Sampling error in the estimation of the mean climatology affects the success of this approach. This is also the case for full-field initialization, although as anomaly initialisation is affected to a smaller degree by the drift, the sampling error is assumed to be smaller (Hazeleger et al., 2013a). The relative merits of anomaly versus full initialization are being quantified (Hazeleger et al., 2013a; Magnusson et al., 2013; Smith et al., 2013a), although no initialization method was found to be definitely better in terms of forecast quality. Another less widely explored alternative is dynamic bias correction in which multi-year monthly mean analysis increments are added during the integration of the ocean model (Wang et al., 2013). Figure 11.2 includes predictions performed with both full and anomaly initialization systems.

11.2.3.2 Forecast Quality Assessment

The quality of a forecast system is assessed by estimating, among others, the accuracy, skill and reliability of a set of hindcasts (Jolliffe and Stephenson, 2011). These three terms—accuracy, skill and reliability—are used here in a strict technical sense. A suite of measures needs to be considered, particularly when a forecast system are compared. The accuracy of a forecast system refers to the average distance/error between forecasts and observations. The skill score is a relative measure of the quality of the forecasting system compared to some benchmark or reference forecast (e.g., climatology or persistence). The reliability, which is a property of the specific forecast system, measures the trustworthiness of the predictions. Reliability measures how well the predicted probability distribution matches the observed relative frequency of the forecast event. Accuracy and reliability are aspects of forecast quality that can be improved by improving the individual forecast systems or by combining several of them into a multi-model prediction. The reliability can be improved by *a posteriori* corrections to model spread. Forecast quality can also be improved by unequal weighting (Weigel et al., 2010; DelSole et al., 2013), although this option has not been explored in decadal prediction to date, because a long training sample is required to obtain robust weights.

The assessment of forecast quality depends on the quantities of greatest interest to those who use the information. World Meteorological Organization (WMO)'s Standard Verification System (SVS) for Long-Range Forecasts (LRF) (WMO, 2002) outlines specifications for long-range (sub-seasonal to seasonal) forecast quality assessment. These measures are also described in Jolliffe and Stephenson (2011) and Wilks (2006). A recommendation for a deterministic metric for decadal climate predictions is the mean square skill score (MSSS), and for a probabilistic metric, the continuous ranked probability skill score (CRPSS) as described in Goddard et al. (2013) and Meehl et al. (2013d). For dynamical ensemble systems, a useful measure of the characteristics of an ensemble forecast system is spread. The relative spread can be described in terms of the ratio between the mean spread around the ensemble mean and the root mean square error (RMSE) of the ensemble-mean prediction, or spread-to-RMSE ratio. A ratio of 1 is considered a desirable feature for a Gaussian-distributed variable of a well-calibrated (i.e., reliable) prediction system (Palmer et al., 2006). The importance of using statistical inference in forecast quality assessments has been recently emphasized (Garcia-Serrano and Doblas-Reyes, 2012; Goddard et al., 2013). This is even more important when there are only small samples available (Kumar, 2009) and a small number of degrees of freedom (Gangstø et al., 2013). Confidence intervals for the scores are typically computed using either parametric or bootstrap methods (Lanzante, 2005; Jolliffe, 2007; Hanlon et al., 2013).

The skill of seasonal predictions can vary from generation to generation (Power et al. 1999) and from one generation of forecast systems to the next (Balmaseda et al., 1995). This highlights the possibility that the skill of decadal predictions might also vary from one period to another. Certain initial conditions might precede more predictable near-term states than other initial conditions, and this has the potential to be reflected in predictive skill assessments. However, the short length of the period available to initialize and verify the predictions makes the analysis of the variations in skill very difficult.

11.2.3.3 Pre-CMIP5 Decadal Prediction Experiments

Early decadal prediction studies found little additional predictive skill from initialization, over that due to changes in radiative forcing (RF), on global (Pierce et al., 2004) and regional scales (Troccoli and Palmer, 2007). However, neither of these studies considered more than two start dates. More comprehensive tests, which considered at least nine different start dates indicated temperature skill (Smith et al., 2007; Keenlyside et al., 2008; Pohlmann et al., 2009; Sugiura et al., 2009; Mochizuki et al., 2010; Smith et al., 2010; Doblas-Reyes et al., 2011; Garcia-Serrano and Doblas-Reyes, 2012; Garcia-Serrano et al., 2012; Kroger et al., 2012; Matei et al., 2012b; van Oldenborgh et al., 2012; Wu and Zhou, 2012; MacLeod et al., 2013). Moreover, this skill was enhanced by initialization (local increase in correlation of 0.1 to 0.3, depending on the system) mostly over the ocean, in particular over the North Atlantic and subtropical Pacific oceans. Regions with skill improvements from initialization for precipitation are small and rarely statistically significant (Goddard et al., 2013).

11.2.3.4 Coupled Model Intercomparison Project Phase 5 Decadal Prediction Experiments

Indices of global mean temperature, the Atlantic Multi-decadal Variability (AMV; (Trenberth and Shea, 2006)) and the Inter-decadal Pacific Oscillation (IPO; Power et al., 1999) or Pacific Decadal Oscillation (PDO) are used as benchmarks to assess the ability of decadal forecast systems to predict multi-annual averages of climate variability (Kim et al., 2012; van Oldenborgh et al., 2012; Doblas-Reyes et al., 2013; Goddard et al., 2013; see also Figure 11.3). Initialized predictions of global mean surface air temperature (GMST) for the following year are now being performed in almost-real time (Folland et al., 2013).

Non-initialized predictions (or projections) of the global mean temperature are statistically significantly skilful for most of the forecast ranges considered (*high confidence*), due to the almost monotonic increase in temperature, pointing to the importance of the time-varying RF (Murphy et al., 2010; Kim et al., 2012). This leads to a high (above 0.9) correlation of the ensemble mean prediction that varies very as a function of forecast lead time. This holds whether the changes in the external forcing (i.e., changes in natural and/or anthropogenic atmospheric composition) are specified (i.e., CMIP5) or are projected (ENSEMBLES). The skill of the multi-annual global mean surface temperature improves with initialization, although this is mainly evidenced when the accuracy is measured in terms of the RMSE (Doblas-Reyes et al., 2013). An improved prediction of global mean surface temperature is evidenced by the closer fit of the initialized predictions during the 21st century (Figure 11.3; Meehl and Teng, 2012; Doblas-Reyes et al., 2013; Guemas et al., 2013; Box 9.2). The impact of initialization is seen as a better representation of the phase of the internal variability, in particular in increasing the upper ocean heat content (Meehl et al., 2011) and in terms of a correction of the model's forced response.

The AMV (Chapter 14) has important impacts on temperature and precipitation over land (Li and Bates, 2007; Li et al., 2008; Semenov et al., 2010). The AMV index shows a large fraction of its variability on decadal time scales and has multi-year predictability (Murphy et al., 2010; Garcia-Serrano and Doblas-Reyes, 2012). The AMV has been connected to multi-decadal variability of Atlantic tropical cyclones (Goldenberg et al., 2001; Zhang and Delworth, 2006; Smith et al., 2010; Dunstone et al., 2011). Figure 11.3 shows that the CMIP5 multi-model ensemble mean has skill on multi-annual time scales, the skill being generally larger than for the single-model forecast systems (Garcia-Serrano and Doblas-Reyes, 2012; Kim et al., 2012). The skill of the AMV index improves with initialization (*high confidence*) for the early forecast ranges. In particular, the RMSE is substantially reduced (indicating improved skill) with initialization for the AMV. The positive correlation of the non-initialized AMV predictions is consistent with the view that part of the recent variability is due to external forcings (Evan et al., 2009; Ottera et al., 2010; Chang et al., 2011; Booth et al., 2012; Garcia-Serrano et al., 2012; Terray, 2012; Villarini and Vecchi, 2012; Doblas-Reyes et al., 2013).

Pacific decadal variability is associated with potentially important climate impacts, including rainfall over America, Asia, Africa and Australia (Power et al., 1999; Deser et al., 2004; Seager et al., 2008; Zhu et al., 2011; Li et al., 2012). The combination of Pacific and Atlantic variability and climate change is an important driver of multi-decadal USA drought (McCabe et al., 2004; Burgman et al., 2010) including key events like the American dustbowl of the 1930s (Schubert et al., 2004). van Oldenborgh et al. (2012) reported weak skill in hindcasting the IPO in the ENSEMBLES multi-model. Doblas-Reyes et al. (2013) show that the ensemble-mean skill of the ENSEMBLES multi-model IPO is not statistically significant at the 95% level and shows no clear impact of the initialization, in agreement with the predictability study of Meehl et al. (2010). On the other hand, case studies suggest that there might be some initial states that can produce skill in predicting IPO-related decadal variability for some time periods (e.g., Chikamoto et al., 2012b; Meehl and Arblaster, 2012; Meehl et al., 2013a).

The higher AMV and global mean temperature skill of the CMIP5 predictions with respect to the ENSEMBLES hindcasts (van Oldenborgh et al., 2012; Goddard et al., 2013) might be partly due to the CMIP5 multi-model using specified instead of projected aerosol loading (especially the volcanic aerosol) and solar irradiance variations during the simulations. As these forcings cannot be specified in a real forecast setting, ENSEMBLES offers an estimate of the skill closer to what could be expected from a real-time forecast system such as the one described in (Smith et al., 2013a). The use of correct forcings nevertheless allows a more powerful test of the effect of initialization on the ability of models to reproduce past observations.

Near-term prediction systems have significant skill for temperature over large regions (Figure 11.4), especially over the oceans (Smith et al., 2010; Doblas-Reyes et al., 2011; Kim et al., 2012; Matei et al., 2012b; van Oldenborgh et al., 2012; Hanlon et al., 2013). It has been shown that a large part of the skill corresponds to the correct representation of the long-term trend (*high confidence*) as the skill decreases substantially after an estimate of the long-term trend is removed from both the predictions and the observations (e.g., Corti et al., 2012; van Oldenborgh et al., 2012; MacLeod et al., 2013). Robust skill increase due to initialization (Figure 11.4) is limited to areas of the North Atlantic, the Indian Ocean and the southeast Pacific (*high confidence*) (Doblas-Reyes et al., 2013), in agreement with previous results (Pohlmann et al., 2009; Smith et al., 2010; Mochizuki et al., 2012) and predictability estimates (Branstator and Teng, 2012). Similar results have been found in several individual forecast systems (e.g., Muller et al., 2012; Bellucci et al., 2013). However, the impact of initialization on the skill in those regions, though robust (as shown by the agreement between the different CMIP5 systems) is small and not statistically significant with 90% confidence.

The improvement in retrospective North Atlantic variability predictions from initialization (Smith et al., 2010; Dunstone et al., 2011; Garcia-Serrano et al., 2012; Hazeleger et al., 2013b) suggests that internal variability was important to North Atlantic variability during the past few decades. However, the interpretation of the results is complicated by the fact that the impact on skill varies slightly with the forecast quality measure used (Figure 11.3; Doblas-Reyes et al., 2013). This has been attributed to, among other things, the different impact of the predicted local trends on the scores used (Goddard et al., 2013). Skill in hindcasts of subpolar Atlantic temperature, which is evident in Figure 11.4, is

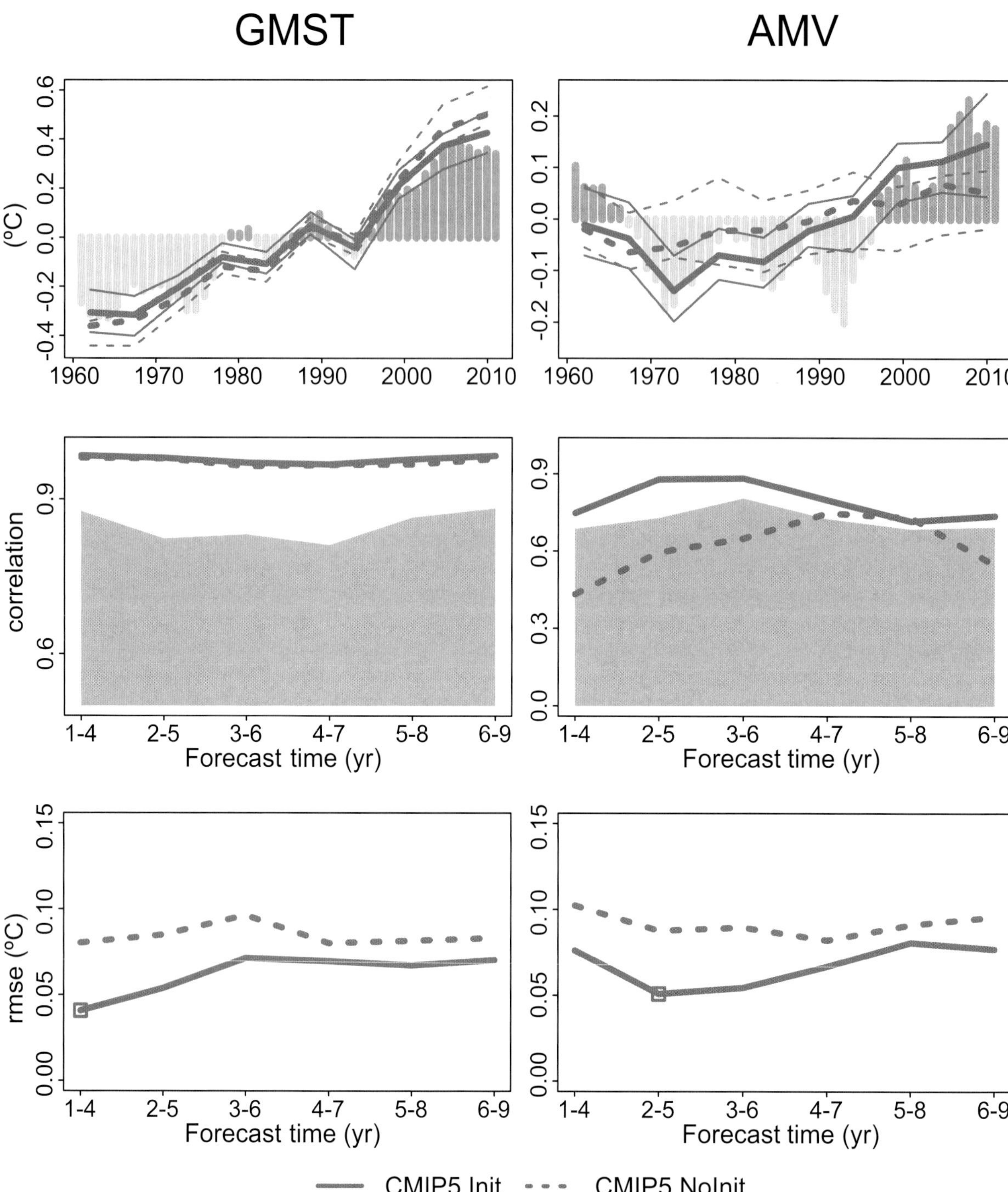

11

Figure 11.3 | Decadal prediction forecast quality of two climate indices. (Top row) Time series of the 2- to 5-year average ensemble-mean initialized hindcast anomalies and the corresponding non-initialized experiments for two climate indices: global mean surface temperature (GMST, left) and the Atlantic multi-decadal variability (AMV, right). The observational time series, Goddard Institute of Space Studies (GISS) GMST and Extended Reconstructed Sea Surface Temperature (ERSST) for the AMV, are represented with dark grey (positive anomalies) and light grey (negative anomalies) vertical bars, where a 4-year running mean has been applied for consistency with the time averaging of the predictions. Predicted time series are shown for the CMIP5 Init (solid) and NoInit (dotted) simulations with hindcasts started every 5 years over the period 1960–2005. The lower and upper quartile of the multi-model ensemble are plotted using thin lines. The AMV index was computed as the SST anomalies averaged over the region Equator to 60°N and 80°W to 0°W minus the SST anomalies averaged over 60°S to 60°N. Note that the vertical axes are different for each time series. (Middle row) Correlation of the ensemble mean prediction with the observational reference along the forecast time for 4-year averages of the three sets of CMIP5 hindcasts for Init (solid) and NoInit (dashed). The one-sided 95% confidence level with a *t* distribution is represented in grey. The effective sample size has been computed taking into account the autocorrelation of the observational time series. A two-sided *t* test (where the effective sample size has been computed taking into account the autocorrelation of the observational time series) has been used to test the differences between the correlation of the initialized and non-initialized experiments, but no differences where found statistically significant with a confidence equal or higher than 90%. (Bottom row) Root mean square error (RMSE) of the ensemble mean prediction along the forecast time for 4-year averages of the CMIP5 hindcasts for Init (solid) and NoInit (dashed). A two-sided *F* test (where the effective sample size has been computed taking into account the autocorrelation of the observational time series) has been used to test the ratio between the RMSE of the Init and NoInit, and those forecast times with differences statistically significant with a confidence equal or higher than 90% are indicated with an open square. (Adapted from Doblas-Reyes et al., 2013.)

improved more by initialization than is skill in hindcasting sub-tropical Atlantic temperature (Garcia-Serrano et al., 2012; Robson et al., 2012; Hazeleger et al., 2013b). This is relevant because the sub-polar branch of the AMV is a source of skill for multi-year North Atlantic tropical storm frequency predictions (Smith et al., 2010). Vecchi et al. (2013) argued that the nominal improvement in multi-year forecasts of North Atlantic hurricane frequency was mainly due to persistence.

Sugiura et al. (2009) reported on skill in hindcasting the Pacific Decadal Oscillation (PDO) in their forecast system. They ascribed the skill to the interplay between Rossby waves and a clockwise propagation of ocean heat content anomalies along the Kuroshio–Oyashio extension and subtropical subduction pathway. However, as Figure 11.4 shows, the Pacific Ocean has the lowest temperature skill overall, with no consistent impact from initialization. The central North Pacific has zero or negative skill, which may be due to the relatively large amplitude of the interannual variability when compared to the long-term trend; the overall failure to predict the largest warming events (Guémas et al., 2012) beyond a few months; and differences (compared to AMV) in how surface temperature and upper ocean heat content interact for the PDO (Mochizuki et al., 2010; Chikamoto et al., 2012a; Mochizuki et al., 2012). There is a robust loss of skill due to initialization in the CMIP5 predictions over the equatorial Pacific (Doblas-Reyes et al., 2013) that has not been adequately explained.

The AMV is thought to be related to the AMOC (Knight et al., 2005). An assessment of the impact of observing systems on AMOC predictability indicates that the recent dense observations of oceanic temperature and salinity are crucial to constraining the AMOC in one model Zhang et al. (2007a). The observing system representative of the pre-2000s was not as effective, indicating that inadequate observations in the past might also limit the impact of initialization on the predictions. This has been confirmed by Pohlmann et al. (2013) using decadal predictions, where they also find a positive impact from initialization that agrees with Hazeleger et al. (2013b). Assessments of the skill of prediction systems to hindcast past variability in the AMOC have been attempted (Pohlmann et al., 2013; Swingedouw et al., 2013) although direct measures of the AMOC are far too short to underpin a reliable estimate of skill, and longer histories are poorly known (Matei et al., 2012a; Vecchi et al., 2012). There is *very low confidence* in current estimates of the skill of the AMOC hindcasts. Sustained ocean observations, such as Argo, a broad global array of temperature/salinity profiling floats, and Rapid Climate Change-Meridional Overturning Circulation and Heatflux Array (RAPID-MOCHA), will be needed to build a capability to reliably predict the AMOC (Srokosz et al., 2012).

Climate prediction is, by nature, probabilistic. Probabilistic predictions are expected to be skilful, but also reliable. Decadal predictions should be evaluated on the basis of whether they give an accurate estimation of the relative frequency of the predicted outcome. This question can be addressed using, among other tools, attributes diagrams (Mason, 2004). They measure how closely the forecast probabilities of an event correspond to the mean probability of observing the event. They are based on a discrete binning of many forecast probabilities taken over a given geographical region. Figure 11.5 illustrates the CMIP5 multi-model Init and NoInit attributes diagrams for predictions of both the global and North Atlantic SSTs to be in the lower tercile (where the tercile threshold has been estimated separately for the predictions and the observations). The diagrams are constructed using predictions for each grid point over the corresponding area. For perfect reliability the forecast probability and the frequency of occurrence should be equal, and the plotted points should lie on the diagonal (solid black line in the figure). When the line joining the bullets (the reliability curve) has positive slope it indicates that as the forecast probability of the event increases, so does the chance of observing the event. The predictions therefore can be considered as moderately reliable. However, if the slope of the curve is less than the slope of the diagonal, then the forecast system is overconfident. If the reliability curve is mainly horizontal, then the frequency of occurrence of the event does not depend on the forecast probabilities and the predictions contain no more information than a random guess. An ideal forecast should have a good resolution whilst retaining reliability, that is, probability forecasts should be both sharp and reliable.

In agreement with Corti et al. (2012), CMIP5 multi-model surface temperature predictions are more reliable for the North Atlantic than when considered over the global oceans, and have a tendency to be overconfident particularly for the global oceans (*medium confidence*). This means that the multi-model ensemble spread should not be considered as a robust measure of the actual uncertainty, at least for multi-annual averages. The attributes diagrams already take into account the systematic error in the simulated variability by estimating separately the event thresholds for the predictions and the observational reference. For the North Atlantic, initialization improves the reliability of the predictions, which translates into an increase of the Brier skill score, the probabilistic skill measure with respect to a naïve climatological prediction (which is reliable, but not skilful) used to aggregate the information in the attributes diagram. However, the uncertainty associated with these estimates is not negligible. This is due mainly to the small sample of start dates, which has the consequence that the number of predictions with a given probability is small to give a robust estimate of the observed relative frequency (Brocker and Smith, 2007). In addition to this, there are biases in the reliability diagram itself (Ferro and Fricker, 2012). These results suggest that the multi-model ensemble should be used with care when estimating probability forecasts or the uncertainty of the mean predictions. Given that the models used for the dynamical predictions are the same as those used for the projections, this verification also provides useful information for the assessment of the projections (cf. Box 11.2).

The skill in hindcasting precipitation over land (Figure 11.6) is much lower than the skill in hindcasting temperature over land. This is consistent with predictability studies discussed previously (e.g., Box 11.1) (*high confidence*). Several regions, especially in the Northern Hemisphere (NH) and West Africa (Gaetani and Mohino, 2013), have skill but these regions are not statistically significant with a 95% confidence level. The positive skill in hindcasting precipitation can be attributed mostly to variable RF (*high confidence*) as initialization improves the skill very little (Goddard et al., 2013). The areas with positive skill agree with those where the precipitation trends of multi-annual averages are the largest (Doblas-Reyes et al., 2013). The skill in areas like West Africa might be associated with the positive AMV skill, as the AMV drives interannual variability in precipitation over this region (van Oldenborgh et al., 2012).

The small amount of statistically significant differences found between the initialized and non-initialized experiments does not necessarily mean that the impact of the initialization does not have a physical basis. A comparison of the global mean temperature and AMV forecast quality using 1- and 5-year intervals between start dates (Garcia-Serrano et al., 2012) suggests that, although a five-year interval sampling allows an estimate of the level of skill, local maxima as a function of forecast time might well be due to poor sampling of the start dates (Garcia-Serrano and Doblas-Reyes, 2012; Kharin et al., 2012; Doblas-Reyes et al., 2013; Goddard et al., 2013). Several signals, such as the

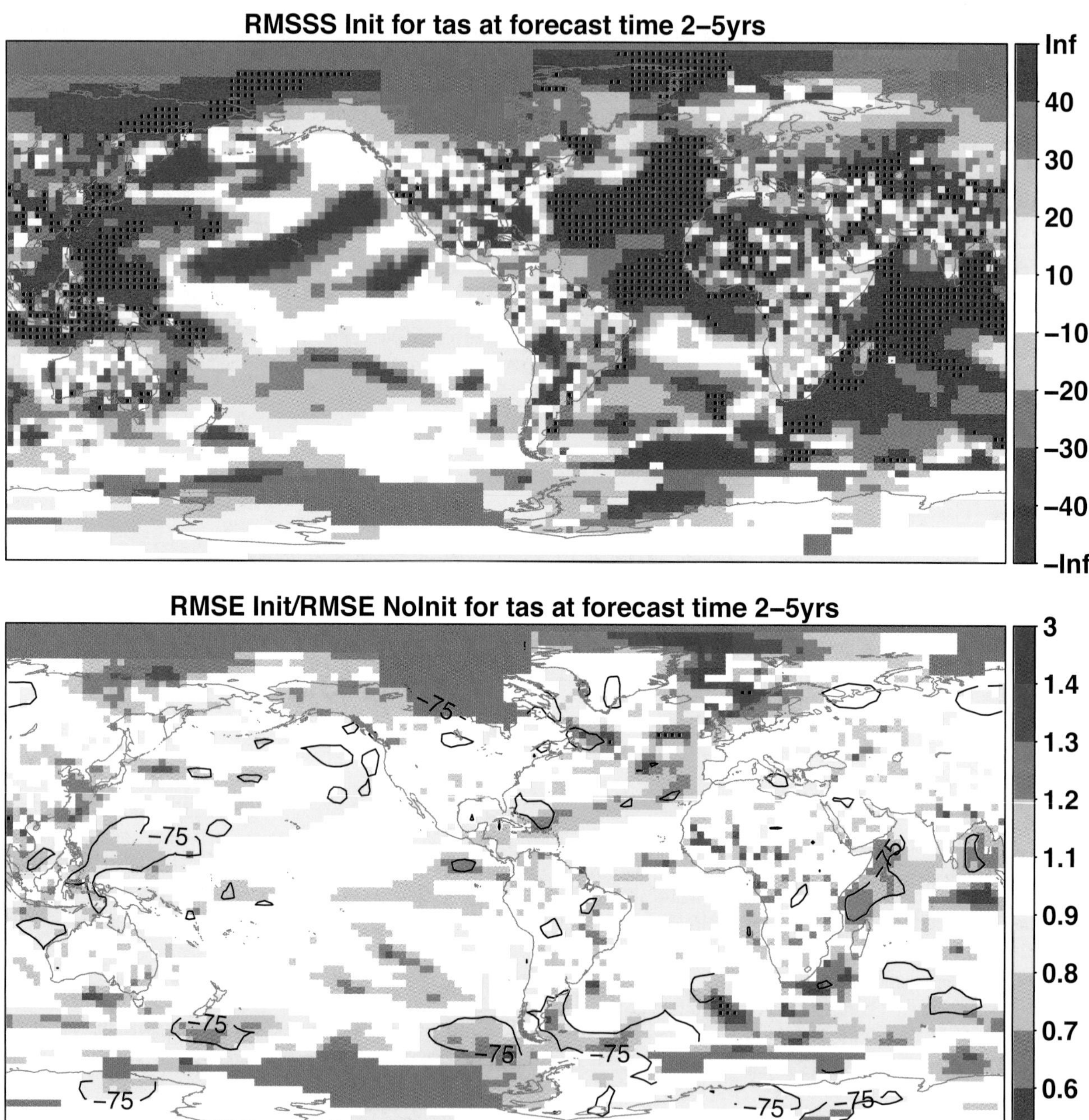

Figure 11.4 | (a) Root mean square skill score of the near surface air temperature forecast quality for the forecast time 2 to 5 years from the multi-model ensemble mean of the CMIP5 Init experiment with 5-year interval between start dates over the period 1960–2005. A combination of temperatures from Global Historical Climatology Network/Climate Anomaly Monitoring System (GHCN/CAMS) air temperature over land, Extended Reconstructed Sea Surface Temperature (ERSST) and Goddard Institute of Space Studies Surface Temperature Analysis (GISTEMP) 1200 over the polar areas is used as a reference. Black dots correspond to the points where the skill score is statistically significant with 95% confidence using a one-sided *F*-test taking into account the autocorrelation of the observation minus prediction time series. (b) Ratio between the root mean square error of the ensemble mean of Init and NoInit. Dots are used for the points where the ratio is significantly above or below 1, with 90% confidence using a two-sided *F*-test taking into account the autocorrelation of the observation minus prediction time series. Contours are used for areas where the ratio of at least 75% of the single forecast systems is either above or below one agreeing with the value of the ratio in the multi-model ensemble. Poorly observationally sampled areas are masked in grey. The original model data have been bilinearly interpolated to the observational grid. The ensemble mean of each forecast system has been estimated before computing the multi-model ensemble mean. (Adapted from Doblas-Reyes et al., 2013.)

skill improvement for temperature over the North Atlantic, are robust in the sense that it is found in more than 75% of forecast system. However, it is difficult to obtain statistical significance with these limited samples. The low start date sampling frequency is one of the limitations of the core CMIP5 near-term prediction experiment, the other one being the short length of the period of study, limited by the availability of observational data. Results estimated with yearly start dates are more robust than with a 5-year start date frequency. However, even with 1-year start date frequency, the impact of the initialization is similar. The spatial distribution of the skill does not change substantially with the different start date frequency. The skill and the initialization impact are both slightly reduced in the results with yearly start dates, but at the same time the spatial variability is substantially reduced.

The CMIP5 multi-model overestimates the spread of the multi-annual average temperature (Doblas-Reyes et al., 2013). Figure 11.7 shows the ratio of the spread around the ensemble mean prediction and the RMSE of the ensemble mean prediction of Init and NoInit, which in a well-calibrated system is expected to be close to 1. However, the ratio is overestimated over the North Atlantic, the Indian Ocean and the Arctic, and underestimated over the North Pacific and most continental areas, suggesting that the CMIP5 systems do not discriminate

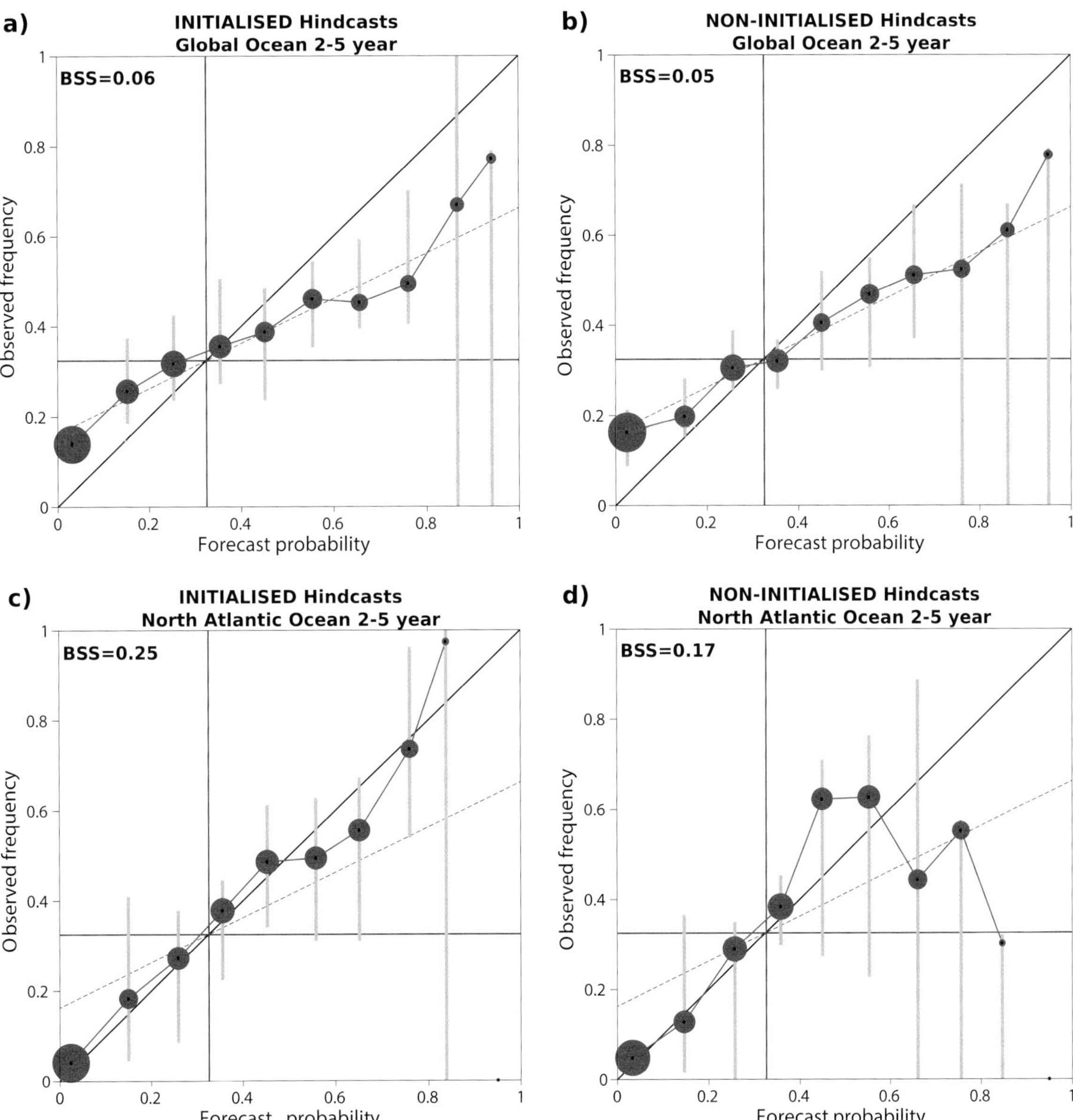

Figure 11.5 | Attributes diagram for the CMIP5 multi-model decadal initialized (a and c) and non-initialized (b and d) hindcasts for the event 'surface air temperature anomalies below the lower tercile over (a) and (b) the global oceans (60°N to 60°S) and (c) and (d) the North Atlantic (87.5°N to 30°N, 80°W to 10°W) for the forecast time 2 to 5 years. The red bullets in the figure correspond to the number of probability bins (10 in this case) used to estimate forecast probabilities. The size of the bullets represents the number of forecasts in a specific probability category and is a measure of the sharpness (or variance of the forecast probabilities) of the predictions. The blue horizontal and vertical lines indicate the climatological frequency of the event in the observations and the mean forecast probability, respectively. Grey vertical bars indicate the uncertainty in the observed frequency for each probability category estimated at 95% level of confidence with a bootstrap resampling procedure based on 1000 samples. The longer the bars, the more the vertical position of the bullets may change as new hindcasts become available. The black dashed line separates skilful from unskilled regions in the diagram in the Brier skill score sense. The Brier skill score with respect to the climatological forecast is drawn in the top left corner of each panel. (Adapted from Corti et al., 2012.)

between the regions where the spread should be reduced according to the RMSE level in the area. These results are found for both the Init and NoInit ensembles and agree with the overconfidence of the probability forecasts shown in Figure 11.6 (Corti et al., 2012). The spread overestimation also agrees with the results found for the indices illustrate in Figure 11.3 (Doblas-Reyes et al., 2013). The spread overestimation points to the need for a careful interpretation of current ensemble and probabilistic climate information for climate adaptation and services.

The skill of extreme daily temperature and precipitation in multi-annual time scales has also been assessed (Eade et al., 2012; Hanlon et al., 2013). There is little improvement in skill with the initialization beyond

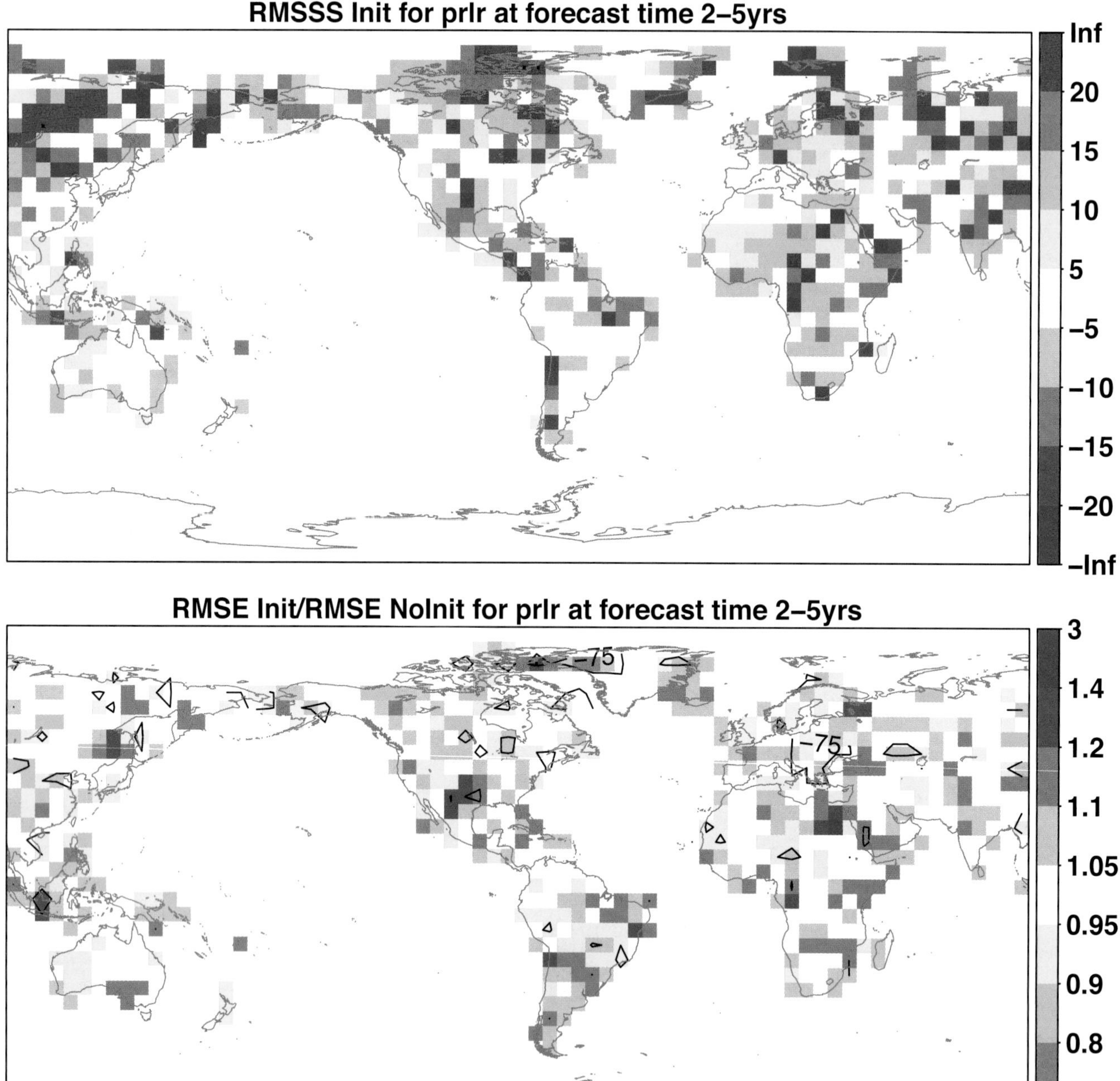

Figure 11.6 | (a) Root mean square skill score for precipitation hindcasts for the forecast time 2 to 5 years from the multi-model ensemble mean of the CMIP5 Init experiment with 5-year interval between start dates over the period 1960–2005. Global Precipitation Climatology Centre (GPCC) precipitation is used as a reference. Black dots correspond to the points where the skill score is statistically significant with 95% confidence using a one-sided *F*-test taking into account the autocorrelation of the observation minus prediction time series. (b) Ratio between the root mean square error of the ensemble mean of Init and NoInit. Dots are used for the points where the ratio is significantly above or below one with 90% confidence using a two-sided *F*-test taking into account the autocorrelation of the observation minus prediction time series. Contours are used for areas where the ratio of at least 75% of the single forecast systems is either above or below 1, agreeing with the value of the ratio in the multi-model ensemble. The model original data have been bilinearly interpolated to the observational grid. The ensemble mean of each forecast system has been estimated before computing the multi-model ensemble mean. (Adapted from Doblas-Reyes et al., 2013.)

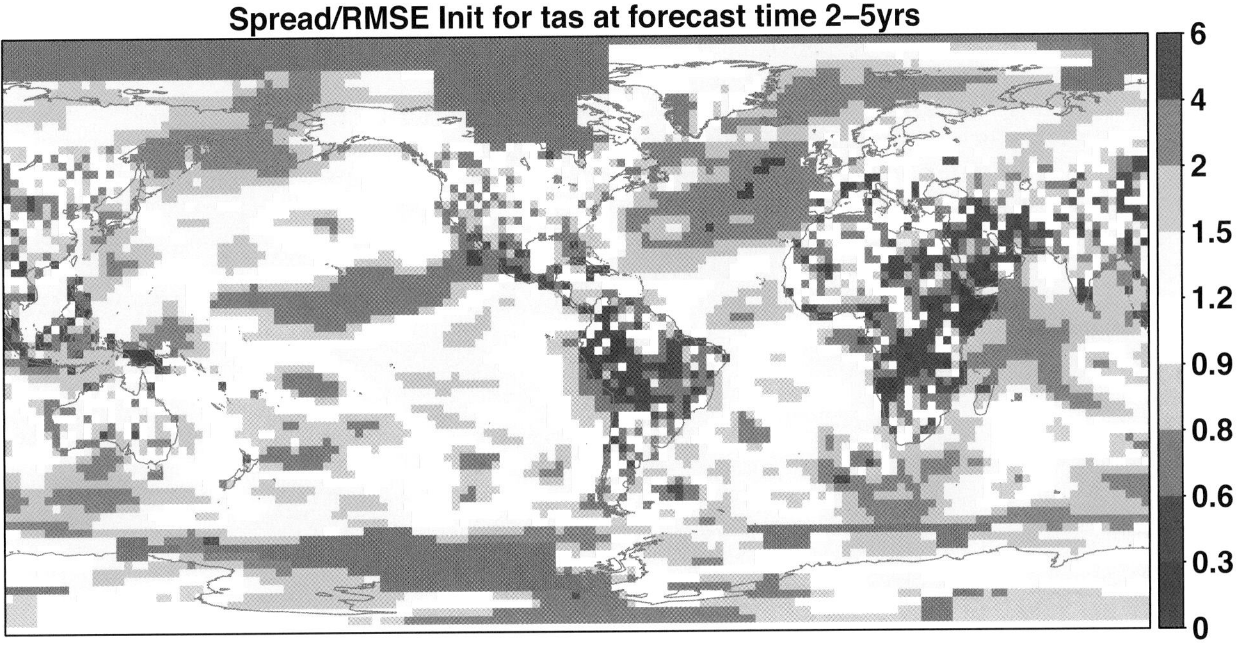

Figure 11.7 | Ratio between the surface temperature spread around the ensemble mean and the root mean square error (RMSE) of the ensemble-mean prediction of Init and NoInit for the forecast time 2 to 5 years with 5-year interval between start dates over the period 1960–2005. A combination of temperatures from Global Historical Climatology Network/Climate Anomaly Monitoring System (GHCN/CAMS) air temperature over land, Extended Reconstructed Sea Surface Temperature (ERSST) v3b over sea and Goddard Institute of Space Studies Surface Temperature Analysis (GISTEMP) 1200 over the polar areas is used as a reference to compute the RMSE. (Adapted from Doblas-Reyes et al., 2013.)

the first year, suggesting that skill then arises largely from the varying external forcing. The skill for extremes is generally similar to, but slightly lower than, that for the mean.

Responding to the increases in decadal skill in certain regions due to initialization, a coordinated quasi-operational decadal prediction initiative has been organized (Smith et al., 2013b). The forecast systems participating in the initiative are based on those of CMIP5 and have been evaluated for forecast quality. Statistical predictions are also included in the initiative. The most recent forecast shows (compared to the projections) substantial warming of the north Atlantic subpolar gyre, cooling of the north Pacific throughout the next decade and cooling over most land and ocean regions and in the global average out to several years ahead. However, in the absence of explosive or frequent volcanic eruptions, global surface temperature is predicted to continue to rise and, to a certain degree, recover from the reduced rate of warming (see Box 9.2).

11.2.3.5 Realizing Potential

Although idealized model experiments show considerable promise for predicting internal variability, realizing this potential is a challenging task. There are three main hurdles: (1) the limited availability of data to initialize and verify predictions, (2) limited progress in initialization techniques for decadal predictions and (3) dynamical model shortcomings that require validating how the simulated variance compares with the observed variance.

It is expected that the availability of temperature and salinity data in the top 2 km of the ocean through the enhanced global deployment of Argo floats will give a step change in our ability to initialize and predict ocean heat and density anomalies (Zhang et al., 2007a; Dunstone and Smith, 2010). Another important advancement is the availability of highly accurate altimetry data, made especially useful after the launching of TOPography EXperiment (TOPEX)/Poseidon in 1992. Argo and altimeter data became available only in 2000 and 1992 respectively, so an accurate estimate of their impact on real forecasts has to wait (Dunstone and Smith, 2010). In all cases, both the length of the observational data sets and the reduced coverage of the data available, especially before 2000, are serious limitations to obtain robust estimates of forecast quality.

Improved initialization of other aspects such as sea ice, snow cover, frozen soil and soil moisture, may also have potential to contribute to predictive skill beyond the seasonal time scale. This could be investigated, for example by using measurements of soil moisture from the Soil Moisture and Ocean Salinity (SMOS) satellite launched in 2009, or by initializing sea ice thickness with observations from the CryoSat-2 satellite launched in 2010. Along the same line, understanding the links between the initialization and the correct prediction of both the internal and external variability should help improving forecast quality (Solomon et al., 2011).

Many of the current decadal prediction systems use relatively simple initialization schemes and do not adopt fully coupled initialization/ensemble generation schemes. Assimilation schemes offer opportunities for fully coupled initialization including assimilation of variables such as sea ice, snow cover and soil moisture, although they present technically and scientifically challenging problems. This approach has been tested in schemes like four-dimensional variational data assimilation (4DVAR; Sugiura et al., 2008) and the ensemble Kalman filter (Keppenne et al., 2005; Zhang et al., 2007a).

Bias correction is used to reduce the effects of model drift, but the nonlinearity in the climate system (e.g., Power (1995) might limit the effectiveness of bias correction and thereby reduce forecast quality. Understanding and reducing both drift and systematic errors is important (Palmer and Weisheimer, 2011), as it is also for seasonal-to-interannual climate prediction and for climate change projections. While improving models is the highest priority, efforts to quantify the degree of interference between model bias and predictive signals should not be overlooked.

11.3 Near-term Projections

11.3.1 Introduction

In this section the outlook for global and regional climate up to mid-century is assessed, based on climate model projections. In contrast to the predictions discussed in Section 11.2, these projections are not initialized using observations; instead, they are initialized from historical simulations of the evolution of climate from pre-industrial conditions up to the present. The historical simulations are forced by estimates of past anthropogenic and natural climate forcing agents, and the projections are obtained by forcing the models with scenarios for future climate forcing agents. Major use is made of the CMIP5 model experiments forced by the Representative Concentration Pathway (RCP) scenarios discussed in Chapters 1 and 8. Projections of climate change in this and subsequent chapters are expressed relative to the reference period: 1986–2005. In this chapter most emphasis is given to the period 2016–2035, but some information on changes projected before and after this period (up to mid-century) is also provided. Longer-term projections are assessed in Chapters 12 and 13.

Key assessment questions addressed in this section are: *What is the externally forced signal of near-term climate change, and how large is it compared to natural internal variability?* From the point of view of climate impacts, the absolute magnitude of climate change may in some instances be less important than the magnitude relative to the local level of natural internal variability. Because many systems are naturally adapted to a background level of variability, it may be changes that move outside of this range that are most likely to trigger impacts that are unprecedented in the recent past (e.g., Lobell and Burke (2008) for crops).

An important conclusion of the AR4 (Section 10.3.1) was that near-term climate projections are not very sensitive to plausible alternative non-mitigation scenarios for GHG concentrations (specifically the Special Report on Emission Scenarios (SRES) scenarios; comparison with RCP scenarios is discussed in Chapter 1), that is, in the near term, different scenarios give rise to similar magnitudes and patterns of climate change. (Note, however, that some impacts may be more sensitive.) For this reason, most of the projections presented in this chapter are based on one specific RCP scenario, RCP4.5. RCP4.5 was chosen because of its intermediate GHG forcing. However, there is greater sensitivity to other forcing agents, in particular anthropogenic aerosols (e.g., Chalmers et al., 2012). Consequently, a further question addressed in this section (especially in Section 11.3.6.1) is: *To what extent are near-term climate projections sensitive to alternative scenarios for anthropogenic forcing?* Note finally that a great deal of additional information on near-term projections is provided in Annex I.

11.3.1.1 Uncertainty in Near-term Climate Projections

As discussed in Chapters 1 (Section 1.4) and 12 (Section 12.2), climate projections are subject to several sources of uncertainty. Here three main sources are distinguished. The first arises from natural *internal variability*, which is intrinsic to the climate system, and includes phenomena such as variability in the mid-latitude storm tracks and the ENSO. The existence of internal variability places fundamental limits on the precision with which future climate variables can be projected. The second is uncertainty concerning the past, present and future *forcing* of the climate system by natural and anthropogenic forcing agents such as GHGs, aerosols, solar forcing and land use change. Forcing agents may be specified in various ways, for example, as *emissions* or as *concentrations* (see Section 12.2). The third is uncertainty related to the *response* of the climate system to the specified forcing agents.

Quantifying the uncertainty that arises from each of the three sources is an important challenge. For projections, no attempt is made to predict the evolution of the internal variability. Instead, the statistics of this variability are included as a component of the uncertainty associated with a projection. The magnitude of internal variability can be estimated from observations (Chapters 2, 3 and 4) or from climate models (Chapter 9). Challenges arise in estimating the variability on decadal and longer time scales, and for rare events such as extremes, as observational records are often too short to provide robust estimates.

Uncertainty concerning the past forcing of the climate system arises from a lack of direct or proxy observations, and from observational errors. This uncertainty can influence future projections of some variables (particularly large-scale ocean variables) for years or even decades ahead (e.g., Meehl and Hu, 2006; Stenchikov et al., 2009; Gregory, 2010). Uncertainty about future forcing arises from the inability to predict future anthropogenic emissions and land use change, and natural forcings (e.g., volcanoes), and from uncertainties concerning carbon cycle and other biogeochemical feedbacks (Chapters 6, 12 and Annex II.4.1). The uncertainties in future anthropogenic forcing are typically investigated through the development of specific scenarios (e.g., for emissions or concentrations), such as the RCP scenarios (Chapters 1 and 8). Different scenarios give rise to different climate projections, and the spread of such projections is commonly described as *scenario uncertainty.* The sensitivity of climate projections to alternative scenarios for future anthropogenic emissions is discussed especially in Section 11.3.6.1

To project the climate response to specified forcing agents, climate models are required. The term *model uncertainty* describes uncertainty about the extent to which any particular climate model provides an accurate representation of the real climate system. This uncertainty arises from approximations required in the development of models. Such approximations affect the representation of all aspects of the climate including natural internal variability and the response to external forcings. As discussed in Chapter 1 (Section 1.4.2), the term *model uncertainty* is sometimes used in a narrower sense to describe the spread between projections generated using different models or model

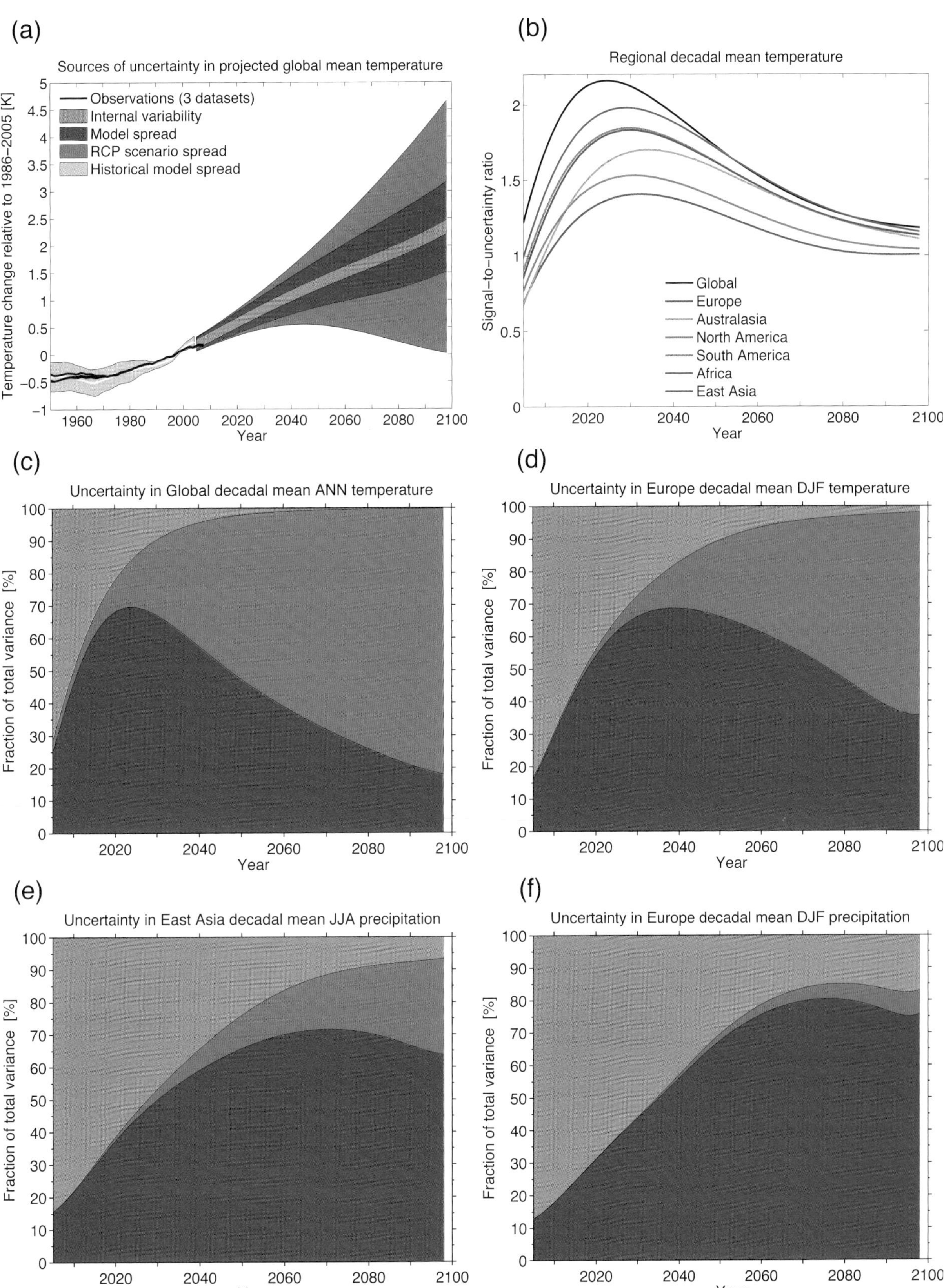

Figure 11.8 | Sources of uncertainty in climate projections as a function of lead time based on an analysis of CMIP5 results. (a) Projections of global mean decadal mean surface air temperature to 2100 together with a quantification of the uncertainty arising from internal variability (orange), model spread (blue) and RCP scenario spread (green). (b) Signal-to-uncertainty ratio for various global and regional averages. The signal is defined as the simulated multi-model mean change in surface air temperature relative to the simulated mean surface air temperature in the period 1986–2005, and the uncertainty is defined as the total uncertainty. (c–f) The fraction of variance explained by each source of uncertainty for: global mean decadal and annual mean temperature (c), European (30°N to 75°N, 10°W to 40°E) decadal mean boreal winter (December to February) temperature (d) and precipitation (f), and East Asian (5°N to 45°N, 67.5°E to 130°E) decadal mean boreal summer (June to August) precipitation (e). See text and Hawkins and Sutton (2009) and Hawkins and Sutton (2011) for further details.

versions; however, such a measure is crude as it takes no account of factors such as model quality (Chapter 9) or model independence. The term *model response uncertainty* is used here to describe the dimension of model uncertainty that is directly related to the response to external forcings. To obtain projections of extreme events such as tropical cyclones, or regional phenomena such as orographic rainfall, it is sometimes necessary to employ a dynamical or statistical downscaling procedure. Such downscaling introduces an additional dimension of model uncertainty (e.g., Alexandru et al., 2007).

The relative importance of the different sources of uncertainty depends on the variable of interest, the space and time scales involved (Section 10.5.4.3 of Meehl et al. (2007b)), and the lead-time of the projection. Figure 11.8 provides an illustration of these dependencies based on an analysis of CMIP5 projections (following Hawkins and Sutton, 2009, 2011;Yip et al., 2011). In this example, the forcing-related uncertainty is estimated using the spread of projections for different RCP scenarios (i.e., scenario uncertainty), while the spread among different models for individual RCP scenarios is used as a measure of the model response uncertainty. Internal variability is estimated from the models as in Hawkins and Sutton (2009). Key points are: (1) the uncertainty in *near-term* projections is dominated by internal variability and model spread. This finding provides some of the rationale for considering near-term projections separately from long-term projections. Note, however, that the RCP scenarios do not sample the full range of uncertainty in future anthropogenic forcing, and that uncertainty in aerosol forcings in particular may be more important than is suggested by Figure 11.8 (see Section 11.3.6.1); (2) internal variability becomes increasingly important on smaller space and time scales; (3) for projections of precipitation, scenario uncertainty is less important and (on regional scales) internal variability is generally more important than for projections of surface air temperature; (4) the full model uncertainty may well be larger or smaller than the model spread due to common errors or unrealistic models.

A key quantity for any climate projection is the signal-to-noise (S/N) ratio (Christensen et al., 2007), where the 'signal' is a measure of the amplitude of the projected climate change, and the noise is a measure of the uncertainty in the projection. Higher S/N ratios indicate more robust projections of change and/or changes that are large relative to background levels of variability. Depending on the purpose, it may be useful to identify the noise with the total uncertainty, or with a specific component such as the internal variability. The evolution of the S/N ratio with lead time depends on whether the signal grows more rapidly than the noise, or vice versa. Figure 11.8 (top right) shows that, when the noise is identified with the total uncertainty, the S/N ratio for surface air temperature is typically higher at lower latitudes and has a maximum at a lead time of a few decades (Cox and Stephenson, 2007; Hawkins and Sutton, 2009). The former feature is primarily a consequence of the greater amplitude of internal variability in mid-latitudes. The latter feature arises because over the first few decades, when scenario uncertainty is small, the signal grows most rapidly, but subsequently, the contribution from scenario uncertainty grows more rapidly than does the signal, so the S/N ratio falls. See Hawkins and Sutton (2009, 2011) for further details.

11.3.2 Near-term Projected Changes in the Atmosphere and Land Surface

11.3.2.1 Surface Temperature

11.3.2.1.1 Global mean surface air temperature

Figure 11.9 (a) and (b) show CMIP5 projections of global mean surface air temperature under RCP4.5. The 5 to 95% range for the projected anomaly for the period 2016–2035, relative to the reference period 1986–2005, is 0.47°C to 1.00°C (see also Table 12.2). However, as discussed in Section 11.3.1.1, this range provides only a very crude measure of uncertainty, and there is no guarantee that the real world must lie within this range. Obtaining better estimates is an important challenge. One approach involves initializing climate models using observations, as discussed in Section 11.2. Figure 11.9 (b) compares multi-model initialized climate predictions (8 models from Smith et al., 2013b), initialized in 2011; 14 CMIP5 decadal prediction experiment models following the methodology of Meehl and Teng (2012), initialized in 2006 with the 'raw' uninitialized CMIP5 projections. The 5 to 95% range for both sets of initialized predictions is cooler (by about 15% for the median values) than the corresponding range for the raw projections, particularly at the upper end. The differences are partly a consequence of initializing the models in a state that is cool (in comparison to the median of the raw projections) as a result of the recent hiatus in global mean surface temperature rise (see Box 9.2). However, it is not yet possible to attribute all of the reasons with confidence because the raw projections are based on a different, and larger, set of models than the initialized predictions, and because of uncertainties related to the bias adjustment of the initialized predictions (Goddard et al., 2013; Meehl et al., 2013d)

Another approach to making projections involves weighting models according to some measure of their quality (see Chapter 9). A specific approach of this type, known as Allen, Stott and Kettleborough (ASK) (Allen et al., 2000; Stott and Kettleborough, 2002), is based on the use of results from detection and attribution studies (Chapter 10), in which the fit between observations and model simulations of the past is used to scale projections of the future. ASK requires specific simulations to be carried out with individual forcings (e.g., anthropogenic GHG forcing alone), and only some of the centres participating in CMIP5 have carried out the necessary integrations. Biases in ASK-derived projections may arise from errors in the specified forcings, or in the simulated patterns of response, and/or from nonlinearities in the responses to forcings.

Figure 11.9c shows the projected range of global mean surface air temperature change derived using the ASK approach for RCP4.5 (Stott and G. Jones, 2012; Stott et al., 2013) applied to six models and compares this with the range derived from the 42 CMIP5 models. In this case decadal means are shown. The 5 to 95% confidence interval for the projected temperature anomaly for the period 2016–2035, based on the ASK method, is 0.39°C to 0.87°C. As for the initialized predictions shown in Figure 11.9b, both the lower and upper values are below the corresponding values obtained from the raw CMIP5 results, although there is substantial overlap between the two ranges. The relative cooling of the ASK results is directly related to evidence presented in

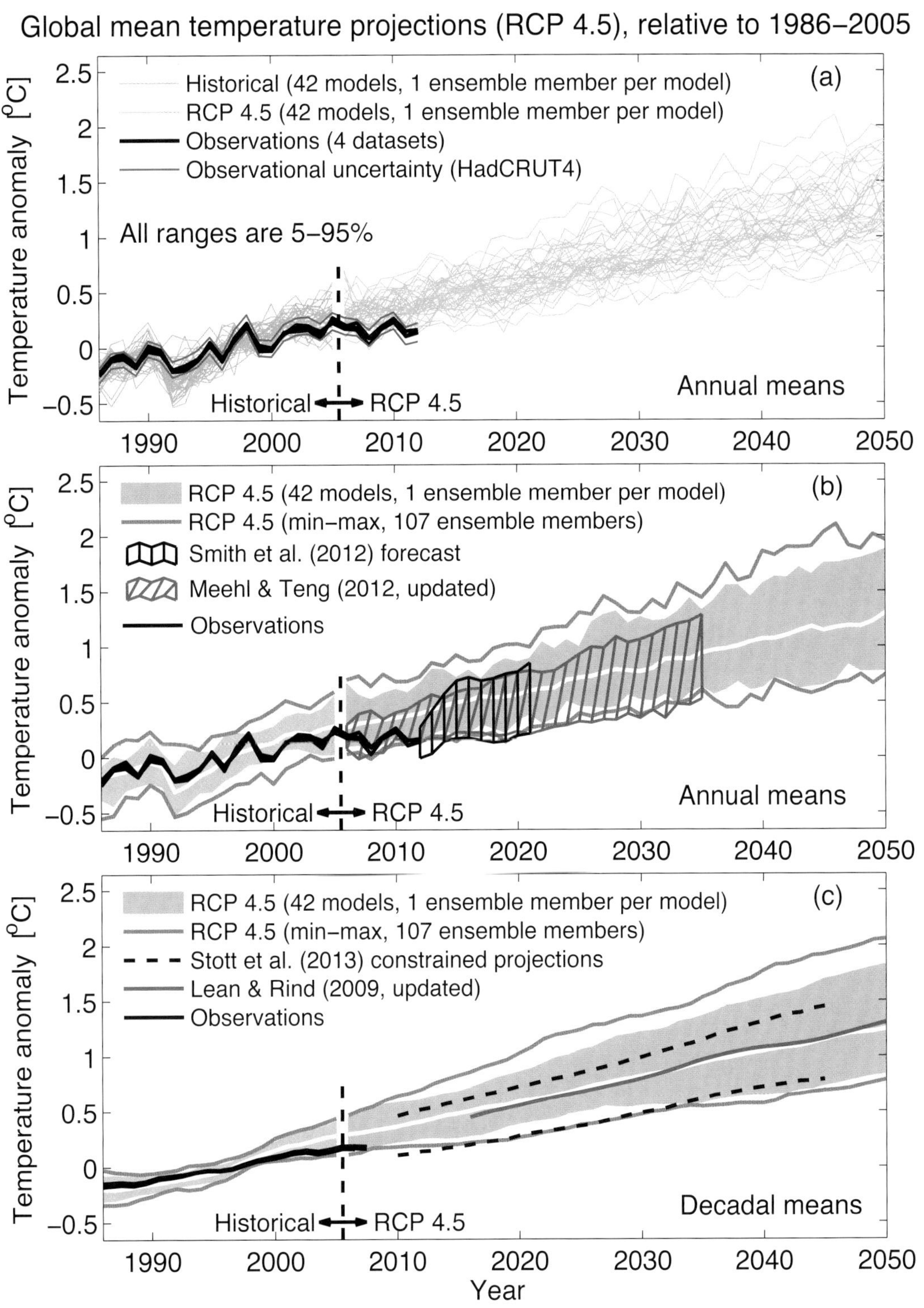

Figure 11.9 | (a) Projections of global mean, annual mean surface air temperature 1986–2050 (anomalies relative to 1986–2005) under RCP4.5 from CMIP5 models (blue lines, one ensemble member per model), with four observational estimates: Hadley Centre/Climate Research Unit gridded surface temperature data set 3 (HadCRUT3: Brohan et al., 2006); European Centre for Medium range Weather Forecast (ECMWF) interim reanalysis of the global atmosphere and surface conditions (ERA-Interim: Simmons et al., 2010); Goddard Institute of Space Studies Surface Temperature Analysis (GISTEMP: Hansen et al., 2010); National Oceanic and Atmospheric Administration (NOAA: Smith et al. (2008) for the period 1986–2011 (black lines). (b) As in (a) but showing the 5 to 95% range (grey and blue shades, with the multi-model median in white) of annual mean CMIP5 projections using one ensemble member per model from RCP4.5 scenario, and annual mean observational estimates (solid black line). The maximum and minimum values from CMIP5 are shown by the grey lines. Red hatching shows 5 to 95% range for predictions initialized in 2006 for 14 CMIP5 models applying the Meehl and Teng (2012) methodology. Black hatching shows the 5 to 95% range for predictions initialized in 2011 for eight models from Smith et al. (2013b). (c) As (a) but showing the 5 to 95% range (grey and blue shades, with the multi-model median in white) of decadal mean CMIP5 projections using one ensemble member per model from RCP4.5 scenario, and decadal mean observational estimates (solid black line). The maximum and minimum values from CMIP5 are shown by the grey lines. The dashed black lines show an estimate of the projected 5 to 95% range for decadal mean global mean surface air temperature for the period 2016–2040 derived using the ASK methodology applied to six CMIP5 GCMs. (From Stott et al., 2013.) The red line shows a statistical prediction based on the method of Lean and Rind (2009), updated for RCP4.5.

Chapter 10 (Section 10.3.1) that 'This provides evidence that some CMIP5 models have a higher transient response to GHGs and a larger response to other anthropogenic forcings (dominated by the effects of aerosols) than the real world (*medium confidence*).' The ASK results and the initialised predictions both suggest that those CMIP5 models that warm most rapidly over the period (1986–2005) to (2016–2035) may be inconsistent with the observations. This possibility is also suggested by comparing the models with the observed rate of warming since 1986—see Box 9.2 for a full discussion of this comparison. Lastly, Figure 11.9 also shows a statistical prediction for global mean surface air temperature, using the method of Lean and Rind (2009), which uses multiple linear regression to decompose observed temperature variations into distinct components. This prediction is very similar to the CMIP5 multi-model median.

The projections shown in Figure 11.9 assume the RCP4.5 scenario and use the 1986–2005 reference period. In Section 11.3.6 additional uncertainties associated with future forcing, climate responses and sensitivity to the choice of reference period, are discussed. An overall assessment of the *likely* range for future global mean surface air temperature is provided in Section 11.3.6.3.

For the remaining projections in this chapter the spread among the CMIP5 models is used as a simple, but crude, measure of uncertainty. The extent of agreement between the CMIP5 projections provides rough guidance about the likelihood of a particular outcome. But—as partly illustrated by the discussion above—it must be kept firmly in mind that the real world could fall outside of the range spanned by these particular models. See Section 11.3.6 for further discussion.

11.3.2.1.2 Regional and seasonal patterns of surface warming

The geographical pattern of near-term surface warming simulated by the CMIP5 models (Figure 11.10) is consistent with previous IPCC reports in a number of key aspects, although weaknesses in the ability of current models to capture observed regional trends (Box 11.2) must be kept in mind. First, temperatures over land increase more rapidly than over sea (e.g., Manabe et al., 1991; Sutton et al., 2007). Processes that contribute to this land–sea warming contrast include different local feedbacks over ocean and land and changes in atmospheric energy transport from ocean to land regions (e.g., Lambert and Chiang, 2007; Vidale et al., 2007; Shimpo and Kanamitsu, 2009; Fasullo, 2010; Boer, 2011; Joshi et al., 2011).

Second, the projected warming in wintertime shows a pronounced polar amplification in the NH (see Box 5.1). This feature is found in virtually all coupled model projections, but the CMIP3 simulations generally appeared to underestimate this effect in comparison to

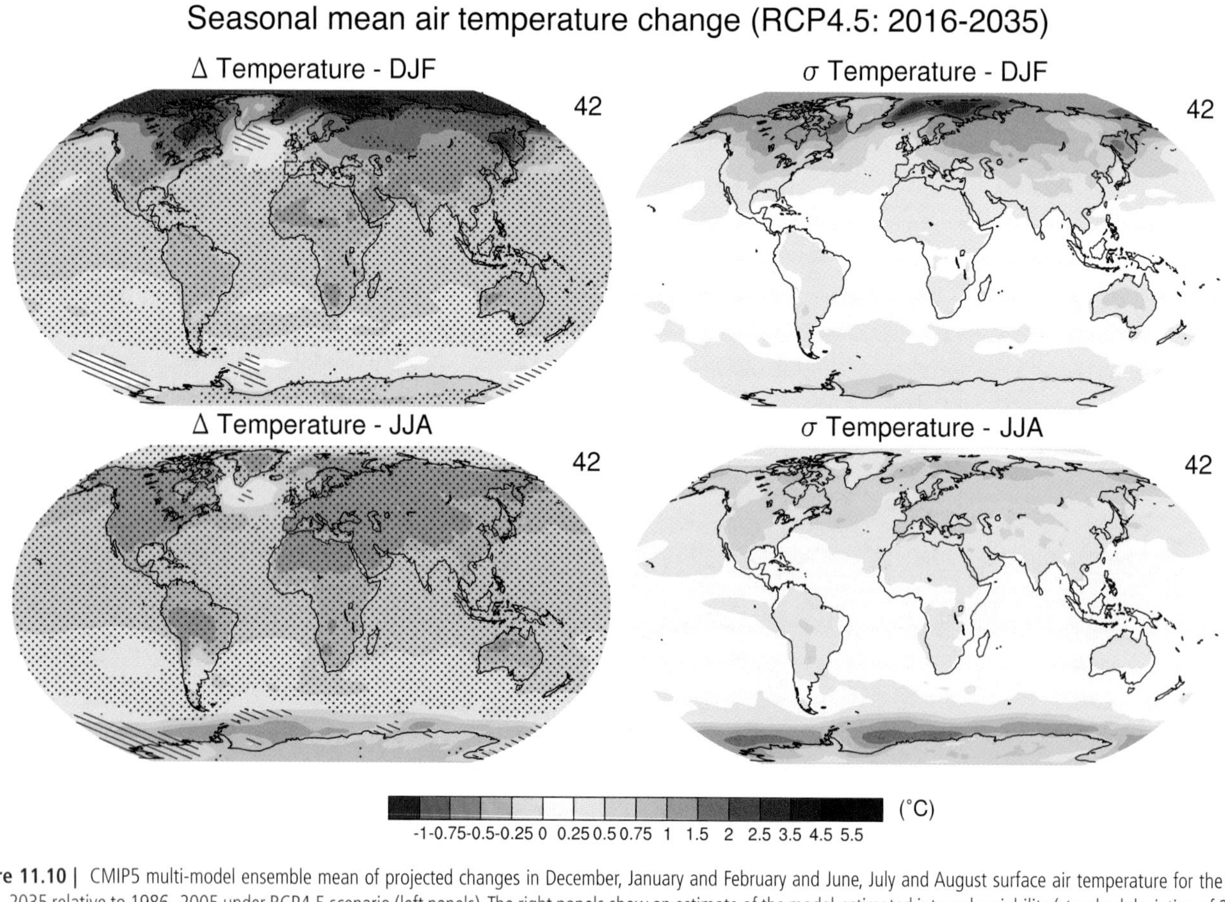

Figure 11.10 | CMIP5 multi-model ensemble mean of projected changes in December, January and February and June, July and August surface air temperature for the period 2016–2035 relative to 1986–2005 under RCP4.5 scenario (left panels). The right panels show an estimate of the model-estimated internal variability (standard deviation of 20-year means). Hatching in left-hand panels indicates areas where projected changes are small compared to the internal variability (i.e., smaller than one standard deviation of estimated internal variability), and stippling indicates regions where the multi-model mean projections deviate significantly from the simulated 1986–2005 period (by at least two standard deviations of internal variability) and where at least 90% of the models agree on the sign of change. The number of models considered in the analysis is listed in the top-right portion of the panels; from each model one ensemble member is used. See Box 12.1 in Chapter 12 for further details and discussion. Technical details are in Annex I.

observations (Stroeve et al., 2007; Screen and Simmonds, 2010). Several studies have isolated mechanisms behind this amplification, which include reductions in snow cover and retreat of sea ice (e.g., Serreze et al., 2007; Comiso et al., 2008); changes in atmospheric and oceanic circulations (Chylek et al., 2009, 2010; Simmonds and Keay, 2009); presence of anthropogenic soot in the Arctic environment (Flanner et al., 2007; Quinn et al., 2008; Jacobson, 2010; Ramana et al., 2010); and increases in cloud cover and water vapour (Francis, 2007; Schweiger et al., 2008). Most studies argue that changes in sea ice are central to the polar amplification—see Section 11.3.4.1 for further discussion. Further information about the regional changes in surface air temperature projected by the CMIP5 models is presented in Annex I.

As discussed in Sections 11.1 and 11.3.1, the signal of climate change is emerging against a background of natural internal variability. The concept of 'emergence' describes the magnitude of the climate change signal relative to this background variability, and may be useful for some climate impact assessments (e.g., AR4, Chapter 11, Table 11.1; Mahlstein et al., 2011; Hawkins and Sutton, 2012; see also FAQ 10.2). However, it is important to recognize that there is no single metric of emergence. It depends on user-driven choices of variable, space and time scale, of the baseline relative to which changes are measured (e.g., pre-industrial versus recent climate) and of the threshold at which emergence is defined.

Figure 11.11 quantifies the 'Time of Emergence' (ToE) of the mean warming signal relative to the recent past (1986–2005), based on the CMIP5 RCP4.5 projections, using a spatial resolution of 2.5° latitude × 2.5° longitude, the standard deviation of interannual variations as the measure of internal variability, and a signal-to-noise threshold of 1. Because of the dependence on user-driven choices, the most important information in Figure 11.11 is the geographical and seasonal variation in ToE, seen in the maps, and the variation in ToE between models, shown in the histograms. Consistent with Mahlstein et al. (2011), the earliest ToE is found in the tropics, with ToE in mid-latitudes typically a decade or so later. Over North Africa and Asia, earlier ToE is found for the warm half-year (April to September) than for the cool half-year. Earlier ToE is generally found for larger space and time scales, because the variance of natural internal variability decreases with averaging (Section 11.3.1.1 and AR4, Section 10.5.4.3). This tendency can be seen in Figure 11.11 by comparing the median value of the histograms for area averages with the area average of the median ToE inferred from the maps (e.g., for Region 2). The large range of values for ToE implied by different CMIP5 models, which can be as much as 30 years, is a consequence of differences in both the magnitude of the warming signal simulated by the models (i.e., uncertainty in the climate response, see Section 11.3.1.1) and in the amplitude of simulated natural internal variability (Hawkins and Sutton, 2012).

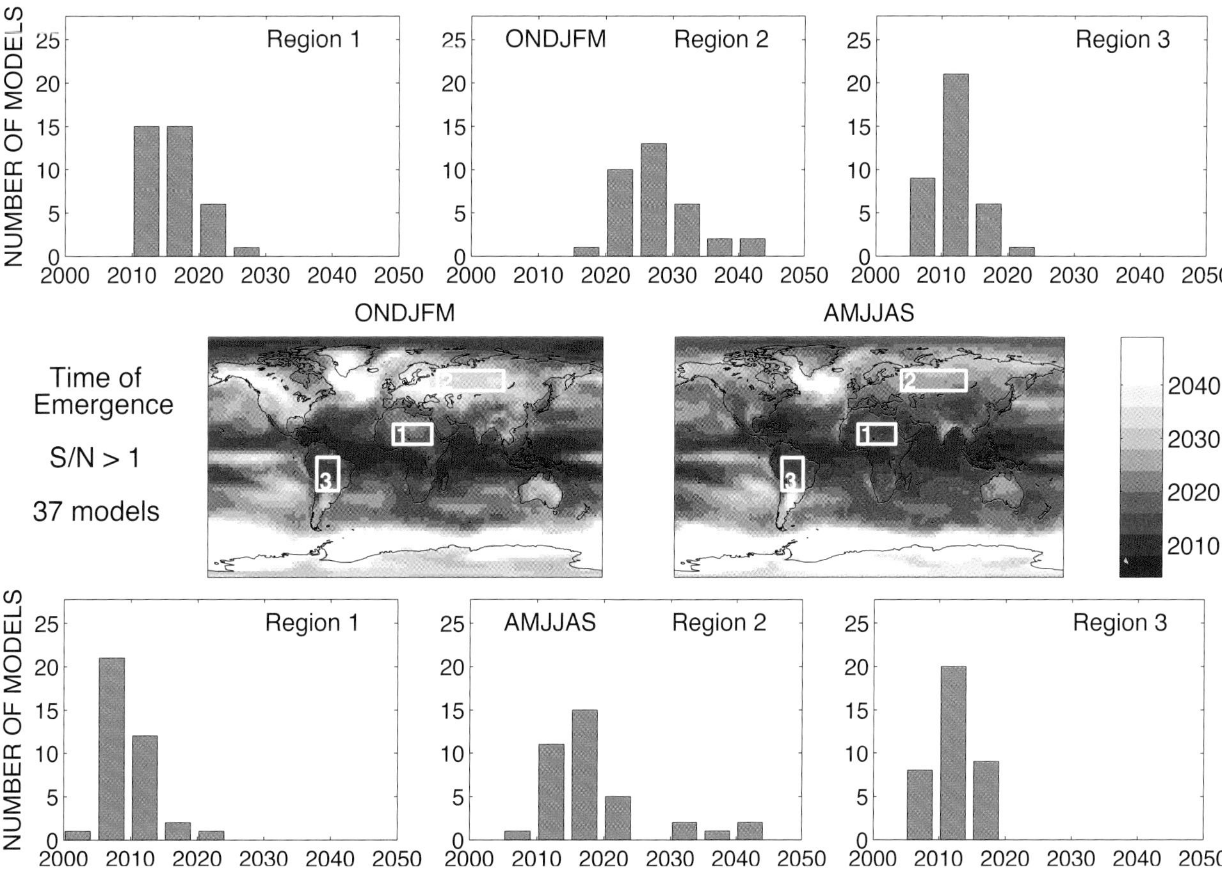

Figure 11.11 | Time of Emergence (ToE) of significant local warming derived from 37 CMIP5 models under the RCP4.5 scenario. Warming is quantified as the half-year mean temperature anomaly relative to 1986–2005, and the noise as the standard deviation of half-year mean temperature derived from a control simulation of the relevant model. Central panels show the median time at which the signal-to-noise ratio exceeds a threshold value of 1 for (left) the October to March half year and (right) the April to September half year, using a spatial resolution of 2.5° × 2.5°. Histograms show the distribution of ToE for area averages over the regions indicated obtained from the different CMIP5 models. Full details of the methodology may be found in Hawkins and Sutton (2012).

In summary, it is *very likely* that anthropogenic warming of surface air temperature over the next few decades will proceed more rapidly over land areas than over oceans, and that the warming over the Arctic in winter will be greater than the global mean warming over the same period. Relative to background levels of natural internal variability, near-term increases in seasonal mean and annual mean temperatures are expected to occur more rapidly in the tropics and subtropics than in mid-latitudes (*high confidence*).

11.3.2.2 Free Atmospheric Temperature

Changes in zonal mean temperature for the near-term period (2016–2035 compared to the base period 1986–2005) for the multi-model CMIP5 ensemble show a pattern similar to that in the CMIP3, with warming in the troposphere and cooling in the stratosphere of a couple of degrees that is significant even in the near term period. There is relatively greater warming in the tropical upper troposphere and northern high latitudes. A more detailed assessment of observed and simulated changes in free atmospheric temperatures can be found in Sections 10.3.1.2.1 and 12.4.3.2.

11.3.2.3 The Water Cycle

As discussed in the AR4 (Section 10.3.6; Meehl et al., 2007b), the IPCC Technical Paper on Climate Change and Water (Bates et al., 2008) and the Special Report on Managing the Risks of Extreme Events and Disasters to Advance Climate Change Adaptation (Seneviratne et al., 2012), a general intensification of the global hydrological cycle, and of precipitation extremes, are expected for a future warmer climate (e.g., (Huntington, 2006; Williams et al., 2007; Wild et al., 2008; Chou et al., 2009; Dery et al., 2009; O'Gorman and Schneider, 2009; Lu and Fu, 2010; Seager et al., 2010; Wu et al., 2010; Kao and Ganguly, 2011; Muller et al., 2011; Durack et al., 2012). In this section, projected changes in the time-mean hydrological cycle are discussed; changes in extremes, are presented in Section 11.3.2.5 while processes underlying precipitation changes are treated in Chapter 7.

11.3.2.3.1 Changes in precipitation

AR4 projections of the spatial patterns of precipitation change in response to GHG forcing (Chapter 10, Section 10.3.2) showed consistency between models on the largest scales (i.e., zonal means) but large uncertainty on smaller scales. The consistent pattern was characterized by increases at high latitudes and in wet regions (including the maxima in mean precipitation found in the tropics), and decreases in dry regions (including large parts of the subtropics). Large uncertainties in the sign of projected change were seen especially in regions located on the borders between regions of increases and regions of decreases. More recent research has highlighted the fact that if models agree that the projected change is small in some sense relative to internal variability, then agreement on the sign of the change is not expected (Tebaldi et al., 2011; Power et al., 2012). This recognition led to the identification of subregions within the border regions, where models agree that projected changes are either zero or small (Power et al., 2012). This, and other considerations, also led to the realization that the consensus among models on precipitation projections is more widespread than might have been inferred on the basis of the projections described in the AR4 (Power et al., 2012). Information on the reliability of near-term projections can also be obtained from verification of past regional trends (Räisänen (2007); Box 11.2)

Since the AR4 there has also been considerable progress in understanding the factors that govern the spatial pattern of change in precipitation (*P*), precipitation minus evaporation (*P* – *E*), and inter-model differences in these patterns. The general pattern of wet-get-wetter (also referred to as 'rich-get-richer', e.g., Held and Soden, 2006; Chou et al., 2009; Allan et al., 2010) and dry-get-drier has been confirmed, although with deviations in some dry regions at present that are projected to become wetter by some models, e.g., Northeast Brazil in austral summer and East Africa (see Annex I). It has been demonstrated that the wet-get-wetter pattern implies an enhanced seasonal precipitation range between wet and dry seasons in the tropics, and enhanced inter-hemispheric precipitation gradients (Chou et al., 2007).

It has recently been proposed that analysis of the energy budget, previously applied only to the global mean, may provide further insights into the controls on regional changes in precipitation (Levermann et al., 2009; Muller and O'Gorman, 2011; O'Gorman et al., 2012). Muller and O'Gorman (2011) argue in particular that changes in radiative and surface sensible heat fluxes provide a guide to the local precipitation response over land. Projected and observed patterns of oceanic precipitation change in the tropics tend to follow patterns of SST change because of local changes in atmospheric stability, such that regions warming more than the tropics as a whole tend to exhibit an increase in local precipitation, while regions warming less tend to exhibit reduced precipitation (Johnson and Xie, 2010; Xie et al., 2010).

AR4 (Section 10.3.2 and Chapter 11) showed that, especially in the near term, and on regional or smaller scales, the magnitude of projected changes in mean precipitation was small compared to the magnitude of natural internal variability (Christensen et al., 2007). Recent work has confirmed this result, and provided more quantification (e.g., Hawkins and Sutton, 2011; Hoerling et al., 2011; Rowell, 2011; Deser et al., 2012; Power et al., 2012). Hawkins and Sutton (2011) presented further analysis of CMIP3 results and found that, on spatial scales of the order of 1000 km, internal variability contributes 50 to 90% of the total uncertainty in all regions for projections of decadal and seasonal mean precipitation change for the next decade, and is the most important source of uncertainty for many regions for lead times up to three decades ahead (Figure 11.8). Thereafter, response uncertainty is generally dominant. Forcing uncertainty (except for that relating to aerosols, see Section 11.4.7) is generally negligible for near-term projections. The S/N ratio for projected changes in seasonal mean precipitation is highest in the subtropics and at high latitudes. Rowell (2011) found that the contribution of response uncertainty to the total uncertainty (response plus internal variability) in local precipitation change is highest in the deep tropics, particularly over South America, Africa, the east and central Pacific, and the Atlantic. Over tropical land and summer mid-latitude continents the representation of SST changes, atmospheric processes, land surface processes, and the terrestrial carbon cycle all contribute to the uncertainty in projected changes in rainfall.

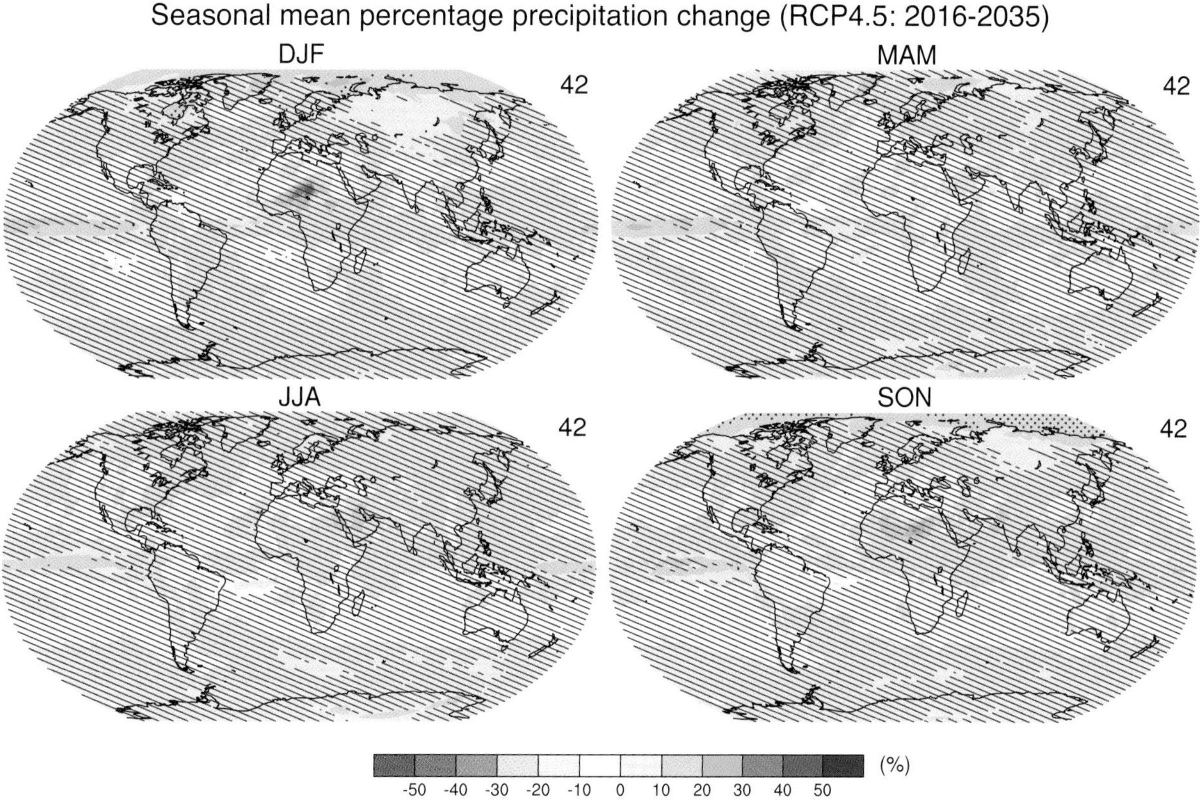

Figure 11.12 | CMIP5 multi-model ensemble mean of projected changes (%) in precipitation for 2016–2035 relative to 1986–2005 under RCP4.5 for the four seasons. The number of CMIP5 models used is indicated in the upper right corner. Hatching and stippling as in Figure 11.10.

In addition to the response to GHG forcing, forcing from natural and anthropogenic aerosols may exert significant impacts on regional patterns of precipitation change as well as on global mean temperature (Bollasina et al., 2011; Yue et al., 2011; Fyfe et al., 2012). Precipitation changes may arise as a consequence of temperature and stratification changes driven by aerosol-induced radiative effects, and/or as indirect aerosol effects on cloud microphysics (Chapter 7). Future emissions of aerosols and aerosol precursors are subject to large uncertainty, and further large uncertainties arise in assessing the responses to these emissions. These issues are discussed in Section 11.3.6.

Figures 11.12 and 11.13a present projections of near-term changes in precipitation from CMIP5. Regional maps and time series are presented in Annex I. The basic pattern of wet regions tending to get wetter and dry regions tending to get dryer is apparent, although with some regional deviations as mentioned previously. However, the

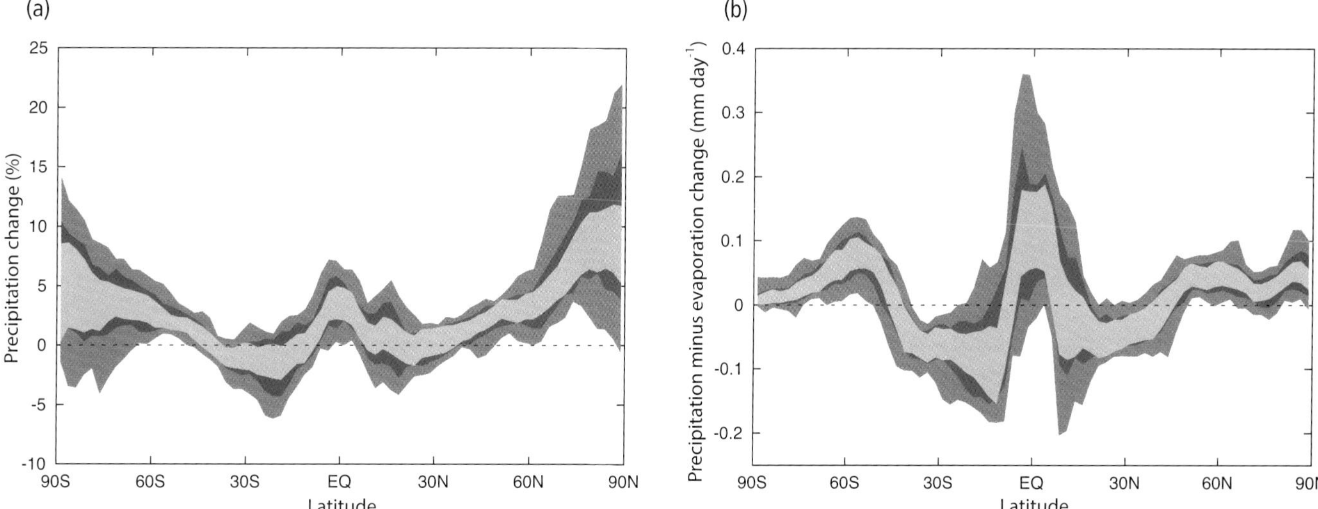

Figure 11.13 | CMIP5 multi-model projections of changes in annual and zonal mean (a) precipitation (%) and (b) precipitation minus evaporation (mm day^{-1}) for the period 2016–2035 relative to 1986–2005 under RCP4.5. The light blue denotes the 5 to 95% range, the dark blue the 17 to 83% range of model spread. The grey indicates the 1σ range of natural variability derived from the pre-industrial control runs (see Annex I for details).

large response uncertainty is evident in the substantial spread in the magnitude of projected change simulated by different climate models (Figure 11.13a). In addition, it is important to recognize—as discussed in previous sections—that models may agree and still be in error (e.g., Power et al. 2012). In particular, there is some evidence from comparing observations with simulations of the recent past that climate models might be underestimating the magnitude of changes in precipitation in many regions (Pincus et al., 2008; Liepert and Previdi, 2009; Schaller et al., 2011; Joetzjer et al., 2012) This evidence is discussed in detail in Chapter 9 (Section 9.4.1) and Box 11.2, and could imply that projected changes in precipitation are underestimated by current models. However, the magnitude of any underestimation has yet to be quantified, and is subject to considerable uncertainty.

Figures 11.12 and 11.13a also highlight the large amplitude of the natural internal variability of mean precipitation. On regional scales, mean projected changes are almost everywhere smaller than the estimated standard deviation of natural internal variability. The only exceptions are the northern high latitudes and the equatorial Pacific Ocean (Figure 11.12). For zonal means (Figure 11.13a) and at high latitudes only, the projected changes relative to the recent past exceed the estimated standard deviation of internal variability.

Overall, zonal mean precipitation will *very likely* increase in high and some of the mid latitudes, and will *more likely than not* decrease in the subtropics. At more regional scales precipitation changes may be influenced by anthropogenic aerosol emissions and will be strongly influenced by natural internal variability.

11

11.3.2.3.2 Changes in evaporation, evaporation minus precipitation, runoff, soil moisture, relative humidity and specific humidity

Because the variability of the atmospheric moisture storage is negligible, global mean increases in evaporation are required to balance increases in precipitation in response to anthropogenic forcing (Meehl et al., 2007a; Trenberth et al., 2007; Bates et al., 2008; Lu and M. Cai, 2009). The global atmospheric water content is constrained by the Clausius–Clapeyron equation to increase at around 7% K^{-1}; however, both the global precipitation and evaporation in global warming simulations increase at 1 to 3% K^{-1} (Lambert and Webb, 2008; Lu and M.Cai, 2009).

Changes in evapotranspiration over land are influenced not only by the response to RF, but also by the vegetation response to elevated CO_2 concentrations. Physiological effects of CO_2 may involve both the stomatal response, which acts to restrict transpiration (Field et al., 1995; Hungate et al., 2002; Cao et al., 2009, 2010; Lammertsma et al., 2011), and an increase in plant growth and leaf area, which acts to increase evapotranspiration (El Nadi, 1974; Bounoua et al., 2010). Simulation of the latter process requires the inclusion of vegetation models that allow spatial and temporal variability in the amount of active biomass, either by changes in the phenological cycle or changes in the biome structure.

In response to GHG forcing, dry land areas tend to show a reduction of evaporation and often precipitation, accompanied by a drying of the soil and an increase of surface temperature, in response to decreases in latent heat fluxes from the surface (e.g., Fischer et al., 2007; Seneviratne et al., 2010). Jung et al. (2010) use a mixture of observations and models to illustrate a recent global mean decline in land surface evaporation due to soil-moisture limitations. Accompanying precipitation effects are more subtle, as there are significant uncertainties and large geographical variations regarding the soil-moisture precipitation feedback (Hohenegger et al., 2009; Taylor et al., 2011). AR4 projections (Meehl et al. (2007b) of annual mean soil moisture changes for the 21st century showed a tendency for decreases in the subtropics, southern South America and the Mediterranean region, and increases in limited areas of east Africa and central Asia. Changes seen in other regions were mostly not consistent or statistically significant.

AR4 projections of 21st century runoff changes (Meehl et al., 2007b) showed consistency in sign among models indicating annual mean reductions in southern Europe and increases in Southeast Asia and at high northern latitudes. Projected changes in global mean runoff associated with the physiological effects of doubled CO_2 concentrations show increases of 6 to 8% relative to pre-industrial levels, an increase that is comparable to that simulated in response to RF changes (11% ± 6%) (Betts et al., 2007; Cao et al., 2010). Gosling et al. (2011) assess the projected impacts of climate change on river runoff from global and basin-scale hydrological models obtaining increased runoff with global warming in the Liard (Canada), Rio Grande (Brazil) and Xiangxi (China) basins and decrease for the Okavango (southwest Africa).

Consideration of hydrological drought conditions employs a range of different dryness indicators, such as soil moisture or other drought indices that integrate precipitation and evaporation effects (Seneviratne et al., 2012). There are large uncertainties in regional drought projections (Burke and Brown, 2008), and very few studies have addressed the near-term future (Sheffield and Wood, 2008; Dai, 2011). In order to provide an indication of future changes of water availability, Figure 11.13b presents zonal mean changes in precipitation minus evaporation ($P - E$) from CMIP5. As in the case of precipitation (Figure 11.13a), the uncertainty is dominated by model differences as opposed to natural variability (compare blue versus grey shading). The results are consistent with the wet-get-wetter and dry-get-drier pattern (e.g., Held and Soden 2006): In the high latitudes and the tropics, most of the models project zonal-mean increases in $P - E$, which over land would need to be compensated by increases in runoff (see next paragraph). In contrast, zonal mean projected changes in the subtropics are negative, indicating decreases in water availability. Although this pattern is evident in most or all of the models, and although several studies project drought increases in the near term future (Sheffield and Wood, 2008; Dai, 2011), the assessment is debated in the literature based on discrepancies in the recent past and due to natural variability (Seneviratne et al., 2012; Sheffield et al., 2012).

The global distribution of the 2016–2035 changes in annual mean evaporation, evaporation minus precipitation ($E - P$), surface runoff, soil moisture, relative humidity and surface-level specific humidity from the CMIP5 multi-model ensemble under RCP4.5 are shown in Figure 11.14. Changes in evaporation over land (Figure11.14a), are mostly positive with the largest values at northern high latitudes, in agreement with projected temperature increases (Figure 11.10). Over the oceans, evaporation is also projected to increase in most regions. Projected changes are larger than the estimated standard deviation of internal variability only at high latitudes and over the tropical oceans. Decreases in evaporation over land (i.e., Australia, southern Africa, northeastern South America and Mexico) and oceans are smaller than the estimated standard deviation of internal variability; the only exception is the western North Atlantic, although the model agreement is low in that region. Projected changes in $(E - P)$ over land (Figure 11.14b) are generally consistent with the zonal mean changes shown in Figure 11.13b. In the high northern latitudes and the tropics, $(E - P)$ changes are mostly negative as dominated by precipitation increases (Figure 11.12), while in

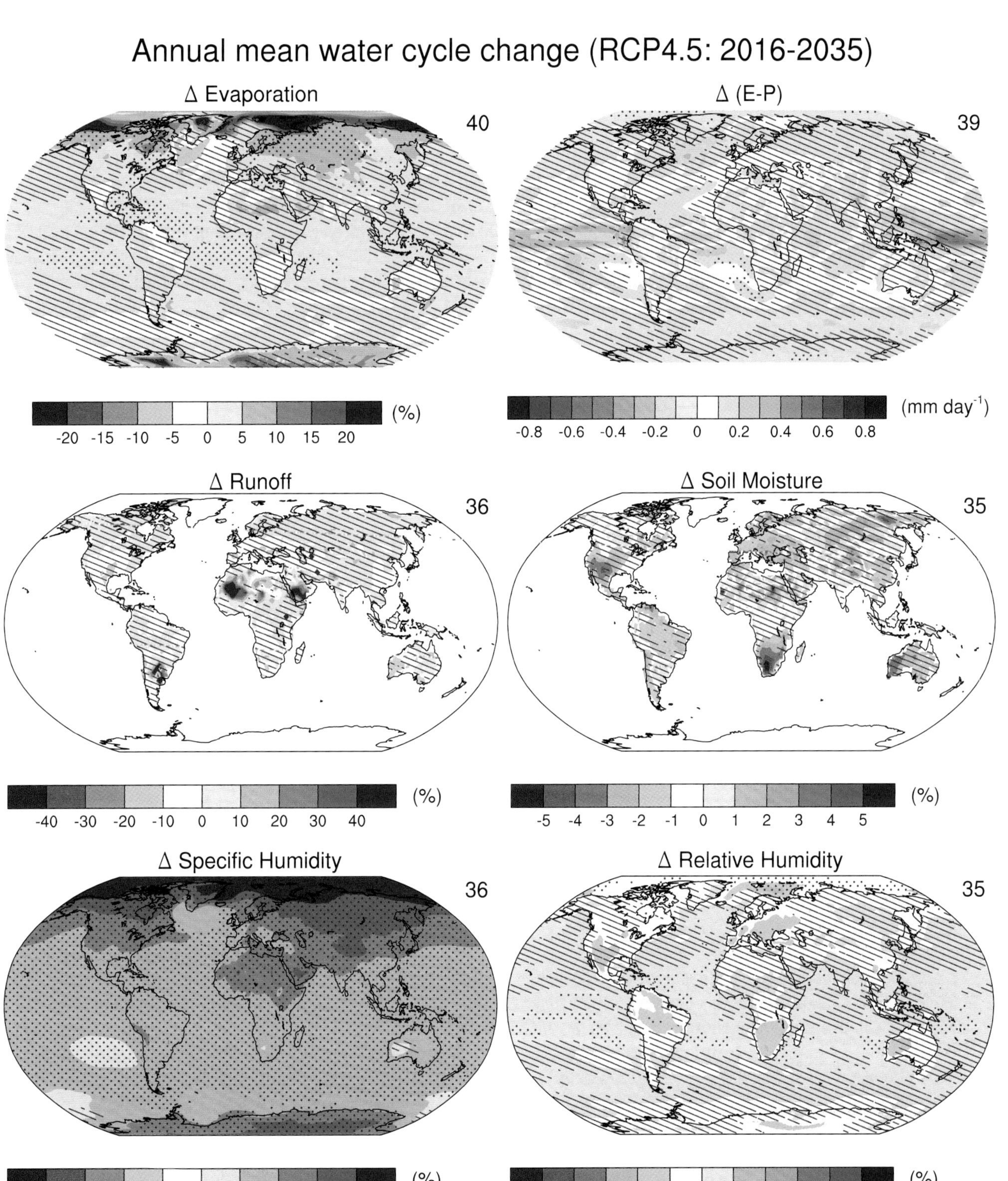

Figure 11.14 | CMIP5 multi-model annual mean projected changes for the period 2016–2035 relative to 1986–2005 under RCP4.5 for: (a) evaporation (%), (b) evaporation minus precipitation ($E - P$, mm day^{-1}), (c) total runoff (%), (d) soil moisture in the top 10 cm (%), (e) relative change in specific humidity (%), and (f) absolute change in relative humidity (%). The number of CMIP5 models used is indicated in the upper right corner of each panel. Hatching and stippling as in Figure 11.10.

11

the subtropics several areas exhibit increases in ($E - P$), in particular in Europe, western Australia and central-western USA. However, in most locations changes are smaller than internal variability.

Annual mean shallow soil moisture (Figure 11.14d) shows decreases in most subtropical regions (except La Plata basin in South America) and in central Europe, and increases in northern mid-to-high latitudes. Projected changes are larger than the estimated internal variability only in southern Africa, the Amazon region and Europe. Projected changes in runoff (Figure 11.14c) show decreases in northern Africa, western Australia, southern Europe and southwestern USA and increases larger than the internal variability in northwestern Africa, southern Arabia and southeastern South America associated to the projected changes in precipitation (Figure 11.12). Owing to the simplified hydrological models in many CMIP5 climate models, the projections of soil moisture and runoff have large model uncertainties.

Changes in near-surface specific humidity are positive, with the largest values at northern high latitudes when expressed in percentage terms (Figure 11.14e). This is consistent with the projected increases in temperature when assuming constant relative humidity. These changes are larger than the estimated standard deviation of internal variability almost everywhere: the only exceptions are oceanic regions such as the northern North Atlantic and around Antarctica. In comparison, absolute changes in near-surface relative humidity (Figure 11.14f) are much smaller, on the order of a few percent, with general decreases over most land areas, and small increases over the oceans. Significant decreases relative to natural variability are projected in the Amazonia, southern Africa and Europe, although the model agreement in these regions is low.

Over the next few decades projected increases in near-surface specific humidity are *very likely*, and projected increases in evaporation are *likely* in many land regions. There is *low confidence* in projected changes in soil moisture and surface runoff.

11.3.2.4 Atmospheric Circulation

11.3.2.4.1 Northern Hemisphere extratropical circulation

In the NH extratropics, some Atmosphere–Ocean General Circulation Models (AOGCMs) indicate changes to atmospheric circulation from anthropogenic forcing by the mid-21st century, including a poleward shift of the jet streams and associated zonal mean storm tracks (Miller et al., 2006; Pinto et al., 2007; Paeth and Pollinger, 2010) and a strengthening of the Atlantic storm track (Pinto et al., 2007), Figure 11.15. Consistent with this, the CMIP5 AOGCMs exhibit an ensemble mean increase in the North Atlantic Oscillation (NAO) and Northern Annular Model (NAM) indices by 2050, especially in autumn and winter (Gillett et al., 2013).

However, there are reasons to be cautious over these near-term projections. Although models simulate the broad features of the large-scale circulation well, there remain quite significant biases in many models (see Sections 9.4.1.4.3 and 9.5.3.2). The response of the NH circulation can be sensitive to small changes in model formulation (Sigmond et al., 2007), and to features that are known to be poorly simulated in many climate models. These features include high- and low-latitude physics (Rind, 2008; Woollings, 2010), ocean circulation (Woollings and Blackburn, 2012), tropical circulation (Haarsma and Selten, 2012) and stratospheric dynamics (Huebener et al., 2007; Morgenstern et al., 2010; Scaife et al., 2012). As a result, there is considerable model uncertainty in the response of the NH storm track position (Ulbrich et al., 2008), stationary waves (Brandefelt and Kornich, 2008) and the jet streams (Miller et al., 2006; Ihara and Kushnir, 2009; Woollings and Blackburn, 2012). Further, CMIP5 models show that the response of NH extratropical circulation to even strong GHG forcing remains weak compared to recent multidecadal variability and a recent detection and attribution study suggests that tropospheric ozone and aerosol changes may have been a key driver to NH extratropical circulation changes (Gillett et al., 2013). Some AOGCMs simulate multi-decadal NAO variability as large as that recently observed with no external forcing (Selten et al., 2004; Semenov et al., 2008). This suggests that internal variability could dominate the anthropogenically forced response in the near term (Deser et al., 2012).

Some studies have predicted a shift to the negative phase of the Atlantic Multi-decadal Oscillation (AMO) over the coming few decades, with potential impacts on atmospheric circulation around the Atlantic sector (Knight et al., 2005; Sutton and Hodson, 2005; Folland et al., 2009). It has also been suggested that there may be significant changes in solar forcing over the next few decades, which could have an influence on NAO-related atmospheric circulation (Lockwood et al., 2011), although these predictions are highly uncertain (see Section 11.3.6.2.2).

There is only *medium confidence* in near-term projections of a northward shift of NH storm track and westerlies, and an increase of the NAO/NAM because of the large response uncertainty and the potentially large influence of internal variability.

11.3.2.4.2 Southern Hemisphere extratropical circulation

Increases in GHGs, and related dynamical processes, are projected to lead to poleward shifts in the annual mean position of Southern Hemisphere (SH) extratropical storm tracks and winds (Figure 11.17; Chapters 10 and 12). A key issue in projections of near-term SH extratropical circulation change is the extent to which changes driven by stratospheric ozone recovery will counteract changes driven by increasing GHGs. Several observational and modeling studies (Gillett and Thompson, 2003; Shindell and Schmidt, 2004; Arblaster and Meehl, 2006; Roscoe and Haigh, 2007; Fogt et al., 2009; Polvani et al., 2011a; Gillett et al., 2013) indicate that, over the late 20th and early 21st centuries, the observed summertime poleward shift of the westerly jet (a positive Southern Annular Mode (SAM)) has been caused primarily by the depletion of stratospheric ozone, with increasing GHGs contributing only a smaller fraction to the observed trends. The latest generation of climate models project substantially smaller poleward trends in SH atmospheric circulation in austral summer over the coming half century compared to those over the late 20th century, as the recovery of stratospheric ozone will oppose the effects of continually increasing GHGs (Arblaster et al., 2011; McLandress et al., 2011; Polvani et al., 2011a; Eyring et al., 2013). Locally, internal variability may be a dominant contributor to near-term changes in lower-tropospheric zonal winds (Figure 11.17). The average 2016–2035 SH extratropical storm tracks and zonal winds are *likely* to shift poleward relative to 1986–2005. However, even though a full recovery of the ozone hole is not expected until the 2060s to 2070s (Table 5.4; WMO, 2010; see Chapter 12), it is *likely* that over the near term there will be a reduced rate in the austral summertime poleward shift of the SH circumpolar trough, SH extratropical storm tracks and winds compared to its movement over the past 30 years, including the possibility of no detectable shift.

11.3.2.4.3 Tropical circulation

Increases in GHGs are expected to lead to a poleward shift of the Hadley Circulation (Lu et al., 2007; Chapter 12, Figure 11.18). Relative to the late 20th century, the tendency towards a poleward expansion of the Hadley Circulation will start to emerge by the mid-2030s, with certain intra-model consensus in the SH expansion, despite the counteracting effect of ozone recovery (Figure 11.18). As with near-term changes in SH extratropical circulation, a key for near-term projections of the structure of the SH Hadley Circulation is the extent to which future stratospheric ozone recovery will counteract the impact of GHGs. The poleward expansion of the Hadley Circulation, particularly of the SH branch during austral summer, during the later decades of the 20th century has been largely attributed to the combined impact of stratospheric ozone depletion (Thompson and Solomon, 2002; Son et al., 2008, 2009a, 2009b; Polvani et al., 2011a, 2011b; Min and Son, 2013) and the concurrent increase in GHGs (Arblaster and Meehl, 2006; Arblaster et al., 2011) as discussed in the previous section. The poleward expansion of the Hadley Circulation driven by the response of the atmosphere to increasing GHGs (Lu et al., 2007; Kang et al., 2011; Staten et al., 2011; Butler et al., 2012) would be counteracted in the SH by reduced stratospheric ozone depletion but depends on the rate of ozone recovery (UNEP and WMO, 2011). Increases in the incoming solar radiation can lead to a widening of the Hadley Cell (Haigh, 1996; Haigh et al., 2005) and large volcanic eruption to

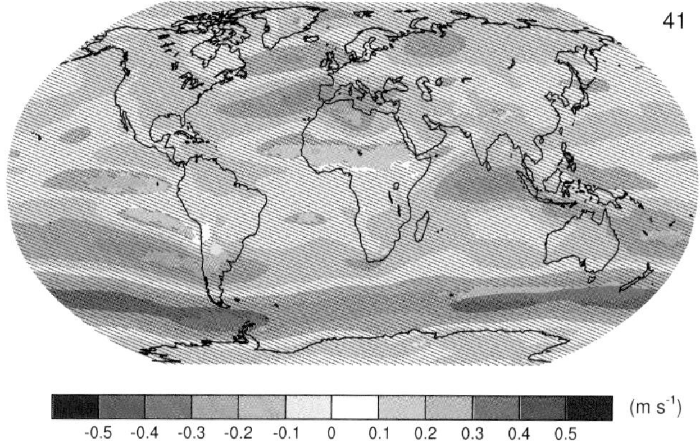

Figure 11.15 | CMIP5 multi-model ensemble mean of projected changes (m s^{-1}) in zonal (west-to-east) wind at 850 hPa for 2016–2035 relative to 1986–2005 under RCP4.5. The number of CMIP5 models used is indicated in the upper right corner. Hatching and stippling as in Figure 11.10.

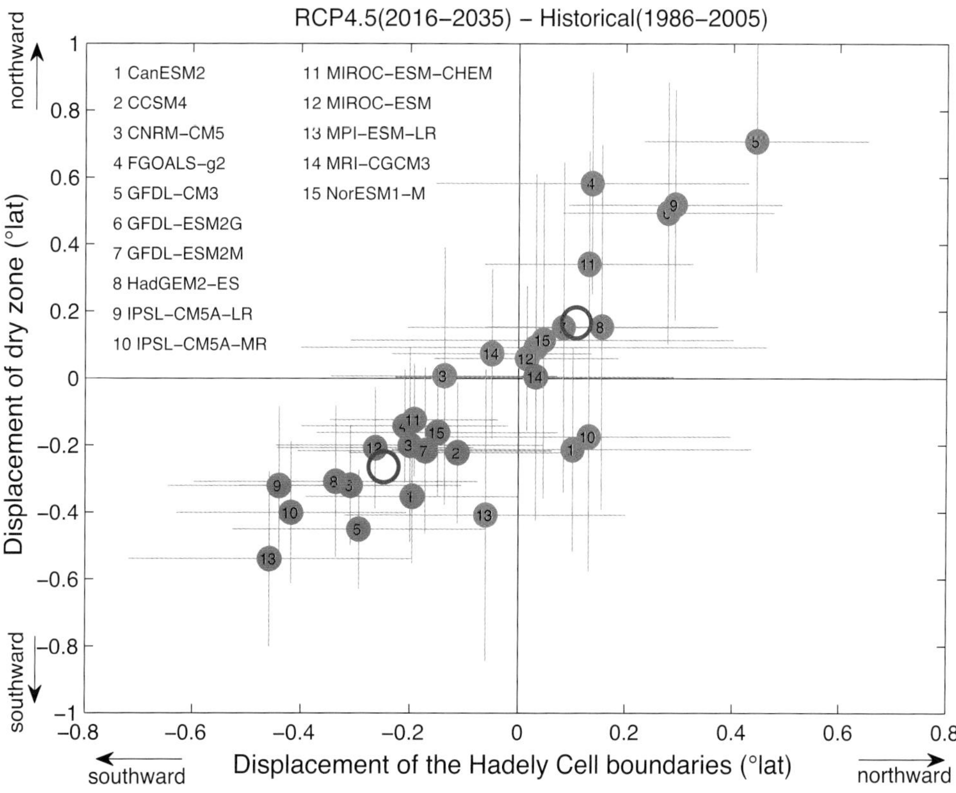

Figure 11.16 | Projected changes in the annual averaged poleward edge of the Hadley Circulation (horizontal axis) and sub-tropical dry zones (vertical axis) based on 15 Atmosphere–Ocean General Circulation Models (AOGCMs) from the CMIP5 (Taylor et al., 2012) multi-model ensemble, under 21st century RCP4.5. Orange symbols show the change in the northern edge of the Hadley Circulation/dry zones, while blue symbols show the change in the southern edge of the Hadley Circulation/dry zones. Open circles indicate the multi-model average, while horizontal and vertical coloured lines indicate the ±1 standard deviation range for internal climate variability estimated from each model. Values referenced to the 1986–2005 climatology. (Figure based on the methodology of Lu et al., 2007.)

contraction of the tropics and the tropical circulation (Lu et al., 2007; Birner, 2010). So future solar variations and volcanic activities could also lead to variations in the width of the Hadley Cell. The poleward extent of the Hadley Circulation and associated dry zones can exhibit substantial internal variability (e.g., Birner, 2010; Davis and Rosenlof, 2012) that can be as large as its near-term projected changes (Figure 11.16). There is also considerable uncertainty in the amplitude of the poleward shift of the Hadley Circulation in response to GHGs across multiple AOGCMs (Lu et al., 2007; Figure 11.16). It is *likely* that the poleward extent of the Hadley Circulation will increase through the mid-21st century. However, because of the counteracting impacts of future changes in stratospheric ozone and GHG concentrations, it is *unlikely* that it will continue to expand poleward in the SH as rapidly as it did in recent decades.

The Hadley Cell expansion in the NH has been largely attributed to the low-frequency variability of the SST (Hu et al., 2013), the increase of black carbon (BC) and tropospheric ozone (Allen and Sherwood, 2011). Internal variability in the poleward edge of the NH Hadley Circulation is large relative the radiatively forced signal (Figure 11.16. Given the complexity in the forcing mechanism of the NH expansion and the uncertainties in future concentrations of tropospheric pollutants, there is *low confidence* in the character of near-term changes to the structure of the NH Hadley Circulation.

Global climate models and theoretical considerations suggest that a warming of the tropics should lead to a weakening of the zonally asymmetric or Walker Circulation (Knutson and Manabe, 1995; Held and Soden, 2006; Vecchi and Soden, 2007; Gastineau et al., 2009). Aerosol forcing can modify both Hadley and Walker Circulations, which—depending on the details of the aerosol forcing—may lead to temporary reversals or enhancements in any GHG-driven weakening of the Walker Circulation (Sohn and Park, 2010; Bollasina et al., 2011; Merrifield, 2011; DiNezio et al., 2013). Meanwhile, the strength and structure of the Walker Circulation are impacted by internal climate variations, such as the ENSO (e.g., Battistiand Sarachik, 1995), the PDO (e.g., Zhang et al. 1997) and the IPO (Power et al., 1999, 2006; Meehl and Hu, 2006; Meehl and Arblaster, 2011; Power and Kociuba, 2011b; Meehl and Arblaster, 2012; Meehl et al., 2013a). Even on time scales of 30 to 100 years, substantial variations in the strength of the Pacific Walker Circulation in the absence of changes in RF are possible (Power et al., 2006; Vecchi et al., 2006). Estimated near-term weakening of the Walker Circulation from CMIP3 models under the A1B scenario (Vecchi and Soden, 2007; Power and Kociuba, 2011a) are *very likely* to be smaller than the impact of internal climate variations over 50-year time scales (Vecchi et al., 2006). There is also considerable response uncertainty in the amplitude of the weakening of Walker Circulation in response to GHG increase across multiple AOGCMs (Vecchi and Soden, 2007; DiNezio et al., 2009; Power and Kociuba, 2011a, 2011b). Thus, there is *low confidence* in projected near-term changes to the Walker Circulation. It is *very likely* that there will be decades in which the Walker Circulation strengthens and weakens due to internal variability through the mid-century as the externally forced change is small compared to internally generated decadal variability.

11.3.2.5 Atmospheric Extremes

Extreme events in a changing climate are the subject of Chapter 3 (Seneviratne et al., 2012) of the IPCC Special Report on Extremes (SREX). This previous IPCC chapter provides an assessment of more than 1000 studies. Here the focus is on near-term aspects and an assessment of more recent studies is provided.

11.3.2.5.1 Temperature extremes

In the AR4 (Meehl et al., 2007b), cold episodes were projected to decrease significantly in a future warmer climate and it was considered *very likely* that heat waves would be more intense, more frequent and last longer towards the end of the 21st century. These conclusions have generally been confirmed in subsequent studies addressing both global scales (Clark et al., 2010; Diffenbaugh and Scherer, 2011; Caesar and Lowe, 2012; Orlowsky and Seneviratne, 2012; Sillmann et al., 2013) and regional scales (e.g., Gutowski et al., 2008; Alexander and Arblaster, 2009; Fischer and Schar, 2009; Marengo et al., 2009; Meehl et al., 2009a; Diffenbaugh and Ashfaq, 2010; Fischer and Schar, 2010; Cattiaux et al., 2012; Wang et al., 2012). In the SREX assessment it is

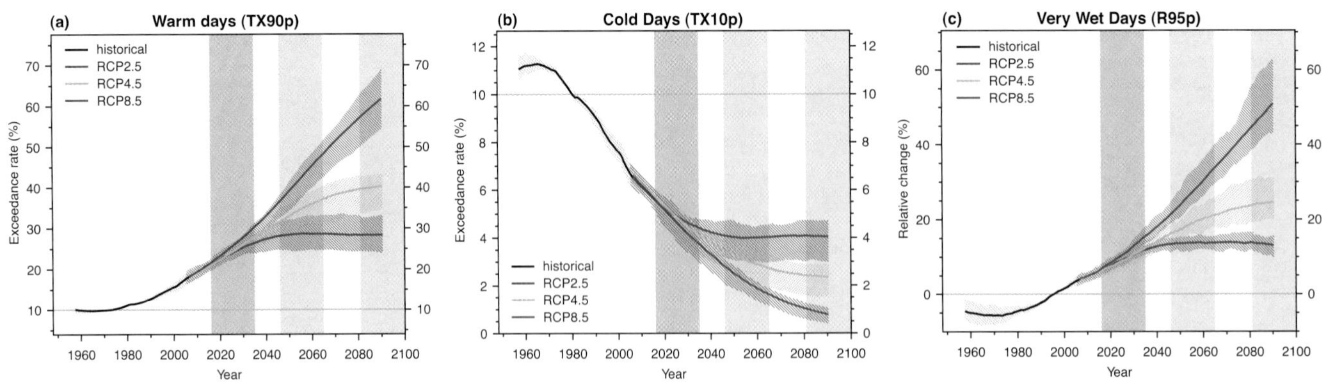

Figure 11.17 | Global projections of the occurrence of (a) warm days (TX90p), (b) cold days (TX10p) and (c) precipitation amount from very wet days (R95p). Results are shown from CMIP5 for the RCP2.6, RCP4.5 and RCP8.5 scenarios. Solid lines indicate the ensemble median and shading indicates the interquartile spread between individual projections (25th and 75th percentiles). The specific definitions of the indices shown are (a) percentage of days annually with daily maximum surface air temperature (T_{max}) exceeding the 90th percentile of T_{max} for 1961–1990, (b) percentage of days with T_{max} below the 10th percentile and (c) percentage change relative to 1986–2005 of the annual precipitation amount from daily events above the 95th percentile. (From Sillmann et al., 2013.)

concluded that increases in the number of warm days and nights and decreases in the number of cold days and nights are *virtually certain* on the global scale.

None of the aforementioned studies specifically addressed the near term. However, detection and attribution studies (see also Chapter 10) show that temperature extremes have already increased in many regions, consistent with climate change projections, and analyses of CMIP5 global projections show that this trend will continue and become more notable. The CMIP5 model ensemble exhibits a significant decrease in the frequency of cold nights, an increase in the frequency of warm days and nights and an increase in the duration of warm spells (Sillmann et al., 2013). These changes are particularly evident in global mean projections (see Figure 11.17). Figure 11.17 shows that for the next few decades—as discussed in the introduction to the current chapter—these changes are remarkably insensitive to the emission scenario considered (Caesar and Lowe, 2012). In most land regions and in the near-term, the frequency of warm days and warm nights will thus *likely* continue to increase, while that of cold days and cold nights will *likely* continue to decrease.

Near-term projections from General Circulation Model–Regional Climate Model (GCM–RCM) model chains (van der Linden and Mitchell, 2009) for Europe are shown in Figure 11.18, displaying near-term changes in mean and extreme temperature (left-hand panels) and precipitation (right-hand panels) relative to the reference period 1986–2005. In terms of mean June, July and August (JJA) temperatures (Figure 11.18a), projections show a warming of 0.6°C to 1.5°C, with highest changes over the land portion of the Mediterranean. The north–south gradient in the projections is consistent with the AR4. Daytime extreme summer temperatures in southern and central Europe are projected to warm substantially faster than mean temperatures (compare Figure 11.18a and b). This difference between changes in mean and extremes can be explained by increases in interannual and/or synoptic variability, or increases in diurnal temperature range (Gregory and Mitchell, 1995; Schar et al., 2004; Fischer and Schar, 2010; Hansen, 2012; Quesada et al., 2012; Seneviratne et al., 2012). There is some evidence, however, that this effect is overestimated in some of the models (Fischer et al., 2012; Stegehuis et al., 2012), leading to a potential overestimation of the projected Mediterranean summer mean warming (Buser et al., 2009; Boberg and Christensen, 2012). With regard to near-term projections of

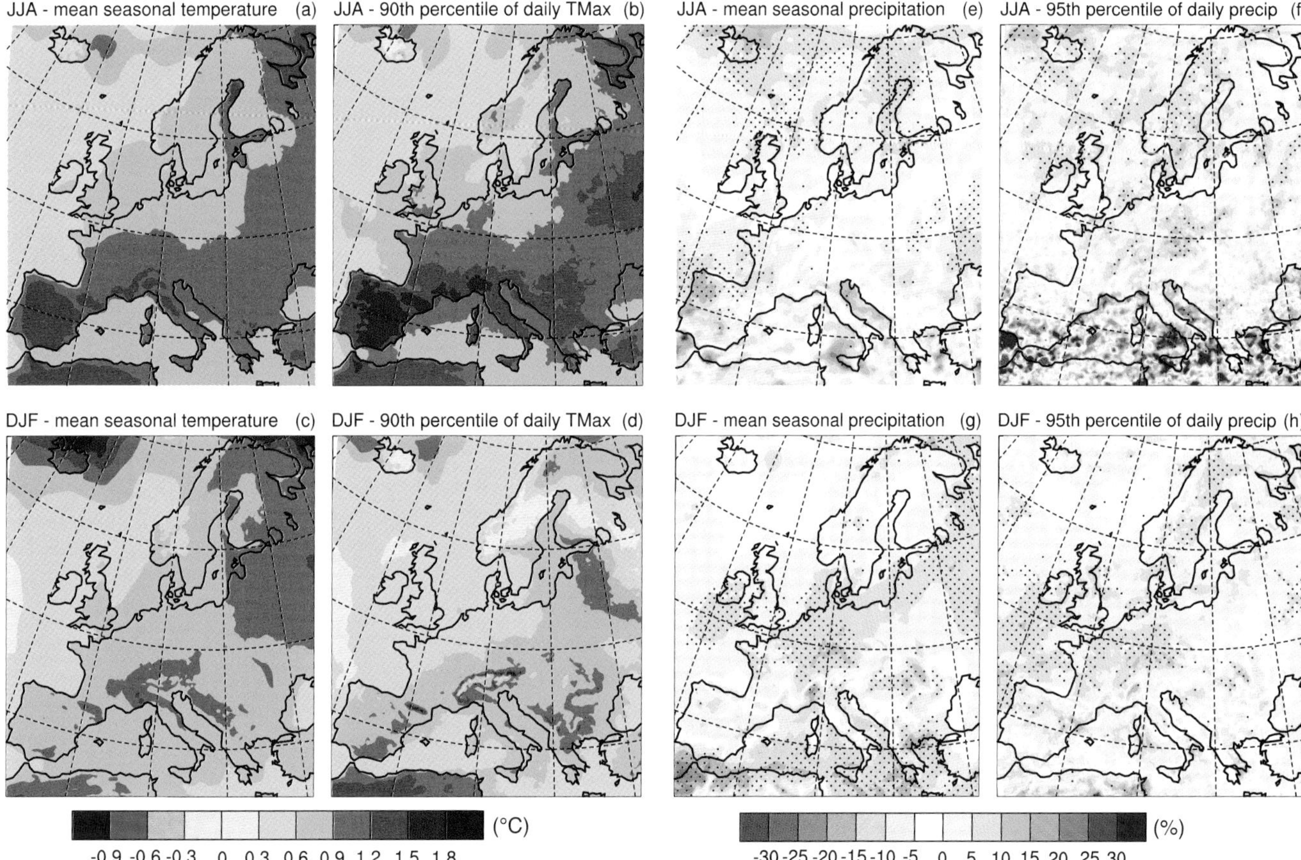

Figure 11.18 | European-scale projections from the ENSEMBLES regional climate modelling project for 2016–2035 relative to 1986–2005, with top and bottom panels applicable to June, July and August (JJA) and December, January, February (DJF), respectively. For temperature, projected changes (°C) are displayed in terms of ensemble mean changes of (a, c) mean seasonal surface temperature, and (b, d) the 90th percentile of daily maximum temperatures. For precipitation, projected changes (%) are displayed in terms of ensemble mean changes of (e, g) mean seasonal precipitation and (f, h) the 95th percentile of daily precipitation. The stippling in (e–h) highlights regions where 80% of the models agree in the sign of the change (for temperature all models agree on the sign of the change). The analysis includes the following 10 GCM-RCM simulation chains for the SRES A1B scenario (naming includes RCM group and GCM simulation): HadRM3Q0-HadCM3Q0, ETHZ-HadCM3Q0, HadRM3Q3-HadCM3Q3, SMHI-HadCM3Q3, HadRM3Q16-HadCM3Q16, SMHI-BCM, DMI-ARPEGE, KNMI-ECHAM5, MPI-ECHAM5, DMI-ECHAM5. (Rajczak et al., 2013.)

record heat compared to record cold (Meehl et al., 2009b) show, for one model, that over the USA the ratio of daily record high temperatures to daily record low temperatures could increase from an early 2000s value of roughly 2 to 1 to a mid-century value of about 20 to 1.

In terms of December, January and February (DJF) temperatures (Figure 11.18c), projections show a warming of 0.3°C to 1.8°C, with the largest changes in the N–NE part of Europe. This characteristic pattern of changes tends to persist to the end of century (van der Linden and Mitchell, 2009). In contrast to JJA temperatures, daytime high-percentile (i.e., warm) winter temperatures are projected to warm slower than mean temperatures (compare Figure 11.18c and Figure 11.18d), while low-percentile (i.e., cold) winter temperatures warm faster than the mean. This behaviour is indicative of reductions in internal variability, which may be linked to changes in storm track activity, reductions in diurnal temperature range and changes in snow cover (e.g., Colle et al. 2013; Dutra et al., 2011).

11.3.2.5.2 Heavy precipitation events

For the 21st century, the AR4 and the SREX concluded that heavy precipitation events were *likely* to increase in many areas of the globe (IPCC, 2007). Since AR4, a larger number of additional studies have been published using global and regional climate models (Fowler et al., 2007; Gutowski et al., 2007; Sun et al., 2007; Im et al., 2008; O'Gorman and Schneider, 2009; Xu et al., 2009; Hanel and Buishand, 2011; Heinrich and Gobiet, 2011; Meehl et al., 2012b). For the near term, CMIP5 global projections (Figure 11.17c) confirm a clear tendency for increases in heavy precipitation events in the global mean, but there are significant variations across regions (Sillmann et al., 2013). Past observations have also shown that interannual and decadal variability in mean and heavy precipitation are large, and are in addition strongly affected by internal variability (e.g., El Niño), volcanic forcing and anthropogenic aerosol loads (see Section 2.3.1). In general models have difficulties in representing these variations, particularly in the tropics (see Section 9.5.4.2). Thus the frequency and intensity of heavy precipitation events will *likely* increase over many land areas in the near term, but this trend will not be apparent in all regions, because of natural variability and possible influences of anthropogenic aerosols.

Simulations with regional climate models demonstrate that the response in terms of heavy precipitation events to anthropogenic climate change may become evident in some but not all regions in the near term. For instance, ENSEMBLES projections for Europe (see Figure 11.18e–h) confirm the previous IPCC results that changes in mean precipitation as well as heavy precipitation events are characterized by a pronounced north–south gradient in the extratropics, especially in the winter season, with precipitation increases in the higher latitudes and decreases in the subtropics. Although this pattern starts to emerge in the near term, the projected changes are statistically significant only in a fraction of the domain. The results are affected by both changes in water vapour content as induced by large-scale warming and large-scale circulation changes. Figure 11.18e–h also shows that mid- and high-latitude projections for changes in DJF extremes and means are qualitatively similar in the near term, at least for the event size considered.

Previous work reviewed in AR4 has established that extreme precipitation events may increase substantially stronger than mean precipitation amounts. More specifically, extreme events may increase with the atmospheric water vapour content, that is, up to the rate of the Clausius–Clapeyron (CC) relationship (e.g., Allen and Ingram, 2002). More recent work suggests that increases beyond this threshold may occur for short-term events associated with thunderstorms (Lenderink and Van Meijgaard, 2008; Lenderink and Meijgaard, 2010) and tropical convection (O'Gorman, 2012). A number of studies showed strong dependencies on location and season, but confirm the existence of significant deviations from the CC scaling (e.g., Lenderink et al., 2011; Mishra et al., 2012; Berg et al., 2013). Studies with cloud-resolving models generally support the existence of temperature-precipitation relations that are close to or above (up to about twice) the CC relation (Muller et al., 2011; Singleton and Toumi, 2012).

11.3.2.5.3 Tropical cyclones

The projected response of tropical cyclones (TCs) at the end of the 21st century is summarized in Section 14.6.1 and the IPCC Special Report on Extremes (SREX) (Seneviratne et al., 2012). Relative to the number of studies focussing on projections of TC activity at the end of the 21st century (Section 14.6.1; Knutson et al., 2010; Seneviratne et al., 2012 there are fewer studies that have explored near-term projections of TC activity (Table 11.2); the North Atlantic (NA) stands out as the basin with most studies. In the NA, there are mixed projections for basin-wide TC frequency, suggesting significant decreases (Knutson et al., 2013a) or non-significant changes (Villarini et al., 2011; Villarini and Vecchi, 2012). Multi-model mean projected NA TC frequency changes based on CMIP3 and CMIP5 over the first half of the 21st century were smaller than the overall uncertainty estimated from the Coupled General Circulation Models (CGCMs), with internal climate variability being a leading source of uncertainty through the mid-21st century (Villarini et al., 2011; Villarini and Vecchi, 2012). Therefore, based on the limited literature available, the conflicting near-term projections in basins with more than one study, the large influence of internal variability, the lack of confidently detected/attributed changes in TC activity (Chapter 10) and the conflicting projections for basin-wide TC frequency even at the end of the 21st century (Chapter 14), there is currently *low confidence* in basin-scale and global projections of trends in tropical cyclone frequency to the mid-21st century.

Exploring different hurricane intensity measures, two studies project near-term increases of NA hurricane intensity (Knutson et al., 2013a; Villarini and Vecchi, 2013), driven in large part by projected reductions in NA tropospheric aerosols in CMIP5 future forcing scenarios. Studies project near-term increases in the frequency Category 4–5 TCs in the NA (Knutson et al., 2013a) and southwest Pacific (Leslie et al., 2007). Published studies agree in the sign of projected mid-century intensity change (intensification), but the only basin with more than one study exploring intensity is the NA. For the NA, an estimate of the time scale of emergence of projected changes in intense TC frequency exceeds 60 years (Bender et al., 2010), although that estimate depends crucially on the amplitude of internal climate variations of intense hurricane frequency (e.g., Emanuel, 2011), which remains poorly constrained at the moment. Therefore, there is *low confidence* in near-term TC intensity projections in all TC basins.

Table 11.2 | Summary of studies exploring near-term projections of tropical cyclone (TC) activity. First column lists the TC basin explored, the second column summarizes the changes in TC activity reported in each study, the third column presents notes on the methodology and the fourth column provides a reference to the study.

TC Basin Explored	Projected Change in TC Activity Reported	Notes	Reference
Global	Reduced global, Northern Hemisphere and Southern Hemisphere frequency 2016–2035 relative to 1986–2005.	High-resolution atmospheric model forced by CMIP3 SRES A1B multi-model SST change 2004–2099.	Sugi and Yoshimura (2012)
N.W. Pacific	Over first half of 21st century: Reduced Activity over South China Sea, Increased Activity near subtropical Asia	Statistical downscale of five CMIP3 models under SRES A1B.	Wang et al. (2011)
N.W. Pacific	Over 2001–2040, a decrease in TC frequency in the East China Sea, and a frequency decrease and increase in intensity of Yangze River Basin landfalling typhoons.	Statistical downscaling of CGCM forced by CMIP3 SRES A1B scenario.	Orlowsky and Seneviratne (2012)
S.W. Pacific	Differences of 2000–2050 with 1970–2000. Negligible change in overall frequency. Significant (~15%) increase in number of Category 4–5 TCs.	Dynamical regional downscale of coupled AOGCM forced with IPCC IS92a increasing CO_2 scenario.	Leslie et al. (2007)
N. Atlantic	Linear trend in TC frequency 2001–2050: Ensemble-mean non-significant decrease in TC frequency (–5%). Ensemble range of –50% to +30%.	Statistical downscaling of CMIP3 models under A1B scenario.	Villarini et al. (2011)
N. Atlantic	TC frequency averaged 2016–2035 minus 1986–2005: Ensemble-mean non-significant increase for RCP2.6 (4%), non-significant decrease for RCP4.5 (–2%) and RCP8.5 (–1%). Ensemble range of –30% to 27% across all scenarios/models.	Statistical downscaling of CMIP5 RCP2.6, RCP4.5 and RCP8.5	Villarini and Vecchi (2012)
N. Atlantic	Power Dissipation Index averaged 2016–2035 minus 1986–2005: Ensemble mean significant increase for RCP2.6 (23%) and RCP8.5 (17%), non-significant increase for RCP4.5 (10%). Ensemble range of –43% to 78% across all scenarios/models.	Statistical downscaling of CMIP5 RCP2.6, RCP4.5 and RCP4.5	Villarini and Vecchi (2013)
N. Atlantic	Difference 2016–2035 minus 1986–2005 averages: Significant decrease (–20%) to overall TC and hurricane frequency. Significant increase (+45%) in number of Category 4–5 TCs. Significant increase in precipitation of hurricanes (11%) and tropical storms (18%).	Double dynamical refinement of CMIP5 RCP4.5 multi-model ensemble projections.	Knutson et al. (2013a)

Modes of climate variability that in the past have led to variations in the intensity, frequency and structure of tropical cyclones across the globe—such as the ENSO (e.g., Zhang and Delworth, 2006; Wang et al., 2007; Callaghan and Power, 2011; Chapter 14)—are *very likely* to continue influencing TC activity through the mid-21st century. Therefore, it is *very likely* that over the next few decades tropical cyclone frequency, intensity and spatial distribution globally, and in individual basins, will vary from year to year and decade to decade.

11.3.3 Near-term Projected Changes in the Ocean

11.3.3.1 Temperature

Globally averaged surface and near-surface ocean temperatures are projected by AOGCMs to warm over the early 21st century, in response to both present day atmospheric concentrations of GHGs ('committed warming'; e.g., Meehl et al., 2006) and projected future changes in RF (Figure 11.19). Globally averaged SST shows substantial year-to-year and decade-to-decade variability (e.g., Knutson et al., 2006; Meehl et al., 2011), whereas the variability of depth-averaged ocean temperatures is much less (e.g., Meehl et al., 2011; Palmer et al., 2011). The rate at which globally averaged surface and depth-averaged temperatures rise in response to a given scenario for RF shows a considerable spread between models (an example of response uncertainty; see Section 11.2), due to differences in climate sensitivity and ocean heat uptake (e.g., Gregory and Forster, 2008). In the CMIP5 models under all RCP forcing scenarios, globally averaged SSTs are projected to be warmer over the near term relative to 1986–2005 (Figure 11.20).

A key uncertainty in the future evolution of globally averaged oceanic temperature are possible future large volcanic eruptions, which could impact the radiative balance of the planet for 2 to 3 years after their eruption and act to reduce oceanic temperature for decades into the future (Delworth et al., 2005; Stenchikov et al., 2009; Gregory, 2010). An estimate using the GFDL-CM2.1 coupled AOGCM (Stenchikov et al., 2009) suggests that a single Tambora (1815)-like volcano could erase the projected global ocean depth-averaged temperature increase for many years to a decade. A Pinatubo (1991)-like volcano could erase the projected increase for 2 to 10 years. See Section 11.3.6 for further discussion.

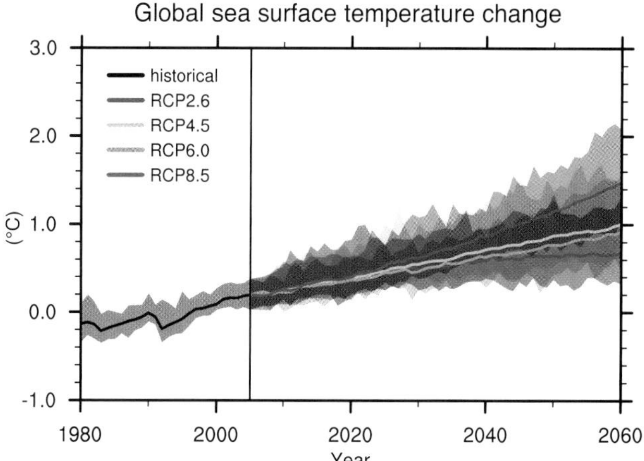

Figure 11.19 | Projected changes in annual averaged, globally averaged, surface ocean temperature based on 12 Atmosphere–Ocean General Circulation Models (AOGCMs) from the CMIP5 (Meehl et al., 2007b) multi-model ensemble, under 21st century scenarios RCP2.6, RCP4.5, RCP6.0 and RCP8.5. Shading indicates the 90% range of projected annual global mean surface temperature anomalies. Anomalies computed against the 1986–2005 average from the historical simulations of each model.

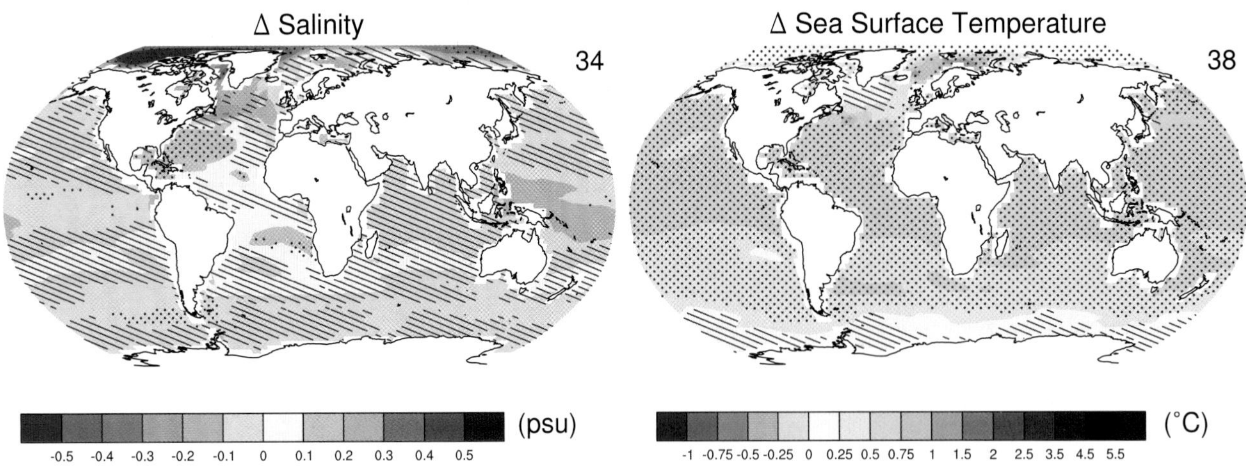

Figure 11.20 | CMIP5 multi-model ensemble mean of projected changes in sea surface temperature (right panel; °C) and sea surface salinity (left panel; practical salinity units) for 2016–2035 relative to 1986–2005 under RCP4.5. The number of CMIP5 models used is indicated in the upper right corner. Hatching and stippling as in Figure 11.10.

In the absence of multiple major volcanic eruptions (see Section 11.3.6.2), it is *very likely* that globally averaged surface and depth-averaged temperatures averaged 2016–2035 will be warmer than those averaged over 1986–2005.

There are regional variations in the projected amplitude of ocean temperature change (Figure 11.20) which are influenced by ocean circulation as well as surface heating (Timmermann et al., 2007; Vecchi and Soden, 2007; DiNezio et al., 2009; Yin et al., 2009; Xie et al., 2010; Yin et al., 2010), including changes in tropospheric aerosol concentrations (e.g., Booth et al., 2012; Villarini and Vecchi, 2012). Inter-decadal variability of upper ocean temperatures is larger in mid-latitudes, particularly in the NH, than in the tropics. A consequence of this contrast is that it will take longer in the mid-latitudes than in the tropics for the anthropogenic warming signal to emerge from the noise of internal variability (Wang et al., 2010).

Projected changes to thermal structure of the tropical Indo-Pacific are strongly dependent on the future behaviour of the Walker Circulation (Vecchi and Soden, 2007; DiNezio et al., 2009; Timmermann et al., 2010), in addition to changes in heat transport and changes in surface heat fluxes. It is *likely* that internal climate variability will be a dominant contributor to changes in the depth and tilt of the equatorial thermocline, and the strength of the east–west gradient of SST across the Pacific through the mid-21st century; thus it is *likely* there will be multi-year periods with increases or decreases in these measures.

11.3.3.2 Salinity

Changes in sea surface salinity are expected in response to changes in precipitation, evaporation and runoff (see Section 11.3.2.3), as well as ocean circulation; in general (but not in every region), salty regions are expected to become saltier and fresh regions fresher (e.g., Durack et al. 2012; Terray et al. 2012; Figure 11.20). As discussed in Chapter 10 (Section 10.4.2), observation-based and attribution studies have found some evidence of an emerging anthropogenic signal in salinity change (Section 10.4.2), in particular increases in surface salinity in the subtropical North Atlantic, and decreases in the west Pacific warm pool region (Stott et al., 2008; Cravatte et al., 2009; Durack and Wijffels, 2010; Durack et al., 2012; Pierce et al., 2012; Terray et al., 2012). Models generally predict increases in salinity in the tropical and (especially) subtropical Atlantic, and decreases in the western tropical Pacific over the next few decades (Figure 11.20) (Durack et al., 2012; Terray et al., 2012). These projected decreases in the Atlantic and in the western tropical Pacific are considered *likely*.

Projected near-term increases in freshwater flux into the Arctic Ocean produce a fresher surface layer and increased transport of fresh water into the North Atlantic (Holland et al., 2006; Holland et al., 2007; Vavrus et al., 2012). Such contributions to decreased density of the ocean surface layer in the North Atlantic could act to reduce deep ocean convection there and contribute to a near-term reduction of strength of Atlantic Meridional Ocean Circulation (AMOC). However, the strength of the AMOC can also be modulated by changes in temperature, such as those from changing RF (Delworth and Dixon, 2006).

11.3.3.3 Circulation

As discussed in previous assessment reports, the AMOC is generally projected to weaken over the next century in response to increase in atmospheric GHG. However, the rate and magnitude of weakening is very uncertain. Response uncertainty is a major contributor in the near term, but the influence of anthropogenic aerosols and natural RFs (solar, volcanic) cannot be neglected, and could be as important as the influence of GHGs (e.g., Delworth and Dixon, 2006; Stenchikov et al., 2009). For example, the rate of weakening of the AMOC in two models with different climate sensitivities is quite different, with the less sensitive model (CCSM4) showing less weakening and a more rapid recovery than the more sensitive model (Community Earth System Model 1/Community Atmosphere Model 5 (CESM1/CAM5; Meehl et al., 2013c). In addition, the natural variability of the AMOC on decadal time scales is poorly known and poorly understood, and could dominate any anthropogenic response in the near term (Drijfhout and Hazeleger, 2007). The AMOC is known to play an important role in the decadal variability of the North Atlantic Ocean, but climate models show large differences in their simulation of both the amplitude and spectrum of AMOC variability (e.g., Bryan et al., 2006; Msadek et al., 2010). In some AOGCMs changes in SH surface winds influence the evolution of the AMOC on time scales of many decades (Delworth and Zeng, 2008), so the delayed response to SH wind changes, driven by the historical reduction in stratospheric ozone along with its projected recovery, could be an additional confounding issue (Section 11.3.2.3). Overall, it is *likely* that there will be some decline in the AMOC by 2050, but decades during which the AMOC increases are also to be expected. There is *low confidence* in projections of when an anthropogenic influence on the AMOC might be detected (Baehr et al., 2008; Roberts and Palmer, 2012).

Projected changes to oceanic circulation in the Indo-Pacific are strongly dependent on future response of the Walker Circulation (Vecchi and Soden, 2007; DiNezio et al., 2009), the near-term projected weakening of which is smaller than the expected variability on time scales of decades to years (Section 11.3.2.4.3). Taking variability into account, there is *medium confidence* in a weakening of equatorial Pacific circulation, including equatorial upwelling and the shallow subtropical overturning in the Pacific, and the Indonesian Throughflow over the coming decades.

11.3.4 Near-term Projected Changes in the Cryosphere

This section assesses projected near-term changes of elements of the cryosphere. These consist of sea ice, snow cover and near-surface permafrost (frozen ground), changes to the Arctic Ocean and possible abrupt changes involving the cryosphere. Glaciers and ice sheets are addressed in Chapter 13. Here near-term changes in the geographical coverage of sea ice, snow cover and near-surface permafrost are assessed.

Trends due to changes in external forcing exist alongside considerable interannual and decadal variability. This complicates our ability to make specific, precise short-term projections, and delays the emergence of a forced signal above the noise.

11.3.4.1 Sea Ice

Though most of the CMIP5 models project a nearly ice-free Arctic (sea ice extent less than 1×10^6 km^2 for at least 5 consecutive years) at the end of summer by 2100 in the RCP8.5 scenario (see Section 12.4.6.1), some show large changes in the near term as well. Some previous models project an ice-free summer period in the Arctic Ocean by 2040 (Holland et al., 2006), and even as early as the late 2030s using a criterion of 80% sea ice area loss (e.g., Zhang, 2010). By scaling six CMIP3 models to recent observed September sea ice changes, a nearly ice-free Arctic in September is projected to occur by 2037, reaching the first quartile of the distribution for timing of September sea ice loss by 2028 (Wang and Overland, 2009). However, a number of models that have fairly thick Arctic sea ice produce a slower near-term decrease in sea ice extent compared to observations (Stroeve et al., 2007). Based on a linear extrapolation into the future of the recent sea ice volume trend from a hindcast simulation conducted with a regional model of the Arctic sea ice–ocean system (Maslowski et al., 2012) projected that it would take only until about 2016 to reach a nearly ice-free Arctic Ocean in summer. However, such an approach not only neglects the effect of year-to-year or longer-term variability (Overland and Wang, 2013) but also ignores the negative feedbacks that can occur when the sea ice cover becomes thin (Notz, 2009). Mahlstein and Knutti (2012) estimated the annual mean global surface warming threshold for nearly ice-free Arctic conditions in September to be ~2°C above the present derived from both CMIP3 models and observations.

An analysis of CMIP3 model simulations indicates that for near-term predictions the dominant factor for decreasing sea ice is increased ice melt, and reductions in ice growth play a secondary role (Holland et al., 2010). Arctic sea ice has larger volume loss when there is thicker ice initially across the CMIP3 models, with a projected accumulated mass loss of about 0.5 m by 2020, and roughly 1.0 m by 2050, with considerable model spread (Holland et al., 2010). The CMIP3 models tended to under-estimate the observed rapid decline of summer Arctic sea ice during the satellite era, but these recent trends are more accurately simulated in the CMIP5 models (see Section 12.4.6.1). For CMIP3 models, results indicate that the changes in Arctic sea ice mass budget over the 21st century are related to the late 20th century mean sea ice thickness distribution (Holland et al., 2010), average sea ice thickness (Bitz, 2008; Hodson et al., 2012), fraction of thin ice cover (Boe et al., 2009) and oceanic heat transport to the Arctic (Mahlstein et al., 2011). Acceleration of sea ice drift observed over the last three decades, underestimated in CMIP3 projections (Rampal et al., 2011), and the presence of fossil-fuel and biofuel soot in the Arctic environment (Jacobson, 2010), could also contribute to ice-free late summer conditions over the Arctic in the near term. Details on the transition to an ice-free summer over the Arctic are presented in Chapter 12 (Sections 12.4.6.1 and 12.5.5.7).

The discussion in Section 12.4.6.1 makes the case for assessing near-term projections of Arctic sea ice by weighting/recalibrating the models based on their present-day Arctic sea ice simulations, with a credible underlying physical basis in order to increase confidence in the results, and accounting for the potentially large imprint of natural variability on both observations and model simulations (see Section 9.8.3). A subselection of a set of CMIP5 models that fits those criteria, following the methodology proposed by Massonnet et al. (2012), is applied in Chapter 12 (Section 12.4.6.1) to the full set of models that provided the CMIP5 database with sea ice output. Among the five selected models, four project a nearly ice-free Arctic Ocean in September (sea ice extent less than 1×10^6 km^2 for at least 5 consecutive years) before 2050 for RCP8.5, the earliest and latest years of near disappearance of the sea ice pack being about 2040 and about 2060, respectively. The potential irreversibility of the Arctic sea ice loss and the possibility of an abrupt transition toward an ice-free Arctic Ocean are discussed in Section 12.5.5.7.

In light of all these results and others discussed in greater detail in Section 12.4.6.1, it is *very likely* that the Arctic sea ice cover will continue to shrink and thin all year round during the 21st century as the annual mean global surface temperature rises. It is also *likely* that the Arctic Ocean will become nearly ice-free in September before the middle of the century for high GHG emissions such as those corresponding to RCP8.5 (*medium confidence*).

In early 21st century simulations, Antarctic sea ice cover is projected to decrease in the CMIP5 models, though CMIP3 and CMIP5 models simulate recent decreases in Antarctic sea ice extent compared to slight increases in the observations (Section 12.4.6.1). However, there is the possibility that melting of the Antarctic ice sheet could be changing the vertical ocean temperature stratification around Antarctica and encourage sea ice growth (Bintanja et al., 2013). This and other evidence discussed in Section 12.4.6.1 leads to the assessment that there is *low confidence* in Antarctic sea ice model projections that show near-term decreases of sea ice cover because of the wide range of model responses and the inability of almost all of the models to reproduce the mean seasonal cycle, interannual variability and overall increase of the Antarctic sea ice areal coverage observed during the satellite era (see Section 9.4.3).

11.3.4.2 Snow Cover

Decreases of snow cover extent (SCE, defined over ice-free land areas) are strongly connected to a shortening of seasonal snow cover duration (Brown and Mote, 2009) and are related to both precipitation and temperature changes (see Section 12.4.6.2). This has implications for snow on sea ice where loss of sea ice area in autumn delays snowfall accumulation, with CMIP5 multi-model mean values of snow depth in April north of 70°N reduced from about 28 cm to roughly 18 cm for the 2031–2050 period compared to the 1981–2000 average (Hezel et al., 2012). The snow accumulation season by mid-century in one model is projected to begin later in autumn, with the melt season initiated earlier in the spring (Lawrence and Slater, 2010). As discussed in greater detail in Section 12.4.6.2, projected increases in snowfall across much of the northern high latitudes act to increase snow amounts, but warming reduces the fraction of precipitation that falls as snow. In addition, the reduction of Arctic sea ice also provides an increased moisture source for snowfall (Liu et al., 2012). Whether the average SCE decreases or increases by mid-century depends on the balance between these competing factors. The dividing line where models transition from simulating increasing or decreasing maximum snow water equivalent roughly coincides with the –20°C isotherm in the mid-20th century November to March mean surface air temperature (Raisanen, 2008). The projected change of SCE over some regions is inconsistent with that of extreme snowfall, a major contributor to SCE. For instance, SCE is projected to decrease over northern China by the mid-21st century (Shi et al., 2011), while the extreme snowfall events over the region are projected to increase (Sun et al., 2010).

Time series of projected changes in relative SCE (for NH ice-free land areas) are shown in Figure 12.32. Multi-model averages from the CMIP5 archive (Brutel-Vuilmet et al., 2013) show percentage decreases of NH SCE ± 1 standard deviation for the 2016–2035 time period for a March to April average using a 15% extent threshold for the four RCP scenarios as follows: RCP2.6: –5.2% ± 1.9% (21 models); RCP4.5: –5.3% ± 1.5% (24 models); RCP6.0: –4.5% ± 1.2% (16 models); RCP8.5: –6.0% ± 2.0% (24 models).

11.3.4.3 Near Surface Permafrost

Virtually all near-term projections indicate a substantial amount of near-surface permafrost degradation (typically taking place in the upper 2 to 3 m; see Callaghan et al. (2011) and see glossary for detailed definition), and thaw depth deepening over much of the permafrost area (Sushama et al., 2006; Lawrence et al., 2008; Guo and Wang, 2012). As discussed in more detail in Section 12.4.6.2, these projections have increased credibility compared to the previous generation of models assessed in the AR4 because current climate models represent permafrost more accurately (Alexeev et al., 2007; Nicolsky et al., 2007; Lawrence et al., 2008). The reduction in annual mean near-surface permafrost area for the 2016–2035 time period compared to the 1986–2005 reference period for the CMIP5 models (Slater and Lawrence, 2013) for the NH for the four RCP scenarios is 21% ± 5% (RCP2.6), 18% ± 6% (RCP4.5), 18% ± 3% (RCP6.0) and 20% ± 5% (RCP8.5).

11.3.5 Projections for Atmospheric Composition and Air Quality to 2100

The future evolution of atmospheric composition is determined by the chemical–physical processes in the atmosphere, forced primarily by anthropogenic and natural emissions and by interactions with the biosphere and ocean (Chapters 2, 6, 7, 8 and 12). Twenty-first century projections of the chemically reactive GHGs, including methane (CH_4), nitrous oxide (N_2O) and ozone (O_3), as well as aerosols, are assessed here (Section 11.3.5.1). Future air pollution, specifically ground-level O_3 and $PM_{2.5}$ (particulate matter with a diameter of less than 2.5 µm, a measure of aerosol concentration), is also assessed here (Section 11.3.5.2). The impact of changes in natural emissions and deposition through altered land use (Heald et al., 2008; Chen et al., 2009a; Cook et al., 2009; Wu et al., 2012) and production of food or biofuels (Chapter 6) on atmospheric composition and air quality are not assessed here. Projected CO_2 abundances are discussed in Chapters 6 and 12.

Projections for the 21st century are based predominantly on the CMIP5 models that included atmospheric chemistry and the related ACCMIP (Atmospheric Chemistry and Climate Model Intercomparison Project) models, driven by the RCP emission and climate scenarios. These and the earlier SRES scenarios include only direct anthropogenic emissions. Natural emissions may also change with biosphere feedbacks in response to climate or land use change (Chapters 6, 8). Emphasis is placed on evaluating the 21st-century RCP scenarios from emissions to abundance, summarized in tables in Annex II. For the well-mixed greenhouse gases (WMGHGs), the effective radiative forcing (ERF) in both RCP and SRES scenarios increases similarly before 2040 with little spread (±16% in ERF; see Tables AII.6.1 to AII.6.10), but by 2050 the RCP2.6 scenario diverges, falling well below the envelope containing both the SRES and other RCP scenarios.

National and regional regulations implemented on emissions contributing to ground-level ozone and $PM_{2.5}$ pollution influence global atmospheric chemistry and climate (NRC, 2009; HTAP, 2010a), as was recognized in the TAR (Jacob et al., 1993; Penner et al., 1993; Johnson et al., 1999; Prather et al., 2001). Ozone and aerosols are radiatively active species (Chapters 7 and 8) and many of their precursors serve as indirect GHGs (e.g., nitrogen oxides (NO_x), carbon monoxide (CO), Non Methane Volatile Organic Compounds (NMVOC)) by changing the atmospheric oxidative capacity, and thereby the lifetimes and abundances of CH_4, hydrofluorocarbons (HFCs) and tropospheric O_3 (Chapter 8). Consequently their evolution can influence near-term climate both regionally and globally (Section 11.3.6.1 and FAQ 8.2). The RCP and SRES scenarios differ greatly in terms of the short-lived air pollutants and aerosol climate forcing. The CMIP3 climate simulations driven by the SRES scenarios projected a wide range of future air pollutant trajectories, including unconstrained growth that resulted in very large tropospheric O_3 increases (Prather et al., 2003). Subsequently, the near-term projections of current legislation (CLE) and maximum feasible reductions (MFR) emissions illustrated the impacts of air pollution control strategies on air quality, global atmospheric chemistry and near-term climate (Dentener et al., 2005, 2006; Stevenson et al., 2006). The RCP scenarios applied in the CMIP5 climate models all assume a continuation of current trends in air pollution policies (van Vuuren et al., 2011) and thus do not cover the range of future pollutant emissions found in the literature, specifically those with higher pollutant emissions (Dentener et al., 2005; Kloster et al., 2008; Pozzer et al., 2012); see Chapter 8.

The new RCP emissions are compared to the older SRES and other published emission scenarios in Annex II (Tables AII.2.1 to AII.2.22) and Figures 8.2 and 8.SM.1. By 2030 the RCP aerosol and ozone precursor emissions are smaller than SRES by factors of 1.2 to 3. For these short-lived air pollutants, the spread across RCPs by 2030 is much smaller than the range between the CLE and MFR scenarios: ±12% vs. ±31% for nitrogen oxides; ±17% vs. ±60% for sulphate; ±5% vs. ±11% for carbon monoxide. BC aerosol emissions also vary little across the RCPs: ±4% range in 2030; ±15% in 2100. Most of this spread is due to uncertain projections for the rapidly industrializing nations. From 2000 to 2030, sulphur dioxide (SO_2) emissions decline in the RCPs by –15% to –8% per decade, within the range of the MFR and CLE scenarios (–23% to +2% per decade), but far below the SRES range (+4% to +21% per decade). Evaluation of recent trends in SO_2 emissions shows a trend similar to the near-term RCP projections (Smith et al., 2011; Klimont et al., 2013), but independent estimates for recent trends in other aerosol species are not available. The RCP trend in NO_x emissions (–5% to +2% per decade) is likewise within the CLE-MFR range, but far below the SRES trends (+10% to +30% per decade). For OC and BC emissions, the RCP trend lies between the SRES B1/A2 range. A simple sum of the main four aerosol emissions (N, S, OC, BC; Tables AII.2.18 to AII.2.22) in the SRES vs. RCP scenarios indicates that the CMIP3 simulations driven by the SRES scenarios have about 40% more aerosols in 2000 than the CMIP5 simulations driven by the RCP scenarios. On average, these aerosols increase by 9% per decade in the SRES scenarios but decrease by 5% per decade in the RCP scenarios over the near term. By 2030, the CMIP3 models thus include up to three times more anthropogenic aerosols under the SRES scenarios than the CMIP5 models driven by the RCP scenarios (*high confidence*).

11.3.5.1 Reactive Greenhouse Gases and Aerosols

The IPCC has assessed previous emission-based scenarios for future GHGs and aerosols in the SAR (IS92) and TAR/AR4 (SRES). The new RCP scenarios are different in that they embed a simple, parametric model of atmospheric chemistry and biogeochemistry that maps emissions onto atmospheric abundances (the 'concentration pathways') (Lamarque et al., 2011; Meinshausen et al., 2011a, 2011b; van Vuuren et al., 2011). As an integrated product, the RCP-prescribed emissions, abundances and RF used in the CMIP5 model ensembles do not reflect the current best understanding of natural and anthropogenic emissions, atmospheric chemistry and biogeochemistry and RF of climate (Chapters 2, 6 and 8) (see, e.g., Dlugokencky et al., 2011; Prather et al., 2012; Lamarque et al., 2013; Stevenson et al., 2013; Voulgarakis et al., 2013; Young et al., 2013). Rather, the best estimates of atmospheric abundances and associated RF include a more complete atmospheric chemistry description and a fuller set of uncertainties than considered in the RCPs provided to the CMIP5 models. While this widens the range of climate forcing for each individual scenario, this uncertainty generally remains smaller than the range across the four RCP scenarios.

11.3.5.1.1 Methane, nitrous oxide and the fluorinated gases

Kyoto GHG abundances projected to year 2100 are given in Annex II (Tables AII.4.1–AII.4.15) as both RCP published values (Meinshausen et al., 2011b) and derived from the RCP anthropogenic emissions pathways. The latter includes current best estimates of atmospheric chemistry and natural sources, with uncertainties (denoted RCP[&]). Emissions of CH_4 and N_2O, primarily from the agriculture, forestry and other land use sectors (AFOLU) are uncertain, typically by 25% or more (Prather et al., 2009; NRC, 2010). Following the method of Prather et al. (2012) a best estimate and uncertainty range for the year 2011 anthropogenic and natural emissions of CH_4 and N_2O are derived using updated AR5 values (see Chapters 2, 5 and 6). The re-scaled RCP[&] anthropogenic-only emissions of CH_4 and N_2O are given in Tables AII.2.2 and AII.2.3 and differ from the published RCPs by a single scale factor for each species. An uncertainty range for 2011 values (*likely*, ±1 standard deviation in %, based on Prather et al. 2012) is applied to all subsequent years. Abundances are then integrated using these rescaled RCP[&] anthropogenic emissions, the best estimate for natural emissions, and a model projecting changes in tropospheric OH (see Holmes et al., 2013; for details). Similar scaling to match current observational constraints (harmonization) was done for the SRES emissions (Prather et al., 2001) and the RCPs (Meinshausen et al., 2011b). However, these earlier harmonizations used older values for lifetimes and natural sources, and did not provide estimates of uncertainty.

Combining CH_4 observations, lifetime estimates for the present day, the ACCMIP studies, plus estimated limits on changing natural sources, gives a year 2011 total anthropogenic CH_4 emission of 354 ± 45 $Tg(CH_4)\ yr^{-1}$ (Montzka et al., 2011; Prather et al., 2012) (Chapters 2, 6 and 8). The RCP total emission lies within 10% of this value, and thus the scaling factor between the RCP[&] and RCP total emission, is small (Table AII.2.2). Projection of the tropospheric OH lifetime of CH_4 (AII.5.8) is based on the ACCMIP simulations of the RCPs for 2100 time slice simulations (Voulgarakis et al., 2013), other modelling studies (Stevenson et al., 2006; John et al., 2012) and multi-model sensitivity analyses of key factors (Holmes et al., 2013) that includes uncertainties in emissions from agricultural, forest and land use sources, in atmospheric lifetimes, and in chemical feedbacks and loss. Lifetimes, and thus future CH_4 abundances, decrease slowly under RCP2.6 and RCP4.5, remain almost constant under RCP6.0 and increase slowly under RCP8.5. Future changes in natural sources of CH_4 due to land use and climate change are included in a few CMIP5 models and may alter future CH_4 abundances (Chapter 6), but there is limited evidence, and thus these changes are not included in the RCP[&] projections.

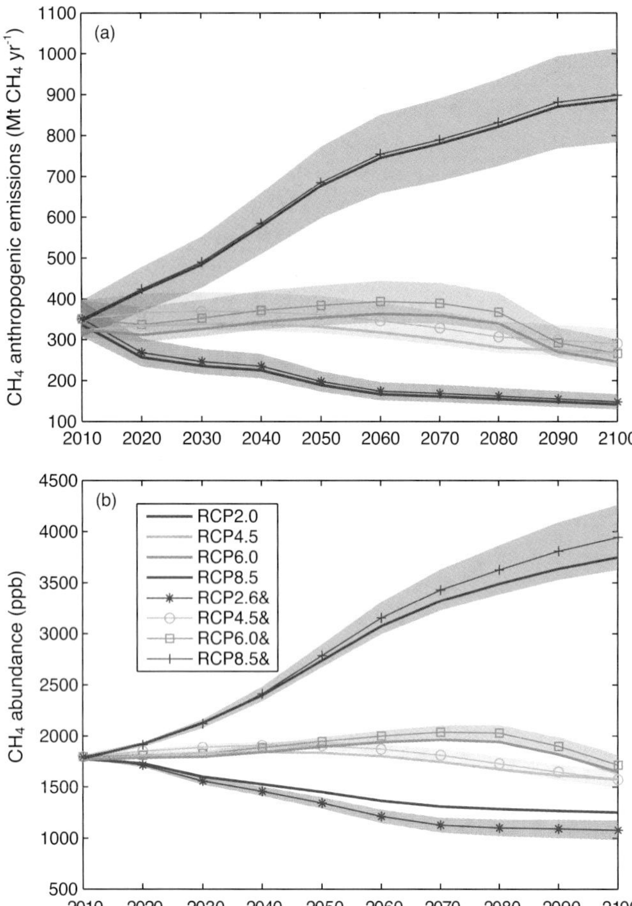

Figure 11.21 | Projections for CH_4 (a) anthropogenic emissions ($MtCH_4$ yr^{-1}) and (b) atmospheric abundances (ppb) for the four RCP scenarios (2010–2100). Natural emissions in 2010 are estimated to be 202 ± 35 $MtCH_4$ yr^{-1} (see Chapter 8). The thick solid lines show the published RCP2.6 (light blue), RCP4.5 (dark blue), RCP6.0 (orange) and RCP8.5 (red) values. Thin lines with markers show values from this assessment (denoted as RCPn.n&, following methods of Prather et al. (2012) and Holmes et al. (2013): red plus, RCP8.5; orange square, RCP6.0; light blue circle, RCP4.5; dark blue asterisk, RCP2.6. The shaded region shows the *likely* range from the Monte Carlo calculations that consider uncertainties, including in current anthropogenic emissions.

The resulting best estimates of total CH_4 anthropogenic emissions and abundances (RCP&) are compared with RCP values in Figure 11.21. For RCP2.6, the CH_4 abundance is projected to decline continuously over the century by about 30%, whereas in RCP 4.5 and 6.0 it peaks mid-century and then declines to below the year 2011 abundance by the end of the century. Throughout the century, the uncertainty in CH_4 abundance for an individual scenario is less than range from RCP2.6 to RCP8.5. For example, by year 2020 the spread in CH_4 abundance across the RCPs is already large, 1720 to 1920 ppb, with uncertainty in each scenario estimated at only ±20 ppb. The *likely* range for RCP& CH_4 is 30% wider than that in the RCP CH_4 abundances used to force the CMIP5 models (Figure 11.21): by year 2100 the *likely* range of RCP8.5& CH_4 abundance extends 520 ppb above the single-valued RCP8.5 CH_4 abundance, and RCP2.6& CH_4 extends 230 ppb below RCP2.6 CH_4.

Substantial effort has gone into identifying and quantifying individual sources of N_2O (see Chapter 6) but less into evaluating its lifetime and chemical feedbacks. Recent multi-model, chemistry–climate studies (CCMVal) project a more vigorous stratospheric overturning by 2100 that is expected to shorten the N_2O lifetime (Oman et al., 2010; Strahan et al., 2011), but no evaluation of the lifetime is reported. Here we combine observations of N_2O (pre-industrial, present, and present trends; Chapter 2), with two modern studies of the lifetime (Hsu and Prather, 2010; Fleming et al., 2011), and a Monte Carlo method (Prather et al., 2012) to estimate a year 2011 total anthropogenic emission of 6.7 ± 1.3 $TgN(N_2O)$ yr^{-1} (Table AII.2.3). All RCP N_2O (anthropogenic) emissions are reduced by 20% so that year 2011 values are consistent with an observationally constrained budget using a longer lifetime than adopted by the RCPs (Table AII.2.3). The N_2O lifetime (Table AII.5.9) is projected to decrease by 2 to 4% by year 2100, due to changing circulation and chemistry in the stratosphere (Fleming et al., 2011) and to the negative chemical feedback on its own lifetime (Prather and Hsu, 2010). In the near term, the spread in N_2O across RCP&s is small: 330 to 332 ± 4 ppb in year 2020; 346 to 365 ± 11 ppb in year 2050. By year 2100, the range of best-estimate N_2O concentrations across the RCP&s (354–425 ppb) is 20% smaller than that across the RCPs (344–435 ppb), but the *likely* range in RCP&s encompasses the RCP range.

Recent measurements show some discrepancies with bottom-up inventories of the industrially produced, synthetic fluorinated (F) gases (AII.2.4 to AII.2.15). European HFC-23 emissions are greatly under-reported (Keller et al., 2011) while HFC-125 and 152a are roughly consistent with emissions inventories (Brunner et al., 2012). Globally, HFC-365mfc and HFC-245fa emissions are overestimated (Vollmer et al., 2011) while SF_6 appears to be under-reported (Levin et al., 2010). For HFC-134a, combining current measurements and lifetimes (Table 2.1, Chapter 8; WMO, 2010; Prather et al., 2012) gives an estimate of 2010 emissions (~150 Gg yr^{-1}) that is consistent with the RCP range (139 to 153 Gg yr^{-1}). Without clear guidance on how to correct or place uncertainty on the RCP F-gas emissions, the RCP emissions are reported without uncertainty estimates in Annex II Tables AII.2.4 to AII.2.15. For the very long-lived SF_6 and perfluorocarbons (CF_4, C_2F_6, C_6F_{10}) uncertainty in lifetimes does not significantly affect the projected abundances over the 21st century (AII.4.4 to AII.4.7). Projected HFC abundances depend on the changes in tropospheric OH, which determines their atmospheric lifetime (Chapter 8). The relative change in hydroxyl radical (OH), as indicated by the projected OH lifetime of CH_4 (AII.5.8), is used to project HFCs including uncertainties (*likely* range) (AII.4.8 to AII.4.15) (Prather et al., 2012).

Scenarios for the ozone-depleting GHG under control of the Montreal Protocol (chlorofluorocarbons (CFCs), HCFCs, halons in AII.4.16) follow scenario A1 of the 2010 WMO Ozone Assessment (WMO, 2010; Table 5-A3). All CFC abundances decline throughout the century, but some HCFC abundances increase to 2030 before their phase-out and decline. The summed ERF of all these F-gases is approximately constant (0.35 to 0.39 W m^{-2}) up to year 2040 for all RCPs but declines thereafter. In RCP8.5, the drop in ERF from the Montreal Protocol gases is nearly made up by the growth in HFCs (Tables AII.6.4 to AII.6.6, Chapter 8).

11.3.5.1.2 Tropospheric and stratospheric O_3

Projected O_3 changes are broken into tropospheric and stratospheric columns (Dobson Unit (DU); see AII.5.1 and AII.5.2) because each has different driving factors and RF efficiencies (Chapter 8). Tropospheric

O_3 changes are driven by anthropogenic emissions of CH_4, NO_x, CO, NMVOC (AII.2.2.16 to AII.2.2.18). Small changes (<10%) are projected over the next few decades. By 2100 tropospheric O_3 decreases in RCP2.6, 4.5 and 6.0 but increases in RCP8.5 due to CH_4 increases. Higher tropospheric temperatures and humidity drive a decline in tropospheric O_3, but stratospheric O_3 recovery and increased stratosphere–troposphere exchange can counter that (Shindell et al., 2006; Zeng et al., 2008, 2010; Kawase et al., 2011; Lamarque et al., 2011). The latter effect is difficult to quantify but it is included in some of the ACCMIP and CMIP5 models used to project tropospheric O_3. Changes in natural emissions of NO_x, particularly soil and lightning NO_x, and biogenic NMVOC may also alter tropospheric O_3 abundances (Wild, 2007; Wu et al., 2007). However, global estimates of their change with climate (e.g., Kesik et al., 2006; Monson et al., 2007; Butterbach-Bahl et al., 2009; Price, 2013) remain highly uncertain.

Best estimates for projected tropospheric O_3 change following the RCP scenarios (Table AII.5.2) are based on ACCMIP time slice simulations for 2030 and 2100 with chemistry–climate models (Young et al., 2013) and the CMIP5 simulations (Eyring et al., 2013). There is *high confidence* in these results because similar estimates are obtained when projections are made using the response of tropospheric O_3 to key forcing factors that vary across scenarios (Prather et al., 2001; Stevenson et al., 2006; Oman et al., 2010; Wild et al., 2012). The ACCMIP models show a wide range in tropospheric O_3 burden changes from 2000 to 2100: –5 DU (–15%) in RCP2.6 to +5 DU in RCP8.5. The CMIP5 results are similar but not identical: –3 DU (–9%) to +10 DU (+30%). The 2030 and 2100 multi-model mean estimates are more robust for ACCMIP which includes 5 to 11 models (range depends on time slice and scenario) than for CMIP5 (4 models). Tropospheric O_3 changes in the near term (2030–2040) are small (±2 DU), except for RCP8.5 (>3 DU), which shows continued growth through to 2100 driven primarily by CH_4 increases. The ERF from tropospheric O_3 changes (AII.6.7b) parallels the O_3 burden change (Stevenson et al., 2013).

Stratospheric O_3 is being driven by declining chlorine levels, changing N_2O and CH_4, cooler temperatures from increased CO_2, and a more vigorous overturning circulation in the stratosphere driven by more wave propagation under climate change (Butchart et al., 2006; Eyring et al., 2010; Oman et al., 2010). Overall stratospheric O_3 is expected to increase in the coming decades, reversing the majority of the loss that occurred between 1980 and 2000. Best estimates for global mean stratospheric O_3 change under the RCP scenarios (Table AII.5.1) are taken from the CMIP5 results (Eyring et al., 2013). By 2100 stratospheric O_3 columns show a 5 to 7% increase above 2000 levels for all RCPs, recovering to within 1% of the pre-ozone hole 1980 levels by 2050, but with latitudinal differences.

11.3.5.1.3 Aerosols

Aerosol species can be emitted directly (mineral dust, sea salt, BC and some organic carbon (OC)) or indirectly through precursor gases (SO_2, ammonia, nitrogen oxides, hydrocarbons); see Chapter 7. CMIP5 models (Lamarque et al., 2011; Shindell et al., 2013) have projected changes in aerosol burden (Tg) and aerosol optical depth (AOD) to year 2100 using RCP emissions for anthropogenic source (Tables AII.5.3 to AII.5.8). Total AOD is dominated by dust and sea salt, but absorbing aerosol optical depth (AAOD) is primarily of anthropogenic origin (Chapter 7). Uniformly, anthropogenic aerosols decrease under RCPs as expected from the declining emissions (11.3.5, Figure 8.2, AII.2.17 to AII.2.22). From years 2010 to 2030 the aerosol burdens decrease across the RCPs but at varied rates: for sulphate from 6% (RCP8.5) to 23% (RCP2.6); for BC from 5% (RCP4.5) to 15% (RCP2.6), and for OC from 0% (RCP6.0) to 11% (RCP4.5). The summed aerosol loading of these three anthropogenic components drop from year 2010 to year 2030 by 5% to 12% (across RCPs), and by year 2100 this drop is 24% to 39% (Tables AII.5.5 to AII.5.7). These evolving aerosol loadings reduce the magnitude of the negative aerosol forcing (Chapter 8; Table AII.6.9) even in the near term (11.3.6.1).

11.3.5.2 Projections of Air Quality for the 21st Century

Future air quality depends on anthropogenic emissions (local, regional and global), natural biogenic emissions and the physical climate (e.g., Steiner et al., 2006, 2010; Meleux et al., 2007; Tao et al., 2007; Wu et al., 2008; Doherty et al., 2009; Carlton et al., 2010; Tai et al., 2010; Hoyle et al., 2011). This assessment focuses on O_3 and $PM_{2.5}$ in surface air, reflecting the preponderance of published literature and multi-model assessments for these air pollutants (e.g., HTAP, 2010a) plus the chemistry–climate CMIP5 and ACCMIP model simulations. Nitrogen and acid deposition is addressed in Chapter 6. Toxic atmospheric species such as mercury and persistent organic pollutants are outside this assessment (Jacob and Winner, 2009; NRC, 2009; HTAP, 2010b, 2010c).

The global and continental-scale surface O_3 and $PM_{2.5}$ changes assessed here include (1) the impact of climate change (Section 11.3.5.2.1), and (2) the impact of changing global and regional anthropogenic emissions (Section 11.3.5.2.2). Changes in local emissions within a metropolitan region or surrounding air basin on local air quality projections are not assessed here. Anthropogenic emissions of O_3 precursors include NO_x, CH_4, CO, and NMVOC; $PM_{2.5}$ is both directly emitted (OC, BC) and produced photochemically from precursor emissions (NO_x, NH_3, SO_2, NMVOC) (see Tables AII.2.2,16-22). Recent reviews describe the impact of temperature-driven processes on O_3 and $PM_{2.5}$ air quality from observational and modelling evidence (Isaksen et al., 2009; Jacob and Winner, 2009; Fiore et al., 2012). Projecting future air quality empirically from a mean surface warming using the observed correlation with temperature is problematic, as there is little evidence that future pollution episodes can be simply modelled as all else being equal except for a uniform temperature shift. Air quality relationships with synoptic conditions may be more robust (e.g., Dharshana et al., 2010; Appelhans et al., 2012; Tai et al., 2012a, 2012b), but require the ability to project changes in key conditions such as blocking and stagnation episodes. The response of blocking frequency to global warming is complex, with summertime increases possible over some regions, but models are generally biased compared to observed blocking statistics, and indicate even larger uncertainty in projecting changes in blocking intensity and persistence (Box 14.2).

11.3.5.2.1 Climate-driven changes

Projecting regional air quality faces the challenge of simulating first the changes in regional climate and then the feedbacks from atmospheric chemistry and the biosphere. The air pollution response to climate-driven changes in the biosphere is uncertain as to sign because of competing effects: for example, plants currently emit more NMVOC with warmer temperatures; with higher CO_2 and water stress plants may emit less; with a warmer climate the vegetation types may shift to emit either more or less NMVOC; shifting vegetation types may also alter surface uptake of ozone and aerosols; and our understanding of chemical oxidation pathways for biogenic emissions is incomplete (e.g., Monson et al., 2007; Carlton et al., 2009; Hallquist et al., 2009; Ito et al., 2009; Pacifico et al., 2009, 2012; Paulot et al., 2009). Although studies have split the cause of air quality changes into climate versus emissions, these attributions are difficult to assess for several reasons: the global-to-regional down-scaling of meteorology that is model dependent (see Chapters 9 and 14; also Manders et al., 2012), the brief simulations that preclude clear separation of climate change from climate variability (Nolte et al., 2008; Fiore et al., 2012; Langner et al., 2012a), and the lack of systematically explored standard scenarios for local anthropogenic emissions, land use change and biogenic emissions.

Ozone

Globally, a warming climate decreases baseline surface O_3 almost everywhere but increases O_3 levels in some polluted regions and seasons. The surface ozone response to climate change alone between 2000 and 2030 is shown in Figure 11.22 (CLIMATE), where the ranges reflect multi-model differences in spatial averages (solid green lines) and spatial variability within a single model (dashed green lines). There is *high confidence* that in unpolluted regions, higher water vapour abundances and temperatures enhance O_3 destruction, leading to lower baseline O_3 levels in a warmer climate (e.g., global average in Figure 11.22). Higher CH_4 levels such as in RCP8.5 can offset this climate-driven decrease in baseline O_3. Other large-scale factors that could increase baseline O_3 in a warming climate include increased lightning NO_x and stratospheric influx of O_3 (see Section 11.3.5.1). Evidence and agreement are limited regarding the impact of climate change on long-range transport of pollutants (Wu et al., 2008; HTAP, 2010a; Doherty et al., 2013). The global chemistry-climate models assessed here (Figures 11.22, 11.23ab) include most of these feedback processes, but a systematic evaluation of their relative impacts is lacking.

In polluted regions, observations show that high-O_3 episodes correlate with high temperatures (e.g., Lin et al., 2001; Bloomer et al., 2009; Rasmussen et al., 2012), but these episodes also coincide with cloud-free enhanced photochemistry and with air stagnation that concentrates pollution near the surface (e.g., AR4 Box 7.4). Other temperature-related factors, such as biogenic emissions from vegetation and soils, volatilization of NMVOC, thermal decomposition of organic nitrates to NO_x and wildfire frequency may increase with a warming climate and are expected to increase surface O_3 (e.g., Doherty et al., 2013; Skjøth and Geels, 2013; and as reviewed by Isaksen et al. (2009), Jacob and Winner (2009) and Fiore et al. (2012)), although some of these processes are known to have optimal temperature ranges (e.g., Sillman and Samson, 1995; Guenther et al., 2006; Steiner et al., 2010). Overall, the integrated effect of these processes on O_3 remains poorly understood, and they have been implemented with varying levels of complexity in the models assessed here.

Models show that a warmer atmosphere can lead to local O_3 increases during the peak pollution season (e.g., by 2 to 6 ppb within Central Europe by 2030; green dashed line for Europe in Figure 11.22). Regional models projecting summer daytime statistics tend to simulate a wider range of climate-driven changes (e.g., Zhang et al., 2008; Avise et al., 2012; Kelly et al., 2012), with most studies focusing on 2050 (Fiore et al., 2012) or beyond. For example, summer temperature extremes over parts of Europe are projected to warm more than the corresponding mean local temperatures due to enhanced variability at interannual to intraseasonal time scales (see Section 12.4.3.3). Several modelling studies note a longer season for O_3 pollution in a warmer world (Nolte et al., 2008; Racherla and Adams, 2008). For some regions, models agree on the sign of the O_3 response to a warming climate (e.g., increases in northeastern USA and southern Europe; decreases in northern Europe), but they often disagree (e.g., the midwest, southeast, and western USA (Jacob and Winner, 2009; Weaver et al., 2009; Langner et al., 2012a; Langner et al., 2012b; Manders et al., 2012)). Several studies have suggested a role for changing synoptic meteorology on future air pollution levels (Leibensperger et al., 2008; Jacob and Winner, 2009; Weaver et al., 2009; Lang and Waugh, 2011; Tai et al., 2012a, 2012b; Turner et al., 2013), but projected regional changes in synoptic conditions are uncertain (see Sections 11.3.2.4, 12.4.3.3 and Box 14.2). Observational and modelling evidence together indicate that, all else being equal, a warming climate is expected to increase surface O_3 in polluted regions (*medium confidence*), although a systematic evaluation of all the factors driving extreme pollution episodes is lacking.

Aerosols

Evaluations as to whether climate change will worsen or improve aerosol pollution are model-dependent. Assessments are confounded by opposing influences on the individual species contributing to total $PM_{2.5}$ and large interannual variability caused by the small-scale meteorology (e.g., convection and precipitation) that controls aerosol concentrations (Mahmud et al., 2010). For a full discussion, see Chapter 7. Higher temperatures generally decrease nitrate aerosol through enhanced volatility but increase sulphate aerosol through faster production, although observed $PM_{2.5}$–temperature correlations also reflect humidity and synoptic meteorology (e.g., Aw and Kleeman, 2003; Liao et al., 2006; Racherla and Adams, 2006; Unger et al., 2006a; Hedegaard et al., 2008; Jacobson, 2008; Kleeman, 2008; Pye et al., 2009; Tai et al., 2012b). Natural aerosols may increase with temperature, particularly carbonaceous aerosol from wildfires, mineral dust, and biogenic secondary organic aerosol (SOA; Section 7.3.5; Mahowald and Luo, 2003; Tegen et al., 2004; Jickells et al., 2005; Woodward et al., 2005; Mahowald et al., 2006; Liao et al., 2007; Mahowald, 2007; Tagaris et al., 2007; Heald et al., 2008; Spracklen et al., 2009; Jiang et al., 2010; Yue et al., 2010; Carvalho et al., 2011; Fiore et al., 2012). SOA formation also depends on anthropogenic emissions and atmospheric oxidizing capacity (Carlton et al., 2010; Jiang et al., 2010).

Aerosols are scavenged from the atmosphere by precipitation and direct deposition (see Chapter 7). Hence most components of $PM_{2.5}$ are anti-correlated with precipitation (Tai et al., 2010), and aerosol burdens are expected to decrease on average where precipitation increases (Racherla and Adams, 2006; Liao et al., 2007; Tagaris et al., 2007; Zhang et al., 2008; Avise et al., 2009; Pye et al., 2009). However, a shift in the frequency and type of precipitation may be as important as the change in mean precipitation (see Chapter 7). Seasonal and regional

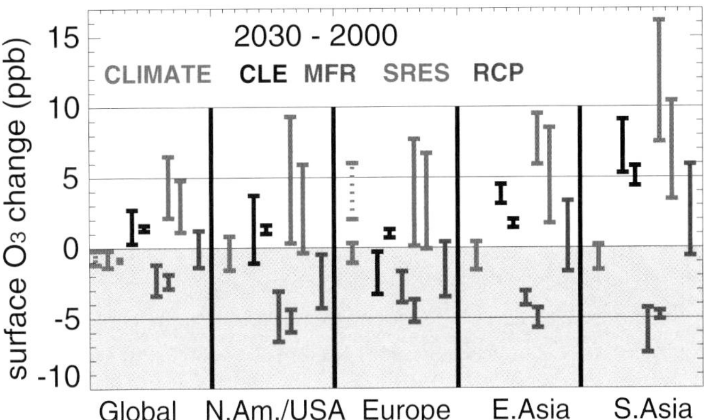

Figure 11.22 | Changes in surface O_3 (ppb) between year 2000 and 2030 driven by climate alone (CLIMATE, green) or driven by emissions alone, following current legislation (CLE, black), maximum feasible reductions (MFR, grey), SRES (blue) and RCP (red) emission scenarios. Results are reported globally and for the four northern mid-latitude source regions used by the Task Force on Hemispheric Transport of Air Pollution (HTAP, 2010a). Where two vertical bars are shown (CLE, MFR, SRES), they represent the multi-model standard deviation of the annual mean based on (left bar; SRES includes A2 only) the Atmospheric Composition Change: a European Network (ACCENT)/Photocomp study (Dentener et al., 2006) and (right bar) the parametric HTAP ensemble (Wild et al., 2012; four SRES and RCP scenarios included). Under Global, the leftmost (dashed green) vertical bar denotes the spatial range in climate-only changes from one model (Stevenson et al., 2005) while the green square shows global annual mean climate-only changes in another model (Unger et al., 2006b). Under Europe, the dashed green bar denotes the range of climate-only changes in summer daily maximum O_3 in one model (Forkel and Knoche 2006). (Adapted from Figure 3 of Fiore et al., 2012.)

differences in aerosol burdens versus precipitation further preclude a simple scaling of aerosol response to precipitation changes (Kloster et al., 2010; Fang et al., 2011). Climate-driven changes in the frequency of drizzle and the mixing depths or ventilation of the surface layer also influence projected changes in $PM_{2.5}$ (e.g., Kleeman, 2008; Dawson et al., 2009; Jacob and Winner, 2009; Mahmud et al., 2010), and aerosols in turn can influence locally clouds, precipitation and scavenging (e.g., Zhang et al., 2010b; see Section 7.6).

While $PM_{2.5}$ is expected to decrease in regions where precipitation increases, the climate variability at these scales results in only *low confidence* for projections at best. Further, consensus is lacking on the other factors including climate-driven changes in biogenic and mineral dust aerosols, leading to *no confidence level* being attached to the overall impact of climate change on $PM_{2.5}$ distributions.

11.3.5.2.2 Changes driven by regional and global anthropogenic pollutant emissions

Projections for annual-mean surface O_3 and $PM_{2.5}$ for 2000 through 2100 are shown in Figures 11.23a and 11.25b, respectively. Changes are spatially averaged over selected world (land-only) regions and include the combined effects of emission and climate changes under the RCPs. Results are taken from the ACCMIP models and a subset of the CMIP5 models that included atmospheric chemistry. Large interannual variations are evident in the CMIP5 transient simulations, and large regional variations occur in both the CMIP5 and the ACCMIP decadal time slice simulations (see Lamarque et al., (2013) for ACCMIP overview).

The largest surface O_3 changes under the RCP scenarios are much smaller than those projected under the older SRES scenarios (Figures 11.22 and 11.23a; Table AII.7; Lamarque et al., 2011; Wild et al., 2012). By 2100, global annual multi-model mean surface O_3 rises by 12 ppb in SRES A2, but by only 3 ppb in RCP8.5. Much larger O_3 decreases are projected to occur by 2030 under the MFR scenario (Figure 11.22), which assumes that existing control technologies are applied uniformly across the globe (Dentener et al., 2006).

For RCP2.6, RCP4.5 and RCP6.0, the CMIP5/ACCMIP models project that continental-scale spatially averaged near-term surface O_3 decreases or changes little (–4 to +1 ppb) from 2000 to 2030 for all regions except South Asia, whereas the long-term change to 2100 is a consistent decrease (–14 to –3 ppb) for all regions (Figure 11.23a; and Table AII.7.3). For RCP8.5, the CMIP5/ACCMIP models project continental-scale spatial average surface O_3 increases of up to +5 ppb for both 2030 and 2100 (Figure 11.23a; Table AII.7.3). The increases under RCP8.5 reflect the prominent rise in methane abundances (Kawase et al., 2011; Lamarque et al., 2011; Wild et al., 2012), which by 2100 raise background O_3 levels by 5 to 14 ppb over continental-scale regions, and on average by about 8 ppb (25% above current levels) above RCP4.5 and RCP6.0 which include more stable methane pathways over the 21st century (*high confidence*). Earlier studies have shown that rising CH_4 abundances (and global NO_x emissions) increase baseline O_3, and can offset aggressive local emission reductions and lengthen the O_3 pollution season (Jacob et al., 1999; Prather et al., 2001, 2003; Fiore et al., 2002, 2009; Hogrefe et al., 2004; Granier et al., 2006; Szopa et al., 2006; Tao et al., 2007; Huang et al., 2008; Lin et al., 2008; Wu et al., 2008; Avise et al., 2009; Chen et al., 2009b; HTAP, 2010a; Wild et al., 2012; Lei et al., 2013).

The O_3 changes driven by the RCP emissions scenarios with fixed, present-day climate (Figure 11.22; Wild et al., 2012) are similar to the changes estimated with the full chemistry–climate models (Figure 11.23a). Although the regions considered are not identical, the evidence supports a major role for global emissions in determining near-term O_3 concentrations. Overall, the multi-model ranges associated with the influence of near-term climate change on global and regional O_3 air quality are smaller than those across emission scenarios (Figure 11.22; HTAP, 2010a; Wild et al., 2012).

Aerosol changes driven by anthropogenic emissions depend somewhat on oxidant levels (e.g., Unger et al., 2006a; Kleeman, 2008; Leibensperger et al., 2011a), but generally sulphate follows SO_2 emissions and carbonaceous aerosols follow the primary elemental and OC emissions. Competition between sulphate and nitrate for ammonium (see Chapter 7) means that reducing SO_2 emissions while increasing NH_3 emissions as in the RCPs (Tables AII.2.19 and AII.2.20) would lead to near-term nitrate aerosol levels equal to or higher than those of sulphate in some regions; see Section 7.3.5.2 (Bauer et al., 2007; Pye et al., 2009; Bellouin et al., 2011; Henze et al., 2012).

Regional $PM_{2.5}$ in the CMIP5 and ACCMIP chemistry–climate models following the RCP scenarios generally declines over the 21st century, with little difference across the individual scenarios except for the South and East Asia regions (Figure 11.23b). The noisy projections over Africa, the Middle East and to some extent Australia, reflect dust sources and their strong dependence on interannual meteorological variability. Over the two Asian regions, different $PM_{2.5}$ levels between the RCPs are due to (1) OC emission trajectories over South Asia and (2) combined changes in carbonaceous aerosol and SO_2 over East Asia (Fiore et al., 2012) (Figure 8.SM.1).

Global emissions of aerosols and precursors can contribute to high-PM events. For example, dust trans-oceanic transport events are observed to increase aerosols in downwind regions (Prospero, 1999; Grousset et al., 2003; Chin et al., 2007; Fairlie et al., 2007; Huang et al., 2008; Liu et al., 2009; Ramanathan and Feng, 2009; HTAP, 2010a). The balance between regional and global anthropogenic emissions versus climate-driven changes for $PM_{2.5}$ will vary regionally with future changes in precipitation, wildfires, dust and biogenic emissions.

In summary, lower air pollution levels are projected following the RCP emissions as compared to the SRES emissions in the TAR and AR4, reflecting implementation of air pollution control measures (*high confidence*). The range in projections of air quality is driven primarily by emissions (including CH_4) rather than by physical climate change (*medium confidence*). The total emission-driven range in air quality—including the CLE and MFR scenarios—is larger than that spanned by the RCPs (see Section 11.3.5.1 for comparison of RCPs and SRES).

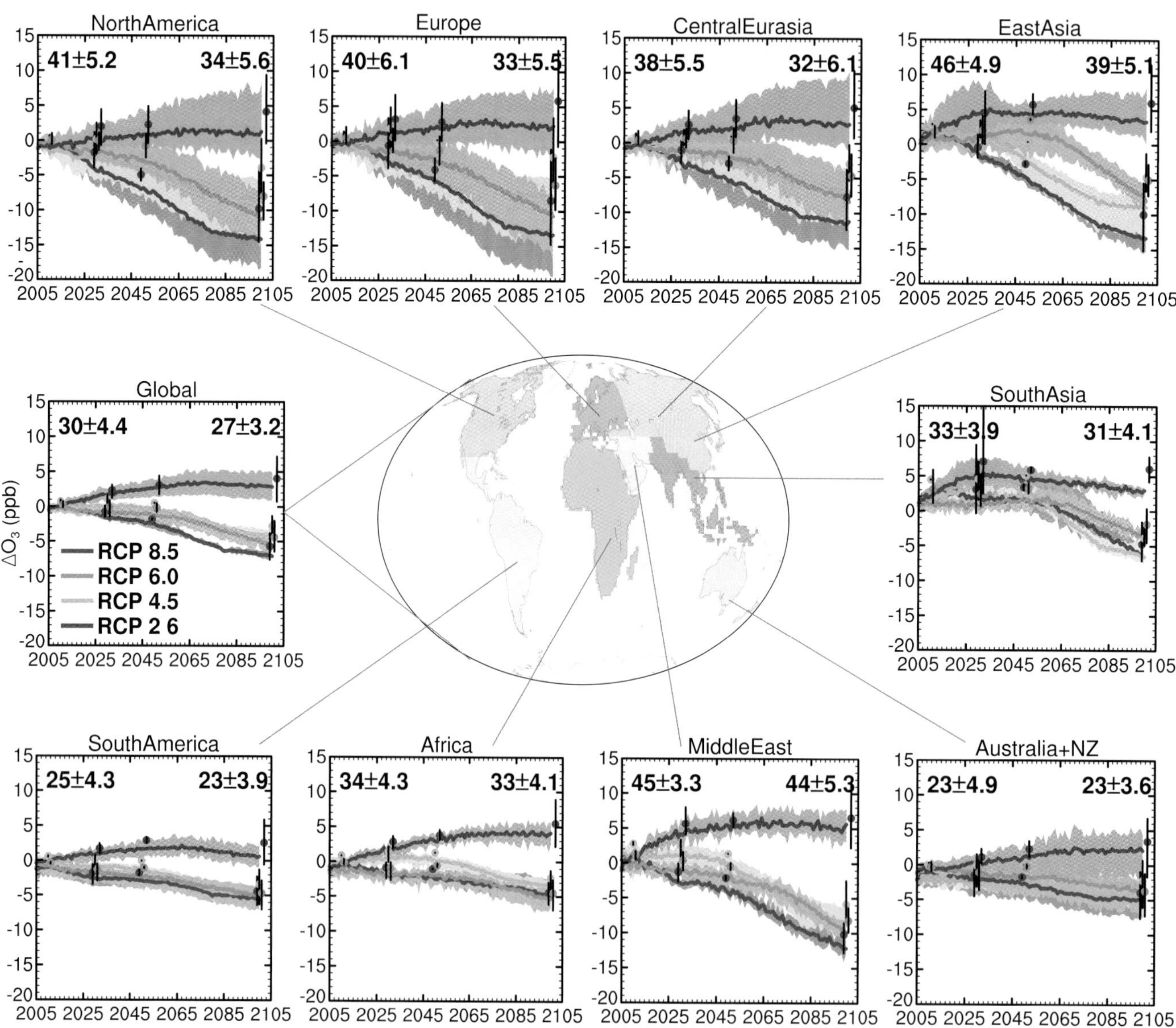

Figure 11.23a | Projected changes in annual mean surface O_3 (ppb mole fraction) from 2000 to 2100 following the RCP scenarios (8.5, red; 6.0, orange; 4.5, light blue; 2.6, dark blue). Results in each box are averaged over the designated coloured land regions. Continuous coloured lines and shading denote the average and full range of four chemistry–climate models (GFDL-CM3, GISS-E2-R, and NCAR-CAM3.5 from CMIP5 plus LMDz-ORINCA). Coloured dots and vertical black bars denote the average and full range of the ACCMIP models (CESM-CAM-superfast, CICERO-OsloCTM2, CMAM, EMAC-DLR, GEOSCCM, GFDL-AM3, HadGEM2, MIROC-CHEM, MOCAGE, NCAR-CAM3.5, STOC-HadAM3, UM-CAM) for decadal time slices centred on 2010, 2030, 2050 and 2100. Participation in the decadal slices ranges from 2 to 12 models (see (Lamarque et al., 2013)). Changes are relative to the 1986–2005 reference period for the CMIP5 transient simulations, and relative to the average of the 1980 and 2000 decadal time slices for the ACCMIP ensemble. The average value and model standard deviation for the reference period is shown in the top of each panel for CMIP5 models (left) and ACCMIP models (right). In cases where multiple ensemble members are available from a single model, they are averaged prior to inclusion in the multi-model mean. (Adapted from Fiore et al., 2012.)

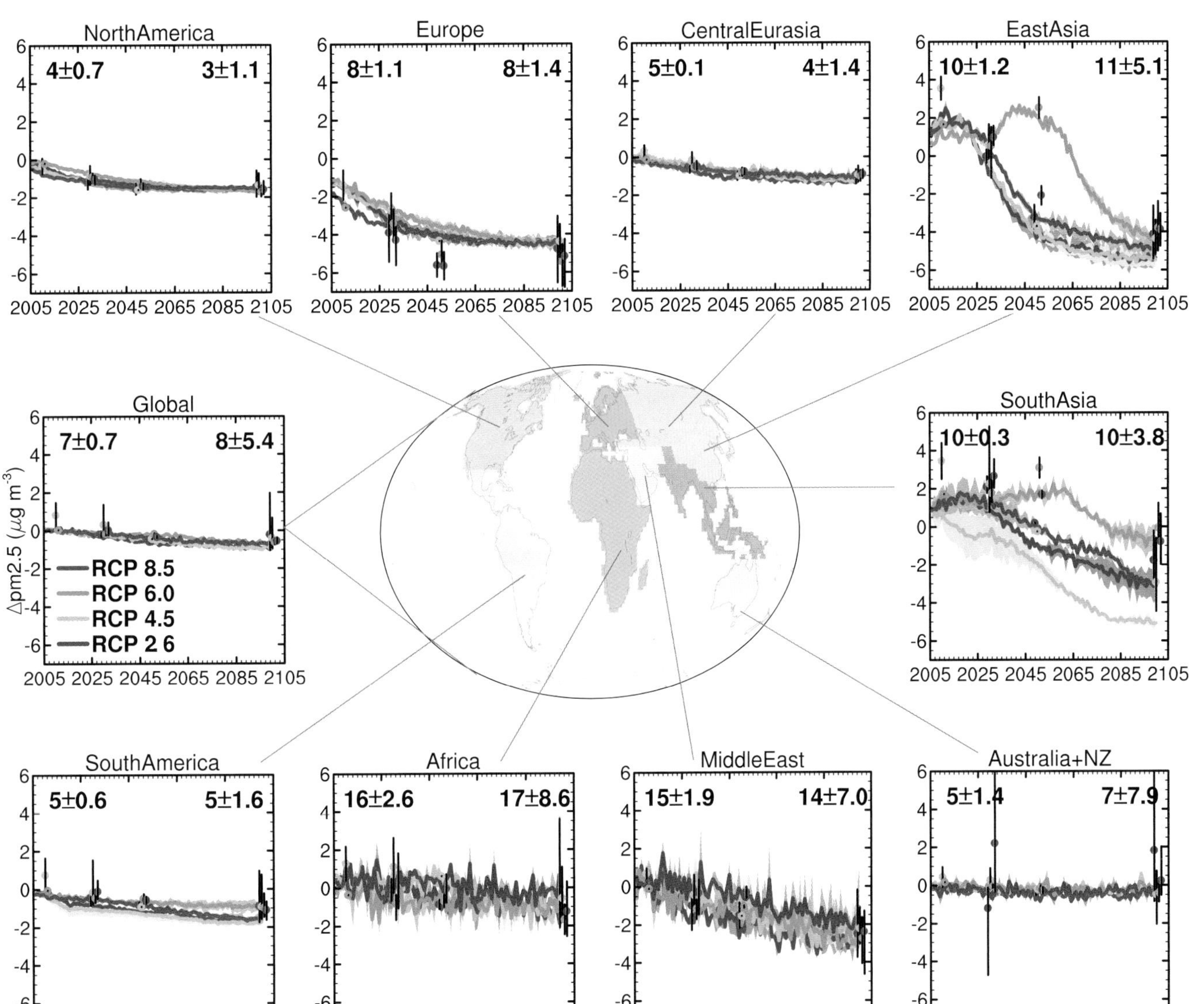

Figure 11.23b | Projected changes in annual mean surface $PM_{2.5}$ (micrograms per cubic metre of aerosols with diameter less than 2.5 µm) from 2000 to 2100 following the RCP scenarios (8.5 red, 6.0 orange, 4.5 light blue, 2.6 dark blue). $PM_{2.5}$ values are calculated as the sum of individual aerosol components (black carbon + organic carbon + sulphate + secondary organic aerosol + 0.1*dust + 0.25*sea salt). Nitrate was not reported for most models and is not included here. See Figure 11.23a for details, but note that fewer models contribute: GISS-E2-R and GFDL-CM3 from CMIP5; CICERO-OsloCTM2, GEOSCCM, GFDL-AM3, HadGEM2, MIROC-CHEM, and NCAR-CAM3.5 from ACCMIP. (Adapted from Fiore et al., 2012.)

11.3.5.2.3 Extreme weather and air pollution

Extreme air quality episodes are associated with changing weather patterns, such as heat waves and stagnation episodes (Logan, 1989; Vukovich, 1995; Cox and Chu, 1996; Mickley et al., 2004; Stott et al., 2004). Heat waves are generally associated with poor air quality (Ordóñez et al., 2005; Vautard et al., 2005; Lee et al., 2006b; Struzewska and Kaminski, 2008; Tressol et al., 2008; Vieno et al., 2010; Hodnebrog et al., 2012). Although anthropogenic climate change has increased the near-term risk of such heat waves (Stott et al., 2004; Clark et al., 2010; Diffenbaugh and Ashfaq, 2010; Chapter 10; Section 11.3.2.5.1), projected changes in the frequency of regional air stagnation events, which are largely driven by blocking events, remain difficult to assess: the frequency of blocking events with persistent high pressure is projected to decrease in a warming climate but increases may occur in some regions, and projected changes in their intensity and duration remain uncertain (Chapters 9 and 14; Box 14.2). Projections in regional air pollution extremes are necessarily conditioned on projected changes in these weather patterns. The severity of extreme pollution events also depends on local emissions (see references in Fiore et al., 2012). Feedbacks from vegetation (higher biogenic NMVOC emissions, lower stomatal uptake of O_3 with higher temperatures) can combine with similar positive feedbacks via dust and wildfires to worsen air pollution and its impacts during heat waves (Lee et al., 2006a; Jiang et al., 2008; Royal Society, 2008; Flannigan et al., 2009; Andersson and Engardt, 2010; Vieno et al., 2010; Hodnebrog et al., 2012; Jaffe and Wigder, 2012; Mues et al., 2012).

There is high agreement across numerous modelling studies projecting increases in extreme O_3 pollution events over the USA and Europe, but the projections do not consistently agree at the regional level (Kleeman, 2008; Jacob and Winner, 2009; Jacobson and Streets, 2009; Weaver et al., 2009; Huszar et al., 2011; Katragkou et al., 2011; Langner et al., 2012b) because they depend on accurate projections of local emissions, regional climate and poorly understood biospheric feedbacks. Although observational evidence clearly demonstrates a strong statistical correlation between extreme temperatures (heat waves) and pollution events, this temperature correlation reflects in part the coincident occurrence of stagnation events and clear skies that also drive extreme pollution. Mechanistic understanding of biogenic emissions, deposition and atmospheric chemistry is consistent with a temperature-driven increase in pollution extremes in already polluted regions, although these processes may not scale simply with mean temperature under a changing climate (see Section 11.3.5.2.1), and better projections of the changing meteorology at regional scales are needed. Assuming all else is equal (e.g., local anthropogenic emissions) this collective evidence indicates that uniformly higher temperatures in polluted environments will trigger regional feedbacks during air stagnation episodes that will increase peak pollution (*medium confidence*).

11.3.6 Additional Uncertainties in Projections of Near-term Climate

11

As discussed in Section 11.3.1, most of the projections presented in Sections 11.3.2 to 11.3.4 are based on the RCP4.5 scenario and rely on the spread among the CMIP5 ensemble of opportunity as an *ad hoc* measure of uncertainty. It is possible that the real world might follow a path outside (above or below) the range projected by the CMIP5 models. Such an eventuality could arise if there are processes operating in the real world that are missing from, or inadequately represented in, the models. Two main possibilities must be considered: (1) Future radiative and other forcings may diverge from the RCP4.5 scenario and, more generally, could fall outside the range of *all* the RCP scenarios; (2) The response of the real climate system to radiative and other forcing may differ from that projected by the CMIP5 models. A third possibility is that internal fluctuations in the real climate system are inadequately simulated in the models. The fidelity of the CMIP5 models in simulating internal climate variability is discussed in Chapter 9.

Future changes in RF will be caused by anthropogenic and natural processes. The consequences for near-term climate of uncertainties in anthropogenic emissions and land use are discussed in Section 11.3.6.1. The uncertainties in natural RF that are most important for near-term climate are those associated with future volcanic eruptions and variations in the radiation received from the Sun (solar output), and are discussed in Section 11.3.6.2. In addition, carbon cycle and other biogeochemical feedbacks in a warming climate could potentially lead to abundances of CO_2 and CH_4 (and hence RF) outside the range of the RCP scenarios, but these feedbacks are not expected to play a major role in near term climate—see Chapters 6 and 12 for further discussion.

The response of the climate system to radiative and other forcing is influenced by a very wide range of processes, not all of which are adequately simulated in the CMIP5 models (Chapter 9). Of particular concern for projections are mechanisms that could lead to major 'surprises' such as an abrupt or rapid change that affects global-to-continental scale climate. Several such mechanisms are discussed in this assessment report; these include: rapid changes in the Arctic (Section 11.3.4 and Chapter 12), rapid changes in the ocean's overturning circulation (Chapter 12), rapid change of ice sheets (Chapter 13) and rapid changes in regional monsoon systems and hydrological climate (Chapter 14). Additional mechanisms may also exist as synthesized in Chapter 12. These mechanisms have the potential to influence climate in the near term as well as in the long term, albeit the likelihood of substantial impacts increases with global warming and is generally lower for the near term. Section 11.3.6.3 provides an overall assessment of projections for global mean surface air temperature, taking into account all known quantifiable uncertainties.

11.3.6.1 Uncertainties in Future Anthropogenic Forcing and the Consequences for Near-term Climate

Climate projections for periods prior to year 2050 are not very sensitive to available alternative scenarios for anthropogenic CO_2 emissions (see Section 11.3.2.1.1; Stott and Kettleborough, 2002; Meehl et al., 2007b). Near-term projections, however, may be sensitive to changes in emissions of climate forcing agents with lifetimes shorter than CO_2, particularly the GHGs CH_4 (lifetime of a decade), tropospheric O_3 (lifetime of weeks), and tropospheric aerosols (lifetime of days). Although the RCPs and SRES scenarios span a similar range of total effective radiative forcing (ERF, see Section 7.5, Figure 7.3, Chapter 8), they include different ranges of ERF from aerosol, CH_4, and tropospheric O_3 (see Section 11.3.5.1, Tables AII.6.2 and AII.6.7 to AII.6.10). From years 2000 to 2030 the change in ERF across the RCPs ranges from –0.05 to +0.14 W m^{-2} for CH_4 and from –0.04 to +0.08 W m^{-2} for tropospheric O_3 (Tables AII.6.2 and AII.6.7; Stevenson et al., 2013). From years 2000 to 2030 the total aerosol ERF becomes less negative, increasing by +0.26 W m^{-2} for RCP8.5 (only RCP evaluated; for ACCMIP results see Table AII.6.9; Shindell et al., 2013). Total ERF change across scenarios derived from the CMIP5 ensemble can be compared only beginning in 2010. For the period 2010 to 2030, total ERF in the CMIP5 decadal averages increases by +0.5 to +1.0 W m^{-2} (RCP2.6 and RCP6.0 to RCP8.5; Table AII.6.10) while total ERF from the published RCPs increases by +0.7 to +1.1 W m^{-2} (RCP2.6 and RCP6.0 to RCP8.5, Table AII.6.8). Here we re-examine the near-term temperature increases projected from the RCPs (see Section 11.3.2.1.1) and assess the potential for changes in near-term anthropogenic forcing to induce climate responses that fall outside these scenarios.

For the different RCP pathways the increase in global mean surface temperature by 2026–2035 relative to the reference period 1986-2005 ranges from 0.74°C (RCP2.6 and RCP6.0) to 0.94°C (RCP8.5) (median of CMIP5 models, see Figure 11.24, Table AII.7.5). This inter-scenario range of 0.20°C is smaller than the inter-model spread for an individual scenario: 0.33°C to 0.52°C (defined as the 17 to 83% range of the decadal means of the models). This RCP inter-scenario spread may be too narrow as discussed in Section 11.3.5.1. The temperature increase of the most rapidly warming scenario (RCP8.5) emerges from inter-model spread (i.e., becomes greater than two times the 17 to 83% range) by about 2040, due primarily to increasing CH_4 and CO_2. By 2050 the inter-scenario spread is 0.8°C whereas the model spread

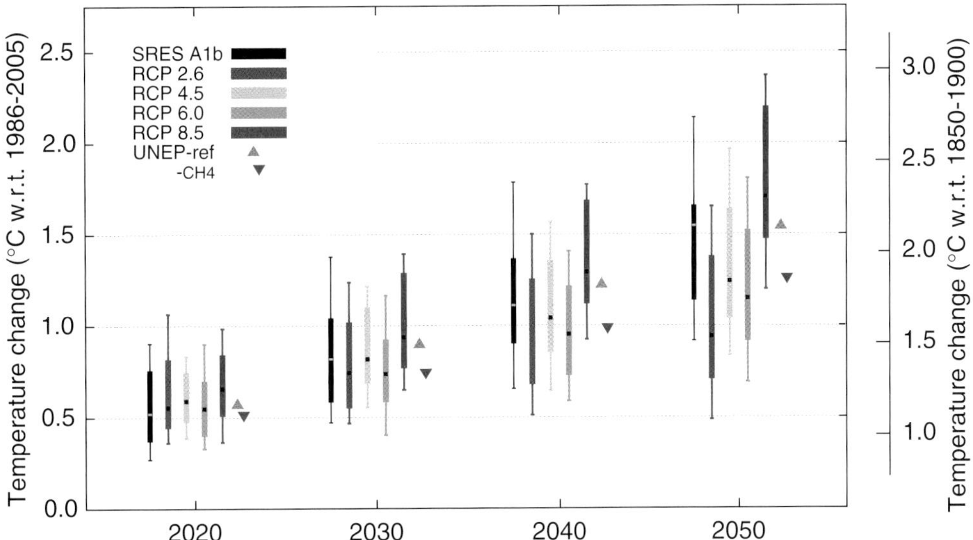

Figure 11.24a | Near-term increase in global mean surface air temperatures (°C) across scenarios. Increases in 10-year mean (2016–2025, 2026–2035, 2036–2045 and 2046–2055) relative to the reference period (1986–2005) of the globally averaged surface air temperatures. Results are shown for the CMIP5 model ensembles (see Annex I for listing of models included) for RCP2.6 (dark blue), RCP4.5 (light blue), RCP6.0 (orange), and RCP8.5 (red) and the CMIP3 model ensemble (22 models) for SRES A1b (black). The multi-model median (square), 17 to 83% range (wide boxes), 5 to 95% range (whiskers) across all models are shown for each decade and scenario. Values are provided in Table AII.7.5. Also shown are best estimates for a UNEP scenario (UNEP-ref, grey upward triangles) and one that implements technological controls on methane emissions (UNEP CH4, red downward-pointing triangles) (UNEP and WMO, 2011; Shindell et al., 2012a). Both UNEP scenarios are adjusted to reflect the 1986–2005 reference period. The right-hand floating axis shows increases in global mean surface air temperature relative to the early instrumental period (0.61°C), defined from the difference between 1850–1900 and 1986–2005 in the Hadley Centre/Climate Research Unit gridded surface temperature data set 4 (HadCRUT4) global mean temperature analysis (Chapter 2 and Table AII.1.3). Note that uncertainty remains on how to match the 1986–2005 reference period in observations with that in CMIP5 results. See discussion of Figure 11.25.

for each scenario is only 0.6°C. At 2040 the ERF in the published RCPs ranges from 2.6 (RCP2.6) to 3.6 (RCP8.5) W m^{-2}, and about 40% of this difference is due to the steady increases in CH_4 and tropospheric O_3 found only in RCP8.5. RCP6.0 has the lowest ERF and thus warms less rapidly than other RCPs up to 2030 (Table AII.6.8).

In terms of geographic patterns of warming, differences between RCP8.5 and RCP2.6 are within ±0.5°C over most of the globe for both summer and winter seasons for 2016–2035 (Figure 11.24b), but by 2036–2055 RCP8.5 is projected to be warmer than RCP2.6 by 0.5°C to 1.0°C over most continents, and by more than 1.0°C over the Arctic in winter. Although studies suggest that the Arctic response is complex and particularly sensitive to BC aerosols (Flanner et al., 2007; Quinn et al., 2008; Jacobson, 2010; Ramana et al., 2010; Bond et al., 2013; Sand et al., 2013), the difference in ERF between RCP2.6 and RCP8.5 is dominated by the GHGs, as the BC atmospheric burden is decreasing through the century with little difference across the RCPs (Table AII.5.7).

Large changes in emissions of the well-mixed greenhouse gases (WMGHGs) produce only modest changes in the near term because these gases are long lived: For example, a 50% cut in Kyoto-gas emissions beginning in 1990 offsets the warming that otherwise would have occurred by only –0.11°C ± 0.03°C after 12 years (Prather et al., 2009). In contrast, many studies have noted the large potential for air pollutant emission reductions to influence near-term climate because RF from these species responds almost immediately to changes in emissions. Decreases in sulphate aerosol have occurred through mitigation of both air pollution and fossil-fuel emissions, and are expected to produce a near-term rise in surface temperatures (e.g., Jacobson and Streets, 2009; Raes and Seinfeld, 2009; Wigley et al., 2009; Kloster et al., 2010; Makkonen et al., 2012).

Because global mean aerosol forcing decreases in all RCP scenarios (AII.5.3 to AII.5.7, AII.6.9; see Section 11.3.5), the potential exists for a systematic difference between the CMIP3 models forced with the SRES scenarios and the CMIP5 models forced with the RCP scenarios. One study directly addressed the impacts of aerosols on climate under the RCP4.5 scenario, and found that the aerosol emission reductions induce about a 0.2°C warming in the near term compared with fixed 2005 aerosol levels (more indicative of the SRES CMIP3 aerosols) (Levy et al., 2013). The cooling over the period 1951–2010 that is attributed to non-WMGHG anthropogenic forcing in the CMIP5 models (Figures 10.4 and 10.5) has a *likely* range of –0.25°C ± 0.35°C compared to +0.9°C ± 0.4°C for WMGHG. The non-WMGHG forcing generally includes the influence of non-aerosol warming agents over the historical period such as tropospheric ozone, and a simple correction would give an aerosol-only cooling that is about 50% larger in magnitude (see ERF components, Chapter 8). The near-term reductions in total aerosol emissions, however, even under the MFR scenario, are at most about 50% (AII.2.17 to AII.2.22), indicating a maximum near-term temperature response of about half that induced by the addition of aerosols over the last century. Hence, the evidence indicates that differences in aerosol loading from the SRES (conservatively assuming roughly constant aerosols) to the RCP scenarios can increase warming in the CMIP5 models relative to the CMIP3 models by up to 0.2°C in the near term for the same WMGHG forcing (*medium confidence*).

Many studies show that air pollutants influence climate and identify approaches to mitigate both air pollution and global warming by

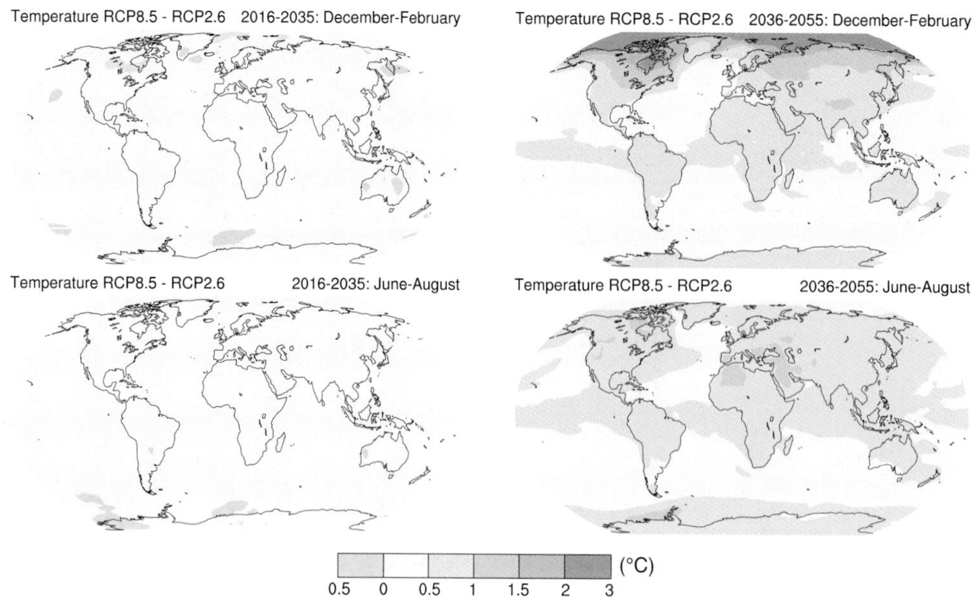

Figure 11.24b | Global maps of near-term differences in surface air temperature across the RCP scenarios. Differences between (RCP8.5) and low (RCP2.6) scenarios for the CMIP5 model ensemble (31 models) are shown for averages over 2016–2035 (left) and 2036–2055 (right) in boreal winter (December, January and February; top row) and summer (June, July and August; bottom row).

decreasing CH_4, tropospheric O_3 and absorbing aerosols, particularly BC (e.g., Hansen et al., 2000; Fiore et al., 2002, 2008, 2009; Dentener et al., 2005; West et al., 2006; Royal Society, 2008; Jacobson, 2010; Penner et al., 2010; UNEP and WMO, 2011; Anenberg et al., 2012; Shindell et al., 2012b; Unger, 2012; Bond et al., 2013). An alternative set of technologically based scenarios (UNEP and WMO, 2011) that examined controls on CH_4 and BC emissions designed to reduce tropospheric CH_4, O_3 and BC also included reductions of co-emitted species (e.g., CO, OC, NO_x). These reductions were applied in two CMIP5 models, and then those model responses were combined with the AR4 best estimates for the range of climate sensitivity and for uncertainty estimates for each component of RF (Shindell et al., 2012a). This approach provided a near-term best estimate and range of global mean temperature change for the reference (UNEP-ref) and CH_4-mitigation (UNEP-CH4) scenarios (Figure 11.24a, adjusted to reflect the 1986–2005 reference period). Under UNEP-CH4, anthropogenic CH_4 emissions decrease by 24% from 2010 to 2030, and global warming is reduced by 0.16°C (best estimate) at 2030 and by 0.28°C at 2050. A third UNEP scenario (UNEP-BC+CH4; not shown) adds reductions in BC by 78% onto CH_4 mitigation and reduces warming by an additional 0.12°C (best estimate) at 2030. However, it greatly increases the uncertainty owing to poor understanding of associated cloud adjustments (i.e., semi-direct and indirect effects) as well as of the ratio of BC to co-emitted reflective OC aerosols, their size distributions and mixing states (see Chapter 7, Section 7.5). Corresponding BC reductions in the RCPs are only 4 to 11%.

Beyond global mean temperature, shifting magnitudes and geographic patterns of emissions may induce aerosol-specific changes in regional atmospheric circulation and precipitation. See Chapter 7, especially Sections 7.6.2 and 7.6.4, for assessment of this work (Roeckner et al., 2006; Menon and et al., 2008; Ming et al., 2010, 2011; Ott et al., 2010; Randles and Ramaswamy, 2010; Allen and Sherwood, 2011; Bollasina et al., 2011; Leibensperger et al., 2011b;Fyfe et al., 2012; Ganguly et al., 2012; Rotstayn et al., 2012; Shindell et al., 2012b; Teng et al., 2012; Bond et al., 2013). Recent trends in aerosol–fog interactions and snowpack decline are implicated in more rapid regional warming in Europe (van Oldenborgh et al., 2010; Ceppi et al., 2012; Scherrer et al., 2012), and coupling of aerosols and soil moisture could increase near-term local warming in the eastern USA (Mickley et al., 2011). Major changes in the tropical circulation and rainfall have been attributed to increasing aerosols, but studies often disagree in sign (see Section 11.3.2.4.3, Chapters 10 and 14). The lack of standardization (e.g., different regions, different mixtures of reflecting and absorbing aerosols) and agreement across studies prevents generalization of these findings to project aerosol-induced changes in regional atmospheric circulation or precipitation in the near term.

Land use and land cover change (LULCC; see Chapter 6), including deforestation, forest degradation and agricultural expansion for bioenergy (Georgescu et al., 2009; Anderson-Teixeira et al., 2012), can alter global climate forcing through changing surface albedo (assessed as ERF; Chapter 8), the hydrological cycle, GHGs (for CO_2, see Chapters 6 and 12), or aerosols. The shift from forest to grassland in many places since the pre-industrial era has been formally attributed as a cause of regionally lower mean and extreme temperatures (Christidis et al., 2013). RCP CO_2 and CH_4 anthropogenic emissions include land use changes (Hurtt et al., 2011) that vary with the underlying storylines and differ across RCPs. These global-scale changes in crop and pasture land projected over the near term (+2% for RCP2.6 and RCP8.5; –4% for RCP4.5and RCP6.0) are smaller in magnitude than the 1950–2000 change (+6%) (see Figure 6.23). Overall LULCC has had small impact on ERF (–0.15 W m^{-2}; see AII.1.2) and thus as projected is not a major factor in near-term climate change on global scales.

Land use changes can also lead to sustained near-term changes in regional climate through modification of the biogeophysical properties that alter the water and energy cycles. Local- and regional-scale climate responses to LULCC can exceed those associated with global mean warming (Baidya Roy and Avissar, 2002; Findell et al., 2007; Pitman et al., 2009, 2012; Pielke et al., 2011; Boisier et al., 2012; de Noblet-Ducoudre et al., 2012; Lee and Berbery, 2012). Examples of LULCC-driven changes include: Brazilian conversion to sugarcane induces seasonal shifts of 1 to 2°C (Georgescu et al., 2013); European forested areas experience less severe heat waves (Teuling et al., 2010); and deforested regions over the Amazon lack deep convective clouds (Wang et al., 2009). Systematic assessment of near-term, local-to-regional climate change is beyond the scope here.

In summary, climate projections for the near term are not very sensitive to the range in anthropogenic emissions of CO_2 and other WMGHGs. By the 2040s the CMIP5 median for global mean temperature ranges from a low of +0.9°C (RCP2.6 and RCP6.0) to a high of +1.3°C (RCP8.5) above the CMIP5 reference period (Figure 11.24a; Table AII.7.5). See discussion below regarding possible offsets between the observed and CMIP5 reference periods. Alternative CH_4 scenarios incorporating large emission reductions outside the RCP range would offset near-term warming by –0.2°C (*medium confidence*). Aerosols remain a major source of uncertainty in near-term projections, on both global and regional scales. Removal of half of the sulphate aerosol, as projected before 2030 in the MFR scenario and by 2050 in most RCPs, would increase warming by up to +0.2°C (*medium confidence*). Actions to reduce BC aerosol could reduce warming, but the magnitude is highly uncertain, depending on co-emitted (reflective) aerosols and aerosol-cloud interactions (Chapter 7; Section 7.5). In addition, near-term climate change, including extremes and precipitation, may be driven locally by land use change and shifting geographic patterns of aerosols; and these regional climatic effects may exceed those induced by the global ERF.

11.3.6.2 Uncertainties in Future Natural Radiative Forcing and the Consequences for Near-term Climate

11.3.6.2.1 The effects of future volcanic eruptions

As discussed in Chapters 8 and 10, explosive volcanic eruptions are the major cause of natural variations in RF on interannual to decadal time scales. Most important are large tropical and subtropical eruptions that inject substantial amounts of SO_2 directly into the stratosphere. The subsequent formation of sulphate aerosols leads to a negative RF of several watts per metre squared, with a typical lifetime of a year (Robock, 2000). The eruption of Mt Pinatubo in 1991 was one of the largest in recent times, with a return period of about three times per century, but dwarfed by Tambora in 1815 (Gao et al., 2008). Mt Pinatubo caused a rapid drop in a global mean surface air temperature of several tenths of a degree Celsius over the following year, but this signal disappeared over the next five years (Hansen et al., 1992; Soden et al., 2002; Bender et al., 2010). In addition to global mean cooling, there are effects on the hydrological cycle (e.g., Trenberth and Dai, 2007), atmosphere and ocean circulation (e.g., Stenchikov et al., 2006; Ottera et al., 2010). The surface climate response typically persists for a few years, but the subsurface ocean response can persist for decades or centuries, with consequences for sea level rise (Delworth et al., 2005; Stenchikov et al., 2009; Gregory, 2010; Timmreck, 2012).

Although it is possible to detect when various existing volcanoes become more active, or are more likely to erupt, the precise timing of an eruption, the amount of SO_2 emitted and its distribution in the stratosphere are not predictable until after the eruption. Eruptions comparable to Mt Pinatubo can be expected to cause a short-term cooling of the climate with related effects on surface climate that persist for a few years before a return to warming trajectories discussed in Section 11.3.2. Larger eruptions, or several eruptions occurring close together in time, would lead to larger and/or more persistent effects.

11.3.6.2.2 The effects of future changes in solar forcing

Some of the future CMIP5 climate simulations using the RCP scenarios include an 11-year variation in total solar irradiance (TSI) but no underlying trend beyond 2005. Chapter 10 noted that there has been little observed trend in TSI during a time period of rapid global warming since the late 1970s, but that the 11-year solar cycle does introduce a significant and measurable pattern of response in the troposphere (Section 10.3.1.1.3). As discussed in Chapter 8 (Section 8.4.1.3), the Sun has been in a 'grand solar maximum' of magnetic activity on the multi-decadal time scale. However, the most recent solar minimum was the lowest and longest since 1920, and some studies (e.g., Lockwood, 2010) suggest there could be a continued decline towards a much quieter period in the coming decades, but there is *low confidence* in these projections (Section 8.4.1.3). Nevertheless, if there is such a reduction in solar activity, there is *high confidence* that the variations in TSI RF will be much smaller than the projected increased forcing due to GHGs (Section 8.4.1.3). In addition, studies that have investigated the effect of a possible decline in TSI on future climate have shown that the associated decrease in global mean surface temperature is much smaller than the warming expected from increases in anthropogenic GHGs (Feulner and Rahmstorf, 2010; Jones et al., 2012; Meehl et al., 2013b) However, regional impacts could be more significant (Xoplaki et al., 2001; Mann et al., 2009; Gray et al., 2010; Ineson et al., 2011).

As discussed in Section 8.4.1, a recent satellite measurement (Harder et al., 2009) found much greater than expected reduction at ultraviolet (UV) wavelengths in the recent declining solar cycle phase. Changes in solar UV drive stratospheric O_3 chemistry and can change RF. Haigh et al. (2010) show that if these observations are correct, they imply the opposite relationship between solar RF and solar activity over that period than has hitherto been assumed. These new measurements therefore increase uncertainty in estimates of the sign of solar RF, but they are not expected to alter estimates of the maximum absolute magnitude of the solar contribution to RF, which remains small (Chapter 8). However, they do suggest the possibility of a much larger impact of solar variations on the stratosphere than previously thought, and some studies have suggested that this may lead to significant regional impacts on climate (as discussed in Section 10.3.1.1.3) that are not necessarily reflected by the RF metric (see Section 8.4.1).

In summary, possible future changes in solar irradiance could influence the rate at which global mean surface air temperature increases, but there is *high confidence* that this influence will be small in comparison to the influence of increasing concentrations of GHGs in the atmosphere. Understanding of the impacts of changes in solar irradiance on continental and sub-continental scale climate remains low.

Frequently Asked Questions

FAQ 11.2 | How Do Volcanic Eruptions Affect Climate and Our Ability to Predict Climate?

Large volcanic eruptions affect the climate by injecting sulphur dioxide gas into the upper atmosphere (also called stratosphere), which reacts with water to form clouds of sulphuric acid droplets. These clouds reflect sunlight back to space, preventing its energy from reaching the Earth's surface, thus cooling it, along with the lower atmosphere. These upper atmospheric sulphuric acid clouds also locally absorb energy from the Sun, the Earth and the lower atmosphere, which heats the upper atmosphere (see FAQ 11.2, Figure 1). In terms of surface cooling, the 1991 Mt Pinatubo eruption in the Philippines, for example, injected about 20 million tons of sulphur dioxide (SO_2) into the stratosphere, cooling the Earth by about 0.5°C for up to a year. Globally, eruptions also reduce precipitation, because the reduced incoming shortwave at the surface is compensated by a reduction in latent heating (i.e., in evaporation and hence rainfall).

For the purposes of predicting climate, an eruption causing significant global surface cooling and upper atmospheric heating for the next year or so can be expected. The problem is that, while a volcano that has become more active can be detected, the precise timing of an eruption, or the amount of SO_2 injected into the upper atmosphere and how it might disperse cannot be predicted. This is a source of uncertainty in climate predictions.

Large volcanic eruptions produce lots of particles, called ash or tephra. However, these particles fall out of the atmosphere quickly, within days or weeks, so they do not affect the global climate. For example, the 1980 Mount St. Helens eruption affected surface temperatures in the northwest USA for several days but, because it emitted little SO_2 into the stratosphere, it had no detectable global climate impacts. If large, high-latitude eruptions inject sulphur into the stratosphere, they will have an effect only in the hemisphere where they erupted, and the effects will only last a year at most, as the stratospheric cloud they produce only has a lifetime of a few months.

Tropical or subtropical volcanoes produce more global surface or tropospheric cooling. This is because the resulting sulphuric acid cloud in the upper atmosphere lasts between one and two years, and can cover much of the globe. However, their regional climatic impacts are difficult to predict, because dispersion of stratospheric sulphate aerosols depends heavily on atmospheric wind conditions at the time of eruption. Furthermore, the surface cooling effect is typically not uniform: because continents cool more than the ocean, the summer monsoon can weaken, reducing rain over Asia and Africa. The climatic response is complicated further by the fact that upper atmospheric clouds from tropical eruptions also absorb sunlight and heat from the Earth, which produces more upper atmosphere warming in the tropics than at high latitudes.

The largest volcanic eruptions of the past 250 years stimulated scientific study. After the 1783 Laki eruption in Iceland, there were record warm summer temperatures in Europe, followed by a very cold winter. Two large eruptions, an unidentified one in 1809, and the 1815 Tambora eruption caused the 'Year Without a Summer' in 1816. Agricultural failures in Europe and the USA that year led to food shortages, famine and riots.

The largest eruption in more than 50 years, that of Agung in 1963, led to many modern studies, including observations and climate model calculations. Two subsequent large eruptions, El Chichón in 1982 and Pinatubo in 1991, inspired the work that led to our current understanding of the effects of volcanic eruptions on climate.

(continued on next page)

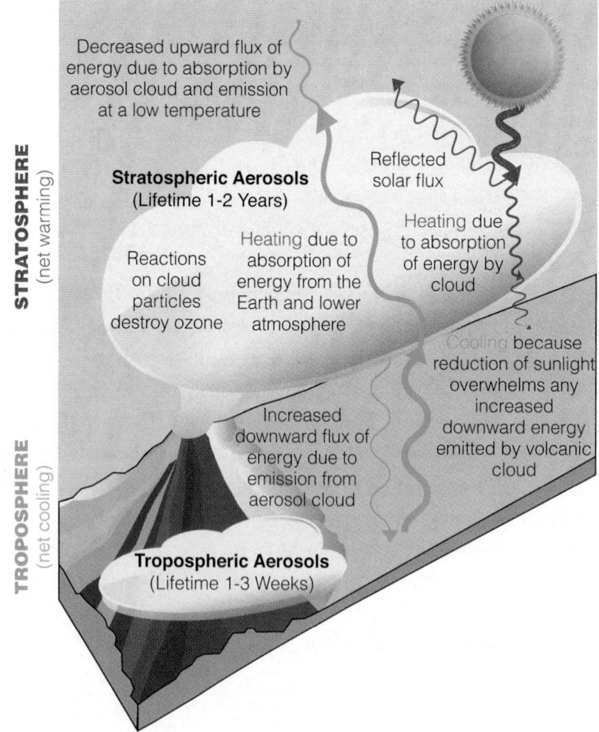

FAQ 11.2, Figure 1 | Schematic of how large tropical or sub-tropical volcanoes impact upper atmospheric (stratospheric) and lower atmospheric (tropospheric) temperatures.

FAQ 11.2 (continued)

Volcanic clouds remain in the stratosphere only for a couple of years, so their impact on climate is correspondingly short. But the impacts of consecutive large eruptions can last longer: for example, at the end of the 13th century there were four large eruptions—one every ten years. The first, in 1258 CE, was the largest in 1000 years. That sequence of eruptions cooled the North Atlantic Ocean and Arctic sea ice. Another period of interest is the three large, and several lesser, volcanic events during 1963–1991 (see Chapter 8 for how these eruptions affected atmospheric composition and reduced shortwave radiation at the ground.

Volcanologists can detect when a volcano becomes more active, but they cannot predict whether it will erupt, or if it does, how much sulphur it might inject into the stratosphere. Nevertheless, volcanoes affect the ability to predict climate in three distinct ways. First, if a violent eruption injects significant volumes of sulphur dioxide into the stratosphere, this effect can be included in climate predictions. There are substantial challenges and sources of uncertainty involved, such as collecting good observations of the volcanic cloud, and calculating how it will move and change during its lifetime. But, based on observations, and successful modelling of recent eruptions, some of the effects of large eruptions can be included in predictions.

The second effect is that volcanic eruptions are a potential source of uncertainty in our predictions. Eruptions cannot be predicted in advance, but they will occur, causing short-term climatic impacts on both local and global scales. In principle, this potential uncertainty can be accounted for by including random eruptions, or eruptions based on some scenario in our near-term ensemble climate predictions. This area of research needs further exploration. The future projections in this report do not include future volcanic eruptions.

Third, the historical climate record can be used, along with estimates of observed sulphate aerosols, to test the fidelity of our climate simulations. While the climatic response to explosive volcanic eruptions is a useful analogue for some other climatic forcings, there are limitations. For example, successfully simulating the impact of one eruption can help validate models used for seasonal and interannual predictions. But in this way not all the mechanisms involved in global warming over the next century can be validated, because these involve long term oceanic feedbacks, which have a longer time scale than the response to individual volcanic eruptions.

11.3.6.3 Synthesis of Near-term Projections of Global Mean Surface Air Temperature

Figure 11.25 provides a synthesis of near-term projections of global mean surface air temperature (GMST) from CMIP5, CMIP3 and studies that have attempted to use observations to quantify projection uncertainty (see Section 11.3.2.1). On the basis of this evidence, an attempt is made here to assess a *likely* range for GMST in the period 2016–2035. Such an overall assessment is not straightforward. The following points must be taken into account:

1. No likelihoods are associated with the different RCP scenarios. For this reason, previous IPCC Assessment Reports have only presented projections that are conditional on specific scenarios. Here we attempt a broader assessment across all four RCP scenarios. This is possible only because, as discussed in Section 11.3.6.1, *near-term* projections of GMST are not especially sensitive to these different scenarios.

2. In the near term it is expected that increases in GMST will be driven by past and future increases in GHG concentrations and future decreases in anthropogenic aerosols, as found in all the RCP scenarios. Figure 11.25c shows that in the near term the CMIP3 projections based on the SRES scenarios are generally cooler than the CMIP5 projections based on the RCP scenarios. This difference is at least partly attributable to higher aerosol concentrations in the SRES scenarios (see Section 11.3.6.1).

3. The CMIP3 and CMIP5 projections are ensembles of opportunity, and it is explicitly recognized that there are sources of uncertainty not simulated by the models. Evidence of this can be seen by comparing the Rowlands et al. (2012) projections for the A1B scenario, which were obtained using a very large ensemble in which the physics parameterizations were perturbed in a single climate model, with the corresponding raw multi-model CMIP3 projections. The former exhibit a substantially larger *likely* range than the latter. A pragmatic approach to addressing this issue, which was used in the AR4 and is also used in Chapter 12, is to consider the 5 to 95% CMIP3/5 range as a '*likely*' rather than '*very likely*' range.

4. As discussed in Section 11.3.6.2, the RCP scenarios assume no underlying trend in total solar irradiance and no future volcanic eruptions. Future volcanic eruptions cannot be predicted and there is *low confidence* in projected changes in solar irradiance (Chapter 8). Consequently the possible effects of future changes in natural forcings are excluded from the assessment here.

5. As discussed in Section 11.3.2.1.1 observationally constrained 'ASK' projections (Gillett et al., 2013; Stott et al., 2013) are 10 to 15% cooler (median values for RCP4.5; 6–10% cooler for RCP8.5), and have a narrower range, than the corresponding 'raw' (uninitialized) CMIP5 projections. The reduced rate of warming in the ASK projections is related to evidence from Chapter 10 (Section 10.3.1) that 'some CMIP5 models have a higher transient response to GHGs and a larger response to other anthropogenic forcings (dominated by the effects of aerosols) than the real world (*medium confidence*).' These models may warm too rapidly as GHGs increase and aerosols decline.

6. Over the last two decades the observed rate of increase in GMST has been at the lower end of rates simulated by CMIP5 models (Figure 11.25a). This hiatus in GMST rise is discussed in detail in Box 9.2 (Chapter 9), where it is concluded that the hiatus is attributable, in roughly equal measure, to a decline in the rate of increase in ERF and a cooling contribution from internal variability (expert judgment, *medium confidence*). The decline in the rate of increase in ERF is attributed primarily to natural (solar and volcanic) forcing but there is *low confidence* in quantifying the role of forcing trend in causing the hiatus, because of uncertainty in the magnitude of the volcanic forcing trend and *low confidence* in the aerosol forcing trend. Concerning the higher rate of warming in CMIP5 simulations it is concluded that there is a substantial contribution from internal variability but that errors in ERF and in model responses may also contribute. There is *low confidence* in this assessment because of uncertainties in aerosol forcing in particular.

   The observed hiatus has important implications for near-term projections of GMST. A basic issue concerns the sensitivity of projections to the choice of reference period. Figure 11.25b and c shows the 5 to 95% ranges for CMIP5 projections using a 1986–2005 reference period (light grey), and the same projections using a 2006–2012 reference period (dark grey). The latter projections are cooler, and the effect of using a more recent reference period appears similar to the effect of initialization (discussed in Section 11.3.2.1.1 and shown in Figure 11.25c for RCP4.5). Using this more recent reference period, the 5 to 95% range for the mean GMST in 2016–2035 relative to 1986–2005 is 0.36°C to 0.79°C (using all RCP scenarios, weighted to ensure equal weights per model and using an estimate of the observed GMST anomaly for (2006–2012)–(1986–2005) of 0.16°C). This range may be compared with the range of 0.48°C to 1.15°C obtained from the CMIP5 models using the original 1986–2005 reference period.

7. In view of the sensitivity of projections to the reference period it is helpful to consider the possible rate of change of GMST in the near term. The CMIP5 5 to 95% ranges for GMST trends in the period 2012–2035 are 0.11°C to 0.41°C per decade. This range is similar to, though slightly narrower than, the range found by Easterling and Wehner (2009) for the CMIP3 SRES A2 scenario over the longer period 2000–2050. It may also be compared with recent rates in the observational record (e.g., ~0.26°C per decade for 1984–1998 and ~0.04°C per decade for hiatus period 1998–2012; See Box 9.2). The RCP scenarios project that ERF will increase more rapidly in the near term than occurred over the hiatus period (see Box 9.2 and Annex II), which is consistent with more rapid warming. In addition, Box 9.2 includes an assessment that internal variability is *more likely than not* to make a positive contribution to the increase in GMST in the near term. Internal variability is included in the CMIP5 projections, but because most of the CMIP5 simulations do not reproduce the observed reduction in global mean surface warming over the last 10 to 15 years, the distribution of CMIP5 near-term trends will not reflect this assessment and might, as a result, be biased low. This uncertainty, however, is somewhat counter balanced by the evidence of point 5, which suggests a high bias in the distribution of near-term trends. A further projection of GMST for the period 2016–2035 may be obtained by starting from the observed GMST for 2012 (0.14°C relative to 1986–2005) and projecting increases at rates between the 5 to 95% CMIP5 range of 0.11°C to 0.41°C per decade. The resulting range of 0.29°C to 0.69°C, relative to 1986–2005, is shown on Figure 11.25(c).

Overall, in the absence of major volcanic eruptions—which would cause significant but temporary cooling—and, assuming no significant future long term changes in solar irradiance, it is *likely* (>66% probability) that the GMST anomaly for the period 2016–2035, relative to the reference period of 1986–2005 will be in the range 0.3°C to 0.7°C (expert assessment, to one significant figure; *medium confidence*). This range is consistent, to one significant figure, with the range obtained by using CMIP5 5 to 95% model trends for 2012–2035. It is also consistent with the CMIP5 5 to 95% range for all four RCP scenarios of 0.36°C to 0.79°C, using the 2006–2012 reference period, after the upper and lower bounds are reduced by 10% to take into account the evidence noted under point 5 that some models may be too sensitive to anthropogenic forcing. The 0.3°C to 0.7°C range includes the *likely* range of the ASK projections and initialized predictions for RCP4.5. It corresponds to a rate of change of GMST between 2012 and 2035 in the range 0.12°C to 0.42°C per decade. The higher rates of change can be associated with a significant positive contribution from internal variability (Box 9.2) and/or high rates of increase in ERF (e.g., as found in RCP8.5). Note that an upper limit of 0.8°C on the 2016–2035 GMST corresponds to a rate of change over the period 2012–2035 of 0.49°C per decade, which is considered *unlikely*. The assessed rates of change are consistent with the AR4 SPM statement that 'For the next two decades, a warming of about 0.2°C per decade is projected for a range of SRES emission scenarios'. However, the implied rates of warming over the period from 1986–2005 to 2016–2035 are lower as a result of the hiatus: 0.10°C to 0.23°C per decade, suggesting the AR4 assessment was near the upper end of current expectations for this specific time interval.

The assessment here provides only a *likely* range for GMST. Possible reasons why the real world might depart from this range include: RF departs significantly from the RCP scenarios, due to either natural (e.g., major volcanic eruptions, changes in solar irradiance) or anthropogenic (e.g., aerosol or GHG emissions) causes; processes that are poorly simulated in the CMIP5 models exert a significant influence on GMST. The latter class includes: a possible strong 'recovery' from the recent hiatus in GMST; the possibility that models might underestimate decadal variability (but see Section 9.5.3.1); the possibility that model sensitivity to anthropogenic forcing may differ from that of the real world (see point

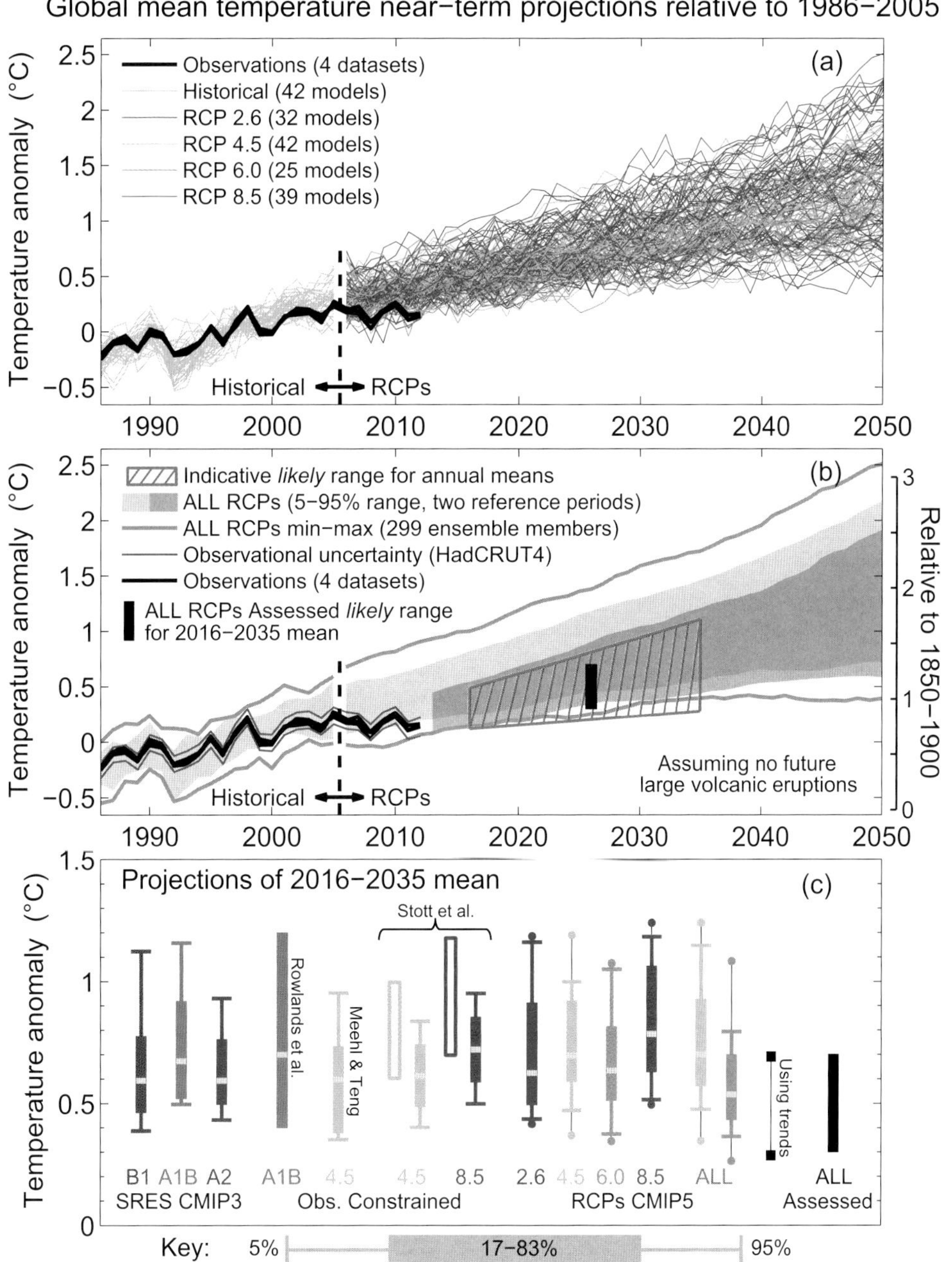

Figure 11.25 | Synthesis of near-term projections of global mean surface air temperature (GMST). (a) Simulations and projections of annual mean GMST 1986–2050 (anomalies relative to 1986–2005). Projections under all RCPs from CMIP5 models (grey and coloured lines, one ensemble member per model), with four observational estimates (Hadley Centre/Climate Research Unit gridded surface temperature data set 4 (HadCRUT4): Morice et al., 2012); European Centre for Medium range Weather Forecast (ECMWF) interim reanalysis of the global atmosphere and surface conditions (ERA-Interim): Simmons et al., 2010); Goddard Institute of Space Studies Surface Temperature Analysis (GISTEMP): Hansen et al., 2010); National Oceanic and Atmospheric Administration (NOAA): Smith et al., 2008)) for the period 1986–2012 (black lines). (b) As (a) but showing the 5 to 95% range of annual mean CMIP5 projections (using one ensemble member per model) for all RCPs using a reference period of 1986–2005 (light grey shade) and all RCPs using a reference period of 2006–2012, together with the observed anomaly for (2006–2012) to (1986–2005) of 0.16°C (dark grey shade). The percentiles for 2006 onwards have been smoothed with a 5-year running mean for clarity. The maximum and minimum values from CMIP5 using all ensemble members and the 1986–2005 reference period are shown by the grey lines (also smoothed). Black lines show annual mean observational estimates. The red hatched region shows the indicative *likely* range for annual mean GMST during the period 2016–2035 based on the 'ALL RCPs Assessed' *likely* range for the 20-year mean GMST anomaly for 2016–2035, which is shown as a black bar in both (b) and (c) (see text for details). The temperature scale on the right hand side shows changes relative to a reference period of 1850-1900, assuming a warming of GMST between 1850-1900 and 1986-2005 of 0.61°C estimated from HadCRUT4.The temperature scale relative to the 1850-1900 period on the right-hand side assumes a warming of GMST prior to 1986–2005 of 0.61°C estimated from HadCRUT4. (c) A synthesis of projections for the mean GMST anomaly for 2016–2035 relative to 1986–2005. The box and whiskers represent the 66% and 90% ranges. Shown are unconstrained SRES CMIP3 and RCP CMIP5 projections; observationally constrained projections: Rowlands et al. (2012) for SRES A1B scenario, updated to remove simulations with large future volcanic eruptions; Meehl and Teng (2012) for RCP4.5 scenario, updated to include 14 CMIP5 models; Stott et al. (2013), based on six CMIP5 models with unconstrained 66% ranges for these six models shown as unfilled boxes; unconstrained projections for all four RCP scenarios using two reference periods as in panel b (light grey and dark grey shades, consistent with panel b); 90% range estimated using CMIP5 trends for the period 2012–2035 and the observed GMST anomaly for 2012; an overall *likely* (>66%) assessed range for all RCP scenarios. The dots for the CMIP5 estimates show the maximum and minimum values using all ensemble members. The medians (or maximum likelihood estimate for Rowlands et al. 2012) are indicated by a grey band.

Table 11.3 | Percentage of CMIP5 models for which the projected change in global mean surface air temperature, relative to 1850-1900, crosses the specified temperature levels, by the specified time periods and assuming the specified RCP scenarios. The projected temperature change relative to the mean temperature in the period 1850-1900 is calculated using the models' projected temperature change relative to 1986–2005 plus the observed temperature change between 1850–1900 and 1986–2005 of 0.61°C estimated from the Hadley Centre/Climate Research Unit gridded surface temperature data set 4 (HadCRUT4; Morice et al., 2012). The percentages in brackets use an alternative reference period for the model projections of 2006–2012, together with the observed temperature difference between 1986–2005 and 2006–2012 of 0.16°C. The definition of crossing is that the 20-year mean exceeds the specified temperature level. Note that these percentages should *not* be interpreted as likelihoods because there are other sources of uncertainty (see discussion in Section 11.3.6.3).

Scenario	Early (2016–2035)	Mid (2046–2065)
	Temperature +1.0°C	
RCP 2.6	100% (84%)	100% (94%)
RCP 4.5	98% (93%)	100% (100%)
RCP 6.0	96% (80%)	100% (100%)
RCP 8.5	100% (100%)	100% (100%)
	Temperature +1.5°C	
RCP 2.6	22% (0%)	56% (28%)
RCP 4.5	17% (0%)	95% (86%)
RCP 6.0	12% (0%)	92% (88%)
RCP 8.5	33% (5%)	100% (100%)
	Temperature +2.0°C	
RCP 2.6	0% (0%)	16% (3%)
RCP 4.5	0% (0%)	43% (29%)
RCP 6.0	0% (0%)	32% (20%)
RCP 8.5	0% (0%)	95% (90%)
	Temperature +3.0°C	
RCP 2.6	0% (0%)	0% (0%)
RCP 4.5	0% (0%)	0% (0%)
RCP 6.0	0% (0%)	0% (0%)
RCP 8.5	0% (0%)	21% (5%)

5); and the possibility of abrupt changes in climate (see introduction to Sections 11.3.6 and 12.5.5).

The assessment here has focused on 20-year mean values of GMST for the period 2016–2035. There is no unique method to derive a *likely* range for annual mean values from the range for 20-year means, so such calculations necessarily involve additional uncertainties (beyond those outlined in the previous paragraph), and lower confidence. Nevertheless, it is useful to attempt to estimate a range for annual mean values, which may be compared with raw model projections and, in the future, with observations. To do so, the following simple approach is used: (1) Starting in 2009 from the observed GMST anomaly for 2006–2012 of 0.16°C (relative to 1986–2005), linear trends are projected over the period 2009–2035 with maximum and minimum gradients selected to be consistent with the 0.3°C to 0.7°C range for the mean GMST in the period 2016–2035; 2). To take into account the expected year-to-year variability of annual mean values, the resulting linear trends are offset by ±0.1°C. The value of 0.1°C is based on the standard deviation of annual means in CMIP5 control runs (to one significant figure). These calculations provide an indicative *likely* range for annual mean GMST, which is shown as the red hatched area in Figure 11.25b. Note that this range does not take into account the expected impact of any future volcanic eruptions.

The assessed *likely* range for GMST in the period 2016–2035 may also be used to assess the likelihood that GMST will cross policy-relevant levels, relative to earlier time periods (Joshi et al., 2011). Using the 1850–1900 period, and the observed temperature rise between 1850–1900 and 1986–2005 of 0.61°C (estimated from the HadCRUT4 data set (Morice et al., 2012) gives a *likely* range for the GMST anomaly in 2016–2035 of 0.91°C–1.31°C, and supports the following conclusions: it is *more likely than not* that the mean GMST for the period 2016–2035 will be more than 1°C above the mean for 1850–1900, and *very unlikely* that it will be more than 1.5°C above the 1850–1900 mean (expert assessment, *medium confidence*). Additional information about the possibility of GMST crossing specific temperature levels is provided in Table 11.3, which shows the percentage of CMIP5 models for which the projected change in GMST exceeds specific temperature levels, under each RCP scenario, in two time periods (early century: 2016–2035 and mid-century: 2046–2065), and also using the two different reference periods discussed under point 6 and illustrated in Figure 11.25. However, these percentages should *not* be interpreted as likelihoods because—as discussed in this section—there are sources of uncertainty not captured by the CMIP5 ensemble. Note finally that it is *very likely* that specific temperature levels will be crossed temporarily in individual years before a permanent crossing is established (Joshi et al., 2011), but Table 11.3 is based on 20-year mean values.

Box 11.2 | Ability of Climate Models to Simulate Observed Regional Trends

The ability of models to simulate past climate change on regional scales can be used to investigate whether the multi-model ensemble spread covers the forcing and model uncertainties. Agreement between observed and simulated regional trends, taking natural variability and model spread into account, would build confidence in near-term projections. Although large-scale features are simulated well (see Chapter 10), on sub-continental and smaller scales the observed trends are, in general, more often in the tails of the distribution of modelled trends than would be expected by chance fluctuations (Bhend and Whetton, 2012; Knutson et al., 2013b; van Oldenborgh et al., 2013). Natural variability and model spread are larger at smaller scales (Stott et al., 2010), but this is not enough to bridge the gap between models and observations. Downscaling with Regional Climate Models (RCMs) does not affect seasonal mean trends except near mountains or coastlines in Europe (van Oldenborgh et al., 2009; van Haren et al., 2012). These results hold for both observed and modelled estimates of natural variability and for various analyses of the observations. Given the statistical nature of the comparisons, it is currently not possible to say in which regions observed discrepancies are due to coincidental natural variability and in which regions they are due to forcing or model deficiencies. These results show that in general the Coupled Model Intercomparison Project Phase 5 (CMIP5) ensemble cannot be taken as a reliable regional probability forecast, but that the true uncertainty can be larger than the model spread indicated in the maps in this chapter and Annex I.

Temperature

Räisänen (2007) and Yokohata et al. (2012) compared regional linear temperature trends during 1955–2005 (1961–2000) with corresponding trends in the CMIP3 ensemble. They found that the range of simulated trends captured the observed trend in nearly all locations. Using another metric, Knutson et al., (2013b) found that CMIP5 models did slightly better than CMIP3 in reproducing linear trends (see also Figure 10.2, Section 10.3.1.1.2). The linear CMIP5 temperature trends are compared with the observed trends in Box 11.2, Figure 1a–h. The rank histograms show the warm bias in global mean temperature (see Chapter 10) and some overconfidence, but within the inter-model spread. However, the apparent agreement appears to be for the wrong reason. Many of the models that appear to correctly simulate observed high regional trends do so because they have a high climate response (i.e., the global temperature rises quickly) and do not simulate the observed spatial pattern of trends (Kumar et al., 2013). To address this, Bhend and Whetton (2012) and van Oldenborgh et al. (2013) use another definition of the local trend: the regression of the local temperature on the (low-pass filtered) global mean temperature. This definition separates the local temperature response pattern from the global mean climate response. They find highly significant discrepancies between the CMIP3 and CMIP5 trend patterns and a variety of estimates of observed trend estimates. These discrepancies are defined relative to an error model that includes the (modelled or observed) natural variability, model spread and spatial autocorrelations. In the following, areas where the observed and modelled trends show marked differences are noted. Areas of agreement are covered in Section 10.3.1.1.4.

In December to February the observed Arctic amplification extends further south than modelled in Central Asia and northwestern North America. In June to August southern Europe and North Africa have warmed significantly faster than both CMIP3 and CMIP5 models simulated (van Oldenborgh et al., 2009); this also holds for the Middle East. The observed Indo-Pacific warm pool trend is significantly higher than the modelled trend year-round (Shin and Sardeshmukh, 2011; Williams and Funk, 2011), and the North Pacific and the southeastern USA and adjoining ocean trends were lower. Direct causes for many of these discrepancies are known (e.g., December to February circulation trends that differ between the observation and the models (Gillett et al., 2005; Gillett and Stott, 2009; van Oldenborgh et al., 2009; Bhend and Whetton, 2012) or teleconnections from other areas with trend biases (Deser and Phillips, 2009; Meehl et al., 2012a), but the causes of the underlying discrepancies are often unknown. Possibilities include observational uncertainties (note, however, that the areas where the observations warm more than the models do not correspond to areas of increased urbanization or irrigation; cf. Section 2.4.1.3), an underestimation of the low-frequency variability (Knutson et al. (2013b) show evidence that this is probably not the case for temperature outside the tropics), unrealistic local forcing (e.g., aerosols (Ruckstuhl and Norris, 2009)), or missing or misrepresented processes in models (e.g., fog (Vautard et al., 2009; Ceppi et al., 2012)).

Precipitation

In spite of the larger variability relative to the trends and observational uncertainties (cf. Section 2.5.1.2), annual mean regional linear precipitation trends have been found to differ significantly between observations and CMIP3 models, both in the zonal mean (Allan and Soden, 2007; Zhang et al., 2007b) and regionally (Räisänen, 2007). The comparison is shown in Box 11.2, Figure 1i–p for the CMIP5 half-year seasons used in Annex I, following van Oldenborgh et al. (2013). In both half years the observations fall more often in the highest and lowest 5% than expected by chance fluctuations within the ensemble (grey area). The differences larger than the difference between the CRU and GPCC analyses (cf. Figure 2.29) are noted below. *(continued on next page)*

Box 11.2 (continued)

In Europe there are large-scale differences between observed trends and trends, both in General Circulation Models (GCMs) and RCMs (Bhend and von Storch, 2008), which are ascribed to circulation change discrepancies in winter and in summer sea surface temperature (SST) trend biases (Lenderink et al., 2009; van Haren et al., 2012) and the misrepresentation of Summer North Atlantic Oscillation (NAO) teleconnections (Bladé et al., 2012). Central North America has become much wetter over 1950–2012, especially in winter, which is not simulated by the CMIP5 models. Larger observed northwest Australian rainfall increases than in CMIP3 in summer are driven by ozone forcings in two climate models (Kang et al., 2011) and aerosols in another (Rotstayn et al., 2012). The Guinea Coast has become drier in the observations than in the models. The CMIP5 patterns seem to reproduce the observed patterns somewhat better than the CMIP3 patterns (Bhend and Whetton, 2012), but the remaining discrepancies imply that CMIP5 projections cannot be used as reliable precipitation forecasts.

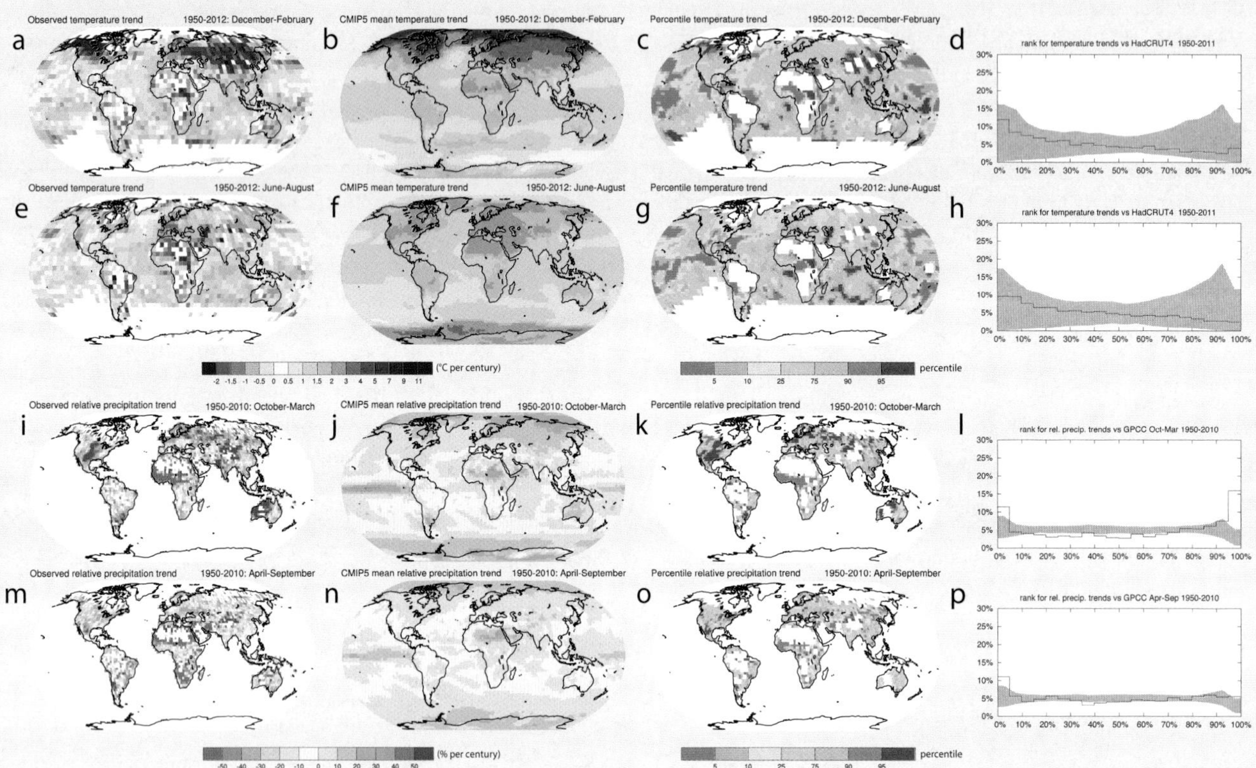

Box 11.2, Figure 1 | (a) Observed linear December to February temperature trend 1950–2012 (Hadley Centre/Climate Research Unit gridded surface temperature data set 4.1.1.0 (HadCRUT4.1.1.0, °C per century). (b) The equivalent CMIP5 ensemble mean trend. (c) Quantile of the observed trend in the ensemble, and (d) the corresponding rank histogram, the grey band denotes the 90% band of intermodel fluctuations (following Annan and Hargreaves, 2010). (e–h) Same for June to August. (i–l) Same for October to March precipitation (Global Precipitation Climatology Centre (GPCC) v7) 1950–2010, % per century). (m–p) Precipitation in April to September. Grid boxes where less than 50% of the years have observations are left white. (Based on Räisänen (2007) and van Oldenborgh et al. (2013).)

Acknowledgements

The authors thank Ed Hawkins (U. Reading, UK) for extensive input to discussions on the assessment of near-term global temperature and his work on key synthesis figures, and Jan Sedlacek (ETH, Switzerland) for his outstanding work on the production of numerous figures in this chapter.

References

Alexander, L. V., and J. M. Arblaster, 2009: Assessing trends in observed and modelled climate extremes over Australia in relation to future projections. *Int. J. Climatol.*, **29**, 417–435.

Alexander, M. A., L. Matrosova, C. Penland, J. D. Scott, and P. Chang, 2008: Forecasting Pacific SSTs: Linear inverse model predictions of the PDO. *J. Clim.*, **21**, 385–402.

Alexandru, A., R. de Elia, and R. Laprise, 2007: Internal variability in regional climate downscaling at the seasonal scale. *Mon. Weather Rev.*, **135**, 3221–3238.

Alexeev, V. A., D. J. Nicolsky, V. E. Romanovsky, and D. M. Lawrence, 2007: An evaluation of deep soil configurations in the CLM3 for improved representation of permafrost. *Geophys. Res. Lett.*, **34**, L09502.

Allan, R. P., and B. J. Soden, 2007: Large discrepancy between observed and simulated precipitation trends in the ascending and descending branches of the tropical circulation. *Geophys. Res. Lett.*, **34**, L18705.

Allan, R. P., B. J. Soden, V. O. John, W. Ingram, and P. Good, 2010: Current changes in tropical precipitation. *Environ. Res. Lett.*, **5**, 025205.

Allen, M. R., and W. J. Ingram, 2002: Constraints on future changes in climate and the hydrologic cycle. *Nature*, **419**, 224–232.

Allen, M. R., P. A. Stott, J. F. B. Mitchell, R. Schnur, and T. L. Delworth, 2000: Quantifying the uncertainty in forecasts of anthropogenic climate change. *Nature*, **407**, 617–620.

Allen, R., and S. Sherwood, 2011: The impact of natural versus anthropogenic aerosols on atmospheric circulation in the Community Atmosphere Model. *Clim. Dyn.*, **36**, 1959–1978.

Anderson-Teixeira, K., P. Snyder, T. Twine, S. Cuadra, M. Costa, and E. DeLucia, 2012: Climate-regulation services of natural and agricultural ecoregions of the Americas. *Nature Clim. Change*, doi:10.1038/nclimate1346.

Andersson, C., and M. Engardt, 2010: European ozone in a future climate: Importance of changes in dry deposition and isoprene emissions. *J. Geophys. Res.*, **115**, D02303.

Anenberg, S. C., et al., 2012: Global air quality and health co-benefits of mitigating near-term climate change through methane and black carbon emission controls. *Environ. Health Perspect.*, **120**, 831–839.

Annan, J. D., and J. C. Hargreaves, 2010: Reliability of the CMIP3 ensemble. *Geophys. Res. Lett.*, **37**, L02703.

Appelhans, T., A. Sturman, and P. Zawar-Reza, 2012: Synoptic and climatological controls of particulate matter pollution in a Southern Hemisphere coastal city. *Int. J. Climatol., 33, 463-479.*

Arblaster, J. M., and G. A. Meehl, 2006: Contributions of external forcings to southern annular mode trends. *J. Clim.*, **19**, 2896–2905.

Arblaster, J. M., G. A. Meehl, and D. J. Karoly, 2011: Future climate change in the Southern Hemisphere: Competing effects of ozone and greenhouse gases. *Geophys. Res. Lett.*, **38**, L02701.

Avise, J., R. G. Abraham, S. H. Chung, J. Chen, and B. Lamb, 2012: Evaluating the effects of climate change on summertime ozone using a relative response factor approach for policymakers. *J. Air Waste Manage. Assoc.*, **62**, 1061–1074.

Avise, J., J. Chen, B. Lamb, C. Wiedinmyer, A. Guenther, E. Salathé©, and C. Mass, 2009: Attribution of projected changes in summertime US ozone and PM2.5 concentrations to global changes. *Atmos. Chem. Phys.*, **9**, 1111–1124.

Aw, J., and M. J. Kleeman, 2003: Evaluating the first-order effect of intraannual temperature variability on urban air pollution. *J. Geophys. Res.*, **108**, 4365.

Baehr, J., K. Keller, and J. Marotzke, 2008: Detecting potential changes in the meridional overturning circulation at 26°N in the Atlantic. *Clim. Change*, **91**, 11–27.

Baidya Roy, S., and R. Avissar, 2002: Impact of land use/land cover change on regional hydrometeorology in Amazonia,. *J. Geophys. Res.*, **107(D20)**, LBA 4-1-LBA 4-12. DOI: 10.1029/2000JD000266.

Balmaseda, M., and D. Anderson, 2009: Impact of initialization strategies and observations on seasonal forecast skill. *Geophys. Res. Lett.*, **36**, L01701.

Balmaseda, M. A., M. K. Davey, and D. L. T. Anderson, 1995: Decadal and seasonal dependence of ENSO prediction skill. *J. Clim.*, **8**, 2705–2715.

Bates, B. C., Z. W. Kundzewicz, S. Wu, and J. P. Palutikof, 2008: *Climate Change and Water*. Technical Paper of the Intergovernmental Panel on Climate Change. IPCC, 210 pp.

Battisti, D., and E. Sarachik, 1995: Understanding and predicting ENSO. *Rev. Geophys.*, 1367–1376.

Bauer, S. E., D. Koch, N. Unger, S. M. Metzger, D. T. Shindell, and D. G. Streets, 2007: Nitrate aerosols today and in 2030: A global simulation including aerosols and tropospheric ozone. *Atmos. Chem. Phys.*, **7**, 5043–5059.

Bellouin, N., J. G. L. Rae, A. Jones, C. E. Johnson, J. M. Haywood, and O. Boucher, 2011: Aerosol forcing in the CMIP5 simulations by Hadgem2-ES and the role of ammonium nitrate. *J. Geophys. Res. Atmos.*, doi:10.1029/2011JD016074.

Bellucci, A., et al., 2013: Decadal climate predictions with a coupled OAGCM initialized with oceanic reanalyses. *Clim. Dyn.*, **40**, 1483–1497.

Bender, F. A. M., A. M. L. Ekman, and H. Rodhe, 2010: Response to the eruption of Mount Pinatubo in relation to climate sensitivity in the CMIP3 models. *Clim. Dyn.*, **35**, 875–886.

Berg, P., C. Moseley, and J.O. Haerter, 2013: Strong increase in convective precipitation in response to higher temperatures. *Nature Geosci.*, **6**, 181-185, DOI: 10.1038/ngeo1731.

Berner, J., F. J. Doblas-Reyes, T. N. Palmer, G. Shutts, and A. Weisheimer, 2008: Impact of a quasi-stochastic cellular automaton backscatter scheme on the systematic error and seasonal prediction skill of a global climate model. *Philos. Trans. R. Soc. London A*, **366**, 2561–2579.

Betts, R. A., et al., 2007: Projected increase in continental runoff due to plant responses to increasing carbon dioxide. *Nature*, **448**, 1037–1041, DOI 10.1038/nature06045.

Bhend, J., and H. von Storch, 2008: Consistency of observed winter precipitation trends in northern Europe with regional climate change projections. *Clim. Dyn.*, **31**, 17–28.

Bhend, J., and P. Whetton, 2012: Consistency of simulated and observed regional changes in temperature, sea level pressure and precipitation. *Clim. Change*, doi:10.1007/s10584-012-0691-2.

Bintanja, R., G.J. van Oldenborgh, S.S. Drijfhout, B. Wouters, and C. A. Katsman, 2013: Important role for ocean warming and increased ice-shelf melt in Antarctic sea-ice expansion. *Nature Geosci*, **6**, 376–379.

Birner, T., 2010: Recent widening of the tropical belt from global tropopause statistics: Sensitivities. *J. Geophys. Res. Atmos.*, **115**, DOI:10.1029/2010JD014664.

Bitz, C., 2008: Some aspects of uncertainty in predicting sea ice thinning. In: *Arctic Sea Ice Decline: Observations, Projections, Mechanisms, and Implications*. Geophysical Monographs, 180. American Geophysical Union, Washington, DC, pp. 63–76.

Bladé, I., D. Fortuny, G. J. van Oldenborgh, and B. Liebmann, 2012: The summer North Atlantic Oscillation in CMIP3 models and related uncertainties in projected summer drying in Europe. *J. Geophys. Res.*, **116**, D16104.

Bloomer, B. J., J. W. Stehr, C. A. Piety, R. J. Salawitch, and R. R. Dickerson, 2009: Observed relationships of ozone air pollution with temperature and emissions. *Geophys. Res. Lett.*, **36**, L09803.

Boberg, F., and J. H. Christensen, 2012: Overestimation of Mediterranean summer temperature projections due to model deficiencies. *Nature Clim. Change*, **2(6)**, 433–436.

Boe, J. L., A. Hall, and X. Qu, 2009: September sea-ice cover in the Arctic Ocean projected to vanish by 2100. *Nature Geosci.*, **2**, 341–343.

Boer, G. J., 2000: A study of atmosphere-ocean predictability on long time scales. *Clim. Dyn.*, **16**, 469–477.

Boer, G. J., 2004: Long time-scale potential predictability in an ensemble of coupled climate models. *Clim. Dyn.*, **23**, 29–44.

Boer, G. J., 2011: Decadal potential predictability of twenty-first century climate. *Clim. Dyn.*, **36**, 1119–1133.

Boer, G. J., and S. J. Lambert, 2008: Multi-model decadal potential predictability of precipitation and temperature. *Geophys. Res. Lett.*, **35**, L05706.

Boer, G. J., V. V. Kharin, and W. J. Merryfield, 2013: Decadal predictability and forecast skill. *Clim. Dyn.*, doi:10.1007/s00382-013-1705-0.

Boisier, J. P., et al., 2012: Attributing the impacts of land-cover changes in temperate regions on surface temperature and heat fluxes to specific causes: Results from the first LUCID set of simulations. *J. Geophys. Res. Atmos.*, **117**.

Bollasina, M. A., Y. Ming, and V. Ramaswamy, 2011: Anthropogenic aerosols and the weakening of the South Asian summer monsoon. *Science*, doi:10.1126/science.1204994.

Bond, T. C., et al., 2013: Bounding the role of black carbon in the climate system: A scientific assessment. *J. Geophys. Res.*, doi:10.1002/jgrd.50171.

11

Booth, B. B. B., N. J. Dunstone, P. R. Halloran, T. Andrews, and N. Bellouin, 2012: Aerosols implicated as a prime driver of twentieth-century North Atlantic climate variability. *Nature*, **485**, 534–534.

Bounoua, L., F. G. Hall, P. J. Sellers, A. Kumar, G. J. Collatz, C. J. Tucker, and M. L. Imhoff, 2010: Quantifying the negative feedback of vegetation to greenhouse warming: A modeling approach. *Geophys. Res. Lett.*, **27**, L23701.

Brandefelt, J., and H. Kornich, 2008: Northern Hemisphere stationary waves in future climate projections. *J. Clim.*, doi: 10.1175/2008JCLI2373.1, 6341-6353.

Branstator, G., and H. Y. Teng, 2010: Two limits of initial-value decadal predictability in a CGCM. *J. Clim.*, **23**, 6292–6311.

Branstator, G., and H. Y. Teng, 2012: Potential impact of initialization on decadal predictions as assessed for CMIP5 models. *Geophys. Res. Lett.*, **39**, L12703.

Branstator, G., H. Y. Teng, G. A. Meehl, M. Kimoto, J. R. Knight, M. Latif, and A. Rosati, 2012: Systematic estimates of initial-value decadal predictability for six AOGCMs. *J. Clim.*, **25**, 1827–1846.

Brocker, J., and L. A. Smith, 2007: Increasing the reliability of reliability diagrams. *Weather Forecast.*, **22**, 651–661.

Brohan, P., J. J. Kennedy, I. Harris, S. F. B. Tett, and P. D. Jones, 2006: Uncertainty estimates in regional and global observed temperature changes: A new data set from 1850. *J. Geophys. Res. Atmos.*, **111**, DOI 10.1029/2005JD006548.

Brown, R. D., and P. W. Mote, 2009: The response of Northern Hemisphere snow cover to a changing climate. *J. Clim.*, **22**, 2124–2145.

Brunner, D., S. Henne, C. A. Keller, S. Reimann, M. K. Vollmer, S. O'Doherty, and M. Maione, 2012: An extended Kalman-filter for regional scale inverse emission estimation. *Atmos. Chem. Phys.*, **12**, 3455–3478.

Brutel-Vuilmet, C., M. Menegoz, and G. Krinner, 2013: An analysis of present and future seasonal Northern Hemisphere land snow cover simulated by CMIP5 coupled climate models. *Cryosphere*, **7**, 67–80.

Bryan, F. O., G. Danabasoglu, N. Nakashiki, Y. Yoshida, D. H. Kim, J. Tsutsui, and S. C. Doney, 2006: Response of the North Atlantic thermohaline circulation and ventilation to increasing carbon dioxide in CCSM3. *J. Clim.*, **19**, 2382–2397.

Burgman, R., R. Seager, A. Clement, and C. Herweijer, 2010: Role of tropical Pacific SSTs in global medieval hydroclimate: A modeling study. *Geophys. Res. Lett.*, **37**, L06705.

Burke, E., and S. Brown, 2008: Evaluating uncertainties in the projection of future drought. *J. Hydrometeor*, **9**, 292–299.

Buser, C. M., H.R. Künsch, D. Lüthi, M. Wild, and C. Schär, 2009: Bayesian multi-model projection of climate: Bias assumptions and interannual variability. *Clim. Dyn.*, **33(6)**, 849-868, DOI:10.1007/s00382-009-0588-6.

Butchart, N., et al., 2006: Simulations of anthropogenic change in the strength of the Brewer-Dobson circulation. *Clim. Dyn.*, **27**, 727–741.

Butler, T. M., Z. S. Stock, M. R. Russo, H. A. C. Denier van der Gon, and M. G. Lawrence, 2012: Megacity ozone air quality under four alternative future scenarios. *Atmos. Chem. Phys.*, **12**, 4413–4428.

Butterbach-Bahl, K., M. Kahl, L. Mykhayliv, C. Werner, R. Kiese, and C. Li, 2009: A European-wide inventory of soil NO emissions using the biogeochemical models DNDC/Forest-DNDC. *Atmos. Environ.*, **43**, 1392–1402.

Caesar, J., and J. A. Lowe, 2012: Comparing the impacts of mitigation versus non-intervention scenarios on future temperature and precipitation extremes in the HadGEM2 climate model. *J. Geophys. Res.*, **117**, D15109.

Callaghan, J., and S. B. Power, 2011: Variability and decline in the number of severe tropical cyclones making land-fall over eastern Australia since the late nineteenth century. *Clim. Dyn.*, **37**, 647–662.

Callaghan, T. V., M. Johansson, O. Anisimov, H. H. Christiansen, A. Instanes, V. Romanovsky, and S. Smith, 2011: Changing permafrost and its impacts. In: *Snow, Water, Ice and Permafrost in the Arctic (SWIPA)*. Arctic Monitoring and Assessment Program (AMAP).

Cao, L., G. Bala, K. Caldeira, R. Nemani, and G. Ban-Weiss, 2009: Climate response to physiological forcing of carbon dioxide simulated by the coupled Community Atmosphere Model (CAM3.1) and Community Land Model (CLM3.0). *Geophys. Res. Lett.*, **36**, L10402.

Cao, L., G. Bala, K. Caldeira, R. Nemani, and G. Ban-Weiss, 2010: Importance of carbon dioxide physiological forcing to future climate change. *Proc. Natl. Acad. Sci. U.S.A.*, **107**, 9513–9518.

Carlton, A. G., C. Wiedinmyer, and J. H. Kroll, 2009: A review of Secondary Organic Aerosol (SOA) formation from isoprene. *Atmos. Chem. Phys.*, **9**, 4987–5005.

Carlton, A. G., R. W. Pinder, P. V. Bhave, and G. A. Pouliot, 2010: To what extent can biogenic SOA be controlled? *Environ. Sci. Technol.*, **44**, 3376–3380.

Carvalho, A., A. Monteiro, M. Flannigan, S. Solman, A. I. Miranda, and C. Borrego, 2011: Forest fires in a changing climate and their impacts on air quality. *Atmos. Environ.*, **45**, 5545–5553.

Cattiaux, J., P. Yiou, and R. Vautard, 2012: Dynamics of future seasonal temperature trends and extremes in Europe: A multi-model analysis from CMIP. *Clim. Dyn.*, **38(9–10)**, 1949-1964, DOI: 10.1007/s00382-001-1211-1.

Ceppi, P., S. C. Scherrer, A. M. Fischer, and C. Appenzeller, 2012: Revisiting Swiss temperature trends 1959–2008. *Int. J. Climatol.*, **32**, 203–213.

Chalmers, N., E. J. Highwood, E. Hawkins, R. Sutton, and L. J. Wilcox, 2012: Aerosol contribution to the rapid warming of near-term climate under RCP 2.6. *Geophys. Res. Lett.*, **39**, L18709.

Chang, C. Y., J. C. H. Chiang, M. F. Wehner, A. R. Friedman, and R. Ruedy, 2011: Sulfate aerosol control of tropical Atlantic climate over the twentieth century. *J. Clim.*, **24**, 2540–2555.

Chen, J., J. Avise, A. Guenther, C. Wiedinmyer, E. Salathe, R. B. Jackson, and B. Lamb, 2009a: Future land use and land cover influences on regional biogenic emissions and air quality in the United States. *Atmos. Environ.*, **43**, 5771–5780.

Chen, J., et al., 2009b: The effects of global changes upon regional ozone pollution in the United States. *Atmos. Chem. Phys.*, **9**, 1125–1141.

Chevallier, M., and D. Salas-Melia, 2012: The role of sea ice thickness distribution in the Arctic sea ice potential predictability: A diagnostic approach with a coupled GCM. *J. Clim.*, **25**, 3025–3038.

Chikamoto, Y., M. Kimoto, M. Watanabe, M. Ishii, and T. Mochizuki, 2012a: Relationship between the Pacifc and Atlantic stepwise climate change during the 1990s. *Geophys. Res. Lett.*, **39**, L21710.

Chikamoto, Y., et al., 2012b: Predictability of a stepwise shift in Pacific climate during the late 1990s in hindcast experiments using MIROC. *J. Meteorol. Soc. Jpn.*, **90A**, 1–21.

Chin, M., T. Diehl, P. Ginoux, and W. Malm, 2007: Intercontinental transport of pollution and dust aerosols: Implications for regional air quality. *Atmos. Chem. Phys.*, **7**, 5501–5517.

Chou, C., J. Y. Tu, and P. H. Tan, 2007: Asymmetry of tropical precipitation change under global warming. *Geophys. Res. Lett.*, **34**, L17708.

Chou, C., J. D. Neelin, C. A. Chen, and J. Y. Tu, 2009: Evaluating the "rich-get-richer" mechanism in tropical precipitation change under global warming. *J. Clim.*, **22**, 1982–2005.

Christensen, J. H., et al., 2007: Regional climate projections. In: *Climate Change 2007: The Physical Science Basis. Contribution of Working Group I to the Fourth Assessment Report of the Intergovernmental Panel on Climate Change* [Solomon, S., D. Qin, M. Manning, Z. Chen, M. Marquis, K. B. Averyt, M. Tignor and H. L. Miller (eds.)] Cambridge University Press, Cambridge, United Kingdom and New York, NY, USA, pp. 847–940.

Christidis, N., P. A. Stott, G. C. Hegerl, and R. A. Betts, 2013: The role of land use change in the recent warming of daily extreme temperatures. *Geophys. Res. Lett.*, **40**, 589–594.

Chylek, P., C. K. Folland, G. Lesins, and M. K. Dubey, 2010: Twentieth century bipolar seesaw of the Arctic and Antarctic surface air temperatures. *Geophys. Res. Lett.*, **37**, L08703.

Chylek, P., C. K. Folland, G. Lesins, M. K. Dubey, and M. Y. Wang, 2009: Arctic air temperature change amplification and the Atlantic Multidecadal Oscillation. *Geophys. Res. Lett.*, **36**, L14801.

Clark, R. T., J. M. Murphy, and S. J. Brown, 2010: Do global warming targets limit heatwave risk? *Geophys. Res. Lett.*, **37**, L17703.

Colle, B. A., Z. Zhang, K.A. Lombardo, E. Chang, P. Liu, M. Zhang, and S. Hameed, 2013: Historical and future predictions of eastern North America and western Atlantic extratropical cyclones in CMIP5 during the cool Season. *J. Clim.*, doi:10.1175/JCLI-D-12-00498.1.

Collins, M., 2002: Climate predictability on interannual to decadal time scales: The initial value problem. *Clim. Dyn.*, **19**, 671–692.

Collins, M., and B. Sinha, 2003: Predictability of decadal variations in the thermohaline circulation and climate. *Geophys. Res. Lett.*, **30**, 1306.

Collins, M., et al., 2006: Interannual to decadal climate predictability in the North Atlantic: A multimodel-ensemble study. *J. Clim.*, **19**, 1195–1203.

Comiso, J. C., C. L. Parkinson, R. Gersten, and L. Stock, 2008: Accelerated decline in the Arctic Sea ice cover. *Geophys. Res. Lett.*, **35**, L01703.

Cook, B. I., R. L. Miller, and R. Seager, 2009: Amplification of the North American "Dust Bowl" drought through human-induced land degradation. *Proc. Natl Acad. Sci. U.S.A.*, **106**, 4997–5001.

Corti, S., A. Weisheimer, T.N. Palmer, F. J. Doblas-Reyes, and L. Magnusson, 2012: Reliability of decadal predictions. *Geophys. Res. Lett.*, doi:10.1029/2012GL053354.

Cox, P., and D. Stephenson, 2007: Climate change - A changing climate for prediction. *Science*, **317**, 207–208.

Cox, W., and S. Chu, 1996: Assessment of interannual ozone variation in urban areas from a climatological perspective. *Atmos. Environ.*, **30**, 2615–2625.

Cravatte, S., T. Delcroix, D. Zhang, M. McPhaden, and J. Leloup, 2009: Observed freshening and warming of the western Pacific Warm Pool. *Clim. Dyn.*, **33**, 565–589.

Dai, A., 2011: Drought under global warming: A review. *WIREs Clim. Change*, **2**, 45–65.

Davis, S. M., and K. H. Rosenlof, 2012: A Multidiagnostic intercomparison of tropical-width time series using reanalyses and satellite observations. *J. Clim.*, **25**, 1061–1078.

Dawson, J. P., P. N. Racherla, B. H. Lynn, P. J. Adams, and S. N. Pandis, 2009: Impacts of climate change on regional and urban air quality in the eastern United States: Role of meteorology. *J. Geophys. Res.*, **114**, D05308.

de Noblet-Ducoudre, N., et al., 2012: Determining robust impacts of land-use-induced land cover changes on surface climate over North America and Eurasia: Results from the first set of LUCID experiments. *J. Clim.*, **25**, 3261–3281.

DelSole, T., and X. Feng, 2013: The "Shukla–Gutzler" method for estimating potential seasonal predictability. *Mon. Weather Rev.*, **141**, 822–832.

DelSole, T., X. S. Yang, and M. K. Tippett, 2013: Is unequal weighting significantly better than equal weighting for multi-model forecasting? *Q. J. R. Meteorol. Soc.*, **139**, 176–183.

Delworth, T., and K. Dixon, 2006: Have anthropogenic aerosols delayed a greenhouse gas-induced weakening of the North Atlantic thermohaline circulation? *Geophys. Res. Lett.*, doi:10.1029/2005GL024980, L02606.

Delworth, T., V. Ramaswamy, and G. Stenchikov, 2005: The impact of aerosols on simulated ocean temperature and heat content in the 20th century. *Geophys. Res. Lett.*, **32**, L24709, doi: 10.1029/2005GL024457.

Delworth, T. L., and F. Zeng, 2008: Simulated impact of altered Southern Hemisphere winds on the Atlantic Meridional Overturning Circulation. *Geophys. Res. Lett.*, **35**, L20708, doi: 10.1029/2008GL035166.

Dentener, F., et al., 2005: The impact of air pollutant and methane emission controls on tropospheric ozone and radiative forcing: CTM calculations for the period 1990–2030. *Atmos. Chem. Phys.*, **5**, 1731–1755.

Dentener, F., et al., 2006: The global atmospheric environment for the next generation. *Environ. Sci. Technol.*, **40**, 3586–3594.

Dery, S. J., M. A. Hernandez-Henriquez, J. E. Burford, and E. F. Wood, 2009: Observational evidence of an intensifying hydrological cycle in northern Canada. *Geophys. Res. Lett.*, **36**, L13402.

Deser, C., and A. S. Phillips, 2009: Atmospheric circulation trends, 1950–2000: The relative roles of sea surface temperature forcing and direct atmospheric radiative forcing. *J. Clim.*, **22**, 396–413.

Deser, C., A. S. Phillips, and J. W. Hurrell, 2004: Pacific interdecadal climate variability: Linkages between the tropics and the North Pacific during boreal winter since 1900. *J. Clim.*, **17**, 3109–3124.

Deser, C., A. Phillips, V. Bourdette, and H. Y. Teng, 2012: Uncertainty in climate change projections: The role of internal variability. *Clim. Dyn.*, **38**, 527–546.

Dharshana, K. G. T., S. Kravtsov, and J. D. W. Kahl, 2010: Relationship between synoptic weather disturbances and particulate matter air pollution over the United States. *J. Geophys. Res. Atmos.*, **115**, D24219, doi:10.1029/2010JD014852.

Diffenbaugh, N. S., and M. Ashfaq, 2010: Intensification of hot extremes in the United States. *Geophys. Res. Lett.*, **37**, L15701.

Diffenbaugh, N. S., and M. Scherer, 2011: Observational and model evidence of global emergence of permanent, unprecedented heat in the 20th and 21st centuries. *Clim. Change*, **107(3–4)**, 615–624.

DiNezio, P., G. A. Vecchi, and A. Clement, 2013: Detectability of changes in the Walker Circulation in response to global warming. *J. Clim.*, doi:10.1175/JCLI-D-12-00531.1.

DiNezio, P., A. Clement, G. Vecchi, B. Soden, and B. Kirtman, 2009: Climate response of the equatorial Pacific to global warming. *J. Clim.*, doi: 10.1175/2009JCLI2982.1, 4873–4892.

Dlugokencky, E. J., E. G. Nisbet, R. Fisher, and D. Lowry, 2011: Global atmospheric methane: Budget, changes and dangers. *Philos. Trans. R. Soc. London A*, **369**, 2058–2072.

Doblas-Reyes, F. J., M. A. Balmaseda, A. Weisheimer, and T. N. Palmer, 2011: Decadal climate prediction with the ECMWF coupled forecast system: Impact of ocean observations. *J. Geophys. Res. Atmos*, **116**, D19111.

Doblas-Reyes, F. J., et al., 2009: Addressing model uncertainty in seasonal and annual dynamical ensemble forecasts. *Q. J. R. Meteorol. Soc.*, **135**, 1538–1559.

Doblas-Reyes, F. J., et al., 2013: Initialized near-term regional climate change prediction. *Nature Commun.*, **4**, 1715.

Doherty, R., et al., 2009: Current and future climate- and air pollution-mediated impacts on human health. *Environ. Health*, **8**, doi: 10.1186/1476-069X-8-S1-S8.

Doherty, R. M., et al., 2013: Impacts of climate change on surface ozone and intercontinental ozone pollution: A multi-model study. *J. Geophys. Res. Atmos.*, doi:10.1002/jgrd.50266.

Drijfhout, S. S., and W. Hazeleger, 2007: Detecting Atlantic MOC changes in an ensemble of climate change simulations. *J. Clim.*, **20**, 1571–1582.

Du, H., F. J. Doblas-Reyes, J. Garcia-Serrano, V. Guemas, Y. Soufflet, and B. Wouters, 2012: Sensitivity of decadal predictions to the initial atmospheric and oceanic perturbations. *Clim. Dyn.*, **39**, 2013–2023.

Dunstone, N. J., and D. M. Smith, 2010: Impact of atmosphere and sub-surface ocean data on decadal climate prediction. *Geophys. Res. Lett.*, **37**, L02709.

Dunstone, N. J., D. M. Smith, and R. Eade, 2011: Multi-year predictability of the tropical Atlantic atmosphere driven by the high latitude North Atlantic Ocean. *Geophys. Res. Lett.*, **38**, L14701.

Durack, P. J., and S. E. Wijffels, 2010: Fifty-year trends in global ocean salinities and their relationship to broad-scale warming. *J. Clim.*, **23**, 4342–4362.

Durack, P. J., S. E. Wijffels, and R. J. Matear, 2012: Ocean salinities reveal strong global water cycle intensification during 1950 to 2000. *Science*, **336**, 455-458, doi: 10.1126/science.1212222.

Dutra, E., C. Schar, P. Viterbo, and P. M. A. Miranda, 2011: Land-atmosphere coupling associated with snow cover. *Geophys. Res. Lett.*, **38**, L15707.

Eade, R., E. Hamilton, D. M. Smith, R. J. Graham, and A. A. Scaife, 2012: Forecasting the number of extreme daily events out to a decade ahead. *J. Geophys. Res.*, **117**, D21110, doi:10.1029/2012JD018015.

Easterling, D. R., and M. F. Wehner, 2009: Is the climate warming or cooling? *Geophys. Res. Lett.*, **36**, L08706.

El Nadi, A. H., 1974: The significance of leaf area in evapotranspiration. *Ann. Bot*, **38(3)**, 607–611.

Emanuel, K., 2011: Global warming effects on U.S. hurricane damage. *Weather Clim. Soc.*, **3**, 261–268.

Evan, A. T., D. J. Vimont, A. K. Heidinger, J. P. Kossin, and R. Bennartz, 2009: The role of aerosols in the evolution of tropical North Atlantic Ocean Temperature anomalies. *Science*, **324**, 778–781.

Eyring, V., et al., 2013: Long-term changes in tropospheric and stratospheric ozone and associated climate impacts in CMIP5 simulations. *J. Geophys., Res.*, **118**. 5029-5060, doi:10.1002/jgrd.50316.

Eyring, V., et al., 2010: Multi-model assessment of stratospheric ozone return dates and ozone recovery in CCMVal-2 models. *Atmos. Chem. Phys.*, **10**, 9451–9472.

Fairlie, T. D., D. J. Jacob, and R. J. Park, 2007: The impact of transpacific transport of mineral dust in the United States. *Atmos. Environ.*, **41**, 1251–1266.

Fang, Y., et al., 2011: The impacts of changing transport and precipitation on pollutant distributions in a future climate. *J. Geophys. Res.*, **116**, D18303.

Fasullo, J. T., 2010: Robust land-ocean contrasts in energy and water cycle feedbacks. *J. Clim.*, **23**, 4677–4693.

Ferro, C. A. T., and T. E. Fricker, 2012: A bias-corrected decomposition of the Brier score. *Q. J. R. Meteorol. Soc.*, **138**, 1954–1960.

Feulner, G., and S. Rahmstorf, 2010: On the effect of a new grand minimum of solar activity on the future climate on Earth. *Geophys. Res. Lett.*, **37**, L05707, doi: 10.1029/2010GL042710.

Field, C. B., R. B. Jackson, and H. A. Mooney, 1995: Stomatal responses to increased CO_2—Implications from the plant to the global-scale. *Plant Cell Environ.*, **18**, 1214–1225.

Findell, K. L., E. Shevliakova, P. C. D. Milly, and R. J. Stouffer, 2007: Modeled impact of anthropogenic land cover change on climate. *J. Clim.*, **20**, 3621–3634.

Fiore, A. M., J. J. West, L. W. Horowitz, V. Naik, and M. D. Schwarzkopf, 2008: Characterizing the tropospheric ozone response to methane emission controls and the benefits to climate and air quality. *J. Geophys. Res.*, **113**, D08307.

Fiore, A. M., D. J. Jacob, B. D. Field, D. G. Streets, S. D. Fernandes, and C. Jang, 2002: Linking ozone pollution and climate change: The case for controlling methane. *Geophys. Res. Lett.*, **29**, 1919.

11

Fiore, A. M., et al., 2012: Global air quality and climate. *Chem. Soc. Rev.*, **41**, 6663–6683.

Fiore, A. M., et al., 2009: Multimodel estimates of intercontinental source-receptor relationships for ozone pollution. *J. Geophys. Res.*, **114**, D04301.

Fischer, E. M., and C. Schar, 2009: Future changes in daily summer temperature variability: Driving processes and role for temperature extremes. *Clim. Dyn.*, **33**, 917–935.

Fischer, E. M., and C. Schar, 2010: Consistent geographical patterns of changes in high-impact European heatwaves. *Nature Geosci.*, **3**, 398–403.

Fischer, E. M., J. Rajczak, and C. Schär, 2012: Changes in European summer temperature variability revisited. *Geophys. Res. Lett.*, **6**, L19702.

Fischer, E. M., S. I. Seneviratne, P. L. Vidale, D. Luthi, and C. Schar, 2007: Soil moisture–atmosphere interactions during the 2003 European summer heat wave. *J. Clim.*, **20**, 5081–5099.

Flanner, M. G., C. S. Zender, J. T. Randerson, and P. J. Rasch, 2007: Present-day climate forcing and response from black carbon in snow. *J. Geophys. Res.*, **112**, D11202, doi: 10.1029/2006JD008003.

Flannigan, M. D., M. A. Krawchuk, W. J. de Groot, B. M. Wotton, and L. M. Gowman, 2009: Implications of changing climate for global wildland fire. *Int. J. Wildland Fire*, **18**, 483–507.

Fleming, E., C. Jackman, R. Stolarski, and A. Douglass, 2011: A model study of the impact of source gas changes on the stratosphere for 1850–2100. *Atmos. Chem. Phys.*, **11**, 8515–8541.

Fogt, R. L., J. Perlwitz, A. J. Monaghan, D. H. Bromwich, J. M. Jones, and G. J. Marshall, 2009: Historical SAM variability. Part II: Twentieth-century variability and trends from reconstructions, observations, and the IPCC AR4 Models. *J. Clim.*, **22**, 5346–5365.

Folland, C., J. Knight, H. Linderholm, D. Fereday, S. Ineson, and J. Hurrell, 2009: The summer North Atlantic Oscillation: Past, present, and future. *J. Clim.*, doi: 10.1175/2008JCLI2459.1, 1082–1103.

Folland, C. K., A.W. Colman, D.M. Smith, O. Boucher, D. E. Parker, and J.-P. Vernier, 2013: High predictive skill of global surface temperature a year ahead. *Geophys. Res. Lett.*, **40**, 761–767.

Forkel, R. and R. Knoche, 2006: Regional climate change and its impact on photooxidant concentrations in

southern Germany: Simulations with a coupled regional climate-chemistry model. J. Geophys. Res., 2006, 111, D12302.

Fowler, H. J., M. Ekstrom, S. Blenkinsop, and A. P. Smith, 2007: Estimating change in extreme European precipitation using a multimodel ensemble. *J. Geophys. Res. Atmos.*, **112**, D18104, doi: 10.1029/2007JD008619.

Francis, J. A. H., E, 2007: Changes in the fabric of the Arctic's greenhouse blanket. *Environ. Res. Lett.*, doi:10.1088/1748-9326/2/4/045011.

Fyfe, J. C., N. P. Gillett, and G. J. Marshall, 2012: Human influence on extratropical Southern Hemisphere summer precipitation. *Geophys. Res. Lett.*, **39**, L23711.

Fyfe, J. C., W. J. Merryfield, V. Kharin, G. J. Boer, W. S. Lee, and K. von Salzen, 2011: Skillful predictions of decadal trends in global mean surface temperature. *Geophys. Res. Lett.*, **38**, L22801.

Gaetani, M., and E. Mohino, 2013: Decadal prediction of the Sahelian precipitation in CMIP5 simulations. *J. Clim.*, doi:10.1175/JCLI-D-12-00635.1.

Gangstø, R., A. P. Weigel, M. A. Liniger, and C. Appenzeller, 2013: Comments on the evaluation of decadal predictions. *Clim. Res.*, **55**, 181–200.

Ganguly, D., P. J. Rasch, H. Wang, and J.-H. Yoon, 2012: Climate response of the South Asian monsoon system to anthropogenic aerosols. *J. Geophys. Res.*, **117**, D13209.

Gao, C. C., A. Robock, and C. Ammann, 2008: Volcanic forcing of climate over the past 1500 years: An improved ice core-based index for climate models. *J. Geophys. Res. Atmos.*, **113**, D23111, doi: 10.1029/2008JD010239.

Garcia-Serrano, J., and F. J. Doblas-Reyes, 2012: On the assessment of near-surface global temperature and North Atlantic multi-decadal variability in the ENSEMBLES decadal hindcast. *Clim. Dyn.*, **39**, 2025–2040.

Garcia-Serrano, J., F. J. Doblas-Reyes, and C. A. S. Coelho, 2012: Understanding Atlantic multi-decadal variability prediction skill. *Geophys. Res. Lett.*, **39**, L18708, doi:10.1029/2012GL053283.

Gastineau, G., L. Li, and H. Le Treut, 2009: The Hadley and Walker Circulation changes in global warming conditions described by idealized atmospheric simulations. *J. Clim.*, **22**, 3993–4013.

Georgescu, M., D. B. Lobell, and C. B. Field, 2009: Potential impact of U.S. biofuels on regional climate. *Geophys. Res. Lett.*, **36**, L21806.

Georgescu, M., D. B. Lobell, C. B. Field, and A. Mahalov, 2013: Simulated hydroclimatic impacts of projected Brazilian sugarcane expansion. *Geophys. Res. Lett.*, **40**, 972-977, doi:10.1002/grl,50206.

Gillett, N., R. Allan, and T. Ansell, 2005: Detection of external influence on sea level pressure with a multi-model ensemble. *Geophys. Res. Lett.*, **32**, L19714, doi: 10.1029/2005GL023640.

Gillett, N., V. Arora, D. Matthews, and M. Allen, 2013: Constraining the ratio of global warming to cumulative CO_2 emissions using CMIP5 simulations. *J. Clim.*, doi:10.1175/JCLI-D-12-00476.1.

Gillett, N. P., and D. W. J. Thompson, 2003: Simulation of recent Southern Hemisphere climate change. *Science*, **302**, 273–275.

Gillett, N. P., and P. A. Stott, 2009: Attribution of anthropogenic influence on seasonal sea level pressure. *Geophys. Res. Lett.*, **36**, L23709.

Goddard, L., et al., 2013: A verification framework for interannual-to-decadal predictions experiments. *Clim. Dyn.*, **40**, 245–272.

Goldenberg, S. B., C. W. Landsea, A. M. Mestas-Nunez, and W. M. Gray, 2001: The recent increase in Atlantic hurricane activity: Causes and implications. *Science*, **293**, 474–479.

Gosling, S. N., R. G. Taylor, N. W. Arnell, and M. C. Todd, 2011: A comparative analysis of projected impacts of climate change on river runoff from global and catchment-scale hydrological models. *Hydrol. Earth Syst. Sci.*, **15**, 279–294.

Granier, C., et al., 2006: Ozone pollution from future ship traffic in the Arctic northern passages. *Geophys. Res. Lett.*, **33**, L13807.

Gray, L., et al., 2010: Solar Influences on climate. *Rev. Geophys.*, **48**, RG4001, doi: 10.1029/2009/RG000282.

Gregory, J., 2010: Long-term effect of volcanic forcing on ocean heat content. *Geophys. Res. Lett.*, doi:10.1029/2010GL045507, L22701.

Gregory, J. M., and J. F. B. Mitchell, 1995: Simulation of daily variability of surface-temperature and precipitation over Europe in the current and 2xco(2) climates using the Ukmo Climate Model. *Q. J. R. Meteorol. Soc.*, **121**, 1451–1476.

Gregory, J. M., and P. M. Forster, 2008: Transient climate response estimated from radiative forcing and observed temperature change. *J. Geophys. Res. Atmos.*, **113**, D23105, doi:10.1029/2008JD010405.

Griffies, S. M., and K. Bryan, 1997: A predictability study of simulated North Atlantic multidecadal variability. *Clim. Dyn.*, **13**, 459–487.

Grotzner, A., M. Latif, A. Timmermann, and R. Voss, 1999: Interannual to decadal predictability in a coupled ocean-atmosphere general circulation model. *J. Clim.*, **12**, 2607–2624.

Grousset, F. E., P. Ginoux, A. Bory, and P. E. Biscaye, 2003: Case study of a Chinese dust plume reaching the French Alps. *Geophys. Res. Lett.*, **30**, L22701.

Guemas, V., F. J. Doblas-Reyes, I. Andreu-Burillo, and M. Asif, 2013: Retrospective prediction of the global warming slowdown in the last decade. *Nature Clim. Change*, doi:10.1038/nclimate1863.

Guémas, V., F.J. Doblas-Reyes, F. Lienert, Y. Soufflet, and H. Du, 2012: Identifying the causes of the poor decadal climate prediction skill over the North Pacific. *J. Geophys. Res.*, **117**, D20111.

Guenther, A., T. Karl, P. Harley, C. Wiedinmyer, P. I. Palmer, and C. Geron, 2006: Estimates of global terrestrial isoprene emissions using MEGAN (Model of Emissions of Gases and Aerosols from Nature). *Atmos. Chem. Phys.*, **6**, 3181–3210.

Guo, D. L., and H. Wang, 2012: A projection of permafrost degradation on the Tibetan Plateau during the 21st century. *J. Geophys. Res.*, 117, D05106, doi:10.1029/2011JD016545.

Gutowski, W. J., K. A. Kozak, R. W. Arritt, J. H. Christensen, J. C. Patton, and E. S. Takle, 2007: A possible constraint on regional precipitation intensity changes under global warming. *J. Hydrometeorol.*, **8**, 1382–1396.

Gutowski, W. J., et al., 2008: Causes of observed changes in extremes and projections of future changes. In: *Weather and Climate Extremes in a Changing Climate. Regions of Focus: North America, Hawaii, Caribbean, and U.S. Pacific Islands* [T. R. Karl, G. A. Meehl, D. M. Christopher, S. J. Hassol, A. M. Waple and W. L. Murray (eds.)]. U.S. Climate Change Science Program and the Subcommittee on Global Change Research.

Haarsma, R. J., and F. M. Selten, 2012: Anthropogenic changes in the Walker Circulation and their impact on the extra-tropical planetary wave structure in the Northern Hemisphere. *Clim. Dyn.*, doi: 10.1007/s00382-012-1308-1.

Haigh, J., A. Winning, R. Toumi, and J. Harder, 2010: An influence of solar spectral variations on radiative forcing of climate. *Nature*, **467**, 696–699.

Haigh, J. D., 1996: The impact of solar variability on climate. *Science*, **272**, 981–984.

Haigh, J. D., M. Blackburn, and R. Day, 2005: The response of tropospheric circulation to perturbations in lower-stratospheric temperature. *J. Clim.*, **18**, 3672–3685.

Hallquist, M., et al., 2009: The formation, properties and impact of secondary organic aerosol: Current and emerging issues. *Atmos. Chem. Phys.*, **9**, 5155–5236.

Hanel, M., and T. A. Buishand, 2011: Analysis of precipitation extremes in an ensemble of transient regional climate model simulations for the Rhine basin. *Clim. Dyn.*, **36**, 1135–1153.

Hanlon, H. M., G. C. Hegerl, S. F. B. Tett, and D. M. Smith, 2013: Can a decadal forecasting system predict temperature extreme indices? *J. Clim.*, doi:10.1175/JCLI-D-12-00512.1.

Hansen, J., A. Lacis, R. Ruedy, and M. Sato, 1992: Potential climate impact of Mount-Pinatubo eruption. *Geophys. Res. Lett.*, **19**, 215–218.

Hansen, J., R. Ruedy, M. Sato, and K. Lo, 2010: Global surface temperature change. *Rev. Geophys.*, **48**.

Hansen, J., M. Sato, R. Ruedy, A. Lacis, and V. Oinas, 2000: Global warming in the twenty-first century: An alternative scenario. *Proc. Natl. Acad. Sci. U.S.A.*, **97**, 9875–9880.

Hansen, J., M. Sato and R. Ruedy, 2012: Perception of climate change. *Proc. Natl. Acad. Sci. U.S.A.*, **109(37)**, E2415–E2423.

Harder, J., J. Fontenla, P. Pilewskie, E. Richard, and T. Woods, 2009: Trends in solar spectral irradiance variability in the visible and infrared. *Geophys. Res. Lett.*, **36**, L07801, doi: 10.1029/2008GL036797.

Hawkins, E., and R. Sutton, 2009: The potential to narrow uncertainty in regional climate predictions. *Bull. Am. Meteorol. Soc.*, 90, 1095-1107, doi: 10.1175/2009BAMS2607.1.

Hawkins, E., and R. Sutton, 2011: The potential to narrow uncertainty in projections of regional precipitation change. *Clim. Dyn.*, **37**, 407–418.

Hawkins, E., and R. Sutton, 2012: Time of emergence of climate signals. *Geophys. Res. Lett.*, doi:10.1029/2011GL050087.

Hawkins, E., J. Robson, R. Sutton, D. Smith, and N. Keenlyside, 2011: Evaluating the potential for statistical decadal predictions of SSTs with a perfect model approach. *Clim. Dyn.*, **37**, 2495.

Hazeleger, W., et al., 2013a: Multiyear climate predictions using two initialisation strategies. *Geophys. Res. Lett.*, doi::10.1002/grl.50355.

Hazeleger, W., et al., 2013b: Predicting multi-year North Atlantic Ocean variability. *J. Geophys. Res.*, doi:10.1002/grl.50355.

Heald, C. L., et al., 2008: Predicted change in global secondary organic aerosol concentrations in response to future climate, emissions, and land use change. *J. Geophys. Res.*, **113**, D05211.

Hedegaard, G. B., J. Brandt, J. H. Christensen, L. M. Frohn, C. Geels, K. M. Hansen, and M. Stendel, 2008: Impacts of climate change on air pollution levels in the Northern Hemisphere with special focus on Europe and the Arctic. *Atmos. Chem. Phys.*, **8**, 3337–3367.

Heinrich, G., and A. Gobiet, 2011: The future of dry and wet spells in Europe: A comprehensive study based on the ENSEMBLES regional climate models. *Int. J. Climatol*, doi:658 10.1002/joc.2421.

Held, I., and B. Soden, 2006: Robust responses of the hydrological cycle to global warming. *J. Clim.*, 5686–5699.

Henze, D. K., et al., 2012: Spatially Refined Aerosol Direct Radiative Forcing Efficiencies. *Environ. Sci. Technol.*, **46**, 9511–9518.

Hermanson, L., and R. T. Sutton, 2010: Case studies in interannual to decadal climate predictability. *Clim. Dyn.*, **35**, 1169–1189.

Hezel, P. j., X. Zhang, C.M. Bitz, and B. P. Kelly, 2012: Projected decline in snow depth on Arctic sea ice casued by progressively later autumn open ocean freeze-up this century. *Geophys. Res. Lett.*, **39**, L17505, doi:10.1029/2012GL052794.

Ho, C. K., Hawkins, Shaffrey, and Underwood, 2012a: Statistical decadal predictions for sea surface temperatures: A benchmark for dynamical GCM predictions. *Clim. Dyn.*, doi:10.1007/s00382-012-1531-9.

Ho, C. K., D. B. Stephenson, M. Collins, C. A. T. Ferro, and S. J. Brown, 2012b: Calibration strategies: A source of additional uncertainty in climate change projections. *Bull. Am. Meteorol. Soc.*, **93**, 21–26.

Hodnebrog, Ø., et al., 2012: Impact of forest fires, biogenic emissions and high temperatures on the elevated Eastern Mediterranean ozone levels during the hot summer of 2007. *Atmos. Chem. Phys.*, **12**, 8727–8750.

Hodson, D. L. R., S.P.E. Keeley, A. West, J. Ridley, E. Hawkins, and H. T. Hewitt, 2012: Identifying uncertainties in Arctic climate change projections. *Clim. Dyn.*, doi:10.1007/s00382-012-1512-z.

Hoerling, M., et al., 2011: On North American decadal climate for 2011–20. *J. Clim.*, **24**, 4519–4528.

Hogrefe, C., et al., 2004: Simulating changes in regional air pollution over the eastern United States due to changes in global and regional climate and emissions. *J. Geophys. Res.*, **109**, D22301.

Hohenegger, C., P. Brockhaus, C. S. Bretherton, and C. Schar, 2009: The soil moisture-precipitation feedback in simulations with explicit and parameterized convection. *J. Clim.*, **22**, 5003–5020.

Holland, M. M., J. Finnis, and M. C. Serreze, 2006: Simulated Arctic Ocean freshwater budgets in the twentieth and twenty-first centuries. *J. Clim.*, **19**, 6221–6242.

Holland, M. M., M.C. Serreze, and J. Stroeve, 2010: The sea ice mass budget of the Arctic and its future change as simulated by coupled climate models. *Clim. Dyn.*, **34**, 185-200, doi: 10.1007/s00382-008-0493-4.

Holland, M. M., J. Finnis, A. P. Barrett, and M. C. Serreze, 2007: Projected changes in arctic ocean freshwater budgets. *J. Geophys. Res.*, **112**, G04S55, doi: 10.1029/2006/JG000354.

Holmes, C. D., M. J. Prather, O. A. Søvde, and G. Myhre, 2013: Future methane, hydroxyl, and their uncertainties: Key climate and emission parameters for future predictions. *Atmos Chem Phys*, **13**, 285–302.

Hoyle, C. R., et al., 2011: A review of the anthropogenic influence on biogenic secondary organic aerosol. *Atmos. Chem. Phys.*, **11**, 321–343.

Hsu, J., and M. Prather, 2010: Global long-lived chemical modes excited in a 3-D chemistry transport model: Stratospheric N_2O, NO_y, O_3 and CH_4 chemistry. *Geophys. Res. Lett.*, **37**, L07805.

HTAP, 2010a: *Hemispheric Transport of Air Pollution 2010, Part A: Ozone and Particulate Matter*. Air Pollution Studies No. 17. United Nations, New York, NY, USA, and Geneva, Swtzerland, 278 pp.

HTAP, 2010a, 2010b: *Hemispheric Transport of Air Pollution 2010, Part C: Persistent Organic Pollutants*. Air Pollution Studies No. 19. United Nations, New York, NY, USA, and Geneva, Switzerland, 278 pp.

HTAP, 2010a, 2010c: *Hemispheric Transport of Air Pollution 2010, Part B: Mercury*. Air Pollution Studies No. 18. United Nations, New York, NY, USA, and Geneva, Switzerland, 278 pp.

Hu, Y., L. Tao, and J. Liu, 2013: Poleward expansion of the Hadley Circulation in CMIP5 simulations. *Adv. Atmos. Sci.*, **30**, 790–795.

Huang, H.-C., et al., 2008: Impacts of long-range transport of global pollutants and precursor gases on U.S. air quality under future climatic conditions. *J. Geophys. Res.*, **113**, D19307.

Huebener, H., U. Cubasch, U. Langematz, T. Spangehl, F. Niehorster, I. Fast, and M. Kunze, 2007: Ensemble climate simulations using a fully coupled ocean-troposphere-stratosphere general circulation model. *Philos. Trans. R. Soc. London A*, doi: 10.1098/rsta.2007.2078, 2089-2101.

Hungate, B. A., et al., 2002: Evapotranspiration and soil water content in a scrub-oak woodland under carbon dioxide enrichment. *Global Change Biol.*, **8**, 289–298.

Huntington, T. G., 2006: Evidence for intensification of the global water cycle: Review and synthesis. *J. Hydrol.*, **319**, 83–95.

Hurtt, G. C., et al., 2011: Harmonization of land-use scenarios for the period 1500–2100: 600 years of global gridded annual land-use transitions, wood harvest, and resulting secondary lands. *Clim. Change*, **109**, 117–161.

Huszar, P., et al., 2011: Effects of climate change on ozone and particulate matter over Central and Eastern Europe. *Clim. Res.*, **50**, 51–68.

Ihara, C., and Y. Kushnir, 2009: Change of mean midlatitude westerlies in 21st century climate simulations. *Geophys. Res. Lett.*, doi:10.1029/2009GL037674, L13701.

Im, E. S., W. J. Gutowski, and F. Giorgi, 2008: Consistent changes in twenty-first century daily precipitation from regional climate simulations for Korea using two convection parameterizations. *Geophys. Res. Lett.*, **35**, L14706.

Ineson, S., A. Scaife, J. Knight, J. Manners, N. Dunstone, L. Gray, and J. Haigh, 2011: Solar forcing of winter climate variability in the Northern Hemisphere. *Nature Geosci.*, **4**, 753–757.

ICPO, 2011: *Decadal and Bias Correction for Decadal Climate Predictions*. CLIVAR Publication Series No.150, International CLIVAR Project Office. 6 pp.

IPCC, 2007: *Climate Change 2007: The Physical Science Basis. Contribution of Working Group I to the Fourth Assessment Report of the Intergovernmental Panel on Climate Change* [Solomon, S., D. Qin, M. Manning, Z. Chen, M. Marquis, K. B. Averyt, M. Tignor and H. L. Miller (eds.)] Cambridge University Press, Cambridge, United Kingdom and New York, NY, USA, 996 pp.

Isaksen, I. S. A., et al., 2009: Atmospheric composition change: Climate-chemistry interactions. *Atmos. Environ.*, **43**, 5138–5192.

Ishii, M., and M. Kimoto, 2009: Reevaluation of historical ocean heat content variations with an XBT depth bias correction. *J. Oceanogr.*, **65**, 287–299.

Ishii, M., M. Kimoto, K. Sakamoto, and S. Iwasaki, 2006: Steric sea level changes estimated from historical ocean subsurface temperature and salinity analyses. *J. Oceanogr.*, **62**, 155–170.

Ito, A., S. Sillman, and J. E. Penner, 2009: Global chemical transport model study of ozone response to changes in chemical kinetics and biogenic volatile organic compounds emissions due to increasing temperatures: Sensitivities to isoprene nitrate chemistry and grid resolution. *J. Geophys. Res.*, **114**, D09301.

Jacob, D. J., and D. A. Winner, 2009: Effect of climate change on air quality. *Atmos. Environ.*, **43**, 51–63.

Jacob, D. J., J. A. Logan, and P. P. Murti, 1999: Effect of rising Asian emissions on surface ozone in the United States. *Geophys. Res. Lett.*, **26**, 2175–2178.

Jacob, D. J., et al., 1993: Factors regulating ozone over the United-States and its export to the global atmosphere. *J. Geophys. Res. Atmos.*, **98**, 14817–14826.

Jacobson, M., 2008: Effects of wind-powered hydrogen fuel cell vehicles on stratospheric ozone and global climate. *Geophys. Res. Lett.*, doi:10.1029/2008GL035102, L14706.

Jacobson, M., 2010: Short-term effects of controlling fossil-fuel soot, biofuel soot and gases, and methane on climate, Arctic ice, and air pollution health. *J. Geophys. Res.*, D14209, doi:10.1029/2009JD013795.

Jacobson, M., and D. Streets, 2009: Influence of future anthropogenic emissions on climate, natural emissions, and air quality. *J. Geophys. Res.*, D08118, doi:10.1029/2008JD011476.

Jaffe, D. A., and N. L. Wigder, 2012: Ozone production from wildfires: A critical review. *Atmos. Environ.*, **51**, 1–10.

Jai, L., and T. DelSole, 2012: Multi-year predictability of temperature and precipitation in multiple climate models. *Geophys. Res. Lett.*, **39**, L17705.

Jiang, X., Z.-L. Yang, H. Liao, and C. Wiedinmyer, 2010: Sensitivity of biogenic secondary organic aerosols to future climate change at regional scales: An online coupled simulation. *Atmos. Environ.*, **44**, 4891–4907.

Jiang, X., C. Wiedinmyer, F. Chen, Z.-L. Yang, and J. C.-F. Lo, 2008: Predicted impacts of climate and land use change on surface ozone in the Houston, Texas, area. *J. Geophys. Res.*, **113**, D20312.

Jickells, T. D., et al., 2005: Global iron connections between desert dust, ocean biogeochemistry, and climate. *Science*, **308**, 67–71.

Joetzjer, E., H. Douville, C. Delire, and P. Ciais, 2012: Present-day and future Amazonian precipitation in global climate models: CMIP5 versus CMIP3. *Clim. Dyn.*, doi:10.1007/s00382-012-1644-1.

John, J. G., A. M. Fiore, V. Naik, L. W. Horowitz, and J. P. Dunne, 2012: Climate versus emission drivers of methane lifetime from 1860–2100. *Atmos. Chem. Phys.*, **12**, 12021-12036, doi: 10.5194/acp-12-12021-2012.

Johnson, C. E., W. J. Collins, D. S. Stevenson, and R. G. Derwent, 1999: Relative roles of climate and emissions changes on future tropospheric oxidant concentrations. *J. Geophys. Res.*, **104**, 18631–18645.

Johnson, N. C., and S. P. Xie, 2010: Changes in the sea surface temperature threshold for tropical convection. *Nature Geosci.*, **3**, 842–845.

Jolliffe, I. T., 2007: Uncertainty and inference for verification measures. *Weather Forecast.*, **22**, 637–650.

Jolliffe, I. T., and D. B. Stephenson, 2011: *Forecast Verification: A Practitioner's Guide in Atmospheric Science,* 2nd ed. John Wiley & Sons, Hoboken, NJ, USA.

Jones, G., M. Lockwood, and P. Stott, 2012: What influence will future solar activity changes over the 21st century have on projected global near-surface temperature changes? *J. Geophys. Res.*, **117**, D05103, doi.1029/2011JD17013.

Joshi, M., E. Hawkins, R. Sutton, J. Lowe, and D. Frame, 2011: Projections of when temperature change will exceed 2 degrees C above pre-industrial levels. *Nature Clim. Change*, **1**, 407–412.

Jung, M., et al., 2010: Recent decline in the global land evapotranspiration trend due to limited moisture supply. *Nature*, **467**, 951–954.

Kang, S. M., L. M. Polvani, J. C. Fyfe, and M. Sigmond, 2011: Impact of polar ozone depletion on subtropical precipitation. *Science*, **332**, 951–954.

Kao, S. C., and A. R. Ganguly, 2011: Intensity, duration, and frequency of precipitation extremes under 21st-century warming scenarios. *J. Geophys. Res.*, **116**, D16119, doi:10.1029/2010JD015529.

Katragkou, E., P. Zanis, I. Kioutsioukis, I. Tegoulias, D. Melas, B. C. Kruger, and E. Coppola, 2011: Future climate change impacts on summer surface ozone from regional climate-air quality simulations over Europe. *J. Geophys. Res.*, **116**, D22307, doi:10.1029/2011JD015899.

Kawase, H., T. Nagashima, K. Sudo, and T. Nozawa, 2011: Future changes in tropospheric ozone under Representative Concentration Pathways (RCPs). *Geophys. Res. Lett.*, **38**, L05801.

Keenlyside, N. S., and J. Ba, 2010: Prospects for decadal climate prediction. *WIREs Clim. Change*, **1**, 627–635.

Keenlyside, N. S., M. Latif, J. Jungclaus, L. Kornblueh, and E. Roeckner, 2008: Advancing decadal-scale climate prediction in the North Atlantic sector. *Nature*, **453**, 84–88.

Keller, C. A., D. Brunner, S. Henne, M. K. Vollmer, S. O'Doherty, and S. Reimann, 2011: Evidence for under-reported western European emissions of the potent greenhouse gas HFC-23. *Geophys. Res. Lett.*, **38**, L15808.

Kelly, J., P. A. Makar, and D. A. Plummer, 2012: Projections of mid-century summer air-quality for North America: Effects of changes in climate and precursor emissions. *Atmos Chem Phys*, **12**, 5367–5390.

Keppenne, C. L., M. M. Rienecker, N. P. Kurkowski, and D. A. Adamec, 2005: Ensemble Kalman filter assimilation of temperature and altimeter data with bias correction and application to seasonal prediction. *Nonlin. Process. Geophys.*, **12**, 491–503.

Kesik, M., et al., 2006: Future scenarios of N2O and NO emissions from European forest soils. *J. Geophys., Res.*, **111**, G02018.

Kharin, V. V., G. J. Boer, W. J. Merryfield, J. F. Scinocca, and W. S. Lee, 2012: Statistical adjustment of decadal predictions in a changing climate. *Geophys. Res. Lett.*, **39**, L19705.

Kim, H. M., P. J. Webster, and J. A. Curry, 2012: Evaluation of short-term climate change prediction in multi-model CMIP5 decadal hindcasts. *Geophys. Res. Lett.*, **39**, L10701.

Kleeman, M. J., 2008: A preliminary assessment of the sensitivity of air quality in California to global change. *Clim. Change*, **87(Suppl 1)**, S273–S292.

Kleeman, R., Y. M. Tang, and A. M. Moore, 2003: The calculation of climatically relevant singular vectors in the presence of weather noise as applied to the ENSO problem. *J. Atmos. Sci.*, **60**, 2856–2868.

Klimont, Z., S. J. Smith, and J. Cofala, 2013: The last decade of global anthropogenic sulfur dioxide: 2000–2011 emissions. *Environ. Res. Lett.*, **8**, 014003, doi:10.1088/1748-9326/8/1/014003.

Kloster, S., F. Dentener, J. Feichter, F. Raes, U. Lohmann, E. Roeckner, and I. Fischer-Bruns, 2010: A GCM study of future climate response to aerosol pollution reductions. *Clim. Dyn.*, **34**, 1177–1194.

Kloster, S., et al., 2008: Influence of future air pollution mitigation strategies on total aerosol radiative forcing. *Atmos. Chem. Phys.*, **8**, 6405–6437.

Knight, J., R. Allan, C. Folland, M. Vellinga, and M. Mann, 2005: A signature of persistent natural thermohaline circulation cycles in observed climate. *Geophys. Res. Lett.*, doi:10.1029/2005GL024233, L20708.

Knutson, T., and S. Manabe, 1995: Time-mean response over the tropical Pacific to increased CO_2 in a coupled ocean-atmosphere model. *J. Clim.*, **8**, 2181–2199.

Knutson, T.R., and coauthors, 2013a: Dynamical Downscaling Projections of Late 21st Century Atlantic Hurricane Activity CMIP3 and CMIP5 Model-based Scenarios. *J. Climate,* doi:10.1175/JCLI-D-12-00539.1

Knutson, T. R., F. Zeng, and A. T. Wittenberg 2013b: Multimodel Assessment of Regional Surface Temperature Trends: CMIP3 and CMIP5 Twentieth-Century Simulations. *J. Clim.*, **26**, 4168–4185.

Knutson, T. R., et al., 2006: Assessment of Twentieth-Century regional surface temperature trends using the GFDL CM2 coupled models. *J. Clim.*, **19**, 1624-1651, doi: 10.1175/JCLI3709.1.

Knutson, T. R., et al., 2010: Tropical cyclones and climate change. *Nature Geosci*, **3**, 157–163.

Knutti, R., D. Masson, and A. Gettelman, 2013: Climate model genealogy: Generation CMIP5 and how we got there. *Geophys. Res. Lett.*, doi:10.1002/grl.50256.

Koster, R. D., et al., 2010: Contribution of land surface initialization to subseasonal forecast skill: First results from a multi-model experiment. *Geophys. Res. Lett.*, **37**, L02402.

Kroger, J., W. A. Muller, and J. S. von Storch, 2012: Impact of different ocean reanalyses on decadal climate prediction. *Clim. Dyn.*, **39**, 795–810.

Krueger, O., and J.-S. von Storch, 2011: A simple empirical model for decadal climate prediction. *J. Clim.*, **24**, 1276–1283.

Kumar, A., 2009: Finite samples and uncertainty estimates for skill measures for seasonal prediction. *Mon. Weather Rev.*, **137**, 2622–2631.

Kumar, S., V. Merwade, D. Niyogi, and J. L. Kinter III, 2013: Evaluation of temperature and precipitation trends and long-term persistence in CMIP5 20th century climate simulations. *J. Clim.*, doi:10.1175/JCLI-D-12-00259.1.

Laepple, T., S. Jewson, and K. Coughlin, 2008: Interannual temperature predictions using the CMIP3 multi-model ensemble mean. *Geophys. Res. Lett.*, **35**, L10701.

Lamarque, J.-F., et al., 2011: Global and regional evolution of short-lived radiatively-active gases and aerosols in the Representative Concentration Pathways. *Clim. Change*, doi:10.1007/s10584-011-0155-0, 1–22.

Lamarque, J. F., et al., 2013: The Atmospheric Chemistry and Climate Model Intercomparison Project (ACCMIP): Overview and description of models, simulations and climate diagnostics. *Geosci. Model Dev.*, **6**, 179-206, doi 10.5194/gmf-6-179-2013.

Lambert, F. H., and J. C. H. Chiang, 2007: Control of land-ocean temperature contrast by ocean heat uptake. *Geophys. Res. Lett.*, **34**, L13704, doi: 10.1029/2007GL029755.

Lambert, F. H., and M. J. Webb, 2008: Dependency of global mean precipitation on surface temperature. *Geophys. Res. Lett.*, **35**, L23803.

Lammertsma, E. I., H. J. de Boer, S. C. Dekker, D. L. Dilcher, A. F. Lotter, and F. Wagner-Cremer, 2011: Global CO2 rise leads to reduced maximum stomatal conductance in Florida vegetation. *Proc. Acad. Sci. U.S.A.*, **108**, 4035–4040.

Lang, C., and D. W. Waugh, 2011: Impact of climate change on the frequency of Northern Hemisphere summer cyclones. *J. Geophys. Res.*, **116**, D04103, doi: 10.1029/2010JD014300.

Langner, J., M. Engardt, and C. Andersson, 2012a: European summer surface ozone 1990–2100. *Atmos. Chem. Phys*.,**12**, 10097–10105.

Langner, J., M. Engardt, A. Baklanov, J. H. Christensen, M. Gauss, C. Geels, G. B. Hedegaard, R. Nuterman, D. Simpson, J. Soares, M. Sofiev, P. Wind, and A. Zakey, 2012b: A multi-model study of impacts of climate change on surface ozone in Europe. *Atmos. Chem. Phys.*, **12**, 10423-10440.

Lanzante, J. R., 2005: A cautionary note on the use of error bars. *J. Clim.*, **18**, 3699–3703.

Latif, M., M. Collins, H. Pohlmann, and N. Keenlyside, 2006: A review of predictability studies of Atlantic sector climate on decadal time scales. *J. Clim.*, **19**, 5971–5987.

Latif, M., C. W. Boning, J. Willebrand, A. Biastoch, A. Alvarez-Garcia, N. Keenlyside, and H. Pohlmann, 2007: Decadal to multidecadal variability of the Atlantic MOC: Mechanisms and predictability. In: *Ocean Circulation: Mechanisms and Impacts - Past and Future Changes of Meridional Overturning.* AGU Monograph 173. [A. Schmittner, J. C. H. Chiang and S. R. Hemming (eds.)]. American Geophysical Union, Washington, DC, pp. 149–166.

Lawrence, D. M., and A. G. Slater, 2010: The contribution of snow condition trends to future ground climate. *Clim. Dyn.*, **34**, 969–981.

Lawrence, D. M., A. G. Slater, V. E. Romanovsky, and D. J. Nicolsky, 2008: Sensitivity of a model projection of near-surface permafrost degradation to soil column depth and representation of soil organic matter. *J. Geophys. Res.*, **113**, F02011, doi:10.1029/2007JF000883/

Lean, J. L., and D. H. Rind, 2009: How will Earth's surface temperature change in future decades? *Geophys. Res. Lett.*, **36**, L15708.

Lee, J. D., et al., 2006a: Ozone photochemistry and elevated isoprene during the UK heatwave of august 2003. *Atmos. Environ.*, **40**, 7598-7613.

Lee, S.-J., and E. H. Berbery, 2012: Land cover change effects on the climate of the La Plata Basin. *J. Hydrometeorol.*, **13**, 84–102.

Lee, T. C. K., F. W. Zwiers, X. B. Zhang, and M. Tsao, 2006b: Evidence of decadal climate prediction skill resulting from changes in anthropogenic forcing. *J. Clim.*, **19**, 5305–5318.

Lei, H., D. J. Wuebbles, X.-Z. Liang, and S. Olsen, 2013: Domestic versus international contributions on 2050 ozone air quality: How much is convertible by regional control? *Atmos. Environ.*, **68**, 315–325.

Leibensperger, E. M., L. J. Mickley, and D. J. Jacob, 2008: Sensitivity of US air quality to mid-latitude cyclone frequency and implications of 1980–2006 climate change. *Atmos. Chem. Phys.*, **8**, 7075–7086.

Leibensperger, E. M., L. J. Mickley, D. J. Jacob, and S. R. H. Barrett, 2011a: Intercontinental influence of NOx and CO emissions on particulate matter air quality. *Atmos. Environ.*, **45**, 3318–3324.

Leibensperger, E. M., L. J. Mickley, D. J. Jacob, W.-T. Chen, J. H. Seinfeld, A. Nenes, P. J. Adams, D. G. Streets, N. Kumar, and D. Rind, 2012: Climatic effects of 1950–2050 changes in US anthropogenic aerosols – Part 2: Climate response. *Atmos. Chem. Phys.*, **12**, 3349-3362.

Lenderink, G., and E. Van Meijgaard, 2008: Increase in hourly precipitation extremes beyond expectations from temperature changes. *Nature Geosci.*, **1**, 511–514.

Lenderink, G., and E. v. Meijgaard, 2010: Linking increases in hourly precipitation extremes to atmospheric temperature and moisture changes. *Environ. Res. Lett.*, **5(2)**, 025208.

Lenderink, G., E. van Meijgaard, and F. Selten, 2009: Intense coastal rainfall in the Netherlands in response to high sea surface temperatures: Analysis of the event of August 2006 from the perspective of a changing climate. *Clim. Dyn.*, **32**, 19–33.

Lenderink, G., H.Y. Mok, T.C. Lee, and G. J. v. Oldenborgh, 2011: Scaling and trends of hourly precipitation extremes in two different climate zones—Hong Kong and the Netherlands. *Hydrol. Earth Syst. Sci.*, **15(9)**, 3033–3041.

Leslie, L. M., D. J. Karoly, M. Leplastrier, and B. W. Buckley, 2007: Variability of tropical cyclones over the southwest Pacific Ocean using a high-resolution climate model. *Meteorol. Atmos. Phys.*, **97**, 171–180.

Levermann, A., J. Schewe, V. Petoukhov, and H. Held, 2009: Basic mechanism for abrupt monsoon transitions. *Proc. Natl. Acad. Sci. U.S.A.*, **106**, 20572–20577.

Levin, I., et al., 2010: The global SF6 source inferred from long-term high precision atmospheric measurements and its comparison with emission inventories. *Atmos. Chem. Phys.*, **10**, 2655–2662.

Levy, H., L. W. Horowitz, Daniel Schwarzkopf, M. M., G. Y., N. J.-C., and V. Ramaswamy, 2013: The roles of aerosol direct and indirect effects in past and future climate change. *J. Geophys. Res.*, doi:10.1002/jgrd.50192.

Li, H. L., H. J. Wang, and Y. Z. Yin, 2012: Interdecadal variation of the West African summer monsoon during 1979–2010 and associated variability. *Clim. Dyn.*, doi:10.1007/s00382-012-1426-9.

Li, S. L., and G. T. Bates, 2007: Influence of the Atlantic multidecadal oscillation on the winter climate of East China. *Adv. Atmos. Sci.*, **24**, 126–135.

Li, S. L., J. Perlwitz, X. W. Quan, and M. P. Hoerling, 2008: Modelling the influence of North Atlantic multidecadal warmth on the Indian summer rainfall. *Geophys. Res. Lett.*, **35**, L05804.

Liao, H., W.-T. Chen, and J. H. Seinfeld, 2006: Role of climate change in global predictions of future tropospheric ozone and aerosols. *J. Geophys. Res.*, **111**, D12304.

Liao, K.-J., et al., 2007: Sensitivities of ozone and fine Particulate matter formation to emissions under the impact of potential future climate change. *Environ. Sci. Technol.*, **41**, 8355–8361.

Liepert, B. G., and M. Previdi, 2009: Do models and observations disagree on the rainfall response to global warming? *J. Clim.*, **22**, 3156–3166.

Lin, C. Y. C., D. J. Jacob, and A. M. Fiore, 2001: Trends in exceedances of the ozone air quality standard in the continental United States, 1980–1998. *Atmos. Environ.*, **35**, 3217–3228.

Lin, J.-T., D. J. Wuebbles, and X.-Z. Liang, 2008: Effects of intercontinental transport on surface ozone over the United States: Present and future assessment with a global model. *Geophys. Res. Lett.*, **35**, L02805.

Liu, J., D. L. Mauzerall, L. W. Horowitz, P. Ginoux, and A. M. Fiore, 2009: Evaluating inter-continental transport of fine aerosols: (1) Methodology, global aerosol distribution and optical depth. *Atmos. Environ.*, **43**, 4327–4338.

Liu, J., J. A. Curry, H. Wang, M. Song, and R. M. Horton, 2012: Impact of declining Arctic sea ice on winter snowfall. *Proc. Natl. Acad. Sci. U.S.A.*, **109**, 4074–4079.

Lobell, D. B., and M. B. Burke, 2008: Why are agricultural impacts of climate change so uncertain? The importance of temperature relative to precipitation. *Environ. Res. Lett.*, **3**, L05804.

Lockwood, M., 2010: Solar change and climate: An update in the light of the current exceptional solar minimum. *Proc. R. Soc. London A*, **466**, 303–329.

Lockwood, M., R. G. Harrison, M. J. Owens, L. Barnard, T. Woollings, and F. Steinhilber, 2011: The solar influence on the probability of relatively cold UK winters in the future. *Environ. Res. Lett.*, **6**, 034004, doi:10.1088/1748-9326/6/3/034004.

Logan, J. A., 1989: Ozone in rural areas of the United States. *J. Geophys. Res.*, **94**, 8511–8532.

Lu, J., and M.Cai, 2009: Stabilization of the atmospheric boundary layer and the muted global hydrological cycle response to global warming. *J. Hydrometeor*, **10**, 347-352, doi: 10.1175/2008JHM1058.1.

Lu, J., G. Vecchi, and T. Reichler, 2007: Expansion of the Hadley Cell under global warming. *Geophys. Res. Lett.*, doi:10.1029/2006GL028443, L06805.

Lu, R. Y., and Y. H. Fu, 2010: Intensification of East Asian summer rainfall interannual variability in the twenty-first century simulated by 12 CMIP3 coupled models. *J. Clim.*, **23**, 3316–3331.

MacLeod, D. A., C Caminade, and A. P. Morse, 2013: Useful decadal climate prediction at regional scales? A look at the ENSEMBLES stream 2 decadal hindcasts. *Environ. Res. Lett.*, **7**, 044012.

Magnusson, L., M. Balmaseda, S. Corti, F. Molteni, and T. Stockdale, 2013: Evaluation of forecast strategies for seasonal and decadal forecasts in presence of systematic model errors. *Clim. Dyn.*, doi:10.1007/s00382-012-1599-2.

Mahlstein, I., and R. Knutti, 2012: September Arctic sea ice predicted to disappear near 2C global warming above present. *J. Geophys. Res.*, **117**, D06104.

Mahlstein, I., R. Knutti, S. Solomon, and R. W. Portmann, 2011: Early onset of significant local warming in low latitude countries. *Environ. Res. Lett.*, **6**, L06805.

Mahmud, A., M. Hixson, J. Hu, Z. Zhao, S. H. Chen, and M. J. Kleeman, 2010: Climate impact on airborne particulate matter concentrations in California using seven year analysis periods. *Atmos. Chem. Phys.*, **10**, 11097–11114.

Mahowald, N. M., 2007: Anthropocene changes in desert area: Sensitivity to climate model predictions. *Geophys. Res. Lett.*, **34**, L18817.

Mahowald, N. M., and C. Luo, 2003: A less dusty future? *Geophys. Res. Lett.*, **30**, 1903.

Mahowald, N. M., D. R. Muhs, S. Levis, P. J. Rasch, M. Yoshioka, C. S. Zender, and C. Luo, 2006: Change in atmospheric mineral aerosols in response to climate: Last glacial period, preindustrial, modern, and doubled carbon dioxide climates. *J. Geophys. Res.*, **111**, D10202.

Makkonen, R., A. Asmi, V. M. Kerminen, M. Boy, A. Arneth, P. Hari, and M. Kulmala, 2012: Air pollution control and decreasing new particle formation lead to strong climate warming. *Atmos. Chem. Phys.*, **12**, 1515–1524.

Manabe, S., R. J. Stouffer, M. J. Spelman, and K. Bryan, 1991: Transient responses of a coupled Ocean Atmosphere Model to gradual changes of atmospheric CO_2 .1. Annual mean response. *J. Clim.*, **4**, 785–818.

Manders, A. M. M., E. van Meijgaard, A. C. Mues, R. Kranenburg, L. H. van Ulft, and M. Schaap, 2012: The impact of differences in large-scale circulation output from climate models on the regional modeling of ozone and PM. *Atmos Chem Phys*, **12**, 9441–9458.

Mann, M., et al., 2009: Global signatures and dynamical origins of the Little Ice Age and Medieval Climate Anomaly. *Science*, **326**, 1256–1260.

Marengo, J. A., R. Jones, L. M. Alves, and M. C. Valverde, 2009: Future change of temperature and precipitation extremes in South America as derived from the PRECIS regional climate modeling system. *Int. J. Climatol.*, **29**, 2241–2255.

Maslowski, W., J. C. Kinney, M. Higgens, and A. Roberts, 2012: The future of Arctic sea ice. *Annu. Rev. Earth Planet. Sci.*, **40**, 625–654.

Mason, S. J., 2004: On using "climatology" as a reference strategy in the Brier and ranked probability skill scores. *Mon. Weather Rev.*, **132**, 1891–1895.

Massonnet, T. Fichefet, H. Goosse, C. M. Bitz, G. Philippon-Berthier, M. M. Holland, and P.-Y. Barriat, 2012: Constraining projections of summer Arctic sea ice. *Cryosphere*, **6**, 1383–1394.

Matei, D., J. Baehr, J. H. Jungclaus, H. Haak, W. A. Muller, and J. Marotzke, 2012a: Multiyear prediction of monthly mean Atlantic Meridional Overturning Circulation at 26.5 degrees N. *Science*, **335**, 76–79.

Matei, D., H. Pohlmann, J. Jungclaus, W. Muller, H. Haak, and J. Marotzke, 2012b: Two tales of initializing decadal climate prediction experiments with the ECHAM5/MPI-OM Model. *J. Clim.*, **25**, 8502–8523.

McCabe, G. J., M. A. Palecki, and J. L. Betancourt, 2004: Pacific and Atlantic Ocean influences on multidecadal drought frequency in the United States. *Proc. Natl. Acad.Sci. U.S.A.*, **101**, 4136 – 4141.

McLandress, C., T. G. Shepherd, J. F. Scinocca, D. A. Plummer, M. Sigmond, A. I. Jonsson, and M. C. Reader, 2011: Separating the dynamical effects of climate change and ozone depletion. Part II. Southern Hemisphere troposphere. *J. Clim.*, **24**, 1850–1868.

Meehl, G., et al., 2006: Climate change projections for the twenty-first century and climate change commitment in the CCSM3. *J. Clim.*, 2597–2616.

Meehl, G. A., and A. X. Hu, 2006: Megadroughts in the Indian monsoon region and southwest North America and a mechanism for associated multidecadal Pacific sea surface temperature anomalies. *J. Clim.*, **19**, 1605–1623.

Meehl, G. A., and J. M. Arblaster, 2011: Decadal variability of Asian-Australian monsoon-ENSO-TBO relationships. *J. Clim.*, **24**, 4925–4940.

Meehl, G. A., and J. M. Arblaster, 2012: Relating the strength of the tropospheric biennial oscillation (TBO) to the phase of the Interdecadal Pacific Oscillation (IPO). *Geophys. Res. Lett.*, **39**, L20716.

Meehl, G. A., and H. Y. Teng, 2012: Case studies for initialized decadal hindcasts and predictions for the Pacific region. *Geophys. Res. Lett.*, **39**, L22705.

Meehl, G. A., A. X. Hu, and C. Tebaldi, 2010: Decadal prediction in the Pacific region. *J. Clim.*, **23**, 2959–2973.

Meehl, G. A., J. M. Arblaster, and G. Branstator, 2012a: Mechanisms contributing to the warming hole and the consequent U.S. east-west differential of heat extremes. *J. Clim.*, **25**, 6394–6408.

Meehl, G.A., A. Hu, J.M. Arblaster, J. Fasullo, and K.E. Trenberth, 2013a: Externally forced and internally generated decadal climate variability associated with the Interdecadal Pacific Oscillation, J. Climate, 26, 7298-7310, doi: http://dx.doi.org/10.1175/JCLI-D-12-00548.1

Meehl, G. A., J.M. Arblaster, and D. R. Marsh, 2013b: Could a future "Grand Solar Minimum" like the Maunder Minimum stop global warming? *Geophys. Res. Lett.*, doi: 10.1002/grl.50361.

Meehl, G. A., C. Tebaldi, G. Walton, D. Easterling, and L. McDaniel, 2009a: Relative increase of record high maximum temperatures compared to record low minimum temperatures in the U. S. *Geophys. Res. Lett.*, **36**, L08703.

Meehl, G. A., J. M. Arblaster, J. T. Fasullo, A. Hu, and K. E. Trenberth, 2011: Model-based evidence of deep-ocean heat uptake during surface-temperature hiatus periods. *Nature Clim. Change*, **1**, 360–364.

Meehl, G. A., et al., 2007a: The WCRP CMIP3 multimodel dataset - A new era in climate change research. *Bull. Am. Meteorol. Soc.*, **88**, 1383-1394.

Meehl, G. A., et al., 2013c: Climate change projections in CESM1(CAM5) compared to CCSM4. *J. Clim.*, doi:10.1175/JCLI-D-12-00572.1.

Meehl, G. A., et al., 2012b: Climate system response to external forcings and climate change projections in CCSM4. *J. Clim.*, **25**, 3661–3683.

Meehl, G. A., et al., 2007b: Global climate projections. In: *Climate Change 2007: The Physical Science Basis. Contribution of Working Group I to the Fourth Assessment Report of the Intergovernmental Panel on Climate Change* [Solomon, S., D. Qin, M. Manning, Z. Chen, M. Marquis, K. B. Averyt, M. Tignor and H. L. Miller (eds.)] Cambridge University Press, Cambridge, United Kingdom and New York, NY, USA, pp. 747–846.

Meehl, G. A., et al., 2009b: Decadal prediction: Can it be skillful? *Bull. Am. Meteorol. Soc.*, **90**, 1467–1485.

Meehl, G. A., et al., 2013d: Decadal climate prediction: An update from the trenches. *Bull. Am. Meteorol. Soc.*, doi:10.1175/BAMS-D-12-00241.1.

Meinshausen, M., T. M. L. Wigley, and S. C. B. Raper, 2011a: Emulating atmosphere-ocean and carbon cycle models with a simpler model, MAGICC6—Part 2: Applications. *Atmos Chem Phys*, **11**, 1457–1471.

Meinshausen, M., S. J. Smith, K. Calvin, and J. Daniel, 2011b: The RCP greenhouse gas concentrations and their extensions from 1765 to 2300. *Clim. Change*, doi: 10.1007/s10584-011-0156–z.

Meleux, F., F. Solmon, and F. Giorgi, 2007: Increase in summer European ozone amounts due to climate change. *Atmos. Environ.*, **41**, 7577–7587.

Menon, S., and et al., 2008: Aerosol climate effects and air quality impacts from 1980 to 2030. *Environ. Res. Lett.*, **3**, 024004.

Merrifield, M. A., 2011: A shift in western tropical Pacific sea level trends during the 1990s. *J. Clim.*, **24**, 4126–4138.

Merryfield, W. J., et al., 2013: The Canadian Seasonal to Interannual Prediction System. Part I: Models and initialization. *Mon. Weather Rev.*, doi:10.1175/MWR-D-12-00216.1.

Mickley, L. J., D. J. Jacob, B. D. Field, and D. Rind, 2004: Effects of future climate change on regional air pollution episodes in the United States. *Geophys. Res. Lett.*, **31**, L24103.

Mickley, L. J., E. M. Leibensperger, D. J. Jacob, and D. Rind, 2011: Regional warming from aerosol removal over the United States: Results from a transient 2010–2050 climate simulation. *Atmos. Environ.*, doi:10.1016/j.atmosenv.2011.07.030.

Miller, R., G. Schmidt, and D. Shindell, 2006: Forced annular variations in the 20th century intergovernmental panel on climate change fourth assessment report models. *J. Geophys. Res.*,D18101, doi:10.1029/2005JD006323.

Min, S.-K., and S.-K. Son, 2013: Multi-model attribution of the Southern Hemisphere Hadley Cell widening: Major role of ozone depletion. *J. Geophys. Res.*, **118**, 3007–3015.

Ming, Y., V. Ramaswamy, and G. Persad, 2010: Two opposing effects of absorbing aerosols on global-mean precipitation. *Geophys. Res. Lett.*, **37**, L13701.

Ming, Y., V. Ramaswamy, and G. Chen, 2011: A model investigation of aerosol-induced changes in boreal winter extratropical circulation. *J. Clim.*, doi:10.1175/2011jcli4111.1.

Mishra, V., J. M. Wallace, and D. P. Lettenmaier, 2012: Relationship between hourly extreme precipitation and local air temperature in the United States. *Geophys. Res. Lett.*, **39**, L16403.

Mochizuki, T., et al., 2012: Decadal prediction using a recent series of MIROC global climate models. *J. Meteorol. Soc. Jpn.*, **90A**, 373–383.

Mochizuki, T., et al., 2010: Pacific decadal oscillation hindcasts relevant to near-term climate prediction. *Proc, Natl. Acad. Sci. U.S.A.*, **107**, 1833–1837.

Monson, R. K., et al., 2007: Isoprene emission from terrestrial ecosystems in response to global change: Minding the gap between models and observations. *Philos. Trans.R. Soc. A*, **365**, 1677–1695.

Montzka, S., M. Krol, E. Dlugokencky, B. Hall, P. Jockel, and J. Lelieveld, 2011: Small interannual variability of global atmospheric hydroxyl. *Science*, **331**, 67–69.

Morgenstern, O., et al., 2010: Anthropogenic forcing of the Northern Annular Mode in CCMVal-2 models. *J. Geophys. Res.*, D00M03, doi:10.1029/2009JD013347.

Morice, C. P., J. J. Kennedy, N. A. Rayner, and P. D. Jones, 2012: Quantifying uncertainties in global and regional temperature change using an ensemble of observational estimates: The HadCRUT4 data set. *J. Geophys. Res.*, **117**, D08101, doi: 10.1029/2011JD017187.

Msadek, R., K. Dixon, T. Delworth, and W. Hurlin, 2010: Assessing the predictability of the Atlantic meridional overturning circulation and associated fingerprints. *Geophys. Res. Lett.*, doi:10.1029/2010GL044517, L19608.

Mues, A., A. Manders, M. Schaap, A. Kerschbaumer, R. Stern, and P. Builtjes, 2012: Impact of the extreme meteorological conditions during the summer 2003 in Europe on particulate matter concentrations. *Atmos. Environ.*, **55**, 377–391.

Muller, C. J., and P. A. O'Gorman, 2011: An energetic perspective on the regional response of precipitation to climate change. *Nature Clim. Change*, **1**, 266–271.

Muller, C. J., P. A. O'Gorman, and L. E. Back, 2011: Intensification of precipitation extremes with warming in a cloud-resolving model. *J. Clim.*, **24**, 2784–2800.

Muller, W. A., et al., 2012: Forecast skill of multi-year seasonal means in the decadal prediction system of the Max Planck Institute for Meteorology. *Geophys. Res. Lett.*, **39**, L22707.

Murphy, J., et al., 2010: Towards prediction of decadal climate variability and change. *Proced. Environ. Sci.*, **1**, *287–304*.

Murphy, J. M., B. B. B. Booth, M. Collins, G. R. Harris, D. M. H. Sexton, and M. J. Webb, 2007: A methodology for probabilistic predictions of regional climate change from perturbed physics ensembles. *Philos. Trans. R. Soc. A*, **365**, 1993–2028.

Newman, M., 2007: Interannual to decadal predictability of tropical and North Pacific sea surface temperatures. *J. Clim.*, **20**, 2333–2356.

Newman, M., 2013: An empirical benchmark for decadal forecasts of global surface temperature anomalies. *J. Clim.*, doi:10.1175/JCLI-D-12-00590.1.

Nicolsky, D. J., V. E. Romanovsky, V. A. Alexeev, and D. M. Lawrence, 2007: Improved modeling of permafrost dynamics in a GCM land-surface scheme. *Geophys. Res. Lett.*, **34**, L08501.

Nolte, C. G., A. B. Gilliland, C. Hogrefe, and L. J. Mickley, 2008: Linking global to regional models to assess future climate impacts on surface ozone levels in the United States. *J. Geophys. Res.*, **113**, D14307.

Notz, D., 2009: The future of ice sheets and sea ice: Between reversible retreat and unstoppable loss. *Proc. Natl. Acad. Sci. U.S.A.*, **106**, 20590–20595.

NRC, 2009: *Global Sources of Local Pollution:An Assessment of Long-Range Transport of Key Air Pollutants to and from the United States.* The National Academies Press, Washington, DC.

NRC, 2010: *Greenhouse Gas Emissions: Methods to Support International Climate Agreements.* National Research Council, Washington, DC.

O'Gorman, P. A., 2012: Sensitivity of tropical precipitation extremes to climate change. *Nature Geosci.*, **5(10)**, 697–700.

O'Gorman, P. A., and T. Schneider, 2009: The physical basis for increases in precipitation extremes in simulations of 21st-century climate change. *Proc. Natl. Acad. Sci. U.S.A.*, **106**, 14773–14777.

O'Gorman, P. A., R. P. Allan, M. P. Byrne, and M. Previdi, 2012: Energetic constraints on precipitation under climate change. *Surv. Geophys.*, **33**, 585–608.

Oman, L., et al., 2010: Multimodel assessment of the factors driving stratospheric ozone evolution over the 21st century. *J. Geophys. Res.*, **115**, D24306, doi: 10.1029/2010JD014362.

Ordóñez, C., H. Mathis, M. Furger, S. Henne, C. Hüglin, J. Staehelin, and A. S. H. Prévôt, 2005: Changes of daily surface ozone maxima in Switzerland in all seasons from 1992 to 2002 and discussion of summer 2003. *Atmos. Chem. Phys.*, **5**, 1187–1203.

Orlowsky, B., and S. I. Seneviratne, 2012: Global changes in extremes events: Regional and seasonal dimension. Climatic Change. *Climate Change*, **110(3–4)**, 669–696.

Ott, L., et al., 2010: Influence of the 2006 Indonesian biomass burning aerosols on tropical dynamics studied with the GEOS-5 AGCM. *J. Geophys. Res.*, **115**, D14121.

Ottera, O. H., M. Bentsen, H. Drange, and L. L. Suo, 2010: External forcing as a metronome for Atlantic multidecadal variability. *Nature Geosci.*, **3**, 688–694.

Overland, J. E., and M. Wang, 2013: When will the summer Arctic be nearly ice free? *Geophys. Res. Lett.*, doi:10.1002/grl.50316.

Pacifico, F., S. P. Harrison, C. D. Jones, and S. Sitch, 2009: Isoprene emissions and climate. *Atmos. Environ.*, **43**, 6121–6135.

Pacifico, F., G. A. Folberth, C. D. Jones, S. P. Harrison, and W. J. Collins, 2012: Sensitivity of biogenic isoprene emissions to past, present, and future environmental conditions and implications for atmospheric chemistry. *J. Geophys. Res. Atmos.*, **117**, D22302.

Paeth, H., and F. Pollinger, 2010: Enhanced evidence in climate models for changes in extratropical atmospheric circulation. *Tellus A*, **62**, 647–660.

Palmer, M. D., D. J. McNeall, and N. J. Dunstone, 2011: Importance of the deep ocean for estimating decadal changes in Earth's radiation balance. *Geophys. Res. Lett.*, **38**, L13707.

Palmer, T. N., and A. Weisheimer, 2011: Diagnosing the causes of bias in climate models—why is it so hard? *Geophys. Astrophys. Fluid Dyn.*, **105**, 351–365.

Palmer, T. N., R. Buizza, R. Hagedon, A. Lawrence, M. Leutbecher, and L. Smith, 2006: Ensemble prediction: A pedagogical perspective. *ECMWF Newslett.*, **106**, 10–17.

Palmer, T. N., et al., 2004: Development of a European multimodel ensemble system for seasonal-to-interannual prediction (DEMETER). *Bull. Am. Meteorol. Soc.*, **85**, 853-872.

Paolino, D. A., J. L. Kinter, B. P. Kirtman, D. H. Min, and D. M. Straus, 2012: The Impact of Land Surface and Atmospheric Initialization on Seasonal Forecasts with CCSM. *J. Clim.*, **25**, 1007–1021.

Paulot, F., J. D. Crounse, H. G. Kjaergaard, J. H. Kroll, J. H. Seinfeld, and P. O. Wennberg, 2009: Isoprene photooxidation: New insights into the production of acids and organic nitrates. *Atmos. Chem. Phys.*, **9**, 1479–1501.

Penner, J. E., H. Eddleman, and T. Novakov, 1993: Towards the development of a global inventory for black carbon emissions. *Atmos. Environ. A*, **27**, 1277–1295.

Penner, J. E., M. J. Prather, I. S. A. Isaksen, J. S. Fuglestvedt, Z. Klimont, and D. S. Stevenson, 2010: Short-lived uncertainty? *Nature Geosci*, **3(9)**, 587–588.

Persechino, A., J. Mignot, D. Swingedouw, S. Labetoulle, and E. Guilyardi, 2012: Decadal predictability of the Atlantic Meridional Overturning Circulation and climate in the IPSL-CM5A-LR model. *Clim. Dyn*, doi: 10.1007/s00382-012-1466-1.

Pielke, R. A., et al., 2011: Land use/land cover changes and climate: Modeling analysis and observational evidence. *WIREs Clim. Change*, **2**, 828–850.

Pierce, D. W., P. J. Gleckler, T. P. Barnett, B. D. Santer, and P. J. Durack, 2012: The fingerprint of human-induced changes in the ocean's salinity and temperature fields. *Geophys. Res. Lett.*, **39**, L21704, doi:10.1029/2012GL053389.

Pierce, D. W., T. P. Barnett, R. Tokmakian, A. Semtner, M. Maltrud, J. A. Lysne, and A. Craig, 2004: The ACPI Project, Element 1: Initializing a coupled climate model from observed conditions. *Clim. Change*, **62**, 13–28.

Pincus, R., C. P. Batstone, R. J. P. Hofmann, K. E. Taylor, and P. J. Glecker, 2008: Evaluating the present day simulation of clouds, precipitation, and radiation in climate models. *J. Geophys. Res.*, **113**, D14209.

Pinto, J., U. Ulbrich, G. Leckebusch, T. Spangehl, M. Reyers, and S. Zacharias, 2007: Changes in storm track and cyclone activity in three SRES ensemble experiments with the ECHAM5/MPI-OM1 GCM. *Clim. Dyn.*, doi: 10.1007/s00382-007-0230-4, 195–210.

Pitman, A. J., et al., 2012: Effects of land cover change on temperature and rainfall extremes in multi-model ensemble simulations. *Earth Syst. Dyn.*, **3**, 213–231.

Pitman, A. J., et al., 2009: Uncertainties in climate responses to past land cover change: First results from the LUCID intercomparison study. *Geophys. Res. Lett.*, **36**, L14814.

Pohlmann, H., J. H. Jungclaus, A. Köhl, D. Stammer, and J. Marotzke, 2009: Initializing decadal climate predictions with the GECCO oceanic synthesis: Effects on the North Atlantic. *J. Clim.*, **22**, 3926–3938.

Pohlmann, H., M. Botzet, M. Latif, A. Roesch, M. Wild, and P. Tschuck, 2004: Estimating the decadal predictability of a coupled AOGCM. *J. Clim.*, **17**, 4463–4472.

Pohlmann, H., et al., 2013: Predictability of the mid-latitude Atlantic meridional overturning circulation in a multi-model system. *Clim. Dyn.*, doi:10.1007/s00382-013-1663-6.

Polvani, L. M., M. Previdi, and C. Deser, 2011a: Large cancellation, due to ozone recovery, of future Southern Hemisphere atmospheric circulation trends. *Geophys. Res. Lett.*, **38**, L04707.

Polvani, L. M., D. W. Waugh, G. J. P. Correa, and S.-W. Son, 2011b: Stratospheric ozone depletion: The main driver of twentieth-century atmospheric circulation changes in the Southern Hemisphere. *J. Clim.*, **24**, 795–812.

Power, S., and R. Colman, 2006: Multi-year predictability in a coupled general circulation model. *Clim. Dyn.*, **26**, 247–272.

Power, S., T. Casey, C. Folland, A. Colman, and V. Mehta, 1999: Inter-decadal modulation of the impact of ENSO on Australia. *Clim. Dyn.*, **15**, 319–324.

Power, S. B., 1995: Climate drift in a global ocean General-Circulation Model. *J. Phys. Oceanogr.*, **25**, 1025–1036.

Power, S. B., and G. Kociuba, 2011a: The impact of global warming on the Southern Oscillation Index. *Clim. Dyn.*, **37**, 1745–1754.

Power, S. B., and G. Kociuba, 2011b: What caused the observed twentieth-century weakening of the Walker Circulation? *J. Clim.*, **24**, 6501–6514.

Power, S. B., M. Haylock, R. Colman, and X. Wang, 2006: The predictability of interdecadal changes in ENSO and ENSO teleconnections. *J. Clim.*, **19**, 4755–4771.

Power, S. B., F. Delage, R. Colman, and A. Moise, 2012: Consensus on 21st century rainfall projections in climate models more widespread than previously thought. *J. Clim.*, doi::10.1175/JCLI-D-11-00354.1.

Pozzer, A., et al., 2012: Effects of business-as-usual anthropogenic emissions on air quality. *Atmos Chem Phys*, **12**, 6915–6937.

Prather, M., et al., 2001: Atmospheric chemistry and greenhouse gases. In: *Climate Change 2001: The Scientific Basis. Contribution of Working Group I to the Third Assessment Report of the Intergovernmental Panel on Climate Change* [J. T. Houghton, Y. Ding, D. J. Griggs, M. Noquer, P. J. van der Linden, X. Dai, K. Maskell and C. A. Johnson (eds.)]. Cambridge University Press, Cambridge, United Kingdom and New York, NY, USA, pp. 239–287.

Prather, M., et al., 2003: Fresh air in the 21st century? *Geophys. Res. Lett.*, **30**, 1100.

Prather, M. J., and J. Hsu, 2010: Coupling of nitrous oxide and methane by global atmospheric chemistry. *Science*, **330**, 952–954.

Prather, M. J., C. D. Holmes, and J. Hsu, 2012: Reactive greenhouse gas scenarios: Systematic exploration of uncertainties and the role of atmospheric chemistry. *Geophys. Res. Lett.*, **39**, L09803.

Prather, M. J., et al., 2009: Tracking uncertainties in the causal chain from human activities to climate. *Geophys. Res. Lett.*, **36**, L05707.

Price, C., 2013: Lightning applications in weather and climate. *Surv. Geophys.*, doi: 10.1007/s10712-012-9218-7.

Prospero, J. M., 1999: Long-term measurements of the transport of African mineral dust to the southeastern United States: Implications for regional air quality. *J. Geophys. Res. Atmos.*, **104**, 15917–15927.

Pye, H. O. T., H. Liao, S. Wu, L. J. Mickley, D. J. Jacob, D. K. Henze, and J. H. Seinfeld, 2009: Effect of changes in climate and emissions on future sulfate-nitrate-ammonium aerosol levels in the United States. *J. Geophys. Res.*, **114**, D01205.

Quesada, B., R. Vautard, P. Yiou, M. Hirschi, and S. Seneviratne, 2012: Asymmetric European summer heat predictability from wet and dry southern winters and springs. *Nature Clim. Change*, **2 (10)**, 736–741.

Quinn, P. K., et al., 2008: Short-lived pollutants in the Arctic: Their climate impact and possible mitigation strategies. *Atmos. Chem. Phys.*, **8**, 1723–1735.

Racherla, P. N., and P. J. Adams, 2006: Sensitivity of global tropospheric ozone and fine particulate matter concentrations to climate change. *J. Geophys. Res.*, **111**, D24103.

Racherla, P. N., and P. J. Adams, 2008: The response of surface ozone to climate change over the eastern United States. *Atmos. Chem. Phys.*, **8**, 871–885.

Raes, F., and J. H. Seinfeld, 2009: New directions: Climate change and air pollution abatement: A bumpy road. *Atmos. Environ.*, **43**, 5132–5133.

Räisänen, J., 2008: Warmer climate: Less or more snow? *Clim. Dyn.*, **30**, 307–319.

Räisänen, J., 2007: How reliable are climate models? *Tellus A*, **59**, 2–29.

Räisänen, J., and L. Ruokolainen, 2006: Probabilistic forecasts of near-term climate change based on a resampling ensemble technique. *Tellus A*, **58**, 461–472.

Rajczak, J., P. Pall, and C. Schär, 2013: Projections of extreme precipitation events in regional climate simulations for the European and Alpine regions. *J. Geophys. Res.*, doi:10.1002/jgrd.50297.

Ramana, M. V., V. Ramanathan, Y. Feng, S. C. Yoon, S. W. Kim, G. R. Carmichael, and J. J. Schauer, 2010: Warming influenced by the ratio of black carbon to sulphate and the black-carbon source. *Nature Geosci*, **3**, 542–545.

Ramanathan, V., and Y. Feng, 2009: Air pollution, greenhouse gases and climate change: Global and regional perspectives. *Atmos. Environ.*, **43**, 37–50.

Rampal, P., J. Weiss, C. Dubois, and J. M. Campin, 2011: IPCC climate models do not capture Arctic sea ice drift acceleration: Consequences in terms of projected sea ice thinning and decline. *J. Geophys. Res.*, **116**, C00D07, doi: 10.1029/2011JC007110.

Randles, C. A., and V. Ramaswamy, 2010: Direct and semi-direct impacts of absorbing biomass burning aerosol on the climate of southern Africa: A Geophysical Fluid Dynamics Laboratory GCM sensitivity study. *Atmos. Chem. Phys.*, **10**, 9819–9831.

Rasmussen, D. J., A. M. Fiore, V. Naik, L. W. Horowitz, S. J. McGinnis, and M. G. Schultz, 2012: Surface ozone-temperature relationships in the eastern US: A monthly climatology for evaluating chemistry-climate models. *Atmos. Environ.*, doi:10.1016/j.atmosenv.2011.11.021.

Rind, D., 2008: The consequences of not knowing low-and high-latitude climate sensitivity. *Bull. Am. Meteorol. Soc.*, doi: 10.1175/2007BAMS2520.1, 855–864.

Roberts, C. D., and M. D. Palmer, 2012: Detectability of changes to the Atlantic meridional overturning circulation in the Hadley Centre Climate Models. *Clim. Dyn.*, **39**, 2533-2546, doi: 10.1007/s00382-012-1306-3.

Robock, A., 2000: Volcanic eruptions and climate. *Rev. Geophys.*, **38**, 191–219.

Robson, J. I., R. T. Sutton, and D. M. Smith, 2012: Initialized decadal predictions of the rapid warming of the North Atlantic ocean in the mid 1990s. *Geophys. Res. Lett.*, **39**, L19713, doi: 10.1029/2012GL053370.

Roeckner, E., P. Stier, J. Feichter, S. Kloster, M. Esch, and I. Fischer-Bruns, 2006: Impact of carbonaceous aerosol emissions on regional climate change. *Clim. Dyn.*, **27**, 553–571.

Roscoe, H. K., and J. D. Haigh, 2007: Influences of ozone depletion, the solar cycle and the QBO on the Southern Annular Mode. *Q. J. R. Meteorol. Soc.*, **133**, 1855–1864.

Rotstayn, L. D., S. J. Jeffrey, M. A. Collier, S. M. Dravitzki, A. C. Hirst, J. I. Syktus, and K. K. Wong, 2012: Aerosol- and greenhouse gas-induced changes in summer rainfall and circulation in the Australasian region: A study using single-forcing climate simulations. *Atmos. Chem. Phys.*, **12**, 6377–6404.

Rowell, D. P., 2011: Sources of uncertainty in future change in local precipitation. *Clim. Dyn.*, doi:10.1007/s00382-011-1210-2.

Rowlands, D. J., et al., 2012: Broad range of 2050 warming from an observationally constrained large climate model ensemble. *Nature Geosci.*, **5**, 256–260.

Royal Society, 2008: *Ground-Level Ozone in the 21st Century: Future Trends, Impacts and Policy Implications*.The Royal Society, London, United Kingdom.

Ruckstuhl, C., and J. R. Norris, 2009: How do aerosol histories affect solar "dimming" and "brightening" over Europe?: IPCC-AR4 models versus observations. *J. Geophys. Res. Atmos.*, **114**, D00D04, doi: 1029/2008JD011066.

Saha, S., et al., 2010: The NCEP Climate Forecast System Reanalysis. *Bull. Am. Meteorol. Soc.*, **91**, 1015–1057.

Sand, T., K. Berntsen, J. E. Kay, J. F. Lamarque, Ø. Seland, and A. Kirkevåg, 2013: The Arctic response to remote and local forcing of black carbon. *Atmos Chem Phys*, **13**, 211–224.

Scaife, A. A., et al., 2012: Climate change projections and stratosphere-troposphere interaction. *Clim. Dyn.*, **38**, 2089–2097.

Schaller, N., I. Mahlstein, J. Cermak, and R. Knutti, 2011: Analyzing precipitation projections: A comparison of different approaches to climate model evaluation. *J. Geophys. Res.*, **116**, D10118.

Schar, C., P. L. Vidale, D. Luthi, C. Frei, C. Haberli, M. A. Liniger, and C. Appenzeller, 2004: The role of increasing temperature variability in European summer heatwaves. *Nature*, **427**, 332–336.

Scherrer, S. C., P. Ceppi, M. Croci-Maspoli, and C. Appenzeller, 2012: Snow-albedo feedback and Swiss spring temperature trends. *Theor. Appl. Climatol.*, **110**, 509–516.

Schneider, E. K., B. Huang, Z. Zhu, D. G. DeWitt, J. L. Kinter, K. B.P., and J. Shukla, 1999: Ocean data assimilation, initialization and predictions of ENSO with a coupled GCM. *Mon. Weather Rev.*, **127**, 1187–1207.

Schneider, N., and A. J. Miller, 2001: Predicting western North Pacific Ocean climate. *J. Clim.*, **14**, 3997–4002.

Schubert, S., M. J. Suarez, P. J. Pegion, R. D. Koster, and J. T. Bacmeister, 2004: On the cause of the 1930s Dust Bowl. *Science*, **303**, 1855–1859.

Schweiger, A. J., R. W. Lindsay, S. Vavrus, and J. A. Francis, 2008: Relationships between Arctic sea ice and clouds during autumn. *J. Clim.*, **21**, 4799–4810.

Screen, J. A., and I. Simmonds, 2010: The central role of diminishing sea ice in recent Arctic temperature amplification. *Nature*, **464**, 1334–1337.

Seager, R., Y. Kushnir, M. Ting, M. Cane, N. Naik, and J. Miller, 2008: Would advance knowledge of 1930s SSTs have allowed prediction of the dust bowl drought? *J. Clim.*, **21**, 3261–3281.

Seager, R., N. Naik, W. Baethgen, A. Robertson, Y. Kushnir, J. Nakamura, and S. Jurburg, 2010: Tropical oceanic causes of interannual to multidecadal precipitation variability in southeast South America over the past century. *J. Clim.*, **23**, 5517–5539.

Selten, F., G. Branstator, H. Dijkstra, and M. Kliphuis, 2004: Tropical origins for recent and future Northern Hemisphere climate change. *Geophys. Res. Lett.*, doi:10.1029/2004GL020739, L21205.

Semenov, V., M. Latif, J. Jungclaus, and W. Park, 2008: Is the observed NAO variability during the instrumental record unusual? *Geophys. Res. Lett.*, doi:10.1029/2008GL033273, L11701.

Semenov, V. A., M. Latif, D. Dommenget, N. S. Keenlyside, A. Strehz, T. Martin, and W. Park, 2010: The impact of North Atlantic-Arctic multidecadal variability on Northern Hemisphere surface air temperature. *J. Clim.*, **23**, 5668–5677.

Seneviratne, S. I., et al., 2010: Investigating soil moisture-climate interactions in a changing climate: A review. *Earth Sci. Rev.*, **99**, 125–161.

Seneviratne, S. I., et al., 2012: Changes in climate extremes and their impacts on the natural physical environment. In: *IPCC Special Report on Extreme Events and Disasters (SREX)*. World Meteorological Organization, Geneva, Switzerland, pp.

Serreze, M. C., A. P. Barrett, A. G. Slater, M. Steele, J. L. Zhang, and K. E. Trenberth, 2007: The large-scale energy budget of the Arctic. *J. Geophys. Res.*, **112**, D11122, doi: 10.1029/2006JD008230.

Sevellec, F., and A. Fedorov, 2012: Model bias reduction and the limits of oceanic decadal predictability: Importance of the deep ocean. *J. Clim.*, doi:10.1175/JCLI-D-12-00199.1.

Sheffield, J., and E. F. Wood, 2008: Projected changes in drought occurrence under future global warming from multi-model, multi-scenario, IPCC AR4 simulations. *Clim. Dyn.*, **31**, 79–105.

Sheffield, J., E. F. Wood, and M. L. Roderick, 2012: Little change in global drought over the past 60 years. *Nature*, **491**, 435–440.

Shi, Y., X. J. Gao, J. Wu, and F. Giorgi, 2011: Changes in snow cover over China in the 21st century as simulated by a high resolution regional climate model. *Environ. Res. Lett.*, **6**, 045401, doi: 10.1088/1748-9326/6/4/045401.

Shimpo, A., and M. Kanamitsu, 2009: Planetary scale land-ocean contrast of near-surface air temperature and precipitation forced by present and future SSTs. *J. Meteorol. Soc. Jpn.*, **87**, 877–894.

Shin, S.-I., and P. D. Sardeshmukh, 2011: Critical influence of the pattern of tropical ocean warming on remote climate trends. *Clim. Dyn.*, **36**, 1577–1591.

Shindell, D., et al., 2013: Radiative forcing in the ACCMIP historical and future climate simulations. *Atmos. Chem. Phys.*, **13**, 2939-2974.

Shindell, D., et al., 2012a: Simultaneously mitigating near-term climate change and improving human health and food security. *Science*, **335**, 183–189.

Shindell, D. T., and G. A. Schmidt, 2004: Southern Hemisphere climate response to ozone changes and greenhouse gas increases. *Geophys. Res. Lett.*, **31**, L18209, doi:10.1029/2004GL020724.

Shindell, D. T., A. Voulgarakis, G. Faluvegi, and G. Milly, 2012: Precipitation response to regional radiative forcing. *Atmos. Chem. Phys*., **12**, 6969–6982.

Shindell, D. T., et al., 2006: Simulations of preindustrial, present-day, and 2100 conditions in the NASA GISS composition and climate model G-PUCCINI. *Atmos. Chem. Phys.*, **6**, 4427–4459.

Sigmond, M., P. Kushner, and J. Scinocca, 2007: Discriminating robust and non-robust atmospheric circulation responses to global warming. *J. Geophys. Res.*, D20121, doi:10.1029/2006JD008270.

Sillman, S., and P. J. Samson, 1995: Impact of temperature on oxidant photochemistry in urban, polluted rural and remote environments. *J. Geophys. Res.*, **100**, 11497–11508.

Sillmann, J., V.V. Kharin, F. W. Zwiers, and X. Zhang, 2013: Climate extreme indices in the CMIP5 multi-model ensemble. Part 2: Future climate projections. *J. Geophys. Res.*, **118**, 1–21.

Simmonds, I., and K. Keay, 2009: Extraordinary September Arctic sea ice reductions and their relationships with storm behavior over 1979–2008. *Geophys. Res. Lett.*, **36**, L19715.

Simmons, A. J., K. M. Willett, P. D. Jones, P. W. Thorne, and D. P. Dee, 2010: Low-frequency variations in surface atmospheric humidity, temperature, and precipitation: Inferences from reanalyses and monthly gridded observational data sets. *J. Geophys. Res. Atmos.*, **115**, D01110, doi:10.1029/2009JD012442.

Singleton, A., and R. Toumi, 2012: Super-Clausius-Clapeyron scaling of rainfall in a model squall line. *Q. J. R. Meteorol. Soc.*, **139**, 334–339.

Skjøth, C. A., and C. Geels, 2013: The effect of climate and climate change on ammonia emissions in Europe. *Atmos Chem Phys*, **13**, 117–128.

Slater, A. G., and D. M. Lawrence, 2013: Diagnosing present and future permafrost from climate models. *J. Clim.*, doi:10.1175/JCLI-D-12-00341.1.

Smith, D. M., and J. M. Murphy, 2007: An objective ocean temperature and salinity analysis using covariances from a global climate model. *J. Geophys. Res.*, **112**, C02022.

Smith, D. M., A. A. Scaife, and B. P. Kirtman, 2012: What is the current state of scientific knowledge with regard to seasonal and decadal forecasting? *Environ. Res. Lett.*, **7**, 015602.

Smith, D. M., R. Eade, and H. Pohlmann, 2013a: A comparison of full-field and anomaly initialization for seasonal to decadal climate prediction. *Clim. Dyn.*, doi:10.1007/s00382-013-1683-2.

Smith, D. M., S. Cusack, A. W. Colman, C. K. Folland, G. R. Harris, and J. M. Murphy, 2007: Improved surface temperature prediction for the coming decade from a global climate model. *Science*, **317**, 796–799.

Smith, D. M., R. Eade, N. J. Dunstone, D. Fereday, J. M. Murphy, H. Pohlmann, and A. A. Scaife, 2010: Skilful multi-year predictions of Atlantic hurricane frequency. *Nature Geosci.*, **3**, 846–849.

Smith, D. M., et al., 2013b: Real-time multi-model decadal climate predictions. *Clim. Dyn.*, doi:10.1007/s00382-012-1600–0.

Smith, S. J., J. van Aardenne, Z. Klimont, R. J. Andres, A. Volke, and S. Delgado Arias, 2011: Anthropogenic sulfur dioxide emissions: 1850–2005. *Atmos Chem Phys*, **11**, 1101–1116.

Smith, T. M., R. W. Reynolds, T. C. Peterson, and J. Lawrimore, 2008: Improvements to NOAA's historical merged land-ocean surface temperature analysis (1880–2006). *J. Clim.*, **21**, 2283–2296.

Soden, B. J., R. T. Wetherald, G. L. Stenchikov, and A. Robock, 2002: Global cooling after the eruption of Mount Pinatubo: A test of climate feedback by water vapor. *Science*, **296**, 727–730.

Sohn, B. J., and S.-C. Park, 2010: Strengthened tropical circulations in past three decades inferred from water vapor transport. *J. Geophys. Res.*, **115**, D15112.

Solomon, A., et al., 2011: Distinguishing the roles of natural and anthropogenically forced decadal climate variability. *Bull. Am. Meteorol. Soc.*, **92**, 141–156.

Son, S., N. Tandon, L. Polvani, and D. Waugh, 2009a: Ozone hole and Southern Hemisphere climate change. *Geophys. Res. Lett.*, doi:10.1029/2009GL038671, L15705.

Son, S., et al., 2009b: The impact of stratospheric ozone recovery on tropopause height trends. *J. Clim.*, doi: 10.1175/2008JCLI2215.1, 429–445.

Son, S. W., et al., 2008: The impact of stratospheric ozone recovery on the Southern Hemisphere westerly jet. *Science*, **320**, 1486–1489.

Spracklen, D. V., L. J. Mickley, J. A. Logan, R. C. Hudman, R. Yevich, M. D. Flannigan, and A. L. Westerling, 2009: Impacts of climate change from 2000 to 2050 on wildfire activity and carbonaceous aerosol concentrations in the western United States. *J. Geophys. Res.*, **114**, D20301.

Srokosz, M., et al., 2012: Past, present, and future changes in the Atlantic Meridional Overturning Circulation. *Bull. Am. Meteorol. Soc.*, **93**, 1663–1676.

Stainforth, D. A., et al., 2005: Uncertainty in predictions of the climate response to rising levels of greenhouse gases. *Nature*, **433**, 403–406.

Stammer, D., 2006: *Report of the First CLIVAR Workshop on Ocean Reanalysis*. WCRP Informal Publication No. 9/2006. ICPO Publication Series No. 93. World Climate Research Programme, World Meteorological Organization, Geneva, Switzerland.

Stan, C., and B. P. Kirtman, 2008: The influence of atmospheric noise and uncertainty in ocean initial conditions on the limit of predictability in a coupled GCM. *J. Clim.*, **21**, 3487–3503.

Staten, P. W., J. J. Rutz, T. Reichler, and J. Lu, 2011: Breaking down the tropospheric circulation response by forcing. *Clim. Dyn.*, doi:10.1007/s00382-011-1267-y.

Stegehuis, A. I., R. Vautard, P. Ciais, R. Teuling, M. Jung, and P. Yiou, 2012: Summer temperatures in Europe and land heat fluxes in observation-based data and regional climate model simulations. *Clim. Dyn.*, doi:10.1007/s00382-012-1559-x.

Steiner, A. L., S. Tonse, R. C. Cohen, A. H. Goldstein, and R. A. Harley, 2006: Influence of future climate and emissions on regional air quality in California. *J. Geophys. Res.*, **111**, D18303.

Steiner, A. L., A. J. Davis, S. Sillman, R. C. Owen, A. M. Michalak, and A. M. Fiore, 2010: Observed suppression of ozone formation at extremely high temperatures due to chemical and biophysical feedbacks. *Proc. Natl. Acad. Sci. U.S.A.*, doi:10.1073/pnas.1008336107.

Stenchikov, G., T. Delworth, V. Ramaswamy, R. Stouffer, A. Wittenberg, and F. Zeng, 2009: Volcanic signals in oceans. *J. Geophys. Res. Atmos.*, doi:ARTN D16104, 10.1029/2008JD011673, -.

Stenchikov, G., K. Hamilton, R. Stouffer, A. Robock, V. Ramaswamy, B. Santer, and H. Graf, 2006: Arctic Oscillation response to volcanic eruptions in the IPCC AR4 climate models. *J. Geophys. Res.*, **111**, D07107, doi.1029/2005JD006286.

Stevenson, D., R. Doherty, M. Sanderson, C. Johnson, B. Collins, and D. Derwent, 2005: Impacts of climate change and variability on tropospheric ozone and its precursors. *Faraday Discuss.*, **130**, 41–57.

Stevenson, D. S., et al., 2013: Tropospheric ozone changes, radiative forcing and attribution to emissions in the Atmospheric Chemistry and Climate Model Intercomparison Project (ACCMIP). *Atmos. Chem. Phys.*, **13**, 3063-2085. doi:10.5194/acp-13-3063-2013.

Stevenson, D. S., et al., 2006: Multimodel ensemble simulations of present-day and near-future tropospheric ozone. *J. Geophys. Res.*, **111**, D08301.

Stockdale, T. N., 1997: Coupled ocean–atmosphere forecasts in the presence of climate drift. *Mon. Weather Rev.*, **125**, 809–818.

Stockdale, T. N., D. L. T. Anderson, J. O. S. Alves, and M. A. Balmaseda, 1998: Global seasonal rainfall forecasts using a coupled ocean-atmosphere model. *Nature*, **392**, 370–373.

Stott, P., D. Stone, and M. Allen, 2004: Human contribution to the European heatwave of 2003. *Nature*, **432**, 610–614.

Stott, P., R. Sutton, and D. Smith, 2008: Detection and attribution of Atlantic salinity changes. *Geophys. Res. Lett.*, doi:10.1029/2008GL035874, L21702.

Stott, P., P. Good, G. Jones, N. Gillet, and E. Hawkins, 2013: Upper range of climate warming projections are inconsistent with past warming. *Environ. Res. Lett.*, **8**, 014024, doi:10.1088/1748-9326/8/1/014024.

Stott, P., N. Gillett, G. Hegerl, D. Karoly, D. Stone, X. Zhang, and F. Zwiers, 2010: Detection and attribution of climate change: A regional perspective. *WIREs Clim. Change*, **1**, 192–211.

Stott, P. A., and J. A. Kettleborough, 2002: Origins and estimates of uncertainty in predictions of twenty-first century temperature rise. *Nature*, **416**, 723–726.

Stott, P. A., and G. Jones, 2012: Observed 21st century temperatures further constrain decadal predictions of future warming. *Atmos. Sci. Lett.*, **13**, 151–156.

Strahan, S., et al., 2011: Using transport diagnostics to understand chemistry climate model ozone simulations. *J. Geophys. Res. Atmos.*, **116**, D17302, doi:10.1029/2010/JD015360.

Stroeve, J., M. M. Holland, W. Meier, T. Scambos, and M. Serreze, 2007: Arctic sea ice decline: Faster than forecast. *Geophys. Res. Lett.*, **34**, L09501.

Struzewska, J., and J. W. Kaminski, 2008: Formation and transport of photooxidants over Europe during the July 2006 heat wave - observations and GEM-AQ model simulations. *Atmos. Chem. Phys.*, **8**, 721–736.

Sugi, M., and J. Yoshimura, 2012: Decreasing trend of tropical cyclone frequency in 228-year high-resolution AGCM simulations. *Geophys. Res. Lett.*, **39**, L19805, doi: 10.1029/2012GL053360.

Sugiura, N., et al., 2008: Development of a four-dimensional variational coupled data assimilation system for enhanced analysis and prediction of seasonal to interannual climate variations. *J. Geophys. Res. C*, **113**, C10017.

Sugiura, N., et al., 2009: Potential for decadal predictability in the North Pacific region. *Geophys. Res. Lett.*, **36**, L20701.

Sun, J., and H. Wang, 2006: Relationship between Arctic Oscillation and Pacific Decadal Oscillation on decadal timescales. *Chin. Sci. Bull.*, **51**, 75–79.

Sun, J., H. Wang, W. Yuan, and H. Chen, 2010: Spatial-temporal features of intense snowfall events in China and their possible change. *J. Geophys. Res.*, **115**, D16110, doi: 10.1029/2009JD013541.

Sun, Y., S. Solomon, A. Dai, and R. W. Portmann, 2007: How often will it rain? *J. Clim.*, **20(19)**, 4801–4818.

Sushama, L., R. Laprise, and M. Allard, 2006: Modeled current and future soil thermal regime for northeast Canada. *J. Geophys. Res. Atmos.*, **111**, D18111, doi: 10.1029/20005JD007027.

Sutton, R., and D. Hodson, 2005: Atlantic Ocean forcing of North American and European summer climate. *Science*, doi: 10.1126/science.1109496, 115–118.

Sutton, R. T., B. W. Dong, and J. M. Gregory, 2007: Land/sea warming ratio in response to climate change: IPCC AR4 model results and comparison with observations. *Geophys. Res. Lett.*, **34**, L02701.

Swingedouw, D., J. Mignot, S. Labetoulle, E. Guilyardi, and G. Madec, 2013: Initialisation and predictability of the AMOC over the last 50 years in a climate model. *Clim. Dyn.*, doi:10.1007/s00382-012-1516-8.

Szopa, S., D. A. Hauglustaine, R. Vautard, and L. Menut, 2006: Future global tropospheric ozone changes and impact on European air quality. *Geophys. Res. Lett.*, **33**, L14805.

Tagaris, E., et al., 2007: Impacts of global climate change and emissions on regional ozone and fine particulate matter concentrations over the United States. *J. Geophys. Res.*, **112**, D14312.

Tai, A. P. K., L. J. Mickley, and D. J. Jacob, 2010: Correlations between fine particulate matter (PM2.5) and meteorological variables in the United States: Implications for the sensitivity of PM2.5 to climate change. *Atmos. Environ.*, **44**, 3976–3984.

Tai, A. P. K., L. J. Mickley, and D. J. Jacob, 2012a: Impact of 2000–2050 climate change on fine particulate matter (PM2.5) air quality inferred from a multi-model analysis of meteorological modes. *Atmos. Chem. Phys.*, **12**, 11329-11337, doi: 10.5194/acp-12-11329-2012.

Tai, A. P. K., L. J. Mickley, D. J. Jacob, E. M. Leibensperger, L. Zhang, J. A. Fisher, and H. O. T. Pye, 2012b: Meteorological modes of variability for fine particulate matter (PM2.5) air quality in the United States: Implications for PM2.5 sensitivity to climate change. *Atmos. Chem. Phys.*, **12**, 3131–3145.

Tao, Z., A. Williams, H.-C. Huang, M. Caughey, and X.-Z. Liang, 2007: Sensitivity of U.S. surface ozone to future emissions and climate changes. *Geophys. Res. Lett.*, **34**, L08811.

Tatebe, H., et al., 2012: Initialization of the climate model MIROC for decadal prediction with hydographic data assimilation. *J. Meteorol. Soc. Jpn.*, **90A**, 275–294.

Taylor, C. M., A. Gounou, F. Guichard, P. P. Harris, R. J. Ellis, F. Couvreux, and M. De Kauwe, 2011: Frequency of Sahelian storm initiation enhanced over mesoscale soil-moisture patterns. *Nature Geosci.*, **4**, 430–433.

Taylor, K. E., R. J. Stouffer, and G. A. Meehl, 2012: An overview of Cmip5 and the experiment design. *Bull. Am. Meteorol. Soc.*, **93**, 485–498.

Tebaldi, C., J. M. Arblaster, and R. Knutti, 2011: Mapping model agreement on future climate projections. *Geophys. Res. Lett.*, **38**, L23701.

Tegen, I., M. Werner, S. P. Harrison, and K. E. Kohfeld, 2004: Relative importance of climate and land use in determining present and future global soil dust emission. *Geophys. Res. Lett.*, **31**, L05105.

Teng, H., W. M. Washington, G. Branstator, G. A. Meehl, and J.-F. Lamarque, 2012: Potential impacts of Asian carbon aerosols on future US warming. *Geophys. Res. Lett.*, **39**, L11703.

Teng, H. Y., G. Branstator, and G. A. Meehl, 2011: Predictability of the Atlantic Overturning Circulation and associated surface patterns in two CCSM3 climate change ensemble experiments. *J. Clim.*, **24**, 6054–6076.

Terray, L., 2012: Evidence for multiple drivers of North Atlantic multi-decadal climate variability. *Geophys. Res. Lett.*, **39**, L19712.

Terray, L., L. Corre, S. Cravatte, T. Delcroix, G. Reverdin, and A. Ribes, 2012: Near-surface salinity as nature's rain gauge to detect human influence on the tropical water cycle. *J. Clim.*, **25**, 958–977.

Teuling, A. J., et al., 2010: Contrasting response of European forest and grassland energy exchange to heatwaves. *Nature Geosci.*, **3**, 722–727.

Thompson, D. W. J., and S. Solomon, 2002: Interpretation of recent Southern Hemisphere climate change. *Science*, **296**, 895–899.

Timmermann, A., S. McGregor, and F. Jin, 2010: Wind effects on past and future regional sea level trends in the Southern Indo-Pacific. *J. Clim.*, doi: 10.1175/2010JCLI3519.1, 4429-4437.

Timmermann, A., et al., 2007: The influence of a weakening of the Atlantic Meridional Overturning Circulation on ENSO. *J. Clim.*, **20**, 4899–4919.

Timmreck, C., 2012: Modeling the climatic effects of large explosive volcanic eruptions. *WIREs Clim. Change*, **3**, 545–564.

Toyoda, T., et al., 2011: Impact of the assimilation of sea ice concentration data on an atmosphere-ocean-sea ice coupled simulation of the Arctic Ocean climate. *SOLA*, **7**, 37–40.

Trenberth, K., and A. Dai, 2007: Effects of Mount Pinatubo volcanic eruption on the hydrological cycle as an analog of geoengineering. *Geophys. Res. Lett.*, **34**, L15702, doi: 10.1029/2007GL030524.

Trenberth, K. E., and D. J. Shea, 2006: Atlantic hurricanes and natural variability in 2005. *Geophys. Res. Lett.*, **33**, L12704.

Trenberth, K. E., et al., 2007: Observations: Atmospheric surface and climate change. In: *Climate Change 2007: The Physical Science Basis. Contribution of Working Group I to the Fourth Assessment Report of the Intergovernmental Panel on Climate Change* [Solomon, S., D. Qin, M. Manning, Z. Chen, M. Marquis, K. B. Averyt, M. Tignor and H. L. Miller (eds.)] Cambridge University Press, Cambridge, United Kingdom and New York, NY, USA, pp. 235–336.

Tressol, M., et al., 2008: Air pollution during the 2003 European heat wave as seen by MOZAIC airliners. *Atmos. Chem. Phys.*, **8**, 2133–2150.

Troccoli, A., and T. N. Palmer, 2007: Ensemble decadal predictions from analysed initial conditions. *Philos. Trans. R. Soc. A*, **365**, 2179–2191.

Turner, A. J., A. M. Fiore, L. W. Horowitz, and M. Bauer, 2013: Summertime cyclones over the Great Lakes Storm Track from 1860–2100: Variability, trends, and association with ozone pollution. *Atmos. Chem. Phys.*, **13**, 565–578.

Tziperman, E., L. Zanna, and C. Penland, 2008: Nonnormal thermohaline circulation dynamics in a coupled ocean-atmosphere GCM. *J. Phys. Oceanogr.*, **38**, 588–604.

Ulbrich, U., J. Pinto, H. Kupfer, G. Leckebusch, T. Spangehl, and M. Reyers, 2008: Changing Northern Hemisphere storm tracks in an ensemble of IPCC climate change simulations. *J. Clim.*, doi: 10.1175/2007JCLI1992.1, 1669–1679.

UNEP and WMO, 2011: Integrated Assessment of Black Carbon and Tropospheric Ozone. United Nations Environment Programme & World Meteorological Organization [Available at http://www.unep.org/dewa/Portals/67/pdf/BlackCarbon_SDM.pdf]

Unger, N., 2012: Global climate forcing by criteria air pollutants. *Annu. Rev. Environ. Resour.*, **37**, 1-24.

Unger, N., D. T. Shindell, D. M. Koch, and D. G. Streets, 2006a: Cross influences of ozone and sulfate precursor emissions changes on air quality and climate. *Proc. Natl. Acad. Sci. U.S.A.*, **103**, 4377–4380.

Unger, N., D. T. Shindell, D. M. Koch, M. Amann, J. Cofala, and D. G. Streets, 2006b: Influences of man-made emissions and climate changes on tropospheric ozone, methane, and sulfate at 2030 from a broad range of possible futures. *J. Geophys. Res. Atmos.*, **111**, D12313, doi: 10.1029/2005JD006518.

van der Linden, P., and J. F. B. Mitchell, 2009: ENSEMBLES: Climate change and its impacts. Summary of research and results from the ENSEMBLES project [Available from the Met Office Hadley Centre, Fitzroy Road, Exeter EX1 3PB, United Kingdom].

van Haren, R., G.J. van Oldenborgh, G. Lenderink, M. Collins, and W. Hazeleger, 2012: SST and circulation trend biases cause an underestimation of European precipitation trends precipitation trends. *Clim. Dyn.*, **40**, 1–20.

van Oldenborgh, G. J., P. Yiou, and R. Vautard, 2010: On the roles of circulation and aerosols in the decline of mist and dense fog in Europe over the last 30 years. *Atmos. Chem. Phys.*, **10**, 4597–4609.

van Oldenborgh, G. J., F. J. Doblas-Reyes, B. Wouters, and W. Hazeleger, 2012: Decadal prediction skill in a multi-model ensemble. *Clim. Dyn.*, **38**, 1263–1280.

van Oldenborgh, G. J., F.J. Doblas-Reyes, S. S. Drijfhout, and E. Hawkins, 2013: Reliability of regional climate model trends. *Environ. Res. Lett.*, **8**, 014055.

van Oldenborgh, G. J., et al., 2009: Western Europe is warming much faster than expected. *Clim. Past*, **5**, 1–12.

van Vuuren, D., et al., 2011: The representative concentration pathways: An overview. *Clim. Change*, doi:10.1007/s10584-011-0148-z, 1-27.

Vautard, R., P. Yiou, and G. van Oldenborgh, 2009: Decline of fog, mist and haze in Europe over the past 30 years. *Nature Geosci.*, **2**, 115–119.

Vautard, R., C. Honoré, M. Beekmann, and L. Rouil, 2005: Simulation of ozone during the August 2003 heat wave and emission control scenarios. *Atmos. Environ.*, **39**, 2957–2967.

Vavrus, S. J., M. M. Holland, A. Jahn, D. A. Bailey, and B. A. Blazey, 2012: Twenty-first-century Arctic climate change in CCSM4. *J. Clim.*, **25**, 2696–2710.

Vecchi, G., and B. Soden, 2007: Global warming and the weakening of the tropical circulation. *J. Clim.*, doi: 10.1175/JCLI4258.1, 4316–4340.

Vecchi, G., B. Soden, A. Wittenberg, I. Held, A. Leetmaa, and M. Harrison, 2006: Weakening of tropical Pacific atmospheric circulation due to anthropogenic forcing. *Nature*, doi: 10.1038/nature04744, 73–76.

Vecchi, G. A., et al., 2012: Technical comment on "Multiyear prediction of monthly mean Atlantic meridional overturning circulation at 26.5°N. *Science*, **338**, 604.

Vecchi, G.A., R. Msadek, W. Anderson, Y.-S. Chang, T. Delworth, K. Dixon, R. Gudgel, A. Rosati, W. Stern, G. Villarini, A. Wittenberg, X. Yang, F. Zeng, R. Zhang and S. Zhang (2013): Multi-year Predictions of North Atlantic Hurricane Frequency: Promise and Limitations. J. Climate, doi:10.1175/JCLI-D-12-00464.1

Vidale, P. L., D. Luethi, R. Wegmann, and C. Schaer, 2007: European summer climate variability in a heterogeneous multi-model ensemble. *Clim. Change*, **81**, 209–232.

Vieno, M., et al., 2010: Modelling surface ozone during the 2003 heat-wave in the UK. *Atmos. Chem. Phys.*, **10**, 7963–7978.

Vikhliaev, Y., B. Kirtman, and P. Schopf, 2007: Decadal North Pacific bred vectors in a coupled GCM. *J. Clim.*, **20**, 5744–5764.

Villarini, G., and G. A. Vecchi, 2012: 21st century projections of North Atlantic tropical storms from CMIP5 models. *Nature Clim. Change*, doi:Nature Climate Change :10:1038/NCLIMATE1530.

Villarini, G., and G. A. Vecchi, 2013: Projected increases in North Atlantic tropical cyclone intensity from CMIP5 models. *J. Clim.*, **26**, 3231–3240.

Villarini, G., G. A. Vecchi, T. R. Knutson, M. Zhao, and J. A. Smith, 2011: North Atlantic tropical storm frequency response to anthropogenic forcing: Projections and sources of uncertainty. *J. Clim.*, **24**, 3224–3238.

Vollmer, M. K., et al., 2011: Atmospheric histories and global emissions of the anthropogenic hydrofluorocarbons HFC-365mfc, HFC-245fa, HFC-227ea, and HFC-236fa. *J. Geophys. Res.*, **116**, D08304.

Voulgarakis, A., et al., 2013: Analysis of present day and future OH and methane lifetime in the ACCMIP simulations. *Atmos. Chem. Phys.*, **13**, 2563–2587.

Vukovich, F. M., 1995: Regional-scale boundary layer ozone variations in the eastern United States and their association with meteorological variations. *Atmos. Environ.*, **29**, 2259–2273.

Wang, B., et al., 2013: Preliminary evaluations on skills of FGOALS-g2 in decadal predictions. *Adv. Atmos. Sci.*, **30(3)**, 674–683.

Wang, H. J., J. Q. Sun, and K. Fan, 2007: Relationships between the North Pacific Oscillation and the typhoon/hurricane frequencies. *Sci. China D*, **50**, 1409–1416.

Wang, H. J., et al., 2012: Extreme climate in China: Facts, simulation and projection. *Meteorol. Z.*, **21(3)**, 279–304.

Wang, J., F. , et al., 2009: Impact of deforestation in the Amazon Basin on cloud climatology. *Proc. Natl. Acad. Sci.*, **106**, 3670–3674.

Wang, M., J. Overland, and N. Bond, 2010: Climate projections for selected large marine ecosystems. *J. Mar. Syst.*, doi: 10.1016/j.jmarsys.2008.11.028, 258–266.

Wang, M. Y., and J. E. Overland, 2009: A sea ice free summer Arctic within 30 years? *Geophys. Res. Lett.*, **36**, L07502, doi: 10.1029/2009GL037820.

Wang, R. F., L. G. Wu, and C. Wang, 2011: Typhoon track changes associated with global warming. *J. Clim.*, **24**, 3748–3752.

Weaver, C. P., et al., 2009: A preliminary synthesis of modeled climate change impacts on U.S. regional ozone concentrations. *Bull. Am. Meteorol. Soc.*, **90**, 1843–1863.

Weigel, A. P., R. Knutti, M. A. Liniger, and C. Appenzeller, 2010: Risks of model weighting in multimodel climate projections. *J. Clim.*, **23**, 4175–4191.

Weisheimer, A., T. N. Palmer, and F. J. Doblas-Reyes, 2011: Assessment of representations of model uncertainty in monthly and seasonal forecast ensembles. *Geophys. Res. Lett.*, **38**, L16703.

West, J. J., A. M. Fiore, L. W. Horowitz, and D. L. Mauzerall, 2006: Global health benefits of mitigating ozone pollution with methane emission controls. *Proc. Natl. Acad. Sci. U.S.A.*, **103**, 3988–3993.

Wigley, T., et al., 2009: Uncertainties in climate stabilization. *Clim. Change*, **97**, 85–121.

Wild, M., J. Grieser, and C. Schaer, 2008: Combined surface solar brightening and increasing greenhouse effect support recent intensification of the global land-based hydrological cycle. *Geophys. Res. Lett.*, **35**, L17706.

Wild, O., 2007: Modelling the global tropospheric ozone budget: Exploring the variability in current models. *Atmos. Chem. Phys.*, **7**, 2643–2660.

Wild, O., et al., 2012: Modelling future changes in surface ozone: A parameterized approach. *Atmos. Chem. Phys.*, **12**, 2037-2054, doi: 10.5194/acp-12-2037-2012.

Wilks, D. S., 2006: *Statistical Methods in the Atmospheric Sciences*, Vol. 91. Academic Press, Elsevier, San Diego, CA, USA, 627 pp.

Williams, A., and C. Funk, 2011: A westward extension of the warm pool leads to a westward extension of the Walker circulation, drying eastern Africa. *Clim. Dyn.*, **37**, 2417–2435.

Williams, P. D., E. Guilyardi, R. Sutton, J. Gregory, and G. Madec, 2007: A new feedback on climate change from the hydrological cycle. *Geophys. Res. Lett.*, **34**, L08706.

Woodward, S., D. L. Roberts, and R. A. Betts, 2005: A simulation of the effect of climate change; induced desertification on mineral dust aerosol. *Geophys. Res. Lett.*, **32**, L18810.

Woollings, T., 2010: Dynamical influences on European climate: An uncertain future. *Philos. Trans. R. Soc. A*, doi: 10.1098/rsta.2010.0040, 3733–3756.

Woollings, T., and M. Blackburn, 2012: The North Atlantic jet stream under climate change and its relation to the NAO and EA patterns. *J. Clim.*, **25**, 886–902.

WMO, 2002: Standardised Verification System (SVS) for Long-Range Forecasts (LRF). *New Attachment II-9 to the Manual on the GDPS (WMO-No. 485)* [W. SVS-LRF (ed.)]. World Meteorological Organization, Geneva, Switzerland.

WMO, 2010: *Scientific Assessment of Ozone Depletion: 2010*. Global Ozone Research and Monitoring Project-Report No. 52. 516. World Meteorological Organization, Geneva, Switzerland.

Wu, B., and T. J. Zhou, 2012: Prediction of decadal variability of sea surface temperature by a coupled global climate model FGOALS_gl developed in LASG/IAP. *Chin. Sci. Bull.*, **57**, 2453–2459.

Wu, P. L., R. Wood, J. Ridley, and J. Lowe, 2010: Temporary acceleration of the hydrological cycle in response to a CO_2 rampdown. *Geophys. Res. Lett.*, **37**, L12705.

Wu, S., L. J. Mickley, J. O. Kaplan, and D. J. Jacob, 2012: Impacts of changes in land use and land cover on atmospheric chemistry and air quality over the 21st century. *Atmos. Chem. Phys.*, **12**, 1597–1609.

Wu, S., L. J. Mickley, D. J. Jacob, D. Rind, and D. G. Streets, 2008: Effects of 2000–2050 changes in climate and emissions on global tropospheric ozone and the policy-relevant background surface ozone in the United States. *J. Geophys. Res.*, **113**, D18312.

Wu, S., L. J. Mickley, D. J. Jacob, J. A. Logan, R. M. Yantosca, and D. Rind, 2007: Why are there large differences between models in global budgets of tropospheric ozone? *J. Geophys. Res.*, **112**, D05302.

Xie, S., C. Deser, G. Vecchi, J. Ma, H. Teng, and A. Wittenberg, 2010: Global warming pattern formation: Sea surface temperature and rainfall. *J. Clim.*, doi: 10.1175/2009JCLI3329.1, 966–986.

Xin, X. G., T. W. Wu, and J. Zhang, 2013: Introduction of CMIP5 experiments carried out with the Climate System Models of Beijing Climate Center. *Adv. Clim. Change Res*, **4**, 41–49.

Xoplaki, E., P. Maheras, and J. Luterbacher, 2001: Variability of climate in Meridional Balkans during the periods 1675–1715 and 1780–1830 and its impact on human life. *Clim. Change*, **48**, 581–615.

Xu, Y., C.H. Xu, X.J. Gao, and Y. Luo, 2009: Projected changes in temperature and precipitation extremes over the Yangtze River Basin of China in the 21st century. *Quat. Int.*, **208**, 44-52.

Yang, X., et al., 2013: A predictable AMO-like pattern in GFDL's fully-coupled ensemble initialization and decadal forecasting system. *J. Clim.*, **26(2)**, 650-661.

Yeager, S., A. Karspeck, G. Danabasoglu, J. Tribbia, and H. Teng, 2012: A decadal prediction case study: Late 20th century North Atlantic ocean heat content. *J. Clim.*, **25**, 5173–5189.

Yin, J. J., M. E. Schlesinger, and R. J. Stouffer, 2009: Model projections of rapid sea-level rise on the northeast coast of the United States. *Nature Geosci.*, **2**, 262–266.

Yin, J. J., S. M. Griffies, and R. J. Stouffer, 2010: Spatial variability of sea level rise in twenty-first century projections. *J. Clim.*, **23**, 4585–4607.

Yip, S., C. A. T. Ferro, D. B. Stephenson, and E. Hawkins, 2011: A simple, coherent framework for partitioning uncertainty in climate predictions. *J. Clim.*, **24**, 4634–4643.

Yokohata, T., J. D. Annan, M. Collins, C. S. Jackson, M. Tobis, M. J. Webb, and J. C. Hargreaves, 2012: Reliability of multi-model and structurally different single-model ensembles. *Clim. Dyn.*, **39**, 599–616.

Young, P. J., et al., 2013: Pre-industrial to end 21st century projections of tropospheric ozone from the Atmospheric Chemistry and Climate Model Intercomparison Project (ACCMIP). *Atmos. Chem. Phys.*, **13**, 2063–2090.

Yue, X., H. J. Wang, H. Liao, and K. Fan, 2010: Simulation of dust aerosol radiative feedback using the GMOD: 2. Dust-climate interactions. *J. Geophys. Res. Atmos.*, **115**, D04201, doi: 10.1029/2009JD012063.

Yue, X., H. Liao, H. Wang, S. Li, and J. Tang, 2011: Role of sea surface temperature responses in simulation of the climatic effect of mineral dust aerosol,. *Atmos. Chem. Phys.*, **11**, 6049-6069, doi: 10.5194/acp-11-6049-2011.

Zanna, L., 2012: Forecast skill and predictability of observed Atlantic sea surface temperatures. *J. Clim.*, **25**, 5047–5056.

Zeng, G., J. A. Pyle, and P. J. Young, 2008: Impact of climate change on tropospheric ozone and its global budgets. *Atmos. Chem. Phys.*, **8**, 369–387.

Zeng, G., O. Morgenstern, P. Braesicke, and J. A. Pyle, 2010: Impact of stratospheric ozone recovery on tropospheric ozone and its budget. *Geophys. Res. Lett.*, **37**, L09805.

Zhang, R., and T. L. Delworth, 2006: Impact of Atlantic multidecadal oscillations on India/Sahel rainfall and Atlantic hurricanes. *Geophys. Res. Lett.*, **33**, L17712.

Zhang, S., A. Rosati, and T. Delworth, 2010a: The adequacy of observing systems in monitoring the Atlantic Meridional Overturning Circulation and North Atlantic Climate. *J. Clim.*, **23**, 5311–5324.

Zhang, S., M. J. Harrison, A. Rosati, and A. A. Wittenberg, 2007a: System design and evaluation of coupled ensemble data assimilation for global oceanic climate studies. *Mon. Weather Rev.*, **135**, 3541–3564.

Zhang, X., et al., 2007b: Detection of human influence on twentieth-century precipitation trends. *Nature*, **448**, 461–465.

Zhang, X. D., 2010: Sensitivity of arctic summer sea ice coverage to global warming forcing: Towards reducing uncertainty in arctic climate change projections. *Tellus A*, **62**, 220–227.

Zhang, Y., J. Wallace, and D. Battisti, 1997: ENSO-like interdecadal variability: 1900–93. *J. Clim.*, 1004–1020.

Zhang, Y., X. Y. Wen, and C. J. Jang, 2010b: Simulating chemistry-aerosol-cloud-radiation-climate feedbacks over the continental US using the online-coupled Weather Research Forecasting Model with chemistry (WRF/Chem). *Atmos. Environ.*, **44**, 3568–3582.

Zhang, Y., X.-M. Hu, L. R. Leung, and W. I. Gustafson, Jr., 2008: Impacts of regional climate change on biogenic emissions and air quality. *J. Geophys. Res.*, **113**, D18310.

Zhu, Y. L., H. J. Wang, W. Zhou, and J. H. Ma, 2011: Recent changes in the summer precipitation pattern in East China and the background circulation. *Clim. Dyn.*, **36**, 1463–1473.

12

Long-term Climate Change: Projections, Commitments and Irreversibility

Coordinating Lead Authors:
Matthew Collins (UK), Reto Knutti (Switzerland)

Lead Authors:
Julie Arblaster (Australia), Jean-Louis Dufresne (France), Thierry Fichefet (Belgium), Pierre Friedlingstein (UK/Belgium), Xuejie Gao (China), William J. Gutowski Jr. (USA), Tim Johns (UK), Gerhard Krinner (France/Germany), Mxolisi Shongwe (South Africa), Claudia Tebaldi (USA), Andrew J. Weaver (Canada), Michael Wehner (USA)

Contributing Authors:
Myles R. Allen (UK), Tim Andrews (UK), Urs Beyerle (Switzerland), Cecilia M. Bitz (USA), Sandrine Bony (France), Ben B.B. Booth (UK), Harold E. Brooks (USA), Victor Brovkin (Germany), Oliver Browne (UK), Claire Brutel-Vuilmet (France), Mark Cane (USA), Robin Chadwick (UK), Ed Cook (USA), Kerry H. Cook (USA), Michael Eby (Canada), John Fasullo (USA), Erich M. Fischer (Switzerland), Chris E. Forest (USA), Piers Forster (UK), Peter Good (UK), Hugues Goosse (Belgium), Jonathan M. Gregory (UK), Gabriele C. Hegerl (UK/Germany), Paul J. Hezel (Belgium/USA), Kevin I. Hodges (UK), Marika M. Holland (USA), Markus Huber (Switzerland), Philippe Huybrechts (Belgium), Manoj Joshi (UK), Viatcheslav Kharin (Canada), Yochanan Kushnir (USA), David M. Lawrence (USA), Robert W. Lee (UK), Spencer Liddicoat (UK), Christopher Lucas (Australia), Wolfgang Lucht (Germany), Jochem Marotzke (Germany), François Massonnet (Belgium), H. Damon Matthews (Canada), Malte Meinshausen (Germany), Colin Morice (UK), Alexander Otto (UK/Germany), Christina M. Patricola (USA), Gwenaëlle Philippon-Berthier (France), Prabhat (USA), Stefan Rahmstorf (Germany), William J. Riley (USA), Joeri Rogelj (Switzerland/Belgium), Oleg Saenko (Canada), Richard Seager (USA), Jan Sedláček (Switzerland), Len C. Shaffrey (UK), Drew Shindell (USA), Jana Sillmann (Canada), Andrew Slater (USA/Australia), Bjorn Stevens (Germany/USA), Peter A. Stott (UK), Robert Webb (USA), Giuseppe Zappa (UK/Italy), Kirsten Zickfeld (Canada/Germany)

Review Editors:
Sylvie Joussaume (France), Abdalah Mokssit (Morocco), Karl Taylor (USA), Simon Tett (UK)

This chapter should be cited as:
Collins, M., R. Knutti, J. Arblaster, J.-L. Dufresne, T. Fichefet, P. Friedlingstein, X. Gao, W.J. Gutowski, T. Johns, G. Krinner, M. Shongwe, C. Tebaldi, A.J. Weaver and M. Wehner, 2013: Long-term Climate Change: Projections, Commitments and Irreversibility. In: *Climate Change 2013: The Physical Science Basis. Contribution of Working Group I to the Fifth Assessment Report of the Intergovernmental Panel on Climate Change* [Stocker, T.F., D. Qin, G.-K. Plattner, M. Tignor, S.K. Allen, J. Boschung, A. Nauels, Y. Xia, V. Bex and P.M. Midgley (eds.)]. Cambridge University Press, Cambridge, United Kingdom and New York, NY, USA.

Table of Contents

12

Executive Summary

This chapter assesses long-term projections of climate change for the end of the 21st century and beyond, where the forced signal depends on the scenario and is typically larger than the internal variability of the climate system. Changes are expressed with respect to a baseline period of 1986–2005, unless otherwise stated.

Scenarios, Ensembles and Uncertainties

The Coupled Model Intercomparison Project Phase 5 (CMIP5) presents an unprecedented level of information on which to base projections including new Earth System Models with a more complete representation of forcings, new Representative Concentration Pathways (RCP) scenarios and more output available for analysis. The four RCP scenarios used in CMIP5 lead to a total radiative forcing (RF) at 2100 that spans a wider range than that estimated for the three Special Report on Emission Scenarios (SRES) scenarios (B1, A1B, A2) used in the Fourth Assessment Report (AR4), RCP2.6 being almost 2 W m^{-2} lower than SRES B1 by 2100. The magnitude of future aerosol forcing decreases more rapidly in RCP scenarios, reaching lower values than in SRES scenarios through the 21st century. Carbon dioxide (CO_2) represents about 80 to 90% of the total anthropogenic forcing in all RCP scenarios through the 21st century. The ensemble mean total effective RFs at 2100 for CMIP5 concentration-driven projections are 2.2, 3.8, 4.8 and 7.6 W m^{-2} for RCP2.6, RCP4.5, RCP6.0 and RCP8.5 respectively, relative to about 1850, and are close to corresponding Integrated Assessment Model (IAM)-based estimates (2.4, 4.0, 5.2 and 8.0 W m^{-2}). {12.2.1, 12.3, Table 12.1, Figures 12.1, 12.2, 12.3, 12.4}

New experiments and studies have continued to work towards a more complete and rigorous characterization of the uncertainties in long-term projections, but the magnitude of the uncertainties has not changed significantly since AR4. There is overall consistency between the projections based on CMIP3 and CMIP5, for both large-scale patterns and magnitudes of change. Differences in global temperature projections are largely attributable to a change in scenarios. Model agreement and confidence in projections depend on the variable and spatial and temporal averaging. The well-established stability of large-scale geographical patterns of change during a transient experiment remains valid in the CMIP5 models, thus justifying pattern scaling to approximate changes across time and scenarios under such experiments. Limitations remain when pattern scaling is applied to strong mitigation scenarios, to scenarios where localized forcing (e.g., aerosols) are significant and vary in time and for variables other than average temperature and precipitation. {12.2.2, 12.2.3, 12.4.2, 12.4.9, Figures 12.10, 12.39, 12.40, 12.41}

Projections of Temperature Change

Global mean temperatures will continue to rise over the 21st century if greenhouse gas (GHG) emissions continue unabated. Under the assumptions of the concentration-driven RCPs, global mean surface temperatures for 2081–2100, relative to 1986–2005 will *likely*[1] be in the 5 to 95% range of the CMIP5 models; 0.3°C to 1.7°C (RCP2.6), 1.1°C to 2.6°C (RCP4.5), 1.4°C to 3.1°C (RCP6.0), 2.6°C to 4.8°C (RCP8.5). Global temperatures averaged over the period 2081–2100 are projected to *likely* exceed 1.5°C above 1850-1900 for RCP4.5, RCP6.0 and RCP8.5 (*high confidence*), are *likely* to exceed 2°C above 1850-1900 for RCP6.0 and RCP8.5 (*high confidence*) and are *more likely than not* to exceed 2°C for RCP4.5 (*medium confidence*). Temperature change above 2°C under RCP2.6 is *unlikely* (*medium confidence*). Warming above 4°C by 2081–2100 is *unlikely* in all RCPs (*high confidence*) except for RCP8.5, where it is *about as likely as not* (*medium confidence*). {12.4.1, Tables 12.2, 12.3, Figures 12.5, 12.8}

Temperature change will not be regionally uniform. There is *very high confidence*[2] that globally averaged changes over land will exceed changes over the ocean at the end of the 21st century by a factor that is *likely* in the range 1.4 to 1.7. In the absence of a strong reduction in the Atlantic Meridional Overturning, the Arctic region is projected to warm most (*very high confidence*). This polar amplification is not found in Antarctic regions due to deep ocean mixing, ocean heat uptake and the persistence of the Antarctic ice sheet. Projected regional surface air temperature increase has minima in the North Atlantic and Southern Oceans in all scenarios. One model exhibits marked cooling in 2081–2100 over large parts of the Northern Hemisphere (NH), and a few models indicate slight cooling locally in the North Atlantic. Atmospheric zonal mean temperatures show warming throughout the troposphere, especially in the upper troposphere and northern high latitudes, and cooling in the stratosphere. {12.4.2, 12.4.3, Table 12.2, Figures 12.9, 12.10, 12.11, 12.12}

It is *virtually certain* that, in most places, there will be more hot and fewer cold temperature extremes as global mean temperatures increase. These changes are expected for events defined as extremes on both daily and seasonal time scales. Increases in the frequency, duration and magnitude of hot extremes along with heat stress are expected; however, occasional cold winter extremes will continue to occur. Twenty-year return values of low temperature events are projected to increase at a rate greater than winter mean temperatures in most regions, with the largest changes in the return values of low temperatures at high latitudes. Twenty-year return values for high temperature events are projected to increase at a rate similar to or greater than the rate of increase of summer mean temperatures in most regions. Under RCP8.5 it is *likely* that, in most land regions, a current 20-year high temperature event will occur more frequently by the end of the 21st

12

[1] In this Report, the following terms have been used to indicate the assessed likelihood of an outcome or a result: Virtually certain 99–100% probability, Very likely 90–100%, Likely 66–100%, About as likely as not 33–66%, Unlikely 0–33%, Very unlikely 0–10%, Exceptionally unlikely 0–1%. Additional terms (Extremely likely: 95–100%, More likely than not >50–100%, and Extremely unlikely 0–5%) may also be used when appropriate. Assessed likelihood is typeset in italics, e.g., *very likely* (see Section 1.4 and Box TS.1 for more details).

[2] In this Report, the following summary terms are used to describe the available evidence: limited, medium, or robust; and for the degree of agreement: low, medium, or high. A level of confidence is expressed using five qualifiers: very low, low, medium, high, and very high, and typeset in italics, e.g., *medium confidence*. For a given evidence and agreement statement, different confidence levels can be assigned, but increasing levels of evidence and degrees of agreement are correlated with increasing confidence (see Section 1.4 and Box TS.1 for more details).

century (at least doubling its frequency, but in many regions becoming an annual or 2-year event) and a current 20-year low temperature event will become exceedingly rare. {12.4.3, Figures 12.13, 12.14}

Changes in Atmospheric Circulation

Mean sea level pressure is projected to decrease in high latitudes and increase in the mid-latitudes as global temperatures rise. In the tropics, the Hadley and Walker Circulations are *likely* to slow down. Poleward shifts in the mid-latitude jets of about 1 to 2 degrees latitude are *likely* at the end of the 21st century under RCP8.5 in both hemispheres (*medium confidence*), with weaker shifts in the NH. In austral summer, the additional influence of stratospheric ozone recovery in the Southern Hemisphere opposes changes due to GHGs there, though the net response varies strongly across models and scenarios. Substantial uncertainty and thus *low confidence* remains in projecting changes in NH storm tracks, especially for the North Atlantic basin. The Hadley Cell is *likely* to widen, which translates to broader tropical regions and a poleward encroachment of subtropical dry zones. In the stratosphere, the Brewer–Dobson circulation is *likely* to strengthen. {12.4.4, Figures 12.18, 12.19, 12.20}

Changes in the Water Cycle

It is *virtually certain* that, in the long term, global precipitation will increase with increased global mean surface temperature. Global mean precipitation will increase at a rate per degree Celsius smaller than that of atmospheric water vapour. It will *likely* increase by 1 to 3% °C^{-1} for scenarios other than RCP2.6. For RCP2.6 the range of sensitivities in the CMIP5 models is 0.5 to 4% °C^{-1} at the end of the 21st century. {12.4.1, Figures 12.6, 12.7}

Changes in average precipitation in a warmer world will exhibit substantial spatial variation. Some regions will experience increases, other regions will experience decreases and yet others will not experience significant changes at all. There is *high confidence* that the contrast of annual mean precipitation between dry and wet regions and that the contrast between wet and dry seasons will increase over most of the globe as temperatures increase. The general pattern of change indicates that high latitude land masses are *likely* to experience greater amounts of precipitation due to the increased specific humidity of the warmer troposphere as well as increased transport of water vapour from the tropics by the end of this century under the RCP8.5 scenario. Many mid-latitude and subtropical arid and semi-arid regions will *likely* experience less precipitation and many moist mid-latitude regions will *likely* experience more precipitation by the end of this century under the RCP8.5 scenario. Globally, for short-duration precipitation events, a shift to more intense individual storms and fewer weak storms is *likely* as temperatures increase. Over most of the mid-latitude land-masses and over wet tropical regions, extreme precipitation events will *very likely* be more intense and more frequent in a warmer world. The global average sensitivity of the 20-year return value of the annual maximum daily precipitation increases ranges from 4% °C^{-1} of local temperature increase (average of CMIP3 models) to 5.3% °C^{-1} of local temperature increase (average of CMIP5 models) but regionally there are wide variations. {12.4.5, Figures 12.10, 12.22, 12.26, 12.27}

Annual surface evaporation is projected to increase as global temperatures rise over most of the ocean and is projected to change over land following a similar pattern as precipitation. Decreases in annual runoff are *likely* in parts of southern Europe, the Middle East, and southern Africa by the end of the 21st century under the RCP8.5 scenario. Increases in annual runoff are *likely* in the high northern latitudes corresponding to large increases in winter and spring precipitation by the end of the 21st century under the RCP8.5 scenario. Regional to global-scale projected decreases in soil moisture and increased risk of agricultural drought are *likely* in presently dry regions and are projected with *medium confidence* by the end of the 21st century under the RCP8.5 scenario. Prominent areas of projected decreases in evaporation include southern Africa and north western Africa along the Mediterranean. Soil moisture drying in the Mediterranean, southwest USA and southern African regions is consistent with projected changes in Hadley Circulation and increased surface temperatures, so surface drying in these regions as global temperatures increase is *likely* with *high confidence* by the end of this century under the RCP8.5 scenario. In regions where surface moistening is projected, changes are generally smaller than natural variability on the 20-year time scale. {12.4.5, Figures 12.23, 12.24, 12.25}

Changes in Cryosphere

It is *very likely* that the Arctic sea ice cover will continue shrinking and thinning year-round in the course of the 21st century as global mean surface temperature rises. At the same time, in the Antarctic, a decrease in sea ice extent and volume is expected, but with *low confidence*. Based on the CMIP5 multi-model ensemble, projections of average reductions in Arctic sea ice extent for 2081–2100 compared to 1986–2005 range from 8% for RCP2.6 to 34% for RCP8.5 in February and from 43% for RCP2.6 to 94% for RCP8.5 in September (*medium confidence*). A nearly ice-free Arctic Ocean (sea ice extent less than 1×10^6 km^{2}for at least 5 consecutive years) in September before mid-century is *likely* under RCP8.5 (*medium confidence*), based on an assessment of a subset of models that most closely reproduce the climatological mean state and 1979–2012 trend of the Arctic sea ice cover. Some climate projections exhibit 5- to 10-year periods of sharp summer Arctic sea ice decline—even steeper than observed over the last decade—and it is *likely* that such instances of rapid ice loss will occur in the future. There is little evidence in global climate models of a tipping point (or critical threshold) in the transition from a perennially ice-covered to a seasonally ice-free Arctic Ocean beyond which further sea ice loss is unstoppable and irreversible. In the Antarctic, the CMIP5 multi-model mean projects a decrease in sea ice extent that ranges from 16% for RCP2.6 to 67% for RCP8.5 in February and from 8% for RCP2.6 to 30% for RCP8.5 in September for 2081–2100 compared to 1986–2005. There is, however, *low confidence* in those values as projections because of the wide inter-model spread and the inability of almost all of the available models to reproduce the mean annual cycle, interannual variability and overall increase of the Antarctic sea ice areal coverage observed during the satellite era. {12.4.6, 12.5.5, Figures 12.28, 12.29, 12.30, 12.31}

It is *very likely* that NH snow cover will reduce as global temperatures rise over the coming century. A retreat of permafrost extent with rising global temperatures is *virtually certain*. Snow cover changes result from precipitation and ablation changes, which are sometimes opposite. Projections of the NH spring snow covered area by the end of the 21st century vary between a decrease of 7% (RCP2.6) and a decrease of 25% (RCP8.5), with a pattern that is fairly consistent between models. The projected changes in permafrost are a response not only to warming but also to changes in snow cover, which exerts a control on the underlying soil. By the end of the 21st century, diagnosed near-surface permafrost area is projected to decrease by between 37% (RCP2.6) and 81% (RCP8.5) (*medium confidence*). {12.4.6, Figures 12.32, 12.33}

Changes in the Ocean

The global ocean will warm in all RCP scenarios. The strongest ocean warming is projected for the surface in subtropical and tropical regions. At greater depth the warming is projected to be most pronounced in the Southern Ocean. Best estimates of ocean warming in the top one hundred meters are about 0.6°C (RCP2.6) to 2.0°C (RCP8.5), and about 0.3°C (RCP2.6) to 0.6°C (RCP8.5) at a depth of about 1 km by the end of the 21st century. For RCP4.5 by the end of the 21st century, half of the energy taken up by the ocean is in the uppermost 700 m and 85% is in the uppermost 2000 m. Due to the long time scales of this heat transfer from the surface to depth, ocean warming will continue for centuries, even if GHG emissions are decreased or concentrations kept constant. {12.4.7, 12.5.2–12.5.4, Figure 12.12}

It is *very likely* that the Atlantic Meridional Overturning Circulation (AMOC) will weaken over the 21st century but it is *very unlikely* that the AMOC will undergo an abrupt transition or collapse in the 21st century. Best estimates and ranges for the reduction from CMIP5 are 11% (1 to 24%) in RCP2.6 and 34% (12 to 54%) in RCP8.5. There is *low confidence* in assessing the evolution of the AMOC beyond the 21st century. {12.4.7, Figure 12.35}

Carbon Cycle

When forced with RCP8.5 CO_2 emissions, as opposed to the RCP8.5 CO_2 concentrations, 11 CMIP5 Earth System Models with interactive carbon cycle simulate, on average, a 50 ppm (min to max range –140 to +210 ppm) larger atmospheric CO_2 concentration and 0.2°C (min to max range –0.4 to +0.9°C) larger global surface temperature increase by 2100. {12.4.8, Figures 12.36, 12.37}

Long-term Climate Change, Commitment and Irreversibility

Global temperature equilibrium would be reached only after centuries to millennia if RF were stabilized. Continuing GHG emissions beyond 2100, as in the RCP8.5 extension, induces a total RF above 12 W m^{-2} by 2300. Sustained negative emissions beyond 2100, as in RCP2.6, induce a total RF below 2 W m^{-2} by 2300. The projected warming for 2281–2300, relative to 1986–2005, is 0.0°C to 1.2°C for RCP2.6 and 3.0°C to 12.6°C for RCP8.5 (*medium confidence*). In much the same way as the warming to a rapid increase of forcing is delayed, the cooling after a decrease of RF is also delayed. {12.5.1, Figures 12.43, 12.44}

A large fraction of climate change is largely irreversible on human time scales, unless net anthropogenic CO_2 emissions were strongly negative over a sustained period. For scenarios driven by CO_2 alone, global average temperature is projected to remain approximately constant for many centuries following a complete cessation of emissions. The positive commitment from CO_2 may be enhanced by the effect of an abrupt cessation of aerosol emissions, which will cause warming. By contrast, cessation of emission of short-lived GHGs will contribute a cooling. {12.5.3, 12.5.4, Figures 12.44, 12.45, 12.46, FAQ 12.3}

Equilibrium Climate Sensitivity and Transient Climate Response

Estimates of the equilibrium climate sensitivity (ECS) based on observed climate change, climate models and feedback analysis, as well as paleoclimate evidence indicate that ECS is *likely* in the range 1.5°C to 4.5°C with *high confidence*, *extremely unlikely* less than 1°C (*high confidence*) and *very unlikely* greater than 6°C (*medium confidence*). The transient climate response (TCR) is *likely* in the range 1°C to 2.5°C and *extremely unlikely* greater than 3°C, based on observed climate change and climate models. {Box 12.2, Figures 1, 2}

Climate Stabilization

The principal driver of long-term warming is total emissions of CO_2 and the two quantities are approximately linearly related. The global mean warming per 1000 PgC (transient climate response to cumulative carbon emissions (TCRE)) is *likely* between 0.8°C to 2.5°C per 1000 PgC, for cumulative emissions less than about 2000 PgC until the time at which temperatures peak. To limit the warming caused by anthropogenic CO_2 emissions alone to be *likely* less than 2°C relative to the period 1861-1880, total CO_2 emissions from all anthropogenic sources would need to be limited to a cumulative budget of about 1000 PgC since that period. About half [445 to 585 PgC] of this budget was already emitted by 2011. Accounting for projected warming effect of non-CO_2 forcing, a possible release of GHGs from permafrost or methane hydrates, or requiring a higher likelihood of temperatures remaining below 2°C, all imply a lower budget. {12.5.4, Figures 12.45, 12.46, Box 12.2}

Some aspects of climate will continue to change even if temperatures are stabilized. Processes related to vegetation change, changes in the ice sheets, deep ocean warming and associated sea level rise and potential feedbacks linking for example ocean and the ice sheets have their own intrinsic long time scales and may result in significant changes hundreds to thousands of years after global temperature is stabilized. {12.5.2 to 12.5.4}

Abrupt Change

Several components or phenomena in the climate system could potentially exhibit abrupt or nonlinear changes, and some are known to have done so in the past. Examples include the AMOC, Arctic sea ice, the Greenland ice sheet, the Amazon forest and monsoonal circulations. For some events, there is information on potential consequences, but in general there is *low confidence* and little consensus on the likelihood of such events over the 21st century. {12.5.5, Table 12.4}

12.1 Introduction

Projections of future climate change are not like weather forecasts. It is not possible to make deterministic, definitive predictions of how climate will evolve over the next century and beyond as it is with short-term weather forecasts. It is not even possible to make projections of the frequency of occurrence of all possible outcomes in the way that it might be possible with a calibrated probabilistic medium-range weather forecast. Projections of climate change are uncertain, first because they are dependent primarily on scenarios of future anthropogenic and natural forcings that are uncertain, second because of incomplete understanding and imprecise models of the climate system and finally because of the existence of internal climate variability. The term climate projection tacitly implies these uncertainties and dependencies. Nevertheless, as greenhouse gas (GHG) concentrations continue to rise, we expect to see future changes to the climate system that are greater than those already observed and attributed to human activities. It is possible to understand future climate change using models and to use models to characterize outcomes and uncertainties under specific assumptions about future forcing scenarios.

This chapter assesses climate projections on time scales beyond those covered in Chapter 11, that is, beyond the mid-21st century. Information from a range of different modelling tools is used here; from simple energy balance models, through Earth System Models of Intermediate Complexity (EMICs) to complex dynamical climate and Earth System Models (ESMs). These tools are evaluated in Chapter 9 and, where possible, the evaluation is used in assessing the validity of the projections. This chapter also summarizes some of the information on leading-order measures of the sensitivity of the climate system from other chapters and discusses the relevance of these measures for climate projections, commitments and irreversibility.

Since the AR4 (Meehl et al., 2007b) there have been a number of advances:

- New scenarios of future forcings have been developed to replace the Special Report on Emissions Scenarios (SRES). The Representative Concentration Pathways (RCPs, see Section 12.3) (Moss et al., 2010), have been designed to cover a wide range of possible magnitudes of climate change in models rather than being derived sequentially from storylines of socioeconomic futures. The aim is to provide a range of climate responses while individual socioeconomic scenarios may be derived, scaled and interpolated (some including explicit climate policy). Nevertheless, many studies that have been performed since AR4 have used SRES and, where appropriate, these are assessed. Simplified scenarios of future change, developed strictly for understanding the response of the climate system rather than to represent realistic future outcomes, are also synthesized and the understanding of leading-order measures of climate response such as the equilibrium climate sensitivity (ECS) and the transient climate response (TCR) are assessed.

- New models have been developed with higher spatial resolution, with better representation of processes and with the inclusion of more processes, in particular processes that are important in simulating the carbon cycle of the Earth. In these models, emissions of GHGs may be specified and these gases may be chemically active in the atmosphere or be exchanged with pools in terrestrial and oceanic systems before ending up as an airborne concentration (see Figure 10.1 of AR4).

- New types of model experiments have been performed, many coordinated by the Coupled Model Intercomparison Project Phase 5 (CMIP5) (Taylor et al., 2012), which exploit the addition of these new processes. Models may be driven by emissions of GHGs, or by their concentrations with different Earth System feedback loops cut. This allows the separate assessment of different feedbacks in the system and of projections of physical climate variables and future emissions.

- Techniques to assess and quantify uncertainties in projections have been further developed but a full probabilistic quantification remains difficult to propose for most quantities, the exception being global, temperature-related measures of the system sensitivity to forcings, such as ECS and TCR. In those few cases, projections are presented in the form of probability density functions (PDFs). We make the distinction between the spread of a multi-model ensemble, an *ad hoc* measure of the possible range of projections and the quantification of uncertainty that combines information from models and observations using statistical algorithms. Just like climate models, different techniques for quantifying uncertainty exist and produce different outcomes. Where possible, different estimates of uncertainty are compared.

Although not an advance, as time has moved on, the baseline period from which climate change is expressed has also moved on (a common baseline period of 1986–2005 is used throughout, consistent with the 2006 start-point for the RCP scenarios). Hence climate change is expressed as a change with respect to a recent period of history, rather than a time before significant anthropogenic influence. It should be borne in mind that some anthropogenically forced climate change had already occurred by the 1986–2005 period (see Chapter 10).

The focus of this chapter is on global and continental/ocean basin-scale features of climate. For many aspects of future climate change, it is possible to discuss generic features of projections and the processes that underpin them for such large scales. Where interesting or unique changes have been investigated at smaller scales, and there is a level of agreement between different studies of those smaller-scale changes, these may also be assessed in this chapter, although where changes are linked to climate phenomena such as El Niño, readers are referred to Chapter 14. Projections of atmospheric composition, chemistry and air quality for the 21st century are assessed in Chapter 11, except for CO_2 which is assessed in this chapter. An innovation for AR5 is Annex I: Atlas of Global and Regional Climate Projections, a collection of global and regional maps of projected climate changes derived from model output. A detailed commentary on each of the maps presented in Annex I is not provided here, but some discussion of generic features is provided.

Projections from regional models driven by boundary conditions from global models are not extensively assessed but may be mentioned in this chapter. More detailed regional information may be found in Chapter 14 and is also now assessed in the Working Group II report, where it can more easily be linked to impacts.

12.2 Climate Model Ensembles and Sources of Uncertainty from Emissions to Projections

12.2.1 The Coupled Model Intercomparison Project Phase 5 and Other Tools

Many of the figures presented in this chapter and in others draw on data collected as part of CMIP5 (Taylor et al., 2012). The project involves the worldwide coordination of ESM experiments including the coordination of input forcing fields, diagnostic output and the hosting of data in a distributed archive. CMIP5 has been unprecedented in terms of the number of modelling groups and models participating, the number of experiments performed and the number of diagnostics collected. The archive of model simulations began being populated by mid-2011 and continued to grow during the writing of AR5. The production of figures for this chapter draws on a fixed database of simulations and variables that was available on 15 March 2013 (the same as the cut-off date for the acceptance of the publication of papers). Different figures may use different subsets of models and there are unequal numbers of models that have produced output for the different RCP scenarios. Figure 12.1 gives a summary of which output was available from which model for which scenario. Where multiple runs are performed with exactly the same model but with different initial conditions, we choose only one ensemble member (usually the first but in cases where that was not available, the first available member is chosen) in order not to weight models with more ensemble members than others unduly in the multi-model synthesis. Rather than give an exhaustive account of which models were used to make which figures, this summary information is presented as a guide to readers.

In addition to output from CMIP5, information from a coordinated set of simulations with EMICs is also used (Zickfeld et al., 2013) to investigate long-term climate change beyond 2100. Even more simplified energy balance models or emulation techniques are also used, mostly to estimate responses where ESM experiments are not available (Meinshausen et al., 2011a; Good et al., 2013). An evaluation of the models used for projections is provided in Chapter 9 of this Report.

12.2.2 General Concepts: Sources of Uncertainties

The understanding of the sources of uncertainty affecting future climate change projections has not substantially changed since AR4, but many experiments and studies since then have proceeded to explore and characterize those uncertainties further. A full characterization,

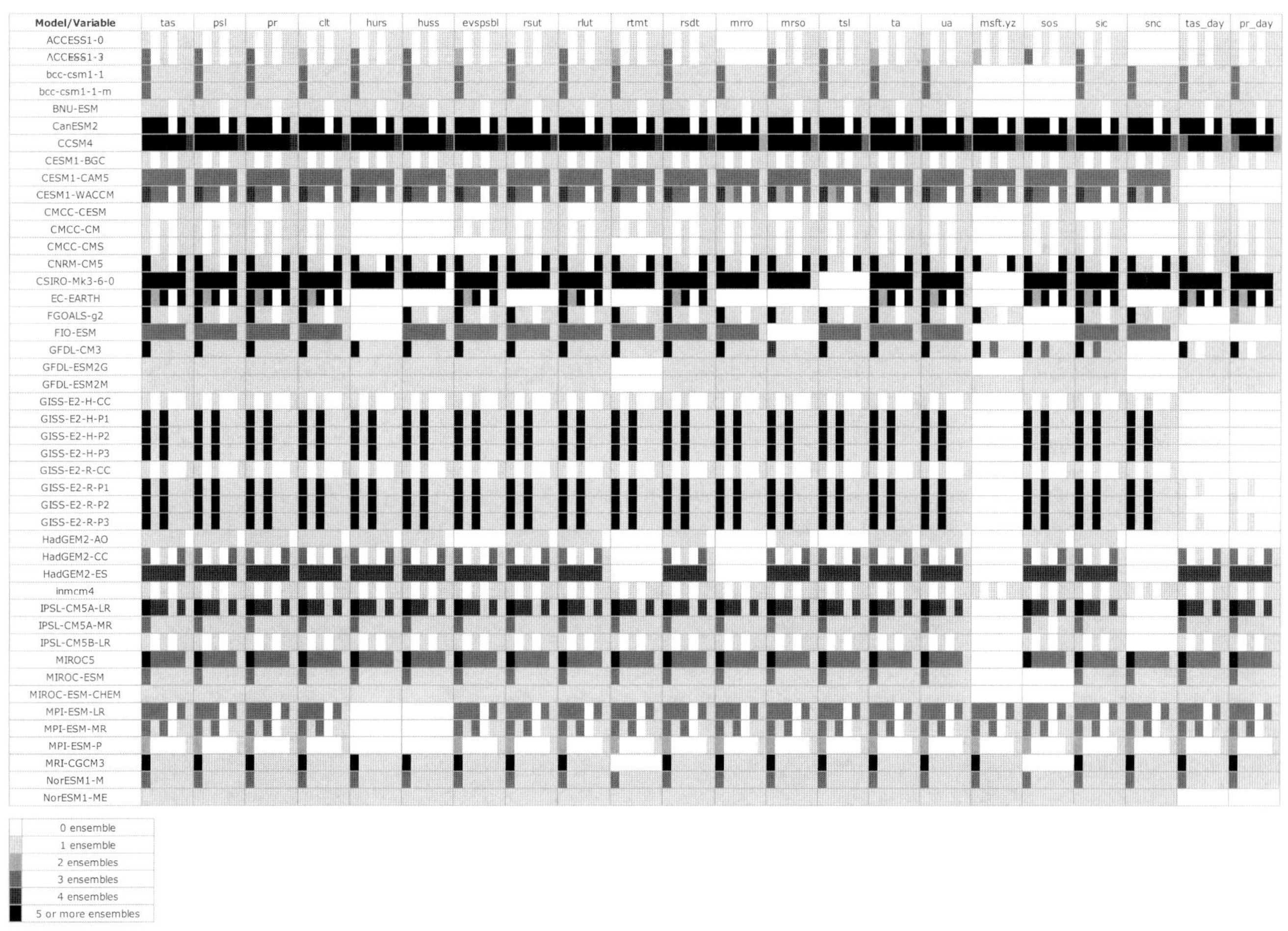

Figure 12.1 | A summary of the output used to make the CMIP5 figures in this chapter (and some figures in Chapter 11). The climate variable names run along the horizontal axis and use the standard abbreviations in the CMIP5 protocol (Taylor et al., 2012, and online references therein). The climate model names run along the vertical axis. In each box the shading indicates the number of ensemble members available for historical, RCP2.6, RCP4.5, RCP6.0, RCP8.5 and pre-industrial control experiments, although only one ensemble member per model is used in the relevant figures.

Frequently Asked Questions
FAQ 12.1 | Why Are So Many Models and Scenarios Used to Project Climate Change?

Future climate is partly determined by the magnitude of future emissions of greenhouse gases, aerosols and other natural and man-made forcings. These forcings are external to the climate system, but modify how it behaves. Future climate is shaped by the Earth's response to those forcings, along with internal variability inherent in the climate system. A range of assumptions about the magnitude and pace of future emissions helps scientists develop different emission scenarios, upon which climate model projections are based. Different climate models, meanwhile, provide alternative representations of the Earth's response to those forcings, and of natural climate variability. Together, ensembles of models, simulating the response to a range of different scenarios, map out a range of possible futures, and help us understand their uncertainties.

Predicting socioeconomic development is arguably even more difficult than predicting the evolution of a physical system. It entails predicting human behaviour, policy choices, technological advances, international competition and cooperation. The common approach is to use scenarios of plausible future socioeconomic development, from which future emissions of greenhouse gases and other forcing agents are derived. It has not, in general, been possible to assign likelihoods to individual forcing scenarios. Rather, a set of alternatives is used to span a range of possibilities. The outcomes from different forcing scenarios provide policymakers with alternatives and a range of possible futures to consider.

Internal fluctuations in climate are spontaneously generated by interactions between components such as the atmosphere and the ocean. In the case of near-term climate change, they may eclipse the effect of external perturbations, like greenhouse gas increases (see Chapter 11). Over the longer term, however, the effect of external forcings is expected to dominate instead. Climate model simulations project that, after a few decades, different scenarios of future anthropogenic greenhouse gases and other forcing agents—and the climate system's response to them—will differently affect the change in mean global temperature (FAQ 12.1, Figure 1, left panel). Therefore, evaluating the consequences of those various scenarios and responses is of paramount importance, especially when policy decisions are considered.

Climate models are built on the basis of the physical principles governing our climate system, and empirical understanding, and represent the complex, interacting processes needed to simulate climate and climate change, both past and future. Analogues from past observations, or extrapolations from recent trends, are inadequate strategies for producing projections, because the future will not necessarily be a simple continuation of what we have seen thus far.

Although it is possible to write down the equations of fluid motion that determine the behaviour of the atmosphere and ocean, it is impossible to solve them without using numerical algorithms through computer model simulation, similarly to how aircraft engineering relies on numerical simulations of similar types of equations. Also, many small-scale physical, biological and chemical processes, such as cloud processes, cannot be described by those equations, either because we lack the computational ability to describe the system at a fine enough resolution to directly simulate these processes or because we still have a partial scientific understanding of the mechanisms driving these processes. Those need instead to be approximated by so-called parameterizations within the climate models, through which a mathematical relation between directly simulated and approximated quantities is established, often on the basis of observed behaviour.

There are various alternative and equally plausible numerical representations, solutions and approximations for modelling the climate system, given the limitations in computing and observations. This diversity is considered a healthy aspect of the climate modelling community, and results in a range of plausible climate change projections at global and regional scales. This range provides a basis for quantifying uncertainty in the projections, but because the number of models is relatively small, and the contribution of model output to public archives is voluntary, the sampling of possible futures is neither systematic nor comprehensive. Also, some inadequacies persist that are common to all models; different models have different strength and weaknesses; it is not yet clear which aspects of the quality of the simulations that can be evaluated through observations should guide our evaluation of future model simulations. *(continued on next page)*

FAQ 12.1 (continued)

Models of varying complexity are commonly used for different projection problems. A faster model with lower resolution, or a simplified description of some climate processes, may be used in cases where long multi-century simulations are required, or where multiple realizations are needed. Simplified models can adequately represent large-scale average quantities, like global average temperature, but finer details, like regional precipitation, can be simulated only by complex models.

The coordination of model experiments and output by groups such as the Coupled Model Intercomparison Project (CMIP), the World Climate Research Program and its Working Group on Climate Models has seen the science community step up efforts to evaluate the ability of models to simulate past and current climate and to compare future climate change projections. The 'multi-model' approach is now a standard technique used by the climate science community to assess projections of a specific climate variable.

FAQ 12.1, Figure 1, right panels, shows the temperature response by the end of the 21st century for two illustrative models and the highest and lowest RCP scenarios. Models agree on large-scale patterns of warming at the surface, for example, that the land is going to warm faster than ocean, and the Arctic will warm faster than the tropics. But they differ both in the magnitude of their global response for the same scenario, and in small scale, regional aspects of their response. The magnitude of Arctic amplification, for instance, varies among different models, and a subset of models show a weaker warming or slight cooling in the North Atlantic as a result of the reduction in deepwater formation and shifts in ocean currents.

There are inevitable uncertainties in future external forcings, and the climate system's response to them, which are further complicated by internally generated variability. The use of multiple scenarios and models have become a standard choice in order to assess and characterize them, thus allowing us to describe a wide range of possible future evolutions of the Earth's climate.

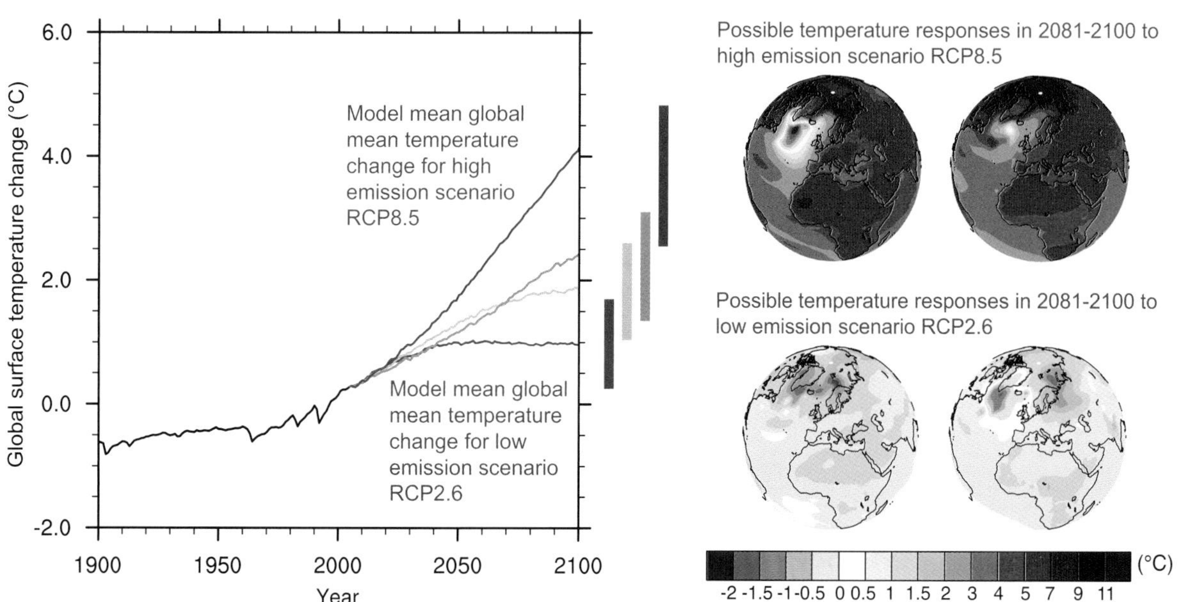

FAQ 12.1, Figure 1 | Global mean temperature change averaged across all Coupled Model Intercomparison Project Phase 5 (CMIP5) models (relative to 1986–2005) for the four Representative Concentration Pathway (RCP) scenarios: RCP2.6 (dark blue), RCP4.5 (light blue), RCP6.0 (orange) and RCP8.5 (red); 32, 42, 25 and 39 models were used respectively for these 4 scenarios. *Likely* ranges for global temperature change by the end of the 21st century are indicated by vertical bars. Note that these ranges apply to the difference between two 20-year means, 2081–2100 relative to 1986–2005, which accounts for the bars being centred at a smaller value than the end point of the annual trajectories. For the highest (RCP8.5) and lowest (RCP2.6) scenario, illustrative maps of surface temperature change at the end of the 21st century (2081–2100 relative to 1986–2005) are shown for two CMIP5 models. These models are chosen to show a rather broad range of response, but this particular set is not representative of any measure of model response uncertainty.

qualitative and even more so quantitative, involves much more than a measure of the range of model outcomes, because additional sources of information (e.g., observational constraints, model evaluation, expert judgement) lead us to expect that the uncertainty around the future climate state does not coincide straightforwardly with those ranges. In fact, in this chapter we highlight wherever relevant the distinction between model uncertainty evaluation, which encompasses the understanding that models have intrinsic shortcoming in fully and accurately representing the real system, and cannot all be considered independent of one another (Knutti et al., 2013), and a simpler descriptive quantification, based on the range of outcomes from the ensemble of models.

Uncertainty affecting mid- to long-term projections of climatic changes stems from distinct but possibly interacting sources. Figure 12.2 shows a schematic of the chain from scenarios, through ESMs to projections. Uncertainties affecting near-term projections of which some aspect are also relevant for longer-term projections are discussed in Section 11.3.1.1 and shown in Figure 11.8.

Future anthropogenic emissions of GHGs, aerosol particles and other forcing agents such as land use change are dependent on socioeconomic factors including global geopolitical agreements to control those emissions. Systematic studies that attempt to quantify the likely ranges of anthropogenic emission have been undertaken (Sokolov et al., 2009) but it is more common to use a scenario approach of different but plausible—in the sense of technically feasible—pathways, leading to the concept of *scenario uncertainty*. AR4 made extensive use of the SRES scenarios (IPCC, 2000) developed using a sequential approach, that is, socioeconomic factors feed into emissions scenarios which are then used either to directly force the climate models or to determine concentrations of GHGs and other agents required to drive these models. This report also assesses outcomes of simulations that use the new RCP scenarios, developed using a parallel process (Moss et al., 2010) whereby different targets in terms of RF at 2100 were selected (2.6, 4.5, 6.0 and 8.5 W m^{-2}) and GHG and aerosol emissions consistent with those targets, and their corresponding socioeconomic drivers were developed simultaneously (see Section 12.3). Rather than being identified with one socioeconomic storyline, RCP scenarios are consistent with many possible economic futures (in fact, different combinations of GHG and aerosol emissions can lead to the same RCP). Their development was driven by the need to produce scenarios that could be input to climate model simulations more expediently while corresponding socioeconomic scenarios would be developed in parallel, and to produce a wide range of model responses that may be scaled and interpolated to estimate the response under other scenarios, involving different measures of adaptation and mitigation.

In terms of the uncertainties related to the RCP emissions scenarios, the following issues can be identified:

- No probabilities or likelihoods have been attached to the alternative RCP scenarios (as was the case for SRES scenarios). Each of them should be considered plausible, as no study has questioned their technical feasibility (see Chapter 1).

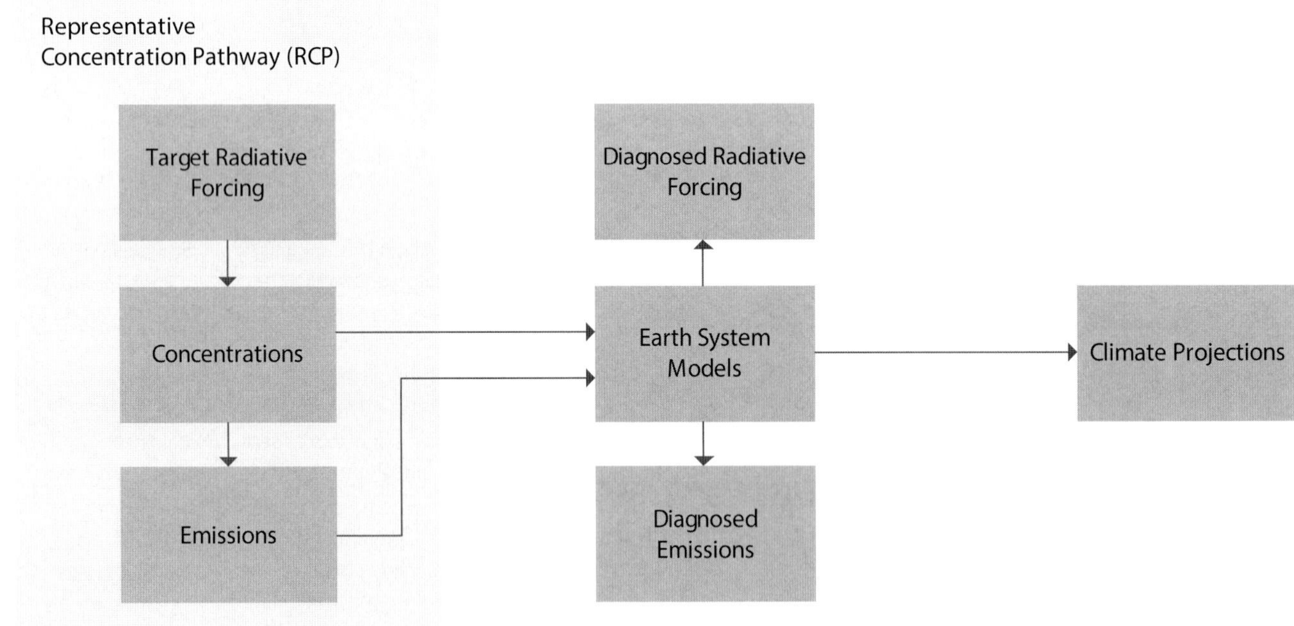

Figure 12.2 | Links in the chain from scenarios, through models to climate projections. The Representative Concentration Pathways (RCPs) are designed to sample a range of radiative forcing (RF) of the climate system at 2100. The RCPs are translated into both concentrations and emissions of greenhouse gases using Integrated Assessment Models (IAMs). These are then used as inputs to dynamical Earth System Models (ESMs) in simulations that are either concentration-driven (the majority of projection experiments) or emissions-driven (only for RCP8.5). Aerosols and other forcing factors are implemented in different ways in each ESM. The ESM projections each have a potentially different RF, which may be viewed as an output of the model and which may not correspond to precisely the level of RF indicated by the RCP nomenclature. Similarly, for concentration-driven experiments, the emissions consistent with those concentrations diagnosed from the ESM may be different from those specified in the RCP (diagnosed from the IAM). Different models produce different responses even under the same RF. Uncertainty propagates through the chain and results in a spread of ESM projections. This spread is only one way of assessing uncertainty in projections. Alternative methods, which combine information from simple and complex models and observations through statistical models or expert judgement, are also used to quantify that uncertainty.

- Despite the naming of the RCPs in terms of their target RF at 2100 or at stabilization (Box 1.1), climate models translate concentrations of forcing agents into RF in different ways due to their different structural modelling assumptions. Hence a model simulation of RCP6.0 may not attain exactly a RF of 6 W m^{-2}; more accurately, an RCP6.0 forced model experiment may not attain exactly the same RF as was intended by the specification of the RCP6.0 forcing inputs. Thus in addition to the scenario uncertainty there is RF uncertainty in the way the RCP scenarios are implemented in climate models.

- Some model simulations are concentration-driven (GHG concentrations are specified) whereas some models, which have Earth Systems components, convert emission scenarios into concentrations and are termed emissions-driven. Different ESMs driven by emissions may produce different concentrations of GHGs and aerosols because of differences in the representation and/or parameterization of the processes responsible for the conversion of emissions into concentrations. This aspect may be considered a facet of forcing uncertainty, or may be compounded in the category of model uncertainty, which we discuss below. Also, aerosol loading and land use changes are not dictated intrinsically by the RCP specification. Rather, they are a result of the Integrated Assessment Model that created the emission pathway for a given RCP.

SRES and RCPs account for future changes only in anthropogenic forcings. With regard to solar forcing, the 1985–2005 solar cycle is repeated. Neither projections of future deviations from this solar cycle, nor future volcanic RF and their uncertainties are considered.

Any climate projection is subject to sampling uncertainties that arise because of internal variability. In this chapter, the prediction of, for example, the amplitude or phase of some mode of variability that may be important on long time scales is not addressed (see Sections 11.2 and 11.3). Any climate variable projection derived from a single simulation of an individual climate model will be affected by internal variability (stemming from the chaotic nature of the system), whether it be a variable that involves a long time average (e.g., 20 years), a snapshot in time or some more complex diagnostic such as the variance computed from a time series over many years. No amount of time averaging can reduce internal variability to zero, although for some EMICs and simplified models, which may be used to reproduce the results of more complex model simulations, the representation of internal variability is excluded from the model specification by design. For different variables, and different spatial and time scale averages, the relative importance of internal variability in comparison with other sources of uncertainty will be different. In general, internal variability becomes more important on shorter time scales and for smaller scale variables (see Section 11.3 and Figure 11.2). The concept of signal-to-noise ratio may be used to quantify the relative magnitude of the forced response (signal) versus internal variability (noise). Internal variability may be sampled and estimated explicitly by running ensembles of simulations with slightly different initial conditions, designed explicitly to represent internal variability, or can be estimated on the basis of long control runs where external forcings are held constant. In the case of both multi-model and perturbed physics ensembles (see below), there is an implicit perturbation in the initial state of each run considered, which means that these ensembles sample both modelling uncertainty and internal variability jointly.

The ability of models to mimic nature is achieved by simplification choices that can vary from model to model in terms of the fundamental numeric and algorithmic structures, forms and values of parameterizations, and number and kinds of coupled processes included. Simplifications and the interactions between parameterized and resolved processes induce 'errors' in models, which can have a leading-order impact on projections. It is possible to characterize the choices made when building and running models into structural—indicating the numerical techniques used for solving the dynamical equations, the analytic form of parameterization schemes and the choices of inputs for fixed or varying boundary conditions—and parametric—indicating the choices made in setting the parameters that control the various components of the model. The community of climate modellers has regularly collaborated in producing coordinated experiments forming multi-model ensembles (MMEs), using both global and regional model families, for example, CMIP3/5 (Meehl et al., 2007a), ENSEMBLES (Johns et al., 2011) and Chemistry–Climate Model Validation 1 and 2 (CCM-Val-1 and 2; Eyring et al., 2005), through which structural uncertainty can be at least in part explored by comparing models, and perturbed physics ensembles (PPEs, with e.g., Hadley Centre Coupled Model version 3 (HadCM3; Murphy et al., 2004), Model for Interdiciplinary Research On Climate (MIROC; Yokohata et al., 2012), Community Climate System Model 3 (CCSM3; Jackson et al., 2008; Sanderson, 2011)), through which uncertainties in parameterization choices can be assessed in a given model. As noted below, neither MMEs nor PPEs represent an adequate sample of all the possible choices one could make in building a climate model. Also, current models may exclude some processes that could turn out to be important for projections (e.g., methane clathrate release) or produce a common error in the representation of a particular process. For this reason, it is of critical importance to distinguish two different senses in which the uncertainty terminology is used or misused in the literature (see also Sections 1.4.2, 9.2.2, 9.2.3, 11.2.1 and 11.2.2). A narrow interpretation of the concept of model uncertainty often identifies it with the range of responses of a model ensemble. In this chapter this type of characterization is referred as model range or model spread. A broader concept entails the recognition of a fundamental uncertainty in the representation of the real system that these models can achieve, given their necessary approximations and the limits in the scientific understanding of the real system that they encapsulate. When addressing this aspect and characterizing it, this chapter uses the term model uncertainty.

The relative role of the different sources of uncertainty—model, scenario and internal variability—as one moves from short- to mid- to long-term projections and considers different variables at different spatial scales has to be recognized (see Section 11.3). The three sources exchange relevance as the time horizon, the spatial scale and the variable change. In absolute terms, internal variability is generally estimated, and has been shown in some specific studies (Hu et al., 2012) to remain approximately constant across the forecast horizon, with model ranges and scenario/forcing variability increasing over time. For forecasts of global temperatures after mid-century, scenario and model ranges dominate the amount of variation due to internally generated variability, with scenarios accounting for the largest source of uncertainty in projections by the end of the century. For global average precipitation projections, scenario uncertainty has a much smaller role even by the end of the 21st century and model range maintains the largest share across all projection horizons. For temperature and precipitation projections at smaller spatial scales, internal variability may remain a significant source of uncertainty up until middle of the 21st century in some regions (Hawkins and Sutton, 2009, 2011; Rowell, 2012; Knutti and Sedláček, 2013). Within single model experiments, the persistently significant role of internally generated variability for regional projections even beyond short- and mid-term horizons has been documented by analyzing relatively large ensembles sampling initial conditions (Deser et al., 2012a, 2012b).

12.2.3 From Ensembles to Uncertainty Quantification

Ensembles like CMIP5 do not represent a systematically sampled family of models but rely on self-selection by the modelling groups. This opportunistic nature of MMEs has been discussed, for example, in Tebaldi and Knutti (2007) and Knutti et al. (2010a). These ensembles are therefore not designed to explore uncertainty in a coordinated manner, and the range of their results cannot be straightforwardly interpreted as an exhaustive range of plausible outcomes, even if some studies have shown how they appear to behave as well calibrated probabilistic forecasts for some large-scale quantities (Annan and Hargreaves, 2010). Other studies have argued instead that the tail of distributions is by construction undersampled (Räisänen, 2007). In general, the difficulty in producing quantitative estimates of uncertainty based on multiple model output originates in their peculiarities as a statistical sample, neither random nor systematic, with possible dependencies among the members (Jun et al., 2008; Masson and Knutti, 2011; Pennell and Reichler, 2011; Knutti et al., 2013) and of spurious nature, that is, often counting among their members models with different degrees of complexities (different number of processes explicitly represented or parameterized) even within the category of general circulation models.

Agreement between multiple models can be a source of information in an uncertainty assessment or confidence statement. Various methods have been proposed to indicate regions where models agree on the projected changes, agree on no change or disagree. Several of those methods are compared in Box 12.1. Many figures use stippling or hatching to display such information, but it is important to note that confidence cannot be inferred from model agreement alone.

Perturbed physics experiments (PPEs) differ in their output interpretability for they can be, and have been, systematically constructed and as such lend themselves to a more straightforward treatment through statistical modelling (Rougier, 2007; Sanso and Forest, 2009). Uncertain parameters in a single model to whose values the output is known to be sensitive are targeted for perturbations. More often it is the parameters in the atmospheric component of the model that are varied (Collins et al., 2006a; Sanderson et al., 2008), and to date have in fact shown to be the source of the largest uncertainties in large-scale response, but lately, with much larger computing power expense, also parameters within the ocean component have been perturbed (Collins et al., 2007; Brierley et al., 2010). Parameters in the land surface schemes have also been subject to perturbation studies (Fischer et al., 2011; Booth et al., 2012; Lambert et al., 2012). Ranges of possible values are explored and often statistical models that fit the relationship between parameter values and model output, that is, emulators, are trained on the ensemble and used to predict the outcome for unsampled parameter value combinations, in order to explore the parameter space more thoroughly that would otherwise be computationally affordable (Rougier et al., 2009). The space of a single model simulations (even when filtered through observational constraints) can show a large range of outcomes for a given scenario (Jackson et al., 2008). However, multi-model ensembles and perturbed physics ensembles produce modes and distributions of climate responses that can be different from one another, suggesting that one type of ensemble cannot be used as an analogue for the other (Murphy et al., 2007; Sanderson et al., 2010; Yokohata et al., 2010; Collins et al., 2011).

Many studies have made use of results from these ensembles to characterize uncertainty in future projections, and these will be assessed and their results incorporated when describing specific aspects of future climate responses. PPEs have been uniformly treated across the different studies through the statistical framework of analysis of computer experiments (Sanso et al., 2008; Rougier et al., 2009; Harris et al., 2010) or, more plainly, as a thorough exploration of alternative responses reweighted by observational constraints (Murphy et al., 2004; Piani et al., 2005; Forest et al., 2008; Sexton et al., 2012). In all cases the construction of a probability distribution is facilitated by the systematic nature of the experiments. MMEs have generated a much more diversified treatment (1) according to the choice of applying weights to the different models on the basis of past performance or not (Weigel et al., 2010) and (2) according to the choice between treating the different models and the truth as indistinguishable or treating each model as a version of the truth to which an error has been added (Annan and Hargreaves, 2010; Sanderson and Knutti, 2012). Many studies can be classified according to these two criteria and their combination, but even within each of the four resulting categories different studies produce different estimates of uncertainty, owing to the preponderance of *a priori* assumptions, explicitly in those studies that approach the problem through a Bayesian perspective, or only implicit in the choice of likelihood models, or weighting. This makes the use of probabilistic and other results produced through statistical inference necessarily dependent on agreeing with a particular set of assumptions (Sansom et al., 2013), given the lack of a full exploration of the robustness of probabilistic estimates to varying these assumptions.

In summary, there does not exist at present a single agreed on and robust formal methodology to deliver uncertainty quantification estimates of future changes in all climate variables (see also Section 9.8.3 and Stephenson et al., 2012). As a consequence, in this chapter, statements using the calibrated uncertainty language are a result of the expert judgement of the authors, combining assessed literature results with an evaluation of models demonstrated ability (or lack thereof) in simulating the relevant processes (see Chapter 9) and model consensus (or lack thereof) over future projections. In some cases when a significant relation is detected between model performance and reliability of its future projections, some models (or a particular parametric configuration) may be excluded (e.g., Arctic sea ice; Section 12.4.6.1 and Joshi et al., 2010) but in general it remains an open research question to find significant connections of this kind that justify some form of weighting across the ensemble of models and produce aggregated

Box 12.1 | Methods to Quantify Model Agreement in Maps

The climate change projections in this report are based on ensembles of climate models. The ensemble mean is a useful quantity to characterize the average response to external forcings, but does not convey any information on the robustness of this response across models, its uncertainty and/or likelihood or its magnitude relative to unforced climate variability. In the IPCC AR4 WGI contribution (IPCC, 2007) several criteria were used to indicate robustness of change, most prominently in Figure SPM.7. In that figure, showing projected precipitation changes, stippling marked regions where at least 90% of the CMIP3 models agreed on the sign of the change. Regions where less than 66% of the models agreed on the sign were masked white. The resulting large white area was often misinterpreted as indicating large uncertainties in the different models' response to external forcings, but recent studies show that, for the most part, the disagreement in sign among models is found where projected changes are small and still within the modelled range of internal variability, that is, where a response to anthropogenic forcings has not yet emerged locally in a statistically significant way (Tebaldi et al., 2011; Power et al., 2012).

A number of methods to indicate model robustness, involving an assessment of the significance of the change when compared to internal variability, have been proposed since AR4. The different methods share the purpose of identifying regions with large, significant or robust changes, regions with small changes, regions where models disagree or a combination of those. They do, however, use different assumptions about the statistical properties of the model ensemble, and therefore different criteria for synthesizing the information from it. Different methods also differ in the way they estimate internal variability. We briefly describe and compare several of these methods here.

Method (a): The default method used in Chapters 11,12 and 14 as well as in the Annex I (hatching only) is shown in Box 12.1, Figure 1a, and is based on relating the climate change signal to internal variability in 20-year means of the models as a reference[3]. Regions where the multi-model mean change exceeds two standard deviations of internal variability and where at least 90% of the models agree on the sign of change are stippled and interpreted as 'large change with high model agreement'. Regions where the model mean is less than one standard deviation of internal variability are hatched and interpreted as 'small signal or low agreement of models'. This can have various reasons: (1) changes in individual models are smaller than internal variability, or (2) although changes in individual models are significant, they disagree about the sign and the multi-model mean change remains small. Using this method, the case where all models scatter widely around zero and the case where all models agree on near zero change therefore are both hatched (e.g., precipitation change over the Amazon region by the end of the 21st century, which the following methods mark as 'inconsistent model response').

Method (b): Method (a) does not distinguish the case where all models agree on no change and the case where, for example, half of the models show a significant increase and half a decrease. The distinction may be relevant for many applications and a modification of method (a) is to restrict hatching to regions where there is high agreement among the models that the change will be 'small', thus eliminating the ambiguous interpretation 'small or low agreement' in (a). In contrast to method (a) where the model mean is compared to variability, this case (b) marks regions where at least 80% of the individual models show a change smaller than two standard deviations of variability with hatching. Grid points where many models show significant change but don't agree are no longer hatched (Box 12.1, Figure 1b).

Method (c): Knutti and Sedláček (2013) define a dimensionless robustness measure, R, which is inspired by the signal-to-noise ratio and the ranked probability skill score. It considers the natural variability and agreement on magnitude and sign of change. A value of $R = 1$ implies perfect model agreement; low or negative values imply poor model agreement (note that by definition R can assume any negative value). Any level of R can be chosen for the stippling. For illustration, in Box 12.1, Figure 1c, regions with $R > 0.8$ are marked with small dots, regions with $R > 0.9$ with larger dots and are interpreted as 'robust large change'. This yields similar results to method (a) for the end of the century, but with some areas of moderate model robustness ($R > 0.8$) already for the near-term projections, even though the signal is still within the noise. Regions where at least 80% of the models individually show no significant change are hatched and interpreted as 'changes unlikely to emerge from variability'[4].There is less hatching in this method than in method (a),

(continued on next page)

[3] The internal variability in this method is estimated using pre-industrial control runs for each of the models which are at least 500 years long. The first 100 years of the pre-industrial are ignored. Variability is calculated for every grid point as the standard deviation of non-overlapping 20-year means, multiplied by the square root of 2 to account for the fact that the variability of a difference in means is of interest. A quadratic fit as a function of time is subtracted from these at every grid point to eliminate model drift. This is by definition the standard deviation of the difference between two independent 20-year averages having the same variance and estimates the variation of that difference that would be expected due to unforced internal variability. The median across all models of that quantity is used.

[4] Variability in methods b–d is estimated from interannual variations in the base period within each model.

12

Box 12.1 (continued)

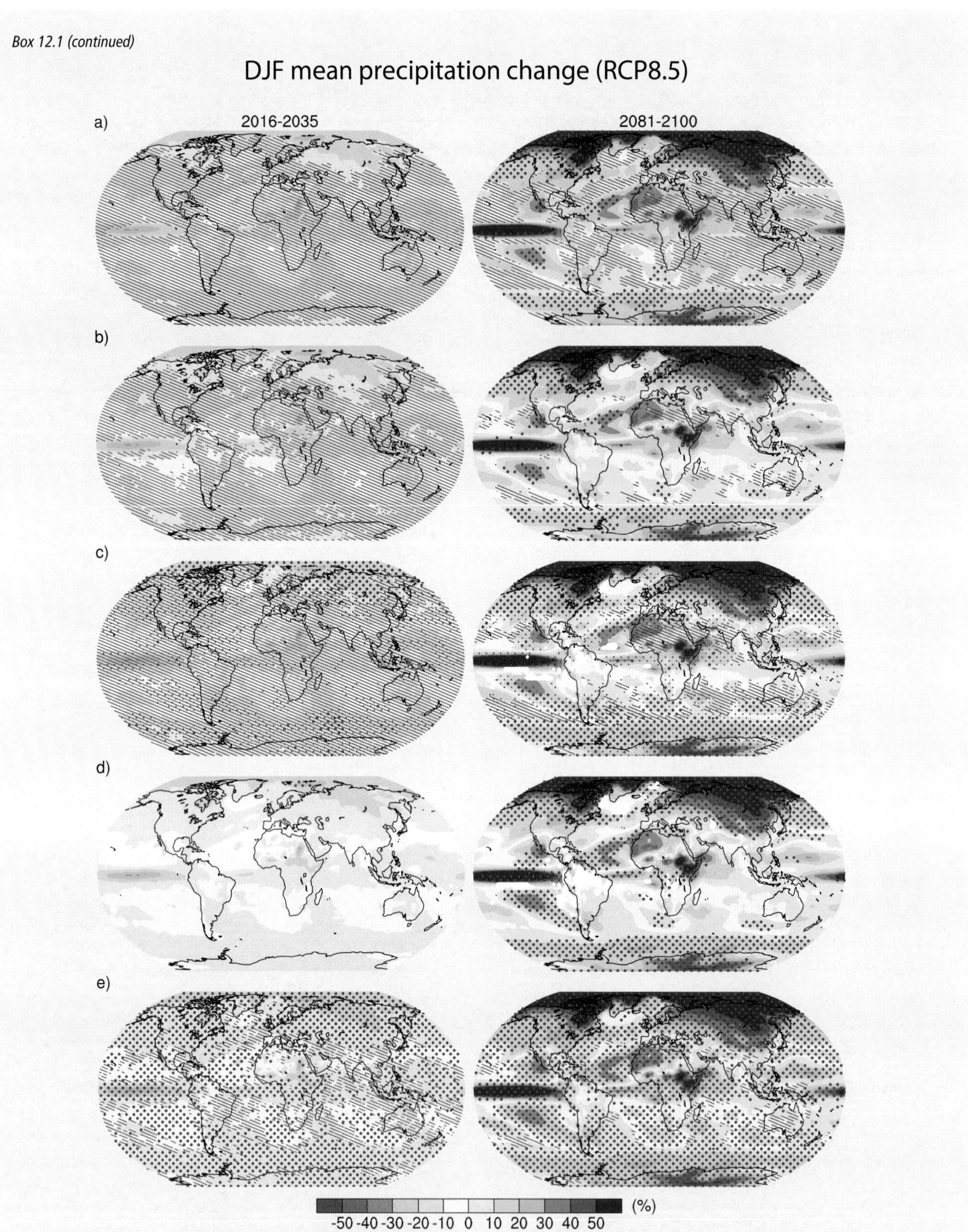

Box 12.1, Figure 1 | Projected change in December to February precipitation for 2016–2035 and 2081–2100, relative to 1986–2005 from CMIP5 models. The choice of the variable and time frames is just for illustration of how the different methods compare in cases with low and high signal-to-noise ratio (left and right column, respectively). The colour maps are identical along each column and only stippling and hatching differ on the basis of the different methods. Different methods for stippling and hatching are shown determined (a) from relating the model mean to internal variability, (b) as in (a) but hatching here indicates high agreement for 'small change', (c) by the robustness measure by Knutti and Sedláček (2013), (d) by the method proposed by Tebaldi et al. (2011) and (e) by the method by Power et al. (2012). Detailed technical explanations for each method are given in the text. 39 models are used in all panels.

Box 12.1 (continued)

because it requires 80% of the models to be within variability, not just the model average. Regions where at least 50% of the models show significant change but $R < 0.5$ are masked as white to indicate 'models disagreeing on the projected change projections' (Box 12.1, Figure 1c).

Method (d): Tebaldi et al. (2011) start from IPCC AR4 SPM7 but separate lack of model agreement from lack of signal (Box 12.1, Figure 1e). Grid points are stippled and interpreted as 'robust large change' when more than 50% of the models show significant change and at least 80% of those agree on the sign of change. Grid points where more than 50% of the models show significant change but less than 80% of those agree on the sign of change are masked as white and interpreted as 'unreliable'. The results are again similar to the methods above. No hatching was defined in that method (Box 12.1 Figure 1d). (See also Neelin et al., 2006 for a similar approach applied to a specific regional domain.)

Method (e): Power et al. (2012) identify three distinct regions using various methods in which projections can be very loosely described as either: 'statistically significant', 'small (relative to temporal variability) or zero, but not statistically significant' or 'uncertain'. The emphasis with this approach is to identify robust signals taking the models at face value and to address the questions: (1) What will change? (2) By how much? and (3) What will not change? The underlying consideration here is that statistical testing under the assumption of model independence provides a worthwhile, albeit imperfect, line of evidence that needs to be considered in conjunction with other evidence (e.g., degree of interdependence, ability of models to simulate the past), in order to assess the degree of confidence one has in a projected change.

The examples given here are not exhaustive but illustrate the main ideas. Other methods include simply counting the number of models agreeing on the sign (Christensen et al., 2007), or varying colour hue and saturation to indicate magnitude of change and robustness of change separately (Kaye et al., 2012). In summary, there are a variety of ways to characterize magnitude or significance of change, and agreement between models. There is also a compromise to make between clarity and richness of information. Different methods serve different purposes and a variety of criteria can be justified to highlight specific properties of multi-model ensembles. Clearly only a subset of information regarding robust and uncertain change can be conveyed in a single plot. The methods above convey some important pieces of this information, but obviously more information could be provided if more maps with additional statistics were provided. In fact Annex I provides more explicit information on the range of projected changes evident in the models (e.g., the median, and the upper and lower quartiles). For most of the methods there is a necessity to choose thresholds for the level of agreement that cannot be identified objectively, but could be the result of individual, application-specific evaluations. Note also that all of the above methods measure model agreement in an ensemble of opportunity, and it is impossible to derive a confidence or likelihood statement from the model agreement or model spread alone, without considering consistency with observations, model dependence and the degree to which the relevant processes are understood and reflected in the models (see Section 12.2.3).

The method used by Power et al. (2012) differs from the other methods in that it tests the statistical significance of the ensemble mean rather than a single simulation. As a result, the area where changes are significant increases with an increasing number of models. Already for the period centred on 2025, most of the grid points when using this method show significant change in the ensemble mean whereas in the other methods projections for this time period are classified as changes not exceeding internal variability. The reason is that the former produces a statement about the mean of the distribution being significantly different from zero, equivalent to treating the ensemble as 'truth plus error', that is, assuming that the models are independent and randomly distributed around reality. Methods a–d, on the other hand, use an 'indistinguishable' interpretation, in which each model and reality are drawn from the same distribution. In that case, the stippling and hatching characterize the likelihood of a single member being significant or not, rather than the ensemble mean. There is some debate in the literature on how the multi-model ensembles should be interpreted statistically. This and past IPCC reports treat the model spread as some measure of uncertainty, irrespective of the number of models, which implies an 'indistinguishable' interpretation. For a detailed discussion readers are referred to the literature (Tebaldi and Knutti, 2007; Annan and Hargreaves, 2010; Knutti et al., 2010a, 2010b; Annan and Hargreaves, 2011a; Sanderson and Knutti, 2012).

future projections that are significantly different from straightforward one model–one vote (Knutti, 2010) ensemble results. Therefore, most of the analyses performed for this chapter make use of all available models in the ensembles, with equal weight given to each of them unless otherwise stated.

12.2.4 Joint Projections of Multiple Variables

While many of the key processes relevant to the simulation of single variables are understood, studies are only starting to focus on assessing projections of joint variables, especially when extremes or variability in the individual quantities are of concern. A few studies have addressed projected changes in joint variables, for example, by combining mean temperature and precipitation (Williams et al., 2007; Tebaldi and Lobell, 2008; Tebaldi and Sanso, 2009; Watterson, 2011; Watterson and Whetton, 2011a; Sexton et al., 2012), linking soil moisture, precipitation and temperature mean and variability (Seneviratne et al., 2006; Fischer and Schär, 2009; Koster et al., 2009b, 2009c), combining temperature and humidity (Diffenbaugh et al., 2007; Fischer and Schär, 2010; Willett and Sherwood, 2012), linking summertime temperature and soil moisture to prior winter snowpack (Hall et al., 2008) or linking precipitation change to circulation, moisture and moist static energy budget changes (Neelin et al., 2003; Chou and Neelin, 2004; Chou et al., 2006, 2009). Models may have difficulties simulating all relevant interactions between atmosphere and land surface and the water cycle that determine the joint response, observations to evaluate models are often limited (Seneviratne et al., 2010), and model uncertainties are therefore large (Koster et al., 2006; Boé and Terray, 2008; Notaro, 2008; Fischer et al., 2011). In some cases, correlations between, for example, temperature and precipitation or accumulated precipitation and temperature have found to be too strong in climate models (Trenberth and Shea, 2005; Hirschi et al., 2011). The situation is further complicated by the fact that model biases in one variable affect other variables. The standard method for model projections is to subtract model biases derived from control integrations (assuming that the bias remains constant in a future scenario integration). Several studies note that this may be problematic when a consistent treatment of biases in multiple variables is required (Christensen et al., 2008; Buser et al., 2009), but there is no consensus at this stage for a methodology addressing this problem (Ho et al., 2012). More generally the existence of structural errors in models according to which an unavoidable discrepancy (Rougier, 2007) between their simulations and reality cannot be avoided is relevant here, as well as for univariate projections. In the recent literature an estimate of this discrepancy has been proposed through the use of MMEs, using each model in turn as a surrogate for reality, and measuring the distance between it and the other models of the ensemble. Some summary statistic of these measures is then used to estimate the distance between models and the real world (Sexton and Murphy, 2012; Sexton et al., 2012; Sanderson, 2013). Statistical frameworks to deal with multivariate projections are challenging even for just two variables, as they have to address a trade-off between modelling the joint behavior at scales that are relevant for impacts—that is, fine spatial and temporal scales, often requiring complex spatio-temporal models—and maintaining computational feasibility. In one instance (Tebaldi and Sanso, 2009) scales were investigated at the seasonal and sub-continental level, and projections of the forced response of temperature and precipitation at those scales did not show significant correlations, likely because of the heterogeneity of the relation between the variables within those large averaged regions and seasons. In Sexton et al. (2012) the spatial scale focussed on regions of Great Britain and correlation emerged as more significant, for example, between summer temperatures and precipitation amounts. Fischer and Knutti (2013) estimated strong relationships between variables making up impact relevant indices (e.g., temperature and humidity) and showed how in some cases, uncertainties across models are larger for a combined variable than if the uncertainties in the individual underlying variables were treated independently (e.g., wildfires), whereas in other cases the uncertainties in the combined variables are smaller than in the individual ones (e.g., heat stress for humans).

Even while recognizing the need for joint multivariate projections, the above limitations at this stage prevent a quantitative assessment for most cases. A few robust qualitative relationships nonetheless emerge from the literature and these are assessed, where appropriate, in the rest of the chapter. For applications that are sensitive to relationships between variables, but still choose to use the multi-model framework to determine possible ranges for projections, sampling from univariate ranges may lead to unrealistic results when significant correlations exist. IPCC assessments often show model averages as best estimates, but such averages can underestimate spatial variability, and more in general they neither represent any of the actual model states (Knutti et al., 2010a) nor do they necessarily represent the joint best estimate in a multivariate sense. Impact studies usually need temporally and spatially coherent multivariate input from climate model simulations. In those cases, using each climate model output individually and feeding it into the impact model, rather than trying to summarise a multivariate distribution from the MME and sample from it, is likely to be more consistent, assuming that the climate model itself correctly captures the spatial covariance, the temporal co-evolution and the relevant feedbacks.

12.3 Projected Changes in Forcing Agents, Including Emissions and Concentrations

The experiments that form the basis of global future projections discussed in this chapter are extensions of the simulations of the observational record discussed in Chapters 9 and 10. The scenarios assessed in AR5, introduced in Chapter 1, include four new scenarios designed to explore a wide range of future climate characterized by representative trajectories of well-mixed greenhouse gas (WMGHG) concentrations and other anthropogenic forcing agents. These are described further in Section 12.3.1. The implementation of forcing agents in model projections, including natural and anthropogenic aerosols, ozone and land use change are discussed in Section 12.3.2, with a strong focus on CMIP5 experiments. Global mean emissions, concentrations and RFs applicable to the historical record simulations assessed in Chapters 8, 9 and 10, and the future scenario simulations assessed here, are listed in Annex II. Global mean RF for the 21st century consistent with these scenarios, derived from CMIP5 and other climate model studies, is discussed in Section 12.3.3.

12.3.1 Description of Scenarios

Long-term climate change projections reflect how human activities or natural effects could alter the climate over decades and centuries. In this context, defined scenarios are important, as using specific time series of emissions, land use, atmospheric concentrations or RF across multiple models allows for coherent climate model intercomparisons and synthesis. Some scenarios present a simple stylized future (not accompanied by a socioeconomic storyline) and are used for process understanding. More comprehensive scenarios are produced by Integrated Assessment Models (IAMs) as internally consistent sets of emissions and socioeconomic assumptions (e.g., regarding population and socioeconomic development) with the aim of presenting several plausible future worlds (see Section 1.5.2 and Box 1.1). In general it is these scenarios that are used for policy relevant climate change, impact, adaptation and mitigation analysis. It is beyond the scope of this report to consider the full range of currently published scenarios and their implications for mitigation policy and climate targets—that is covered by the Working Group III contribution to the AR5. Here, we focus on the RCP scenarios used within the CMIP5 intercomparison exercise (Taylor et al. 2012) along with the SRES scenarios (IPCC, 2000) developed for the IPCC Third Assessment Report (TAR) but still widely used by the climate community.

12.3.1.1 Stylized Concentration Scenarios

A 1% per annum compound increase of atmospheric CO_2 concentration until a doubling or a quadrupling of its initial value has been widely used since the second phase of CMIP (Meehl et al., 2000) and the Second Assessment Report (Kattenberg et al., 1996). This stylized scenario is a useful benchmark for comparing coupled model climate sensitivity, climate feedback and transient climate response, but is not used directly for future projections. The exponential increase of CO_2 concentration induces approximately a linear increase in RF due to a 'saturation effect' of the strong absorbing bands (Augustsson and Ramanathan, 1977; Hansen et al., 1988; Myhre et al., 1998). Thus, a linear ramp function in forcing results from these stylized pathways, adding to their suitability for comparative diagnostics of the models' climate feedbacks and inertia. The CMIP5 intercomparison project again includes such a stylized pathway, in which the CO_2 concentration reaches twice the initial concentration after 70 years and four times the initial concentration after 140 years. The corresponding RFs are 3.7 W m^{-2} (Ramaswamy et al., 2001) and 7.4 W m^{-2} respectively with a range of ±20% accounting for uncertainties in radiative transfer calculations and rapid adjustments (see Section 8.3.2.1), placing them within the range of the RFs at the end of the 21st century for the future scenarios presented below. The CMIP5 project also includes a second stylized experiment in which the CO_2 concentration is quadrupled instantaneously, which allows a distinction between effective RFs and longer-term climate feedbacks (Gregory et al., 2004).

12.3.1.2 The Socioeconomic Driven Scenarios from the Special Report on Emission Scenarios

The climate change projections undertaken as part of CMIP3 and discussed in AR4 were based primarily on the SRES A2, A1B and B1 scenarios (IPCC, 2000). These scenarios were developed using IAMs and resulted from specific socioeconomic scenarios, that is, from storylines about future demographic and economic development, regionalization, energy production and use, technology, agriculture, forestry, and land use. All SRES scenarios assumed that no climate mitigation policy would be undertaken. Based on these SRES scenarios, global climate models were then forced with corresponding WMGHG and aerosol concentrations, although the degree to which models implemented these forcings differed (Meehl et al., 2007b, Table 10.1). The resulting climate projections, together with the socioeconomic scenarios on which they are based, have been widely used in further analysis by the impact, adaptation and vulnerability research communities.

12.3.1.3 The New Concentration Driven Representative Concentration Pathway Scenarios, and Their Extensions

As introduced in Box 1.1 and mentioned in Section 12.1, a new parallel process for scenario development was proposed in order to facilitate the interactions between the scientific communities working on climate change, adaptation and mitigation (Hibbard et al., 2007; Moss et al., 2008, 2010; van Vuuren et al., 2011). These new scenarios, Representative Concentration Pathways, are referred to as pathways in order to emphasize that they are not definitive scenarios, but rather internally consistent sets of time-dependent forcing projections that could potentially be realized with more than one underlying socioeconomic scenario. The primary products of the RCPs are concentrations but they also provide gas emissions. They are representative in that they are one of several different scenarios, sampling the full range of published scenarios (including mitigation scenarios) at the time they were defined, that have similar RF and emissions characteristics. They are identified by the approximate value of the RF (in W m^{-2}) at 2100 or at stabilization after 2100 in their extensions, relative to pre-industrial (Moss et al., 2008; Meinshausen et al., 2011c). RCP2.6 (the lowest of the four, also referred to as RCP3-PD) peaks at 3.0 W m^{-2} and then declines to 2.6 W m^{-2} in 2100, RCP4.5 (medium-low) and RCP6.0 (medium-high) stabilize after 2100 at 4.2 and 6.0 W m^{-2} respectively, while RCP8.5 (highest) reaches 8.3 W m^{-2} in 2100 on a rising trajectory (see also Figure 12.3a which takes into account the efficacies of the various anthropogenic forcings). The primary objective of these scenarios is to provide all the input variables necessary to run comprehensive climate models in order to reach a target RF (Figure 12.2). These scenarios were developed using IAMs that provide the time evolution of a large ensemble of anthropogenic forcings (concentration and emission of gas and aerosols, land use changes, etc.) and their individual RF values (Moss et al., 2008, 2010; van Vuuren et al., 2011). Note that due to the substantial uncertainties in RF, these forcing values should be understood as comparative 'labels', not as exact definitions of the forcing that is effective in climate models. This is because concentrations or emissions, rather than the RF itself, are prescribed in the CMIP5 climate model runs. The forcing as manifested in climate models is discussed in Section 12.3.3.

Various steps were necessary to turn the selected 'raw' RCP scenarios from the IAMs into data sets usable by the climate modelling community. First, harmonization with historical data was performed for emissions of reactive gases and aerosols (Lamarque et al., 2010; Granier et al., 2011; Smith et al., 2011), land use (Hurtt et al., 2011), and for GHG emissions and concentrations (Meinshausen et al., 2011c). Then

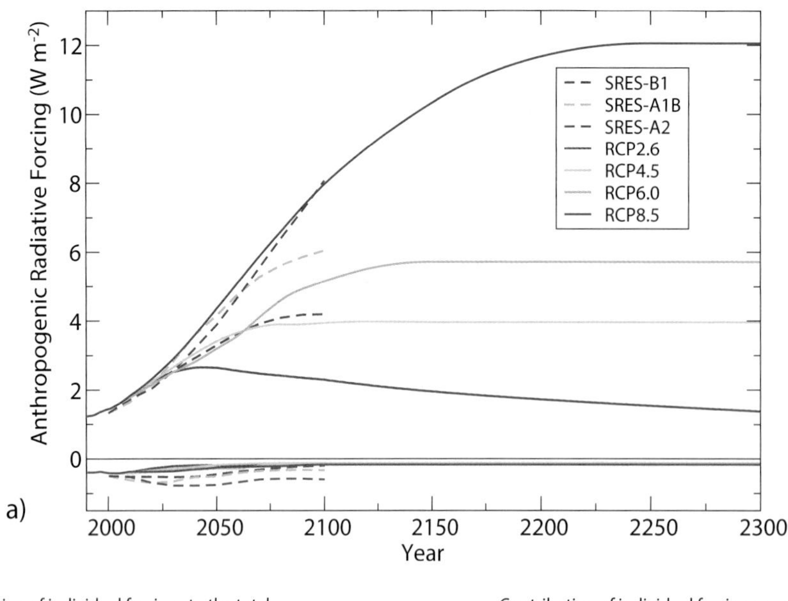

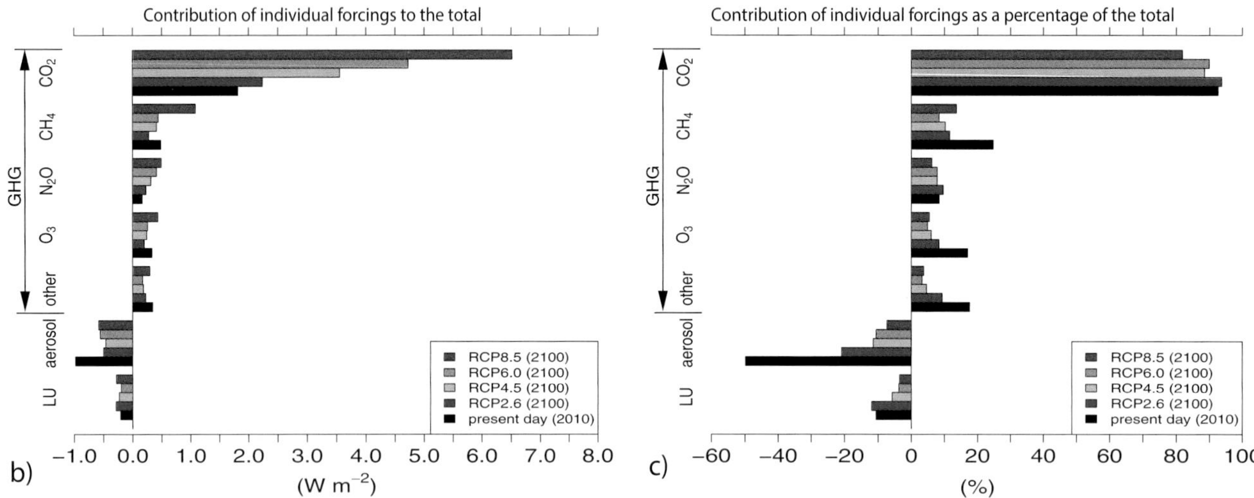

Figure 12.3 | (a) Time evolution of the total anthropogenic (positive) and anthropogenic aerosol (negative) radiative forcing (RF) relative to pre-industrial (about 1765) between 2000 and 2300 for RCP scenarios and their extensions (continuous lines), and SRES scenarios (dashed lines) as computed by the Integrated Assessment Models (IAMs) used to develop those scenarios. The four RCP scenarios used in CMIP5 are: RCP2.6 (dark blue), RCP4.5 (light blue), RCP6.0 (orange) and RCP8.5 (red). The three SRES scenarios used in CMIP3 are: B1 (blue, dashed), A1B (green, dashed) and A2 (red, dashed). Positive values correspond to the total anthropogenic RF. Negative values correspond to the forcing from all anthropogenic aerosol–radiation interactions (i.e., direct effects only). The total RF of the SRES and RCP families of scenarios differs in 2000 because the number of forcings represented and our knowledge about them have changed since the TAR. The total RF of the RCP family is computed taking into account the efficacy of the various forcings (Meinshausen et al., 2011a). (b) Contribution of the individual anthropogenic forcings to the total RF in year 2100 for the four RCP scenarios and at present day (year 2010). The individual forcings are gathered into seven groups: carbon dioxide (CO_2), methane (CH_4), nitrous oxide (N_2O), ozone (O_3), other greenhouse gases, aerosol (all effects unlike in (a), i.e., aerosol–radiation and aerosol–cloud interactions, aerosol deposition on snow) and land use (LU). (c) As in (b), but the individual forcings are relative to the total RF (i.e., RF_x/RF_{tot}, in %, with RF_x individual RFs and RF_{tot} total RF). Note that the RFs in (b) and (c) are not efficacy adjusted, unlike in (a). The values shown in (a) are summarized in Table AII.6.8. The values shown in (b) and (c) have been directly extracted from data files (hosted at http://tntcat.iiasa.ac.at:8787/RcpDb/) compiled by the four modelling teams that developed the RCP scenarios and are summarized in Tables AII.6.1 to AII.6.3 for CO_2, CH_4 and N_2O respectively.

12

atmospheric chemistry runs were performed to estimate ozone and aerosol distributions (Lamarque et al., 2011). Finally, a single carbon cycle model with a representation of carbon–climate feedbacks was used in order to provide consistent values of CO_2 concentration for the CO_2 emission provided by a different IAM for each of the scenarios. This methodology was used to produce consistent data sets across scenarios but does not provide uncertainty estimates for them. After these processing steps, the final RCP data sets comprise land use data, harmonized GHG emissions and concentrations, gridded reactive gas and aerosol emissions, as well as ozone and aerosol abundance fields. These data are used as forcings in individual climate models. The number and type of forcings included primarily depend on the experiment. For instance, while the CO_2 concentration is prescribed in most experiments, CO_2 emissions are prescribed in some others (see Box 6.4 and Section 12.3.2.1). Which of these forcings are included in individual CMIP5 models, and variations in their implementation, is described in Section 12.3.2.2.

During this development process, the total RF and the RF of individual forcing agents have been estimated by the IAMs and made available via the RCP database (Meinshausen et al., 2011c). Each individual anthropogenic forcing varies from one scenario to another. They have been aggregated into a few groups in Figure 12.3b and c. The total anthropogenic RF estimated by the IAMs in 2010 is about 0.15 W m^{-2} lower than Chapter 8's best estimate of ERF in 2010 (2.2 W m^{-2}), the difference arising from a revision of the RF due to aerosols and land use in the current assessment compared to AR4. All the other individual forcings are consistent to within 0.02 W m^{-2}. The change in CO_2 concentration is the main cause of difference in the total RF among the scenarios (Figure 12.3b). The relative contribution[5] of CO_2 to the total anthropogenic forcing is currently (year 2010) about 80 to 90% and does not vary much across the scenarios (Figure 12.3c), as was also the case for SRES scenarios (Ramaswamy et al., 2001). Aerosols have a large negative contribution to the total forcing (about –40 to –50% in 2010), but this contribution decreases (in both absolute and relative terms) in the future for all the RCPs scenarios. This means that while anthropogenic aerosols have had a cooling effect in the past, their decrease in all RCP scenarios relative to current levels is expected to have a net warming effect in the future (Levy II et al., 2013; see also Figure 8.20). The 21st century decrease in the magnitude of future aerosol forcing was not as large and as rapid in the SRES scenarios (Figure 12.3a). However, even in the SRES scenarios, aerosol effects were expected to have a diminishing role in the future compared to GHG forcings, mainly because of the accumulation of GHG in the atmosphere (Dufresne et al., 2005). Other forcings do not change much in the future, except CH_4 which increases in the RCP8.5 scenario. Note that the estimates of all of these individual RFs are subject to many uncertainties (see Sections 7.5, 8.5 and 11.3.6). In this section and in Table AII.6.8, the RF values for RCP scenarios are derived from published equivalent-CO_2 (CO_2eq) concentration data that aggregates all anthropogenic forcings including GHGs and aerosols. The conversion to RF uses the formula: RF = $3.71/\ln(2) \cdot \ln(CO_2eq/278)$ W m^{-2}, where CO_2eq is in ppmv.

The four RCPs (Meinshausen et al., 2011c) are based on IAMs up to the end of the 21st century only. In order to investigate longer-term climate change implications, these RCPs were also extended until 2300. The extensions, formally named Extended Concentration Pathways (ECPs) but often simply referred to as RCP extensions, use simple assumptions about GHG and aerosol emissions and concentrations beyond 2100 (such as stabilization or steady decline) and were designed as hypothetical 'what-if' scenarios, not as an outcome of an IAM assuming socioeconomic considerations beyond 2100 (Meinshausen et al., 2011c) (see Box 1.1). In order to continue to investigate a broad range of possible climate futures, RCP2.6 assumes small constant net negative emissions after 2100 and RCP8.5 assumes stabilization with high emissions between 2100 and 2150, then a linear decrease until 2250. The two middle RCPs aim for a smooth stabilization of concentrations by 2150. RCP8.5 stabilizes concentrations only by 2250, with CO_2 concentrations of approximately 2000 ppmv, nearly seven times the pre-industrial level. As RCP2.6 implies net negative CO_2 emissions after around 2070 and throughout the extension, CO_2 concentrations slowly reduce towards 360 ppmv by 2300.

12.3.1.4 Comparison of Special Report on Emission Scenarios and Representative Concentration Pathway Scenarios

The four RCP scenarios used in CMIP5 lead to RF values that range from 2.3 to 8.0 W m^{-2} at 2100, a wider range than that of the three SRES scenarios used in CMIP3 which vary from 4.2 to 8.1 W m^{-2} at 2100 (see Table AII.6.8 and Figure 12.3). The SRES scenarios do not assume any policy to control climate change, unlike the RCP scenarios. The RF of RCP2.6 is hence lower by 1.9 W m^{-2} than the three SRES scenarios and very close to the ENSEMBLES E1 scenario (Johns et al., 2011). RCP4.5 and SRES B1 have similar RF at 2100, and comparable time evolution (within 0.2 W m^{-2}). The RF of SRES A2 is lower than RCP8.5 throughout the 21st century, mainly due to a faster decline in the radiative effect of aerosols in RCP8.5 than SRES A2, but they converge to within 0.1 W m^{-2} at 2100. RCP6.0 lies in between SRES B1 and SRES A1B. Results obtained with one General Circulation Model (GCM) (Dufresne et al., 2013) and with a reduced-complexity model (Rogelj et al., 2012) confirm that the differences in temperature responses are consistent with the differences in RFs estimates. RCP2.6, which assumes strong mitigation action, yields a smaller temperature increase than any SRES scenario. The temperature increase with the RCP4.5 and SRES B1 scenarios are close and the temperature increase is larger with RCP8.5 than with SRES A2. The spread of projected global mean temperature for the RCP scenarios (Section 12.4.1) is considerably larger (at both the high and low response ends) than for the three SRES scenarios used in CMIP3 (B1, A1B and A2) as a direct consequence of the larger range of RF across the RCP scenarios compared to that across the three SRES scenarios (see analysis of SRES versus RCP global temperature projections in Section 12.4.9 and Figure 12.40).

12.3.2 Implementation of Forcings in Coupled Model Intercomparison Project Phase 5 Experiments

The CMIP5 experimental protocol for long-term transient climate experiments prescribes a common basis for a comprehensive set of anthropogenic forcing agents acting as boundary conditions in three experimental phases—historical, RCPs and ECPs (Taylor et al., 2012). To permit common implementations of this set of forcing agents in CMIP5 models, self-consistent forcing data time series have been computed and provided to participating models (see Sections 9.3.2.2 and 12.3.1.3) comprising emissions or concentrations of GHGs and related compounds, ozone and atmospheric aerosols and their chemical precursors, and land use change.

The forcing agents implemented in Atmosphere–Ocean General Circulation Models (AOGCMs) and ESMs used to make long-term climate projections in CMIP5 are summarized in Table 12.1. The number of CMIP5 models listed here is about double the number of CMIP3 models listed in Table 10.1 of AR4 (Meehl et al., 2007b).

Natural forcings (arising from solar variability and aerosol emissions via volcanic activity) are also specified elements in the CMIP5 experimental protocol, but their future time evolutions are not prescribed

5 The range of the relative contribution of CO_2 and aerosols to the total anthropogenic forcing is derived here from the RF values given by the IAMs and the best estimate assessed in Chapter 8.

Table 12.1 | Radiative forcing agents in the CMIP5 multi-model global climate projections. See Table 9.A.1 for descriptions of the models and main model references. Earth System Models (ESMs) are highlighted in bold. Numeric superscripts indicate model-specific references that document forcing implementations. Forcing agents are mostly implemented in close conformance with standard prescriptions (Taylor et al., 2012) and recommended data sets (Lamarque et al., 2010; Cionni et al., 2011; Lamarque et al., 2011; Meinshausen et al., 2011c) provided for CMIP5. Variations in forcing implementations are highlighted with superscripts and expanded in the table footnotes. Entries mean: n.a.: Forcing agent is not included in either the historical or future scenario simulations; Y: Forcing agent included (via prescribed concentrations, distributions or time series data); E: Concentrations of forcing agent calculated interactively driven by prescribed emissions or precursor emissions; Es: Concentrations of forcing agent calculated interactively constrained by prescribed surface concentrations. For a more detailed classification of ozone chemistry and ozone forcing implementations in CMIP5 models see Eyring et al. (2013).

Model	Forcing Agents																
	Greenhouse Gases						*Aerosols*									*Other*	
	CO_2 [ce]	CH_4	N_2O	Trop O_3	Strat O_3	CFCs	SO_4	Black carbon	Organic carbon	Nitrate	Cloud albedo effect [ac]	Cloud lifetime effect [ac]	Dust	Volcanic	Sea salt	Land use	Solar
ACCESS-1.0 [1]	Y [p]	Y	Y	Y [b]	Y [b]	Y	E	E	E	n.a.	Y	Y	Y [pd]	Y [v5]	Y [pd]	n.a.	Y
ACCESS-1.3 [1]	Y [p]	Y	Y	Y [b]	Y [b]	Y	E	E	E	n.a.	Y	Y	n.a.	Y [v5]	Y [pd]	n.a.	Y
BCC-CSM1.1 [2]	Y/E [p]	Y	Y	Y [b]	Y [b]	Y	Y [a]	Y [a]	Y [a]	n.a.	n.a.	n.a.	Y [a]	Y [v0]	Y [a]	n.a.	Y
BCC-CSM1.1(m) [2]	Y/E [p]	Y	Y	Y [b]	Y [b]	Y	Y [a]	Y [a]	Y [a]	n.a.	n.a.	n.a.	Y [a]	Y [v0]	Y [a]	n.a.	Y
BNU-ESM	Y/E [p]	Y	Y	Y [a]	Y [a]	Y	Y [a]	Y [a]	Y [a]	n.a.	n.a.	n.a.	Y [a]	Y [v0]	Y [a]	n.a.	Y
CanCM4	Y	Y	Y	Y [b]	Y [b]	Y	E	E	E	n.a.	Y [so]	n.a.	Y [pd]	Y/E [st,v0]	Y [pd]	n.a.	Y
CanESM2	Y/E [p]	Y	Y	Y [b]	Y [b]	Y	E	E	E	n.a.	Y [so]	n.a.	Y [pd]	Y/E [st,v0]	Y [pd]	Y [cr]	Y
CCSM4 [3]	Y [p]	Y	Y	Y [a]	Y [a]	Y	Y [a]	Y [a]	Y [a]	n.a.	n.a.	n.a.	Y [a]	Y [v0]	Y [a]	Y	Y
CESM1(BGC) [4]	Y/E [p]	Y	Y	Y [a]	Y [a]	Y	Y [a]	Y [a]	Y [a]	n.a.	n.a.	n.a.	Y [a]	Y [v0]	Y [a]	Y	Y
CESM1(CAM5) [5]	Y [p]	Y	Y	Y [a]	Y [a]	Y	E	E	E	n.a.	Y	Y	E	Y [v0]	E	Y	Y
CESM1(CAM5.1,FV2) [5]	Y [p]	Y	Y	Y [a]	Y [a]	Y	E	E	E	n.a.	Y	Y	E	Y [v0]	E	Y	Y
CESM1(FASTCHEM)	Y [p]	Y [a]	Y	E	E	Y	E	Y [a]	Y [a]	n.a.	n.a.	n.a.	Y [a]	Y [v0]	Y [a]	Y	Y
CESM1(WACCM) [6]	Es [p]	Es	Es	E/Es [op]	E/Es [op]	Es	Y	Y	Y	n.a.	n.a.	n.a.	Y [a]	Y [v0]	Y [a]	Y	Y
CMCC-CESM [7]	Y	Y	Y	Y [b]	Y [b]	Y	Y [a]	n.a.	n.a.	n.a.	Y [so]	n.a.	Y [fx]	n.a.	Y [fx]	n.a.	Y [or]
CMCC-CM	Y	Y	Y	Y [b]	Y [b]	Y	Y [a]	n.a.	n.a.	n.a.	Y [so]	n.a.	Y [fx]	n.a.	Y [fx]	n.a.	Y [or]
CMCC-CMS	Y	Y	Y	Y [b]	Y [b]	Y	Y [a]	n.a.	n.a.	n.a.	Y [so]	n.a.	Y [fx]	n.a.	Y [fx]	n.a.	Y [or]
CNRM-CM5 [8]	Y	Y	Y	Y [c]	Y [c]	Y	Y [e]	Y [e]	Y [e]	n.a.	Y [so,ic]	n.a.	Y [e]	Y [v1]	Y [e]	n.a.	Y
CSIRO-Mk3.6.0 [9]	Y	Y	Y	Y [b]	Y [b]	Y	E	E	E	n.a.	Y	Y	Y [pd]	Y [v0]	Y [pd]	n.a.	Y
EC-EARTH [10]	Y	Y	Y	Y [b]	Y [b]	Y	Y [a]	Y [a]	Y [a]	n.a.	n.a.	n.a.	Y [a]	Y [v1]	Y [a]	Y	Y
FGOALS-g2 [11]	Y	Y	Y	Y [b]	Y [b]	Y	Y [a]	Y [a]	Y [a]	n.a.	Y	Y	Y [a]	n.a.	Y [a]	n.a.	Y
FGOALS-s2 [12]	Y/E	Y	Y	Y [b]	Y [b]	Y	Y [a]	Y [a]	Y [a]	n.a.	n.a.	n.a.	Y [a]	Y [v0]	Y [a]	n.a.	Y
FIO-ESM	Y/E	Y	Y	Y [a]	Y [a]	Y	Y [a]	Y [a]	Y [a]	n.a.	n.a.	n.a.	Y [a]	Y [v0]	Y [a]	n.a.	Y
GFDL-CM3 [13]	Y [p]	Y/Es [rc]	Y/Es [rc]	E	E	Y/Es [rc]	E	E	E	n.a./E [rc]	Y	Y	E [pd]	Y/E [st,v0]	E [pd]	Y	Y
GFDL-ESM2G	Y/E [p]	Y	Y	Y [b]	Y [b]	Y	Y [a]	Y [a]	Y [a]	n.a.	n.a.	n.a.	Y [fx]	Y [v0]	Y [fx]	Y	Y
GFDL-ESM2M	Y/E [p]	Y	Y	Y [b]	Y [b]	Y	Y [a]	Y [a]	Y [a]	n.a.	n.a.	n.a.	Y [fx]	Y [v0]	Y [fx]	Y	Y
GISS-E2-p1 [14]	Y	Y	Y	Y [d]	Y [d]	Y	Y	Y	Y	Y	Y	n.a.	Y [fx]	Y [v4]	Y [fx]	Y	Y [or]
GISS-E2-p2 [14]	Y	Es/E [hf]	Es	E	E	Es/E [hf]	E	E	E	E	Y	n.a.	Y [pd]	Y [v4]	Y [pd]	Y	Y [or]

(continued on next page)

Table 12.1 (continued)

Model	Forcing Agents																
	Greenhouse Gases						*Aerosols*									*Other*	
	CO_2 [ce]	CH_4	N_2O	Trop O_3	Strat O_3	CFCs	SO_4	Black carbon	Organic carbon	Nitrate	Cloud albedo effect [ac]	Cloud lifetime effect [ac]	Dust	Volcanic	Sea salt	Land use	Solar
GISS-E2-p3 [14]	Y	Es/E [hf]	Es	E	E	Es/E [hf]	E	E	E	E	Y	n.a.	Y [pd]	Y [v4]	Y [pd]	Y	Y [or]
HadCM3	Y [p]	Y	Y	Y [b]	Y [b]	Y	E	n.a.	n.a.	n.a.	Y [so]	n.a.	n.a.	Y [v2]	n.a.	n.a.	Y
HadGEM2-AO [15]	Y [p]	Y	Y	Y [b]	Y [b]	Y	E	E	E	n.a.	Y	Y	Y [pd]	Y [v2]	Y [pd]	Y	Y
HadGEM2-CC [16,17]	Y [p]	Y	Y	Y [b]	Y [b]	Y	E	E	E	n.a.	Y	Y	Y [pd]	Y [v2]	Y [pd]	Y	Y
HadGEM2-ES [16]	Y/E [p]	Es	Y	E	Y [b]	Y	E	E	E	n.a.	Y	Y	Y [pd]	Y [v2]	Y [pd]	Y	Y
INM-CM4	Y/E	Y	Y	Y [b]	Y [b]	n.a.	Y [fx]	n.a.	n.a.	n.a.	Y [so]	n.a.	n.a.	Y [v0]	n.a.	Y	Y
IPSL-CM5A-LR [18]	Y/E [p]	Y	Y	Y [e]	Y [e]	Y	Y [e]	Y [e]	Y [e]	n.a.	Y	n.a.	Y [e]	Y [v1]	Y [e]	Y	Y
IPSL-CM5A-MR [18]	Y/E [p]	Y	Y	Y [e]	Y [e]	Y	Y [e]	Y [e]	Y [e]	n.a.	Y	n.a.	Y [e]	Y [v1]	Y [e]	Y	Y
IPSL-CM5B-LR [18]	Y [p]	Y	Y	Y [e]	Y [e]	Y	Y [e]	Y [e]	Y [e]	n.a.	Y	n.a.	Y [e]	Y [v1]	Y [e]	Y	Y
MIROC-ESM [19]	Y/E [p]	Y	Y	Y [f]	Y [f]	Y	E	E	E	n.a.	Y [ic]	Y [ic]	Y [pd]	Y [v3]	Y [pd]	Y	Y
MIROC-ESM-CHEM [19]	Y [p]	Y	Y	E	E	Y	E	E	E	n.a.	Y [ic]	Y [ic]	Y [pd]	Y [v3]	Y [pd]	Y	Y
MIROC4h [20]	Y [p]	Y	Y	Y [g]	Y [g]	Y	E	E	E	n.a.	Y	Y	Y [pd]	Y [v3]	Y [pd]	Y [cr]	Y
MIROC5 [20]	Y [p]	Y	Y	Y [f]	Y [f]	Y	E	E	E	n.a.	Y [ic]	Y [ic]	Y [pd]	Y [v3]	Y [pd]	Y [cr]	Y
MPI-ESM-LR	Y/E [p]	Y	Y	Y [b]	Y [b]	Y	Y [h]	Y [h]	Y [h]	Y [h]	n.a.	n.a.	Y [h]	Y [v0]	Y [h]	Y	Y [or]
MPI-ESM-MR	Y [p]	Y	Y	Y [b]	Y [b]	Y	Y [h]	Y [h]	Y [h]	Y [h]	n.a.	n.a.	Y [h]	Y [v0]	Y [h]	Y	Y [or]
MPI-ESM-P	Y [p]	Y	Y	Y [b]	Y [b]	Y	Y [h]	Y [h]	Y [h]	Y [h]	n.a.	n.a.	Y [h]	Y [v0]	Y [h]	Y	Y [or]
MRI-CGCM3 [21]	Y	Y	Y	Y [b]	Y [b]	Y	E	E	E	n.a.	Y [ic]	Y [ic]	E [pd]	E [v0]	E [pd]	Y	Y
MRI-ESM1 [22]	E	Y	Y	E	E	Es	E	E	E	n.a.	Y [ic]	Y [ic]	E [pd]	E [v0]	E [pd]	Y	Y
NorESM1-M [23]	Y [p]	Y	Y	Y [a]	Y [a]	Y	E	E	E	n.a.	Y	Y	E	Y/E [st,v1]	E [pd]	Y	Y
NorESM1-ME [23]	Y/E [p]	Y	Y	Y [a]	Y [a]	Y	E	E	E	n.a.	Y	Y	E	Y/E [st,v1]	E [pd]	Y	Y

Notes:

Model-specific references relating to forcing implementations:

1 Dix et al. (2013)
2 Wu et al. (2013); Xin et al. (2013a, 2013b)
3 Meehl et al. (2012); Gent et al. (2011)
4 Long et al. (2013); Meehl et al. (2012)
5 Meehl et al. (2013)
6 Calvo et al. (2012); Meehl et al. (2012)
7 Cagnazzo et al. (2013)
8 Voldoire et al. (2013)
9 Rotstayn et al. (2012)
10 Hazeleger et al. (2013)
11 Li et al. (2013c)
12 Bao et al. (2013)
13 Levy II et al. (2013)
14 Shindell et al. (2013a). GISS-E2-R and GISS-E2-H model variants are forced similarly and both represented here as GISS-E2. Both -R and -H model versions have three variants: in physics version 1 (p1) aerosols and ozone are specified from pre-computed transient aerosol and ozone fields, in physics version 2 (p2) aerosols and atmospheric chemistry are calculated online as a function of atmospheric state and transient emissions inventories, while in physics version 3 (p3) atmospheric composition is calculated as for p2 but the aerosol impacts on clouds (and hence the aerosol indirect effect) is calculated interactively. In p1 and p2 variants the aerosol indirect effect is parameterized following Hansen et al. (2005b).
15 HadGEM2-AO is forced in a similar way to HadGEM2-ES and HadGEM2-CC following Jones et al. (2011), but tropospheric ozone, stratospheric ozone and land cover are prescribed.
16 Jones et al. (2011)
17 Hardiman et al. (2012)
18 Dufresne et al. (2013)
19 Watanabe et al. (2011)
20 Komuro et al. (2012)
21 Yukimoto et al. (2012)
22 Adachi et al. (2013)
23 Iversen et al. (2013); Kirkevåg et al. (2013); Tjiputra et al. (2013)

(continued on next page)

Table 12.1 (continued)

Additional notes:

[ce] Separate entries for CO_2 denote 'concentration-driven' and 'emissions-driven' experiments as indicated.

[ac] 'Cloud albedo effect' and 'Cloud lifetime effect' are classical terms (as used in AR4) to describe indirect effects of radiative forcing associated with aerosols. They relate to the revised terminologies defined in Chapter 7 and used in AR5: 'Radiative forcing from aerosol–cloud interactions (RFaci)' and 'Effective radiative forcing from aerosol–cloud interactions (ERFaci)'. RFaci equates to cloud albedo effect, while ERFaci is the effective forcing resulting from cloud albedo effect plus cloud lifetime effect, including all rapid adjustments to cloud lifetime and thermodynamics (Section 7.1.3, Figure 7.3).

[p] Physiological forcing effect of CO_2 via plant stomatal response and evapotranspiration (Betts et al., 2007) included.

[rc] Separate entries denote different treatments used for radiation and chemistry respectively.

[hf] Separate entries denote treatment for historical and future (RCPs) respectively.

[a] Three-dimensional tropospheric ozone, stratospheric ozone, methane, and/or aerosol distributions specified as monthly 10-year mean concentrations, computed off-line using CAM-Chem – a modified version of CAM3.5 with interactive chemistry – driven with specified emissions for the historical period (Lamarque et al., 2010) and RCPs (Lamarque et al., 2011) with sea surface temperature and sea ice boundary conditions based on CCSM3's projections for the closest corresponding AR4 scenarios.

[b] Ozone prescribed using the original or slightly modified IGAC/SPARC ozone data set (Cionni et al., 2011); in some models this data set is modified to add a future solar cycle and in some models tropospheric ozone is zonally averaged.

[c] Linearized 2D ozone chemistry scheme (Cariolle and Teyssedre, 2007) including transport and photochemistry, reactive to stratospheric chlorine concentrations but not tropospheric chemical emissions.

[d] Ozone prescribed using the data set described in Hansen et al. (2007), with historical tropospheric ozone being calculated by a CCM and stratospheric ozone taken from Randel and Wu (2007) in the past. Tropospheric ozone is held constant from 1990 onwards, while stratospheric ozone is constant from 1997 to 2003 and then returned linearly to its 1979 value over the period 2004 to 2050.

[e] For IPSL-CM5 model versions, ozone and aerosol concentrations are calculated semi-offline with the atmospheric general circulation model including interactive chemistry and aerosol, following the four RCPs in the future (Dufresne et al., 2013; Szopa et al., 2013). The same aerosol concentration fields (but not ozone) are also prescribed for the CNRM-CM5 model.

[f] Ozone concentrations computed off-line by Kawase et al. (2011) using a CCM forced with CMIP5 emissions.

[g] Ozone concentrations computed off-line by Sudo et al. (2003) for the historical period and Kawase et al. (2011) for the future.

[h] Time dependent climatology based on simulations and observations; aerosols are distinguished only with respect to coarse and fine mode, and anthropogenic and natural origins, not with respect to composition.

[op] Separate entries denote different ozone chemistry precursors.

[so] RFaci from sulphate aerosol only.

[st] Separate entries denote stratosphere and troposphere respectively.

[ic] Radiative effects of aerosols on ice clouds are represented.

[pd] Prognostic or diagnostic scheme for dust/sea salt aerosol with emissions/concentrations determined by the model state rather than externally prescribed.

[fx] Fixed prescribed climatology of dust/sea salt aerosol concentrations with no year-to-year variability.

[v0] Explosive volcanic aerosol returns rapidly in future to zero (or near-zero) background, like that in the pre-industrial control experiment.

[v1] Explosive volcanic aerosol returns rapidly in future to constant (average volcano) background, the same as in the pre-industrial control experiment.

[v2] Explosive volcanic aerosol returns slowly in future (over several decades) to constant (average volcano) background like that in the pre-industrial control experiment.

[v3] Explosive volcanic aerosol returns rapidly in future to near-zero background, below that in the pre-industrial control experiment.

[v4] Explosive volcanic aerosol set to zero in future, but constant (average volcano) background in the pre-industrial control experiment.

[v5] Explosive volcanic aerosol returns slowly in future (over several decades) to constant (average volcano) background, but zero background in the pre-industrial control experiment.

[cr] Land use change represented via crop change only.

[or] Realistic time-varying orbital parameters for solar forcing (in historical period only for GISS-E2).

very precisely. A repeated 11-year cycle for total solar irradiance (Lean and Rind, 2009) is suggested for future projections but the periodicity is not specified precisely as solar cycles vary in length. Some models include the effect of orbital variations as well, but most do not. For volcanic eruptions, no specific CMIP5 prescription is given for future emissions or concentration data, the general recommendation being that volcanic aerosols should either be omitted entirely both from the control experiment and future projections or the same background volcanic aerosols should be prescribed in both. This provides a consistent framework for model intercomparison given a lack of knowledge of when future large eruptions will occur. In general models have adhered to this guidance, but there are variations in the background volcanic aerosol levels chosen (zero or an average volcano background in general) and some cases, for example, Australian Community Climate and Earth System Simulator (ACCESS)1.0 and ACCESS1.3 (Dix et al., 2013), where the background volcanic aerosol in future differs significantly from that in the control experiment, with a small effect on future RF.

For the other natural aerosols (dust, sea-salt, etc.), no emission or concentration data are recommended. The emissions are potentially computed interactively by the models themselves and may change with climate, or prescribed from separate model simulations carried out in the implementation of CMIP5 experiments, or simply held constant. Natural aerosols (mineral dust and sea salt) are in a few cases prescribed with no year-to-year variation (giving no transient forcing effect), in some cases prescribed from data sets computed off-line as described above, and in other cases calculated interactively via prognostic or diagnostic calculations. The degree to which natural aerosol emissions are interactive is effectively greater in some such models than others, however, as mineral dust emissions are more constrained when land vegetation cover is specified (e.g., as in Commonwealth Scientific and Industrial Research Organisation (CSIRO)-Mk3.6.0) (Rotstayn et al., 2012) than when vegetation is allowed to evolve dynamically (e.g., as in Hadley Centre new Global Environmental Model 2-ES (HadGEM2-ES)) (Jones et al., 2011) (Table 9.A.1).

12.3.2.1 'Emissions-Driven' versus 'Concentration Driven' Experiments

A novel feature within the CMIP5 experimental design is that experiments with prescribed anthropogenic emissions are included in addition to classical experiments with prescribed concentration pathways for WMGHGs (Taylor et al., 2012). The essential features of these two classes of experiment are described in Box 6.4. The CMIP5 protocol includes experiments in which 'ESMs' (models possessing at least a carbon cycle, allowing for interactive calculation of atmospheric CO_2 or compatible emissions) and AOGCMs (that do not possess such an interactive carbon cycle) are both forced with WMGHG concentration pathways to derive a range of climate responses consistent with those pathways from the two types of model. The range of climate responses including climate–carbon cycle feedbacks can additionally be explored in ESMs driven with emissions rather than concentrations, analogous to Coupled Climate Carbon Cycle Model Intercomparison Project (C^4MIP) experiments (Friedlingstein et al., 2006)—see Box 6.4. Results from the two types of experiment cannot be compared directly, but they provide complementary information. Uncertainties in the forward climate response driven with specified emissions or concentrations can be derived from all participating models, while concentration-driven ESM experiments also permit a policy-relevant diagnosis of the range of anthropogenic carbon emissions compatible with the imposed concentration pathways (Hibbard et al., 2007; Moss et al., 2010).

WMGHG forcing implementations in CMIP5 concentration-driven experiments conform closely in almost all cases to the standard protocol (Table 12.1; CO_2, CH_4, N_2O, chlorofluorocarbons (CFCs)), imposing an effective control over the RF due to WMGHGs across the multi-model ensemble, apart from the model spread arising from radiative transfer codes (Collins et al., 2006b; Meehl et al., 2007b). The ability of ESMs to determine their own WMGHG concentrations in emissions-driven experiments means that RF due to WMGHGs is less tightly controlled in such experiments. Even in concentration-driven experiments, many models implement some emissions-driven forcing agents (more often aerosols, but also ozone in some cases), leading to a potentially greater spread in both the concentrations and hence RF of those emissions-driven agents.

12.3.2.2 Variations Between Model Forcing Implementations

Apart from the distinction between concentration-driven and emissions-driven protocols, a number of variations are present in the implementation of forcing agents listed in Table 12.1, which generally arise due to constraining characteristics of the model formulations, various computational efficiency considerations or local implementation decisions. In a number of models, off-line modelling using an aerosol chemistry climate model has been used to convert emissions into concentrations compatible with the specific model formulation or characteristics. As a result, although detailed prescriptions are given for the forcing agents in CMIP5 experiments in emissions terms, individual modelling approaches lead to considerable variations in their implementations and consequential RFs. This was also the case in the ENSEMBLES multi-model projections, in which similar forcing agents to CMIP5 models were applied but again with variations in the implementation of aerosol, ozone and land use forcings, prescribing the SRES A1B and E1 scenarios in a concentration-driven protocol (Johns et al., 2011) akin to the CMIP5 protocol.

Methane, nitrous oxide and CFCs (typically with some aggregation of the multiple gases) are generally prescribed in CMIP5 models as well-mixed concentrations following the forcing data time series provided for the given scenarios. In a number of models (CESM1(WACCM), GFDL-CM3, GISS-E2-p2, GISS-E2-p3, HadGEM2-ES and MRI-ESM1) the three-dimensional concentrations in the atmosphere of some species evolve interactively driven by the full emissions/sinks cycle (in some cases constrained by prescribed concentrations at the surface, e.g., HadGEM2-ES for methane). In cases where the full emissions/sinks cycle is modelled, the radiation scheme is usually passed the time-varying 3-D concentrations, but some models prescribe different concentrations for the purpose of radiation.

Eyring et al. (2013) document, in greater detail than Table 12.1, the implementations of tropospheric and stratospheric ozone in CMIP5 models, including their ozone chemistry schemes and modifications applied to reference data sets in models driven by concentrations. In most models that prescribe ozone, concentrations are based on the original or slightly modified CMIP5 standard ozone data set computed as part of the International Global Atmospheric Chemistry/Stratospheric Processes and their Role in Climate (IGAC/SPARC) activity (Cionni et al., 2011). In the stratosphere, this data set is based on observations of the past (Randel and Wu, 2007) continued into the future with the multi-model mean of 13 chemistry–climate models (CCMs) projections following the SRES A1B (IPCC, 2000) and SRES A1 adjusted halogen scenario (WMO, 2007). The stratospheric zonal mean ozone field is merged with a 3-D tropospheric ozone time series generated as the mean of two CCMs (Goddard Institute of Space Studies-Physical Understanding of Composition-Climate Interactions and Impacts (GISS-PUCCINI), Shindell et al., 2006; CAM3.5, Lamarque et al., 2010) in the past and continued by one CCM (CAM3.5) in the future. Some CMIP5 models (MIROC-ESM, MIROC4h, MIROC5 and GISS-E2-p1) prescribe ozone concentrations using different data sets but again following just one GHG scenario in the future for the projection of stratospheric ozone. In other models (e.g., Institut Pierre Simon Laplace (IPSL)-CM5, CCSM4) ozone is again prescribed, but supplied as concentrations from off-line computations using a related CCM. Some models determine ozone interactively from specified emissions via on-line atmospheric chemistry (CESM1(FASTCHEM), CESM1(WACCM), CNRM-CM5, GFDL-CM3, GISS-E2-p2, GISS-E2-p3, MIROC-ESM-CHEM, MRI-ESM1; and HadGEM2-ES for tropospheric ozone only). Computing ozone concentrations interactively allows the fast coupling between chemistry and climate to be captured, but modelling of chemistry processes is sometimes simplified (CNRM-CM5, CESM(FASTCHEM)) in comparison with full complexity CCMs to reduce the computational cost. Compared to CMIP3, in which all models prescribed ozone and around half of them used a fixed ozone climatology, this leads to substantial improvement to ozone forcings in CMIP5, although differences remain among the models with interactive chemistry.

For atmospheric aerosols, either aerosol precursor emissions-driven or concentration-driven forcings are applied depending on individual model characteristics (see Sections 7.3 and 7.4 for an assessment of aerosols processes including aerosol–radiation and aerosol–cloud interactions). A larger fraction of models in CMIP5 than CMIP3 prescribe aerosol precursor emissions rather than concentrations. Many still prescribe concentrations pre-computed either using a directly related aerosol CCM or from output of another, complex, emissions-driven aerosol chemistry model within the CMIP5 process. As for ozone, aerosol concentrations provided from off-line simulations help to reduce the computational burden of the projections themselves. For several of the concentration-driven models (CCSM4, IPSL-CM5A variants, MPI-ESM-LR, MPI-ESM-MR), additional emissions-driven simulations have been undertaken to tailor the prescribed concentrations closely to the model's individual aerosol–climate characteristics. Lamarque et al. (2010, 2011) provided the recommended CMIP5 aerosols data set which has been used in several of the models driven by concentrations. Compared with the CMIP3 models, a much larger fraction of CMIP5 models now incorporate black and organic carbon aerosol forcings. Also, a larger fraction of CMIP5 than CMIP3 models now includes a range of processes that combine in the effective RF from aerosol–cloud interactions (ERFaci; see Section 7.1.3 and Figure 7.3). Previously such processes were generally termed aerosol indirect effects, usually separated into cloud albedo (or first indirect) effect and cloud lifetime (or second indirect) effect. Many CMIP5 models only include the interaction between sulphate aerosol and cloud, and the majority of them only model the effect of aerosols on cloud albedo rather than cloud lifetime (Table 12.1). No CMIP5 models represent urban aerosol pollution explicitly so that is not listed in Table 12.1 (see Section 11.3.5.2 for discussion of future air quality). Only one model (GISS-E2) explicitly includes nitrate aerosol as a separate forcing, though it is also included within the total aerosol mixture in the Max Planck Institute-Earth System Model (MPI-ESM) model versions.

Land use change is typically applied by blending anthropogenic land surface disturbance via crop and pasture fraction changes with underlying land cover maps of natural vegetation, but model variations in the underlying land cover maps and biome modelling mean that the land use forcing agent is impossible to impose in a completely common way at present (Pitman et al., 2009). Most CMIP5 models represent crop and pasture disturbance separately, while some (Canadian Earth System Model (CanESM2), MIROC4h, MIROC5) represent crop but not pasture. Some models (e.g., HadGEM2-ES, MIROC-ESM and MPI-ESM versions) allow a dynamical representation of natural vegetation changes alongside anthropogenic disturbance (see also Sections 9.4.4.3 and 9.4.4.4).

Treatment of the CO_2 emissions associated with land cover changes is also model dependent. Some models do not account for land cover changes at all, some simulate the biophysical effects but are still forced externally by land cover change induced CO_2 emissions (in emissions-driven simulations), while the most advanced ESMs simulate both biophysical effects of land cover changes and their associated CO_2 emissions.

12.3.3 Synthesis of Projected Global Mean Radiative Forcing for the 21st Century

Quantification of future global mean RF is of interest as it is directly related to changes in the global energy balance of the climate system and resultant climate change. Chapter 8 discusses RF concepts and methods for computing it that form the basis of analysis directly from the output of model projections.

We assess three related estimates of projected global mean forcing and its range through the 21st century in the context of forcing estimated for the recent past (Figure 12.4). The estimates used are: the total forcings for the defined RCP scenarios, harmonized to RF in the past (Meinshausen et al., 2011a; Meinshausen et al., 2011c); the total effective radiative forcing (ERF) estimated from CMIP5 models through the 21st century for the four RCP experiments (Forster et al., 2013); and that estimated from models in the Atmospheric Chemistry and Climate Model Intercomparison Project (ACCMIP; Lamarque et al., 2013—see Section 8.2.2) for RCP time-slice experiments (Shindell et al., 2013b). Methodological differences mean that whereas CMIP5 estimates include both natural and anthropogenic forcings based entirely on ERF, ACCMIP estimates anthropogenic composition forcing only (neglecting forcing changes due to natural, i.e., solar and volcanic, and land use factors) based on a combination of ERF for aerosols and RF for WMGHG (see Section 8.5.3). Note also that total forcing for the defined RCP scenarios is based on Meinshausen et al. (2011c) but combining total anthropogenic ERF (allowing for efficacies of the various anthropogenic forcings as in Figure 12.3) with natural (solar and volcanic) RF.

The CMIP5 multi-model ensemble mean ERF at 2100 (relative to an 1850–1869 base period) is 2.2, 3.8, 4.8 and 7.6 W m^{-2} respectively for RCP2.6, RCP4.5, RCP6.0 and RCP8.5 concentration-driven projections, with a 1-σ range based on annual mean data for year 2100 of about ±0.5 to 1.0 W m^{-2} depending on scenario (lowest for RCP2.6 and highest for RCP8.5). The CMIP5-based ERF estimates are close to the total forcing at 2100 (relative to an 1850–1859 base period) of 2.4, 4.0, 5.2 and 8.0 W m^{-2} as defined for the four RCPs.

The spread in ERF indicated from CMIP5 model results with specified GHG concentration pathways is broadly consistent with that found for CMIP3 models for the A1B scenario using the corresponding method (Forster and Taylor, 2006). As for CMIP3 models, part of the forcing spread in CMIP5 models (Forster et al., 2013) is consistent with differences in GHG forcings arising from the radiative transfer codes (Collins et al., 2006b). Aerosol forcing implementations in CMIP5 models also vary considerably, however (Section 12.3.2), leading to a spread in aerosol concentrations and forcings which contributes to the overall model spread. A further small source of spread in CMIP5 results possibly arises from an underlying ambiguity in the CMIP5 experimental design regarding the volcanic forcing offset between the historical experiment versus the pre-industrial control experiment. Most models implement zero volcanic forcing in the control experiment but some use constant negative forcing equal to the time-mean of historical volcanic forcing (see Table 12.1 and Section 12.3.2). The effect of this volcanic forcing offset persists into the future projections.

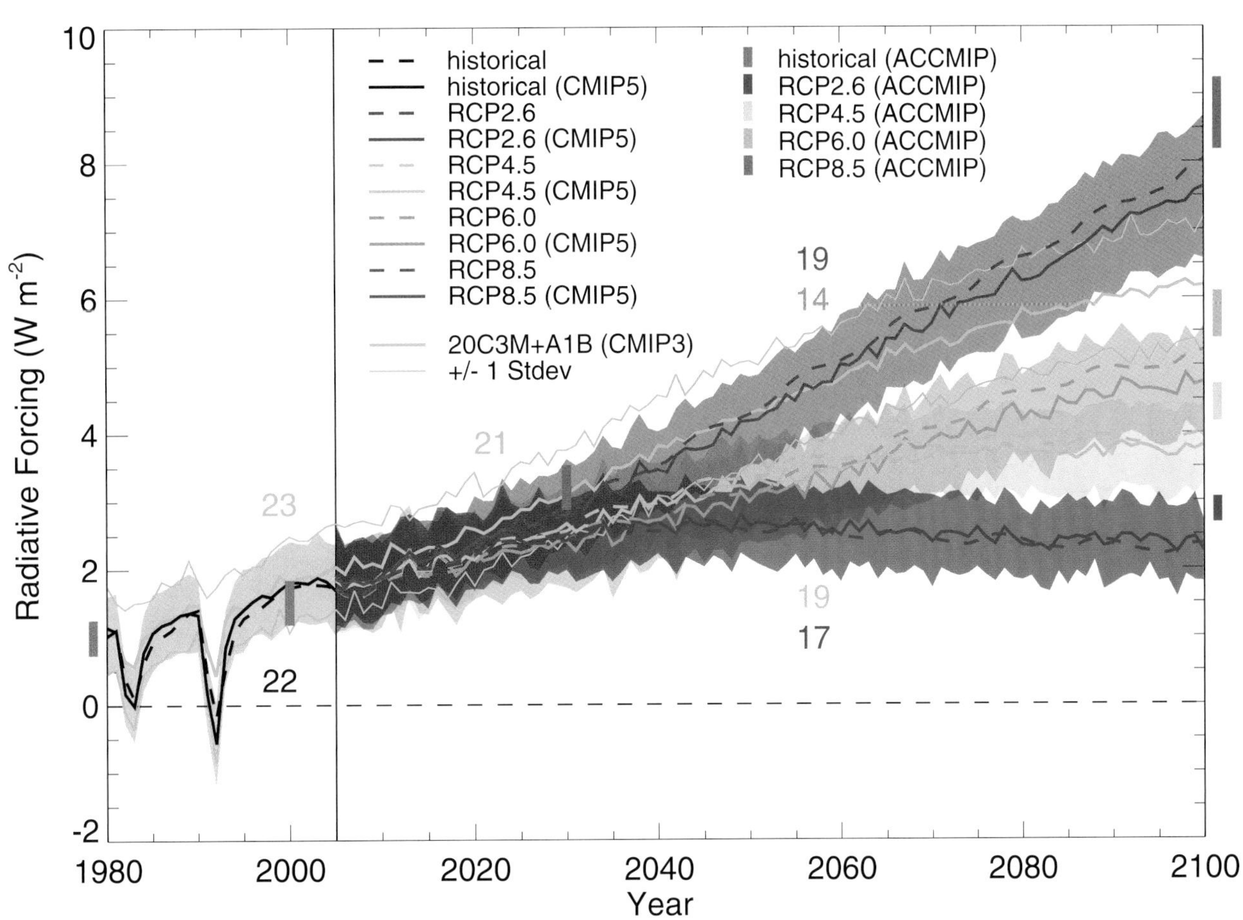

Figure 12.4 | Global mean radiative forcing (RF, W m^{-2}) between 1980 and 2100 estimated by alternative methods. The baseline is circa 1850 but dependent on the methods. Dashed lines indicate the total anthropogenic plus natural (solar and volcanic) RF for the RCP scenarios as defined by Meinshausen et al. (2011c), taking into account the efficacies of the various anthropogenic forcings (Meinshausen et al., 2011a), normalized by the mean between 1850 and 1859. Solid lines are multi-model mean effective radiative forcing (ERF) realized in a subset of CMIP5 models for the concentration-driven historical experiment and RCP scenarios, normalized either with respect to the 1850–1869 base period or with respect to the pre-industrial control simulation (Forster et al., 2013). (The subset of CMIP5 models included is defined by Table 1 of Forster et al. (2013) but omitting the FGOALS-s2 (Flexible Global Ocean-Atmosphere-Land System) model, the historical and RCP simulations of which were subsequently withdrawn from the CMIP5 archive.) This CMIP5-based estimate assumes each model has an invariant climate feedback parameter, calculated from abrupt 4 × CO_2 experiments using the method of Gregory et al. (2004). Each individual CMIP5 model's forcing estimate is an average over all available ensemble members, and a 1-σ inter-model range around the multi-model mean is shaded in light colour. Grey or coloured vertical bars illustrate the 1-σ range (68% confidence interval) of anthropogenic composition forcing (excluding natural and land use change forcings, based on ERF for aerosols combined with RF for WMGHG) estimated in ACCMIP models (Shindell et al., 2013b) for time slice experiments at 1980, 2000, 2030 (RCP8.5 only) and 2100 (all RCPs). The ACCMIP ranges plotted have been converted from the 5 to 95% ranges given in Shindell et al. (2013b) (Table 8) to a 1-σ range. Note that the ACCMIP bars at 1980 and 2100 are shifted slightly to aid clarity. The mean ERF diagnosed from 21 CMIP3 models for the SRES A1B scenario, as in Forster and Taylor (2006), is also shown (thick green line) with a 1-σ range (thinner green lines). The number of models included in CMIP3 and CMIP5 ensemble means is shown colour coded. (See Tables AII.6.8 to AII.6.10. Note that the CMIP5 model ranges given in Table AII.6.10 are based on decadal averages and therefore differ slightly from the ranges based on annual data shown in this figure.)

ACCMIP projected forcing at 2030 (for RCP8.5) and 2100 (all RCPs) is systematically higher than corresponding CMIP5 ERF, although with some overlap between 1-σ ranges. CMIP5 and ACCMIP comprise different sets of models and they are related in many but not all cases (Section 8.2.2). Confining analysis to a subset of closely related models also gives higher forcing estimates from ACCMIP compared to CMIP5 so the discrepancy in multi-model ensemble mean forcings appears unrelated to the different model samples associated with the two methods of estimation. The discrepancy is thought to originate mostly from differences in the underlying methodologies used to estimate RF, but is not yet well understood (see also Section 8.5.3).

There is *high confidence* in projections from ACCMIP models (Shindell et al., 2013b) based on the GISS-E2 CMIP5 simulations (Shindell et al., 2013a) and an earlier study with a version of the HadGEM2-ES model related to that used in CMIP5 (Bellouin et al., 2011), consistent with understanding of the processes controlling nitrate formation (Adams et al., 2001), that nitrate aerosols (which provide a negative forcing) will increase substantially over the 21st century under the RCPs (Section 8.5.3, Figure 8.20). The magnitude of total aerosol-related forcing (also negative in sign) will therefore tend to be underestimated in the CMIP5 multi-model mean ERF, as nitrate aerosol has been omitted as a forcing from almost all CMIP5 models.

Natural RF variations are, by their nature, difficult to project reliably (see Section 8.4). There is *very high confidence* that Industrial Era natural forcing has been a small fraction of the (positive) anthropogenic forcing except for brief periods following large volcanic eruptions (Sections 8.5.1 and 8.5.2). Based on that assessment and the assumption that variability in natural forcing remains of a similar magnitude and character to that over the Industrial Era, total anthropogenic forcing relative to pre-industrial, for any of the RCP scenarios through the 21st century, is *very likely* to be greater in magnitude than changes in natural (solar plus volcanic) forcing on decadal time scales.

In summary, global mean forcing projections derived from climate models exhibit a substantial range for the given RCP scenarios in concentration-driven experiments, contributing to the projected global mean temperature range (Section 12.4.1). Forcings derived from ACCMIP models for 2100 are systematically higher than those estimated from CMIP5 models for reasons that are not fully understood but are partly due to methodological differences. The multi-model mean estimate of combined anthropogenic plus natural forcing from CMIP5 is consistent with indicative RCP forcing values at 2100 to within 0.2 to 0.4 W m^{-2}.

12.4 Projected Climate Change over the 21st Century

12.4.1 Time-Evolving Global Quantities

12.4.1.1 Projected Changes in Global Mean Temperature and Precipitation

A consistent and robust feature across climate models is a continuation of global warming in the 21st century for all the RCP scenarios (Figure 12.5 showing changes in concentration-driven model simulations). Temperature increases are almost the same for all the RCP scenarios during the first two decades after 2005 (see Figure 11.25). At longer time scales, the warming rate begins to depend more on the specified GHG concentration pathway, being highest (>0.3°C per decade) in the highest RCP8.5 and significantly lower in RCP2.6, particularly after about 2050 when global surface temperature response stabilizes (and declines thereafter). The dependence of global temperature rise on GHG forcing at longer time scales has been confirmed by several studies (Meehl et al., 2007b). In the CMIP5 ensemble mean, global warming under RCP2.6 stays below 2°C above 1850-1900 levels throughout the 21st century, clearly demonstrating the potential of mitigation policies (note that to translate the anomalies in Figure 12.5 into anomalies with respect to that period, an assumed 0.61°C of observed warming since 1850–1900, as discussed in Section 2.4.3, should be added). This is in agreement with previous studies of aggressive mitigation scenarios (Johns et al., 2011; Meehl et al., 2012). Note, however, that some individual ensemble members do show warming exceeding 2°C above 1850-1900 (see Table 12.3). As for the other pathways, global warming exceeds 2°C within the 21st century under RCP4.5, RCP6.0 and RCP8.5, in qualitative agreement with previous studies using the SRES A1B and A2 scenarios (Joshi et al., 2011). Global mean temperature increase exceeds 4°C under RCP8.5 by 2100. The CMIP5 concentration-driven global temperature projections are broadly similar to CMIP3 SRES scenarios discussed in AR4 (Meehl et al., 2007b) and Section 12.4.9, although the overall range of the former is larger primarily because of the low-emission mitigation pathway RCP2.6 (Knutti and Sedláček, 2013).

The multi-model global mean temperature changes under different RCPs are summarized in Table 12.2. The relationship between cumulative anthropogenic carbon emissions and global temperature is assessed in Section 12.5 and only concentration-driven models are

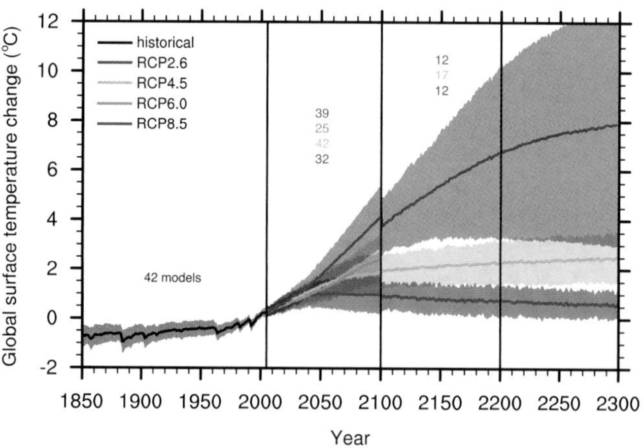

Figure 12.5 | Time series of global annual mean surface air temperature anomalies (relative to 1986–2005) from CMIP5 concentration-driven experiments. Projections are shown for each RCP for the multi-model mean (solid lines) and the 5 to 95% range (±1.64 standard deviation) across the distribution of individual models (shading). Discontinuities at 2100 are due to different numbers of models performing the extension runs beyond the 21st century and have no physical meaning. Only one ensemble member is used from each model and numbers in the figure indicate the number of different models contributing to the different time periods. No ranges are given for the RCP6.0 projections beyond 2100 as only two models are available.

Table 12.2 | CMIP5 annual mean surface air temperature anomalies (°C) from the 1986–2005 reference period for selected time periods, regions and RCPs. The multi-model mean ±1 standard deviation ranges across the individual models are listed and the 5 to 95% ranges from the models' distribution (based on a Gaussian assumption and obtained by multiplying the CMIP5 ensemble standard deviation by 1.64) are given in brackets. Only one ensemble member is used from each model and the number of models differs for each RCP (see Figure 12.5) and becomes significantly smaller after 2100. No ranges are given for the RCP6.0 projections beyond 2100 as only two models are available. Using Hadley Centre/Climate Research Unit gridded surface temperature data set 4 (HadCRUT4) and its uncertainty estimate (5 to 95% confidence interval), the observed warming to the 1986–2005 reference period (see Section 2.4.3) is 0.61°C ± 0.06°C (1850–1900), 0.30°C ± 0.03°C (1961–1990), 0.11°C ± 0.02°C (1980–1999). Decadal values are provided in Table AII.7.5, but note that percentiles of the CMIP5 distributions cannot directly be interpreted in terms of calibrated language.

		RCP2.6 (ΔT in °C)	RCP4.5 (ΔT in °C)	RCP6.0 (ΔT in °C)	RCP8.5 (ΔT in °C)
Global:	2046–2065	1.0 ± 0.3 (0.4, 1.6)	1.4 ± 0.3 (0.9, 2.0)	1.3 ± 0.3 (0.8, 1.8)	2.0 ± 0.4 (1.4, 2.6)
	2081–2100	1.0 ± 0.4 (0.3, 1.7)	1.8 ± 0.5 (1.1, 2.6)	2.2 ± 0.5 (1.4, 3.1)	3.7 ± 0.7 (2.6, 4.8)
	2181–2200	0.7 ± 0.4 (0.1, 1.3)	2.3 ± 0.5 (1.4, 3.1)	3.7 ± 0.7 (-,-)	6.5 ± 2.0 (3.3, 9.8)
	2281–2300	0.6 ± 0.3 (0.0, 1.2)	2.5 ± 0.6 (1.5, 3.5)	4.2 ± 1.0 (-,-)	7.8 ± 2.9 (3.0, 12.6)
Land: 2081–2100		1.2 ± 0.6 (0.3, 2.2)	2.4 ± 0.6 (1.3, 3.4)	3.0 ± 0.7 (1.8, 4.1)	4.8 ± 0.9 (3.4, 6.2)
Ocean: 2081–2100		0.8 ± 0.4 (0.2, 1.4)	1.5 ± 0.4 (0.9, 2.2)	1.9 ± 0.4 (1.1, 2.6)	3.1 ± 0.6 (2.1, 4.0)
Tropics: 2081–2100		0.9 ± 0.3 (0.3, 1.4)	1.6 ± 0.4 (0.9, 2.3)	2.0 ± 0.4 (1.3, 2.7)	3.3 ± 0.6 (2.2, 4.4)
Polar: Arctic: 2081–2100		2.2 ± 1.7 (-0.5, 5.0)	4.2 ± 1.6 (1.6, 6.9)	5.2 ± 1.9 (2.1, 8.3)	8.3 ± 1.9 (5.2, 11.4)
Polar: Antarctic: 2081–2100		0.8 ± 0.6 (-0.2, 1.8)	1.5 ± 0.7 (0.3, 2.7)	1.7 ± 0.9 (0.2, 3.2)	3.1 ± 1.2 (1.1, 5.1)

included here. Warming in 2046–2065 is slightly larger under RCP4.5 compared to RCP6.0, consistent with its greater total anthropogenic forcing at that time (see Table A.II.6.12). For all other periods the magnitude of global temperature change increases from RCP2.6 to RCP8.5. Beyond 2100, RCP2.6 shows a decreasing trend whereas under all other RCPs warming continues to increase. Also shown in Table 12.2 are projected changes at 2081–2100 averaged over land and ocean separately as well as area-weighted averages over the Tropics (30°S to 30°N), Arctic (67.5°N to 90°N) and Antarctic (90°S to 55°S) regions. Surface air temperatures over land warm more than over the ocean, and northern polar regions warm more than the tropics. The excess of land mass in the Northern Hemisphere (NH) in comparison with the Southern Hemisphere (SH), coupled with the greater uptake of heat by the Southern Ocean in comparison with northern ocean basins means that the NH generally warms more than the SH. Arctic warming is much greater than in the Antarctic, due to the presence of the Antarctic ice sheet and differences in local responses in snow and ice. Mechanisms behind these features of warming are discussed in Section 12.4.3. Maps and time series of regional temperature changes are displayed in Annex I and regional averages are discussed in Section 14.8.1.

Global annual multi-model mean temperature changes above 1850-1900 are listed in Table 12.3 for the 2081–2100 period (assuming 0.61°C warming since 1850–1900 as discussed in Section 2.4.3) along with the percentage of 2081–2100 projections from the CMIP5 models exceeding policy-relevant temperature levels under each RCP. These complement a similar discussion for the near-term projections in Table 11.3 which are based on the CMIP5 ensemble as well as evidence (discussed in Sections 10.3.1, 11.3.2.1.1 and 11.3.6.3) that some CMIP5 models have a higher sensitivity to GHGs and a larger response to other anthropogenic forcings (dominated by the effects of aerosols) than the real world (*medium confidence*). The percentage calculations for the long-term projections in Table 12.3 are based solely on the CMIP5 ensemble, using one ensemble member for each model. For these long-term projections, the 5 to 95% ranges of the CMIP5 model ensemble are considered the *likely* range, an assessment based on the fact that the 5 to 95% range of CMIP5 models' TCR coincides with the assessed *likely* range of the TCR (see Section 12.4.1.2 below and Box 12.2). Based on this assessment, global mean temperatures averaged in the period 2081–2100 are projected to *likely* exceed 1.5°C above 1850-1900 for RCP4.5, RCP6.0 and RCP8.5 (*high confidence*). They are also *likely* to exceed 2°C above 1850-1900 for RCP6.0 and RCP8.5 (*high confidence*) and *more likely than not* to exceed 2°C for RCP4.5 (*medium confidence*). Temperature change above 2°C under RCP2.6 is *unlikely* but is assessed only with *medium confidence* as some CMIP5 ensemble members do produce a global mean temperature change above 2°C. Warming above 4°C by 2081–2100 is *unlikely* in all RCPs (*high confidence*) except RCP8.5. Under the latter, the 4°C global temperature level is exceeded in more than half of ensemble members, and is assessed to be *about as likely as not* (*medium confidence*). Note that the likelihoods of exceeding specific temperature levels show some sensitivity to the choice of reference period (see Section 11.3.6.3).

CMIP5 models on average project a gradual increase in global precipitation over the 21st century: change exceeds 0.05 mm day^{-1} (~2% of global precipitation) and 0.15 mm day^{-1} (~5% of global precipitation) by 2100 in RCP2.6 and RCP8.5, respectively. The relationship between global precipitation and global temperature is approximately linear (Figure 12.6). The precipitation sensitivity, that is, the change of global precipitation with temperature, is about 1 to 3% °C^{-1} in most models, tending to be highest for RCP2.6 and RCP4.5 (Figure 12.7; note that only global values are discussed in this section, ocean and land changes are discussed in Section 12.4.5.2). These behaviours are consistent with previous studies, including CMIP3 model projections for SRES scenarios and AR4 constant composition commitment experiments (Meehl et al., 2007b), and ENSEMBLES multi-model results for SRES A1B and E1 scenarios (Johns et al., 2011).

The processes that govern global precipitation changes are now well understood and have been presented in Section 7.6. They are briefly summarized here and used to interpret the long-term projected changes. The precipitation sensitivity (about 1 to 3% °C^{-1}) is very different from the water vapour sensitivity (~7% °C^{-1}) as the main physical

Table 12.3 | CMIP5 global annual mean temperature changes above 1850-1900 for the 2081–2100 period of each RCP scenario (mean, ±1 standard deviation and 5 to 95% ranges based on a Gaussian assumption and obtained by multiplying the CMIP5 ensemble standard deviation by 1.64), assuming 0.61°C warming has occurred prior to 1986–2005 (second column). For a number of temperature levels (1°C, 1.5°C, 2°C, 3°C and 4°C), the proportion of CMIP5 model projections for 2081–2100 above those levels under each RCP scenario are listed. Only one ensemble member is used for each model.

	ΔT (°C) 2081–2100	$\Delta T > +1.0$°C	$\Delta T > +1.5$°C	$\Delta T > +2.0$°C	$\Delta T > +3.0$°C	$\Delta T > +4.0$°C
RCP2.6	1.6 ± 0.4 (0.9, 2.3)	94%	56%	22%	0%	0%
RCP4.5	2.4 ± 0.5 (1.7, 3.2)	100%	100%	79%	12%	0%
RCP6.0	2.8 ± 0.5 (2.0, 3.7)	100%	100%	100%	36%	0%
RCP8.5	4.3 ± 0.7 (3.2, 5.4)	100%	100%	100%	100%	62%

laws that drive these changes also differ. Water vapour increases are primarily a consequence of the Clausius–Clapeyron relationship associated with increasing temperatures in the lower troposphere (where most atmospheric water vapour resides). In contrast, future precipitation changes are primarily the result of changes in the energy balance of the atmosphere and the way that these later interact with circulation, moisture and temperature (Mitchell et al., 1987; Boer, 1993; Vecchi and Soden, 2007; Previdi, 2010; O'Gorman et al., 2012). Indeed, the radiative cooling of the atmosphere is balanced by latent heating (associated with precipitation) and sensible heating. Since AR4, the changes in heat balance and their effects on precipitation have been analyzed in detail for a large variety of forcings, simulations and models (Takahashi, 2009a; Andrews et al., 2010; Bala et al., 2010; Ming et al., 2010; O'Gorman et al., 2012; Bony et al., 2013).

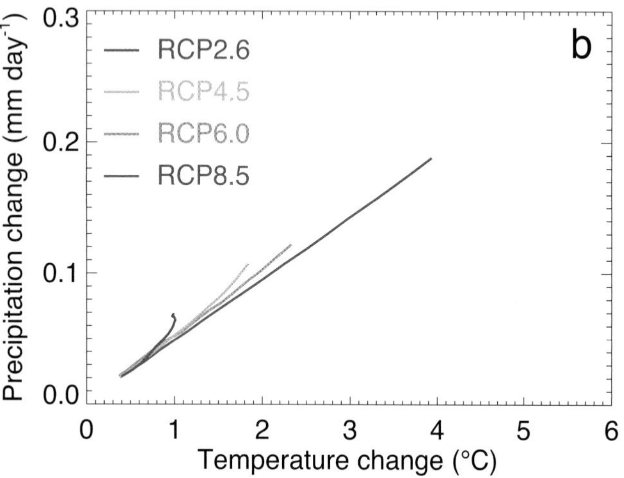

Figure 12.6 | Global mean precipitation (mm day^{-1}) versus temperature (°C) changes relative to 1986–2005 baseline period in CMIP5 model concentrations-driven projections for the four RCPs for (a) means over decadal periods starting in 2006 and overlapped by 5 years (2006–2015, 2011–2020, up to 2091–2100), each line representing a different model (one ensemble member per model) and (b) corresponding multi-model means for each RCP.

An increase of CO_2 decreases the radiative cooling of the troposphere and reduces precipitation (Andrews et al., 2010; Bala et al., 2010). On longer time scales than the fast hydrological adjustment time scale (Andrews et al., 2010; Bala et al., 2010; Cao et al., 2012; Bony et al., 2013), the increase of CO_2 induces a slow increase of temperature and water vapour, thereby enhancing the radiative cooling of the atmosphere and increasing global precipitation (Allen and Ingram, 2002; Yang et al., 2003; Held and Soden, 2006). Even after the CO_2 forcing stabilizes or begins to decrease, the ocean continues to warm, which then drives up global temperature, evaporation and precipitation. In addition, nonlinear effects also affect precipitation changes (Good et al., 2012). These different effects explain the steepening of the precipitation versus temperature relationship in RCP2.6 and RCP4.5 scenarios (Figure 12.6), as RF stabilizes and/or declines from the mid-century (Figure 12.4). In idealized CO_2 ramp-up/ramp-down experiments, this effect produces an hydrological response overshoot (Wu et al., 2010). An increase of absorbing aerosols warms the atmosphere and reduces precipitation, and the surface temperature response may be too small to compensate this decrease (Andrews et al., 2010; Ming et al., 2010; Shiogama et al., 2010a). Change in scattering aerosols or incoming solar radiation modifies global precipitation mainly via the response of the surface temperature (Andrews et al., 2009; Bala et al., 2010).

The main reasons for the inter-model spread of the precipitation sensitivity estimate among GCMs have not been fully understood. Nevertheless, spread in the changes of the cloud radiative effect has been shown to have an impact (Previdi, 2010), although the effect is less important for precipitation than it is for the climate sensitivity estimate (Lambert and Webb, 2008). The lapse rate plus water vapour feedback and the response of the surface heat flux (Previdi, 2010; O'Gorman et al., 2012), the shortwave absorption by water vapour (Takahashi, 2009b) or by aerosols, have been also identified as important factors.

Global precipitation sensitivity estimates from observations are very sensitive to the data and the time period considered. Some

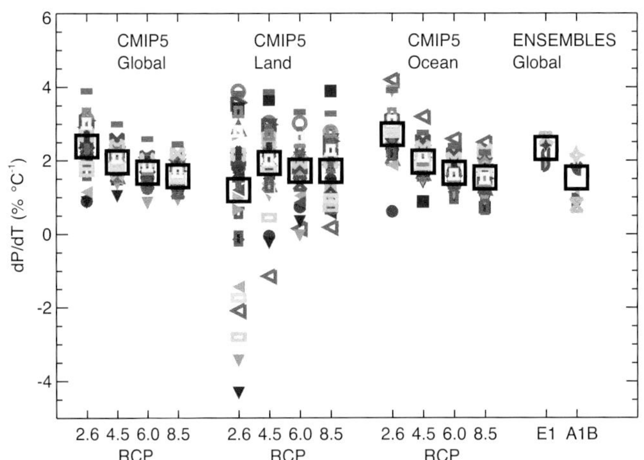

Figure 12.7 | Percentage changes over the 21st century in global, land and ocean precipitation per degree Celsius of global warming in CMIP5 model concentration-driven projections for the four RCP scenarios. Annual mean changes are calculated for each year between 2006 and 2100 from one ensemble member per model relative to its mean precipitation and temperature for the 1986–2005 baseline period, and the gradient of a least-squares fit through the annual data is derived. Land and ocean derived values use global mean temperature in the denominator of $\delta P/\delta T$. Each coloured symbol represents a different model, the same symbol being used for the same model for different RCPs and larger black squares being the multi-model mean. Also shown for comparison are global mean results for ENSEMBLES model concentrations-driven projections for the E1 and A1B scenarios (Johns et al., 2011), in this case using a least-squares fit derived over the period 2000–2099 and taking percentage changes relative to the 1980–1999 baseline period. Changes of precipitation over land and ocean are discussed in Section 12.4.5.2.

observational studies suggest precipitation sensitivity values higher than model estimates (Wentz et al., 2007; Zhang et al., 2007), although more recent studies suggest consistent values (Adler et al., 2008; Li et al., 2011b).

12.4.1.2 Uncertainties in Global Quantities

Uncertainties in global mean quantities arise from variations in internal natural variability, model response and forcing pathways. Table 12.2 gives two measures of uncertainty in the CMIP5 model projections, the standard deviation and the 5 to 95% range across the ensemble's distribution. Because CMIP5 was not designed to explore fully the uncertainty range in projections (see Section 12.2), neither its standard deviation nor its range can be interpreted directly as an uncertainty statement about the corresponding real quantities, and other techniques and arguments to assess uncertainty in future projections must be considered. Figure 12.8 summarizes the uncertainty ranges in global mean temperature changes at the end of the 21st century under the various scenarios quantified by various methods. Individual CMIP5 models are shown by red crosses. Red bars indicate mean and 5 to 95% percentiles based on assuming a normal distribution for the CMIP5 sample (i.e., ±1.64 standard deviations). Estimates from the simple climate carbon cycle Model for the Assessment of Greenhouse Gas-Induced Climate Change (MAGICC; Meinshausen et al., 2011a; Meinshausen et al., 2011b) calibrated to C[4]MIP (Friedlingstein et al., 2006) carbon cycle models, assuming a PDF for climate sensitivity that corresponds to the assessment of IPCC AR4 (Meehl et al., 2007b, Box 10.2), are given as yellow bars (Rogelj et al., 2012). Note that not all models have simulated all scenarios. To test the effect of undersampling, and to generate a consistent set of uncertainties across scenarios, a step response method that estimates the total warming as sum of responses to small forcing steps (Good et al., 2011a) is used to emulate 23 CMIP5 models under the different scenarios (those 23 models that supplied the necessary simulations to compute the emulators, i.e., CO_2 step change experiments). This provides means and ranges (5 to 95%) that are comparable across scenarios (blue). See also Section 12.4.9 for a discussion focussed on the differences between CMIP3 and CMIP5 projections of global average temperature changes.

For the CO_2 concentration-driven simulations (Figure 12.8a), the dominant driver of uncertainty in projections of global temperature for the higher RCPs beyond 2050 is the transient climate response (TCR), for RCP2.6, which is closer to equilibrium by the end of the century, it is both the TCR and the equilibrium climate sensitivity (ECS). In a transient situation, the ratio of temperature to forcing is approximately constant and scenario independent (Meehl et al., 2007b, Appendix 10.A.1; Gregory and Forster, 2008; Knutti et al., 2008b; Good et al., 2013). Therefore, the uncertainty in TCR maps directly into the uncertainty in global temperature projections for the RCPs other than RCP2.6. The assessed *likely* range of TCR based on various lines of evidence (see Box 12.2) is similar to the 5 to 95% percentile range of TCR in CMIP5. In addition, the assessed *likely* range of ECS is also consistent with the CMIP5 range (see Box 12.2). There is little evidence that the CMIP5 models are significantly over- or underestimating the RF. The RF uncertainty is small compared to response uncertainty (see Figure 12.4), and is considered by treating the 5 to 95% as a *likely* rather than *very likely* range. Kuhlbrodt and Gregory (2012) suggest that models might be overestimating ocean heat uptake, as previously suggested by Forest et al. (2006), but observationally constrained estimates of TCR are unaffected by that. The ocean heat uptake efficiency does not contribute much to the spread of TCR (Knutti and Tomassini, 2008; Kuhlbrodt and Gregory, 2012).

Therefore, for global mean temperature projections only, the 5 to 95% range (estimated as 1.64 times the sample standard deviation) of the CMIP5 projections can also be interpreted as a *likely* range for future temperature change between about 2050 and 2100. *Confidence* in this assessment is *high* for the end of the century because the warming then is dominated by CO_2 and the TCR. *Confidence* is only *medium* for mid-century when the contributions of RF and initial conditions to the total temperature response uncertainty are larger. The *likely* ranges are an expert assessment, taking into account many lines of evidence, in much the same way as in AR4 (Figure SPM.5), and are not probabilistic. The *likely* ranges for 2046–2065 do not take into account the possible influence of factors that lead to near-term (2016–2035) projections of global mean surface temperature (GMST) that are somewhat cooler than the 5 to 95% model ranges (see Section 11.3.6), because the influence of these factors on longer term projections cannot be quantified. A few recent studies indicate that some of the models with the strongest transient climate response might overestimate the near term warming (Otto et al., 2013; Stott et al., 2013) (see Sections 10.8.1, 11.3.2.1.1), but there is little evidence of whether and how much that affects the long-term warming response. One perturbed physics ensemble combined with observations indicates warming that exceeds the AR4 at the top end but used a relatively short time period of warming

(50 years) to constrain the models' projections (Rowlands et al., 2012) (see Sections 11.3.2.1.1 and 11.3.6.3). GMSTs for 2081–2100 (relative to 1986–2005) for the CO_2 concentration driven RCPs is therefore assessed to *likely* fall in the range 0.3°C to 1.7°C (RCP2.6), 1.1°C to 2.6°C (RCP4.5), 1.4°C to 3.1°C (RCP6.0), and 2.6°C to 4.8°C (RCP8.5) estimated from CMIP5. Beyond 2100, the number of CMIP5 simulations is insufficient to estimate a *likely* range. Uncertainties before 2050 are assessed in Section 11.3.2.1.1. The assessed *likely* range is very similar to the range estimated by the pulse response model, suggesting that the different sample of models for the different RCPs are not strongly affecting the result, and providing further support that this pulse response technique can be used to emulate temperature and ocean heat uptake in Chapter 13 and Section 12.4.9. The results are consistent with the probabilistic results from MAGICC, which for the lower RCPs have a slightly narrower range due to the lack of internal variability in the simple model, and the fact that non-CO_2 forcings are treated more homogeneously than in CMIP5 (Meinshausen et al., 2011a, 2011b). This is particularly pronounced for RCP2.6 where the CMIP5 range is substantially larger, partly due to the larger fraction of non-CO_2 forcings in that scenario.

The uncertainty estimate in AR4 for the SRES scenarios was –40% to +60% around the CMIP3 means (shown here in grey for comparison). That range was asymmetric and wider for the higher scenarios because it included the uncertainty in carbon cycle climate feedbacks. The SRES scenarios are based on the assumption of prescribed emissions, which then translates to uncertainties in concentrations that propagate through to uncertainties in the temperature response. The RCP scenarios assume prescribed concentrations. For scenarios that stabilize (RCP2.6) that approach of constant fractional uncertainty underestimates the uncertainty and is no longer applicable, mainly because internal variability has a larger relative contribution to the total uncertainty (Good et al., 2013; Knutti and Sedláček, 2013). For the RCPs, the carbon cycle climate feedback uncertainty is not included because the simulations are driven by concentrations. Furthermore, there is no clear evidence that distribution of CMIP5 global temperature changes deviates from a normal distribution. For most other variables the shape of the distribution is unclear, and standard deviations are simply used as an indication of model spread, not representing a formal uncertainty assessment.

Simulations with prescribed CO_2 emissions rather than concentrations are only available for RCP8.5 (Figure 12.8b) and from MAGICC. The projected temperature change in 2100 is slightly higher and the uncertainty range is wider as a result of uncertainties in the carbon cycle climate feedbacks. The CMIP5 range is consistent with the uncertainty range given in AR4 for SRES A2 in 2100. Further details about emission versus concentration driven simulations are given in Section 12.4.8.

In summary, the projected changes in global temperature for 2100 in the RCP scenarios are very consistent with those obtained by CMIP3 for SRES in IPCC AR4 (see Section 12.4.9) when taking into account the differences in scenarios. The *likely* uncertainty ranges provided here are similar for RCP4.5 and RCP6.0 but narrower for RCP8.5 compared to AR4. There was no scenario as low as RCP2.6 in AR4. The uncertainties in global temperature projections have not decreased significantly in CMIP5 (Knutti and Sedláček, 2013), but the assessed ranges cannot be compared between AR4 and AR5. The main reason is that uncertainties in carbon cycle feedbacks are not considered in the concentration driven RCPs. In contrast, the *likely* range in AR4 included those. The assessed *likely* ranges are therefore narrower for the high RCPs. The differences in the projected warming are largely attributable to the difference in scenarios (Knutti and Sedláček, 2013), and the change in the future and reference period, rather than to developments in modelling since AR4. A detailed comparison between the SRES and RCP scenarios and the CMIP3 and CMIP5 models is given in Section 12.4.9.

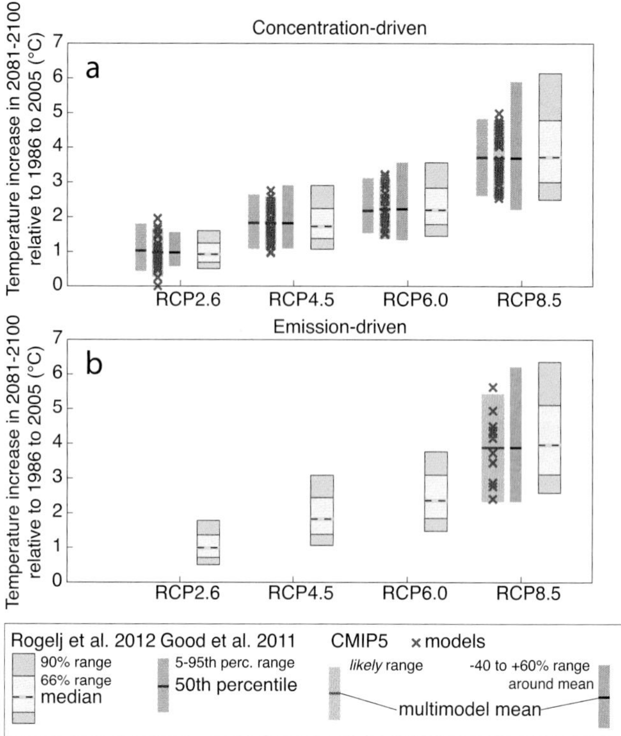

Figure 12.8 | Uncertainty estimates for global mean temperature change in 2081–2100 with respect to 1986–2005. Red crosses mark projections from individual CMIP5 models. Red bars indicate mean and 5 to 95% ranges based on CMIP5 (1.64 standard deviations), which are considered as a *likely* range. Blue bars indicate 5 to 95% ranges from the pulse response emulation of 21 models (Good et al., 2011a). Grey bars mark the range from the mean of CMIP5 minus 40% to the mean +60%, assessed as *likely* in AR4 for the SRES scenarios. The yellow bars show the median, 17 to 83% range and 5 to 95% range based on Rogelj et al. (2012). See also Figures 12.39 and 12.40.

12.4.2 Pattern Scaling

12.4.2.1 Definition and Use

In this chapter we show geographical patterns of projected changes in climate variables according to specific scenarios and time horizons. Alternative scenarios and projection times can be inferred from those shown by using some established approximation methods. This is especially the case for large-scale regional patterns of average temperature and—with additional caveats—precipitation changes. In fact, 'pattern scaling' is an approximation that has been explicitly suggested in the description of the RCPs (Moss et al., 2010) as a method for deriving impact-relevant regional projections for scenarios that have not been simulated by global and regional climate models. It was first proposed by Santer et al. (1990) and revisited later by numerous studies (e.g., Huntingford and Cox, 2000). It relies on the existence of robust geographical patterns of change, emerging at the time when the response to external forcings emerges from the noise, and persisting across the length of the simulation, across different scenarios, and even across models, modulated by the corresponding changes in global average temperature. The robustness of temperature change patterns has been amply documented from the original paper onward. An example is given in Figure 12.9 for surface air temperature from each of the CMIP5 models highlighting both similarities and differences between the responses of different models. The precipitation pattern was shown to scale linearly with global average temperature to a sufficient accuracy in CMIP3 models (Neelin et al., 2006) for this to be useful for projections related to the hydrological cycle. Shiogama et al. (2010b) find similar results with the caution that in the early stages of warming aerosols modify the pattern. A more mixed evaluation can be found in

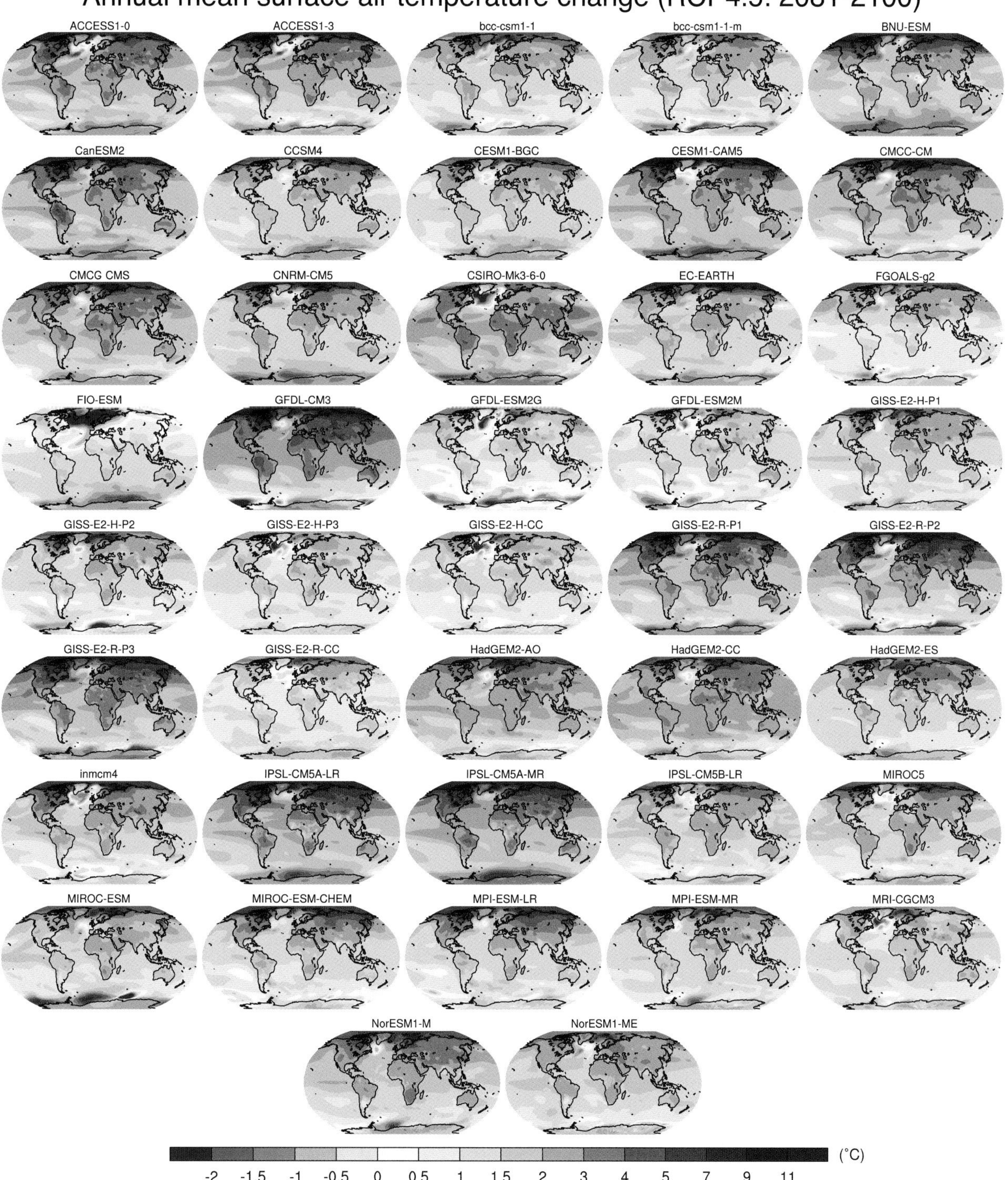

Figure 12.9 | Surface air temperature change in 2081–2100 displayed as anomalies with respect to 1986–2005 for RCP4.5 from one ensemble member of each of the concentration-driven models available in the CMIP5 archive.

12

Good et al. (2012), where some land areas in the low latitudes exhibit a nonlinear relation to global average temperature, but, largely, average precipitation change over the remaining regions can be well approximated by a grid-point specific linear function of global average temperature change. It is in the latter quantity that the dependence of the evolution of the change in time on the model (e.g., its climate sensitivity) and the forcing (e.g., the emission scenario) is encapsulated.

In analytical terms, it is assumed that the following relation holds:

$$C\ (t,\xi) = T_G(t)\ \chi(\xi) + R\ (t,\ \xi)$$

where the symbol ξ identifies the geographic location (model grid point or other spatial coordinates) and possibly the time of year (e.g., a June–July–August average). The index t runs along the length of the forcing scenario of interest. $T_G(t)$ indicates global average temperature change at time t under this scenario; $\chi(\xi)$ is the time-invariant geographic pattern of change per 1°C global surface temperature change for the variable of interest (which represents the forced component of the change) and $C\ (t,\xi)$ is the actual field of change for that variable at the specific time t under this scenario. The $R\ (t,\ \xi)$ is a residual term and highlights the fact that pattern scaling cannot reconstruct model behaviour with complete accuracy due to both natural variability and because of limitations of the methodology discussed below. This way, regionally and temporally differentiated results under different scenarios or climate sensitivities can be approximated by the product of a spatial pattern, constant over time, scenario and model characteristics, and a time evolving global mean change in temperature. Model and scenario dependence are thus captured through the global mean temperature response, and simple climate models calibrated against fully coupled climate models can be used to simulate the latter, at a great saving in computational cost. The spatial pattern can be estimated through the available coupled model simulations under the assumption that it does not depend on the specific scenario(s) used.

The choice of the pattern in the studies available in the literature can be as simple as the ensemble average field of change (across models and/or across scenarios, for the coupled experiments available), normalized by the corresponding change in global average temperature, choosing a segment of the simulations when the signal has emerged from the noise of natural variability from a baseline of reference (e.g., the last 20 years of the 21st century compared to pre-industrial or current climate) and taking the difference of two multi-decadal means. Similar properties and results have been obtained using more sophisticated multivariate procedures that optimize the variance explained by the pattern (Holden and Edwards, 2010). The validity of this approximation is discussed by Mitchell et al. (1999) and Mitchell (2003). Huntingford and Cox (2000) evaluate the quality of the approximation for numerous variables, showing that the technique performs best for temperature, downward longwave radiation, relative humidity, wind speeds and surface pressure while showing relatively larger limitations for rainfall rate anomalies. Joshi et al. (2013) have recently shown that the accuracy of the approximation, especially across models, is improved by adding a second term, linear in the land–sea surface warming ratio, another quantity that can be easily estimated from existing coupled climate model simulations. There exist of course differences between the patterns generated by different GCMs (documented for example for CMIP3 in Watterson and Whetton, 2011b), but uncertainty can be characterized, for example, by the inter-model spread in the pattern $\chi(\xi)$. Recent applications of the methodology to probabilistic future projections have in fact sought to fully quantify errors introduced by the approximation, on the basis of the available coupled model runs (Harris et al., 2006).

Pattern scaling and its applications have been documented in IPCC WGI Reports before (IPCC, 2001, Section 13.5.2.1; Meehl et al., 2007b, Section 10.3.2). It has been used extensively for regional temperature and precipitation change projections, for example, Murphy et al. (2007), (Watterson, 2008), Giorgi (2008), Harris et al. (2006, 2010), May (2008a), Ruosteenoja et al. (2007), Räisänen and Ruokolainen (2006), Cabre et al. (2010) and impact studies, for example, as described in Dessai et al. (2005) and Fowler et al. (2007b). Recent studies have focussed on patterns linked to warming at certain global average temperature change thresholds (e.g., May, 2008a; Sanderson et al., 2011) and patterns derived under the RCPs (Ishizaki et al., 2012).

There are basic limitations to this approach, besides a degradation of its performance as the regional scale of interest becomes finer and in the presence of regionally specific forcings. Recent work with MIROC3.2 (Shiogama et al., 2010a; Shiogama et al., 2010b) has revealed a dependence of the precipitation sensitivity (global average precipitation change per 1°C of global warming—see Figure 12.6) on the scenario, due to the precipitation being more sensitive to carbon aerosols than WMGHGs. In fact, there are significant differences in black and organic carbon aerosol forcing between the emission scenarios investigated by Shiogama et al. (2010a; 2010b). Levy II et al. (2013) confirm that patterns of precipitation change are spatially correlated with the sources of aerosol emissions, in simulations where the indirect effect is represented. This is a behaviour that is linked to a more general limitation of pattern scaling, which breaks down if aerosol forcing is significant. The effects of aerosols have a regional nature and are thus dependent on the future sources of pollution which are likely to vary geographically in the future and are difficult to predict (May, 2008a). For example, Asian and North American aerosol production are likely to have different time histories and future projections. Schlesinger et al. (2000) extended the methodology of pattern scaling by isolating and recombining patterns derived by dedicated experiments with a coupled climate model where sulphate aerosols were increased for various regions in turn. More recently, in an extension of pattern scaling into a probabilistic treatment of model, scenario and initial condition uncertainties, Frieler et al. (2012) derived joint probability distributions for regionally averaged temperature and precipitation changes as linear functions of global average temperature and additional predictors including regionally specific sulphate aerosol and black carbon emissions.

Pattern scaling is less accurate for strongly mitigated stabilization scenarios. This has been shown recently by May (2012), comparing patterns of temperature change under a scenario limiting global warming since pre-industrial times to 2°C and patterns produced by a scenario that reaches 4.5°C of global average temperature change. The limitations of pattern scaling in approximating changes while the climate system approaches equilibrium have found their explanation in Manabe and Wetherald (1980) and Mitchell et al. (1999). Both studies point out that as the temperatures of the deep oceans reach equilibrium (over multiple centuries) the geographical distribution of warming changes as well, for example, showing a larger warming of the high latitudes in the SH than in the earlier periods of the transient response, relative to the global mean warming. More recently, Held et al. (2010) showed how this slow warming pattern is in fact present during the initial transient response of the system as well, albeit with much smaller amplitude. Further, Gillett et al. (2011) show how in a simulation in which emissions cease, regional temperatures and precipitation patterns exhibit ongoing changes, even though global mean temperature remains almost constant. Wu et al. (2010) showed that the global precipitation response shows a nonlinear response to strong mitigation scenarios, with the hydrological cycle continuing to intensify even after atmospheric CO_2 concentration, and thus global average temperature, start decreasing. Regional nonlinear responses to mitigation scenarios of precipitation and sea surface temperatures (SSTs) are shown by Chadwick et al. (2013).

Other areas where pattern scaling shows a lack of robustness are the edges of polar ice caps and sea ice extent, where at an earlier time in the simulation ice melts and regions of sharp gradient surface, while later in the simulation, in the absence of ice, the gradient will become less steep. Different sea ice representations in models also make the location of such regions much less robust across the model ensembles and the scenarios.

Pattern scaling has not been as thoroughly explored for quantities other than average temperature and precipitation. Impact relevant extremes, for example, seem to indicate a critical dependence on the scale at which their changes are evaluated, with studies showing that some aspects of their statistics change in a close-to-linear way with mean temperature (Kharin et al., 2007; Lustenberger et al., 2013) while others have documented the dependence of their changes on moments of their statistical distribution other than the mean (Ballester et al., 2010a), which would make pattern scaling inadequate.

12.4.2.2 Coupled Model Intercomparison Project Phase 5 Patterns Scaled by Global Average Temperature Change

On the basis of CMIP5 simulations, we show geographical patterns (Figure 12.10) of warming and precipitation change and indicate measures of their variability across models and across RCPs. The patterns are scaled to 1°C global mean surface temperature change above the reference period 1986–2005 for 2081–2100 (first row) and for a period of approximate stable temperature, 2181–2200 (thus excluding RCP8.5, which does not stabilize by that time) (second row). Spatial correlation of fields of temperature and precipitation change range from 0.93 to 0.99 when considering ensemble means under different RCPs. The lower values are found when computing correlation between RCP2.6 and the higher RCPs, and may be related to the high mitigation

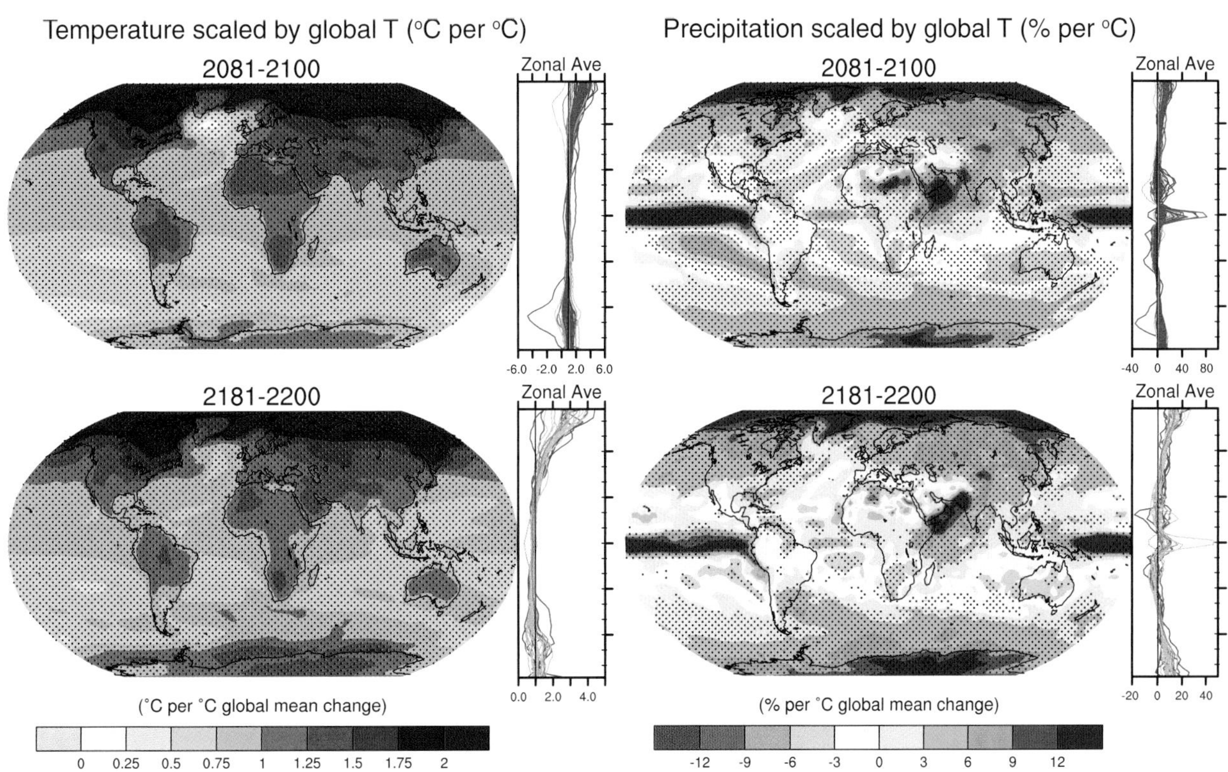

Figure 12.10 | Temperature (left) and precipitation (right) change patterns derived from transient simulations from the CMIP5 ensembles, scaled to 1°C of global mean surface temperature change. The patterns have been calculated by computing 20-year averages at the end of the 21st (top) and 22nd (bottom) centuries and over the period 1986–2005 for the available simulations under all RCPs, taking their difference (percentage difference in the case of precipitation) and normalizing it, grid-point by grid-point, by the corresponding value of global average temperature change for each model and scenario. The normalized patterns have then been averaged across models and scenarios. The colour scale represents degrees Celsius (in the case of temperature) and percent (in the case of precipitation) per 1°C of global average temperature change. Stippling indicates where the mean change averaged over all realizations is larger than the 95% percentile of the distribution of models. Zonal means of the geographical patterns are shown for each individual model for RCP2.6 (blue), 4.5 (light blue), 6.0 (orange) and 8.5 (red). RCP8.5 is excluded from the stabilization figures. The RCP2.6 simulation of the FIO-ESM (First Institute of Oceanography) model was excluded because it did not show any warming by the end of the 21st century, thus not complying with the method requirement that the pattern be estimated at a time when the temperature change signal from CO_2 increase has emerged.

enacted under RCP2.6 from early in the 21st century. Pattern correlation varies between 0.91 and 0.98 for temperature and between 0.91 and 0.96 for precipitation when comparing patterns computed by averaging and normalizing changes at the end of the 21st, 22nd and 23rd centuries, with the largest value representing the correlation between the patterns at the end of the 22nd and 23rd centuries, the lowest representing the correlation between the pattern at the end of the 21st and the pattern at the end of the 23rd century. The zonal means shown to the side of each plot represent each model by one line, colour coding the four different scenarios. They show good agreement of models and scenarios over low and mid-latitudes for temperature, but higher spread across models and especially across scenarios for the areas subject to polar amplification, for which the previous discussion about the sensitivity of the patterns to the sea ice edge may be relevant. A comparison of the mean of the lines to their spread indicates overall the presence of a strong mean signal with respect to the spread of the ensemble. Precipitation shows an opposite pattern of inter-model spread, with larger variations in the low latitudes and around the equator, and smaller around the high latitudes. Precipitation has also a lower signal-to-noise ratio (measured as above by comparing the ensemble mean change magnitude to the spread across models and scenarios of these zonal mean averages).

As already mentioned, although we do not explicitly use pattern scaling in the sections that follow, we consider it a useful approximation when the need emerges to interpolate or extrapolate results to different scenarios or time periods, noting the possibility that the scaling may break down at higher levels of global warming, and that the validity of the approximation is limited to broad patterns of change, as opposed to local scales. An important caveat is that pattern scaling only applies to the climate response that is externally forced. The actual response is a combination of forced change and natural variability, which is not and should not be scaled up or down by the application of this technique, which becomes important on small spatial scales and shorter time scales, and whose relative magnitude compared to the forced component also depends on the variable (Hawkins and Sutton, 2009, 2011; Mahlstein et al., 2011; Deser et al., 2012a, 2012b; Mahlstein et al., 2012) (see Section 11.2). One approach to produce projections that include both components is to estimate natural variability separately, scale the forced response and add the two.

12.4.3 Changes in Temperature and Energy Budget

12.4.3.1 Patterns of Surface Warming: Land–Sea Contrast, Polar Amplification and Sea Surface Temperatures

Patterns of surface air temperature change for various RCPs show widespread warming during the 21st century (Figure 12.11; see Annex I for seasonal patterns). A key feature that has been present throughout the history of coupled modelling is the larger warming over land compared to oceans, which occurs in both transient and equilibrium climate change (e.g., Manabe et al., 1990). The degree to which warming is larger over land than ocean is remarkably constant over time under transient warming due to WMGHGs (Lambert and Chiang, 2007; Boer, 2011; Lambert et al., 2011) suggesting that heat capacity differences between land and ocean do not play a major role in the land–sea warming contrast (Sutton et al., 2007; Joshi et al., 2008, 2013). The phenomenon is predominantly a feature of the surface and lower atmosphere (Joshi et al., 2008). Studies have found it occurs due to contrasts in surface sensible and latent fluxes over land (Sutton et al., 2007), land–ocean contrasts in boundary layer lapse rate changes (Joshi et al., 2008), boundary layer relative humidity and associated low-level cloud cover changes over land (Doutriaux-Boucher et al., 2009; Fasullo, 2010) and soil moisture reductions (Dong et al., 2009; Clark et al., 2010) under climate change. The land–sea warming contrast is also sensitive to aerosol forcing (Allen and Sherwood, 2010; Joshi et al., 2013). Globally averaged warming over land and ocean is identified separately in Table 12.2 for the CMIP5 models and the ratio of land to ocean warming is *likely* in the range of 1.4 to 1.7, consistent with previous studies (Lambert et al., 2011). The CMIP5 multi-model mean ratio is approximately constant from 2020 through to 2100 (based on an update of Joshi et al., 2008 from available CMIP5 models).

Amplified surface warming in Arctic latitudes is also a consistent feature in climate model integrations (e.g., Manabe and Stouffer, 1980). This is often referred to as polar amplification, although numerous studies have shown that under transient forcing, this is primarily an Arctic phenomenon (Manabe et al., 1991; Meehl et al., 2007b). The lack of an amplified transient warming response in high Southern polar latitudes has been associated with deep ocean mixing, strong ocean heat uptake and the persistence of the vast Antarctic ice sheet. In equilibrium simulations, amplified warming occurs in both polar regions.

On an annual average, and depending on the forcing scenario (see Table 12.2), the CMIP5 models show a mean Arctic (67.5°N to 90°N) warming between 2.2 and 2.4 times the global average warming for 2081–2100 compared to 1986–2005. Similar polar amplification factors occurred in earlier coupled model simulations (e.g., Holland and Bitz, 2003; Winton, 2006a). This factor in models is slightly higher than the observed central value, but it is within the uncertainty of the best estimate from observations of the recent past (Bekryaev et al., 2010). The uncertainty is large in the observed factor because station records are short and sparse (Serreze and Francis, 2006) and the forced signal is contaminated by the noise of internal variability. By contrast, model trends in surface air temperature are 2.5 to 5 times higher than observed over Antarctica, but here also the observational estimates have a very large uncertainty, so, for example, the CMIP3 ensemble mean is consistent with observations within error estimates (Monaghan et al., 2008). Moreover, recent work suggests more widespread current West Antarctic surface warming than previously estimated (Bromwich et al., 2013).

The amplified Arctic warming in models has a distinct seasonal character (Manabe and Stouffer, 1980; Rind, 1987; Holland and Bitz, 2003; Lu and Cai, 2009; Kumar et al., 2010). Arctic amplification (defined as the 67.5 N° to 90°N warming compared to the global average warming for 2081–2100 versus 1986–2005) peaks in early winter (November to December) with a CMIP5 RCP4.5 multi-model mean warming for 67.5°N to 90°N exceeding the global average by a factor of more than 4. The warming is smallest in summer when excess heat at the Arctic surface goes into melting ice or is absorbed by the ocean, which has a relatively large thermal inertia. Simulated Arctic warming also has a consistent vertical structure that is largest in the lower troposphere

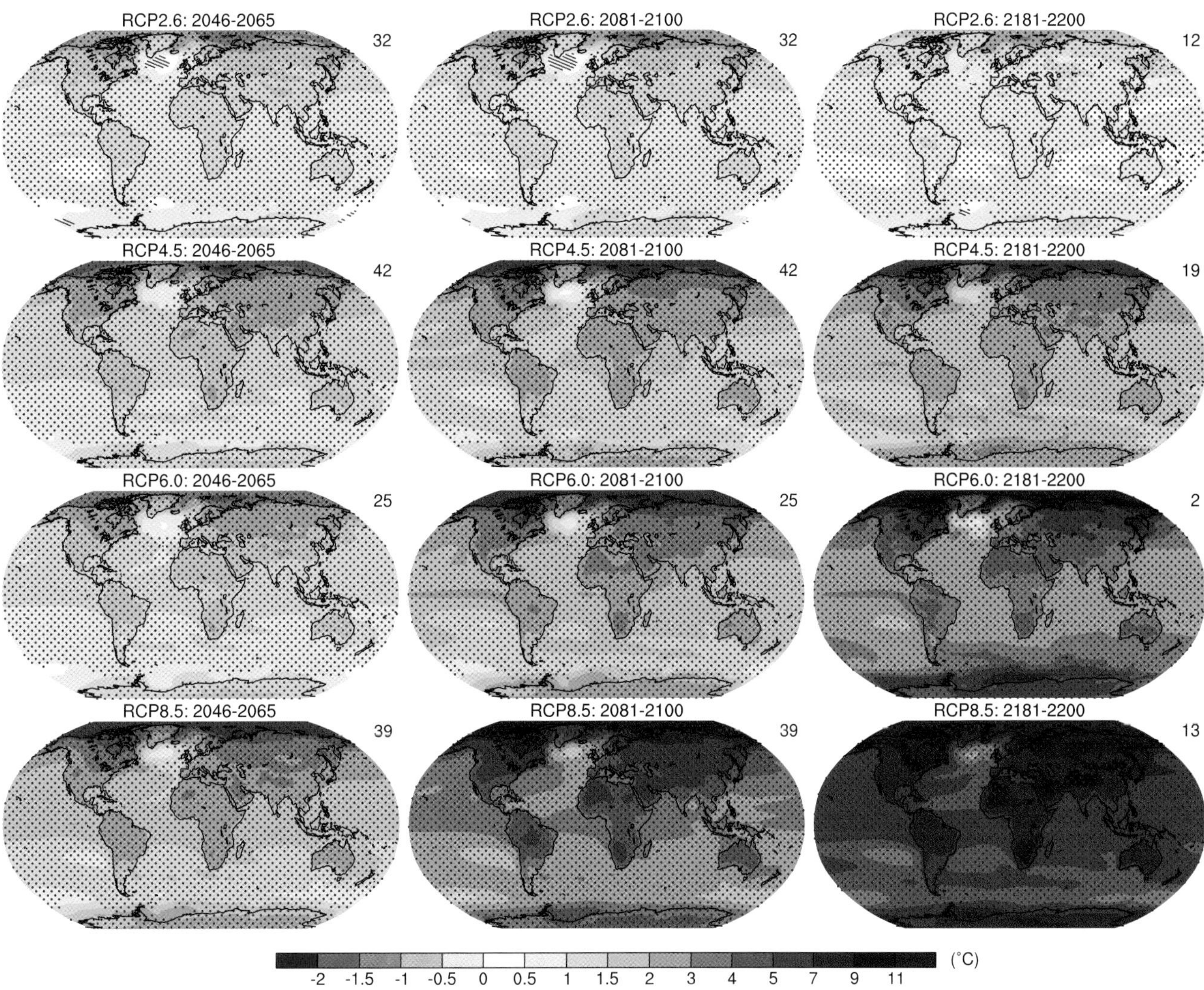

Figure 12.11 | Multi-model ensemble average of surface air temperature change (compared to 1986–2005 base period) for 2046–2065, 2081–2100, 2181–2200 for RCP2.6, 4.5, 6.0 and 8.5. Hatching indicates regions where the multi-model mean change is less than one standard deviation of internal variability. Stippling indicates regions where the multi-model mean change is greater than two standard deviations of internal variability and where at least 90% of the models agree on the sign of change (see Box 12.1). The number of CMIP5 models used is indicated in the upper right corner of each panel.

(e.g., Manabe et al., 1991; Kay et al., 2012). This is in agreement with recent observations (Serreze et al., 2009; Screen and Simmonds, 2010) but contrary to an earlier study that suggested a larger warming aloft (Graversen et al., 2008). The discrepancy in observed vertical structure may reflect inadequacies in data sets (Bitz and Fu, 2008; Grant et al., 2008; Thorne, 2008) and sensitivity to the time period used for averaging (see also Box 2.3).

As also discussed in Box 5.1, there are many mechanisms that contribute to Arctic amplification, some of which were identified in early modelling studies (Manabe and Stouffer, 1980). Feedbacks associated with changes in sea ice and snow amplify surface warming near the poles (Hall, 2004; Soden et al., 2008; Graversen and Wang, 2009; Kumar et al., 2010). The longwave radiation changes in the top of the atmosphere associated with surface warming opposes surface warming at all latitudes, but less so in the Arctic (Winton, 2006a; Soden et al., 2008). Rising temperature globally is expected to increase the horizontal latent heat transport by the atmosphere into the Arctic (Flannery, 1984; Alexeev et al., 2005; Cai, 2005; Langen and Alexeev, 2007; Kug et al., 2010), which warms primarily the lower troposphere. On average, CMIP3 models simulate enhanced latent heat transport (Held and Soden, 2006), but north of about 65°N, the sensible heat transport declines enough to more than offset the latent heat transport increase (Hwang et al., 2011). Increased atmospheric heat transport into the Arctic and subsidence warming has been associated with a teleconnection driven by enhanced convection in the tropical western Pacific (Lee et al., 2011). Ocean heat transport plays a role in the simulated Arctic amplification, with both large late 20th century transport (Mahlstein and Knutti, 2011) and increases over the 21st century (Hwang et al., 2011; Bitz et al., 2012) associated with higher amplification. As noted by Held and Soden (2006), Kay et al. (2012), and Alexeev and Jackson (2012), diagnosing the role of various factors in amplified warming is complicated by coupling in the system in which local feedbacks interact with poleward heat transports.

Although models consistently exhibit Arctic amplification as global mean temperatures rise, the multitude of physical processes described above mean that they differ considerably in the magnitude. Previous work has implicated variations across climate models in numerous factors including inversion strength (Boé et al., 2009a), ocean heat transport (Holland and Bitz, 2003; Mahlstein and Knutti, 2011), albedo feedback (Winton, 2006a), longwave radiative feedbacks (Winton, 2006a) and shortwave cloud feedback (Crook et al., 2011; Kay et al., 2012) as playing a role in the across-model scatter in polar amplification. The magnitude of amplification is generally higher in models with less extensive late 20th century sea ice in June, suggesting that the initial ice state influences the 21st century Arctic amplification. The pattern of simulated Arctic warming is also associated with the initial ice state, and in particular with the location of the winter sea ice edge (Holland and Bitz, 2003; Räisänen, 2007; Bracegirdle and Stephenson, 2012). This relationship has been suggested as a constraint on projected Arctic warming (Abe et al., 2011; Bracegirdle and Stephenson, 2012), although, in general, the ability of models to reproduce observed climate and its trends is not a sufficient condition for attributing *high confidence* to the projection of future trends (see Section 9.8).

Minima in surface warming occur in the North Atlantic and Southern Oceans under transient forcing in part due to deep ocean mixed layers in those regions (Manabe et al., 1990; Xie et al., 2010). Trenberth and Fasullo (2010) find that the large biases in the Southern Ocean energy budget in CMIP3 coupled models negatively correlate with equilibrium climate sensitivity (see Section 12.5.3), suggesting that an improved mean state in the Southern Ocean is needed before warming there can be understood. In the equatorial Pacific, warming is enhanced in a narrow band which previous assessments have described as 'El Niño-like', as may be expected from the projected decrease in atmospheric tropical circulations (see Section 12.4.4). However, DiNezio et al. (2009) highlight that the tropical Pacific warming in the CMIP3 models is not 'El Niño-like' as the pattern of warming and associated teleconnections (Xie et al., 2010; Section 12.4.5.2) is quite distinct from that of an El Niño event. Instead the pattern is of enhanced equatorial warming and is due to a meridional minimum in evaporative damping on the equator (Liu et al., 2005) and ocean dynamical changes that can be decoupled from atmospheric changes (DiNezio et al., 2009) (see also further discussion in Section 12.4.7).

In summary, there is robust evidence over multiple generations of models and *high confidence* in these large-scale warming patterns. In the absence of a strong reduction in the Atlantic Meridional Overturning Circulation (AMOC), there is *very high confidence* that the Arctic region is projected to warm most.

12.4.3.2 Zonal Average Atmospheric Temperature

Zonal temperature changes at the end of the 21st century show warming throughout the troposphere and, depending on the scenario, a mix of warming and cooling in the stratosphere (Figure 12.12). The maximum warming in the tropical upper troposphere is consistent with theoretical explanations and associated with a decline in the moist adiabatic lapse rate of temperature in the tropics as the climate warms (Bony et al., 2006). The northern polar regions also experience large warming in the lower atmosphere, consistent with the mechanisms discussed in Section 12.4.3.1. The tropospheric patterns are similar to those in the TAR and AR4 with the RCP8.5 changes being up to several degrees warmer in the tropics compared to the A1B changes appearing in the AR4. Similar tropospheric patterns appear in the RCP 2.6 and 4.5 changes, but with reduced magnitudes, suggesting some degree of scaling with forcing change in the troposphere, similar to behaviour discussed in the AR4 and Section 12.4.2. The consistency of tropospheric patterns over multiple generations of models indicates *high confidence* in these projected changes.

In the stratosphere, the models show similar tropical patterns of change, with magnitudes differing according to the degree of climate forcing. Substantial differences appear in polar regions. In the north, RCP8.5 and 4.5 yield cooling, though it is more significant in the RCP8.5 ensemble. In contrast, RCP2.6 shows warming, albeit weak and with little significance. In the southern polar region, RCP 2.6 and 4.5 both show significant warming, and RCP8.5 is the outlier, with significant cooling. The polar stratospheric warming, especially in the SH, is similar to that found by Butchart et al. (2010) and Meehl et al. (2012) in GCM simulations that showed effects of ozone recovery in determining the patterns (Baldwin et al., 2007; Son et al., 2010). Eyring et al. (2013) find behaviour in the CMIP5 ensemble both for models with and without interactive chemistry that supports the contention that the polar stratospheric changes in Figure 12.12 are strongly influenced by ozone recovery. Overall, the stratospheric temperature changes do not exhibit pattern scaling with global temperature change and are dependent on ozone recovery.

Away from the polar stratosphere, there is physical and pattern consistency in temperature changes between different generations of models assessed here and in the TAR and AR4. The consistency is especially clear in the northern high latitudes and, coupled with physical understanding, indicates that some of the greatest warming is *very likely* to occur here. There is also consistency across generations of models in relatively large warming in the tropical upper troposphere. Allen and Sherwood (2008) and Johnson and Xie (2010) have presented dynamic and thermodynamic arguments, respectively, for the physical robustness of the tropical behaviour. However, there remains uncertainty about the magnitude of warming simulated in the tropical upper troposphere because large observational uncertainties and contradictory analyses limit a confident assessment of model accuracy in simulating temperature trends in the tropical upper troposphere (Section 9.4.1.4.2). The combined evidence indicates that relatively large warming in the tropical upper troposphere is *likely*, but with *medium confidence*.

12.4.3.3 Temperature Extremes

As the climate continues to warm, changes in several types of temperature extremes have been observed (Donat et al., 2013), and are expected to continue in the future in concert with global warming (Seneviratne et al., 2012). Extremes occur on multiple time scales, from a single day or a few consecutive days (a heat wave) to monthly and seasonal events. Extreme temperature events are often defined by indices (see Box 2.4 for the common definitions used), for example, percentage of days in a year when maximum temperature is above the 90th percentile of a present day distribution or by long period return values. Although changes in temperature extremes are a very robust

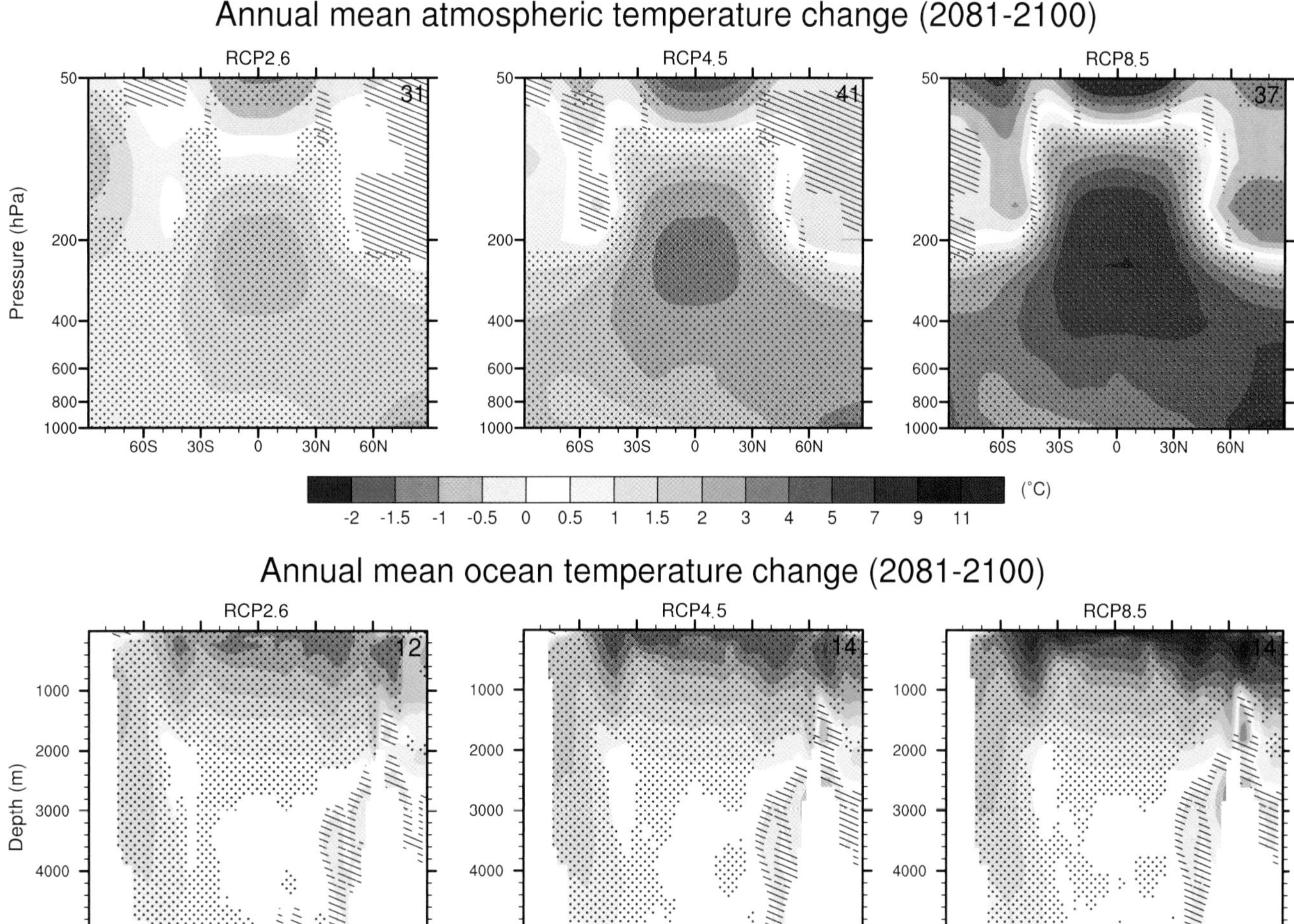

Figure 12.12 | CMIP5 multi-model changes in annual mean zonal mean temperature in the atmosphere and ocean relative to 1986–2005 for 2081–2100 under the RCP2.6 (left), RCP4.5 (centre) and RCP8.5 (right) forcing scenarios. Hatching indicates regions where the multi-model mean change is less than one standard deviation of internal variability. Stippling indicates regions where the multi-model change mean is greater than two standard deviations of internal variability and where at least 90% of the models agree on the sign of change (see Box 12.1).

signature of anthropogenic climate change (Seneviratne et al., 2012), the magnitude of change and consensus among models varies with the characteristics of the event being considered (e.g., time scale, magnitude, duration and spatial extent) as well as the definition used to describe the extreme.

Since the AR4 many advances have been made in establishing global observed records of extremes (Alexander et al., 2006; Perkins et al., 2012; Donat et al., 2013) against which models can be evaluated to give context to future projections (Sillmann and Roeckner, 2008; Alexander and Arblaster, 2009). Numerous regional assessments of future changes in extremes have also been performed and a comprehensive summary of these is given in Seneviratne et al. (2012). Here we summarize the key findings from this report and assess updates since then.

It is *virtually certain* that there will be more hot and fewer cold extremes as global temperature increases (Caesar and Lowe, 2012; Orlowsky and Seneviratne, 2012; Sillmann et al., 2013), consistent with previous assessments (Solomon et al., 2007; Seneviratne et al., 2012). Figure 12.13 shows multi-model mean changes in the absolute temperature indices of the coldest day of the year and the hottest day of the year and the threshold-based indices of frost days and tropical nights from the CMIP5 ensemble (Sillmann et al., 2013). A robust increase in warm temperature extremes and decrease in cold temperature extremes is found at the end of the 21st century, with the magnitude of the changes increasing with increased anthropogenic forcing. The coldest night of the year undergoes larger increases than the hottest day in the globally averaged time series (Figure 12.13b and d). This tendency is consistent with the CMIP3 model results shown in Figure 12.13, which use different models and the SRES scenarios (see Seneviratne et al. (2012) for earlier CMIP3 results). Similarly, increases in the frequency of warm nights are greater than increases in the frequency of warm days (Sillmann et al., 2013). Regionally, the largest increases in the coldest night of the year are projected in the high latitudes of the NH under the RCP8.5 scenario (Figure 12.13a). The subtropics and mid-latitudes exhibit the greatest projected changes in the hottest day of the year, whereas changes in tropical nights and the frequency of warm days and warm nights are largest in the tropics (Sillmann et al., 2013). The number of frost days declines in all regions while significant increases in tropical nights are seen in southeastern North America, the Mediterranean and central Asia.

It is *very likely* that, on average, there will be more record high than record cold temperatures in a warmer average climate. For example, Meehl et al. (2009) find that the current ratio of 2 to 1 for record daily high maxima to low minima over the USA becomes approximately 20 to 1 by the mid-21st century and 50 to 1 by late century in their model simulation of the SRES A1B scenario. However, even at the end of the century daily record low minima continue to be broken, if in a small number, consistent with Kodra et al. (2011), who conclude that cold extremes will continue to occur in a warmer climate, even though their frequency will decline.

It is also *very likely* that heat waves, defined as spells of days with temperature above a threshold determined from historical climatology, will occur with a higher frequency and duration, mainly as a direct consequence of the increase in seasonal mean temperatures (Barnett et al., 2006; Ballester et al., 2010a, 2010b; Fischer and Schär, 2010). Changes in the absolute value of temperature extremes are also *very likely* and expected to regionally exceed global temperature increases by far, with substantial changes in hot extremes projected even for moderate (<2.5°C above present day) average warming levels (Clark et al., 2010; Diffenbaugh and Ashfaq, 2010). These changes often differ from the mean temperature increase, as a result of changes in variability and shape of the temperature distribution (Hegerl et al., 2004; Meehl and Tebaldi, 2004; Clark et al., 2006). For example, summer temperature extremes over central and southern Europe are projected to warm substantially more than the corresponding mean local temperatures as a result of enhanced temperature variability at interannual to intraseasonal time scales (Schär et al., 2004; Clark et al., 2006; Kjellstrom et al., 2007; Vidale et al., 2007; Fischer and Schär, 2009, 2010; Nikulin et al., 2011; Fischer et al., 2012a). Several recent studies have also argued that the probability of occurrence of a Russian heat wave at least as severe as the one in 2010 increases substantially (by a factor of 5 to 10 by the mid-century) along with increasing mean temperatures and enhanced temperature variability (Barriopedro et al., 2011; Dole et al., 2011).

Since the AR4, an increased understanding of mechanisms and feedbacks leading to projected changes in extremes has been gained (Seneviratne et al., 2012). Climate models suggest that hot extremes are amplified by soil moisture-temperature feedbacks (Seneviratne et al., 2006; Diffenbaugh et al., 2007; Lenderink et al., 2007; Vidale et al., 2007; Fischer and Schär, 2009; Fischer et al., 2012a) in northern mid-latitude regions as the climate warms, consistent with previous assessments. Changes in temperature extremes may also be impacted by changes in land–sea contrast, with Watterson et al. (2008) showing an amplification of southern Australian summer warm extremes over the mean due to anomalous temperature advection from warmer continental interiors. The largest increases in the magnitude of warm extremes are simulated over mid-latitude continental areas, consistent with the drier conditions, and the associated reduction in evaporative cooling from the land surface projected over these areas (Kharin et al., 2007). The representation of the latter constitutes a major source of model uncertainty for projections of the absolute magnitude of temperature extremes (Clark et al., 2010; Fischer et al., 2011).

Winter cold extremes also warm more than the local mean temperature over northern high latitudes (Orlowsky and Seneviratne, 2012; Sillmann et al., 2013) as a result of reduced temperature variability related to declining snow cover (Gregory and Mitchell, 1995; Kjellstrom et al., 2007; Fischer et al., 2011) and decreases in land–sea contrast (de Vries et al., 2012). Changes in atmospheric circulation, induced by remote surface heating can also modify the temperature distribution (Haarsma et al., 2009). Sillmann and Croci-Maspoli (2009) note that cold winter extremes over Europe are in part driven by atmospheric blocking and changes to these blocking patterns in the future lead to changes in the frequency and spatial distribution of cold temperature extremes as global temperatures increase. Occasional cold winters will continue to occur (Räisänen and Ylhaisi, 2011).

Human discomfort, morbidity and mortality during heat waves depend not only on temperature but also specific humidity. Heat stress, defined as the combined effect of temperature and humidity, is expected to increase along with warming temperatures and dominates the local decrease in summer relative humidity due to soil drying (Diffenbaugh et al., 2007; Fischer et al., 2012b; Dunne et al., 2013). Areas with abundant atmospheric moisture availability and high present-day temperatures such as Mediterranean coastal regions are expected to experience the greatest heat stress changes because the heat stress response scales with humidity which thus becomes increasingly important to heat stress at higher temperatures (Fischer and Schär, 2010; Sherwood and Huber, 2010; Willett and Sherwood, 2012). For some regions, simulated heat stress indicators are remarkably robust, because those models with stronger warming simulate a stronger decrease in atmospheric relative humidity (Fischer and Knutti, 2013).

Changes in rare temperature extremes can be assessed using extreme value theory based techniques (Seneviratne et al., 2012). Kharin et al. (2007), in an analysis of CMIP3 models, found large increases in the 20-year return values of the annual maximum and minimum daily averaged surface air temperatures (i.e., the size of an event that would be expected on average once every 20 years, or with a 5% chance every year) with larger changes over land than ocean. Figure 12.14 displays the end of 21st century change in the magnitude of these rare events from the CMIP5 models in the RCP2.6, 4.5 and 8.5 scenarios (Kharin et al., 2013). Comparison to the changes in summer mean temperature shown in Figure AI.5 and A1.7 of Annex I Supplementary Material reveals that rare high temperature events are projected to change at rates similar to or slightly larger than the summertime mean temperature in many land areas. However, in much of Northern Europe 20-year return values of daily high temperatures are projected to increase 2°C or more than JJA mean temperatures under RCP8.5, consistent with previous studies (Sterl et al., 2008; Orlowsky and Seneviratne, 2012). Rare low temperature events are projected to experience significantly larger increases than the mean in most land regions, with a pronounced effect at high latitudes. Twenty-year return values of cold extremes increase significantly more than

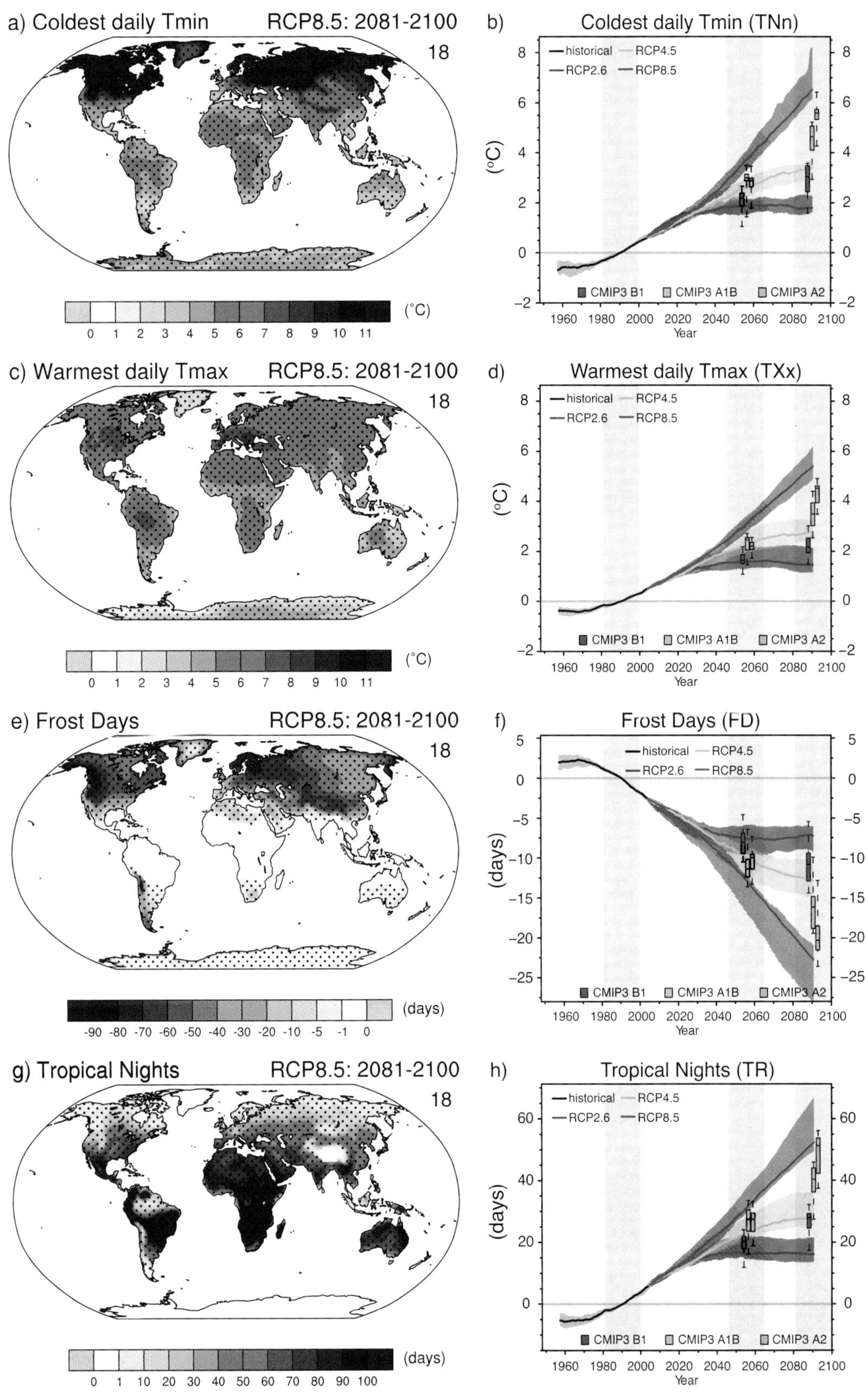

Figure 12.13 | CMIP5 multi-model mean geographical changes (relative to a 1981–2000 reference period in common with CMIP3) under RCP8.5 and 20-year smoothed time series for RCP2.6, RCP4.5 and RCP8.5 in the (a, b) annual minimum of daily minimum temperature, (c, d) annual maximum of daily maximum temperature, (e, f) frost days (number of days below 0°C) and (g, h) tropical nights (number of days above 20°C). White areas over land indicate regions where the index is not valid. Shading in the time series represents the interquartile ensemble spread (25th and 75th quantiles). The box-and-whisker plots show the interquartile ensemble spread (box) and outliers (whiskers) for 11 CMIP3 model simulations of the SRES scenarios A2 (orange), A1B (cyan), and B1 (purple) globally averaged over the respective future time periods (2046–2065 and 2081–2100) as anomalies from the 1981–2000 reference period. Stippling indicates grid points with changes that are significant at the 5% level using a Wilcoxon signed-ranked test. (Updated from Sillmann et al. (2013), excluding the FGOALS-s2 model.)

winter mean temperature changes, particularly over parts of North America and Europe. Kharin et al. (2013) concluded from the CMIP5 models that it is *likely* that in most land regions a current 20 year maximum temperature event is projected to become a one-in-two-year event by the end of the 21st century under the RCP4.5 and RCP8.5 scenarios, except for some regions of the high latitudes of the NH where it is *likely* to become a one-in-five-year event (see also Seneviratne et al. (2012) Figure 3.5). Current 20-year minimum temperature events are projected to become exceedingly rare, with return periods *likely* increasing to more than 100 years in almost all locations under RCP8.5 (Kharin et al., 2013). Section 10.6.1.1 notes that a number of detection and attribution studies since SREX suggest that the model changes may tend to be too large for warm extremes and too small for cold extremes and thus these likelihood statements are somewhat less strongly stated than a direct interpretation of model output and its ranges. The CMIP5 analysis shown in Figure 12.14 reinforces this assessment of large changes in the frequency of rare events, particularly in the RCP8.5 scenario (Kharin et al., 2013).

There is high consensus among models in the sign of the future change in temperature extremes, with recent studies confirming this conclusion from the previous assessments (Tebaldi et al., 2006; Meehl et al., 2007b; Orlowsky and Seneviratne, 2012; Seneviratne et al., 2012; Sillmann et al., 2013). However, the magnitude of the change remains uncertain owing to scenario and model (both structural and parameter) uncertainty (Clark et al., 2010) as well as internal variability. These uncertainties are much larger than corresponding uncertainties in the magnitude of mean temperature change (Barnett et al., 2006; Clark et al., 2006; Fischer and Schär, 2010; Fischer et al., 2011).

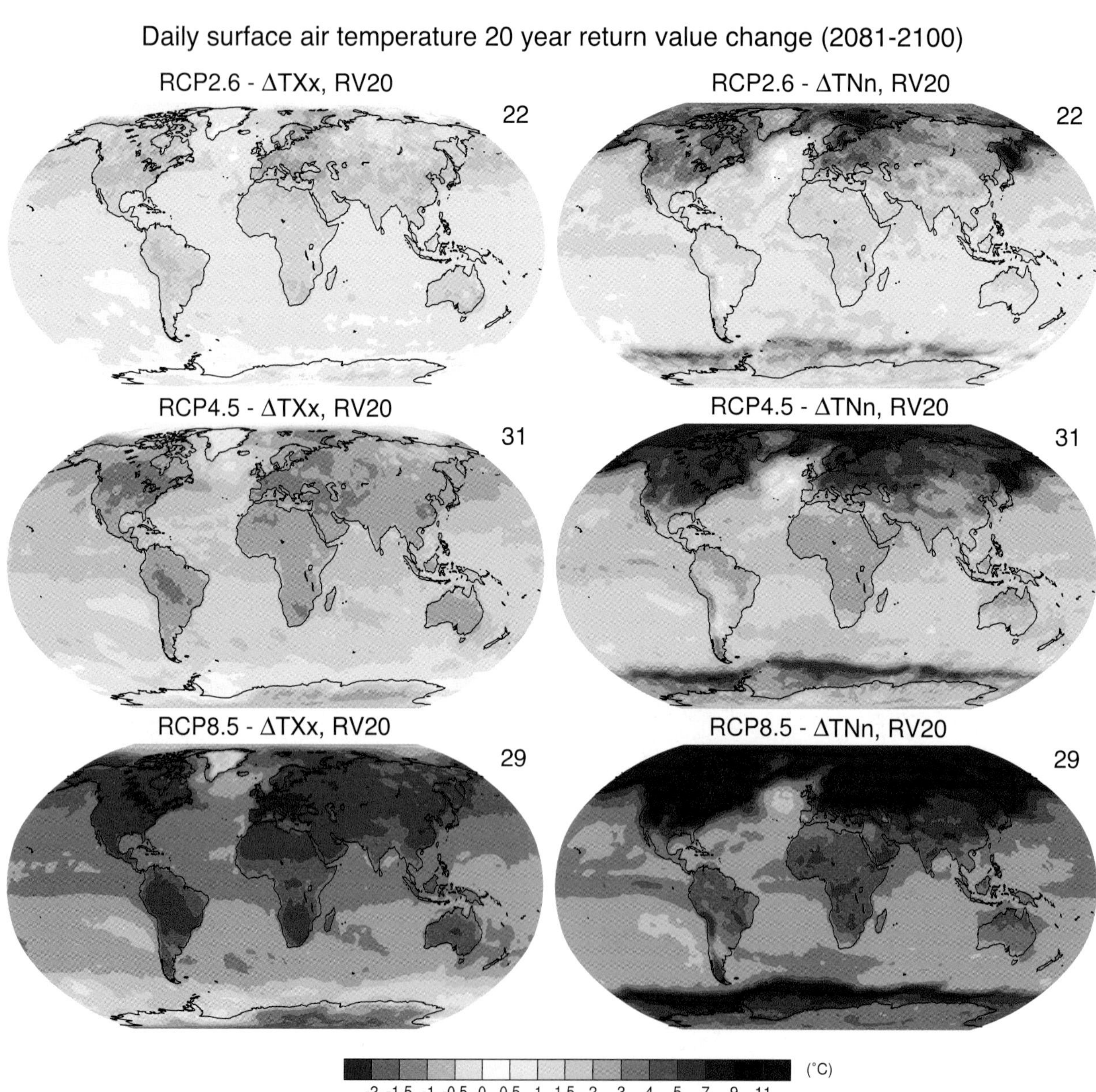

Figure 12.14 | The CMIP5 multi-model median change in 20-year return values of annual warm temperature extremes (left-hand panels) and cold temperature extremes (right-hand panels) as simulated by CMIP5 models in 2081–2100 relative to 1986–2005 in the RCP2.6 (top), RCP4.5 (middle panels), and RCP8.5 (bottom) experiments.

12

12.4.3.4 Energy Budget

Anthropogenic or natural perturbations to the climate system produce RFs that result in an imbalance in the global energy budget at the top of the atmosphere (TOA) and affect the global mean temperature (Section 12.3.3). The climate responds to a change in RF on multiple time scales and at multiyear time scales the energy imbalance (i.e., the energy heating or cooling the Earth) is very close to the ocean heat uptake due to the much lower thermal inertia of the atmosphere and the continental surfaces (Levitus et al., 2005; Knutti et al., 2008a; Murphy et al., 2009; Hansen et al., 2011). The radiative responses of the fluxes at TOA are generally analysed using the forcing-feedback framework and are presented in Section 9.7.2.

CMIP5 models simulate a small increase of the energy imbalance at the TOA over the 20th century (see Box 3.1, Box 9.2 and Box 13.1). The future evolution of the imbalance is very different depending on the scenario (Figure 12.15a): for RCP8.5 it continues to increase rapidly, much less for RCP6.0, it is almost constant for RCP4.5 and decreases for RCP2.6. This latter negative trend reveals the quasi-stabilization characteristic of RCP2.6. (In a transient scenario simulation, the TOA imbalance is always less than the RF because of the slow rate of ocean heat uptake.)

The rapid fluctuations that are simulated during the 20th century originate from volcanic eruptions that are prescribed in the models (see Section 12.3.2). These aerosols reflect solar radiation and thus decrease the amount of SW radiation absorbed by the Earth (Figure 12.15c). The minimum of shortwave (SW) radiation absorbed by the Earth during the period 1960–2000 is due mainly to two factors: a sequence of volcanic eruptions and an increase of the reflecting aerosol burden due to human activities (see Sections 7.5, 8.5 and 9.4.6). During the 21st century, the absorbed SW radiation monotonically increases for the RCP8.5 scenario, and increases and subsequently stabilizes for the other scenarios, consistent with what has been previously obtained with CMIP3 models and SRES scenarios (Trenberth and Fasullo, 2009). The two main contributions to the SW changes are the change of clouds (see Section 12.4.3.5) and the change of the cryosphere (see Section 12.4.6) at high latitudes. In the longwave (LW) domain (Figure 12.15b), the net flux at TOA represents the opposite of the flux that is emitted by the Earth's surface and atmosphere toward space, i.e., a negative anomaly represents an increase of the emitted

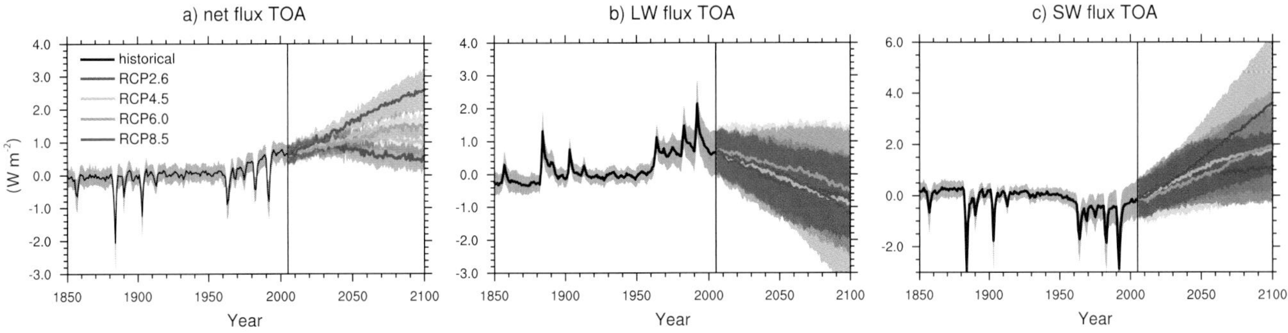

Figure 12.15 | Time series of global and annual multi-model mean (a) net total radiation anomaly at the top of the atmosphere (TOA), (b) net longwave radiation anomaly at the TOA and (c) net shortwave radiation anomaly at the TOA from the CMIP5 concentration-driven experiments for the historical period (black) and the four RCP scenarios. All the fluxes are positive downward and units are W m^{-2}. The anomalies are calculated relative to the 1900–1950 base period as this is a common period to all model experiments with few volcanic eruptions and relatively small trends. One ensemble member is used for each individual CMIP5 model and the ± standard deviation across the distribution of individual models is shaded.

12

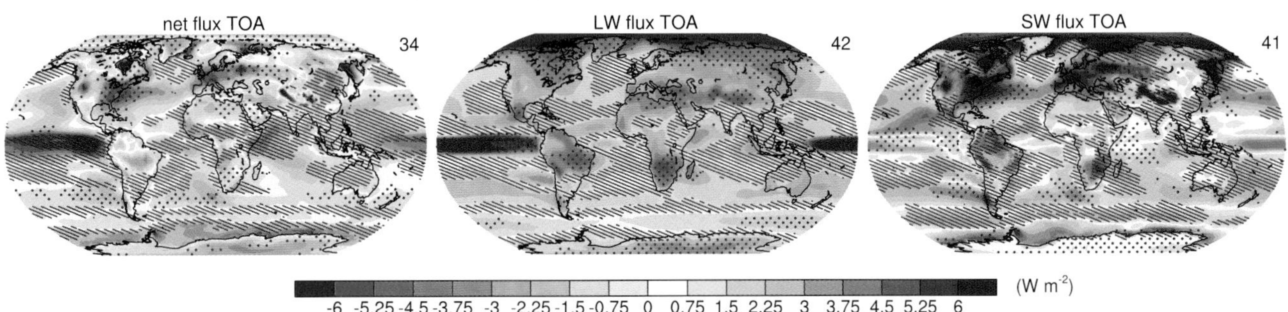

Figure 12.16 | Multi-model CMIP5 average changes in annual mean (left) net total radiation anomaly at the top of the atmosphere (TOA), (middle) net longwave radiation anomaly at the TOA and (right) net shortwave radiation anomaly at the TOA for the RCP4.5 scenario averaged over the periods 2081–2100. All fluxes are positive downward, units are W m^{-2}. The net radiation anomalies are computed with respect to the 1900–1950 base period. Hatching indicates regions where the multi-model mean change is less than one standard deviation of internal variability. Stippling indicates regions where the multi-model mean change is greater than two standard deviations of internal variability and where at least 90% of models agree on the sign of change (see Box 12.1).

LW radiation. The LW net flux depends mainly on two factors: the surface temperature and the magnitude of the greenhouse effect of the atmosphere. During the 20th century, the rapid fluctuations of LW radiation are driven by volcanic forcings, which decrease the absorbed SW radiation, surface temperature, and the LW radiation emitted by the Earth toward space. During the period 1960–2000, the fast increase of GHG concentrations also decreases the radiation emitted by the Earth. In response to this net heating of the Earth, temperatures warm and thereby increase emitted LW radiation although the change of the temperature vertical profile, water vapour, and cloud properties modulate this response (e.g., Bony et al., 2006; Randall et al., 2007).

12.4.3.5 Clouds

This section provides a summary description of future changes in clouds and their feedbacks on climate. A more general and more precise description and assessment of the role of clouds in the climate system is provided in Chapter 7, in particular Section 7.2 for cloud processes and feedbacks and Section 7.4 for aerosol–cloud interactions. Cloud feedbacks and adjustments are presented in Section 7.2.5 and a synthesis is provided in Section 7.2.6. Clouds are a major component of the climate system and play an important role in climate sensitivity (Cess et al., 1990; Randall et al., 2007), the diurnal temperature range (DTR) over land (Zhou et al., 2009), and land–sea contrast (see Section 12.4.3.1). The observed global mean cloud RF is about –20 W m^{-2} (Loeb et al., 2009) (see Section 7.2.1), that is, clouds have a net cooling effect. Current GCMs simulate clouds through various complex parameterizations (see Section 7.2.3), and cloud feedback is a major source of the spread of the climate sensitivity estimate (Soden and Held, 2006; Randall et al., 2007; Dufresne and Bony, 2008) (see Section 9.7.2).

Under future projections the multi-model pattern of total cloud amount shows consistent decreases in the subtropics, in conjunction with a decrease of the relative humidity there, and increases at high latitudes. Another robust pattern is an increase in cloud cover at all latitudes in the vicinity of the tropopause, a signature of the increase of the altitude of high level clouds in convective regions (Wetherald and Manabe, 1988; Meehl et al., 2007b; Soden and Vecchi, 2011; Zelinka et al., 2012). Low-level clouds were identified as a primary cause of inter-model spread in cloud feedbacks in CMIP3 models (Bony and Dufresne, 2005; Webb et al., 2006; Wyant et al., 2006). Since AR4, these results have been confirmed along with the positive feedbacks due to high level clouds in the CMIP3 or CFMIP models (Zelinka and Hartmann, 2010; Soden and Vecchi, 2011; Webb et al., 2013) and CMIP5 models (Vial et al., 2013). Since AR4, the response of clouds has been partitioned in a direct or 'rapid' response of clouds to CO_2 and a 'slow' response of clouds to the surface temperature increase (i.e., the usual feedback response) (Gregory and Webb, 2008). The radiative effect of clouds depends mainly on their fraction, optical depth and temperature. The contribution of these variables to the cloud feedback has been quantified for the multi-model CMIP3 (Soden and Vecchi, 2011) and CFMIP1 database (Zelinka et al., 2012). These findings are consistent with the radiative changes obtained with the CMIP5 models (Figure 12.16) and may be summarized as follows (see Section 7.2.5 for more details).

The dominant contributor to the SW cloud feedback is the change in cloud fraction. The reduction of cloud fraction between 50°S and 50°N, except along the equator and the eastern part of the ocean basins (Figure 12.17), contributes to an increase in the absorbed solar radiation (Figure 12.16c). Physical mechanisms and the role of different parameterizations have been proposed to explain this reduction of low-level clouds (Zhang and Bretherton, 2008; Caldwell and Bretherton, 2009; Brient and Bony, 2013; Webb et al., 2013). Poleward of 50°S, the cloud fraction and the cloud optical depth increases, thereby increasing cloud reflectance. This leads to a decrease of solar absorption around Antarctica where the ocean is nearly ice free in summer (Figure 12.16c). However, there is *low confidence* in this result because GCMs do not reproduce the nearly 100% cloud cover observed there and the negative feedback could be overestimated (Trenberth and Fasullo, 2010) or, at the opposite, underestimated because the cloud optical depth simulated by models is biased high there (Zelinka et al., 2012).

In the LW domain, the tropical high cloud changes exert the dominant effect. A lifting of the cloud top with warming is simulated consistently across models (Meehl et al., 2007b) which leads to a positive feedback whereby the LW emissions from high clouds decrease as they cool (Figure 12.16b). The dominant driver of this effect is the increase of tropopause height and physical explanations have been proposed (Hartmann and Larson, 2002; Lorenz and DeWeaver, 2007; Zelinka

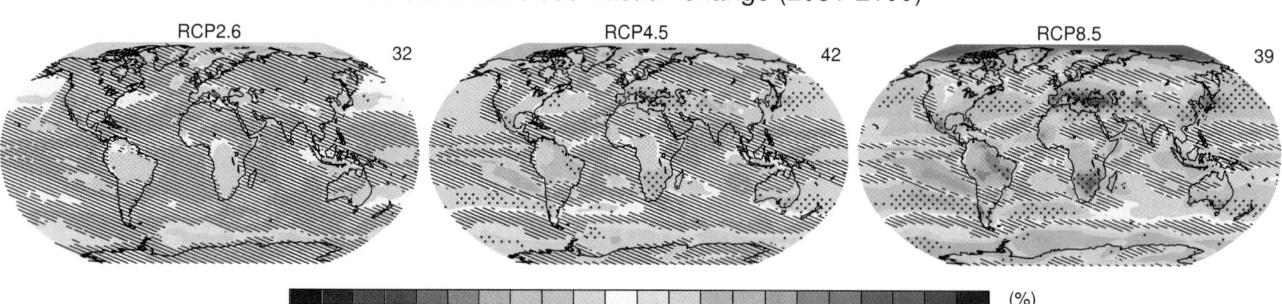

Figure 12.17 | CMIP5 multi-model changes in annual mean total cloud fraction (in %) relative to 1986–2005 for 2081–2100 under the RCP2.6 (left), RCP4.5 (centre) and RCP8.5 (right) forcing scenarios. Hatching indicates regions where the multi-model mean change is less than one standard deviation of internal variability. Stippling indicates regions where the multi-model mean change is greater than two standard deviations of internal variability and where 90% of the models agree on the sign of change (see Box 12.1). The number of CMIP5 models used is indicated in the upper right corner of each panel.

and Hartmann, 2010). Although the decrease in cloudiness generally increases outgoing longwave radiation and partly offsets the effect of cloud rising, the net effect is a consistent positive global mean LW cloud feedback across CMIP and CFMIP models. Global mean SW cloud feedbacks range from slightly negative to strongly positive (Soden and Vecchi, 2011; Zelinka et al., 2012), with an inter-model spread in net cloud feedback being mainly attributable to low-level cloud changes.

In summary, both the multi-model mean and the inter-model spread of the cloud fraction and radiative flux changes simulated by the CMIP5 models are consistent with those previously obtained by the CMIP3 models. These include decreases in cloud amount in the subtropics, increases at high latitudes and increases in the altitude of high level clouds in convective regions. Many of these changes have been understood primarily as responses to large-scale circulation changes (see Section 7.2.6).

12.4.4 Changes in Atmospheric Circulation

Projected changes in energy and water cycles couple with changes in atmospheric circulation and mass distribution. Understanding this coupling is necessary to assess physical behaviour underlying projected changes, particularly at regional scales, revealing why changes occur and the realism of the changes. The focus in this section is on atmospheric circulation behaviour that CMIP5 GCMs resolve well. Thus, the section includes discussion of extratropical cyclones but not tropical cyclones: extratropical cyclones are fairly well resolved by most CMIP5 GCMs, whereas tropical cyclones are not, requiring resolutions finer than used by the large majority of CMIP5 GCMs (see Section 9.5.4.3). Detailed discussion of tropical cyclones appears in Section 14.6.1 (see also Section 11.3.2.5.3 for near term changes and Section 3.4.4 in Seneviratne et al. (2012)). Regional detail concerning extratropical storm tracks, including causal processes, appears in Section 14.6.2 (see also Section 11.3.2.4 for near-term changes and Seneviratne et al. (2012) for an assessment of projected changes related to weather and climate extremes).

12.4.4.1 Mean Sea Level Pressure and Upper-Air Winds

Sea level pressure gives an indication of surface changes in atmospheric circulation (Figure 12.18). As in previous assessments, a robust feature of the pattern of change is a decrease in high latitudes and increases in the mid-latitudes, associated with poleward shifts in the SH mid-latitude storm tracks (Section 12.4.4.3) and positive trends in the annular modes (Section 14.5) as well as an expansion of the Hadley Cell (Section 12.4.4.2). Similar patterns of sea level pressure change are found in observed trends over recent decades, suggesting an already detectable change (Gillett and Stott, 2009; Section 10.3.3.4), although the observed patterns are influenced by both natural and anthropogenic forcing as well as internal variability and the relative importance of these influences is likely to change in the future. Internal variability has been found to play a large role in uncertainties of future sea level pressure projections, particularly at higher latitudes (Deser et al., 2012a).

In boreal winter, decreases of sea level pressure over NH high latitudes are slightly weaker in the CMIP5 ensemble compared to previous assessments, consistent with Scaife et al. (2012) and Karpechko and Manzini (2012), who suggest that improvements in the representation of the stratosphere can influence this pattern. In austral summer, the SH projections are impacted by the additional influence of stratospheric ozone recovery (see Section 11.3.2.4.2) which opposes changes due to GHGs. Under the weaker GHG emissions of RCP2.6, decreases in sea level pressure over the SH mid-latitudes and increases over SH high latitudes are consistent with expected changes from ozone recovery (Arblaster et al., 2011; McLandress et al., 2011; Polvani et al., 2011). For

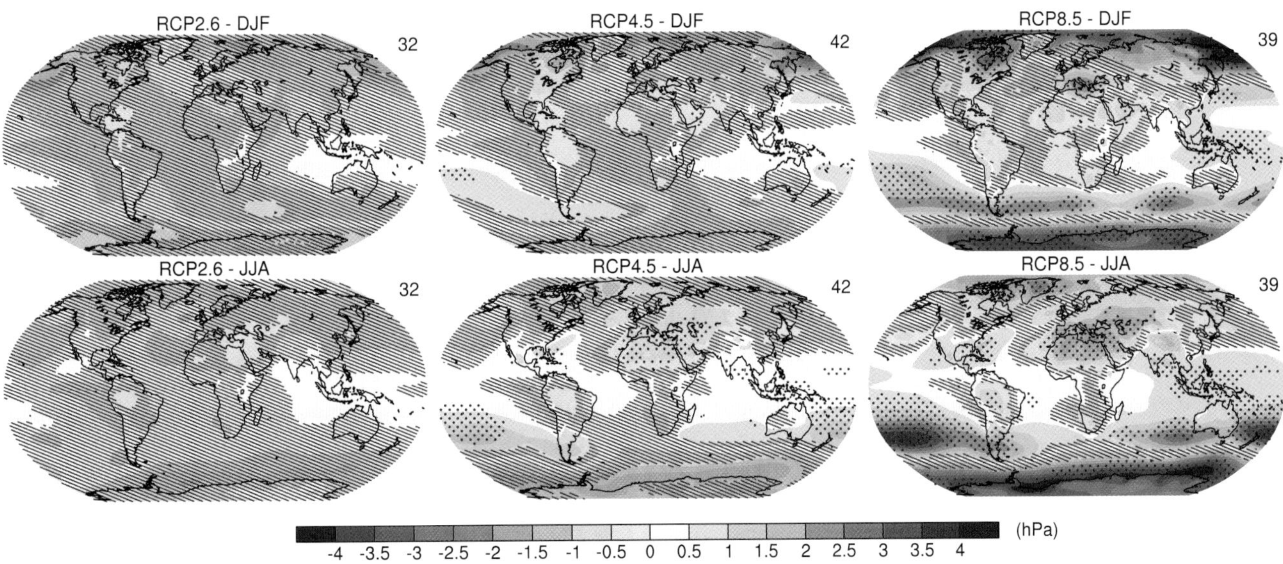

Figure 12.18 | CMIP5 multi-model ensemble average of December, January and February (DJF, top row) and June, July and August (JJA, bottom row) mean sea level pressure change (2081–2100 minus 1986–2005) for, from left to right, RCP2.6, 4.5 and 8.5. Hatching indicates regions where the multi-model mean change is less than one standard deviation of internal variability. Stippling indicates regions where the multi-model mean change is greater than two standard deviations of internal variability and where at least 90% of models agree on the sign of change (see Box 12.1).

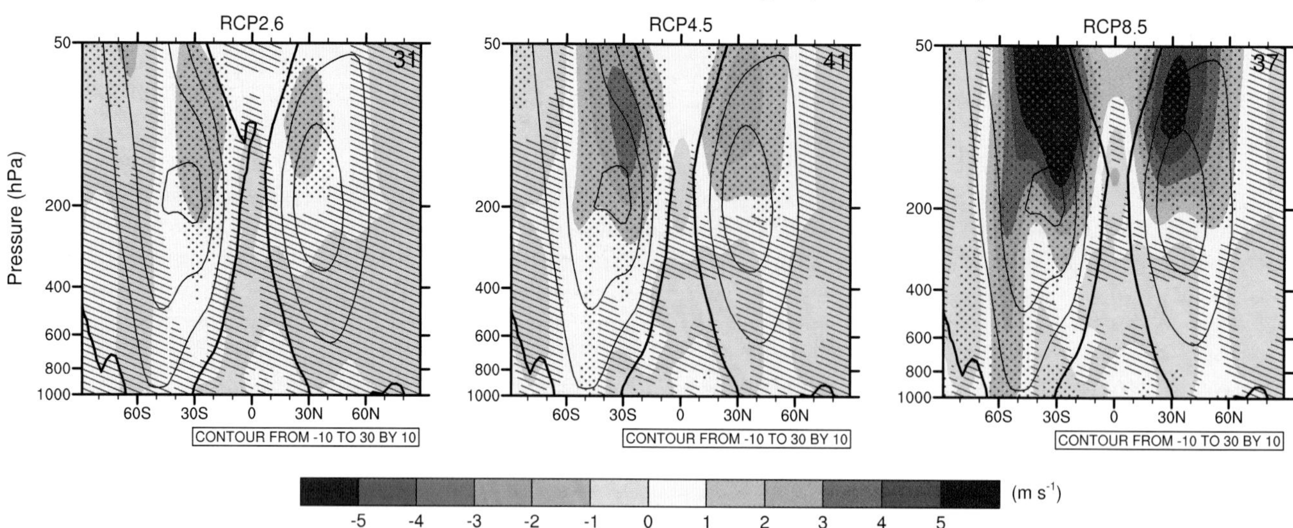

Figure 12.19 | Coupled Model Intercomparison Project Phase 5 (CMIP5) multi-model ensemble average of zonal and annual mean wind change (2081–2100 minus 1986–2005) for, from left to right, Representative Concentration Pathway 2.6 (RCP2.6), 4.5 and 8.5. Black contours represent the multi-model average for the 1986–2005 base period. Hatching indicates regions where the multi-model mean change is less than one standard deviation of internal variability. Stippling indicates regions where the multi-model mean change is greater than two standard deviations of internal variability and where at least 90% of models agree on the sign of change (see Box 12.1).

all other RCPs, the magnitude of SH extratropical changes scales with the RF, as found in previous model ensembles (Paeth and Pollinger, 2010; Simpkins and Karpechko, 2012).

Large increases in seasonal sea level pressure are also found in regions of sub-tropical drying such as the Mediterranean and northern Africa in DJF and Australia in JJA. Projected changes in the tropics are less consistent across the models; however, a decrease in the eastern equatorial Pacific and increase over the maritime continent, associated with a weakening of the Walker Circulation (Vecchi and Soden, 2007; Power and Kociuba, 2011b), is found in all RCPs.

Future changes in zonal and annual mean zonal winds (Figure 12.19) are seen throughout the atmosphere with stronger changes in higher RCPs. Large increases in winds are evident in the tropical stratosphere and a poleward shift and intensification of the SH tropospheric jet is seen under RCP4.5 and RCP8.5, associated with an increase in the SH upper tropospheric meridional temperature gradient (Figure 12.12) (Wilcox et al., 2012). In the NH, the response of the tropospheric jet is weaker and complicated by the additional thermal forcing of polar amplification (Woollings, 2008). Barnes and Polvani (2013) evaluate changes in the annual mean mid-latitude jets in the CMIP5 ensemble, finding consistent poleward shifts in both hemispheres under RCP8.5 for the end of the 21st century. In the NH, the poleward shift is ~1°, similar to that found for the CMIP3 ensemble (Woollings and Blackburn, 2012). In the SH, the annual mean mid-latitude jet shifts poleward by ~2° under RCP8.5 at the end of the 21st century in the CMIP5 multi-model mean (Barnes and Polvani, 2013), with a similar shift of 1.5° in the surface westerlies (Swart and Fyfe, 2012). A strengthening of the SH surface westerlies is also found under all RCPs except RCP2.6 (Swart and Fyfe, 2012), with largest changes in the Pacific basin (Bracegirdle et al., 2013). In austral summer, ozone recovery offsets changes in GHGs to some extent, with a weak reversal of the jet shift found in the multi-model mean under the low emissions scenario of RCP2.6 (Swart and Fyfe, 2012) and weak or poleward shifts in other RCPs (Swart and Fyfe, 2012; Wilcox et al., 2012). Eyring et al. (2013) note the sensitivity of the CMIP5 SH summertime circulation changes to both the strength of the ozone recovery (simulated by some models interactively) and the rate of GHG increases.

Although the poleward shift of the tropospheric jets are robust across models and *likely* under increased GHGs, the dynamical mechanisms behind these projections are still not completely understood and have been explored in both simple and complex models (Chen et al., 2008; Lim and Simmonds, 2009; Butler et al., 2010). The shifts are associated with a strengthening in the upper tropospheric meridional temperature gradient (Wilcox et al., 2012) and hypotheses for associated changes in planetary wave activity and/or synoptic eddy characteristics that impact on the position of the jet have been put forward (Gerber et al., 2012). Equatorward biases in the position of the SH jet (Section 9.5.3.2), while somewhat improved over similar biases in the CMIP3 models (Kidston and Gerber, 2010) still remain, limiting our confidence in the magnitude of future changes.

In summary, poleward shifts in the mid-latitude jets of about 1 to 2 degrees latitude are *likely* at the end of the 21st century under RCP8.5 in both hemispheres (*medium confidence*) with weaker shifts in the NH and under lower emission scenarios. Ozone recovery will *likely* weaken the GHG-induced changes in the SH extratropical circulation in austral summer.

12.4.4.2 Planetary-Scale Overturning Circulations

Large-scale atmospheric overturning circulations and their interaction with other atmospheric mechanisms are significant in determining tropical climate and regional changes in response to enhanced RF. Observed changes in tropical atmospheric circulation are assessed in Section 2.7.5, while Section 10.3.3 discusses attribution of these observed changes to anthropogenic forcing. Evidence is inconclusive on recent trends in the strength of the Hadley (Stachnik and Schumacher, 2011) and Walker Circulations (Vecchi et al., 2006; Sohn and Park, 2010; Merrifield, 2011; Luo et al., 2012; Tokinaga et al., 2012), though there is *medium confidence* of an anthropogenic influence on the observed widening of the Hadley Circulation (Hu and Fu, 2007; Johanson and Fu, 2009; Davis and Rosenlof, 2012). In the projections, there are indications of a weakening of tropical overturning of air as the climate warms (Held and Soden, 2006; Vecchi and Soden, 2007; Gastineau et al., 2008, 2009; Chou and Chen, 2010; Chadwick et al., 2012; Bony et al., 2013). In the SRES A1B scenario, CMIP3 models show a remarkable agreement in simulating a weakening of the tropical atmospheric overturning circulation (Vecchi and Soden, 2007). CMIP5 models also show a consistent weakening (Chadwick et al., 2012). Along the ascending branches of tropical overturning cells, a reduction in convective mass flux from the boundary layer to the free atmosphere is implied by the differential response to global warming of the boundary-layer moisture content and surface evaporation. This weakening of vertical motion along the ascending regions of both the tropical meridional and near-equatorial zonal cells is associated with an imbalance in the rate of atmospheric moisture increase and that of global mean precipitation (Held and Soden, 2006). A reduction in the compensating climatological subsidence along the downward branches of overturning circulations, where the rate of increase of static stability exceeds radiative cooling, is implied.

Several mechanisms have been suggested for the changes in the intensity of the tropical overturning circulation. The weakening of low-level convective mass flux along ascending regions of tropical overturning cells has been ascribed to changes in the hydrologic cycle (Held and Soden, 2006; Vecchi and Soden, 2007). Advection of dry air from subsidence regions towards the ascending branches of large-scale tropical circulation has been suggested to be a feasible mechanism weakening ascent along the edges of convection regions (Chou et al., 2009). A deepening of the tropical troposphere in response to global warming increases the vertical extent of convection, which has been shown to increase the atmosphere's moist stability and thus also weakening overturning cells (Chou and Chen, 2010). An imbalance between the increase in diabatic heating of the troposphere and in static stability whereby the latter increases more rapidly has also been thought to play a role in weakening tropical ascent (Lu et al., 2008). Mean advection of enhanced vertical stratification under GHG forcing which involves cooling of convective regions and warming of subsidence regions has been shown to slow down tropical cells (Ma et al., 2012). The latest findings using CMIP5 models reveal that an increase in GHGs (particularly CO_2) contributes significantly to weakening tropical overturning cells by reducing radiative cooling in the upper atmosphere (Bony et al., 2013). SST gradients have also been found to play a role in altering the strength of tropical cells (Tokinaga et al., 2012; Ma and Xie, 2013). Evidence has been provided suggesting that the SH Hadley Cell may strengthen in response to meridional SST gradients featuring reduced warming in the SH subtropical oceans relative to the NH, particularly over the Pacific and Indian Oceans (Ma and Xie, 2013). The north-to-south SST warming gradients are a source of intermodel differences in their projections of changes in the SH Hadley Circulation.

Apart from changes in Hadley Circulation strength, a robust feature in 21st century climate model simulations is an increase in the cell's depth and width (Mitas and Clement, 2006; Frierson et al., 2007; Lu et al., 2007; Lu et al., 2008), with the latter change translating to a broadening of tropical regions (Seidel and Randel, 2007; Seidel et al., 2008) and a poleward displacement of subtropical dry zones (Lu et al., 2007; Scheff and Frierson, 2012). The increase in the cell's depth is consistent with a tropical tropopause rise. The projected increase in the height of the tropical tropopause and the associated increase in meridional temperature gradients close to the tropopause slope have been proposed to be an important mechanism behind the Hadley cell expansion and the poleward displacement of the subtropical westerly jet (Lu et al., 2008; Johanson and Fu, 2009). An increase in subtropical and mid-latitude static stability has been found to be an important factor widening the Hadley Cell by shifting baroclinic eddy activity and the associated eddy-driven jet and subsidence poleward (Mitas and Clement, 2006; Lu et al., 2008). The projected widening of the Hadley Cell is consistent with late 20th century observations, where ~2° to 5° latitude expansion was found (Fu et al., 2006; Johanson and Fu, 2009). The consistency of simulated changes in CMIP3 and CMIP5 models and the consistency of Hadley Cell changes with the projected tropopause rise and increase in subtropical and mid-latitude static stability indicate that a widening and weakening of the NH Hadley Cell by the late 21st century is *likely*.

The zonally asymmetric Walker Circulation is projected to weaken under global warming (Power and Kociuba, 2011a, 2011b), more than the Hadley Circulation (Lu et al., 2007; Vecchi and Soden, 2007). The consistency of the projected Walker Circulation slowdown from CMIP3 to CMIP5 suggests that its change is robust (Ma and Xie, 2013). Almost everywhere around the equatorial belt, changes in the 500 hPa vertical motion oppose the climatological background motion, notably over the maritime continent (Vecchi and Soden, 2007; Shongwe et al., 2011). Around the Indo-Pacific warm pool, in response to a spatially uniform SST warming, the climatological upper tropospheric divergence weakens (Ma and Xie, 2013). Changes in the strength of the Walker Circulation also appear to be linked to differential warming between the Indian and Pacific Ocean warming at low latitudes (Luo et al., 2012). Over the equatorial Pacific Ocean, where mid-tropospheric ascent is projected to strengthen, changes in zonal SST and hence sea level pressure gradients induce low-level westerly wind anomalies that act to weaken the low-level branch of the Pacific Walker Circulation. These projected changes in the tropical Pacific circulation are already occurring (Zhang and Song, 2006). However, the projected weakening of the Pacific Walker Cell does not imply an increase in the frequency and/or magnitude of El Niño events (Collins et al., 2010). The consistency of simulated changes in CMIP3 and CMIP5 models and the consistency of Walker Cell changes with equatorial SST and pressure-gradient changes that are already observed indicate that a weakening of the Walker Cell by the late 21st century is *likely*.

In the upper atmosphere, a robust feature of projected stratospheric circulation change is that the Brewer–Dobson circulation will *likely* strengthen in the 21st century (Butchart et al., 2006, 2010; Li et al., 2008; McLandress and Shepherd, 2009; Shepherd and McLandress, 2011). In a majority of model experiments, the projected changes in the large-scale overturning circulation in the stratosphere feature an intensification of tropical upward mass flux, which may extend to the upper stratosphere. The proposed driver of the increase in mass flux at the tropical lower stratosphere is the enhanced propagation of wave activity, mainly resolved planetary waves, associated with a positive trend in zonal wind structure (Butchart and Scaife, 2001; Garcia and Randel, 2008). In the 21st century, increases in wave excitation from diabatic heating in the upper tropical troposphere could reinforce the wave forcing on the tropical upwelling branch of the stratospheric mean meridional circulation (Calvo and Garcia, 2009). Parameterized orographic gravity waves that result from strengthening of subtropical westerly jets and cause more waves to propagate into the lower stratosphere also play a role (Sigmond et al., 2004; Butchart et al., 2006). The projected intensification in tropical upwelling is counteracted by enhanced mean extratropical/polar lower stratospheric subsidence. In the NH high latitudes, the enhanced downwelling is associated with an increase in stationary planetary wave activities (McLandress and Shepherd, 2009). The intensification of the stratospheric meridional residual circulation has already been reported in studies focussing on the last decades of the 20th century (Garcia and Randel, 2008; Li et al., 2008; Young et al., 2012). The projected increase in troposphere-to-stratosphere mass exchange rate (Butchart et al., 2006) and stratospheric mixing associated with the strengthening of the Brewer–Dobson circulation will *likely* result in a decrease in the mean age of air in the lower stratosphere. In the mid-latitude lower stratosphere, quasi-horizontal mixing is a significant contributor to reducing the lifetimes of air. There are some suggestions that the changes in stratospheric overturning circulation could lead to a reduction in tropical ozone concentrations and an increase at high latitudes (Jiang et al., 2007) and an increase in the amplitude of the annual cycle of stratospheric ozone (Randel et al., 2007).

12.4.4.3 Extratropical Storms: Tracks and Influences on Planetary-Scale Circulation and Transports

Since the AR4, there has been continued evaluation of changes in extratropical storm tracks under projected warming using both CMIP3 and, more recently, CMIP5 simulations, as well as supporting studies using single models or idealized simulations. CMIP3 analyses use a variety of methods for diagnosing storm tracks, but diagnosis of changes in the tracks appears to be relatively insensitive to methods used (Ulbrich et al., 2013). Analyses of SH storm tracks generally agree with earlier studies, showing that extratropical storm tracks will tend to shift poleward (Bengtsson et al., 2009; Gastineau et al., 2009; Gastineau and Soden, 2009; Perrie et al., 2010; Schuenemann and Cassano, 2010; Chang et al., 2012b). The behaviour is consistent with a *likely* trend in observed storm-track behaviour (see Section 2.7.6). Similar behaviour appears in CMIP5 simulations for the SH (Figure 12.20c, d). In SH winter there is a clear poleward shift in storm tracks of several degrees and a reduction in storm frequency of only a few percent (not shown). The poleward shift at the end of the century is consistent with a poleward shift in the SH of the latitudes with strongest tropospheric jets (Figure 12.19). This appears to coincide with shifts in baroclinic dynamics governing extratropical storms (Frederiksen et al., 2011), though the degree of jet shift appears to be sensitive to bias in a model's contemporary-climate storm tracks (Chang et al., 2012a, 2012b). Although there is thus some uncertainty in the degree of shift, the consistency of behaviour with observation-based trends, consistency between CMIP5 and CMIP3 projections under a variety of diagnostics and the physical consistency of the storm response with other climatic changes gives *high confidence* that a poleward shift of several degrees in SH storm tracks is *likely* by the end of the 21st century under the RCP8.5 scenario.

In the NH winter (Figure 12.20a, b), the CMIP5 multi-model ensemble shows an overall reduced frequency of storms and less indication of a poleward shift in the tracks. The clearest poleward shift in the NH winter at the end of the 21st century occurs in the Asia-Pacific storm track, where intensification of the westerly jet promotes more intense cyclones in an ensemble of CMIP5 models (Mizuta, 2012). Otherwise, changes in winter storm-track magnitude, as measured by band-pass sea level pressure fluctuations, show only small change relative to interannual and inter-decadal variability by the end of the 21st century in SRES A1B and RCP4.5 simulations for several land areas over the NH (Harvey et al., 2012). Consistency in CMIP3 and CMIP5 changes seen in the SH are absent in the NH (Chang et al., 2012a). Factors identified that affect changes in the North Atlantic basin's storm track include horizontal resolution (Colle et al., 2013) and how models simulate changes in the Atlantic's meridional overturning circulation (Catto et al., 2011; Woollings et al., 2012), the zonal jet and Hadley Circulation (Mizuta, 2012; Zappa et al., 2013) and subtropical upper troposphere temperature (Haarsma et al., 2013). Substantial uncertainty and thus *low confidence* remains in projecting changes in NH winter storm tracks, especially for the North Atlantic basin.

Additional analyses of CMIP3 GCMs have determined other changes in properties of extratropical storms. Most analyses find that the frequency of storms decreases in projected climates (Finnis et al., 2007; Favre and Gershunov, 2009; Dowdy et al., 2013), though the occurrence of strong storms may increase in some regions (Pinto et al., 2007; Bengtsson et al., 2009; Ulbrich et al., 2009; Zappa et al., 2013). Many studies focus on behaviour of specific regions, and results of these studies are detailed in Section 14.6.2.

Changes in extratropical storms in turn may influence other large-scale climatic changes. Kug et al. (2010) in a set of time-slice simulations show that a poleward shift of storm tracks in the NH could enhance polar warming and moistening. The Arctic Oscillation (AO) is sensitive to synoptic eddy vorticity flux, so that projected changes in storm tracks can alter the AO (Choi et al., 2010). The net result is that changes in extratropical storms alter the climate in which they are embedded, so that links between surface warming, extratropical storms and their influence on climate are more complex than simple responses to changes in baroclinicity (O'Gorman, 2010).

12.4.5 Changes in the Water Cycle

The water cycle consists of water stored on the Earth in all its phases, along with the movement of water through the Earth's climate system. In the atmosphere, water occurs primarily as gaseous water vapour, but it also occurs as solid ice and liquid water in clouds. The ocean is primarily liquid water, but is partly covered by ice in polar regions. Terrestrial water in liquid form appears as surface water (lakes, rivers), soil moisture and groundwater. Solid terrestrial water occurs in ice sheets, glaciers, frozen lakes, snow and ice on the surface and permafrost.

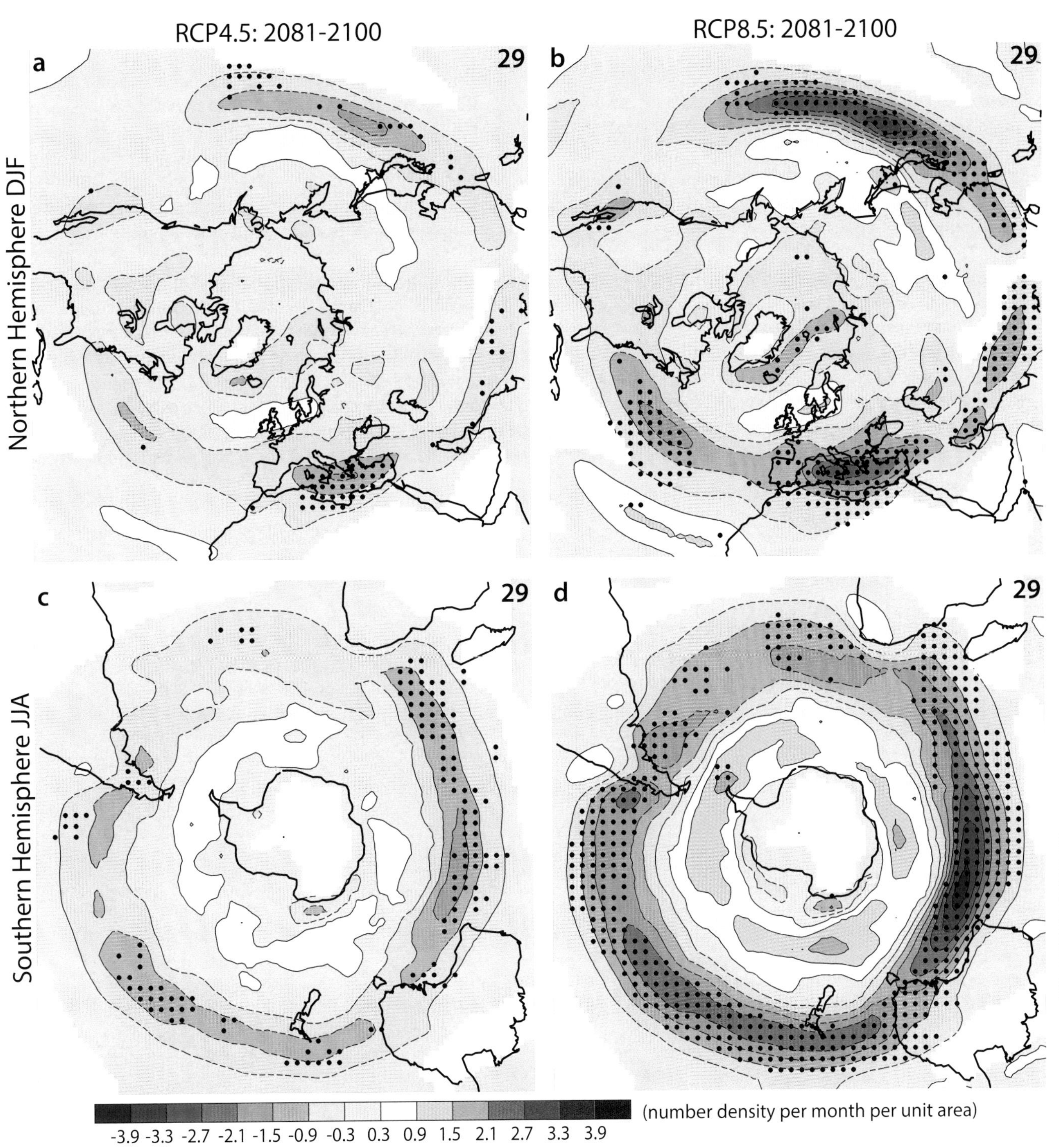

Figure 12.20 | Change in winter, extratropical storm track density (2081–2100) – (1986–2005) in CMIP5 multi-model ensembles: (a) RCP4.5 Northern Hemisphere December, January and February (DJF) and (b) RCP8.5 Northern Hemisphere DJF, (c) RCP4.5 Southern Hemisphere June, July and August (JJA) and (d) RCP8.5 Southern Hemisphere JJA. Storm-track computation uses the method of Bengtsson et al. (2006, their Figure 13a) applied to 6-hourly 850 hPa vorticity computed from horizontal winds in the CMIP5 archive. The number of models used appears in the upper right of each panel. DJF panels include data for December 1985 and 2080 and exclude December 2005 and December 2100 for in-season continuity. Stippling marks locations where at least 90% of the models agree on the sign of the change; note that this criterion differs from that used for many other figures in this chapter, due to the small number of models providing sufficient data to estimate internal variability of 20-year means of storm-track statistics. Densities have units (number density per month per unit area), where the unit area is equivalent to a 5° spherical cap (~10^6 km^2). Locations where the scenario or contemporary-climate ensemble average is below 0.5 density units are left white.

Projections of future changes in the water cycle are inextricably connected to changes in the energy cycle (Section 12.4.3) and atmospheric circulation (Section 12.4.4).

Saturation vapour pressure increases with temperature, but projected future changes in the water cycle are far more complex than projected temperature changes. Some regions of the world will be subject to decreases in hydrologic activity while others will be subject to increases. There are important local seasonal differences among the responses of the water cycle to climate change as well.

At first sight, the water cycles simulated by CMIP3/5 models may appear to be inconsistent, particularly at regional scales. Anthropogenic changes to the water cycle are superimposed on complex naturally varying modes of the climate (such as El Niño-Southern Oscillation (ENSO), AO, Pacific Decadal Oscillation (PDO), etc.) aggravating the differences between model projections. However, by careful consideration of the interaction of the water cycle with changes in other aspects of the climate system, the mechanisms of change are revealed, increasing confidence in projections.

12.4.5.1 Atmospheric Humidity

Atmospheric water vapour is the primary GHG in the atmosphere. Its changes affect all parts of the water cycle. However, the amount of water vapour is dominated by naturally occurring processes and not significantly affected directly by human activities. A common experience from past modelling studies is that relative humidity (RH) remains approximately constant on climatological time scales and planetary space scales, implying a strong constraint by the Clausius–Clapeyron relationship on how specific humidity will change. The AR4 stated that 'a broad-scale, quasi-unchanged RH response [to climate change] is uncontroversial' (Randall et al., 2007). However, underlying this fairly straightforward behaviour are changes in RH that can influence changes in cloud cover and atmospheric convection (Sherwood, 2010). More recent analysis provides further detail and insight on RH changes. Analysis of CMIP3 and CMIP5 models shows near-surface RH decreasing over most land areas as temperatures increase with the notable exception of parts of tropical Africa (O'Gorman and Muller, 2010) (Figure 12.21). The prime contributor to these decreases in RH over land is the larger temperature increases over land than over ocean in the RCP scenarios (Joshi et al., 2008; Fasullo, 2010; O'Gorman and Muller, 2010). The specific humidity of air originating over more slowly warming oceans will be governed by saturation temperatures of oceanic air. As this air moves over land and is warmed, its relative humidity drops as any further moistening of the air over land is insufficient to maintain constant RH, a behaviour Sherwood et al. (2010) term a last-saturation-temperature constraint. The RH decrease over most land areas by the end of the 21st century is consistent with a last-saturation-temperature constraint and with observed behaviour during the first decade of the current century (Section 2.5.5; Simmons et al., 2010). Land–ocean differences in warming are projected to continue through the 21st century, and although the CMIP5 projected changes are small, they are consistent with a last-saturation constraint, indicating with *medium confidence* that reductions in near-surface RH over many land areas are *likely*.

12.4.5.2 Patterns of Projected Average Precipitation Changes

Global mean precipitation changes have been presented in Section 12.4.1.1. The processes that govern large-scale changes in precipitation are presented in Section 7.6, and are used here to interpret the

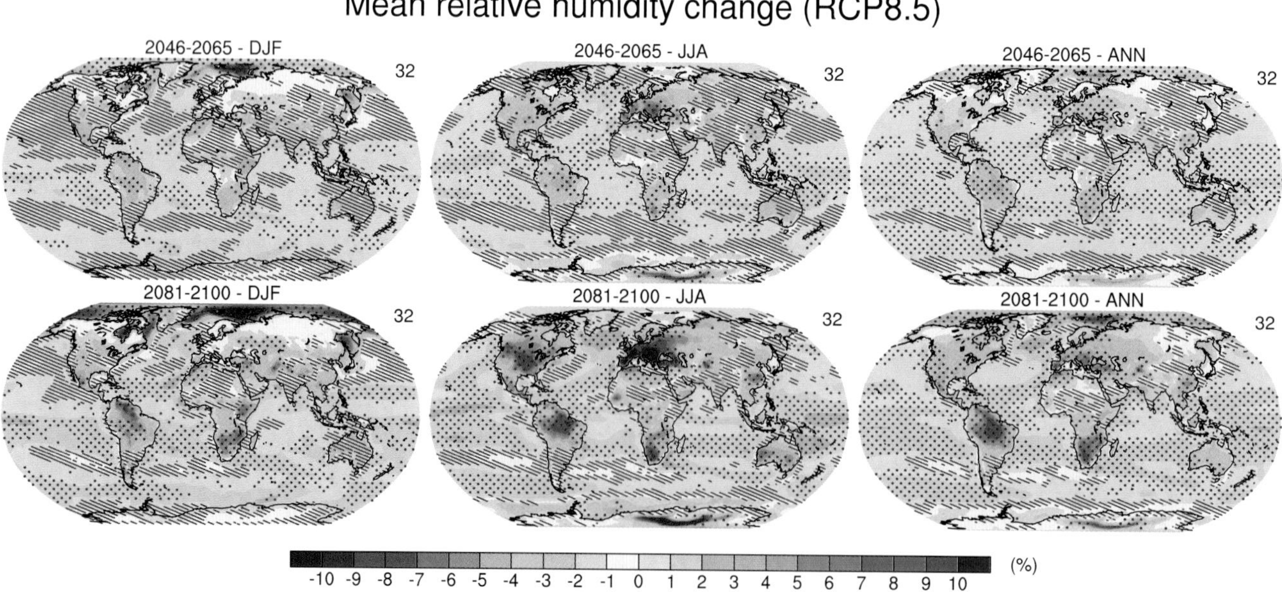

Figure 12.21 | Projected changes in near-surface relative humidity from the CMIP5 models under RCP8.5 for the December, January and February (DJF, left), June, July and August (JJA, middle) and annual mean (ANN, right) averages relative to 1986–2005 for the periods 2046–2065 (top row), 2081–2100 (bottom row). The changes are differences in relative humidity percentage (as opposed to a fractional or relative change). Hatching indicates regions where the multi-model mean change is less than one standard deviation of internal variability. Stippling indicates regions where the multi-model mean change is greater than two standard deviations of internal variability and where at least 90% of models agree on the sign of change (see Box 12.1).

projected changes in RCP scenarios. Changes in precipitation extremes are presented in Section 12.4.5.5. Further discussion of regional changes, in particular the monsoon systems, is presented in Chapter 14.

Figure 12.22 shows the CMIP5 multi-model average percentage change in seasonal mean precipitation in the middle of the 21st century, at the end of the 21st century and at the end of the 22nd century for the RCP8.5 scenario relative to the 1986–2005 average. Precipitation changes for all the scenarios are shown in Annex I Supplementary Material and scale approximately with the global mean temperature (Section 12.4.2). In many regions, changes in precipitation exhibit strong seasonal characteristics so that, in regions where the sign of the precipitation changes varies with the season, the annual mean values (Figure 12.10) may hide some of these seasonal changes, resulting in weaker confidence than seasonal mean values (Chou et al., 2013; Huang et al., 2013).

The patterns of multi-model precipitation changes displayed in Figure 12.22 tend to smooth and decrease the spatial contrast of precipitation changes simulated by each model, in particular over regions where model results disagree. Thus the amplitude of the multi-model ensemble mean precipitation response significantly underestimates the median amplitude computed from each individual model (Neelin et al., 2006; Knutti et al., 2010a). The CMIP3/5 multi-model ensemble precipitation projections must be interpreted in the context of uncertainty. Multi-model projections are not probabilistic statements about the likelihood of changes. Maps of multi-model projected changes are smoothly varying but observed changes are and will continue to be much more granular.

To analyze the patterns of projected precipitation changes, a useful framework consists in decomposing them into a part that is related to atmospheric circulation changes and a part that is related mostly to water vapour changes, referred to as dynamical and thermodynamical components, respectively. However, the definition of these two components may differ among studies. At the time of the AR4, the robust changes of the difference between precipitation and evaporation ($P - E$) were interpreted as a wet-get-wetter and dry-get-drier type of response (Mitchell et al., 1987; Chou and Neelin, 2004; Held and Soden, 2006). The theoretical background, which is more relevant over oceans than over land, is that the lower-tropospheric water vapour increase with temperature enhances the moisture transported by the circulation. This leads to additional moisture convergence within the convergence zones and to additional moisture divergence in the descent zones, increasing the contrast in precipitation minus evaporation values between moisture convergence and divergence regions. A weakening of the tropical overturning circulation (see Section 12.4.4.2) partially opposes this thermodynamic response (Chou and Neelin, 2004; Held and Soden, 2006; Vecchi and Soden, 2007; Chou et al., 2009; Seager et al., 2010; Allan, 2012; Bony et al., 2013). At the regional scale the dynamic response may be larger than the thermodynamic response, and this has been analyzed in more detail since the AR4 (Chou et al., 2009; Seager et al., 2010; Xie et al., 2010; Muller and O'Gorman, 2011; Chadwick et al., 2012; Scheff and Frierson, 2012; Bony et al., 2013; Ma and Xie, 2013). Over continents, this simple wet-get-wetter and dry-get-drier type of response fails for some important regions such as the Amazon. At the global scale, the net water vapour transport from oceans to land increases, and therefore the average $P - E$ over continents also increases (Liepert and Previdi, 2012).

In the mid and high latitudes, a common feature across generations of climate models is a simulated increased precipitation. The thermodynamical component explains most of the projected increase (Emori and Brown, 2005; Seager et al., 2010). This is consistent with theoretical explanations assuming fixed atmospheric flow patterns but increased water vapour in the lower troposphere (Held and Soden, 2006). In addition to this thermodynamical effect, water transport may be modified by the poleward shift of the storm tracks and by the increase of their intensity (Seager et al., 2010; Wu et al., 2011b), although confidence in such changes in storm tracks may not be high (see Section 12.4.4). On seasonal time scales, the minimum and maximum values of precipitation both increase, with a larger increase of the maximum and therefore an increase of the annual precipitation range (Seager et al., 2010; Chou and Lan, 2012). In particular, the largest changes over northern Eurasia and North America are projected to occur during winter. At high latitudes of the NH, the precipitation increase may lead to an increase of snowfall in the colder regions and a decrease of snowfall in the warmer regions due to the decreased number of freezing days (see Section 12.4.6.2).

Most models simulate a large increase of the annual mean precipitation over the equatorial ocean and an equatorward shift of the Intertropical Convergence Zone (ITCZ), in both summer and winter seasons, that are mainly explained by atmospheric circulation changes (Chou et al., 2009; Seager et al., 2010; Sobel and Camargo, 2011). The changes of the atmospheric circulation have different origins. Along the margins of the convection zones, spatial inhomogeneities, including local convergence feedback or the rate at which air masses from dry regions tend to flow into the convection zone, can yield a considerable sensitivity in precipitation response (Chou et al., 2006; Neelin et al., 2006). Along the equator, atmosphere–ocean interactions yield to a maximum of SST warming and a large precipitation increase there (Xie et al., 2010; Ma and Xie, 2013). Model studies with idealized configurations suggest that tropical precipitation changes should be interpreted as responses to changes of the atmospheric energy budget rather than responses to changes of SST (Kang and Held, 2012). All of these atmospheric circulation changes, and therefore precipitation changes, can differ considerably from model to model. This is the case over both ocean and land. For instance, the spread of model projections in the Sahel region, West Africa, is large in both the CMIP3 and CMIP5 multi-model data base (Roehrig et al., 2013).

In the subtropical dry regions, there is a robust decrease of $P - E$ that is accounted for by the thermodynamic contribution (Chou and Neelin, 2004; Held and Soden, 2006; Chou et al., 2009; Seager et al., 2010; Bony et al., 2013). Over ocean, the spatial heterogeneity of temperature increase impacts the lower-tropospheric water vapour increase, which impacts both the thermodynamic and the dynamic responses (Xie et al., 2010; Ma and Xie, 2013). In addition, the pattern of precipitation changes in dry regions may be different from that of $P - E$ because the contribution of evaporation changes can be as large (but of opposite sign) as the moisture transport changes (Chou and Lan, 2012; Scheff and Frierson, 2012; Bony et al., 2013). This is especially the case over the subsidence regions during the warm season over land where the agreement between models is the smallest (Chou et al., 2009; Allan, 2012). A robust feature is the decline of precipitation on the poleward flanks of the subtropical dry zones as a consequence of the Hadley Cell expansion, with possible additional decrease from a poleward shift of the mid latitude storm tracks (Seager et al., 2010; Scheff and Frierson, 2012). On seasonal time scales, the minimum and the maximum values of precipitation both increase, with a larger increase of the maximum and therefore an increase of the annual precipitation range (Sobel and Camargo, 2011; Chou and Lan, 2012).

Long-term precipitation changes are driven mainly by the increase of the surface temperature, as presented above, but other factors also contribute to them. Recent studies suggest that CO_2 increase has a significant direct influence on atmospheric circulation, and therefore on global and tropical precipitation changes (Andrews et al., 2010; Bala et al., 2010; Cao et al., 2012; Bony et al., 2013). Over the ocean, the positive RF from increased atmospheric CO_2 reduces the radiative cooling of the troposphere and the large scale rising motion and hence reduces precipitation in the convective regions. Over large landmasses, the direct effect of CO_2 on precipitation is the opposite owing to the small thermal inertia of land surfaces (Andrews et al., 2010; Bala et al., 2010; Cao et al., 2012; Bony et al., 2013). Regional precipitation changes are also influenced by aerosol and ozone (Ramanathan et al., 2001; Allen et al., 2012; Shindell et al., 2013a) through both local and large-scale processes, including changes in the circulation. Stratospheric ozone depletion contributes to the poleward expansion of the Hadley Cell and the related change of precipitation in the SH (Kang et al., 2011) whereas black carbon and tropospheric ozone increases are major contributors in the NH (Allen et al., 2012). Regional precipitation changes depend on regional forcings and on how models simulate their local and remote effects. Based on CMIP3 results, the inter-model spread of the estimate of precipitation changes over land is larger than the inter-scenario spread except in East Asia (Frieler et al., 2012).

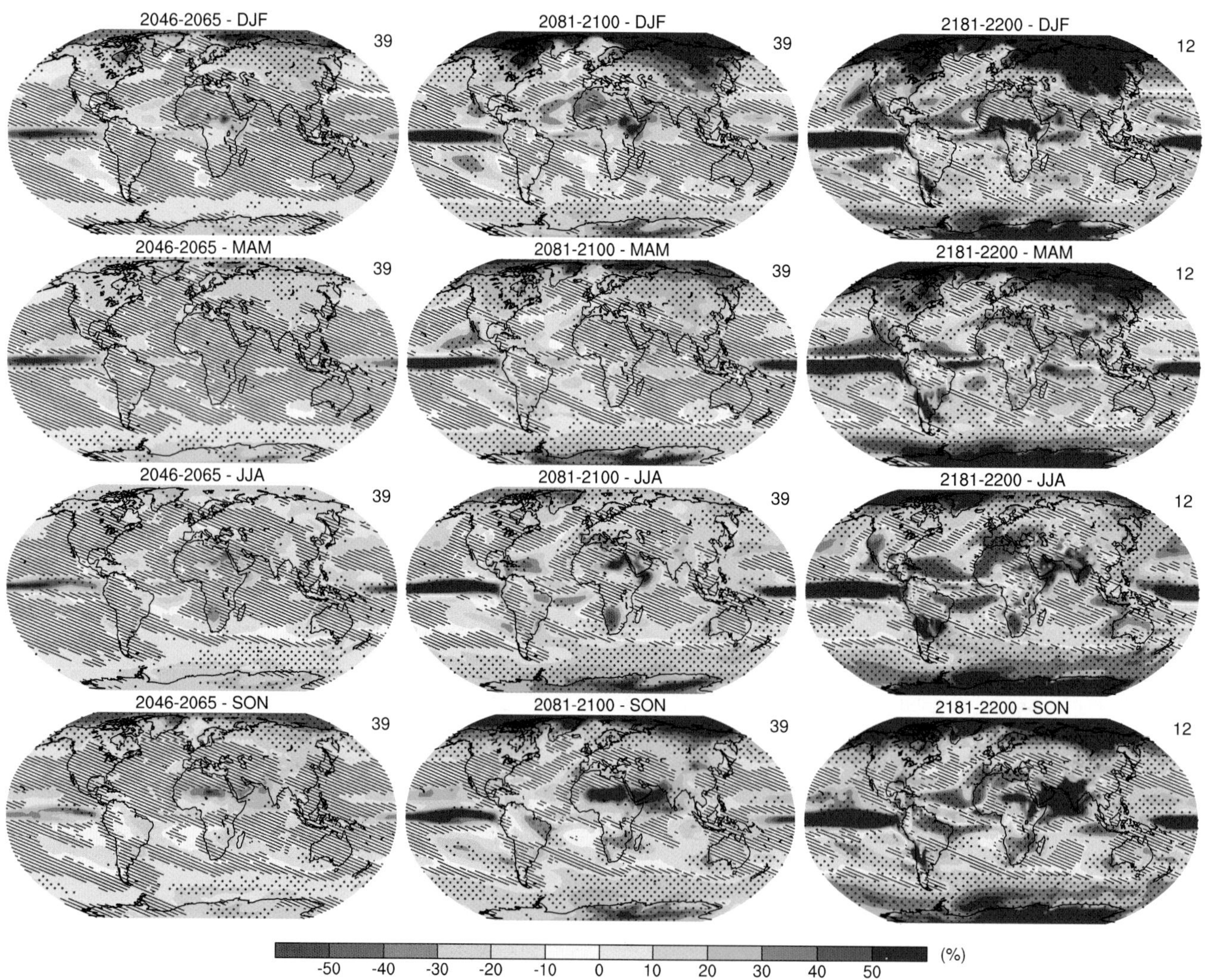

Figure 12.22 | Multi-model CMIP5 average percentage change in seasonal mean precipitation relative to the reference period 1986–2005 averaged over the periods 2045–2065, 2081–2100 and 2181–2200 under the RCP8.5 forcing scenario. Hatching indicates regions where the multi-model mean change is less than one standard deviation of internal variability. Stippling indicates regions where the multi-model mean change is greater than two standard deviations of internal variability and where at least 90% of models agree on the sign of change (see Box 12.1).

12

Projected precipitation changes vary greatly between models, much more so than for temperature projections. Part of this variance is due to genuine differences between the models including their ability to replicate observed precipitation patterns (see Section 9.4.1.1). However, a large part of it is also the result of the small ensemble size from each model (Rowell, 2012). This is especially true for regions of small projected changes located between two regions: one experiencing significant increases while the other experiences significant decreases. Individual climate model realizations will differ in their projection of future precipitation changes in these regions simply owing to their internal variability (Deser et al., 2012b; Deser et al., 2012a). Multi-model projections containing large numbers of realizations would tend to feature small changes in these regions, and hatching in Figure 12.22 indicates regions where the projected multi-model mean change is less than one standard deviation of internal variability (method (a), Box 12.1). Confidence in projections in regions of limited or no change in precipitation may be more difficult to obtain than confidence in regions of large projected changes. However, Power et al. (2012) and Tebaldi et al. (2011) show that for some of the regions featuring small multi-model average projected changes, effective consensus in projections may be better than the metrics reported in AR4 would imply.

Since the AR4, progress has been made in the understanding of the processes that control large scale precipitation changes. There is *high confidence* that the contrast of seasonal mean precipitation between dry and wet regions will increase in a warmer climate over most of the globe although there may be regional exceptions to this general pattern. This response is particularly robust when considering $P - E$ changes as a function of atmospheric dynamical regimes. However, it is important to note that significant exceptions can occur in specific regions especially along the equator and on the poleward edges of the subtropical dry zone. In these regions, atmospheric circulation changes lead to shifts of the precipitation patterns. There is *high confidence* that the contrast between wet and dry seasons will increase over most of the globe as temperatures increase. Over the mid- and high-latitude regions, projected precipitation increases in winter are larger than in summer. Over most of the subtropical oceans, projected precipitation increases in summer are larger than in winter.

The changes in precipitation shown in Figure 12.22 exhibit patterns that become more pronounced and confidence in them increases as temperatures increase. More generally, the spatial and temporal changes in precipitation between two scenarios or within two periods of a given scenario exhibit the pattern scaling behavior and limitations described in Section 12.4.2. The patterns and the associated multi-model spreads in CMIP5 for the RCP scenarios are very similar to those in CMIP3 for the SRES scenarios discussed in the AR4, with the projections in CMIP5 being slightly more consistent over land than those from CMIP3 (Knutti and Sedláček, 2013). The largest percentage changes are at the high latitudes. By the end of the 21st century, over the large northern land masses, increased precipitation is *likely* under the RCP8.5 scenario in the winter and spring poleward of 50°N. The robustness across scenarios, the magnitude of the projected changes versus natural variability and physical explanations described above yield *high confidence* that the projected changes would be larger than natural 20-year variations (see Box 12.1). In the tropics, precipitation changes exhibit strong regional contrasts, with increased precipitation over the equatorial Pacific and Indian Oceans and decreases over much of the subtropical ocean. However, decreases are not projected to be larger than natural 20-year variations anywhere until the end of this century under the RCP8.5 scenario. Decreased precipitation in the Mediterranean, Caribbean and Central America, southwestern United States and South Africa is *likely* under the RCP8.5 scenario and is projected with *medium confidence* to be larger than natural variations by the end of the 22nd century in some seasons (Box 12.1). The CMIP3 models' historical simulations of zonal mean precipitation trends were shown to underestimate observed trends (Gillett et al., 2004; Lambert et al., 2005; Zhang et al., 2007; Liepert and Previdi, 2009) (see Section 10.3.2.2). Therefore it is *more likely than not* that the magnitude of the projected future changes in Figure 12.22 based on the multi-model mean is underestimated. Observational uncertainties including limited global coverage and large natural variability, in addition to challenges in precipitation modelling, limit confidence in assessment of climatic changes in precipitation.

12.4.5.3 Soil Moisture

Near-surface soil moisture is the net result of a suite of complex processes (e.g., precipitation evapotranspiration, drainage, overland flow, infiltration), and heterogeneous and difficult-to-characterize aboveground and belowground system properties (e.g., slope, soil texture). As a result, regional to global-scale simulations of soil moisture and drought remain relatively uncertain (Burke and Brown, 2008; Henderson-Sellers et al., 2008). The AR4 (Section 8.2.3.2) discussed the lack of assessments of global-scale models in their ability to simulate soil moisture, and this problem appears to have persisted (Section 9.4.4.2). Furthermore, consistent multi-model projections of total soil moisture are difficult to make owing to substantial differences between climate models in the depth of their soil. However, Koster et al. (2009a) argued that once climatological statistics affecting soil moisture were accounted for, different models tend to agree on soil moisture projections.

The AR4 summarized multi-model projections of 21st century annual mean soil moisture changes as decreasing in the subtropics and Mediterranean region, and increasing in east Africa and central Asia. Dai (2013) found similar changes in an ensemble of 11 CMIP5 GCMs under RCP4.5. Figure 12.23 shows projected changes in surface soil moisture (upper 10 cm) in the CMIP5 ensemble at the end of the 21st century under the RCPs 2.6, 4.5, 6.0 and 8.5. We focus on this new CMIP5 specification because it describes soil moisture at a consistent depth across all CMIP5 models. The broad patterns are moderately consistent across the RCPs, with the changes tending to become stronger as the strength of the forcing change increases. The agreement among CMIP5 models and the consistency with other physical features of climate change indicate *high confidence* in certain regions where surface soils are projected to dry. There is little-to-no confidence anywhere in projections of moister surface soils. Under RCP8.5, with the largest projected change, individual ensemble members (not shown) show consistency across the ensemble for drying in the Mediterranean region, northeast and southwest South America, southern Africa, and southwestern USA. However, ensemble members show disagreement on the sign of change in large regions such as central Asia or the high northern latitudes. The Mediterranean, southwestern USA, northeast South America and southern African drying regions are consistent with

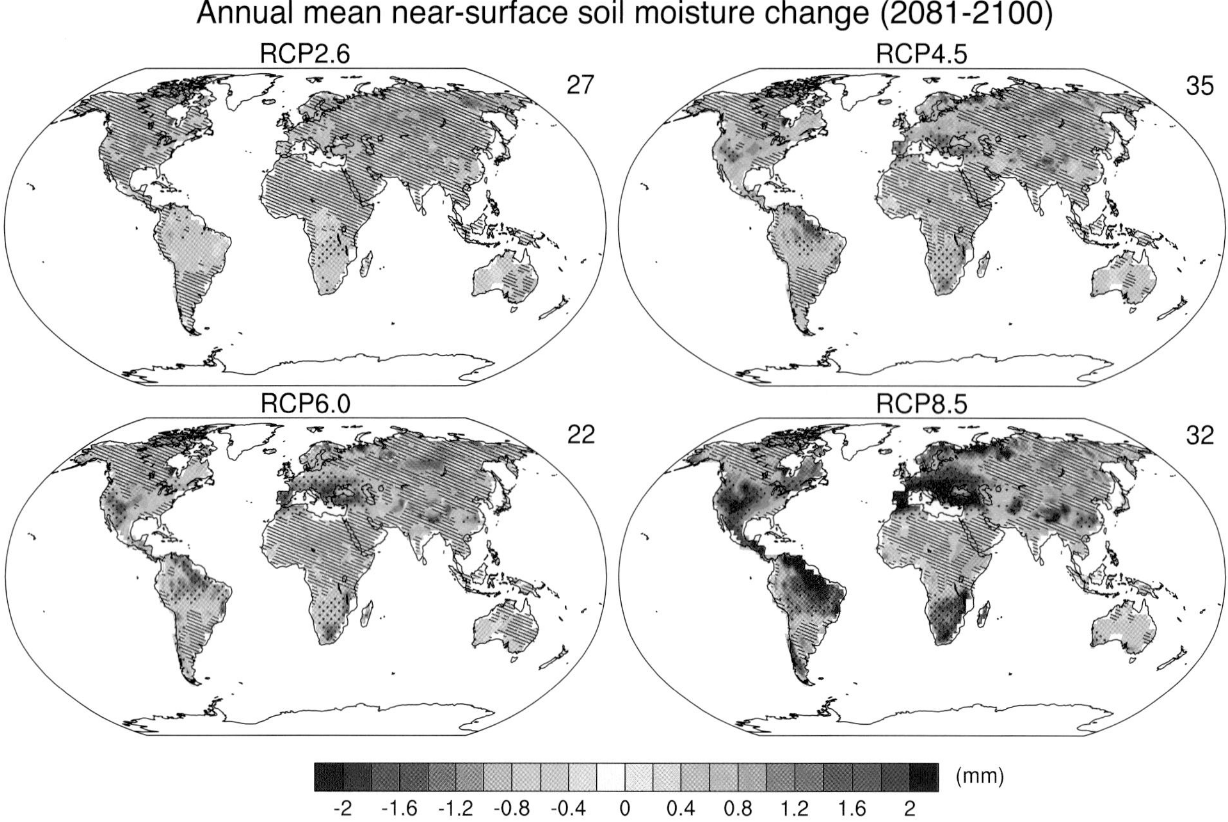

Figure 12.23 | Change in annual mean soil moisture (mass of water in all phases in the uppermost 10 cm of the soil) (mm) relative to the reference period 1986–2005 projected for 2081–2100 from the CMIP5 ensemble. Hatching indicates regions where the multi-model mean change is less than one standard deviation of internal variability. Stippling indicates regions where the multi-model mean change is greater than two standard deviations of internal variability and where at least 90% of models agree on the sign of change (see Box 12.1). The number of CMIP5 models used is indicated in the upper right corner of each panel.

projected widening of the Hadley Circulation that shifts downwelling, thus inhibiting precipitation in these regions. The large-scale drying in the Mediterranean, southwest USA, and southern Africa appear across generations of projections and climate models and is deemed *likely* as global temperatures rise and will increase the risk of agricultural drought. In addition, an analysis of CMIP3 and CMIP5 projections of soil moisture in five drought-prone regions indicates that the differences in future forcing scenarios are the largest source of uncertainty in such regions rather than differences between model responses (Orlowsky and Seneviratne, 2012).

Other recent assessments include multi-model ensemble approaches, dynamical downscaling, and regional climate models applied around the globe and illustrate the variety of issues influencing soil moisture changes. Analyses of the southwestern USA using CMIP3 models (Christensen and Lettenmaier, 2007; Seager et al., 2007) show consistent projections of drying, primarily due to a decrease in winter precipitation. In contrast, Kellomaki et al. (2010) find that SRES A2 projections for Finland yield decreased snow depth, but soil moisture generally increases, consistent with the general increase in precipitation occurring in high northern latitudes. Kolomyts and Surova (2010), using projections from the CMIP3 models, GISS and HadCM2, under the SRES A2 forcing, show that vegetation type has substantial influence on the development of pronounced drying over the 21st century in Middle Volga Region forests.

Projected changes in soil moisture from the CMIP3/5 models also show substantial seasonal variation. For example, soil moisture changes in the North American midlatitudes, coupled with projected warming, increases the strength of land–atmosphere coupling during spring and summer in 15 GCMs under RCP8.5 (Dirmeyer et al., 2013). For the Cline River watershed in western Canada, Kienzle et al. (2012) find decreases in summer soil moisture content, but annual increases averaging 2.6% by the 2080s using a suite of CMIP3 GCMs simulating B1, A1B and A2 scenarios to drive a regional hydrology model. Hansen et al. (2007), using dynamical downscaling of one GCM running the A2 scenario, find summer soil moisture decreases in Mongolia of up to 6% due to increased potential evaporation in a warming climate and decreased precipitation and decreased precipitation.

Soil moisture projections in high latitude permafrost regions are critically important for assessing future climate feedbacks from trace-gas emissions (Zhuang et al., 2004; Riley et al., 2011) and vegetation changes (Chapin et al., 2005). In addition to changes in precipitation, snow cover and evapotranspiration, future changes in high-latitude soil moisture also will depend on permafrost degradation, thermokarst evolution, rapid changes in drainage (Smith et al., 2005), and changes in plant communities and their water demands. Current understanding of these interacting processes at scales relevant to climate is poor, so that full incorporation in current GCMs is lacking.

12.4.5.4 Runoff and Evaporation

In the AR4, 21st century model-projected runoff consistently showed decreases in southern Europe, the Middle East, and southwestern USA and increases in Southeast Asia, tropical East Africa and at high northern latitudes. The same general features appear in the CMIP5 ensemble of GCMs for all four RCPs shown in Figure 12.24, with the areas of most robust change typically increasing with magnitude of forcing change. However, the robustness of runoff decreases in the southwestern USA is less in the CMIP5 models compared to the AR4. The large decreases in runoff in southern Europe and southern Africa are consistent with changes in the Hadley Circulation and related precipitation decreases and warming-induced evapotranspiration increases. The high northern latitude runoff increases are *likely* under RCP8.5 and consistent with the projected precipitation increases (Figure 12.22). The consistency of changes across different generations of models and different forcing scenarios, together with the physical consistency of change indicates that decreases are also *likely* in runoff in southern Europe, the Middle East, and southern Africa in this scenario.

A number of reports since the AR4 have updated findings from CMIP3 models and analyzed a large set of mechanisms affecting runoff. Several studies have focussed on the Colorado River basin in the United States (Christensen and Lettenmaier, 2007; McCabe and Wolock, 2007; Barnett and Pierce, 2008; Barnett et al., 2008) showing that runoff reductions that do happen under global warming occur through a combination of evapotranspiration increases and precipitation decreases, with the overall reduction in river flow exacerbated by human water demands on the basin's supply.

A number of CMIP3 analyses have examined trends and seasonal shifts in runoff. For example, Kienzle et al. (2012) studied climate change scenarios over the Cline River watershed in western Canada and projected (1) spring runoff and peak streamflow up to 4 weeks earlier than in 1961–1990; (2) significantly higher streamflow between October and June; and (3) lower streamflow between July and September. For the Mediterranean basin, an ensemble of regional climate models driven by several GCMs using the A1B scenario have a robust decrease in runoff emerging only after 2050 (Sanchez-Gomez et al., 2009).

Annual mean surface evaporation in the models assessed in AR4 showed increases over most of the ocean and increases or decreases over land with largely the same pattern over land as increases and decreases in precipitation. Similar behaviour occurs in an ensemble of CMIP5 models (Figure 12.25). Evaporation increases over most of the ocean and land, with prominent areas of decrease over land occurring in southern Africa and northwestern Africa along the Mediterranean. The areas of decrease correspond to areas with reduced precipitation. There is some uncertainty about storm-track changes over Europe (see Sections 12.4.3 and 14.6.2). However, the consistency of the decreases across different generations of models and different forcing scenarios along with the physical basis for the precipitation decrease

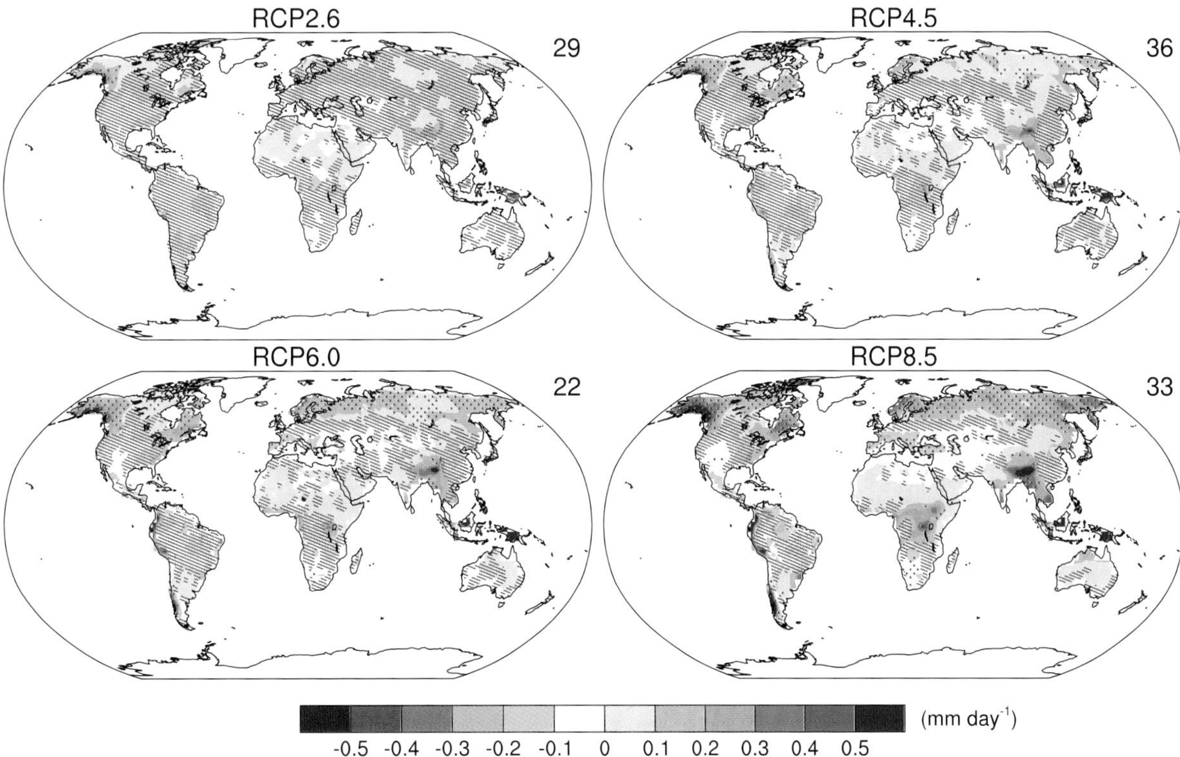

Figure 12.24 | Change in annual mean runoff relative to the reference period 1986–2005 projected for 2081–2100 from the CMIP5 ensemble. Hatching indicates regions where the multi-model mean change is less than one standard deviation of internal variability. Stippling indicates regions where the multi-model mean change is greater than two standard deviations of internal variability and where at least 90% of models agree on the sign of change (see Box 12.1). The number of CMIP5 models used is indicated in the upper right corner of each panel.

12

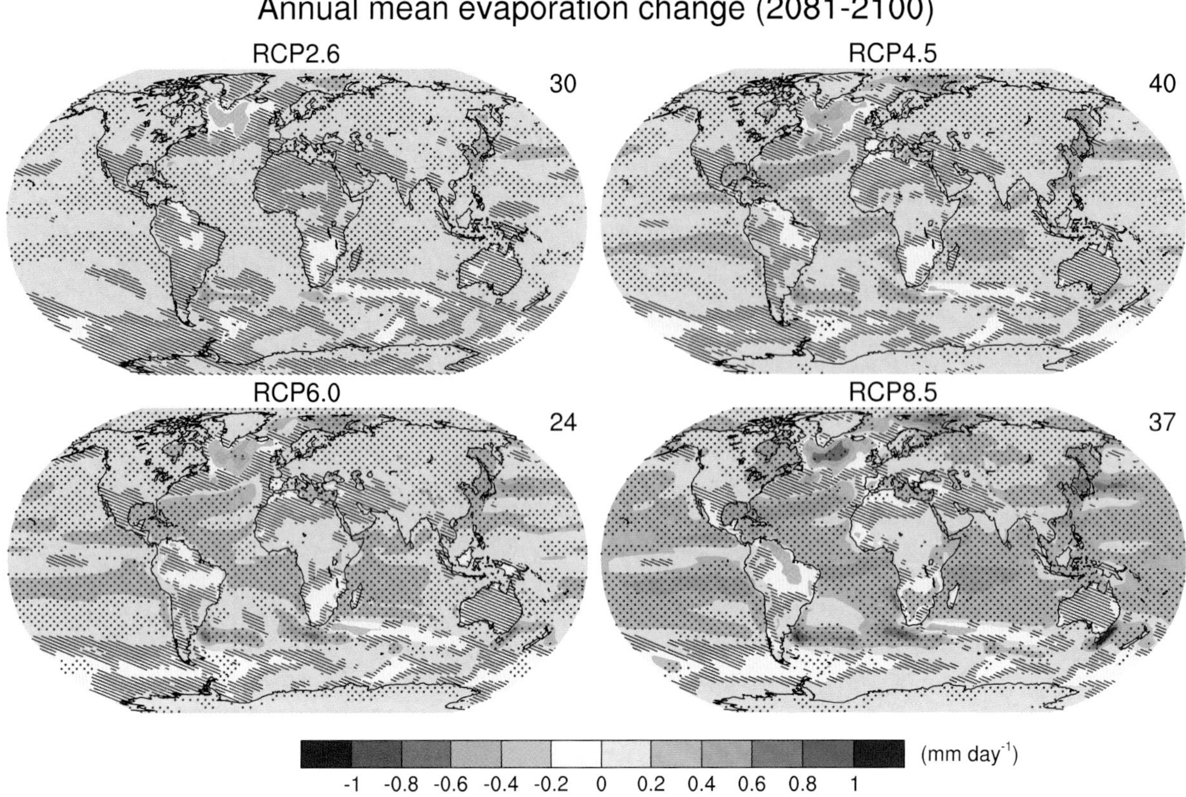

Figure 12.25 | Change in annual mean evaporation relative to the reference period 1986–2005 projected for 2081–2100 from the CMIP5 ensemble. Hatching indicates regions where the multi-model mean change is less than one standard deviation of internal variability. Stippling indicates regions where the multi-model mean change is greater than two standard deviations of internal variability and where at least 90% of models agree on the sign of change (see Box 12.1). The number of CMIP5 models used is indicated in the upper right corner of each panel.

indicates that these decreases in annual mean evaporation are *likely* under RCP8.5, but with *medium confidence.* Annual mean evaporation increases over land in the northern high latitudes are consistent with the increase in precipitation and the overall warming that would increase potential evaporation. For the northern high latitudes, the physical consistency and the similar behaviour across multiple generations and forcing scenarios indicates that annual mean evaporation increases there are *likely*, with *high confidence*.

Evapotranspiration changes partly reflect changes in precipitation. However, some changes might come from altered biological processes. For example, increased atmospheric CO_2 promotes stomatal closure and reduced transpiration (Betts et al., 2007; Cruz et al., 2010) which can potentially yield increased runoff. There is potential for substantial feedback between vegetation changes and regional water cycles, though the impact of such feedback remains uncertain at this point due to limitations on modelling crop and other vegetation processes in GCMs (e.g., Newlands et al., 2012) and uncertainties in plant response, ecosystem shifts and land management changes.

12.4.5.5 Extreme Events in the Water Cycle

In addition to the changes in the seasonal pattern of mean precipitation described above, the distribution of precipitation events is projected to undergo profound changes (Gutowski et al., 2007; Sun et al., 2007; Boberg et al., 2010). At daily to weekly scales, a shift to more intense individual storms and fewer weak storms is projected (Seneviratne et al., 2012). At seasonal or longer time scales, increased evapotranspiration over land can lead to more frequent and more intense periods of agricultural drought.

A general relationship between changes in total precipitation and extreme precipitation does not exist (Seneviratne et al., 2012). Two possible mechanisms controlling short-term extreme precipitation amounts are discussed at length in the literature and are similar to the thermodynamic and dynamical mechanisms detailed above for changes in average precipitation.

The first considers that extreme precipitation events occur when most of the available atmospheric water vapour rapidly precipitates out in a single storm. The maximum amount of water vapour in air (saturation) is determined by the Clausius–Clapeyron relationship. As air temperature increases, this saturated amount of water also increases (Allen and Ingram, 2002; Pall et al., 2007; Allan and Soden, 2008; Kendon et al., 2010). Kunkel et al. (2013) examined the CMIP5 model RCP4.5 and 8.5 projections for changes in maximum water vapour concentrations, a principal factor controlling the probable bound on maximum precipitation, concluding that maximum water vapour changes are comparable to mean water vapour changes but that the potential for changes in dynamical factors is less compelling. Such increases in atmospheric water vapour are expected to increase the intensity of individual precipitation events, but have less impact on their frequency. As a result projected increases in extreme precipitation may be more reliable than similar projections of changes in mean precipitation in some regions (Kendon et al., 2010).

A second mechanism for extreme precipitation put forth by O'Gorman and Schneider (2009a, 2009b) is that such events are controlled by anomalous horizontal moisture flux convergence and associated convective updrafts which would change in a more complicated fashion in a warmer world (Sugiyama et al., 2010). Emori and Brown (2005) showed that the thermodynamic mechanism dominated over the dynamical mechanism nearly everywhere outside the tropical warm pool. However, Utsumi et al. (2011) used gridded observed daily data to find that daily extreme precipitation monotonically increases with temperature only at high latitudes, with the opposite behaviour in the tropics and a mix in the mid-latitudes. Li et al. (2011a) found that both mechanisms contribute to extreme precipitation in a high-resolution aquaplanet model with updrafts as the controlling element in the tropics and air temperature controlling the mid-latitudes consistent with the results by Chou et al. (2012). Using a high-resolution regional model, Berg et al. (2009) found a seasonal dependence in Europe with the Clausius–Clapeyron relationship providing an upper limit to daily precipitation intensity in winter but water availability rather than storage capacity is the controlling factor in summer. Additionally, Lenderink and Van Meijgaard (2008) found that very short (sub-daily) extreme precipitation events increase at a rate twice the amount predicted by Clausius–Clapeyron scaling in a very high-resolution model over Europe suggesting that both mechanisms can interact jointly. Gastineau and Soden (2009) found in the CMIP3 models that the updrafts associated with the most extreme tropical precipitation events actually weaken despite an increase in the frequency of the heaviest rain rates further complicating simple mechanistic explanations. See also Sections 7.6.5 and 11.3.2.5.2.

Projections of changes in future extreme precipitation may be larger at the regional scales than for future mean precipitation, but natural variability is also larger causing a tendency for signal-to-noise ratios to decrease when considering increasingly extreme metrics. However, mechanisms of natural variability still are a large factor in assessing the robustness of projections (Kendon et al., 2008). In addition, large-scale circulation changes, which are uncertain, could dominate over the above mechanisms depending on the rarity and type of events considered. However, analysis of CMIP3 models suggests circulation changes are potentially insufficient to offset the influence of increasing atmospheric water vapour on extreme precipitation change over Europe at least on large spatial scales (Kendon et al., 2010). An additional shift of the storm track has been shown in models with a better representation of the stratosphere, and this is found to lead to an enhanced increase in extreme rainfall over Europe in winter (Scaife et al., 2012).

Similar to temperature extremes (Section 12.4.3.3), the definition of a precipitation extreme depends very much on context and is often used in discussion of particular climate-related impacts (Seneviratne et al. (2012), Box 3.1). Consistently, climate models project future episodes of more intense precipitation in the wet seasons for most of the land areas, especially in the NH and its higher latitudes, and the monsoon regions of the world, and at a global average scale. The actual magnitude of the projected change is dependent on the model used,

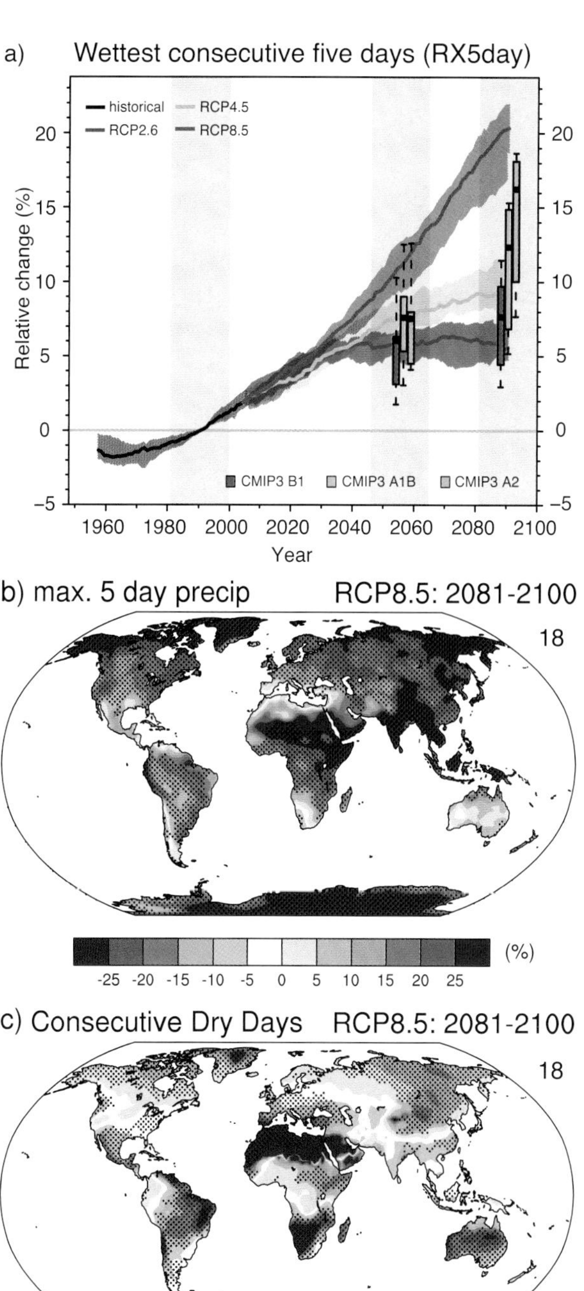

Figure 12.26 | (a, b) Projected percent changes (relative to the 1981–2000 reference period in common with CMIP3) from the CMIP5 models in RX5day, the annual maximum five-day precipitation accumulation. (a) Global average percent change over land regions for the RCP2.6, RCP4.5 and RCP8.5 scenarios. Shading in the time series represents the interquartile ensemble spread (25th and 75th quantiles). The box-and-whisker plots show the interquartile ensemble spread (box) and outliers (whiskers) for 11 CMIP3 model simulations of the SRES scenarios A2 (orange), A1B (cyan) and B1 (purple) globally averaged over the respective future time periods (2046–2065 and 2081–2100) as anomalies from the 1981–2000 reference period. (b) Percent change over the 2081–2100 period in the RCP8.5 scenario. (c) Projected change in annual CDD, the maximum number of consecutive dry days when precipitation is less than 1 mm, over the 2081–2100 period in the RCP8.5 scenario (relative to the 1981–2000 reference period) from the CMIP5 models. Stippling indicates gridpoints with changes that are significant at the 5% level using a Wilcoxon signed-ranked test. (Updated from Sillmann et al. (2013), excluding the FGOALS-s2 model.)

Frequently Asked Questions

FAQ 12.2 | How Will the Earth's Water Cycle Change?

The flow and storage of water in the Earth's climate system are highly variable, but changes beyond those due to natural variability are expected by the end of the current century. In a warmer world, there will be net increases in rainfall, surface evaporation and plant transpiration. However, there will be substantial differences in the changes between locations. Some places will experience more precipitation and an accumulation of water on land. In others, the amount of water will decrease, due to regional drying and loss of snow and ice cover.

The water cycle consists of water stored on the Earth in all its phases, along with the movement of water through the Earth's climate system. In the atmosphere, water occurs primarily as a gas—water vapour—but it also occurs as ice and liquid water in clouds. The ocean, of course, is primarily liquid water, but the ocean is also partly covered by ice in polar regions. Terrestrial water in liquid form appears as surface water—such as lakes and rivers—soil moisture and groundwater. Solid terrestrial water occurs in ice sheets, glaciers, snow and ice on the surface and in permafrost and seasonally frozen soil.

Statements about future climate sometimes say that the water cycle will accelerate, but this can be misleading, for strictly speaking, it implies that the cycling of water will occur more and more quickly with time and at all locations. Parts of the world will indeed experience intensification of the water cycle, with larger transports of water and more rapid movement of water into and out of storage reservoirs. However, other parts of the climate system will experience substantial depletion of water, and thus less movement of water. Some stores of water may even vanish.

As the Earth warms, some general features of change will occur simply in response to a warmer climate. Those changes are governed by the amount of energy that global warming adds to the climate system. Ice in all forms will melt more rapidly, and be less pervasive. For example, for some simulations assessed in this report, summer Arctic sea ice disappears before the middle of this century. The atmosphere will have more water vapour, and observations and model results indicate that it already does. By the end of the 21st century, the average amount of water vapour in the atmosphere could increase by 5 to 25%, depending on the amount of human emissions of greenhouse gases and radiatively active particles, such as smoke. Water will evaporate more quickly from the surface. Sea level will rise due to expansion of warming ocean waters and melting land ice flowing into the ocean (see FAQ 13.2).

These general changes are modified by the complexity of the climate system, so that they should not be expected to occur equally in all locations or at the same pace. For example, circulation of water in the atmosphere, on land and in the ocean can change as climate changes, concentrating water in some locations and depleting it in others. The changes also may vary throughout the year: some seasons tend to be wetter than others. Thus, model simulations assessed in this report show that winter precipitation in northern Asia may increase by more than 50%, whereas summer precipitation there is projected to hardly change. Humans also intervene directly in the water cycle, through water management and changes in land use. Changing population distributions and water practices would produce further changes in the water cycle.

Water cycle processes can occur over minutes, hours, days and longer, and over distances from metres to kilometres and greater. Variability on these scales is typically greater than for temperature, so climate changes in precipitation are harder to discern. Despite this complexity, projections of future climate show changes that are common across many models and climate forcing scenarios. Similar changes were reported in the AR4. These results collectively suggest well understood mechanisms of change, even if magnitudes vary with model and forcing. We focus here on changes over land, where changes in the water cycle have their largest impact on human and natural systems.

Projected climate changes from simulations assessed in this report (shown schematically in FAQ 12.2, Figure 1) generally show an increase in precipitation in parts of the deep tropics and polar latitudes that could exceed 50% by the end of the 21st century under the most extreme emissions scenario. In contrast, large areas of the subtropics could have decreases of 30% or more. In the tropics, these changes appear to be governed by increases in atmospheric water vapour and changes in atmospheric circulation that further concentrate water vapour in the tropics and thus promote more tropical rainfall. In the subtropics, these circulation changes simultaneously promote less rainfall despite warming in these regions. Because the subtropics are home to most of the world's deserts, these changes imply increasing aridity in already dry areas, and possible expansion of deserts. *(continued on next page)*

FAQ 12.2 (continued)

Increases at higher latitudes are governed by warmer temperatures, which allow more water in the atmosphere and thus, more water that can precipitate. The warmer climate also allows storm systems in the extratropics to transport more water vapour into the higher latitudes, without requiring substantial changes in typical wind strength. As indicated above, high latitude changes are more pronounced during the colder seasons.

Whether land becomes drier or wetter depends partly on precipitation changes, but also on changes in surface evaporation and transpiration from plants (together called evapotranspiration). Because a warmer atmosphere can have more water vapour, it can induce greater evapotranspiration, given sufficient terrestrial water. However, increased carbon dioxide in the atmosphere reduces a plant's tendency to transpire into the atmosphere, partly counteracting the effect of warming.

In the tropics, increased evapotranspiration tends to counteract the effects of increased precipitation on soil moisture, whereas in the subtropics, already low amounts of soil moisture allow for little change in evapotranspiration. At higher latitudes, the increased precipitation generally outweighs increased evapotranspiration in projected climates, yielding increased annual mean runoff, but mixed changes in soil moisture. As implied by circulation changes in FAQ 12.2, Figure 1, boundaries of high or low moisture regions may also shift.

A further complicating factor is the character of rainfall. Model projections show rainfall becoming more intense, in part because more moisture will be present in the atmosphere. Thus, for simulations assessed in this report, over much of the land, 1-day precipitation events that currently occur on average every 20 years could occur every 10 years or even more frequently by the end of the 21st century. At the same time, projections also show that precipitation events overall will tend to occur less frequently. These changes produce two seemingly contradictory effects: more intense downpours, leading to more floods, yet longer dry periods between rain events, leading to more drought.

At high latitudes and at high elevation, further changes occur due to the loss of frozen water. Some of these are resolved by the present generation of global climate models (GCMs), and some changes can only be inferred because they involve features such as glaciers, which typically are not resolved or included in the models. The warmer climate means that snow tends to start accumulating later in the fall, and melt earlier in the spring. Simulations assessed in this report show March to April snow cover in the Northern Hemisphere is projected to decrease by approximately 10 to 30% on average by the end of this century, depending on the greenhouse gas scenario. The earlier spring melt alters the timing of peak springtime flow in rivers receiving snowmelt. As a result, later flow rates will decrease, potentially affecting water resource management. These features appear in GCM simulations.

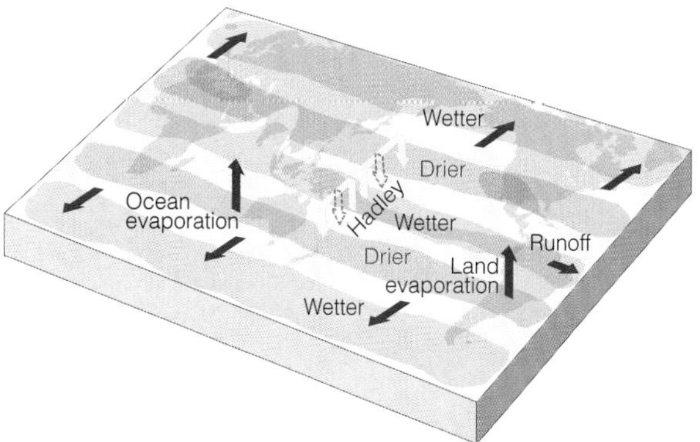

FAQ 12.2, Figure 1 | Schematic diagram of projected changes in major components of the water cycle. The blue arrows indicate major types of water movement changes through the Earth's climate system: poleward water transport by extratropical winds, evaporation from the surface and runoff from the land to the oceans. The shaded regions denote areas more likely to become drier or wetter. Yellow arrows indicate an important atmospheric circulation change by the Hadley Circulation, whose upward motion promotes tropical rainfall, while suppressing subtropical rainfall. Model projections indicate that the Hadley Circulation will shift its downward branch poleward in both the Northern and Southern Hemispheres, with associated drying. Wetter conditions are projected at high latitudes, because a warmer atmosphere will allow greater precipitation, with greater movement of water into these regions.

Loss of permafrost will allow moisture to seep more deeply into the ground, but it will also allow the ground to warm, which could enhance evapotranspiration. However, most current GCMs do not include all the processes needed to simulate well permafrost changes. Studies analysing soils freezing or using GCM output to drive more detailed land models suggest substantial permafrost loss by the end of this century. In addition, even though current GCMs do not explicitly include glacier evolution, we can expect that glaciers will continue to recede, and the volume of water they provide to rivers in the summer may dwindle in some locations as they disappear. Loss of glaciers will also contribute to a reduction in springtime river flow. However, if annual mean precipitation increases—either as snow or rain—then these results do not necessarily mean that annual mean river flow will decrease.

12

but there is strong agreement across the models over the direction of change (Tebaldi et al., 2006; Goubanova and Li, 2007; Chen and Knutson, 2008; Haugen and Iversen, 2008; May, 2008b; Kysely and Beranova, 2009; Min et al., 2011; Sillmann et al., 2013). Regional details are less robust in terms of the relative magnitude of changes but remain in good accord across models in terms of the sign of the change and the large-scale geographical patterns (Meehl et al., 2005a; CCSP, 2008a). In semi-arid regions of the midlatitudes and subtropics such as the Mediterranean, the southwest USA, southwestern Australia, southern Africa and a large portion of South America, the tendency manifested in the majority of model simulations is for longer dry periods and is consistent with the average decreases shown in Figure 12.22. Figure 12.26 shows projected percent changes in RX5day, the annual maximum of consecutive 5-day precipitation over land regions obtained from the CMIP5 models (Box 2.4, Table 1). Globally averaged end of 21st century changes over land range from 5% (RCP2.6) to 20% (RCP8.5) more precipitation during very wet 5-day periods. Results from the CMIP3 models are shown for comparison (see Section 12.4.9). Locally, the few regions where this index of extreme precipitation decreases in the late 21st century RCP8.5 projection coincide with areas of robust decreases in the mean precipitation of Figure 12.22.

Drought is discussed extensively in the SREX report (Seneviratne et al., 2012) and the conclusions about future drought risk described there based on CMIP3 models are reinforced by the CMIP5 models. As noted in the SREX reports, assessments of changes in drought characteristics with climate change should be made in the context of specific impacts questions. The risk of future agricultural drought episodes is increased in the regions of robust soil moisture decrease described in Section 12.4.5.3 and shown in Figure 12.23. Other measures in the literature of future agricultural drought are largely focussed on the Palmer Drought Severity Index (Wehner et al., 2011; Schwalm et al., 2012; Dai, 2013) and project 'extreme' drought as the normal climatological state by the end of the 21st century under the high emission scenarios in many mid-latitude locations. However, this measure of agricultural drought has been criticized as overly sensitive to increased temperatures due to a simplified soil moisture model (Hoerling et al., 2012). The consecutive dry-day index (CDD) is the length of the longest period of consecutive days with precipitation less than 1 mm (Box 2.4, Table 1). CMIP5 projected changes in CDD over the 2081–2100 period under the RCP8.5 scenario (relative to the 1981–2000 reference period in common with CMIP3) from the CMIP5 models are shown in Figure 12.26c and exhibit patterns similar to projected changes in both precipitation and soil moisture (Sillmann et al., 2013). Substantial increases in this measure of meteorological drought are projected in the Mediterranean, Central America, Brazil, South Africa and Australia while decreases are projected in high northern latitudes.

Truly rare precipitation events can cause very significant impacts. The statistics of these events at the tails of the precipitation distribution are well described by Extreme Value (EV) Theory although there are significant biases in the direct comparison of gridded model output and actual station data (Smith et al., 2009). There is also strong evidence that model resolution plays a key role in replicating EV quantities estimated from gridded observational data, suggesting that high-resolution models may provide somewhat more confidence in projection of changes in rare precipitation events (Fowler et al., 2007a; Wehner et al., 2011). Figure 12.27 shows the late 21st century changes per degree Celsius in local warming in 20-year return values of annual maximum daily precipitation relative to the late 20th century (left) and the associated return periods of late 20th century 20-year return values at the end of the 21st century from the CMIP5 models. Across future emission scenarios, the global average of the CMIP5 multi-model median return value sensitivity is an increase of 5.3% °C^{-1} (Kharin et al., 2013). The CMIP5 land average is close to the CMIP3 value of 4% °C^{-1} reported by Min et al. (2011) for a subset of CMIP3 models. Corresponding with this change, the global average of return periods of late 20th century 20-year return values is reduced from 20 years to 14 years for a 1°C local warming. Return periods are projected to be reduced by about 10 to 20% °C^{-1} over the most of the mid-latitude land masses with larger reductions over wet tropical regions (Kharin et al., 2013). Hence, extreme precipitation events will *very likely* be more intense

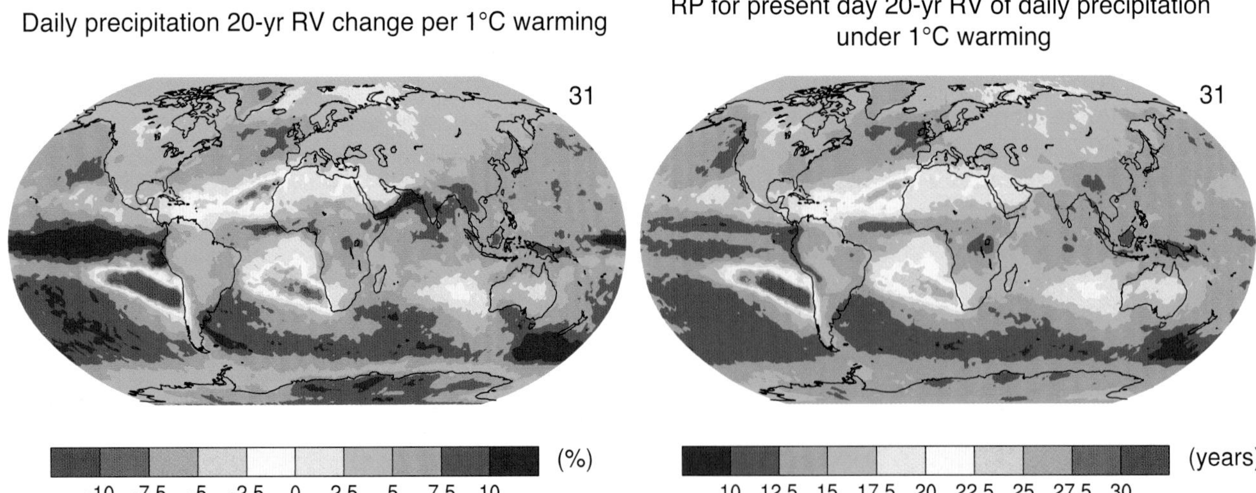

Figure 12.27 | (Left) The CMIP5 2081–2100 multi-model ensemble median percent change in 20-year return values of annual maximum daily precipitation per 1°C of local warming relative to the 1986–2005 reference period. (Right) The average 2081–2100 CMIP5 multi-model ensemble median of the return periods (years) of 1986–2005 20-year return values of annual maximum daily precipitation corresponding to 1°C of local warming. Regions of no change would have return periods of 20 years.

12

and more frequent in these regions in a warmer climate. Reductions in return values (or equivalently, increases in return period) are confined to convergent oceanic regions where circulation changes have reduced the available water vapour.

Severe thunderstorms, associated with large hail, high winds, and tornadoes, are another example of extreme weather associated with the water cycle. The large-scale environments in which they occur are characterized by large Convective Available Potential Energy (CAPE) and deep tropospheric wind shear (Brooks et al., 2003; Brooks, 2009). Del Genio et al. (2007), Trapp et al. (2007, 2009), and Van Klooster and Roebber (2009) found a general increase in the energy and decrease in the shear terms from the late 20th century to the late 21st century over the USA using a variety of regional model simulations embedded in global model SRES scenario simulations. The relative change between these two competing factors would tend to favour more environments that would support severe thunderstorms, providing storms are initiated. Trapp et al. (2009), for example, found an increase in favourable thunderstorm conditions for all regions of the USA east of the Rocky Mountains. Large variability in both the energy and shear terms means that statistical significance is not reached until late in the 21st century under high forcing scenarios. One way of assessing the possibility of a change in the frequency of future thunderstorms is to look at historical records of observed tornado, hail and wind occurrence with respect to the environmental conditions (Brooks, 2013). This indicates that an increase in the fraction of severe thunderstorms containing non-tornadic winds would be consistent with the model projections of increased energy and decreased shear, but there has not been enough research to make a firm conclusion regarding future changes in frequency or magnitude.

Less work has been done on projected changes outside of the USA. Marsh et al. (2009) found that mean energy decreased in the warm season in Europe while it increased in the cool season. Even though the energy decreases in the warm season, the number of days with favourable environments for severe thunderstorms increases because of an increasing number of days with relatively large values of available energy. For Europe, with the Mediterranean Sea and Sahara Desert to the south, questions remain about changes in boundary layer moisture, a main driver of the energy term. Niall and Walsh (2005) examined changes in CAPE, which may be associated with hailstorm occurrence in southeastern Australia using a global model, and found little change under warmer conditions. Leslie et al. (2008) reconsidered the southeastern Australia hail question by nesting models with 1 km horizontal grid spacing and using sophisticated microphysical parameterizations and found an increase in the frequency of large hail by 2050 under the SRES A1B scenario, but with extremely large internal variability in the environments and hail size.

Overall, for all parts of the world studied, the results are suggestive of a trend toward environments favouring more severe thunderstorms, but the small number of analyses precludes any likelihood estimate of this change.

12.4.6 Changes in Cryosphere

12.4.6.1 Changes in Sea Ice Cover

Based on the analysis of CMIP3 climate change simulations (e.g., Arzel et al., 2006; Zhang and Walsh, 2006), the AR4 concludes that the Arctic and Antarctic sea ice covers are projected to shrink in the 21st century under all SRES scenarios, with a large range of model responses (Meehl et al., 2007b). It also stresses that, in some projections, the Arctic Ocean becomes almost entirely ice-free in late summer during the second half of the 21st century. These conclusions were confirmed by further analyses of the CMIP3 archives (e.g., Stroeve et al., 2007; Bracegirdle et al., 2008; Lefebvre and Goosse, 2008; Boé et al., 2009b; Sen Gupta et al., 2009; Wang and Overland, 2009; Zhang, 2010b; NRC, 2011; Körper et al., 2013). Figures 12.28 and 12.29 and the studies of Maksym et al. (2012), Massonnet et al. (2012), Stroeve et al. (2012) and Wang and Overland (2012) show that the CMIP5 AOGCMs/ESMs as a group also project decreases in sea ice extent through the end of this century in both hemispheres under all RCPs. However, as in the case of CMIP3, the inter-model spread is considerable.

In the NH, in accordance with CMIP3 results, the absolute rate of decrease of the CMIP5 multi-model mean sea ice areal coverage is greatest in September. The reduction in sea ice extent between the time periods 1986–2005 and 2081–2100 for the CMIP5 multi-model average ranges from 8% for RCP2.6 to 34% for RCP8.5 in February and from 43% for RCP2.6 to 94% for RCP8.5 in September. *Medium confidence* is attached to these values as projections of sea ice extent decline in the real world due to errors in the simulation of present-day sea ice extent (mean and trends—see Section 9.4.3) and because of the large spread of model responses. About 90% of the available CMIP5 models reach nearly ice-free conditions (sea ice extent less than 1×10^6 km^2 for at least 5 consecutive years) during September in the Arctic before 2100 under RCP8.5 (about 45% under RCP4.5). By the end of the 21st century, the decrease in multi-model mean sea ice volume ranges from 29% for RCP2.6 to 73% for RCP8.5 in February and from 54% for RCP2.6 to 96% for RCP8.5 in September. *Medium confidence* is attached to these values as projections of the real world sea ice volume. In February, these percentages are much higher than the corresponding ones for sea ice extent, which is indicative of a substantial sea ice thinning.

A frequent criticism of the CMIP3 models is that, as a group, they strongly underestimate the rapid decline in summer Arctic sea ice extent observed during the past few decades (e.g., Stroeve et al., 2007; Winton, 2011), which suggests that the CMIP3 projections of summer Arctic sea ice areal coverage might be too conservative. As shown in Section 9.4.3 and Figure 12.28b, the magnitude of the CMIP5 multi-model mean trend in September Arctic sea ice extent over the satellite era is more consistent with, but still underestimates, the observed one (see also Massonnet et al., 2012; Stroeve et al., 2012; Wang and Overland, 2012; Overland and Wang, 2013). Owing to the shortness of the observational record, it is difficult to ascertain the relative influence of natural variability on this trend. This hinders the comparison between modelled and observed trends, and hence the estimate of the sensitivity of the September Arctic sea ice extent to global surface temperature change (i.e., the decrease in sea ice extent per degree global warming) (Kay et al., 2011; Winton, 2011; Mahlstein and Knutti, 2012). This sensitivity may be crucial for determining future sea ice losses. Indeed, a clear relationship exists at longer than decadal time scales in climate change simulations between the annual mean or September mean Arctic sea ice extent and the annual mean global surface temperature change for ice extents larger than ~1 × 10^6 km^2 (e.g., Ridley et al., 2007; Zhang, 2010b; NRC, 2011; Winton, 2011; Mahlstein and Knutti, 2012). This relationship is illustrated in Figure 12.30 for both CMIP3 and CMIP5 models. From this figure, it can be seen that the sea ice sensitivity varies significantly from model to model and is generally larger and in better agreement among models in CMIP5.

A complete and detailed explanation for what controls the range of Arctic sea ice responses in models over the 21st century remains elusive, but the Arctic sea ice provides an example where process-based constraints can be used to reduce the spread of model projections (Overland et al., 2011; Collins et al., 2012; Hodson et al., 2012). For CMIP3 models, results indicate that the changes in Arctic sea ice mass budget over the 21st century are related to the late 20th century mean sea ice thickness distribution (Holland et al., 2010), average sea ice thickness (Bitz, 2008; Hodson et al., 2012), fraction of thin ice cover (Boé et al., 2009b) and oceanic heat transport to the Arctic (Mahlstein and Knutti, 2011). For CMIP5 models, Massonnet et al. (2012) showed that the time needed for the September Arctic sea ice areal coverage to drop below a certain threshold is highly correlated with the September sea ice extent and annual mean sea ice volume averaged over the past several decades (Figure 12.31a, b). The timing of a seasonally ice-free Arctic Ocean or the fraction of remaining sea ice in September at any time during the 21st century were also found to correlate with the past trend in September Arctic sea ice extent and the amplitude of the mean seasonal cycle of sea ice extent (Boé et al., 2009b; Collins et al., 2012; Massonnet et al., 2012) (Figure 12.31c, d). All these empirical

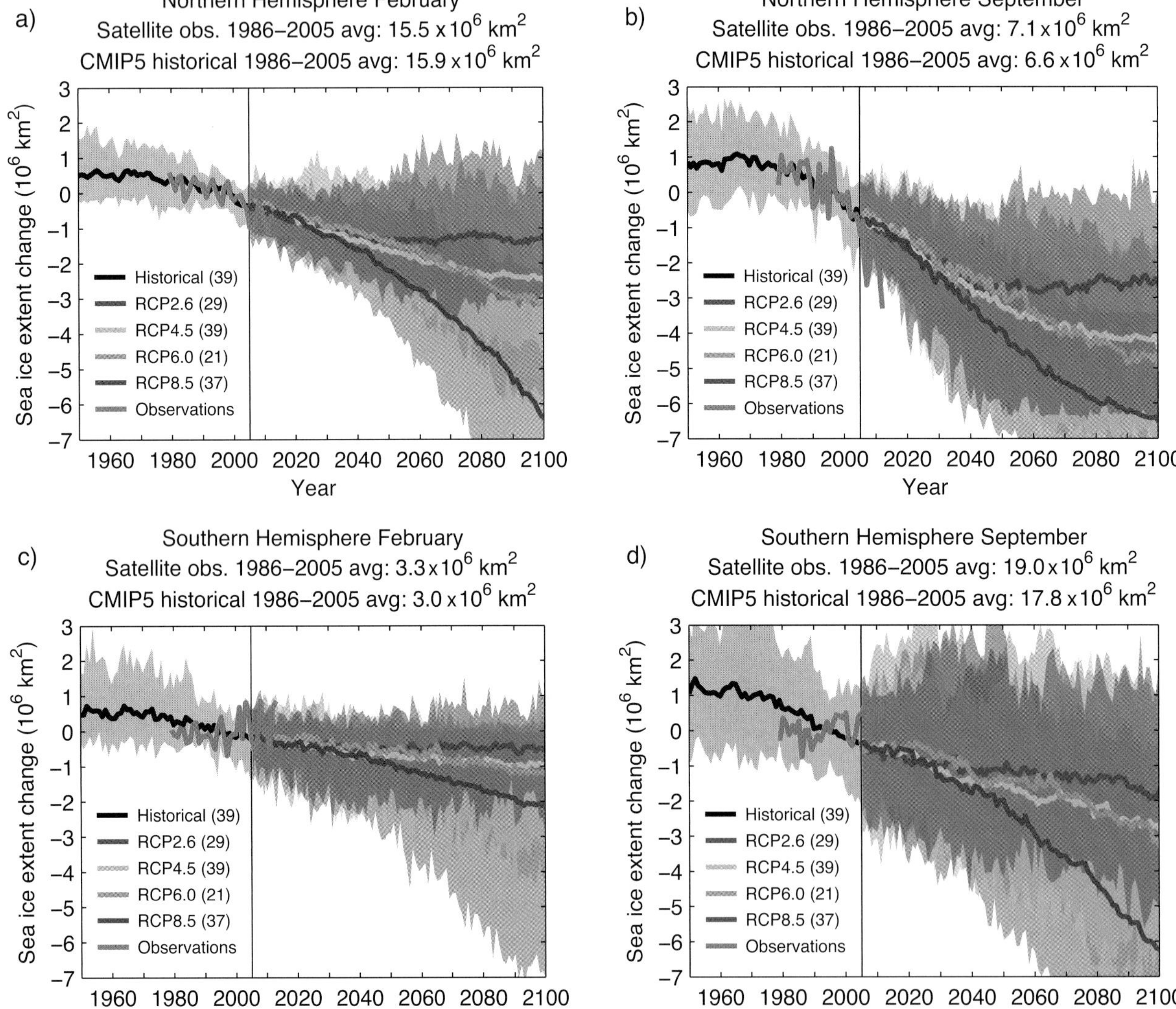

Figure 12.28 | Changes in sea ice extent as simulated by CMIP5 models over the second half of the 20th century and the whole 21st century under RCP2.6, RCP4.5, RCP6.0 and RCP8.5 for (a) Northern Hemisphere February, (b) Northern Hemisphere September, (c) Southern Hemisphere February and (d) Southern Hemisphere September. The solid curves show the multi-model means and the shading denotes the 5 to 95% range of the ensemble. The vertical line marks the end of CMIP5 historical climate change simulations. One ensemble member per model is taken into account in the analysis. Sea ice extent is defined as the total ocean area where sea ice concentration exceeds 15% and is calculated on the original model grids. Changes are relative to the reference period 1986–2005. The number of models available for each RCP is given in the legend. Also plotted (solid green curves) are the satellite data of Comiso and Nishio (2008, updated 2012) over 1979–2012.

relationships can be understood on simple physical grounds (see the aforementioned references for details).

These results lend support for weighting/recalibrating the models based on their present-day Arctic sea ice simulations. Today, the optimal approach for constraining sea ice projections from climate models is unclear, although one notes that these methods should have a credible underlying physical basis in order to increase confidence in their results (see Section 12.2). In addition, they should account for the potentially large imprint of natural variability on both observations and model simulations when these two sources of information are to be compared (see Section 9.8.3). This latter point is particularly critical if the past sea ice trend or sensitivity is used in performance metrics given the relatively short observational period (Kay et al., 2011; Overland et al., 2011; Mahlstein and Knutti, 2012; Massonnet et al., 2012; Stroeve et al., 2012). A number of studies have applied such metrics to the CMIP3 and CMIP5 models. Stroeve et al. (2007) and Stroeve et al. (2012) rejected several CMIP3 and CMIP5 models, respectively, on

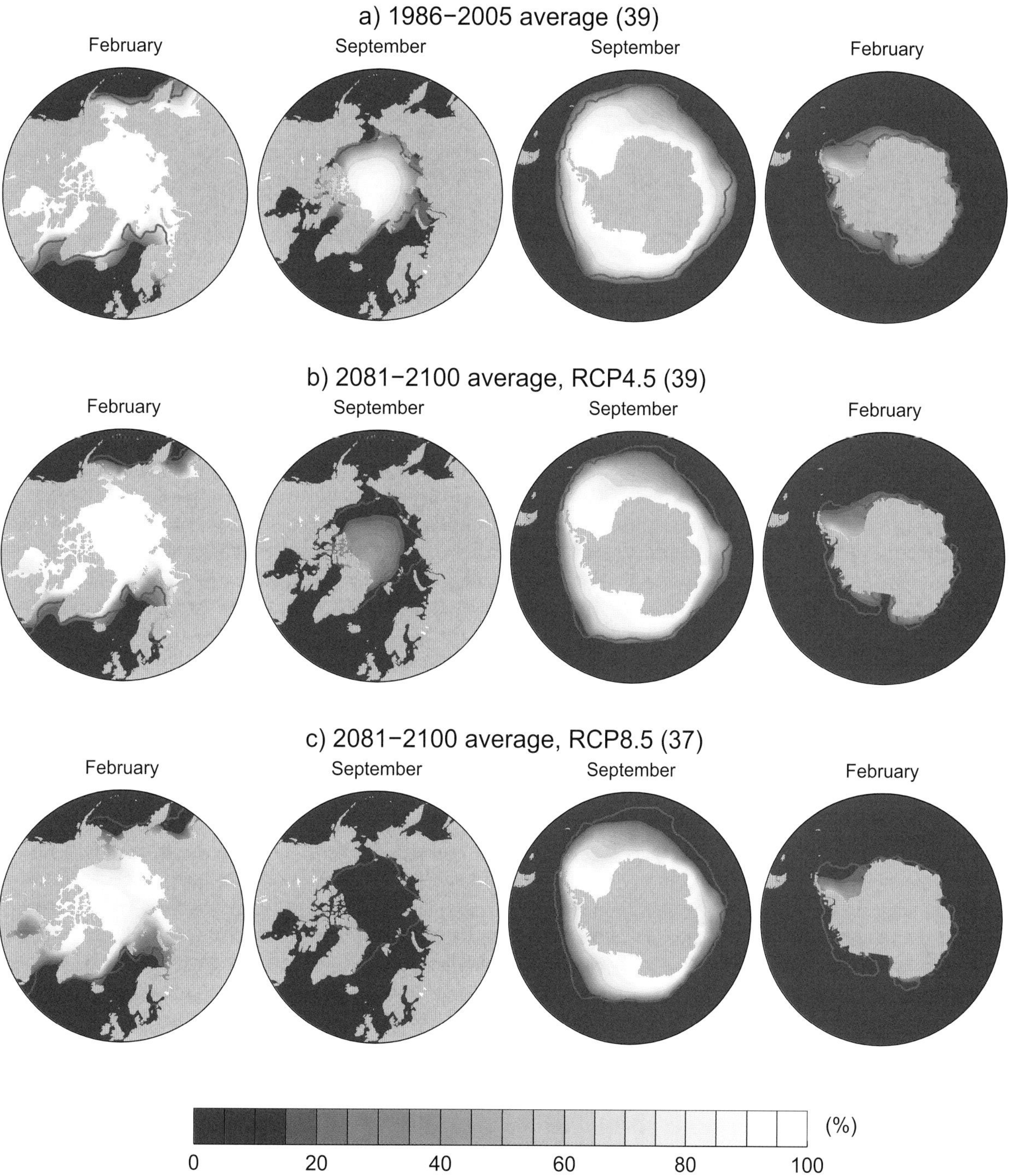

Figure 12.29 | February and September CMIP5 multi-model mean sea ice concentrations (%) in the Northern and Southern Hemispheres for the periods (a) 1986–2005, (b) 2081–2100 under RCP4.5 and (c) 2081–2100 under RCP8.5. The model sea ice concentrations are interpolated onto a 1° × 1° regular grid. One ensemble member per model is taken into account in the analysis, and the multi-model mean sea ice concentration is shown where it is larger than 15%. The number of models available for each RCP is given in parentheses. The pink lines indicate the observed 15% sea ice concentration limits averaged over 1986–2005 (Comiso and Nishio, 2008, updated 2012).

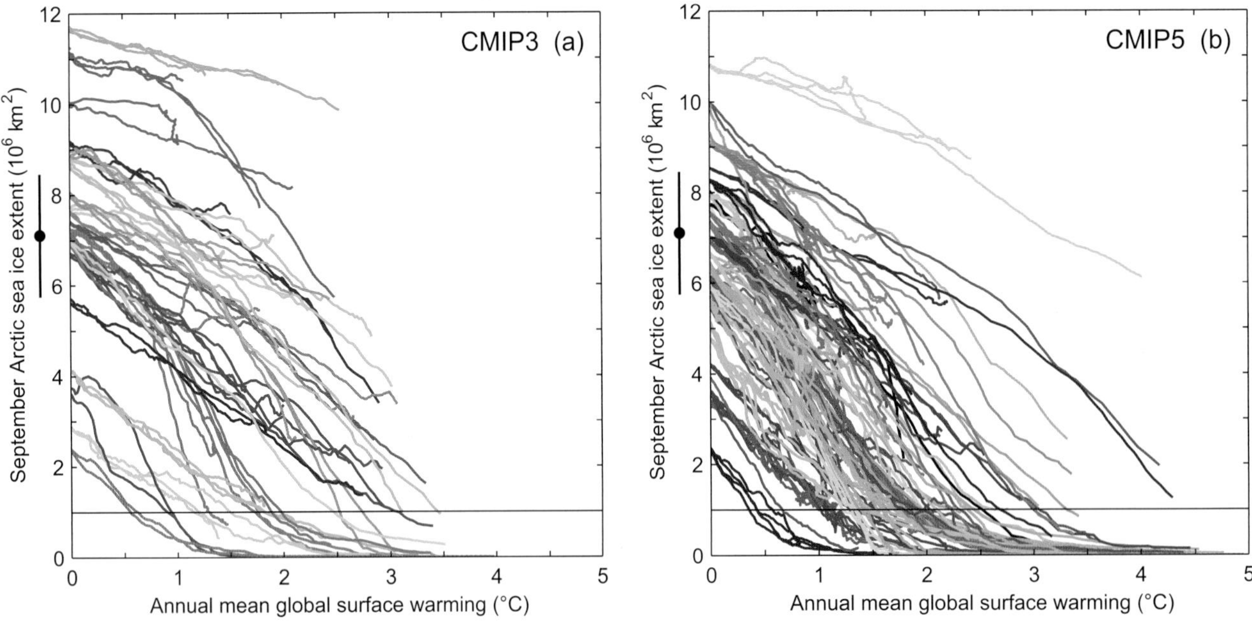

Figure 12.30 | September Arctic sea ice extent as a function of the annual mean global surface warming relative to the period 1986–2005 for (a) CMIP3 models (all SRES scenarios) and (b) CMIP5 models (all RCPs). The ice extents and global temperatures are computed on a common latitude-longitude grid for CMIP3 and on the original model grids for CMIP5. One ensemble member per model is taken into account in the analysis. A 21-year running mean is applied to the model output. The full black circle and vertical bar on the left-hand side of the *y*-axis indicate the mean and ±2 standard deviations about the mean of the observed September Arctic sea ice extent over 1986–2005, respectively (Comiso and Nishio, 2008, updated 2012). The horizontal line corresponds to a nearly ice-free Arctic Ocean in September.

the basis of their simulated late 20th century mean September Arctic sea ice extent. Wang and Overland (2009) selected a subset of CMIP3 models (and Wang and Overland (2012) did the same for the CMIP5 models) based on their fidelity to the observed mean seasonal cycle of Arctic sea ice extent in the late 20th century and then scaled the chosen models to the recently observed September sea ice extent. Zhang (2010b) retained a number of CMIP3 models based on the regression between summer sea ice loss and Arctic surface temperature change. Boé et al. (2009b) and Mahlstein and Knutti (2012) did not perform a model selection but rather recalibrated the CMIP3 Arctic sea ice projections on available observations of September Arctic sea ice trend and sensitivity to global surface temperature change, respectively. Finally, Massonnet et al. (2012) selected a subset of CMIP5 models on the basis of the four relationships illustrated in Figure 12.31a–d.

12

These various methods all suggest a faster rate of summer Arctic sea ice decline than the multi-model mean. Although they individually provide a reduced range for the year of near disappearance of the September Arctic sea ice compared to the original CMIP3/CMIP5 multi-model ensemble, they lead to different timings (Overland and Wang, 2013). Consequently, the time interval obtained when combining all these studies remains wide: 2020–2100+ (2100+ = not before 2100) for the SRES A1B scenario and RCP4.5 (Stroeve et al., 2007, 2012; Boé et al., 2009b; Wang and Overland, 2009, 2012; Zhang, 2010b; Massonnet et al., 2012) and 2020–2060 for RCP8.5 (Massonnet et al., 2012; Wang and Overland, 2012). The method proposed by Massonnet et al. (2012) is applied here to the full set of models that provided the CMIP5 database with sea ice output. The natural variability of each of the four diagnostics shown in Figure 12.31a–d is first estimated by averaging over all available models with more than one ensemble member the diagnostic standard deviations derived from the model ensemble members. Then, for each model, a ±2 standard deviation interval is constructed around the ensemble mean or single realization of the diagnostic considered. A model is retained if, for each diagnostic, either this interval overlaps a ±20% interval around the observed/reanalysed value of the diagnostic or at least one ensemble member from that model gives a value for the diagnostic that falls within ±20% of the observational/reanalysed data. The outcome is displayed in Figure 12.31e for RCP8.5. Among the five selected models (ACCESS1.0, ACCESS1.3, GFDL-CM3, IPSL-CM5A-MR, MPI-ESM-MR), four project a nearly ice-free Arctic Ocean in September before 2050 (2080) for RCP8.5 (RCP4.5), the earliest and latest years of near disappearance of the sea ice pack being about 2040 and about 2060 (about 2040 and 2100+), respectively. It should be mentioned that Maslowski et al. (2012) projected that it would take only until about 2016 to reach a nearly ice-free Arctic Ocean in summer, based on a linear extrapolation into the future of the recent sea ice volume trend from a hindcast simulation conducted with a regional model of the Arctic sea ice–ocean system. However, such an extrapolation approach is problematic as it ignores the negative feedbacks that can occur when the sea ice cover becomes thin (e.g., Bitz and Roe, 2004; Notz, 2009) and neglects the effect of year-to-year or longer-term variability (Overland and Wang, 2013). Mahlstein and Knutti (2012) encompassed the dependence of sea ice projections on the forcing scenario by determining the annual mean global surface warming threshold for nearly ice-free conditions in September. Their best estimate of ~2°C above the present derived from both CMIP3 models and observations is consistent with the 1.6 to 2.1°C range (mean value: 1.9°C) obtained from the CMIP5 model subset shown in Figure 12.31e (see also Figure 12.30b). The reduction in September Arctic sea ice extent by the end of the 21st century, averaged over this subset of models, ranges from 56% for RCP2.6 to 100% for RCP8.5.

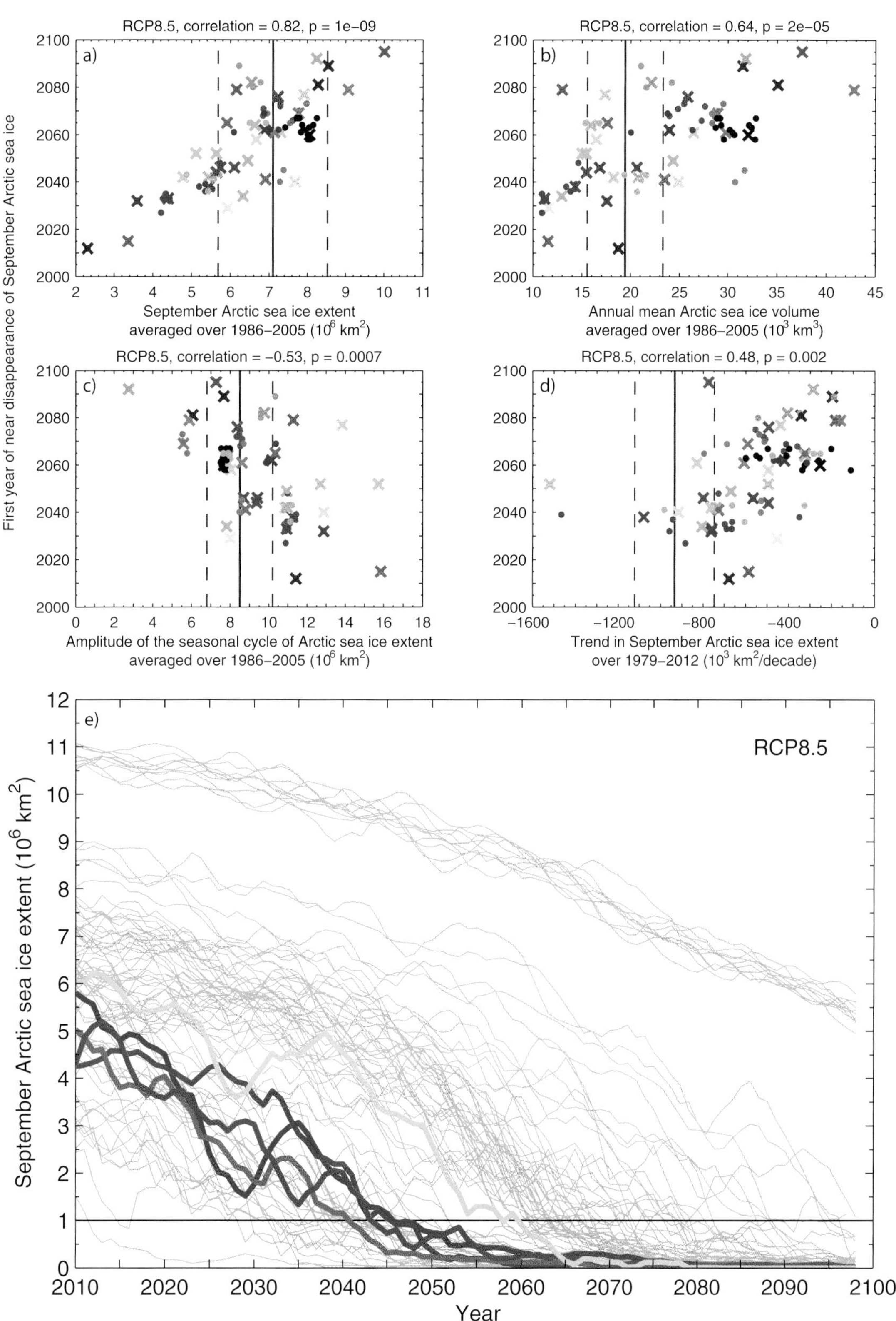

Figure 12.31 | (a–d) First year during which the September Arctic sea ice extent falls below 1 × 10^6 km^2 in CMIP5 climate projections (37 models, RCP8.5) as a function of (a) the September Arctic sea ice extent averaged over 1986–2005, (b) the annual mean Arctic sea ice volume averaged over 1986–2005, (c) the amplitude of the 1986–2005 mean seasonal cycle of Arctic sea ice extent and (d) the trend in September Arctic sea ice extent over 1979–2012. The sea ice diagnostics displayed are calculated on the original model grids. The correlations and one-tailed p-values are computed from the multi-member means for models with several ensemble members (coloured crosses), but the ensemble members of individual models are also depicted (coloured dots). The vertical solid and dashed lines show the corresponding observations or bias-adjusted PIOMAS (Pan-Arctic Ice-Ocean Modelling and Assimilation System) reanalysis data (a, c and d: Comiso and Nishio, 2008, updated 2012; b: Schweiger et al., 2011) and the ±20% interval around these data, respectively. (e) Time series of September Arctic sea ice extent (5-year running mean) as simulated by all CMIP5 models and their ensemble members under RCP8.5 (thin curves). The thick, coloured curves correspond to a subset of five CMIP5 models selected on the basis of panels a–d following Massonnet et al. (2012) (see text for details). Note that each of these models provides only one ensemble member for RCP8.5.

In light of all these results, it is *very likely* that the Arctic sea ice cover will continue to shrink and thin all year round during the 21st century as the annual mean global surface temperature rises. It is also *likely* that the Arctic Ocean will become nearly ice-free in September before the middle of the century for high GHG emissions such as those corresponding to RCP8.5 (*medium confidence*). The potential irreversibility of the Arctic sea ice loss and the possibility of an abrupt transition toward an ice-free Arctic Ocean are discussed in Section 12.5.5.7.

In the SH, the decrease in sea ice extent between 1986–2005 and 2081–2100 projected by the CMIP5 models as a group varies from 16% for RCP2.6 to 67% for RCP8.5 in February and from 8% to 30% in September. In contrast with the NH, the absolute rate of decline is greatest in wintertime. Eisenman et al. (2011) argue that this hemispheric asymmetry in the seasonality of sea ice loss is fundamentally related to the geometry of coastlines. For each forcing scenario, the relative changes in multi-model mean February and September Antarctic sea ice volumes by the end of the century are of the same order as the corresponding ones for sea ice extent. About 75% of the available CMIP5 models reach a nearly ice-free state in February within this century under RCP8.5 (about 60% under RCP4.5). For RCP8.5, only small portions of the Weddell and Ross Seas stay ice-covered in February during 2081–2100 in those models that do not project a seasonally ice-free Southern Ocean (see Figure 12.29c). Nonetheless, there is *low confidence* in these Antarctic sea ice projections because of the wide range of model responses and the inability of almost all of the models to reproduce the mean seasonal cycle, interannual variability and overall increase of the Antarctic sea ice areal coverage observed during the satellite era (see Section 9.4.3; Maksym et al., 2012; Turner et al., 2013; Zunz et al., 2013).

12.4.6.2 Changes in Snow Cover and Frozen Ground

12

Excluding ice sheets and glaciers, analyses of seasonal snow cover changes generally focus on the NH, where the configuration of the continents on the Earth induces a larger maximum seasonal snow cover extent (SCE) and a larger sensitivity of SCE to climate changes. Seasonal snow cover extent and snow water equivalent (SWE) respond to both temperature and precipitation. At the beginning and the end of the snow season, SCE decreases are closely linked to a shortening of the seasonal snow cover duration, while SWE is more sensitive to snowfall amount (Brown and Mote, 2009). Future widespread reductions of SCE, particularly in spring, are simulated by the CMIP3 models (Roesch, 2006; Brown and Mote, 2009) and confirmed by the CMIP5 ensemble (Brutel-Vuilmet et al., 2013). The NH spring (March-April average) snow cover area changes are coherent in the CMIP5 models although there is considerable scatter. Relative to the 1986–2005 reference period, the CMIP5 models simulate a weak decrease of about 7 ± 4% (one-σ inter-model dispersion) for RCP2.6 during the last two decades of the 21st century, while SCE decreases of about 13 ± 4% are simulated for RCP4.5, 15 ± 5% for RCP6.0, and 25 ± 8% for RCP8.5 (Figure 12.32). There is *medium confidence* in these numbers because of the considerable inter-model scatter mentioned above and because snow processes in global climate models are strongly simplified.

Projections for the change in annual maximum SWE are more mixed. Warming decreases SWE both by reducing the fraction of precipitation that falls as snow and by increasing snowmelt, but projected increases in precipitation over much of the northern high latitudes during winter months act to increase snow amounts. Whether snow covering the ground will become thicker or thinner depends on the balance between these competing factors. Both in the CMIP3 (Räisänen, 2008) and in the CMIP5 models (Brutel-Vuilmet et al., 2013), annual maximum SWE tends to increase or only marginally decrease in the coldest

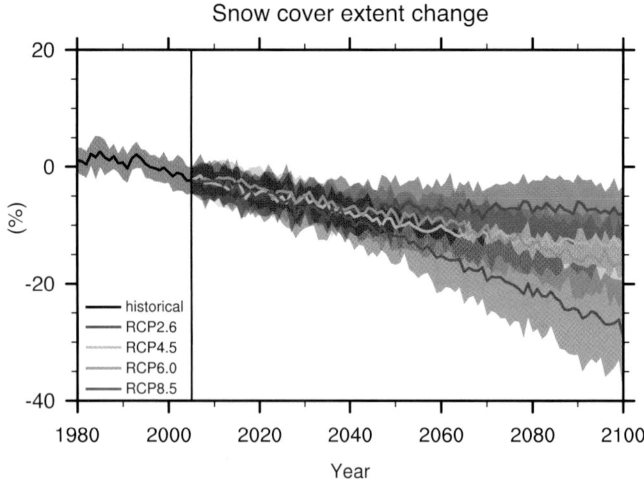

Figure 12.32 | Northern Hemisphere spring (March to April average) snow cover extent change (in %) in the CMIP5 ensemble, relative to the simulated extent for the 1986–2005 reference period. Thick lines mark the multi-model average, shading indicates the inter-model spread (one standard deviation). The observed March to April average snow cover extent for the 1986–2005 reference period is $32.6 \cdot 10^6$ km^2 (Brown and Robinson, 2011).

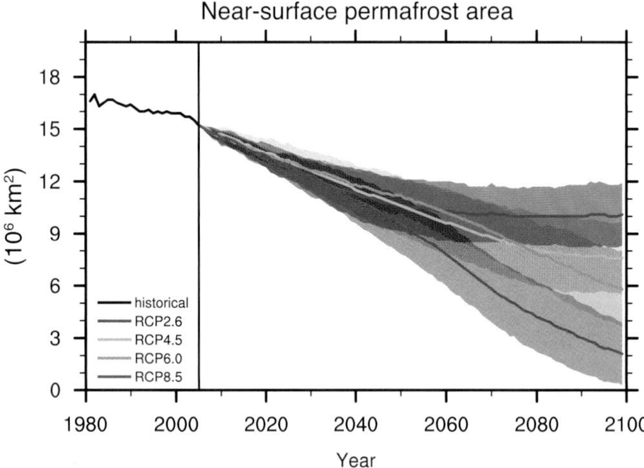

Figure 12.33 | Northern Hemisphere near-surface permafrost area, diagnosed for the available CMIP5 models by Slater and Lawrence (2013) following Nelson and Outcalt (1987) and using 20-year average bias-corrected monthly surface air temperatures and snow depths. Thick lines: multi-model average. Shading and thin lines indicate the inter-model spread (one standard deviation). The black line for the historical period is diagnosed from the average of the European Centre for Medium range Weather Forecast (ECMWF) reanalysis of the global atmosphere and surface conditions (ERA), Japanese ReAnalysis (JRA), Modern Era Retrospective-analysis for Research and Applications (MERRA) and Climate Forecast System Reanalysis and Reforecast (CFSRR) reanalyses (Slater and Lawrence, 2013). Estimated present permafrost extent is between 12 and 17 million km^2 (Zhang et al., 2000).

regions, while annual maximum SWE decreases are strong closer to the southern limit of the seasonally snow-covered area.

It is thus *very likely* (*high confidence*) that by the end of the 21st century, NH spring snow cover extent will be substantially lower than today if anthropogenic climate forcing is similar to the stronger scenarios considered here. Conversely, there is only *medium confidence* in the latitudinal pattern of annual maximum SWE changes (increase or little change in the coldest regions, stronger decrease further to the South) because annual maximum SWE is influenced by competing factors (earlier melt onset, higher solid precipitation rates in some regions).

The strong projected warming across the northern high latitudes in climate model simulations has implications for frozen ground. Recent projections of the extent of near-surface permafrost (see Glossary) degradation continue to vary widely depending on the underlying climate forcing scenario and model physics, but virtually all of them indicate substantial near-surface permafrost degradation and thaw depth deepening over much of the permafrost area (Saito et al., 2007; Lawrence et al., 2008a, 2012; Koven et al., 2011, 2013; Eliseev et al., 2013; Slater and Lawrence, 2013). Permafrost at greater depths is less directly relevant to the surface energy and water balance, and its degradation naturally occurs much more slowly (Delisle, 2007). Climate models are beginning to represent permafrost physical processes and properties more accurately (Alexeev et al., 2007; Nicolsky et al., 2007; Lawrence et al., 2008a; Rinke et al., 2008; Koven et al., 2009; Gouttevin et al., 2012), but there are large disagreements in the calculation of current frozen soil extent and active layer depth due to differences in the land model physics in the CMIP5 ensemble (Koven et al., 2013). The projected changes in permafrost are a response not only to warming, but also to changes in snow conditions because snow properties and their seasonal evolution exert significant control on soil thermal state (Zhang, 2005; Lawrence and Slater, 2010; Shkolnik et al., 2010; Koven et al., 2013). Applying the surface frost index method (Nelson and Outcalt, 1987) to coupled climate model anomalies from the CMIP5 models (Slater and Lawrence, 2013) yields a reduction of the diagnosed 2080–2099 near-surface permafrost area (continuous plus discontinuous near-surface permafrost) by 37 ± 11% (RCP2.6), 51 ± 13% (RCP4.5), 58 ± 13% (RCP6.0), and 81±12% (RCP8.5), compared to the 1986–2005 diagnosed near-surface permafrost area, with *medium confidence* in the numbers as such because of the strongly simplified soil physical processes in current-generation global climate models (Figure 12.33). The uncertainty range given here is the 1-σ inter-model dispersion. Applying directly the model output to diagnose permafrost extent and its changes over the 21st century yields similar relative changes (Koven et al., 2013). In summary, based on high agreement across CMIP5 and older model projections, fundamental process understanding, and paleoclimatic evidence (e.g., Vaks et al., 2013), it appears *virtually certain* (*high confidence*) that near-surface permafrost extent will shrink as global climate warms. However, the amplitude of the projected reductions of near-surface permafrost extent not only depends on the emission scenario and the global climate model response, but also very much on the permafrost-related soil processes taken into account in the models.

12.4.7 Changes in the Ocean

12.4.7.1 Sea Surface Temperature, Salinity and Ocean Heat Content

Projected increase of SST and heat content over the next two decades is relatively insensitive to the emissions trajectory. However, projected outcomes diverge as the 21st century progresses. When SSTs increase as a result of external forcing, the interior water masses respond to the integrated signal at the surface, which is then propagated down to greater depth (Gleckler et al., 2006; Gregory, 2010). Changes in globally averaged ocean heat content currently account for about 90% of the change in global energy inventory since 1970 (see Box 3.1). Heat is transported within the interior of the ocean by its large-scale general circulation and by smaller-scale mixing processes. Changes in transports lead to redistribution of existing heat content and can cause local cooling even though the global mean heat content is rising (Banks and Gregory, 2006; Lowe and Gregory, 2006; Xie and Vallis, 2012).

Figure 12.12 shows the multi-model mean projections of zonally averaged ocean temperature change under three emission scenarios. The differences in projected ocean temperature changes for different RCPs manifest themselves more markedly as the century progresses. The largest warming is found in the top few hundred metres of the subtropical gyres, similar to the observed pattern of ocean temperature changes (Levitus et al., 2012, see also Section 3.2). Surface warming varies considerably between the emission scenarios ranging from about 1°C (RCP2.6) to more than 3°C in RCP8.5. Mixing and advection processes gradually transfer the additional heat to deeper levels of about 2000 m at the end of the 21st century. Depending on the emission scenario, global ocean warming between 0.5°C (RCP2.6) and 1.5°C (RCP8.5) will reach a depth of about 1 km by the end of the century. The strongest warming signal is found at the surface in subtropical and tropical regions. At depth the warming is most pronounced in the Southern Ocean. From an energy point of view, for RCP4.5 by the end of the 21st century, half of the energy taken up by the ocean is in the uppermost 700 m, and 85% is in the uppermost 2000 m.

In addition to the upper-level warming, the patterns are further characterized by a slight cooling in parts of the northern mid- and high latitudes below 1000 m and a pronounced heat uptake in the deep Southern Ocean at the end of the 21st century. The cooling may be linked to the projected decrease of the strength of the AMOC (see Section 12.4.7.2; 13.4.1; Banks and Gregory, 2006).

The response of ocean temperatures to external forcing comprises mainly two time scales: a relatively fast adjustment of the ocean mixed layer and the slow response of the deep ocean (Hansen et al., 1985; Knutti et al., 2008a; Held et al., 2010). Simulations with coupled ocean–atmosphere GCMs suggest time-scales of several millennia until the deep ocean is in equilibrium with the external forcing (Stouffer, 2004; Hansen et al., 2011; Li et al., 2013a). Thus, the long time-scale of the ocean response to external forcing implies an additional commitment to warming for many centuries when GHG emissions are decreased or concentrations kept constant (see Section 12.5.2). Further assessment of ocean heat uptake and its relationship to projections of sea level rise is presented in Section 13.4.1.

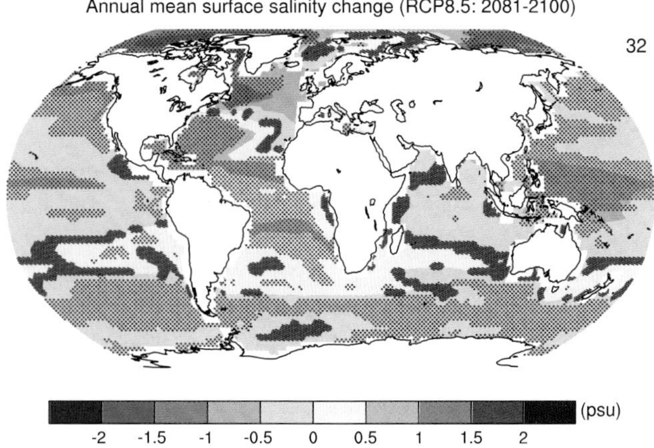

Figure 12.34 | Projected sea surface salinity differences 2081–2100 for RCP8.5 relative to 1986–2005 from CMIP5 models. Hatching indicates regions where the multi-model mean change is less than one standard deviation of internal variability. Stippling indicates regions where the multi-model mean change is greater than two standard deviations of internal variability and where at least 90% of the models agree on the sign of change (see Box 12.1). The number of CMIP5 models used is indicated in the upper right corner.

Durack and Wijffels (2010) and Durack et al. (2012) examined trends in global sea surface salinity (SSS) changes over the period 1950–2008. Their analysis revealed strong, spatially coherent trends in SSS over much of the global ocean, with a pattern that bears striking resemblance to the climatological SSS field and is associated with an intensification of the global water cycle (see Sections 3.3.2.1, 10.4.2 and 12.4.5). The CMIP5 climate model projections available suggest that high SSS subtropical regions that are dominated by net evaporation are typically getting more saline; lower SSS regions at high latitudes are typically getting fresher. They also suggest a continuation of this trend in the Atlantic where subtropical surface waters become more saline as the century progresses (Figure 12.34) (see also Terray et al., 2012). At the same time, the North Pacific is projected to become less saline.

12.4.7.2 Atlantic Meridional Overturning

Almost all climate model projections reveal an increase of high latitude temperature and high latitude precipitation (Meehl et al., 2007b). Both of these effects tend to make the high latitude surface waters lighter and hence increase their stability. As seen in Figure 12.35, all models show a weakening of the AMOC over the course of the 21st century (see Section 12.5.5.2 for further analysis). Projected changes in the strength of the AMOC at high latitudes appear stronger in Geophysical Fluid Dynamics Laboratory (GFDL) CM2.1 when density is used as a vertical coordinate instead of depth (Zhang, 2010a). Once the RF is stabilized, the AMOC recovers, but in some models to less than its pre-industrial level. The recovery may include a significant overshoot (i.e., a weaker circulation may persist) if the anthropogenic RF is eliminated (Wu et al., 2011a). Gregory et al. (2005) found that for all eleven models

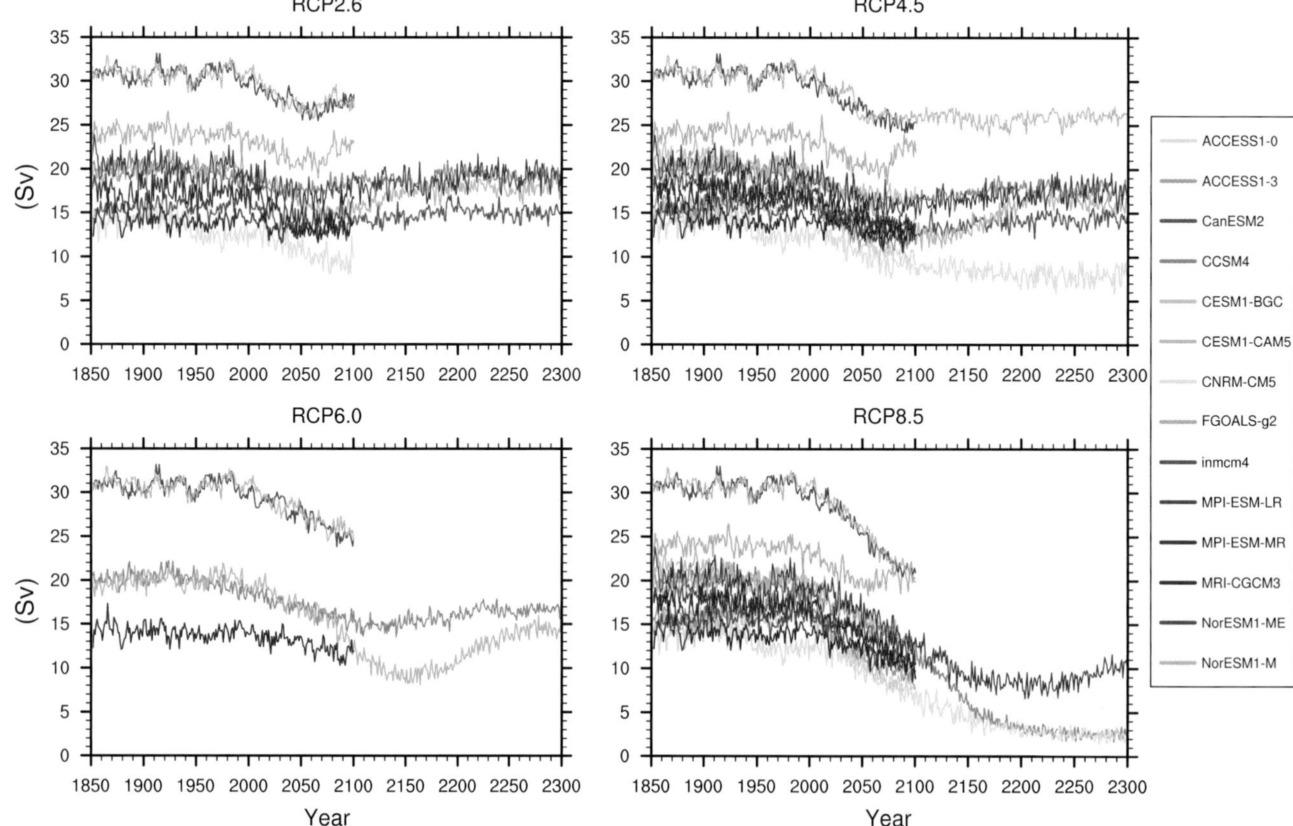

Figure 12.35 | Multi-model projections of Atlantic Meridional Overturning Circulation (AMOC) strength at 30°N from 1850 through to the end of the RCP extensions. Results are based on a small number of CMIP5 models available. Curves show results from only the first member of the submitted ensemble of experiments.

analysed (six from CMIP2/3 and five EMICs), the AMOC reduction was caused more by changes in surface heat flux than changes in surface freshwater flux. They further found that models with a stronger AMOC in their control run exhibited a larger weakening (see also Gregory and Tailleux, 2011).

Based on the assessment of the CMIP5 RCP simulations and on our understanding gleaned from analysis of CMIP3 models, observations and our understanding of physical mechanisms, it is *very likely* that the AMOC will weaken over the 21st century. Best estimates and ranges for the reduction from CMIP5 are 11% (1 to 24%) in RCP2.6 and 34% (12 to 54%) in RCP8.5. There is *low confidence* in assessing the evolution of the AMOC beyond the 21st century.

12.4.7.3 Southern Ocean

A dominant and robust feature of the CMIP3 climate projections assessed in AR4 is the weaker surface warming at the end of the 21st century in the Southern Ocean area compared to the global mean. Furthermore, the Antarctic Circumpolar Current (ACC) moves southward in most of the climate projections analysed in response to the simulated southward shift and strengthening of the SH mid-latitude westerlies (Meehl et al., 2007b).

The additional analyses of the CMIP3 model output performed since the release of AR4 confirm and refine the earlier findings. The displacement and intensification of the mid-latitude westerlies contribute to a large warming between 40°S and 60°S from the surface to mid-depths (Fyfe et al., 2007; Sen Gupta et al., 2009). Part of this warming has been attributed to the southward translation of the Southern Ocean current system (Sen Gupta et al., 2009). Moreover, the wind changes influence the surface temperature through modifications of the latent and sensible heat fluxes and force a larger northward Ekman transport of relatively cold polar surface water (Screen et al., 2010). This also leads to a stronger upwelling that brings southward and upward relatively warm and salty deep water, resulting in a subsurface salinity increase at mid-depths south of 50°S (Sen Gupta et al., 2009; Screen et al., 2010).

Overall, CMIP3 climate projections exhibit a decrease in mixed layer depth at southern mid- and high latitudes by the end of the 21st century. This feature is a consequence of the enhanced stratification resulting from surface warming and freshening (Lefebvre and Goosse, 2008; Sen Gupta et al., 2009; Capotondi et al., 2012). Despite large inter-model differences, there is a robust weakening of Antarctic Bottom Water production and its northward outflow, which is consistent with the decrease in surface density and is manifest as a warming signal close to the Antarctic margin that reaches abyssal depths (Sen Gupta et al., 2009).

In the vicinity of the Antarctic ice sheet, CMIP3 models project an average warming of ~0.5C° at depths of 200–500 m in 2091–2100 compared to 1991–2000 for the SRES A1B scenario, which has the potential to impact the mass balance of ice shelves (Yin et al., 2011). More detailed regional modelling using the SRES A1B scenario indicates that a redirection of the coastal current into the cavities underlying the Filchner-Ronne ice shelf during the second half of the 21st century might enhance the average basal melting rate there from 0.2 m yr^{-1} to almost 4 m yr^{-1} (Hellmer et al., 2012; see Section 13.4.4.2).

There are very few published analyses of CMIP5 climate projections focusing on the Southern Ocean. Meijers et al. (2012) found a wide variety of ACC responses to climate warming scenarios across CMIP5 models. Models show a high correlation between the changes in ACC strength and position, with a southward (northward) shift of the ACC core as the ACC gets stronger (weaker). No clear relationship between future changes in wind stress and ACC strength was identified, while the weakening of the ACC transport simulated at the end of the 21st century by many models was found to correlate with the strong decrease in the surface heat and freshwater fluxes in the ACC region (Meijers et al., 2012; Downes and Hogg, 2013). In agreement with the CMIP3 assessment (Sen Gupta et al., 2009), subtropical gyres generally strengthen under RCP4.5 and RCP8.5 and all expand southward, inducing a southward shift of the northern boundary of the ACC at most longitudes in the majority of CMIP5 models (Meijers et al., 2012). As in CMIP3 climate projections, an overall shallowing of the deep mixed layers that develop on the northern edge of the ACC in winter is observed, with larger shallowing simulated by models with deeper mixed layers during 1976–2005 (Sallée et al., 2013a). Sallée et al. (2013b) reported a warming of all mode, intermediate and deep water masses in the Southern Ocean. The largest temperature increase is found in mode and intermediate water layers. Consistently with CMIP3 projections (Downes et al., 2010), these water layers experience a freshening, whereas bottom water becomes slightly saltier. Finally, Sallée et al. (2013b) noted an enhanced upwelling of circumpolar deep water and an increased subduction of intermediate water that are nearly balanced by interior processes (diapycnal fluxes).

A number of studies suggest that oceanic mesoscale eddies might influence the response of the Southern Ocean circulation, meridional heat transport and deep water formation to changes in wind stress and surface buoyancy flux (Böning et al., 2008; Farneti et al., 2010; Downes et al., 2011; Farneti and Gent, 2011; Saenko et al., 2012; Spence et al., 2012). These eddies are not explicitly resolved in climate models and their role in future circulation changes still needs to be precisely quantified. Some of the CMIP5 models have output the meridional overturning due to the Eulerian mean circulation and that induced by parameterized eddies, thus providing a quantitative estimate of the role of the mesoscale circulation in a warming climate. On this basis, Downes and Hogg (2013) found that, under RCP8.5, the strengthening (weakening) of the upper (lower) Eulerian mean meridional overturning cell in the Southern Ocean is significantly correlated with the increased overlying wind stress and surface warming and is partly compensated at best by changes in eddy-induced overturning.

None of the CMIP3 and CMIP5 models include an interactive ice sheet component. When climate–ice sheet interactions are accounted for in an EMIC under a 4 × CO_2 scenario, the meltwater flux from the Antarctic ice sheet further reduces the surface density close to Antarctica and the rate of Antarctic Bottom Water formation. This ultimately results in a smaller surface warming at high southern latitudes compared to a simulation in which the freshwater flux from the melting ice sheet is not taken into account (Swingedouw et al., 2008). Nevertheless, in this study, this effect becomes significant only after more than one century.

12.4.8 Changes Associated with Carbon Cycle Feedbacks and Vegetation Cover

Climate change may affect the global biogeochemical cycles changing the magnitude of the natural sources and sinks of major GHGs. Numerous studies investigated the interactions between climate change and the carbon cycle (e.g., Friedlingstein et al., 2006), methane cycle (e.g., O'Connor et al., 2010), ozone (Cionni et al., 2011) or aerosols (e.g., Carslaw et al., 2010). Many CMIP5 ESMs now include a representation of the carbon cycle as well as atmospheric chemistry, allowing interactive projections of GHGs (mainly CO_2 and O_3) and aerosols. With such models, projections account for the imposed changes in anthropogenic emissions, but also for changes in natural sources and sinks as they respond to changes in climate and atmospheric composition. If included in ESMs, the impact on projected concentration, RF and hence on climate can be quantified. Climate-induced changes on the carbon cycle are assessed below, while changes in natural emissions of CH_4 are assessed in Chapter 6, changes in atmospheric chemistry in Chapter 11, and climate–aerosol interactions are assessed in Chapter 7.

12.4.8.1 Carbon Dioxide

As presented in Section 12.3, the CMIP5 experimental design includes, for the RCP8.5 scenario, experiments driven either by prescribed anthropogenic CO_2 emissions or concentration. The historical and 21st century emission-driven simulations allow evaluating the climate response of the Earth system when atmospheric CO_2 and the climate response are interactively being calculated by the ESMs. In such ESMs, the atmospheric CO_2 is calculated as the difference between the imposed anthropogenic emissions and the sum of land and ocean carbon uptakes. As most of these ESMs account for land use changes and their CO_2 emissions, the only external forcing is fossil fuel CO_2 emissions (along with all non-CO_2 forcings as in the C-driven RCP8.5 simulations). For a given ESM, the emission driven and concentration driven simulations would show different climate projections if the simulated atmospheric CO_2 in the emission driven run is significantly different from the one prescribed for the concentration driven runs. This would happen if the ESMs carbon cycle is different from the one simulated by MAGICC6, the model used to calculate the CMIP5 GHGs concentrations from the emissions for the four RCPs (Meinshausen et al., 2011c). When driven by CO_2 concentration, the ESMs can calculate the fossil fuel CO_2 emissions that would be compatible with the prescribed atmospheric CO_2 trajectory, allowing comparison with the set of CO_2 emissions initially estimated by the IAMs (Arora et al., 2011; Jones et al., 2013) (see Section 6.4.3, Box 6.4).

Figure 12.36 shows the simulated atmospheric CO_2 and global average surface air temperature warming (relative to the 1986–2005 reference period) for the RCP8.5 emission driven simulations from the CMIP5 ESMs, compared to the concentration driven simulations from the same models. Most (seven out of eleven) of the models estimate a larger CO_2 concentration than the prescribed one. By 2100, the multi-model average CO_2 concentration is 985 ± 97 ppm (full range 794 to 1142 ppm), while the CO_2 concentration prescribed for the RCP8.5 is 936 ppm. Figure 12.36 also shows the range of atmospheric CO_2 projections when the MAGICC6 model, used to provide the RCP concentrations, is tuned to emulate combinations of climate sensitivity uncertainty taken from 19 CMIP3 models and carbon cycle feedbacks uncertainty taken from 10 C^{4}MIP models, generating 190 model simulations (Meinshausen et al., 2011c; Meinshausen et al., 2011b). The emulation of the CMIP3/C^{4}MIP models shows for the RCP8.5, a range of simulated CO_2 concentrations of 794 to 1149 ppm (90% confidence level), extremely similar to what is obtained with the CMIP5 ESMs, with atmospheric concentration as high as 1150 ppm by 2100, that is, more than 200 ppm above the prescribed CO_2 concentration.

Global warming simulated by the E-driven runs show higher upper ends than when atmospheric CO_2 concentration is prescribed. For the models assessed here, the global surface temperature change (2081–2100 average relative to 1986–2005 average) ranges between 2.6°C and 4.7°C, with a multi-model average of 3.7°C ± 0.7°C for the concentration driven simulations, while the emission driven simulations give a range of 2.5°C to 5.6°C, with a multi-model average of 3.9°C ± 0.9°C, that is, 5% larger than for the concentration driven runs. The models that simulate the largest CO_2 concentration by 2100 have the largest warming amplification in the emission driven simulations, with an additional warming of more than 0.5°C.

The uncertainty on the carbon cycle has been shown to be of comparable magnitude to the uncertainty arising from physical climate processes (Gregory et al., 2009). Huntingford et al. (2009) used a simple model to characterize the relative role of carbon cycle and climate sensitivity uncertainties in contributing to the range of future temperature changes, concluding that the range of carbon cycle processes represent about 40% of the physical feedbacks. Perturbed parameter ensembles systematically explore land carbon cycle parameter uncertainty and illustrate that a wide range of carbon cycle responses are consistent with the same underlying model structures and plausible parameter ranges (Booth et al., 2012; Lambert et al., 2012). Figure 12.37 shows how the comparable range of future climate change (SRES A1B) arises from parametric uncertainty in land carbon cycle and atmospheric feedbacks. The same ensemble shows that the range of atmospheric CO_2 in the land carbon cycle ensemble is wider than the full SRES concentration range (B1 to A1FI scenario).

The CMIP5 ESMs described above do not include the positive feedback arising from the carbon release from high latitudes permafrost thawing under a warming scenario, which could further increase the atmospheric CO_2 concentration and the warming. Two recent studies investigated the climate–permafrost feedback from simulations with models of intermediate complexity (EMICs) that accounts for a permafrost carbon module (MacDougall et al., 2012; Schneider von Deimling et al., 2012). Burke et al. (2012) also estimated carbon loss from permafrost, from a diagnostic of the present-day permafrost carbon store and future soil warming as simulated by CMIP5 models. However, this last study did not quantify the effect on global temperature. Each of these studies found that the range of additional warming due to the permafrost carbon loss is quite large, because of uncertainties in future high latitude soil warming, amount of carbon stored in permafrost soils, vulnerability of freshly thawed organic material, the proportion of soil carbon that might be emitted as carbon dioxide via aerobic decomposition or as methane via anaerobic decomposition (Schneider von Deimling et al., 2012). For the RCP8.5, the additional warming from permafrost ranges between 0.04°C and 0.69°C by 2100 although

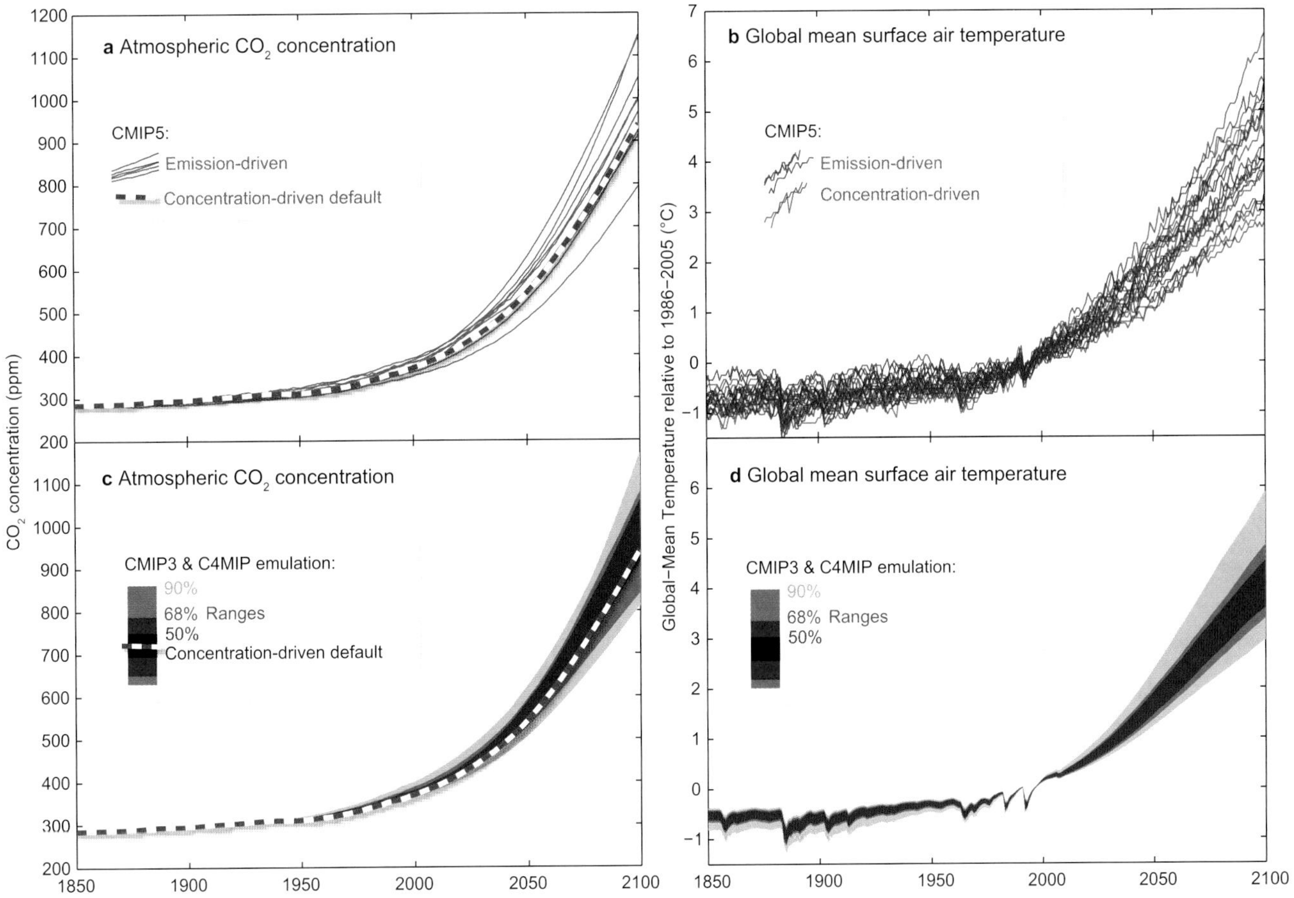

Figure 12.36 | Simulated changes in (a) atmospheric CO_2 concentration and (b) global averaged surface temperature (°C) as calculated by the CMIP5 Earth System Models (ESMs) for the RCP8.5 scenario when CO_2 emissions are prescribed to the ESMs as external forcing (blue). Also shown (b, in red) is the simulated warming from the same ESMs when directly forced by atmospheric CO_2 concentration (a, red white line). Panels (c) and (d) show the range of CO_2 concentrations and global average surface temperature change simulated by the Model for the Assessment of Greenhouse Gas-Induced Climate Change 6 (MAGICC6) simple climate model when emulating the CMIP3 models climate sensitivity range and the Coupled Climate Carbon Cycle Model Intercomparison Project (C^4MIP) models carbon cycle feedbacks. The default line in (c) is identical to the one in (a).

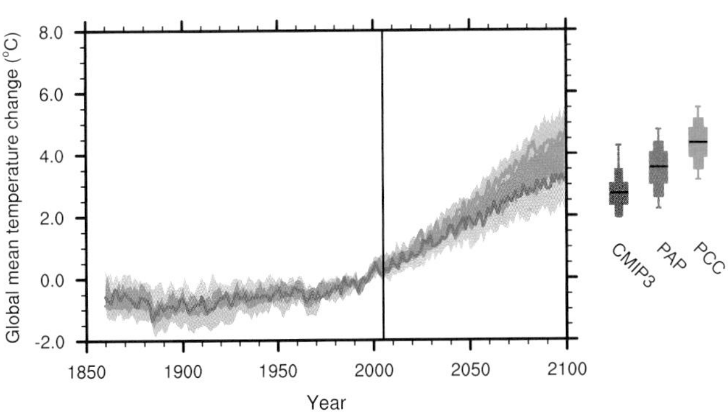

Figure 12.37 | Uncertainty in global mean temperature from Met Office Hadley Centre climate prediction model 3 (HadCM3) results exploring atmospheric physics and terrestrial carbon cycle parameter perturbations under the SRES A1B scenario (Murphy et al., 2004; Booth et al., 2012). Relative uncertainties in the Perturbed Carbon Cycle (PCC, green plume) and Perturbed Atmospheric Processes (PAP, blue plume) on global mean anomalies of temperature (relative to the 1986–2005 period). The standard simulations from the two ensembles, HadCM3 (blue solid) and HadCM3C (green solid) are also shown. Three bars are shown on the right illustrating the 2100 temperature anomalies associated with the CMIP3/AR4 ensemble (black) the PAP ensemble (blue) and PCC ensemble (green). The ranges indicate the full range, 10th to 90th, 25th to 75th and 50th percentiles.

there is *medium confidence* in these numbers as are the ones on the amount of carbon released (see Section 12.5.5.4) (MacDougall et al., 2012; Schneider von Deimling et al., 2012).

12.4.8.2 Changes in Vegetation Cover

Vegetation cover can also be affected by climate change, with forest cover potentially being decreasing (e.g., in the tropics) or increasing (e.g., in high latitudes). In particular, the Amazon forest has been the subject of several studies, generally agreeing that future climate change would increase the risk tropical Amazon forest being replaced by seasonal forest or even savannah (Huntingford et al., 2008; Jones et al., 2009; Malhi et al., 2009). Increase in atmospheric CO_2 would partly reduce such risk, through increase in water efficiency under elevated CO_2 (Lapola et al., 2009; Malhi et al., 2009). Recent multi-model estimates based on different CMIP3 climate scenarios and different dynamic global vegetation models predict a moderate risk of tropical forest reduction in South America and even lower risk for African and Asian tropical forests (see also Section 12.5.5.6) (Gumpenberger et al., 2010; Huntingford et al., 2013).

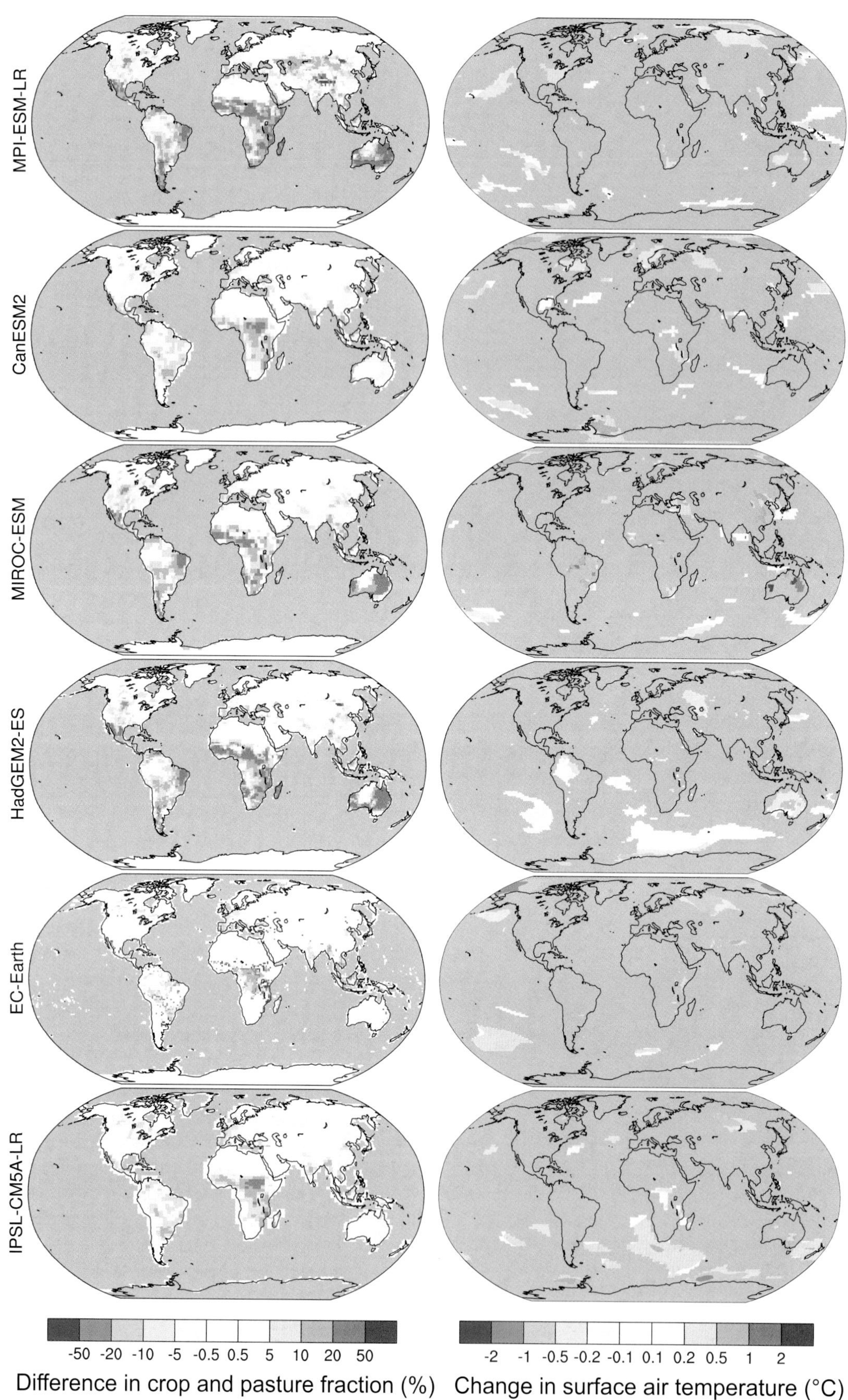

Figure 12.38 | Impact of land use change on surface temperature. LUCID-CMIP5 experiments where six ESMs were forced either with or without land use change beyond 2005 under the RCP8.5 scenario. Left maps of changes in total crop and pasture fraction (%) in the RCP8.5 simulations between 2006 and 2100 as implemented in each ESM. Right maps show the differences in surface air temperature (averaged over the 2071–2100 period) between the simulations with and without land use change beyond 2005. Only statistically significant changes ($p < 0.05$) are shown.

ESMs simulations with interactive vegetation confirmed known biophysical feedback associated with large-scale changes in vegetation. In the northern high latitudes, warming-induced vegetation expansion reduces surface albedo, enhancing the warming over these regions (Falloon et al., 2012; Port et al., 2012), with potentially larger amplification due to ocean and sea ice response (Swann et al., 2010). Over tropical forest, reduction of forest coverage would reduce evapotranspiration, also leading to a regional warming (Falloon et al., 2012; Port et al., 2012).

CMIP5 ESMs also include human induced land cover changes (deforestation, reforestation) affecting the climate system through changes in land surface physical properties (Hurtt et al., 2011). Future changes in land cover will have an impact on the climate system through biophysical and biogeochemical processes (e.g., Pongratz et al., 2010). Biophysical processes include changes in surface albedo and changes in partitioning between latent and sensible heat, while biogeochemical feedbacks essentially include change in CO_2 sources and sinks but could potentially also include changes in N_2O or CH_4 emissions. The biophysical response to future land cover changes has been investigated within the SRES scenarios. Using the SRES A2 2100 land cover, Davin et al. (2007) simulated a global cooling of 0.14 K relatively to a simulation with present-day land cover, the cooling being largely driven by change in albedo. Regional analyses have been performed in order to quantify the biophysical impact of biofuels plantation generally finding a local to regional cooling when annual crops are replaced by bioenergy crops, such as sugar cane (Georgescu et al., 2011; Loarie et al., 2011). However, some energy crops require nitrogen inputs for their production, leading inevitably to nitrous oxide (N_2O) emissions, potentially reducing the direct cooling effect and the benefit of biofuels as an alternative to fossil fuel emissions. Such emission estimates are still uncertain, varying strongly for different crops, management methods, soil types and reference systems (St. Clair et al., 2008; Smeets et al., 2009).

In the context of the Land-Use and Climate, IDentification of robust impacts (LUCID) project (Pitman et al., 2009) ESMs performed additional CMIP5 simulations in order to separate the biophysical from the biogeochemical effects of land use changes in the RCP scenarios. The LUCID–CMIP5 experiments were designed to complement RCP8.5 and RCP2.6 simulations of CMIP5, both of which showing an intensification of land use change over the 21st century. The LUCID–CMIP5 analysis was focussed on a difference in climate and land-atmosphere fluxes between the average of ensemble of simulations with and without land use changes by the end of 21st century (Brovkin et al., 2013). Due to different interpretation of land use classes, areas of crops and pastures were specific for each ESM (Figure 12.38, left). On the global scale, simulated biophysical effects of land use changes projected in the CMIP5 experiments with prescribed CO_2 concentrations were not significant. However, these effects were significant for regions with land use changes >10%. Only three out of six participating models, CanESM2, HadGEM2-ES and MIROC-ESM, reveal statistically significant changes in regional mean annual mean surface air temperature for the RCP8.5 scenario (Figure 12.38, right). However, there is *low confidence* on the overall effect as there is no agreement among the models on the sign of the global average temperature change due to the biophysical effects of land use changes (Brovkin et al., 2013). Changes in land surface albedo, available energy, latent and sensible heat fluxes were relatively small but significant in most of ESMs for regions with substantial land use changes. The scale of climatic effects reflects a small magnitude of land use changes in both the RCP2.6 and 8.5 scenarios and their limitation mainly to the tropical and subtropical regions where differences between biophysical effects of forests and grasslands are less pronounced than in mid- and high latitudes. LUCID-CMIP5 did not perform similar simulations for the RCP4.5 or RCP6.0 scenarios. As these two scenarios show a global decrease of land use area, one might expect their climatic impact to be different from the one seen in the RC2.6 and RCP8.5.

12.4.9 Consistency and Main Differences Between Coupled Model Intercomparison Project Phase 3/ Coupled Model Intercomparison Project Phase 5 and Special Report on Emission Scenarios/ Representative Concentration Pathways

In the experiments collected under CMIP5, both models and scenario have changed with respect to CMIP3 making a comparison with earlier results and the scientific literature they generated (on which some of this chapter's content is still based) complex. The set of models used in AR4 (the CMIP3 models) have been superseded by the new CMIP5 models (Table 12.1; Chapter 9) and the SRES scenarios have been replaced by four RCPs (Section 12.3.1). In addition, the baseline period used to compute anomalies has advanced 6 years, from 1980–1999 to 1986–2005.

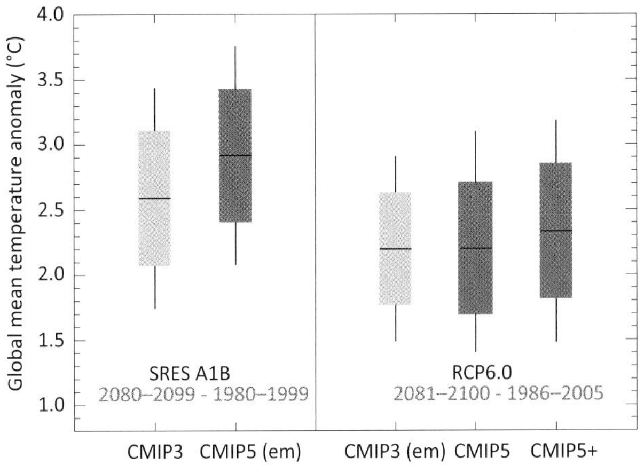

Figure 12.39 | Global mean temperature anomalies at the end of the 21st century from General Circulation Model (GCM) experiments and emulators comparing CMIP3/ CMIP5 responses under SRES A1B and RCP6.0. The boxes and whiskers indicate the 5th percentile, mean value – 1 standard deviation, mean, mean value + 1 standard deviation and 95th percentile of the distributions. The first box-and-whiskers on the left is computed directly from the CMIP3 ensemble and corresponds to the numbers quoted in AR4. The emulated SRES A1B projections (second from left) of CMIP5 are obtained by the method of Good et al. (2011a) and are calculated for the period 2080-2099 expressed with respect to the AR4 baseline period of 1980–1999. Because of the method, the subset of CMIP5 that are emulated are restricted to those with pre-industrial control, abrupt 4 × CO_2, historical, RCP4.5 and RCP8.5 simulations. The emulated RCP6.0 projections of CMIP3 (third from left, see also Figure 12.8) are from Knutti and Sedláček (2013) obtained using the method of Meinshausen et al. (2011b; 2011c) and are calculated for the slightly different future period 2081–2100 to be consistent with the rest of this chapter, and are expressed with respect to the AR5 baseline period of 1986–2005. The box-and-whiskers fourth from the left are a graphical representation of the numbers shown in Table 12.2. The final box-and-whiskers on the right is a combination of CMIP5 model output and emulation of CMIP5 RCP6.0 numbers for those models that did not run RCP6.0.

It would be extremely costly computationally to rerun the full CMIP3 ensemble under the new RCPs and/or the full CMIP5 ensemble under the old SRES scenarios in order to separate model and scenario effects. In the absence of a direct comparison, we rely on simplified modelling frameworks to emulate CMIP3/5 SRES/RCP model behaviour and compare them. Figure 12.39 shows an emulation of the global mean temperature response at the end of the 21st century that one would expect from the CMIP5 models if they were run under SRES A1B. In this case, anomalies are computed with respect to 1980–1999 for direct comparison with the values reported in AR4 (Meehl et al., 2007b) which used that baseline. The method used to emulate the SRES A1B response of the CMIP5 is documented by Good et al. (2011a; 2013). Ensemble-mean A1B RF was computed from CMIP3 projections using the Forster and Taylor (2006) method, scaled to ensure consistency with the forcing required by the method. The simple model is only used to predict the temperature difference between A1B and RCP8.5, and between A1B and RCP4.5 separately for each model. These differences are then added to CMIP5 GCM simulations of RCP8.5 and RCP4.5 respectively, and averaged to give a single A1B estimate. The emulated CMIP5 SRES A1B results show a slightly larger mean response than the actual CMIP3 models, with a similar spread (±1 standard deviation is used in this case). The main reason for this is the slightly larger mean transient climate response (TCR) in the subset of CMIP5 models available in comparison with the AR4 CMIP3 models. An alternative emulation is presented by Knutti and Sedláček (2013) who use the simplified MAGICC models with parameters chosen to emulate the response of the CMIP3 models to RCP6.0 forcing, with anomalies expressed with respect to the 1986–2005 baseline period (Figure 12.39). They too find a larger mean response in the CMIP5 case but also a larger spread (±1 standard deviation) in CMIP5. Uncertainties in the different approaches to emulating climate model simulations, for example estimating the non-GHG RF, and the small sample sizes of CMIP3 and CMIP5 make it difficult to draw conclusions on the statistical significance of the differences displayed in Figure 12.39, but the same uncertainties lead us to conclude that on the basis of these analyses there appears to be no fundamental difference between the behaviour of the CMIP5 ensemble, in comparison with CMIP3.

Meinshausen et al. (2011a; 2011b) tuned MAGICC6 to emulate 19 GCMs from CMIP3. The results are temperature projections and their uncertainties (based on the empirical distribution of the ensemble) under each of the RCPs, extended to year 2500 (under constant emissions for the lowest RCP and constant concentrations for the remaining three). In the same paper, an ensemble produced by combining carbon cycle parameter calibration to nine C[4]MIP models with the 19 CMIP3 model parameter calibrations is also used to estimate the emissions implied by the various concentration pathways, had the CMIP3 models included a carbon cycle component. Rogelj et al. (2012) used the same tool but performed a fully probabilistic analysis of the SRES and RCP scenarios using a parameter space that is consistent with

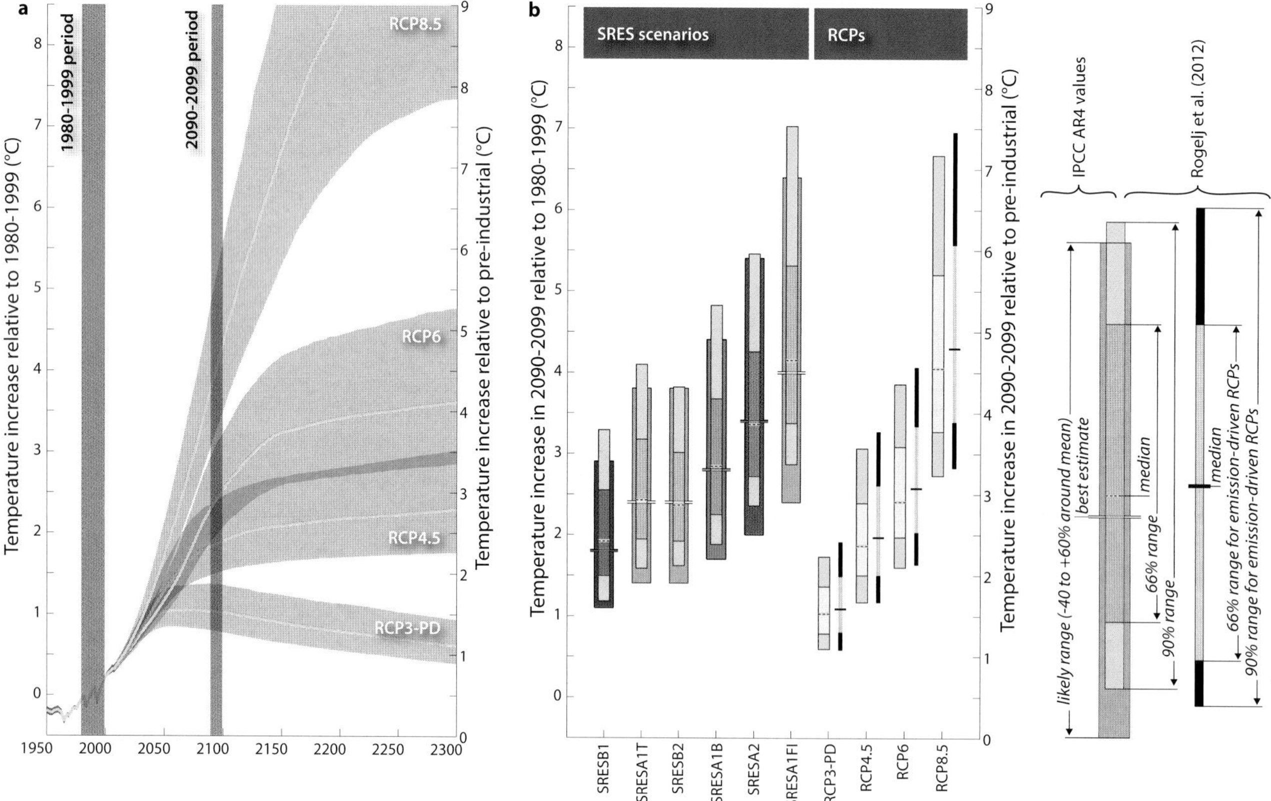

Figure 12.40 | Temperature projections for SRES scenarios and the RCPs. (a) Time-evolving temperature distributions (66% range) for the four RCP scenarios computed with the ECS distribution from Rogelj et al. (2012) and a model setup representing closely the carbon-cycle and climate system uncertainty estimates of the AR4 (grey areas). Median paths are drawn in yellow. Red shaded areas indicate time periods referred to in panel b. (b) Ranges of estimated average temperature increase between 2090 and 2099 for SRES scenarios and the RCPs respectively. Note that results are given both relative to 1980–1999 (left scale) and relative to pre-industrial (right scale). Yellow ranges indicate results obtained by Rogelj et al. (2012). Colour-coding of AR4 ranges is chosen to be consistent with AR4 (Meehl et al., 2007b). RCP2.6 is labelled as RCP3-PD here.

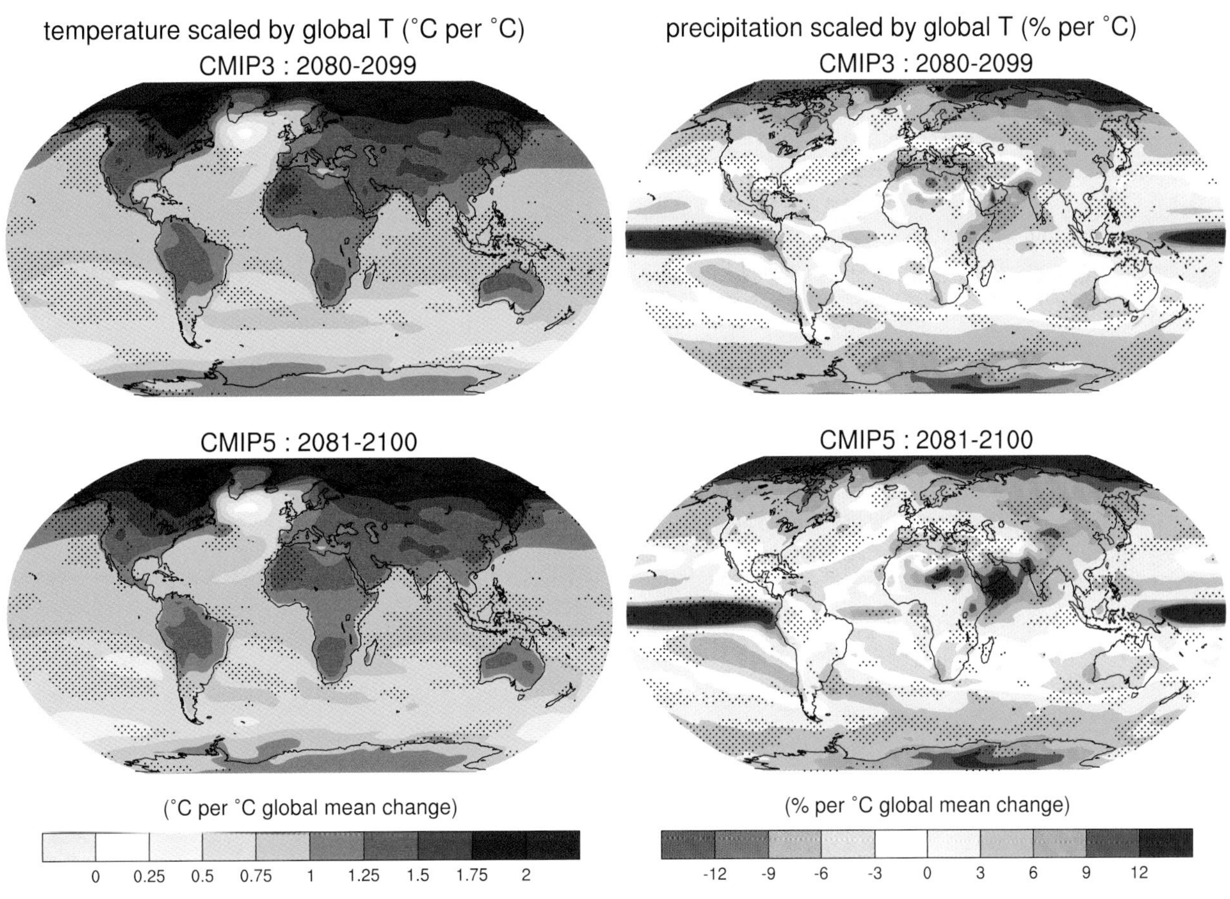

Figure 12.41 | Patterns of temperature (left column) and percent precipitation change (right column) for the CMIP3 models average (first row) and CMIP5 models average (second row), scaled by the corresponding global average temperature changes. The patterns are computed in both cases by taking the difference between the averages over the last 20 years of the 21st century experiments (2080–2099 for CMIP3 and 2081–2100 for CMIP5) and the last twenty years of the historic experiments (1980–1999 for CMIP3, 1986–2005 for CMIP5) and rescaling each difference by the corresponding change in global average temperature. This is done first for each individual model, and then the results are averaged across models. For the CMIP5 patterns, the RCP2.6 simulation of the FIO-ESM model was excluded because it did not show any warming by the end of the 21st century, thus not complying with the method requirement that the pattern be estimated at a time when the temperature change signal from CO_2 increase has emerged. Stippling indicates a measure of significance of the difference between the two corresponding patterns obtained by a bootstrap exercise. Two subsets of the pooled set of CMIP3 and CMIP5 ensemble members of the same size as the original ensembles, but without distinguishing CMIP3 from CMIP5 members, were randomly sampled 500 times. For each random sample we compute the corresponding patterns and their difference, then the true difference is compared, grid-point by grid-point, to the distribution of the bootstrapped differences, and only grid-points at which the value of the difference falls in the tails of the bootstrapped distribution (less than the 2.5 percentiles or the 97.5 percentiles) are stippled.

CMIP3/C[4]MIP but a more general uncertainty characterization for key quantities like equilibrium climate sensitivity, similarly to the approach utilized by Meinshausen et al. (2009). Observational or other historical constraints are also used in this study and the analysis is consistent with the overall assessment of sources and ranges of uncertainties for relevant quantities (equilibrium climate sensitivity above all) from AR4 (Meehl et al., 2007b , Box 10.2). Figure 12.40 summarizes results of this probabilistic comparison for global temperature. The RCPs span a large range of stabilization, mitigation and non-mitigation pathways and the resulting range of temperature changes are larger than those produced under SRES scenarios, which do not consider mitigation options. The SRES results span an interval between just above 1.0°C and 6.5°C when considering the respective *likely* ranges of all scenarios, including B1 as the lowest and A1FI as the highest. Emissions under RCP8.5 are highest and the resulting temperature changes *likely* range from 4.0°C to 6.1°C by 2100. The lowest RCP2.6 assumes significant mitigation and the global temperature change *likely* remains below 2°C.

Similar temperature change projections by the end of the 21st century are obtained under RCP8.5 and SRES A1FI, RCP6 and SRES B2 and RCP4.5 and SRES B1. There remain large differences though in the transient trajectories, with rates of change slower or faster for the different pairs. These differences can be traced back to the interplay of the (negative) short-term effect of sulphate aerosols and the (positive) effect of long-lived GHGs. Impact studies may be sensitive to the differences in these temporal profiles so care should be taken in approximating SRES with RCPs and vice versa.

While simple models can separate the effect of the scenarios and the model response, no studies are currently available that allow an attribution of the CMIP3-CMIP5 differences to changes in the transient climate response, the carbon cycle, and the inclusion of new processes (chemistry, land surface, vegetation). The fact that these sets of CMIP3 and CMIP5 experiments do not include emission-driven runs would suggest that differences in the representation of the carbon cycle are very unlikely to explain differences in the simulations, since the only effect of changes in the carbon cycle representation would affect the land surface, and thus would have only a minor effect on the climate response at the global scale.

Figure 12.41 shows a comparison of the patterns of warming and precipitation change from CMIP3 (using 23 models and three SRES scenarios) and CMIP5 (using 46 models and four RCPs), utilizing the pattern scaling methodology (Section 12.4.2). The geographic patterns of mean change are very similar across the two ensembles of models, with pattern correlations of 0.98 for temperature and 0.90 for precipitation changes. However there exist significant differences in the absolute values of the patterns, if not in their geographic shapes. A simple bootstrapping exercise that pooled together all models and scenarios and resampled 500 times the same numbers of models/scenarios divided into two groups, but without distinguishing CMIP3 from CMIP5 (and thus SRES from RCPs) allows to compute a measure of significance of the actual differences in the patterns. Stippling in Figure 12.41 marks the large regions where the difference is significant for temperature and precipitation patterns. The temperature pattern from CMIP5 shows significantly larger warming per degree Celsius of global mean temperature change in the NH and less warming per degree Celsius in the SH compared to the corresponding pattern from CMIP3. For precipitation patterns, CMIP5 shows significantly larger increases per degree Celsius in the NH and significantly larger decreases per degree Celsius in the SH compared to CMIP3. Even in this case we do not have studies that allow tracing the source of these differences to specific changes in models' configurations, processes represented or scenarios run.

Knutti and Sedláček (2013) attempt to identify or rule out at least some of these sources. Differences in model projections spread or its counterpart, robustness, between CMIP3 and CMIP5 are discussed, and it is shown that by comparing the behaviour of only a subset of 11 models, contributed to the two CMIPs by the same group of institutions, the robustness of CMIP5 versus that of CMIP3 actually decreases slightly. This would suggest that the enhanced robustness of CMIP5 is not clearly attributable to advances in modelling, and may be a result of the fact that the CMIP5 ensemble contains different versions of the same model that are counted as independent in this measure of robustness.

A comparison of CMIP3 and CMIP5 results for extreme indices is provided in Sections 12.4.3.3 and Figure 12.13 for temperature extremes, and Section 12.4.5.5 and Figure 12.26 for extremes in the water cycle.

12.5 Climate Change Beyond 2100, Commitment, Stabilization and Irreversibility

This section discusses the long term (century to millennia) climate change based on the RCP scenario extensions and idealized scenarios, the commitment from current atmospheric composition and from past emissions, the concept of cumulative carbon and the resulting constraints on emissions for various temperature targets. The term irreversibility is used in various ways in the literature. This report defines a perturbed state as irreversible on a given time scale if the recovery time scale from this state due to natural processes is significantly longer than the time it takes for the system to reach this perturbed state (see Glossary), for example, the climate change resulting from the long residence time of a CO_2 perturbation in the atmosphere. These results are discussed in Sections 12.5.2 to 12.5.4. Aspects of irreversibility in the context of abrupt change, multiple steady states and hysteresis are discussed in Section 12.5.5 and in Chapter 13 for ice sheets and sea level rise.

12.5.1 Representative Concentration Pathway Extensions

The CMIP5 intercomparison project includes simulations extending the four RCP scenarios to the year 2300 (see Section 12.3.1). This allows exploring the longer-term climate response to idealized GHG and aerosols forcings (Meinshausen et al., 2011c). Continuing GHG emissions beyond 2100 as in the RCP8.5 extension induces a total RF above 12 W m^{-2} by 2300, while sustaining negative emissions beyond 2100, as in the RCP2.6 extension, induces a total RF below 2 W m^{-2} by 2300. The projected warming for 2281–2300, relative to 1986–2005, is 0.6°C (range 0.0°C to 1.2°C) for RCP2.6, 2.5°C (range 1.5°C to 3.5°C) for RCP4.5, and 7.8°C (range 3.0°C to 12.6°C) for RCP8.5 (*medium confidence*, based on a limited number of CMIP5 simulations) (Figures 12.3 and 12.5, Table 12.2).

EMICs simulations have been performed following the same CMIP5 protocol for the historical simulation and RCP scenarios extended to 2300 (Zickfeld et al., 2013). These scenarios have been prolonged beyond 2300 to investigate longer-term commitment and irreversibility (see below). Up to 2300, projected warming and the reduction of the AMOC as simulated by the EMICs are similar to those simulated by the CMIP5 ESMs (Figures 12.5 and 12.42).

12.5.2 Climate Change Commitment

Climate change commitment, the fact that the climate will change further after the forcing or emissions have been eliminated or held constant, has attracted increased attention by scientists and policymakers shortly before the completion of IPCC AR4 (Hansen et al., 2005a; Meehl et al., 2005b, 2006; Wigley, 2005) (see also AR4 Section 10.7.1). However, the argument that the surface response would lag the RF due to the large thermal reservoir of the ocean in fact goes back much longer (Bryan et al., 1982; Hansen et al., 1984, 1985; Siegenthaler and Oeschger, 1984; Schlesinger, 1986; Mitchell et al., 2000; Wetherald et al., 2001). The discussion in this section is framed largely in terms of temperature change, but other changes in the climate system (e.g., precipitation) are closely related to changes in temperature (see Sections 12.4.1.1 and 12.4.2). A summary of how past emissions relate to future warming is also given in FAQ 12.3.

The Earth system has multiple response time scales related to different thermal reservoirs (see also Section 12.5.3). For a step change in forcing (instantaneous increase in the magnitude of the forcing and constant forcing after that), a large fraction of the total of the surface temperature response will be realized within years to a few decades (Brasseur and Roeckner, 2005; Knutti et al., 2008a; Murphy et al., 2009; Hansen et al., 2011). The remaining response, realized over centuries, is controlled by the slow mixing of the energy perturbation into the ocean (Stouffer, 2004). The response time scale depends on the amount of ocean mixing

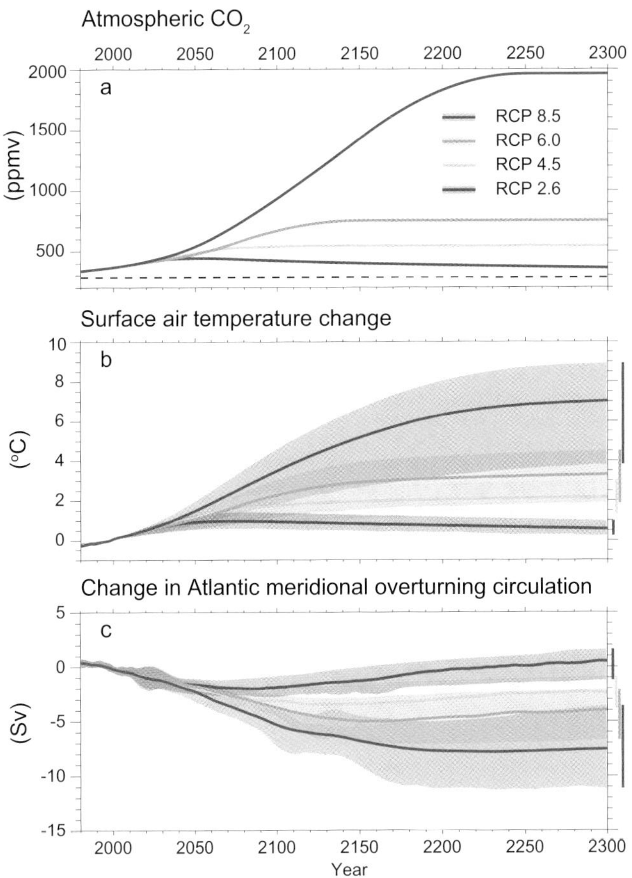

Figure 12.42 | (a) Atmospheric CO_2, (b) projected global mean surface temperature change and (c) projected change in the Atlantic meridional overturning circulation, as simulated by EMICs for the four RCPs up to 2300 (Zickfeld et al., 2013). A 10-year smoothing was applied. Shadings and bars denote the minimum to maximum range. The dashed line on (a) indicates the pre-industrial CO_2 concentration.

and the strength of climate feedbacks, and is longer for higher climate sensitivity (Hansen et al., 1985; Knutti et al., 2005). The transient climate response is therefore smaller than the equilibrium response, in particular for high climate sensitivities. This can also be interpreted as the ocean heat uptake being a negative feedback (Dufresne and Bony, 2008; Gregory and Forster, 2008). Delayed responses can also occur due to processes other than ocean warming, for example, vegetation change (Jones et al., 2009) or ice sheet melt that continues long after the forcing has been stabilized (see Section 12.5.3).

Several forms of commitment are often discussed in the literature. The most common is the 'constant composition commitment', the warming that would occur after stabilizing all radiative constituents at a given year (for example year 2000) levels. For year 2000 commitment, AOGCMs estimated a most likely value of about 0.6°C for 2100 (relative to 1980–1999, AR4 Section 10.7.1). A present-day composition commitment simulation is not part of CMIP5, so direct comparison with CMIP3 is not possible. However, the available CMIP5 results based on the RCP4.5 extension with constant RF (see Section 12.5.1) are consistent with those numbers, with an additional warming of about 0.5°C 200 years after stabilization of the forcing (Figures 12.5 and 12.42).

A measure of constant composition commitment is the fraction of realized warming which can be estimated as the ratio of the warming at a given time to the long-term equilibrium warming (e.g., Stouffer, 2004; Meehl et al., 2007b, Section 10.7.2; Eby et al., 2009; Solomon et al., 2009). EMIC simulations have been performed with RCPs forcing up to 2300 prolonged until the end of the millennium with a constant forcing set at the value reached by 2300 (Figure 12.43). When the forcing stabilizes, the fraction of realized warming is significantly below unity. However, the fraction of realized warming depends on the history of the forcing. For the RCP4.5 and RCP6.0 extension scenarios with early stabilization, it is about 75% at the time of forcing stabilization; while for RCP8.5, with stabilization occurring later, it is about 85% (see Figure 12.43); but for a 1% yr^{-1} CO_2 increase to 2 × CO_2 or 4 × CO_2 and constant forcing thereafter, the fraction of realized warming is much smaller, about 40 to 70% at the time when the forcing is kept constant. The fraction of realized warming rises typically by 10% over the century following the stabilization of forcing. Due to the long time scales in the deep ocean, full equilibrium is reached only after hundreds to thousands of years (Hansen et al., 1985; Gregory et al., 2004; Stouffer, 2004; Meehl et al., 2007b, Section 10.7.2; Knutti et al., 2008a; Danabasoglu and Gent, 2009; Held et al., 2010; Hansen et al., 2011; Li et al., 2013a).

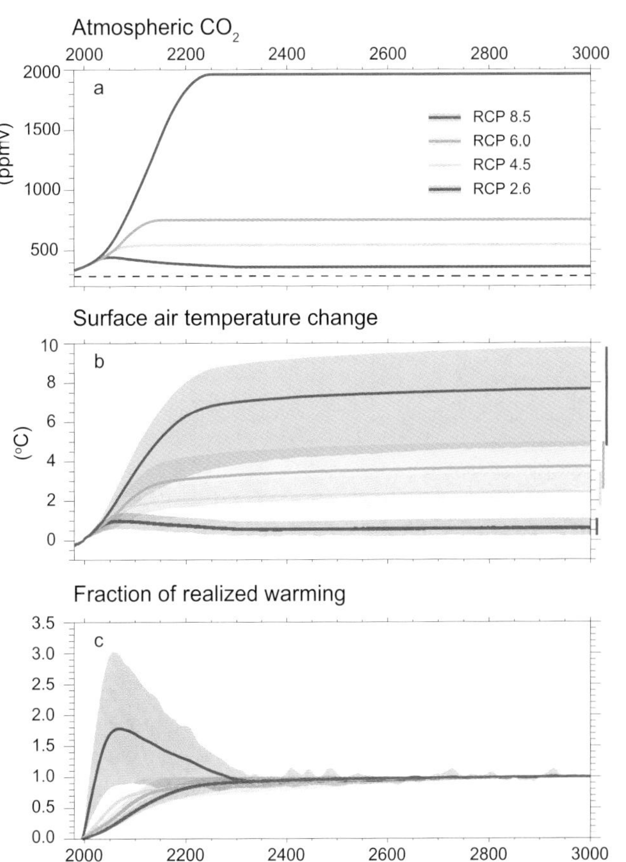

Figure 12.43 | (a) Atmospheric CO_2, (b) projected global mean surface temperature change and (c) fraction of realized warming calculated as the ratio of global temperature change at a given time to the change averaged over the 2980–2999 time period, as simulated by Earth System Models of Intermediate Complexity (EMICs) for the 4 RCPs up to 2300 followed by a constant (year 2300 level) radiative forcing up to the year 3000 (Zickfeld et al., 2013). A 10-year smoothing was applied. Shadings and bars denote the minimum to maximum range. The dashed line on (a) indicates the pre-industrial CO_2 concentration.

12

'Constant emission commitment' is the warming that would result from maintaining annual anthropogenic emissions at the current level. Few studies exist but it is estimated to be about 1°C to 2.5°C by 2100 assuming constant (year 2010) emissions in the future, based on the MAGICC model calibrated to CMIP3 and C[4]MIP models (Meinshausen et al., 2011a; Meinshausen et al., 2011b) (see FAQ 12.3). Such a scenario is different from non-intervention economic scenarios, and it does not stabilize global temperature, as any plausible emission path after 2100 would cause further warming. It is also different from a constant cumulative emission scenario which implies zero emissions in the future.

Another form of commitment involves climate change when anthropogenic emissions are set to zero ('zero emission commitment'). Results from a variety of models ranging from EMICs (Meehl et al., 2007b; Weaver et al., 2007; Matthews and Caldeira, 2008; Plattner et al., 2008; Eby et al., 2009; Solomon et al., 2009; Friedlingstein et al., 2011) to ESMs (Frölicher and Joos, 2010; Gillett et al., 2011; Gillett et al., 2013) show that abruptly setting CO_2 emissions to zero (keeping other forcings constant if accounted for) results in approximately constant global temperature for several centuries onward. Those results indicate that past emissions commit us to persistent warming for hundreds of years, continuing at about the level of warming that has been realized. On near equilibrium time scales of a few centuries to about a millennium, the temperature response to CO_2 emissions is controlled by climate sensitivity (see Box 12.2) and the cumulative airborne fraction of CO_2 over these time scales. After about a thousand years (i.e., near thermal equilibrium) and cumulative CO_2 emissions less than about 2000 PgC, approximately 20 to 30% of the cumulative anthropogenic carbon emissions still remain in the atmosphere (Montenegro et al., 2007; Plattner et al., 2008; Archer et al., 2009; Frölicher and Joos, 2010; Joos et al., 2013) (see Box 6.1) and maintain a substantial temperature response long after emissions have ceased (Friedlingstein and Solomon, 2005; Hare and Meinshausen, 2006; Weaver et al., 2007; Matthews and Caldeira, 2008; Plattner et al., 2008; Eby et al., 2009; Lowe et al., 2009; Solomon et al., 2009, 2010; Frölicher and Joos, 2010; Zickfeld et al., 2012). In the transient phase, on a 100- to 1000-year time scale, the approximately constant temperature results from a compensation between delayed commitment warming (Meehl et al., 2005b; Wigley, 2005) and the reduction in atmospheric CO_2 resulting from ocean and land carbon uptake as well as from the nonlinear dependence of RF on atmospheric CO_2 (Meehl et al., 2007b; Plattner et al., 2008; Solomon et al., 2009; Solomon et al., 2010). The commitment associated with past emissions depends, as mentioned above, on the value of climate sensitivity and cumulative CO_2 airborne fraction, but it also depends on the choices made for other RF constituents. In a CO_2 only case and for equilibrium climate sensitivities near 3°C, the warming commitment (i.e., the warming relative to the time when emissions are stopped) is near zero or slightly negative. For high climate sensitivities, and in particular if aerosol emissions are eliminated at the same time, the commitment from past emission can be significantly positive, and is a superposition of a fast response to reduced aerosols emissions and a slow response associated with high climate sensitivities (Brasseur and Roeckner, 2005; Hare and Meinshausen, 2006; Armour and Roe, 2011; Knutti and Plattner, 2012; Matthews and Zickfeld, 2012) (see FAQ 12.3). In the real world, the emissions of CO_2 and non-CO_2 forcing agents are of course coupled. All of the above studies support the conclusion that temperatures would decrease only very slowly (if at all), even for strong reductions or complete elimination of CO_2 emissions, and might even increase temporarily for an abrupt reduction of the short-lived aerosols (FAQ 12.3). The implications of this fact for climate stabilization are discussed in Section 12.5.4.

New EMIC simulations with pre-industrial CO_2 emissions and zero non-CO_2 forcings after 2300 (Zickfeld et al., 2013) confirm this behaviour (Figure 12.44) seen in many earlier studies (see above). Switching off anthropogenic CO_2 emissions in 2300 leads to a continuous slow decline of atmospheric CO_2, to a significantly slower decline of global temperature and to a continuous increase in ocean thermal expansion

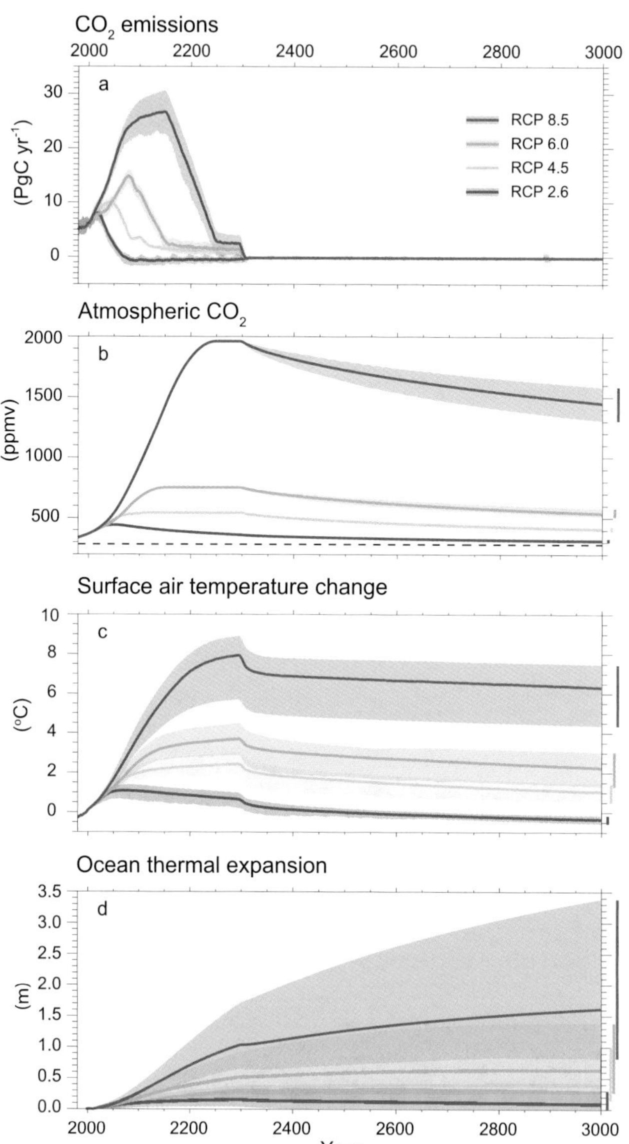

Figure 12.44 | (a) Compatible anthropogenic CO_2 emissions up to 2300, followed by zero emissions after 2300, (b) prescribed atmospheric CO_2 concentration up to 2300 followed by projected CO_2 concentration after 2300, (c) global mean surface temperature change and (d) ocean thermal expansion as simulated by Earth System Models of Intermediate Complexity (EMICs) for the four concentration driven RCPs with all forcings included (Zickfeld et al., 2013). A 10-year smoothing was applied. The drop in temperature in 2300 is a result of eliminating all non-CO_2 forcings along with CO_2 emissions. Shadings and bars denote the minimum to maximum range. The dashed line on (b) indicates the pre-industrial CO_2 concentration.

12

over the course of the millennium. Larger forcings induce longer delays before the Earth system would reach equilibrium. For RCP8.5, by year 3000 (700 years after emissions have ceased) global temperature has decreased only by 1°C to 2°C (relative to its peak value by 2300) and ocean thermal expansion has almost doubled (relative to 2300) and is still increasing (Zickfeld et al., 2013).

The previous paragraph discussed climate change commitment from GHGs that have already been emitted. Another form of commitment refers to climate change associated with heat and carbon that has gone into the land surface and oceans. This would be relevant to the consequences of a one-time removal of all of the excess CO_2 in the atmosphere and is computed by taking a transient simulation and instantaneously setting atmospheric CO_2 concentrations to initial (pre-industrial) values (Cao and Caldeira, 2010). In such an extreme case, there would be a net flux of CO_2 from the ocean and land surface to the atmosphere, releasing an amount of CO_2 representing about 30% of what was removed from the atmosphere, i.e., the airborne fraction applies equally to positive and negative emissions, and it depends on the emissions history. A related form of experiment investigates the consequences of an initial complete removal followed by sustained removal of any CO_2 returned to the atmosphere from the land surface and oceans, and is computed by setting atmospheric CO_2 concentrations to pre-industrial values and maintaining this concentration (Cao and Caldeira, 2010). In this case, only about one-tenth of the pre-existing temperature perturbation persists for more than half of a century. A similar study performed with a GFDL AOGCM where forcing was instantaneously returned to its pre-industrial value, found larger residual warming, up to 30% of the pre-existing warming (Held et al., 2010).

Several studies on commitment to past emissions have demonstrated that the persistence of warming is substantially longer than the lifetime of anthropogenic GHGs themselves, as a result of nonlinear absorption effects as well as the slow heat transfer into and out of the ocean. In much the same way as the warming to a step increase of forcing is delayed, the cooling after setting RF to zero is also delayed. Loss of excess heat from the ocean will lead to a positive surface air temperature anomaly for decades to centuries (Held et al., 2010; Solomon et al., 2010; Bouttes et al., 2013).

A more general form of commitment is the question of how much warming we are committed to as a result of inertia and hence commitments related to the time scales for energy system transitions and other societal, economic and technological aspects (Grubb, 1997; Washington et al., 2009; Davis et al., 2010). For example, Davis et al. (2010) estimated climate commitment of 1.3°C (range 1.1°C to 1.4°C, relative to pre-industrial) from existing CO_2-emitting devices under specific assumptions regarding their lifetimes. These forms of commitment, however, are strongly based on political, economic and social assumptions that are outside the domain of IPCC WGI and are not further considered here.

12.5.3 Forcing and Response, Time Scales of Feedbacks

Equilibrium climate sensitivity (ECS), transient climate response (TCR) and climate feedbacks are useful concepts to characterize the response of a model to an external forcing perturbation. However, there are limitations to the concept of RF (Joshi et al., 2003; Shine et al., 2003; Hansen et al., 2005b; Stuber et al., 2005), and the separation of forcings and fast (or rapid) responses (e.g., clouds changing almost instantaneously as a result of CO_2-induced heating rates rather than as a response to the slower surface warming) is sometimes difficult (Andrews and Forster, 2008; Gregory and Webb, 2008). Equilibrium warming also depends on the type of forcing (Stott et al., 2003; Hansen et al., 2005b; Davin et al., 2007). ECS is time or state dependent in some models (Senior and Mitchell, 2000; Gregory et al., 2004; Boer et al., 2005; Williams et al., 2008; Colman and McAvaney, 2009; Colman and Power, 2010), and in some but not all models climate sensitivity from a slab ocean version differs from that of coupled models or the effective climate sensitivity (see Glossary) diagnosed from a transient coupled integration (Gregory et al., 2004; Danabasoglu and Gent, 2009; Li et al., 2013a). The computational cost of coupled AOGCMs is often prohibitively large to run simulations to full equilibrium, and only a few models have performed those (Manabe and Stouffer, 1994; Voss and Mikolajewicz, 2001; Gregory et al., 2004; Danabasoglu and Gent, 2009; Li et al., 2013a). Because of the time dependence of effective climate sensitivity, fitting simple models to AOGCMs over the first few centuries may lead to errors when inferring the response on multi-century time scales. In the HadCM3 case the long-term warming would be underestimated by 30% if extrapolated from the first century (Gregory et al., 2004), in other models the warming of the slab and coupled model is almost identical (Danabasoglu and Gent, 2009). The assumption that the response to different forcings is approximately additive appears to be justified for large-scale temperature changes but limited for other climate variables (Boer and Yu, 2003; Sexton et al., 2003; Gillett et al., 2004; Meehl et al., 2004; Jones et al., 2007). A more complete discussion of the concept of ECS and the limitations is given in Knutti and Hegerl (2008). The CMIP5 model estimates of ECS and TCR are also discussed in Section 9.7. Despite all limitations, the ECS and TCR remain key concepts to characterize the transient and near equilibrium warming as a response to RF on time scales of centuries. Their overall assessment is given in Box 12.2.

A number of recent studies suggest that equilibrium climate sensitivities determined from AOGCMs and recent warming trends may significantly underestimate the true Earth system sensitivity (see Glossary) which is realized when equilibration is reached on millennial time scales (Hansen et al., 2008; Rohling et al., 2009; Lunt et al., 2010; Pagani et al., 2010; Rohling and Members, 2012). The argument is that slow feedbacks associated with vegetation changes and ice sheets have their own intrinsic long time scales and are not represented in most models (Jones et al., 2009). Additional feedbacks are mostly thought to be positive but negative feedbacks of smaller magnitude are also simulated (Swingedouw et al., 2008; Goelzer et al., 2011). The climate sensitivity of a model may therefore not reflect the sensitivity of the full Earth system because those feedback processes are not considered (see also Sections 10.8, 5.3.1 and 5.3.3.2; Box 5.1). Feedbacks determined in very different base state (e.g., the Last Glacial Maximum) differ from those in the current warm period (Rohling and Members, 2012), and relationships between observables and climate sensitivity are model dependent (Crucifix, 2006; Schneider von Deimling et al., 2006; Edwards et al., 2007; Hargreaves et al., 2007, 2012). Estimates of climate sensitivity based on paleoclimate archives (Hansen

Frequently Asked Questions

FAQ 12.3 | What Would Happen to Future Climate if We Stopped Emissions Today?

Stopping emissions today is a scenario that is not plausible, but it is one of several idealized cases that provide insight into the response of the climate system and carbon cycle. As a result of the multiple time scales in the climate system, the relation between change in emissions and climate response is quite complex, with some changes still occurring long after emissions ceased. Models and process understanding show that as a result of the large ocean inertia and the long lifetime of many greenhouse gases, primarily carbon dioxide, much of the warming would persist for centuries after greenhouse gas emissions have stopped.

When emitted in the atmosphere, greenhouse gases get removed through chemical reactions with other reactive components or, in the case of carbon dioxide (CO_2), get exchanged with the ocean and the land. These processes characterize the lifetime of the gas in the atmosphere, defined as the time it takes for a concentration pulse to decrease by a factor of e (2.71). How long greenhouse gases and aerosols persist in the atmosphere varies over a wide range, from days to thousands of years. For example, aerosols have a lifetime of weeks, methane (CH_4) of about 10 years, nitrous oxide (N_2O) of about 100 years and hexafluoroethane (C_2F_6) of about 10,000 years. CO_2 is more complicated as it is removed from the atmosphere through multiple physical and biogeochemical processes in the ocean and the land; all operating at different time scales. For an emission pulse of about 1000 PgC, about half is removed within a few decades, but the remaining fraction stays in the atmosphere for much longer. About 15 to 40% of the CO_2 pulse is still in the atmosphere after 1000 years.

As a result of the significant lifetimes of major anthropogenic greenhouse gases, the increased atmospheric concentration due to past emissions will persist long after emissions are ceased. Concentration of greenhouse gases would not return immediately to their pre-industrial levels if emissions were halted. Methane concentration would return to values close to pre-industrial level in about 50 years, N_2O concentrations would need several centuries, while CO_2 would essentially never come back to its pre-industrial level on time scales relevant for our society. Changes in emissions of short-lived species like aerosols on the other hand would result in nearly instantaneous changes in their concentrations.

The climate system response to the greenhouse gases and aerosols forcing is characterized by an inertia, driven mainly by the ocean. The ocean has a very large capacity of absorbing heat and a slow mixing between the surface and the deep ocean. This means that it will take several centuries for the whole ocean to warm up and to reach equilibrium with the altered radiative forcing. The surface ocean (and hence the continents) will continue to warm until it reaches a surface temperature in equilibrium with this new radiative forcing. The AR4 showed that if concentration of greenhouse gases were held constant at present day level, the Earth surface would still continue to warm by about 0.6°C over the 21st century relative to the year 2000. This is the climate commitment to current concentrations (or constant composition commitment), shown in grey in FAQ 12.3, Figure 1. Constant emissions at current levels would further increase the atmospheric concentration and result in much more warming than observed so far (FAQ 12.3, Figure 1, red lines).

Even if anthropogenic greenhouses gas emissions were halted now, the radiative forcing due to these long-lived greenhouse gases concentrations would only slowly decrease in the future, at a rate determined by the lifetime of the gas (see above). Moreover, the

(continued on next page)

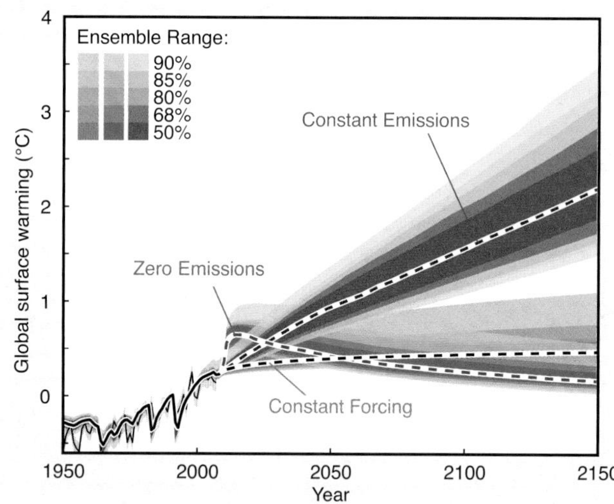

FAQ 12.3, Figure 1 | Projections based on the energy balance carbon cycle model Model for the Assessment of Greenhouse Gas-Induced Climate Change (MAGICC) for constant atmospheric composition (constant forcing, grey), constant emissions (red) and zero future emissions (blue) starting in 2010, with estimates of uncertainty. Figure adapted from Hare and Meinshausen (2006) based on the calibration of a simple carbon cycle climate model to all Coupled Model Intercomparison Project Phase 3 (CMIP3) and Coupled Climate Carbon Cycle Model Intercomparison Project (C4MIP) models (Meinshausen et al., 2011a; Meinshausen et al., 2011b). Results are based on a full transient simulation starting from pre-industrial and using all radiative forcing components. The thin black line and shading denote the observed warming and uncertainty.

FAQ 12.3 (continued)

climate response of the Earth System to that radiative forcing would be even slower. Global temperature would not respond quickly to the greenhouse gas concentration changes. Eliminating CO_2 emissions only would lead to near constant temperature for many centuries. Eliminating short-lived negative forcings from sulphate aerosols at the same time (e.g., by air pollution reduction measures) would cause a temporary warming of a few tenths of a degree, as shown in blue in FAQ 12.3, Figure 1. Setting all emissions to zero would therefore, after a short warming, lead to a near stabilization of the climate for multiple centuries. This is called the commitment from past emissions (or zero future emission commitment). The concentration of GHG would decrease and hence the radiative forcing as well, but the inertia of the climate system would delay the temperature response.

As a consequence of the large inertia in the climate and carbon cycle, the long-term global temperature is largely controlled by total CO_2 emissions that have accumulated over time, irrespective of the time when they were emitted. Limiting global warming below a given level (e.g., 2°C above pre-industrial) therefore implies a given budget of CO_2, that is, higher emissions earlier implies stronger reductions later. A higher climate target allows for a higher CO_2 concentration peak, and hence larger cumulative CO_2 emissions (e.g., permitting a delay in the necessary emission reduction).

Global temperature is a useful aggregate number to describe the magnitude of climate change, but not all changes will scale linearly global temperature. Changes in the water cycle for example also depend on the type of forcing (e.g., greenhouse gases, aerosols, land use change), slower components of the Earth system such as sea level rise and ice sheet would take much longer to respond, and there may be critical thresholds or abrupt or irreversible changes in the climate system.

et al., 2008; Rohling et al., 2009; Lunt et al., 2010; Pagani et al., 2010; Schmittner et al., 2011; Rohling and Members, 2012), most but not all based on climate states colder than present, are therefore not necessarily representative for an estimate of climate sensitivity today (see also Sections 5.3.1, 5.3.3.2, Box 5.1). Also it is uncertain on which time scale some of those Earth system feedbacks would become significant.

Equilibrium climate sensitivity undoubtedly remains a key quantity, useful to relate a change in GHGs or other forcings to a global temperature change. But the above caveats imply that estimates based on past climate states very different from today, estimates based on time scales different than those relevant for climate stabilization (e.g., estimates based on climate response to volcanic eruptions), or based on forcings other than GHGs (e.g., spatially non-uniform land cover changes, volcanic eruptions or solar forcing) may differ from the climate sensitivity measuring the climate feedbacks of the Earth system today, and this measure, in turn, may be slightly different from the sensitivity of the Earth in a much warmer state on time scales of millennia. The TCR and the transient climate response to cumulative carbon emissions (TCRE) are often more directly relevant to evaluate short term changes and emission reductions needed for stabilization (see Section 12.5.4).

12.5.4 Climate Stabilization and Long-term Climate Targets

This section discusses the relation between emissions and climate targets, in the context of the uncertainties characterizing both the transient and the equilibrium climate responses to emissions. 'Climate targets' considered here are both stabilizing temperature at a specified value and avoiding a warming beyond a predefined threshold. The latter idea of limiting peak warming is a more general concept than stabilization of temperature or atmospheric CO_2, and one that is more realistic than an exact climate stabilization which would require perpetual non-zero positive emissions to counteract the otherwise unavoidable long-term slow decrease in global temperature (Matsuno et al., 2012a) (Figure 12.44).

12.5.4.1 Background

The concept of stabilization is strongly linked to the ultimate objective of the UNFCCC, which is 'to achieve [...] stabilization of greenhouse gas concentrations in the atmosphere at a level that would prevent dangerous anthropogenic interference with the climate system'. Recent policy discussions focussed on a global temperature increase, rather than on GHG concentrations. The most prominent target currently discussed is the 2°C temperature target, that is, to limit global temperature increase relative to pre-industrial times to below 2°C. The 2°C target has been used first by the European Union as a policy target in 1996 but can be traced further back (Jaeger and Jaeger, 2010; Randalls, 2010). Climate impacts however are geographically diverse (Joshi et al., 2011) and sector specific, and no objective threshold defines when dangerous interference is reached. Some changes may be delayed or irreversible, and some impacts are likely to be beneficial. It is thus not possible to define a single critical threshold without value judgments and without assumptions on how to aggregate current and future costs and benefits. Targets other than 2°C have been proposed (e.g., 1.5°C global warming relative to pre-industrial), or targets based on CO_2 concentration levels, for example, 350 ppm (Hansen et al., 2008). The rate of change may also be important (e.g., for adaptation). This section does not advocate or defend any threshold, nor does it judge the economic or political feasibility of such goals, but simply assesses the implications of different illustrative climate targets on allowed carbon emissions, based on our current understanding of climate and carbon cycle feedbacks.

12.5.4.2 Constraints on Cumulative Carbon Emissions

The current RF from GHGs maintained indefinitely (i.e., the commitment from constant greenhouse gas concentrations) would correspond to approximately 2°C warming. That, however, does not imply that the commitment from past emissions has already exceeded 2°C. Part of the positive RF from GHGs is currently compensated by negative aerosol forcing, and stopping GHG emissions would lead to a decrease in the GHG forcing. Actively removing CO_2 from the atmosphere, for example by the combined use of biomass energy and carbon capture and storage, would further accelerate the decrease in GHG forcing.

The total amount of anthropogenic CO_2 released in the atmosphere (often termed cumulative carbon emission) is a good indicator of the atmospheric CO_2 concentration and hence of the global warming response to CO_2. The ratio of global temperature change to total cumulative anthropogenic CO_2 emissions (TCRE) is relatively constant over time and independent of the scenario, but is model dependent as it depends on the model cumulative airborne fraction of CO_2 and ECS/TCR (Matthews and Caldeira, 2008; Allen et al., 2009; Gregory et al., 2009; Matthews et al., 2009; Meinshausen et al., 2009; Zickfeld et al., 2009; Bowerman et al., 2011; Knutti and Plattner, 2012; Zickfeld et al., 2012, 2013). This is consistent with an earlier study indicating that the global warming potential of CO_2 is approximately independent of the scenario (Caldeira and Kasting, 1993). The concept of a constant ratio of cumulative emissions of CO_2 to temperature holds well only until temperatures peak (see Figure 12.45e) and only for smoothly varying cumulative CO_2 emissions (Gillett et al., 2013). It does not hold for stabilization on millennial time scales or for non-CO_2 forcings, and there is limited evidence for its applicability for cumulative emissions exceeding 2000 PgC owing to limited simulations available (Plattner et al., 2008; Hajima et al., 2012; Matsuno et al., 2012b; Gillett et al., 2013; Zickfeld et al., 2013). For non-CO_2 forcings with shorter atmospheric life times than CO_2 the rate of emissions at the time of peak warming is more important than the cumulative emissions over time (Smith et al., 2012).

Assuming constant climate sensitivity and fixed carbon cycle feedbacks, long-term (several centuries to millennium) stabilization of global temperatures requires eventually the stabilization of atmospheric concentrations (or decreasing concentrations if the temperature should be stabilized more quickly). This requires decreasing emissions to near-zero (Jones et al., 2006; Meehl et al., 2007b; Weaver et al., 2007; Matthews and Caldeira, 2008; Plattner et al., 2008; Allen et al., 2009; Matthews et al., 2009; Meinshausen et al., 2009; Zickfeld et al., 2009; Friedlingstein et al., 2011; Gillett et al., 2011; Roeckner et al., 2011; Knutti and Plattner, 2012; Matsuno et al., 2012a).

The relationships between cumulative emissions and temperature for various studies are shown in Figure 12.45. Note that some lines mark the evolution of temperature as a function of emissions over time while other panels show peak temperatures for different simulations. Also some models prescribe only CO_2 emissions while others use multi gas scenarios, and the time horizons differ. The warming is usually larger if non-CO_2 forcings are considered, since the net effect of the non-CO_2 forcings is positive in most scenarios (Hajima et al., 2012). Not all numbers are therefore directly comparable. Matthews et al. (2009) estimated the TCRE as 1°C to 2.1°C per 1000 PgC (TtC, or 10^{12} metric tonnes of carbon) (5 to 95%) based on the C^{4}MIP model range (Figure 12.45a). The ENSEMBLES E1 show a range of 1°C to 4°C per 1000 PgC (scaled from 0.5°C to 2°C for 500 PgC, Figure 12.45d) (Johns et al., 2011). Rogelj et al. (2012) estimate a 5 to 95% range of about 1°C to 2°C per 1000 PgC (Figure 12.45e) based on the MAGICC model calibrated to the C^{4}MIP model range and the *likely* range of 2°C to 4.5°C for climate sensitivity given in AR4. Allen et al. (2009) used a simple model and found 1.3°C to 3.9°C per 1000 PgC (5 to 95%) for peak warming (Figure 12.45g) and 1.4°C to 2.5°C for TCRE. The EMICs TCRE simulations suggest a range of about 1.4°C to 2.5°C per 1000 PgC and a mean of 1.9°C per 1000 PgC (Zickfeld et al., 2013) (Figure 12.45h). The results of Meinshausen et al. (2009) confirm the approximate linearity between temperature and CO_2 emissions (Figure 12.45b). Their results are difficult to compare owing to the shorter time period considered, but the model was found to be consistent with that of Allen et al. (2009). Zickfeld et al. (2009), using an EMIC, find a best estimate of about 1.5°C per 1000 PgC. Gillett et al. (2013) find a range of 0.8°C to 2.4°C per 1000 PgC in 15 CMIP5 models and derive an observationally constrained range of 0.7°C to 2.0°C per 1000 PgC. Results from much earlier model studies support the near linear relationship of cumulative emissions and global temperature, even though these studies did not discuss the linear relationship. An example is given in Figure 12.45c based on data shown in IPCC TAR Figure 13.3 (IPCC, 2001) and IPCC AR4 Figure 10.35 (Meehl et al., 2007b). The relationships between cumulative CO_2 emissions and temperature in CMIP5 are shown in Figure 12.45f for the 1% yr^{-1} CO_2 increase scenarios and in Figure 12.45i for the RCP8.5 emission driven ESM simulations (Gillett et al., 2013). Compatible emissions from concentration driven CMIP5 ESMs are discussed in Section 6.4.3.3.

Expert judgement based on the available evidence therefore suggests that the TCRE is *likely* between 0.8°C to 2.5°C per 1000 PgC, for cumulative CO_2 emissions less than about 2000 PgC until the time at which temperature peaks. Under these conditions, and for low to medium estimates of climate sensitivity, the TCRE is nearly identical to the peak climate response to cumulative carbon emissions. For high climate sensitivity, strong carbon cycle climate feedbacks or large cumulative emissions, the peak warming can be delayed and the peak response may be different from TCRE, but is often poorly constrained by models and observations. The range of TCRE assessed here is consistent with other recent attempts to synthesize the available evidence (NRC, 2011; Matthews et al., 2012). The results by Schwartz et al. (2010, 2012) imply a much larger warming for the carbon emitted over the historical period and have been questioned by Knutti and Plattner (2012) for neglecting the relevant response time scales and combining a transient airborne fraction with an equilibrium climate sensitivity.

The TCRE can be compared to the temperature response to emissions on a time scale of about 1000 years after emissions cease. This can be estimated from the *likely* range of equilibrium climate sensitivity (1.5°C to 4.5°C) and a cumulative CO_2 airborne fraction after about

1000 years of about 25 ± 5% (Archer et al., 2009; Joos et al., 2013). Again combining the extreme values would suggest a range of 0.6°C to 2.7°C per 1000 PgC, and 1.5°C per 1000 PgC for an ECS of 3°C and a cumulative airborne fraction of 25%. However, this equilibrium estimate is based on feedbacks estimated for the present day climate. Climate and carbon cycle feedbacks may increase substantially on long time scales and for high cumulative CO_2 emissions (see Section 12.5.3), introducing large uncertainties in particular on the upper bound. Based on paleoclimate data and an analytical model, Goodwin et al. (2009) estimate a long term RF of 1.5 W m^{-2} for an emission of 1000 PgC. For an equilibrium climate sensitivity of 3°C this corresponds to a warming of 1.2°C on millennial time scales, consistent with the climate carbon cycle models results discussed above.

The uncertainty in TCRE is caused by the uncertainty in the physical feedbacks and ocean heat uptake (reflected in TCR) and uncertainties in carbon cycle feedbacks (affecting the cumulative airborne fraction of CO_2). TCRE only characterizes the warming due to CO_2 emissions, and contributions from non-CO_2 gases need to be considered separately when estimating likelihoods to stay below a temperature limit. Warming as a function of cumulative CO_2 emissions is similar in the four RCP scenarios, and larger than that due to CO_2 alone, since non-CO_2 forcings contribute warming in these scenarios (compare Figure 12.45 f, i) (Hajima et al., 2012).

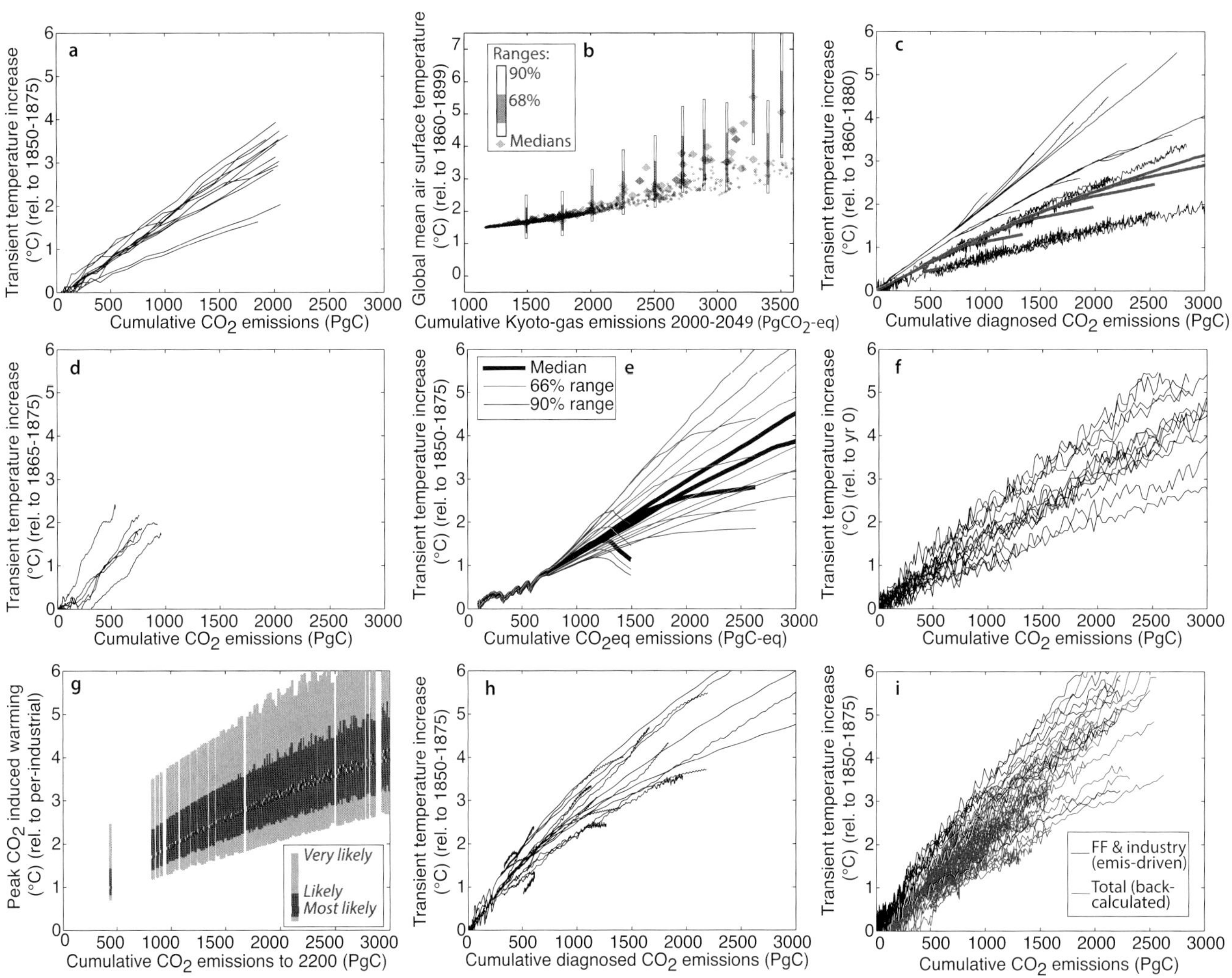

Figure 12.45 | Global temperature change vs. cumulative carbon emissions for different scenarios and models. (a) Transient global temperature increase vs. cumulative CO_2 emissions for Coupled Climate Carbon Cycle Model Intercomparison Project (C[4]MIP) (Matthews et al., 2009). (b) Maximum temperature increase until 2100 vs. cumulative Kyoto-gas emissions (CO_2 equivalent; note that all other panels are given in C equivalent) (Meinshausen et al., 2009). (c) Transient temperature increase vs. cumulative CO_2 emissions for IPCC TAR models (red, IPCC TAR Figure 13.3) and IPCC AR4 Earth System Models of Intermediate Complexity (EMICs, black: IPCC AR4 Figure 10.35). (d) As in (a) but for the ENSEMBLES E1 scenario (Johns et al., 2011). (e) Transient temperature increase for the RCP scenarios based on the Model for the Assessment of Greenhouse Gas-Induced Climate Change (MAGICC) model constrained to C[4]MIP, observed warming, and the IPCC AR4 climate sensitivity range (Rogelj et al., 2012). (f) Transient temperature change from the CMIP5 1% yr^{-1} concentration driven simulations. (g) Peak CO_2 induced warming vs. cumulative CO_2 emissions to 2200 (Allen et al., 2009; Bowerman et al., 2011). (h) Transient temperature increase from the new EMIC RCP simulations (Zickfeld et al., 2013). (i) Transient temperature change from the CMIP5 historical and RCP8.5 emission driven simulations (black) and transient temperature change in all concentration-driven CMIP5 RCP simulations with back-calculated emissions (red). Note that black lines in panel (i) do not include land use CO_2 and that warming in (i) is higher than in (f) due to additional non-CO_2 forcings.

12

Box 12.2 | Equilibrium Climate Sensitivity and Transient Climate Response

Equilibrium climate sensitivity (ECS) and transient climate response (TCR) are useful metrics summarizing the global climate system's temperature response to an externally imposed radiative forcing (RF). ECS is defined as the equilibrium change in annual mean global surface temperature following a doubling of the atmospheric CO_2 concentration (see Glossary), while TCR is defined as the annual mean global surface temperature change at the time of CO_2 doubling following a linear increase in CO_2 forcing over a period of 70 years (see Glossary). Both metrics have a broader application than these definitions imply: ECS determines the eventual warming in response to stabilization of atmospheric composition on multi-century time scales, while TCR determines the warming expected at a given time following any steady increase in forcing over a 50- to 100-year time scale.

ECS and TCR can be estimated from various lines of evidence. The estimates can be based on the values of ECS and TCR diagnosed from climate models (Section 9.7.1; Table 9.5), or they can be constrained by analysis of feedbacks in climate models (see Section 9.7.2), patterns of mean climate and variability in models compared to observations (Section 9.7.3.3), temperature fluctuations as reconstructed from paleoclimate archives (Sections 5.3.1 and 5.3.3.2; Box 5.1), observed and modelled short-term perturbations of the energy balance like those caused by volcanic eruptions (Section 10.8), and the observed surface and ocean temperature trends since pre-industrial (see Sections 10.8.1 and 10.8.2; Figure 10.20). For many applications, the limitations of the forcing-feedback analysis framework and the dependence of feedbacks on time scales and the climate state (see Section 12.5.3) must be kept in mind. Some studies estimate the TCR as the ratio of global mean temperature change to RF (Section 10.8.2.2) (Gregory and Forster, 2008; Padilla et al., 2011; Schwartz, 2012). Those estimates are scaled by the RF of 2 × CO_2 (3.7 W m^{-2}; Myhre et al., 1998) to be comparable to TCR in the following discussion.

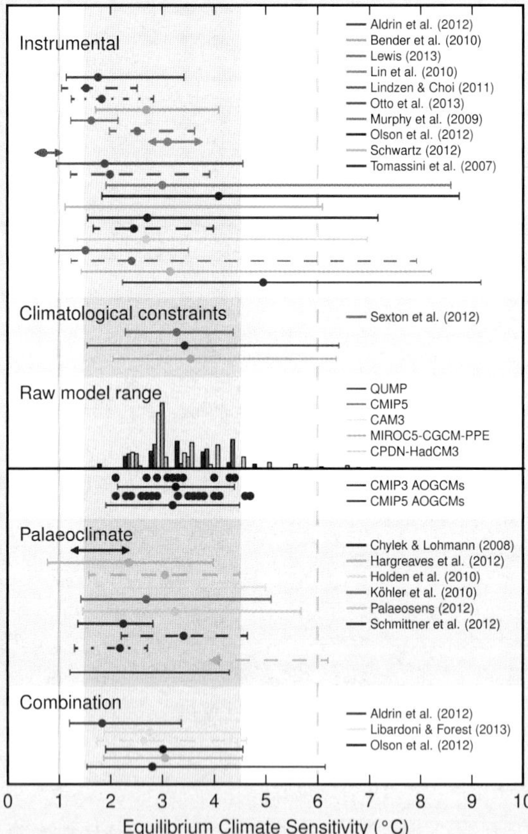

Box 12.2, Figure 1 | Probability density functions, distributions and ranges for equilibrium climate sensitivity, based on Figure 10.20b plus climatological constraints shown in IPCC AR4 (Meehl et al., 2007b; Box 10.2, Figure 1), and results from CMIP5 (Table 9.5). The grey shaded range marks the *likely* 1.5°C to 4.5°C range, and the grey solid line the *extremely unlikely* less than 1°C, the grey dashed line the *very unlikely* greater than 6°C. See Figure 10.20b and Chapter 10 Supplementary Material for full caption and details. Labels refer to studies since AR4. Full references are given in Section 10.8.

Newer studies of constraints based on the observed warming since pre-industrial, analysed using simple and intermediate complexity models, improved statistical methods, and several different and newer data sets, are assessed in detail in Section 10.8.2. Together with results from feedback analysis and paleoclimate constraints (Sections 5.3.1 and 5.3.3.2; Box 5.1), but without considering the CMIP based evidence, these studies show ECS is *likely* between 1.5°C to 4.5°C (*medium confidence*) and *extremely unlikely* less than 1.0°C (see Section 10.8.2). A few studies argued for very low values of climate sensitivity, but many of them have received criticism in the literature (see Section 10.8.2). Estimates based on AOGCMs and feedback analysis indicate a range of 2°C to 4.5°C, with the CMIP5 model mean at 3.2°C, similar to CMIP3. A summary of published ranges and PDFs of ECS is given in Box 12.2, Figure 1. Distributions and ranges for the TCR are shown in Box 12.2, Figure 2.

Simultaneously imposing different constraints from the observed warming trends, volcanic eruptions, model climatology, and paleoclimate, for example, by using a distribution obtained from the Last Glacial Maximum as a prior for the 20th century analysis, yields a more narrow range for climate sensitivity (see Figure 10.20; Section 10.8.2.5) (e.g., Annan and Hargreaves, 2006, 2011b; Hegerl et al., 2006; Aldrin et al., 2012). However, such methods are sensitive to assumptions of independence of the various lines of evidence, which might have shared biases (Lemoine, 2010), and the assumption that each individual line of evidence is unbiased and its uncertainties are captured completely. Expert elicitations for PDFs of climate sensitivity exist (Morgan and Keith, 1995; Zickfeld et al., 2010), but have also received some criticism (Millner et al., 2013). They are not used formally here because the experts base their opinion on the same studies as we assess. The peer-reviewed literature provides no consensus on a

(continued on next page)

Box 12.2 (continued)

formal statistical method to combine different lines of evidence. All methods in general are sensitive to the assumed prior distributions. These limitations are discussed in detail in Section 10.8.2.

Based on the combined evidence from observed climate change including the observed 20th century warming, climate models, feedback analysis and paleoclimate, ECS is *likely* in the range 1.5°C to 4.5°C with *high confidence*. The combined evidence increases the confidence in this final assessment compared to that based on the observed warming and paleoclimate only. ECS is positive, *extremely unlikely* less than 1°C (*high confidence*), and *very unlikely* greater than 6°C (*medium confidence*). The upper limit of the *likely* range is unchanged compared to AR4. The lower limit of the *likely* range of 1.5°C is less than the lower limit of 2°C in AR4. This change reflects the evidence from new studies of observed temperature change, using the extended records in atmosphere and ocean. These studies suggest a best fit to the observed surface and ocean warming for ECS values in the lower part of the *likely* range. Note that these studies are not purely observational, because they require an estimate of the response to RF from models. In addition, the uncertainty in ocean heat uptake remains substantial (see Section 3.2, Box 13.1). Accounting for short term variability in simple models remains challenging, and it is important not to give undue weight to any short time period that might be strongly affected by internal variability (see Box 9.2). On the other hand, AOGCMs show very good agreement with observed climatology with ECS values in the upper part of the 1.5°C to 4.5°C range (Section 9.7.3.3), but the simulation of key feedbacks like clouds remains challenging in those models. The estimates from the observed warming, paleoclimate, and from climate models are consistent within their uncertainties, each is supported by many studies and multiple data sets, and in combination they provide *high confidence* for the assessed *likely* range. Even though this assessed range is similar to previous reports (Charney, 1979; IPCC, 2001), confidence today is much higher as a result of high quality and longer observational records with a clearer anthropogenic signal, better process understanding, more and better understood evidence from paleoclimate reconstructions, and better climate models with higher resolution that capture many more processes more realistically. Box 12.2 Figure 1 illustrates that all these lines of evidence individually support the assessed *likely* range of 1.5°C to 4.5°C.

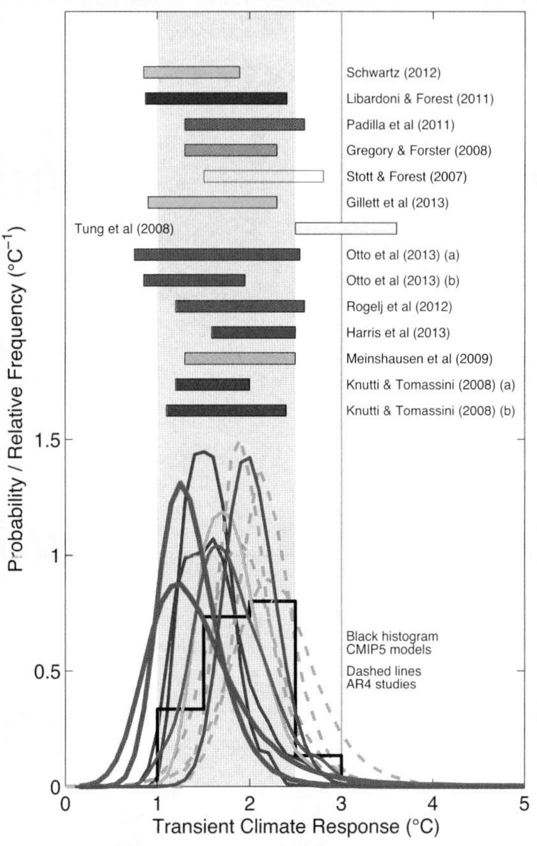

Box 12.2, Figure 2 | Probability density functions, distributions and ranges (5 to 95%) for the transient climate response from different studies, based on Figure 10.20a, and results from CMIP5 (black histogram; Table 9.5). The grey shaded range marks the *likely* 1°C to 2.5°C range, and the grey solid line marks the *extremely unlikely* greater than 3°C. See Figure 10.20a and Chapter 10 Supplementary Material for full caption and details. Full references are given in Section 10.8.

The tails of the ECS distribution are now better understood. Multiple lines of evidence provide *high confidence* that an ECS value less than 1°C is *extremely unlikely*. The assessment that ECS is *very unlikely* greater than 6°C is an expert judgment informed by several lines of evidence. First, the comprehensive climate models used in the CMIP5 exercise produce an ECS range of 2.1°C to 4.7°C (Table 9.5), very similar to CMIP3. Second, comparisons of perturbed-physics ensembles against the observed climate find that models with ECS values in the range 3°C to 4°C show the smallest errors for many fields (Section 9.7.3.3). Third, there is increasing evidence that the aerosol RF of the 20th century is not strongly negative, which makes it unlikely that the observed warming was caused by a very large ECS in response to a very small net forcing. Fourth, multiple and at least partly independent observational constraints from the satellite period, instrumental period and palaeoclimate studies continue to yield very low probabilities for ECS larger than 6°C, particularly when including most recent ocean and atmospheric data (see Box 12.2, Figure 1).

Analyses of observations and simulations of the instrumental period are estimating the effective climate sensitivity (a measure of the strengths of the climate feedbacks today, see Glossary), rather than ECS directly. In some climate models ECS tends to be higher than the effective climate sensitivity (see Section 12.5.3), because the feedbacks that are represented in the models (water vapour, lapse

(continued on next page)

Box 12.2 (continued)

rate, albedo and clouds) vary with the climate state. On time scales of many centuries, additional feedbacks with their own intrinsic time scales (e.g., vegetation, ice sheets; see Sections 5.3.3 and 12.5.3) (Jones et al., 2009; Goelzer et al., 2011) may become important but are not usually modelled. The resulting Earth system sensitivity is less well constrained but likely to be larger than ECS (Hansen et al., 2008; Rohling et al., 2009; Lunt et al., 2010; Pagani et al., 2010; Rohling and Members, 2012), implying that lower atmospheric CO_2 concentrations are needed to meet a given temperature target on multi-century time scales. A number of caveats, however, apply to those studies (see Section 12.5.3). Those long-term feedbacks have their own intrinsic time scales, and are less likely to be proportional to global mean temperature change.

For scenarios of increasing RF, TCR is a more informative indicator of future climate than ECS (Frame et al., 2005; Held et al., 2010). This assessment concludes with *high confidence* that the TCR is *likely* in the range 1°C to 2.5°C, close to the estimated 5 to 95% range of CMIP5 (1.2°C to 2.4°C; see Table 9.5), is positive and *extremely unlikely* greater than 3°C. As with the ECS, this is an expert-assessed range, supported by several different and partly independent lines of evidence, each based on multiple studies, models and data sets. TCR is estimated from the observed global changes in surface temperature, ocean heat uptake and RF, the detection/attribution studies identifying the response patterns to increasing GHG concentrations (Section 10.8.1), and the results of CMIP3 and CMIP5 (Section 9.7.1). Estimating TCR suffers from fewer difficulties in terms of state- or time-dependent feedbacks (see Section 12.5.3), and is less affected by uncertainty as to how much energy is taken up by the ocean. Unlike ECS, the ranges of TCR estimated from the observed warming and from AOGCMs agree well, increasing our confidence in the assessment of uncertainties in projections over the 21st century.

Another useful metric relating directly CO_2 emissions to temperature is the transient climate response to cumulative carbon emission (TCRE) (see Sections 12.5.4 and 10.8.4). This metric is useful to determine the allowed cumulative carbon emissions for stabilization at a specific global temperature. TCRE is defined as the annual mean global surface temperature change per unit of cumulated CO_2 emissions, usually 1000 PgC, in a scenario with continuing emissions (see Glossary). It considers physical and carbon cycle feedbacks and uncertainties, but not additional feedbacks associated for example with the release of methane hydrates or large amounts of carbon from permafrost. The assessment based on climate models as well as the observed warming suggests that the TCRE is *likely* between 0.8°C to 2.5°C per 1000 PgC (10^{12} metric tons of carbon), for cumulative CO_2 emissions less than about 2000 PgC until the time at which temperatures peak. Under these conditions, and for low to medium estimates of climate sensitivity, the TCRE gives an accurate estimate of the peak global mean temperature response to cumulated carbon emissions. TCRE has the advantage of directly relating global mean surface temperature change to CO_2 emissions, but as a result of combining the uncertainty in both TCR and the carbon cycle response, it is more uncertain. It also ignores non-CO_2 forcings and the fact that other components of the climate system (e.g., sea level rise, ice sheets) have their own intrinsic time scales, resulting in climate change not avoided by limiting global temperature change.

12.5.4.3 Conclusions and Limitations

One difficulty with the concepts of climate stabilization and targets is that stabilization of global temperature does not imply stabilization for all aspects of the climate system. For example, some models show significant hysteresis behaviour in the global water cycle, because global precipitation depends on both atmospheric CO_2 and temperature (Wu et al., 2010). Processes related to vegetation changes (Jones et al., 2009) or changes in the ice sheets (Charbit et al., 2008; Ridley et al., 2010) as well as ocean acidification, deep ocean warming and associated sea level rise (Meehl et al., 2005b; Wigley, 2005; Zickfeld et al., 2013) (see Figure 12.44d), and potential feedbacks linking, for example, ocean and the ice sheets (Gillett et al., 2011; Goelzer et al., 2011), have their own intrinsic long time scales. Those will result in significant changes hundreds to thousands of years after global temperature is stabilized. Thermal expansion, in contrast to global mean temperature, also depends on the evolution of surface temperature (Stouffer and Manabe, 1999; Bouttes et al., 2013; Zickfeld et al., 2013).

The simplicity of the concept of a cumulative carbon emission budget makes it attractive for policy (WBGU, 2009). The principal driver of long term warming is the total cumulative emission of CO_2 over time. To limit warming caused by CO_2 emissions to a given temperature target, cumulative CO_2 emissions from all anthropogenic sources therefore need to be limited to a certain budget. Higher emissions in earlier decades simply imply lower emissions by the same amount later on. This is illustrated in the RCP2.6 scenario in Figure 12.46a/b. Two idealized emission pathways with initially higher emissions (even sustained at high level for a decade in one case) eventually lead to the same warming if emissions are then reduced much more rapidly. Even a stepwise emission pathway with levels constant at 2010 and zero near mid-century would eventually lead to a similar warming as they all have identical cumulative emissions.

However, several aspects related to the concept of a cumulative carbon emission budget should be kept in mind. The ratio of global temperature and cumulative carbon is only approximately constant. It is the result of an interplay of several compensating carbon cycle and climate feedback processes operating on different time scales (a cancellation of variations in the increase in RF per ppm of CO_2, the ocean heat uptake efficiency and the airborne fraction) (Gregory et al., 2009; Matthews et al., 2009; Solomon et al., 2009). It depends on the modelled climate sensitivity and carbon cycle feedbacks. Thus, the allowed emissions for a given temperature target are uncertain (see Figure 12.45) (Matthews et al., 2009; Zickfeld et al., 2009; Knutti and Plattner, 2012). Nevertheless, the relationship is nearly linear in all models. Most models do not consider the possibility that long term feedbacks (Hansen et al., 2007; Knutti and Hegerl, 2008) may be different (see Section 12.5.3). Despite the fact that stabilization refers to equilibrium, the results assessed here are primarily relevant for the next few centuries and may differ for millennial scales. Notably, many of these limitations apply similarly to other policy targets, for example, stabilizing the atmospheric CO_2 concentration.

Non-CO_2 forcing constituents are important, which requires either assumptions on how CO_2 emission reductions are linked to changes in other forcings (Meinshausen et al., 2006; Meinshausen et al., 2009; McCollum et al., 2013), or separate emission budgets and climate modelling for short-lived and long-lived gases. So far, many studies ignored non-CO_2 forcings altogether. Those that consider them find significant effects, in particular warming of several tenths of a degree for abrupt reductions in emissions of short-lived species, like aerosols (Brasseur and Roeckner, 2005; Hare and Meinshausen, 2006; Zickfeld et al., 2009; Armour and Roe, 2011; Tanaka and Raddatz, 2011) (see also FAQ 12.3). Other studies, which model reductions that explicitly target warming from short-lived non-CO_2 species only, find important short-term cooling benefits shortly after the reduction of these species (Shindell et al., 2012), but do not extend beyond 2030.

The concept of cumulative carbon also implies that higher initial emissions can be compensated by a faster decline in emissions later or by negative emissions. However, in the real world short-term and long-term goals are not independent and mitigation rates are limited by economic constraints and existing infrastructure (Rive et al., 2007; Mignone et al., 2008; Meinshausen et al., 2009; Davis et al., 2010; Friedlingstein et al., 2011; Rogelj et al., 2013). An analysis of 193 published emission pathways with an energy balance model (UNEP, 2010; Rogelj et al., 2011) is shown in Figure 12.46c, d. Those emission pathways that *likely* limit warming below 2°C (above pre-industrial) by 2100 show emissions of about 31 to 46 Pg(CO_2-eq) yr^{-1} and 17 to 23 Pg(CO_2-eq) yr^{-1} by 2020 and 2050, respectively. Median 2010 emissions of all models are 48 Pg(CO_2-eq) yr^{-1}. Note that, as opposed to Figure 12.46a, b, many scenarios still have positive emissions in 2100. As these will not be zero immediately after 2100, they imply that the warming may exceed the target after 2100.

The aspects discussed above do not limit the robustness of the overall scientific assessment, but highlight factors that need to be considered when determining cumulative CO_2 emissions consistent with a given temperature target. In conclusion, taking into account the available information from multiple lines of evidence (observations, models and process understanding), the near linear relationship between cumulative CO_2 emissions and peak global mean temperature is well established in the literature and robust for cumulative total CO_2 emissions up to about 2000 PgC. It is consistent with the relationship inferred from past cumulative CO_2 emissions and observed warming, is supported by process understanding of the carbon cycle and global energy balance, and emerges as a robust result from the entire hierarchy of models.

Using a best estimate for the TCRE would provide a most likely value for the cumulative CO_2 emissions compatible with stabilization at a given temperature. However, such a budget would imply about 50% probability for staying below the temperature target. Higher probabilities for staying below a temperature or concentration target require significantly lower budgets (Knutti et al., 2005; Meinshausen et al., 2009; Rogelj et al., 2012). Based on the assessment of TCRE (assuming a normal distribution with a ±1 standard deviation range of 0.8-2.5°C per 1000 PgC), limiting the warming caused by anthropogenic CO_2 emissions alone (i.e., ignoring other radiative forcings) to less than 2°C since the period 1861–1880 with a probability of >33%, >50% and >66%, total CO_2 emissions from all anthropogenic sources would need to be below a cumulative budget of about 1570 PgC, 1210 PgC and 1000 PgC since 1870, respectively. An amount of 515 [445 to 585] PgC was emitted between 1870 and 2011. Accounting for non-CO_2 forcings contributing to peak warming, or requiring a higher likelihood of temperatures remaining below 2°C, both imply lower cumulative CO_2 emissions. A possible release of GHGs from permafrost or methane hydrates, not accounted for in current models, would also further reduce the anthropogenic CO_2 emissions compatible with a given temperature target. When accounting for the non-CO_2 forcings as in the RCP scenarios, compatible carbon emissions since 1870 are reduced to about 900 PgC, 820 PgC and 790 PgC to limit warming to less than 2°C since the period 1861–1880 with a probability of >33%, >50%, and >66%, respectively. These estimates were derived by computing the fraction of CMIP5 ESMs and EMICs that stay below 2°C for given cumulative emissions following RCP8.5, as shown in TFE.8 Figure 1c. The non-CO_2 forcing in RCP8.5 is higher than in RCP2.6. Because all likelihood statements in calibrated IPCC language are open intervals, the provided estimates are thus both conservative and consistent choices valid for non-CO_2 forcings across all RCP scenarios. There is no RCP scenario which limits warming to 2°C with probabilities of >33% or >50%, and which could be used to directly infer compatible cumulative emissions. For a probability of >66% RCP2.6 can be used as a comparison. Combining the average back-calculated fossil fuel carbon emissions for RCP2.6 between 2012 and 2100 (270 PgC) with the average historical estimate of 515 PgC gives a total of 785 PgC, i.e., 790 PgC when rounded to 10 PgC. As the 785 PgC estimate excludes an explicit assessment of future land-use change emissions, the 790 PgC value also remains a conservative estimate consistent with the overall likelihood assessment. The ranges of emissions for these three likelihoods based on the RCP scenarios are rather narrow, as they are based on a single scenario and on the limited sample of models available (TFE.8 Figure 1c). In contrast to TCRE they do not include observational constraints or account for sources of uncertainty not sampled by the models. The concept of a fixed cumulative CO_2 budget holds not just for 2°C, but for any temperature level explored with models so far (up to about 5°C; see Figures 12.44 to 12.46), with higher temperature levels implying larger budgets.

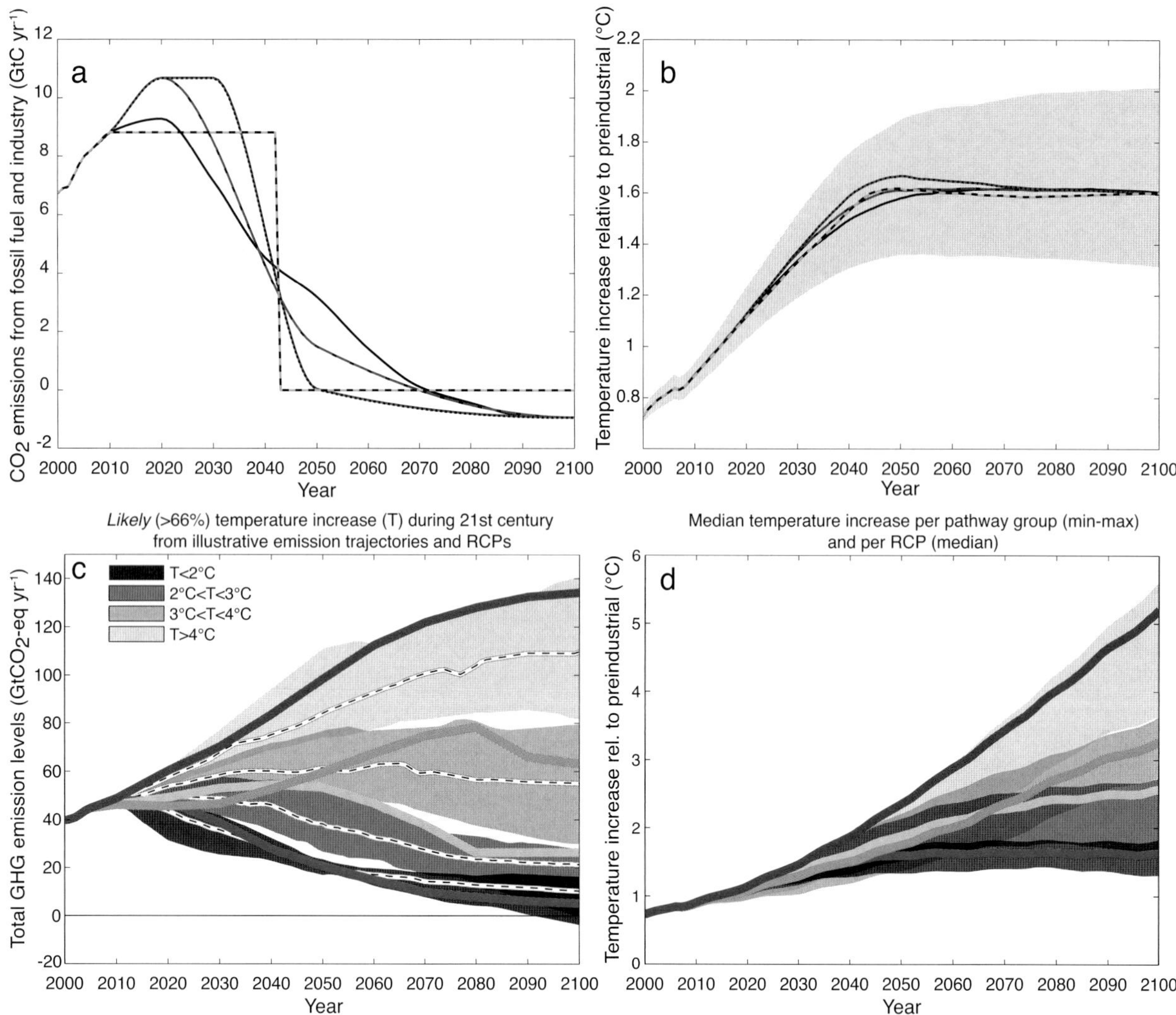

Figure 12.46 | (a) CO_2 emissions for the RCP2.6 scenario (black) and three illustrative modified emission pathways leading to the same warming. (b) Global temperature change relative to pre-industrial for the pathways shown in panel (a). (c) Grey shaded bands show Integrated Assessment Model (IAM) emission pathways over the 21st century. The pathways were grouped based on ranges of *likely* avoided temperature increase in the 21st century. Pathways in the darkest three bands *likely* stay below 2°C, 3°C, 4°C by 2100, respectively (see legend), while those in the lightest grey band are higher than that. Emission corridors were defined by, at each year, identifying the 15th to 85th percentile range of emissions and drawing the corresponding bands across the range. Individual scenarios that follow the upper edge of the bands early on tend to follow the lower edge of the band later on. Black-white lines show median paths per range. (d) Global temperature relative to pre-industrial for the pathways in (c). (Data in (c) and (d) based on Rogelj et al. (2011).) Coloured lines in (c) and (d) denote the four RCP scenarios.

12

12.5.5 Potentially Abrupt or Irreversible Changes

12.5.5.1 Introduction

This report adopts the definition of abrupt climate change used in Synthesis and Assessment Product 3.4 of the U.S. Climate Change Science Program CCSP (CCSP, 2008b). We define *abrupt climate change* as a large-scale change in the climate system that takes place over a few decades or less, persists (or is anticipated to persist) for at least a few decades, and causes substantial disruptions in human and natural systems (see Glossary). Other definitions of abrupt climate change exist. For example, in the AR4 climate change was defined as abrupt if it occurred faster than the typical time scale of the responsible forcing.

A number of components or phenomena within the Earth system have been proposed as potentially possessing critical thresholds (sometimes referred to as tipping points (Lenton et al., 2008)), beyond which abrupt or nonlinear transitions to a different state ensues. The term irreversibility is used in various ways in the literature. The AR5 report defines a perturbed state as *irreversible* on a given time scale if the recovery time scale from this state due to natural processes is significantly longer than the time it takes for the system to reach this perturbed state (see Glossary). In that context, most aspects of the climate change resulting from CO_2 emissions are irreversible, due to the long residence time of the CO_2 perturbation in the atmosphere and the resulting warming (Solomon et al., 2009). These results are discussed in Sections 12.5.2 to 12.5.4. Here, we also assess aspects of irreversibility in the context of abrupt change, multiple steady states and hysteresis, i.e., the question whether a change (abrupt or not) would be reversible if the forcing was reversed or removed (e.g., Boucher et al., 2012). Irreversibility of ice sheets and sea level rise are also assessed in Chapter 13.

Table 12.4 | Components in the Earth system that have been proposed in the literature as potentially being susceptible to abrupt or irreversible change. Column 2 defines whether or not a potential change can be considered to be abrupt under the AR5 definition. Column 3 states whether or not the process is irreversible in the context of abrupt change, and also gives the typical recovery time scales. Column 4 provides an assessment, if possible, of the likelihood of occurrence of abrupt change in the 21st century for the respective components or phenomena within the Earth system, for the scenarios considered in this chapter.

Change in climate system component	Potentially abrupt (AR5 definition)	Irreversibility if forcing reversed	Projected likelihood of 21st century change in scenarios considered
Atlantic MOC collapse	Yes	Unknown	*Very unlikely* that the AMOC will undergo a rapid transition (*high confidence*)
Ice sheet collapse	No	Irreversible for millennia	*Exceptionally unlikely* that either Greenland or West Antarctic Ice sheets will suffer near-complete disintegration (*high confidence*)
Permafrost carbon release	No	Irreversible for millennia	Possible that permafrost will become a net source of atmospheric greenhouse gases (*low confidence*)
Clathrate methane release	Yes	Irreversible for millennia	*Very unlikely* that methane from clathrates will undergo catastrophic release (*high confidence*)
Tropical forests dieback	Yes	Reversible within centuries	*Low confidence* in projections of the collapse of large areas of tropical forest
Boreal forests dieback	Yes	Reversible within centuries	*Low confidence* in projections of the collapse of large areas of boreal forest
Disappearance of summer Arctic sea ice	Yes	Reversible within years to decades	*Likely* that the Arctic Ocean becomes nearly ice-free in September before mid-century under high forcing scenarios such as RCP8.5 (*medium confidence*)
Long-term droughts	Yes	Reversible within years to decades	*Low confidence* in projections of changes in the frequency and duration of megadroughts
Monsoonal circulation	Yes	Reversible within years to decades	*Low confidence* in projections of a collapse in monsoon circulations

In this section we examine the main components or phenomena within the Earth system that have been proposed in the literature as potentially being susceptible to abrupt or irreversible change (see Table 12.4). Abrupt changes that arise from nonlinearities within the climate system are inherently difficult to assess and their timing, if any, of future occurrences is difficult to predict. Nevertheless, progress is being made exploring the potential existence of early warning signs for abrupt climate change (see e.g., Dakos et al., 2008; Scheffer et al., 2009).

12.5.5.2 The Atlantic Meridional Overturning

EMICs for which the stability has been systematically assessed by suitably designed hysteresis experiments robustly show a threshold beyond which the Atlantic thermohaline circulation cannot be sustained (Rahmstorf et al., 2005). This is also the case for one low-resolution ESM (Hawkins et al., 2011). However, proximity to this threshold is highly model dependent and influenced by factors that are currently poorly understood. There is some indication that the CMIP3 climate models may generally overestimate the stability of the Atlantic Ocean circulation (Hofmann and Rahmstorf, 2009; Drijfhout et al., 2010). In particular, De Vries and Weber (2005), Dijkstra (2007), Weber et al. (2007), Huisman et al. (2010), Drijfhout et al. (2010) and Hawkins et al. (2011) suggest that the sign of net freshwater flux into the Atlantic transported through its southern boundary via the overturning circulation determines whether or not the AMOC is in a mono-stable or bi-stable state. For the pre-industrial control climate of most of the CMIP3 models, Drijfhout et al. (2010) found that the salt flux was negative (implying a positive freshwater flux), indicating that they were in a mono-stable regime. However, this is not the case in the CMIP5 models where Weaver et al. (2012) found that the majority of the models were in a bi-stable regime during RCP integrations. Observations suggest that the present day ocean is in a bi-stable regime, thereby allowing for multiple equilibria and a stable 'off' state of the AMOC (Bryden et al., 2011; Hawkins et al., 2011).

In addition to the main threshold for a complete breakdown of the circulation, others may exist that involve more limited changes, such as a cessation of Labrador Sea deep water formation (Wood et al., 1999). Rapid melting of the Greenland ice sheet causes increases in freshwater runoff, potentially weakening the AMOC. None of the CMIP5 simulations include an interactive ice sheet component. However, Jungclaus et al. (2006), Mikolajewicz et al. (2007), Driesschaert et al. (2007) and Hu et al. (2009) found only a slight temporary effect of increased melt water fluxes on the AMOC, that was either small compared to the effect of enhanced poleward atmospheric moisture transport or only noticeable in the most extreme scenarios.

Although many more model simulations have been conducted since the AR4 under a wide range of forcing scenarios, projections of the AMOC behaviour have not changed. Based on the available CMIP5 models, EMICs and the literature, it remains *very likely* that the AMOC will weaken over the 21st century relative to pre-industrial. Best estimates and ranges for the reduction from CMIP5 are 11% (1 to 24%) in RCP2.6 and 34% (12 to 54%) in RCP8.5 (Weaver et al., 2012) (see Section 12.4.7.2, Figure 12.35). But there is *low confidence* in the magnitude of the weakening. Drijfhout et al. (2012) show that the AMOC decrease per degree global mean temperature rise varies from 1.5 to 1.9 Sv ($10^6\,m^3\,s^{-1}$) for the CMIP5 multi-model ensemble members they considered depending on the scenario, but that the standard deviation in this regression is almost half the signal.

The FIO-ESM model shows cooling over much of the NH that may be related to a strong reduction of the AMOC in all RCP scenarios (even RCP2.6), but the limited output available from the model precludes an assessment of the response and realism of this response. Hence it is not included the overall assessment of the likelihood of abrupt changes.

It is *unlikely* that the AMOC will collapse beyond the end of the 21st century for the scenarios considered but a collapse beyond the 21st century for large sustained warming cannot be excluded.There is *low confidence* in assessing the evolution of the AMOC beyond the 21st century. Two of the CMIP5 models revealed an eventual slowdown of the AMOC to an off state (Figure 12.35). But this did not occur abruptly.

As assessed by Delworth et al. (2008), for an abrupt transition of the AMOC to occur, the sensitivity of the AMOC to forcing would have to be far greater that seen in current models. Alternatively, significant ablation of the Greenland ice sheet greatly exceeding even the most aggressive of current projections would be required (Swingedouw et al., 2007; Hu et al., 2009). While neither possibility can be excluded entirely, it is *unlikely* that the AMOC will collapse beyond the end of the 21st century because of global warming based on the models and range of scenarios considered.

12.5.5.3 Ice Sheets

As detailed in Section 13.4.3, all available modelling studies agree that the Greenland ice sheet will significantly decrease in area and volume in a warmer climate as a consequence of increased melt rates not compensated for by increased snowfall rates and amplified by positive feedbacks. Conversely, the surface mass balance of the Antarctic ice sheet is projected to increase in most projections because increased snowfall rates outweigh melt increase (see Section 13.4.4).

Irreversibility of ice sheet volume and extent changes can arise because of the surface-elevation feedback that operates when a decrease of the elevation of the ice sheet induces a decreased surface mass balance (generally through increased melting), and therefore essentially applies to Greenland. As detailed in Section 13.4.3.3, several stable states of the Greenland ice sheet might exist (Charbit et al., 2008; Ridley et al., 2010; Langen et al., 2012; Robinson et al., 2012; Solgaard and Langen, 2012), and the ice sheet might irreversibly shrink to a stable smaller state once a warming threshold is crossed for a certain amount of time, with the critical duration depending on how far the temperature threshold has been exceeded. Based on the available evidence (see Section 13.4.3.3), an irreversible decrease of the Greenland ice sheet due to surface mass balance changes appears *very unlikely* in the 21st century but *likely* on multi-centennial to millennial time scales in the strongest forcing scenarios.

In theory (Weertman, 1974; Schoof, 2007) ice sheet volume and extent changes can be abrupt because of the grounding line instability that can occur in coastal regions where bedrock is retrograde (i.e., sloping towards the interior of the ice sheet) and below sea level (see Section 4.4.4 and Box 13.2). This essentially applies to West Antarctica, but also to parts of Greenland and East Antarctica. Furthermore, ice shelf decay induced by oceanic or atmospheric warming might lead to abruptly accelerated ice flow further inland (De Angelis and Skvarca, 2003). Because ice sheet growth is usually a slow process, such changes could also be irreversible in the definition adopted here. The available evidence (see Section 13.4) suggests that it is *exceptionally unlikely* that the ice sheets of either Greenland or West Antarctica will suffer a near-complete disintegration during the 21st century. More generally, the potential for abrupt and/or irreversible ice sheet changes (or the initiation thereof) during the 21st century and beyond is discussed in detail in Sections 13.4.3 and 13.4.4.

12.5.5.4 Permafrost Carbon Storage

Since the IPCC AR4, estimates of the amount of carbon stored in permafrost have been significantly revised upwards (Tarnocai et al., 2009), putting the permafrost carbon stock to an equivalent of twice the atmospheric carbon pool (Dolman et al., 2010). Because of low carbon input at high latitudes, permafrost carbon is to a large part of Pleistocene (Zimov et al., 2006) or Holocene (Smith et al., 2004) origin, and its potential vulnerability is dominated by decomposition (Eglin et al., 2010). The conjunction of a long carbon accumulation time scale on one hand and potentially rapid permafrost thawing and carbon decomposition under warmer climatic conditions (Zimov et al., 2006; Schuur et al., 2009; Kuhry et al., 2010) on the other hand suggests potential irreversibility of permafrost carbon decomposition (leading to an increase of atmospheric CO_2 and/or CH_4 concentrations) on time scales of hundreds to thousands of years in a warming climate. Indeed, recent observations (Dorrepaal et al., 2009; Kuhry et al., 2010) suggest that this process, induced by widespread permafrost warming and thawing (Romanovsky et al., 2010), might be already occurring. However, the existing modelling studies of permafrost carbon balance under future warming that take into account at least some of the essential permafrost-related processes (Khvorostyanov et al., 2008; Wania et al., 2009; Koven et al., 2011; Schaefer et al., 2011; MacDougall et al., 2012; Schneider von Deimling et al., 2012) do not yield coherent results beyond the fact that present-day permafrost might become a net emitter of carbon during the 21st century under plausible future warming scenarios (*low confidence*). This also reflects an insufficient understanding of the relevant soil processes during and after permafrost thaw, including processes leading to stabilization of unfrozen soil carbon (Schmidt et al., 2011), and precludes a firm assessment of the amplitude of irreversible changes in the climate system potentially related to permafrost degassing and associated global feedbacks at this stage (see also Sections 6.4.3.4 and 6.4.7.2 and FAQ 6.1).

12.5.5.5 Atmospheric Methane from Terrestrial and Oceanic Clathrates

Model simulations (Fyke and Weaver, 2006; Reagan and Moridis, 2007; Lamarque, 2008; Reagan and Moridis, 2009) suggest that clathrate deposits in shallow regions (in particular at high latitude regions and in the Gulf of Mexico) are susceptible to destabilization via ocean warming. However, concomitant sea level rise due to changes in ocean mass enhances clathrate stability in the ocean (Fyke and Weaver, 2006). A recent assessment of the potential for a future abrupt release of methane was undertaken by the U.S. Climate Change Science Program (Synthesis and Assessment Product 3.4 see Brooke et al., 2008). They concluded that it was *very unlikely* that such a catastrophic release would occur this century. However, they argued that anthropogenic warming will *very likely* lead to enhanced methane emissions from both terrestrial and oceanic clathrates (Brooke et al., 2008). Although difficult to formally assess, initial estimates of the 21st century positive feedback from methane clathrate destabilization are small but not insignificant (Fyke and Weaver, 2006; Archer, 2007; Lamarque, 2008). Nevertheless, on multi-millennial time scales, the positive feedback to anthropogenic warming of such methane emissions is potentially larger (Archer and Buffett, 2005; Archer, 2007; Brooke et al., 2008). Once more, due to the difference between release and accumulation time scales, such emissions are irreversible. See also FAQ 6.1.

12.5.5.6 Tropical and Boreal Forests

12.5.5.6.1 Tropical forests

In today's climate, the strongest growth in the Amazon rainforest occurs during the dry season when strong insolation is combined with water drawn from underground aquifers that store the previous wet season's rainfall (Huete et al., 2006). AOGCMs do not agree about how the dry season length in the Amazon may change in the future under the SRES A1 scenario (Bombardi and Carvalho, 2009), but simulations with coupled regional climate/potential vegetation models are consistent in simulating an increase in dry season length, a 70% reduction in the areal extent of the rainforest by the end of the 21st century using the SRES A2 scenario, and an eastward expansion of the Caatinga vegetation (Cook and Vizy, 2008; Sorensson et al., 2010). In addition, some models have demonstrated the existence of multiple equilibria of the tropical South American climate–vegetation system (e.g., Oyama and Nobre, 2003). The transition could be abrupt when the dry season becomes too long for the vegetation to survive, although the resilience of the vegetation to a longer dry period may be increased by the CO_2 fertilization effect (Zelazowski et al., 2011). Deforestation may also increase dry season length (Costa and Pires, 2010) and drier conditions increase the likelihood of wildfires that, combined with fire ignition associated with human activity, can undermine the forest's resiliency to climate change (see also Section 6.4.8.1). If climate change brings drier conditions closer to those supportive of seasonal forests rather than rainforest, fire can act as a trigger to abruptly and irreversibly change the ecosystem (Malhi et al., 2009). However, the existence of refugia is an important determinant of the potential for the re-emergence of the vegetation (Walker et al., 2009).

Analysis of projected change in the climate–biome space of current vegetation distributions suggest that the risk of Amazonian forest dieback is small (Malhi et al., 2009), a finding supported by modelling when strong carbon dioxide fertilization effects on Amazonian vegetation are assumed (Rammig et al., 2010). However, the strength of CO_2 fertilization on tropical vegetation is poorly known (see Box 6.3). Uncertainty concerning the existence of critical thresholds in the Amazonian and other tropical rainforests purely driven by climate change therefore remains high, and so the possibility of a critical threshold being crossed in precipitation volume cannot be ruled out (Nobre and Borma, 2009; Good et al., 2011b, 2011c). Nevertheless, there is still some question as to whether a transition of the Amazonian or other tropical rainforests into a lower biomass state could result from the combined effects of limits to carbon fertilization, climate warming, potential precipitation decline in interaction with the effects of human land use.

12.5.5.6.2 Boreal forest

Evidence from field observations and biogeochemical modelling make it scientifically conceivable that regions of the boreal forest could tip into a different vegetation state under climate warming, but uncertainties on the likelihood of this occurring are very high (Lenton et al., 2008; Allen et al., 2010). This is mainly due to large gaps in knowledge concerning relevant ecosystemic and plant physiological responses to warming (Niinemets, 2010). The main response is a potential advancement of the boreal forest northward and the potential transition from a forest to a woodland or grassland state on its dry southern edges in the continental interiors, leading to an overall increase in herbaceous vegetation cover in the affected parts of the boreal zone (Lucht et al., 2006). The proposed potential mechanisms for decreased forest growth and/or increased forest mortality are: increased drought stress under warmer summer conditions in regions with low soil moisture (Barber et al., 2000; Dulamsuren et al., 2009, 2010); desiccation of saplings with shallow roots due to summer drought periods in the top soil layers, causing suppression of forest reproduction (Hogg and Schwarz, 1997); leaf tissue damage due to high leaf temperatures during peak summer temperatures under strong climate warming; and increased insect, herbivory and subsequent fire damage in damaged or struggling stands (Dulamsuren et al., 2008). The balance of effects controlling standing biomass, fire type and frequency, permafrost thaw depth, snow volume and soil moisture remains uncertain. Although the existence of, and the thresholds controlling, a potential critical threshold in the boreal forest are extremely uncertain, its existence cannot at present be ruled out.

12.5.5.7 Sea Ice

Several studies based on observational data or model hindcasts suggest that the rapidly declining summer Arctic sea ice cover might reach or might already have passed a tipping point (Lindsay and Zhang, 2005; Wadhams, 2012; Livina and Lenton, 2013). Identifying Arctic sea ice tipping points from the short observational record is difficult due to high interannual and decadal variability. In some climate projections, the decrease in summer Arctic sea ice areal coverage is not gradual but is instead punctuated by 5- to10- year periods of strong ice loss (Holland et al., 2006; Vavrus et al., 2012; Döscher and Koenigk, 2013). Still, these abrupt reductions do not necessarily require the existence of a tipping point in the system or further imply an irreversible behaviour (Amstrup et al., 2010; Lenton, 2012). The 5- to 10-year events discussed by Holland et al. (2006) arise when large natural climate variability in the Arctic reinforces the anthropogenically-forced change (Holland et al., 2008). Positive trends on the same time scale also occur when internal variability counteracts the forced change until the middle of the 21st century (Holland et al., 2008; Kay et al., 2011; Vavrus et al., 2012).

Further work using single-column energy-balance models (Merryfield et al., 2008; Eisenman and Wettlaufer, 2009; Abbot et al., 2011) yielded mixed results about the possibility of tipping points and bifurcations in the transition from perennial to seasonal sea ice cover. Thin ice and snow covers promote strong longwave radiative loss to space and high ice growth rates (e.g., Bitz and Roe, 2004; Notz, 2009; Eisenman, 2012). These stabilizing negative feedbacks can be large enough to overcome the positive surface–albedo feedback and/or cloud feedback, which act to amplify the forced sea ice response. In such low-order models, the emergence of multiple stable states with increased climate forcing is a parameter-dependent feature (Abbot et al., 2011; Eisenman, 2012). For example, Eisenman (2012) showed with a single-column energy-balance model that certain parameter choices that cause thicker ice or warmer ocean under a given climate forcing make the model more prone to bifurcations and hence irreversible behaviour.

The reversibility of sea ice loss with respect to global or hemispheric mean surface temperature change has been directly assessed in AOGCMs/ESMs by first raising the CO_2 concentration until virtually all sea ice disappears year-round and then lowering the CO_2 level at the same rate as during the ramp-up phase until it reaches again the initial value (Armour et al., 2011; Boucher et al., 2012; Ridley et al., 2012; Li et al., 2013b). None of these studies show evidence of a bifurcation leading to irreversible changes in Arctic sea ice. AOGCMs have also been used to test summer Arctic sea ice recovery after either sudden or very rapid artificial removal, and all had sea ice return within a few years (Schröder and Connolley, 2007; Sedláček et al., 2011; Tietsche et al., 2011). In the Antarctic, as a result of the strong coupling between the Southern Ocean's surface and the deep ocean, the sea ice areal coverage in some of the models integrated with ramp-up and ramp-down atmospheric CO_2 concentration exhibits a significant lag relative to the global or hemispheric mean surface temperature (Ridley et al., 2012; Li et al., 2013b), so that its changes may be considered irreversible on centennial time scales.

Diagnostic analyses of a few global climate models have shown abrupt sea ice losses in the transition from seasonal to year-round Arctic ice-free conditions after raising CO_2 to very high levels (Winton, 2006b; Ridley et al., 2008; Li et al., 2013b), but without evidence for irreversible changes. Winton (2006b, 2008) hypothesized that the small ice cap instability (North, 1984) could cause such an abrupt transition. With a low-order Arctic sea ice model, Eisenman and Wettlaufer (2009) also found an abrupt change behaviour in the transition from seasonal ice to year-round ice-free conditions, accompanied by an irreversible bifurcation to a new stable, annually ice-free state. They concluded that the cause is a loss of the stabilizing effect of sea ice growth when the ice season shrinks in time. The Arctic sea ice may thus experience a sharp transition to annually ice-free conditions, but the irreversible nature of this transition seems to depend on the model complexity and structure.

In conclusion, rapid summer Arctic sea ice losses are *likely* to occur in the transition to seasonally ice-free conditions. These abrupt changes might have consequences throughout the climate system as noted by Vavrus et al. (2011) for cloud cover and Lawrence et al. (2008b) for the high-latitude ground state. Furthermore, the interannual-to-decadal variability in the summer Arctic sea ice extent is projected to increase in response to global warming (Holland et al., 2008; Goosse et al., 2009). These studies suggest that large anomalies in Arctic sea ice areal coverage, like the ones that occurred in 2007 and 2012, might become increasingly frequent. However, there is little evidence in global climate models of a tipping point (or critical threshold) in the transition from a perennially ice-covered to a seasonally ice-free Arctic Ocean beyond which further sea ice loss is unstoppable and irreversible.

12.5.5.8 Hydrologic Variability: Long-Term Droughts and Monsoonal Circulation

12.5.5.8.1 Long-term Droughts

As noted in Section 5.5.5, long-term droughts (often called megadroughts, see Glossary) are a recurring feature of Holocene paleoclimate records in North America, East and South Asia, Europe, Africa and India. The transitions into and out of the long-term droughts take many years. Because the long-term droughts all ended, they are not irreversible. Nonetheless transitions over years to a decade into a state of long-term drought would have impacts on human and natural systems.

AR4 climate model projections (Milly et al., 2008) and CMIP5 ensembles (Figure 12.23) both suggest widespread drying and drought across most of southwestern North America and many other subtropical regions by the mid to late 21st century (see Section 12.4.5), although without abrupt change. Some studies suggest that this subtropical drying may have already begun in southwestern North America (Seager et al., 2007; Seidel and Randel, 2007; Barnett et al., 2008; Pierce et al., 2008). More recent studies (Hoerling et al., 2010; Seager and Vecchi, 2010; Dai, 2011; Seager and Naik, 2012) suggest that regional reductions in precipitation are due primarily to internal variability and that the anthropogenic forced trends are currently weak in comparison.

While previous long-term droughts in southwest North America arose from natural causes, climate models project that this region will undergo progressive aridification as part of a general drying and poleward expansion of the subtropical dry zones driven by rising GHGs (Held and Soden, 2006; Seager et al., 2007; Seager and Vecchi, 2010). The models project the aridification to intensify steadily as RF and global warming progress without abrupt changes. Because of the very long lifetime of the anthropogenic atmospheric CO_2 perturbation, such drying induced by global warming would be largely irreversible on millennium time scale (Solomon et al., 2009; Frölicher and Joos, 2010; Gillett et al., 2011) (see Sections 12.5.2 and 12.5.4). For example, Solomon et al. (2009) found in a simulation where atmospheric CO_2 increases to 600 ppm followed by zero emissions, that the 15% reduction in precipitation in areas such as southwest North America, southern Europe and western Australia would persist long after emissions ceased. This, however, is largely a consequence of the warming persisting for centuries after emissions cease rather than an irreversible behaviour of the water cycle itself.

12.5.5.8.2 Monsoonal circulation

Climate model simulations and paleo-reconstructions provide evidence of past abrupt changes in Saharan vegetation, with the 'green Sahara' conditions (Hoelzmann et al., 1998) of the African Humid Period (AHP) during the mid-Holocene serving as the most recent example. However, Mitchell (1990) and Claussen et al. (2003) note that the mid-Holocene is not a direct analogue for future GHG-induced climate change since the forcings are different: a increased shortwave forcing in the NH summer versus a globally and seasonally uniform atmospheric CO_2 increase, respectively. Paleoclimate examples suggest that a strong radiative or SST forcing is needed to achieve a rapid climate change, and that the rapid changes are reversible when the forcing is withdrawn. Both the abrupt onset and termination of the AHP were triggered when northern African summer insolation was 4.2% higher than present day, representing a local increase of about 19 W m^{-2} (deMenocal et al., 2000). Note that the globally averaged radiative anthropogenic forcing from 1750 to 2011 (Table 8.6) is small compared to this local increase in insolation. A rapid Saharan greening has been simulated in a climate model of intermediate complexity forced by a rapid increase in atmospheric CO_2, with the overall extent of greening depending on the equilibrium atmospheric CO_2 level reached (Claussen et al., 2003).

Abrupt Saharan vegetation changes of the Younger Dryas are linked with a rapid AMOC weakening which is considered *very unlikely* during the 21st century and *unlikely* beyond that as a consequence of global warming.

Studies with conceptual models (Zickfeld et al., 2005; Levermann et al., 2009) have shown that the Indian summer monsoon can operate in two stable regimes: besides the 'wet' summer monsoon, a stable state exists which is characterized by low precipitation over India. These studies suggest that any perturbation of the radiative budget that tends to weaken the driving pressure gradient has the potential to induce abrupt transitions between these two regimes.

Numerous studies with coupled ocean–atmosphere models have explored the potential impact of anthropogenic forcing on the Indian monsoon (see also Section 14.2). When forced with anticipated increases in GHG concentrations, the majority of these studies show an intensification of the rainfall associated with the Indian summer monsoon (Meehl and Washington, 1993; Kitoh et al., 1997; Douville et al., 2000; Hu et al., 2000; May, 2002; Ueda et al., 2006; Kripalani et al., 2007; Stowasser et al., 2009; Cherchi et al., 2010). Despite the intensification of precipitation, several of these modelling studies show a weakening of the summer monsoon circulation (Kitoh et al., 1997; May, 2002; Ueda et al., 2006; Kripalani et al., 2007; Stowasser et al., 2009; Cherchi et al., 2010). The net effect is nevertheless an increase of precipitation due to enhanced moisture transport into the Asian monsoon region (Ueda et al., 2006). In recent years, studies with GCMs have also explored the direct effect of aerosol forcing on the Indian monsoon (Lau et al., 2006; Meehl et al., 2008; Randles and Ramaswamy, 2008; Collier and Zhang, 2009). Considering absorbing aerosols (black carbon) only, Meehl et al. (2008) found an increase in pre-monsoonal precipitation, but a decrease in summer monsoon precipitation over parts of South Asia. In contrast, Lau et al. (2006) found an increase in May–June–July precipitation in that region. If an increase in scattering aerosols only is considered, the monsoon circulation weakens and precipitation is inhibited (Randles and Ramaswamy, 2008). More recently, Bollasina et al. (2011) showed that anthropogenic aerosols played a fundamental role in driving the recent observed weakening of the summer monsoon. Given that the effect of increased atmospheric regional loading of aerosols is opposed by the concomitant increases in GHG concentrations, it is *unlikely* that an abrupt transition to the dry summer monsoon regime will be triggered in the 21st century.

Acknowledgements

We especially acknowledge the input of Contributing Authors Urs Beyerle for maintaining the database of CMIP5 output, Jan Sedláček for producing a large number of CMIP5 figures, and Joeri Rogelj for preparing synthesis figures. Chapter technical assistants Oliver Stebler, Franziska Gerber and Barbara Aellig, provided great help in assembling the chapter and Sébastien Denvil and Jérôme Raciazek provided technical assistance in downloading the CMIP5 data.

References

Abbot, D. S., M. Silber, and R. T. Pierrehumbert, 2011: Bifurcations leading to summer Arctic sea ice loss. *J. Geophys. Res.*, **116**, D19120.

Abe, M., H. Shiogama, T. Nozawa, and S. Emori, 2011: Estimation of future surface temperature changes constrained using the future-present correlated modes in inter-model variability of CMIP3 multimodel simulations. *J. Geophys. Res.*, **116**, D18104.

Adachi, Y., et al., 2013: Basic performance of a new earth system model of the Meteorological Research Institute (MRI-ESM1). *Papers Meteorol. Geophys.*, doi:10.2467/mripapers.64.

Adams, P. J., J. H. Seinfeld, D. Koch, L. Mickley, and D. Jacob, 2001: General circulation model assessment of direct radiative forcing by the sulfate-nitrate-ammonium-water inorganic aerosol system. *J. Geophys. Res.*, **106**, 1097–1111.

Adler, R. F., G. J. Gu, J. J. Wang, G. J. Huffman, S. Curtis, and D. Bolvin, 2008: Relationships between global precipitation and surface temperature on interannual and longer timescales (1979–2006). *J. Geophys. Res.*, **113**, D22104.

Aldrin, M., M. Holden, P. Guttorp, R. B. Skeie, G. Myhre, and T. K. Berntsen, 2012: Bayesian estimation of climate sensitivity based on a simple climate model fitted to observations of hemispheric temperatures and global ocean heat content. *Environmetrics*, **23**, 253–271.

Alexander, L. V., and J. M. Arblaster, 2009: Assessing trends in observed and modelled climate extremes over Australia in relation to future projections. *Int. J. Climatol.*, **29**, 417–435.

Alexander, L. V., et al., 2006: Global observed changes in daily climate extremes of temperature and precipitation. *J. Geophys. Res.*, **111**, D05109.

Alexeev, V., and C. Jackson, 2012: Polar amplification: Is atmospheric heat transport important? *Clim. Dyn.*, doi:10.1007/s00382-012-1601-z.

Alexeev, V., D. Nicolsky, V. Romanovsky, and D. Lawrence, 2007: An evaluation of deep soil configurations in the CLM3 for improved representation of permafrost. *Geophys. Res. Lett.*, **34**, L09502.

Alexeev, V. A., P. L. Langen, and J. R. Bates, 2005: Polar amplification of surface warming on an aquaplanet in "ghost forcing" experiments without sea ice feedbacks. *Clim. Dyn.*, **24**, 655–666.

Allan, R., and B. Soden, 2008: Atmospheric warming and the amplification of precipitation extremes. *Science*, **321**, 1481–1484.

Allan, R. P., 2012: Regime dependent changes in global precipitation. *Clim. Dyn.*, doi:10.1007/s00382-011-1134-x.

Allen, C., et al., 2010: A global overview of drought and heat-induced tree mortality reveals emerging climate change risks for forests. *Forest Ecol. Manage.*, **259**, 660–684.

Allen, M. R., and W. J. Ingram, 2002: Constraints on future changes in climate and the hydrologic cycle. *Nature*, **419**, 224–232.

Allen, M. R., D. J. Frame, C. Huntingford, C. D. Jones, J. A. Lowe, M. Meinshausen, and N. Meinshausen, 2009: Warming caused by cumulative carbon emissions towards the trillionth tonne. *Nature*, **458**, 1163–1166.

Allen, R. J., and S. C. Sherwood, 2008: Warming maximum in the tropical upper troposphere deduced from thermal winds. *Nature Geosci.*, **1**, 399–403.

Allen, R. J., and S. C. Sherwood, 2010: Aerosol-cloud semi-direct effect and land-sea temperature contrast in a GCM. *Geophys. Res. Lett.*, **37**, L07702.

Allen, R. J., S. C. Sherwood, J. R. Norris, and C. S. Zender, 2012: Recent Northern Hemisphere tropical expansion primarily driven by black carbon and tropospheric ozone. *Nature*, **485**, 350–354.

Amstrup, S., E. DeWeaver, D. Douglas, B. Marcot, G. Durner, C. Bitz, and D. Bailey, 2010: Greenhouse gas mitigation can reduce sea-ice loss and increase polar bear persistence. *Nature*, **468**, 955–958.

Andrews, T., and P. M. Forster, 2008: CO_2 forcing induces semi-direct effects with consequences for climate feedback interpretations. *Geophys. Res. Lett.*, **35**, L04802.

Andrews, T., P. M. Forster, and J. M. Gregory, 2009: A surface energy perspective on climate change. *J. Clim.*, **22**, 2557–2570.

Andrews, T., P. Forster, O. Boucher, N. Bellouin, and A. Jones, 2010: Precipitation, radiative forcing and global temperature change. *Geophys. Res. Lett.*, **37**, L14701.

Annan, J. D., and J. C. Hargreaves, 2006: Using multiple observationally-based constraints to estimate climate sensitivity. *Geophys. Res. Lett.*, **33**, L06704.

Annan, J. D., and J. C. Hargreaves, 2010: Reliability of the CMIP3 ensemble. *Geophys. Res. Lett.*, **37**, L02703.

Annan, J. D., and J. C. Hargreaves, 2011a: Understanding the CMIP3 multi-model ensemble. *J. Clim.*, **24**, 4529–4538.

Annan, J. D., and J. C. Hargreaves, 2011b: On the generation and interpretation of probabilistic estimates of climate sensitivity. *Clim. Change*, **104**, 423–436.

Arblaster, J. M., G. A. Meehl, and D. J. Karoly, 2011: Future climate change in the Southern Hemisphere: Competing effects of ozone and greenhouse gases. *Geophys. Res. Lett.*, **38**, L02701.

Archer, D., 2007: Methane hydrate stability and anthropogenic climate change. *Biogeosciences*, **4**, 521–544.

Archer, D., and B. Buffett, 2005: Time-dependent response of the global ocean clathrate reservoir to climatic and anthropogenic forcing. *Geochem. Geophys. Geosyst.*, **6**, Q03002.

Archer, D., et al., 2009: Atmospheric lifetime of fossil fuel carbon dioxide. *Annu. Rev. Earth Planet. Sci.*, **37**, 117–134.

Armour, K., and G. Roe, 2011: Climate commitment in an uncertain world. *Geophys. Res. Lett.*, **38**, L01707.

Armour, K., I. Eisenman, E. Blanchard-Wrigglesworth, K. McCusker, and C. Bitz, 2011: The reversibility of sea ice loss in a state-of-the-art climate model. *Geophys. Res. Lett.*, **38**, L16705.

Arora, V. K., et al., 2011: Carbon emission limits required to satisfy future representative concentration pathways of greenhouse gases. *Geophys. Res. Lett.*, **38**, L05805.

Arzel, O., T. Fichefet, and H. Goosse, 2006: Sea ice evolution over the 20th and 21st centuries as simulated by current AOGCMs. *Ocean Model.*, **12**, 401–415.

Augustsson, T., and V. Ramanathan, 1977: Radiative-convective model study of CO_2 climate problem. *J. Atmos. Sci.*, **34**, 448–451.

Bala, G., K. Caldeira, and R. Nemani, 2010: Fast versus slow response in climate change: Implications for the global hydrological cycle. *Clim. Dyn.*, **35**, 423–434.

Baldwin, M. P., M. Dameris, and T. G. Shepherd, 2007: Atmosphere—How will the stratosphere affect climate change? *Science*, **316**, 1576–1577.

Ballester, J., F. Giorgi, and X. Rodo, 2010a: Changes in European temperature extremes can be predicted from changes in PDF central statistics. *Clim. Change*, **98**, 277–284.

Ballester, J., X. Rodo, and F. Giorgi, 2010b: Future changes in Central Europe heat waves expected to mostly follow summer mean warming. *Clim. Dyn.*, **35**, 1191–1205.

Banks, H. T., and J. M. Gregory, 2006: Mechanisms of ocean heat uptake in a coupled climate model and the implications for tracer based predictions of ocean heat uptake. *Geophys. Res. Lett.*, **33**, L07608.

Bao, Q., et al., 2013: The Flexible Global Ocean-Atmosphere-Land system model, Spectral Version 2: FGOALS-s2. *Adv. Atmos. Sci.*, **30**, 561–576.

Barber, V., G. Juday, and B. Finney, 2000: Reduced growth of Alaskan white spruce in the twentieth century from temperature-induced drought stress. *Nature*, **405**, 668–673.

Barnes, E. A., and L. M. Polvani, 2013: Response of the midlatitude jets and of their variability to increased greenhouse gases in the CMIP5 models. *J. Clim.*, doi:10.1175/JCLI-D-12-00536.1.

Barnett, D. N., S. J. Brown, J. M. Murphy, D. M. H. Sexton, and M. J. Webb, 2006: Quantifying uncertainty in changes in extreme event frequency in response to doubled CO_2 using a large ensemble of GCM simulations. *Clim. Dyn.*, **26**, 489–511.

Barnett, T., and D. Pierce, 2008: When will Lake Mead go dry? *Water Resour. Res.*, **44**, W03201.

Barnett, T. P., et al., 2008: Human-induced changes in the hydrology of the western United States. *Science*, **319**, 1080–1083.

Barriopedro, D., E. M. Fischer, J. Luterbacher, R. Trigo, and R. Garcia-Herrera, 2011: The hot summer of 2010: Redrawing the temperature record map of Europe. *Science*, **332**, 220–224.

Bekryaev, R. V., I. V. Polyakov, and V. A. Alexeev, 2010: Role of polar amplification in long-term surface air temperature variations and modern Arctic warming. *J. Clim.*, **23**, 3888–3906.

Bellouin, N., J. Rae, A. Jones, C. Johnson, J. Haywood, and O. Boucher, 2011: Aerosol forcing in the Hadley Centre CMIP5 simulations and the role of ammonium nitrate. *J. Geophys. Res.*, **116**, D20206.

Bengtsson, L., K. I. Hodges, and E. Roeckner, 2006: Storm tracks and climate change. *J. Clim.*, **19**, 3518–3543.

Bengtsson, L., K. I. Hodges, and N. Keenlyside, 2009: Will extratropical storms intensify in a warmer climate? *J. Clim.*, **22**, 2276–2301.

Berg, P., J. O. Haerter, P. Thejll, C. Piani, S. Hagemann, and J. H. Christensen, 2009: Seasonal characteristics of the relationship between daily precipitation intensity and surface temperature. *J. Geophys. Res.*, **114**, D18102.

Betts, R., et al., 2007: Projected increase in continental runoff due to plant responses to increasing carbon dioxide. *Nature*, **448**, 1037–1041.

Bitz, C., and G. Roe, 2004: A mechanism for the high rate of sea ice thinning in the Arctic Ocean. *J. Clim.*, **17**, 3623–3632.

Bitz, C., and Q. Fu, 2008: Arctic warming aloft is data set dependent. *Nature*, **455**, E3–E4.

Bitz, C. M., 2008: Some aspects of uncertainty in predicting sea ice thinning. In: *Arctic Sea Ice Decline: Observations, Projections, Mechanisms, and Implications* [E. T. DeWeaver, C. M. Bitz and L. B. Tremblay (eds.)]. American Geophysical Union, Washington, DC, pp. 63–76.

Bitz, C. M., J. K. Ridley, M. M. Holland, and H. Cattle, 2012: Global climate models and 20th and 21st century Arctic climate change. In: *Arctic Climate Change – The ACSYS Decade and Beyond* [P. Lemke (ed.)]. Springer Science+Business Media, Dordrecht, Netherlands, pp. 405–436.

Boberg, F., P. Berg, P. Thejll, W. Gutowski, and J. Christensen, 2010: Improved confidence in climate change projections of precipitation evaluated using daily statistics from the PRUDENCE ensemble. *Clim. Dyn.*, **35**, 1097–1106.

Boé, J., and L. Terray, 2008: Uncertainties in summer evapotranspiration changes over Europe and implications for regional climate change. *Geophys. Res. Lett.*, **35**, L05702.

Boé, J., A. Hall, and X. Qu, 2009a: Current GCMs' unrealistic negative feedback in the Arctic. *J. Clim.*, **22**, 4682–4695.

Boé, J. L., A. Hall, and X. Qu, 2009b: September sea-ice cover in the Arctic Ocean projected to vanish by 2100. *Nature Geosci.*, **2**, 341–343.

Boer, G. J., 1993: Climate change and the regulation of the surface moisture and energy budgets. *Clim. Dyn.*, **8**, 225–239.

Boer, G. J., 2011: The ratio of land to ocean temperature change under global warming. *Clim. Dyn.*, **37**, 2253–2270.

Boer, G. J., and B. Yu, 2003: Climate sensitivity and response. *Clim. Dyn.*, **20**, 415–429.

Boer, G. J., K. Hamilton, and W. Zhu, 2005: Climate sensitivity and climate change under strong forcing. *Clim. Dyn.*, **24**, 685–700.

Bollasina, M. A., Y. Ming, and V. Ramaswamy, 2011: Anthropogenic aerosols and the weakening of the South Asian summer monsoon. *Science*, **334**, 502–505.

Bombardi, R., and L. Carvalho, 2009: IPCC global coupled model simulations of the South America monsoon system. *Clim. Dyn.*, **33**, 893–916.

Böning, C., A. Dispert, M. Visbeck, S. Rintoul, and F. Schwarzkopf, 2008: The response of the Antarctic Circumpolar Current to recent climate change. *Nature Geosci.*, **1**, 864–869.

Bony, S., and J. L. Dufresne, 2005: Marine boundary layer clouds at the heart of tropical cloud feedback uncertainties in climate models. *Geophys. Res. Lett.*, **32**, L20806.

Bony, S., G. Bellon, D. Klocke, S. Sherwood, S. Fermepin, and S. Denvil, 2013: Robust direct effect of carbon dioxide on tropical circulation and regional precipitation. *Nature Geosci.*, doi:10.1038/ngeo1799.

Bony, S., et al., 2006: How well do we understand and evaluate climate change feedback processes? *J. Clim.*, **19**, 3445–3482.

Booth, B. B. B., et al., 2012: High sensitivity of future global warming to land carbon cycle processes. *Environ. Res. Lett.*, **7**, 024002.

Boucher, O., et al., 2012: Reversibility in an Earth System model in response to CO_2 concentration changes. *Environ. Res. Lett.*, **7**, 024013.

Bouttes, N., J. M. Gregory, and J. A. Lowe, 2013: The reversibility of sea level rise. *J. Clim.*, **26**, 2502–2513.

Bowerman, N., D. Frame, C. Huntingford, J. Lowe, and M. Allen, 2011: Cumulative carbon emissions, emissions floors and short-term rates of warming: Implications for policy. *Philos. Trans. R. Soc. A*, **369**, 45–66.

Bracegirdle, T., and D. Stephenson, 2012: Higher precision estimates of regional polar warming by ensemble regression of climate model projections. *Clim. Dyn.*, **39**, 2805–2821.

Bracegirdle, T., W. Connolley, and J. Turner, 2008: Antarctic climate change over the twenty first century. *J. Geophys. Res.*, **113**, D03103.

Bracegirdle, T. J., et al., 2013: Assessment of surface winds over the Atlantic, Indian, and Pacific Ocean sectors of the Southern Ocean in CMIP5 models: Historical bias, forcing response, and state dependence. *J. Geophys. Res.*, **118**, 547–562.

Brasseur, G., and E. Roeckner, 2005: Impact of improved air quality on the future evolution of climate. *Geophys. Res. Lett.*, **32**, L23704.

Brient, F., and S. Bony, 2013: Interpretation of the positive low-cloud feedback predicted by a climate model under global warming. *Clim. Dyn.*, **40**, 2415–2431.

Brierley, C. M., M. Collins, and A. J. Thorpe, 2010: The impact of perturbations to ocean-model parameters on climate and climate change in a coupled model. *Clim. Dyn.*, **34**, 325–343.

Bromwich, D. H., J. P. Nicolas, A. J. Monaghan, M. A. Lazzara, L. M. Keller, G. A. Weidner, and A. B. Wilson, 2013: Central West Antarctica among the most rapidly warming regions on Earth. *Nature Geosci.*, **6**, 139–145.

Brooke, E., D. Archer, E. Dlugokencky, S. Frolking, and D. Lawrence, 2008: Potential for abrupt changes in atmospheric methane. *Abrupt Climate Change: A Report by the U.S. Climate Change Science Program and the Subcommittee on Global Change Research*. U.S. Geological Survey, Washington, DC, pp. 163–201.

Brooks, H. E., 2009: Proximity soundings for severe convection for Europe and the United States from reanalysis data. *Atmos. Res.*, **93**, 546–553.

Brooks, H. E., 2013: Severe thunderstorms and climate change. *Atmos. Res.*, **123**, 129–138.

Brooks, H. E., J. W. Lee, and J. P. Craven, 2003: The spatial distribution of severe thunderstorm and tornado environments from global reanalysis data. *Atmos. Res.*, **67–68**, 73–94.

Brovkin, V., et al., 2013: Effect of anthropogenic land-use and land cover changes on climate and land carbon storage in CMIP5 projections for the 21st century. *J. Clim.*, doi:10.1175/JCLI-D-12-00623.1.

Brown, R., and P. Mote, 2009: The response of Northern Hemisphere snow cover to a changing climate. *J. Clim.*, **22**, 2124–2145.

Brown, R. D., and D. A. Robinson, 2011: Northern Hemisphere spring snow cover variability and change over 1922–2010 including an assessment of uncertainty. *Cryosphere*, **5**, 219–229.

Brutel-Vuilmet, C., M. Menegoz, and G. Krinner, 2013: An analysis of present and future seasonal Northern Hemisphere land snow cover simulated by CMIP5 coupled climate models. *Cryosphere*, **7**, 67–80.

Bryan, K., F. G. Komro, S. Manabe, and M. J. Spelman, 1982: Transient climate response to increasing atmospheric carbon-dioxide. *Science*, **215**, 56–58.

Bryden, H. L., B. A. King, and G. D. McCarthy, 2011: South Atlantic overturning circulation at 24S. *J. Mar. Res.*, **69**, 38–55.

Burke, E., and S. Brown, 2008: Evaluating uncertainties in the projection of future drought. *J. Hydrometeorol.*, **9**, 292–299.

Burke, E. J., C. D. Jones, and C. D. Koven, 2012: Estimating the permafrost-carbon-climate response in the CMIP5 climate models using a simplified approach. *J. Clim.*, doi:10.1175/JCLI-D-12-00550.1.

Buser, C. M., H. R. Kunsch, D. Luthi, M. Wild, and C. Schär, 2009: Bayesian multi-model projection of climate: Bias assumptions and interannual variability. *Clim. Dyn.*, **33**, 849–868.

Butchart, N., and A. A. Scaife, 2001: Removal of chlorofluorocarbons by increased mass exchange between the stratosphere and troposphere in a changing climate. *Nature*, **410**, 799–802.

Butchart, N., et al., 2006: Simulations of anthropogenic change in the strength of the Brewer-Dobson circulation. *Clim. Dyn.*, **27**, 727–741.

Butchart, N., et al., 2010: Chemistry-climate model simulations of twenty-first century stratospheric climate and circulation changes. *J. Clim.*, **23**, 5349–5374.

Butler, A. H., D. W. J. Thompson, and R. Heikes, 2010: The steady-state atmospheric circulation response to climate change-like thermal forcings in a simple General Circulation Model. *J. Clim.*, **23**, 3474–3496.

Cabre, M. F., S. A. Solman, and M. N. Nunez, 2010: Creating regional climate change scenarios over southern South America for the 2020's and 2050's using the pattern scaling technique: Validity and limitations. *Clim. Change*, **98**, 449–469.

Caesar, J., and J. A. Lowe, 2012: Comparing the impacts of mitigation versus non-intervention scenarios on future temperature and precipitation extremes in the HadGEM2 climate model. *J. Geophys. Res.*, **117**, D15109.

Cagnazzo, C., E. Manzini, P. G. Fogli, M. Vichi, and P. Davini, 2013: Role of stratospheric dynamics in the ozone–carbon connection in the Southern Hemisphere. *Clim. Dyn.*, doi:10.1007/s00382-013-1745-5.

Cai, M., 2005: Dynamical amplification of polar warming. *Geophys. Res. Lett.*, **32**, L22710.

Caldeira, K., and J. F. Kasting, 1993: Insensitivity of global warming potentials to carbon-dioxide emission scenarios. *Nature*, **366**, 251–253.

Caldwell, P., and C. S. Bretherton, 2009: Response of a subtropical stratocumulus-capped mixed layer to climate and aerosol changes. *J. Clim.*, **22**, 20–38.

Calvo, N., and R. R. Garcia, 2009: Wave forcing of the tropical upwelling in the lower stratosphere under increasing concentrations of greenhouse gases. *J. Atmos. Sci.*, **66**, 3184–3196.

Calvo, N., R. R. Garcia, D. R. Marsh, M. J. Mills, D. E. Kinnison, and P. J. Young, 2012: Reconciling modeled and observed temperature trends over Antarctica. *Geophys. Res. Lett.*, **39**, L16803.

Cao, L., and K. Caldeira, 2010: Atmospheric carbon dioxide removal: Long-term consequences and commitment. *Environ. Res. Lett.*, **5**, 024011.

Cao, L., G. Bala, and K. Caldeira, 2012: Climate response to changes in atmospheric carbon dioxide and solar irradiance on the time scale of days to weeks. *Environ. Res. Lett.*, **7**, 034015.

Capotondi, A., M. Alexander, N. Bond, E. Curchitser, and J. Scott, 2012: Enhanced upper ocean stratification with climate change in the CMIP3 models. *J. Geophys. Res.*, **117**, C04031.

Cariolle, D., and H. Teyssedre, 2007: A revised linear ozone photochemistry parameterization for use in transport and general circulation models: Multi-annual simulations. *Atmos. Chem. Phys.*, **7**, 2183–2196.

Carslaw, K., O. Boucher, D. Spracklen, G. Mann, J. Rae, S. Woodward, and M. Kulmala, 2010: A review of natural aerosol interactions and feedbacks within the Earth system. *Atmos. Chem. Phys.*, **10**, 1701–1737.

Catto, J. L., L. C. Shaffrey, and K. I. Hodges, 2011: Northern Hemisphere extratropical cyclones in a warming climate in the HiGEM high-resolution climate model. *J. Clim.*, **24**, 5336–5352.

CCSP, 2008a: *Weather and Climate Extremes in a Changing Climate: A Report by the U.S. Climate Change Science Program and the Subcommittee on Global Change Research*. Department of Commerce, NOAA's National Climatic Data Center, College Park, MD, 164 pp.

CCSP, 2008b: *Abrupt Climate Change. A Report by the U.S. Climate Change Science Program and the Subcommittee on Global Change Research*. U.S. Geological Survey, Washington, DC, 459 pp.

Cess, R., et al., 1990: Intercomparison and interpretation of climate feedback processes in 19 atmospheric general-circulation models. *J. Geophys. Res.*, **95**, 16601–16615.

Chadwick, R., I. Boutle, and G. Martin, 2012: Spatial patterns of precipitation change in CMIP5: Why the rich don't get richer in the Tropics. *J. Clim.*, doi:10.1175/JCLI-D-12-00543.1.

Chadwick, R., P. Wu, P. Good, and T. Andrews, 2013: Asymmetries in tropical rainfall and circulation patterns in idealised CO_2 removal experiments. *Clim. Dyn.*, **40**, 295–316.

Chang, E. K. M., Y. Guo, and X. Xia, 2012a: CMIP5 multimodel ensemble projection of storm track change under global warming. *J. Geophys. Res.*, **117**, D23118.

Chang, E. K. M., Y. Guo, X. Xia, and M. Zheng, 2012b: Storm track activity in IPCC AR4/CMIP3 model simulations. *J. Clim.*, **26**, 246–260.

Chapin, F., et al., 2005: Role of land-surface changes in Arctic summer warming. *Science*, **310**, 657–660.

Charbit, S., D. Paillard, and G. Ramstein, 2008: Amount of CO_2 emissions irreversibly leading to the total melting of Greenland. *Geophys. Res. Lett.*, **35**, L12503.

Charney, J. G., 1979: *Carbon Dioxide and Climate: A Scientific Assessment.* National Academies of Science Press, Washington, DC, 22 pp.

Chen, C. T., and T. Knutson, 2008: On the verification and comparison of extreme rainfall indices from climate models. *J. Clim.*, **21**, 1605–1621.

Chen, G., J. Lu, and D. M. W. Frierson, 2008: Phase speed spectra and the latitude of surface westerlies: Interannual variability and global warming trend. *J. Clim.*, **21**, 5942–5959.

Cherchi, A., A. Alessandri, S. Masina, and A. Navarra, 2010: Effect of increasing CO_2 levels on monsoons. *Clim. Dyn.*, **37**, 83–101.

Choi, D. H., J. S. Kug, W. T. Kwon, F. F. Jin, H. J. Baek, and S. K. Min, 2010: Arctic Oscillation responses to greenhouse warming and role of synoptic eddy feedback. *J. Geophys. Res. Atmos.*, **115**, D17103.

Chou, C., and J. D. Neelin, 2004: Mechanisms of global warming impacts on regional tropical precipitation. *J. Clim.*, **17**, 2688–2701.

Chou, C., and C. Chen, 2010: Depth of convection and the weakening of tropical circulation in global warming. *J. Clim.*, **23**, 3019–3030.

Chou, C., and C.-W. Lan, 2012: Changes in the annual range of precipitation under global warming. *J. Clim.*, **25**, 222–235.

Chou, C., J. D. Neelin, J. Y. Tu, and C. T. Chen, 2006: Regional tropical precipitation change mechanisms in ECHAM4/OPYC3 under global warming. *J. Clim.*, **19**, 4207–4223.

Chou, C., J. D. Neelin, C. A. Chen, and J. Y. Tu, 2009: Evaluating the "Rich-Get-Richer" mechanism in tropical precipitation change under global warming. *J. Clim.*, **22**, 1982–2005.

Chou, C., C. Chen, P.-H. Tan, and K.-T. Chen, 2012: Mechanisms for global warming impacts on precipitation frequency and intensity. *J. Clim.*, **25**, 3291–3306.

Chou, C., J. C. H. Chiang, C.-W. Lan, C.-H. Chung, Y.-C. Liao, and C.-J. Lee, 2013: Increase in the range between wet and dry season precipitation. *Nature Geosci.*, **6**, 263–267.

Christensen, J. H., F. Boberg, O. B. Christensen, and P. Lucas-Picher, 2008: On the need for bias correction of regional climate change projections of temperature and precipitation. *Geophys. Res. Lett.*, **35**, L20709.

Christensen, J. H., et al., 2007: Regional climate projections. In: *Climate Change 2007: The Physical Science Basis. Contribution of Working Group I to the Fourth Assessment Report of the Intergovernmental Panel on Climate Change* [Solomon, S., D. Qin, M. Manning, Z. Chen, M. Marquis, K. B. Averyt, M. Tignor and H. L. Miller (eds.)] Cambridge University Press, Cambridge, United Kingdom and New York, NY, USA, pp. 847–940.

Christensen, N., and D. Lettenmaier, 2007: A multimodel ensemble approach to assessment of climate change impacts on the hydrology and water resources of the Colorado River Basin. *Hydrol. Earth Syst. Sci.*, **11**, 1417–1434.

Cionni, I., et al., 2011: Ozone database in support of CMIP5 simulations: Results and corresponding radiative forcing. *Atmos. Chem. Phys.*, **11**, 11267–11292.

Clark, R. T., S. J. Brown, and J. M. Murphy, 2006: Modeling Northern Hemisphere summer heat extreme changes and their uncertainties using a physics ensemble of climate sensitivity experiments. *J. Clim.*, **19**, 4418–4435.

Clark, R. T., J. M. Murphy, and S. J. Brown, 2010: Do global warming targets limit heatwave risk? *Geophys. Res. Lett.*, **37**, L17703.

Claussen, M., V. Brovkin, A. Ganopolski, C. Kubatzki, and V. Petoukhov, 2003: Climate change in northern Africa: The past is not the future. *Clim. Change*, **57**, 99–118.

Colle, B. A., Z. Zhang, K. A. Lombardo, E. Chang, P. Liu, and M. Zhang, 2013: Historical evaluation and future prediction of eastern North America and western Atlantic extratropical cyclones in the CMIP5 models during the cool season. *J. Clim.*, doi:10.1175/JCLI-D-12-00498.1.

Collier, J., and G. Zhang, 2009: Aerosol direct forcing of the summer Indian monsoon as simulated by the NCAR CAM3. *Clim. Dyn.*, **32**, 313–332.

Collins, M., C. M. Brierley, M. MacVean, B. B. B. Booth, and G. R. Harris, 2007: The sensitivity of the rate of transient climate change to ocean physics perturbations. *J. Clim.*, **20**, 2315–2320.

Collins, M., B. B. B. Booth, G. Harris, J. M. Murphy, D. M. H. Sexton, and M. J. Webb, 2006a: Towards quantifying uncertainty in transient climate change. *Clim. Dyn.*, **27**, 127–147.

Collins, M., R. E. Chandler, P. M. Cox, J. M. Huthnance, J. Rougier, and D. B. Stephenson, 2012: Quantifying future climate change. *Nature Clim. Change*, **2**, 403–409.

Collins, M., B. Booth, B. Bhaskaran, G. Harris, J. Murphy, D. Sexton, and M. Webb, 2011: Climate model errors, feedbacks and forcings: A comparison of perturbed physics and multi-model ensembles. *Clim. Dyn.*, **36**, 1737–1766.

Collins, M., et al., 2010: The impact of global warming on the tropical Pacific ocean and El Nino. *Nature Geosci.*, **3**, 391–397.

Collins, W. D., et al., 2006b: Radiative forcing by well-mixed greenhouse gases: Estimates from climate models in the Intergovernmental Panel on Climate Change (IPCC) Fourth Assessment Report (AR4). *J. Geophys. Res.*, **111**, D14317.

Colman, R., and B. McAvaney, 2009: Climate feedbacks under a very broad range of forcing. *Geophys. Res. Lett.*, **36**, L01702.

Colman, R., and S. Power, 2010: Atmospheric radiative feedbacks associated with transient climate change and climate variability. *Clim. Dyn.*, **34**, 919–933.

Comiso, J. C., and F. Nishio, 2008: Trends in the sea ice cover using enhanced and compatible AMSR-E, SSM/I, and SMMR data. *J. Geophys. Res.*, **113**, C02S07.

Cook, K., and E. Vizy, 2008: Effects of twenty-first-century climate change on the Amazon rain forest. *J. Clim.*, **21**, 542–560.

Costa, M., and G. Pires, 2010: Effects of Amazon and Central Brazil deforestation scenarios on the duration of the dry season in the arc of deforestation. *Int. J. Climatol.*, **30**, 1970–1979.

Crook, J. A., P. M. Forster, and N. Stuber, 2011: Spatial patterns of modeled climate feedback and contributions to temperature response and polar amplification. *J. Clim.*, **24**, 3575–3592.

Crucifix, M., 2006: Does the Last Glacial Maximum constrain climate sensitivity? *Geophys. Res. Lett.*, **33**, L18701.

Cruz, F. T., A. J. Pitman, J. L. McGregor, and J. P. Evans, 2010: Contrasting regional responses to increasing leaf-level atmospheric carbon dioxide over Australia. *J. Hydrometeorol.*, **11**, 296–314.

Dai, A., 2011: Drought under global warming: A review. *WIREs Clim. Change*, **2**, 45–65.

Dai, A., 2013: Increasing drought under global warming in observations and models. *Nature Clim. Change*, **3**, 52–58.

Dakos, V., M. Scheffer, E. H. van Nes, V. Brovkin, V. Petoukhov, and H. Held, 2008: Slowing down as an early warning signal for abrupt climate change. *Proc. Natl. Acad. Sci. U.S.A.*, **105**, 14308–14312.

Danabasoglu, G., and P. Gent, 2009: Equilibrium climate sensitivity: Is it accurate to use a slab ocean model? *J. Clim.*, **22**, 2494–2499.

Davin, E. L., N. de Noblet-Ducoudre, and P. Friedlingstein, 2007: Impact of land cover change on surface climate: Relevance of the radiative forcing concept. *Geophys. Res. Lett.*, **34**, L13702.

Davis, S., K. Caldeira, and H. Matthews, 2010: Future CO_2 emissions and climate change from existing energy infrastructure. *Science*, **329**, 1330–1333.

Davis, S. M., and K. H. Rosenlof, 2012: A multidiagnostic intercomparison of tropical-width time series using reanalyses and satellite observations. *J. Clim.*, **25**, 1061–1078.

De Angelis, H., and P. Skvarca, 2003: Glacier surge after ice shelf collapse. *Science*, **299**, 1560–1562.

de Vries, H., R. J. Haarsma, and W. Hazeleger, 2012: Western European cold spells in current and future climate. *Geophys. Res. Lett.*, **39**, L04706.

de Vries, P., and S. Weber, 2005: The Atlantic freshwater budget as a diagnostic for the existence of a stable shut down of the meridional overturning circulation. *Geophys. Res. Lett.*, **32**, L09606.

Del Genio, A. D., M.-S. Yao, and J. Jonas, 2007: Will moist convection be stronger in a warmer climate? *Geophys. Res. Lett.*, **34**, L16703.

Delisle, G., 2007: Near-surface permafrost degradation: How severe during the 21st century? *Geophys. Res. Lett.*, **34**, L09503.

Delworth, T. L., et al., 2008: The potential for abrupt change in the Atlantic meridional overturning circulation. In: *Abrupt Climate Change: A Report by the U.S. Climate Change Science Program and the Subcommittee on Global Change Research*, U.S. Geological Survey, Washington, DC, pp. 258–359.

deMenocal, P., J. Ortiz, T. Guilderson, J. Adkins, M. Sarnthein, L. Baker, and M. Yarusinsky, 2000: Abrupt onset and termination of the African Humid Period: Rapid climate responses to gradual insolation forcing. *Quaternary Science Reviews*, **19**, 347–361.

Deser, C., A. Phillips, V. Bourdette, and H. Teng, 2012a: Uncertainty in climate change projections: The role of internal variability. *Clim. Dyn.*, **38**, 527–546.

Deser, C., R. Knutti, S. Solomon, and A. S. Phillips, 2012b: Communication of the role of natural variability in future North American climate. *Nature Clim. Change*, **2**, 775–779.

Dessai, S., X. F. Lu, and M. Hulme, 2005: Limited sensitivity analysis of regional climate change probabilities for the 21st century. *J. Geophys. Res. Atmos.*, **110**, D19108.

Diffenbaugh, N. S., and M. Ashfaq, 2010: Intensification of hot extremes in the United States. *Geophys. Res. Lett.*, **37**, L15701.

Diffenbaugh, N. S., J. S. Pal, F. Giorgi, and X. J. Gao, 2007: Heat stress intensification in the Mediterranean climate change hotspot. *Geophys. Res. Lett.*, **34**, L11706.

Dijkstra, H., 2007: Characterization of the multiple equilibria regime in a global ocean model. *Tellus A*, **59**, 695–705.

DiNezio, P. N., A. C. Clement, G. A. Vecchi, B. J. Soden, and B. P. Kirtman, 2009: Climate response of the equatorial Pacific to global warming. *J. Clim.*, **22**, 4873–4892.

Dirmeyer, P. A., Y. Jin, B. Singh, and X. Yan, 2013: Evolving land-atmosphere interactions over North America from CMIP5 simulations. *J. Clim.*, doi:10.1175/JCLI-D-12-00454.1.

Dix, M., et al., 2013: The ACCESS Coupled Model: Documentation of core CMIP5 simulations and initial results. *Aust. Meteorol. Oceanogr. J.*, **63**, 83-199.

Dole, R., et al., 2011: Was there a basis for anticipating the 2010 Russian heat wave? *Geophys. Res. Lett.*, **38**, L06702.

Dolman, A., G. van der Werf, M. van der Molen, G. Ganssen, J. Erisman, and B. Strengers, 2010: A carbon cycle science update since IPCC AR-4. *Ambio*, **39**, 402–412.

Donat, M. G., et al., 2013: Updated analyses of temperature and precipitation extreme indices since the beginning of the twentieth century: The HadEX2 dataset. *J. Geophys. Res.*, **118**, 2098–2118.

Dong, B. W., J. M. Gregory, and R. T. Sutton, 2009: Understanding land-sea warming contrast in response to increasing greenhouse gases. Part I: Transient adjustment. *J. Clim.*, **22**, 3079–3097.

Dorrepaal, E., S. Toet, R. van Logtestijn, E. Swart, M. van de Weg, T. Callaghan, and R. Aerts, 2009: Carbon respiration from subsurface peat accelerated by climate warming in the subarctic. *Nature*, **460**, 616–619.

Döscher, R., and T. Koenigk, 2013: Arctic rapid sea ice loss events in regional coupled climate scenario experiments. *Ocean Sci.*, **9**, 217–248.

Doutriaux-Boucher, M., M. J. Webb, J. M. Gregory, and O. Boucher, 2009: Carbon dioxide induced stomatal closure increases radiative forcing via a rapid reduction in low cloud. *Geophys. Res. Lett.*, **36**, L02703.

Douville, H., J. F. Royer, J. Polcher, P. Cox, N. Gedney, D. B. Stephenson, and P. J. Valdes, 2000: Impact of CO_2 doubling on the Asian summer monsoon: Robust versus model-dependent responses. *J. Meteorol. Soc. Jpn.*, **78**, 421–439.

Dowdy, A. J., G. A. Mills, B. Timbal, and Y. Wang, 2013: Changes in the risk of extratropical cyclones in Eastern Australia. *J. Clim.*, **26**, 1403–1417.

Downes, S., A. Budnick, J. Sarmiento, and R. Farneti, 2011: Impacts of wind stress on the Antarctic Circumpolar Current fronts and associated subduction. *Geophys. Res. Lett.*, **38**, L11605.

Downes, S. M., and A. M. Hogg, 2013: Southern Ocean circulation and eddy compensation in CMIP5 models. *J. Clim.*, doi:10.1175/JCLI-D-12-00504.1.

Downes, S. M., N. L. Bindoff, and S. R. Rintoul, 2010: Changes in the subduction of Southern Ocean water masses at the end of the twenty-first century in eight IPCC models. *J. Clim.*, **23**, 6526–6541.

Driesschaert, E., et al., 2007: Modeling the influence of Greenland ice sheet melting on the Atlantic meridional overturning circulation during the next millennia. *Geophys. Res. Lett.*, **34**, L10707.

Drijfhout, S., G. J. van Oldenborgh, and A. Cimatoribus, 2012: Is a decline of AMOC causing the warming hole above the North Atlantic in observed and modeled warming patterns? *J. Clim.*, **25**, 8373–8379.

Drijfhout, S. S., S. Weber, and E. van der Swaluw, 2010: The stability of the MOC as diagnosed from the model projections for the pre industrial, present and future climate. *Clim. Dyn.*, **37**, 1575–1586.

Dufresne, J.-L., et al., 2013: Climate change projections using the IPSL-CM5 Earth system model: From CMIP3 to CMIP5. *Clim. Dyn.*, **40**, 2123–2165.

Dufresne, J., J. Quaas, O. Boucher, S. Denvil, and L. Fairhead, 2005: Contrasts in the effects on climate of anthropogenic sulfate aerosols between the 20th and the 21st century. *Geophys. Res. Lett.*, **32**, L21703.

Dufresne, J. L., and S. Bony, 2008: An assessment of the primary sources of spread of global warming estimates from coupled atmosphere-ocean models. *J. Clim.*, **21**, 5135–5144.

Dulamsuren, C., M. Hauck, and M. Muhlenberg, 2008: Insect and small mammal herbivores limit tree establishment in northern Mongolian steppe. *Plant Ecol.*, **195**, 143–156.

Dulamsuren, C., M. Hauck, and C. Leuschner, 2010: Recent drought stress leads to growth reductions in *Larix sibirica* in the western Khentey, Mongolia. *Global Change Biol.*, **16**, 3024–3035.

Dulamsuren, C., et al., 2009: Water relations and photosynthetic performance in *Larix sibirica* growing in the forest-steppe ecotone of northern Mongolia. *Tree Physiol.*, **29**, 99–110.

Dunne, J. P., R. J. Stouffer, and J. G. John, 2013: Reductions in labour capacity from heat stress under climate warming. *Nature Clim. Change*, doi:10.1038/nclimate1827.

Durack, P., and S. Wijffels, 2010: Fifty-year trends in global ocean salinities and their relationship to broad-scale warming. *J. Clim.*, **23**, 4342–4362.

Durack, P. J., S. E. Wijffels, and R. J. Matear, 2012: Ocean salinities reveal strong global water cycle intensification during 1950 to 2000. *Science*, **336**, 455–458.

Eby, M., K. Zickfeld, A. Montenegro, D. Archer, K. Meissner, and A. Weaver, 2009: Lifetime of anthropogenic climate change: Millennial time scales of potential CO_2 and surface temperature perturbations. *J. Clim.*, **22**, 2501–2511.

Edwards, T., M. Crucifix, and S. Harrison, 2007: Using the past to constrain the future: How the palaeorecord can improve estimates of global warming. *Prog. Phys. Geogr.*, **31**, 481–500.

Eglin, T., et al., 2010: Historical and future perspectives of global soil carbon response to climate and land-use changes. *Tellus B*, **62**, 700–718.

Eisenman, I., 2012: Factors controlling the bifurcation structure of sea ice retreat. *J. Geophys. Res.*, **117**, D01111.

Eisenman, I., and J. Wettlaufer, 2009: Nonlinear threshold behavior during the loss of Arctic sea ice. *Proc. Natl. Acad. Sci. U.S.A.*, **106**, 28–32.

12

Eisenman, I., T. Schneider, D. S. Battisti, and C. M. Bitz, 2011: Consistent changes in the sea ice seasonal cycle in response to global warming. *J. Clim.*, **24**, 5325–5335.

Eliseev, A., P. Demchenko, M. Arzhanov, and I. Mokhov, 2013: Transient hysteresis of near-surface permafrost response to external forcing. *Clim. Dyn.*, doi:10.1007/s00382-013-1672-5.

Emori, S., and S. Brown, 2005: Dynamic and thermodynamic changes in mean and extreme precipitation under changed climate. *Geophys. Res. Lett.*, **32**, L17706.

Eyring, V., et al., 2005: A strategy for process-oriented validation of coupled chemistry-climate models. *Bull. Am. Meteorol. Soc.*, **86**, 1117–1133.

Eyring, V., et al., 2013: Long-term ozone changes and associated climate impacts in CMIP5 simulations. *J. Geophys. Res.*, doi:10.1002/jgrd.50316.

Falloon, P. D., R. Dankers, R. A. Betts, C. D. Jones, B. B. B. Booth, and F. H. Lambert, 2012: Role of vegetation change in future climate under the A1B scenario and a climate stabilisation scenario, using the HadCM3C Earth system model. *Biogeosciences*, **9**, 4739–4756.

Farneti, R., and P. Gent, 2011: The effects of the eddy-induced advection coefficient in a coarse-resolution coupled climate model. *Ocean Model.*, **39**, 135–145.

Farneti, R., T. Delworth, A. Rosati, S. Griffies, and F. Zeng, 2010: The role of mesoscale eddies in the rectification of the Southern Ocean response to climate change. *J. Phys. Oceanogr.*, **40**, 1539–1557.

Fasullo, J. T., 2010: Robust land-ocean contrasts in energy and water cycle feedbacks. *J. Clim.*, **23**, 4677–4693.

Favre, A., and A. Gershunov, 2009: North Pacific cyclonic and anticyclonic transients in a global warming context: Possible consequences for Western North American daily precipitation and temperature extremes. *Clim. Dyn.*, **32**, 969–987.

Finnis, J., M. M. Holland, M. C. Serreze, and J. J. Cassano, 2007: Response of Northern Hemisphere extratropical cyclone activity and associated precipitation to climate change, as represented by the Community Climate System Model. *J. Geophys. Res.*, **112**, G04S42.

Fischer, E. M., and C. Schär, 2009: Future changes in daily summer temperature variability: Driving processes and role for temperature extremes. *Clim. Dyn.*, **33**, 917–935.

Fischer, E. M., and C. Schär, 2010: Consistent geographical patterns of changes in high-impact European heatwaves. *Nature Geosci.*, **3**, 398–403.

Fischer, E. M., and R. Knutti, 2013: Robust projections of combined humidity and temperature extremes. *Nature Clim. Change*, **3**, 126–130.

Fischer, E. M., D. M. Lawrence, and B. M. Sanderson, 2011: Quantifying uncertainties in projections of extremes—A perturbed land surface parameter experiment. *Clim. Dyn.*, **37**, 1381–1398.

Fischer, E. M., J. Rajczak, and C. Schär, 2012a: Changes in European summer temperature variability revisited. *Geophys. Res. Lett.*, **39**, L19702.

Fischer, E. M., K. W. Oleson, and D. M. Lawrence, 2012b: Contrasting urban and rural heat stress responses to climate change. *Geophys. Res. Lett.*, **39**, L03705.

Flannery, B. P., 1984: Energy-balance models incorporating transport of thermal and latent energy. *J. Atmos. Sci.*, **41**, 414–421.

Forest, C. E., P. H. Stone, and A. P. Sokolov, 2006: Estimated PDFs of climate system properties including natural and anthropogenic forcings. *Geophys. Res. Lett.*, **33**, L01705.

Forest, C. E., P. H. Stone, and A. P. Sokolov, 2008: Constraining climate model parameters from observed 20th century changes. *Tellus A*, **60**, 911–920.

Forster, P., and K. Taylor, 2006: Climate forcings and climate sensitivities diagnosed from coupled climate model integrations. *J. Clim.*, **19**, 6181–6194.

Forster, P. M., T. Andrews, P. Good, J. M. Gregory, L. S. Jackson, and M. Zelinka, 2013: Evaluating adjusted forcing and model spread for historical and future scenarios in the CMIP5 generation of climate models. *J. Geophys. Res.*, **118**, 1139–1150.

Fowler, H., M. Ekstrom, S. Blenkinsop, and A. Smith, 2007a: Estimating change in extreme European precipitation using a multimodel ensemble. *J. Geophys. Res.*, **112**, D18104.

Fowler, H. J., S. Blenkinsop, and C. Tebaldi, 2007b: Linking climate change modelling to impacts studies: Recent advances in downscaling techniques for hydrological modelling. *Int. J. Climatol.*, **27**, 1547–1578.

Frame, D., B. Booth, J. Kettleborough, D. Stainforth, J. Gregory, M. Collins, and M. Allen, 2005: Constraining climate forecasts: The role of prior assumptions. *Geophys. Res. Lett.*, **32**, L09702.

Frederiksen, C. S., J. S. Frederiksen, J. M. Sisson, and S. L. Osbrough, 2011: Australian winter circulation and rainfall changes and projections. *Int. J. Clim. Change Strat. Manage.*, **3**, 170–188.

Friedlingstein, P., and S. Solomon, 2005: Contributions of past and present human generations to committed warming caused by carbon dioxide. *Proc. Natl. Acad. Sci. U.S.A.*, **102**, 10832–10836.

Friedlingstein, P., S. Solomon, G. Plattner, R. Knutti, P. Ciais, and M. Raupach, 2011: Long-term climate implications of twenty-first century options for carbon dioxide emission mitigation. *Nature Clim. Change*, **1**, 457–461.

Friedlingstein, P., et al., 2006: Climate-carbon cycle feedback analysis: Results from the C[4]MIP model intercomparison. *J. Clim.*, **19**, 3337–3353.

Frieler, K., M. Meinshausen, M. Mengel, N. Braun, and W. Hare, 2012: A scaling approach to probabilistic assessment of regional climate. *J. Clim.*, **25**, 3117–3144.

Frierson, D., J. Lu, and G. Chen, 2007: Width of the Hadley cell in simple and comprehensive general circulation models. *Geophys. Res. Lett.*, **34**, L18804.

Frölicher, T., and F. Joos, 2010: Reversible and irreversible impacts of greenhouse gas emissions in multi-century projections with the NCAR global coupled carbon cycle-climate model. *Clim. Dyn.*, **35**, 1439–1459.

Fu, Q., C. M. Johanson, J. M. Wallace, and T. Reichler, 2006: Enhanced mid-latitude tropospheric warming in satellite measurements. *Science*, **312**, 1179–1179.

Fyfe, J., O. Saenko, K. Zickfeld, M. Eby, and A. Weaver, 2007: The role of poleward-intensifying winds on Southern Ocean warming. *J. Clim.*, **20**, 5391–5400.

Fyke, J., and A. Weaver, 2006: The effect of potential future climate change on the marine methane hydrate stability zone. *J. Clim.*, **19**, 5903–5917.

Garcia, R. R., and W. J. Randel, 2008: Acceleration of the Brewer-Dobson circulation due to increases in greenhouse gases. *J. Atmos. Sci.*, **65**, 2731–2739.

Gastineau, G., and B. J. Soden, 2009: Model projected changes of extreme wind events in response to global warming. *Geophys. Res. Lett.*, **36**, L10810.

Gastineau, G., H. Le Treut, and L. Li, 2008: Hadley circulation changes under global warming conditions indicated by coupled climate models. *Tellus A*, **60**, 863–884.

Gastineau, G., L. Li, and H. Le Treut, 2009: The Hadley and Walker circulation changes in global warming conditions described by idealized atmospheric simulations. *J. Clim.*, **22**, 3993–4013.

Gent, P. R., et al., 2011: The Community Climate System Model Version 4. *J. Clim.*, **24**, 4973–4991.

Georgescu, M., D. Lobell, and C. Field, 2011: Direct climate effects of perennial bioenergy crops in the United States. *Proc. Natl. Acad. Sci. U.S.A.*, **109**, 4307–4312.

Gerber, E. P., et al., 2012: Assessing and understanding the impact of stratospheric dynamics and variability on the Earth system. *Bull. Am. Meteorol. Soc.*, **93**, 845–859.

Gillett, N., M. Wehner, S. Tett, and A. Weaver, 2004: Testing the linearity of the response to combined greenhouse gas and sulfate aerosol forcing. *Geophys. Res. Lett.*, **31**, L14201.

Gillett, N. P., and P. A. Stott, 2009: Attribution of anthropogenic influence on seasonal sea level pressure. *Geophys. Res. Lett.*, **36**, L23709.

Gillett, N. P., V. K. Arora, D. Matthews, and M. R. Allen, 2013: Constraining the ratio of global warming to cumulative CO_2 emissions using CMIP5 simulations. *J. Clim.*, doi:10.1175/JCLI-D-12-00476.1.

Gillett, N. P., V. K. Arora, K. Zickfeld, S. J. Marshall, and A. J. Merryfield, 2011: Ongoing climate change following a complete cessation of carbon dioxide emissions. *Nature Geosci.*, **4**, 83–87.

Giorgi, F., 2008: A simple equation for regional climate change and associated uncertainty. *J. Clim.*, **21**, 1589–1604.

Gleckler, P. J., K. AchutaRao, J. M. Gregory, B. D. Santer, K. E. Taylor, and T. M. L. Wigley, 2006: Krakatoa lives: The effect of volcanic eruptions on ocean heat content and thermal expansion. *Geophys. Res. Lett.*, **33**, L17702.

Goelzer, H., P. Huybrechts, M. Loutre, H. Goosse, T. Fichefet, and A. Mouchet, 2011: Impact of Greenland and Antarctic ice sheet interactions on climate sensitivity. *Clim. Dyn.*, **37**, 1005–1018.

Good, P., J. M. Gregory, and J. A. Lowe, 2011a: A step-response simple climate model to reconstruct and interpret AOGCM projections. *Geophys. Res. Lett.*, **38**, L01703.

Good, P., J. M. Gregory, J. A. Lowe, and T. Andrews, 2013: Abrupt CO_2 experiments as tools for predicting and understanding CMIP5 representative concentration pathway projections. *Clim. Dyn.*, **40**, 1041–1053.

Good, P., C. Jones, J. Lowe, R. Betts, B. Booth, and C. Huntingford, 2011b: Quantifying environmental drivers of future tropical forest extent. *J. Clim.*, **24**, 1337–1349.

Good, P., et al., 2012: A step-response approach for predicting and understanding non-linear precipitation changes. *Clim. Dyn.*, **39**, 2789–2803.

Good, P., et al., 2011c: A review of recent developments in climate change science. Part I: Understanding of future change in the large-scale climate system. *Prog. Phys. Geogr.*, **35**, 281–296.

Goodwin, P., R. Williams, A. Ridgwell, and M. Follows, 2009: Climate sensitivity to the carbon cycle modulated by past and future changes in ocean chemistry. *Nature Geosci.*, **2**, 145–150.

Goosse, H., O. Arzel, C. Bitz, A. de Montety, and M. Vancoppenolle, 2009: Increased variability of the Arctic summer ice extent in a warmer climate. *Geophys. Res. Lett.*, **36**, L23702.

Goubanova, K., and L. Li, 2007: Extremes in temperature and precipitation around the Mediterranean basin in an ensemble of future climate scenario simulations. *Global Planet. Change*, **57**, 27–42.

Gouttevin, I., G. Krinner, P. Ciais, J. Polcher, and C. Legout, 2012: Multi-scale validation of a new soil freezing scheme for a land-surface model with physically-based hydrology. *Cryosphere*, **6**, 407–430.

Granier, C., et al., 2011: Evolution of anthropogenic and biomass burning emissions at global and regional scales during the 1980–2010 period. *Clim. Change*, **109**, 163–190.

Grant, A., S. Brönnimann, and L. Haimberger, 2008: Recent Arctic warming vertical structure contested. *Nature*, **455**, E2–E3.

Graversen, R., and M. Wang, 2009: Polar amplification in a coupled climate model with locked albedo. *Clim. Dyn.*, **33**, 629–643.

Graversen, R., T. Mauritsen, M. Tjernstrom, E. Kallen, and G. Svensson, 2008: Vertical structure of recent Arctic warming. *Nature*, **541**, 53–56.

Gregory, J., and M. Webb, 2008: Tropospheric adjustment induces a cloud component in CO_2 forcing. *J. Clim.*, **21**, 58–71.

Gregory, J., and P. Forster, 2008: Transient climate response estimated from radiative forcing and observed temperature change. *J. Geophys. Res.*, **113**, D23105.

Gregory, J. M., 2010: Long-term effect of volcanic forcing on ocean heat content. *Geophys. Res. Lett.*, **37**, L22701.

Gregory, J. M., and J. F. B. Mitchell, 1995: Simulation of daily variability of surface-temperature and precipitation over Europe in the current and $2xCO_2$ climates using the UKMO climate model. *Q. J. R. Meteorol. Soc.*, **121**, 1451–1476.

Gregory, J. M., and R. Tailleux, 2011: Kinetic energy analysis of the response of the Atlantic meridional overturning circulation to CO_2-forced climate change. *Clim. Dyn.*, **37**, 893–914.

Gregory, J. M., C. D. Jones, P. Cadule, and P. Friedlingstein, 2009: Quantifying carbon cycle feedbacks. *J. Clim.*, **22**, 5232–5250.

Gregory, J. M., et al., 2004: A new method for diagnosing radiative forcing and climate sensitivity. *Geophys. Res. Lett.*, **31**, L03205.

Gregory, J. M., et al., 2005: A model intercomparison of changes in the Atlantic thermohaline circulation in response to increasing atmospheric CO_2 concentration. *Geophys. Res. Lett.*, **32**, L12703.

Grubb, M., 1997: Technologies, energy systems and the timing of CO_2 emissions abatement—An overview of economic issues. *Energy Policy*, **25**, 159–172.

Gumpenberger, M., et al., 2010: Predicting pan-tropical climate change induced forest stock gains and losses-implications for REDD. *Environ. Res. Lett.*, **5**, 014013.

Gutowski, W., K. Kozak, R. Arritt, J. Christensen, J. Patton, and E. Takle, 2007: A possible constraint on regional precipitation intensity changes under global warming. *J. Hydrometeorol.*, **8**, 1382–1396.

Haarsma, R. J., F. Selten, and G. J. van Oldenborgh, 2013: Anthropogenic changes of the thermal and zonal flow structure over Western Europe and Eastern North Atlantic in CMIP3 and CMIP5 models. *Clim. Dyn.*, doi:10.1007/s00382-013-1734-8.

Haarsma, R. J., F. Selten, B. V. Hurk, W. Hazeleger, and X. L. Wang, 2009: Drier Mediterranean soils due to greenhouse warming bring easterly winds over summertime central Europe. *Geophys. Res. Lett.*, **36**, L04705.

Hajima, T., T. Ise, K. Tachiiri, E. Kato, S. Watanabe, and M. Kawamiya, 2012: Climate change, allowable emission, and Earth system response to representative concentration pathway scenarios. *J. Meteorol. Soc. Jpn.*, **90**, 417–433.

Hall, A., 2004: The role of surface albedo feedback in climate. *J. Clim.*, **17**, 1550–1568.

Hall, A., X. Qu, and J. Neelin, 2008: Improving predictions of summer climate change in the United States. *Geophys. Res. Lett.*, **35**, L01702.

Hansen, J., M. Sato, P. Kharecha, and K. von Schuckmann, 2011: Earth's energy imbalance and implications. *Atmos. Chem. Phys.*, **11**, 13421–13449.

Hansen, J., G. Russell, A. Lacis, I. Fung, D. Rind, and P. Stone, 1985: Climate response-times—Dependence on climate sensitivity and ocean mixing. *Science*, **229**, 857–859.

Hansen, J., M. Sato, P. Kharecha, G. Russell, D. Lea, and M. Siddall, 2007: Climate change and trace gases. *Philos. Trans. R. Soc. A*, **365**, 1925–1954.

Hansen, J., et al., 1984: Climate sensitivity: Analysis of feedback mechanisms. In: *Climate Processes and Climate Sensitivity* [J. Hansen and T. Takahashi (eds.)]. American Geophysical Union, Washington, DC, pp. 130–163.

Hansen, J., et al., 1988: Global climate changes as forecast by Goddard Institute for Space Studies 3-dimensional model. *J. Geophys. Res. Atmos.*, **93**, 9341–9364.

Hansen, J., et al., 2008: Target atmospheric CO_2: Where should humanity aim? *Open Atmos. Sci. J.*, **2**, 217–231.

Hansen, J., et al., 2005a: Earth's energy imbalance: Confirmation and implications. *Science*, **308**, 1431–1435.

Hansen, J., et al., 2005b: Efficacy of climate forcings. *J. Geophys. Res.*, **110**, D18104.

Hardiman, S., N. Butchart, T. Hinton, S. Osprey, and L. Gray, 2012: The effect of a well resolved stratosphere on surface climate: Differences between CMIP5 simulations with high and low top versions of the Met Office climate model. *J. Clim.*, **35**, 7083–7099.

Hare, B., and M. Meinshausen, 2006: How much warming are we committed to and how much can be avoided? *Clim. Change*, **75**, 111–149.

Hargreaves, J. C., A. Abe-Ouchi, and J. D. Annan, 2007: Linking glacial and future climates through an ensemble of GCM simulations. *Clim. Past*, **3**, 77–87.

Hargreaves, J. C., J. D. Annan, M. Yoshimori, and A. Abe-Ouchi, 2012: Can the Last Glacial Maximum constrain climate sensitivity? *Geophys. Res. Lett.*, **39**, L24702.

Harris, G. R., M. Collins, D. M. H. Sexton, J. M. Murphy, and B. B. B. Booth, 2010: Probabilistic projections for 21st century European climate. *Nat. Hazards Earth Syst. Sci.*, **10**, 2009–2020.

Harris, G. R., D. M. H. Sexton, B. B. B. Booth, M. Collins, J. M. Murphy, and M. J. Webb, 2006: Frequency distributions of transient regional climate change from perturbed physics ensembles of general circulation model simulations. *Clim. Dyn.*, **27**, 357–375.

Hartmann, D. L., and K. Larson, 2002: An important constraint on tropical cloud-climate feedback. *Geophys. Res. Lett.*, **29**, 1951.

Harvey, B. J., L. C. Shaffrey, T. J. Woollings, G. Zappa, and K. I. Hodges, 2012: How large are projected 21st century storm track changes? *Geophys. Res. Lett.*, **39**, L18707.

Haugen, J., and T. Iversen, 2008: Response in extremes of daily precipitation and wind from a downscaled multi-model ensemble of anthropogenic global climate change scenarios. *Tellus A*, **60**, 411–426.

Hawkins, E., and R. Sutton, 2009: The potential to narrow uncertainty in regional climate predictions. *Bull. Am. Meteorol. Soc.*, **90**, 1095–1107.

Hawkins, E., and R. Sutton, 2011: The potential to narrow uncertainty in projections of regional precipitation change. *Clim. Dyn.*, **37**, 407–418.

Hawkins, E., R. Smith, L. Allison, J. Gregory, T. Woollings, H. Pohlmann, and B. de Cuevas, 2011: Bistability of the Atlantic overturning circulation in a global climate model and links to ocean freshwater transport. *Geophys. Res. Lett.*, **38**, L16699.

Hazeleger, W., et al., 2013: Multiyear climate predictions using two initialisation strategies. *Geophys. Res. Lett.*, doi:10.1002/grl.50355.

Hegerl, G., T. Crowley, W. Hyde, and D. Frame, 2006: Climate sensitivity constrained by temperature reconstructions over the past seven centuries. *Nature*, **440**, 1029–1032.

Hegerl, G. C., F. W. Zwiers, P. A. Stott, and V. V. Kharin, 2004: Detectability of anthropogenic changes in annual temperature and precipitation extremes. *J. Clim.*, **17**, 3683–3700.

Held, I., and B. Soden, 2006: Robust responses of the hydrological cycle to global warming. *J. Clim.*, **19**, 5686–5699.

Held, I. M., M. Winton, K. Takahashi, T. Delworth, F. R. Zeng, and G. K. Vallis, 2010: Probing the fast and slow components of global warming by returning abruptly to preindustrial forcing. *J. Clim.*, **23**, 2418–2427.

Hellmer, H. H., F. Kauker, R. Timmermann, J. Determann, and J. Rae, 2012: Twenty-first-century warming of a large Antarctic ice-shelf cavity by a redirected coastal current. *Nature*, **484**, 225–228.

Henderson-Sellers, A., P. Irannejad, and K. McGuffie, 2008: Future desertification and climate change: The need for land-surface system evaluation improvement. *Global and Planetary Change*, **64**, 129–138.

Hibbard, K. A., G. A. Meehl, P. A. Cox, and P. Friedlingstein, 2007: A strategy for climate change stabilization experiments. *EOS Transactions AGU*, **88**, 217–221.

Hirschi, M., et al., 2011: Observational evidence for soil-moisture impact on hot extremes in southeastern Europe. *Nature Geosci.*, **4**, 17–21.

Ho, C. K., D. B. Stephenson, M. Collins, C. A. T. Ferro, and S. J. Brown, 2012: Calibration strategies: A source of additional uncertainty in climate change projections. *Bull. Am. Meteorol. Soc.*, **93**, 21–26.

Hodson, D. L. R., S. P. E. Keeley, A. West, J. Ridley, E. Hawkins, and H. T. Hewitt, 2012: Identifying uncertainties in Arctic climate change projections. *Clim. Dyn.*, doi:10.1007/s00382-012-1512-z.

Hoelzmann, P., D. Jolly, S. Harrison, F. Laarif, R. Bonnefille, and H. Pachur, 1998: Mid-Holocene land-surface conditions in northern Africa and the Arabian Peninsula: A data set for the analysis of biogeophysical feedbacks in the climate system. *Global Biogeochem. Cycles*, **12**, 35–51.

Hoerling, M., J. Eischeid, and J. Perlwitz, 2010: Regional precipitation trends: Distinguishing natural variability from anthropogenic forcing. *J. Clim.*, **23**, 2131–2145.

Hoerling, M. P., J. K. Eischeid, X.-W. Quan, H. F. Diaz, R. S. Webb, R. M. Dole, and D. R. Easterling, 2012: Is a transition to semipermanent drought conditions imminent in the US Great Plains? *J. Clim.*, **25**, 8380–8386.

Hofmann, M., and S. Rahmstorf, 2009: On the stability of the Atlantic meridional overturning circulation. *Proc. Natl. Acad. Sci. U.S.A.*, **106**, 20584–20589.

Hogg, E., and A. Schwarz, 1997: Regeneration of planted conifers across climatic moisture gradients on the Canadian prairies: Implications for distribution and climate change. *J. Biogeogr.*, **24**, 527–534.

Holden, P. B., and N. R. Edwards, 2010: Dimensionally reduced emulation of an AOGCM for application to integrated assessment modelling. *Geophys. Res. Lett.*, **37**, L21707.

Holland, M., C. Bitz, and B. Tremblay, 2006: Future abrupt reductions in the summer Arctic sea ice. *Geophys. Res. Lett.*, **33**, L23503.

Holland, M., M. Serreze, and J. Stroeve, 2010: The sea ice mass budget of the Arctic and its future change as simulated by coupled climate models. *Clim. Dyn.*, **34**, 185–200.

Holland, M. M., and C. M. Bitz, 2003: Polar amplification of climate change in coupled models. *Clim. Dyn.*, **21**, 221–232.

Holland, M. M., C. M. Bitz, B. Tremblay, and D. A. Bailey, 2008: The role of natural versus forced change in future rapid summer Arctic ice loss. In: *Arctic Sea Ice Decline: Observations, Projections, Mechanisms, and Implications* [E. T. DeWeaver, C. M. Bitz and L. B. Tremblay (eds.)]. American Geophysical Union, Washington, DC, pp. 133–150.

Hu, A., G. Meehl, W. Han, and J. Yin, 2009: Transient response of the MOC and climate to potential melting of the Greenland ice sheet in the 21st century. *Geophys. Res. Lett.*, **36**, L10707.

Hu, Y., and Q. Fu, 2007: Observed poleward expansion of the Hadley circulation since 1979. *Atmos. Chem. Phys.*, **7**, 5229–5236.

Hu, Z.-Z., M. Latif, E. Roeckner, and L. Bengtsson, 2000: Intensified Asian summer monsoon and its variability in a coupled model forced by increasing greenhouse gas concentrations. *Geophys. Res. Lett.*, **27**, 2681–2684.

Hu, Z. Z., A. Kumar, B. Jha, and B. H. Huang, 2012: An analysis of forced and internal variability in a warmer climate in CCSM3. *J. Clim.*, **25**, 2356–2373.

Huang, P., S.-P. Xie, K. Hu, G. Huang, and R. Huang, 2013: Patterns of the seasonal response of tropical rainfall to global warming. *Nature Geosci.*, **6**, 357–361.

Huete, A. R., et al., 2006: Amazon rainforests green-up with sunlight in dry season. *Geophys. Res. Lett.*, **33**, L06405.

Huisman, S., M. den Toom, H. Dijkstra, and S. Drijfhout, 2010: An indicator of the multiple equilibria regime of the Atlantic meridional overturning circulation. *J. Phys. Oceanogr.*, **40**, 551–567.

Huntingford, C., and P. M. Cox, 2000: An analogue model to derive additional climate change scenarios from existing GCM simulations. *Clim. Dyn.*, **16**, 575–586.

Huntingford, C., J. Lowe, B. Booth, C. Jones, G. Harris, L. Gohar, and P. Meir, 2009: Contributions of carbon cycle uncertainty to future climate projection spread. *Tellus B*, **61**, 355–360.

Huntingford, C., et al., 2008: Towards quantifying uncertainty in predictions of Amazon 'dieback'. *Philos. Trans. R. Soc. B*, **363**, 1857–1864.

Huntingford, C., et al., 2013: Simulated resilience of tropical rainforests to CO_2-induced climate change. *Nature Geosci.*, **6**, 268–273.

Hurtt, G., et al., 2011: Harmonization of land-use scenarios for the period 1500–2100: 600 years of global gridded annual land-use transitions, wood harvest, and resulting secondary lands. *Clim. Change*, **109**, 117–161.

Hwang, Y.-T., D. M. W. D.M.W. Frierson, B. J. Soden, and I. M. Held, 2011: Corrigendum for Held and Soden (2006). *J. Clim.*, **24**, 1559–1560.

IPCC, 2000: *IPCC Special Report on Emissions Scenarios. Prepared by Working Group III of the Intergovernmental Panel on Climate Change.* Cambridge University Press, Cambridge, United Kingdom, and New York, NY, USA.

IPCC, 2001: *Climate Change 2001: The Scientific Basis. Contribution of Working Group I to the Third Assessment Report of the Intergovernmental Panel on Climate Change* [J. T. Houghton, Y. Ding, D. J. Griggs, M. Noquer, P. J. van der Linden, X. Dai, K. Maskell and C. A. Johnson (eds.)]. Cambridge University Press, Cambridge, United Kingdom and New York, NY, USA, 881 pp.

IPCC, 2007: *Climate Change 2007: The Physical Science Basis. Contribution of Working Group I to the Fourth Assessment Report of the Intergovernmental Panel on Climate Change* [Solomon, S., D. Qin, M. Manning, Z. Chen, M. Marquis, K. B. Averyt, M. Tignor and H. L. Miller (eds.)]. Cambridge University Press, Cambridge, United Kingdom and New York, NY, USA, 996 pp.

Ishizaki, Y., et al., 2012: Temperature scaling pattern dependence on representative concentration pathway emission scenarios. *Clim. Change*, **112**, 535–546.

Iversen, T., et al., 2013: The Norwegian Earth System Model, NorESM1-M – Part 2: Climate response and scenario projections. *Geosci. Model Dev.*, **6**, 389–415.

Jackson, C. S., M. K. Sen, G. Huerta, Y. Deng, and K. P. Bowman, 2008: Error reduction and convergence in climate prediction. *J. Clim.*, **21**, 6698–6709.

Jaeger, C., and J. Jaeger, 2010: Three views of two degrees. *Clim. Change Econ.*, **3**, 145–166.

Jiang, X., S. J. Eichelberger, D. L. Hartmann, R. Shia, and Y. L. Yung, 2007: Influence of doubled CO_2 on ozone via changes in the Brewer-Dobson circulation. *J. Atmos. Sci.*, **64**, 2751–2755.

Johanson, C. M., and Q. Fu, 2009: Hadley Cell widening: Model simulations versus observations. *J. Clim.*, **22**, 2713–2725.

Johns, T. C., et al., 2011: Climate change under aggressive mitigation: The ENSEMBLES multi-model experiment. *Clim. Dyn.*, **37**, 1975–2003.

Johnson, N. C., and S.-P. Xie, 2010: Changes in the sea surface temperature threshold for tropical convection. *Nature Geosci.*, **3**, 842–845.

Jones, A., J. Haywood, and O. Boucher, 2007: Aerosol forcing, climate response and climate sensitivity in the Hadley Centre climate model. *J. Geophys. Res.*, **112**, D20211.

Jones, C., P. Cox, and C. Huntingford, 2006: Climate-carbon cycle feedbacks under stabilization: Uncertainty and observational constraints. *Tellus B*, **58**, 603–613.

Jones, C., J. Lowe, S. Liddicoat, and R. Betts, 2009: Committed terrestrial ecosystem changes due to climate change. *Nature Geosci.*, **2**, 484–487.

Jones, C. D., et al., 2013: 21st Century compatible CO_2 emissions and airborne fraction simulated by CMIP5 Earth System models under 4 Representative Concentration Pathways. *J. Clim.*, doi:10.1175/JCLI-D-12-00554.1.

Jones, C. D., et al., 2011: The HadGEM2–ES implementation of CMIP5 centennial simulations. *Geosci. Model Dev.*, **4**, 543–570.

Joos, F., et al., 2013: Carbon dioxide and climate impulse response functions for the computation of greenhouse gas metrics: A multi-model analysis. *Atmos. Chem. Phys.*, **13**, 2793–2825.

Joshi, M., E. Hawkins, R. Sutton, J. Lowe, and D. Frame, 2011: Projections of when temperature change will exceed 2°C above pre-industrial levels. *Nature Clim. Change*, **1**, 407–412.

Joshi, M., K. Shine, M. Ponater, N. Stuber, R. Sausen, and L. Li, 2003: A comparison of climate response to different radiative forcings in three general circulation models: Towards an improved metric of climate change. *Clim. Dyn.*, **20**, 843–854.

Joshi, M. M., F. H. Lambert, and M. J. Webb, 2013: An explanation for the difference between twentieth and twenty-first century land–sea warming ratio in climate models. *Clim. Dyn.*, doi:10.1007/s00382-013-1664-5.

Joshi, M. M., M. J. Webb, A. C. Maycock, and M. Collins, 2010: Stratospheric water vapour and high climate sensitivity in a version of the HadSM3 climate model. *Atmos. Chem. Phys.*, **10**, 7161–7167.

Joshi, M. M., J. M. Gregory, M. J. Webb, D. M. H. Sexton, and T. C. Johns, 2008: Mechanisms for the land/sea warming contrast exhibited by simulations of climate change. *Clim. Dyn.*, **30**, 455–465.

Jun, M., R. Knutti, and D. W. Nychka, 2008: Spatial analysis to quantify numerical model bias and dependence: How many climate models are there? *J. Am. Stat. Assoc. Appl. Case Stud.*, **103**, 934–947.

Jungclaus, J., H. Haak, M. Esch, E. Röckner, and J. Marotzke, 2006: Will Greenland melting halt the thermohaline circulation? *Geophys. Res. Lett.*, **33**, L17708.

Kang, S. M., and I. M. Held, 2012: Tropical precipitation, SSTs and the surface energy budget: A zonally symmetric perspective. *Clim. Dyn.*, **38**, 1917–1924.

Kang, S. M., L. M. Polvani, J. C. Fyfe, and M. Sigmond, 2011: Impact of polar ozone depletion on subtropical precipitation. *Science*, **332**, 951–954.

Karpechko, A. Y., and E. Manzini, 2012: Stratospheric influence on tropospheric climate change in the Northern Hemisphere. *J. Geophys. Res.*, **117**, D05133.

Kattenberg, A., et al., 1996: Climate models—Projections of future climate. In: *Climate Change 1995: The Science of Climate Change. Contribution of WGI to the Second Assessment Report of the Intergovernmental Panel on Climate Change* [J. T. Houghton, L. G. Meira . A. Callander, N. Harris, A. Kattenberg and K. Maskell (eds.)]. Cambridge University Press, Cambridge, United Kingdom, and New York, NY, USA, pp. 285–357.

Kawase, H., T. Nagashima, K. Sudo, and T. Nozawa, 2011: Future changes in tropospheric ozone under Representative Concentration Pathways (RCPs). *Geophys. Res. Lett.*, **38**, L05801.

Kay, J., M. Holland, and A. Jahn, 2011: Inter-annual to multi-decadal Arctic sea ice extent trends in a warming world. *Geophys. Res. Lett.*, **38**, L15708.

Kay, J. E., M. M. Holland, C. Bitz, E. Blanchard-Wrigglesworth, A. Gettelman, A. Conley, and D. Bailey, 2012: The influence of local feedbacks and northward heat transport on the equilibrium Arctic climate response to increased greenhouse gas forcing in coupled climate models. *J. Clim.*, **25**, 5433–5450.

Kaye, N., A. Hartley, and D. Hemming, 2012: Mapping the climate: Guidance on appropriate techniques to map climate variables and their uncertainty. *Geosci. Model Dev.*, **5**, 245–256.

Kellomaki, S., M. Maajarvi, H. Strandman, A. Kilpelainen, and H. Peltola, 2010: Model computations on the climate change effects on snow cover, soil moisture and soil frost in the boreal conditions over Finland. *Silva Fennica*, **44**, 213–233.

Kendon, E., D. Rowell, and R. Jones, 2010: Mechanisms and reliability of future projected changes in daily precipitation. *Clim. Dyn.*, **35**, 489–509.

Kendon, E., D. Rowell, R. Jones, and E. Buonomo, 2008: Robustness of future changes in local precipitation extremes. *J. Clim.*, **17**, 4280–4297.

Kharin, V. V., F. W. Zwiers, X. B. Zhang, and G. C. Hegerl, 2007: Changes in temperature and precipitation extremes in the IPCC ensemble of global coupled model simulations. *J. Clim.*, **20**, 1419–1444.

Kharin, V. V., F. W. Zwiers, X. Zhang, and M. Wehner, 2013: Changes in temperature and precipitation extremes in the CMIP5 ensemble. *Clim. Change*, doi:10.1007/s10584-013-0705-8.

Khvorostyanov, D., P. Ciais, G. Krinner, and S. Zimov, 2008: Vulnerability of east Siberia's frozen carbon stores to future warming. *Geophys. Res. Lett.*, **35**, L10703.

Kidston, J., and E. P. Gerber, 2010: Intermodel variability of the poleward shift of the austral jet stream in the CMIP3 integrations linked to biases in 20th century climatology. *Geophys. Res. Lett.*, **37**, L09708.

Kienzle, S., M. Nemeth, J. Byrne, and R. MacDonald, 2012: Simulating the hydrological impacts of climate change in the upper North Saskatchewan River basin, Alberta, Canada. *J. Hydrol.*, **412**, 76–89.

Kirkevåg, K., et al., 2013: Aerosol–climate interactions in the Norwegian Earth System Model – NorESM1–M. *Geosci. Model Dev.*, **6**, 207–244.

Kitoh, A., S. Yukimoto, A. Noda, and T. Motoi, 1997: Simulated changes in the Asian summer monsoon at times of increased atmospheric CO_2. *J. Meteorol. Soc. Jpn.*, **75**, 1019–1031.

Kjellstrom, E., L. Barring, D. Jacob, R. Jones, G. Lenderink, and C. Schär, 2007: Modelling daily temperature extremes: Recent climate and future changes over Europe. *Clim. Change*, **81**, 249–265.

Knutti, R., 2010: The end of model democracy? *Clim. Change*, **102**, 395–404.

Knutti, R., and G. C. Hegerl, 2008: The equilibrium sensitivity of the Earth's temperature to radiation changes. *Nature Geosci.*, **1**, 735–743.

Knutti, R., and L. Tomassini, 2008: Constraints on the transient climate response from observed global temperature and ocean heat uptake. *Geophys. Res. Lett.*, **35**, L09701.

Knutti, R., and G.-K. Plattner, 2012: Comment on 'Why hasn't Earth warmed as much as expected?' by Schwartz et al. 2010. *J. Clim.*, **25**, 2192–2199.

Knutti, R., and J. Sedláček, 2013: Robustness and uncertainties in the new CMIP5 climate model projections. *Nature Clim. Change*, **3**, 369–373.

Knutti, R., D. Masson, and A. Gettelman, 2013: Climate model genealogy: Generation CMIP5 and how we got there. *Geophys. Res. Lett.*, **40**, 1194–1199.

Knutti, R., S. Krähenmann, D. Frame, and M. Allen, 2008a: Comment on "Heat capacity, time constant, and sensitivity of Earth's climate system" by S. E. Schwartz. *J. Geophys. Res.*, **113**, D15103.

Knutti, R., F. Joos, S. Müller, G. Plattner, and T. Stocker, 2005: Probabilistic climate change projections for CO_2 stabilization profiles. *Geophys. Res. Lett.*, **32**, L20707.

Knutti, R., R. Furrer, C. Tebaldi, J. Cermak, and G. A. Meehl, 2010a: Challenges in combining projections from multiple climate models. *J. Clim.*, **23**, 2739–2758.

Knutti, R., G. Abramowitz, M. Collins, V. Eyring, P. J. Gleckler, B. Hewitson, and L. Mearns, 2010b: Good practice guidance paper on assessing and combining multi model climate projections. *Meeting Report of the Intergovernmental Panel on Climate Change Expert Meeting on Assessing and Combining Multi-Model Climate Projections*. IPCC Working Group I Technical Support Unit, University of Bern, Bern, Switzerland.

Knutti, R., et al., 2008b: A review of uncertainties in global temperature projections over the twenty-first century. *J. Clim.*, **21**, 2651–2663.

Kodra, E., K. Steinhaeuser, and A. R. Ganguly, 2011: Persisting cold extremes under 21st-century warming scenarios. *Geophys. Res. Lett.*, **38**, L08705.

Kolomyts, E., and N. Surova, 2010: Predicting the impact of global warming on soil water resources in marginal forests of the middle Volga region. *Water Resour.*, **37**, 89–101.

Komuro, Y., et al., 2012: Sea-ice in twentieth-century simulations by new MIROC coupled models: A comparison between models with high resolution and with ice thickness distribution. *J. Meteorol. Soc. Jpn.*, **90A**, 213–232.

Körper, J., et al., 2013: The effects of aggressive mitigation on steric sea level rise and sea ice changes. *Clim. Dyn.*, **40**, 531–550.

Koster, R., Z. Guo, R. Yang, P. Dirmeyer, K. Mitchell, and M. Puma, 2009a: On the nature of soil moisture in land surface models. *J. Clim.*, **22**, 4322–4335.

Koster, R., et al., 2006: GLACE: The Global Land-Atmosphere Coupling Experiment. Part I: Overview. *J. Hydrometeorol.*, **7**, 590–610.

Koster, R. D., S. D. Schubert, and M. J. Suarez, 2009b: Analyzing the concurrence of meteorological droughts and warm periods, with implications for the determination of evaporative regime. *J. Clim.*, **22**, 3331–3341.

Koster, R. D., H. L. Wang, S. D. Schubert, M. J. Suarez, and S. Mahanama, 2009c: Drought-induced warming in the continental United States under different SST regimes. *J. Clim.*, **22**, 5385–5400.

Koven, C., P. Friedlingstein, P. Ciais, D. Khvorostyanov, G. Krinner, and C. Tarnocai, 2009: On the formation of high-latitude soil carbon stocks: Effects of cryoturbation and insulation by organic matter in a land surface model. *Geophys. Res. Lett.*, **36**, L21501.

Koven, C. D., W. J. Riley, and A. Stern, 2013: Analysis of permafrost thermal dynamics and response to climate change in the CMIP5 Earth system models. *J. Clim.*, **26**, 1877–1900.

Koven, C. D., et al., 2011: Permafrost carbon-climate feedbacks accelerate global warming. *Proc. Natl. Acad. Sci. U.S.A.*, **108**, 14769–14774.

Kripalani, R., J. Oh, A. Kulkarni, S. Sabade, and H. Chaudhari, 2007: South Asian summer monsoon precipitation variability: Coupled climate model simulations and projections under IPCC AR4. *Theor. Appl. Climatol.*, **90**, 133–159.

Kug, J., D. Choi, F. Jin, W. Kwon, and H. Ren, 2010: Role of synoptic eddy feedback on polar climate responses to the anthropogenic forcing. *Geophys. Res. Lett.*, **37**, L14704.

Kuhlbrodt, T., and J. M. Gregory, 2012: Ocean heat uptake and its consequences for the magnitude of sea level rise and climate change. *Geophys. Res. Lett.*, **39**, L18608.

Kuhry, P., E. Dorrepaal, G. Hugelius, E. Schuur, and C. Tarnocai, 2010: Potential remobilization of belowground permafrost carbon under future global warming. *Permafr. Periglac. Process.*, **21**, 208–214.

Kumar, A., et al., 2010: Contribution of sea ice loss to Arctic amplification. *Geophys. Res. Lett.*, **37**, L21701.

Kunkel, K. E., T. R. Karl, D. R. Easterling, K. Redmond, J. Young, X. Yin, and P. Hennon, 2013: Probable Maximum Precipitation (PMP) and climate change. *Geophys. Res. Lett.*, **40**, 1402–1408.

Kysely, J., and R. Beranova, 2009: Climate-change effects on extreme precipitation in central Europe: Uncertainties of scenarios based on regional climate models. *Theor. Appl. Climatol.*, **95**, 361–374.

Lamarque, J.-F., et al., 2011: Global and regional evolution of short-lived radiatively-active gases and aerosols in the Representative Concentration Pathways. *Clim. Change*, **109**, 191–212.

Lamarque, J., 2008: Estimating the potential for methane clathrate instability in the 1%-CO_2 IPCC AR-4 simulations. *Geophys. Res. Lett.*, **35**, L19806.

Lamarque, J., et al., 2010: Historical (1850–2000) gridded anthropogenic and biomass burning emissions of reactive gases and aerosols: Methodology and application. *Atmos. Chem. Phys.*, **10**, 7017–7039.

Lamarque, J. F., et al., 2013: The Atmospheric Chemistry and Climate Model Intercomparison Project (ACCMIP): Overview and description of models, simulations and climate diagnostics. *Geosci. Model Dev.*, **6**, 179–206.

12

Lambert, F., and M. Webb, 2008: Dependency of global mean precipitation on surface temperature. *Geophys. Res. Lett.*, **35**, L16706.

Lambert, F. H., and J. C. H. Chiang, 2007: Control of land-ocean temperature contrast by ocean heat uptake. *Geophys. Res. Lett.*, **34**, L13704.

Lambert, F. H., M. J. Webb, and M. J. Joshi, 2011: The relationship between land-ocean surface temperature contrast and radiative forcing. *J. Clim.*, **24**, 3239–3256.

Lambert, F. H., N. P. Gillett, D. A. Stone, and C. Huntingford, 2005: Attribution studies of observed land precipitation changes with nine coupled models. *Geophys. Res. Lett.*, **32**, L18704.

Lambert, F. H., G. R. Harris, M. Collins, J. M. Murphy, D. M. H. Sexton, and B. B. B. Booth, 2012: Interactions between perturbations to different Earth system components simulated by a fully-coupled climate model. *Clim. Dyn.*, doi:10.1007/s00382-012-1618-3.

Langen, P. L., and V. A. Alexeev, 2007: Polar amplification as a preferred response in an idealized aquaplanet GCM. *Clim. Dyn.*, **29**, 305–317.

Langen, P. L., A. M. Solgaard, and C. S. Hvidberg, 2012: Self-inhibiting growth of the Greenland Ice Sheet. *Geophys. Res. Lett.*, **39**, L12502.

Lapola, D. M., M. D. Oyama, and C. A. Nobre, 2009: Exploring the range of climate biome projections for tropical South America: The role of CO_2 fertilization and seasonality. *Global Biogeochem. Cycles*, **23**, GB3003.

Lau, K., M. Kim, and K. Kim, 2006: Asian summer monsoon anomalies induced by aerosol direct forcing: The role of the Tibetan Plateau. *Clim. Dyn.*, **26**, 855–864.

Lawrence, D., and A. Slater, 2010: The contribution of snow condition trends to future ground climate. *Clim. Dyn.*, **34**, 969–981.

Lawrence, D., A. Slater, and S. Swenson, 2012: Simulation of present-day and future permafrost and seasonally frozen ground conditions in CCSM4. *J. Clim.*, **25**, 2207–2225.

Lawrence, D., A. Slater, V. Romanovsky, and D. Nicolsky, 2008a: Sensitivity of a model projection of near-surface permafrost degradation to soil column depth and representation of soil organic matter. *J. Geophys. Res. Earth Surface*, **113**, F02011.

Lawrence, D., A. Slater, R. Tomas, M. Holland, and C. Deser, 2008b: Accelerated Arctic land warming and permafrost degradation during rapid sea ice loss. *Geophys. Res. Lett.*, **35**, L11506.

Lean, J., and D. Rind, 2009: How will Earth's surface temperature change in future decades? *Geophys. Res. Lett.*, **36**, L15708.

Lee, S., T. Gong, N. Johnson, S. B. Feldstein, and D. Pollard, 2011: On the possible link between tropical convection and the Northern Hemisphere Arctic surface air temperature change between 1958 and 2001. *J. Clim.*, **24**, 4350–4367.

Lefebvre, W., and H. Goosse, 2008: Analysis of the projected regional sea-ice changes in the Southern Ocean during the twenty-first century. *Clim. Dyn.*, **30**, 59–76.

Lemoine, D. M., 2010: Climate sensitivity distributions dependence on the possibility that models share biases. *J. Clim.*, **23**, 4395–4415.

Lenderink, G., and E. Van Meijgaard, 2008: Increase in hourly precipitation extremes beyond expectations from temperature changes. *Nature Geosci.*, **1**, 511–514.

Lenderink, G., A. van Ulden, B. van den Hurk, and E. van Meijgaard, 2007: Summertime inter-annual temperature variability in an ensemble of regional model simulations: Analysis of the surface energy budget. *Clim. Change*, **81**, 233–247.

Lenton, T., H. Held, E. Kriegler, J. Hall, W. Lucht, S. Rahmstorf, and H. Schellnhuber, 2008: Tipping elements in the Earth's climate system. *Proc. Natl. Acad. Sci. U.S.A.*, **105**, 1786–1793.

Lenton, T. M., 2012: Arctic climate tipping points. *Ambio*, **41**, 10–22.

Leslie, L. M., M. Leplastrier, and B. W. Buckley, 2008: Estimating future trends in severe hailstorms over the Sydney Basin: A climate modelling study. *Atmos. Res.*, **87**, 37–51.

Levermann, A., J. Schewe, V. Petoukhov, and H. Held, 2009: Basic mechanism for abrupt monsoon transitions. *Proc. Natl. Acad. Sci. U.S.A.*, **106**, 20572–20577.

Levitus, S., J. Antonov, and T. Boyer, 2005: Warming of the world ocean, 1955–2003. *Geophys. Res. Lett.*, **32**, L02604.

Levitus, S., et al., 2012: World ocean heat content and thermosteric sea level change (0–2000 m), 1955–2010. *Geophys. Res. Lett.*, **39**, L10603.

Levy II, H., L. W. Horowitz, M. D. Schwarzkopf, Y. Ming, J.-C. Golaz, V. Naik, and V. Ramaswamy, 2013: The roles of aerosol direct and indirect effects in past and future climate change. *J. Geophys. Res.*, doi:10.1002/jgrd.50192.

Li, C., J. S. von Storch, and J. Marotzke, 2013a: Deep-ocean heat uptake and equilibrium climate response. *Clim. Dyn.*, **40**, 1071–1086.

Li, C., D. Notz, S. Tietsche, and J. Marotzke, 2013b: The transient versus the equilibrium response of sea ice to global warming. *J. Clim.*, doi:10.1175/JCLI-D-12-00492.1.

Li, F., J. Austin, and J. Wilson, 2008: The strength of the Brewer-Dobson circulation in a changing climate: Coupled chemistry-climate model simulations. *J. Clim.*, **21**, 40–57.

Li, F., W. Collins, M. Wehner, D. Williamson, J. Olson, and C. Algieri, 2011a: Impact of horizontal resolution on simulation of precipitation extremes in an aqua-planet version of Community Atmospheric Model (CAM3). *Tellus*, **63**, 884–892.

Li, L., X. Jiang, M. Chahine, E. Olsen, E. Fetzer, L. Chen, and Y. Yung, 2011b: The recycling rate of atmospheric moisture over the past two decades (1988–2009). *Environ. Res. Lett.*, **6**, 034018.

Li, L. J., et al., 2013c: The Flexible Global Ocean-Atmosphere-Land System Model: Grid-point Version 2: FGOALS-g2. *Adv. Atmos. Sci.*, **30**, 543–560.

Liepert, B. G., and M. Previdi, 2009: Do models and observations disagree on the rainfall response to global warming? *J. Clim.*, **22**, 3156–3166.

Liepert, B. G., and M. Previdi, 2012: Inter-model variability and biases of the global water cycle in CMIP3 coupled climate models. *Environ. Res. Lett.*, **7**, 014006.

Lim, E. P., and I. Simmonds, 2009: Effect of tropospheric temperature change on the zonal mean circulation and SH winter extratropical cyclones. *Clim. Dyn.*, **33**, 19–32.

Lindsay, R., and J. Zhang, 2005: The thinning of Arctic sea ice, 1988–2003: Have we passed a tipping point? *J. Clim.*, **18**, 4879–4894.

Liu, Z., S. J. Vavrus, F. He, N. Wen, and Y. Zhong, 2005: Rethinking tropical ocean response to global warming: The enhanced equatorial warming. *J. Clim.*, **18**, 4684–4700.

Livina, V. N., and T. M. Lenton, 2013: A recent tipping point in the Arctic sea-ice cover: Abrupt and persistent increase in the seasonal cycle since 2007. *Cryosphere*, **7**, 275–286.

Loarie, S. R., D. B. Lobell, G. P. Asner, Q. Z. Mu, and C. B. Field, 2011: Direct impacts on local climate of sugar-cane expansion in Brazil. *Nature Clim. Change*, **1**, 105–109.

Loeb, N. G., et al., 2009: Toward optimal closure of the Earth's Top-of-Atmosphere radiation budget. *J. Clim.*, **22**, 748–766.

Long, M. C., K. Lindsay, S. Peacock, J. K. Moore, and S. C. Doney, 2013: Twentieth-century oceanic carbon uptake and storage in CESM1(BGC). *J. Clim.*, doi:10.1175/JCLI-D-12-00184.1.

Lorenz, D. J., and E. T. DeWeaver, 2007: Tropopause height and zonal wind response to global warming in the IPCC scenario integrations. *J. Geophys. Res. Atmos.*, **112**, D10119.

Lowe, J., C. Huntingford, S. Raper, C. Jones, S. Liddicoat, and L. Gohar, 2009: How difficult is it to recover from dangerous levels of global warming? *Environ. Res. Lett.*, **4**, 014012.

Lowe, J. A., and J. M. Gregory, 2006: Understanding projections of sea level rise in a Hadley Centre coupled climate model. *J. Geophys. Res.*, **111**, C11014.

Lu, J., and M. Cai, 2009: Seasonality of polar surface warming amplification in climate simulations. *Geophys. Res. Lett.*, **36**, L16704.

Lu, J., G. Vecchi, and T. Reichler, 2007: Expansion of the Hadley cell under global warming. *Geophys. Res. Lett.*, **34**, L06805.

Lu, J., G. Chen, and D. Frierson, 2008: Response of the zonal mean atmospheric circulation to El Niño versus global warming. *J. Clim.*, **21**, 5835–5851.

Lucht, W., S. Schaphoff, T. Ebrecht, U. Heyder, and W. Cramer, 2006: Terrestrial vegetation redistribution and carbon balance under climate change. *Carbon Balance Manage.*, **1**, 1-6.

Lunt, D., A. Haywood, G. Schmidt, U. Salzmann, P. Valdes, and H. Dowsett, 2010: Earth system sensitivity inferred from Pliocene modelling and data. *Nature Geosci.*, **3**, 60–64.

Luo, J. J., W. Sasaki, and Y. Masumoto, 2012: Indian Ocean warming modulates Pacific climate change. *Proc. Natl. Acad. Sci. U.S.A.*, **109**, 18701–18706.

Lustenberger, A., R. Knutti, and E. M. Fischer, 2013: The potential of pattern scaling for projecting temperature-related extreme indices. *Int. J. Climatol.*, doi:10.1002/joc.3659.

Ma, J., and S.-P. Xie, 2013: Regional patterns of sea surface temperature change: A source of uncertainty in future projections of precipitation and atmospheric circulation. *J. Clim.*, **26**, 2482–2501.

Ma, J., S.-P. Xie, and Y. Kosaka, 2012: Mechanisms for tropical tropospheric circulation change in response to global warming. *J. Clim.*, **25**, 2979–2994.

MacDougall, A. H., C. A. Avis, and A. J. Weaver, 2012: Significant contribution to climate warming from the permafrost carbon feedback. *Nature Geosci.*, **5**, 719–721.

Mahlstein, I., and R. Knutti, 2011: Ocean heat transport as a cause for model uncertainty in projected Arctic warming. *J. Clim.*, **24**, 1451–1460.

Mahlstein, I., and R. Knutti, 2012: September Arctic sea ice predicted to disappear near 2°C global warming above present. *J. Geophys. Res.*, **117**, D06104.
Mahlstein, I., R. Knutti, S. Solomon, and R. W. Portmann, 2011: Early onset of significant local warming in low latitude countries. *Environ. Res. Lett.*, **6**, 034009.
Mahlstein, I., R. W. Portmann, J. S. Daniel, S. Solomon, and R. Knutti, 2012: Perceptible changes in regional precipitation in a future climate. *Geophys. Res. Lett.*, **39**, L05701.
Maksym, T., S. E. Stammerjohn, S. Ackley, and R. Massom, 2012: Antarctic sea ice—A polar opposite? *Oceanography*, **25**, 140–151.
Malhi, Y., et al., 2009: Exploring the likelihood and mechanism of a climate-change-induced dieback of the Amazon rainforest. *Proc. Natl. Acad. Sci. U.S.A.*, **106**, 20610–20615.
Manabe, S., and R. Stouffer, 1980: Sensitivity of a global climate model to an increase of CO_2 concentration in the atmosphere. *J. Geophys. Res.*, **85**, 5529–5554.
Manabe, S., and R. T. Wetherald, 1980: Distribution of climate change resulting from an increase in CO_2 content of the atmosphere. *J. Atmos. Sci.*, **37**, 99–118.
Manabe, S., and R. Stouffer, 1994: Multiple-century response of a coupled ocean-atmosphere model to an increase of atmospheric carbon-dioxide. *J. Clim.*, **7**, 5–23.
Manabe, S., K. Bryan, and M. J. Spelman, 1990: Transient-response of a global ocean atmosphere model to a doubling of atmospheric carbon-dioxide. *J. Phys. Oceanogr.*, **20**, 722–749.
Manabe, S., R. J. Stouffer, M. J. Spelman, and K. Bryan, 1991: Transient responses of a coupled ocean atmosphere model to gradual changes of atmospheric CO_2. Part I: Annual mean response. *J. Clim.*, **4**, 785–818.
Marsh, P. T., H. E. Brooks, and D. J. Karoly, 2009: Preliminary investigation into the severe thunderstorm environment of Europe simulated by the Community Climate System Model 3. *Atmos. Res.*, **93**, 607–618.
Maslowski, W., J. C. Kinney, M. Higgins, and A. Roberts, 2012: The future of Arctic sea ice. In: *Annual Review of Earth and Planetary Sciences* [R. Jeanloz (ed.)]. Annual Reviews, Palo Alto, CA, USA, pp. 625–654.
Masson, D., and R. Knutti, 2011: Climate model genealogy. *Geophys. Res. Lett.*, **38**, L08703.
Massonnet, F., T. Fichefet, H. Goosse, C. M. Bitz, G. Philippon-Berthier, M. Holland, and P. Y. Barriat, 2012: Constraining projections of summer Arctic sea ice. *Cryosphere*, **6**, 1383–1394.
Matsuno, T., K. Maruyama, and J. Tsutsui, 2012a: Stabilization of atmospheric carbon dioxide via zero emissions-An alternative way to a stable global environment. Part 1: Examination of the traditional stabilization concept. *Proc. Jpn. Acad. B*, **88**, 368–384.
Matsuno, T., K. Maruyama, and J. Tsutsui, 2012b: Stabilization of atmospheric carbon dioxide via zero emissions-An alternative way to a stable global environment. Part 2: A practical zero-emissions scenario. *Proc. Jpn. Acad. B*, **88**, 385–395.
Matthews, H., and K. Caldeira, 2008: Stabilizing climate requires near-zero emissions. *Geophys. Res. Lett.*, **35**, L04705.
Matthews, H., N. Gillett, P. Stott, and K. Zickfeld, 2009: The proportionality of global warming to cumulative carbon emissions. *Nature*, **459**, 829–832.
Matthews, H. D., and K. Zickfeld, 2012: Climate response to zeroed emissions of greenhouse gases and aerosols. *Nature Clim. Change*, **2**, 338–341.
Matthews, H. D., S. Solomon, and R. Pierrehumbert, 2012: Cumulative carbon as a policy framework for achieving climate stabilization. *Philos. Trans. R. Soc. A*, **370**, 4365–4379.
May, W., 2002: Simulated changes of the Indian summer monsoon under enhanced greenhouse gas conditions in a global time-slice experiment. *Geophys. Res. Lett.*, **29**, 1118.
May, W., 2008a: Climatic changes associated with a global "2°C-stabilization" scenario simulated by the ECHAM5/MPI-OM coupled climate model. *Clim. Dyn.*, **31**, 283–313.
May, W., 2008b: Potential future changes in the characteristics of daily precipitation in Europe simulated by the HIRHAM regional climate model. *Clim. Dyn.*, **30**, 581–603.
May, W., 2012: Assessing the strength of regional changes in near-surface climate associated with a global warming of 2°C. *Clim. Change*, **110**, 619–644.
McCabe, G., and D. Wolock, 2007: Warming may create substantial water supply shortages in the Colorado River basin. *Geophys. Res. Lett.*, **34**, L22708.
McCollum, D., V. Krey, K. Riahi, P. Kolp, A. Grubler, M. Makowski, and N. Nakicenovic, 2013: Climate policies can help resolve energy security and air pollution challenges. *Clim. Change*, doi:10.1007/s10584-013-0710-y.
McLandress, C., and T. G. Shepherd, 2009: Simulated anthropogenic changes in the Brewer-Dobson circulation, including its extension to high latitudes. *J. Clim.*, **22**, 1516–1540.
McLandress, C., T. G. Shepherd, J. F. Scinocca, D. A. Plummer, M. Sigmond, A. I. Jonsson, and M. C. Reader, 2011: Separating the dynamical effects of climate change and ozone depletion. Part II: Southern Hemisphere troposphere. *J. Clim.*, **24**, 1850–1868.
Meehl, G., and W. Washington, 1993: South Asian summer monsoon variability in a model with doubled atmospheric carbon-dioxide concentration. *Science*, **260**, 1101–1104.
Meehl, G., J. Arblaster, and C. Tebaldi, 2005a: Understanding future patterns of increased precipitation intensity in climate model simulations. *Geophys. Res. Lett.*, **32**, L18719.
Meehl, G., J. Arblaster, and W. Collins, 2008: Effects of black carbon aerosols on the Indian monsoon. *J. Clim.*, **21**, 2869–2882.
Meehl, G., et al., 2012: Climate system response to external forcings and climate change projections in CCSM4. *J. Clim.*, **25**, 3661–3683.
Meehl, G. A., and C. Tebaldi, 2004: More intense, more frequent, and longer lasting heat waves in the 21st century. *Science*, **305**, 994–997.
Meehl, G. A., G. J. Boer, C. Covey, M. Latif, and R. J. Stouffer, 2000: The Coupled Model Intercomparison Project (CMIP). *Bull. Am. Meteorol. Soc.*, **81**, 313–318.
Meehl, G. A., C. Tebaldi, G. Walton, D. Easterling, and L. McDaniel, 2009: Relative increase of record high maximum temperatures compared to record low minimum temperatures in the U. S. *Geophys. Res. Lett.*, **36**, L23701.
Meehl, G. A., W. M. Washington, C. M. Ammann, J. M. Arblaster, T. M. L. Wigley, and C. Tebaldi, 2004: Combinations of natural and anthropogenic forcings in twentieth-century climate. *J. Clim.*, **17**, 3721–3727.
Meehl, G. A., et al., 2005b: How much more global warming and sea level rise? *Science*, **307**, 1769–1772.
Meehl, G. A., et al., 2007a: The WCRP CMIP3 multimodel dataset - A new era in climate change research. *Bull. Am. Meteorol. Soc.*, **88**, 1383–1394.
Meehl, G. A., et al., 2006: Climate change projections for the twenty-first century and climate change commitment in the CCSM3. *J. Clim.*, **19**, 2597–2616.
Meehl, G. A., et al., 2013: Climate change projections in CESM1(CAM5) compared to CCSM4. *J. Clim.*, doi:10.1175/JCLI-D-12-00572.1.
Meehl, G. A., et al., 2007b: Global climate projections. In: *Climate Change 2007: The Physical Science Basis. Contribution of Working Group I to the Fourth Assessment Report of the Intergovernmental Panel on Climate Change* [Solomon, S., D. Qin, M. Manning, Z. Chen, M. Marquis, K. B. Averyt, M. Tignor and H. L. Miller (eds.)] Cambridge University Press, Cambridge, United Kingdom and New York, NY, USA, pp. 747–846.
Meijers, A. J. S., E. Shuckburgh, N. Bruneau, J.-B. Sallee, T. J. Bracegirdle, and Z. Wang, 2012: Representation of the Antarctic Circumpolar Current in the CMIP5 climate models and future changes under warming scenarios. *J. Geophys. Res.*, **117**, C12008.
Meinshausen, M., S. Raper, and T. Wigley, 2011a: Emulating coupled atmosphere-ocean and carbon cycle models with a simpler model, MAGICC6–Part 1: Model description and calibration. *Atmos. Chem. Phys.*, **11**, 1417–1456.
Meinshausen, M., T. Wigley, and S. Raper, 2011b: Emulating atmosphere-ocean and carbon cycle models with a simpler model, MAGICC6–Part 2: Applications. *Atmos. Chem. Phys.*, **11**, 1457–1471.
Meinshausen, M., B. Hare, T. Wigley, D. Van Vuuren, M. Den Elzen, and R. Swart, 2006: Multi-gas emissions pathways to meet climate targets. *Clim. Change*, **75**, 151–194.
Meinshausen, M., et al., 2009: Greenhouse-gas emission targets for limiting global warming to 2°C. *Nature*, **458**, 1158–1162.
Meinshausen, M., et al., 2011c: The RCP greenhouse gas concentrations and their extensions from 1765 to 2300. *Clim. Change*, **109**, 213–241.
Merrifield, M. A., 2011: A shift in western tropical Pacific sea level trends during the 1990s. *J. Clim.*, **24**, 4126–4138.
Merryfield, W. J., M. M. Holland, and A. H. Monahan, 2008: Multiple equilibria and abrupt transitions in Arctic summer sea ice extent. In: *Arctic Sea Ice Decline: Observations, Projections, Mechanisms, and Implications*. American Geophysical Union, Washington, DC, pp. 151–174.
Mignone, B., R. Socolow, J. Sarmiento, and M. Oppenheimer, 2008: Atmospheric stabilization and the timing of carbon mitigation. *Clim. Change*, **88**, 251–265.
Mikolajewicz, U., M. Vizcaino, J. Jungclaus, and G. Schurgers, 2007: Effect of ice sheet interactions in anthropogenic climate change simulations. *Geophys. Res. Lett.*, **34**, L18706.

Millner, A., R. Calel, D. A. Stainforth, and G. MacKerron, 2013: Do probabilistic expert elicitations capture scientists' uncertainty about climate change? *Clim. Change*, **116**, 427–436.

Milly, P., J. Betancourt, M. Falkenmark, R. Hirsch, Z. Kundzewicz, D. Lettenmaier, and R. Stouffer, 2008: Stationarity is dead: Whither water management? *Science*, **319**, 573–574.

Min, S., X. Zhang, F. Zwiers, and G. Hegerl, 2011: Human contribution to more-intense precipitation extremes. *Nature*, **470**, 378–381.

Ming, Y., V. Ramaswamy, and G. Persad, 2010: Two opposing effects of absorbing aerosols on global-mean precipitation. *Geophys. Res. Lett.*, **37**, L13701.

Mitas, C., and A. Clement, 2006: Recent behavior of the Hadley cell and tropical thermodynamics in climate models and reanalyses. *Geophys. Res. Lett.*, **33**, L01810.

Mitchell, J. F. B., 1990: Is the Holocene a good analogue for greenhouse warming? *J. Clim.*, **3**, 1177–1192.

Mitchell, J. F. B., C. A. Wilson, and W. M. Cunnington, 1987: On CO_2 climate sensitivity and model dependence of results. *Q. J. R. Meteorol. Soc.*, **113**, 293–322.

Mitchell, J. F. B., T. C. Johns, W. J. Ingram, and J. A. Lowe, 2000: The effect of stabilising atmospheric carbon dioxide concentrations on global and regional climate change. *Geophys. Res. Lett.*, **27**, 2977–2980.

Mitchell, J. F. B., T. C. Johns, M. Eagles, W. J. Ingram, and R. A. Davis, 1999: Towards the construction of climate change scenarios. *Clim. Change*, **41**, 547–581.

Mitchell, T. D., 2003: Pattern scaling - An examination of the accuracy of the technique for describing future climates. *Clim. Change*, **60**, 217–242.

Mizuta, R., 2012: Intensification of extratropical cyclones associated with the polar jet change in the CMIP5 global warming projections. *Geophys. Res. Lett.*, **39**, L19707.

Monaghan, A., D. Bromwich, and D. Schneider, 2008: Twentieth century Antarctic air temperature and snowfall simulations by IPCC climate models. *Geophys. Res. Lett.*, **35**, L07502.

Montenegro, A., V. Brovkin, M. Eby, D. Archer, and A. Weaver, 2007: Long term fate of anthropogenic carbon. *Geophys. Res. Lett.*, **34**, L19707.

Morgan, M. G., and D. W. Keith, 1995: Climate-change - Subjective judgments by climate experts. *Environ. Sci. Technol.*, **29**, A468–A476.

Moss, R. H., et al., 2010: The next generation of scenarios for climate change research and assessment. *Nature*, **463**, 747–756.

Moss, R. H., et al., 2008: Towards new scenarios for analysis of emissions, climate change, impacts, and response strategies. In: *IPCC Expert Meeting Report: Towards New Scenarios*. Intergovernmental Panel on Climate Change, Geneva, Switzerland, 132 pp.

Muller, C. J., and P. A. O'Gorman, 2011: An energetic perspective on the regional response of precipitation to climate change. *Nature Clim. Change*, **1**, 266–271.

Murphy, D. M., S. Solomon, R. W. Portmann, K. H. Rosenlof, P. M. Forster, and T. Wong, 2009: An observationally based energy balance for the Earth since 1950. *J. Geophys. Res.*, **114**, D17107.

Murphy, J., D. Sexton, D. Barnett, G. Jones, M. Webb, and M. Collins, 2004: Quantification of modelling uncertainties in a large ensemble of climate change simulations. *Nature*, **430**, 768–772.

Murphy, J. M., B. B. B. Booth, M. Collins, G. R. Harris, D. M. H. Sexton, and M. J. Webb, 2007: A methodology for probabilistic predictions of regional climate change from perturbed physics ensembles. *Philos. Trans. R. Soc. A*, **365**, 1993–2028.

Myhre, G., E. Highwood, K. Shine, and F. Stordal, 1998: New estimates of radiative forcing due to well mixed greenhouse gases. *Geophys. Res. Lett.*, **25**, 2715–2718.

Neelin, J. D., C. Chou, and H. Su, 2003: Tropical drought regions in global warming and El Niño teleconnections. *Geophys. Res. Lett.*, **30**, 2275.

Neelin, J. D., M. Munnich, H. Su, J. E. Meyerson, and C. E. Holloway, 2006: Tropical drying trends in global warming models and observations. *Proc. Natl. Acad. Sci. U.S.A.*, **103**, 6110–6115.

Nelson, F., and S. Outcalt, 1987: A computational method for prediction and regionalization of permafrost. *Arct. Alpine Res.*, **19**, 279–288.

Newlands, N. K., G. Espino-Hernández, and R. S. Erickson, 2012: Understanding crop response to climate variability with complex agroecosystem models. *Int. J. Ecol.*, **2012**, 756242.

Niall, S., and K. Walsh, 2005: The impact of climate change on hailstorms in southeastern Australia. *Int. J. Climatol.*, **25**, 1933–1952.

Nicolsky, D., V. Romanovsky, V. Alexeev, and D. Lawrence, 2007: Improved modeling of permafrost dynamics in a GCM land-surface scheme. *Geophys. Res. Lett.*, **34**, L08501.

Niinemets, U., 2010: Responses of forest trees to single and multiple environmental stresses from seedlings to mature plants: Past stress history, stress interactions, tolerance and acclimation. *Forest Ecol. Manage.*, **260**, 1623–1639.

Nikulin, G., E. Kjellstrom, U. Hansson, G. Strandberg, and A. Ullerstig, 2011: Evaluation and future projections of temperature, precipitation and wind extremes over Europe in an ensemble of regional climate simulations. *Tellus A*, **63**, 41–55.

Nobre, C., and L. Borma, 2009: 'Tipping points' for the Amazon forest. *Curr. Opin. Environ. Sustain.*, **1**, 28–36.

North, G., 1984: The small ice cap instability in diffuse climate models. *J. Atmos. Sci.*, **41**, 3390–3395.

Notaro, M., 2008: Statistical identification of global hot spots in soil moisture feedbacks among IPCC AR4 models. *J. Geophys. Res.*, **113**, D09101.

Notz, D., 2009: The future of ice sheets and sea ice: Between reversible retreat and unstoppable loss. *Proc. Natl. Acad. Sci. U.S.A.*, **106**, 20590–20595.

NRC, 2011: *Climate Stabilization Targets: Emissions, Concentrations, and Impacts over Decades to Millennia.* National Academies Press, Washington, DC, 298 pp.

O'Connor, F., et al., 2010: Possible role of wetlands, permafrost, and methane hydrates in the methane cycle under future climate change: A review. *Rev. Geophys.*, **48**, RG4005.

O'Gorman, P., and T. Schneider, 2009a: Scaling of precipitation extremes over a wide range of climates simulated with an idealized GCM. *J. Clim.*, **22**, 5676–5685.

O'Gorman, P., and T. Schneider, 2009b: The physical basis for increases in precipitation extremes in simulations of 21st-century climate change. *Proc. Natl. Acad. Sci. U.S.A.*, **106**, 14773–14777.

O'Gorman, P., R. Allan, M. Byrne, and M. Previdi, 2012: Energetic constraints on precipitation under climate change. *Surv. Geophys.*, **33**, 585–608.

O'Gorman, P. A., 2010: Understanding the varied response of the extratropical storm tracks to climate change. *Proc. Natl. Acad. Sci. U.S.A.*, **107**, 19176–19180.

O'Gorman, P. A., and C. J. Muller, 2010: How closely do changes in surface and column water vapor follow Clausius-Clapeyron scaling in climate change simulations? *Environ. Res. Lett.*, **5**, 025207.

Orlowsky, B., and S. I. Seneviratne, 2012: Global changes in extreme events: Regional and seasonal dimension. *Clim. Change*, **110**, 669–696.

Otto, A., et al., 2013: Energy budget constraints on climate response. *Nature Geosci.*, **6**, 415-416.

Overland, J. E., and M. Wang, 2013: When will the summer Arctic be nearly sea ice free? *Geophys. Res. Lett.*, doi:10.1002/grl.50316.

Overland, J. E., M. Wang, N. A. Bond, J. E. Walsh, V. M. Kattsov, and W. L. Chapman, 2011: Considerations in the selection of global climate models for regional climate projections: The Arctic as a case study. *J. Clim.*, **24**, 1583–1597.

Oyama, M. D., and C. A. Nobre, 2003: A new climate-vegetation equilibrium state for Tropical South America. *Geophys. Res. Lett.*, **30**, 2199.

Padilla, L., G. Vallis, and C. Rowley, 2011: Probabilistic estimates of transient climate sensitivity subject to uncertainty in forcing and natural variability. *J. Clim.*, **24**, 5521–5537.

Paeth, H., and F. Pollinger, 2010: Enhanced evidence in climate models for changes in extratropical atmospheric circulation. *Tellus A*, **62**, 647–660.

Pagani, M., Z. Liu, J. LaRiviere, and A. Ravelo, 2010: High Earth-system climate sensitivity determined from Pliocene carbon dioxide concentrations. *Nature Geosci.*, **3**, 27–30.

Pall, P., M. Allen, and D. Stone, 2007: Testing the Clausius-Clapeyron constraint on changes in extreme precipitation under CO_2 warming. *Clim. Dyn.*, **28**, 351–363.

Pennell, C., and T. Reichler, 2011: On the effective number of climate models. *J. Clim.*, **24**, 2358–2367.

Perkins, S. E., L. V. Alexander, and J. R. Nairn, 2012: Increasing frequency, intensity and duration of observed global heatwaves and warm spells. *Geophys. Res. Lett.*, **39**, L20714.

Perrie, W., Y. H. Yao, and W. Q. Zhang, 2010: On the impacts of climate change and the upper ocean on midlatitude northwest Atlantic landfalling cyclones. *J. Geophys. Res.*, **115**, D23110.

Piani, C., D. J. Frame, D. A. Stainforth, and M. R. Allen, 2005: Constraints on climate change from a multi-thousand member ensemble of simulations. *Geophys. Res. Lett.*, **32**, L23825.

Pierce, D., et al., 2008: Attribution of declining Western US snowpack to human effects. *J. Clim.*, **21**, 6425–6444.

Pinto, J. G., U. Ulbrich, G. C. Leckebusch, T. Spangehl, M. Reyers, and S. Zacharias, 2007: Changes in storm track and cyclone activity in three SRES ensemble experiments with the ECHAM5/MPI-OM1 GCM. *Clim. Dyn.*, **29**, 195–210.

Pitman, A., et al., 2009: Uncertainties in climate responses to past land cover change: First results from the LUCID intercomparison study. *Geophys. Res. Lett.*, **36**, L14814.

Plattner, G.-K., et al., 2008: Long-term climate commitments projected with climate-carbon cycle models. *J. Clim.*, **21**, 2721– 2751.

Polvani, L. M., M. Previdi, and C. Deser, 2011: Large cancellation, due to ozone recovery, of future Southern Hemisphere atmospheric circulation trends. *Geophys. Res. Lett.*, **38**, L04707.

Pongratz, J., C. Reick, T. Raddatz, and M. Claussen, 2010: Biogeophysical versus biogeochemical climate response to historical anthropogenic land cover change. *Geophys. Res. Lett.*, **37**, L08702.

Port, U., V. Brovkin, and M. Claussen, 2012: The influence of vegetation dynamics on anthropogenic climate change. *Earth Syst. Dyn.*, **3**, 233–243.

Power, S., and G. Kociuba, 2011a: The impact of global warming on the Southern Oscillation Index. *Clim. Dyn.*, **37**, 1745–1754.

Power, S., F. Delage, R. Colman, and A. Moise, 2012: Consensus on twenty-first-century rainfall projections in climate models more widespread than previously thought. *J. Clim.*, **25**, 3792–3809.

Power, S. B., and G. Kociuba, 2011b: What caused the observed twentieth-century weakening of the Walker circulation? *J. Clim.*, **24**, 6501–6514.

Previdi, M., 2010: Radiative feedbacks on global precipitation. *Environ. Res. Lett.*, **5**, 025211.

Rahmstorf, S., et al., 2005: Thermohaline circulation hysteresis: A model intercomparison. *Geophys. Res. Lett.*, **32**, L23605.

Räisänen, J., 2007: How reliable are climate models? *Tellus A*, **59**, 2–29.

Räisänen, J., 2008: Warmer climate: Less or more snow? *Clim. Dyn.*, **30**, 307–319.

Räisänen, J., and L. Ruokolainen, 2006: Probabilistic forecasts of near-term climate change based on a resampling ensemble technique. *Tellus A*, **58**, 461–472.

Räisänen, J., and J. S. Ylhaisi, 2011: Cold months in a warming climate. *Geophys. Res. Lett.*, **38**, L22704.

Ramanathan, V., P. J. Crutzen, J. T. Kiehl, and D. Rosenfeld, 2001: Aerosols, climate, and the hydrologic cycle. *Science*, **294**, 2119–2124.

Ramaswamy, V., et al., 2001: Radiative forcing of climate change. In: *Climate Change 2001: The Scientific Basis. Contribution of Working Group I to the Third Assessment Report of the Intergovernmental Panel on Climate Change* [J. T. Houghton, Y. Ding, D. J. Griggs, M. Noquer, P. J. van der Linden, X. Dai, K. Maskell and C. A. Johnson (eds.)]. Cambridge University Press, Cambridge, United Kingdom and New York, NY, USA pp. 349-416.

Rammig, A., et al., 2010: Estimating the risk of Amazonian forest dieback. *New Phytologist*, **187**, 694–706.

Randall, D. A., et al., 2007: Climate models and their evaluation. In: *Climate Change 2007: The Physical Science Basis. Contribution of Working Group I to the Fourth Assessment Report of the Intergovernmental Panel on Climate Change* [Solomon, S., D. Qin, M. Manning, Z. Chen, M. Marquis, K. B. Averyt, M. Tignor and H. L. Miller (eds.)] Cambridge University Press, Cambridge, United Kingdom and New York, NY, USA, pp. 589–662.

Randalls, S., 2010: History of the 2°C climate target. *WIREs Climate Change*, **1**, 598–605.

Randel, W., and F. Wu, 2007: A stratospheric ozone profile data set for 1979–2005: Variability, trends, and comparisons with column ozone data. *J. Geophys. Res.*, **112**, D06313.

Randel, W. J., M. Park, F. Wu, and N. Livesey, 2007: A large annual cycle in ozone above the tropical tropopause linked to the Brewer-Dobson circulation. *J. Atmos. Sci.*, **64**, 4479–4488.

Randles, C., and V. Ramaswamy, 2008: Absorbing aerosols over Asia: A Geophysical Fluid Dynamics Laboratory general circulation model sensitivity study of model response to aerosol optical depth and aerosol absorption. *J. Geophys. Res.*, **113**, D21203.

Reagan, M., and G. Moridis, 2007: Oceanic gas hydrate instability and dissociation under climate change scenarios. *Geophys. Res. Lett.*, **34**, L22709.

Reagan, M., and G. Moridis, 2009: Large-scale simulation of methane hydrate dissociation along the West Spitsbergen Margin. *Geophys. Res. Lett.*, **36**, L23612.

Ridley, J., J. Lowe, and D. Simonin, 2008: The demise of Arctic sea ice during stabilisation at high greenhouse gas concentrations. *Clim. Dyn.*, **30**, 333–341.

Ridley, J., J. Lowe, C. Brierley, and G. Harris, 2007: Uncertainty in the sensitivity of Arctic sea ice to global warming in a perturbed parameter climate model ensemble. *Geophys. Res. Lett.*, **34**, L19704.

Ridley, J., J. Gregory, P. Huybrechts, and J. Lowe, 2010: Thresholds for irreversible decline of the Greenland ice sheet. *Clim. Dyn.*, **35**, 1049–1057.

Ridley, J. K., J. A. Lowe, and H. T. Hewitt, 2012: How reversible is sea ice loss? *Cryosphere*, **6**, 193–198.

Riley, W. J., et al., 2011: Barriers to predicting changes in global terrestrial methane fluxes: Analyses using CLM4ME, a methane biogeochemistry model integrated in CESM. *Biogeosciences*, **8**, 1925–1953.

Rind, D., 1987: The doubled CO_2 climate - Impact of the sea-surface temperature-gradient. *J. Atmos. Sci.*, **44**, 3235–3268.

Rinke, A., P. Kuhry, and K. Dethloff, 2008: Importance of a soil organic layer for Arctic climate: A sensitivity study with an Arctic RCM. *Geophys. Res. Lett.*, **35**, L13709.

Rive, N., A. Torvanger, T. Berntsen, and S. Kallbekken, 2007: To what extent can a long-term temperature target guide near-term climate change commitments? *Clim. Change*, **82**, 373–391.

Robinson, A., R. Calov, and A. Ganopolski, 2012: Multistability and critical thresholds of the Greenland ice sheet. *Nature Clim. Change*, **2**, 429–432.

Roeckner, E., M. A. Giorgetta, T. Crueger, M. Esch, and J. Pongratz, 2011: Historical and future anthropogenic emission pathways derived from coupled climate-carbon cycle simulations. *Clim. Change*, **105**, 91–108.

Roehrig, R., D. Bouniol, F. Guichard, F. Hourdin, and J.-L. Redelsperger, 2013: The present and future of the West African monsoon: A process-oriented assessment of CMIP5 simulations along the AMMA transect. *J. Clim.*, doi:10.1175/JCLI-D-12-00505.1.

Roesch, A., 2006: Evaluation of surface albedo and snow cover in AR4 coupled climate models. *J. Geophys. Res.*, **111**, D15111.

Rogelj, J., M. Meinshausen, and R. Knutti, 2012: Global warming under old and new scenarios using IPCC climate sensitivity range estimates. *Nature Clim. Change*, **2**, 248–253.

Rogelj, J., D. L. McCollum, B. C. O'Neill, and K. Riahi, 2013: 2020 emissions levels required to limit warming to below 2°C. *Nature Clim. Change*, **3**, 405–412.

Rogelj, J., et al., 2011: Emission pathways consistent with a 2°C global temperature limit. *Nature Clim. Change*, **1**, 413–418.

Rohling, E., and P. P. Members, 2012: Making sense of palaeoclimate sensitivity. *Naturc*, **491**, 683–691.

Rohling, E., K. Grant, M. Bolshaw, A. Roberts, M. Siddall, C. Hemleben, and M. Kucera, 2009: Antarctic temperature and global sea level closely coupled over the past five glacial cycles. *Nature Geosci.*, **2**, 500–504.

Romanovsky, V. E., S. L. Smith, and H. H. Christiansen, 2010: Permafrost thermal state in the polar Northern Hemisphere during the international polar year 2007–2009: A synthesis. *Permafr. Periglac. Process.*, **21**, 106–116.

Rotstayn, L. D., S. J. Jeffrey, M. A. Collier, S. M. Dravitzki, A. C. Hirst, J. I. Syktus, and K. K. Wong, 2012: Aerosol- and greenhouse gas-induced changes in summer rainfall and circulation in the Australasian region: A study using single-forcing climate simulations. *Atmos. Chem. Phys.*, **12**, 6377–6404.

Rougier, J., 2007: Probabilistic inference for future climate using an ensemble of climate model evaluations. *Clim. Change*, **81**, 247–264.

Rougier, J., D. M. H. Sexton, J. M. Murphy, and D. Stainforth, 2009: Analyzing the climate sensitivity of the HadSM3 climate model using ensembles from different but related experiments. *J. Clim.*, **22**, 3540–3557.

Rowell, D. P., 2012: Sources of uncertainty in future changes in local precipitation. *Clim. Dyn.*, doi:10.1007/s00382-011-1210-2.

Rowlands, D. J., et al., 2012: Broad range of 2050 warming from an observationally constrained large climate model ensemble. *Nature Geosci.*, **5**, 256–260.

Ruosteenoja, K., H. Tuomenvirta, and K. Jylha, 2007: GCM-based regional temperature and precipitation change estimates for Europe under four SRES scenarios applying a super-ensemble pattern-scaling method. *Clim. Change*, **81**, 193–208.

Saenko, O. A., A. S. Gupta, and P. Spence, 2012: On challenges in predicting bottom water transport in the Southern Ocean. *J. Clim.*, **25**, 1349–1356.

Saito, K., M. Kimoto, T. Zhang, K. Takata, and S. Emori, 2007: Evaluating a high-resolution climate model: Simulated hydrothermal regimes in frozen ground regions and their change under the global warming scenario. *J. Geophys. Res.*, **112**, F02S11.

Sallée, J.-B., E. Shuckburgh, N. Bruneau, A. J. S. Meijers, T. Bracegirdle, and Z. Wang, 2013a: Assessment of Southern Ocean mixed-layer depths in CMIP5 models: Historical bias and forcing response. *J. Geophys. Res.*, doi:10.1002/jgrc.20157.

Sallée, J.-B., E. Shuckburgh, N. Bruneau, A. J. S. Meijers, T. J. Bracegirdle, Z. Wang, and T. Roy, 2013b: Assessment of Southern Ocean water mass circulation and characteristics in CMIP5 models: Historical bias and forcing response. *J. Geophys. Res.*, doi:10.1002/jgrc.20135.

Sanchez-Gomez, E., S. Somot, and A. Mariotti, 2009: Future changes in the Mediterranean water budget projected by an ensemble of regional climate models. *Geophys. Res. Lett.*, **36**, L21401.

Sanderson, B. M., 2011: A multimodel study of parametric uncertainty in predictions of climate response to rising greenhouse gas concentrations. *J. Clim.*, **25**, 1362–1377.

Sanderson, B. M., 2013: On the estimation of systematic error in regression-based predictions of climate sensitivity. *Clim. Change*, doi:10.1007/s10584-012-0671-6.

Sanderson, B. M., and R. Knutti, 2012: On the interpretation of constrained climate model ensembles. *Geophys. Res. Lett.*, **39**, L16708.

Sanderson, B. M., K. M. Shell, and W. Ingram, 2010: Climate feedbacks determined using radiative kernels in a multi-thousand member ensemble of AOGCMs. *Clim. Dyn.*, **35**, 1219–1236.

Sanderson, B. M., et al., 2008: Constraints on model response to greenhouse gas forcing and the role of subgrid-scale processes. *J. Clim.*, **21**, 2384–2400.

Sanderson, M. G., D. L. Hemming, and R. A. Betts, 2011: Regional temperature and precipitation changes under high-end (>= 4°C) global warming. *Philos. Trans. R. Soc. A*, **369**, 85–98.

Sanso, B., and C. Forest, 2009: Statistical calibration of climate system properties. *J. R. Stat. Soc. C*, **58**, 485–503.

Sanso, B., C. E. Forest, and D. Zantedeschi, 2008: Inferring climate system properties using a computer model. *Bayes. Anal.*, **3**, 1–37.

Sansom, P. G., D. B. Stephenson, C. A. T. Ferro, G. Zappa, and L. Shaffrey, 2013: Simple uncertainty frameworks for selecting weighting schemes and interpreting multi-model ensemble climate change experiments. *J. Clim.*, doi:10.1175/JCLI-D-12-00462.1.

Santer, B. D., T. M. L. Wigley, M. E. Schlesinger, and J. F. B. Mitchell, 1990: *Developing Climate Scenarios from Equilibrium GCM Results.* Max-Planck-Institut-für-Meteorologie Report. Max-Planck-Institut-für-Meteorologie, Hamburg, Germany, 29 pp.

Scaife, A. A., et al., 2012: Climate change projections and stratosphere-troposphere interaction. *Clim. Dyn.*, **38**, 2089–2097.

Schaefer, K., T. Zhang, L. Bruhwiler, and A. Barrett, 2011: Amount and timing of permafrost carbon release in response to climate warming. *Tellus B*, **63**, 165–180.

Schär, C., P. L. Vidale, D. Lüthi, C. Frei, C. Häberli, M. A. Liniger, and C. Appenzeller, 2004: The role of increasing temperature variability in European summer heatwaves. *Nature*, **427**, 332–336.

Scheff, J., and D. M. W. Frierson, 2012: Robust future precipitation declines in CMIP5 largely reflect the poleward expansion of model subtropical dry zones. *Geophys. Res. Lett.*, **39**, L18704.

Scheffer, M., et al., 2009: Early-warning signals for critical transitions. *Nature*, **461**, 53–59.

Schlesinger, M., 1986: Equilibrium and transient climatic warming induced by increased atmospheric CO_2. *Clim. Dyn.*, **1**, 35–51.

Schlesinger, M., et al., 2000: Geographical distributions of temperature change for scenarios of greenhouse gas and sulfur dioxide emissions. *Technol. Forecast. Soc. Change*, **65**, 167–193.

Schmidt, M. W. I., et al., 2011: Persistence of soil organic matter as an ecosystem property. *Nature*, **478**, 49–56.

Schmittner, A., et al., 2011: Climate sensitivity estimated from temperature reconstructions of the Last Glacial Maximum. *Science*, **334**, 1385–1388.

Schneider von Deimling, T., H. Held, A. Ganopolski, and S. Rahmstorf, 2006: Climate sensitivity estimated from ensemble simulations of glacial climate. *Clim. Dyn.*, **27**, 149–163.

Schneider von Deimling, T., M. Meinshausen, A. Levermann, V. Huber, K. Frieler, D. Lawrence, and V. Brovkin, 2012: Estimating the near-surface permafrost-carbon feedback on global warming. *Biogeosciences*, **9**, 649–665.

Schoof, C., 2007: Ice sheet grounding line dynamics: Steady states, stability, and hysteresis. *J. Geophys. Res.*, **112**, F03S28.

Schröder, D., and W. M. Connolley, 2007: Impact of instantaneous sea ice removal in a coupled general circulation model. *Geophys. Res. Lett.*, **34**, L14502.

Schuenemann, K. C., and J. J. Cassano, 2010: Changes in synoptic weather patterns and Greenland precipitation in the 20th and 21st centuries: 2. Analysis of 21st century atmospheric changes using self-organizing maps. *J. Geophys. Res.*, **115**, D05108.

Schuur, E., J. Vogel, K. Crummer, H. Lee, J. Sickman, and T. Osterkamp, 2009: The effect of permafrost thaw on old carbon release and net carbon exchange from tundra. *Nature*, **459**, 556–559.

Schwalm, C. R., et al., 2012: Reduction in carbon uptake during turn of the century drought in western North America. *Nature Geosci.*, **5**, 551–556.

Schwartz, S., R. Charlson, R. Kahn, J. Ogren, and H. Rodhe, 2010: Why hasn't Earth warmed as much as expected? *J. Clim.*, **23**, 2453–2464.

Schwartz, S., R. Charlson, R. Kahn, J. Ogren, and H. Rodhe, 2012: Reply to "Comments on 'Why hasn't Earth warmed as much as expected?'". *J. Clim.*, **25**, 2200–2204.

Schwartz, S. E., 2012: Determination of Earth's transient and equilibrium climate sensitivities from observations over the twentieth century: Strong dependence on assumed forcing. *Surv. Geophys.*, **33**, 745–777.

Schweiger, A., R. Lindsay, J. Zhang, M. Steele, H. Stern, and R. Kwok, 2011: Uncertainty in modeled Arctic sea ice volume. *J. Geophys. Res.*, **116**, C00D06.

Screen, J., and I. Simmonds, 2010: The central role of diminishing sea ice in recent Arctic temperature amplification. *Nature*, **464**, 1334–1337.

Screen, J. A., N. P. Gillett, A. Y. Karpechko, and D. P. Stevens, 2010: Mixed layer temperature response to the Southern Annular Mode: Mechanisms and model representation. *J. Clim.*, **23**, 664–678.

Seager, R., and G. A. Vecchi, 2010: Greenhouse warming and the 21st century hydroclimate of the southwestern North America. *Proc. Natl. Acad. Sci. U.S.A.*, **107**, 21277–21282.

Seager, R., and N. Naik, 2012: A mechanisms-based approach to detecting recent anthropogenic hydroclimate change. *J. Clim.*, **25**, 236–261.

Seager, R., N. Naik, and G. A. Vecchi, 2010: Thermodynamic and dynamic mechanisms for large-scale changes in the hydrological cycle in response to global warming. *J. Clim.*, **23**, 4651–4668.

Seager, R., et al., 2007: Model projections of an imminent transition to a more arid climate in southwestern North America. *Science*, **316**, 1181–1184.

Sedláček, J., R. Knutti, O. Martius, and U. Beyerle, 2011: Impact of a reduced Arctic sea ice cover on ocean and atmospheric properties. *J. Clim.*, **25**, 307–319.

Seidel, D., and W. Randel, 2007: Recent widening of the tropical belt: Evidence from tropopause observations. *J. Geophys. Res.*, **112**, D20113.

Seidel, D. J., Q. Fu, W. J. Randel, and T. J. Reichler, 2008: Widening of the tropical belt in a changing climate. *Nature Geosci.*, **1**, 21–24.

Sen Gupta, A., A. Santoso, A. Taschetto, C. Ummenhofer, J. Trevena, and M. England, 2009: Projected changes to the Southern Hemisphere ocean and sea ice in the IPCC AR4 climate models. *J. Clim.*, **22**, 3047–3078.

Seneviratne, S. I., D. Lüthi, M. Litschi, and C. Schär, 2006: Land-atmosphere coupling and climate change in Europe. *Nature*, **443**, 205–209.

Seneviratne, S. I., et al., 2010: Investigating soil moisture-climate interactions in a changing climate: A review. *Earth Sci. Rev.*, **99**, 125–161.

Seneviratne, S. I., et al., 2012: Changes in climate extremes and their impacts on the natural physical environment. In: *Managing the Risks of Extreme Events and Disasters to Advance Climate Change Adaptation. A Special Report of Working Groups I and II of the Intergovernmental Panel on Climate Change (IPCC)* [C. B. Field, et al. (eds.)]. Cambridge University Press, Cambridge, United Kingdom, and New York, NY, USA, pp. 109–230.

Senior, C. A., and J. F. B. Mitchell, 2000: The time-dependence of climate sensitivity. *Geophys. Res. Lett.*, **27**, 2685–2688.

Serreze, M., A. Barrett, J. Stroeve, D. Kindig, and M. Holland, 2009: The emergence of surface-based Arctic amplification. *Cryosphere*, **3**, 11–19.

Serreze, M. C., and J. A. Francis, 2006: The Arctic amplification debate. *Clim. Change*, **76**, 241–264.

Sexton, D., H. Grubb, K. Shine, and C. Folland, 2003: Design and analysis of climate model experiments for the efficient estimation of anthropogenic signals. *J. Clim.*, **16**, 1320–1336.

Sexton, D. M. H., and J. M. Murphy, 2012: Multivariate probabilistic projections using imperfect climate models. Part II: Robustness of methodological choices and consequences for climate sensitivity. *Clim. Dyn.*, 2543–2558.

Sexton, D. M. H., J. M. Murphy, M. Collins, and M. J. Webb, 2012: Multivariate probabilistic projections using imperfect climate models. Part I: Outline of methodology. *Clim. Dyn.*, 2513–2542.

Shepherd, T. G., and C. McLandress, 2011: A robust mechanism for strengthening of the Brewer-Dobson circulation in response to climate change: Critical-layer control of subtropical wave breaking. *J. Atmos. Sci.*, **68**, 784–797.

Sherwood, S. C., 2010: Direct versus indirect effects of tropospheric humidity changes on the hydrologic cycle. *Environ. Res. Lett.*, **5**, 025206.

Sherwood, S. C., and M. Huber, 2010: An adaptability limit to climate change due to heat stress. *Proc. Natl. Acad. Sci. U.S.A.*, **107**, 9552–9555.

Sherwood, S. C., W. Ingram, Y. Tsushima, M. Satoh, M. Roberts, P. L. Vidale, and P. A. O'Gorman, 2010: Relative humidity changes in a warmer climate. *J. Geophys. Res.*, **115**, D09104.

Shindell, D., et al., 2012: Simultaneously mitigating near-term climate change and improving human health and food security. *Science*, **335**, 183–189.

Shindell, D. T., et al., 2006: Simulations of preindustrial, present-day, and 2100 conditions in the NASA GISS composition and climate model G-PUCCINI. *Atmos. Chem. Phys.*, **6**, 4427–4459.

Shindell, D. T., et al., 2013a: Interactive ozone and methane chemistry in GISS-E2 historical and future climate simulations. *Atmos. Chem. Phys.*, **13**, 2653–2689.

Shindell, D. T., et al., 2013b: Radiative forcing in the ACCMIP historical and future climate simulations. *Atmos. Chem. Phys.*, **13**, 2939–2974.

Shine, K. P., J. Cook, E. J. Highwood, and M. M. Joshi, 2003: An alternative to radiative forcing for estimating the relative importance of climate change mechanisms. *Geophys. Res. Lett.*, **30**, 2047.

Shiogama, H., S. Emori, K. Takahashi, T. Nagashima, T. Ogura, T. Nozawa, and T. Takemura, 2010a: Emission scenario dependency of precipitation on global warming in the MIROC3.2 model. *J. Clim.*, **23**, 2404–2417.

Shiogama, H., et al., 2010b: Emission scenario dependencies in climate change assessments of the hydrological cycle. *Clim. Change*, **99**, 321–329.

Shkolnik, I., E. Nadyozhina, T. Pavlova, E. Molkentin, and A. Semioshina, 2010: Snow cover and permafrost evolution in Siberia as simulated by the MGO regional climate model in the 20th and 21st centuries. *Environ. Res. Lett.*, **5**, 015005.

Shongwe, M. E., G. J. van Oldenborgh, B. van den Hurk, and M. van Aalst, 2011: Projected changes in mean and extreme precipitation in Africa under global warming. Part II: East Africa. *J. Clim.*, **24**, 3718–3733.

Siegenthaler, U., and H. Oeschger, 1984: Transient temperature changes due to increasing CO_2 using simple models. *Ann. Glaciol.*, **5**, 153–159.

Sigmond, M., P. C. Siegmund, E. Manzini, and H. Kelder, 2004: A simulation of the separate climate effects of middle-atmosphere and tropospheric CO_2 doubling. *J. Clim.*, **17**, 2352–2367.

Sillmann, J., and E. Roeckner, 2008: Indices for extreme events in projections of anthropogenic climate change. *Clim. Change*, **86**, 83–104.

Sillmann, J., and M. Croci-Maspoli, 2009: Present and future atmospheric blocking and its impact on European mean and extreme climate. *Geophys. Res. Lett.*, **36**, L10702.

Sillmann, J., V. V. Kharin, F. W. Zwiers, X. Zhang, and D. Bronaugh, 2013: Climate extremes indices in the CMIP5 multimodel ensemble: Part 2. Future climate projections. *J. Geophys. Res.*, **118**, 2473–2493.

Simmons, A. J., K. M. Willett, P. D. Jones, P. W. Thorne, and D. P. Dee, 2010: Low-frequency variations in surface atmospheric humidity, temperature, and precipitation: Inferences from reanalyses and monthly gridded observational data sets. *J. Geophys. Res.*, **115**, D01110.

Simpkins, G. R., and A. Y. Karpechko, 2012: Sensitivity of the southern annular mode to greenhouse gas emission scenarios. *Clim. Dyn.*, **38**, 563–572.

Slater, A. G., and D. M. Lawrence, 2013: Diagnosing present and future permafrost from climate models. *J. Clim.*, doi:10.1175/JCLI-D-12-00341.1.

Smeets, E. M. W., L. F. Bouwmanw, E. Stehfest, D. P. van Vuuren, and A. Posthuma, 2009: Contribution of N_2O to the greenhouse gas balance of first-generation biofuels. *Global Change Biol.*, **15**, 1–23.

Smith, L., Y. Sheng, G. MacDonald, and L. Hinzman, 2005: Disappearing Arctic lakes. *Science*, **308**, 1429–1429.

Smith, L., et al., 2004: Siberian peatlands a net carbon sink and global methane source since the early Holocene. *Science*, **303**, 353–356.

Smith, R. L., C. Tebaldi, D. Nychka, and L. O. Mearns, 2009: Bayesian modeling of uncertainty in ensembles of climate models. *J. Am. Stat. Assoc.*, **104**, 97–116.

Smith, S. J., J. van Aardenne, Z. Klimont, R. J. Andres, A. Volke, and S. Delgado Arias, 2011: Anthropogenic sulfur dioxide emissions: 1850–2005. *Atmos. Chem. Phys.*, **11**, 1101–1116.

Smith, S. M., J. A. Lowe, N. H. A. Bowerman, L. K. Gohar, C. Huntingford, and M. R. Allen, 2012: Equivalence of greenhouse-gas emissions for peak temperature limits. *Nature Clim. Change*, **2**, 535–538.

Sobel, A. H., and S. J. Camargo, 2011: Projected future seasonal changes in tropical summer climate. *J. Clim.*, **24**, 473–487.

Soden, B., I. Held, R. Colman, K. Shell, J. Kiehl, and C. Shields, 2008: Quantifying climate feedbacks using radiative kernels. *J. Clim.*, **21**, 3504–3520.

Soden, B. J., and I. M. Held, 2006: An assessment of climate feedbacks in coupled ocean-atmosphere models. *J. Clim.*, **19**, 3354–3360.

Soden, B. J., and G. A. Vecchi, 2011: The vertical distribution of cloud feedback in coupled ocean-atmosphere models. *Geophys. Res. Lett.*, **38**, L12704.

Sohn, B. J., and S.-C. Park, 2010: Strengthened tropical circulations in past three decades inferred from water vapor transport. *J. Geophys. Res.*, **115**, D15112.

Sokolov, A. P., et al., 2009: Probabilistic forecast for twenty-first-century climate based on uncertainties in emissions (without policy) and climate parameters. *J. Clim.*, **23**, 2230–2231.

Solgaard, A. M., and P. L. Langen, 2012: Multistability of the Greenland ice sheet and the effects of an adaptive mass balance formulation. *Clim. Dyn.*, **39**, 1599–1612.

Solomon, S., G. Plattner, R. Knutti, and P. Friedlingstein, 2009: Irreversible climate change due to carbon dioxide emissions. *Proc. Natl. Acad. Sci. U.S.A.*, **106**, 1704–1709.

Solomon, S., J. Daniel, T. Sanford, D. Murphy, G. Plattner, R. Knutti, and P. Friedlingstein, 2010: Persistence of climate changes due to a range of greenhouse gases. *Proc. Natl. Acad. Sci. U.S.A.*, **107**, 18354–18359.

Solomon, S., et al., 2007: Technical Summary. In: *Climate Change 2007: The Physical Science Basis. Contribution of Working Group I to the Fourth Assessment Report of the Intergovernmental Panel on Climate Change* [Solomon, S., D. Qin, M. Manning, Z. Chen, M. Marquis, K. B. Averyt, M. Tignor and H. L. Miller (eds.)] Cambridge University Press, Cambridge, United Kingdom and New York, NY, USA, pp. 19–92.

Son, S. W., et al., 2010: Impact of stratospheric ozone on Southern Hemisphere circulation change: A multimodel assessment. *J. Geophys. Res.*, **115**, D00M07.

Sorensson, A., C. Menendez, R. Ruscica, P. Alexander, P. Samuelsson, and U. Willen, 2010: Projected precipitation changes in South America: A dynamical downscaling within CLARIS. *Meteorol. Z.*, **19**, 347–355.

Spence, P., O. A. Saenko, C. O. Dufour, J. Le Sommer, and M. H. England, 2012: Mechanisms maintaining Southern Ocean meridional heat transport under projected wind forcing. *J. Phys. Oceanogr.*, **42**, 1923–1931.

St. Clair, S., J. Hillier, and P. Smith, 2008: Estimating the pre-harvest greenhouse gas costs of energy crop production. *Biomass Bioenerg.*, **32**, 442–452.

Stachnik, J. P., and C. Schumacher, 2011: A comparison of the Hadley circulation in modern reanalyses. *J. Geophys. Res.*, **116**, D22102.

Stephenson, D. B., M. Collins, J. C. Rougier, and R. E. Chandler, 2012: Statistical problems in the probabilistic prediction of climate change. *Environmetrics*, **23**, 364–372.

Sterl, A., et al., 2008: When can we expect extremely high surface temperatures? *Geophys. Res. Lett.*, **35**, L14703.

Stott, P., G. Jones, and J. Mitchell, 2003: Do models underestimate the solar contribution to recent climate change? *J. Clim.*, **16**, 4079–4093.

Stott, P., P. Good, G. A. Jones, N. Gillett, and E. Hawkins, 2013: The upper end of climate model temperature projections is inconsistent with past warming. *Environ. Res. Lett.*, **8**, 014024.

Stouffer, R., 2004: Time scales of climate response. *J. Clim.*, **17**, 209–217.

Stouffer, R. J., and S. Manabe, 1999: Response of a coupled ocean-atmosphere model to increasing atmospheric carbon dioxide: Sensitivity to the rate of increase. *J. Clim.*, **12**, 2224–2237.

Stowasser, M., H. Annamalai, and J. Hafner, 2009: Response of the South Asian summer monsoon to global warming: Mean and synoptic systems. *J. Clim.*, **22**, 1014–1036.

Stroeve, J., M. Holland, W. Meier, T. Scambos, and M. Serreze, 2007: Arctic sea ice decline: Faster than forecast. *Geophys. Res. Lett.*, **34**, L09501.

Stroeve, J. C., V. Kattsov, A. Barrett, M. Serreze, T. Pavlova, M. Holland, and W. N. Meier, 2012: Trends in Arctic sea ice extent from CMIP5, CMIP3 and observations. *Geophys. Res. Lett.*, **39**, L16502.

Stuber, N., M. Ponater, and R. Sausen, 2005: Why radiative forcing might fail as a predictor of climate change. *Clim. Dyn.*, **24**, 497–510.

Sudo, K., M. Takahashi, and H. Akimoto, 2003: Future changes in stratosphere-troposphere exchange and their impacts on future tropospheric ozone simulations. *Geophys. Res. Lett.*, **30**, 2256.

Sugiyama, M., H. Shiogama, and S. Emori, 2010: Precipitation extreme changes exceeding moisture content increases in MIROC and IPCC climate models. *Proc. Natl. Acad. Sci. U.S.A.*, **107**, 571–575.

Sun, Y., S. Solomon, A. Dai, and R. W. Portmann, 2007: How often will it rain? *J. Clim.*, **20**, 4801–4818.

Sutton, R. T., B. W. Dong, and J. M. Gregory, 2007: Land/sea warming ratio in response to climate change: IPCC AR4 model results and comparison with observations. *Geophys. Res. Lett.*, **34**, L02701.

Swann, A. L., I. Y. Fung, S. Levis, G. B. Bonan, and S. C. Doney, 2010: Changes in Arctic vegetation amplify high-latitude warming through the greenhouse effect. *Proc. Natl. Acad. Sci. U.S.A.*, **107**, 1295–1300.

Swart, N. C., and J. C. Fyfe, 2012: Observed and simulated changes in the Southern Hemisphere surface westerly wind-stress. *Geophys. Res. Lett.*, **39**, L16711.

Swingedouw, D., P. Braconnot, P. Delecluse, E. Guilyardi, and O. Marti, 2007: Quantifying the AMOC feedbacks during a 2xCO_2 stabilization experiment with land-ice melting. *Clim. Dyn.*, **29**, 521–534.

Swingedouw, D., T. Fichefet, P. Huybrechts, H. Goosse, E. Driesschaert, and M. Loutre, 2008: Antarctic ice-sheet melting provides negative feedbacks on future climate warming. *Geophys. Res. Lett.*, **35**, L17705.

Szopa, S., et al., 2013: Aerosol and ozone changes as forcing for climate evolution between 1850 and 2100. *Clim. Dyn.*, **40**, 2223–2250.

Takahashi, K., 2009a: Radiative constraints on the hydrological cycle in an idealized radiative-convective equilibrium model. *J. Atmos. Sci.*, **66**, 77–91.

Takahashi, K., 2009b: The global hydrological cycle and atmospheric shortwave absorption in climate models under CO_2 forcing. *J. Clim.*, **22**, 5667–5675.

Tanaka, K., and T. Raddatz, 2011: Correlation between climate sensitivity and aerosol forcing and its implication for the "climate trap". *Clim. Change*, **109**, 815–825.

Tarnocai, C., J. Canadell, E. Schuur, P. Kuhry, G. Mazhitova, and S. Zimov, 2009: Soil organic carbon pools in the northern circumpolar permafrost region. *Global Biogeochem. Cycles*, **23**, GB2023.

Taylor, K. E., R. J. Stouffer, and G. A. Meehl, 2012: A summary of the CMIP5 experiment design. *Bull. Am. Meteorol. Soc.*, **93**, 485–498.

Tebaldi, C., and R. Knutti, 2007: The use of the multi-model ensemble in probabilistic climate projections. *Philos. Trans. R. Soc. A*, **365**, 2053–2075.

Tebaldi, C., and D. B. Lobell, 2008: Towards probabilistic projections of climate change impacts on global crop yields. *Geophys. Res. Lett.*, **35**, L08705.

Tebaldi, C., and B. Sanso, 2009: Joint projections of temperature and precipitation change from multiple climate models: A hierarchical Bayesian approach. *J. R. Stat. Soc. A*, **172**, 83–106.

Tebaldi, C., J. M. Arblaster, and R. Knutti, 2011: Mapping model agreement on future climate projections. *Geophys. Res. Lett.*, **38**, L23701.

Tebaldi, C., K. Hayhoe, J. M. Arblaster, and G. A. Meehl, 2006: Going to the extremes. *Clim. Change*, **79**, 185–211.

Terray, L., L. Corre, S. Cravatte, T. Delcroix, G. Reverdin, and A. Ribes, 2012: Near-surface salinity as nature's rain gauge to detect human influence on the tropical water cycle. *J. Clim.*, **25**, 958–977.

Thorne, P., 2008: Arctic tropospheric warming amplification? *Nature*, **455**, E1–E2.

Tietsche, S., D. Notz, J. H. Jungclaus, and J. Marotzke, 2011: Recovery mechanisms of Arctic summer sea ice. *Geophys. Res. Lett.*, **38**, L02707.

Tjiputra, J. F., et al., 2013: Evaluation of the carbon cycle components in the Norwegian Earth System Model (NorESM). *Geosci. Model Dev.*, **6**, 301–325.

Tokinaga, H., S.-P. Xie, C. Deser, Y. Kosaka, and Y. M. Okumura, 2012: Slowdown of the Walker circulation driven by tropical Indo-Pacific warming. *Nature*, **491**, 439–443.

Trapp, R. J., N. S. Diffenbaugh, and A. Gluhovsky, 2009: Transient response of severe thunderstorm forcing to elevated greenhouse gas concentrations. *Geophys. Res. Lett.*, **36**, L01703.

Trapp, R. J., N. S. Diffenbaugh, H. E. Brooks, M. E. Baldwin, E. D. Robinson, and J. S. Pal, 2007: Changes in severe thunderstorm environment frequency during the 21st century caused by anthropogenically enhanced global radiative forcing. *Proc. Natl. Acad. Sci. U.S.A.*, **104**, 19719–19723.

Trenberth, K. E., and D. J. Shea, 2005: Relationships between precipitation and surface temperature. *Geophys. Res. Lett.*, **32**, L14703.

Trenberth, K. E., and J. T. Fasullo, 2009: Global warming due to increasing absorbed solar radiation. *Geophys. Res. Lett.*, **36**, L07706.

Trenberth, K. E., and J. T. Fasullo, 2010: Simulation of present-day and twenty-first-century energy budgets of the Southern Oceans. *J. Clim.*, **23**, 440–454.

Turner, J., T. J. Bracegirdle, T. Phillips, G. J. Marshall, and J. S. Hosking, 2013: An initial assessment of Antarctic sea ice extent in the CMIP5 models. *J. Clim.*, **26**, 1473–1484.

Ueda, H., A. Iwai, K. Kuwako, and M. Hori, 2006: Impact of anthropogenic forcing on the Asian summer monsoon as simulated by eight GCMs. *Geophys. Res. Lett.*, **33**, L06703.

Ulbrich, U., G. C. Leckebusch, and J. G. Pinto, 2009: Extra-tropical cyclones in the present and future climate: A review. *Theor. Appl. Climatol.*, **96**, 117–131.

Ulbrich, U., et al., 2013: Are greenhouse gas signals of Northern Hemisphere winter extra-tropical cyclone activity dependent on the identification and tracking algorithm? *Meteorol. Z.*, **22**, 61–68.

UNEP, 2010: The emissions gap report: Are the Copenhagen Accord pledges sufficient to limit global warming to 2°C or 1.5°C? , 55 pp.

Utsumi, N., S. Seto, S. Kanae, E. E. Maeda, and T. Oki, 2011: Does higher surface temperature intensify extreme precipitation? *Geophys. Res. Lett.*, **38**, L16708.

Vaks, A., et al., 2013: Speleothems reveal 500,000-year history of Siberian permafrost. *Science*, **340**, 183–186.

Van Klooster, S. L., and P. J. Roebber, 2009: Surface-based convective potential in the contiguous United States in a business-as-usual future climate. *J. Clim.*, **22**, 3317–3330.

van Vuuren, D. P., et al., 2011: RCP3–PD: Exploring the possibilities to limit global mean temperature change to less than 2°C. *Clim. Change*, **109**, 95–116.

Vavrus, S., M. Holland, and D. Bailey, 2011: Changes in Arctic clouds during intervals of rapid sea ice loss. *Clim. Dyn.*, **36**, 1475–1489.

Vavrus, S. J., M. M. Holland, A. Jahn, D. A. Bailey, and B. A. Blazey, 2012: Twenty-first-century Arctic climate change in CCSM4. *J. Clim.*, **25**, 2696–2710.

Vecchi, G. A., and B. J. Soden, 2007: Global warming and the weakening of the tropical circulation. *Bull. Am. Meteorol. Soc.*, **88**, 1529–1530.

Vecchi, G. A., B. J. Soden, A. T. Wittenberg, I. M. Held, A. Leetmaa, and M. J. Harrison, 2006: Weakening of tropical Pacific atmospheric circulation due to anthropogenic forcing. *Nature*, **441**, 73–76.

Vial, J., J.-L. Dufresne, and S. Bony, 2013: On the interpretation of inter-model spread in CMIP5 climate sensitivity estimates. *Clim. Dyn.*, doi:10.1007/s00382-013-1725-9.

Vidale, P. L., D. Lüthi, R. Wegmann, and C. Schär, 2007: European summer climate variability in a heterogeneous multi-model ensemble. *Clim. Change*, **81**, 209–232.

Voldoire, A., et al., 2013: The CNRM-CM5.1 global climate model: Description and basic evaluation. *Clim. Dyn.*, **40**, 2091–2121.

Voss, R., and U. Mikolajewicz, 2001: Long-term climate changes due to increased CO_2 concentration in the coupled atmosphere-ocean general circulation model ECHAM3/LSG. *Clim. Dyn.*, **17**, 45–60.

Wadhams, P., 2012: Arctic ice cover, ice thickness and tipping points. *Ambio*, **41**, 23–33.

Walker, R., et al., 2009: Protecting the Amazon with protected areas. *Proc. Natl. Acad. Sci. U.S.A.*, **106**, 10582–10586.

Wang, M., and J. Overland, 2009: A sea ice free summer Arctic within 30 years? *Geophys. Res. Lett.*, **36**, L07502.

Wang, M., and J. E. Overland, 2012: A sea ice free summer Arctic within 30 years: An update from CMIP5 models. *Geophys. Res. Lett.*, **39**, L18501.

Wania, R., I. Ross, and I. Prentice, 2009: Integrating peatlands and permafrost into a dynamic global vegetation model: 2. Evaluation and sensitivity of vegetation and carbon cycle processes. *Global Biogeochem. Cycles*, **23**, GB3015.

Washington, W., et al., 2009: How much climate change can be avoided by mitigation? *Geophys. Res. Lett.*, **36**, L08703.

Watanabe, S., et al., 2011: MIROC-ESM 2010: Model description and basic results of CMIP5-20c3m experiments. *Geosci. Model Dev.*, **4**, 845–872.

Watterson, I. G., 2008: Calculation of probability density functions for temperature and precipitation change under global warming. *J. Geophys. Res.*, **113**, D12106.

Watterson, I. G., 2011: Calculation of joint PDFs for climate change with properties matching recent Australian projections. *Aust. Meteorol. Oceanogr. J.*, **61**, 211–219.

Watterson, I. G., and P. H. Whetton, 2011a: Joint PDFs for Australian climate in future decades and an idealized application to wheat crop yield. *Aust. Meteorol. Oceanogr. J.*, **61**, 221–230.

Watterson, I. G., and P. H. Whetton, 2011b: Distributions of decadal means of temperature and precipitation change under global warming. *J. Geophys. Res.*, **116**, D07101.

Watterson, I. G., J. L. McGregor, and K. C. Nguyen, 2008: Changes in extreme temperatures of Australasian summer simulated by CCAM under global warming, and the roles of winds and land-sea contrasts. *Aust. Meteorol. Mag.*, **57**, 195–212.

WBGU, 2009: *Solving the Climate Dilemma: The Budget Approach.* German Advisory Council on Global Change, Berlin, 59 pp.

Weaver, A., K. Zickfeld, A. Montenegro, and M. Eby, 2007: Long term climate implications of 2050 emission reduction targets. *Geophys. Res. Lett.*, **34**, L19703.

Weaver, A. J., et al., 2012: Stability of the Atlantic meridional overturning circulation: A model intercomparison. *Geophys. Res. Lett.*, **39**, L20709.

Webb, M., et al., 2006: On the contribution of local feedback mechanisms to the range of climate sensitivity in two GCM ensembles. *Clim. Dyn.*, **27**, 17–38.

Webb, M. J., F. H. Lambert, and J. M. Gregory, 2013: Origins of differences in climate sensitivity, forcing and feedback in climate models. *Clim. Dyn.*, **40**, 677–707.

Weber, S., et al., 2007: The modern and glacial overturning circulation in the Atlantic Ocean in PMIP coupled model simulations. *Clim. Past*, **3**, 51–64.

Weertman, J., 1974: Stability of the junction of an ice sheet and an ice shelf. J. Glaciol., **13**, 3–11.

Wehner, M., D. Easterling, J. Lawrimore, R. Heim, R. Vose, and B. Santer, 2011: Projections of future drought in the continental United States and Mexico. *J. Hydrometeorol.*, **12**, 1359–1377.

Weigel, A., R. Knutti, M. Liniger, and C. Appenzeller, 2010: Risks of model weighting in multimodel climate projections. *J. Clim.*, **23**, 4175–4191.

Wentz, F., L. Ricciardulli, K. Hilburn, and C. Mears, 2007: How much more rain will global warming bring? *Science*, **317**, 233–235.

Wetherald, R., and S. Manabe, 1988: Cloud feedback processes in a General-Circulation Model. *J. Atmos. Sci.*, **45**, 1397–1415.

Wetherald, R. T., R. J. Stouffer, and K. W. Dixon, 2001: Committed warming and its implications for climate change. *Geophys. Res. Lett.*, **28**, 1535–1538.

Wigley, T. M. L., 2005: The climate change commitment. *Science*, **307**, 1766–1769.

Wilcox, L. J., A. J. Charlton-Perez, and L. J. Gray, 2012: Trends in Austral jet position in ensembles of high- and low-top CMIP5 models. *J. Geophys. Res.*, **117**, D13115.

Willett, K., and S. Sherwood, 2012: Exceedance of heat index thresholds for 15 regions under a warming climate using the wet-bulb globe temperature. *Int. J. Climatol.*, **32**, 161–177.

Williams, J. W., S. T. Jackson, and J. E. Kutzbach, 2007: Projected distributions of novel and disappearing climates by 2100 AD. *Proc. Natl. Acad. Sci. U.S.A.*, **104**, 5738–5742.

Williams, K. D., W. J. Ingram, and J. M. Gregory, 2008: Time variation of effective climate sensitivity in GCMs. *J. Clim.*, **21**, 5076–5090.

Winton, M., 2006a: Amplified Arctic climate change: What does surface albedo feedback have to do with it? *Geophys. Res. Lett.*, **33**, L03701.

Winton, M., 2006b: Does the Arctic sea ice have a tipping point? *Geophys. Res. Lett.*, **33**, L23504.

Winton, M., 2008: Sea ice-albedo feedback and nonlinear Arctic climate change. In: *Arctic Sea Ice Decline: Observations, Projections, Mechanisms, and Implications* [E. T. DeWeaver, C. M. Bitz and L. B. Tremblay (eds.)]. American Geophysical Union, Washington, DC, pp. 111–131.

Winton, M., 2011: Do climate models underestimate the sensitivity of Northern Hemisphere sea ice cover? *J. Clim.*, **24**, 3924–3934.

WMO, 2007: Scientific assessment of ozone depletion. In: *2006, Global Ozone Research and Monitoring Project*. World Meteorological Organization, Geneva, Switzerland, 572 pp.

Wood, R., A. Keen, J. Mitchell, and J. Gregory, 1999: Changing spatial structure of the thermohaline circulation in response to atmospheric CO_2 forcing in a climate model. *Nature*, **399**, 572–575.

Woollings, T., 2008: Vertical structure of anthropogenic zonal-mean atmospheric circulation change. *Geophys. Res. Lett.*, **35**, L19702.

Woollings, T., and M. Blackburn, 2012: The North Atlantic jet stream under climate change and its relation to the NAO and EA patterns. *J. Clim.*, **25**, 886–902.

Woollings, T., J. M. Gregory, J. G. Pinto, M. Reyers, and D. J. Brayshaw, 2012: Response of the North Atlantic storm track to climate change shaped by ocean-atmosphere coupling. *Nature Geosci.*, **5**, 313–317.

Wu, P., R. Wood, J. Ridley, and J. Lowe, 2010: Temporary acceleration of the hydrological cycle in response to a CO_2 rampdown. *Geophys. Res. Lett.*, **37**, L12705.

Wu, P., L. Jackson, A. Pardaens, and N. Schaller, 2011a: Extended warming of the northern high latitudes due to an overshoot of the Atlantic meridional overturning circulation. *Geophys. Res. Lett.*, **38**, L24704.

Wu, T., et al., 2013: Global carbon budgets simulated by the Beijing Climate Center Climate System Model for the last century. *J. Geophys. Res.*, doi:10.1002/jgrd.50320.

Wu, Y., M. Ting, R. Seager, H.-P. Huang, and M. A. Cane, 2011b: Changes in storm tracks and energy transports in a warmer climate simulated by the GFDL CM2.1 model. *Clim. Dyn.*, **37**, 53–72.

Wyant, M. C., et al., 2006: A comparison of low-latitude cloud properties and their response to climate change in three AGCMs sorted into regimes using mid-tropospheric vertical velocity. *Clim. Dyn.*, **27**, 261–279.

Xie, P., and G. Vallis, 2012: The passive and active nature of ocean heat uptake in idealized climate change experiments. *Clim. Dyn.*, **38**, 667–684.

Xie, S. P., C. Deser, G. A. Vecchi, J. Ma, H. Y. Teng, and A. T. Wittenberg, 2010: Global warming pattern formation: Sea surface temperature and rainfall. *J. Clim.*, **23**, 966–986.

Xin, X., L. Zhang, J. Zhang, T. Wu, and Y. Fang, 2013a: Climate change projections over East Asia with BCC_CSM1.1 climate model under RCP scenarios. *J. Meteorol. Soc. Jpn.*, **4**, 413-429.

Xin, X., T. Wu, J. Li, Z. Wang, W. Li, and F. Wu, 2013b: How well does BCC_CSM1.1 reproduce the 20th century climate change in China? *Atmos. Ocean. Sci. Lett.*, **6**, 21–26.

Yang, F. L., A. Kumar, M. E. Schlesinger, and W. Q. Wang, 2003: Intensity of hydrological cycles in warmer climates. *J. Clim.*, **16**, 2419–2423.

Yin, J., J. Overpeck, S. Griffies, A. Hu, J. Russell, and R. Stouffer, 2011: Different magnitudes of projected subsurface ocean warming around Greenland and Antarctica. *Nature Geosci.*, **4**, 524–528.

Yokohata, T., M. Webb, M. Collins, K. Williams, M. Yoshimori, J. Hargreaves, and J. Annan, 2010: Structural similarities and differences in climate responses to CO_2 increase between two perturbed physics ensembles. *J. Clim.*, **23**, 1392–1410.

Yokohata, T., J. D. Annan, M. Collins, C. S. Jackson, M. Tobis, M. Webb, and J. C. Hargreaves, 2012: Reliability of multi-model and structurally different single-model ensembles. *Clim. Dyn.*, **39**, 599–616.

Young, P. J., K. H. Rosenlof, S. Solomon, S. C. Sherwood, Q. Fu, and J.-F. Lamarque, 2012: Changes in stratospheric temperatures and their implications for changes in the Brewer Dobson circulation, 1979–2005. *J. Clim.*, **25**, 1759–1772.

Yukimoto, S., et al., 2012: A new global climate model of the Meteorological Research Institute: MRI-CGCM3–Model description and basic performance. *J. Meteorol. Soc. Jpn.*, **90A**, 23–64.

Zappa, G., L. C. Shaffrey, K. I. Hodges, P. G. Sansom, and D. B. Stephenson, 2013: A multi-model assessment of future projections of North Atlantic and European extratropical cyclones in the CMIP5 climate models. *J. Clim.*, doi:10.1175/JCLI-D-12-00573.1.

Zelazowski, P., Y. Malhi, C. Huntingford, S. Sitch, and J. Fisher, 2011: Changes in the potential distribution of humid tropical forests on a warmer planet. *Philos. Trans. R. Soc. A*, **369**, 137–160.

Zelinka, M., and D. Hartmann, 2010: Why is longwave cloud feedback positive? *J. Geophys. Res.*, **115**, D16117.

Zelinka, M., S. Klein, and D. Hartmann, 2012: Computing and partitioning cloud feedbacks using Cloud property histograms. Part II: Attribution to changes in cloud amount, altitude, and optical depth. *J. Clim.*, **25**, 3736–3754.

Zhang, M., and H. Song, 2006: Evidence of deceleration of atmospheric vertical overturning circulation over the tropical Pacific. *Geophys. Res. Lett.*, **33**, L12701.

Zhang, M. H., and C. Bretherton, 2008: Mechanisms of low cloud-climate feedback in idealized single-column simulations with the Community Atmospheric Model, version 3 (CAM3). *J. Clim.*, **21**, 4859–4878.

Zhang, R., 2010a: Northward intensification of anthropogenically forced changes in the Atlantic meridional overturning circulation (AMOC). *Geophys. Res. Lett.*, **37**, L24603.

Zhang, T., 2005: Influence of the seasonal snow cover on the ground thermal regime: An overview. *Rev. Geophys.*, **43**, RG4002.

Zhang, T., J. A. Heginbottom, R. G. Barry, and J. Brown, 2000: Further statistics on the distribution of permafrost and ground ice in the Northern Hemisphere 1. *Polar Geogr.*, **24**, 126–131.

Zhang, X., 2010b: Sensitivity of Arctic summer sea ice coverage to global warming forcing: Towards reducing uncertainty in arctic climate change projections. *Tellus A*, **62**, 220–227.

Zhang, X., and J. Walsh, 2006: Toward a seasonally ice-covered Arctic Ocean: Scenarios from the IPCC AR4 model simulations. *J. Clim.*, **19**, 1730–1747.

Zhang, X. B., et al., 2007: Detection of human influence on twentieth-century precipitation trends. *Nature*, **448**, 461–U464.

Zhou, L. M., R. E. Dickinson, P. Dirmeyer, A. Dai, and S. K. Min, 2009: Spatiotemporal patterns of changes in maximum and minimum temperatures in multi-model simulations. *Geophys. Res. Lett.*, **36**, L02702.

Zhuang, Q., et al., 2004: Methane fluxes between terrestrial ecosystems and the atmosphere at northern high latitudes during the past century: A retrospective analysis with a process-based biogeochemistry model. *Global Biogeochem. Cycles*, **18**, GB3010.

Zickfeld, K., V. K. Arora, and N. P. Gillett, 2012: Is the climate response to CO_2 emissions path dependent? *Geophys. Res. Lett.*, **39**, L05703.

Zickfeld, K., B. Knopf, V. Petoukhov, and H. Schellnhuber, 2005: Is the Indian summer monsoon stable against global change? *Geophys. Res. Lett.*, **32**, L15707.

Zickfeld, K., M. Eby, H. Matthews, and A. Weaver, 2009: Setting cumulative emissions targets to reduce the risk of dangerous climate change. *Proc. Natl. Acad. Sci. U.S.A.*, **106**, 16129–16134.

Zickfeld, K., M. Morgan, D. Frame, and D. Keith, 2010: Expert judgments about transient climate response to alternative future trajectories of radiative forcing. *Proc. Natl. Acad. Sci. U.S.A.*, **107**, 12451–12456.

Zickfeld, K., et al., 2013: Long-term climate change commitment and reversibility: An EMIC intercomparison. *J. Clim.*, doi:10.1175/JCLI-D-12-00584.1.

Zimov, S., S. Davydov, G. Zimova, A. Davydova, E. Schuur, K. Dutta, and F. Chapin, 2006: Permafrost carbon: Stock and decomposability of a globally significant carbon pool. *Geophys. Res. Lett.*, **33**, L20502.

Zunz, V., H. Goosse, and F. Massonnet, 2013: How does internal variability influence the ability of CMIP5 models to reproduce the recent trend in Southern Ocean sea ice extent? *Cryosphere*, **7**, 451–468.

Sea Level Change

Coordinating Lead Authors:
John A. Church (Australia), Peter U. Clark (USA)

Lead Authors:
Anny Cazenave (France), Jonathan M. Gregory (UK), Svetlana Jevrejeva (UK), Anders Levermann (Germany), Mark A. Merrifield (USA), Glenn A. Milne (Canada), R. Steven Nerem (USA), Patrick D. Nunn (Australia), Antony J. Payne (UK), W. Tad Pfeffer (USA), Detlef Stammer (Germany), Alakkat S. Unnikrishnan (India)

Contributing Authors:
David Bahr (USA), Jason E. Box (Denmark/USA), David H. Bromwich (USA), Mark Carson (Germany), William Collins (UK), Xavier Fettweis (Belgium), Piers Forster (UK), Alex Gardner (USA), W. Roland Gehrels (UK), Rianne Giesen (Netherlands), Peter J. Gleckler (USA), Peter Good (UK), Rune Grand Graversen (Sweden), Ralf Greve (Japan), Stephen Griffies (USA), Edward Hanna (UK), Mark Hemer (Australia), Regine Hock (USA), Simon J. Holgate (UK), John Hunter (Australia), Philippe Huybrechts (Belgium), Gregory Johnson (USA), Ian Joughin (USA), Georg Kaser (Austria), Caroline Katsman (Netherlands), Leonard Konikow (USA), Gerhard Krinner (France), Anne Le Brocq (UK), Jan Lenaerts (Netherlands), Stefan Ligtenberg (Netherlands), Christopher M. Little (USA), Ben Marzeion (Austria), Kathleen L. McInnes (Australia), Sebastian H. Mernild (USA), Didier Monselesan (Australia), Ruth Mottram (Denmark), Tavi Murray (UK), Gunnar Myhre (Norway), J.P. Nicholas (USA), Faezeh Nick (Norway), Mahé Perrette (Germany), David Pollard (USA), Valentina Radić (Canada), Jamie Rae (UK), Markku Rummukainen (Sweden), Christian Schoof (Canada), Aimée Slangen (Australia/Netherlands), Jan H. van Angelen (Netherlands), Willem Jan van de Berg (Netherlands), Michiel van den Broeke (Netherlands), Miren Vizcaíno (Netherlands), Yoshihide Wada (Netherlands), Neil J. White (Australia), Ricarda Winkelmann (Germany), Jianjun Yin (USA), Masakazu Yoshimori (Japan), Kirsten Zickfeld (Canada)

Review Editors:
Jean Jouzel (France), Roderik van de Wal (Netherlands), Philip L. Woodworth (UK), Cunde Xiao (China)

This chapter should be cited as:
Church, J.A., P.U. Clark, A. Cazenave, J.M. Gregory, S. Jevrejeva, A. Levermann, M.A. Merrifield, G.A. Milne, R.S. Nerem, P.D. Nunn, A.J. Payne, W.T. Pfeffer, D. Stammer and A.S. Unnikrishnan, 2013: Sea Level Change. In: *Climate Change 2013: The Physical Science Basis. Contribution of Working Group I to the Fifth Assessment Report of the Intergovernmental Panel on Climate Change* [Stocker, T.F., D. Qin, G.-K. Plattner, M. Tignor, S.K. Allen, J. Boschung, A. Nauels, Y. Xia, V. Bex and P.M. Midgley (eds.)]. Cambridge University Press, Cambridge, United Kingdom and New York, NY, USA.

Table of Contents

Supplementary Material

Supplementary Material is available in online versions of the report.

Executive Summary

This chapter considers changes in global mean sea level, regional sea level, sea level extremes, and waves. Confidence in projections of global mean sea level rise has increased since the Fourth Assessment Report (AR4) because of the improved physical understanding of the components of sea level, the improved agreement of process-based models with observations, and the inclusion of ice-sheet dynamical changes.

Past Sea Level Change

Paleo sea level records from warm periods during the last 3 million years indicate that global mean sea level has exceeded 5 m above present (*very high confidence*)[1] when global mean temperature was up to 2°C warmer than pre-industrial (*medium confidence*). There is *very high confidence* that maximum global mean sea level during the last interglacial period (~129 to 116 ka) was, for several thousand years, at least 5 m higher than present and *high confidence* that it did not exceed 10 m above present, implying substantial contributions from the Greenland and Antarctic ice sheets. This change in sea level occurred in the context of different orbital forcing and with high latitude surface temperature, averaged over several thousand years, at least 2°C warmer than present (*high confidence*) {5.3.4, 5.6.1, 5.6.2, 13.2.1}

Proxy and instrumental sea level data indicate a transition in the late 19th century to the early 20th century from relatively low mean rates of rise over the previous two millennia to higher rates of rise (*high confidence*). It is *likely*[2] that the rate of global mean sea level rise has continued to increase since the early 20th century, with estimates that range from 0.000 [–0.002 to 0.002] mm yr^{-2} to 0.013 [0.007 to 0.019] mm yr^{-2}. It is *very likely* that the global mean rate was 1.7 [1.5 to 1.9] mm yr^{-1} between 1901 and 2010 for a total sea level rise of 0.19 [0.17 to 0.21] m. Between 1993 and 2010, the rate was *very likely* higher at 3.2 [2.8 to 3.6] mm yr^{-1}; similarly high rates *likely* occurred between 1920 and 1950. {3.7.2, 3.7.4, 5.6.3, 13.2.1, 13.2.2, Figure 13.3}

Understanding of Sea Level Change

Ocean thermal expansion and glacier melting have been the dominant contributors to 20th century global mean sea level rise. Observations since 1971 indicate that thermal expansion and glaciers (excluding Antarctic glaciers peripheral to the ice sheet) explain 75% of the observed rise (*high confidence*). The contribution of the Greenland and Antarctic ice sheets has increased since the early 1990s, partly from increased outflow induced by warming of the immediately adjacent ocean. Natural and human-induced land water storage changes have made only a small contribution; the rate of groundwater depletion has increased and now exceeds the rate of reservoir impoundment. Since 1993, when observations of all sea level components are available, the sum of contributions equals the observed global mean sea level rise within uncertainties (*high confidence*). {Chapters 3, 4, 13.3.6, Figure 13.4, Table 13.1}

There is *high confidence* in projections of thermal expansion and Greenland surface mass balance, and *medium confidence* in projections of glacier mass loss and Antarctic surface mass balance. There has been substantial progress in ice-sheet modelling, particularly for Greenland. Process-based model calculations of contributions to past sea level change from ocean thermal expansion, glacier mass loss and Greenland ice-sheet surface mass balance are consistent with available observational estimates of these contributions over recent decades. Ice-sheet flowline modelling is able to reproduce the observed acceleration of the main outlet glaciers in the Greenland ice sheet, thus allowing estimates of the 21st century dynamical response (*medium confidence*). Significant challenges remain in the process-based projections of the dynamical response of marine-terminating glaciers and marine-based sectors of the Antarctic ice sheet. Alternative means of projection of the Antarctic ice-sheet contribution (extrapolation within a statistical framework and informed judgement) provide *medium confidence* in a *likely* range. There is currently *low confidence* in projecting the onset of large-scale grounding line instability in the marine-based sectors of the Antarctic ice sheet. {13.3.1 to 13.3.3, 13.4.3, 13.4.4}

The sum of thermal expansion simulated by Coupled Model Intercomparison Project phase 5 (CMIP5) Atmosphere–Ocean General Circulation Models (AOGCMs), glacier mass loss computed by global glacier models using CMIP5 climate change simulations, and estimates of land water storage explain 65% of the observed global mean sea level rise for 1901–1990 and 90% for 1971–2010 and 1993–2010 (*high confidence*). When observed climate parameters are used, the glacier models indicate a larger Greenland peripheral glacier contribution in the first half of the 20th century such that the sum of thermal expansion, glacier mass loss and changes in land water storage and a small ongoing Antarctic ice-sheet contribution are within 20% of the observations throughout the 20th century. Model-based estimates of ocean thermal expansion and glacier contributions indicate that the greater rate of global mean sea level rise since 1993 is a response to radiative forcing (RF, both anthropogenic and natural) and increased loss of ice-sheet mass and not part of a natural oscillation (*medium confidence*). {13.3.6, Figures 13.4, 13.7, Table 13.1}

1 In this Report, the following summary terms are used to describe the available evidence: limited, medium, or robust; and for the degree of agreement: low, medium, or high. A level of confidence is expressed using five qualifiers: very low, low, medium, high, and very high, and typeset in italics, e.g., *medium confidence*. For a given evidence and agreement statement, different confidence levels can be assigned, but increasing levels of evidence and degrees of agreement are correlated with increasing confidence (see Section 1.4 and Box TS.1 for more details).

2 In this Report, the following terms have been used to indicate the assessed likelihood of an outcome or a result: Virtually certain 99–100% probability, Very likely 90–100%, Likely 66–100%, About as likely as not 33–66%, Unlikely 0–33%, Very unlikely 0–10%, Exceptionally unlikely 0–1%. Additional terms (Extremely likely: 95–100%, More likely than not >50–100%, and Extremely unlikely 0–5%) may also be used when appropriate. Assessed likelihood is typeset in italics, e.g., *very likely* (see Section 1.4 and Box TS.1 for more details).

The Earth's Energy Budget

Independent estimates of effective RF of the climate system, the observed heat storage, and surface warming combine to give an energy budget for the Earth that is closed within uncertainties (*high confidence*), and is consistent with the *likely* range of climate sensitivity. The largest increase in the storage of heat in the climate system over recent decades has been in the oceans; this is a powerful observation for the detection and attribution of climate change. {Boxes 3.1, 13.1}

Global Mean Sea Level Rise Projections

It is *very likely* that the rate of global mean sea level rise during the 21st century will exceed the rate observed during 1971–2010 for all Representative Concentration Pathway (RCP) scenarios due to increases in ocean warming and loss of mass from glaciers and ice sheets. Projections of sea level rise are larger than in the AR4, primarily because of improved modeling of land-ice contributions. For the period 2081–2100, compared to 1986–2005, global mean sea level rise is *likely* (*medium confidence*) to be in the 5 to 95% range of projections from process-based models, which give 0.26 to 0.55 m for RCP2.6, 0.32 to 0.63 m for RCP4.5, 0.33 to 0.63 m for RCP6.0, and 0.45 to 0.82 m for RCP8.5. For RCP8.5, the rise by 2100 is 0.52 to 0.98 m with a rate during 2081–2100 of 8 to 16 mm yr^{-1}. We have considered the evidence for higher projections and have concluded that there is currently insufficient evidence to evaluate the probability of specific levels above the assessed *likely* range. Based on current understanding, only the collapse of marine-based sectors of the Antarctic ice sheet, if initiated, could cause global mean sea level to rise substantially above the *likely* range during the 21st century. This potential additional contribution cannot be precisely quantified but there is *medium confidence* that it would not exceed several tenths of a meter of sea level rise during the 21st century. {13.5.1, Table 13.5, Figures 13.10, 13.11}

Some semi-empirical models project a range that overlaps the process-based *likely* range while others project a median and 95th percentile that are about twice as large as the process-based models. In nearly every case, the semi-empirical model 95th percentile is higher than the process-based *likely* range. Despite the successful calibration and evaluation of semi-empirical models against the observed 20th century sea level record, there is no consensus in the scientific community about their reliability, and consequently *low confidence* in projections based on them. {13.5.2, 13.5.3, Figure 13.12}

13

It is *virtually certain* that global mean sea level rise will continue beyond 2100, with sea level rise due to thermal expansion to continue for many centuries. The amount of longer term sea level rise depends on future emissions. The few available process-based models that go beyond 2100 indicate global mean sea level rise above the pre-industrial level to be less than 1 m by 2300 for greenhouse gas concentrations that peak and decline and remain below 500 ppm CO_2-eq, as in scenario RCP2.6. For a radiative forcing that corresponds to above 700 ppm CO_2-eq but below 1500 ppm, as in the scenario RCP8.5, the projected rise is 1 m to more than 3 m (*medium confidence*). This assessment is based on *medium confidence* in the modelled contribution from thermal expansion and *low confidence* in the modelled contribution from ice sheets. The amount of ocean thermal expansion increases with global warming (0.2 to 0.6 m $°C^{-1}$) but the rate of the glacier contribution decreases over time as their volume (currently 0.41 m sea level equivalent) decreases. Sea level rise of several meters could result from long-term mass loss by ice sheets (consistent with paleo data observations of higher sea levels during periods of warmer temperatures), but there is *low confidence* in these projections. Sea level rise of 1 to 3 m per degree of warming is projected if the warming is sustained for several millennia (*low confidence*). {13.5.4, Figures 13.4.3, 13.4.4}

The available evidence indicates that sustained global warming greater than a certain threshold above pre-industrial would lead to the near-complete loss of the Greenland ice sheet over a millennium or more, causing a global mean sea level rise of about 7 m. Studies with fixed ice-sheet topography indicate the threshold is greater than 2°C but less than 4°C (*medium confidence*) of global mean surface temperature rise with respect to pre-industrial. The one study with a dynamical ice sheet suggests the threshold is greater than about 1°C (*low confidence*) global mean warming with respect to pre-industrial. We are unable to quantify a *likely* range. Whether or not a decrease in the Greenland ice sheet mass loss is irreversible depends on the duration and degree of exceedance of the threshold. Abrupt and irreversible ice loss from a potential instability of marine-based sectors of the Antarctic ice sheet in response to climate forcing is possible, but current evidence and understanding is insufficient to make a quantitative assessment. {5.8, 13.3, 13.4 }

Regional Sea Level Change Projections

It is *very likely* that in the 21st century and beyond, sea level change will have a strong regional pattern, with some places experiencing significant deviations of local and regional sea level change from the global mean change. Over decadal periods, the rates of regional sea level change as a result of climate variability can differ from the global average rate by more than 100% of the global average rate. By the end of the 21st century, it is *very likely* that over about 95% of the world ocean, regional sea level rise will be positive, and most regions that will experience a sea level fall are located near current and former glaciers and ice sheets. About 70% of the global coastlines are projected to experience a relative sea level change within 20% of the global mean sea level change. {13.6.5, Figures 13.18 to 13.22}

Projections of 21st Century Sea Level Extremes and Surface Waves

It is *very likely* that there will be a significant increase in the occurrence of future sea level extremes in some regions by 2100, with a *likely* increase in the early 21st century. This increase will primarily be the result of an increase in mean sea level (*high confidence*), with the frequency of a particular sea level extreme increasing by an order of magnitude or more in some regions by the end of the 21st century. There is *low confidence* in region-specific projections of storminess and associated storm surges. {13.7.2, Figure 13.25}

It is *likely* (*medium confidence*) that annual mean significant wave heights will increase in the Southern Ocean as a result of enhanced wind speeds. Southern Ocean generated swells are *likely* to affect heights, periods, and directions of waves in adjacent basins. It is *very likely* that wave heights and the duration of the wave season will increase in the Arctic Ocean as a result of reduced sea-ice extent. In general, there is *low confidence* in region-specific projections due to the *low confidence* in tropical and extratropical storm projections, and to the challenge of downscaling future wind fields from coarse-resolution climate models. {13.7.3; Figure 13.26}

13.1 Components and Models of Sea Level Change

13.1.1 Introduction and Chapter Overview

Changes in sea level occur over a broad range of temporal and spatial scales, with the many contributing factors making it an integral measure of climate change (Milne et al., 2009; Church et al., 2010). The primary contributors to contemporary sea level change are the expansion of the ocean as it warms and the transfer of water currently stored on land to the ocean, particularly from land ice (glaciers and ice sheets) (Church et al., 2011a). Observations indicate the largest increase in the storage of heat in the climate system over recent decades has been in the oceans (Section 3.2) and thus sea level rise from ocean warming is a central part of the Earth's response to increasing greenhouse gas (GHG) concentrations.

The First IPCC Assessment Report (FAR) laid the groundwork for much of our current understanding of sea level change (Warrick and Oerlemans, 1990). This included the recognition that sea level had risen during the 20th century, that the rate of rise had increased compared to the 19th century, that ocean thermal expansion and the mass loss from glaciers were the main contributors to the 20th century rise, that during the 21st century the rate of rise was projected to be faster than during the 20th century, that sea level will not rise uniformly around the world, and that sea level would continue to rise well after GHG emissions are reduced. They also concluded that no major dynamic response of the ice sheets was expected during the 21st century, leaving ocean thermal expansion and the melting of glaciers as the main contributors to the 21st century rise. The Second Assessment Report (SAR) came to very similar conclusions (Warrick et al., 1996).

By the time of the Third Assessment Report (TAR), coupled Atmosphere–Ocean General Circulation Models (AOGCMs) and ice-sheet models largely replaced energy balance climate models as the primary techniques supporting the interpretation of observations and the projections of sea level (Church et al., 2001). This approach allowed consideration of the regional distribution of sea level change in addition to the global average change. By the time of the Fourth Assessment Report (AR4), there were more robust observations of the variations in the rate of global average sea level rise for the 20th century, some understanding of the variability in the rate of rise, and the satellite altimeter record was long enough to reveal the complexity of the time-variable spatial distribution of sea level (Bindoff et al., 2007). Nevertheless, three central issues remained. First, the observed sea level rise over decades was larger than the sum of the individual contributions estimated from observations or with models (Rahmstorf et al., 2007, 2012a), although in general the uncertainties were large enough that there was no significant contradiction. Second, it was not possible to make confident projections of the regional distribution of sea level rise. Third, there was insufficient understanding of the potential contributions from the ice sheets. In particular, the AR4 recognized that existing ice-sheet models were unable to simulate the recent observations of ice-sheet accelerations and that understanding of ice-sheet dynamics was too limited to assess the likelihood of continued acceleration or to provide a best estimate or an upper bound for their future contributions.

Despite changes in the scenarios between the four Assessments, the sea level projections for 2100 (compared to 1990) for the full range of scenarios were remarkably similar, with the reduction in the upper end in more recent reports reflecting the smaller increase in radiative forcing (RF) in recent scenarios due to smaller GHG emissions and the inclusion of aerosols, and a reduction in uncertainty in projecting the contributions: 31 to 110 cm in the FAR, 13 to 94 cm in the SAR, 9 to 88 cm in the TAR and 18 to 59 cm in AR4 (not including a possible additional allowance for a dynamic ice-sheet response).

Results since the AR4 show that for recent decades, sea level has continued to rise (Section 3.7). Improved and new observations of the ocean (Section 3.7) and the cryosphere (Chapter 4) and their representation in models have resulted in better understanding of 20th century sea level rise and its components (this chapter). Records of past sea level changes constrain long-term land-ice response to warmer climates as well as extend the observational record to provide a longer context for current sea level rise (Section 5.6).

This chapter provides a synthesis of past and contemporary sea level change at global and regional scales. Drawing on the published refereed literature, including as summarized in earlier chapters of this Assessment, we explain the reasons for contemporary change and assess confidence in and provide global and regional projections of *likely* sea level change for the 21st century and beyond. We discuss the primary factors that cause regional sea level to differ from the global average and how these may change in the future. In addition, we address projected changes in surface waves and the consequences of sea level and climate change for extreme sea level events.

13.1.2 Fundamental Definitions and Concepts

The height of the ocean surface at any given location, or sea level, is measured either with respect to the surface of the solid Earth (relative sea level (RSL)) or a geocentric reference such as the reference ellipsoid (geocentric sea level). RSL is the more relevant quantity when considering the coastal impacts of sea level change, and it has been measured using tide gauges during the past few centuries (Sections 13.2.2 and 3.7) and estimated for longer time spans from geological records (Sections 13.2.1 and 5.6). Geocentric sea level has been measured over the past two decades using satellite altimetry (Sections 13.2.2 and 3.7).

A temporal average for a given location, known as Mean Sea Level (MSL; see Glossary), is applied to remove shorter period variability. Apart from Section 13.7, which considers high-frequency changes in ocean surface height, the use of 'sea level' elsewhere in this chapter refers to MSL. It is common to average MSL spatially to define global mean sea level (GMSL; see Glossary). In principle, integrating RSL change over the ocean area gives the change in ocean water volume, which is directly related to the processes that dominate sea level change (changes in ocean temperature and land-ice volume). In contrast, a small correction (–0.15 to –0.5 mm yr^{-1}) needs to be subtracted from altimetry observations to estimate ocean water volume change (Tamisiea, 2011). Local RSL change can differ significantly from GMSL because of spatial variability in changes of the sea surface and ocean floor height (see FAQ 13.1 and Section 13.6).

13.1.3 Processes Affecting Sea Level

This chapter focusses on processes within the ocean, atmosphere, land ice, and hydrological cycle that are climate sensitive and are expected to contribute to sea level change at regional to global scales in the coming decades to centuries (Figure 13.1). Figure 13.2 is a navigation aid for the different sections of this chapter and sections of other chapters that are relevant to sea level change.

Changes in ocean currents, ocean density and sea level are all tightly coupled such that changes at one location impact local sea level and sea level far from the location of the initial change, including changes in sea level at the coast in response to changes in open-ocean temperature (Landerer et al., 2007; Yin et al., 2010). Although both temperature and salinity changes can contribute significantly to regional sea level change (Church et al., 2010), only temperature change produces a significant contribution to global average ocean volume change due to thermal expansion or contraction (Gregory and Lowe, 2000). Regional atmospheric pressure anomalies also cause sea level to vary through atmospheric loading (Wunsch and Stammer, 1997). All of these climate-sensitive processes cause sea level to vary on a broad range of space and time scales from relatively short-lived events, such as waves and storm surges, to sustained changes over several decades or centuries that are associated with atmospheric and ocean modes of climate variability (White et al., 2005; Miller and Douglas, 2007; Zhang and Church, 2012).

Water and ice mass exchange between the land and the oceans leads to a change in GMSL. A signal of added mass to the ocean propagates rapidly around the globe such that all regions experience a sea level change within days of the mass being added (Lorbacher et al., 2012). In addition, an influx of freshwater changes ocean temperature and salinity and thus changes ocean currents and local sea level (Stammer, 2008; Yin et al., 2009), with signals taking decades to propagate around the global ocean. The coupled atmosphere–ocean system can also adjust to temperature anomalies associated with surface freshwater anomalies through air–sea feedbacks, resulting in dynamical adjustments of sea level (Okumura et al., 2009; Stammer et al., 2011). Water mass exchange between land and the ocean also results in patterns of sea level change called 'sea level fingerprints' (Clark and Lingle, 1977; Conrad and Hager, 1997; Mitrovica et al., 2001) due to change in the gravity field and vertical movement of the ocean floor associated with visco-elastic Earth deformation (Farrell and Clark, 1976). These changes in mass distribution also affect the Earth's inertia tensor and therefore rotation, which produces an additional sea level response (Milne and Mitrovica, 1998).

There are other processes that affect sea level but are not associated with contemporary climate change. Some of these result in changes that are large enough to influence the interpretation of observational records and sea level projections at regional and global scales. In particular, surface mass transfer from land ice to oceans during the last deglaciation contributes significantly to present-day sea level change due to the ongoing visco-elastic deformation of the Earth and the corresponding changes of the ocean floor height and gravity (referred to as glacial isostatic adjustment (GIA)) (Lambeck and Nakiboglu, 1984; Peltier and Tushingham, 1991). Ice sheets also have long response times and so continue to respond to past climate change (Section 13.1.5).

Anthropogenic processes that influence the amount of water stored in the ground or on its surface in lakes and reservoirs, or cause changes in land surface characteristics that influence runoff or evapotranspiration rates, will perturb the hydrological cycle and cause sea level change (Sahagian, 2000; Wada et al., 2010). Such processes include water impoundment (dams, reservoirs), irrigation schemes, and groundwater depletion (Section 13.4.5).

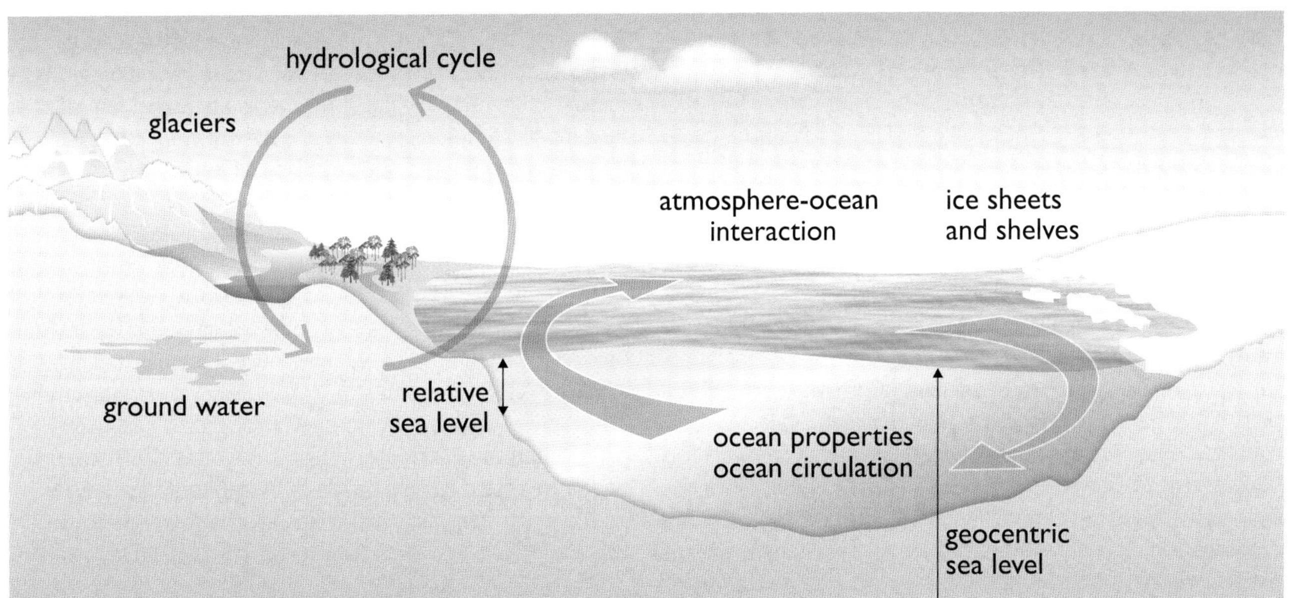

Figure 13.1 | Climate-sensitive processes and components that can influence global and regional sea level and are considered in this chapter. Changes in any one of the components or processes shown will result in a sea level change. The term 'ocean properties' refers to ocean temperature, salinity and density, which influence and are dependent on ocean circulation. Both relative and geocentric sea level vary with position. Note that the geocenter is not shown.

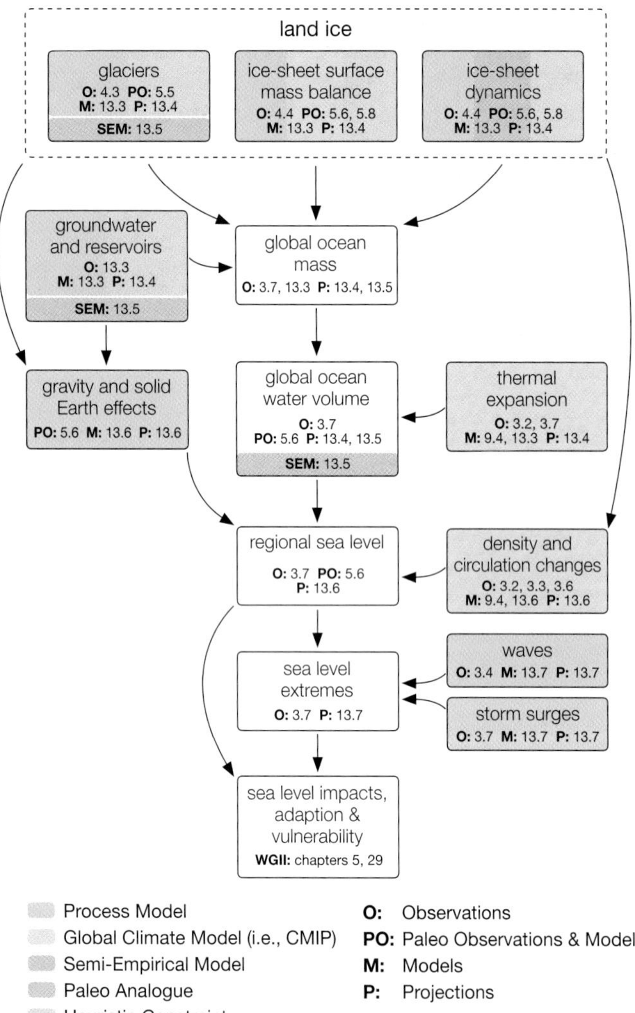

Figure 13.2 | Schematic representation of key linkages between processes and components that contribute to sea level change and are considered in this report. Colouring of individual boxes indicates the types of models and approaches used in projecting the contribution of each process or component to future sea level change. The diagram also serves as an index to the sections in this Assessment that are relevant to the assessment of sea level projections via the section numbers given at the bottom of each box. Note gravity and solid Earth effects change the shape of the ocean floor and surface and thus are required to infer changes in ocean water volume from both relative and geocentric sea level observations.

Sea level changes due to tectonic and coastal processes are beyond the scope of this chapter. With the exception of earthquakes, which can cause rapid local changes and tsunamis (Broerse et al., 2011) and secular RSL changes due to post-seismic deformation (Watson et al., 2010), tectonic processes cause, on average, relatively low rates of sea level change (order 0.1 mm yr^{-1} or less; Moucha et al., 2008). Sediment transfer and compaction (including from ground water depletion) in the coastal zone are particularly important in deltaic regions (Blum and Roberts, 2009; Syvitski et al., 2009). Although they can dominate sea level change in these localized areas, they are less important as a source of sea level change at regional and global scales and so are not considered further in this chapter (see discussion in Working Group II, Chapter 5). Estimates of sediment delivery to the oceans (Syvitski and Kettner, 2011) suggest a contribution to GMSL rise of order 0.01 mm yr^{-1}.

13.1.4 Models Used to Interpret Past and Project Future Changes in Sea Level

AOGCMs have components representing the ocean, atmosphere, land, and cryosphere, and simulate sea surface height changes relative to the geoid resulting from the natural forcings of volcanic eruptions and changes in solar irradiance, and from anthropogenic increases in GHGs and aerosols (Chapter 9). AOGCMs also exhibit internally generated climate variability, including such modes as the El Niño-Southern Oscillation (ENSO), the Pacific Decadal Oscillation (PDO), the North Atlantic Oscillation (NAO) and others that affect sea level (White et al., 2005; Zhang and Church, 2012). Critical components for global and regional changes in sea level are changes in surface wind stress and air–sea heat and freshwater fluxes (Lowe and Gregory, 2006; Timmermann et al., 2010; Suzuki and Ishii, 2011) and the resultant changes in ocean density and circulation, for instance in the strength of the Atlantic Meridional Overturning Circulation (AMOC) (Yin et al., 2009; Lorbacher et al., 2010; Pardaens et al., 2011a). As in the real world, ocean density, circulation and sea level are dynamically connected in AOGCMs and evolve together. Offline models are required for simulating glacier and ice-sheet changes (Section 13.1.4.1).

Geodynamic surface-loading models are used to simulate the RSL response to past and contemporary changes in surface water and land-ice mass redistribution and contemporary atmospheric pressure changes. The sea surface height component of the calculation is based solely on water mass conservation and perturbations to gravity, with no considerations of ocean dynamic effects. Application of these models has focussed on annual and interannual variability driven by contemporary changes in the hydrological cycle and atmospheric loading (Clarke et al., 2005; Tamisiea et al., 2010), and on secular trends associated with past and contemporary changes in land ice and hydrology (Lambeck et al., 1998; Mitrovica et al., 2001; Peltier, 2004; Riva et al., 2010).

Semi-empirical models (SEMs) project sea level based on statistical relationships between observed GMSL and global mean temperature (Rahmstorf, 2007a; Vermeer and Rahmstorf, 2009; Grinsted et al., 2010) or total RF (Jevrejeva et al., 2009, 2010). The form of this relationship is motivated by physical considerations, and the parameters are determined from observational data—hence the term 'semi-empirical' (Rahmstorf et al., 2012b). SEMs do not explicitly simulate the underlying processes, and they use a characteristic response time that could be considerably longer than the time scale of interest (Rahmstorf, 2007a) or one that is explicitly determined by the model (Grinsted et al., 2010).

Storm-surge and wave-projection models are used to assess how changes in storminess and MSL impact sea level extremes and wave climates. The two main approaches involve dynamical (Lowe et al., 2010) and statistical models (Wang et al., 2010). The dynamical models are forced by near-surface wind and mean sea level pressure fields derived from regional or global climate models (Lowe et al., 2010).

In this chapter, we use the term 'process-based models' (see Glossary) to refer to sea level and land-ice models (Section 13.1.4.1) that aim to simulate the underlying processes and interactions, in contrast to 'semi-empirical models' which do not. Although these two approaches are distinct, semi-empirical methods are often employed in components of the process-based models (e.g., glacier models in which surface mass balance is determined by a degree-day method (Braithwaite and Olesen, 1989)).

13.1.4.1 Models Used to Project Changes in Ice Sheets and Glaciers

The representation of glaciers and ice sheets within AOGCMs is not yet at a stage where projections of their changing mass are routinely available. Additional process-based models use output from AOGCMs to evaluate the consequences of projected climate change on these ice masses.

The overall contribution of an ice mass to sea level involves changes to either its surface mass balance (SMB) or changes in the dynamics of ice flow that affect outflow (i.e., solid ice discharge) to the ocean. SMB is primarily the difference between snow accumulation and the melt and sublimation of snow and ice (ablation). An assessment of observations related to this mass budget can be found in Section 4.4.2. Although some ice-sheet models used in projections incorporate both effects, most studies have focussed on either SMB or flow dynamics. It is assumed that the overall contribution can be found by summing the contributions calculated independently for these two sources, which is valid if they do not interact significantly. Although this can be addressed using a correction term to SMB in ice-sheet projections over the next century, such interactions become more important on longer time scales when, for example, changes in ice-sheet topography may significantly affect SMB or dynamics.

Projecting the sea level contribution of land ice requires comparing the model results with a base state that assumes no significant sea level contribution. This base state is taken to be either the pre-industrial period or, because of our scant knowledge of the ice sheets before the advent of satellites, the late 20th century. In reality, even at these times, the ice sheets may have been contributing to sea level change (Huybrechts et al., 2011; Box and Colgan, 2013) and this contribution, although difficult to quantify, should be included in the observed sea level budget (Gregory et al., 2013b).

Regional Climate Models (RCMs), which incorporate or are coupled to sophisticated representations of the mass and energy budgets associated with snow and ice surfaces, are now the primary source of ice-sheet SMB projections. A major source of uncertainty lies in the ability of these schemes to adequately represent the process of internal refreezing of melt water within the snowpack (Bougamont et al., 2007; Fausto et al., 2009). These models require information on the state of the atmosphere and ocean at their lateral boundaries, which are derived from reanalysis data sets or AOGCMs for past climate, or from AOGCM projections of future climate.

Models of ice dynamics require a fairly complete representation of stresses within an ice mass in order to represent the response of ice flow to changes at the marine boundary and the governing longitudinal stresses (Schoof, 2007a). For Antarctica, there is also a need to employ high spatial resolution (<1 km) to capture the dynamics of grounding line migration robustly so that results do not depend to an unreasonable extent on model resolution (Durand et al., 2009; Goldberg et al., 2009; Morlighem et al., 2010; Cornford et al., 2013; Pattyn et al., 2013). One-dimensional flowline models have been developed to the stage that modelled iceberg calving is comparable with observations (Nick et al., 2009). The success of this modelling approach relies on the ability of the model's computational grid to evolve to continuously track the migrating calving front. Although this is relatively straightforward in a one-dimensional model, this technique is difficult to incorporate into three-dimensional ice-sheet models that typically use a computational grid that is fixed in space.

The main challenge faced by models attempting to assess sea level change from glaciers is the small number of glaciers for which mass budget observations are available (about 380) (Cogley, 2009a) (see Sections 4.3.1 and 4.3.4) as compared to the total number (the Randolph Glacier Inventory contains more than 170,000) (Arendt et al., 2012). Statistical techniques are used to derive relations between observed SMB and climate variables for the small sample of surveyed glaciers, and then these relations are used to upscale to regions of the world. These techniques often include volume–area scaling to estimate glacier volume from their more readily observable areas. Although tidewater glaciers may also be affected by changes in outflow related to calving, the complexity of the associated processes means that most studies limit themselves to assessing the effects of SMB changes.

13.2 Past Sea Level Change

13.2.1 The Geological Record

Records of past sea level change provide insight into the sensitivity of sea level to past climate change as well as context for understanding current changes and evaluating projected changes. Since the AR4, important progress has been made in understanding the amplitude and variability of sea level during past intervals when climate was warmer than pre-industrial, largely through better accounting of the effects of proxy uncertainties and GIA on coastal sequences (Kopp et al., 2009, 2013; Raymo et al., 2011; Dutton and Lambeck, 2012; Lambeck et al., 2012; Raymo and Mitrovica, 2012) (Chapter 5). Here we summarize the constraints provided by the record of past sea level variations during times when global temperature was similar to or warmer than today.

13.2.1.1 The Middle Pliocene

There is *medium confidence* that during the warm intervals of the middle Pliocene (3.3 to 3.0 Ma), global mean surface temperatures were 2°C to 3.5°C warmer than for pre-industrial climate (Section 5.3.1). There are multiple lines of evidence that GMSL during these middle Pliocene warm periods was higher than today, but low agreement on how high it reached (Section 5.6.1). The most robust lines of evidence come from proximal sedimentary records that suggest periodic deglaciation of the West Antarctic ice sheet (WAIS) and parts of the East Antarctic ice sheet (EAIS) (Naish et al., 2009; Passchier, 2011) and from ice-sheet models that suggest near-complete deglaciation of the Greenland ice sheet, WAIS and partial deglaciation of the EAIS

(Pollard and DeConto, 2009; Hill et al., 2010; Dolan et al., 2011). The assessment by Chapter 5 suggests that GMSL was above present, but that it did not exceed 20 m above present, during the middle Pliocene warm periods (*high confidence*).

13.2.1.2 Marine Isotope Stage 11

During marine isotope stage 11 (MIS 11; 401 to 411 ka), Antarctic ice core and tropical Pacific paleo temperature estimates suggest that global temperature was 1.5°C to 2.0°C warmer than pre-industrial (*low confidence*) (Masson-Delmotte et al., 2010). Studies of the magnitude of sea level highstands from raised shorelines attributed to MIS 11 have generated highly divergent estimates. Since the AR4, studies have accounted for GIA effects (Raymo and Mitrovica, 2012) or reported elevations from sites where the GIA effects are estimated to be small (Muhs et al., 2012; Roberts et al., 2012). From this evidence, our assessment is that MIS 11 GMSL reached 6 to 15 m higher than present (*medium confidence*), requiring a loss of most or all of the present Greenland ice sheet and WAIS plus a reduction in the EAIS of up to 5 m equivalent sea level if sea level rise was at the higher end of the range.

13.2.1.3 The Last Interglacial Period

New data syntheses and model simulations since the AR4 indicate that during the Last Interglacial Period (LIG, ~129 to 116 ka), global mean annual temperature was 1°C to 2°C warmer than pre-industrial (*medium confidence*) with peak global annual sea surface temperatures (SSTs) that were 0.7°C ± 0.6°C warmer (*medium confidence*) (Section 5.3.4). High latitude surface temperature, averaged over several thousand years, was at least 2°C warmer than present (*high confidence*) (Section 5.3.4). There is robust evidence and high agreement that under the different orbital forcing and warmer climate of the LIG, sea level was higher than present. There have been a large number of estimates of the magnitude of LIG GMSL rise from localities around the globe, but they are generally from a small number of RSL reconstructions, and do not consider GIA effects, which can be substantial (Section 5.6.2). Since the AR4, two approaches have addressed GIA effects in order to infer LIG sea level from RSL observations at coastal sites. Kopp et al. (2009, 2013) obtained a probabilistic estimate of GMSL based on a large and geographically broadly distributed database of LIG sea level indicators. Their analysis accounted for GIA effects (and their uncertainties) as well as uncertainties in geochronology, the interpretation of sea level indicators, and regional tectonic uplift and subsidence. Kopp et al. (2013) concluded that GMSL was 6.4 m (95% probability) and 7.7 m (67% probability) higher than present, and with a 33% probability that it exceeded 8.8 m. The other approach, taken by Dutton and Lambeck (2012), used data from far-field sites that are tectonically stable. Their estimate of 5.5 to 9 m LIG GMSL is consistent with the probabilistic estimates made by Kopp et al. (2009, 2013). Chapter 5 thus concluded there is *very high confidence* that the maximum GMSL during the LIG was at least 5 m higher than present and *high confidence* it did not exceed 10 m. The best estimate is 6 m higher than present. Chapter 5 also concluded from ice-sheet model simulations and elevation changes derived from a new Greenland ice core that the Greenland ice sheet *very likely* contributed between 1.4 and 4.3 m sea level equivalent. This implies with *medium confidence* a contribution from the Antarctic ice sheet to the global mean sea level during the last interglacial period, but this is not yet supported by observational and model evidence.

There is *medium confidence* for a sea level fluctuation of up to 4 m during the LIG, but regional sea level variability and uncertainties in sea level proxies and their ages cause differences in the timing and amplitude of the reported fluctuation (Kopp et al., 2009, 2013; Thompson et al., 2011). For the time interval during the LIG in which GMSL was above present, there is *high confidence* that the maximum 1000-year average rate of GMSL rise associated with the sea level fluctuation exceeded 2 m kyr^{-1} but that it did not exceed 7 m kyr^{-1} (Chapter 5) (Kopp et al., 2013). Faster rates lasting less than a millennium cannot be ruled out by these data. Therefore, there is *high confidence* that there were intervals when rates of GMSL rise during the LIG exceeded the 20th century rate of 1.7 [1.5 to 1.9] mm yr^{-1}.

13.2.1.4 The Late Holocene

Since the AR4, there has been significant progress in resolving the sea level history of the last 7000 years. RSL records indicate that from ~7 to 3 ka, GMSL *likely* rose 2 to 3 m to near present-day levels (Chapter 5). Based on local sea level records spanning the last 2000 years, there is *medium confidence* that fluctuations in GMSL during this interval have not exceeded ~ ±0.25 m on time scales of a few hundred years (Section 5.6.3, Figure 13.3a). The most robust signal captured in salt marsh records from both Northern and Southern Hemispheres supports the AR4 conclusion for a transition from relatively low rates of change during the late Holocene (order tenths of mm yr^{-1}) to modern rates (order mm yr^{-1}) (Section 5.6.3, Figure 13.3b). However, there is variability in the magnitude and the timing (1840–1920) of this increase in both paleo and instrumental (tide gauge) records (Section 3.7). By combining paleo sea level records with tide gauge records at the same localities, Gehrels and Woodworth (2013) concluded that sea level began to rise above the late Holocene background rate between 1905 and 1945, consistent with the conclusions by Lambeck et al. (2004).

13.2.2 The Instrumental Record (~1700–2012)

The instrumental record of sea level change is mainly comprised of tide gauge measurements over the past two to three centuries (Figures 13.3b and 13.3c) and, since the early 1990s, of satellite-based radar altimeter measurements (Figure 13.3d).

13.2.2.1 The Tide Gauge Record (~1700–2012)

The number of tide gauges has increased since the first gauges at some northern European ports were installed in the 18th century; Southern Hemisphere (SH) measurements started only in the late 19th century. Section 3.7 assesses 20th century sea level rise estimates from tide gauges (Douglas, 2001; Church and White, 2006, 2011; Jevrejeva et al., 2006, 2008; Holgate, 2007; Ray and Douglas, 2011), and concludes that even though different strategies were developed to account for inhomogeneous tide gauge data coverage in space and time, and to correct for vertical crustal motions (also sensed by tide gauges, in addition to sea level change and variability), it is *very likely*

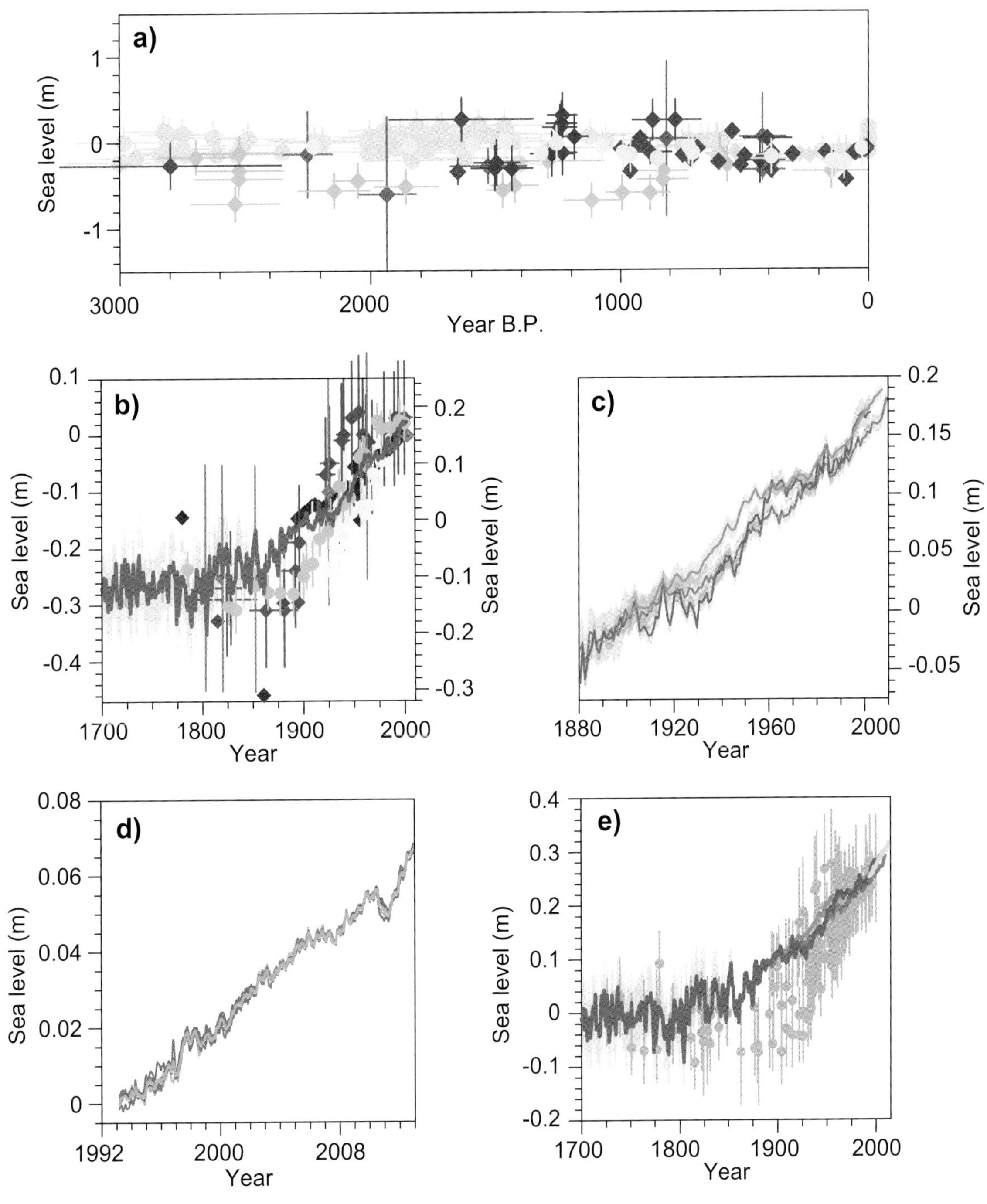

Figure 13.3 | (a) Paleo sea level data for the last 3000 years from Northern and Southern Hemisphere sites. The effects of glacial isostatic adjustment (GIA) have been removed from these records. Light green = Iceland (Gehrels et al., 2006), purple = Nova Scotia (Gehrels et al., 2005), bright blue = Connecticut (Donnelly et al., 2004), blue = Nova Scotia (Gehrels et al., 2005), red = United Kingdom (Gehrels et al., 2011), green = North Carolina (Kemp et al., 2011), brown = New Zealand (Gehrels et al., 2008), grey = mid-Pacific Ocean (Woodroffe et al., 2012). (b) Paleo sea level data from salt marshes since 1700 from Northern and Southern Hemisphere sites compared to sea level reconstruction from tide gauges (blue time series with uncertainty) (Jevrejeva et al., 2008). The effects of GIA have been removed from these records by subtracting the long-term trend (Gehrels and Woodworth, 2013). Ordinate axis on the left corresponds to the paleo sea level data. Ordinate axis on the right corresponds to tide gauge data. Green and light green = North Carolina (Kemp et al., 2011), orange = Iceland (Gehrels et al., 2006), purple = New Zealand (Gehrels et al., 2008), dark green = Tasmania (Gehrels et al., 2012), brown = Nova Scotia (Gehrels et al., 2005). (c) Yearly average global mean sea level (GMSL) reconstructed from tide gauges by three different approaches. Orange from Church and White (2011), blue from Jevrejeva et al. (2008), green from Ray and Douglas (2011) (see Section 3.7). (d) Altimetry data sets from five groups (University of Colorado (CU), National Oceanic and Atmospheric Administration (NOAA), Goddard Space Flight Centre (GSFC), Archiving, Validation and Interpretation of Satellite Oceanographic (AVISO), Commonwealth Scientific and Industrial Research Organisation (CSIRO)) with mean of the five shown as bright blue line (see Section 3.7). (e) Comparison of the paleo data from salt marshes (purple symbols, from (b)), with tide gauge and altimetry data sets (same line colours as in (c) and (d)). All paleo data were shifted by mean of 1700–1850 derived from the Sand Point, North Carolina data. The Jevrejeva et al. (2008) tide gauge data were shifted by their mean for 1700–1850; other two tide gauge data sets were shifted by the same amount. The altimeter time series has been shifted vertically upwards so that their mean value over the 1993–2007 period aligns with the mean value of the average of all three tide gauge time series over the same period.

Frequently Asked Questions

FAQ 13.1 | Why Does Local Sea Level Change Differ from the Global Average?

Shifting surface winds, the expansion of warming ocean water, and the addition of melting ice can alter ocean currents which, in turn, lead to changes in sea level that vary from place to place. Past and present variations in the distribution of land ice affect the shape and gravitational field of the Earth, which also cause regional fluctuations in sea level. Additional variations in sea level are caused by the influence of more localized processes such as sediment compaction and tectonics.

Along any coast, vertical motion of either the sea or land surface can cause changes in sea level relative to the land (known as relative sea level). For example, a local change can be caused by an increase in sea surface height, or by a decrease in land height. Over relatively short time spans (hours to years), the influence of tides, storms and climatic variability—such as El Niño—dominates sea level variations. Earthquakes and landslides can also have an effect by causing changes in land height and, sometimes, tsunamis. Over longer time spans (decades to centuries), the influence of climate change—with consequent changes in volume of ocean water and land ice—is the main contributor to sea level change in most regions. Over these longer time scales, various processes may also cause vertical motion of the land surface, which can also result in substantial changes in relative sea level.

Since the late 20th century, satellite measurements of the height of the ocean surface relative to the center of the Earth (known as geocentric sea level) show differing rates of geocentric sea level change around the world (see FAQ 13.1, Figure 1). For example, in the western Pacific Ocean, rates were about three times greater than the global mean value of about 3 mm per year from 1993 to 2012. In contrast, those in the eastern Pacific Ocean are lower than the global mean value, with much of the west coast of the Americas experiencing a fall in sea surface height over the same period. *(continued on next page)*

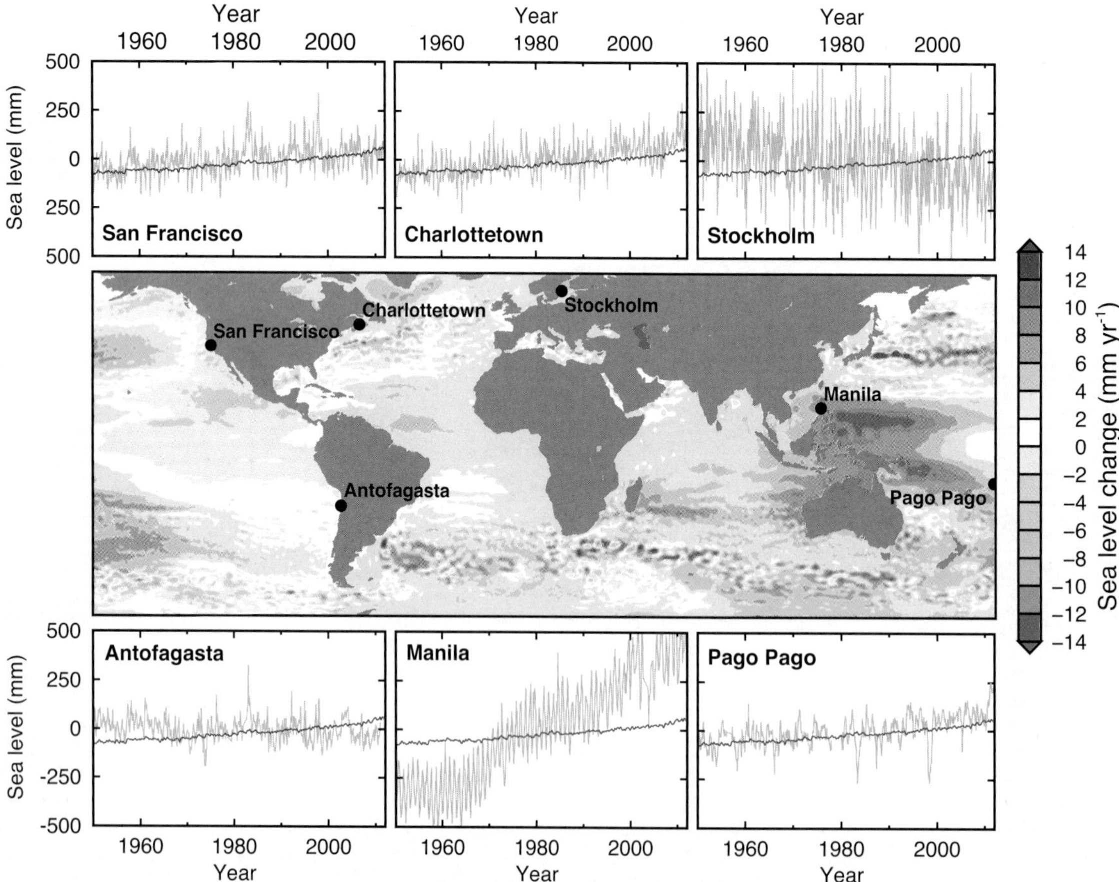

FAQ13.1, Figure 1 | Map of rates of change in sea surface height (geocentric sea level) for the period 1993–2012 from satellite altimetry. Also shown are relative sea level changes (grey lines) from selected tide gauge stations for the period 1950–2012. For comparison, an estimate of global mean sea level change is also shown (red lines) with each tide gauge time series. The relatively large, short-term oscillations in local sea level (grey lines) are due to the natural climate variability described in the main text. For example, the large, regular deviations at Pago Pago are associated with the El Niño-Southern Oscillation.

FAQ 13.1 (continued)

Much of the spatial variation shown in FAQ 13.1, Figure 1 is a result of natural climate variability—such as El Niño and the Pacific Decadal Oscillation—over time scales from about a year to several decades. These climate variations alter surface winds, ocean currents, temperature and salinity, and hence affect sea level. The influence of these processes will continue during the 21st century, and will be superimposed on the spatial pattern of sea level change associated with longer term climate change, which also arises through changes in surface winds, ocean currents, temperature and salinity, as well as ocean volume. However, in contrast to the natural variability, the longer term trends accumulate over time and so are expected to dominate over the 21st century. The resulting rates of geocentric sea level change over this longer period may therefore exhibit a very different pattern from that shown in FAQ 13.1, Figure 1.

Tide gauges measure relative sea level, and so they include changes resulting from vertical motion of both the land and the sea surface. Over many coastal regions, vertical land motion is small, and so the long-term rate of sea level change recorded by coastal and island tide gauges is similar to the global mean value (see records at San Francisco and Pago Pago in FAQ 13.1, Figure 1). In some regions, vertical land motion has had an important influence. For example, the steady fall in sea level recorded at Stockholm (FAQ 13.1, Figure 1) is caused by uplift of this region after the melting of a large (>1 km thick) continental ice sheet at the end of the last Ice Age, between ~20,000 and ~9000 years ago. Such ongoing land deformation as a response to the melting of ancient ice sheets is a significant contributor to regional sea level changes in North America and northwest Eurasia, which were covered by large continental ice sheets during the peak of the last Ice Age.

In other regions, this process can also lead to land subsidence, which elevates relative sea levels, as it has at Charlottetown, where a relatively large increase has been observed, compared to the global mean rate (FAQ 13.1, Figure 1). Vertical land motion due to movement of the Earth's tectonic plates can also cause departures from the global mean sea level trend in some areas—most significantly, those located near active subduction zones, where one tectonic plate slips beneath another. For the case of Antofagasta (FAQ 13.1, Figure 1) this appears to result in steady land uplift and therefore relative sea level fall.

In addition to regional influences of vertical land motion on relative sea level change, some processes lead to land motion that is rapid but highly localized. For example, the greater rate of rise relative to the global mean at Manila (FAQ 13.1, Figure 1) is dominated by land subsidence caused by intensive groundwater pumping. Land subsidence due to natural and anthropogenic processes, such as the extraction of groundwater or hydrocarbons, is common in many coastal regions, particularly in large river deltas.

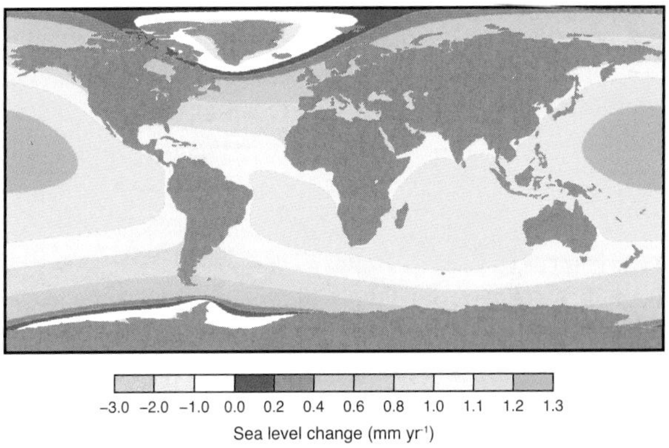

FAQ13.1, Figure 2 | Model output showing relative sea level change due to melting of the Greenland ice sheet and the West Antarctic ice sheet at rates of 0.5 mm yr^{-1} each (giving a global mean value for sea level rise of 1 mm yr^{-1}). The modelled sea level changes are less than the global mean value in areas near the melting ice but enhanced further afield. (Adapted from Milne et al., 2009)

It is commonly assumed that melting ice from glaciers or the Greenland and Antarctic ice sheets would cause globally uniform sea level rise, much like filling a bath tub with water. In fact, such melting results in regional variations in sea level due to a variety of processes, including changes in ocean currents, winds, the Earth's gravity field and land height. For example, computer models that simulate these latter two processes predict a regional fall in relative sea level around the melting ice sheets, because the gravitational attraction between ice and ocean water is reduced, and the land tends to rise as the ice melts (FAQ 13.1, Figure 2). However, further away from the ice sheet melting, sea level rise is enhanced, compared to the global average value.

In summary, a variety of processes drive height changes of the ocean surface and ocean floor, resulting in distinct spatial patterns of sea level change at local to regional scales. The combination of these processes produces a complex pattern of total sea level change, which varies through time as the relative contribution of each process changes. The global average change is a useful single value that reflects the contribution of climatic processes (e.g., land-ice melting and ocean warming), and represents a good estimate of sea level change at many coastal locations. At the same time, however, where the various regional processes result in a strong signal, there can be large departures from the global average value.

that the long-term trend estimate in GMSL is 1.7 [1.5 to 1.9] mm yr^{-1} between 1901 and 2010 for a total sea level rise of 0.19 [0.17 to 0.21] m (Figure 13.3c). Interannual and decadal-scale variability is superimposed on the long-term MSL trend, and Chapter 3 noted that discrepancies between the various published MSL records are present at these shorter time scales.

Section 3.7 also concludes that it is *likely* that the rate of sea level rise increased from the 19th century to the 20th century. Taking this evidence in conjunction with the proxy evidence for a change of rate (Sections 5.6.3 and 13.2.1; Figure 13.3b), there is *high confidence* that the rate of sea level rise has increased during the last two centuries, and it is *likely* that GMSL has accelerated since the early 1900's. Because of the presence of low-frequency variations (e.g., multi-decadal variations seen in some tide gauge records; Chambers et al. (2012)), sea level acceleration results are sensitive to the choice of the analysis time span. When a 60-year oscillation is modelled along with an acceleration term, the estimated acceleration in GMSL (twice the quadratic term) computed over 1900–2010 ranges from 0.000 [–0.002 to 0.002] mm yr^{-2} in the Ray and Douglas (2011) record, to 0.013 [0.007 to 0.019] mm yr^{-2} in the Jevrejeva et al. (2008) record, and 0.012 [0.009 to 0.015] mm yr^{-2} in the Church and White (2011) record. For comparison, Church and White (2011) estimated the acceleration term to be 0.009 [0.004 to 0.014] mm yr^{-2} over the 1880–2009 time span when the 60-year cycle is not considered.

13.2.2.2 The Satellite Altimeter Record (1993–2012)

The high-precision satellite altimetry record started in 1992 and provides nearly global (±66°) sea level measurements at 10-day intervals. Ollivier et al. (2012) showed that Envisat, which observes to ±82° latitude, provides comparable GMSL estimates. Although there are slight differences at interannual time scales in the altimetry-based GMSL time series produced by different groups (Masters et al., 2012), there is very good agreement on the 20-year long GMSL trend (Figure 13.3d). After accounting for the ~ –0.3 mm yr^{-1} correction related to the increasing size of the global ocean basins due to GIA (Peltier, 2009), a GMSL rate of 3.2 [2.8 to 3.6] mm yr^{-1} over 1993–2012 is found by the different altimetry data processing groups. The current level of precision is derived from assessments of all source of errors affecting the altimetric measurements (Ablain et al., 2009) and from tide gauge comparisons (Beckley et al., 2010; Nerem et al., 2010). Chapter 3 concludes that the GMSL trend since 1993 is *very likely* higher compared to the mean rates over the 20th century, and that it is *likely* that GMSL rose between 1920 and 1950 at a rate comparable to that observed since 1993. This recent higher rate is also seen in tide gauge data over the same period, but on the basis of observations alone it does not necessarily reflect a recent acceleration, considering the previously reported multi-decadal variations of mean sea level. The rapid increase in GMSL since 2011 is related to the recovery from the 2011 La Niña event (Section 13.3.5) (Boening et al., 2012).

13.3 Contributions to Global Mean Sea Level Rise During the Instrumental Period

In order to assess our understanding of the causes of observed changes and our confidence in projecting future changes we compare observational estimates of contributions with results derived from AOGCM experiments, beginning in the late 19th century, forced with estimated past time-dependent anthropogenic changes in atmospheric composition and natural forcings due to volcanic aerosols and variations in solar irradiance (Section 10.1). This period and these simulations are often referred to as "historical."

13.3.1 Thermal Expansion Contribution

13.3.1.1 Observed

Important progress has been realized since AR4 in quantifying the observed thermal expansion component of global mean sea level rise. This progress reflects (1) the detection of systematic time-dependent depth biases affecting historical expendable bathythermograph data (Gouretski and Koltermann, 2007) (Chapter 3), (2) the newly available Argo Project ocean (temperature and salinity) data with almost global coverage (not including ice-covered regions and marginal seas) of the oceans down to 2000 m since 2004–2005, and (3) estimates of the deep-ocean contribution using ship-based data collected during the World Ocean Circulation Experiment and revisit cruises (Johnson and Gruber, 2007; Johnson et al., 2007; Purkey and Johnson, 2010; Kouketsu et al., 2011).

For the period 1971–2010, the rate for the 0 to 700 m depth range is 0.6 [0.4 to 0.8] mm yr^{-1} (Section 3.7.2 and Table 3.1). Including the deep-ocean contribution for the same period increases the value to 0.8 [0.5 to 1.1] mm yr^{-1} (Table 13.1). Over the altimetry period (1993–2010), the rate for the 0 to 700 m depth range is 0.8 [0.5 to 1.1] mm yr^{-1} and 1.1 [0.8 to 1.4] mm yr^{-1} when accounting for the deep ocean (Section 3.7.2, Table 3.1, Table 13.1).

13.3.1.2 Modelled

GMSL rise due to thermal expansion is approximately proportional to the increase in ocean heat content (Section 13.4.1). Historical GMSL rise due to thermal expansion simulated by CMIP5 models is shown in Table 13.1 and Figure 13.4a. The model spread is due to uncertainty in RF and modelled climate response (Sections 8.5.2, 9.4.2.2, 9.7.2.5 and 13.4.1).

In the time mean of several decades, there is a negative volcanic forcing if there is more volcanic activity than is typical of the long term, and a positive forcing if there is less. In the decades after major volcanic eruptions, the rate of expansion is temporarily enhanced, as the ocean recovers from the cooling caused by the volcanic forcing (Church et al., 2005; Gregory et al., 2006) (Figure 13.4a). During 1961–1999, a period when there were several large volcanic eruptions, the CMIP3 simulations with both natural and anthropogenic forcing have substantially smaller increasing trends in the upper 700 m than those with anthropogenic forcing only (Domingues et al., 2008) because the natural volcanic forcing tends to cool the climate system, thus reducing ocean

Table 13.1 | Global mean sea level budget (mm yr^{-1}) over different time intervals from observations and from model-based contributions. Uncertainties are 5 to 95%. The Atmosphere–Ocean General Circulation Model (AOGCM) historical integrations end in 2005; projections for RCP4.5 are used for 2006–2010. The modelled thermal expansion and glacier contributions are computed from the CMIP5 results, using the model of Marzeion et al. (2012a) for glaciers. The land water contribution is due to anthropogenic intervention only, not including climate-related fluctuations.

Source	1901–1990	1971–2010	1993–2010
Observed contributions to global mean sea level (GMSL) rise			
Thermal expansion	–	0.8 [0.5 to 1.1]	1.1 [0.8 to 1.4]
Glaciers except in Greenland and Antarctica[a]	0.54 [0.47 to 0.61]	0.62 [0.25 to 0.99]	0.76 [0.39 to 1.13]
Glaciers in Greenland[a]	0.15 [0.10 to 0.19]	0.06 [0.03 to 0.09]	0.10 [0.07 to 0.13][b]
Greenland ice sheet	–	–	0.33 [0.25 to 0.41]
Antarctic ice sheet	–	–	0.27 [0.16 to 0.38]
Land water storage	–0.11 [–0.16 to –0.06]	0.12 [0.03 to 0.22]	0.38 [0.26 to 0.49]
Total of contributions	–	–	**2.8 [2.3 to 3.4]**
Observed GMSL rise	**1.5 [1.3 to 1.7]**	**2.0 [1.7 to 2.3]**	**3.2 [2.8 to 3.6]**
Modelled contributions to GMSL rise			
Thermal expansion	0.37 [0.06 to 0.67]	0.96 [0.51 to 1.41]	1.49 [0.97 to 2.02]
Glaciers except in Greenland and Antarctica	0.63 [0.37 to 0.89]	0.62 [0.41 to 0.84]	0.78 [0.43 to 1.13]
Glaciers in Greenland	0.07 [–0.02 to 0.16]	0.10 [0.05 to 0.15]	0.14 [0.06 to 0.23]
Total including land water storage	**1.0 [0.5 to 1.4]**	**1.8 [1.3 to 2.3]**	**2.8 [2.1 to 3.5]**
Residual[c]	**0.5 [0.1 to 1.0]**	**0.2 [–0.4 to 0.8]**	**0.4 [–0.4 to 1.2]**

Notes:

[a] Data for all glaciers extend to 2009, not 2010.

[b] This contribution is not included in the total because glaciers in Greenland are included in the observational assessment of the Greenland ice sheet.

[c] Observed GMSL rise – modelled thermal expansion – modelled glaciers – observed land water storage.

heat uptake (Levitus et al., 2001). The models including natural forcing are closer to observations, though with a tendency to underestimate the trend by about 10% (Sections 9.4.2.2 and 10.4.1).

Gregory (2010) and Gregory et al. (2013a) proposed that AOGCMs underestimate ocean heat uptake in their historical simulations because their control experiments usually omit volcanic forcing, so the imposition of historical volcanic forcing on the simulated climate system represents a time mean negative forcing relative to the control climate. The apparent long persistence of the simulated oceanic cooling following the 1883 eruption of Krakatau (Delworth et al., 2005; Gleckler et al., 2006a, 2006b; Gregory et al., 2006) is a consequence of this bias, which also causes a model-dependent underestimate of up to 0.2 mm yr^{-1} of thermal expansion on average during the 20th century (Gregory et al., 2013a, 2013b). This implies that CMIP5 results may be similarly underestimated, depending on the details of the individual model control runs. Church et al. (2013) proposed a correction of 0.1 mm yr^{-1} to the model mean rate, which we apply in the sea level budget in Table 13.1 and Figure 13.7. The corrected CMIP5 model mean rate for 1971–2010 is close to the central observational estimate; the model mean rate for 1993–2010 exceeds the central observational estimate but they are not statistically different given the uncertainties (Table 13.1 and Figure 13.4a). This correction is not made to projections of thermal expansion because it is very small compared with the projected increase in the rate (Section 13.5.1).

In view of the improvement in observational estimates of thermal expansion, the good agreement of historical model results with observational estimates, and their consistency with understanding of the energy budget and RF of the climate system (Box 13.1), we have *high confidence* in the projections of thermal expansion using AOGCMs.

13.3.2 Glaciers

13.3.2.1 Observed

'Glaciers' are defined here as all land-ice masses, including those peripheral to (but not including) the Greenland and Antarctic ice sheets. The term 'glaciers and ice caps' was applied to this category in the AR4. Changes in aggregate glacier volume have conventionally been determined by various methods of repeat mapping of surface elevation to detect elevation (and thus volume) change. Mass changes are determined by compilation and upscaling of limited direct observations of surface mass balance (SMB). Since 2003, gravity observations from Gravity Recovery and Climate Experiment (GRACE) satellites have been used to detect mass change of the world's glaciers.

The combined records indicate that a net decline of global glacier volume began in the 19th century, before significant anthropogenic RF had started, and was probably the result of warming associated with the termination of the Little Ice Age (Crowley, 2000; Gregory et al., 2006, 2013b). Global rates of glacier volume loss did not increase significantly during much of the 20th century (Figure 4.12). In part this may have been because of an enhanced rate of loss due to unforced high-latitude variability early in the century, while anthropogenic warming was still comparatively small (Section 13.3.2.2). It is *likely* that anthropogenic forcing played a statistically significant role in acceleration of global glacier losses in the latter decades of the 20th

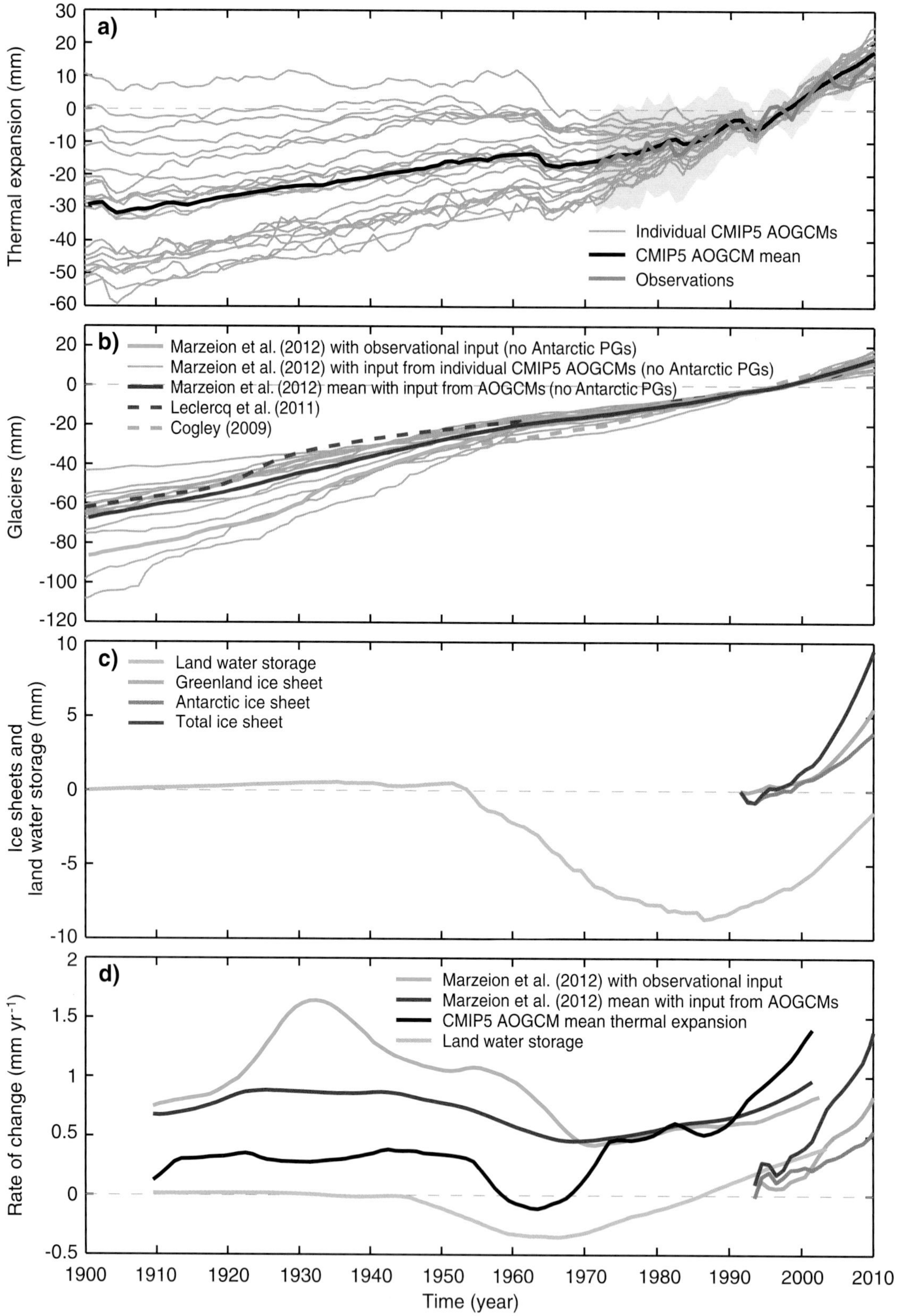

Figure 13.4 | Comparison of modelled and observed components of global mean sea level change since 1900. Changes in glaciers, ice sheets and land water storage are shown as positive sea level rise when mass is added to the ocean. (a) Ocean thermal expansion. Individual CMIP5 Atmosphere–Ocean General Circulation Model (AOGCM) simulations are shown in grey, the AOGCM average is black, observations in teal with the 5 to 95% uncertainties shaded. (b) Glaciers (excluding Antarctic peripheral glaciers). Model simulations by Marzeion et al. (2012a) with input from individual AOGCMs are shown in grey with the average of these results in bright purple. Model simulations by Marzeion et al. (2012a) forced by observed climate are shown in light blue. The observational estimates by Cogley (2009b) are shown in green (dashed) and by Leclercq et al. (2011) in red (dashed). (c) Changes in land water storage (yellow/orange, the sum of groundwater depletion and reservoir storage) start at zero in 1900. The Greenland ice sheet (green), the Antarctic ice sheet (blue) and the sum of the ice sheets (red), start at zero at the start of the record in 1991. (d) The rate of change (19-year centred trends) for the terms in (a)–(c), and for the ice sheets (5-year centred trends). All curves in (a) and (b) are shown with zero time-mean over the period 1986–2005 and the colours in (d) are matched to earlier panels. (Updated from Church et al., 2013)

century relative to rates in the 19th century (Section 10.5.2.2). It is also *likely* that, during the 20th century, the progressive loss of glacier area significantly restricted the rate of mass loss (Gregory et al., 2013b).

The earliest sea level assessments recognized that glaciers have been significant contributors to GMSL rise (Meier, 1984). As assessed in Chapter 4, observations, improved methods of analysis and a new, globally complete inventory indicate that glaciers, including those around the ice-sheet peripheries, *very likely* continue to be significant contributors to sea level, but are also highly variable on annual to decadal time scales. It is assumed that all glacier losses contribute to sea level rise, but the potential role of terrestrial interception of runoff, either in lakes formed following future ice retreat or in groundwater, has yet to be evaluated. For the period 2003–2009, the sea level contribution of all glaciers globally, including those glaciers surrounding the periphery of the two ice sheets, is 0.71 [0.64 to 0.79] mm yr^{-1} sea level equivalent (SLE) (Section 4.3.3, Table 4.4). Depending on the method used, however, loss-rate measurements of the two ice sheets can be very difficult to separate from losses from the peripheral glaciers. To avoid double counting, total cryospheric losses are determined by adding estimates of glacier losses excluding the peripheral glaciers to losses from the ice sheets including their peripheral glaciers. The sea level contribution of all glaciers *excluding* those glaciers surrounding the periphery of the two ice sheets was 0.54 [0.47-0.61] mm yr^{-1} SLE for 1901-1990, 0.62 [0.25-0.99] mm yr^{-1} SLE for 1971-2009, 0.76 [0.39-1.13] mm yr^{-1} SLE for 1993-2009, and 0.83 [0.46-1.20] mm yr^{-1} SLE for 2005-2009 (Section 4.3.3.4, Table 13.1).

13.3.2.2 Modelled

Global glacier mass balance models are calibrated using data from the few well-observed glaciers. Approximately 100 glacier mass balance records are available in any given year over the past half-century; only 17 glaciers exist with records of 30 years or more (Dyurgerov and Meier, 2005; Kaser et al., 2006; Cogley, 2012). Confidence in these models for projections of future change (Section 13.4.2) depends on their ability to reproduce past observed glacier change using corresponding climate observations as the forcing (Raper and Braithwaite, 2005; Meier et al., 2007; Bahr et al., 2009; Radić and Hock, 2011; Marzeion et al., 2012b; 2012a; Giesen and Oerlemans, 2013). Model validation is challenging owing to the scarcity of independent observations (unused in model calibration), but uncertainties have been evaluated by methods such as cross validation of hindcast projections for individual glaciers drawn from the sample of glacier observations averaged for calibration (Marzeion et al., 2012a; Radić et al., 2013).

Confidence in the use of AOGCM climate simulations as input to glacier projections is gained from the agreement since the mid-20th century of glacier models forced by AOGCM simulations with glacier models forced by observations (Marzeion et al., 2012a) (Figure 13.4b). In the earlier 20th century, around the 1930s, glaciers at high northern latitudes lost mass at an enhanced rate (Oerlemans et al., 2011; Leclercq et al., 2012); in the model, observed forcings produced larger glacier losses than did AOGCM forcings (Marzeion et al., 2012a) (Figure 13.4d). This is judged *likely* to be due to an episode of unforced, regionally variable warming around Greenland (Box, 2002; Chylek et al., 2004) rather than to RF of the climate system, and is consequently not reproduced by AOGCM experiments (Section 10.2). In our analysis of the budget of GMSL rise (Section 13.3.6), we take the difference between the simulations using AOGCM forcing and the simulation using observations as an estimate of the influence of unforced climate variability on global glacier mass balance (Figure 13.4b).

There is *medium confidence* in the use of glacier models to make global projections based on AOGCM results. The process-based understanding of glacier surface mass balance, the consistency of models and observations of glacier changes, and the evidence that AOGCM climate simulations can provide realistic input all give confidence, which on the other hand is limited because the set of well-observed glaciers is a very small fraction of the total.

13.3.3 Greenland and Antarctic Ice Sheets

13.3.3.1 Observed Mass Balance

The Greenland ice sheet's mass balance is comprised of its surface mass balance and outflow, whereas Antarctica's mass budget is dominated by accumulation and outflow in the form of calving and ice flow into floating (and therefore sea level neutral) ice shelves. Knowledge of the contribution of the Greenland and Antarctic ice sheets to observed sea level changes over the last two decades comes primarily from satellite and airborne surveys. Three main techniques are employed: the mass budget method, repeat altimetry, and gravimetric methods that measure temporal variations in the Earth's gravity field (Section 4.4.2).

Observations indicate that the Greenland contribution to GMSL has *very likely* increased from 0.09 [–0.02 to 0.20] mm yr^{-1} for 1992–2001 to 0.59 [0.43 to 0.76] mm yr^{-1} for 2002–2011 (Section 4.4.3, Figure 13.4). The average rate of the Antarctica contribution to sea level rise *likely* increased from 0.08 [–0.10 to 0.27] mm yr^{-1} for 1992–2001 to 0.40 [0.20 to 0.61] mm yr^{-1} for 2002–2011 (Section 4.4.3). For the budget period 1993–2010, the combined contribution of the ice sheets is 0.60 [0.42 to 0.78] mm yr^{-1}. For comparison, the AR4's assessment for the period 1993–2003 was 0.21 ± 0.07 mm yr^{-1} for Greenland and 0.21 ± 0.35 mm yr^{-1} for Antarctica.

13.3.3.2 Modelled Surface Mass Balance

Projections of changes in the SMB of the Antarctic and Greenland ice sheets are obtained from RCM or downscaled AOGCM simulations (Sections 13.4.3.1 and 13.4.4.1). A spatial resolution of a few tens kilometres or finer is required in order to resolve the strong gradients in SMB across the steep slopes of the ice-sheet margins. Although simulations of SMB at particular locations may have errors of 5 to 20% compared with *in situ* observations, there is good agreement between methods involving RCMs and observational methods of evaluating ice-sheet mass balance (Shepherd et al., 2012). In the present climate, for both Greenland and Antarctica, the mean SMB over the ice-sheet area is positive, giving a negative number when expressed as sea level equivalent (SLE).

In Greenland, the average and standard deviation of accumulation (precipitation minus sublimation) estimates for 1961–1990 is –1.62 ± 0.21 mm yr^{-1} SLE from the models in Table 13.2, agreeing with published observation-based accumulation maps, for example −1.42 ± 0.11 mm yr^{-1} SLE by Bales et al. (2009) and −1.63 ± 0.23 mm yr^{-1} SLE by Burgess et al. (2010). For SMB (accumulation minus runoff, neglecting drifting snow erosion, which is small), the models give −0.92 ± 0.26 mm yr^{-1} SLE for 1961–1990 (Table 13.2).

All of these models indicate that Greenland ice sheet SMB showed no significant trend from the 1960s to the 1980s, then started becoming less positive (becoming less negative expressed as SLE) in the early 1990s, on average by 3% yr^{-1}. This results in a statistically significant and increasing (i.e., becoming more positive) contribution to the rate of GMSL rise (SMB trend column of Table 13.2, Figure 13.5). The largest trends are found in models with coupled snow and atmosphere simulations using the Regional Atmospheric Climate Model 2 (RACMO2) and the Modèle Atmosphérique Régional (MAR). Van den Broeke et al. (2009) concluded that the mass loss during 2000–2008 is equally split between SMB and dynamical change. Rignot et al. (2011) indicated that SMB change accounts for about 60% of the mass loss since 1992 and Sasgen et al. (2012) showed that SMB change, simulated by RACMO2 (Ettema et al., 2009, an earlier version of the model in Table 13.2), accounts for about 60% of the observed rate of mass loss during 2002–2010, with an observational estimate of the increase in ice outflow accounting for the remainder. This satisfactory consistency, within uncertainties, in estimates for the Greenland ice-sheet mass budget gives confidence in SMB simulations of the past, and hence also in the similar models used for projections of SMB changes (Section 13.4.3.1).

This recent trend towards increasingly less positive SMB is caused almost entirely by increased melting and subsequent runoff, with variability in accumulation being comparatively small (Sasgen et al., 2012; Vernon et al., 2013). This tendency is related to pronounced regional warming, which may be attributed to some combination of anthropogenic climate change and anomalous regional variability in recent years (Hanna et al., 2008; 2012; Fettweis et al., 2013). Greenland SMB models forced by boundary conditions from AOGCM historical simulations (Rae et al., 2012; Fettweis et al., 2013) do not show statistically significant trends towards increasing contributions to GMSL, implying that the dominant contribution is internally generated regional climate variability, which is not expected to be reproduced by AOGCM historical simulations (Section 10.2). We have *high confidence* in projections of future warming in Greenland because of the agreement of models in predicting amplified warming at high northern latitudes (Sections 12.4.3.1, 14.8.2) for well-understood physical reasons, although there remains uncertainty in the size of the amplification, and we have *high confidence* in projections of increasing surface melting (Section 13.4.3.1) because of the sensitivity to warming demonstrated by SMB models of the past.

All Greenland SMB simulations for the first half of the 20th century depend on reconstructions of meteorological variability over the ice sheet made using empirical relationships based on observations from coastal stations and estimates of accumulation from ice cores. Despite the similar input data sets in all cases, the various climate reconstruction and SMB methods used have led to a range of results (Fettweis et al., 2008; Wake et al., 2009; Hanna et al., 2011; Box, 2013; Box and Colgan, 2013; Box et al., 2013; Gregory et al., 2013b). For 1901–1990, Hanna et al. (2011) have a time-mean GMSL contribution of −0.3 mm yr^{-1}, while Box and Colgan (2013) have a weakly positive contribution and the others are about zero. In all cases, there is substantial variability associated with regional climate fluctuations, in particular the warm episode in the 1930s, during which glaciers retreated in southeastern Greenland (Bjork et al., 2012). Chylek et al. (2004) argued that this episode was associated with the NAO rather than with global climate change.

In Antarctica, accumulation (precipitation minus sublimation) approximates SMB because surface melting and runoff are negligible in the present climate (Section 4.4.2.1.1). There are uncertainties in model- and observation-based estimates of Antarctic SMB. Global climate models do not account for snow hydrology or for drifting snow processes which remove an estimated 7% of the accumulated snow (Lenaerts et al., 2012), and the ice sheet's steep coastal slopes are not well captured by coarse-resolution models. Observation-based estimates rely on sparse accumulation measurements with very little coverage in high-accumulation areas. For the Antarctic ice sheet and ice shelves

Table 13.2 | Surface mass balance (SMB) and rates of change of SMB of the Greenland ice sheet, calculated from ice-sheet SMB models using meteorological observations and reanalyses as input, expressed as sea level equivalent (SLE). A negative SLE number for SMB indicates that accumulation exceeds runoff. A positive SLE for SMB anomaly indicates that accumulation has decreased, or runoff has increased, or both. Uncertainties are one standard deviation. Uncertainty in individual model results reflects temporal variability (1 standard deviations of annual mean values indicated); the uncertainty in the model average is 1 standard deviation of variation across models.

Reference and Model[a]	Time-Mean SMB 1961–1990 mm yr^{-1} SLE	Rate of Change of SMB 1991–2010 mm yr^{-2} SLE	Time-Mean SMB Anomaly (With Respect to 1961–1990 Time-Mean SMB)[b] mm yr^{-1} SLE		
			1971–2010	1993–2010	2005–2010
RACMO2, Van Angelen et al. (2012), 11 km RCM	−1.13 ± 0.30	0.04 ± 0.01	0.07 ± 0.33	0.23 ± 0.30	0.47 ± 0.24
MAR, Fettweis et al. (2011), 25 km RCM	−1.17 ± 0.31	0.05 ± 0.01	0.12 ± 0.38	0.36 ± 0.33	0.64 ± 0.22
PMM5, Box et al. (2009), 25 km RCM	−0.98 ± 0.18	0.02 ± 0.01	0.00 ± 0.19	0.10 ± 0.22	0.23 ± 0.21
ECMWFd, Hanna et al. (2011), 5 km PDD	−0.77 ± 0.27	0.02 ± 0.01	0.02 ± 0.28	0.12 ± 0.27	0.24 ± 0.19
SnowModel, Mernild and Liston (2012), 5 km EBM	−0.54 ± 0.21	0.03 ± 0.01	0.09 ± 0.25	0.19 ± 0.24	0.36 ± 0.23
Model Average	−0.92 ± 0.26	0.03 ± 0.01	0.06 ± 0.05	0.20 ± 0.10	0.39 ± 0.17

Notes:

[a] The approximate spatial resolution is stated and the model type denoted by PDD = positive degree day, EBM = Energy Balance Model, RCM = Regional Climate Model.

[b] Difference from the time-mean SMB of 1961–1990. This difference equals the sea level contribution from Greenland SMB changes if the ice sheet is assumed to have been near zero mass balance during 1961–1990 (Hanna et al., 2005; Sasgen et al., 2012).

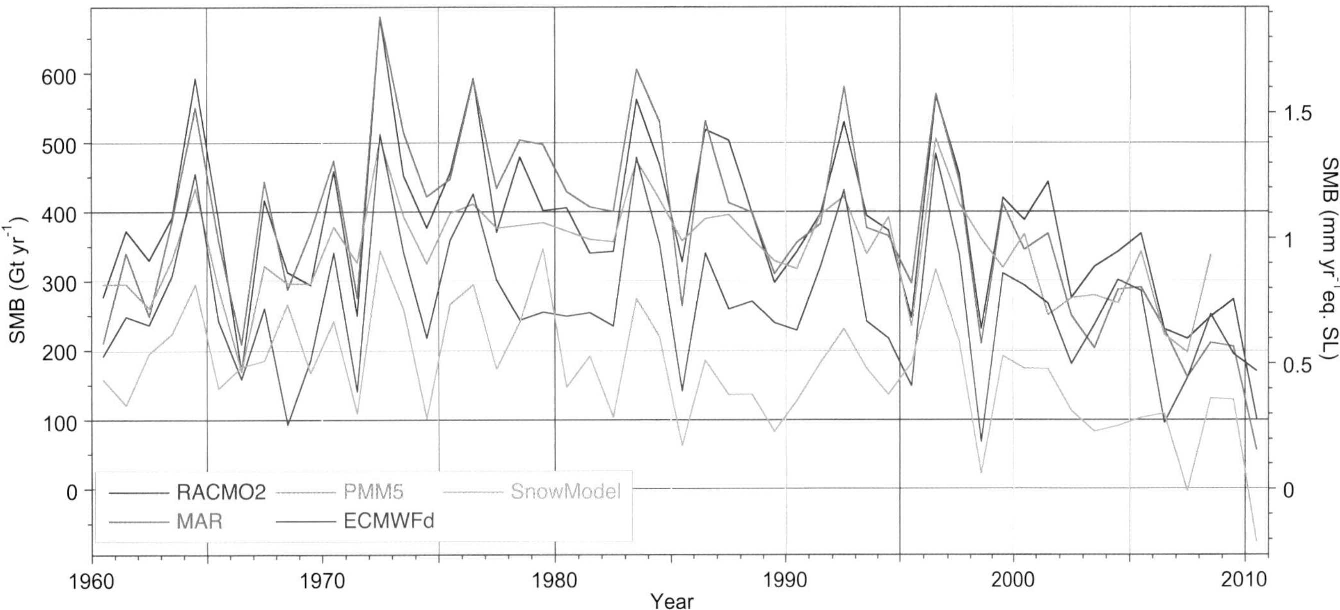

Figure 13.5 | Annual mean surface mass balance (accumulation minus ablation) for the Greenland ice sheet, simulated by five regional climate models for the period 1960–2010.

together, CMIP3 AOGCMs simulate SMB for 1979–2000 of –7.1 ± 1.5 mm yr^{-1} SLE (Connolley and Bracegirdle, 2007; Uotila et al., 2007), the mean being about 10% larger in magnitude than observation-based estimates, for instance –6.3 mm yr^{-1} SLE from Vaughan et al. (1999). For the SMB of the grounded ice sheet alone, four global reanalysis models, with resolutions of 38 to 125 km (Bromwich et al., 2011), give –5.2 ± 0.5 mm yr^{-1} SLE for 1979–2010, which compares well with an observational estimate of –4.9 ± 0.1 mm yr^{-1} SLE for 1950–2000 (Arthern et al., 2006). Because of higher accumulation near the coast, the regional climate model RACMO2 gives the somewhat larger value of –5.5 ± 0.3 mm yr^{-1} SLE for 1979–2000 (Lenaerts et al., 2012). This relatively good agreement, combined with the similarity of the geographical distribution of modelled and observed SMB, give *medium confidence* in the realism of the RCM SMB simulation.

Some global reanalyses have been shown to contain spurious trends in various quantities in the SH related to changes in the observing systems, for example, new satellite observations (Bromwich et al., 2007; 2011). In the RCMs and in global reanalyses that are not affected by spurious trends, no significant trend is present in accumulation since 1980 (Section 4.4.2.3). This agrees with observation-based studies (Monaghan et al., 2006; Anschütz et al., 2009) (Chapter 4) and implies that Antarctic SMB change has not contributed significantly to recent changes in the rate of GMSL rise. Likewise, CMIP3 historical simulations do not exhibit any systematic trend in Antarctic precipitation during the late 20th century (Uotila et al., 2007). No observational assessments have been made of variability in SMB for the whole ice sheet for the earlier part of the 20th century, or of its longer term mean.

General Circulation Model (GCM) and Regional Circulation Model (RCM) projections consistently indicate significant Antarctic warming and concomitant increase in precipitation. We have *high confidence* in expecting a relationship between these quantities on physical grounds (Section 13.4.4.1) and from ice core evidence (Van Ommen et al., 2004; Lemieux-Dudon et al., 2010; Stenni et al., 2011). The absence of a significant trend in Antarctic precipitation up to the present is not inconsistent with the expected relationship, because observed temperature trends over the majority of the continent are weak (Section 10.5.2.1) and trends in Antarctic precipitation simulated for recent decades are much smaller than interannual variability (van den Broeke et al., 2006; Uotila et al., 2007). Taking all these considerations together, we have *medium confidence* in model projections of a future Antarctic SMB increase, implying a negative contribution to GMSL rise (see also Sections 13.4.4.1, 13.5.3 and 14.8.15).

13.3.4 Contributions from Water Storage on Land

Changes in water storage on land in response to climate change and variability (i.e., water stored in rivers, lakes, wetlands, the vadose zone, aquifers and snow pack at high latitudes and altitudes) and from direct human-induced effects (i.e., storage of water in reservoirs and groundwater pumping) have the potential to contribute to sea level change. Based on satellite observations of the Northern Hemisphere (NH) snowpack, Biancamaria et al. (2011) found no significant trend in the contribution of snow to sea level. Estimates of climate-related changes in land water storage over the past few decades rely on global hydrological models because corresponding observations are inadequate (Milly et al., 2010). In assessing the relation between terrestrial water storage and climate, Milly et al. (2003) and Ngo-Duc et al. (2005) found no long-term climatic trend in total water storage, but documented interannual to decadal fluctuations, equivalent to several millimetres of sea level. Recent studies have shown that interannual variability in observed GMSL correlates with ENSO indices (Nerem et al., 2010) and is inversely related to ENSO-driven changes of terrestrial water storage, especially in the tropics (Llovel et al., 2011). During El Niño events, sea level (and ocean mass) tends to be higher because ocean precipitation increases and land precipitation decreases in the tropics (Cazenave et al., 2012). The reverse happens during La Niña events, as seen during 2010–2011, when there was a decrease in GMSL due to a temporary increase in water storage on the land, especially in Australia, northern South America, and southeast Asia (Boening et al., 2012) (Section 13.3.5).

Direct human interventions on land water storage also induce sea level changes (Sahagian, 2000; Gornitz, 2001; Huntington, 2008; Lettenmaier and Milly, 2009). The largest contributions come from impoundment in reservoirs and groundwater withdrawal. Over the past half-century, storage in tens of thousands of reservoirs has offset some of the sea level rise that would otherwise have occurred. Chao et al. (2008) estimated that the nearly 30,000 reservoirs built during the 20th century resulted in nominal reservoir storage up to 2007 equivalent to ~23 mm of sea level fall (mostly since 1940), with a stabilization in recent years. Chao et al. further assumed that the reservoirs were 85% full, and by including seepage into groundwater as estimated from a model, they obtained a total of 30 mm of sea level fall (equivalent to a rate of sea level fall of 0.55 mm yr^{-1} from 1950 to 2000). Their seepage estimate was argued to be unrealistically large, however, because it assumes aquifers are infinite and have no interfering boundary conditions (Lettenmaier and Milly, 2009; Konikow, 2013). Chao et al. (2008) argued that sedimentation of reservoirs does not reduce their sea level contribution, but their argument is disputed (Gregory et al., 2013b). Lettenmaier and Milly (2009) suggested a loss of capacity due to sedimentation at 1% yr^{-1}. Given the uncertainty about them, neither the seepage nor the effect of sedimentation is included in the budget (Section 13.3.6). Here the (negative) GMSL contribution from reservoir storage is estimated as 85% [70 to 100%] of the nominal capacity (with the lower limit coming from Pokhrel et al. (2012)).

Konikow (2011) estimated that human-induced groundwater depletion contributed 0.26 ± 0.07 mm yr^{-1} to GMSL rise over 1971–2008 and 0.34 ± 0.07 mm yr^{-1} over 1993–2008 (based mostly on observational methods), whereas Wada et al. (2012) estimated values of 0.42 ± 0.08 mm yr^{-1} over 1971–2008 and 0.54 ± 0.09 mm yr^{-1} over 1993–2008 (based on modelling of water fluxes). The average of these two series with the difference as a measure of the uncertainty is used in the sea level budget (Section 13.3.6). Pokhrel et al. (2012) estimated a larger groundwater depletion, but Konikow (2013) (disputed by Pohkrel et al. (2013)) argued that their underlying assumptions of defining depletion as equivalent to groundwater use, and allowing unlimited extraction to meet water demand, led to substantial overestimates of depletion.

In summary, climate-related changes in water and snow storage on land do not show significant long-term trends for the recent decades. However, direct human interventions in land water storage (reservoir impoundment and groundwater depletion) have each contributed at least several tenths of mm yr^{-1} of sea level change (Figure 13.4, Table 13.1). Reservoir impoundment exceeded groundwater depletion for the majority of the 20th century but groundwater depletion has increased and now exceeds current rates of impoundment, contributing to an increased rate of GMSL rise. The net contribution for the 20th century is estimated by adding the average of the two groundwater depletion estimates to the reservoir storage term (Figure 13.4c). The trends are -0.11 [-0.16 to -0.06] mm yr^{-1} for 1901-1990, 0.12 [0.03 to 0.22] mm yr^{-1} for 1971 to 2010 and 0.38 [0.26 to 0.49] mm yr^{-1} for 1993 to 2010 (Table 13.1).

13.3.5 Ocean Mass Observations from Gravity Recovery and Climate Experiment

As discussed in Chapter 3, it has been possible to directly estimate changes in ocean mass using satellite gravity data from GRACE since 2002 (Chambers et al., 2004, 2010; Chambers, 2006; Cazenave et al., 2009; Leuliette and Miller, 2009; Llovel et al., 2010). These measurements represent the sum of total land ice plus land water components, and thus provide an independent assessment of these contributions. However, GRACE is also sensitive to mass redistribution associated with GIA and requires that this effect (on the order of –0.7 to –1.3 mm yr^{-1} when averaged over the ocean domain) (Paulson et al., 2007; Peltier, 2009; Chambers et al., 2010; Tamisiea, 2011) be removed before estimating the ocean-mass component. Most recent estimates (Leuliette and Willis, 2011; von Schuckmann and Le Traon, 2011) report a global mean ocean mass increase of 1.8 [1.4 to 2.2] mm yr^{-1} over 2003–2012 after correcting for the GIA component. The associated error results from the low signal-to-noise ratio over the ocean domain and uncertainty in the model-based GIA correction (Quinn and Ponte, 2010).

Chapter 3 notes that, in terms of global averages, the sum of the contribution to GMSL due to change in global ocean mass (the barystatic contribution), measured by GRACE, and the contribution due to global ocean thermal expansion (the thermosteric contribution), measured by the Argo Project, agrees within uncertainties with the GMSL change observed by satellite altimetry (Leuliette and Willis, 2011; von Schuckmann and Le Traon, 2011), although there is still a missing contribution from expansion in the deep ocean below 2000 m. These data sets have allowed an investigation of the cause of variability in sea level over the last few years (Figure 13.6). In particular, Boening et al. (2012) concluded that the decrease in GMSL over 2010–2011 followed by a rapid increase since 2011 was related to the 2011 La Niña event, whereby changes in land/ocean precipitation patterns caused a temporary increase in water storage on the land (and corresponding decrease in GMSL) during the La Niña event, especially in Australia, northern South America and southeast Asia (Boening et al., 2012).

13.3.6 Budget of Global Mean Sea Level Rise

Drawing on Sections 13.3.1 to 13.3.5, the budget of GMSL rise (Table 13.1, Figure 13.7) is analysed using models and observations for the periods 1901–1990 (the 20th century, excluding the period after 1990 when ice-sheet contributions to GMSL rise have increased; Sections 4.4 and 13.3.3.1), since 1971 (when significantly more ocean data became available and systematic glacier reconstructions began), and since 1993 (when precise satellite sea level altimetry began). The 2005–2010 period when Argo and GRACE data are available is short and strongly affected by interannual climate variability, as discussed in the previous subsection (Section 13.3.5 and Figure 13.6). Such variability is not externally forced and is therefore not expected to be reproduced in AOGCM historical experiments. For the contribution from land water storage (Figure 13.4c) we use the estimated effect of human intervention and neglect effects from climate-related variation, which

13

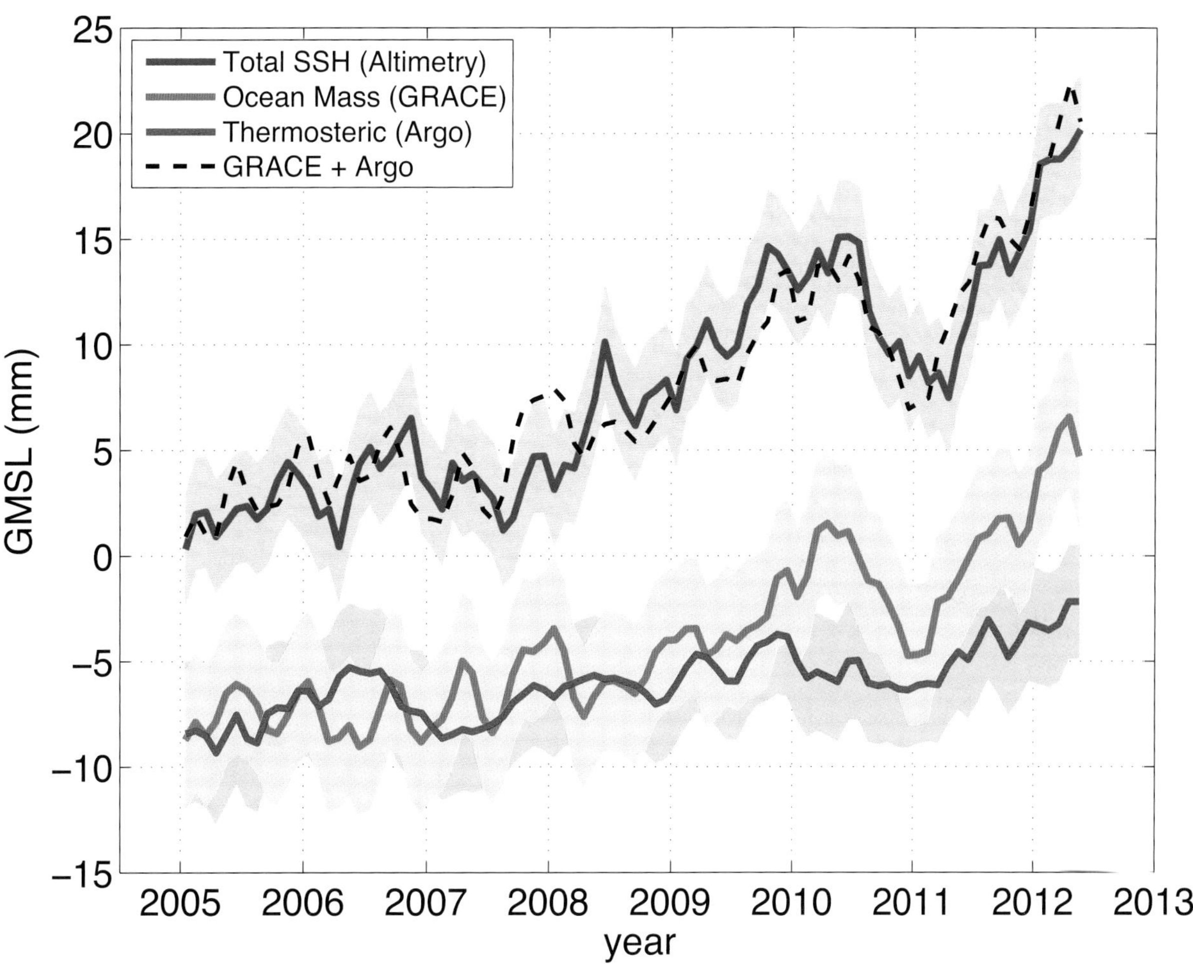

Figure 13.6 | Global mean sea level from altimetry from 2005 to 2012 (blue line). Ocean mass changes are shown in green (as measured by Gravity Recovery and Climate Experiment (GRACE)) and thermosteric sea level changes (as measured by the Argo Project) are shown in red. The black line shows the sum of the ocean mass and thermosteric contributions. (Updated from Boening et al., 2012)

are unimportant on multi-decadal time scales (Section 13.3.4). Contributions due to runoff from thawed permafrost, change in atmospheric moisture content, and sedimentation in the ocean are not considered in the budget because they are negligible compared with observed GMSL rise and the uncertainties.

For 1993–2010, allowing for uncertainties, the observed GMSL rise is consistent with the sum of the observationally estimated contributions (*high confidence*) (Table 13.1, Figure 13.7e). The two largest terms are ocean thermal expansion (accounting for about 35% of the observed GMSL rise) and glacier mass loss (accounting for a further 25%, not including that from Greenland and Antarctica). Observations indicate an increased ice-sheet contribution over the last two decades (Sections 4.4.2.2, 4.4.2.3 and 13.3.3.1) (Shepherd et al., 2012). The closure of the observational budget since 1993, within uncertainties, represents a significant advance since the AR4 in physical understanding of the causes of past GMSL change, and provides an improved basis for critical evaluation of models of these contributions in order to assess their reliability for making projections.

The observational budget cannot be rigorously assessed for 1901–1990 or 1971–2010 because there is insufficient observational information to estimate ice-sheet contributions with *high confidence* before the 1990s, and ocean data sampling is too sparse to permit an estimate of global mean thermal expansion before the 1970s. However, a closed observational GMSL budget since the 1970s can be demonstrated with reasonable estimates of ice-sheet contributions (Church et al., 2011a; Moore et al., 2011) (Table 13.1, Figure 13.7). For 1971–2010, the observed contributions from thermal expansion and mass loss from glaciers (not including those in Antarctica) alone explain about 75% of the observed GMSL (*high confidence*).

AOGCM-based estimates of thermal expansion, which agree well with observations since 1971, observational estimates of the glacier contribution, and the estimated change in land water storage (Figure 13.4c), which is relatively small, can all be made from the start of the 20th century (Sections 13.3.1.2, 13.3.2.2 and 13.3.4, Table 13.1). Model estimates of Greenland ice-sheet SMB changes give an uncertain but relatively small contribution during most of the 20th century, increasing since the early 1990s (Section 13.3.3.2). There could be a small constant contribution from the Antarctic ice sheet (Huybrechts et al., 2011; Gregory et al., 2013b) due to long-term adjustment to climate change in previous millennia. Any secular rate of sea level rise in the late Holocene was small (order of few tenths mm yr^{-1}) (Section

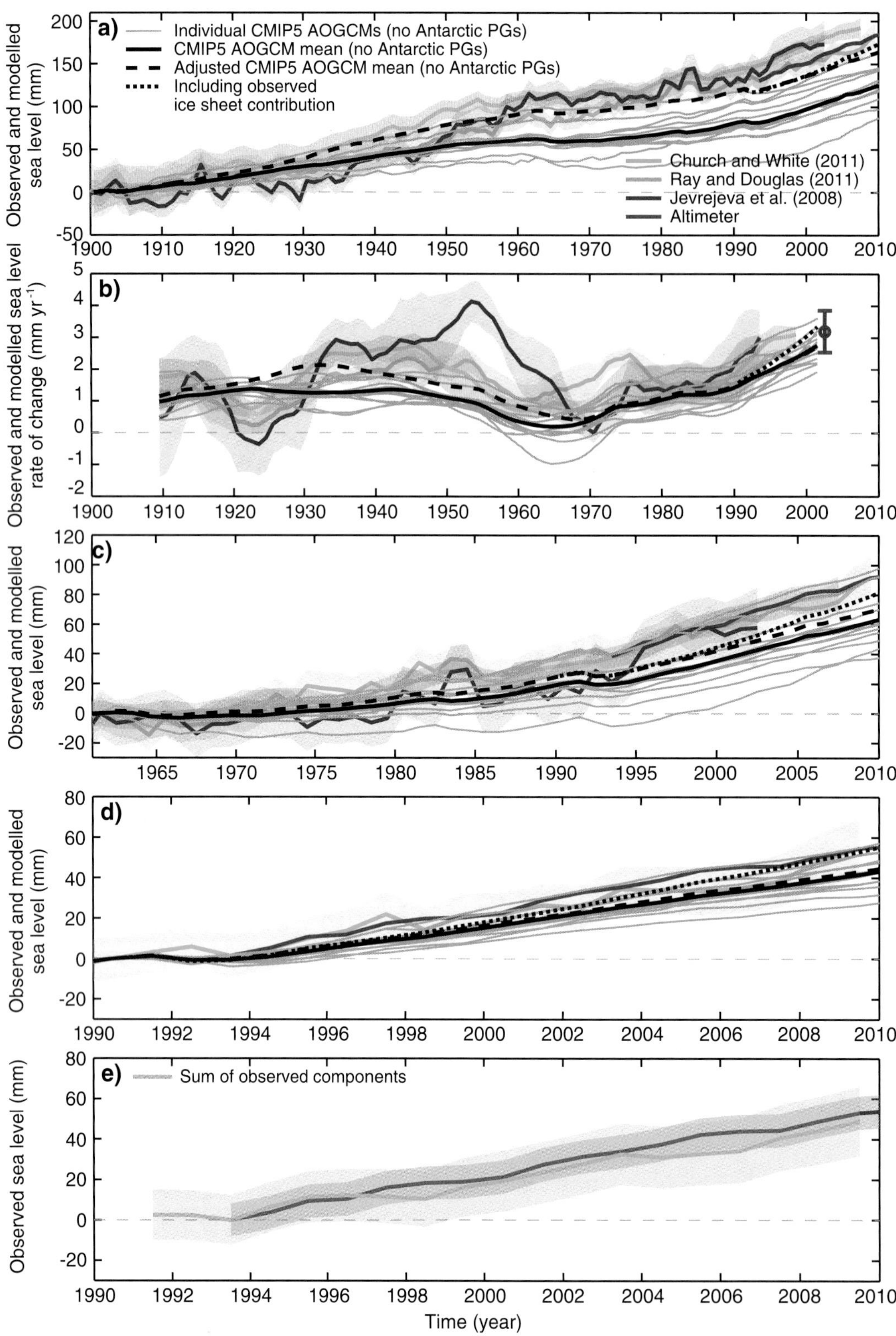

Figure 13.7 | (a) The observed and modelled sea level for 1900 to 2010. (b) The rates of sea level change for the same period, with the satellite altimeter data shown as a red dot for the rate. (c) The observed and modelled sea level for 1961 to 2010. (d) The observed and modelled sea level for 1990 to 2010. Panel (e) compares the sum of the observed contributions (orange) and the observed sea level from the satellite altimeter data (red). The estimates of global mean sea level are from Jevrejeva et al. (2008), Church and White (2011), and Ray and Douglas (2011), with the shading indicating the uncertainty estimates (two standard deviations). The satellite altimeter data since 1993 are shown in red. The grey lines in panels (a)-(d) are the sums of the contributions from modelled ocean thermal expansion and glaciers (excluding glaciers peripheral to the Antarctic ice sheet; from Marzeion et al., 2012a), plus changes in land-water storage (see Figure 13.4). The black line is the mean of the grey lines plus a correction of thermal expansion for the omission of volcanic forcing in the AOGCM control experiments (see Section 13.3.1.2). The dashed black line (adjusted model mean) is the sum of the corrected model mean thermal expansion, the change in land water storage, the Marzeion et al. (2012a) glacier estimate using observed (rather than modelled) climate (see Figure 13.4), and an illustrative long-term ice-sheet contribution (of 0.1 mm yr^{-1}). The dotted black line is the adjusted model mean but now including the observed ice-sheet contributions, which begin in 1993. Because the observational ice-sheet estimates include the glaciers peripheral to the Greenland and Antarctic ice sheets (from Section 4.4), the contribution from glaciers to the adjusted model mean excludes the peripheral glaciers to avoid double counting. (Figure and caption updated from Church et al., 2013).

13.2.1.4), probably less than 0.2 mm yr^{-1} (see discussion in Gregory et al., (2013b). Including these ice-sheet contributions (but omitting Antarctic SMB variations, for which no observationally based information for the ice sheet as a whole is available for the majority of the 20th century), GMSL rise during the 20th century can be accounted for within uncertainties, including the observation that the linear trend of GMSL rise during the last 50 years is little larger than for the 20th century, despite the increasing anthropogenic forcing (Gregory et al., 2013b). Model-based attribution of sea level change to RFs is discussed in Section 10.4.3.

The sum of CMIP5 AOGCM thermal expansion (Section 13.3.1.2), glacier model results with CMIP5 AOGCM input (not including glaciers in Antarctica; Section 13.3.2.2; Marzeion et al., (2012a)), and anthropogenic intervention in land water storage (Section 13.3.4) accounts for about 65% of the observed rate of GMSL rise for 1901–1990, and 90% for 1971–2010 and 1993–2010 (*high confidence*) (Table 13.1; Figure 13.7). In all periods, the residual is small enough to be attributed to the ice sheets (Section 13.3.3.2).

The unusually warm conditions in the Arctic during the 1930s (Chylek et al., 2004), which are attributed to unforced climate variability (Delworth and Knutson, 2000) and are therefore not expected to be simulated by AOGCMs, *likely* produced a greater mass loss by glaciers in high northern latitudes (Section 13.3.2.2). The difference between the glacier mass loss calculated with the Marzeion et al. (2012a) model when it is forced with observed climate rather than AOGCM simulated climate (the purple and blue curves in Figure 13.4b) is an estimate of this effect.

If the glacier model results for observational input are used (Marzeion et al. (2012a), not including glaciers in Antarctica) and an illustrative value of 0.1 mm yr^{-1} is included for a long-term Antarctic contribution, the model mean is within 20% of the observed GMSL rise for the 20th century (Figure 13.7a,c, dashed line), and 10% since 1993 (Figure 13.7d, dashed line; Church et al. (2013)). When the observed ice-sheet contributions since 1992 are included as well, the sum is almost equivalent to the observed rise (dotted line in Figure 13.7). Both observations and models have a maximum rate of rise in the 1930–1950 period, a minimum rate in the 1960s and a maximum rate over the last two decades (Figure 13.7b). This agreement provides evidence that the larger rate of rise since 1990, with a significant component of ocean thermal expansion (Figure 13.4d), results from increased RF (both natural and anthropogenic) and increased ice-sheet discharge, rather than a natural oscillation (*medium confidence*) (Church et al., 2013).

In summary, the evidence now available gives a clearer account of observed GMSL change than in previous IPCC assessments, in two respects. First, reasonable agreement can be demonstrated throughout the period since 1900 between GMSL rise as observed and as calculated from the sum of contributions. From 1993, all contributions can be estimated from observations; for earlier periods, a combination of models and observations is needed. Second, when both models and observations are available, they are consistent within uncertainties. These two advances give confidence in the 21st century sea level projections. The ice-sheet contributions have the potential to increase substantially due to rapid dynamical change (Sections 13.1.4.1, 13.4.3.2 and 13.4.4.2) but have been relatively small up to the present (Sections 4.4 and 13.3.3.2). Therefore, the closure of the sea level budget to date does not test the reliability of ice-sheet models in projecting future rapid dynamical change; we have only *medium confidence* in these models, on the basis of theoretical and empirical understanding of the relevant processes and observations of changes up to the present (13.4.3.2, 13.4.4.2).

Box 13.1 | The Global Energy Budget

The global energy balance is a fundamental aspect of the Earth's climate system. At the top of the atmosphere (TOA), the boundary of the climate system, the balance involves shortwave radiation received from the Sun, and shortwave radiation reflected and longwave radiation emitted by the Earth (Section 1.2.2). The rate of storage of energy in the Earth system must be equal to the net downward radiative flux at the TOA.

The TOA fluxes (Section 2.3) have been measured by the Earth Radiation Budget Experiment (ERBE) satellites from 1985 to 1999 (Wong et al., 2006) and the Cloud and the Earth's Radiant Energy System (CERES) satellites from March 2000 to the present (Loeb et al., 2009). The TOA radiative flux measurements are highly precise, allowing identification of changes in the Earth's net energy budget from year to year within the ERBE and CERES missions (Kato, 2009; Stackhouse et al., 2010; Loeb et al., 2012), but the absolute calibration of the instruments is not sufficiently accurate to allow determination of the absolute TOA energy flux or to provide continuity across missions (Loeb et al., 2009).

The ocean has stored more than 90% of the increase in energy in the climate system over recent decades (Box 3.1), resulting in ocean thermal expansion and hence sea level rise (Sections 3.7, 9.4 and 13.3.1). Thus the energy and sea level budgets are linked and must be consistent (Church et al., 2011b). This Box focusses on the Earth's global energy budget since 1970 when better global observational data coverage is available. The RFs (from Chapter 8), the global averaged surface temperatures (Hadley Centre/Climate Research Unit gridded surface temperature data set 4 (HadCRUT4) (Morice et al., 2012), and the rate of energy storage are relative to the time mean of 1860 to 1879. Otto et al. (2013) used an energy imbalance over this reference period of 0.08 ± 0.03 W m^{-2}, which is subtracted from the observed energy storage. *(continued on next page)*

13

Box 13.1 (continued)

Since 1970, the effective radiative forcing (ERF) of the climate system has been positive as a result of increased greenhouse gas (GHG) concentrations (well-mixed and short-lived GHGs, tropospheric and stratospheric ozone, and stratospheric water vapour) and a small increase in solar irradiance (Box 13.1, Figure 1a). This positive ERF has been partly compensated by changes in tropospheric aerosols which predominantly reflect sunlight and modify cloud properties and structure in ways that tend to reinforce the negative ERF, although black carbon produces positive forcing. Explosive volcanic eruptions (such as El Chichón in Mexico in 1982 and Mt. Pinatubo in the Philippines in 1991) can inject sulphur dioxide into the stratosphere, giving rise to stratospheric aerosol, which persists for several years. This reflects some of the incoming solar radiation, and thus gives a further negative forcing. Changes in surface albedo from land-use change have also led to a greater reflection of shortwave radiation back to space and hence a negative forcing. Since 1970, the net ERF of the climate system (including black carbon on snow and combined contrails and contrail-induced cirrus, not shown) has increased (Chapter 8), resulting in a cumulative total energy inflow (Box 13.1, Figure 1a). From 1971 to 2010, the total energy inflow (relative to the reference period 1860-1879) is estimated to be 790 [105 to 1,370] ZJ (1 ZJ = 10^{21} J).

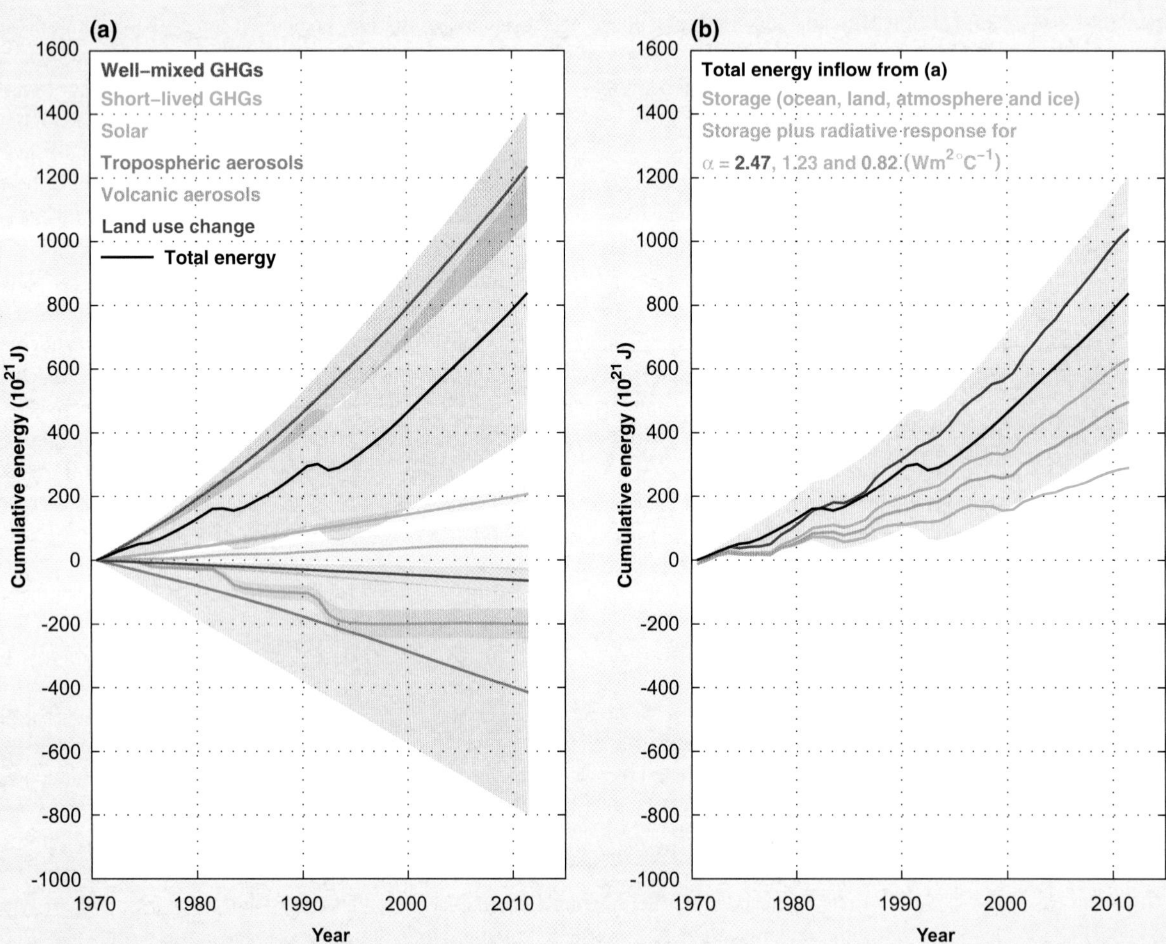

Box 13.1, Figure 1 | The Earth's energy budget from 1970 through 2011. (a) The cumulative energy flux into the Earth system from changes in well-mixed and short-lived greenhouse gases, solar forcing, changes in tropospheric aerosol forcing, volcanic forcing and surface albedo (relative to 1860–1879) are shown by the coloured lines and these are added to give the cumulative energy inflow (black; including black carbon on snow and combined contrails and contrail-induced cirrus, not shown separately). (b) The cumulative total energy inflow from (a, black) is balanced by the sum of the warming of the Earth system (blue; energy absorbed in warming the ocean, the atmosphere and the land and in the melting of ice) and an increase in outgoing radiation inferred from changes in the global averaged surface temperature. The sum of these two terms is given for a climate feedback parameter α of 0.82, 1.23 and 2.47 W m^{-2} °C^{-1} (corresponding to an equilibrium climate sensitivity of 4.5, 3.0 and 1.5°C, respectively). The energy budget would be closed for a particular value of α if that line coincided with the total energy inflow. For clarity, all uncertainties (shading) shown are for a *likely* range.

If the ERF were fixed, the climate system would eventually warm sufficiently that the radiative response would balance the ERF, and there would be zero net heat flux into the system. As the ERF is increasing, the ocean's large capacity to store heat means the climate system is not in equilibrium (Hansen et al., 2005), and continues to store energy (Box 3.1 and Box 13.1, Figure 1b). This storage provides

(continued on next page)

13

Box 13.1 (continued)

strong evidence of a changing climate. The majority of this additional heat is in the upper 700 m of the ocean but there is also warming in the deep and abyssal ocean (Box 3.1). The associated thermal expansion of the ocean has contributed about 40% of the observed sea level rise since 1971 (Sections 13.3.1, 13.3.6; Church et al., (2011b)). A small amount of additional heat has been used to warm the continents, warm and melt glacial and sea ice, and warm the atmosphere. The estimated increase in energy in the Earth system between 1971 and 2010 is 274 [196 to 351] ZJ (Box 3.1).

As the climate system warms, energy is lost to space through increased outgoing radiation. This radiative response by the system is predominantly due to increased thermal grey-body radiation emitted by the atmosphere and surface, but is modified by climate feedbacks, such as changes in water vapour, surface albedo and clouds, which affect both outgoing longwave and reflected shortwave radiation. Following Murphy et al. (2009), Box 13.1, Figure 1b relates the cumulative total energy inflow to the Earth system to the change in energy storage and the cumulative outgoing radiation. Calculation of the latter is based on the observed globally averaged surface temperature change ΔT relative to a reference temperature for which the Earth system would be in radiative balance. This temperature change is multiplied by the climate feedback parameter α, which in turn is related to the equilibrium climate sensitivity. For equilibrium climate sensitivities of 4.5°C, 3.0°C to 1.5°C (Box 12.2) and an ERF for a doubled CO_2 concentration of 3.7 ± 0.74 W m^{-2} (Sections 8.1, 8.3), the corresponding estimates of the climate feedback parameter α are 0.82, 1.23 and 2.47 W m^{-2} °C^{-1}.

In addition to these forced variations in the Earth's energy budget, there is also internal variability on decadal time scales. Observations and models indicate that because of the comparatively small heat capacity of the atmosphere, a decade of steady or even decreasing surface temperature can occur in a warming world (Easterling and Wehner, 2009; Palmer et al., 2011). General Circulation Model simulations indicate that these periods are associated with a transfer of heat from the upper to the deeper ocean, of order 0.1 W m^{-2} (Katsman and van Oldenborgh, 2011; Meehl et al., 2011), with a near steady (Meehl et al., 2011) or an increased radiation to space (Katsman and van Oldenborgh, 2011), again of order 0.1 W m^{-2}. Although these natural fluctuations represent a large amount of heat, they are significantly smaller than the anthropogenic forcing of the Earth's energy budget (Huber and Knutti, 2012), particularly when looking at time scales of several decades or more (Santer et al., 2011).

These independent estimates of ERF, observed heat storage, and surface warming combine to give an energy budget for the Earth that is consistent with the assessed *likely* range of climate sensitivity (1.5°C to 4.5°C; Box 12.2) to within estimated uncertainties (*high confidence*). Quantification of the terms in the Earth's energy budget and verification that these terms balance over recent decades provides strong evidence for our understanding of anthropogenic climate change. Changes in the Earth's energy storage are a powerful observation for the detection and attribution of climate change (Section 10.3) (Gleckler et al., 2012; Huber and Knutti, 2012).

13.4 Projected Contributions to Global Mean Sea Level

13.4.1 Ocean Heat Uptake and Thermal Expansion

More than 90% of the net energy increase of the climate system on multiannual time scales is stored in the ocean (Box 3.1). GMSL rise due to thermal expansion is approximately proportional to the increase in ocean heat content. The constant of proportionality is 0.11 ± 0.01 m per 10^{24} J for the ensemble of CMIP5 models (Kuhlbrodt and Gregory, 2012); it depends on the vertical and latitudinal distribution of warming in the ocean, because the expansion of sea water per degree Celsius of warming is greater at higher temperature and higher pressure (Russell et al., 2000; Hallberg et al., 2012; Körper et al., 2013; Perrette et al., 2013).

For the early decades of the 21st century, the upper ocean dominates the ocean heat uptake, and ocean heat content rises roughly linearly with global mean surface air temperature (SAT) change (Pardaens et al., 2011b; Körper et al., 2013). On multi-decadal time scales under scenarios of steadily increasing RF, the rate of increase of ocean heat content is approximately proportional to the global mean SAT change from equilibrium (Gregory, 2000; Meehl et al., 2007; Rahmstorf, 2007a; Gregory and Forster, 2008; Katsman et al., 2008; Schwartz, 2012), with the constant of proportionality (in W m^{-2} °C^{-1}) being the ocean heat uptake efficiency κ.

The ocean heat uptake efficiency quantifies the effect of ocean heat uptake on moderating time-dependent climate change; neglecting the small fraction of heat stored elsewhere in the climate system, the surface warming can be approximated as $F/(\alpha+\kappa)$, where F is the RF and α is the climate feedback parameter (Raper et al., 2002), and hence the rate of ocean heat uptake is approximately $\kappa F/(\alpha+\kappa)$. In CMIP3 and CMIP5, the model spread in projections of surface warming is dominated by the spread in F and α, but the spread in κ accounts for a substantial part of the spread in projections of ocean heat uptake (Dufresne and Bony, 2008; Gregory and Forster, 2008; Knutti and Tomassini, 2008; Geoffroy et al., 2012; Sriver et al., 2012; Forster et al., 2013).

The spread in κ relates to differences among models in heat-transport processes within the ocean. The warming spreads downwards from the surface over time, and the greatest increases in projected ocean heat content occur where the warming penetrates most deeply, in the Southern Ocean and the North Atlantic (Figure 12.12; Section 12.4.7.1) (Kuhlbrodt and Gregory, 2012). Changes in convection and the large-scale vertical circulation are particularly important to heat uptake in the North Atlantic (Banks and Gregory, 2006; Rugenstein et al., 2013). Heat is also transported vertically by eddies, especially in the Southern Ocean, and by turbulent mixing. These processes are parameterized in models when they occur at unresolved scales. Observed ocean heat uptake has been used in conjunction with observed global SAT change to constrain the ocean effective thermal diffusivity representing all unresolved vertical transports in simple climate models and EMICs (Forest et al., 2008; Knutti and Tomassini, 2008; Marčelja, 2010; Sokolov et al., 2010). The simulated ocean vertical temperature profile and the depth of penetration of the warming in AOGCMs have also been evaluated by comparison with observations, and both bear a relationship to κ (Hallberg et al., 2012; Kuhlbrodt and Gregory, 2012). Such comparisons suggest that model projections might be biased towards overestimating ocean heat uptake and thermal expansion for a given surface warming (Sections 9.4.2.2, 10.8.3 and 13.3.1.2). The physical causes of this tendency are unclear. Although the simulated vertical temperature profile is affected by the model representation of vertical heat transport processes, Brierley et al. (2010) found only a small effect on κ from variation of model parameters that influence interior heat transport.

Because the ocean integrates the surface heat flux, thermal expansion projections following different scenarios do not significantly diverge for several decades. Scenarios assuming strong mitigation of GHG emissions begin to show a reduced rate of thermal expansion beyond about 2040; the amount by 2100 is about one third less than in a non-mitigation scenario (Washington et al., 2009; Pardaens et al., 2011b; Körper et al., 2013), and half as much in RCP2.6 as in RCP8.5 (Yin, 2012) (Section 13.5.1). The integrating effect means that thermal expansion depends not only on the cumulative total, but also on the pathway of CO_2 emissions; reducing emissions earlier rather than later, for the same cumulative total, leads to a larger mitigation of sea level rise due to thermal expansion (Zickfeld et al., 2012; Bouttes et al., 2013). The integrating effect also means that annual time series of global ocean thermal expansion show less interannual variability than time series of global SAT. For the present assessment of GMSL rise, projections of ocean heat uptake and thermal expansion up to 2100 have been derived from the CMIP5 AOGCMs (Yin, 2012). Methods are described in Section 13.5.1 and the Supplementary Material and the results for ocean heat uptake are shown in Figure 13.8, and for thermal expansion in Table 13.5 and Figures 13.10 and 13.11.

Ocean heat uptake efficiency is not constant on time scales of many decades or in scenarios of stable or decreasing RF (Rahmstorf, 2007a; Schewe et al., 2011; Bouttes et al., 2013). A good representation of AOGCM behaviour is obtained by distinguishing a shallow layer, which is associated with surface temperature variations on decadal time scales, from a deep layer, which has the majority of the heat capacity (Hansen et al., 1985; Knutti et al., 2008; Held et al., 2010; Olivié et al., 2012; Schwartz, 2012; Bouttes et al., 2013; Geoffroy et al., 2013). Ocean heat uptake and thermal expansion take place not only while atmospheric GHG concentrations are rising, but continue for many centuries to millennia after stabilization of RF, at a rate which declines on a centennial time scale (Stouffer, 2004; Meehl et al., 2005; 2007; Solomon et al., 2009; Hansen et al., 2011; Meehl et al., 2012; Schwartz, 2012; Bouttes et al., 2013; Li et al., 2013; Zickfeld et al., 2013). This is because the time scale for warming the deep ocean is much longer than for the shallow ocean (Gregory, 2000; Held et al., 2010).

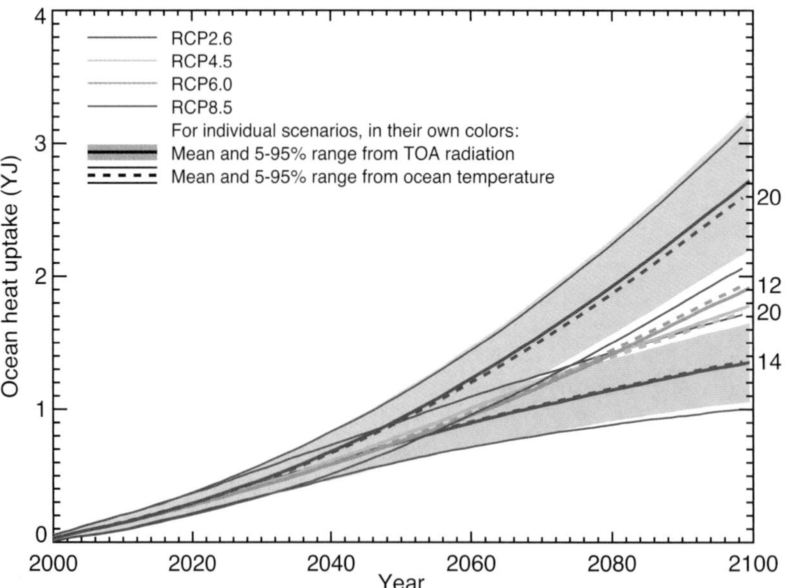

Figure 13.8 | Heat uptake by the climate system during the 21st century relative to 1986–2005 projected by CMIP5 Atmosphere–Ocean General Circulation Models (AOGCMs) under RCP scenarios (1 YJ = 10^{24} J). The heat uptake is diagnosed by two different methods. The thick solid lines, and the coloured ranges for RCP2.6 and RCP8.5, are the time- and global integral of the net downward radiative flux perturbation at the top of the atmosphere, from the 21 AOGCMs used to make the global mean sea level projections (in some cases estimated from other scenarios, as described in the Supplementary Material). The broken solid lines, and the thin solid lines delimiting ranges for RCP2.6 and RCP8.5, are the global volume integral of ocean temperature change, in a smaller and different set of AOGCMs for each scenario. The difference between the two diagnoses is due partly to the different sets of models (which is a consequence of diagnostics available in the CMIP5 data set), and partly to heat uptake in other parts of the simulated climate system than the ocean water. In both methods, climate drift in the pre-industrial control run has been subtracted.

The rate and the stabilization time scale for thermal expansion depend on the GHG stabilization level. For the highest scenario (RCP8.5), GMSL rise due to thermal expansion can exceed 2 m above the pre-industrial level by the year 2500 (Section 12.5.2, Figure 12.44, Figure 13.14a), and is still rising at that time. Changes in ocean circulation, particularly due to a reduction in deep water formation, can also have a large effect on global ocean heat uptake, and may relate nonlinearly to global surface warming (Levermann et al., 2005; Fluckiger et al., 2006; Vellinga and Wood, 2008). As the rate of ocean heat uptake decreases, the surface warming approaches the level determined by the equilibrium climate sensitivity.

On a multi-millennial time scale, the range from Earth System Models of Intermediate Complexity suggests that thermal expansion contributes between 0.20 to 0.63 m per degree Celsius of global mean temperature increase (Meehl et al., 2007; Zickfeld et al., 2013) (Section 12.5.2 and Figure 13.14a). The median of the six models of 0.42 m $°C^{-1}$ is consistent with a thermal expansion of 0.38 m $°C^{-1}$ that would result from a uniform increase in ocean temperature from the presently observed temperature and salinity distribution (Levitus et al., 2009). Uncertainty arises due to the different spatial distribution of the warming in models and the dependence of the expansion on local temperature and salinity.

13.4.2 Glaciers

The 21st century sea level contribution from glaciers presented in the AR4 assessment ranged from 0.06 to 0.15 m SLE by 2100 across a range of scenarios (Meehl et al., 2007). The Randolph Glacier Inventory (RGI) (Arendt et al., 2012) has improved projections of glacier contribution to sea level rise by providing the first globally complete accounting of glacier location, area, and area-elevation distribution (hypsometry). Several analyses of scenario-dependent SMB glacier projections (referred to here as process-based models) have been produced using the RGI, including Marzeion et al. (2012a), Giesen and Oerlemans (2013), and Radić et al. (2013). The Marzieon and Radić approaches each used different suites of CMIP5 AOGCM models to calculate SMB terms from RCP forcings, and the model by Slangen and van de Wal (2011) was used to calculate SMB terms from RCP forcings (Supplementary Material 13.SM.1). Giesen and Oerlemans (2013) used CRU forcing but calculated SMB from three different combinations of variations in modelled temperature, precipitation, and atmospheric transmissivity. Only their results for varying temperature are shown here. Machguth et al. (2013) is also included in Table 13.3, but this projection represents changes in Greenland peripheral glaciers only, and is not included in the global glacier summaries. Although these details differ among the models, all share a generally common time-evolving structure, with SMB initially determined by model-generated climate forcing applied to a subset of global glaciers, the ensuing volume change converted to area change via volume-area scaling, and this result upscaled to a new global distribution and hypsometry to create initial conditions for the subsequent time step. These methods are described further in Section13.5.1 and in the Supplementary Material. Related results are shown in Table 13.5 and Figures 13.10 and 13.11.

Although the peripheral glaciers surrounding the ice sheets are included with the ice sheets in assessment of present-day changes (Table 13.1), future projections should ideally assess the peripheral glaciers separately, as these are too small and dynamically responsive to be modelled adequately with coarse-grid, non-dynamic ice-sheet SMB models. The peripheral glaciers surrounding both the Greenland and Antarctic ice sheets are thus included in the process-based models described above, but for projections shown in Table 13.5, the Antarctic peripheral glaciers are included with the Antarctic ice sheet whereas the Greenland peripheral glaciers are included with the remaining world's glaciers. Projected losses from glaciers peripheral to both ice sheets are listed separately in Table 13.3.

Several glacier loss projections derived from model types other than process-based models have been published since 2007; their projections range from 0.08 to 0.39 m SLE by 2100 (Table 13.3). These used methods of projecting future losses from glaciers developed in response to the absence of a global compilation of glacier observations after 2005 and the absence of a globally complete glacier inventory to provide geographic boundary conditions for conventional modelling. These methods include extrapolation from observed rates (Meier et al., 2007), semi-empirical methods applied to sea level change components (Jevrejeva et al., 2012b), kinematic (or 'limit seeking') projections (Pfeffer et al., 2008), and power-law scaling estimates based on re-establishing equilibrium accumulation-area ratios (AARs) from initial non-equilibrium AARs (Bahr et al., 2009). Strengths of these approaches include the fact that observations used to calibrate extrapolation and semi-empirical projection partially account for future dynamically forced losses, that semi-empirical methods use modelled future forcings as guidance for projections, and that AAR equilibration has strong physical and theoretical underpinnings and gives generalized but robust projections. These strengths partially offset the weaknesses of these models, which include, in the case of extrapolation and semi-empirical projection, an assumption of statistical stationarity that may not be valid, while the AAR equilibration approach gives only a final steady state value, so that rates or cumulative losses at any intermediate time must be estimated by area-response time scaling. However, these alternate methods are valuable because of their construction on fundamental and robust principles together with their use of the limited available information to produce projections that, although imprecise, are transparent, and require less detailed input information or knowledge of details of complex processes in comparison to process-based models.

Published results from process-based models are shown in Table 13.3. Glacier contributions at 2100, expressed as SLE, range between 0.04 and 0.11 m for Special Report on Emission Scenarios (SRES) A1B, 0.07 and 0.17 m for RCP2.6, between 0.07 and 0.20 m for RCP4.5, between 0.07 and 0.20 m for RCP6.0, and between 0.12 and 0.26 m for RCP8.5.

The projections derived from alternative models are also shown in Table 13.3; the mean and range of these models listed here is 0.24 [0.08 to 0.39] m SLE, consistent with the process-based models. Results from the process-based models, plotted as time series and grouped by forcing scenario, are shown in Figure 13.9. See Table 13.3 for specific start/end dates for each projection.

Unresolved uncertainties in the projection of glacier contributions to sea level rise include the potential for near-term dynamic response from marine-terminating glaciers and interception of terrestrial runoff. Of the about 734,000 km² of global glacier area exclusive of that peripheral to the Greenland and Antarctic ice sheets, 280,500 km² (38%) drains through marine-terminating outlets (Gardner et al., 2013). Although the long-term potential for dynamic discharge from glaciers (as opposed to ice sheets) is limited by their small total volume, dynamic losses may be an important component of total sea level rise on the decade-to-century scale. In Alaska, Columbia Glacier lost 7.65 Gt yr^{-1} between 1996 and 2007, with 94% of that loss coming from rapid tidewater retreat (Rasmussen et al., 2011); the loss from this single 1000 km² glacier is 1.3% of the global cryospheric component of sea level rise during 1993–2010 (Table 13.1) and 0.7% of total sea level rise. The observations required to estimate the potential for similar dynamic response from other glacier regions do not exist at this time, but the dynamic contribution could be large on the century time scale. If the basin-wide thinning rate observed at Columbia Glacier

Table 13.3 | Twenty-first century sea level rise projections for global glaciers, from process-based surface mass balance models, and from alternate model strategies. Dates for beginning and end of model period are as shown; mean and 5% to 95% confidence sea level equivalents are shown in metres. Process-based models all use variations on Atmosphere–Ocean General Circulation Model (AOGCM) mass balance forcing applied to inventoried glacier hypsometries on a subset of global glaciers and upscaling by power-law techniques to the global total. Calving and rapid dynamic response are not included in any of the models except for Jevrejeva et al. (2012b), where calving losses are present to a limited degree in input data, and NRC (2012), where calving is explicitly included in future losses. Other model details are discussed in the text.

				Contribution to Global Mean Sea Level Rise (SLR)		Peripheral Glacier (PG) Contribution	
Reference	**Model**	**Starting Date**	**End Date**	**Projected SLR (m) from Glaciers *except* Antarctic PGs**		**Greenland Ice Sheet PG (m)**	**Antarctic Ice Sheet PG (m)**
Process-based Surface Mass Balance (SMB) Models				**Mean**	**[5 to 95%] confidence**	**5 to 95% confidence**	**5 to 95% confidence**
Scenario RCP2.6							
Marzeion et al. (2012a)		1986–2005 Mean	2099	0.12	[0.07–0.17]	0.007–0.02	0.02–0.04
Slangen and van de Wal (2011)		2000	2099	0.10	[0.07–0.13]	0.004–0.007	0.02–0.03
Scenario RCP4.5							
Marzeion et al. (2012a)		1986–2005 Mean	2099	0.14	[0.08–0.20]	0.009–0.022	0.02–0.04
Radic et al. (2013)		2006	2099	0.13	[0.07–0.20]	0.0–0.024	0.02
Slangen and van de Wal (2011)		2000	2099	0.12	[0.07–0.17]	0.005–0.01	0.03–0.04
Scenario RCP6.0							
Marzeion et al. (2012a)		1986–2005 Mean	2099	0.15	[0.09–0.20]	0.01–0.022	0.02–0.04
Slangen and van de Wal (2011)		2000	2099	0.14	[0.07–0.20]	0.006–0.01	0.04
Scenario RCP8.5							
Marzeion et al. (2012a)		1986–2005 Mean	2099	0.18	[0.12–0.25]	0.015–0.025	0.02–0.05
Radic et al. (2013)		2006	2099	0.19	[0.12–0.26]	0.009–0.031	0.02–0.03
Slangen and van de Wal (2011)		2000	2099	0.18	[0.12–0.24]	0.008–0.015	0.04–0.06
Scenario A1B							
Giesen and Oerlemans (2013)		2012	2099	0.08	[0.04–0.11]	0.004–0.021	0.01–0.04
Scenario A1B and RCP4.5							
Machguth et al. (2013)[a]		2000	2098			0.006–0.011	
Alternate Models							
Meier et al. (2007)	Extrapolation with fixed rate	2006	2100	0.3	[0.08–0.13]		
	Extrapolation with fixed acceleration	2006	2100	0.24	[0.11–0.37]		
Pfeffer et al. (2008)	Low-range projection	2007	2100	0.17			
	High-range projection	2007	2100	0.24			
Bahr et al. (2009)	AAR fixed at present values	Find equilibrium value		0.18	[0.15–0.27]		
	AAR declines at current rate	Find equilibrium value		0.38	[0.35–0.39]		
National Research Council (2012)	Generalized linear model extrapolation, variable rate	2010	2100	0.14	[0.13–0.16]		
Jeverjeva et al. (2012b)	Semi-empirical projection of components of sea level rise, forced by radiation	2009	2100	0.26			

Notes

[a] This projection represents changes in Greenland peripheral glaciers only, and is not included in the global glacier summaries.

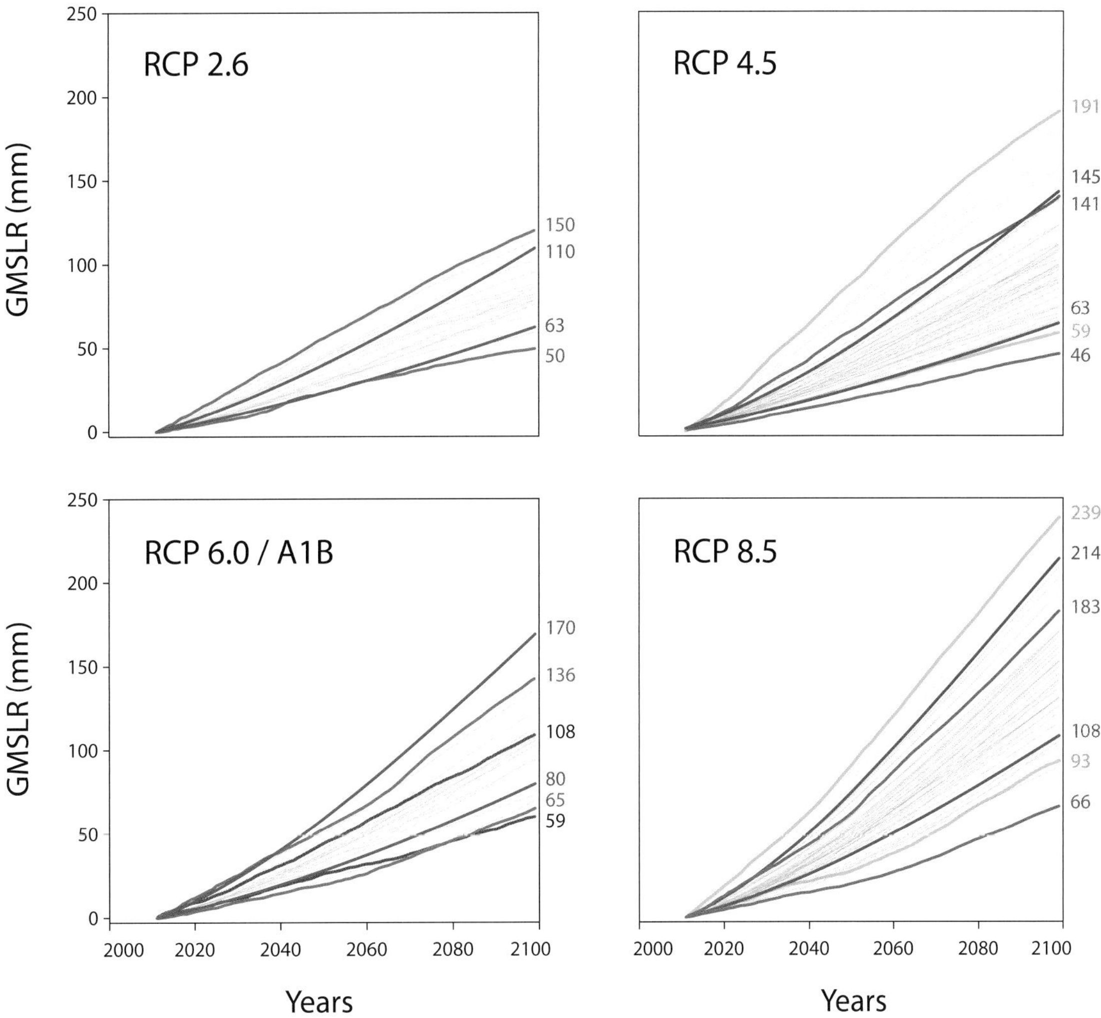

Figure 13.9 | Time series plots for process-based model projections of sea level contributions from global glaciers (in mm), including peripheral glaciers surrounding the Greenland ice sheet but excluding the glaciers surrounding the Antarctic ice sheet. Projections are grouped by forcing scenario as indicated on the plots. Results are plotted for a common time interval of 2011 to 2099. Colours correspond to particular model analyses: red = Marzeion et al. (2012a); blue = Slangen and van de Wal (2011); green = Radić et al. (2013); black = Giesen and Oerlemans (2013). Individual Atmosphere–Ocean General Circulation Model (AOGCM) projections are plotted for each analysis, so the ranges of the curves at 2099 are different than those listed in Table 13.3, where 5 to 95% confidence limits are shown. In the panel showing results for RCP6.0 and A1B forcings, only Geisen and Oerlemans (black lines) use the A1B forcing.

over the past 25 years (about 5 m yr^{-1}) were to occur over the area of global glaciers draining through marine outlets (280,500 km^2) during the next 89 years (2011–2100), the sea level contribution would be approximately 30 cm SLE, compatible with Jeverajeva et al's (2012b) projected loss of 26 cm SLE from glaciers. Although this is a rough calculation and an upper bound, because drainage through marine outlets does not guarantee tidewater instability, it indicates that the potential for a significant sea level response to dynamic retreat of glaciers cannot be rejected *a priori*.

Completion of the global glacier inventory has allowed large improvements in assessment and modelling, but further uncertainties related to the inventory remain to be resolved, including those arising from the size cutoff decided for the inventory (Bahr and Radić, 2012). Another source of uncertainty is interception of glacier runoff by land hydrology. Despite rapidly growing knowledge of changes in terrestrial water storage, especially through increased reliability of GRACE observations, glacier mass loss is still generally assumed to flow directly to the ocean, with no delay or interception by surface or aquifer storage. Although this probably will not apply to discharge from glaciers located near coasts (e.g., Canadian Arctic, Patagonia, Alaska, ice-sheet peripheries), runoff from interior regions (e.g., Alps, High Mountain Asia) may be significantly intercepted before reaching the ocean. Whether terrestrial interception has any significant effect on the net glacier contribution to sea level rise is undetermined at this time.

13.4.3 Greenland Ice Sheet

13.4.3.1 Surface Mass Balance Change

Greenland SMB is positive in the present climate but shows a decreasing trend (Section 13.3.3.2), which implies an increasing contribution to GMSL rise. Like the AR4, all recent studies have indicated that the future sea level contribution from Greenland SMB change will be increasingly positive because the increase in ablation (mostly runoff) outweighs that in accumulation (mostly snowfall), and that scenarios of greater RF lead to a larger sea level contribution. Precipitation is projected to increase at about 5% per °C of annual-mean warming over Greenland, but the increase in snowfall is smaller because the fraction of rain increases as temperature rises (Gregory and Huybrechts, 2006; Fettweis et al., 2013).

We compare post-AR4 studies of Greenland SMB change using time-dependent simulations of the 21st century by CMIP3 AOGCMs for scenario SRES A1B and CMIP5 AOGCMs for scenario RCP4.5 (Table 13.4). The time-integral of the Greenland SMB anomaly with respect to a reference period is interpreted as a contribution to GMSL rise, on the assumption that the ice sheet was in approximate mass balance during the reference period (see discussion in Sections 13.1.4.1 and 13.3.3.2); this assumption can be avoided only if ice-sheet outflow is also modelled. Making this assumption, the Greenland SMB contribution lies in the range 0.00 to 0.13 m for these two scenarios.

The spread in the magnitude and patterns of Greenland climate change projected by the AOGCMs causes a large part of the spread in the projected contribution to GMSL rise (Table 13.4). Yoshimori and Abe-Ouchi (2012) found that the inter-model spread in global mean SAT change accounts for about 60% of the spread in the change of projected Greenland ablation. Two important contributions to the remaining spread are the weakening of the AMOC, which affects the magnitude of warming over Greenland, and the SAT of Greenland in the model control climate, which affects the sensitivity of melting to warming (Yoshimori and Abe-Ouchi, 2012; Fettweis et al., 2013).

Ablation is computed using either an EBM, which may be stand-alone or part of a regional climate model, or from surface air temperature using an empirical temperature index method, mostly commonly the positive-degree-day (PDD) method, in which melting is proportional to the time-integral of temperature above the freezing point. Meltwater production increases faster than linearly with temperature increase because of reduced albedo due to refreezing of meltwater in the snowpack and expansion of the area of bare ice (van Angelen et al., 2012; Fettweis et al., 2013; Franco et al., 2013). The simulation of this positive albedo feedback on mass loss depends sensitively on the model

Table 13.4 | Contribution to sea level rise from change in the surface mass balance of the Greenland ice sheet during the 21st century. Where given, ranges are 5 to 95% estimated from the published results and indicate the uncertainty due to the climate change modelling by Atmosphere–Ocean General Circulation Model (AOGCMs), except where noted otherwise.

Reference	Model[a]	Contribution to Global Mean Sea Level Rise			
		starting from	up to	amount (m)[b]	rate (mm yr^{-1})[b]
Scenario SRES A1B, CMIP3 AOGCMs					
AR4 (Meehl et al., 2007)[c]	20 km PDD	1990	2090–2099	0.01–0.08[d]	0.3–1.9[d]
Bengtsson et al. (2011)[e]	60 km (T213) EBM	1959–1989	2069–2099	—	1.4
Fettweis et al. (2008)[f]	TI from 25 km EBM	1970–1999	2090–2099	0.03–0.05	0.3–1.0
Graversen et al. (2011)	10 km PDD	2000	2100	0.02–0.08 0.00–0.17[g]	 0.0–2.1[g]
Mernild et al. (2010)	25 km EBM	1980–1999	2070–2079	0.02	0.5
Rae et al. (2012)[h]	25 km EBM	1980–1999	2090–2099	0.01, 0.04, 0.06	0.3,1.2,1.5
Seddik et al. (2012)[i]	10 km[e] PDD	2004	2104	0.02, 0.04	—
Yoshimori and Abe-Ouchi (2012)	1–2 km TI	1980–1999	2090–2099	0.02–0.13	0.2–2.0
Scenario RCP4.5, CMIP5 AOGCMs					
Fettweis et al. (2013)[c]	25 km RCM	1980–1999	2100	0.02–0.11	0.1–1.2 in 2080–2099
Gregory and Huybrechts (2006)[c,j]	20 km PDD	1980–1999	2100	0.00–0.06	0.0–0.8 in 2080–2099
Van Angelen et al. (2012)[k]	11 km RCM	1960–1990	2100	0.11[l]	1.7[l] in 2079–2098
Yoshimori and Abe-Ouchi (2012)[j]	1–2 km TI	1980–1999	2090–2099	0.00–0.11	0.0–1.8

Notes:

[a] The spatial resolution is stated and the surface mass balance (SMB) method denoted by TI = temperature index, PDD = positive degree day, EBM = Energy Balance Model.

[b] The amount of sea level rise is the time-integral of the SMB anomaly from the period or date labelled 'starting from' to the one labelled 'up to'. Unless otherwise indicated, the SMB anomaly is calculated relative to the mean SMB for the 'starting from' period, and the rate of sea level rise is the SMB anomaly in the 'up to' period.

[c] These results are estimated from global mean surface air temperature (SAT) change, using formulae fitted to results from a Greenland SMB model.

[d] The SMB anomaly is relative to the late 19th century.

[e] This experiment used time-slices, with boundary conditions from the European Centre for Medium range Weather Forecasts (ECMWF) and Hamburg 5 (ECHAM5) GCM, rather than a simulation of the complete century; thus, results are not available for the amount.

[f] Fettweis et al. (2008) and Franco et al. (2011) used a hybrid approach: they derived a regression relationship from simulations of the recent past using a Regional Climate Model (RCM), incorporating an EBM, between annual anomalies in Greenland climate and in Greenland SMB, then applied this relationship to project future SMB from projected future climate anomalies. The method assumes that a relationship derived from past variability will also hold for future forced climate change.

[g] Range including uncertainty in choice of emission scenario (B1, A1B or A2), SMB modelling and ice-sheet dynamical modelling, as well as uncertainty in climate modelling.

[h] Results are given for the Hadley Centre Regional Model 3P (HadRM3P), High-Resolution Hamburg climate model 5 (HIRHAM5) and the Modèle Atmosphérique Régional (MAR) RCMs driven with the same boundary conditions from the ECHAM5/MPI-OM AOGCM.

[i] Results are given for two ice sheet models (Elmer/Ice, SImulation COde for POLythermal Ice Sheets (SICOPOLIS)) using the same AOGCM climate boundary conditions. The resolution given is for SICOPOLIS; Elmer/Ice has variable resolution.

[j] Results calculated from CMIP5 AOGCMs by the same method as used in the paper.

[k] These results were obtained from the model of Van Angelen et al. (2012) using boundary conditions from the HadGEM2-ES AOGCM and are shown by Fettweis et al. (2013).

[l] With respect to 1992–2011 as a reference period, during which there is a significant simulated trend in SMB (Section 13.3.3.2), the amount is 0.07 m and the rate 1.4 mm yr^{-1}.

snow-albedo parameterization (Rae et al., 2012). Goelzer et al. (2013) projected 14 to 31% more runoff during the 21st century when using an EBM than when using a PDD method, mainly because of the omission of the albedo feedback in the latter. However, other studies using temperature index methods (Graversen et al., 2011; Yoshimori and Abe-Ouchi, 2012) have ranges extending to higher values than those from EBMs, indicating that this is not the only difference between these classes of methods (Table 13.4).

SMB simulations are also particularly sensitive to the treatment of meltwater refreezing (Bougamont et al., 2007; Rae et al., 2012). The pore space in the present-day percolation zone could accommodate 1 to 5 mm SLE of refrozen meltwater over the next several decades (Harper et al., 2012), and the importance of meltwater refreezing will become greater as melting becomes prevalent in areas where it has previously been rare. On the other hand, refreezing will be restricted, and runoff consequently increased, by the expansion of the area of bare ice (Fettweis et al., 2013).

Another source of model spread is the representation of topography, which is lower when represented at coarser resolution. This allows precipitation to spread further inland because of reduced topographic barriers (Bengtsson et al., 2011), and enhances ablation because there is more area at lower, warmer altitudes (Bengtsson et al., 2011; Seddik et al., 2012). Most of the models in Table 13.4 use a fixed Greenland topography, and thus cannot simulate the positive feedback on ablation that can be expected as the ice-sheet surface becomes lower. Dynamical models are required to simulate this effect (Section 13.4.3.2).

For the present assessment of GMSL rise, changes in Greenland ice sheet SMB up to 2100 have been computed from global mean SAT change projections derived from the CMIP5 AOGCMs, following methods described in Section 13.5.1 and the Supplementary Material. The distribution of results, shown in Table 13.5 and Figures 13.10 and 13.11, covers the ranges obtained using the methods of Fettweis et al. (2013), Gregory and Huybrechts (2006), and Yoshimori and Abe-Ouchi (2012).

On multi-centennial to millennial time scales, feedbacks between regional climate and the ice sheet become increasingly relevant, especially under strong climate change scenarios, thus requiring coupled climate ice-sheet models to capture potential feedbacks beyond the year 2100. These models apply a reduced spatial resolution in order to be computationally efficient enough to evaluate longer time scales and to combine the different climatic components. Consistent with regional climate models for the 21st century, they project an increasingly negative mass balance for the Greenland ice sheet for all warming scenarios which is mainly due to a decreasing SMB (Ridley et al., 2005; Winguth et al., 2005; Driesschaert et al., 2007; Mikolajewicz et al., 2007a; Swingedouw et al., 2008; Vizcaíno et al., 2008, 2010; Huybrechts et al., 2011; Goelzer et al., 2013). The main feedbacks between climate and the ice sheet arise from changes in ice elevation, atmospheric and ocean circulation, and sea-ice distribution.

Comparing the different feedbacks, *high confidence* can be assigned to the models' ability to capture the feedback between SMB and surface elevation. As a consequence, a nonlinear increase in ice loss from Greenland with increasing regional RF is found across different scenarios (Driesschaert et al., 2007). This nonlinearity arises from the increase in both the length of the ablation season and the daily amount of melting as the ice-sheet surface lowers. This SMB-surface elevation feedback is also the main reason for the threshold behaviour of the Greenland ice sheet on multi-millennial time scales (Section 13.4.3.3).

Medium-to-*low confidence* is assigned to the models' representation of the atmospheric and ocean circulation and sea-ice changes. On multi-centennial time scales, Swingedouw et al. (2008) found enhanced ice loss from Greenland in a coupled simulation (compared to the uncoupled version) in which ice topography and meltwater flux influence the ocean and atmospheric circulation as well as sea-ice distribution. Vizcaíno et al. (2010) found reduced ice loss due to the coupling, mainly caused by the effect of topographic changes on the surface temperature, but less pronounced in amplitude compared with Swingedouw et al. (2008). Both the atmospheric circulation and the ocean currents, especially in coastal areas, are poorly resolved by these models. It is therefore *likely* that the time scales associated with ocean transport processes are distorted and there is *low confidence* that these feedbacks, although existent, can be quantified accurately by the applied models.

The AMOC exerts a strong influence on regional climate around the Greenland ice sheet and consequently its SMB. Most CMIP5 models show a reduction of the AMOC under future warming during the 21st century and beyond (Section 12.4.7.2). Although coupled climate–ice sheet models show some influence of meltwater from Greenland on the AMOC, the uncertainty between models with respect to the AMOC response to warming is significantly larger than the difference between simulations with or without this feedback within one model.

In the coupled climate–ice sheet model applied by Mikolajewicz et al. (2007a) and Vizcaíno et al. (2008), the AMOC shows a strongly nonlinear response to global warming. A weak AMOC reduction is found for 1%-per-year-CO_2-increase scenarios up to 560 and 840 ppm, and a near-complete cessation of the AMOC for 1120 ppm. As a consequence, after 600 years of integration, the sea level contribution for the 1120 ppm scenario is similar to that of the 560 ppm scenario, but doubles for the medium scenario, which stabilizes at 840 ppm. In the most recent model version (Vizcaíno et al., 2010), the AMOC shows a strong weakening of the AMOC in all scenarios (~60% reduction in 560 ppm scenario; ~80% for 1120 ppm). The total sea level contribution from Greenland, including the effect of the AMOC weakening, is ~1 m (corresponding to an average rate of 1.7 mm yr^{-1}) for 560 ppm CO_2 and ~3 m (5 mm yr^{-1}) for 1120 ppm CO_2.

Even though the AMOC weakening in the model by Huybrechts et al. (2011) is less pronounced (10 to 25%), the ice loss through melting is significantly weaker in this model. During the first 1000 years of integration, the Greenland ice sheet contributes 0.36 m (corresponding to an average rate of 0.36 mm yr^{-1}) for 560 ppm CO_2 and 2.59 m (2.59 mm yr^{-1}) for 1120 ppm CO_2. In Huybrechts et al. (2011), the respective increases in global mean SAT are 2.4°C (2 × CO_2) and 6.3°C (4 × CO_2) after 1000 years with respect to pre-industrial. This relatively weak warming response to GHG forcing compared to CMIP5 models and the climate model used in Vizcaíno et al. (2010) explains the relatively small sea level response.

Using the same model as Huybrechts et al. (2011), albeit with a slightly higher polar warming, Goelzer et al. (2012) computed temperatures and sea level under the SRES scenarios B1, A1B and A2, with subsequent GHG stabilization after the year 2100. As in Huybrechts et al. (2011), the ice-sheet evolution is dominated by the SMB. They find sea level contributions of 1.4, 2.6 and 4.2 m in the year 3000 for the scenarios B1, A1B and B2, which correspond to mean rates of sea level rise of 1.4 mm yr^{-1}, 2.6 mm yr^{-1}, and 4.2 mm yr^{-1}, respectively.

In summary, coupled climate-ice sheet models consistently show an increasingly negative mass balance of the Greenland ice sheet due mainly to a decreasing SMB under warming scenarios on centennial time scales beyond 2100. On multi-millennial time scales, these models show a threshold temperature beyond which the melting of the Greenland ice sheet self-amplifies and the ice volume is reduced to less than 30% of its present volume (Section 13.4.3.3).

13.4.3.2 Dynamical Change

Observations suggest three main mechanisms by which climate change can affect the dynamics of ice flow in Greenland (Sections 4.4.3 and 4.4.4): by directly affecting ice loss (outflow) through the calving of icebergs and marine melt from marine-terminating outlet glaciers; by altering basal sliding through the interaction of surface melt water with the glacier bed; and indirectly through the interaction between SMB and ice flow. We assess the consequences of each of these processes.

Section 4.4.3.2 presents the observational basis on which concerns about increased ice loss by calving and marine melt are based. In particular, recent increases in loss are thought to be linked to the migration of subtropical water masses around the coast of Greenland (Holland et al., 2008) and its occupation of coastal fjords (Straneo et al., 2010; Christoffersen et al., 2011). Output from 19 AOGCMs under scenario A1B showed warming of 1.7°C to 2.0°C around Greenland over the course of the 21st century (Yin et al., 2011), suggesting that the trend towards increased outflow triggered by warming coastal waters will continue.

Although projections of outflow are at a fairly early stage, literature now exists to make an assessment. Flowline modelling has successfully simulated the observed retreat and associated acceleration of the main outlet glaciers of the Greenland ice sheet (Helheim and Petermann Glaciers (Nick et al., 2009, 2012); Jakobshavn Isbræ (Vieli and Nick, 2011)). The same model has been used to project mass loss from these glaciers (Nick et al., 2013), as well as Kangerdlugssuaq Glacier, using ocean and atmosphere forcing based on scenarios A1B and RCP8.5. At 2100, total projected SLR spans 8 to 13 mm for A1B and 11 to 17 mm for RCP8.5. These figures generalize to 40 to 63 mm and 57 to 85 mm, respectively, for the whole ice sheet based on a simple scaling between modelled and total ice-sheet area (a factor of ~5). Price et al. (2011) modelled the century-scale response of the ice sheet to the observed recent retreat of three outlet glaciers (Jakobshavn Isbræ, and Helheim and Kangerdlugssuaq Glaciers). At 2100, the projected SLR associated with the three modelled outlet glaciers is 0.6 to 1.4 mm, which equates to SLR of 4 to 8 mm after scaling (by a factor of ~6) to all outlet glaciers based on observed mass loss (van den Broeke et al., 2009). Total projected SLR then varies between 10 and 45 mm at 2100 if successive retreats are specified with a notional repeat interval between 50 and 10 years.

Goelzer et al. (2013) implemented the Nick et al. (2013) retreat chronology within a 5-km resolution ice-sheet model along with their own generalization for including unsampled outlet glaciers. Associated SLR at 2100 is projected to vary between 8 and 18 mm. Graversen et al. (2011) attempted to capture the effect of increased outflow by enhancing basal sliding and generated SLR of 9 to 24 mm at 2100.

Two estimates of the effect of dynamical change on Greenland's contribution to SLR by 2100 have been made on the basis of physical intuition. Pfeffer et al. (2008) developed a low scenario by assuming a first-decade doubling of outlet glacier velocity throughout the ice sheet that equates to 93 mm SLR, while a high scenario that assumes an order of magnitude increase on the same time scale contributes 467 mm. Katsman et al. (2011) used a similar methodology to obtain an estimate of 100 mm SLR.

Based primarily on Nick et al. (2013), we assess the upper limit of the *likely* range of this dynamical effect to be 85 mm for RCP8.5 and 63 mm for all other RCP scenarios for the year 2100. We have *medium confidence* in this as an upper limit because it is compatible with Katsman et al. (2011), the low scenario of Pfeffer et al. (2008), and Price et al. (2011) in the probable event of a sub-decadal recurrence interval. Although the *likely* upper limit is less than the high scenario of Pfeffer et al. (2008), process modelling gives no support to the order of magnitude increase in flow on which this scenario is based. It is higher than the contributions found by Goelzer et al. (2013) and Graversen et al. (2011) for which there are two potential explanations. First, the generalization used to extrapolate from the modelled sample to all outlet glaciers differs. Nick et al. (2013) used a scaling similar to the independently derived value of Price et al. (2011), while the implied scaling used by Goelzer et al. (2013) is substantially lower. Second, Goelzer et al. (2013) suggested that surface ice melt and calving each remove marginal ice (see below), implying that by not including surface melt, overall mass loss by dynamics may be over predicted by the flowline model of Nick et al. (2013). At present, these studies cannot be reconciled and we therefore use the more inclusive range.

The lower limit of the *likely* range is assessed as 20 mm for RCP8.5 and 14 mm for all other RCP scenarios. This reflects the individual outlet glacier projections of Nick et al. (2013) but uses a lower generalization more similar to that found by Goelzer et al. (2013). This assessment of the lower limit is compatible with Price et al. (2011) and Graversen et al. (2011).

Section 4.4.3.2 assesses understanding of the link between abundant summer meltwater, lubrication of the ice-sheet base, and enhanced ice flow. Although this mechanism appears important in modulating present-day ice flow, it is not supported as the cause of recent mass loss. Goelzer et al. (2013) incorporated a parameterization of this effect based on field observations, which results in less than a millimetre SLR by 2100 in their projections. Bindschadler et al. (2013) reported a suite of experiments assessing this effect in an eight-model ensemble, but their parameterization appears overly simplistic and may well exaggerate the importance of the effect. These projections do not incorporate the effect on ice flow of the latent heat released by increased future quantities of melt water within the ice sheet (Phillips et al., 2010; 2013), for which no projections are currently available. Basal lubrication is therefore assessed as making an insignificant contribution to the *likely* range of SLR over the next century and is omitted in the remainder of the assessment. We have *medium confidence* in projections of this effect primarily because recent improvements in process-based understanding show that it has little contribution to mass loss (Section 4.4.3.2); the potential of latent-heat effects in the future limits a higher level of confidence.

Finally, we assess the level of interaction between SMB change and ice flow. In AR4, this effect is assessed as 0 ± 10% (*likely* range) of SMB, based on Huybrechts and de Wolde (1999) and Gregory and Huybrechts (2006). This assessment included both the positive feedback between SMB and the height of the ice sheet, and a countering negative feedback involving ice flow and depletion effects. The latter effect is partly included in our assessment of the direct impacts of climate change on ice flow, and we therefore limit our assessment to the SMB-height feedback. Few studies explicitly determine this effect, but Goelzer et al. (2013) reported that it amounts to 5 to 15% of SMB over the course of the 21st century, which we extend slightly (0 to 15%) to reflect possible interaction with mass loss by calving (Goelzer et al., 2013).

Goelzer et al. (2013) and Gillet-Chaulet et al. (2012) suggested that SMB and ice dynamics cannot be assessed separately because of the strong interaction between ice loss and climate due to, for instance, calving and SMB. The current assessment has by necessity separated these effects because the type of coupled ice sheet-climate models needed to make a full assessment do not yet exist. These interactions may well combine to reduce SLR in comparison to the assessed range because of the mass-depletion effect of retreating outlet glaciers. Another source of uncertainty is the bedrock topography of Greenland, although recent improvements in data coverage (Bamber et al., 2013) suggest that the majority of the ice sheet rests on bedrock above sea level and the number of deep bedrock troughs penetrating into the interior of Greenland are limited, thus limiting the potential for marine ice-sheet instability (see Box 13.2).

Although not strictly comparable because they contain a different balance of ice-dynamical effects, the assessment is consistent with Bindschadler et al. (2013), who reported an extensive model inter-comparison exercise in which standardized experiments are combined to represent the impact of climate change under RCP8.5 on the Greenland ice sheet. The resultant projection included contributions from lubrication, marine melt and SMB-coupling and generated a mean SLR at 2100 of 162 mm over five models, or 53 mm if an outlier with anomalously high response is removed (including SMB results in SLR at 2100 of 223 and 114 mm for five- and four-model means, respectively). This comparison provides further weight to our confidence.

In summary, dynamical change within the Greenland ice sheet is *likely* (*medium confidence*) to lead to SLR during the next century with a range of 20 to 85 mm for RCP8.5, and 14 to 63 mm for all other scenarios by year 2100. The latter are assumed to have uniform SLR in the absence of literature allowing these scenarios to be assessed individually, although dependency on scenario is expected to exist. In addition, mass loss associated with SMB-height feedback is *likely* to contribute a further 0 to 15% of SMB (in itself scenario dependent). This equates to, for example, 0 to 14 mm by 2100 based on the central estimate of RCP8.5. The peripheral glaciers of Greenland are not included here but are in the assessment of global glaciers' contribution to SLR (Section 13.4.2). All the available literature suggests that this dynamical contribution to sea level rise will continue well beyond 2100.

13.4.3.3 Possible Irreversibility of Greenland Ice Loss and Associated Temperature Threshold

A number of model results agree in showing that the Greenland ice sheet, like other climatic subsystems (Lenton et al., 2008; Levermann et al., 2012) (see Section 12.5.5), exhibits a strongly nonlinear and potentially irreversible response to surface warming. The mechanism of this threshold behaviour is the SMB-height feedback (Section 13.4.3.2); that is, as the surface is lowered due to ice loss, the associated warming of the near surface increases ablation, leading to further ice loss. This feedback is small but not negligible in the 21st century (Section 13.4.3.2) and becomes important for projections for the 22nd century (Goelzer et al. 2013) and beyond. This nonlinear behaviour may be accelerated by a reduced surface albedo caused by surface melting which tends to further decrease the surface mass balance (Box et al., 2012) (Section 13.4.3.1).

Although the mean SMB of the Greenland ice sheet is positive, in a steady state it must be balanced by ice outflow, so the ice sheet must extend to the coast. In a warmer climate, the mean SMB is reduced (Section 13.4.3.1) and the steady-state ice sheet will have a lower surface and volume. Models show a threshold in surface warming beyond which self-amplifying feedbacks result in a partial or near-complete ice loss on Greenland (Greve, 2000; Driesschaert et al., 2007; Charbit et al., 2008; Ridley et al., 2010; Robinson et al., 2012). If a temperature above this threshold is maintained over a multi-millennial time period, the majority of the Greenland ice sheet will be lost by changes in SMB on a millennial to multi-millennial time scale (equivalent to a sea level rise of about 7 m; Table 4.1). During the Middle Pliocene warm intervals, when global mean temperature was 2°C to 3.5°C higher than pre-industrial, ice-sheet models suggest near-complete deglaciation of Greenland (Hill et al., 2010).

A simplifying assumption is that the threshold is the warming required with the current ice-sheet topography to reduce the mean SMB to zero, on the argument that the ice sheet margin must then retreat from the coast. Using this criterion, Gregory and Huybrechts (2006) estimated that the SMB threshold occurs for a GMST increase of 3.1 [1.9 to 4.6] °C (4.5 [3.0 to 6.0] °C for Greenland surface temperature) above pre-industrial (assumed to be a steady state). More recent studies have found thresholds below or in the lower part of this range. In a coupled ice sheet–climate model of intermediate complexity, Huybrechts et al. (2011) found this threshold at 2.5°C for annual average Greenland SAT. Comparing three regional climate models, Rae et al. (2012) found a strong dependence of the threshold on the model formulation of the SMB. Based on the model's performance against observations and the physical detail of its surface scheme, MAR is considered the most realistic model, and yields a threshold value 2.8 [2.1 to 3.4] °C for changes in Greenland annual average temperature compared to pre-industrial. Using MAR driven with output from various CMIP5 AOGCMs, Fettweis et al. (2013) evaluated the threshold as ~3°C in GMST above 1980–1999 (hence about 3.5°C relative to pre-industrial), and found that it is not exceeded in the 21st century under the RCP4.5 scenario but is reached around 2070 under the RCP8.5 scenario.

Some of the uncertainty in the threshold results from the value assumed for the steady state ice-sheet SMB (see Table 13.2), and whether this is assumed to be pre-industrial or a more recent period. For 400 Gt yr^{-1} (Fettweis et al., 2013), the parametrization of Greenland ice sheet SMB used for present assessment of 21st century changes (Section 13.4.3.1, Supplementary Material) gives a global warming threshold of 3.0 [2.1 to 4.1] °C with respect to 1860–1879 (the reference period used in Box 13.1); for 225 Gt yr^{-1} (Gregory and Huybrechts, 2006, following Church et al., 2001), the threshold is 2.1 [1.5 to 3.0] °C.

Although a negative SMB is a sufficient condition for passing the threshold, it will overestimate the value of the threshold quantitatively, because the SMB–height feedback (even without passing the threshold) means that the steady-state SMB is reduced by more than is calculated assuming fixed topography. The actual SMB change will depend on the dynamical response of the ice sheet that determines its topography. Constraining simulations with a dynamic ice-sheet model to changes during the last interglacial, Robinson et al. (2012) estimated the threshold as 1.6 [0.9 to 2.8] °C global averaged temperature above pre-industrial. In these simulations, they find that the threshold is passed when southeastern Greenland has a negative SMB. The near-complete ice loss then occurs through ice flow and SMB.

The complete loss of the ice sheet is not inevitable because it has a long time scale (tens of millennia near the threshold and a millennium or more for temperatures a few degrees above the threshold). If the surrounding temperatures decline before the ice sheet is eliminated, the ice sheet might regrow. In the context of future GHG emissions, the time scale of ice loss is competing with the time scale of temperature decline after a reduction of GHG emissions (Allen et al., 2009; Solomon et al., 2009; Zickfeld et al., 2009). The outcome therefore depends on both the CO_2 concentration and on how long it is sustained. Charbit et al. (2008) found that loss of the ice sheet is inevitable for cumulative emissions above about 3000 GtC, but a partial loss followed by regrowth occurs for cumulative emissions less than 2500 GtC. Ridley et al. (2010) identified three steady states of the ice sheet. If the CO_2 concentration is returned to pre-industrial when more than 20 to 40% of the ice sheet has been lost, it will regrow only to 80% of its original volume due to a local climate feedback in one region; if 50% or more, it regrows to 20 to 40% of the original. Similar states with ice volume around 20%, 60 to 80% and 100% of the initial ice volume are also found in other models (Langen et al., 2012; Robinson et al., 2012). If all the ice is lost, temperatures must decline to below a critical threshold for regrowth of the ice sheet (Robinson et al., 2012; Solgaard and Langen, 2012).

On the evidence of paleo data and modelling (Section 5.6.2.3, 13.2.1), it is *likely* that during the LIG, when global mean temperature never exceeded 2°C pre-industrial, the Greenland ice sheet contributed no more than ~4 m to GMSL. This could indicate that the threshold for near-complete deglaciation had not been passed, or that it was not greatly exceeded so that the rate of mass loss was low; however, the forcing responsible for the LIG warming was orbital rather than from CO_2 (van de Berg et al., 2011), so it is not a direct analogue and the applicable threshold may be different. Studies with fixed-topography ice sheets indicate a threshold of 2°C or above of global warming with respect to pre-industrial for near-complete loss of the Greenland ice sheet, while the one study (and therefore *low confidence*) presently available with a dynamical ice sheet suggests that the threshold could be as low as about 1°C (Robinson et al. 2012). Recent studies with fixed-topography ice sheets indicate that the threshold is less than about 4°C (*medium confidence* because of multiple studies). With currently available information, we do not have sufficient confidence to assign a *likely* range for the threshold. If the threshold is exceeded temporarily, an irreversible loss of part or most of the Greenland ice sheet could result, depending on the duration and amount that the threshold is exceeded.

13.4.4 Antarctic Ice Sheet

13.4.4.1 Surface Mass Balance Change

Because the ice loss from Antarctica due to surface melt and runoff is about 1% of the total mass gain from snowfall, most ice loss occurs through solid ice discharge into the ocean. In the 21st century, ablation is projected to remain small on the Antarctic ice sheet because low surface temperatures inhibit surface melting, except near the coast and on the Antarctic Peninsula, and meltwater and rain continue to freeze in the snowpack (Ligtenberg et al., 2013). Projections of Antarctic SMB changes over the 21st century thus indicate a negative contribution to sea level because of the projected widespread increase in snowfall associated with warming air temperatures (Krinner et al., 2007; Uotila et al., 2007; Bracegirdle et al., 2008). Several studies (Krinner et al., 2007; Uotila et al., 2007; Bengtsson et al., 2011) have shown that the precipitation increase is directly linked to atmospheric warming via the increased moisture holding capacity of warmer air, and is therefore larger for scenarios of greater warming. The relationship is exponential, resulting in an increase of SMB as a function of Antarctic SAT change evaluated in various recent studies with high-resolution (~60 km) models as 3.7% $°C^{-1}$ (Bengtsson et al., 2011), 4.8% $°C^{-1}$ (Ligtenberg et al., 2013) and ~7% $°C^{-1}$ (Krinner et al., 2007). These agree well with the sensitivity of 5.1 ± 1.5% $°C^{-1}$ (one standard deviation) of CMIP3 AOGCMs (Gregory and Huybrechts, 2006).

The effect of atmospheric circulation changes on continental-mean SMB is an order of magnitude smaller than the effect of warming, but circulation changes can have a large influence on regional changes in accumulation, particularly near the ice-sheet margins (Uotila et al., 2007) where increased accumulation may induce additional ice flow across the grounding line (Huybrechts and De Wolde, 1999; Gregory and Huybrechts, 2006; Winkelmann et al., 2012). Simulated SMB is strongly and nonlinearly influenced by ocean surface temperature and sea-ice conditions (Swingedouw et al., 2008). This dependence means that the biases in the model-control climate may distort the SMB sensitivity to climate change, suggesting that more accurate predictions may be obtained from regional models by using boundary conditions constructed by combining observed present-day climate with projected climate change (Krinner et al., 2008). There is a tendency for higher resolution models to simulate a stronger future precipitation increase because of better representation of coastal and orographic precipitation processes (Genthon et al., 2009).

For the present assessment of GMSL rise, changes in Antarctic ice-sheet SMB up to 2100 have been computed from global mean SAT change projections derived from the CMIP5 AOGCMs, using the range of sensitivities of precipitation increase to atmospheric warming summarized above, and the ratio of Antarctic to global warming evaluated from CMIP3 AOGCMs by Gregory and Huybrechts (2006) (see also Section 13.5.1 and Supplementary Material). The results are shown in Table 13.5 and Figures 13.10 and 13.11. The projected change in ice outflow is affected by the SMB because of the influence of topography on ice dynamics (Section 13.4.4.2 and Supplementary Material). Ozone recovery, through its influence on atmospheric circulation at high southern latitudes (Section 10.3.3.3), may offset some effects of GHG increase in the 21st century, but Antarctic precipitation is nonetheless projected to increase (Polvani et al., 2011). Bintanja et al. (2013) suggested that Antarctic warming and precipitation increase may be suppressed in the future by expansion of Antarctic sea ice, promoted by freshening of the surface ocean, caused by basal melting of ice shelves, and they conducted an AOGCM sensitivity test of this hypothesis. We consider these possibilities in Section 13.5.3.

Beyond the year 2100, regional climate simulations run at high spatial resolution (5 to 55 km) but without climate-ice sheet feedbacks included show a net ice gain until the year 2200 (Ligtenberg et al., 2013). During the 22nd century, the ice gain is equivalent to an average rate of sea level fall of 1.2 mm yr^{-1} for the A1B scenario and 0.46 mm yr^{-1} for the E1 scenario.

For multi-centennial to multi-millennial projections, feedbacks between the ice sheet and regional climate need to be accounted for. This is currently done using ice-sheet models coupled to climate models of intermediate complexity, which have a significantly lower spatial resolution in the atmospheric component than regional climate models used to assess future SMB within the 21st century. These coarser resolution models capture the increase in snowfall under future warming, but the regional distribution is represented less accurately. Accordingly, there is *low confidence* in their ability to model spatial melting and accumulation patterns accurately. In contrast, *medium confidence* can be assigned to the models' projection of total accumulation on Antarctica, as it is controlled by the large-scale moisture transport toward the continent.

In idealized scenarios of 1% increase of CO_2 yr^{-1} up to 560 ppm with subsequent stabilization, Vizcaíno et al. (2010) and Huybrechts et al. (2011) found an initial increase of ice volume due to additional snowfall during the first 600 years of integration. In both models, the changes in SMB dominate the mass changes during and beyond the first 100 years. After 600 years of integration, Vizcaíno et al. (2010) found a mass gain corresponding to a sea level fall of 0.15 m (–0.25 mm yr^{-1} on average). For the same experiment and the same period, Huybrechts et al. (2011) found a sea level fall of 0.08 m (–0.13 mm yr^{-1} on average). In a similar experiment but allowing GHG concentrations to reach 1120 ppm CO_2 before being stabilized, both models show a net positive sea level contribution after 600 years of integration. Huybrechts et al. (2011) found a weak sea level contribution during the first 500 years of integration followed by a stronger and relatively constant long-term average rate of ~2 mm yr^{-1} after 1000 years of integration up to a total contribution of ~4 m SLE after 3000 years of integration. Although they found some grounding line retreat due to basal ice-shelf melt, the multi-centennial evolution of the ice sheet is dominated by changes in SMB whereas the solid-ice discharge after an initial increase shows a significant decrease during the scenario.

For the same scenario, Vizcaíno et al. (2010) found that the initial mass gain is followed by a weak mass loss. After 250 years of integration, the contribution is stronger and relatively constant at a rate of about 3 mm yr^{-1}, corresponding to a net contribution of 1.2 m to global mean sea level rise after 600 years.

The same model as in Huybrechts et al. (2011), although with a slightly stronger polar amplification, was applied to the three SRES scenarios used in the AR4 (B1, A1B, A2) with stabilization in the year 2100 (Goelzer et al., 2012). For the lowest scenario (B1), they found practically no net Antarctic contribution to sea level in the year 3000. Under the medium scenario (A1B), the ice sheet contributes 0.26 m, and under the highest scenario (A2), it contributes 0.94 m SLE in the year 3000.

These simulations include a negative feedback on the regional climate by ice-sheet melt through which summer temperatures can be significantly reduced over Antarctica (Swingedouw et al., 2008). However, due to the coarse resolution and the high polar amplification, there is *low confidence* in the model's representation of oceanic circulation changes around Antarctica.

In both models (Vizcaíno et al., 2010; Huybrechts et al., 2011), the ice sheets are not equilibrated with the surrounding climate after the integration period under the 1120 ppm CO_2 forcing. Though GHG concentrations were stabilized after 120 years of integration, the Antarctic ice sheet continues to contribute to sea level rise at a relatively constant rate for another 480 years in Vizcaíno et al. (2010) and 2880 years in Huybrechts et al. (2011). This is consistent with a positive sea level contribution from Antarctica during past warmer climates (Sections 13.2.1 and 13.5.4).

In summary, both coupled ice sheet-climate models consistently show that for high-emission scenarios, the surface melt increases and leads to an ice loss on multi-centennial time scales. The long time period over which the Antarctic ice sheet continues to lose mass indicates a potential role of the feedback between climate and ice sheet. Consistent with regional climate models for the 21st and 22nd centuries, both coarse-resolution coupled models show a positive SMB change for most of the first 100 years of climate change. Due to the inertia in the climate system, regional temperatures continue to rise after that. Together with enhanced solid ice discharge, this results in mass loss of the ice sheet. The corresponding decline in surface elevation increases the surface temperature and leads to additional ice loss.

13.4.4.2 Dynamical Change

The Antarctic ice sheet represents the largest potential source of future SLR; the West Antarctic ice sheet alone has the potential to raise sea level by ~4.3 m (Fretwell et al., 2013). The rate at which this reservoir will be depleted and cause sea level to rise, however, is not easily quantifiable. In this section, we focus on dynamical changes (i.e., those related to the flow of the ice sheet) that affect SLR by altering the flux of ice across the grounding line (or outflow) that separates ice resting on bedrock (some of which does not currently displace ocean water) from floating ice shelves (which already displace ocean water and have only a negligible effect on sea level) (Jenkins and Holland, 2007).

Issues associated with the inability of models to reproduce recently observed changes in the dynamics of the Antarctic ice sheet prevented the AR4 from quantifying the effect of these changes on future sea level. Since the AR4, progress has been made in understanding the observations (Sections 4.4.3 and 4.4.4), and projections are becoming available. It must be stressed, however, that this field has yet to reach the same level of development that exists for modelling many other components of the Earth system. There is an underlying concern that observations presage the onset of large-scale grounding line retreat in what is termed the Marine Ice Sheet Instability (MISI; Box 13.2), and much of the research assessed here attempts to understand the applicability of this theoretical concept to projected SLR from Antarctica.

There are three distinct processes that could link climate change to dynamical change of the Antarctic ice sheet and potentially trigger increased outflow. These may operate directly through the increased potential for melt ponds to form on the upper surface of ice shelves, which may destabilize them, or by increases in submarine melt experienced by ice shelves as a consequence of oceanic warming, which leads to their thinning, as well as indirectly by coupling between SMB and ice flow. Section 4.4.3.2 presents the observational basis on which understanding of these processes is based, while their potential future importance is assessed here. Literature on the two mechanisms directly linked to climate change will be assessed first, followed by their relation to outflow change and lastly SMB coupling.

There is strong evidence that regional warming and increased melt water ponding in the Antarctic Peninsula led to the collapse of ice shelves along the length of peninsula (Cook and Vaughan, 2010), most notably the Larsen B ice shelf (MacAyeal et al., 2003). Substantial local warming (~5 to 7 °C) would, however, be required before the main Antarctic ice shelves (the Ross and Filchner-Ronne ice shelves) would become threatened (Joughin and Alley, 2011). An assessment of the AR4 AOGCM ensemble under scenario A1B yielded an Antarctic continental-average warming rate of 0.034 ± 0.01°C yr^{-1} (Bracegirdle et al., 2008), suggesting that the required level of warming may not be approached by the end of the 21st century. Using an intermediate complexity model with scenario A2, Fyke et al. (2010) found that melt starts to reach significant levels over these ice shelves around 2100 to 2300. Barrand et al. (2013) made a process-based assessment of the effect of ice-shelf collapse on outflow from the Antarctic Peninsula, which yields a range of SLR at 2100 between 10 and 20 mm, with a bounding maximum of 40 mm.

There is good evidence linking the focus of current Antarctic mass loss in the Amundsen Sea sector of the WAIS (containing Pine Island and Thwaites Glaciers) (Shepherd and Wingham, 2007; Rignot et al., 2008; Pritchard et al., 2009) to the presence of relatively warm Circumpolar Deep Water on the continental shelf (Thoma et al., 2008; Jenkins et al., 2010). However, it is not possible to determine whether this upwelling was related directly or indirectly to a rise in global mean temperature. Yin et al. (2011) assessed output from 19 AOGCMs under scenario A1B to determine how subsurface temperatures are projected to evolve around the West and East Antarctic ice sheets. They showed decadal-mean warming of 0.4°C to 0.7°C and 0.4°C to 0.9°C around West and East Antarctica, respectively (25th to 75th percentiles of ensemble) by the end of the 21st century. More detailed regional modelling using scenario A1B illustrates the potential for warm water to invade the ocean cavity underlying the Filchner-Ronne ice shelf in the second half of the 21st century, with an associated 20-fold increase in melt (Hellmer et al., 2012). Based on the limited literature, there is *medium confidence* that oceanic processes may potentially trigger further dynamical change particularly in the latter part of the 21st century, while there is also *medium confidence* that atmospheric change will not affect dynamics outside of the Antarctic Peninsula during the 21st century.

Several process-based projections of the future evolution of Pine Island Glacier have now been made, and some of the issues that this modelling faced (such as the need for sub-kilometre resolution to ensure consistent results; Cornford et al. (2013), Durand et al. (2009)) are being resolved (Pattyn et al., 2013). In experiments using an idealized increase in marine melt, Joughin et al. (2010) demonstrated only limited (~25 km) retreat of the grounding line before a new equilibrium position was established. Gladstone et al. (2012) used a flowline model forced with ocean-model output (Hellmer et al., 2012) to identify two modes of retreat: one similar to that identified by Joughin et al. (2010), and a second characterized by complete collapse from 2150 onwards. More sophisticated ice-flow modelling (albeit with idealized forcing) suggests grounding line retreat of ~100 km in 50 years (Cornford et al., 2013). These studies support the theoretical finding of Gudmundsson et al. (2012) that grounding line retreat, if triggered, does not inevitably lead to MISI but may halt if local buttressing from ice rises or channel sidewalls is sufficient. Parizek et al. (2013) used a flowline model to study Thwaites Glacier and found that grounding line retreat is possible only under extreme ocean forcing. It is also thought that considerably less back pressure is exerted by Thwaites' ice shelf in comparison to Pine Island's (Rignot, 2001; 2008), which may make it less sensitive to forcing by submarine melt (Schoof, 2007a; Goldberg et al., 2012). Based on this literature, there is *high confidence* that the retreat of Pine Island Glacier (if it occurs) can be characterized by a SLR measured in centimetres by 2100, although there is *low confidence* in the models' ability to determine the probability or timing of any such retreat. There is also *medium confidence* (in the light of the limited literature) that Thwaites Glacier is probably less prone to undergo ocean-driven grounding line retreat than its neighbour in the 21st century. No process-based modelling is available on which to be base projections of EAIS glaciers currently losing mass, such as Totten and Cook Glaciers.

Bindschadler et al. (2013) reported a model inter-comparison exercise on the impact of climate change under RCP8.5 on the Antarctic ice sheet. The resultant projection includes contributions from increased marine melt in the Amundsen Sea and Amery sectors, and generated a mean SLR at 2100 of ~100 mm over four models (with overall SLR of 81 mm when SMB change was included). There is, however, *low confidence* in the projection because of the unproven ability of many of the contributing models to simulate grounding line motion. Bindschadler et al. (2013) also reported idealized experiments in which ice-shelf melt is increased by 2, 20 and 200 m yr^{-1}. The resulting five-model mean SLR of 69, 693 and 3477 mm by 2100, respectively, can be considered only as a general indication because of the shortcomings of the contributing models (e.g., two do not include ice shelves) which may be offset by the use of a multi-model mean. Although grounding line migration in the 20 m yr^{-1} experiment is extensive in some models and consistent with what might be expected under MISI (Bindschadler et al., 2013), the 200 m yr^{-1} experiment is unrealistic, even if used as a proxy for the improbable atmosphere-driven collapse of the major ice shelves, and is not considered further.

We now assess two alternatives to process-based modelling that exist in the literature: the development of plausible high-end projections based on physical intuition (Pfeffer et al., 2008; Katsman et al., 2011) and the use of a probabilistic framework for extrapolating current observations of the ice sheet's mass budget (Little et al., 2013a; 2013b). Pfeffer et al. (2008) postulated a possible but extreme scenario of 615 mm SLR based on vastly accelerated outflow in the Amundsen Sea sector and East Antarctica, whereas a more plausible scenario involving reduced acceleration in the Amundsen Sea sector yields 136 mm. Katsman et al. (2011) used similar assumptions in a 'modest' scenario that generates SLR of 70 to 150 mm, and a 'severe' scenario that attempts to capture the consequences of the collapse of the WAIS through the MISI and has a SLR contribution of 490 mm. The NRC (2012) extrapolated mass-budget observations of the ice sheet to generate a projection of 157 to 323 mm (including future SMB change), with an additional 77 to 462 mm accounting for 21st-century increases in outflow (summing as 234 to 785 mm).

Little et al. (2013a) applied a range of linear growth rates to present-day SMB and outflow observations of Antarctic sectors (Rignot et al., 2008; Shuman et al., 2011; Zwally and Giovinetto, 2011), which are then weighted using a continental-scale observational synthesis (Shepherd et al., 2012) (consistent with the assessment of Chapter 4). In the case of Pine Island Glacier, growth rates are based on the process-based modelling of Joughin et al. (2010). Within this framework, SLR at 2100 has a 5 to 95% range of –20 to 185 mm for dynamical change only, and –86 to 133 mm when SMB change is included (based on Uotila et al. (2007)). Projections for the Antarctic Peninsula are consistent with the process-based modelling of Barrand et al. (2013). Further, Little et al. (2013a) found that the upper (95%) limit of the projected range can only approach 400 mm under scenarios expected for MISI (such as the immediate collapse of Pine Island and Thwaites Glaciers or all marine-based sectors experiencing the same rates of mass loss as Pine Island Glacier).

Our assessment of the *likely* range of SLR is based on the weighted 5-95% range (-20 to 185 mm) of Little et al. (2013), which is consistent with the lower scenarios of Katsman et al. (2011) (70 to 150 mm) and Pfeffer et al. (2008) (136 mm), and with the RCP8.5 projection and low-melt experiment of Bindschadler et al. (2013) (~100 and 69 mm, respectively). The base projection of the NRC (2012) (157 to 323 mm including future SMB change), however, is less compatible. This moderate level of consistency across a range of techniques suggests *medium confidence* in this assessment. We assess this as the *likely* (as opposed to *very likely*) range because it is based primarily on perturbations of the ice sheet's present-day state of mass imbalance and does not include the potentially large increases in outflow that may be associated with the MISI discussed below.

The probability of extensive grounding line retreat being both triggered and continuing to the extent that it contributes to significant SLR in the 21st century is very poorly constrained, as the results of a recent expert elicitation indicate (Bamber and Aspinall, 2013). We have *medium confidence*, however, that this probability lies outside of the *likely* range of SLR. Five arguments support this assessment. First, the partial loss of Pine Island Glacier is included by Little et al. (2013a) in their range and the full loss of the ice stream (if it were to occur) is thought to raise sea level by centimetres only (consistent with the use of the Little et al. (2013a) 5 to 95% range as the assessed *likely* range). Second, the current grounding line position of the neighbouring Thwaites Glacier appears to be more stable than that of Pine Island Glacier (Parizek et al., 2013) so that its potentially large contribution to SLR by 2100 is assessed to have a significantly lower probability. Third, there is a low probability that atmospheric warming in the 21st century will lead to extensive grounding line retreat outside of the Antarctic Peninsula because summer air temperatures will not rise to the level where significant surface melt and ponding occur. Fourth, although this retreat may be triggered by oceanic warming during the 21st century (in particular, under the Filchner-Ronne ice shelf), current literature suggests that this may occur late in the century (Hellmer et al., 2012), reducing the time over which enhanced outflow could affect end-of-century SLR. Finally, there are theoretical grounds to believe that grounding line retreat may stabilize (Gudmundsson et al., 2012) so that MISI (and associated high SLR) is not inevitable.

We next assess the magnitude of potential SLR at 2100 in the event that MISI affects the Antarctic ice sheet. Bindschadler et al. (2013), Katsman et al. (2011), the NRC (2012), and Pfeffer et al. (2008) presented contrasting approaches that can be used to make this assessment, which are upper-end estimates of 693, 490, 785 and 615 mm, respectively. Together this literature suggests with *medium confidence* that this contribution would be several tenths of a metre. The literature does not offer a means of assessing the probability of this contribution, however, other than our assessment (above) that it lies above the *likely* range.

Literature investigating the relation between the SLR generated by dynamical change and emission scenario does not currently exist. There is also a lack of literature on the relation between emission scenario and the intrusions of warm water into ice-shelf cavities thought to be important in triggering observed mass loss (Jacobs et al., 2011) and potentially important in the future (Hellmer et al., 2012). It is therefore premature to attach a scenario dependence to projections of dynamical change, even though such a dependency is expected to exist.

Likely increases in snowfall over the next century (Section 13.4.4.1) will affect the amount of ice lost by outflow across the grounding line because of local changes in ice thickness and stress regime (Huybrechts and De Wolde, 1999). This effect was incorporated in AR4 projections for 2100 as a compensatory mass loss amounting to 0 to 10% of the SMB mass gain (Gregory and Huybrechts, 2006). Winkelmann et al. (2012) re-evaluated the effect and reported a range of 15 to 35% for the next century (30 to 65% after 500 years). The two studies are difficult to compare because of differences in model physics and experimental design so that the use of their joint range (0 to 35%) is an appropriate assessment of the *likely* range of this effect. This range is supported by Barrand et al. (2013), who quantified the effect for the Antarctic Peninsula as ~15% of SMB. Moreover, because this contribution relies on similar physics to the grounding line migration discussed above, it is appropriate to assume that their uncertainties are correlated. Winkelmann et al. (2012) showed that the fractional size of this compensatory effect is independent of scenario. Accounting for this effect equates to SLR of 0 to 32 mm by 2100 based on the SMB range over all emission-scenario projections in Section 13.5.1.1.

Beyond the 21st century, only projections with coarse-resolution ice sheet–climate models are available (Vizcaíno et al., 2010; Huybrechts et al., 2011). *Confidence* in the ability of these two models to capture both change in the oceanic circulation around Antarctica and the response of the ice sheet to these changes, especially a potential grounding line retreat, is *low*. The model applied by Vizcaíno et al. (2010) lacks a dynamic representation of ice shelves. Because dynamic ice discharge from Antarctica occurs predominately through ice shelves, it is *likely* that the projections using this model considerably underestimate the Antarctic contribution.

In summary, it is *likely* that dynamical change within the Antarctic ice sheet will lead to SLR during the next century with a range of –20 to 185 mm. SLR beyond the *likely* range is poorly constrained and considerably larger increases are possible (the underlying probability distribution is asymmetric towards larger rise), which will probably be associated with the MISI (Box 13.2). Although the likelihood of such SLR cannot yet be assessed more precisely than falling above the *likely* range, literature suggests (with *medium confidence*) that its potential magnitude is several tenths of a metre. We are unable to assess SLR as a function of emission scenario, although a dependency of SLR on scenario is expected to exist. In addition, coupling between SMB and dynamical change is *likely* to make a further contribution to SLR of 0 to 35% of the SMB. All the available literature suggests that this dynamical contribution to sea level rise will continue well beyond 2100.

13.4.4.3 Possible Irreversibility of Ice Loss from West Antarctica

Due to relatively weak snowfall on Antarctica and the slow ice motion in its interior, it can be expected that the WAIS would take at least several thousand years to regrow if it was eliminated by dynamic ice discharge. Consequently any significant ice loss from West Antarctic that occurs within the next century will be irreversible on a multi-centennial to millennial time scale. We discuss here the possibility of abrupt dynamic ice loss from West Antarctica (see Section 12.5.5 for definition of abrupt).

Information on the ice and bed topography of WAIS suggests that it has about 3.3 m of equivalent global sea level grounded on areas with downward sloping bedrock (Bamber et al., 2009). As detailed in Box 13.2, large areas of the WAIS may therefore be subject to potential ice loss via the MISI. As it is the case for other potential instabilities within the climate system (Section 12.5.5), there are four lines of evidence to assess the likelihood of a potential occurrence: theoretical understanding, present-day observations, numerical simulations, and paleo records.

The MISI is based on a number of studies that indicated the theoretical existence of the instability (Weertman, 1961; Schoof, 2007a) (see also Box 13.2). The most fundamental derivation, that is, starting from a first-principle ice equation, states that in one-dimensional ice flow the grounding line between grounded ice sheet and floating ice shelf cannot be stable on a landward sloping bed. The limitation of the one-dimensional case disregards possible stabilizing effects of the ice shelves (Dupont and Alley, 2005). Although it is clear that ice shelves that are laterally constrained by embayments inhibit ice flow into the ocean, the effect has not been quantified against the MISI. The same is true for other potentially stabilizing effects such as sedimentation near the grounding line (Alley et al., 2007) or the influence of large-scale bedrock roughness (i.e., topographic pinning points) on ice flow. Although these stabilizing effects need to be further investigated and quantified against the destabilizing effect of the MISI, no studies are available that would allow dismissing the MISI on theoretical grounds.

Although direct observations of ice dynamics are available, they are neither detailed enough nor cover a sufficiently long period to allow the monitoring of the temporal evolution of an MISI. Most Antarctic ice loss that has been detected during the satellite period has come from the WAIS (Rignot et al., 2008; Pritchard et al., 2012). Some studies have found an acceleration of ice loss (Rignot et al., 2011) as well as enhanced basal ice-shelf melting (Pritchard et al., 2012), but the short period of observations does not allow one to either dismiss or confirm that these changes are associated with destabilization of WAIS.

Paleo records suggest that WAIS may have deglaciated several times during warm periods of the last 5 million years, but they contain no information on rates (Naish et al., 2009). Although coarse-resolution models are in principle capable of modelling the MISI, there is *medium confidence* in their ability to simulate the correct response time to external perturbations on decadal to centennial time scales (Pattyn et al., 2013). One of these models (Pollard and DeConto, 2009) reproduced paleo records of deglaciation with a forced ice-sheet model at 40 km resolution and parameterized ice flow across the grounding line according to Schoof (2007a). These simulations showed a sea level rise of about 7 m over time spans of 1000 to 7000 years with approximately equal contributions from West and East Antarctica. However, no available model results or paleo records have indicated the possibility of self-accelerated ice discharge from these regions.

In summary, ice-dynamics theory, numerical simulations, and paleo records indicate that the existence of a marine-ice sheet instability associated with abrupt and irreversible ice loss from the Antarctic ice sheet is possible in response to climate forcing. However, theoretical considerations, current observations, numerical models, and paleo records currently do not allow a quantification of the timing of the onset of such an instability or of the magnitude of its multi-century contribution.

Box 13.2 | History of the Marine Ice-Sheet Instability Hypothesis

Marine ice sheets rest on bedrock that is submerged below sea level (often by 2 to 3 km). The most well-researched marine ice sheet is the West Antarctic ice sheet (WAIS) where approximately 75% of the ice sheet's area currently rests on bedrock below sea level. The East Antarctic ice sheet (EAIS), however, also has appreciable areas grounded below sea level (~35%), in particular around the Totten and Cook Glaciers.

These ice sheets are fringed by floating ice shelves, which are fed by flow from grounded ice across a grounding line (GL). The GL is free to migrate both seawards and landwards as a consequence of the local balance between the weight of ice and displaced ocean water. Depending on a number of factors, which include ice-shelf extent and geometry, ice outflow to the ocean generally (but not always) increases with ice thickness at the GL. Accordingly, when the ice sheet rests on a bed that deepens towards the ice-sheet interior (see Box 13.2, Figure 1a), the ice outflow to the ocean will generally increase as the GL retreats. It is this feature that gives rise to the Marine Ice-Sheet Instability (MISI), which states that a GL cannot remain stable on a landward-deepening slope. Even if snow accumulation and outflow were initially in balance (Box 13.2, Figure 1b), natural fluctuations in climate cause the GL to fluctuate slightly (Box 13.2, Figure 1c). In the case of a retreat, the new GL position is then associated with deeper bedrock and thicker ice, so that outflow increases (Box 13.2, Figure 1d). This increased outflow leads to further, self-sustaining retreat until a region of shallower, seaward-sloping bedrock is reached. Stable configurations can therefore exist only where the GL rests on slopes that deepen towards the ocean. A change in climate can therefore potentially force a large-scale retreat of the GL from one bedrock ridge to another further inland. *(continued on next page)*

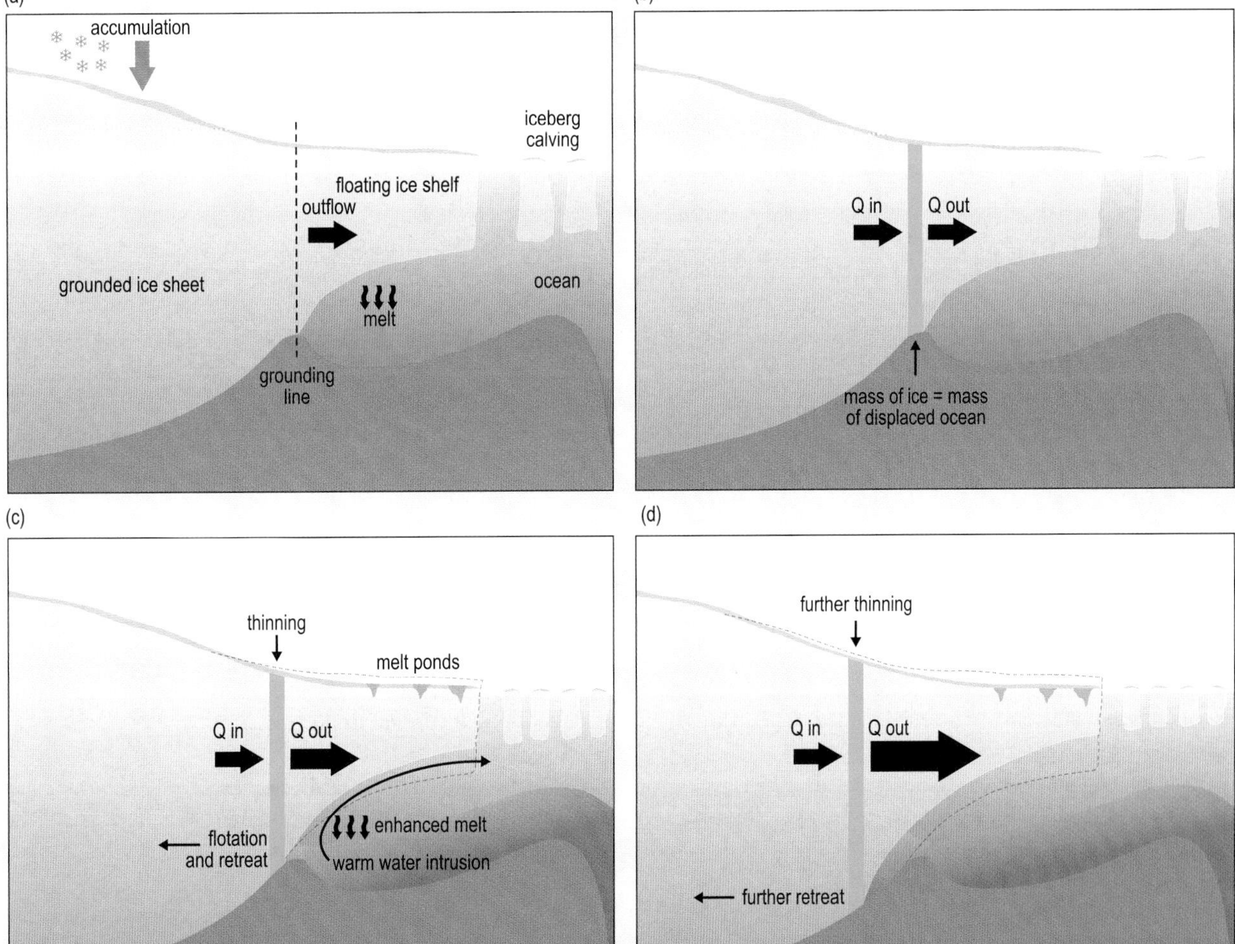

Box 13.2, Figure 1 | Schematic of the processes leading to the potentially unstable retreat of a grounding line showing (a) geometry and ice fluxes of a marine ice sheet, (b) the grounding line in steady state, (c) climate change triggering mass outflow from the ice sheet and the start of grounding line retreat and (d) self-sustained retreat of the grounding line.

Box 13.2 (continued)

The MISI has a long history based on theoretical discussions that were started by Weertman (1974) and Mercer (1978), and has seen many refinements over the subsequent years. The advent of satellite-based observations has given fresh impetus to this debate, in particular work on the GL retreat and associated thinning of Pine Island (PIG), Thwaites (TG) and Smith Glaciers (all part of the WAIS), which are collectively responsible for most of Antarctica's present mass loss (Rignot et al., 2008). These observations highlighted the need to develop a better understanding of the MISI to make more accurate projections of the ice sheet's future contribution to sea level rise.

Early studies of the MISI were not based on a formal derivation from the basic laws of mechanics thought to control ice-sheet flow and the robustness of their results was therefore difficult to assess. An open question was the expected impact of changes at the GL on the ice-sheet flow (Hindmarsh, 1993). Recently, however, a more complete analysis from first principles has been developed that suggests that the fundamental relation between thickness and flux at the GL exists and has a power of ~5 (i.e., that a 10% increase in thickness leads to a 60% increase in flux) (Schoof, 2007b, 2011). This analysis, however, does not include ice shelves that occupy laterally constrained embayments, which is often the case (for instance at PIG). In such situations, drag from ice-shelf sidewalls may suppress the positive feedback between increasing ice thickness and ice flux at the GL (Dupont and Alley, 2005; Goldberg et al., 2009; Gudmundsson et al., 2012). Other factors that could suppress the instability include a sea level fall adjacent to the GL resulting from the isostatic and gravitational effects of ice loss (Gomez et al., 2010b).

Two processes that could trigger GL retreat are particularly relevant to contemporary polar climate change. The first is the presence of warmer ocean water under ice shelves, which leads to enhanced submarine ice-shelf melt (Jacobs et al., 2011). The second is the presence of melt water ponds on the surface of the ice shelf, which can cause stress concentrations allowing fractures to penetrate the full ice-shelf thickness. This process appears to have been a primary factor in the collapse of the Larsen B Ice Shelf (LBIS) over the course of two months in 2002 (MacAyeal et al., 2003). The collapse of the LBIS provided a natural demonstration of the linkage between the structural integrity of an ice shelf and the flow of grounded ice draining into it. Following the breakup of LBIS, the speeds of the glaciers feeding the collapsed portion of the shelf increased two- to eightfold, while the flow of glaciers draining into a surviving sector was unaltered (Rignot et al., 2004; Scambos et al., 2004; Rott et al., 2011). This indicates that a mechanical link does indeed exist between shelf and sheet, and has important implications for the future evolution of the far more significant PIG and TG systems of the WAIS.

The recent strides made in placing MISI on a sound analytical footing are, however, limited to the analysis of steady states. Numerical modelling is needed to simulate the GL retreat rates that are required to make accurate SLR projections. There are major challenges in designing models whose results are not controlled by the details of their numerical design. Problems arise at the GL because, in addition to flotation, basal traction is dramatically reduced as the ice loses contact with the underlying bedrock (Pattyn et al., 2006). This is a topic of active research, and a combination of more complete modelling of the GL stress regime (Favier et al., 2012) and the use of high-resolution (subkilometre) models (Durand et al., 2009; Cornford et al., 2013) shows promise towards resolving these problems. Much progress has also been made by using model inter-comparison as a means of understanding these effects (Pattyn et al., 2013).

13.4.5 Anthropogenic Intervention in Water Storage on Land

The potential future effects that human activities have on changing water storage on land, thus affecting sea level, have been little studied in the published peer-reviewed scientific literature. For depletion of groundwater arising from extraction (for agriculture and other uses), we consider two possibilities. The first assumes that this contribution to GMSL rise continues throughout the 21st century at the rate of 0.40 ± 0.11 mm yr^{-1} (mean ± SD) assessed for 2001–2008 by Konikow (2011), amounting to 38 [21 to 55] mm by 2081–2100 relative to 1986–2005. The second uses results from land surface hydrology models (Wada et al., 2012) with input from climate and socioeconomic projections for SRES scenarios, yielding 70 [51 to 90] mm for the same time interval. Because of the improved treatment of groundwater recharge by Wada et al. (2012), this is less than Rahmstorf et al. (2012b) obtained by assuming that the groundwater extraction estimates of Wada et al. (2010) can be scaled up in the future with global population. These two possibilities indicate a range of about 20 to 90 mm for the contribution of groundwater depletion to GMSL rise.

For the rate of impoundment of water in reservoirs, we evaluate two possibilities. The first assumes it will continue throughout the 21st century (e.g., Lempérière, 2006) at the average rate of –0.2 ± 0.05 mm yr^{-1} SLE (mean ± SD) estimated for 1971–2010 using data updated from Chao et al. (2008), giving a negative contribution to GMSL rise of –19 [–11 to –27] mm by 2081–2100 relative to 1986–2005. The second assumes it will be zero after 2010 (i.e., no further net impoundment), as shown for the 1990s and 2000s by Lettenmaier and Milly (2009) (see Section 13.3.4 for discussion). A zero contribution implies a balance between further construction of reservoir capacity and reduction of storage volume by sedimentation, each of which could plausibly

Frequently Asked Questions

FAQ 13.2: Will the Greenland and Antarctic Ice Sheets Contribute to Sea Level Change over the Rest of the Century?

The Greenland, West and East Antarctic ice sheets are the largest reservoirs of freshwater on the planet. As such, they have contributed to sea level change over geological and recent times. They gain mass through accumulation (snowfall) and lose it by surface ablation (mostly ice melt) and outflow at their marine boundaries, either to a floating ice shelf, or directly to the ocean through iceberg calving. Increases in accumulation cause global mean sea level to fall, while increases in surface ablation and outflow cause it to rise. Fluctuations in these mass fluxes depend on a range of processes, both within the ice sheet and without, in the atmosphere and oceans. Over the course of this century, however, sources of mass loss appear set to exceed sources of mass gain, so that a continuing positive contribution to global sea level can be expected. This FAQ summarizes current research on the topic and provides indicative magnitudes for the various end-of-century (2081-2100 with respect to 1986-2005) sea level contributions from the full assessment, which are reported as the two-in-three probability level across all emission scenarios.

Over millennia, the slow horizontal flow of an ice sheet carries mass from areas of net accumulation (generally, in the high-elevation interior) to areas of net loss (generally, the low-elevation periphery and the coastal perimeter). At present, Greenland loses roughly half of its accumulated ice by surface ablation, and half by calving. Antarctica, on the other hand, loses virtually all its accumulation by calving and submarine melt from its fringing ice shelves. Ice shelves are floating, so their loss has only a negligible direct effect on sea level, although they can affect sea level indirectly by altering the mass budget of their parent ice sheet (see below).

In East Antarctica, some studies using satellite radar altimetry suggest that snowfall has increased, but recent atmospheric modelling and satellite measurements of changes in gravity find no significant increase. This apparent disagreement may be because relatively small long-term trends are masked by the strong interannual variability of snowfall. Projections suggest a substantial increase in 21st century Antarctic snowfall, mainly because a warmer atmosphere would be able to carry more moisture into polar regions. Regional changes in atmospheric circulation probably play a secondary role. For the whole of the Antarctic ice sheet, this process is projected to contribute between 0 and 70 mm to sea level fall.

Currently, air temperatures around Antarctica are too cold for substantial surface ablation. Field and satellite-based observations, however, indicate enhanced outflow—manifested as ice-surface lowering—in a few localized coastal regions. These areas (Pine Island and Thwaites Glaciers in West Antarctica, and Totten and Cook Glaciers in East Antarctica) all lie within kilometre-deep bedrock troughs towards the edge of Antarctica's continental shelf. The increase in outflow is thought to have been triggered by regional changes in ocean circulation, bringing warmer water in contact with floating ice shelves.

On the more northerly Antarctic Peninsula, there is a well-documented record of ice-shelf collapse, which appears to be related to the increased surface melting caused by atmospheric warming over recent decades. The subsequent thinning of glaciers draining into these ice shelves has had a positive—but minor—effect on sea level, as will any further such events on the Peninsula. Regional projections of 21st century atmospheric temperature change suggest that this process will probably not affect the stability of the large ice shelves of both the West and East Antarctica, although these ice shelves may be threatened by future oceanic change (see below).

Estimates of the contribution of the Antarctic ice sheets to sea level over the last few decades vary widely, but great strides have recently been made in reconciling the observations. There are strong indications that enhanced outflow (primarily in West Antarctica) currently outweighs any increase in snow accumulation (mainly in East Antarctica), implying a tendency towards sea level rise. Before reliable projections of outflow over the 21st century can be made with greater confidence, models that simulate ice flow need to be improved, especially of any changes in the grounding line that separates floating ice from that resting on bedrock and of interactions between ice shelves and the ocean. The concept of 'marine ice-sheet instability' is based on the idea that the outflow from an ice sheet resting on bedrock below sea level increases if ice at the grounding line is thicker and, therefore, faster flowing. On bedrock that slopes downward towards the ice-sheet interior, this creates a vicious cycle of increased outflow, causing ice at the grounding line to thin and go afloat. The grounding line then retreats down slope into thicker ice that, in turn, drives further increases in outflow. This feedback could potentially result in the rapid loss of parts of the ice sheet, as grounding lines retreat along troughs and basins that deepen towards the ice sheet's interior.

FAQ 13.2 (continued)

Future climate forcing could trigger such an unstable collapse, which may then continue independently of climate. This potential collapse might unfold over centuries for individual bedrock troughs in West Antarctica and sectors of East Antarctica. Much research is focussed on understanding how important this theoretical concept is for those ice sheets. Sea level could rise if the effects of marine instability become important, but there is not enough evidence at present to unambiguously identify the precursor of such an unstable retreat. Change in outflow is projected to contribute between –20 (i.e., fall) and 185 mm to sea level rise by year 2100, although the uncertain impact of marine ice-sheet instability could increase this figure by several tenths of a metre. Overall, increased snowfall seems set to only partially offset sea level rise caused by increased outflow.

In Greenland, mass loss through more surface ablation and outflow dominates a possible recent trend towards increased accumulation in the interior. Estimated mass loss due to surface ablation has doubled since the early 1990s. This trend is expected to continue over the next century as more of the ice sheet experiences surface ablation for longer periods. Indeed, projections for the 21st century suggest that increasing mass loss will dominate over weakly increasing accumulation. The refreezing of melt water within the snow pack high up on the ice sheet offers an important (though perhaps temporary) dampening effect on the relation between atmospheric warming and mass loss.

Although the observed response of outlet glaciers is both complex and highly variable, iceberg calving from many of Greenland's major outlet glaciers has increased substantially over the last decade, and constitutes an appreciable additional mass loss. This seems to be related to the intrusion of warm water into the coastal seas around Greenland, but it is not clear whether this phenomenon is related to inter-decadal variability, such as the North Atlantic

(continued on next page)

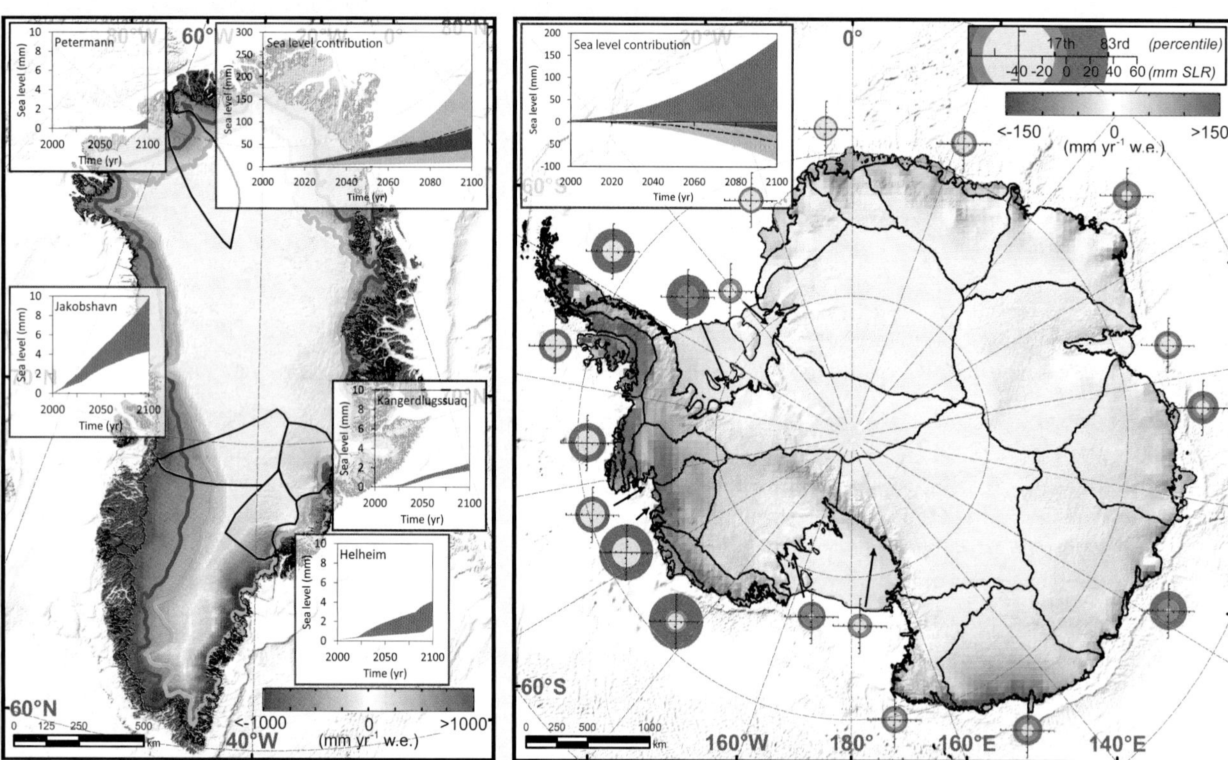

FAQ 13.2, Figure 1 | Illustrative synthesis of projected changes in SMB and outflow by 2100 for (a) Greenland and (b) Antarctic ice sheets. Colours shown on the maps refer to projected SMB change between the start and end of the 21st century using the RACMO2 regional atmospheric climate model under future warming scenarios A1B (Antarctic) and RCP4.5 (Greenland). For Greenland, average equilibrium line locations during both these time periods are shown in purple and green, respectively. Ice-sheet margins and grounding lines are shown as black lines, as are ice-sheet sectors. For Greenland, results of flowline modelling for four major outlet glaciers are shown as inserts, while for Antarctica the coloured rings reflect projected change in outflow based on a probabilistic extrapolation of observed trends. The outer and inner radius of each ring indicate the upper and lower bounds of the two-thirds probability range of the contribution, respectively (scale in upper right); red refers to mass loss (sea level rise) while blue refers to mass gain (sea level fall). Finally, the sea level contribution is shown for each ice sheet (insert located above maps) with light grey referring to SMB (model experiment used to generate the SMB map is shown as a dashed line) and dark grey to outflow. All projections refer to the two-in-three probability range across all scenarios.

13

FAQ 13.2 (continued)

Oscillation, or a longer term trend associated with greenhouse gas–induced warming. Projecting its effect on 21st century outflow is therefore difficult, but it does highlight the apparent sensitivity of outflow to ocean warming. The effects of more surface melt water on the lubrication of the ice sheet's bed, and the ability of warmer ice to deform more easily, may lead to greater rates of flow, but the link to recent increases in outflow is unclear. Change in the net difference between surface ablation and accumulation is projected to contribute between 10 and 160 mm to sea level rise in 2081-2100 (relative to 1986-2005), while increased outflow is projected to contribute a further 10 to 70 mm (Table 13.5).

The Greenland ice sheet has contributed to a rise in global mean sea level over the last few decades, and this trend is expected to increase during this century. Unlike Antarctica, Greenland has no known large-scale instabilities that might generate an abrupt increase in sea level rise over the 21st century. A threshold may exist, however, so that continued shrinkage might become irreversible over multi-centennial time scales, even if the climate were to return to a pre-industrial state over centennial time scales. Although mass loss through the calving of icebergs may increase in future decades, this process will eventually end when the ice margin retreats onto bedrock above sea level where the bulk of the ice sheet resides.

have a rate of about 1% yr^{-1} of existing capacity (Lempérière, 2006; Lettenmaier and Milly, 2009). These two possibilities together indicate a range of about 0 to 30 mm of GMSL fall for the contribution of reservoir impoundment.

Our assessment thus leads to a range of –10 to +90 mm for the net contribution to GMSL rise from anthropogenic intervention in land water storage by 2081–2100 relative to 1986–2005. This range includes the range of 0 to 40 mm assumed by Katsman et al. (2008). Because of the limited information available, we do not have sufficient confidence to give ranges for individual RCP scenarios.

13.5 Projections of Global Mean Sea Level Rise

Process-based projections for GMSL rise during the 21st century, given in Section 13.5.1, are the sum of contributions derived from models that were evaluated by comparison with observations in Section 13.3 and used to project the contributions in Section 13.4. Projections of GMSL rise by semi-empirical models (SEMs) are given in Section 13.5.2. We compare these two and other approaches in Section 13.5.3 and assess the level of confidence that we can place in each approach. Longer term projections are discussed in Section 13.5.4.

13.5.1 Process-Based Projections for the 21st Century

The process-based projections of GMSL rise for each RCP scenario are based on results from 21 CMIP5 AOGCMs from which projections of SAT change and thermal expansion are available (see Section 13.4.1). Where CMIP5 results were not available for a particular AOGCM and scenario, they were estimated (Good et al., 2011; 2013) (Section 12.4.1.2; Supplementary Material). The projections of thermal expansion do not include an adjustment for the omission of volcanic forcing in AOGCM spin-up (Section 13.3.4.2), as this is uncertain and relatively small (about 10 mm during the 21st century). Changes in glacier and ice-sheet SMB are calculated from the global mean SAT projections using parameterizations derived from the results of process-based models of these components (note that glaciers on Antarctica are covered by the Antarctic ice-sheet SMB projection, and are therefore not included in the glacier projections) (Sections 13.4.2, 13.4.3.1, 13.4.4.1 and Supplementary Material). According to the assessment in Section 12.4.1.2, global mean SAT change is *likely* to lie within the 5 to 95% range of the projections of CMIP5 models. Following this assessment, the 5 to 95% range of model results for each of the GMSL rise contributions that is projected on the basis of CMIP5 results is interpreted as the *likely* range.

Possible ice-sheet dynamical changes by 2100 are assessed from the published literature (Sections 13.4.3.2 and 13.4.4.2), which as yet provides only a partial basis for making projections related to particular scenarios. They are thus treated as independent of scenario, except that a higher rate of change is used for Greenland ice sheet outflow under RCP8.5. Projections of changes in land water storage due to human intervention are also treated as independent of emissions scenario, because we do not have sufficient information to give ranges for individual scenarios. The scenario-independent treatment does not imply that the contributions concerned will not depend on the scenario followed, only that the current state of knowledge does not permit a quantitative assessment of the dependence. For each of these contributions, our assessment of the literature provides a 5-95% range for the late 21st century (2100 for Greenland and Antarctic ice-sheet dynamics, 2081-2100 for land water storage). For consistency with the treatment of the CMIP5-derived results, we interpret this range as the *likely* range. We assume that each of these contributions begins from its present-day rate and that the rate increases linearly in time, in order to interpolate from the present day to the late 21st century (see Supplementary Material for details).

The *likely* range of GMSL rise given for each RCP combines the uncertainty in global climate change, represented by the CMIP5 ensemble (Section 12.4.1.2), with the uncertainties in modelling the contributions to GMSL. The part of the uncertainty related to the magnitude of global climate change is correlated among all the scenario-dependent contributions, while the methodological uncertainties are treated as independent (see also Supplementary Material).

The sum of the projected contributions gives the *likely* range for future GMSL rise. The median projections for GMSL in all scenarios lie within a range of 0.05 m until the middle of the century (Figure 13.11), because the divergence of the climate projections has a delayed effect owing to the time-integrating characteristic of sea level. By the late 21st century (over an interval of 95 years, between the 20-year mean of 2081–2100 and the 20-year mean of 1986–2005), they have a spread of about 0.25 m, with RCP2.6 giving the least amount of rise (0.40 [0.26 to 0.55] m) (*likely* range) and RCP8.5 giving the most (0.63 [0.45 to 0.82] m). RCP4.5 and RCP6.0 are very similar at the end of the century (0.47 [0.32 to 0.63] m and 0.48 [0.33 to 0.63]] m respectively), but RCP4.5 has a greater rate of rise earlier in the century than RCP6.0 (Figure 13.10 and Table 13.5). At 2100, the *likely* ranges are 0.44 [0.28–0.61] m (RCP2.6), 0.53 [0.36–0.71] m (RCP4.5), 0.55 [0.38–0.73] m (RCP6.0), and 0.74 [0.52–0.98] m (RCP8.5).

In all scenarios, the rate of rise at the start of the RCP projections (2007–2013) is about 3.7 mm yr^{-1}, slightly above the observational range of 3.2 [2.8 to 3.6] mm yr^{-1} for 1993–2010, because the modelled contributions for recent years, although consistent with observations for 1993–2010 (Section 13.3), are all in the upper part of the observational ranges, perhaps related to the simulated rate of climatic warming being greater than has been observed (Box 9.2). In the projections, the rate of rise initially increases. In RCP2.6 it becomes roughly constant (central projection 4.5 mm yr^{-1}) before the middle of the century, and subsequently declines slightly. The rate of rise becomes roughly constant in RCP4.5 and RCP6.0 by the end of the century, whereas acceleration continues throughout the century in RCP8.5, reaching 11 [8 to 16] mm yr^{-1} in 2081–2100.

In all scenarios, thermal expansion is the largest contribution, accounting for about 30 to 55% of the projections. Glaciers are the next largest, accounting for 15-35% of the projections. By 2100, 15 to 55% of the present volume of glaciers outside Antarctica is projected to be eliminated under RCP2.6, and 35 to 85% under RCP8.5 (Table 13.SM.2). SMB change on the Greenland ice sheet makes a positive contribution, whereas SMB change in Antarctica gives a negative contribution (Sections 13.4.3.1 and 13.4.4.1). The positive contribution due to rapid dynamical changes that result in increased ice outflow from both ice sheets together has a *likely* range of 0.03 to 0.20 m in RCP8.5 and 0.03 to 0.19 m in the other RCPs. There is a relatively small positive contribution from human intervention in land water storage, predominantly due to increasing extraction of groundwater.

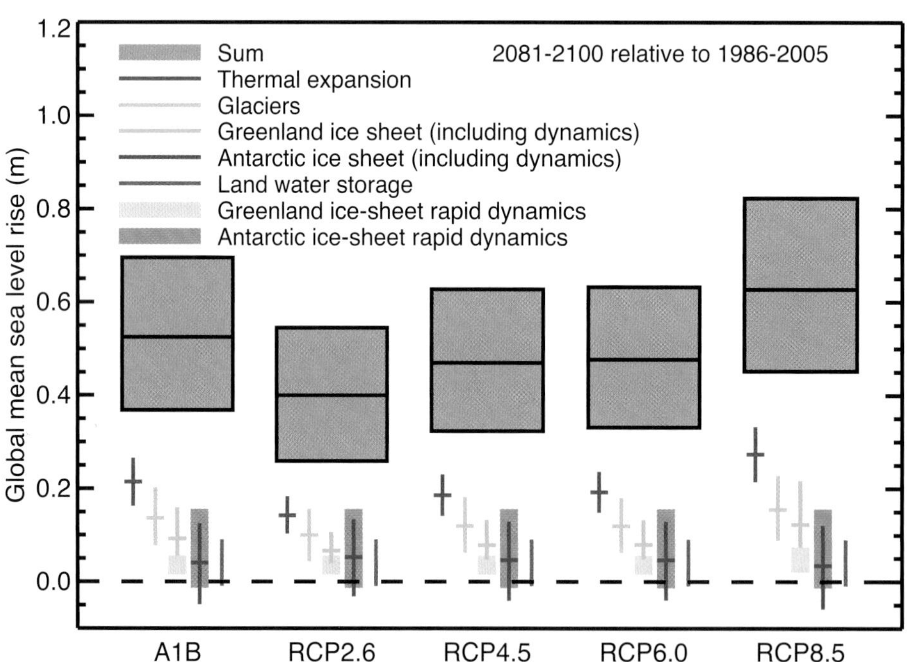

Figure 13.10 | Projections from process-based models with *likely* ranges and median values for global mean sea level rise and its contributions in 2081–2100 relative to 1986–2005 for the four RCP scenarios and scenario SRES A1B used in the AR4. The contributions from ice sheets include the contributions from ice-sheet rapid dynamical change, which are also shown separately. The contributions from ice-sheet rapid dynamical change and anthropogenic land water storage are treated as having uniform probability distributions, and as independent of scenario (except that a higher rate of change is used for Greenland ice-sheet outflow under RCP8.5). This treatment does not imply that the contributions concerned will not depend on the scenario followed, only that the current state of knowledge does not permit a quantitative assessment of the dependence. See discussion in Sections 13.5.1 and 13.5.3 and Supplementary Material for methods. Only the collapse of the marine-based sectors of the Antarctic ice sheet, if initiated, could cause global mean sea level (GMSL) to rise substantially above the *likely* range during the 21st century. This potential additional contribution cannot be precisely quantified but there is *medium confidence* that it would not exceed several tenths of a meter of sea level rise.

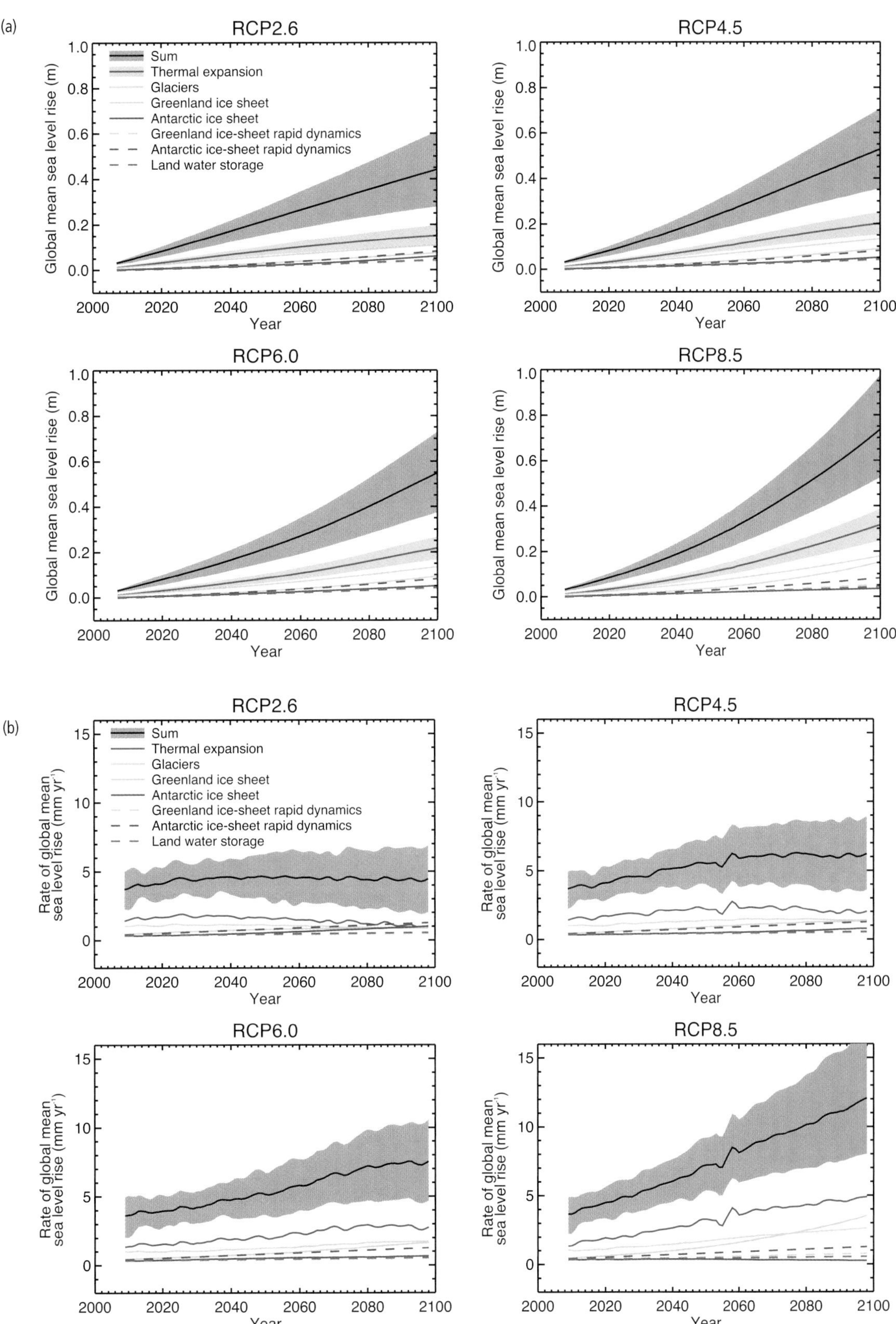

Figure 13.11 | Projections from process-based models of (a) global mean sea level (GMSL) rise relative to 1986–2005 and (b) the rate of GMSL rise and its contributions as a function of time for the four RCP scenarios and scenario SRES A1B. The lines show the median projections. For GMSL rise and the thermal expansion contribution, the *likely* range is shown as a shaded band. The contributions from ice sheets include the contributions from ice-sheet rapid dynamical change, which are also shown separately. The time series for GMSL rise plotted in (a) are tabulated in Annex II (Table AII.7.7), and the time series of GMSL rise and all of its contributions are available in the Supplementary Material. The rates in (b) are calculated as linear trends in overlapping 5-year periods. Only the collapse of the marine-based sectors of the Antarctic ice sheet, if initiated, could cause GMSL to rise substantially above the *likely* range during the 21st century. This potential additional contribution cannot be precisely quantified but there is *medium confidence* that it would not exceed several tenths of a metre of sea level rise.

Table 13.5 | Median values and *likely* ranges for projections of global mean sea level (GMSL) rise and its contributions in metres in 2081–2100 relative to 1986–2005 for the four RCP scenarios and SRES A1B, GMSL rise in 2046–2065 and 2100, and rates of GMSL rise in mm yr^{-1} in 2081–2100. See Section 13.5.1 concerning how the *likely* range is defined. Because some of the uncertainties in modelling the contributions are treated as uncorrelated, the sum of the lower bound of contributions does not equal the lower bound of the sum, and similarly for the upper bound (see Supplementary Material). Because of imprecision from rounding, the sum of the medians of contributions may not exactly equal the median of the sum. The net contribution (surface mass balance (SMB) + dynamics) for each ice sheet, and the contribution from rapid dynamical change in both ice sheets together, are shown as additional lines below the sum; they are not contributions in addition to those given above the sum. The contributions from ice-sheet rapid dynamical change and anthropogenic land water storage are treated as having uniform probability distributions, uncorrelated with the magnitude of global climate change (except for the interaction between Antarctic ice sheet SMB and outflow), and as independent of scenario (except that a higher rate of change is used for Greenland ice sheet outflow under RCP8.5). This treatment does not imply that the contributions concerned will not depend on the scenario followed, only that the current state of knowledge does not permit a quantitative assessment of the dependence. Regional sea level change is expected in general to differ from the global mean (see Section 13.6).

	SRES A1B	RCP2.6	RCP4.5	RCP6.0	RCP8.5
Thermal expansion	0.21 [0.16 to 0.26]	0.14 [0.10 to 0.18]	0.19 [0.14 to 0.23]	0.19 [0.15 to 0.24]	0.27 [0.21 to 0.33]
Glaciers[a]	0.14 [0.08 to 0.21]	0.10 [0.04 to 0.16]	0.12 [0.06 to 0.19]	0.12 [0.06 to 0.19]	0.16 [0.09 to 0.23]
Greenland ice-sheet SMB[b]	0.05 [0.02 to 0.12]	0.03 [0.01 to 0.07]	0.04 [0.01 to 0.09]	0.04 [0.01 to 0.09]	0.07 [0.03 to 0.16]
Antarctic ice-sheet SMB[c]	–0.03 [–0.06 to –0.01]	–0.02 [–0.04 to –0.00]	–0.02 [–0.05 to –0.01]	–0.02 [–0.05 to –0.01]	–0.04 [–0.07 to –0.01]
Greenland ice-sheet rapid dynamics	0.04 [0.01 to 0.06]	0.04 [0.01 to 0.06]	0.04 [0.01 to 0.06]	0.04 [0.01 to 0.06]	0.05 [0.02 to 0.07]
Antarctic ice-sheet rapid dynamics	0.07 [–0.01 to 0.16]	0.07 [–0.01 to 0.16]	0.07 [–0.01 to 0.16]	0.07 [–0.01 to 0.16]	0.07 [–0.01 to 0.16]
Land water storage	0.04 [–0.01 to 0.09]	0.04 [–0.01 to 0.09]	0.04 [–0.01 to 0.09]	0.04 [–0.01 to 0.09]	0.04 [–0.01 to 0.09]
Global mean sea level rise in 2081–2100	0.52 [0.37 to 0.69]	0.40 [0.26 to 0.55]	0.47 [0.32 to 0.63]	0.48 [0.33 to 0.63]	0.63 [0.45 to 0.82]
Greenland ice sheet	0.09 [0.05 to 0.15]	0.06 [0.04 to 0.10]	0.08 [0.04 to 0.13]	0.08 [0.04 to 0.13]	0.12 [0.07 to 0.21]
Antarctic ice sheet	0.04 [–0.05 to 0.13]	0.05 [–0.03 to 0.14]	0.05 [–0.04 to 0.13]	0.05 [–0.04 to 0.13]	0.04 [–0.06 to 0.12]
Ice-sheet rapid dynamics	0.10 [0.03 to 0.19]	0.10 [0.03 to 0.19]	0.10 [0.03 to 0.19]	0.10 [0.03 to 0.19]	0.12 [0.03 to 0.20]
Rate of global mean sea level rise	8.1 [5.1 to 11.4]	4.4 [2.0 to 6.8]	6.1 [3.5 to 8.8]	7.4 [4.7 to 10.3]	11.2 [7.5 to 15.7]
Global mean sea level rise in 2046–2065	0.27 [0.19 to 0.34]	0.24 [0.17 to 0.32]	0.26 [0.19 to 0.33]	0.25 [0.18 to 0.32]	0.30 [0.22 to 0.38]
Global mean sea level rise in 2100	0.60 [0.42 to 0.80]	0.44 [0.28 to 0.61]	0.53 [0.36 to 0.71]	0.55 [0.38 to 0.73]	0.74 [0.52 to 0.98]
Only the collapse of the marine-based sectors of the Antarctic ice sheet, if initiated, could cause GMSL to rise substantially above the *likely* range during the 21st century. This potential additional contribution cannot be precisely quantified but there is *medium confidence* that it would not exceed several tenths of a meter of sea level rise.					

Notes:

[a] Excluding glaciers on Antarctica but including glaciers peripheral to the Greenland ice sheet.

[b] Including the height–SMB feedback.

[c] Including the interaction between SMB change and outflow.

13.5.2 Semi-Empirical Projections for the 21st Century

The development of semi-empirical models (SEMs) was motivated by two problems. First, process-based modelling was incomplete in the AR4 because of the unavailability of ice-sheet dynamical models which could be used to simulate the observed recent accelerations in ice flow and make projections with confidence (Meehl et al., 2007) (Sections 13.1.4.1, 13.4.3.2 and 13.4.4.2). Second, in all previous IPCC assessments, observed GMSL rise during the 20th century could not be completely accounted for by the contributions to GMSL from thermal expansion, glaciers and ice sheets. For example, the AR4 assessed the mean observational rate for 1961–2003 as 1.8 ± 0.5 mm yr^{-1}, and the sum of contributions as 1.1 ± 0.5 mm yr^{-1} (Bindoff et al., 2007; Hegerl et al., 2007). With the central estimates, only about 60% of observed sea level rise was thus explained, and the potential implication was that projections using process-based models which reproduce only those known contributions would underestimate future sea level rise (Rahmstorf, 2007a; Jevrejeva et al., 2009; Grinsted et al., 2010). SEMs do not aim to solve the two problems that motivated their development, but instead provide an alternative approach for projecting GMSL.

The semi-empirical approach regards a change in sea level as an integrated response of the entire climate system, reflecting changes in the dynamics and thermodynamics of the atmosphere, ocean and cryosphere; it does not explicitly attribute sea level rise to its individual physical components. SEMs use simple physically motivated relationships, with various analytical formulations and parameters determined from observational time series, to predict GMSL for the 21st century (Figure 13.12 and Table 13.6) and beyond, from either global mean SAT (Rahmstorf, 2007a; Horton et al., 2008; Vermeer and Rahmstorf, 2009; Grinsted et al., 2010; Rahmstorf et al., 2012b) or RF (Jevrejeva et al., 2009; 2010, 2012a).

SEMs are designed to reproduce the observed sea level record over their period of calibration, as this provides them with model parameters needed to make projections (Rahmstorf, 2007a; Jevrejeva et al., 2009; Vermeer and Rahmstorf, 2009; Grinsted et al., 2010). A test of the predictive skill of the models requires simulating a part of the observed record that has not been used for calibration. For instance, Rahmstorf (2007b) calibrated for 1880–1940 and predicted 1940–2000, obtaining results within 0.02 m of observed. Jevrejeva et al. (2012b) calibrated

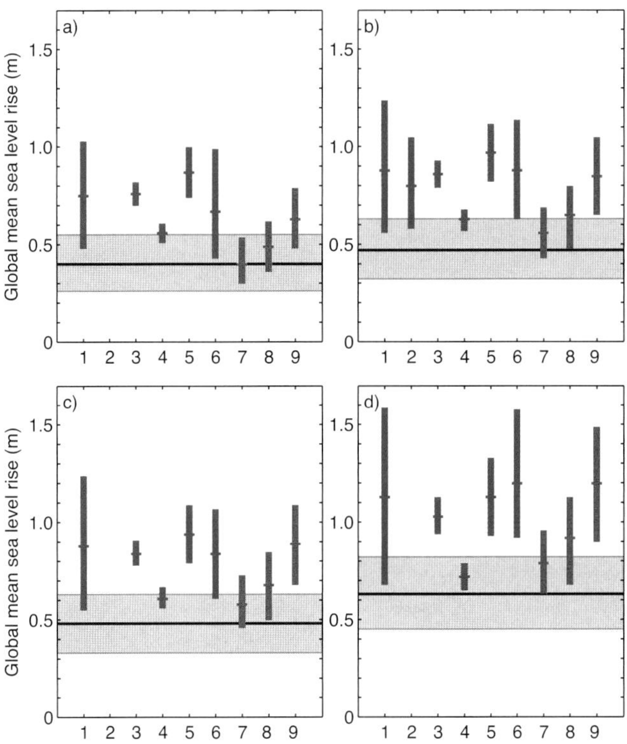

Figure 13.12 | Median and range (5 to 95%) for projections of global mean sea level rise (metres) in 2081–2100 relative to 1986–2005 by semi-empirical models for (a) RCP2.6, (b) RCP4.5, (c) RCP6.0 and (d) RCP8.5. Blue bars are results from the models using RCP temperature projections, red bars are using RCP radiative forcing (RF). The numbers on the horizontal axis refer to the literature source of the projection and the sea level reconstruction used for calibration (for studies using RCP temperature projections) or reconstruction of RF (for studies using RCP RF). (1) Rahmstorf et al. (2012b), with Kemp et al. (2011); (2) Schaeffer et al. (2012); (3) Rahmstorf et al. (2012b), with Church and White (2006); (4) Rahmstorf et al. (2012b), with Church and White (2011); (5) Rahmstorf et al. (2012b), with Jevrejeva et al. (2008); (6) Grinsted et al. (2010), with Moberg et al. (2005); (7) Jevrejeva et al. (2012a), with Goosse et al. (2005); (8) Jevrejeva et al. (2012a), with Crowley et al. (2003); (9) Jevrejeva et al. (2012a) with Tett et al. (2007). Also shown for comparison is the median (thick black line) and *likely* range (horizontal grey bar) (as defined in Section 13.5.1) from the process-based projections (Table 13.5), which are assessed as having *medium confidence*, in contrast to SEMs, which are assessed as having *low confidence* (Section 13.5.3).

up to 1950 and predicted 0.03 m (about 25%) less than observed for 1950–2009, and 3.8 mm yr^{-1} for 1993–2010, which is about 20% more than observed.

The GMSL estimates used for calibrating the SEMs are based on the existing sparse network of long tide-gauge records, and are thus uncertain, especially before the late 19th century; these uncertainties are reflected in the observational estimates of the rate of GMSL rise (Sections 3.7 and 13.2.2). Consequently, the projections may be sensitive to the statistical treatment of the temporal variability in the instrumental record of sea level change (Holgate et al., 2007; Rahmstorf, 2007b; Schmith et al., 2007). Rahmstorf et al. (2012b) reported that GMSL projections for the RCP4.5 scenario for 2100 (Table 13.6) varied by ±0.04 m when the embedding dimension used for temporal smoothing during the calibration was varied within a range of 0 to 25 years.

Furthermore, there is some sensitivity to the choice of data sets used for calibration. For instance, when calibrated up to 1960 and used to predict 1961–2003, the model of Bittermann et al. (2013) overestimates the GMSL data set of Jevrejeva et al. (2008) by 75%, but makes an accurate estimate for the Church and White (2011) data set, although these two data sets have similar rates of sea level rise in the predicted period. The central projections of Rahmstorf et al. (2012b) for 2100 under RCP4.5 (Table 13.6) for calibration with the GMSL data set of Church and White (2006) are about 0.2 m more than for calibration with the Church and White (2011) data set, although the two Church and White (2006, 2011) data sets differ at all times by less than one standard deviation. The ranges of the projections by Grinsted et al. (2010) and Jevrejeva et al. (2010, 2012a, 2012b) allow for the uncertainty in the GMSL reconstructions through the use of an uncertainty covariance matrix in determining the model parameters. Grinsted et al. (2010) also investigated the sensitivity to the temperature data set used as predictor, and Jevrejeva et al. (2010) investigated the sensitivity to RF as predictor (Table 13.6). In the latter case, three data sets gave median projections under RCP4.5 for 2100 within a range of about ±0.20 m.

SEM projections will be biased unless contributions to past GMSL rise which correlate with but are not physically related to contemporary changes in the predictor variable (either global mean SAT change or RF) are subtracted from the observational sea level record before the calibration (Vermeer and Rahmstorf, 2009; Jevrejeva et al., 2012b; Rahmstorf et al., 2012b; Orlić and Pasarić, 2013). These include groundwater depletion due to anthropogenic intervention and storage of water by dams (Section 13.3.4), ongoing adjustment of the Greenland and Antarctic ice sheets to climate change in previous centuries and millennia (Section 13.3.6), and the effects of internally generated regional climate variability on glaciers (Marzeion et al., 2012a; Church et al., 2013, Sections 13.3.2.2 and 13.3.6) and ice sheets (Section 13.3.3.2). For instance, Jevrejeva et al. (2012b) found that their median projections for 2100 were reduced by 0.02 to 0.10 m by excluding some such contributions.

Making projections with a SEM assumes that sea level change in the future will have the same relationship as it has had in the past to RF or global mean temperature change. The appropriate choice for the formulation of the SEM may depend on the nature of the climate forcing and the time scale, and potentially nonlinear physical processes may not scale in the future in ways which can be calibrated from the past (von Storch et al., 2008; Vermeer and Rahmstorf, 2009; Rahmstorf et al., 2012b; Orlić and Pasarić, 2013). Two such effects that could lead to overestimated or underestimated projections by SEMs have been discussed in the literature.

First, AOGCMs indicate that the ocean heat uptake efficiency tends to decline as warming continues and heat penetrates more deeply (Gregory and Forster, 2008). A linear scaling of the rate of global ocean heat uptake with global SAT determined from the past, as proposed by Rahmstorf (2007a), will thus overestimate future time-integrated heat content change and the consequent global ocean thermal expansion on a century time scale (Orlić and Pasarić, 2013). Rahmstorf (2007a) found that the linear scaling overestimated by 0.12 m (about 30%) the thermal expansion simulated by a climate model with a 3D ocean from 1990 to 2100 under scenario SRES A1FI. Furthermore, the Rahmstorf (2007a) model is inadequate for simulating sea level variations of the last millennium (von Storch et al., 2008), which arise predominantly from episodic volcanic forcing, rather than the sustained forcing on multi-decadal time scales for which it was intended. In both applications, the AOGCM behaviour is more accurately reproduced by taking into account the vertical profile of warming, at least by distinguishing the upper (mixed layer) and lower (thermocline) layers (Vermeer and Rahmstorf, 2009; Held et al., 2010) (Section 13.4.1), or by introducing a relaxation time scale for sea level rise (Jevrejeva et al., 2012b).

Second, the sensitivity of glaciers to warming will tend to decrease as the area most prone to ablation and the remaining volume decrease, partly counteracted by lowering of the surface due to thinning (Huss et al., 2012) Section 13.4.2). On the other hand, glaciers at high latitudes that currently have negligible surface melting will begin to ablate as the climate becomes warmer, tending to give an increase in sensitivity (Rahmstorf et al., 2012b) (Section 13.4.2). Estimating the balance of these two effects will require detailed modelling of glacier SMB. The absence of a multidecadal acceleration in the rate of glacier mass loss in observations of the 20th and simulations of the 21st centuries (Section 4.3.3) (Radic and Hock, 2010; Marzeion et al., 2012a), despite rising global temperatures, suggests that the reduction in sensitivity may dominate (Gregory et al., 2013b).

13.5.3 Confidence in *Likely* Ranges and Bounds

The AR4 (Meehl et al., 2007) presented process-model-based projections of GMSL rise for the end of the 21st century, but did not provide a best estimate or *likely* range principally because scientific

Table 13.6 | Global mean sea level (GMSL) rise (metres) projected by semi-empirical models and compared with the IPCC AR4 and AR5 projections. In each case the results have a probability distribution whose 5th, 50th and 95th percentiles are shown in the columns as indicated. The AR5 5 to 95% process-based model range is interpreted as a *likely* range (*medium confidence*) (Section 13.5.1).

	From	To	5%	50%	95%
Scenario SRES A1B					
IPCC AR4[a]	1990	2100	0.22	0.37	0.50
IPCC AR4[a,b]	1990	2100	0.22	0.43	0.65
IPCC AR5 (also in Table 13.5)	1996	2100	0.42	0.60	0.80
Rahmstorf (2007a)[c]	1990	2100	—	0.85	—
Horton et al. (2008)[d]	2000	2100	0.62	0.74	0.88
Vermeer and Rahmstorf (2009)	1990	2100	0.98	1.24	1.56
Grinsted et al. (2010) with Brohan et al. (2006) temperature for calibration	1990	2100	0.32	0.83	1.34
Grinsted et al. (2010) with Moberg et al. (2005) temperature for calibration	1990	2100	0.91	1.12	1.32
Jevrejeva et al. (2010) with Crowley et al. (2003) forcing for calibration	1990	2100	0.63	0.86	1.06
Jevrejeva et al. (2010) with Goosse et al. (2005) forcing for calibration	1990	2100	0.60	0.75	1.15
Jevrejeva et al. (2010) with Tett et al. (2007) forcing for calibration	1990	2100	0.87	1.15	1.40
Scenario RCP4.5					
IPCC AR5 (also in Table 13.5)	1986–2005	2081–2100	0.32	0.47	0.63
Grinsted et al. (2010) calibrated with Moberg et al. (2005) temperature	1986–2005	2081–2100	0.63	0.88	1.14
Rahmstorf et al. (2012b) calibrated with Church and White (2006) GMSL	1986–2005	2081–2100	0.79	0.86	0.93
Rahmstorf et al. (2012b) calibrated with Church and White (2011) GMSL	1986–2005	2081–2100	0.57	0.63	0.68
Rahmstorf et al. (2012b) calibrated with Jevrejeva et al. (2008) GMSL	1986–2005	2081–2100	0.82	0.97	1.12
Rahmstorf et al. (2012b) calibrated with proxy data	1986–2005	2081–2100	0.56	0.88	1.24
Jevrejeva et al. (2012a) calibrated with Goosse et al. (2005) radiative forcing	1986–2005	2081–2100	0.43	0.56	0.69
Jevrejeva et al. (2012a) calibrated with Crowley et al. (2003) radiative forcing	1986–2005	2081–2100	0.48	0.65	0.80
Jevrejeva et al. (2012a) calibrated with Tett et al. (2007) radiative forcing	1986–2005	2081–2100	0.65	0.85	1.05
Schaeffer et al. (2012)	1986–2005	2081–2100	0.58	0.80	1.05

Notes:

[a] Extrapolated to 2100 using the projected rates of sea level rise for 2090–2099 in Table 10.7 of Meehl et al. (2007).

[b] Including scaled-up ice-sheet discharge given in Table 10.7 of Meehl et al. (2007) and extrapolated to 2100 as an illustration of the possible magnitude of this effect.

[c] Uncertainty range not given.

[d] The mean value and the range are shown for semi-empirical model projections based on results from 11 GCMs.

understanding at the time was not sufficient to allow an assessment of the possibility of future rapid changes in ice-sheet dynamics (on time scales of a few decades, Section 4.4.5). Future rapid changes in ice-sheet outflow were consequently not included in the ranges given by the AR4. For the SRES A1B scenario, the AR4 range was 0.21 to 0.48 m, and for the highest emissions scenario, A1FI, it was 0.26 to 0.59 m. The AR4 also noted that if ice-sheet outflow increased linearly with global mean surface air temperature, the AR4 maximum projections would be raised by 0.1 to 0.2 m. The AR4 was unable to exclude larger values or to assess their likelihood.

Since the publication of the AR4, upper bounds of up to 2.4 m for GMSL rise by 2100 have been estimated by other approaches, namely SEMs (Section 13.5.2), evidence from past climates (Section 13.2.1) and physical constraints on ice-sheet dynamics (Sections 13.4.3.2 and 13.4.4.2). The broad range of values reflects the different methodologies for obtaining the upper bound, involving different constraining factors and sources of evidence. In particular, the upper bound is strongly affected by the choice of probability level, which in some approaches is unknown because the probability of the underlying assumptions is not quantified (Little et al., 2013b).

The confidence that can be placed in projections of GMSL rise and its upper bound by the various approaches must be considered. Confidence arises from the nature, quantity, quality and consistency of the evidence.

The first approach is based on process-based projections, which use the results from several models for each contribution (Sections 13.4 and 13.5.1; Table 13.5). There is medium evidence in support of this approach, arising from our understanding of the modelled physical processes, the consistency of the models with wider physical understanding of those processes as elements of the climate system (e.g., Box 13.1), the consistency of modelled and observed contributions (Sections 13.3.1 to 13.3.5), the consistency of observed and modelled GMSL (Section 13.3.6), and the consistency of process-based projections based on the CMIP5 ensemble of AOGCMs, which have a range of 50 to 60% of the ensemble mean under a given scenario (Table 13.5). Considering this evidence, we have *medium confidence* in the process-based projections.

The second approach uses SEMs (Section 13.5.2, Table 13.6), which make projections by calibrating a physically motivated relationship between GMSL and some other parameter of the climate system in the past and applying it to the future, without quantifying the contributory physical processes. If we had no physical understanding of the causes of sea level rise, the semi-empirical approach to projections would be the only possible one, but extrapolation beyond the range of calibration implies uncertainty that is difficult to quantify, owing to the assumption that sea level change in the future will have the same relationship as it has had in the past to RF or global mean temperature change (Section 13.5.2). As a result, there is low agreement and no consensus in the scientific community about the reliability of SEM projections, despite their successful calibration and evaluation against the observed 20th century sea level record.

For a given RCP, some SEMs project a range that overlaps the process-based *likely* range while others project a median and 95-percentile that are about twice as large as the process-based models. In nearly every case, the SEM 95-percentile is above the process-based *likely* range (Figure 13.12). Two physical explanations have been suggested for the higher projections. First, the contribution from accelerated calving of tidewater glaciers may be substantial and included in SEMs but not process-based models (Jevrejeva et al., 2012b); however, this could account for only 0.1 to 0.2 m of additional GMSL rise. Second, SEMs may allow for rapid ice-sheet dynamical change (Section 4.4.4) in response to future climate change (Grinsted et al., 2010; Little et al., 2013a). In order for large ice-sheet dynamical changes to be predictable by SEMs, two conditions must be met. First, these changes must have contributed substantially to sea level rise during the period of calibration. This is *very unlikely* to be the case, because it is *very likely* that dynamical changes have contributed only a small part of the observed sea level rise during the 20th century, rising to about 15% during 1993–2010 (Section 13.3.6). Second, the changes must have a link to global surface temperature or RF. Current understanding of recent dynamical changes in Greenland and Antarctica is that they have been triggered by local changes in ocean temperature (Holland et al., 2008; Thoma et al., 2008; Jacobs et al., 2011), but a link has not been demonstrated between these changes and global climate change or its drivers. Consequently there is great uncertainty regarding whether recent ice-sheet dynamical changes indicate a long-term trend or instead arise from internal variability (Bamber and Aspinall, 2013). Hence there is no evidence that ice-sheet dynamical change is the explanation for the higher GMSL rise projections of SEMs, implying that either there is some other contribution which is presently unidentified or underestimated by process-based models, or that the projections of SEMs are overestimates (cf. Section 13.5.2). Because of the limited or medium evidence supporting SEMs, and the low agreement about their reliability, we have *low confidence* in their projections.

The third approach uses paleo records of sea level change that show that rapid GMSL rise has occurred during glacial terminations, at rates that averaged about 10 mm yr^{-1} over centuries, with at least one instance (Meltwater Pulse 1A) that exceeded 40 mm yr^{-1} (Section 5.6.3), but this rise was primarily from much larger ice-sheet sources that no longer exist. Contributions from these vanished ice sheets could have continued even after sea level and climate had reached interglacial states, if the Greenland and Antarctic ice sheets contracted during the termination to smaller sizes than at present. During past interglacial periods, only the Greenland and Antarctic ice sheets were present. For the time interval during the LIG in which GMSL was above present, there is *high confidence* that the maximum 1000-year average rate of GMSL rise during these periods exceeded 2 m kyr^{-1} but did not exceed 7 m kyr^{-1} (Kopp et al., 2013) (Sections 5.6.2 and 13.2.1.3). Because climate variations during interglacial periods had different forcings from anthropogenic climate change, they give only a limited basis for predictions of the future, and we do not consider that they provide upper bounds for GMSL rise during the 21st century.

The fourth approach is concerned particularly with the contribution from ice-sheet dynamical change, for which it considers kinematic limits. Pfeffer et al. (2008) argued that scenarios of GMSL rise exceeding 2 m by 2100 are physically untenable, ruling out, for example, the heuristic argument of Hansen et al. (2007) giving 5 m by 2100. Pfeffer et al. (2008) constructed scenarios of 0.8 m and 2.0 m, and Katsman et al. (2011) of 1.15 m, for GMSL rise by 2100, including ice-sheet rapid dynamical acceleration. Although these authors considered their scenarios to be physically possible, they are unable to quantify their likelihood, because the probability of the assumptions on which they depend cannot be estimated from observations of the response of the Greenland and Antarctic ice sheets to climate change or variability on century time scales. These scenarios involve contributions of ~0.5 m from Antarctica. This is much greater than any process-based projections of dynamical ice-sheet change (Section 13.4.4.2), and would require either a sustained high increase in outflow in all marine-based sectors or the localized collapse of the ice sheet in the Amundsen Sea sector (Little et al., 2013a).

In summary, we have greater confidence in the process-based projections than in the other approaches, and our assessment is that GMSL rise during the 21st century for each RCP scenario is *likely* (*medium confidence*) to lie within the 5 to 95% range given by the process-based projections (Section 13.5.1 and Table 13.5; see Section 13.5.4 for following centuries), which are consistent with the *likely* ranges projected for global mean surface air temperature change (Section 12.4.1.2). We are not able to assess a *very likely* range on the same basis, because there is no assessment available of the *very likely* range for global mean SAT change, and because we cannot robustly quantify the probability of ice-sheet dynamical changes which would give rise to greater values.

Under the RCP8.5 scenario, which has the highest RF, the *likely* range reaches 0.98 m by 2100 relative to 1986–2005. Observations do not show an increase in Antarctic precipitation, which is projected by models and makes a negative contribution to the projected GMSL rise (Table 13.5). The recovery of Antarctic stratospheric ozone concentration and increased basal melting of Antarctic ice shelves have both been suggested as giving rise to mechanisms whereby the Antarctic warming and precipitation increase might be suppressed with respect to CMIP5 projections (Section 13.4.4.1). If the Antarctic precipitation increase is omitted from the process-based projections, the *likely* range for RCP8.5 at 2100 reaches 1.03 m (assuming uncorrelated errors). Higher values for 2100 are given in the scientific literature on the basis of various approaches: 1.15 m (Katsman et al., 2011), 1.21 m (Schaeffer et al., 2012) (for RCP4.5), 1.40 m (National Research Council, 2012), 1.65 m (Jevrejeva et al., 2012b) (for RCP8.5), 1.79 m (Vermeer and Rahmstorf, 2009) (for SRES A1FI), 1.90 m (Rahmstorf et al., 2012b) (with proxy calibration, for RCP8.5), 2.0 m (Pfeffer et al., 2008), 2.25 m (Sriver et al., 2012), and 2.4 m (Nicholls et al., 2011). Considering this inconsistent evidence, we conclude that the probability of specific levels above the *likely* range cannot be reliably evaluated.

Only the collapse of marine-based sectors of the Antarctic ice sheet could cause GMSL rise substantially above the *likely* range during the 21st century. Expert estimates of contributions from this source have a wide spread (Bamber and Aspinall, 2013), indicating a lack of consensus on the probability for such a collapse. The potential additional contribution to GMSL rise also cannot be precisely quantified, but there is *medium confidence* that, if a collapse were initiated, it would not exceed several tenths of a metre during the 21st century (Section 13.4.4.2).

The time mean rate of GMSL rise during the 21st century is *very likely* to exceed the rate of 2.0 [1.7 to 2.3] mm yr^{-1} observed during 1971–2010, because the process-based GMSL projections indicate a significantly greater rate even under the RCP2.6 scenario, which has the lowest RF. It has been asserted that the acceleration of GMSL rise implied by the IPCC AR4 projections is inconsistent with the observed magnitude of acceleration during the 20th century (Boretti, 2011, 2012b, 2012a, 2012c, 2013a, 2013b. 2013c; Boretti and Watson, 2012; Parker, 2013a, 2013b, 2013c). Refuting this argument, Hunter and Brown (2013) show that the acceleration projected in the AR4 is consistent with observations since 1990s. Present understanding of the contributions to GMSL rise (Section 13.3) gives an explanation of the rate of 20th century GMSL rise and confidence in the process-based projections, which indicate a greater rate of rise in the 21st century because of increasing forcing.

The improved agreement of process-based models with observations and physical understanding represents progress since the AR4, in which there was insufficient confidence to give *likely* ranges for 21st century GMSL rise, as we have done here. For scenario SRES A1B, which was assessed in the AR4, the *likely* range on the basis of science assessed in the AR5 is 0.60 [0.42 to 0.80] m by 2100 relative to 1986–2005, and 0.57 [0.40 to 0.76] m by 2090–2099 relative to 1990. Compared with the AR4 projection of 0.21 to 0.48 m for the same scenario and period, the largest increase is from the inclusion of rapid changes in Greenland and Antarctic ice sheet outflow, for which the combined *likely* range is 0.03 to 0.21 m by 2091–2100 (assuming uncorrelated uncertainties). These terms were omitted in the AR4 because a basis to make projections was not available in published literature at that time. The contribution from thermal expansion is similar to the AR4 projection and has smaller uncertainty. The contribution from glaciers is larger than in the AR4 primarily because of the greater estimate of the present glacier volume in new inventories (although the glacier area estimate is similar, Table 4.1), and the Greenland SMB contribution is larger because of recent improvement in models of relevant surface processes. Further progress on modelling each of the contributions is still needed in order to attain *high confidence* in GMSL projections, in particular concerning the probability distribution of GMSL above the *likely* ranges.

13.5.4 Long-term Scenarios

Less information is available on climate change beyond the year 2100 than there is up to the year 2100. However, the ocean and ice sheets will continue to respond to changes in external forcing on multi-centennial to multi-millennial time scales. For the period up to the year 2500, available physical model projections discussed in Sections 13.4.1-4 are combined into an assessment of future sea level rise. Paleo simulations are combined with paleo data to estimate the sea level commitment on a multi-millennial time scale beyond 2500 for different levels of sustained increases in global mean temperature.

The RCPs, as applied in Chapter 12 and Sections 13.4 and 13.5.1, are defined up to the year 2100. Their extension up to the year 2300 is used to project long-term climate change (Section 12.3.1.3) (Meinshausen et al., 2011), but they are not directly derived from integrated assessment models. In simulations that are reported here up to the year 2500, the RF has generally been kept constant at the 2300 level except for RCP2.6, in which the forcing continues to decline at the 2300 rate. Some model simulations of ice sheets and ocean warming assessed here have used scenarios different from the RCP scenarios. Because of the limited number of available simulations, sea level projections beyond the year 2100 have thus been grouped into three categories according to their GHG concentration in the 22nd century: *low scenarios* in which atmospheric GHG concentrations peak and decline and do not exceed values that are equivalent to 500 ppm CO_2, *medium scenarios* with concentrations between 500 and 700 ppm CO_2-eq, and *high scenarios* above 700 ppm. As a consequence, the model spread shown in Figure 13.13 and Table 13.8 combines different scenarios and is not merely due to different model physics. The low scenarios include RCP2.6, SRES B1 and scenarios with 0.5 and 2% yr^{-1} increases in CO_2 followed by no emissions after 450 ppm has been reached, and the commitment scenarios, CC, in Goelzer et al. (2013) which stabilize CO_2 at present-day levels. In a number of the low scenarios, the global mean temperature peaks during the 21st century and declines thereafter. These peak-and-decline scenarios include RCP2.6 as well as all scenarios with no GHG emissions after a specified year. Even in these scenarios sea level continues to rise up to the year 2500 in accordance with the multi-millennial sea level commitment of about 2 m $°C^{-1}$ as discussed in Section 13.5.4.2. The medium scenarios include RCP4.5 as well as scenarios with 1% yr^{-1} increase in CO_2 up to 560 ppm and SRES-B1 and SRES-A1B. The high scenarios include RCP6.0 and RCP8.5 as well as 1120 ppm scenarios and SRES A2. Also included are scenarios with 0.5 and 2% increase in CO_2 and a SRES A2 scenario with zero emissions after 1200 and 1120 ppm have been reached, respectively.

13.5.4.1 Multi-centennial Projections

The multi-centennial sea level contributions from ocean expansion and the cryospheric components are discussed in Sections 13.4.1 to 13.4.4. A synthesis of these contributions is provided in Table 13.8 and Figure 13.13 for the end of each century until the year 2500. Thermal expansion contributions (dark blue bars, Figure 13.13) were obtained from coarse-resolution coupled climate models (Vizcaíno et al., 2008; Solomon et al., 2009; Gillett et al., 2011; Schewe et al., 2011; Zickfeld et al., 2013). For comparison, the full model spread of the CMIP5 models which were integrated beyond 2100 is provided in Table 13.7 and as light blue bars in Figure 13.13. Even though the models used for the long-term projections (Table 13.8) are less complex compared to the CMIP5 models, their model spread for the different periods and scenarios encompasses the CMIP5 spread, which provides *medium confidence* in the application of the less complex models beyond 2300.

Contributions from the Greenland and Antarctic ice sheets were obtained with climate models of comparable complexity coupled to ice-sheet models (Vizcaíno et al., 2010; Huybrechts et al., 2011; Goelzer et al., 2012). Glacier projections were obtained by application of the method by Marzeion et al. (2012a) to the CMIP5 model output for scenarios and models that were integrated up to the year 2300. For 2400 and 2500, the same model spread as for 2300 is shown. This is probably underestimating the glacier's sea level contribution beyond 2300.

The ranges of sea level contributions provided in Figure 13.13 and Table 13.8 only represent the model spread and cannot be interpreted as uncertainty ranges. An uncertainty assessment cannot be provided beyond the year 2100 because of the small number of available simulations, the fact that different scenarios were combined within one scenario group, and the overall *low confidence* in the ability of the coarse-resolution ice-sheet models to capture the dynamic ice discharge from Greenland and Antarctica, as discussed below. The range for the total sea level change was obtained by taking the sum of contributions that result in the lowest and the highest sea level rise and thereby covers the largest possible model spread.

Except for the glacier models (Section 13.4.2), the models used here for the period beyond 2100 are different from the models used for the 21st century (Sections 13.4.1, 13.4.3, 13.4.4, and 13.5.1). Generally, the model spread for the total sea level contribution in 2100 is slightly lower than the *likely* range provided in Section 13.5.1 (light red bars in Figure 13.13). This is due to the ice-sheet models, particularly of the Antarctic ice sheet, as coarse-resolution model results for thermal expansion cover the range of the CMIP5 projections (light blue vertical lines in Figure 13.13 and Table 13.7.) and the glacier contribution is the same.

Projections beyond 2100 show positive contributions to sea level from thermal expansion, glaciers and changes in Greenland ice sheet SMB. Due to enhanced accumulation under warming, the Antarctic ice sheet SMB change makes a negative contribution to sea level in scenarios below 700 ppm CO_2-eq. These results were obtained with fully coupled climate–ice sheet models which need to apply a relatively low spatial resolution. In light of the discussion in Section 13.3.3.2 and the assessment of the 21st century changes in Section 13.4.4.1, there is *low confidence* in this result. For scenarios above 700 ppm CO_2-eq, Antarctic SMB change is contributing positively to GMSL.

As discussed in Sections 13.4.3.2 and 13.4.4.2, there is *medium confidence* in the ability of coupled ice sheet–climate models to project sea level contributions from dynamic ice-sheet changes in Greenland and Antarctica for the 21st century. In Greenland, dynamic mass loss is limited by topographically defined outlets regions. Furthermore, solid ice discharge induced from interaction with the ocean is self-limiting because retreat of the ice sheet results in less contact with the ocean and less mass loss by iceberg calving (Pfeffer et al., 2008; Graversen et al., 2011; Price et al., 2011). By contrast, the bedrock topography of Antarctica is such that parts of the retreating ice sheet will remain in contact with the ocean. In particular, due to topography that is sloping landward, especially in West Antarctica, enhanced rates of mass loss are expected as the ice retreats.

Although the model used by Huybrechts et al. (2011) is in principle capable of capturing grounding line motion of marine ice sheets (see Box 13.2), *low confidence* is assigned to the model's ability to capture the associated time scale and the perturbation required to initiate a retreat (Pattyn et al., 2013). The model used by Vizcaino et al. (2010) does not represent ice-shelf dynamics and is thus lacking a fundamental process that can trigger the instability. As stated by the authors, *low confidence* is thus also assigned to the model's ability to project future solid ice discharge from Antarctica. It is thus *likely* that the values depicted in Figure 13.13 systematically underestimate Antarctica's future contribution. As detailed in Section 13.5.4.2, simulations of the last 5 Myr (Pollard and DeConto, 2009) indicate that on

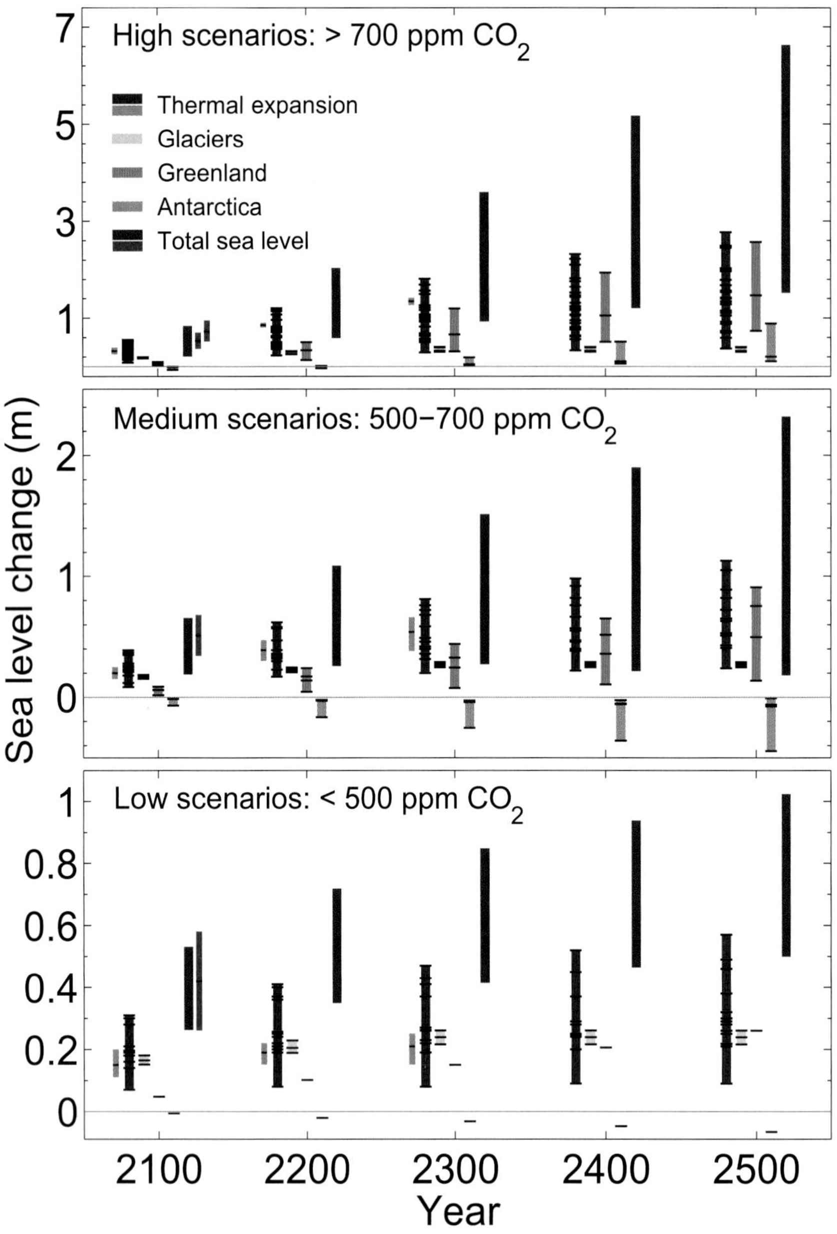

Figure 13.13 | Sea level projections beyond the year 2100 are grouped into three categories according to the concentration of GHG concentration (in CO_2-eq) in the year 2100 (upper panel: >700 ppm including RCP6.0 and RCP8.5; middle panel: 500–700 ppm including RCP4.5; lower panel: <500 ppm including RCP2.6). Colored bars show the full model spread. Horizontal lines provide the specific model simulations. The different contributions are given from left to right as thermal expansion from the CMIP5 simulations up to 2300 (as used for the 21st century projections, section 13.5.1, light blue, with the median indicated by the horizontal bar), thermal expansion for the models considered in this section (dark blue), glaciers (light green), Greenland ice sheet (dark green), Antarctic ice sheet (orange), and the total contribution (red). The range provided for the total sea level change represents the maximum possible spread that can be obtained from the four different contributions. Light red-shaded bars show the *likely* range for the 21st century total sea level projection of the corresponding scenarios from Figure 13.10 with the median as the horizontal line. In the upper panel, the left light red bar corresponds to RCP6.0 and the right light red bar corresponds to RCP8.5.

multi-millennial time scales, the Antarctic ice sheet loses mass for elevated temperatures, in contrast to the projections until the year 2500 for the low and medium scenarios.

The model spread of total sea level change in 2300 ranges from 0.41 to 0.85 m for the low scenario (Table 13.8). Using an SEM, Schaeffer et al. (2012) obtained a significantly larger 90% confidence range of 1.3 to 3.3 m for the RCP2.6 scenario. The RCP4.5 scenario, for which they obtained a range of 2.3 to 5.5 m, is categorized here as a medium scenario, and is also significantly higher than the range of 0.27 to 1.51 m computed by the process-based models. Using a different semi-empirical approach, Jevrejeva et al. (2012a) obtained a 90% confidence range of 0.13 to 1.74 m for RCP2.6 in the year 2500, which encloses the model spread of 0.50 to 1.02 m for the low scenario from the process-based models. For the medium and high scenarios, however, they obtained ranges of 0.72 to 4.3 m and 1.0 to 11.5 m, respectively, which are significantly higher than the corresponding process-based model spread of 0.18 to 2.32 m and 1.51 to 6.63 m (Table 13.8). Because projections of land water storage are not available for years beyond 2100 these were not included here.

The higher estimates from the SEMs than the process-based models used here for the long-term projections are consistent with the relation between the two modelling approaches for the 21st century (Figure 13.12). Section 13.5.3 concluded that the limited or medium evidence supporting SEMs, and the low agreement about their reliability, provides *low confidence* in their projections for the 21st century. We note here that the confidence in the ability of SEMs is further reduced with the length of the extrapolation period and the deviation of the future forcing from the forcing of the learning period (Schaeffer et al., 2012), thus decreasing confidence over the long time frames considered here.

For increasing global mean SAT, sea level is *virtually certain* to continue to rise beyond the year 2500 as shown by available process-based model simulations of thermal expansion and ice sheets that were computed beyond 2500 (Rahmstorf and Ganopolski, 1999; Ridley et al., 2005; Winguth et al., 2005; Driesschaert et al., 2007; Mikolajewicz et al., 2007b; Swingedouw et al., 2008; Vizcaíno et al., 2008; Solomon et al., 2009; Vizcaíno et al., 2010; Gillett et al., 2011; Goelzer et al., 2011; Huybrechts et al., 2011; Schewe et al., 2011).

13.5.4.2 Multi-Millennial Projections

Here sea level commitment in response to a global mean temperature increase on a multi-millennial time scale is assessed. Figure 13.14 shows the sea level rise after several millennia of constant global mean temperature increase above pre-industrial. The thermal expansion of the ocean was taken from 1000-year integrations with six coupled climate models as used in the AR4 (models Bern2D, CGoldstein, CLIMate and BiosphERe-2 (CLIMBER-2), Massachusetts Institute of Technology (MIT), MoBidiC, and Loch-Vecode-Ecbilt-CLio-agIsm Model (LOVECLIM) in Figure 10.34 in Meehl et al. (2007)). These yield a rate of sea level change in the range of 0.20 to 0.63 m °C^{-1} (Figure 13.14a). For reference, a spatially uniform increase of ocean temperature yields a global mean sea level rise of 0.38 m °C^{-1} when added to observed data (Levitus et al., 2009) (black dots in Figure 13.14a). Uncertainty arises due to the different spatial distribution of the warming in models and the dependence of the expansion on local temperature and salinity. The contribution for glaciers was obtained with the models from Mazeion et al. (2012a) and Radic and Hock (2011) by integration with fixed boundary conditions corresponding to different global mean SAT levels for 3000 years.

As detailed in Sections 13.4.3.2 and 13.4.4.2, there is *low confidence* in the ability of current Antarctic ice-sheet models to capture the temporal response to changes in external forcing on a decadal to centennial time scale. On multi-centennial to multi-millennial time scales, however, these models can be validated against paleo sea level records. The contributions from the Greenland ice sheet were computed with a dynamic ice-sheet model coupled to an energy-moisture balance model for the SMB (Robinson et al., 2012). The model's parameters were constrained by comparison with SMB estimates and topographic data for the present day and with estimated summit-elevation changes from ice-core records for the Last Interglacial period (LIG), in order to ensure that the coupled model ensemble has a realistic sensitivity to climatic change. The parameter spread leads to a spread in ice-sheet responses (dark green lines in Figure 13.14c). The contribution to sea level commitment from the Greenland ice sheet is relatively weak (on average 0.18 m °C^{-1} up to 1°C and 0.34 m °C^{-1} between 2°C and 4°C) apart from the abrupt threshold of ice loss between 0.8°C and 2.2°C above pre-industrial (90% confidence interval in the particular model calculations reported here) (Figure 13.14c). This represents a change from a fully ice-covered Greenland to an essentially ice-free state, reducing the ice sheet to around 10% of present-day volume and raising sea level by over 6 m (Ridley et al., 2005; Ridley et al., 2010). The threshold temperature is lower than estimates obtained from the assumption that the threshold coincides with a negative total SMB of the Greenland ice sheet (see Section 13.4.3.3 for a more complete discussion).

The Antarctic ice sheet contribution comes from a simulation of the last 5 million years (Pollard and DeConto, 2009), which is in good agreement with regional paleo records (Naish et al., 2009). The sensitivity of the ice sheet was extracted from this model simulation by correlating the ice volume with the global mean temperature which forces the simulation. The standard deviation of the resulting scatter is used as a measure of uncertainty (Figure 13.14d). Uncertainty arises from uncertainty in the forcing data, the ice physics representation, and from the time-dependent nature of the simulation. For example, the existence of hysteresis behavior on the sub-continental scale can lead to different contributions for the same temperature increase. The Antarctic ice sheet shows a relatively constant commitment of 1.2 m °C^{-1}. Paleorecords indicate that a potential hysteresis behaviour of East Antarctica requires a temperature increase above 4°C and is thereby outside of the scope discussed here (Foster and Rohling, 2013).

In order to compare the model results with past sea level anomalies for the temperature range up to 4°C, we focus on the three previous periods of warmer climates and higher sea levels than pre-industrial that were assessed in Sections 5.6.1, 5.6.2 and 13.2.1: the middle Pliocene, MIS 11, and the LIG (Figure 13.14e). In each case, there is reasonable agreement between the model result of a long-term sea level response for a given temperature with the information from the paleo record.

The ability of the physical models to reproduce paleo sea level records on a multi-millennial time scale provides confidence in applying them to millennial time frames. After 2000 years, the sea level contribution will be largely independent of the exact warming path during the first century. As can be seen from Figure 10.34 of AR4, the oceanic heat content will be largely equilibrated after 2000 years; the same is true for the glacier component. The situation for Antarctica is slightly more complicated, but as can be inferred from Pollard and DeConto (2009), much of the retreat of the West Antarctic ice sheet will have already occurred by 2000 years, especially if the warming occurs on a decadal to centennial time scale. The opposite and smaller trend in East Antarctic ice volume due to increased snowfall in a warmer environment will also have largely equilibrated (Uotila et al., 2007; Winkelmann et al., 2012).

The most significant difference arises from the contribution of the Greenland ice sheet. Consistent with previous estimates (Huybrechts et al., 2011; Goelzer et al., 2012), the rate of the sea level contribution from Greenland increases with temperature. The transient simulations for an instantaneous temperature increase show a quasi-quadratic dependence of the sea level contribution on this temperature increase after 2000 years (Figure 13.14h) (Robinson et al. 2012). The results are quantitatively consistent with previous estimates on a millennial time scale (Huybrechts et al., 2011; Goelzer et al., 2012). The sea level contribution of the Greenland ice sheet after 2000 years of integration at 560 ppm was plotted against the average Greenland temperature divided by the standard polar amplification of 1.5 between global mean and Greenland mean temperature increase (Gregory and Huybrechts, 2006, black dot in Figure 13.14h). Taken together, these results imply that a sea level rise of 1 to 3 m °C^{-1} is expected if the warming is sustained for several millennia (*low confidence*) (Figure 13.14e, 13.14j).

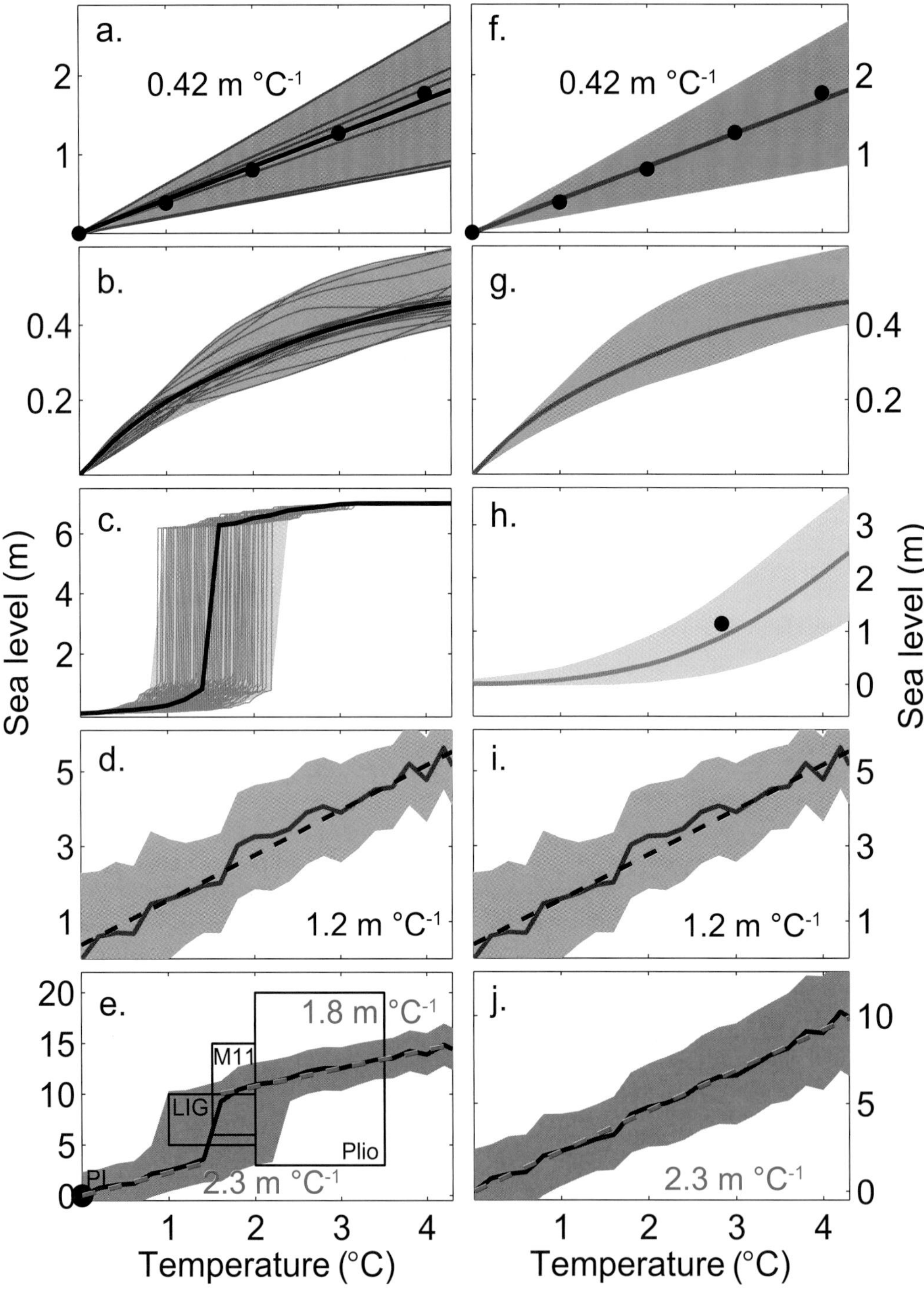

Figure 13.14 | (Left column) Multi-millennial sea level commitment per degree Celsius of warming as obtained from physical model simulations of (a) ocean warming, (b) mountain glaciers and (c) the Greenland and (d) the Antarctic ice sheets. (e) The corresponding total sea level commitment, compared to paleo estimates from past warm periods (PI = pre-industrial, LIG = last interglacial period, M11 = Marine Isotope Stage 11, Plio = Mid-Pliocene). Temperatures are relative to pre-industrial. Dashed lines provide linear approximations in (d) and (e) with constant slopes of 1.2, 1.8 and 2.3 m °C^{-1}. Shading as well as the vertical line represents the uncertainty range as detailed in the text. (Right column) 2000-year-sea level commitment. The difference in total sea level commitment (j) compared to the fully equilibrated situation (e) arises from the Greenland ice sheet which equilibrates on tens of thousands of years. After 2000 years one finds a nonlinear dependence on the temperature increase (h) consistent with coupled climate–ice sheet simulations by Huybrechts et al. (2011) (black dot). The total sea level commitment after 2000 years is quasi-linear with a slope of 2.3 m °C^{-1}.

Table 13.7 | Median and model spread of the thermal expansion of CMIP5 comprehensive climate models. RCP2.6 belongs to the low scenarios as shown in Figure 13.13 and Table 13.8; RCP4.5 is a 'medium scenario' and RCP8.5 a 'high scenario'. The model spread in Table 13.8 encloses the CMIP5 model spread for all scenarios. Sea level contributions are provided in metres.

	Mean 2191–2200			Mean 2291–2300		
Scenario	**No. of Models**	**Median**	**Model Spread**	**No. of Models**	**Median**	**Model Spread**
RCP2.6	3	0.19 m	0.15–0.22 m	3	0.21 m	0.15–0.25 m
RCP4.5	7	0.39 m	0.30–0.47 m	6	0.54 m	0.38–0.66 m
RCP8.5	2	0.85m	0.80–0.90 m	2	1.34 m	1.26–1.41 m

Table 13.8 | Model spread of sea level contribution and total sea level change for low, medium and high scenarios as defined in the text and shown in Figure 13.13. As detailed in the text, there is *low confidence* in the ice-sheet models' ability to project rapid dynamical change in the Antarctic ice sheet, which may result in a systematic underestimation of the ice-sheet contributions. The unit of all sea level contributions is metres.

Contribution	Scenario	2100	2200	2300	2400	2500
Thermal expansion	Low	0.07 to 0.31 m	0.08 to 0.41 m	0.08 to 0.47 m	0.09 to 0.52 m	0.09 to 0.57 m
Glaciers	Low	0.15 to 0.18 m	0.19 to 0.23 m	0.22 to 0.26 m	0.22 to 0.26 m[b]	0.22 to 0.26 m[b]
Greenland ice sheet	Low	0.05 m[a]	0.10 m[a]	0.15 m[a]	0.21 m[a]	0.26 m[a]
Antarctic ice sheet	Low	–0.01 m[a]	–0.02 m[a]	–0.03 m[a]	–0.05 m[a]	–0.07 m[a]
Total	**Low**	**0.26 to 0.53 m**	**0.35 to 0.72 m**	**0.41 to 0.85 m**	**0.46 to 0.94 m**	**0.50 to 1.02 m**
Thermal expansion	Medium	0.09 to 0.39 m	0.17 to 0.62 m	0.20 to 0.81 m	0.22 to 0.98 m	0.24 to 1.13 m
Glaciers	Medium	0.15 to 0.19 m	0.21 to 0.25 m	0.25 to 0.29 m	0.25 to 0.29 m[b]	0.25 to 0.29 m[b]
Greenland ice sheet	Medium	0.02 to 0.09 m	0.05 to 0.24 m	0.08 to 0.44 m	0.11 to 0.65 m	0.14 to 0.91 m
Antarctic ice sheet	Medium	–0.07 to –0.01 m	–0.17 to –0.02 m	–0.25 to –0.03 m	–0.36 to –0.02 m	–0.45 to –0.01 m
Total	**Medium**	**0.19 to 0.66 m**	**0.26 to 1.09 m**	**0.27 to 1.51 m**	**0.21 to 1.90 m**	**0.18 to 2.32 m**
Thermal expansion	High	0.08 to 0.55 m	0.23 to 1.20 m	0.29 to 1.81 m	0.33 to 2.32 m	0.37 to 2.77 m
Glaciers	High	0.17 to 0.19 m	0.25 to 0.32 m	0.30 to 0.40 m	0.30 to 0.40 m[b]	0.30 to 0.40 m[b]
Greenland ice sheet	High	0.02 to 0.09 m	0.13 to 0.50 m	0.31 to 1.19 m	0.51 to 1.94 m	0.73 to 2.57 m
Antarctic ice sheet	High	–0.07 to –0.00 m	–0.04 to 0.01 m	0.02 to 0.19 m	0.06 to 0.51 m	0.11 to 0.88 m
Total	**High**	**0.21 to 0.83 m**	**0.58 to 2.03 m**	**0.92 to 3.59 m**	**1.20 to 5.17 m**	**1.51 to 6.63 m**

Notes:

[a] The value is based on one simulation only.

[b] Owing to lack of available simulations the same interval used as for the year 2300.

13.6 Regional Sea Level Changes

Regional sea level changes may differ substantially from a global average, showing complex spatial patterns which result from ocean dynamical processes, movements of the sea floor, and changes in gravity due to water mass redistribution (land ice and other terrestrial water storage) in the climate system. The regional distribution is associated with natural or anthropogenic climate modes rather than factors causing changes in the global average value, and include such processes as a dynamical redistribution of water masses and a change of water mass properties caused by changes in winds and air pressure, air–sea heat and freshwater fluxes and ocean currents. Because the characteristic time scales of all involved processes are different, their relative contribution to net regional sea level variability or change will depend fundamentally on the time scale considered.

13.6.1 Regional Sea Level Changes, Climate Modes and Forced Sea Level Response

As discussed in Chapter 3, most of the regional sea level changes observed during the recent altimetry era or reconstructed during past decades from tide gauges appear to be steric (Levitus et al., 2005, 2009; Lombard et al., 2005a, 2005b; Ishii and Kimoto, 2009; Stammer et al., 2013). Moreover, steric changes observed during the altimetry era appear to be primarily thermosteric in nature, although halosteric effects, which can reduce or enhance thermosteric changes, are also important in some regions (e.g., Atlantic Ocean, Bay of Bengal). Ocean models and ocean reanalysis-based results (Carton et al., 2005; Wunsch and Heimbach, 2007; Stammer et al., 2011) as well as ocean circulation models without data assimilation (Lombard et al., 2009) confirm these results.

Observations and ocean reanalysis (Stammer et al., 2011; 2013) also agree in showing that steric spatial patterns over the last half of the 20th century fluctuate in space and time as part of modes of the coupled ocean–atmosphere system such as ENSO, the NAO, and the PDO (Levitus et al., 2005; Lombard et al., 2005a; Di Lorenzo et al., 2010; Lozier et al., 2010; Zhang and Church, 2012). In these cases, regional sea level variability is associated with changing wind fields and resulting changes in the ocean circulation (Kohl and Stammer, 2008). For example, the large rates of sea level rise in the western tropical Pacific and of sea level fall in the eastern Pacific over the period 1993–2010 correspond to an increase in the strength of the trade winds in the central and eastern tropical Pacific over the same period (Timmermann et al., 2010; Merrifield and Maltrud, 2011; Nidheesh et al., 2012). The long-term sea level trend from 1958 to 2001 in the tropical Pacific can also be explained as the ocean's dynamical response to variations in the wind forcing (Qiu and Chen, 2006; Timmermann et al., 2010).

Spatial variations in trends in regional sea level may also be specific to a particular sea or ocean basin. For example, a sea level rise of 5.4 ± 0.3 mm yr^{-1} in the region between Japan and Korea from 1993 to 2001 is nearly two times the GMSL trend, with more than 80% of this rise being thermosteric (Kang et al., 2005). Han et al. (2010) found that regional changes of sea level in the Indian Ocean that have emerged since the 1960s are driven by changing surface winds associated with a combined enhancement of Hadley and Walker Cells.

13.6.2 Coupled Model Intercomparison Project Phase 5 General Circulation Model Projections on Decadal to Centennial Time Scales

CMIP5 projections of regional sea level provide information primarily about dynamical sea level changes resulting from increased heat uptake and changes in the wind forcing. On decadal time scales, the CMIP5 model ensemble identifies strong interannual variability (up to 8 cm, root-mean square (RMS)) associated with ENSO and dynamics of the equatorial current system in the tropical Pacific and Indian Oceans (Figure 13.15a). Similar variability in the amplitude of sea level change but due to other climate modes is also apparent in the North Atlantic Current and in parts of the Southern Ocean.

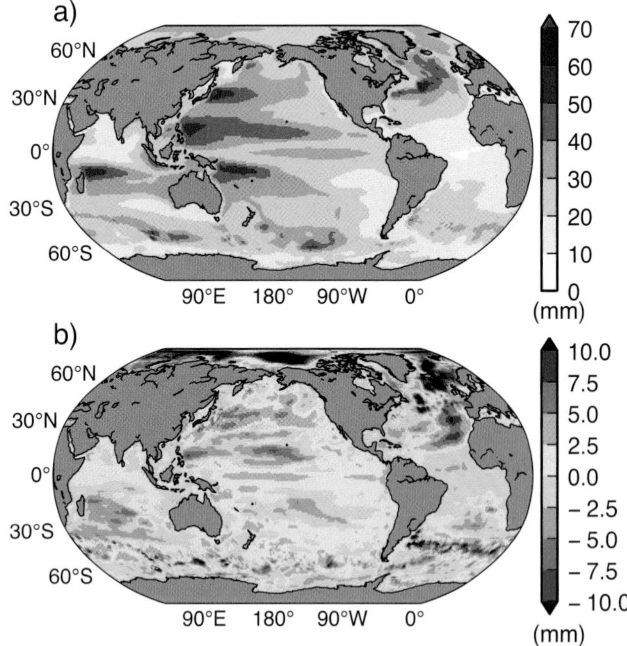

Figure 13.15 | (a) Root-mean square (RMS) interannual dynamic sea level variability (millimetres) in a CMIP5 multi-model ensemble (21 models), built from the historically forced experiments during the period 1951–2005. (b) Changes in the ensemble average interannual dynamic sea level variability (standard deviation (SD), in millimetres) evaluated over the period 2081–2100 relative to the reference period 1986–2005. The projection data (2081–2100) are from the CMIP5 RCP4.5 experiment.

Toward the end of the 21st century, the CMIP5 results indicate that it is possible that the interannual to decadal variability of dynamic sea level can weaken in some parts of the world ocean, for example, the western low-latitude Pacific and parts of the Indian Ocean, whereas it could be amplified in other parts, for example, the North Pacific, the eastern tropical Pacific, the eastern subtropical Atlantic and the Arctic (Figure 13.15b).

Longer-than-decadal-time-scale regional sea level changes can increasingly be expected to result from long-term changes in the wind field, changes in the regional and global ocean heat and freshwater content and the associated dynamical adjustment (with associated redistribution of ocean properties), and (to a lesser extent) from atmospheric pressure. The CMIP5 projections of steric sea level changes toward the end of the 21st century reveal a clear regional pattern in dynamical sea level change (Figure 13.16), in which the Southern Ocean shows a net decline relative to the global mean, while the remaining global ocean displays complex ridge-and-trough structures superimposed on a generally rising sea level (Yin, 2012). For example, in the North Atlantic, the largest sea level rise is along and north of the North Atlantic Current, but less so further to the south in the center of the warmer subtropical gyre. A similar dipole pattern was observed in CMIP3 results there due to a weakening of the AMOC which leads to a local steric sea level rise east of North America, resulting in more water on the shelf and directly impacting northeastern North America (Levermann et al., 2005; Landerer et al., 2007; Yin et al., 2010). A similar pattern can be observed in the North Pacific, but here and in other parts of the world ocean (e.g., Southern Ocean), regional sea level patterns are largely the result of changes in wind forcing, associated changes in the circulation, and an associated redistribution of heat and freshwater. Some regional changes can also be expected to result from modifications in the expansion coefficient due to changes in the ocean's regional heat content (Kuhlbrodt and Gregory, 2012).

The CMIP5 ensemble indicates that regions showing an enhanced sea level toward the end of the 21st century coincide with those showing the largest uncertainty (Figure 13.16b). Although this also appeared in the earlier CMIP3 SRES A1B results, the CMIP5 results, by comparison, show a general reduction in the ensemble spread, especially in high latitudes. On a global average, this reduction is from 5.7 cm to 2.1 cm, RMS.

The contribution of changes of global ocean heat storage to regional steric sea level anomalies is *virtually certain* to increase with time as the climate warming signal increasingly penetrates into the deep ocean (Pardaens et al., 2011a). For the last three decades of the 21st century, the AR4 climate model ensemble mean shows a significant heat storage increase (Yin et al., 2010), about half of which is stored in the ocean below 700 m depth. Recent detection of ongoing changes in the ocean salinity structure (Durack and Wijffels, 2010) (Section 3.3.2) may also contribute to future regional steric sea level changes. Halosteric effects can dominate in some regions, especially in regions of high-latitude water mass formation where long-term heat and freshwater changes are expected to occur (e.g., in the subpolar North Atlantic, the Arctic, the Southern Ocean) (Yin et al., 2010; Pardaens et al., 2011a). Because of an anticipated increase in atmospheric moisture transport from low to high latitudes (Pardaens et al., 2003), halosteric anomalies are positive in the Arctic Ocean and dominate regional sea level

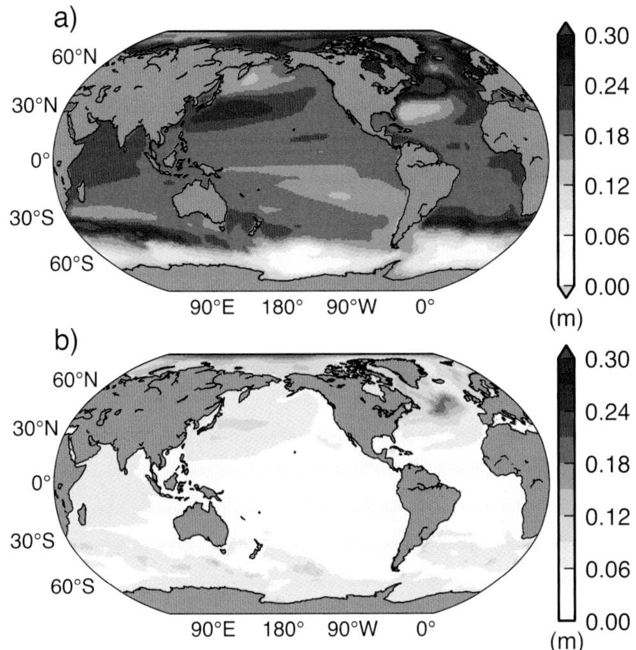

Figure 13.16 | (a) Ensemble mean projection of the time-averaged dynamic and steric sea level changes for the period 2081–2100 relative to the reference period 1986–2005, computed from 21 CMIP5 climate models (in metres), using the RCP4.5 experiment. The figure includes the globally averaged steric sea level increase of 0.18 ± 0.05 m. (b) Root-mean square (RMS) spread (deviation) of the individual model result around the ensemble mean (metres). Note that the global mean is different from the value in Table 13.5, by less than 0.01 m, because a slightly different set of CMIP5 models was used (see the Supplementary Material).

anomalies there (Yin et al., 2010). It is *likely* that future thermosteric changes will dominate the steric variations in the Southern Ocean, and strong compensation between thermosteric and halosteric change will characterize the Atlantic (Pardaens et al., 2011a).

13.6.3 Response to Atmospheric Pressure Changes

Regional sea level also adjusts to regional changes in atmospheric sea level pressure relative to its instantaneous mean over the ocean. Over time scales longer than a few days, the adjustment is nearly isostatic. Sea level pressure is projected to increase over the subtropics and mid-latitudes (depressing sea level) and decrease over high latitudes (raising sea level), especially over the Arctic (order several millibars), by the end of the 21st century associated with a poleward expansion of the Hadley Circulation and a poleward shift of the storm tracks of several degrees latitude (Section 12.4.4) (Held and Soden, 2006). These changes may therefore contribute positively to the sea level rise in the Arctic in the range of up to 1.5 cm and about 2.5 cm for RCP4.5 and RCP8.5, respectively (Yin et al., 2010) (Figure 13.17). In contrast, air pressure changes oppose sea level rise in mid- and low latitudes albeit with small amplitudes. Air pressure may also influence regional sea level elsewhere, as demonstr ated by sea level changes in the Mediterranean in the second half of the 20th century (Tsimplis et al., 2005).

13.6.4 Response to Freshwater Forcing

Enhanced freshwater fluxes derived from an increase in ice-sheet meltwater at high latitudes results in a regional pattern of sea level rise

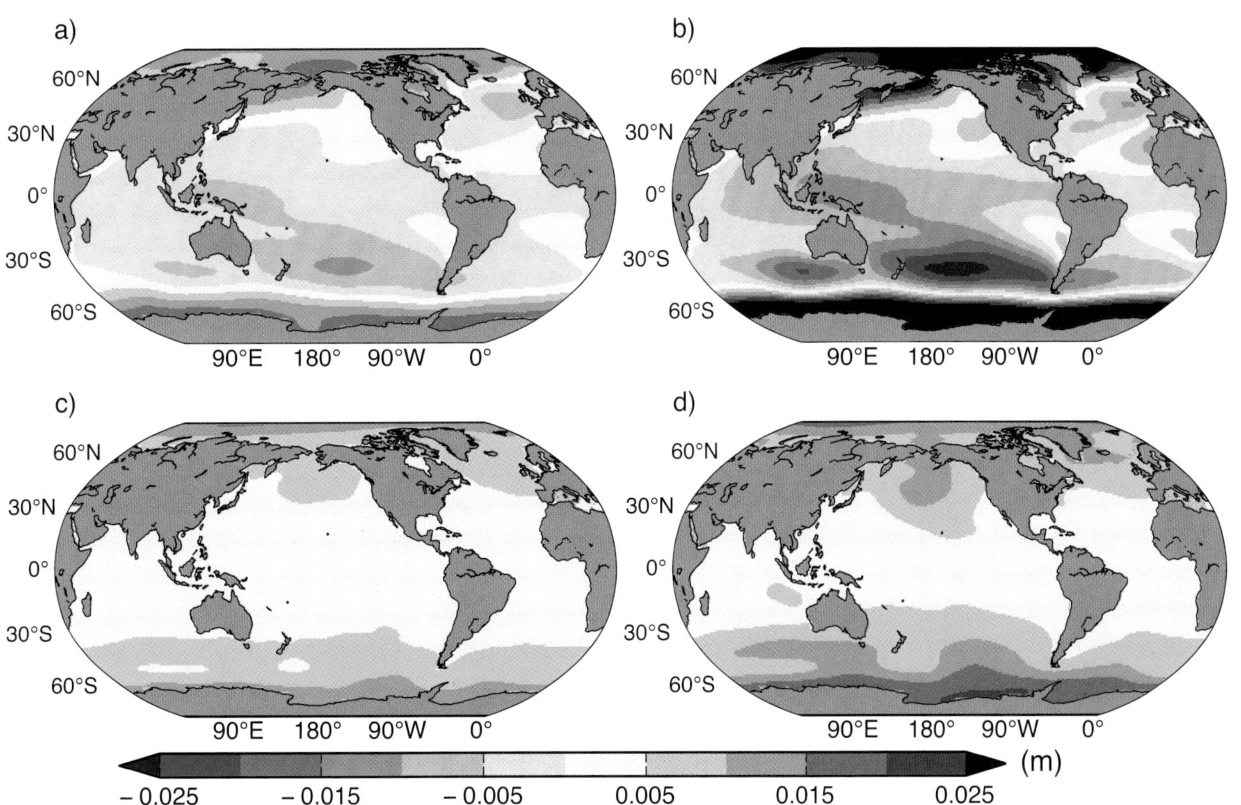

Figure 13.17 | Projected ensemble mean sea level change (metres) due to changes in atmospheric pressure loading over the period from 1986–2005 to 2081–2100 for (a) RCP4.5 and (b) RCP8.5 (contour interval is 0.005 m). Standard deviation of the model ensemble due to the atmospheric pressure loading for (c) RCP4.5 and (d) RCP8.5 (contour interval is 0.005 m).

originating from adjustments in ocean dynamics and in the solid earth. Neither effect is included in CMIP5 models, although the latter adjustment is computed off line here.

13.6.4.1 Dynamic Ocean Response to Cryospheric Freshwater Forcing

The addition of freshwater from glaciers and ice sheets to the ocean leads to an instantaneous increase in global mean sea level, but because it is communicated around the ocean basins via a dynamical adjustment, it is not instantaneously globally uniform (Kawase, 1987; Cane, 1989). For the addition of mass, the barotropic adjustment of the ocean takes place in a few days (Gower, 2010; Lorbacher et al., 2012). The addition of freshwater to the ocean from melting of the Greenland ice sheet results in an additional basin-wide steric response of the North Atlantic within months and is communicated to the global ocean via boundary waves, equatorial Kelvin waves, and westward propagating baroclinic Rossby waves on decadal time scales (Stammer, 2008). A similar response but with a different pattern can be observed from Antarctic meltwater input. In both cases, an associated complete baroclinic adjustment of the global ocean might take as long as several centuries. The adjustment of the ocean to high-latitude meltwater input also involves atmospheric teleconnections; such a response to Greenland meltwater pulses could lead to sea level changes in the Pacific within months (Stammer et al., 2011). On longer-than-decadal time scales, the freshwater input to the North Atlantic raises sea level in the Arctic Ocean and leads to an anomalous southward Bering Strait throughflow, transporting colder, fresher water from the Arctic Ocean into the North Pacific (Hu et al., 2010) and causing North Pacific cooling (Okumura et al., 2009). Meltwater forcing in the subpolar North Atlantic also causes changes of the AMOC (Section 12.4.7.2), which in turn causes dynamical changes of sea level in the North Atlantic, particularly in its northwestern region (Lorbacher et al., 2010). The combination of this dynamic sea level rise and the global mean sea level rise makes the northeastern North American coast vulnerable to some of the fastest and largest sea level rises during this century (Yin et al., 2009).

13.6.4.2 Earth and Gravitational Response to Contemporary Surface Water Mass Redistribution

Deformational, rotational and gravitational responses to mass redistribution between the cryosphere, the land and the oceans produce distinctive regional departures from GMSL, referred to as sea level fingerprints (Mitrovica et al., 2001, 2009; Gomez et al., 2010a; Riva et al., 2010) (Section 13.1, FAQ 13.1). Many existing studies of these effects have not defined a specific rate of ice-sheet mass loss (Mitrovica et al., 2001) or are based on end-member scenarios of ice retreat, such as from the WAIS (Bamber et al., 2009; Mitrovica et al., 2009; Gomez et al., 2010a) and marine-based parts of the East Antarctic ice sheet (Gomez et al., 2010a). Bamber and Riva (2010) calculated the sea level fingerprint of all contemporary land-ice melt and each of its major components. Spada et al. (2013) examined the regional sea level pattern from future ice melt based on the A1B scenario.

As can be seen from Figure 13.18, a characteristic of the sea level fingerprints is that regions adjacent to the source of the mass loss are subject to relative sea level fall of about an order of magnitude greater than the equivalent GMSL rise from these mass contributions, whereas in the far field the sea level rise is larger (up to about 30%) than the global average rise (Mitrovica et al., 2001, 2009; Gomez et al., 2010a). Gomez et al. (2010a) and Mitrovica et al. (2011) showed that differences in the maximum predicted rise (relative to the global mean) between published results is due to the accuracy with which water expulsion from the deglaciated marine basins is calculated. These changes are in addition to the ongoing response to past changes (e.g., glacial isostatic adjustment in response to the last deglaciation). Mitrovica et al. (2001) suggested that the lower rates of sea level change inferred from tide gauge records at European sites relative to the global average were consistent with 20th century melting from Greenland. Similarly, Gehrels and Woodworth (2013) suggested that the larger magnitude of the early 20th century sea level acceleration observed in Australia and New Zealand, as compared with the North Atlantic, may represent a fingerprint of the increased melt contributions of Greenland and Arctic glaciers in the 1930s. Nevertheless, current rates of ice-sheet melting are difficult to distinguish from dynamic variability (Kopp et al., 2010; Hay et al., 2013), but it is *likely* that with further ice-sheet melting they will begin to dominate the regional patterns of sea level change toward the end of the 21st century, especially under climate forcing conditions for which ice-sheet melting contributes more than 20-cm equivalent sea level rise (Kopp et al., 2010). These changes are in addition to the ongoing response to past changes (e.g., GIA in response to the last deglaciation; Figure 13.18a).

Water mass redistributions associated with land hydrology changes other than those from land ice may also produce spatially variable fingerprints in sea level (Fiedler and Conrad, 2010). In particular, regional changes in the terrestrial storage of water can lead to a sea level response on interannual and longer time scales, specifically near large river basins (Riva et al., 2010).

13.6.5 Regional Relative Sea Level Changes

Regional relative sea level change projections can be estimated from a combination of the various contributions to sea level change described above, emerging from the ocean, atmospheric pressure loading and the solid Earth.

Over the next few decades, regional relative sea level changes over most parts of the world are *likely* to be dominated by dynamical changes (mass redistribution and steric components) resulting from natural variability, although exceptions are possible at sites near rapidly melting ice sheets where static effects could become large. However, towards the end of the 21st century, regional patterns in sea level from all other contributions will progressively emerge and eventually dominate over the natural variability.

Ensemble mean estimates of relative sea level change during the period 2081–2100 relative to 1986–2000 resulting from GIA and from glacier and ice-sheet melting for RCP4.5 and RCP8.5 scenarios (Figure 13.18) suggest that for the 21st century, past, present and future loss of land-ice mass will *very likely* be an important contributor to spatial patterns in relative sea level change, leading to rates of maximum rise at low-to-mid latitudes. Hu et al. (2011) and Sallenger et al. (2012) also

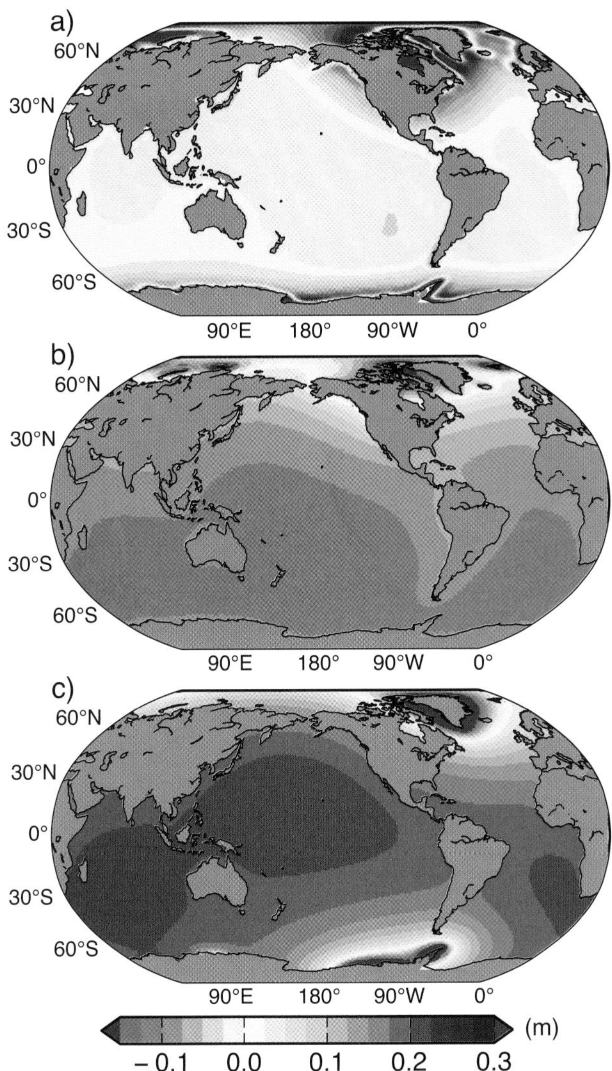

Figure 13.18 | Ensemble mean regional contributions to sea level change (metres) from (a) glacial isostatic adjustment (GIA), (b) glaciers and (c) ice-sheet surface mass balance (SMB). Panels (b) and (c) are based on information available from scenario RCP4.5. All panels represent changes between the periods 1986–2000 and 2081–2100.

suggested that steric and dynamical sea level changes can potentially increase the sea level near the northeastern coast of North America and in the western Pacific. Considerable uncertainties remain, however, in both the sea level budget and in the regional expression of sea level rise. In addition, local sea level rise can also partly be compensated by vertical land movement resulting from GIA, especially in some formerly glaciated high-latitude regions where high rates of land uplift may lead to a decrease of relative sea level. For example, Johansson et al. (2014) reported a 29 cm sea level rise in the Gulf of Finland and 27 cm fall in the Bay of Bothnia.

The ensemble mean regional relative sea level change between 1986–2005 and 2081–2100 for the RCP4.5 scenario (not including the dynamic ocean contribution in response to the influx of freshwater associated with land-ice loss and changes in terrestrial ground water) reveals that many regions are *likely* to experience regional sea level changes that differ substantially from the global mean (Figure 13.19).

Figure 13.20 shows ensemble mean regional relative sea level change between 1986–2005 and 2081–2100 for RCPs 2.6, 6.0 and 8.5.

It is *very likely* that over about 95% of the world ocean, regional relative sea level rise will be positive, while most regions that will experience a sea level fall are located near current and former glaciers and ice sheets. Figure 13.21b shows that over most of the oceans (except for limited regions around western Antarctica, Greenland, and high Arctic regions), estimated regional sea level changes are significant at the 90% confidence limit. Local sea level changes deviate more than 10% and 25% from the global mean projection for as much as 30% and 9% of the ocean area, respectively, indicating that spatial

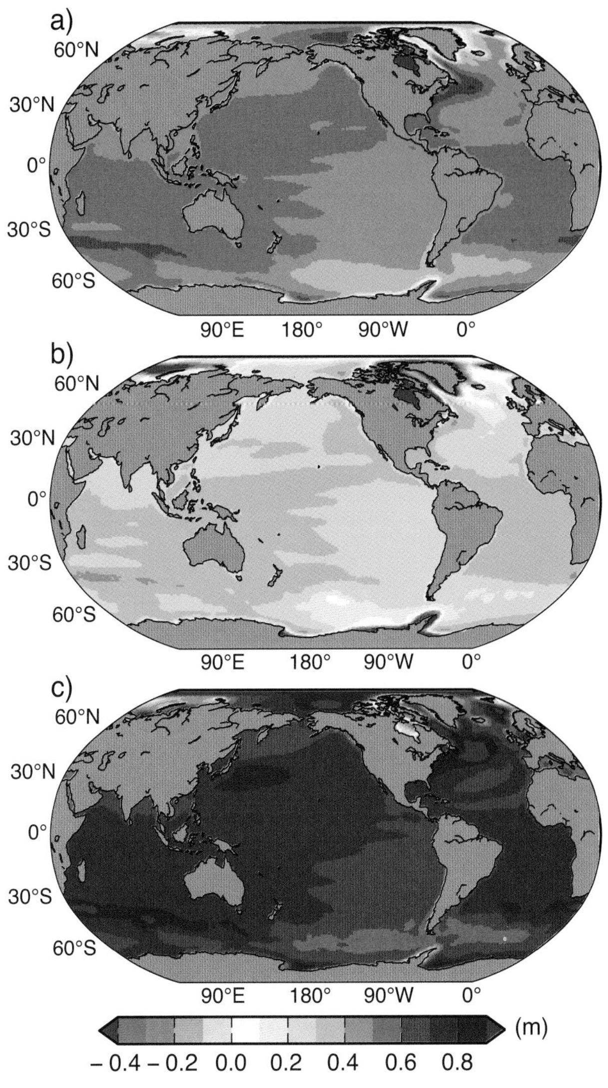

Figure 13.19 | (a) Ensemble mean regional relative sea level change (m) evaluated from 21 models of the CMIP5 scenario RCP 4.5, including atmospheric loading, plus land-ice, GIA and terrestrial water sources, between 1986–2005 and 2081–2100. Global mean is 0.48 m, with a total range of -1.74 to +0.71 m. (b) The local, lower 90% uncertainty bound (p=0.05) for RCP4.5 scenario sea level rise (plus non-scenario components). (c) The local, upper 90% uncertainty bound (p=0.95) for RCP4.5 scenario sea level rise (plus non-scenario components). Note that the global mean is different from the value in Table 13.5, by less than 0.01 m, because a slightly different set of CMIP5 models was used (see the Supplementary Material) and that panels (b) and (c) contain local uncertainties not present in global uncertainties.

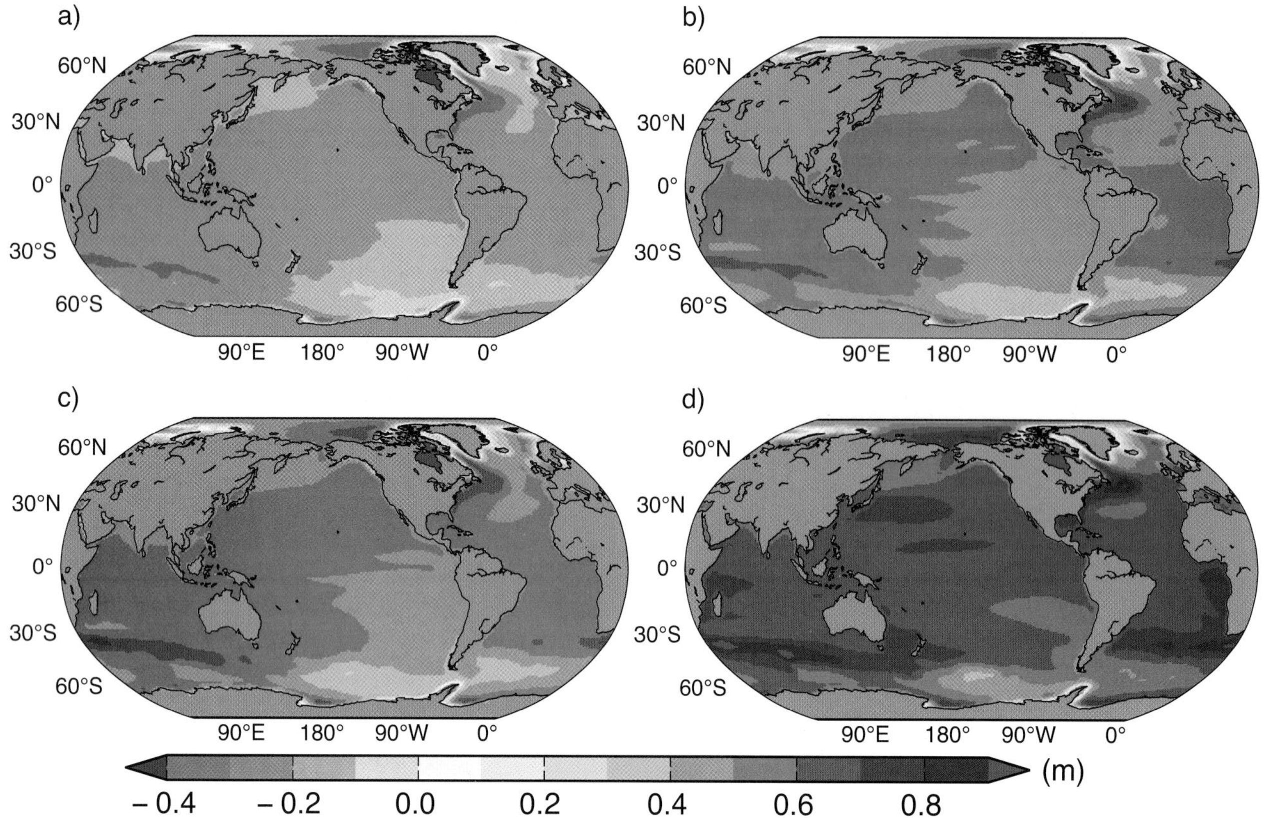

Figure 13.20 | Ensemble mean regional relative sea level change (metres) evaluated from 21 CMIP5 models for the RCP scenarios (a) 2.6, (b) 4.5, (c) 6.0 and (d) 8.5 between 1986–2005 and 2081–2100. Each map includes effects of atmospheric loading, plus land ice, glacial isostatic adjustment (GIA) and terrestrial water sources.

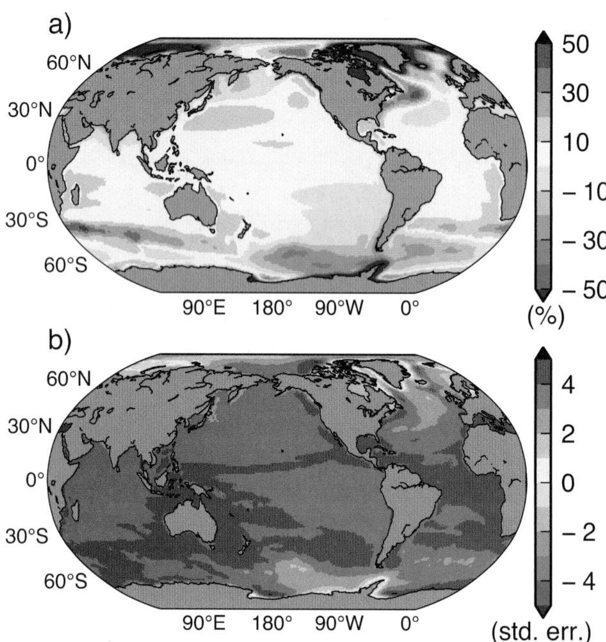

Figure 13.21 | (a) Percentage of the deviation of the ensemble mean regional relative sea level change between 1986–2005 and 2081–2100 from the global mean value. The figure was computed for RCP4.5, but to first order is representative for all RCPs. (b) Total RCP4.5 sea level change (plus all other components) divided by the combined standard error of all components (see Supplementary Material Section 13.SM.2). Assuming a normal distribution, or a *t*-distribution given the number of models as an approximation of the number of degrees of freedom, a region passes the 90% confidence level where the change is greater than 2 standard errors, which is most of the ocean except for limited regions around western Antarctica, Greenland and high Arctic regions.

variations can be large. Regional changes in sea level reach values of up to 30% above the global mean value in the Southern Ocean and around North America, between 10 and 20% in equatorial regions and up to 50% below the global mean in the Arctic region and some regions near Antarctica (Figure 13.21a).

Figure 13.22 shows that, between 1986–2005 and 2081–2100, sea level changes along the world's coastlines associated with the RCP4.5 and RCP8.5 scenarios have a substantially skewed non-Gaussian distribution, with significant coastal deviations from the global mean. When the coastlines around Antarctica and Greenland are excluded (Figure 13.22b), many negative changes disappear, but the general structures of the global histograms remain. In general, changes along the coastlines will range from about 30 cm to 55 cm for an RCP 4.5 scenario, peaking near 50 cm, and from about 40 cm to more than 80 cm under a RCP 8.5 scenario, peaking near 65 cm. About 68% and 72% of the coastlines will experience a relative sea level change within ±20% of the GMSL change for RCP4.5 and RCP8.5, respectively. In both cases, the maximum of the histogram is slightly higher than the GMSL, whereas the arithmetic mean is lower. Only some coastlines will experience a sea level rise of up to about 40% above GMSL change.

Figure 13.23 shows the combination of the natural variability (annual mean) and the CMIP5 projected sea level rise for the RCP4.5 scenario for a number of locations distributed around the world. For example, at Pago Pago (14°S,195°E) in the western equatorial Pacific, the

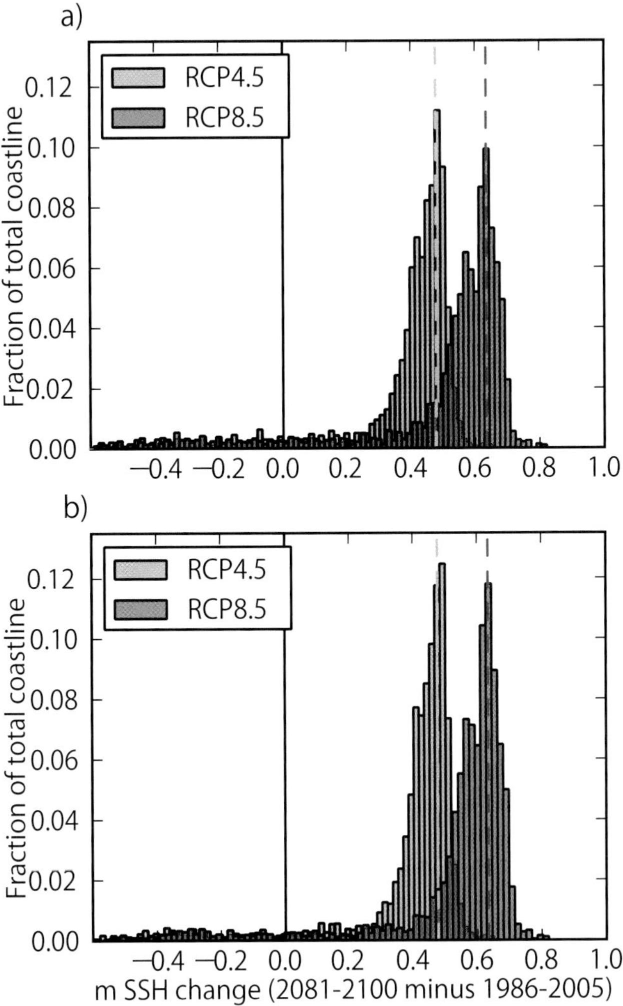

Figure 13.22 | (a) Histograms of the deviation of the ensemble mean regional relative sea level change (Figure 13.20) along all coastlines (represented by the closest model grid point) between 1986–2005 and 2081–2100 from the global mean value. Shown are results for RCP4.5 (blue) and RCP8.5 (pink), respectively. (b) Same as in (a) but excluding Antarctic and Greenland coastlines. Vertical dashed lines represent global mean sea level changes for the two RCPs.

historical record indicates that annual variability in mean sea level has been about 21 cm (5 to 95% range). Projections by individual climate models indicate that it is *very likely* that a similar range of natural variability will continue through the 21st century (Figure 13.15b). However, by 2100, the average projected sea level for the RCP4.5 scenario of 0.52 [0.32 to 0.70] m is greater than any observations of annual mean sea level in the instrumental record. Of all the examples shown, the greatest sea level increase will be in New York, which is representative of the enhanced sea level rise there due to ocean processes and GIA in the region (compare Figures 13.16 and 13.18). The figure also reveals the large spatial inhomogeneity of interannual to decadal variability. In each case, monthly variability and extreme sea levels from winds and waves associated with weather phenomena (Section 13.7) need to be considered in addition to these projections of regional sea level.

13.6.6 Uncertainties and Sensitivity to Ocean/Climate Model Formulations and Parameterizations

Uncertainties of climate models are discussed in detail in Chapter 9. Sea level is a property of the ocean connected to nearly all dynamical and thermodynamical processes over the full ocean column, from the surface fluxes to the ocean bottom. Although many of the processes are to first order correctly simulated in climate models, differences between models (Figure 13.24) indicate that uncertainties in simulated and projected steric sea level (globally and regionally) remain poorly understood. Moreover, the spread in ocean heat uptake efficiency among models is responsible for 50% of the spread in heat uptake (Kuhlbrodt and Gregory, 2012). In addition, some processes are not part of the CMIP5 simulations, such as the dynamical response of the ocean to meltwater input or the GIA/rotational/gravitational processes associated with this ice mass loss. Stammer and Hüttemann (2008) showed that coupled climate models that do not include the effect of changes in atmospheric moisture content on sea level pressure will underestimate future regional atmospheric pressure loading effects by up to 2 cm. Other uncertainties result from GIA/rotational/gravitational effects as well as from uncertainties in air–sea fluxes.

Improvements in the skill of a sea level projection require (1) better parameterizations of unresolved physical processes, (2) improved numerical algorithms for such processes as temperature and salinity advection, (3) refined grid resolution to better represent such features as boundary currents and mesoscale eddies, and (4) the elimination of obsolete assumptions that have a direct impact on sea level (Griffies and Greatbatch, 2012). Among the many limiting approximations made in ocean models, the Boussinesq approximation has been found to only marginally impact regional patterns (i.e., deviations from global mean) when directly compared to non-Boussinesq simulations (Losch et al., 2004), thus lending greater confidence in Boussinesq models for addressing questions of regional sea level change. Furthermore, for global sea level, the now-standard *a posteriori* adjustment (Greatbatch, 1994; Griffies and Greatbatch, 2012) accurately incorporates the missing global steric effect. The representation of dense overflows can also affect sea level simulations, and is particularly problematic in many ocean models used for climate studies, with direct impacts on the simulated vertical patterns of ocean heat uptake (Legg et al., 2009).

Coarse-resolution ocean–climate simulations require a parameterization of mesoscale and smaller eddies, but the parameterizations as well as the details of their numerical implementations can greatly impact the simulation. As shown by Hallberg and Gnanadesikan (2006) and Farneti et al. (2010), coarse-resolution climate models may be overestimating the Antarctic Circumpolar Current response to wind changes. Better implementations of eddy parameterizations reduce such biases (Farneti and Gent, 2011; Gent and Danabasoglu, 2011), and they form the basis for some, but not all, of the CMIP5 simulations. Moreover, Vinogradov and Ponte (2011) suggested that as one considers regional sea level variability and its relevant dynamics and forcing, mesoscale ocean features become important factors on a sub-decadal time scale. Suzuki et al. (2005) compared changes in mean dynamic sea level in 2080–2100 relative to 1980–2000 as obtained from a low- and a high-resolution ocean component of a coupled model and concluded that although changes are comparable between runs, the high-resolution model captures enhanced details owing to resolving ocean eddy dynamics.

Even with a perfect ocean model, skill in sea level projections depends on skill of the coupled climate model in which errors impacting sea level may originate from non-ocean components. Furthermore, initialization is fundamental to the prediction problem, particularly for simulation of low-frequency climate variability modes (Meehl et al., 2010). Projections of land-ice melting and the resultant sea level rise patterns also have large uncertainties, with additional uncertainties arising from GIA models such as the mantle viscosity structure. Each of the many uncertainties and errors results in considerable spread in the projected patterns of sea level change (Figure 13.24) (Pardaens et al., 2011a; Slangen et al., 2012). In addition to ocean–climate model formulations and parameterizations, uncertainty in predictions of sea level change may be associated with specified freshwater forcing. Whether or not an ocean model is coupled with an ice-sheet model, the forcing should distinguish between runoff and iceberg flux. Martin and Adcroft (2010) reported the only attempt thus far to explicitly represent iceberg drift and melting in a fully coupled climate model.

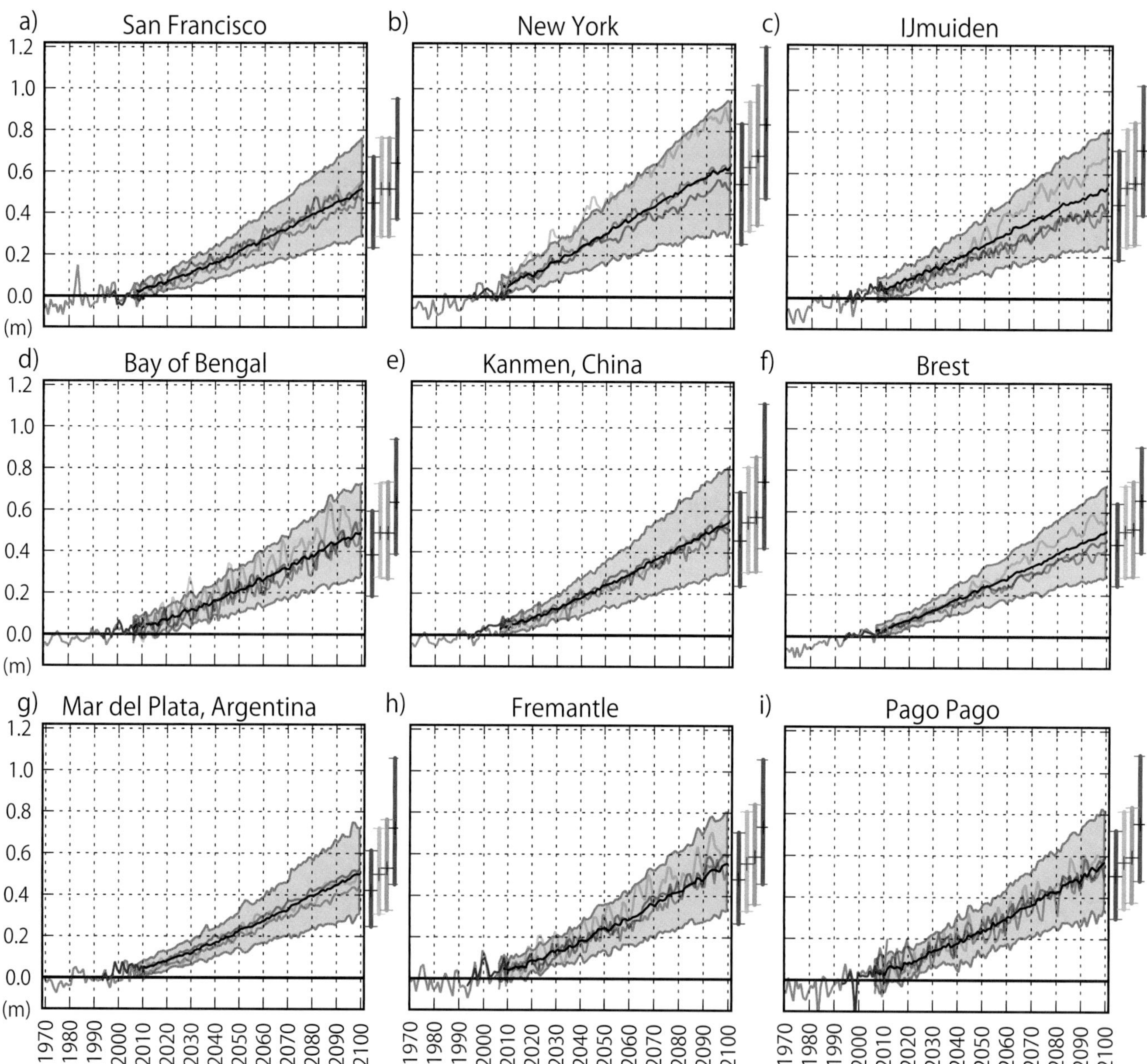

Figure 13.23 | Observed and projected relative sea level change (compare Figure 13.20) near nine representative coastal locations for which long tide-gauge measurements are available. The observed *in situ* relative sea level records from tide gauges (since 1970) are plotted in yellow, and the satellite record (since 1993) is provided as purple lines. The projected range from 21 CMIP5 RCP4.5 scenario runs (90% uncertainty) is shown by the shaded region for the period 2006–2100, with the bold line showing the ensemble mean. Coloured lines represent three individual climate model realizations drawn randomly from three different climate models used in the ensemble. Station locations of tide gauges are: (a) San Francisco: 37.8°N, 122.5°W; (b) New York: 40.7°N, 74.0°W; (c) Ijmuiden: 52.5°N, 4.6°E; (d) Haldia: 22.0°N, 88.1°E; (e) Kanmen, China: 28.1°N, 121.3°E; (f) Brest: 48.4°N, 4.5°W; (g) Mar del Plata, Argentina: 38.0°S, 57.5°W; (h) Fremantle: 32.1°S, 115.7°E; (i) Pago Pago: 14.3°S, 170.7°W. Vertical bars at the right sides of each panel represent the ensemble mean and ensemble spread (5 to 95%) of the *likely* (*medium confidence*) sea level change at each respective location at the year 2100 inferred from RCPs 2.6 (dark blue), 4.5 (light blue), 6.0 (yellow) and 8.5 (red).

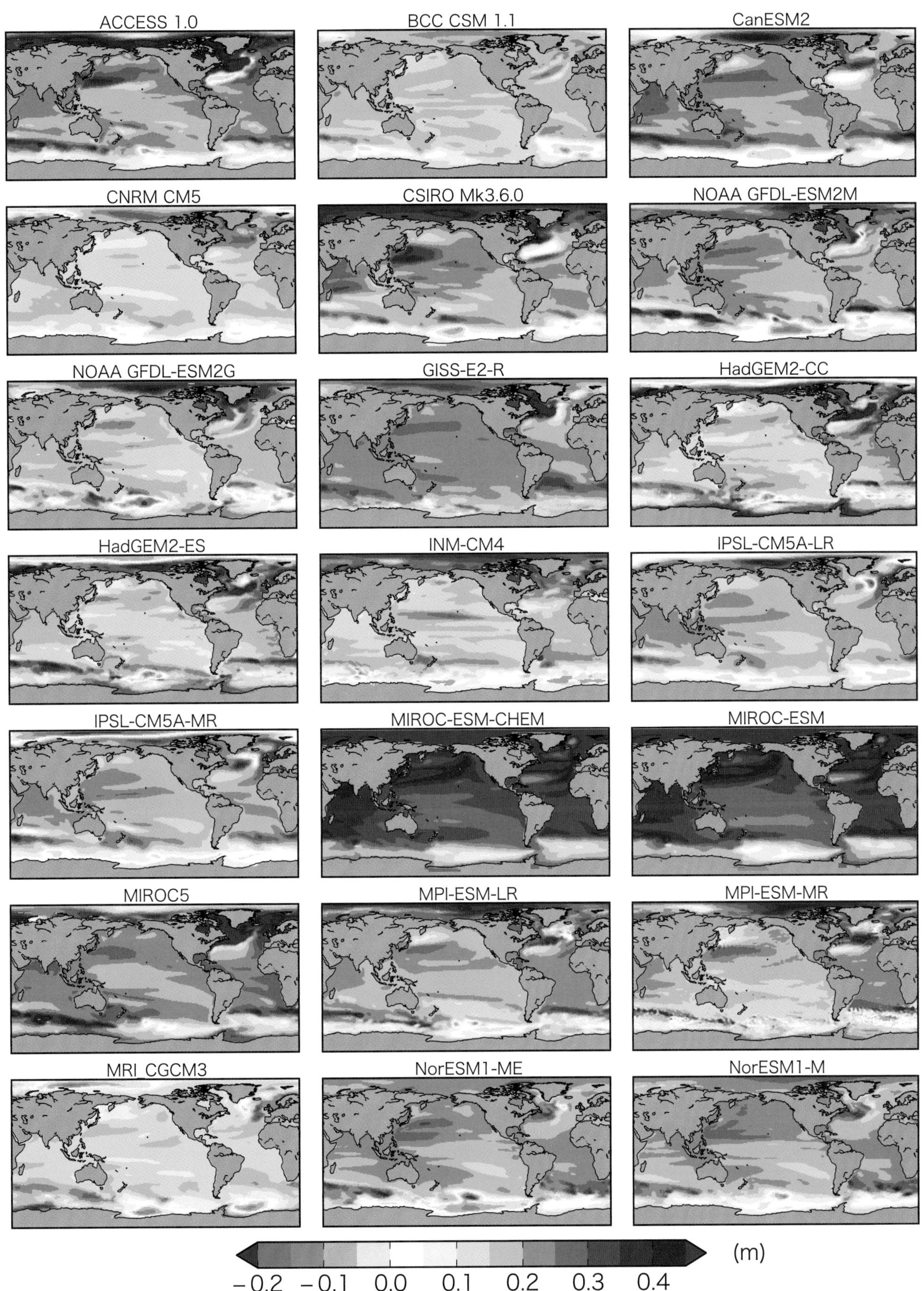

Figure 13.24 | Projected relative sea level change (in m) from the combined global steric plus dynamic topography and glacier contributions for the RCP4.5 scenario over the period from 1986–2005 to 2081–2100 for each individual climate model used in the production of Figure 13.16a.

13.7 Projections of 21st Century Sea Level Extremes and Waves

Climate change will affect sea levels extremes and ocean waves in two principal ways. First, because extratropical and tropical storms are one of the key drivers of sea level extremes and waves, future changes in intensity, frequency, duration, and path of these storms will impact them. Second, sea level rise adds to the heights of sea level extremes, regardless of any changes in the storm-related component. MSL change may also accentuate the threat of coastal inundation due to changes in wave runup. Observations of changes in sea level extremes and waves are discussed in Chapter 3. Sea level extremes at the coast occur mainly in the form of storm surges and tsunamis, but because the latter are not climate driven, we assess only projections for sea level extremes based on estimates of future storminess and MSL change.

13.7.1 Observed Changes in Sea Level Extremes

As discussed in the AR4 (Bindoff et al., 2007) and confirmed by more recent studies (Menéndez and Woodworth, 2010), statistical analyses of tide-gauge observations have shown an increase in observed sea level extremes worldwide that are caused primarily by an increase in MSL (Chapter 3). Dominant modes of climate variability, particularly ENSO and NAO, also have a measureable influence on sea level extremes in many regions (Lowe et al., 2010; Walsh et al., 2011). These impacts are due to sea level anomalies associated with climate modes, as well as mode-related changes in storminess. There has been some indication that the amplitude and phase of major tidal constituents have exhibited long-term change (Jay, 2009; Muller et al., 2011), but their impacts on extreme sea level are not well understood. Using particle size analysis of cores collected in the Mackenzie Delta in the Arctic region, Vermaire et al. (2013) inferred increased storm surge activity in the region during the last approximately 150 years, which they related to the annual mean temperature anomaly in the NH and a decrease in summer sea-ice extent.

13.7.2 Projections of Sea Level Extremes

13.7.2.1 Recent Assessments of Projections of Sea Level Extremes

The AR4 assessed projections of storm surges for a few regions (Europe, Australia, the Bay of Bengal) based on a limited number of dynamical modelling studies (Christensen et al., 2007). Although these results generally indicated higher magnitude surges in future scenarios, there was *low confidence* in these projections because of the wide spread in underlying AOGCM and RCM projections.

13

Studies since the AR4 have further assessed the relative contributions of sea level rise and storminess on projected sea level extremes. Lowe et al. (2010) concluded that the increases in the observed sea level extremes in the 20th century occurred primarily through an increase in MSL, and that the same applies to projections for the 21st century. The IPCC Special Report on Managing the Risks of Extreme Events and Disasters to Advance Climate Change Adaptation (SREX) assessment concluded that it is *very likely* that MSL rise will contribute to an increase in future sea level extremes (Seneviratne et al., 2012). It noted that changes in storminess may also affect sea level extremes but the limited geographical coverage of studies and uncertainties associated with storminess changes prevent a general assessment. The global tropical cyclone frequency will *likely* decrease or remain roughly constant, but it is *more likely than not* that the frequency of the most intense storms will increase in some ocean basins (Chapter 14). Uncertainties in projections of cyclone frequency and tracks make it difficult to project how these changes will impact particular regions. Similarly, while the SREX and the current assessment (Chapter 14) find that it is *likely* that there has been a poleward shift in the main northern and southern extra-tropical cyclone tracks during the last 50 years, and that regional changes may be substantial, there is only *low confidence* in region-specific projections.

13.7.2.2 Projections Based on Dynamical and Statistical Approaches

Projected changes in storm surges (relative to MSL) have been assessed by applying climate–model forcing to storm-surge models. Return periods of sea level extremes (see Glossary) exceeding a given threshold level, referred to as return levels, are used in quantifying projected changes. Using three regionally downscaled GCMs for A2, B2 and A1B scenarios, Debernard and Roed (2008) found an 8 to 10% increase in the 99th percentile surge heights between 1961–1990 and 2071–2100, mainly during the winter season, along the coastlines of the eastern North Sea and the northwestern British Isles, and decreases south of Iceland. Using a downscaled GCM under an A1B scenario, Wang et al. (2008) projected a significant increase in wintertime storm surges around most of Ireland between 1961–1990 and 2031–2060. Sterl et al. (2009) concatenated the output from a 17-member ensemble of A1B simulations from a GCM over the periods 1950–2000 and 2050–2100 into a single longer time series to estimate 10,000-year return levels of surge heights along the Dutch coastline. No statistically significant change in this value was projected for the 21st century because projected wind speed changes were not associated with the maximum surge-generating northerlies. Using an ensemble of three climate models under A2 simulations, Colberg and McInnes (2012) found that changes in the 95th percentile sea level height (with respect to mean sea level) across the southern Australian coast in 2081–2100 compared to 1981–2000 were small (±0.1 m), mostly negative, and despite some inter-model differences, resembled the changes in wind patterns simulated by the climate models (McInnes et al., 2011). These studies demonstrate that the results are sensitive to the particular choice of GCM or RCM, therefore identifying uncertainties associated with the projections. For the tropical east coast of Australia, Harper et al. (2009) found that a 10% increase in tropical cyclone intensity for 2050 led to increases in the 100-year return level (including tides) that at most locations were smaller than 0.1 m with respect to mean sea level.

Several regional storm-surge studies have considered the relative contribution of the two main causative factors on changes in future sea level extremes (e.g., McInnes et al. (2009, 2013) for the southeastern coast of Australia; Brown et al. (2010) for the eastern Irish Sea; Woth et al. (2006) for the North Sea; Lowe et al. (2009) for the United Kingdom coast). They concluded that sea level rise has a greater potential than meteorological changes to increase sea level extremes by the end of the 21st century in these locations. Unnikrishnan et al. (2011) used

RCM simulations to force a storm-surge model for the Bay of Bengal and found that the combined effect of MSL rise of 4 mm yr^{-1} and RCM projections for the A2 scenario (2071–2100) gave an increase in 100-year return levels of total sea level (including tides) between 0.40 to 0.67 m (about 15 to 20%) along the northern part of the east coast of India, except around the head of the bay, compared to those in the base line (1961–1990) scenario.

Using six hypothetical hurricanes producing approximate 100-year return levels, Smith et al. (2010) found that in the regions of large surges on the southeastern Louisiana coast, the effect of MSL rise added linearly to the simulated surges. However, in the regions of moderate surges (2–3 m), particularly in wetland-fronted areas, the increase in surge height was 1–3 m larger than the increase in mean sea level rise. They showed that sea level rise alters the speed of propagation of surges and their amplification in different regions of the coast. For the Gulf of Mexico, Mousavi et al. (2011) developed a simple relationship between hurricane-induced storm surges, sea level rise and hurricane intensification through increased SSTs for three modelled major historical cyclones, concluding that the dynamic interaction of surge and sea level rise lowered or amplified the surge at different points within a shallow coastal bay.

Higher mean sea levels can significantly decrease the return period for exceeding given threshold levels. For a network of 198 tide gauges covering much of the globe, Hunter (2012) determined the factor by which the frequency of sea levels exceeding a given height would be increased for a MSL rise of 0.5 m (Figure 13.25a). These calculations have been repeated here (Figure 13.25b) using regional RSL projections and their uncertainty using the RCP4.5 scenario (Section 13.6, Figure 13.19a). This multiplication factor depends exponentially on the inverse of the Gumbel scale parameter (a factor that describes the statistics of sea level extremes caused by the combination of tides and storm surges) (Coles and Tawn, 1990). The scale parameter is generally

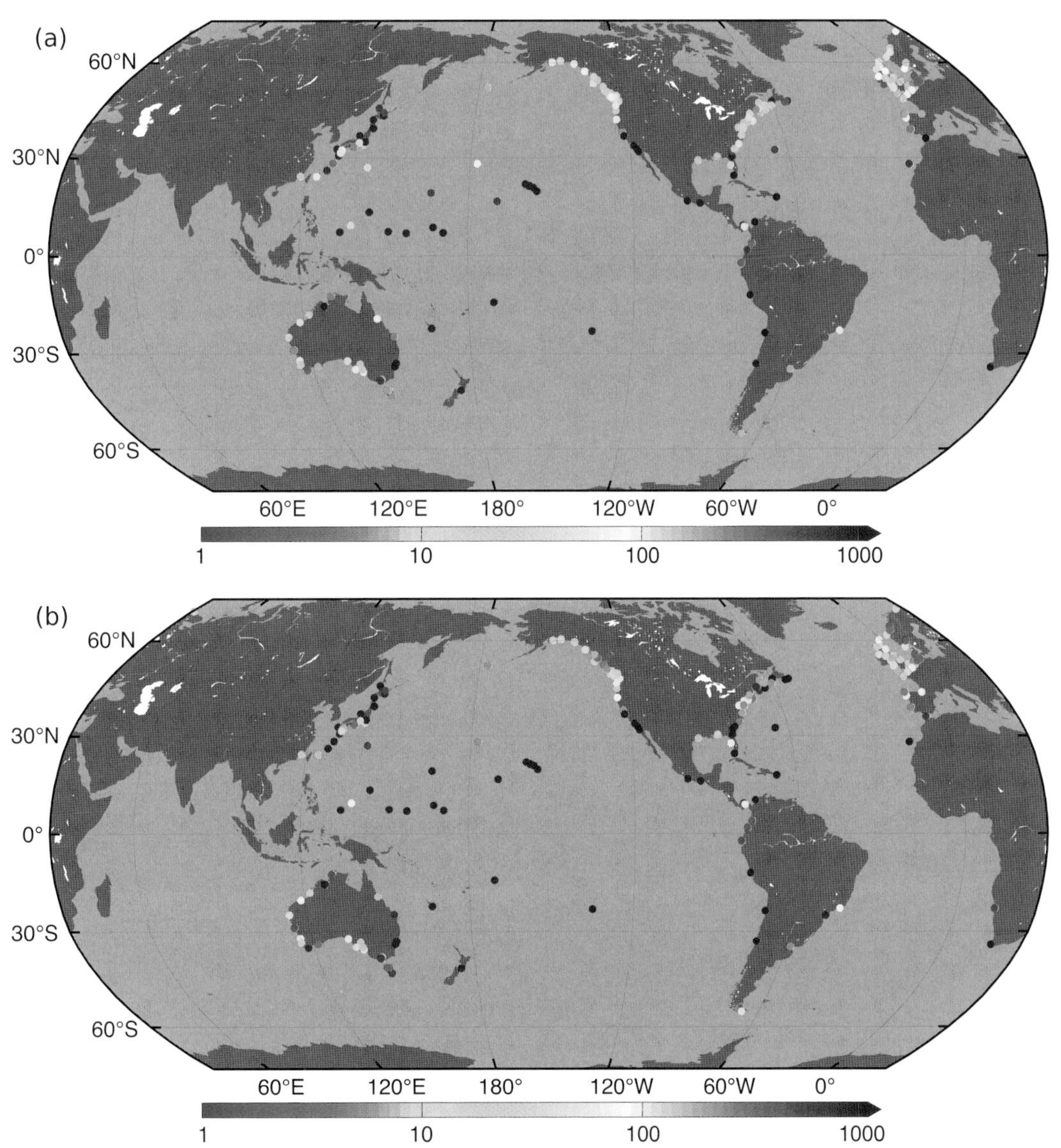

Figure 13.25 | The estimated multiplication factor (shown at tide gauge locations by colored dots), by which the frequency of flooding events of a given height increase for (a) a mean sea level (MSL) rise of 0.5 m (b) using regional projections of MSL for the RCP4.5 scenario, shown in Figure13.19a.

13

large where tides and/or storm surges are large, leading to a small multiplication factor, and vice versa. Figure 13.25a shows that a 0.5 m MSL rise would *likely* result in the frequency of sea level extremes increasing by an order of magnitude or more in some regions. The multiplication factors are found to be similar or slightly higher, in general, when accounting for regional MSL projections (Figure 13.25b). Specifically, in regions having higher regional projections of MSL, such as the east coast of Canada and the USA (where GIA results in a larger sea level rise) and/or in regions of large uncertainty (e.g. in regions near the former Laurentide ice sheet where the GIA uncertainty is large), the multiplication factor is higher, whereas in regions having lower regional projections of MSL, such as the northwest region of North America (where the land is rising due to present changes in glaciers and icecaps), the multiplication factor is lower. In another study, large increases in the frequency of sea level extremes for 2050 were found for a network of sites around the USA coastline based on semi-empirical MSL rise projections and 20th century statistics of extremes (Tebaldi et al., 2012). Using projected time series of tides, MSL rise, components of sea level fluctuations from projected MSLP and wind stress fields, and a contribution for ENSO variability through projected SSTs for the 21st century, Cayan et al. (2008) showed that for high-end scenarios of MSL rise, the frequency and magnitude of extremes along the California coast increases considerably relative to those experienced in the 20th century.

In summary, dynamical and statistical methods on regional scales show that it is *very likely* that there will be an increase in the occurrence of future sea level extremes in some regions by 2100, with a *likely* increase in the early 21st century. The combined effects of MSL rise and changes in storminess will affect future extremes. There is *high confidence* that extremes will increase with MSL rise yet there is *low confidence* in region-specific projections in storminess and storm surges.

13.7.3 Projections of Ocean Waves

Changes in ocean wave conditions are determined by changes in the major wind systems, especially in the main areas affected by tropical and extra-tropical storms. Based on *in situ* and satellite altimeter observations and wave–model hindcasts, it is *likely* that mean significant wave heights (SWH, defined as the average of the highest one third of wave heights) have increased in regions of the North Pacific and the North Atlantic over the past half century, and in the Southern Ocean since the mid 1980s (Chapter 3, Seneviratne et al., 2012). The limited observational wave record makes it difficult to separate long-term trends from multi decadal variability (Young et al., 2011). A number of studies have related changes in wind–wave climatologies to modes of climate variability such as ENSO (Allan and Komar, 2006; Adams et al., 2008; Menéndez et al., 2008), the NAO (Woolf et al., 2002; Izaguirre et al., 2010), and the Southern Annular Mode (SAM) (Hemer et al., 2010; Izaguirre et al., 2011). Although anthropogenic influences have been considered (Wang et al., 2009), it is *likely* that reported SWH trends over the past half-century largely reflect natural variations in wind forcing. Recent reductions in summer sea ice extent have resulted in enhanced wave activity in the Arctic Ocean due to increased fetch area and longer duration of the open-water season (Francis et al., 2011; Overeem et al., 2011).

In general, there is *low confidence* in projections of future storm conditions (Chapters 12 and 14) and hence in projections of ocean waves. Nevertheless, there has been continued progress in translating climate model outputs into wind–wave projections. In the AR4, projected changes in global SWHs were based on a single statistical model (Wang and Swail, 2006). The projected conditions were consistent with increased wind speeds associated with mid-latitude storms, but they considered only a limited five-member ensemble for a single future emission scenario (SRES A2); wave parameters other than SWH were not considered.

Since the AR4, global wave–climate projections for the end of the 21st century have been made by dynamically downscaling CMIP3 AOGCM results. A multi-model ensemble based on dynamical models forced with various GHG emission scenarios (SRES A1B: Mori et al. (2010), Fan et al. (2013), Semedo et al. (2013); SRES A2: Hemer et al. (2012a), as well as the statistical model of Wang and Swail (2006) forced with emission scenarios IS92a and SRES A2 and B2, has been constructed as part of the Coordinated Ocean Wave Climate Project (COWCLIP) (Hemer et al., 2013). In general, the ensemble projected changes of annual mean SWH (Figure 13.26a) resemble the statistical projections of Wang and Swail (2006) under an A2 scenario. The largest change is projected to be in the Southern Ocean, where mean SWHs at the end of the 21st century are approximately 5 to 10% higher than the present-day mean. SWH increase in this region reflects the projected strengthening of the westerlies over the Southern Ocean, particularly during austral winter (Figure 13.26c). Another region of SWH increase in the ensembles is in the tropical South Pacific associated with a projected strengthening of austral winter easterly trade winds in the CMIP3 multi-model data set (Figure 13.26c). Negligible change or a mean SWH decrease is projected for all other ocean basins, with decreases identified in the trade wind region of the North Pacific, the mid-latitude westerlies in all basins, and in the trade and monsoon wind regions of the Indian Ocean. Hemer et al. (2013) found that variance of wave–climate projections associated with wave downscaling methodology dominated other sources of variance within the projections such as the climate scenario or climate model uncertainties. Mori et al. (2013) reported similar findings.

Three CMIP3-based model projections (Mori et al., 2010; Hemer et al., 2012b; Fan et al., 2013) were used to compare projections of wave direction and period (Hemer et al., 2013). Wave direction (Figure 13.26d) exhibits clockwise rotation in the tropics, consistent with a higher contribution from northward propagating swell from the Southern Ocean. Wave period (Figure 13.26e) shows an increase over the eastern Pacific, which is also attributed to enhanced wave generation in the Southern Ocean and northward swell propagation. A projected decrease in wave periods in the North Atlantic and western and central North Pacific is symptomatic of weaker wind forcing in these regions.

SWH projections based on CMIP5 winds for emission scenarios RCP4.5 and RCP8.5 (Dobrynin et al., 2012) exhibit similar regional patterns for the end of the 21st century to the CMIP3 results presented in Figure 13.26A. Dobrynin et al. (2012) reported SWH increases in the Arctic Ocean, an area not considered by Hemer et al. (2013), and in basins connected to the Southern Ocean, particularly for RCP8.5. The probability of extreme wave heights is projected to increase in the SH, the

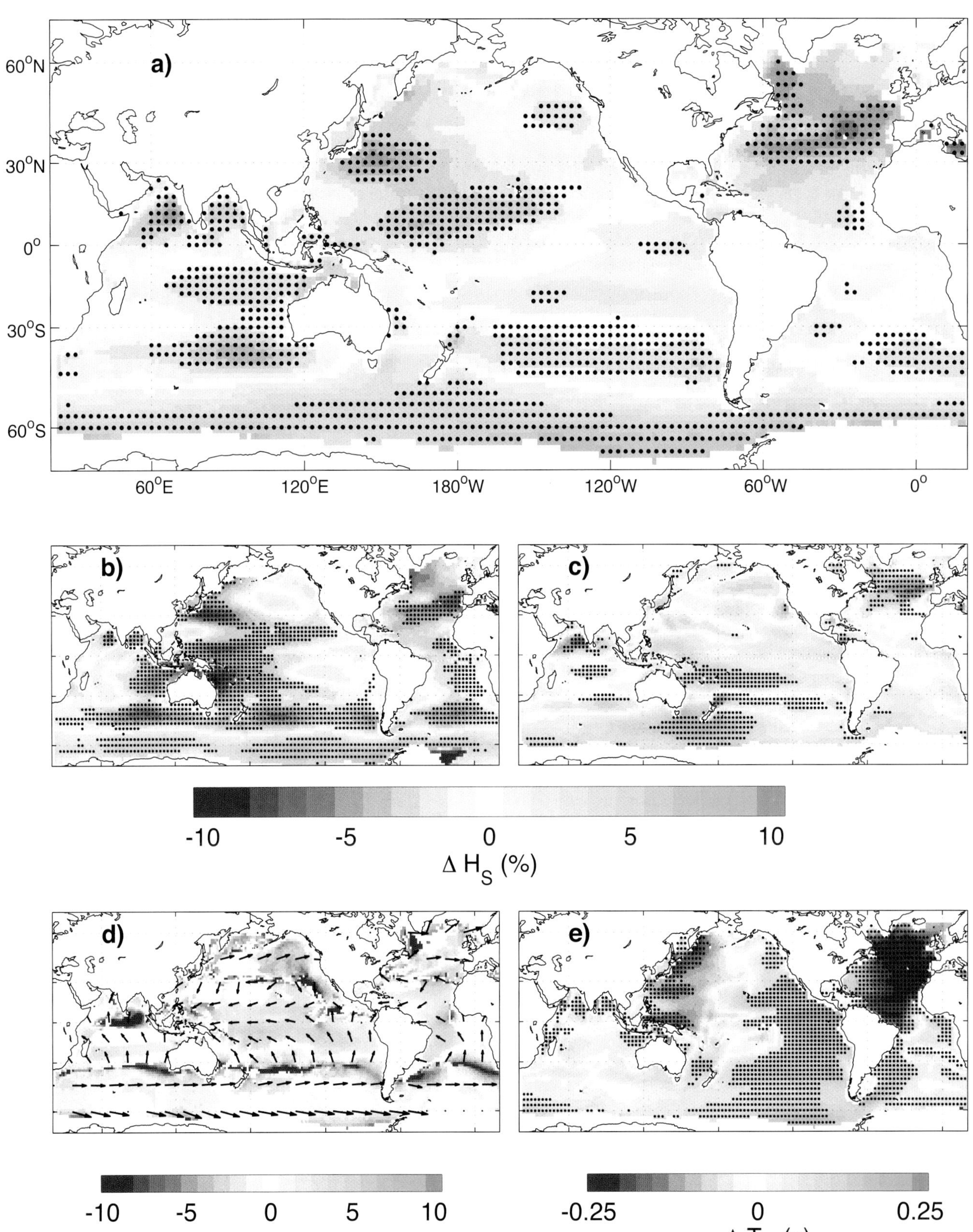

Figure 13.26 | Projected changes in wind–wave conditions (~2075–2100 compared with ~1980–2009) derived from the Coordinated Ocean Wave Climate Projection (COWCLIP) Project (Hemer et al., 2013). (a) Percentage difference in annual mean significant wave height. (b) Percentage difference in means of January to March significant wave height. (c) Percentage difference in means of July to September significant wave height. Hashed regions indicate projected change is greater than the 5-member ensemble standard deviation. (d) As for (a), but displaying absolute changes in mean wave direction, with positive values representing projected clockwise rotation relative to displayed vectors, and colours shown only where ensemble members agree on sign of change. (e) As for (a), but displaying absolute changes in mean wave period. The symbol ~ is used to indicate that the reference periods differ slightly for the various model studies considered.

Arctic and Indian Oceans, but decrease in the North and Equatorial Atlantic and in the Pacific. In addition to wind changes, the projected loss of summer sea ice extent in the Arctic Ocean is *very likely* to increase overall wave activity there (Manson and Solomon, 2007; Overeem et al., 2011).

Model intercomparisons are starting to identify common features of global wave projections but in general there is *low confidence* in wave model projections because of uncertainties regarding future wind states, particularly storm geography, the limited number of model simulations used in the ensemble averages, and the different methodologies used to downscale climate model results to regional scales (Hemer et al., 2012a). Despite these uncertainties, it appears *likely* (*medium confidence*) that enhanced westerly surface winds in the SH (discussed in Chapter 12) will lead to enhanced wave generation in that region by the end of the 21st century.

A number of dynamical wave projection studies have been carried out with a regional focus. For the Mediterranean Sea, Lionello et al. (2008; 2010) projected a widespread shift of the wave height distribution to lower values by the mid-21st century under an SRES A1B scenario, implying a decrease in mean and extreme wave heights. Caires et al. (2008) and Debernard and Røed (2008) reported a decrease (4 to 6% of present values) in the annual 99th percentile SWH south of Iceland by the end of the 21st century, and an increase (6 to 8%) along the North Sea east coast (SRES A2, B2, A1B scenarios). Grabemann and Weisse (2008) found increases (up to 18% of present values) in annual 99th percentile SWH in the North Sea by the end of the 21st century, with an increase in the frequency of extreme wave events over large areas of the southern and eastern North Sea (SRES A2, B2 scenarios). Charles et al. (2012) projected a general decrease in wave heights in the Bay of Biscay by the end of the 21st century (SRES A2, A1B, B1 scenarios), accompanied by clockwise rotations in winter swell (attributed to a projected northward shift in North Atlantic storm tracks) and summer sea and intermediate waves (attributed to a projected slackening of westerly winds). Along the Portuguese coast, Andrade et al. (2007) found little projected change in SWH and a tendency for a more northerly wave direction than present (SRES A2 scenario).

In the Pacific, multi-model projections by Graham et al. (2013) (SRES A2 scenario) indicate a decrease in boreal winter upper-quantile SWHs over the mid-latitude North Pacific by the end of the 21st century associated with a projected decrease in wind speeds along the southern flank of the main westerlies. There is a less robust tendency for higher extreme waves at higher latitudes. On the southeastern Australian coast, Hemer et al. (2012b) used multi-model projections (SRES A2 and B1 scenarios) to identify a decrease in mean SWH (<0.2 m) by the end of the 21st century compared to present due to a projected decrease in regional storm wave energy, and a shift to a more southerly wave direction, consistent with a projected southward shift of the subtropical ridge in the forcing fields.

13.8 Synthesis and Key Uncertainties

There has been significant progress in our understanding of sea level change since the AR4. Paleo data now provide *high confidence* that sea levels were substantially higher when GHG concentrations were higher or surface temperatures were warmer than pre-industrial. The combination of paleo sea level data and long tide gauge records confirms that the rate of rise has increased from low rates of change during the late Holocene (order tenths of mm yr^{-1}) to rates of almost 2 mm yr^{-1} averaged over the 20th century, with a *likely* continuing acceleration during the 20th century (Figure 13.27). Since 1993, the sum of observed contributions to sea level rise is in good agreement with the observed rise.

Understanding of the components that contribute to total sea level rise has improved significantly. For the 20th century, the range from an ensemble of such process-based models encompasses the observed rise when allowances are made for lack of inclusion of volcanic forcing in AOGCM control simulations, natural climate variability, and a possible small long-term ice-sheet contribution. Ice-sheet contributions to the 20th century sea level rise were small, however, and this agreement is thus not an evaluation of ice-sheet models. Nevertheless, there has been significant improvement in accounting for important physical processes in ice-sheet models, particularly of the dynamical response of individual glacier systems to warmer ocean waters in the immediate vicinity of the outlet glaciers. Although there are as yet no complete simulations of regional ocean temperature changes near ice sheets and of the ice-sheet response to realistic climate change forcing, the publications to date have allowed an assessment of the *likely* range of sea level rise for the 21st century (Figure 13.27).

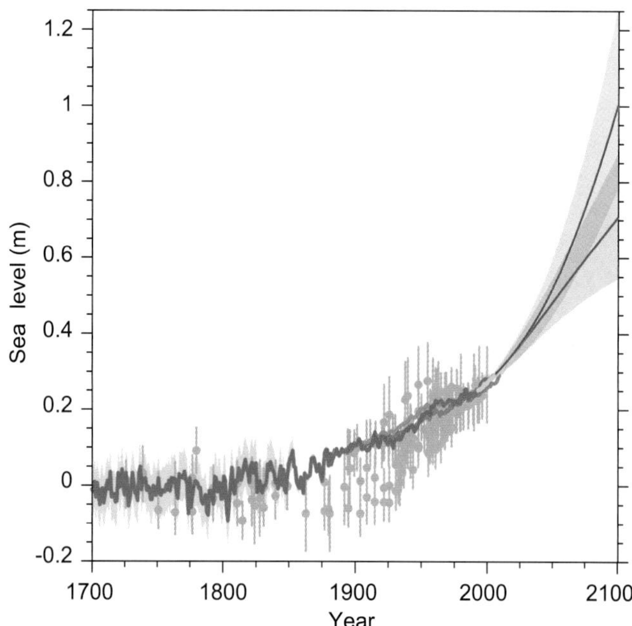

Figure 13.27 | Compilation of paleo sea level data, tide gauge data, altimeter data (from Figure 13.3), and central estimates and *likely* ranges for projections of global mean sea level rise for RCP2.6 (blue) and RCP8.5 (red) scenarios (Section 13.5.1), all relative to pre-industrial values.

These observations, together with our current scientific understanding and projections of future climate and sea level, imply that it is *virtually certain* that sea level will continue to rise during the 21st century and beyond. For the first few decades of the 21st century, regional sea level change will be dominated by climate variability superimposed on the climate change signal. For all scenarios, the rate of 21st century GMSL rise is *very likely* to exceed the average rate during the 20th century. For the RCP8.5 scenario, the projected rate of GMSL rise by the end of the 21st century will approach average rates experienced during the deglaciation of the Earth after the Last Glacial Maximum. These rates imply a significant transfer of mass from the ice sheets to the oceans and associated regional departures of sea level rise from the global average, in addition to the regional patterns from changing atmosphere–ocean interactions.

Sea level rise has already led to a significant increase in the return frequency of sea level extremes at many locations, and it is *very likely* that this will continue during the 21st century, although there is *low confidence* in projections of changes in storminess. The first assessment of surface waves indicates a *likely* (*medium confidence*) increase in the height of waves in the Southern Ocean.

Despite this progress, significant uncertainties remain, particularly related to the magnitude and rate of the ice-sheet contribution for the 21st century and beyond, the regional distribution of sea level rise, and the regional changes in storm frequency and intensity. For coastal planning, sea level rise needs to be considered in a risk management framework, requiring knowledge of the frequency of sea level variability (from climate variability and extreme events) in future climates, projected changes in mean sea level, and the uncertainty of the sea level projections (Hunter, 2010, 2012), as well as local issues such as the compaction of sediments in deltaic regions and the changing supply of these sediments to maintain the height of the deltas (Syvitski et al., 2009). Although improved understanding has allowed the projection of a *likely* range of sea level rise during the 21st century, it has not been possible to quantify a *very likely* range or give an upper bound to future rise. The potential collapse of ice shelves, as observed on the Antarctic Peninsula (Rignot et al., 2004; Scambos et al., 2004; Rott et al., 2011), could lead to a larger 21st century rise of up to several tenths of a metre.

Sea level will continue to rise for centuries, even if GHG concentrations are stabilized, with the amount of rise dependent on future GHG emissions. For higher emission scenarios and warmer temperatures, surface melting of the Greenland ice sheet is projected to exceed accumulation, leading to its long-term decay and a sea level rise of metres, consistent with paleo sea level data.

Acknowledgements

We thank Lea Crosswell and Louise Bell for their assistance in drafting a number of diagrams in this chapter and Jorie Clark for assistance with managing chapter references.

References

Ablain, M., A. Cazenave, G. Valladeau, and S. Guinehut, 2009: A new assessment of the error budget of global mean sea level rate estimated by satellite altimetry over 1993–2008. *Ocean Sci.*, **5**, 193–201.

Adams, P. N., D. L. Inman, and N. E. Graham, 2008: Southern California deep-water wave climate: Characterization and application to coastal processes. *J. Coast. Res.*, **24**, 1022–1035.

Allan, J. C., and P. D. Komar, 2006: Climate controls on US West Coast erosion processes. *J. Coast. Res.*, **22**, 511–529.

Allen, M. R., D. J. Frame, C. Huntingford, C. D. Jones, J. A. Lowe, M. Meinshausen, and N. Meinshausen, 2009: Warming caused by cumulative carbon emissions towards the trillionth tonne. *Nature*, **458**, 1163–1166.

Alley, R. B., S. Anandakrishnan, T. K. Dupont, B. R. Parizek, and D. Pollard, 2007: Effect of sedimentation on ice-sheet grounding-line stability. *Science*, **315**, 1838–1841.

Andrade, C., H. O. Pires, R. Taborda, and M. C. Freitas, 2007: Projecting future changes in wave climate and coastal response in Portugal by the end of the 21st century. *J. Coast. Res.*, **SI 50**, 263–257.

Anschütz, H., et al., 2009: Revisiting sites of the South Pole Queen Maud Land Traverses in East Antarctica: Accumulation data from shallow firn cores. *J. Geophys. Res. Atmos.*, **114**, D012204.

Arendt, A., et al., 2012: Randolph Glacier Inventory [v2.0]: A dataset of global glacier outlines. Global Land Ice Measurements from Space, Boulder CO, USA. Digital Media.

Arthern, R., D. P. Winebrenner, and D. G. Vaughan, 2006: Antarctic snow accumulation mapped using polarization of 4.3-cm wavelength microwave emission. *J. Geophys. Res. Atmos.*, **111**, D06107.

Bahr, D. B., and V. Radić, 2012: Significant contribution to total mass from very small glaciers. *Cryosphere*, **6**, 763–770.

Bahr, D. B., M. Dyurgerov, and M. F. Meier, 2009: Sea-level rise from glaciers and ice caps: A lower bound. *Geophys. Res. Lett.*, **36**, L03501.

Bales, R. C., et al., 2009: Annual accumulation for Greenland updated using ice core data developed during 2000–2006 and analysis of daily coastal meteorological data. *J. Geophys. Res. Atmos.*, **114**, D06116.

Bamber, J., and R. Riva, 2010: The sea level fingerprint of recent ice mass fluxes. *Cryosphere*, **4**, 621–627.

Bamber, J. L., and W. P. Aspinall, 2013: An expert judgement assessment of future sea level rise from the ice sheets. *Nature Clim. Change*, **3**, 424–427.

Bamber, J. L., R. E. M. Riva, B. L. A. Vermeersen, and A. M. LeBrocq, 2009: Reassessment of the potential sea-level rise from a collapse of the West Antarctic Ice Sheet. *Science*, **324**, 901–903.

Bamber, J. L., et al., 2013: A new bed elevation dataset for Greenland. *Cryosphere*, **7**, 499–510.

Banks, H. T., and J. M. Gregory, 2006: Mechanisms of ocean heat uptake in a coupled climate model and the implications for tracer based predictions of ocean heat uptake. *Geophys. Res. Lett.*, **33**, L07608.

Barrand, N. E., et al., 2013: Computing the volume response of the Antarctic Peninsula ice sheet to warming scenarios to 2200. *J. Glaciol.*, **55**, 397–409.

Beckley, B. D., et al., 2010: Assessment of the Jason-2 Extension to the TOPEX/Poseidon, Jason-1 sea-surface height time series for global mean sea level monitoring. *Mar. Geodesy*, **33**, 447–471.

Bengtsson, L., S. Koumoutsaris, and K. Hodges, 2011: Large-scale surface mass balance of ice sheets from a comprehensive atmosphere model. *Surv. Geophys.*, **32**, 459–474.

Biancamaria, S., A. Cazenave, N. M. Mognard, W. Llovel, and F. Frappart, 2011: Satellite-based high latitude snow volume trend, variability and contribution to sea level over 1989/2006. *Global Planet. Change*, **75**, 99–107.

Bindoff, N. L., et al., 2007: Observations: Oceanic climate change and sea level. In: *Climate Change 2007: The Physical Science Basis. Contribution of Working Group I to the Fourth Assessment Report of the Intergovernmental Panel on Climate Change* [Solomon, S., D. Qin, M. Manning, Z. Chen, M. Marquis, K. B. Averyt, M. Tignor and H. L. Miller (eds.)] Cambridge University Press, Cambridge, United Kingdom and New York, NY, USA, pp. 385–432.

Bindschadler, R. A., et al., 2013: Ice-sheet model sensitivities to environmental forcing and their use in projecting future sea level (The SeaRISE Project). *J. Glaciol.*, **59**, 195–224.

Bintanja, R., G. J. van Oldenborgh, S. S. Drijfhout, B. Wouters, and C. A. Katsman, 2013: Important role for ocean warming and increased ice-shelf melt in Antarctic sea-ice expansion. *Nature Geosci.*, **6**, 376–379.

Bittermann, K., S. Rahmstorf, M. Perrette, and M. Vermeer, 2013: Predictability of 20th century sea-level rise from past data. *Environ. Res. Lett.*, **8**, 014013.

Bjork, A. A., et al., 2012: An aerial view of 80 years of climate-related glacier fluctuations in southeast Greenland. *Nature Geosci.*, **5**, 427–432.

Blum, M. D., and H. H. Roberts, 2009: Drowning of the Mississippi Delta due to insufficient sediment supply and global sea-level rise. *Nature Geosci.*, **2**, 488–491.

Boening, C., J. K. Willis, F. W. Landerer, R. S. Nerem, and J. Fasullo, 2012: The 2011 La Niña: So strong, the oceans fell. *Geophys. Res. Lett.*, **39**, L19602.

Boretti, A., 2011: The measured rate of rise of sea levels is not increasing and climate models should be revised to match the experimental evidence. *R. Soc. Publish. eLett.*, **July 12, 2011**.

Boretti, A., 2012a: Short term comparison of climate model predictions and satellite altimeter measurements of sea levels. *Coast. Eng.*, **60**, 319–322.

Boretti, A., 2012b: Is there any support in the long term tide gauge data to the claims that parts of Sydney will be swamped by rising sea levels? *Coast. Eng.*, **64**, 161–167.

Boretti, A. A., 2012c: Discussion of Natalya N. Warner, Philippe E. Tissot, "Storm flooding sensitivity to sea level rise for Galveston Bay, Texas", *Ocean Eng.* 44 (2012), 23–32. *Ocean Eng.*, **55**, 235–237.

Boretti, A., and T. Watson, 2012: The inconvenient truth: Ocean levels are not accelerating in Australia or over the world. *Energy Environ.*, **23**, 801–817.

Boretti, A., 2013a: Discussion of Christine C. Shepard, Vera N. Agostini, Ben Gilmer, Tashya Allen, Jeff Stone, William Brooks and Michael W. Beck. Reply: Evaluating alternative future sea-level rise scenarios, *Nat. Hazards*, 2012 doi:10.1007/s11069-012-0160-2. *Nat. Hazards*, **65**, 967–975.

Boretti, A., 2013b: Discussion of J.A.G. Cooper, C. Lemckert, Extreme sea level rise and adaptation options for coastal resort cities: A qualitative assessment from the Gold Coast, Australia. *Ocean Coast. Manage.*, **78**, 132–135.

Boretti, A. A.,, 2013c: Discussion of "Dynamic System Model to Predict Global Sea-Level Rise and Temperature Change" by Mustafa M. Aral, Jiabao Guan, and Biao Chang. *J. Hydrol. Eng.*, **18**, 370–372.

Bougamont, M., et al., 2007: The impact of model physics on estimating the surface mass balance of the Greenland ice sheet. *Geophys. Res. Lett.*, **34**, L17501.

Bouttes, N., J. M. Gregory, and J. A. Lowe, 2013: The reversibility of sea-level rise. *J. Clim.*, **26**, 2502–2513.

Box, J. E., 2002: Survey of Greenland instrumental temperature records: 1873–2001. *Int. J. Climatol.*, **22**, 1829–1847.

Box, J. E., 2013: Greenland ice sheet mass balance reconstruction. Part II: Surface mass balance (1840–2010). *J. Clim.*, **26**, 6974-6989.

Box, J. E., and W. Colgan, 2013: Greenland ice sheet mass balance reconstruction. Part III: Marine ice loss and total mass balance (1840–2010). *J. Clim.*, **26**, 6990-7002.

Box, J. E., L. Yang, D. H. Bromwich, and L. S. Bai, 2009: Greenland ice sheet surface air temperature variability: 1840–2007. *J. Clim.*, **22**, 4029–4049.

Box, J. E., X. Fettweis, J. C. Stroeve, M. Tedesco, D. K. Hall, and K. Steffen, 2012: Greenland ice sheet albedo feedback: Thermodynamics and atmospheric drivers. *Cryosphere*, **6**, 821–839.

Box, J. E., et al., 2013: Greenland ice sheet mass balance reconstruction. Part I: Net snow accumulation (1600–2009). *J. Clim.*, **26**, 3919-3934.

Bracegirdle, T. J., W. M. Connolley, and J. Turner, 2008: Antarctic climate change over the twenty first century. *J. Geophys. Res. Atmos.*, **113**, D03103.

Braithwaite, R. J., and O. B. Olesen, 1989: Calculation of glacier ablation from air temperature, West Greenland. In: *Glacier Fluctuations and Climatic Change* [J. Oerlemans (ed.)]. Kluwer Academic, Dordrecht, Netherlands, pp. 219–233.

Brierley, C., M. Collins, and A. Thorpe, 2010: The impact of perturbations to ocean-model parameters on climate and climate change in a coupled model. *Clim. Dyn.*, **34**, 325–343.

Broerse, D. B. T., L. L. A. Vermeersen, R. E. M. Riva, and W. van der Wal, 2011: Ocean contribution to co-seismic crustal deformation and geoid anomalies: Application to the 2004 December 26 Sumatra-Andaman earthquake. *Earth Planet. Sci. Lett.*, **305**, 341–349.

Brohan, P., J. J. Kennedy, I. Harris, S. F. B. Tett, and P. D. Jones, 2006: Uncertainty estimates in regional and global observed temperature changes: A new data set from 1850. *J. Geophys. Res. Atmos.*, **111**, D12106.

Bromwich, D. H., J. P. Nicolas, and A. J. Monaghan, 2011: An assessment of precipitation changes over Antarctica and the Southern Ocean since 1989 in contemporary global reanalyses. *J. Clim.*, **24**, 4189–4209.

Bromwich, D. H., R. L. Fogt, K. I. Hodges, and J. E. Walsh, 2007: A tropospheric assessment of the ERA-40, NCEP, and JRA-25 global reanalyses in the polar regions. *J. Geophys. Res. Atmos.*, **112**, D10111.

Brown, J., A. Souza, and J. Wolf, 2010: Surge modelling in the eastern Irish Sea: Present and future storm impact. *Ocean Dyn.*, **60**, 227–236.

Burgess, E. W., R. R. Forster, J. E. Box, E. Mosley-Thompson, D. H. Bromwich, R. C. Bales, and L. C. Smith, 2010: A spatially calibrated model of annual accumulation rate on the Greenland Ice Sheet (1958–2007). *Journal of Geophys. Res. Earth Surf.*, **115**, F02004.

Caires, S., J. Groeneweg, and A. Sterl, 2008: Past and future changes in North Sea extreme waves. In: *Proceedings of the 31st International Conference on Coastal Engineering*, Vols. 1-5 [J.M. Smith (ed.)], World Scientific Publishing Company, Singapore, pp. 547–559.

Cane, M. A., 1989: A mathmatical note on Kawase study of deep ocean. *J. Phys. Oceanogr.*, **19**, 548–550.

Carton, J. A., B. S. Giese, and S. A. Grodsky, 2005: Sea level rise and the warming of the oceans in the Simple Ocean Data Assimilation (SODA) ocean reanalysis. *J. Geophys. Res. Oceans*, **110**, C09006.

Cayan, D., P. Bromirski, K. Hayhoe, M. Tyree, M. Dettinger, and R. Flick, 2008: Climate change projections of sea level extremes along the California coast. *Clim. Change*, **87**, 57–73.

Cazenave, A., et al., 2009: Sea level budget over 2003–2008: A reevaluation from GRACE space gravimetry, satellite altimetry and Argo. *Global Planet. Change*, **65**, 83–88.

Cazenave, A., et al., 2012: Estimating ENSO influence on the global mean sea level, 1993–2010. *Mar. Geodesy*, **35 (SI1)**, 82–97.

Chambers, D. P., 2006: Evaluation of new GRACE time-variable gravity data over the ocean. *Geophys. Res. Lett.*, **33**, L17603.

Chambers, D. P., J. Wahr, and R. S. Nerem, 2004: Preliminary observations of global ocean mass variations with GRACE. *Geophys. Res. Lett.*, **31**, L13310.

Chambers, D. P., M. A. Merrifield, and R. S. Nerem, 2012: Is there a 60-year oscillation in global mean sea level? *Geophys. Res. Lett.*, **39**, L18607.

Chambers, D. P., J. M. Wahr, M. Tamisiea, and R. S. Nerem, 2010: Ocean mass from GRACE and glacial isostatic adjustment. *J. Geophys. Res.*, **115**, B11415.

Chao, B. F., Y. H. Wu, and Y. S. Li, 2008: Impact of artificial reservoir water impoundment on global sea level. *Science*, **320**, 212–214.

Charbit, S., D. Paillard, and G. Ramstein, 2008: Amount of CO_2 emissions irreversibly leading to the total melting of Greenland. *Geophys. Res. Lett.*, **35**, L12503.

Charles, E. D., D. Idier, P. Delecluse, M. Deque, and G. Le Cozannet, 2012: Climate change impact on waves in the Bay of Biscay, France. *Ocean Dyn.*, **62**, 831–848.

Christensen, J. H., et al., 2007: Regional climate projections. In: *Climate Change 2007: The Physical Science Basis. Contribution of Working Group I to the Fourth Assessment Report of the Intergovernmental Panel on Climate Change* [Solomon, S., D. Qin, M. Manning, Z. Chen, M. Marquis, K. B. Averyt, M. Tignor and H. L. Miller (eds.)] Cambridge University Press, Cambridge, United Kingdom and New York, NY, USA, pp. 849–925.

Christoffersen, P., et al., 2011: Warming of waters in an East Greenland fjord prior to glacier retreat: Mechanisms and connection to large-scale atmospheric conditions. *Cryosphere*, **5**, 701–714.

Church, J. A., and N. J. White, 2006: A 20th century acceleration in global sea-level rise. *Geophys. Res. Lett.*, **33**, L01602.

Church, J. A., and N. J. White, 2011: Sea-level rise from the late 19th to the early 21st century. *Surv. Geophys.*, **32**, 585–602.

Church, J. A., N. J. White, and J. M. Arblaster, 2005: Significant decadal-scale impact of volcanic eruptions on sea level and ocean heat content. *Nature*, **438**, 74–77.

Church, J. A., P. L. Woodworth, T. Aarup, and W. S. Wilson, (eds.) 2010: Understanding Sea-Level Rise and Variability. Wiley-Blackwell, Hoboken, NJ, USA, 428 pp.

Church, J. A., D. Monselesan, J. M. Gregory, and B. Marzeion, 2013: Evaluating the ability of process based models to project sea-level change. *Environ. Res. Lett.*, **8**, 015051.

Church, J. A., J. M. Gregory, N. J. White, S. M. Platten, and J. X. Mitrovica, 2011a: Understanding and projecting sea level change. *Oceanography*, **24**, 130–143.

Church, J. A., et al., 2001: Changes in sea level. *Climate Change 2001: The Scientific Basis. Contribution of Working Group I to the Third Assessment Report of the Intergovernmental Panel on Climate Change* [J. T. Houghton, Y. Ding, D. J. Griggs, M. Noquer, P. J. van der Linden, X. Dai, K. Maskell and C. A. Johnson (eds.)]. Cambridge University Press, Cambridge, United Kingdom and New York, NY, USA, pp. 639–693.

Church, J. A., et al., 2011b: Revisiting the Earth's sea-level and energy budgets from 1961 to 2008. *Geophys. Res. Lett.*, **38**, L18601.

Chylek, P., J. E. Box, and G. Lesins, 2004: Global warming and the Greenland ice sheet. *Clim. Change*, **63**, 201–221.

Clark, J. A., and C. S. Lingle, 1977: Future sea-level changes due to West Antarctic ice sheet fluctuations. *Nature*, **269**, 206–209.

Clarke, P. J., D. A. Lavallee, G. Blewitt, T. M. van Dam, and J. M. Wahr, 2005: Effect of gravitational consistency and mass conservation on seasonal surface mass loading models. *Geophys. Res. Lett.*, **32**, L08306.

Cogley, G., 2009a: Geodetic and direct mass-balance measurements: Comparison and joint analysis. *Ann. Glaciol.*, **50**, 96–100.

Cogley, G., 2012: The future of the world's glaciers. In: *Future Climates of the World*, 2nd ed. [A. Henderson-Sellers and K. McGuffie (eds.)]. Elsevier, Amsterdam, Netherlands, and Philadelphia, PA, USA, pp. 197–222.

Cogley, J. G., 2009b: A more complete version of the World Glacier Inventory. *Ann. Glaciol.*, **50**, 32–38.

Colberg, F., and K. L. McInnes, 2012: The impact of future changes in weather patterns on extreme sea levels over southern Australia. *J. Geophys. Res. Oceans*, **117**, C08001.

Coles, S. G., and J. A. Tawn, 1990: Statistics of coastal flood prevention. *Philos. Trans. R. Soc. London A*, **332**, 457–476.

Connolley, W. M., and T. J. Bracegirdle, 2007: An Antarctic assessment of IPCC AR4 coupled models. *Geophys. Res. Lett.*, **34**, L22505.

Conrad, C. P., and B. H. Hager, 1997: Spatial variations in the rate of sea level rise caused by the present-day melting of glaciers and ice sheets. *Geophys. Res. Lett.*, **24**, 1503–1506.

Cook, A. J., and D. G. Vaughan, 2010: Overview of areal changes of the ice shelves on the Antarctic Peninsula over the past 50 years. *Cryosphere*, **4**, 77–98.

Cornford, S. L., et al., 2013: Adaptive mesh, finite volume modeling of marine ice sheets. *J. Comput. Phys.*, **232**, 529–549.

Crowley, T., 2000: Causes of climate change over the past 1000 years. *Science*, **289**, 270–277.

Crowley, T. J., S. K. Baum, K.-Y. Kim, G. C. Hegerl, and W. T. Hyde, 2003: Modeling ocean heat content changes during the last millennium. *Geophys. Res. Lett.*, **30**, 1932.

Debernard, J. B., and L. P. Røed, 2008: Future wind, wave and storm surge climate in the Northern Seas: A revisit. *Tellus A*, **60**, 427–438.

Delworth, T. L., and T. R. Knutson, 2000: Simulation of early 20th century global warming. *Science*, **287**, 2246–2250.

Delworth, T. L., V. Ramaswamy, and G. L. Stenchikov, 2005: The impact of aerosols on simulated ocean temperature and heat content in the 20th century. *Geophys. Res. Lett.*, **32**, L24709.

Di Lorenzo, E., et al., 2010: Central Pacific El Niño and decadal climate change in the North Pacific Ocean. *Nature Geosci.*, **3**, 762–765.

Dobrynin, M., J. Murawsky, and S. Yang, 2012: Evolution of the global wind wave climate in CMIP5 experiments. *Geophys. Res. Lett.*, **39**, L18606.

Dolan, A. M., A. M. Haywood, D. J. Hill, H. J. Dowsett, S. J. Hunter, D. J. Lunt, and S. J. Pickering, 2011: Sensitivity of Pliocene ice sheets to orbital forcing. *Palaeogeogr. Palaeoclimatol. Palaeoecol.*, **309**, 98–110.

Domingues, C. M., J. A. Church, N. J. White, P. J. Gleckler, S. E. Wijffels, P. I. M. Barker, and J. R. Dunn, 2008: Improved estimates of upper-ocean warming and multi-decadal sea-level rise. *Nature*, **453**, 1090–1093.

Donnelly, J. P., P. Cleary, P. Newby, and R. Ettinger, 2004: Coupling instrumental and geological records of sea-level change: Evidence from southern New England of an increase in the rate of sea-level rise in the late 19th century. *Geophys. Res. Lett.*, **31**, L05203.

Douglas, B. C., 2001: Sea level change in the era of the recording tide gauge. In: *Sea Level Rise, History and Consequences*. International Geophysics Series, Volume 75 [B. Douglas, M. S. Kearney and S. P. Leatherman (eds.)]. Academic Press, San Diego, CA, USA, pp. 37–64.

Driesschaert, E., et al., 2007: Modeling the influence of Greenland ice sheet melting on the Atlantic meridional overturning circulation during the next millennia. *Geophys. Res. Lett.*, **34**, L10707.

13

Dufresne, J. L., and S. Bony, 2008: An assessment of the primary sources of spread of global warming estimates from coupled atmosphere-ocean models. *J. Clim.*, **21**, 5135–5144.

Dupont, T. K., and R. B. Alley, 2005: Assessment of the importance of ice-shelf buttressing to ice-sheet flow. *Geophys. Res. Lett.*, **32**, L04503.

Durack, P. J., and S. E. Wijffels, 2010: Fifty-year trends in global ocean salinities and their relationship to broad-scale warming. *J. Clim.*, **23**, 4342–4362.

Durand, G., O. Gagliardini, T. Zwinger, E. Le Meur, and R. C. A. Hindmarsh, 2009: Full Stokes modeling of marine ice sheets: Influence of the grid size. *Ann. Glaciol.*, **50**, 109–114.

Dutton, A., and K. Lambeck, 2012: Ice volume and sea level during the last interglacial. *Science*, **337**, 216–219.

Dyurgerov, M. B., and M. F. Meier, 2005: Glaciers and the changing Earth system: A 2004 snapshot. Occasional Paper. Institute of Arctic and Alpine Research, University of Colorado, Boulder,CO, USA.

Easterling, D. R., and M. F. Wehner, 2009: Is the climate warming or cooling? *Geophys. Res. Lett.*, **36**, L08706.

Ettema, J., M. R. van den Broeke, E. van Meijgaard, W. J. van de Berg, J. L. Bamber, J. E. Box, and R. C. Bales, 2009: Higher surface mass balance of the Greenland ice sheet revealed by high-resolution climate modeling. *Geophys. Res. Lett.*, **36**, L12501.

Fan, Y., I. M. Held, S. J. Lin, and X. and Wang, 2013: Ocean warming effect on surface gravity wave climate change for the end of the 21st century. *J. Clim.*, **26**, 6046-6066.

Farneti, R., and P. R. Gent, 2011: The effects of the eddy-induced advection coefficient in a coarse-resolution coupled climate model. *Ocean Model.*, **39**, 135–145.

Farneti, R., T. L. Delworth, A. J. Rosati, S. M. Griffies, and F. Zeng, 2010: The role of mesoscale eddies in the rectification of the Southern Ocean response to climate change. *J. Phys. Oceanogr.*, **40**, 1539–1557.

Farrell, W. E., and J. A. Clark, 1976: On postglacial sea level. *Geophys. J. R. Astron. Soc.*, **46**, 647–667.

Fausto, R. S., A. P. Ahlstrom, D. van As, S. J. Johnsen, P. L. Langen, and K. Steffen, 2009: Improving surface boundary conditions with focus on coupling snow densification and meltwater retention in large-scale ice-sheet models of Greenland. *J. Glaciol.*, **55**, 869–878.

Favier, L., O. Gagliardini, G. Durand, and T. Zwinger, 2012: A three-dimensional full Stokes model of the grounding line dynamics: Effect of a pinning point beneath the ice shelf. *Cryosphere*, **6**, 101–112.

Fettweis, X., E. Hanna, H. Gallee, P. Huybrechts, and M. Erpicum, 2008: Estimation of the Greenland ice sheet surface mass balance for the 20th and 21st centuries. *Cryosphere*, **2**, 117–129.

Fettweis, X., A. Belleflame, M. Erpicum, B. Franco, and S. Nicolay, 2011: Estimation of the sea level rise by 2100 resulting from changes in the surface mass balance of the Greenland ice sheet. *Clim. Change Geophys. Found. Ecol. Effects* [J. Blanco and H. Kheradmand (eds.)]. Croatia: Intech, pp. 503–520.

Fettweis, X., B. Franco, M. Tedesco, J. H. van Angelen, J. T. M. Lenaerts, M. R. van den Broeke, and H. Gallee, 2013: Estimating Greenland ice sheet surface mass balance contribution to future sea level rise using the regional atmospheric model MAR. *Cryosphere*, **7**, 469–489.

Fiedler, J. W., and C. P. Conrad, 2010: Spatial variability of sea level rise due to water impoundment behind dams. *Geophys. Res. Lett.*, **37**, L12603.

Fluckiger, J., R. Knutti, and J. W. C. White, 2006: Oceanic processes as potential trigger and amplifying mechanisms for Heinrich events. *Paleoceanography*, **21**, PA2014.

Forest, C. E., P. H. Stone, and A. P. Sokolov, 2008: Constraining climate model parameters from observed 20th century changes. *Tellus A*, **60**, 911–920.

Forster, P. M., T. Andrews, I. Goodwin, J. M. Gregory, L. S. Jackson, and M. Zelinka, 2013: Evaluating adjusted forcing and model spread for historical and future scenarios in the CMIP5 generation of climate models. *J. Geophys. Res.*, **118**, 1139–1150.

Foster, G. L., and E. J. Rohling, 2013: Relationship between sea level and climate forcing by CO_2 on geological timescales. *Proc. Natl. Acad. Sci. U.S.A.*, **110**, 1209–1214.

Francis, O. P., G. G. Panteleev, and D. E. Atkinson, 2011: Ocean wave conditions in the Chukchi Sea from satellite and in situ observations. *Geophys. Res. Lett.*, **38**, L24610.

Franco, B., X. Fettweis, and M. Erpicum, 2013: Future projections of the Greenland ice sheet energy balance driving the surface melt. *Cryosphere*, **7**, 1–18.

Franco, B., X. Fettweis, M. Erpicum, and S. Nicolay, 2011: Present and future climates of the Greenland ice sheet according to the IPCC AR4 models. *Clim. Dyn.*, **36**, 1897–1918.

Fretwell, P., et al., 2013: Bedmap2: Improved ice bed, surface and thickness datasets for Antarctica. *Cryosphere*, **7**, 375–393.

Fyke, J. G., L. Carter, A. Mackintosh, A. J. Weaver, and K. J. Meissner, 2010: Surface melting over ice shelves and ice sheets as assessed from modeled surface air temperatures. *J. Clim.*, **23**, 1929–1936.

Gardner, A. S., et al., 2013: A reconciled estimate of glacier contributions to sea level rise: 2003 to 2009. *Science*, **340**, 852-857.

Gehrels, R., and P. L. Woodworth, 2013: When did modern rates of sea-level rise start? *Global Planet. Change*, **100**, 263–277.

Gehrels, W. R., B. Hayward, R. M. Newnham, and K. E. Southall, 2008: A 20th century acceleration of sea-level rise in New Zealand. *Geophys. Res. Lett.*, **35**, L02717.

Gehrels, W. R., D. A. Dawson, J. Shaw, and W. A. Marshall, 2011: Using Holocene relative sea-level data to inform future sea-level predictions: An example from southwest England. *Global Planet. Change*, **78**, 116–126.

Gehrels, W. R., et al., 2005: Onset of recent rapid sea-level rise in the western Atlantic Ocean. *Quat. Sci. Rev.*, **24**, 2083–2100.

Gehrels, W. R., et al., 2006: Rapid sea-level rise in the North Atlantic Ocean since the first half of the nineteenth century. *Holocene*, **16**, 949–965.

Gehrels, W. R., et al., 2012: Nineteenth and twentieth century sea-level changes in Tasmania and New Zealand. *Earth Planet. Sci. Lett.*, **315**, 94–102.

Gent, P. R., and G. Danabasoglu, 2011: Response to increasing southern hemisphere winds in CCSM4. *J. Clim.*, **24**, 4992–4998.

Genthon, C., G. Krinner, and H. Castebrunet, 2009: Antarctic precipitation and climate-change predictions: Horizontal resolution and margin vs plateau issues. *Ann. Glaciol.*, **50**, 55–60.

Geoffroy, O., D. Saint-Martin, and A. Ribes, 2012: Quantifying the source of spread in climate change experiments. *Geophys. Res. Lett.*, **39**, L24703.

Geoffroy, O., D. Saint-Martin, D. J. L. Olivie, A. Voldoire, G. Belon, and S. Tyteca, 2013: Transient climate response in a two-box energy-balance model. Part I: Analytical solution and parameter calibration using CMIP5. *J. Clim.*, **26**, 1841-1857.

Giesen, R. H., and J. Oerlemans, 2013: Climate-model induced differences in the 21st century global and regional glacier contributions to sea-level rise. *Clim. Dyn.*, **41**, 3283–3300.

Gillet-Chaulet, F., et al., 2012: Greenland ice sheet contribution to sea-level rise from a new-generation ice-sheet model. *Cryosphere*, **6**, 1561–1576.

Gillett, N. P., V. K. Arora, K. Zickfeld, S. J. Marshall, and A. J. Merryfield, 2011: Ongoing climate change following a complete cessation of carbon dioxide emissions. *Nature Geosci.*, **4**, 83–87.

Gladstone, R. M., et al., 2012: Calibrated prediction of Pine Island Glacier retreat during the 21st and 22nd centuries with a coupled flowline model. *Earth Planet. Sci. Lett.*, **333**, 191–199.

Gleckler, P. J., K. AchutaRao, J. M. Gregory, B. D. Santer, K. E. Taylor, and T. M. L. Wigley, 2006a: Krakatoa lives: The effect of volcanic eruptions on ocean heat content and thermal expansion. *Geophys. Res. Lett.*, **33**, L17702.

Gleckler, P. J., T. M. L. Wigley, B. D. Santer, J. M. Gregory, K. AchutaRao, and K. E. Taylor, 2006b: Volcanoes and climate: Krakatoa's signature persists in the ocean. *Nature*, **439**, 675.

Gleckler, P. J., et al., 2012: Human-induced global ocean warming on multidecadal timescales. *Nature Clim. Change*, **2**, 524–529.

Goelzer, H., P. Huybrechts, M. F. Loutre, H. Goosse, T. Fichefet, and A. Mouchet, 2011: Impact of Greenland and Antarctic ice sheet interactions on climate sensitivity. *Clim. Dyn.*, **37**, 1005–1018.

Goelzer, H., P. Huybrechts, S. C. B. Raper, M. F. Loutre, H. Goosse, and T. Fichefet, 2012: Millennial total sea-level commitments projected with the Earth system model of intermediate complexity LOVECLIM. *Environ. Res. Lett.*, **7**, 045401.

Goelzer, H., et al., 2013: Sensitivity of Greenland ice sheet projections to model formulations. *J. Glaciol.*, **59**, 733-749.

Goldberg, D., D. M. Holland, and C. Schoof, 2009: Grounding line movement and ice shelf buttressing in marine ice sheets. *J. Geophys. Res. Earth Surf.*, **114**, F04026.

Goldberg, D. N., C. M. Little, O. V. Sergienko, A. Gnanadesikan, R. Hallberg, and M. Oppenheimer, 2012: Investigation of land ice-ocean interaction with a fully coupled ice-ocean model: 2. Sensitivity to external forcings. *J. Geophys. Res. Earth Surf.*, **117**, F02038.

Gomez, N., J. X. Mitrovica, P. Huybers, and P. U. Clark, 2010a: Sea level as a stabilizing factor for marine-ice-sheet grounding lines. *Nature Geosci.*, **3**, 850–853.

Gomez, N., J. X. Mitrovica, M. E. Tamisiea, and P. U. Clark, 2010b: A new projection of sea level change in response to collapse of marine sectors of the Antarctic Ice Sheet. *Geophys. J. Int.*, **180**, 623–634.

Good, P., J. M. Gregory, and J. A. Lowe, 2011: A step-response simple climate model to reconstruct and interpret AOGCM projections. *Geophys. Res. Lett.*, **38**, L01703.

Good, P., J. M. Gregory, J. A. Lowe, and T. Andrews, 2013: Abrupt CO2 experiments as tools for predicting and understanding CMIP5 representative concentration pathway projections. *Clim. Dyn.*, **40**, 1041–1053.

Goosse, H., H. Renssen, A. Timmermann, and R. S. Bradley, 2005: Internal and forced climate variability during the last millennium: A model-data comparison using ensemble simulations. *Quat. Sci. Rev.*, **24**, 1345–1360.

Gornitz, V., 2001: Impoundment, groundwater mining, and other hydrologic transformations: Impacts on global sea level rise. *Sea Level Rise, History and Consequences*. International Geophysics Series, Volume 75 [B. Douglas, M. S. Kearney and S. P. Leatherman (eds.)]. Academic Press, San Diego, CA, USA, pp. 97–119.

Gouretski, V., and K. P. Koltermann, 2007: How much is the ocean really warming? *Geophys. Res. Lett.*, **34**, L01610.

Gower, J. F. R., 2010: Comment on "Response of the global ocean to Greenland and Antarctic ice melting" by D. Stammer. *J. Geophys. Res. Oceans*, **115**, C10009.

Grabemann, I., and R. Weisse, 2008: Climate change impact on extreme wave conditions in the North Sea: An ensemble study. *Ocean Dyn.*, **58**, 199–212.

Graham, N. E., D. R. Cayan, P. Bromirski, and R. Flick, 2013: Multi-model projections of 21st century North Pacific winter wave climate under the IPCC A2 scenario. *Clim. Dyn.*, **40**, 1335–1360.

Graversen, R. G., S. Drijfhout, W. Hazeleger, R. van de Wal, R. Bintanja, and M. Helsen, 2011: Greenland's contribution to global sea level rise by the end of the 21st century. *Clim. Dyn.*, **37**, 1427–1442.

Greatbatch, R. J., 1994: A note on the representation of steric sea-levels in models that conserve volume rather than mass. *J. Geophys. Res. Oceans*, **99**, 12767–12771.

Gregory, J. M., 2000: Vertical heat transports in the ocean and their effect on time-dependent climate change. *Clim. Dyn.*, **16**, 501–515.

Gregory, J. M., 2010: Long-term effect of volcanic forcing on ocean heat content. *Geophys. Res. Lett.*, **37**, L22701.

Gregory, J. M., and J. A. Lowe, 2000: Predictions of global and regional sea-level rise using AOGCMs with and without flux adjustment. *Geophys. Res. Lett.*, **27**, 3069–3072.

Gregory, J. M., and P. Huybrechts, 2006: Ice-sheet contributions to future sea-level change. *Philos. R. Soc. London A*, **364**, 1709–1731.

Gregory, J. M., and P. M. Forster, 2008: Transient climate response estimated from radiative forcing and observed temperature change. *J. Geophys. Res. Atmos.*, **113**, D23105.

Gregory, J. M., J. A. Lowe, and S. F. B. Tett, 2006: Simulated global-mean sea-level changes over the last half-millennium. *J. Clim.*, **19**, 4576–4591.

Gregory, J. M., et al., 2013a: Climate models without pre-industrial volcanic forcing underestimate historical ocean thermal expansion. *Geophys. Res. Lett.*, **40**, 1–5.

Gregory, J. M., et al., 2013b: Twentieth-century global-mean sea level rise: Is the whole greater than the sum of the parts? *J. Clim.*, **26**, 4476-4499.

Greve, R., 2000: On the response of the Greenland ice sheet to greenhouse climate change. *Clim. Change*, **46**, 289–303.

Griffies, S. M., and R. J. Greatbatch, 2012: Physical processes that impact the evolution of global mean sea level in ocean climate models. *Ocean Model.*, **51**, 37–72.

Grinsted, A., J. C. Moore, and S. Jevrejeva, 2010: Reconstructing sea level from paleo and projected temperatures 200 to 2100 AD. *Clim. Dyn.*, **34**, 461–472.

Gudmundsson, G. H., J. Krug, G. Durand, F. L., and O. Gagliardini, 2012: The stability of grounding lines on retrograde slopes. *Cryosphere Discuss.*, **6**, 2597–2619.

Hallberg, R., and A. Gnanadesikan, 2006: The role of eddies in determining the structure and response of the wind-driven southern hemisphere overturning: Results from the Modeling Eddies in the Southern Ocean (MESO) project. *J. Phys. Oceanogr.*, **36**, 2232–2252.

Hallberg, R., A. Adcroft, J. Dunne, J. Krasting, and R. J. Stouffer, 2013: Sensitivity of 21st century global-mean steric sea level rise to ocean model formulation. *J. Clim.*, **26**, 2947-2956.

Han, W. Q., et al., 2010: Patterns of Indian Ocean sea-level change in a warming climate. *Nature Geosci.*, **3**, 546–550.

Hanna, E., S. H. Mernild, J. Cappelen, and K. Steffen, 2012: Recent warming in Greenland in a long-term instrumental (1881–2012) climatic context: I. Evaluation of surface air temperature records. *Environ. Res. Lett.*, **7**, 045404.

Hanna, E., P. Huybrechts, I. Janssens, J. Cappelen, K. Steffen, and A. Stephens, 2005: Runoff and mass balance of the Greenland ice sheet: 1958–2003. *J. Geophys. Res. Atmos.*, **110**, D13108.

Hanna, E., et al., 2008: Increased runoff from melt from the Greenland Ice Sheet: A response to global warming. *J. Clim.*, **21**, 331–341.

Hanna, E., et al., 2011: Greenland Ice Sheet surface mass balance 1870 to 2100 based on twentieth century reanalysis, and links with global climate forcing. *J. Geophys. Res.*, **116**, D24121.

Hansen, J., M. Sato, P. Kharecha, and K. von Schuckmann, 2011: Earth's energy imbalance and implications. *Atmos. Chem. Phys.*, **11**, 13421–13449.

Hansen, J., G. Russell, A. Lacis, I. Fung, D. Rind, and P. Stone, 1985: Climate response times—dependence on climate sensitivity and ocean mixing. *Science*, **229**, 857–859.

Hansen, J., M. Sato, P. Kharecha, G. Russell, D. Lea, and M. Siddall, 2007: Climate change and trace gases. *Philos. Trans. R. Soc. London A*, **365**, 1925–1954.

Hansen, J., et al., 2005: Earth's energy imbalance: Confirmation and implications. *Science*, **308**, 1431–1435.

Harper, B., T. Hardy, L. Mason, and R. Fryar, 2009: Developments in storm tide modelling and risk assessment in the Australian region. *Nat. Hazards*, **51**, 225–238.

Harper, J., N. Humphrey, W. T. Pfeffer, J. Brown, and X. Fettweis, 2012: Greenland ice-sheet contribution to sea-level rise buffered by meltwater storage in firn. *Nature*, **491**, 240-243.

Hay, C. C., E. Morrow, R. E. Kopp, and J. X. Mitrovica, 2013: Estimating the sources of global sea level rise with data assimilation techniques. *Proc. Natl. Acad. Sci. U.S.A.*, **110**, 3692–3699.

Hegerl, G. C., et al., 2007: Understanding and attributing climate change. In: *Climate Change 2007: The Physical Science Basis. Contribution of Working Group I to the Fourth Assessment Report of the Intergovernmental Panel on Climate Change* [Solomon, S., D. Qin, M. Manning, Z. Chen, M. Marquis, K. B. Averyt, M. Tignor and H. L. Miller (eds.)]. Cambridge University Press, Cambridge, United Kingdom and New York, NY, USA, pp. 663–745.

Held, I. M., and B. J. Soden, 2006: Robust responses of the hydrological cycle to global warming. *J. Clim.*, **19**, 5686–5699.

Held, I. M., M. Winton, K. Takahashi, T. Delworth, F. R. Zeng, and G. K. Vallis, 2010: Probing the fast and slow components of global warming by returning abruptly to preindustrial forcing. *J. Clim.*, **23**, 2418–2427.

Hellmer, H. H., F. Kauker, R. Timmermann, J. Determann, and J. Rae, 2012: Twenty-first-century warming of a large Antarctic ice-shelf cavity by a redirected coastal current. *Nature*, **485**, 225–228.

Hemer, M. A., J. A. Church, and J. R. Hunter, 2010: Variability and trends in the directional wave climate of the Southern Hemisphere. *Int. J. Climatol.*, **30**, 475–491.

Hemer, M. A., J. Katzfey, and C. Trenham, 2012a: Global dynamical projections of surface ocean wave climate for a future high greenhouse gas emission scenario. *Ocean Model.*, **70**, 221-245.

Hemer, M. A., K. L. McInnes, and R. Ranasinghe, 2012b: Projections of climate change-driven variations in the offshore wave climate off southeastern Australia. *Int. J. Climatol.*, **33**, 1615-1632.

Hemer, M. A., Y. Fan, N. Mori, A. Semedo, and X. L. Wang, 2013: Projected future changes in wind-wave climate in a multi-model ensemble. *Nature Clim. Change*, **3**, 471–476.

Hill, D. J., A. M. Dolan, A. M. Haywood, S. J. Hunter, and D. K. Stoll, 2010: Sensitivity of the Greenland ice sheet to Pliocene sea surface temperatures. *Stratigraphy*, **7**, 111–121.

Hindmarsh, R. C. A., 1993: Qualitative dynamics of marine ice sheets. *Ice Clim. Syst. I*, **12**, 68–99.

Holgate, S., S. Jevrejeva, P. Woodworth, and S. Brewer, 2007: Comment on "A semi-empirical approach to projecting future sea-level rise". *Science*, **317**, 2.

Holgate, S. J., 2007: On the decadal rates of sea level change during the twentieth century. *Geophys. Res. Lett.*, **34**, L01602.

Holland, D. M., R. H. Thomas, B. De Young, M. H. Ribergaard, and B. Lyberth, 2008: Acceleration of Jakobshavn Isbrae triggered by warm subsurface ocean waters. *Nature Geosci.*, **1**, 659–664.

Horton, R., C. Herweijer, C. Rosenzweig, J. P. Liu, V. Gornitz, and A. C. Ruane, 2008: Sea level rise projections for current generation CGCMs based on the semi-empirical method. *Geophys. Res. Lett.*, **35**, L02715.

Hu, A., G. A. Meehl, W. Han, and J. Yin, 2011: Effect of the potential melting of the Greenland Ice Sheet on the Meridional Overturning Circulation and global climate in the future. *Deep-Sea Res. Pt. Ii*, **58**, 1914–1926.

Hu, A. X., et al., 2010: Influence of Bering Strait flow and North Atlantic circulation on glacial sea-level changes. *Nature Geosci.*, **3**, 118–121.

Huber, M., and R. Knutti, 2012: Anthropogenic and natural warming inferred from changes in earth's energy balance. *Nature Geosci.*, **5**, 31–36.

Hunter, J., 2010: Estimating sea-level extremes under conditions of uncertain sea-level rise. *Clim. Change*, **99**, 331–350.

Hunter, J., 2012: A simple technique for estimating an allowance for uncertain sea-level rise. *Clim. Change*, **113**, 239–252.

Hunter, J. R., and M. J. I. Brown, 2013: Discussion of Boretti, A., 'Is there any support in the long term tide gauge data to the claims that parts of Sydney will be swamped by rising sea levels?'. *Coastal Eng.*, **75**, 1–3.

Huntington, T. G., 2008: Can we dismiss the effect of changes in land-based water storage on sea-level rise? *Hydrol. Proc.*, **22**, 717–723.

Huss, M., R. Hock, A. Bauder, and M. Funk, 2012: Conventional versus reference-surface mass balance. *J. Glaciol.*, **58**, 278–286.

Huybrechts, P., and J. De Wolde, 1999: The dynamic response of the Greenland and Antarctic ice sheets to multiple-century climatic warming. *J. Clim.*, **12**, 2169–2188.

Huybrechts, P., H. Goelzer, I. Janssens, E. Driesschaert, T. Fichefet, H. Goosse, and M. F. Loutre, 2011: Response of the Greenland and Antarctic ice sheets to multi-millennial greenhouse warming in the earth system model of intermediate complexity LOVECLIM. *Surv. Geophys.*, **32**, 397–416.

Ishii, M., and M. Kimoto, 2009: Reevaluation of historical ocean heat content variations with time-varying XBT and MBT depth bias corrections. *J. Oceanogr.*, **65**, 287–299.

Izaguirre, C., F. J. Méndez, M. Menéndez, and I. J. Losada, 2011: Global extreme wave height variability based on satellite data. *Geophys. Res. Lett.*, **38**, L10607.

Izaguirre, C., F. J. Méndez, M. Menéndez, A. Luceño, and I. J. Losada, 2010: Extreme wave climate variability in southern Europe using satellite data. *J. Geophys. Res. Oceans*, **115**, C04009.

Jacobs, S. S., A. Jenkins, C. F. Giulivi, and P. Dutrieux, 2011: Stronger ocean circulation and increased melting under Pine Island Glacier ice shelf. *Nature Geosci.*, **4**, 519–523.

Jay, D. A., 2009: Evolution of tidal amplitudes in the eastern Pacific Ocean. *Geophys. Res. Lett.*, **36**, L04603.

Jenkins, A., and D. Holland, 2007: Melting of floating ice and sea level rise. *Geophys. Res. Lett.*, **34**, L16609.

Jenkins, A., P. Dutrieux, S. S. Jacobs, S. D. McPhail, J. R. Perrett, A. T. Webb, and D. White, 2010: Observations beneath Pine Island Glacier in West Antarctica and implications for its retreat. *Nature Geosci.*, **3**, 468–472.

Jevrejeva, S., A. Grinsted, and J. C. Moore, 2009: Anthropogenic forcing dominates sea level rise since 1850. *Geophys. Res. Lett.*, **36**, L20706.

Jevrejeva, S., J. C. Moore, and A. Grinsted, 2010: How will sea level respond to changes in natural and anthropogenic forcings by 2100? *Geophys. Res. Lett.*, **37**, L07703.

Jevrejeva, S., J. C. Moore, and A. Grinsted, 2012a: Sea level projections to AD 2500 with a new generation of climate change scenarios. *Global Planet. Change*, **80–81**, 14–20.

Jevrejeva, S., J. C. Moore, and A. Grinsted, 2012b: Potential for bias in 21st century semiempirical sea level projections. *J. Geophys. Res.*, **117**, D20116.

Jevrejeva, S., A. Grinsted, J. C. Moore, and S. Holgate, 2006: Nonlinear trends and multiyear cycles in sea level records. *J. Geophys. Res. Oceans*, **111**, C09012.

Jevrejeva, S., J. C. Moore, A. Grinsted, and P. L. Woodworth, 2008: Recent global sea level acceleration started over 200 years ago? *Geophys. Res. Lett.*, **35**, L08715.

Johansson, M. M., H. Pellikka, K. K. Kahma, and K. Ruosteenoja, 2013: Global sea level rise scenarios adapted to the Finnish coast. *J. Mar. Syst.*, **129**, 35–46.

Johnson, G. C., and N. Gruber, 2007: Decadal water mass variations along 20 W in the Northeastern Atlantic Ocean. *Prog. Oceanogr.*, **73**, 277–295.

Johnson, G. C., S. Mecking, B. M. Sloyan, and S. E. Wijffels, 2007: Recent bottom water warming in the Pacific Ocean. *J. Clim.*, **20**, 5365–5375.

Joughin, I., and R. B. Alley, 2011: Stability of the West Antarctic ice sheet in a warming world. *Nature Geosci.*, **4**, 506–513.

Joughin, I., B. E. Smith, and D. M. Holland, 2010: Sensitivity of 21st century sea level to ocean-induced thinning of Pine Island Glacier, Antarctica. *Geophys. Res. Lett.*, **37**, L20502.

Kang, S. K., J. Y. Cherniawsky, M. G. G. Foreman, H. S. Min, C. H. Kim, and H. W. Kang, 2005: Patterns of recent sea level rise in the East/Japan Sea from satellite altimetry and in situ data. *J. Geophys. Res. Oceans*, **110**, C07002.

Kaser, G., J. G. Cogley, M. B. Dyurgerov, M. F. Meier, and A. Ohmura, 2006: Mass balance of glaciers and ice caps: Consensus estimates for 1961–2004. *Geophys. Res. Lett.*, **33**, L19501.

Kato, S., 2009: Interannual variability of the global radiation budget. *J. Clim.*, **22**, 4893–4907.

Katsman, C., W. Hazeleger, S. Drijfhout, G. Oldenborgh, and G. Burgers, 2008: Climate scenarios of sea level rise for the northeast Atlantic Ocean: A study including the effects of ocean dynamics and gravity changes induced by ice melt. *Clim. Change*, **91**, 351–374.

Katsman, C. A., and G. J. van Oldenborgh, 2011: Tracing the upper ocean's missing heat. *Geophys. Res. Lett.*, **38**, L14610.

Katsman, C. A., et al., 2011: Exploring high-end scenarios for local sea level rise to develop flood protection strategies for a low-lying delta - the Netherlands as an example. *Clim. Dyn.*, **109**, 617–645.

Kawase, M., 1987: Establishment of deep ocean circulation driven by deep-water. *J. Phys. Oceanogr.*, **17**, 2294–2317.

Kemp, A. C., B. P. Horton, J. P. Donnelly, M. E. Mann, M. Vermeer, and S. Rahmstorf, 2011: Climate related sea-level variations over the past two millennia. *Proc. Natl. Acad. Sci. U.S.A.*, **108**, 11017–11022.

Knutti, R., and L. Tomassini, 2008: Constraints on the transient climate response from observed global temperature and ocean heat uptake. *Geophys. Res. Lett.*, **35**, L09701.

Knutti, R., S. Krahenmann, D. J. Frame, and M. R. Allen, 2008: Comment on "Heat capacity, time constant, and sensitivity of Earth's climate system" by S. E. Schwartz. *J. Geophys. Res. Atmos.*, **113**, D15103.

Kohl, A., and D. Stammer, 2008: Decadal sea level changes in the 50-year GECCO ocean synthesis. *J. Clim.*, **21**, 1876–1890.

Konikow, L. F., 2011: Contribution of global groundwater depletion since 1900 to sea-level rise. *Geophys. Res. Lett.*, **38**, L17401.

Konikow, L. F., 2013: Comment on "Model estimates of sea-level change due to anthropogenic impacts on terrestrial water storage" by Pokhrel et al. *Nature Geosci.*, **6**, 2.

Kopp, R. E., F. J. Simons, J. X. Mitrovica, A. C. Maloof, and M. Oppenheimer, 2009: Probabilistic assessment of sea level during the last interglacial stage. *Nature*, **462**, 863–868.

Kopp, R. E., F. J. Simons, J. X. Mitrovica, A. C. Maloof, and M. Oppenheimer 2013: A probabilistic assessment of sea level variations within the last interglacial stage. *Geophys. J. Int.*, **193**, 711–716.

Kopp, R. E., J. X. Mitrovica, S. M. Griffies, J. J. Yin, C. C. Hay, and R. J. Stouffer, 2010: The impact of Greenland melt on local sea levels: A partially coupled analysis of dynamic and static equilibrium effects in idealized water-hosing experiments. *Clim. Change*, **103**, 619–625.

Körper, J., et al., 2013: The effect of aggressive mitigation on sea level rise and sea ice changes. *Clim. Dyn.*, **40**, 531–550.

Kouketsu, S., et al., 2011: Deep ocean heat content changes estimated from observation and reanalysis product and their influence on sea level change. *J. Geophys. Res. Oceans*, **116**, C03012.

Krinner, G., O. Magand, I. Simmonds, C. Genthon, and J. L. Dufresne, 2007: Simulated Antarctic precipitation and surface mass balance at the end of the twentieth and twenty-first centuries. *Clim. Dyn.*, **28**, 215–230.

Krinner, G., B. Guicherd, K. Ox, C. Genthon, and O. Magand, 2008: Influence of oceanic boundary conditions in simulations of Antarctic climate and surface mass balance change during the coming century. *J. Clim.*, **21**, 938–962.

Kuhlbrodt, T., and J. M. Gregory, 2012: Ocean heat uptake and its consequences for the magnitude of sea level rise and climate change. *Geophys. Res. Lett.*, **39**, L18608.

Lambeck, K., and S. M. Nakiboglu, 1984: Recent global changes in sea level. *Geophys. Res. Lett.*, **11**, 959–961.

Lambeck, K., C. Smither, and M. Ekman, 1998: Tests of glacial rebound models for Fennoscandinavia based on instrumented sea- and lake-level records. *Geophys. J. Int.*, **135**, 375–387.

Lambeck, K., A. Purcell, and A. Dutton, 2012: The anatomy of interglacial sea levels: The relationship between sea levels and ice volumes during the Last Interglacial. *Earth Planet. Sci. Lett.*, **315**, 4–11.

Lambeck, K., M. Anzidei, F. Antonioli, A. Benini, and A. Esposito, 2004: Sea level in Roman time in the Central Mediterranean and implications for recent change. *Earth Planet. Sci. Lett.*, **224**, 563–575.

Landerer, F. W., J. H. Jungclaus, and J. Marotzke, 2007: Regional dynamic and steric sea level change in response to the IPCC-A1B scenario. *J. Phys. Oceanogr.*, **37**, 296–312.

Langen, P. L., A. M. Solgaard, and C. S. Hvidberg, 2012: Self-inhibiting growth of the Greenland Ice Sheet. *Geophys. Res. Lett.*, **39**, L12502.

Leclercq, P. W., J. Oerlemans, and J. G. Cogley, 2011: Estimating the glacier contribution to sea-level rise for the period 1800–2005. *Surv. Geophys.*, **32**, 519–535.

Leclercq, P. W., A. Weidick, F. Paul, T. Bolch, M. Citterio, and J. Oerlemans, 2012: Brief communication "Historical glacier length changes in West Greenland". *Cryosphere*, **6**, 1339–1343.

Legg, S., et al., 2009: Improving oceanic overflow representation in climate models. *Bull. Am. Meteorol. Soc.*, **90**, 657–670.

Lemieux-Dudon, B., et al., 2010: Consistent dating for Antarctic and Greenland ice cores. *Quat. Sci. Rev.*, **29**, 8–20.

Lempérière, F., 2006: The role of dams in the XXI century: Achieving a sustainable development target. *Int. J. Hydropower Dams*, **13**, 99–108.

Lenaerts, J. T. M., M. R. van den Broeke, W. J. van de Berg, E. van Meijgaard, and P. Kulpers Munneke, 2012: A new high-resolution surface mass balance map of Antarctica (1989–2009) based on regional atmospheric climate modeling. *Geophys. Res. Lett.*, **39**, L04501.

Lenton, T. M., H. Held, E. Kriegler, J. W. Hall, W. Lucht, S. Rahmstorf, and H. J. Schellnhuber, 2008: Tipping elements in the Earth's climate system. *Proc. Natl. Acad. Sci. U.S.A.*, **105**, 1786–1793.

Lettenmaier, D. P., and P. C. D. Milly, 2009: Land waters and sea level. *Nature Geosci.*, **2**, 452–454.

Leuliette, E. W., and L. Miller, 2009: Closing the sea level rise budget with altimetry, Argo, and GRACE. *Geophys. Res. Lett.*, **36**, L04608.

Leuliette, E. W., and J. K. Willis, 2011: Balancing the sea level budget. *Oceanography*, **24**, 122–129.

Levermann, A., A. Griesel, M. Hofmann, M. Montoya, and S. Rahmstorf, 2005: Dynamic sea level changes following changes in the thermohaline circulation. *Clim. Dyn.*, **24**, 347–354.

Levermann, A., et al., 2012: Potential climatic transitions with profound impact on Europe Review of the current state of six 'tipping elements of the climate system'. *Clim. Change*, **110**, 845–878.

Levitus, S., J. Antonov, and T. Boyer, 2005: Warming of the world ocean, 1955–2003. *Geophys. Res. Lett.*, **32**, L02604.

Levitus, S., J. I. Antonov, J. L. Wang, T. L. Delworth, K. W. Dixon, and A. J. Broccoli, 2001: Anthropogenic warming of Earth's climate system. *Science*, **292**, 267–270.

Levitus, S., J. I. Antonov, T. P. Boyer, R. A. Locarnini, H. E. Garcia, and A. V. Mishonov, 2009: Global ocean heat content 1955–2008 in light of recently revealed instrumentation problems. *Geophys. Res. Lett.*, **36**, L07608.

Li, C., J. S. von Storch, and J. Marotzke, 2013: Deep-ocean heat uptake and equilibrium climate response. *Clim. Dyn.*, **40**, 1071–1086.

Ligtenberg, S. R. M., W. J. van de Berg, M. R. van den Broeke, J. G. L. Rae, and E. van Meijgaard, 2013: Future surface mass balance of the Antarctic ice sheet and its influence on sea level change, simulated by a regional atmospheric climate model. *Clim. Dyn.*, **41**, 867–884.

Lionello, P., M. B. Galati, and E. Elvini, 2010: Extreme storm surge and wind wave climate scenario simulations at the Venetian littoral. *Phys. Chem. Earth Pts. A/B/C*, **40–41**, 86–92.

Lionello, P., S. Cogo, M. B. Galati, and A. Sanna, 2008: The Mediterranean surface wave climate inferred from future scenario simulations. *Global Planet. Change*, **63**, 152–162.

Little, C. M., M. Oppenheimer, and N. M. Urban, 2013a: Upper bounds on twenty-first-century Antarctic ice loss assessed using a probabilistic framework. *Nature Clim. Change*, **7**, 654-659.

Little, C. M., N. M. Urban, and M. Oppenheimer, 2013b: Probabilistic framework for assessing the ice sheet contribution to sea level change. *Proc. Natl. Acad. Sci. U.S.A.*, **110**, 3264–3269.

Llovel, W., S. Guinehut, and A. Cazenave, 2010: Regional and interannual variability in sea level over 2002–2009 based on satellite altimetry, Argo float data and GRACE ocean mass. *Ocean Dyn.*, **60**, 1193–1204.

Llovel, W., et al., 2011: Terrestrial waters and sea level variations on interannual time scale. *Global Planet. Change*, **75**, 76–82.

Loeb, N. G., et al., 2009: Toward optimal closure of the earth's top-of-atmosphere radiation budget. *J. Clim.*, **22**, 748–766.

Loeb, N. G., et al., 2012: Observed changes in top-of-the-atmosphere radiation and upper-ocean heating consistent within uncertainty. *Nature Geosci.*, **5**, 110–113.

Lombard, A., G. Garric, and T. Penduff, 2009: Regional patterns of observed sea level change: Insights from a 1/4A degrees global ocean/sea-ice hindcast. *Ocean Dyn.*, **59**, 433–449.

Lombard, A., A. Cazenave, P. Y. Le Traon, and M. Ishii, 2005a: Contribution of thermal expansion to present-day sea-level change revisited. *Global Planet. Change*, **47**, 1–16.

Lombard, A., A. Cazenave, K. DoMinh, C. Cabanes, and R. S. Nerem, 2005b: Thermosteric sea level rise for the past 50 years: Comparison with tide gauges and inference on water mass contribution. *Global Planet. Change*, **48**, 303–312.

Lorbacher, K., J. Dengg, C. W. Boning, and A. Biastoch, 2010: Regional patterns of sea level change related to interannual variability and multidecadal trends in the Atlantic meridional overturning circulation. *J. Clim.*, **23**, 4243–4254.

Lorbacher, K., S. J. Marsland, J. A. Church, S. M. Griffies, and D. Stammer, 2012: Rapid barotrophic sea-level rise from ice-sheet melting scenarios. *J. Geophys. Res.*, **117**, C06003.

Losch, M., A. Adcroft, and J. M. Campin, 2004: How sensitive are coarse general circulation models to fundamental approximations in the equations of motion? *J. Phys. Oceanogr.*, **34**, 306–319.

Lowe, J. A., and J. M. Gregory, 2006: Understanding projections of sea level rise in a Hadley Centre coupled climate model. *J. Geophys. Res. Oceans*, **111**, C11014.

Lowe, J. A., et al., 2009: UK Climate Projections science report: Marine and coastal projections. M. O. H. Centre, Ed.

Lowe, J. A., et al., 2010: Past and future changes in extreme sea levels and waves. In: *Understanding Sea-Level Rise and Variability* [J. A. Church, P. L. Woodworth, T. Aarup and W. S. Wilson (eds.)]. Wiley-Blackwell, Hoboken, NJ, USA, pp. 326– 375.

Lozier, M. S., V. Roussenov, M. S. C. Reed, and R. G. Williams, 2010: Opposing decadal changes for the North Atlantic meridional overturning circulation. *Nature Geosci.*, **3**, 728–734.

MacAyeal, D. R., T. A. Scambos, C. L. Hulbe, and M. A. Fahnestock, 2003: Catastrophic ice-shelf break-up by an ice-shelf-fragment- capsize mechanism. *J. Glaciol.*, **49**, 22–36.

Machguth, H., et al., 2013: The future sea-level rise contribution of Greenland's glaciers and ice caps. *Environ. Res. Lett.*, **8**, 025005.

Manson, G. K., and S. M. Solomon, 2007: Past and future forcing of Beaufort sea coastal change. *Atmos. Ocean*, **45**, 107–122.

Marčelja, S., 2010: The timescale and extent of thermal expansion of the global ocean due to climate change. *Ocean Sci.*, **6**, 179–184.

Martin, T., and A. Adcroft, 2010: Parameterizing the fresh-water flux from land ice to ocean with interactive icebergs in a coupled climate model. *Ocean Model.*, **34**, 111–124.

Marzeion, B., A. H. Jarosch, and M. Hofer, 2012a: Past and future sea-level changes from the surface mass balance of glaciers. *Cryosphere*, **6**, 1295–1322.

Marzeion, B., M. Hofer, A. H. Jarosch, G. Kaser, and T. Mölg, 2012b: A minimal model for reconstructing interannual mass balance variability of glaciers in the European Alps. *Cryosphere*, **6**, 71–84.

Masson-Delmotte, V., et al., 2010: EPICA Dome C record of glacial and interglacial intensities. *Quat. Sci. Rev.*, **29**, 113–128.

Masters, D., R. S. Nerem, C. Choe, E. Leuliette, B. Beckley, N. White, and M. Ablain, 2012: Comparison of global mean sea level time series from TOPEX/Poseidon, Jason-1, and Jason-2. *Mar. Geodesy*, **35**, 20–41.

McInnes, K., I. Macadam, G. Hubbert, and J. O'Grady, 2009: A modelling approach for estimating the frequency of sea level extremes and the impact of climate change in southeast Australia. *Nat. Hazards*, **51**, 115–137.

McInnes, K., I. Macadam, G. Hubbert, and J. G. O'Grady, 2013: An assessment of current and future vulnerability to coastal inundation due to sea level extremes in Victoria, southeast Australia. *Int. J. Climatol.*, **33**, 33–47.

McInnes, K. L., T. A. Erwin, and J. M. Bathols, 2011: Global climate model projected changes in 10 m wind due to anthropogenic climate change. *Atmos. Sci. Lett.*, **12**, 325–333.

Meehl, G. A., A. X. Hu, and C. Tebaldi, 2010: Decadal Prediction in the Pacific Region. *J. Clim.*, **23**, 2959–2973.

Meehl, G. A., J. M. Arblaster, J. T. Fasullo, A. Hu, and K. E. Trenberth, 2011: Model-based evidence of deep-ocean heat uptake during surface-temperature hiatus periods. *Nature Clim. Change*, **1**, 360–364.

Meehl, G. A., et al., 2005: How much more global warming and sea level rise? *Science*, **307**, 1769–1772.

Meehl, G. A., et al., 2007: Global climate projections.In: *Climate Change 2007: The Physical Science Basis. Contribution of Working Group I to the Fourth Assessment Report of the Intergovernmental Panel on Climate Change* [Solomon, S., D. Qin, M. Manning, Z. Chen, M. Marquis, K. B. Averyt, M. Tignor and H. L. Miller (eds.)]. Cambridge University Press, Cambridge, United Kingdom and New York, NY, USA, pp. 755–828.

Meehl, G. A., et al., 2012: Relative outcomes of climate change mitigation related to global temperature versus sea-level rise. *Nature Clim. Change*, **2**, 576–580.

Meier, M. F., 1984: Contribution of small glaciers to global sea level. *Science*, **226**, 1418–1421.

Meier, M. F., et al., 2007: Glaciers dominate eustatic sea-level rise in the 21st century. *Science*, **317**, 1064–1067.

Meinshausen, M., et al., 2011: The RCP greenhouse gas concentrations and their extensions from 1765 to 2300. *Clim. Change*, **109**, 213–241.

Menéndez, M., and P. L. Woodworth, 2010: Changes in extreme high water levels based on a quasi-global tide-gauge data set. *J. Geophys. Res. Oceans*, **115**, C10011.

Menéndez, M., F. J. Méndez, I. J. Losada, and N. E. Graham, 2008: Variability of extreme wave heights in the northeast Pacific Ocean based on buoy measurements. *Geophys. Res. Lett.*, **35**, L22607.

Mercer, J. H., 1978: West Antarctic ice sheet and CO_2 greenhouse effect: A threat of disaster. *Nature*, **271**, 321–325.

Mernild, S. H., and G. E. Liston, 2012: Greenland freshwater runoff. Part II: Distribution and trends, 1960–2010. *J. Clim.*, **25**, 6015.

Mernild, S. H., G. E. Liston, C. A. Hiemstra, and J. H. Christensen, 2010: Greenland ice sheet surface mass-balance modeling in a 131-yr perspective, 1950–2080. *J. Hydrometeorol.*, **11**, 3–25.

Merrifield, M. A., and M. E. Maltrud, 2011: Regional sea level trends due to a Pacific trade wind intensification. *Geophys. Res. Lett.*, **38**, L21605.

Mikolajewicz, U., M. Vizcaíno, J. Jungclaus, and G. Schurgers, 2007a: Effect of ice sheet interactions in anthropogenic climate change simulations. *Geophys. Res. Lett.*, **34**, L18706.

Mikolajewicz, U., M. Groger, E. Maier-Reimer, G. Schurgers, M. Vizcaíno, and A. Winguth, 2007b: Long-term effects of anthropogenic CO_2 emissions simulated with a complex earth system model. *Clim. Dyn.*, **28**, 599–634.

Miller, L., and B. C. Douglas, 2007: Gyre-scale atmospheric pressure variations and their relation to 19th and 20th century sea level rise. *Geophys. Res. Lett.*, **34**, L16602.

Milly, P. C. D., A. Cazenave, and M. C. Gennero, 2003: Contribution of climate-driven change in continental water storage to recent sea-level rise. *Proc. Natl. Acad. Sci. U.S.A.*, **100**, 13158–13161.

Milly, P. C. D., et al., 2010: Terrestrial water-storage contributions to sea-level rise and variability. In: *Understanding Sea-Level Rise and Variability* [J. A. Church, P. L. Woodworth, T. Aarup and W. S. Wilson (eds.)]. Wiley-Blackwell, Hoboken, NJ, USA, pp. 226–255.

Milne, G. A., and J. X. Mitrovica, 1998: Postglacial sea-level change on a rotating Earth. *Geophys. J. Int.*, **133**, 1–19.

Milne, G. A., W. R. Gehrels, C. W. Hughes, and M. E. Tamisiea, 2009: Identifying the causes of sea-level change. *Nature Geosci.*, **2**, 471–478.

Mitrovica, J. X., N. Gomez, and P. U. Clark, 2009: The sea-level fingerprint of West Antarctic collapse. *Science*, **323**, 753–753.

Mitrovica, J. X., M. E. Tamisiea, J. L. Davis, and G. A. Milne, 2001: Recent mass balance of polar ice sheets inferred from patterns of global sea-level change. *Nature*, **409**, 1026–1029.

Mitrovica, J. X., N. Gomez, E. Morrow, C. Hay, K. Latychev, and M. E. Tamisiea, 2011: On the robustness of predictions of sea-level fingerprints. *Geophys. J. Int.*, **187**, 729–742.

Moberg, A., D. M. Sonechkin, K. Holmgren, N. M. Datsenko, and W. Karlen, 2005: Highly variable Northern Hemisphere temperatures reconstructed from low- and high-resolution proxy data. *Nature*, **433**, 613–617.

Monaghan, A. J., et al., 2006: Insignificant change in Antarctic snowfall since the International Geophysical Year. *Science*, **313**, 827–831.

Moore, J. C., S. Jevrejeva, and A. Grinsted, 2011: The historical global sea level budget. *Ann. Glaciol.*, **52**, 8–14.

Mori, N., T. Shimura, T. Yasuda, and H. Mase, 2013: Multi-model climate projections of ocean surface variables under different climate scenarios—Future change of waves, sea level, and wind. *Ocean Eng.*, **71**, 122-129.

Mori, N., T. Yasuda, H. Mase, T. Tom, and Y. Oku, 2010: Projection of extreme wave climate change under global warming. *Hydrol. Res. Lett.*, **4**, 15–19.

Morice, C. P., J. J. Kennedy, N. A. Rhayner, and P. D. Jones, 2012: Quantifying uncertainties in global and regional temperature change using an ensemble of observational estimates: The HadCRUT4 data set. *J. Geophys. Res. Atmos.*, **117**, D08101.

Morlighem, M., E. Rignot, H. Seroussi, E. Larour, and H. Ben Dhia, 2010: Spatial patterns of basal drag inferred using control methods from a full-Stokes and simpler models for Pine Island Glacier, West Antarctica. *Geophys. Res. Lett.*, **37**, L14502.

Moucha, R., A. M. Forte, J. X. Mitrovica, D. B. Rowley, S. Quere, N. A. Simmons, and S. P. Grand, 2008: Dynamic topography and long-term sea-level variations: There is no such thing as a stable continental platform. *Earth Planet. Sci. Lett.*, **271**, 101–108.

Mousavi, M., J. Irish, A. Frey, F. Olivera, and B. Edge, 2011: Global warming and hurricanes: The potential impact of hurricane intensification and sea level rise on coastal flooding. *Clim. Change*, **104**, 575–597.

Muhs, D. R., J. M. Pandolfi, K. R. Simmons, and R. R. Schumann, 2012: Sea-level history of past interglacial periods from uranium-series dating of corals, Curacao, Leeward Antilles islands. *Quat. Res.*, **78**, 157–169.

Muller, M., B. K. Arbic, and J. X. Mitrovica, 2011: Secular trends in ocean tides: Observations and model results. *J. Geophys. Res. Oceans*, **116**, C05013.

Murphy, D. M., S. Solomon, R. W. Portmann, K. H. Rosenlof, P. M. Forster, and T. Wong, 2009: An observationally based energy balance for the Earth since 1950. *J. Geophys. Res. Oceans*, **114**, D17107.

Naish, T., et al., 2009: Obliquity-paced Pliocene West Antarctic ice sheet oscillations. *Nature*, **458**, 322–328.

National Research Council, 2012: *Sea-Level Rise for the Coasts of California, Oregon, and Washington: Past, Present, and Future.* The National Academies Press, Washington, DC.

Nerem, R. S., D. P. Chambers, C. Choe, and G. T. Mitchum, 2010: Estimating mean sea level change from the TOPEX and Jason altimeter missions. *Mar. Geodesy*, **33**, 435–446.

Ngo-Duc, T., K. Laval, J. Polcher, A. Lombard, and A. Cazenave, 2005: Effects of land water storage on global mean sea level over the past half century. *Geophys. Res. Lett.*, **32**, L09704.

Nicholls, R. J., et al., 2011: Sea-level rise and its possible impacts given a 'beyond 4 degrees C world' in the twenty-first century. *Philos. Trans. R. Soc. London A*, **369**, 161–181.

Nick, F. M., A. Vieli, I. M. Howat, and I. Joughin, 2009: Large-scale changes in Greenland outlet glacier dynamics triggered at the terminus. *Nature Geosci.*, **2**, 110–114.

Nick, F. M., et al., 2012: The response of Petermann Glacier, Greenland, to large calving events, and its future stability in the context of atmospheric and oceanic warming. *J. Glaciol.*, **58**, 229–239.

Nick, F. M., et al., 2013: Future sea-level rise from Greenland's major outlet glaciers in a warming climate. *Nature*, **497**, 235–238.

Nidheesh, A. G., M. Lengaine, J. Vialard, A. S. Unnikrishnam, and H. Dayan, 2013: Decadal and long-term sea level variability in the tropical Indo-Pacific Ocean. *Clim. Dyn.*, **41**, 381–402.

Oerlemans, J., J. Jania, and L. Kolondra, 2011: Application of a minimal glacier model to Hansbreen, Svalbard. *Cryosphere*, **5**, 1–11.

Okumura, Y. M., C. Deser, A. Hu, A. Timmermann, and S. P. Xie, 2009: North Pacific climate response to freshwater forcing in the Subarctic North Atlantic: Oceanic and atmospheric pathways. *J. Clim.*, **22**, 1424–1445.

Olivié, D. J. L., G. P. Peters, and D. Saint-Martin, 2012: Atmosphere Response Time Scales Estimated from AOGCM Experiments. *J. Clim.*, **25**, 7956–7972.

Ollivier, A., Y. Faugere, N. Picot, M. Ablain, P. Femenias, and J. Benveniste, 2012: Envisat Ocean Altimeter becoming relevant for mean sea level trend studies. *Mar. Geodesy*, **35**, 118–136.

Orlić, M., and Z. Pasarić, 2013: Semi-empirical versus process-based sea-level projections for the twenty-first century. *Nature Clim. Change*, **8**, 735-738.

Otto, A., et al., 2013: Energy budget contraints on climate response. *Nature Geosci.*, **6**, 415–416.

Overeem, I., R. S. Anderson, C. W. Wobus, G. D. Clow, F. E. Urban, and N. Matell, 2011: Sea ice loss enhances wave action at the Arctic coast. *Geophys. Res. Lett.*, **38**, L17503.

Palmer, M. D., D. J. McNeall, and N. J. Dunstone, 2011: Importance of the deep ocean for estimating decadal changes in Earth's radiation balance. *Geophys. Res. Lett.*, **38**, L13707.

Pardaens, A., J. M. Gregory, and J. Lowe, 2011a: A model study of factors influencing projected changes in regional sea level over the twenty-first century. *Clim. Dyn.*, **36**, 2015–2033.

Pardaens, A. K., H. T. Banks, J. M. Gregory, and P. R. Rowntree, 2003: Freshwater transports in HadCM3. *Clim. Dyn.*, **21**, 177–195.

Pardaens, A. K., J. A. Lowe, S. Brown, R. J. Nicholls, and D. de Gusmao, 2011b: Sea-level rise and impacts projections under a future scenario with large greenhouse gas emission reductions. *Geophys. Res. Lett.*, **38**, L12604.

Parizek, B. R., et al., 2013: Dynamic (in)stability of Thwaites Glacier, West Antarctica. *J. Geophys. Res. Earth Surf.*, **118**, 638–655.

Parker, A., 2013a: Comment to M Lichter and D Felsenstein, Assessing the costs of sea-level rise and extreme flooding at the local level: A GIS-based approach. *Ocean Coast. Manage.*, **78**, 138–142.

Parker, A., 2013b: Sea level trends at locations of the United States with more than 100 years of recording. *Nat. Hazards*, **65**, 1011–1021.

Parker, A., 2013c: Comment to Shepard, CC, Agostini, VN, Gilmer, B., Allen, T., Stone, J., Brooks, W., Beck, MW: Assessing future risk: Quantifying the effects of sea level rise on storm surge risk for the southern shores of Long Island. *Nat. Hazards*, **65**, 977–980.

Passchier, S., 2011: Linkages between East Antarctic Ice Sheet extent and Southern Ocean temperatures based on a Pliocene high-resolution record of ice-rafted debris off Prydz Bay, East Antarctica. *Paleoceanography*, **26**, Pa4204.

Pattyn, F., A. Huyghe, S. De Brabander, and B. De Smedt, 2006: Role of transition zones in marine ice sheet dynamics. *J. Geophys. Res. Earth Surf.*, **111**, F02004.

Pattyn, F., et al., 2013: Grounding-line migration in plan-view marine ice-sheet models: Results of the ice2sea MISMIP3d intercomparison. *J. Glaciol.*, **59**, 410–422.

Paulson, A., S. J. Zhong, and J. Wahr, 2007: Inference of mantle viscosity from GRACE and relative sea level data. *Geophys. J. Int.*, **171**, 497–508.

Peltier, W. R., 2004: Global glacial isostasy and the surface of the ice-age earth: The ICE-5G (VM2) model and GRACE. *Annu. Rev. Earth Planet. Sci.*, **32**, 111–149.

Peltier, W. R., 2009: Closure of the budget of global sea level rise over the GRACE era: The importance and magnitudes of the required corrections for global glacial isostatic adjustment. *Quat. Sci. Rev.*, **28**, 1658–1674.

Peltier, W. R., and A. M. Tushingham, 1991: Influence of glacial isostatic-adjustment on tide gauge measurements of secular sea-level change. *J. Geophys. Res. Solid Earth Planets*, **96**, 6779–6796.

Perrette, M., F. W. Landerer, R. Riva, K. Frieler, and M. Meinshausen, 2013: A scaling approach to project regional sea level rise and its uncertainties. *Earth Syst. Dyn.*, **4**, 11–29.

Pfeffer, W. T., J. T. Harper, and S. O'Neel, 2008: Kinematic constraints on glacier contributions to 21st-century sea-level rise. *Science*, **321**, 1340–1343.

Phillips, T., H. Rajaram, and K. Steffen, 2010: Cryo-hydrologic warming: A potential mechanism for rapid thermal response of ice sheets. *Geophys. Res. Lett.*, **37**, L20503.

Phillips, T., H. Rajaram, W. Colgan, K. Steffen, and W. Abdalati, 2013: Evaluation of cryo-hydrologic warming as an explanation for increased ice velocities in the wet snow zone, Sermeq Avannarleq, West Greenland. *J. Geophys. Res.*, **118**, 1241–1256.

Pokhrel, Y. N., N. Hanasaki, P. J. F. Yeh, T. J. Yamada, S. Kanae, and T. Oki, 2012: Model estimates of sea-level change due to anthropogenic impacts on terrestrial water storage. *Nature Geosci.*, **5**, 389–392.

Pokhrel, Y. N., N. Hanasaki, P. J. F. Yeh, T. J. Yamada, S. Kanae, and T. Oki, 2013: Overestimated water storage Reply. *Nature Geosci.*, **6**, 2–3.

Pollard, D., and R. M. DeConto, 2009: Modelling West Antarctic ice sheet growth and collapse through the past five million years. *Nature*, **458**, 329–332.

Polvani, L. M., M. Previdi, and C. Deser, 2011: Large cancellation, due to ozone recovery, of future Southern Hemisphere atmospheric circulation trends. *Geophys. Res. Lett.*, **38**, L04707.

Price, S. F., A. J. Payne, I. M. Howat, and B. E. Smith, 2011: Committed sea-level rise for the next century from Greenland ice sheet dynamics during the past decade. *Proc. Natl. Acad. Sci. U.S.A.*, **108**, 8978–8983.

Pritchard, H. D., R. J. Arthern, D. G. Vaughan, and L. A. Edwards, 2009: Extensive dynamic thinning on the margins of the Greenland and Antarctic ice sheets. *Nature*, **461**, 971–975.

Pritchard, H. D., S. R. M. Ligtenberg, H. A. Fricker, D. G. Vaughan, M. R. van den Broeke, and L. Padman, 2012: Antarctic ice-sheet loss driven by basal melting of ice shelves. *Nature*, **484**, 502–505.

Purkey, S. G., and G. C. Johnson, 2010: Warming of global abyssal and deep southern ocean waters between the 1990s and 2000s: Contributions to global heat and sea level rise budgets. *J. Clim.*, **23**, 6336–6351.

Qiu, B., and S. M. Chen, 2006: Decadal variability in the large-scale sea surface height field of the South Pacific Ocean: Observations and causes. *J. Phys. Oceanogr.*, **36**, 1751–1762.

Quinn, K. J., and R. M. Ponte, 2010: Uncertainty in ocean mass trends from GRACE. *Geophys. J. Int.*, **181**, 762–768.

Radic, V., and R. Hock, 2010: Regional and global volumes of glaciers derived from statistical upscaling of glacier inventory data. *J. Geophys. Res. Earth Surf.*, **115**, F01010.

Radić, V., and R. Hock, 2011: Regionally differentiated contribution of mountain glaciers and ice caps to future sea-level rise. *Nature Geosci.*, **4**, 91–94.

Radić, V., A. Bliss, A. D. Beedlow, R. Hock, E. Miles, and J. G. Cogley, 2013: Regional and global projections of the 21st century glacier mass changes in response to climate scenarios from global climate models. *Clim. Dyn.*, doi:10.1007/s00382-013-1719-7.

Rae, J. G. L., et al., 2012: Greenland ice sheet surface mass balance: Evaluating simulations and making projections with regional climate models. *Cryosphere*, **6**, 1275–1294.

Rahmstorf, S., 2007a: A semi-empirical approach to projecting future sea-level rise. *Science*, **315**, 368–370.

Rahmstorf, S., 2007b: Response to comments on "A semi-empirical approach to projecting future sea-level rise". *Science*, **317**, 1866.

Rahmstorf, S., and A. Ganopolski, 1999: Long-term global warming scenarios computed with an efficient coupled climate model. *Clim. Change*, **43**, 353–367.

Rahmstorf, S., M. Perrette, and M. Vermeer, 2012a: Testing the robustness of semi-empirical sea level projections. *Clim. Dyn.*, **39**, 861–875.

Rahmstorf, S., G. Foster, and A. Cazenave, 2012b: Comparing climate projections to observations up to 2011. *Environ. Res. Lett.*, **7**, 044035.

Rahmstorf, S., A. Cazenave, J. A. Church, J. E. Hansen, R. F. Keeling, D. E. Parker, and R. C. J. Somerville, 2007: Recent climate observations compared to projections. *Science*, **316**, 709–709.

Raper, S. C. B., and R. J. Braithwaite, 2005: The potential for sea level rise: New estimates from glacier and ice cap area and volume distributions. *Geophys. Res. Lett.*, **32**, L05502.

Raper, S. C. B., J. M. Gregory, and R. J. Stouffer, 2002: The role of climate sensitivity and ocean heat uptake on AOGCM transient temperature response. *J. Clim.*, **15**, 124–130.

Rasmussen, L. A., H. Conway, R. M. Krimmel, and R. Hock, 2011: Surface mass balance, thinning and iceberg production, Columbia Glacier, Alaska, 1948–2007. *J. Glaciol.*, **57**, 431–440.

Ray, R. D., and B. C. Douglas, 2011: Experiments in reconstructing twentieth-century sea levels. *Prog. Oceanogr.*, **91**, 495–515.

Raymo, M. E., and J. X. Mitrovica, 2012: Collapse of polar ice sheets during the stage 11 interglacial. *Nature*, **483**, 453–456.

Raymo, M. E., J. X. Mitrovica, M. J. O'Leary, R. M. DeConto, and P. L. Hearty, 2011: Departures from eustasy in Pliocene sea-level records. *Nature Geosci.*, **4**, 328–332.

Ridley, J., J. M. Gregory, P. Huybrechts, and J. Lowe, 2010: Thresholds for irreversible decline of the Greenland ice sheet. *Clim. Dyn.*, **35**, 1065–1073.

Ridley, J. K., P. Huybrechts, J. M. Gregory, and J. A. Lowe, 2005: Elimination of the Greenland ice sheet in a high CO_2 climate. *J. Clim.*, **18**, 3409–3427.

Rignot, E., 2001: Evidence for rapid retreat and mass loss of Thwaites Glacier, West Antarctica. *J. Glaciol.*, **47**, 213–222.

Rignot, E., 2008: Changes in West Antarctic ice stream dynamics observed with ALOS PALSAR data. *Geophys. Res. Lett.*, **35**, L12505.

Rignot, E., I. Velicogna, M. R. van den Broeke, A. Monaghan, and J. Lenaerts, 2011: Acceleration of the contribution of the Greenland and Antarctic ice sheets to sea level rise. *Geophys. Res. Lett.*, **38**, L05503.

Rignot, E., G. Casassa, P. Gogineni, W. Krabill, A. Rivera, and R. Thomas, 2004: Accelerated ice discharge from the Antarctic Peninsula following the collapse of Larsen B ice shelf. *Geophys. Res. Lett.*, **31**, L18401.

Rignot, E., J. L. Bamber, M. R. Van Den Broeke, C. Davis, Y. H. Li, W. J. Van De Berg, and E. Van Meijgaard, 2008: Recent Antarctic ice mass loss from radar interferometry and regional climate modelling. *Nature Geosci.*, **1**, 106–110.

Riva, R. E. M., J. L. Bamber, D. A. Lavallee, and B. Wouters, 2010: Sea-level fingerprint of continental water and ice mass change from GRACE. *Geophys. Res. Lett.*, **37**, L19605.

Roberts, D. L., P. Karkanas, Z. Jacobs, C. W. Marean, and R. G. Roberts, 2012: Melting ice sheets 400,000 yr ago raised sea level by 13 m: Past analogue for future trends. *Earth Planet. Sci. Lett.*, **357**, 226–237.

Robinson, A., R. Calov, and A. Ganopolski, 2012: Multistability and critical thresholds of the Greenland ice sheet. *Nature Clim. Change*, **2**, 429–432.

Rott, H., F. Muller, T. Nagler, and D. Floricioiu, 2011: The imbalance of glaciers after disintegration of Larsen-B ice shelf, Antarctic Peninsula. *Cryosphere*, **5**, 125–134.

Rugenstein, M., M. Winton, R. J. Stouffer, S. M. Griffies, and R. W. Hallberg, 2013: Northern high latitude heat budget decomposition and transient warming. *J. Clim.*, **26**, 609-621.

Russell, G. L., V. Gornitz, and J. R. Miller, 2000: Regional sea-level changes projected by the NASA/GISS atmosphere-ocean model. *Clim. Dyn.*, **16**, 789–797.

Sahagian, D., 2000: Global physical effects of anthropogenic hydrological alterations: Sea level and water redistribution. *Global Planet. Change*, **25**, 39–48.

Sallenger, A. H., K. S. Doran, and P. A. Howd, 2012: Hotspot of accelerated sea-level rise on the Atlantic coast of North America. *Nature Clim. Change*, **2**, 884-888.

Santer, B. D., et al., 2011: Separating signal and noise in atmospheric temperature changes: The importance of timescale. *J. Geophys. Res. Atmos.*, **116**, D22105.

Sasgen, I., et al., 2012: Timing and origin of recent regional ice-mass loss in Greenland. *Earth Planet. Sci. Lett.*, **333**, 293–303.

Scambos, T. A., J. A. Bohlander, C. A. Shuman, and P. Skvarca, 2004: Glacier acceleration and thinning after ice shelf collapse in the Larsen B embayment, Antarctica. *Geophys. Res. Lett.*, **31**, L18402.

Schaeffer, M., W. Hare, S. Rahmstorf, and M. Vermeer, 2012: Long-term sea-level rise implied by 1.5°C and 2°C warming levels. *Nature Clim. Change*, **2**, 867–870.

Schewe, J., A. Levermann, and M. Meinshausen, 2011: Climate change under a scenario near 1.5 °C of global warming: Monsoon intensification, ocean warming and steric sea level rise. *Earth Syst. Dyn.*, **2**, 25–35.

Schmith, T., S. Johansen, and P. Thejll, 2007: Comment on "A semi-empirical approach to projecting future sea-level rise". *Science*, **317**, 1866.

Schoof, C., 2007a: Marine ice-sheet dynamics. Part 1. the case of rapid sliding. *J. Fluid Mech.*, **573**, 27–55.

Schoof, C., 2007b: Ice sheet grounding line dynamics: Steady states, stability, and hysteresis. *J. Geophys. Res. Earth Surf.*, **112**, F03S28.

Schoof, C., 2011: Marine ice sheet dynamics. Part 2. A Stokes flow contact problem. *J. Fluid Mech.*, **679**, 122–155.

Schwartz, S. E., 2012: Determination of Earth's transient and equilibrium climate sensitivities from observations over the twentieth century: Strong dependence on assumed forcing. *Surv. Geophys.*, **33**, 745–777.

Seddik, H., R. Greve, T. Zwinger, F. Gillet-Chaulet, and O. Gagliardini, 2012: Simulations of the Greenland ice sheet 100 years into the future with the full Stokes model Elmer/Ice. *J. Glaciol.*, **58**, 427–440.

Semedo, A., R. Weisse, A. Beherens, A. Sterl, L. Bengtson, and H. Gunther, 2013: Projection of global wave climate change towards the end of the 21st century. *J. Clim.*, **26**, 8269-8288.

Seneviratne, S.I., N. Nicholls, D. Easterling, C.M. Goodess, S. Kanae, J. Kossin, Y. Luo, J. Marengo, K. McInnes, M. Rahimi, M. Reichstein, A. Sorteberg, C. Vera, and X. Zhang, 2012: Changes in climate extremes and their impacts on the natural physical environment. In: Managing the Risks of Extreme Events and Disasters to Advance Climate Change Adaptation [Field, C.B., V. Barros, T.F. Stocker, D. Qin, D.J. Dokken, K.L. Ebi, M.D. Mastrandrea, K.J. Mach, G.-K. Plattner, S.K. Allen, M. Tignor, and P.M. Midgley (eds.)]. A Special Report of Working Groups I and II of the Intergovernmental Panel on Climate Change (IPCC). Cambridge University Press, Cambridge, UK, and New York, NY, USA, pp. 109-230.

Shepherd, A., and D. Wingham, 2007: Recent sea-level contributions of the Antarctic and Greenland ice sheets. *Science*, **315**, 1529–1532.

Shepherd, A., et al., 2012: A reconciled estimate of ice-sheet mass balance. *Science*, **338**, 1183–1189.

Shuman, C. A., E. Berthier, and T. A. Scambos, 2011: 2001–2009 elevation and mass losses in the Larsen A and B embayments, Antarctic Peninsula. *J. Glaciol.*, **57**, 737–754.

Slangen, A. B. A., and R. S. W. van de Wal, 2011: An assessment of uncertainties in using volume-area modelling for computing the twenty-first century glacier contribution to sea-level change. *Cryosphere*, **5**, 673–686.

Slangen, A. B. A., C. A. Katsman, R. S. W. van de Wal, L. L. A. Vermeersen, and R. E. M. Riva, 2012: Towards regional projections of twenty-first century sea-level change based on IPCC SRES scenarios. *Clim. Dyn.*, **38**, 1191–1209.

Smith, J. M., M. A. Cialone, T. V. Wamsley, and T. O. McAlpin, 2010: Potential impact of sea level rise on coastal surges in southeast Louisiana. *Ocean Eng.*, **37**, 37–47.

Sokolov, A. P., C. E. Forest, and P. H. Stone, 2010: Sensitivity of climate change projections to uncertainties in the estimates of observed changes in deep-ocean heat content. *Clim. Dyn.*, **34**, 735–745.

Solgaard, A. M., and P. L. Langen, 2012: Multistability of the Greenland ice sheet and the effects of an adaptive mass balance formulation. *Clim. Dyn.*, **39**, 1599–1612.

Solomon, S., G.-K. Plattner, R. Knutti, and P. Friedlingstein, 2009: Irresversible climate change due to carbon dioxide emissions. *Proc. Natl. Acad. Sci. U.S.A.*, **106**, 1704–1709.

Spada, G., J. L. Bamber, and R. T. W. L. Hurkmans, 2013: The gravitationally consistent sea-level fingerprint of future terrestrial ice loss. *Geophys. Res. Lett.*, **40**, 482–486.

Sriver, R. L., N. M. Urban, R. Olson, and K. Keller, 2012: Toward a physically plausible upper bound of sea-level rise projections. *Clim. Change*, **115**, 893–902.

Stackhouse, P. W., T. Wong, N. G. Loeb, D. P. Kratz, A. C. Wilber, D. R. Doelling, and L. C. Nguyen, 2010: Earth radiation budget at top-of-atmosphere. *Bull. Am. Meteorol. Soc.*, **90**, S33–S34.

Stammer, D., 2008: Response of the global ocean to Greenland and Antarctic ice melting. *J. Geophys. Res. Oceans*, **113**, C06022.

Stammer, D., and S. Huttemann, 2008: Response of regional sea level to atmospheric pressure loading in a climate change scenario. *J. Clim.*, **21**, 2093–2101.

Stammer, D., A. Cazenave, R. M. Ponte, and M. E. Tamisiea, 2013: Causes for contemporary regional sea level changes. In: *Annual Review of Marine Science*, Vol. 5 [C. A. Carlson and S. J. Giovannoni (eds.)]. Annual Reviews, Palo Alto, CA, USA, pp. 21–46.

Stammer, D., N. Agarwal, P. Herrmann, A. Kohl, and C. R. Mechoso, 2011: Response of a coupled ocean-atmosphere model to Greenland ice melting. *Surv. Geophys.*, **32**, 621–642.

Stenni, B., et al., 2011: Expression of the bipolar see-saw in Antarctic climate records during the last deglaciation. *Nature Geosci.*, **4**, 46–49.

Sterl, A., H. van den Brink, H. de Vries, R. Haarsma, and E. van Meijgaard, 2009: An ensemble study of extreme North Sea storm surges in a changing climate. *Ocean Sci. Discuss.*, **6**, 1031–1059.

Stouffer, R. J., 2004: Time scales of climate response. *J. Clim.*, **17**, 209–217.

Straneo, F., et al., 2010: Rapid circulation of warm subtropical waters in a major glacial fjord in East Greenland. *Nature Geosci.*, **3**, 182–186.

Suzuki, T., and M. Ishii, 2011: Regional distribution of sea level changes resulting from enhanced greenhouse warming in the Model for Interdisciplinary Research on Climate version 3.2. *Geophys. Res. Lett.*, **38**, L02601.

Suzuki, T., et al., 2005: Projection of future sea level and its variability in a high-resolution climate model: Ocean processes and Greenland and Antarctic ice-melt contributions. *Geophys. Res. Lett.*, **32**, L19706.

Swingedouw, D., T. Fichefet, P. Huybrechts, H. Goosse, E. Driesschaert, and M. F. Loutre, 2008: Antarctic ice-sheet melting provides negative feedbacks on future climate warming. *Geophys. Res. Lett.*, **35**, L17705.

Syvitski, J. P. M., and A. Kettner, 2011: Sediment flux and the Anthropocene. *Philos. Trans. R. Soc. London A*, **369**, 957–975.

Syvitski, J. P. M., et al., 2009: Sinking deltas due to human activities. *Nature Geosci.*, **2**, 681–686.

Tamisiea, M. E., 2011: Ongoing glacial isostatic contributions to observations of sea level change. *Geophys. J. Int.*, **186**, 1036–1044.

Tamisiea, M. E., E. M. Hill, R. M. Ponte, J. L. Davis, I. Velicogna, and N. T. Vinogradova, 2010: Impact of self attraction and loading on the annual cycle in sea level. *J. Geophys. Res.*, **115**, C07004.

Tebaldi, C., B. H. Strauss, and C. E. Zervas, 2012: Modelling sea level rise impacts on storm surges along US coasts. *Environ. Res. Lett.*, **7**, 2–11.

Tett, S. F. B., et al., 2007: The impact of natural and anthropogenic forcings on climate and hydrology since 1550. *Clim. Dyn.*, **28**, 3–34.

Thoma, M., A. Jenkins, D. Holland, and S. Jacobs, 2008: Modelling circumpolar deep water intrusions on the Amundsen Sea continental shelf, Antarctica. *Geophys. Res. Lett.*, **35**, L18602.

Thompson, W. G., H. A. Curran, M. A. Wilson, and B. White, 2011: Sea-level oscillations during the last interglacial highstand recorded by Bahamas corals. *Nature Geosci.*, **4**, 684–687.
Timmermann, A., S. McGregor, and F. F. Jin, 2010: Wind effects on past and future regional sea level trends in the Southern Indo-Pacific. *J. Clim.*, **23**, 4429–4437.
Tsimplis, M., E. Alvarez-Fanjul, D. Gomis, L. Fenoglio-Marc, and B. Perez, 2005: Mediterranean Sea level trends: Atmospheric pressure and wind contribution. *Geophys. Res. Lett.*, **32**, L20602.
Unnikrishnan, A. S., M. R. R. Kumar, and B. Sindhu, 2011: Tropical cyclones in the Bay of Bengal and extreme sea-level projections along the east coast of India in a future climate scenario. *Curr. Sci. (India)*, **101**, 327–331.
Uotila, P., A. H. Lynch, J. J. Cassano, and R. I. Cullather, 2007: Changes in Antarctic net precipitation in the 21st century based on Intergovernmental Panel on Climate Change (IPCC) model scenarios. *J. Geophys. Res. Atmos.*, **112**, D10107.
van Angelen, J. H., et al., 2012: Sensitivity of Greenland ice sheet surface mass balance to surface albedo parameterization: A study with a regional climate model. *Cryosphere*, **6**, 1531–1562.
van de Berg, W. J., M. van den Broeke, J. Ettema, E. van Meijgaard, and F. Kaspar, 2011: Significant contribution of insolation to Eemian melting of the Greenland ice sheet. *Nature Geosci.*, **4**, 679–683.
van den Broeke, M., W. J. van de Berg, and E. van Meijgaard, 2006: Snowfall in coastal West Antarctica much greater than previously assumed. *Geophys. Res. Lett.*, **33**, L02505.
van den Broeke, M., et al., 2009: Partitioning recent Greenland mass loss. *Science*, **326**, 984–986.
Van Ommen, T. D., V. Morgan, and M. A. J. Curran, 2004: Deglacial and Holocene changes in accumulation at Law Dome, East Antarctica. *Ann. Glaciol.*, **39**, 395–365.
Vaughan, D. G., J. L. Bamber, M. Giovinetto, J. Russell, and A. P. R. Cooper, 1999: Reassessment of net surface mass balance in Antarctica. *J. Clim.*, **12**, 933–946.
Vellinga, M., and R. Wood, 2008: Impacts of thermohaline circulation shutdown in the twenty-first century. *Clim. Change*, **91**, 43–63.
Vermaire, J. C., M. F. J. Pisaric, J. R. Thienpont, C. J. C. Mustaphi, S. V. Kokelj, and J. P. Smol, 2013: Arctic climate warming and sea ice declines lead to increased storm surge activity. *Geophys. Res. Lett.*, **40**, 1386-1390.
Vermeer, M., and S. Rahmstorf, 2009: Global sea level linked to global temperature. *Proc. Natl. Acad. Sci. U.S.A.*, **106**, 21527–21532.
Vernon, C. L., J. L. Bamber, J. E. Box, M. R. van den Broeke, X. Fettweis, E. Hanna, and P. Huybrechts, 2013: Surface mass balance model intercomparison for the Greenland ice sheet. *Cryosphere*, **7**, 599–614.
Vieli, A., and F. M. Nick, 2011: Understanding and modelling rapid dynamic changes of tidewater outlet glaciers: Issues and implications. *Surv. Geophys.*, **32**, 437–458.
Vinogradov, S. V., and R. M. Ponte, 2011: Low-frequency variability in coastal sea level from tide gauges and altimetry. *J. Geophys. Res. Oceans*, **116**, C07006.
Vizcaíno, M., U. Mikolajewicz, J. Jungclaus, and G. Schurgers, 2010: Climate modification by future ice sheet changes and consequences for ice sheet mass balance. *Clim. Dyn.*, **34**, 301–324.
Vizcaíno, M., U. Mikolajewicz, M. Groger, E. Maier-Reimer, G. Schurgers, and A. M. E. Winguth, 2008: Long-term ice sheet-climate interactions under anthropogenic greenhouse forcing simulated with a complex Earth System Model. *Clim. Dyn.*, **31**, 665–690.
von Schuckmann, K., and P. Y. Le Traon, 2011: How well can we derive Global Ocean Indicators from Argo data? *Ocean Sci.*, **7**, 783–791.
von Storch, H., E. Zorita, and J. F. Gonzalez-Rouco, 2008: Relationship between global mean sea-level and global mean temperature in a climate simulation of the past millennium. *Ocean Dyn.*, **58**, 227–236.
Wada, Y., L. P. H. van Beek, C. M. van Kempen, J. W. T. M. Reckman, S. Vasak, and M. F. P. Bierkens, 2010: Global depletion of groundwater resources. *Geophys. Res. Lett.*, **37**, L20402.
Wada, Y., L. P. H. van Beek, F. C. S. Weiland, B. F. Chao, Y. H. Wu, and M. F. P. Bierkens, 2012: Past and future contribution of global groundwater depletion to sea-level rise. *Geophys. Res. Lett.*, **39**, L09402.
Wake, L. M., P. Huybrechts, J. E. Box, E. Hanna, I. Janssens, and G. A. Milne, 2009: Surface mass-balance changes of the Greenland ice sheet since 1866. *Ann. Glaciol.*, **50**, 178–184.
Walsh, K. J. E., K. McInnes, and J. L. McBride, 2011: Climate change impacts on tropical cyclones and extreme sea levels in the South Pacific—a regional assessment. *Global Planet. Change*, **80–81**, 149–164.
Wang, S., R. McGrath, J. Hanafin, P. Lynch, T. Semmler, and P. Nolan, 2008: The impact of climate change on storm surges over Irish waters. *Ocean Model.*, **25**, 83–94.
Wang, X. L., and V. R. Swail, 2006: Climate change signal and uncertainty in projections of ocean wave heights. *Clim. Dyn.*, **26**, 109–126.
Wang, X. L., V. R. Swail, and A. Cox, 2010: Dynamical versus statistical downscaling methods for ocean wave heights. *Int. J. Climatol.*, **30**, 317–332.
Wang, X. L., V. R. Swail, F. Zwiers, X. Zhang, and Y. Feng, 2009: Detection of external influence on trends of atmospheric storminess and northern oceans wave heights. *Clim. Dyn.*, **32**, 189–203.
Warrick, R. A., and J. Oerlemans, 1990: Sea level rise. In: *Climate Change: The IPCC Scientific Assessment* [J. T. Houghton, G. J. Jenkins and J. J. Ephraum (eds.)]. Cambridge University Press, Cambridge, United Kingdom, and New York, NY, USA, pp. 260–281.
Warrick, R. A., C. Le Provost, M. F. Meier, J. Oerlemans, and P. L. Woodworth, 1996: Changes in sea level. In: *Climate Change 1995: The Science of Climate Change. Contribution of WGI to the Second Assessment Report of the Intergovernmental Panel on Climate Change* [J. T. Houghton, L. G. Meira . A. Callander, N. Harris, A. Kattenberg and K. Maskell (eds.)]. Cambridge University Press, Cambridge, United Kingdom, and New York, NY, USA, pp. 359–405.
Washington, W. M., et al., 2009: How much climate change can be avoided by mitigation? *Geophys. Res. Lett.*, **36**, L08703.
Watson, C., et al., 2010: Twentieth century constraints on sea level change and earthquake deformation at Macquarie Island. *Geophys. J. Int.*, **182**, 781–796.
Weertman, J., 1961: Stability of Ice-Age ice sheets. *J. Geophys. Res.*, **66**, 3783–3792.
Weertman, J., 1974: Stability of the junction of an ice sheet and an ice shelf. *J. Glaciol.*, **13**, 3–11.
White, N. J., J. A. Church, and J. M. Gregory, 2005: Coastal and global averaged sea level rise for 1950 to 2000. *Geophys. Res. Lett.*, **32**, L01601.
Winguth, A., U. Mikolajewicz, M. Groger, E. Maier-Reimer, G. Schurgers, and M. Vizcaíno, 2005: Centennial-scale interactions between the carbon cycle and anthropogenic climate change using a dynamic Earth system model. *Geophys. Res. Lett.*, **32**, L23714.
Winkelmann, R., A. Levermann, M. A. Martin, and K. Frieler, 2012: Increased future ice discharge from Antarctica owing to higher snowfall. *Nature*, **492**, 239–242.
Wong, T., B. A. Wielecki, R. B. I. Lee, G. L. Smith, K. A. Bush, and J. K. Willis, 2006: Reexamination of the observed decadal variability of the earth radiation budget using altitude-corrected ERBE/ERBS nonscanner WFOV data. *J. Clim.*, **19**, 4028–4040.
Woodroffe, C. D., H. V. McGregor, K. Lambeck, S. G. Smithers, and D. Fink, 2012: Mid-Pacific microatolls record sea-level stability over the past 5000 yr. *Geology*, **40**, 951–954.
Woolf, D. K., P. G. Challenor, and P. D. Cotton, 2002: Variability and predictability of the North Atlantic wave climate. *J. Geophys. Res. Oceans*, **107**, C103145.
Woth, K., R. Weisse, and H. von Storch, 2006: Climate change and North Sea storm surge extremes: An ensemble study of storm surge extremes expected in a changed climate projected by four different regional climate models. *Ocean Dyn.*, **56**, 3–15.
Wunsch, C., and D. Stammer, 1997: Atmospheric loading and the oceanic "inverted barometer" effect. *Rev. Geophys.*, **35**, 79–107.
Wunsch, C., and P. Heimbach, 2007: Practical global oceanic state estimation. *Physica D*, **230**, 197–208.
Yin, J., 2012: Century to multi-century sea level rise projections from CMIP5 models. *Geophys. Res. Lett.*, **39**, L17709.
Yin, J. J., M. E. Schlesinger, and R. J. Stouffer, 2009: Model projections of rapid sea-level rise on the northeast coast of the United States. *Nature Geosci.*, **2**, 262–266.
Yin, J. J., S. M. Griffies, and R. J. Stouffer, 2010: Spatial variability of sea level rise in twenty-first century projections. *J. Clim.*, **23**, 4585–4607.
Yin, J. J., J. T. Overpeck, S. M. Griffies, A. X. Hu, J. L. Russell, and R. J. Stouffer, 2011: Different magnitudes of projected subsurface ocean warming around Greenland and Antarctica. *Nature Geosci.*, **4**, 524–528.
Yoshimori, M., and A. Abe-Ouchi, 2012: Sources of spread in multi-model projections of the Greenland ice-sheet surface mass balance. *J. Clim.*, **25**, 1157–1175.
Young, I. R., S. Zieger, and A. V. Babanin, 2011: Global trends in wind speed and wave height. *Science*, **332**, 451–455.
Zhang, X. B., and J. A. Church, 2012: Sea level trends, interannual and decadal variability in the Pacific Ocean. *Geophys. Res. Lett.*, **39**, L21701.
Zickfeld, K., V. K. Arora, and N. P. Gillett, 2012: Is the climate response to CO_2 emissions path dependent? *Geophys. Res. Lett.*, **39**, L05703.

Zickfeld, K., M. Eby, H. D. Matthews, and A. J. Weaver, 2009: Setting cumulative emissions targets to reduce the risk of dangerous climate change. *Proc. Natl. Acad. Sci. U.S.A.*, **106**, 16129–16134.

Zickfeld, K., et al., 2013: Long-term climate change commitment and reversibility: An EMIC intercomparison. *J. Clim.*, **26**, 5782-5809.

Zwally, H. J., and M. B. Giovinetto, 2011: Overview and assessment of Antarctic Ice-Sheet mass balance estimates: 1992–2009. *Surv. Geophys.*, **32**, 351–376.

13

14

Climate Phenomena and their Relevance for Future Regional Climate Change

Coordinating Lead Authors:
Jens Hesselbjerg Christensen (Denmark), Krishna Kumar Kanikicharla (India)

Lead Authors:
Edvin Aldrian (Indonesia), Soon-Il An (Republic of Korea), Iracema Fonseca Albuquerque Cavalcanti (Brazil), Manuel de Castro (Spain), Wenjie Dong (China), Prashant Goswami (India), Alex Hall (USA), Joseph Katongo Kanyanga (Zambia), Akio Kitoh (Japan), James Kossin (USA), Ngar-Cheung Lau (USA), James Renwick (New Zealand), David B. Stephenson (UK), Shang-Ping Xie (USA), Tianjun Zhou (China)

Contributing Authors:
Libu Abraham (Qatar), Tércio Ambrizzi (Brazil), Bruce Anderson (USA), Osamu Arakawa (Japan), Raymond Arritt (USA), Mark Baldwin (UK), Mathew Barlow (USA), David Barriopedro (Spain), Michela Biasutti (USA), Sébastien Biner (Canada), David Bromwich (USA), Josephine Brown (Australia), Wenju Cai (Australia), Leila V. Carvalho (USA/Brazil), Ping Chang (USA), Xiaolong Chen (China), Jung Choi (Republic of Korea), Ole Bøssing Christensen (Denmark), Clara Deser (USA), Kerry Emanuel (USA), Hirokazu Endo (Japan), David B. Enfield (USA), Amato Evan (USA), Alessandra Giannini (USA), Nathan Gillett (Canada), Annamalai Hariharasubramanian (USA), Ping Huang (China), Julie Jones (UK), Ashok Karumuri (India), Jack Katzfey (Australia), Erik Kjellström (Sweden), Jeff Knight (UK), Thomas Knutson (USA), Ashwini Kulkarni (India), Koteswara Rao Kundeti (India), William K. Lau (USA), Geert Lenderink (Netherlands), Chris Lennard (South Africa), Lai-yung Ruby Leung (USA), Renping Lin (China), Teresa Losada (Spain), Neil C. Mackellar (South Africa), Victor Magaña (Mexico), Gareth Marshall (UK), Linda Mearns (USA), Gerald Meehl (USA), Claudio Menéndez (Argentina), Hiroyuki Murakami (USA/Japan), Mary Jo Nath (USA), J. David Neelin (USA), Geert Jan van Oldenborgh (Netherlands), Martin Olesen (Denmark), Jan Polcher (France), Yun Qian (USA), Suchanda Ray (India), Katharine Davis Reich (USA), Belén Rodriguez de Fonseca (Spain), Paolo Ruti (Italy), James Screen (UK), Jan Sedláček (Switzerland) Silvina Solman (Argentina), Martin Stendel (Denmark), Samantha Stevenson (USA), Izuru Takayabu (Japan), John Turner (UK), Caroline Ummenhofer (USA), Kevin Walsh (Australia), Bin Wang (USA), Chunzai Wang (USA), Ian Watterson (Australia), Matthew Widlansky (USA), Andrew Wittenberg (USA), Tim Woollings (UK), Sang-Wook Yeh (Republic of Korea), Chidong Zhang (USA), Lixia Zhang (China), Xiaotong Zheng (China), Liwei Zou (China)

Review Editors:
John Fyfe (Canada), Won-Tae Kwon (Republic of Korea), Kevin Trenberth (USA), David Wratt (New Zealand)

This chapter should be cited as:
Christensen, J.H., K. Krishna Kumar, E. Aldrian, S.-I. An, I.F.A. Cavalcanti, M. de Castro, W. Dong, P. Goswami, A. Hall, J.K. Kanyanga, A. Kitoh, J. Kossin, N.-C. Lau, J. Renwick, D.B. Stephenson, S.-P. Xie and T. Zhou, 2013: Climate Phenomena and their Relevance for Future Regional Climate Change. In: *Climate Change 2013: The Physical Science Basis. Contribution of Working Group I to the Fifth Assessment Report of the Intergovernmental Panel on Climate Change* [Stocker, T.F., D. Qin, G.-K. Plattner, M. Tignor, S.K. Allen, J. Boschung, A. Nauels, Y. Xia, V. Bex and P.M. Midgley (eds.)]. Cambridge University Press, Cambridge, United Kingdom and New York, NY, USA.

Table of Contents

Supplementary Material is available in online versions of the report.

14

Executive Summary

This chapter assesses the scientific literature on projected changes in major climate phenomena and more specifically their relevance for future change in regional climates, contingent on global mean temperatures continue to rise.

Regional climates are the complex result of processes that vary strongly with location and so respond differently to changes in global-scale influences. The following large-scale climate phenomena are increasingly well simulated by climate models and so provide a scientific basis for understanding and developing credibility in future regional climate change. A phenomenon is considered relevant to regional climate change if there is confidence that it has influence on the regional climate and there is confidence that the phenomenon will change, particularly under the Representative Concentration Pathway 4.5 (RCP4.5) or higher end scenarios. {Table 14.3}

Monsoon Systems

There is growing evidence of improved skill of climate models in reproducing climatological features of the global monsoon. Taken together with identified model agreement on future changes, the global monsoon, aggregated over all monsoon systems, is *likely*[1] to strengthen in the 21st century with increases in its area and intensity, while the monsoon circulation weakens. Monsoon onset dates are *likely* to become earlier or not to change much and monsoon retreat dates are *likely* to delay, resulting in lengthening of the monsoon season in many regions. {14.2.1}

Future increase in precipitation extremes related to the monsoon is *very likely* in South America, Africa, East Asia, South Asia, Southeast Asia and Australia. Lesser model agreement results in *medium confidence*[2] that monsoon-related interannual precipitation variability will increase in the future. {14.2.1, 14.8.5, 14.8.7, 14.8.9, 14.8.11, 14.8.12, 14.8.13}

Model skill in representing regional monsoons is lower compared to the global monsoon and varies across different monsoon systems. There is *medium confidence* that overall precipitation associated with the Asian-Australian monsoon will increase but with a north–south asymmetry: Indian and East Asian monsoon precipitation is projected to increase, while projected changes in Australian summer monsoon precipitation are small. There is *medium confidence* that the Indian summer monsoon circulation will weaken, but this is compensated by increased atmospheric moisture content, leading to more precipitation. For the East Asian summer monsoon, both monsoon circulation and precipitation are projected to increase. There is *medium confidence* that the increase of the Indian summer monsoon rainfall and its extremes throughout the 21st century will be the largest among all monsoons. {14.2.2, 14.8.9, 14.8.11, 14.8.13}

There is *low confidence* in projections of changes in precipitation amounts for the North American and South American monsoons, but *medium confidence* that the North American monsoon will arrive and persist later in the annual cycle, and *high confidence* in expansion of the South American monsoon area. {14.2.3, 14.8.3, 14.8.4, 14.8.5}

There is *low confidence* in projections of a small delay in the development of the West African rainy season and an intensification of late-season rains. Model limitations in representing central features of the West African monsoon result in *low confidence* in future projections. {14.2.4, 14.8.7}

Tropical Phenomena

Based on models' ability to reproduce general features of the Indian Ocean Dipole and agreement on future projections, the tropical Indian Ocean is *likely* to feature a zonal (east–west) pattern of change in the future with reduced warming and decreased precipitation in the east, and increased warming and increased precipitation in the west, directly influencing East Africa and Southeast Asia precipitation. {14.3, 14.8.7, 14.8.12}

A newly identified robust feature in model simulations of tropical precipitation over oceans gives *medium confidence* that annual precipitation change follows a 'warmer-get-wetter' pattern, increasing where warming of sea surface temperature exceeds the tropical mean and vice versa. There is *medium confidence* in projections showing an increase in seasonal mean precipitation on the equatorial flank of the Inter-Tropical Convergence Zone (ITCZ) affecting parts of Central America, the Caribbean, South America, Africa and West Asia despite shortcomings in many models in simulating the ITCZ. There is *medium confidence* that the frequency of zonally oriented South Pacific Convergence Zone events will increase, with the South Pacific Convergence Zone (SPCZ) lying well to the northeast of its average position, a feature commonly reproduced in models that simulate the SPCZ realistically, resulting in reduced precipitation over many South Pacific island nations. Similarly there is *medium confidence* that the South Atlantic Convergence Zone will shift southwards, leading to an increase in precipitation over southeastern South America and a reduction immediately north thereof. {14.3, 14.8.4, 14.8.5, 14.8.7, 14.8.11, 14.8.14}

---

1 In this Report, the following terms have been used to indicate the assessed likelihood of an outcome or a result: Virtually certain 99–100% probability, Very likely 90–100%, Likely 66–100%, About as likely as not 33–66%, Unlikely 0–33%, Very unlikely 0–10%, Exceptionally unlikely 0–1%. Additional terms (Extremely likely: 95–100%, More likely than not >50–100%, and Extremely unlikely 0–5%) may also be used when appropriate. Assessed likelihood is typeset in italics, e.g., *very likely* (see Section 1.4 and Box TS.1 for more details).

2 In this Report, the following summary terms are used to describe the available evidence: limited, medium, or robust; and for the degree of agreement: low, medium, or high. A level of confidence is expressed using five qualifiers: very low, low, medium, high, and very high, and typeset in italics, e.g., *medium confidence*. For a given evidence and agreement statement, different confidence levels can be assigned, but increasing levels of evidence and degrees of agreement are correlated with increasing confidence (see Section 1.4 and Box TS.1 for more details).

There is *low confidence* in projections of future changes in the Madden–Julian Oscillation owing to poor ability of the models to simulate it and its sensitivity to ocean warming patterns. The implications for future projections of regional climate extremes in West Asia, South Asia, Southeast Asia and Australia are therefore highly uncertain when associated with the Madden–Julian Oscillation. {14.3, 14.8.10, 14.8.11, 14.8.12, 14.8.13}

There is *low confidence* in the projections of future changes for the tropical Atlantic, both for the mean and interannual modes, because of systematic errors in model simulations of current climate. The implications for future changes in Atlantic hurricanes and tropical South American and West African precipitation are therefore uncertain. {14.3, 14.6.1, 14.8.5, 14.8.7 }

The realism of the representation of El Niño-Southern Oscillation (ENSO) in climate models is increasing and models simulate ongoing ENSO variability in the future. Therefore there is *high confidence* that ENSO *very likely* remains as the dominant mode of interannual variability in the future and due to increased moisture availability, the associated precipitation variability on regional scales *likely* intensifies. An eastward shift in the patterns of temperature and precipitation variations in the North Pacific and North America related to El Niño and La Niña (teleconnections), a feature consistently simulated by models, is projected for the future, but with *medium confidence,* while other regional implications including those in Central and South America, the Caribbean, Africa, most of Asia, Australia and most Pacific Islands are more uncertain. However, natural modulations of the variance and spatial pattern of ENSO are so large in models that *confidence* in any specific projected change in its variability in the 21st century remains *low*. {14.4, 14.8.3, 14.8.4, 14.8.5, 14.8.7, 14.8.9, 14.8.11, 14.8.12, 14.8.13, 14.8.14}

Cyclones

Based on process understanding and agreement in 21st century projections, it is *likely* that the global frequency of occurrence of tropical cyclones will either decrease or remain essentially unchanged, concurrent with a *likely* increase in both global mean tropical cyclone maximum wind speed and precipitation rates. The future influence of climate change on tropical cyclones is *likely* to vary by region, but the specific characteristics of the changes are not yet well quantified and there is *low confidence* in region-specific projections of frequency and intensity. However, better process understanding and model agreement in specific regions provide *medium confidence* that precipitation will be more extreme near the centres of tropical cyclones making landfall in North and Central America; East Africa; West, East, South and Southeast Asia as well as in Australia and many Pacific islands. Improvements in model resolution and downscaling techniques increase confidence in projections of intense storms, and the frequency of the most intense storms will *more likely than not* increase substantially in some basins. {14.6, 14.8.3, 14.8.4, 14.8.7, 14.8.9, 14.8.10, 14.8.11, 14.8.12, 14.8.13, 14.8.14}

Despite systematic biases in simulating storm tracks, most models and studies are in agreement on the future changes in the number of extratropical cyclones (ETCs). The global number of ETCs is *unlikely* to decrease by more than a few percent. A small poleward shift is *likely* in the Southern Hemisphere (SH) storm track. It is *more likely than not*, based on projections with *medium confidence*, that the North Pacific storm track will shift poleward. However, it is *unlikely* that the response of the North Atlantic storm track is a simple poleward shift. There is *low confidence* in the magnitude of regional storm track changes, and the impact of such changes on regional surface climate. It is *very likely* that increases in Arctic, Northern European, North American and SH winter precipitation by the end of the 21st century (2081–2100) will result from more precipitation in ETCs associated with enhanced extremes of storm-related precipitation. {14.6, 14.8.2, 14.8.3, 14.8.5, 14.8.6, 14.8.13, 14.8.15}

Blocking

Increased ability in simulating blocking in models and higher agreement on projections indicate that there is *medium confidence* that the frequency of Northern and Southern Hemisphere blocking will not increase, while trends in blocking intensity and persistence remain uncertain. The implications for blocking-related regional changes in North America, Europe and Mediterranean and Central and North Asia are therefore also uncertain. {14.8.3, 14.8.6, 14.8.8, Box 14.2}

Annular and Dipolar Modes of Variability

Models are generally able to simulate gross features of annular and dipolar modes. Model agreement in projections indicates that future boreal wintertime North Atlantic Oscillation is *very likely* to exhibit large natural variations and trend of similar magnitude to that observed in the past and is *likely* to become slightly more positive on average, with some, but not well documented, implications for winter conditions in the Arctic, North America and Eurasia. The austral summer/autumn positive trend in Southern Annular Mode is *likely* to weaken considerably as stratospheric ozone recovers through the mid-21st century with some, but not well documented, implications for South America, Africa, Australia, New Zealand and Antarctica. {14.5.1, 14.5.2, 14.8.2, 14.8.3, 14.8.5, 14.8.6, 14.8.7, 14.8.8, 14.8.13, 14.8.15}

Atlantic Multi-decadal Oscillation

Multiple lines of evidence from paleo reconstructions and model simulations indicate that the Atlantic Multi-decadal Oscillation (AMO) is *unlikely* to change its behaviour in the future as the mean climate changes. However, natural fluctuations in the AMO over the coming few decades are *likely* to influence regional climates at least as strongly as will human-induced changes, with implications for Atlantic major hurricane frequency, the West African wet season, North American and European summer conditions. {14.7.6, 14.2.4, 14.6.1, 14.8.3, 14.8.6}

Pacific South American Pattern

Understanding of underlying physical mechanisms and the projected sea surface temperatures in the equatorial Indo-Pacific regions gives *medium confidence* that future changes in the mean atmospheric circulation for austral summer will project on this pattern, thereby influencing the South American Convergence Zone and precipitation over southeastern South America. {14.7.2, 14.8.5}

14.1 Introduction

Regional climates are the complex outcome of local physical processes and the non-local responses to large-scale phenomena such as the El Niño-Southern Oscillation (ENSO) and other dominant modes of climate variability. The dynamics of regional climates are determined by local weather systems that control the net transport of heat, moisture and momentum into a region. Regional climate is interpreted in the widest sense to mean the whole joint probability distribution of climate variables for a region including the time mean state, the variance and co-variance and the extremes.

This chapter assesses the physical basis of future regional climate change in the context of changes in the following types of phenomena: *monsoons and tropical convergence zones,* large-scale *modes of climate variability* and *tropical and extratropical cyclones*. Assessment of future changes in these phenomena is based on climate model projections (e.g., the Coupled Model Intercomparison Project Phase 3 (CMIP3) and CMIP5 multi-model ensembles described in Chapter 12) and an understanding of how well such models represent the key processes in these phenomena. More generic processes relevant to regional climate change, such as thermodynamic processes and land–atmosphere feedback processes, are assessed in Chapter 12. Local processes such as snow–albedo feedback, moisture feedbacks due to local vegetation, effects of steep complex terrain etc. can be important for changes but are in general beyond the scope of this chapter. The main focus here is on large-scale atmospheric phenomena rather than more local feedback processes or impacts such as floods and droughts.

Sections 14.1.1 to 14.1.3 introduce the three main classes of phenomena addressed in this Assessment and then Section 14.1.4 summarizes their main impacts on precipitation and surface temperature. Specific climate phenomena are then addressed in Sections 14.2 to 14.7, which build on key findings from the Fourth Assessment Report, AR4 (IPCC, 2007a), and provide an assessment of process understanding and how well models simulate the phenomenon and an assessment of future projections for the phenomena. In Section 14.8, future regional climate changes are assessed, and where possible, interpreted in terms of future changes in phenomena. In particular, the relevance of the various phenomena addressed in this chapter for future climate change in the regions covered in Annex I are emphasized. The regions are those defined in previous regional climate change assessments (IPCC, 2007a, 2007b, 2012). Regional Climate Models (RCMs) and other downscaling tools required for local impact assessments are assessed in Section 9.6 and results from these studies are used where such supporting information adds additional relevant details to the assessment.

14.1.1 Monsoons and Tropical Convergence Zones

The major monsoon systems are associated with the seasonal movement of convergence zones over land, leading to profound seasonal changes in local hydrological cycles. Section 14.2 assesses current understanding of monsoonal behaviour in the present and future climate, how monsoon characteristics are influenced by the large-scale tropical modes of variability and their potential changes and how the monsoons in turn affect regional extremes. Convergence zones over the tropical oceans not only play a fundamental role in determining regional climates but also influence the global atmospheric circulation. Section 14.3 presents an assessment of these and other important tropical phenomena.

14.1.2 Modes of Climate Variability

Regional climates are strongly influenced by modes of climate variability (see Box 14.1 for definitions of mode, regime and teleconnection). This chapter assesses major modes such as El Niño-Southern Oscillation (ENSO, Section 14.4), the North Atlantic Oscillation/Northern Annular Mode (NAO/NAM) and Southern Annular Mode (SAM) in the extratropics (Section 14.5) and various other well-known modes such as the Pacific North American (PNA) pattern, Pacific Decadal Oscillation (PDO), Atlantic Multi-decadal Oscillation (AMO), etc. (Section 14.7). Many of these modes are described in previous IPCC reports (e.g., Section 3.6 of AR4 WG1). Chapter 2 gives operational definitions of mode indices (Box 2.5, Table 1) and an assessment of observed historical behaviour (Section 2.7.8). Climate models are generally able to simulate the gross features of many of the modes of variability (Section 9.5), and so provide useful tools for understanding how modes might change in the future (Müller and Roeckner, 2008; Handorf and Dethloff, 2009).

Modes and regimes provide a simplified description of variations in the climate system. In the simplest paradigm, variations in climate variables are described by linear projection onto a set of mode indices (Baldwin et al., 2009; Baldwin and Thompson, 2009; Hurrell and Deser, 2009). For example, a large fraction of interannual variance in Northern Hemisphere (NH) sea level pressure is accounted for by linear combinations of the NAM and the PNA modes (Quadrelli and Wallace, 2004). Alternatively, the nonlinear regime paradigm considers the probability distribution of local climate variables to be a multi-modal mixture of distributions related to a discrete set of regimes/types (Palmer, 1999; Cassou and Terray, 2001; Monahan et al., 2001).

There is ongoing debate on the relevance of the different paradigms (Stephenson et al., 2004; Christiansen, 2005; Ambaum, 2008; Fereday et al., 2008), and care is required when interpreting these constructs (Monahan et al., 2009; Takahashi et al., 2011).

Modes of climate variability may respond to climate change in one or more of the following ways:

- *No change*—the modes will continue to behave as they have done in the recent past.

- *Index changes*—the probability distributions of the mode indices may change (e.g., shifts in the mean and/or variance, or more complex changes in shape such as changes in local probability density, e.g., frequency of regimes).

- *Spatial changes*—the climate patterns associated with the modes may change spatially (e.g., new flavours of ENSO; see Section 14.4 and Supplementary Material) or the local amplitudes of the climate patterns may change (e.g., enhanced precipitation for a given change in index (Bulic and Kucharski, 2012)).

- *Structural changes*—the types and number of modes and their mutual dependencies may change; completely new modes could in principle emerge.

An assessment of changes in modes of variability can be problematic for several reasons. First, interpretation depends on how one separates modes of variability from forced changes in the time mean or variations in the annual cycle (Pezzulli et al., 2005; Compo and Sardeshmukh, 2010). Modes of variability are generally defined using indices based on either detrended anomalies (Deser et al., 2010b) or anomalies obtained by removing the time mean over a historical reference period (see Box 2.5). The mode index in the latter approach will include changes in the mean, whereas by definition there is no trend in a mode index when it is based on detrended anomalies. Second, it can be difficult to separate natural variations from forced responses, for example, warming trends in the N. Atlantic during the 20th century that may be due to trends in aerosol and other forcings rather than natural internal variability (see Sections 14.6.2 and 14.7.1). Finally, modes of climate variability are nonlinearly related to one another (Hsieh et al., 2006) and this relationship can change in time (e.g., trends in correlation between ENSO and NAO indices).

Even when the change in a mode of variability index does not contribute greatly to mean regional climate change, a climate mode may still play an important role in regional climate variability and extremes. Natural variations, such as those due to modes of variability, are a major source of uncertainty in future projections of mean regional climate (Deser et al., 2012). Furthermore, changes in the extremes of regional climate are *likely* to be sensitive to small changes in variance or shape of the distribution of the mode indices or the mode spatial patterns (Coppola et al., 2005; Scaife et al., 2008).

14.1.3 Tropical and Extratropical Cyclones

Tropical and extratropical cyclones (TCs and ETCs) are important weather phenomena intimately linked to regional climate phenomena and modes of climate variability. Both types of cyclone can produce extreme wind speeds and precipitation (see Section 3.4, IPCC Special Report on Managing the Risks of Extreme Events and Disasters to Advance Climate Change Adaptation (SREX; IPCC, 2012)). Sections 14.6.1 and 14.6.2 assess the recent progress in scientific understanding of how these important weather systems are *likely* to change in the future.

14.1.4 Summary of Climate Phenomena and their Impact on Regional Climate

Box 14.1, Figure 1 illustrates the large-scale climate phenomena assessed in this chapter. Many of the climate phenomena are evident in the map of annual mean rainfall (central panel). The most abundant annual rainfall occurs in the tropical convergence zones: Inter-Tropical Convergence Zone (ITCZ) over the Pacific, Atlantic and African equatorial belt (see Section 14.3.1.1), South Pacific Convergence Zone (SPCZ) over central South Pacific (see Section 14.3.1.2) and South Atlantic Convergence Zone (SACZ) over Southern South America and Southern Atlantic (see Section 14.3.1.3). In the global monsoon domain (white contours on the map), large amounts of precipitation occur but only in certain seasons (see Section 14.2.1). Local maxima in precipitation are also apparent over the major storm track regions in mid-latitudes (see Section 14.7.2). Box 14.1 Figure 1 also illustrates surface air temperature (left panels) and precipitation (right panels) teleconnection patterns for ENSO (in December to February and June to August; see Section 14.4), NAO (in December to February; see Section 14.5.1) and SAM (in September to November; see Section 14.5.2). The teleconnection patterns were obtained by taking the correlation between monthly gridded temperature and precipitation anomalies and indices for the modes (see Box 14.1 definitions). It can be seen that all three modes have far-reaching effects on temperature and precipitation in many parts of the world. Box 14.1, Table 1 briefly summarizes the main regional impacts of different well-known modes of climate variability.

Box 14.1 | Conceptual Definitions and Impacts of Modes of Climate Variability

This box briefly defines key concepts used to interpret modes of variability (below) and summarizes regional impacts associated with well-known modes (Box 14.1, Table 1 and Box 14.1, Figure 1). The terms below are used to describe variations in time series variables reported at a set of geographically fixed spatial locations, for example, a set of observing stations or model grid points (based on the more complete statistical and dynamical interpretation in Stephenson et al. (2004)).

Climate indices
Time series constructed from climate variables that provides an aggregate summary of the state of the climate system. For example, the difference between sea level pressure in Iceland and the Azores provides a simple yet useful historical NAO index (see Section 14.5 and Box 2.5 for definitions of this and other well-known observational indices). Because of their maximum variance properties, climate indices are often defined using *principal components*.

Principal component
A linear combination of a set of time series variables that has maximum variance subject to certain normalization constraints. Principal components are widely used to define optimal *climate indices* from gridded datasets (e.g., the Arctic Oscillation (AO) index, defined as the leading principal component of NH sea level pressure; Section 14.5). *(continued on next page)*

Box 14.1 (continued)

Climate pattern
A set of coefficients obtained by 'projection' (regression) of climate variables at different spatial locations onto a *climate index* time series.

Empirical Orthogonal Function
The *climate pattern* obtained if the *climate index* is a *principal component*. It is an eigenvector of the covariance matrix of gridded climate data.

Teleconnection
A statistical association between climate variables at widely separated, geographically fixed spatial locations. Teleconnections are caused by large spatial structures such as basin-wide coupled modes of ocean–atmosphere variability, Rossby wave-trains, mid-latitude jets and storm-tracks, etc.

Teleconnection pattern
A correlation map obtained by calculating the correlation between variables at different spatial locations and a climate index. It is the special case of a *climate pattern* obtained for standardized variables and a standardized *climate index*, that is, the variables and index are each centred and scaled to have zero mean and unit variance. One-point teleconnection maps are made by choosing a variable at one of the locations to be the climate index. *(continued on next page)*

Box 14.1, Table 1 | Regional climate impacts of fundamental modes of variability.

Mode	Regional Climate Impacts
ENSO	Global impact on interannual variability in global mean temperature. Influences severe weather and tropical cyclone activity worldwide. The diverse El Niño flavours present different teleconnection patterns that induce large impacts in numerous regions from polar to tropical latitudes (Section 14.4).
PDO	Influences surface air temperature and precipitation over the entire North American continent and extratropical North Pacific. Modulates ENSO rainfall teleconnections, e.g., Australian climate (Section 14.7.3).
IPO	Modulates decadal variability in Australian rainfall, and ENSO teleconnections to rainfall, surface temperature, river flow and flood risk over Australia, New Zealand and the SPCZ (Section 14.7.3).
NAO	Influences the N. Atlantic jet stream, storm tracks and blocking and thereby affects winter climate in over the N. Atlantic and surrounding landmasses. The summer NAO (SNAO) influences Western Europe and Mediterranean basin climates in the season (Section 14.5.1).
NAM	Modulates the intensity of mid-latitude storms throughout the Northern Hemisphere and thereby influences North America and Eurasia climates as well as sea ice distribution across the Arctic sea (Section 14.5.1).
NPO	Influences winter air temperature and precipitation over much of western North America as well as Arctic sea ice in the Pacific sector (Section 14.5.1).
SAM	Influences temperature over Antarctica, Australia, Argentina, Tasmania and the south of New Zealand and precipitation over southern South America, New Zealand, Tasmania, Australia and South Africa (Section 14.5.2).
PNA	Influences the jet stream and storm tracks over the Pacific and North American sectors, exerting notable influences on the temperature and precipitation in these regions on intraseasonal and interannual time scales (Section 14.7.2).
PSA	Influences atmospheric circulation over South America and thereby has impacts on precipitation over the continent (Section 14.7.1).
AMO	Influences air temperatures and rainfall over much of the Northern Hemisphere, in particular, North America and Europe. It is associated with multidecadal variations in Indian, East Asian and West African monsoons, the North African Sahel and northeast Brazil rainfall, the frequency of North American droughts and Atlantic hurricanes (Section 14.7.6).
AMM	Influences seasonal hurricane activity in the tropical Atlantic on both decadal and interannual time scales. Its variability is influenced by other modes, particularly ENSO and NAO (Section 14.3.4).
AN	Affects the West African Monsoon, the oceanic forcing of Sahel rainfall on both decadal and interannual time-scales and the spatial extension of drought in South Africa (Section 14.3.4).
IOB	Associated with the intensity of Northwest Pacific monsoon, the tropical cyclone activity over the Northwest Pacific and anomalous rainfall over East Asia (Section 14.3.3).
IOD	Associated with droughts in Indonesia, reduced rainfall over Australia, intensified Indian summer monsoon, floods in East Africa, hot summers over Japan, and anomalous climate in the extratropical Southern Hemisphere (Section 14.3.3).
TBO	Modulates the strength of the Indian and West Pacific monsoons. Affects droughts and floods over large areas of south Asia and Australia (Section 14.7.4).
MJO	Modulates the intensity of monsoon systems around the globe and tropical cyclone activity in the Indian, Pacific and Atlantic Oceans. Associated with enhanced rainfall in Western North America, northeast Brazil, Southeast Africa and Indonesia during boreal winter and Central America/Mexico and Southeast Asia during boreal summer (Section 14.3.2).
QBO	Strongly affects the strength of the northern stratospheric polar vortex as well as the extratropical troposphere circulation, occurring preferentially in boreal winter (Section 14.7.5).
BLC	Associated with cold air outbreaks, heat-waves, floods and droughts in middle and high latitudes of both hemispheres (Box 14.2).

Notes:
AMM: Atlantic Meridional Mode
AMO: Atlantic Multi-decadal Oscillation
AN: Atlantic Niño pattern
BLC: Blocking events
ENSO: El Niño-Southern Oscillation
IOB: Indian Ocean Basin pattern
IOD: Indian Ocean Dipole pattern
IPO: Interdecadal Pacific Oscillation
MJO: Madden-Julian Oscillation
NAM: Northern Annular Mode
NAO: North Atlantic Oscillation
NPO: North Pacific Oscillation
PDO: Pacific Decadal Oscillation
PNA: Pacific North America pattern
PSA: Pacific South America pattern
QBO: Quasi-Biennial Oscillation
SAM: Southern Annular Mode
TBO: Tropospheric Biennial Oscillation

Box 14.1 (continued)

Mode of climate variability
Underlying space–time structure with preferred spatial pattern and temporal variation that helps account for the gross features in variance and for *teleconnections*. A mode of variability is often considered to be the product of a spatial *climate pattern* and an associated *climate index* time series.

Climate regime
A set of similar states of the climate system that occur more frequently than nearby states due to either more persistence or more often recurrence. In other words, a cluster in climate state space associated with a local maximum in the probability density function.

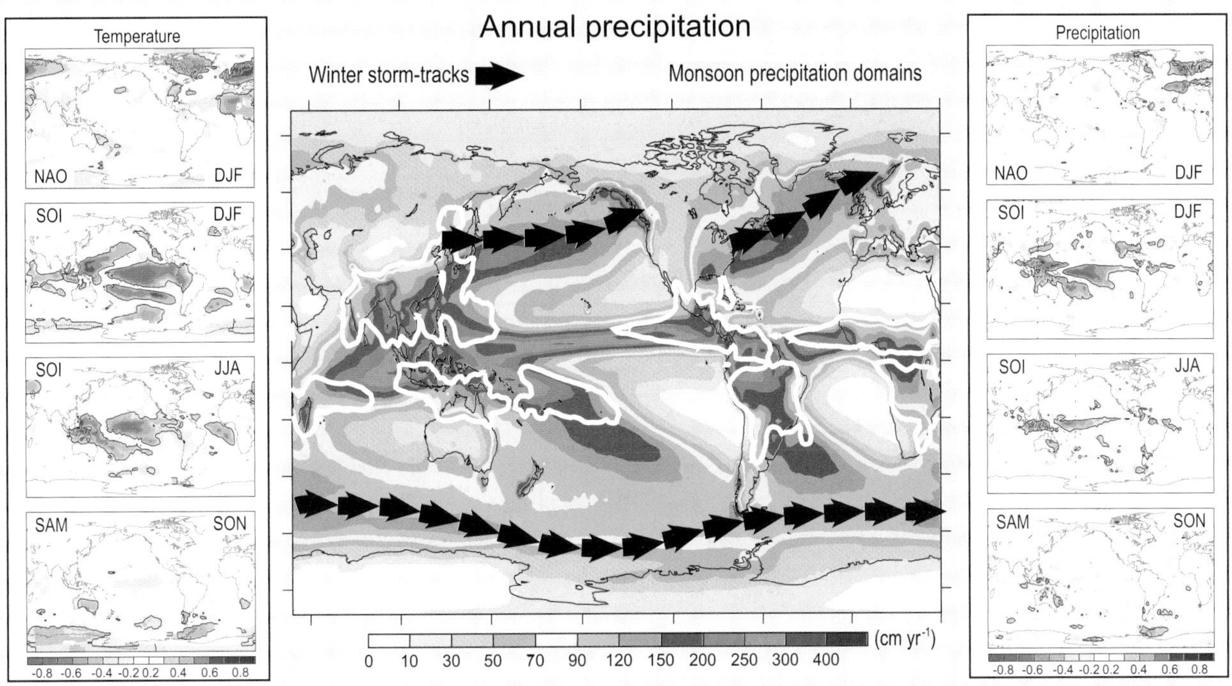

Box 14.1, Figure 1 | Global distribution of average annual rainfall (in cm/year) from 1979–2010 Global Precipitation Climatology Project (GPCP) database, monsoon precipitation domain (white contours) as defined in Section 14.2.1, and winter storm-tracks in both hemispheres (black arrows). In left (right) column seasonal correlation maps of North Atlantic Oscillation (NAO), Southern Oscillation Index (SOI, the atmospheric component of El Niño-Southern Oscillation (ENSO)) and Southern Annular Mode (SAM) mode indexes vs. monthly temperature (precipitation) anomalies in boreal winter (December, January and February (DJF)), austral winter (June, July and August (JJA)) and austral spring (September, October and November (SON)). Black contours indicate a 99% significance level. The mode indices were taken from National Oceanic and Atmospheric Administration (NOAA, http://www.esrl.noaa.gov/psd/data/climateindices/list/), global temperatures from NASA Goddard Institute of Space Studies Surface Temperature Analysis (GISTEMP, http://data.giss.nasa.gov/gistemp/) and global precipitations from GPCP (http://www.esrl.noaa.gov/psd/data/gridded/data.gpcp.html).

14.2 Monsoon Systems

Monsoons are a seasonal phenomenon responsible for producing the majority of wet season rainfall within the tropics. The precipitation characteristics over the Asian-Australian, American and African monsoons can be viewed as an integrated global monsoon system, associated with a global-scale atmospheric overturning circulation (Trenberth et al., 2000). In Section 14.2.1, changes in precipitation of the global monsoon system are assessed. Changes in regional monsoons are assessed in Sections 14.2.2 to 14.2.4.

14.2.1 Global Overview

The global *land* monsoon precipitation displays a decreasing trend over the last half-century, with primary contributions from the weakened summer monsoon systems in the NH (Wang and Ding, 2006). The combined global ocean–land monsoon precipitation has intensified during 1979–2008, mainly due to an upward trend in the NH summer oceanic monsoon precipitation (Zhou et al., 2008b; Hsu et al., 2011; Wang et al., 2012b). Because the fractional increase in monsoon area is greater than that in total precipitation, the ratio of the latter to the former (a measure of the global monsoon intensity) exhibits a decreasing trend (Hsu et al., 2011). CMIP5 models generally reproduce the observed global monsoon domain, but the disparity between the best and poorest models is large (Section 9.5.2.4).

In the CMIP5 models the global monsoon area (GMA), the global monsoon total precipitation (GMP) and the global monsoon precipitation intensity (GMI) are projected to increase by the end of the 21st century (2081–2100, Hsu et al., 2013; Kitoh et al., 2013; Figure 14.1). See Supplementary Material Section 14.SM.1.2 for the definitions of GMA, GMP and GMI. The CMIP5 model projections show an expansion of GMA mainly over the central to eastern tropical Pacific, the southern Indian Ocean and eastern Asia. In all RCP scenarios, GMA is *very likely* to increase, and GMI is *likely* to increase, resulting in a *very likely* increase in GMP, by the end of the 21st century (2081–2100). The 100-year median changes in GMP are +5%, +8%, +10%, and +16% in RCP2.6, RCP4.5, RCP6.0, and RCP8.5 scenarios, respectively. Indices of precipitation extremes such as simple daily precipitation intensity index (SDII), defined as the total precipitation divided by the number of days with precipitation greater than or equal to 1 mm, annual maximum 5-day precipitation total (R5d) and consecutive dry days (CDD) all indicate that intense precipitation will increase at larger rates than those of mean precipitation (Figure 14.1). The standard deviation of interannual variability in seasonal average precipitation (Psd) is projected to increase by many models but some models show a decrease in Psd. This is related to uncertainties in projections of future changes in tropical sea surface temperature (SST). Regarding seasonality, CMIP5 models project that monsoon onset dates will come earlier or not change much while monsoon retreat dates will delay, resulting in a lengthening of the monsoon season in many regions.

CMIP5 models show a decreasing trend of lower-troposphere wind convergence (dynamical factor) throughout the 20th and 21st centuries (Figure 14.2d). With increased moisture (see also Section 12.4), the moisture flux convergence shows an increasing trend from 1980 through the 21st century (Figure 14.2c). Surface evaporation shows a similar trend (Figure 14.2b) associated with warmer SSTs. Therefore, the global monsoon precipitation increases (Figure 14.2a) due to increases in moisture flux convergence and surface evaporation despite

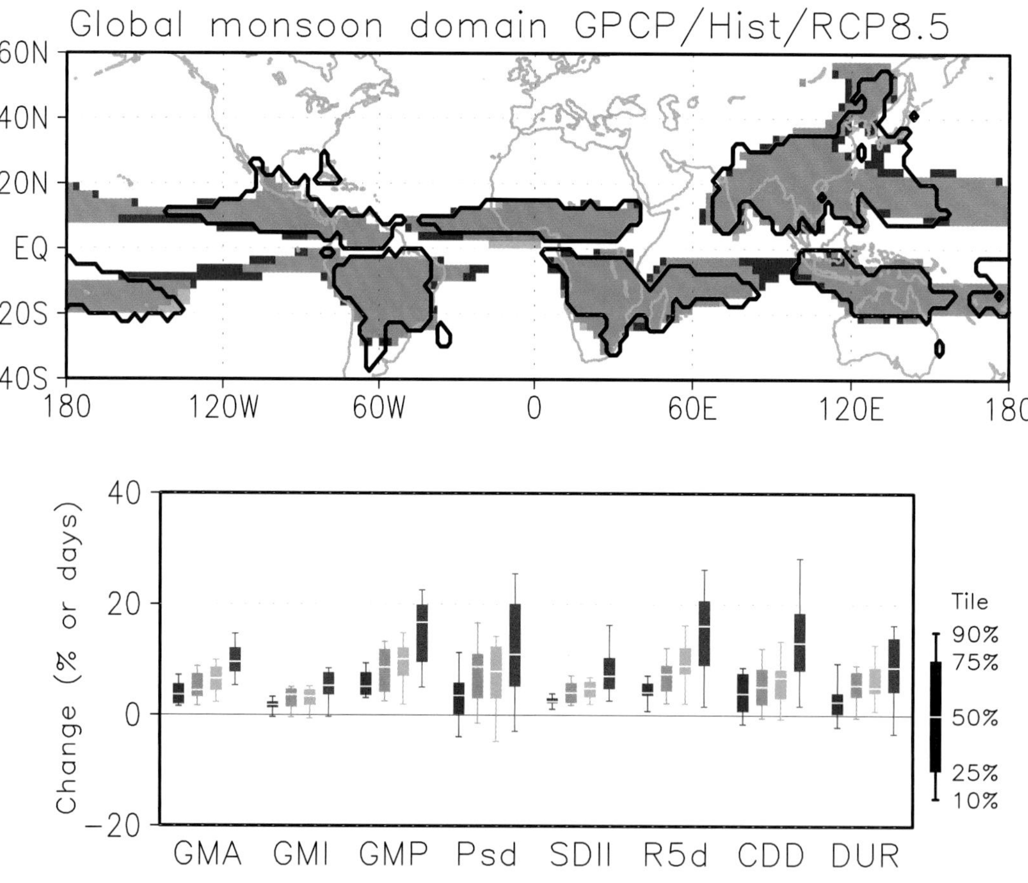

Figure 14.1 | (Upper) Observed (thick contour) and simulated (shading) global monsoon domain, based on the definition of Wang et al. (2011). The observations are based on GPCP v2.2 data (Huffman et al., 2009), and the simulations are based on 26 CMIP5 multi-model mean precipitation with a common 2.5 by 2.5 degree grid in the present day (1986–2005) and the future (2080–2099; RCP8.5 scenario). Orange (dark blue) shading shows monsoon domain only in the present day (future). Light blue shading shows monsoon domain in both periods. (Lower) Projected changes for the future (2080–2099) relative to the present day (1986-2005) in the global monsoon area (GMA) and global monsoon intensity (GMI), global monsoon total precipitation (GMP), standard deviation of interannual variability in seasonal average precipitation (Psd), simple daily precipitation intensity index (SDII), seasonal maximum 5-day precipitation total (R5d), seasonal maximum consecutive dry days (CDD) and monsoon season duration (DUR), under the RCP2.6 (dark blue; 18 models), RCP4.5 (light blue; 24 models), RCP6.0 (orange; 14 models) and RCP8.5 scenarios (red; 26 models). Units are % except for DUR (days). Box-and-whisker plots show the 10th, 25th, 50th, 75th and 90th percentiles. All of the indices are calculated for the summer season (May to September in the Northern Hemisphere; November to March in the Southern Hemisphere). The indices of Psd, SDII, R5d and CDD calculated for each model's original grid, and then averaged over the monsoon domains determined by each model at the present-day. The indices of DUR are calculated for seven regional monsoon domains based on the criteria proposed by Wang and LinHo (2002) using regionally averaged climatological cycles of precipitation, and then their changes are averaged with weighting based on their area at the present day.

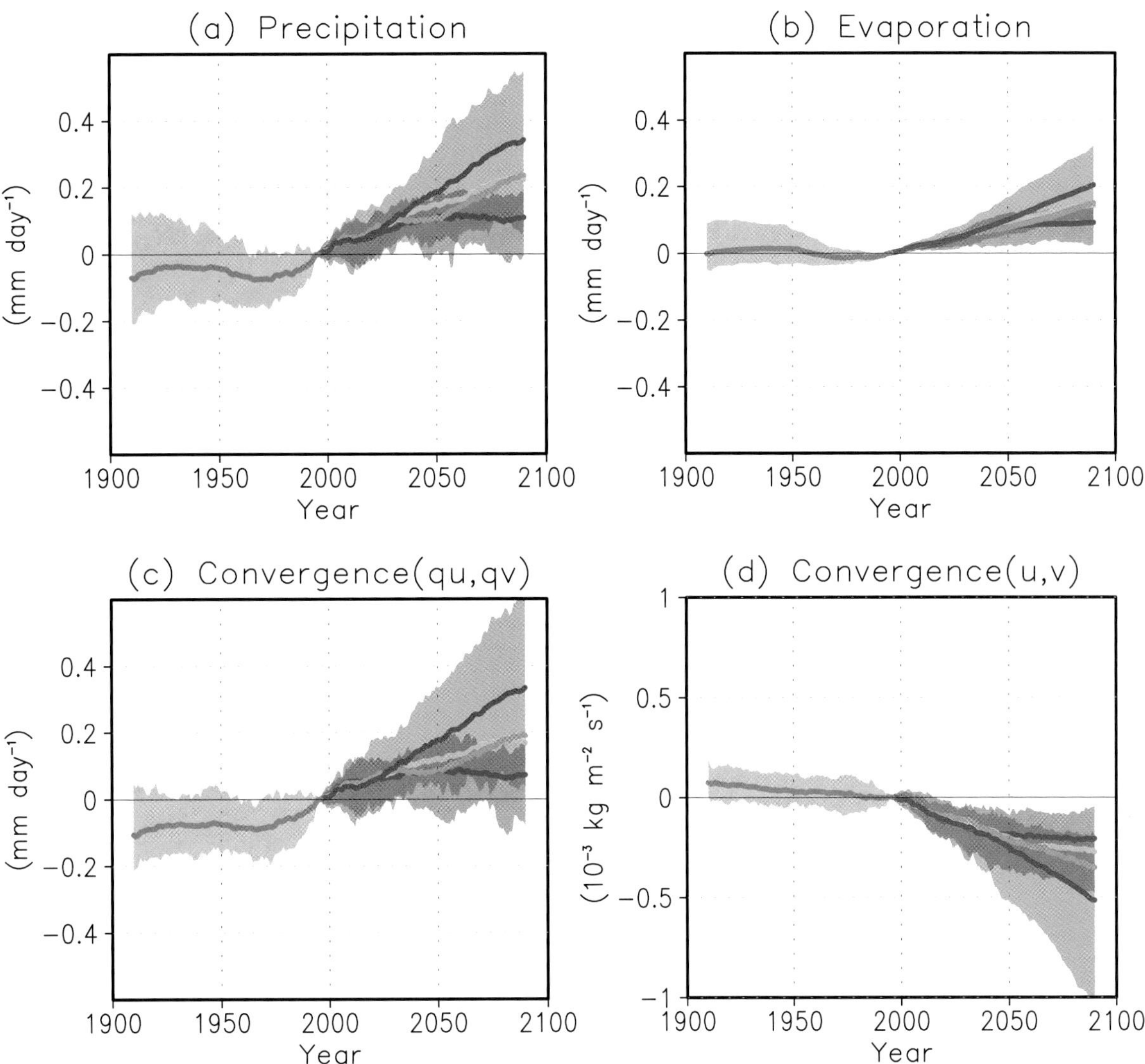

Figure 14.2 | Time series of simulated anomalies, smoothed with a 20-year running mean over the global land monsoon domain for (a) precipitation (mm day^{-1}), (b) evaporation (mm day^{-1}), (c) water vapour flux convergence in the lower (below 500 hPa) troposphere (mm day^{-1}), and (d) wind convergence in the lower troposphere (10^{-3} kg m^{-2} s^{-1}), relative to the present-day (1986–2005), based on CMIP5 multi-model monthly outputs. Historical (grey; 29 models), RCP2.6 (dark blue; 20 models), RCP4.5 (light blue; 24 models), RCP6.0 (orange; 16 models), and RCP8.5 (red; 24 models) simulations are shown in the 10th and 90th percentile (shading), and in all model averages (thick lines).

a weakened monsoon circulation. Besides greenhouse gases (GHGs), monsoons are affected by changes in aerosol loadings (Ramanathan et al., 2005). The aerosol direct forcing may heat the atmosphere but cools the surface, altering atmospheric stability and inducing horizontal pressure gradients that modulate the large-scale circulation and hence monsoon rainfall (Lau et al., 2008). However, the representation of aerosol forcing differs among models, and remains an important source of uncertainty (Chapter 7 and Section 12.2.2), particularly in some regional monsoon systems.

14.2.2 Asian-Australian Monsoon

The seasonal variation in the thermal contrast between the large Eurasian landmass and the Pacific-Indian Oceans drives the powerful Asian-Australian monsoon (AAM) system (Figure 14.3), which consists of five major subsystems: Indian (also known as South Asian), East Asian, Maritime Continent, Australian, and Western North Pacific monsoons. More than 85% of CMIP5 models show an increase in mean precipitation of the East Asian summer (EAS) monsoon, while more than 95% of models project an increase in heavy precipitation events (Figure 14.4). All models and all scenarios project an increase in both the mean and extreme precipitation in the Indian summer monsoon (referred to as SAS in Figures 14.3 and 14.4) . In these two regions, the interannual standard deviation of seasonal mean precipitation also increases. Over the Australian-Maritime Continent (AUSMC) monsoon region, agreement among models is low. Figure 14.5 shows the time-series of circulation indices representing EAS, Indian (IND), Western North Pacific (WNP) and Australian (AUS) summer monsoon systems. The Indian monsoon circulation index is *likely* to decrease in the 21st century, while a slight increase in the East Asian monsoon circulation is projected. Scatter among models is large for the western North Pacific and Australian monsoon circulation change.

Factors that limit the confidence in quantitative assessment of monsoon changes include sensitivity to model resolution (Cherchi and Navarra, 2007; Klingaman et al., 2011), model biases (Levine and

Frequently Asked Questions
FAQ 14.1 | How is Climate Change Affecting Monsoons?

Monsoons are the most important mode of seasonal climate variation in the tropics, and are responsible for a large fraction of the annual rainfall in many regions. Their strength and timing is related to atmospheric moisture content, land–sea temperature contrast, land cover and use, atmospheric aerosol loadings and other factors. Overall, monsoonal rainfall is projected to become more intense in future, and to affect larger areas, because atmospheric moisture content increases with temperature. However, the localized effects of climate change on regional monsoon strength and variability are complex and more uncertain.

Monsoon rains fall over all tropical continents: Asia, Australia, the Americas and Africa. The monsoon circulation is driven by the difference in temperature between land and sea, which varies seasonally with the distribution of solar heating. The duration and amount of rainfall depends on the moisture content of the air, and on the configuration and strength of the atmospheric circulation. The regional distribution of land and ocean also plays a role, as does topography. For example, the Tibetan Plateau—through variations in its snow cover and surface heating—modulates the strength of the complex Asian monsoon systems. Where moist on-shore winds rise over mountains, as they do in southwest India, monsoon rainfall is intensified. On the lee side of such mountains, it lessens.

Since the late 1970s, the East Asian summer monsoon has been weakening and not extending as far north as it used to in earlier times , as a result of changes in the atmospheric circulation. That in turn has led to increasing drought in northern China, but floods in the Yangtze River Valley farther south. In contrast, the Indo-Australian and Western Pacific monsoon systems show no coherent trends since the mid-20th century, but are strongly modulated by the El Niño-Southern Oscillation (ENSO). Similarly, changes observed in the South American monsoon system over the last few decades are strongly related to ENSO variability. Evidence of trends in the North American monsoon system is limited, but a tendency towards heavier rainfalls on the northern side of the main monsoon region has been observed. No systematic long-term trends have been observed in the behaviour of the Indian or the African monsoons.

The land surface warms more rapidly than the ocean surface, so that surface temperature contrast is increasing in most regions. The tropical atmospheric overturning circulation, however, slows down on average as the climate warms due to energy balance constraints in the tropical atmosphere. These changes in the atmospheric circulation lead to regional changes in monsoon intensity, area and timing. There are a number of other effects as to how

(continued on next page)

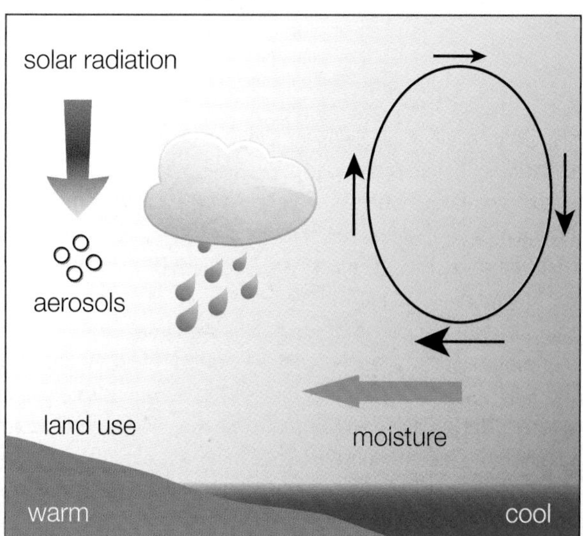

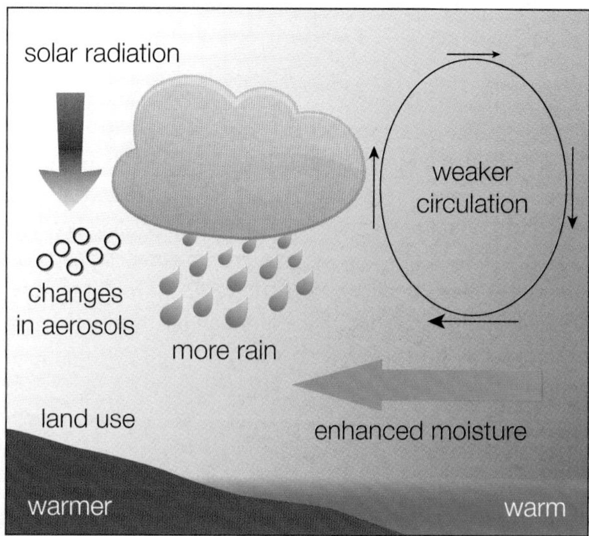

FAQ 14.1, Figure 1 | Schematic diagram illustrating the main ways that human activity influences monsoon rainfall. As the climate warms, increasing water vapour transport from the ocean into land increases because warmer air contains more water vapour. This also increases the potential for heavy rainfalls. Warming-related changes in large-scale circulation influence the strength and extent of the overall monsoon circulation. Land use change and atmospheric aerosol loading can also affect the amount of solar radiation that is absorbed in the atmosphere and land, potentially moderating the land–sea temperature difference.

FAQ 14.1 (continued)

climate change can influence monsoons. Surface heating varies with the intensity of solar radiation absorption, which is itself affected by any land use changes that alter the reflectivity (albedo) of the land surface. Also, changing atmospheric aerosol loadings, such as air pollution, affect how much solar radiation reaches the ground, which can change the monsoon circulation by altering summer solar heating of the land surface. Absorption of solar radiation by aerosols, on the other hand, warms the atmosphere, changing the atmospheric heating distribution.

The strongest effect of climate change on the monsoons is the increase in atmospheric moisture associated with warming of the atmosphere, resulting in an increase in total monsoon rainfall even if the strength of the monsoon circulation weakens or does not change.

Climate model projections through the 21st century show an increase in total monsoon rainfall, largely due to increasing atmospheric moisture content. The total surface area affected by the monsoons is projected to increase, along with the general poleward expansion of the tropical regions. Climate models project from 5% to an approximately 15% increase of global monsoon rainfall depending on scenarios. Though total tropical monsoon rainfall increases, some areas will receive less monsoon rainfall, due to weakening tropical wind circulations. Monsoon onset dates are *likely* to be early or not to change much and the monsoon retreat dates are *likely* to delay, resulting in lengthening of the monsoon season.

Future regional trends in monsoon intensity and timing remain uncertain in many parts of the world. Year-to-year variations in the monsoons in many tropical regions are affected by ENSO. How ENSO will change in future—and how its effects on monsoon will change—also remain uncertain. However, the projected overall increase in monsoon rainfall indicates a corresponding increase in the risk of extreme rain events in most regions.

Turner, 2012; Bollasina and Ming, 2013), poor skill in simulating the Madden–Julian Oscillation (MJO; Section 9.1.3.3) and uncertainties in projected ENSO changes (Collins et al., 2010; Section 14.4) and in the representation of aerosol effects (Section 9.4.6).

14.2.2.1 Indian Monsoon

The Indian summer monsoon is known to have undergone abrupt shifts in the past millennium, giving rise to prolonged and intense droughts (Meehl and Hu, 2006; Sinha et al., 2011; see also Chapter 2). The observed recent weakening tendency in seasonal rainfall and the regional re-distribution has been partially attributed to factors such as changes in black carbon and/or sulphate aerosols (Chung and Ramanathan, 2006; Lau et al., 2008; Bollasina et al., 2011), land use (Niyogi et al., 2010; see also Chapter 10) and SSTs (Annamalai et al., 2013). An increase in extreme rainfall events occurred at the expense of weaker rainfall events (Goswami et al., 2006) over the central Indian region, and in many other areas (Krishnamurthy et al., 2009). With a declining number of monsoon depressions (Krishnamurthy and Ajayamohan, 2010), the upward trend in extreme rainfall events may be due to enhanced moisture content (Goswami et al., 2006) or warmer SSTs in the tropical Indian Ocean (Rajeevan et al., 2008).

CMIP3 projections show suppressed rainfall over the equatorial Indian Ocean (Cai et al., 2011e; Turner and Annamalai, 2012), and an increase in seasonal mean rainfall over India (Ueda et al., 2006; Annamalai

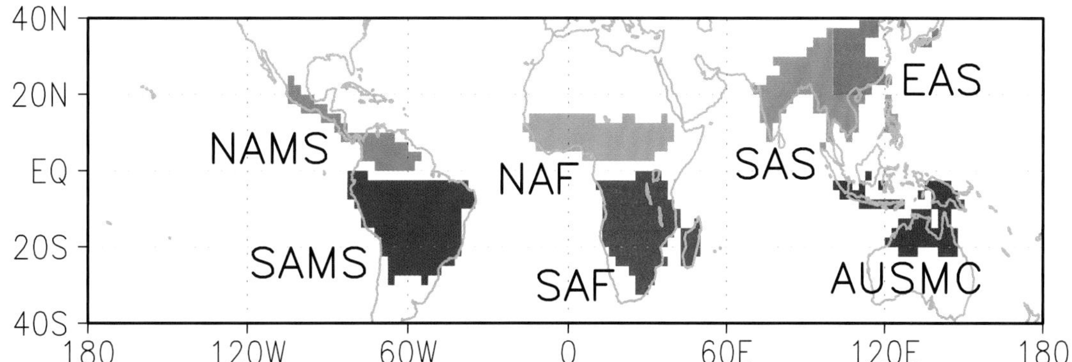

Figure 14.3 | Regional land monsoon domain based on 26 CMIP5 multi-model mean precipitation with a common 2.5° × 2.5° grid in the present-day (1986–2005). For regional divisions, the equator separates the northern monsoon domains (North America Monsoon System (NAMS), North Africa (NAF), Southern Asia (SAS) and East Asian summer (EAS)) from the southern monsoon domains (South America Monsoon System (SAMS), South Africa (SAF), and Australian-Maritime Continent (AUSMC)), 60°E separates NAF from SAS, and 20°N and 100°E separates SAS from EAS. All the regional domains are within 40°S to 40°N.

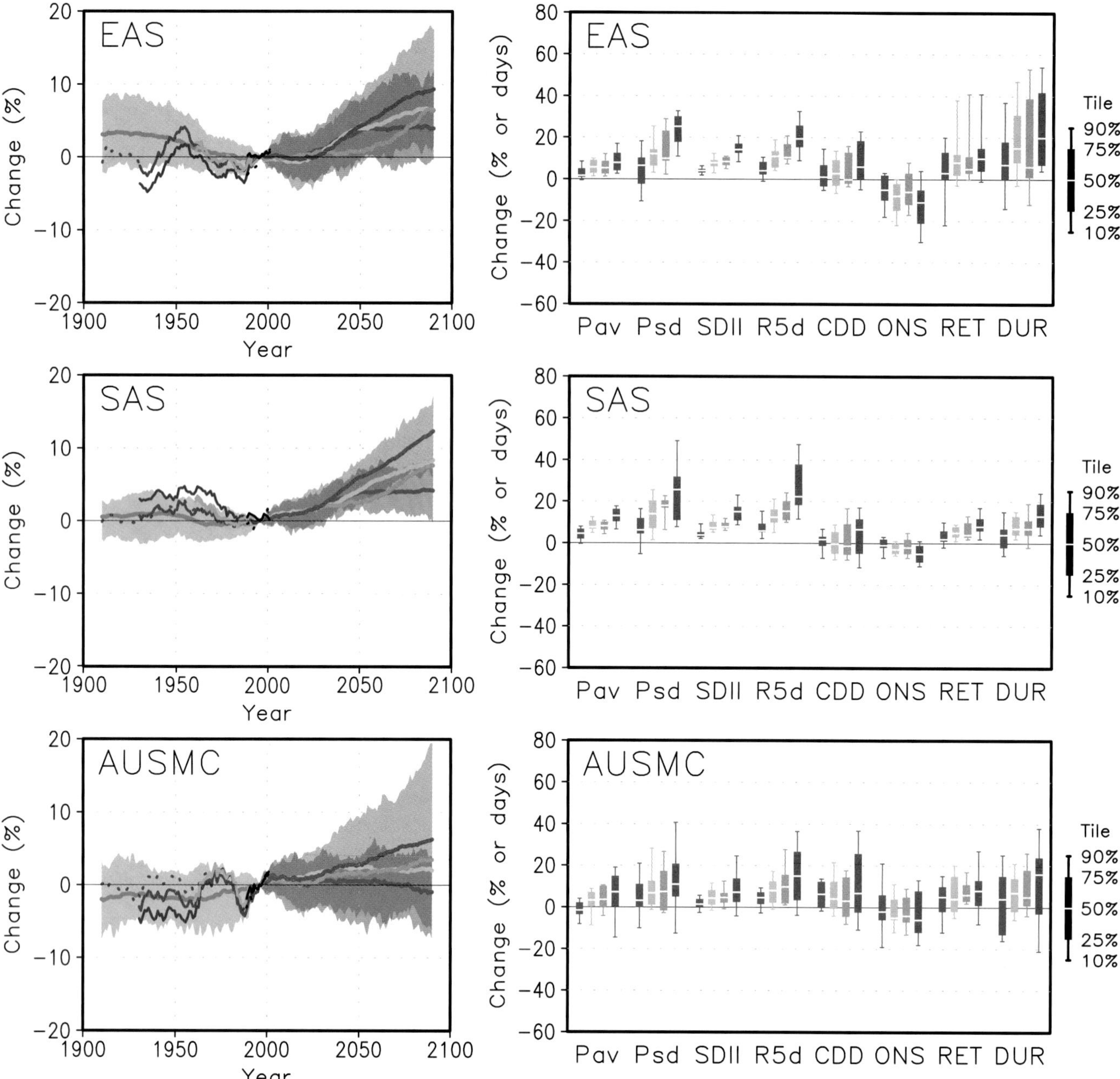

Figure 14.4 | Changes in precipitation indices over the regional land monsoon domains of (upper) East Asian summer (EAS), (middle) Southern Asia (SAS), and (lower) Australian-Maritime Continent (AUSMC) based on CMIP5 multi-models. (Left) Time series of observed and model-simulated summer precipitation anomalies (%) relative to the present-day average. All the time series are smoothed with a 20-year running mean. For the time series of simulations, all model averages are shown by thick lines for the historical (grey; 40 models), RCP2.6 (dark blue; 24 models), RCP4.5 (light blue; 34 models), RCP6.0 (orange; 20 models), and RCP8.5 scenarios (red; 32 models). Their intervals between 10th and 90th percentiles are shown by shading for RCP2.6 and RCP8.5 scenarios. For the time series of observations, Climate Research Unit (CRU) TS3.2 (update from Mitchell and Jones, 2005; dark blue), Global Precipitation Climatology Centre (GPCC) v6 (Becker et al., 2013; deep green), GPCC Variability Analysis of Surface Climate Observations (VASClimO; Beck et al., 2005; light green), Highly Resolved Observational Data Integration Towards the Evaluation of Water Resources (APHRODITE) v1101 (Yatagai et al., 2012; only for EAS and SAS regions; light blue), Global Precipitation Climatology Project (GPCP) v2.2 (updated from Huffman et al., 2009; black), and Climate Prediction Center (NOAA) Merged Analysis of Precipitation (CMAP) v1201 (updated from Xie and Arkin, 1997; black with dots) are shown. GPCC v6 with dot line, GPCC VASClimO, GPCP v2.2 and CMAP v1201 are calculated using all grids for the period of 1901–2010, 1951–2000, 1979–2010, 1979–2010, respectively. CRU TS3.2, GPCC v6 with solid line, and APHRODITE v1101, are calculated using only grid boxes (2.5° in longitude/latitude) where at least one observation site exists for more than 80% of the period of 1921–2005, 1921–2005, and 1951–2005, respectively. (Right) Projected changes for the future (2080-2099) relative to the present-day average in averaged precipitation (Pav), standard deviation of interannual variability in seasonal average precipitation (Psd), simple precipitation daily intensity index (SDII), seasonal maximum 5-day precipitation total (R5d), seasonal maximum consecutive dry days (CDD), monsoon onset date (ONS), retreat date (RET), and duration (DUR), under the RCP2.6 (18 models), RCP4.5 (24 models), RCP6.0 (14 models) and RCP8.5 scenarios (26 models). Units are % in Pav, Psd, SDII, R5d, and CDD; days in ONS, RET, and DUR. Box-whisker plots show the 10th, 25th, 50th, 75th and 90th percentiles. All of the indices are calculated for the summer season (May to September in the Northern Hemisphere; November to March in the Southern Hemisphere). The indices of Pav, Psd, SDII, R5d and CDD are calculated for each model's original grid, and then averaged over the monsoon domains determined by each model at the present day. The indices of ONS, RET and DUR are calculated based on the criteria proposed by Wang and LinHo (2002) using regionally averaged climatological cycles of precipitation.

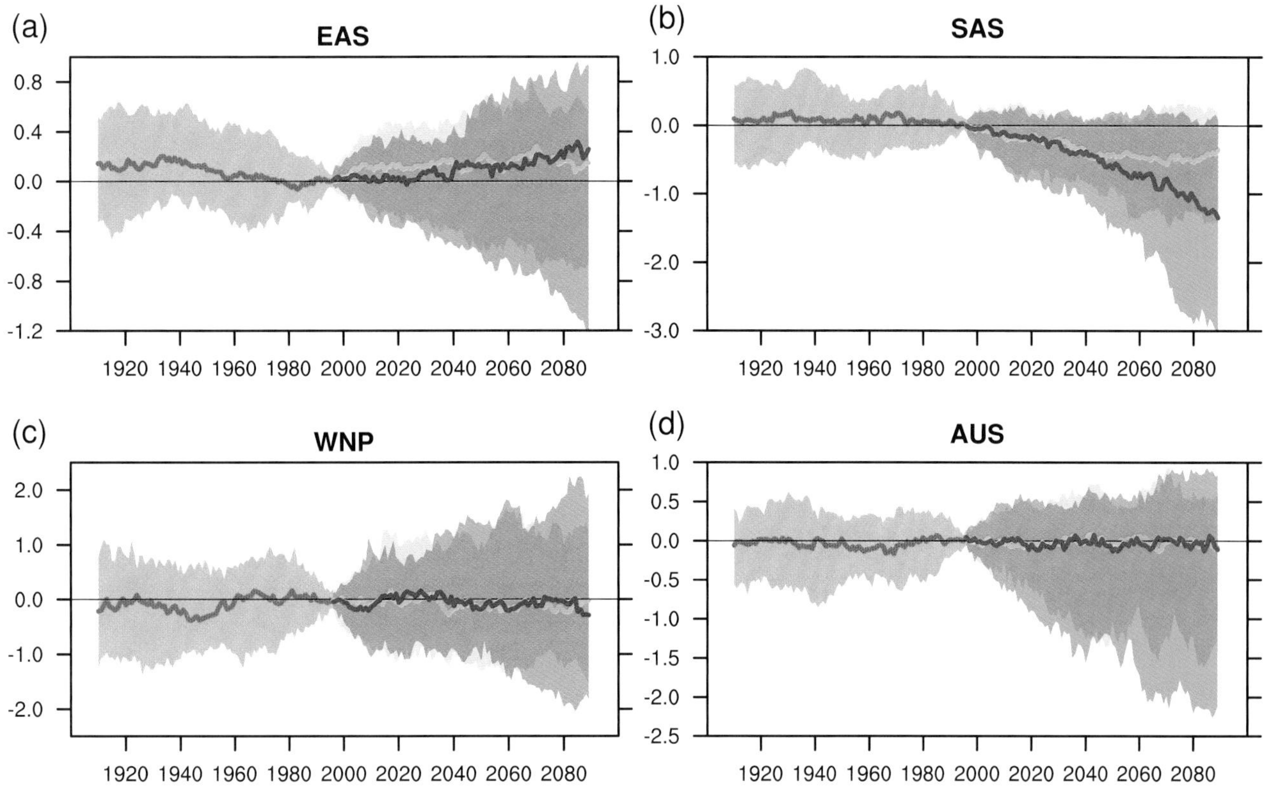

Figure 14.5 | Time series of summer monsoon indices (21-year running mean) relative to the base period average (1986–2005). Historical (gray), RCP4.5 (light blue) and RCP8.5 (red) simulations by 39 CMIP5 model ensembles are shown in 10th and 90th (shading), and 50th (thick line) percentiles. (a) East Asian summer monsoon (defined as June, July and August (JJA) sea level pressure difference between 160°E and 110°E from 10°N to 50°N), (b) Indian summer monsoon (defined as meridional differences of the JJA 850 hPa zonal winds averaged over 5°N to 15°N, 40°E to 80°E and 20°N to 30°N, 60°E to 90°E), (c) western North Pacific summer monsoon (defined as meridional differences of the JJA 850 hPa zonal winds averaged over 5°N to 15°N, 100°E to 130°E and 20°N to 30°N, 110°E to 140°E), (d) Australian summer monsoon (defined as December, January and February (DJF) 850 hPa zonal wind anomalies averaged over 10°S to 0°, 120°E to 150°E). (See Wang et al. (2004) and Zhou et al. (2009c) for indices definitions.)

et al., 2007; Turner et al., 2007a; Kumar et al., 2011b; Sabade et al., 2011). These results are generally confirmed by CMIP5 projections (Chaturvedi et al., 2012). The projected changes in Indian monsoon rainfall increase with the anthropogenic forcing among RCPs (May, 2011; see Figure 14.4; SAS).

In a suite of models that realistically simulate ENSO–monsoon relationships, normal monsoon years are *likely* to become less frequent in the future, but there is no clear consensus about the occurrence of extreme monsoon years (Turner and Annamalai, 2012). CMIP3 models indicate ENSO–monsoon relationships to persist in the future (Kumar et al., 2011b), but there is *low confidence* in the projection of ENSO variability (Section 14.4). Sub-seasonal scale monsoon variability is linked to the MJO but again the *confidence* in the future projection of MJO remains *low* (Section 14.3.2).

CMIP5 models project an increase in mean precipitation as well as its interannual variability and extremes (Figure 14.4; SAS). All models project an increase in heavy precipitation events but disagree on CDD changes. Regarding seasonality, model agreement is high on an earlier onset and later retreat, and hence longer duration. The monsoon circulation weakens in the future (Figure 14.5; IND) but the precipitation increases. Like the global monsoon (Section 14.2.1), the precipitation increase is largely due to the increased moisture flux from ocean to land.

14.2.2.2 East Asian Monsoon

The East Asian monsoon is characterized by a wet season and southerly flow in summer and by dry cold northerly flow in winter. The East Asian summer (EAS) monsoon circulation has experienced an inter-decadal weakening from the 1960s to the 1980s (Hori et al., 2007; Li et al., 2010a), associated with deficient rainfall in North China and excessive rainfall in central East China along 30°N (Hu, 1997; Wang, 2001; Gong and Ho, 2002; Yu et al., 2004). The summer monsoon circulation has begun to recover in recent decades (Liu et al., 2012a; Zhu et al., 2012). The summer rainfall amount over East Asia shows no clear trend during the 20th century (Zhang and Zhou, 2011), although significant trends may be found in local station records (Wang et al., 2006). The winter monsoon circulation weakened significantly after the 1980s (Wang et al., 2009a; Wang and Chen, 2010). See Supplementary Material Sections 14.SM.1.3 to 14.SM.1.7 for additional discussions of natural variability.

CMIP3 models show reasonable skill in simulating large-scale circulation of the EAS monsoon (Boo et al., 2011), but their performance is poor in reproducing the monsoon rainband (Lin et al., 2008a; Li and Zhou, 2011). Only a few CMIP3 models reproduce the Baiu rainband (Ninomiya, 2012) and high-resolution models (Kitoh and Kusunoki, 2008) show better performance than low resolution CMIP3 type models in simulating the monsoon rainband (Kitoh and Kusunoki,

2008). CMIP3 models show large uncertainties in projections of monsoon precipitation and circulation (Ding et al., 2007; Kripalani et al., 2007a) but the simulation of interannual variability of the EAS monsoon circulation has improved from CMIP3 to CMIP5 (Sperber et al., 2012). Climate change may bring a change in the position of the monsoon rain band (Li et al., 2010a).

CMIP5 projections indicate a *likely* increase in both the circulation (Figure 14.5) and rainfall of the EAS monsoon (Figure 14.4) throughout the 21st century. This is different from other Asian-Australian monsoon subsystems, where the increase in precipitation (Figure 14.4) is generally associated with weakening monsoon circulation (Figure 14.5). Interannual variability of seasonal mean rainfall is *very likely* to increase except for RCP2.6 (Figure 14.4). Heavy precipitation events (SDII and R5d) are also *very likely* to increase. CMIP5 models project an earlier monsoon onset and longer duration but the spread among models is large (Figure 14.4).

14.2.2.3 Maritime Continent Monsoon

Interaction between land and water characterizes the Maritime Continent region located between the Asian continent and Australia. It provides a land bridge along which maximum convection marches from the Asian summer monsoon regime (generally peaking in June, July and August) to the Australian summer monsoon system (generally peaking in December, January and February).

Phenomena such as the MJO (Tangang et al., 2008; Section 14.3.2; Hidayat and Kizu, 2010; Salahuddin and Curtis, 2011), and ENSO (Aldrian and Djamil, 2008; Moron et al., 2010; Section 14.4) influence Maritime Continent Monsoon variability. Rainfall extremes in the Maritime Continent are strongly influenced by diurnal rainfall variability (Qian, 2008; Qian et al., 2010a; Robertson et al., 2011; Ward et al., 2011) as well as the MJO. There have been no obvious trends in extreme rainfall indices in Indonesia, except evidence of a decrease in some areas in annual rainfall and an increase in the ratio of the wet to dry season rainfall (Aldrian and Djamil, 2008).

Modelling the Maritime Continent monsoon is a challenge because of the coarse resolution of contemporary large-scale coupled climate models (Aldrian and Djamil, 2008; Qian, 2008). Most CMIP3 models tend to simulate increasing precipitation in the tropical central Pacific but declining trends over the Maritime Continent for June to August (Ose and Arakawa, 2011), consistent with a decreasing zonal SST gradient across the equatorial Pacific and a weakening Walker Circulation (Collins et al., 2010). Projections of CMIP5 models are consistent with those of CMIP3 models, with decreasing precipitation during boreal summer and increasing precipitation during boreal winter, but model agreement is not high (Figures AI.66-67; Figure 12.22, but see also Figure 14.27).

14.2.2.4 Australian Monsoon

Some indices of the Australian summer monsoon (Wang et al., 2004; Li et al., 2012a) show a clear post-1980 reduction, but another index by Kajikawa et al. (2010) does not fully exhibit this change. Over northwest Australia, summer rainfall has increased by more than 50% (Rotstayn et al., 2007; Shi et al., 2008b; Smith et al., 2008), whereas over northeast Australia, summer rainfall has decreased markedly since around 1980 (Li et al., 2012a).

Models in general show skill in representing the gross spatial characteristics of Australian monsoon summer precipitation (Moise et al., 2005). Further, atmospheric General Circulation Models (GCMs) forced by SST anomalies can skilfully reproduce monsoon-related zonal wind variability over recent decades (Zhou et al., 2009a). Recent analysis of the skill of a suite of CMIP3 models showed a good representation in the ensemble mean, but a very large range of biases across individual models (more than a factor of 6; Colman et al., 2011). Most CMIP models have biases in monsoon seasonality, but CMIP5 models generally perform better than CMIP3 (Jourdain et al., 2013).

In climate change projections, overall changes in tropical Australian rainfall are small, with substantial uncertainties (Figure 14.4; Moise et al., 2012; see also Figure 14.27). Using a group of CMIP5 models that exhibit a realistic present-day climatology, most projections using the RCP8.5 scenario produced 5% to 20% more monsoon rainfall by the late 21st century compared to the pre-industrial period (Jourdain et al., 2013). Most CMIP3 model projections suggest delayed monsoon onset and reduced monsoon duration over northern Australia. Weaker model agreement is seen over the interior of the Australian continent, where ensembles show an approximate 7-day delay of both the onset and retreat with little change in duration (Zhang et al., 2013a). CMIP5 model agreement in changes of monsoon precipitation seasonality is low (Figure 14.4).

14.2.2.5 Western North Pacific Monsoon

The western North Pacific summer monsoon (WNPSM) occupies a broad oceanic region of the South China and Philippine Seas, featuring a monsoon trough and a subtropical anticyclonic ridge to the north (Zhang and Wang, 2008).

The western North Pacific monsoon does not show any trend during 1950–1999. Since the late 1970s, the correlation has strengthened between interannual variability in the western North Pacific monsoon and ENSO (Section 14.4), a change mediated by Indian Ocean SST (Huang et al., 2010; Xie et al., 2010a). This occurred despite a weakening of the Indian monsoon–ENSO correlation in this period (Wang et al., 2008a).

CMIP5 models project little change in western North Pacific monsoon circulation (Figure 14.5) but enhanced precipitation (Figures AI.66-67; Figure 12.22; but see also Figure 14.24) due to increased moisture convergence (Chapter 12).

14.2.3 American Monsoons

The American monsoons, the North America Monsoon System (NAMS) and the South America Monsoon System (SAMS), are associated with large inter-seasonal differences in precipitation, humidity and atmospheric circulation (Vera et al., 2006; Marengo et al., 2010a). NAMS and SAMS indices are often, though not always, defined in terms of precipitation characteristics (Wang and LinHo, 2002).

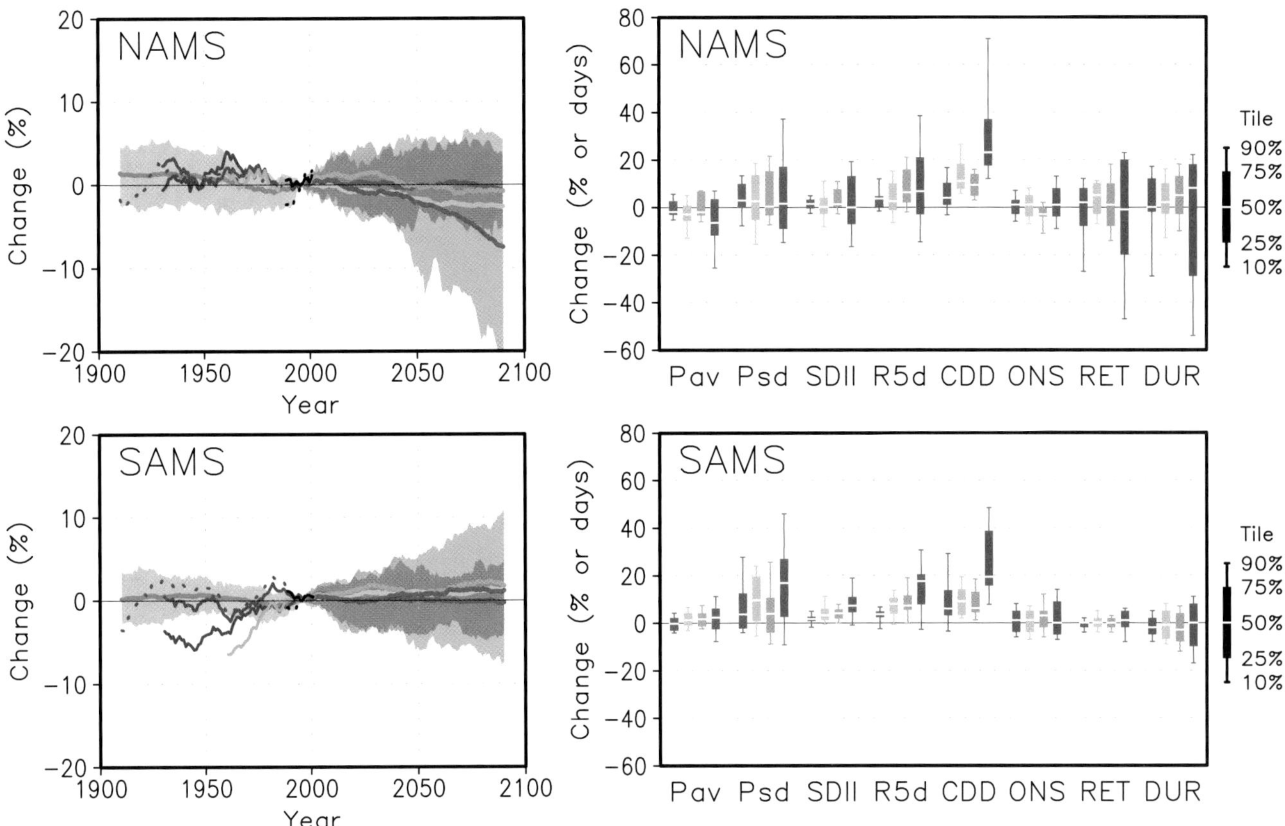

Figure 14.6 | As in Figure 14.4, except for (upper) North America Monsoon System (NAMS) and (lower) South America Monsoon System (SAMS).

14.2.3.1 North America Monsoon System

The warm season precipitation in northern Mexico and the southwestern USA is strongly influenced by the NAMS. It has been difficult to simulate many important NAMS-related phenomenon in global climate models (Castro et al., 2007; Lin et al., 2008b; Cerezo-Mota et al., 2011), though the models capture gross-scale features associated with the NAMS seasonal cycle (Liang et al., 2008b; Gutzler, 2009). See Supplementary Material Section 14.SM.1.8 for a more detailed discussion of NAMS dynamics.

In the NAMS core region, no distinct precipitation trends have been seen over the last half of the 20th century (Anderson et al., 2010; Arriaga-Ramirez and Cavazos, 2010), due to countervailing trends in increasing intensity and decreasing frequency of events, as well as the decreasing length of the monsoon season itself (Englehart and Douglas, 2006). However, monsoonal stream flow in western Mexico has been decreasing, possibly as a result of changing precipitation characteristics or antecedent hydrological conditions rather than overall precipitation amounts (Gochis et al., 2007). There has also been a systematic delay in monsoon onset, peak and termination (Grantz et al., 2007) as well as an increase in extreme precipitation events associated with land-falling hurricanes (Cavazos et al., 2008). Finally, positive trends in NAMS precipitation amounts have been detected in the northern fringes of the core area, that is, Arizona and western New Mexico (Anderson et al., 2010), consistent with northward NAMS expansion during relatively warm periods in the Holocene (Petersen, 1994; Mock and Brunelle-Daines, 1999; Harrison et al., 2003; Poore et al., 2005; Metcalfe et al., 2010).

Over the coming century, CMIP5 simulations generally project a precipitation reduction in the core zone of the monsoon (Figures AI.27 and Figure 14.6), but this signal is not particularly consistent across models, even under the RCP8.5 scenario (Cook and Seager, 2013). Thus *confidence* in projections of monsoon precipitation changes is currently *low*. CMIP5 models have no consensus on future changes of monsoon timing (Figure 14.6). Temperature increases are consistently projected in all models (Annex I). This will *likely* increase the frequency of extreme summer temperatures (Diffenbaugh and Ashfaq, 2010; Anderson, 2011; Duffy and Tebaldi, 2012), together with projected increase in consecutive dry days (Figure 14.6).

14.2.3.2 South America Monsoon System

The SAMS mainly influences precipitation in the South American tropics and subtropics (Figure 14.1). The main characteristics of SAMS onset are increased humidity flux from the Atlantic Ocean over northern South America, an eastward shift of the subtropical high, strong northwesterly moisture flux east of the tropical Andes, and establishment of the Bolivian High (Raia and Cavalcanti, 2008; Marengo et al., 2010a; Silva and Kousky, 2012). Recent SAMS indices have been calculated based on different variables, such as a large scale index (Silva and Carvalho, 2007), moisture flux (Raia and Cavalcanti, 2008), and wind (Gan et al., 2006), in addition to precipitation (Nieto-Ferreira and

Rickenbach, 2010; Seth et al., 2010; Kitoh et al., 2013). As seen below, conclusions regarding SAMS changes can depend on the index chosen.

SAMS duration and amplitude obtained from the observed large-scale index have both increased in the last 32 years (Jones and Carvalho, 2013). Increase of extreme precipitation and consecutive dry days have been observed in the SAMS region from 1969 to 2009 (Skansi et al., 2013). The overall annual cycle of precipitation in the SAMS region, including SAMS onset and demise, is generally well represented by models (Bombardi and Carvalho, 2009; Seth et al., 2010; Kitoh et al., 2013). Extreme precipitation indices in SAMS region are also well simulated by CMIP5 models (Kitoh et al., 2013). CMIP5 models subjected to historical forcing show increases in SAMS amplitude, earlier onset and later demise during the 1951–2005 period (Jones and Carvalho, 2013). Using a precipitation based index, precipitation increases in austral summer but decreases in austral spring, indicating delayed SAMS onset in the CMIP3 projections (Seth et al., 2011). CMIP5 projections based on the global precipitation index (Section 14.2.1) consistently show small precipitation increases and little change in onset and retreat (Kitoh et al., 2013; Figure 14.6). On the other hand, when using a different index, earlier onsets and later demises and thus, longer duration of the SAMS by the end of the 21st century (2081–2100) has been found (Jones and Carvalho, 2013). Thus there is *medium confidence* that SAMS overall precipitation will remain unchanged. The different estimates of changes in timing underscores potential uncertainties related to SAMS timing due to differences in SAMS indices. The models do show significant and robust increases in extreme precipitation indices in the SAMS region, such as seasonal maximum 5-day precipitation total and number of consecutive dry days (Figure 14.6), leading to *medium confidence* in projections of these characteristics.

14.2.4 African Monsoon

In Africa, monsoon circulation affects precipitation in West Africa where notable upper air flow reversals are observed. East and south African precipitation is generally described by variations in the tropical convergence zone rather than as a monsoon feature. This section covers the West African monsoon, and Section 14.8.7 also covers the latter two regions.

The West African monsoon develops during northern spring and summer, with a rapid northward jump of the rainfall belt from along the Gulf of Guinea at 5°N in May to June to the Sahel at 10°N in July to August. Factors influencing the West African monsoon include interannual to decadal variations, land processes and the direct response to radiative forcing. Cross-equatorial tropical Atlantic SST patterns influence the monsoon flow and moistening of the boundary layer, so that a colder northern tropical Atlantic induces negative rainfall anomalies (Biasutti et al., 2008; Giannini et al., 2008; Xue et al., 2010; Rowell, 2011).

In CMIP3 simulations, rainfall is projected to decrease in the early part but increase towards the end of the rainy season, implying a small delay in the monsoon season and an intensification of late-season rains (Biasutti and Sobel, 2009; Biasutti et al., 2009; Seth et al., 2010). CMIP5 models, on the other hand, simulate the variability of tropical Atlantic SST patterns with little credibility (Section 9.4.2.5.2) and model resolution is known to limit the ability to capture the mesoscale 'squall line' systems that form a central element in the maintenance of the rainy season (Ruti and Dell'Aquila, 2010; see also Section 14.8.7). Therefore, projections of the West African monsoon rainfall appear to be uncertain, reflected by considerable model deficiencies and spread in the projections (Figure 14.7). Note that this figure is based on a somewhat eastward extended area for the West African monsoon (NAF in Figure 14.3), seen as a component of the global monsoon system (Section 14.2.1). The limitations of model simulations in the region arising from the lack of convective organization (Kohler et al., 2010) leading to the underestimation of interannual variability (Scaife et al., 2009) imply that *confidence* in projections of the African monsoon is *low*.

The limited information that could be deduced from CMIP3 has not improved much in CMIP5. Figure 14.7 largely confirms the findings based on CMIP3. The CMIP5 model ensemble projects a modest change in the onset date (depending on the scenario) and a small delay in the retreat date, leading to a small increase in the duration of the rainy season. The delay in the monsoon retreat is larger in the high-end emission scenarios. The interannual variance and the 5-day rain intensity show a robust increase, while a small increase in dry day periods is less significant.

14.2.5 Assessment Summary

It is projected that global monsoon precipitation will *likely* strengthen in the 21st century with increase in its area and intensity while the monsoon circulation weakens. Precipitation extremes including precipitation intensity and consecutive dry days are *likely* to increase at higher rates than those of mean precipitation. Overall, CMIP5 models project that the monsoon onset will be earlier or not change much and the monsoon retreat dates will delay, resulting in a lengthening of the monsoon season. Such features are *likely* to occur in most of Asian-Australian Monsoon regions.

There is *medium confidence* that overall precipitation associated with the Asian-Australian monsoon will increase but with a north-south asymmetry: Indian and East Asian monsoon precipitation is projected to increase, while projected changes in Australian summer monsoon precipitation are small. There is *medium confidence* that the Indian summer monsoon circulation will weaken, but this is compensated by increased atmospheric moisture content, leading to more precipitation. For the East Asian summer monsoon, both monsoon circulation and precipitation are projected to increase. There is *low confidence* that over the Maritime Continent boreal summer rainfall will decrease and boreal winter rainfall will increase. There is *low confidence* that changes in the tropical Australian monsoon rainfall are small. There is *low confidence* that Western North Pacific summer monsoon circulation changes are small, but with increased rainfall due to enhanced moisture. There is *medium confidence* in an increase of Indian summer monsoon rainfall and its extremes throughout the 21st century under all RCP scenarios. Their percentage change ratios are the largest and model agreement is highest among all monsoon regions.

There is *low confidence* in projections of American monsoon precipitation changes but there is *high confidence* in increases of precipitation extremes, of wet days and consecutive dry days. There is *medium*

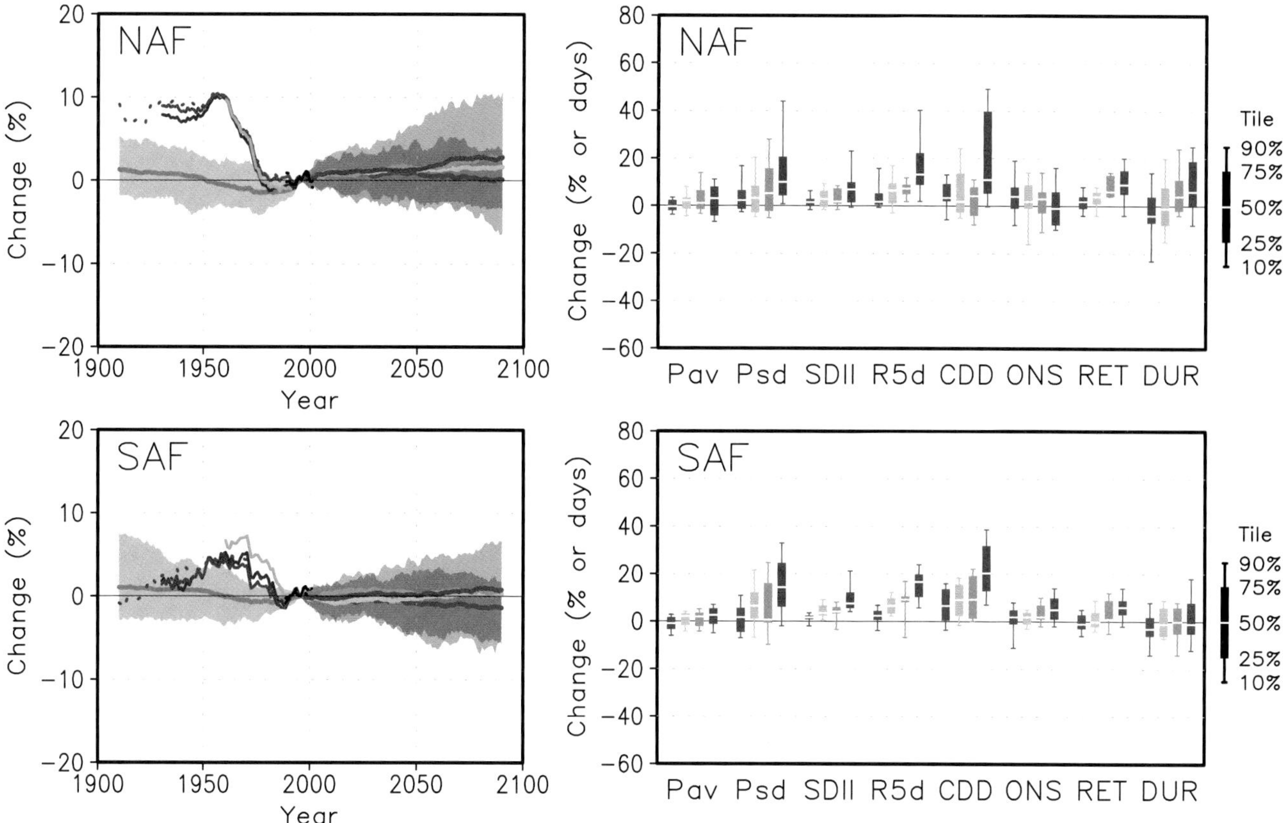

Figure 14.7 | As in Figure 14.4, except for (upper) North Africa (NAF) and (lower) South Africa (SAF).

confidence in precipitation associated with the NAMS will arrive later in the annual cycle, and persist longer. Projections of changes in the timing and duration of the SAMS remain uncertain. There is *high confidence* in the expansion of SAMS, resulting from increased temperature and humidity.

Based on how models represent known drivers of the West African monsoon, there is *low confidence* in projections of its future development based on CMIP5. *Confidence* is *low* in projections of a small delay in the onset of the West African rainy season with an intensification of late-season rains.

14.3 Tropical Phenomena

14.3.1 Convergence Zones

Section 7.6 presents a radiative perspective of changes in convection (including the differences between GHG and aerosol forcings), and Section 12.4.5.2 discusses patterns of precipitation change on the global scale. The emphasis here is on regional aspects of tropical changes. Tropical convection over the oceans, averaged for a month or longer, is organized into long and narrow convergence zones, often anchored by SST structures. Latent heat release in convection drives atmospheric circulation and affects global climate. In model experiments where spatially uniform SST warming is imposed, precipitation increases in these tropical convergence zones (Xie et al., 2010b), following the 'wet-get-wetter' paradigm (Held and Soden, 2006). On the flanks of a convergence zone, rainfall may decrease because of the increased horizontal gradient in specific humidity and the resultant increase in dry advection into the convergence zone (Neelin et al., 2003).

Although these arguments based on moist atmospheric dynamics call for changes in tropical convection to be organized around the climatological rain band, studies since AR4 show that such changes in a warmer climate also depend on the spatial pattern of SST warming. As a result of the SST pattern effect, rainfall change does not generally project onto the climatological convergence zones, especially for the annual mean. In CMIP3/5 model projections, annual rainfall change over tropical oceans follows a 'warmer-get-wetter' pattern, increasing where the SST warming exceeds the tropical mean and vice versa (Figure 14.8, Xie et al., 2010b; Sobel and Camargo, 2011; Chadwick et al., 2013). Differences among models in the SST warming pattern are an important source of uncertainty in rainfall projections, accounting for a third of inter-model variability in annual precipitation change in the tropics (Ma and Xie, 2013).

Figure 14.8 presents selected indices for several robust patterns of SST warming for RCP8.5. They include greater warming in the NH than in the Southern Hemisphere (SH), a pattern favouring rainfall increase at locations north of the equator and decreases to the south (Friedman et al., 2013); enhanced equatorial warming (Liu et al., 2005) that anchors a pronounced rainfall increase in the equatorial Pacific; reduced warming in the subtropical Southeast Pacific that weakens convection there; decreased zonal SST gradient across the equatorial Pacific (see Section 14.4) and increased westward SST gradient across the equatorial

Indian Ocean (see Section 14.3.3) that together contribute to the weakened Walker cells.

Changes in tropical convection affect the pattern of SST change (Chou et al., 2005) and such atmospheric and oceanic perturbations are inherently coupled. The SST pattern effect dominates the annual rainfall change while the wet-get-wetter effect becomes important for seasonal mean rainfall in the summer hemisphere (Huang et al., 2013). This is equivalent to an increase in the annual range of precipitation in a warmer climate (Chou et al., 2013). Given uncertainties in SST warming pattern, the confidence is generally higher for seasonal than annual mean changes in tropical rainfall.

14.3.1.1 Inter-Tropical Convergence Zone

The Inter-Tropical Convergence Zone (ITCZ) is a zonal band of persistent low-level convergence, atmospheric convection, and heavy rainfall. Over the Atlantic and eastern half of the Pacific, the ITCZ is displaced north of the equator due to ocean–atmosphere interaction (Xie et al., 2007) and extratropical influences (Kang et al., 2008; Fučkar et al., 2013). Many models show an unrealistic double-ITCZ pattern over the tropical Pacific and Atlantic, with excessive rainfall south of the equator (Section 9.4.2.5.1). This bias needs to be kept in mind in assessing ITCZ changes in model projections, especially for boreal spring when the model biases are largest.

The global zonal mean ITCZ migrates back and forth across the equator following the sun. In CMIP5, seasonal mean rainfall is projected to increase on the equatorward flank of the ITCZ (Figure 14.9). The co-migration of rainfall increase with the ITCZ is due to the wet-get-wetter effect while the equatorward displacement is due to the SST pattern effect (Huang et al., 2013).

14.3.1.2 South Pacific Convergence Zone

The South Pacific Convergence Zone (SPCZ, Widlansky et al., 2011) extends southeastward from the tropical western Pacific to French Polynesia and the SH mid-latitudes, contributing most of the yearly rainfall

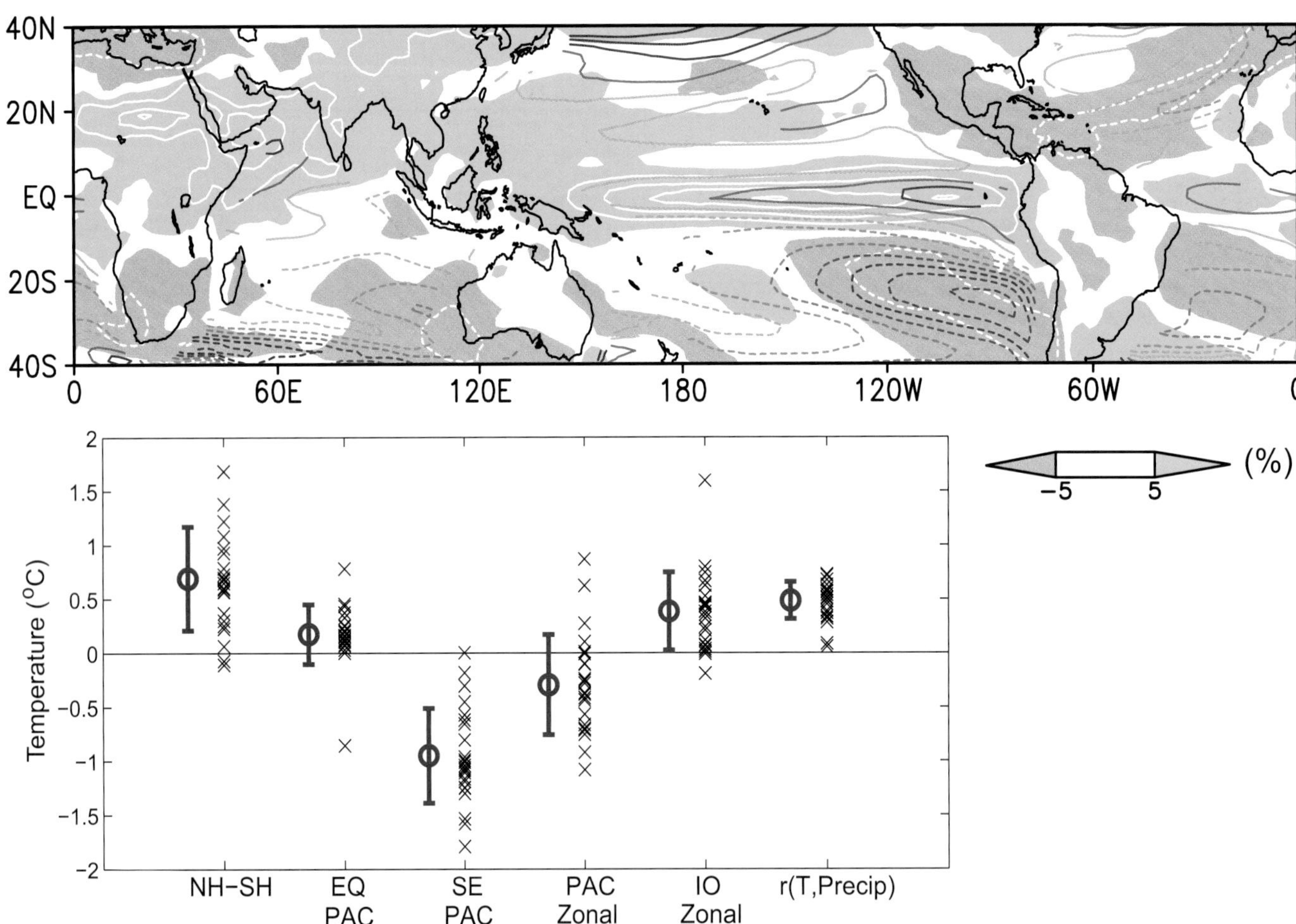

Figure 14.8 | (Upper panel) Annual mean precipitation percentage change ($\delta P/P$ in green/gray shade and white contours at 20% intervals), and relative SST change (colour contours at intervals of 0.2°C; negative dashed) to the tropical (20°S to 20°N) mean warming in RCP8.5 projections, shown as 23 CMIP5 model ensemble mean. (Lower panel) Sea surface temperature (SST) warming pattern indices in the 23-model RCP8.5 ensemble, shown as the 2081–2100 minus 1986–2005 difference. From left: Northern (EQ to 60°N) minus Southern (60°S to EQ) Hemisphere; equatorial (120°E to 60°W, 5°S to 5°N) and Southeast (130°W to 70°W, 30°S to 15°S) Pacific relative to the tropical mean warming; zonal SST gradient in the equatorial Pacific (120°E to 180°E minus 150°W to 90°W, 5°S to 5°N) and Indian (50°E to 70°E, 10°S to 10°N minus 90°E to 110°S, 10°S to EQ) Oceans. (Rightmost) Spatial correlation (*r*) between relative SST change and precipitation percentage change ($\delta P/P$) in the tropics (20°S to 20°N) in each model. (The spatial correlation for the multi-model ensemble mean fields in the upper panel is 0.63). The circle and error bar indicate the ensemble mean and ±1 standard deviation, respectively. The upper panel is a CMIP5 update of Ma and Xie (2013), and see text for indices in the lower panel.

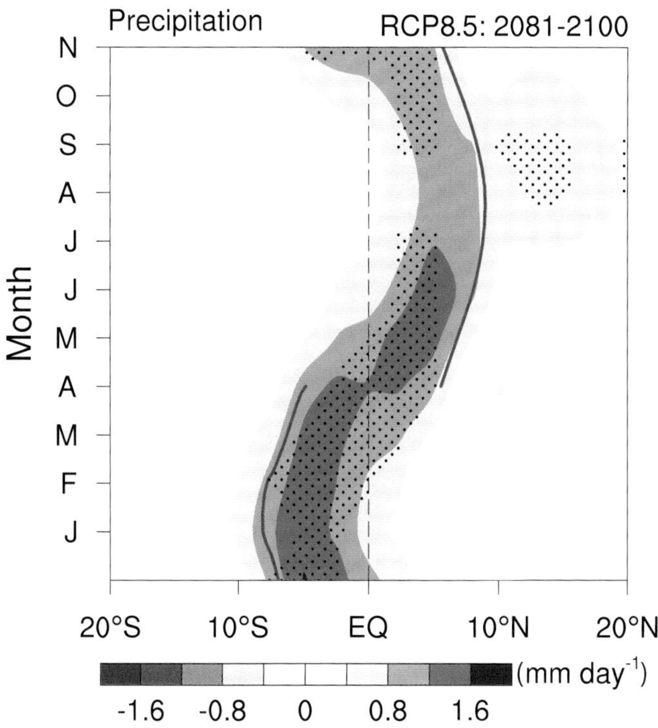

Figure 14.9 | Seasonal cycle of zonal mean tropical precipitation change (2081–2100 in RCP8.5 minus 1986–2005) in CMIP5 multi-model ensemble (MME) mean. Eighteen CMIP5 models were used. Stippling indicates that more than 90% models agree on the sign of MME change. The red curve represents the meridional maximum of the climatological rainfall. (Adapted from Huang et al., 2013.)

to the many South Pacific island nations under its influence. The SPCZ is most pronounced during austral summer (December, January and February (DJF)).

Zonal and meridional SST gradients, trade wind strength, and subsidence over the eastern Pacific are important mechanisms for SPCZ orientation and variability (Takahashi and Battisti, 2007; Lintner and Neelin, 2008; Vincent et al., 2011; Widlansky et al., 2011). Many GCMs simulate the SPCZ as lying east–west, giving a 'double-ITCZ' structure and missing the southeastward orientation (Brown et al., 2012a).

The majority of CMIP models simulate increased austral summer mean precipitation in the SPCZ, with decreased precipitation at the eastern edge of the SPCZ (Brown et al., 2012a; Brown et al., 2012b). The position of the SPCZ varies on interannual to decadal time scales, shifting northeast in response to El Niño (Folland et al., 2002; Vincent et al., 2011). Strong El Niño events induce a zonally oriented SPCZ located well northeast of its average position, while more moderate ENSO (Section 14.4) events are associated with movement of the SPCZ to the northeast or southwest, without a change in its orientation.

Models from both CMIP3 and CMIP5 that simulate the SPCZ well show a consistent tendency towards much more frequent zonally oriented SPCZ events in future (Cai et al., 2012b). The mechanism appears to be associated with a reduction in near-equatorial meridional SST gradient, a robust feature of modelled SST response to anthropogenic forcing (Widlansky et al., 2013). An increased frequency of zonally oriented SPCZ events would have major implications for regional climate, possibly leading to longer dry spells in the southwest Pacific.

14.3.1.3 South Atlantic Convergence Zone

The South Atlantic Convergence Zone (SACZ) extends from the Amazon region through southeastern Brazil towards the Atlantic Ocean during austral summer (Cunningham and Cavalcanti, 2006; Carvalho et al., 2011; de Oliveira Vieira et al., 2013). Floods or dry conditions in southeastern Brazil are often related to SACZ variability (Muza et al., 2009; Lima et al., 2010; Vasconcellos and Cavalcanti, 2010). A subset of CMIP models simulate the SACZ (Vera and Silvestri, 2009; Seth et al., 2010; Yin et al., 2012) and its variability as a dipolar structure (Junquas et al., 2012; Cavalcanti and Shimizu, 2012).

A southward displacement of SACZ and intensification of the southern centre of the precipitation dipole are suggested in projections of CMIP3 and CMIP5 models (Seth et al., 2010; Junquas et al., 2012; Cavalcanti and Shimizu, 2012). This displacement is consistent with the increased precipitation over southeastern South America, south of 25°S, projected for the second half of the 21st century, in CMIP3, CMIP5 and regional models (Figure AI.34, Figure 14.21). It is also consistent with the southward displacement of the Atlantic subtropical high (Seth et al., 2010) related to the southward expansion of the Hadley Cell (Lu et al., 2007). Pacific SST warming and the strengthening of the Pacific–South American (PSA)-like wave train (Section 14.6.2) are potential mechanisms for changes in the dipolar pattern resulting in SACZ change (Junquas et al., 2012). This change is also supported by the intensification and increased frequency of the low level jet over South America in future projections (Soares and Marengo, 2009; Seth et al., 2010).

14.3.2 Madden–Julian Oscillation

The MJO is the dominant mode of tropical intraseasonal (20 to 100 days) variability (Zhang, 2005). The MJO modulates tropical cyclone activity (Frank and Roundy, 2006), contributes to intraseasonal fluctuations of the monsoons (Maloney and Shaman, 2008), and excites teleconnection patterns outside the tropics (L'Heureux and Higgins, 2008; Lin et al., 2009). Simulation of the MJO by GCMs remains challenging, but with some improvements made in recent years (Section 9.5.2.3).

Possible changes in the MJO in a future warmer climate have just begun to be explored with models that simulate the phenomenon. In the Max Planck Institute Earth System Model, MJO variance increases appreciably with increasing warming (Schubert et al., 2013). The change in MJO variance is highly sensitive to the spatial pattern of SST warming (Maloney and Xie, 2013). In light of the low skill in simulating MJO, and its sensitive to SST warming pattern, which in itself is subject to large uncertainties, it is currently not possible to assess how the MJO will change in a warmer climate.

14.3.3 Indian Ocean Modes

The tropical Indian Ocean SST exhibits two modes of interannual variability (Schott et al., 2009; Deser et al., 2010b): the Indian Ocean Basin (IOB) mode featuring a basin-wide structure of the same sign, and the Indian Ocean Dipole (IOD) mode with largest amplitude in the eastern

Indian Ocean off Indonesia, and weaker anomalies of the opposite polarity over the rest of the basin (Box 2.5). Both modes are statistically significantly correlated with ENSO (Section 14.4). CMIP models simulate both modes well (Section 9.5.3.4.2).

The formation of IOB is linked to ENSO via an atmospheric bridge and surface heat flux adjustment (Klein et al., 1999; Alexander et al., 2002). Ocean–atmosphere interactions within the Indian Ocean are important for the long persistence of this mode (Izumo et al., 2008; Wu et al., 2008; Du et al., 2009). The basin mode affects the termination of ENSO events (Kug and Kang, 2006), it induces coherent atmospheric anomalies in the summer following El Niño (Xie et al., 2009), including supressed convection (Wang et al., 2003) and reduced tropical cyclone activity (Du et al., 2011) over the Northwest Pacific and anomalous rainfall over East Asia (Huang et al., 2004).

IOD develops in July to November and involves Bjerknes feedback between zonal SST gradient, zonal wind and thermocline tilt along the equator (Saji et al., 1999; Webster et al., 1999). A positive IOD event (with negative SST anomalies off Sumatra) is associated with droughts in Indonesia, reduced rainfall over Australia, intensified Indian summer monsoon, increased precipitation in East Africa and anomalous conditions in the extratropical SH (Yamagata et al., 2004). Most CMIP3 models are able to reproduce the general features of the IOD, including its phase lock onto the July to November season, while detailed analysis of CMIP5 simulations are not yet available (Section 9.5.3.4.2)

Basin-mean SST has risen steadily for much of the 20th century, a trend captured by CMIP3 20th century simulations (Alory et al., 2007). The SST increase over the North Indian Ocean since about 1930 is noticeably weaker than for the rest of the basin. This spatial pattern is suggestive of the effects of reduced surface solar radiation due to Asian brown clouds (Chung and Ramanathan, 2006) and it affects Arabian Sea cyclones (Evan et al., 2011b). In the equatorial Indian Ocean, coral isotope records off Indonesia indicate a reduced SST warming and/or increased salinity during the 20th century (Abram et al., 2008). From ship-borne surface measurements, an easterly wind change especially during July to October has been observed over the past six decades, a result consistent with a reduction of marine cloudiness in the east and a decreasing precipitation trend over the maritime continent (Tokinaga et al., 2012). Atmospheric reanalysis products have difficulty representing these changes (Han et al., 2010).

The projected changes over the equatorial Indian Ocean include easterly wind anomalies, a shoaling thermocline (Vecchi and Soden, 2007a; Du and Xie, 2008) and reduced SST warming in the east (Stowasser et al., 2009), a result confirmed by CMIP5 multi-model analysis (Zheng et al., 2013; Figure 14.10). The change in zonal SST gradient, in turn, reinforces the easterly wind change, indicative of a positive feedback between them as envisioned by Bjerknes (1969). This coupled pattern is most pronounced during July to November, and is broadly consistent with the observed changes in the equatorial Indian Ocean.

In one CMIP3 model, the IOB mode and its capacitor effect persist longer, through summer into early fall towards the end of the century (2081–2100, Zheng et al., 2011). This increased persistence intensifies ENSO's influence on the Northwest Pacific summer monsoon. The *confidence* level of this relationship is *low* due to the lack of multi-model studies.

The IOD variability in SST remains nearly unchanged in future projections of CMIP3 and CMIP5 (Ihara et al., 2009; Figure 14.11a) despite the easterly wind change that lifts the thermocline (Figure 14.10b) and intensifies thermocline feedback on SST in the eastern equatorial Indian Ocean. The global increase in atmospheric dry static stability weakens the atmospheric response to zonal SST gradient changes,

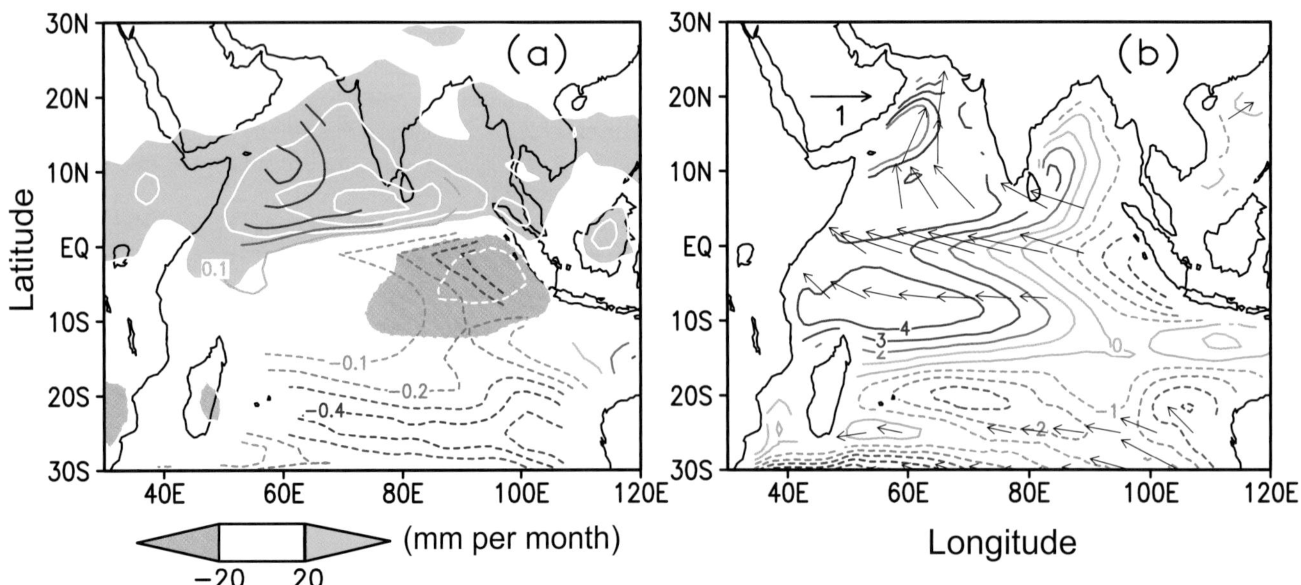

Figure 14.10 | September to November changes in a 22-model CMIP5 ensemble (2081–2100 in RCP8.5 minus 1986–2005 in historical run). (a) Sea surface temperature (SST, colour contours at 0.1°C intervals) relative to the tropical mean (20°S to 20°N), and precipitation (shading and white contours at 20 mm per month intervals). (b) Surface wind velocity (m s^{-1}), and sea surface height deviation from the global mean (contours, centimetres). Over the equatorial Indian Ocean, ocean–atmospheric changes form Bjerknes feedback, with the reduced SST warming and suppressed convection in the east. (Updated with CMIP5 from Xie et al., 2010b.)

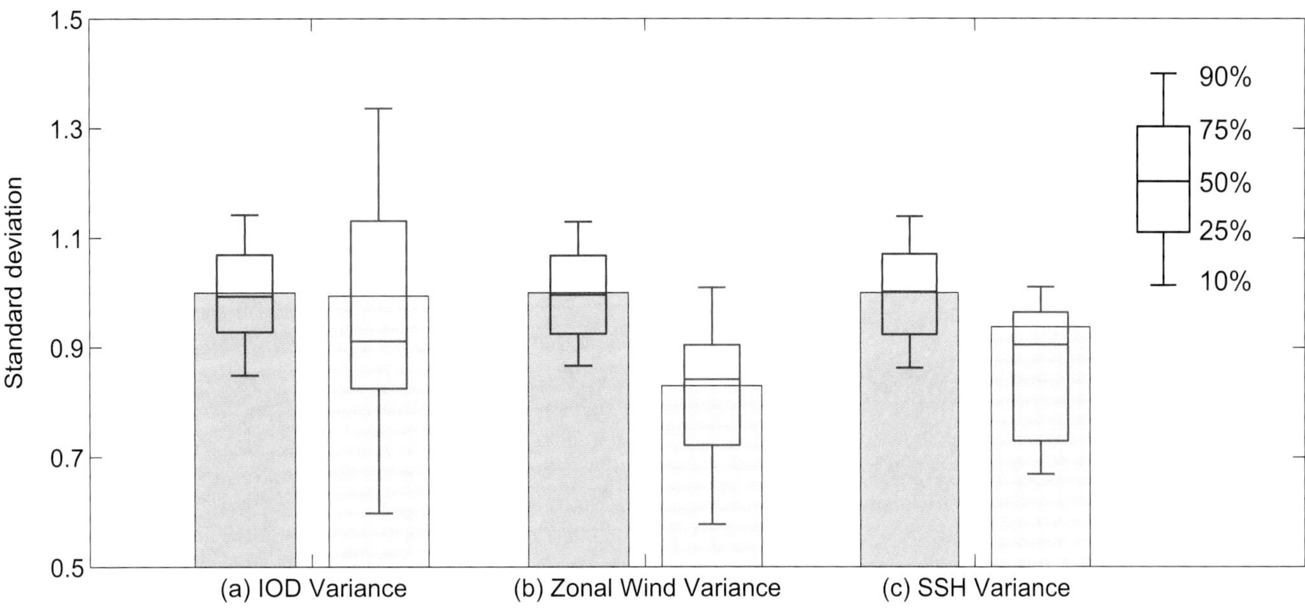

Figure 14.11 | CMIP5 multi-model ensemble mean standard deviations of interannual variability for September to November in pre-industrial (PiControl; blue bars) and RCP8.5 (red) runs: (a) the Indian Ocean dipole index defined as the western (50°E to 70°E, 10°S to 10°N) minus eastern (90°E to 110°E, 10°S to 0°) SST difference; (b) zonal wind in the central equatorial Indian Ocean (70°E to 90°E, 5°S to 5°N); and (c) sea surface height in the eastern equatorial Indian Ocean (90°E to 110°E, 10°S to 0°). The standard deviation is normalized by the pre-industrial (PiControl) value for each model before ensemble average. Blue box-and-whisker plots show the 10th, 25th, 50th, 75th and 90th percentiles of 51-year windows for PiControl, representing natural variability. Red box-and-whisker plots represent inter-model variability for RCP8.5, based on the nearest rank. (Adapted from Zheng et al., 2013.)

countering the enhanced thermocline feedback (Zheng et al., 2010). The weakened atmospheric feedback is reflected in a decrease in IOD variance in both zonal wind and the thermocline depth (Zheng et al., 2013; Figure 14.11b, c).

14.3.4 Atlantic Ocean Modes

The Atlantic features a northward-displaced ITCZ (Section 14.3.1.1), and a cold tongue that develops in boreal summer. Climate models generally fail to simulate these characteristics of tropical Atlantic climate (Section 9.5.3.3). The biases severely limit model skill in simulating modes of Atlantic climate variability and in projecting future climate change in the Atlantic sector. In-depth analysis of the CMIP5 projections of Atlantic Ocean Modes has not yet been fully explored, but see Section 12.4.3.

The inter-hemispheric SST gradient displays pronounced interannual to decadal variability (Box 2.5, Figure 2), referred to as the Atlantic meridional mode (AMM; Servain et al., 1999; Chiang and Vimont, 2004; Xie and Carton, 2004). A thermodynamic feedback between surface winds, evaporation and SST (WES; Xie and Philander, 1994) is fundamental to the AMM (Chang et al., 2006). This mode affects precipitation in northeastern Brazil by displacing the ITCZ (Servain et al., 1999; Chiang and Vimont, 2004; Xie and Carton, 2004), and Atlantic hurricane activity (Vimont and Kossin, 2007; Smirnov and Vimont, 2011).

The Atlantic Niño mode represents interannual variability in the equatorial cold tongue, akin to ENSO (Box 2.5, Figure 2). Bjerknes feedback is considered important for energizing the mode (Zebiak, 1993; Carton and Huang, 1994; Keenlyside and Latif, 2007). This mode affects the West Africa Monsoon (Vizy and Cook, 2002; Giannini et al., 2003).

Over the past century, the Atlantic has experienced a pronounced and persistent warming trend. The warming has brought detectable changes in atmospheric circulation and rainfall patterns in the region. In particular, the ITCZ has shifted southward and land precipitation has increased over the equatorial Amazon, equatorial West Africa, and along the Guinea coast, while it has decreased over the Sahel (Deser et al., 2010a; Tokinaga and Xie, 2011; see also Sections 2.5 and 2.7). Atlantic Niño variability has weakened by 40% in amplitude from 1960 to 1999, associated with a weakening of the equatorial cold tongue (Tokinaga and Xie, 2011).

The CMIP3 20th century climate simulations generally capture the warming trend of the basin-averaged SST over the tropical Atlantic. A majority of the models also seem to capture the secular trend in the tropical Atlantic SST inter-hemispheric gradient and, as a result, the southward shift of the Atlantic ITCZ over the past century (Chang et al., 2011).

Many CMIP3 model simulations with the A1B emission scenario show only minor changes in the SST variance associated with the AMM. However, the few models that give the best AMM simulation over the 20th century project a weakening in future AMM activity (Breugem et al., 2006), possibly due to the northward shift of the ITCZ (Breugem et al., 2007). At present, model projections of future change in AMM activity is considered highly uncertain because of the poorly simulated Atlantic ITCZ. In fact, uncertainty in projected changes in Atlantic meridional SST gradient limits the confidence in regional climate projections surrounding the tropical Atlantic Ocean (Good et al., 2008).

A majority of CMIP3 models forced with the A1B emission scenario project no major change in Atlantic Niño activity in the 21st century, while a few models project a sizable decrease in future activity (Breugem et al., 2006).

CMIP5 projections show an accelerated SST warming over much of the tropical Atlantic (Figure 12.11). RCP8.5 projections of the inter-hemispheric SST gradient change within the basin, however, are not consistent among CMIP5 models as future GHG increase dominates over the anthropogenic aerosol effect.

14.3.5 Assessment Summary

There is *medium confidence* that annual rainfall changes over tropical oceans follow a 'warmer-get-wetter' pattern, increasing where the SST warming exceeds the tropical mean and vice versa. One third of inter-model differences in precipitation projection are due to those in SST pattern. The SST pattern effect on precipitation change is a new finding since AR4.

The wet-get-wetter effect is more obvious in the seasonal than annual rainfall change in the tropics. Confidence is generally higher in seasonal than in annual mean changes in tropical precipitation. There is *medium confidence* that seasonal rainfall will increase on the equatorward flank of the current ITCZ; that the frequency of zonally oriented SPCZ events will increase, with the SPCZ lying well to the northeast of its average position during those events; and that the SACZ shifts southwards, in conjunction with the southward displacement of the South Atlantic subtropical high, leading to an increase in precipitation over southeastern South America.

Owing to models' ability to reproduce general features of IOD and agreement on future projections, it is *likely* that the tropical Indian Ocean will feature a zonal pattern with reduced (enhanced) warming and decreased (increased) rainfall in the east (west), a pattern especially pronounced during August to November. The Indian Ocean dipole mode will *very likely* remain active, with interannual variability unchanged in SST but decreasing in thermocline depth. There is *low confidence* in changes in the summer persistence of the Indian Ocean SST response to ENSO and in ENSO's influence on summer climate over the Northwest Pacific and East Asia.

The observed SST warming in the tropical Atlantic represents a reduction in spatial variation in climatology: the warming is weaker north than south of the equator; and the equatorial cold tongue weakens both in the mean and interannual variability. There is *low confidence* in projected changes over the tropical Atlantic, both for the mean and interannual modes, because of large errors in model simulations of current climate.

There is *low confidence* in how MJO will change in the future due to the poor skill of models in simulating MJO and the sensitivity of its change to SST warming patterns that are themselves subject to large uncertainties in the projections.

14.4 El Niño-Southern Oscillation

The ENSO is a coupled ocean–atmosphere phenomenon naturally occurring at the interannual time scale over the tropical Pacific (see Box 2.5, Supplementary Material Section 14.SM.2, and Figure 14.12).

14.4.1 Tropical Pacific Mean State

SST in the western tropical Pacific has increased by up to 1.5°C per century, and the warm pool has expanded (Liu and Huang, 2000; Huang and Liu, 2001; Cravatte et al., 2009). Studies disagree on how the east–west SST gradient along the equator has changed, some showing a strengthening (Cane et al., 1997; Hansen et al., 2006; Karnauskas et al., 2009; An et al., 2011) and others showing a weakening (Deser et al., 2010a; Tokinaga et al., 2012), because of observational uncertainties associated with limited data sampling, changing measurement techniques, and analysis procedures. Most CMIP3 and CMIP5 models also disagree on the response of zonal SST gradient across the equatorial Pacific (Yeh et al., 2012).

The Pacific Ocean warms more near the equator than in the subtropics in CMIP3 and CMIP5 projections (Liu et al., 2005; Gastineau and Soden, 2009; Widlansky et al., 2013; Figure 14.12) because of the difference in evaporative damping (Xie et al., 2010b). Other oceanic changes include a basin-wide thermocline shoaling (Vecchi and Soden, 2007a; DiNezio et al., 2009; Collins et al., 2010; Figure 14.12), a weakening of surface currents, and a slight upward shift and strengthening of the equatorial undercurrent (Luo and Rothstein, 2011; Sen Gupta et al., 2012). A weakening of tropical atmosphere circulation during the 20th century was documented in observations and reanalyses (Vecchi et al., 2006; Zhang and Song, 2006; Vecchi and Soden, 2007a; Bunge and Clarke, 2009; Karnauskas et al., 2009; Yu and Zwiers, 2010; Tokinaga et al., 2012) and in CMIP models (Vecchi and Soden, 2007a; Gastineau and Soden, 2009). The Pacific Walker Circulation, however, intensified during the most recent two decades (Mitas and Clement, 2005; Liu and Curry, 2006; Mitas and Clement, 2006; Sohn and Park, 2010; Li and Ren, 2012; Zahn and Allan, 2011; Zhang et al., 2011a), illustrating the effects of natural variability.

14.4.2 El Niño Changes over Recent Decades and in the Future

The amplitude modulation of ENSO at longer time scales has been observed in reconstructed instrumental records (Gu and Philander, 1995; Wang, 1995; Mitchell and Wallace, 1996; Wang and Wang, 1996; Power et al., 1999; An and Wang, 2000; Yeh and Kirtman, 2005; Power and Smith, 2007; Section 5.4.1), in proxy records (Cobb et al., 2003; Braganza et al., 2009; Li et al., 2011c; Yan et al., 2011), and is also simulated by coupled GCMs (Lau et al., 2008; Wittenberg, 2009). Some studies have suggested that the modulation was due to changes in mean climate conditions in the tropical Pacific (An and Wang, 2000; Fedorov and Philander, 2000; Wang and An, 2001, 2002; Li et al., 2011c), as observed since the 1980s (An and Jin, 2000; An and Wang, 2000; Fedorov and Philander, 2000; Kim and An, 2011). With three events during 2000-2010, which meets intensity in Nino4 being larger than in Nino3, two events during 1990-2000 and only two events are found for 1950-1990 the maximum SST warming during El Niño now

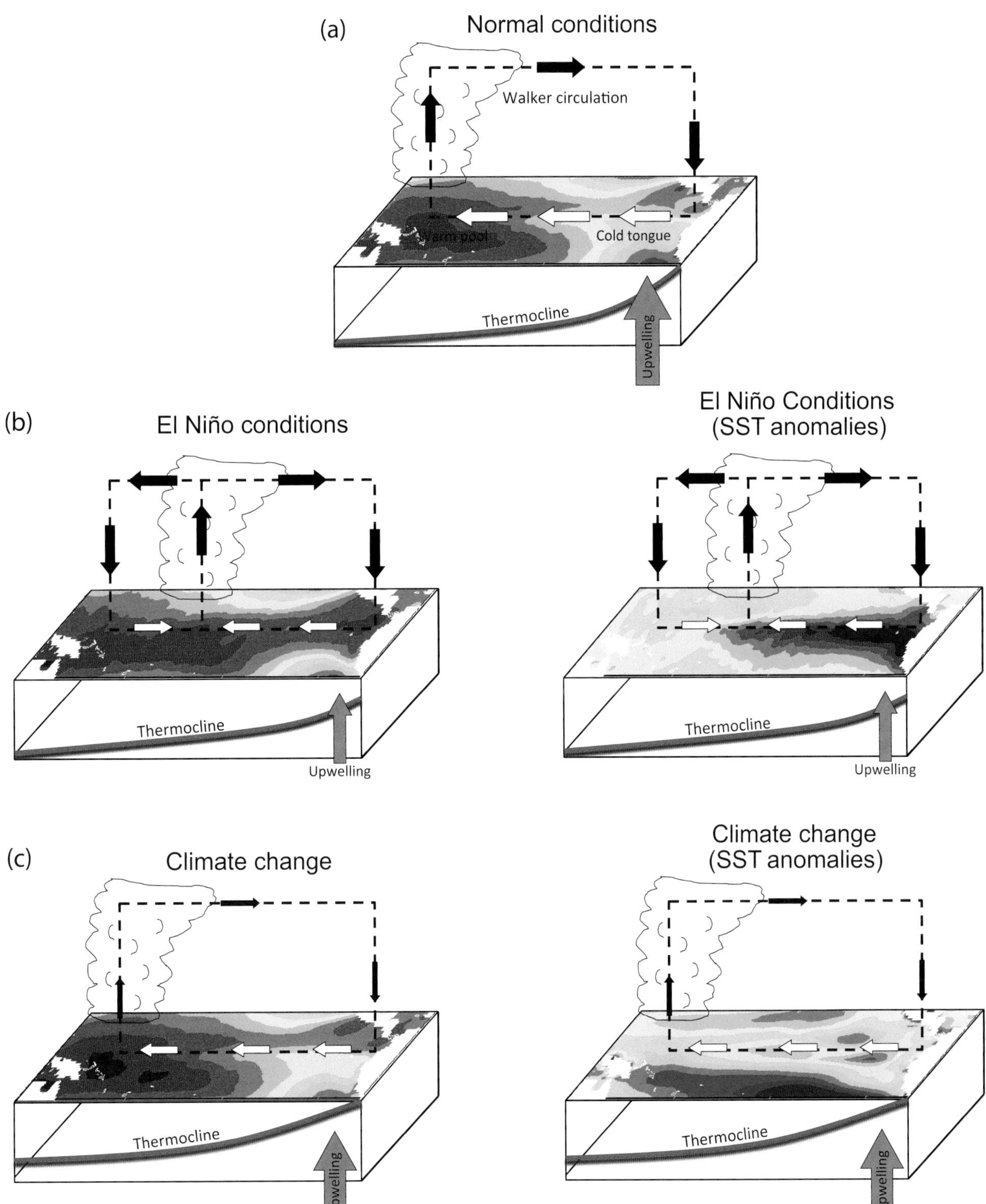

Figure 14.12 | Idealized schematic showing atmospheric and oceanic conditions of the tropical Pacific region and their interactions during normal conditions, El Niño conditions, and in a warmer world. (a) Mean climate conditions in the tropical Pacific, indicating sea surface temperatures (SSTs), surface wind stress and associated Walker Circulation, the mean position of convection and the mean upwelling and position of the thermocline. (b) Typical conditions during an El Niño event. SSTs are anomalously warm in the east; convection moves into the central Pacific; the trade winds weaken in the east and the Walker Circulation is disrupted; the thermocline flattens and the upwelling is reduced. (c) The likely mean conditions under climate change derived from observations, theory and coupled General Circulation Models (GCMs). The trade winds weaken; the thermocline flattens and shoals; the upwelling is reduced although the mean vertical temperature gradient is increased; and SSTs (shown as anomalies with respect to the mean tropical-wide warming) increase more on the equator than off. Diagrams with absolute SST fields are shown on the left, diagrams with SST anomalies are shown on the right. For the climate change fields, anomalies are expressed with respect to the basin average temperature change so that blue colours indicate a warming smaller than the basin mean, not a cooling (Collins et al., 2010).

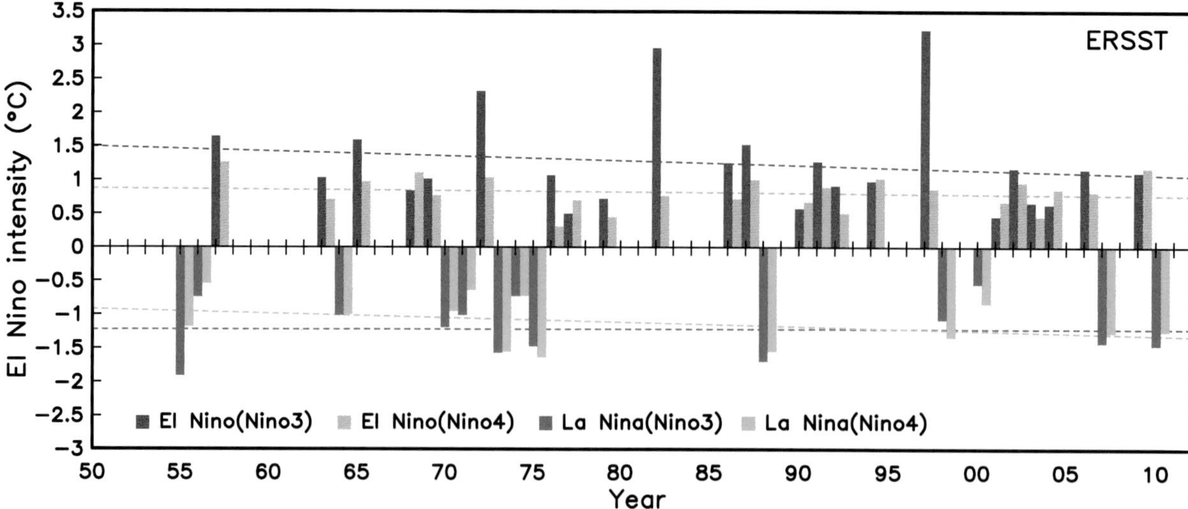

Figure 14.13 | Intensities of El Niño and La Niña events for the last 60 years in the eastern equatorial Pacific (Niño3 region) and in the central equatorial Pacific (Niño4 region), and the estimated linear trends, obtained from Extended Reconstructed Sea Surface Temperature v3 (ERSSTv3).

appears to occur more often in the central Pacific (Figure 14.13; Ashok et al., 2007; Kao and Yu, 2009; Kug et al., 2009; Section 9.5.3.4.1 and Supplementary Material Section 14.SM.2; Yeh et al., 2009), with global impacts that are distinct from 'standard' El Niño events where the maximum warming is over the eastern Pacific (Kumar et al., 2006a; Ashok et al., 2007; Kao and Yu, 2009; Hu et al., 2012b). During the past century, an increasing trend in ENSO amplitude was also observed (Li et al., 2011c; Vance et al., 2012), possibly caused by a warming climate (Zhang et al., 2008; Kim and An, 2011) although other reconstructions in this data-sparse region dispute this trend (Giese and Ray, 2011).

Long coupled GCM simulations show that decadal-to-centennial modulations of ENSO can be generated without any change in external forcing (Wittenberg, 2009; Yeh et al., 2011), with multi-decadal epochs of anomalous ENSO behaviour. The modulations result from nonlinear processes in the tropical climate system (Timmermann et al., 2003), the interaction with the mean climate state (Ye and Hsieh, 2008; Choi et al., 2009, 2011, 2012), or from random changes in ENSO activity triggered by chaotic atmospheric variability (Power and Colman, 2006; Power et al., 2006). There is little consensus as to whether the decadal modulations of ENSO properties (amplitude and spatial pattern) during recent decades are due to anthropogenic effects or natural variability. Instrumental SST records are available back to the 1850s, but good observations of the coupled air–sea feedbacks that control ENSO behaviour—including subsurface temperature and current fluctuations, and air–sea exchanges of heat, momentum and water—are available only after the late 1970s, making observed historical variations in ENSO feedbacks highly uncertain (Chen, 2003; Wittenberg, 2004).

CMIP5 models show some improvement compared to CMIP3, especially in ENSO amplitude (Section 9.5.3.4.1). Selected CMIP5 models that simulate well strong El Niño events show a gradual increase of El Niño intensity, especially over the central Pacific (Kim and Yu, 2012). CMIP3 models suggested a westward shift of SST variability in future projections (Boer, 2009; Yeh et al., 2009). Generally, however, future changes in El Niño intensity in CMIP5 models are model dependent (Guilyardi et al., 2012; Kim and Yu, 2012; Stevenson et al., 2012), and not significantly distinguished from natural modulations (Stevenson, 2012; Figure 14.14). Because the change in tropical mean conditions (especially the zonal gradient) in a warming climate is model dependent (Section 14.4.1), changes in ENSO intensity for the 21st century (Solomon and Newman, 2011; Hu et al., 2012a) are uncertain (Figure 14.14). Future changes in ENSO depend on competing changes in coupled ocean–atmospheric feedback (Philip and Van Oldenborgh, 2006; Collins et al., 2010; Vecchi and Wittenberg, 2010), and on the dynamical regime a given model is in. There is *high confidence*, however, that ENSO will remain the dominant mode of natural climate variability in the 21st century (Collins et al., 2010; Guilyardi et al. 2012; Kim and Yu 2012; Stevenson 2012).

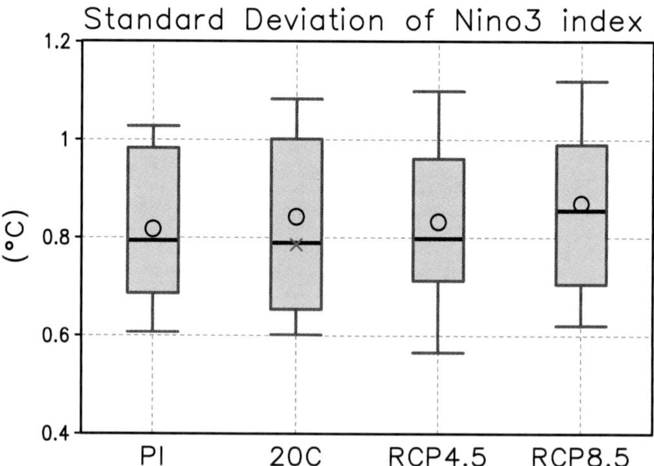

Figure 14.14 | Standard deviation in CMIP5 multi-model ensembles of sea surface temperature variability over the eastern equatorial Pacific Ocean (Nino3 region: 5°S-5°N, 150°W-90°W), a measure of El Nino amplitude, for the pre-industrial (PI) control and 20th century (20C) simulations, and 21st century projections using RCP4.5 and RCP8.5. Thirty-one models are used for the ensemble average. Open circles indicate multi-model ensemble means, and the red cross symbol is the observed standard deviation for January 1870 – December 2011 obtained from HadISSTv1. The linear trend and climatological mean of seasonal cycle have been removed. Box-whisker plots show the 16th, 25th, 50th, 75th, and 84th percentiles.

14.4.3 Teleconnections

There is little improvement in the CMIP5 ensemble relative to CMIP3 in the amplitude and spatial correlation metrics of precipitation teleconnections in response to ENSO, in particular within regions of strong observed precipitation teleconnections (equatorial South America, the western equatorial Pacific and a southern section of North America; Langenbrunner and Neelin, 2013). Scenario projections in CMIP3 and CMIP5 showed a systematic eastward shift in both El Niño- and La Niña-induced teleconnection patterns over the extratropical NH (Meehl and Teng, 2007; Stevenson et al., 2012; Figure 14.15), which might be due to the eastward migration of tropical convection centres associated with the expansion of the warm pool in a warm climate (Muller and Roeckner, 2006; Müller and Roeckner, 2008; Cravatte et al., 2009; Kug et al., 2010), or changes in the mid-latitude mean circulation (Meehl and Teng, 2007). Some models produced an intensified ENSO teleconnection pattern over the North Atlantic region in a warmer climate (Müller and Roeckner, 2008; Bulic et al., 2012) and a weakened teleconnection pattern over the North Pacific (Stevenson, 2012). It is unclear whether the eastward shift of tropical convection is related to longitudinal shifts in El Niño maximum SST anomalies (see Supplementary Material Section 14.SM.2) or to changes in the mean state in the tropical Pacific. Some coupled GCMs, which do not show an increase in the central Pacific warming during El Nino in response to a warming climate, do not produce a substantial change in the longitudinal location of tropical convection (Müller and Roeckner, 2008; Yeh et al., 2009).

In a warmer climate, the increase in atmospheric moisture intensifies temporal variability of precipitation even if atmospheric circulation variability remains the same (Trenberth 2011; Section 12.4.5). This applies to ENSO-induced precipitation variability but the possibility of changes in ENSO teleconnections complicates this general conclusion, making it somewhat regional-dependent (Seager et al. 2012)

14.4.4 Assessment Summary

ENSO shows considerable inter-decadal modulations in amplitude and spatial pattern within the instrumental record. Models without changes in external forcing display similar modulations, and there is little consensus on whether the observed changes in ENSO are due to external forcing or natural variability (see also Section 10.3.3 for an attribution discussion).

There is *high confidence* that ENSO will remain the dominant mode of interannual variability with global influences in the 21st century, and due to changes in moisture availability ENSO-induced rainfall variability on regional scales will intensify. There is *medium confidence* that ENSO-induced teleconnection patterns will shift eastward over the North Pacific and North America. There is *low confidence* in changes in the intensity and spatial pattern of El Niño in a warmer climate.

14.5 Annular and Dipolar Modes

The North Atlantic Oscillation (NAO), the North Pacific Oscillation (NPO) and the Northern and Southern Annular Modes (NAM and SAM) are dominant modes of variability in the extratropics. These modes are the focus of much research attention, especially in impact studies, where they are often used as aggregate descriptors of past regional climate

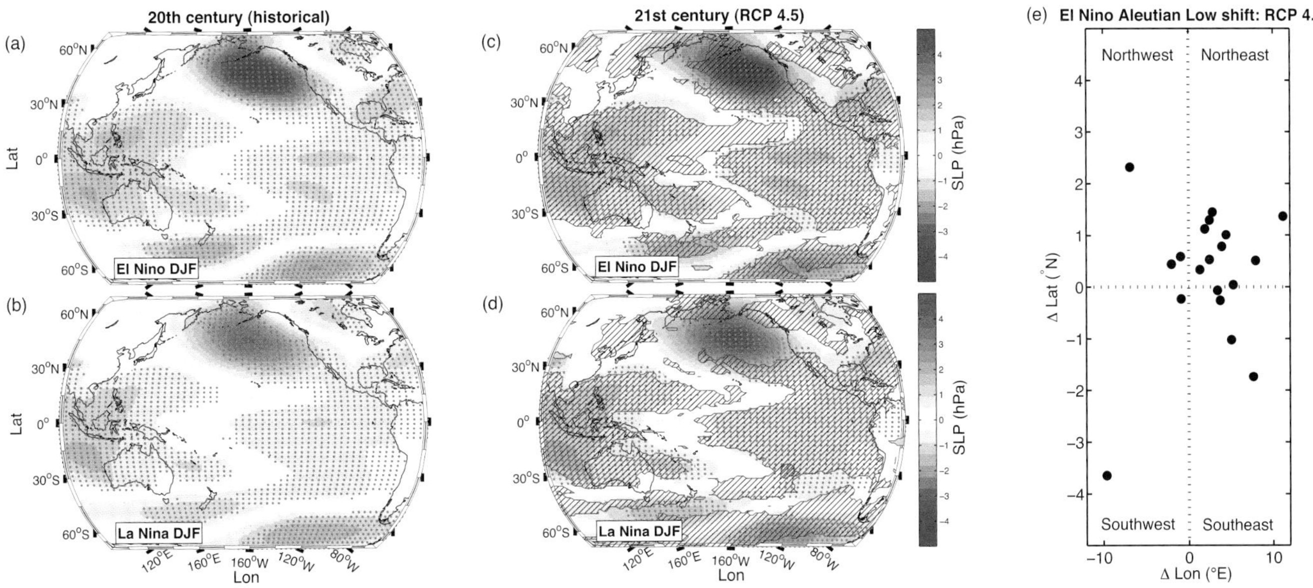

Figure 14.15 | Changes to sea level pressure (SLP) teleconnections during December, January and February (DJF) in the CMIP5 models. (a) SLP anomalies for El Niño during the 20th century. (b) SLP anomalies for La Nina during the 20th century. (c) SLP anomalies for El Niño during RCP4.5. (d) SLP anomalies for La Niña during RCP4.5. Maps in (a)–(d) are stippled where more than two thirds of models agree on the sign of the SLP anomaly ((a),(b): 18 models; (c),(d): 12 models), and hatched where differences between the RCP4.5 multi-model mean SLP anomaly exceed the 60th percentile (red-bordered regions) or are less than the 40th percentile (blue-bordered regions) of the distribution of 20th century ensemble means. In all panels, El Niño (La Niña) periods are defined as years having DJF Nino3.4 SST above (below) one standard deviation relative to the mean of the detrended time series. For ensemble mean calculations, all SLP anomalies have been normalized to the standard deviation of the ensemblemember detrended Nino3.4 SST. (e) Change in the 'centre of mass' of the Aleutian Low SLP anomaly, RCP4.5–20th century. The Aleutian Low SLP centre of mass is a vector with two elements (lat, lon), and is defined as the sum of (lat, lon) weighted by the SLP anomaly, over all points in the region 180°E to 120°E, 40°N to 60°N having a negative SLP anomaly during El Niño.

trends and variations over many parts of the world. For example, since IPCC (2007a) more than 2000 scientific articles have been published, which include NAO, AO, or NAM in either the title or abstract. This assessment focusses on recent research on these modes that is most relevant for future regional climate change. Past behaviour of these modes inferred from observations is assessed in Section 2.7.8.

14.5.1 Northern Modes

The NAO is a well-established dipolar mode of climate variability having opposite variations in sea level pressure between the Atlantic subtropical high and the Iceland/Arctic low (Wanner et al., 2001; Hurrell et al., 2003; Budikova, 2009). It is strongly associated with the tropospheric jet, storms (see Section 14.6.2), and blocking that determine the weather and climate over the North Atlantic and surrounding continents (Hurrell and Deser, 2009; Box 14.2). The NAO exists in boreal summer as well as in boreal winter, albeit with different physical characteristics (Sun et al., 2008; Folland et al., 2009).

Over the North Pacific, there is a similar wintertime dipolar mode known as the NPO associated with north–south displacements of the Asian-Pacific jet stream and the Pacific storm track. The NPO influences winter air temperature and precipitation over much of western North America as well as sea ice over the Pacific sector of the Arctic, more so than either ENSO (Section 14.4) or the PNA (Linkin and Nigam, 2008).

These dipolar modes have been interpreted as the regional manifestation of an annular mode in sea level pressure known as the Arctic Oscillation (AO; Thompson and Wallace, 1998) or the Northern Annular Mode (NAM; Thompson and Wallace, 2000). The AO (NAM at 1000 hPa) index and the NAO index (see Box 2.5, Table 1) are strongly correlated but the AO spatial pattern is more zonally symmetric and so differs from the NAO over the N. Pacific (Ambaum et al., 2001; Feldstein and Franzke, 2006). Hereafter, the term NAO is used to denote NAO, AO and NAM in boreal winter unless further distinction is required.

Climate models are generally able to simulate the gross features of NAO and NPO (see Section 9.5.3.2). It has been argued that these modes may be a preferred pattern of response to climate change (Gerber et al., 2008). However, this is not supported by a detailed examination of the vertical structure of the simulated global warming response (Woollings, 2008). Hori et al. (2007) noted that NAO variability did not change substantially in the Special Report on Emission Scenarios (SRES)-A1B and 20th century scenarios and so concluded that the trend in the NAO index (defined relative to a historical mean state) is a result of an anthropogenic trend in the basic mean state rather than due to changes in NAO variability. However, other research indicates that there is a coherent two-way interaction between the trend in the mean state and the NAO-like modes of variability—the mode and/or regime structure change due to changes in the mean state (Branstator and Selten, 2009 ; Barnes and Polvani, 2013). Section 14.6.2 assesses the jet and storm track changes associated with the projected responses.

Model simulations have underestimated the magnitude of the large positive trend from 1960-2000 in winter NAO observations, which now appears to be more likely due to natural variability rather than anthropogenic influences (see Section 10.3.3.2). Some studies have even considered NAO to be a source of natural variability that needs to be removed before detection and attribution of anthropogenic changes (Zhang et al., 2006). Detection of regional surface air temperature response to anthropogenic forcing has been found to be robust to the exclusion of model-simulated AO and PNA changes (Wu and Karoly, 2007). Model projections of wintertime European precipitation have been shown to become more consistent with observed trends after removal of trends due to NAO (Bhend and von Storch, 2008). Underestimation of trends in NAO can lead to biases in projections of regional climate, for example, Arctic sea ice (Koldunov et al., 2010).

Underestimation of NAO long-term variability may be due to missing or poorly represented processes in climate models. Recent observational and modelling studies have helped to confirm that the lower stratosphere plays an important role in explaining recent more negative NAO winters and long-term trends in NAO (Scaife et al., 2005; Dong et al., 2011; Ouzeau et al., 2011; Schimanke et al., 2011). This is supported by evidence that seasonal forecasts of NAO can be improved by inclusion of the stratospheric Quasi-Biennial Oscillation (QBO; Boer and Hamilton, 2008; Marshall and Scaife, 2010). Other studies have found that observed changes in stratospheric water vapour changes from 1965–1995 led to an impact on NAO simulated by a model, and have suggested that changes in stratospheric water vapour may be another possible pathway for communicating tropical forcing to the extratropics (Joshi et al., 2006; Bell et al., 2009). There is growing evidence that future NAO projections are sensitive to how climate models resolve stratospheric processes and troposphere–stratosphere interactions (Sigmond and Scinocca, 2010; Scaife et al., 2011a; Karpechko and Manzini, 2012).

Several recent studies of historical data have found a positive association between solar activity and NAO (Haigh and Roscoe, 2006; Kodera et al., 2008; Lockwood et al., 2010), while other studies have found little imprint of solar and volcanic forcing on NAO (Casty et al., 2007). Positive associations between NAO and solar forcing have been reproduced in recent modelling studies (Lee et al., 2008; Ineson et al., 2011) but no significant changes were found in CMIP5 projections of NAO due to changes in solar irradiance or aerosol forcing.

Observational studies have noted weakening of NAO during periods of reduced Arctic sea ice (Strong et al., 2009; Wu and Zhang, 2010). Several modelling studies have also shown a negative NAO response to the partial removal of sea ice in the Arctic or high latitudes (Kvamsto et al., 2004; Magnusdottir et al., 2004; Seierstad and Bader, 2009; Deser et al., 2010c; Screen et al., 2012). However, the strength and timing of the response to sea ice loss varies considerably between studies, and can be hard to separate from common responses to warming of the troposphere and from natural climate variability. The impact of sea ice loss in individual years on NAO is small and hard to detect (Bluthgen et al., 2012). Reviews of the emerging literature on this topic can be found in Budikova (2009) and Bader et al. (2011).

The NPO contributes to the excitation of ENSO events via the 'Seasonal Footprinting Mechanism' (SFM; Anderson, 2003; Vimont et al., 2009; Alexander et al., 2010). Some studies indicate that warm events in the central tropical Pacific Ocean may in turn excite the NPO (Di Lorenzo et al., 2009).

Recent multi-model studies of NAO (Hori et al., 2007; Karpechko, 2010; Zhu and Wang, 2010; Gillett and Fyfe, 2013) reconfirm the small positive response of boreal winter NAO indices to GHG forcing noted in earlier studies reported in AR4 (Kuzmina et al., 2005; Miller et al., 2006; Stephenson et al., 2006). Projected trends in wintertime NAO indices are generally found to have small amplitude compared to natural internal variations (Deser et al., 2012). Furthermore, there is substantial variation in NAO projections from different climate models. For example, one study found no significant NAO trends in two simulations with the ECHAM4/OPYC3 model (Fischer-Bruns et al., 2009), whereas another study found a strong positive trend in NAO in the ECHAM5/MPI-OM SRES A1B simulations (Müller and Roeckner, 2008). The model dependence of the response is an important source of uncertainty in the regional climate change response (Karpechko, 2010). A multi-model study of 24 climate model projections suggests that there are no major changes in the NPO due to greenhouse warming (Furtado et al., 2011).

Figures 14.16a, b summarize the wintertime NAO and NAM indices simulated by models participating in the CMIP5 experiment (Gillett and Fyfe, 2013). The multi-model mean of the NAO and NAM indices are similar and exhibit small linear trends in agreement with those shown for the NAM index in AR4 (AR4, Figure 10.17a). The multi-model mean projected increase of around 1 to 2 hPa from 1850 to 2100 is smaller than the spread of around 2 to 4 hPa between model simulations (Figure 14.16).

Some differences in model projections can be accounted for by changes in the NAO spatial pattern, for example, northeastward shifts in NAO centres of action have been found to be important for estimating the trend in the NAO index (Ulbrich and Christoph, 1999; Hu and Wu, 2004). Individual model simulations have shown the spatial extent of NAO influence decreases with GHG forcing (Fischer-Bruns et al., 2009), a positive feedback between jet and storm tracks that enhances a poleward shift in the NAO pattern (Choi et al., 2010), and changes in the NAO pattern but with no changes in the propagation conditions for Rossby waves (Brandefelt, 2006). One modelling study found a trend in the correlation between NAO and ENSO during the 21st century (Muller and Roeckner, 2006). Such changes in the structure of NAO and/or its interaction with other modes of variability would could lead to important regional climate impacts.

14.5.2 Southern Annular Mode

The Southern Annular Mode (SAM, also known as Antarctic Oscillation, AAO), is the leading mode of climate variability in the SH extratropics, describing fluctuations in the latitudinal position and strength of the mid-latitude eddy-driven westerly jet (see Box 2.5; Section 9.5.3.2). SAM variability has a major influence on the climate of Antarctica, Australasia, southern South America and South Africa (Watterson, 2009; Thompson et al., 2011 and references therein).

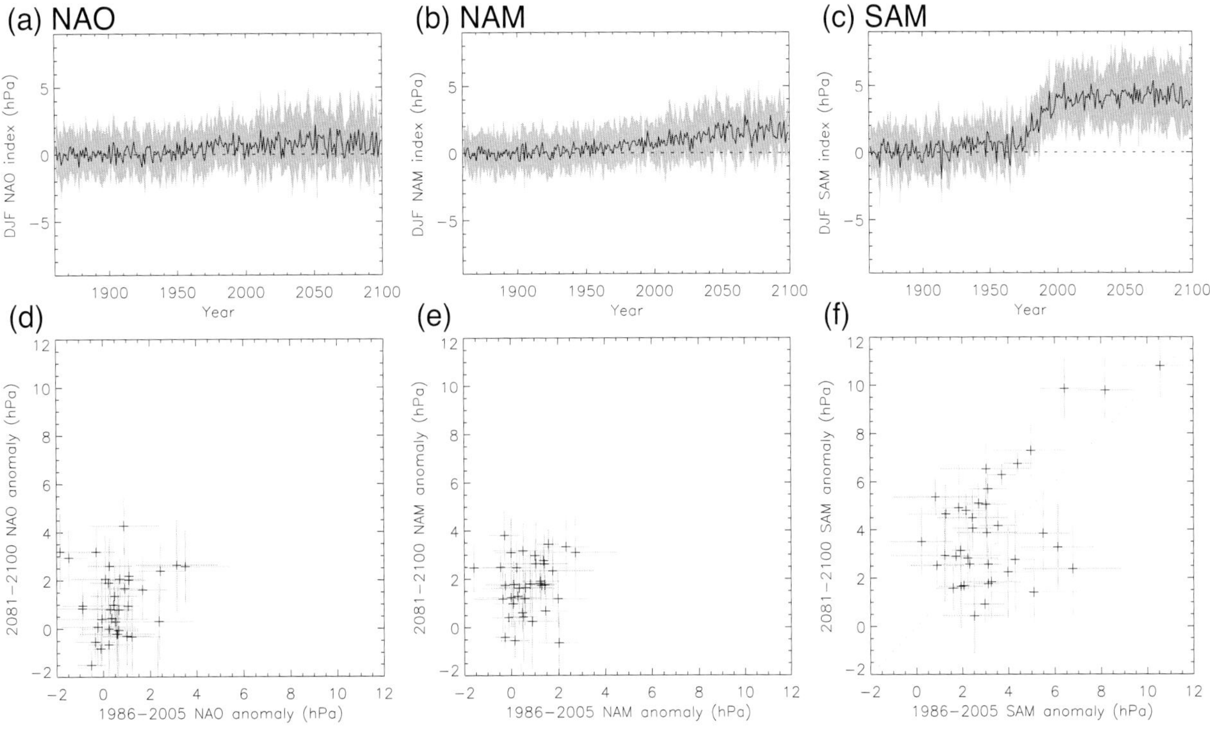

Figure 14.16 | Summary of multi-model ensemble simulations of wintertime (December to February) mean North Atlantic Oscillation (NAO), Northern Annular Mode (NAM) and Southern Annular Mode (SAM) sea level pressure (SLP) indices for historical and RCP4.5 scenarios produced by 39 climate models participating in CMIP5. Panels (a)–(c) show time series of the ensemble mean (black line) and inter-quartile range (grey shading) of the mean index for each model. Panels (d)–(f) show scatter plots of individual model 2081–2100 time means versus 1986–2005 time means (black crosses) together with (–2,+2) standard error bars. The NAO index is defined here as the difference of regional averages: (90°W to 60°E, 20°N to 55°N) minus (90°W to 60°E, 55°N to 90°N) (see Stephenson et al., 2006). The NAM and SAM are defined as zonal indices: NAM as the difference in zonal mean SLP at 35°N and 65°N (Li and Wang, 2003) and SAM as the difference in zonal mean SLP at 40°S and 65°S (Gong and Wang, 1999). All indices have been centred to have zero time mean from 1861–1900. Comparison of simulated and observed trends from 1961–2011 is shown in Figure 10.13.

The physical mechanisms of the SAM are well understood, and the SAM is well represented in climate models, although the detailed spatial and temporal characteristics vary between models (Raphael and Holland, 2006). In the past few decades the SAM index has exhibited a positive trend in austral summer and autumn (Figure 14.16, Marshall, 2007; Jones et al., 2009b), a change attributed to the effects of ozone depletion and, to a lesser extent, the increase in GHGs (Thompson et al., 2011, see also Section 10.3.3.3). It is *likely* that these two factors will continue to be the principal drivers into the future, but as the ozone hole recovers they will be competing to push the SAM in opposite directions (Arblaster et al., 2011; Thompson et al., 2011; Bracegirdle et al., 2013), at least during late austral spring and summer, when ozone depletion has had its greatest impact on the SAM. The SAM is also influenced by teleconnections to the tropics, primarily associated with ENSO (Carvalho et al., 2005; L'Heureux and Thompson, 2006). Changes to the tropical circulation, and to such teleconnections, as the climate warms could further affect SAM variability (Karpechko et al., 2010). See Supplementary Material Section 14.SM.3.1 for further details on the observed variability of SAM.

The CMIP3 models projected a continuing positive trend in the SAM in both summer and winter (Miller et al., 2006). However, those models generally had poor simulations of stratospheric ozone, and tended to underestimate natural variability and to misrepresent observed trends in the SAM, indicating that care should be taken in interpretation of their future SAM projections (Fogt et al., 2009). Arblaster et al. (2011) showed that there can be large differences in the sensitivity of these models to CO_2 increases, which affects their projected trends in the SAM.

Since the AR4 a number of chemistry–climate models (CCMs) have been run that have fully interactive stratospheric chemistry, although unlike coupled atmosphere ocean models they are usually not coupled to the oceans (see also Sections 9.1.3.2.8 and 9.4.6.2). The majority of CCMs and coupled models, which generally compare well to reanalyses (Gerber et al., 2010) although many exhibit biases in their placement of the SH eddy-driven jet (Wilcox et al., 2012; Bracegirdle et al., 2013), indicate that through to at least the mid-21st century the current observed SAM changes are neutralized or reversed during austral summer (Perlwitz et al., 2008; Son et al., 2010; Polvani et al., 2011; Bracegirdle et al., 2013). Figure 14.16 shows the projected ensemble-mean future SAM index evolution during DJF from a suite of CMIP5 models, suggesting that the recent positive trend will weaken considerably as stratospheric ozone concentrations recover over southern high latitudes.

Projected 21st century changes in the SAM, and the closely associated SH eddy-driven jet position, vary by season (Gillett and Fyfe, 2013), and are sensitive to the rate of ozone recovery (Son et al., 2010; Eyring et al., 2013) and to GHG emissions scenario (Swart and Fyfe, 2012; Eyring et al., 2013). In the RCP2.6 scenario, with small increases in GHGs, ozone recovery may dominate in austral summer giving a small projected equatorward jet shift (Eyring et al., 2013) with little change in the annual mean jet position (Swart and Fyfe, 2012). In RCP8.5 large GHG increases are expected to dominate, giving an ongoing poleward shift of the SH jet in all seasons (Swart and Fyfe, 2012; Eyring et al., 2013). In RCP4.5 the influences of ozone recovery and GHG increases are expected to approximately balance in austral summer, with an ongoing poleward jet shift projected in the other seasons (Swart and Fyfe, 2012; Eyring et al., 2013; Gillett and Fyfe, 2013).

14.5.3 Assessment Summary

Future boreal wintertime NAO is *very likely* to exhibit large natural variations and trend of similar magnitude to that observed in the past; is *very likely* to be differ quantitatively from individual climate model projections; is *likely* to become slightly more positive (on average) due to increases in GHGs. The austral summer/autumn positive trend in SAM is *likely* to weaken considerably as ozone depletion recovers through to the mid-21st century. There is *medium confidence* from recent studies that projected changes in NAO and SAM are sensitive to boundary processes, which are not yet well represented in many climate models currently used for projections, for example, stratosphere-troposphere interaction, ozone chemistry, solar forcing and atmospheric response to Arctic sea ice loss. There is *low confidence* in projections of other modes such as the NPO due to the small number of modelling studies.

Box 14.2 | Blocking

Atmospheric blocking is associated with persistent, slow-moving high-pressure systems that interrupt the prevailing westerly winds of middle and high latitudes and the normal eastward progress of extratropical storm systems. Overall, blocking activity is more frequent at the exit zones of the jet stream and shows appreciable seasonal variability in both hemispheres, reaching a maximum in winter–spring and a minimum in summer–autumn (e.g., Wiedenmann et al., 2002). In the Northern Hemisphere (NH), the preferred locations for winter blocking are the North Atlantic and North Pacific, whereas continental blocks are relatively more frequent in summer (Tyrlis and Hoskins, 2008; Barriopedro et al., 2010). Southern Hemisphere (SH) blocking is less frequent than in the NH, and it tends to be concentrated over the Southeast Pacific and the Indian Ocean (Berrisford et al., 2007).

Blocking is a complex phenomenon that involves large- and small-scale components of the atmospheric circulation, and their mutual interactions. Although there is not a widely accepted blocking theory, transient eddy activity is considered to play an important role in blocking occurrence and maintenance through feedbacks between the large-scale flow and synoptic eddies (e.g., Yamazaki and Itoh, 2009). *(continued on next page)*

14

Box 14.2 (continued)

Blocking is an important component of intraseasonal variability in the extratropics and causes climate anomalies over large areas of Europe (Trigo et al., 2004; Masato et al., 2012), North America (Carrera et al., 2004), East Asia (e.g., Wang et al., 2010; Cheung et al., 2012), high-latitude regions of the SH (Mendes et al., 2008) and Antarctica (Massom et al., 2004; Scarchilli et al., 2011). Blocking can also be responsible for extreme events (e.g., Buehler et al., 2011; Pfahl and Wernli, 2012), such as cold spells in winter (e.g., 2008 in China, Zhou et al., 2009d; or 2010 in Europe, Cattiaux, 2010) and summer heat waves in the NH (e.g., 2010 in Russia, Matsueda, 2011; Lupo et al., 2012) and in southern Australia (Pezza et al., 2008).

At interannual time scales, there are statistically significant relationships between blocking activity and several dominant modes of atmospheric variability, such as the NAO (Section 14.5.1) and wintertime blocking in the Euro-Atlantic sector (Croci-Maspoli et al., 2007a; Luo et al., 2010), the winter PNA (Section 14.7.1) and blocking frequency in the North Pacific (Croci-Maspoli et al., 2007a), or the SAM (Section 14.5.2) and winter blocking activity near the New Zealand sector (Berrisford et al., 2007). Multi-decadal variability in winter blocking over the North Atlantic and the North Pacific seem to be related, respectively, with the Atlantic Meridional Overturning Circulation (AMOC; Häkkinen et al., 2011; Section 14.7.6) and the Pacific Decadal Oscillation (PDO; Chen and Yoon, 2002; Section 14.7.3), although this remains an open question.

Other important scientific issues related to the blocking phenomenon include the mechanisms of blocking onset and maintenance, two-way interactions between blocking and stratospheric processes (e.g., Martius et al., 2009; Woollings et al., 2010), influence on blocking of slowly varying components of the climate system (sea surface temperature (SST), sea ice, etc., Liu et al., 2012b), and external forcings.

The most consistent long-term observed trends in blocking for the second half of the 20th century are the reduced winter activity over the North Atlantic (e.g., Croci-Maspoli et al., 2007b), which is consistent with the observed increasing North Atlantic Oscillation (NAO) trend from the 1960s to the mid-1990s (Section 2.7.8), as well as an eastward shift of intense winter blocking over the Atlantic and Pacific Oceans (Davini et al., 2012). The apparent decreasing trend in SH blocking activity (e.g., Dong et al., 2008) seems to be in agreement with the upward trend in the SAM.

The AR4 (Section 8.4.5) reported a tendency for General Circulation Models (GCMs) to underestimate NH blocking frequency and persistence, although most models were able to capture the preferred locations for blocking occurrence and their seasonal distributions. Several intercomparison studies based on a set of CMIP3 models (Scaife et al., 2010; Vial and Osborn, 2012) revealed some progress in the simulation of NH blocking activity, mainly in the North Pacific, but only modest improvements in the North Atlantic. In the SH, blocking frequency and duration was also underestimated, particularly over the Australia–New Zealand sector (Matsueda et al., 2010). CMIP5 models still show a general blocking frequency underestimation over the Euro-Atlantic sector, and some tendency to overestimate North Pacific blocking (Section 9.5.2.2), with considerable inter-model spread (Box 14.2, Figure 1).

Model biases in the mean flow, rather than in variability, can explain a large part of the blocking underestimation and they are usually evidenced as excessive zonality of the flow or systematic shifts in the latitude of the jet stream (Matsueda et al., 2010; Scaife et al., 2011b; Barnes and Hartmann, 2012; Vial and Osborn, 2012; Anstey et al., 2013; Dunn-Sigouin and Son, 2013). Increasing the horizontal resolution in atmospheric GCMs with prescribed SSTs has been shown to significantly reduce blocking biases, particularly in the Euro-Atlantic sector and Australasian sectors (e.g., Matsueda et al., 2010; Jung et al., 2011; Dawson et al., 2012; Berckmans et al., 2013), while North Pacific blocking could be more sensitive to systematic errors in tropical SSTs (Hinton et al., 2009). Also blocking biases are smaller in those CMIP5 models with higher horizontal and vertical resolution (Anstey et al., 2013). However, the improvement of blocking simulation with increasing horizontal resolution is less clear in coupled models than in atmospheric GCMs with prescribed SSTs, indicating that both SSTs and the relative coarse resolution in OGCM (Scaife et al., 2011b) are important causes of blocking biases.

Most CMIP3 models projected significant reductions in NH annual blocking frequency (Barnes et al., 2012), particularly during winter, but CMIP5 models seem to indicate weaker decreases in the future (Dunn-Sigouin and Son, 2013) and a more complex response than that reported for CMIP3 models, including possible regional increases of blocking frequency in summer (Cattiaux et al., 2013; Masato et al., 2013). There is high agreement that winter blocking frequency over the North Atlantic and North Pacific will not increase under enhanced GHG concentrations (Barnes et al., 2012; Dunn-Sigouin and Son, 2013). Future strengthening of the zonal wind and meridional jet displacements may partially account for some of the projected changes in blocking frequency over the ocean basins of both hemispheres (Matsueda et al., 2010; Barnes and Hartmann, 2012; Dunn-Sigouin and Son, 2013). Future trends in blocking intensity and persistence are even more uncertain, with no clear signs of significant changes. How the location and frequency of blocking events will evolve in future are both critically important for understanding regional climate change in particular with respect to extreme conditions (e.g., Sillmann et al., 2011; de Vries et al., 2013). *(continued on next page)*

Box 14.2 (continued)

In summary, the increased ability in simulating blocking in some models indicate that there is *medium confidence* that the frequency of NH and SH blocking will not increase, while trends in blocking intensity and persistence remain uncertain. The implications for blocking related regional changes in North America, Europe and Mediterranean and Central and North Asia are therefore also uncertain [Box 14.2 and 14.8.3, 14.8.6, 14.8.8]

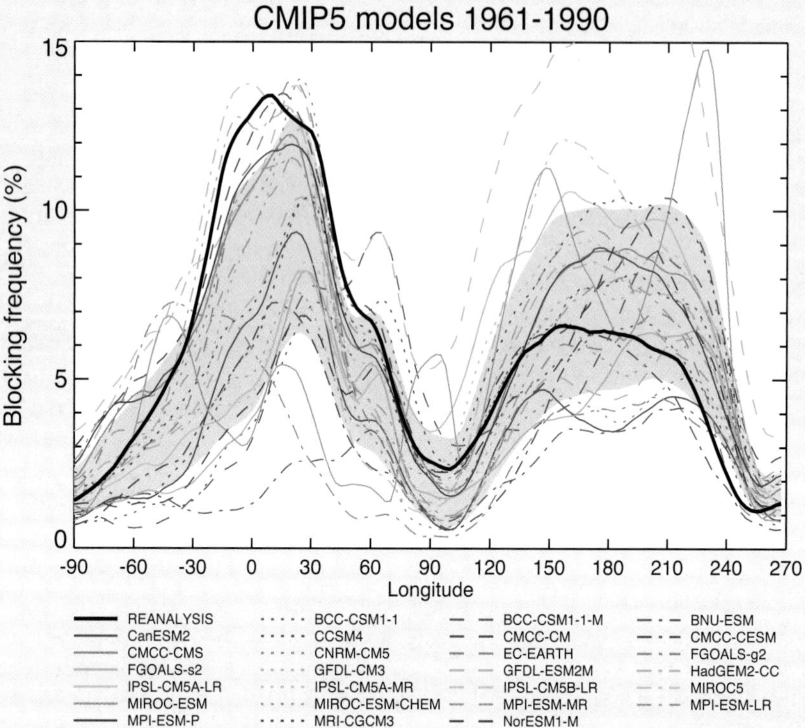

Box 14.2, Figure 1 | Annual mean blocking frequency in the NH (expressed in % of time, that is, 1% means about 4 days per year) as simulated by a set of CMIP5 models (colour lines) for the 1961–1990 period of one run of the historical simulation. Grey shading shows the mean model result plus/minus one standard deviation. Black thick line indicates the observed blocking frequency derived from the National Centers for Environmental Prediction/National Center for Atmospheric Research (NCEP/NCAR) reanalysis. Only CMIP5 models with available 500 hPa geopotential height daily data at http://pcmdi3.llnl.gov/esgcet/home.htm have been used. Blocking is defined as in Barriopedro et al. (2006), which uses a modified version of the(Tibaldi and Molteni, 1990) index. Daily data was interpolated to a common regular 2.5° × 2.5° longitude–latitude grid before detecting blocking.

14.6 Large-scale Storm Systems

14.6.1 Tropical Cyclones

The potential for regional changes in future tropical cyclone frequency, track and intensity is of great interest, not just because of the associated negative effects, but also because tropical cyclones can play a major role in maintaining regional water resources (Jiang and Zipser, 2010; Lam et al., 2012; Prat and Nelson, 2012). Past and projected increases in human exposure to tropical cyclones in many regions (Peduzzi et al., 2012) heightens the interest further.

14.6.1.1 Understanding the Causes of Past and Projected Regional Changes

Detection of past trends in measures of tropical cyclone activity is constrained by the quality of historical records and uncertain quantification of natural variability in these measures (Knutson et al., 2010; Lee et al., 2012; Seneviratne et al., 2012). Observed regional climate variability generally represents a complex convolution of natural and anthropogenic factors, and the response of tropical cyclones to each factor is not yet well understood (see also Section 10.6.1.5 and Supplementary Material Section 14.SM.4.1.2). For example, the steady long-term increase in tropical Atlantic SST due to increasing GHGs can be dominated by shorter-term decadal variability forced by both external and internal factors (Mann and Emanuel, 2006; Baines and Folland, 2007; Evan et al., 2009, 2011a; Ting et al., 2009; Zhang and Delworth, 2009; Chang et al., 2011; Solomon and Newman, 2011; Booth et al., 2012; Camargo et al., 2012; Villarini and Vecchi, 2012). Similarly, tropical upper-tropospheric temperatures, which modulate tropical cyclone potential intensity (Emanuel, 2010), can be forced by slowly evolving changes in the stratospheric circulation of ozone (Brewer–Dobson circulation) due to climate change with occasional large amplitude and persistent changes forced by volcanic eruptions (Thompson and

Solomon, 2009; Evan, 2012). This convolution of anthropogenic and natural factors, as represented in a climate model, has also been shown to be useful in prediction of Atlantic tropical storm frequency out to a few years (Smith et al., 2010).

In addition to greenhouse warming scenarios, tropical cyclones can also respond to anthropogenic forcing via different and possibly unexpected pathways. For example, increasing anthropogenic emissions of black carbon and other aerosols in South Asia has been linked to a reduction of SST gradients in the Northern Indian Ocean (Chung and Ramanathan, 2006; Meehl et al., 2008), which has in turn been linked to a weakening of the vertical wind shear in the region. Evan et al. (2011b) linked the reduced wind shear to the observed increase in the number of very intense storms in the Arabian Sea, including five very severe cyclones that have occurred since 1998, but the fundamental cause of this proposed linkage is not yet certain (Evan et al., 2012; Wang et al., 2012a). Furthermore, it is possible that a substantial part of the multi-decadal variability of North Atlantic SST is radiatively forced, via the cloud albedo effect, by what are essentially pollution aerosols emitted from North America and Europe (Baines and Folland, 2007; Booth et al., 2012), although the relative contribution of this forcing to the observed variability has been questioned (Zhang et al., 2013b). Note that in the North Atlantic, the evidence suggests that the *reduction* of pollution aerosols is linked to tropical SST increases, while in the northern Indian Ocean, *increases* in aerosol pollution have been linked to reduced vertical wind shear. Both of these effects (increasing SST and reduced shear) have been observed to be related to increased tropical cyclone activity.

Finally, in addition to interannual-to-multi-decadal forcing of tropical Atlantic SST via radiative dimming (Evan et al., 2009; Evan et al., 2011a), dust aerosols have a large and more immediate *in situ* effect on the regional thermodynamic and kinematic environment (Dunion and Marron, 2008; Dunion, 2011), and Saharan dust storms—whose frequency has been linked to atmospheric CO_2 concentration (Mahowald, 2007)—have also been linked to reduced strengthening of tropical cyclones (Dunion and Velden, 2004; Wu, 2007). Direct *in situ* relationships have also been identified between aerosol pollution concentrations and tropical cyclone structure and intensity (Khain et al., 2008, 2010; Rosenfeld et al., 2011). Thus, when assessing changes in tropical cyclone activity, it is clear that detection and attribution aimed simply at long-term linear trends forced by increasing well-mixed GHGs is not adequate to provide a complete picture of the potential anthropogenic contributions to the changes in tropical cyclone activity that have been observed (Section 10.6).

14.6.1.2 Regional Numerical Projections

Similar to observational analyses, confidence in numerical simulations of tropical cyclone activity (Supplementary Material Tables 14.SM.1 to 14.SM.4) is reduced when model spatial domain is reduced from global to region-specific (IPCC SREX Box 3.2; see also Section 9.5.4.3). The assessment provided by Knutson et al. (2010) of projections based on the SRES A1B scenario concluded that it is *likely* that the global frequency of tropical cyclones will either decrease or remain essentially unchanged while mean intensity (as measured by maximum wind speed) increases by +2 to +11% and tropical cyclone rainfall rates increase by about 20% within 100 km of the cyclone centre. However, inter-model differences in regional projections lead to lower confidence in basin-specific projections, and confidence is particularly low for projections of frequency within individual basins. For example, a recent study by Ying et al. (2012) showed that numerical projections of 21st century changes in tropical cyclone frequency in the western North Pacific range broadly from –70% to +60%, while there is better model agreement in measures of mean intensity and precipitation, which are projected to change in the region by –3% to +18% and +5% to +30%, respectively. The available modelling studies that are capable of producing very strong cyclones typically project substantial increases in the frequency of the most intense cyclones and it is *more likely than not* that this increase will be larger than 10% in some basins (Emanuel et al., 2008; Bender et al., 2010; Knutson et al., 2010, 2013; Yamada et al., 2010; Murakami et al., 2012). It should be emphasized that this metric is generally more important to physical and societal impacts than overall frequency or mean intensity.

As seen in Tables 14.SM.1 to 14.SM.4 of the Supplementary Material, as well as the previous assessments noted above, model projections often vary in the details of the models and the experiments performed, and it is difficult to objectively assess their combined results to form a consensus, particularly by region. It is useful to do this after normalizing the model output using a combination of objective and subjective expert judgements. The results of this are shown in Figure 14.17, and are based on a subjective normalization of the model output to four common metrics under a common future scenario projected through the 21st century. The global assessment is essentially the same as Knutson et al. (2010) and the assessment of projections in the western North Pacific is essentially unchanged from Ying et al. (2012). The annual frequency of tropical cyclones is generally projected to decrease or remain essentially unchanged in the next century in most regions although as noted above, the confidence in the projections is lower in specified regions than global projections. The decrease in storm frequency is apparently related to a projected decrease of upward deep convective mass flux and increase in the saturation deficit of the middle troposphere in the tropics associated with global warming (Bengtsson et al., 2007; Emanuel et al., 2008, 2012; Zhao et al., 2009; Held and Zhao, 2011; Murakami et al., 2012; Sugi et al., 2012; Sugi and Yoshimura, 2012).

A number of experiments that are able to simulate intense tropical cyclones project increases in the frequency of these storms in some regions, although there are presently only limited studies to assess and there is insufficient data to draw from in most regions to make a confident assessment (Figure 14.17). Confidence is somewhat better in the North Atlantic and western North Pacific basins where an increase in the frequency of the strongest storms is *more likely than not*. The models generally project an increase in mean lifetime-maximum intensity of simulated storms (Supplementary Material Table 14.SM.3), which is consistent with a projected increase in the frequency in the more intense storms (Elsner et al., 2008). The projected increase in intensity concurrent with a projected decrease in frequency can be argued to result from a difference in scaling between projected changes in surface enthalpy fluxes and the Clausius–Clapeyron relationship associated with the moist static energy of the middle troposphere (Emanuel et al., 2008). The increase in rainfall rates associated with

14

tropical cyclones is a consistent feature of the numerical models under greenhouse warming as atmospheric moisture content in the tropics and tropical cyclone moisture convergence is projected to increase (Trenberth et al., 2005, 2007a; Allan and Soden, 2008; Knutson et al., 2010, 2013). Although no broad-scale, detectable long-term changes in tropical cyclone rainfall rates have been reported, preliminary evidence for a detectable anthropogenic increase in atmospheric moisture content over ocean regions has been reported by Santer et al. (2007).

A number of studies since the AR4 have attempted to project future changes in tropical cyclone tracks and genesis at inter- or intra-basin scale (Leslie et al., 2007; Vecchi and Soden, 2007b; Emanuel et al., 2008; Yokoi and Takayabu, 2009; Zhao et al., 2009; Li et al., 2010b; Murakami and Wang, 2010; Lavender and Walsh, 2011; Murakami et al., 2011a, 2013). These studies suggest that projected changes in tropical cyclone activity are strongly correlated with projected changes in the spatial pattern of tropical SST (Sugi et al., 2009; Chauvin and Royer, 2010; Murakami et al., 2011b; Zhao and Held, 2012) and associated weakening of the Pacific Walker Circulation (Vecchi and Soden, 2007a), indicating that reliable projections of regional tropical cyclone activity depend critically on the reliability of the projected pattern of SST changes. However, assessing changes in regional tropical cyclone frequency is still limited because confidence in projections critically depend on the performance of control simulations (Murakami and Sugi, 2010), and current climate models still fail to simulate observed temporal and spatial variations in tropical cyclone frequency (Walsh et al., 2012). As noted above, tropical cyclone genesis and track variability is modulated in most regions by known modes of atmosphere–ocean variability. The details of the relationships vary by region (e.g., El Niño events tend to cause track shifts in western North Pacific typhoons and tend to suppress Atlantic storm genesis and development). Similarly, it has been demonstrated that accurate modelling of tropical cyclone activity fundamentally depends on the model's ability to reproduce these modes of variability (e.g., Yokoi and Takayabu, 2009). Reliable projections of future tropical cyclone activity, both global and regional, will then depend critically on reliable projections of the behaviour of these modes of variability (e.g., ENSO) under global warming, as well as an adequate understanding of their physical links with tropical cyclones. At present there is still uncertainty in their projected behaviours (e.g., Section 14.4).

The reduction in signal-to-noise ratio that accompanies changing focus from global to regional scales also lengthens the emergence time scale (i.e., the time required for a trend signal to rise above the natural variability in a statistically significant way). Based on changes in tropical cyclone intensity predicted by idealized numerical simulations with carbon dioxide (CO_2)-induced tropical SST warming, Knutson and Tuleya (2004) suggested that clearly detectable increases may not be manifest for decades to come. The more recent high-resolution dynamical downscaling study of Bender et al. (2010) supports this argument and suggests that the predicted increases in the frequency of the strongest Atlantic storms may not emerge as a statistically significant signal until the latter half of the 21st century under the SRES A1B emissionscenario. However, regional forcing by agents other than GHGs,

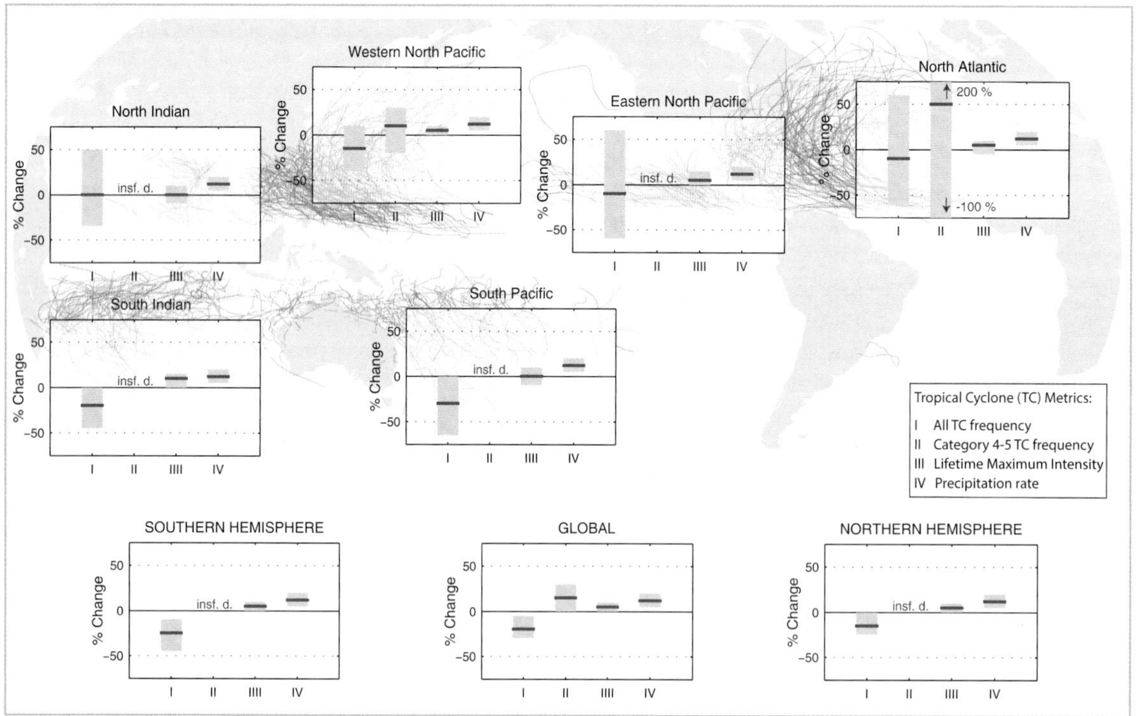

Figure 14.17 | General consensus assessment of the numerical experiments described in Supplementary Material Tables 14.SM.1 to 14.SM.4. All values represent expected percent change in the average over period 2081–2100 relative to 2000–2019, under an A1B-like scenario, based on expert judgement after subjective normalization of the model projections. Four metrics were considered: the percent change in (I) the total annual frequency of tropical storms, (II) the annual frequency of Category 4 and 5 storms, (III) the mean Lifetime Maximum Intensity (LMI; the maximum intensity achieved during a storm's lifetime) and (IV) the precipitation rate within 200 km of storm centre at the time of LMI. For each metric plotted, the solid blue line is the best guess of the expected percent change, and the coloured bar provides the 67% (*likely*) confidence interval for this value (note that this interval ranges across –100% to +200% for the annual frequency of Category 4 and 5 storms in the North Atlantic). Where a metric is not plotted, there are insufficient data (denoted 'insf. d.') available to complete an assessment. A randomly drawn (and coloured) selection of historical storm tracks are underlain to identify regions of tropical cyclone activity.

such as anthropogenic aerosols, is known to affect the regional climatic conditions differently (e.g., Zhang and Delworth, 2009), and there is evidence that tropical cyclone activity may have changed in some regions because of effects from anthropogenic aerosol pollution. The fidelity of the emergence time scales projected under A1B warming thus depends on the fidelity of A1B aerosol projections, which are known to be uncertain (Forster et al., 2007; Haerter et al., 2009).

14.6.2 Extratropical Cyclones

Some agreement on the response of storm tracks to anthropogenic forcing started to emerge in the climate model projections from CMIP3, with many models projecting a poleward shift of the storm tracks (Yin, 2005) and an expansion of the tropics (Lu et al., 2007). As stated in AR4 (Meehl et al., 2007) this response is particularly clear in the SH, but less clear in the NH. Although clearer in zonal mean responses (Yin, 2005), regional responses at different longitudes differ widely from this in many models (Ulbrich et al., 2008). There is a strong two-way coupling between storm tracks and the large-scale circulation (Lorenz and Hartmann, 2003; Robinson, 2006; Gerber and Vallis, 2007), which results in associated shifts in storm tracks and westerly jet streams (Raible, 2007; Athanasiadis et al., 2010).

14.6.2.1 Process Understanding of Future Changes

Future storm track change is the result of several different competing dynamical influences (Held, 1993; O'Gorman, 2010; Woollings, 2010). It is becoming apparent that the relatively modest storm track response in many models reflects the partial cancelling of opposing tendencies (Son and Lee, 2005; Lim and Simmonds, 2009; Butler et al., 2010).

One of the most important factors is the change in the meridional temperature gradient from which ETCs draw most of their energy. This gradient is projected to increase in the upper troposphere due to tropical amplification and decrease in the lower troposphere due to polar amplification, and it is still unclear whether this will lead to an overall increase or decrease in ETC activity. The projected response can involve an increase in eddy activity at upper levels and a decrease at lower levels (Hernandez-Deckers and von Storch, 2010), although in other models changes in low level eddy activity are more in line with the upper level wind changes (Mizuta et al., 2011; Wu et al., 2011; Mizuta, 2012). The projected warming pattern also changes vertical temperature gradients leading to increased stability at low latitudes and decreased stability at higher latitudes, and there is some modelling evidence that this may be a strong factor in the response (Lu et al., 2008, 2010; Kodama and Iwasaki, 2009; Lim and Simmonds, 2009). Increasing depth of the troposphere might also be important for future changes (Lorenz and DeWeaver, 2007).

Uncertainties in the projections of large-scale warming contribute to uncertainty in the storm track response (Rind, 2008). Several mechanisms have been proposed to explain how the storm tracks respond to the large-scale changes, including changes in eddy phase speed (Chen et al., 2007, 2008; Lu et al., 2008), eddy source regions (Lu et al., 2010) and eddy length scales (Kidston et al., 2011) with a subsequent effect on wave-breaking characteristics (Riviere, 2011). Furthermore, changes in the large-scale warming might also be partly due to changes in the storm tracks due to the two-way coupling between storm tracks and the large-scale flow. However, there is evidence that the amplitude of the tropical and polar warming are largely determined by atmospheric poleward heat fluxes set by local processes (Hwang and Frierson, 2010).

Local processes could prove important for the storm track response in certain regions, for example, sea ice loss (Kvamsto et al., 2004; Seierstad and Bader, 2009; Deser et al., 2010c; Bader et al., 2011) and spatial changes in SSTs (Graff and LaCasce, 2012). Local land–sea contrast in warming also has a local influence on baroclinicity along the eastern continental coastlines (Long et al., 2009; McDonald, 2011). It is not clear how the storm track responds to multiple forcings with some studies suggesting a linear response (Lim and Simmonds, 2009) while others suggest more complex interaction (Butler et al., 2010).

The projected increase in moisture content in a warmer atmosphere is also *likely* to have competing effects. Latent heating has been shown to play a role in invigorating individual ETCs, especially in the downstream development over eastern ocean (Dacre and Gray, 2009; Fink et al., 2009, 2012). However, there is evidence that the overall effect of moistening is to weaken ETCs by improving the efficiency of poleward heat transport and hence reducing the dry baroclinicity (Frierson et al., 2007; O'Gorman and Schneider, 2008; Schneider et al., 2010; Lucarini and Ragone, 2011). Consistent with this, studies have shown that precipitation is projected to increase in ETCs despite no increase in wind speed intensity of ETCs (Bengtsson et al., 2009; Zappa et al., 2013b).

14.6.2.2 Regional Projections

Large-scale projections of ETCs are assessed in Section 12.4.4.3. This section complements this by presenting a more detailed assessment of regional changes.

Individual model projections of regional storm track changes are often comparable with the magnitude of interannual natural variability and so the changes are expected to be relevant for regional climate. However, the magnitude of the response is model dependent at any given location, especially over land (Harvey et al., 2012). There is also disagreement between different cyclone/storm track identification methods, even when applied to the same data (Raible et al., 2008; Ulbrich et al., 2009), although in the response to anthropogenic forcing, these differences appear mainly in the statistics of weak cyclones (Ulbrich et al., 2013). Conversely, when the same method is applied to different models the spread between the model responses is often larger than the ensemble mean response, especially in the NH (Ulbrich et al., 2008; Laine et al., 2009).

The poleward shift of the SH storm track remains one of the most reproducible projections, yet even here there is considerable quantitative uncertainty. This is partly associated with the varied model biases in jet latitude (Kidston and Gerber, 2010) although factors such as the varied cloud response may play a role (Trenberth and Fasullo, 2010). Many models project a similar poleward shift in the North Pacific (Bengtsson et al., 2006; Ulbrich et al., 2008; Catto et al., 2011), although this is often weaker compared to natural variability and often varies considerably between ensemble members (Pinto et al., 2007; McDonald, 2011). Poleward shifts are generally less clear at the surface than in the upper troposphere (Yin, 2005; McDonald, 2011; Chang et al., 2012), which reduces their relevance for regional impacts. However, a shift in extreme surface winds is still detectable in the zonal mean, especially in the subtropics and the southern high latitudes (Gastineau and Soden, 2009). A weakening of the Mediterranean storm track is a particularly robust response (Pinto et al., 2007; Loeptien et al., 2008; Ulbrich et al., 2009; Donat et al., 2011) for which increasing static stability is important (Raible et al., 2010). In general, the storm track response in summer is weaker than in winter with less consistency between models (Lang and Waugh, 2011).

The response of the North Atlantic storm track is more complex than a poleward shift in many models, with an increase in storm activity and a downstream extension of the storm track into Europe (Bengtsson et al., 2006; Pinto et al., 2007; Ulbrich et al., 2008; Catto et al., 2011; McDonald, 2011). In some models this regional response is important (Ulbrich et al., 2009), with storm activity over Western Europe increasing by 50% (McDonald, 2011) or by an amount comparable to the natural variability (Pinto et al., 2007; Woollings et al., 2012). The return periods of intense cyclones are shortened (Della-Marta and Pinto, 2009) with impact on potential wind damage (Leckebusch et al., 2007; Donat et al., 2011) and economic losses (Pinto et al., 2012). This response is related to the local minimum in warming in the North Atlantic ocean, which serves to increase the meridional temperature gradient on its southern side (Laine et al., 2009; Catto et al., 2011). The minimum in warming arises due to the weakening of northward ocean heat transports by the Atlantic Meridional Overturning Circulation (AMOC), and the varying AMOC responses of the models can account for a large fraction of the variance in the Atlantic storm track projections (Woollings et al., 2012). CMIP5 models show a similar, albeit weaker extension of the storm track towards Europe, flanked by reductions in cyclone activity on both the northern and southern sides (Harvey et al., 2012; Zappa et al., 2013b). Despite large biases in the mean state, the model responses were found to agree with one another within sampling variation caused by natural variability (Sansom et al., 2013). Colle et al. (2013) noted similar reductions but also found that the higher resolution CMIP5 models gave more realistic ETC performance in the historical period. The best 7 models were found to give projections of increased 10 to 20% increase in cyclone track density over the eastern USA, including 10 to 40% more intense (<980 hPa) cyclones.

There is general agreement that there will be a small global reduction in ETC numbers (Ulbrich et al., 2009). In individual regions there can be much larger changes which are comparable to natural variations, but these changes are not reproduced by the majority of the models (e.g., Donat et al., 2011). ETC intensities are particularly sensitive to the method and quantity used to define them, so there is little consensus on changes in intensity (Ulbrich et al., 2009). While there are indications that the absolute values of pressure minima deepen in future scenario simulations (Lambert and Fyfe, 2006), this is often associated with large-scale pressure changes rather than changes in the pressure gradients or winds associated with ETCs (Bengtsson et al., 2009; Ulbrich et al., 2009; McDonald, 2011). The CMIP5 model projections show little evidence of change in the intensity of winds associated with ETCs (Zappa et al., 2013b).

There are systematic storm track biases common to many models, which might have some influence on the projected storm track response to forcing (Chang et al., 2012). Some models with improved representation of the stratosphere have shown a markedly different circulation response in the NH, with consequences for Atlantic/European storm activity in particular (Scaife et al., 2011a). Concerns over the skill of many models in representing both the stratosphere and the ocean mean that *confidence* in NH storm track projections remains *low*. Higher horizontal resolution can improve ETC representation, yet there are still relatively few high-resolution global models which have been used for storm track projections (Geng and Sugi, 2003; Bengtsson et al., 2009; Catto et al., 2011; Colle et al., 2013; Zappa et al., 2013a). Several studies have used RCMs to simulate storms at high resolution in particular regions. In multi-model experiments over Europe, the ETC response is more sensitive to the choice of driving GCM than the choice of RCM (Leckebusch et al., 2006; Donat et al., 2011), highlighting the importance of large-scale circulation uncertainties. There has been little work on potential changes to mesoscale storm systems, although it has been suggested that polar lows may reduce in frequency due to an increase in static stability (Zahn and von Storch, 2010). Higher resolution runs of one climate model also suggest an increase in intensity of autumn ETCs due to increased transitioning of Atlantic hurricanes (Haarsma et al., 2013).

14.6.3 Assessment Summary

The influence of past and future climate change on tropical cyclones is *likely* to vary by region, but the specific characteristics of the changes are not yet well understood, and the substantial influence of ENSO and other known climate modes on global and regional tropical cyclone activity emphasizes the need for more reliable assessments of future changes in the characteristics of these modes. Recent advances in understanding and phenomenological evidence for shorter-term effects on tropical cyclones from aerosol forcing are providing increasingly greater confidence that anthropogenic forcing has had a measurable effect on tropical cyclone activity in certain regions. Shorter term increases such as those observed in the Atlantic over the past 30 to 40 years appear to be robust and have been hypothesized to be related, in part, to regional external forcing by GHGs and aerosols, but the more steady century-scale trends that may be expected from CO_2 forcing alone are much more difficult to assess given the data uncertainty in the available tropical cyclone records.

Although projections under 21st century greenhouse warming indicate that it is *likely* that the global frequency of tropical cyclones will either decrease or remain essentially unchanged, concurrent with a *likely* increase in both global mean tropical cyclone maximum wind speed and rainfall rates, there is *low confidence* in region-specific projections of frequency and intensity. Still, based on high-resolution modelling studies, the frequency of the most intense storms, which are associated with particularly extensive physical effects, will *more likely than not* increase substantially in some basins under projected 21st century warming and there is *medium confidence* that tropical cyclone rainfall rates will increase in every affected region.

The global number of ETCs is *unlikely* to decrease by more than a few percent due to anthropogenic change. A small poleward shift is *likely* in the SH storm track, but the magnitude is model dependent. There is only *medium confidence* in projections of storm track shifts in the

Northern Hemisphere. Nevertheless, model results suggests that it is *more likely than not* that the N. Pacific storm track will shift poleward, and that it is *unlikely* that the N. Atlantic storm track will respond with a simple poleward shift. There is *low confidence* in the magnitude of regional storm track changes, and the impact of such changes on regional surface climate.

14.7 Additional Phenomena of Relevance

14.7.1 Pacific–South American Pattern

The PSA pattern is a teleconnection from the tropics to SH extratropics through Rossby wave trains (Box 2.5). This pattern induces atmospheric circulation anomalies over South America, affecting extreme precipitation (Drumond and Ambrizzi, 2005; Cunningham and Cavalcanti, 2006; Muza et al., 2009; Vasconcellos and Cavalcanti, 2010). PSA trends for recent decades depend on the choice of indices and are hence uncertain (Section 2.7.8). The PSA pattern is reproduced in many model simulations (Solman and Le Treut, 2006; Vera and Silvestri, 2009; Bates, 2010; Rodrigues et al., 2011; Cavalcanti and Shimizu, 2012).

The intensification and westward displacement of the PSA wave pattern in projections of CMIP3 may be related to the increase in frequency and intensity of positive SST anomalies in the tropical Pacific by the end of the 21st century (2081–2100, Junquas et al., 2012). These perturbations of the PSA characteristics are linked with changes in SACZ dipole precipitation and affect South America precipitation (Section 14.3.1.3). The PSA pattern occurrence and implications for precipitation increase in the southeastern South America have been associated with the zonally asymmetric part of the global SST change in the equatorial Indian–western Pacific Oceans (Junquas et al., 2013).

Process understanding of the formation of the PSA gives *medium confidence* that future changes in the mean atmospheric circulation for austral summer will project on the PSA pattern thereby influencing the SACZ and precipitation over southeastern South America. However, the literature is not sufficient to assess more general changes in PSA, and *confidence* is *low* in its future projections.

14.7.2 Pacific–North American Pattern

The PNA pattern, as defined in Box 2.5, Table 1 and portrayed in Box 2.5, Figure 2 is a prominent mode of atmospheric variability over the North Pacific and the North American land mass. This phenomenon exerts notable influences on the temperature and precipitation variability in these regions (e.g., Nigam, 2003). The PNA pattern is related to ENSO events in the tropical Pacific (see Section 14.4), and also serves as a bridge linking ENSO and NAO variability (see Li and Lau, 2012). The data records indicate a significant positive trend in the wintertime PNA index over the past 60 years (see Table 2.14 and Box 2.5, Figure 1).

Stoner et al. (2009) have assessed the capability of 22 CMIP3 GCMs in replicating the essential aspects of the observed PNA pattern. Their results indicate that a majority of the models overestimate the fraction of variance explained by the PNA pattern, and that the spatial characteristics of PNA patterns simulated in 14 of the 22 models are in good agreement with the observations. The model-projected future evolution of the PNA pattern has not yet been fully assessed and therefore *confidence* in its future development is *low*.

14.7.3 Pacific Decadal Oscillation/Inter-decadal Pacific Oscillation

The Pacific Decadal Oscillation (PDO, Box 2.5) refers to the leading Empirical Orthogonal Function (EOF) of monthly SST anomalies over the North Pacific (north of 20°N) from which the globally averaged SST anomalies have been subtracted (Mantua et al., 1997). It exhibits anomalies of one sign along the west coast of North America and of opposite sign in the western and central North Pacific (see also Section 9.5.3.6 and Chapter 11). The PDO is closely linked to fluctuations in the strength of the wintertime Aleutian Low Pressure System. The time scale of the PDO is around 20 to 30 years, with changes of sign between positive and negative polarities in the 1920s, the late 1940s, the late 1970s and around 2000.

The extension of the PDO to the whole Pacific basin is known as the Inter-decadal Pacific Oscillation (IPO, Power et al., 1999). The IPO is nearly identical in form to the PDO in the NH but is defined globally, as a leading EOF (principal component) of 13-year lowpass-filtered global SST anomalies (Parker et al., 2007) and has substantial amplitude in the tropical and southern Pacific. The time series of the PDO and IPO correlate highly on an annual basis. The PDO/IPO pattern is considered to be the result of internal climate variability (Schneider and Cornuelle, 2005; Alexander, 2010) and has not been observed to exhibit a long-term trend. The PDO/IPO is associated with ENSO modulations, with more El Niño activity during the positive PDO/IPO and more La Niña activity during the negative PDO/IPO.

At the time of the AR4, little had been published on modelling of the PDO/IPO or of its evolution in future. In a recent study, Furtado et al. (2011) found that the PDO/IPO did not exhibit major changes in spatial or temporal characteristics under GHG warming in most of the 24 CMIP3 models used, although some models indicated a weak shift toward more occurrences of the negative phase of the PDO/IPO by the end of the 21st century (2081–2100, Lapp et al., 2012). However, given that the models strongly underestimate the PDO/IPO connection with tropical Indo-Pacific SST variations (Furtado et al., 2011; Lienert et al., 2011), the credibility of the projections remains uncertain. Furthermore, internal variability is so high that it is hard to detect any forced changes in the Aleutian Low for the next half a century (Deser et al., 2012; Oshima et al., 2012). Therefore *confidence* is *low* in projections of future changes in PDO/IPO.

14.7.4 Tropospheric Biennial Oscillation

The Tropospheric Biennial Oscillation (TBO; Meehl, 1997) is a proposed mechanism for the biennial tendency in large-scale drought and floods of south Asia and Australia. Multiple studies imply that TBO involves the Asian-Australian monsoon, the IOD and ENSO (Sections 14.2.2, 14.3.3 and 14.4; see also Supplementary Material Section 14.SM.5.2).

The IPO (Section 14.7.3) affects the decade-to-decade strength of the TBO. A major contributor to recent change in the TBO comes from increase of SST in the Indian Ocean that contributes to stronger trade winds in the Pacific, one of the processes previously identified with strengthening the TBO (Meehl and Arblaster, 2012). Thus, prediction of decadal variability assessed in Chapter 11 that can be associated, for example, with the IPO (e.g., Meehl et al., 2010) can influence the accuracy of shorter-term predictions of the TBO across the entire Indo-Pacific region (Turner et al., 2011), but the relevance for longer time scales is uncertain.

Since AR4, little work has been done to document the ability of climate models to simulate the TBO. However, Li et al. (2012b) showed that there is an overall improvement in the seasonality of monsoons rainfall related to changes in the TBO from CMIP3 to CMIP5, with most CMIP5 models better simulating both the monsoon timing and the very low rainfall rates outside of the monsoon season (see also Section 14.2.2). In addition they concluded that the India-Australia link seems to be robust in all models.

With regard to possible future behaviour of the TBO, no analysis using multiple GCMs has been made since the AR4. In models that more accurately simulate the TBO in the present-day climate, the TBO strengthens in a future warmer climate (Nanjundiah et al., 2005). However, as with ENSO (Section 14.4) and IOD (Section 14.3.3), internally generated decadal variability complicates the interpretation of such future changes. Therefore, it remains unclear how future changes in the TBO will emerge and how this may influence regional climate in the future. *Confidence* in the projected future changes in TBO remains *low*.

14.7.5 Quasi-Biennial Oscillation

The QBO is a near-periodic, large-amplitude, downward propagating oscillation in zonal winds in the equatorial stratosphere (Baldwin et al., 2001). It is driven by vertically propagating internal waves that are generated in the tropical troposphere (Plumb, 1977). The QBO has substantial effects on the global stratospheric circulation, in particular the strength of the northern stratospheric polar vortex as well as the extratropical troposphere (Boer, 2009; Marshall and Scaife, 2009; Garfinkel and Hartmann, 2011). These extratropical effects occur primarily in winter when the stratosphere and troposphere are strongly coupled (Anstey and Shepherd, 2008; Garfinkel and Hartmann, 2011).

It has been unclear how the QBO will respond to future climate change related to GHG increase and recovery of stratospheric ozone. Climate models assessed in the AR4 did not simulate the QBO as they lacked the necessary vertical resolution (Kawatani et al., 2011). Recent model studies without using gravity wave parameterization (Kawatani et al., 2011; Kawatani et al., 2012) showed that the QBO period and amplitude may become longer and weaker, and the downward penetration into the lowermost stratosphere may be more curtailed in a warmer climate. This finding is attributed to the effect of increased equatorial upwelling (stronger Brewer–Dobson circulation; Butchart et al., 2006; Garcia and Randel, 2008; McLandress and Shepherd, 2009; Okamoto et al., 2011) dominating the effect of increased wave forcing (more convective activity). Two studies with gravity wave parameterization, however, gave conflicting results depending on the simulated changes in the intensity of the Brewer–Dobson circulation (Watanabe and Kawatani, 2012).

There are limited published results on the future behaviour of the QBO, using CMIP5 models. On the basis of the recent literature, it is uncertain how the period or amplitude of the QBO may change in future and *confidence* in the projections remains *low*.

14.7.6 Atlantic Multi-decadal Oscillation

The AMO (Box 2.5; see also Section 9.5.3.3.2) is a fluctuation seen in the instrumental SST record throughout the North Atlantic Ocean and is related to variability in the thermohaline circulation (Knight et al., 2005). Area-mean North Atlantic SST shows variations with a range of about 0.4°C (see Box 2.5) and a warming of a similar magnitude since 1870. The AMO has a quasi-periodicity of about 70 years, although the approximately 150-year instrumental record possesses only a few distinct phases—warm during approximately 1930–1965 and after 1995, and cool between 1900–1930 and 1965–1995. The phenomenon has also been referred to as 'Atlantic Multidecadal Variability' to avoid the implication of temporal regularity. Along with secular trends and Pacific variability, the AMO is one of the principal features of multidecadal variability in the instrumental climate record.

The AR4 highlighted a number of important links between the AMO and regional climates. Subsequent research using observational and paleoclimate records, and climate models, has confirmed and expanded upon these connections, such as West African monsoon and Sahel rainfall (Mohino et al., 2011; Section 14.2.4), summer climate in North America (Seager et al., 2008; Section 14.8.3; Feng et al., 2011) and Europe (Folland et al., 2009; Ionita et al., 2012; Section 14.8.6), the Arctic (Chylek et al., 2009; Mahajan et al., 2011), and Atlantic major hurricane frequency (Chylek and Lesins, 2008; Zhang and Delworth, 2009; Section 14.6.1). Further, the list of AMO influences around the globe has been extended to include decadal variations in many other regions (e.g., Zhang and Delworth, 2006; Kucharski et al., 2009a, 2009b; Huss et al., 2010; Marullo et al., 2011; Wang et al., 2011).

Paleo reconstructions of Atlantic temperatures show AMO-like variability well before the instrumental era, as noted in the AR4 (Chapter 6; see also Section 5.4.2). Recent analyses confirm this, and suggest potential for intermittency in AMO variability (Saenger et al., 2009; Zanchettin et al., 2010; Chylek et al., 2012). Control simulations of climate models run for hundreds or thousands of years also show long-lived Atlantic multi-decadal variability (Menary et al., 2012). These lines of evidence suggest that AMO variability will continue into the future. No fundamental changes in the characteristics of North Atlantic multi-decadal variability in the 21st century are seen in CMIP3 models (Ting et al., 2011).

Many studies have diagnosed a trend towards a warm North Atlantic in recent decades additional to that implied by global climate forcings (Knight, 2009; Polyakov et al., 2010). It is unclear exactly when the current warm phase of the AMO will terminate, but may occur within the next few decades, leading to a cooling influence in the North Atlantic and offsetting some of the effects characterizing global warming (Keenlyside et al., 2008; see also Section 11.3.3.3).

Some similarity in the shape of the instrumental time series of global and NH mean surface temperatures and the AMO has long been noted. By removing an estimate of the effect of interannual variability phenomena like ENSO (Section 14.4), AMO transitions have been shown to have the potential to produce large and abrupt changes in hemispheric temperatures (Thompson et al., 2010). Estimates of the AMO's contribution to recent climate change are uncertain, however, as attribution of the observed AMO requires a model (physical or conceptual) whose assumptions are nearly always difficult to verify (Knight, 2009).

14.7.7 Assessment Summary

Literature is generally insufficient to assess future changes in behaviour of the PNA, PSA, TBO, QBO and PDO/IPO. *Confidence* in the projections of changes in these modes is therefore *low*. However, process understanding of the formation of the PSA gives *medium confidence* that future changes in the mean atmospheric circulation for austral summer will project on the PSA pattern thereby influencing the SACZ and precipitation over Southeastern South America.

Paleoclimate reconstructions and long model simulations indicate that the AMO is *unlikely* to change its behaviour in the future as the mean climate changes. However, natural fluctuations in the AMO over the coming few decades are *likely* to influence regional climates at least as strongly as will human-induced changes.

14.8 Future Regional Climate Change

14.8.1 Overview

The following sections assess future climate projections for several regions, and relate them, where possible, to projected changes in the major *climate phenomena* assessed in Sections 14.2 to 14.7. The regional climate change assessments are mainly of mean surface air temperature and mean precipitation based primarily on multi-model ensemble projections from general circulation models. Reference is made to the appropriate projection maps from CMIP5 (Taylor et al., 2011c) presented in Annex I: Atlas of Global and Regional Climate Projections. Annex I uses smaller sub-regions similar to those introduced by SREX (Seneviratne et al., 2012). Table 14.1 presents a quantitative summary of the regional area averages over three projection periods (2016–2035, 2046–2065 and 2081–2100 with respect to the reference period 1986–2005, representing near future, middle century and end of century) for the RCP4.5 scenario. The 26 land regions assessed here are presented in Seneviratne et al., 2012, page 12 and the coordinates can be found from their online Appendix 3.A. Added to this are six additional regions containing the two polar regions, the Caribbean, Indian Ocean and Pacific Island States (see Annex I for further details). Table 14.1 identifies the smaller sub-domains grouped within the somewhat large regions that are discussed in Sections 14.8.2 to 14.8.15. Tables for RCP2.6, RCP6.0 and RCP8.5 scenarios are presented in Supplementary Material Tables 14.SM.1a to 14.SM.1c. For continental-scale regions, projected changes in mean precipitation between (2081–2100) and (1986–2005) are compared in two generations of models forced under two comparable emission scenarios: RCP4.5 in CMIP5 versus A1B in CMIP3. In contrast to the Annex, the seasons here are chosen differently for each region so as to best capture the regional features such as monsoons. Downscaling issues are illustrated in panels showing results from an ensemble of high-resolution time-slice experiments with the Meteorological Research Institute (MRI) model (Endo et al., 2012; Mizuta et al., 2012). To facilitate a direct comparison across the scenarios, the precipitation changes are normalized by the global annual mean surface air temperature changes in each scenario. Published results using other downscaling methods are also assessed when found essential to illustrate issues related to regional climate change.

Regional climate projections are generally more uncertain than projections of global mean temperature but the sources of uncertainty are similar (see Chapters 8, 11, and 12) yet differ in relative importance. For example, natural variability (Deser et al., 2012), aerosol forcing (Chapter 7) and land use/cover changes (DeFries et al., 2002; Moss et al., 2010) all become more important sources of uncertainty on a regional scale. Regional climate assessments incur additional uncertainty due to the cascade of uncertainty through the hierarchy of models needed to generate local information (cf. downscaling in Section 9.6). Calibration (bias correction) of model output to match local observations is an additional important source of uncertainty in regional climate projections (e.g., Ho et al., 2012), which should be considered when interpreting the regional projections. Therefore, the model spread shown in Annex 1 should not be interpreted as the final uncertainty in the observable regional climate change response.

Table 14.2 summarizes the assessed confidence in the ability of CMIP5 models to represent regional scale present-day climate (temperature and precipitation, based on Chapter 9), the main controlling phenomena for weather and climate in that region and the assessed resulting confidence in the future projections. There is generally less confidence in projections of precipitation than of temperature. For example, in Annex I, the temperature projections for 2081–2100 are almost always above the model estimates of natural variability, whereas the precipitation projections less frequently rise above natural variability. Although some projections are robust for reasons that are well understood (e.g., the projected increase in precipitation at high latitudes), many other regions have precipitation projections that vary in sign and magnitude across the models. These issues are further discussed in Section 12.4.5.2. Details on how the confidence table is constructed are found in the Supplementary Material.

Credibility in regional climate change projections is increased if it is possible to find key drivers of the change that are known to be well-simulated and well-projected by climate models. Table 14.3 summarizes the assessment of how major climate phenomena might be relevant for future regional climate change. For each entry in the table, the relevance is based on an assessment of confidence in future change in the phenomenon and the confidence in how the phenomenon influences regional climate. For example, NAO is assigned high relevance (red) for the Arctic region because NAO is known to influence the Arctic and there is *high confidence* that the NAO index will increase in response to anthropogenic forcing. If there is *low confidence* in how a phenomenon might change (e.g., ENSO) but *high confidence* that it has a strong regional impact, then the cell in the table is assigned medium relevance (yellow). It can be seen from the table that there are many cases where major phenomena are assessed to have high (red) or medium (yellow) relevance for future regional climate change. See Supplementary Material Section 14.SM.6.1 for more details on how this relevance table was constructed.

Frequently Asked Questions

FAQ 14.2 | How Are Future Projections in Regional Climate Related to Projections of Global Means?

The relationship between regional climate change and global mean change is complex. Regional climates vary strongly with location and so respond differently to changes in global-scale influences. The global mean change is, in effect, a convenient summary of many diverse regional climate responses.

Heat and moisture, and changes in them, are not evenly distributed across the globe for several reasons:

- External forcings vary spatially (e.g., solar radiation depends on latitude, aerosol emissions have local sources, land use changes regionally, etc.).
- Surface conditions vary spatially, for example, land/sea contrast, topography, sea surface temperatures, soil moisture content.
- Weather systems and ocean currents redistribute heat and moisture from one region to another.

Weather systems are associated with regionally important climate phenomena such as monsoons, tropical convergence zones, storm tracks and important modes of climate variability (e.g., El Niño-Southern Oscillation (ENSO), North Atlantic Oscillation (NAO), Southern Annular Mode (SAM), etc.). In addition to modulating regional warming, some climate phenomena are also projected to change in the future, which can lead to further impacts on regional climates (see Table 14.3).

Projections of change in surface temperature and precipitation show large regional variations (FAQ 14.2, Figure 1). Enhanced surface warming is projected to occur over the high-latitude continental regions and the Arctic ocean,

(continued on next page)

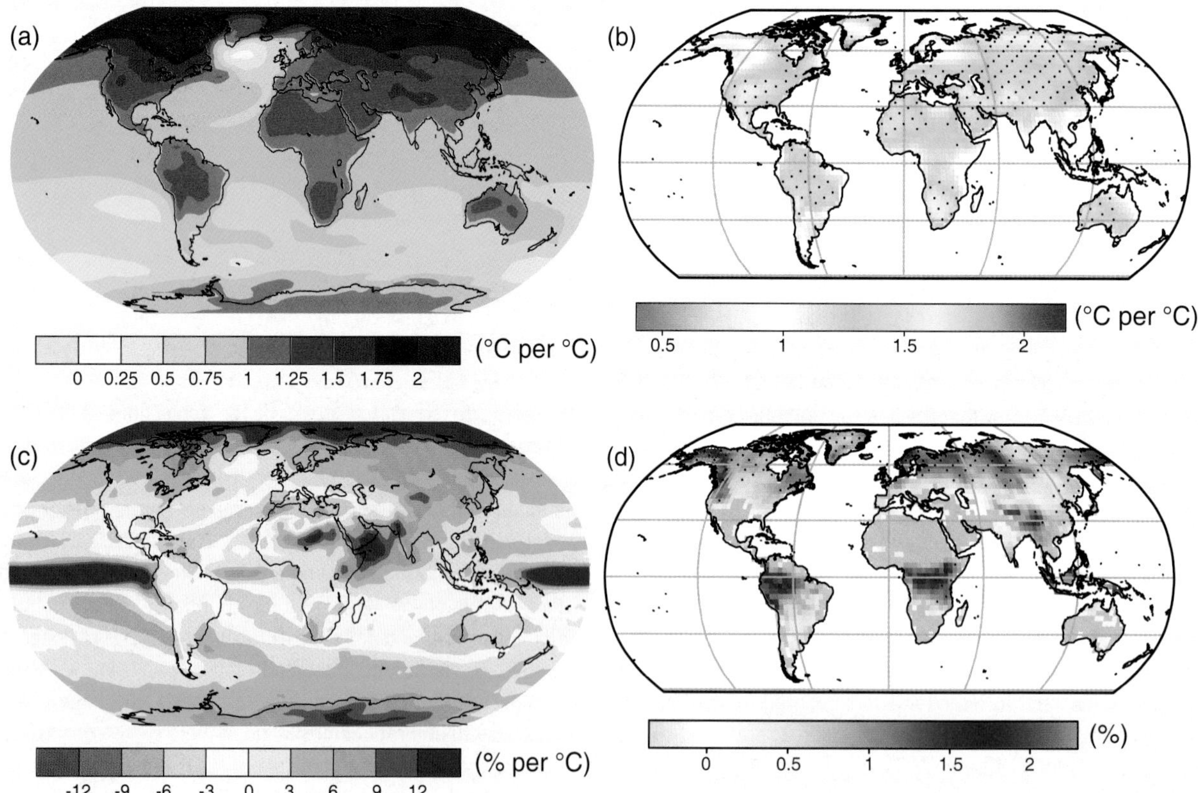

FAQ 14.2, Figure 1 | Projected 21st century changes in annual mean and annual extremes (over land) of surface air temperature and precipitation: (a) mean surface temperature per °C of global mean change, (b) 90th percentile of daily maximum temperature per °C of global average maximum temperature, (c) mean precipitation (in % per °C of global mean temperature change), and (d) fraction of days with precipitation exceeding the 95th percentile. Sources: Panels (a) and (c) projected changes in means between 1986–2005 and 2081–2100 from CMIP5 simulations under RCP4.5 scenario (see Chapter 12, Figure 12.41); Panels (b) and (d) projected changes in extremes over land between 1980–1999 and 2081–2100 (adapted from Figures 7 and 12 of Orlowsky and Seneviratne, 2012).

FAQ 14.2 (continued)

while over other oceans and lower latitudes changes are closer to the global mean (FAQ 14.2, Figure 1a). For example, warming near the Great Lakes area of North America is projected to be about 50% greater than that of the global mean warming. Similar large regional variations are also seen in the projected changes of more extreme temperatures (FAQ 14.2, Figure 1b). Projected changes in precipitation are even more regionally variable than changes in temperature (FAQ 14.2, Figure 1c, d), caused by modulation from climate phenomena such as the monsoons and tropical convergence zones. Near-equatorial latitudes are projected to have increased mean precipitation, while regions on the poleward edges of the subtropics are projected to have reduced mean precipitation. Higher latitude regions are projected to have increased mean precipitation and in particular more extreme precipitation from extratropical cyclones.

Polar regions illustrate the complexity of processes involved in regional climate change. Arctic warming is projected to increase more than the global mean, mostly because the melting of ice and snow produces a regional feedback by allowing more heat from the Sun to be absorbed. This gives rise to further warming, which encourages more melting of ice and snow. However, the projected warming over the Antarctic continent and surrounding oceans is less marked in part due to a stronger positive trend in the Southern Annular Mode. Westerly winds over the mid-latitude southern oceans have increased over recent decades, driven by the combined effect of loss of stratospheric ozone over Antarctica, and changes in the atmosphere's temperature structure related to increased greenhouse gas concentrations. This change in the Southern Annular Mode is well captured by climate models and has the effect of reducing atmospheric heat transport to the Antarctic continent. Nevertheless, the Antarctic Peninsula is still warming rapidly, because it extends far enough northwards to be influenced by the warm air masses of the westerly wind belt.

14.8.2 Arctic

This section is concerned with temperature and precipitation dimensions of Arctic climate change, and their links to climate phenomena. The reader is referred elsewhere for information on sea ice loss (Sections 4.2.2, 5.5.2 and Chapter 10), and projections of sea ice change (Sections 9.4.3, 9.8.3 and Chapters 11 and 12).

Arctic climate is affected by three modes of variability: NAO (Section 14.5.1), PDO (Section 14.7.3) and AMO (Section 14.7.6). The NAO index correlates positively with temperatures in the northeastern Eurasian sector, and correlates negatively with temperatures in the Baffin Bay and Canadian Archipelago, but exhibits little relationship with central Arctic temperatures (Polyakov et al., 2003). The PDO plays a role in temperature variability of Alaska and the Yukon (Hartmann and Wendler, 2005). The AMO is positively associated with SST throughout the Arctic (Chylek et al., 2009; Levitus et al., 2009; Chylek et al., 2010) (Mahajan et al., 2011). ETCs are also mainly responsible for winter precipitation in the region (see Table 14.3).

The surface and lower troposphere in the Arctic and surrounding land areas show regional warming over the past three decades of about 1 °C per decade—significantly greater than the global mean trend (Figures 2.22 and 2.25). According to temperature reconstructions, this signal is highly unusual: Temperatures averaged over the Arctic over the past few decades are significantly higher than any seen over the past 2000 years (Kaufman et al., 2009). Temperatures 11 ka were greater than the 20th century mean, but this is probably a strongly forced signal, since summer solar radiation was 9% greater than present (Miller et al., 2010). Finally, warmer temperatures have been sustained in pan-Arctic land areas where a declining NAO over the past decade ought to have caused cooling (Semenov, 2007; Turner et al., 2007b). Since AR4, evidence has also emerged that precipitation has trended upward in most pan-Arctic land areas over the past few decades (e.g., Pavelsky and Smith, 2006; Rawlins et al., 2010), though the evidence remains mixed (e.g., Dai et al., 2009). Increasing ETC activity over the Canadian Arctic has also been observed (Section 2.6.4).

Since AR4, there has been progress in adapting RCMs for polar applications (Wilson et al., 2012). These models have been evaluated with regard to their ability to simulate Arctic clouds, surface heat fluxes, and boundary layer processes (Tjernstrom et al., 2004; Inoue et al., 2006; Rinke et al., 2006). They have been used to improve simulations of Arctic-specific climate processes, such as glacial mass balance (Zhang et al., 2007). A few regional models have been used for Arctic climate change projections (e.g., Zahn and von Storch, 2010; Koenigk et al., 2011; Döscher and Koenigk, 2012). For information on GCM quality in the Arctic, see Chapter 9 and the brief summary of assessed confidence in the CMIP5 models in Table 14.2.

The CMIP5 model simulations exhibit an ensemble-mean polar amplified warming, especially in winter, similar to CMIP3 model simulations (Bracegirdle and Stephenson, 2012; see also Box 5.1). For RCP4.5, ensemble-mean winter warming rises to 5.0°C over pan-Arctic land areas by the end of the 21st century (2081–2100), and about 7.0°C over the Arctic Sea (Table14.1). Throughout the century, the warming exceeds simulated estimates of internal variability (Figure AI.8). The RCP4.5 ensemble-mean warming is more modest in JJA (Table 14.1), reaching about 2.2°C by century's end over pan-Arctic land areas, and 1.5°C over the Arctic Sea. The summer warming exceeds variability estimates by about mid-century (Figure AI.9). These simulated anthropogenic seasonal warming patterns match qualitatively the observed warming patterns over the past six decades (AMAP, 2011), and the observed warming patterns are *likely* to be at least partly anthropogenic in origin (Section 10.3.1.1.4). Given the magnitude of future projected changes relative to variability, and the presence of anthropogenic signals already, it is *likely* future Arctic surface temperature changes will continue to be strongly influenced by the anthropogenic forcing over the coming decades.

The CMIP5 models robustly project precipitation increases in the pan-Arctic (both land and sea) region over the 21st century, as did their CMIP3 counterparts (Kattsov et al., 2007; Rawlins et al., 2010). Under the RCP4.5 scenario, the cold season, ensemble mean precipitation increases about 25% by the century's end (Table 14.1), due to enhanced precipitation in ETCs (Table 14.3). However, this signal does not rise consistently above the noise of simulated variability until mid 21st century (Figure AI.10). During the warm season, precipitation increases are smaller, about 15% (Table 14.1), though these signals also rise above variability by mid 21st century (Figure AI.11). The inter-model spread in the precipitation increase is generally as large as the ensemble mean signal itself (similar to CMIP3 model behaviour, Holland and Webster, 2007), so the magnitude of the future increase is uncertain. However, since nearly all models project a large precipitation increase rising above the variability year-round, it is *likely* the pan-Arctic region will experience a statistically significant increase in precipitation by mid-century (see also Table 14.2). The small projected increase in the NAO is *likely* to affect Arctic precipitation (and temperature) patterns in the coming century (Section 14.5.1; Table 14.3), though the importance of these signals relative to anthropogenic signals described here is unclear.

In summary: It is *likely* Arctic surface temperature changes will be strongly influenced by anthropogenic forcing over the coming decades dominating natural variability such as induced by NAO. It is *likely* the pan-Arctic region will experience a significant increase in precipitation by mid-century due mostly to enhanced precipitation in ETCs.

14.8.3 North America

The climate of North America (NA) is affected by the following phenomena: NAO (Section 14.5.1), ENSO (Section 14.4), PNA (Section 14.7.2), PDO (Section 14.7.3), NAMS (Section 14.2.3), TCs and ETCs (Section 14.6). The NAO affects temperature and precipitation over Eastern NA during winter (Hurrell et al., 2003). Positive PNA brings warmer temperatures to northern Western NA and Alaska in winter, cooler temperatures to the southern part of Eastern NA, and dry conditions to much of Eastern NA (Nigam, 2003). The PNA can also be excited by ENSO-related SST anomalies (Horel and Wallace, 1981; Nigam, 2003). The PDO is linked to decadal climate anomalies resembling those of the PNA. The NAMS brings excess summer rainfall to Central America and Mexico and the southern portion of Western NA (Gutzler, 2004). TCs also impact the Gulf Coast and Eastern NA (see Section 14.6.1). The AMM and AMO may affect their frequency and intensity (Landsea et al., 1999; Goldenberg et al., 2001; Cassou et al., 2007; Emanuel, 2007; Vimont and Kossin, 2007; Smirnov and Vimont, 2011). ETCs are also mainly responsible for winter precipitation, especially in the northern half of NA. See Table 14.3 for a summary of this information.

A general surface warming over NA has been documented over the last century (see Section 2.4). It is particularly large over Alaska and northern Western NA during winter and spring and the northern part of Eastern NA during summer (Zhang et al., 2011b). There is also a cooling tendency over Central and Eastern NA (i.e., the 'warming hole' discussed in Section 2.4.1) during spring, though it is absent in lower tropospheric temperature (cf. Figure 2.25). The warming has coincided with a general decline in NA snow extent and depth (Brown and Mote, 2009; McCabe and Wolock, 2010; Kapnick and Hall, 2012). Consistent with surface temperature trends, temperature extremes also exhibit secular changes. Cold days and nights have decreased in the last half century, while warm days and nights have increased (see Chapter 2). These changes are especially apparent for nightly extremes (Vincent et al., 2007). It is unclear whether there have been mean precipitation trends over the last 50 years (Section 2.5.1; Zhang et al., 2011b). However, precipitation extremes increased, especially over Central and Eastern NA (see Section 2.6.2 and Seneviratne et al., 2012).

Table 14.2 provides an assessment of GCM quality for simulations of temperature, precipitation, and main phenomena in NA's regions. Regarding regional modelling experiments since AR4, biases have decreased somewhat as resolutions increase. The North American Regional Climate Change Assessment Program created a simulation suite for NA at 50-km resolution. When forced by reanalyses, this suite generally reproduces climate variability within observational error (Leung and Qian, 2009; Wang et al., 2009b; Gutowski, 2010; Mearns et al., 2012).Other regional modelling experiments covering parts or all of NA have shown improvements as resolution increases (Liang et al., 2008a; Lim et al., 2011; Yeung et al., 2011), including for extremes (Kawazoe and Gutowski, 2013). Bias reductions are large for snowpack in topographically complex Western NA, as revealed by 2- to 20-km resolution regional simulations (Qian et al., 2010b; Salathe Jr. et al., 2010; Pavelsky et al., 2011; Rasmussen et al., 2011). Thus there has been substantial progress since AR4 in understanding the value of regional modelling in simulating NA climate. The added value of using regional models to simulate climate change is discussed in Section 9.6.6.

NA warming patterns in RCP4.5 CMIP5 projections are generally similar to those of CMIP3 (Figures AI.4 and AI.5, Table 14.1). In winter, warming is greatest in Alaska, Canada, and Greenland (Figures AI.12 and AI.16), while in summer, maximum warming shifts south, to Western, Central, and Eastern NA. Examining near-term (2016–2035) CMIP5 projections of the less sensitive models (25th percentile, i.e., upper left maps in Figures AI.12, AI.13, AI.16, AI.17, AI.20 AI.21, AI.24 and AI.25), the warming generally exceeds natural variability estimates. Exceptions are Alaska, parts of Western, Central, and Eastern NA, and Canada and Greenland during winter, when natural variability linked to wintertime storms is particularly large. By 2046–2065, warming in all regions exceeds the natural variability estimate for all models. Thus it is *very likely* the warming signal will be large compared to natural variability in all NA regions throughout the year by mid-century. This warming generally leads to a two- to four fold increase in simulated heat wave frequency over the 21st century (e.g., Lau and Nath, 2012).

Anthropogenic climate change may also bring systematic cold-season precipitation changes. As with previous models, CMIP5 projections generally agree in projecting a winter precipitation increase over the northern half of NA (Figure 14.18 and AI.19). This is associated with increased atmospheric moisture, increased moisture convergence, and a poleward shift in ETC activity (Section 14.6.2 and Table 14.3). The change is consistent with CMIP3 model projections of positive NAO trends (Table 14.3; Hori et al., 2007; Karpechko, 2010; Zhu and Wang, 2010). Winter precipitation increases extend southward into the USA (northern portions of SREX regions 3 to 5; Neelin et al., 2013) but with decreasing strength relative to natural variability. This behaviour is qualitatively reproduced in higher resolution simulations (Figure 14.18).

Warm-season precipitation also exhibits significant increases in Alaska, northern Canada, and Eastern NA by century's end (Figures 14.18, AI.19, AI.22). However, CMIP5 models disagree on the sign of the precipitation change over the rest of NA (Figures AI.26 and AI.27), consistent with CMIP3 results (Figure 14.18; Neelin et al., 2006; Rauscher et al., 2008; Seth et al., 2010). One set of high resolution simulatons (Endo et al., 2012) shows a tendency towards more precipitation than either CMIP3 or CMIP5 models (Figure 14.18), suggesting the simulated warm-season precipitation change in the region may be resolution-dependent. Future precipitation changes associated with the NAMS are likewise uncertain, though there is *medium confidence* the phenomenon will move to later in the annual cycle (Section 14.2.3, Table 14.3). As there is *medium confidence* tropical cyclones will be associated with greater rainfall rates, the Gulf and East coasts of NA may be impacted by greater precipitation when tropical cyclones occur (Table 14.3).

CMIP3 models showed a 21st century precipitation decrease across much of southwestern NA, accompanied by a robust evaporation increase characteristic of mid-latitude continental warming (Seager et al., 2007; Seager and Vecchi, 2010) and an increase in drought frequency (Sheffield and Wood, 2008; Gutzler and Robbins, 2011). When downscaled, CMIP3 models showed less drying in the region (Gao et al., 2012c) and an extreme precipitation increase, despite overall drying (Dominguez et al., 2012). CMIP5 models do not consistently show such a precipitation decrease in this region (Neelin et al., 2013). This is one of the few emerging differences between the two ensembles in climate projections over NA. However, the CMIP5 models still show a strong decrease in soil moisture here (Dai, 2013), due to increasing evaporation.

In summary, it is *very likely* that by mid-century the anthropogenic warming signal will be large compared to natural variability such as that stemming from the NAO, ENSO, PNA, PDO, and the NAMS in all

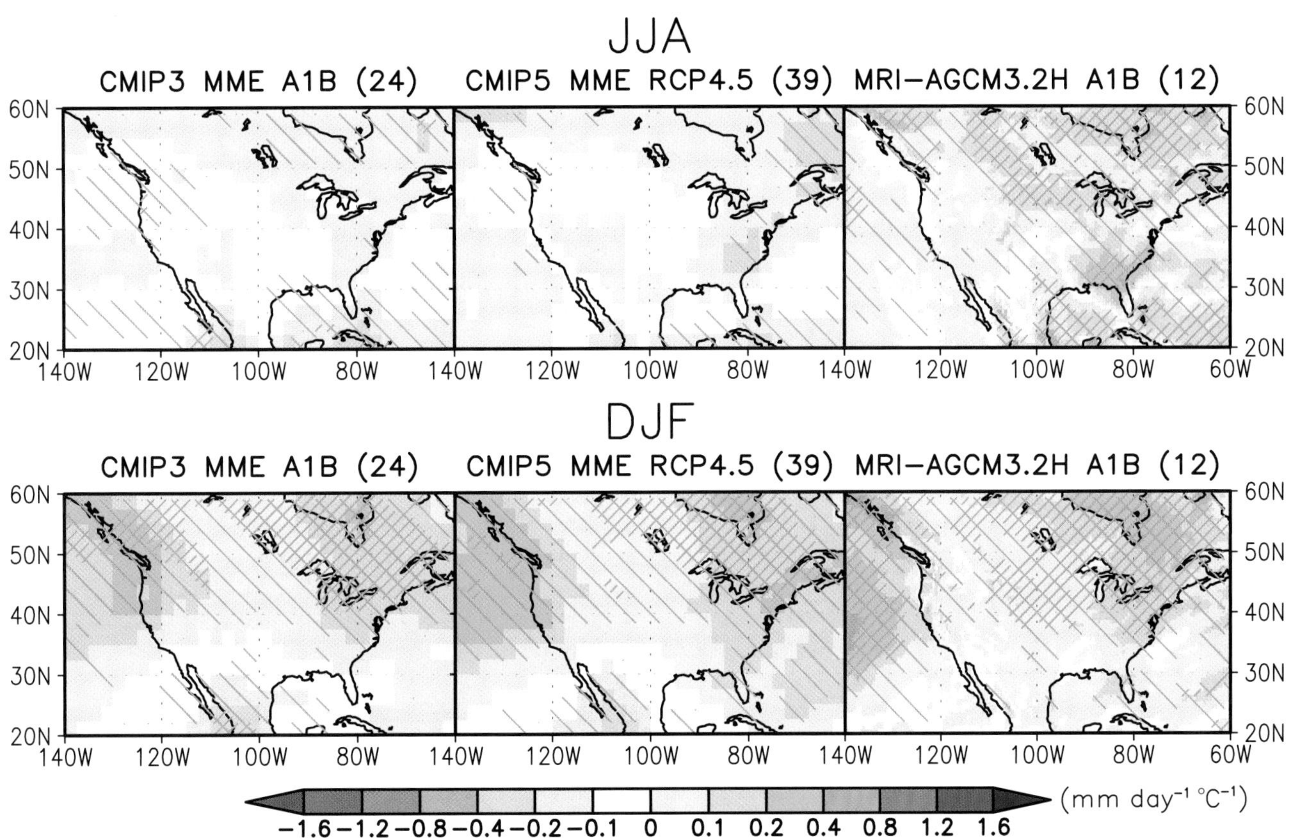

Figure 14.18 | Maps of precipitation changes for North America in 2080–2099 with respect to 1986–2005 in June, July and August (above) and December to February (below) in the SRES A1B scenario with 24 CMIP3 models (left), and in the RCP4.5 scenario with 39 CMIP5 models (middle). Right figures are the precipitation changes in 2075–2099 with respect to 1979–2003 in the SRES A1B scenario with the 12 member 60 km mesh Meteorological Research Institute (MRI)-Atmospheric General Circulation Model 3.2 (AGCM3.2) multi-physics, multi-SST ensembles (Endo et al., 2012). Precipitation changes are normalized by the global annual mean surface air temperature changes in each scenario. Light hatching denotes where more than 66% of models (or members) have the same sign with the ensemble mean changes, while dense hatching denotes where more than 90% of models (or members) have the same sign with the ensemble mean changes.

NA regions throughout the year. It is *likely* that the northern half of NA will experience an increase in precipitation over the 21st century, due in large part to a precipitation increase within ETCs.

14.8.4 Central America and Caribbean

The Central America and the Caribbean (CAC) region is affected by several phenomena, including the ITCZ (Section 14.3.1.1), NAMS (Section 14.2.3.1), ENSO (Section 14.4) and TCs (Section 14.6.1; Table 14.3; also Gamble and Curtis, 2008). The annual cycle results from air–sea interactions over the Western Hemisphere warm pool in the tropical eastern north Pacific and the Intra Americas Seas (Amador et al., 2006; Wang et al., 2007). The Caribbean Low Level Jet is a key element of the region's summer climate (Cook and Vizy, 2010) and is controlled by the size and intensity of the Western Hemisphere warm pool (Wang et al., 2008b). It is also modulated by SST gradients between the eastern equatorial Pacific and tropical Atlantic (Taylor et al., 2011d). ENSO is the main driver of climate variability, with El Niño being associated with dry conditions and La Niña with wet conditions (Karmalkar et al., 2011). Other teleconnection patterns, such as the NAO (Section 14.5.1) and the strength of boreal winter convection over the Amazon, influence trade winds over the Tropical North Atlantic and can combine with ENSO to modulate the summer Western Hemisphere warm pool (e.g., Enfield et al., 2006). Table 14.3 summarizes the main phenomena and their relevance to climate change over the CAC.

Because inter-decadal climate variations can be large in the CAC region, precipitation trends must be interpreted carefully. From 1950 to 2003, negative trends were seen in several data sets in the Caribbean region and parts of Central America (Neelin et al., 2006). However, regarding secular trends (1901–2005), this signal was identified only in the Caribbean region (Trenberth et al., 2007b). Prolonged dry or wet periods are related to decadal variability of the adjacent Pacific and Atlantic (Mendoza et al., 2007; Seager et al., 2009; Mendez and Magaña, 2010), and the intensity of easterlies over the region. For instance, increased easterly surface winds over Puerto Rico from 1950 to 2000 disrupted a pattern of inland moisture convergence, leading to a dramatic precipitation decrease (Comarazamy and Gonzalez, 2011).

Table 14.2 provides an overall assessment of GCM quality for simulations of temperature, precipitation and main phenomena in the CAC sub-regions. Annual cycles of temperature and precipitation are well

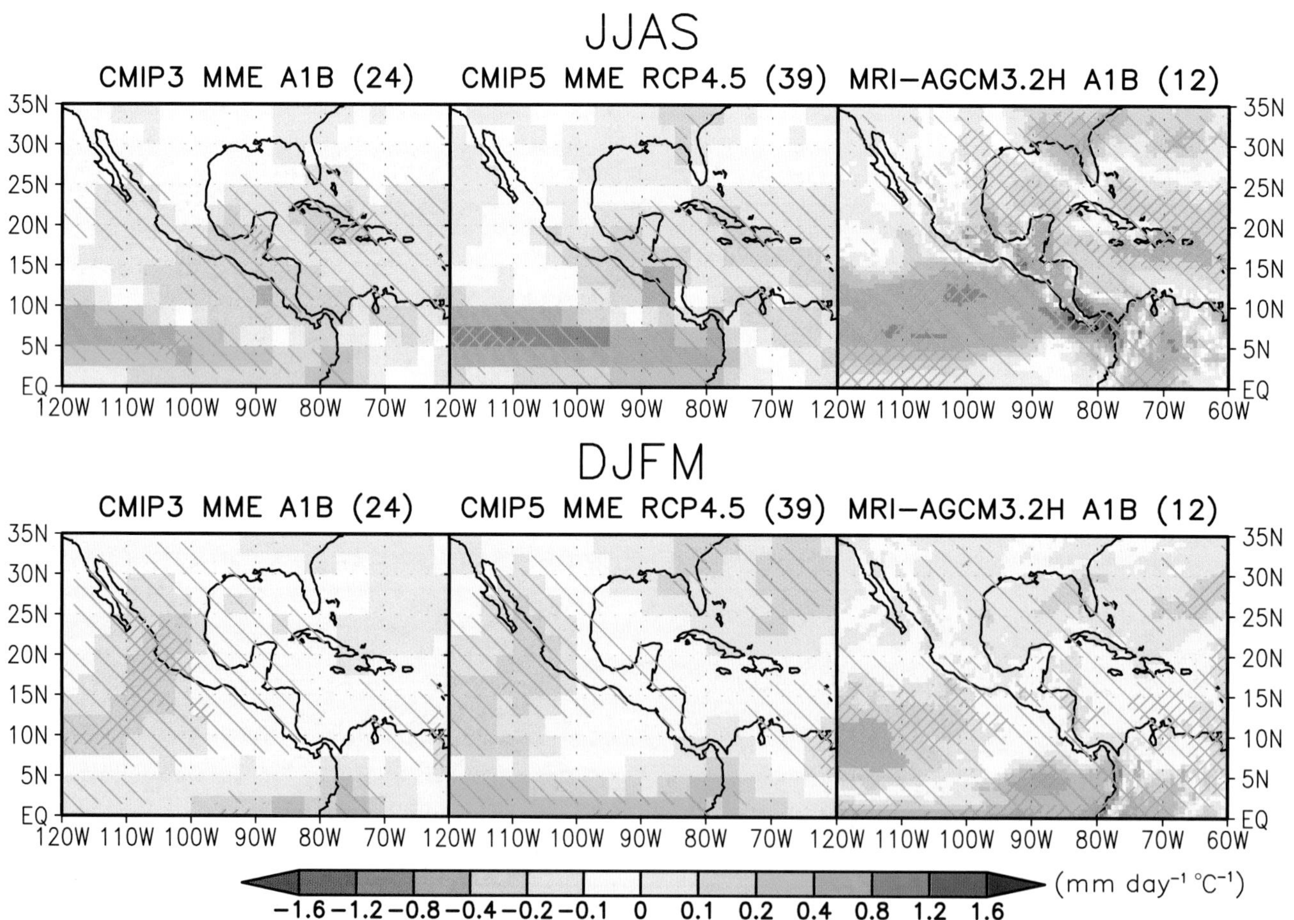

Figure 14.19 | Maps of precipitation changes for Central America and Caribbean in 2080–2099 with respect to 1986–2005 in June to September (above) and December to March (below) in the SRES A1B scenario with 24 CMIP3 models (left), and in the RCP4.5 scenario with 39 CMIP5 models (middle). Right figures are the precipitation changes in 2075–2099 with respect to 1979–2003 in the SRES A1B scenario with the 12 member 60 km mesh Meteorological Research Institute (MRI)-Atmospheric General Circulation Model 3.2 (AGCM3.2) multi-physics, multi-SST ensembles (Endo et al., 2012). Precipitation changes are normalized by the global annual mean surface air temperature changes in each scenario. Light hatching denotes where more than 66% of models (or members) have the same sign with the ensemble mean changes, while dense hatching denotes where more than 90% of models (or members) have the same sign with the ensemble mean changes.

simulated by CMIP5 models, though precipitation from June to October is underestimated (Figure 9.38). Regional models also simulate temperature and precipitation climatologies, and the magnitude and annual cycle of the Caribbean Low-Level Jet reasonably well (Campbell et al., 2010; Taylor et al., 2013).

CMIP3 models generally projected a precipitation reduction over much of the Caribbean region, consistent with the observed negative trend since 1950 (Neelin et al., 2006; Rauscher et al., 2008). The subtropics are generally expected to dry as global climate warms (Held and Soden, 2006), but in both CMIP3 and CMIP5 models the CAC region shows the greatest drying. Future drying may also be related to strengthening of the Caribbean Low-Level Jet (Taylor et al., 2013) and subsidence over the Caribbean region associated with warmer SSTs in the tropical Pacific than Atlantic (Taylor et al., 2011d). A high-resolution regional Ocean GCM using a CMIP3 ensemble for boundary conditions confirms that the Intra American Seas circulation weakens by similar rate as the reduction in Atlantic Meridional Overturning (Liu et al., 2012c). This weakening causes the Gulf of Mexico to warm less than other oceans.

Downscaling experiments for the region have shown a mid-21st century warming between 2°C and 3°C (Vergara et al., 2007; Rauscher et al., 2008; Karmalkar et al., 2011). Precipitation decreases over most of the CAC region, similar to the signal in driving global models (Campbell et al., 2010; Hall et al., 2012). However, only a few downscaling studies took into account key elements of the region's climate, such as easterly wave activity, TCs, or interannual variability mechanisms linked to ENSO (Karmalkar et al., 2011).

By century's end, CMIP5 models project greatest warming in the CAC region in JJA. Warming is projected to be larger over Central America than the Caribbean in summer and winter (Figures AI.24, AI.2, Table 14.1). From October to March, ensemble mean projections indicate precipitation decrease in northern Central America, including Mexico. In the Caribbean precipitation is projected to decrease in the south (consistent with the observed trends) but to increase in the north (Figure AI.26). From April to September, the projected zone of precipitation reduction expands over the entire CAC region, and this signal is generally larger than the models' estimates of natural variability (Figure AI.27). Precipitation changes projected by CMIP3, CMIP5 and a high-resolution model show a similar reduction in parts of Mexico and the southern Caribbean in DJFM, and in Central America and the Caribbean in JJAS (Figure 14.19). The CMIP5 ensemble shows greater agreement in the DJFM precipitation increase in the northern Caribbean sector than CMIP3. These projected changes are also reflected in Table 14.1. Figures AI.26, AI.27 and Figure 14.19 suggest an intensification and southward displacement of the East Pacific ITCZ, which can contribute to drying in southern Central America (Karmalkar et al., 2011).

ENSO will continue to influence CAC climate, but changes in ENSO frequency or intensity remain uncertain (Section 14.4). Projected drier conditions may also be related to decreased frequency of TCs, though the associated rainfall rate of these systems are higher in future projections (Section 14.6.1).

In summary, owing to model agreement on projections and the degree of consistency with observed trends, it is *likely* warm-season precipitation will decrease in the Caribbean region, over the coming century. However, there is only *medium confidence* that Central America will experience a decrease in precipitation.

14.8.5 South America

South America (SA) is affected by several climate phenomena. ENSO (Section 14.4) and Atlantic Ocean modes (Section 14.3.4) have a role in interannual variability of many regions. The SAMS (Section 14.2.3.2) is responsible for rainfall over large areas, while the SACZ (Section 14.3.1.3) and Atlantic ITCZ (Section 14.3.1.1) also affect precipitation. Teleconnections such as the PSA (Section 14.7.1), the SAM (Section 14.5.2) with related ETCs (Section 14.6.2) and the IOD (Section 14.3.3) also influence climate variability. Table 14.3 summarizes the main phenomena and their assessed relevance to climate change over SA.

Positive minimum temperature trends have been observed in SA (Alexander et al., 2006; Marengo and Camargo, 2008; Rusticucci and Renom, 2008; Marengo et al., 2009; Seneviratne et al., 2012; Skansi et al., 2013). Glacial retreat in the tropical Andes was observed in the last three decades (Vuille et al., 2008; Rabatel et al., 2013). In contrast to the warming over the continental interior, a prominent but localized coastal cooling was detected during the past 30 to 50 years, extending from central Peru (Gutiérrez et al., 2011) to northern (Schulz et al., 2012) and central Chile (Falvey and Garreaud, 2009). Observed precipitation changes include a significant increase in precipitation during the 20th century over the southern sector of southeastern SA, a negative trend in SACZ continental area (Section 2.5.1; Barros et al., 2008), a negative trend in mean precipitation and precipitation extremes in central-southern Chile, and a positive trend in southern Chile (Haylock et al., 2006; Quintana and Aceituno, 2012). Other detected changes include positive extreme precipitation trends in southeastern SA, central-northern Argentina and northwestern Peru and Ecuador (Section 2.6.2; Haylock et al., 2006; Dufek et al., 2008; Marengo et al., 2009; Re and Barros, 2009; Skansi et al., 2013).

Table 14.2 provides an overall assessment of GCM quality for simulations of temperature, precipitation and main phenomena in the sub-regions of SA. In general, GCM results are consistent with observed temperature tendencies (e.g., Haylock et al., 2006). Trends toward warmer nights in CMIP3 models (Marengo et al., 2010b; Rusticucci et al., 2010) are consistent with observed trends. CMIP3 models, however, do not simulate the cooling ocean and warming land trends observed in the last 30 years along subtropical western SA noted above. The number of warm nights in SA is well represented in CMIP5 simulations (Sillmann et al., 2013). CMIP5 models reproduce the annual cycle of precipitation over SA, though the multi-model mean underestimates rainfall over some areas (Figure 9.38). In tropical SA, rainy season precipitation is better reproduced in the CMIP5 ensemble than CMIP3 (Figure 9.39). CMIP3 models were able to simulate extreme precipitation indices over SA (Rusticucci et al., 2010), but CMIP5 models improved them globally (Sillmann et al., 2013). CMIP5 also improved simulations of precipitation indices in the SAMS region (Kitoh et al., 2013; Section 14.2.3.2). The main precipitation features are well represented by regional models in several areas of SA (Solman et al., 2008; Alves and Marengo, 2010; Chou et al., 2012; Solman et al., 2013). However, regional models underestimate daily precipitation intensity in the La

Plata Basin and in eastern Northeastern SA in DJF and almost over the whole continent in JJA (Carril et al., 2012).

Regarding future projections, CMIP5 models indicate higher temperatures over all of SA, with the largest changes in southeastern Amazonia by century's end. (Figures AI.28, AI.29, Table 14.1). Temperature changes projected by RCMs forced by a suite of CMIP3 models agree that the largest warming occurs over the southern Amazon during austral winter. Regional models project a greater frequency of warm nights over SA, except in parts of Argentina, and a reduction of cold nights over the whole continent (Marengo et al., 2009). CMIP5 projections confirm the results of CMIP3 in AR4 and SREX (see Section 12.4.9).

CMIP5 results confirm precipitation changes projected by CMIP3 models in the majority of SA regions, with increased confidence, as more models agree in the changes (AI.30, AI.34, AI31, AI35). Inter-model spread in precipitation also decreased in some SA regions from CMIP3 to CMIP5 (Blázquez and Nuñez, 2012). CMIP5 precipitation

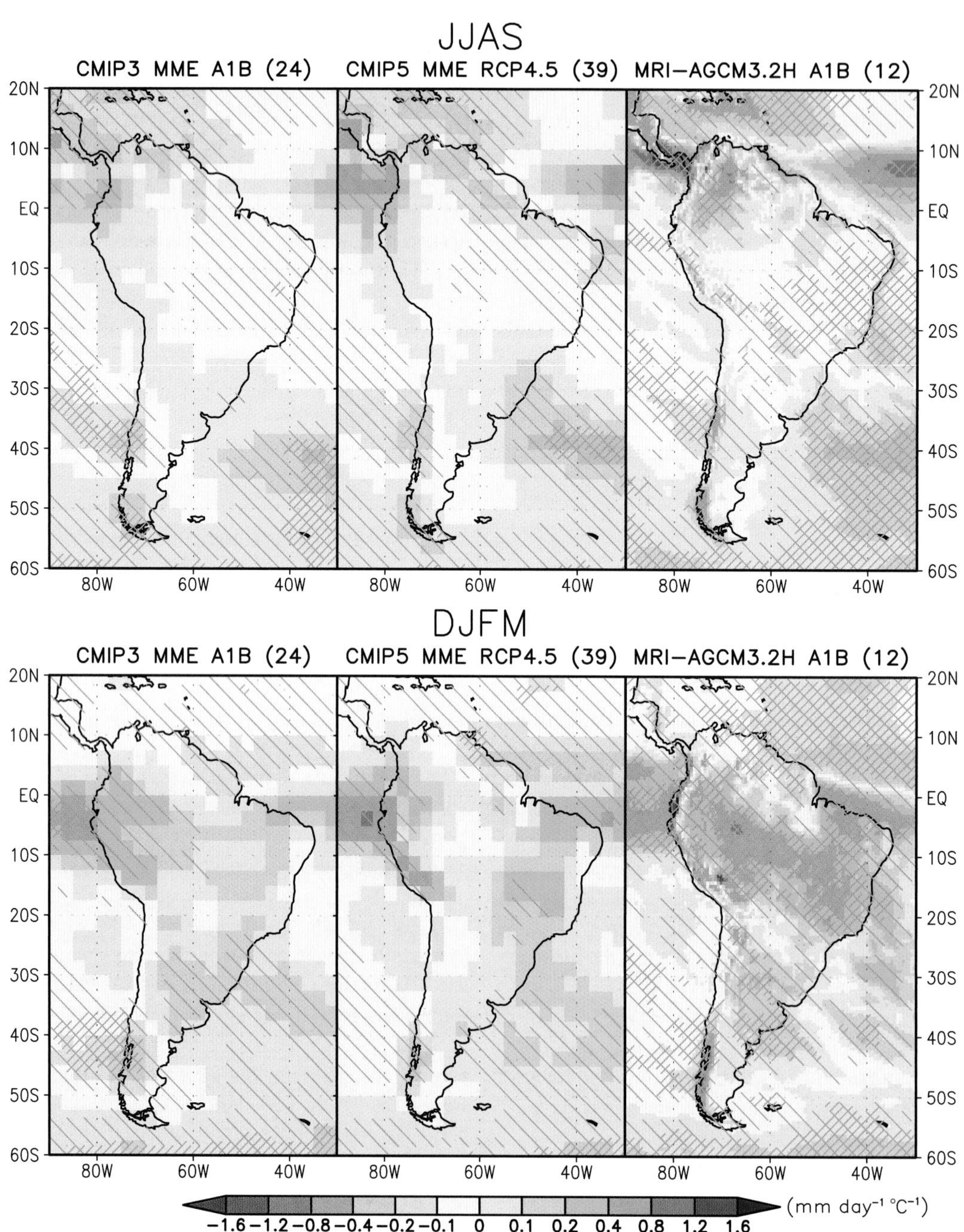

Figure 14.20 | Maps of precipitation changes for South America in 2080–2099 with respect to 1986–2005 in June to September (above) and December to March (below) in the SRES A1B scenario with 24 CMIP3 models (left), and in the RCP4.5 scenario with 39 CMIP5 models (middle). Right figures are the precipitation changes in 2075–2099 with respect to 1979–2003 in the SRES A1B scenario with the 12-member 60-km mesh Meteorological Research Institute (MRI)- Atmospheric General Circulation Model 3.2 (AGCM3.2) multi-physics, multi-SST ensembles (Endo et al., 2012). Precipitation changes are normalized by the global annual mean surface air temperature changes in each scenario. Light hatching denotes where more than 66% of models (or members) have the same sign with the ensemble mean changes, while dense hatching denotes where more than 90% of models (or members) have the same sign with the ensemble mean changes.

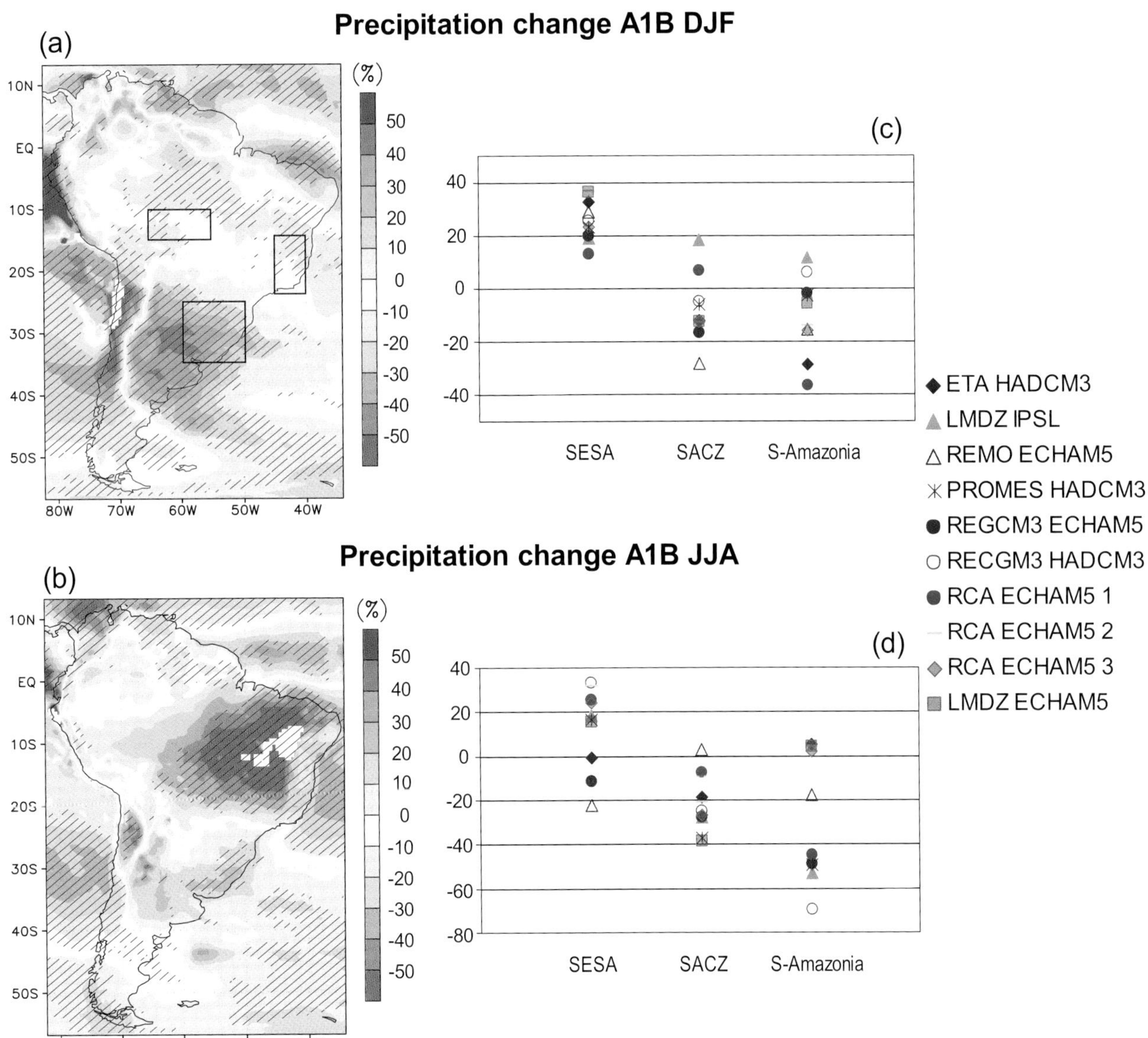

Figure 14.21 | (a) December, January and February (DJF) and (b) June, July and August (JJA) relative precipitation change in 2071–2100 with respect to 1961–1990 in the A1B scenario from an ensemble of 10 Regional Climate Models (RCMs) participating in the Europe–South America Network for Climate Change Assessment and Impact Studies-La Plata Basin (CLARIS-LPB) Project. Hatching denotes areas where 8 out of 10 RCMs agree in the sign of the relative change. (c) DJF and (d) JJA dispersion among regional model projections of precipitation changes averaged over land grid points in Southeastern South America (SESA, 35°S to 25°S, 60°W to 50°W), South Atlantic Convergence Zone (SACZ, 25°S to 15°S, 45°W to 40°W) and southern Amazonia (15°S to 10°S, 65°W to 55°W), indicated by the boxes in (a).

projections for the end of the twenty-first century (2081–2100) show a precipitation increase from October to March over the southern part of Southeast Brazil and the La Plata Basin, the extreme south of Chile, the northwest coast of SA, and the Atlantic ITCZ, extending to a small area of the northeastern Brazil coast (Figures AI.30 and AI.34). Reduced October to March rainfall is projected in the extreme northern region of SA, eastern Brazil, and central Chile. In eastern Amazonia and northeastern Brazil, CMIP5 models show both drying and moistening. This uncertainty can also be seen in Table 14.1.

Figure 14.20 confirms that changes in northwestern, southwestern and southeastern SA are consistent among CMIP3 and CMIP5 ensembles and a high-resolution model ensemble, which gives more confidence in these results. However, in eastern Amazonia and northeast and eastern Brazil, there is less agreement. Results from high-resolution or regional models forced by CMIP3 models provides further indication the projected changes are robust. A high-resolution regional model ensemble projects precipitation increases during austral summer over the La Plata Basin region, northwestern SA and southernmost Chile, and a decrease over northern SA, eastern Amazonia, eastern Brazil, central Chile and the Altiplano (Figure 14.21). Other regional models also project a precipitation increase over the Peruvian coast and Ecuador and a reduction in the Amazon Basin (Marengo et al., 2010b; Marengo et al., 2012).

From April to September, the CMIP5 ensemble projects precipitation increases over the La Plata Basin and northwestern SA near the coast (Figures AI31, AI35). In contrast, a reduction is projected for northeast

Brazil and eastern Amazonia. Precipitation is projected to decrease in Central Chile, but to increase over extreme southern areas. In CMIP3 models, a precipitation reduction in the central Andes resulted from a moisture transport decrease from the continental interior to the Altiplano (Minvielle and Garreaud, 2011). CMIP3 and CMIP5 models are consistent in projecting drier conditions in eastern Amazonia during the dry season and wetter conditions in western Amazonia (Malhi et al., 2008; Cook et al., 2011). The Amazon forest's future is discussed in Section 12.5.5.6.1. Areas of maximum change in CMIP5 are consistent with those of CMIP3 in JJA, agreeing also with a high-resolution model ensemble (Figure 14.20). Increased precipitation in southeastern SA is projected by a high-resolution model ensemble in all four seasons (Blázquez et al., 2012). The austral winter precipitation increase over the La Plata Basin and southern Chile, and the reduction in eastern Amazonia and northeast Brazil, are also projected by RCMs (Figure 14.21) as in CMIP5 models. A relevant result from a RCM is the precipitation decrease over most of SA north of 20°S in austral spring, suggesting a longer dry season (Sörensson et al., 2010; see also Section 14.2.3.2). Note that average CMIP5 spatial values in Table 14.1 are consistent with changes seen in the maps, unless for the west coast of SA, where there are spatial variations within the area and the values do not reflect the changes.

Regional model projections and a high-resolution model ensemble indicate an increase in the number of consecutive dry days in northeastern SA (Marengo et al., 2009; Kitoh et al., 2011). An increase in heavy precipitation events over almost the entire continent, especially Amazonia, southern Brazil and northern Argentina, is projected by a high-resolution model ensemble (Kitoh et al., 2011) and in subtropical areas of South America by regional models (Marengo et al., 2009). Seneviratne et al. (2012) indicated *low* to *medium confidence* in CMIP3 SA precipitation trends. However, the increased ability of CMIP5 models to represent extremes (Kitoh et al., 2013) provides higher confidence in the signals discussed above (Section 14.2.3.2), consistent with global changes in land areas (Section 12.4.5).

Precipitation changes projected over SA are consistent with El Niño influences, for example, rainfall increase over southeastern and northwestern SA and decrease over eastern Amazonia. However, CMIP3 models could not represent certain features of ENSO well (Roxy et al., 2013) and there is no consensus about future ENSO behaviour (Coelho and Goddard, 2009; Collins et al., 2010) even with CMIP5 results (Section 14.4). As the various types of ENSO produce different impacts on SA (Ashok et al., 2007; Hill et al., 2011; Tedeschi et al., 2013), future ENSO effects remain uncertain. It is *very likely* that ENSO remains the dominant mode of interannual variability in the future (Section 14.4.2). Therefore, regions in SA currently influenced by Pacific SST will continue to experience ENSO effects on precipitation and temperature.

Projected precipitation increases in the southern sector of southeastern SA are consistent with changes in the SACZ dipole (Section 14.3.1.3) and PSA (Section 14.7.1). Increased precipitation in this region may also have a contribution from a more frequent and intense Low Level Jet (Nuñez et al., 2009; Soares and Marengo, 2009). CMIP3 model analyses show little impact on extreme precipitation from SAM changes toward century's end, except in Patagonia (Menendez and Carril, 2010). However, the southward shift of stormtracks associated with the SAM's projected positive trend (Reboita et al., 2009; Section 14.5.2) impacts zones of cyclogenesis off the southeast SA coast (Kruger et al., 2011; Section 12.4.4).

In summary, it is *very likely* temperatures will increase over the whole continent, with greatest warming projected in southern Amazonia. It is *likely* there will be an increase (reduction) in frequency of warm (cold) nights in most regions. It is *very likely* precipitation will increase in the southern sector of southeastern and northwestern SA, and decrease in Central Chile and extreme north of the continent. It is *very likely* that less rainfall will occur in eastern Amazonia, northeast and eastern Brazil during the dry season. However, in the rainy season there is *medium confidence* in the precipitation changes over these regions. There is *high confidence* in an increase of precipitation extremes.

14.8.6 Europe and Mediterranean

This section assesses regional climate change in Europe and the North African and West Asian rims of the Mediterranean basin. Area-average summaries are presented for the three sub-regions of Northern Europe (NEU), Central Europe (CEU) and Mediterranean (MED) (cf. Tables 14.1 to 14.2).

The most relevant climate phenomena for this region are NAO (Section 14.5.1), ETCs (Section 14.6.2) and blocking (Box 14.2, Folland et al., 2009; Feliks et al., 2010; Dole et al., 2011; Mariotti and Dell'Aquila, 2012). These phenomena also interact with longer time-scale North Atlantic ocean-atmosphere phenomena such as the AMO (Section 14.7.6, Mariotti and Dell'Aquila, 2012; Sutton and Dong, 2012). Other phenomena have minor influence in limited sectors of the region (see Supplementary Material Section 14.SM.6.3).

Recent 1981-2012 trends in annual mean temperature in each sub-region exceed the global mean land trend as can be inferred from Figure 2.22. Consistent with previous AR4 conclusions (Section 11.3), recent studies of extreme events (Section 2.6.1) point to a *very likely* increase of the number of warm days and nights, and decrease of the number of cold days and nights, since 1950 in Europe. Heat waves can be amplified by drier soil conditions resulting from warming (Vautard et al., 2007; Seneviratne et al., 2010; Hirschi et al., 2011). Several studies (Section 2.6.2.1) also indicate general increases in the intensity and frequency of extreme precipitation especially in winter during the last four decades however there are inconsistencies between studies, regions and seasons.

The ability of climate models to simulate the climate in this region has improved in many important aspects since AR4 (see Figure 9.38). Particularly relevant for this region are increased model resolution and a better representation of the land surface processes in many of the models that participated in the recent CMIP5 experiment. Table 14.2 provides an assessment of the CMIP5 quality for simulations of temperature, precipitation, and main phenomena in the region. The CMIP5 projections reveal warming in all seasons for the three sub-regions, while precipitation projections are more variable across sub-regions and seasons. In the winter half year (October to March), NEU and CEU are projected to have increased mean precipitation associated with increased atmospheric moisture, increased moisture convergence and intensification in ETC activity (Section 14.6.2 and Table 14.3) and no change or a moderate reduction in the MED. In the summer half year (April to September) , NEU and CEU mean precipitation are projected to have only small changes whereas there is a notable reduction in MED (see Table 14.1, Figures AI.36 to AI.37 and AI.42-AI.43). Figure 14.22 illustrates that the precipitation changes are broadly consistent with the findings CMIP3.

High-resolution projections from the Japanese high-resolution model ensemble also agree with these findings and are consistent with downscaling results from coordinated multi-model GCM/RCM experiments (e.g., ENSEMBLES, Déqué et al., 2012). In general, regional climate change amplitudes for temperature and precipitation follow the global warming amplitude although modulated both by changes in the large-scale circulation and by regional feedback processes (Kjellstrom et al., 2011), which confirms assessments in AR4 (Christensen et al., 2007).

Some new investigations have focussed on the uncertainties associated with model projections. A large ensemble of RCM-GCM shows that the temperature response is robust in spite of a considerable uncertainty related to choice of model combination (GCM/RCM) and sampling (natural variability), even for the 2021–2050 time frame (Déqué et al., 2012). Other studies based on CMIP3 projections suggest that GHG-forced changes in the MED are *likely* to become distinguishable from the 'noise' created by internal decadal variations in decades beyond 2020–2030 (Giorgi and Bi, 2009). It has been also shown using an ensemble of RCM simulations that the removal of NAO-related variability leads to an earlier emergence of change in seasonal mean temperatures for some regions in Europe (Kjellström et al., 2013). Hence, in the near term, decadal predictability is *likely* to be critically dependent on the regional impacts of modes of variability 'internal' to the climate system (Section 11.3). However, it has been shown that NAO trends do not account for a large fraction of the long-term future change in mean temperature or precipitation (Stephenson et al., 2006) and that large-scale atmospheric circulation changes in CMIP5 models are not the main driver of the warming projected in Europe by the end of the century (2081–2100; Cattiaux et al., 2013). Therefore, changes in climate phenomena contribute to the uncertainty in the near-term projections rather than long-term changes in this region (Table 14.3), further supporting the credibility in model projection (Table 14.2).

Recent studies have clearly identified a possible amplification of temperature extremes by changes in soil moisture (Jaeger and Seneviratne,

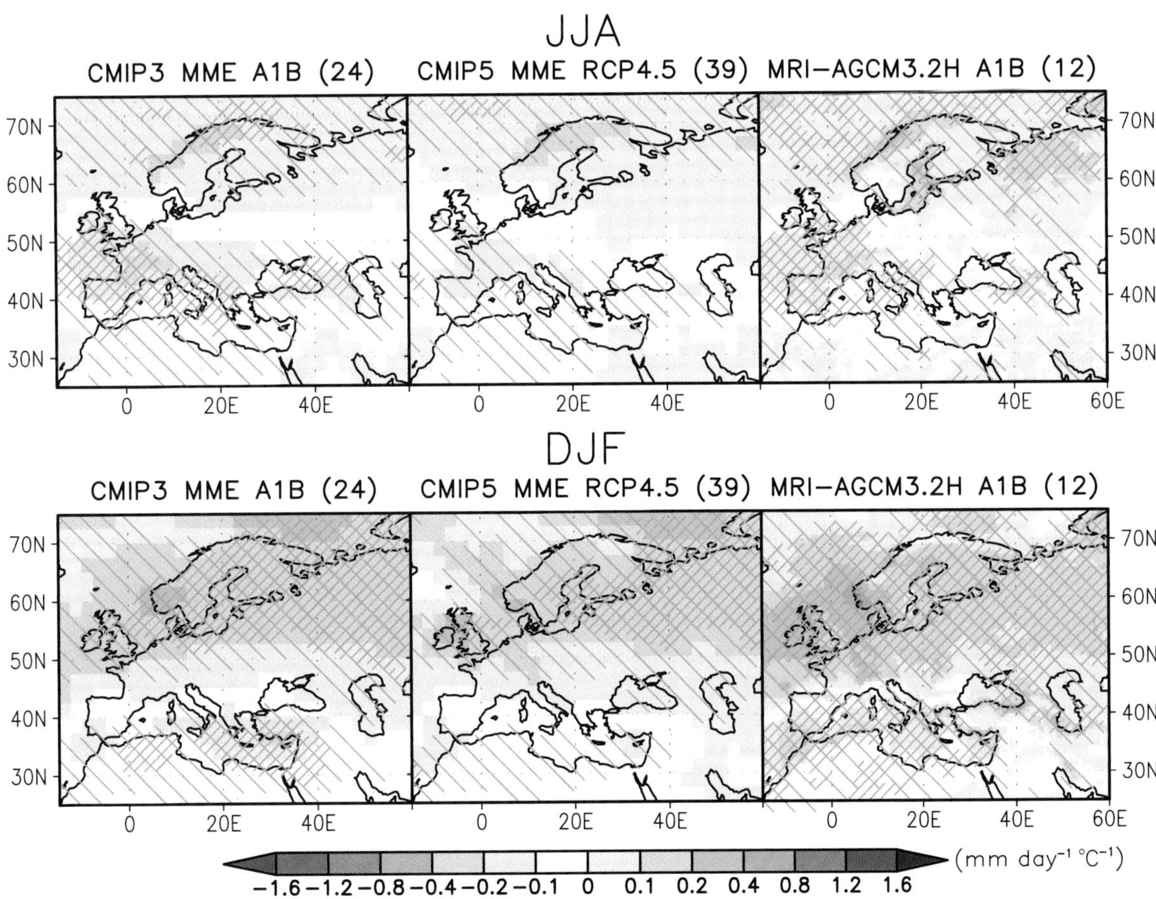

Figure 14.22 | Maps of precipitation changes for Europe and Mediterranean in 2080–2099 with respect to 1986–2005 in June to August (above) and December to February (below) in the SRES A1B scenario with 24 CMIP3 models (left), and in the RCP4.5 scenario with 39 CMIP5 models (middle). Right figures are the precipitation changes in 2075–2099 with respect to 1979–2003 in the SRES A1B scenario with the 12 member 60 km mesh Meteorological Research Institute (MRI)-Atmospheric General Circulation Model 3.2 (AGCM3.2) multi-physics, multi-sea surface temperature (SST) ensembles (Endo et al., 2012). Precipitation changes are normalized by the global annual mean surface air temperature changes in each scenario. Light hatching denotes where more than 66% of models (or members) have the same sign with the ensemble mean changes, while dense hatching denotes where more than 90% of models (or members) have the same sign with the ensemble mean changes.

2010; Hirschi et al., 2011), acting as a mechanism that further magnifies the intensity and frequency of heat waves given the projected enhance of summer drying conditions. This is in line with the assessed results presented in SREX (Seneviratne et al., 2012). At the other end of the spectrum, studies indicate that European winter variability may be related to sea ice reductions in the Barents-Kara Sea (Petoukhov and Semenov, 2010) and CMIP5 models in projections for the future in general exhibit a similar relation until the summer sea ice has almost disappeared (Yang and Christensen, 2012). Although the mechanism behind this relation remains unclear this suggests that cold winters in Europe will continue to occur in coming decades, despite an overall warming.

Although climate models have improved fidelity in simulating aspects of regional climates over Europe and the Mediterranean, the spread in projections is still substantial, partly due to large amounts of natural variability in this region (particularly NAO and AMO), besides the inherent model deficiencies .

In summary, there is *high confidence* in model projections of mean temperature in this region. It is *very likely* that temperatures will continue to increase throughout the 21st century over all of Europe and the Mediterranean region. It is *likely* that winter mean temperature will rise more in NEU than in CEU or MED, whereas summer warming will *likely* be more intense in MED and CEU than in NEU. The length, frequency, and/or intensity of warm spells or heat waves are assessed to be *very likely* to increase throughout the whole region. There is *medium confidence* in an annual mean precipitation increase in NEU and CEU, while a decrease is *likely* in MED summer mean precipitation.

14.8.7 Africa

The African continent encompasses a variety of climatic zones. Here the continent is divided into four major sub-regions: Sahara (SAH), Western Africa (WAF), Eastern Africa (EAF) and Southern Africa (SAF). A fifth Mediterranean region to the north of Sahara is discussed in Section 14.8.6. In tropical latitudes, rainfall follows insolation (this simplified picture is modified by the presence of orography, especially in the Great Horn of Africa, the geography of the coastline, and by the oceans). The most relevant phenomena affecting climate variability are the monsoons (Section 14.2.4), ENSO (Section 14.4), Indian and Atlantic Ocean SSTs (IOD, Section 14.3.3; AMM Section 14.3.4; AMO Section 14.7.6) and the atmospheric Walker Circulation (Section 2.7.5). Tropical cyclones impact East African and Madagascan coastal regions (Section 14.6.1) and ETCs clearly impact southern Africa (Section 14.6.2).

Sub-Saharan Sahelian climate is dominated by the monsoonal system that brings rainfall to the region during only one season (Polcher et al., 2011). Most of the rain between May/June and September comes from mesoscale 'squall line' systems that travel short distances in their lifetime (~1000 to 2000 km), and whose distribution is somewhat modified by the synoptic scale African Easterly Wave (Ruti and Dell'Aquila, 2010). The onset of the rainy season in West Africa is a key parameter triggering changes in the vegetation and surface properties, that implies feedbacks to the local atmospheric heat and moisture cycle. The length and frequency of dry spells as well as the length or cumulated rainfall of the season also affect this. All are affected by a large interannual variability (Janicot et al., 2011). When evaluating models their ability to reproduce such characteristics of the African monsoon is essential. A large effect of natural multi-decadal SST and warming of the oceans on Sahel rainfall is *very likely* (Hoerling et al., 2006; Ting et al., 2009, 2011; Mohino et al., 2011; Rodriguez-Fonseca et al., 2011).

East Africa experiences a semi-annual rainfall cycle, driven by the ITCZ movement across the equator. Direct links between the region's rainfall and ENSO have been demonstrated (Giannini et al., 2008) and references therein), but variations in Indian Ocean SST (phases of the IOD) are recognized as the dominant driver of east African rainfall variability (Marchant et al., 2007). This feature acts to enhance rainfall through either anomalous low-level easterly flow of moist air into the continent (Shongwe et al., 2011), or a weakening of the low-level westerly flow over the northern Indian Ocean that transports moisture away from the continent (Black et al., 2003). Although the effect of the IOD is evident in the short rainy season, Shongwe et al. (2011) do not find a similar relationship for the long rains. Williams and Funk (2011), however, argue for a reduction in the long rains over Kenya and Ethiopia in response to warmer Indian Ocean SSTs.

Variability in southern Africa's climate is strongly influenced by its adjacent oceans (Rouault et al., 2003; Hansingo and Reason, 2008, 2009; Hermes and Reason, 2009) as well as by ENSO (Vigaud et al., 2009; Pohl et al., 2010). Although it is generally observed that El Niño events correspond to conditions of below-average rainfall over much of southern Africa (Mason, 2001; Giannini et al., 2008; Manatsa et al., 2008) the ENSO teleconnection is not linear, but rather has complex influence in which a number of regimes of local rainfall response can be identified (Fauchereau et al., 2009). The extreme southwestern parts of southern Africa receive rainfall in austral winter brought by mid-latitude frontal systems mostly associated with passing ETCs, but the majority of the region experiences a single summer rainfall season occurring between November and April. A semi-permanent zone of sub-tropical convergence is a major contributor to summer rainfall in sub-tropical southern Africa (Fauchereau et al., 2009; Vigaud et al., 2012).

Because of its exceptional magnitude and its clear link to global SST, 20th century decadal rainfall variability in the Sahel is a test of GCMs ability to produce realistic long-term changes in tropical precipitation. Despite biases in the region (Cook and Vizy, 2006) the CMIP3 coupled models overall can capture the observed correlation between Sahel rainfall and basin-wide area averaged SST variability (Biasutti et al., 2008) even though individual models may fail, especially at interannual time scales (Lau et al., 2006; Joly et al., 2007). Recently, Ackerley et al. (2011) used a perturbed physics ensemble and reached a similar result for the role of atmospheric sulphate, confirming previous results (Rotstayn and Lohmann, 2002; Held et al., 2005). Since AR4, only limited information about improved performance has been documented and only in WAF have major efforts been focussing on relating model behaviour with ability to simulate local climate processes in such details. However, in a comparative study of the ability of CMIP3 and CMIP5 to simulate multiple SST–Africa teleconnections, Rowell (2013) found varying degrees of success in simulating these. In particular, no clear indication of an improvement in the CMIP5 models vs. the CMI3 models was identified.

In projections of the 21st century, the CMIP3 models produced both significant drying and significant moistening (Held et al., 2005; Biasutti and Giannini, 2006; Cook and Vizy, 2006; Lau et al., 2006), and the mechanisms by which a model dries or wets the Sahel are not fully understood (Cook, 2008). At least qualitatively, the CMIP3 ensemble simulates a more robust response during the pre-onset and the demise portion of the rainy season (Biasutti and Sobel, 2009; Seth et al., 2011). Rainfall is projected to decrease in the early phase of the seasons—implying a small delay in the main rainy season, but is projected to increase at the end of the season—implying an intensification of late-season rains (d'Orgeval et al., 2006), although this appears to be less robust in the CMIP5 models (Section 14.2.4). Projections of a change in the timing of the rains is common to other monsoon regions (Li et al., 2006; Biasutti and Sobel, 2009; Seth et al., 2011), including southern Africa.

The relevance of a local effect is supported by several lines of evidence. There is observational evidence that local soil moisture gradients can trigger convective systems and that these surface contrasts are as important as topography for generating these systems, which bring most of the rain to the region (Taylor et al., 2011a, 2011b). Additional evidence comes from simulations of future rainfall changes in West Africa by RCMs subject to coupled model-derived boundary conditions (Patricola and Cook, 2010), documenting a wetting response of the Sahel to increased GHG in the absence of other forcings. But the relative importance of this effect versus the response to SST trends is not well quantified, mostly due to the limitation of using a single RCM.

An evaluation of six GCMs over East Africa by Conway et al. (2007) reveals no clear multi-model trend in mean annual rainfall by the 2080s, but some indications of increased SON and decreased March,

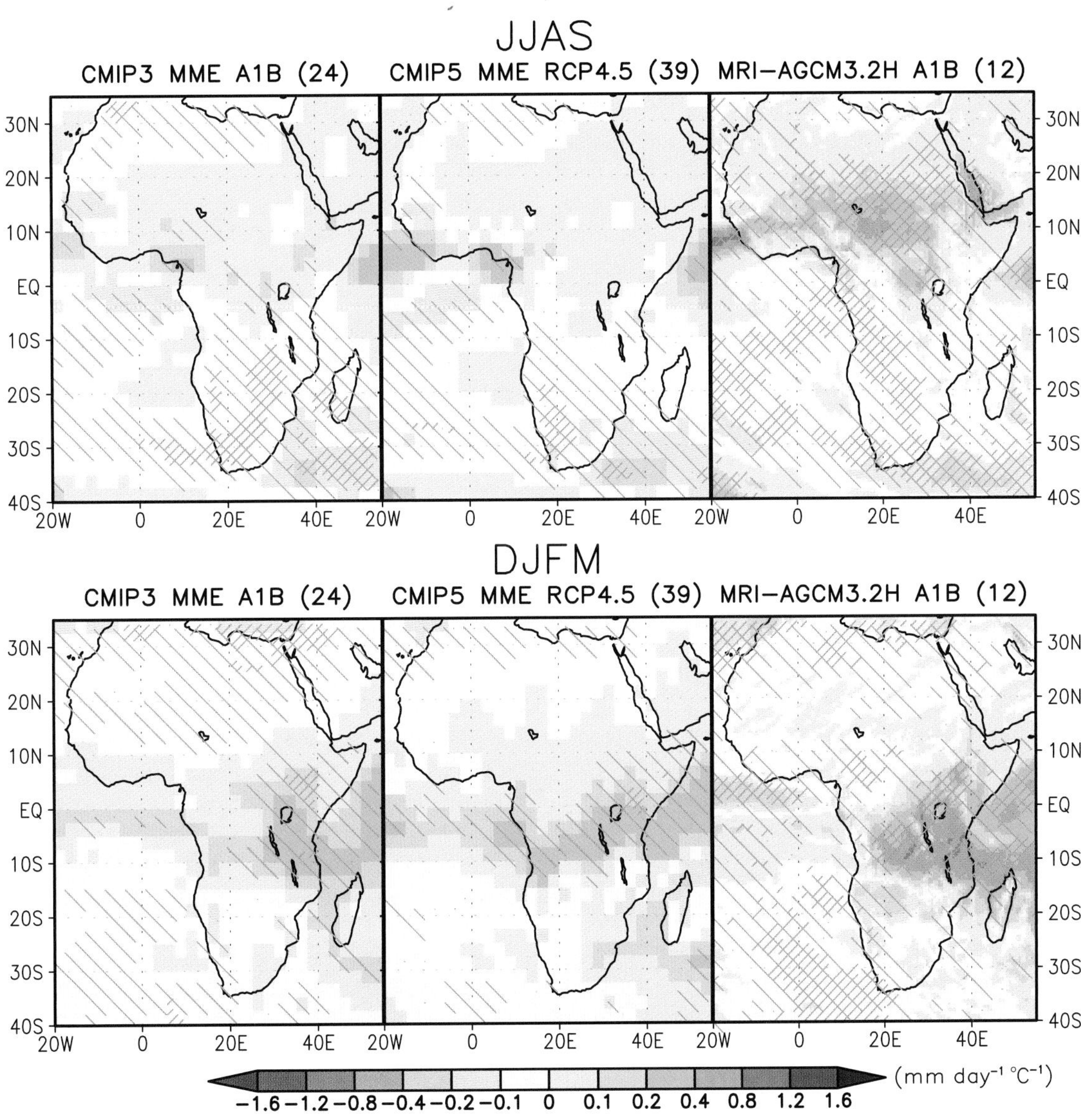

Figure 14.23 | Maps of precipitation changes for Africa in 2080–2099 with respect to 1986–2005 in June to September (above) and December to March (below) in the SRES A1B scenario with 24 CMIP3 models (left), and in the RCP4.5 scenario with 39 CMIP5 models (middle). Right figures are the precipitation changes in 2075–2099 with respect to 1979–2003 in the SRES A1B scenario with the 12-member 60-km mesh Meteorological Research Institute (MRI)-Atmospheric Generl Circulation Model 3.2 (AGCM3.2) multi-physics, multi-sea surface temperature (SST) ensembles (Endo et al., 2012). Precipitation changes are normalized by the global annual mean surface air temperature changes in each scenario. Light hatching denotes where more than 66% of models (or members) have the same sign with the ensemble mean changes, while dense hatching denotes where more than 90% of models (or members) have the same sign with the ensemble mean changes.

April and May (MAM) rainfall are noted. They found inconsistency in how the models represent changes in the IOB and consequent changes in rainfall over East Africa. Shongwe et al. (2011) analysed an ensemble of 12 CMIP3 GCMs (forced with A1B emissions). They found widespread increases in short season (OND) rainfall including extreme precipitation across the region, with statistically significant ensemble mean increases. For the long rains (MAM), similar changes in the sign and magnitude of mean and extreme seasonal rainfall were seen, but model skill in simulating the MAM season is relatively poor. The changes shown for the short rains are consistent with a differential warming of 21st century Indian Ocean SSTs, which leads to a positive IOD-like state (see Section 14.3.3). The atmospheric consequence of this is a weakening of the descending branch of the East African Walker Cell and enhancement of low-level moisture convergence over east Africa (Vecchi and Soden, 2007a; Shongwe et al., 2011).

In an assessment of 19 CMIP3 models run with the A1B emissions forcing, Giannini et al. (2008) note a tendency toward a persistent El Niño-like pattern (see Section 14.4) in the equatorial Pacific along with a decrease in rainfall over southern Africa. Dynamical downscaling of a single GCM by Engelbrecht et al. (2011) shows—for the austral winter—an intensification of the southern edge of the subtropical high pressure belt resulting in southward displacement of the mid-latitude systems that bring frontal rain to the south western parts of the continent, thus resulting in decrease in rainfall. The decrease in summer rainfall is consistent with high-resolution (18 km) RCM simulations done by Haensler et al. (2011) which indicate widespread reductions in rainfall over southern Africa under the A1B scenario.

Shongwe et al. (2009) identified reduction in spring (SON) rainfall throughout the eastern parts of southern Africa. There is good consensus amongst the models used, with the spring anomalies indicating a trend toward later onset of the summer rainy season. Autumn (MAM) reductions are shown for most of southern Africa while eastern South Africa experiences no change and eastern parts of southern Africa show a small increase.

Table 14.2 provides an overall assessment of CMIP5 quality for simulations of temperature, precipitation, and main phenomena in the different sub-regions of Africa. Overall, *confidence* in the projected precipitation changes is at best *medium*. This is owing to the overall modest ability of models to capture the most important phenomena having a strong control on African climates (Table 14.3).

The ability of climate models to simulate historical climate, its change, and its variability, has improved in many aspects since the AR4 (see Section 9.6.1). But for Africa there is no clear evidence that the modest increase in resolution and a better representation of the land surface processes in many CMIP5 models have resulted in marked improvements (e.g., Figure 9.39). The CMIP5 models projection for this century is further warming in all seasons in the considered four sub-regions, while precipitation show some distinct sub-regional and seasonally dependent changes. In the October to March half year all four regions are projected to receive practically unaltered precipitation amounts by 2081–2100, although somewhat elevated in RCP8.5. In the April to September half year SAH, WAF and EAF will experience little change but a quite notable reduction in SAF is projected (see Table 14.1, Figures AI.40 to AI.51). This is consistent with the results from CMIP3 as depicted in Figure 14.23 in the West African monsoon wet season and austral summer. High resolution information provided by the Japanese high-resolution model ensemble also matches this finding.

In summary, given models' ability to capture local processes, large scale climate evolution and their linkages, it is *very likely* that all of Africa will continue to warm during the 21st century. The overall quality of the CMIP5 models imply, that SAH already very dry is *very likely* to remain very dry. But there is *low confidence* in projection statements about drying or wetting of WAF. Owing to models' ability to capture the overall monsoonal behaviour, there is *medium confidence* in projections of a small delay in the rainy season with an increase at the end of the season. There is *medium confidence* in projections showing little change in mean precipitation in EAF and reduced precipitation in the Austral winter in SAF, as models tend to represent Indian Ocean SST developments with credibility. Likewise, increasing rainfall in EAF is *likely* for the short rainy season, but *low confidence* exists in projections regarding drying or wetting in the long rainy season.

14.8.8 Central and North Asia

This area mostly covering the interior of a large continent extending from the Tibetan plateau to the Arctic is mainly influenced by weather systems coming from the west or south, giving some dependency on the AAM (Section 14.2.2) on the one hand and NAO/NAM (Section 14.5.1) on the other, with associated atmospheric blocking as an additional phenomenon of influence related to the latter (Box 14.2). In particular, the variability and long-term change of the climate system in central Asia and northern Asia are closely related to variations of the NAO and NAM (Takaya and Nakamura, 2005; Knutson et al., 2006; Popova and Shmakin, 2010; Sung et al., 2010; Table 14.3).

As a part of the polar amplification, large warming trends in recent decades are observed in the northern Asian sector (e.g., Figure 2.22). The warming trend was particularly strong in the cold season (November to March), with an increase of 2.4°C per 50 years in the mid-latitude semi-arid area of Asia, where the annual rainfall is within the range of 200 to 600 mm over the period of 1901–2009 (Huang et al., 2012). The observations indicate some increasing trends of heavy precipitation events in northern Asia, but no spatially coherent trends in central Asia (Seneviratne et al., 2012).

The CMIP5 models generally have difficulties in representing the mean climate expressed as the climatological means of both temperature and precipitation (Table 14.2) for the sub-regions represented in this area, which is partly related to the poor resolution unable to resolve the complex mountainous terrain dominating this region. But the scarceness of observational data and issues related to how these best can be compared with coarse resolution models adds to the uncertainty regarding model quality.

The model projections presented in AR4 (Section 11.4) indicated strong warming in northern Asia during winter and in central Asia during summer. Precipitation was projected to increase throughout the year in northern Asia with the largest fractional increase during winter. For central Asia, a majority of the CMIP3 models projected decreasing

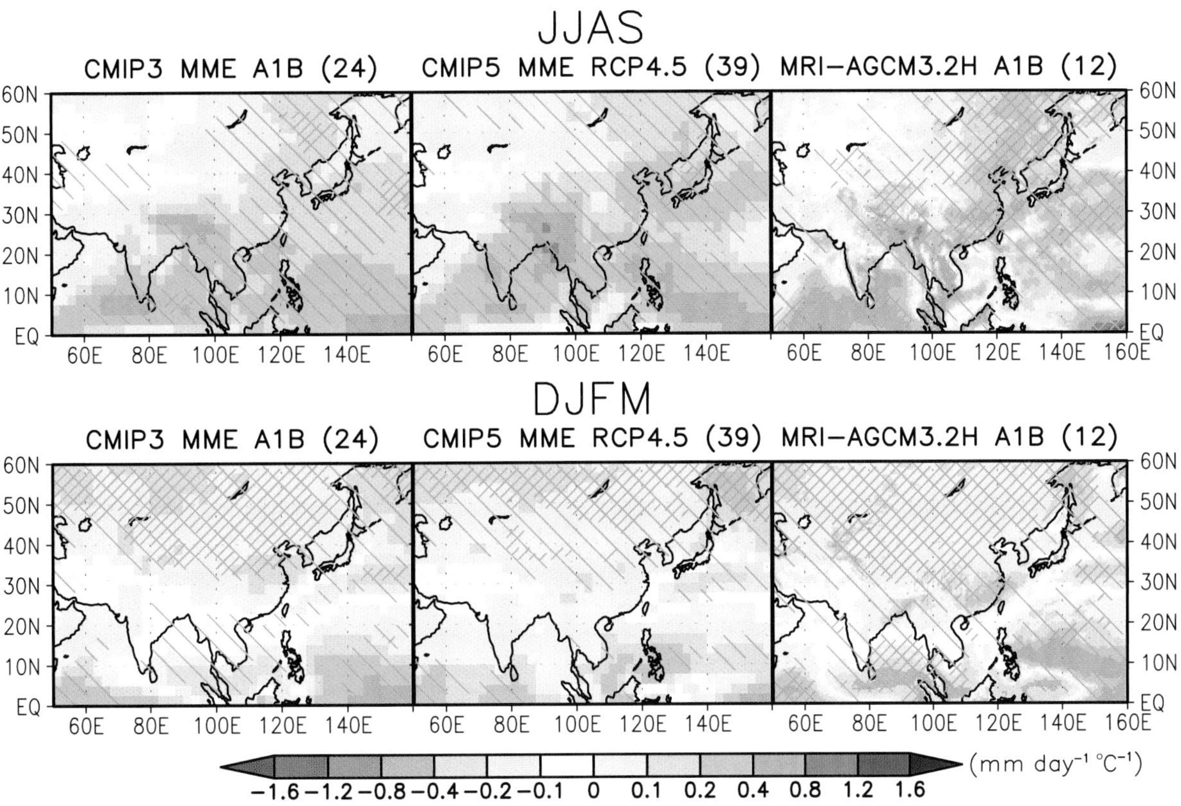

Figure 14.24 | Maps of precipitation changes for Central, North, East and South Asia in 2080–2099 with respect to 1986–2005 in June to September (above) and December to March (below) in the SRES A1B scenario with 24 CMIP3 models (left), and in the RCP4.5 scenario with 39 CMIP5 models (middle). Right figures are the precipitation changes in 2075–2099 with respect to 1979–2003 in the SRES A1B scenario with the 12-member 60-km mesh Meteorological Research Institute (MRI)-Atmospheric General Circulation Model 3.2 (AGCM3.2) multi-physics, multi-sea surface temperature (SST) ensembles (Endo et al., 2012). Precipitation changes are normalized by the global annual mean surface air temperature changes in each scenario. Light hatching denotes where more than 66% of models (or members) have the same sign with the ensemble mean changes, while dense hatching denotes where more than 90% of models (or members) have the same sign with the ensemble mean changes.

precipitation during spring and summer. Seneviratne et al. (2012) indicate increases in all precipitation extreme indices for northern Asia and in the 20-year return value of annual maximum daily precipitation for central Asia. These projections are supported by output from CMIP5 models subject to various RCP scenarios (see Annex I). CMIP5 projected temperature increase in Central Asia of comparable magnitude in both JJA and in DJF. In North Asia, temperatures rise more in DJF than in JJA, while less annual variation is found over Central Asia and the Tibetan Plateau (Table 14.1, Figures AI.12 to AI.13, AI.52 to AI.55 and AI.56 to AI.57).

With an RCM Sato et al. (2007) projected precipitation decreases over northern Mongolia and increases over southern Mongolia in July. Soil moisture over Mongolia decreases in July as a result of the combined effect of decreased precipitation and increased potential evaporation due to rising surface temperature. In North Asia, all CMIP5 models projects an increase in precipitation in the winter half year, and summer half year precipitation is also projected to increase (Table 14.1; Figures AI.14 to AI.15). In Central Asia and the Tibetan Plateau, model agreement is lower on changes both for winter and summer precipitation (Figure 14.24; Table 14.1; Figures AI.54 to AI.55 and AI.58 to AI.59). The ability of these CMIP5 models to simulate precipitation over this region varies (Table 14.3). The reasonable level of agreement in projections of precipitation to be positive and significantly above the 20-year natural variability (Table 14.2), and therefore suggests that *confidence* in the sign of the projected change in future precipitation is *medium*.

In summary, all the areas are projected to warm, a stronger than global mean warming trend is projected for northern Asia during winter. For central Asia, warming magnitude is similar between winter and summer. Precipitation in northern Asia will *very likely* increase, whereas the precipitation over central Asia is *likely* to increase. Extreme precipitation events will *likely* increase in both regions.

14.8.9 East Asia

Summer is the rainy season for East Asia. The Meiyu-Changma-Baiu rain band is the defining feature of East Asian summer climate, extending from eastern China through central Japan (Ding and Chan, 2005; Zhou et al., 2009b). The summer rain band is anchored by the subtropical westerly jet (Sampe and Xie, 2010), and located on the northwestern flank of the western North Pacific subtropical high (Zhou and Yu, 2005). The wintertime circulation is characterized by monsoonal northerlies between the Siberian High and the Aleutian Low.

Both the East Asian summer and winter monsoon circulations have experienced an inter-decadal scale weakening after the 1970s due to natural variability of the coupled climate system, leading to enhanced

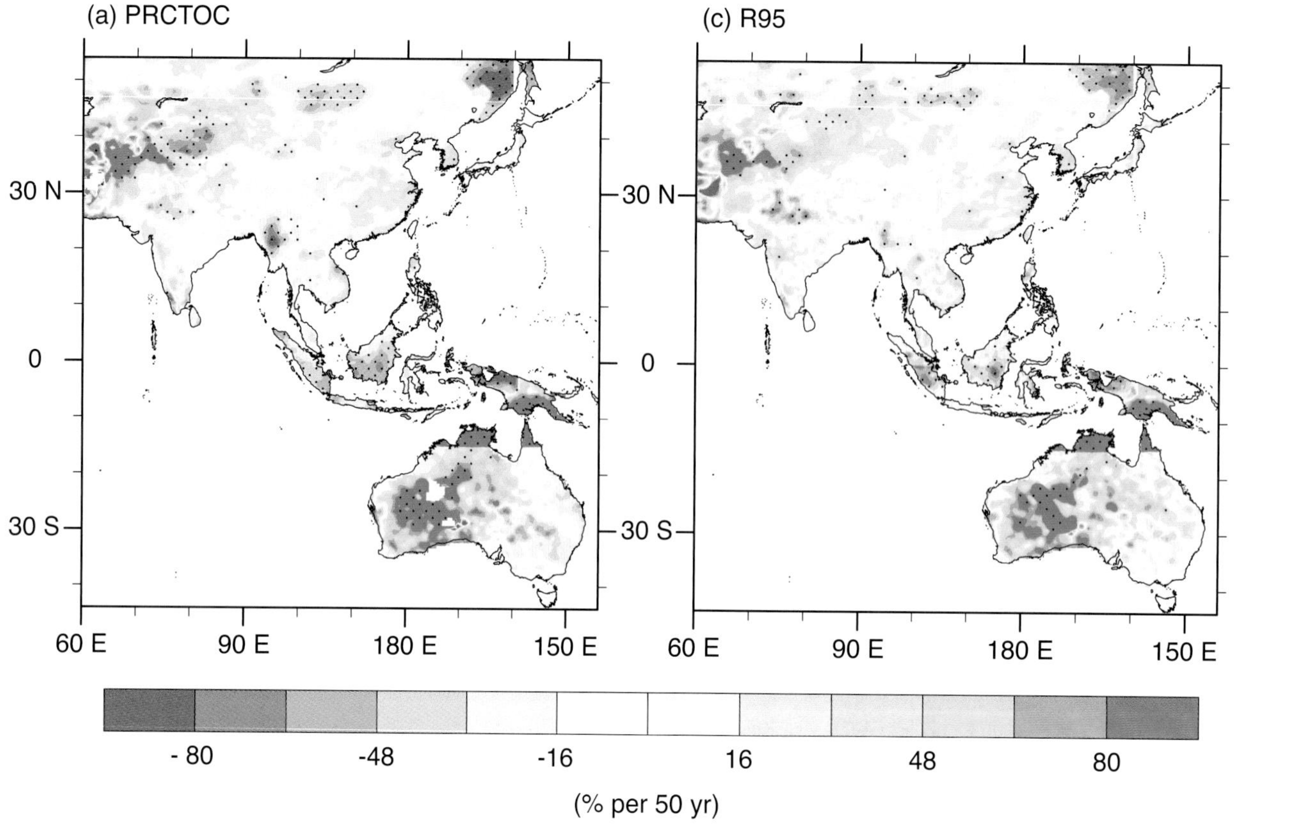

Figure 14.25 | Linear trend for local summer (a) total precipitation and (b) R95 (summer total precipitation when PR >95th percentile) during 1961–2006. The unit is % per 50-years. The trends statistically significant at the 5% level are dotted. The daily precipitation data over Australia and China are produced by the Australian Water Availability Project (AWAP, Jones et al., 2009a) and National Climate Centre China of China Meteorological Administration (Wu and Gao, 2013), respectively, while that over the other area is compiled by the Highly Resolved Observational Data Integration Towards the Evaluation of Water Resources (APHRODITE) project (Yatagai et al., 2012). The resolution of precipitation data set is 0.5° × 0.5°. Local summer is defined as June, July and August in the Northern Hemisphere, and December, January and February in the Southern Hemisphere.

mean and extreme precipitation along the Yangtze River Valley (30°N) but deficient mean precipitation in North China in summer (Figure 14.25), and a warmer climate in winter. The observed monsoon circulation changes are partly reproduced by GCMs driven by PDO-related SST patterns but the quality of precipitation simulation is poor (Zhou et al., 2008a; Li et al., 2010a; Zhou and Zou, 2010).

In AR4, the regional warming is projected to be above the global mean in East Asia (Christensen et al., 2007). "It is *very likely* that heat waves/hot spells in summer will be of longer duration, more intense and more frequent, but very cold days are *very likely* to decrease in frequency. The precipitation is *likely* to increase in both boreal winter and summer, while the frequency of intense precipitation events is *very likely* to increase. Extreme rainfall and winds associated with tropical cyclones are *likely* to increase". CMIP5 results support many of these assessments.

More recent analysis suggested that CMIP3 models projected increased summer precipitation in amount and intensity over East Asia (Figure 14.24 for SRES A1B scenario) due to enhanced moisture convergence in a warmer climate (Ding et al., 2007; Sun and Ding, 2010; Chen et al., 2011; Kusunoki and Arakawa, 2012), along with an increase in interannual variability (Lu and Fu, 2010). CMIP5 projections for RCP4.5 support those from AR4 for summer (Figure 14.24), with 90% of the models projecting a precipitation increase in the winter half year (see Table 14.1 and Figures AI.56 to AI.59). CMIP3 models projections indicated a decrease of winter precipitation extending northeastward from South China Sea to south of Japan under SRES A1B scenario, changes seen in CMIP5 projections but with smaller spatial coverage (Figure 14.24).

An increase of extreme precipitation is projected over East Asia in a warmer climate (Jiang et al., 2011; Lee et al., 2011; Li et al., 2011a, 2011b). A high-resolution model projects an increase of Meiyu precipitation in May through July, Changma precipitation over Korean peninsula in May, and Baiu precipitation over Japan in July (Kusunoki and Mizuta, 2008), and an increase of heavy precipitation over East Asia under SRES A1B scenario (Kusunoki and Mizuta, 2008; Endo, 2012). CMIP3 models project a late withdrawal of Baiu (Kitoh and Uchiyama, 2006), as has been observed in eastern and western Japan (Endo, 2010). There is a significant increase in mean, daily maximum and minimum temperatures in southeastern China, associated with a decrease in the number of frost days and an increase in the heat wave duration under SRES A2 scenario (Chen et al., 2011). The CMIP5 model projections also indicate an increase of temperature in both boreal winter and summer over East Asia for RCP4.5 (Table 14.1). A decrease of the annual and seasonal maximum wind speeds is found under SRES A2 scenario due to both the reduced intensity of cold waves and the reduced intensity of the winter monsoons (Jiang and Zhao, 2013).

The future warming patterns simulated by RCMs essentially follow those of the driving GCMs (e.g., Dairaku et al., 2008). For summer precipitation, however, RCM downscaling usually shows different regional details due to more realistic topographic forcing than in GCMs (Gao et al., 2008, 2012a). The uncertainty of precipitation projection in eastern China is larger than that in western China (Gao et al., 2012b). RCM downscaling indicates that both the seasonal mean summer rainfall and extreme precipitation around Japan Islands are projected to increase (Im et al., 2008; Iizumi et al., 2012).

Projections with a 5-km RCM show that the heaviest hourly precipitation is projected to increase even in the near future (2030s) when temperature increase is modest (Kitoh et al., 2009). A southwest expansion of the subtropical anticyclone over the northwestern Pacific Ocean associated with El Niño-like mean state changes in the Pacific and a dry air intrusion in the mid-troposphere from the Asian continent gives a favourable condition for intense precipitation in the Baiu season in Japan (Kanada et al., 2010). Increased water vapour supply from the south of the Baiu front and an intensified frontal zone with intense mean updrafts contribute to the increased occurrence of intense daily precipitation during the late Baiu season (Kanada et al., 2012).

In summary, based on CMIP5 model projections, there is *medium confidence* that with an intensified East Asian summer monsoon, summer precipitation over East Asia will increase (Table 14.3). Under RCP4.5 scenario, precipitation increase is *likely* over East Asia during the Meiyu-Changma-Baiu season in May to July, and precipitation extremes are *very likely* to increase over the eastern Asian continent in all seasons and over Japan in summer. However, there is only *low confidence* in more specific details of the projected changes due to the limited skill of CMIP5 models in simulating monsoon features such as the East Asian monsoon rainband (Table 14.2).

14.8.10 West Asia

This region extends from the Mediterranean to the western fringes of South Asia, covering the Middle East and the Arabian Peninsula and includes large areas of barren desert. The climate over this region varies from arid to semi-arid and precipitation is primarily received in the cold season.

The western part of the region is on the margin of Atlantic and Mediterranean influences, primarily the NAO (Section 14.5.1) during winter months, and indirectly the monsoon heat low (Section 14.2.2.1) in the summer months. Precipitation in this region comes largely from passing ETCs (Section 14.6.2). Land-falling TCs (Section 14.6.1) that occasionally influence the eastern part of the Arabian Peninsula are notable extreme events. Pacific Ocean variability, associated with ENSO (Section 14.2.4), and the ITCZ (Section 14.3.1) are also known to impact weather and climate in different parts of West Asia.

In recent decades, there appears to be a weak but non-significant downward trend in mean precipitation (Zhang et al., 2005; Alpert et al., 2008; AlSarmi and Washington, 2011; Tanarhte et al., 2012), although intense weather events appear to be increasing (Alpert et al., 2002; Yosef et al., 2009). In contrast, upward temperature trends are notable and robust (Alpert et al., 2008; AlSarmi and Washington, 2011; Tanarhte et al., 2012).

The ability of climate models to simulate historical climate, its change and its variability, has improved in many important aspects since the AR4 (see Figure 9.39 in Chapter 9). CMIP5 models tend to be able to reproduce the basic climate state of the region as well as the main phenomena affecting it with some fidelity (Table 14.2), but the region is at the fringes of the influence of different drivers of European, Asian and African climates and remains poorly analysed in the peer-reviewed literature with respect to climate model performances.

The CMIP5 model projections for this century are for further warming in all seasons, while precipitation shows some distinct sub-regional and seasonally dependent changes, characterized by model scatter. In both winter (October to March) and summer (April to September) precipitation in general is projected to decrease, (see Table 14.1, Figures AI.52 to AI.55). However, the various interacting dynamical influences on precipitation of the region (that models have varying success in capturing in the current climate) results in uncertainty in both the patterns and magnitude of future precipitation change. Indeed, while the overall pattern of change has remained the same between CMIP3 and CMIP5, the confidence has decreased somewhat and the boundary between the Mediterranean decreases and the general mid-latitude increase to the north has shifted closer to the region (Figures 14.26 and AI.54 to AI.55). So, although the Mediterranean side still appears *likely* to become drier, the *likely* precipitation changes for the interior land masses are less clear and the intensified and northward shifting ITCZ may imply an increase in precipitation in the most southern part of the Arabian Peninsula. Overall, the projections by the end of the century (2081–2100) indicates little overall change, although with a tendency for reduced precipitation, particular in the high end scenarios (Figures AI.5 to AI.55). However, regardless of the sign of precipitation change in the high mountain regions of the interior, the influence of warming on the snow pack will *very likely* cause important changes in the timing and amount of the spring melt (Diffenbaugh et al., 2013).

Recent downscaling results (Lionello et al., 2008; Evans, 2009; Jin et al., 2010; Dai, 2011) suggest that the eastern Mediterranean will experience a decrease in precipitation during the rainy season due to a northward displacement of the storm tracks (Section 14.6.2). A northward shift in the ITCZ results in more precipitation in the southern part, not previously being seriously affected by it. A moderate change in the annual cycle of precipitation has also been simulated by some models. Precipitation and temperature statistics in RCMs for an area consisting of the western part of the Arab Peninsula was assessed by Black (2009) and Onol and Semazzi (2009) confirming GCM-based findings. Increased drought duration has been projected (Kim and Byun, 2009). Inland from the Mediterranean coastal areas, resolution of the terrain becomes more important and, while downscaled results (Evans, 2008; Marcella and Eltahir, 2011; Lelieveld et al., 2012) broadly agree with GCM projections, higher resolution results in some differences associated with mountain barrier jets (Evans, 2008; see also Figure 14.26).

In summary, since AR4 climate models appear to have only modestly improved fidelity in simulating aspects of large-scale climate

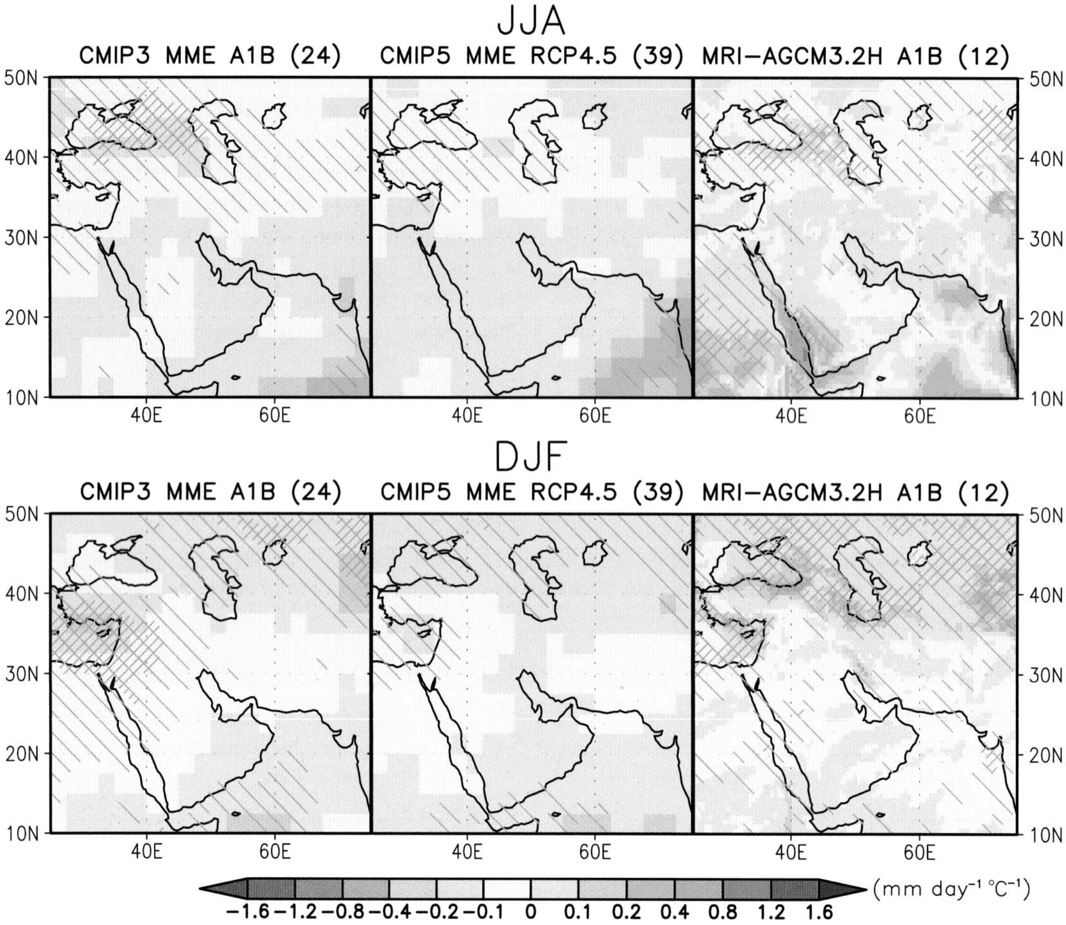

Figure 14.26 | Maps of precipitation changes for West Asia in 2080–2099 with respect to 1986–2005 in June, July and August (above) and December, January and February (below) in the SRES A1B scenario with 24 CMIP3 models (left), and in the RCP4.5 scenario with 39 CMIP5 models (middle). The figures on the right are the precipitation changes in 2075–2099 with respect to 1979–2003 in the SRES A1B scenario with the 12-member 60-km mesh Meteorological Research Institute (MRI)-Atmospheric General Circulation Model 3.2 (AGCM3.2) multi-physics, multi-sea surface temperature (SST) ensembles (Endo et al., 2012). Precipitation changes are normalized by the global annual mean surface air temperature changes in each scenario. Light hatching denotes where more than 66% of models (or members) have the same sign with the ensemble mean changes, while dense hatching denotes where more than 90% of models (or members) have the same sign with the ensemble mean changes.

phenomena influencing regional climates over West Asia. Model agreement, however, indicates that it is *very likely* that temperatures will continue to increase. But at the same time, model agreement on projected precipitation changes have reduced, resulting in *medium confidence* in projections showing an overall reduction in precipitation.

14.8.11 South Asia

From June through September, the Indian summer monsoon (Section 14.2.2.1) dominates South Asia, while the northeast winter monsoon contributes substantially to annual rainfall over southeastern India and Sri Lanka. The winter weather systems are also important in northern parts of South Asia, that is, the western Himalayas.

Seasonal mean rainfall shows interdecadal variability, noticeably a declining trend with more frequent deficit monsoons (Kulkarni, 2012). There are regional inhomogeneities: precipitation decreased over central India along the monsoon trough (Figure 14.25) thought to be due to a number of factors (Section 14.2.2) including black carbon, sulphate aerosols (Chung and Ramanathan, 2007; Bollasina et al., 2011), land use changes (Niyogi et al., 2010) and SST rise over the Indo-Pacific warm pool (Annamalai et al., 2013). The increase in the number of monsoon break days over India (Dash et al., 2009), and the decline in the number of monsoon depressions (Krishnamurthy and Ajayamohan, 2010), are consistent with the overall decrease in seasonal mean rainfall. The frequency of heavy precipitation events is increasing (Rajeevan et al., 2008; Krishnamurthy et al., 2009; Sen Roy, 2009; Pattanaik and Rajeevan, 2010), while light rain events are decreasing (Goswami et al., 2006).

CMIP models reasonably simulate the annual cycle of precipitation and temperature over South Asia (Table 14.2; Figure 9.38) but are limited in simulating fine structures of rainfall variability on sub-seasonal and sub-regional scales (Turner and Annamalai, 2012). CMIP5 models show improved skill in simulating monsoon variability compared to CMIP3 (Sperber et al., 2012; Section 14.2.2).

Summer precipitation changes in South Asia are consistent overall between CMIP3 and CMIP5 (Figure 14.24), but model scatter is large in winter precipitation change (Figures 14.24 and AI.62). Changes in the summer monsoon dominate annual rainfall (see Section 14.2.2). The CMIP3 multi-model ensemble shows an increase in summer precipitation (Kumar et al., 2011a; May, 2011; Sabade et al., 2011), although there are wide variations among model projections (Annamalai et al., 2007; Kripalani et al., 2007b). Spatially, the rainfall increase is stronger over northern parts of South Asia, Bangladesh and Sri Lanka, with a weak decrease over Pakistan (Turner and Annamalai, 2012). In RCP6.0 and RCP8.5 scenarios, frequency of extreme precipitation days shows consistent increasing trends in 2060 and beyond (Chaturvedi et al., 2012; Figure AI.63). In six CMIP3 models, precipitation anomalies during Indian summer monsoon breaks strengthen in a warmer climate, but changes in the timing and duration of active/break spells are variable among models (Mandke et al., 2007). Note that the active/break spells of the monsoon are related to the MJO (see Section 14.3.2), a phenomenon that models simulate poorly (Section 9.5.2.3; Lin et al., 2008a; Sperber and Annamalai, 2008).

High-resolution RCM and GCM projections showed an overall increase of precipitation over a large area of peninsular India (Rupa Kumar et al., 2006; Stowasser et al., 2009; Kumar et al., 2011a), but a significant reduction in orographic rainfall in both seasonal mean and extreme events on west coasts of India (Rajendran and Kitoh, 2008; Ashfaq et al., 2009; Kumar et al., 2013). Such spatial variations in projected precipitation near orography are noticeable in Figure 14.24 on the background of the overall increase.

CMIP5 models project a clear increase in temperature over India especially in winter (Figures AI.60 to AI.61), with enhanced warming during night than day (Kumar et al., 2011a) and over northern India (Kulkarni, 2012). In summer, extremely hot days and nights are projected to increase. Table 14.1 summarizes the projected temperature and precipitation changes for SAS in the RCP4.5 scenario based on CMIP5.

In summary, there is *high confidence* in projected rise in temperature. There is *medium confidence* in summer monsoon precipitation increase in the future over South Asia. Model projections diverge on smaller regional scales.

14.8.12 Southeast Asia

Southeast Asia features a complex range of terrains and land–sea contrasts. Across the region, temperature has been increasing at a rate of 0.14°C to 0.20°C per decade since the 1960s (Tangang et al., 2007), coupled with a rising number of hot days and warm nights, and a decline in cooler weather (Manton et al., 2001; Caesar et al., 2011). A positive trend in the occurrence of heavy (top 10% by rain amount) and light (bottom 5%) rain events and a negative trend in moderate (25 to 75%) rain events has been observed (Lau and Wu, 2007). Annual total wet-day rainfall has increased by 22 mm per decade, while rainfall from extreme rain days has increased by 10 mm per decade (Alexander et al., 2006; Caesar et al., 2011).

Several large-scale phenomena influence the climate of this region. While ENSO (Section 14.4) influence is predominant in East Malaysia and areas east of it, Maritime continent monsoon (Section 14.2.3) influences the climate in Peninsular Malaya. The impact of the IOD (Section 14.3.3) is more prominent in eastern Indonesia. Thus climate variability and trends differ vastly across the region and between seasons. Between 1955 and 2005 the ratio of rainfall in the wet to the dry seasons increased (Aldrian and Djamil, 2008). This appears to be at least in part consistent with an upward trend of the IOD. While an increasing frequency of extreme events has been reported in the northern parts of South East Asia, decreasing trends in such events are reported in Myanmar (Chang, 2011); see also Figure 14.25.

For a given region, strong seasonality in change is observed. In Peninsular Malaya during the southwest monsoon season, total rainfall and the frequency of wet days decreased, but rainfall intensity increased in much of the region (Deni et al., 2010). During the northeast monsoon, total rainfall, the frequency of extreme rainfall events, and rainfall intensity all increased over the peninsula (Suhaila et al., 2010).

High-resolution model simulations are necessary to resolve complex terrain such as in Southeast Asia (Nguyen et al., 2012; Section 14.2.2.4). In a RCM downscaling simulation using the A1B emission scenario (Chotamonsak et al., 2011), regional average rainfall was projected to increase, consistent with a combination of the 'warmer getting wetter' mechanism (Section 14.3.1), an increase in summer monsoon, though there is a lack of consensus on future ENSO changes. The spatial pattern of change is similar to that projected in the AR4 (Christensen et al., 2007, Section 11.4).

The median increase in temperature over land ranges from 0.8°C in RCP2.6 to 3.2°C in RCP8.5 by the end of this century (2081–2100). A moderate increase in precipitation is projected for the region: 1% in RCP2.6 increasing to 8% in RCP8.5 by 2100 (Table 14.1, Supplementary Material Table 14.SM.1a to 14.SM.1c, Figures 14.27 and AI.64 to AI.65). On islands neighbouring the southeast tropical Indian Ocean, rainfall is projected to decrease during July to November (the IOD prevalent season), consistent with a slower oceanic warming in the east than in the west tropical Indian Ocean, despite little change projected in the IOD (Section 14.3.3).

In summary, warming is *very likely* to continue with substantial sub-regional variations. There is *medium confidence* in a moderate increase in rainfall, except on Indonesian islands neighbouring the southeast Indian Ocean. Strong regional variations are expected because of terrain.

14.8.13 Australia and New Zealand

The climate of Australia is a mix of tropical and extratropical influences. Northern Australia lies in the tropics and is strongly affected by the Australian monsoon circulation (Section 14.2.2) and ENSO (Section 14.4). Southern Australia extends into the extratropical westerly circulation and is also affected by the middle latitude storm track (Section 14.6.2), the SAM (Section 14.5.2), mid-latitude transient wave propagation, and remotely by the IOD (Section 14.3.3) and ENSO.

Eastern–northeastern Australian rainfall is strongly influenced by the ENSO cycle, with La Niña years typically associated with wet conditions and more frequent and intense tropical cyclones in summer, and El Niño years with drier than normal conditions, most notably in spring. The SAM plays a significant role in modulating southern Australian rainfall, the positive SAM being associated with generally above-normal rainfall during summer (Hendon et al., 2007; Thompson et al., 2011), but in winter with reduced rainfall, particularly in Southwest

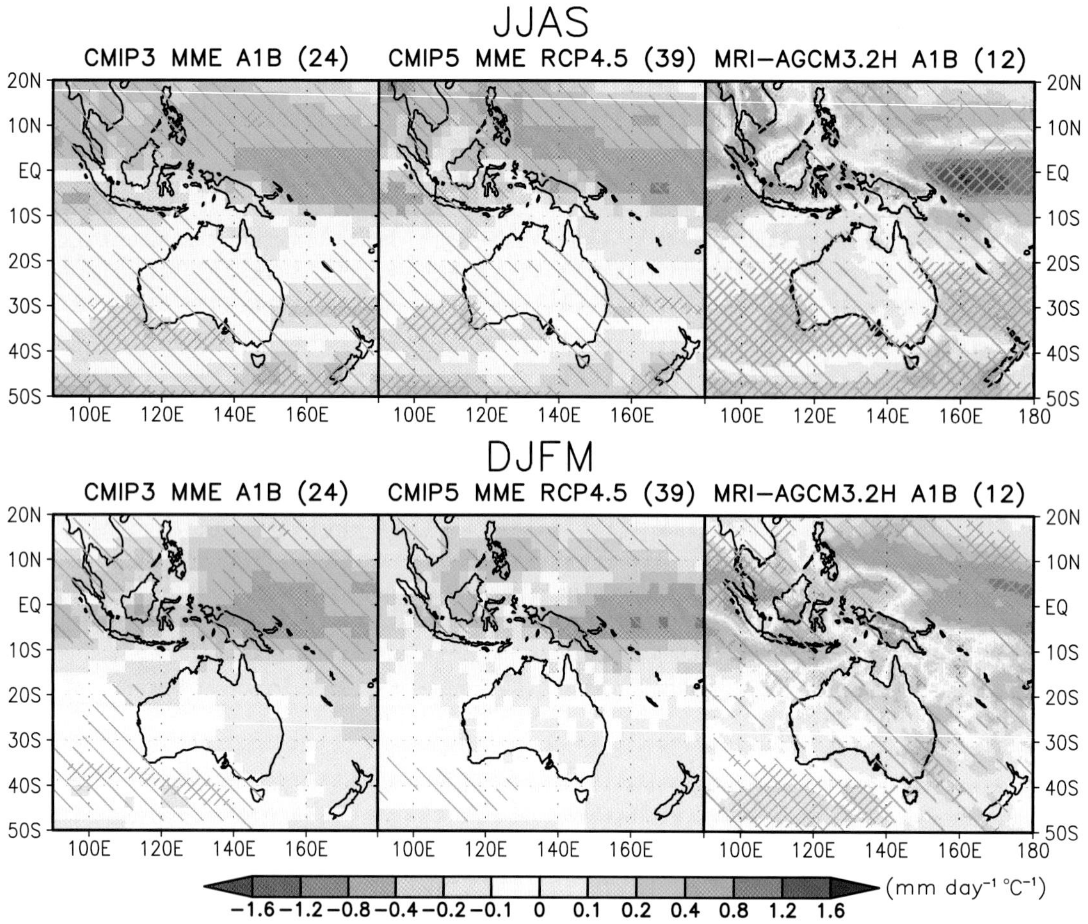

Figure 14.27 | Maps of precipitation changes for Southeast Asia, Australia and New Zealand in 2080–2099 with respect to 1986–2005 in June to September (above) and December to March (below) in the SRES A1B scenario with 24 CMIP3 models (left), and in the RCP4.5 scenario with 39 CMIP5 models (middle). Right figures are the precipitation changes in 2075–2099 with respect to 1979–2003 in the SRES A1B scenario with the 12-member 60- km mesh Meteorological Research Institute (MRI)-Atmospheric General Circulation Model 3.2 (AGCM3.2) multi-physics, multi-sea surface temperature (SST) ensembles (Endo et al., 2012). Precipitation changes are normalized by the global annual mean surface air temperature changes in each scenario. Light hatching denotes where more than 66% of models (or members) have the same sign with the ensemble mean changes, while dense hatching denotes where more than 90% of models (or members) have the same sign with the ensemble mean changes.

Western Australia (Hendon et al., 2007; Meneghini et al., 2007; Pezza et al., 2008; Risbey et al., 2009; Cai et al., 2011c). Rossby wavetrains induced by tropical convective anomalies associated with the IOD (Cai et al., 2009), and associated with ENSO through its coherence with the IOD (Cai et al., 2011b) also have a strong impact, leading to lower winter and spring rainfall particularly over Southeastern Australia during positive IOD and El Niño events. Along the eastern seaboard, ETCs (Section 14.6.2) exert a strong influence on the regional climate, while ENSO and other teleconnections play a lesser role (Risbey et al., 2009; Dowdy et al., 2012).

Significant trends have been observed in Australian rainfall over recent decades (Figure 14.25), varying vastly by region and season. Increasing summer rainfall and decreasing temperature trends over northwest Australia have raised the question of whether aerosols originating in the NH play a role (Rotstayn et al., 2007; Shi et al., 2008b; Smith et al., 2008; Rotstayn et al., 2009; Cai et al., 2011d), but there is no consensus at present. By contrast, a prominent rainfall decline has been experienced in austral winter over southwest Western Australia (Cai and Cowan, 2006; Bates et al., 2008) and in mid-to-late autumn over southeastern Australia (Murphy and Timbal, 2008). Over southwest Western Australia, the decrease in winter rainfall since the late 1960s of about 20% have led to an even bigger (~50%) drop in inflow into dams. The rainfall decline has been linked to changes in large-scale mean sea level pressure (Bates et al., 2008), shifts in synoptic systems (Hope et al., 2006), changes in baroclinicity (Frederiksen and Frederiksen, 2007), the SAM (Cai and Cowan, 2006; Meneghini et al., 2007), land cover changes (Timbal and Arblaster, 2006), anthropogenic forcing (Timbal et al., 2006), Indian Ocean warming (England et al., 2006) and teleconnection to Antarctic precipitation (van Ommen and Morgan, 2010).

Over southeastern Australia, the decreasing rainfall trend is largest in autumn with sustained declines during the drought of 1997–2009, especially in May (Cai and Cowan, 2008; Murphy and Timbal, 2008; Cai et al., 2012a). The exact causes remain contentious, and for the decrease in May, may include ENSO variability and long-term Indian Ocean warming (Cai and Cowan, 2008; Ummenhofer et al., 2009b), a weakening of the subtropical storm track due to decreasing baroclinic instability of the subtropical jet (Frederiksen et al., 2010; Frederiksen et al., 2011a, 2011b) and a poleward shift the ocean–atmosphere circulation

(Smith and Timbal, 2012; Cai and Cowan, 2013). The well-documented poleward expansion of the subtropical dry zone (Seidel et al., 2008; Johanson and Fu, 2009; Lucas et al., 2012), particularly in April and May, is shown to account for much of the April–May reduction (Cai et al., 2012a). Rainfall trends over southeastern Australia in spring, far weaker but with a signature in the subtropical ridge (Cai et al., 2011a; Timbal and Drosdowsky, 2012), have been shown to be linked with trends and variability in the IOD (Cai et al., 2009; Ummenhofer et al., 2009b). Antarctic proxy data that capture both eastern Australian rainfall and ENSO variability (Vance et al., 2012) show a predominance of El Niño/drier conditions in the 20th century than was the average over the last millennium.

On seasonal to decadal time scales, New Zealand precipitation is modulated by the SAM (Kidston et al., 2009; Thompson et al., 2011), ENSO (Kidson and Renwick, 2002; Ummenhofer and England, 2007) and the IPO (Griffiths, 2007). Increased westerly flow across New Zealand, associated with negative SAM and with El Niño events, leads to increased rainfall and generally lower than normal temperatures in western regions. The positive SAM and La Niña conditions are generally associated with increased rainfall in the north and east of the country, and warmer than normal conditions. On longer time scales, a drying trend since 1979 across much of New Zealand during austral summer is consistent with recent trends in the SAM and to a lesser extent ENSO and the IPO (Griffiths, 2007; Ummenhofer et al., 2009a). In western regions, however, the drying is accompanied by a trend towards increased heavy rainfall (Griffiths, 2007). Temperatures over New Zealand have risen by just under 1°C over the past century (Dean and Stott, 2009). The upward trend has been modulated by an increase in the frequency of cool southerly wind flows over the country since the 1950s, without which the observed warming is consistent with large-scale anthropogenic forcing (Dean and Stott, 2009).

A recent analysis (Irving et al., 2012; their Figure 9) shows that climate projections over Australia using CMIP5 models, which generally simulate the climate of Australia well (Watterson et al., 2013), are highly consistent with existing CMIP3-derived projections. The projected changes include a further 1.0 to 5.0°C temperature rise by the year 2070 (relative to 1990); a long-term drying over southern areas during winter, particularly in the southwest (Figure 14.27), that is consistent with an upward trend of the SAM (Pitman and Perkins, 2008; Shi et al., 2008a; Cai et al., 2011c); a long-term rainfall decline over southern and eastern areas during spring, in part consistent with a upward trend of the IOD index (Smith and Chandler, 2010; Zheng et al., 2010; Weller and Cai, 2013; Zheng et al., 2013). Precipitation change in northeast Australia remains uncertain (Moise et al., 2012), related to the lack of consensus over how ENSO may change (Collins et al., 2010; Section 14.4). In terms of climate extremes, more frequent hot days and nights and less frequent cold days and nights are projected (Alexander and Arblaster, 2009). Changes in the intensity and frequency of extreme rainfall events generally follow the mean rainfall change (Kharin et al., 2007), although there is an increase in most regions in the intensity of short duration extremes (e.g., Alexander and Arblaster, 2009).

For New Zealand, future climate projections suggest further increases in the westerlies in winter and spring, though model biases in jet latitude in the present climate reduce confidence in the detail of future projections (Barnes et al., 2010). The influence of poleward expansion of the subtropical high-pressure belt is projected to lead to drier conditions in parts of the country (Figure 14.27; Table 14.1), and a decrease in westerly wind strength in northern regions. Such projections imply increased seasonality of rainfall in many regions of New Zealand (Reisinger et al., 2010). Both flood and drought occurrence is projected to approximately double over New Zealand during the 21st century, under the SRES A1B scenario. Temperatures are projected to rise at about 70% of the global rate, because of the buffering effect of the oceans around New Zealand. Temperature rises are projected to be smallest in spring (SON) while the season of greatest warming varies by region around the country. Continued decreases in frost frequency, and increases in the frequency of high-temperature extremes, are expected, but have not been quantified (Reisinger et al., 2010).

In summary, based on understanding of recent trends and on CMIP5 results, it is *likely* that cool season precipitation will decrease over southern Australia associated in part with trends in the SAM, the IOD and a poleward shift and expansion of the subtropical dry zone. It is *very likely* that Australia will continue to warm through the 21st century, at a rate similar to the global land surface mean. The frequency of very warm days is *very likely* to increase through this century, across the whole country.

It is *very likely* that temperatures will continue to rise over New Zealand. Precipitation is *likely* to increase in western regions in winter and spring, but the magnitude of change is *likely* to remain comparable to that of natural climate variability through the rest of the century. In summer and autumn, it is *as likely as not* that precipitation amounts will change.

14.8.14 Pacific Islands Region

The Pacific Islands region includes the northwest tropical Pacific, and the tropical southwest Pacific. North of the Equator, the wet season occurs from May to November. In the south, the wet seasons occurs from November to April.

The phenomena mainly responsible for climate variations in the Pacific Islands are ENSO (Section 14.4), the SPCZ (Section 14.3.1.2), the ITCZ (Section 14.3.1.1) and the WNPSM (Section 14.2.2.5). During El Niño events, the ITCZ and SPCZ move closer to the equator, rainfall decreases in western regions and increases in the central Pacific, and tropical cyclone numbers tend to increase and to occur farther east than normal (Diamond et al., 2012). During La Niña, the western tropical Pacific tends to experience above-average numbers of tropical cyclones (Nicholls et al., 1998; Lavender and Walsh, 2011).

The seasonal evolution of the SPCZ has a strong influence on the seasonality of the climate of the southern tropical Pacific, particularly during the wet season. The SPCZ moves northward during moderate El Niño events and southward during La Niña events (Folland et al., 2002; Vincent et al., 2011). During El Niño events, southwest Pacific Island nations experience an increased occurrence of forest fires and droughts (Salinger et al., 2001; Kumar et al., 2006b), and an increased probability of tropical cyclone damage, as tropical cyclogenesis tends to reside within 6° to 10° south of the SPCZ (Vincent et al., 2011).

Nauru experiences drought during La Niña as the SPCZ and ITCZ move to the west (Brown et al., 2012c). During strong El Niño events (e.g., 1982/1983, 1997/1998) the SPCZ undergoes an extreme swing of up to 10 degrees towards the equator and collapses to a more zonally oriented structure (Vincent et al., 2011; Section 14.3.2). The impacts from these zonal SPCZ events are much more severe than those from moderate El Niño events (Vincent et al., 2011; Cai et al., 2012b), and can induce massive droughts and food shortages (Barnett, 2011).

Temperatures have increased at a rate between 0.1°C and 0.2°C per decade throughout the Pacific Islands during the 20th century (Folland et al., 2003). Changes in temperature extremes have followed those of mean temperatures (Manton et al., 2001; Griffiths et al., 2005). During 1961–2000, locations to the northeast of the SPCZ became wetter, with the largest trends occurring in the eastern Pacific Ocean (east of 160°W), while locations to the southwest of the SPCZ became drier (Griffiths et al., 2003), indicative of a northeastward shift of the SPCZ. Trends in the frequency of rain days were generally similar to those of total annual rainfall (Manton et al., 2001; Griffiths et al., 2003). Since 1980, western Pacific monsoon- and ITCZ-related rain during June to August has decreased (Hennessy et al., 2011).

Future projections for tropical Pacific Island nations are based on direct outputs from a suite of CMIP3 models, updated using CMIP5 wherever available (Brown et al., 2011; Hennessy et al., 2011; Irving et al., 2011; Moise and Delage, 2011; Perkins, 2011; Perkins et al., 2012). These projections carry a large uncertainty, even in the sign of change, as discussed below and as evident in Table 14.1.

Annual average air and sea surface temperature are projected to continue to increase for all tropical Pacific countries. By 2055, under the high A2 emissions scenario, the increase is projected to be 1°C to 2°C. A rise in the number of hot days and warm nights is also projected, and a decline in cooler weather, as already observed (Manton et al., 2001). For a low-emission scenario, the lower range decreases about 0.5°C while the upper range reduces by between 0.2°C and 0.5°C.

To a large extent, the response of the ITCZ, the SPCZ, and the WNPSM to greenhouse warming will determine how rainfall patterns will change in tropical Pacific. In northwestern and near-equatorial regions, rainfall during all seasons is projected to increase in the 21st century. Wet season increases are consistent with the expected intensification of the WNPSM and the ITCZ (Smith et al., 2012a). For the southwestern tropical Pacific, the CMIP3 and CMIP5 ensemble mean change in summer rainfall is far smaller than the inter-model range (Brown et al., 2012b; Widlansky et al., 2013). There is a projected intensification in the western part of the SPCZ and near the equator with little mean change in SPCZ position (Brown et al., 2012a; Brown et al., 2012b). For the southern group of the Cook Islands, the Solomon Islands, and Tuvalu, average rainfall during the wet season is projected to increase; and for Vanuatu, Tonga, Samoa, Niue, Fiji, a decrease in dry season rainfall is accompanied by an increase in the wet season, indicating an intensified seasonal cycle.

Extreme rainfall days are *likely* to occur more often in all regions related to an intensification of the ITCZ and the SPCZ (Perkins, 2011). Although the intensification appears to be reproduced in CMIP5 models (Brown et al., 2012a), it has recently been questioned (Widlansky et al., 2013; see Section 14.3.1). There are two competing mechanisms, the 'wet regions getting wetter' and the 'warmest getting wetter, or coldest getting drier' paradigms. These two mechanisms compete within much of the SPCZ region. Based on a multi-model ensemble of 55 greenhouse warming experiments, in which model biases were corrected, tropical SST changes between 2°C to 3°C resulted in a 5% decrease of austral summer moisture convergence in the current SPCZ region (Widlansky et al., 2013). This projects a diminished rainy season for most Southwest Pacific island nations. In Samoa and neighbouring islands, summer rainfall may decrease on average by 10 to 20% during the 21st century as simulated by the hierarchy of bias-corrected atmospheric model experiments. Less rainfall, combined with increasing surface temperatures and enhanced potential evaporation, could increase the chance for longer-term droughts in the region. Such projections are completely opposite to those based on direct model outputs (Figure 14.27).

Recent downscaling experiments support the above conclusion regarding the impact of biases on the SPCZ change, and suggest that the projected intensification of the ITCZ may have uncertainties of a similar nature (Chapter 7 of Hennessy et al., 2011). In these experiments a bias correction is applied to average sea surface temperatures, and the atmosphere is forced with the 'correct' climatological seasonal cycle together with warming derived from large-scale model outputs. The results show opposite changes in much of the SPCZ and some of the ITCZ regions, resulting in much lower confidence in rainfall projections.

Despite the uncertainty, there is general agreement in model projections regarding an increase in rainfall along the equator (Tables 14.1 and 14.2), and regarding a faster warming rate in the equatorial Pacific than the off-equatorial regions (Xie et al., 2010b). A potential consequence is an increase in the frequency of the zonal SPCZ events (Cai et al., 2012b).

In summary, based on CMIP3 and CMIP5 model projections and recently observed trends, it is *very likely* that temperatures, including the frequency and magnitude of extreme high temperatures, will continue to increase through the 21st century. In equatorial regions, the consistency across model projections suggests that rainfall is *likely* to increase. However, given new model results and physical insights since the AR4, the rainfall outlook is uncertain in regions directly affected by the SPCZ and western portion of the ITCZ.

14.8.15 Antarctica

Much of the climate variability of Antarctica is modulated by the Southern Annular Mode (SAM, Section 14.5.2), the high-latitude atmospheric response to ENSO (Section 14.4) and interactions between the two (Stammerjohn et al., 2008; Fogt et al., 2011; see also Sections 2.7 and 10.3.3). Signatures of the SAM and ENSO in Antarctic temperature, snow accumulation and sea ice have been documented by many observational and modelling studies (Bromwich et al., 2004; Guo et al., 2004; Kaspari et al., 2004; van den Broeke and van Lipzig, 2004; Marshall, 2007).

The positive SAM is associated on average with warmer conditions over the Peninsula and colder conditions over East Antarctica, with a mixed and generally non-significant impact over West Antarctica (Kwok and Comiso, 2002; Thompson and Solomon, 2002; van den Broeke and van Lipzig, 2004). ENSO is associated with circulation anomalies over the southeast Pacific that primarily affect West Antarctica (Bromwich et al., 2004; Guo et al., 2004; Turner, 2004). ENSO variability tends to produce out-of-phase variations between the western and eastern sectors of West Antarctica (Bromwich et al., 2004; Kaspari et al., 2004), in association with the PSA pattern (Section 14.7.1).

The positive summer/autumn trend in the SAM index in recent decades (Section 14.5.2) has been related to the contrasting temperature trend patterns observed in these two seasons, with warming in the east and north of the Antarctic Peninsula and cooling (or no significant temperature change) over much of East Antarctica (Turner et al., 2005; Thompson et al., 2011). The high polarity of the SAM is also consistent with the significant increase in snow accumulation observed in the southern part of the Peninsula (Thomas et al., 2008).

Unlike the eastern Antarctic Peninsula, its western coast shows maximum warming in austral winter (when the SAM does not exhibit any significant trend), which has been attributed to reduced sea ice concentrations in the Bellingshausen Sea. Recent studies have emphasized the role of tropical SST forcing not directly linked to ENSO to explain the prominent spring- and wintertime atmospheric warming in West Antarctica (Ding et al., 2011; Schneider et al., 2012). There is further evidence of tropical SST influence on Antarctic temperatures and precipitation on decadal to inter-decadal time scales (Monaghan and Bromwich, 2008; Okumura et al., 2012).

Modelling of Antarctic climate remains challenging, in part because of the nature of the high-elevation ice sheet in the east Antarctic and its effects on regional climate (Section 9.4.1.1). Moreover, modelling ice properties themselves, for both land ice and sea ice, is an area that is still developing despite improvements in recent years (Vancoppenolle et al., 2009; Picard et al., 2012; Section 9.4.3). Modelling the role of the stratosphere and of ozone recovery is critical for Antarctic climate, as stratospheric change is intimately linked to trends in the SAM (Section 14.5.2).

The projected easing of the positive SAM trend in austral summer (Section 14.5.2) may act to delay future loss of Antarctic sea ice (Bitz and Polvani, 2012; Smith et al., 2012b). It is unclear what effect ENSO will have on future Antarctic climate change as the ENSO response to climate change remains uncertain (see 12.4.4.1 and 14.5.2 for more information). Seasonally, changes in the strength of the circumpolar westerlies are also expected during the 21st century as a result of changes in the semi-annual oscillation caused by alterations in the mid- to high-latitude temperature gradient in the SH. Bracegirdle et al. (2008) considered modelled circulation changes over the Southern Ocean and found a more pronounced strengthening of the autumn peak of the semi-annual oscillation compared with the spring peak.

Future changes in surface temperature over Antarctica are *likely* to be smaller than the global mean, and much smaller than those projected for the Arctic, because of the buffering effect of the southern oceans, and the thermal mass of the east Antarctic ice sheet (Section 12.4.6). Warming is *likely* to bring increased precipitation on average across Antarctica (Bracegirdle et al., 2008), but the spatial pattern of precipitation change remains uncertain.

In summary, consistency across CMIP5 projections suggests it is *very likely* that Antarctic temperatures will increase through the rest of the century, but more slowly than the global mean rate of increase (Table 14.1). SSTs of the oceans around Antarctica are *likely* to rise more slowly than surface air temperature over the Antarctic land mass. As temperatures rise, it is also *likely* that precipitation will increase (Table 14.1), up to 20% or more over the East Antarctic. However, given known difficulties associated with correctly modelling Antarctic climate, and uncertainties associated with future SAM and ENSO trends and the extent of Antarctic sea ice, precipitation projections have only *medium confidence*.

Table 14.1 | Temperature and precipitation projections by the CMIP5 global models. The figures shown are averages over SREX regions (Seneviratne et al., 2012) of the projections by a set of 42 global models for the RCP4.5 scenario. Added to the SREX regions are a six other regions including the two Polar Regions, the Caribbean, Indian Ocean and Pacific Island States (see Annex I for further details). The 26 SREX regions are: Alaska/NW Canada (ALA), Eastern Canada/Greenland/Iceland (CGI), Western North America (WNA), Central North America (CNA), Eastern North America (ENA), Central America/Mexico (CAM), Amazon (AMZ), NE Brazil (NEB), West Coast South America (WSA), Southeastern South America (SSA), Northern Europe (NEU), Central Europe (CEU), Southern Europe/the Mediterranean (MED), Sahara (SAH), Western Africa (WAF), Eastern Africa (EAF), Southern Africa (SAF), Northern Asia (NAS), Western Asia (WAS), Central Asia (CAS), Tibetan Plateau (TIB), Eastern Asia (EAS), Southern Asia (SAS), Southeastern Asia (SEA), Northern Australia (NAS) and Southern Australia/New Zealand (SAU). The area-mean temperature and precipitation responses are first averaged for each model over the 1986–2005 period from the historical simulations and the 2016–2035, 2046–2065 and 2081–2100 periods of the RCP4.5 experiments. Based on the difference between these two periods, the table shows the 25th, 50th and 75th percentiles, and the lowest and highest response among the 42 models, for temperature in degrees Celsius and precipitation as a percent change. Regions in which the middle half (25 to 75%) of this distribution is all of the same sign in the precipitation response are coloured light brown for decreasing precipitation and light green for increasing precipitation. Information is provided for land areas contained in the boxes unless otherwise indicated. The temperature responses are averaged over the boreal winter and summer seasons; December, January and February (DJF) and June, July and August (JJA) respectively. The precipitation responses are averaged over half year periods, boreal winter; October, November, December, January, February and March (ONDJFM) and summer; April, May, June, July, August and September (AMJJAS).

RCP4.5			Temperature (°C)					Precipitation (%)				
REGION	**MONTH[a]**	**Year**	**min**	**25%**	**50%**	**75%**	**max**	**min**	**25%**	**50%**	**75%**	**max**
Arctic												
(land)	DJF	2035	0.6	1.5	1.7	2.2	4.2	3	7	9	11	19
		2065	0.4	3.0	3.4	4.5	8.0	5	14	17	21	37
		2100	–0.9	3.7	5.0	6.2	10.0	–2	18	24	30	50
	JJA	2035	0.3	0.8	1.0	1.2	3.0	–3	4	5	7	20
		2065	0.5	1.3	1.8	2.3	4.8	1	7	10	12	34
		2100	0.3	1.8	2.2	3.0	6.0	–2	10	13	17	39
	Annual	2035	0.4	1.3	1.5	1.7	3.8	1	5	6	8	20
		2065	0.3	2.4	2.8	3.5	6.4	3	11	13	15	35
		2100	–0.4	3.0	3.9	4.7	7.8	–2	14	17	21	43
(sea)	DJF	2035	0.2	2.2	2.8	3.3	6.7	–1	7	9	15	25
		2065	–0.5	4.2	5.1	6.8	11.4	–2	14	18	25	39
		2100	–2.2	5.4	7.0	9.1	14.8	–10	23	26	37	48
	JJA	2035	0.1	0.5	0.6	0.7	1.9	–3	4	6	7	17
		2065	0.0	0.8	1.2	1.4	2.9	–2	9	11	14	23
		2100	–0.3	1.2	1.5	2.1	4.0	–3	12	16	18	29
	Annual	2035	0.2	1.5	2.0	2.3	4.7	0	6	8	9	21
		2065	–0.1	2.9	3.7	4.7	7.4	–1	11	13	20	28
		2100	–1.0	3.7	4.9	6.5	9.3	–7	16	21	26	37
High latitudes												
Canada/ Greenland/ Iceland	DJF	2035	–0.2	1.2	1.7	1.9	3.1	0	4	5	9	14
		2065	0.6	2.8	3.4	3.9	6.6	3	9	12	15	21
		2100	–0.5	3.2	4.6	5.6	8.1	–2	11	15	22	32
	JJA	2035	0.1	0.7	1.0	1.2	3.0	0	2	3	4	8
		2065	0.5	1.3	1.8	2.3	4.5	2	5	6	9	16
		2100	0.2	1.7	2.3	3.0	5.6	1	6	9	12	20
	Annual	2035	0.2	1.1	1.3	1.6	2.9	0	3	4	6	9
		2065	0.4	2.0	2.5	2.9	5.2	3	7	9	11	17
		2100	–0.2	2.6	3.2	4.0	6.4	0	10	11	15	22
North Asia	DJF	2035	0.5	1.1	1.5	2.2	4.0	2	6	8	10	22
		2065	1.2	2.3	3.0	3.6	6.0	5	11	14	18	34
		2100	0.2	3.0	3.8	4.9	7.8	5	13	18	22	44
	JJA	2035	0.1	0.8	1.0	1.4	2.5	1	2	4	6	16
		2065	0.8	1.5	2.0	2.7	4.4	–1	5	8	10	21
		2100	0.8	1.9	2.4	3.5	5.1	–3	6	9	12	30
	Annual	2035	0.4	1.1	1.3	1.6	3.0	1	4	5	7	18
		2065	0.8	2.0	2.4	2.9	4.9	2	8	9	12	25
		2100	0.2	2.5	3.2	3.8	5.8	1	10	12	15	35

(continued on next page)

Table 14.1 (continued)

RCP4.5			Temperature (°C)					Precipitation (%)				
REGION	**MONTH**[a]	**Year**	**min**	**25%**	**50%**	**75%**	**max**	**min**	**25%**	**50%**	**75%**	**max**
North America												
Alaska/	DJF	2035	0.0	1.1	1.7	2.4	3.4	–1	3	5	8	12
NW Canada		2065	1.2	2.8	3.6	4.8	7.4	3	9	11	17	29
		2100	2.3	3.5	4.8	5.9	9.7	7	11	17	21	42
	JJA	2035	0.3	0.7	1.0	1.4	2.8	–1	2	5	7	16
		2065	0.7	1.3	1.8	2.3	4.9	–2	6	10	12	29
		2100	0.9	1.8	2.2	3.1	5.2	–2	9	12	16	34
	Annual	2035	0.4	1.0	1.4	1.8	2.8	0	3	6	7	14
		2065	1.4	2.1	2.7	3.6	5.2	4	8	10	13	28
		2100	1.7	2.5	3.5	4.3	6.7	3	11	14	17	33
West North	DJF	2035	–0.4	0.7	1.1	1.5	2.5	–2	0	3	4	8
America		2065	0.9	1.7	2.2	2.6	4.0	–3	3	4	6	11
		2100	1.3	2.2	2.6	3.4	5.2	–4	4	6	8	17
	JJA	2035	0.3	0.9	1.1	1.3	2.1	–6	–1	1	3	9
		2065	0.8	1.7	2.0	2.6	3.4	–7	–1	1	4	10
		2100	0.9	2.1	2.5	3.4	4.6	–8	–1	2	6	10
	Annual	2035	0.3	0.8	1.0	1.3	1.9	–4	–1	2	3	6
		2065	0.9	1.7	2.0	2.5	3.4	–3	1	3	5	11
		2100	1.1	2.0	2.6	3.4	4.3	–4	2	4	6	14
Central North	DJF	2035	–0.1	0.7	1.1	1.6	2.9	–8	–1	1	5	11
America		2065	0.9	1.6	2.2	2.7	4.2	–7	1	4	7	17
		2100	1.2	2.0	2.7	3.6	4.9	–6	–1	4	9	18
	JJA	2035	0.3	0.8	1.1	1.4	2.3	–7	–2	0	3	9
		2065	0.9	1.7	2.1	2.5	3.5	–16	–1	2	5	12
		2100	1.0	2.1	2.5	3.1	4.6	–13	–1	2	5	13
	Annual	2035	0.4	0.9	1.1	1.3	2.0	–4	–1	1	3	7
		2065	1.0	1.7	2.0	2.4	3.4	–7	0	3	4	14
		2100	1.1	2.0	2.6	3.1	4.3	–4	0	3	6	10
Eastern North	DJF	2035	0.0	0.8	1.1	1.7	2.2	–6	0	3	7	12
America		2065	0.9	1.7	2.4	2.8	4.1	–2	4	7	9	18
		2100	0.7	2.2	2.9	3.8	4.8	–4	6	9	12	20
	JJA	2035	0.1	0.8	1.0	1.2	1.9	–4	0	3	5	9
		2065	0.8	1.5	2.0	2.4	3.9	–6	2	4	6	14
		2100	1.0	2.0	2.5	3.1	4.8	–7	2	5	7	14
	Annual	2035	0.4	0.8	1.1	1.3	1.9	–4	1	3	5	9
		2065	1.0	1.7	2.1	2.4	3.5	–1	3	5	7	14
		2100	1.0	2.1	2.7	3.1	4.2	–2	4	7	9	14

(continued on next page)

Table 14.1 (continued)

RCP4.5			Temperature (°C)					Precipitation (%)				
REGION	MONTH[a]	Year	min	25%	50%	75%	max	min	25%	50%	75%	max
Central America												
Central America	DJF	2035	0.3	0.6	0.8	0.9	1.3	−8	−3	−1	2	10
		2065	0.7	1.2	1.5	1.7	2.1	−15	−4	−2	3	10
		2100	1.0	1.6	1.8	2.4	2.7	−22	−5	0	2	11
	JJA	2035	0.5	0.7	0.8	1.0	1.4	−8	−3	−1	2	7
		2065	1.1	1.3	1.6	1.9	2.5	−15	−6	−2	1	6
		2100	1.1	1.6	2.0	2.5	3.2	−17	−6	−2	1	12
	Annual	2035	0.4	0.7	0.9	0.9	1.3	−8	−3	−1	1	6
		2065	1.0	1.3	1.5	1.8	2.4	−14	−6	−2	1	6
		2100	1.2	1.6	1.9	2.5	3.0	−17	−5	−2	1	9
Caribbean (land and sea)	DJF	2035	0.3	0.5	0.6	0.7	1.0	−13	−4	0	3	8
		2065	0.6	1.0	1.2	1.4	1.8	−14	−6	−1	3	16
		2100	0.7	1.2	1.4	1.9	2.4	−22	−6	0	5	15
	JJA	2035	0.3	0.5	0.6	0.7	1.1	−17	−9	−6	0	11
		2065	0.7	0.9	1.1	1.4	2.0	−25	−16	−11	−4	16
		2100	0.7	1.1	1.3	1.8	2.5	−36	−18	−10	−3	13
	Annual	2035	0.3	0.5	0.6	0.7	1.1	−12	−5	−3	1	8
		2065	0.6	0.9	1.1	1.4	1.9	−19	−11	−5	−2	17
		2100	0.7	1.2	1.4	1.9	2.4	−29	−10	−5	−1	14
South America												
Amazon	DJF	2035	0.4	0.7	0.8	0.9	1.6	−12	−2	0	2	4
		2065	0.8	1.3	1.6	1.9	3.0	−22	−3	−1	2	6
		2100	0.7	1.7	2.0	2.5	3.7	−22	−4	−1	1	8
	JJA	2035	0.5	0.8	1.0	1.1	1.8	−14	−3	0	2	5
		2065	1.0	1.5	1.8	2.1	3.3	−25	−4	−1	2	11
		2100	1.3	1.8	2.2	2.8	4.2	−31	−4	−1	1	9
	Annual	2035	0.4	0.8	0.9	1.0	1.8	−13	−2	0	1	4
		2065	0.9	1.4	1.7	2.1	3.3	−23	−3	−1	1	7
		2100	1.0	1.8	2.1	2.8	4.0	−25	−4	−1	1	7
Northeast Brazil	DJF	2035	0.4	0.6	0.7	0.9	1.3	−10	−2	1	3	17
		2065	0.8	1.3	1.5	1.7	2.3	−15	−5	0	4	21
		2100	0.8	1.6	1.8	2.4	2.9	−17	−5	−1	5	25
	JJA	2035	0.3	0.7	0.8	1.0	1.7	−16	−6	−3	2	15
		2065	0.8	1.4	1.6	1.9	3.0	−29	−10	−5	1	18
		2100	1.1	1.7	1.9	2.5	3.3	−39	−14	−9	−4	27
	Annual	2035	0.4	0.7	0.8	0.9	1.4	−11	−3	0	3	13
		2065	0.8	1.4	1.6	1.8	2.6	−17	−6	−2	3	20
		2100	1.0	1.7	1.9	2.5	3.1	−19	−7	−3	3	26
West Coast South America	DJF	2035	0.5	0.7	0.8	0.9	1.2	−4	−1	1	3	6
		2065	0.9	1.2	1.5	1.7	2.1	−7	−1	1	4	7
		2100	1.0	1.6	1.9	2.2	2.9	−8	0	2	5	9
	JJA	2035	0.5	0.7	0.9	0.9	1.3	−9	−1	0	2	7
		2065	1.1	1.3	1.5	1.8	2.5	−10	−2	−1	2	9
		2100	1.3	1.6	1.9	2.4	3.0	−11	−2	1	4	11
	Annual	2035	0.5	0.7	0.8	0.9	1.2	−4	0	1	2	5
		2065	1.0	1.2	1.5	1.7	2.3	−6	−1	1	2	5
		2100	1.1	1.5	1.8	2.3	2.8	−7	0	2	4	7

(continued on next page)

Table 14.1 (continued)

RCP4.5			Temperature (°C)					Precipitation (%)				
REGION	MONTH[a]	Year	min	25%	50%	75%	max	min	25%	50%	75%	max
Southeastern South America	DJF	2035	0.2	0.6	0.7	0.9	1.4	−7	0	2	4	10
		2065	0.7	1.1	1.3	1.6	2.4	−6	0	3	6	15
		2100	0.6	1.3	1.7	2.2	3.0	−6	1	4	7	18
	JJA	2035	0.0	0.4	0.6	0.8	1.2	−12	−1	2	6	19
		2065	0.4	1.0	1.2	1.5	2.1	−13	1	5	7	17
		2100	0.9	1.3	1.5	1.9	2.7	−18	1	4	8	27
	Annual	2035	0.3	0.5	0.6	0.8	1.3	−6	0	1	4	12
		2065	0.6	1.0	1.3	1.6	2.3	−6	1	3	6	13
		2100	0.7	1.3	1.6	2.2	2.7	−8	1	4	7	17
Europe												
Northern Europe	DJF	2035	−0.3	0.6	1.3	2.3	3.0	−4	2	4	6	12
		2065	−0.5	1.8	2.7	3.5	5.7	−1	3	8	11	24
		2100	−3.2	2.6	3.4	4.4	6.0	2	7	11	14	25
	JJA	2035	0.2	0.6	0.9	1.3	2.6	−6	2	4	6	11
		2065	0.0	1.2	1.8	2.5	3.6	−10	2	3	8	18
		2100	−1.1	1.6	2.2	3.0	4.7	−4	2	5	8	23
	Annual	2035	0.1	0.8	1.1	1.6	2.7	−2	2	3	6	12
		2065	−0.5	1.6	2.0	2.8	3.8	−5	3	5	9	17
		2100	−2.3	2.1	2.7	3.5	4.5	1	5	8	10	24
Central Europe	DJF	2035	−0.4	0.6	1.2	1.7	2.5	−4	0	3	5	11
		2065	0.3	1.4	2.1	2.7	3.6	−3	2	6	10	17
		2100	−0.8	2.0	2.6	3.4	5.1	−4	3	7	11	18
	JJA	2035	0.3	0.9	1.1	1.5	2.4	−8	−3	0	4	9
		2065	0.4	1.7	2.0	2.6	4.3	−13	−4	1	3	8
		2100	0.4	2.0	2.7	3.0	4.6	−16	−6	0	5	13
	Annual	2035	0.3	0.7	1.1	1.4	2.3	−3	−1	2	3	8
		2065	0.4	1.5	1.9	2.4	3.2	−6	0	3	5	9
		2100	−0.3	2.0	2.6	3.1	4.0	−5	0	4	6	14
Southern Europe/ Mediterranean	DJF	2035	−0.1	0.6	0.8	1.0	1.5	−11	−4	−2	2	8
		2065	0.1	1.2	1.5	1.8	2.3	−15	−6	−3	0	7
		2100	−0.2	1.5	2.0	2.4	3.0	−19	−7	−4	−1	9
	JJA	2035	0.6	0.9	1.2	1.4	2.9	−16	−7	−4	−1	5
		2065	1.0	1.9	2.2	2.6	4.3	−24	−12	−9	−4	5
		2100	1.2	2.3	2.8	3.3	5.5	−28	−17	−11	−6	2
	Annual	2035	0.3	0.8	1.0	1.2	2.0	−12	−4	−2	0	3
		2065	0.7	1.5	1.7	2.1	3.1	−14	−8	−5	−2	3
		2100	0.6	2.0	2.3	2.7	4.0	−19	−10	−6	−3	4
Africa												
Sahara	DJF	2035	0.1	0.8	1.0	1.1	1.5	−43	−11	−2	6	33
		2065	0.6	1.5	1.7	2.0	2.5	−29	−15	−7	1	92
		2100	0.7	1.8	2.2	2.6	3.1	−42	−14	−7	4	98
	JJA	2035	0.4	0.9	1.1	1.2	2.0	−25	−5	3	8	45
		2065	0.9	1.7	2.0	2.4	3.5	−31	−11	1	14	70
		2100	1.1	2.2	2.4	3.2	4.5	−28	−15	−1	10	108
	Annual	2035	0.4	0.9	1.0	1.1	1.5	−25	−7	0	7	45
		2065	1.0	1.6	1.8	2.2	2.8	−31	−11	−3	8	57
		2100	1.0	2.0	2.2	2.9	3.8	−27	−14	−6	9	86

(continued on next page)

Table 14.1 (continued)

RCP4.5			Temperature (°C)					Precipitation (%)				
REGION	MONTH[a]	Year	min	25%	50%	75%	max	min	25%	50%	75%	max
West Africa	DJF	2035	0.4	0.8	0.9	1.0	1.3	−5	−1	2	3	9
		2065	0.9	1.4	1.6	1.9	2.7	−10	1	4	5	7
		2100	1.3	1.7	2.0	2.5	3.6	−5	1	4	6	11
	JJA	2035	0.6	0.7	0.8	0.9	1.2	−4	0	1	2	6
		2065	1.0	1.3	1.5	1.9	2.6	−9	−1	2	3	6
		2100	0.9	1.6	1.8	2.6	3.3	−12	0	2	4	9
	Annual	2035	0.6	0.7	0.8	0.9	1.2	−4	−1	1	3	8
		2065	1.1	1.3	1.5	1.9	2.5	−10	0	2	4	6
		2100	1.0	1.6	1.9	2.6	3.2	−8	1	3	4	8
East Africa	DJF	2035	0.4	0.7	0.8	1.0	1.2	−4	−1	1	5	10
		2065	0.8	1.3	1.5	1.8	2.5	−3	−1	3	7	19
		2100	1.0	1.6	1.9	2.4	3.2	−6	−1	5	10	25
	JJA	2035	0.5	0.7	0.9	1.0	1.2	−8	−3	0	2	12
		2065	0.8	1.4	1.6	1.9	2.4	−10	−4	1	3	18
		2100	0.7	1.7	2.0	2.5	3.1	−12	−4	0	5	19
	Annual	2035	0.5	0.7	0.8	0.9	1.2	−5	−2	1	3	10
		2065	1.0	1.3	1.6	1.9	2.4	−6	−2	1	6	17
		2100	1.0	1.6	2.0	2.5	3.1	−7	−2	2	8	21
Southern Africa	DJF	2035	0.6	0.7	0.9	1.1	1.3	−11	−4	−2	0	3
		2065	1.0	1.4	1.7	2.0	2.6	−19	−5	−3	−1	4
		2100	1.1	1.8	2.1	2.7	3.3	−19	−7	−3	1	5
	JJA	2035	0.5	0.8	0.9	1.0	1.5	−18	−9	−4	−1	9
		2065	1.1	1.5	1.7	2.0	2.5	−29	−13	−8	−3	4
		2100	1.4	1.8	2.1	2.6	3.3	−29	−18	−9	−3	12
	Annual	2035	0.6	0.8	0.9	1.0	1.4	−13	−5	−2	0	4
		2065	1.1	1.5	1.7	2.1	2.6	−15	−7	−4	−1	4
		2100	1.4	1.8	2.1	2.7	3.3	−20	−7	−5	−1	5
West Indian Ocean	DJF	2035	0.3	0.5	0.6	0.7	1.0	−10	0	2	3	10
		2065	0.6	1.0	1.1	1.3	1.8	−10	−1	2	5	13
		2100	0.8	1.2	1.4	1.8	2.3	−9	−1	2	6	22
	JJA	2035	0.4	0.5	0.6	0.7	1.0	−5	−1	2	5	12
		2065	0.6	0.9	1.1	1.3	1.8	−7	−1	1	5	12
		2100	0.7	1.2	1.4	1.8	2.3	−7	0	2	5	19
	Annual	2035	0.3	0.5	0.6	0.7	1.0	−5	1	2	3	7
		2065	0.6	1.0	1.1	1.3	1.8	−4	−1	2	4	11
		2100	0.8	1.2	1.4	1.8	2.2	−5	0	2	5	19
Asia												
West Asia	DJF	2035	0.0	0.8	1.1	1.4	1.8	−12	0	3	6	14
		2065	0.5	1.5	1.9	2.3	3.2	−10	−1	2	7	21
		2100	0.6	1.9	2.4	2.9	3.8	−11	−3	4	9	20
	JJA	2035	0.2	0.9	1.1	1.3	2.1	−10	−2	1	5	55
		2065	1.1	1.7	2.1	2.6	4.0	−20	−6	−3	2	51
		2100	1.2	2.0	2.7	3.4	4.7	−29	−6	−1	4	60
	Annual	2035	0.1	0.9	1.0	1.2	1.8	−9	−2	3	4	27
		2065	0.7	1.7	1.9	2.3	3.2	−12	−2	0	4	27
		2100	0.9	2.1	2.5	3.1	4.1	−19	−2	1	6	28

(continued on next page)

Table 14.1 (continued)

RCP4.5			Temperature (°C)					Precipitation (%)				
REGION	**MONTH[a]**	**Year**	**min**	**25%**	**50%**	**75%**	**max**	**min**	**25%**	**50%**	**75%**	**max**
Central Asia	**DJF**	**2035**	−0.1	0.8	1.3	1.6	2.4	−6	0	4	8	19
		2065	0.6	1.7	2.4	2.9	4.0	−9	−2	4	10	17
		2100	1.0	2.3	2.7	3.3	5.4	−12	−1	5	12	25
	JJA	**2035**	0.3	0.9	1.1	1.4	2.1	−13	−3	2	6	17
		2065	1.1	1.7	2.1	2.6	4.3	−22	−5	1	6	16
		2100	0.9	2.1	2.7	3.4	5.0	−17	−3	1	5	18
	Annual	**2035**	0.2	0.8	1.1	1.3	2.0	−6	−1	2	6	13
		2065	0.7	1.7	2.2	2.5	3.6	−13	−2	2	6	16
		2100	0.8	2.2	2.6	3.2	4.8	−12	−4	4	8	18
Eastern Asia	**DJF**	**2035**	0.3	0.8	1.0	1.3	2.3	−9	−1	1	3	7
		2065	0.8	1.6	2.0	2.5	3.4	−5	3	5	9	16
		2100	0.9	2.1	2.7	3.1	4.7	−9	5	9	15	30
	JJA	**2035**	0.4	0.7	0.9	1.1	1.6	−3	0	2	3	6
		2065	0.7	1.4	1.9	2.3	3.1	−2	3	6	8	18
		2100	0.7	1.8	2.2	2.8	3.9	1	4	7	11	24
	Annual	**2035**	0.3	0.9	0.9	1.1	1.7	−3	0	2	3	7
		2065	0.9	1.6	1.9	2.2	3.0	−1	4	6	8	18
		2100	0.7	1.9	2.4	3.0	3.9	−1	5	7	11	21
Tibetan Plateau	**DJF**	**2035**	0.0	0.9	1.2	1.5	2.2	−3	2	4	8	15
		2065	0.9	1.9	2.3	2.9	3.9	−1	6	8	12	17
		2100	1.4	2.3	2.8	3.5	5.5	2	6	11	16	25
	JJA	**2035**	0.4	0.9	1.1	1.3	2.3	−5	1	3	5	12
		2065	1.0	1.7	2.1	2.5	4.4	−3	2	6	9	25
		2100	0.9	2.2	2.5	3.1	5.4	−4	5	9	13	37
	Annual	**2035**	0.3	0.9	1.2	1.4	2.0	−2	1	4	5	11
		2065	1.0	1.8	2.2	2.6	3.6	−1	4	7	9	22
		2100	0.9	2.2	2.6	3.3	4.9	−1	6	9	14	32
South Asia	**DJF**	**2035**	0.1	0.7	1.0	1.1	1.4	−18	−6	−1	4	8
		2065	0.6	1.6	1.8	2.3	2.6	−17	−3	4	7	13
		2100	1.4	2.0	2.3	3.0	3.7	−14	0	8	14	28
	JJA	**2035**	0.3	0.6	0.7	0.9	1.3	−3	2	3	6	9
		2065	0.9	1.1	1.3	1.7	2.6	−3	5	7	11	33
		2100	0.7	1.4	1.7	2.2	3.3	−7	8	10	13	37
	Annual	**2035**	0.2	0.7	0.8	1.0	1.3	−2	1	3	4	7
		2065	0.8	1.4	1.6	1.9	2.5	−2	3	7	9	26
		2100	1.3	1.7	2.1	2.7	3.5	−3	6	10	12	27
North Indian Ocean	**DJF**	**2035**	0.1	0.5	0.6	0.7	1.0	−16	−3	1	7	22
		2065	0.5	1.0	1.2	1.5	1.9	−7	1	5	15	33
		2100	0.8	1.3	1.5	2.0	2.5	−9	5	9	20	41
	JJA	**2035**	0.2	0.5	0.6	0.7	1.0	−8	−1	2	5	16
		2065	0.6	1.0	1.2	1.4	1.9	−7	2	6	9	23
		2100	0.8	1.3	1.4	1.9	2.5	−10	5	8	12	36
	Annual	**2035**	0.2	0.5	0.6	0.7	1.0	−5	0	1	4	12
		2065	0.5	1.0	1.1	1.4	1.9	−4	3	6	9	22
		2100	0.9	1.3	1.5	2.0	2.5	−5	5	9	13	38

(continued on next page)

Table 14.1 (continued)

RCP4.5			Temperature (°C)					Precipitation (%)				
REGION	MONTH[a]	Year	min	25%	50%	75%	max	min	25%	50%	75%	max
Southeast Asia (land)	DJF	2035	0.3	0.5	0.7	0.8	1.1	−2	1	2	4	12
		2065	0.6	1.1	1.3	1.6	2.2	−1	1	3	8	13
		2100	0.8	1.4	1.6	2.2	3.0	−5	2	6	9	19
	JJA	2035	0.3	0.6	0.7	0.8	1.2	−3	0	1	3	7
		2065	0.7	1.1	1.2	1.5	2.2	−2	0	3	7	13
		2100	0.8	1.4	1.5	2.0	2.7	−3	2	4	9	19
	Annual	2035	0.3	0.6	0.7	0.8	1.2	−2	0	1	3	8
		2065	0.7	1.1	1.2	1.6	2.2	−1	1	3	7	13
		2100	0.8	1.4	1.6	2.1	2.7	−2	2	5	10	18
Southeast Asia (sea)	DJF	2035	0.3	0.5	0.6	0.7	1.1	−3	0	2	3	9
		2065	0.6	0.9	1.1	1.3	1.9	−4	0	3	6	10
		2100	0.9	1.2	1.4	1.7	2.5	−5	1	3	6	11
	JJA	2035	0.3	0.5	0.6	0.6	1.0	−4	0	1	2	7
		2065	0.7	0.9	1.1	1.3	1.9	−2	2	3	5	9
		2100	0.9	1.2	1.4	1.7	2.5	−1	2	3	6	16
	Annual	2035	0.3	0.5	0.6	0.7	1.0	−4	0	2	3	8
		2065	0.6	1.0	1.1	1.3	1.9	−2	1	3	5	7
		2100	0.9	1.2	1.4	1.7	2.5	−3	2	4	6	9
Australia												
North Australia	DJF	2035	0.2	0.6	0.9	1.1	1.9	−20	−5	−2	3	8
		2065	0.6	1.2	1.5	2.1	3.4	−18	−6	0	3	12
		2100	1.1	1.6	2.0	2.6	4.0	−31	−8	−4	3	9
	JJA	2035	0.4	0.8	0.9	1.1	1.4	−48	−10	−4	1	15
		2065	0.9	1.4	1.6	1.9	2.3	−53	−15	−7	−1	17
		2100	0.9	1.7	2.0	2.5	2.9	−46	−19	−8	2	11
	Annual	2035	0.3	0.7	0.9	1.1	1.6	−24	−6	−3	1	7
		2065	0.7	1.3	1.6	1.9	2.6	−21	−7	−2	2	11
		2100	1.0	1.7	2.0	2.5	3.4	−33	−9	−4	1	8
South Australia/ New Zealand	DJF	2035	−0.1	0.6	0.8	1.0	1.2	−27	−5	−2	2	7
		2065	0.4	1.2	1.5	1.7	2.2	−18	−4	0	2	11
		2100	0.7	1.5	1.8	2.3	3.0	−17	−6	−2	2	8
	JJA	2035	0.2	0.6	0.7	0.8	1.0	−22	−3	−1	1	4
		2065	0.6	1.1	1.2	1.4	1.6	−21	−6	−3	2	11
		2100	0.7	1.4	1.6	1.8	2.4	−20	−9	−3	2	7
	Annual	2035	0.1	0.6	0.7	0.8	1.0	−24	−3	−2	1	5
		2065	0.6	1.1	1.3	1.5	1.7	−18	−5	−1	1	10
		2100	0.9	1.5	1.8	2.0	2.4	−17	−9	−2	2	7

(continued on next page)

Table 14.1 (continued)

RCP4.5			Temperature (°C)					Precipitation (%)				
REGION	MONTH[a]	Year	min	25%	50%	75%	max	min	25%	50%	75%	max
The Pacific												
Northern Tropical Pacific	DJF	2035	0.2	0.5	0.6	0.7	0.9	−7	−2	0	3	11
		2065	0.7	1.0	1.1	1.4	1.9	−4	−2	1	6	12
		2100	0.9	1.2	1.4	1.7	2.4	−6	−1	1	5	20
	JJA	2035	0.3	0.5	0.6	0.7	1.0	−11	−2	1	3	8
		2065	0.6	0.9	1.0	1.3	2.0	−9	−2	2	5	9
		2100	0.8	1.1	1.4	1.8	2.6	−11	−1	2	4	16
	Annual	2035	0.3	0.5	0.6	0.7	1.0	−8	−2	1	3	7
		2065	0.6	1.0	1.1	1.3	1.9	−7	−1	1	4	9
		2100	0.9	1.2	1.4	1.7	2.4	−8	0	1	4	18
Equatorial Pacific	DJF	2035	0.1	0.5	0.6	0.7	1.2	−9	−1	7	11	44
		2065	0.5	1.0	1.2	1.4	2.5	−4	5	12	19	226
		2100	0.4	1.2	1.5	1.8	3.3	−27	7	16	29	309
	JJA	2035	0.1	0.5	0.6	0.7	1.1	−18	5	10	14	40
		2065	0.7	1.0	1.1	1.4	2.3	0	11	15	25	143
		2100	0.5	1.2	1.5	1.8	2.9	−19	13	23	33	125
	Annual	2035	0.1	0.5	0.7	0.7	1.1	−11	3	7	12	40
		2065	0.7	1.0	1.2	1.4	2.3	−1	7	12	24	194
		2100	0.5	1.2	1.4	1.8	2.9	−23	13	19	29	225
Southern Pacific	DJF	2035	0.3	0.4	0.5	0.6	0.9	−7	−1	1	2	6
		2065	0.6	0.8	1.0	1.2	1.5	−22	0	2	4	6
		2100	0.8	1.0	1.3	1.5	2.0	−24	−1	3	5	8
	JJA	2035	0.3	0.4	0.5	0.6	0.9	−10	0	1	3	8
		2065	0.6	0.8	1.0	1.1	1.6	−18	−1	1	4	7
		2100	0.8	1.0	1.2	1.5	2.1	−17	−2	2	4	10
	Annual	2035	0.3	0.4	0.5	0.6	0.9	−8	0	1	2	7
		2065	0.6	0.8	1.0	1.1	1.6	−21	0	2	3	5
		2100	0.8	1.1	1.2	1.5	2.0	−21	0	2	4	6
Antarctica												
(land)	DJF	2035	0.1	0.5	0.6	0.8	1.3	−3	1	3	4	8
		2065	0.1	1.0	1.3	1.6	2.3	−7	3	5	8	14
		2100	0.5	1.5	1.7	2.1	3.1	−5	4	8	10	17
	JJA	2035	−0.5	0.6	0.8	0.9	1.8	−3	2	5	6	13
		2065	−0.1	1.2	1.4	1.8	2.5	1	6	8	13	16
		2100	−0.3	1.5	1.9	2.4	3.8	−1	9	12	15	23
	Annual	2035	−0.1	0.5	0.7	0.9	1.3	−3	2	4	5	9
		2065	0.0	1.1	1.3	1.7	2.3	−3	4	7	10	14
		2100	0.1	1.5	1.8	2.3	3.2	−3	7	9	13	21
(sea)	DJF	2035	−0.3	0.2	0.4	0.5	0.7	−1	1	3	3	5
		2065	−0.4	0.5	0.6	0.9	1.3	0	3	4	5	8
		2100	−0.3	0.6	0.9	1.2	1.8	0	4	5	7	11
	JJA	2035	−0.7	0.4	0.6	1.0	1.9	0	2	2	4	5
		2065	−0.6	0.7	1.1	1.6	3.3	2	4	5	7	10
		2100	−0.8	1.1	1.4	2.2	3.8	3	5	7	10	13
	Annual	2035	−0.4	0.3	0.5	0.7	1.3	0	2	2	4	5
		2065	−0.5	0.5	0.8	1.2	2.3	2	3	4	6	9
		2100	−0.5	0.8	1.2	1.7	2.6	1	4	6	9	12

Notes:

[a] Precipitation changes cover 6 months; ONDJFM and AMJJAS for winter and summer (Northern Hemisphere).

Table 14.2 | Assessed *confidence* (*high, medium, low*) in climate projections of regional temperature and precipitation change from the multi-model ensemble of CMIP5 models for the RCP4.5 scenario. Column 1 refers to the SREX regions (cf. Seneviratne et al., 2012, page 12. The region's coordinates can be found from their online Appendix 3.A) and six additional regions including the two polar regions, the Caribbean, Indian Ocean and Pacific Island States (see Annex I for further details). Columns 2 to 4 show confidence in models' ability to simulate present-day mean temperature and precipitation as well as the most important phenomena for that region based on Figures 9.39, 9.40, and 9.45. In column 4, the individual phenomena are listed, with associated confidence levels shown below, in the same order as the phenomena. Note that only phenomena assessed in Figure 9.45 are listed. Column 5 is an interpretation of the relevance of the main climate phenomena for future regional climate change, based on Table 14.3. Note that the SREX regions are smaller than the regions listed in Table 14.3. Columns 6 and 7 express confidence in projected temperature and precipitation changes, based solely on model agreement for 2080–2099 vs. 1985–2005, as listed in Table 14.1 and in the maps shown in Annex I. The confidence is assessed for two periods for temperature (DJF and JJA) and two-half year periods for precipitation (October to March and April to September). When the projections are consistent with no significant change, it is marked by an asterisk (*) and the assigned *confidence* is *medium*. Further details on how confidence levels have been assigned are provided in the Supplementary Material (Section 14.SM.6.1).

	Present			Future		
SREX Region	**Temperature**	**Precipitation**	**Main Phenomenon**	**Relevance of Main Phenomena**	**Temperature**	**Precipitation**
1. ALA	M	L	PNA/PDO M/M	H	H/H	H/H
2. CGI	H	M	NAO H	H	H/H	H/H
3. WNA	M	L	PNA/ENSO/PDO/Monsoon M/M/M/M	M-H	H/H	M/M*
4. CNA	L	H	PNA/ENSO M/M	M-H	H/H	M*/M*
5. ENA	H	H	PNA/ENSO/NAO/Monsoon M/M/H/M	M-H	H/H	H/M
6. CAM	H	M	ENSO/TC M/H	M-H	H/H	M*/M*
7. AMZ	H	L	ENSO M	M	H/H	M*/M*
8. NEB	H	M	ENSO M	M-H	H/H	M*/L
9. WSA	M	L	ENSO /SAM M/M	M-H	H/H	L/M*
10. SSA	H	L	ENSO/SAM M/M	M-H	H/H	L/L
11. NEU	M	H	NAO/blocking H/L	H	H/H	H/L
12. CEU	H	M	NAO/blocking H/L	H	H/H	M/M*
13. MED	H	H	NAO/blocking H/L	H	H/H	L/M
14. SAH	M	L	NAO H	H	H/H	M*/M*
15. WAF	M	L	Monsoon/AMO M/M	M	H/H	L/M*
16. EAF	H	L	IOD M	M	H/H	M*/M*
17. SAF	H	L	SAM/TC M/H	H	H/H	M*/L
18. NAS	M	L	NAO/Blocking H/L	M	H/H	H/H
19. WAS	H	L	NAO/IOD/TC H/M/H	M-H	H/H	M*/M*
20. CAS	M	L	N/A	N/A	H/H	M*/M*
21. TIB	M	L	Monsoon M	M	H/H	H/H
22. EAS	M	M	ENSO/Monsoon/TC M/M/H	M-H	H/H	M/H
23. SAS	M	M	Monsoon/IOD/ENSO/TC/MJO M/M/M/H/L	L-H	H/H	M*/H
24. SEA	H	M	Monsoon/IOD/ENSO/TC/MJO M/M/M/H/L	L-H	H/H	M/M

(continued on next page)

Table 14.2 (continued)

	Present			Future		
SREX Region	**Temperature**	**Precipitation**	**Main Phenomenon**	**Relevance of Main Phenomena**	**Temperature**	**Precipitation**
25. NAU	H	M	ENSO/Monsoon/TC/IOD/MJO M/M/H/M/L	L-H	H/H	M*/M*
26. SAU	H	L	SAM M	M-H	H/H	M*/M*
1. Arctic (land)	H	L	NAO H	H	H/H	H/H
2. Arctic (sea)	H	L	NAO H	H	H/H	H/H
3. Antarctic (land)	M	M	SAM M	L-H	H/H	H/H
4. Antarctic (sea)	M	M	SAM M	L-H	H/H	H/H
5. Caribbean	H	L	TC/ENSO H/M	M-H	H/H	M*/M
6. West Indian Ocean	H	M	IOD M	N/A	H/H	M*/M*
7. North Indian Ocean	H	M	Monsoon/MJO M/L	N/A	H/H	L/M
8. SE Asia (sea)	H	M	Monsoon/IOD/ENSO/TC/MJO M/M/M/H/L	L-H	H/H	L/M
9. Northern Tropical Pacific	H	L	ENSO/TC M/H	M-H	H/H	M*/M*
10. Equatorial Tropical Pacific	H	M	ENSO/MJO M/L	M-H	H/H	M/M
11. Southern Tropical Pacific	H	H	ENSO//MJO M/L	M-H	H/H	M*/M*

Table 14.3 | Summary of the relevance of projected changes in major phenomena for mean change in future regional climate. The relevance is classified into high (red), medium (yellow), low (cyan), and 'no obvious relevance' (grey), based on confidence that there will be a change in the phenomena ('HP' for high, 'MP' for medium, 'LP' for low), and confidence in the impact of the phenomena on each region ('HI' for high, 'MI' for medium, 'LI' for low). More information on how these assessments have been constructed is given in the Supplementary Material (Section 14.SM.6.1).

Phenomena Regions	Section	Monsoon Systems *MP—see Section 14.2*	Tropical Phenomena[a] *HP/MP/LP/LP—See Section 14.3*	ENSO *LP—See Section 14.4*	Annular and Dipolar Modes *HP—See Section 14.5*	Tropical Cyclones *MP—See Section 14.6.1*	Extratropical Cyclones[b] *MP/HP—See Section 14.6.2*
Arctic	*14.8.2*				*HP/HI* The small projected increase in NAO is *likely* to contribute to wintertime changes in temperature and precipitation.		*MP/HI* Projected increase in precipitation in extratropical cyclones is *likely* to enhance mean precipitation.
North America	*14.8.3*	MP/HI It is *likely* the number of consecutive dry days will increase, and overall water availability will be reduced.	HP/LI Projected ITCZ shifts unrelated to ENSO changes will impact temperature and precipitation, especially in winter.	LP/HI *Likely* changes in N. American precipitation if ENSO changes.	HP/MI The small projected increase in the NAO index is *likely* to contribute to wintertime temperature and precipitation changes in NE America.	MP/HI Projected increases in extreme precipitation near the centres of tropical cyclones making landfall along the western coast of the USA and Mexico, the Gulf Mexico, and the eastern coast of the USA and Canada.	MP/HI Projected increases in precipitation in extratropical cyclones will lead to large increases in wintertime precipitation over the northern third of the continent.
Central America and Caribbean	*14.8.4*	*MP/HI* Projected reduction in mean precipitation .	*HP/HI* Reduced mean precipitation in southern Central America if there is a southward displacement of the East Pacific ITCZ.	*LP/HI* Reduced mean precipitation if El Niño events become more frequent and/or intense.		*MP/HI* More extreme precipitation near the centres of tropical cyclones making landfall along the eastern and western coasts.	
South America	*14.8.5*	*MP/HI* Projected increase in extreme precipitation and in the extension of monsoon area.	*HP/HI* Projected increase in the mean precipitation in the southeast due to the projected southward displacement of the SACZ.	*LP/HI* Reduced mean precipitation in eastern Amazonia and increased precipitation in the La Plata Basin.	*HP/HI* Poleward shift of storm tracks due to projected positive trend in SAMS phase leads to less precipitation in central Chile and increased precipitation in the southern tip of South America.		*HP/HI* Southward displacement of cyclogenesis activity increases the precipitation in the extreme south.
Europe and Mediterranean	*14.8.6*				*HP/HI* Projected increase in the NAO will lead to enhanced winter warming and precipitation over NW Europe.		*MP/HI* Enhanced extremes of storm-related precipitation and decreased frequency of storm-related precipitation over the E. Mediterranean.
Africa	*14.8.7*	*MP/HI* Projected enhancement of summer precipitation in West Africa.	*HP/LI* Enhanced precipitation in parts of East Africa due to projected shifts in ITCZ. Modified precipitation in West or East Africa according to variations in Atlantic or Indian Ocean SSTs.	*LP/HI* Increased precipitation in East Africa and decreased precipitation and enhanced warming in southern Africa if El Niño events become more frequent and/or intense.	*HP/HI* Enhanced winter warming over southern Africa due to projected increase in SAM.	*MP/HI* Projected increase in extreme precipitation near the centres of tropical cyclones making landfall along the eastern coast (including Madagascar).	*HP/HI* Enhanced extremes of storm-related precipitation and decreased frequency of storm-related precipitation over southwestern Africa.
Central and North Asia	*14.8.8*	*MP/MI* Projected enhancement in summer mean precipitation.			*HP/LI* Projected enhancement in winter warming over North Asia.		
East Asia	*14.8.9*	*MP/MI* Enhanced summer precipitation due to intensification of East Asian summer monsoon circulation.		*LP/HI* Enhanced warming if El Niño events become more frequent and/or intense.		*MP/HI* Projected increase in extreme precipitation near the centres of tropical cyclones making landfall in Japan, along coasts of east China Sea and Sea of Japan.	*MP/MI* Projected reduction in midwinter precipitation.

(continued on next page)

Table 14.3 (continued)

Phenomena Regions	**Section**	**Monsoon Systems** *MP—see Section 14.2*	**Tropical Phenomena**[a] *HP/MP/LP/LP—See Section 14.3*	**ENSO** *LP—See Section 14.4*	**Annular and Dipolar Modes** *HP—See Section 14.5*	**Tropical Cyclones** *MP—See Section 14.6.1*	**Extratropical Cyclones**[b] *MP/HP—See Section 14.6.2*
West Asia	*14.8.10*		*HP/LI* Enhanced precipitation in southern parts of West Asia due to projected northward shift in ITCZ.			*MP/HI* Projected increase in extreme precipitation near the centres of tropical cyclones making landfall on the Arabian Peninsula.	*MP/LI* Projected decrease in mean precipitation due to north-ward shift of storm tracks.
South Asia	*14.8.11*	*MP/MI* Enhanced summer precipitation associated with Indian Monsoon.	*LP/MI* Strengthened break mon-soon precipitation anomalies associated with MJO.	*LP/HI* Enhanced warming and increased summer season rainfall variability due to ENSO.		*MP/HI* Projected increase in extreme precipitation near the centres of tropical cyclones making landfall along coasts of Bay of Bengal and Arabian Sea.	
Southeast Asia	*14.8.12*	*LP/MI* Decrease in precipitation over Maritime continent.	*HP/MI* Projected changes in IOD-like warming pattern will reduce mean precipitation in Indonesia during Jul-Oct.	*LP/HI* Reduction in mean precipitation and enhanced warming if El Niño events become more frequent and/or intense.		*MP/HI* Projected increase in extreme precipitation near the centres of tropical cyclones making landfall along coasts of South China Sea, Gulf of Thailand, and Andaman Sea.	
Australia and New Zealand	*14.8.13*	*MP/LI* Mean monsoon pre-cipitation may increase over northern Australia.	*HP/LI* More frequent zonal SPCZ episodes may reduce pre-cipitation in NE Australia.	*LP/HI* Reduced precipitation in North and East Australia and NZ if El Niño events become more frequent and/or intense.	*HP/MI* Increased warming and reduced precipitation in NZ and South Aust. due to projected positive trend in SAM.	*MP/HI* More extreme precipitation near the centres of tropical cyclones making landfall along the eastern, western, and northern coasts of Australia.	*HP/HI* Projected increase in extremes of storm-related precipitation.
Pacific Islands Region	*14.8.14*		*HP/LI* Increased mean precipitation along equator with ITCZ intensification. More frequent zonal SPCZ episodes leading to reduced precipitation in southwest and increases in east.	*HP/LI* Increased mean precipita-tion in central/east Pacific if El Niño events become more frequent and/or intense.		*HP/HI* More extreme precipitation near the centres of tropical cyclones passing over or near Pacific islands.	
Antarctica	*14.8.15*			*LP/MI* Increased warming over Antarctic Peninsula and reduced across central Pacific if El Niño events become more frequent and/or intense.	*HP/HI* Increased warming over Antarctic Peninsula and west Antarctic related to positive trend projected in SAM.		*HP/MI* Increased precipitation in coastal areas due to projected poleward shift of storm track.

References

Abram, N. J., M. K. Gagan, J. E. Cole, W. S. Hantoro, and M. Mudelsee, 2008: Recent intensification of tropical climate variability in the Indian Ocean. *Nature Geosci.*, **1**, 849–853.

Ackerley, D., B. B. B. Booth, S. H. E. Knight, E. J. Highwood, D. J. Frame, M. R. Allen, and D. P. Rowell, 2011: Sensitivity of twentieth-century Sahel rainfall to sulfate aerosol and CO_2 forcing. *J. Clim.*, **24**, 4999–5014.

Aldrian, E., and Y. S. Djamil, 2008: Spatio-temporal climatic change of rainfall in east Java Indonesia. *Int. J. Climatol.*, **28**, 435–448.

Alexander, L. V., and J. M. Arblaster, 2009: Assessing trends in observed and modelled climate extremes over Australia in relation to future projections. *Int. J. Climatol.*, **29**, 417–435.

Alexander, L. V., et al., 2006: Global observed changes in daily climate extremes of temperature and precipitation. *J. Geophys. Res. Atmos.*, **111**, D05109.

Alexander, M., D. Vimont, P. Chang, and J. Scott, 2010: The impact of extratropical atmospheric variability on ENSO: Testing the seasonal footprinting mechanism using coupled model experiments. *J. Clim.*, **23**, 2885–2901.

Alexander, M., I. Blade, M. Newman, J. Lanzante, N. Lau, and J. Scott, 2002: The atmospheric bridge: The influence of ENSO teleconnections on air-sea interaction over the global oceans. *J. Clim.*, **15**, 2205–2231.

Alexander, M. A., 2010: Extratropical air-sea interaction, SST variability and the Pacific Decadal Oscillation (PDO). In: *Climate Dynamics: Why Does Climate Vary?* [D. S. a. F. Bryan (ed.)]. American Geophysical Union, Washingon, DC, pp. 123–148.

Allan, R., and B. Soden, 2008: Atmospheric warming and the amplification of precipitation extremes. *Science*, **321**, 1481–1484.

Alory, G., S. Wijffels, and G. Meyers, 2007: Observed temperature trends in the Indian Ocean over 1960–1999 and associated mechanisms. *Geophys. Res. Lett.*, **34**, L02606.

Alpert, P., S. Krichak, H. Shafir, D. Haim, and I. Osetinsky, 2008: Climatic trends to extremes employing regional modeling and statistical interpretation over the E. Mediterranean. *Global Planet. Change*, **63**, 163–170.

Alpert, P., et al., 2002: The paradoxical increase of Mediterranean extreme daily rainfall in spite of decrease in total values. *Geophys. Res. Lett.*, **29**, 31–34.

AlSarmi, S., and R. Washington, 2011: Recent observed climate change over the Arabian Peninsula. *J. Geophys. Res. Atmos.*, **116**, D11109.

Alves, L. M., and J. A. Marengo, 2010: Assessment of regional seasonal predictability using the PRECIS regional climate modeling system over South America. *Theor. Appl. Climatol.*, **100**, 337–350.

Amador, J. A., E. J. Alfaro, O. G. Lizano, and V. O. Magana, 2006: Atmospheric forcing of the eastern tropical Pacific: A review. *Prog. Oceanogr.*, **69**, 101–142.

AMAP, 2011: *Snow, Water, Ice and Permafrost in the Arctic (SWIPA): Climate Change and the Cryosphere*, Arctic Monitoring and Assessment Programme, Oslo, Norway, 538 pp.

Ambaum, M., B. Hoskins, and D. Stephenson, 2001: Arctic oscillation or North Atlantic oscillation? *J. Clim.*, **14**, 3495–3507.

Ambaum, M. H. P., 2008: Unimodality of wave amplitude in the Northern Hemisphere. *J. Atmos. Sci.*, **65**, 1077–1086.

An, S.-I., J.-W. Kim, S.-H. Im, B.-M. Kim, and J.-H. Park, 2011: Recent and future sea surface temperature trends in the tropical Pacific warm pool and cold tongue regions. *Clim. Dyn.*, doi:10.1007/s00382-011-1129-7.

An, S. I., and B. Wang, 2000: Interdecadal change of the structure of the ENSO mode and its impact on the ENSO frequency. *J. Clim.*, **13**, 2044–2055.

An, S. I., and F. F. Jin, 2000: An Eigen analysis of the interdecadal changes in the structure and frequency of ENSO mode. *Geophys. Res. Lett.*, **27**, 2573–2576.

Anderson, B., J. Wang, G. Salvucci, S. Gopal, and S. Islam, 2010: Observed trends in summertime precipitation over the southwestern United States. *J. Clim.*, **23**, 1937–1944.

Anderson, B. T., 2003: Tropical Pacific sea surface temperatures and preceding sea level pressure anomalies in the subtropical North Pacific. *J. Geophys. Res. Atmos.*, **108**, 4732.

Anderson, B. T., 2011: Near-term increase in frequency of seasonal temperature extremes prior to the 2 degree C global warming target. *Clim. Change*, **108**, 581–589.

Annamalai, H., K. Hamilton, and K. R. Sperber, 2007: The South Asian summer monsoon and its relationship with ENSO in the IPCC AR4 simulations. *J. Clim.*, **20**, 1071–1092.

Annamalai, H., J. Hafner, K. P. Sooraj, and P. Pillai, 2013: Global warming shifts monsoon circulation, drying South Asia. *J. Clim.*, **26**, 2701–2718.

Anstey, J. A., and T. G. Shepherd, 2008: Response of the northern stratospheric polar vortex to the seasonal alignment of QBO phase transitions. *Geophys. Res. Lett.*, **35**, L22810.

Anstey, J. A., et al., 2013: Multi-model analysis of Northern Hemisphere winter blocking, Part I: Model biases and the role of resolution. *J. Geophys. Res. Atmos.*, **118**, doi: 10.1002/jgrd.50231.

Arblaster, J. M., G. A. Meehl, and D. J. Karoly, 2011: Future climate change in the Southern Hemisphere: Competing effects of ozone and greenhouse gases. *Geophys. Res. Lett.*, **38**, L02701.

Arriaga-Ramirez, S., and T. Cavazos, 2010: Regional trends of daily precipitation indices in northwest Mexico and southwest United States. *J. Geophys. Res.*, **115**, D144111.

Ashfaq, M., S. Ying, T. Wen-wen, R. J. Trapp, G. Xueijie, J. S. Pal, and N. S. Diffenbaugh, 2009: Suppression of South Asian summer monsoon precipitation in the 21st century. *Geophys. Res. Lett.*, doi:10.1029/2008gl036500.

Ashok, K., S. K. Behera, S. A. Rao, H. Y. Weng, and T. Yamagata, 2007: El Nino Modoki and its possible teleconnection. *J. Geophys. Res. Oceans*, **112**, C11007.

Athanasiadis, P. J., J. M. Wallace, and J. J. Wettstein, 2010: Patterns of wintertime jet stream variability and their relation to the storm tracks. *J. Atmos. Sci.*, **67**, 1361–138.

Bader, J., M. D. S. Mesquita, K. I. Hodges, N. Keenlyside, S. Osterhus, and M. Miles, 2011: A review on Northern Hemisphere sea-ice, storminess and the North Atlantic Oscillation: Observations and projected changes. *Atmos. Res.*, **101**, 809–834.

Baines, P. G., and C. K. Folland, 2007: Evidence for a rapid global climate shift across the late 1960s. *J. Clim.*, **20**, 2721–2744.

Baldwin, M., D. Stephenson, and I. Jolliffe, 2009: Spatial weighting and iterative projection methods for EOFs. *J. Clim.*, **22**, 234–243.

Baldwin, M. P., and D. W. J. Thompson, 2009: A critical comparison of stratosphere-troposphere coupling indices. *Q. J. R. Meteorol. Soc.*, **135**, 1661–1672.

Baldwin, M. P., et al., 2001: The quasi-biennial oscillation. *Rev. Geophys.*, **39**, 179–229.

Barnes, E., J. Slingo, and T. Woollings, 2012: A methodology for the comparison of blocking climatologies across indices, models and climate scenarios. *Clim. Dyn.*, **38**, 2467–2481.

Barnes, E. A., and D. L. Hartmann, 2012: Detection of Rossby wave breaking and its response to shifts of the midlatitude jet with climate change. *J. Geophys. Res. Atmos.*, **117**, D09117.

Barnes, E. A., and L. Polvani, 2013: Response of the midlatitude jets and of their variability to increased greenhouse gases in the CMIP5 models. *J. Clim.*, **26**, 7117–7135.

Barnes, E. A., D. L. Hartmann, D. M. W. Frierson, and J. Kidston, 2010: Effect of latitude on the persistence of eddy-driven jets. *Geophys. Res. Lett.*, **37**, L11804.

Barnett, J., 2011: Dangerous climate change in the Pacific Islands: Food production and food security. *Region. Environ. Change*, **11**, S229–S237.

Barriopedro, D., R. Garcia-Herrera, A. R. Lupo, and E. Hernandez, 2006: A climatology of Northern Hemisphere blocking. *J. Clim.*, **19**, 1042–1063.

Barriopedro, D., R. García-Herrera, J. F. González-Rouco, and R. M. Trigo, 2010: Application of blocking diagnosis methods to General Circulation Models. Part II: Model simulations. *Clim. Dyn.*, **35**, 1393–1409.

Barros, V. R., M. Doyle, and I. Camilloni, 2008: Precipitation trends in southeastern South America: Relationship with ENSO phases and the low-level circulation. *Theor. Appl. Climatol.*, **93**, 19–33.

Bates, B., P. Hope, B. Ryan, I. Smith, and S. Charles, 2008: Key findings from the Indian Ocean Climate Initiative and their impact on policy development in Australia. *Clim. Change*, **89**, 339–354.

Bates, S. C., 2010: Seasonal influences on coupled ocean-atmosphere variability in the tropical Atlantic ocean. *J. Clim.*, **23**, 582–604.

Beck, C., J. Grieser, and B. Rudolf, 2005: A new monthly precipitation climatology for the global land areas for the period 1951 to 2000. In: *Climate Status Report 2004*. German Weather Service, Offenbach, Germany, pp. 181–190.

Becker, A., P. Finger, A. Meyer-Christoffer, B. Rudolf, K. Schamm, U. Schneider, and M. Ziese, 2013: A description of the global land-surface precipitation data products of the Global Precipitation Climatology Centre with sample applications including centennial (trend) analysis from 1901–present. *Earth Syst. Sci. Data*, **5**, 71–99.

Bell, C. J., L. J. Gray, A. J. Charlton-Perez, M. M. Joshi, and A. A. Scaife, 2009: Stratospheric communication of El Niño teleconnections to European winter. *J. Clim.*, **22**, 4083–4096.

Bender, M. A., T. R. Knutson, R. E. Tuleya, J. J. Sirutis, G. A. Vecchi, S. T. Garner, and I. M. Held, 2010: Modeled impact of anthropogenic warming on the frequency of intense Atlantic hurricanes. *Science*, **327**, 454–458.

Bengtsson, L., K. I. Hodges, and E. Roeckner, 2006: Storm tracks and climate change. *J. Clim.*, **19**, 3518–3543.

Bengtsson, L., K. I. Hodges, and N. Keenlyside, 2009: Will extratropical storms intensify in a warmer climate? *J. Clim.*, **22**, 2276–2301.

Bengtsson, L., K. I. Hodges, M. Esch, N. Keenlyside, L. Kornblueh, J.-J. Luo, and T. Yamagata, 2007: How may tropical cyclones change in a warmer climate? *Tellus A*, **59**, 539–561.

Berckmans, J., T. Woollings, M.-E. Demory, P.-L. Vidale, and M. Roberts, 2013: Atmospheric blocking in a high resolution climate model: Influences of mean state, orography and eddy forcing. *Atmos. Sci. Lett.*, **14**, 34–40.

Berrisford, P., B. J. Hoskins, and E. Tyrlis, 2007: Blocking and Rossby wave-breaking on the dynamical tropopause in the Southern Hemisphere. *J. Atmos. Sci.*, **64**, 2881–2898.

Bhend, J., and H. von Storch, 2008: Consistency of observed winter precipitation trends in northern Europe with regional climate change projections. *Clim. Dyn.*, **31**, 17–28.

Biasutti, M., and A. Giannini, 2006: Robust Sahel drying in response to late 20th century forcings. *Geophys. Res. Lett.*, **33**, L11706.

Biasutti, M., and A. H. Sobel, 2009: Delayed seasonal cycle and African monsoon in a warmer climate. *Geophys. Res. Lett.*, **36**, L23707.

Biasutti, M., A. H. Sobel, and S. J. Camargo, 2009: The role of the Sahara Low in summertime Sahel rainfall variability and change in the CMIP3 models. *J. Clim.*, **22**, 5755–5771.

Biasutti, M., I. Held, A. Sobel, and A. Giannini, 2008: SST forcings and Sahel rainfall variability in simulations of the twentieth and twenty-first centuries. *J. Clim.*, **21**, 3471–3486.

Bitz, C. M., and L. M. Polvani, 2012: Antarctic climate response to stratospheric ozone depletion in a fine resolution ocean climate model. *Geophys. Res. Lett.*, **39**, L20705.

Bjerknes, J., 1969: Atmospheric teleconnections from the Equatorial Pacific. *Mon. Weather Rev.*, **97**, 163–172.

Black, E., 2009: The impact of climate change on daily precipitation statistics in Jordan and Israel. *Atmos. Sci. Lett.*, **10**, 192–200.

Black, E., J. Slingo, and K. Sperber, 2003: An observational study of the relationship between excessively strong short rains in coastal East Africa and Indian Ocean SST. *Mon. Weather Rev.*, **131**, 74–94.

Blázquez, J., and M. Nuñez, 2012: Analysis of uncertainties in future climate projections for South America: Comparison of WCRP-CMIP3 and WCRP-CMIP5 models. *Clim. Dyn.*, doi:10.1007/s00382-012-1489-7, 1-18.

Blázquez, J., M. N. Nuñez, and S. Kusunoki, 2012: Climate projections and uncertainties over South America from MRI/JMA global model experiments. *Atmos. Clim. Sci.*, **2**, 381–400.

Bluthgen, J., R. Gerdes, and M. Werner, 2012: Atmospheric response to the extreme Arctic sea ice conditions in 2007. *Geophys. Res. Lett.*, **39**, L02707.

Boer, G., 2009: Changes in interannual variability and decadal potential predictability under global warming. *J. Clim.*, **22**, 3098–3109.

Boer, G. J., and K. Hamilton, 2008: QBO influence on extratropical predictive skill. *Clim. Dyn.*, **31**, 987–1000.

Bollasina, M., and Y. Ming, 2013: The general circulation model precipitation bias over the southwestern equatorial Indian Ocean and its implications for simulating the South Asian monsoon. *Clim. Dyn.*, **40**, 823–838.

Bollasina, M. A., Y. Ming, and V. Ramaswamy, 2011: Anthropogenic aerosols and the weakening of the south Asian summer monsoon. *Science*, **334**, 502–505.

Bombardi, R. J., and L. M. V. Carvalho, 2009: IPCC global coupled model simulations of the South America monsoon system. *Clim. Dyn.*, **33**, 893–916.

Boo, K. O., G. Martin, A. Sellar, C. Senior, and Y. H. Byun, 2011: Evaluating the East Asian monsoon simulation in climate models. *J. Geophys. Res.*, **116**, D01109.

Booth, B. B. B., N. J. Dunstone, P. R. Halloran, T. Andrews, and N. Bellouin, 2012: Aerosols implicated as a prime driver of twentieth-century North Atlantic climate variability. *Nature*, **484**, 228–232.

Bracegirdle, T. J., and D. B. Stephenson, 2012: Higher precision estimates of regional polar warming by ensemble regression of climate model projections. *Clim. Dyn.*, **39**, 2805–2821.

Bracegirdle, T. J., W. M. Connolley, and J. Turner, 2008: Antarctic climate change over the twenty first century. *J. Geophys. Res.*, **113**, D03103.

Bracegirdle, T. J., et al., 2013: Assessment of surface winds over the Atlantic, Indian, and Pacific Ocean sectors of the Southern Ocean in CMIP5 models: Historical bias, forcing response, and state dependence. *J. Geophys. Res. Atmos.*, **118**, 547–562.

Braganza, K., J. Gergis, S. Power, J. Risbey, and A. Fowler, 2009: A multiproxy index of the El Niño-Southern Oscillation, AD 1525–1982. *J. Geophys. Res. Atmos.*, **114**, D05106.

Brandefelt, J., 2006: Atmospheric modes of variability in a changing climate. *J. Clim.*, **19**, 5934–5943.

Branstator, G., and F. Selten, 2009: "Modes of Variability" and Climate Change. *J. Clim.*, **22**, 2639–2658.

Breugem, W., W. Hazeleger, and R. Haarsma, 2006: Multimodel study of tropical Atlantic variability and change. *Geophys. Res. Lett.*, doi:10.1029/2006GL027831.

Breugem, W., W. Hazeleger, and R. Haarsma, 2007: Mechanisms of northern tropical Atlantic variability and response to CO2 doubling. *J. Clim.*, doi:DOI 10.1175/JCLI4137.1, 2691–2705.

Bromwich, D. H., A. J. Monaghan, and Z. C. Guo, 2004: Modeling the ENSO modulation of Antarctic climate in the late 1990s with the polar MM5. *J. Clim.*, **17**, 109–132.

Brown, J., A. Moise, and R. Colman, 2012a: The South Pacific Convergence Zone in CMIP5 simulations of historical and future climate. *Clim. Dyn.*, doi:10.1007/s00382-012-1591-x, 1–19.

Brown, J., A. Moise, and F. Delage, 2012b: Changes in the South Pacific Convergence Zone in IPCC AR4 future climate projections. *Clim. Dyn.*, **39**, 1–19.

Brown, J., S. Power, F. Delage, R. Colman, A. Moise, and B. Murphy, 2011: Evaluation of the South Pacific Convergence Zone in IPCC AR4 climate model simulations of the twentieth century. *J. Clim.*, **24**, 1565–1582.

Brown, J., et al., 2012c: Implications of CMIP3 model biases and uncertainties for climate projections in the western tropical Pacific. *Clim. Change*, doi:10.1007/s10584-012-0603-5, 1–15.

Brown, R., and P. Mote, 2009: The response of Northern Hemisphere snow cover to a changing climate. *J. Clim.*, doi:10.1175/2008JCLI2665.1, 2124–2145.

Budikova, D., 2009: Role of Arctic sea ice in global atmospheric circulation: A review. *Global Planet. Change*, **68**, 149–163.

Buehler, T., C. C. Raible, and T. F. Stocker, 2011: The relationship of winter season North Atlantic blocking frequencies to extreme cold and dry spells in the ERA-40. *Tellus A*, **63**, 212–222.

Bulic, I., and F. Kucharski, 2012: Delayed ENSO impact on spring precipitation over the North/Atlantic European region. *Clim. Dyn.*, **38**, 2593–2612.

Bulic, I., C. Brankovic, and F. Kucharski, 2012: Winter ENSO teleconnections in a warmer climate. *Clim. Dyn.*, **38**, 1593–1613.

Bunge, L., and A. J. Clarke, 2009: A verified estimation of the El Nino index Nino-3.4 since 1877. *J. Clim.*, **22**, 3979–3992.

Butchart, N., et al., 2006: Simulations of anthropogenic change in the strength of the Brewer-Dobson circulation. *Clim. Dyn.*, **27**, 727–741.

Butler, A. H., D. W. J. Thompson, and R. Heikes, 2010: The steady-state atmospheric circulation response to climate change-like thermal forcings in a simple General Circulation Model. *J. Clim.*, **23**, 3474–3496.

Caesar, J., et al., 2011: Changes in temperature and precipitation extremes over the Indo-Pacific region from 1971 to 2005. *Int. J. Climatol.*, **31**, 791–801.

Cai, W., and T. Cowan, 2008: Dynamics of late autumn rainfall reduction over southeastern Australia. *Geophys. Res. Lett.*, **35**, L09708.

Cai, W., and T. Cowan, 2013: Southeast Australia autumn rainfall reduction: A climate-change induced poleward shift of ocean-atmosphere circulation. *J. Clim.*, **26**, 189–205.

Cai, W., T. Cowan, and A. Sullivan, 2009: Recent unprecedented skewness towards positive Indian Ocean Dipole occurrences and its impact on Australian rainfall. *Geophys. Res. Lett.*, **36**, L11705.

Cai, W., P. van Rensch, and T. Cowan, 2011a: Influence of global-scale variability on the subtropical ridge over southeast Australia. *J. Clim.*, **24**, 6035–6053.

Cai, W., T. Cowan, and M. Thatcher, 2012a: Rainfall reductions over Southern Hemisphere semi-arid regions: The role of subtropical dry zone expansion. *Sci. Rep.*, **2**, doi: 10.1038/srep00702.

Cai, W., P. van Rensch, T. Cowan, and H. H. Hendon, 2011b: Teleconnection pathways of ENSO and the IOD and the mechanisms for impacts on Australian rainfall. *J. Clim.*, **24**, 3910–3923.

Cai, W., P. van Rensch, S. Borlace, and T. Cowan, 2011c: Does the Southern Annular Mode contribute to the persistence of the multidecade-long drought over southwest Western Australia? *Geophys. Res. Lett.*, **38**, L14712.

Cai, W., T. Cowan, A. Sullivan, J. Ribbe, and G. Shi, 2011d: Are anthropogenic aerosols responsible for the northwest Australia summer rainfall increase? A CMIP3 perspective and implications. *J. Clim.*, **24**, 2556–2564.

Cai, W., et al., 2012b: More extreme swings of the South Pacific convergence zone due to greenhouse warming. *Nature*, **488**, 365–369.

Cai, W. J., and T. Cowan, 2006: SAM and regional rainfall in IPCC AR4 models: Can anthropogenic forcing account for southwest Western Australian winter rainfall reduction? *Geophys. Res. Lett.*, **33**, doi: 10.1029/2006gl028037.

Cai, W. J., A. Sullivan, and T. Cowan, 2011e: Interactions of ENSO, the IOD, and the SAM in CMIP3 Models. *J. Clim.*, **24**, 1688–1704.

Camargo, S., M. Ting, and Y. Kushnir, 2012: Influence of local and remote SST on North Atlantic tropical cyclone potential intensity *Clim. Dyn.*, **40**, 1515–1529.

Campbell, J. D., M. A. Taylor, T. S. Stephenson, R. A. Watson, and F. S. Whyte, 2010: Future climate of the Caribbean from a regional climate model. *Int. J. Climatol.*, **31**, 1866–1878.

Cane, M. A., et al., 1997: Twentieth-century sea surface temperature trends. *Science*, **275**, 957–960.

Carrera, M. L., R. W. Higgins, and V. E. Kousky, 2004: Downstream weather impacts associated with atmospheric blocking over the northeast Pacific. *J. Clim.*, **17**, 4823–4839.

Carril, A. F., et al., 2012: Performance of a multi-RCM ensemble for South Eastern South America. *Clim. Dyn.*, **39**, 2747–2768.

Carton, J., and B. Huang, 1994: Warm events in the tropical Atlantic. *J. Phys. Oceanogr.*, **24**, 888–903.

Carvalho, L. M. V., C. Jones, and T. Ambrizzi, 2005: Opposite phases of the antarctic oscillation and relationships with intraseasonal to interannual activity in the tropics during the austral summer. *J. Clim.*, **18**, 702–718.

Carvalho, L. M. V., A. E. Silva, C. Jones, B. Liebmann, P. L. Silva Dias, and H. R. Rocha, 2011: Moisture transport and intraseasonal variability in the South America Monsoon System. *Clim. Dyn.*, **36**, 1865–1880.

Cassou, C., and L. Terray, 2001: Dual influence of Atlantic and Pacific SST anomalies on the North Atlantic/Europe winter climate. *Geophys. Res. Lett.*, **28**, 3195–3198.

Cassou, C., C. Deser, and M. A. Alexander, 2007: Investigating the impact of reemerging sea surface temperature anomalies on the winter atmospheric circulation over the North Atlantic. *J. Clim.*, **20**, 3510–3526.

Castro, C. L., R. A. Pielke Sr., and J. O. Adegoke, 2007: Investigation of the summer climate of the contiguous United States and Mexico using the Regional Atmospheric Modeling System (RAMS). Part I: Model climatology (1950–2002). *J. Clim.*, **20**, 3844–3865.

Casty, C., C. C. Raible, T. F. Stocker, H. Wanner, and J. Luterbacher, 2007: A European pattern climatology 1766–2000. *Clim. Dyn.*, **29**, 791–805.

Cattiaux, J., H. Douville, and Y. Peings, 2013: European temperatures in CMIP5: Origins of present-day biases and future uncertainties. *Clim. Dyn.*, doi:10.1007/s00382-013-1731-y, 1–19.

Catto, J. L., L. C. Shaffrey, and K. I. Hodges, 2011: Northern Hemisphere extratropical cyclones in a warming climate in the HiGEM High-Resolution Climate Model. *J. Clim.*, **24**, 5336–5352.

Cavalcanti, I. F. A., and M. H. Shimizu, 2012: Climate fields over South America and variability of SACZ and PSA in HadGEM-ES. *Am. J. Clim. Change*, **1**, 132–144.

Cavazos, T., C. Turrent, and D. P. Lettenmaier, 2008: Extreme precipitation trends associated with tropical cyclones in the core of the North American monsoon. *Geophys. Res. Lett.*, **35**, doi: 10.1029/2008GL035832.

Cerezo-Mota, R., M. Allen, and R. Jones, 2011: Mechanisms controlling precipitation in the northern portion of the North American monsoon. *J. Clim.*, **24**, 2771–2783.

Chadwick, R., I. Boutle, and G. Martin, 2013: Spatial patterns of precipitation change in CMIP5: Why the rich don't get richer in the tropics. *J. Clim.*, **26**, 3803–3822.

Chang, C.-H., 2011: Preparedness and storm hazards in a global warming world: Lessons from Southeast Asia. *Nat. Hazards*, **56**, 667–679.

Chang, C., J. Chiang, M. Wehner, A. Friedman, and R. Ruedy, 2011: Sulfate aerosol control of tropical Atlantic climate over the twentieth century. *J. Clim.*, **24**, 2540–2555.

Chang, E. K. M., Y. Guo, and X. Xia, 2012: CMIP5 multimodel ensemble projection of storm track change under global warming. *J. Geophys. Res. Atmos.*, **117**, doi: 10.1029/2012jd018578.

Chang, P., et al., 2006: Climate fluctuations of tropical coupled systems - The role of ocean dynamics. *J. Clim.*, **19**, 5122–5174.

Chaturvedi, R. K., J. Joshi, M. Jayaraman, G. Bala, and N. H. Ravindranath, 2012: Multi-model climate change projections for India under Representative Concentration Pathways (RCPs): A preliminary analysis. *Curr. Sci.*, **103**, 791–802.

Chauvin, F., and J.-F. Royer, 2010: Role of the SST Anomaly structures in response of cyclogenesis to global warming. In: *Hurricanes and Climate Change* [J. B. Elsner, R. E. Hodges, J. C. Malmstadt and K. N. Scheitlin (eds.)]. Springer Science+Business Media, Dordrecht, Netherlands, pp. 39–56.

Chen, D., 2003: A comparison of wind products in the context of ENSO prediction. *Geophys. Res. Lett.*, **30**, **doi:** 10.1029/2002GL016121.

Chen, G., I. M. Held, and W. A. Robinson, 2007: Sensitivity of the latitude of the surface westerlies to surface friction. *J. Atmos. Sci.*, **64**, 2899–2915.

Chen, G., J. Lu, and D. M. W. Frierson, 2008: Phase speed spectra and the latitude of surface westerlies: Interannual variability and global warming trend. *J. Clim.*, **21**, 5942–5959.

Chen, T.-C., and J.-h. Yoon, 2002: Interdecadal variation of the North Pacific wintertime blocking. *Mon. Weather Rev.*, **130**, 3136–3143.

Chen, W., Z. Jiang, L. Li, and P. Yiou, 2011: Simulation of regional climate change under the IPCC A2 scenario in southeast China. *Clim. Dyn.*, **36**, 491–507.

Cherchi, A., and A. Navarra, 2007: Sensitivity of the Asian summer monsoon to the horizontal resolution: Differences between AMIP-type and coupled model experiments. *Clim. Dyn.*, **28**, 273–290.

Cheung, H. N., W. Zhou, H. Y. Mok, and M. C. Wu, 2012: Relationship between Ural–Siberian blocking and the East Asian winter monsoon in relation to the Arctic Oscillation and the El Niño–Southern Oscillation. *J. Clim.*, **25**, 4242–4257.

Chiang, J., and D. Vimont, 2004: Analogous Pacific and Atlantic meridional modes of tropical atmosphere-ocean variability. *J. Clim.*, 4143–4158.

Choi, D. H., J. S. Kug, W. T. Kwon, F. F. Jin, H. J. Baek, and S. K. Min, 2010: Arctic Oscillation responses to greenhouse warming and role of synoptic eddy feedback. *J. Geophys. Res. Atmos.*, **115**, doi: 10.1029/2010jd014160.

Choi, J., S. An, and S. Yeh, 2012: Decadal amplitude modulation of two types of ENSO and its relationship with the mean state. *Clim. Dyn.*, **38**, 2631–2644.

Choi, J., S. I. An, B. Dewitte, and W. W. Hsieh, 2009: Interactive feedback between the Tropical Pacific Decadal Oscillation and ENSO in a Coupled General Circulation Model. *J. Clim.*, **22**, 6597–6611.

Choi, J., S.-I. An, J.-S. Kug, and S.-W. Yeh, 2011: The role of mean state on changes in El Niño's flavor. *Clim. Dyn.*, **37**, 1205–1215.

Chotamonsak, C., E. P. Salathe, Jr., J. Kreasuwan, S. Chantara, and K. Siriwitayakorn, 2011: Projected climate change over Southeast Asia simulated using a WRF regional climate model. *Atmos. Sci. Lett.*, **12**, 213–219.

Chou, C., J. D. Neelin, U. Lohmann, and J. Feichter, 2005: Local and remote impacts of aerosol climate forcing on tropical precipitation. *J. Clim.*, **18**, 4621–4636.

Chou, C., J. C. H. Chiang, C.-W. Lan, C.-H. Chung, Y.-C. Liao, and C.-J. Lee, 2013: Increase in the range between wet and dry season precipitation. *Nature Geosci.*, **6**, 263–267.

Chou, S., et al., 2012: Downscaling of South America present climate driven by 4-member HadCM3 runs. *Clim. Dyn.*, **38**, 635–653.

Christensen, J. H., et al., 2007: Regional climate projections. In: *Climate Change 2007: The Physical Science Basis. Contribution of Working Group I to the Fourth Assessment Report of the Intergovernmental Panel on Climate Change* [Solomon, S., D. Qin, M. Manning, Z. Chen, M. Marquis, K. B. Averyt, M. Tignor and H. L. Miller (eds.)] Cambridge University Press, Cambridge, United Kingdom and New York, NY, USA, pp. 847–940.

Christiansen, B., 2005: The shortcomings of nonlinear principal component analysis in identifying circulation regimes. *J. Clim.*, **18**, 4814–4823.

Chung, C. E., and V. Ramanathan, 2006: Weakening of North Indian SST gradients and the monsoon rainfall in India and the Sahel. *J. Clim.*, **19**, 2036–2045.

Chung, C. E., and V. Ramanathan, 2007: Relationship between trends in land precipitation and tropical SST gradient. *Geophys. Res. Lett.*, **34**, doi: 10.1029/2007gl030491.

Chylek, P., and G. Lesins, 2008: Multidecadal variability of Atlantic hurricane activity: 1851–2007. *J. Geophys. Res.*, **113**, D22106.

Chylek, P., C. K. Folland, G. Lesins, and M. Dubey, 2010: The 20th Century bipolar seesaw of the Arctic and Antarctic surface air temperatures. *Geophys. Res. Lett.*, **37**, doi: 10.1029/2010GL042793.

Chylek, P., C. K. Folland, G. Lesins, M. Dubey, and M. Wang, 2009: Arctic air temperature change amplification and the Atlantic Multidecadal Oscillation. *Geophys. Res. Lett.*, **36**, doi: 10.1029/ 2009GL038777.

Chylek, P., C. Folland, L. Frankcombe, H. Dijkstra, G. Lesins, and M. Dubey, 2012: Greenland ice core evidence for spatial and temporal variability of the Atlantic Multidecadal Oscillation. *Geophys. Res. Lett.*, **39**, L09705.

Cobb, K. M., C. D. Charles, H. Cheng, and R. L. Edwards, 2003: El Nino/Southern Oscillation and tropical Pacific climate during the last millennium. *Nature*, **424**, 271–276.

Coelho, C. A. S., and L. Goddard, 2009: El Nino-induced tropical droughts in climate change projections. *J. Clim.*, **22**, 6456–6476.

Colle, B. A., Z. Zhang, K. A. Lombardo, E. Chang, P. Liu, and M. Zhang, 2013: Historical evaluation and future prediction of eastern North America and western Atlantic extratropical cyclones in the CMIP5 models during the cool season. *J. Clim.*, **26**, 6882–6903.

Collins, M., et al., 2010: The impact of global warming on the tropical Pacific ocean and El Niño. *Nature Geosci.*, **3**, 391–397.

Colman, R. A., A. F. Moise, and L. I. Hanson, 2011: Tropical Australian climate and the Australian monsoon as simulated by 23 CMIP3 models. *J. Geophys. Res. Atmos.*, **116,** doi: 10.1029/2010jd015149.

Comarazamy, D. E., and J. E. Gonzalez, 2011: Regional long-term climate change (1950–2000) in the midtropical Atlantic and its impacts on the hydrological cycle of Puerto Rico. *J. Geophys. Res. Atmos.*, **116**, doi: 10.1029/2010jd015414.

Compo, G. P., and P. D. Sardeshmukh, 2010: Removing ENSO-related variations from the climate record. *J. Clim.*, **23**, 1957–1978.

Conway, D., C. Hanson, R. Doherty, and A. Persechino, 2007: GCM simulations of the Indian Ocean dipole influence on East African rainfall: Present and future. *Geophys. Res. Lett.*, **34**, doi: 10.1029/2006GL027597.

Cook, B., N. Zeng, and J.-H. Yoon, 2011: Will Amazonia dry out? Magnitude and causes of change from IPCC Climate Model Projections. *Earth Interact.*, **16**, 1–27.

Cook, B. I., and R. Seager, 2013: The response of the North American Monsoon to increased greenhouse gas forcing. *J. Geophys. Res.*, **118,**

Cook, K., 2008: Climate science: The mysteries of Sahel droughts. *Nature Geosci.*, **1**, 647–648.

Cook, K. H., and E. K. Vizy, 2006: Coupled model simulations of the west African monsoon system: Twentieth- and twenty-first-century simulations. *J. Clim.*, **19**, 3681–3703.

Cook, K. H., and E. K. Vizy, 2010: Hydrodynamics of the Caribbean Low-Level Jet and its relationship to precipitation. *J. Clim.*, **23**, 1477–1494.

Coppola, E., F. Kucharski, F. Giorgi, and F. Molteni, 2005: Bimodality of the North Atlantic Oscillation in simulations with greenhouse gas forcing. *Geophys. Res. Lett.*, **32**, doi: 10.1029/2005gl024080.

Cravatte, S., T. Delcroix, D. Zhang, M. McPhaden, and J. Leloup, 2009: Observed freshening and warming of the western Pacific Warm Pool. *Clim. Dyn.*, **33**, 565–589.

Croci-Maspoli, M., C. Schwierz, and H. Davies, 2007a: Atmospheric blocking: Space-time links to the NAO and PNA. *Clim. Dyn.*, **29**, 713–725.

Croci-Maspoli, M., C. Schwierz, and H. C. Davies, 2007b: A multifaceted climatology of atmospheric blocking and its recent linear trend. *J. Clim.*, **20**, 633–649.

Cunningham, C. A. C., and I. F. D. Cavalcanti, 2006: Intraseasonal modes of variability affecting the South Atlantic Convergence Zone. *Int. J. Climatol.*, **26**, 1165–1180.

d'Orgeval, T., J. Polcher, and L. Li, 2006: Uncertainties in modelling future hydrological change over West Africa. *Clim. Dyn.*, **26**, 93–108.

Dacre, H. F., and S. L. Gray, 2009: The spatial distribution and evolution characteristics of North Atlantic Cyclones. *Mon. Weather Rev.*, **137**, 99–115.

Dai, A., 2011: Drought under global warming: A review. *WIREs Clim. Change*, **2**, 45–65.

Dai, A., 2013: Increasing drought under global warming in observations and models. *Nature Clim. Change*, **3**, 52–58.

Dai, A., T. Qian, K. E. Trenberth, and J. D. Milliman, 2009: Changes in continental freshwater discharge from 1948 to 2004. *J. Clim.*, **22**, 2773–2792.

Dairaku, K., S. Emori, and T. Nozawa, 2008: Impacts of global warming on hydrological cycles in the Asian monsoon region. *Adv. Atmos. Sci.*, **25**, 960–973.

Dash, S. K., M. A. Kulkarni, U. C. Mohanty, and K. Prasad, 2009: Changes in the characteristics of rain events in India. *J. Geophys. Res. Atmos.*, **114**, D10109.

Davini, P., C. Cagnazzo, S. Gualdi, and A. Navarra, 2012: Bidimensional diagnostics, variability, and trends of Northern Hemisphere blocking. *J. Clim.*, **25**, 6496–6509.

Dawson, A., T. N. Palmer, and S. Corti, 2012: Simulating regime structures in weather and climate prediction models. *Geophys. Res. Lett.*, **39**, L21805.

de Oliveira Vieira, S., P. Satyamurty, and R. V. Andreoli, 2013: On the South Atlantic Convergence Zone affecting southern Amazonia in austral summer. *Atmos. Sci. Lett.*, **14**, 1–6.

de Vries, H., T. Woollings, J. Anstey, R. J. Haarsma, and W. Hazeleger, 2013: Atmospheric blocking and its relation to jet changes in a future climate. *Clim. Dyn.*, doi:10.1007/s00382-013-1699-7, 1–12.

Dean, S., and P. Stott, 2009: The effect of local circulation variability on the detection and attribution of New Zealand temperature trends. *J. Clim.*, **22**, 6217–6229.

DeFries, R., L. Bounoua, and G. Collatz, 2002: Human modification of the landscape and surface climate in the next fifty years. *Global Change Biol.*, **8**, 438–458.

Della-Marta, P. M., and J. G. Pinto, 2009: Statistical uncertainty of changes in winter storms over the North Atlantic and Europe in an ensemble of transient climate simulations. *Geophys. Res. Lett.*, **36**, doi: 10.1029/2009gl038557.

Deni, S. M., J. Suhaila, W. Z. W. Zin, and A. A. Jemain, 2010: Spatial trends of dry spells over Peninsular Malaysia during monsoon seasons. *Theor. Appl. Climatol.*, **99**, 357–371.

Déqué, M., S. Somot, E. Sanchez-Gomez, C. M. Goodess, D. Jacob, G. Lenderink, and O. B. Christensen, 2012: The spread amongst ENSEMBLES regional scenarios: Regional climate models, driving general circulation models and interannual variability. *Clim. Dyn.*, **38**, 951–964.

Deser, C., A. S. Phillips, and M. A. Alexander, 2010a: Twentieth century tropical sea surface temperature trends revisited. *Geophys. Res. Lett.*, **37**, doi: 10.1029/2010gl043321.

Deser, C., M. A. Alexander, S.-P. Xie, and A. S. Phillips, 2010b: Sea surface temperature variability: Patterns and mechanisms. *Annu. Rev. Mar. Sci.*, **2**, 115–143.

Deser, C., R. Tomas, M. Alexander, and D. Lawrence, 2010c: The seasonal atmospheric response to projected Arctic sea ice loss in the late twenty-first century. *J. Clim.*, **23**, 333–351.

Deser, C., A. Phillips, V. Burdette, and H. Teng, 2012: Uncertainty in climate change projections: The role of internal variability. *Clim. Dyn.*, **38**, 527–546.

Di Lorenzo, E., et al., 2009: Nutrient and salinity decadal variations in the central and eastern North Pacific. *Geophys. Res. Lett.*, **36**, doi: 10.1029/2009GL038261.

Diamond, H. J., A. M. Lorrey, and J. A. Renwick, 2012: A southwest Pacific tropical cyclone climatology and linkages to the El Niño–Southern Oscillation. *J. Clim.*, **26**, 3–25.

Diffenbaugh, N. S., and M. Ashfaq, 2010: Intensification of hot extremes in the United States. *Geophys. Res. Lett.*, **37**, L15701.

Diffenbaugh, N. S., M. Scherer, and M. Ashfaq, 2013: Response of snow-dependent hydrologic extremes to continued global warming. *Nature Clim. Change*, **3**, 379–384.

DiNezio, P. N., A. C. Clement, G. A. Vecchi, B. J. Soden, and B. P. Kirtman, 2009: Climate response of the equatorial Pacific to global warming. *J. Clim.*, **22**, 4873–4892.

Ding, Q., E. Steig, D. Battisti, and M. Kuttel, 2011: Winter warming in West Antarctica caused by central tropical Pacific warming. *Nature Geosci.*, **4**, 398–403.

Ding, Y., and J. C. L. Chan, 2005: The East Asian summer monsoon: An overview. *Meteorol. Atmos. Phys.*, **89**, 117–142.

Ding, Y., G. Ren, Z. Zhao, Y. Xu, Y. Luo, Q. Li, and J. Zhang, 2007: Detection, causes and projection of climate change over China: An overview of recent progress. *Adv. Atmos. Sci.*, doi:DOI 10.1007/s00376-007-0954-4, 954–971.

Dole, R., M. Hoerling, J. Perlwitz, J. Eischeid, and P. Pegion, 2011: Was there a basis for anticipating the 2010 Russian heat wave? *Geophys. Res. Lett.*, L06702, doi 10.1029/2010GL046582.

Dominguez, F., E. Rivera, D. P. Lettenmaier, and C. L. Castro, 2012: Changes in winter precipitation extremes for the western United States under a warmer climate as simulated by regional climate models. *Geophys. Res. Lett.*, **39**, L05803.

Donat, M. G., G. C. Leckebusch, S. Wild, and U. Ulbrich, 2011: Future changes in European winter storm losses and extreme wind speeds inferred from GCM and RCM multi-model simulations. *Nat. Hazards Earth Syst. Sci.*, **11**, 1351–1370.

Dong, B., R. T. Sutton, and T. Woollings, 2011: Changes of interannual NAO variability in response to greenhouse gases forcing. *Clim. Dyn.*, **37**, 1621–1641.

Dong, L., T. J. Vogelsang, and S. J. Colucci, 2008: Interdecadal trend and ENSO-related interannual variability in Southern Hemisphere blocking. *J. Clim.*, **21**, 3068–3077.

Döscher, R., and T. Koenigk, 2012: Arctic rapid sea ice loss events in regional coupled climate scenario experiments. *Ocean Sci. Discuss.*, **9**, 2327–2373.

Dowdy, A. J., G. A. Mills, B. Timbal, and Y. Wang, 2012: Changes in the risk of extratropical cyclones in eastern Australia. *J. Clim.*, **26**, 1403–1417.

Drumond, A. R. M., and T. Ambrizzi, 2005: The role of SST on the South American atmospheric circulation during January, February and March 2001. *Clim. Dyn.*, **24**, 781–791.

Du, Y., and S.-P. Xie, 2008: Role of atmospheric adjustments in the tropical Indian Ocean warming during the 20th century in climate models. *Geophys. Res. Lett.*, **35**, doi: 10.1029/2008GL033631.

Du, Y., L. Yang, and S. Xie, 2011: Tropical Indian Ocean influence on Northwest Pacific tropical cyclones in summer following strong El Niño. *J. Clim.*, **24**, 315–322.

Du, Y., S. P. Xie, G. Huang, and K. M. Hu, 2009: Role of air-sea interaction in the long persistence of El Niño-induced north Indian Ocean warming. *J. Clim.*, **22**, 2023–2038.

Dufek, A. S., T. Ambrizzi, and R. P. Rocha, 2008: Are reanalysis data useful for calculating climate indices over South America? *Ann. NY Acad. Sci.*, **1146**, 87–104.

Duffy, P. B., and C. Tebaldi, 2012: Increasing prevalence of extreme summer temperatures in the U.S. *Clim. Change*, **111**, 487–495.

Dunion, J., and C. Velden, 2004: The impact of the Saharan air layer on Atlantic tropical cyclone activity. *Bull. Am. Meteorol. Soc.*, **85**, 353–365.

Dunion, J., and C. Marron, 2008: A reexamination of the Jordan mean tropical sounding based on awareness of the Saharan air layer: Results from 2002. *J. Clim.*, **21**, 5242–5253.

Dunion, J. P., 2011: Rewriting the climatology of the tropical North Atlantic and Caribbean Sea atmosphere. *J. Clim.*, **24**, 893–908.

Dunn-Sigouin, E., and S.-W. Son, 2013: Northern Hemisphere blocking frequency and duration in the CMIP5 models. *J. Geophys. Res. Atmos.*, **118**, 1179–1188.

Elsner, J. B., J. P. Kossin, and T. H. Jagger, 2008: The increasing intensity of the strongest tropical cyclones. *Nature*, **455**, 92–95.

Emanuel, K., 2007: Environmental factors affecting tropical cyclone power dissipation. *J. Clim.*, **20**, 5497–5509.

Emanuel, K., 2010: Tropical cyclone activity downscaled from NOAA-CIRES reanalysis, 1908–1958. *J. Adv. Model. Earth Syst.*, **2**, 12.

Emanuel, K., R. Sundararajan, and J. Williams, 2008: Hurricanes and global warming: Results from downscaling IPCC AR4 simulations. *Bull. Am. Meteorol. Soc.*, **89**, 347–367.

Emanuel, K., S. Solomon, D. Folini, S. Davis, and C. Cagnazzo, 2012: Influence of tropical tropopause layer cooling on Atlantic hurricane activity. *J. Clim.*, **26**, 2288–2301.

Endo, H., 2010: Long-term changes of seasonal progress in Baiu rainfall using 109 years (1901–2009) daily station data. *Sola*, **7**, 5–8.

Endo, H., 2012: Future changes of Yamase bringing unusually cold summers over northeastern Japan in CMIP3 multi-models. *J. Meteorol. Soc. Jpn.*, **90A**, 123-136.

Endo, H., A. Kitoh, T. Ose, R. Mizuta, and S. Kusunoki, 2012: Future changes and uncertainties in Asian precipitation simulated by multiphysics and multi–sea surface temperature ensemble experiments with high-resolution Meteorological Research Institute atmospheric general circulation models (MRI-AGCMs). *J. Geophys. Res.*, **117**, D16118.

Enfield, D., S. K. Lee, and C. Wang, 2006: How are large Western Hemisphere warm pools formed? *Prog. Oceanogr.*, **70**, 346–365.

Engelbrecht, C. J., F. A. Engelbrecht, and L. L. Dyson, 2011: High-resolution model-projected changes in mid-tropospheric closed-lows and extreme rainfall events over southern Africa. *Int. J. Climatol.*, **33**, 173–187.

England, M. H., C. C. Ummenhofer, and A. Santoso, 2006: Interannual rainfall extremes over southwest Western Australia linked to Indian ocean climate variability. *J. Clim.*, **19**, 1948–1969.

Englehart, P. J., and A. V. Douglas, 2006: Defining intraseasonal rainfall variability within the North American monsoon. *J. Clim.*, **19**, 4243–4253.

Evan, A., 2012: Atlantic hurricane activity following two major volcanic eruptions. *J. Geophys. Res. Atmos.*, **117**, doi: 10.1029/2011JD016716.

Evan, A., G. Foltz, D. Zhang, and D. Vimont, 2011a: Influence of African dust on ocean-atmosphere variability in the tropical Atlantic. *Nature Geosci.*, **4**, 762–765.

Evan, A. T., J. P. Kossin, C. E. Chung, and V. Ramanathan, 2011b: Arabian Sea tropical cyclones intensified by emissions of black carbon and other aerosols. *Nature*, **479**, 94–97.

Evan, A. T., J. P. Kossin, C. Chung, and V. Ramanathan, 2012: Evan et al. reply to Wang et al. (2012), "Intensified Arabian Sea tropical storms". *Nature*, **489**, E2–E3.

Evan, A. T., D. J. Vimont, A. K. Heidinger, J. P. Kossin, and R. Bennartz, 2009: The role of aerosols in the evolution of tropical North Atlantic Ocean temperature anomalies. *Science*, **324**, 778–781.

Evans, J. P., 2008: Changes in water vapor transport and the production of precipitation in the eastern Fertile Crescent as a result of global warming. *J. Hydrometeorol.*, **9**, 1390–1401.

Evans, J. P., 2009: 21st century climate change in the Middle East. *Clim. Change*, **92**, 417–432.

Eyring, V., et al., 2013: Long-term ozone changes and associated climate impacts in CMIP5 simulations. *J. Geophys. Res. Atmos*, doi:10.1002/jgrd.50316.

Falvey, M., and R. D. Garreaud, 2009: Regional cooling in a warming world: Recent temperature trends in the southeast Pacific and along the west coast of subtropical South America (1979–2006). *J. Geophys. Res. Atmos.*, **114**, D04102.

Fauchereau, N., B. Pohl, C. Reason, M. Rouault, and Y. Richard, 2009: Recurrent daily OLR patterns in the Southern Africa/Southwest Indian Ocean region, implications for South African rainfall and teleconnections. *Clim. Dyn.*, **32**, 575–591.

Fedorov, A. V., and S. G. Philander, 2000: Is El Nino changing? *Science*, **288**, 1997–2002.

Feldstein, S. B., and C. Franzke, 2006: Are the North Atlantic Oscillation and the Northern Annular Mode distinguishable? *J. Atmos. Sci.*, **63**, 2915–2930.

Feliks, Y., M. Ghil, and A. W. Robertson, 2010: Oscillatory climate modes in the eastern Mediterranean and their synchronization with the North Atlantic Oscillation. *J. Clim.*, **23**, 4060–4079.

Feng, S., Q. Hu, and R. Oglesby, 2011: Influence of Atlantic sea surface temperatures on persistent drought in North America. *Clim. Dyn.*, **37**, 569–586.

Fereday, D. R., J. R. Knight, A. A. Scaife, C. K. Folland, and A. Philipp, 2008: Cluster analysis of North Atlantic-European circulation types and links with tropical Pacific sea surface temperatures. *J. Clim.*, **21**, 3687–3703.

Fink, A., S. Pohle, J. Pinto, and P. Knippertz, 2012: Diagnosing the influence of diabatic processes on the explosive deepening of extratropical cyclones. *Geophys. Res. Lett.*, **39**, doi: 10.1029/2012GL051025.

Fink, A. H., T. Bruecher, V. Ermert, A. Krueger, and J. G. Pinto, 2009: The European storm Kyrill in January 2007: Synoptic evolution, meteorological impacts and some considerations with respect to climate change. *Nat. Hazards Earth Syst. Sci.*, **9**, 405–423.

Fischer-Bruns, I., D. F. Banse, and J. Feichter, 2009: Future impact of anthropogenic sulfate aerosol on North Atlantic climate. *Clim. Dyn.*, **32**, 511–524.

Fogt, R., D. Bromwich, and K. Hines, 2011: Understanding the SAM influence on the South Pacific ENSO teleconnection. *Clim. Dyn.*, **36**, 1555–1576.

Fogt, R. L., J. Perlwitz, A. J. Monaghan, D. H. Bromwich, J. M. Jones, and G. J. Marshall, 2009: Historical SAM variability. Part II: Twentieth-century variability and Ttrends from reconstructions, observations, and the IPCC AR4 models. *J. Clim.*, **22**, 5346–5365.

Folland, C., M. Salinger, N. Jiang, and N. Rayner, 2003: Trends and variations in South Pacific island and ocean surface temperatures. *J. Clim.*, **16**, 2859–2874.

Folland, C. K., J. A. Renwick, M. J. Salinger, and A. B. Mullan, 2002: Relative influences of the Interdecadal Pacific Oscillation and ENSO on the South Pacific Convergence Zone. *Geophys. Res. Lett.*, **29**, doi: 10.1029/2001GL014201.

Folland, C. K., J. Knight, H. W. Linderholm, D. Fereday, S. Ineson, and J. W. Hurrell, 2009: The summer North Atlantic Oscillation: Past, present, and future. *J. Clim.*, **22**, 1082–1103.

Forster, P., et al., 2007: Changes in atmospheric constituents and in radiative forcing. In: *Climate Change 2007: The Physical Science Basis. Contribution of Working Group I to the Fourth Assessment Report of the Intergovernmental Panel on Climate Change* [Solomon, S., D. Qin, M. Manning, Z. Chen, M. Marquis, K. B. Averyt, M. Tignor and H. L. Miller (eds.)] Cambridge University Press, Cambridge, United Kingdom and New York, NY, USA, pp. 129–234.

Frank, W., and P. Roundy, 2006: The role of tropical waves in tropical cyclogenesis. *Mon. Weather Rev.*, **134**, 2397–2417.

Frederiksen, C. S., J. S. Frederiksen, J. M. Sisson, and S. L. Osbrough, 2011a: Changes and projections in the annual cycle of the Southern Hemisphere circulation, storm tracks and Australian rainfall. *Int. J. Clim. Change Impacts Respons.*, **2**, 143–162.

Frederiksen, C. S., J. S. Frederiksen, J. M. Sisson, and S. L. Osbrough, 2011b: Australian winter circulation and rainfall changes and projections. *Int. J. Clim. Change Strat. Manage.*, **3**, 170–188.

Frederiksen, J. S., and C. S. Frederiksen, 2007: Interdecadal changes in southern hemisphere winter storm track modes. *Tellus A*, **59**, 599–617.

Frederiksen, J. S., C. S. Frederiksen, S. L. Osbrough, and J. M. Sisson, 2010: Causes of changing Southern Hemispheric weather systems. In: *Managing Climate Change* [I. Jupp, P. Holper and W. Cai (eds.)]. CSIRO Publishing, Collingwood, Victoria, Australia, pp. 85–98.

Friedman, A. R., Y. T. Hwang, J. C. H. Chiang, and D. M. W. Frierson, 2013: Interhemispheric temperature asymmetry over the 20th century and in future projections. *J. Clim.*, doi:10.1175/JCLI-D-12-00525.1.

Frierson, D. M. W., I. M. Held, and P. Zurita-Gotor, 2007: A gray-radiation aquaplanet moist GCM. Part II: Energy transports in altered climates. *J. Atmos. Sci.*, **64**, 1680–1693.

Fučkar, N. S., S.-P. Xie, R. Farneti, E. A. Maroon, and D. M. W. Frierson, 2013: Influence of the extratropical ocean circulation on the intertropical convergence zone in an idealized coupled general circulation model. *J. Clim.*, **26**, 4612–4629.

Furtado, J., E. Di Lorenzo, N. Schneider, and N. A. Bond, 2011: North Pacific decadal variability and climate change in the IPCC AR4 models. *J. Clim.*, **24**, 3049–3066.

Gamble, D. W., and S. Curtis, 2008: Caribbean precipitation: Review, model and prospect. *Prog. Phys. Geogr.*, **32**, 265–276.

Gan, M. A., V. B. Rao, and M. C. L. Moscati, 2006: South American monsoon indices. *Atmos. Sci. Lett.*, **6**, 219–223.

Gao, X., Y. Shi, and F. Giorgi, 2012a: A high resolution simulation of climate change over China. *Sci. China Earth Sci.*, **54**, 462–472.

Gao, X., Y. Shi, R. Song, F. Giorgi, Y. Wang, and D. Zhang, 2008: Reduction of future monsoon precipitation over China: Comparison between a high resolution RCM simulation and the driving GCM. *Meteorol. Atmos. Phys.*, **100**, 73–86.

Gao, X., Y. Shi, D. Zhang, J. Wu, F. Giorgi, Z. Ji, and Y. Wang, 2012b: Uncertainties in monsoon precipitation projections over China: Results from two high-resolution RCM simulations. *Clim. Res.*, **2**, 213.

Gao, Y., L. R. Leung, E. P. Salathé, F. Dominguez, B. Nijssen, and D. P. Lettenmaier, 2012c: Moisture flux convergence in regional and global climate models: Implications for droughts in the southwestern United States under climate change. *Geophys. Res. Lett.*, **39**, L09711.

Garcia, R., and W. J. Randel, 2008: Acceleration of the Brewer–Dobson circulation due to increases in greenhouse gases. *J. Atmos. Sci.*, **65**, 2731–2739.

Garfinkel, C. I., and D. L. Hartmann, 2011: The influence of the Quasi-Biennial Oscillation on the troposphere in wintertime in a hierarchy of models, Part 1: Simplified dry GCMs. *J. Atmos. Sci.*, **68**, 1273–1289.

Gastineau, G., and B. J. Soden, 2009: Model projected changes of extreme wind events in response to global warming. *Geophys. Res. Lett.*, **36**, doi: 10.1029/2009gl037500.

Geng, Q. Z., and M. Sugi, 2003: Possible change of extratropical cyclone activity due to enhanced greenhouse gases and sulfate aerosols - Study with a high-resolution AGCM. *J. Clim.*, **16**, 2262–2274.

Gerber, E. P., and G. K. Vallis, 2007: Eddy-zonal flow interactions and the persistence of the zonal index. *J. Atmos. Sci.*, **64**, 3296–3311.

Gerber, E. P., L. M. Polvani, and D. Ancukiewicz, 2008: Annular mode time scales in the Intergovernmental Panel on Climate Change Fourth Assessment Report models. *Geophys. Res. Lett.*, **35**, doi: 10.1029/2008gl035712.

Gerber, E. P., et al., 2010: Stratosphere-troposphere coupling and annular mode variability in chemistry-climate models. *J. Geophys. Res.*, **115**, doi: 10.1029/2009jd013770.

Giannini, A., R. Saravanan, and P. Chang, 2003: Oceanic forcing of Sahel rainfall on interannual to interdecadal time scales. *Science*, **302**, 1027–1030.

Giannini, A., M. Biasutti, I. Held, and A. Sobel, 2008: A global perspective on African climate. *Clim. Change*, **90**, 359–383.

Giese, B., and S. Ray, 2011: El Niño variability in simple ocean data assimilation (SODA), 1871–2008. *J. Geophys. Res. Oceans*, **116**, 10.1029/2010JC006695.

Gillett, N. P., and J. C. Fyfe, 2013: Annular mode changes in the CMIP5 simulations. *Geophys. Res. Lett.*, **40**, .

Giorgi, F., and X. Bi, 2009: Time of emergence (TOE) of GHG-forced precipitation change hot-spots. *Geophys. Res. Lett.*, **36**, doi:10.1029/2009GL037593.

Gochis, D. J., L. Castillo-Brito, and J. Shuttleworth, 2007: Correlations between sea-surface temperatures and warm season streamflow in northwest Mexico. *Int. J. Climatol.*, **27**, 883–901.

Goldenberg, S. B., C. Landsea, A. M. Mestas-Nunez, and W. M. Gray, 2001: The recent increase in Atlantic hurricane activity: Causes and implications. *Science*, **293**, 474–479.

Gong, D. Y., and S. W. Wang, 1999: Definition of Antarctic Oscillation Index. *Geophys. Res. Lett.*, **26**, 459–462.

Gong, D. Y., and C. H. Ho, 2002: The Siberian High and climate change over middle to high latitude Asia. *Theor. Appl. Climatol.*, **72**, 1–9.

Good, P., J. A. Lowe, M. Collins, and W. Moufouma-Okia, 2008: An objective tropical Atlantic sea surface temperature gradient index for studies of south Amazon dry-season climate variability and change. *Philos. Trans. R. Soc. London B*, **363**, 1761–1766.

Goswami, B. N., V. Venugopal, D. Sengupta, M. S. Madhusoodanan, and P. K. Xavier, 2006: Increasing trend of extreme rain events over India in a warming environment. *Science*, **314**, 1442–1445.

Graff, L., and J. LaCasce, 2012: Changes in the extratropical storm tracks in response to changes in SST in an AGCM. *J. Clim.*, **25**, 1854–1870.

Grantz, K., B. Rajagopalan, M. Clark, and E. Zagona, 2007: Seasonal shifts in the North American monsoon. *J. Clim.*, **20**, 1923–1935.

Griffiths, G., M. Salinger, and I. Leleu, 2003: Trends in extreme daily rainfall across the South Pacific and relationship to the South Pacific Convergence Zone. *Int. J. Climatol.*, **23**, 847–869.

Griffiths, G., et al., 2005: Change in mean temperature as a predictor of extreme temperature change in the Asia-Pacific region. *Int. J. Climatol.*, **25**, 1301–1330.

Griffiths, G. M., 2007: Changes in New Zealand daily rainfall extremes 1930 - 2004. *Weather Clim.*, **27**, 3–44.

Gu, D. F., and S. G. H. Philander, 1995: Secular changes of annual and interannual variability in the tropics during the past century. *J. Clim.*, **8**, 864–876.

Guilyardi, E., H. Bellenger, M. Collins, S. Ferrett, W. Cai, and A. Wittenberg, 2012: A first look at ENSO in CMIP5. *CLIVAR Exchanges,* **58**, 29-32.

Guo, Z. C., D. H. Bromwich, and K. M. Hines, 2004: Modeled antarctic precipitation. Part II: ENSO modulation over West Antarctica. *J. Clim.*, **17**, 448–465.

Gutiérrez, D., et al., 2011: Coastal cooling and increased productivity in the main upwelling zone off Peru since the mid-twentieth century. *Geophys. Res. Lett.*, **38**, L07603.

Gutowski, W. J. et al., 2010: Regional, extreme monthly precipitation simulated by NARCCAP RCMs. *J. Hydrometeorol.*, **11**, 1373–1379.

Gutzler, D. S., 2004: An index of interannual precipitation variability in the core of the North American monsoon region. *J. Clim.*, **17**, 4473–4480.

Gutzler, D. S., and T. O. Robbins, 2011: Climate variability and projected change in the western United States: Regional downscaling and drought statistics. *Clim. Dyn.*, **37**, 835–849.

Gutzler, D. S., L. N. Long, J. Schemm, S. B. Roy, M. Bosilovich, J. C. Collier, M. Kanamitsu, P. Kelly, D. Lawrence, M. I. Lee, R. L. Sánchez, B. Mapes, K. Mo, A. Nunes, E. A. Ritchie, J. Roads, S. Schubert, H. Wei, and G. J. Zhang, 2009: Simulations of the 2004 North American Monsoon: NAMAP2. *J. Climate*, **22**, 6716-6740.

Haarsma, R. J., et al., 2013: More hurricanes to hit Western Europe due to global warming. *Geophys. Res. Lett.*, doi:10.1002/grl.50360.

Haensler, A., S. Hagemann, and D. Jacob, 2011: The role of the simulation setup in a long-term high-resolution climate change projection for the southern African region. *Theor. Appl. Climatol.*, **106**, 153–169.

Haerter, J., E. Roeckner, L. Tomassini, and J. von Storch, 2009: Parametric uncertainty effects on aerosol radiative forcing. *Geophys. Res. Lett.*, **36**, doi: 10.1029/2009GL039050.

Haigh, J. D., and H. K. Roscoe, 2006: Solar influences on polar modes of variability. *Meteorol. Z.*, **15**, 371–378.

Häkkinen, S., P. B. Rhines, and D. L. Worthen, 2011: Atmospheric blocking and atlantic multidecadal ocean variability. *Science*, **334**, 655–659.

Hall, T., A. Sealy, T. Stephenson, S. Kusunoki, M. Taylor, A. A. Chen, and A. Kitoh, 2012: Future climate of the Caribbean from a super-high-resolution atmospheric general circulation model. *Theor. Appl. Climatol.*, doi:10.1007/s00704-012-0779-7, 1–17.

Han, W., et al., 2010: Patterns of Indian Ocean sea-level change in a warming climate. *Nature Geosci.*, **3**, 546–550.

Handorf, K., and Dethloff, 2009: Atmospheric teleconnections and flow regimes under future climate projections. 237–255.

Hansen, J., M. Sato, R. Ruedy, K. Lo, D. W. Lea, and M. Medina-Elizade, 2006: Global temperature change. *Proc. Natl. Acad. Sci. U.S.A.*, **103**, 14288–14293.

Hansingo, K., and C. Reason, 2008: Modelling the atmospheric response to SST dipole patterns in the South Indian Ocean with a regional climate model. *Meteorol. Atmos. Phys.*, **100**, 37–52.

Hansingo, K., and C. Reason 2009: Modelling the atmospheric response over southern Africa to SST forcing in the southeast tropical Atlantic and southwest subtropical Indian Oceans. *Int. J. Climatol.*, **29**, 1001–1012.

Harrison, S. P., et al., 2003: Mid-Holocene climates of the Americas: A dynamical response to changed seasonality. *Clim. Dyn.*, **20**, 663–688.

Hartmann, B., and G. Wendler, 2005: The Significance of the 1976 Pacific climate shift in the climatology of Alaska. *J. Clim.*, **18**, 4824–4839.

Harvey, B. J., L. C. Shaffrey, T. J. Woollings, G. Zappa, and K. I. Hodges, 2012: How large are projected 21st century storm track changes? *Geophys. Res. Lett.*, **39**, L18707.

Haylock, M. R., et al., 2006: Trends in total and extreme South American rainfall in 1960–2000 and links with sea surface temperature. *J. Clim.*, **19**, 1490–1512.

Held, I., and M. Zhao, 2011: The response of tropical cyclone statistics to an increase in CO_2 with fixed sea surface temperatures. *J. Clim.*, **24**, 5353–5364.

Held, I., T. Delworth, J. Lu, K. Findell, and T. Knutson, 2005: Simulation of Sahel drought in the 20th and 21st centuries. *Proc. Natl. Acad. Sci. U.S.A.*, **102**, 17891–17896.

Held, I. M., 1993: Large-scale dynamics and global warming. *Bull. Am. Meteorol. Soc.*, **74**, 228–241.

Held, I. M., and B. J. Soden, 2006: Robust responses of the hydrological cycle to global warming. *J. Clim.*, **19**, 5686–5699.

Hendon, H. H., D. W. J. Thompson, and M. C. Wheeler, 2007: Australian rainfall and surface temperature variations associated with the Southern Hemisphere annular mode. *J. Clim.*, **20**, 2452–2467.

Hennessy, K., S. Power, and G. Cambers, Eds., 2011: *Climate change in the Pacific: Scientific Assessment and New Research. Regional Overview (Volume 1) and Country Reports (Volume 2).* Australian Bureau of Meteorology (BoM) and Commonwealth Scientific and Industrial Organisation (CSIRO), Melbourne, Australia.

Hermes, J., and C. Reason, 2009: Variability in sea-surface temperature and winds in the tropical south-east Atlantic Ocean and regional rainfall relationships. *Int. J. Climatol.*, **29**, 11–21.

Hernandez-Deckers, D., and J.-S. von Storch, 2010: Energetics responses to increases in greenhouse gas concentration. *J. Clim.*, **23**, 3874–3887.

Hidayat, R., and S. Kizu, 2010: Influence of the Madden-Julian Oscillation on Indonesian rainfall variability in austral summer. *Int. J. Climatol.*, **30**, 1816–1825.

Hill, K. J., A. S. Taschetto, and M. H. England, 2011: Sensitivity of South American summer rainfall to tropical Pacific Ocean SST anomalies. *Geophys. Res. Lett.*, **38**, L01701.

Hinton, T. J., B. J. Hoskins, and G. M. Martin, 2009: The influence of tropical sea surface temperatures and precipitation on North Pacific atmospheric blocking. *Climate Dynamics*, **33**, 549-563.

Hirschi, M., et al., 2011: Observational evidence for soil-moisture impact on hot extremes in southeastern Europe. *Nature Geosci.*, **4**, 17–21.

Ho, C. K., D. B. Stephenson, M. Collins, C. A. T. Ferro, and S. J. Brown, 2012: Calibration strategies: A source of additional uncertainty in climate change projections. *Bull. Am. Meteorol. Soc.*, **93**, 21–26.

Hoerling, M., J. Hurrell, J. Eischeid, and A. Phillips, 2006: Detection and attribution of twentieth-century northern and southern African rainfall change. *J. Clim.*, **19**, 3989–4008.

Holland, G. J., and P. J. Webster, 2007: Heightened tropical cyclone activity in the North Atlantic: Natural variability or climate trend? *Philos. Trans. R. Soc. London A*, **365**, 2695–2716.

Hope, P. K., W. Drosdowsky, and N. Nicholls, 2006: Shifts in the synoptic systems influencing southwest Western Australia. *Clim. Dyn.*, **26**, 751–764.

Horel, J. D., and J. M. Wallace, 1981: Planetary-scale atmospheric phenomena associated with the Southern Oscillation. *Mon. Weather Rev.*, **109**, 813–829.

Hori, M. E., D. Nohara, and H. L. Tanaka, 2007: Influence of Arctic Oscillation towards the Northern Hemisphere surface temperature variability under the global warming scenario. *J. Meteorol. Soc. Jpn.*, **85**, 847–859.

Hsieh, W. W., A. Wu, and A. Shabbar, 2006: Nonlinear atmospheric teleconnections. *Geophys. Res. Lett.*, **33**, doi: 10.1029/2005gl025471.

Hsu, P.-C., T. Li, and B. Wang, 2011: Trends in global monsoon area and precipitation in the past 30 years. *Geophys. Res. Lett.*, **38**, doi: 10.1029/2011GL046893.

Hsu, P.-C., T. Li, H. Murakami, and A. Kitoh, 2013: Future change of the global monsoon revealed from 19 CMIP5 models. *J. Geophys. Res. Atmos.*, **118**, doi: 10.1002/jgrd.50145.

Hu, Z., A. Kumar, B. Jha, and B. Huang, 2012a: An Analysis of Forced and internal variability in a warmer climate in CCSM3. *J. Clim.*, **25**, 2356–2373.

Hu, Z., A. Kumar, B. Jha, W. Wang, B. Huang, and B. Huang, 2012b: An analysis of warm pool and cold tongue El Niños: Air-sea coupling processes, global influences, and recent trends. *Clim. Dyn.*, **38**, 2017–2035.

Hu, Z. Z., 1997: Interdecadal variability of summer climate over East Asia and its association with 500 hPa height and global sea surface temperature. *J. Geophys. Res. Atmos.*, **102**, 19403–19412.

Hu, Z. Z., and Z. H. Wu, 2004: The intensification and shift of the annual North Atlantic Oscillation in a global warming scenario simulation. *Tellus A*, **56**, 112–124.

Huang, B., and Z. Liu, 2001: Temperature trend of the last 40 yr in the upper Pacific Ocean. *J. Clim.*, **14**, 3738–3750.

Huang, G., K. M. Hu, and S. P. Xie, 2010: Strengthening of tropical Indian Ocean teleconnection to the northwest Pacific since the mid-1970s: An atmospheric GCM study. *J. Clim.*, **23**, 5294–5304.

Huang, J., X. Guan, and F. Ji, 2012: Enhanced cold-season warming in semi-arid regions. *Atmos. Chem. Phys. Discuss.*, **12**, 4627–4653.

Huang, P., S.-P. Xie, K. Hu, G. Huang, and R. Huang, 2013: Patterns of the seasonal response of tropical rainfall to global warming. *Nature Geosci.*, **6**, 357–361.

Huang, R., W. Chen, B. Yang, and R. Zhang, 2004: Recent advances in studies of the interaction between the east Asian winter and summer monsoons and ENSO cycle. *Adv. Atmos. Sci.*, **21**, 407–424.

Huffman, G. J., R. F. Adler, D. T. Bolvin, and G. Gu, 2009: Improving the global precipitation record: GPCP Version 2.1. *Geophys. Res. Lett.*, **36**, L17808.

Hurrell, J. W., and C. Deser, 2009: North Atlantic climate variability: The role of the North Atlantic Oscillation. *J. Mar. Syst.*, **78**, 28–41.

Hurrell, J. W., Y. Kushnir, G. Ottersen, and M. Visbeck, 2003: An overview of the North Atlantic Oscillation. In: *The North Atlantic Oscillation: Climate Significance and Environmental Impact* [J. W. Hurrell, Y. Kushnir, M. Visbeck and G. Ottersen (eds.)]. American Geophysical Union, Washington, DC, pp. 1–35.

Huss, M., R. Hock, A. Bauder, and M. Funk, 2010: 100-year mass changes in the Swiss Alps linked to the Atlantic Multidecadal Oscillation. *Geophys. Res. Lett.*, **37**, doi: 10.1029/2010GL042616.

Hwang, Y.-T., and D. M. W. Frierson, 2010: Increasing atmospheric poleward energy transport with global warming. *Geophys. Res. Lett.*, **37**, doi: 10.1029/2010GL045440.

Ihara, C., Y. Kushnir, M. Cane, and V. de la Pena, 2009: Climate Change over the Equatorial Indo-Pacific in Global Warming. *J. Clim.*, **22**, 2678–2693.

Iizumi, T., F. Uno, and M. Nishimori, 2012: Climate downscaling as a source of uncertainty in projecting local climate change impacts. *J. Meteorol. Soc. Jpn.*, **90B**, 83–90.

Im, E. S., J. B. Ahn, W. T. Kwon, and F. Giorgi, 2008: Multi-decadal scenario simulation over Korea using a one-way double-nested regional climate model system. Part 2: Future climate projection (2021–2050). *Clim. Dyn.*, **30**, 239–254.

Ineson, S., A. A. Scaife, J. R. Knight, J. C. Manners, N. J. Dunstone, L. J. Gray, and J. D. Haigh, 2011: Solar forcing of winter climate variability in the Northern Hemisphere. *Nature Geosci.*, **4**, 753–757.

Inoue, J., J. Liu, and J. A. Curry, 2006: Intercomparison of arctic regional climate models: Modeling clouds and radiation for SHEBA in May 1998. *J. Clim.*, **19**, 4167–4178.

Ionita, M., G. Lohmann, N. Rimbu, S. Chelcea, and M. Dima, 2012: Interannual to decadal summer drought variability over Europe and its relationship to global sea surface temperature. *Clim. Dyn.*, **38**, 363–377.

IPCC, 2007a: *Climate Change 2007: The Physical Science Basis. Contribution of Working Group I to the Fourth Assessment Report of the Intergovernmental Panel on Climate Change* [Solomon, S., D. Qin, M. Manning, Z. Chen, M. Marquis, K. B. Averyt, M. Tignor and H. L. Miller (eds.)] Cambridge University Press, Cambridge, United Kingdom and New York, NY, USA,996 pp.

IPCC, 2007b: *Climate Change 2007: Impacts, Adaptation and Vulnerability. Contribution of Working Group II to the Fourth Assessment Report of the Intergovernmental Panel on Climate Change (IPCC)* [M. L. Parry, O. F. Canziani, J. P. Palutikof, P. J. van der Linden and C. E. Hanson (eds.)]. Cambridge University Press, Cambridge, United Kingdom and New York, NY, USA, 976 pp.

IPCC, 2012: *Managing the Risks of Extreme Events and Disasters to Advance Climate Change Adaptation. A Special Report of Working Groups I and II of the Intergovernmental Panel on Climate Change [*C. B. Field, V. Barros, T. F. Stocker, D. Qin, D. J. Dokken, K. L. Ebi, M. D. Mastrandrea, K. J. Mach, G.-K. Plattner, S. K. Allen, M. Tignor and P.M. Midgley (eds.)]. Cambridge University Press, Cambridge, United Kingdom and New York, NY, USA, 582 pp.

Irving, D., P. Whetton, and A. Moise, 2012: Climate projections for Australia: A first glance at CMIP5. *Aust. Mereorol. Oceanogr. J.*, **62**, 211–225.

Irving, D., et al., 2011: Evaluating global climate models for the Pacific island region. *Clim. Res.*, **49**, 169–187.

Izumo, T., C. D. Montegut, J. J. Luo, S. K. Behera, S. Masson, and T. Yamagata, 2008: The role of the western Arabian Sea upwelling in Indian monsoon rainfall variability. *J. Clim.*, **21**, 5603–5623.

Jaeger, E. B., and S. I. Seneviratne, 2010: Impact of soil moisture–atmosphere coupling on European climate extremes and trends in a regional climate model. *Clim. Dyn.*, **36**, 1919–1939.

Janicot, S., et al., 2011: Intraseasonal variability of the West African monsoon. *Atmos. Sci. Lett.*, **12**, 58–66.

Jiang, H., and E. Zipser, 2010: Contribution of tropical cyclones to the global precipitation from eight seasons of TRMM data: Regional, seasonal, and interannual variations. *J. Clim.*, **23**, 1526–1543.

Jiang, Y. L., and Z. Zhao, 2013: Maximum wind speed changes over China. *Acta Meteorol. Sin.*, **27**, 63–74.

Jiang, Z., J. Song, L. Li, W. Chen, Z. Wang, and J. Wang, 2011: Extreme climate events in China: IPCC-AR4 model evaluation and projection. *Clim. Change*, **110**, 385–401.

Jin, F., A. Kitoh, and P. Alpert, 2010: Water cycle changes over the Mediterranean: A comparison study of a super-high-resolution global model with CMIP3. *Philos. Trans. R. Soc. London A*, **68**, 5137–5149.

Johanson, C. M., and Q. Fu, 2009: Hadley cell widening: Model simulations versus observations. *J. Clim.*, **22**, 2713–2725.

Joly, M., A. Voldoire, H. Douville, P. Terray, and J.-F. Royer, 2007: African monsoon teleconnections with tropical SSTs: Validation and evolution in a set of IPCC4 simulations. *Clim. Dyn.*, **29**, 1–20.

Jones, C., and L. M. V. Carvalho, 2013: Climate change in the South American Monsoon System: Present climate and CMIP5 projections. *J. Clim.*, doi:10.1175/JCLI-D-12-00412.1.

Jones, D. A., W. Wang, and R. Fawcett, 2009a: High-quality spatial climate data-sets for Australia. *Aust. Meteorol. Oceanogr. J.*, **58**, 233–248.

Jones, J. M., R. L. Fogt, M. Widmann, G. J. Marshall, P. D. Jones, and M. Visbeck, 2009b: Historical SAM variability. Part I: Century-ength seasonal reconstructions. *J. Clim.*, **22**, 5319–5345.

Joshi, M. M., A. J. Charlton, and A. A. Scaife, 2006: On the influence of stratospheric water vapor changes on the tropospheric circulation. *Geophys. Res. Lett.*, **33**, doi: 10.1029/2006gl025983.

Jourdain, N., A. Gupta, A. Taschetto, C. Ummenhofer, A. Moise, and K. Ashok, 2013: The Indo-Australian monsoon and its relationship to ENSO and IOD in reanalysis data and the CMIP3/CMIP5 simulations. *Clim. Dyn.*, doi:10.1007/s00382-013-1676-1, 1–30.

Jung, T., et al., 2011: High-resolution global climate simulations with the ECMWF model in project Athena: Experimental design, model climate, and seasonal forecast skill. *J. Clim.*, **25**, 3155–3172.

Junquas, C., C. Vera, L. Li, and H. Treut, 2012: Summer precipitation variability over Southeastern South America in a global warming scenario. *Climate Dynamics*, **38**, 1867-1883.

Junquas, C., C. S. Vera, L. Li, and H. Treut, 2013: Impact of projected SST changes on summer rainfall in southeastern South America. *Clim. Dyn.*, **40**, 1569–1589.

Kajikawa, Y., B. Wang, and J. Yang, 2010: A multi-time scale Australian monsoon index. *Int. J. Climatol.*, **30**, 1114–1120.

Kanada, S., M. Nakano, and T. Kato, 2010: Changes in mean atmospheric structures around Japan during July due to global warming in regional climate experiments using a cloud resolving model. *Hydrol. Res. Lett.*, **4**, 11–14.

Kanada, S., M. Nakano, and T. Kato, 2012: Projections of future changes in precipitation and the vertical structure of the frontal zone during the Baiu season in the vicinity of Japan using a 5-km-mesh regional climate model. *J. Meteorol. Soc. Jpn.*, **90A**, 65–86.

Kang, S., I. Held, D. Frierson, and M. Zhao, 2008: The response of the ITCZ to extratropical thermal forcing: Idealized slab-ocean experiments with a GCM. *J. Clim.*, **21**, 3521–3532.

Kao, H. Y., and J. Y. Yu, 2009: Contrasting Eastern-Pacific and Central-Pacific types of ENSO. *J. Clim.*, **22**, 615–632.

Kapnick, S., and A. Hall, 2012: Causes of recent changes in western North American snowpack. *Clim. Dyn.*, **38**, 1885–1899.

Karmalkar, A. V., R. S. Bradley, and H. F. Diaz, 2011: Climate change in Central America and Mexico: Regional climate model validation and climate change projections. *Clim. Dyn.*, **37**, 605–629.

Karnauskas, K. B., R. Seager, A. Kaplan, Y. Kushnir, and M. A. Cane, 2009: Observed strengthening of the zonal sea surface temperature gradient across the equatorial Pacific Ocean. *J. Clim.*, **22**, 4316–4321.

Karpechko, A. Y., 2010: Uncertainties in future climate attributable to uncertainties in future Northern Annular Mode trend. *Geophys. Res. Lett.*, **37**, doi: 10.1029/2010gl044717.

Karpechko, A. Y., and E. Manzini, 2012: Stratospheric influence on tropospheric climate change in the Northern Hemisphere. *J. Geophys. Res.*, **117**, doi: 10.1029/2011JD017036.

Karpechko, A. Y., N. P. Gillett, L. J. Gray, and M. Dall'Amico, 2010: Influence of ozone recovery and greenhouse gas increases on Southern Hemisphere circulation. *J. Geophys. Res.*, **115**, D22117.

Kaspari, S., P. A. Mayewski, D. A. Dixon, V. B. Spikes, S. B. Sneed, M. J. Handley, and G. S. Hamilton, 2004: Climate variability in West Antarctica derived from annual accumulation-rate records from ITASE firn/ice cores. *Annals of Glaciology*, **39**, 585–594.

Kattsov, V. M., J. E. Walsh, W. L. Chapman, V. A. Govorkova, T. V. Pavlova, and X. D. Zhang, 2007: Simulation and projection of arctic freshwater budget components by the IPCC AR4 global climate models. *J. Hydrometeorol.*, **8**, 571–589.

Kaufman, D. S., et al., 2009: Recent warming reverses long-term Arctic cooling. *Science*, **325**, 1236–1239.

Kawatani, Y., K. Hamilton, and S. Watanabe, 2011: The Quasi-Biennial Oscillation in a double CO_2 climate. *J. Atmos. Sci.*, **68**, 265–283.

Kawatani, Y., K. Hamilton, and A. Noda, 2012: The effects of changes in sea surface temperature and CO_2 concentration on the Quasi-Biennial Oscillation. *J. Atmos. Sci.*, **69**, 1734–1749.

Kawazoe, S., and W. Gutowski, 2013: Regional, very heavy daily precipitation in NARCCAP simulations. *J. Hydrometeorol.*, doi:10.1175/jhm-d-12-068.1.

Keenlyside, N., and M. Latif, 2007: Understanding equatorial Atlantic interannual variability. *J. Clim.*, **20**, 131–142.

Keenlyside, N., M. Latif, J. Jungclaus, L. Kornblueh, and E. Roeckner, 2008: Advancing decadal-scale climate prediction in the North Atlantic sector. *Nature*, **453**, 84–88.

Khain, A., B. Lynn, and J. Dudhia, 2010: Aerosol effects on intensity of landfalling hurricanes as seen from simulations with the WRF model with spectral bin microphysics. *J. Atmos. Sci.*, **67**, 365–384.

Khain, A., N. Cohen, B. Lynn, and A. Pokrovsky, 2008: Possible aerosol effects on lightning activity and structure of hurricanes. *J. Atmos. Sci.*, **65**, 3652–3677.

Kharin, V. V., F. W. Zwiers, X. Zhang, and G. C. Hegerl, 2007: Changes in temperature and precipitation extremes in the IPCC ensemble of global coupled model simulations. *J. Clim.*, **20**, 1419–1444.

Kidson, J. W., and J. A. Renwick, 2002: Patterns of convection in the tropical Pacific and their influence on New Zealand weather. *Int. J. Climatol.*, **22**, 151–174.

Kidston, J., and E. P. Gerber, 2010: Intermodel variability of the poleward shift of the austral jet stream in the CMIP3 integrations linked to biases in 20th century climatology. *Geophys. Res. Lett.*, **37**.

Kidston, J., J. A. Renwick, and J. McGregor, 2009: Hemispheric-scale seasonality of the Southern Annular Mode and impacts on the climate of New Zealand. *J. Clim.*, **22**, 4759–4770.

Kidston, J., G. K. Vallis, S. M. Dean, and J. A. Renwick, 2011: Can the increase in the eddy length scale under global warming cause the poleward shift of the jet streams? *J. Clim.*, **24**, 3764–3780.

Kim, B. M., and S. I. An, 2011: Understanding ENSO regime behavior upon an Increase in the warm-pool temperature using a simple ENSO model. *J. Clim.*, **24**, 1438–1450.

Kim, D., and H. Byun, 2009: Future pattern of Asian drought under global warming scenario. *Theor. Appl. Climatol.*, **98**, 137–150.

Kim, S. T., and J.-Y. Yu, 2012: The two types of ENSO in CMIP5 models. *Geophys. Res. Lett.*, doi:10.1029/2012GL052006.

Kitoh, A., and T. Uchiyama, 2006: Changes in onset and withdrawal of the East Asian summer rainy season by multi-model global warming experiments. *J. Meteorol. Soc. Jpn.*, **84**, 247–258.

Kitoh, A., and S. Kusunoki, 2008: East Asian summer monsoon simulation by a 20-km mesh AGCM. *Clim. Dyn.*, **31**, 389–401.

Kitoh, A., S. Kusunoki, and T. Nakaegawa, 2011: Climate change projections over South America in the late 21st century with the 20 and 60 km mesh Meteorological Research Institute atmospheric general circulation model (MRI-AGCM). *J. Geophys. Res. Atmos.*, **116**, D06105.

Kitoh, A., T. Ose, K. Kurihara, S. Kusunoki, M. Sugi, and KAKUSHIN Team-3 Modeling Group, 2009: Projection of changes in future weather extremes using super-high-resolution global and regional atmospheric models in the KAKUSHIN Program: Results of preliminary experiments. *Hydrol. Res. Lett.*, **3**, 49–53.

Kitoh, A., H. Endo, K. Krishna Kumar, I. F. A. Cavalcanti, P. Goswami, and T. Zhou, 2013: Monsoons in a changing world regional perspective in a global context. *J. Geophys. Res. Atmos.*, **118**, doi: 10.1002/jgrd.50258.

Kjellstrom, E., G. Nikulin, U. Hansson, G. Strandberg, and A. Ullerstig, 2011: 21st century changes in the European climate: Uncertainties derived from an ensemble of regional climate model simulations. *Tellus A*, **63**, 24–40.

Kjellström, E., P. Thejll, M. Rummukainen, J. H. Christensen, F. Boberg, C. O. B, and C. Fox Maule, 2013: Emerging regional climate change signals for Europe under varying large-scale circulation conditions. *Clim. Res.*, **56**, 103–119.

Klein, S. A., B. J. Soden, and N.-C. Lau, 1999: Remote sea surface temperature variations during ENSO: Evidence for a tropical atmospheric bridge. *J. Clim.*, **12**, 917–932.

Klingaman, N. P., S. J. Woolnough, H. Weller, and J. M. Slingo, 2011: The impact of finer-resolution air-sea coupling on the Intraseasonal Oscillation of the Indian monsoon. *J. Clim.*, **24**, 2451–2468.

Knight, J., 2009: The Atlantic Multidecadal Oscillation inferred from the forced climate response in coupled general ciculation models. *J. Clim.*, **22**, 1610–1625.

Knight, J. R., R. J. Allan, C. K. Folland, M. Vellinga, and M. E. Mann, 2005: A signature of persistent natural thermohaline circulation cycles in observed climate. *Geophys. Res. Lett.*, **32**, L20708.

Knutson, T. R., and R. E. Tuleya, 2004: Impact of CO_2-induced warming on simulated hurricane intensity and precipitation: Sensitivity to the choice of climate model and convective parameterization. *J. Clim.*, **17**, 3477–3495.

Knutson, T. R., et al., 2006: Assessment of twentieth-century regional surface temperature trends using the GFDL CM2 coupled models. *J. Clim.*, **19**, 1624–1651.

Knutson, T. R., et al., 2010: Tropical cyclones and climate change. *Nature Geosci.*, **3**, 157–163.

Knutson, T. R., et al., 2013: Dynamical downscaling projections of 21st century Atlantic hurricane activity: CMIP3 and CMIP5 model-based scenarios. *J. Clim.*, **26**, 6591–6617.

Kodama, C., and T. Iwasaki, 2009: Influence of the SST rise on baroclinic instability wave activity under an aquaplanet condition. *J. Atmos. Sci.*, **66**, 2272–2287.

Kodera, K., M. E. Hori, S. Yukimoto, and M. Sigmond, 2008: Solar modulation of the Northern Hemisphere winter trends and its implications with increasing CO2. *Geophys. Res. Lett.*, **35**, doi: 10.1029/2007gl031958.

Koenigk, T., R. Döscher, and G. Nikulin, 2011: Arctic future scenario experiments with a coupled regional climate model. *Tellus A*, **63**, 69–86.

Kohler, M., N. Kalthoff, and C. Kottmeier, 2010: The impact of soil moisture modifications on CBL characteristics in West Africa: A case-study from the AMMA campaign. *Q. J. R. Meteorol. Soc.*, **136**, 442–455.

Koldunov, N. V., D. Stammer, and J. Marotzke, 2010: Present-day Arctic sea ice variability in the Coupled ECHAM5/MPI-OM Model. *J. Clim.*, **23**, 2520–2543.

Kripalani, R., J. Oh, and H. Chaudhari, 2007a: Response of the East Asian summer monsoon to doubled atmospheric CO2: Coupled climate model simulations and projections under IPCC AR4. *Theor. Appl. Climatol.*, **87**, 1–28.

Kripalani, R. H., J. H. Oh, A. Kulkarni, S. S. Sabade, and H. S. Chaudhari, 2007b: South Asian summer monsoon precipitation variability: Coupled climate model simulations and projections under IPCC AR4. *Theor. Appl. Climatol.*, **90**, 133–159.

Krishnamurthy, C. K. B., U. Lall, and H. H. Kwon, 2009: Changing frequency and intensity of rainfall extremes over India from 1951 to 2003. *J. Clim.*, **22**, 4737–4746.

Krishnamurthy, V., and R. S. Ajayamohan, 2010: Composite structure of monsoon low pressure systems and its relation to Indian rainfall. *J. Clim.*, **23**, 4285–4305.

Kruger, L. F., R. P. da Rocha, M. S. Reboita, and T. Ambrizzi, 2011: RegCM3 nested in the HadAM3 scenarios A2 and B2: projected changes in cyclogeneses, temperature and precipitation over South Atlantic Ocean. *Clim. Change*, **113**, 599–621.

Kucharski, F., A. Bracco, J. Yoo, A. Tompkins, L. Feudale, P. Ruti, and A. Dell'Aquila, 2009a: A Gill-Matsuno-type mechanism explains the tropical Atlantic influence on African and Indian monsoon rainfall. *Q. J. R. Meteorol. Soc.*, **135**, 569–579.

Kucharski, F., et al., 2009b: The CLIVAR C20C project: Skill of simulating Indian monsoon rainfall on interannual to decadal timescales. Does GHG forcing play a role? *Clim. Dyn.*, **33**, 615–627.

Kug, J.-S., and I.-S. Kang, 2006: Interactive Feedback between ENSO and the Indian Ocean. *J. Clim.*, **19**, 1784–1801.

Kug, J.-S., F.-F. Jin, and S.-I. An, 2009: Two types of El Nino events: Cold tongue El Niño and warm pool El Niño. *J. Clim.*, **22**, 1499–1515.

Kug, J. S., S. I. An, Y. G. Ham, and I. S. Kang, 2010: Changes in El Niño and La Niña teleconnections over North Pacific-America in the global warming simulations. *Theor. Appl. Climatol.*, **100**, 275–282.

Kulkarni, A., 2012: Weakening of Indian summer monsoon rainfall in warming environment. *Theor. Appl. Climatol.*, doi:10.1007/s00704-012-0591-4.

Kumar, K., S. Patwardhan, A. Kulkarni, K. Kamala, K. Rao, and R. Jones, 2011a: Simulated projections for summer monsoon climate over India by a high-resolution regional climate model (PRECIS). *Curr. Sci.*, **101**, 312–326.

Kumar, K., et al., 2011b: The once and future pulse of Indian monsoonal climate. *Clim. Dyn.*, **36**, 2159–2170.

Kumar, K. K., B. Rajagopalan, M. Hoerling, G. Bates, and M. Cane, 2006a: Unraveling the mystery of Indian Monsoon failure during El Niño. *Science*, **314**, 115–119.

Kumar, P., et al., 2013: Downscaled climate change projections with uncertainty assessment over India using a high resolution multi-model approach. *Sci. Total Environ.*, doi:10.1016/j.scitotenv.2013.01.051.

Kumar, V., R. Deo, and V. Ramachandran, 2006b: Total rain accumulation and rain-rate analysis for small tropical Pacific islands: A case study of Suva, Fiji. *Atmos. Sci. Lett.*, **7**, 53–58.

Kusunoki, S., and R. Mizuta, 2008: Future changes in the Baiu rain band projected by a 20-km mesh global atmospheric model: Sea surface temperature dependence. *Sola*, **4**, 85–88.

Kusunoki, S., and O. Arakawa, 2012: Change in the precipitation intensity of the East Asian summer monsoon projected by CMIP3 models. *Clim. Dyn.*, **38**, 2055–2072.

Kuzmina, S. I., L. Bengtsson, O. M. Johannessen, H. Drange, L. P. Bobylev, and M. W. Miles, 2005: The North Atlantic Oscillation and greenhouse-gas forcing. *Geophys. Res. Lett.*, **32**, doi: 10.1029/2004gl021064.

Kvamsto, N., P. Skeie, and D. Stephenson, 2004: Impact of labrador sea-ice extent on the North Atlantic oscillation. *Int. J. Climatol.*, **24**, 603–612.

Kwok, R., and J. C. Comiso, 2002: Southern ocean climate and sea ice anomalies associated with the Southern Oscillation. *J. Clim.*, **15**, 487–501.

L'Heureux, M. L., and D. W. J. Thompson, 2006: Observed relationships between the El Niño–Southern Oscillation and the extratropical zonal-mean circulation. *J. Clim.*, **19**, 276–287.

L'Heureux, M. L., and R. W. Higgins, 2008: Boreal winter links between the Madden-Julian oscillation and the Arctic oscillation. *J. Clim.*, **21**, 3040–3050.

Laine, A., M. Kageyama, D. Salas-Melia, G. Ramstein, S. Planton, S. Denvil, and S. Tyteca, 2009: An energetics study of wintertime Northern Hemisphere storm tracks under 4 × CO(2) conditions in two ocean-atmosphere coupled models. *J. Clim.*, **22**, 819–839.

Lam, H., M. H. Kok, and K. K. Y. Shum, 2012: Benefits from typhoons—the Hong Kong perspective. *Weather*, **67**, 16–21.

Lambert, S. J., and J. C. Fyfe, 2006: Changes in winter cyclone frequencies and strengths simulated in enhanced greenhouse warming experiments: Results from the models participating in the IPCC diagnostic exercise. *Clim. Dyn.*, **26**, 713–728.

Landsea, C. W., R. A. Pielke, A. Mestas-Nunez, and J. A. Knaff, 1999: Atlantic basin hurricanes: Indices of climatic changes. *Clim. Change*, **42**, 89–129.

Lang, C., and D. W. Waugh, 2011: Impact of climate change on the frequency of Northern Hemisphere summer cyclones. *J. Geophys. Res. Atmos.*, **116**, D04103.

Langenbrunner, B., and J. D. Neelin, 2013: Analyzing ENSO teleconnections in CMIP models as a measure of model fidelity in simulating precipitation. *J. Clim.*, doi:10.1175/jcli-d-12-00542.1.

Lapp, S. L., J. M. St. Jacques, E. M. Barrow, and D. J. Sauchyn, 2012: GCM projections for the Pacifi Decadal Oscillation under greenhouse forcing for the early 21st century. *International Journal of Climatology*, **32**, 1423–1442.

Lau, K., S. Shen, K. Kim, and H. Wang, 2006: A multimodel study of the twentieth-century simulations of Sahel drought from the 1970s to 1990s. *J. Geophys. Res. Atmos.*, **111**.

Lau, K., et al., 2008: The Joint Aerosol-Monsoon Experiment—A new challenge for monsoon climate research. *Bull. Am. Meteorol. Soc.*, doi:10.1175/BAMS-89-3-369, 369–383.

Lau, K. M., and H. T. Wu, 2007: Detecting trends in tropical rainfall characteristics, 1979–2003. *Int. J. Climatol.*, **27**, 979–988.

Lau, N.-C., and M. J. Nath, 2012: A model study of heat waves over North America: Meteorological aspects and projections for the 21st Century. *J. Clim.*, **25**, 4761–4784.

Lavender, S., and K. Walsh, 2011: Dynamically downscaled simulations of Australian region tropical cyclones in current and future climates. *Geophys. Res. Lett.*, **38**, doi: 10.1029/2011GL047499.

Leckebusch, G. C., U. Ulbrich, L. Froehlich, and J. G. Pinto, 2007: Property loss potentials for European midlatitude storms in a changing climate. *Geophys. Res. Lett.*, **34**, doi: 10.1029/2006gl027663.

Leckebusch, G. C., B. Koffi, U. Ulbrich, J. G. Pinto, T. Spangehl, and S. Zacharias, 2006: Analysis of frequency and intensity of European winter storm events from a multi-model perspective, at synoptic and regional scales. *Clim. Res.*, **31**, 59–74.

Lee, J. N., S. Hameed, and D. T. Shindell, 2008: The northern annular mode in summer and its relation to solar activity variations in the GISS ModelE. *J. Atmos. Sol. Terres. Phys.*, **70**, 730–741.

Lee, T.-C., K.-Y. Chan, H.-S. Chan, and M.-H. Kok, 2011: Projections of extreme rainfall in Hong Kong in the 21st century. *Acta Meteorol. Sin.*, **25**, 691–709.

Lee, T.-C., T. R. Knutson, H. Kamahori, and M. Ying, 2012: Impacts of climate change on tropical cyclones in the western North Pacific basin. Part I: Past observations. *Trop. Cyclone Res. Rev.*, **1**, 213–230.

Lelieveld, J., et al., 2012: Climate change and impacts in the Eastern Mediterranean and the Middle East. *Clim. Change*, **114**, 667–687.

Leslie, L., D. Karoly, M. Leplastrier, and B. Buckley, 2007: Variability of tropical cyclones over the southwest Pacific Ocean using a high-resolution climate model. *Meteorol. Atmos. Phys.*, **97**, 171–180.

Leung, L. R., and Y. Qian, 2009: Atmospheric rivers induced heavy precipitation and flooding in the western U.S. simulated by the WRF regional climate model. *Geophys. Res. Lett.*, **36**, L03820.

Levine, R. C., and A. G. Turner, 2012: Dependence of Indian monsoon rainfall on moisture fluxes across the Arabian Sea and the impact of coupled model sea surface temperature biases. *Clim. Dyn.*, **38,** 2167-2190.

Levitus, S., G. Matishov, D. Seidov, and I. Smolyar, 2009: Barents Sea multidecadal variability. *Geophys. Res. Lett.*, **36**, L19604.

Li, B., and T. J. Zhou, 2011: El Nino-Southern Oscillation-related principal interannual variability modes of early and late summer rainfall over East Asia in sea surface temperature-driven atmospheric general circulation model simulations. *J. Geophys. Res. Atmos.*, **116**, 15.

Li, G., and B. Ren, 2012: Evidence for strengthening of the tropical Pacific ocean surface wind speed during 1979–2001. *Theor. Appl. Climatol.*, doi:10.1007/s00704-0110-463-3.

Li, H., A. Dai, T. Zhou, and J. Lu, 2010a: Responses of East Asian summer monsoon to historical SST and atmospheric forcing during 1950–2000. *Clim. Dyn.*, **34**, 501–514.

Li, H. M., L. Feng, and T. J. Zhou, 2011a: Multi-model projection of July-August climate extreme changes over China under CO_2 doubling. Part II: Temperature. *Adv. Atmos. Sci.*, **28**, 448–463.

Li, H. M., L. Feng, and T. J. Zhou, 2011b: Multi-model projection of July-August climate extreme changes over China under CO_2 doubling. Part I: precipitation. *Adv. Atmos. Sci.*, **28**, 433–447.

Li, J., and J. Wang, 2003: A modified zonal index and its physical sense. *Geophys. Res. Lett.*, **30**, doi: 10.1029/2003GL017441.

Li, J., J. Feng, and Y. Li, 2012a: A possible cause of decreasing summer rainfall in northeast Australia. *Int. J. Climatol.*, **32**, 995–1005.

Li, J. B., et al., 2011c: Interdecadal modulation of El Nino amplitude during the past millennium. *Nature Clim. Change*, **1**, 114–118.

Li, T., P. Liu, X. Fu, B. Wang, and G. Meehl, 2006: Spatiotemporal structures and mechanisms of the tropospheric biennial oscillation in the Indo-Pacific warm ocean regions. *J. Clim.*, **19**, 3070–3087.

Li, T., M. Kwon, M. Zhao, J. Kug, J. Luo, and W. Yu, 2010b: Global warming shifts Pacific tropical cyclone location. *Geophys. Res. Lett.*, **37**, doi: 10.1029/2010GL045124.

Li, Y., and N. Lau, 2012: Impact of ENSO on the atmospheric variability over the north Atlantic in late winter—Role of transient eddies. *J. Clim.*, **25**, 320–342.

Li, Y., N. C. Jourdain, A. S. Taschetto, C. C. Ummenhofer, K. Ashok, and A. Sen Gupta, 2012b: Evaluation of monsoon seasonality and the tropospheric biennial oscillation transitions in the CMIP models. *Geophys. Res. Lett.*, **39**, L20713.

Liang, X.-Z., K. E. Kunkel, G. A. Meehl, R. G. Jones, and J. X. L. Wang, 2008a: Regional climate models downscaling analysis of general circulation models present climate biases propagation into future change projections. *Geophys. Res. Lett.*, **35**, L08709.

Liang, X.-Z., J. Zhu, K. E. Kunkel, M. Ting, and J. X. L. Wang, 2008b: Do GCMs simulate the North American monsoon precipitation seasonal-interannual variability? *J. Clim.*, **21**, 4424–4448.

Lienert, F., J. C. Fyfe, and W. J. Marryfield, 2011: Do climate models capture the tropical influences on North Pacific sea surface temperature variability? *J. Clim.*, **24**, 6203–6209.

Lim, E.-P., and I. Simmonds, 2009: Effect of tropospheric temperature change on the zonal mean circulation and SH winter extratropical cyclones. *Clim. Dyn.*, **33**, 19–32.

Lim, Y.-K., L. B. Stefanova, S. C. Chan, S. D. Schubert, and J. J. O'Brien, 2011: High-resolution subtropical summer precipitation derived from dynamical downscaling of the NCEP/DOE reanalysis:how much small-scale information is added by a regional model? *Clim. Dyn.*, **37**, 1061–1080.

Lima, K., P. Satyamurty, and J. Fernández, 2010: Large-scale atmospheric conditions associated with heavy rainfall episodes in Southeast Brazil. *Theor. Appl. Climatol.*, **101**, 121–135.

Lin, H., G. Brunet, and J. Derome, 2009: An observed connection between the North Atlantic Oscillation and the Madden-Julian Oscillation. *J. Clim.*, **22**, 364–380.

Lin, J. L., et al., 2008a: Subseasonal variability associated with Asian summer monsoon simulated by 14 IPCC AR4 coupled GCMs. *J. Clim.*, **21**, 4541–4567.

Lin, J. L., et al., 2008b: North American monsoon and convectively coupled equatorial waves simulated by IPCC AR4 coupled GCMs. *J. Clim.*, **21**, 2919–2937.

Linkin, M., and S. Nigam, 2008: The North Pacific Oscillation-West Pacific teleconnection pattern: Mature-phase structure and winter impacts. *J. Clim.*, **21**, 1979–1997.

Lintner, B., and J. Neelin, 2008: Eastern margin variability of the South Pacific Convergence Zone. *Geophys. Res. Lett.*, **35**, doi: 10.1029/2008gl034298.

Lionello, P., S. Planton, and X. Rodo, 2008: Preface: Trends and climate change in the Mediterranean region. *Global Planet. Change*, **63**, 87–89.

Liu, H. W., T. J. Zhou, Y. X. Zhu, and Y. H. Lin, 2012a: The strengthening East Asia summer monsoon since the early 1990s. *Chinese Science Bulletin*, **57**, 1553–1558.

Liu, J., J. A. Curry, H. Wang, M. Song, and R. M. Horton, 2012b: Impact of declining Arctic sea ice on winter snowfall. *Proc. Natl. Acad. Sci. U.S.A.*, **109**, 4074–4079.

Liu, J. P., and J. A. Curry, 2006: Variability of the tropical and subtropical ocean surface latent heat flux during 1989–2000. *Geophys. Res. Lett.*, **33**, doi: 10.1029/2005gl024809.

Liu, Y., S.-K. Lee, B. A. Muhling, J. T. Lamkin, and D. B. Enfield, 2012c: Significant reduction of the Loop Current in the 21st century and its impact on the Gulf of Mexico. *J. Geophys. Res.*, **117**, C05039.

Liu, Z., and B. Huang, 2000: Cause of tropical Pacific warming trend. *Geophys. Res. Lett.*, **27**, 1935–1938.

Liu, Z., S. Vavrus, F. He, N. Wen, and Y. Zhong, 2005: Rethinking tropical ocean response to global warming: The enhanced equatorial warming. *J. Clim.*, **18**, 4684–4700.

Lockwood, M., R. G. Harrison, T. Woollings, and S. K. Solanki, 2010: Are cold winters in Europe associated with low solar activity? *Environ. Res. Lett.*, **5, doi:** 10.1088/1748-9326/5/2/024001.

Loeptien, U., O. Zolina, S. Gulev, M. Latif, and V. Soloviov, 2008: Cyclone life cycle characteristics over the Northern Hemisphere in coupled GCMs. *Clim. Dyn.*, **31**, 507–532.

Long, Z., W. Perrie, J. Gyakum, R. Laprise, and D. Caya, 2009: Scenario changes in the climatology of winter midlatitude cyclone activity over eastern North America and the Northwest Atlantic. *J. Geophys. Res. Atmos.*, **114**, doi: 10.1029/2008jd010869.

Lorenz, D. J., and D. L. Hartmann, 2003: Eddy-zonal flow feedback in the Northern Hemisphere winter. *J. Clim.*, **16**, 1212–1227.

Lorenz, D. J., and E. T. DeWeaver, 2007: Tropopause height and zonal wind response to global warming in the IPCC scenario integrations. *J. Geophys. Res. Atmos.*, **112**, doi: 10.1029/2006jd008087.

Lu, J., G. A. Vecchi, and T. Reichler, 2007: Expansion of the Hadley cell under global warming. *Geophys. Res. Lett.*, **34**, doi: 10.1029/2006gl028443.

Lu, J., G. Chen, and D. M. W. Frierson, 2008: Response of the zonal mean atmospheric circulation to El Nino versus global warming. *J. Clim.*, **21**, 5835–5851.

Lu, J., G. Chen, and D. M. W. Frierson, 2010: The position of the mid latitude storm track and eddy-driven westerlies in Aquaplanet AGCMs. *J. Atmos. Sci.*, **67**, 3984–4000.

Lu, R., and Y. Fu, 2010: Intensification of East Asian summer rainfall interannual variability in the twenty-first century simulated by 12 CMIP3 coupled models. *J. Clim.*, doi:10.1175/2009JCLI3130.1, 3316–3331.

Lucarini, V., and F. Ragone, 2011: Energetics of climate models: Net energy balance and meridional enthalpy transport. *Rev. Geophys.*, **49**, doi: 10.1029/2009RG000323.

Lucas, C., H. Nguyen, and B. Timbal, 2012: An observational analysis of southern hemisphere tropical expansion. *J. Geophys. Res.*, **117**, doi: 10.1029/2011JD017033.

Luo, D., W. Zhou, and K. Wei, 2010: Dynamics of eddy-driven North Atlantic Oscillations in a localized shifting jet: Zonal structure and downstream blocking. *Clim. Dyn.*, **34**, 73–100.
Luo, Y., and L. M. Rothstein, 2011: Response of the Pacific ocean circulation to climate change. *Atmosphere-ocean*, **49**, 235–244.
Lupo, A. R., I. I. Mokhov, M. G. Akperov, A. V. Chernokulsky, and H. Athar, 2012: A dynamic analysis of the role of the planetary- and synoptic-scale in the summer of 2010 blocking episodes over the European part of Russia. *Adv. Meteorol.*, **2012**, 11.
Ma, J., and S.-P. Xie, 2013: Regional patterns of sea surface temperature change: A source of uncertainty in future projections of precipitation and atmospheric circulation. *J. Clim.*, **26**, 2482–2501.
Magnusdottir, G., C. Deser, and R. Saravanan, 2004: The effects of North Atlantic SST and sea ice anomalies on the winter circulation in CCM3. Part I: Main features and storm track characteristics of the response. *J. Clim.*, **17**, 857–876.
Mahajan, S., R. Zhang, and T. L. Delworth, 2011: Impact of the Atlantic meridional overturning circulation (AMOC) on Arctic surface air temperature and sea ice variability. *J. Clim.*, **24**, 6573–6581.
Mahowald, N., 2007: Anthropocene changes in desert area: Sensitivity to climate model predictions. *Geophys. Res. Lett.*, **34**, doi: 10.1029/2007GL030472.
Malhi, Y., J. T. Roberts, R. A. Betts, T. J. Killeen, W. Li, and C. A. Nobre, 2008: Climate change, deforestation, and the fate of the Amazon. *Science*, **319**, 169–172.
Maloney, E. D., and J. Shaman, 2008: Intraseasonal variability of the West African monsoon and Atlantic ITCZ. *J. Clim.*, **21**, 2898–2918.
Maloney, E. D., and S.-P. Xie, 2013: Sensitivity of MJO activity to the pattern of climate warming. *J. Adv. Model. Earth Syst.*, **5**, 32–47.
Manatsa, D., W. Chingombe, H. Matsikwa, and C. H. Matarira, 2008: The superior influence of Darwin Sea level pressure anomalies over ENSO as a simple drought predictor for Southern Africa. *Theor. Appl. Climatol.*, **92**, 1–14.
Mandke, S. K., A. K. Sahai, M. A. Shinde, S. Joseph, and R. Chattopadhyay, 2007: Simulated changes in active/break spells during the Indian summer monsoon due to enhanced CO_2 concentrations: Assessment from selected coupled atmosphere-ocean global climate models. *Int. J. Climatol.*, **27**, 837–859.
Mann, M. E., and K. A. Emanuel, 2006: Atlantic hurricane trends linked to climate change. *Eos Trans.*, **87**, 233–241.
Manton, M. J., et al., 2001: Trends in extreme daily rainfall and temperature in Southeast Asia and the South Pacific: 1961–1998. *Int. J. Climatol.*, **21**, 269–284.
Mantua, N. J., S. R. Hare, Y. Zhang, J. M. Wallace, and R. C. Francis, 1997: A Pacific interdecadal climate oscillation with impacts on salmon production. *Bull. Am. Meteorol. Soc.*, **78**, 1069–1079.
Marcella, M. P., and E. A. B. Eltahir, 2011: Modeling the summertime climate of Southwest Asia: The role of land surface processes in shaping the climate of semiarid regions. *J. Clim.*, **25**, 704–719.
Marchant, R., C. Mumbi, S. Behera, and T. Yamagata, 2007: The Indian Ocean dipole—the unsung driver of climatic variability in East Africa. *Afr. J. Ecol.*, **45**, 4–16.
Marengo, J., et al., 2010a: Recent developments on the South American Monsoon system. *Int. J. Climatol.*, **32**, 1–21.
Marengo, J., et al., 2012: Development of regional future climate change scenarios in South America using the Eta CPTEC/HadCM3 climate change projections: Climatology and regional analyses for the Amazon, São Francisco and the Paraná River basins. *Clim. Dyn.*, **38**, 1829–1848.
Marengo, J. A., and C. C. Camargo, 2008: Surface air temperature trends in Southern Brazil for 1960–2002. *Int. J. Climatol.*, **28**, 893–904.
Marengo, J. A., R. Jones, L. M. Alves, and M. C. Valverde, 2009: Future change of temperature and precipitation extremes in South America as derived from the PRECIS regional climate modeling system. *Int. J. Climatol.*, **29**, 2241–2255.
Marengo, J. A., M. Rusticucci, O. Penalba, and M. Renom, 2010b: An intercomparison of observed and simulated extreme rainfall and temperature events during the last half of the twentieth century: Part 2: Historical trends. *Clim. Change*, **98**, 509–529.
Mariotti, A., and A. Dell'Aquila, 2012: Decadal climate variability in the Mediterranean region: Roles of large-scale forcings and regional processes. *Clim. Dyn.*, **38**, 1129–1145.
Marshall, A. G., and A. A. Scaife, 2009: Impact of the QBO on surface winter climate. *J. Geophys. Res.*, **114**, doi: 10.1029/ 2009jd011737.
Marshall, A. G., and A. A. Scaife, 2010: Improved predictability of stratospheric sudden warming events in an atmospheric general circulation model with enhanced stratospheric resolution. *J. Geophys. Res. Atmos.*, **115**, doi: 10.1029/2009jd012643.
Marshall, G. J., 2007: Half-century seasonal relationships between the Southern Annular Mode and Antarctic temperatures. *Int. J. Climatol.*, **27**, 373–383.
Martius, O., L. M. Polvani, and H. C. Davies, 2009: Blocking precursors to stratospheric sudden warming events. *Geophys. Res. Lett.*, **36**, L14806.
Marullo, S., V. Artale, and R. Santoleri, 2011: The SST multidecadal variability in the Atlantic–Mediterranean region and its relation to AMO. *J. Clim.*, **24**, 4385–4401.
Masato, G., B. J. Hoskins, and T. J. Woollings, 2012: Wave-breaking characteristics of midlatitude blocking. *Q. J. R. Meteorol. Soc.*, **138**, 1285–1296.
Masato, G., B. J. Hoskins, and T. Woollings, 2013: Winter and summer Northern Hemisphere blocking in CMIP5 models. *J. Clim.*, doi:10.1175/jcli-d-12-00466.1.
Mason, S., 2001: El Nino, climate change, and Southern African climate. *Environmetrics*, **12**, 327–345.
Massom, R. A., M. J. Pook, J. C. Comiso, N. Adams, J. Turner, T. Lachlan-Cope, and T. T. Gibson, 2004: Precipitation over the interior East Antarctic ice sheet related to midlatitude blocking-high activity. *J. Clim.*, **17**, 1914–1928.
Matsueda, M., 2011: Predictability of Euro-Russian blocking in summer of 2010. *Geophys. Res. Lett.*, **38**, L06801.
Matsueda, M., H. Endo, and R. Mizuta, 2010: Future change in Southern Hemisphere summertime and wintertime atmospheric blockings simulated using a 20-km-mesh AGCM. *Geophys. Res. Lett.*, **37**, L02803.
May, W., 2011: The sensitivity of the Indian summer monsoon to a global warming of 2 degrees C with respect to pre-industrial times. *Clim. Dyn.*, **37**, 1843–1868.
McCabe, G., and D. Wolock, 2010: Long-term variability in Northern Hemisphere snow cover and associations with warmer winters. *Clim. Change*, doi:10.1007/s10584-009-9675-2, 141–153.
McDonald, R. E., 2011: Understanding the impact of climate change on Northern Hemisphere extra-tropical cyclones. *Clim. Dyn.*, **37**, 1399–1425.
McLandress, C., and T. G. Shepherd, 2009: Simulated anthropogenic changes in the Brewer–Dobson circulation, including its extension to high latitudes. *J. Clim.*, **22**, 1516–1540.
Mearns, L. O., R. Arritt, S. Biner, M. Bukovsky, S. Stain, and et al., 2012: The North American regional climate change assessment program: Overview of phase I results. *Bull. Am. Meteorol. Soc.*, **93**, 1337–1362.
Meehl, G., and H. Teng, 2007: Multi-model changes in El Nino teleconnections over North America in a future warmer climate. *Clim. Dyn.*, **29**, 779–790.
Meehl, G., J. Arblaster, and W. Collins, 2008: Effects of Black Carbon Aerosols on the Indian Monsoon. *J. Clim.*, **21**, 2869–2882.
Meehl, G., A. Hu, and C. Tebaldi, 2010: Decadal Prediction in the Pacific Region. *J. Clim.*, **23**, 2959–2973.
Meehl, G. A., 1997: The south Asian monsoon and the tropospheric biennial oscillation. *J. Clim.*, **10**, 1921–1943.
Meehl, G. A., and A. Hu, 2006: Megadroughts in the Indian monsoon region and southwest North America and a mechanism for associated multidecadal Pacific sea surface temperature anomalies. *J. Clim.*, **19**, 1605–1623.
Meehl, G. A., and J. M. Arblaster, 2012: Relating the strength of the tropospheric biennial oscillation (TBO) to the phase of the Interdecadal Pacific Oscillation (IPO). *Geophys. Res. Lett.*, **39**, L20716.
Meehl, G. A., et al., 2007: Global climate projections. In: *Climate Change 2007: The Physical Science Basis. Contribution of Working Group I to the Fourth Assessment Report of the Intergovernmental Panel on Climate Change* [Solomon, S., D. Qin, M. Manning, Z. Chen, M. Marquis, K. B. Averyt, M. Tignor and H. L. Miller (eds.)] Cambridge University Press, Cambridge, United Kingdom and New York, NY, USA, pp. 747–846.
Menary, M., W. Park, K. Lohmann, M. Vellinga, M. Palmer, M. Latif, and J. Jungclaus, 2012: A multimodel comparison of centennial Atlantic meridional overturning circulation variability. *Clim. Dyn.*, **38**, 2377–2388.
Mendes, M. C. D., R. M. Trigo, I. F. A. Cavalcanti, and C. C. Da Camara, 2008: Blocking episodes in the Southern Hemisphere: Impact on the climate of adjacent continental areas. *Pure Appl. Geophys.*, **165**, 1941–1962.
Mendez, M., and V. Magana, 2010: Regional aspects of prolonged meteorological droughts over Mexico and Central America. *J. Clim.*, **23**, 1175–1188.
Mendoza, B., V. Garcia-Acosta, V. Velasco, E. Jauregui, and R. Diaz-Sandoval, 2007: Frequency and duration of historical droughts from the 16th to the 19th centuries in the Mexican Maya lands, Yucatan Peninsula. *Clim. Change*, **83**, 151–168.

Meneghini, B., I. Simmonds, and I. N. Smith, 2007: Association between Australian rainfall and the Southern Annular Mode. *Int. J. Climatol.*, **27**, 109–121.

Menendez, C. G., and A. Carril, 2010: Potential changes in extremes and links with the Southern Annular Mode as simulated by a multi-model ensemble. *Clim. Change*, **98**, 359–377.

Metcalfe, S. E., M. D. Jones, S. J. Davies, A. Noren, and A. MacKenzie, 2010: Climate variability over the last two millennia in the North American Monsoon, recorded in laminated lake sediments from Laguna de Juanacatlan, Mexico. *Holocene*, **20**, 1195–1206.

Miller, G. H., et al., 2010: Temperature and precipitation history of the Arctic. *Q. Sci. Rev.*, **29**, 1679–1715.

Miller, R. L., G. A. Schmidt, and D. T. Shindell, 2006: Forced annular variations in the 20th century intergovernmental panel on climate change fourth assessment report models. *J. Geophys. Res. Atmos.*, **111**, doi: 10.1029/2005jd006323.

Minvielle, M., and R. D. Garreaud, 2011: Projecting rainfall changes over the South American altiplano. *J. Clim.*, **24**, 4577–4583.

Mitas, C. M., and A. Clement, 2005: Has the Hadley cell been strengthening in recent decades? *Geophys. Res. Lett.*, **32**, doi: 10.1029/2004gl021765.

Mitas, C. M., and A. Clement, 2006: Recent behavior of the Hadley cell and tropical thermodynamics in climate models and reanalyses. *Geophys. Res. Lett.*, **33**, doi: 10.1029/2005gl024406.

Mitchell, T. D., and P. D. Jones, 2005: An improved method of constructing a database of monthly climate observations and associated high-resolution grids. *Int. J. Climatol.*, **25**, 693–712.

Mitchell, T. P., and J. M. Wallace, 1996: ENSO seasonality: 1950–78 versus 1979–92. *J. Clim.*, **9**, 3149–3161.

Mizuta, R., 2012: Intensification of extratropical cyclones associated with the polar jet change in the CMIP5 global warming projections. *Geophys. Res. Lett.*, **39**, doi: 10.1029/2012GL053032.

Mizuta, R., M. Matsueda, H. Endo, and S. Yukimoto, 2011: Future change in extratropical cyclones associated with change in the upper troposphere. *J. Clim.*, **24**, 6456–6470.

Mizuta, R., et al., 2012: Climate simulations using MRI-AGCM3.2 with 20-km grid. *J. Meteorol. Soc. Jpn.*, **90A**, 233–258.

Mock, C. J., and A. R. Brunelle-Daines, 1999: A modern analogue of western United States summer palaeoclimate at 6000 years before present. *Holocene*, **9**, 541–545.

Mohino, E., S. Janicot, and J. Bader, 2011: Sahel rainfall and decadal to multi-decadal sea surface temperature variability. *Clim. Dyn.*, **37**, 419–440.

Moise, A., and F. Delage, 2011: New climate model metrics based on object-orientated pattern matching of rainfall. *J. Geophys. Res. Atmos.*, **116**, doi: 10.1029/2010JD015318.

Moise, A. F., R. A. Colman, and J. R. Brown, 2012: Behind uncertainties in projections of Australian tropical climate: Analysis of 19 CMIP3 models. *J. Geophys. Res. Atmos.*, **117**, doi: 10.1029/2011jd017365.

Moise, A. F., R. A. Colman, and H. Zhang, 2005: Coupled model simulations of current Australian surface climate and its changes under greenhouse warming: An analysis of 18 CMIP2 models. *Aust. Meteorol. Mag.*, **54**, 291–307.

Monaghan, A. J., and D. H. Bromwich, 2008: Advances in describing recent Antarctic climate variablity. *Bull. Am. Meteorol. Soc.*, **89**, 1295–1306.

Monahan, A. H., L. Pandolfo, and J. C. Fyfe, 2001: The preferred structure of variability of the Northern Hemisphere atmospheric circulation. *Geophys. Res. Lett.*, **28**, 1019–1022.

Monahan, A. H., J. C. Fyfe, M. H. P. Ambaum, D. B. Stephenson, and G. R. North, 2009: Empirical Orthogonal Functions: The medium is the message. *J. Clim.*, **22**, 6501–6514.

Moron, V., A. W. Robertson, and J.-H. Qian, 2010: Local versus regional-scale characteristics of monsoon onset and post-onset rainfall over Indonesia. *Clim. Dyn.*, **34**, 281–299.

Moss, R. H., et al., 2010: The next generation of scenarios for climate change research and assessment. *Nature*, **463**, 747–756.

Muller, W. A., and E. Roeckner, 2006: ENSO impact on midlatitude circulation patterns in future climate change projections. *Geophys. Res. Lett.*, **33**, doi: 10.1029/2005gl025032.

Müller, W. A., and E. Roeckner, 2008: ENSO teleconnections in projections of future climate in ECHAM5/MPI-OM. *Clim. Dyn.*, **31**, 533–549.

Murakami, H., and B. Wang, 2010: Future change of North Atlantic tropical cyclone tracks: Projection by a 20–km-mesh global atmospheric model. *J. Clim.*, **23**, 2699–2721.

Murakami, H., and M. Sugi, 2010: Effect of model resolution on tropical cyclone climate projections. *Sola*, **6**, 73–76.

Murakami, H., B. Wang, and A. Kitoh, 2011a: Future change of Western North Pacific typhoons: Projections by a 20-km-mesh global atmospheric model. *J. Clim.*, **24**, 1154–1169.

Murakami, H., R. Mizuta, and E. Shindo, 2011b: Future changes in tropical cyclone activity projected by multi-physics and multi-SST ensemble experiments using the 60-km-mesh MRI-AGCM. *Clim. Dyn.*, doi:10.1007/s00382-011-1223-x.

Murakami, H., M. Sugi, and A. Kitoh, 2013: Future changes in tropical cyclone activity in the North Indian Ocean projected by high-resolution MRI-AGCMs. *Clim. Dyn.*, **40**, 1949–1968.

Murakami, H., et al., 2012: Future changes in tropical cyclone activity projected by the new high-resolution MRI-AGCM. *J. Clim.*, **25**, 3237–3260.

Murphy, B. F., and B. Timbal, 2008: A review of recent climate variability and climate change in southeastern Australia. *Int. J. Climatol.*, **28**, 859–879.

Muza, M. N., L. M. V. Carvalho, C. Jones, and B. Liebmann, 2009: Intraseasonal and interannual variability of extreme dry and wet events over southeastern South America and the subtropical Atlantic during austral summer. *J. Clim.*, **22**, 1682–1699.

Nanjundiah, R., V. Vidyunmala, and J. Srinivasan, 2005: The impact of increase in CO_2 on the simulation of tropical biennial oscillations (TBO) in 12 coupled general circulation models. *Atmos. Sci. Lett.*, **6**, 183–191.

Neelin, J., C. Chou, and H. Su, 2003: Tropical drought regions in global warming and El Nino teleconnections. *Geophys. Res. Lett.*, **30**, doi: 10.1029/2003GL018625.

Neelin, J. D., M. Munnich, H. Su, J. E. Meyerson, and C. E. Holloway, 2006: Tropical drying trends in global warming models and observations. *Proc. Natl. Acad. Sci.*, **103**, 6110–6115.

Neelin, J. D., B. Langenbrunner, J. E. Meyerson, A. Hall, and N. Berg, 2013: California winter precipitation change under global warming in CMIP5 models. *J. Clim.*, **26**, 6238–6256.

Nguyen, K., J. Katzfey, and J. McGregor, 2012: Global 60 km simulations with CCAM: Evaluation over the tropics. *Clim. Dyn.*, **39**, 637–654.

Nicholls, N., C. Landsea, and J. Gill, 1998: Recent trends in Australian region tropical cyclone activity. *Meteorol. Atmos. Phys.*, **65**, 197–205.

Nieto-Ferreira, R., and T. Rickenbach, 2010: Regionality of monsoon onset in South America: A three-stage conceptual model. *Int. J. Climatol.*, **31**, 1309–1321.

Nigam, S., 2003: Teleconnections. In: *Encyclopedia of Atmospheric Sciences* [J. A. P. J. R. Holton and J. A. Curry (eds.)]. Academic Press, San Diego, CA, USA, pp. 2243–2269.

Ninomiya, K., 2012: Characteristics of intense rainfalls over southwestern Japan in the Baiu season in the CMIP3 20th century simulation and 21st century projection. *J. Meteorol. Soc. Jpn.*, **90A**, 327–338.

Niyogi, D., C. Kishtawal, S. Tripathi, and R. S. Govindaraju, 2010: Observational evidence that agricultural intensification and land use change may be reducing the Indian summer monsoon rainfall. *Water Resources Research*, **46,** W03533, doi: 03510.01029/02008wr007082.

Nuñez, M. N., S. A. Solman, and M. F. Cabre, 2009: Regional climate change experiments over southern South America. II: Climate change scenarios in the late twenty-first century. *Clim. Dyn.*, **32**, 1081–1095.

O'Gorman, P. A., 2010: Understanding the varied response of the extratropical storm tracks to climate change. *Proc. Natl. Acad. Sci. U.S.A.*, **107**, 19176–19180.

O'Gorman, P. A., and T. Schneider, 2008: Energy of midlatitude transient eddies in idealized simulations of changed climates. *J. Clim.*, **21**, 5797–5806.

Okamoto, K., K. Sato, and H. Akiyoshi, 2011: A study on the formation and trend of the Brewer-Dobson circulation. *J. Geophys. Res.*, **116**, doi: 10.1029/2010JD014953.

Okumura, Y. M., D. Schneider, C. Deser, and R. Wilson, 2012: Decadal-interdecadal climate variability over Antarctica and linkages to the tropics: Analysis of ice core, instrumental, and tropical proxy data. *J. Clim.*, **25**, 7421–7441.

Onol, B., and F. Semazzi, 2009: Regionalization of climate change simulations over the Eastern Mediterranean. *J. Clim.*, **22**, 1944–1961.

Orlowsky, B., and S. Seneviratne, 2012: Global changes in extreme events: Regional and seasonal dimension. *Clim. Change*, **110**, 669–696.

Ose, T., and O. Arakawa, 2011: Uncertainty of future precipitation change due to global warming associated with sea surface temperature change in the tropical Pacific. *J. Meteorol. Soc. Jpn.*, **89**, 539–552.

Oshima, K., Y. Tanimoto, and S. P. Xie, 2012: Regional patterns of wintertime SLP change over the North Pacific and their uncertainty in CMIP3 multi-model projections. *J. Meteorol. Soc. Jpn.*, **90**, 385–396.

Ouzeau, G., J. Cattiaux, H. Douville, A. Ribes, and D. Saint-Martin, 2011: European cold winter 2009–2010: How unusual in the instrumental record and how reproducible in the ARPEGE-Climat model? *Geophys. Res. Lett.*, **38**, 6.

Palmer, T. N., 1999: A nonlinear dynamical perspective on climate prediction. *J. Clim.*, **12**, 575–591.

Parker, D., C. Folland, A. Scaife, J. Knight, A. Colman, P. Baines, and B. Dong, 2007: Decadal to multidecadal variability and the climate change background. *J. Geophys. Res.*, **112**, D18115.

Patricola, C., and K. Cook, 2010: Northern African climate at the end of the twenty-first century: An integrated application of regional and global climate models. *Clim. Dyn.*, **35**, 193–212.

Pattanaik, D. R., and M. Rajeevan, 2010: Variability of extreme rainfall events over India during southwest monsoon season. *Meteorol. Appl.*, **17**, 88–104.

Pavelsky, T., S. Kapnick, and A. Hall, 2011: Accumulation and melt dynamics of snowpack from a multiresolution regional climate model in the central Sierra Nevada, California. *J. Geophys. Res. Atmos.*, **116**, D16115.

Pavelsky, T. M., and L. C. Smith, 2006: Intercomparison of four global precipitation data sets and their correlation with increased Eurasian river discharge to the Arctic Ocean. *J. Geophys. Res. Atmos.*, **111**, D21112.

Peduzzi, P., et al., 2012: Global trends in tropical cyclone risk. *Nature Clim. Change*, **2**, 289–294.

Perkins, S., 2011: Biases and model agreement in projections of climate extremes over the tropical Pacific. *Earth Interactions*, **15**, 1-36.

Perkins, S., D. Irving, J. Brown, S. Power, A. Moise, R. Colman, and I. Smith, 2012: CMIP3 ensemble climate projections over the western tropical Pacific based on model skill. *Clim. Res.*, **51**, 35–58.

Perlwitz, J., S. Pawson, R. L. Fogt, J. E. Nielsen, and W. D. Neff, 2008: Impact of stratospheric ozone hole recovery on Antarctic climate. *Geophys. Res. Lett.*, **35**, doi: 10.1029/2008gl033317.

Petersen, K. L., 1994: A warm and wet Little Climatic Optimum and a cold and dry Little Ice Age in the southern Rocky Mountains, U.S.A. *Clim. Change*, **26**, 243–269.

Petoukhov, V., and V. A. Semenov, 2010: A link between reduced Barents-Kara sea ice and cold winter extremes over northern continents. *J. Geophys. Res. Atmos.*, **115**, D21111.

Pezza, A. B., T. Durrant, I. Simmonds, and I. Smith, 2008: Southern Hemisphere synoptic behavior in extreme phases of SAM, ENSO, sea ice extent, and southern Australia rainfall. *J. Clim.*, **21**, 5566–5584.

Pezzulli, S., D. Stephenson, and A. Hannachi, 2005: The variability of seasonality. *J. Clim.*, **18**, 71–88.

Pfahl, S., and H. Wernli, 2012: Quantifying the relevance of atmospheric blocking for co-located temperature extremes in the Northern Hemisphere on (sub-)daily time scales. *Geophys. Res. Lett.*, doi:10.1029/2012GL052261.

Philip, S., and G. Van Oldenborgh, 2006: Shifts in ENSO coupling processes under global warming. *Geophys. Res. Lett.*, **33**, doi: 10.1029/2006GL026196.

Picard, G., F. Domine, G. Krinner, L. Arnaud, and E. Lefebvre, 2012: Inhibition of the positive snow-albedo feedback by precipitation in interior Antarctica *Nature Clim. Change*, doi:10.1038/NCLIMATE1590.

Pinto, J. G., M. K. Karreman, K. Born, P. M. Della-Marta, and M. Klawa, 2012: Loss potentials associated with European windstorms under future climate conditions. *Clim. Res.*, **54**, 1–20.

Pinto, J. G., U. Ulbrich, G. C. Leckebusch, T. Spangehl, M. Reyers, and S. Zacharias, 2007: Changes in storm track and cyclone activity in three SRES ensemble experiments with the ECHAM5/MPI-OM1 GCM. *Clim. Dyn.*, **29**, 195–210.

Pitman, A. J., and S. E. Perkins, 2008: Regional projections of future seasonal and annual changes in rainfall and temperature over Australia based on skill-selected AR4 models. *Earth Interact.*, **12**, 1–50.

Plumb, R. A., 1977: The interaction of two internal waves with the mean flow: Implications for the theory of the quasi-biennial oscillation. *J. Atmos. Sci.*, **34**, 1847–1858.

Pohl, B., N. Fauchereau, C. Reason, and M. Rouault, 2010: Relationships between the Antarctic Oscillation, the Madden - Julian Oscillation, and ENSO, and Consequences for Rainfall Analysis. *J. Clim.*, **23**, 238–254.

Polcher, J., et al., 2011: AMMA's contribution to the evolution of prediction and decision-making systems for West Africa. *Atmos. Sci. Lett.*, **12**, 2–6.

Polvani, L. M., M. Previdi, and C. Deser, 2011: Large cancellation, due to ozone recovery, of future Southern Hemisphere atmospheric circulation trends. *Geophys. Res. Lett.*, **38**, doi: 10.1029/2011gl046712.

Polyakov, I., V. Alexeev, U. Bhatt, E. Polyakova, and X. Zhang, 2010: North Atlantic warming: Patterns of long-term trend and multidecadal variability. *Clim. Dyn.*, **34**, 439–457.

Polyakov, I. V., et al., 2003: Variability and trends of air temperature and pressure in the maritime Arctic, 1875–2000. *J. Clim.*, **16**, 2067–2077.

Poore, R. Z., M. J. Pavich, and H. D. Grissino-Mayer, 2005: Record of the North American southwest monsoon from Gulf of Mexico sediment cores. *Geology*, **33**, 209–212.

Popova, V. V., and A. B. Shmakin, 2010: Regional structure of surface-air temperature fluctuatoons in Northern Eurasia in the latter half of the 20th and early 21st centuries. *Izvestiya Atmos. Ocean. Phys.*, **46**, 144–158.

Power, S., and R. Colman, 2006: Multi-year predictability in a coupled general circulation model. *Clim. Dyn.*, **26**, 247–272.

Power, S., M. Haylock, R. Colman, and X. Wang, 2006: The predictability of interdecadal changes in ENSO activity and ENSO teleconnections. *J. Clim.*, **19**, 4755–4771.

Power, S., T. Casey, C. Folland, A. Colman, and V. Mehta, 1999: Inter-decadal modulation of the impact of ENSO on Australia. *Clim. Dyn.*, **15**, 319–324.

Power, S. B., and I. N. Smith, 2007: Weakening of the Walker Circulation and apparent dominance of El Nino both reach record levels, but has ENSO really changed? *Geophys. Res. Lett.*, **34**, L18702.

Prat, O. P., and B. R. Nelson, 2012: Precipitation contribution of tropical cyclones in the Southeastern United States from 1998 to 2009 using TRMM satellite data. *J. Clim.*, **26**, 1047–1062.

Qian, J.-H., 2008: Why precipitation is mostly concentrated over islands in the Maritime Continent. *J. Atmos. Sci.*, **65**, 1428–1441.

Qian, J.-H., A. W. Robertson, and V. Moron, 2010a: Interactions among ENSO, the Monsoon, and Diurnal Cycle in rainfall variability over Java, Indonesia. *J. Atmos. Sci.*, **67**, 3509–3524.

Qian, Y., S. J. Ghan, and L. R. Leung, 2010b: Downscaling hydroclimate changes over the Western US based on CAM subgrid scheme and WRF regional climate simulations. *Int. J. Climatol.*, **30**, 675–693.

Quadrelli, R., and J. M. Wallace, 2004: A simplified linear framework for interpreting patterns of Northern Hemisphere wintertime climate variability. *J. Clim.*, **17**, 3728–3744.

Quintana, J. M., and P. Aceituno, 2012: Changes in the rainfall regime along the extratropical west coast of South America (Chile): 30–43ºS. *Atmosfera*, **25**, 1–22.

Rabatel, A., et al., 2013: Current state of glaciers in the tropical Andes: A multi-century perspective on glacier evolution and climate change. *Cryosphere*, **7**, 81–102.

Raia, A., and I. F. A. Cavalcanti, 2008: The life cycle of the South American Monsoon System. *J. Clim.*, **21**, 6227–6246.

Raible, C., 2007: On the relation between extremes of midlatitude cyclones and the atmospheric circulation using ERA40. *Geophys. Res. Lett.*, **34**, doi: 10.1029/2006GL029084.

Raible, C. C., B. Ziv, H. Saaroni, and M. Wild, 2010: Winter synoptic-scale variability over the Mediterranean Basin under future climate conditions as simulated by the ECHAM5. *Clim. Dyn.*, **35**, 473–488.

Raible, C. C., P. M. Della-Marta, C. Schwierz, H. Wernli, and R. Blender, 2008: Northern Hemisphere extratropical cyclones: A comparison of detection and tracking methods and different reanalyses. *Mon. Weather Rev.*, **136**, 880–897.

Rajeevan, M., J. Bhate, and A. K. Jaswal, 2008: Analysis of variability and trends of extreme rainfall events over India using 104 years of gridded daily rainfall data. *Geophys. Res. Lett.*, **35**, doi: 10.1029/2008gl035143.

Rajendran, K., and A. Kitoh, 2008: Indian summer monsoon in future climate projection by a super high-resolution global model. *Curr. Sci.*, **95**, 1560–1569.

Ramanathan, V., et al., 2005: Atmospheric brown clouds: Impacts on South Asian climate and hydrological cycle. *Proc. Natl. Acad. Sci. U.S.A.*, doi: 10.1073/pnas.0500656102, 5326–5333.

Raphael, M. N., and M. M. Holland, 2006: Twentieth century simulation of the southern hemisphere climate in coupled models. Part 1: Large scale circulation variability. *Clim. Dyn.*, **26**, 217–228.

Rasmussen, R., et al., 2011: High-resolution coupled climate runoff simulations of seasonal snowfall over Colorado; A process study of current and warmer climate. *J. Clim.*, **24**, 3015–3048.

Rauscher, S. A., F. Giorgi, N. S. Diffenbaugh, and A. Seth, 2008: Extension and Intensification of the Meso-American mid-summer drought in the twenty-first century. *Clim. Dyn.*, **31**, 551–571.

Rawlins, M. A., et al., 2010: Analysis of the Arctic system for freshwater cycle intensification: Observations and expectations. *J. Clim.*, **23**, 5715–5737.

Re, M., and V. Barros, 2009: Extreme rainfalls in SE South America. *Clim. Change*, **96**, 119–136.

Reboita, M. S., T. Ambrizzi, and R. P. da Rocha, 2009: Relationship between the southern annular mode and southern hemisphere atmospheric systems. *Rev. Brasil. Meteorol.*, **24**, doi: 10.1590/S0102-77862009000100005.

Reisinger, A., A. B. Mullan, M. Manning, D. Wratt, and R. Nottage, 2010: Global and local climate change scenarios to support adaptation in New Zealand. In: *Climate Change Adaptation in New Zealand: Future Scenarios and Some Sectoral Perspectives* [R. A. C. Nottage, D. S. Wratt, J. F. Bornman, and K. Jones (eds.)] VUW Press, Wellington, New Zealand, pp. 26–43.

Rind, D., 2008: The consequences of not knowing low-and high-latitude climate sensitivity. *Bull. Am. Meteorol. Soc.*, **89**, 855–864.

Rinke, A., et al., 2006: Evaluation of an ensemble of Arctic regional climate models: Spatiotemporal fields during the SHEBA year. *Clim. Dyn.*, **26**, 459–472.

Risbey, J. S., M. J. Pook, P. C. McIntosh, M. C. Wheeler, and H. H. Hendon, 2009: On the remote drivers of rainfall variability in Australia. *Mon. Weather Rev.*, **137**, 3233–3253.

Riviere, G., 2011: A dynamical interpretation of the poleward shift of the jet streams in global warming scenarios. *J. Atmos. Sci.*, **68**, 1253–1272.

Robertson, A. W., et al., 2011: The Maritime Continent monsoon. In: *The Global Monsoon System: Research and Forecast*, 2nd ed. [C. P. Chang, Y. Ding, N. C. Lau, R. H. Johnson, B. Wang and T. Yasunari (eds.)] World Scientific Singapore, pp. 85–98.

Robinson, W. A., 2006: On the self-maintenance of midlatitude jets. *J. Atmos. Sci.*, **63**, 2109–2122.

Rodrigues, R. R., R. J. Haarsma, E. J. D. Campos, and T. Ambrizzi, 2011: The impacts of inter–El Niño variability on the tropical Atlantic and northeast Brazil climate. *J. Clim.*, **24**, 3402–3422.

Rodriguez-Fonseca, B., et al., 2011: Interannual and decadal SST-forced responses of the West African monsoon. *Atmos. Sci. Lett.*, **12**, 67–74.

Rosenfeld, D., M. Clavner, and R. Nirel, 2011: Pollution and dust aerosols modulating tropical cyclones intensities. *Atmos. Res.*, **102**, 66–76.

Rotstayn, L., and U. Lohmann, 2002: Tropical rainfall trends and the indirect aerosol effect. *J. Clim.*, **15**, 2103–2116.

Rotstayn, L. D., et al., 2007: Have Australian rainfall and cloudiness increased due to the remote effects of Asian anthropogenic aerosols? *J. Geophys. Res. Atmos.*, **112**, D09202.

Rotstayn, L. D., et al., 2009: Improved simulation of Australian climate and ENSO-related climate variability in a GCM with an interactive aerosol treatment. *Int. J. Climatol.*, doi:10.1002/joc.1952.

Rouault, M., P. Florenchie, N. Fauchereau, and C. Reason, 2003: South East tropical Atlantic warm events and southern African rainfall. *Geophys. Res. Lett.*, **30**, doi: 10.1029/2002GL014840.

Rowell, D. P., 2011: Sources of uncertainty in future changes in local precipitation. *Clim. Dyn.*, **39**, 1929–1950.

Rowell, D. P., 2013: Simulating SST teleconnections to Africa: What is the state of the art? *J. Clim.*, doi:10.1175/jcli-d-12–00761.1.

Roxy, M., N. Patil, K. Ashok, and K. Aparna, 2013: Revisiting the Indian summer monsoon-ENSO links in the IPCC AR4 projections: A cautionary outlook. *Global Planet. Change*, doi:10.1016/j.gloplacha.2013.02.003, early on-line release.

Rupa Kumar, K., et al., 2006: High-resolution climate change scenarios for India for the 21st century. *Curr. Sci.*, **90**, 334–345.

Rusticucci, M., and M. Renom, 2008: Variability and trends in indices of quality-controlled daily temperature extremes in Uruguay. *Int. J. Climatol.*, **28**, 1083–1095.

Rusticucci, M., J. Marengo, O. Penalba, and M. Renom, 2010: An intercomparison of model-simulated in extreme rainfall and temperature events during the last half of the twentieth century. Part 1: Mean values and variability. *Clim. Change*, **98**, 493–508.

Ruti, P., and A. Dell'Aquila, 2010: The twentieth century African easterly waves in reanalysis systems and IPCC simulations, from intra-seasonal to inter-annual variability. *Clim. Dyn.*, **35**, 1099–1117.

Sabade, S., A. Kulkarni, and R. Kripalani, 2011: Projected changes in South Asian summer monsoon by multi-model global warming experiments. *Theor. Appl. Climatol.*, **103**, 543–565.

Saenger, C., A. Cohen, D. Oppo, R. Halley, and J. Carilli, 2009: Surface-temperature trends and variability in the low-latitude North Atlantic since 1552. *Nature Geosci.*, **2**, 492–495.

Saji, N. H., B. N. Goswami, P. N. Vinayachandran, and T. Yamagata, 1999: A dipole mode in the tropical Indian Ocean. *Nature*, **401**, 360–363.

Salahuddin, A., and S. Curtis, 2011: Climate extremes in Malaysia and the equatorial South China Sea. *Global Planet. Change*, **78**, 83–91.

Salathe Jr, E. P., L. R. Leung, Y. Qian, and Y. Zhang, 2010: Regional climate model projections for the State of Washington. *Clim. Change*, **102**, 51–75.

Salinger, M. J., J. A. Renwick, and A. B. Mullan, 2001: Interdecadal Pacific Oscillation and South Pacific climate. *Int. J. Climatol.*, **21**, 1705–1722.

Sampe, T., and S.-P. Xie, 2010: Large-scale dynamics of the Meiyu-Baiu rainband: Environmental forcing by the westerly jet. *J. Climate*, **23**, 113–134.

Sansom, P. G., D. B. Stephenson, C. A. T. Ferro, G. Zappa, and L. Shaffrey, 2013: Simple uncertainty frameworks for selecting weighting schemes and interpreting multi-model ensemble climate change experiments. *J. Clim.*, doi:10.1175/JCLI-D-12–00462.1.

Santer, B. D., et al., 2007: Identification of human-induced changes in atmospheric moisture content. *Proc. Natl. Acad. Sci. U.S.A.*, **104**, 15248–15253.

Sato, T., F. Kimura, and A. Kitoh, 2007: Projection of global warming onto regional precipitation over Mongolia using a regional climate model. *J. Hydrol.*, **333**, 144–154.

Scaife, A., et al., 2011a: Climate change projections and stratosphere–troposphere interaction. *Clim. Dyn.*, **38**, 2089–2097.

Scaife, A., et al., 2009: The CLIVAR C20C project: Selected twentieth century climate events. *Clim. Dyn.*, **33**, 603–614.

Scaife, A. A., J. R. Knight, G. K. Vallis, and C. K. Folland, 2005: A stratospheric influence on the winter NAO and North Atlantic surface climate. *Geophys. Res. Lett.*, **32**, doi: 10.1029/2005gl023226.

Scaife, A. A., C. K. Folland, L. V. Alexander, A. Moberg, and J. R. Knight, 2008: European climate extremes and the North Atlantic Oscillation. *J. Clim.*, **21**, 72–83.

Scaife, A. A., T. Wollings, J. Knight, G. Martin, and T. Hinton, 2010: Atmospheric blocking and mean biases in 18 climate models. *Journal of Climate*, **23**, 6143-6152.

Scaife, A. A., et al., 2011b: Improved Atlantic winter blocking in a climate model. *Geophys. Res. Lett.*, **38**, L23703.

Scarchilli, C., M. Frezzotti, and P. Ruti, 2011: Snow precipitation at four ice core sites in East Antarctica: Provenance, seasonality and blocking factors. *Clim. Dyn.*, **37**, 2107–2125.

Schimanke, S., J. Koerper, T. Spangehl, and U. Cubasch, 2011: Multi-decadal variability of sudden stratospheric warmings in an AOGCM. *Geophys. Res. Lett.*, **38**, L01801.

Schneider, D., C. Deser, and Y. Okumura, 2012: An assessment and interpretation of the observed warming of West Antarctica in the austral spring. *Clim. Dyn.*, **38**, 323–347.

Schneider, N., and B. Cornuelle, 2005: The forcing of the Pacific decadal oscillation. *J. Clim.*, **18**, 4355–4373.

Schneider, T., P. A. O'Gorman, and X. J. Levine, 2010: Water vapor and the dynamics of climate changes. *Rev. Geophys.*, **48**, RG3001.

Schott, F. A., S.-P. Xie, and J. P. McCreary, 2009: Indian Ocean circulation and climate variability. *Rev. Geophys.*, **47**, RG1002.

Schubert, J. J., B. Stevens, and T. Crueger, 2013: The Madden-Julian Oscillation as simulated by the MPI Earth System Model: Over the last and into the next millennium. *J. Adv. Model. Earth Syst.*, **5**, 71–84.

Schulz, N., J. P. Boisier, and P. Aceituno, 2012: Climate change along the arid coast of northern Chile. *Int. J. Climatol.*, **32**, 1803–1814.

Screen, J. A., I. Simmonds, C. Deser, and R. Tomas, 2012: The atmospheric response to three decades of observed Arctic sea ice loss. *J. Clim.*, **26**, 1230–1248.

Seager, R., and G. Vecchi, 2010: Greenhouse warming and the 21st century hydroclimate of southwestern North America. *Proc. Natl. Acad. Sci. U.S.A.*, **107**, 21277–21282.

Seager, R., Y. Kushnir, M. Ting, M. Cane, N. Naik, and J. Miller, 2008: Would advance knowledge of 1930s SSTs have allowed prediction of the dust bowl drought? *J. Clim.*, **21**, 3261–3281.

Seager, R., N. Naik, and L. Vogel, 2012: Does Global Warming Cause Intensified Interannual Hydroclimate Variability? *J. Clim.*, **25**, 3355-3372

Seager, R., et al., 2007: Model projections of an imminent transition to a more arid climate in southwestern North America. *Science*, **316**, 1181–1184.

Seager, R., et al., 2009: Mexican drought: An observational modeling and tree ring study of variability and climate change. *Atmosfera*, **22**, 1–31.

Seidel, D. J., Q. Fu, W. J. Randel, and T. J. Reichler, 2008: Widening of the tropical belt in a changing climate. *Nature Geosci.*, **1**, 21–24.

Seierstad, I. A., and J. Bader, 2009: Impact of a projected future Arctic Sea Ice reduction on extratropical storminess and the NAO. *Clim. Dyn.*, **33**, 937–943.

Semenov, V. A., 2007: Structure of temperature variability in the high latitudes of the Northern Hemisphere. *Izvestiya Atmos. Ocean. Phys.*, **43**, 687–695.

Sen Gupta, A., A. Ganachaud, S. McGregor, J. N. Brown, and L. Muir, 2012: Drivers of the projected changes to the Pacific Ocean equatorial circulation. *Geophys. Res. Lett.*, **39**, L09605.

Sen Roy, S., 2009: A spatial analysis of extreme hourly precipitation patterns in India. *Int. J. Climatol.*, **29**, 345–355.

Seneviratne, S., et al., 2010: Investigating soil moisture-climate interactions in a changing climate: A review. *Earth Sci. Rev.*, **95**, 125–161.

Seneviratne, S. I., et al., 2012: Changes in climate extremes and their impacts on the natural physical environment. In: *Managing the Risks of Extreme Events and Disasters to Advance Climate Change Adaptation*. *A Special Report of Working Groups I and II of the Intergovernmental Panel on Climate Change (IPCC)* [C. B. Field, V. Barros, T. F. Stocker, D. Qin, D. J. Dokken, K. L. Ebi, M. D. Mastrandrea, K. J. Mach, G. -K. Plattner, S. K. Allen, M. Tignor and P. M. Midgley (eds.)]. Cambridge University Press, Cambridge, United Kingdom, and New York, NY, USA, pp. 109–230.

Servain, J., I. Wainer, J. McCreary, and A. Dessier, 1999: Relationship between the equatorial and meridional modes of climatic variability in the tropical Atlantic. *Geophys. Res. Lett.*, **26**, 485–488.

Seth, A., M. Rojas, and S. A. Rauscher, 2010: CMIP3 projected changes in the annual cycle of the South American Monsoon. *Clim. Change*, **98**, 331–357.

Seth, A., S. A. Rauscher, M. Rojas, A. Giannini, and S. J. Camargo, 2011: Enhanced spring convective barrier for monsoons in a warmer world? *Clim. Change*, **104**, 403–414.

Sheffield, J., and E. F. Wood, 2008: Projected changes in drought occurrence under future global warming from multi-model, multi-scenario, IPCC AR4 simulations. *Clim. Dyn.*, **31**, 79–105.

Shi, G., J. Ribbe, W. Cai, and T. Cowan, 2008a: An interpretation of Australian rainfall projections. *Geophys. Res. Lett.*, **35**, L02702.

Shi, G., W. Cai, T. Cowan, J. Ribbe, L. Rotstayn, and M. Dix, 2008b: Variability and trend of North West Australia rainfall: Observations and coupled climate modeling. *J. Clim.*, **21**, 2938–2959.

Shongwe, M., G. van Oldenborgh, B. van den Hurk, and M. van Aalst, 2011: Projected changes in mean and extreme precipitation in Africa under global warming. Part II: East Africa. *J. Clim.*, **24**, 3718–3733.

Shongwe, M. E., G. J. van Oldenborgh, B. van den Hurk, B. de Boer, C. A. S. Coelho, and M. K. van Aalst, 2009: Projected changes in mean and extreme precipitation in Africa under global warming. Part I: Southern Africa. *J. Clim.*, **22**, 3819–3837.

Sigmond, M., and J. F. Scinocca, 2010: The influence of the basic state on the Northern Hemisphere circulation response to climate change. *J. Clim.*, **23**, 1434–1446.

Sillmann, J., M. Croci-Maspoli, M. Kallache, and R. W. Katz, 2011: Extreme cold winter temperatures in Europe under the influence of North Atlantic atmospheric blocking. *J. Clim.*, **24**, 5899–5913.

Sillmann, J., V. V. Kharin, X. Zhang, F. W. Zwiers, and D. Bronaugh, 2013: Climate extremes indices in the CMIP5 multimodel ensemble: Part 1. Model evaluation in the present climate. *J. Geophys. Res. Atmos.*, **118**, 1716–1733.

Silva, A. E., and L. M. V. Carvalho, 2007: Large-scale index for South America Monsoon (LISAM). *Atmos. Sci. Lett.*, **8**, 51–57.

Silva, V. B. S., and V. E. Kousky, 2012: The South American Monsoon System: Climatology and variability. Chapter 5 in: *Modern Climatology* [S.-Y. Wang (ed.)], pp 123-152.

Sinha, A., et al., 2011: A global context for megadroughts in monsoon Asia during the past millennium. *Quat. Sci. Rev.*, **30**, 47–62.

Skansi, M. d. l. M., et al., 2013: Warming and wetting signals emerging from analysis of changes in climate extreme indices over South America. *Global Planet. Change*, **100**, 295–307.

Smirnov, D., and D. Vimont, 2011: Variability of the Atlantic Meridional Mode during the Atlantic hurricane season. *J. Clim.*, **24**, 1409–1424.

Smith, D. M., R. Eade, N. J. Dunstone, D. Fereday, J. M. Murphy, H. Pohlmann, and A. A. Scaife, 2010: Skilful multi-year predictions of Atlantic hurricane frequency. *Nature Geosci*, **3**, 846–849.

Smith, I., and E. Chandler, 2010: Refining rainfall projections for the Murray Darling Basin of south-east Australia—the effect of sampling model results based on performance. *Clim. Change*, **102**, 377–393.

Smith, I. N., and B. Timbal, 2012: Links between tropical indices and southern Australian rainfall. *Int. J. Climatol.*, **32**, 33–40.

Smith, I. N., L. Wilson, and R. Suppiah, 2008: Characteristics of the northern Australian rainy season. *J. Clim.*, **21**, 4298–4311.

Smith, I. N., A. F. Moise, and R. Colman, 2012a: Large scale circulation features in the tropical Western Pacific and their representation in climate models. *J. Geophys. Res.*, **117**, doi: 10.1029/2011JD016667.

Smith, K. L., L. M. Polvani, and D. R. Marsh, 2012b: Mitigation of 21st century Antarctic sea ice loss by stratospheric ozone recovery. *Geophys. Res. Lett.*, **39**, doi: 10.1029/2012GL053325.

Soares, W. R., and J. A. Marengo, 2009: Assessments of moisture fluxes east of the Andes in South America in a global warming scenario. *Int. J. Climatol.*, **29**, 1395–1414.

Sobel, A., and S. Camargo, 2011: Projected future seasonal changes in tropical summer climate. *J. Clim.*, **24**, 473–487.

Sohn, B., and S. Park, 2010: Strengthened tropical circulations in past three decades inferred from water vapor transport. *J. Geophys. Res. Atmos.*, **115**, doi: 10.1029/2009JD013713.

Solman, S., M. Nuñez, and M. Cabré, 2008: Regional climate change experiments over southern South America. I: Present climate. *Clim. Dyn.*, **30**, 533–552.

Solman, S., et al., 2013: Evaluation of an ensemble of regional climate model simulations over South America driven by the ERA-Interim reanalysis: Model performance and uncertainties. *Clim. Dyn.*, doi:10.1007/s00382-013-1667-2, 1–19.

Solman, S. A., and H. Le Treut, 2006: Climate change in terms of modes of atmospheric variability and circulation regimes over southern South America. *Clim. Dyn.*, **26**, 835–854.

Solomon, A., and M. Newman, 2011: Decadal predictability of tropical Indo-Pacific Ocean temperature trends due to anthropogenic forcing in a coupled climate model. *Geophys. Res. Lett.*, **38**, doi: 10.1029/2010GL045978.

Son, S. W., and S. Y. Lee, 2005: The response of westerly jets to thermal driving in a primitive equation model. *J. Atmos. Sci.*, **62**, 3741–3757.

Son, S. W., et al., 2010: Impact of stratospheric ozone on Southern Hemisphere circulation change: A multimodel assessment. *J. Geophys. Res.*, **115**, D00M07.

Sörensson, A. A., C. Menéndez, R. Ruscica, P. Alexander, P. Samuelsson, and U. Willén, 2010: Projected precipitation changes in South America: A dynamical downscaling within CLARIS. . *Meteorol. Z.*, **19**, 347–355.

Sperber, K., and H. Annamalai, 2008: Coupled model simulations of boreal summer intraseasonal (30–50 day) variability, Part 1: Systematic errors and caution on use of metrics. *Clim. Dyn.*, **31**, 345–372.

Sperber, K. R., et al., 2012: The Asian summer monsoon: An intercomparison of CMIP5 vs. CMIP3 simulations of the late 20th century. *Clim. Dyn.*, doi:10.1007/s00382-012-1607-6, 1–34.

Stammerjohn, S. E., D. G. Martinson, R. C. Smith, X. Yuan, and D. Rind, 2008: Trends in Antarctic annual sea ice retreat and advance and their relation to El Niño-Southern Oscillation and Southern Annular Mode variability. *J. Geophys. Res.*, **113**, C03S90.

Stephenson, D., A. Hannachi, and A. O'Neill, 2004: On the existence of multiple climate regimes. *Q. J. R. Meteorol. Soc.*, **130**, 583–605.

Stephenson, D., V. Pavan, M. Collins, M. Junge, and R. Quadrelli, 2006: North Atlantic Oscillation response to transient greenhouse gas forcing and the impact on European winter climate: A CMIP2 multi-model assessment. *Clim. Dyn.*, **27**, 401–420.

Stevenson, S., B. Fox-Kemper, M. Jochum, R. Neale, C. Deser, and G. Meehl, 2012: Will there be a significant change to El Nino in the twenty-first century? *J. Clim.*, **25**, 2129–2145.

Stevenson, S. L., 2012: Significant changes to ENSO strength and impacts in the twenty-first century: Results from CMIP5. *Geophys. Res. Lett.*, doi:10.1029/2012GL052759.

Stoner, A. M. K., K. Hayhoe, and D. J. Wuebbles, 2009: Assessing General Circulation Model simulations of atmospheric teleconnection patterns. *J. Clim.*, **22**, 4348–4372.

Stowasser, M., H. Annamalai, and J. Hafner, 2009: Response of the South Asian summer monsoon to global warming: Mean and synoptic systems. *J. Clim.*, **22**, 1014–1036.

Strong, C., G. Magnusdottir, and H. Stern, 2009: Observed feedback between winter sea ice and the North Atlantic Oscillation. *J. Clim.*, **22**, 6021–6032.

Sugi, M., and J. Yoshimura, 2012: Decreasing trend of tropical cyclone frequency in 228-year high-resolution AGCM simulations. *Geophys. Res. Lett.*, **39**, L19805.

Sugi, M., H. Murakami, and J. Yoshimura, 2009: A reduction in global tropical cyclone frequency due to global warming. *Sola*, **5**, 164–167.

Sugi, M., H. Murakami, and J. Yoshimura, 2012: On the mechanism of tropical cyclone frequency changes due to global warming. *J. Meteorol. Soc. Jpn.*, **90A**, 397–408.

Suhaila, J., S. M. Deni, W. Z. W. Zin, and A. A. Jemain, 2010: Spatial patterns and trends of daily rainfall regime in Peninsular Malaysia during the southwest and northeast monsoons: 1975–2004. *Meteorol. Atmos. Phys.*, **110**, 1–18.

Sun, J., H. Wang, and W. Yuan, 2008: Decadal variations of the relationship between the summer North Atlantic Oscillation and middle East Asian air temperature. *J. Geophys. Res. Atmos.*, **113**, D15107.

Sun, Y., and Y. H. Ding, 2010: A projection of future changes in summer precipitation and monsoon in East Asia. *Science China Earth Sciences*, **53**, 284–300.

Sung, M.-K., G.-H. Lim, and J.-S. Kug, 2010: Phase asymmetric downstream development of the North Atlantic Oscillation and its impact on the East Asian winter monsoon. *J. Geophys. Res.*, **115**, doi: 10.1029/2009JD013153.

Sutton, R. T., and B. Dong, 2012: Atlantic Ocean influence on a shift in European climate in the 1990s. *Nature Geosci.*, **5**, 788–792.

Swart, N. C., and J. C. Fyfe, 2012: Observed and simulated changes in the Southern Hemisphere surface westerly wind-stress. *Geophys. Res. Lett.*, **39**, L16711.

Takahashi, K., and D. S. Battisti, 2007: Processes controlling the mean tropical Pacific precipitation pattern. Part II: The SPCZ and the southeast Pacific dry zone. *J. Clim.*, **20**, 5696–5706.

Takahashi, K., A. Montecinos, K. Goubanova, and B. Dewitte, 2011: ENSO regimes: Reinterpreting the canonical and Modoki El Nino. *Geophys. Res. Lett.*, **38**, doi: 10.1029/2011gl047364.

Takaya, K., and H. Nakamura, 2005: Mechanisms of intraseasonal amplification of the cold Siberian high. *J. Atmos. Sci.*, **62**, 4423–4440.

Tanarhte, M., P. Hadjinicolaou, and J. Lelieveld, 2012: Intercomparison of temperature and precipitation data sets based on observations in the Mediterranean and the Middle East. *J. Geophys. Res. Atmos.*, **117**, doi: 10.1029/2011JD017293.

Tangang, F. T., L. Juneng, and S. Ahmad, 2007: Trend and interannual variability of temperature in Malaysia: 1961–2002. *Theor. Appl. Climatol.*, **89**, 127–141.

Tangang, F. T., et al., 2008: On the roles of the northeast cold surge, the Borneo vortex, the Madden-Julian Oscillation, and the Indian Ocean Dipole during the extreme 2006/2007 flood in southern Peninsular Malaysia. *Geophys. Res. Lett.*, **35**, L14S07.

Taylor, C., A. Gounou, F. Guichard, P. Harris, R. Ellis, F. Couvreux, and M. De Kauwe, 2011a: Frequency of Sahelian storm initiation enhanced over mesoscale soil-moisture patterns. *Nature Geosci.*, **4**, 430–433.

Taylor, C., et al., 2011b: New perspectives on land-atmosphere feedbacks from the African Monsoon Multidisciplinary Analysis. *Atmos. Sci. Lett.*, **12**, 38–44.

Taylor, K. E., R. J. Stouffer, and G. A. Meehl, 2011c: An overview of CMIP5 and the experiment design. *Bull. Am. Meteorol. Soc.*, **93**, 485–498.

Taylor, M. A., F. S. Whyte, T. S. Stephenson, and C. J.D, 2013: Why dry? Investigating the future evolution of the Caribbean Low Level Jet to explain projected Caribbean drying. *Int. J. Climatol.*, **33**, 784–792.

Taylor, M. A., T. S. Stephenson, A. Owino, A. A. Chen, and J. D. Campbell, 2011d: Tropical gradient influences on Caribbean rainfall. *J. Geophys. Res.*, **116**, D00Q08.

Tedeschi, R. G., I. F. A. Cavalcanti, and A. M. Grimm, 2013: Influences of two types of ENSO on South American precipitation. *Int. J. Climatol.*, **33**, 1382–1400.

Thomas, E. R., G. J. Marshall, and J. R. McConnell, 2008: A doubling in snow accumulation in the western Antarctic Peninsula since 1850. *Geophys. Res. Lett.*, **35**, L01706.

Thompson, D., and S. Solomon, 2009: Understanding recent stratospheric climate change. *J. Clim.*, **22**, 1934–1943.

Thompson, D. W. J., and J. M. Wallace, 1998: The Arctic Oscillation signature in the wintertime geopotential height and temperature fields. *Geophys. Res. Lett.*, **25**, 1297–1300.

Thompson, D. W. J., and J. M. Wallace, 2000: Annular modes in the extratropical circulation. Part I: Month-to-month variability. *J. Clim.*, **13**, 1000–1016.

Thompson, D. W. J., and S. Solomon, 2002: Interpretation of recent Southern Hemisphere climate change. *Science*, **296**, 895–899.

Thompson, D. W. J., J. M. Wallace, J. J. Kennedy, and P. D. Jones, 2010: An abrupt drop in Northern Hemisphere sea surface temperature around 1970. *Nature*, **467**, 444–447.

Thompson, D. W. J., S. Solomon, P. J. Kushner, M. H. England, K. M. Grise, and D. J. Karoly, 2011: Signatures of the Antarctic ozone hole in Southern Hemisphere surface climate change. *Nature Geosci.*, **4**, 741–749.

Timbal, B., and J. M. Arblaster, 2006: Land cover change as an additional forcing to explain the rainfall decline in the south west of Australia. *Geophys. Res. Lett.*, **33**, L07717.

Timbal, B., and W. Drosdowsky, 2012: The relationship between the decline of South-eastern Australian rainfall and the strengthening of the subtropical ridge. *Int. J. Climatol.*, doi:10.1002/joc.3492.

Timbal, B., J. M. Arblaster, and S. Power, 2006: Attribution of the late-twentieth-century rainfall decline in southwest Australia. *J. Clim.*, **19**, 2046–2062.

Timmermann, A., F. F. Jin, and J. Abshagen, 2003: A nonlinear theory for El Nino bursting. *J. Atmos. Sci.*, **60**, 152–165.

Ting, M., Y. Kushnir, R. Seager, and C. Li, 2009: Forced and internal twentieth-century SST trends in the north Atlantic. *J. Clim.*, **22**, 1469–1481.

Ting, M., Y. Kushnir, R. Seager, and C. Li, 2011: Robust features of Atlantic multi-decadal variability and its climate impacts. *Geophys. Res. Lett.*, **38**, L17705.

Tjernstrom, M., et al., 2004: Modeling the Arctic boundary layer: An evalutation of six ARCMIP regional-scale models with data from the SHEBA project. *Bound. Layer Meteorol.*, **117**, 337–381.

Tokinaga, H., and S. P. Xie, 2011: Weakening of the equatorial Atlantic cold tongue over the past six decades. *Nature Geosci.*, **4**, 222–226.

Tokinaga, H., S. Xie, A. Timmermann, S. McGregor, T. Ogata, H. Kubota, and Y. Okumura, 2012: Regional patterns of tropical Indo-Pacific climate change: Evidence of the Walker Circulation weakening. *J. Clim.*, **25**, 1689–1710.

Trenberth, K.E., 2011: Changes in precipitation with climate change. *Climate Res.*, **47**, 123-138.

Trenberth, K., and J. Fasullo, 2010: Simulation of present-day and twenty-first-century energy budgets of the southern oceans. *J. Clim.*, **23**, 440–454.

Trenberth, K., J. Fasullo, and L. Smith, 2005: Trends and variability in column-integrated atmospheric water vapor. *Clim. Dyn.*, **24**, 741–758.

Trenberth, K., C. Davis, and J. Fasullo, 2007a: Water and energy budgets of hurricanes: Case studies of Ivan and Katrina. *J. Geophys. Res. Atmos.*, **112**, doi: 10.1029/2006JD008303.

Trenberth, K. E., D. P. Stepaniak, and J. M. Caron, 2000: The global monsoon as seen through the divergent atmospheric circulation. *J. Clim.*, **13**, 3969–3993.

Trenberth, K. E., et al., 2007b: Observations: Surface and atmospheric climate change. In: *Climate Change 2007: The Physical Science Basis. Contribution of Working Group I to the Fourth Assessment Report of the Intergovernmental Panel on Climate Change* [Solomon, S., D. Qin, M. Manning, Z. Chen, M. Marquis, K. B. Averyt, M. Tignor and H. L. Miller (eds.)] Cambridge University Press, Cambridge, United Kingdom and New York, NY, USA, pp. 235–336.

Trigo, R. M., I. F. Trigo, C. C. DaCamara, and T. J. Osborn, 2004: Climate impact of the European winter blocking episodes from the NCEP/NCAR Reanalyses. *Clim. Dyn.*, **23**, 17–28.

Turner, A., K. Sperber, J. Slingo, G. A. Meehl, C. R. Mechoso, M. Kimoto, and A. Giannini, 2011: Modelling monsoons: Understanding and predicting current and future behaviour. World Scientific Series on Asia-Pacific Weather and Climate, Vol. 5. *The Global Monsoon System: Research and Forecast*, 2nd ed. [C. P. Chang, Y. Ding, N.-C. Lau, R. H. Johnson, B. Wang and T. Yasunari (eds.)]. World Scientific Publication Company, Singapore, 608 pp.

Turner, A. G., and H. Annamalai, 2012: Climate change and the South Asian summer monsoon. *Nature Clim. Change*, **2**, 587–595.

Turner, A. G., P. M. Inness, and J. M. Slingo, 2007a: The effect of doubled CO_2 and model basic state biases on the monsoon-ENSO system. I: Mean response and interannual variability. *Q. J. R. Meteorol. Soc.*, **133**, 1143–1157.

Turner, J., 2004: The El Niño–southern oscillation and Antarctica. *Int. J. Climatol.*, **24**, 1–31.

Turner, J., J. E. Overland, and J. E. Walsh, 2007b: An Arctic and Antarctic perspective on recent climate change. *Int. J. Climatol.*, **27**, 277–293.

Turner, J., et al., 2005: Antarctic climate change during the last 50 years. *Int. J. Climatol.*, **25**, 279–294.

Tyrlis, E., and B. J. Hoskins, 2008: Aspects of a Northern Hemisphere atmospheric blocking climatology. *J. Atmos. Sci.*, **65**, 1638–1652.

Ueda, H., A. Iwai, K. Kuwako, and M. E. Hori, 2006: Impact of anthropogenic forcing on the Asian summer monsoon as simulated by eight GCMs. *Geophys. Res. Lett.*, **33**, doi: 10.1029/2005gl025336.

Ulbrich, U., and M. Christoph, 1999: A shift of the NAO and increasing storm track activity over Europe due to anthropogenic greenhouse gas forcing. *Clim. Dyn.*, **15**, 551–559.

Ulbrich, U., G. C. Leckebusch, and J. G. Pinto, 2009: Extra-tropical cyclones in the present and future climate: A review. *Theor. Appl. Climatol.*, **96**, 117–131.

Ulbrich, U., J. G. Pinto, H. Kupfer, G. C. Leckebusch, T. Spangehl, and M. Reyers, 2008: Changing northern hemisphere storm tracks in an ensemble of IPCC climate change simulations. *J. Clim.*, **21**, 1669–1679.

Ulbrich, U., et al., 2013: Are Greenhouse Gas Signals of Northern Hemisphere winter extra-tropical cyclone activity dependent on the identification and tracking methodology? *Meteorol. Z.*, **22**, 61-68.

Ummenhofer, C. C., and M. H. England, 2007: Interannual extremes in New Zealand precipitation linked to modes of Southern Hemisphere climate variability. *J. Clim.*, **20**, 5418–5440.

Ummenhofer, C. C., A. Sen Gupta, and M. H. England, 2009a: Causes of late twentieth-century trends in New Zealand precipitation. *J. Clim.*, **22**, 3–19.

Ummenhofer, C. C., M. H. England, P. C. McIntosh, G. A. Meyers, M. J. Pook, J. S. Risbey, A. S. Gupta, and A. S. Taschetto, 2009b: What causes southeast Australia's worst droughts? *Geophys. Res. Lett.*, **36**, doi: 10.1029/2008gl036801.

van den Broeke, M. R., and N. P. M. van Lipzig, 2004: Changes in Antarctic temperature, wind and precipitation in response to the Antarctic Oscillation. *Ann. Glaciol.*, **39**, 119–126.

van Ommen, T. D., and V. Morgan, 2010: Snowfall increase in coastal East Antarctica linked with southwest Western Australian drought. *Nature Geosci*, **3**, 267–272.

Vance, T. R., T. D. van Ommen, M. A. J. Curran, C. T. Plummer, and A. D. Moy, 2012: A millennial proxy record of ENSO and eastern Australian rainfall from the Law Dome ice core, East Antarctica. *J. Clim.*, **26**, 710–725.

Vancoppenolle, M., T. Fichefet, H. Goosse, S. Bouillon, G. Madec, and M. A. M. Maqueda, 2009: Simulating the mass balance and salinity of arctic and antarctic sea ice. 1. Model description and validation. *Ocean Model.*, **27**, 33–53.

Vasconcellos, F. C., and I. F. A. Cavalcanti, 2010: Extreme precipitation over Southeastern Brazil in the austral summer and relations with the Southern Hemisphere annular mode. *Atmos. Sci. Lett.*, **11**, 21–26.

Vautard, R., et al., 2007: Summertime European heat and drought waves induced by wintertime Mediterranean rainfall deficit. *Geophys. Res. Lett.*, **34**, doi: 10.1029/2006GL028001.

Vecchi, G., and A. Wittenberg, 2010: El Nino and our future climate: Where do we stand? *WIREs Clim Change*, **1**, 260–270.

Vecchi, G. A., and B. J. Soden, 2007a: Global warming and the weakening of the tropical circulation. *J. Clim.*, **20**, 4316–4340.

Vecchi, G. A., and B. J. Soden, 2007b: Increased tropical Atlantic wind shear in model projections of global warming. *Geophys. Res. Lett.*, **34**, L08702.

Vecchi, G. A., B. J. Soden, A. T. Wittenberg, I. M. Held, A. Leetmaa, and M. J. Harrison, 2006: Weakening of tropical Pacific atmospheric circulation due to anthropogenic forcing. *Nature*, **441**, 73–76.

Vera, C., and G. Silvestri, 2009: Precipitation interannual variability in South America from the WCRP-CMIP3 multi-model dataset. *Clim. Dyn.*, **32**, 1003–1014.

Vera, C., et al., 2006: Toward a unified view of the American Monsoon Systems. *J. Clim.*, **19**, 4977–5000.

Vergara, W., et al., 2007: Visualizing future climate in Latin America: Results from the application of the Earth Simulator. In: *Latin America and Caribbean Region Sustainable Development Working Paper No. 30.* The World Bank, Washington, DC, 82 pp.

Vial, J., and T. Osborn, 2012: Assessment of atmosphere-ocean general circulation model simulations of winter northern hemisphere atmospheric blocking. *Clim. Dyn.*, **39**, 95–112.

Vigaud, N., B. Pohl, and J. Crétat, 2012: Tropical-temperate interactions over southern Africa simulated by a regional climate model. *Clim. Dyn.*, doi:10.1007/s00382-012-1314-3, 1–22.

Vigaud, N., Y. Richard, M. Rouault, and N. Fauchereau, 2009: Moisture transport between the South Atlantic Ocean and southern Africa: Relationships with summer rainfall and associated dynamics. *Clim. Dyn.*, **32**, 113–123.

Villarini, G., and G. A. Vecchi, 2012: Twenty-first-century projections of North Atlantic tropical storms from CMIP5 models. *Nature Clim. Change*, **2**, 604–607.

Vimont, D., M. Alexander, and A. Fontaine, 2009: Midlatitude excitation of tropical variability in the Pacific: The role of thermodynamic coupling and seasonality. *J. Clim.*, **22**, 518–534.

Vimont, D. J., and J. P. Kossin, 2007: The Atlantic Meridional Mode and hurricane activity. *Geophys. Res. Lett.*, **34**, L07709.

Vincent, E., M. Lengaigne, C. Menkes, N. Jourdain, P. Marchesiello, and G. Madec, 2011: Interannual variability of the South Pacific Convergence Zone and implications for tropical cyclone genesis. *Clim. Dyn.*, **36**, 1881–1896.

Vincent, L. A., W. A. van Wijngaarden, and R. Ropkinson, 2007: Surface temperature and humidity trends in Canda for 1953–2005. *J. Clim.*, **20**, 5100–5113.

Vizy, E., and K. Cook, 2002: Development and application of a mesoscale climate model for the tropics: Influence of sea surface temperature anomalies on the West African monsoon. *J. Geophys. Res. Atmos.*, **107**, ACL 2-1-ACL 2–22.

Vuille, M., B. Francou, P. Wagnon, I. Juen, G. Kaser, B. G. Mark, and R. S. Bradley, 2008: Climate change and tropical Andean glaciers: Past, present and future. *Earth Sci. Rev.*, **89**, 79–96.

Walsh, K., K. McInnes, and J. McBride, 2012: Climate change impacts on tropical cyclones and extreme sea levels in the South Pacific - A regional assessment. *Global Planet. Change*, **80–81**, 149–164.

Wang, B., 1995: Interdecadal changes in El-Nino onset in the last four decades. *J. Clim.*, **8**, 267–285.

Wang, B., and Y. Wang, 1996: Temporal structure of the Southern Oscillation as revealed by waveform and wavelet analysis. *J. Clim.*, **9**, 1586–1598.

Wang, B., and S. I. An, 2001: Why the properties of El Nino changed during the late 1970s. *Geophys. Res. Lett.*, **28**, 3709–3712.

Wang, B., and S. I. An, 2002: A mechanism for decadal changes of ENSO behavior: Roles of background wind changes. *Clim. Dyn.*, **18**, 475–486.

Wang, B., and LinHo, 2002: Rainy season of the Asian-Pacific summer monsoon. *J. Clim.*, **15**, 386–398.

Wang, B., and Q. Ding, 2006: Changes in global monsoon precipitation over the past 56 years. *Geophys. Res. Lett.*, **33**, L06711.

Wang, B., R. G. Wu, and T. Li, 2003: Atmosphere-warm ocean interaction and its impacts on Asian-Australian monsoon variation. *J. Clim.*, **16**, 1195–1211.

Wang, B., I. S. Kang, and J. Y. Lee, 2004: Ensemble simulations of Asian-Australian monsoon variability by 11 AGCMs. *J. Clim.*, **17**, 803–818.

Wang, B., Q. Ding, and J. Jhun, 2006: Trends in Seoul (1778–2004) summer precipitation. *Geophys. Res. Lett.*, **33**, L15803.

Wang, B., J. Yang, and T. J. Zhou, 2008a: Interdecadal changes in the major modes of Asian-Australian monsoon variability: Strengthening relationship with ENSO since the late 1970s. *J. Clim.*, **21**, 1771–1789.

Wang, B., H.-J. Kim, K. Kikuchi, and A. Kitoh, 2011: Diagnostic metrics for evaluation of annual and diurnal cycles. *Clim. Dyn.*, **37**, 941–955.

Wang, B., S. Xu, and L. Wu, 2012a: Intensified Arabian Sea tropical storms. *Nature*, **489**, E1–E2.

Wang, B., J. Liu, H.-J. Kim, P. J. Webster, and S.-Y. Yim, 2012b: Recent change of the global monsoon precipitation (1979–2008). *Clim. Dyn.*, **39**, 1123–1135.

Wang, C., S. K. Lee, and D. B. Enfield, 2007: Impact of the Atlantic warm pool on the summer climate of the Western Hemisphere. *J. Clim.*, **20**, 5021–5040.

Wang, C., S. K. Lee, and D. B. Enfield, 2008b: Climate response to anomalously large and small Atlantic warm pools during the summer. *J. Clim.*, **21**, 2437–2450.

Wang, H., 2001: The weakening of the Asian monsoon circulation after the end of 1970's. *Adv. Atmos. Sci.*, 376–386.

Wang, L., and W. Chen, 2010: How well do existing indices measure the strength of the East Asian winter monsoon? *Adv. Atmos. Sci.*, **27**, 855–870.

Wang, L., R. Huang, L. Gu, W. Chen, and L. Kang, 2009a: Interdecadal variations of the east Asian winter monsoon and their association with quasi-stationary planetary wave activity. *J. Clim.*, **22**, 4860–4872.

Wang, L., W. Chen, W. Zhou, J. C. L. Chan, D. Barriopedro, and R. Huang, 2010: Effect of the climate shift around mid 1970s on the relationship between wintertime Ural blocking circulation and East Asian climate. *Int. J. Climatol.*, **30**, 153–158.

Wang, S. Y., R. R. Gillies, E. S. Takle, and W. J. Gutowski, 2009b: Evaluation of precipitation in the Intermountain Region as simulated by the NARCCAP regional climate models. *Geophys. Res. Lett.*, **36**, L11704.

Wang, X., C. Z. Wang, W. Zhou, D. X. Wang, and J. Song, 2011: Teleconnected influence of North Atlantic sea surface temperature on the El Niño onset. *Clim. Dyn.*, **37**, 663–676.

Wanner, H., et al., 2001: North Atlantic Oscillation—Concepts and studies. *Surveys in Geophysics*, **22**, 321–382.

Ward, P., M. Marfai, Poerbandono, and E. Aldrian, 2011: Climate adaptation in the city of Jakarta. Chapter 13 in: *Climate Adaptation and Flood Risk in Coastal Cities* [J. Aerts, W. Botzen, M. Bowman, P. Ward and P. Dircke (eds.)]. Routledge Earthscan, Amsterdam, Netherlands, 330 pp.

Watanabe, S., and Y. Kawatani, 2012: Sensitivity of the QBO to mean tropical upwelling under a changing climate simulated with an Earth System Model. *J. Meteorol. Soc. Jpn. II*, **90A**, 351–360.

Watterson, I., A. C. Hirst, and L. D. Rotstayn, 2013: A skill-score based evaluation of simulated Australian climate. *Australian Meteorol. Oceanogr. J.*, **63**, 181-190.

Watterson, I. G., 2009: Components of precipitation and temperature anomalies and change associated with modes of the Southern Hemisphere. *Int. J. Climatol.*, **29**, 809–826.

Webster, P. J., A. M. Moore, J. P. Loschnigg, and R. R. Leben, 1999: Coupled ocean-atmosphere dynamics in the Indian Ocean during 1997–98. *Nature*, **401**, 356–360.

Weller, E., and W. Cai, 2013: Realism of the Indian Ocean Dipole in CMIP5 models: The implication for climate projections. *J. Clim.*, **26**, 6649–6659.

Widlansky, M., P. Webster, and C. Hoyos, 2011: On the location and orientation of the South Pacific Convergence Zone. *Clim. Dyn.*, **36**, 561–578.

Widlansky, M. J., et al., 2013: Changes in South Pacific rainfall bands in a warming climate. *Nature Clim. Change*, **3**, 417–423.

Wiedenmann, J. M., A. R. Lupo, I. I. Mokhov, and E. A. Tikhonova, 2002: The climatology of blocking anticyclones for the Northern and Southern Hemispheres: Block intensity as a diagnostic. *J. Clim.*, **15**, 3459–3473.

Wilcox, L. J., A. J. Charlton-Perez, and L. J. Gray 2012: Trends in Austral jet position in ensembles of high- and low-top CMIP5 models. *J. Geophys. Res.*, doi:10.1029/2012JD017597.

Williams, A., and C. Funk, 2011: A westward extension of the warm pool leads to a westward extension of the Walker circulation, drying eastern Africa. *Clim. Dyn.*, **37**, 2417–2435.

Wilson, A. B., D. H. Bromwich, and K. M. Hines, 2012: Evaluation of Polar WRF forecasts on the Arctic System Reanalysis domain:2. Atmopsheric hydrologic cycle. *J. Geophys. Res.*, **17**, D04107.

Wittenberg, A., 2004: Extended wind stress analyses for ENSO. *J. Clim.*, **17**, 2526–2540.

Wittenberg, A. T., 2009: Are historical records sufficient to constrain ENSO simulations? *Geophys. Res. Lett.*, **36**, L12702.

Woollings, T., 2008: Vertical structure of anthropogenic zonal-mean atmospheric circulation change. *Geophys. Res. Lett.*, **35**, L19702.

Woollings, T., 2010: Dynamical influences on European climate: An uncertain future. *Philos. Trans. R. Soc. London A,* **368**, 3733–3756.

Woollings, T., A. Charlton-Perez, S. Ineson, A. G. Marshall, and G. Masato, 2010: Associations between stratospheric variability and tropospheric blocking. *J. Geophys. Res. Atmos.*, **115**, D06108.

Woollings, T., J. Gregory, J. Pinto, M. Reyers, and D. Brayshaw, 2012: Response of the North Atlantic storm track to climate change shaped by ocean-atmosphere coupling. *Nature Geosci.*, **5**, 313–317.

Wu, J., and X. J. Gao, 2013: A gridded daily observation dataset over China region and comparison with the other datasets. *Chin. J. Geophys (in Chinese)*, **56**, 1102–1111.

Wu, L. G., 2007: Impact of Saharan air layer on hurricane peak intensity. *Geophys. Res. Lett.*, **34**, doi: 10.1029/2007GL029564.

Wu, Q., and X. Zhang, 2010: Observed forcing-feedback processes between Northern Hemisphere atmospheric circulation and Arctic sea ice coverage. *J. Geophys. Res. Atmos.*, **115**., doi: 10.1029/2009jd013574.

Wu, Q. G., and D. J. Karoly, 2007: Implications of changes in the atmospheric circulation on the detection of regional surface air temperature trends. *Geophys. Res. Lett.*, **34**, L08703.

Wu, R., B. P. Kirtman, and V. Krishnamurthy, 2008: An asymmetric mode of tropical Indian Ocean rainfall variability in boreal spring. *J. Geophys. Res. Atmos.*, **113**, D05104.

Wu, Y., M. Ting, R. Seager, H.-P. Huang, and M. A. Cane, 2011: Changes in storm tracks and energy transports in a warmer climate simulated by the GFDL CM2.1 model. *Clim. Dyn.*, **37**, 53–72.

Xie, P., and P. A. Arkin, 1997: Global precipitation: A 17-year monthly analysis based on gauge observations, satellite estimates, and numerical model outputs. *Bull. Am. Meteorol. Soc.*, **78**, 2539–2558.

Xie, S.-P., et al., 2007: A regional ocean–atmosphere model for Eastern Pacific climate: Toward reducing tropical biases. *J. Clim.*, **20**, 1504–1522.

Xie, S. P., and S. G. H. Philander, 1994: A coupled ocean-atmosphere model of relevance to the ITCZ in the eastern Pacific. *Tellus A*, **46**, 340–350.

Xie, S. P., and J. A. Carton, 2004: Tropical Atlantic variability: Patterns, mechanisms, and impacts. *Earth Clim. Ocean-Atmos. Interact.*, American Geophysical Union, 121–142.

Xie, S. P., K. Hu, J. Hafner, H. Tokinaga, Y. Du, G. Huang, and T. Sampe, 2009: Indian Ocean capacitor effect on Indo-western Pacific climate during the summer following El Niño. *J. Clim.*, **22**, 730–747.

Xie, S. P., Y. Du, G. Huang, X. T. Zheng, H. Tokinaga, K. M. Hu, and Q. Y. Liu, 2010a: Decadal shift in El Niño influences on Indo-western Pacific and east Asian climate in the 1970s. *J. Clim.*, **23**, 3352–3368.

Xie, S. P. D., C. Deser, G. A. Vecchi, J. Ma, H. Teng, and A. T. Wittenberg, 2010b: Global warming pattern formation: Sea surface temperature and rainfall. *J. Clim.*, **23**, 966–986.

Xu, Y., X.-J. Gao, and F. Giorgi, 2009: Regional variability of climate change hot-spots in East Asia. *Adv. Atmos. Sci.*, **26**, 783–792.

Xue, Y., et al., 2010: Intercomparison and analyses of the climatology of the West African Monsoon in the West African Monsoon Modeling and Evaluation project (WAMME) first model intercomparison experiment. *Clim. Dyn.*, **35**, 3–27.

Yamada, Y., K. Oouchi, M. Satoh, H. Tomita, and W. Yanase, 2010: Projection of changes in tropical cyclone activity and cloud height due to greenhouse warming: Global cloud-system-resolving approach. *Geophys. Res. Lett.*, **37**, L07709.

Yamagata, T., S. K. Behera, J.-J. Luo, S. Masson, M. Jury, and S. A. Rao, 2004: Coupled ocean-atmosphere variability in the tropical Indian Ocean. *Earth Clim. Ocean-Atmos. Interact.*, American Geophysical Union, 189–212.

Yamazaki, A., and H. Itoh, 2009: Selective absorption mechanism for the maintenance of blocking. *Geophys. Res. Lett.*, **36**, L05803.

Yan, H., L. G. Sun, Y. H. Wang, W. Huang, S. C. Qiu, and C. Y. Yang, 2011: A record of the Southern Oscillation Index for the past 2,000 years from precipitation proxies. *Nature Geosci.*, **4**, 611–614.

Yang, S., and J. H. Christensen, 2012: Arctic sea ice reduction and European cold winters in CMIP5 climate change experiments. *Geophys. Res. Lett.*, **39**, L20707.

Yatagai, A., K. Kamiguchi, O. Arakawa, A. Hamada, N. Yasutomi, and A. Kitoh, 2012: APHRODITE: Constructing a long-term daily gridded precipitation dataset for Asia based on a dense network of rain gauges. *Bull. Am. Meteorol. Soc.*, **93**, 1401–1415.

Ye, Z. Q., and W. W. Hsieh, 2008: Changes in ENSO and associated overturning circulations from enhanced greenhouse gases by the end of the twentieth century. *J. Clim.*, **21**, 5745–5763.

Yeh, S.-W., Y.-G. Ham, and J.-Y. Lee, 2012: Changes in the tropical Pacific SST Trend from CMIP3 to CMIP5 and its implication of ENSO. *J. Clim.*, **25**, 7764–7771.

Yeh, S.-W., B. P. Kirtman, J.-S. Kug, W. Park, and M. Latif, 2011: Natural variability of the central Pacific El Nino event on multi-centennial timescales. *Geophys. Res. Lett.*, **38**, L02704.

Yeh, S. W., and B. P. Kirtman, 2005: Pacific decadal variability and decadal ENSO amplitude modulation. *Geophys. Res. Lett.*, **32**, L05703.

Yeh, S. W., J. S. Kug, B. Dewitte, M. H. Kwon, B. P. Kirtman, and F. F. Jin, 2009: El Nino in a changing climate. *Nature*, **461**, 511–515.

Yeung, J. K., J. A. Smith, G. Villarini, A. A.N., M. L. Baeck, and W. F. Krajewski, 2011: Analyses of the warm season rainfall climatology of the northeastern US using regional climate model simulations and radar rainfall fields. *Adv. Water Resour.*, **34**, 184–204.

Yin, J. H., 2005: A consistent poleward shift of the storm tracks in simulations of 21st century climate. *Geophys. Res. Lett.*, **32**, 4.

Yin, L., R. Fu, E. Shevliakova, and R. Dickinson, 2012: How well can CMIP5 simulate precipitation and its controlling processes over tropical South America? *Clim. Dyn.*, doi:10.1007/s00382-012-1582–y.

Ying, M., T. R. Knutson, H. Kamahori, and T.-C. Lee, 2012: Impacts of climate change on tropical cyclones in the Western North Pacific Basin. Part II: Late twenty-first century projections. *Trop. Cyclone Res. Rev.*, **1**, 231–241.

Yokoi, S., and Y. Takayabu, 2009: Multi-model projection of global warming impact on tropical cyclone genesis frequency over the western north Pacific. *J. Meteorol. Soc. Jpn.*, **87**, 525–538.

Yosef, Y., H. Saaroni, and P. Alpert, 2009: Trends in daily rainfall intensity over Israel 1950/1–2003/4. *Open Atmos. Sci. J.*, **3**, 196–203.

Yu, B., and F. W. Zwiers, 2010: Changes in equatorial atmospheric zonal circulations in recent decades. *Geophys. Res. Lett.*, **37**, L05701.

Yu, R. C., B. Wang, and T. J. Zhou, 2004: Tropospheric cooling and summer monsoon weakening trend over East Asia. *Geophys. Res. Lett.*, **31**, L22212.

Zahn, M., and H. von Storch, 2010: Decreased frequency of North Atlantic polar lows associated with future climate warming. *Nature*, **467**, 309–312.

Zahn, M., and R. Allan, 2011: Changes in water vapor transports of the ascending branch of the tropical circulation. *J. Geophys. Res. Atmos.*, **116**, doi: 10.1029/2011JD016206.

Zanchettin, D., A. Rubino, and J. Jungclaus, 2010: Intermittent multidecadal-to-centennial fluctuations dominate global temperature evolution over the last millennium. *Geophys. Res. Lett.*, **37**, L14702.

Zappa, G., L. C. Shaffrey, and K. I. Hodges, 2013a: The ability of CMIP5 models to simulate North Atlantic extratropical cyclones. *J. Clim.*, doi:10.1175/jcli-d-12-00501.1.

Zappa, G., L. C. Shaffrey, K. I. Hodges, P. G. Sansom, and D. B. Stephenson, 2013b: A multi-model assessment of future projections of North Atlantic and European extratropical cyclones in the CMIP5 climate models. *J. Clim.*, doi:10.1175/jcli-d-12-00573.1.

Zebiak, S. E., 1993: Air–sea interaction in the equatorial Atlantic region. *J. Clim.*, **6**, 1567–1586.

Zhang, C., 2005: Madden-Julian Oscillation. *Rev. Geophys.*, **43**, RG2003.

Zhang, H., P. Liang, A. Moise, and L. Hanson, 2013a: The response of summer monsoon onset/retreat in Sumatra-Java and tropical Australia region to global warming in CMIP3 models. *Clim. Dyn.*, **40**, 377–399.

Zhang, J., U. S. Bhatt, W. V. Tangborn, and C. S. Lingle, 2007: Climate downscaling for estimating glacier mass balances in northwestern North America: Validation with a USGS benchmark glacier. *Geophys. Res. Lett.*, **34**, L21505.

Zhang, L., L. Wu, and L. Yu, 2011a: Oceanic origin of a recent La Nia-like trend in the tropical Pacific. *Adv. Atmos. Sci.*, **28**, 1109–1117.

Zhang, L. X., and T. J. Zhou, 2011: An assessment of monsoon precipitation changes during 1901–2001. *Clim. Dyn.*, **37**, 279–296.

Zhang, M. H., and H. Song, 2006: Evidence of deceleration of atmospheric vertical overturning circulation over the tropical Pacific. *Geophys. Res. Lett.*, **33**, L12701.

Zhang, Q., Y. Guan, and H. Yang, 2008: ENSO amplitude change in observation and coupled models. *Adv. Atmos. Sci.*, **25**, 361–366.

Zhang, R., and T. L. Delworth, 2006: Impact of Atlantic multidecadal oscillations on India/Sahel rainfall and Atlantic hurricanes. *Geophys. Res. Lett.*, **33**, L17712.

Zhang, R., and T. L. Delworth, 2009: A new method for attributing climate variations over the Atlantic Hurricane Basin's main development region. *Geophys. Res. Lett.*, **36**, L06701.

Zhang, R., et al., 2013b: Have aerosols caused the observed Atlantic multidecadal variability? *J. Atmos. Sci.*, **70**, 1135–1144.

Zhang, S., and B. Wang, 2008: Global summer monsoon rainy seasons. *Int. J. Climatol.*, **28**, 1563–1578.

Zhang, X., R. Brown, L. Vincent, W. Skinner, Y. Feng, and E. Mekis, 2011b: Canadian climate trends, 1950–2007. Canadian Biodiversity: Ecosystem Status and Trends 2012, Technical Thematic Report No. 5. Canadian Councils of Resource Ministers, Ottowa, iv + 21p.

Zhang, X., et al., 2005: Trends in Middle East climate extreme indices from 1950 to 2003. *J. Geophys. Res. Atmos.*, **110**, doi: 10.1029/2005JD006181.

Zhang, X. B., F. W. Zwiers, and P. A. Stott, 2006: Multimodel multisignal climate change detection at regional scale. *J. Clim.*, **19**, 4294–4307.

Zhao, M., and I. Held, 2012: TC-permitting GCM simulations of hurricane frequency response to sea surface temperature anomalies projected for the late twenty-first century. *J. Clim.*, **25**, 2995–3009.

Zhao, M., I. M. Held, S. J. Lin, and G. A. Vecchi, 2009: Simulations of global hurricane climatology, interannual variability, and response to global warming using a 50-km resolution GCM. *J. Clim.*, **22**, 6653–6678.

Zheng, X.-T., S.-P. Xie, and Q. Liu, 2011: Response of the Indian Ocean basin mode and its capacitor effect to global warming. *J. Clim.*, **24**, 6146–6164.

Zheng, X.-T., Y. Du, L. Liu, G. Huang, and Q. Liu, 2013: Indian Ocean Dipole response to global warming in the CMIP5 multi-model ensemble. *J. Clim.*, **26**, 6067–6080.

Zheng, X. T., S. P. Xie, G. A. Vecchi, Q. Y. Liu, and J. Hafner, 2010: Indian Ocean Dipole response to global warming: Analysis of ocean-atmospheric feedbacks in a coupled model. *J. Clim.*, **23**, 1240–1253.

Zhou, T., B. Wu, and B. Wang, 2009a: How well do atmospheric general circulation models capture the leading modes of the interannual variability of the Asian-Australian monsoon? *J. Clim.*, **22**, 1159–1173.

Zhou, T., R. Yu, H. Li, and B. Wang, 2008a: Ocean forcing to changes in global monsoon precipitation over the recent half-century. *J. Clim.*, **21**, 3833–3852.

Zhou, T. J., and R. C. Yu, 2005: Atmospheric water vapor transport associated with typical anomalous summer rainfall patterns in China. *J. Geophys. Res. Atmos.*, **110**, D08104.

Zhou, T. J., and L. W. Zou, 2010: Understanding the predictability of East Asian summer monsoon from the reproduction of land-sea thermal contrast change in AMIP-type simulation. *J. Clim.*, **23**, 6009–6026.

Zhou, T. J., L. X. Zhang, and H. M. Li, 2008b: Changes in global land monsoon area and total rainfall accumulation over the last half century. *Geophys. Res. Lett.*, **35**, L16707.

Zhou, T. J., D. Y. Gong, J. Li, and B. Li, 2009b: Detecting and understanding the multi-decadal variability of the East Asian Summer Monsoon—Recent progress and state of affairs. *Meteorol. Z.*, **18**, 455–467.

Zhou, T. J., et al., 2009c: The CLIVAR C20C project: Which components of the Asian-Australian monsoon circulation variations are forced and reproducible? *Clim. Dyn.*, **33**, 1051–1068.

Zhou, W., J. C. L. Chan, W. Chen, J. Ling, J. G. Pinto, and Y. Shao, 2009d: Synoptic-scale controls of persistent low temperature and icy weather over southern China in January 2008. *Mon. Weather Rev.*, **137**, 3978–3991.

Zhu, C., B. Wang, W. Qian, and B. Zhang, 2012: Recent weakening of northern East Asian summer monsoon: A possible response to global warming. *Geophys. Res. Lett.*, **39**, doi: 10.1029/2012GL051155.

Zhu, Y. L., and H. J. Wang, 2010: The Arctic and Antarctic Oscillations in the IPCC AR4 Coupled Models. *Acta Meteorol. Sin.*, **24**, 176–188.

Annexes

Annex I: Atlas of Global and Regional Climate Projections

Editorial Team:
Geert Jan van Oldenborgh (Netherlands), Matthew Collins (UK), Julie Arblaster (Australia), Jens Hesselbjerg Christensen (Denmark), Jochem Marotzke (Germany), Scott B. Power (Australia), Markku Rummukainen (Sweden), Tianjun Zhou (China)

Advisory Board:
David Wratt (New Zealand), Francis Zwiers (Canada), Bruce Hewitson (South Africa)

Review Editor Team:
Pascale Delecluse (France), John Fyfe (Canada), Karl Taylor (USA)

This annex should be cited as:
IPCC, 2013: Annex I: Atlas of Global and Regional Climate Projections [van Oldenborgh, G.J., M. Collins, J. Arblaster, J.H. Christensen, J. Marotzke, S.B. Power, M. Rummukainen and T. Zhou (eds.)]. In: *Climate Change 2013: The Physical Science Basis. Contribution of Working Group I to the Fifth Assessment Report of the Intergovernmental Panel on Climate Change* [Stocker, T.F., D. Qin, G.-K. Plattner, M. Tignor, S.K. Allen, J. Boschung, A. Nauels, Y. Xia, V. Bex and P.M. Midgley (eds.)]. Cambridge University Press, Cambridge, United Kingdom and New York, NY, USA.

Table of Contents

Introduction and Scope

This Annex presents a series of figures showing global and regional patterns of climate change computed from global climate model output gathered as part of the Coupled Model Intercomparison Project Phase 5 (CMIP5; Taylor et al., 2012). Maps of surface air temperature change and relative precipitation change (i.e., change expressed as a percentage of mean precipitation) in different seasons are presented for the globe and for a number of different sub-continental-scale regions. Twenty-year average changes for the near term (2016–2035), for the mid term (2046–2065) and for the long term (2081–2100) are given, relative to a reference period of 1986–2005. Time series for temperature and relative precipitation changes are shown for global land and sea averages, the 26 sub-continental SREX (IPCC Special Report on Managing the Risks of Extreme Events and Disasters to Advance Climate Change Adaptation) regions (IPCC, 2012) augmented with polar regions and the Caribbean, two Indian Ocean and three Pacific Ocean regions. In total this Annex gives projections for 35 regions, 2 variables and 2 seasons. The projections are made under the Representative Concentration Pathway (RCP) scenarios, which are introduced in Chapter 1 with more technical detail given in Section 12.3 (also note the discussion of near-term biases in Sections 11.3.5.1 and 11.3.6.1). Maps are shown only for the RCP4.5 scenario; however, the time series presented show how the area-average response varies among the RCP2.6, RCP4.5, RCP6.0 and RCP8.5 scenarios. Spatial maps for the other RCP scenarios and additional seasons are presented in the Annex I Supplementary Material. Figures AI.1 and AI.2 give a graphical explanation of aspects of both the time series plots and the spatial maps. While some of the background to the information presented is given here, discussion of the maps and time series, as well as important additional background, is provided in Chapters 9, 11, 12 and 14. Figure captions on each page of the Atlas reference the specific sub-sections in the report relevant to the regions considered on that page.

The projection of future climate change involves the careful evaluation of models, taking into account uncertainties in observations and consideration of the physical basis of the findings, in order to characterize the credibility of the projections and assess their sensitivity to uncertainties. As discussed in Chapter 9, different climate models have varying degrees of success in simulating past climate variability and mean state when compared to observations. Verification of regional trends is discussed in Box 11.2 and provides further information on the credibility of model projections. The information presented in this Annex is based entirely on all available CMIP5 model output with equal weight given to each model or version with different parameterizations.

Complementary methods for making quantitative projections, in which model output is combined with information about model performance using statistical techniques, exist and should be considered in impacts studies (see Sections 9.8.3, 11.3.1 and 12.2.2 to 12.2.3). Although results from the application of such methods can be assessed alongside the projections from CMIP5 presented here, it is beyond the scope of this Annex. Nor do the simple maps provided represent a robust estimate of the uncertainty associated with the projections. Here the range of model spread is provided as a simple, albeit imperfect, guide to the range of possible futures (including the effect of natural variability). Alternative approaches used to estimate projection uncertainty are discussed in Sections 11.3.1 and 12.2.2 to 12.2.3. The reliability of past trends is assessed in Box 11.2, which concludes that the time series and maps cannot be interpreted literally as probability density functions. They should not be interpreted as 'forecasts'.

Projections of future climate change are conditional on assumptions of climate forcing, affected by shortcomings of climate models and inevitably also subject to internal variability when considering specific periods. Projected patterns of climate change may differ from one climate model generation to the next due to improvements in models. Some model-inadequacies are common to all models, but so are many patterns of change across successive generations of models, which gives some confidence in projections. The information presented is intended to be only a starting point for anyone interested in more detailed information on projections of future climate change and complements the assessment in Chapters 11, 12 and 14.

Technical Notes

Data and Processing: The figures have been constructed using the CMIP5 model output available at the time of the AR5 cut-off for accepted papers (15 March 2013). This data set comprises 32/42/25/39 scenario experiments for RCP2.6/4.5/6.0/8.5 from 42 climate models (Table AI.1). Only concentration-driven experiments are used (i.e., those in which concentrations rather than emissions of greenhouse gases are prescribed) and only one ensemble member from each model is selected, even if multiple realizations exist with different initial conditions and different realizations of natural variability. Hence each model is given equal weight. Maps from only one scenario (RCP4.5) are shown but time series are included from all four RCPs. Maps from other RCPs are presented in the Annex I Supplementary Material.

Reference Period: Projections are expressed as anomalies with respect to the reference period of 1986–2005 for both time series and spatial maps (i.e., differences between the future period and the reference period). Thus the changes are relative to the climate change that has already occurred since the pre-industrial period and which is discussed in Chapters 2 and 10. For quantities where the trend is larger than the natural variability such as large-area temperature changes, a more recent reference period would give better estimates (see Section 11.3.6.1); for quantities where the natural variability is much larger than the trend a longer reference period would be preferable.

Equal Model Weighting: Model evaluation uses a multitude of techniques (see Chapter 9) and there is no consensus in the community about how to use this information to assign likelihood to different model projections. Consequently, the different CMIP5 models used for the projections in the Atlas are all considered to give equally likely projections in the sense of 'one model, one vote'. Models with variations in physical parameterization schemes are treated as distinct models.

Variables: Two variables have been plotted: surface air temperature change and relative precipitation change. The relative precipitation change is defined as the percentage change from the 1986–2005 reference period in each ensemble member. For the time series, the variables are first averaged over the domain and then the changes from the reference period are computed. This implies that in regions with large climatological precipitation gradients, the change is generally dominated by the areas with the most precipitation.

Seasons: For temperature, the standard meteorological seasons June to August and December to February are shown, as these often correspond roughly with the warmest and coldest seasons. The annual mean and remaining seasons, March to May and September to October can be found in the Annex I Supplementary Material. For precipitation, the half-years April to September and October to March are shown so that in most monsoon areas the local rain seasons are entirely contained within the seasonal range plotted. Because the seasonal average is computed first, followed by the percentile change, these numbers are dominated by the rainy months within the half-year. The annual means are included in the Supplementary Material.

Regions: In addition to the global maps, the areas defined in the SREX (IPCC, 2012) are plotted with the addition of six regions containing the Caribbean, Indian Ocean and Pacific Island States and land and sea areas of the two polar regions. For regions containing large land-areas, averages are computed only over land grid points only. For ocean regions, averages are computed over both land and ocean grid points (see figure captions). A grid box is considered land if the land fraction is larger than 50% and sea if it is smaller than this. SREX regions with long coastlines (west coast of South America, North Europe, Southeast Asia) therefore include some influence of the ocean. Note that temperature and precipitation over islands may be very different from those over the surrounding sea.

Time Series: For each of the resulting areas the areal mean is computed on the original model grid using land, sea or all points, depending on the definition of the region (see above). As an indication of the model uncertainty and natural variability, the time series of each model and scenario over the common period 1900–2100 are shown on the top of the page as anomalies relative to 1986–2005 (the seasons December to February and October to March are counted towards the second year in the interval). The multi-model ensemble means are also shown. Finally, for the period 2081–2100, the 20-year means are computed and the box-and-whisker plots show the 5th, 25th, 50th (median), 75th and 95th percentiles sampled over the distribution of the 20-year means of the model time series indicated in Table AI.1, including both natural variability and model spread. In the 20-year means the natural variability is suppressed relative to the annual values in the time series whereas the model uncertainty is the same. Note that owing to a smaller number of models, the box-and-whisker plots for the RCP2.6 scenario and especially the RCP6.0 scenario are less certain than those for RCP4.5 and RCP8.5.

Spatial Maps: The maps in the Atlas show, for an area encompassing two or three regions, the difference between the periods 2016–2035, 2046–2065 and 2081–2100 and the reference period 1986–2005. As local projections of climate change are uncertain, a measure of the range of model projections is shown in addition to the median response of the model ensemble interpolated to a common 2.5° grid (the interpolation was done bilinearly for surface air temperature and first order conservatively for precipitation). It should again be emphasized (see above) that this range does not represent the full uncertainty in the projection. On the left, the 25th percentile of the distribution of ensemble members is shown, on the right the 75th percentile. The median is shown in the middle (different from similar plots in Chapters 11 and 12 and the time series which show the multi-model mean). The distribution combines the effects of natural variability and model spread. The colour scale is kept constant over all maps.

Hatching: Hatching indicates regions where the magnitude of the change of the 20-year mean is less than 1 standard deviation of model-estimated present-day natural variability of 20-year mean differences. The natural variability is estimated using all pre-industrial control runs which are at least 500 years long. The first 100 years of the pre-industrial are ignored. The natural variability is then calculated for every grid point as the standard deviation of non-overlapping 20-year means after a quadratic fit is subtracted at every grid point to eliminate model drift. This is multiplied by the square root of 2, a factor that arises as the comparison is between two distributions of numbers. The median across all models of that quantity is used. This characterizes the typical difference between two 20-year averages that would be expected due to unforced internal variability. The hatching is applied to all maps so, for example, if the 25th percentile of the distribution of model projections is less than 1 standard deviation of natural variability, it is hatched.

The hatching can be interpreted as some indication of the strength of the future anomalies from present-day climate, when compared to the strength of present day internal 20-year variability. It either means that the change is relatively small or that there is little agreement between models on the sign of the change. It is presented only as a guide to assessing the strength of change as the difference between two 20-year intervals. Using other measures of natural variability would give smaller or larger hatched areas, but the colours underneath the hatching would not be very different. Other methods of hatching and stippling are possible (see Box 12.1) and, in cases where such information is critical, it is recommended that thorough attention is paid to assessing significance using a statistical test appropriate to the problem being considered.

Scenarios: Spatial patterns of changes for scenarios other than RCP4.5 can be found in the Annex I Supplementary Material.

References

IPCC, 2012: *Managing the Risks of Extreme Events and Disasters to Advance Climate Change Adaptation.* A Special Report of Working Groups I and II of the Intergovernmental Panel on Climate Change [C. B. Field, V. Baros, T. F. Stocker, D. Qin, D. J. Dokken, K. L. Ebi, M. D. Mastrandrea, K .J. Mach, G.-K. Plattner, S. K. Allen, M. Tignor and P. M. Midgley (eds.)]. Cambridge University Press, Cambridge, United Kingdom, and New York, NY, USA, 582 pp.

Taylor, K. E., R. J. Stouffer, and G. A. Meehl, 2012: A summary of the CMIP5 experiment design. *Bull. Am. Meteorol. Soc.*, **93**, 485–498.

Table AI.1 | The CMIP5 models used in this Annex for each of the historical and RCP scenario experiments. A number in each column is the identifier of the single ensemble member from that model that is used. A blank indicates no run was used, usually because that scenario run was not available. For the pre-industrial control column (piControl), a 'tas' indicates that those control simulations are used in the estimate of internal variability of surface air temperature and a 'pr' indicates that those control simulations are used in the estimate of precipitation internal variability.

CMIP5 Model Name	piControl	Historical	RCP2.6	RCP4.5	RCP6.0	RCP8.5
ACCESS1-0	tas/pr	1		1		1
ACCESS1-3	tas/pr	1		1		1
bcc-csm1-1	tas/pr	1	1	1	1	1
bcc-csm1-1-m		1	1	1	1	
BNU-ESM	tas/pr	1	1	1		1
CanESM2	tas/pr	1	1	1		1
CCSM4	tas/pr	1	1	1	1	1
CESM1-BGC	tas/pr	1		1		1
CESM1-CAM5		1	1	1	1	1
CMCC-CM		1		1		1
CMCC-CMS	tas/pr	1		1		1
CNRM-CM5	tas/pr	1	1	1		1
CSIRO-Mk3-6-0	tas/pr	1	1	1	1	1
EC-EARTH		8	8	8		8
FGOALS-g2	tas/pr	1	1	1		1
FIO-ESM	tas/pr	1	1	1	1	1
GFDL-CM3	tas/pr	1	1	1	1	1
GFDL-ESM2G	tas/pr	1	1	1	1	1
GFDL-ESM2M	tas/pr	1	1	1	1	1
GISS-E2-H p1		1	1	1	1	1
GISS-E2-H p2	tas/pr	1	1	1	1	1
GISS-E2-H p3	tas/pr	1	1	1	1	1
GISS-E2-H-CC		1		1		
GISS-E2-R p1		1	1	1	1	1
GISS-E2-R p2	pr	1	1	1	1	1
GISS-E2-R p3	pr	1	1	1	1	1
GISS-E2-R-CC		1		1		
HadGEM2-AO		1	1	1	1	1
HadGEM2-CC		1		1		1
HadGEM2-ES		2	2	2	2	2
inmcm4	tas/pr	1		1		1
IPSL-CM5A-LR	tas/pr	1	1	1	1	1
IPSL-CM5A-MR		1	1	1	1	1
IPSL-CM5B-LR		1		1		1
MIROC5	tas/pr	1	1	1	1	1
MIROC-ESM	tas/pr	1	1	1	1	1
MIROC-ESM-CHEM		1	1	1	1	1
MPI-ESM-LR	tas/pr	1	1	1		1
MPI-ESM-MR	tas/pr	1	1	1		1
MPI-ESM-P	tas/pr					
MRI-CGCM3	tas/pr	1	1	1	1	1
NorESM1-M	tas/pr	1	1	1	1	1
NorESM1-ME		1	1	1	1	1
Number of models		42	32	42	25	39

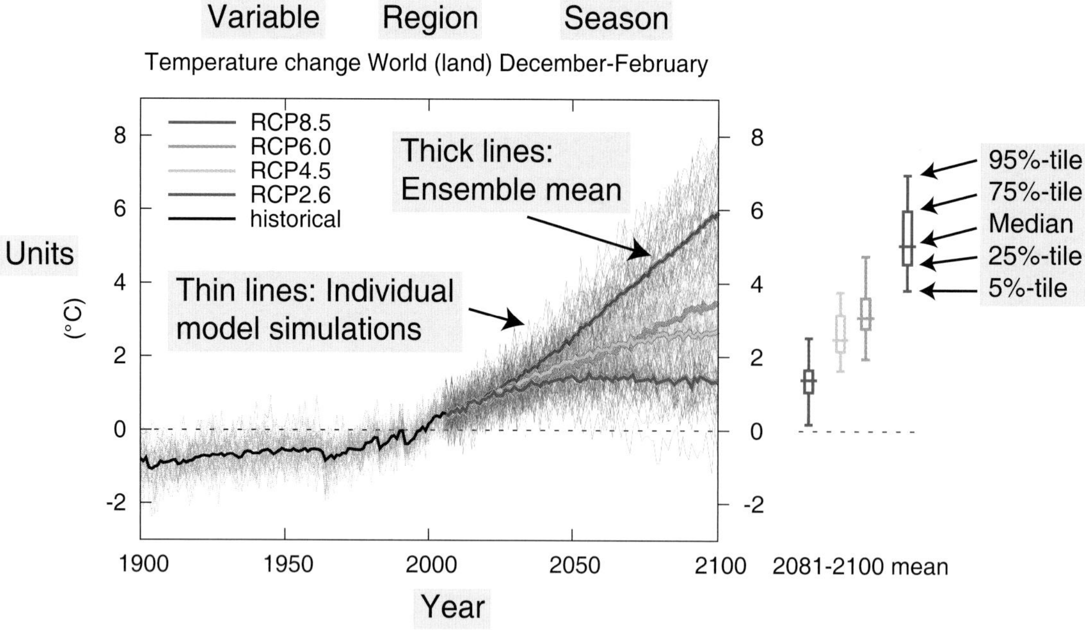

Figure AI.1 | Explanation of the features of a typical time series figure presented in Annex I.

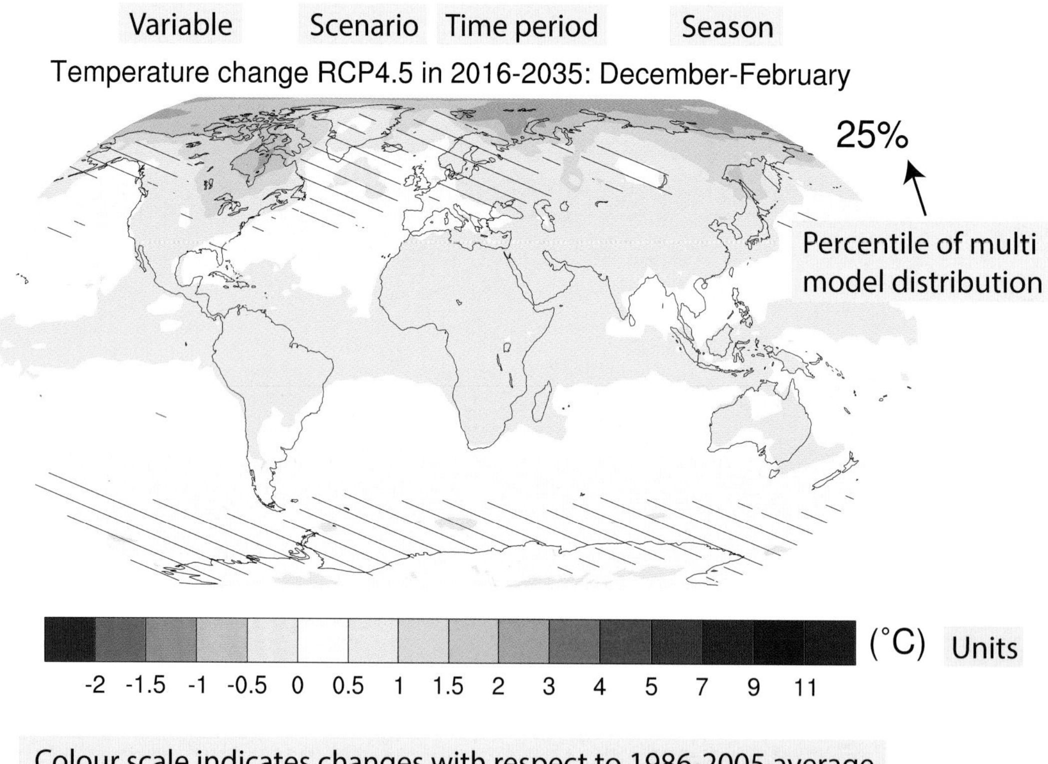

Figure AI.2 | Explanation of the features of a typical spatial map presented in Annex I. Hatching indicates regions where the magnitude of the 25th, median or 75th percentile of the 20-year mean change is less than 1 standard deviation of model-estimated natural variability of 20-year mean differences.

Atlas

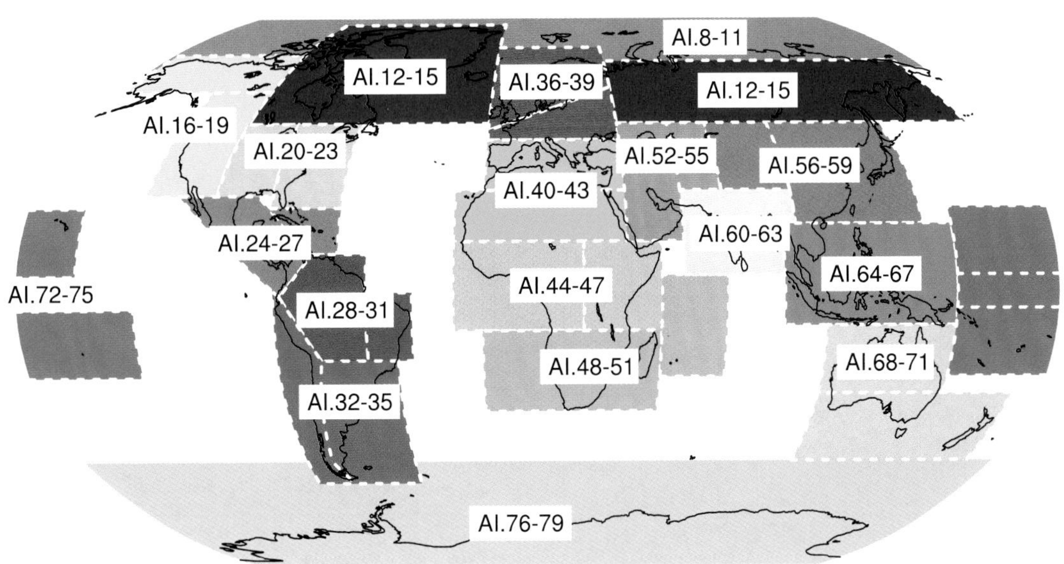

Figure AI.3 | Overview of the SREX, ocean and polar regions used.

Figures AI.4 to AI.7: World
Figures AI.8 to AI.11: Arctic
Figures AI.12 to AI.15: High latitudes
Figures AI.16 to AI.19: North America (West)
Figures AI.20 to AI.23: North America (East)
Figures AI.24 to AI.27: Central America and Caribbean
Figures AI.28 to AI.31: Northern South America
Figures AI.32 to AI.35: Southern South America
Figures AI.36 to AI.39: North and Central Europe
Figures AI.40 to AI.43: Mediterranean and Sahara
Figures AI.44 to AI.47: West and East Africa
Figures AI.48 to AI.51: Southern Africa and West Indian Ocean
Figures AI.52 to AI.55: West and Central Asia
Figures AI.56 to AI.59: Eastern Asia and Tibetan Plateau
Figures AI.60 to AI.63: South Asia
Figures AI.64 to AI.67: Southeast Asia
Figures AI.68 to AI.71: Australia and New Zealand
Figures AI.72 to AI.75: Pacific Islands region
Figures AI.76 to AI.79: Antarctica

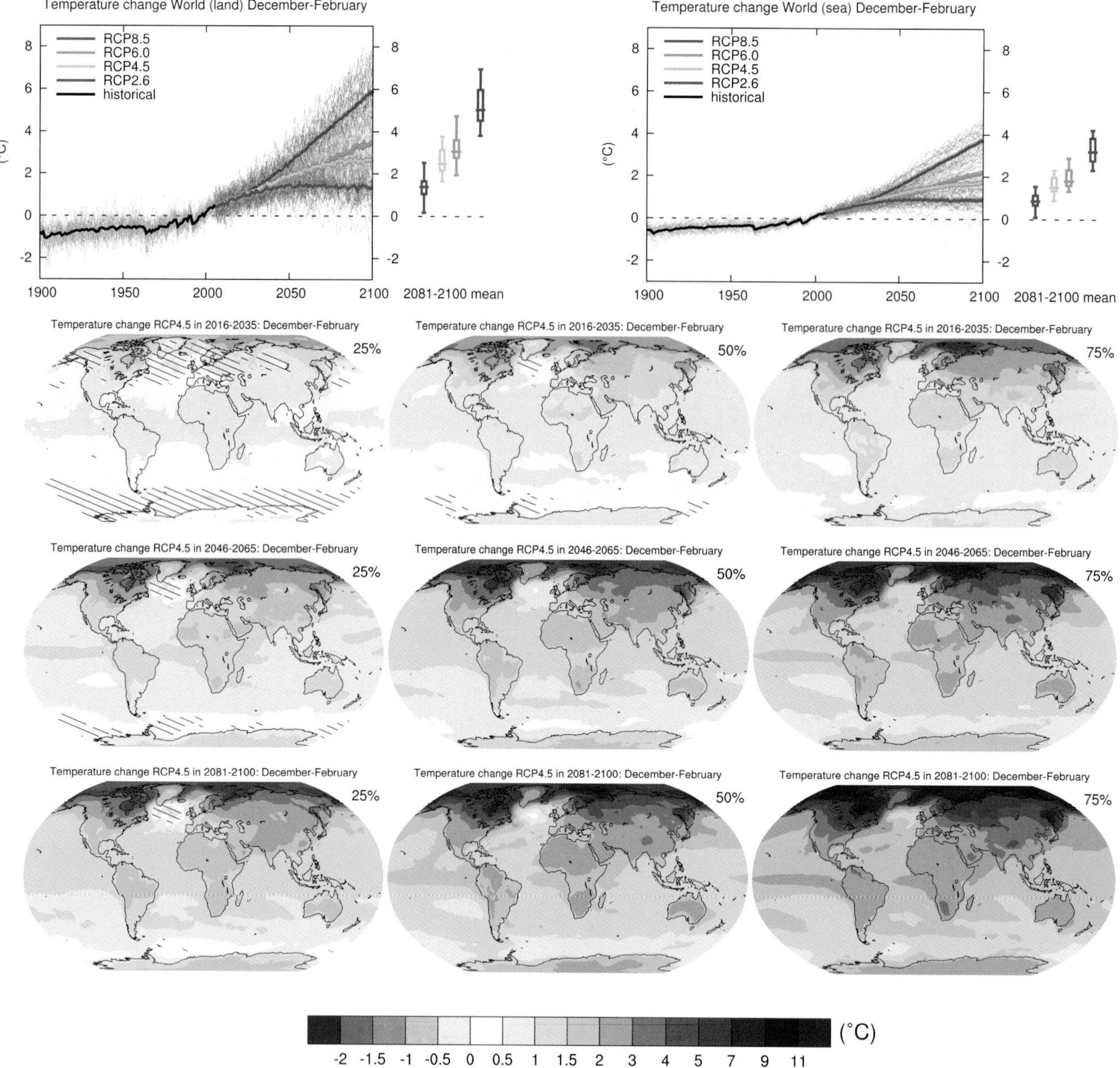

Figure AI.4 | (Top left) Time series of temperature change relative to 1986–2005 averaged over land grid points over the globe in December to February. (Top right) Same for sea grid points. Thin lines denote one ensemble member per model, thick lines the CMIP5 multi-model mean. On the right-hand side the 5th, 25th, 50th (median), 75th and 95th percentiles of the distribution of 20-year mean changes are given for 2081–2100 in the four RCP scenarios.

(Below) Maps of temperature changes in 2016–2035, 2046–2065 and 2081–2100 with respect to 1986–2005 in the RCP4.5 scenario. For each point, the 25th, 50th and 75th percentiles of the distribution of the CMIP5 ensemble are shown; this includes both natural variability and inter-model spread. Hatching denotes areas where the 20-year mean differences of the percentiles are less than the standard deviation of model-estimated present-day natural variability of 20-year mean differences.

Sections 9.4.1.1, 9.6.1.1, 10.3.1.1.4, 11.3.2.1.2, 11.3.3.1, Box 11.2, 12.4.3.1 and 12.4.7 contain relevant information regarding the evaluation of models in this region, the model spread in the context of other methods of projecting changes and the role of modes of variability and other climate phenomena.

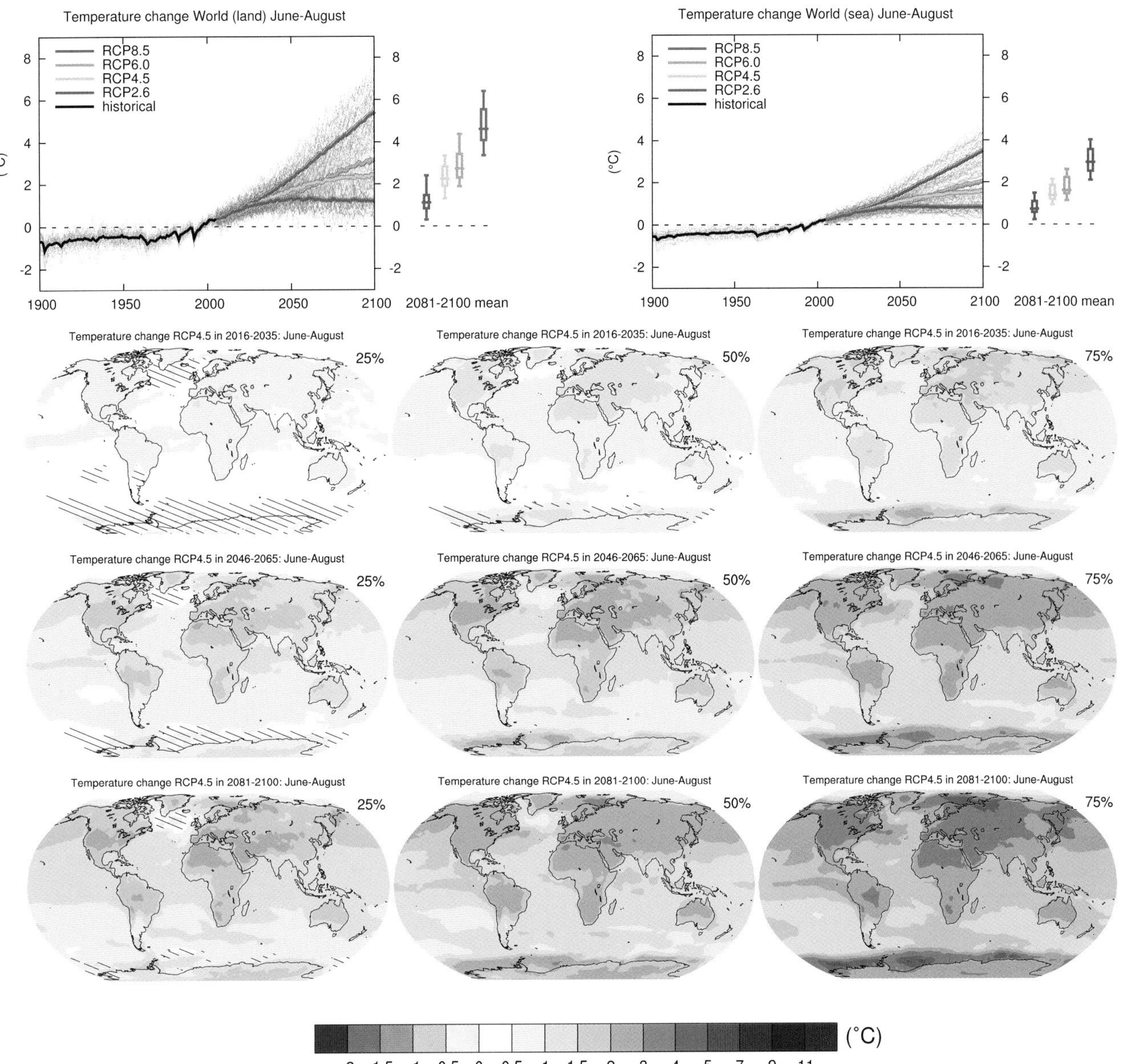

Figure AI.5 | (Top left) Time series of temperature change relative to 1986–2005 averaged over land grid points over the globe in June to August. (Top right) Same for sea grid points. Thin lines denote one ensemble member per model, thick lines the CMIP5 multi-model mean. On the right-hand side the 5th, 25th, 50th (median), 75th and 95th percentiles of the distribution of 20-year mean changes are given for 2081–2100 in the four RCP scenarios.

(Below) Maps of temperature changes in 2016–2035, 2046–2065 and 2081–2100 with respect to 1986–2005 in the RCP4.5 scenario. For each point, the 25th, 50th and 75th percentiles of the distribution of the CMIP5 ensemble are shown; this includes both natural variability and inter-model spread. Hatching denotes areas where the 20-year mean differences of the percentiles are less than the standard deviation of model-estimated present-day natural variability of 20-year mean differences.

Sections 9.4.1.1, 9.6.1.1, 10.3.1.1.4, 11.3.2.1.2, 11.3.3.1, Box 11.2, 12.4.3.1 and 12.4.7 contain relevant information regarding the evaluation of models in this region, the model spread in the context of other methods of projecting changes and the role of modes of variability and other climate phenomena.

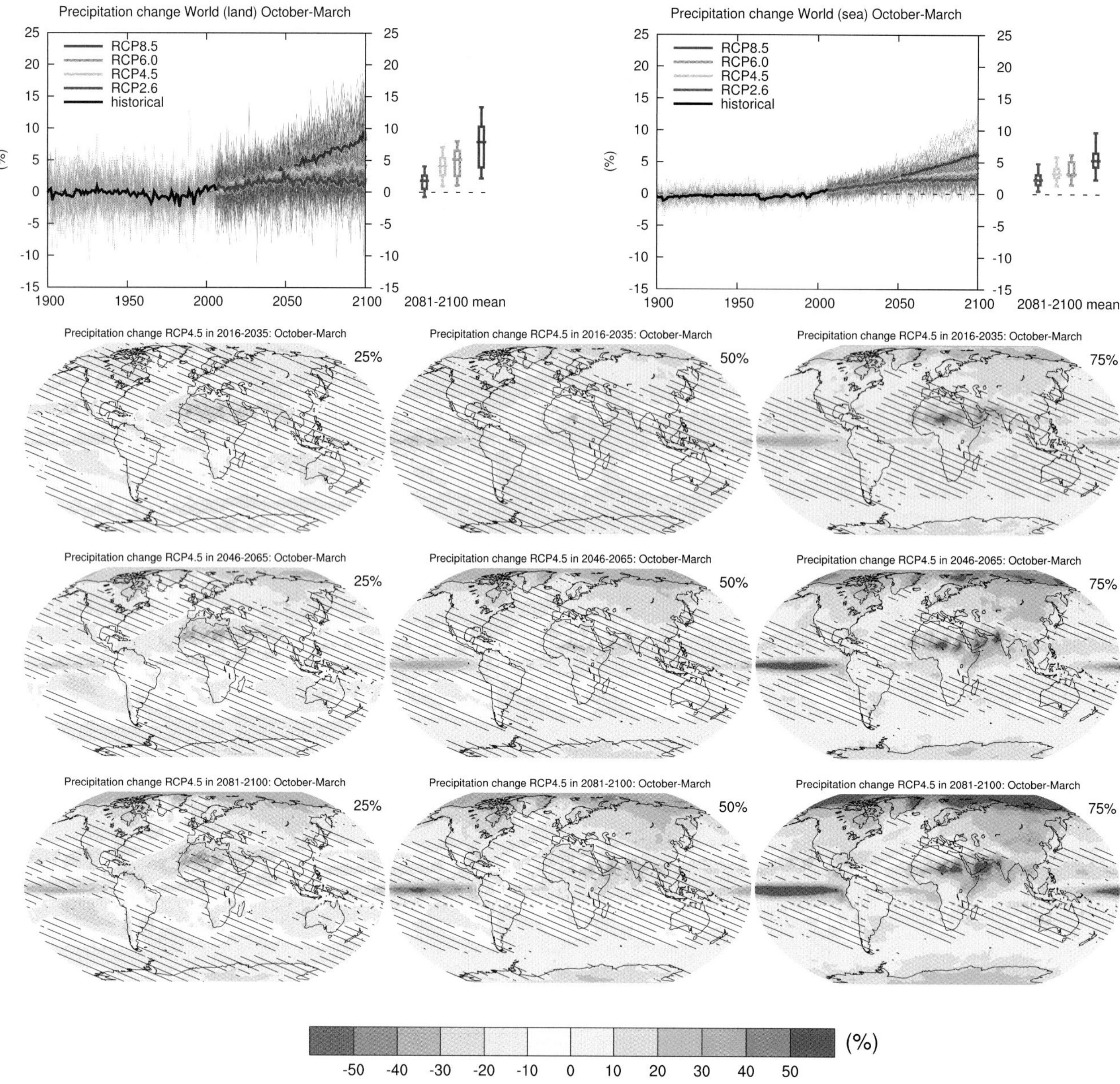

Figure AI.6 | (Top left) Time series of relative change relative to 1986–2005 in precipitation averaged over land grid points over the globe in October to March. (Top right) Same for sea grid points. Thin lines denote one ensemble member per model, thick lines the CMIP5 multi-model mean. On the right-hand side the 5th, 25th, 50th (median), 75th and 95th percentiles of the distribution of 20-year mean changes are given for 2081–2100 in the four RCP scenarios.

(Below) Maps of precipitation changes in 2016–2035, 2046–2065 and 2081–2100 with respect to 1986–2005 in the RCP4.5 scenario. For each point, the 25th, 50th and 75th percentiles of the distribution of the CMIP5 ensemble are shown; this includes both natural variability and inter-model spread. Hatching denotes areas where the 20-year mean differences of the percentiles are less than the standard deviation of model-estimated present-day natural variability of 20-year mean differences.

Sections 9.4.1.1, 9.6.1.1, 10.3.2.2, 11.3.2.3.1, Box 11.2, 12.4.5.2, 14.2 contain relevant information regarding the evaluation of models in this region, the model spread in the context of other methods of projecting changes and the role of modes of variability and other climate phenomena.

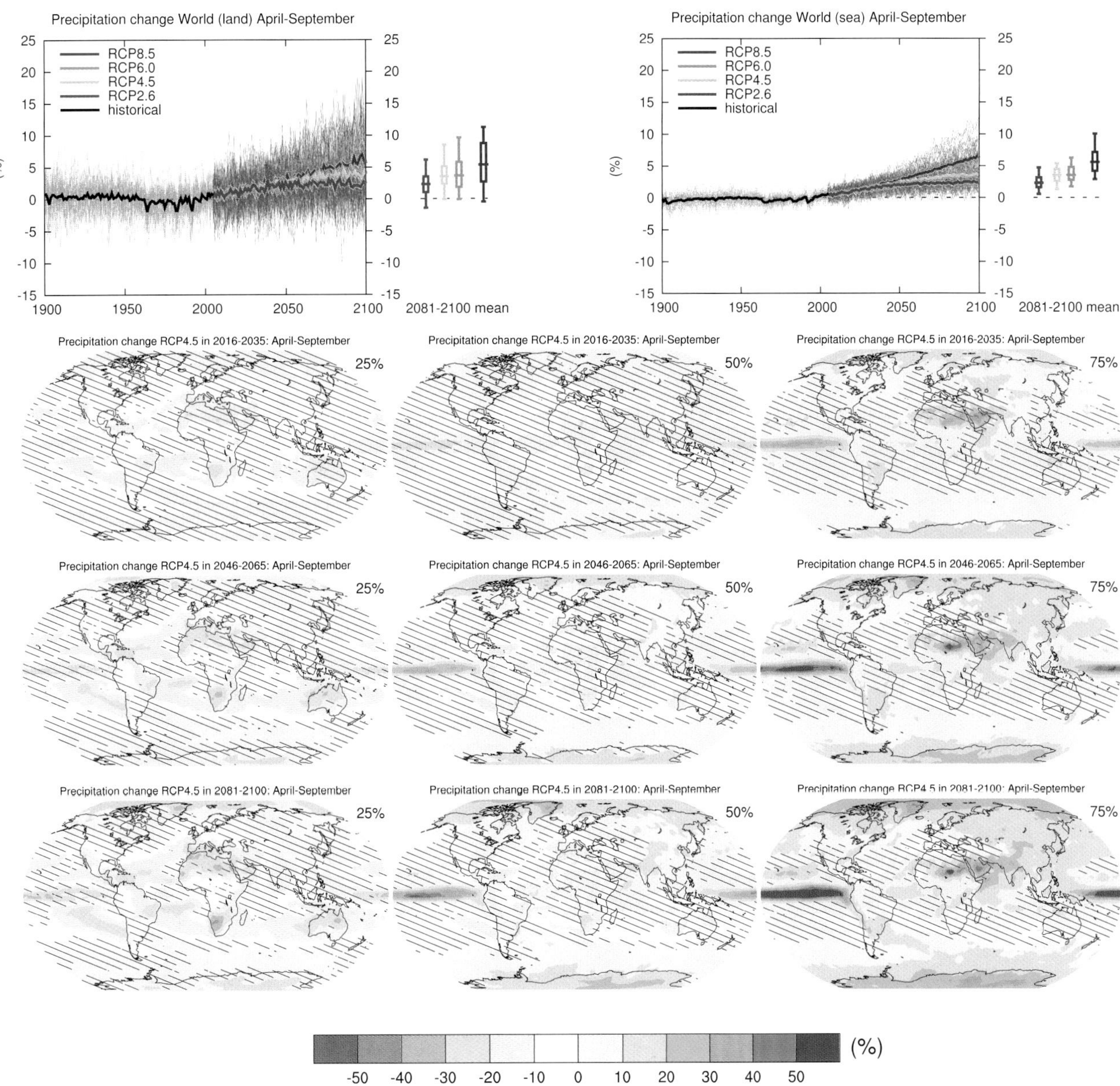

Figure AI.7 | (Top left) Time series of relative change relative to 1986–2005 in precipitation averaged over land grid points over the globe in April to September. (Top right) Same for sea grid points. Thin lines denote one ensemble member per model, thick lines the CMIP5 multi-model mean. On the right-hand side the 5th, 25th, 50th (median), 75th and 95th percentiles of the distribution of 20-year mean changes are given for 2081–2100 in the four RCP scenarios.

(Below) Maps of precipitation changes in 2016–2035, 2046–2065 and 2081–2100 with respect to 1986–2005 in the RCP4.5 scenario. For each point, the 25th, 50th and 75th percentiles of the distribution of the CMIP5 ensemble are shown; this includes both natural variability and inter-model spread. Hatching denotes areas where the 20-year mean differences of the percentiles are less than the standard deviation of model-estimated present-day natural variability of 20-year mean differences.

Sections 9.4.1.1, 9.6.1.1, 10.3.2.2, 11.3.2.3.1, Box 11.2, 12.4.5.2, 14.2 contain relevant information regarding the evaluation of models in this region, the model spread in the context of other methods of projecting changes and the role of modes of variability and other climate phenomena.

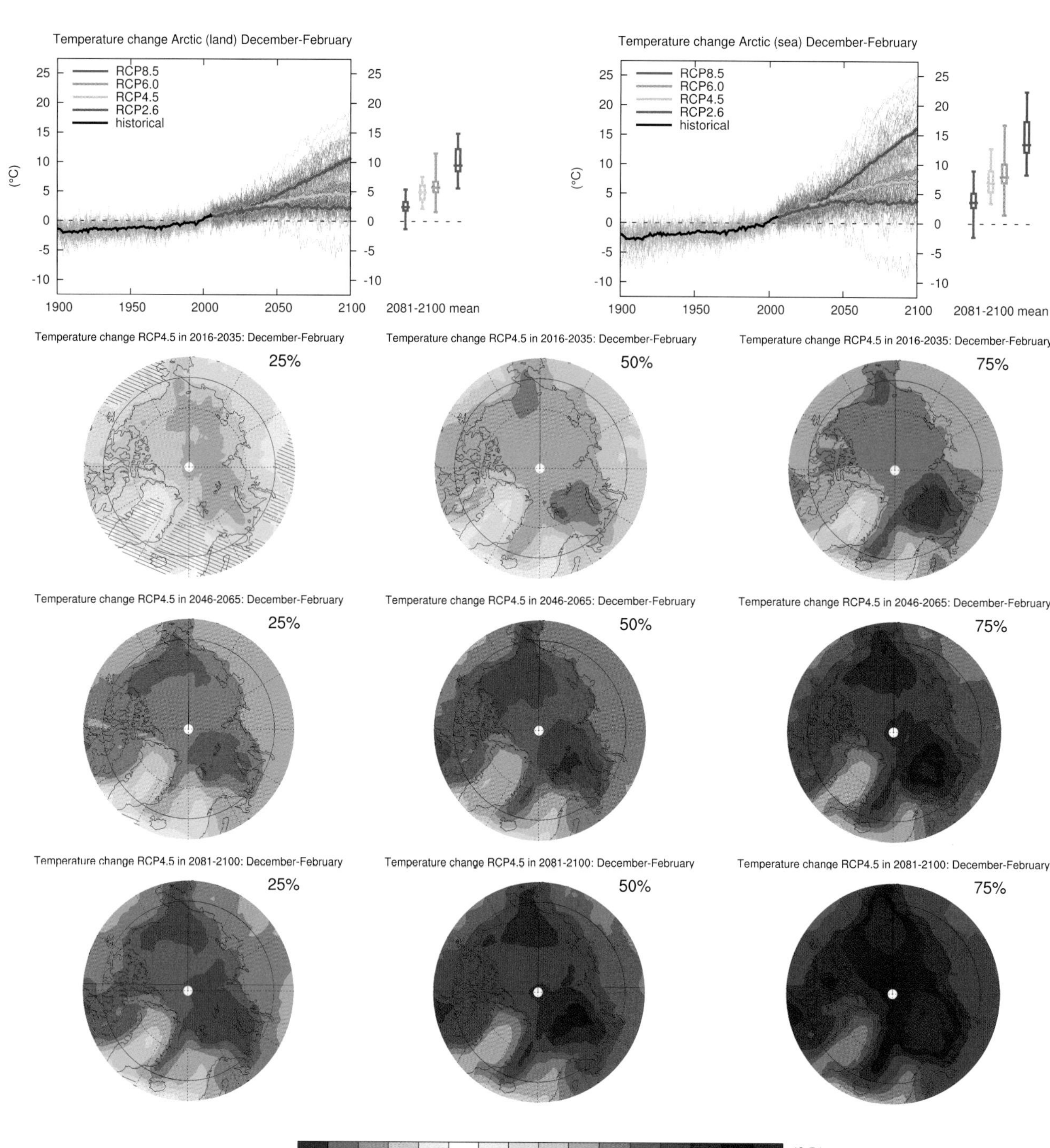

Figure AI.8 | (Top left) Time series of temperature change relative to 1986–2005 averaged over land grid points in the Arctic (67.5°N to 90°N) in December to February. (Top right) Same for sea grid points. Thin lines denote one ensemble member per model, thick lines the CMIP5 multi-model mean. On the right-hand side the 5th, 25th, 50th (median), 75th and 95th percentiles of the distribution of 20-year mean changes are given for 2081–2100 in the four RCP scenarios.

(Below) Maps of temperature changes in 2016–2035, 2046–2065 and 2081–2100 with respect to 1986–2005 in the RCP4.5 scenario. For each point, the 25th, 50th and 75th percentiles of the distribution of the CMIP5 ensemble are shown; this includes both natural variability and inter-model spread. Hatching denotes areas where the 20-year mean differences of the percentiles are less than the standard deviation of model-estimated present-day natural variability of 20-year mean differences.

Sections 9.4.1.1, 9.6.1.1, 10.3.1.1.4, 11.3.2.1.2, Box 11.2, 12.4.3.1, 14.8.2 contain relevant information regarding the evaluation of models in this region, the model spread in the context of other methods of projecting changes and the role of modes of variability and other climate phenomena.

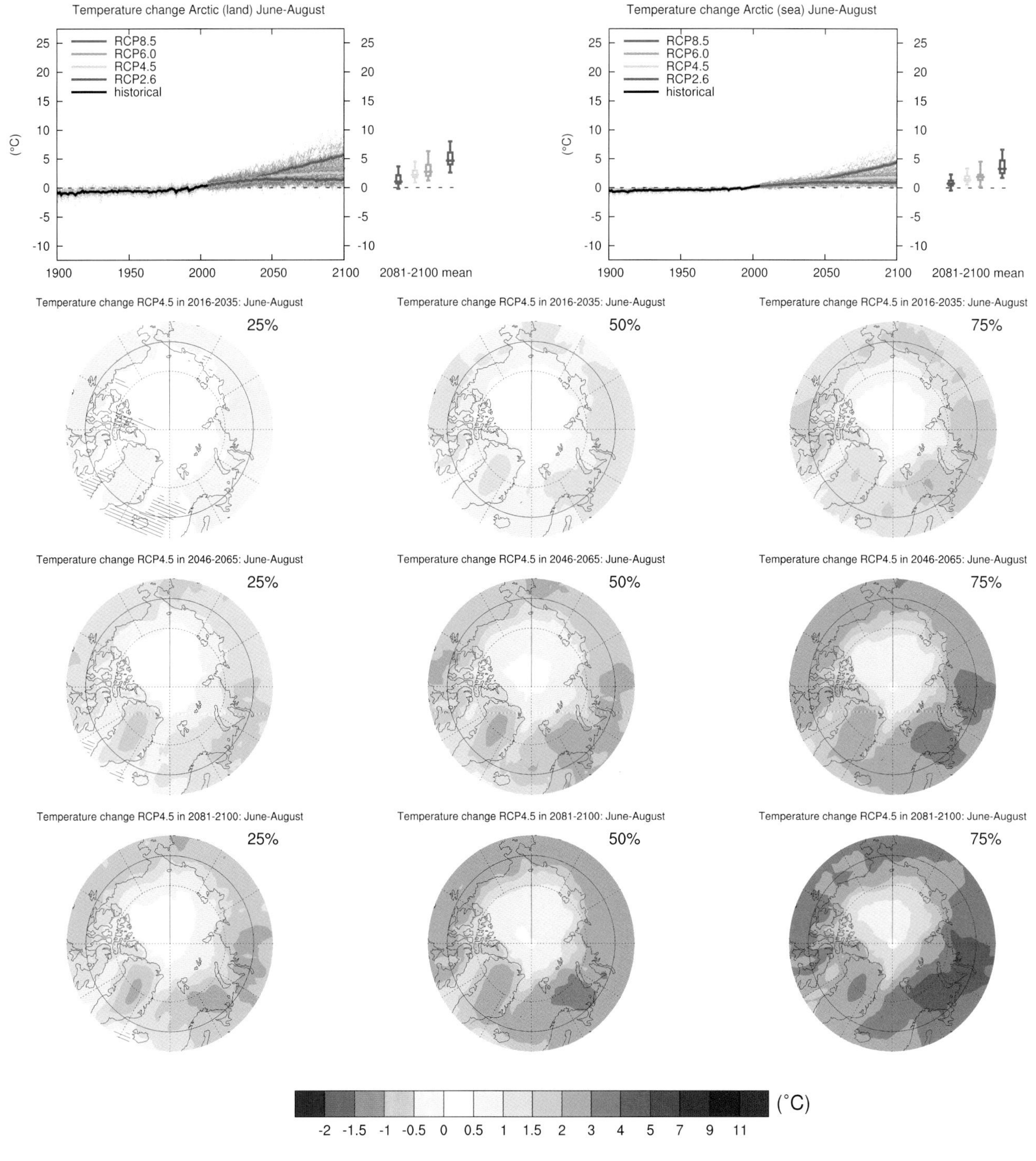

Temperature change Arctic (land) June-August

Temperature change Arctic (sea) June-August

RCP8.5
RCP6.0
RCP4.5
RCP2.6
historical

(°C)

1900 1950 2000 2050 2100 2081-2100 mean

Temperature change RCP4.5 in 2016-2035: June-August 25%

Temperature change RCP4.5 in 2016-2035: June-August 50%

Temperature change RCP4.5 in 2016-2035: June-August 75%

Temperature change RCP4.5 in 2046-2065: June-August 25%

Temperature change RCP4.5 in 2046-2065: June-August 50%

Temperature change RCP4.5 in 2046-2065: June-August 75%

Temperature change RCP4.5 in 2081-2100: June-August 25%

Temperature change RCP4.5 in 2081-2100: June-August 50%

Temperature change RCP4.5 in 2081-2100: June-August 75%

-2 -1.5 -1 -0.5 0 0.5 1 1.5 2 3 4 5 7 9 11 (°C)

Figure AI.9 | (Top left) Time series of temperature change relative to 1986–2005 averaged over land grid points in the Arctic (67.5°N to 90°N) in June to August. (Top right) Same for sea grid points. Thin lines denote one ensemble member per model, thick lines the CMIP5 multi-model mean. On the right-hand side the 5th, 25th, 50th (median), 75th and 95th percentiles of the distribution of 20-year mean changes are given for 2081–2100 in the four RCP scenarios.

(Below) Maps of temperature changes in 2016–2035, 2046–2065 and 2081–2100 with respect to 1986–2005 in the RCP4.5 scenario. For each point, the 25th, 50th and 75th percentiles of the distribution of the CMIP5 ensemble are shown; this includes both natural variability and inter-model spread. Hatching denotes areas where the 20-year mean differences of the percentiles are less than the standard deviation of model-estimated present-day natural variability of 20-year mean differences.

Sections 9.4.1.1, 9.6.1.1, 10.3.1.1.4, 11.3.2.1.2, Box 11.2, 12.4.3.1, 14.8.2 contain relevant information regarding the evaluation of models in this region, the model spread in the context of other methods of projecting changes and the role of modes of variability and other climate phenomena.

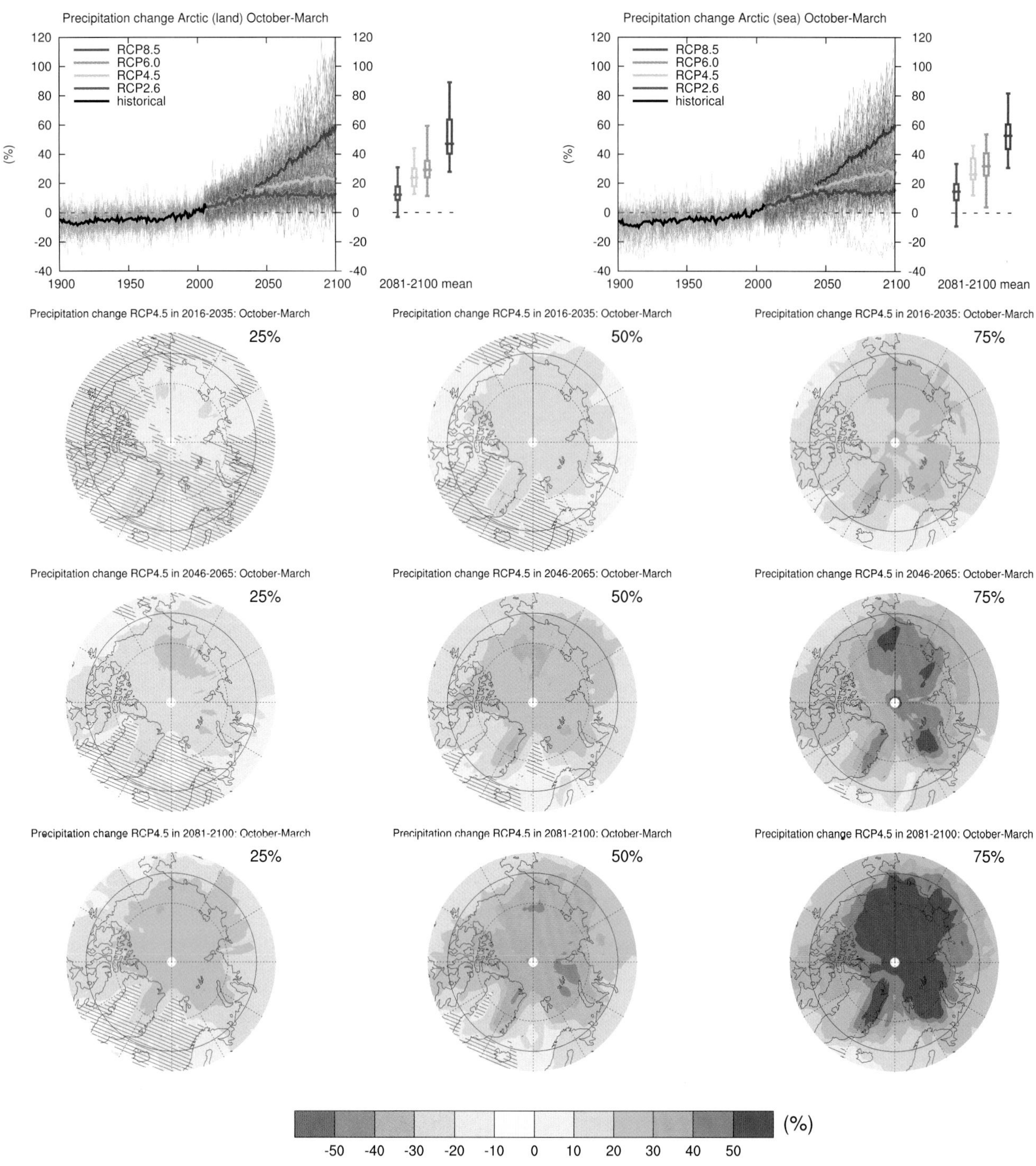

Figure AI.10 | (Top left) Time series of relative change relative to 1986–2005 in precipitation averaged over land grid points in the Arctic (67.5°N to 90°N) in October to March. (Top right) Same for sea grid points. Thin lines denote one ensemble member per model, thick lines the CMIP5 multi-model mean. On the right-hand side the 5th, 25th, 50th (median), 75th and 95th percentiles of the distribution of 20-year mean changes are given for 2081–2100 in the four RCP scenarios.

(Below) Maps of precipitation changes in 2016–2035, 2046–2065 and 2081–2100 with respect to 1986–2005 in the RCP4.5 scenario. For each point, the 25th, 50th and 75th percentiles of the distribution of the CMIP5 ensemble are shown; this includes both natural variability and inter-model spread. Hatching denotes areas where the 20-year mean differences of the percentiles are less than the standard deviation of model-estimated present-day natural variability of 20-year mean differences.

Sections 9.4.1.1, 9.6.1.1, 11.3.2.3.1, Box 11.2, 12.4.5.2, 14.8.2 contain relevant information regarding the evaluation of models in this region, the model spread in the context of other methods of projecting changes and the role of modes of variability and other climate phenomena.

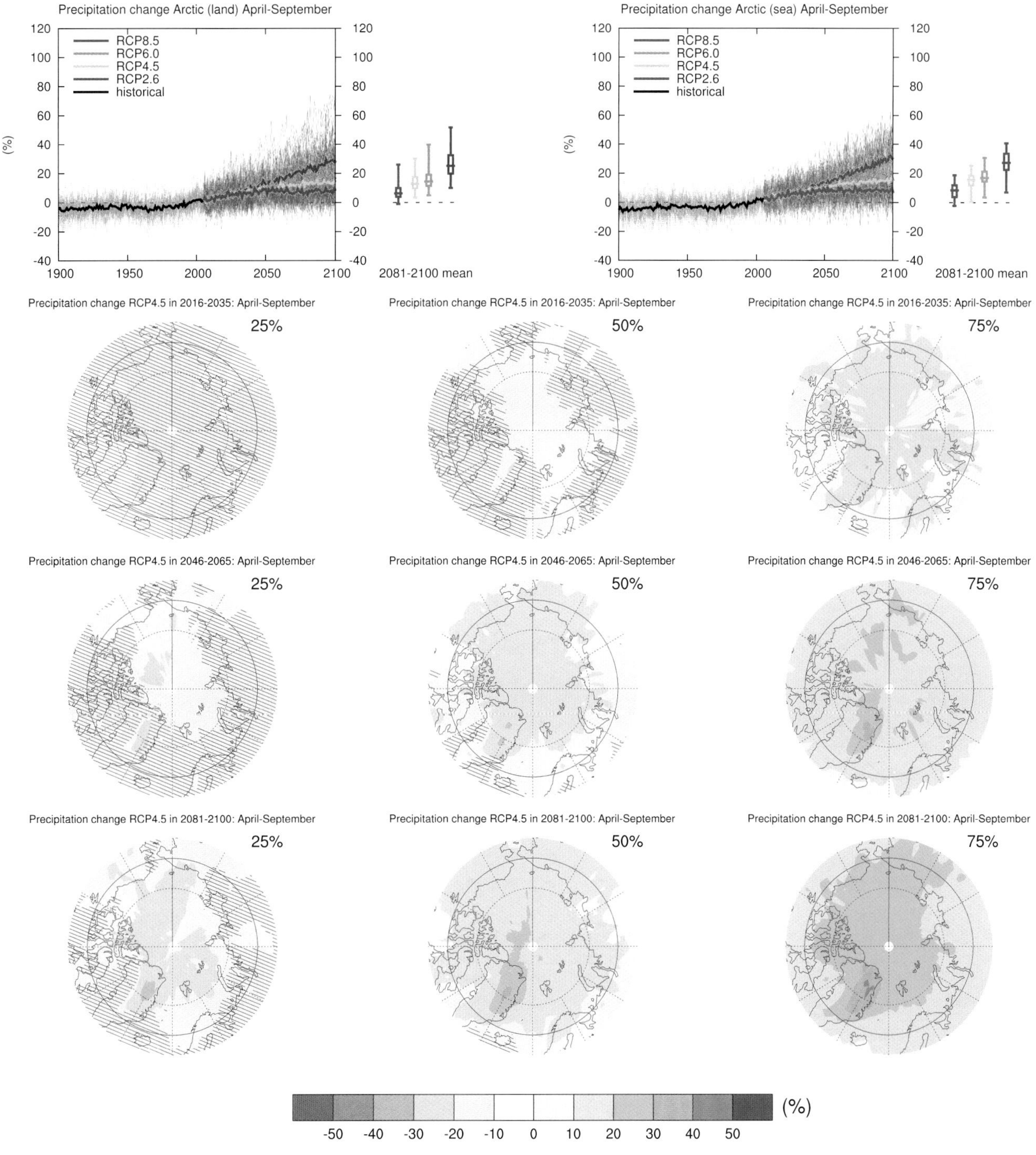

Figure AI.11 | (Top left) Time series of relative change relative to 1986–2005 in precipitation averaged over land grid points in the Arctic (67.5°N to 90°N) in April to September. (Top right) Same for sea grid points. Thin lines denote one ensemble member per model, thick lines the CMIP5 multi-model mean. On the right-hand side the 5th, 25th, 50th (median), 75th and 95th percentiles of the distribution of 20-year mean changes are given for 2081–2100 in the four RCP scenarios.

(Below) Maps of precipitation changes in 2016–2035, 2046–2065 and 2081–2100 with respect to 1986–2005 in the RCP4.5 scenario. For each point, the 25th, 50th and 75th percentiles of the distribution of the CMIP5 ensemble are shown; this includes both natural variability and inter-model spread. Hatching denotes areas where the 20-year mean differences of the percentiles are less than the standard deviation of model-estimated present-day natural variability of 20-year mean differences.

Sections 9.4.1.1, 9.6.1.1, 11.3.2.3.1, Box 11.2, 12.4.5.2, 14.8.2 contain relevant information regarding the evaluation of models in this region, the model spread in the context of other methods of projecting changes and the role of modes of variability and other climate phenomena.

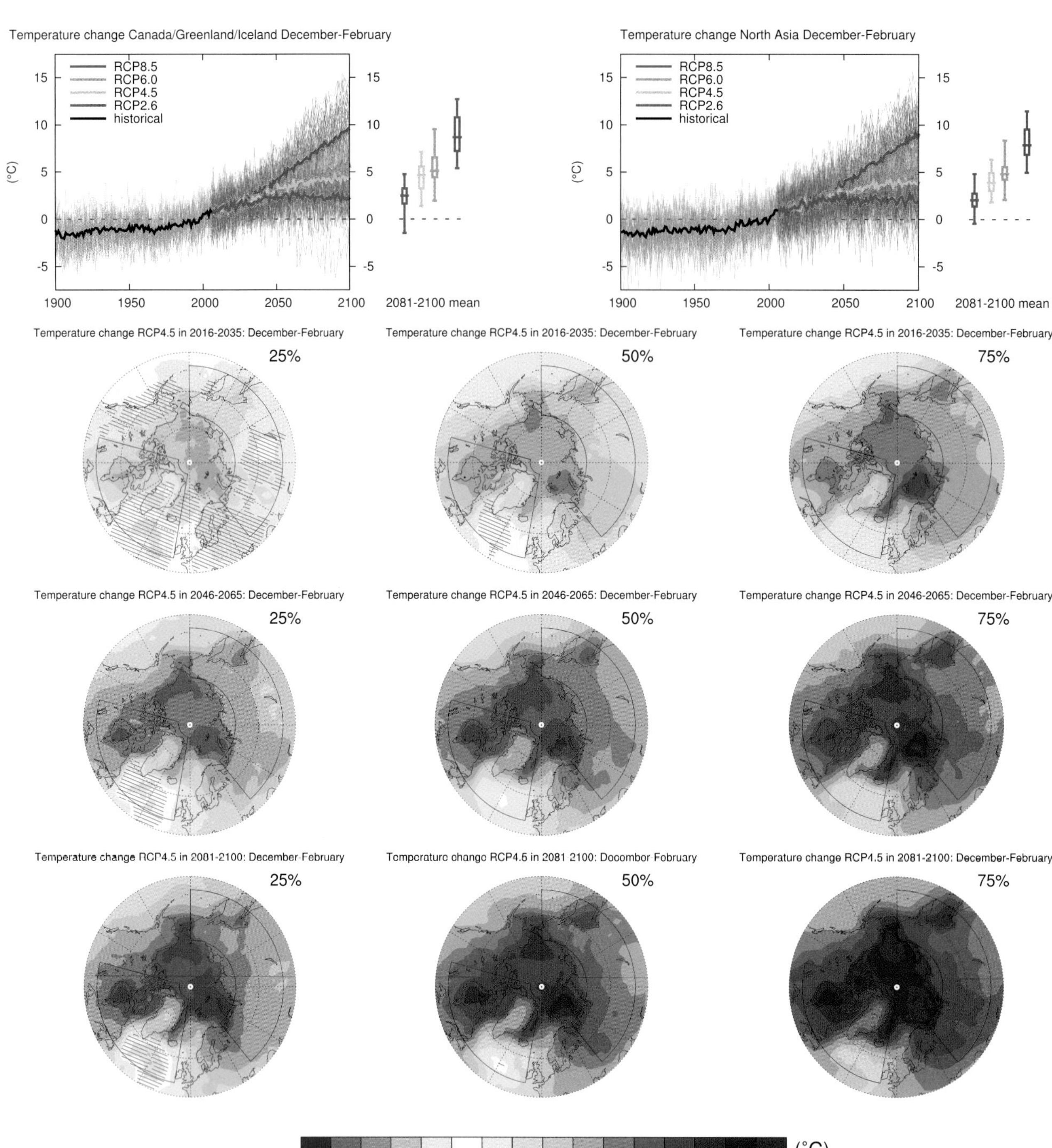

Figure AI.12 | (Top left) Time series of temperature change relative to 1986–2005 averaged over land grid points in Canada/Greenland/Iceland (50°N to 85°N, 105°W to 10°W) in December to February. (Top right) Same for land grid points in North Asia (50°N to 70°N, 40°E to 180°E). Thin lines denote one ensemble member per model, thick lines the CMIP5 multi-model mean. On the right-hand side the 5th, 25th, 50th (median), 75th and 95th percentiles of the distribution of 20-year mean changes are given for 2081–2100 in the four RCP scenarios.

(Below) Maps of temperature changes in 2016–2035, 2046–2065 and 2081–2100 with respect to 1986–2005 in the RCP4.5 scenario. For each point, the 25th, 50th and 75th percentiles of the distribution of the CMIP5 ensemble are shown; this includes both natural variability and inter-model spread. Hatching denotes areas where the 20-year mean differences of the percentiles are less than the standard deviation of model-estimated present-day natural variability of 20-year mean differences.

Sections 9.4.1.1, 9.6.1.1, 10.3.1.1.4, 11.3.2.1.2, Box 11.2, 14.8.2, 14.8.8 contain relevant information regarding the evaluation of models in this region, the model spread in the context of other methods of projecting changes and the role of modes of variability and other climate phenomena.

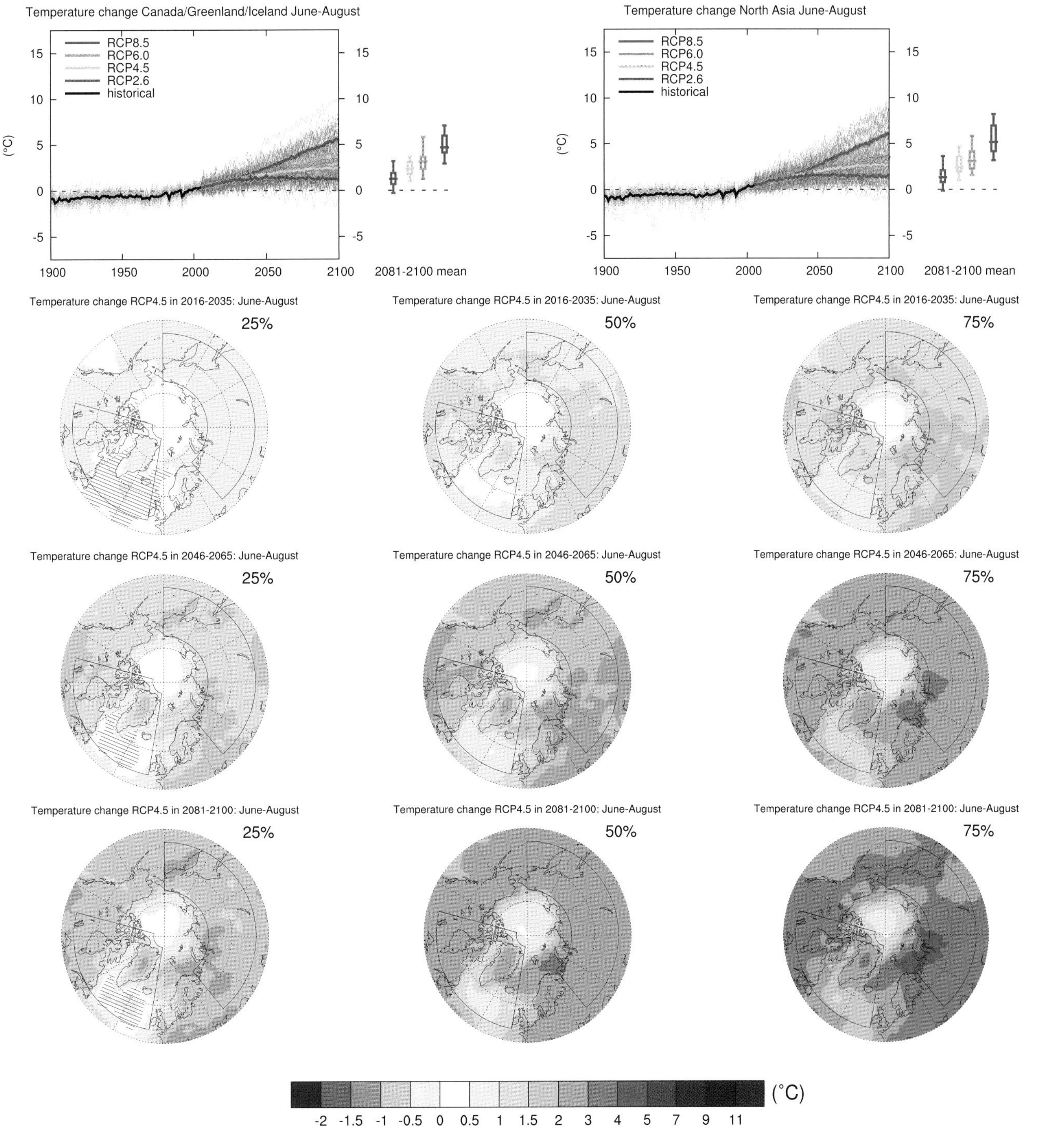

Temperature change Canada/Greenland/Iceland June-August

Temperature change North Asia June-August

RCP8.5
RCP6.0
RCP4.5
RCP2.6
historical

(°C)

1900 1950 2000 2050 2100 2081-2100 mean

Temperature change RCP4.5 in 2016-2035: June-August 25%

Temperature change RCP4.5 in 2016-2035: June-August 50%

Temperature change RCP4.5 in 2016-2035: June-August 75%

Temperature change RCP4.5 in 2046-2065: June-August 25%

Temperature change RCP4.5 in 2046-2065: June-August 50%

Temperature change RCP4.5 in 2046-2065: June-August 75%

Temperature change RCP4.5 in 2081-2100: June-August 25%

Temperature change RCP4.5 in 2081-2100: June-August 50%

Temperature change RCP4.5 in 2081-2100: June-August 75%

-2 -1.5 -1 -0.5 0 0.5 1 1.5 2 3 4 5 7 9 11 (°C)

Figure AI.13 | (Top left) Time series of temperature change relative to 1986–2005 averaged over land grid points in Canada/Greenland/Iceland (50°N to 85°N, 105°W to 10°W) in June to August. (Top right) Same for land grid points in North Asia (50°N to 70°N, 40°E to 180°E). Thin lines denote one ensemble member per model, thick lines the CMIP5 multi-model mean. On the right-hand side the 5th, 25th, 50th (median), 75th and 95th percentiles of the distribution of 20-year mean changes are given for 2081–2100 in the four RCP scenarios.

(Below) Maps of temperature changes in 2016–2035, 2046–2065 and 2081–2100 with respect to 1986–2005 in the RCP4.5 scenario. For each point, the 25th, 50th and 75th percentiles of the distribution of the CMIP5 ensemble are shown; this includes both natural variability and inter-model spread. Hatching denotes areas where the 20-year mean differences of the percentiles are less than the standard deviation of model-estimated present-day natural variability of 20-year mean differences.

Sections 9.4.1.1, 9.6.1.1, 10.3.1.1.4, 11.3.2.1.2, Box 11.2, 14.8.2, 14.8.8 contain relevant information regarding the evaluation of models in this region, the model spread in the context of other methods of projecting changes and the role of modes of variability and other climate phenomena.

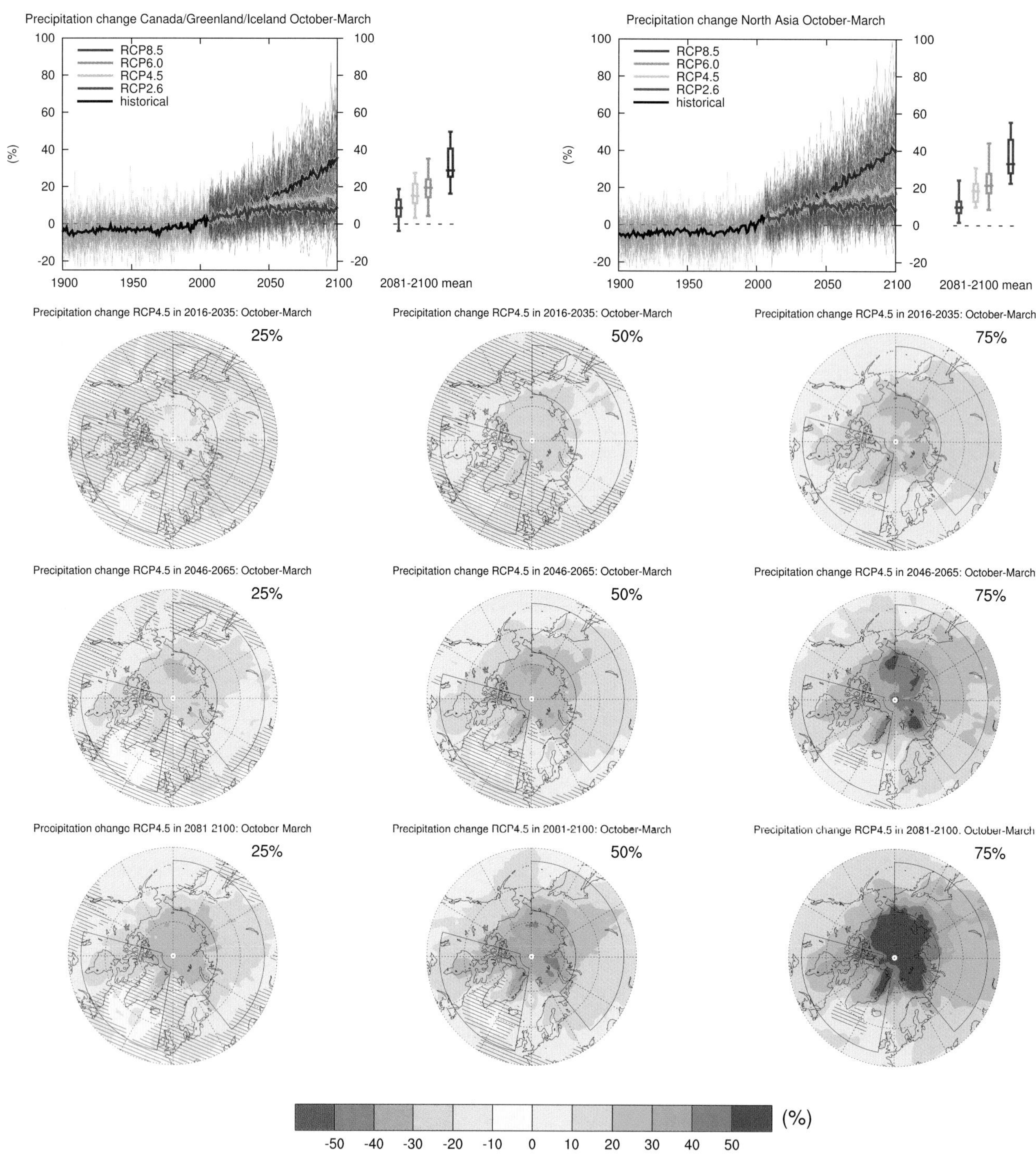

Figure AI.14 | (Top left) Time series of relative change relative to 1986–2005 in precipitation averaged over land grid points in Canada/Greenland/Iceland (50°N to 85°N, 105°W to 10°W) in October to March. (Top right) Same for land grid points in North Asia (50°N to 70°N, 40°E to 180°E). Thin lines denote one ensemble member per model, thick lines the CMIP5 multi-model mean. On the right-hand side the 5th, 25th, 50th (median), 75th and 95th percentiles of the distribution of 20-year mean changes are given for 2081–2100 in the four RCP scenarios.

(Below) Maps of precipitation changes in 2016–2035, 2046–2065 and 2081–2100 with respect to 1986–2005 in the RCP4.5 scenario. For each point, the 25th, 50th and 75th percentiles of the distribution of the CMIP5 ensemble are shown; this includes both natural variability and inter-model spread. Hatching denotes areas where the 20-year mean differences of the percentiles are less than the standard deviation of model-estimated present-day natural variability of 20-year mean differences.

Sections 9.4.1.1, 9.6.1.1, 10.3.2.2, 11.3.2.3.1, Box 11.2, 12.4.5.2, 14.8.2, 14.8.8 contain relevant information regarding the evaluation of models in this region, the model spread in the context of other methods of projecting changes and the role of modes of variability and other climate phenomena.

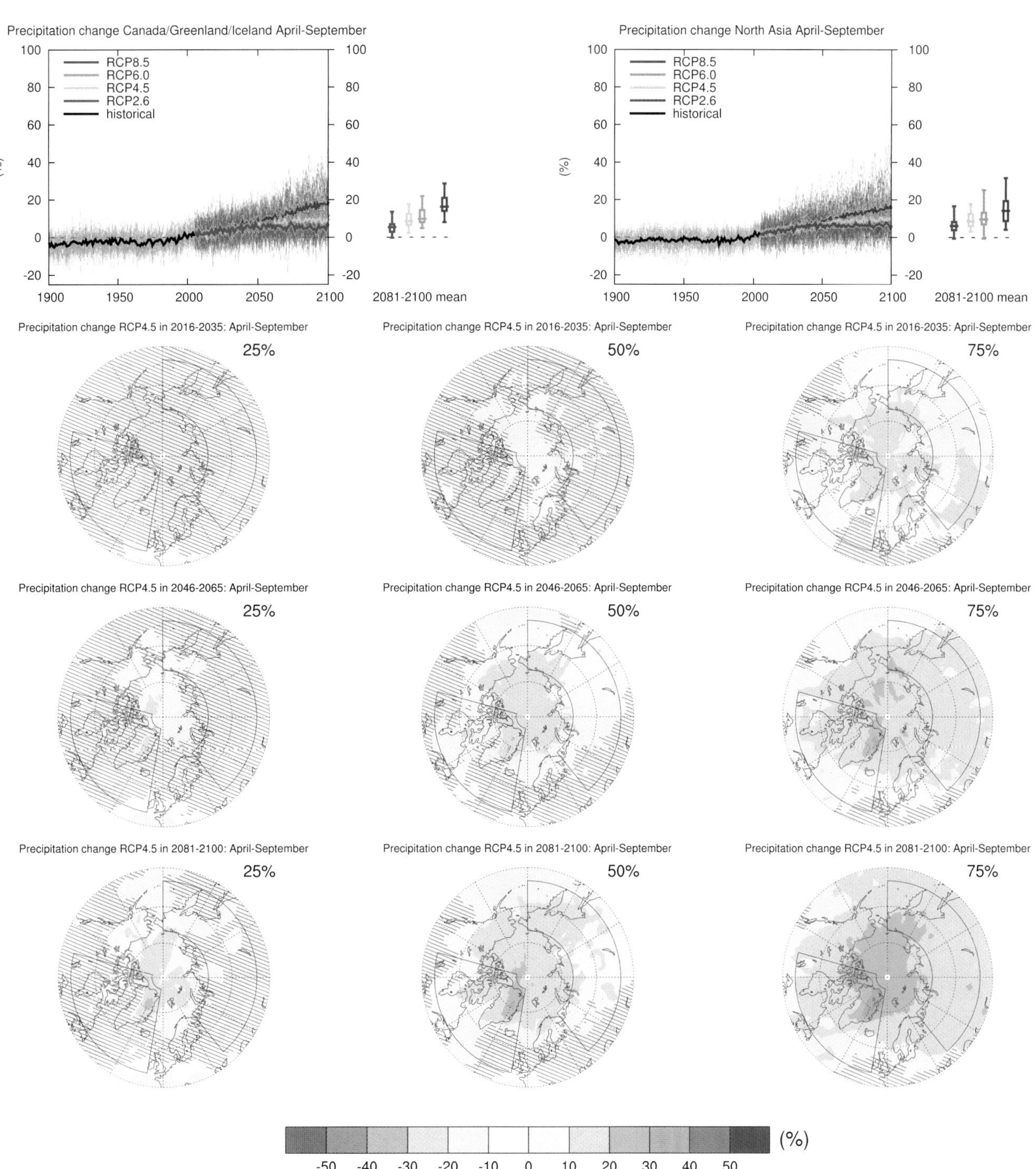

Figure AI.15 | (Top left) Time series of relative change relative to 1986–2005 in precipitation averaged over land grid points in Canada/Greenland/Iceland (50°N to 85°N, 105°W to 10°W) in April to September. (Top right) Same for land grid points in North Asia (50°N to 70°N, 40°E to 180°E). Thin lines denote one ensemble member per model, thick lines the CMIP5 multi-model mean. On the right-hand side the 5th, 25th, 50th (median), 75th and 95th percentiles of the distribution of 20-year mean changes are given for 2081–2100 in the four RCP scenarios.

(Below) Maps of precipitation changes in 2016–2035, 2046–2065 and 2081–2100 with respect to 1986–2005 in the RCP4.5 scenario. For each point, the 25th, 50th and 75th percentiles of the distribution of the CMIP5 ensemble are shown; this includes both natural variability and inter-model spread. Hatching denotes areas where the 20-year mean differences of the percentiles are less than the standard deviation of model-estimated present-day natural variability of 20-year mean differences.

Sections 9.4.1.1, 9.6.1.1, 10.3.2.2, 11.3.2.3.1, Box 11.2, 12.4.5.2, 14.8.2, 14.8.8 contain relevant information regarding the evaluation of models in this region, the model spread in the context of other methods of projecting changes and the role of modes of variability and other climate phenomena.

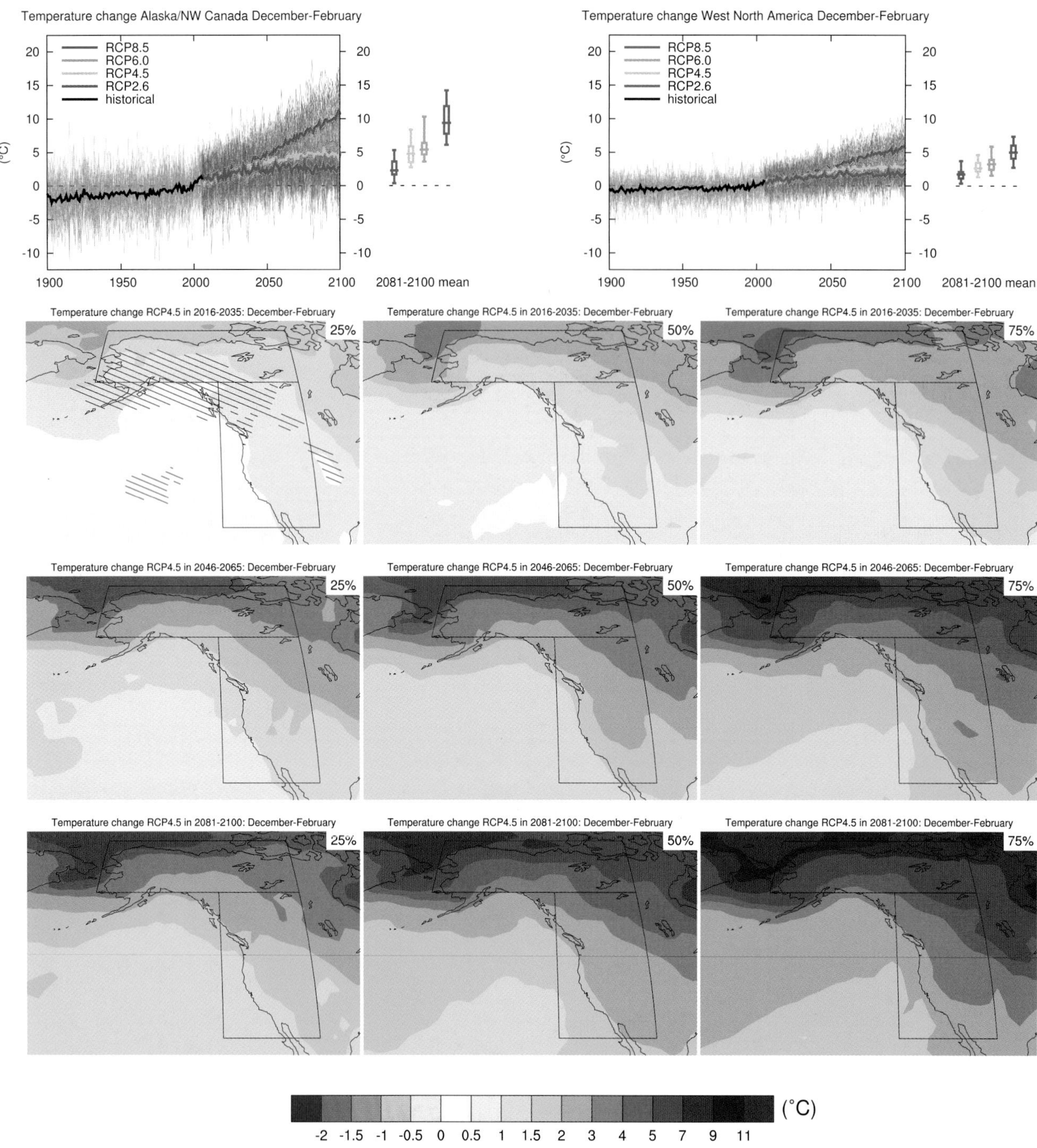

Figure AI.16 | (Top left) Time series of temperature change relative to 1986–2005 averaged over land grid points in Alaska/NW Canada (60°N to 72.6°N, 168°W to 105°W) in December to February. (Top right) Same for land grid points in West North America (28.6°N to 60°N, 130°W to 105°W). Thin lines denote one ensemble member per model, thick lines the CMIP5 multi-model mean. On the right-hand side the 5th, 25th, 50th (median), 75th and 95th percentiles of the distribution of 20-year mean changes are given for 2081–2100 in the four RCP scenarios.

(Below) Maps of temperature changes in 2016–2035, 2046–2065 and 2081–2100 with respect to 1986–2005 in the RCP4.5 scenario. For each point, the 25th, 50th and 75th percentiles of the distribution of the CMIP5 ensemble are shown; this includes both natural variability and inter-model spread. Hatching denotes areas where the 20-year mean differences of the percentiles are less than the standard deviation of model-estimated present-day natural variability of 20-year mean differences.

Sections 9.4.1.1, 9.6.1.1, 10.3.1.1.4, Box 11.2, 14.8.3 contain relevant information regarding the evaluation of models in this region, the model spread in the context of other methods of projecting changes and the role of modes of variability and other climate phenomena.

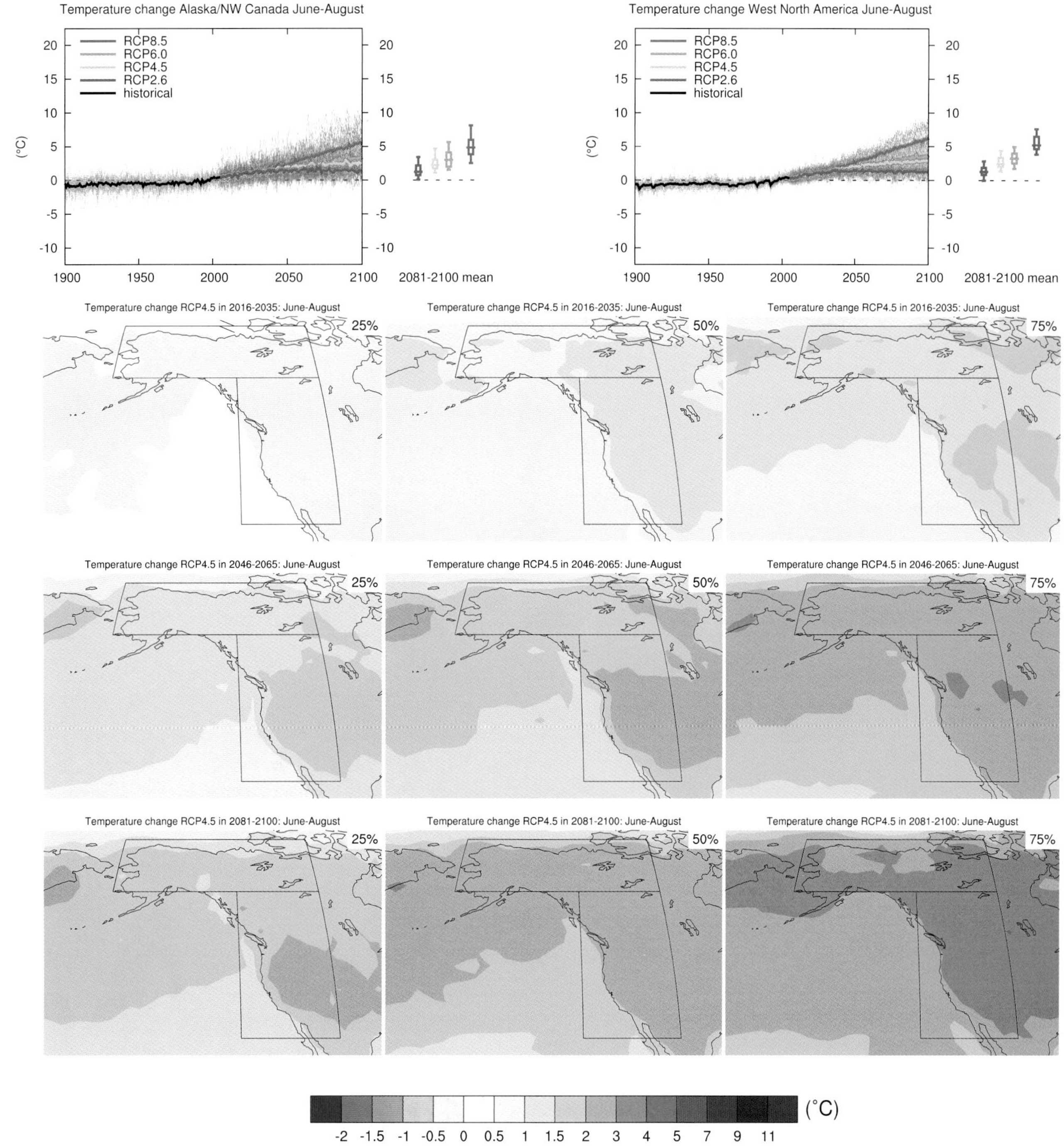

Figure AI.17 | (Top left) Time series of temperature change relative to 1986–2005 averaged over land grid points in Alaska/NW Canada (60°N to 72.6°N, 168°W to 105°W) in June to August. (Top right) Same for land grid points in West North America (28.6°N to 60°N, 130°W to 105°W). Thin lines denote one ensemble member per model, thick lines the CMIP5 multi-model mean. On the right-hand side the 5th, 25th, 50th (median), 75th and 95th percentiles of the distribution of 20-year mean changes are given for 2081–2100 in the four RCP scenarios.

(Below) Maps of temperature changes in 2016–2035, 2046–2065 and 2081–2100 with respect to 1986–2005 in the RCP4.5 scenario. For each point, the 25th, 50th and 75th percentiles of the distribution of the CMIP5 ensemble are shown; this includes both natural variability and inter-model spread. Hatching denotes areas where the 20-year mean differences of the percentiles are less than the standard deviation of model-estimated present-day natural variability of 20-year mean differences.

Sections 9.4.1.1, 9.6.1.1, 10.3.1.1.4, Box 11.2, 14.8.3 contain relevant information regarding the evaluation of models in this region, the model spread in the context of other methods of projecting changes and the role of modes of variability and other climate phenomena.

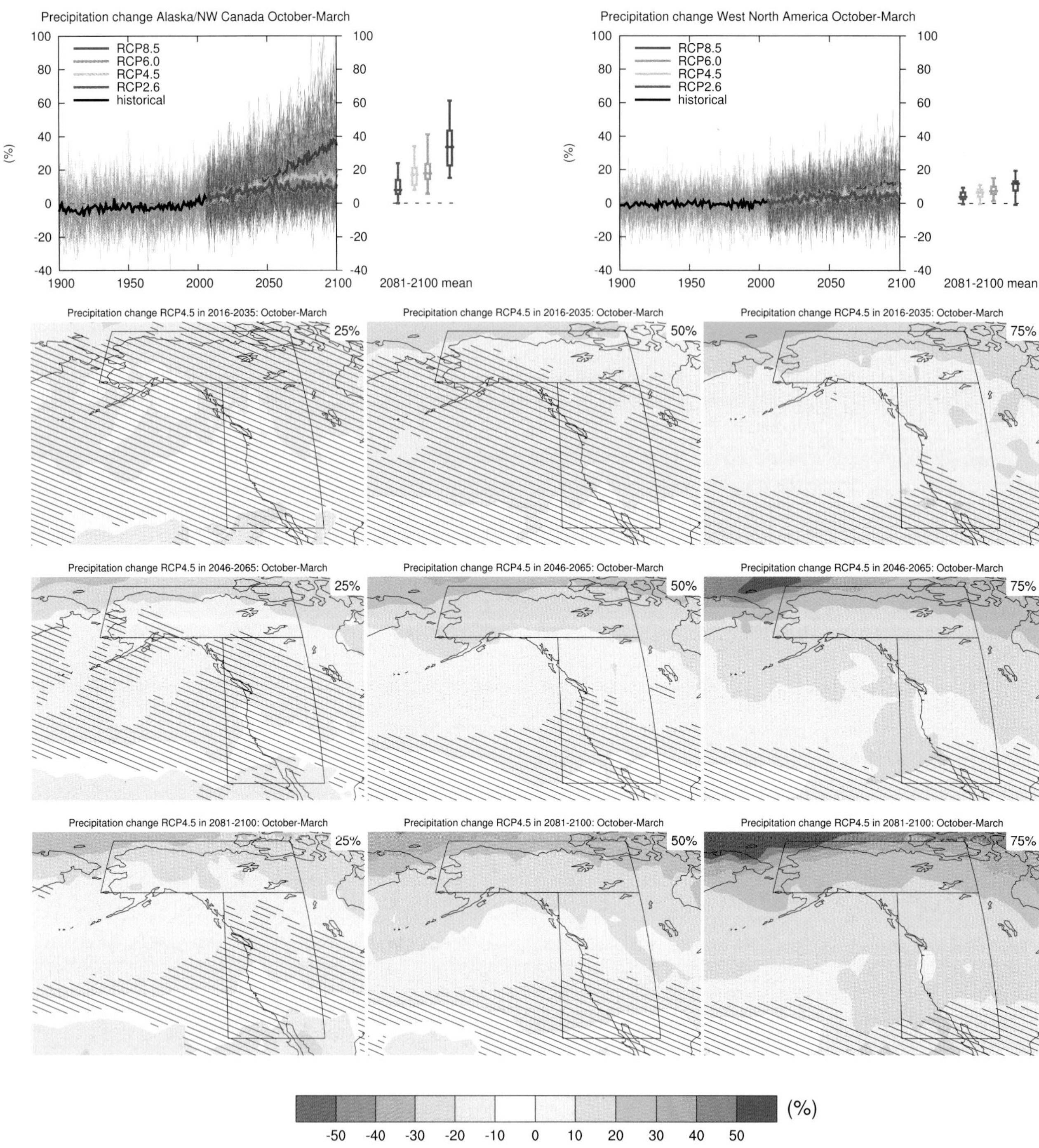

Figure AI.18 | (Top left) Time series of relative change relative to 1986–2005 in precipitation averaged over land grid points in Alaska/NW Canada (60°N to 72.6°N, 168°W to 105°W) in October to March. (Top right) Same for land grid points in West North America (28.6°N to 60°N, 130°W to 105°W). Thin lines denote one ensemble member per model, thick lines the CMIP5 multi-model mean. On the right-hand side the 5th, 25th, 50th (median), 75th and 95th percentiles of the distribution of 20-year mean changes are given for 2081–2100 in the four RCP scenarios.

(Below) Maps of precipitation changes in 2016–2035, 2046–2065 and 2081–2100 with respect to 1986–2005 in the RCP4.5 scenario. For each point, the 25th, 50th and 75th percentiles of the distribution of the CMIP5 ensemble are shown; this includes both natural variability and inter-model spread. Hatching denotes areas where the 20-year mean differences of the percentiles are less than the standard deviation of model-estimated present-day natural variability of 20-year mean differences.

Sections 9.4.1.1, 9.6.1.1, Box 11.2, 12.4.5.2, 14.2.3.1, 14.8.3 contain relevant information regarding the evaluation of models in this region, the model spread in the context of other methods of projecting changes and the role of modes of variability and other climate phenomena.

AI

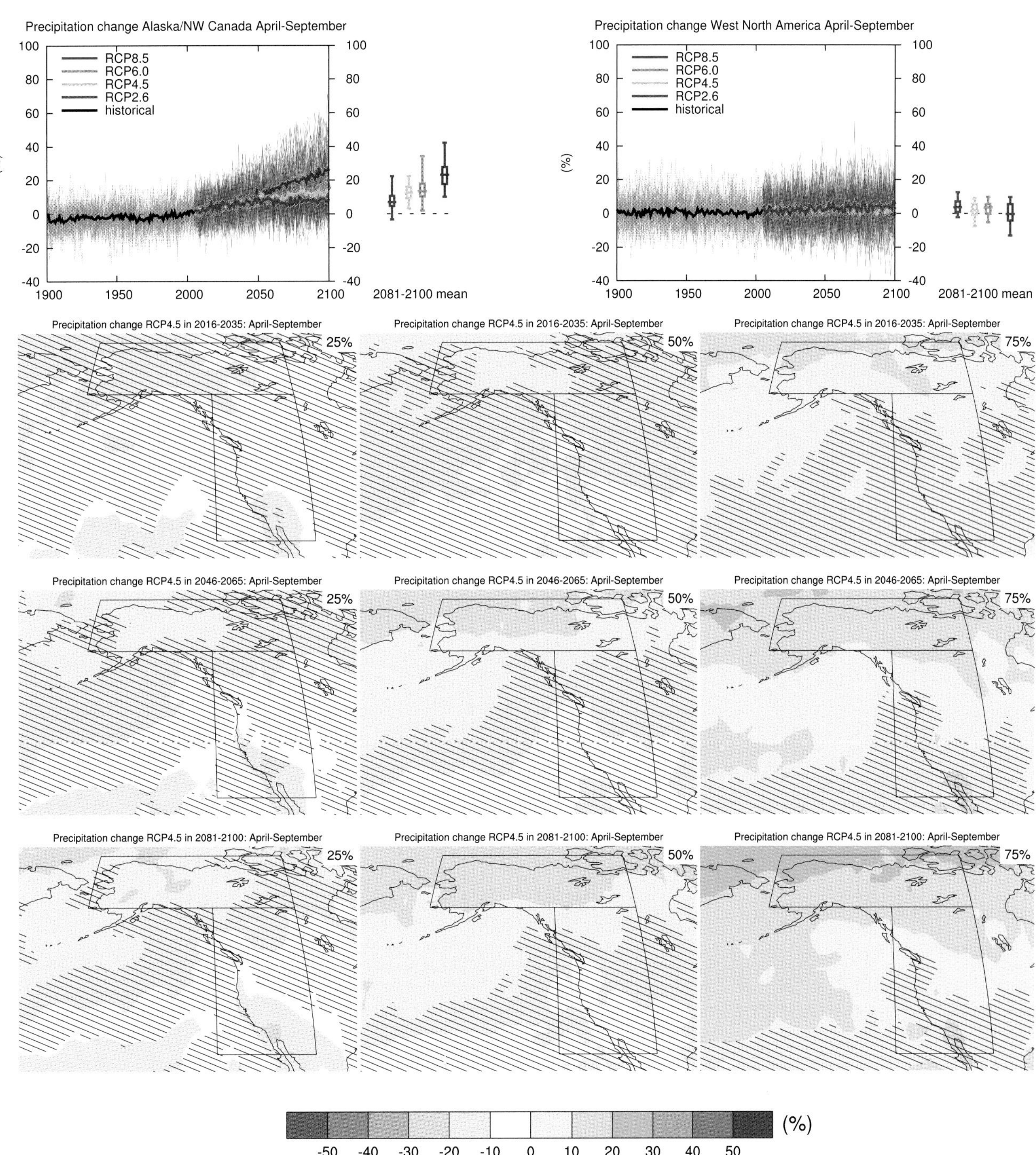

Figure AI.19 | (Top left) Time series of relative change relative to 1986–2005 in precipitation averaged over land grid points in Alaska/NW Canada (60°N to 72.6°N, 168°W to 105°W) in April to September. (Top right) Same for land grid points in West North America (28.6°N to 60°N, 130°W to 105°W). Thin lines denote one ensemble member per model, thick lines the CMIP5 multi-model mean. On the right-hand side the 5th, 25th, 50th (median), 75th and 95th percentiles of the distribution of 20-year mean changes are given for 2081–2100 in the four RCP scenarios.

(Below) Maps of precipitation changes in 2016–2035, 2046–2065 and 2081–2100 with respect to 1986–2005 in the RCP4.5 scenario. For each point, the 25th, 50th and 75th percentiles of the distribution of the CMIP5 ensemble are shown; this includes both natural variability and inter-model spread. Hatching denotes areas where the 20-year mean differences of the percentiles are less than the standard deviation of model-estimated present-day natural variability of 20-year mean differences.

Sections 9.4.1.1, 9.6.1.1, Box 11.2, 12.4.5.2, 14.2.3.1, 14.8.3 contain relevant information regarding the evaluation of models in this region, the model spread in the context of other methods of projecting changes and the role of modes of variability and other climate phenomena.

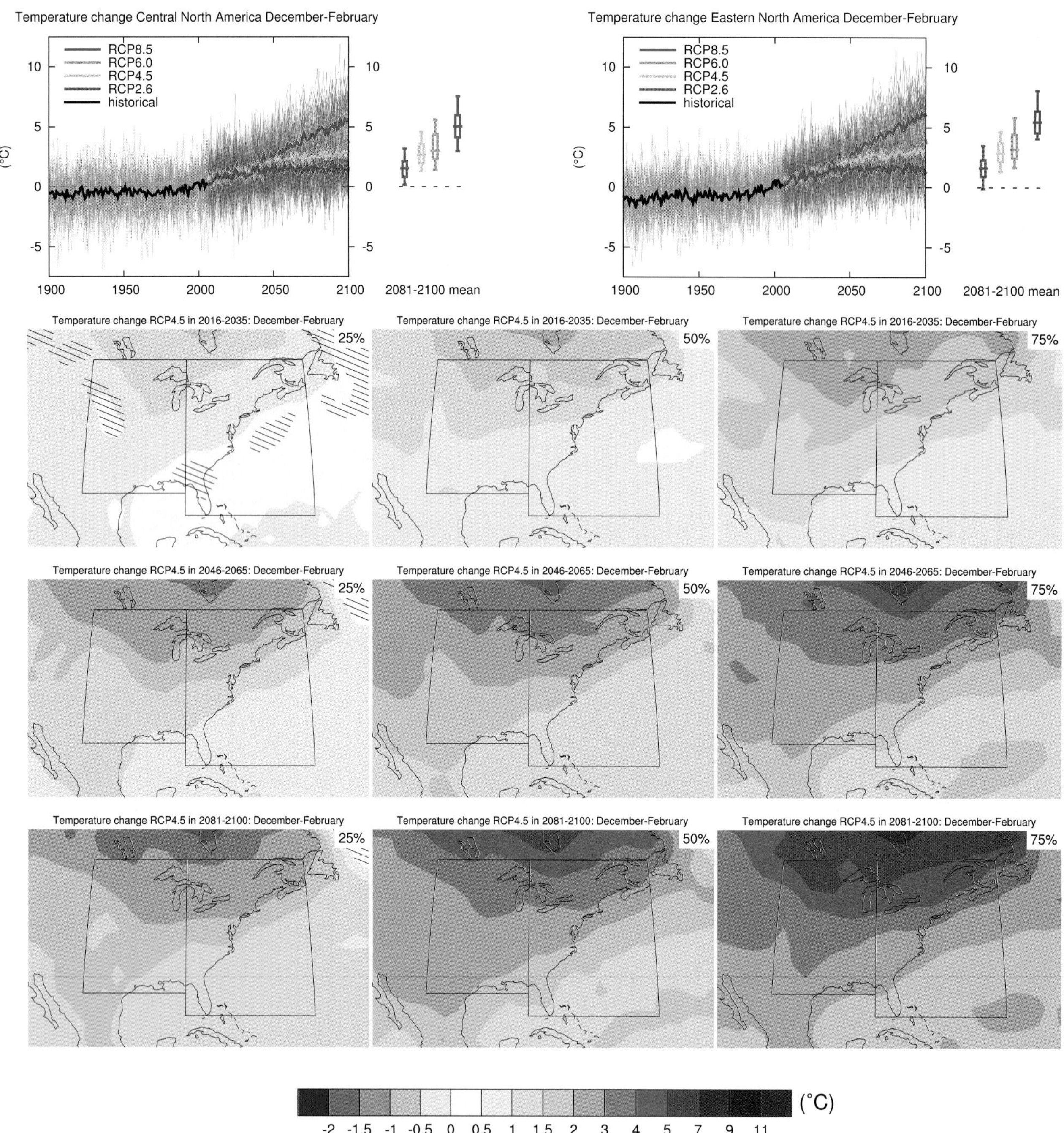

Figure AI.20 | (Top left) Time series of temperature change relative to 1986–2005 averaged over land grid points in Central North America (28.6°N to 50°N, 105°W to 85°W) in December to February. (Top right) Same for land grid points in Eastern North America (25°N to 50°N, 85°W to 60°W). Thin lines denote one ensemble member per model, thick lines the CMIP5 multi-model mean. On the right-hand side the 5th, 25th, 50th (median), 75th and 95th percentiles of the distribution of 20-year mean changes are given for 2081–2100 in the four RCP scenarios.

(Below) Maps of temperature changes in 2016–2035, 2046–2065 and 2081–2100 with respect to 1986–2005 in the RCP4.5 scenario. For each point, the 25th, 50th and 75th percentiles of the distribution of the CMIP5 ensemble are shown; this includes both natural variability and inter-model spread. Hatching denotes areas where the 20-year mean differences of the percentiles are less than the standard deviation of model-estimated present-day natural variability of 20-year mean differences.

Sections 9.4.1.1, 9.6.1.1, 10.3.1.1.4, Box 11.2, 14.8.3 contain relevant information regarding the evaluation of models in this region, the model spread in the context of other methods of projecting changes and the role of modes of variability and other climate phenomena.

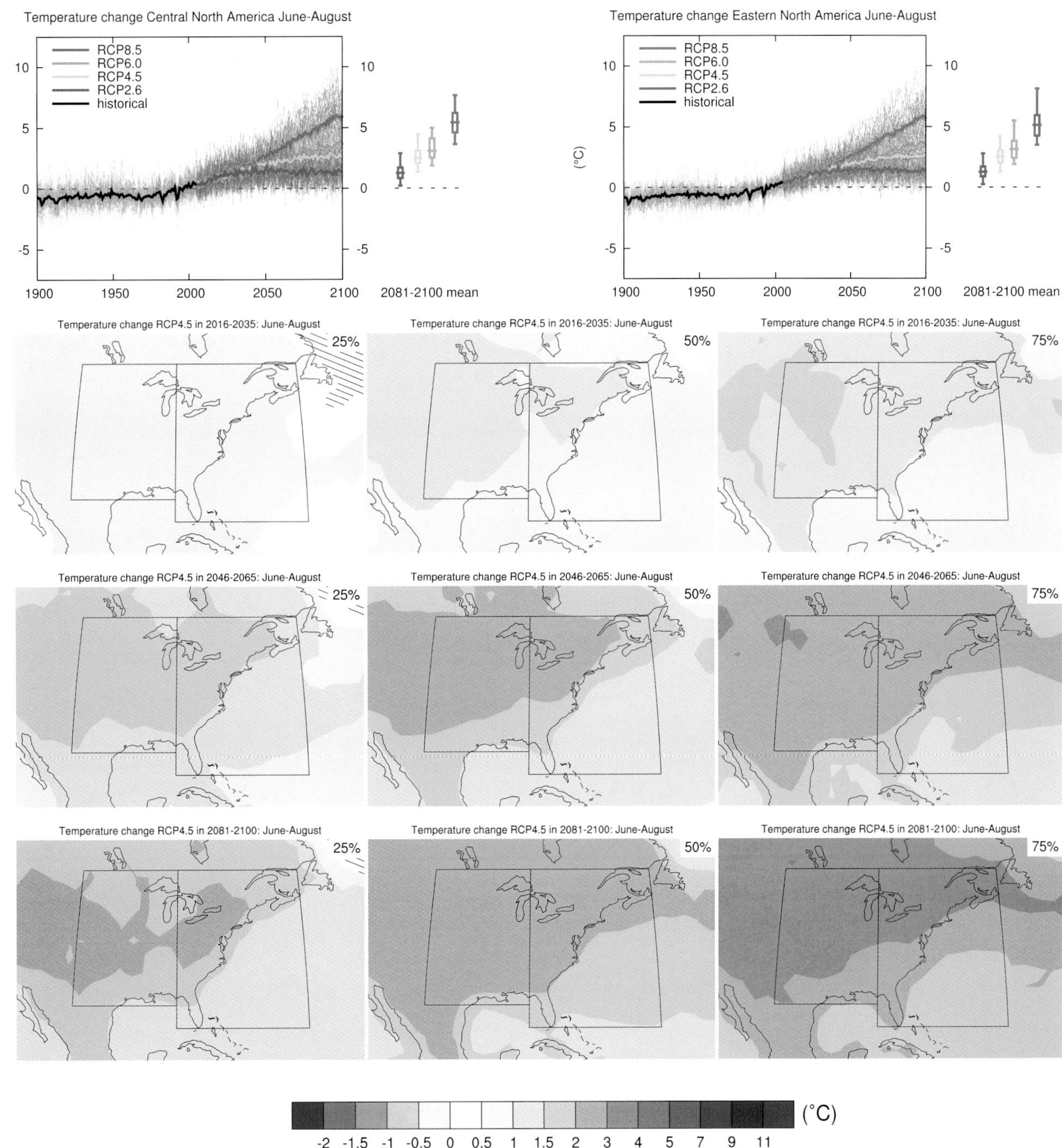

Figure AI.21 | (Top left) Time series of temperature change relative to 1986–2005 averaged over land grid points in Central North America (28.6°N to 50°N, 105°W to 85°W) in June to August. (Top right) Same for land grid points in Eastern North America (25°N to 50°N, 85°W to 60°W). Thin lines denote one ensemble member per model, thick lines the CMIP5 multi-model mean. On the right-hand side the 5th, 25th, 50th (median), 75th and 95th percentiles of the distribution of 20-year mean changes are given for 2081–2100 in the four RCP scenarios.

(Below) Maps of temperature changes in 2016–2035, 2046–2065 and 2081–2100 with respect to 1986–2005

in the RCP4.5 scenario. For each point, the 25th, 50th and 75th percentiles of the distribution of the CMIP5 ensemble are shown; this includes both natural variability and inter-model spread. Hatching denotes areas where the 20-year mean differences of the percentiles are less than the standard deviation of model-estimated present-day natural variability of 20-year mean differences.

Sections 9.4.1.1, 9.6.1.1, 10.3.1.1.4, Box 11.2, 14.8.3 contain relevant information regarding the evaluation of models in this region, the model spread in the context of other methods of projecting changes and the role of modes of variability and other climate phenomena.

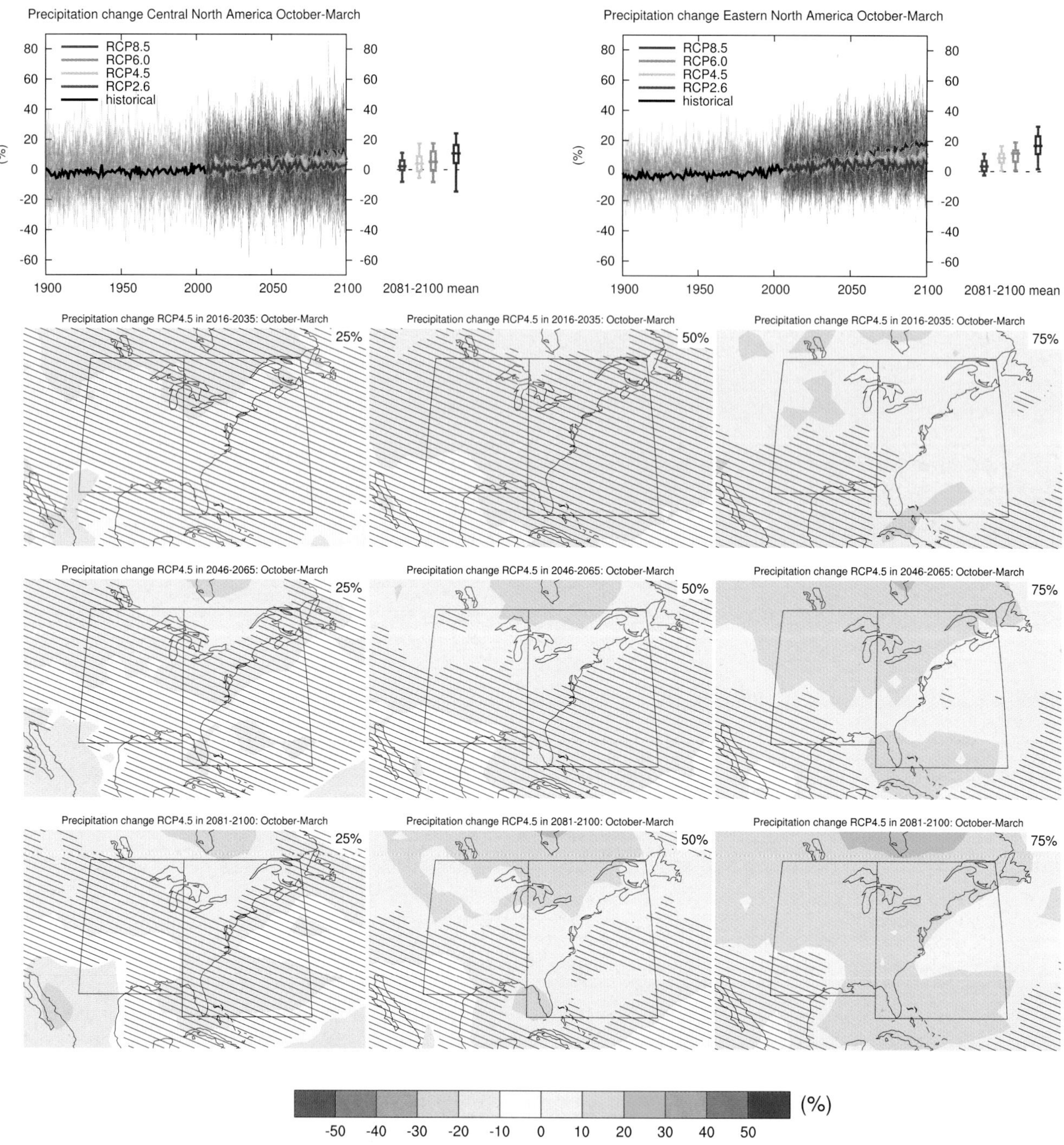

Figure AI.22 | (Top left) Time series of relative change relative to 1986–2005 in precipitation averaged over land grid points in Central North America (28.6°N to 50°N, 105°W to 85°W) in October to March. (Top right) Same for land grid points in Eastern North America (25°N to 50°N, 85°W to 60°W). Thin lines denote one ensemble member per model, thick lines the CMIP5 multi-model mean. On the right-hand side the 5th, 25th, 50th (median), 75th and 95th percentiles of the distribution of 20-year mean changes are given for 2081–2100 in the four RCP scenarios.

(Below) Maps of precipitation changes in 2016–2035, 2046–2065 and 2081–2100 with respect to 1986–2005 in the RCP4.5 scenario. For each point, the 25th, 50th and 75th percentiles of the distribution of the CMIP5 ensemble are shown; this includes both natural variability and inter-model spread. Hatching denotes areas where the 20-year mean differences of the percentiles are less than the standard deviation of model-estimated present-day natural variability of 20-year mean differences.

Sections 9.4.1.1, 9.6.1.1, Box 11.2, 14.8.3 contain relevant information regarding the evaluation of models in this region, the model spread in the context of other methods of projecting changes and the role of modes of variability and other climate phenomena.

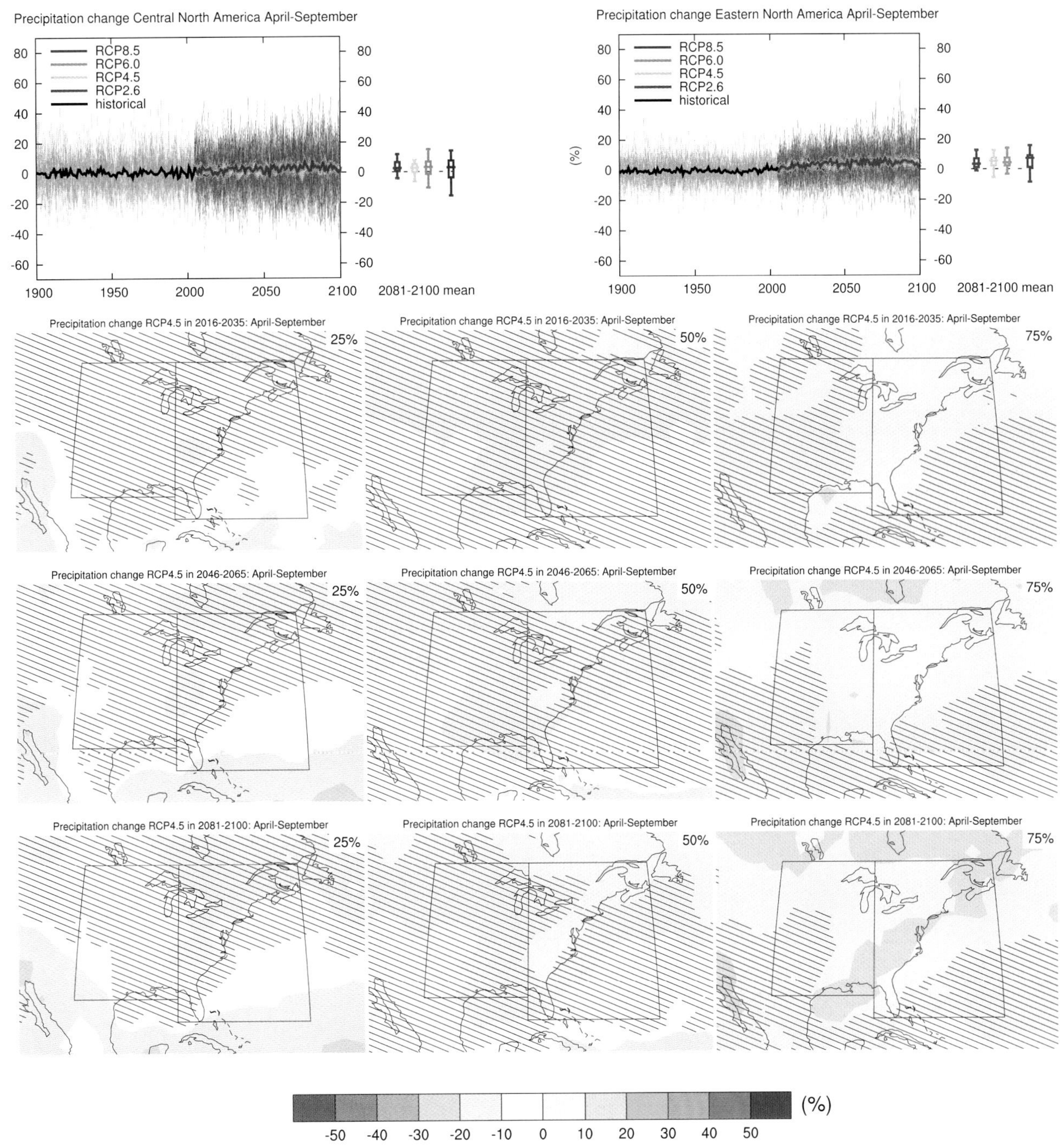

Figure AI.23 | (Top left) Time series of relative change relative to 1986–2005 in precipitation averaged over land grid points in Central North America (28.6°N to 50°N, 105°W to 85°W) in April to September. (Top right) Same for land grid points in Eastern North America (25°N to 50°N, 85°W to 60°W). Thin lines denote one ensemble member per model, thick lines the CMIP5 multi-model mean. On the right-hand side the 5th, 25th, 50th (median), 75th and 95th percentiles of the distribution of 20-year mean changes are given for 2081–2100 in the four RCP scenarios.

(Below) Maps of precipitation changes in 2016–2035, 2046–2065 and 2081–2100 with respect to 1986–2005 in the RCP4.5 scenario. For each point, the 25th, 50th and 75th percentiles of the distribution of the CMIP5 ensemble are shown; this includes both natural variability and inter-model spread. Hatching denotes areas where the 20-year mean differences of the percentiles are less than the standard deviation of model-estimated present-day natural variability of 20-year mean differences.

Sections 9.4.1.1, 9.6.1.1, Box 11.2, 14.8.3 contain relevant information regarding the evaluation of models in this region, the model spread in the context of other methods of projecting changes and the role of modes of variability and other climate phenomena.

Temperature change Central America December-February

Temperature change Caribbean (land and sea) December-February

RCP8.5
RCP6.0
RCP4.5
RCP2.6
historical

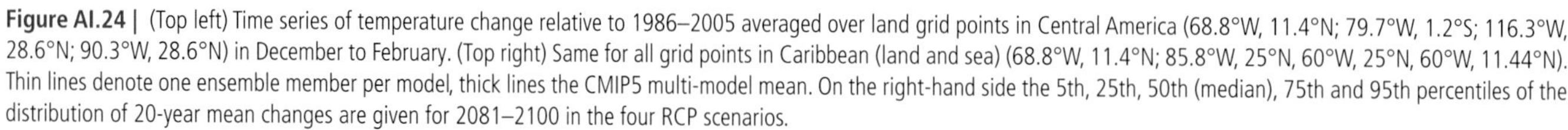

Figure AI.24 | (Top left) Time series of temperature change relative to 1986–2005 averaged over land grid points in Central America (68.8°W, 11.4°N; 79.7°W, 1.2°S; 116.3°W, 28.6°N; 90.3°W, 28.6°N) in December to February. (Top right) Same for all grid points in Caribbean (land and sea) (68.8°W, 11.4°N; 85.8°W, 25°N, 60°W, 25°N, 60°W, 11.44°N). Thin lines denote one ensemble member per model, thick lines the CMIP5 multi-model mean. On the right-hand side the 5th, 25th, 50th (median), 75th and 95th percentiles of the distribution of 20-year mean changes are given for 2081–2100 in the four RCP scenarios.

(Below) Maps of temperature changes in 2016–2035, 2046–2065 and 2081–2100 with respect to 1986–2005 in the RCP4.5 scenario. For each point, the 25th, 50th and 75th percentiles of the distribution of the CMIP5 ensemble are shown; this includes both natural variability and inter-model spread. Hatching denotes areas where the 20-year mean differences of the percentiles are less than the standard deviation of model-estimated present-day natural variability of 20-year mean differences.

Sections 9.4.1.1, 9.6.1.1, 10.3.1.1.4, Box 11.2, 14.8.4 contain relevant information regarding the evaluation of models in this region, the model spread in the context of other methods of projecting changes and the role of modes of variability and other climate phenomena.

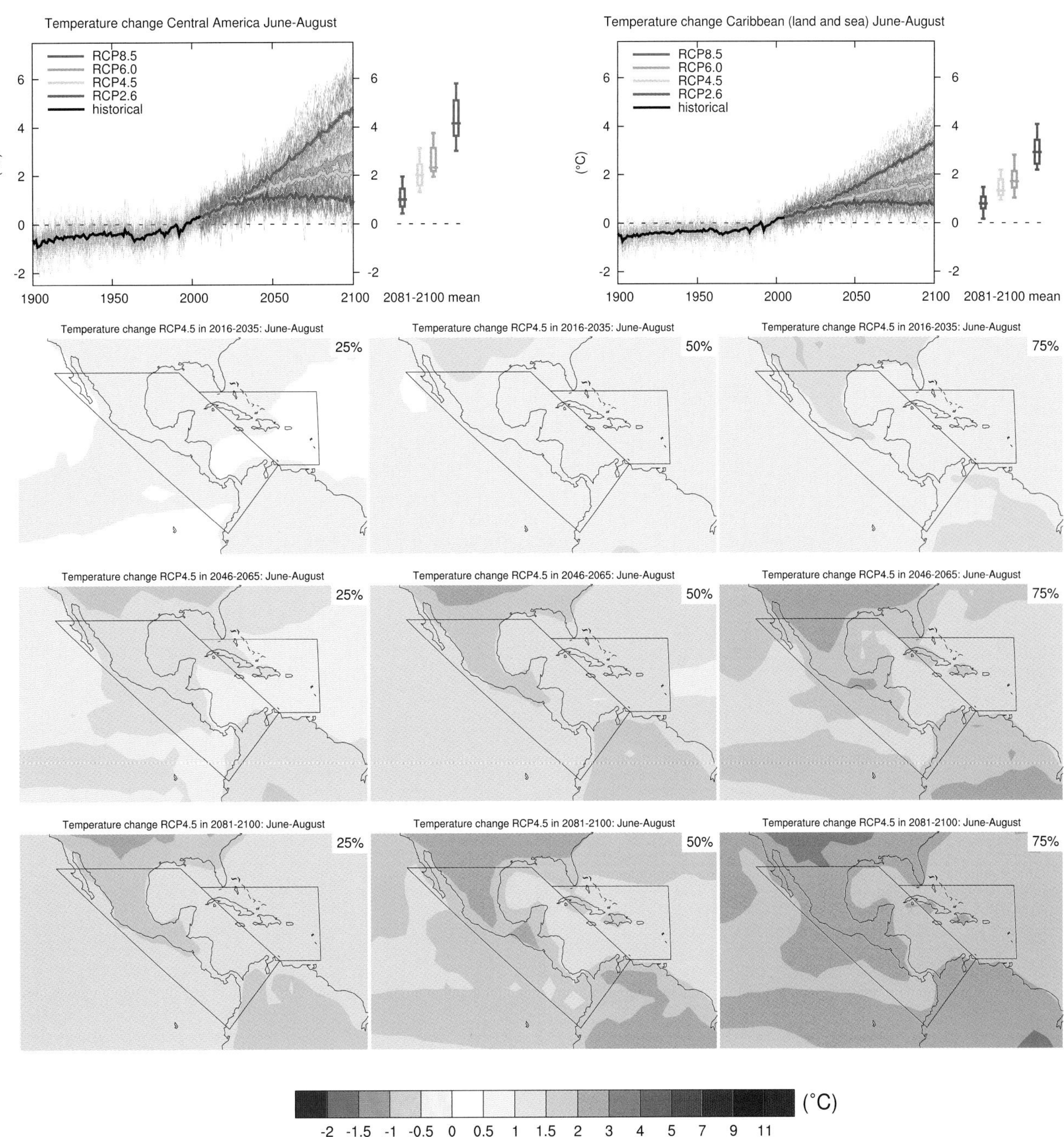

Figure AI.25 | (Top left) Time series of temperature change relative to 1986–2005 averaged over land grid points in Central America (68.8°W, 11.4°N; 79.7°W, 1.2°S; 116.3°W, 28.6°N; 90.3°W, 28.6°N) in June to August. (Top right) Same for all grid points in Caribbean (land and sea) (68.8°W, 11.4°N; 85.8°W, 25°N, 60°W, 25°N, 60°W, 11.44°N). Thin lines denote one ensemble member per model, thick lines the CMIP5 multi-model mean. On the right-hand side the 5th, 25th, 50th (median), 75th and 95th percentiles of the distribution of 20-year mean changes are given for 2081–2100 in the four RCP scenarios.

(Below) Maps of temperature changes in 2016–2035, 2046–2065 and 2081–2100 with respect to 1986–2005 in the RCP4.5 scenario. For each point, the 25th, 50th and 75th percentiles of the distribution of the CMIP5 ensemble are shown; this includes both natural variability and inter-model spread. Hatching denotes areas where the 20-year mean differences of the percentiles are less than the standard deviation of model-estimated present-day natural variability of 20-year mean differences.

Sections 9.4.1.1, 9.6.1.1, 10.3.1.1.4, Box 11.2, 14.8.4 contain relevant information regarding the evaluation of models in this region, the model spread in the context of other methods of projecting changes and the role of modes of variability and other climate phenomena.

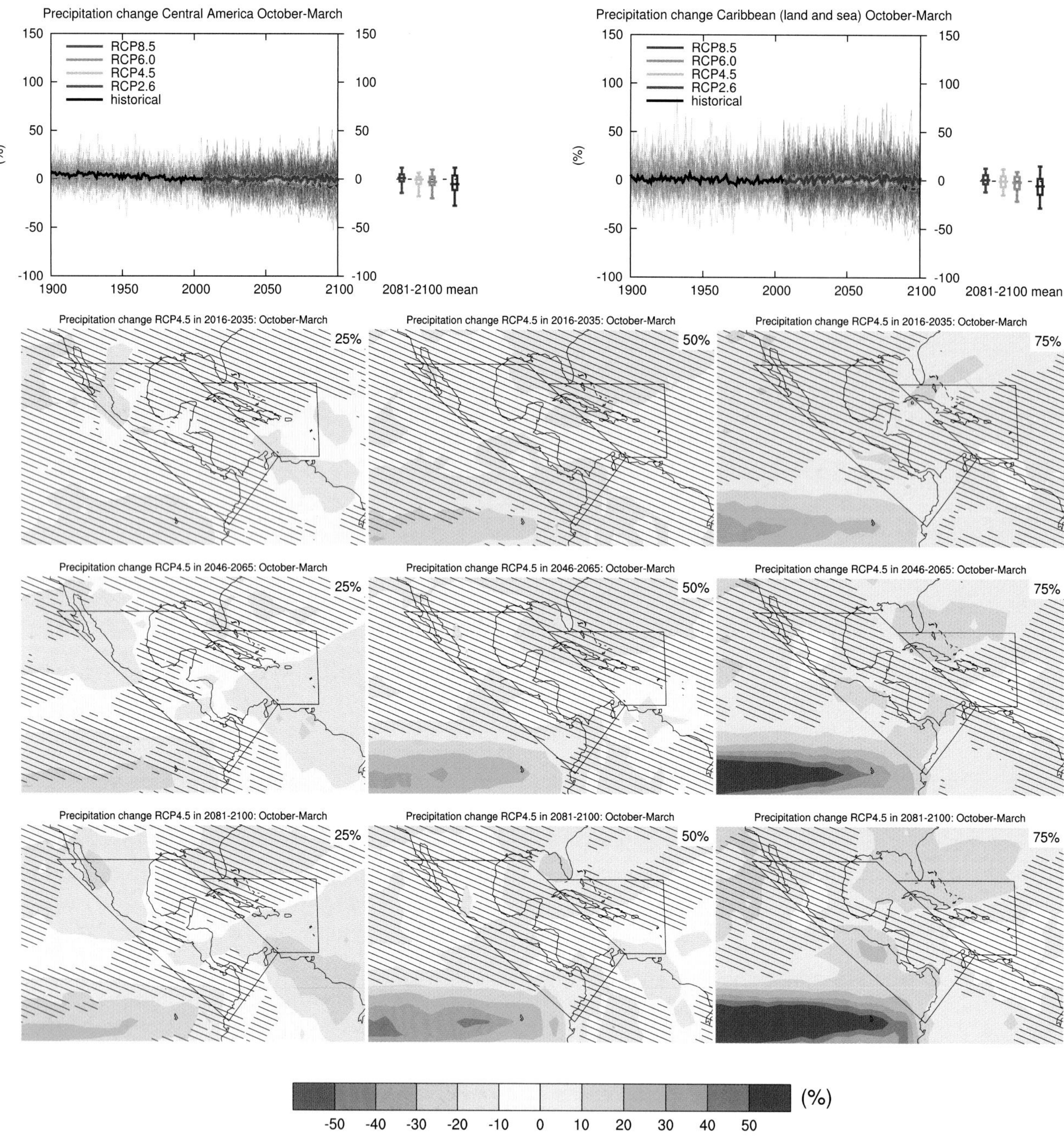

Figure AI.26 | (Top left) Time series of relative change relative to 1986–2005 in precipitation averaged over land grid points in Central America (68.8°W,11.4°N; 79.7°W, 1.2°S; 116.3°W,28.6°N; 90.3°W,28.6°N) in October to March. (Top right) Same for all grid points in Caribbean (land and sea) (68.8°W, 11.4°N; 85.8°W, 25°N, 60°W, 25°N, 60°W, 11.44°N). Thin lines denote one ensemble member per model, thick lines the CMIP5 multi-model mean. On the right-hand side the 5th, 25th, 50th (median), 75th and 95th percentiles of the distribution of 20-year mean changes are given for 2081–2100 in the four RCP scenarios.

(Below) Maps of precipitation changes in 2016–2035, 2046–2065 and 2081–2100 with respect to 1986–2005 in the RCP4.5 scenario. For each point, the 25th, 50th and 75th percentiles of the distribution of the CMIP5 ensemble are shown; this includes both natural variability and inter-model spread. Hatching denotes areas where the 20-year mean differences of the percentiles are less than the standard deviation of model-estimated present-day natural variability of 20-year mean differences.

Sections 9.4.1.1, 9.6.1.1, Box 11.2, 12.4.5.2, 14.2.3.1, 14.8.4 contain relevant information regarding the evaluation of models in this region, the model spread in the context of other methods of projecting changes and the role of modes of variability and other climate phenomena.

Precipitation change Central America April-September

Precipitation change Caribbean (land and sea) April-September

Figure AI.27 | (Top left) Time series of relative change relative to 1986–2005 in precipitation averaged over land grid points in Central America (68.8°W, 11.4°N; 79.7°W, 1.2°S; 116.3°W, 28.6°N; 90.3°W, 28.6°N) in April to September. (Top right) Same for all grid points in Caribbean (land and sea) (68.8°W, 11.4°N; 85.8°W, 25°N, 60°W, 25°N, 60°W, 11.44°N). Thin lines denote one ensemble member per model, thick lines the CMIP5 multi-model mean. On the right-hand side the 5th, 25th, 50th (median), 75th and 95th percentiles of the distribution of 20-year mean changes are given for 2081–2100 in the four RCP scenarios.

(Below) Maps of precipitation changes in 2016–2035, 2046–2065 and 2081–2100 with respect to 1986–2005 in the RCP4.5 scenario. For each point, the 25th, 50th and 75th percentiles of the distribution of the CMIP5 ensemble are shown; this includes both natural variability and inter-model spread. Hatching denotes areas where the 20-year mean differences of the percentiles are less than the standard deviation of model-estimated present-day natural variability of 20-year mean differences.

Sections 9.4.1.1, 9.6.1.1, Box 11.2, 12.4.5.2, 14.2.3.1, 14.8.4 contain relevant information regarding the evaluation of models in this region, the model spread in the context of other methods of projecting changes and the role of modes of variability and other climate phenomena.

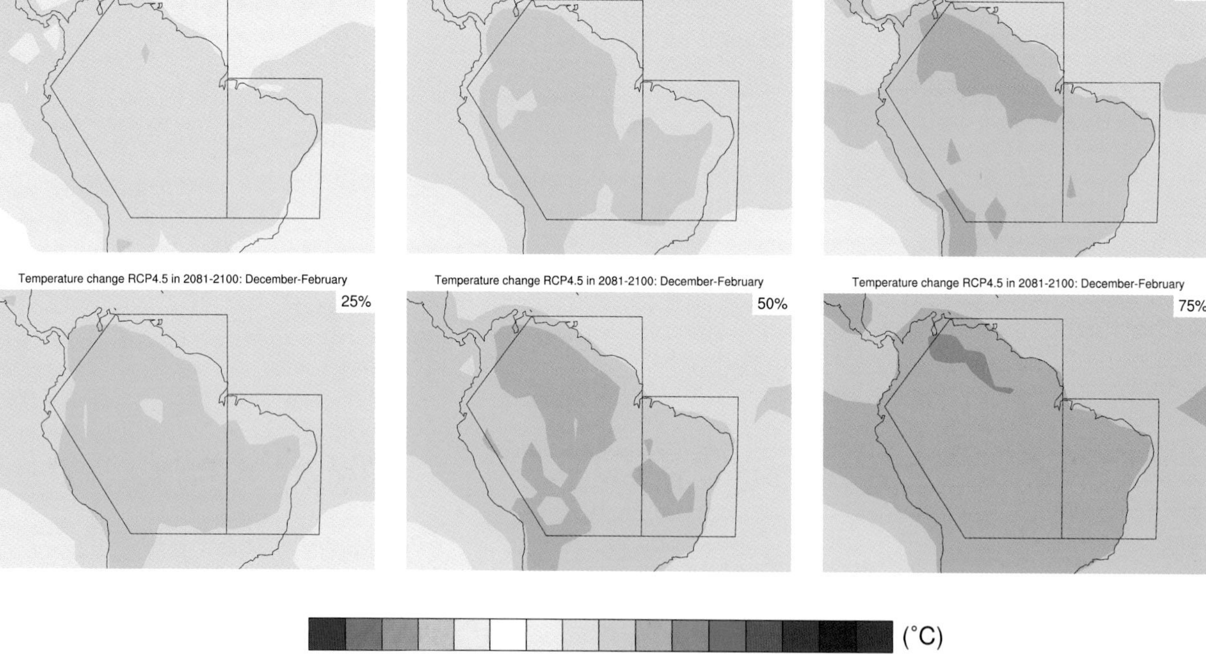

Figure AI.28 | (Top left) Time series of temperature change relative to 1986–2005 averaged over land grid points in the Amazon (20°S, 66.4°W; 1.24°S, 79.7°W; 11.44°N, 68.8°W; 11.44°N, 50°W; 20°S, 50°W) in December-February. (Top right) Same for land grid points in northeast Brazil (20°S to EQ, 50°W to 34°W). Thin lines denote one ensemble member per model, thick lines the CMIP5 multi-model mean. On the right-hand side the 5th, 25th, 50th (median), 75th and 95th percentiles of the distribution of 20-year mean changes are given for 2081–2100 in the four RCP scenarios.

(Below) Maps of temperature changes in 2016–2035, 2046–2065 and 2081–2100 with respect to 1986–2005 in the RCP4.5 scenario. For each point, the 25th, 50th and 75th percentiles of the distribution of the CMIP5 ensemble are shown; this includes both natural variability and inter-model spread. Hatching denotes areas where the 20-year mean differences of the percentiles are less than the standard deviation of model-estimated present-day natural variability of 20-year mean differences.

Sections 9.4.1.1, 9.6.1.1, 10.3.1.1.4, Box 11.2, 14.8.5 contain relevant information regarding the evaluation of models in this region, the model spread in the context of other methods of projecting changes and the role of modes of variability and other climate phenomena.

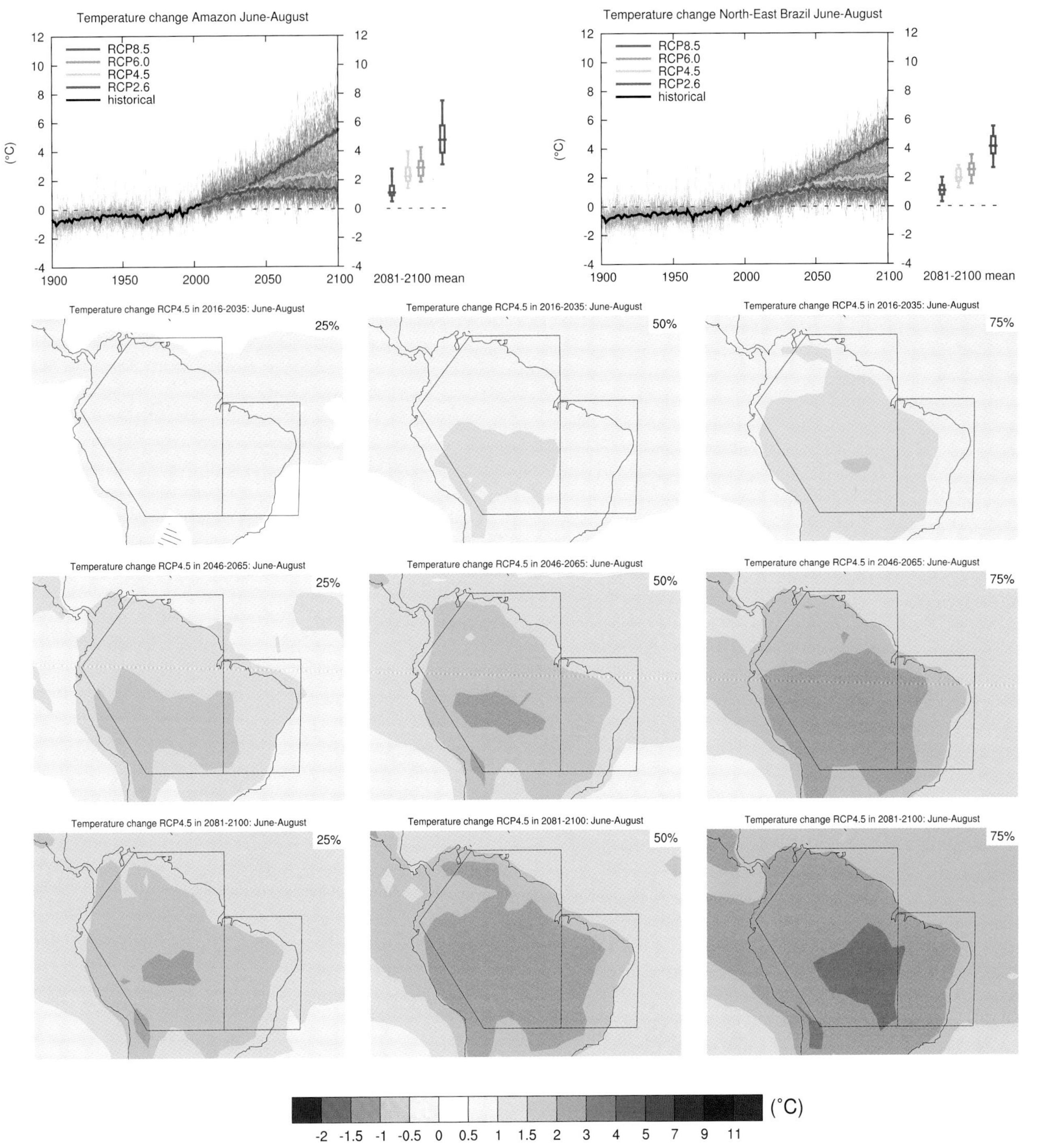

Figure AI.29 | (Top left) Time series of temperature change relative to 1986–2005 averaged over land grid points in the Amazon (20°S, 66.4°W; 1.24°S, 79.7°W; 11.44°N, 68.8°W; 11.44°N, 50°W; 20°S, 50°W) in June to August. (Top right) Same for land grid points in northeast Brazil (20°S to EQ, 50°W to 34°W). Thin lines denote one ensemble member per model, thick lines the CMIP5 multi-model mean. On the right-hand side the 5th, 25th, 50th (median), 75th and 95th percentiles of the distribution of 20-year mean changes are given for 2081–2100 in the four RCP scenarios.

(Below) Maps of temperature changes in 2016–2035, 2046–2065 and 2081–2100 with respect to 1986–2005 in the RCP4.5 scenario. For each point, the 25th, 50th and 75th percentiles of the distribution of the CMIP5 ensemble are shown; this includes both natural variability and inter-model spread. Hatching denotes areas where the 20-year mean differences of the percentiles are less than the standard deviation of model-estimated present-day natural variability of 20-year mean differences.

Sections 9.4.1.1, 9.6.1.1, 10.3.1.1.4, Box 11.2, 14.8.5 contain relevant information regarding the evaluation of models in this region, the model spread in the context of other methods of projecting changes and the role of modes of variability and other climate phenomena.

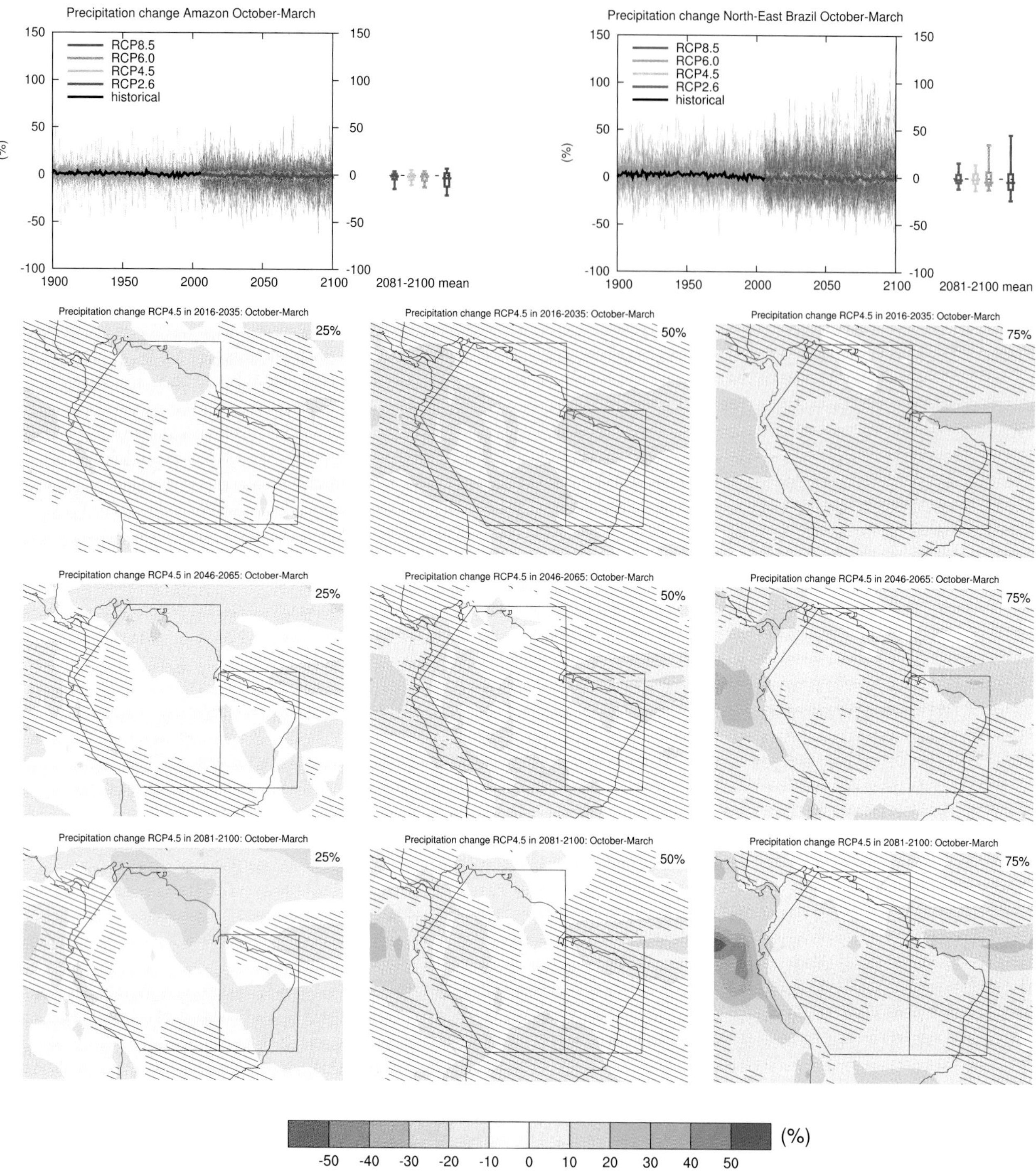

Figure AI.30 | (Top left) Time series of relative change relative to 1986–2005 in precipitation averaged over land grid points in the Amazon (20°S, 66.4°W; 1.24°S, 79.7°W; 11.44°N, 68.8°W; 11.44°N, 50°W; 20°S, 50°W) in October to March. (Top right) Same for land grid points in northeast Brazil (20°S to EQ, 50°W to 34°W). Thin lines denote one ensemble member per model, thick lines the CMIP5 multi-model mean. On the right-hand side the 5th, 25th, 50th (median), 75th and 95th percentiles of the distribution of 20-year mean changes are given for 2081–2100 in the four RCP scenarios.

(Below) Maps of precipitation changes in 2016–2035, 2046–2065 and 2081–2100 with respect to 1986–2005 in the RCP4.5 scenario. For each point, the 25th, 50th and 75th percentiles of the distribution of the CMIP5 ensemble are shown; this includes both natural variability and inter-model spread. Hatching denotes areas where the 20-year mean differences of the percentiles are less than the standard deviation of model-estimated present-day natural variability of 20-year mean differences.

Sections 9.4.1.1, 9.6.1.1, 11.3.2.1.2, Box 11.2, 14.2.3.2, 14.8.5 contain relevant information regarding the evaluation of models in this region, the model spread in the context of other methods of projecting changes and the role of modes of variability and other climate phenomena.

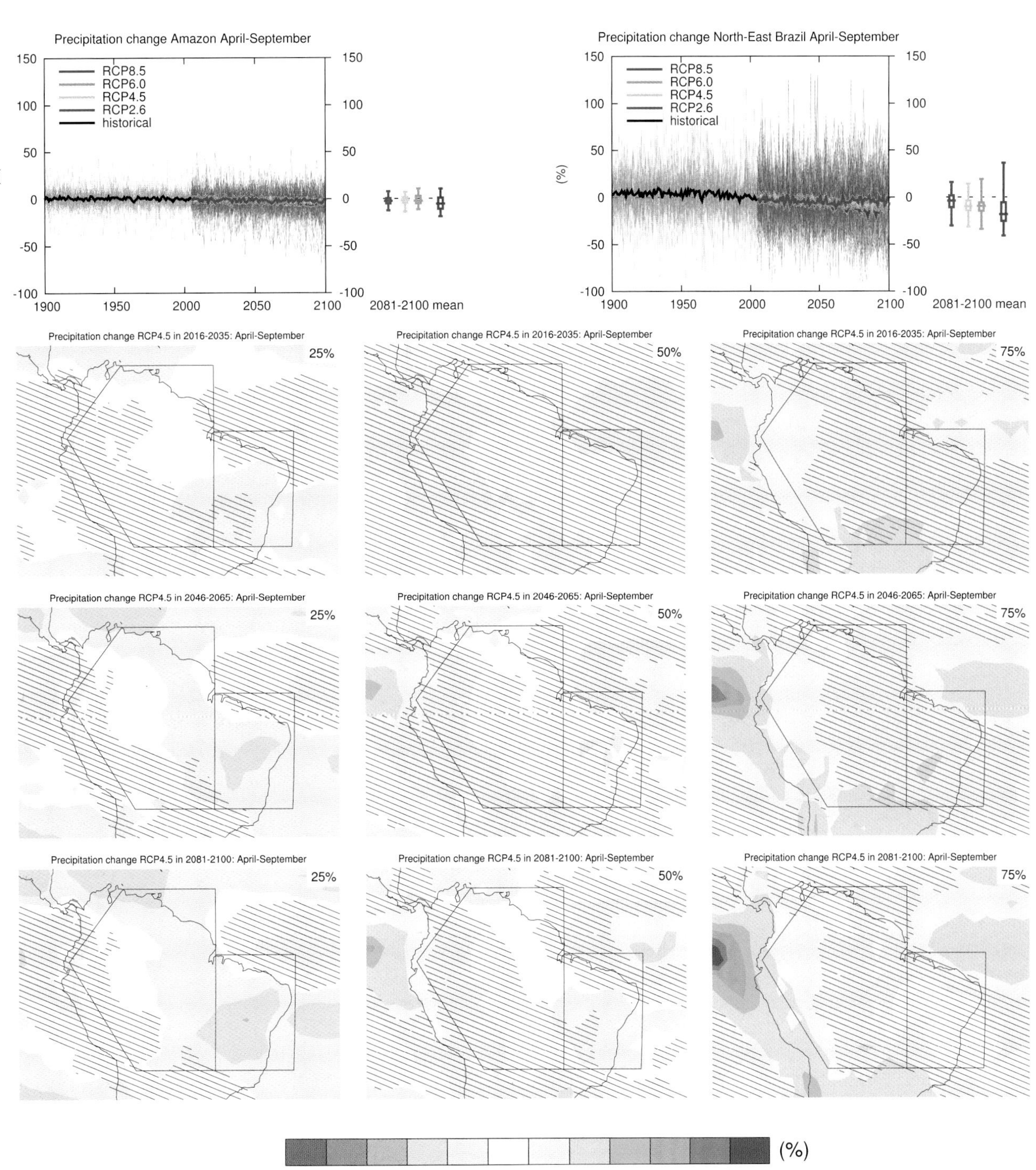

Figure AI.31 | (Top left) Time series of relative change relative to 1986–2005 in precipitation averaged over land grid points in the Amazon (20°S, 66.4°W; 1.24°S, 79.7°W; 11.44°N, 68.8°W; 11.44°N, 50°W; 20°S, 50°W) in April to September. (Top right) Same for land grid points in northeast Brazil (20°S to EQ, 50°W to 34°W). Thin lines denote one ensemble member per model, thick lines the CMIP5 multi-model mean. On the right-hand side the 5th, 25th, 50th (median), 75th and 95th percentiles of the distribution of 20-year mean changes are given for 2081–2100 in the four RCP scenarios.

(Below) Maps of precipitation changes in 2016–2035, 2046–2065 and 2081–2100 with respect to 1986–2005 in the RCP4.5 scenario. For each point, the 25th, 50th and 75th percentiles of the distribution of the CMIP5 ensemble are shown; this includes both natural variability and inter-model spread. Hatching denotes areas where the 20-year mean differences of the percentiles are less than the standard deviation of model-estimated present-day natural variability of 20-year mean differences.

Sections 9.4.1.1, 9.6.1.1, 11.3.2.1.2, Box 11.2, 14.2.3.2, 14.8.5 contain relevant information regarding the evaluation of models in this region, the model spread in the context of other methods of projecting changes and the role of modes of variability and other climate phenomena.

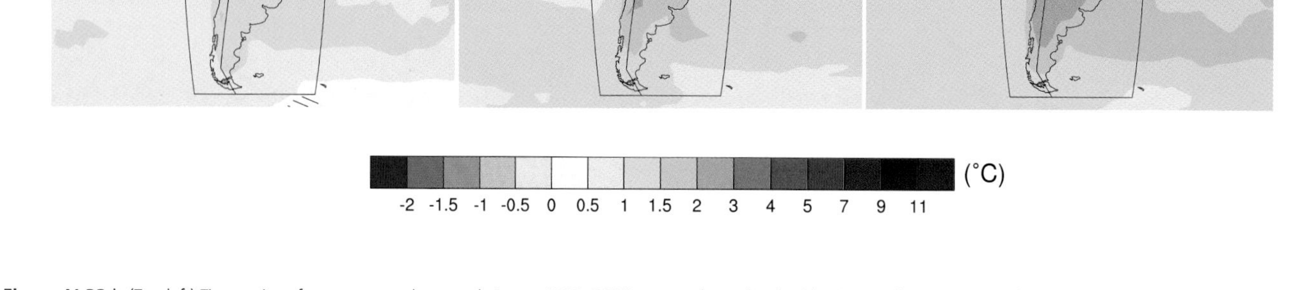

Figure AI.32 | (Top left) Time series of temperature change relative to 1986–2005 averaged over land grid points in the west coast of South America (79.7°W, 1.2°S; 66.4°W, 20°S; 72.1°W, 50°S; 67.3°W, 56.7°S; 82.0°W, 56.7°S; 82.2°W, 0.5°N) in December to February. (Top right) Same for land grid points in southeastern South America (39.4°W, 20°S; 39.4°W, 56.6°S; 67.3°W, 56.7°S; 72.1°W, 50°S; 66°W, 20°S). Thin lines denote one ensemble member per model, thick lines the CMIP5 multi-model mean. On the right-hand side the 5th, 25th, 50th (median), 75th and 95th percentiles of the distribution of 20-year mean changes are given for 2081–2100 in the four RCP scenarios.

(Below) Maps of temperature changes in 2016–2035, 2046–2065 and 2081–2100 with respect to 1986–2005 in the RCP4.5 scenario. For each point, the 25th, 50th and 75th percentiles of the distribution of the CMIP5 ensemble are shown; this includes both natural variability and inter-model spread. Hatching denotes areas where the 20-year mean differences of the percentiles are less than the standard deviation of model-estimated present-day natural variability of 20-year mean differences.

Sections 9.4.1.1, 9.6.1.1, 10.3.1.1.4, Box 11.2, 14.8.5 contain relevant information regarding the evaluation of models in this region, the model spread in the context of other methods of projecting changes and the role of modes of variability and other climate phenomena.

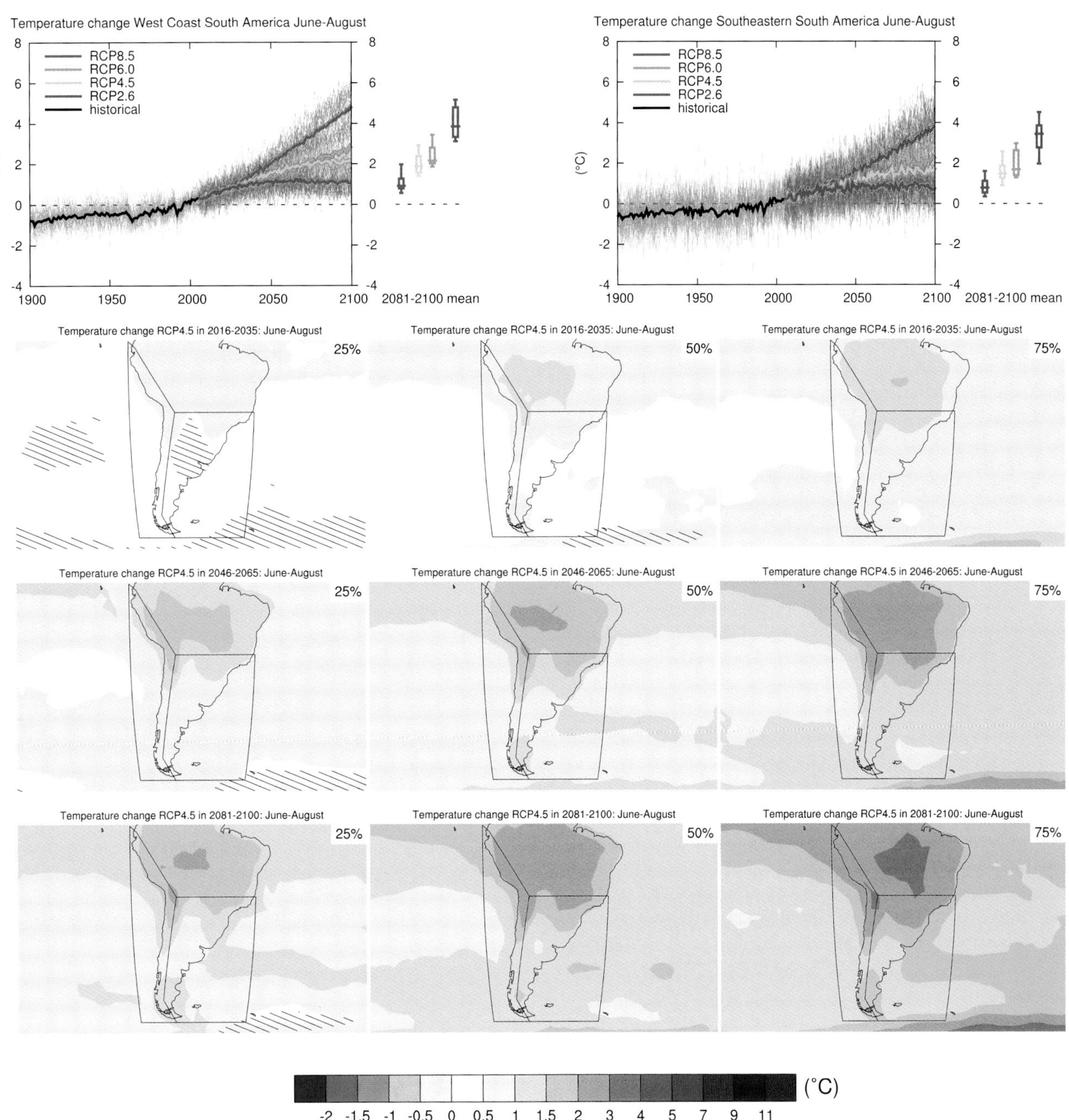

Figure AI.33 | (Top left) Time series of temperature change relative to 1986–2005 averaged over land grid points in the west coast of South America (79.7°W, 1.2°S; 66.4°W, 20°S; 72.1°W, 50°S; 67.3°W, 56.7°S; 82.0°W, 56.7°S; 82.2°W, 0.5°N) in June to August. (Top right) Same for land grid points in southeastern South America (39.4°W, 20°S; 39.4°W, 56.6°S; 67.3°W, 56.7°S; 72.1°W, 50°S; 66°W, 20°S). Thin lines denote one ensemble member per model, thick lines the CMIP5 multi-model mean. On the right-hand side the 5th, 25th, 50th (median), 75th and 95th percentiles of the distribution of 20-year mean changes are given for 2081–2100 in the four RCP scenarios.

(Below) Maps of temperature changes in 2016–2035, 2046–2065 and 2081–2100 with respect to 1986–2005 in the RCP4.5 scenario. For each point, the 25th, 50th and 75th percentiles of the distribution of the CMIP5 ensemble are shown; this includes both natural variability and inter-model spread. Hatching denotes areas where the 20-year mean differences of the percentiles are less than the standard deviation of model-estimated present-day natural variability of 20-year mean differences.

Sections 9.4.1.1, 9.6.1.1, 10.3.1.1.4, Box 11.2, 14.8.5 contain relevant information regarding the evaluation of models in this region, the model spread in the context of other methods of projecting changes and the role of modes of variability and other climate phenomena.

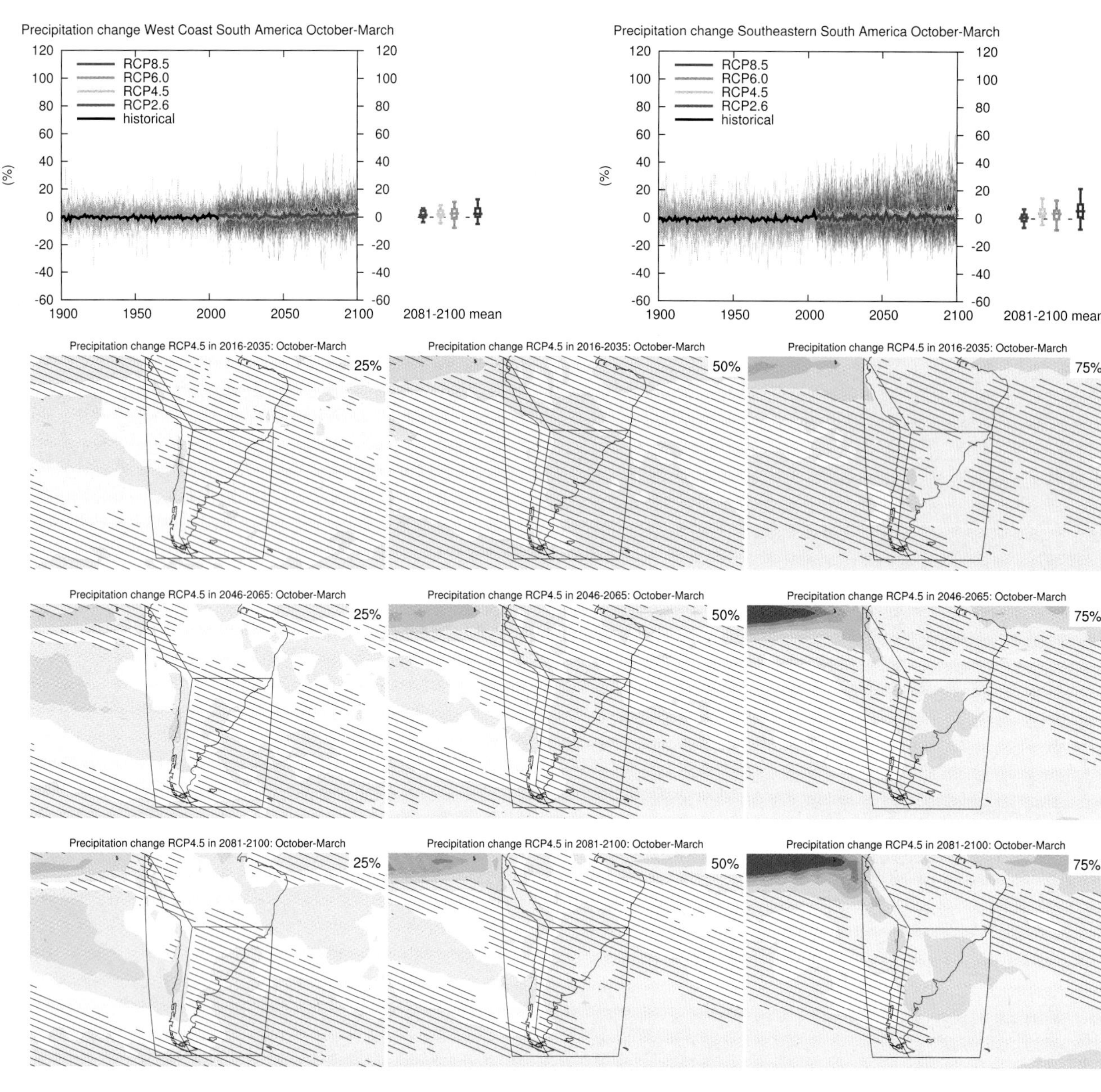

Figure AI.34 | (Top left) Time series of relative change relative to 1986–2005 in precipitation averaged over land grid points in the west coast of South America (79.7°W, 1.2°S; 66.4°W, 20°S; 72.1°W, 50°S; 67.3°W, 56.7°S; 82.0°W, 56.7°S; 82.2°W, 0.5°N) in October to March. (Top right) Same for land grid points in southeastern South America (39.4°W, 20°S; 39.4°W, 56.6°S; 67.3°W, 56.7°S; 72.1°W, 50°S; 66°W, 20°S). Thin lines denote one ensemble member per model, thick lines the CMIP5 multi-model mean. On the right-hand side the 5th, 25th, 50th (median), 75th and 95th percentiles of the distribution of 20-year mean changes are given for 2081–2100 in the four RCP scenarios.

(Below) Maps of precipitation changes in 2016–2035, 2046–2065 and 2081–2100 with respect to 1986–2005 in the RCP4.5 scenario. For each point, the 25th, 50th and 75th percentiles of the distribution of the CMIP5 ensemble are shown; this includes both natural variability and inter-model spread. Hatching denotes areas where the 20-year mean differences of the percentiles are less than the standard deviation of model-estimated present-day natural variability of 20-year mean differences.

Sections 9.4.1.1, 9.6.1.1, Box 11.2, 12.4.5.2, 14.8.5 contain relevant information regarding the evaluation of models in this region, the model spread in the context of other methods of projecting changes and the role of modes of variability and other climate phenomena.

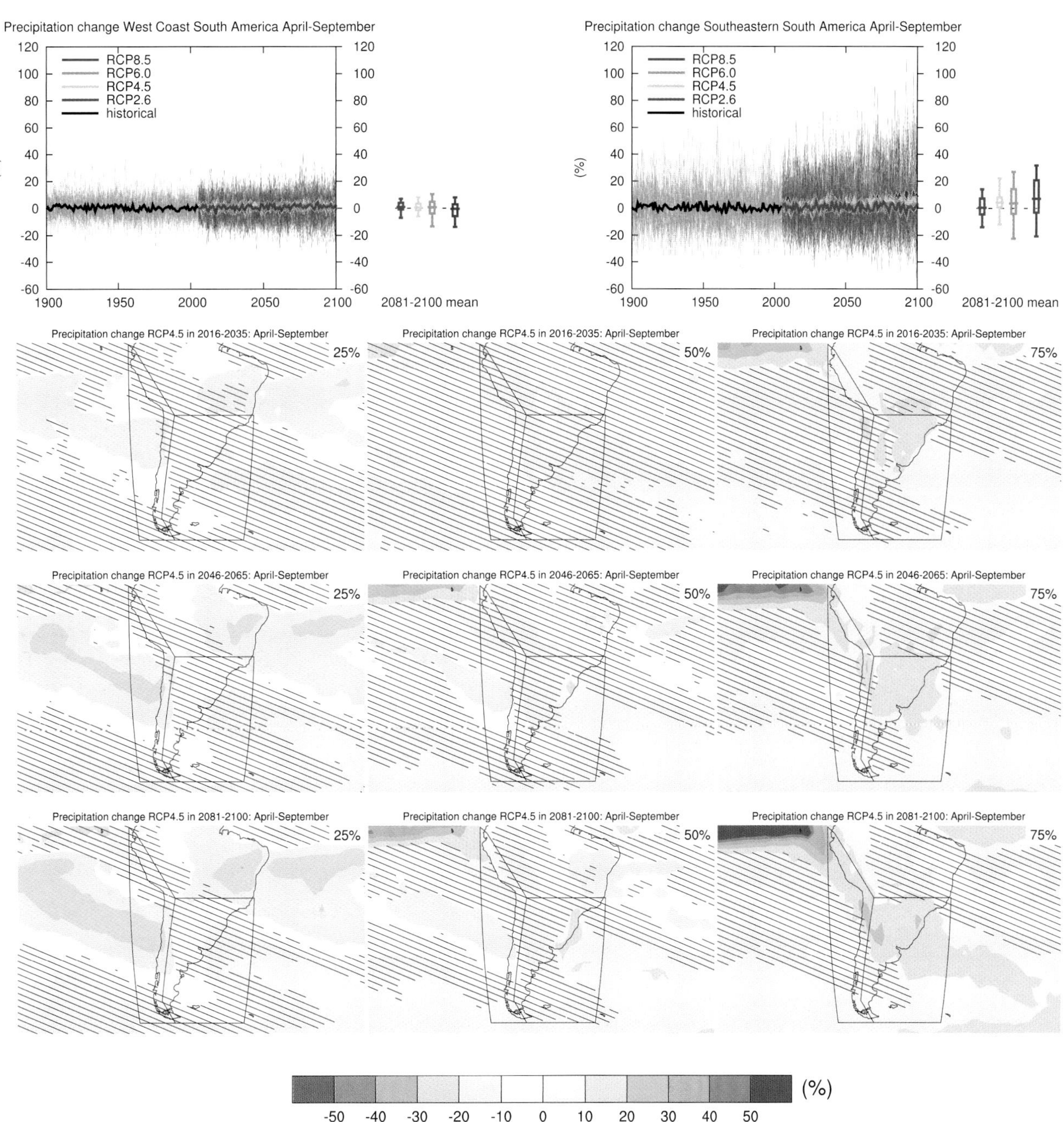

Figure AI.35 | (Top left) Time series of relative change relative to 1986–2005 in precipitation averaged over land grid points in the west coast of South America (79.7°W, 1.2°S; 66.4°W, 20°S; 72.1°W, 50°S; 67.3°W, 56.7°S; 82.0°W, 56.7°S; 82.2°W, 0.5°N) in April to September. (Top right) Same for land grid points in southeastern South America (39.4°W, 20°S; 39.4°W, 56.6°S; 67.3°W, 56.7°S; 72.1°W, 50°S; 66°W, 20°S). Thin lines denote one ensemble member per model, thick lines the CMIP5 multi-model mean. On the right-hand side the 5th, 25th, 50th (median), 75th and 95th percentiles of the distribution of 20-year mean changes are given for 2081–2100 in the four RCP scenarios.

(Below) Maps of precipitation changes in 2016–2035, 2046–2065 and 2081–2100 with respect to 1986–2005 in the RCP4.5 scenario. For each point, the 25th, 50th and 75th percentiles of the distribution of the CMIP5 ensemble are shown; this includes both natural variability and inter-model spread. Hatching denotes areas where the 20-year mean differences of the percentiles are less than the standard deviation of model-estimated present-day natural variability of 20-year mean differences.

Sections 9.4.1.1, 9.6.1.1, Box 11.2, 12.4.5.2, 14.8.5 contain relevant information regarding the evaluation of models in this region, the model spread in the context of other methods of projecting changes and the role of modes of variability and other climate phenomena.

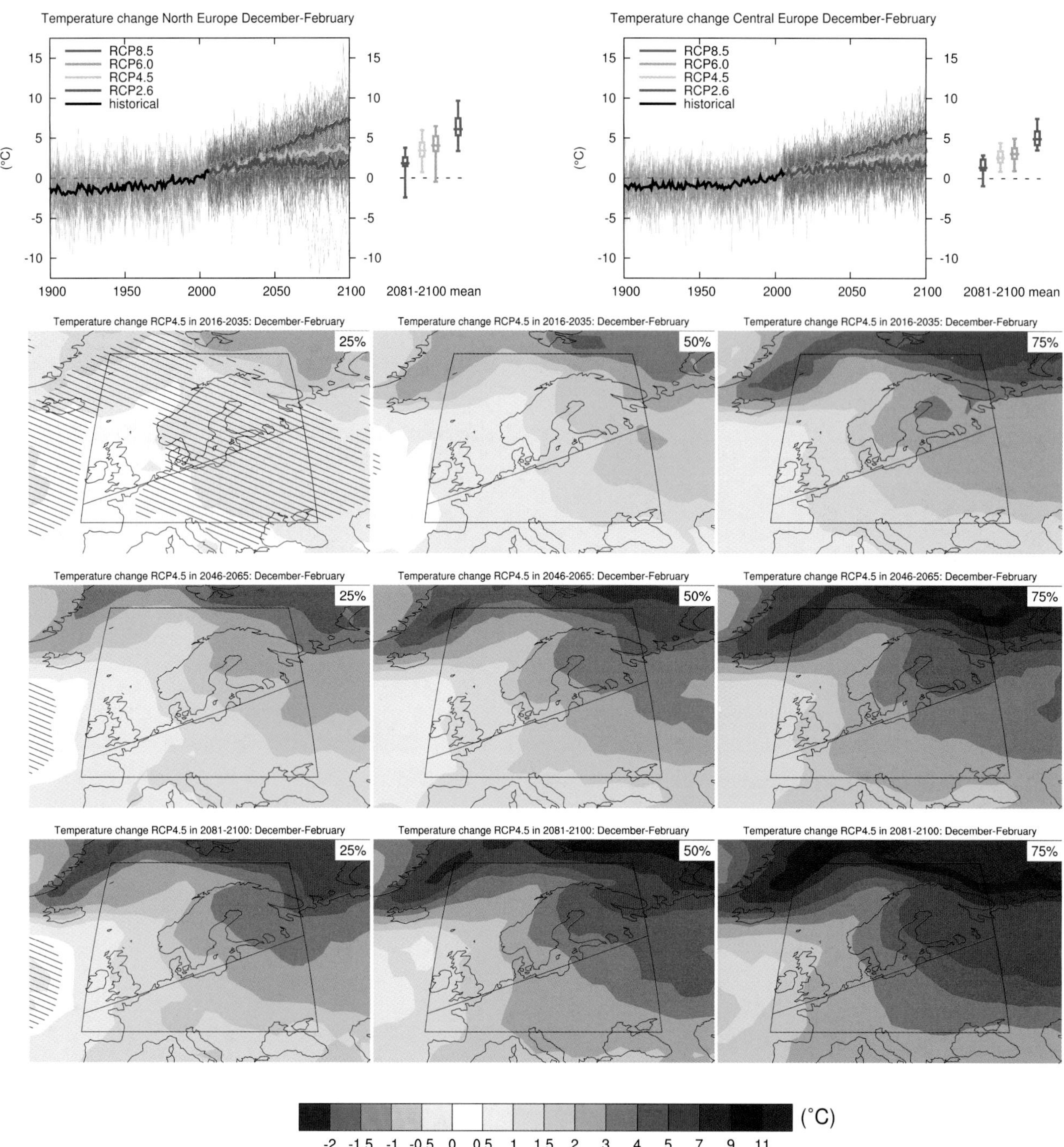

Figure AI.36 | (Top left) Time series of temperature change relative to 1986–2005 averaged over land grid points in North Europe (10°W, 48°N; 10°W, 75°N; 40°E, 75°N; 40°E, 61.3°N) in December to February. (Top right) Same for land grid points in Central Europe (10°W, 45°N; 10°W, 48°N; 40°E, 61.3°N; 40°E, 45°N). Thin lines denote one ensemble member per model, thick lines the CMIP5 multi-model mean. On the right-hand side the 5th, 25th, 50th (median), 75th and 95th percentiles of the distribution of 20-year mean changes are given for 2081–2100 in the four RCP scenarios.

(Below) Maps of temperature changes in 2016–2035, 2046–2065 and 2081–2100 with respect to 1986–2005 in the RCP4.5 scenario. For each point, the 25th, 50th and 75th percentiles of the distribution of the CMIP5 ensemble are shown; this includes both natural variability and inter-model spread. Hatching denotes areas where the 20-year mean differences of the percentiles are less than the standard deviation of model-estimated present-day natural variability of 20-year mean differences.

Sections 9.4.1.1, 9.6.1.1, 10.3.1.1.4, 10.3, Box 11.2, 14.8.6 contain relevant information regarding the evaluation of models in this region, the model spread in the context of other methods of projecting changes and the role of modes of variability and other climate phenomena.

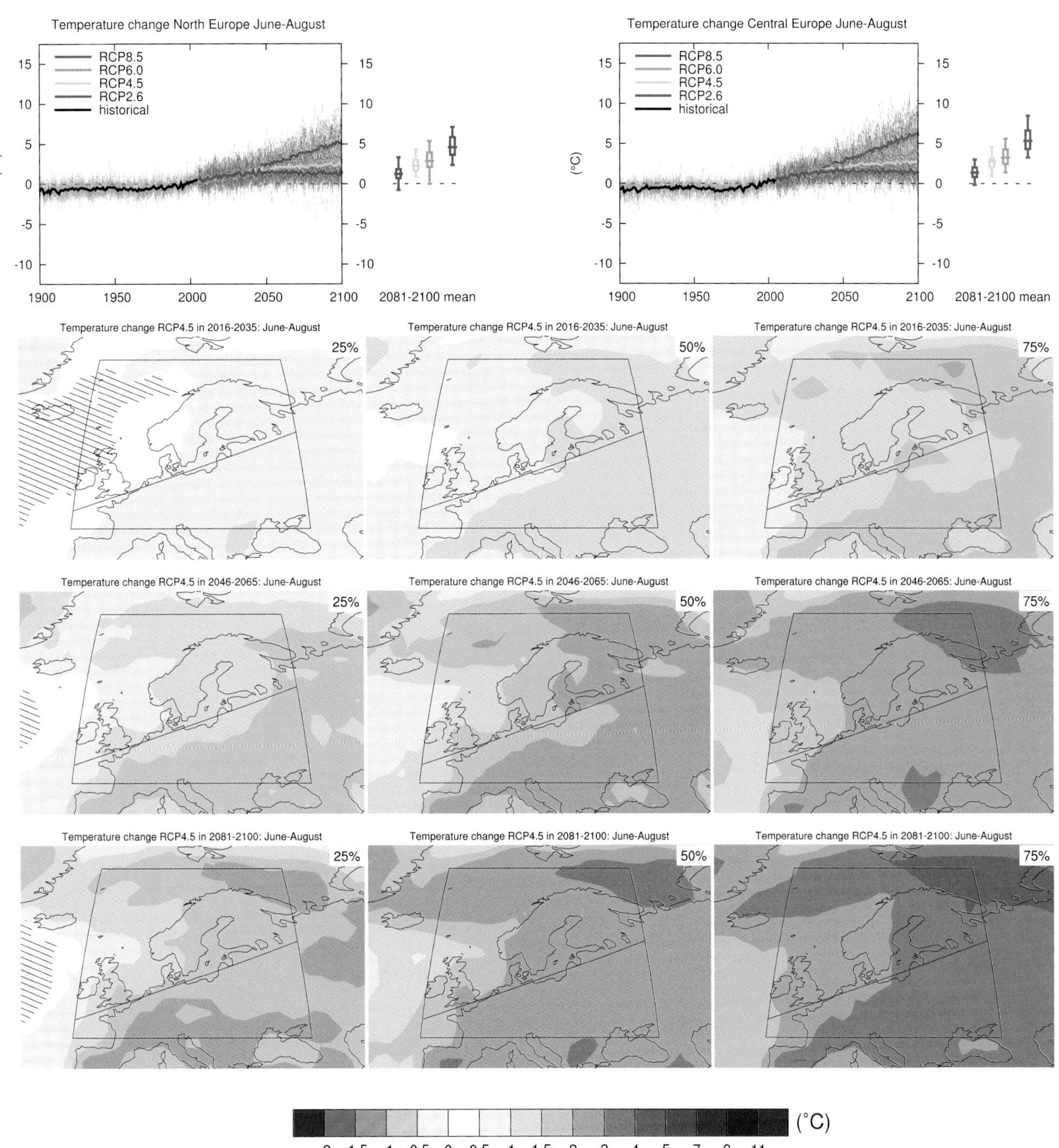

Figure AI.37 | (Top left) Time series of temperature change relative to 1986–2005 averaged over land grid points in North Europe (10°W, 48°N; 10°W, 75°N; 40°E, 75°N; 40°E, 61.3°N) in June to August. (Top right) Same for land grid points in Central Europe (10°W, 45°N; 10°W, 48°N; 40°E, 61.3°N; 40°E, 45°N). Thin lines denote one ensemble member per model, thick lines the CMIP5 multi-model mean. On the right-hand side the 5th, 25th, 50th (median), 75th and 95th percentiles of the distribution of 20-year mean changes are given for 2081–2100 in the four RCP scenarios.

(Below) Maps of temperature changes in 2016–2035, 2046–2065 and 2081–2100 with respect to 1986–2005 in the RCP4.5 scenario. For each point, the 25th, 50th and 75th percentiles of the distribution of the CMIP5 ensemble are shown; this includes both natural variability and inter-model spread. Hatching denotes areas where the 20-year mean differences of the percentiles are less than the standard deviation of model-estimated present-day natural variability of 20-year mean differences.

Sections 9.4.1.1, 9.6.1.1, 10.3.1.1.4, 10.3, Box 11.2, 14.8.6 contain relevant information regarding the evaluation of models in this region, the model spread in the context of other methods of projecting changes and the role of modes of variability and other climate phenomena.

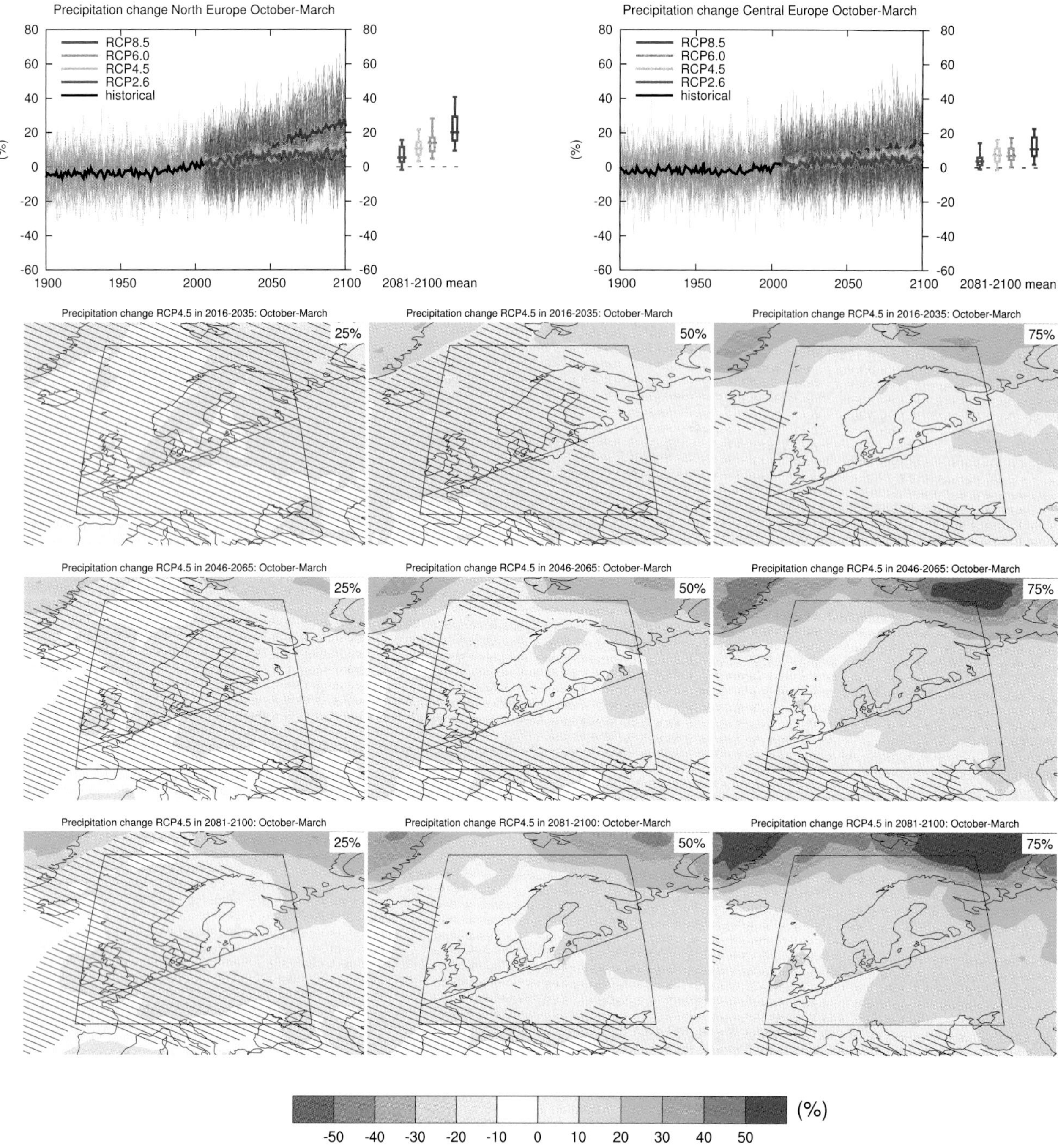

Figure AI.38 | (Top left) Time series of relative change relative to 1986–2005 in precipitation averaged over land grid points in North Europe (10°W, 48°N; 10°W, 75°N; 40°E, 75°N; 40°E, 61.3°N) in October to March. (Top right) Same for land grid points in Central Europe (10°W, 45°N; 10°W, 48°N; 40°E, 61.3°N; 40°E, 45°N). Thin lines denote one ensemble member per model, thick lines the CMIP5 multi-model mean. On the right-hand side the 5th, 25th, 50th (median), 75th and 95th percentiles of the distribution of 20-year mean changes are given for 2081–2100 in the four RCP scenarios.

(Below) Maps of precipitation changes in 2016–2035, 2046–2065 and 2081–2100 with respect to 1986–2005 in the RCP4.5 scenario. For each point, the 25th, 50th and 75th percentiles of the distribution of the CMIP5 ensemble are shown; this includes both natural variability and inter-model spread. Hatching denotes areas where the 20-year mean differences of the percentiles are less than the standard deviation of model-estimated present-day natural variability of 20-year mean differences.

Sections 9.4.1.1, 9.6.1.1, Box 11.2, 12.4.5.2, 14.8.6 contain relevant information regarding the evaluation of models in this region, the model spread in the context of other methods of projecting changes and the role of modes of variability and other climate phenomena.

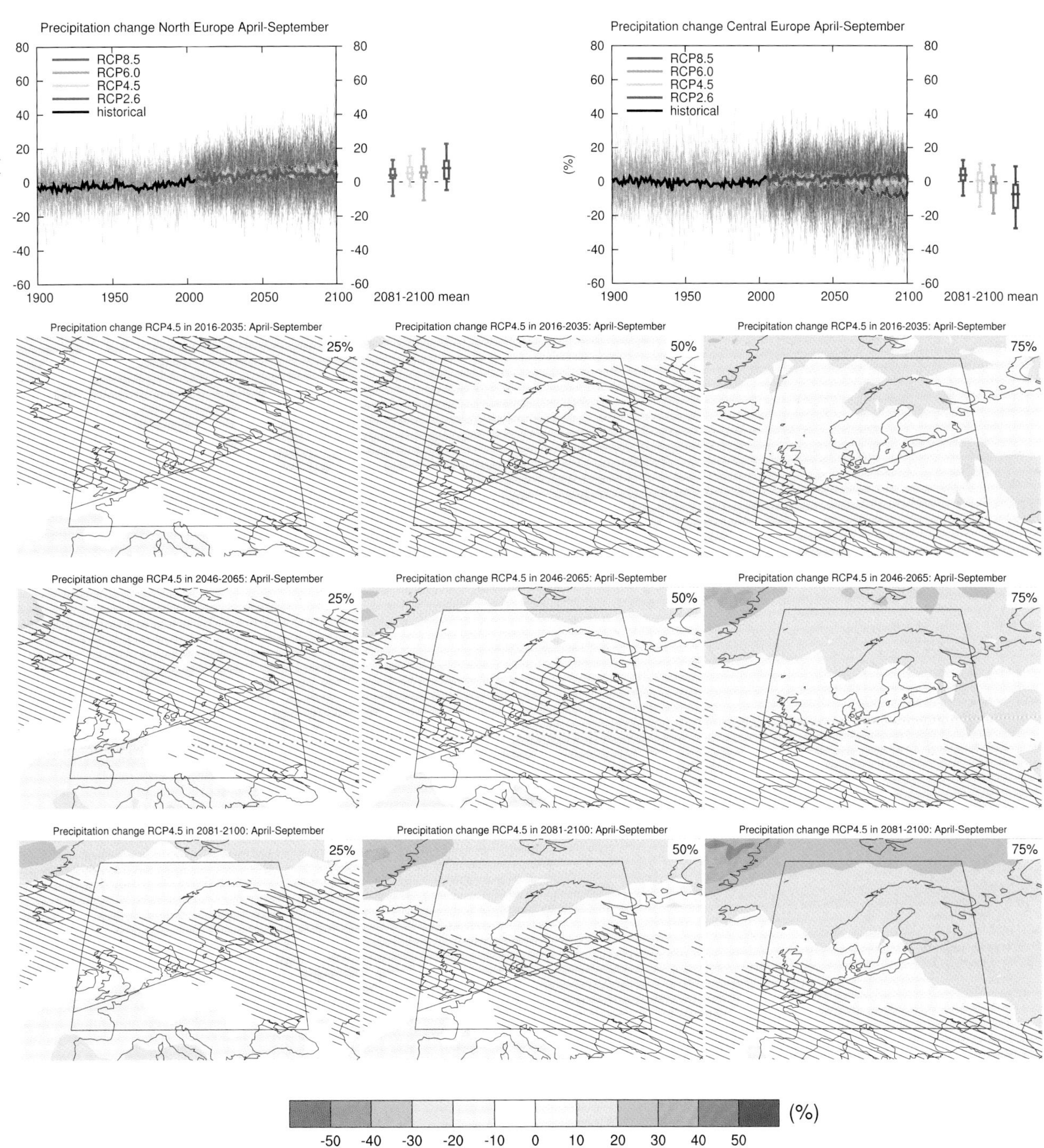

Figure AI.39 | (Top left) Time series of relative change relative to 1986–2005 in precipitation averaged over land grid points in North Europe (10°W, 48°N; 10°W, 75°N; 40°E, 75°N; 40°E, 61.3°N) in April to September. (Top right) Same for land grid points in Central Europe (10°W, 45°N; 10°W, 48°N; 40°E, 61.3°N; 40°E, 45°N). Thin lines denote one ensemble member per model, thick lines the CMIP5 multi-model mean. On the right-hand side the 5th, 25th, 50th (median), 75th and 95th percentiles of the distribution of 20-year mean changes are given for 2081–2100 in the four RCP scenarios.

(Below) Maps of precipitation changes in 2016–2035, 2046–2065 and 2081–2100 with respect to 1986–2005 in the RCP4.5 scenario. For each point, the 25th, 50th and 75th percentiles of the distribution of the CMIP5 ensemble are shown; this includes both natural variability and inter-model spread. Hatching denotes areas where the 20-year mean differences of the percentiles are less than the standard deviation of model-estimated present-day natural variability of 20-year mean differences.

Sections 9.4.1.1, 9.6.1.1, Box 11.2, 12.4.5.2, 14.8.6 contain relevant information regarding the evaluation of models in this region, the model spread in the context of other methods of projecting changes and the role of modes of variability and other climate phenomena.

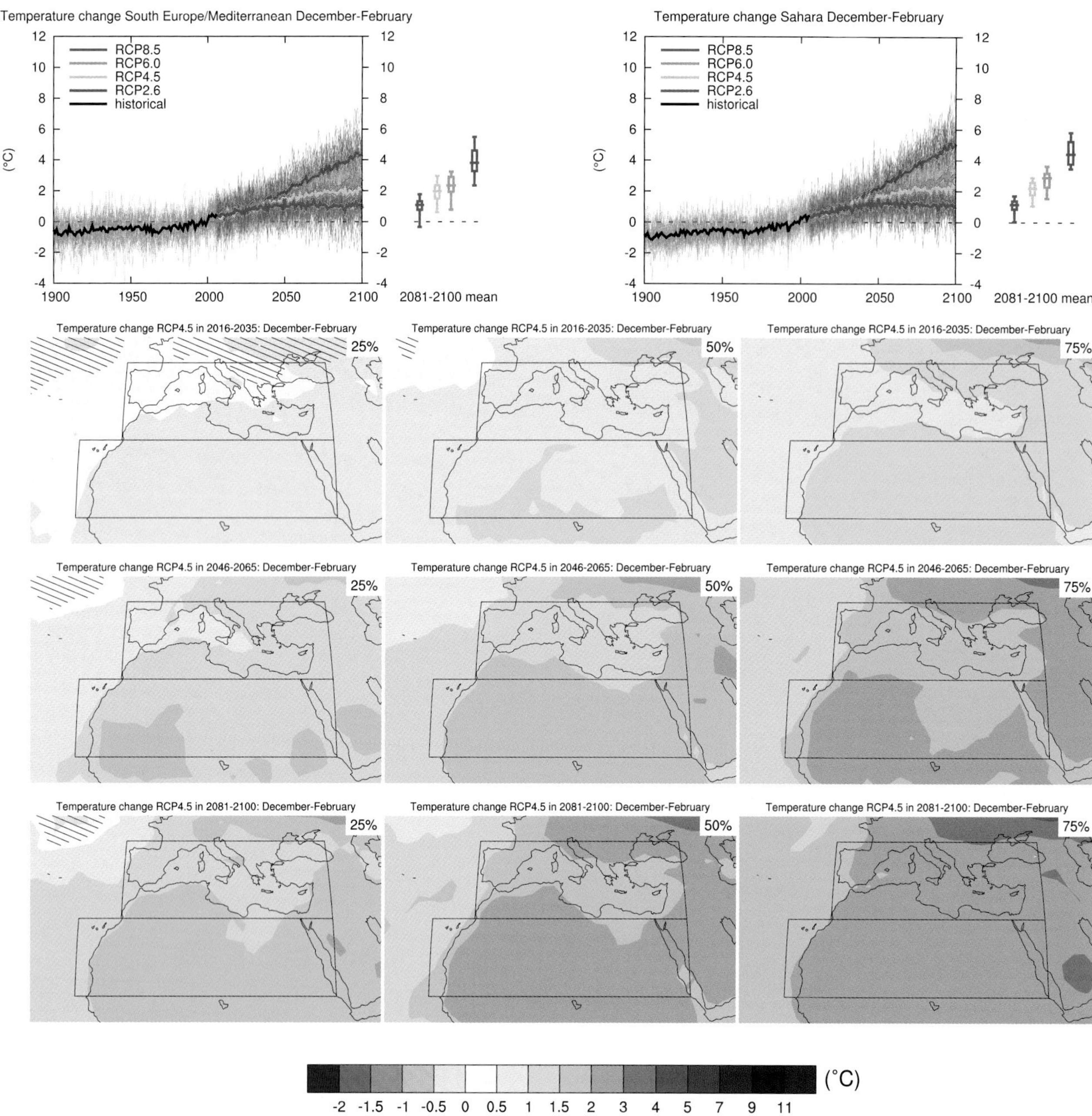

Figure AI.40 | (Top left) Time series of temperature change relative to 1986–2005 averaged over land grid points in the region South Europe/Mediterranean (30°N to 45°N, 10°W to 40°E) in December to February. (Top right) Same for land grid points in the Sahara (15°N to 30°N, 20°W to 40°E). Thin lines denote one ensemble member per model, thick lines the CMIP5 multi-model mean. On the right-hand side the 5th, 25th, 50th (median), 75th and 95th percentiles of the distribution of 20-year mean changes are given for 2081–2100 in the four RCP scenarios.

(Below) Maps of temperature changes in 2016–2035, 2046–2065 and 2081–2100 with respect to 1986–2005 in the RCP4.5 scenario. For each point, the 25th, 50th and 75th percentiles of the distribution of the CMIP5 ensemble are shown; this includes both natural variability and inter-model spread. Hatching denotes areas where the 20-year mean differences of the percentiles are less than the standard deviation of model-estimated present-day natural variability of 20-year mean differences.

Sections 9.4.1.1, 9.6.1.1, 10.3.1.1.4, Box 11.2, 14.8.6, 14.8.7 contain relevant information regarding the evaluation of models in this region, the model spread in the context of other methods of projecting changes and the role of modes of variability and other climate phenomena.

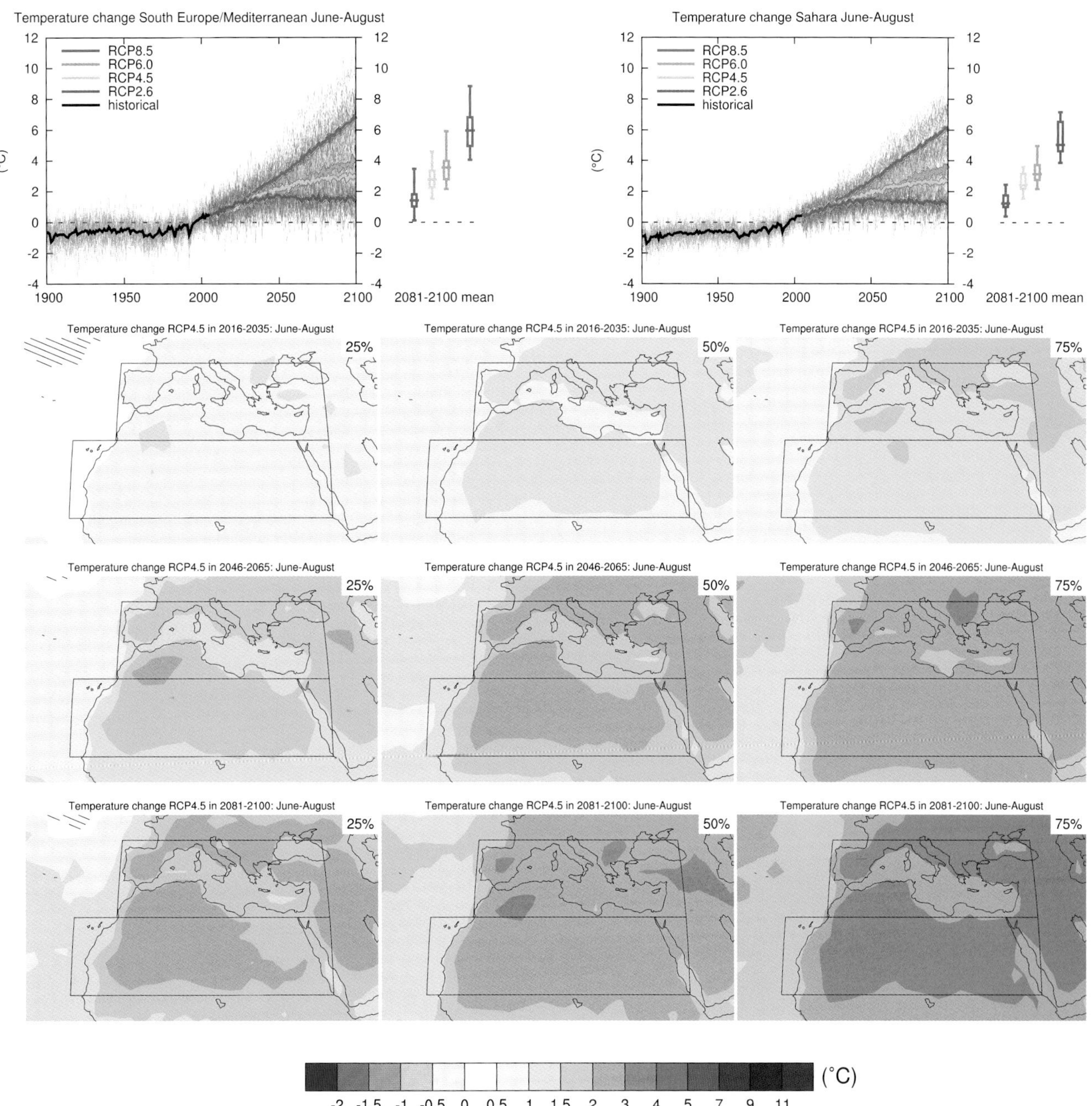

Figure AI.41 | (Top left) Time series of temperature change relative to 1986–2005 averaged over land grid points in the region South Europe/Mediterranean (30°N to 45°N, 10°W to 40°E) in June to August. (Top right) Same for land grid points in the Sahara (15°N to 30°N, 20°W to 40°E). Thin lines denote one ensemble member per model, thick lines the CMIP5 multi-model mean. On the right-hand side the 5th, 25th, 50th (median), 75th and 95th percentiles of the distribution of 20-year mean changes are given for 2081–2100 in the four RCP scenarios.

(Below) Maps of temperature changes in 2016–2035, 2046–2065 and 2081–2100 with respect to 1986–2005 in the RCP4.5 scenario. For each point, the 25th, 50th and 75th percentiles of the distribution of the CMIP5 ensemble are shown; this includes both natural variability and inter-model spread. Hatching denotes areas where the 20-year mean differences of the percentiles are less than the standard deviation of model-estimated present-day natural variability of 20-year mean differences.

Sections 9.4.1.1, 9.6.1.1, 10.3.1.1.4, Box 11.2, 14.8.6, 14.8.7 contain relevant information regarding the evaluation of models in this region, the model spread in the context of other methods of projecting changes and the role of modes of variability and other climate phenomena.

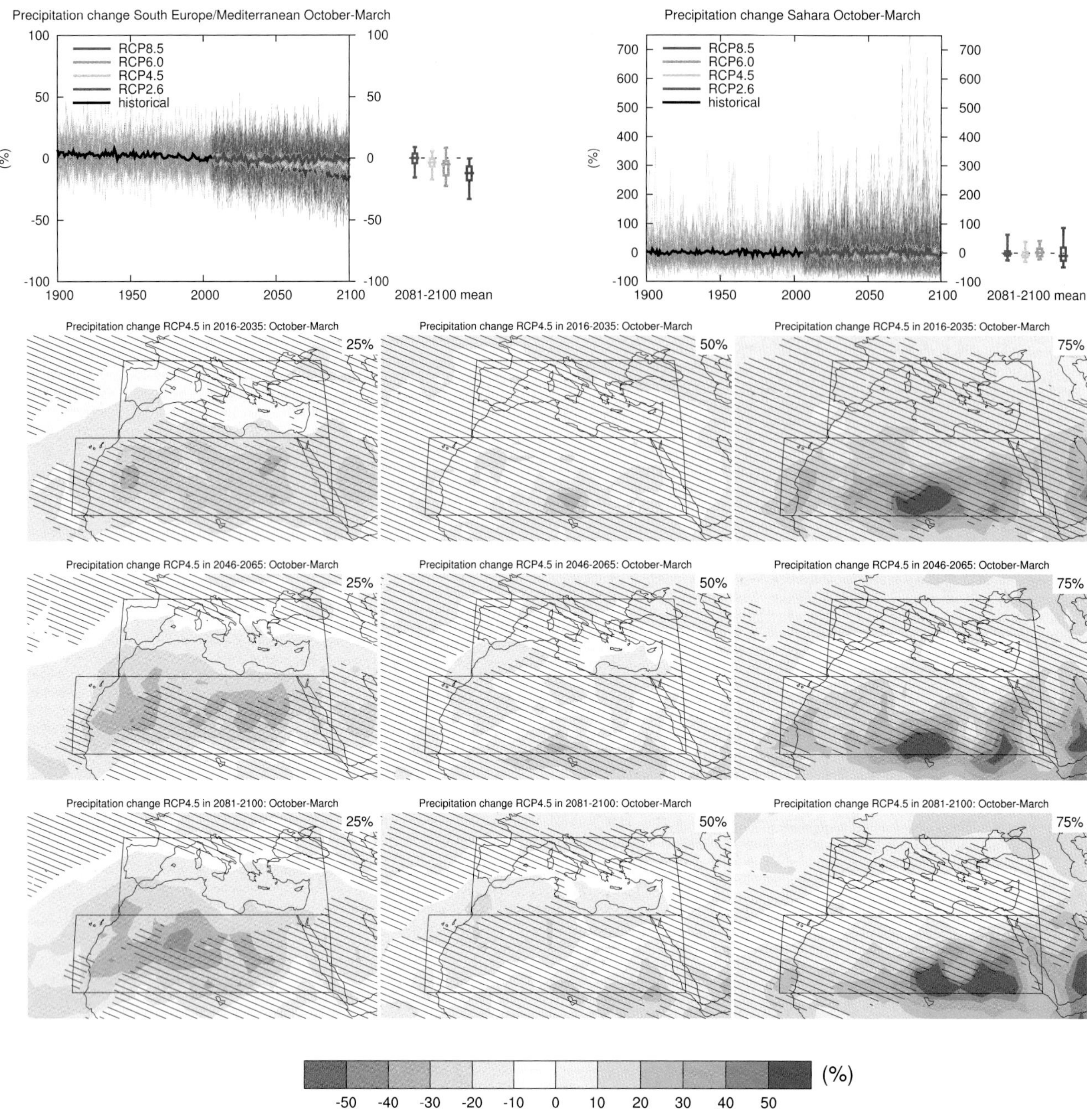

Figure AI.42 | (Top left) Time series of relative change relative to 1986–2005 in precipitation averaged over land grid points in the region South Europe/Mediterranean (30°N to 45°N, 10°W to 40°E) in October to March. (Top right) Same for land grid points in the Sahara (15°N to 30°N, 20°W to 40°E). Thin lines denote one ensemble member per model, thick lines the CMIP5 multi-model mean. On the right-hand side the 5th, 25th, 50th (median), 75th and 95th percentiles of the distribution of 20-year mean changes are given for 2081–2100 in the four RCP scenarios. Note different scales.

(Below) Maps of precipitation changes in 2016–2035, 2046–2065 and 2081–2100 with respect to 1986–2005 in the RCP4.5 scenario. For each point, the 25th, 50th and 75th percentiles of the distribution of the CMIP5 ensemble are shown; this includes both natural variability and inter-model spread. Hatching denotes areas where the 20-year mean differences of the percentiles are less than the standard deviation of model-estimated present-day natural variability of 20-year mean differences.

Sections 9.4.1.1, 9.6.1.1, Box 11.2, 12.4.5.2, 14.8.6, 14.8.7 contain relevant information regarding the evaluation of models in this region, the model spread in the context of other methods of projecting changes and the role of modes of variability and other climate phenomena.

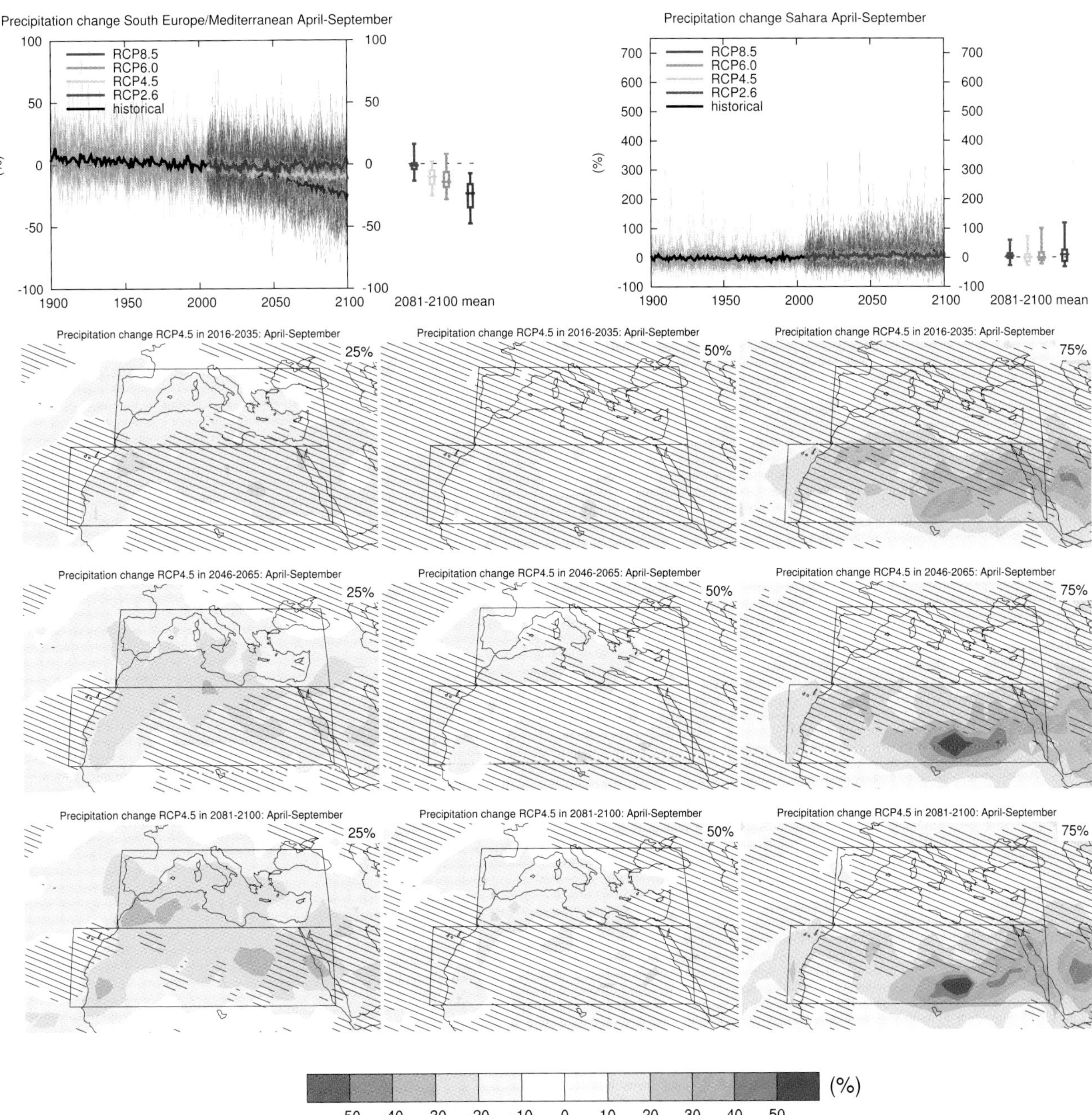

Figure AI.43 | (Top left) Time series of relative change relative to 1986–2005 in precipitation averaged over land grid points in the region South Europe/Mediterranean (30°N to 45°N, 10°W to 40°E) in April to September. (Top right) Same for land grid points in the Sahara (15°N to 30°N, 20°W to 40°E). Thin lines denote one ensemble member per model, thick lines the CMIP5 multi-model mean. On the right-hand side the 5th, 25th, 50th (median), 75th and 95th percentiles of the distribution of 20-year mean changes are given for 2081–2100 in the four RCP scenarios. Note different scales.

(Below) Maps of precipitation changes in 2016–2035, 2046–2065 and 2081–2100 with respect to 1986–2005 in the RCP4.5 scenario. For each point, the 25th, 50th and 75th percentiles of the distribution of the CMIP5 ensemble are shown; this includes both natural variability and inter-model spread. Hatching denotes areas where the 20-year mean differences of the percentiles are less than the standard deviation of model-estimated present-day natural variability of 20-year mean differences.

Sections 9.4.1.1, 9.6.1.1, Box 11.2, 12.4.5.2, 14.8.6, 14.8.7 contain relevant information regarding the evaluation of models in this region, the model spread in the context of other methods of projecting changes and the role of modes of variability and other climate phenomena.

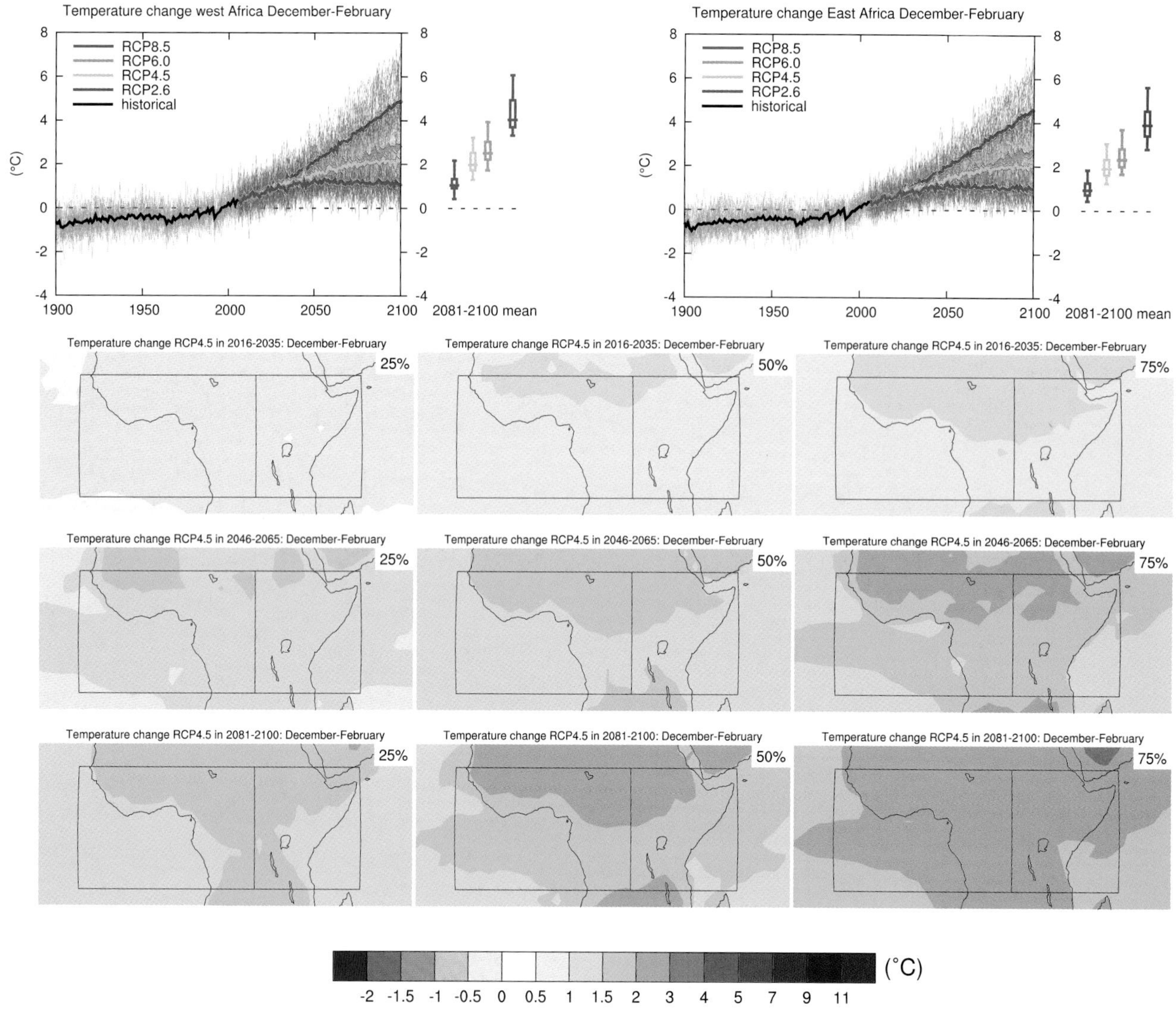

Figure AI.44 | (Top left) Time series of temperature change relative to 1986–2005 averaged over land grid points in West Africa (11.4°S to 15°N, 20°W to 25°E) in December to February. (Top right) Same for land grid points in East Africa (11.3°S to 15°N, 25°E to 52°E). Thin lines denote one ensemble member per model, thick lines the CMIP5 multi-model mean. On the right-hand side the 5th, 25th, 50th (median), 75th and 95th percentiles of the distribution of 20-year mean changes are given for 2081–2100 in the four RCP scenarios.

(Below) Maps of temperature changes in 2016–2035, 2046–2065 and 2081–2100 with respect to 1986–2005 in the RCP4.5 scenario. For each point, the 25th, 50th and 75th percentiles of the distribution of the CMIP5 ensemble are shown; this includes both natural variability and inter-model spread. Hatching denotes areas where the 20-year mean differences of the percentiles are less than the standard deviation of model-estimated present-day natural variability of 20-year mean differences.

Sections 9.4.1.1, 9.6.1.1, 10.3.1.1.4, Box 11.2, 14.8.7 contain relevant information regarding the evaluation of models in this region, the model spread in the context of other methods of projecting changes and the role of modes of variability and other climate phenomena.

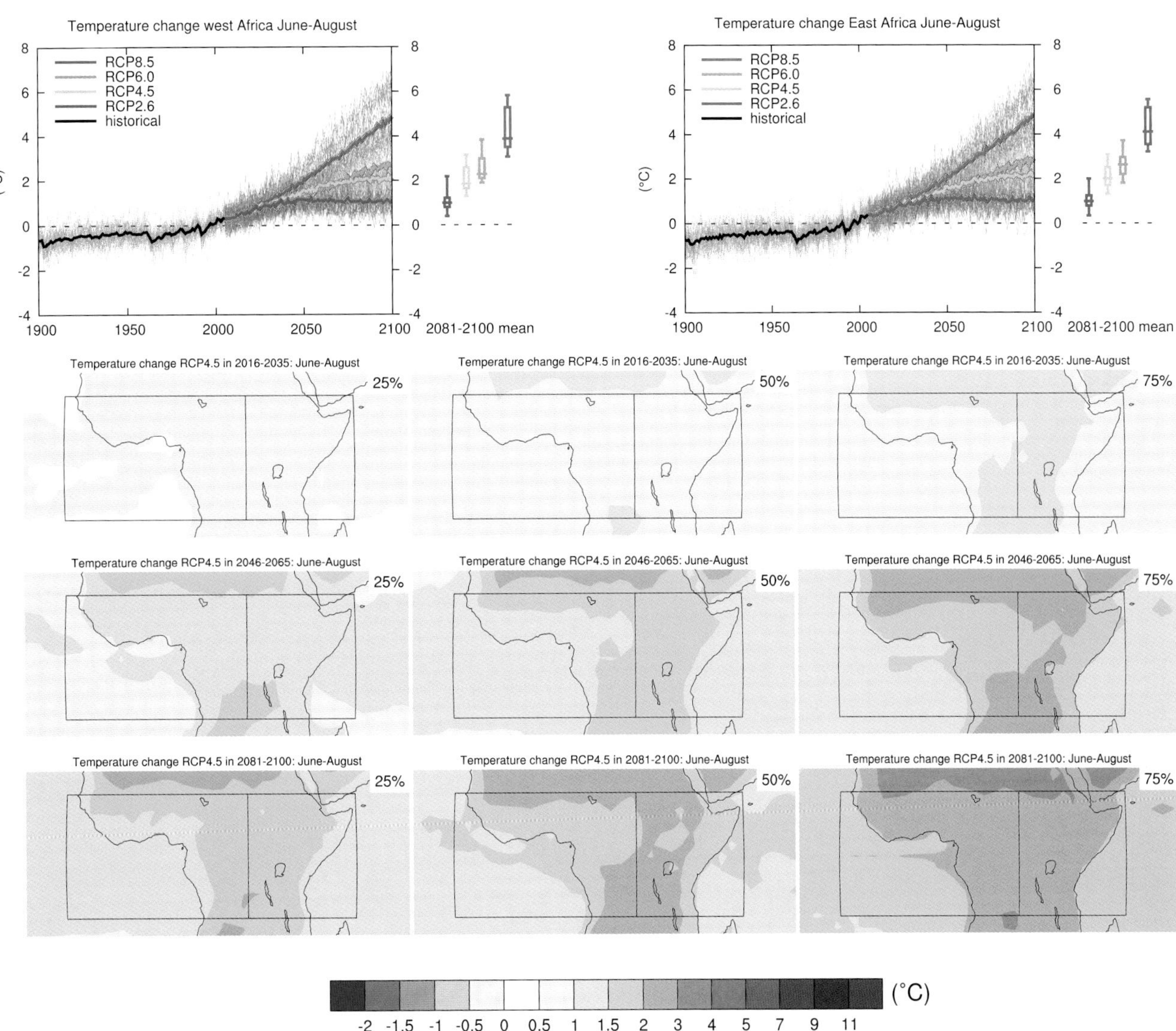

Figure AI.45 | (Top left) Time series of temperature change relative to 1986–2005 averaged over land grid points in West Africa (11.4°S to 15°N, 20°W to 25°E) in June to August. (Top right) Same for land grid points in East Africa (11.3°S to 15°N, 25°E to 52°E). Thin lines denote one ensemble member per model, thick lines the CMIP5 multi-model mean. On the right-hand side the 5th, 25th, 50th (median), 75th and 95th percentiles of the distribution of 20-year mean changes are given for 2081–2100 in the four RCP scenarios.

(Below) Maps of temperature changes in 2016–2035, 2046–2065 and 2081–2100 with respect to 1986–2005 in the RCP4.5 scenario. For each point, the 25th, 50th and 75th percentiles of the distribution of the CMIP5 ensemble are shown; this includes both natural variability and inter-model spread. Hatching denotes areas where the 20-year mean differences of the percentiles are less than the standard deviation of model-estimated present-day natural variability of 20-year mean differences.

Sections 9.4.1.1, 9.6.1.1, 10.3.1.1.4, Box 11.2, 14.8.7 contain relevant information regarding the evaluation of models in this region, the model spread in the context of other methods of projecting changes and the role of modes of variability and other climate phenomena.

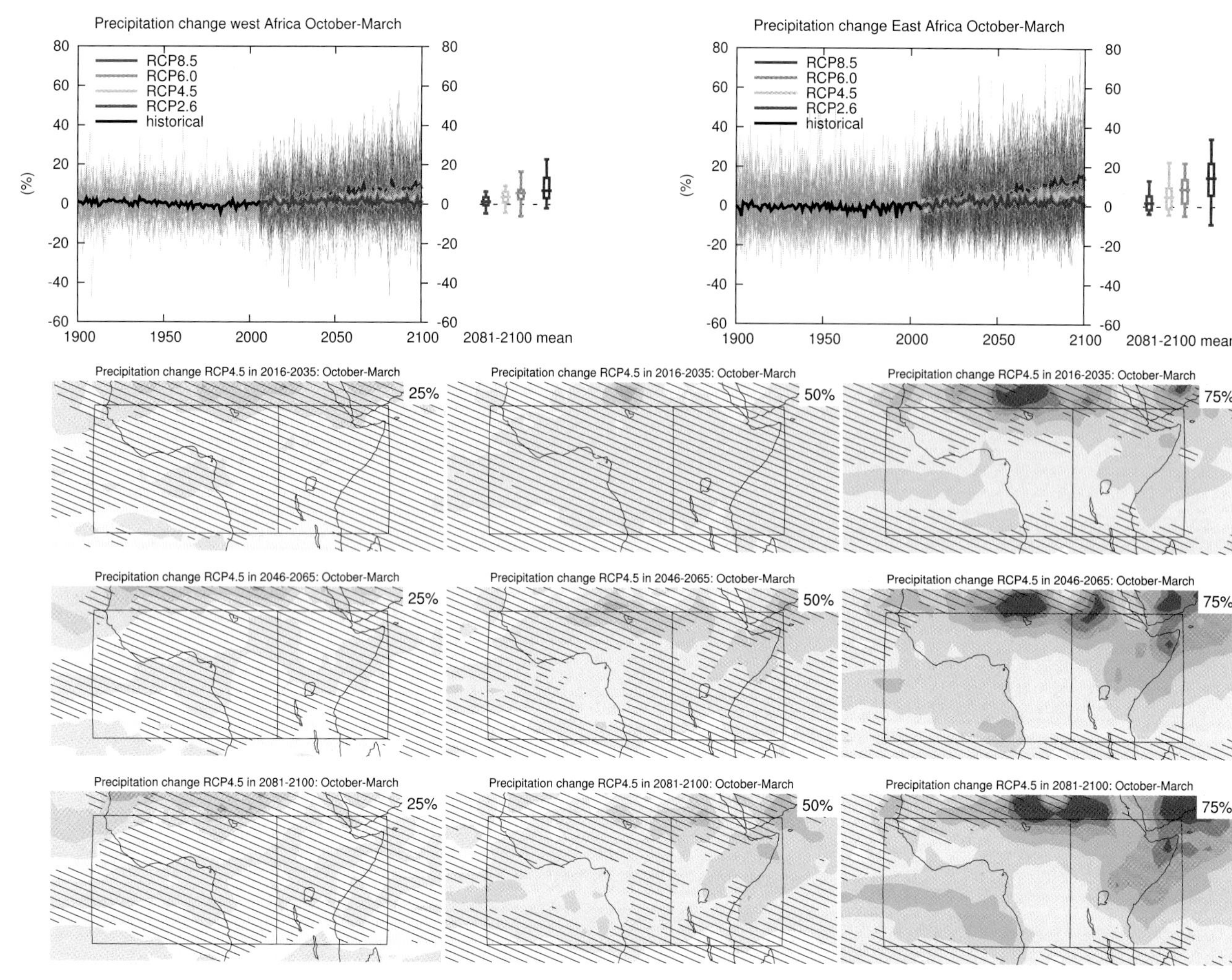

Figure AI.46 | (Top left) Time series of relative change relative to 1986–2005 in precipitation averaged over land grid points in West Africa (11.4°S to 15°N, 20°W to 25°E) in October to March. (Top right) Same for land grid points in East Africa (11.3°S to 15°N, 25°E to 52°E). Thin lines denote one ensemble member per model, thick lines the CMIP5 multi-model mean. On the right-hand side the 5th, 25th, 50th (median), 75th and 95th percentiles of the distribution of 20-year mean changes are given for 2081–2100 in the four RCP scenarios.

(Below) Maps of precipitation changes in 2016–2035, 2046–2065 and 2081–2100 with respect to 1986–2005 in the RCP4.5 scenario. For each point, the 25th, 50th and 75th percentiles of the distribution of the CMIP5 ensemble are shown; this includes both natural variability and inter-model spread. Hatching denotes areas where the 20-year mean differences of the percentiles are less than the standard deviation of model-estimated present-day natural variability of 20-year mean differences.

Sections 9.4.1.1, 9.6.1.1, 11.3.2.1.2, Box 11.2, 12.4.5.2, 14.2.4, 14.8.7 contain relevant information regarding the evaluation of models in this region, the model spread in the context of other methods of projecting changes and the role of modes of variability and other climate phenomena.

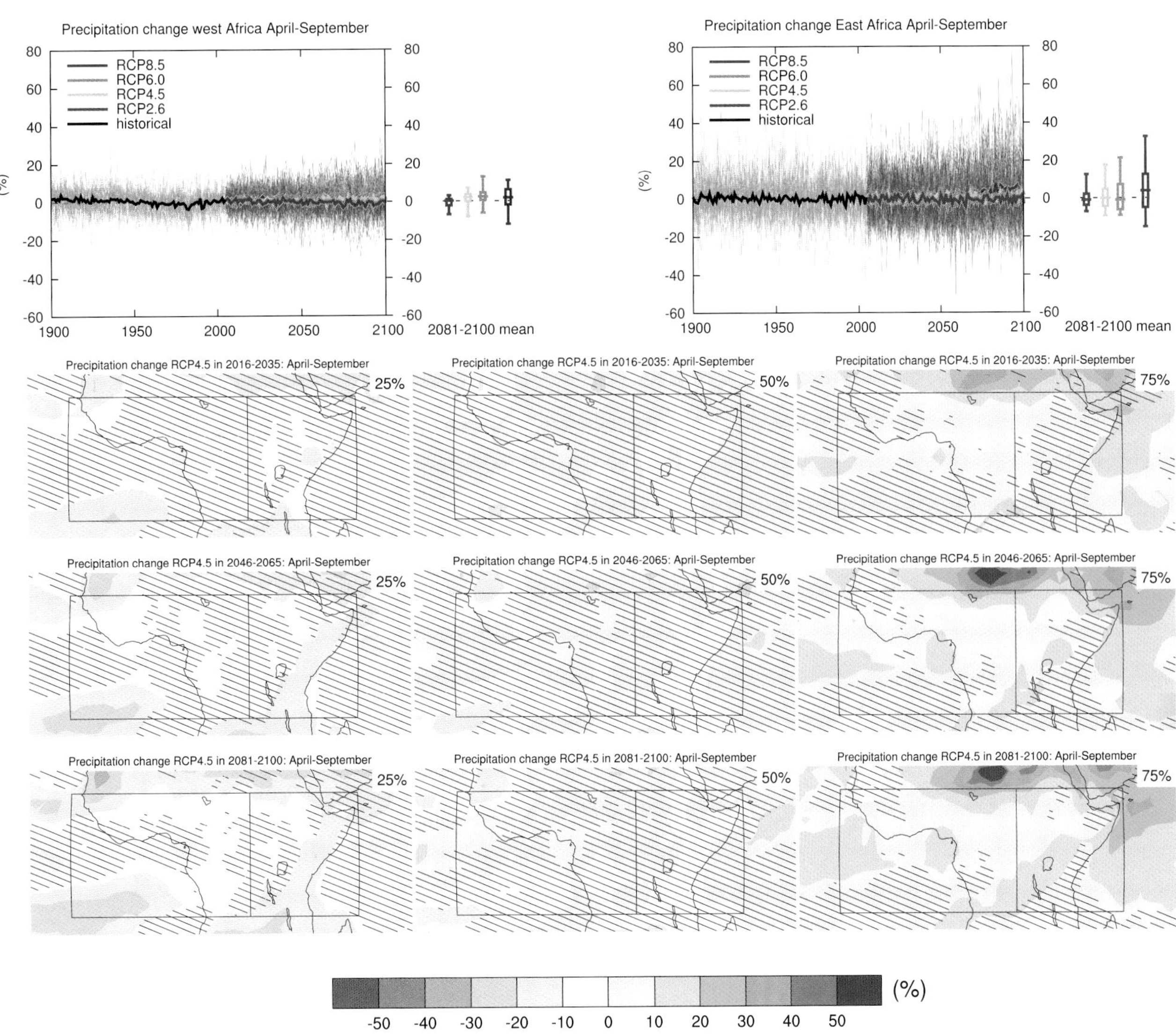

Figure AI.47 | (Top left) Time series of relative change relative to 1986–2005 in precipitation averaged over land grid points in West Africa (11.4°S to 15°N, 20°W to 25°E) in April to September. (Top right) Same for land grid points in East Africa (11.3°S to 15°N, 25°E to 52°E). Thin lines denote one ensemble member per model, thick lines the CMIP5 multi-model mean. On the right-hand side the 5th, 25th, 50th (median), 75th and 95th percentiles of the distribution of 20-year mean changes are given for 2081–2100 in the four RCP scenarios.

(Below) Maps of precipitation changes in 2016–2035, 2046–2065 and 2081–2100 with respect to 1986–2005 in the RCP4.5 scenario. For each point, the 25th, 50th and 75th percentiles of the distribution of the CMIP5 ensemble are shown; this includes both natural variability and inter-model spread. Hatching denotes areas where the 20-year mean differences of the percentiles are less than the standard deviation of model-estimated present-day natural variability of 20-year mean differences.

Sections 9.4.1.1, 9.6.1.1, 11.3.2.1.2, Box 11.2, 12.4.5.2, 14.2.4, 14.8.7 contain relevant information regarding the evaluation of models in this region, the model spread in the context of other methods of projecting changes and the role of modes of variability and other climate phenomena.

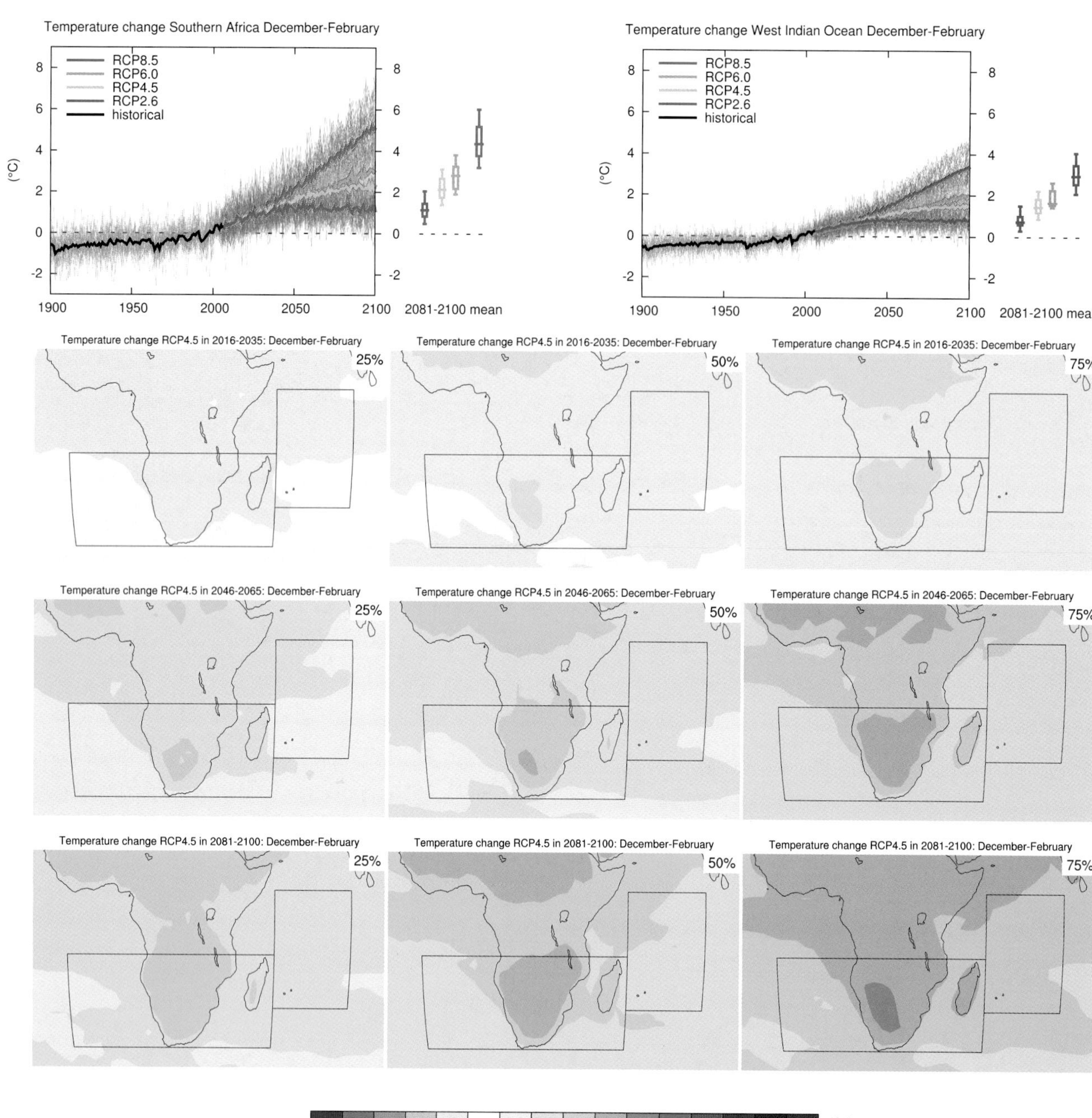

Figure AI.48 | (Top left) Time series of temperature change relative to 1986–2005 averaged over land grid points in Southern Africa (35°S to 11.4°S, 10°W to 52°E) in December to February. (Top right) Same for sea grid points in the West Indian Ocean (25°S to 5°N, 52°E to 75°E). Thin lines denote one ensemble member per model, thick lines the CMIP5 multi-model mean. On the right-hand side the 5th, 25th, 50th (median), 75th and 95th percentiles of the distribution of 20-year mean changes are given for 2081–2100 in the four RCP scenarios.

(Below) Maps of temperature changes in 2016–2035, 2046–2065 and 2081–2100 with respect to 1986–2005 in the RCP4.5 scenario. For each point, the 25th, 50th and 75th percentiles of the distribution of the CMIP5 ensemble are shown; this includes both natural variability and inter-model spread. Hatching denotes areas where the 20-year mean differences of the percentiles are less than the standard deviation of model-estimated present-day natural variability of 20-year mean differences.

Sections 9.4.1.1, 9.6.1.1, 10.3.1.1.4, Box 11.2, 14.8.7 contain relevant information regarding the evaluation of models in this region, the model spread in the context of other methods of projecting changes and the role of modes of variability and other climate phenomena.

Temperature change Southern Africa June-August

RCP8.5
RCP6.0
RCP4.5
RCP2.6
historical

(°C)

1900 1950 2000 2050 2100 2081-2100 mean

Temperature change West Indian Ocean June-August

RCP8.5
RCP6.0
RCP4.5
RCP2.6
historical

(°C)

1900 1950 2000 2050 2100 2081-2100 mean

Temperature change RCP4.5 in 2016-2035: June-August 25%

Temperature change RCP4.5 in 2016-2035: June-August 50%

Temperature change RCP4.5 in 2016-2035: June-August 75%

Temperature change RCP4.5 in 2046-2065: June-August 25%

Temperature change RCP4.5 in 2046-2065: June-August 50%

Temperature change RCP4.5 in 2046-2065: June-August 75%

Temperature change RCP4.5 in 2081-2100: June-August 25%

Temperature change RCP4.5 in 2081-2100: June-August 50%

Temperature change RCP4.5 in 2081-2100: June-August 75%

Figure AI.49 | (Top left) Time series of temperature change relative to 1986–2005 averaged over land grid points in Southern Africa (35°S to 11.4°S, 10°W to 52°E) in June to August. (Top right) Same for sea grid points in the West Indian Ocean (25°S to 5°N, 52°E to 75°E). Thin lines denote one ensemble member per model, thick lines the CMIP5 multi-model mean. On the right-hand side the 5th, 25th, 50th (median), 75th and 95th percentiles of the distribution of 20-year mean changes are given for 2081–2100 in the four RCP scenarios.

(Below) Maps of temperature changes in 2016–2035, 2046–2065 and 2081–2100 with respect to 1986–2005 in the RCP4.5 scenario. For each point, the 25th, 50th and 75th percentiles of the distribution of the CMIP5 ensemble are shown; this includes both natural variability and inter-model spread. Hatching denotes areas where the 20-year mean differences of the percentiles are less than the standard deviation of model-estimated present-day natural variability of 20-year mean differences.

Sections 9.4.1.1, 9.6.1.1, 10.3.1.1.4, Box 11.2, 14.8.7 contain relevant information regarding the evaluation of models in this region, the model spread in the context of other methods of projecting changes and the role of modes of variability and other climate phenomena.

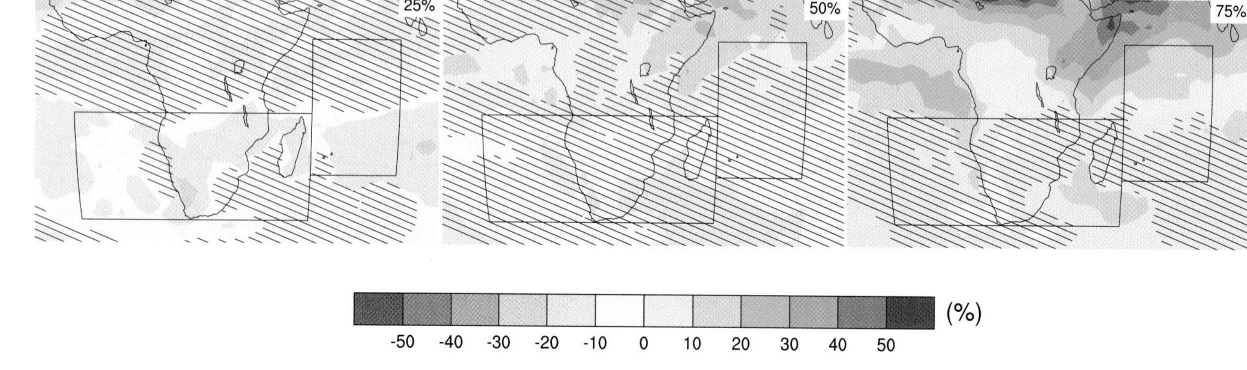

Figure AI.50 | (Top left) Time series of relative change relative to 1986–2005 in precipitation averaged over land grid points in Southern Africa (35°S to 11.4°S, 10°W to 52°E) in October to March. (Top right) Same for sea grid points in the West Indian Ocean (25°S to 5°N, 52°E to 75°E). Thin lines denote one ensemble member per model, thick lines the CMIP5 multi-model mean. On the right-hand side the 5th, 25th, 50th (median), 75th and 95th percentiles of the distribution of 20-year mean changes are given for 2081–2100 in the four RCP scenarios.

(Below) Maps of precipitation changes in 2016–2035, 2046–2065 and 2081–2100 with respect to 1986–2005 in the RCP4.5 scenario. For each point, the 25th, 50th and 75th percentiles of the distribution of the CMIP5 ensemble are shown; this includes both natural variability and inter-model spread. Hatching denotes areas where the 20-year mean differences of the percentiles are less than the standard deviation of model-estimated present-day natural variability of 20-year mean differences.

Sections 9.4.1.1, 9.6.1.1, Box 11.2, 12.4.5.2, 14.8.7 contain relevant information regarding the evaluation of models in this region, the model spread in the context of other methods of projecting changes and the role of modes of variability and other climate phenomena.

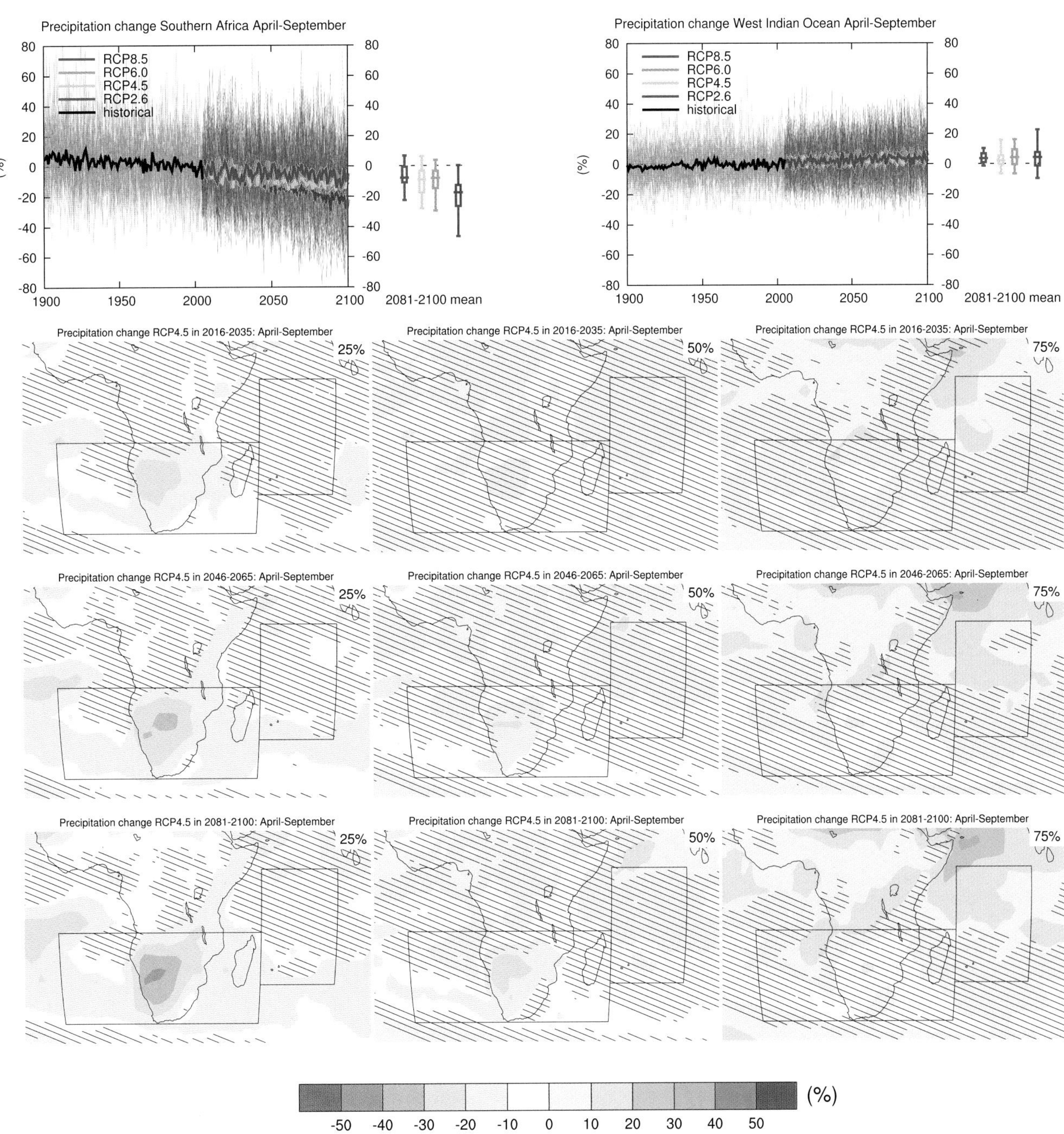

Figure AI.51 | (Top left) Time series of relative change relative to 1986–2005 in precipitation averaged over land grid points in Southern Africa (35°S to 11.4°S, 10°W to 52°E) in April to September. (Top right) Same for sea grid points in the West Indian Ocean (25°S to 5°N, 52°E to 75°E). Thin lines denote one ensemble member per model, thick lines the CMIP5 multi-model mean. On the right-hand side the 5th, 25th, 50th (median), 75th and 95th percentiles of the distribution of 20-year mean changes are given for 2081–2100 in the four RCP scenarios.

(Below) Maps of precipitation changes in 2016–2035, 2046–2065 and 2081–2100 with respect to 1986–2005 in the RCP4.5 scenario. For each point, the 25th, 50th and 75th percentiles of the distribution of the CMIP5 ensemble are shown; this includes both natural variability and inter-model spread. Hatching denotes areas where the 20-year mean differences of the percentiles are less than the standard deviation of model-estimated present-day natural variability of 20-year mean differences.

Sections 9.4.1.1, 9.6.1.1, Box 11.2, 12.4.5.2, 14.8.7 contain relevant information regarding the evaluation of models in this region, the model spread in the context of other methods of projecting changes and the role of modes of variability and other climate phenomena.

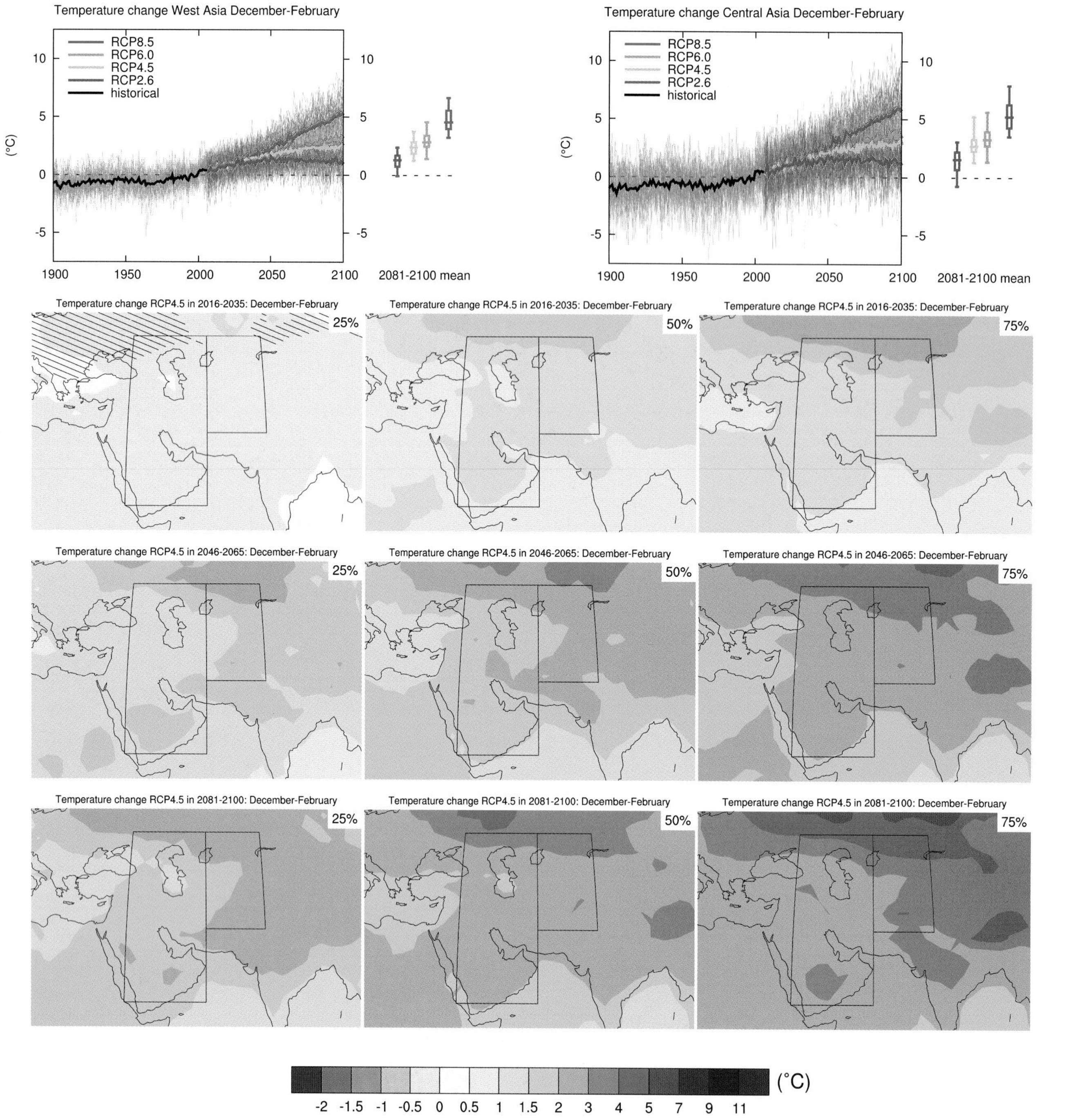

Figure AI.52 | (Top left) Time series of temperature change relative to 1986–2005 averaged over land grid points in West Asia (15°N to 50°N, 40°E to 60°E) in December to February. (Top right) Same for land grid points in Central Asia (30°N to 50°N, 60°E to 75°E). Thin lines denote one ensemble member per model, thick lines the CMIP5 multi-model mean. On the right-hand side the 5th, 25th, 50th (median), 75th and 95th percentiles of the distribution of 20-year mean changes are given for 2081–2100 in the four RCP scenarios.

(Below) Maps of temperature changes in 2016–2035, 2046–2065 and 2081–2100 with respect to 1986–2005 in the RCP4.5 scenario. For each point, the 25th, 50th and 75th percentiles of the distribution of the CMIP5 ensemble are shown; this includes both natural variability and inter-model spread. Hatching denotes areas where the 20-year mean differences of the percentiles are less than the standard deviation of model-estimated present-day natural variability of 20-year mean differences.

Sections 9.4.1.1, 9.6.1.1, 10.3.1.1.4, Box 11.2, 14.8.8, 14.8.10 contain relevant information regarding the evaluation of models in this region, the model spread in the context of other methods of projecting changes and the role of modes of variability and other climate phenomena.

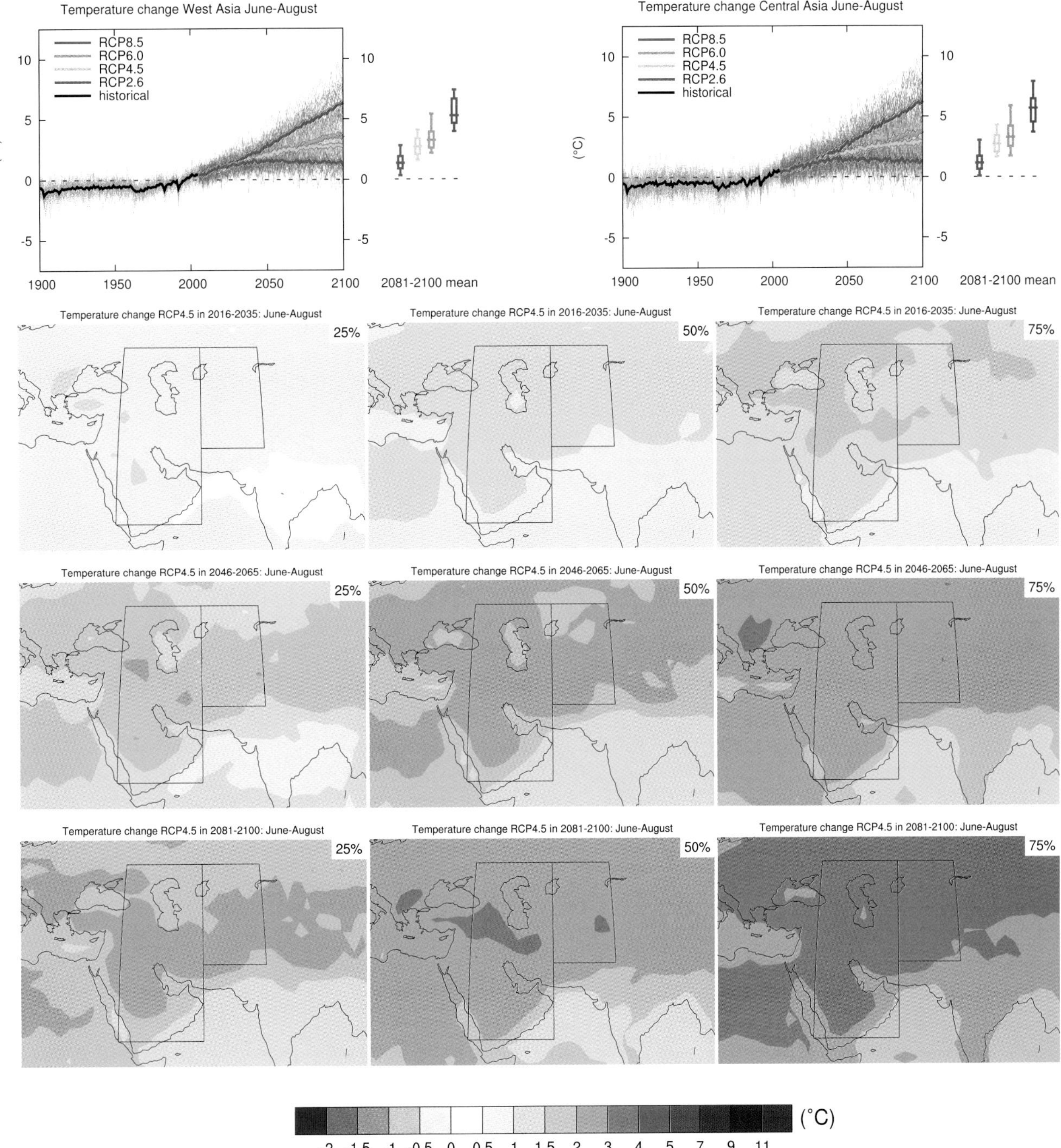

Figure AI.53 | (Top left) Time series of temperature change relative to 1986–2005 averaged over land grid points in West Asia (15°N to 50°N, 40°E to 60°E) in June to August. (Top right) Same for land grid points in Central Asia (30°N to 50°N, 60°E to 75°E). Thin lines denote one ensemble member per model, thick lines the CMIP5 multi-model mean. On the right-hand side the 5th, 25th, 50th (median), 75th and 95th percentiles of the distribution of 20-year mean changes are given for 2081–2100 in the four RCP scenarios.

(Below) Maps of temperature changes in 2016–2035, 2046–2065 and 2081–2100 with respect to 1986–2005 in the RCP4.5 scenario. For each point, the 25th, 50th and 75th percentiles of the distribution of the CMIP5 ensemble are shown; this includes both natural variability and inter-model spread. Hatching denotes areas where the 20-year mean differences of the percentiles are less than the standard deviation of model-estimated present-day natural variability of 20-year mean differences.

Sections 9.4.1.1, 9.6.1.1, 10.3.1.1.4, Box 11.2, 14.8.8, 14.8.10 contain relevant information regarding the evaluation of models in this region, the model spread in the context of other methods of projecting changes and the role of modes of variability and other climate phenomena.

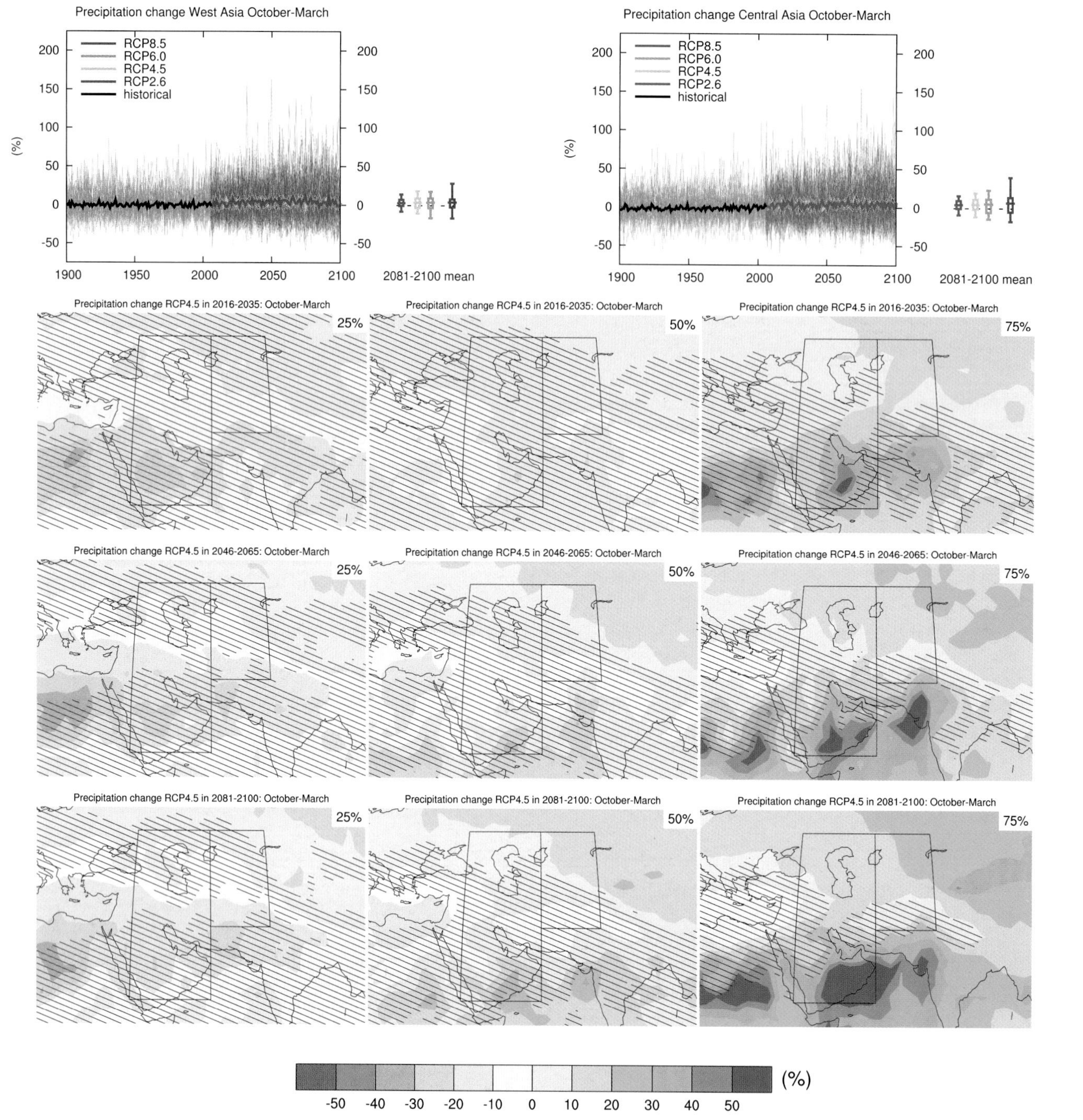

Figure AI.54 | (Top left) Time series of relative change relative to 1986–2005 in precipitation averaged over land grid points in West Asia (15°N to 50°N, 40°E to 60°E) in October to March. (Top right) Same for land grid points in Central Asia (30°N to 50°N, 60°E to 75°E). Thin lines denote one ensemble member per model, thick lines the CMIP5 multi-model mean. On the right-hand side the 5th, 25th, 50th (median), 75th and 95th percentiles of the distribution of 20-year mean changes are given for 2081–2100 in the four RCP scenarios.

(Below) Maps of precipitation changes in 2016–2035, 2046–2065 and 2081–2100 with respect to 1986–2005 in the RCP4.5 scenario. For each point, the 25th, 50th and 75th percentiles of the distribution of the CMIP5 ensemble are shown; this includes both natural variability and inter-model spread. Hatching denotes areas where the 20-year mean differences of the percentiles are less than the standard deviation of model-estimated present-day natural variability of 20-year mean differences.

Sections 9.4.1.1, 9.6.1.1, Box 11.2, 12.4.5.2, 14.8.8, 14.8.10 contain relevant information regarding the evaluation of models in this region, the model spread in the context of other methods of projecting changes and the role of modes of variability and other climate phenomena.

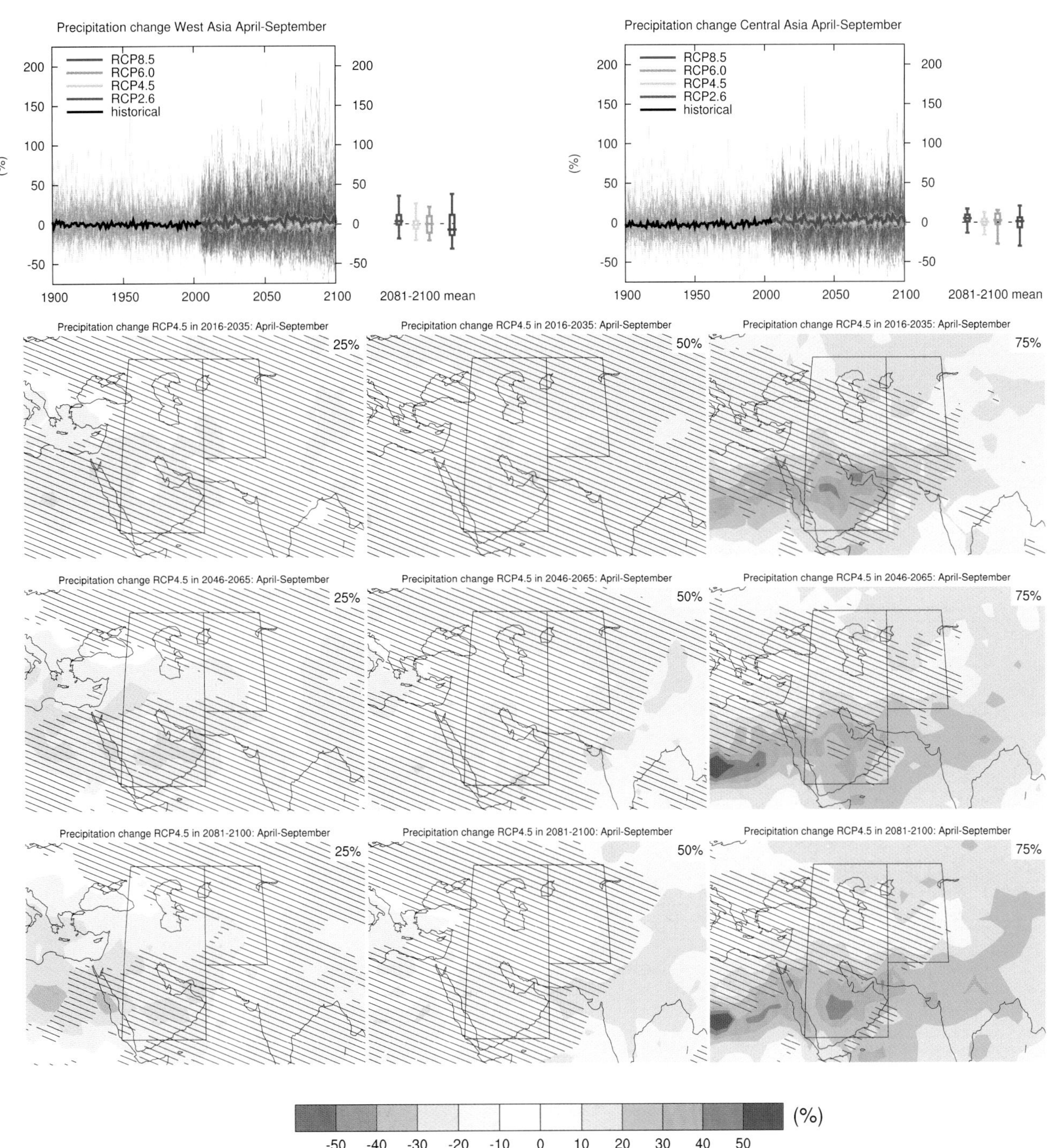

Figure AI.55 | (Top left) Time series of relative change relative to 1986–2005 in precipitation averaged over land grid points in West Asia (15°N to 50°N, 40°E to 60°E) in April to September. (Top right) Same for land grid points in Central Asia (30°N to 50°N, 60°E to 75°E). Thin lines denote one ensemble member per model, thick lines the CMIP5 multi-model mean. On the right-hand side the 5th, 25th, 50th (median), 75th and 95th percentiles of the distribution of 20-year mean changes are given for 2081–2100 in the four RCP scenarios.

(Below) Maps of precipitation changes in 2016–2035, 2046–2065 and 2081–2100 with respect to 1986–2005 in the RCP4.5 scenario. For each point, the 25th, 50th and 75th percentiles of the distribution of the CMIP5 ensemble are shown; this includes both natural variability and inter-model spread. Hatching denotes areas where the 20-year mean differences of the percentiles are less than the standard deviation of model-estimated present-day natural variability of 20-year mean differences.

Sections 9.4.1.1, 9.6.1.1, Box 11.2, 12.4.5.2, 14.8.8, 14.8.10 contain relevant information regarding the evaluation of models in this region, the model spread in the context of other methods of projecting changes and the role of modes of variability and other climate phenomena.

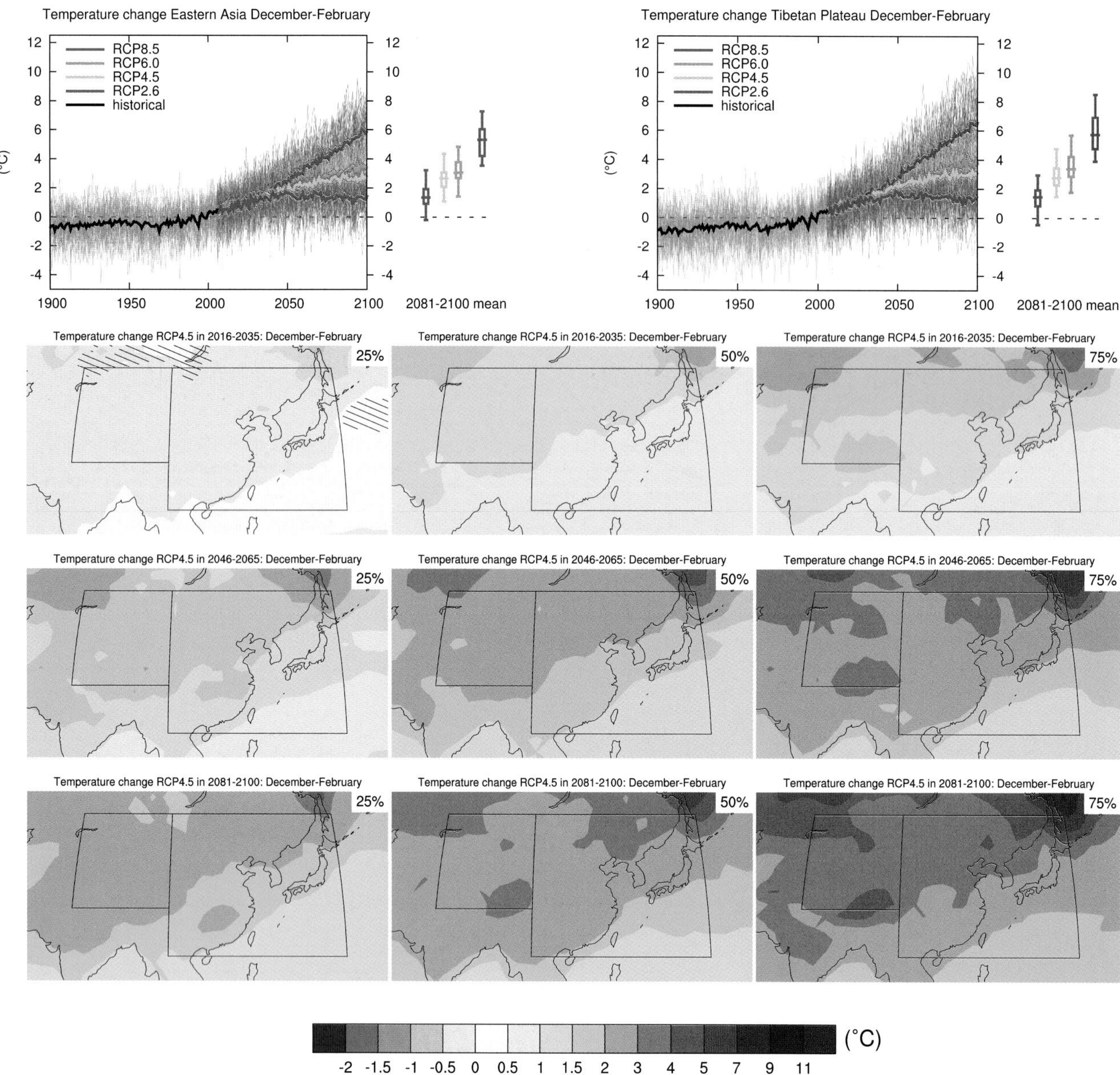

Figure AI.56 | (Top left) Time series of temperature change relative to 1986–2005 averaged over land grid points in Eastern Asia (20°N to 50°N, 100°E to 145°E) in December to February. (Top right) Same for land grid points on the Tibetan Plateau (30°N to 50°N, 75°E to 100°E). Thin lines denote one ensemble member per model, thick lines the CMIP5 multi-model mean. On the right-hand side the 5th, 25th, 50th (median), 75th and 95th percentiles of the distribution of 20-year mean changes are given for 2081–2100 in the four RCP scenarios.

(Below) Maps of temperature changes in 2016–2035, 2046–2065 and 2081–2100 with respect to 1986–2005 in the RCP4.5 scenario. For each point, the 25th, 50th and 75th percentiles of the distribution of the CMIP5 ensemble are shown; this includes both natural variability and inter-model spread. Hatching denotes areas where the 20-year mean differences of the percentiles are less than the standard deviation of model-estimated present-day natural variability of 20-year mean differences.

Sections 9.4.1.1, 9.6.1.1, 10.3.1.1.4, Box 11.2, 14.8.8, 14.8.9 contain relevant information regarding the evaluation of models in this region, the model spread in the context of other methods of projecting changes and the role of modes of variability and other climate phenomena.

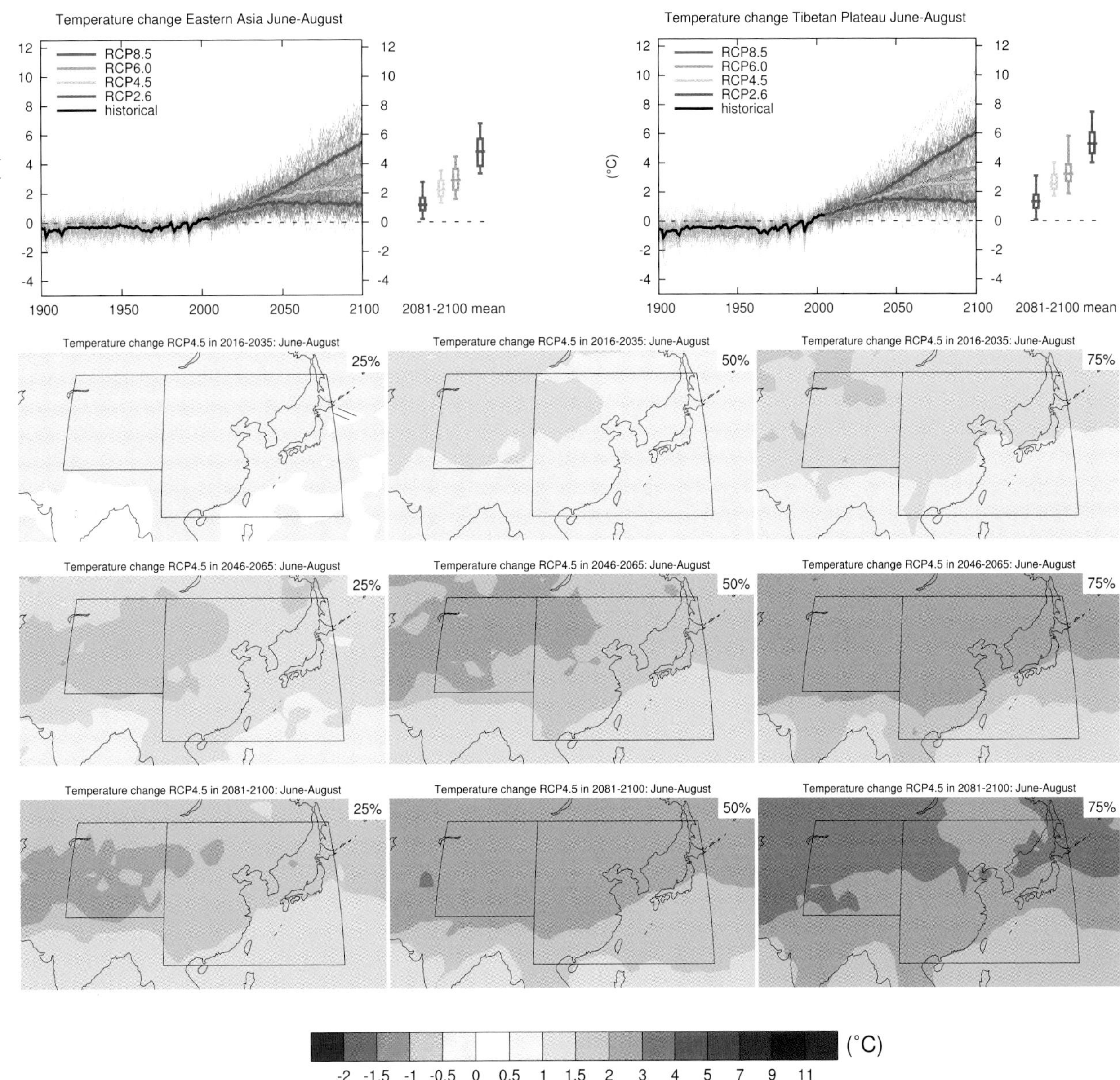

Figure AI.57 | (Top left) Time series of temperature change relative to 1986–2005 averaged over land grid points in Eastern Asia (20°N to 50°N, 100°E to 145°E) in June to August. (Top right) Same for land grid points on the Tibetan Plateau (30°N to 50°N, 75°E to 100°E). Thin lines denote one ensemble member per model, thick lines the CMIP5 multi-model mean. On the right-hand side the 5th, 25th, 50th (median), 75th and 95th percentiles of the distribution of 20-year mean changes are given for 2081–2100 in the four RCP scenarios.

(Below) Maps of temperature changes in 2016–2035, 2046–2065 and 2081–2100 with respect to 1986–2005 in the RCP4.5 scenario. For each point, the 25th, 50th and 75th percentiles of the distribution of the CMIP5 ensemble are shown; this includes both natural variability and inter-model spread. Hatching denotes areas where the 20-year mean differences of the percentiles are less than the standard deviation of model-estimated present-day natural variability of 20-year mean differences.

Sections 9.4.1.1, 9.6.1.1, 10.3.1.1.4, Box 11.2, 14.8.8, 14.8.9 contain relevant information regarding the evaluation of models in this region, the model spread in the context of other methods of projecting changes and the role of modes of variability and other climate phenomena.

Precipitation change Eastern Asia October-March

Precipitation change Tibetan Plateau October-March

RCP8.5
RCP6.0
RCP4.5
RCP2.6
historical

(%)

2081-2100 mean

Precipitation change RCP4.5 in 2016-2035: October-March

Precipitation change RCP4.5 in 2046-2065: October-March

Precipitation change RCP4.5 in 2081-2100: October-March

25% 50% 75%

-50 -40 -30 -20 -10 0 10 20 30 40 50 (%)

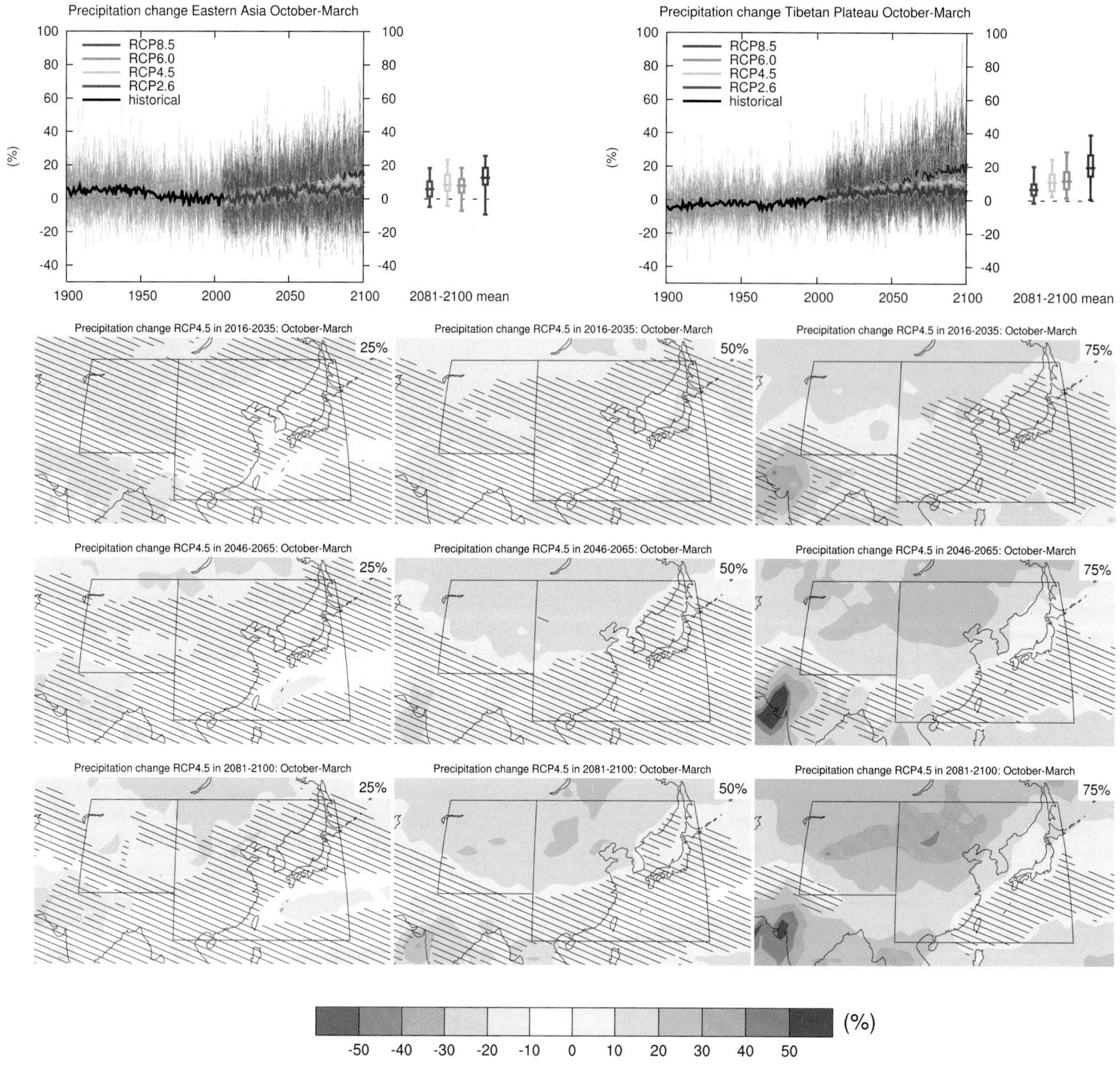

Figure AI.58 | (Top left) Time series of relative change relative to 1986–2005 in precipitation averaged over land grid points in Eastern Asia (20°N to 50°N, 100°E to 145°E) in October to March. (Top right) Same for land grid points on the Tibetan Plateau (30°N to 50°N, 75°E to 100°E). Thin lines denote one ensemble member per model, thick lines the CMIP5 multi-model mean. On the right-hand side the 5th, 25th, 50th (median), 75th and 95th percentiles of the distribution of 20-year mean changes are given for 2081–2100 in the four RCP scenarios.

(Below) Maps of precipitation changes in 2016–2035, 2046–2065 and 2081–2100 with respect to 1986–2005 in the RCP4.5 scenario. For each point, the 25th, 50th and 75th percentiles of the distribution of the CMIP5 ensemble are shown; this includes both natural variability and inter-model spread. Hatching denotes areas where the 20-year mean differences of the percentiles are less than the standard deviation of model-estimated present-day natural variability of 20-year mean differences.

Sections 9.4.1.1, 9.6.1.1, Box 11.2, 14.2.2.2, 14.8.8, 14.8.9 contain relevant information regarding the evaluation of models in this region, the model spread in the context of other methods of projecting changes and the role of modes of variability and other climate phenomena.

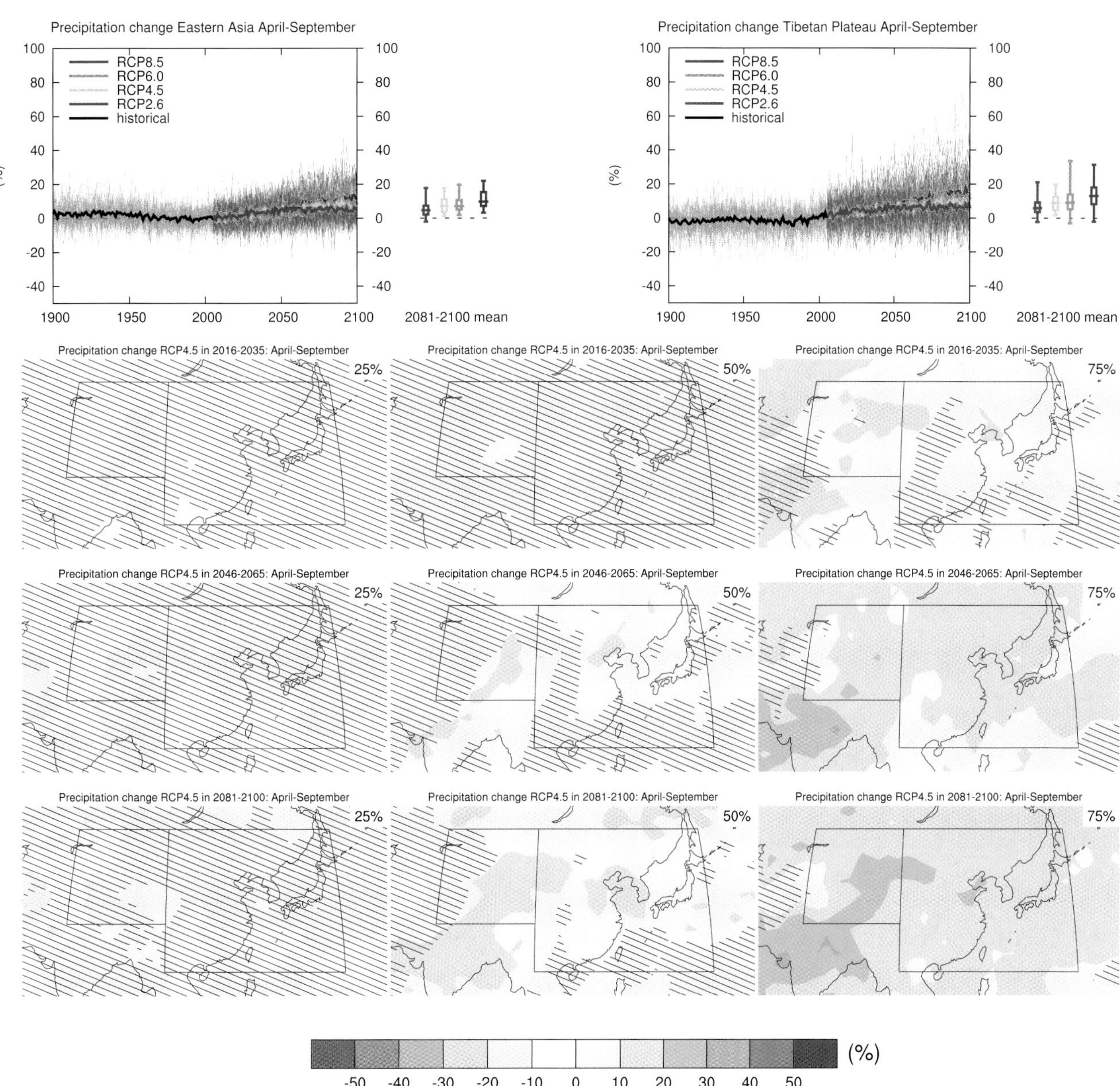

Figure AI.59 | (Top left) Time series of relative change relative to 1986–2005 in precipitation averaged over land grid points in Eastern Asia (20°N to 50°N, 100°E to 145°E) in April to September. (Top right) Same for land grid points on the Tibetan Plateau (30°N to 50°N, 75°E to 100°E). Thin lines denote one ensemble member per model, thick lines the CMIP5 multi-model mean. On the right-hand side the 5th, 25th, 50th (median), 75th and 95th percentiles of the distribution of 20-year mean changes are given for 2081–2100 in the four RCP scenarios.

(Below) Maps of precipitation changes in 2016–2035, 2046–2065 and 2081–2100 with respect to 1986–2005 in the RCP4.5 scenario. For each point, the 25th, 50th and 75th percentiles of the distribution of the CMIP5 ensemble are shown; this includes both natural variability and inter-model spread. Hatching denotes areas where the 20-year mean differences of the percentiles are less than the standard deviation of model-estimated present-day natural variability of 20-year mean differences.

Sections 9.4.1.1, 9.6.1.1, Box 11.2, 14.2.2.2, 14.8.8, 14.8.9 contain relevant information regarding the evaluation of models in this region, the model spread in the context of other methods of projecting changes and the role of modes of variability and other climate phenomena.

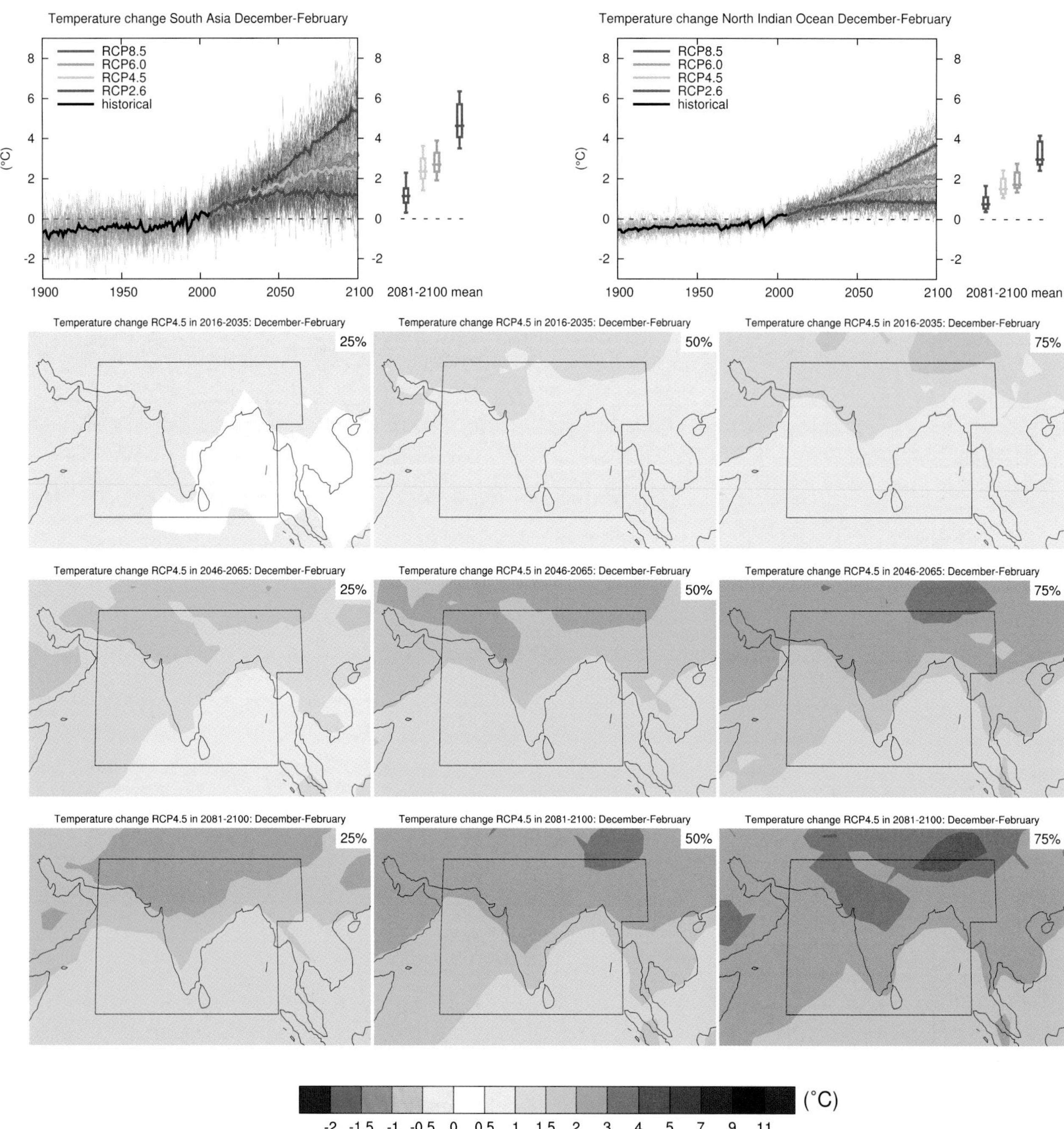

Figure AI.60 | (Top left) Time series of temperature change relative to 1986–2005 averaged over land grid points in South Asia (60°E, 5°N; 60°E, 30°N; 100°E, 30°N; 100°E, 20°E; 95°E, 20°N; 95°E, 5°N) in December to February. (Top right) Same for sea grid points in the North Indian Ocean (5°N to 30°N, 60°E to 95°E). Thin lines denote one ensemble member per model, thick lines the CMIP5 multi-model mean. On the right-hand side the 5th, 25th, 50th (median), 75th and 95th percentiles of the distribution of 20-year mean changes are given for 2081–2100 in the four RCP scenarios.

(Below) Maps of temperature changes in 2016–2035, 2046–2065 and 2081–2100 with respect to 1986–2005 in the RCP4.5 scenario. For each point, the 25th, 50th and 75th percentiles of the distribution of the CMIP5 ensemble are shown; this includes both natural variability and inter-model spread. Hatching denotes areas where the 20-year mean differences of the percentiles are less than the standard deviation of model-estimated present-day natural variability of 20-year mean differences.

Sections 9.4.1.1, 9.6.1.1, 10.3.1.1.4, Box 11.2, 14.8.11 contain relevant information regarding the evaluation of models in this region, the model spread in the context of other methods of projecting changes and the role of modes of variability and other climate phenomena.

Temperature change South Asia June-August

RCP8.5
RCP6.0
RCP4.5
RCP2.6
historical

(°C)

1900 1950 2000 2050 2100 2081-2100 mean

Temperature change North Indian Ocean June-August

RCP8.5
RCP6.0
RCP4.5
RCP2.6
historical

(°C)

1900 1950 2000 2050 2100 2081-2100 mean

Temperature change RCP4.5 in 2016-2035: June-August 25%

Temperature change RCP4.5 in 2016-2035: June-August 50%

Temperature change RCP4.5 in 2016-2035: June-August 75%

Temperature change RCP4.5 in 2046-2065: June-August 25%

Temperature change RCP4.5 in 2046-2065: June-August 50%

Temperature change RCP4.5 in 2046-2065: June-August 75%

Temperature change RCP4.5 in 2081-2100: June-August 25%

Temperature change RCP4.5 in 2081-2100: June-August 50%

Temperature change RCP4.5 in 2081-2100: June-August 75%

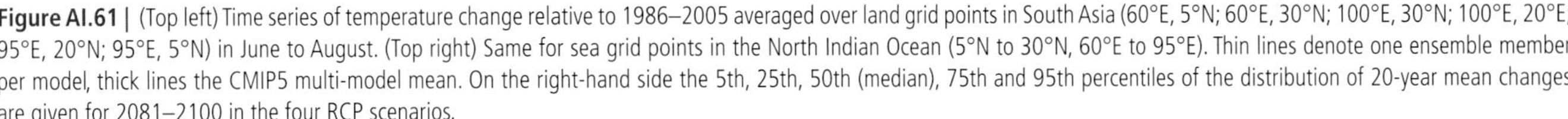

Figure AI.61 | (Top left) Time series of temperature change relative to 1986–2005 averaged over land grid points in South Asia (60°E, 5°N; 60°E, 30°N; 100°E, 30°N; 100°E, 20°E; 95°E, 20°N; 95°E, 5°N) in June to August. (Top right) Same for sea grid points in the North Indian Ocean (5°N to 30°N, 60°E to 95°E). Thin lines denote one ensemble member per model, thick lines the CMIP5 multi-model mean. On the right-hand side the 5th, 25th, 50th (median), 75th and 95th percentiles of the distribution of 20-year mean changes are given for 2081–2100 in the four RCP scenarios.

(Below) Maps of temperature changes in 2016–2035, 2046–2065 and 2081–2100 with respect to 1986–2005 in the RCP4.5 scenario. For each point, the 25th, 50th and 75th percentiles of the distribution of the CMIP5 ensemble are shown; this includes both natural variability and inter-model spread. Hatching denotes areas where the 20-year mean differences of the percentiles are less than the standard deviation of model-estimated present-day natural variability of 20-year mean differences.

Sections 9.4.1.1, 9.6.1.1, 10.3.1.1.4, Box 11.2, 14.8.11 contain relevant information regarding the evaluation of models in this region, the model spread in the context of other methods of projecting changes and the role of modes of variability and other climate phenomena.

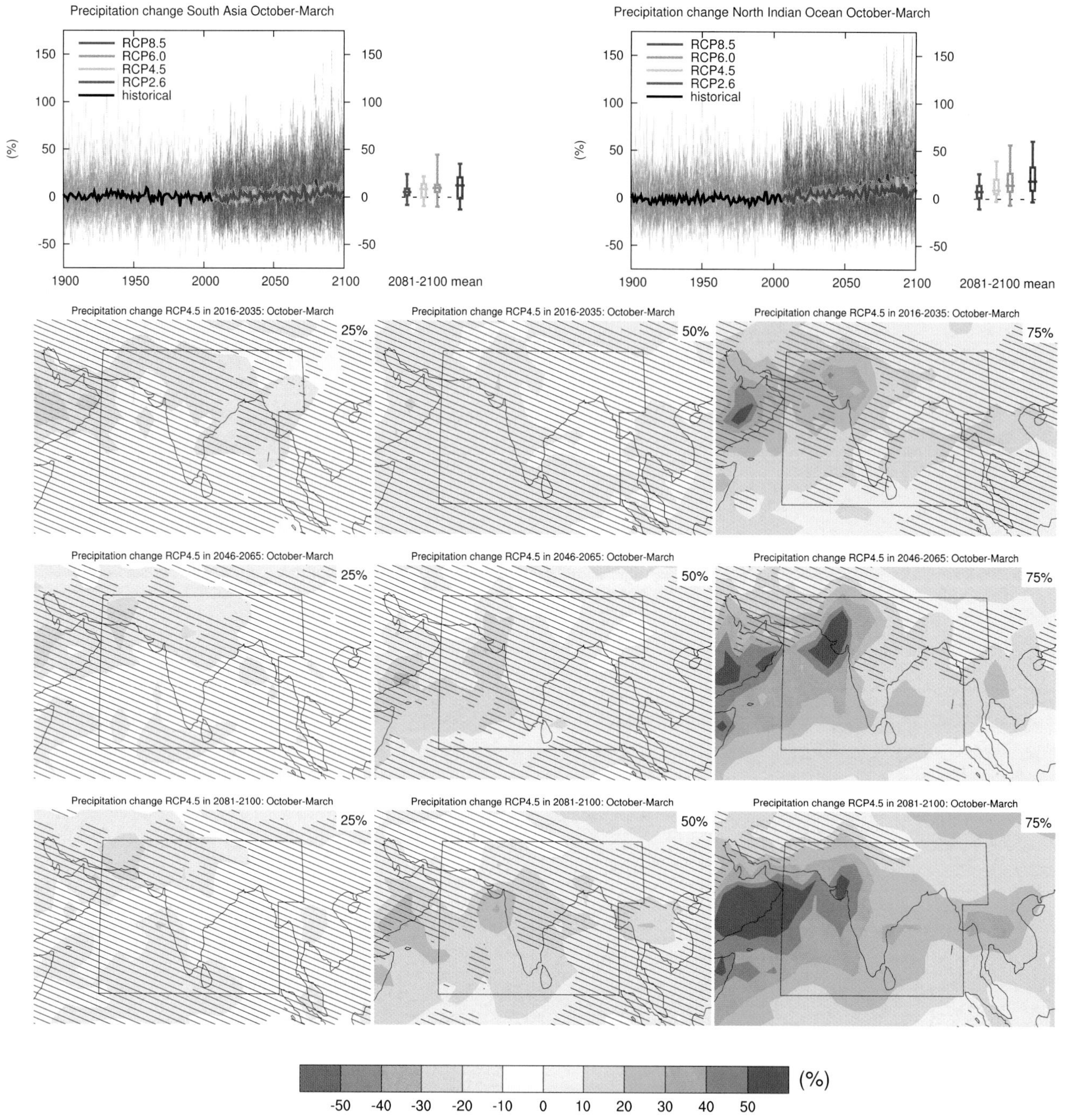

Precipitation change South Asia October-March

RCP8.5
RCP6.0
RCP4.5
RCP2.6
historical

Precipitation change North Indian Ocean October-March

Precipitation change RCP4.5 in 2016-2035: October-March 25% 50% 75%

Precipitation change RCP4.5 in 2046-2065: October-March 25% 50% 75%

Precipitation change RCP4.5 in 2081-2100: October-March 25% 50% 75%

Figure AI.62 | (Top left) Time series of relative change relative to 1986–2005 in precipitation averaged over land grid points in South Asia (60°E, 5°N; 60°E, 30°N; 100°E, 30°N; 100°E, 20°E; 95°E, 20°N; 95°E, 5°N) in October to March. (Top right) Same for sea grid points in the North Indian Ocean (5°N to 30°N, 60°E to 95°E). Thin lines denote one ensemble member per model, thick lines the CMIP5 multi-model mean. On the right-hand side the 5th, 25th, 50th (median), 75th and 95th percentiles of the distribution of 20-year mean changes are given for 2081–2100 in the four RCP scenarios.

(Below) Maps of precipitation changes in 2016–2035, 2046–2065 and 2081–2100 with respect to 1986–2005 in the RCP4.5 scenario. For each point, the 25th, 50th and 75th percentiles of the distribution of the CMIP5 ensemble are shown; this includes both natural variability and inter-model spread. Hatching denotes areas where the 20-year mean differences of the percentiles are less than the standard deviation of model-estimated present-day natural variability of 20-year mean differences.

Sections 9.4.1.1, 9.6.1.1, Box 11.2, 14.2.2.1, 14.8.11 contain relevant information regarding the evaluation of models in this region, the model spread in the context of other methods of projecting changes and the role of modes of variability and other climate phenomena.

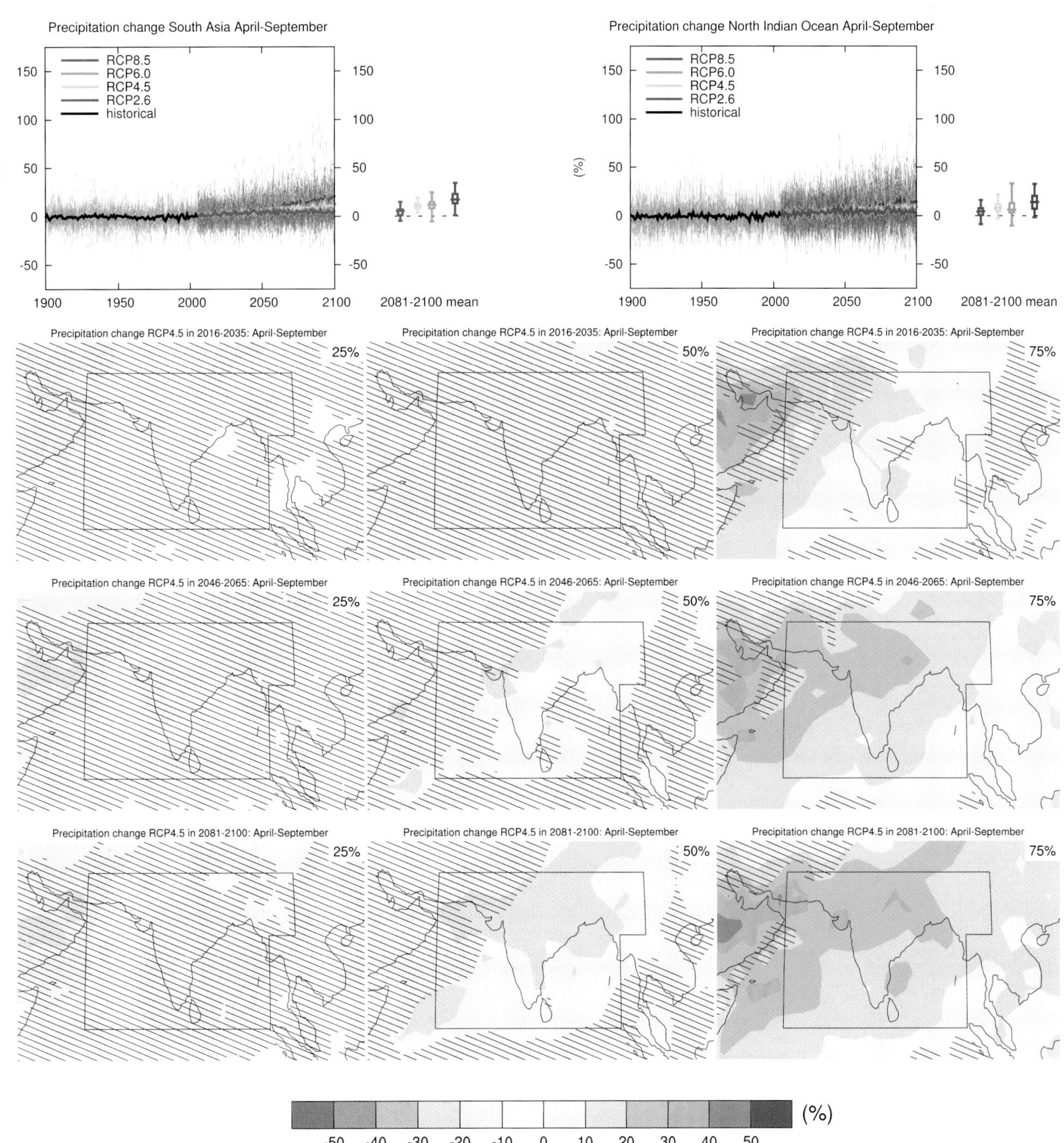

Figure AI.63 | (Top left) Time series of relative change relative to 1986–2005 in precipitation averaged over land grid points in South Asia (60°E, 5°N; 60°E, 30°N; 100°E, 30°N; 100°E, 20°E; 95°E, 20°N; 95°E, 5°N) in April to September. (Top right) Same for sea grid points in the North Indian Ocean (5°N to 30°N, 60°E to 95°E). Thin lines denote one ensemble member per model, thick lines the CMIP5 multi-model mean. On the right-hand side the 5th, 25th, 50th (median), 75th and 95th percentiles of the distribution of 20-year mean changes are given for 2081–2100 in the four RCP scenarios.

(Below) Maps of precipitation changes in 2016–2035, 2046–2065 and 2081–2100 with respect to 1986–2005 in the RCP4.5 scenario. For each point, the 25th, 50th and 75th percentiles of the distribution of the CMIP5 ensemble are shown; this includes both natural variability and inter-model spread. Hatching denotes areas where the 20-year mean differences of the percentiles are less than the standard deviation of model-estimated present-day natural variability of 20-year mean differences.

Sections 9.4.1.1, 9.6.1.1, Box 11.2, 14.2.2.1, 14.8.11 contain relevant information regarding the evaluation of models in this region, the model spread in the context of other methods of projecting changes and the role of modes of variability and other climate phenomena.

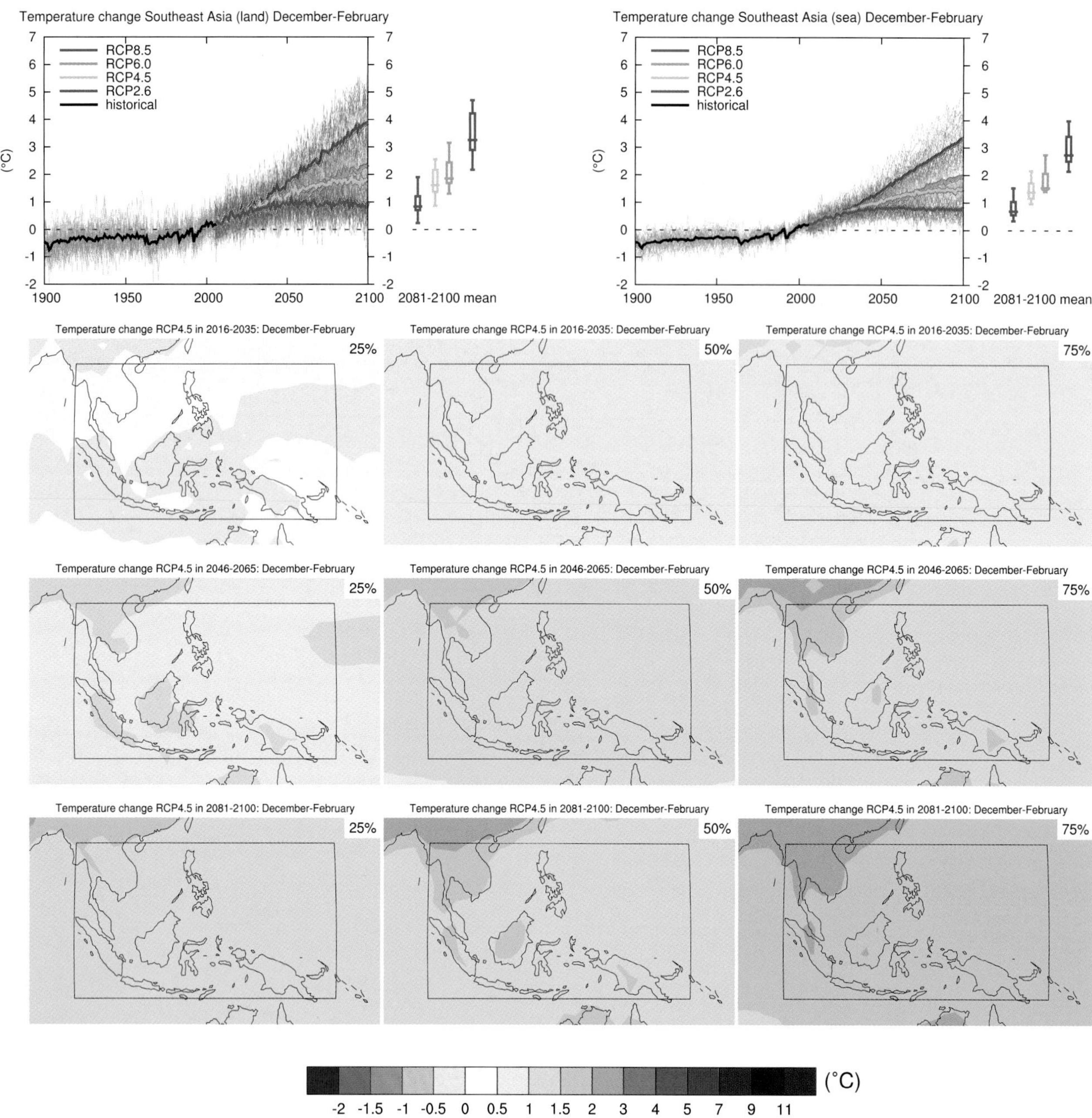

Figure AI.64 | (Top left) Time series of temperature change relative to 1986–2005 averaged over land grid points in Southeast Asia (10°S to 20°N, 95°E to 155°E) in December to February. (Top right) Same for sea grid points. Thin lines denote one ensemble member per model, thick lines the CMIP5 multi-model mean. On the right-hand side the 5th, 25th, 50th (median), 75th and 95th percentiles of the distribution of 20-year mean changes are given for 2081–2100 in the four RCP scenarios.

(Below) Maps of temperature changes in 2016–2035, 2046–2065 and 2081–2100 with respect to 1986–2005 in the RCP4.5 scenario. For each point, the 25th, 50th and 75th percentiles of the distribution of the CMIP5 ensemble are shown; this includes both natural variability and inter-model spread. Hatching denotes areas where the 20-year mean differences of the percentiles are less than the standard deviation of model-estimated present-day natural variability of 20-year mean differences.

Sections 9.4.1.1, 9.6.1.1, 10.3.1.1.4, Box 11.2, 14.8.12 contain relevant information regarding the evaluation of models in this region, the model spread in the context of other methods of projecting changes and the role of modes of variability and other climate phenomena.

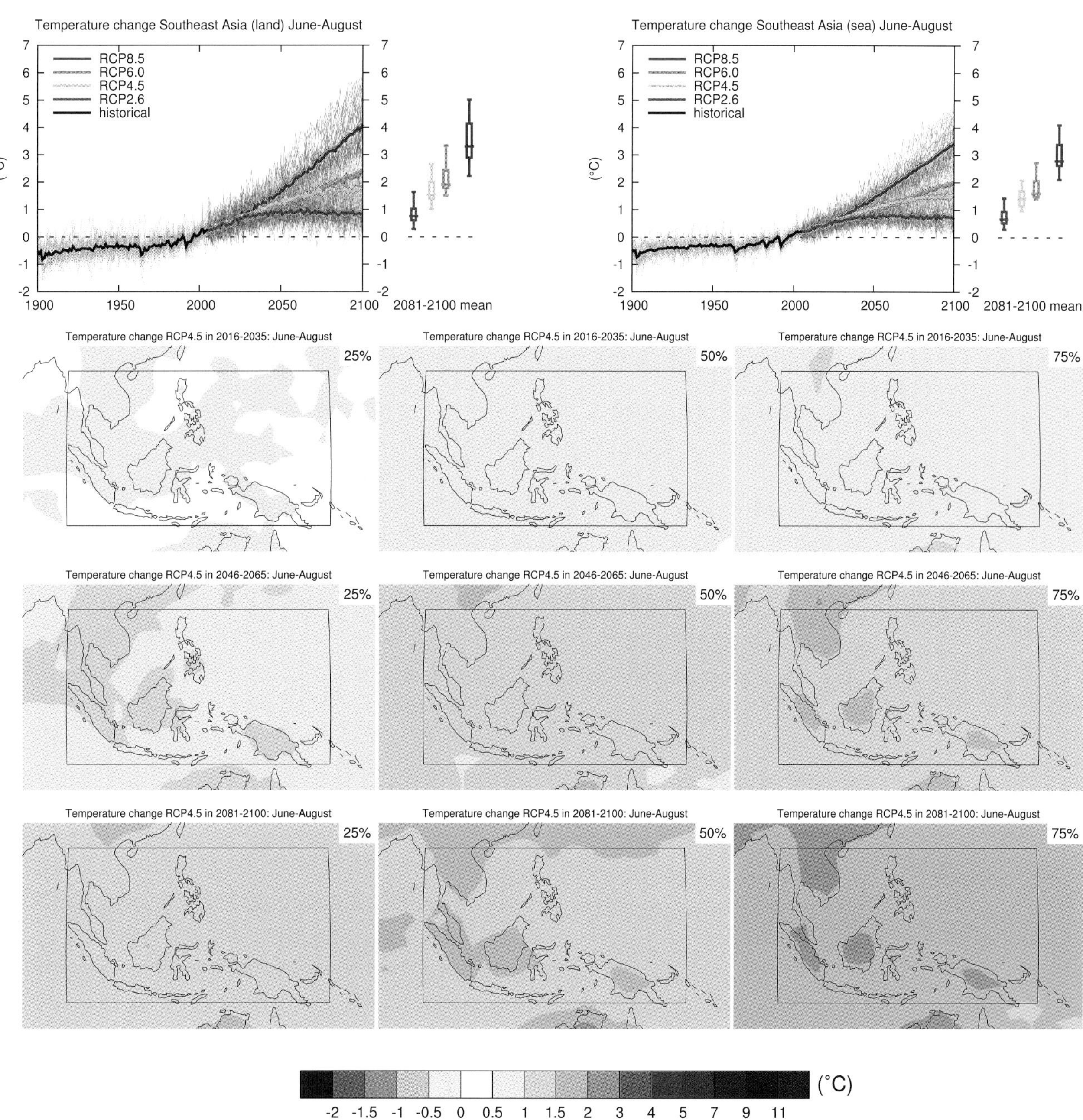

Figure AI.65 | (Top left) Time series of temperature change relative to 1986–2005 averaged over land grid points in Southeast Asia (10°S to 20°N, 95°E to 155°E) in June to August. (Top right) Same for sea grid points. Thin lines denote one ensemble member per model, thick lines the CMIP5 multi-model mean. On the right-hand side the 5th, 25th, 50th (median), 75th and 95th percentiles of the distribution of 20-year mean changes are given for 2081–2100 in the four RCP scenarios.

(Below) Maps of temperature changes in 2016–2035, 2046–2065 and 2081–2100 with respect to 1986–2005 in the RCP4.5 scenario. For each point, the 25th, 50th and 75th percentiles of the distribution of the CMIP5 ensemble are shown; this includes both natural variability and inter-model spread. Hatching denotes areas where the 20-year mean differences of the percentiles are less than the standard deviation of model-estimated present-day natural variability of 20-year mean differences.

Sections 9.4.1.1, 9.6.1.1, 10.3.1.1.4, Box 11.2, 14.8.12 contain relevant information regarding the evaluation of models in this region, the model spread in the context of other methods of projecting changes and the role of modes of variability and other climate phenomena.

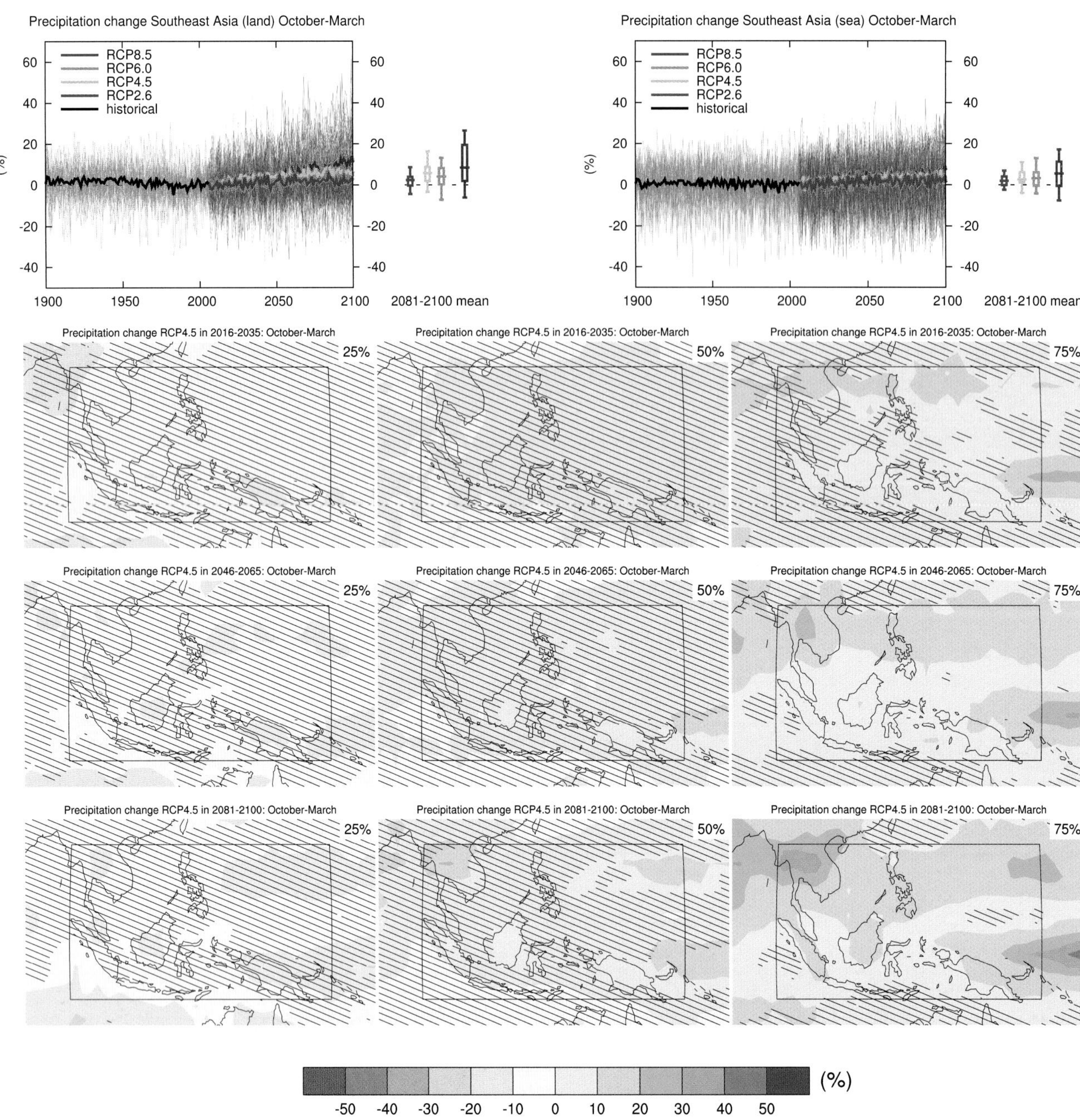

Figure AI.66 | (Top left) Time series of relative change relative to 1986–2005 in precipitation averaged over land grid points in Southeast Asia (10°S to 20°N, 95°E to 155°E) in October to March. (Top right) Same for sea grid points. Thin lines denote one ensemble member per model, thick lines the CMIP5 multi-model mean. On the right-hand side the 5th, 25th, 50th (median), 75th and 95th percentiles of the distribution of 20-year mean changes are given for 2081–2100 in the four RCP scenarios.

(Below) Maps of precipitation changes in 2016–2035, 2046–2065 and 2081–2100 with respect to 1986–2005 in the RCP4.5 scenario. For each point, the 25th, 50th and 75th percentiles of the distribution of the CMIP5 ensemble are shown; this includes both natural variability and inter-model spread. Hatching denotes areas where the 20-year mean differences of the percentiles are less than the standard deviation of model-estimated present-day natural variability of 20-year mean differences.

Sections 9.4.1.1, 9.6.1.1, Box 11.2, 14.2.2.3, 14.2.2.5, 14.8.12 contain relevant information regarding the evaluation of models in this region, the model spread in the context of other methods of projecting changes and the role of modes of variability and other climate phenomena.

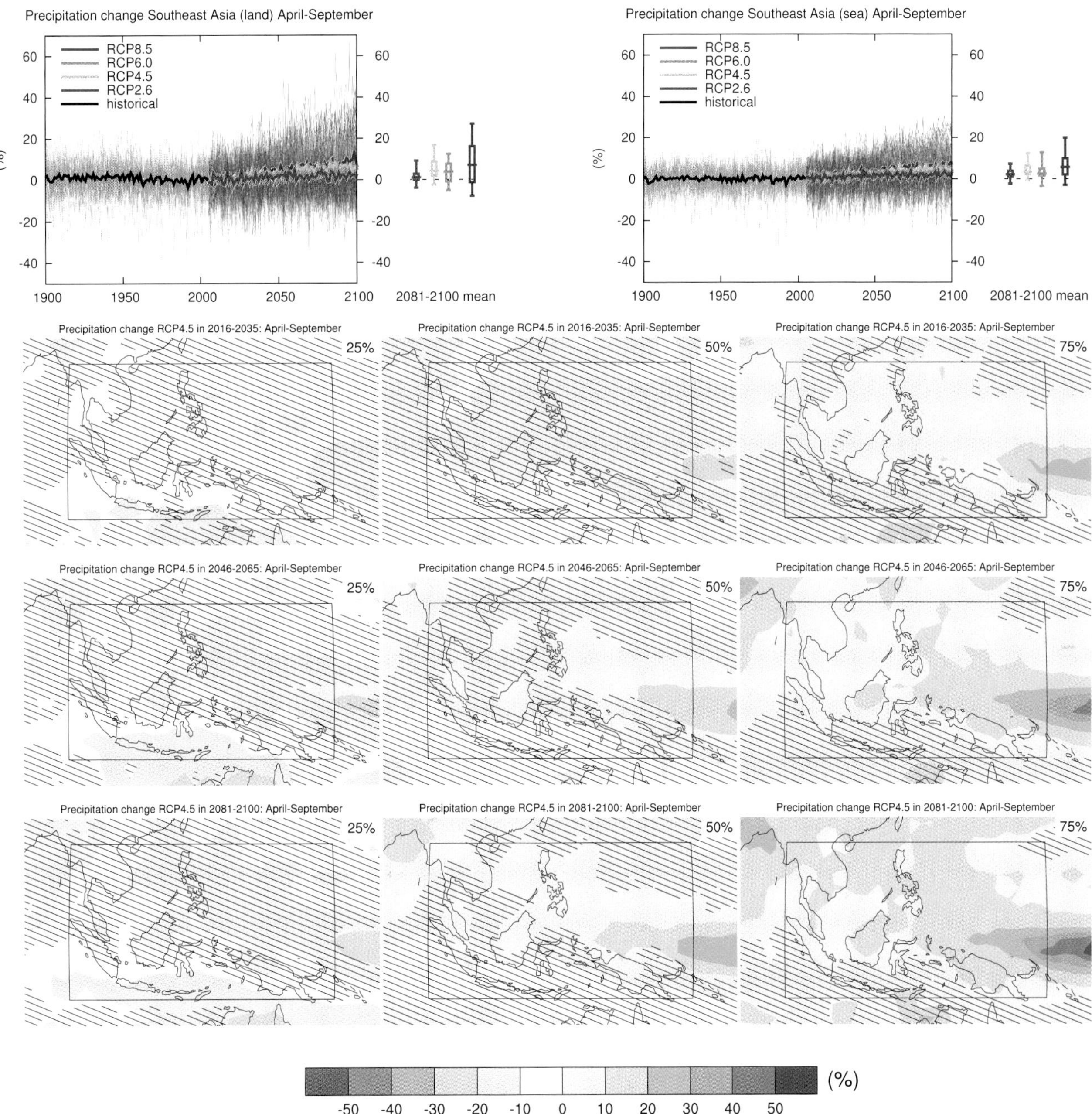

Figure AI.67 | (Top left) Time series of relative change relative to 1986–2005 in precipitation averaged over land grid points in Southeast Asia (10°S to 20°N, 95°E to 155°E) in April to September. (Top right) Same for sea grid points. Thin lines denote one ensemble member per model, thick lines the CMIP5 multi-model mean. On the right-hand side the 5th, 25th, 50th (median), 75th and 95th percentiles of the distribution of 20-year mean changes are given for 2081–2100 in the four RCP scenarios.

(Below) Maps of precipitation changes in 2016–2035, 2046–2065 and 2081–2100 with respect to 1986–2005 in the RCP4.5 scenario. For each point, the 25th, 50th and 75th percentiles of the distribution of the CMIP5 ensemble are shown; this includes both natural variability and inter-model spread. Hatching denotes areas where the 20-year mean differences of the percentiles are less than the standard deviation of model-estimated present-day natural variability of 20-year mean differences.

Sections 9.4.1.1, 9.6.1.1, Box 11.2, 14.2.2.3, 14.2.2.5, 14.8.12 contain relevant information regarding the evaluation of models in this region, the model spread in the context of other methods of projecting changes and the role of modes of variability and other climate phenomena.

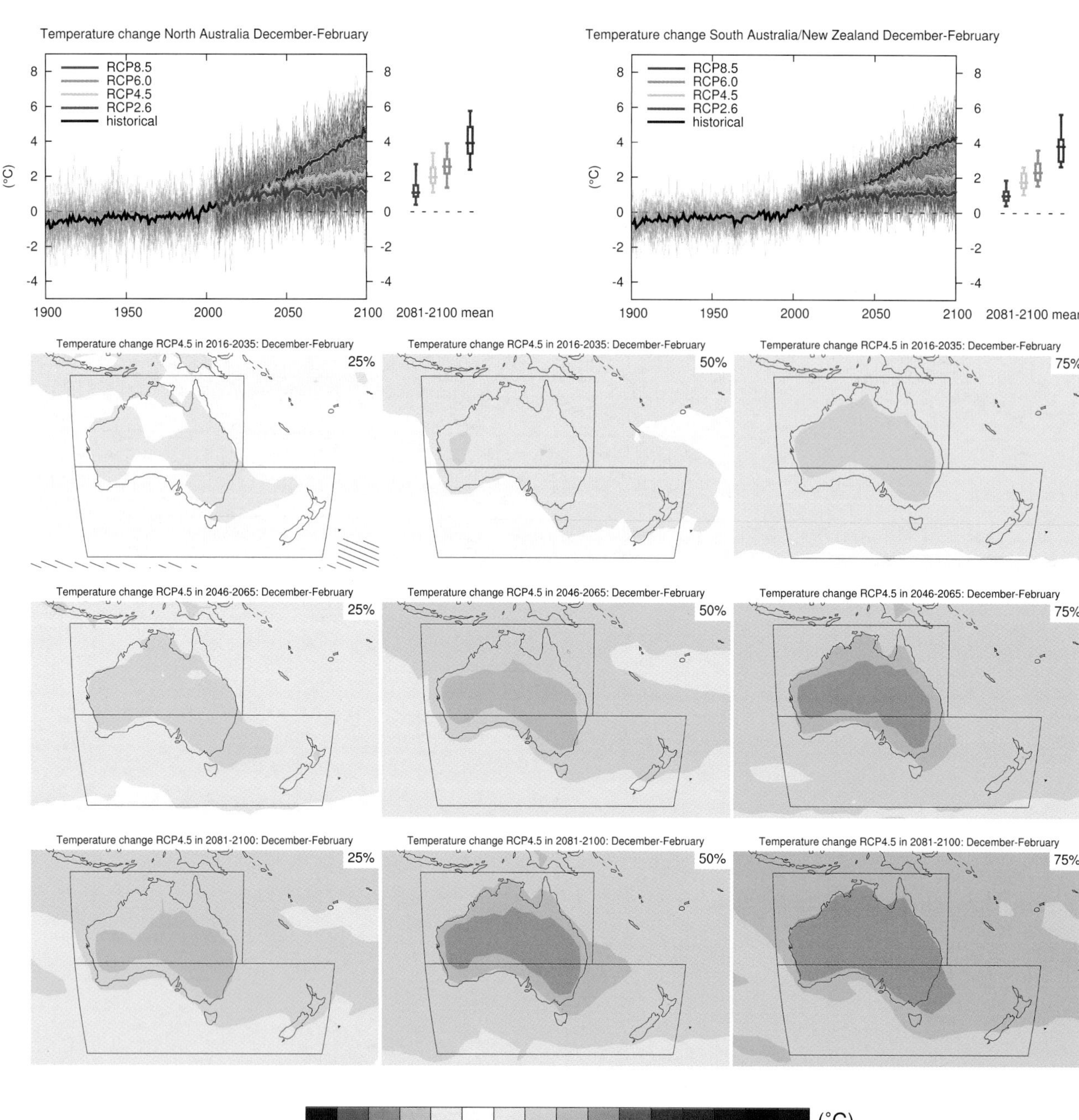

Figure AI.68 | (Top left) Time series of temperature change relative to 1986–2005 averaged over land grid points in North Australia (30°S to 10°S, 110°E to 155°E) in December to February. (Top right) Same for land grid points in South Australia/New Zealand (50°S to 30°S, 110°E to 180°E). Thin lines denote one ensemble member per model, thick lines the CMIP5 multi-model mean. On the right-hand side the 5th, 25th, 50th (median), 75th and 95th percentiles of the distribution of 20-year mean changes are given for 2081–2100 in the four RCP scenarios.

(Below) Maps of temperature changes in 2016–2035, 2046–2065 and 2081–2100 with respect to 1986–2005 in the RCP4.5 scenario. For each point, the 25th, 50th and 75th percentiles of the distribution of the CMIP5 ensemble are shown; this includes both natural variability and inter-model spread. Hatching denotes areas where the 20-year mean differences of the percentiles are less than the standard deviation of model-estimated present-day natural variability of 20-year mean differences.

Sections 9.4.1.1, 9.6.1.1, 10.3.1.1.4, Box 11.2, 14.8.13 contain relevant information regarding the evaluation of models in this region, the model spread in the context of other methods of projecting changes and the role of modes of variability and other climate phenomena.

AI

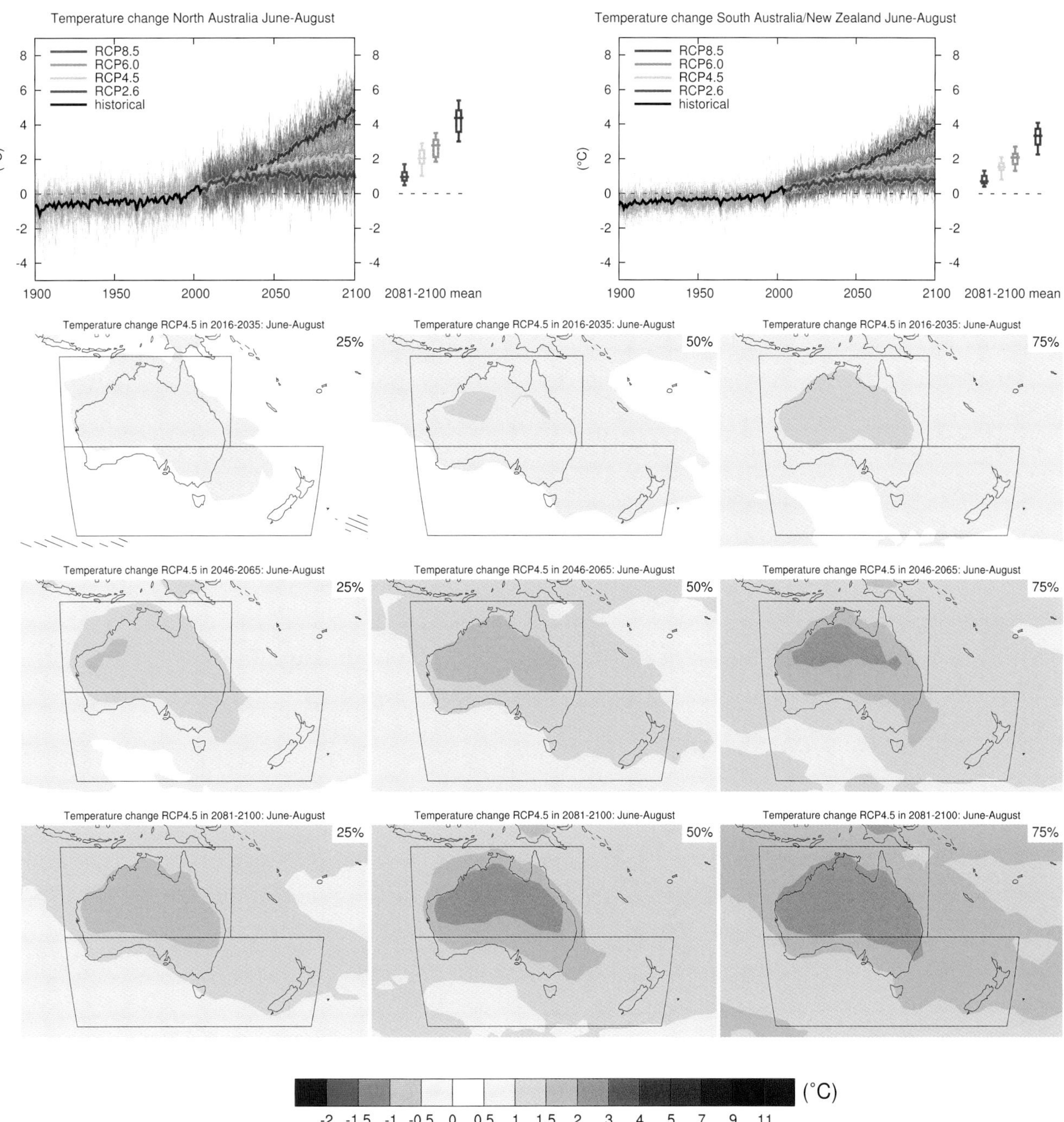

Figure AI.69 | (Top left) Time series of temperature change relative to 1986–2005 averaged over land grid points in North Australia (30°S to 10°S, 110°E to 155°E) in June to August. (Top right) Same for land grid points in South Australia/New Zealand (50°S to 30°S, 110°E to 180°E). Thin lines denote one ensemble member per model, thick lines the CMIP5 multi-model mean. On the right-hand side the 5th, 25th, 50th (median), 75th and 95th percentiles of the distribution of 20-year mean changes are given for 2081–2100 in the four RCP scenarios.

(Below) Maps of temperature changes in 2016–2035, 2046–2065 and 2081–2100 with respect to 1986–2005 in the RCP4.5 scenario. For each point, the 25th, 50th and 75th percentiles of the distribution of the CMIP5 ensemble are shown; this includes both natural variability and inter-model spread. Hatching denotes areas where the 20-year mean differences of the percentiles are less than the standard deviation of model-estimated present-day natural variability of 20-year mean differences.

Sections 9.4.1.1, 9.6.1.1, 10.3.1.1.4, Box 11.2, 14.8.13 contain relevant information regarding the evaluation of models in this region, the model spread in the context of other methods of projecting changes and the role of modes of variability and other climate phenomena.

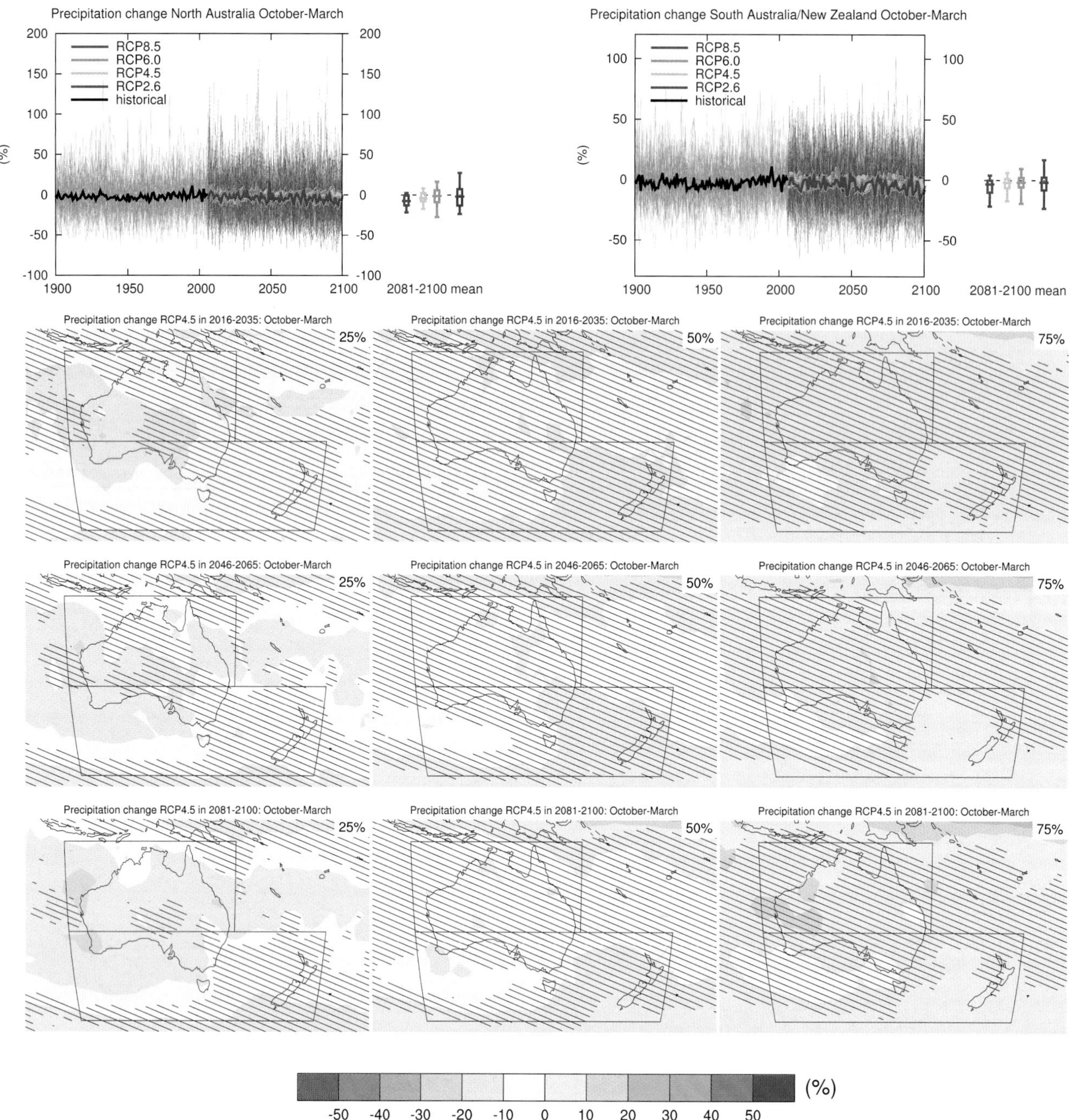

Figure AI.70 | (Top left) Time series of relative change relative to 1986–2005 in precipitation averaged over land grid points in North Australia (30°S to 10°S, 110°E to 155°E) in October to March. (Top right) Same for land grid points in South Australia/New Zealand (50°S to 30°S, 110°E to 180°E). Thin lines denote one ensemble member per model, thick lines the CMIP5 multi-model mean. On the right-hand side the 5th, 25th, 50th (median), 75th and 95th percentiles of the distribution of 20-year mean changes are given for 2081–2100 in the four RCP scenarios. Note different scales.

(Below) Maps of precipitation changes in 2016–2035, 2046–2065 and 2081–2100 with respect to 1986–2005 in the RCP4.5 scenario. For each point, the 25th, 50th and 75th percentiles of the distribution of the CMIP5 ensemble are shown; this includes both natural variability and inter-model spread. Hatching denotes areas where the 20-year mean differences of the percentiles are less than the standard deviation of model-estimated present-day natural variability of 20-year mean differences.

Sections 9.4.1.1, 9.6.1.1, Box 11.2, 14.2.2.4, 14.8.13 contain relevant information regarding the evaluation of models in this region, the model spread in the context of other methods of projecting changes and the role of modes of variability and other climate phenomena.

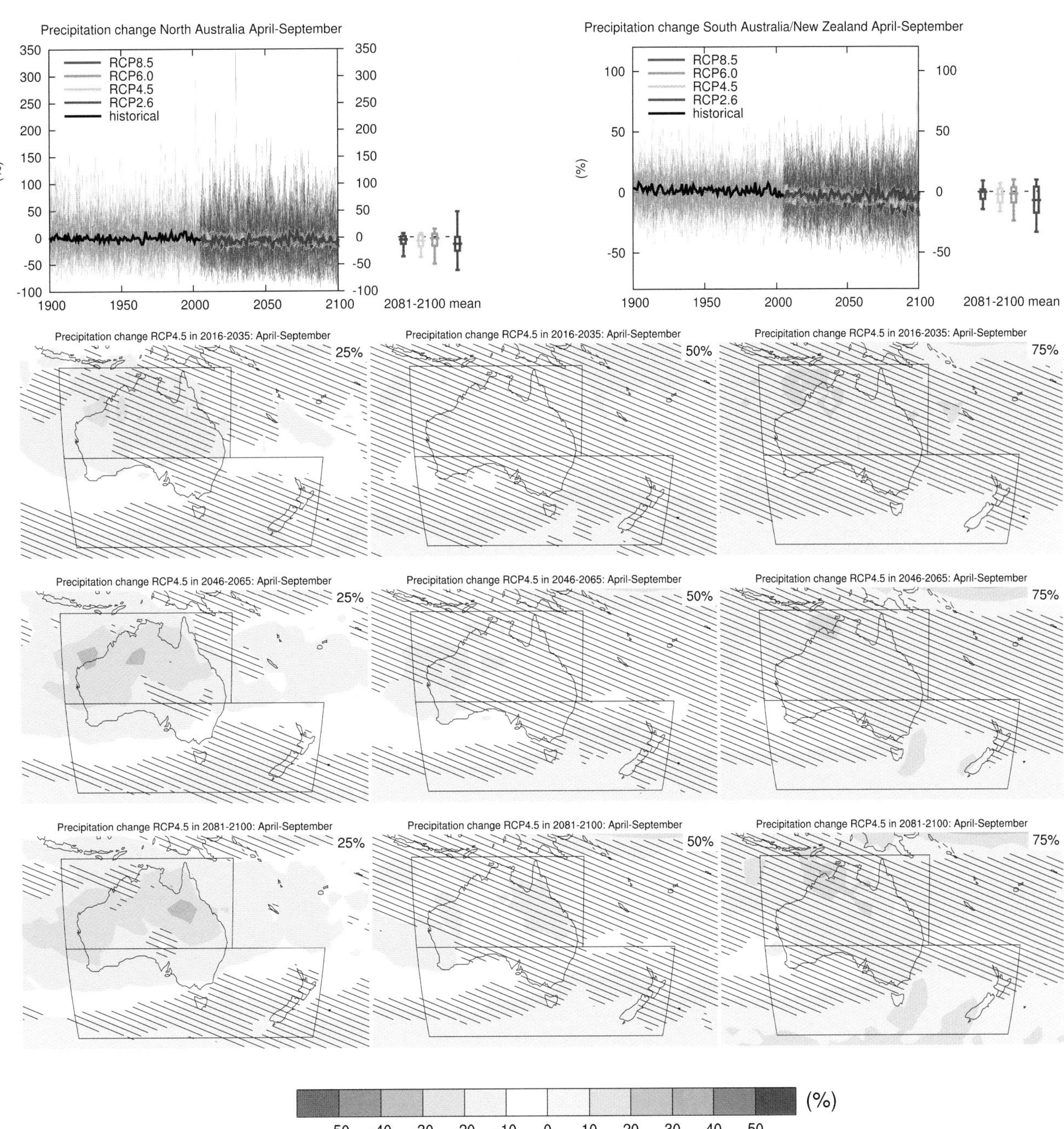

Figure AI.71 | (Top left) Time series of relative change relative to 1986–2005 in precipitation averaged over land grid points in North Australia (30°S to 10°S, 110°E to 155°E) in April to September. (Top right) Same for land grid points in South Australia/New Zealand (50°S to 30°S, 110°E to 180°E). Thin lines denote one ensemble member per model, thick lines the CMIP5 multi-model mean. On the right-hand side the 5th, 25th, 50th (median), 75th and 95th percentiles of the distribution of 20-year mean changes are given for 2081–2100 in the four RCP scenarios. Note different scales.

(Below) Maps of precipitation changes in 2016–2035, 2046–2065 and 2081–2100 with respect to 1986–2005 in the RCP4.5 scenario. For each point, the 25th, 50th and 75th percentiles of the distribution of the CMIP5 ensemble are shown; this includes both natural variability and inter-model spread. Hatching denotes areas where the 20-year mean differences of the percentiles are less than the standard deviation of model-estimated present-day natural variability of 20-year mean differences.

Sections 9.4.1.1, 9.6.1.1, Box 11.2, 14.2.2.4, 14.8.13 contain relevant information regarding the evaluation of models in this region, the model spread in the context of other methods of projecting changes and the role of modes of variability and other climate phenomena.

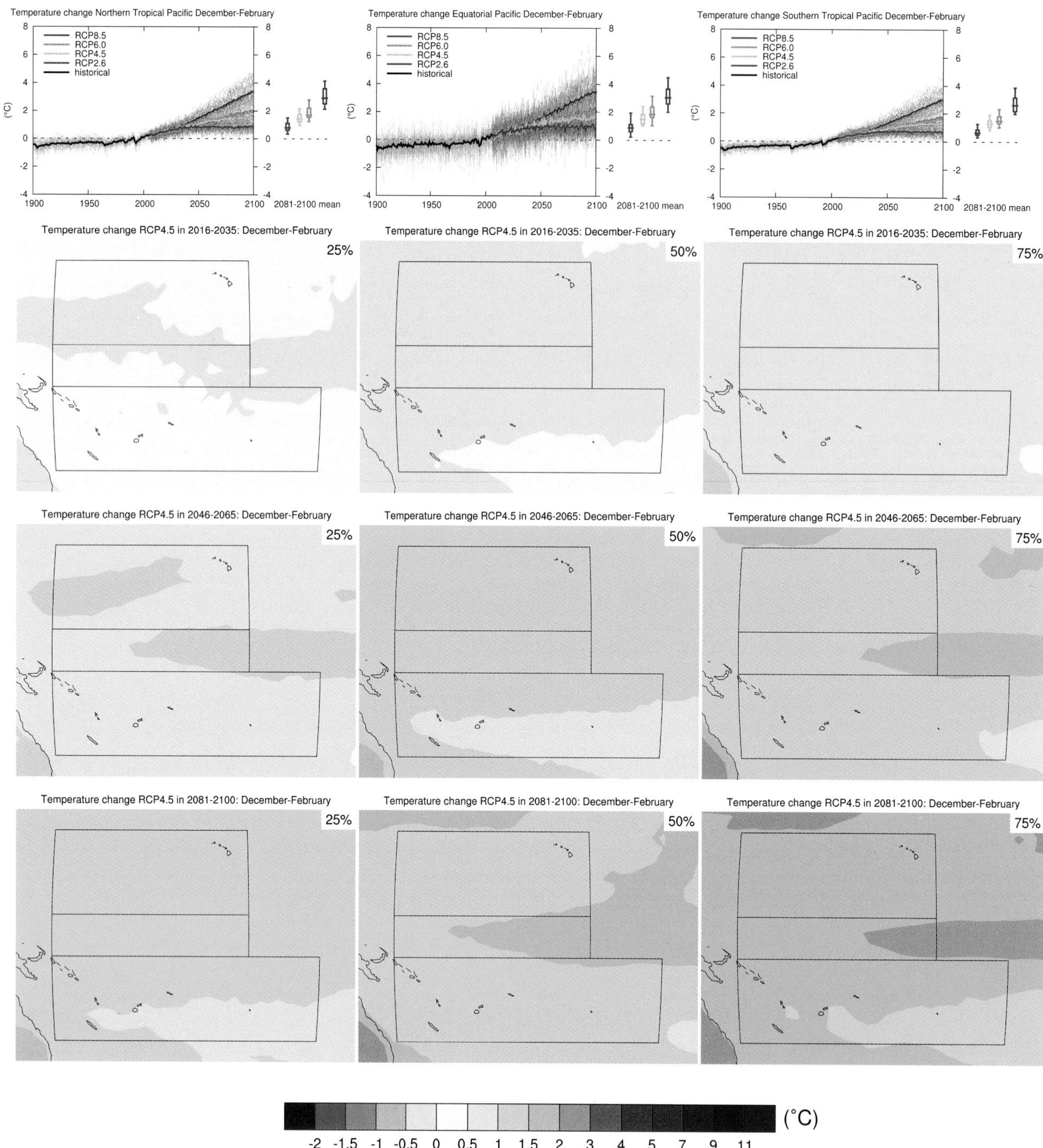

Figure AI.72 | (Top left) Time series of temperature change relative to 1986–2005 averaged over all grid points in the Northern Tropical Pacific (5°N to 25°N, 155°E to 150°W) in December to February. Top middle: same for all grid points in the Equatorial Pacific (5°S to 5°N, 155°E to 150°W). (Top right) Same for all grid points in the Southern Tropical Pacific (5°S to 5°N, 155°E to 150°W). Thin lines denote one ensemble member per model, thick lines the CMIP5 multi-model mean. On the right-hand side the 5th, 25th, 50th (median), 75th and 95th percentiles of the distribution of 20-year mean changes are given for 2081–2100 in the four RCP scenarios.

(Below) Maps of temperature changes in 2016–2035, 2046–2065 and 2081–2100 with respect to 1986–2005 in the RCP4.5 scenario. For each point, the 25th, 50th and 75th percentiles of the distribution of the CMIP5 ensemble are shown; this includes both natural variability and inter-model spread. Hatching denotes areas where the 20-year mean differences of the percentiles are less than the standard deviation of model-estimated present-day natural variability of 20-year mean differences.

Sections 9.4.1.1, 9.6.1.1, 10.3.1.1.4, Box 11.2, 12.4.3.1, 14.4.1, 14.8.14 contain relevant information regarding the evaluation of models in this region, the model spread in the context of other methods of projecting changes and the role of modes of variability and other climate phenomena.

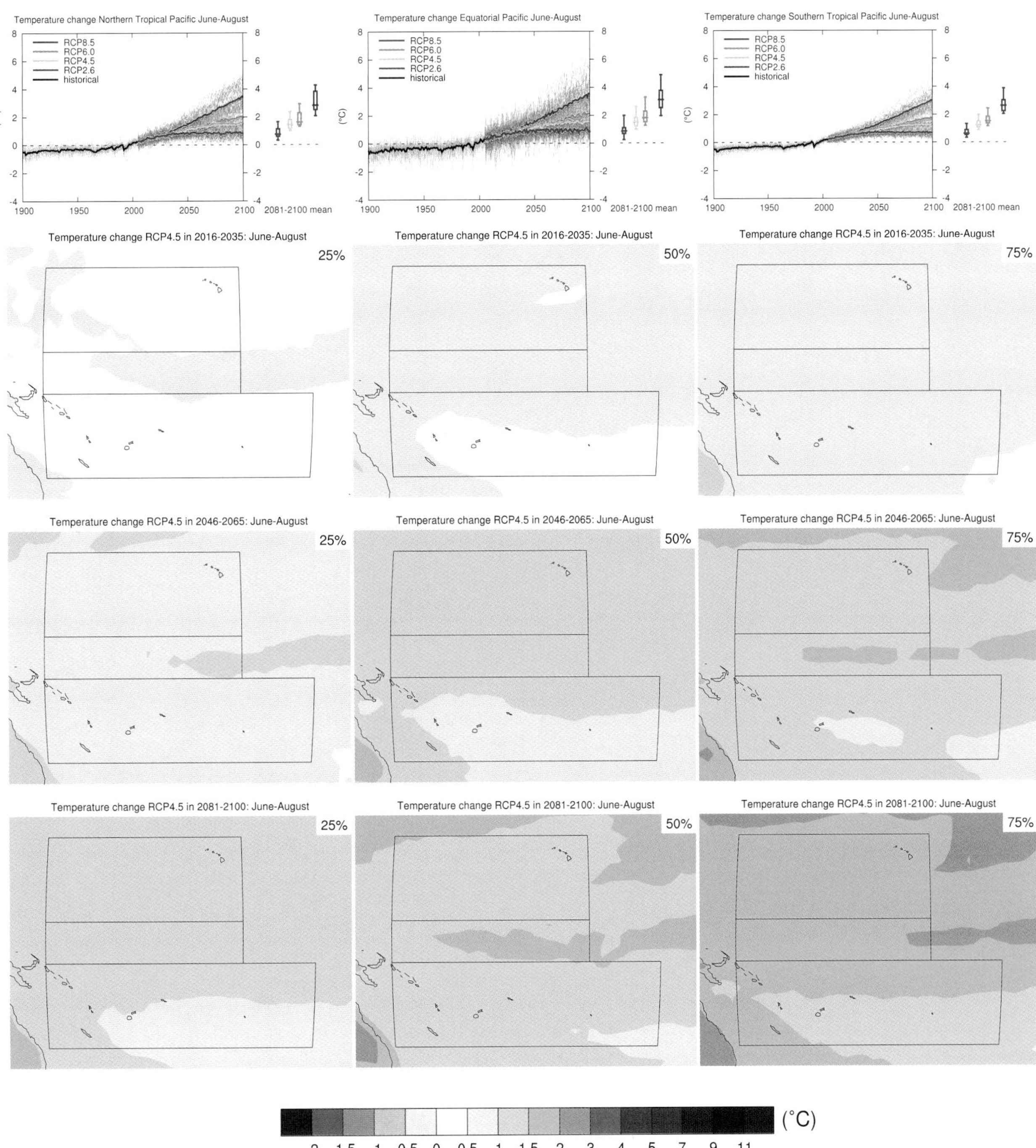

Figure AI.73 | (Top left) Time series of temperature change relative to 1986–2005 averaged over all grid points in the Northern Tropical Pacific (5°N to 25°N, 155°E to 150°W) in June to August. Top middle: same for all grid points in the Equatorial Pacific (5°S to 5°N, 155°E to 150°W). (Top right) Same for all grid points in the Southern Tropical Pacific (5°S to 5°N, 155°E to 150°W). Thin lines denote one ensemble member per model, thick lines the CMIP5 multi-model mean. On the right-hand side the 5th, 25th, 50th (median), 75th and 95th percentiles of the distribution of 20-year mean changes are given for 2081–2100 in the four RCP scenarios.

(Below) Maps of temperature changes in 2016–2035, 2046–2065 and 2081–2100 with respect to 1986–2005 in the RCP4.5 scenario. For each point, the 25th, 50th and 75th percentiles of the distribution of the CMIP5 ensemble are shown; this includes both natural variability and inter-model spread. Hatching denotes areas where the 20-year mean differences of the percentiles are less than the standard deviation of model-estimated present-day natural variability of 20-year mean differences.

Sections 9.4.1.1, 9.6.1.1, 10.3.1.1.4, Box 11.2, 12.4.3.1, 14.4.1, 14.8.14 contain relevant information regarding the evaluation of models in this region, the model spread in the context of other methods of projecting changes and the role of modes of variability and other climate phenomena.

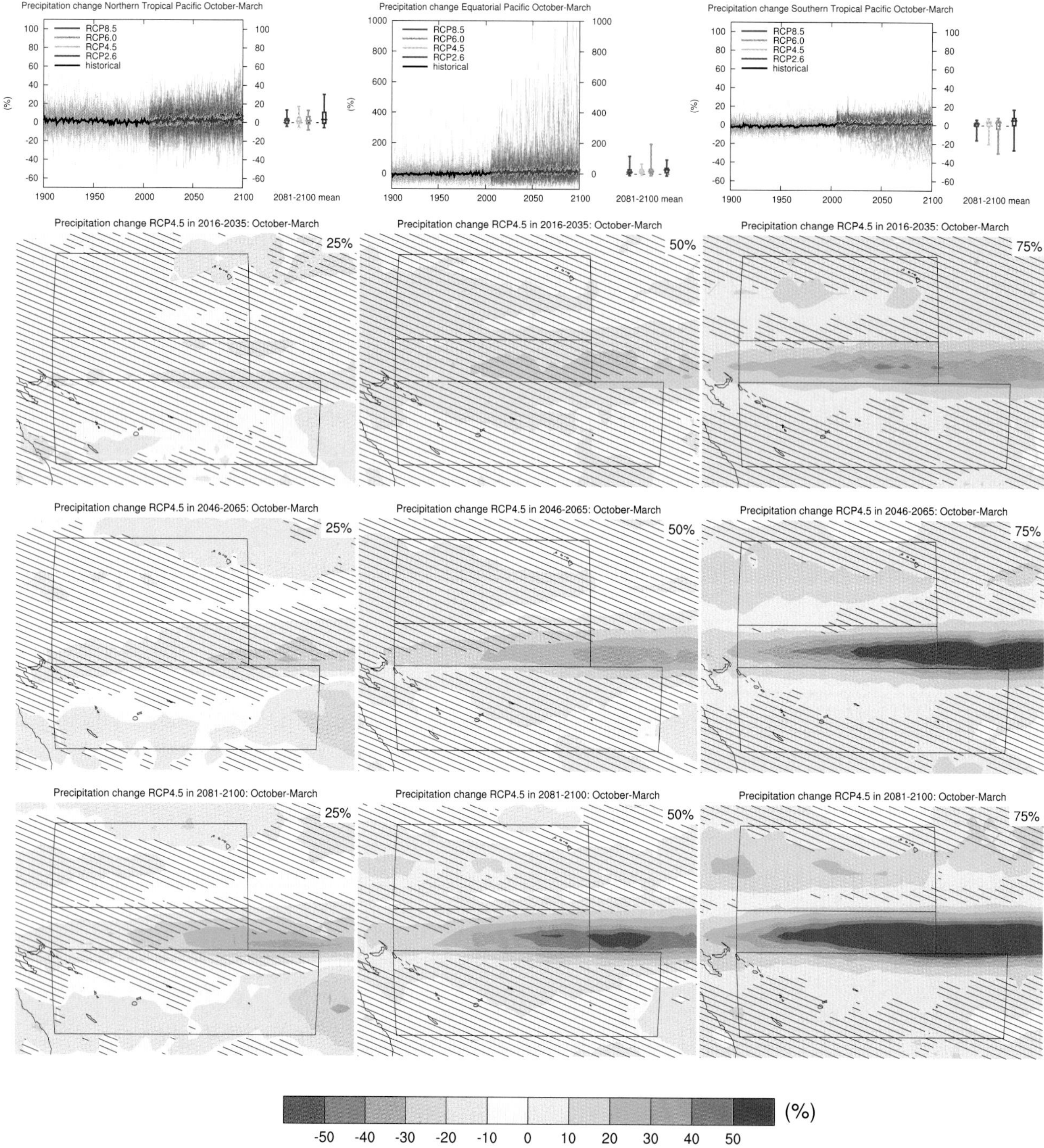

Figure AI.74 | (Top left) Time series of relative change relative to 1986–2005 in precipitation averaged over all grid points in the Northern Tropical Pacific (5°N to 25°N, 155°E to 150°W) in October to March. Top middle: same for all grid points in the Equatorial Pacific (5°S to 5°N, 155°E to 150°W). (Top right) Same for all grid points in the Southern Tropical Pacific (5°S to 5°N, 155°E to 150°W). Thin lines denote one ensemble member per model, thick lines the CMIP5 multi-model mean. On the right-hand side the 5th, 25th, 50th (median), 75th and 95th percentiles of the distribution of 20-year mean changes are given for 2081–2100 in the four RCP scenarios. Note different scales.

(Below) Maps of precipitation changes in 2016–2035, 2046–2065 and 2081–2100 with respect to 1986–2005 in the RCP4.5 scenario. For each point, the 25th, 50th and 75th percentiles of the distribution of the CMIP5 ensemble are shown; this includes both natural variability and inter-model spread. Hatching denotes areas where the 20-year mean differences of the percentiles are less than the standard deviation of model-estimated present-day natural variability of 20-year mean differences.

Sections 9.4.1.1, 9.6.1.1, 11.3.2.1.2, Box 11.2, 12.4.5.2, 14.8.14 contain relevant information regarding the evaluation of models in this region, the model spread in the context of other methods of projecting changes and the role of modes of variability and other climate phenomena.

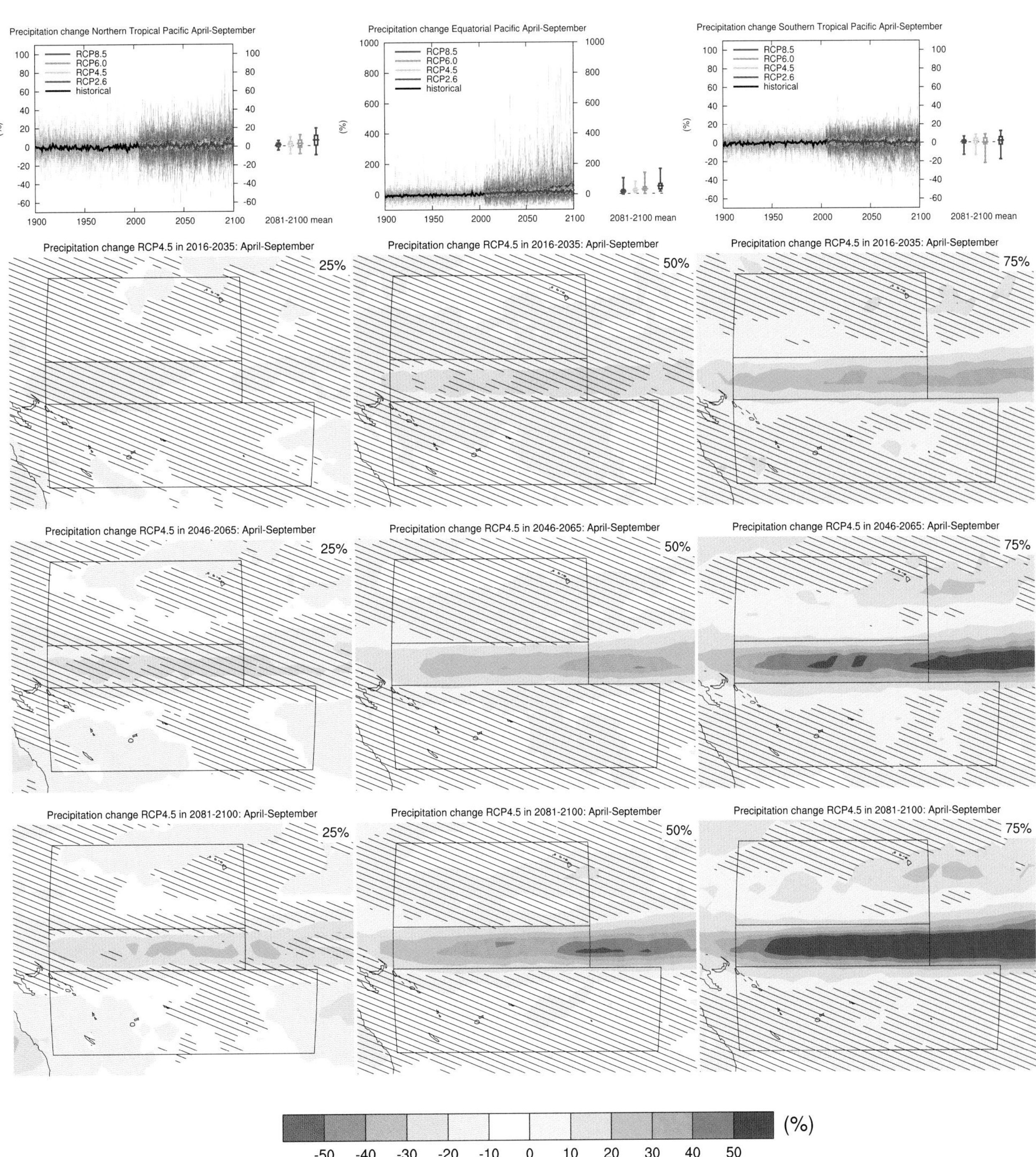

Figure AI.75 | (Top left) Time series of relative change relative to 1986–2005 in precipitation averaged over all grid points in the Northern Tropical Pacific (5°N to 25°N, 155°E to 150°W) in April to September. Top middle: same for all grid points in the Equatorial Pacific (5°S to 5°N, 155°E to 150°W). (Top right) Same for all grid points in the Southern Tropical Pacific (5°S to 5°N, 155°E to 150°W). Thin lines denote one ensemble member per model, thick lines the CMIP5 multi-model mean. On the right-hand side the 5th, 25th, 50th (median), 75th and 95th percentiles of the distribution of 20-year mean changes are given for 2081–2100 in the four RCP scenarios. Note different scales.

(Below) Maps of precipitation changes in 2016–2035, 2046–2065 and 2081–2100 with respect to 1986–2005 in the RCP4.5 scenario. For each point, the 25th, 50th and 75th percentiles of the distribution of the CMIP5 ensemble are shown; this includes both natural variability and inter-model spread. Hatching denotes areas where the 20-year mean differences of the percentiles are less than the standard deviation of model-estimated present-day natural variability of 20-year mean differences.

Sections 9.4.1.1, 9.6.1.1, 11.3.2.1.2, Box 11.2, 12.4.5.2, 14.8.14 contain relevant information regarding the evaluation of models in this region, the model spread in the context of other methods of projecting changes and the role of modes of variability and other climate phenomena.

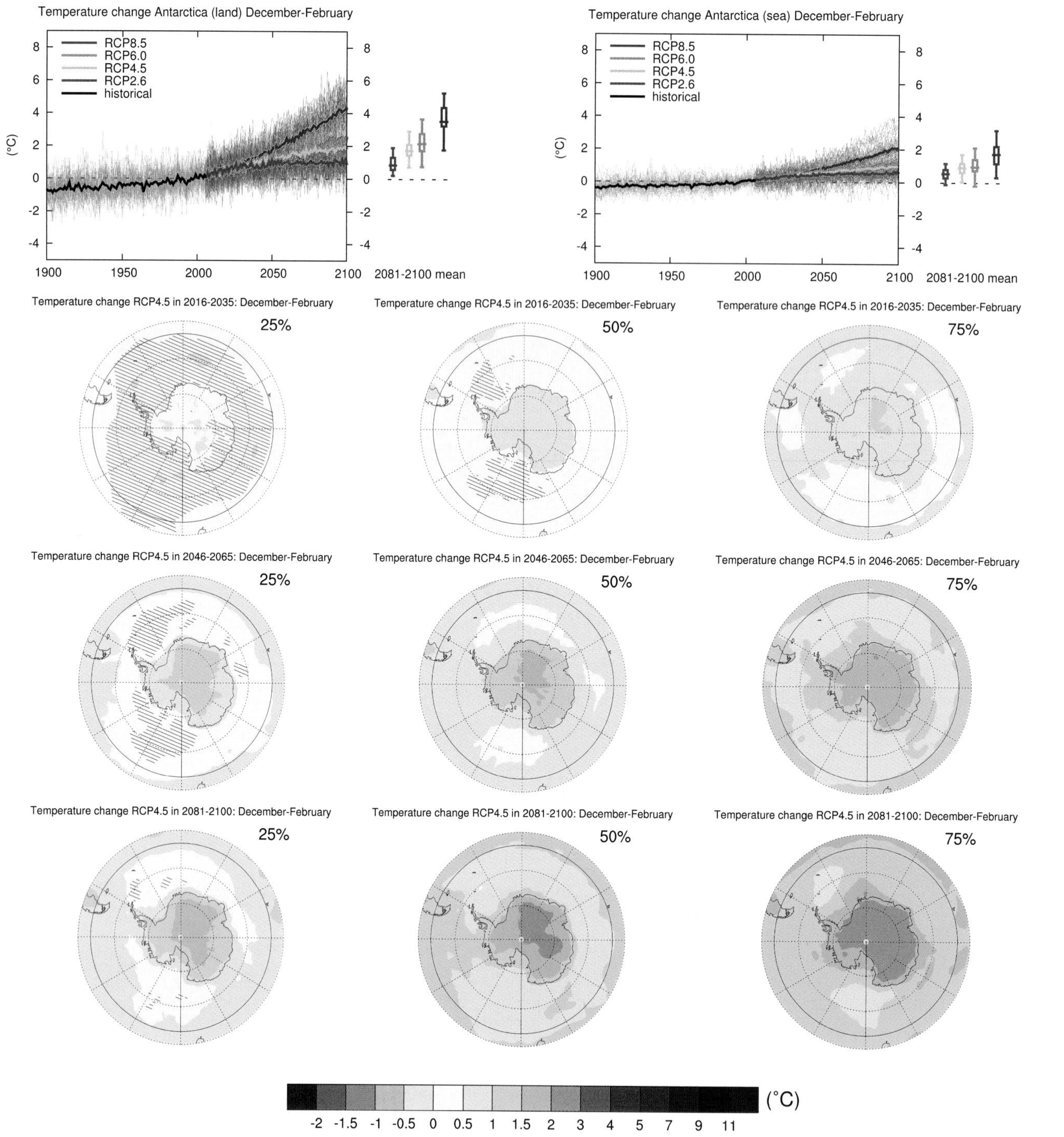

Temperature change Antarctica (land) December-February

Temperature change Antarctica (sea) December-February

RCP8.5
RCP6.0
RCP4.5
RCP2.6
historical

(°C)

1900 1950 2000 2050 2100 2081-2100 mean

Temperature change RCP4.5 in 2016-2035: December-February 25%

Temperature change RCP4.5 in 2016-2035: December-February 50%

Temperature change RCP4.5 in 2016-2035: December-February 75%

Temperature change RCP4.5 in 2046-2065: December-February 25%

Temperature change RCP4.5 in 2046-2065: December-February 50%

Temperature change RCP4.5 in 2046-2065: December-February 75%

Temperature change RCP4.5 in 2081-2100: December-February 25%

Temperature change RCP4.5 in 2081-2100: December-February 50%

Temperature change RCP4.5 in 2081-2100: December-February 75%

-2 -1.5 -1 -0.5 0 0.5 1 1.5 2 3 4 5 7 9 11 (°C)

Figure AI.76 | (Top left) Time series of temperature change relative to 1986–2005 averaged over land grid points in Antarctica (90°S to 50°S) in December to February. (Top right) Same for sea grid points. Thin lines denote one ensemble member per model, thick lines the CMIP5 multi-model mean. On the right-hand side the 5th, 25th, 50th (median), 75th and 95th percentiles of the distribution of 20-year mean changes are given for 2081–2100 in the four RCP scenarios.

(Below) Maps of temperature changes in 2016–2035, 2046–2065 and 2081–2100 with respect to 1986–2005 in the RCP4.5 scenario. For each point, the 25th, 50th and 75th percentiles of the distribution of the CMIP5 ensemble are shown; this includes both natural variability and inter-model spread. Hatching denotes areas where the 20-year mean differences of the percentiles are less than the standard deviation of model-estimated present-day natural variability of 20-year mean differences.

Sections 9.4.1.1, 9.6.1.1, 10.3.1.1.4, Box 11.2, 12.4.3.1, 14.8.15 contain relevant information regarding the evaluation of models in this region, the model spread in the context of other methods of projecting changes and the role of modes of variability and other climate phenomena.

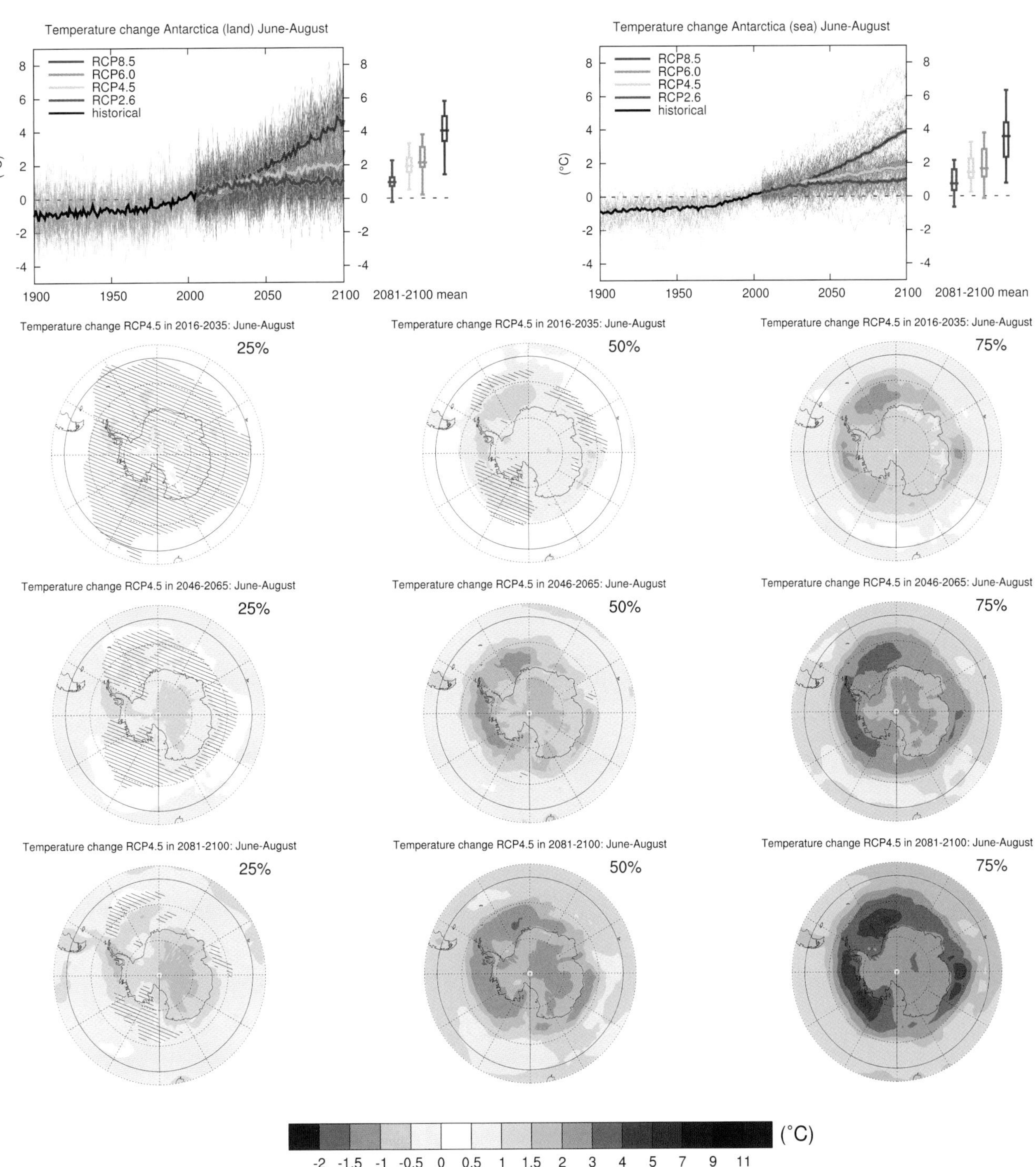

Figure AI.77 | (Top left) Time series of temperature change relative to 1986–2005 averaged over land grid points in Antarctica (90°S to 50°S) in June to August. (Top right) Same for sea grid points. Thin lines denote one ensemble member per model, thick lines the CMIP5 multi-model mean. On the right-hand side the 5th, 25th, 50th (median), 75th and 95th percentiles of the distribution of 20-year mean changes are given for 2081–2100 in the four RCP scenarios.

(Below) Maps of temperature changes in 2016–2035, 2046–2065 and 2081–2100 with respect to 1986–2005 in the RCP4.5 scenario. For each point, the 25th, 50th and 75th percentiles of the distribution of the CMIP5 ensemble are shown; this includes both natural variability and inter-model spread. Hatching denotes areas where the 20-year mean differences of the percentiles are less than the standard deviation of model-estimated present-day natural variability of 20-year mean differences.

Sections 9.4.1.1, 9.6.1.1, 10.3.1.1.4, Box 11.2, 12.4.3.1, 14.8.15 contain relevant information regarding the evaluation of models in this region, the model spread in the context of other methods of projecting changes and the role of modes of variability and other climate phenomena.

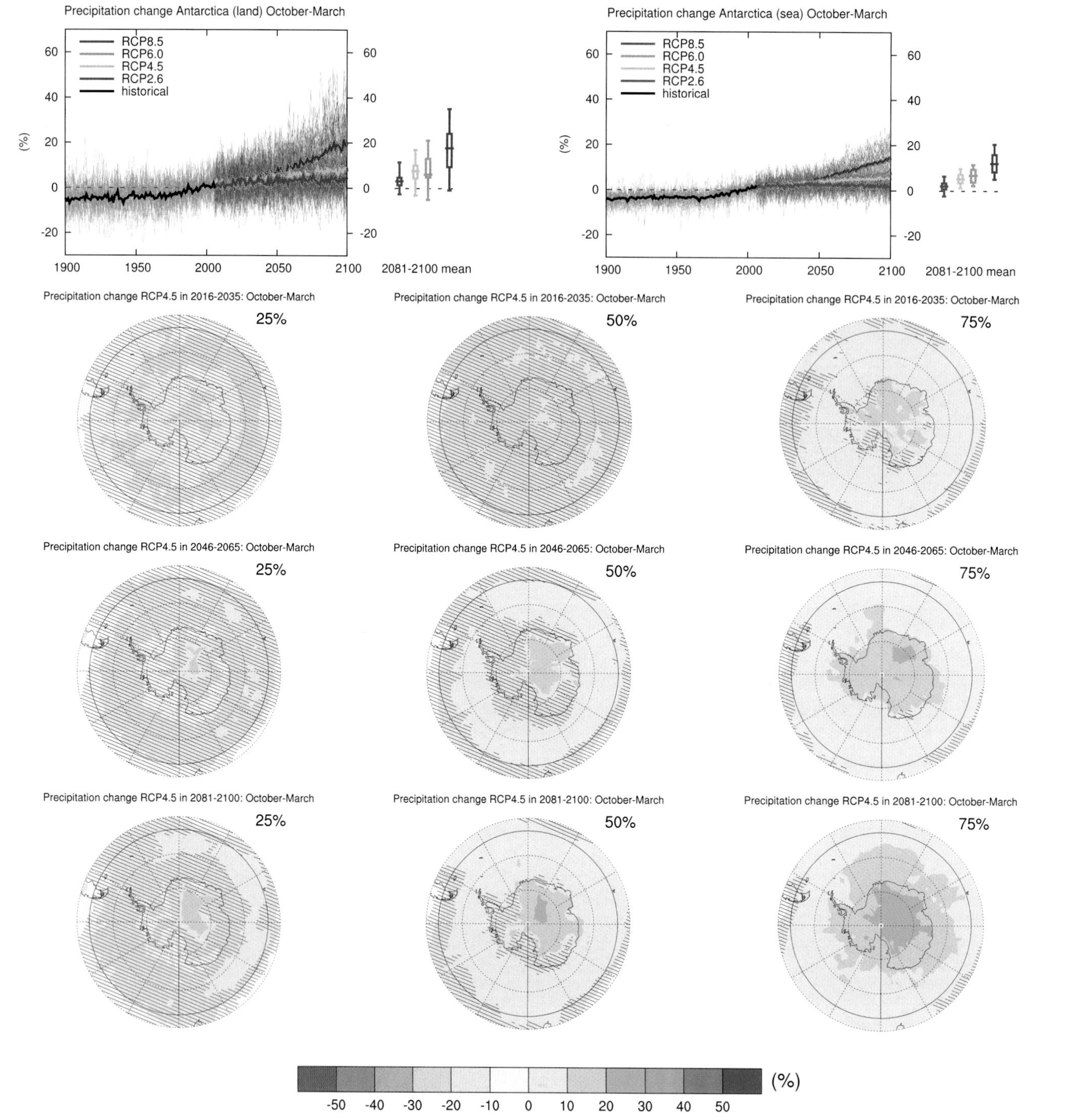

Precipitation change Antarctica (land) October-March

Precipitation change Antarctica (sea) October-March

RCP8.5
RCP6.0
RCP4.5
RCP2.6
historical

(%)

2081-2100 mean

Precipitation change RCP4.5 in 2016-2035: October-March 25%

Precipitation change RCP4.5 in 2016-2035: October-March 50%

Precipitation change RCP4.5 in 2016-2035: October-March 75%

Precipitation change RCP4.5 in 2046-2065: October-March 25%

Precipitation change RCP4.5 in 2046-2065: October-March 50%

Precipitation change RCP4.5 in 2046-2065: October-March 75%

Precipitation change RCP4.5 in 2081-2100: October-March 25%

Precipitation change RCP4.5 in 2081-2100: October-March 50%

Precipitation change RCP4.5 in 2081-2100: October-March 75%

-50 -40 -30 -20 -10 0 10 20 30 40 50 (%)

Figure AI.78 | (Top left) Time series of relative change relative to 1986–2005 in precipitation averaged over land grid points in Antarctica (90°S to 50°S) in October to March. (Top right) Same for sea grid points. Thin lines denote one ensemble member per model, thick lines the CMIP5 multi-model mean. On the right-hand side the 5th, 25th, 50th (median), 75th and 95th percentiles of the distribution of 20-year mean changes are given for 2081–2100 in the four RCP scenarios.

(Below) Maps of precipitation changes in 2016–2035, 2046–2065 and 2081–2100 with respect to 1986–2005 in the RCP4.5 scenario. For each point, the 25th, 50th and 75th percentiles of the distribution of the CMIP5 ensemble are shown; this includes both natural variability and inter-model spread. Hatching denotes areas where the 20-year mean differences of the percentiles are less than the standard deviation of model-estimated present-day natural variability of 20-year mean differences.

Sections 9.4.1.1, 9.6.1.1, 10.3.2.2, Box 11.2, 12.4.5.2, 14.8.15 contain relevant information regarding the evaluation of models in this region, the model spread in the context of other methods of projecting changes and the role of modes of variability and other climate phenomena.

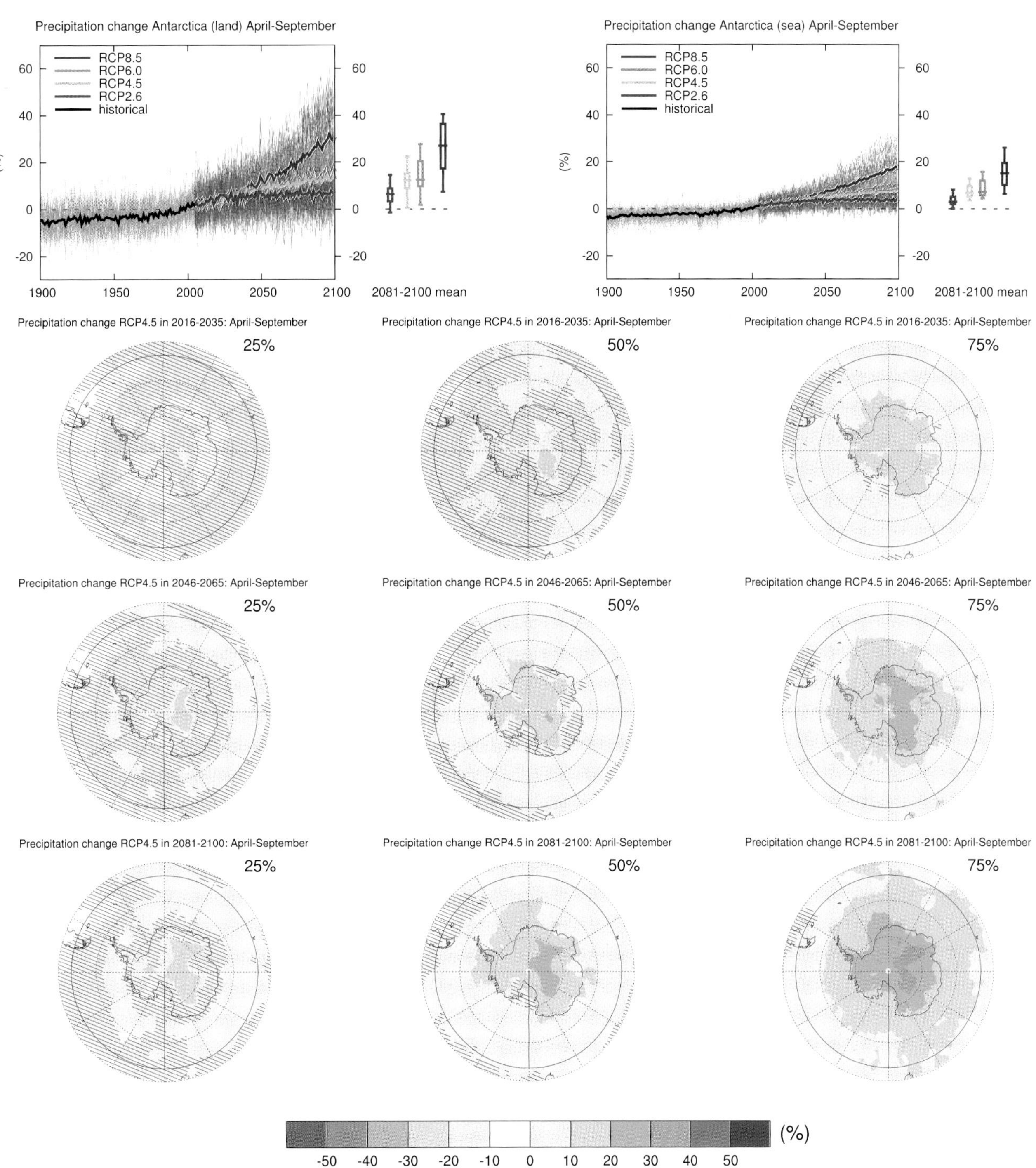

Figure AI.79 | (Top left) Time series of relative change relative to 1986–2005 in precipitation averaged over land grid points in Antarctica (90°S to 50°S) in April to September. (Top right) Same for sea grid points. Thin lines denote one ensemble member per model, thick lines the CMIP5 multi-model mean. On the right-hand side the 5th, 25th, 50th (median), 75th and 95th percentiles of the distribution of 20-year mean changes are given for 2081–2100 in the four RCP scenarios.

(Below) Maps of precipitation changes in 2016–2035, 2046–2065 and 2081–2100 with respect to 1986–2005 in the RCP4.5 scenario. For each point, the 25th, 50th and 75th percentiles of the distribution of the CMIP5 ensemble are shown; this includes both natural variability and inter-model spread. Hatching denotes areas where the 20-year mean differences of the percentiles are less than the standard deviation of model-estimated present-day natural variability of 20-year mean differences.

Sections 9.4.1.1, 9.6.1.1, 10.3.2.2, Box 11.2, 12.4.5.2, 14.8.15 contain relevant information regarding the evaluation of models in this region, the model spread in the context of other methods of projecting changes and the role of modes of variability and other climate phenomena.

Annex II: Climate System Scenario Tables

Editorial Team:
Michael Prather (USA), Gregory Flato (Canada), Pierre Friedlingstein (UK/Belgium), Christopher Jones (UK), Jean-François Lamarque (USA), Hong Liao (China), Philip Rasch (USA)

Contributors:
Olivier Boucher (France), François-Marie Bréon (France), Tim Carter (Finland), William Collins (UK), Frank J. Dentener (EU/Netherlands), Edward J. Dlugokencky (USA), Jean-Louis Dufresne (France), Jan Willem Erisman (Netherlands), Veronika Eyring (Germany), Arlene M. Fiore (USA), James Galloway (USA), Jonathan M. Gregory (UK), Ed Hawkins (UK), Chris Holmes (USA), Jasmin John (USA), Tim Johns (UK), Fiona Lo (USA), Natalie Mahowald (USA), Malte Meinshausen (Germany), Colin Morice (UK), Vaishali Naik (USA/India), Drew Shindell (USA), Steven J. Smith (USA), David Stevenson (UK), Peter W. Thorne (USA/Norway/UK), Geert Jan van Oldenborgh (Netherlands), Apostolos Voulgarakis (UK/Greece), Oliver Wild (UK), Donald Wuebbles (USA), Paul Young (UK)

This annex should be cited as:
IPCC, 2013: Annex II: Climate System Scenario Tables [Prather, M., G. Flato, P. Friedlingstein, C. Jones, J.-F. Lamarque, H. Liao and P. Rasch (eds.)]. In: *Climate Change 2013: The Physical Science Basis. Contribution of Working Group I to the Fifth Assessment Report of the Intergovernmental Panel on Climate Change* [Stocker, T.F., D. Qin, G.-K. Plattner, M. Tignor, S.K. Allen, J. Boschung, A. Nauels, Y. Xia, V. Bex and P.M. Midgley (eds.)]. Cambridge University Press, Cambridge, United Kingdom and New York, NY, USA.

Table of Contents

Introduction

Annex II presents, in tabulated form, data related to historical and projected changes in the climate system that are assessed in the chapters of this report (see Section 1.6). It also includes some comparisons with the Third Assessment Report (TAR) and Fourth Assessment Report (AR4) results. These data include values for emissions into the atmosphere, atmospheric abundances and burdens (integrated abundance), effective radiative forcing (ERF; includes adjusted forcing from aerosols, see Chapters 7 and 8), and global mean surface temperatures and sea level. Projections from 2010 to 2100 focus on the RCP scenarios (Moss et al., 2010; Lamarque et al., 2010; 2011; Meinshausen et al., 2011a; van Vuuren et al., 2011; see also Chapters 1, 6, 8, 11, 12 and 13). Projections also include previous IPCC scenarios (IPCC Scenarios 1992a (IS92a), Special Report on Emission Scenarios (SRES) A2 and B1, TAR Appendix II) and some alternative near-term scenarios for methane (CH_4) and short-lived pollutants that impact climate or air quality. Emissions from biomass burning are included as anthropogenic. ERF from land use change is also included in some tables.

Where uncertainties or ranges are presented here, they are noted in each table as being a recommended value or model ensemble mean/median with a 68% confidence interval (16 to 84%, $\pm1\sigma$ for a normal distribution) or 90% confidence interval (5 to 95%, $\pm1.645\sigma$ for a normal distribution) or statistics (standard deviation, percentiles, or minimum/maximum) of an ensemble of models. In some cases these are a formal evaluation of uncertainty as assessed in the chapters, but in other cases (specifically Tables AII.2.1, 3.1, 4.1, 5.1, 6.10, 7.1 to 7.5) they just describe the statistical results from the available models, and the referenced chapters must be consulted for the assessed uncertainty or confidence level of these results. In the case of Table AII.7.5, for example, the global mean surface temperature change (°C) relative to 1986–2005 is a statistical summary of the spread in the Coupled Model Intercomparison Project (CMIP) ensembles for each of the scenarios: model biases and model dependencies are not accounted for; the percentiles do not correspond to the assessed uncertainty derived in Chapters 11 (Section 11.3.6.3) and 12 (Section 12.4.1); and statistical spread across models cannot be interpreted in terms of calibrated language (Section 12.2).

The Representative Concentration Pathway (RCP) scenarios for emissions include only anthropogenic sources and use a single model to project from emissions to abundances to radiative forcing to climate change (Meinshausen et al., 2011a; 2011b). We include projected changes in natural carbon dioxide (CO_2) sources and sinks for 2010–2100 based on this assessment (Chapters 6 and 12). Present-day natural and anthropogenic emissions of CH_4 and nitrous oxide (N_2O) are assessed and used to scale the RCP anthropogenic emissions to be consistent with these best estimates (Chapters 6 and 11). Current model evaluations of atmospheric chemistry and the carbon cycle, including results from the CMIP5 and Atmospheric Chemistry and Climate Model Intercomparison Project (ACCMIP) projects, are used to project future composition and ERF separately from the RCP model (see Sections 6.4.3, 11.3.5 and 12.3). Thus, projected changes in greenhouse gases (GHGs), aerosols and ERF evaluated in this report may differ from the published RCPs and from what was used in the CMIP5 runs, and these are denoted RCP$^{\&}$. The CMIP5 climate projections used for the most part the RCP concentration pathways for well-mixed greenhouse gases (WMGHG) and the emissions pathways for ozone (O_3) and aerosol precursors. Such differences are discussed in the relevant chapters and noted in the tables.

For each species, the abundances (given as dry air mole fraction: ppm = micromoles per mole (10^{-6}); ppb = nanomoles per mole (10^{-9}); and ppt = picomoles per mole (10^{-12})), burdens (global total in grams, 1 Tg = 10^{12} g), average column amount (1 Dobson Unit (DU) = 2.687×10^{16} molecules per cm^2), AOD (mean aerosol optical depth at 550 nm), ERF (effective radiative forcing, W m^{-2}), and other climate system quantities are calculated for scenarios using methodologies based on the latest climate chemistry and climate carbon models (see Chapters 2, 6, 7, 8, 10, 11 and 12). Results are shown for individual years (e.g., 2010 = year 2010) and decadal averages (e.g., 2020^d = average of years 2016 through 2025), although some 10-year periods are different, see table notes. Year 2011 is the last year for observed quantities (denoted 2011* or 2011^{obs}). Results are shown as global mean values except for environmental data focussing on air quality (Tables AII.7.1–AII.7.4), which give regional mean surface abundances of O_3 and fine particulate matter with diameter less than 2.5 µm ($PM_{2.5}$). Results for global mean surface temperature (Tables AII.7.5 and AII.7.6) show only raw CMIP5 data or data from previous assessments. For best estimates of near-term and long-term temperature change see Chapters 11 and 12, respectively. Results for global mean sea level rise (Table AII.7.7) are assessed values with uncertainties described in Chapter 13.

AII

Chemical Abbreviations and Symbols

Well Mixed Greenhouse Gases (WMGHG)

CO_2	carbon dioxide (KP, Kyoto Protocol gas)
CH_4	methane (KP)
N_2O	nitrous oxide (KP)
HFC	hydrofluorocarbon[1] (a class of compounds: HFC-32, HFC-134a, …) (KP)
PFC	perfluorocarbon (a class of compounds: CF_4, C_2F_6, …) (KP)
SF_6	sulphur hexafluoride (KP)
NF_3	nitrogen trifluoride (KP)
CFC	chlorofluorocarbon (a class of compounds: $CFCl_3$, CF_2Cl_2, …) (MP, Montreal Protocol gas)
HCFC	hydrochlorofluorocarbon[1] (a class of compounds: HCFC-22, HCFC-141b, …) (MP)
CCl_4	carbon tetrachloride (MP)
CH_3CCl_3	methyl chloroform (MP)

[1] A few HFCs and HCFCs are very short lived in the atmosphere and therefore not well mixed.

Ozone and Aerosols, and their Precursors

O_3	ozone (both stratospheric and tropospheric)
NO_x	sum of NO (nitric oxide) and NO_2 (nitrogen dioxide)
NH_3	ammonia
CO	carbon monoxide
NMVOC	a class of compounds comprising all non-methane volatile organic compounds (i.e., hydrocarbons that may also contain oxygen, also known as biogenic VOC or NMHC)
OH	hydroxyl radical
$PM_{2.5}$	any aerosols with diameter less than 2.5 μm
BC	black carbon aerosol
OC	organic carbon aerosol
SO_2	sulphur dioxide, a gas
SO_x	oxidized sulphur in gaseous form, including SO_2
$SO_4^=$	sulphate ion, usually as sulphuric acid or ammonium sulphate in aerosol

List of Tables

AII.7: Environmental Data

Table AII.7.1: Global mean surface O_3 change (ppb)
Table AII.7.2: Surface O_3 change (ppb) for HTAP regions
Table AII.7.3: Surface O_3 change (ppb) from CMIP5/ACCMIP for continental regions
Table AII.7.4: Surface particulate matter change ($\log_{10}$[$PM_{2.5}$ (microgram/m^3)]) from CMIP5/ACCMIP for continental regions
Table AII.7.5: CMIP5 (RCP) and CMIP3 (SRES A1B) global mean surface temperature change (°C) relative to 1986–2005 reference period
Table AII.7.6: Global mean surface temperature change (°C) relative to 1990 from the TAR
Table AII.7.7: Global mean sea level rise (m) with respect to 1986–2005 at 1 January on the years indicated

References

Calvin, K., et al., 2012: The role of Asia in mitigating climate change: Results from the Asia modeling exercise. *Energy Econ*., **34**, S251–S260.

Cionni, I., V. Eyring, J. Lamarque, W. Randel, D. Stevenson, F. Wu, G. Bodeker, T. Shepherd, D. Shindell, and D. Waugh, 2011: Ozone database in support of CMIP5 simulations: Results and corresponding radiative forcing. *Atmos. Chem. Phys*., **11**, 11267–11292.

Cofala, J., M. Amann, Z. Klimont, K. Kupiainen, and L. Hoglund-Isaksson, 2007: Scenarios of global anthropogenic emissions of air pollutants and methane until 2030. *Atmos. Environ*., **41**, 8486–8499.

Dentener, F., D. Stevenson, J. Cofala, R. Mechler, M. Amann, P. Bergamaschi, F. Raes, and R. Derwent, 2005: The impact of air pollutant and methane emission controls on tropospheric ozone and radiative forcing: CTM calculations for the period 1990-2030. *Atmos. Chem. Phys*., **5**, 1731–1755.

Dentener, F., et al., 2006: The global atmospheric environment for the next generation. *Environ. Sci. Technol*., **40**, 3586–3594.

Douglass, A. and V. Fioletov, 2010: *Stratospheric Ozone and Surface Ultraviolet Radiation in Scientific Assessment of Ozone Depletion: 2010.* Global Ozone Research and Monitoring Project-Report No. 52.World Meteorological Organization, Geneva, Switzerland.

Erisman, J. W., M. A. Sutton, J. Galloway, Z. Klimont, and W. Winiwarter, 2008: How a century of ammonia synthesis changed the world. *Nature Geosci*., **1**, 636–639.

Eyring, V., et al., 2013: Long-term ozone changes and associated climate impacts in CMIP5 simulations. *J. Geophys. Res*., doi:10.1002/jgrd.50316.

Fiore, A. M., et al., 2012: Global air quality and climate. *Chem. Soc. Rev*., **41**, 6663–6683.

Fleming, E., C. Jackman, R. Stolarski and A. Douglass, 2011: A model study of the impact of source gas changes on the stratosphere for 1850-2100. *Atmos. Chem. Phys*., **11**, 8515–8541.

Forster, P. M., T. Andrews, P. Good, J. M. Gregory, L. S. Jackson, and M. Zelinka, 2013: Evaluating adjusted forcing and model spread for historical and future scenarios in the CMIP5 generation of climate models. *J. Geophys. Res.*, **118**, 1139–1150.

Friedlingstein, P., et al., 2006: Climate-carbon cycle feedback analysis: Results from the C4MIP model intercomparison. *J. Clim.*, **19**, 3337–3353.

Holmes, C. D., M. J. Prather, A.O. Søvde, and G. Myhre, 2013: Future methane, hydroxyl, and their uncertainties: Key climate and emission parameters for future predictions. *Atmos. Chem. Phys*., **13**, 285–302.

HTAP, 2010. *Hemispheric Transport of Air Pollution 2010, Part A: Ozone and Particulate Matter*. United Nations, Geneva, Switzerland.

Jones, C. D., et al., 2013: 21st Century compatible CO2 emissions and airborne fraction simulated by CMIP5 Earth System models under 4 Representative Concentration Pathways. *J. Clim*., doi:10.1175/JCLI-D-12-00554.1.

Lamarque, J. F., G. P. Kyle, M. Meinshausen, K. Riahi, S. J. Smith, D. P. Van Vuuren, A. J. Conley, and F. Vitt, 2011: Global and regional evolution of short-lived radiatively-active gases and aerosols in the Representative Concentration Pathways. *Clim. Change*, **109**, 191–212.

Lamarque, J. F., et al., 2010: Historical (1850-2000) gridded anthropogenic and biomass burning emissions of reactive gases and aerosols: methodology and application. *Atmos. Chem. Phys*., **10**, 7017–7039.

Lamarque, J. F., et al., 2013: The Atmospheric Chemistry and Climate Model Intercomparison Project (ACCMIP): Overview and description of models, simulations and climate diagnostics. *Geosci. Model Dev*., **6**, 179–206.

Meinshausen, M., T. M. L. Wigley, and S. C. B. Raper, 2011b: Emulating atmosphere-ocean and carbon cycle models with a simpler model, MAGICC6-Part 2: Applications. *Atmos. Chem. Phys*., **11**, 1457–1471.

Meinshausen, M., et al., 2011a: The RCP greenhouse gas concentrations and their extensions from 1765 to 2300. *Clim. Change*, **109**, 213–241.

Moss, R. H., et al., 2010: The next generation of scenarios for climate change research and assessment. *Nature*, **463**, 747–756.

Prather, M., et al., 2001: Atmospheric chemistry and greenhouse gases. In: *Climate Change 2001: The Scientific Basis. Contribution of Working Group I to the Third Assessment Report of the Intergovernmental Panel on Climate Change* [J. T. Houghton, Y. Ding, D. J. Griggs, M. Noquer, P. J. van der Linden, X. Dai, K. Maskell and C. A. Johnson (eds.)]. Cambridge University Press, Cambridge, United Kingdom and New York, NY, USA, pp. 239–287.

Prather, M., et al., 2003: Fresh air in the 21st century? *Geophys. Res. Lett*., **30**, 1100.

Prather, M. J., C. D. Holmes, and J. Hsu, 2012: Reactive greenhouse gas scenarios: Systematic exploration of uncertainties and the role of atmospheric chemistry. *Geophys. Res. Lett*., **39**, L09803.

Rogelj, J., et al., 2011: Emission pathways consistent with a 2°C global temperature limit. *Nature Clim. Change*, **1**, 413–418.

Shindell, D.T., J.-F. Lamarque, M. Schulz, M. Flanner, et al., 2013: Radiative forcing in the ACCMIP historical and future climate simulations. *Atmos. Chem. Phys*., **13**, 2939–2974.

Stevenson, D. S., et al., 2013: Tropospheric ozone changes, radiative forcing and attribution to emissions in the Atmospheric Chemistry and Climate Model Intercomparison Project (ACCMIP). *Atmos. Chem. Phys*., **13**, 3063–3085.

van Vuuren, D. P., et al., 2008: Temperature increase of 21st century mitigation scenarios. *Proc. Natl. Acad. Sci. U.S.A.*, **105**, 15258–15262.

van Vuuren, D., et al., 2011: The representative concentration pathways: An overview. *Clim. Change*, **109**, 5–31.

Voulgarakis, A., et al., 2013: Analysis of present day and future OH and methane lifetime in the ACCMIP simulations. 21 *Atmos. Chem. Phys*., **13**, 2563–2587.

Wild, O., A.M. Fiore et al., 2012: Modelling future changes in surface ozone: A parameterized approach. *Atmos. Chem. Phys.*, **12**, 2037–2054.

WMO. 2010. *Scientific Assessment of Ozone Depletion: 2010*. Global Ozone Research and Monitoring Project—Report No. 52. World Meteorological Organization, Geneva, Switzerland.

Young, P. J., et al., 2013: Pre-industrial to end 21st century projections of tropospheric ozone from the Atmospheric Chemistry and Climate Model Intercomparison Project (ACCMIP). Atmos. Chem. Phys., **13**, 2063–2090.

Tables

AII.1: Historical Climate System Data

Table AII.1.1a | Historical abundances of the Kyoto greenhouse gases

Year	CO_2 (ppm)	CH_4 (ppb)	N_2O (ppb)
PI*	278 ± 2	722 ± 25	270 ± 7
1755	276.7	723	272.8
1760	276.5	726	274.1
1765	276.6	730	274.2
1770	277.3	733	273.7
1775	278.0	736	273.1
1780	278.2	739	272.4
1785	278.6	742	271.9
1790	280.0	745	271.8
1795	281.4	748	272.1
1800	282.6	751	272.6
1805	283.6	755	272.1
1810	284.2	760	271.4
1815	284.0	765	271.5
1820	283.3	769	272.9
1825	283.1	774	274.1
1830	283.8	779	273.7
1835	283.9	784	270.5
1840	284.1	789	269.6
1845	285.8	795	270.3
1850	286.8	802	270.4
1855	286.4	808	270.6
1860	286.1	815	271.7
1865	286.3	823	272.3
1870	288.0	831	273.0
1875	289.4	839	274.7
1880	289.8	847	275.8
1885	290.9	856	277.2
1890	293.1	866	278.3
1895	295.4	877	277.7
1900	296.2	891	277.3
1905	297.4	912	279.2
1910	299.3	935	280.8
1915	301.1	961	282.7
1920	303.3	990	285.1
1925	304.7	1020	284.3
1930	306.6	1049	284.9
1935	308.4	1077	286.6
1940	310.4	1102	287.7
1945	310.9	1129	288.0
1950	311.2	1162	287.6
1955	313.4	1207	289.6
1956	314.0	1217	290.4
1957	314.6	1228	291.2
1958	315.3	1239	291.7

Year	CO_2 (ppm)	CH_4 (ppb)	N_2O (ppb)
PI*	278 ± 2	722 ± 25	270 ± 7
1959	316.0	1251	292.1
1960	316.7	1263	292.4
1961	317.4	1275	292.5
1962	318.0	1288	292.5
1963	318.5	1301	292.6
1964	319.0	319.0	292.6
1965	319.7	1328	292.7
1966	320.6	1343	292.9
1967	321.5	1357	293.3
1968	322.5	1372	293.8
1969	323.5	1388	294.4
1970	324.6	1403	295.2
1971	325.6	1419	296.0
1972	326.8	1435	296.9
1973	328.0	1451	297.8
1974	329.2	1467	298.4
1975	330.2	1483	299.0
1976	331.3	1500	299.4
1977	332.7	1516	299.8
1978	334.3	1532	300.2
1979	336.2	1549	300.7
1980	338.0	1567	301.3
1981	339.3	1587	302.0
1982	340.5	1607	303.0
1983	342.1	1626	303.9
1984	343.7	1643	304.5
1985	345.2	1657	305.5
1986	346.6	1670	305.9
1987	348.4	1682	306.3
1988	350.5	1694	306.7
1989	352.2	1704	307.8
1990	353.6	1714	308.7
1991	354.8	1725	309.3
1992	355.7	1733	309.8
1993	356.6	1738	310.1
1994	358.0	1743	310.4
1995	359.9	1747	311.0
1996	361.4	1751	311.8
1997	363.1	1757	312.7
1998	365.2	1765	313.7
1999	367.2	1771	314.7
2000	368.7	1773	315.6
2001	370.2	1773	316.3
2002	372.3	1774	317.0

(continued on next page)

Table AII.1.1a *(continued)*

Year	CO_2 (ppm)	CH_4 (ppb)	N_2O (ppb)
PI*	**278 ± 2**	**722 ± 25**	**270 ± 7**
2003	374.5	1776	317.6
2004	376.6	1776	318.3
2005	378.7	1776	319.1
2006	380.8	1776	319.8
2007	382.7	1781	320.6
2008	384.6	1787	321.4
2005	378.7	1776	319.1
2006	380.8	1776	319.8
2007	382.7	1781	320.6
2008	384.6	1787	321.4
2009	386.4	1792	322.3
2010	388.4	1798	323.2
2011*	390.5 ± 0.3	1803 ± 4	324 ± 1

Year	SF_6 (ppt)	CF_4 (ppt)	C_2F_6 (ppt)	C_6F_{14} (ppt)	NF_3 (ppt)
PI*	**0**	**35**	**0**	**0**	
1900	0	35	0	0	
1910	0	35	0.1	0	
1920	0	35	0.1	0	
1930	0	36	0.2	0	
1940	0	37	0.3	0	
1950	0	39	0.5	0	
1960	0.1	43	0.6	0	
1970	0.3	51	0.8	0	
1980	0.8	60	1.2	0	
1990	2.4	68	1.9	0	
2000	4.5	76	2.9	0	
2005	5.6	75	3.7	0	0.3
2010	7.0	78.3	4.1	0	
2011*	7.3 ± 0.1	79.0	4.2	0	0.6

Year	HFC-23 (ppt)	HFC-32 (ppt)	HFC-125 (ppt)	HFC-134a (ppt)	HFC-143a (ppt)	HFC-227ea (ppt)	HFC-245fa (ppt)	HFC-43-10mee (ppt)
PI*	**0**	**0**	**0**	**0**	**0**	**0**	**0**	**0**
1940	0.1	0	0	0	0	0	0	0
1950	0.3	0	0	0	0	0	0	0
1960	0.7	0	0	0	0	0	0	0
1970	1.6	0	0	0	0	0	0	0
1980	3.7	0	0	0	0.2	0	0	0
1990	7.9	0	0.1	0	0.6	0	0	0
2000	14.8	0	1.3	14	3.1	0.1	0	0
2010	23.2	4.1	8.2	58	10.9	0.6	1.1	0
2011*	24.0	4.9	9.6	63 ± 1	12.0	0.65	1.24	0

Notes:

Abundances are mole fraction of dry air for the lower, well-mixed atmosphere (ppm = micromoles per mole, ppb = nanomoles per mole, ppt = picomoles per mole). Values refer to single-year average. Uncertainties (5 to 95% confidence intervals) are given for 2011 only when more than one laboratory reports global data. Pre-industrial (PI*, taken to be 1750 for GHG) and present day (2011*) abundances are from Chapter 2, Tables 2.1 and 2.SM.1; see also Chapter 6 for Holocene variability (10 ppm CO_2, 40 ppb CH_4, 10 ppb N_2O). Intermediate data for CO_2, CH_4 and N_2O are from Chapters 2 and 8, Figure 8.6. See also Appendix 1.A. Intermediate data for the F-gases are taken from Meinshausen et al. (2011).

Table AII.1.1b | Historical abundances of the Montreal Protocol greenhouse gases (all ppt)

Year	CFC-11	CFC-12	CFC-113	CFC-114	CFC-115	CCl_4	CH_3CCl_3	HCFC-22
PI*	**0**	**0**	**0**	**0**	**0**	**0**	**0**	**0**
1960	9.5	29.5	1.9	3.8	0.0	52.1	1.5	2.1
1965	23.5	58.8	3.1	5.0	0.0	64.4	4.7	4.9
1970	52.8	114.3	5.5	6.5	0.2	75.9	16.2	12.1
1975	106.1	203.1	10.4	8.3	0.6	85.5	40.0	23.8
1980	161.9	297.4	19.0	10.7	1.3	93.3	81.6	42.5
1985	205.4	381.2	37.3	12.9	2.8	99.6	106.1	62.7
1990	256.2	477.5	67.6	15.4	4.7	106.5	127.2	88.2
1995	267.4	523.8	83.6	16.1	6.8	103.2	110.3	113.6
2000	261.7	541.0	82.3	16.5	7.9	98.6	49.7	139.5
2005	251.6	542.7	78.8	16.6	8.3	93.7	20.1	165.5
2010	240.9	532.5	75.6	16.4	8.4	87.6	8.3	206.8
2011*	238 ± 1	528±2	74.3±0.5	15.8	8.4	86±2	6.4±0.4	213±2

Year	HCFC-141b	HCFC-142b	Halon 1211	Halon 1202	Halon 1301	Halon 2402	CH_3Br	CH_3Cl
PI*	**0**	**0**	**0**	**0**	**0**	**0**		
1960	0.0	0.0	0.00	0.00	0.00	0.00	6.5	510
1965	0.0	0.0	0.00	0.00	0.00	0.00	6.7	528
1970	0.0	0.0	0.02	0.00	0.00	0.02	0.0	540
1975	0.0	0.2	0.12	0.01	0.04	0.06	7.4	546
1980	0.0	0.4	0.42	0.01	0.24	0.15	7.7	548
1985	0.0	0.7	1.04	0.02	0.74	0.26	8.2	549
1990	0.0	1.2	2.27	0.03	1.66	0.41	8.6	550
1995	2.7	6.3	3.34	0.04	2.63	0.52	9.2	550
2000	11.8	11.4	4.02	0.04	2.84	0.50	8.9	550
2005	17.5	15.1	4.26	0.02	3.03	0.48	7.9	550
2010	20.3	20.5	4.07	0.00	3.20	0.46	7.2	550
2011*	21.4±0.5	21.2±0.5	4.07	0.00	3.23	0.45	7.1	534

Notes:

See Table AII.1.1a. For present-day (2011*) see Chapter 2. Intermediate years are from Scenario A1, WMO Ozone Assessment (WMO, 2010).

Table AII.1.2 | Historical effective radiative forcing (ERF) (W m^{-2}), including land use change (LUC)

Year	CO_2	GHG Other*	O_3 (Trop)	O_3 (Strat)	Aerosol (Total)	LUC	H_2O (Strat)	BC Snow	Con–trails	Solar	Volcano
1750	0.000	0.000	0.000	0.000	0.000	0.000	0.000	0.000	0.000	0.000	–0.001
1751	–0.023	0.004	0.000	0.000	–0.002	0.000	0.000	0.000	0.000	–0.014	0.000
1752	–0.024	0.006	0.001	0.000	–0.004	–0.001	0.000	0.000	0.000	–0.029	0.000
1753	–0.024	0.007	0.001	0.000	–0.005	–0.001	0.000	0.000	0.000	–0.033	0.000
1754	–0.025	0.008	0.002	0.000	–0.007	–0.002	0.000	0.001	0.000	–0.043	0.000
1755	–0.026	0.010	0.002	0.000	–0.009	–0.002	0.000	0.001	0.000	–0.054	–0.664
1756	–0.026	0.011	0.003	0.000	–0.011	–0.002	0.000	0.001	0.000	–0.055	0.000
1757	–0.027	0.013	0.003	0.000	–0.013	–0.003	0.000	0.001	0.000	–0.048	0.000
1758	–0.028	0.014	0.003	0.000	–0.014	–0.003	0.000	0.001	0.000	–0.050	0.000
1759	–0.028	0.015	0.004	0.000	–0.016	–0.004	0.000	0.001	0.000	–0.102	0.000
1760	–0.029	0.016	0.004	0.000	–0.018	–0.004	0.000	0.001	0.000	–0.112	–0.060
1761	–0.029	0.017	0.005	0.000	–0.020	–0.004	0.000	0.002	0.000	–0.016	–1.093
1762	–0.029	0.017	0.005	0.000	–0.021	–0.005	0.001	0.002	0.000	–0.007	–0.300
1763	–0.029	0.018	0.006	0.000	–0.023	–0.005	0.001	0.002	0.000	–0.018	–0.093
1764	–0.028	0.018	0.006	0.000	–0.025	–0.006	0.001	0.002	0.000	–0.022	–0.021
1765	–0.026	0.018	0.006	0.000	–0.027	–0.006	0.001	0.002	0.000	–0.054	–0.003
1766	–0.024	0.018	0.007	0.000	–0.029	–0.006	0.001	0.002	0.000	–0.048	0.000
1767	–0.022	0.018	0.007	0.000	–0.030	–0.007	0.001	0.003	0.000	–0.036	0.000
1768	–0.020	0.018	0.008	0.000	–0.032	–0.007	0.001	0.003	0.000	0.016	0.000
1769	–0.017	0.018	0.008	0.000	–0.034	–0.008	0.001	0.003	0.000	0.050	0.000
1770	–0.014	0.018	0.009	0.000	–0.036	–0.008	0.001	0.003	0.000	0.081	0.000
1771	–0.011	0.018	0.009	0.000	–0.038	–0.008	0.001	0.003	0.000	0.055	0.000
1772	–0.008	0.018	0.009	0.000	–0.039	–0.009	0.001	0.003	0.000	0.052	–0.070
1773	–0.005	0.018	0.010	0.000	–0.041	–0.009	0.001	0.003	0.000	0.016	–0.020
1774	–0.003	0.018	0.010	0.000	–0.043	–0.010	0.001	0.004	0.000	–0.002	–0.005
1775	–0.001	0.018	0.011	0.000	–0.045	–0.010	0.001	0.004	0.000	–0.038	–0.001
1776	0.001	0.018	0.011	0.000	–0.046	–0.010	0.001	0.004	0.000	–0.045	0.000
1777	0.002	0.018	0.011	0.000	–0.048	–0.011	0.001	0.004	0.000	–0.036	0.000
1778	0.003	0.018	0.012	0.000	–0.050	–0.011	0.001	0.004	0.000	0.017	–0.067
1779	0.003	0.018	0.012	0.000	–0.052	–0.012	0.001	0.004	0.000	–0.034	–0.071
1780	0.003	0.018	0.013	0.000	–0.054	–0.012	0.002	0.004	0.000	–0.069	–0.018
1781	0.004	0.018	0.013	0.000	–0.055	–0.012	0.002	0.005	0.000	–0.057	–0.004
1782	0.004	0.018	0.014	0.000	–0.057	–0.013	0.002	0.005	0.000	–0.028	–0.001
1783	0.006	0.018	0.014	0.000	–0.059	–0.013	0.002	0.005	0.000	–0.065	–7.857
1784	0.009	0.018	0.014	0.000	–0.061	–0.014	0.002	0.005	0.000	–0.059	–0.522
1785	0.012	0.018	0.015	0.000	–0.062	–0.014	0.002	0.005	0.000	–0.046	–0.121
1786	0.017	0.018	0.015	0.000	–0.064	–0.014	0.002	0.005	0.000	–0.022	–0.027
1787	0.021	0.018	0.016	0.000	–0.066	–0.015	0.002	0.005	0.000	–0.001	–0.002
1788	0.027	0.018	0.016	0.000	–0.068	–0.015	0.002	0.006	0.000	0.034	–0.133
1789	0.033	0.019	0.017	0.000	–0.070	–0.016	0.002	0.006	0.000	–0.033	–0.041
1790	0.038	0.019	0.017	0.000	–0.071	–0.016	0.002	0.006	0.000	–0.058	–0.009
1791	0.044	0.019	0.017	0.000	–0.073	–0.016	0.002	0.006	0.000	–0.056	–0.001
1792	0.050	0.020	0.018	0.000	–0.075	–0.017	0.002	0.006	0.000	–0.051	0.000
1793	0.055	0.020	0.018	0.000	–0.077	–0.017	0.002	0.006	0.000	–0.065	0.000
1794	0.060	0.021	0.019	0.000	–0.079	–0.018	0.002	0.006	0.000	–0.064	–0.157
1795	0.066	0.022	0.019	0.000	–0.080	–0.018	0.002	0.007	0.000	–0.027	0.000
1796	0.070	0.023	0.020	0.000	–0.082	–0.018	0.002	0.007	0.000	–0.033	–0.781
1797	0.075	0.023	0.020	0.000	–0.084	–0.019	0.002	0.007	0.000	–0.043	–0.071
1798	0.079	0.024	0.020	0.000	–0.086	–0.019	0.002	0.007	0.000	–0.045	–0.016

Table AII.1.2 | *(continued)*

Year	CO_2	GHG Other*	O_3 (Trop)	O_3 (Strat)	Aerosol (Total)	LUC	H_2O (Strat)	BC Snow	Con-trails	Solar	Volcano
1799	0.084	0.025	0.021	0.000	–0.087	–0.020	0.003	0.007	0.000	–0.047	–0.002
1800	0.088	0.025	0.021	0.000	–0.089	–0.020	0.003	0.007	0.000	–0.055	0.000
1801	0.092	0.026	0.022	0.000	–0.091	–0.020	0.003	0.007	0.000	–0.021	–0.154
1802	0.096	0.026	0.022	0.000	–0.093	–0.021	0.003	0.008	0.000	–0.010	–0.048
1803	0.099	0.026	0.023	0.000	–0.095	–0.021	0.003	0.008	0.000	–0.033	–0.011
1804	0.103	0.026	0.023	0.000	–0.096	–0.022	0.003	0.008	0.000	–0.040	–0.230
1805	0.106	0.026	0.023	0.000	–0.098	–0.022	0.003	0.008	0.000	–0.046	–0.070
1806	0.109	0.026	0.024	0.000	–0.100	–0.022	0.003	0.008	0.000	–0.036	–0.016
1807	0.112	0.026	0.024	0.000	–0.102	–0.023	0.003	0.008	0.000	–0.057	–0.002
1808	0.114	0.026	0.025	0.000	–0.104	–0.023	0.003	0.008	0.000	–0.065	0.000
1809	0.116	0.026	0.025	0.000	–0.105	–0.024	0.003	0.009	0.000	–0.065	–6.947
1810	0.117	0.026	0.025	0.000	–0.107	–0.024	0.003	0.009	0.000	–0.070	–2.254
1811	0.118	0.027	0.026	0.000	–0.109	–0.024	0.003	0.009	0.000	–0.072	–0.836
1812	0.119	0.027	0.026	0.000	–0.111	–0.025	0.003	0.009	0.000	–0.072	–0.308
1813	0.118	0.028	0.027	0.000	–0.112	–0.025	0.004	0.009	0.000	–0.069	–0.109
1814	0.117	0.029	0.027	0.000	–0.114	–0.026	0.004	0.009	0.000	–0.064	0.000
1815	0.115	0.030	0.028	0.000	–0.116	–0.026	0.004	0.009	0.000	–0.062	–11.629
1816	0.113	0.031	0.028	0.000	–0.118	–0.026	0.004	0.010	0.000	–0.052	–4.553
1817	0.110	0.032	0.028	0.000	–0.120	–0.027	0.004	0.010	0.000	–0.048	–2.419
1818	0.107	0.034	0.029	0.000	–0.121	–0.027	0.004	0.010	0.000	–0.053	–0.915
1819	0.104	0.035	0.029	0.000	–0.123	–0.028	0.004	0.010	0.000	–0.054	–0.337
1820	0.101	0.037	0.030	0.000	–0.125	–0.028	0.004	0.010	0.000	–0.059	–0.039
1821	0.099	0.038	0.030	0.000	–0.127	–0.028	0.004	0.010	0.000	–0.065	0.000
1822	0.097	0.040	0.031	0.000	–0.128	–0.029	0.004	0.010	0.000	–0.066	0.000
1823	0.096	0.041	0.031	0.000	–0.130	–0.029	0.004	0.011	0.000	–0.068	0.000
1824	0.097	0.042	0.031	0.000	–0.132	–0.030	0.004	0.011	0.000	–0.059	0.000
1825	0.098	0.043	0.032	0.000	–0.134	–0.030	0.005	0.011	0.000	–0.052	0.000
1826	0.100	0.044	0.032	0.000	–0.136	–0.030	0.005	0.011	0.000	–0.044	0.000
1827	0.103	0.045	0.033	0.000	–0.137	–0.031	0.005	0.011	0.000	–0.018	0.000
1828	0.106	0.045	0.033	0.000	–0.139	–0.031	0.005	0.011	0.000	–0.008	0.000
1829	0.109	0.045	0.034	0.000	–0.141	–0.032	0.005	0.011	0.000	–0.006	0.000
1830	0.111	0.045	0.034	0.000	–0.143	–0.032	0.005	0.012	0.000	0.002	0.000
1831	0.113	0.044	0.034	0.000	–0.145	–0.032	0.005	0.012	0.000	0.002	–1.538
1832	0.114	0.043	0.035	0.000	–0.146	–0.033	0.005	0.012	0.000	–0.020	–1.229
1833	0.114	0.041	0.035	0.000	–0.148	–0.033	0.005	0.012	0.000	–0.035	–0.605
1834	0.114	0.039	0.036	0.000	–0.150	–0.034	0.005	0.012	0.000	–0.038	–0.223
1835	0.113	0.037	0.036	0.000	–0.152	–0.034	0.005	0.012	0.000	–0.033	–4.935
1836	0.112	0.036	0.037	0.000	–0.153	–0.034	0.005	0.012	0.000	0.017	–1.445
1837	0.112	0.035	0.037	0.000	–0.155	–0.035	0.006	0.013	0.000	0.055	–0.523
1838	0.112	0.035	0.037	0.000	–0.157	–0.035	0.006	0.013	0.000	0.051	–0.192
1839	0.114	0.036	0.038	0.000	–0.159	–0.036	0.006	0.013	0.000	0.028	–0.069
1840	0.117	0.037	0.038	0.000	–0.161	–0.036	0.006	0.013	0.000	0.027	–0.047
1841	0.121	0.038	0.039	0.000	–0.162	–0.036	0.006	0.013	0.000	0.007	–0.013
1842	0.127	0.040	0.039	0.000	–0.164	–0.037	0.006	0.013	0.000	–0.006	–0.003
1843	0.135	0.041	0.039	0.000	–0.166	–0.037	0.006	0.013	0.000	–0.013	–0.052
1844	0.142	0.042	0.040	0.000	–0.168	–0.038	0.006	0.014	0.000	–0.024	–0.014
1845	0.149	0.043	0.040	0.000	–0.169	–0.038	0.006	0.014	0.000	–0.026	–0.003
1846	0.155	0.044	0.041	0.000	–0.171	–0.038	0.006	0.014	0.000	–0.024	–0.071
1847	0.160	0.044	0.041	0.000	–0.173	–0.039	0.007	0.014	0.000	–0.062	–0.020
1848	0.163	0.045	0.042	0.000	–0.175	–0.039	0.007	0.014	0.000	–0.018	–0.005

Table AII.1.2 | *(continued)*

Year	CO_2	GHG Other*	O_3 (Trop)	O_3 (Strat)	Aerosol (Total)	LUC	H_2O (Strat)	BC Snow	Con–trails	Solar	Volcano
1849	0.166	0.046	0.042	0.000	–0.177	–0.040	0.007	0.014	0.000	0.043	–0.001
1850	0.167	0.046	0.042	0.000	–0.178	–0.040	0.007	0.014	0.000	0.024	–0.100
1851	0.167	0.047	0.043	0.000	–0.180	–0.040	0.007	0.015	0.000	0.016	–0.075
1852	0.166	0.048	0.044	0.000	–0.182	–0.041	0.007	0.015	0.000	0.020	–0.025
1853	0.164	0.049	0.045	0.000	–0.184	–0.041	0.007	0.015	0.000	0.011	–0.025
1854	0.162	0.050	0.046	0.000	–0.185	–0.041	0.007	0.016	0.000	–0.010	0.000
1855	0.160	0.051	0.047	0.000	–0.187	–0.042	0.007	0.016	0.000	–0.027	–0.050
1856	0.158	0.052	0.048	0.000	–0.189	–0.042	0.007	0.016	0.000	–0.037	–0.975
1857	0.156	0.054	0.049	0.000	–0.191	–0.042	0.008	0.016	0.000	–0.037	–1.500
1858	0.155	0.055	0.050	0.000	–0.192	–0.043	0.008	0.017	0.000	–0.020	–0.725
1859	0.154	0.057	0.050	0.000	–0.194	–0.043	0.008	0.017	0.000	–0.007	–0.275
1860	0.154	0.058	0.051	0.000	–0.196	–0.043	0.008	0.017	0.000	0.029	–0.125
1861	0.153	0.060	0.052	0.000	–0.198	–0.044	0.008	0.018	0.000	0.036	–0.075
1862	0.153	0.061	0.053	0.000	–0.199	–0.044	0.008	0.018	0.000	0.013	–0.350
1863	0.154	0.062	0.054	0.000	–0.201	–0.044	0.008	0.018	0.000	0.006	–0.250
1864	0.156	0.063	0.055	0.000	–0.203	–0.045	0.008	0.018	0.000	–0.017	–0.125
1865	0.158	0.064	0.056	0.000	–0.205	–0.045	0.009	0.019	0.000	–0.018	–0.050
1866	0.162	0.066	0.057	0.000	–0.206	–0.045	0.009	0.019	0.000	–0.021	–0.025
1867	0.167	0.067	0.058	0.000	–0.208	–0.046	0.009	0.019	0.000	–0.037	0.000
1868	0.173	0.068	0.059	0.000	–0.210	–0.046	0.009	0.020	0.000	–0.039	0.000
1869	0.180	0.070	0.059	0.000	–0.212	–0.046	0.009	0.020	0.000	–0.005	–0.025
1870	0.188	0.071	0.060	0.000	–0.213	–0.047	0.009	0.020	0.000	–0.028	–0.025
1871	0.195	0.073	0.061	0.000	–0.215	–0.047	0.009	0.020	0.000	0.025	–0.025
1872	0.202	0.075	0.062	0.000	–0.217	–0.047	0.009	0.021	0.000	0.012	–0.025
1873	0.208	0.077	0.063	0.000	–0.219	–0.048	0.010	0.021	0.000	0.015	–0.075
1874	0.212	0.079	0.064	0.000	–0.220	–0.048	0.010	0.021	0.000	0.000	–0.050
1875	0.215	0.081	0.065	0.000	–0.222	–0.049	0.010	0.022	0.000	–0.015	–0.025
1876	0.218	0.083	0.066	0.000	–0.224	–0.049	0.010	0.022	0.000	–0.029	–0.150
1877	0.219	0.084	0.067	0.000	–0.226	–0.049	0.010	0.022	0.000	–0.033	–0.125
1878	0.219	0.086	0.067	0.000	–0.227	–0.050	0.010	0.022	0.000	–0.041	–0.075
1879	0.221	0.088	0.068	0.000	–0.229	–0.050	0.010	0.023	0.000	–0.044	–0.050
1880	0.222	0.089	0.069	0.000	–0.231	–0.050	0.011	0.023	0.000	–0.039	–0.025
1881	0.224	0.091	0.070	0.000	–0.233	–0.051	0.011	0.023	0.000	–0.007	–0.025
1882	0.228	0.092	0.071	0.000	–0.234	–0.051	0.011	0.024	0.000	–0.019	–0.025
1883	0.232	0.094	0.072	0.000	–0.236	–0.052	0.011	0.024	0.000	–0.031	–1.175
1884	0.238	0.096	0.073	0.000	–0.238	–0.052	0.011	0.024	0.000	0.018	–3.575
1885	0.244	0.098	0.074	0.000	–0.240	–0.053	0.011	0.024	0.000	0.002	–1.575
1886	0.250	0.100	0.075	0.000	–0.241	–0.053	0.011	0.025	0.000	–0.014	–0.900
1887	0.258	0.102	0.075	0.000	–0.243	–0.053	0.012	0.025	0.000	–0.033	–0.925
1888	0.266	0.104	0.076	0.000	–0.245	–0.054	0.012	0.025	0.000	–0.037	–0.550
1889	0.274	0.106	0.077	0.000	–0.247	–0.054	0.012	0.026	0.000	–0.041	–0.725
1890	0.283	0.107	0.078	0.000	–0.248	–0.055	0.012	0.026	0.000	–0.041	–0.975
1891	0.293	0.108	0.079	0.000	–0.250	–0.055	0.012	0.026	0.000	–0.020	–0.750
1892	0.302	0.109	0.080	0.000	–0.252	–0.056	0.012	0.026	0.000	0.004	–0.550
1893	0.311	0.110	0.081	0.000	–0.254	–0.056	0.013	0.027	0.000	0.035	–0.225
1894	0.319	0.111	0.082	0.000	–0.255	–0.057	0.013	0.027	0.000	0.072	–0.100
1895	0.325	0.111	0.083	0.000	–0.257	–0.057	0.013	0.027	0.000	0.052	–0.025
1896	0.330	0.112	0.083	0.000	–0.259	–0.058	0.013	0.028	0.000	0.023	–0.450
1897	0.334	0.113	0.084	0.000	–0.261	–0.058	0.013	0.028	0.000	–0.003	–0.425
1898	0.336	0.114	0.085	0.000	–0.262	–0.059	0.014	0.028	0.000	–0.012	–0.300

Table AII.1.2 | *(continued)*

Year	CO_2	GHG Other*	O_3 (Trop)	O_3 (Strat)	Aerosol (Total)	LUC	H_2O (Strat)	BC Snow	Con-trails	Solar	Volcano
1899	0.337	0.115	0.086	0.000	–0.264	–0.059	0.014	0.028	0.000	–0.017	–0.125
1900	0.339	0.117	0.087	0.000	–0.266	–0.060	0.014	0.029	0.000	–0.028	–0.050
1901	0.341	0.120	0.088	0.000	–0.268	–0.061	0.014	0.029	0.000	–0.043	–0.025
1902	0.344	0.123	0.089	0.000	–0.270	–0.061	0.015	0.030	0.000	–0.048	–0.500
1903	0.349	0.127	0.090	0.000	–0.272	–0.062	0.015	0.030	0.000	–0.036	–1.800
1904	0.355	0.130	0.091	0.000	–0.274	–0.062	0.015	0.031	0.000	0.011	–0.800
1905	0.362	0.134	0.092	0.000	–0.276	–0.063	0.016	0.032	0.000	–0.016	–0.325
1906	0.369	0.138	0.092	0.000	–0.278	–0.063	0.016	0.032	0.000	0.028	–0.175
1907	0.376	0.141	0.093	0.000	–0.280	–0.064	0.016	0.033	0.000	–0.001	–0.225
1908	0.383	0.145	0.094	0.000	–0.282	–0.064	0.017	0.033	0.000	0.020	–0.250
1909	0.389	0.148	0.095	–0.001	–0.284	–0.065	0.017	0.034	0.000	–0.002	–0.100
1910	0.395	0.151	0.096	–0.001	–0.286	–0.065	0.017	0.035	0.000	–0.006	–0.075
1911	0.400	0.155	0.097	–0.001	–0.288	–0.066	0.018	0.035	0.000	–0.032	–0.050
1912	0.406	0.159	0.098	–0.001	–0.289	–0.066	0.018	0.035	0.000	–0.045	–0.475
1913	0.412	0.163	0.100	–0.001	–0.290	–0.067	0.019	0.035	0.000	–0.042	–0.600
1914	0.419	0.167	0.101	–0.001	–0.291	–0.068	0.019	0.035	0.000	–0.033	–0.250
1915	0.427	0.171	0.102	–0.001	–0.292	–0.068	0.019	0.035	0.000	0.013	–0.100
1916	0.436	0.175	0.103	–0.001	–0.293	–0.069	0.020	0.035	0.000	0.068	–0.075
1917	0.445	0.180	0.104	–0.001	–0.294	–0.069	0.020	0.035	0.000	0.086	–0.050
1918	0.453	0.185	0.105	–0.001	–0.296	–0.070	0.021	0.035	0.000	0.121	–0.050
1919	0.460	0.189	0.107	–0.002	–0.297	–0.071	0.021	0.035	0.000	0.073	–0.050
1920	0.466	0.193	0.108	–0.001	–0.298	–0.071	0.022	0.035	0.000	0.039	–0.225
1921	0.472	0.196	0.109	–0.001	–0.302	–0.072	0.022	0.036	0.000	0.012	–0.200
1922	0.476	0.199	0.110	–0.002	–0.305	–0.073	0.022	0.036	0.000	–0.013	–0.075
1923	0.481	0.201	0.111	–0.002	–0.309	–0.073	0.023	0.036	0.000	–0.025	–0.025
1924	0.486	0.203	0.113	–0.002	–0.313	–0.074	0.023	0.036	0.000	–0.029	–0.075
1925	0.491	0.205	0.114	–0.002	–0.317	–0.075	0.024	0.036	0.000	–0.015	–0.075
1926	0.497	0.207	0.115	–0.002	–0.321	–0.076	0.024	0.036	0.000	0.020	–0.050
1927	0.503	0.210	0.116	–0.002	–0.325	–0.076	0.025	0.036	0.000	0.063	–0.050
1928	0.510	0.214	0.117	–0.002	–0.328	–0.077	0.025	0.037	0.000	0.033	–0.125
1929	0.517	0.218	0.119	–0.002	–0.332	–0.078	0.025	0.037	0.000	0.028	–0.250
1930	0.523	0.222	0.120	–0.003	–0.336	–0.079	0.026	0.037	0.000	0.048	–0.150
1931	0.530	0.226	0.122	–0.003	–0.338	–0.080	0.026	0.037	0.000	0.009	–0.125
1932	0.536	0.230	0.124	–0.003	–0.340	–0.081	0.027	0.038	0.000	–0.016	–0.200
1933	0.542	0.234	0.126	–0.003	–0.341	–0.081	0.027	0.038	0.000	–0.029	–0.175
1934	0.548	0.237	0.128	–0.003	–0.343	–0.082	0.027	0.039	0.000	–0.027	–0.100
1935	0.555	0.241	0.130	–0.003	–0.345	–0.083	0.028	0.039	0.000	–0.008	–0.100
1936	0.563	0.244	0.133	–0.003	–0.347	–0.084	0.028	0.040	0.000	0.068	–0.075
1937	0.570	0.247	0.135	–0.003	–0.349	–0.085	0.029	0.040	0.000	0.089	–0.075
1938	0.577	0.251	0.137	–0.003	–0.350	–0.086	0.029	0.040	0.000	0.080	–0.125
1939	0.584	0.254	0.139	–0.004	–0.352	–0.087	0.029	0.041	0.000	0.094	–0.100
1940	0.590	0.257	0.141	–0.004	–0.354	–0.088	0.030	0.041	0.000	0.070	–0.075
1941	0.595	0.261	0.143	–0.004	–0.358	–0.089	0.030	0.042	0.000	0.057	–0.050
1942	0.598	0.264	0.146	–0.004	–0.362	–0.090	0.030	0.042	0.000	0.030	–0.100
1943	0.599	0.267	0.148	–0.004	–0.366	–0.092	0.031	0.043	0.000	–0.005	–0.100
1944	0.599	0.270	0.150	–0.004	–0.370	–0.093	0.031	0.043	0.001	–0.011	–0.050
1945	0.599	0.273	0.152	–0.004	–0.374	–0.094	0.032	0.043	0.001	0.019	–0.050
1946	0.599	0.276	0.154	–0.005	–0.378	–0.095	0.032	0.044	0.001	0.025	–0.050
1947	0.598	0.279	0.156	–0.005	–0.382	–0.096	0.032	0.044	0.002	0.093	–0.050
1948	0.598	0.283	0.158	–0.005	–0.386	–0.097	0.033	0.045	0.002	0.146	–0.050

Table AII.1.2 | *(continued)*

Year	CO_2	GHG Other*	O_3 (Trop)	O_3 (Strat)	Aerosol (Total)	LUC	H_2O (Strat)	BC Snow	Con–trails	Solar	Volcano
1949	0.601	0.287	0.161	–0.005	–0.390	–0.099	0.033	0.045	0.002	0.123	–0.075
1950	0.604	0.291	0.163	–0.005	–0.394	–0.100	0.034	0.046	0.002	0.110	–0.075
1951	0.608	0.296	0.168	–0.005	–0.409	–0.102	0.034	0.046	0.002	0.037	–0.050
1952	0.615	0.302	0.173	–0.006	–0.424	–0.103	0.035	0.047	0.002	0.045	–0.100
1953	0.623	0.308	0.178	–0.006	–0.439	–0.105	0.036	0.047	0.003	0.025	–0.075
1954	0.631	0.315	0.183	–0.006	–0.455	–0.106	0.036	0.048	0.003	0.003	–0.100
1955	0.641	0.323	0.188	–0.006	–0.470	–0.108	0.037	0.048	0.003	0.015	–0.050
1956	0.651	0.332	0.193	–0.007	–0.485	–0.109	0.038	0.049	0.003	0.064	–0.025
1957	0.662	0.341	0.198	–0.007	–0.500	–0.111	0.038	0.050	0.004	0.129	–0.025
1958	0.673	0.349	0.203	–0.007	–0.515	–0.112	0.039	0.050	0.004	0.194	0.000
1959	0.685	0.358	0.208	–0.008	–0.530	–0.114	0.040	0.051	0.004	0.159	0.000
1960	0.698	0.366	0.213	–0.008	–0.546	–0.116	0.041	0.051	0.004	0.151	–0.125
1961	0.709	0.374	0.218	–0.008	–0.563	–0.117	0.041	0.051	0.004	0.110	–0.275
1962	0.719	0.383	0.223	–0.009	–0.580	–0.119	0.042	0.051	0.004	0.051	–0.325
1963	0.727	0.392	0.228	–0.009	–0.598	–0.120	0.043	0.050	0.005	0.038	–1.150
1964	0.735	0.402	0.233	–0.010	–0.615	–0.122	0.044	0.050	0.005	0.019	–1.800
1965	0.748	0.412	0.239	–0.011	–0.632	–0.123	0.045	0.050	0.005	0.008	–1.075
1966	0.762	0.424	0.244	–0.011	–0.650	–0.125	0.046	0.050	0.006	0.012	–0.575
1967	0.778	0.437	0.249	–0.012	–0.667	–0.126	0.047	0.049	0.007	0.055	–0.375
1968	0.794	0.451	0.254	–0.013	–0.684	–0.127	0.048	0.049	0.008	0.086	–0.675
1969	0.811	0.466	0.259	–0.014	–0.701	–0.129	0.049	0.049	0.009	0.077	–0.850
1970	0.828	0.483	0.264	–0.014	–0.719	–0.130	0.050	0.049	0.009	0.092	–0.425
1971	0.846	0.500	0.270	–0.016	–0.722	–0.131	0.050	0.049	0.009	0.082	–0.150
1972	0.865	0.519	0.277	–0.017	–0.725	–0.132	0.051	0.049	0.009	0.076	–0.100
1973	0.885	0.538	0.284	–0.018	–0.728	–0.134	0.052	0.049	0.010	0.044	–0.200
1974	0.904	0.558	0.290	–0.019	–0.732	–0.135	0.053	0.050	0.010	0.023	–0.325
1975	0.920	0.578	0.297	–0.021	–0.735	–0.136	0.054	0.050	0.010	0.006	–0.750
1976	0.938	0.598	0.304	–0.022	–0.738	–0.137	0.055	0.050	0.010	–0.003	–0.350
1977	0.960	0.617	0.310	–0.024	–0.741	–0.138	0.056	0.050	0.011	0.040	–0.125
1978	0.987	0.636	0.317	–0.026	–0.745	–0.138	0.057	0.051	0.011	0.129	–0.200
1979	1.018	0.656	0.324	–0.027	–0.748	–0.139	0.058	0.051	0.012	0.167	–0.225
1980	1.046	0.675	0.330	–0.029	–0.751	–0.140	0.059	0.051	0.012	0.150	–0.125
1981	1.066	0.696	0.335	–0.031	–0.763	–0.141	0.061	0.051	0.012	0.147	–0.125
1982	1.085	0.717	0.339	–0.033	–0.775	–0.141	0.062	0.050	0.012	0.094	–1.325
1983	1.110	0.737	0.343	–0.035	–0.788	–0.142	0.063	0.050	0.012	0.091	–1.875
1984	1.136	0.757	0.348	–0.037	–0.800	–0.143	0.064	0.049	0.013	0.016	–0.750
1985	1.158	0.776	0.352	–0.038	–0.812	–0.143	0.065	0.049	0.014	0.011	–0.325
1986	1.180	0.795	0.356	–0.040	–0.824	–0.144	0.065	0.049	0.015	0.012	–0.350
1987	1.208	0.813	0.360	–0.042	–0.836	–0.144	0.066	0.048	0.016	0.015	–0.250
1988	1.240	0.832	0.365	–0.044	–0.848	–0.145	0.067	0.048	0.017	0.095	–0.200
1989	1.266	0.853	0.369	–0.046	–0.861	–0.145	0.067	0.047	0.018	0.151	–0.150
1990	1.287	0.872	0.373	–0.048	–0.873	–0.146	0.068	0.047	0.019	0.118	–0.150
1991	1.305	0.888	0.375	–0.050	–0.878	–0.146	0.068	0.046	0.019	0.126	–1.350
1992	1.318	0.900	0.376	–0.052	–0.883	–0.146	0.069	0.045	0.020	0.137	–3.025
1993	1.332	0.909	0.378	–0.054	–0.888	–0.147	0.069	0.045	0.022	0.063	–1.225
1994	1.354	0.916	0.379	–0.055	–0.893	–0.147	0.069	0.044	0.024	0.027	–0.500
1995	1.381	0.923	0.380	–0.056	–0.897	–0.147	0.070	0.043	0.025	0.020	–0.250
1996	1.404	0.930	0.382	–0.057	–0.902	–0.148	0.070	0.043	0.027	0.003	–0.175
1997	1.428	0.937	0.383	–0.057	–0.907	–0.148	0.070	0.042	0.028	0.016	–0.125
1998	1.459	0.944	0.385	–0.057	–0.912	–0.148	0.071	0.041	0.029	0.062	–0.075

Table AII.1.2 | *(continued)*

Year	CO_2	GHG Other*	O_3 (Trop)	O_3 (Strat)	Aerosol (Total)	LUC	H_2O (Strat)	BC Snow	Con-trails	Solar	Volcano
1999	1.489	0.952	0.386	−0.056	−0.917	−0.148	0.071	0.041	0.031	0.104	−0.050
2000	1.510	0.957	0.388	−0.056	−0.922	−0.149	0.071	0.040	0.033	0.127	−0.050
2001	1.532	0.961	0.389	−0.055	−0.920	−0.149	0.071	0.040	0.033	0.114	−0.050
2002	1.563	0.965	0.390	−0.055	−0.918	−0.149	0.071	0.040	0.033	0.108	−0.050
2003	1.594	0.969	0.391	−0.054	−0.916	−0.149	0.071	0.040	0.034	0.042	−0.075
2004	1.624	0.973	0.393	−0.053	−0.913	−0.149	0.071	0.040	0.038	0.012	−0.050
2005	1.654	0.976	0.394	−0.053	−0.911	−0.149	0.071	0.040	0.040	−0.011	−0.075
2006	1.684	0.981	0.395	−0.052	−0.909	−0.150	0.071	0.040	0.042	−0.016	−0.100
2007	1.711	0.986	0.396	−0.052	−0.907	−0.150	0.071	0.040	0.044	−0.017	−0.100
2008	1.736	0.992	0.398	−0.051	−0.904	−0.150	0.072	0.040	0.046	−0.025	−0.100
2009	1.762	0.999	0.399	−0.051	−0.902	−0.150	0.072	0.040	0.044	−0.027	−0.125
2010	1.789	1.005	0.400	−0.050	−0.900	−0.150	0.072	0.040	0.048	0.001	−0.100
2011	1.816	1.015	0.400	−0.050	−0.900	−0.150	0.073	0.040	0.050	0.030	−0.125

Notes:

See Figure 8.18, also Sections 8.1 and 11.3.6.1. To get the total ERF (effective radiative forcing) all components can be summed. Small negative values for CO_2 prior to 1800 are due to uncertainty in PI values. GHG other* includes only WMGHG. Aerosol is the sum of direct and indirect effects. LUC includes land use land cover change. Contrails combines aviation contrails (~20% of total) and contrail-induced cirrus. Values are annual average.

Table AII.1.3 | Historical global decadal mean global surface air temperature (°C) relative to 1961–1990 average

Year	HadCRUT4			GISS	NCDC
	Lower (5%)	Median (50%)	Upper (95%)	Median (50%)	Median (50%)
1850[d]	−0.404	−0.320	−0.243		
1860[d]	−0.413	−0.335	−0.263		
1870[d]	−0.326	−0.258	−0.195		
1880[d]	−0.363	−0.297	−0.237	−0.296	−0.291
1890[d]	−0.430	−0.359	−0.299	−0.361	−0.370
1900[d]	−0.473	−0.410	−0.353	−0.418	−0.434
1910[d]	−0.448	−0.387	−0.334	−0.435	−0.430
1920[d]	−0.297	−0.242	−0.193	−0.311	−0.311
1930[d]	−0.166	−0.116	−0.070	−0.172	−0.161
1940[d]	−0.047	−0.002	+0.042	−0.085	−0.063
1950[d]	−0.106	−0.061	−0.017	−0.134	−0.136
1960[d]	−0.093	−0.054	−0.014	−0.104	−0.086
1970[d]	−0.113	−0.077	−0.041	−0.058	−0.060
1980[d]	+0.052	+0.095	+0.135	+0.118	+0.109
1990[d]	+0.221	+0.270	+0.318	+0.275	+0.272
2000[d]	+0.400	+0.453	+0.508	+0.472	+0.450
1986–2005 minus 1850–1900		+0.61 ± 0.06		N/A	N/A
1986–2005 minus 1886–1905		+0.66 ± 0.06		+0.66	+0.66
1986–2005 minus 1961–1990		+0.30 ± 0.03		+0.31	+0.30
1986–2005 minus 1980–1999		+0.11 ± 0.02		+0.11	+0.11
1946–2012 minus 1880–1945		+0.38 ± 0.04		+0.40	+0.39

Notes:

Decadal average (1990[d] = 1990–1999) median global surface air temperatures from HadCRUT4, GISS and NCDC analyses. See Chapter 2, Sections 2.4.3 and 2.SM.4.3.3, Table 2.7, Figures 2.19, 2.20, 2.21 and 2.22, and also Figure 11.24a. Confidence intervals (5 to 95% for HadCRUT4 only) take into account measurement, sampling, bias and coverage uncertainties. Also shown are temperature increases between the CMIP5 reference period (1986–2005) and four earlier averaging periods, where 1850–1900 is the early instrumental temperature record. Uncertainties in these temperature differences are 5 to 95% confidence intervals.

AII.2: Anthropogenic Emissions

See discussion of Figure 8.2 and Section 11.3.5.

Table AII.2.1a | Anthropogenic CO_2 emissions from fossil fuels and other industrial sources (FF) (PgC yr^{-1})

Year	RCP2.6	RCP4.5	RCP6.0	RCP8.5	A2	B1	IS92a	RCP2.6[&]	RCP4.5[&]	RCP6.0[&]	RCP8.5[&]
2000[d]	6.82	6.82	6.82	6.82	6.90	6.90	7.10	6.92 ± 0.80	6.98 ± 0.81	6.76 ± 0.71	6.98 ± 0.81
2010[d]	8.61	8.54	8.39	8.90	8.46	8.50	8.68	8.38 ± 1.03	8.63 ± 1.07	7.66 ± 1.64	8.27 ± 1.68
2020[d]	9.00	9.79	8.99	11.38	11.01	10.00	10.26	8.46 ± 1.38	10.24±1.69	8.33 ± 1.82	10.30 ± 1.87
2030[d]	7.21	10.83	9.99	13.79	13.53	11.20	11.62	6.81 ± 1.49	10.93±1.83	9.20 ± 1.55	12.36 ± 2.25
2040[d]	4.79	11.25	11.47	16.69	15.01	12.20	12.66	4.61 ± 1.60	11.82±1.84	10.04 ± 1.42	15.09 ± 2.15
2050[d]	3.21	10.91	13.00	20.03	16.49	11.70	13.70	2.96 ± 1.80	11.37±1.84	11.14 ± 1.55	18.15 ± 2.56
2060[d]	1.55	9.42	14.73	23.32	18.49	10.20	14.68	1.77 ± 1.06	9.96 ± 2.17	13.22 ± 2.05	21.49 ± 2.42
2070[d]	0.26	7.17	16.33	25.75	20.49	8.60	15.66	0.75 ± 0.90	7.86 ± 1.94	14.57 ± 1.88	23.62 ± 2.43
2080[d]	−0.39	4.62	16.87	27.28	22.97	7.30	17.00	−0.09 ± 0.99	5.17 ± 1.77	15.51 ± 2.29	24.47 ± 2.70
2090[d]	−0.81	4.19	14.70	28.24	25.94	6.10	18.70	−0.30 ± 1.09	5.13 ± 1.53	14.24 ± 1.81	25.30 ± 2.86
2100[d]	−0.92	4.09	13.63	28.68	28.91	5.20	20.40	−0.63 ± 1.17	4.64 ± 1.34	12.78 ± 1.35	25.28 ± 2.73

Notes:

Decadal mean values (2010[d] = average of 2005–2014) are used for emissions because linear interpolation between decadal means conserves total emissions. Data are taken from RCP database (Meinshausen et al., 2011a; http://www.iiasa.ac.at/web-apps/tnt/RcpDb) and may be different from yearly snapshots; for 2100 the average (2095–2100) is used. SRES A2 and B1 and IS92a are taken from TAR Appendix II. RCPn.n[&] values are inferred from ESMs used in CMIP5. The model mean and standard deviation is shown. ESM fossil emissions are taken from 14 models as described in Jones et al. (2013) although not every model has performed every scenario. See Chapter 6, Sections 6.4.3, and 6.4.3.3, and Figure 6.25.

Table AII.2.1b | Anthropogenic CO_2 emissions from agriculture, forestry, land use (AFOLU) (PgC yr^{-1})

Year	RCP2.6	RCP4.5	RCP6.0	RCP8.5	SRES-A2	SRES-B1	IS92a
2000[d]	1.21	1.21	1.21	1.21	1.07	1.07	1.30
2010[d]	1.09	0.94	0.93	1.08	1.12	0.78	1.22
2020[d]	0.97	0.41	0.38	0.91	1.25	0.63	1.14
2030[d]	0.79	0.23	−0.43	0.74	1.19	−0.09	1.04
2040[d]	0.51	0.21	−0.67	0.65	1.06	−0.48	0.92
2050[d]	0.29	0.23	−0.48	0.58	0.93	−0.41	0.80
2060[d]	0.55	0.19	−0.27	0.50	0.67	−0.46	0.54
2070[d]	0.55	0.11	−0.04	0.42	0.40	−0.42	0.28
2080[d]	0.55	0.02	0.20	0.31	0.25	−0.60	0.12
2090[d]	0.59	0.03	0.24	0.20	0.21	−0.78	0.06
2100[d]	0.50	0.04	0.18	0.09	0.18	−0.97	−0.10

Notes:
See Table AII.2.1a.

Table AII.2.1c | Anthropogenic total CO_2 emissions (PgC yr^{-1})

Year	RCP2.6	RCP4.5	RCP6.0	RCP8.5
2000[d]	8.03	8.03	8.03	8.03
2010[d]	9.70	9.48	9.32	9.98
2020[d]	9.97	10.20	9.37	12.28
2030[d]	8.00	11.06	9.57	14.53
2040[d]	5.30	11.46	10.80	17.33
2050[d]	3.50	11.15	12.52	20.61
2060[d]	2.10	9.60	14.46	23.83
2070[d]	0.81	7.27	16.29	26.17
2080[d]	0.16	4.65	17.07	27.60
2090[d]	−0.23	4.22	14.94	28.44
2100[d]	−0.42	4.13	13.82	28.77

Notes:
See Table AII.2.1a.

Table AII.2.2 | Anthropogenic CH_4 emissions (Tg yr^{-1})

Year	RCP2.6	RCP4.5	RCP6.0	RCP8.5	A2	B1	IS92a	RCP2.6[&]	RCP4.5[&]	RCP6.0[&]	RCP8.5[&]
PI								202 ± 28	202 ± 28	202 ± 28	202 ± 28
2010[total]								554 ± 56	554 ± 56	554 ± 56	554 ± 56
2010[anthrop]								352 ± 45	352 ± 45	352 ± 45	352 ± 45
2010[d]	322	322	321	345	370	349	433	352 ± 45	352 ± 45	352 ± 45	352 ± 45
2020[d]	267	334	315	415	424	377	477	268 ± 34	366 ± 47	338 ± 43	424 ± 54
2030[d]	238	338	326	484	486	385	529	246 ± 31	370 ± 47	354 ± 45	490 ± 63
2040[d]	223	337	343	573	542	381	580	235 ± 30	368 ± 47	373 ± 47	585 ± 75
2050[d]	192	331	354	669	598	359	630	198 ± 25	361 ± 46	385 ± 49	685 ± 88
2060[d]	169	318	362	738	654	342	654	174 ± 22	346 ± 44	395 ± 50	754 ± 96
2070[d]	161	301	359	779	711	324	678	169 ± 22	328 ± 42	390 ± 50	790 ±101
2080[d]	155	283	336	820	770	293	704	162 ± 21	306 ± 39	369 ± 47	832 ±106
2090[d]	149	274	278	865	829	266	733	155 ± 20	298 ± 38	293 ± 37	882 ±113
2100[d]	143	267	250	885	889	236	762	148 ± 19	290 ± 37	267 ± 34	899 ±115

Year	MFR	CLE	MFR*	CLE*	Rog[L]	Rog[U]	AME[L]	AME[U]
2000[d]	366	366	303	303				
2010[d]			193	335			332	333
2020[d]			208	383	240	390	294	350
2030[d]	339	478	229	443	217	428	293	376
2040[d]							295	404
2050[d]					178	454	291	426
2060[d]							275	434
2070[d]							254	436
2080[d]							201	430
2090[d]							183	417
2100[d]					121	385	167	406

Notes:

For all anthropogenic emissions see Box 1.1 (Figure 4), Section 8.2.2, Figure 8.2, Sections 11.3.5.1.1 to 3, 11.3.5.2, 11.3.6.1. Ten-year average values (2010[d] = average of 2005–2014; but 2100[d] = average of 2095–2100) are given for RCP-based emissions, but single-year emissions are shown for other scenarios. RCPn.n = harmonized anthropogenic emissions as reported. SRES A2 and B1 and IS92a are from TAR Appendix II. AR5 RCPn.n[&] emissions have ± 1-σ (16 to 84% confidence) uncertainties and are based on the methodology of Prather et al. (2012) updated with CMIP5 results (Holmes et al., 2013; Voulgarakis et al., 2013). Projections of CH_4 lifetimes are harmonized based on PI (1750) and PD (2010) budgets that include uncertainties in lifetimes and abundances. All projected RCP abundances for CH_4 and N_2O (Tables AII.4.2 to AII.4.3) rescale each of the RCP emissions by a fixed factor equal to the ratio of RCP to AR5 anthropogenic emissions at year 2010 to ensure harmonization between total emissions, lifetimes and observed abundances. Natural emissions are kept constant but included as additional uncertainty. Independent emission estimates are shown as follows: MFR/CLE are the maximum feasible reduction and current legislation scenarios from Dentener et al. (2005; 2006), while MFR*/CLE* are the similarly labeled scenarios from Cofala et al. (2007). REF[L]/REF[U] are lower/upper bounds from the reference scenario of van Vuuren et al. (2008), while POL[L]/POL[U] are the lower/upper bounds from their policy scenario. AME[L]/AME[U] are lower/upper bounds from Calvin et al. (2012). Rog[L]/Rog[U] are lower/upper bounds from Rogelj et et. (2011).

Table AII.2.3 | Anthropogenic N_2O emissions (TgN yr^{-1})

Year	RCP2.6	RCP4.5	RCP6.0	RCP8.5	A2	B1	IS92a	RCP2.6[&]	RCP4.5[&]	RCP6.0[&]	RCP8.5[&]
PI								9.1 ± 1.0	9.1 ± 1.0	9.1 ± 1.0	9.1 ± 1.0
2010[total]								15.7 ± 1.1	15.7 ± 1.1	15.7 ± 1.1	15.7 ± 1.1
2010[anthrop]								6.5 ± 1.3	6.5 ± 1.3	6.5 ± 1.3	6.5 ± 1.3
2010[d]	7.7	7.8	8.0	8.25	8.1	7.5	6.2	6.5 ± 1.3	6.5 ± 1.3	6.5 ± 1.3	6.5 ± 1.3
2020[d]	7.4	8.2	8.1	9.5	9.6	8.1	7.1	6.1 ± 1.2	6.8 ± 1.3	6.3 ± 1.2	7.7 ± 1.5
2030[d]	7.3	8.5	8.8	10.7	10.7	8.2	7.7	6.1 ± 1.2	7.1 ± 1.4	7.0 ± 1.4	8.6 ± 1.7
2040[d]	7.1	8.7	9.7	11.9	11.3	8.3	8.0	6.0 ± 1.2	7.2 ± 1.4	7.8 ± 1.5	9.6 ± 1.9
2050[d]	6.3	8.6	10.5	12.7	12.0	8.3	8.3	5.2 ± 1.0	7.1 ± 1.4	8.4 ± 1.6	10.3 ± 2.0
2060[d]	5.8	8.5	11.3	13.4	12.9	7.7	8.3	4.8 ± 0.9	7.1 ± 1.4	9.1 ± 1.8	10.8 ± 2.1
2070[d]	5.7	8.4	12.0	13.9	13.9	7.4	8.4	4.8 ± 0.9	7.0 ± 1.3	9.6 ± 1.9	11.2 ± 2.2
2080[d]	5.6	8.2	12.3	14.5	14.8	7.0	8.5	4.7 ± 0.9	6.8 ± 1.3	9.9 ± 1.9	11.7 ± 2.3
2090[d]	5.5	8.1	12.4	15.2	15.7	6.4	8.6	4.6 ± 0.9	6.8 ± 1.3	9.9 ± 1.9	12.3 ± 2.4
2100[d]	5.3	8.1	12.2	15.7	16.5	5.7	8.7	4.4 ± 0.9	6.7 ± 1.3	9.8 ± 1.9	12.6 ± 2.4

Notes:

See Table AII.2.2.

Table AII.2.4 | Anthropogenic SF_6 emissions (Gg yr^{-1})

Year	RCP2.6	RCP4.5	RCP6.0	RCP8.5	A2	B1
2000[d]	5.70	5.70	5.70	5.70	6.20	6.20
2010[d]	6.14	5.68	7.43	6.93	7.60	5.60
2020[d]	2.87	3.02	9.19	8.12	9.70	5.70
2030[d]	1.96	2.89	9.58	9.83	11.60	7.20
2040[d]	1.53	3.32	9.68	11.14	13.70	8.90
2050[d]	0.76	3.77	9.78	12.07	16.00	10.40
2060[d]	0.51	4.28	9.92	13.69	18.80	10.90
2070[d]	0.42	4.87	10.05	13.72	19.80	9.50
2080[d]	0.32	5.53	10.00	14.79	20.70	7.10
2090[d]	0.19	5.99	9.86	15.96	23.40	6.50
2100[d]	0.07	6.25	9.37	16.79	25.20	6.50

Notes:

For this and all following emissions tables, see Table AII.2.2. RCPn.n = harmonized anthropogenic emissions as reported by RCPs (Lamarque et al., 2010; 2011; Meinshausen et al., 2011a). SRES A2 and B1 and IS92a from TAR Appendix II.

Table AII.2.5 | Anthropogenic CF_4 emissions (Gg yr^{-1})

Year	RCP2.6	RCP4.5	RCP6.0	RCP8.5	A2	B1
2000[d]	11.62	11.62	11.62	11.62	12.60	12.60
2010[d]	13.65	10.69	19.10	11.04	20.30	14.50
2020[d]	12.07	8.77	22.84	11.67	25.20	15.70
2030[d]	7.36	8.47	23.46	12.29	31.40	16.60
2040[d]	5.06	8.68	23.77	12.22	37.90	18.50
2050[d]	2.95	9.04	23.73	12.37	45.60	20.90
2060[d]	2.24	8.95	23.70	11.89	56.00	23.10
2070[d]	2.07	9.04	23.45	11.81	63.60	22.50
2080[d]	1.52	9.51	22.91	11.58	73.20	21.30
2090[d]	1.22	10.50	21.98	11.14	82.80	22.50
2100[d]	1.11	11.05	20.56	10.81	88.20	22.20

Table AII.2.6 | Anthropogenic C_2F_6 emissions (Gg yr^{-1})

Year	RCP2.6	RCP4.5	RCP6.0	RCP8.5	A2	B1
2000[d]	2.43	2.43	2.43	2.43	1.30	1.30
2010[d]	4.29	2.34	2.62	2.50	2.00	1.50
2020[d]	4.98	1.76	2.66	2.61	2.50	1.60
2030[d]	2.33	1.80	2.69	2.75	3.10	1.70
2040[d]	1.15	1.94	2.63	2.74	3.80	1.80
2050[d]	0.55	2.03	2.56	2.79	4.60	2.10
2060[d]	0.34	2.03	2.49	2.71	5.60	2.30
2070[d]	0.26	1.99	2.50	2.74	6.40	2.20
2080[d]	0.16	1.93	2.36	2.74	7.30	2.10
2090[d]	0.10	1.97	2.26	2.68	8.30	2.20
2100[d]	0.09	2.01	2.09	2.63	8.80	2.20

AII

Table AII.2.7 | Anthropogenic C_6F_{14} emissions (Gg yr^{-1})

Year	RCP2.6	RCP4.5	RCP6.0	RCP8.5
2000[d]	0.213	0.213	0.213	0.213
2010[d]	0.430	0.430	0.429	0.430
2020[d]	0.220	0.220	0.220	0.220
2030[d]	0.123	0.123	0.123	0.123
2040[d]	0.112	0.112	0.112	0.112
2050[d]	0.109	0.109	0.109	0.109
2060[d]	0.108	0.108	0.108	0.108
2070[d]	0.106	0.106	0.106	0.106
2080[d]	0.103	0.103	0.103	0.103
2090[d]	0.097	0.097	0.097	0.097
2100[d]	0.090	0.088	0.088	0.090

Table AII.2.8 | Anthropogenic HFC-23 emissions (Gg yr^{-1})

Year	RCP2.6	RCP4.5	RCP6.0	RCP8.5	A2	B1
2000[d]	10.4	10.4	10.4	10.4	13.0	13.0
2010[d]	9.1	9.1	9.1	9.1	15.0	15.0
2020[d]	2.4	2.4	2.4	2.4	5.0	5.0
2030[d]	0.7	0.7	0.7	0.7	2.0	2.0
2040[d]	0.4	0.4	0.4	0.4	2.0	2.0
2050[d]	0.3	0.3	0.3	0.3	1.0	1.0
2060[d]	0.1	0.1	0.1	0.1	1.0	1.0
2070[d]	0.1	0.1	0.1	0.1	1.0	1.0
2080[d]	0.0	0.0	0.0	0.0	1.0	1.0
2090[d]	0.0	0.0	0.0	0.0	1.0	1.0
2100[d]	0.0	0.0	0.0	0.0	1.0	1.0

Table AII.2.9 | Anthropogenic HFC-32 emissions (Gg yr[−1])

Year	RCP2.6	RCP4.5	RCP6.0	RCP8.5	A2	B1
2000[d]	3.5	3.5	3.5	3.5	0.0	0.0
2010[d]	20.1	20.1	20.1	20.1	4.0	3.0
2020[d]	55.4	55.4	55.4	55.4	6.0	6.0
2030[d]	71.2	71.2	71.2	71.2	9.0	8.0
2040[d]	78.8	78.8	78.8	78.8	11.0	10.0
2050[d]	76.5	76.5	76.5	76.5	14.0	14.0
2060[d]	83.6	83.6	83.6	83.6	17.0	14.0
2070[d]	92.7	92.7	92.7	92.7	20.0	14.0
2080[d]	95.4	95.4	95.4	95.4	24.0	14.0
2090[d]	91.0	91.0	91.0	91.0	29.0	14.0
2100[d]	82.7	82.7	82.7	82.7	33.0	13.0

Table AII.2.10 | Anthropogenic HFC-125 emissions (Gg yr[−1])

Year	RCP2.6	RCP4.5	RCP6.0	RCP8.5	A2	B1	IS92a
2000[d]	8	8	8	8	0	0	0
2010[d]	29	18	10	32	11	11	1
2020[d]	82	29	9	63	21	21	9
2030[d]	108	32	9	79	29	29	46
2040[d]	122	31	10	99	35	36	111
2050[d]	122	30	10	115	46	48	175
2060[d]	138	27	11	128	56	48	185
2070[d]	157	24	11	139	66	48	194
2080[d]	165	24	12	144	79	48	199
2090[d]	161	23	12	147	94	46	199
2100[d]	150	23	12	148	106	44	199

Table AII.2.11 | Anthropogenic HFC-134a emissions (Gg yr[−1])

Year	RCP2.6	RCP4.5	RCP6.0	RCP8.5	A2	B1	IS92a
2000[d]	72	72	72	72	80	80	148
2010[d]	146	140	139	153	166	163	290
2020[d]	173	184	153	255	252	249	396
2030[d]	193	208	159	331	330	326	557
2040[d]	209	229	163	402	405	414	738
2050[d]	203	248	167	461	506	547	918
2060[d]	225	246	172	506	633	550	969
2070[d]	252	260	175	553	758	544	1020
2080[d]	263	299	177	602	915	533	1047
2090[d]	256	351	175	651	1107	513	1051
2100[d]	239	400	171	696	1260	486	1055

Table AII.2.12 | Anthropogenic HFC-143a emissions (Gg yr^{-1})

Year	RCP2.6	RCP4.5	RCP6.0	RCP8.5	A2	B1
2000[d]	7.5	7.5	7.5	7.5	0.0	0.0
2010[d]	23.1	14.0	7.0	23.2	9.0	8.0
2020[d]	59.1	17.4	5.4	34.1	16.0	15.0
2030[d]	74.7	20.3	6.0	38.5	22.0	21.0
2040[d]	81.8	23.1	6.6	45.1	27.0	26.0
2050[d]	79.0	25.6	7.1	49.8	35.0	35.0
2060[d]	86.1	25.9	7.7	52.3	43.0	35.0
2070[d]	94.2	28.2	8.3	54.1	51.0	35.0
2080[d]	95.1	33.5	8.7	52.7	61.0	35.0
2090[d]	88.7	39.6	9.0	50.2	73.0	34.0
2100[d]	79.2	45.1	9.1	47.3	82.0	32.0

Table AII.2.13 | Anthropogenic HFC-227ea emissions (Gg yr^{-1})

Year	RCP2.6	RCP4.5	RCP6.0	RCP8.5	A2	B1
2000[d]	1.7	1.7	1.7	1.7	0.0	0.0
2010[d]	7.0	5.3	6.9	8.5	12.0	13.0
2020[d]	2.6	1.4	2.5	2.7	17.0	18.0
2030[d]	0.9	0.3	0.8	0.7	21.0	24.0
2040[d]	0.8	0.2	0.7	0.7	26.0	30.0
2050[d]	0.4	0.1	0.3	0.4	32.0	39.0
2060[d]	0.2	0.0	0.1	0.2	40.0	40.0
2070[d]	0.1	0.0	0.1	0.1	48.0	39.0
2080[d]	0.1	0.0	0.1	0.1	58.0	38.0
2090[d]	0.1	0.0	0.0	0.1	70.0	36.0
2100[d]	0.1	0.0	0.0	0.1	80.0	34.0

Table AII.2.14 | Anthropogenic HFC-245fa emissions (Gg yr^{-1})

Year	RCP2.6	RCP4.5	RCP6.0	RCP8.5	A2	B1
2000[d]	11	11	11	11	0	0
2010[d]	42	46	53	74	59	60
2020[d]	32	86	65	143	79	80
2030[d]	7	95	67	186	98	102
2040[d]	0	97	68	181	121	131
2050[d]	0	95	69	163	149	173
2060[d]	0	87	70	150	190	173
2070[d]	0	82	71	138	228	170
2080[d]	0	80	70	129	276	166
2090[d]	0	81	68	123	334	159
2100[d]	0	83	65	130	388	150

Table AII.2.15 | Anthropogenic HFC-43-10mee emissions (Gg yr^{-1})

Year	RCP2.6	RCP4.5	RCP6.0	RCP8.5	A2	B1
2000[d]	0.6	0.6	0.6	0.6	0.0	0.0
2010[d]	5.6	5.6	5.6	5.6	7.0	6.0
2020[d]	7.2	7.2	7.2	7.2	8.0	7.0
2030[d]	8.1	8.1	8.1	8.1	8.0	8.0
2040[d]	9.4	9.4	9.4	9.1	9.0	9.0
2050[d]	10.8	10.8	10.8	10.4	11.0	11.0
2060[d]	11.1	11.1	11.1	12.1	12.0	11.0
2070[d]	11.0	11.0	11.0	13.9	14.0	11.0
2080[d]	11.0	11.0	10.9	16.2	16.0	11.0
2090[d]	10.7	10.7	10.7	18.9	19.0	11.0
2100[d]	10.5	10.5	10.5	21.4	22.0	10.0

Table AII.2.16 | Anthropogenic CO emissions (Tg yr^{-1})

Year	RCP2.6	RCP4.5	RCP6.0	RCP8.5	A2	B1	IS92a
2000[d]	1071	1071	1071	1071	877	877	1048
2010[d]	1035	1041	1045	1054	977	789	1096
2020[d]	984	997	1028	1058	1075	751	1145
2030[d]	930	986	1030	1019	1259	603	1207
2040[d]	879	948	1046	960	1344	531	1282
2050[d]	825	875	1033	907	1428	471	1358
2060[d]	779	782	996	846	1545	459	1431
2070[d]	718	678	939	799	1662	456	1504
2080[d]	668	571	879	759	1842	426	1576
2090[d]	638	520	835	721	2084	399	1649
2100[d]	612	483	798	694	2326	363	1722

Year	MFR	CLE	REF[L]	REF[U]	POL[L]	POL[U]
2000[d]	977	977	708	1197	706	1197
2010[d]			771	1408	769	1408
2020[d]			755	1629	705	1611
2030[d]	729	904	707	1865	592	1803
2040[d]			695	2165	620	2002
2050[d]			591	2487	482	2218
2060[d]			504	2787	363	2409
2070[d]			450	3052	328	2558
2080[d]			438	3279	268	2635
2090[d]			410	3510	259	2714
2100[d]			363	3735	253	2796

Table AII.2.17 | Anthropogenic NMVOC emissions (Tg yr^{-1})

Year	RCP2.6	RCP4.5	RCP6.0	RCP8.5	A2	B1	IS92a	CLE	MFR
2000[d]	213	213	213	213	141	141	126	147	147
2010[d]	216	209	215	217	155	141	142		
2020[d]	213	197	214	224	179	140	158		
2030[d]	202	201	217	225	202	131	173	146	103
2040[d]	192	201	222	218	214	123	188		
2050[d]	179	191	220	209	225	116	202		
2060[d]	167	180	214	202	238	111	218		
2070[d]	152	167	204	194	251	103	234		
2080[d]	140	152	193	189	275	99	251		
2090[d]	132	145	182	182	309	96	267		
2100[d]	126	141	174	177	342	87	283		

Table AII.2.18 | Anthropogenic NO_X emissions (TgN yr^{-1})

Year	RCP2.6	RCP4.5	RCP6.0	RCP8.5	CLE	MFR
2000[d]	38.5	38.5	38.5	38.5	53.4	53.4
2010[d]	43.5	42.4	43.1	43.5		
2020[d]	47.5	43.5	43.3	48.1		
2030[d]	50.8	45.2	46.2	52.1	69.8	69.8
2040[d]	53.2	46.3	49.8	55.6		
2050[d]	55.5	46.4	53.0	58.4		
2060[d]	58.4	46.0	56.5	60.6		
2070[d]	61.2	45.2	59.5	62.4		
2080[d]	63.3	44.3	60.9	63.8		
2090[d]	65.2	43.9	62.1	65.3		
2100[d]	67.0	43.6	61.8	66.9		

Year	MFR	CLE	REF[L]	REF[U]	POL[L]	POL[U]
2000[d]	38.0	38.0	29.1	41.6	29.1	41.6
2010[d]			26.0	50.2	23.9	50.1
2020[d]			26.3	60.4	21.6	59.2
2030[d]	23.1	42.9	24.4	71.8	16.5	67.4
2040[d]			21.5	86.3	14.1	75.3
2050[d]			17.0	101.7	11.6	83.3
2060[d]			13.2	115.7	11.4	89.8
2070[d]			12.0	127.5	10.5	94.6
2080[d]			11.5	137.2	9.6	97.2
2090[d]			12.0	146.2	8.8	100.1
2100[d]			13.0	155.0	8.0	104.0

Notes:
Odd nitrogen (NO_x) emissions occur as NO or NO_2, measured here as Tg of N.

Table AII.2.19 | Anthropogenic NH_3 emissions (TgN yr^{-1})

Year	RCP2.6	RCP4.5	RCP6.0	RCP8.5	CLE	MFR
2000[d]	38.5	38.5	38.5	38.5	53.4	53.4
2010[d]	43.5	42.4	43.1	43.5		
2020[d]	47.5	43.5	43.3	48.1		
2030[d]	50.8	45.2	46.2	52.1	69.8	69.8
2040[d]	53.2	46.3	49.8	55.6		
2050[d]	55.5	46.4	53.0	58.4		
2060[d]	58.4	46.0	56.5	60.6		
2070[d]	61.2	45.2	59.5	62.4		
2080[d]	63.3	44.3	60.9	63.8		
2090[d]	65.2	43.9	62.1	65.3		
2100[d]	67.0	43.6	61.8	66.9		

AII

Table AII.2.20 | Anthropogenic SO_X emissions (TgS yr^{-1})

Year	RCP2.6	RCP4.5	RCP6.0	RCP8.5	A2	B1	IS92a
2000[d]	55.9	55.9	55.9	55.9	69.0	69.0	79.0
2010[d]	54.9	54.8	55.8	51.9	74.7	73.9	95.0
2020[d]	44.5	50.3	49.9	47.6	99.5	74.6	111.0
2030[d]	30.8	43.2	42.7	42.3	112.5	78.2	125.8
2040[d]	20.9	35.0	41.9	33.5	109.0	78.5	139.4
2050[d]	16.0	26.5	37.8	26.8	105.4	68.9	153.0
2060[d]	13.8	21.0	34.0	23.0	89.6	55.8	151.8
2070[d]	11.9	16.7	23.5	20.3	73.7	44.3	150.6
2080[d]	9.9	13.2	15.9	18.3	64.7	36.1	149.4
2090[d]	8.0	12.0	12.7	14.9	62.5	29.8	148.2
2100[d]	6.7	11.4	10.8	13.1	60.3	24.9	147.0

Year	MFR	CLE	REF[L]	REF[U]	POL[L]	POL[U]
2000[d]	55.6	55.6	50.6	76.4	50.6	76.4
2010[d]			53.1	81.8	52.7	78.7
2020[d]			56.9	84.8	47.7	77.8
2030[d]	17.9	58.8	60.1	86.7	29.8	76.3
2040[d]			52.5	82.9	19.0	72.0
2050[d]			44.2	72.3	12.4	61.7
2060[d]			32.8	73.9	9.5	52.9
2070[d]			30.5	77.7	7.8	49.8
2080[d]			29.6	81.1	6.2	50.5
2090[d]			22.8	84.5	5.1	52.5
2100[d]			18.0	88.0	4.0	54.0

Notes:

Anthropogenic sulphur emissions as SO_2, measured here as Tg of S.

Table AII.2.21 | Anthropogenic OC aerosols emissions (Tg yr^{-1})

Year	RCP2.6	RCP4.5	RCP6.0	RCP8.5	A2	B1	IS92a	MFR*	CLE*
2000[d]	35.6	35.6	35.6	35.6	81.4	81.4	81.4	35.0	35.0
2010[d]	36.6	34.6	36.2	35.6	89.3	74.5	85.2	29.2	34.6
2020[d]	36.6	30.8	36.1	34.5	97.0	71.5	89.0	28.6	32.6
2030[d]	35.3	29.2	36.0	33.2	111.4	59.9	93.9	27.9	30.9
2040[d]	32.3	28.0	36.4	31.6	118.1	54.2	99.8		
2050[d]	30.3	26.8	36.5	30.1	124.7	49.5	105.8		
2060[d]	29.6	25.0	35.7	28.5	133.9	48.6	111.5		
2070[d]	28.2	22.8	34.4	27.4	143.1	48.3	117.2		
2080[d]	27.0	20.7	33.4	26.4	157.2	46.0	122.9		
2090[d]	26.4	19.9	32.7	25.1	176.2	43.8	128.6		
2100[d]	25.5	19.5	32.2	24.1	195.2	41.0	134.4		

Notes:
For both MFR* and CLE* 23 Tg is added to Cofala et al. (2007) values to include biomass burning.

Table AII.2.22 | Anthropogenic BC aerosols emissions (Tg yr^{-1})

Year	RCP2.6	RCP4.5	RCP6.0	RCP8.5	A2	B1	IS92a	MFR*	CLE*
2000[d]	7.88	7.88	7.88	7.88	12.40	12.40	12.40	7.91	7.91
2010[d]	8.49	8.13	8.13	8.06	13.60	11.30	13.00	6.31	8.01
2020[d]	8.27	7.84	7.77	7.66	14.80	10.90	13.60	5.81	7.41
2030[d]	7.03	7.36	7.53	7.04	17.00	9.10	14.30	5.41	7.01
2040[d]	5.80	6.81	7.39	6.22	18.00	8.30	15.20		
2050[d]	5.00	6.21	7.07	5.67	19.00	7.50	16.10		
2060[d]	4.46	5.56	6.48	5.22	20.40	7.40	17.00		
2070[d]	3.99	4.88	5.75	4.88	21.80	7.40	17.90		
2080[d]	3.70	4.23	5.15	4.66	24.00	7.00	18.70		
2090[d]	3.55	4.01	4.70	4.43	26.80	6.70	19.60		
2100[d]	3.39	3.88	4.41	4.27	29.70	6.20	20.50		

Notes:
For both MFR* and CLE* 2.6 Tg added to Cofala et al. (2007) values to include biomass burning.

Table AII.2.23 | Anthropogenic nitrogen fixation (Tg-N yr^{-1})

Year	Historical	SRES A1 + Biofuel	SRES A2	SRES B1	SRES B2	FAO2000 Baseline[a]	FAO2000 Improved[a]	Tilman 2001[a]	Tubiello 2007[a]
1910	0.0								
1920	0.2								
1925	0.6								
1930	0.9								
1935	1.3								
1940	2.2								
1950	3.7								
1955	6.8								
1960	9.5								
1965	18.7								
1970	31.6								
1971	33.3								
1972	36.2								
1973	39.1								
1974	38.6								
1975	43.7								

Table AII.2.23 *(continued)*

Year	Historical	SRES A1 + Biofuel	SRES A2	SRES B1	SRES B2	FAO2000 Baseline[a]	FAO2000 Improved[a]	Tilman 2001[a]	Tubiello 2007[a]
1975	43.7								
1976	46.4								
1977	49.9								
1978	53.8								
1979	57.4								
1980	60.6								
1981	60.3								
1982	61.3								
1983	67.1								
1984	70.9								
1985	70.2								
1986	72.5								
1987	75.8								
1988	79.5								
1989	78.9								
1990	77.1								
1991	75.5								
1992	73.7								
1993	72.3								
1994	72.4								
1995	78.5								
1996	82.6					77.8	77.8		
1997	81.4								
1998	82.8								
1999	84.9								
2000	82.1							87.0	
2001	82.9								
2002	85.2								
2003	90.2								
2004	91.7								
2005	94.2								
2007	98.4								
2010		104.1	101.9	101.7	96.5				
2015		–	–	–	–	106.8	88.0		
2020		122.6	110.7	111.2	100.9			135.0	
2030		141.1	117.6	118.4	103.3	124.5	96.2		
2040		153.3	130.7	122.2	103.5				
2050		165.5	131.1	123.2	101.9			236.0	
2060		171.3	134.0	121.4	99.2				
2070		177.0	132.1	117.5	95.6				
2080		180.1	138.1	111.6	91.5				205
2090		186.0	146.5	108.8	91.3				
2100		192.5	149.8	104.1	91.0				

Notes:
(a) See Chapter 6, Figure 6.30 and Erisman et al. (2008) for details and sources.

AII.3: Natural Emissions

Table AII.3.1a | Net land (natural and land use) CO_2 emissions (PgC yr^{-1})

Year	RCP2.6[&]	RCP4.5[&]	RCP6.0[&]	RCP8.5[&]
2000[d]	−1.02 ± 0.87	−1.14 ± 0.87	−0.92 ± 0.93	−1.14 ± 0.87
2010[d]	−1.49 ± 1.02	−1.85 ± 0.96	−1.03 ± 1.65	−1.30 ± 1.64
2020[d]	−1.24 ± 1.35	−2.83 ± 1.47	−1.79 ± 1.95	−1.43 ± 1.82
2030[d]	−1.28 ± 1.53	−2.84 ± 1.59	−2.37 ± 1.54	−1.76 ± 2.22
2040[d]	−1.21 ± 1.33	−3.25 ± 1.58	−2.27 ± 1.46	−2.15 ± 2.13
2050[d]	−1.00 ± 1.53	−3.07 ± 1.54	−1.98 ± 1.57	−2.35 ± 2.45
2060[d]	−0.76 ± 0.83	−2.80 ± 1.83	−2.46 ± 2.01	−2.71 ± 2.38
2070[d]	−0.68 ± 0.84	−2.59 ± 1.73	−2.40 ± 2.06	−2.57 ± 2.42
2080[d]	−0.15 ± 0.81	−2.04 ± 1.48	−2.22 ± 2.12	−1.96 ± 2.64
2090[d]	−0.03 ± 0.99	−2.12 ± 1.38	−2.77 ± 1.96	−1.63 ± 2.70
2100[d]	0.36 ± 0.95	−1.54 ± 1.25	−2.13 ± 1.32	−1.27 ± 2.90

Notes:

Ten-year average values are shown (2010[d] = average of 2005–2014). CO_2 emissions are inferred from ESMs used in CMIP5 (Jones et al., 2013). See notes Table AII.2.1a and Chapter 6, Sections 6.4.3 and 6.4.3.3 and Figure 6.24.

Table AII.3.1b | Net ocean CO_2 emissions (PgC yr^{-1})

Year	RCP2.6[&]	RCP4.5[&]	RCP6.0[&]	RCP8.5[&]
2000[d]	−2.09 ± 0.19	−2.14 ± 0.32	−2.10 ± 0.17	−2.14 ± 0.32
2010[d]	−2.44 ± 0.22	−2.50 ± 0.42	−2.44 ± 0.20	−2.53 ± 0.43
2020[d]	−2.70 ± 0.26	−2.75 ± 0.46	−2.59 ± 0.22	−3.02 ± 0.51
2030[d]	−2.59 ± 0.30	−2.98 ± 0.52	−2.69 ± 0.22	−3.47 ± 0.54
2040[d]	−2.22 ± 0.32	−3.16 ± 0.56	−2.88 ± 0.27	−3.96 ± 0.67
2050[d]	−1.83 ± 0.33	−3.22 ± 0.60	−3.16 ± 0.31	−4.47 ± 0.76
2060[d]	−1.52 ± 0.30	−3.12 ± 0.63	−3.52 ± 0.36	−4.92 ± 0.84
2070[d]	−1.23 ± 0.23	−2.82 ± 0.61	−3.79 ± 0.41	−5.24 ± 0.97
2080[d]	−0.99 ± 0.27	−2.46 ± 0.59	−4.02 ± 0.44	−5.40 ± 1.14
2090[d]	−0.85 ± 0.26	−2.22 ± 0.53	−3.96 ± 0.43	−5.45 ± 1.18
2100[d]	−0.77 ± 0.26	−2.14 ± 0.47	−3.84 ± 0.42	−5.44 ± 1.22

Notes:

See Table AII.3.1.a.

AII.4: Abundances of the Well-Mixed Greenhouse Gases

Table AII.4.1 | CO_2 abundance (ppm)

Year	Observed	RCP2.6	RCP4.5	RCP6.0	RCP8.5	A2	B1	IS92a	Min	RCP8.5[&]	Max
PI	**278 ± 2**	**278**	**278**	**278**	**278**	**278**	**278**	**278**			
2011[obs]	**390.5 ± 0.3**										
2000		368.9	368.9	368.9	368.9	368	368	368			
2005		378.8	378.8	378.8	378.8					378.8	
2010		389.3	389.1	389.1	389.3	388	387	388	366	394	413
2020		412.1	411.1	409.4	415.8	416	411	414	386	425	449
2030		430.8	435.0	428.9	448.8	448	434	442	412	461	496
2040		440.2	460.8	450.7	489.4	486	460	472	443	504	555
2050		442.7	486.5	477.7	540.5	527	485	504	482	559	627
2060		441.7	508.9	510.6	603.5	574	506	538	530	625	713
2070		437.5	524.3	549.8	677.1	628	522	575	588	703	810
2080		431.6	531.1	594.3	758.2	690	534	615	651	790	914
2090		426.0	533.7	635.6	844.8	762	542	662	722	885	1026
2100		420.9	538.4	669.7	935.9	846	544	713	794	985 ± 97	1142

Notes:

For observations (2011[obs]) see Chapter 2; and for projections see Box 1.1 (Figure 2), Sections 6.4.3.1, 11.3.1.1, 11.3.5.1.1. RCPn.n refers to values taken directly from the published RCP scenarios using the MAGICC model (Meinshausen et al., 2011a; 2011b). These are harmonized to match observations up to 2005 (378.8 ppm) and project future abundances thereafter. RCP8.5[&] shows the average and assessed 90% confidence interval for year 2100, plus the min-max full range derived from the CMIP5 archive for all years (P. Friedlingstein, based on Friedlingstein et al., 2006). 11 ESMs participated (BCC-CSM-1, CanESM2, CESM1-BGC, GFDL-ESM2G, HadGem-2ES, INMCM4, IPSLCM5-LR, MIROC-ESM, MPI-ESM-LR, MRI-ESM1, and Nor-ESM1-ME), running the RCP8.5 anthropogenic emission scenario forced by the RCP8.5 climate change scenario (see Figure 12.36). All abundances are mid-year. Projected values for SRES A2 and B1 and IS92 are the average of reference models taken from the TAR Appendix II.

Table AII.4.2 | CH_4 abundance (ppb)

Year	RCP2.6	RCP4.5	RCP6.0	RCP8.5	A2	B1	IS92a	RCP2.6[&]	RCP4.5[&]	RCP6.0[&]	RCP8.5[&]
PI	**720**	**720**	**720**	**720**				**722 ± 25**	**722 ± 25**	**722 ± 25**	**722 ± 25**
2011[obs]								**1803 ± 4**	**1803 ± 4**	**1803 ± 4**	**1803 ± 4**
2000	1751	1751	1751	1751	1760	1760	1760				
2010	1773	1767	1769	1779	1861	1827	1855	1795 ± 18	1795 ± 18	1795 ± 18	1795 ± 18
2020	1731	1801	1786	1924	1997	1891	1979	1716 ± 23	1847 ± 21	1811 ± 22	1915 ± 25
2030	1600	1830	1796	2132	2163	1927	2129	1562 ± 38	1886 ± 28	1827 ± 28	2121 ± 44
2040	1527	1842	1841	2399	2357	1919	2306	1463 ± 50	1903 ± 37	1880 ± 36	2412 ± 74
2050	1452	1833	1895	2740	2562	1881	2497	1353 ± 60	1899 ± 47	1941 ± 48	2784 ± 116
2060	1365	1801	1939	3076	2779	1836	2663	1230 ± 71	1872 ± 59	1994 ± 61	3152 ± 163
2070	1311	1745	1962	3322	3011	1797	2791	1153 ± 78	1824 ± 72	2035 ± 77	3428 ± 208
2080	1285	1672	1940	3490	3252	1741	2905	1137 ± 88	1756 ± 87	2033 ± 94	3624 ± 250
2090	1268	1614	1819	3639	3493	1663	3019	1135 ± 98	1690 ± 100	1908 ± 111	3805 ± 293
2100	1254	1576	1649	3751	3731	1574	3136	1127 ± 106	1633 ± 110	1734 ± 124	3938 ± 334

Notes:

RCPn.n refers to values taken directly from the published RCP scenarios using the MAGICC model (Meinshausen et al., 2011b) and initialized in year 2005 at 1754 ppb. Values for SRES A2 and B1 and IS92 are from the TAR Appendix II. RCPn.n[&] values are best estimates with uncertainties (68% confidence intervals) from Chapter 11 (Section 11.3.5) based on Holmes et al. (2013) and using RCP[&] emissions and uncertainties tabulated above. For RCP[&] the PI, year 2011 and year 2010 values are based on observations. RCP models used slightly different PI abundances than recommended here (Table AII.1.1, Chapter 2).

Table AII.4.3 | N_2O abundance (ppb)

Year	RCP2.6	RCP4.5	RCP6.0	RCP8.5	A2	B1	IS92a	RCP2.6&	RCP4.5&	RCP6.0&	RCP8.5&
PI	272	272	272	272				270 ± 7	270 ± 7	270 ± 7	270 ± 7
2011obs								324 ± 1	324 ± 1	324 ± 1	324 ± 1
2000	316	316	316	316	316	316	316				
2010	323	323	323	323	325	324	324	323 ± 3	323 ± 3	323 ± 3	323 ± 3
2020	329	330	330	332	335	333	333	330 ± 4	331 ± 4	331 ± 4	332 ± 4
2030	334	337	337	342	347	341	343	336 ± 5	339 ± 5	338 ± 5	342 ± 6
2040	339	344	345	354	360	349	353	342 ± 6	346 ± 7	346 ± 7	353 ± 8
2050	342	351	355	367	373	357	363	346 ± 8	353 ± 9	355 ± 9	365 ± 11
2060	343	356	365	381	387	363	372	349 ± 9	360 ± 10	364 ± 11	377 ± 13
2070	344	361	376	394	401	368	381	351 ± 10	365 ± 12	374 ± 13	389 ± 16
2080	344	366	386	408	416	371	389	352 ± 11	370 ± 13	384 ± 15	401 ± 18
2090	344	369	397	421	432	374	396	353 ± 11	374 ± 14	393 ± 17	413 ± 21
2100	344	372	406	435	447	375	403	354 ± 12	378 ± 16	401 ± 19	425 ± 24

Notes:
See notes Table AII.4.2.

AII

Table AII.4.4 | SF_6 abundance (ppt)

Year	RCP2.6	RCP4.5	RCP6.0	RCP8.5	A2	B1	Obs
2011obs							7.3 ± 0.1
2010	7.0	6.9	7.0	7.0	7	7	
2020	8.9	8.7	10.3	9.9	11	9	
2030	9.7	9.7	14.1	13.4	15	12	
2040	10.4	10.9	17.9	17.6	20	15	
2050	10.8	12.3	21.7	22.1	26	19	
2060	11.0	13.8	25.6	27.2	32	23	
2070	11.2	15.6	29.5	32.6	40	27	
2080	11.3	17.6	33.4	38.1	48	30	
2090	11.4	19.9	37.3	44.1	56	33	
2100	11.4	22.3	41.0	50.5	65	35	

Notes:
Projected SF_6 and PFC abundances (Tables AII.4.4 to AII.4.7) taken directly from RCPs (Meinshausen et al., 2011a). Observed values shown for year 2011.

Table AII.4.5 | CF_4 abundance (ppt)

Year	RCP2.6	RCP4.5	RCP6.0	RCP8.5	A2	B1	Obs
2011 obs							79.0
2010	84	83	85	83	92	91	
2020	93	90	99	91	107	101	
2030	99	95	115	99	125	111	
2040	103	101	130	107	148	122	
2050	106	107	146	115	175	135	
2060	108	113	162	123	208	150	
2070	109	119	177	131	246	164	
2080	110	125	193	138	291	179	
2090	111	131	207	146	341	193	
2100	112	138	222	153	397	208	

AII

Table AII.4.6 | C_2F_6 abundance (ppt)

Year	RCP2.6	RCP4.5	RCP6.0	RCP8.5	A2	B1	Obs
2011 obs							4.2
2010	4.1	3.9	3.9	3.9	4	4	
2020	6.2	4.8	5.0	5.0	5	4	
2030	7.9	5.5	6.2	6.1	6	5	
2040	8.6	6.3	7.3	7.2	7	6	
2050	8.9	7.1	8.4	8.4	9	7	
2060	9.1	7.9	9.4	9.6	11	8	
2070	9.2	8.8	10.5	10.7	14	8	
2080	9.3	9.6	11.5	11.8	17	9	
2090	9.3	10.4	12.5	13.0	20	10	
2100	9.3	11.3	13.4	14.1	23	11	

Table AII.4.7 | C_6F_{14} abundance (ppt)

Year	RCP2.6	RCP4.5	RCP6.0	RCP8.5
2010	0.07	0.07	0.07	0.07
2020	0.13	0.13	0.13	0.13
2030	0.16	0.16	0.16	0.16
2040	0.18	0.18	0.18	0.18
2050	0.20	0.20	0.20	0.20
2060	0.21	0.21	0.21	0.21
2070	0.23	0.23	0.23	0.23
2080	0.25	0.25	0.25	0.25
2090	0.27	0.27	0.27	0.27
2100	0.28	0.28	0.28	0.28

Table AII.4.8 | HFC-23 abundance (ppt)

Year	RCP2.6	RCP4.5	RCP6.0	RCP8.5	A2	B1	RCP2.6&	RCP4.5&	RCP6.0&	RCP8.5&
2011obs							24.0	24.0	24.0	24.0
2010	22.9	22.9	22.9	22.9	26	26	23.2 ± 1	23.2 ± 1	23.2 ± 1	23.2 ± 1
2020	27.2	27.2	27.2	27.2	33	33	26.6 ± 1	26.6 ± 1	26.6 ± 1	26.6 ± 1
2030	27.0	27.0	27.1	27.1	35	35	26.3 ± 1	26.3 ± 1	26.3 ± 1	26.3 ± 1
2040	26.5	26.5	26.6	26.6	35	35	25.7 ± 1	25.8 ± 1	25.8 ± 1	25.8 ± 1
2050	25.8	25.9	25.9	26.0	35	35	24.9 ± 1	25.0 ± 1	25.1 ± 1	25.1 ± 1
2060	25.0	25.1	25.1	25.3	35	34	24.0 ± 1	24.2 ± 1	24.3 ± 1	24.4 ± 1
2070	24.1	24.2	24.4	24.6	34	34	23.0 ± 1	23.4 ± 1	23.4 ± 1	23.6 ± 1
2080	23.3	23.3	23.5	23.8	34	33	22.1 ± 1	22.5 ± 1	22.6 ± 1	22.8 ± 1
2090	22.4	22.5	22.7	23.0	34	33	21.2 ± 1	21.6 ± 1	21.8 ± 1	22.1 ± 1
2100	21.6	21.6	21.9	22.3	33	32	20.3 ± 1	20.8 ± 1	21.0 ± 1	21.3 ± 1

Notes:

RCPn.n HFC abundances (Tables AII.4.8 to AII.4.15) are as reported (Meinshausen et al., 2011a). SRES A2 and B1 and IS92a (where available) are taken from TAR Appendix II. Observed values are shown for 2011 (see Chapter 2, and Table AII.1.1). The AR5 RCPn.n& abundances are calculated starting with observed abundances (adopted for 2010) and future tropospheric OH changes using the methodology of Prather et al. (2012), updated for uncertainty in lifetime and scenario changes in OH using Holmes et al. (2013) and ACCMIP results (Stevenson et al., 2013; Voulgarakis et al., 2013). Projected RCP& abundances are best estimates with 68% confidence range as uncertainties. See also notes Tables AII.4.2 and AII.5.9.

Table AII.4.9 | HFC-32 abundance (ppt)

Year	RCP2.6	RCP4.5	RCP6.0	RCP8.5	A2	B1	RCP2.6&	RCP4.5&	RCP6.0&	RCP8.5&
2011obs							**4.9**	**4.9**	**4.9**	**4.9**
2010	5.7	5.7	5.7	5.7	1	1	4.1 ± 0	4.1 ± 0	4.1 ± 0	4.1 ± 0
2020	21.0	21.0	21.1	21.1	3	3	23.8 ± 2	24.0 ± 2	24.0 ± 2	24.0 ± 2
2030	34.7	35.2	35.5	35.8	4	4	38.1 ± 5	39.1 ± 5	39.1 ± 5	39.2 ± 5
2040	41.1	41.9	42.4	43.6	6	5	44.7 ± 6	46.7 ± 6	46.9 ± 6	47.8 ± 6
2050	41.9	42.8	43.9	46.2	7	7	44.3 ± 7	47.6 ± 7	48.2 ± 7	50.3 ± 8
2060	43.1	43.8	45.6	48.8	9	8	45.0 ± 7	49.6 ± 8	50.6 ± 8	53.8 ± 8
2070	47.9	48.1	50.7	54.7	11	8	49.4 ± 8	54.9 ± 8	56.8 ± 9	60.3 ± 9
2080	51.3	50.5	54.0	58.6	14	8	53.8 ± 9	58.2 ± 9	61.4 ± 10	64.7 ± 10
2090	51.0	49.6	52.8	58.2	17	8	54.0 ± 9	56.9 ±10	60.6 ± 10	64.4 ± 11
2100	47.5	45.6	47.4	53.8	20	8	50.5 ± 9	51.8 ± 9	55.2 ± 10	59.6 ± 11

Table AII.4.10 | HFC-125 abundance (ppt)

Year	RCP2.6	RCP4.5	RCP6.0	RCP8.5	A2	B1	IS92a	RCP2.6&	RCP4.5&	RCP6.0&	RCP8.5&
2011obs								**9.6**	**9.6**	**9.6**	**9.6**
2010	7.1	6.4	5.7	7.7	2	2	0	8.2 ± 1	8.2 ± 1	8.2 ± 1	8.2 ± 1
2020	27.4	14.3	7.6	25.7	8	8	2	30.9 ± 1	16.3 ± 1	9.6 ± 1	27.6 ± 1
2030	60.0	23.2	9.2	48.5	16	16	12	64.1 ± 3	25.2 ± 2	10.9 ± 1	51.0 ± 3
2040	90.5	29.7	10.6	72.0	24	24	40	95.5 ± 7	31.9 ± 3	12.2 ± 1	75.9 ± 5
2050	114.5	34.0	11.8	97.6	34	33	87	119.5 ± 11	36.6 ± 4	13.3 ± 2	103 ± 8
2060	133.4	36.0	12.9	122.9	45	43	137	139.0 ± 15	39.0 ± 5	14.4 ± 2	130 ± 12
2070	154.8	35.8	13.9	147.1	58	49	177	160.8 ± 20	39.4 ± 6	15.5 ± 2	156 ± 16
2080	176.2	34.8	14.8	168.7	72	54	210	183.2 ± 24	39.1 ± 6	16.6 ± 2	180 ± 20
2090	192.3	34.0	15.5	185.8	89	57	236	200.9 ± 29	38.7 ± 7	17.4 ± 3	199 ± 25
2100	200.2	33.2	15.8	198.9	107	58	255	210.5 ± 34	38.1 ± 7	18.0 ± 3	215 ± 30

Table AII.4.11 | HFC-134a abundance (ppt)

Year	RCP2.6	RCP4.5	RCP6.0	RCP8.5	A2	B1	IS92a	RCP2.6&	RCP4.5&	RCP6.0&	RCP8.5&
2011								**63 ± 1**	**63 ± 1**	**63 ± 1**	**63 ± 1**
2010	56	56	56	56	55	55	94	58 ± 3	58 ± 3	58 ± 3	58 ± 3
2020	96	95	90	112	111	108	183	97 ± 5	98 ± 5	91 ± 5	117 ± 5
2030	122	129	109	180	170	165	281	123 ± 9	132 ± 9	110 ± 8	184 ± 11
2040	142	154	121	245	231	223	401	143 ± 12	157 ± 12	122 ± 10	249 ± 17
2050	153	175	129	311	299	293	537	150 ± 15	178 ± 16	130 ± 12	314 ± 24
2060	160	187	135	370	382	352	657	155 ± 16	192 ± 19	137 ± 14	373 ± 32
2070	175	193	141	423	480	380	743	168 ± 18	200 ± 21	143 ± 15	427 ± 39
2080	191	205	144	471	594	391	807	184 ± 21	216 ± 23	148 ± 16	476 ± 47
2090	200	229	144	517	729	390	850	193 ± 23	242 ± 26	150 ± 18	524 ± 56
2100	199	262	141	561	877	379	878	192 ± 25	275 ± 30	148 ± 19	570 ± 64

Table AII.4.12 | HFC-143a abundance (ppt)

Year	RCP2.6	RCP4.5	RCP6.0	RCP8.5	A2	B1	RCP2.6&	RCP4.5&	RCP6.0&	RCP8.5&
2011							**12.0**	**12.0**	**12.0**	**12.0**
2010	10.2	9.4	8.4	10.8	3	2	11 ± 1	11 ± 1	11 ± 1	11 ± 1
2020	33.9	17.8	10.1	28.2	10	9	37 ± 1	19 ± 1	12 ± 1	29 ± 1
2030	72.1	26.8	12.1	46.8	20	18	75 ± 2	28 ± 1	14 ± 1	48 ± 1
2040	109.9	36.0	14.0	65.6	32	29	13 ± 4	38 ± 1	16 ± 1	67 ± 2
2050	142.1	45.4	16.0	85.7	45	43	144 ± 6	47 ± 2	18 ± 1	88 ± 3
2060	168.6	54.0	18.1	105.2	62	57	170 ± 8	56 ± 3	20 ± 1	107 ± 4
2070	196.1	61.4	20.1	123.2	81	68	197 ± 11	64 ± 3	22 ± 1	126 ± 6
2080	222.2	69.7	22.2	138.7	103	77	223 ± 14	73 ± 4	24 ± 2	142 ± 8
2090	242.0	80.2	24.0	150.2	129	85	243 ± 17	85 ± 5	26 ± 2	154 ± 9
2100	252.9	92.6	25.6	157.9	157	90	254 ± 20	98 ± 6	28 ± 2	163 ± 11

Table AII.4.13 | HFC-227ea abundance (ppt)

Year	RCP2.6	RCP4.5	RCP6.0	RCP8.5	A2	B1	RCP2.6&	RCP4.5&	RCP6.0&	RCP8.5&
2011							**0.65**	**0.65**	**0.65**	**0.65**
2010	1.43	1.28	1.42	1.56	2	2	0.6 ± 0.1	0.6 ± 0.1	0.6 ± 0.1	0.6 ± 0.1
2020	2.81	2.10	2.78	3.30	5	6	2.0 ± 0.1	1.5 ± 0.1	2.0 ± 0.1	2.4 ± 0.1
2030	2.48	1.71	2.44	2.77	10	10	2.0 ± 0.1	1.3 ± 0.1	2.0 ± 0.1	2.2 ± 0.1
2040	2.09	1.35	2.04	2.29	14	15	1.8 ± 0.1	1.1 ± 0.1	1.8 ± 0.1	2.0 ± 0.2
2050	1.74	1.06	1.68	1.92	19	21	1.6 ± 0.2	1.0 ± 0.1	1.6 ± 0.2	1.8 ± 0.2
2060	1.35	0.81	1.31	1.55	25	27	1.3 ± 0.2	0.8 ± 0.1	1.3 ± 0.2	1.5 ± 0.2
2070	1.04	0.61	1.01	1.23	32	31	1.1 ± 0.2	0.6 ± 0.1	1.1 ± 0.2	1.3 ± 0.2
2080	0.81	0.45	0.78	0.99	40	34	0.9 ± 0.2	0.5 ± 0.1	0.9 ± 0.2	1.1 ± 0.2
2090	0.63	0.34	0.59	0.79	49	35	0.8 ± 0.2	0.4 ± 0.1	0.8 ± 0.2	0.9 ± 0.2
2100	1.43	1.28	1.42	1.56	2	2	0.6 ± 0.2	0.3 ± 0.1	0.6 ± 0.2	0.8 ± 0.2

Table AII.4.14 | HFC-245fa abundance (ppt)

Year	RCP2.6	RCP4.5	RCP6.0	RCP8.5	A2	B1	RCP2.6&	RCP4.5&	RCP6.0&	RCP8.5&
2011							**1.24**	**1.24**	**1.24**	**1.24**
2010	7.5	7.3	8.2	9.5	8	8	1 ± 0.2	1 ± 0.2	1 ± 0.2	1 ± 0.2
2020	12.1	19.3	18.1	31.5	17	17	10.2 ± 1	18.9 ± 2	16.4 ± 2	31.0 ± 4
2030	7.4	28.2	21.3	51.2	23	23	6.6 ± 1.5	29.2 ± 4	21.6 ± 3	53.1 ± 8
2040	2.3	31.2	22.6	61.7	29	29	2.2 ± 1.0	33.0 ± 6	23.7 ± 4	63.8 ± 10
2050	0.6	31.9	23.3	62.0	36	38	0.7 ± 0.5	34.1 ± 7	24.6 ± 5	64.4 ± 12
2060	0.2	30.6	23.8	59.1	46	43	0.2 ± 0.2	32.9 ± 7	25.3 ± 5	61.7 ± 13
2070	0.0	28.2	24.2	55.3	58	44	0.1 ± 0.1	30.8 ± 7	25.9 ± 5	58.1 ± 13
2080	0.0	26.4	24.3	51.5	72	43	0.0 ± 0.1	29.3 ± 7	26.4 ± 6	54.4 ± 12
2090	0.0	25.8	23.6	48.0	88	42	0.0 ± 0.0	28.6 ± 6	26.0 ± 6	51.0 ± 12
2100	0.0	26.0	22.3	47.3	105	40	0.0 ± 0.0	28.6 ± 6	24.9 ± 6	50.6 ± 11

Table AII.4.15 | HFC-43-10mee abundance (ppt)

Year	RCP2.6	RCP4.5	RCP6.0	RCP8.5	A2	B1	RCP2.6&	RCP4.5&	RCP6.0&	RCP8.5&
2011							—	—	—	—
2010	0.52	0.52	0.52	0.52	1	1	0.0 ± 0.0	0.0 ± 0.0	0.0 ± 0.0	0.0 ± 0.0
2020	1.46	1.46	1.46	1.47	2	1	1.2 ± 0.1	1.2 ± 0.1	1.2 ± 0.1	1.2 ± 0.1
2030	2.09	2.11	2.12	2.14	2	2	2.0 ± 0.2	2.1 ± 0.2	2.1 ± 0.2	2.1 ± 0.2
2040	2.61	2.64	2.66	2.68	3	2	2.7 ± 0.3	2.8 ± 0.3	2.8 ± 0.3	2.8 ± 0.3
2050	3.13	3.17	3.22	3.23	3	3	3.3 ± 0.4	3.4 ± 0.4	3.4 ± 0.4	3.4 ± 0.4
2060	3.56	3.61	3.70	3.83	4	3	3.7 ± 0.6	3.9 ± 0.6	4.0 ± 0.6	4.1 ± 0.6
2070	3.78	3.81	3.96	4.52	4	4	3.9 ± 0.7	4.3 ± 0.7	4.3 ± 0.7	4.9 ± 0.7
2080	3.89	3.88	4.08	5.27	5	4	4.1 ± 0.8	4.4 ± 0.8	4.6 ± 0.8	5.8 ± 0.9
2090	3.93	3.87	4.10	6.14	6	4	4.2 ± 0.8	4.5 ± 0.8	4.7 ± 0.9	6.7 ± 1.0
2100	3.91	3.81	3.99	7.12	7	4	4.2 ± 0.9	4.4 ± 0.9	4.6 ± 0.9	7.9 ± 1.2

Table AII.4.16 | Montreal Protocol greenhouse gas abundances (ppt)

Year	CFC-11	CFC-12	CFC-113	CFC-114	CFC-115	CCl_4	CH_3CCl_3	HCFC-22
2011*	**238 ± 1**	**528 ± 2**	**74.5 ± 0.5**	**15.8**	**8.4**	**86 ± 2**	**6.4 ± 0.4**	**213 ± 2**
2010	240.9	532.5	75.6	16.4	8.4	87.6	8.3	206.8
2020	213.0	492.8	67.4	15.8	8.4	70.9	1.5	301.8
2030	182.6	448.0	59.9	15.1	8.4	54.4	0.2	265.4
2040	153.5	405.8	53.3	14.4	8.4	40.3	0.0	151.0
2050	127.2	367.3	47.4	13.6	8.4	29.2	0.0	71.1
2060	104.4	332.4	42.1	12.9	8.3	20.0	0.0	31.5
2070	85.2	300.7	37.4	12.3	8.3	13.6	0.0	13.7
2080	69.1	272.1	33.3	11.6	8.2	9.3	0.0	5.9
2090	55.9	246.2	29.6	11.1	8.2	6.3	0.0	2.6
2100	45.1	222.8	26.3	10.5	8.1	4.3	0.0	1.1

Year	HCFC-141b	HCFC-142b	Halon 1211	Halon 1202	Halon 1301	Halon 2402	CH_3Br	CH_3Cl
2011*	**21.4 ± 0.5**	**21.2 ± 0.5**	**4.07**	**0.00**	**3.23**	**0.45**	**7.1**	**534**
2010	20.3	20.5	4.07	0.00	3.20	0.46	7.2	550
2020	30.9	30.9	3.08	0.00	3.29	0.38	7.1	550
2030	34.4	31.2	2.06	0.00	3.19	0.27	7.1	550
2040	27.9	23.3	1.30	0.00	2.97	0.18	7.1	550
2050	19.3	14.9	0.78	0.00	2.71	0.12	7.1	550
2060	12.4	9.0	0.46	0.00	2.43	0.07	7.1	550
2070	7.7	5.2	0.26	0.00	2.16	0.05	7.1	550
2080	4.7	3.0	0.15	0.00	1.90	0.03	7.1	550
2090	2.9	1.7	0.08	0.00	1.66	0.02	7.1	550
2100	1.7	0.9	0.05	0.00	1.44	0.01	7.1	550

Notes:

Present day (2011*) is from Chapter 2; projections are from Scenario A1, WMO Ozone Assessment (WMO 2010).

AII.5: Column Abundances, Burdens, and Lifetimes

Table AII.5.1 | Stratospheric O_3 column changes (DU)

Year	Obs	RCP2.6	RCP4.5	RCP6.0	RCP8.5
1850		17	17	17	17
1980	11	15	15	15	15
2000	**269 ± 8**	**276 ± 9**	**276 ± 9**	**276 ± 9**	**276 ± 9**
2010	0	2	–1	1	–2
2020		4	0	3	2
2030		8	4	7	5
2040		9	7	10	9
2050		12	10	13	12
2060		13	12	14	15
2070		13	11	15	16
2080		12	11	16	15
2090		13	12	16	18
2100		15	13	17	20

Notes:

Observed O_3 columns and trends taken from WMO (Douglass and Filetov, 2010), subtracting tropospheric column O_3 (Table AII.5.2) with uncertainty estimates driven by polar variability. CMIP5 RCP results are from Eyring et al. (2013). The multi-model mean is derived from the CMIP5 models with predictive (interactive or semi-offline) stratospheric and tropospheric ozone chemistry. The absolute value is shown for year 2000. All other years are differences relative to (minus) year 2000. The multi-model standard deviation is shown only for year 2000; it does not change much over time; and, representing primarily the spread in absolute O_3 column, it is larger than the standard deviation of the changes (not evaluated here). All models used the same projections for ozone-depleting substances. Near-term differences in projected O_3 columns across scenarios reflect model sampling (i.e., different sets of models contributing to each RCP), while long-term changes reflect changes in N_2O, CH_4 and climate. See Section 11.3.5.1.2.

Table AII.5.2 | Tropospheric O_3 column changes (DU)

Year	CMIP5				ACCMIP			
	RCP2.6	RCP4.5	RCP6.0	RCP8.5	RCP2.6	RCP4.5	RCP6.0	RCP8.5
1850	–10.2	–10.2	–10.2	–10.2	–8.9	–8.9	–8.9	–8.9
1980	–2.0	–2.0	–2.0	–2.0	–1.3	–1.3	–1.3	–1.3
2000	**31.1 ± 3.3**	**31.1 ± 3.3**	**31.1 ± 3.3**	**31.1 ± 3.3**	**30.8 ± 2.1**	**30.8 ± 2.1**	**30.8 ± 2.1**	**30.8 ± 2.1**
2010	1.1	0.6	0.8	0.8				
2020	1.0	0.9	1.0	2.1				
2030	0.6	1.5	1.4	3.5	–1.3	1.0	–0.1	1.8
2040	0.5	1.6	2.1	4.5				
2050	0.0	1.7	2.4	5.7				
2060	–0.7	1.3	2.6	7.1				
2070	–1.6	0.5	2.3	8.1				
2080	–2.5	–0.1	2.0	8.9				
2090	–2.8	–0.4	1.5	9.5				
2100	–3.1	–0.5	1.1	10.2	–5.4	–2.2	–2.6	5.3

(continued on next page)

Table AII.5.2 *(continued)*

Year	A2	B1	IS92a	CLE	MFR
1850					
1980					
2000	**34.0**	**34.0**	**34.0**	**32.6**	**32.6**
2010	1.7	0.8	1.5		
2020	4.2	1.6	3.1		
2030	6.8	1.9	4.7	1.5 ± 0.8	–1.4 ± 0.4
2040	8.6	1.8	6.1		
2050	10.2	1.0	7.6		
2060	11.7	0.0	8.9		
2070	13.2	–0.9	10.0		
2080	15.3	–1.9	11.1		
2090	18.0	–2.8	12.1		
2100	20.8	–3.9	13.2		

Notes:

RCP results from CMIP5 (Eyring et al., 2013) and ACCMIP (Young et al., 2013). For ACCMIP all models have interactive tropospheric ozone chemistry and are included, in contrast to the CMIP5 multi-model mean which includes only those models with predictive (interactive or semi-offline) stratospheric and tropospheric ozone chemistry. The absolute value is shown for year 2000. All other years are differences relative to (minus) year 2000. The multi-model standard deviation is shown only for year 2000; it does not change much over time; and, representing primarily the spread in absolute O_3 columns, it is larger than the standard deviation of the changes across individual models (not evaluated here). SRES values are from TAR Appendix II. CLE/MFR scenarios are from Dentener et al. (2005, 2006): CLE includes climate change, MFR does not. See Section 11.3.5.1.2.

Table AII.5.3 | Total aerosol optical depth (AOD)

Year	(Min)	Historical	(Max)	RCP2.6	RCP4.5	RCP6.0	RCP8.5
1860[d]	0.056	0.101	0.161	0.094	0.101	0.092	0.100
1870[d]	0.058	0.102	0.162	0.095	0.102	0.094	0.101
1180[d]	0.058	0.102	0.163	0.095	0.102	0.094	0.101
1890[d]	0.059	0.104	0.164	0.098	0.104	0.096	0.103
1900[d]	0.058	0.105	0.166	0.099	0.105	0.097	0.104
1910[d]	0.059	0.107	0.169	0.101	0.107	0.099	0.106
1920[d]	0.060	0.108	0.170	0.102	0.108	0.100	0.107
1930[d]	0.061	0.110	0.173	0.104	0.110	0.101	0.109
1940[d]	0.061	0.111	0.175	0.105	0.111	0.103	0.110
1950[d]	0.060	0.115	0.181	0.108	0.115	0.106	0.113
1960[d]	0.064	0.122	0.192	0.116	0.122	0.113	0.120
1970[d]	0.065	0.130	0.204	0.123	0.130	0.120	0.128
1980[d]	0.066	0.135	0.221	0.127	0.135	0.124	0.133
1990[d]	0.068	0.138	0.231	0.129	0.138	0.126	0.135
2000[d]	0.068	0.136	0.232	0.127	0.136	0.124	0.134
2010[d]				0.127	0.137	0.124	0.133
2020[d]				0.123	0.134	0.122	0.132
2030[d]				0.117	0.130	0.119	0.130
2040[d]				0.111	0.126	0.118	0.126
2050[d]				0.108	0.123	0.117	0.124
2060[d]				0.106	0.119	0.116	0.121
2070[d]				0.105	0.116	0.110	0.120
2080[d]				0.103	0.114	0.107	0.118
2090[d]				0.102	0.112	0.106	0.118
2100[d]				0.101	0.111	0.105	0.117
Number of models		21		15	21	13	19

Notes:

Multi-model decadal global means (2030[d] = 2025–2034, 2100[d] = 2095–2100) from CMIP5 models reporting AOD. The numbers of models for each experiment are indicated in the bottom row. The full range of models (given only for historical period for AOD and AAOD) is large and systematic in that models tend to scale relative to one another. Historical estimates for different RCPs vary because of the models included. RCP4.5 included the full set of CMIP5 models contributing aerosol results (21). The standard deviation of the models is 28% (AOD) and 62% (AAOD) (N. Mahowald, CMIP5 archive; Lamarque et al., 2013; Shindell et al., 2013). See Sections 11.3.5.1.3 and 11.3.6.1.

Table AII.5.4 | Absorbing aerosol optical depth (AAOD)

Year	(Min)	Historical	(Max)	RCP2.6	RCP4.5	RCP6.0	RCP8.5
1860[d]	0.00050	0.0035	0.0054	0.0033	0.0035	0.0031	0.0035
1870[d]	0.00060	0.0035	0.0054	0.0033	0.0035	0.0032	0.0036
1180[d]	0.00060	0.0036	0.0054	0.0034	0.0036	0.0032	0.0036
1890[d]	0.00060	0.0036	0.0055	0.0035	0.0036	0.0033	0.0037
1900[d]	0.00070	0.0037	0.0056	0.0035	0.0037	0.0033	0.0038
1910[d]	0.00070	0.0038	0.0057	0.0036	0.0038	0.0034	0.0038
1920[d]	0.00070	0.0038	0.0058	0.0036	0.0038	0.0034	0.0039
1930[d]	0.00070	0.0038	0.0057	0.0036	0.0038	0.0034	0.0038
1940[d]	0.00070	0.0038	0.0057	0.0036	0.0038	0.0034	0.0039
1950[d]	0.00070	0.0038	0.0058	0.0036	0.0038	0.0034	0.0039
1960[d]	0.00080	0.0040	0.0059	0.0038	0.0040	0.0036	0.0040
1970[d]	0.00090	0.0042	0.0065	0.0040	0.0042	0.0038	0.0043
1980[d]	0.00100	0.0046	0.0073	0.0044	0.0046	0.0042	0.0046
1990[d]	0.00110	0.0049	0.0079	0.0047	0.0049	0.0044	0.0049
2000[d]	0.00120	0.0050	0.0084	0.0048	0.0050	0.0045	0.0051
2010[d]				0.0050	0.0051	0.0046	0.0051
2020[d]				0.0050	0.0050	0.0045	0.0050
2030[d]				0.0047	0.0049	0.0045	0.0049
2040[d]				0.0043	0.0048	0.0044	0.0047
2050[d]				0.0041	0.0046	0.0044	0.0046
2060[d]				0.0039	0.0044	0.0043	0.0045
2070[d]				0.0037	0.0042	0.0041	0.0044
2080[d]				0.0037	0.0040	0.0039	0.0043
2090[d]				0.0036	0.0039	0.0038	0.0043
2100[d]				0.0036	0.0039	0.0038	0.0042
Number of models		14		11	14	10	12

Notes:

See notes Table AII.5.3.

Table AII.5.5 | Sulphate aerosol atmospheric burden (TgS)

Year	(Min)	Historical	(Max)	RCP2.6	RCP4.5	RCP6.0	RCP8.5
1860[d]	0.09	0.61	1.42	0.60	0.61	0.57	0.60
1870[d]	0.10	0.62	1.45	0.62	0.62	0.59	0.61
1180[d]	0.12	0.65	1.49	0.64	0.65	0.61	0.64
1890[d]	0.16	0.68	1.57	0.67	0.68	0.64	0.66
1900[d]	0.21	0.73	1.65	0.73	0.73	0.70	0.72
1910[d]	0.23	0.79	1.80	0.79	0.79	0.76	0.78
1920[d]	0.23	0.83	1.84	0.83	0.83	0.80	0.81
1930[d]	0.24	0.87	1.94	0.88	0.87	0.85	0.86
1940[d]	0.25	0.93	2.05	0.95	0.93	0.91	0.92
1950[d]	0.27	1.03	2.21	1.05	1.03	1.01	1.01
1960[d]	0.31	1.25	2.67	1.29	1.25	1.24	1.23
1970[d]	0.35	1.48	3.14	1.52	1.48	1.45	1.47
1980[d]	0.37	1.58	3.33	1.62	1.58	1.54	1.58
1990[d]	0.37	1.59	3.31	1.63	1.59	1.55	1.60
2000[d]	0.37	1.55	3.17	1.59	1.55	1.53	1.56

(continued on next page)

Table AII.5.5 | *(continued)*

Year	(Min)	Historical	(Max)	RCP2.6	RCP4.5	RCP6.0	RCP8.5
2010[d]				1.57	1.59	1.52	1.54
2020[d]				1.43	1.54	1.43	1.51
2030[d]				1.21	1.44	1.33	1.44
2040[d]				1.03	1.31	1.34	1.31
2050[d]				0.94	1.16	1.29	1.20
2060[d]				0.90	1.05	1.24	1.13
2070[d]				0.86	0.96	1.06	1.08
2080[d]				0.81	0.88	0.92	1.05
2090[d]				0.76	0.85	0.86	0.98
2100[d]				0.71	0.83	0.80	0.94
Number of models		18		12	18	10	16

Notes:
See notes Table AII.5.3. The standard deviation of the models is about 50% for sulphate, OC and BC aerosol loadings (N. Mahowald, CMIP5 archive; Lamarque et al., 2013; Shindell et al., 2013).

Table AII.5.6 | OC aerosol atmospheric burden (Tg)

Year	(Min)	Historical	(Max)	RCP2.6	RCP4.5	RCP6.0	RCP8.5
1860[d]	0.34	1.08	2.7	1.09	1.08	1.13	1.12
1870[d]	0.35	1.09	2.7	1.10	1.09	1.14	1.13
1180[d]	0.36	1.09	2.7	1.11	1.09	1.15	1.14
1890[d]	0.35	1.10	2.8	1.12	1.10	1.16	1.15
1900[d]	0.36	1.11	2.8	1.12	1.11	1.16	1.15
1910[d]	0.33	1.10	2.8	1.11	1.10	1.15	1.15
1920[d]	0.34	1.08	2.7	1.09	1.08	1.12	1.13
1930[d]	0.33	1.07	2.6	1.07	1.07	1.11	1.12
1940[d]	0.33	1.07	2.6	1.07	1.07	1.11	1.12
1950[d]	0.36	1.08	2.6	1.08	1.08	1.11	1.12
1960[d]	0.41	1.13	2.7	1.13	1.13	1.17	1.17
1970[d]	0.46	1.20	2.9	1.22	1.20	1.26	1.24
1980[d]	0.54	1.28	3.1	1.32	1.28	1.36	1.33
1990[d]	0.53	1.38	3.3	1.44	1.38	1.48	1.43
2000[d]	0.53	1.41	3.5	1.47	1.41	1.52	1.46
2010[d]				1.59	1.21	1.55	1.29
2020[d]				1.59	1.12	1.56	1.26
2030[d]				1.56	1.08	1.55	1.25
2040[d]				1.47	1.06	1.57	1.22
2050[d]				1.41	1.04	1.57	1.20
2060[d]				1.40	1.01	1.56	1.17
2070[d]				1.36	0.96	1.55	1.14
2080[d]				1.33	0.92	1.55	1.13
2090[d]				1.32	0.90	1.54	1.10
2100[d]				1.30	0.89	1.55	1.09
Number of models		19		12	19	10	17

Notes:
See notes Table AII.5.5.

Table AII.5.7 | BC aerosol atmospheric burden (Tg)

Year	(Min)	Historical	(Max)	RCP2.6	RCP4.5	RCP6.0	RCP8.5
1860[d]	0.037	0.059	0.127	0.058	0.059	0.057	0.059
1870[d]	0.039	0.063	0.133	0.062	0.063	0.061	0.064
1180[d]	0.040	0.068	0.139	0.066	0.068	0.065	0.069
1890[d]	0.043	0.075	0.149	0.070	0.075	0.070	0.076
1900[d]	0.045	0.082	0.156	0.076	0.082	0.075	0.083
1910[d]	0.048	0.089	0.167	0.081	0.089	0.081	0.091
1920[d]	0.049	0.092	0.167	0.083	0.092	0.082	0.095
1930[d]	0.049	0.090	0.161	0.082	0.090	0.081	0.092
1940[d]	0.051	0.091	0.162	0.082	0.091	0.082	0.093
1950[d]	0.053	0.094	0.165	0.085	0.094	0.085	0.096
1960[d]	0.061	0.102	0.179	0.094	0.102	0.094	0.105
1970[d]	0.071	0.115	0.201	0.107	0.115	0.107	0.117
1980[d]	0.088	0.141	0.245	0.130	0.141	0.130	0.144
1990[d]	0.098	0.157	0.274	0.146	0.157	0.145	0.161
2000[d]	0.101	0.164	0.293	0.153	0.164	0.152	0.169
2010[d]				0.170	0.174	0.157	0.170
2020[d]				0.169	0.174	0.152	0.164
2030[d]				0.144	0.166	0.147	0.153
2040[d]				0.120	0.155	0.144	0.138
2050[d]				0.103	0.141	0.138	0.127
2060[d]				0.091	0.126	0.127	0.118
2070[d]				0.081	0.110	0.113	0.110
2080[d]				0.075	0.094	0.101	0.106
2090[d]				0.071	0.087	0.092	0.102
2100[d]				0.068	0.084	0.087	0.099
Number of models		19		13	19	11	17

Notes:

See notes Table AII.5.5.

Table AII.5.8 | CH_4 atmospheric lifetime (yr) against loss by tropospheric OH

Year	RCP2.6[&]	RCP4.5[&]	RCP6.0[&]	RCP8.5[&]	RCP2.6^	RCP4.5^	RCP6.0^	RCP8.5^
2000	11.2 ± 1.3	11.2 ± 1.3	11.2 ± 1.3	11.2 ± 1.3	11.2 ± 1.3	11.2 ± 1.3	11.2 ± 1.3	11.2 ± 1.3
2010	11.2 ± 1.3	11.2 ± 1.3	11.2 ± 1.3	11.2 ± 1.3				
2020	11.0 ± 1.3	11.2 ± 1.3	11.2 ± 1.3	11.2 ± 1.3				
2030	10.8 ± 1.3	11.3 ± 1.4	11.3 ± 1.4	11.4 ± 1.4	10.6 ± 1.4	11.4 ± 2.1	11.1 ± 1.4	11.2 ± 1.4
2040	10.6 ± 1.3	11.3 ± 1.4	11.4 ± 1.4	11.8 ± 1.4				
2050	10.2 ± 1.3	11.3 ± 1.4	11.5 ± 1.4	12.2 ± 1.5				
2060	9.9 ± 1.3	11.2 ± 1.4	11.6 ± 1.4	12.6 ± 1.6				
2070	9.9 ± 1.4	11.2 ± 1.5	11.8 ± 1.5	12.6 ± 1.7				
2080	10.4 ± 1.5	11.1 ± 1.5	11.9 ± 1.6	12.6 ± 1.8				
2090	10.4 ± 1.6	10.9 ± 1.6	11.7 ± 1.7	12.6 ± 1.8				
2100	10.6 ± 1.6	10.7 ± 1.6	11.4 ± 1.8	12.5 ± 1.9	10.7 ± 1.6	10.1 ± 1.5	11.1 ± 1.8	12.1 ± 2.0

Notes:

RCPn.n[&] lifetimes based on best estimate with uncertainty for 2000–2010 (Prather et al., 2012) and then projecting changes in key factors (Holmes et al., 2013). All uncertainties are 68% confidence intervals. RCPn.n^ lifetimes are from ACCMIP results (Voulgarakis et al., 2013) scaled to 11.2 ± 1.3 yr for year 2000; the ACCMIP mean and standard deviation in 2000 are 9.8 ± 1.5 yr. Projected ACCMIP values combine the present day uncertainty with the model standard deviation of future change. Note that the total atmospheric lifetime of CH_4 must include other losses (e.g., stratosphere, surface, tropospheric chlorine), and for 2010 it is 9.1 ± 0.9 yr, see Chapter 8, Section 11.3.5.1.1.

Table AII.5.9 | N_2O atmospheric lifetime (yr)

Year	RCP2.6[&]	RCP4.5[&]	RCP6.0[&]	RCP8.5[&]
2010	131 ± 10	131 ± 10	131 ± 10	131 ± 10
2020	130 ± 10	131 ± 10	131 ± 10	131 ± 10
2030	130 ± 10	130 ± 10	130 ± 10	130 ± 10
2040	130 ± 10	130 ± 10	130 ± 10	129 ± 10
2050	129 ± 10	129 ± 10	129 ± 10	129 ± 10
2060	129 ± 10	129 ± 10	129 ± 10	128 ± 10
2070	129 ± 11	128 ± 11	128 ± 10	128 ± 11
2080	128 ± 11	128 ± 11	128 ± 11	127 ± 11
2090	128 ± 11	128 ± 11	127 ± 11	127 ± 11
2100	128 ± 11	127 ± 11	127 ± 11	126 ± 11

Notes:
RCPn.n[&] lifetimes based on projections from Fleming et al. (2011) and Prather et al. (2012). All uncertainties are 68% confidence intervals.

AII.6: Effective Radiative Forcing

Table AII.6.1 | ERF from CO_2 (W m^{-2})

Year	RCP2.6	RCP4.5	RCP6.0	RCP8.5	A2	B1	IS92a
2000	1.51	1.51	1.51	1.51	1.50	1.50	1.50
2010	1.80	1.80	1.80	1.80	1.78	1.77	1.78
2020	2.11	2.09	2.07	2.15	2.16	2.09	2.13
2030	2.34	2.40	2.32	2.56	2.55	2.38	2.48
2040	2.46	2.70	2.58	3.03	2.99	2.69	2.83
2050	2.49	2.99	2.90	3.56	3.42	2.98	3.18
2060	2.48	3.23	3.25	4.15	3.88	3.20	3.53
2070	2.43	3.39	3.65	4.76	4.36	3.37	3.89
2080	2.35	3.46	4.06	5.37	4.86	3.49	4.25
2090	2.28	3.49	4.42	5.95	5.39	3.57	4.64
2100	2.22	3.54	4.70	6.49	5.95	3.59	5.04

Notes:
RCPn.n ERF based on RCP published projections (Tables AII.4.1 to AII.4.3) and TAR formula for RF. See Chapter 8, Figure 8.18, Section 11.3.5, 11.3.6.1, Figure 12.3. SRES A2 and B1 and IS92a calculated from abundances in Tables AII.4.1 to AII.4.3.

Table AII.6.2 | ERF from CH_4 (W m^{-2})

Year	RCP2.6	RCP4.5	RCP6.0	RCP8.5	A2	B1	IS92a
2000	0.47	0.47	0.47	0.47	0.48	0.48	0.48
2010	0.48	0.48	0.48	0.48	0.51	0.50	0.51
2020	0.47	0.49	0.49	0.54	0.56	0.53	0.56
2030	0.42	0.50	0.49	0.61	0.62	0.54	0.61
2040	0.39	0.51	0.51	0.70	0.68	0.54	0.67
2050	0.36	0.50	0.53	0.80	0.75	0.52	0.73
2060	0.32	0.49	0.54	0.90	0.81	0.51	0.78
2070	0.30	0.47	0.55	0.97	0.88	0.49	0.82
2080	0.29	0.44	0.54	1.01	0.95	0.47	0.85
2090	0.28	0.42	0.50	1.05	1.01	0.44	0.88
2100	0.27	0.41	0.44	1.08	1.07	0.41	0.92

Notes:
See notes Table AII.6.1.

Table AII.6.3 | ERF from N_2O (W m^{-2})

Year	RCP2.6	RCP4.5	RCP6.0	RCP8.5	A2	B1	IS92a
2000	0.15	0.15	0.15	0.15	0.15	0.15	0.15
2010	0.17	0.17	0.17	0.17	0.17	0.17	0.17
2020	0.19	0.19	0.19	0.19	0.20	0.20	0.20
2030	0.20	0.21	0.21	0.23	0.24	0.22	0.23
2040	0.22	0.23	0.24	0.26	0.28	0.25	0.26
2050	0.23	0.25	0.26	0.30	0.32	0.27	0.29
2060	0.23	0.27	0.29	0.34	0.36	0.29	0.32
2070	0.23	0.28	0.33	0.38	0.40	0.30	0.34
2080	0.23	0.30	0.36	0.42	0.44	0.31	0.37
2090	0.23	0.31	0.39	0.46	0.49	0.32	0.39
2100	0.23	0.32	0.41	0.49	0.53	0.32	0.41

AII

Notes:

See notes Table AII.6.1.

Table AII.6.4 | ERF from all HFCs (W m^{-2})

Year	Historical	RCP2.6	RCP4.5	RCP6.0	RCP8.5
2011*	0.019				
2010		0.019	0.019	0.019	0.020
2020		0.038	0.034	0.030	0.044
2030		0.056	0.046	0.036	0.069
2040		0.071	0.055	0.040	0.091
2050		0.083	0.061	0.042	0.110
2060		0.092	0.064	0.044	0.128
2070		0.104	0.066	0.046	0.144
2080		0.116	0.069	0.047	0.159
2090		0.124	0.074	0.047	0.171
2100		0.126	0.080	0.046	0.182

Notes:

See Table 8.3, 8.A.1, Section 11.3.5.1.1. ERF is calculated from RCP published abundances (Meinshausen et al., 2011a; http://www.iiasa.ac.at/web-apps/tnt/RcpDb) and AR5 radiative efficiencies (Chapter 8).

Table AII.6.5 | ERF from all PFCs and SF_6 (W m^{-2})

Year	Historical	RCP2.6	RCP4.5	RCP6.0	RCP8.5
2011*	0.009				
2010		0.009	0.009	0.010	0.009
2020		0.012	0.011	0.013	0.012
2030		0.014	0.013	0.017	0.015
2040		0.015	0.014	0.021	0.019
2050		0.015	0.016	0.025	0.022
2060		0.016	0.017	0.029	0.026
2070		0.016	0.019	0.033	0.031
2080		0.016	0.021	0.038	0.035
2090		0.016	0.023	0.042	0.039
2100		0.016	0.026	0.045	0.044

Notes:

See notes Table AII.6.4.

Table AII.6.6 | ERF from Montreal Protocol greenhouse gases (W m^{-2})

Year	Historical	WMO A1
2011*	0.328	
2020		0.33 ± 0.01
2030		0.29 ± 0.01
2040		0.24 ± 0.01
2050		0.20 ± 0.01
2060		0.17 ± 0.02
2070		0.15 ± 0.02
2080		0.13 ± 0.02
2090		0.11 ± 0.02
2100		0.10 ± 0.02

Notes:

See Table 8.3, 8.A.1. ERF is calculated from AR5 radiative efficiency and projected abundances in Scenario A1 of WMO/UNEP assessment (WMO 2010). The 68% confidence interval shown is approximated by combining uncertainty in the radiative efficiency of each gas (±6.1%) and the decay of each gas since 2010 from Table AII.4.16 (±15%). All sources of uncertainty are assumed to be independent (see Chapters 2 and 8).

Table AII.6.7a | ERF from stratospheric O_3 changes since 1850 (W m^{-2})

Year	AR5	CCMVal-2
1960		0.0
1980		–0.033
2000		–0.079
2011*	–0.05	
2050		–0.055
2100		–0.075

Notes:

AR5 results are from Chapter 8, see also Sections 11.3.5.1.2, 11.3.6.1. CCMVal-2 results (Cionni et al. 2011) are the multi-model average (13 chemistry–climate models) running a single scenario for stratospheric change: REF-B2 scenario of CCMVal-2 with SRES A1B climate scenario.

AII

Table AII.6.7b | ERF from tropospheric O_3 changes since 1850 (W m^{-2})

Year	AR5	RCP2.6	RCP4.5	RCP6.0	RCP8.5
1980		0.31 ± 0.05	0.31 ± 0.05	0.31 ± 0.05	0.31 ± 0.05
2000		0.36	0.36	0.36	0.36
2011*	0.40				
2030		0.32	0.38	0.36	0.44
2100		0.17	0.27	0.27	0.60 ± 0.11

Notes:

AR5 results from Chapter 8; see also Sections 11.3.5.1.2, 11.3.6.1. Model mean results from ACCMIP (Stevenson et al., 2013) using a consistent model set (FGKN), which is similar to the all-model mean. Standard deviation across models shown for 1980s decade is similar for all scenarios except for RCP8.5 at 2100, which is twice as large.

Table AII.6.8: Total anthropogenic ERF from published RCPs and SRES (W m^{-2})

Year	RCP2.6	RCP4.5	RCP6.0	RCP8.5	A2	A1B	B1	IS92a	AR5 Historical
1850	0.12	0.12	0.12	0.12					0.06
1990	1.23	1.23	1.23	1.23	1.03	1.03	1.03	1.03	1.60
2000	1.45	1.45	1.45	1.45	1.33	1.33	1.33	1.31	1.87
2010	1.81	1.81	1.78	1.84	1.74	1.65	1.73	1.63	2.25
2020	2.25	2.25	2.15	2.32	2.04	2.16	2.15	2.00	
2030	2.52	2.67	2.52	2.91	2.56	2.84	2.56	2.40	
2040	2.65	3.07	2.82	3.61	3.22	3.61	2.93	2.82	
2050	2.64	3.42	3.20	4.37	3.89	4.16	3.30	3.25	
2060	2.55	3.67	3.58	5.13	4.71	4.79	3.65	3.76	
2070	2.47	3.84	4.11	5.89	5.56	5.28	3.92	4.24	
2080	2.41	3.90	4.60	6.60	6.40	5.62	4.09	4.74	
2090	2.35	3.91	4.93	7.32	7.22	5.86	4.18	5.26	
2100	2.30	3.94	5.15	7.97	8.07	6.05	4.19	5.79	

Notes:

Derived from RCP published CO_2-eq concentrations that aggregate all anthropogenic forcings including greenhouse gases plus aerosols. These results may not be directly comparable to ERF values used in AR5 because of how aerosol indirect effects are included, but results are similar to those derived using ERF in Chapter 12 (see Figure 12.4). Comparisons with the TAR Appendix II (SRES A2 and B1) may not be equivalent because those total RF values (TAR II.3.11) were made using the TAR Chapter 9 Simple Model, not always consistent with the individual components in that appendix (TAR II.3.1 to 9). See Chapter 1, Sections 11.3.6.1, 12.3.1.3 and 12.3.1.4, Figures 1.15 and 12.3. For AR5 Historical, see Table AII.1.2 and Chapter 8.

Table AII.6.9: ERF components relative to 1850 (W m^{-2}) derived from ACCMIP

Year		WMGHG	Ozone	Aerosol	ERF Net
1930		0.58 ± 0.04	0.09 ± 0.03	–0.24 ± 0.06	0.44 ± 0.07
1980		1.56 ± 0.10	0.30 ± 0.10	–0.90 ± 0.22	1.00 ± 0.26
2000		2.30 ± 0.14	0.33 ± 0.11	–1.17 ± 0.28	1.51 ± 0.33
2030	RCP8.5	3.64 ± 0.22	0.43 ± 0.12	–0.91 ± 0.22	3.20 ± 0.33
2100	RCP2.6	2.83 ± 0.17	0.14 ± 0.07	–0.12 ± 0.06*	2.86 ± 0.19
2100	RCP4.5	4.33 ± 0.26	0.23 ± 0.09	–0.12 ± 0.06*	4.44 ± 0.28
2100	RCP6.0	5.60 ± 0.34	0.25 ± 0.05	–0.12 ± 0.06*	5.74 ± 0.35
2100	RCP8.5	8.27 ± 0.50	0.55 ± 0.18	–0.12 ± 0.03	8.71 ± 0.53

Notes:

Radiative forcing and adjusted forcing from the ACCMIP results (Shindell et al., 2013) are given for all well-mixed greenhouse gases (WMGHG), ozone, aerosols, and the net. Original 90% confidence intervals have been reduced to 68% confidence to compare with the CMIP5 model standard deviations in Table AII.6.10. Some uncertainty ranges (*) are estimated from the 2100 RCP8.5 results (see Chapter 12). See Sections 11.3.5.1.3 and 11.3.6.1, Figure 12.4.

Table AII.6.10 | Total anthropogenic plus natural ERF (W m^{-2}) from CMIP5 and CMIP3, including historical

Year	SRES A1B	RCP2.6[&]	RCP4.5[&]	RCP6.0[&]	RCP8.5[&]
1850s[H]	–0.19 ± 0.19	–0.12 ± 0.07			
1986–2005[H]	1.51 ± 0.44	1.34 ± 0.50			
1986–2005	1.51 ± 0.44	1.31 ± 0.47	1.30 ± 0.48	1.29 ± 0.51	1.30 ± 0.47
2010[d]	2.18 ± 0.53	1.97 ± 0.50	1.91 ± 0.53	1.90 ± 0.54	1.96 ± 0.53
2020[d]	2.58 ± 0.57	2.33 ± 0.47	2.27 ± 0.51	2.16 ± 0.55	2.43 ± 0.52
2030[d]	3.15 ± 0.60	2.50 ± 0.51	2.61 ± 0.54	2.41 ± 0.60	2.92 ± 0.57
2040[d]	3.77 ± 0.72	2.64 ± 0.47	2.98 ± 0.55	2.72 ± 0.58	3.52 ± 0.60
2050[d]	4.32 ± 0.73	2.65 ± 0.47	3.25 ± 0.56	3.07 ± 0.61	4.21 ± 0.63
2060[d]	4.86 ± 0.74	2.57 ± 0.50	3.50 ± 0.59	3.40 ± 0.60	4.97 ± 0.68
2070[d]	5.32 ± 0.79	2.51 ± 0.50	3.65 ± 0.58	3.90 ± 0.65	5.70 ± 0.76
2080[d]	5.71 ± 0.81	2.40 ± 0.46	3.71 ± 0.55	4.27 ± 0.69	6.31 ± 0.81
2090[d]	6.00 ± 0.83	2.44 ± 0.49	3.78 ± 0.58	4.64 ± 0.71	7.13 ± 0.89
2081–2100	5.99 ± 0.78	2.40 ± 0.46	3.73 ± 0.56	4.56 ± 0.70	7.02 ± 0.92

Notes:

CMIP5 historical and RCP results (Forster et al., 2013) are shown with CMIP3 SRES A1B results (Forster and Taylor, 2006). The alternative results for 1986–2005 with CMIP5 are derived from: all models contributing historical experiments (1986–2005[H]), and the subsets of models contributing to each RCP experiment (next line, 1986–2005). For SRES A1B the same set of models is used from 1850 to 2100. Values are 10-year averages (2090[d] = 2086–2095) and show multi-model means and standard deviations. See Chapter 12, Section 12.3 and discussion of Figure 12.4, also Sections 8.1, 9.3.2.2, 11.3.6.1 and 11.3.6.3. Due to lack of reporting, for RCP8.5 the 2081–2100 result contains one fewer model than the 2090d decade, and for A1B the 1850s result has just 5 models and the 2081–2100 result has 3 fewer models than the 2090[d] decade.

AII.7: Environmental Data

Table AII.7.1 | Global mean surface O_3 change (ppb)

	HTAP				SRES			
Year	**RCP2.6**	**RCP4.5**	**RCP6.0**	**RCP8.5**	**A2**	**B1**	**CLE**	**MFR**
2000	27.2 ± 2.9	27.2 ± 2.9	27.2 ± 2.9	27.2 ± 2.9	27.2 ± 2.9	27.2 ± 2.9	28.7	28.7
2010	0.1	0.1	0.0	0.1	1.2	0.6		
2020	–0.3	–0.2	–0.2	0.6	2.8	1.1		
2030	–1.1	–0.1	–0.3	1.0	4.4	1.3	0.7 ± 1.4	–2.3 ± 1.1
2040	–1.5	–0.3	–0.3	1.2	5.3	1.3		
2050	–1.9	–0.8	–0.4	1.5	6.2	0.8		
2060	–2.4	–1.3	–0.5	1.8	7.1	0.2		
2070	–3.0	–1.9	–1.0	1.9	8.0	–0.5		
2080	–3.5	–2.5	–1.5	1.9	9.2	–1.1		
2090	–3.8	–2.8	–2.1	1.9	10.6	–1.7		
2100	–4.2	–3.0	–2.8	1.9	11.9	–2.5		

	CMIP5				ACCMIP			
Year	**RCP2.6**	**RCP4.5**	**RCP6.0**	**RCP8.5**	**RCP2.6**	**RCP4.5**	**RCP6.0**	**RCP8.5**
2000	30.0 ± 4.2	30.0 ± 4.2	30.0 ± 4.2	30.0 ± 4.2	28.1 ± 3.1	28.1 ± 3.2	28.1± 3.1	28.1 ± 3.1
2010	–0.4	–0.2	–0.6	–0.1				
2020	–0.9	–0.3	–0.9	0.7				
2030	–1.8	–0.2	–1.1	1.5	–1.4	0.3	–0.6	1.7
2040	–2.3	–0.3	–1.2	2.0				
2050	–2.9	–0.9	–1.5	2.5				
2060	–4.0	–1.7	–1.9	2.9				
2070	–5.4	–2.8	–2.8	3.1				
2080	–6.4	–3.7	–3.9	3.0				
2090	–6.9	–4.1	–4.8	2.8				
2100	–7.2	–4.3	–5.6	2.7	–6.3	–3.5	–4.9	3.4

Notes:

HTAP results are from Wild et al. (2012) and use the published O_3 sensitivities to regional emissions from the HTAP multi-model study (HTAP 2010) and scale those O_3 changes to the RCP emission scenarios. The ±1 standard deviation (68% confidence interval) over the range of 14 parametric models is shown for year 2000 and is similar for all years. Results from the SRES A2 and B1 scenarios are from the TAR OxComp studies diagnosed by Wild (Prather et al., 2001; 2003). CLE and MFR results (Dentener et al., 2005; 2006) include uncertainty (standard deviation of model results) in the change since year 2000, and CLE alone includes climate effects. The CMIP5 and ACCMIP results are from V. Naik and A. Fiore based on Fiore et al. (2012) and include the standard deviation over the models in year 2000, which is similar for following years. This does not necessarily reflect the uncertainty in the projected change, which may be smaller, see Fiore et al. (2012). The difference in year 2000 between CMIP5 (4 models) and ACCMIP (12 models) reflect different model biases. Even though ACCMIP only has three decades (2000, 2030, 2100), the greater number of models (5 to 11 depending on time slice and scenario) makes this a more robust estimate. See Chapter 11, ES, Section 11.3.5.2.2.

Table AII.7.2 | Surface O_3 change (ppb) for HTAP regions

North America						
Year	RCP2.6	RCP4.5	RCP6.0	RCP8.5	A2	B1
2000	36.1 ± 3.2	36.1 ± 3.2	36.1 ± 3.2	36.1 ± 3.2	36.1 ± 3.2	36.1 ± 3.2
2010	–0.8	–1.1	–0.1	–1.5	1.5	0.4
2020	–1.9	–2.3	–0.9	–1.4	3.6	0.5
2030	–3.7	–2.7	–1.5	–1.1	5.3	–0.1
2040	–4.6	–3.2	–1.9	–1.1	6.2	–0.8
2050	–5.6	–3.9	–2.4	–0.9	6.9	–1.9
2060	–6.5	–4.6	–3.0	–0.7	7.9	–2.9
2070	–7.5	–5.3	–4.0	–0.7	8.8	–3.8
2080	–8.2	–6.1	–4.9	–0.7	10.3	–4.5
2090	–8.5	–6.4	–5.7	–0.8	12.2	–5.2
2100	–8.9	–6.6	–6.7	–0.9	13.9	–6.1

Europe						
Year	RCP2.6	RCP4.5	RCP6.0	RCP8.5	A2	B1
2000	37.8 ± 3.7	37.8 ± 3.7	37.8 ± 3.7	37.8 ± 3.7	37.8 ± 3.7	37.8 ± 3.7
2010	–0.5	–0.3	–0.1	–0.7	1.5	0.3
2020	–1.4	–1.3	–0.7	–0.2	3.7	0.6
2030	–3.0	–1.4	–1.1	0.1	5.7	0.2
2040	–3.8	–1.9	–1.5	0.1	6.7	–0.3
2050	–4.6	–2.7	–2.0	0.3	7.7	–1.2
2060	–5.6	–3.5	–2.6	0.4	8.8	–2.1
2070	–6.6	–4.3	–3.3	0.4	9.8	–3.0
2080	–7.5	–5.1	–4.2	0.2	11.3	–3.8
2090	–8.0	–5.6	–5.2	–0.1	13.4	–4.6
2100	–8.5	–6.0	–6.4	–0.2	15.1	–5.6

South Asia						
Year	RCP2.6	RCP4.5	RCP6.0	RCP8.5	A2	B1
2000	39.6 ± 3.4	39.6 ± 3.4	39.6 ± 3.4	39.6 ± 3.4	39.6 ± 3.4	39.6 ± 3.4
2010	1.5	1.4	0.3	1.4	2.7	1.8
2020	1.6	2.2	0.0	3.9	6.1	3.3
2030	0.5	3.4	–0.6	5.0	8.9	3.9
2040	0.3	3.5	–0.1	5.5	10.4	4.1
2050	0.2	2.9	0.0	5.2	11.7	2.9
2060	–0.1	1.1	0.4	5.1	12.7	1.5
2070	–1.0	–1.2	–0.2	4.9	13.6	–0.1
2080	–2.6	–3.9	–1.7	4.9	14.5	–1.5
2090	–4.4	–5.0	–3.0	4.1	15.1	–3.0
2100	–6.8	–6.0	–4.7	4.0	15.0	–4.6

(continued on next page)

Table AII.7.2 | *(continued)*

	East Asia					
Year	RCP2.6	RCP4.5	RCP6.0	RCP8.5	A2	B1
2000	35.6 ± 2.7	35.6 ± 2.7	35.6 ± 2.7	35.6 ± 2.7	35.6 ± 2.7	35.6 ± 2.7
2010	1.0	0.6	0.5	1.3	2.0	1.1
2020	0.5	0.6	0.4	2.5	4.6	1.9
2030	–1.4	0.2	0.6	2.8	6.8	2.1
2040	–2.7	–0.8	1.4	1.8	8.0	2.0
2050	–3.8	–2.5	1.4	1.4	9.1	0.9
2060	–4.8	–3.6	0.9	1.4	10.2	–0.3
2070	–6.0	–4.6	–0.7	1.2	11.2	–1.4
2080	–6.9	–5.5	–2.2	1.0	12.5	–2.4
2090	–7.4	–5.8	–3.5	0.7	13.9	–3.4
2100	–8.0	–6.0	–4.9	0.5	14.9	–4.6

Notes:
HTAP results from Wild et al. (2012); see Table AII.7.1.

Table AII.7.3 | Surface O_3 change (ppb) from CMIP5/ACCMIP for continental regions

	Africa							
	CMIP5				ACCMIP			
Year	RCP2.6	RCP4.5	RCP6.0	RCP8.5	RCP2.6	RCP4.5	RCP6.0	RCP8.5
2000	33.8 ± 4.3	33.8 ± 4.3	33.8 ± 4.3	33.8 ± 4.3	33.1 ± 4.1	33.1 ± 4.1	33.1 ± 4.1	33.1 ± 4.1
2010	–0.7	–0.1	–1.2	–0.2				
2020	–1.0	0.2	–1.5	0.9				
2030	–1.9	0.5	–1.8	1.7	–1.4	0.9	–1.3	2.4
2040	–2.0	0.6	–1.8	2.6				
2050	–2.3	0.2	–2.0	3.2				
2060	–2.6	–0.3	–2.2	3.7				
2070	–3.2	–1.2	–2.8	4.0				
2080	–3.6	–2.3	–3.7	4.1				
2090	–4.1	–3.0	–4.5	4.1				
2100	–4.8	–3.3	–5.2	4.1	–4.9	–2.9	–4.9	5.0

	Australia							
	CMIP5				ACCMIP			
Year	RCP2.6	RCP4.5	RCP6.0	RCP8.5	RCP2.6	RCP4.5	RCP6.0	RCP8.5
2000	23.3 ± 4.6	23.3 ± 4.6	23.3 ± 4.6	23.3 ± 4.6	23.7 ± 3.5	23.7 ± 3.5	23.7 ± 3.5	23.7 ± 3.5
2010	–1.3	–1.1	–0.8	–0.9				
2020	–1.7	–1.4	–1.0	–0.6				
2030	–2.3	–1.3	–1.4	0.0	–1.8	–0.4	–1.4	0.9
2040	–2.6	–1.2	–1.7	0.5				
2050	–3.0	–1.5	–1.9	0.9				
2060	–3.7	–1.9	–2.0	1.5				
2070	–4.4	–2.4	–2.5	1.8				
2080	–5.0	–2.9	–3.1	1.9				
2090	–5.0	–3.1	–3.5	1.9				
2100	–5.2	–3.2	–4.0	2.0	–4.3	–2.5	–4.0	3.1

(continued on next page)

AII

Table AII.7.3 | *(continued)*

Central Eurasia								
	CMIP5				ACCMIP			
Year	RCP2.6	RCP4.5	RCP6.0	RCP8.5	RCP2.6	RCP4.5	RCP6.0	RCP8.5
2000	38.7 ± 5.3	38.7 ± 5.3	38.7 ± 5.3	38.7 ± 5.3	32.5 ± 6.2	32.5 ± 6.2	32.5 ± 6.2	32.5 ± 6.2
2010	−0.6	−0.6	−0.6	−0.5				
2020	−1.6	−1.2	−1.2	0.5				
2030	−3.2	−1.3	−1.4	1.4	−1.9	−0.1	−0.3	1.8
2040	−4.5	−1.9	−1.7	1.6				
2050	−5.7	−2.9	−2.2	1.8				
2060	−7.2	−4.2	−3.0	2.8				
2070	−9.1	−5.4	−4.3	3.0				
2080	−10.6	−6.5	−6.0	2.9				
2090	−11.2	−6.8	−7.2	2.6				
2100	−11.5	−7.0	−8.1	2.6	−8.5	−3.8	−5.6	4.3

Europe								
	CMIP5				ACCMIP			
Year	RCP2.6	RCP4.5	RCP6.0	RCP8.5	RCP2.6	RCP4.5	RCP6.0	RCP8.5
2000	40.4 ± 6.0	40.4 ± 6.0	40.4 ± 6.0	40.4 ± 6.0	33.6 ± 5.2	33.6 ± 5.2	33.6 ± 5.2	33.6 ± 5.2
2010	−0.4	−0.5	−0.5	−0.4				
2020	−1.5	−1.3	−1.2	0.3				
2030	−3.2	−1.7	−1.7	1.1	−1.6	0.6	−0.4	2.3
2040	−4.6	−2.4	−2.3	1.4				
2050	−6.1	−3.5	−3.0	1.8				
2060	−8.0	−4.9	−4.1	2.4				
2070	−10.4	−6.3	−5.8	2.6				
2080	−12.2	−7.6	−7.6	2.3				
2090	−13.0	−8.0	−9.2	2.1				
2100	−13.4	−8.1	−10.3	2.0	−9.4	−3.5	−7.2	4.9

East Asia								
	CMIP5				ACCMIP			
Year	RCP2.6	RCP4.5	RCP6.0	RCP8.5	RCP2.6	RCP4.5	RCP6.0	RCP8.5
2000	46.3 ± 4.9	46.3 ± 4.9	46.3 ± 4.9	46.3 ± 4.9	41.0 ± 5.5	41.0 ± 5.5	41.0 ± 5.5	41.0 ± 5.5
2010	0.8	0.6	0.1	1.1				
2020	−0.1	0.8	−0.1	2.7				
2030	−2.3	0.5	0.4	3.8	−1.8	1.0	0.4	3.2
2040	−3.9	−0.9	1.1	3.8				
2050	−5.8	−3.3	1.0	3.7				
2060	−8.0	−5.4	0.2	3.9				
2070	−10.2	−7.3	−1.6	3.6				
2080	−12.1	−8.8	−4.0	3.3				
2090	−13.2	−9.4	−6.3	2.9				
2100	−13.9	−9.6	−8.0	2.8	−11.4	−5.9	−6.6	4.6

(continued on next page)

AII

Table AII.7.3 | *(continued)*

Middle East								
	CMIP5				ACCMIP			
Year	RCP2.6	RCP4.5	RCP6.0	RCP8.5	RCP2.6	RCP4.5	RCP6.0	RCP8.5
2000	45.9 ± 3.1	45.9 ± 3.1	45.9 ± 3.1	45.9 ± 3.1	45.7 ± 5.4	45.7 ± 5.4	45.7 ± 5.4	45.7 ± 5.4
2010	−0.4	0.5	−0.7	0.5				
2020	−1.5	0.4	−1.4	2.5				
2030	−3.3	0.6	−1.6	3.8	−2.8	0.9	−1.1	4.1
2040	−3.6	0.2	−2.0	4.4				
2050	−4.6	−0.9	−2.6	4.7				
2060	−6.0	−2.7	−3.5	5.2				
2070	−8.1	−4.9	−4.2	5.1				
2080	−9.9	−7.1	−5.9	5.1				
2090	−11.3	−8.4	−8.2	4.8				
2100	−12.4	−9.0	−9.9	4.6	−11.7	−7.5	−9.8	5.0

North America								
	CMIP5				ACCMIP			
Year	RCP2.6	RCP4.5	RCP6.0	RCP8.5	RCP2.6	RCP4.5	RCP6.0	RCP8.5
2000	40.7 ± 5.1	40.7 ± 5.1	40.7 ± 5.1	40.7 ± 5.1	34.3 ± 5.5	34.3 ± 5.5	34.3 ± 5.5	34.3 ± 5.5
2010	−0.9	−1.2	−0.6	−1.0				
2020	−2.1	−2.4	−1.4	−0.5				
2030	−4.3	−2.8	−1.8	0.1	−2.5	−0.7	−0.8	1.3
2040	−5.7	−3.6	−2.5	0.3				
2050	−7.2	−4.6	−3.1	0.6				
2060	−9.1	−5.8	−4.4	1.0				
2070	−11.4	−7.1	−6.2	1.2				
2080	−13.2	−8.3	−8.1	1.2				
2090	−13.8	−8.5	−9.6	1.0				
2100	−14.1	−8.8	−10.9	0.9	−10.5	−4.7	−8.7	3.4

South America								
	CMIP5				ACCMIP			
Year	RCP2.6	RCP4.5	RCP6.0	RCP8.5	RCP2.6	RCP4.5	RCP6.0	RCP8.5
2000	25.3 ± 4.2	25.3 ± 4.2	25.3 ± 4.2	25.3 ± 4.2	23.7 ± 3.9	23.7 ± 3.9	23.7 ± 3.9	23.7 ± 3.9
2010	−1.4	−0.6	−1.2	−0.3				
2020	−2.1	−1.2	−1.8	0.3				
2030	−2.9	−1.2	−2.1	0.6	−2.3	−0.6	−1.8	1.2
2040	−2.9	−1.3	−2.3	1.1				
2050	−3.2	−1.7	−2.6	1.3				
2060	−3.6	−2.5	−2.9	1.5				
2070	−4.3	−3.6	−3.5	1.5				
2080	−5.1	−4.5	−4.2	1.1				
2090	−5.5	−5.0	−4.7	0.7				
2100	−5.7	−5.2	−5.3	0.4	−5.0	−4.0	−5.2	2.0

(continued on next page)

AII

Table AII.7.3 | *(continued)*

	South Asia							
	CMIP5				ACCMIP			
Year	**RCP2.6**	**RCP4.5**	**RCP6.0**	**RCP8.5**	**RCP2.6**	**RCP4.5**	**RCP6.0**	**RCP8.5**
2000	34.4 ± 3.9	34.4 ± 3.9	34.4 ± 3.9	34.4 ± 3.9	33.7 ± 4.6	33.7 ± 4.6	33.7 ± 4.6	33.7 ± 4.6
2010	1.3	0.9	−0.1	1.3				
2020	1.4	1.6	−0.2	3.1				
2030	0.7	2.7	−0.1	3.9	0.6	2.3	−0.4	4.6
2040	0.6	2.8	0.3	4.0				
2050	0.4	1.6	0.4	3.6				
2060	−0.5	−0.7	0.3	3.2				
2070	−2.0	−3.2	−0.5	2.9				
2080	−3.9	−5.7	−2.0	2.7				
2090	−5.7	−6.7	−3.3	2.2				
2100	−7.1	−7.3	−4.5	1.9	−7.2	−6.1	−4.5	3.6

Notes:
See notes for Table AII.7.1. For definition of regions, see Figure 11.23 and Fiore et al. (2012).

Table AII.7.4 | Surface particulate matter change ($\log_{10}$[$PM_{2.5}$ (microgram/m^3)]) from CMIP5/ACCMIP for continental regions

	Africa			
Year	**RCP2.6**	**RCP4.5**	**RCP6.0**	**RCP8.5**
2000	1.17 ± 0.23			
2030	0.00	0.04	−0.01	0.01
2050	−0.02		−0.02	0.01
2100	0.00	−0.01	−0.03	−0.02

	Australia			
Year	**RCP2.6**	**RCP4.5**	**RCP6.0**	**RCP8.5**
2000	0.65 ± 0.32			
2030	−0.04	0.03	−0.01	0.01
2050	−0.06		−0.02	−0.04
2100	0.00	0.00	−0.03	−0.01

	Central Eurasia			
Year	**RCP2.6**	**RCP4.5**	**RCP6.0**	**RCP8.5**
2000	0.59 ± 0.17			
2030	−0.07	−0.01	−0.05	−0.06
2050	−0.12		−0.08	−0.09
2100	−0.13	−0.11	−0.11	−0.12

	Europe			
Year	**RCP2.6**	**RCP4.5**	**RCP6.0**	**RCP8.5**
2000	0.81 ± 0.09			
2030	−0.20	−0.10	−0.13	−0.24
2050	−0.31		−0.25	−0.33
2100	−0.32	−0.28	−0.37	−0.38

(continued on next page)

Table AII.7.4 | *(continued)*

East Asia				
Year	**RCP2.6**	**RCP4.5**	**RCP6.0**	**RCP8.5**
2000	1.04 ± 0.16			
2030	–0.04	–0.02	0.01	0.01
2050	–0.24		0.07	–0.17
2100	–0.31	–0.33	–0.21	–0.30

Middle East				
Year	**RCP2.6**	**RCP4.5**	**RCP6.0**	**RCP8.5**
2000	1.10 ± 0.27			
2030	–0.06	–0.02	–0.05	–0.03
2050	–0.08		–0.06	–0.03
2100	–0.11	–0.11	–0.10	–0.12

North America				
Year	**RCP2.6**	**RCP4.5**	**RCP6.0**	**RCP8.5**
2000	0.51 ± 0.15			
2030	–0.16	–0.10	–0.10	–0.15
2050	–0.20		–0.16	–0.17
2100	–0.20	–0.19	–0.24	–0.21

South America				
Year	**RCP2.6**	**RCP4.5**	**RCP6.0**	**RCP8.5**
2000	0.71 ± 0.11			
2030	–0.05	–0.04	–0.04	–0.03
2050	–0.10		–0.05	–0.07
2100	–0.11	–0.11	–0.09	–0.12

South Asia				
Year	**RCP2.6**	**RCP4.5**	**RCP6.0**	**RCP8.5**
2000	1.02 ± 0.11			
2030	0.04	0.02	0.03	0.05
2050	–0.05		0.07	0.00
2100	–0.16	–0.24	–0.06	–0.11

Notes:

Decadal average of the $log_{10}[PM_{2.5}]$ values are given only where results include at least four models from either ACCMIP or CMIP5. Results are from A. Fiore and V. Naik based on Fiore et al. (2012) using the CMIP5/ACCMIP archive. Due to the very large systematic spread across models, the statistics were calculated for the log values, but Figure 11.23 shows statistics for direct $PM_{2.5}$ values. Owing to the large spatial variations no global average is given. Model mean and standard deviation are shown for year 2000; differences in $log_{10}[PM_{2.5}]$ are shown for 2030, 2050 and 2100. See notes for Table AII.7.3 and Figure 11.23 for regions; see also Chapter 11, ES.

Table AII.7.5 | CMIP5 (RCP) and CMIP3 (SRES A1B) global mean surface temperature change (°C) relative to 1986–2005 reference period. Results here are a statistical summary of the spread in the CMIP ensembles for each of the scenarios. They do not account for model biases and model dependencies, and the percentiles do not correspond to the assessed uncertainty in Chapters 11 (11.3.6.3) and 12 (12.4.1). The statistical spread across models cannot be interpreted as uncertainty ranges or in terms of calibrated language (Section 12.2).

	RCP2.6					RCP4.5				
Years	**5%**	**17%**	**50%**	**83%**	**95%**	**5%**	**17%**	**50%**	**83%**	**95%**
1850–1990			–0.61					–0.61		
1986–2005			0.00					0.00		
2010d	0.19	0.33	0.36	0.52	0.62	0.22	0.26	0.36	0.48	0.59
2020[d]	0.36	0.45	0.55	0.81	1.07	0.39	0.48	0.59	0.74	0.83
2030[d]	0.47	0.56	0.74	1.02	1.24	0.56	0.69	0.82	1.10	1.22
2040[d]	0.51	0.68	0.88	1.25	1.50	0.64	0.86	1.04	1.35	1.57
2050[d]	0.49	0.71	0.94	1.37	1.65	0.84	1.05	1.24	1.63	1.97
2060[d]	0.36	0.69	0.93	1.48	1.71	0.90	1.13	1.44	1.90	2.19
2070[d]	0.20	0.70	0.89	1.49	1.71	0.98	1.20	1.54	2.07	2.32
2080[d]	0.15	0.62	0.94	1.44	1.79	0.98	1.27	1.62	2.25	2.54
2090[d]	0.18	0.58	0.94	1.53	1.79	1.06	1.33	1.68	2.29	2.59

	RCP6.0					RCP8.5				
Years	**5%**	**17%**	**50%**	**83%**	**95%**	**5%**	**17%**	**50%**	**83%**	**95%**
1850–1990			–0.61					–0.61		
1986–2005			0.00					0.00		
2010[d]	0.21	0.26	0.36	0.47	0.64	0.23	0.29	0.37	0.47	0.62
2020[d]	0.33	0.40	0.55	0.70	0.90	0.37	0.51	0.66	0.84	0.99
2030[d]	0.40	0.59	0.74	0.92	1.17	0.65	0.77	0.94	1.29	1.39
2040[d]	0.59	0.73	0.95	1.21	1.41	0.93	1.13	1.29	1.68	1.77
2050[d]	0.69	0.92	1.15	1.52	1.81	1.20	1.48	1.70	2.19	2.37
2060[d]	0.88	1.08	1.32	1.78	2.18	1.55	1.88	2.16	2.74	2.99
2070[d]	1.08	1.28	1.58	2.14	2.52	1.96	2.25	2.60	3.31	3.61
2080[d]	1.33	1.56	1.81	2.58	2.88	2.31	2.65	3.05	3.93	4.22
2090[d]	1.51	1.72	2.03	2.92	3.24	2.63	2.96	3.57	4.45	4.81

	SRES A1B				
Years	**5%**	**17%**	**50%**	**83%**	**95%**
1850–1990			–0.61		
1986–2005			0.00		
2010[d]	0.15	0.22	0.34	0.44	0.62
2020[d]	0.27	0.37	0.52	0.76	0.91
2030[d]	0.47	0.59	0.82	1.04	1.38
2040[d]	0.65	0.90	1.11	1.36	1.79
2050[d]	0.92	1.14	1.55	1.65	2.14
2060[d]	1.12	1.40	1.75	1.98	2.67
2070[d]	1.40	1.60	2.14	2.39	3.12
2080[d]	1.61	1.80	2.30	2.75	3.47
2090[d]	1.76	1.96	2.54	3.05	3.84

Notes:

This spread in the model ensembles (as shown in Figures 11.26a and 12.5, and discussed in Section 11.3.6) is not a measure of uncertainty. For the AR5 assessment of global mean surface temperature changes and uncertainties see: Section 11.3.6.3 and Figure 11.25 for the near-term (2016–2035) temperatures; and Section 12.4.1 and Tables 12.2–3 for the long term (2081–2100). See discussion about uncertainty and ensembles in Section 12.2, which explains how model spread is not equivalent to uncertainty. Results here are shown for the CMIP5 archive (Annex I, frozen as of March 15, 2013) for the RCPs and the similarly current CMIP3 archive for SRES A1B, which is not the same set of models used in AR4 (Figure SPM.5). Ten-year averages are shown (2030[d] = 2026–2035). Temperature changes are relative to the reference period (1986–2005, defined as zero in this table), using CMIP5 for all four RCPs (G. J. van Oldenborgh, http://climexp.knmi.nl/; see Annex I for listing of models included) and CMIP3 for SRES A1B (22 models). The warming from early instrumental record (1850–1900) to the modern reference period (1986–2005) is derived from HadCRUT4 observations as 0.61°C (C. Morice; see Chapter 2 and Table AII.1.3).

Table AII.7.6 | Global mean surface temperature change (°C) relative to 1990 from the TAR

Years	A1B	A1T	A1FI	A2	B1	B2	IS92a	A1B
PI*	–0.33	–0.33	–0.33	–0.33	–0.33	–0.33	–0.33	–0.33
1990	0.00	0.00	0.00	0.00	0.00	0.00	0.00	0.00
2000	0.16	0.16	0.16	0.16	0.16	0.16	0.15	0.16
2010	0.30	0.40	0.32	0.35	0.34	0.39	0.27	0.30
2020	0.52	0.71	0.55	0.50	0.55	0.66	0.43	0.52
2030	0.85	1.03	0.85	0.73	0.77	0.93	0.61	0.85
2040	1.26	1.41	1.27	1.06	0.98	1.18	0.80	1.26
2050	1.59	1.75	1.86	1.42	1.21	1.44	1.00	1.59
2060	1.97	2.04	2.50	1.85	1.44	1.69	1.26	1.97
2070	2.30	2.25	3.10	2.33	1.63	1.94	1.52	2.30
2080	2.56	2.41	3.64	2.81	1.79	2.20	1.79	2.56
2090	2.77	2.49	4.09	3.29	1.91	2.44	2.08	2.77
2100	2.95	2.54	4.49	3.79	1.98	2.69	2.38	2.95

Notes:

Single-year estimates of mean surface air temperature warming relative to the reference period 1990 for the SRES scenarios evaluated in the TAR. The pre-industrial estimates are for 1750, and all results are based on a simple climate model. See TAR Appendix II.

Table AII.7.7 | Global mean sea level rise (m) with respect to 1986–2005 at 1 January on the years indicated. Values shown as median and *likely* range; see Section 13.5.1.

Year	SRES A1B	RCP2.6	RCP4.5	RCP6.0	RCP8.5
2007	0.03 [0.02 to 0.04]	0.03 [0.02 to 0.04]	0.03 [0.02 to 0.04]	0.03 [0.02 to 0.04]	0.03 [0.02 to 0.04]
2010	0.04 [0.03 to 0.05]	0.04 [0.03 to 0.05]	0.04 [0.03 to 0.05]	0.04 [0.03 to 0.05]	0.04 [0.03 to 0.05]
2020	0.08 [0.06 to 0.10]	0.08 [0.06 to 0.10]	0.08 [0.06 to 0.10]	0.08 [0.06 to 0.10]	0.08 [0.06 to 0.11]
2030	0.12 [0.09 to 0.16]	0.13 [0.09 to 0.16]	0.13 [0.09 to 0.16]	0.12 [0.09 to 0.16]	0.13 [0.10 to 0.17]
2040	0.17 [0.13 to 0.22]	0.17 [0.13 to 0.22]	0.17 [0.13 to 0.22]	0.17 [0.12 to 0.21]	0.19 [0.14 to 0.24]
2050	0.23 [0.17 to 0.30]	0.22 [0.16 to 0.28]	0.23 [0.17 to 0.29]	0.22 [0.16 to 0.28]	0.25 [0.19 to 0.32]
2060	0.30 [0.21 to 0.38]	0.26 [0.18 to 0.35]	0.28 [0.21 to 0.37]	0.27 [0.19 to 0.35]	0.33 [0.24 to 0.42]
2070	0.37 [0.26 to 0.48]	0.31 [0.21 to 0.41]	0.35 [0.25 to 0.45]	0.33 [0.24 to 0.43]	0.42 [0.31 to 0.54]
2080	0.44 [0.31 to 0.58]	0.35 [0.24 to 0.48]	0.41 [0.28 to 0.54]	0.40 [0.28 to 0.53]	0.51 [0.37 to 0.67]
2090	0.52 [0.36 to 0.69]	0.40 [0.26 to 0.54]	0.47 [0.32 to 0.62]	0.47 [0.33 to 0.63]	0.62 [0.45 to 0.81]
2100	0.60 [0.42 to 0.80]	0.44 [0.28 to 0.61]	0.53 [0.36 to 0.71]	0.55 [0.38 to 0.73]	0.74 [0.53 to 0.98]

Annex III: Glossary

Editor:
Serge Planton (France)

This annex should be cited as:
IPCC, 2013: Annex III: Glossary [Planton, S. (ed.)]. In: *Climate Change 2013: The Physical Science Basis. Contribution of Working Group I to the Fifth Assessment Report of the Intergovernmental Panel on Climate Change* [Stocker, T.F., D. Qin, G.-K. Plattner, M. Tignor, S.K. Allen, J. Boschung, A. Nauels, Y. Xia, V. Bex and P.M. Midgley (eds.)]. Cambridge University Press, Cambridge, United Kingdom and New York, NY, USA.

This glossary defines some specific terms as the Lead Authors intend them to be interpreted in the context of this report. Red, italicized words indicate that the term is defined in the Glossary.

Abrupt climate change A large-scale change in the *climate system* that takes place over a few decades or less, persists (or is anticipated to persist) for at least a few decades and causes substantial disruptions in human and natural systems.

Active layer The layer of ground that is subject to annual thawing and freezing in areas underlain by *permafrost*.

Adjustment time See *Lifetime*. See also *Response time*.

Advection Transport of water or air along with its properties (e.g., temperature, chemical tracers) by winds or currents. Regarding the general distinction between advection and *convection*, the former describes transport by large-scale motions of the *atmosphere* or ocean, while convection describes the predominantly vertical, locally induced motions.

Aerosol A suspension of airborne solid or liquid particles, with a typical size between a few nanometres and 10 μm that reside in the *atmosphere* for at least several hours. For convenience the term *aerosol*, which includes both the particles and the suspending gas, is often used in this report in its plural form to mean *aerosol particles*. Aerosols may be of either natural or *anthropogenic* origin. Aerosols may influence *climate* in several ways: directly through scattering and absorbing radiation (see *Aerosol–radiation interaction*) and indirectly by acting as *cloud condensation nuclei* or *ice nuclei*, modifying the optical properties and *lifetime* of clouds (see *Aerosol–cloud interaction*).

Aerosol–cloud interaction A process by which a perturbation to *aerosol* affects the microphysical properties and evolution of clouds through the aerosol role as *cloud condensation nuclei* or ice nuclei, particularly in ways that affect radiation or precipitation; such processes can also include the effect of clouds and precipitation on aerosol. The aerosol perturbation can be *anthropogenic* or come from some natural *source*. The *radiative forcing* from such interactions has traditionally been attributed to numerous *indirect aerosol effects*, but in this report, only two levels of radiative forcing (or effect) are distinguished:

Radiative forcing (or effect) due to aerosol–cloud interactions (RFaci) The radiative forcing (or *radiative effect*, if the perturbation is internally generated) due to the change in number or size distribution of cloud droplets or ice crystals that is the proximate result of an aerosol perturbation, with other variables (in particular total cloud water content) remaining equal. In liquid clouds, an increase in cloud droplet concentration and surface area would increase the cloud *albedo*. This effect is also known as the *cloud albedo effect*, *first indirect effect*, or *Twomey effect*. It is a largely theoretical concept that cannot readily be isolated in observations or comprehensive process models due to the rapidity and ubiquity of *rapid adjustments*.

Effective radiative forcing (or effect) due to aerosol–cloud interactions (ERFaci) The final radiative forcing (or effect) from the aerosol perturbation including the rapid adjustments to the initial change in droplet or crystal formation rate. These adjustments include changes in the strength of *convection*, precipitation efficiency, cloud fraction, *lifetime* or water content of clouds, and the formation or suppression of clouds in remote areas due to altered circulations.

The total effective radiative forcing due to both aerosol–cloud and aerosol–radiation interactions is denoted *aerosol effective radiative forcing (ERFari+aci)*. See also *Aerosol–radiation interaction*.

Aerosol–radiation interaction An interaction of *aerosol* directly with radiation produce *radiative effects*. In this report two levels of radiative forcing (or effect) are distinguished:

Radiative forcing (or effect) due to aerosol–radiation interactions (RFari) The *radiative forcing* (or radiative effect, if the perturbation is internally generated) of an aerosol perturbation due directly to aerosol–radiation interactions, with all environmental variables remaining unaffected. Traditionally known in the literature as the *direct aerosol forcing (or effect)*.

Effective radiative forcing (or effect) due to aerosol-radiation interactions (ERFari) The final radiative forcing (or effect) from the aerosol perturbation including the *rapid adjustments* to the initial change in radiation. These adjustments include changes in cloud caused by the impact of the radiative heating on convective or larger-scale atmospheric circulations, traditionally known as *semi-direct aerosol forcing (or effect)*.

The total effective radiative forcing due to both aerosol–cloud and aerosol–radiation interactions is denoted *aerosol effective radiative forcing (ERFari+aci)*. See also *Aerosol–cloud interaction*.

Afforestation Planting of new *forests* on lands that historically have not contained forests. For a discussion of the term *forest* and related terms such as *afforestation*, *reforestation* and *deforestation*, see the IPCC Special Report on Land Use, Land-Use Change and Forestry (IPCC, 2000). See also the report on Definitions and Methodological Options to Inventory Emissions from Direct Human-induced Degradation of Forests and Devegetation of Other Vegetation Types (IPCC, 2003).

Airborne fraction The fraction of total CO_2 emissions (from fossil fuel and land use change) remaining in the *atmosphere*.

Air mass A widespread body of air, the approximately homogeneous properties of which (1) have been established while that air was situated over a particular *region* of the Earth's surface, and (2) undergo specific modifications while in transit away from the source region (AMS, 2000).

Albedo The fraction of *solar radiation* reflected by a surface or object, often expressed as a percentage. Snow-covered surfaces have a high albedo, the albedo of soils ranges from high to low, and vegetation-covered surfaces and oceans have a low albedo. The Earth's planetary albedo varies mainly through varying cloudiness, snow, ice, leaf area and and cover changes.

Alkalinity A measure of the capacity of an aqueous solution to neutralize acids.

Altimetry A technique for measuring the height of the Earth's surface with respect to the geocentre of the Earth within a defined terrestrial reference frame (geocentric sea level).

Annular modes See *Northern Annular Mode (NAM)* and *Southern Annular Mode (SAM)*.

Anthropogenic Resulting from or produced by human activities.

Atlantic Multi-decadal Oscillation/Variability (AMO/AMV) A multi-decadal (65- to 75-year) fluctuation in the North Atlantic, in which *sea surface temperatures* showed warm phases during roughly 1860 to 1880 and 1930 to 1960 and cool phases during 1905 to 1925 and 1970 to 1990 with a range of approximately 0.4°C. See AMO Index, Box 2.5.

Atmosphere The gaseous envelope surrounding the Earth. The dry atmosphere consists almost entirely of nitrogen (78.1% *volume mixing ratio*) and oxygen (20.9% volume mixing ratio), together with a number of trace gases, such as argon (0.93% volume mixing ratio), helium and radiatively active *greenhouse gases* such as *carbon dioxide* (0.035%

volume mixing ratio) and *ozone*. In addition, the atmosphere contains the greenhouse gas water vapour, whose amounts are highly variable but typically around 1% volume mixing ratio. The atmosphere also contains clouds and *aerosols*.

Atmosphere–Ocean General Circulation Model (AOGCM) See *Climate model*.

Atmospheric boundary layer The atmospheric layer adjacent to the Earth's surface that is affected by friction against that boundary surface, and possibly by transport of heat and other variables across that surface (AMS, 2000). The lowest 100 m of the boundary layer (about 10% of the boundary layer thickness), where mechanical generation of turbulence is dominant, is called the *surface boundary layer* or *surface layer*.

Atmospheric lifetime See *Lifetime*.

Attribution See *Detection and attribution*.

Autotrophic respiration *Respiration* by *photosynthetic* (see *photosynthesis*) organisms (e.g., plants and algaes).

Basal lubrication Reduction of friction at the base of an *ice sheet* or *glacier* due to lubrication by meltwater. This can allow the glacier or ice sheet to slide over its base. Meltwater may be produced by pressure-induced melting, friction or geothermal heat, or surface melt may drain to the base through holes in the ice.

Baseline/reference The baseline (or reference) is the state against which change is measured. A *baseline period* is the period relative to which anomalies are computed. The baseline concentration of a trace gas is that measured at a location not influenced by local *anthropogenic* emissions.

Bayesian method/approach A Bayesian method is a method by which a statistical analysis of an unknown or uncertain quantity(ies) is carried out in two steps. First, a prior probability distribution for the uncertain quantity(ies) is formulated on the basis of existing knowledge (either by eliciting expert opinion or by using existing data and studies). At this first stage, an element of subjectivity may influence the choice, but in many cases, the prior probability distribution can be chosen as neutrally as possible, in order not to influence the final outcome of the analysis. In the second step, newly acquired data are used to update the prior distribution into a posterior distribution. The update is carried out either through an analytic computation or though numeric approximation, using a theorem formulated by and named after the British mathematician Thomas Bayes (1702–1761).

Biological pump The process of transporting carbon from the ocean's surface layers to the deep ocean by the primary production of marine phytoplankton, which converts dissolved inorganic carbon (DIC) and nutrients into organic matter through *photosynthesis*. This natural cycle is limited primarily by the availability of light and nutrients such as phosphate, nitrate and silicic acid, and micronutrients, such as iron. See also *Solubility pump*.

Biomass The total mass of living organisms in a given area or volume; dead plant material can be included as dead biomass. *Biomass burning* is the burning of living and dead vegetation.

Biome A biome is a major and distinct regional element of the *biosphere*, typically consisting of several *ecosystems* (e.g., *forests*, rivers, ponds, swamps within a *region*). Biomes are characterized by typical communities of plants and animals.

Biosphere (terrestrial and marine) The part of the Earth system comprising all *ecosystems* and living organisms, in the *atmosphere*, on land (*terrestrial biosphere*) or in the oceans (*marine biosphere*), including derived dead organic matter, such as litter, soil organic matter and oceanic detritus.

Black carbon (BC) Operationally defined *aerosol* species based on measurement of light absorption and chemical reactivity and/or thermal stability. It is sometimes referred to as *soot*.

Blocking Associated with persistent, slow-moving high-pressure systems that obstruct the prevailing westerly winds in the middle and high latitudes and the normal eastward progress of extratropical transient storm systems. It is an important component of the intraseasonal *climate variability* in the extratropics and can cause long-lived weather conditions such as cold spells in winter and summer *heat waves*.

Brewer–Dobson circulation The meridional overturning circulation of the *stratosphere* transporting air upward in the tropics, poleward to the winter hemisphere, and downward at polar and subpolar latitudes. The Brewer–Dobson circulation is driven by the interaction between upward propagating planetary waves and the mean flow.

Burden The total mass of a gaseous substance of concern in the *atmosphere*.

^{13}C Stable *isotope* of carbon having an atomic weight of approximately 13. Measurements of the ratio of $^{13}C/^{12}C$ in *carbon dioxide* molecules are used to infer the importance of different *carbon cycle* and climate processes and the size of the terrestrial carbon *reservoir*.

^{14}C Unstable *isotope* of carbon having an atomic weight of approximately 14, and a half-life of about 5700 years. It is often used for dating purposes going back some 40 kyr. Its variation in time is affected by the magnetic fields of the Sun and Earth, which influence its production from cosmic rays (see *Cosmogenic radioisotopes*).

AIII

Calving The breaking off of discrete pieces of ice from a *glacier*, *ice sheet* or an *ice shelf* into lake or seawater, producing icebergs. This is a form of mass loss from an ice body. See also *Mass balance/budget (of glaciers or ice sheets)*.

Carbonaceous aerosol *Aerosol* consisting predominantly of organic substances and *black carbon*.

Carbon cycle The term used to describe the flow of carbon (in various forms, e.g., as *carbon dioxide*) through the *atmosphere*, ocean, terrestrial and marine *biosphere* and *lithosphere*. In this report, the reference unit for the global carbon cycle is GtC or equivalently PgC (10^{15}g).

Carbon dioxide (CO_2) A naturally occurring gas, also a by-product of burning fossil fuels from fossil carbon deposits, such as oil, gas and coal, of *burning biomass*, of *land use* changes and of industrial processes (e.g., cement production). It is the principal *anthropogenic greenhouse gas* that affects the Earth's radiative balance. It is the reference gas against which other greenhouse gases are measured and therefore has a *Global Warming Potential* of 1.

Carbon dioxide (CO_2) fertilization The enhancement of the growth of plants as a result of increased atmospheric *carbon dioxide (CO_2)* concentration.

Carbon Dioxide Removal (CDR) Carbon Dioxide Removal methods refer to a set of techniques that aim to remove *CO_2* directly from the *atmosphere* by either (1) increasing natural *sinks* for carbon or (2) using chemical engineering to remove the CO_2, with the intent of reducing the atmospheric CO_2 concentration. CDR methods involve the ocean, land and technical systems, including such methods as *iron fertilization*, large-scale *afforestation* and direct capture of CO_2 from the atmosphere using engineered chemical means. Some CDR methods fall under the category of *geoengineering*, though this may not be the case for others, with the distinction being based on the magnitude, scale, and impact of the particular CDR activities. The boundary between CDR and *mitigation* is not clear and there could be some overlap between the two given current definitions (IPCC, 2012, p. 2). See also *Solar Radiation Management (SRM)*.

CFC See *Halocarbons*.

Chaotic A *dynamical system* such as the *climate system*, governed by nonlinear deterministic equations (see *Nonlinearity*), may exhibit erratic or chaotic behaviour in the sense that very small changes in the initial state of the system in time lead to large and apparently unpredictable changes in its temporal evolution. Such chaotic behaviour limits the *predictability* of the state of a nonlinear dynamical system at specific future times, although changes in its statistics may still be predictable given changes in the system parameters or boundary conditions.

Charcoal Material resulting from charring of *biomass*, usually retaining some of the microscopic texture typical of plant tissues; chemically it consists mainly of carbon with a disturbed graphitic structure, with lesser amounts of oxygen and hydrogen.

Chronology Arrangement of events according to dates or times of occurrence.

Clathrate (methane) A partly frozen slushy mix of *methane* gas and ice, usually found in sediments.

Clausius–Clapeyron equation/relationship The thermodynamic relationship between small changes in temperature and vapour pressure in an equilibrium system with condensed phases present. For trace gases such as water vapour, this relation gives the increase in equilibrium (or saturation) water vapour pressure per unit change in air temperature.

AIII

Climate Climate in a narrow sense is usually defined as the average weather, or more rigorously, as the statistical description in terms of the mean and variability of relevant quantities over a period of time ranging from months to thousands or millions of years. The classical period for averaging these variables is 30 years, as defined by the World Meteorological Organization. The relevant quantities are most often surface variables such as temperature, precipitation and wind. Climate in a wider sense is the state, including a statistical description, of the *climate system*.

Climate–carbon cycle feedback A *climate feedback* involving changes in the properties of land and ocean *carbon cycle* in response to *climate change*. In the ocean, changes in oceanic temperature and circulation could affect the *atmosphere*–ocean CO_2 flux; on the continents, climate change could affect plant *photosynthesis* and soil microbial *respiration* and hence the flux of CO_2 between the atmosphere and the land *biosphere*.

Climate change Climate change refers to a change in the state of the *climate* that can be identified (e.g., by using statistical tests) by changes in the mean and/or the variability of its properties, and that persists for an extended period, typically decades or longer. Climate change may be due to natural internal processes or *external forcings* such as modulations of the *solar cycles*, volcanic eruptions and persistent *anthropogenic* changes in the composition of the *atmosphere* or in *land use*. Note that the *Framework Convention on Climate Change (UNFCCC)*, in its Article 1, defines climate change as: 'a change of climate which is attributed directly or indirectly to human activity that alters the composition of the global atmosphere and which is in addition to natural *climate variability* observed over comparable time periods'. The UNFCCC thus makes a distinction between climate change attributable to human activities altering the atmospheric composition, and climate variability attributable to natural causes. See also *Climate change commitment, Detection and Attribution*.

Climate change commitment Due to the thermal inertia of the ocean and slow processes in the *cryosphere* and land surfaces, the *climate* would continue to change even if the atmospheric composition were held fixed at today's values. Past change in atmospheric composition leads to a *committed climate change*, which continues for as long as a radiative imbalance persists and until all components of the *climate system* have adjusted to a new state. The further change in temperature after the composition of the *atmosphere* is held constant is referred to as the *constant composition temperature commitment* or simply *committed warming* or *warming commitment*. Climate change commitment includes other future changes, for example, in the *hydrological cycle*, in *extreme weather events*, in *extreme climate events*, and in *sea level change*. The *constant emission commitment* is the committed climate change that would result from keeping *anthropogenic* emissions constant and the *zero emission commitment* is the climate change commitment when emissions are set to zero. See also *Climate change*.

Climate feedback An interaction in which a perturbation in one climate quantity causes a change in a second, and the change in the second quantity ultimately leads to an additional change in the first. A negative *feedback* is one in which the initial perturbation is weakened by the changes it causes; a positive feedback is one in which the initial perturbation is enhanced. In this Assessment Report, a somewhat narrower definition is often used in which the climate quantity that is perturbed is the *global mean surface temperature*, which in turn causes changes in the global radiation budget. In either case, the initial perturbation can either be externally forced or arise as part of *internal variability*. See also *Climate Feedback Parameter*.

Climate Feedback Parameter A way to quantify the radiative response of the *climate system* to a *global mean surface temperature* change induced by a *radiative forcing*. It varies as the inverse of the *effective climate sensitivity*. Formally, the Climate Feedback Parameter (α; units: W m^{-2} °C^{-1}) is defined as: $\alpha = (\Delta Q - \Delta F)/\Delta T$, where Q is the global mean radiative forcing, T is the global mean air surface temperature, F is the heat flux into the ocean and Δ represents a change with respect to an unperturbed *climate*.

Climate forecast See *Climate prediction*.

Climate index A time series constructed from climate variables that provides an aggregate summary of the state of the *climate system*. For example, the difference between sea level pressure in Iceland and the Azores provides a simple yet useful historical *NAO* index. Because of their optimal properties, climate indices are often defined using *principal components*—linear combinations of climate variables at different locations that have maximum variance subject to certain normalisation constraints (e.g., the *NAM* and *SAM* indices which are principal components of Northern Hemisphere and Southern Hemisphere gridded pressure anomalies, respectively). See Box 2.5 for a summary of definitions for established observational indices. See also *Climate pattern*.

Climate model (spectrum or hierarchy) A numerical representation of the *climate system* based on the physical, chemical and biological properties of its components, their interactions and *feedback* processes, and accounting for some of its known properties. The climate system can be represented by models of varying complexity, that is, for any one component or combination of components a *spectrum* or *hierarchy* of models can be identified, differing in such aspects as the number of spatial dimensions, the extent to which physical, chemical or biological processes are explicitly represented or the level at which empirical *parametrizations* are involved. Coupled *Atmosphere–Ocean General Circulation Models (AOGCMs)* provide a representation of the climate system that is near or at the most comprehensive end of the spectrum currently available. There is an evolution towards more complex models with interactive chemistry and biology. Climate models are applied as a research tool to study and simulate the *climate*, and for operational purposes, including monthly, seasonal and interannual *climate predictions*. See also *Earth System Model, Earth-System Model of Intermediate Complexity, Energy Balance Model, Process-based Model, Regional Climate Model* and *Semi-empirical model*.

Climate pattern A set of spatially varying coefficients obtained by "projection" (regression) of climate variables onto a *climate index* time series. When the climate index is a principal component, the climate pattern is an eigenvector of the covariance matrix, referred to as an *Empirical Orthogonal Function (EOF)* in climate science.

Climate prediction A climate prediction or *climate forecast* is the result of an attempt to produce (starting from a particular state of the *climate system*) an estimate of the actual evolution of the *climate* in the future, for example, at seasonal, interannual or decadal time scales. Because the future evolution of the climate system may be highly sensitive to initial conditions, such predictions are usually probabilistic in nature. See also *Climate projection*, *Climate scenario*, *Model initialization* and *Predictability*.

Climate projection A climate *projection* is the simulated response of the *climate system* to a *scenario* of future emission or concentration of *greenhouse gases* and *aerosols*, generally derived using *climate models*. Climate projections are distinguished from *climate predictions* by their dependence on the emission/concentration/*radiative forcing* scenario used, which is in turn based on assumptions concerning, for example, future socioeconomic and technological developments that may or may not be realized. See also *Climate scenario*.

Climate regime A state of the *climate system* that occurs more frequently than nearby states due to either more persistence or more frequent recurrence. In other words, a cluster in climate state space associated with a local maximum in the *probability density function*.

Climate response See *Climate sensitivity*.

Climate scenario A plausible and often simplified representation of the future *climate*, based on an internally consistent set of climatological relationships that has been constructed for explicit use in investigating the potential consequences of *anthropogenic climate change*, often serving as input to impact models. *Climate projections* often serve as the raw material for constructing climate scenarios, but climate scenarios usually require additional information such as the observed current climate. A *climate change scenario* is the difference between a climate scenario and the current climate. See also *Emission scenario*, *scenario*.

Climate sensitivity In IPCC reports, *equilibrium climate sensitivity* (units: °C) refers to the equilibrium (steady state) change in the annual *global mean surface temperature* following a doubling of the atmospheric *equivalent carbon dioxide concentration*. Owing to computational constraints, the equilibrium climate sensitivity in a *climate model* is sometimes estimated by running an atmospheric general circulation model coupled to a mixed-layer ocean model, because equilibrium climate sensitivity is largely determined by atmospheric processes. Efficient models can be run to equilibrium with a dynamic ocean. The *climate sensitivity parameter* (units: °C $(W\ m^{-2})^{-1}$) refers to the equilibrium change in the annual global mean surface temperature following a unit change in *radiative forcing*.

The *effective climate sensitivity* (units: °C) is an estimate of the global mean surface temperature response to doubled *carbon dioxide* concentration that is evaluated from model output or observations for evolving non-equilibrium conditions. It is a measure of the strengths of the *climate feedbacks* at a particular time and may vary with forcing history and *climate* state, and therefore may differ from equilibrium climate sensitivity.

The *transient climate response* (units: °C) is the change in the global mean surface temperature, averaged over a 20-year period, centred at the time of atmospheric carbon dioxide doubling, in a climate model simulation in which CO_2 increases at 1% yr^{-1}. It is a measure of the strength and rapidity of the surface temperature response to *greenhouse gas* forcing.

Climate sensitivity parameter See *climate sensitivity*.

Climate system The climate system is the highly complex system consisting of five major components: the *atmosphere*, the *hydrosphere*, the *cryosphere*, the *lithosphere* and the *biosphere*, and the interactions between them. The climate system evolves in time under the influence of its own internal dynamics and because of *external forcings* such as volcanic eruptions, solar variations and *anthropogenic* forcings such as the changing composition of the atmosphere and *land use change*.

Climate variability Climate variability refers to variations in the mean state and other statistics (such as standard deviations, the occurrence of extremes, etc.) of the *climate* on all *spatial and temporal scales* beyond that of individual weather events. Variability may be due to natural internal processes within the *climate system* (*internal variability*), or to variations in natural or *anthropogenic external forcing* (*external variability*). See also *Climate change*.

Cloud condensation nuclei (CCN) The subset of *aerosol* particles that serve as an initial site for the condensation of liquid water, which can lead to the formation of cloud droplets, under typical cloud formation conditions. The main factor that determines which aerosol particles are CCN at a given supersaturation is their size.

Cloud feedback A *climate feedback* involving changes in any of the properties of clouds as a response to a change in the local or *global mean surface temperature*. Understanding cloud feedbacks and determining their magnitude and sign require an understanding of how a change in *climate* may affect the spectrum of cloud types, the cloud fraction and height, the radiative properties of clouds, and finally the Earth's radiation budget. At present, cloud feedbacks remain the largest source of *uncertainty* in *climate sensitivity* estimates. See also *Cloud radiative effect*.

Cloud radiative effect The *radiative effect* of clouds relative to the identical situation without clouds. In previous IPCC reports this was called *cloud radiative forcing*, but that terminology is inconsistent with other uses of the forcing term and is not maintained in this report. See also *Cloud feedback*.

CO_2-equivalent See *Equivalent carbon dioxide*.

Cold days/cold nights Days where maximum temperature, or nights where minimum temperature, falls below the 10th *percentile*, where the respective temperature distributions are generally defined with respect to the 1961–1990 *reference* period. For the corresponding indices, see Box 2.4.

Compatible emissions *Earth System Models* that simulate the land and ocean *carbon cycle* can calculate *CO_2* emissions that are compatible with a given atmospheric CO_2 concentration trajectory. The compatible emissions over a given period of time are equal to the increase of carbon over that same period of time in the sum of the three active *reservoirs*: the *atmosphere*, the land and the ocean.

Confidence The validity of a finding based on the type, amount, quality, and consistency of evidence (e.g., mechanistic understanding, theory, data, models, expert judgment) and on the degree of agreement. Confidence is expressed qualitatively (Mastrandrea et al., 2010). See Figure 1.11 for the levels of confidence and Table 1.1 for the list of *likelihood* qualifiers. See also *Uncertainty*.

Convection Vertical motion driven by buoyancy forces arising from static instability, usually caused by near-surface cooling or increases in salinity in the case of the ocean and near-surface warming or cloud-top radiative cooling in the case of the *atmosphere*. In the atmosphere convection gives rise to cumulus clouds and precipitation and is effective at both scavenging and vertically transporting chemical species. In the ocean convection can carry surface waters to deep within the ocean.

AIII

Cosmogenic radioisotopes Rare radioactive *isotopes* that are created by the interaction of a high-energy cosmic ray particles with atoms nuclei. They are often used as indicator of *solar activity* which modulates the cosmic rays intensity or as tracers of atmospheric transport processes, and are also called *cosmogenic radionuclides*.

Cryosphere All regions on and beneath the surface of the Earth and ocean where water is in solid form, including *sea ice*, lake ice, river ice, snow cover, *glaciers* and *ice sheets*, and *frozen ground* (which includes *permafrost*).

Dansgaard–Oeschger events Abrupt events characterized in Greenland *ice cores* and in *palaeoclimate* records from the nearby North Atlantic by a cold glacial state, followed by a rapid transition to a warmer phase, and a slow cooling back to glacial conditions. Counterparts of Dansgaard–Oeschger events are observed in other regions as well.

Deforestation Conversion of *forest* to non-forest. For a discussion of the term *forest* and related terms such as *afforestation*, *reforestation*, and *deforestation* see the IPCC Special Report on Land Use, Land-Use Change and Forestry (IPCC, 2000). See also the report on Definitions and Methodological Options to Inventory Emissions from Direct Human-induced Degradation of Forests and Devegetation of Other Vegetation Types (IPCC, 2003).

Deglaciation/glacial termination Transitions from full glacial conditions (*ice age*) to warm *interglacials* characterized by global warming and sea level rise due to change in continental ice volume.

AIII

Detection and attribution *Detection of change* is defined as the process of demonstrating that *climate* or a system affected by climate has changed in some defined statistical sense, without providing a reason for that change. An identified change is detected in observations if its *likelihood* of occurrence by chance due to *internal variability* alone is determined to be small, for example, <10%. *Attribution* is defined as the process of evaluating the relative contributions of multiple causal factors to a change or event with an assignment of statistical confidence (Hegerl et al., 2010).

Diatoms Silt-sized algae that live in surface waters of lakes, rivers and oceans and form shells of opal. Their species distribution in ocean cores is often related to past *sea surface temperatures*.

Direct (aerosol) effect See *Aerosol–radiation interaction*.

Direct Air Capture Chemical process by which a pure *CO_2* stream is produced by capturing CO_2 from the ambient air.

Diurnal temperature range The difference between the maximum and minimum temperature during a 24-hour period.

Dobson Unit (DU) A unit to measure the total amount of *ozone* in a vertical column above the Earth's surface (*total column ozone*). The number of Dobson Units is the thickness in units of 10^{-5} m that the ozone column would occupy if compressed into a layer of uniform density at a pressure of 1013 hPa and a temperature of 0°C. One DU corresponds to a column of ozone containing 2.69×10^{20} molecules per square metre. A typical value for the amount of ozone in a column of the Earth's *atmosphere*, although very variable, is 300 DU.

Downscaling Downscaling is a method that derives local- to regional-scale (10 to 100 km) information from larger-scale models or data analyses. Two main methods exist: *dynamical downscaling* and *empirical/statistical downscaling*. The dynamical method uses the output of *regional climate models*, global models with variable spatial *resolution* or high-resolution global models. The empirical/statistical methods develop statistical relationships that link the large-scale atmospheric variables with local/regional climate variables. In all cases, the quality of the driving model remains an important limitation on the quality of the downscaled information.

Drought A period of abnormally dry weather long enough to cause a serious hydrological imbalance. Drought is a relative term; therefore any discussion in terms of precipitation deficit must refer to the particular precipitation-related activity that is under discussion. For example, shortage of precipitation during the growing season impinges on crop production or *ecosystem* function in general (due to *soil moisture* drought, also termed *agricultural drought*), and during the *runoff* and percolation season primarily affects water supplies (*hydrological drought*). Storage changes in soil moisture and groundwater are also affected by increases in actual *evapotranspiration* in addition to reductions in precipitation. A period with an abnormal precipitation deficit is defined as a *meteorological drought*. A *megadrought* is a very lengthy and pervasive drought, lasting much longer than normal, usually a decade or more. For the corresponding indices, see Box 2.4.

Dynamical system A process or set of processes whose evolution in time is governed by a set of deterministic physical laws. The *climate system* is a dynamical system. See also *Abrupt climate change*, *Chaotic*, *Nonlinearity* and *Predictability*.

Earth System Model (ESM) A coupled *atmosphere–ocean general circulation model* in which a representation of the *carbon cycle* is included, allowing for interactive calculation of atmospheric *CO_2* or *compatible emissions*. Additional components (e.g., atmospheric chemistry, *ice sheets*, dynamic vegetation, nitrogen cycle, but also urban or crop models) may be included. See also *Climate model*.

Earth System Model of Intermediate Complexity (EMIC) A *climate model* attempting to include all the most important earth system processes as in ESMs but at a lower *resolution* or in a simpler, more idealized fashion.

Earth System sensitivity The equilibrium temperature response of the coupled *atmosphere*–ocean–*cryosphere*–vegetation–*carbon cycle* system to a doubling of the atmospheric *CO_2* concentration is referred to as Earth System sensitivity. Because it allows slow components (e.g., *ice sheets*, vegetation) of the *climate system* to adjust to the external perturbation, it may differ substantially from the *climate sensitivity* derived from coupled atmosphere–ocean models.

Ecosystem An ecosystem is a functional unit consisting of living organisms, their non-living environment, and the interactions within and between them. The components included in a given ecosystem and its spatial boundaries depend on the purpose for which the ecosystem is defined: in some cases they are relatively sharp, while in others they are diffuse. Ecosystem boundaries can change over time. Ecosystems are nested within other ecosystems, and their scale can range from very small to the entire *biosphere*. In the current era, most ecosystems either contain people as key organisms, or are influenced by the effects of human activities in their environment.

Effective climate sensitivity See *Climate sensitivity*.

Effective radiative forcing See *Radiative forcing*.

Efficacy A measure of how effective a *radiative forcing* from a given *anthropogenic* or natural mechanism is at changing the equilibrium *global mean surface temperature* compared to an equivalent radiative forcing from *carbon dioxide*. A carbon dioxide increase by definition has an efficacy of 1.0. Variations in climate efficacy may result from *rapid adjustments* to the applied forcing, which differ with different forcings.

Ekman pumping Frictional stress at the surface between two fluids (*atmosphere* and ocean) or between a fluid and the adjacent solid surface (the Earth's surface) forces a circulation. When the resulting mass transport is converging, mass conservation requires a vertical flow away from the surface. This is called Ekman pumping. The opposite effect, in case of divergence, is called *Ekman suction*. The effect is important in both the atmosphere and the ocean.

Ekman transport The total transport resulting from a balance between the Coriolis force and the frictional stress due to the action of the wind on the ocean surface. See also *Ekman pumping*.

Electromagnetic spectrum Wavelength or energy range of all electromagnetic radiation. In terms of *solar radiation*, the *spectral irradiance* is the power arriving at the Earth per unit area, per unit wavelength.

El Niño-Southern Oscillation (ENSO) The term *El Niño* was initially used to describe a warm-water current that periodically flows along the coast of Ecuador and Peru, disrupting the local fishery. It has since become identified with a basin-wide warming of the tropical Pacific Ocean east of the dateline. This oceanic event is associated with a fluctuation of a global-scale tropical and subtropical surface pressure pattern called the *Southern Oscillation*. This coupled *atmosphere*–ocean phenomenon, with preferred time scales of two to about seven years, is known as the El Niño-Southern Oscillation (ENSO). It is often measured by the surface pressure anomaly difference between Tahiti and Darwin or the *sea surface temperatures* in the central and eastern equatorial Pacific. During an ENSO event, the prevailing trade winds weaken, reducing upwelling and altering ocean currents such that the sea surface temperatures warm, further weakening the trade winds. This event has a great impact on the wind, sea surface temperature and precipitation patterns in the tropical Pacific. It has climatic effects throughout the Pacific *region* and in many other parts of the world, through global *teleconnections*. The cold phase of ENSO is called *La Niña*. For the corresponding indices, see Box 2.5.

Emission scenario A plausible representation of the future development of emissions of substances that are potentially radiatively active (e.g., *greenhouse gases*, *aerosols*) based on a coherent and internally consistent set of assumptions about driving forces (such as demographic and socioeconomic development, technological change) and their key relationships. *Concentration scenarios*, derived from emission scenarios, are used as input to a *climate model* to compute *climate projections*. In IPCC (1992) a set of emission scenarios was presented which were used as a basis for the climate projections in IPCC (1996). These emission scenarios are referred to as the IS92 scenarios. In the IPCC Special Report on Emission Scenarios (Nakićenović and Swart, 2000) emission scenarios, the so-called *SRES scenarios*, were published, some of which were used, among others, as a basis for the climate projections presented in Chapters 9 to 11 of IPCC (2001) and Chapters 10 and 11 of IPCC (2007). New emission scenarios for *climate change*, the four *Representative Concentration Pathways*, were developed for, but independently of, the present IPCC assessment. See also *Climate scenario* and *Scenario*.

Energy balance The difference between the total incoming and total outgoing energy. If this balance is positive, warming occurs; if it is negative, cooling occurs. Averaged over the globe and over long time periods, this balance must be zero. Because the *climate system* derives virtually all its energy from the Sun, zero balance implies that, globally, the absorbed *solar radiation*, that is, *incoming solar radiation* minus reflected solar radiation at the top of the *atmosphere* and *outgoing longwave radiation* emitted by the climate system are equal. See also *Energy budget*.

Energy Balance Model (EBM) An energy balance model is a simplified model that analyses the *energy budget* of the Earth to compute changes in the *climate*. In its simplest form, there is no explicit spatial dimension and the model then provides an estimate of the changes in globally averaged temperature computed from the changes in radiation. This zero-dimensional energy balance model can be extended to a one-dimensional or two-dimensional model if changes to the energy budget with respect to latitude, or both latitude and longitude, are explicitly considered. See also *Climate model*.

Energy budget (of the Earth) The Earth is a physical system with an energy budget that includes all gains of incoming energy and all losses of outgoing energy. The Earth's energy budget is determined by measuring how much energy comes into the Earth system from the Sun, how much energy is lost to space, and accounting for the remainder on Earth and its *atmosphere*. *Solar radiation* is the dominant source of energy into the Earth system. Incoming solar energy may be scattered and reflected by clouds and *aerosols* or absorbed in the atmosphere. The transmitted radiation is then either absorbed or reflected at the Earth's surface. The average *albedo* of the Earth is about 0.3, which means that 30% of the incident solar energy is reflected into space, while 70% is absorbed by the Earth. Radiant solar or shortwave energy is transformed into sensible heat, latent energy (involving different water states), potential energy, and kinetic energy before being emitted as *infrared radiation*. With the average *surface temperature* of the Earth of about 15°C (288 K), the main outgoing energy flux is in the infrared part of the spectrum. See also *Energy balance*, *Latent heat flux*, *Sensible heat flux*.

Ensemble A collection of model simulations characterizing a *climate prediction* or *projection*. Differences in initial conditions and model formulation result in different evolutions of the modelled system and may give information on *uncertainty* associated with model error and error in initial conditions in the case of *climate forecasts* and on uncertainty associated with model error and with internally generated *climate variability* in the case of climate projections.

Equilibrium and transient climate experiment An *equilibrium climate experiment* is a *climate model* experiment in which the model is allowed to fully adjust to a change in *radiative forcing*. Such experiments provide information on the difference between the initial and final states of the model, but not on the time-dependent response. If the forcing is allowed to evolve gradually according to a prescribed *emission scenario*, the time-dependent response of a climate model may be analysed. Such an experiment is called a *transient climate experiment*. See also *Climate projection*.

Equilibrium climate sensitivity See *Climate sensitivity*.

Equilibrium line The spatially averaged boundary at a given moment, usually chosen as the seasonal *mass budget* minimum at the end of summer, between the region on a *glacier* where there is a net annual loss of ice mass (*ablation* area) and that where there is a net annual gain (*accumulation* area). The altitude of this boundary is referred to as equilibrium line altitude (ELA).

Equivalent carbon dioxide (CO_2) concentration The concentration of *carbon dioxide* that would cause the same *radiative forcing* as a given mixture of carbon dioxide and other forcing components. Those values may consider only *greenhouse gases*, or a combination of greenhouse gases and *aerosols*. Equivalent carbon dioxide concentration is a *metric* for comparing radiative forcing of a mix of different greenhouse gases at a particular time but does not imply equivalence of the corresponding *climate change* responses nor future forcing. There is generally no connection between *equivalent carbon dioxide emissions* and resulting equivalent carbon dioxide concentrations.

Equivalent carbon dioxide (CO_2) emission The amount of *carbon dioxide* emission that would cause the same integrated *radiative forcing*, over a given time horizon, as an emitted amount of a *greenhouse gas* or a mixture of greenhouse gases. The equivalent carbon dioxide emission is obtained by multiplying the emission of a greenhouse gas by its *Global Warming Potential* for the given time horizon. For a mix of greenhouse gases it is obtained by summing the equivalent carbon dioxide emissions of each gas. Equivalent carbon dioxide emission is a common scale for comparing emissions of different greenhouse gases but does not imply equivalence of the corresponding *climate change* responses. See also *Equivalent carbon dioxide concentration*.

Evapotranspiration The combined process of evaporation from the Earth's surface and transpiration from vegetation.

Extended Concentration Pathways See *Representative Concentration Pathways*.

External forcing External forcing refers to a forcing agent outside the *climate system* causing a change in the climate system. Volcanic eruptions, solar variations and *anthropogenic* changes in the composition of the *atmosphere* and *land use change* are external forcings. Orbital forcing is also an external forcing as the *insolation* changes with orbital parameters eccentricity, tilt and precession of the equinox.

Extratropical cyclone A large-scale (of order 1000 km) storm in the middle or high latitudes having low central pressure and fronts with strong horizontal gradients in temperature and humidity. A major cause of extreme wind speeds and heavy precipitation especially in wintertime.

Extreme climate event See *Extreme weather event*.

Extreme sea level See *Storm surge*.

Extreme weather event An extreme weather event is an event that is rare at a particular place and time of year. Definitions of *rare* vary, but an extreme weather event would normally be as rare as or rarer than the 10th or 90th *percentile* of a *probability density function* estimated from observations. By definition, the characteristics of what is called *extreme weather* may vary from place to place in an absolute sense. When a pattern of extreme weather persists for some time, such as a season, it may be classed as an *extreme climate event*, especially if it yields an average or total that is itself extreme (e.g., *drought* or heavy rainfall over a season).

Faculae Bright patches on the Sun. The area covered by faculae is greater during periods of high *solar activity*.

Feedback See *Climate feedback*.

Fingerprint The *climate* response pattern in space and/or time to a specific forcing is commonly referred to as a fingerprint. The spatial patterns of sea level response to melting of *glaciers* or *ice sheets* (or other changes in surface loading) are also referred to as fingerprints. Fingerprints are used to detect the presence of this response in observations and are typically estimated using forced *climate model* simulations.

Flux adjustment To avoid the problem of coupled *Atmosphere–Ocean General Circulation Models (AOGCMs)* drifting into some unrealistic *climate* state, adjustment terms can be applied to the atmosphere-ocean fluxes of heat and moisture (and sometimes the surface stresses resulting from the effect of the wind on the ocean surface) before these fluxes are imposed on the model ocean and atmosphere. Because these adjustments are pre-computed and therefore independent of the coupled model integration, they are uncorrelated with the anomalies that develop during the integration.

Forest A vegetation type dominated by trees. Many definitions of the term *forest* are in use throughout the world, reflecting wide differences in biogeophysical conditions, social structure and economics. For a discussion of the term *forest* and related terms such as *afforestation*, *reforestation* and *deforestation* see the IPCC Report on Land Use, Land-Use Change and Forestry (IPCC, 2000). See also the Report on Definitions and Methodological Options to Inventory Emissions from Direct Human-induced Degradation of Forests and Devegetation of Other Vegetation Types (IPCC, 2003).

Fossil fuel emissions Emissions of *greenhouse gases* (in particular *carbon dioxide*), other trace gases and *aerosols* resulting from the combustion of fuels from fossil carbon deposits such as oil, gas and coal.

Framework Convention on Climate Change See *United Nations Framework Convention on Climate Change (UNFCCC)*.

Free atmosphere The atmospheric layer that is negligibly affected by friction against the Earth's surface, and which is above the *atmospheric boundary layer*.

Frozen ground Soil or rock in which part or all of the *pore water* is frozen. Frozen ground includes *permafrost*. Ground that freezes and thaws annually is called *seasonally frozen ground*.

General circulation The large-scale motions of the *atmosphere* and the ocean as a consequence of differential heating on a rotating Earth. General circulation contributes to the *energy balance* of the system through transport of heat and momentum.

General Circulation Model (GCM) See *Climate model*.

Geoengineering Geoengineering refers to a broad set of methods and technologies that aim to deliberately alter the *climate system* in order to alleviate the impacts of *climate change*. Most, but not all, methods seek to either (1) reduce the amount of absorbed solar energy in the climate system (*Solar Radiation Management*) or (2) increase net carbon sinks from the *atmosphere* at a scale sufficiently large to alter *climate* (*Carbon Dioxide Removal*). Scale and intent are of central importance. Two key characteristics of geoengineering methods of particular concern are that they use or affect the climate system (e.g., atmosphere, land or ocean) globally or regionally and/or could have substantive unintended effects that cross national boundaries. Geoengineering is different from weather modification and ecological engineering, but the boundary can be fuzzy (IPCC, 2012, p. 2).

Geoid The equipotential surface having the same geopotential at each latitude and longitude around the world (geodesists denoting this potential W0) that best approximates the *mean sea level*. It is the surface of reference for measurement of altitude. In practice, several variations of definitions of the geoid exist depending on the way the permanent tide (the zero-frequency gravitational tide due to the Sun and Moon) is considered in geodetic studies.

Geostrophic winds or currents A wind or current that is in balance with the horizontal pressure gradient and the Coriolis force, and thus is outside of the influence of friction. Thus, the wind or current is directly parallel to isobars and its speed is proportional to the horizontal pressure gradient.

Glacial–interglacial cycles Phase of the Earth's history marked by large changes in continental ice volume and global sea level. See also *Ice age* and *Interglacials*.

Glacial isostatic adjustment (GIA) The deformation of the Earth and its gravity field due to the response of the earth–ocean system to changes in ice and associated water loads. It is sometimes referred to as *glacio-hydro isostasy*. It includes vertical and horizontal deformations of the Earth's surface and changes in *geoid* due to the redistribution of mass during the ice–ocean mass exchange.

Glacier A perennial mass of land ice that originates from compressed snow, shows evidence of past or present flow (through internal deformation and/or sliding at the base) and is constrained by internal stress and friction at the base and sides. A glacier is maintained by accumulation of snow at high altitudes, balanced by melting at low altitudes and/or discharge into the sea. An ice mass of the same origin as glaciers, but of continental size, is called an *ice sheet*. For the purpose of simplicity in this Assessment Report, all ice masses other than ice sheets are referred to as glaciers. See also *Equilibrium line* and *Mass balance/budget (of glaciers or ice sheets)*.

Global dimming Global dimming refers to a widespread reduction of *solar radiation* received at the surface of the Earth from about the year 1961 to around 1990.

Global mean surface temperature An estimate of the global mean surface air temperature. However, for changes over time, only anomalies, as departures from a climatology, are used, most commonly based on the area-weighted global average of the *sea surface temperature* anomaly and *land surface air temperature* anomaly.

Global Warming Potential (GWP) An index, based on radiative properties of *greenhouse gases*, measuring the *radiative forcing* following a pulse emission of a unit mass of a given greenhouse gas in the present-day *atmosphere* integrated over a chosen time horizon, relative to that of *carbon dioxide*. The GWP represents the combined effect of the differing times these gases remain in the atmosphere and their relative effectiveness in causing radiative forcing. The *Kyoto Protocol* is based on GWPs from pulse emissions over a 100-year time frame.

Greenhouse effect The infrared *radiative effect* of all infrared-absorbing constituents in the *atmosphere*. *Greenhouse gases*, clouds, and (to a small extent) *aerosols* absorb *terrestrial radiation* emitted by the Earth's surface and elsewhere in the atmosphere. These substances emit *infrared radiation* in all directions, but, everything else being equal, the net amount emitted to space is normally less than would have been emitted in the absence of these absorbers because of the decline of temperature with altitude in the *troposphere* and the consequent weakening of emission. An increase in the concentration of greenhouse gases increases the magnitude of this effect; the difference is sometimes called the enhanced greenhouse effect. The change in a greenhouse gas concentration because of *anthropogenic* emissions contributes to an *instantaneous radiative forcing*. Surface temperature and troposphere warm in response to this forcing, gradually restoring the radiative balance at the top of the atmosphere.

Greenhouse gas (GHG) Greenhouse gases are those gaseous constituents of the *atmosphere*, both natural and *anthropogenic*, that absorb and emit radiation at specific wavelengths within the spectrum of *terrestrial radiation* emitted by the Earth's surface, the atmosphere itself, and by clouds. This property causes the *greenhouse effect.* Water vapour (H_2O), *carbon dioxide (CO_2)*, *nitrous oxide (N_2O)*, *methane (CH_4)* and *ozone* (O_3) are the primary greenhouse gases in the Earth's atmosphere. Moreover, there are a number of entirely human-made greenhouse gases in the atmosphere, such as the *halocarbons* and other chlorine- and bromine-containing substances, dealt with under the *Montreal Protocol.* Beside CO_2, N_2O and CH_4, the *Kyoto Protocol* deals with the greenhouse gases sulphur hexafluoride (SF_6), hydrofluorocarbons (*HFCs*) and perfluorocarbons (PFCs). For a list of *well-mixed greenhouse gases*, see Table 2.A.1.

Gross Primary Production (GPP) The amount of carbon fixed by the autotrophs (e.g. plants and algaes).

Grounding line The junction between a *glacier* or *ice sheet* and *ice shelf*; the place where ice starts to float. This junction normally occurs over a finite zone, rather than at a line.

Gyre Basin-scale ocean horizontal circulation pattern with slow flow circulating around the ocean basin, closed by a strong and narrow (100 to 200 km wide) boundary current on the western side. The subtropical gyres in each ocean are associated with high pressure in the centre of the gyres; the subpolar gyres are associated with low pressure.

Hadley Circulation A direct, thermally driven overturning cell in the *atmosphere* consisting of poleward flow in the upper *troposphere*, subsiding air into the subtropical anticyclones, return flow as part of the trade winds near the surface, and with rising air near the equator in the so-called *Inter-Tropical Convergence Zone*.

Halocarbons A collective term for the group of partially halogenated organic species, which includes the chlorofluorocarbons (*CFCs*), hydrochlorofluorocarbons (*HCFCs*), hydrofluorocarbons (*HFCs*), halons, methyl chloride and methyl bromide. Many of the halocarbons have large *Global Warming Potentials*. The chlorine and bromine-containing halocarbons are also involved in the depletion of the *ozone layer*.

Halocline A layer in the oceanic water column in which salinity changes rapidly with depth. Generally saltier water is denser and lies below less salty water. In some high latitude oceans the surface waters may be colder than the deep waters and the halocline is responsible for maintaining water column stability and isolating the surface waters from the deep waters. See also *Thermocline.*

Halosteric See *Sea level change*.

HCFC See *Halocarbons*.

Heat wave A period of abnormally and uncomfortably hot weather. See also *Warm spell*.

Heterotrophic respiration The conversion of organic matter to *carbon dioxide* by organisms other than autotrophs.

HFC See *Halocarbons*.

Hindcast or retrospective forecast A forecast made for a period in the past using only information available before the beginning of the forecast. A sequence of hindcasts can be used to calibrate the forecast system and/or provide a measure of the average skill that the forecast system has exhibited in the past as a guide to the skill that might be expected in the future.

Holocene The Holocene Epoch is the latter of two epochs in the *Quaternary* System, extending from 11.65 ka (thousand years before 1950) to the present. It is also known as *Marine Isotopic Stage (MIS) 1* or *current interglacial*.

Hydroclimate Part of the *climate* pertaining to the hydrology of a *region*.

Hydrological cycle The cycle in which water evaporates from the oceans and the land surface, is carried over the Earth in atmospheric circulation as water vapour, condenses to form clouds, precipitates over ocean and land as rain or snow, which on land can be intercepted by trees and vegetation, provides *runoff* on the land surface, infiltrates into soils, recharges groundwater, discharges into streams and ultimately flows out into the oceans, from which it will eventually evaporate again. The various systems involved in the hydrological cycle are usually referred to as hydrological systems.

Hydrosphere The component of the *climate system* comprising liquid surface and subterranean water, such as oceans, seas, rivers, fresh water lakes, underground water, etc.

Hypsometry The distribution of land or ice surface as a function of altitude.

Ice age An ice age or *glacial period* is characterized by a long-term reduction in the temperature of the Earth's *climate*, resulting in growth of *ice sheets* and *glaciers*.

Ice–albedo feedback A *climate feedback* involving changes in the Earth's surface *albedo*. Snow and ice have an albedo much higher (up to ~0.8) than the average planetary albedo (~0.3). With increasing temperatures, it is anticipated that snow and ice extent will decrease, the Earth's overall albedo will decrease and more *solar radiation* will be absorbed, warming the Earth further.

Ice core A cylinder of ice drilled out of a *glacier* or *ice sheet*.

Ice sheet A mass of land ice of continental size that is sufficiently thick to cover most of the underlying bed, so that its shape is mainly determined by its dynamics (the flow of the ice as it deforms internally and/or slides at its base). An ice sheet flows outward from a high central ice plateau with a small average surface slope. The margins usually slope more steeply, and most ice is discharged through fast flowing *ice streams* or *outlet glaciers*, in some cases into the sea or into *ice shelves* floating on the sea. There are only two ice sheets in the modern world, one on Greenland and one on Antarctica. During glacial periods there were others.

Ice shelf A floating slab of ice of considerable thickness extending from the coast (usually of great horizontal extent with a very gently sloping surface), often filling embayments in the coastline of an *ice sheet*. Nearly all ice shelves are in Antarctica, where most of the ice discharged into the ocean flows via ice shelves.

Ice stream A stream of ice with strongly enhanced flow that is part of an *ice sheet.* It is often separated from surrounding ice by strongly sheared, crevassed margins. See also *Outlet glacier.*

Incoming solar radiation See *Insolation*.

Indian Ocean Dipole (IOD) Large–scale mode of interannual variability of *sea surface temperature* in the Indian Ocean. This pattern manifests through a zonal gradient of tropical sea surface temperature, which in one extreme phase in boreal autumn shows cooling off Sumatra and warming off Somalia in the west, combined with anomalous easterlies along the equator.

AIII

Indirect aerosol effect See *Aerosol-cloud interaction*.

Industrial Revolution A period of rapid industrial growth with far-reaching social and economic consequences, beginning in Britain during the second half of the 18th century and spreading to Europe and later to other countries including the United States. The invention of the steam engine was an important trigger of this development. The industrial revolution marks the beginning of a strong increase in the use of fossil fuels and emission of, in particular, fossil *carbon dioxide*. In this report the terms *pre-industrial* and *industrial* refer, somewhat arbitrarily, to the periods before and after 1750, respectively.

Infrared radiation See *Terrestrial radiation*.

Insolation The amount of *solar radiation* reaching the Earth by latitude and by season measured in W m^{-2}. Usually *insolation* refers to the radiation arriving at the top of the *atmosphere*. Sometimes it is specified as referring to the radiation arriving at the Earth's surface. See also *Total Solar Irradiance*.

Interglacials or interglaciations The warm periods between *ice age* glaciations. Often defined as the periods at which sea levels were close to present sea level. For the *Last Interglacial (LIG)* this occurred between about 129 and 116 ka (thousand years) before present (defined as 1950) although the warm period started in some areas a few thousand years earlier. In terms of the oxygen *isotope* record interglaciations are defined as the interval between the midpoint of the preceding termination and the onset of the next glaciation. The present interglaciation, the *Holocene*, started at 11.65 ka before present although globally sea levels did not approach their present position until about 7 ka before present.

Internal variability See *Climate variability*.

Inter-Tropical Convergence Zone (ITCZ) The Inter-Tropical Convergence Zone is an equatorial zonal belt of low pressure, strong *convection* and heavy precipitation near the equator where the northeast trade winds meet the southeast trade winds. This band moves seasonally.

Iron fertilization Deliberate introduction of iron to the upper ocean intended to enhance biological productivity which can sequester additional atmospheric *carbon dioxide* into the oceans.

Irreversibility A perturbed state of a *dynamical system* is defined as irreversible on a given timescale, if the recovery timescale from this state due to natural processes is significantly longer than the time it takes for the system to reach this perturbed state. In the context of WGI, the time scale of interest is centennial to millennial. See also *Tipping point*.

Isostatic or Isostasy Isostasy refers to the response of the earth to changes in surface load. It includes the deformational and gravitational response. This response is elastic on short time scales, as in the earth–ocean response to recent changes in mountain glaciation, or viscoelastic on longer time scales, as in the response to the last *deglaciation* following the *Last Glacial Maximum*. See also *Glacial Isostatic Adjustment (GIA)*.

Isotopes Atoms of the same chemical element that have the same the number of protons but differ in the number of neutrons. Some proton–neutron configurations are stable (stable isotopes), others are unstable undergoing spontaneous radioactive decay (*radioisotopes*). Most elements have more than one stable isotope. Isotopes can be used to trace transport processes or to study processes that change the isotopic ratio. Radioisotopes provide in addition time information that can be used for radiometric dating.

Kyoto Protocol The Kyoto Protocol to the *United Nations Framework Convention on Climate Change (UNFCCC)* was adopted in 1997 in Kyoto, Japan, at the Third Session of the Conference of the Parties (COP) to the UNFCCC. It contains legally binding commitments, in addition to those included in the UNFCCC. Countries included in Annex B of the Protocol (most Organisation for Economic Cooperation and Development countries and countries with economies in transition) agreed to reduce their *anthropogenic greenhouse gas* emissions (*carbon dioxide*, *methane*, *nitrous oxide*, hydrofluorocarbons, perfluorocarbons, and sulphur hexafluoride) by at least 5% below 1990 levels in the commitment period 2008–2012. The Kyoto Protocol entered into force on 16 February 2005.

Land surface air temperature The surface air temperature as measured in well-ventilated screens over land at 1.5 m above the ground.

Land use and Land use change *Land use* refers to the total of arrangements, activities and inputs undertaken in a certain land cover type (a set of human actions). The term *land use* is also used in the sense of the social and economic purposes for which land is managed (e.g., grazing, timber extraction and conservation). *Land use change* refers to a change in the use or management of land by humans, which may lead to a change in land cover. Land cover and land use change may have an impact on the surface *albedo*, *evapotranspiration*, *sources* and *sinks* of *greenhouse gases*, or other properties of the *climate system* and may thus give rise to *radiative forcing* and/or other impacts on *climate*, locally or globally. See also the IPCC Report on Land Use, Land-Use Change, and Forestry (IPCC, 2000).

Land water storage Water stored on land other than in *glaciers* and *ice sheets* (that is water stored in rivers, lakes, wetlands, the vadose zone, aquifers, reservoirs, snow and *permafrost*). Changes in land water storage driven by *climate* and human activities contribute to *sea level change*.

La Niña See *El Niño-Southern Oscillation*.

Lapse rate The rate of change of an atmospheric variable, usually temperature, with height. The lapse rate is considered positive when the variable decreases with height.

Last Glacial Maximum (LGM) The period during the last *ice age* when the *glaciers* and *ice sheets* reached their maximum extent, approximately

21 ka ago. This period has been widely studied because the *radiative forcings* and boundary conditions are relatively well known.

Last Interglacial (LIG) See *Interglacials*.

Latent heat flux The turbulent flux of heat from the Earth's surface to the *atmosphere* that is associated with evaporation or condensation of water vapour at the surface; a component of the surface *energy budget*.

Lifetime Lifetime is a general term used for various time scales characterizing the rate of processes affecting the concentration of trace gases. The following lifetimes may be distinguished:

Turnover time (*T*) (also called *global atmospheric lifetime*) is the ratio of the mass *M* of a *reservoir* (e.g., a gaseous compound in the *atmosphere*) and the total rate of removal *S* from the reservoir: $T = M/S$. For each removal process, separate turnover times can be defined. In soil carbon biology, this is referred to as *Mean Residence Time*.

Adjustment time or response time (T_a) is the time scale characterizing the decay of an instantaneous pulse input into the reservoir. The term *adjustment time* is also used to characterize the adjustment of the mass of a reservoir following a step change in the *source* strength. *Half-life* or *decay constant* is used to quantify a first-order exponential decay process. See *Response time* for a different definition pertinent to *climate* variations.

The term *lifetime* is sometimes used, for simplicity, as a surrogate for *adjustment time*.

In simple cases, where the global removal of the compound is directly proportional to the total mass of the reservoir, the adjustment time equals the turnover time: $T = T_a$. An example is *CFC*-11, which is removed from the atmosphere only by photochemical processes in the *stratosphere*. In more complicated cases, where several reservoirs are involved or where the removal is not proportional to the total mass, the equality $T = T_a$ no longer holds. *Carbon dioxide (CO_2)* is an extreme example. Its turnover time is only about 4 years because of the rapid exchange between the atmosphere and the ocean and terrestrial biota. However, a large part of that CO_2 is returned to the atmosphere within a few years. Thus, the adjustment time of CO_2 in the atmosphere is actually determined by the rate of removal of carbon from the surface layer of the oceans into its deeper layers. Although an approximate value of 100 years may be given for the adjustment time of CO_2 in the atmosphere, the actual adjustment is faster initially and slower later on. In the case of *methane (CH_4)*, the adjustment time is different from the turnover time because the removal is mainly through a chemical reaction with the hydroxyl radical (OH), the concentration of which itself depends on the CH_4 concentration. Therefore, the CH_4 removal rate S is not proportional to its total mass *M*.

Likelihood The chance of a specific outcome occurring, where this might be estimated probabilistically. This is expressed in this report using a standard terminology, defined in Table 1.1. See also *Confidence* and *Uncertainty*.

Lithosphere The upper layer of the solid Earth, both continental and oceanic, which comprises all crustal rocks and the cold, mainly elastic part of the uppermost mantle. Volcanic activity, although part of the lithosphere, is not considered as part of the *climate system*, but acts as an *external forcing* factor. See also *Isostatic*.

Little Ice Age (LIA) An interval during the last millennium characterized by a number of extensive expansions of mountain *glaciers* and moderate retreats in between them, both in the Northern and Southern Hemispheres. The timing of glacial advances differs between *regions* and the LIA is, therefore, not clearly defined in time. Most definitions lie in the period 1400 CE and 1900 CE. Currently available *reconstructions* of average Northern Hemisphere temperature indicate that the coolest periods at the hemispheric scale may have occurred from 1450 to 1850 CE.

Longwave radiation See *Terrestrial radiation*.

Madden–Julian Oscillation (MJO) The largest single component of tropical atmospheric intraseasonal variability (periods from 30 to 90 days). The MJO propagates eastwards at around 5 m s^{-1} in the form of a large-scale coupling between atmospheric circulation and deep *convection*. As it progresses, it is associated with large regions of both enhanced and suppressed rainfall, mainly over the Indian and western Pacific Oceans. Each MJO event lasts approximately 30 to 60 days, hence the MJO is also known as the 30- to 60-day wave, or the intraseasonal oscillation.

Marine-based ice sheet An *ice sheet* containing a substantial region that rests on a bed lying below sea level and whose perimeter is in contact with the ocean. The best known example is the West Antarctic ice sheet.

Mass balance/budget (of glaciers or ice sheets) The balance between the mass input to the ice body (*accumulation*) and the mass loss (*ablation* and iceberg *calving*) over a stated period of time, which is often a year or a season. Point mass balance refers to the mass balance at a particular location on the *glacier* or *ice sheet*. Surface mass balance is the difference between surface accumulation and surface ablation. The input and output terms for mass balance are:

Accumulation All processes that add to the mass of a glacier. The main contribution to accumulation is snowfall. Accumulation also includes deposition of hoar, freezing rain, other types of solid precipitation, gain of wind-blown snow, and avalanching.

Ablation Surface processes that reduce the mass of a glacier. The main contributor to ablation is melting with *runoff* but on some glaciers sublimation, loss of wind-blown snow and avalanching are also significant processes of ablation.

Discharge/outflow Mass loss by iceberg calving or ice discharge across the *grounding line* of a floating *ice shelf*. Although often treated as an ablation term, in this report iceberg calving and discharge is considered separately from surface ablation.

Mean sea level The surface level of the ocean at a particular point averaged over an extended period of time such as a month or year. Mean sea level is often used as a national datum to which heights on land are referred.

Medieval Climate Anomaly (MCA) See *Medieval Warm Period*.

Medieval Warm Period (MWP) An interval of relatively warm conditions and other notable *climate* anomalies such as more extensive *drought* in some continental *regions*. The timing of this interval is not clearly defined, with different records showing onset and termination of the warmth at different times, and some showing intermittent warmth. Most definitions lie within the period 900 to 1400 CE. Currently available *reconstructions* of average Northern Hemisphere temperature indicate that the warmest period at the hemispheric scale may have occurred from 950 to 1250 CE. Currently available records and temperature reconstructions indicate that average temperatures during parts of the MWP were indeed warmer in the context of the last 2 kyr, though the warmth may not have been as ubiquitous across seasons and geographical regions as the 20th century warming. It is also called *Medieval Climate Anomaly*.

Meridional Overturning Circulation (MOC) Meridional (north–south) overturning circulation in the ocean quantified by zonal (east–west) sums of mass transports in depth or density layers. In the North Atlantic, away from the subpolar *regions*, the MOC (which is in principle an observable quantity) is often identified with the *thermohaline circulation* (THC), which is a conceptual and incomplete interpretation. It must be borne in mind that the MOC is also driven by wind, and can also include shallower overturning cells such as occur in the upper ocean in the tropics and subtropics, in which warm (light) waters moving poleward are transformed to slightly denser waters and *subducted* equatorward at deeper levels.

Metadata Information about meteorological and climatological data concerning how and when they were measured, their quality, known problems and other characteristics.

Methane (CH_4) Methane is one of the six *greenhouse gases* to be mitigated under the *Kyoto Protocol* and is the major component of natural gas and associated with all hydrocarbon fuels, animal husbandry and agriculture.

Metric A consistent measurement of a characteristic of an object or activity that is otherwise difficult to quantify. Within the context of the evaluation of *climate models*, this is a quantitative measure of agreement between a simulated and observed quantity which can be used to assess the performance of individual models.

Microwave Sounding Unit (MSU) A microwave sounder on National Oceanic and Atmospheric Administration (NOAA) polar orbiter satellites, that estimates the temperature of thick layers of the *atmosphere* by measuring the thermal emission of oxygen molecules from a complex of emission lines near 60 GHz. A series of nine MSUs began making this kind of measurement in late 1978. Beginning in mid 1998, a follow-on series of instruments, the Advanced Microwave Sounding Units (AMSUs), began operation.

AIII

Mineralization/Remineralization The conversion of an element from its organic form to an inorganic form as a result of microbial decomposition. In nitrogen mineralization, organic nitrogen from decaying plant and animal residues (proteins, nucleic acids, amino sugars and urea) is converted to ammonia (NH_3) and ammonium (NH_4^+) by biological activity.

Mitigation A human intervention to reduce the *sources* or enhance the *sinks* of *greenhouse gases*.

Mixing ratio See *Mole fraction*.

Model drift Since model *climate* differs to some extent from observed climate, *climate forecasts* will typically 'drift' from the initial observation-based state towards the model's climate. This drift occurs at different time scales for different variables, can obscure the initial-condition forecast information and is usually removed a posteriori by an empirical, usually linear, adjustment.

Model hierarchy See *Climate model (spectrum or hierarchy)*.

Model initialization A *climate forecast* typically proceeds by integrating a *climate model* forward in time from an initial state that is intended to reflect the actual state of the *climate system*. Available observations of the climate system are 'assimilated' into the model. Initialization is a complex process that is limited by available observations, observational errors and, depending on the procedure used, may be affected by *uncertainty* in the history of climate forcing. The initial conditions will contain errors that grow as the forecast progresses, thereby limiting the time for which the forecast will be useful. See also *Climate prediction*.

Model spread The range or spread in results from *climate models*, such as those assembled for Coupled Model Intercomparison Project Phase 5 (CMIP5). Does not necessarily provide an exhaustive and formal estimate of the *uncertainty* in *feedbacks*, forcing or *projections* even when expressed numerically, for example, by computing a standard deviation of the models' responses. In order to quantify uncertainty, information from observations, physical constraints and expert judgement must be combined, using a statistical framework.

Mode of climate variability Underlying space–time structure with preferred spatial pattern and temporal variation that helps account for the gross features in variance and for *teleconnections*. A mode of variability is often considered to be the product of a spatial *climate pattern* and an associated *climate index* time series.

Mole fraction Mole fraction, or *mixing ratio*, is the ratio of the number of moles of a constituent in a given volume to the total number of moles of all constituents in that volume. It is usually reported for dry air. Typical values for *well-mixed greenhouse gases* are in the order of μmol mol^{-1} (parts per million: *ppm*), nmol mol^{-1} (parts per billion: *ppb*), and fmol mol^{-1} (parts per trillion: *ppt*). Mole fraction differs from *volume mixing ratio*, often expressed in ppmv etc., by the corrections for non-ideality of gases. This correction is significant relative to measurement precision for many greenhouse gases (Schwartz and Warneck, 1995).

Monsoon A monsoon is a tropical and subtropical seasonal reversal in both the surface winds and associated precipitation, caused by differential heating between a continental-scale land mass and the adjacent ocean. Monsoon rains occur mainly over land in summer.

Montreal Protocol The Montreal Protocol on Substances that Deplete the *Ozone Layer* was adopted in Montreal in 1987, and subsequently adjusted and amended in London (1990), Copenhagen (1992), Vienna (1995), Montreal (1997) and Beijing (1999). It controls the consumption and production of chlorine- and bromine-containing chemicals that destroy stratospheric *ozone*, such as chlorofluorocarbons, methyl chloroform, carbon tetrachloride and many others.

Near-surface permafrost A term frequently used in *climate model* applications to refer to *permafrost* at depths close to the ground surface (typically down to 3.5 m). In modelling studies, near-surface permafrost is usually diagnosed from 20 or 30 year climate averages, which is different from the conventional definition of permafrost. Disappearance of near-surface permafrost in a location does not preclude the longer-term persistence of permafrost at greater depth. See also *Active layer*, *Frozen ground* and *Thermokarst*.

Near-term climate forcers (NTCF) Near-term climate forcers (NTCF) refer to those compounds whose impact on *climate* occurs primarily within the first decade after their emission. This set of compounds is primarily composed of those with short *lifetimes* in the atmosphere compared to *well-mixed greenhouse gases*, and has been sometimes referred to as short lived climate forcers or short-lived climate pollutants. However, the common property that is of greatest interest to a climate assessment is the timescale over which their impact on climate is felt. This set of compounds includes *methane*, which is also a well-mixed greenhouse gas, as well as *ozone* and *aerosols*, or their *precursors*, and some halogenated species that are not well-mixed greenhouse gases. These compounds do not accumulate in the atmosphere at decadal to centennial timescales, and so their effect on climate is predominantly in the near term following their emission.

Nitrogen deposition Nitrogen deposition is defined as the nitrogen transferred from the *atmosphere* to the Earth's surface by the processes of wet deposition and dry deposition.

Nitrous oxide (N_2O) One of the six *greenhouse gases* to be mitigated under the *Kyoto Protocol*. The main *anthropogenic source* of nitrous oxide is agriculture (soil and animal manure management), but important contributions also come from sewage treatment, combustion of fossil fuel, and chemical industrial processes. Nitrous oxide is also produced naturally from a wide variety of biological sources in soil and water, particularly microbial action in wet tropical *forests*.

Nonlinearity A process is called *nonlinear* when there is no simple proportional relation between cause and effect. The *climate system* contains many such nonlinear processes, resulting in a system with potentially very complex behaviour. Such complexity may lead to *abrupt climate change*. See also *Chaotic* and *Predictability*.

North Atlantic Oscillation (NAO) The North Atlantic Oscillation consists of opposing variations of surface pressure near Iceland and near the Azores. It therefore corresponds to fluctuations in the strength of the main westerly winds across the Atlantic into Europe, and thus to fluctuations in the embedded *extratropical cyclones* with their associated frontal systems. See NAO Index, Box 2.5.

Northern Annular Mode (NAM) A winter fluctuation in the amplitude of a pattern characterized by low surface pressure in the Arctic and strong mid-latitude westerlies. The NAM has links with the northern polar vortex into the *stratosphere*. Its pattern has a bias to the North Atlantic and its index has a large correlation with the *North Atlantic Oscillation* index. See NAM Index, Box 2.5.

Ocean acidification Ocean acidification refers to a reduction in the *pH* of the ocean over an extended period, typically decades or longer, which is caused primarily by *uptake* of *carbon dioxide* from the *atmosphere*, but can also be caused by other chemical additions or subtractions from the ocean. *Anthropogenic ocean acidification* refers to the component of pH reduction that is caused by human activity (IPCC, 2011, p. 37).

Ocean heat uptake efficiency This is a measure (W m^{-2} $°C^{-1}$) of the rate at which heat storage by the global ocean increases as *global mean surface temperature* rises. It is a useful parameter for *climate change* experiments in which the *radiative forcing* is changing monotonically, when it can be compared with the *Climate Feedback Parameter* to gauge the relative importance of *climate response* and ocean heat *uptake* in determining the rate of climate change. It can be estimated from such an experiment as the ratio of the rate of increase of ocean heat content to the global mean surface air temperature change.

Organic aerosol Component of the *aerosol* that consists of organic compounds, mainly carbon, hydrogen, oxygen and lesser amounts of other elements. See also *Carbonaceous aerosol*.

Outgoing longwave radiation Net outgoing radiation in the infrared part of the spectrum at the top of the *atmosphere*. See also *Terrestrial radiation*.

Outlet glacier A *glacier*, usually between rock walls, that is part of, and drains an *ice sheet*. See also *Ice stream*.

Ozone Ozone, the triatomic form of oxygen (O_3), is a gaseous atmospheric constituent. In the *troposphere*, it is created both naturally and by photochemical reactions involving gases resulting from human activities (*smog*). Tropospheric ozone acts as a *greenhouse gas*. In the *stratosphere*, it is created by the interaction between solar ultraviolet radiation and molecular oxygen (O_2). Stratospheric ozone plays a dominant role in the stratospheric radiative balance. Its concentration is highest in the *ozone layer*.

Ozone hole See *Ozone layer*.

Ozone layer The *stratosphere* contains a layer in which the concentration of *ozone* is greatest, the so-called ozone layer. The layer extends from about 12 to 40 km above the Earth's surface. The ozone concentration reaches a maximum between about 20 and 25 km. This layer has been depleted by human emissions of chlorine and bromine compounds. Every year, during the Southern Hemisphere spring, a very strong depletion of the ozone layer takes place over the Antarctic, caused by *anthropogenic* chlorine and bromine compounds in combination with the specific meteorological conditions of that *region*. This phenomenon is called the *Ozone hole*. See also *Montreal Protocol*.

Pacific Decadal Oscillation (PDO) The pattern and time series of the first empirical orthogonal function of *sea surface temperature* over the North Pacific north of 20°N. The PDO broadened to cover the whole Pacific Basin is known as the Inter-decadal Pacific Oscillation. The PDO and IPO exhibit similar temporal evolution. See also *Pacific Decadal Variability*.

Pacific decadal variability Coupled decadal-to-inter-decadal variability of the atmospheric circulation and underlying ocean in the Pacific Basin. It is most prominent in the North Pacific, where fluctuations in the strength of the winter Aleutian Low pressure system co-vary with North Pacific *sea surface temperatures*, and are linked to decadal variations in atmospheric circulation, sea surface temperatures and ocean circulation throughout the whole Pacific Basin. Such fluctuations have the effect of modulating the *El Niño-Southern Oscillation* cycle. Key measures of Pacific decadal variability are the *North Pacific Index (NPI)*, the *Pacific Decadal Oscillation (PDO)* index and the *Inter-decadal Pacific Oscillation (IPO)* index, all defined in Box 2.5.

Pacific–North American (PNA) pattern An atmospheric large-scale wave pattern featuring a sequence of tropospheric high and low pressure anomalies stretching from the subtropical west Pacific to the east coast of North America. See PNA pattern index, Box 2.5.

Paleoclimate *Climate* during periods prior to the development of measuring instruments, including historic and geologic time, for which only *proxy* climate records are available.

Parameterization In *climate models*, this term refers to the technique of representing processes that cannot be explicitly resolved at the spatial or temporal *resolution* of the model (sub-grid scale processes) by relationships between model-resolved larger-scale variables and the area or time-averaged effect of such subgrid scale processes.

Percentiles The set of partition values which divides the total population of a distribution into 100 equal parts, the 50th percentile corresponding to the *median* of the population.

Permafrost Ground (soil or rock and included ice and organic material) that remains at or below 0°C for at least two consecutive years. See also *Near-surface permafrost*.

pH pH is a dimensionless measure of the acidity of water (or any solution) given by its concentration of hydrogen ions (H^+). pH is measured on a logarithmic scale where pH = $-\log_{10}(H^+)$. Thus, a pH decrease of 1 unit corresponds to a 10-fold increase in the concentration of H^+, or acidity.

Photosynthesis The process by which plants take *carbon dioxide* from the air (or bicarbonate in water) to build carbohydrates, releasing oxygen in the process. There are several pathways of photosynthesis with different responses to atmospheric carbon dioxide concentrations. See also *Carbon dioxide fertilization*.

Plankton Microorganisms living in the upper layers of aquatic systems. A distinction is made between *phytoplankton*, which depend on *photosynthesis* for their energy supply, and *zooplankton*, which feed on phytoplankton.

Pleistocene The Pleistocene Epoch is the earlier of two epochs in the *Quaternary* System, extending from 2.59 Ma to the beginning of the *Holocene* at 11.65 ka.

Pliocene The Plionece Epoch is the last epoch of the *Neogene* System and extends from 5.33 Ma to the beginning of the *Pleistocene* at 2.59 Ma.

Pollen analysis A technique of both relative dating and environmental *reconstruction*, consisting of the identification and counting of pollen types preserved in peat, lake sediments and other deposits. See also *Proxy*.

Precipitable water The total amount of atmospheric water vapour in a vertical column of unit cross-sectional area. It is commonly expressed in terms of the height of the water if completely condensed and collected in a vessel of the same unit cross section.

Precursors Atmospheric compounds that are not *greenhouse gases* or *aerosols*, but that have an effect on greenhouse gas or aerosol concentrations by taking part in physical or chemical processes regulating their production or destruction rates.

Predictability The extent to which future states of a system may be predicted based on knowledge of current and past states of the system. Because knowledge of the *climate system*'s past and current states is generally imperfect, as are the models that utilize this knowledge to produce a *climate prediction*, and because the climate system is inherently *nonlinear* and *chaotic*, predictability of the climate system is inherently limited. Even with arbitrarily accurate models and observations, there may still be limits to the predictability of such a nonlinear system (AMS, 2000).

Prediction quality/skill Measures of the success of a *prediction* against observationally based information. No single measure can summarize all aspects of forecast quality and a suite of *metrics* is considered. Metrics will differ for forecasts given in deterministic and probabilistic form. See also *Climate prediction*.

Pre-industrial See *Industrial Revolution*.

Probability Density Function (PDF) A probability density function is a function that indicates the relative chances of occurrence of different outcomes of a variable. The function integrates to unity over the domain for which it is defined and has the property that the integral over a sub-domain equals the probability that the outcome of the variable lies within that sub-domain. For example, the probability that a temperature anomaly defined in a particular way is greater than zero is obtained from its PDF by integrating the PDF over all possible temperature anomalies greater than zero. Probability density functions that describe two or more variables simultaneously are similarly defined.

Process-based Model Theoretical concepts and computational methods that represent and simulate the behaviour of real-world systems derived from a set of functional components and their interactions with each other and the system environment, through physical and mechanistic processes occurring over time. See also *Climate model*.

Projection A projection is a potential future evolution of a quantity or set of quantities, often computed with the aid of a model. Unlike predictions, projections are conditional on assumptions concerning, for example, future socioeconomic and technological developments that may or may not be realized. See also *Climate prediction* and *Climate projection*.

Proxy A proxy *climate* indicator is a record that is interpreted, using physical and biophysical principles, to represent some combination of climate-related variations back in time. Climate-related data derived in this way are referred to as proxy data. Examples of proxies include *pollen analysis*, *tree ring* records, speleothems, characteristics of corals and various data derived from marine sediments and *ice cores*. Proxy-data can be calibrated to provide quantitative climate information.

Quasi-Biennal Oscillation (QBO) A near-periodic oscillation of the equatorial zonal wind between easterlies and westerlies in the tropical *stratosphere* with a mean period of around 28 months. The alternating wind maxima descend from the base of the mesosphere down to the *tropopause*, and are driven by wave energy that propagates up from the *troposphere*.

Quaternary The Quaternary System is the latter of three systems that make up the *Cenozoic Era* (65 Ma to present), extending from 2.59 Ma to the present, and includes the *Pleistocene* and *Holocene* epochs.

Radiative effect The impact on a radiation flux or heating rate (most commonly, on the downward flux at the top of *atmosphere*) caused by the interaction of a particular constituent with either the *infrared* or *solar radiation* fields through absorption, scattering and emission, relative to an otherwise identical atmosphere free of that constituent. This quantifies the impact of the constituent on the *climate system*. Examples include the *aerosol–radiation interactions*, *cloud radiative effect*, and *greenhouse effect*. In this report, the portion of any top-of-atmosphere radiative effect that is due to *anthropogenic* or other external influences (e.g., volcanic eruptions or changes in the sun) is termed the *instantaneous radiative forcing*.

Radiative forcing Radiative forcing is the change in the net, downward minus upward, radiative flux (expressed in W m^{-2}) at the *tropopause* or top of *atmosphere* due to a change in an external driver of *climate change*, such as, for example, a change in the concentration of *carbon dioxide* or the output of the Sun. Sometimes internal drivers are still treated as forcings even though they result from the alteration in *climate*, for example *aerosol* or *greenhouse gas* changes in *paleoclimates*. The traditional radiative forcing is computed with all tropospheric properties held fixed at their unperturbed values, and after allowing for stratospheric temperatures, if perturbed, to readjust to radiative-dynamical equilibrium. Radiative forcing is called *instantaneous* if no change in stratospheric temperature is accounted for. The radiative forcing once *rapid adjustments* are accounted for is termed the *effective radiative forcing*. For the purposes of this report, radiative forcing is further defined as the change relative to the year 1750 and, unless otherwise noted, refers to a global and annual average value. Radiative forcing is not to be confused with *cloud radiative forcing*, which describes an unrelated measure of the impact of clouds on the radiative flux at the top of the atmosphere.

Rapid adjustment The response to an agent perturbing the *climate system* that is driven directly by the agent, independently of any change in the *global mean surface temperature*. For example, *carbon dioxide* and *aerosols*, by altering internal heating and cooling rates within the *atmosphere*, can each cause changes to cloud cover and other variables thereby producing a *radiative effect* even in the absence of any surface warming or cooling. Adjustments are *rapid* in the sense that they begin to occur right away, before *climate feedbacks* which are driven by warming (although some adjustments may still take significant time to proceed to completion, for example those involving vegetation or *ice sheets*). It is also called the *rapid response* or *fast adjustment*. For further explanation on the concept, see Sections 7.1 and 8.1.

Rapid climate change See *Abrupt climate change*.

Rapid dynamical change (of glaciers or ice sheets) Changes in *glacier* or *ice sheet* mass controlled by changes in flow speed and *discharge* rather than by *accumulation* or *ablation*. This can result in a rate of mass change larger than that due to any imbalance between accumulation and ablation. Rapid dynamical change may be initiated by a climatic trigger, such as incursion of warm ocean water beneath an *ice shelf*, or thinning of a grounded tidewater terminus, which may lead to reactions within the glacier system, that may result in rapid ice loss. See also *Mass balance/budget (of glaciers or ice sheets)*.

Reanalysis Reanalyses are estimates of historical atmospheric temperature and wind or oceanographic temperature and current, and other quantities, created by processing past meteorological or oceanographic data using fixed state-of-the-art weather forecasting or ocean circulation models with data assimilation techniques. Using fixed data assimilation avoids effects from the changing analysis system that occur in operational analyses. Although continuity is improved, global reanalyses still suffer from changing coverage and biases in the observing systems.

Rebound effect When *CO_2* is removed from the *atmosphere*, the CO_2 concentration gradient between atmospheric and land/ocean carbon *reservoirs* is reduced. This leads to a reduction or reversal in subsequent inherent rate of removal of CO_2 from the atmosphere by natural *carbon cycle* processes on land and ocean.

Reconstruction (of climate variable) Approach to reconstructing the past temporal and spatial characteristics of a climate variable from predictors. The predictors can be instrumental data if the reconstruction is used to infill missing data or *proxy* data if it is used to develop *paleoclimate* reconstructions. Various techniques have been developed for this purpose: linear multivariate regression based methods and nonlinear *Bayesian* and analog methods.

Reforestation Planting of *forests* on lands that have previously contained forests but that have been converted to some other use. For a discussion of the term *forest* and related terms such as *afforestation*, *reforestation* and *deforestation*, see the IPCC Report on Land Use, Land-Use Change and Forestry (IPCC, 2000). See also the Report on Definitions and Methodological Options to Inventory Emissions from Direct Human-induced Degradation of Forests and Devegetation of Other Vegetation Types (IPCC, 2003).

Region A region is a territory characterized by specific geographical and climatological features. The *climate* of a region is affected by regional and local scale features like topography, *land use* characteristics and lakes, as well as remote influences from other regions. See also *Teleconnection*.

Regional Climate Model (RCM) A *climate model* at higher *resolution* over a limited area. Such models are used in *downscaling* global *climate* results over specific regional domains.

Relative humidity The relative humidity specifies the ratio of actual water vapour pressure to that at saturation with respect to liquid water or ice at the same temperature. See also *Specific humidity*.

Relative sea level Sea level measured by a *tide gauge* with respect to the land upon which it is situated. See also *Mean sea level* and *Sea level change*.

Representative Concentration Pathways (RCPs) *Scenarios* that include time series of emissions and concentrations of the full suite of *greenhouse gases* and *aerosols* and chemically active gases, as well as *land use*/land cover (Moss et al., 2008). The word *representative* signifies that each RCP provides only one of many possible scenarios that would lead to the specific *radiative forcing* characteristics. The term *pathway* emphasizes that not only the long-term concentration levels are of interest, but also the trajectory taken over time to reach that outcome. (Moss et al., 2010).

RCPs usually refer to the portion of the concentration pathway extending up to 2100, for which Integrated Assessment Models produced corresponding *emission scenarios*. *Extended Concentration Pathways (ECPs)* describe extensions of the RCPs from 2100 to 2500 that were calculated using simple rules generated by stakeholder consultations, and do not represent fully consistent scenarios.

Four RCPs produced from Integrated Assessment Models were selected from the published literature and are used in the present IPCC Assessment as a basis for the *climate predictions* and *projections* presented in Chapters 11 to 14:

RCP2.6 One pathway where radiative forcing peaks at approximately 3 W m^{-2} before 2100 and then declines (the corresponding ECP assuming constant emissions after 2100)

RCP4.5 and RCP6.0 Two intermediate *stabilization pathways* in which radiative forcing is stabilized at approximately 4.5 W m^{-2} and 6.0 W m^{-2} after 2100 (the corresponding ECPs assuming constant concentrations after 2150)

RCP8.5 One high pathway for which radiative forcing reaches greater than 8.5 W m^{-2} by 2100 and continues to rise for some amount of time (the corresponding ECP assuming constant emissions after 2100 and constant concentrations after 2250)

For further description of future scenarios, see Box 1.1.

Reservoir A component of the *climate system*, other than the *atmosphere*, which has the capacity to store, accumulate or release a substance of concern, for example, carbon, a *greenhouse gas* or a *precursor*. Oceans, soils and *forests* are examples of reservoirs of carbon. *Pool* is an equivalent term (note that the definition of pool often includes the atmosphere). The absolute quantity of the substance of concern held within a reservoir at a specified time is called the *stock*.

Resolution In *climate models*, this term refers to the physical distance (metres or degrees) between each point on the grid used to compute the equations. *Temporal resolution* refers to the time step or time elapsed between each model computation of the equations.

Respiration The process whereby living organisms convert organic matter to *carbon dioxide*, releasing energy and consuming molecular oxygen.

Response time The response time or *adjustment time* is the time needed for the *climate system* or its components to re-equilibrate to a new state, following a forcing resulting from external processes. It is very different for various components of the climate system. The response time of the *troposphere* is relatively short, from days to weeks, whereas the *stratosphere* reaches equilibrium on a time scale of typically a few months. Due to their large heat capacity, the oceans have a much longer response time: typically decades, but up to centuries or millennia. The response time of the strongly coupled surface–troposphere system is, therefore, slow compared to that of the stratosphere, and mainly determined by the oceans. The *biosphere* may respond quickly (e.g., to *droughts*), but also very slowly to imposed changes. See *lifetime* for a different definition of response time pertinent to the rate of processes affecting the concentration of trace gases.

Return period An estimate of the average time interval between occurrences of an event (e.g., flood or extreme rainfall) of (or below/above) a defined size or intensity. See also *Return value*.

Return value The highest (or, alternatively, lowest) value of a given variable, on average occurring once in a given period of time (e.g., in 10 years). See also *Return period*.

River discharge See *Streamflow*.

Runoff That part of precipitation that does not evaporate and is not transpired, but flows through the ground or over the ground surface and returns to bodies of water. See also *Hydrological cycle*.

Scenario A plausible description of how the future may develop based on a coherent and internally consistent set of assumptions about key driving forces (e.g., rate of technological change, prices) and relationships. Note that scenarios are neither predictions nor forecasts, but are useful to provide a view of the implications of developments and actions. See also *Climate scenario*, *Emission scenario*, *Representative Concentration Pathways* and *SRES scenarios*.

Sea ice Ice found at the sea surface that has originated from the freezing of seawater. Sea ice may be discontinuous pieces (ice floes) moved on the ocean surface by wind and currents (pack ice), or a motionless sheet attached to the coast (land-fast ice). *Sea ice concentration* is the fraction of the ocean covered by ice. Sea ice less than one year old is called *first-year ice*. *Perennial ice* is sea ice that survives at least one summer. It may be subdivided into *second-year ice* and *multi-year ice*, where multiyear ice has survived at least two summers.

Sea level change Sea level can change, both globally and locally due to (1) changes in the shape of the ocean basins, (2) a change in ocean volume as a result of a change in the mass of water in the ocean, and (3) changes in ocean volume as a result of changes in ocean water density. Global *mean sea level* change resulting from change in the mass of the ocean is called *barystatic*. The amount of barystatic sea level change due to the addition or removal of a mass of water is called its *sea level equivalent (SLE)*. Sea level changes, both globally and locally, resulting from changes in water density are called *steric*. Density changes induced by temperature changes only are called *thermosteric*, while density changes induced by salinity changes are called *halosteric*. Barystatic and steric sea level changes do not include the effect of changes in the shape of ocean basins induced by the change in the ocean mass and its distribution. See also *Relative Sea Level* and *Thermal expansion*.

Sea level equivalent (SLE) The sea level equivalent of a mass of water (ice, liquid or vapour) is that mass, converted to a volume using a density of 1000 kg m^{-3}, and divided by the present-day ocean surface area of 3.625×10^{14} m^2. Thus, 362.5 Gt of water mass added to the ocean will cause 1 mm of global *mean sea level* rise. See also *Sea level change*.

Seasonally frozen ground See *Frozen ground*.

Sea surface temperature (SST) The sea surface temperature is the subsurface bulk temperature in the top few metres of the ocean, measured by ships, buoys and drifters. From ships, measurements of water samples in buckets were mostly switched in the 1940s to samples from engine intake water. Satellite measurements of *skin temperature* (uppermost layer; a fraction of a millimetre thick) in the infrared or the top centimetre or so in the microwave are also used, but must be adjusted to be compatible with the bulk temperature.

AIII

Semi-direct (aerosol) effect See *Aerosol–radiation interaction*.

Semi-empirical model Model in which calculations are based on a combination of observed associations between variables and theoretical considerations relating variables through fundamental principles (e.g., conservation of energy). For example, in sea level studies, semi-empirical models refer specifically to transfer functions formulated to project future global *mean sea level change*, or contributions to it, from future *global mean surface temperature* change or *radiative forcing*.

Sensible heat flux The turbulent or conductive flux of heat from the Earth's surface to the *atmosphere* that is not associated with phase changes of water; a component of the surface *energy budget*.

Sequestration See *Uptake*.

Shortwave radiation See *Solar radiation*.

Significant wave height The average trough-to-crest height of the highest one third of the wave heights (sea and swell) occurring in a particular time period.

Sink Any process, activity or mechanism that removes a *greenhouse gas*, an *aerosol* or a *precursor* of a greenhouse gas or aerosol from the *atmosphere*.

Slab-ocean model A simplified representation in a *climate model* of the ocean as a motionless layer of water with a depth of 50 to 100 m. Climate models with a slab ocean can be used only for estimating the equilibrium response of *climate* to a given forcing, not the transient evolution of climate. See also *Equilibrium and transient climate experiment*.

Snow cover extent The areal extent of snow covered ground.

Snow water equivalent (SWE) The depth of liquid water that would result if a mass of snow melted completely.

Soil moisture Water stored in the soil in liquid or frozen form.

Soil temperature The temperature of the soil. This can be measured or modelled at multiple levels within the depth of the soil.

Solar activity General term describing a variety of magnetic phenomena on the Sun such as *sunspots*, *faculae* (bright areas), and flares (emission of high-energy particles). It varies on time scales from minutes to millions of years. See also *Solar cycle*.

Solar ('11-year') cycle A quasi-regular modulation of *solar activity* with varying amplitude and a period of between 8 and 14 years.

Solar radiation Electromagnetic radiation emitted by the Sun with a spectrum close to the one of a black body with a temperature of 5770 K. The radiation peaks in visible wavelengths. When compared to the *terrestrial radiation* it is often referred to as *shortwave radiation*. See also *Insolation* and *Total solar irradiance (TSI)*.

Solar Radiation Management (SRM) Solar Radiation Management refers to the intentional modification of the Earth's shortwave radiative budget with the aim to reduce *climate change* according to a given *metric* (e.g., *surface temperature*, precipitation, regional impacts, etc). Artificial injection of stratospheric *aerosols* and cloud brightening are two examples of SRM techniques. Methods to modify some fast-responding elements of the longwave radiative budget (such as cirrus clouds), although not strictly speaking SRM, can be related to SRM. SRM techniques do not fall within the usual definitions of *mitigation* and adaptation (IPCC, 2012, p. 2). See also *Solar radiation*, *Carbon Dioxide Removal (CDR)* and *Geoengineering*.

Solubility pump Solubility pump is an important physicochemical process that transports dissolved inorganic carbon from the ocean's surface to its interior. This process controls the inventory of carbon in the ocean. The solubility of gaseous *carbon dioxide* can alter carbon dioxide concentrations in the oceans and the overlying *atmosphere*. See also *Biological pump*.

Source Any process, activity or mechanism that releases a *greenhouse gas*, an *aerosol* or a *precursor* of a greenhouse gas or aerosol into the *atmosphere*.

Southern Annular Mode (SAM) The leading mode of variability of Southern Hemisphere geopotential height, which is associated with shifts in the latitude of the midlatitude jet. See SAM Index, Box 2.5.

Southern Oscillation See *El Niño-Southern Oscillation (ENSO)*.

South Pacific Convergence Zone (SPCZ) A band of low-level convergence, cloudiness and precipitation ranging from the west Pacific warm pool south-eastwards towards French Polynesia, which is one of the most significant features of subtropical Southern Hemisphere *climate*. It shares some characteristics with the *ITCZ*, but is more extratropical in nature, especially east of the Dateline.

Spatial and temporal scales *Climate* may vary on a large range of spatial and temporal scales. Spatial scales may range from local (less than 100 000 km^2), through regional (100 000 to 10 million km^2) to continental (10 to 100 million km^2). Temporal scales may range from seasonal to geological (up to hundreds of millions of years).

Specific humidity The specific humidity specifies the ratio of the mass of water vapour to the total mass of moist air. See also *Relative humidity*.

SRES scenarios SRES scenarios are *emission scenarios* developed by Nakićenović and Swart (2000) and used, among others, as a basis for some of the *climate projections* shown in Chapters 9 to 11 of IPCC (2001) and Chapters 10 and 11 of IPCC (2007). The following terms are relevant for a better understanding of the structure and use of the set of SRES scenarios:

> **Scenario family** Scenarios that have a similar demographic, societal, economic and technical change storyline. Four scenario families comprise the SRES scenario set: A1, A2, B1 and B2.

Illustrative Scenario A scenario that is illustrative for each of the six scenario groups reflected in the Summary for Policymakers of Nakićenović and Swart (2000). They include four revised *marker scenarios* for the scenario groups A1B, A2, B1, B2 and two additional scenarios for the A1FI and A1T groups. All scenario groups are equally sound.

Marker Scenario A scenario that was originally posted in draft form on the SRES website to represent a given scenario family. The choice of markers was based on which of the initial quantifications best reflected the storyline, and the features of specific models. Markers are no more likely than other scenarios, but are considered by the SRES writing team as illustrative of a particular storyline. They are included in revised form in Nakićenović and Swart (2000). These scenarios received the closest scrutiny of the entire writing team and via the SRES open process. Scenarios were also selected to illustrate the other two scenario groups.

Storyline A narrative description of a scenario (or family of scenarios), highlighting the main scenario characteristics, relationships between key driving forces and the dynamics of their evolution.

Steric See *Sea level change*.

Stock See *Reservoir*.

Storm surge The temporary increase, at a particular locality, in the height of the sea due to extreme meteorological conditions (low atmospheric pressure and/or strong winds). The storm surge is defined as being the excess above the level expected from the tidal variation alone at that time and place.

Storm tracks Originally, a term referring to the tracks of individual cyclonic weather systems, but now often generalized to refer to the main *regions* where the tracks of extratropical disturbances occur as sequences of low (cyclonic) and high (anticyclonic) pressure systems.

Stratosphere The highly stratified region of the *atmosphere* above the *troposphere* extending from about 10 km (ranging from 9 km at high latitudes to 16 km in the tropics on average) to about 50 km altitude.

Streamflow Water flow within a river channel, for example expressed in $m^3 s^{-1}$. A synonym for *river discharge*.

Subduction Ocean process in which surface waters enter the ocean interior from the surface mixed layer through *Ekman pumping* and lateral *advection*. The latter occurs when surface waters are advected to a region where the local surface layer is less dense and therefore must slide below the surface layer, usually with no change in density.

Sunspots Dark areas on the Sun where strong magnetic fields reduce the convection causing a temperature reduction of about 1500 K compared to the surrounding regions. The number of sunspots is higher during periods of higher *solar activity*, and varies in particular with the *solar cycle*.

Surface layer See *Atmospheric boundary layer*.

Surface temperature See *Global mean surface temperature*, *Land surface air temperature* and *Sea surface temperature*.

Talik A layer of year-round unfrozen ground that lies in *permafrost* areas.

Teleconnection A statistical association between climate variables at widely separated, geographically-fixed spatial locations. Teleconnections are caused by large spatial structures such as basin-wide coupled modes of ocean–*atmosphere* variability, Rossby wave-trains, mid-latitude jets and *storm tracks*, etc. See also *Teleconnection pattern*.

Teleconnection pattern A correlation map obtained by calculating the correlation between variables at different spatial locations and a *climate index*. It is the special case of a *climate pattern* obtained for standardized variables and a standardized climate index, that is, the variables and index are each centred and scaled to have zero mean and unit variance. One-point teleconnection maps are made by choosing a variable at one of the locations to be the climate index. See also *Teleconnection*.

Terrestrial radiation Radiation emitted by the Earth's surface, the *atmosphere* and the clouds. It is also known as *thermal infrared* or *longwave radiation*, and is to be distinguished from the near-infrared radiation that is part of the solar spectrum. *Infrared radiation*, in general, has a distinctive range of wavelengths (*spectrum*) longer than the wavelength of the red light in the visible part of the spectrum. The spectrum of terrestrial radiation is almost entirely distinct from that of shortwave or *solar radiation* because of the difference in temperature between the Sun and the Earth–atmosphere system. See also *Outgoing longwave radiation*.

Thermal expansion In connection with sea level, this refers to the increase in volume (and decrease in density) that results from warming water. A warming of the ocean leads to an expansion of the ocean volume and hence an increase in sea level. See also *Sea level change*.

Thermocline The layer of maximum vertical temperature gradient in the ocean, lying between the surface ocean and the abyssal ocean. In subtropical regions, its source waters are typically surface waters at higher latitudes that have *subducted* (see *Subduction*) and moved equatorward. At high latitudes, it is sometimes absent, replaced by a *halocline*, which is a layer of maximum vertical salinity gradient.

Thermohaline circulation (THC) Large-scale circulation in the ocean that transforms low-density upper ocean waters to higher-density intermediate and deep waters and returns those waters back to the upper ocean. The circulation is asymmetric, with conversion to dense waters in restricted regions at high latitudes and the return to the surface involving slow upwelling and diffusive processes over much larger geographic regions. The THC is driven by high densities at or near the surface, caused by cold temperatures and/or high salinities, but despite its suggestive though common name, is also driven by mechanical forces such as wind and tides. Frequently, the name THC has been used synonymously with *Meridional Overturning Circulation*.

Thermokarst The process by which characteristic landforms result from the thawing of ice-rich *permafrost* or the melting of massive ground ice.

Thermosteric See *Sea level change*.

Tide gauge A device at a coastal or deep-sea location that continuously measures the level of the sea with respect to the adjacent land. Time averaging of the sea level so recorded gives the observed secular changes of the *relative sea level*.

Tipping point In *climate*, a hypothesized critical threshold when global or regional *climate changes* from one stable state to another stable state. The tipping point event may be irreversible. See also *Irreversibility*.

Total solar irradiance (TSI) The total amount of *solar radiation* in watts per square metre received outside the Earth's *atmosphere* on a surface normal to the incident radiation, and at the Earth's mean distance from the Sun.

Reliable measurements of solar radiation can only be made from space and the precise record extends back only to 1978. The generally accepted value is 1368 W m^{-2} with an accuracy of about 0.2%. It has recently been estimated to 1360.8 ± 0.5 W m^{-2} for the solar minimum of 2008. Variations of a few tenths of a percent are common, usually associated with the passage of *sunspots* across the solar disk. The *solar cycle* variation of TSI is of the order of 0.1% (AMS, 2000). Changes in the ultraviolet part of the spectrum during a solar cycle are comparatively larger (percent) than in TSI. See also *Insolation*.

Transient climate response See *Climate sensitivity*.

Transient climate response to cumulative CO_2 emissions (TCRE) The transient global average *surface temperature* change per unit cumulated *CO_2* emissions, usually 1000 PgC. TCRE combines both information on the *airborne fraction* of cumulated CO_2 emissions (the fraction of the total CO_2 emitted that remains in the *atmosphere*), and on the *transient climate response* (TCR).

Tree rings Concentric rings of secondary wood evident in a cross section of the stem of a woody plant. The difference between the dense, small-celled late wood of one season and the wide-celled early wood of the following spring enables the age of a tree to be estimated, and the ring widths or density can be related to climate parameters such as temperature and precipitation. See also *Proxy*.

Trend In this report, the word *trend* designates a change, generally monotonic in time, in the value of a variable.

Tropopause The boundary between the *troposphere* and the *stratosphere*.

Troposphere The lowest part of the *atmosphere*, from the surface to about 10 km in altitude at mid-latitudes (ranging from 9 km at high latitudes to 16 km in the tropics on average), where clouds and weather phenomena occur. In the troposphere, temperatures generally decrease with height. See also *Stratosphere*.

Turnover time See *Lifetime*.

Uncertainty A state of incomplete knowledge that can result from a lack of information or from disagreement about what is known or even knowable. It may have many types of sources, from imprecision in the data to ambiguously defined concepts or terminology, or uncertain *projections* of human behaviour. Uncertainty can therefore be represented by quantitative measures (e.g., a *probability density function*) or by qualitative statements (e.g., reflecting the judgment of a team of experts) (see Moss and Schneider, 2000; Manning et al., 2004; Mastrandrea et al., 2010). See also *Confidence* and *Likelihood*.

United Nations Framework Convention on Climate Change (UNFCCC) The Convention was adopted on 9 May 1992 in New York and signed at the 1992 Earth Summit in Rio de Janeiro by more than 150 countries and the European Community. Its ultimate objective is the 'stabilisation of *greenhouse gas* concentrations in the *atmosphere* at a level that would prevent dangerous *anthropogenic* interference with the *climate system*'. It contains commitments for all Parties. Under the Convention, Parties included in Annex I (all OECD countries and countries with economies in transition) aim to return greenhouse gas emissions not controlled by the *Montreal Protocol* to 1990 levels by the year 2000. The convention entered in force in March 1994. In 1997, the UNFCCC adopted the *Kyoto Protocol*.

Uptake The addition of a substance of concern to a *reservoir*. The uptake of carbon containing substances, in particular *carbon dioxide*, is often called (carbon) *sequestration*.

Urban heat island (UHI) The relative warmth of a city compared with surrounding rural areas, associated with changes in *runoff*, effects on heat retention, and changes in surface *albedo*.

Ventilation The exchange of ocean properties with the atmospheric *surface layer* such that property concentrations are brought closer to equilibrium values with the *atmosphere* (AMS, 2000), and the processes that propagate these properties into the ocean interior.

Volatile Organic Compounds (VOC) Important class of organic chemical air pollutants that are volatile at ambient air conditions. Other terms used to represent VOCs are *hydrocarbons* (HCs), *reactive organic gases* (ROGs) and *non-methane volatile organic compounds* (NMVOCs). NMVOCs are major contributors (together with NO_x and CO) to the formation of photochemical oxidants such as *ozone*.

Walker Circulation Direct thermally driven zonal overturning circulation in the *atmosphere* over the tropical Pacific Ocean, with rising air in the western and sinking air in the eastern Pacific.

Warm days/warm nights Days where maximum temperature, or nights where minimum temperature, exceeds the 90th *percentile*, where the respective temperature distributions are generally defined with respect to the 1961–1990 *reference* period. For the corresponding indices, see Box 2.4.

Warm spell A period of abnormally hot weather. For the corresponding indices, see Box 2.4. See also *Heat wave*.

Water cycle See *Hydrological cycle*.

Water mass A body of ocean water with identifiable properties (temperature, salinity, density, chemical tracers) resulting from its unique formation process. Water masses are often identified through a vertical or horizontal extremum of a property such as salinity. North Pacific Intermediate Water (NPIW) and Antarctic Intermediate Water (AAIW) are examples of water masses.

Weathering The gradual removal of atmospheric *CO_2* through dissolution of silicate and carbonate rocks. Weathering may involve physical processes (*mechanical weathering*) or chemical activity (*chemical weathering*).

Well-mixed greenhouse gas See *Greenhouse gas*.

Younger Dryas A period 12.85 to 11.65 ka (thousand years before 1950), during the *deglaciation*, characterized by a temporary return to colder conditions in many locations, especially around the North Atlantic.

References

AMS, 2000: *AMS Glossary of Meteorology*, 2nd ed. American Meteorological Society, Boston, MA, http://amsglossary.allenpress.com/glossary/browse.

Hegerl, G. C., O. Hoegh-Guldberg, G. Casassa, M. P. Hoerling, R. S. Kovats, C. Parmesan, D. W. Pierce, and P. A. Stott, 2010: Good practice guidance paper on detection and attribution related to anthropogenic climate change. In: *Meeting Report of the Intergovernmental Panel on Climate Change Expert Meeting on Detection and Attribution of Anthropogenic Climate Change* [T. F. Stocker, C. B. Field, D. Qin, V. Barros, G.-K. Plattner, M. Tignor, P. M. Midgley and K. L. Ebi (eds.)]. IPCC Working Group I Technical Support Unit, University of Bern, Bern, Switzerland.

IPCC, 1992: *Climate Change 1992: The Supplementary Report to the IPCC Scientific Assessment* [J. T. Houghton, B. A. Callander and S. K. Varney (eds.)]. Cambridge University Press, Cambridge, United Kingdom and New York, NY, USA, 116 pp.

IPCC, 1996: *Climate Change 1995: The Science of Climate Change. Contribution of Working Group I to the Second Assessment Report of the Intergovernmental Panel on Climate Change* [J. T. Houghton., L. G. Meira . A. Callander, N. Harris, A. Kattenberg and K. Maskell (eds.)]. Cambridge University Press, Cambridge, United Kingdom and New York, NY, USA, 572 pp.

IPCC, 2000: *Land Use, Land-Use Change, and Forestry. Special Report of the Intergovernmental Panel on Climate Change* [R. T. Watson, I. R. Noble, B. Bolin, N. H. Ravindranath, D. J. Verardo, and D. J. Dokken (eds.)]. Cambridge University Press, Cambridge, United Kingdom and New York, NY, USA, 377 pp.

IPCC, 2001: *Climate Change 2001: The Scientific Basis. Contribution of Working Group I to the Third Assessment Report of the Intergovernmental Panel on Climate Change* [T. Houghton, Y. Ding, D. J. Griggs, M. Noquer, P. J. van der Linden, X. Dai, K. Maskell and C. A. Johnson (eds.)]. Cambridge University Press, Cambridge, United Kingdom and New York, NY, USA, 881 pp.

IPCC, 2003: Definitions and Methodological Options to Inventory Emissions from Direct Human-Induced Degradation of Forests and Devegetation of Other Vegetation Types [Penman, J., M. Gytarsky, T. Hiraishi, T. Krug, D. Kruger, R. Pipatti, L. Buendia, K. Miwa, T. Ngara, K. Tanabe and F. Wagner (eds.)]. The Institute for Global Environmental Strategies (IGES), Japan, 32 pp.

IPCC, 2007: *Climate Change 2007: The Physical Science Basis. Contribution of Working Group I to the Fourth Assessment Report of the Intergovernmental Panel on Climate Change.* [Solomon, S., D. Qin, M. Manning, Z. Chen, M. Marquis, K. B. Averyt, M. Tignor and H. L. Miller (eds.)]. Cambridge University Press, Cambridge, United Kingdom and New York, NY, USA, 996 pp.

IPCC, 2011: *Workshop Report of the Intergovernmental Panel on Climate Change Workshop on Impacts of Ocean Acidification on Marine Biology and Ecosystems* [C. B. Field, V. Barros, T. F. Stocker, D. Qin, K.J. Mach, G.-K. Plattner, M. D. Mastrandrea, M. Tignor and K. L. Ebi (eds.)]. IPCC Working Group II Technical Support Unit, Carnegie Institution, Stanford, CA, USA, 164 pp.

IPCC, 2012: *Meeting Report of the Intergovernmental Panel on Climate Change Expert Meeting on Geoengineering* [O. Edenhofer, R. Pichs-Madruga, Y. Sokona, C. Field, V. Barros, T. F. Stocker, Q. Dahe, J. Minx, K. Mach, G.-K. Plattner, S. Schlömer, G. Hansen and M. Mastrandrea (eds.)]. IPCC Working Group III Technical Support Unit, Potsdam Institute for Climate Impact Research, Potsdam, Germany, 99 pp.

Manning, M., et al., 2004: *IPCC Workshop on Describing Scientific Uncertainties in Climate Change to Support Analysis of Risk of Options*. Workshop Report. IPCC Working Group I Technical Support Unit, Boulder, CO, USA, 138 pp.

Mastrandrea, M. D., C. B. Field, T. F. Stocker, O. Edenhofer, K. L. Ebi, D. J. Frame, H. Held, E. Kriegler, K. J. Mach, P. R. Matschoss, G.-K. Plattner, G. W. Yohe, and F. W. Zwiers, 2010: *Guidance Note for Lead Authors of the IPCC Fifth Assessment Report on Consistent Treatment of Uncertainties*. Intergovernmental Panel on Climate Change (IPCC). http://www.ipcc.ch.

Moss, R., and S. Schneider, 2000: *Uncertainties in the IPCC TAR: Recommendations to Lead Authors for More Consistent Assessment and Reporting*. In: IPCC Supporting Material: Guidance Papers on Cross Cutting Issues in the Third Assessment Report of the IPCC. [Pachauri, R., T. Taniguchi, and K. Tanaka (eds.)]. Intergovernmental Panel on Climate Change, Geneva, pp. 33–51.

Moss, R., et al., 2008: *Towards new scenarios for analysis of emissions, climate change, impacts and response strategies*. Intergovernmental Panel on Climate Change, Geneva, 132 pp.

Moss, R. et al., 2010: The next generation of scenarios for climate change research and assessment. *Nature*, **463**, 747–756.

Nakićenović, N., and R. Swart (eds.), 2000: *Special Report on Emissions Scenarios. A Special Report of Working Group III of the Intergovernmental Panel on Climate Change*. Cambridge University Press, Cambridge, United Kingdom and New York, NY, USA, 599 pp.

Schwartz, S.E., and P. Warneck, 1995: Units for use in atmospheric chemistry. *Pure Appl. Chem.*, **67**, 1377–1406.

Annex IV: Acronyms

This annex should be cited as:

IPCC, 2013: Annex IV: Acronyms. In: *Climate Change 2013: The Physical Science Basis. Contribution of Working Group I to the Fifth Assessment Report of the Intergovernmental Panel on Climate Change* [Stocker, T.F., D. Qin, G.-K. Plattner, M. Tignor, S.K. Allen, J. Boschung, A. Nauels, Y. Xia, V. Bex and P.M. Midgley (eds.)]. Cambridge University Press, Cambridge, United Kingdom and New York, NY, USA.

µmol	Micromole
20C3M	20th Century Climate in Coupled Models
AABW	Antarctic Bottom Water
AAIW	Antarctic Intermediate Water
AAO	Antarctic Oscillation
AATSR	Advanced Along Track Scanning Radiometer
ABA	AMSR Bootstrap Algorithm
ACC	Antarctic Circumpolar Current
ACCENT	Atmospheric Composition Change: a European Network
aci	Aerosol–Cloud Interactions
ACRIM	Active Cavity Radiometer Irradiance Monitor
ACW	Antarctic Circumpolar Wave
AeroCom	Aerosol Model Intercomparison
AERONET	Aerosol Robotic Network
A-FORCE	Aerosol Radiative Forcing in East Asia Aircraft Campaign
AGAGE	Advanced Global Atmospheric Gases Experiment
AGCM	Atmospheric General Circulation Model
AGTP	Absolute Global Temperature Change Potential
AGWP	Absolute Global Warming Potential
AIC	Aircraft-Induced Cirrus
ALOHA	A Long-term Oligotrophic Habitat Assessment
AMIP	Atmospheric Model Intercomparison Project
AMM	Atlantic Meridional Mode
AMO	Atlantic Multi-decadal Oscillation
AMOC	Atlantic Meridional Overturning Circulation
AMSR	Advanced Microwave Scanning Radiometer
AMSU	Advanced Microwave Sounding Unit
AMV	Atlantic Multi-decadal Variability
AO	Arctic Oscillation
AOD	Aerosol Optical Depth
AOGCM	Atmosphere–Ocean General Circulation Model
APHRODITE	Asian Precipitation – Highly Resolved Observational Data Integration Towards Evaluation
AR4	IPCC Fourth Assessment Report
ARCPAC	Aerosol, Radiation, and Cloud Processes affecting Arctic Climate
ARCTAS	Arctic Research of the Composition of the Troposphere from Aircraft and Satellites
ARFI	Aerosol Radiative Forcing over India
ari	Aerosol–Radiation Interactions
ARM	Atmospheric Radiation Measurement
ARTIST	Arctic Radiation and Turbulence Interaction Study
ATL3	Atlantic 3
ATSR	Along Track Scanning Radiometer
AUSMC	Australian-Maritime Continent
AVHRR	Advanced Very High Resolution Radiometer
AVISO	Archiving, Validation and Interpretation of Satellite Oceanographic Data
BATS	Bermuda Atlantic Time Series Study
BC	Black Carbon
BCC	Beijing Climate Center
BCC-CSM	Beijing Climate Center-Climate System Model
BDC	Brewer–Dobson Circulation
BECCS	Bio-Energy with Carbon-Capture and Storage
BMI	Basin Mean Index
BNF	Biological Nitrogen Fixation
BOM	Bureau of Meteorology
C_2Cl_4	Tetrachloroethene
C^4MIP	Coupled Climate Carbon Cycle Model Intercomparison Project
$CaCO_3$	Calcium Carbonate
CALIOP	Cloud-Aerosol Lidar with Orthogonal Polarization
CALIPSO	Cloud-Aerosol Lidar and Infrared Pathfinder Satellite Observations
CAM	Community Atmosphere Model
CAMS	Climate Anomaly Monitoring System
CanESM	Canadian Earth System Model
CASTNET	Clean Air Status and Trends Network
CCCma	Canadian Centre for Climate Modelling and Analysis
CCl_4	Carbon Tetrachloride
CCM	Chemistry–Climate Model
CCMVal	Chemistry–Climate Model Validation
CCN	Cloud Condensation Nuclei
CCR	Carbon Climate Response
CCSM	Community Climate System Model
CCSR	Centre for Climate System Research
CDD	Consecutive Dry Days

CDIAC Carbon Dioxide Information Analysis Center

CDR Carbon Dioxide Removal

CDW Circumpolar Deep Water

CE Common Era

CERES Cloud and the Earth's Radiant Energy System

CESM Community Earth System Model

CESM1–BGC Community Earth System Model 1–Biogeochemical

CF_4 Perfluoromethane

CFC Chlorofluorocarbon

CFC-11 Trichlorofluoromethane ($CFCl_3$)

CFC-113 Trichlorotrifluoroethane ($CF_2ClCFCl_2$)

CFC-12 Dichlorodifluoromethane (CF_2Cl_2)

CFMIP Cloud Feedback Model Intercomparison Project

CFSRR Climate Forecast System Reanalysis and Reforecast

CGCM Coupled General Circulation Model

CH_2Cl_2 Dichloromethane

CH_3Br Bromomethane

CH_3CCl_3 Methyl Chloroform

CH_3Cl Chloromethane

CH_4 Methane

CLIMAP Climate: Long-range Investigation, Mapping, and Prediction

CLIMBER-2 Climate and Biosphere Model

CLIO Coupled Large-scale Ice-Ocean Model

CLM4C Community Land Model for Carbon

CLM4CN Community Land Model for Carbon–Nitrogen

CMAP CPC Merged Analysis of Precipitation

CMDL Climate Monitoring and Diagnostics Laboratory (NOAA)

CMIP3 Coupled Model Intercomparison Project Phase 3

CMIP5 Coupled Model Intercomparison Project Phase 5

CNRM Centre National de Recherches Météorologiques

CO Carbon Monoxide

CO_2 Carbon Dioxide

CO_3^{2-} Carbonate

COADS Comprehensive Ocean–Atmosphere Data Set

COBE-SST Centennial in situ Observation-Based Estimates of Sea Surface Temperature

COCO CCSR Ocean Component Model

COHMAP Cooperative Holocene Mapping Project

CORE Coordinated Ocean-ice Reference Experiments

COWCLIP Coordinated Ocean Wave Climate Project

COWL Cold Ocean/Warm Land

CPC Climate Prediction Center (NOAA)

CPR Cloud Profiling Radar

CRE Cloud Radiative Effect

CRU Climatic Research Unit

CRUTEM4 Climatic Research Unit Gridded Dataset of Global Historical Near-Surface Air TEMperature Anomalies Over Land Version 4

CS Complex Ocean Sediment Model

CSFR Climate Forecast System Reanalysis

CSIRO Commonwealth Scientific and Industrial Research Organisation

CWC Cumulative Warming Commitment

DCESS Danish Center for Earth System Science

DIC Dissolved Inorganic Carbon

DJF December, January and February

DMI Directional Movement Index

DMS Dimethyl Sulphide

DO Dissolved Oxygen; also Dansgaard-Oeschger

DOC Dissolved Organic Carbon

DOE Department of Energy

DTR Diurnal Temperature Range

DU Dobson Units

EAS East Asian Summer

EASM East Asian Summer Monsoon

EBC Equivalent Black Carbon

EBM Energy Balance Model

ECBILT Coupled Atmosphere Ocean Model from de Bilt

ECHAM ECMWF and Hamburg

ECHO-G ECHAM4+HOPE-G

ECMWF European Centre for Medium Range Weather Forecasts

ECS Equilibrium Climate Sensitivity

EDGAR Emission Database for Global Atmospheric Research

EMIC Earth System Model of Intermediate Complexity

ENSO	El Niño-Southern Oscillation
EOF	Empirical Orthogonal Function
ERA-40	ECMWF 40-year ReAnalysis
ERBE	Earth Radiation Budget Experiment
ERBS	Earth Radiation Budget Satellite
ERF	Effective Radiative Forcing
ERFaci	Effective Radiative Forcing due to Aerosol–Cloud Interactions
ERFari	Effective Radiative Forcing due to Aerosol–Radiation Interactions
ERS	European Remote Sensing (Satellite)
ERSST	Extended Reconstructed Sea Surface Temperature
ESA	European Space Agency
ESM	Earth System Model
ESMR	Electrically Scanning Microwave Radiometer
ESRL	Earth System Research Library (NOAA)
ESTOC	European Station for Time Series in the Ocean
ETC	Extratropical Cyclone
FACE	Free-Air CO_2 Enrichment
FAO	Food and Agriculture Organization (UN)
FAR	IPCC First Assessment Report
FGOALS1	Flexible Global Ocean Atmosphere Land System Model Version 1
FIO	First Institute of Oceanography
FLUXNET	Global Network of Flux Towers
FTIR	Fourier Transform Infrared Spectroscopy
FTS	Fourier-Transform Spectrometer
FWCC	Freshwater Content Changes
GCAM	Global Change Assessment Model
GCM	General Circulation Model
GCP	Global Cost Potential
GCRM	Global Cloud-Resolving Models
GEISA	Gestion et Etude des Informations Spectroscopiques Atmosphériques
GENIE-1	Grid Enabled Integrated Earth System Model-1
GeoMIP G1	Geoengineering Model Intercomparison Project G1
GFDL	Geophysical Fluid Dynamics Laboratory
GFED	Global Fire Emissions Database
GHCN	Global Historical Climatology Network
GHCNDEX	Global Historical Climatology Network-Daily Gridded Data Set of Climate Extremes
GHCNv3	Global Historical Climatology Network Version 3
GHG	Greenhouse Gas
GI	Greenland Interstadial
GIA	Glacial Isostatic Adjustment
GIS	Greenland Ice Sheet
GISP	Greenland Ice Sheet Project
GISS	Goddard Institute of Space Studies
GISTEMP	Goddard Institute for Space Studies Surface Temperature Analysis
GL	Grounding Line
GLODAP	Global Ocean Data Analysis Project
GLS	Generalized Least Squares
GMA	Global Monsoon Area
GMD	Global Monitoring Division (NOAA)
GMI	Global Monsoon Precipitation Intensity
GMP	Global Monsoon Total Precipitation
GMSL	Global Mean Sea Level
GMST	Global Mean Surface Temperature
GOCCP	GCM-Oriented CALIPSO Cloud Product
GOGA	Global Ocean Global Atmosphere
GOME	Global Ozone Monitoring Experiment
GOMOS	Global Ozone Monitoring by Occultation of Stars
GOSAT	Greenhouse Gases Observing Satellite
GPCC	Global Precipitation Climatology Centre
GPCP	Global Precipitation Climatology Project
GPH	Geopotential Height
GPP	Gross Primary Productivity
GPS	Global Positioning System
GRACE	Gravity Recovery and Climate Experiment
GRISLI	Grenoble Ice Shelf and Land Ice Model
GS	Greenland Stadial
GSFC	Goddard Space Flight Centre
Gt	Gigatonnes
GTP	Global Temperature Change Potential
GUESS	General Ecosystem Simulator
GWD	Gravity-Wave Drag
GWP	Global Warming Potential

HadAT2	Hadley Centre Atmospheric Temperature Data Set 2
HadCM	Hadley Centre Climate Prediction Models
HadCRUT4	Hadley Centre Climatic Research Unit Gridded Surface Temperature Data Set 4
HadEX	Hadley Centre Gridded Data Set Of Temperature And Precipitation Extremes
HadGEM1	Hadley Centre New Global Environmental Model 1
HadGEM2-ES	Hadley Centre New Global Environmental Model 2-Earth System
HadGHCND	Hadley Centre Gridded Daily Temperatures Data Set
HadISST	Hadley Centre Interpolated SST
HadNMAT2	Hadley Centre Night Marine Air Temperatures Data Set Version 2
HadSLP2r	Hadley Centre Sea Level Pressure Data Set 2r
HadSST3	Hadley Centre Sea Surface Temperature Data Set Version 3
HALOE	Halogen Occultation Experiment
HCFC	Hydrochlorofluorocarbon
HCO_3^-	Bicarbonate Ion
HF	Hickey–Frieden (Radiometer)
HFC	Hydrofluorocarbon
HIPPO	HIAPER Pole-to-Pole Observations
HIRHAM5	High-Resolution Hamburg Climate Model 5
HITRAN	High-Resolution Transmission Molecular Absorption
HOAPS	Hamburg Ocean–Atmosphere Parameters and Fluxes from Satellite
HOT	Hawaii Ocean Time Series
HYDE	History Database of the Environment
HY-INT	Hydroclimatic Intensity
HYLAND	Hybrid Land Terrestrial Ecosystem Model
IAM	Integrated Assessment Model
IASI	Infrared Atmospheric Sounder Interferometer
ICE	Ice Cloud and Land Elevation
ICESat	Ice, Cloud and Land Elevation Satellite
ICOADS	International Comprehensive Ocean-Atmosphere Data Set
IGAC	International Global Atmospheric Chemistry
IMAGE	Integrated Model to Assess the Global Environment
IMBIE	Ice-sheet Mass Balance Intercomparison Experiment
IMPROVE	US Interagency Monitoring of Protected Visual Environments
INMCM4	Institute for Numerical Mathematics Coupled Model 4
IOB	Indian Ocean Basin
IOBM	Indian Ocean Basin Mode
IOD	Indian Ocean Dipole
IODM	Indian Ocean Dipole Mode
IPA	International Permafrost Association
IPO	Inter-decadal Pacific Oscillation
IPSL	Institut Pierre Simon Laplace
IPY	International Polar Year
IR	Infrared
IRF	Impulse Response Function
ISCCP	International Satellite Cloud Climatology Project
ITCZ	Inter-Tropical Convergence Zone
ITF	Indonesian Throughflow
IUK	Iterative Universal Kriging
JIMAR	Joint Institute for Marine and Atmospheric Research
JJA	June, July and August
JMA	Japan Meteorological Agency
JPL	Jet Propulsion Laboratory
ka	1000 Years ago
KCM	Knowledge Capture and Modeling
kyr	1000 Years
LAC	Light-Absorbing Carbon
LBIS	Larsen B Ice Shelf
LBL	Line-by-line (models)
LGM	Last Glacial Maximum
LIA	Little Ice Age
LIG	Last Interglacial
LISAM	Large–scale Index for South America Monsoon
LLGHG	Long-Lived Greenhouse Gas
LMM	Late Maunder Minimum
LNADW	Lower North Atlantic Deep Water
LOSU	Level of Scientific Understanding
LOVECLIM	Loch–Vecode-Ecbilt-Clio-Agism Model

LPB	La Plata Basin
LPJ	Lund-Potsdam-Jena Dynamic Global Model
LRF	Long-Range Forecast
LS	Lower Stratosphere
LSAT	Land-Surface Air Temperature
LSW	Labrador Sea Water
LUC	Land Use and Climate
LUCID	Land Use and Climate, Identification of Robust Impacts
LULC	Land Use and Land Cover
LULCC	Land Use and Land Cover Change
LWCRE	Longwave Cloud Radiative Effect
LWR	Longwave Radiation
MAGICC	Model for the Assessment of Greenhouse Gas Induced Climate Change
MAM	March, April and May
MAR	Modèle Atmosphérique Régional
MARGO	Multiproxy Approach for the Reconstruction of the Glacial Ocean Surface
MAT	Marine Air Temperatures
MBT	Mechanical Bathythermograph
MCA	Medieval Climate Anomaly
MDA	Mineral Dust Aerosol
MDT	Mean Dynamic Topography
MEA	Millennium Ecosystem Assessment
MERRA	Modern Era Reanalysis for Research and Applications
MESSAGE	Model for Energy Supply Strategy Alternatives and their General Environmental Impact
MFR	Maximum Feasible Reduction
MHD	Mace Head
MIP	Model Intercomparison Project
MIPAS	Michelson Interferometer for Passive Atmospheric Sounding
MIROC	Model for Interdisciplinary Research on Climate
MISI	Marine Ice Sheet Instability
MISR	Multi-angle Imaging Spectro-Radiometer
MIT	Massachusetts Institute of Technology
MJO	Madden–Julian Oscillation
MLD	Mixed Layer Depth
MLOST	Merged Land–Ocean Surface Temperature (Analysis)
MLS	Microwave Limb Sounder
MME	Multi-Model Ensemble
MMF	Multiscale Modelling Framework
MMM	Multi-Model Mean
MMTS	Maximum–Minimum Temperature Systems
MOC	Meridional Overturning Circulation
MOCAGE	Modèle de Chimie Atmosphérique à Grande Echelle
MODIS	Moderate Resolution Imaging Spectrometer
MOHC	Met Office Hadley Centre
MOPITT	Measurements of Pollutants in the Troposphere
MPI	Max Planck Institute
MPIOM	Max Planck Institute Ocean Model
MPWP	Mid-Pliocene Warm Period
MRI	Meteorological Research Institute of Japan Meteorological Agency
MSL	Mean Sea Level
MSSS	Mean Square Skill Score
MSU	Microwave Sounding Unit
Mt	Megatonnes
MT	Mid-Tropospheric
MTCO	Mean Temperature of the Coldest Month
MTWA	Mean Temperature of the Warmest Month
MW	Microwave
MXD	Maximum Latewood Density
Ma	Million Years ago
Myr	Million Years
N_2O	Nitrous Oxide
NADW	North Atlantic Deep Water
NAM	Northern Annular Mode
NAMP	National Air Quality Monitoring Programme (India)
NAMS	North American Monsoon System
NAO	North Atlantic Oscillation
NASA	National Aeronautics and Space Administration
NCAR	National Center for Atmospheric Research
NCEP	National Centers for Environmental Prediction
NEC	North Equatorial Current

NEEM	North Greenland Eemian Ice Drilling
NEWS	Global Nutrient Export from WaterSheds
NF_3	Nitrogen Trifluoride
NGRIP	North Greenland Ice Core Project
NH	Northern Hemisphere
NIWA	National Institute of Water and Atmospheric Research
NMAT	Nighttime Marine Air Temperature
NMVOC	Non-Methane Volatile Organic Compound
NNR	NCEP–NCAR
NOAA	National Oceanic and Atmospheric Administration
NODC	National Oceanic Data Center
NorESM	Norwegian Earth System Model
NO_x	Reactive Nitrogen Oxides (the Sum of NO and NO_2)
NPI	North Pacific Index
NPIW	North Pacific Intermediate Water
NPP	Net Primary Productivity
NSIDC	National Snow and Ice Data Center
NT1	National Aeronautics and Space Administration (NASA) Team Algorithm, Version 1
NT2	National Aeronautics and Space Administration (NASA) Team Algorithm, Version 2
NTCF	Near-Term Climate Forcer
$O(^1D)$	Oxygen Radical in the 1D Excited State
O_3	Ozone
OA	Ocean–Atmosphere; also Other Anthropogenic (Forcings)
OAC	Ocean–Atmosphere–Carbon Cycle
OAFlux	Objectively Analyzed Air–Sea Heat Fluxes
OAGCMs	Ocean–Atmosphere General Circulation Models
OAV	Ocean–Atmosphere–Vegetation
OC	Organic Carbon
OCN	Oceanic Carbon and Nutrient Cycling (Model)
ODP	Ocean Drilling Program
ODS	Ozone-Depleting Substance
OH	Hydroxyl Radical
OHC	Ocean Heat Content
OHR	Ocean Heating Rate
OLR	Outgoing Longwave Radiation
OLS	Ordinary Least Squares
OMI	Ozone Monitoring Instrument
ONDJFM	October, November, December, January, February and March
ORC	Oceanic Reservoir Correction
PAGES 2k	Past Global Changes 2k
PARASOL	Polarization and Anisotropy of Reflectances for Atmospheric Sciences Coupled with Observations from Lidar
PATMOS-x	Pathfinder Atmospheres Extended Data Set
PBAPs	Primary Biological Aerosol Particles
PCM	Parallel Climate Model
pCO_2	Partial Pressure of Carbon Dioxide
PDF	Probability Density Function
PDO	Pacific Decadal Oscillation
PDSI	Palmer Drought Severity Index
PETM	Paleocene–Eocene Thermal Maximum
PFC	Perfluorocarbon
PG	Peripheral Glacier
Pg	Petagram
PM_{10}	Particulate Matter with Aerodynamic Diameter <10 μm
$PM_{2.5}$	Particulate Matter with Aerodynamic Diameter <2.5 μm
PMEL	Pacific Marine Environmental Laboratory
PMIP3	Paleoclimate Modelling Intercomparison Project Phase III
PMOD	Physikalisch-Meteorologisches Observatorium Davos
PNA	Pacific–North American (Pattern)
POA	Primary Organic Aerosol
POC	Particulate Organic Carbon
POLDER	Polarization and Directionality of the Earth's Reflectance
PPE	Perturbed-Parameter Ensemble
PRCE	Peak Response to Cumulative Emissions
PREMOS	Precision Monitor Sensor
PSA	Pacific–South American (Pattern)
PSMSL	Permanent Service for Mean Sea Level
PSS	Practical Salinity Scale
PSS78	Practical Salinity Scale 1978

PUCCINI	Physical Understanding of Composition-Climate Interactions and Impacts
QBO	Quasi-Biennial Oscillation
R95p (R99p)	Amount of Precipitation from Days >95th (99th) Percentile
RACMO2	Regional Atmospheric Climate Model 2
RAOBCORE	Radiosone Observation Correction using Reanalyses
RAPID/MOCHA	Rapid Climate Change-Meridional Overturning Circulation and Heatflux Array
RATPAC	Radiosonde Atmospheric Temperature Products for Assessing Climate
RCM	Regional Climate Model
RCP	Representative Concentration Pathway
RE	Radiative Efficiency
REMBO	Regional, Moisture-Balance Orographic Model
REML	Restricted Maximum Likelihood
RF	Radiative Forcing
RFaci	Radiative Forcing from Aerosol–Cloud Interactions
RGI	Randolph Glacier Inventory
RH	Relative Humidity
RICH	Radiosonde Innovation Composite Homogenization
RMIB	Royal Meteorological Institute of Belgium
RMS	Root Mean Square
RMSE	Root Mean Square Error
RO	Radio Occultation
RSCA	Relative Snow-Covered Area
RSL	Relative Sea Level
RSS	Remote Sensing System
RX5day/RX1day	Annual Maximum 5-Day/1-Day Precipitation
S/N	Signal-to-Noise (Ratio)
SACZ	South Atlantic Convergence Zone
SAGE	Stratospheric Aerosol and Gas Experiment or Centre for Sustainability and the Global Environment
SAM	Southern Annular Mode
SAMS	South American Monsoon System
SAMW	Sub-Antarctic Mode Water
SAR	IPCC Second Assessment Report
SAT	Surface Air Temperature
SBA	SSM/I Bootstrap Algorithm
SBUV	Solar Backscatter Ultraviolet
SC	Solar Cycle
SCA	Snow-Covered Area
SCD	Snow Cover Duration
SCE	Snow Cover Extent
SCIA	Scanning Imaging Absorption Spectrometer for Atmospheric Chartography
SCIAMACHY	Scanning Imaging Absorption Spectrometer for Atmospheric Chartography
SD	Snow Depth; also Statistical Downscaling
SDGVM	Sheffield Dynamic Global Vegetation Model
SDII	Simple Daily Precipitation Intensity Index
SeaWiFS	Sea-viewing Wide Field-of-view Sensor
SEM	Semi-Empirical Model
SF_6	Sulphur Hexafluoride
SH	Southern Hemisphere
SICOPOLIS	Simulation Code for Polythermal Ice Sheets
SIM	Spectral Irradiance Monitor
SIO	Scripps Institution of Oceanography
SLE	Sea Level Equivalent
SLP	Sea Level Pressure
SLR	Sea Level Rise
SMB	Surface Mass Balance
SMMR	Scanning Multichannel Microwave Radiometer
SMOS	Soil Moisture and Ocean Salinity
SNO	Simultaneous Nadir Overpass
SO_2	Sulphur Dioxide
SO_2F_2	Sulphuryl Fluoride
SO_4^{2-}	Sulfate
SOA	Secondary Organic Aerosol
SOI	Southern Oscillation Index
SOLSTICE	Solar Stellar Irradiance Comparison Experiment
SON	September, October and November
SORCE	Solar Radiation and Climate Experiment
SPARC	Stratospheric Processes and their Role in Climate Chemistry Climate Model Validation
SPCZ	South Pacific Convergence Zone

SPEI	Standardised Precipitation Evapotranspiration Index
SPI	Standardised Precipitation Index
SPRINTARS	Spectral Radiation-Transport Model for Aerosol Species
SRALT	Satellite Radar Altimetry
SRES	IPCC Special Report on Emission Scenarios
SREX	IPCC Special Report on Managing the Risk of Extreme Events and Disasters to Advance Climate Change Adaptation
SRM	Solar Radiation Management
SSH	Sea Surface Height
SSI	Spectral Solar Irradiance
SSM/I	Special Sensor Microwave/Imager
SSR	Surface Solar Radiation
SSS	Sea Surface Salinity
SST	Sea Surface Temperature
SSU	Stratospheric Sounding Unit
STAR	Center for Satellite Applications and Research
STMW	Subtropical Mode Water
SVS	Standard Verification System (WMO)
SWCRE	Shortwave Cloud Radiative Effect
SWE	Snow Water Equivalent
SWH	Significant Wave Height
SWR	Solar Shortwave Radiation
TBO	Tropospheric Biennial Oscillation
Tg	Teragrams
T/P	TOPEX/Poseidon
TANSO	Thermal and Near Infrared Sensor for Carbon Observation
TAR	IPCC Third Assessment Report
TC	Tropical Cyclone; also Total Carbon
TCCON	Total Carbon Column Observing Network
TCR	Transient Climate Response
TCRE	Transient Climate Response to Cumulative CO_2 Emissions
TES	Tropospheric Emission Spectrometer
TIM	Total Irradiance Monitor
TNI	Trans-Niño Index
TOA	Top of the Atmosphere
TOMS	Total Ozone Mapping Spectrometer
TOPEX	Topography Experiment
TRANSCOM	Atmospheric Tracer Transport Model Intercomparison Project
TRIFFID	Top-down Representation of Interactive Foliage and Flora Including Dynamics
TRUTHS	Traceable Radiometry Underpinning Terrestrial and Helio Studies
TRW	Tree-Ring Width
TSI	Total Solar Irradiance
TTD	Transit Time Distribution
TW	Tidewater
UAH	University of Alabama in Huntsville
UARS	Upper Atmosphere Research Satellite
UCI	University of California, Irvine
UHI	Urban Heat Island
UNADW	Upper North Atlantic Deep Water
UNEP	United Nations Environment Programme
UOHC	Upper (0–700 m) Ocean Heat Content
USHCN	US Historical Climatology Network
UTLS	Upper Troposphere/Lower Stratosphere
UV	Ultraviolet
UVic	University of Victoria
VasClimO	Variability Analyses of Surface Climate Observations
VEGAS	Terrestrial Vegetation and Carbon Model
VIIRS	Visible Infrared Imaging Radiometer Suite
VLM	Vertical Land Motion
VOC	Volatile Organic Compound
VOS	Voluntary Observing Ship
W	Watts
WAIS	West Antarctic Ice Sheet
WASWind	Wave- and Anemometer-Based Sea Surface Wind
WCRP	World Climate Research Programme
WMGHG	Well-Mixed Greenhouse Gas
WMO	World Meteorological Organization
WOCE	World Ocean Circulation Experiment
WSG	Western Subarctic Gyre
XBT	Expendable Bathythermograph

Annex V: Contributors to the IPCC WGI Fifth Assessment Report

This annex should be cited as:

IPCC, 2013: Annex V: Contributors to the IPCC WGI Fifth Assessment Report. In: *Climate Change 2013: The Physical Science Basis. Contribution of Working Group I to the Fifth Assessment Report of the Intergovernmental Panel on Climate Change* [Stocker, T.F., D. Qin, G.-K. Plattner, M. Tignor, S.K. Allen, J. Boschung, A. Nauels, Y. Xia, V. Bex and P.M. Midgley (eds.)]. Cambridge University Press, Cambridge, United Kingdom and New York, NY, USA.

Coordinating Lead Authors, Lead Authors, Review Editors and Contributing Authors are listed alphabetically by surname.

AAMAAS, Borgar
Center for International Climate and Environmental Research Oslo
Norway

ABE-OUCHI, Ayako
University of Tokyo
Japan

ABIODUN, Babatunde
University of Cape Town
South Africa

ABRAHAM, Libu
Qatar Meteorological Department
Qatar

ACHUTARAO, Krishna Mirle
Indian Institute of Technology
India

ADEDOYIN, Akintayo John
University of Botswana
Botswana

ADLER, Robert F.
University of Maryland
USA

AHLSTRÖM, Anders
Lund University
Sweden

ALDRIAN, Edvin
Agency for Meteorology, Climatology and Geophysics
Indonesia

ALDRIN, Magne
Norwegian Computing Center and University of Oslo
Norway

ALEXANDER, Lisa V.
University of New South Wales
Australia

ALLAN, Richard P.
University of Reading
UK

ALLAN, Robert
Met Office Hadley Centre
UK

ALLEN, Myles R.
University of Oxford
UK

ALLEN, Simon K.
IPCC WGI TSU, University of Bern
Switzerland

ALLISON, Ian
Antarctic Climate and Ecosystems Cooperative Research Centre
Australia

AMBRIZZI, Tércio
University of Sao Paulo
Brazil

AN, Soon-Il
Yonsei University
Republic of Korea

ANAV, Alessandro
University of Exeter
UK

ANCHUKAITIS, Kevin
Woods Hole Oceanographic Institution
USA

ANDERSON, Bruce
Boston University
USA

ANDREWS, Oliver
University of East Anglia
UK

ANDREWS, Timothy
Met Office Hadley Centre
UK

AOKI, Shigeru
Hokkaido University
Japan

AOYAMA, Michio
Meteorological Research Institute
Japan

ARAKAWA, Osamu
University of Tsukuba
Japan

ARBLASTER, Julie
Bureau of Meteorology
Australia

ARCHER, David
University of Chicago
USA

ARENDT, Anthony A.
University of Alaska Fairbanks
USA

ARORA, Vivek
Environment Canada
Canada

ARRITT, Raymond
Iowa State University
USA

ARTAXO, Paulo
University of Sao Paulo
Brazil

BAEHR, Johanna
University of Hamburg
Germany

BAHR, David B.
University of Colorado Boulder
USA

BALA, Govindasamy
Indian Institute of Science
India

BALAN SAROJINI, Beena
University of Reading
UK

BALDWIN, Mark
University of Exeter
UK

BAMBER, Jonathan
University of Bristol
UK

BARINGER, Molly
National Oceanic and Atmospheric Administration, Atlantic Oceanographic and Meteorological Laboratory
USA

BARLOW, Mathew
University of Massachusetts
USA

BARRIOPEDRO, David
Universidad Complutense de Madrid
Spain

BARTHOLY, Judit
Eötvös Loránd University
Hungary

BARTLEIN, Patrick J.
University of Oregon
USA

BATES, Nicholas R.
Bermuda Biological Station
Bermuda

BEER, Jürg
Eawag - Swiss Federal Institute of Aquatic Science and Technology
Switzerland

BELLOUIN, Nicolas
University of Reading
UK

BENEDETTI, Angela
European Centre for Medium-Range Weather Forecasts
UK

BENITO, Gerardo
Consejo Superior de Investigaciones Cientificas
Spain

BEYERLE, Urs
ETH Zurich
Switzerland

BIASUTTI, Michela
Columbia University
USA

BINDOFF, Nathaniel L.
University of Tasmania
Australia

BINER, Sébastien
Ouranos Consortium on Regional Climatology and Adaptation to Climate Change
Canada

BITZ, Cecilia M.
University of Washington
USA

BLAKE, Donald R.
University of California Irvine
USA

BODAS-SALCEDO, Alejandro
Met Office Hadley Centre
UK

BOER, George J.
Environment Canada
Canada

BOJARIU, Roxana
National Meteorological Administration
Romania

BONAN, Gordon
National Center for Atmospheric Research
USA

BONY, Sandrine
Laboratoire de Météorologie Dynamique, Institut Pierre Simon Laplace
France

BOOTH, Ben B.B.
Met Office Hadley Centre
UK

BOPP, Laurent
Laboratoire des Sciences du Climat et de l'Environnement, Institut Pierre Simon Laplace
France

BORGES, Alberto Vieira
Université de Liège
Belgium

BOUCHER, Olivier
Laboratoire de Météorologie Dynamique, Institut Pierre Simon Laplace
France

BOUSQUET, Philippe
Laboratoire des Sciences du Climat et de l'Environnement, Institut Pierre Simon Laplace
France

BOUWMAN, Lex
PBL Netherlands Environmental Assessment Agency
Netherlands

BOX, Jason E.
Geological Survey of Denmark and Greenland
Denmark

BOYER, Timothy
National Oceanic and Atmospheric Administration, National Oceanographic Data Center
USA

BRACONNOT, Pascale
Laboratoire des Sciences du Climat et de l'Environnement, Institut Pierre Simon Laplace
France

BRAUER, Achim
GFZ German Research Centre for Geosciences
Germany

BRÉON, François-Marie
Laboratoire des Sciences du Climat et de l'Environnement, Institut Pierre Simon Laplace
France

BRETHERTON, Christopher
University of Washington
USA

BROMWICH, David H.
Ohio State University
USA

BRÖNNIMANN, Stefan
University of Bern
Switzerland

BROOKS, Harold E.
National Oceanic and Atmospheric Administration, National Severe Storms Laboratory
USA

BROVKIN, Victor
Max Planck Institute for Meteorology
Germany

BROWN, Josephine
Bureau of Meteorology
Australia

BROWN, Ross
Environment Canada
Canada

BROWNE, Oliver
University of Edinburgh
UK

BRUHWILER, Lori M.
National Oceanic and Atmospheric Administration, Earth System Research Laboratory
USA

BRUTEL-VUILMET, Claire
Laboratoire de Glaciologie et Géophysique de l'Environnement, Université Joseph Fourier
France

BYRNE, Robert H.
University of South Florida
USA

CAI, Wenju
CSIRO Marine and Atmospheric Research
Australia

CALDEIRA, Kenneth
Carnegie Institution for Science
USA

CAMERON-SMITH, Philip
Lawrence Livermore National Laboratory
USA

CAMILLONI, Ines
Universidad de Buenos Aires
Argentina

CAMPOS, Edmo
University of Sao Paulo
Brazil

CANADELL, Josep
CSIRO Marine and Atmospheric Research
Australia

CANE, Mark
Columbia University
USA

CAO, Long
Zhejiang University
China

CARRASCO, Jorge
Direccion Meteorologica de Chile
Chile

CARSON, Mark
University of Hamburg
Germany

CARTER, Tim
Finnish Environment Institute
Finland

CARVALHO, Leila V.
University of California Santa Barbara
USA

CATTO, Jennifer
Monash University
Australia

CAVALCANTI, Iracema F.A.
National Institute for Space Research
Brazil

CAZENAVE, Anny
Laboratoire d'Etudes en Géophysique et Océanographie Spatiales
France

CHADWICK, Robin
Met Office Hadley Centre
UK

CHAMBERS, Don
University of South Florida
USA

CHANG, Ping
Texas A&M University
USA

CHAPPELLAZ, Jérôme
Laboratoire de Glaciologie et Géophysique de l'Environnement, Université Joseph Fourier
France

CHARABI, Yassine Abdul-Rahman
Sultan Qaboos University
Oman

CHEN, Deliang
University of Gothenburg
Sweden

CHEN, Xiaolong
Institute of Atmospheric Physics, Chinese Academy of Sciences
China

CHEVALLIER, Frédéric
Laboratoire des Sciences du Climat et de l'Environnement, Institut Pierre Simon Laplace
France

CHHABRA, Abha
Indian Space Research Organisation
India

CHIKAMOTO, Yoshimitsu
University of Hawaii
USA

CHOI, Jung
Seoul National University
Republic of Korea

CHOU, Sin Chan
National Institute for Space Research
Brazil

CHRISTENSEN, Jens Hesselbjerg
Danish Meteorological Institute
Denmark

CHRISTENSEN, Ole Bøssing
Danish Meteorological Institute
Denmark

CHRISTIDIS, Nikolaos
Met Office Hadley Centre
UK

CHURCH, John A.
CSIRO Marine and Atmospheric Research
Australia

CIAIS, Philippe
Laboratoire des Sciences du Climat et de l'Environnement, Institut Pierre Simon Laplace
France

CLARK, Peter U.
Oregon State University
USA

CLEVELAND, Cory
University of Montana
USA

CLIFTON, Olivia
Columbia University
USA

COGLEY, J. Graham
Trent University
Canada

COLLINS, Matthew
University of Exeter
UK

COLLINS, William
University of Reading
UK

COLLINS, William
Lawrence Berkeley National Laboratory
USA

COMISO, Josefino C.
National Aeronautics and Space Administration, Goddard Space Flight Center
USA

COOK, Edward
Columbia University
USA

COOK, Kerry H.
University of Texas
USA

COOLEY, Sarah
Woods Hole Oceanographic Institution
USA

COOPER, Owen R.
Cooperative Institute for Research in Environmental Sciences
USA

CORTI, Susanna
Institute of Atmospheric Sciences and Climate
Italy

COX, Peter
University of Exeter
UK

CROWLEY, Thomas
Braeheads Institute
UK

CUBASCH, Ulrich
Freie Universität Berlin
Germany

CUNNINGHAM, Stuart
Scottish Association of Marine Science
UK

DAI, Aiguo
University at Albany
USA

DALSØREN, Stig B.
Center for International Climate and Environmental Research Oslo
Norway

DANIEL, John S.
National Oceanic and Atmospheric Administration, Earth System Research Laboratory
USA

DAVIS, Robert E.
University of Virginia
USA

DAVIS, Sean M.
National Oceanic and Atmospheric Administration, Earth System Research Laboratory
USA

DE CASTRO, Manuel
Universidad de Castilla-La Mancha
Spain

DE DECKKER, Patrick
Australian National University
Australia

DE ELÍA, Ramón
Université du Québec à Montréal and Ouranos Consortium
Canada

DE MENEZES, Viviane Vasconcellos
University of Tasmania
Australia

DE VERNAL, Anne
Université du Québec à Montréal
Canada

DEFRIES, Ruth
Columbia University
USA

DEL GENIO, Anthony
National Aeronautics and Space Administration, Goddard Institute for Space Studies
USA

DELCROIX, Thierry
Laboratoire d'Etudes en Géophysique et Océanographie Spatiales
France

DELECLUSE, Pascale
Météo-France
France

DELMONTE, Barbara
University of Milano-Bicocca
Italy

DELSOLE, Tim
George Mason University
USA

DENTENER, Frank J.
European Commission, Joint Research Center
EU

DESER, Clara
National Center for Atmospheric Research
USA

DINEZIO, Pedro
University of Hawaii
USA

DING, Yihui
National Climate Center, China Meteorological Administration
China

DLUGOKENCKY, Edward J.
National Oceanic and Atmospheric Administration, Earth System Research Laboratory
USA

DOBLAS-REYES, Francisco
Institució Catalana de Recerca i Estudis Avançats and Institut Català de Ciències del Clima
Spain

DOKKEN, Trond
Uni Research Norway
Norway

DOMINGUES, Catia M.
Antarctic Climate and Ecosystems Cooperative Research Centre
Australia

DONAT, Markus G.
University of New South Wales
Australia

DONEY, Scott C.
Woods Hole Oceanographic Institution
USA

DONG, Wenjie
Beijing Normal University
China

DORE, John
Montana State University
USA

DOWSETT, Harry J.
U.S. Geological Survey
USA

DRIOUECH, Fatima
Direction de la Météorologie Nationale
Morocco

DUFRESNE, Jean-Louis
Laboratoire de Météorologie Dynamique, Institut Pierre Simon Laplace
France

DURACK, Paul J.
Lawrence Livermore National Laboratory
USA

EASTERLING, David R.
National Oceanic and Atmospheric Administration, Cooperative Institute for Climate and Satellites
USA

EBY, Michael
University of Victoria
Canada

EDWARDS, R. Lawrence
University of Minnesota
USA

ELISEEV, Alexey
Russian Academy of Sciences
Russian Federation

EMANUEL, Kerry
Massachusetts Institute of Technology
USA

EMORI, Seita
National Institute for Environmental Studies
Japan

ENDO, Hirokazu
Meteorological Research Institute
Japan

ENFIELD, David B.
University of Miami
USA

ERISMAN, Jan Willem
Louis Bolk Institute
Netherlands

EUSKIRCHEN, Eugenie S.
University of Alaska Fairbanks
USA

EVAN, Amato
Scripps Institution of Oceanography
USA

EYRING, Veronika
DLR German Aerospace Center
Germany

FACCHINI, Maria Cristina
Institute of Atmospheric Sciences and Climate
Italy

FASULLO, John
National Center for Atmospheric Research
USA

FEELY, Richard A.
National Oceanic and Atmospheric Administration, Pacific Marine Environmental Laboratory
USA

FEINGOLD, Graham
National Oceanic and Atmospheric Administration, Earth System Research Laboratory
USA

FETTWEIS, Xavier
Université de Liège
Belgium

FICHEFET, Thierry
Université catholique de Louvain
Belgium

FINE, Rana
University of Miami
USA

FIOLETOV, Vitali
Environment Canada
Canada

FIORE, Arlene M.
Columbia University and Lamont-Doherty Earth Observatory
USA

FISCHER, Erich M.
ETH Zurich
Switzerland

FISCHER, Hubertus
University of Bern
Switzerland

FLANNER, Mark
University of Michigan
USA

FLATO, Gregory
Environment Canada
Canada

FLEITMANN, Dominik
University of Reading
UK

FOREST, Chris E.
Pennsylvania State University
USA

FORSTER, Piers
University of Leeds
UK

FOSTER, Gavin
University of Southampton
UK

FRAME, David
Victoria University of Wellington
New Zealand

FREELAND, Howard
Fisheries and Oceans Canada
Canada

FRIEDLINGSTEIN, Pierre
University of Exeter
UK

FRÖHLICH, Claus
Physikalisch-Meteorologisches Observatorium Davos, World Radiation Center
Switzerland

FUGLESTVEDT, Jan
Center for International Climate and Environmental Research Oslo
Norway

FUZZI, Sandro
Institute of Atmospheric Sciences and Climate
Italy

FYFE, John
Environment Canada
Canada

GALLOWAY, James
University of Virginia
USA

GANOPOLSKI, Andrey
Potsdam Institute for Climate Impact Research
Germany

GAO, Xuejie
National Climate Center, China Meteorological Administration
China

GARCÍA-SERRANO, Javier
Institut Català de Ciències del Clima
Spain

GARDNER, Alex S.
Clark University
USA

GARZOLI, Silvia
National Oceanic and Atmospheric Administration, Atlantic Oceanographic and Meteorological Laboratory
USA

GATES, Lydia
Freie Universität Berlin
Germany

GBOBANIYI, Emiola
Swedish Meteorological and Hydrological Institute
Sweden

GEHRELS, W. Roland
University of York
UK

GERLAND, Sebastian
Norwegian Polar Institute
Norway

GHAN, Steven
Pacific Northwest National Laboratory
USA

GIANNINI, Alessandra
Columbia University
USA

GIESEN, Rianne
Utrecht University
Netherlands

GILLETT, Nathan
Environment Canada
Canada

GINOUX, Paul
National Oceanic and Atmospheric Administration, Geophysical Fluid Dynamics Laboratory
USA

GLECKLER, Peter J.
Lawrence Livermore National Laboratory
USA

GONZÁLEZ ROUCO, Jesús Fidel
Universidad Complutense de Madrid
Spain

GONZÁLEZ-DÁVILA, Melchor
Universidad de Las Palmas de Gran Canaria
Spain

GOOD, Peter
Met Office Hadley Centre
UK

GOOD, Simon
Met Office Hadley Centre
UK

GOODESS, Clare
University of East Anglia
UK

GOOSSE, Hugues
Université catholique de Louvain
Belgium

GOSWAMI, Prashant
CSIR Centre for Mathematical Modelling and Computer Simulation
India

GOVIN, Aline
MARUM Center for Marine Environmental Sciences
Germany

GRANIER, Claire
Laboratoire Atmosphères, Milieux, Observations Spatiales, Institut Pierre Simon Laplace
France

GRAVERSON, Rune Grand
Stockholm University
Sweden

GRAY, Lesley
University of Oxford
UK

GREGORY, Jonathan M.
University of Reading and Met Office Hadley Centre
UK

GREVE, Ralf
Hokkaido University
Japan

GRIFFIES, Stephen
National Oceanic and Atmospheric Administration, Geophysical Fluid Dynamics Laboratory
USA

GRUBER, Nicolas
ETH Zurich
Switzerland

GRUBER, Stephan
University of Zurich
Switzerland

GUEMAS, Virginie
Institut Català de Ciències del Clima
Spain

GUILYARDI, Eric
Laboratoire d'Océanographie et du Climat, Institut Pierre Simon Laplace
France

GULEV, Sergey
P.P. Shirshov Institute of Oceanology
Russian Federation

GUPTA, Anil K.
Wadia Institute of Himalayan Geology
India

GURNEY, Kevin
Arizona State University
USA

GUTOWSKI, William J.
Iowa State University
USA

GUTZLER, David
University of New Mexico
USA

HAAS, Christian
York University
Canada

HAGEN, Jon Ove
University of Oslo
Norway

HAIGH, Joanna
Imperial College London
UK

HAIMBERGER, Leopold
University of Vienna
Austria

HALL, Alex
University of California Los Angeles
USA

HANNA, Edward
University of Sheffield
UK

HANSINGO, Kabumbwe
University of Zambia
Zambia

HARGREAVES, Julia
Japan Agency for Marine-Earth Science and Technology
Japan

HARIHARASUBRAMANIAN, Annamalai
University of Hawaii
USA

HARRISON, Sandy
Macquarie University
Australia

HARTMANN, Dennis L.
University of Washington
USA

HAWKINS, Ed
University of Reading
UK

HAYWOOD, Alan
University of Leeds
UK

HEGERL, Gabriele C.
University of Edinburgh
UK

HEIMANN, Martin
Max Planck Institute for Biogeochemistry
Germany

HEINZE, Christoph
University of Bergen
Norway

HELD, Isaac
National Oceanic and Atmospheric Administration, Geophysical Fluid Dynamics Laboratory
USA

HEMER, Mark
CSIRO Marine and Atmospheric Research
Australia

HENSE, Andreas
University of Bonn
Germany

HEWITSON, Bruce
University of Cape Town
South Africa

HEZEL, Paul J.
Université catholique de Louvain
Belgium

HO, Shu-Peng (Ben)
National Center for Atmospheric Research
USA

HOCK, Regine
University of Alaska Fairbanks
USA

HODGES, Kevin I.
University of Reading
UK

HODNEBROG, Øivind
Center for International Climate and Environmental Research Oslo
Norway

HOLGATE, Simon J.
Sea Level Research Foundation
UK

HOLLAND, David
New York University
USA

HOLLAND, Elisabeth A.
University of the South Pacific
Fiji

HOLLAND, Greg
National Center for Atmospheric Research
USA

HOLLAND, Marika M.
National Center for Atmospheric Research
USA

HOLLIS, Chris
GNS Science
New Zealand

HOLMES, Christopher
University of California Irvine
USA

HOOSE, Corinna
Karlsruhe Institute of Technology
Germany

HOPWOOD, Brett
Oak Ridge National Laboratory
USA

HORTON, Ben
Rutgers University
USA

HOUGHTON, Richard A.
Woods Hole Research Center
USA

HOUSE, Joanna I.
University of Bristol
UK

HOUWELING, Sander
Utrecht University
Netherlands

HU, Yongyun
Peking University
China

HUANG, Jianping
Lanzhou University
China

HUANG, Ping
Institute of Atmospheric Physics, Chinese Academy of Sciences
China

HUBER, Markus
ETH Zurich
Switzerland

HUNKE, Elizabeth
Los Alamos National Laboratory
USA

HUNTER, John R.
University of Tasmania
Australia

HUNTER, Stephen
University of Leeds
UK

HURRELL, Jim
National Center for Atmospheric Research
USA

HURTT, George
University of Maryland
USA

HUSS, Matthias
University of Fribourg
Switzerland

HUYBRECHTS, Philippe
Vrije Universiteit Brussel
Belgium

HYDES, David
National Oceanography Centre
UK

ILYINA, Tatiana
Max Planck Institute for Meteorology
Germany

IMBERS QUINTANA, Jara
University of Oxford
UK

INFANTI, Johnna
University of Miami
USA

INGRAM, William
University of Oxford
UK

ISHII, Masayoshi
Meteorological Research Institute
Japan

IVANOVA, Detelina
Lawrence Livermore National Laboratory
USA

JACOB, Daniel
Harvard University
USA

JACOBS, Stanley
Columbia University
USA

JACOBSON, Andrew D.
Northwestern University
USA

JAIN, Atul
University of Illinois
USA

JAIN, Suman
University of Zambia
Zambia

JAKOB, Christian
Monash University
Australia

JANSEN, Eystein
University of Bergen
Norway

JANSSEN, Emily
University of Illinois
USA

JEVREJEVA, Svetlana
National Oceanography Centre
UK

JOHN, Jasmin
National Oceanic and Atmospheric Administration, Geophysical Fluid Dynamics Laboratory
USA

JOHNS, Tim
Met Office Hadley Centre
UK

JOHNSON, Gregory C.
National Oceanic and Atmospheric Administration, Pacific Marine Environmental Laboratory
USA

JONES, Andy
Met Office Hadley Centre
UK

JONES, Christopher
Met Office Hadley Centre
UK

JONES, Julie
University of Sheffield
UK

JOOS, Fortunat
University of Bern
Switzerland

JOSEY, Simon A.
National Oceanography Centre
UK

JOSHI, Manoj
University of East Anglia
UK

JOUGHIN, Ian
University of Washington
USA

JOUSSAUME, Sylvie
Laboratoire des Sciences du Climat et de l'Environnement, Institut Pierre Simon Laplace
France

JOUZEL, Jean
Laboratoire des Sciences du Climat et de l'Environnement, Institut Pierre Simon Laplace
France

JUNGCLAUS, Johann
Max Planck Institute for Meteorology
Germany

KAGEYAMA, Masa
Laboratoire des Sciences du Climat et de l'Environnement, Institut Pierre Simon Laplace
France

KANIKICHARLA, Krishna Kumar
Indian Institute of Tropical Meteorology
India

KANYANGA, Joseph Katongo
Zambia Meteorological Department
Zambia

KANZOW, Torsten
GEOMAR Helmholtz Centre for Ocean Research
Germany

KAPLAN, Alexey
Columbia University
USA

KAPLAN, Jed O.
EPFL Lausanne
Switzerland

KARL, David
University of Hawaii
USA

KARUMURI, Ashok
Indian Institute of Tropical Meteorology
India

KASER, Georg
University of Innsbruck
Austria

KASPAR, Frank
Deutscher Wetterdienst
Germany

KATO, Etsushi
National Institute for Environmental Studies
Japan

KATSMAN, Caroline
Royal Netherlands Meteorological Institute
Netherlands

KATTSOV, Vladimir
Voeikov Main Geophysical Observatory
Russian Federation

KATZFEY, Jack
CSIRO Marine and Atmospheric Research
Australia

KAZMIN, Alexander
P.P. Shirshov Institute of Oceanology
Russian Federation

KEELING, Ralph
Scripps Institution of Oceanography
USA

KENNEDY, John J.
Met Office Hadley Centre
UK

KENT, Elizabeth C.
National Oceanography Centre
UK

KERMINEN, Veli-Matti
Finnish Meteorological Institute
Finland

KEY, Robert M.
Princeton University
USA

KHARIN, Viatcheslav
Environment Canada
Canada

KHATIWALA, Samar
Columbia University
USA

KIMOTO, Masahide
University of Tokyo
Japan

KINNE, Stefan
Max Planck Institute for Meteorology
Germany

KIRSCHKE, Stefanie
Laboratoire des Sciences du Climat et de l'Environnement, Institut Pierre Simon Laplace
France

KIRTMAN, Ben
University of Miami
USA

KITOH, Akio
University of Tsukuba
Japan

KJELLSTRÖM, Erik
Swedish Meteorological and Hydrological Institute
Sweden

KLEIN, Stephen A.
Lawrence Livermore National Laboratory
USA

KLEIN GOLDEWIJK, Kees
Utrecht University and PBL Netherlands Environmental Assessment Agency
Netherlands

KLEIN TANK, Albert M.G.
Royal Netherlands Meteorological Institute
Netherlands

KLEYPAS, Joan
National Center for Atmospheric Research
USA

KLIMONT, Zbigniew
International Institute for Applied Systems Analysis
Austria

KLOSTER, Silvia
Max Planck Institute for Meteorology
Germany

KNIGHT, Jeff
Met Office Hadley Centre
UK

KNUTSON, Thomas
National Oceanic and Atmospheric Administration, Geophysical Fluid Dynamics Laboratory
USA

KNUTTI, Reto
ETH Zurich
Switzerland

KOCH, Dorothy
U.S. Department of Energy
USA

KOIKE, Makoto
University of Tokyo
Japan

KONDO, Yutaka
University of Tokyo
Japan

KONIKOW, Leonard
U.S. Geological Survey
USA

KOPP, Robert
Rutgers University
USA

KÖRPER, Janina
Freie Universität Berlin
Germany

KOSSIN, James P.
National Oceanic and Atmospheric Administration, Cooperative Institute for Meteorological Satellite Studies
USA

KOSTIANOY, Andrey
P.P. Shirshov Institute of Oceanology
Russian Federation

KOVEN, Charles
Lawrence Berkeley National Laboratory
USA

KRAVITZ, Ben
Pacific Northwest National Laboratory
USA

KRINNER, Gerhard
Laboratoire de Glaciologie et Géophysique de l'Environnement, Université Joseph Fourier
France

KROEZE, Carolien
Wageningen University and Open Universiteit Nederland
Netherlands

KULKARNI, Ashwini
Indian Institute of Tropical Meteorology
India

KUNDETI, Koteswara Rao
Indian Institute of Tropical Meteorology
India

KUSHNIR, Yochanan
Columbia University
USA

KWOK, Ronald
National Aeronautics and Space Administration, Jet Propulsion Laboratory
USA

KWON, Won-Tae
National Institute of Meteorological Research
Republic of Korea

LAKEN, Benjamin
Instituto de Astrofisica de Canarias
Spain

LAMARQUE, Jean-François
National Center for Atmospheric Research
USA

LAMBECK, Kurt
Australian National University
Australia

LANDAIS, Amaëlle
Laboratoire des Sciences du Climat et de l'Environnement, Institut Pierre Simon Laplace
France

LANDERER, Felix
National Aeronautics and Space Administration, Jet Propulsion Laboratory
USA

LASSEY, Keith
National Institute of Water and Atmospheric Research
New Zealand

LAU, Ngar-Cheung
National Oceanic and Atmospheric Administration, Geophysical Fluid Dynamics Laboratory
USA

LAU, William K.
National Aeronautics and Space Administration, Goddard Institute for Space Studies
USA

LAW, Rachel M.
CSIRO Marine and Atmospheric Research
Australia

LAWRENCE, David M.
National Center for Atmospheric Research
USA

LE BROCQ, Anne
University of Exeter
UK

LE QUÉRÉ, Corinne
University of East Anglia
UK

LEBSOCK, Matthew
National Aeronautics and Space Administration, Jet Propulsion Laboratory
USA

LEE, David
Manchester Metropolitan University
UK

LEE, Kitack
Pohang University of Science and Technology
Republic of Korea

LEE, Robert W.
University of Reading
UK

LEE, Tong
National Aeronautics and Space Administration, Jet Propulsion Laboratory
USA

LEMKE, Peter
Alfred Wegener Institute for Polar and Marine Research
Germany

LENAERTS, Jan
Utrecht University
Netherlands

LENDERINK, Geert
Royal Netherlands Meteorological Institute
Netherlands

LENNARD, Chris
University of Cape Town
South Africa

LENTON, Andrew
CSIRO Marine and Atmospheric Research
Australia

LEULIETTE, Eric
National Oceanic and Atmospheric Administration, Center for Satellite Applications and Research
USA

LEUNG, Lai-yung Ruby
Pacific Northwest National Laboratory
USA

LEVERMANN, Anders
Potsdam Institute for Climate Impact Research
Germany

LI, Camille
University of Bergen
Norway

LI, Hongmei
Max Planck Institute for Meteorology
Germany

LIAO, Hong
Institute of Atmospheric Physics, Chinese Academy of Sciences
China

LIDDICOAT, Spencer
Met Office Hadley Centre
UK

LIGTENBERG, Stefan
Utrecht University
Netherlands

LIN, Renping
Institute of Atmospheric Physics, Chinese Academy of Sciences
China

LITTLE, Christopher M.
Princeton University
USA

LO, Fiona
Cornell University
USA

LOCKWOOD, Mike
University of Reading
UK

LOEB, Norman G.
National Aeronautics and Space Administration, Langley Research Center
USA

LOHMANN, Ulrike
ETH Zurich
Switzerland

LOMAS, Mark R.
University of Sheffield
UK

LOSADA, Teresa
Universidad de Castilla-La Mancha
Spain

LOTT, Fraser
Met Office Hadley Centre
UK

LU, Jian
George Mason University
USA

LUCAS, Christopher
Bureau of Meteorology
Australia

LUCHT, Wolfgang
Potsdam Institute for Climate Impact Research
Germany

LUNT, Daniel J.
University of Bristol
UK

LUO, Yiqi
University of Oklahoma
USA

LUTERBACHER, Jürg
Justus-Liebig University Giessen
Germany

MACKELLAR, Neil C.
University of Cape Town
South Africa

MAGAÑA, Victor
Universidad Nacional Autonoma de Mexico
Mexico

MAHLSTEIN, Irina
Federal Office of Meteorology and Climatology MeteoSwiss
Switzerland

MAHOWALD, Natalie
Cornell University
USA

MAKI, Takashi
Meteorological Research Institute
Japan

MARENGO, José
National Institute for Space Research
Brazil

MARKUS, Thorsten
National Aeronautics and Space Administration, Goddard Space Flight Center
USA

MARLAND, Gregg
Appalachian State University
USA

MAROTZKE, Jochem
Max Planck Institute for Meteorology
Germany

MARSHALL, Gareth
British Antarctic Survey
UK

MARSTON, George
University of Reading
UK

MARZEION, Ben
University of Innsbruck
Austria

MASSOM, Rob
Australian Antarctic Division
Australia

MASSON-DELMOTTE, Valérie
Laboratoire des Sciences du Climat et de l'Environnement, Institut Pierre Simon Laplace
France

MASSONNET, François
Université catholique de Louvain
Belgium

MATTHEWS, H. Damon
Concordia University
Canada

MAURITZEN, Cecilie
Center for International Climate and Environmental Research Oslo
Norway

MAYORGA, Emilio
University of Washington
USA

MCGREGOR, Shayne
University of New South Wales
Australia

MCINNES, Kathleen L.
CSIRO Marine and Atmospheric Research
Australia

MEARNS, Linda
National Center for Atmospheric Research
USA

MEARS, Carl A.
Remote Sensing Systems
USA

MEEHL, Gerald
National Center for Atmospheric Research
USA

MEINSHAUSEN, Malte
Potsdam Institute for Climate Impact Research
Germany

MELTON, Joe R.
Environment Canada
Canada

MENDOZA, Blanca
Universidad Nacional Autonoma de Mexico
Mexico

MENÉNDEZ, Claudio
Universidad de Buenos Aires
Argentina

MENÉNDEZ, Melisa
Universidad de Cantabria
Spain

MENNE, Matthew
National Oceanic and Atmospheric Administration, National Climatic Data Center
USA

MERCHANT, Christopher J.
University of Edinburgh
UK

MERNILD, Sebastian H.
Los Alamos National Laboratory
USA

MERRIFIELD, Mark A.
University of Hawaii
USA

METZL, Nicolas
Laboratoire d'Océanographie et du Climat, Institut Pierre Simon Laplace
France

MILNE, Glenn A.
University of Ottawa
Canada

MIN, Seung-Ki
Pohang University of Science and Technology
Republic of Korea

MITCHELL, Daniel
University of Oxford
UK

MITROVICA, Jerry X.
Harvard University
USA

MOBERG, Anders
Stockholm University
Sweden

MOHOLDT, Geir
Scripps Institution of Oceanography
USA

MOKHOV, Igor I.
A.M. Obukhov Institute of Atmospheric Physics
Russian Federation

MOKSSIT, Abdalah
Direction de la Météorologie Nationale
Morocco

MÖLG, Thomas
Technical University Berlin
Germany

MONSELESAN, Didier
CSIRO Marine and Atmospheric Research
Australia

MONTZKA, Stephen A.
National Oceanic and Atmospheric Administration, Earth System Research Laboratory
USA

MORAK, Simone
University of Reading
UK

MORDY, Calvin
National Oceanic and Atmospheric Administration, Pacific Marine Environmental Laboratory
USA

MORICE, Colin P.
Met Office Hadley Centre
UK

MOTE, Philip
Oregon State University
USA

MOTTRAM, Ruth
Danish Meteorological Institute
Denmark

MSADEK, Rym
National Oceanic and Atmospheric Administration, Geophysical Fluid Dynamics Laboratory
USA

MUDELSEE, Manfred
Alfred Wegener Institute for Polar and Marine Research
Germany

MÜLLER, Stefanie
Freie Universität Berlin
Germany

MUHS, Daniel R.
U.S. Geological Survey
USA

MULITZA, Stefan
MARUM Center for Marine Environmental Sciences
Germany

MUNHOVEN, Guy
Université de Liège
Belgium

MURAKAMI, Hiroyuki
University of Hawaii
USA

MURPHY, Daniel
National Oceanic and Atmospheric Administration, Earth System Research Laboratory
USA

MURRAY, Tavi
Swansea University
UK

MYHRE, Cathrine Lund
Norwegian Institute for Air Research
Norway

MYHRE, Gunnar
Center for International Climate and Environmental Research Oslo
Norway

MYNENI, Ranga B.
Boston University
USA

NAIK, Vaishali
National Oceanic and Atmospheric Administration, Geophysical Fluid Dynamics Laboratory
USA

NAISH, Tim
Victoria University of Wellington
New Zealand

NAKAJIMA, Teruyuki
University of Tokyo
Japan

NATH, Mary Jo
National Oceanic and Atmospheric Administration, Geophysical Fluid Dynamics Laboratory
USA

NEELIN, J. David
University of California Los Angeles
USA

NEREM, R. Steven
Cooperative Institute for Research in Environmental Sciences
USA

NICHOLAS, J.P.
Ohio State University
USA

NICK, Faezeh
UNIS - The University Centre in Svalbard
Norway

NIELSEN, Claus J.
University of Oslo
Norway

NIWA, Yosuke
Meteorological Research Institute
Japan

NOJIRI, Yukihiro
National Institute for Environmental Studies
Japan

NORBY, Richard J.
Oak Ridge National Laboratory
USA

NORRIS, Joel R.
Scripps Institution of Oceanography
USA

NUNN, Patrick D.
University of New England
Australia

O'CONNOR, Fiona
Met Office Hadley Centre
UK

O'DOWD, Colin
National University of Ireland, Galway
Ireland

O'NEILL, Brian C.
National Center for Atmospheric Research
USA

OLAFSSON, Jon
University of Iceland
Iceland

OLESEN, Martin
Danish Meteorological Institute
Denmark

ORR, James
Laboratoire des Sciences du Climat et de l'Environnement, Institut Pierre Simon Laplace
France

ORSI, Alejandro
Texas A&M University
USA

OSBORN, Timothy
University of East Anglia
UK

OTTO, Alexander
University of Oxford
UK

OTTO, Friederike
University of Oxford
UK

OTTO-BLIESNER, Bette
National Center for Atmospheric Research
USA

OVERDUIN, Pier Paul
Alfred Wegener Institute for Polar and Marine Research
Germany

OVERLAND, James
National Oceanic and Atmospheric Administration, Pacific Marine Environmental Laboratory
USA

PAINTER, Jeff
Lawrence Livermore National Laboratory
USA

PALMER, Tim
University of Oxford
UK

PARK, Geun-Ha
Korea Institute of Ocean Science and Technology
Republic of Korea

PARK, Geun-Ha
National Oceanic and Atmospheric Administration, Atlantic Oceanographic and Meteorological Laboratory
USA

PARKER, David E.
Met Office Hadley Centre
UK

PARRENIN, Frédéric
Laboratoire de Glaciologie et Géophysique de l`Environnement, Université Joseph Fourier
France

PATRA, Prabir
Japan Agency for Marine-Earth Science and Technology
Japan

PATRICOLA, Christina M.
Texas A&M University
USA

PAUL, Frank
University of Zurich
Switzerland

PAVLOVA, Tatiana
Voeikov Main Geophysical Observatory
Russian Federation

PAYNE, Antony J.
University of Bristol
UK

PEARSON, Paul N.
Cardiff University
UK

PENNER, Joyce
University of Michigan
USA

PEREGON, Anna
Laboratoire des Sciences du Climat et de l'Environnement, Institut Pierre Simon Laplace
France

PERLWITZ, Judith
Cooperative Institute for Research in Environmental Sciences
USA

PERRETTE, Mahé
Potsdam Institute for Climate Impact Research
Germany

PETERS, Glen P.
Center for International Climate and Environmental Research Oslo
Norway

PETERS, Wouter
Wageningen University
Netherlands

PETERSCHMITT, Jean-Yves
Laboratoire des Sciences du Climat et de l'Environnement, Institut Pierre Simon Laplace
France

PEYLIN, Philippe
Laboratoire des Sciences du Climat et de l'Environnement, Institut Pierre Simon Laplace
France

PFEFFER, W. Tad
University of Colorado Boulder
USA

PHILIPPON-BERTHIER, Gwenaëlle
Laboratoire des Sciences du Climat et de l'Environnement, Institut Pierre Simon Laplace
France

PIAO, Shilong
Peking University
China

PIERCE, David
Scripps Institution of Oceanography
USA

PIPER, Stephen
Scripps Institution of Oceanography
USA

PITMAN, Andy
University of New South Wales
Australia

PLANTON, Serge
Météo-France
France

PLATTNER, Gian-Kasper
IPCC WGI TSU, University of Bern
Switzerland

POLCHER, Jan
Laboratoire de Météorologie Dynamique, Institut Pierre Simon Laplace
France

POLLARD, David
Pennsylvania State University
USA

POLSON, Debbie
University of Edinburgh
UK

POLYAKOV, Igor
University of Alaska Fairbanks
USA

PONGRATZ, Julia
Max Planck Institute for Meteorology
Germany

POULTER, Benjamin
Laboratoire des Sciences du Climat et de l'Environnement, Institut Pierre Simon Laplace
France

POWER, Scott B.
Bureau of Meteorology
Australia

PRABHAT
Lawrence Berkeley National Laboratory
USA

PRATHER, Michael
University of California Irvine
USA

PROWSE, Terry
Environment Canada
Canada

PURKEY, Sarah G.
University of Washington
USA

QIAN, Yun
Pacific Northwest National Laboratory
USA

QIN, Dahe
Co-Chair IPCC WGI, China Meteorological Administration
China

QIU, Bo
University of Hawaii
USA

QUINN, Terrence
University of Texas
USA

RADIĆ, Valentina
University of British Columbia
Canada

RAE, Jamie
Met Office Hadley Centre
UK

RAHIMZADEH, Fatemeh
Islamic Republic of Iran Meteorological Organization
Iran

RAHMSTORF, Stefan
Potsdam Institute for Climate Impact Research
Germany

RÄISÄNEN, Jouni
University of Helsinki
Finland

RAMASWAMY, Venkatachalam
National Oceanic and Atmospheric Administration, Geophysical Fluid Dynamics Laboratory
USA

RAMESH, Rengaswamy
Physical Research Laboratory
India

RANDALL, David
Colorado State University
USA

RANDEL, William J.
National Center for Atmospheric Research
USA

RASCH, Philip
Pacific Northwest National Laboratory
USA

RAUSER, Florian
Max Planck Institute for Meteorology
Germany

RAVISHANKARA, A.R.
National Oceanic and Atmospheric Administration, Earth System Research Laboratory
USA

RAY, Suchanda
CSIR Centre for Mathematical Modelling and Computer Simulation
India

RAYMOND, Peter A.
Yale University
USA

RAYNAUD, Dominique
Laboratoire de Glaciologie et Géophysique de l`Environnement, Université Joseph Fourier
France

RAYNER, Peter
University of Melbourne
Australia

REASON, Chris
University of Cape Town
South Africa

REICH, Katharine Davis
University of California Los Angeles
USA

REID, Jeffrey
U.S. Naval Research Laboratory
USA

REN, Jiawen
Cold and Arid Regions Environmental and Engineering Research Institute, Chinese Academy of Sciences
China

RENWICK, James
Victoria University of Wellington
New Zealand

REVERDIN, Gilles
Laboratoire d'Océanographie et du Climat, Institut Pierre Simon Laplace
France

RHEIN, Monika
University of Bremen
Germany

RIBES, Aurélien
Météo-France
France

RICHTER, Andreas
University of Bremen
Germany

RICHTER, Carolin
World Meteorological Organization
Switzerland

RIDGWELL, Andy
University of Bristol
UK

RIGBY, Matthew
University of Bristol
UK

RIGNOT, Eric
National Aeronautics and Space Administration, Jet Propulsion Laboratory
USA

RILEY, William J.
Lawrence Berkeley National Laboratory
USA

RINGEVAL, Bruno
Utrecht University
Netherlands

RINTOUL, Stephen R.
CSIRO Marine and Atmospheric Research
Australia

ROBINSON, David
Rutgers University
USA

ROBOCK, Alan
Rutgers University
USA

RÖDENBECK, Christian
Max Planck Institute for Biogeochemistry
Germany

RODRIGUES, Luis R.L.
Institut Català de Ciències del Clima
Spain

RODRÍGUEZ DE FONSECA, Belén
Universidad Complutense de Madrid
Spain

RODWELL, Mark
European Centre for Medium-Range Weather Forecasts
UK

ROEMMICH, Dean
Scripps Institution of Oceanography
USA

ROGELJ, Joeri
ETH Zurich
Switzerland

ROHLING, Eelco
Australian National University
Australia

ROJAS, Maisa
Universidad de Chile
Chile

ROMANOU, Anastasia
Columbia University
USA

ROTH, Raphael
University of Bern
Switzerland

ROTSTAYN, Leon
CSIRO Marine and Atmospheric Research
Australia

RUMMUKAINEN, Markku
Swedish Meteorological and Hydrological Institute
Sweden

RUSTICUCCI, Matilde
Universidad de Buenos Aires
Argentina

RUTI, Paolo
Italian National Agency for New Technologies, Energy and Sustainable Economic Development
Italy

SABINE, Christopher
National Oceanic and Atmospheric Administration, Pacific Marine Environmental Laboratory
USA

SAENKO, Oleg
Environment Canada
Canada

SALZMANN, Ulrich
Northumbria University
UK

SAMSET, Bjørn
Center for International Climate and Environmental Research Oslo
Norway

SANTER, Benjamin D.
Lawrence Livermore National Laboratory
USA

SARR, Abdoulaye
National Meteorological Agency of Senegal
Senegal

SATHEESH, S.K.
Indian Institute of Science
India

SAUNOIS, Marielle
Laboratoire des Sciences du Climat et de l'Environnement, Institut Pierre Simon Laplace
France

SAVARINO, Joël
Laboratoire de Glaciologie et Géophysique de l`Environnement, Université Joseph Fourier
France

SCAIFE, Adam A.
Met Office Hadley Centre
UK

SCHÄR, Christoph
ETH Zurich
Switzerland

SCHMIDT, Hauke
Max Planck Institute for Meteorology
Germany

SCHMIDTKO, Sunke
University of East Anglia
UK

SCHMITT, Raymond
Woods Hole Oceanographic Institution
USA

SCHMITTNER, Andreas
Oregon State University
USA

SCHOOF, Christian
University of British Columbia
Canada

SCHULZ, Jörg
EUMETSAT
Germany

SCHULZ, Michael
MARUM Center for Marine Environmental Sciences
Germany

SCHULZ, Michael
Norwegian Meteorological Institute
Norway

SCHULZWEIDA, Uwe
Max Planck Institute for Meteorology
Germany

SCHURER, Andrew
University of Edinburgh
UK

SCHUUR, Edward
University of Florida
USA

SCINOCCA, John
Environment Canada
Canada

SCREEN, James
University of Exeter
UK

SEAGER, Richard
Columbia University
USA

SEBBARI, Rachid
Direction de la Météorologie Nationale
Morocco

SEDLÁČEK, Jan
ETH Zurich
Switzerland

SEIDEL, Dian J.
National Oceanic and Atmospheric Administration, Air Resources Laboratory
USA

SEMENOV, Vladimir
Russian Academy of Sciences
Russian Federation

SEXTON, David
Met Office Hadley Centre
UK

SHAFFREY, Len C.
University of Reading
UK

SHAKUN, Jeremy
Boston College
USA

SHAO, XueMei
Institute of Geographic Sciences and Natural Resources Research, Chinese Academy of Sciences
China

SHARP, Martin
University of Alberta
Canada

SHEPHERD, Theodore
University of Reading
UK

SHERWOOD, Steven
University of New South Wales
Australia

SHIKLOMANOV, Nikolay
George Washington University
USA

SHIMADA, Koji
Tokyo University of Marine Science and Technology
Japan

SHINDELL, Drew
National Aeronautics and Space Administration, Goddard Institute for Space Studies
USA

SHINE, Keith
University of Reading
UK

SHIOGAMA, Hideo
National Institute for Environmental Studies
Japan

SHONGWE, Mxolisi
South African Weather Service
South Africa

SILLMANN, Jana
Environment Canada
Canada

SIMMONS, Adrian
European Centre for Medium-Range Weather Forecasts
UK

SITCH, Stephen
University of Exeter
UK

SLANGEN, Aimée
CSIRO Marine and Atmospheric Research
Australia

SLATER, Andrew
National Snow and Ice Data Center
USA

SMERDON, Jason
Columbia University
USA

SMIRNOV, Dmitry
Russian Academy of Sciences
Russian Federation

SMITH, Doug
Met Office Hadley Centre
UK

SMITH, Sharon
Natural Resources Canada
Canada

SMITH, Steven J.
Pacific Northwest National Laboratory
USA

SMITH, Thomas M.
National Oceanic and Atmospheric Administration, Center for Satellite Applications and Research
USA

SODEN, Brian J.
University of Miami
USA

SOLMAN, Silvina
Universidad de Buenos Aires
Argentina

SOLOMINA, Olga
Russian Academy of Sciences
Russian Federation

SPAHNI, Renato
University of Bern
Switzerland

SPERBER, Kenneth
Lawrence Livermore National Laboratory
USA

STAMMER, Detlef
University of Hamburg
Germany

STAMMERJOHN, Sharon
University of Colorado Boulder
USA

STEFFEN, Konrad
Swiss Federal Institute for Forest, Snow and Landscape Research WSL
Switzerland

STENDEL, Martin
Danish Meteorological Institute
Denmark

STEPHENS, Graeme
National Aeronautics and Space Administration, Jet Propulsion Laboratory
USA

STEPHENSON, David B.
University of Exeter
UK

STEVENS, Bjorn
Max Planck Institute for Meteorology
Germany

STEVENSON, David S.
University of Edinburgh
UK

STEVENSON, Samantha
University of Hawaii
USA

STIER, Philip
University of Oxford
UK

STÖBER, Uwe
University of Bremen
Germany

STOCKER, Benjamin D.
University of Bern
Switzerland

STOCKER, Thomas F.
Co-Chair IPCC WGI, University of Bern
Switzerland

STORELVMO, Trude
Yale University
USA

STOTT, Peter A.
Met Office Hadley Centre
UK

STRAMMA, Lothar
GEOMAR Helmholtz Centre for Ocean Research
Germany

STUBENRAUCH, Claudia
Laboratoire de Météorologie Dynamique, Institut Pierre Simon Laplace
France

SUGA, Toshio
Tohoku University
Japan

SUTTON, Rowan
University of Reading
UK

SWART, Neil
University of Victoria
Canada

TAKAHASHI, Taro
Columbia University
USA

TAKAYABU, Izuru
Meteorological Research Institute
Japan

TAKEMURA, Toshihiko
Kyushu University
Japan

TALLEY, Lynne D.
Scripps Institution of Oceanography
USA

TANGANG, Fredolin
National University of Malaysia
Malaysia

TANHUA, Toste
GEOMAR Helmholtz Centre for Ocean Research
Germany

TANS, Pieter
National Oceanic and Atmospheric Administration, Earth System Research Laboratory
USA

TARASOV, Pavel
Freie Universität Berlin
Germany

TAYLOR, Karl
Lawrence Livermore National Laboratory
USA

TEBALDI, Claudia
Climate Central, Inc.
USA

TETT, Simon
University of Edinburgh
UK

TEULING, Adriaan J. (Ryan)
Wageningen University
Netherlands

THOMPSON, Rona L.
Norwegian Institute for Air Research
Norway

THORNE, Peter W.
Nansen Environmental and Remote Sensing Center
Norway

THORNTON, Peter
Oak Ridge National Laboratory
USA

TIMMERMANN, Axel
University of Hawaii
USA

TJIPUTRA, Jerry
Uni Research Norway
Norway

TRENBERTH, Kevin
National Center for Atmospheric Research
USA

TÜRKEŞ, Murat
Çanakkale Onsekiz Mart University
Turkey

TURNER, John
British Antarctic Survey
UK

UMMENHOFER, Caroline
Woods Hole Oceanographic Institution
USA

UNNIKRISHNAN, Alakkat S.
National Institute of Oceanography
India

VAN ANGELEN, Jan H.
Utrecht University
Netherlands

VAN DE BERG, Willem Jan
Utrecht University
Netherlands

VAN DE WAL, Roderik
Utrecht University
Netherlands

VAN DEN BROEKE, Michiel
Utrecht University
Netherlands

VAN DEN HURK, Bart
Royal Netherlands Meteorological Institute
Netherlands

VAN DER WERF, Guido
VU University Amsterdam
Netherlands

VAN NOIJE, Twan
Royal Netherlands Meteorological Institute
Netherlands

VAN OLDENBORGH, Geert Jan
Royal Netherlands Meteorological Institute
Netherlands

VAN VUUREN, Detlef
PBL Netherlands Environmental Assessment Agency
Netherlands

VAUGHAN, David G.
British Antarctic Survey
UK

VAUTARD, Robert
Laboratoire des Sciences du Climat et de l'Environnement, Institut Pierre Simon Laplace
France

VAVRUS, Steve
University of Wisconsin
USA

VECCHI, Gabriel
National Oceanic and Atmospheric Administration, Geophysical Fluid Dynamics Laboratory
USA

VELICOGNA, Isabella
University of California Irvine
USA

VERNIER, Jean-Paul
National Aeronautics and Space Administration, Langley Research Center
USA

VESALA, Timo
University of Helsinki
Finland

VINTHER, Bo M.
University of Copenhagen
Denmark

VITERBO, Pedro
Instituto de Meteorologia
Portugal

VIZCAÍNO, Miren
Delft University of Technology
Netherlands

VON SCHUCKMANN, Karina
Institut Français de Recherche pour l'Exploitation de la Mer
France

VON STORCH, Hans
University of Hamburg
Germany

VOULGARAKIS, Apostolos
Imperial College London
UK

WADA, Yoshihide
Utrecht University
Netherlands

WADHAMS, Peter
University of Cambridge
UK

WAELBROECK, Claire
Laboratoire des Sciences du Climat et de l'Environnement, Institut Pierre Simon Laplace
France

WALSH, Kevin
University of Melbourne
Australia

WANG, Bin
University of Hawaii
USA

WANG, Chunzai
National Oceanic and Atmospheric Administration, Atlantic Oceanographic and Meteorological Laboratory
USA

WANG, Fan
Institute of Oceanology, Chinese Academy of Sciences
China

WANG, Hui-Jun
Institute of Atmospheric Physics, Chinese Academy of Sciences
China

WANG, Junhong
National Center for Atmospheric Research
USA

WANG, Muyin
National Oceanic and Atmospheric Administration, Joint Institute for the Study of the Atmosphere and Ocean
USA

WANG, Xiaolan L.
Environment Canada
Canada

WANIA, Rita
Austria

WANNER, Heinz
University of Bern
Switzerland

WANNINKHOF, Rik
National Oceanic and Atmospheric Administration, Atlantic Oceanographic and Meteorological Laboratory
USA

WARD, Daniel S.
Cornell University
USA

WATTERSON, Ian
CSIRO Marine and Atmospheric Research
Australia

WEAVER, Andrew J.
University of Victoria
Canada

WEBB, Mark
Met Office Hadley Centre
UK

WEBB, Robert
National Oceanic and Atmospheric Administration, Earth System Research Laboratory
USA

WEHNER, Michael
Lawrence Berkeley National Laboratory
USA

WEISHEIMER, Antje
University of Oxford
UK

WEISS, Ray F.
Scripps Institution of Oceanography
USA

WHITE, Neil J.
CSIRO Marine and Atmospheric Research
Australia

WIDLANSKY, Matthew
University of Hawaii
USA

WIJFFELS, Susan
CSIRO Marine and Atmospheric Research
Australia

WILD, Martin
ETH Zurich
Switzerland

WILD, Oliver
Lancaster University
UK

WILLETT, Kate M.
Met Office Hadley Centre
UK

WILLIAMS, Keith
Met Office Hadley Centre
UK

WINKELMANN, Ricarda
Potsdam Institute for Climate Impact Research
Germany

WINKER, David
National Aeronautics and Space Administration, Langley Research Center
USA

WINTHER, Jan-Gunnar
Norwegian Polar Institute
Norway

WITTENBERG, Andrew
National Oceanic and Atmospheric Administration, Geophysical Fluid Dynamics Laboratory
USA

WOLF-GLADROW, Dieter
Alfred Wegener Institute for Polar and Marine Research
Germany

WOOD, Simon N.
University of Bath
UK

WOODWORTH, Philip L.
National Oceanography Centre
UK

WOOLLINGS, Tim
University of Reading
UK

WORBY, Anthony
CSIRO Marine and Atmospheric Research
Australia

WRATT, David
National Institute of Water and Atmospheric Research
New Zealand

WUEBBLES, Donald
University of Illinois
USA

WYANT, Matthew
University of Washington
USA

XIAO, Cunde
Chinese Academy of Meteorological Sciences, China Meteorological Administration
China

XIE, Shang-Ping
Scripps Institution of Oceanography
USA

YASHAYAEV, Igor
Bedford Institute of Oceanography
Canada

YASUNARI, Tetsuzo
Nagoya University
Japan

YEH, Sang-Wook
Hanyang University
Republic of Korea

YIN, Jianjun
University of Arizona
USA

YOKOYAMA, Yusuke
University of Tokyo
Japan

YOSHIMORI, Masakazu
University of Tokyo
Japan

YOUNG, Paul
Lancaster University
UK

YU, Lisan
Woods Hole Oceanographic Institution
USA

ZACHOS, James
University of California Santa Cruz
USA

ZAEHLE, Sönke
Max Planck Institute for Biogeochemistry
Germany

ZAPPA, Giuseppe
University of Reading
UK

ZENG, Ning
University of Maryland
USA

ZHAI, Panmao
National Climate Center, China Meteorological Administration
China

ZHANG, Chidong
University of Miami
USA

ZHANG, Hua
National Climate Center, China Meteorological Administration
China

ZHANG, Jianglong
University of North Dakota
USA

ZHANG, Lixia
Institute of Atmospheric Physics, Chinese Academy of Sciences
China

ZHANG, Rong
National Oceanic and Atmospheric Administration, Geophysical Fluid Dynamics Laboratory
USA

ZHANG, Tingjun
Cooperative Institute for Research in Environmental Sciences
USA

ZHANG, Xiao-Ye
Chinese Academy of Meteorological Sciences, China Meteorological Administration
China

ZHANG, Xuebin
Environment Canada
Canada

ZHAO, Lin
Cold and Arid Regions Environmental and Engineering Research Institute, Chinese Academy of Sciences
China

ZHAO, Zong-Ci
National Climate Center, China Meteorological Administration
China

ZHENG, Xiaotong
Ocean University of China
China

ZHOU, Tianjun
Institute of Atmospheric Physics, Chinese Academy of Sciences
China

ZICKFELD, Kirsten
Simon Fraser University
Canada

ZOU, Liwei
Institute of Atmospheric Physics, Chinese Academy of Sciences
China

ZWARTZ, Dan
Victoria University of Wellington
New Zealand

ZWIERS, Francis
University of Victoria
Canada

Annex VI: Expert Reviewers of the IPCC WGI Fifth Assessment Report

This annex should be cited as:

IPCC, 2013: Annex VI: Expert Reviewers of the IPCC WGI Fifth Assessment Report. In: *Climate Change 2013: The Physical Science Basis. Contribution of Working Group I to the Fifth Assessment Report of the Intergovernmental Panel on Climate Change* [Stocker, T.F., D. Qin, G.-K. Plattner, M. Tignor, S.K. Allen, J. Boschung, A. Nauels, Y. Xia, V. Bex and P.M. Midgley (eds.)]. Cambridge University Press, Cambridge, United Kingdom and New York, NY, USA.

AAMAAS, Borgar
Center for International Climate and Environmental Research Oslo
Norway

ABRAHAM, JOHN
University of St. Thomas
USA

ADAM, Hussein
Wad Medani Ahlia College
Sudan

ÅGREN, Göran
Swedish University of Agricultural Sciences
Sweden

ALEXANDER, Lisa
University of New South Wales
Australia

ALEYNIK, Dmitry
Scottish Association for Marine Science
UK

ALLAN, Richard
University of Reading
UK

ALLEN, Simon K.
IPCC WGI TSU, University of Bern
Switzerland

ALLEY, Richard B.
Pennsylvania State University
USA

ALLISON, Ian
Antarctic Climate and Ecosystems Cooperative Research Centre
Australia

ALORY, Gaël
Laboratoire d'Etudes en Géophysique et Océanographie Spatiales
France

ALPERT, Alice
Massachusetts Institute of Technology
USA

AMJAD, Muhammad
Global Change Impact Studies Centre
Pakistan

ANDEREGG, William
Stanford University
USA

ANDERSEN, Bo
Norwegian Space Centre
Norway

ANDREAE, Meinrat O.
Max Planck Institute for Chemistry
Germany

ANDREU-BURILLO, Isabel
Institut Català de Ciències del Clima
Spain

ANDREWS, Oliver David
University of East Anglia
UK

AÑEL CABANELAS, Juan Antonio
University of Oxford
UK

ANENBERG, Susan
U.S. Environmental Protection Agency
USA

ANNAMALAI, H.
University of Hawaii
USA

ANNAN, James
Japan Agency for Marine-Earth Science and Technology
Japan

APITULEY, Arnoud
Royal Netherlands Meteorological Institute
Netherlands

APPENZELLER, Christof
Federal Office of Meteorology and Climatology MeteoSwiss
Switzerland

ARBLASTER, Julie
Bureau of Meteorology
Australia

ARNETH, Almut
Karlsruhe Institute of Technology
Germany

ARORA, Vivek
Environment Canada
Canada

ARTALE, Vincenzo
Italian National Agency for New Technologies, Energy and Sustainable Economic Development
Italy

ARTINANO, Begona
Centro de Investigaciones Energéticas, Medioambientales y Tecnológicas
Spain

ARTUSO, Florinda
Italian National Agency for New Technologies, Energy and Sustainable Economic Development
Italy

ASMI, Ari
University of Helsinki
Finland

AUAD, Guillermo
Bureau of Ocean Energy Management
USA

AUCAMP, Pieter
Ptersa Environmental Management Consultants
South Africa

AZAR, Christian
Chalmers University of Technology
Sweden

BADER, David
Lawrence Livermore National Laboratory
USA

BADIOU, Pascal
Ducks Unlimited Canada
Canada

BAHN, Michael
University of Innsbruck
Austria

BAKAN, Stephan
Max Planck Institute for Meteorology
Germany

BALTENSPERGER, Urs
Paul Scherrer Institute
Switzerland

BAMBER, Jonathan
University of Bristol
UK

BAN-WEISS, George
Lawrence Berkeley National Laboratory and University of Southern California
USA

BARKER, Stephen
Cardiff University
UK

BARNETT, Tim
Scripps Institution of Oceanography
USA

BARRETT, Jack
Imperial College London (retired)
UK

BARRETT, Peter
Victoria University of Wellington
New Zealand

BARRY, Roger
National Snow and Ice Data Center
USA

BATES, J. Ray
University College Dublin
Ireland

BATES, Timothy
National Oceanic and Atmospheric Administration, Pacific Marine Environmental Laboratory
USA

BEKKI, Slimane
Laboratoire Atmosphères, Milieux, Observations Spatiales, Institut Pierre Simon Laplace
France

BELLOUIN, Nicolas
Met Office Hadley Centre
UK

BELTRAN, Catherine
Université Pierre et Marie Curie
France

BENNARTZ, Ralf
University of Wisconsin
USA

BERNHARD, Luzi
Swiss Federal Institute for Forest, Snow and Landscape Research WSL
Switzerland

BERNHARDT, Karl-Heinz
Leibniz Society of Sciences at Berlin
Germany

BERNIER, Pierre
Natural Resources Canada
Canada

BERNTSEN, Terje
University of Oslo
Norway

BERTHIER, Etienne
Laboratoire d'Etudes en Géophysique et Océanographie Spatiales
France

BETTS, Richard
Met Office Hadley Centre
UK

BETZ, Gregor
Karlsruhe Institute of Technology
Germany

BHANDARI, Medani
Syracuse University
USA

BINDOFF, Nathaniel L.
University of Tasmania
Australia

BINTANJA, Richard
Royal Netherlands Meteorological Institute
Netherlands

BLADÉ, Ileana
Universitat de Barcelona
Spain

BLANCO, Juan A.
Universidad Pública de Navarra
Spain

BLATTER, Heinz
ETH Zurich
Switzerland

BLOMQVIST, Sven
Stockholm University
Sweden

BODAS-SALCEDO, Alejandro
Met Office Hadley Centre
UK

BODE, Antonio
Instituto Español de Oceanografia
Spain

BOEHM, Christian Reiner
Imperial College London
UK

BOENING, Carmen
National Aeronautics and Space Administration, Jet Propulsion Laboratory
USA

BOERSMA, Klaas Folkert
Royal Netherlands Meteorological Institute and Eindhoven University of Technology
Netherlands

BOGNER, Jean E.
University of Illinois
USA

BOKO, Michel
Université d'Abomey Calavi
Benin

BOLLASINA, Massimo
National Oceanic and Atmospheric Administration, Geophysical Fluid Dynamics Laboratory
USA

BONNET, Sophie
Université du Québec
Canada

BONY, Sandrine
Laboratoire de Météorologie Dynamique, Institut Pierre Simon Laplace
France

BOOTH, Ben
Met Office Hadley Centre
UK

BOSILOVICH, Michael
National Aeronautics and Space Administration, Goddard Space Flight Center
USA

BOUCHER, Olivier
Laboratoire de Météorologie Dynamique, Institut Pierre Simon Laplace
France

BOULDIN, Jim
University of California Davis
USA

BOURBONNIERE, Richard
Environment Canada
Canada

BOURLES, Bernard
Institut de Recherche pour le Développement
France

BOUSQUET, Philippe
Laboratoire des Sciences du Climat et de l'Environnement, Institut Pierre Simon Laplace
France

BOWEN, Melissa
University of Auckland
New Zealand

BOYER, Timothy
National Oceanic and Atmospheric Administration, National Oceanographic Data Center
USA

BRACEGIRDLE, Thomas
British Antarctic Survey
UK

BRACONNOT, Pascale
Laboratoire des Sciences du Climat et de l'Environnement, Institut Pierre Simon Laplace
France

BRAESICKE, Peter
University of Cambridge
UK

BREGMAN, Abraham
Royal Netherlands Meteorological Institute
Netherlands

BRENDER, Pierre
Laboratoire des Sciences du Climat et de l'Environnement, Institut Pierre Simon Laplace and AgroParisTech
France

BREWER, Michael
National Oceanic and Atmospheric Administration, National Climatic Data Center
USA

BRIERLEY, Christopher
University College London
UK

BRIFFA, Keith
University of East Anglia
UK

BROMWICH, David
Ohio State University
USA

BROOKS, Harold
National Oceanic and Atmospheric Administration, National Severe Storms Laboratory
USA

BROVKIN, Victor
Max Planck Institute for Meteorology
Germany

BROWN, Jaclyn
CSIRO Marine and Atmospheric Research
Australia

BROWN, Josephine
Bureau of Meteorology
Australia

BROWN, Simon
Met Office Hadley Centre
UK

BURKETT, Virginia
U.S. Geological Survey
USA

BURT, Peter
University of Greenwich
UK

BURTON, David
Burton Systems Software
USA

BUTENHOFF, Christopher
Portland State University
USA

BUTLER, James
National Oceanic and Atmospheric Administration, Earth System Research Laboratory
USA

CAESAR, John
Met Office Hadley Centre
UK

CAGNAZZO, Chiara
Institute of Atmospheric Sciences and Climate
Italy

CAI, Rongshuo
Third Institute of Oceanography, State Oceanic Administration
China

CAI, Zucong
Nanjing Normal University
China

CAINEY, Jill
UK

CALVO, Natalia
Universidad Complutense de Madrid
Spain

CAMERON-SMITH, Philip
Lawrence Livermore National Laboratory
USA

CANDELA, Lucila
Universitat Politècnica de Catalunya
Spain

CAO, Jianting
General Institute of Water Resources and Hydropower Planning and Design, Ministry of Water Resources
China

CARDIA SIMÕES, Jefferson
Universidade Federal do Rio Grande do Sul
Brazil

CARDINAL, Damien
Université Pierre et Marie Curie
France

CARTER, Timothy
Finnish Environment Institute
Finland

CASELDINE, Chris
University of Exeter
UK

CASSARDO, Claudio
University of Torino
Italy

CASSOU, Christophe
Centre Européen de Recherche et de Formation Avancée en Calcul Scientifique
France

CEARRETA, Alejandro
Universidad del Pais Vasco
Spain

CERMAK, Jan
Ruhr-Universität Bochum
Germany

CERVARICH, Matthew
University of Illinois
USA

CHADWICK, Robin
Met Office Hadley Centre
UK

CHARLESWORTH, Mark
Keele University
UK

CHARLSON, Robert
University of Washington
USA

CHARPENTIER LJUNGQVIST, Fredrik
Stockholm University
Sweden

CHAUVIN, Fabrice
Météo-France
France

CHAZETTE, Patrick
Laboratoire des Sciences du Climat et de l'Environnement, Institut Pierre Simon Laplace
France

CHE, Tao
Cold and Arid Regions Environmental and Engineering Research Institute, Chinese Academy of Sciences
China

CHEN, Xianyao
First Institute of Oceanography, State Oceanic Administration
China

CHERCHI, Annalisa
Centro Euromediterraneo per i Cambiamenti Climatici and Istituto Nazionale di Geofisica e Vulcanologia
Italy

CHHABRA, Abha
Indian Space Research Organisation
India

CHIKAMOTO, Megumi
University of Hawaii
USA

CHIKAMOTO, Yoshimitsu
University of Hawaii
USA

CHOU, Chia
Academia Sinica
Taiwan, China

CHRISTIAN, James
Fisheries and Oceans Canada
Canada

CHRISTOPHERSEN, Øyvind
Climate and Pollution Agency
Norway

CHRISTY, John
University of Alabama
USA

CHURCH, John
CSIRO Marine and Atmospheric Research
Australia

CHYLEK, Petr
Los Alamos National Laboratory
USA

CIRANO, Mauro
Federal University of Bahia
Brazil

CIURO, Darienne
University of Illinois
USA

CLARK, Robin
Met Office Hadley Centre
UK

CLAUSSEN, Martin
Max Planck Institute for Meteorology
Germany

CLERBAUX, Cathy
Laboratoire Atmosphères, Milieux, Observations Spatiales, Institut Pierre Simon Laplace
France

CLIFT, Peter
Louisiana State University
USA

COAKLEY, James
Oregon State University
USA

COFFEY, Michael
National Center for Atmospheric Research
USA

COGLEY, J. Graham
Trent University
Canada

COLE, Julia
University of Arizona
USA

COLLIER, Mark
CSIRO Marine and Atmospheric Research
Australia

COLLINS, Matthew
University of Exeter
UK

COLLINS, William
University of Reading
UK

COLMAN, Robert
Bureau of Meteorology
Australia

COLOSE, Chris
University at Albany
USA

COOPER, Owen
Cooperative Institute for Research in Environmental Sciences
USA

COPSTEIN WALDEMAR, Celso
Porto Alegre Municipality, Environmental Department
Brazil

CORTESE, Giuseppe
GNS Science
New Zealand

CORTI, Susanna
European Centre for Medium-Range Weather Forecasts and Institute of Atmospheric Sciences and Climate
Italy

COTRIM DA CUNHA, Leticia
Rio de Janeiro State University
Brazil

COUMOU, Dim
Potsdam Institute for Climate Impact Research
Germany

COVEY, Curt
Lawrence Livermore National Laboratory
USA

CRAWFORD, James
USA

CRIMMINS, Allison
U.S. Environmental Protection Agency
USA

CRISTINI, Luisa
University of Hawaii
USA

CROK, Marcel
Netherlands

CURRY, Charles
University of Victoria
Canada

CURTIS, Jeffrey
University of Illinois
USA

DAI, Aiguo
University at Albany and National Center for Atmospheric Research
USA

DAIRAKU, Koji
National Research Institute for Earth Science and Disaster Prevention
Japan

DAMERIS, Martin
DLR German Aerospace Center
Germany

DANIEL, John
National Oceanic and Atmospheric Administration, Earth System Research Laboratory
USA

DANIELS, Emma
Wageningen University
Netherlands

DANIS, François
Laboratoire de Météorologie Dynamique, Institut Pierre Simon Laplace
France

DAUTRAY, Robert
Académie des Sciences
France

DAVIDSON, Eric
Woods Hole Research Center
USA

DAVIES, Michael
Coldwater Consulting Ltd
Canada

DAY, Jonathan
University of Reading
UK

DE ELIA, Ramon
Ouranos Consortium on Regional Climatology and Adaptation to Climate Change
Canada

DE SAEDELEER, Bernard
Université catholique de Louvain
Belgium

DE VRIES, Hylke
Royal Netherlands Meteorological Institute
Netherlands

DEAN, Robert
University of Florida
USA

DEL GENIO, Anthony
National Aeronautics and Space Administration, Goddard Institute for Space Studies
USA

DELPLA, Ianis
Laboratoire d'Etude et de Recherche en Environnement et Santé
France

DELSOLE, Timothy
George Mason University
USA

DELWORTH, Thomas
National Oceanic and Atmospheric Administration, Geophysical Fluid Dynamics Laboratory
USA

DEMORY, Marie-Estelle
University of Reading
UK

DÉQUÉ, Michel
Météo-France
France

DERKSEN, Chris
Environment Canada
Canada

DESIATO, Franco
Institute for Environmental Protection and Research
Italy

DEVARA, Panuganti C.S.
Indian Institute of Tropical Meteorology
India

DEWALS, Benjamin
Université de Liège
Belgium

DEWITT, David G.
Columbia University
USA

DIAZ MOREJON, Cristobal Felix
Ministry of Science, Technology and the Environment
Cuba

DICKENS, Gerald
Rice University
USA

DIEDHIOU, Arona
Institut de Recherche pour le Développement
France

DIMA, Mihai
University of Bucharest
Romania

DING, Yihui
National Climate Center, China Meteorological Administration
China

DING, Yongjian
Cold and Arid Regions Environmental and Engineering Research Institute, Chinese Academy of Sciences
China

DITLEVSEN, Peter
University of Copenhagen
Denmark

DOHERTY, Ruth
University of Edinburgh
UK

DOLE, Randall
National Oceanic and Atmospheric Administration, Earth System Research Laboratory
USA

DOLMAN, Han
VU University Amsterdam
Netherlands

DOMINGUES, Catia M.
Antarctic Climate and Ecosystems Cooperative Research Centre
Australia

DONAHUE, Neil
Carnegie Mellon University
USA

DONNER, Leo
National Oceanic and Atmospheric Administration, Geophysical Fluid Dynamics Laboratory
USA

DOSTAL, Paul
DLR German Aerospace Center
Germany

DOWNES, Stephanie
Australian National University
Australia

DOYLE, Moira Evelina
Universidad de Buenos Aires
Argentina

DRAGONI, Walter
University of Perugia
Italy

DRIJFHOUT, Sybren
Royal Netherlands Meteorological Institute
Netherlands

DU, Enzai
Peking University
China

DUAN, Anmin
Institute of Atmospheric Physics, Chinese Academy of Sciences
China

DUCE, Robert
Texas A&M University
USA

DUDOK DE WIT, Thierry
Université d'Orléans
France

DUNNE, Eimear
Finnish Meteorological Institute
Finland

DUNSTONE, Nick
Met Office Hadley Centre
UK

DURACK, Paul
Lawrence Livermore National Laboratory
USA

DWYER, Ned
University College Cork
Ireland

EASTERBROOK, Don
Western Washington University
USA

EBI, Kristie
Stanford University
USA

EISEN, Olaf
Alfred Wegener Institute for Polar and Marine Research
Germany

EISENMAN, Ian
University of California San Diego
USA

EKHOLM, Tommi
VTT Technical Research Centre of Finland
Finland

ELDEVIK, Tor
University of Bergen
Norway

ELJADID, Ali Geath
Al-Fath University
Libya

EMANUEL, Kerry
Massachusetts Institute of Technology
USA

ENOMOTO, Hiroyuki
National Institute of Polar Research
Japan

ERICKSON, David
Oak Ridge National Laboratory
USA

ESPINOZA, Jhan Carlo
Instituto Geofísico del Perú
Peru

ESSERY, Richard
University of Edinburgh
UK

EVANS, Michael Neil
University of Maryland
USA

EVANS, Wayne
York University
Canada

EXBRAYAT, Jean-François
University of New South Wales
Australia

EYNAUD, Frédérique
Université Bordeaux 1
France

FAHEY, David
National Oceanic and Atmospheric Administration, Earth System Research Laboratory
USA

FAN, Jiwen
Pacific Northwest National Laboratory
USA

FARAGO, Tibor
St. Istvan University
Hungary

FEINGOLD, Graham
National Oceanic and Atmospheric Administration, Earth System Research Laboratory
USA

FEIST, Dietrich
Max Planck Institute for Biogeochemistry
Germany

FERRONE, Andrew
Karlsruhe Institute of Technology
Germany

FESER, Frauke
Helmholtz-Zentrum Geesthacht
Germany

FEULNER, Georg
Potsdam Institute for Climate Impact Research
Germany

FICHEFET, Thierry
Université catholique de Louvain
Belgium

FIELD, Christopher
Carnegie Institution for Science
USA

FISCHER, Andreas
Federal Office of Meteorology and Climatology MeteoSwiss
Switzerland

FISCHER, Hubertus
University of Bern
Switzerland

FISCHLIN, Andreas
ETH Zurich
Switzerland

FISHER, Joshua
National Aeronautics and Space Administration, Jet Propulsion Laboratory
USA

FLORES, José-Abel
Universidad de Salamanca
Spain

FLOSSMANN, Andrea
Université Blaise Pascal
France

FOLBERTH, Gerd
Met Office Hadley Centre
UK

FOLLAND, Christopher
Met Office Hadley Centre
UK

FORBES, Donald
Bedford Institute of Oceanography
Canada

FOREST, Chris
Pennsylvania State University
USA

FORSTER, Piers
University of Leeds
UK

FOSTER, James
National Aeronautics and Space Administration, Goddard Space Flight Center
USA

FOUNTAIN, Andrew
Portland State University
USA

FRANKLIN, James
CLF-Chem Consulting SPRL
Belgium

FRANKS, Stewart
University of Newcastle Australia
Australia

FREDERIKSEN, Carsten
Bureau of Meteorology
Australia

FREDERIKSEN, Jorgen
CSIRO Marine and Atmospheric Research
Australia

FREE, Melissa
National Oceanic and Atmospheric Administration, Air Resources Laboratory
USA

FREELAND, Howard
Fisheries and Oceans Canada
Canada

FREPPAZ, Michele
University of Torino
Italy

FRIEDLINGSTEIN, Pierre
University of Exeter
UK

FROELICHER, Thomas
Princeton University
USA

FRONZEK, Stefan
Finnish Environment Institute
Finland

FRÜH, Barbara
Deutscher Wetterdienst
Germany

FU, Joshua Xiouhua
University of Hawaii
USA

FU, Weiwei
Danish Meteorological Institute
Denmark

FUGLESTVEDT, Jan
Center for International Climate and Environmental Research Oslo
Norway

FUKASAWA, Masao
Japan Agency for Marine-Earth Science and Technology
Japan

FUNG, Inez
University of California Berkeley
USA

FUNK, Martin
ETH Zurich
Switzerland

FYFE, John
Environment Canada
Canada

GAALEMA, Stephen
Black Forest Engineering, LLC
USA

GAGLIARDINI, Olivier
Laboratoire de Glaciologie et Géophysique de l`Environnement, Université Joseph Fourier
France

GAJEWSKI, Konrad
University of Ottawa
Canada

GALDOS, Marcelo
Brazilian Bioethanol Science and Technology Laboratory
Brazil

GALLEGO, David
Universidad Pablo de Olavide
Spain

GANOPOLSKI, Andrey
Potsdam Institute for Climate Impact Research
Germany

GAO, Xuejie
National Climate Center, China Meteorological Administration
China

GARCIA-HERRERA, Ricardo
Universidad Complutense de Madrid
Spain

GARIMELLA, Sarvesh
Massachusetts Institute of Technology
USA

GARREAUD, René
Universidad de Chile
Chile

GATTUSO, Jean-Pierre
Observatoire Océanologique de Villefranche sur Mer, Université Pierre et Marie Curie
France

GAUCI, Vincent
The Open University
UK

GAYO, Eugenia M.
Centro de Investigaciones del Hombre en el Desierto
Chile

GEDNEY, Nicola
Met Office Hadley Centre
UK

GEHRELS, Roland
Plymouth University
UK

GERBER, Stefan
University of Florida
USA

GERLAND, Sebastian
Norwegian Polar Institute
Norway

GERVAIS, François
Université François-Rabelais de Tours
France

GETTELMAN, Andrew
National Center for Atmospheric Research
USA

GHAN, Steven
Pacific Northwest National Laboratory
USA

GHOSH, Sucharita
Swiss Federal Institute for Forest, Snow and Landscape Research WSL
Switzerland

GIFFORD, Roger
CSIRO Plant Industry
Australia

GILBERT, Denis
Fisheries and Oceans Canada
Canada

GILLETT, Nathan
Environment Canada
Canada

GINOUX, Paul
National Oceanic and Atmospheric Administration, Geophysical Fluid Dynamics Laboratory
USA

GIORGETTA, Marco
Max Planck Institute for Meteorology
Germany

GLIKSON, Andrew
Australian National University
Australia

GODIN-BEEKMANN, Sophie
Laboratoire Atmosphères, Milieux, Observations Spatiales, Institut Pierre Simon Laplace
France

GOLAZ, Jean-Christophe
National Oceanic and Atmospheric Administration, Geophysical Fluid Dynamics Laboratory
USA

GONG, Daoyi
Beijing Normal University
China

GONZALEZ, Patrick
U.S. National Park Service
USA

GOOD, Peter
Met Office Hadley Centre
UK

GOOD, Simon
Met Office Hadley Centre
UK

GOODESS, Clare
University of East Anglia
UK

GOOSSE, Hugues
Université catholique de Louvain
Belgium

GORIS, Nadine
University of Bergen and Bjerknes Centre for Climate Research
Norway

GOSWAMI, Santonu
Oak Ridge National Laboratory
USA

GOWER, James
Fisheries and Oceans Canada
Canada

GRAY, Vincent
New Zealand

GREGORY, Jonathan
University of Reading and Met Office Hadley Centre
UK

GREWE, Volker
DLR German Aerospace Center
Germany

GRIFFIES, Stephen
National Oceanic and Atmospheric Administration, Geophysical Fluid Dynamics Laboratory
USA

GRIGGS, David
Monash University
Australia

GRIMM, Alice
Federal University of Parana
Brazil

GRINSTED, Aslak
University of Copenhagen
Denmark

GRUBER, Nicolas
ETH Zurich
Switzerland

GRUBER, Stephan
University of Zurich
Switzerland

GUGLIELMIN, Mauro
University of Insubria
Italy

GUILYARDI, Eric
Laboratoire d'Océanographie et du Climat, Institut Pierre Simon Laplace
France

GUTTORP, Peter
University of Washington and Norwegian Computing Center
USA

GUTZLER, David
University of New Mexico
USA

HAARSMA, Reindert
Royal Netherlands Meteorological Institute
Netherlands

HAEBERLI, Wilfried
University of Zurich
Switzerland

HAFEZ, Yehia
King Abdulaziz University
Saudi Arabia

HAGEN, David L.
AcrossTech
USA

HAGOS, Samson
Pacific Northwest National Laboratory
USA

HAIGH, Joanna
Imperial College London
UK

HAJIMA, Tomohiro
Japan Agency for Marine-Earth Science and Technology
Japan

HALL, Dorothy
National Aeronautics and Space Administration, Goddard Space Flight Center
USA

HALLBERG, Robert
National Oceanic and Atmospheric Administration, Geophysical Fluid Dynamics Laboratory
USA

HALLORAN, Paul
Met Office Hadley Centre
UK

HAN, Dawei
University of Bristol
UK

HANSEN, Bogi
Faroe Marine Research Institute
Faroe Islands

HAO, Aibing
Ministry of Land and Resources
China

HARGREAVES, Julia
Japan Agency for Marine-Earth Science and Technology
Japan

HARNISCH, Jochen
KfW
Germany

HARPER, Joel
University of Montana
USA

HARTMANN, Jens
University of Hamburg
Germany

HASANEAN, Hosny
King Abdulaziz University
Saudi Arabia

HASSLER, Birgit
Cooperative Institute for Research in Environmental Sciences
USA

HAWKINS, Ed
University of Reading
UK

HAYASAKA, Tadahiro
Tohoku University
Japan

HAYWOOD, Jim
Met Office Hadley Centre and University of Exeter
UK

HEGERL, Gabriele
University of Edinburgh
UK

HEIM, Richard
National Oceanic and Atmospheric Administration, National Climatic Data Center
USA

HEINTZENBERG, Jost
Leibniz Institute for Tropospheric Research
Germany

HEINZE, Christoph
University of Bergen and Bjerknes Centre for Climate Research
Norway

HERTWICH, Edgar
Norwegian University of Science and Technology
Norway

HEWITSON, Bruce
University of Cape Town
South Africa

HIGGINS, Paul
American Meteorological Society
USA

HIRST, Anthony
CSIRO Marine and Atmospheric Research
Australia

HISDAL, Hege
Norwegian Water Resources and Energy Directorate
Norway

HOCK, Regine
University of Alaska Fairbanks
USA

HODSON, Dan
University of Reading
UK

HOERLING, Martin
National Oceanic and Atmospheric Administration, Earth System Research Laboratory
USA

HÖGBERG, Peter
Swedish University of Agricultural Sciences
Sweden

HOLGATE, Simon
National Oceanography Centre
UK

HOLLIS, Christopher
GNS Science
New Zealand

HOLTSLAG, Albert A.M.
Wageningen University
Netherlands

HÖNISCH, Bärbel
Columbia University
USA

HOPE, Pandora
Bureau of Meteorology
Australia

HOROWITZ, Larry
National Oceanic and Atmospheric Administration, Geophysical Fluid Dynamics Laboratory
USA

HOURDIN, Frédéric
Laboratoire de Météorologie Dynamique, Institut Pierre Simon Laplace
France

HOUSE, Joanna
University of Bristol
UK

HOUWELING, Sander
Utrecht University
Netherlands

HOVLAND, Martin
University of Bergen
Norway

HOWARD, William
Australian National University
Australia

HREN, Michael
University of Connecticut
USA

HU, Aixue
National Center for Atmospheric Research
USA

HU, Zeng-Zhen
National Oceanic and Atmospheric Administration, National Weather Service
USA

HUANG, Jianping
Lanzhou University
China

HUANG, Lei
National Climate Center, China Meteorological Administration
China

HUANG, Lin
Environment Canada
Canada

HUDSON, James
Desert Research Institute
USA

HUGGEL, Christian
University of Zurich
Switzerland

HUGHES, Malcolm
University of Arizona
USA

HUNTER, John
Antarctic Climate and Ecosystems Cooperative Research Centre
Australia

HURST, Dale
Cooperative Institute for Research in Environmental Sciences
USA

HUYBRECHTS, Philippe
Vrije Universiteit Brussel
Belgium

INCECIK, Selahattin
Istanbul Technical University
Turkey

INGRAM, William
Met Office Hadley Centre and University of Oxford
UK

INOUE, Toshiro
University of Tokyo
Japan

IRVINE, Peter
Institute for Advanced Sustainability Studies
Germany

ISE, Takeshi
University of Hyogo
Japan

ISHII, Masao
Meteorological Research Institute
Japan

ISHIZUKA, Shigehiro
Forestry and Forest Products Institute
Japan

ITO, Akihiko
National Institute for Environmental Studies
Japan

ITO, Takamitsu
Georgia Institute of Technology
USA

ITOH, Kiminori
Yokohama National University
Japan

IVERSEN, Trond
European Centre for Medium-Range Weather Forecasts, UK and Norwegian Meteorological Institute
Norway

JACKSON, Laura
Met Office Hadley Centre
UK

JACOBEIT, Jucundus
University of Augsburg
Germany

JACOBSON, Mark Z.
Stanford University
USA

JAENICKE, Ruprecht
Johannes Gutenberg University Mainz
Germany

JAIN, Sharad K.
Indian Institute of Technology Roorkee
India

JEONG, Myeong-Jae
Gangneung-Wonju National University
Republic of Korea

JIANG, Dabang
Institute of Atmospheric Physics,
Chinese Academy of Sciences
China

JIANG, Jonathan
National Aeronautics and Space
Administration, Jet Propulsion Laboratory
USA

JOHANSSON, Daniel
Chalmers University of Technology
Sweden

JOHN, Jasmin
National Oceanic and Atmospheric
Administration, Geophysical
Fluid Dynamics Laboratory
USA

JOHNS, Tim
Met Office Hadley Centre
UK

JOHNSON, Jennifer
Stanford University
USA

JOHNSON, Nathaniel
University of Hawaii
USA

JONES, Christopher
Met Office Hadley Centre
UK

JONES, Gareth S.
Met Office Hadley Centre
UK

JONES, Philip
University of East Anglia
UK

JOOS, Fortunat
University of Bern
Switzerland

JOSEY, Simon
National Oceanography Centre
UK

JOSHI, Manoj
University of East Anglia
UK

JOUGHIN, Ian
University of Washington
USA

JOUSSAUME, Sylvie
Laboratoire des Sciences du Climat et de
l'Environnement, Institut Pierre Simon Laplace
France

JOYCE, Terrence
Woods Hole Oceanographic Institution
USA

JUCKES, Martin
Science and Technologies Facility Council
UK

JYLHÄ, Kirsti
Finnish Meteorological Institute
Finland

KÄÄB, Andreas
University of Oslo
Norway

KAGEYAMA, Masa
Laboratoire des Sciences du
Climat et de l'Environnement,
Institut Pierre Simon Laplace
France

KAHN, Brian
National Aeronautics and Space
Administration, Jet Propulsion Laboratory
USA

KAHN, Ralph
National Aeronautics and Space
Administration, Goddard Space Flight Center
USA

KALESCHKE, Lars
University of Hamburg
Germany

KANAKIDOU, Maria
University of Crete
Greece

KANAYA, Yugo
Japan Agency for Marine-Earth
Science and Technology
Japan

KANDEL, Robert
Laboratoire de Météorologie Dynamique,
Institut Pierre Simon Laplace
France

KANG, Shichang
Institute of Tibetan Plateau Research,
Chinese Academy of Sciences
China

KANG, Sok Kuh
Korea Ocean Research and
Development Institute
Republic of Korea

KARLSSON, Per Erik
Swedish Environmental Research Institute
Sweden

KAROLY, David
University of Melbourne
Australia

KARPECHKO, Alexey
Finnish Meteorological Institute
Finland

KATBEH-BADER, Nedal
Ministry of Environment Affairs
Palestine

KATO, Etsushi
National Institute for Environmental Studies
Japan

KAUFMANN, Robert
Boston University
USA

KAUPPINEN, Jyrki
University of Turku
Finland

KAVANAGH, Christopher
International Atomic Energy Agency
Monaco

KAWAI, Hiroyasu
Port and Airport Research Institute
Japan

KAWAMIYA, Michio
Japan Agency for Marine-Earth Science
and Technology
Japan

KAWAMURA, Kenji
National Institute of Polar Research
Japan

KAYE, Neil
Met Office Hadley Centre
UK

KEELING, Ralph
Scripps Institution of Oceanography
USA

KEEN, Richard
University of Colorado Boulder (retired)
USA

KEENLYSIDE, Noel
University of Bergen and Bjerknes
Centre for Climate Research
Norway

KELLER, Charles
Los Alamos National Laboratory (retired)
USA

KENDON, Elizabeth
Met Office Hadley Centre
UK

KENNEDY, John
Met Office Hadley Centre
UK

KENT, Elizabeth
National Oceanography Centre
UK

KESKIN, Siddik Sinan
Marmara University
Turkey

KHALIL, Mohammad Aslam Khan
Portland State University
USA

KHESHGI, Haroon
ExxonMobil Research and Engineering
USA

KHMELINSKII, Igor
Universidade do Algarve
Portugal

KHOSRAWI, Farahnaz
Stockholm University
Sweden

KILADIS, George
National Oceanic and Atmospheric Administration, Earth System Research Laboratory
USA

KIM, Daehyun
Columbia University
USA

KIM, Seong-Joong
Korea Polar Research Institute
Republic of Korea

KINDLER, Pascal
University of Geneva
Switzerland

KING, Andrew
University of New South Wales
Australia

KING, Matt
University of Tasmania and Newcastle University
Australia

KINTER, James
Institute of Global Environment and Society, Inc.
USA

KIRCHENGAST, Gottfried
University of Graz
Austria

KIRKEVÅG, Alf
Norwegian Meteorological Institute
Norway

KITOH, Akio
Meteorological Research Institute
Japan

KJELLSTRÖM, Erik
Swedish Meteorological and Hydrological Institute
Sweden

KLEIN TANK, Albert
Royal Netherlands Meteorological Institute
Netherlands

KLINGER, Lee
USA

KLOTZBACH, Philip
Colorado State University
USA

KNUTSON, Thomas
National Oceanic and Atmospheric Administration, Geophysical Fluid Dynamics Laboratory
USA

KNUTTI, Reto
ETH Zurich
Switzerland

KOBASHI, Takuro
National Institute of Polar Research
Japan

KOBAYASHI, Shigeki
Toyota Central R&D Labs., Inc.
Japan

KOBAYASHI, Taiyo
Japan Agency for Marine-Earth Science and Technology
Japan

KOH, Tieh-Yong
Nanyang Technological University
Singapore

KÖHLER, Peter
Alfred Wegener Institute for Polar and Marine Research
Germany

KOMEN, Gerbrand
Royal Netherlands Meteorological Institute and Utrecht University (retired)
Netherlands

KONDO, Yutaka
University of Tokyo
Japan

KONFIRST, Matthew
American Association for the Advancement of Science and National Science Foundation
USA

KONOVALOV, Vladimir
Russian Academy of Sciences
Russian Federation

KOPP, Robert
Rutgers University
USA

KORTELAINEN, Pirkko
Finnish Environment Institute
Finland

KREASUWUN, Jiemjai
Chiang Mai University
Thailand

KREIENKAMP, Frank
Climate & Environment Consulting Potsdam GmbH
Germany

KRINNER, Gerhard
Laboratoire de Glaciologie et Géophysique de l'Environnement, Université Joseph Fourier
France

KRIPALANI, Ramesh
Indian Institute of Tropical Meteorology
India

KRISTJÁNSSON, Jón Egill
University of Oslo
Norway

KRIVOVA, Natalie
Max Planck Institute for Solar System Research
Germany

KUHN, Nikolaus J.
University of Basel
Switzerland

KULSHRESTHA, Umesh
Jawaharlal Nehru University
India

KUSANO, Kanya
Nagoya University
Japan

KUSUNOKI, Shoji
Meteorological Research Institute
Japan

LAGERLOEF, Gary
Earth & Space Research
USA

LAKEN, Benjamin
Instituto de Astrofisíca de Canarias
Spain

LAMBERT, Fabrice
Korea Institute of Ocean
Science and Technology
Republic of Korea

LAMBERT, Francis Hugo
University of Exeter
UK

LANDUYT, William
ExxonMobil Research and Engineering
USA

LANE, Tracy
International Hydropower Association
UK

LANG, Herbert
ETH Zurich
Switzerland

LARTER, Robert
British Antarctic Survey
UK

LAW, Beverly
Oregon State University
USA

LAW, Katharine
Laboratoire Atmosphères, Milieux,
Observations Spatiales, Institut
Pierre Simon Laplace
France

LAWRENCE, Judy
Victoria University of Wellington
New Zealand

LAWRENCE, Mark
Institute for Advanced Sustainability Studies
Germany

LAXON, Seymour
University College London
UK

LE QUÉRÉ, Corinne
University of East Anglia
UK

LEAITCH, Warren Richard
Environment Canada
Canada

LECK, Caroline
Stockholm University
Sweden

LECLERCQ, Paul
Utrecht University
Netherlands

LEE, Arthur
Chevron Corporation
USA

LEE, Jae Hak
Korea Institute of Ocean
Science and Technology
Republic of Korea

LEE, Sai Ming
Hong Kong Observatory
China

LEE, Seoung Soo
National Oceanic and Atmospheric
Administration, Earth System
Research Laboratory
USA

LEE, Tsz-Cheung
Hong Kong Observatory
China

LEGG, Sonya
Princeton University
USA

LEMKE, Peter
Alfred Wegener Institute for
Polar and Marine Research
Germany

LENAERTS, Jan
Utrecht University
Netherlands

LENDERINK, Geert
Royal Netherlands Meteorological Institute
Netherlands

LEROY, Suzanne
Brunel University
UK

LEVIN, Ingeborg
University of Heidelberg
Germany

LEVITUS, Sydney
National Oceanic and Atmospheric
Administration, National
Oceanographic Data Center
USA

LEVY, Julian
Levy Environmental Consulting, Ltd.
USA

LEVY II, Hiram
National Oceanic and Atmospheric
Administration, Geophysical Fluid
Dynamics Laboratory (retired)
USA

LEWIS, Nicholas
UK

LEWITT, Martin
American Geophysical Union
USA

LI, Can
University of Maryland and National
Aeronautics and Space Administration,
Goddard Space Flight Center
USA

LI, Jui-Lin (Frank)
National Aeronautics and Space
Administration, Jet Propulsion Laboratory
USA

LI, Qingxiang
National Meteorological Information Center,
China Meteorological Administration
China

LI, Shenggong
Institute of Geographic Sciences
and Natural Resources Research,
Chinese Academy of Sciences
China

LI, Shuanglin
Institute of Atmospheric Physics,
Chinese Academy of Sciences
China

LI, Weiping
National Climate Center, China
Meteorological Administration
China

LI, Weiwei
University of Illinois
USA

LI, Yueqing
Institute of Plateau Meteorology, China
Meteorological Administration
China

LI, Zhanqing
University of Maryland
USA

LIAO, Hong
Institute of Atmospheric Physics,
Chinese Academy of Sciences
China

LIN, Hai
Environment Canada
Canada

LIN, Jialin
Ohio State University
USA

LINDERHOLM, Hans W.
University of Gothenburg
Sweden

LINDSAY, Ron
University of Washington
USA

LIOU, Kuo-Nan
University of California Los Angeles
USA

LITTLE, Christopher
Princeton University
USA

LIU, Hongyan
Peking University
China

LIU, Ke Xiu
National Marine Data and
Information Service
China

LIU, Qiyong
China CDC
China

LIU, Shaw
Academia Sinica
Taiwan, China

LIU, Xiaohong
Pacific Northwest National Laboratory
USA

LJUNGQVIST, Fredrik
Stockholm University
Sweden

LLOYD, Philip
Cape Peninsula University of Technology
South Africa

LO, Yueh-Hsin
National Taiwan University
Taiwan, China

LOBELL, David
Stanford University
USA

LOEB, Norman
National Aeronautics and Space
Administration, Langley Research Center
USA

LOEW, Alexander
Max Planck Institute for Meteorology
Germany

LOFGREN, Brent
National Oceanic and Atmospheric
Administration, Great Lakes
Environmental Research Laboratory
USA

LOHMANN, Gerrit
Alfred Wegener Institute for
Polar and Marine Research
Germany

LOHMANN, Ulrike
ETH Zurich
Switzerland

LOOKYAT TAYLOR, Helen
World Science Data Base
USA

LÓPEZ MORENO, Juan Ignacio
Instituto Pirenaico de Ecología
Spain

LOUGH, Janice
Australian Institute of Marine Science
Australia

LUCE, Charles
U.S. Forest Service
USA

LÜTHI, Martin
ETH Zurich
Switzerland

LUNT, Daniel
University of Bristol
UK

LUO, Jing-Jia
Bureau of Meteorology
Australia

LUPO, Anthony
University of Missouri
USA

MA, Zhuguo
Institute of Atmospheric Physics,
Chinese Academy of Sciences
China

MACCRACKEN, Michael
Climate Institute
USA

MACGREGOR, Joseph
University of Texas
USA

MÄDER, Claudia
Federal Environment Agency
Germany

MAGGI, Valter
University of Milano-Bicocca
Italy

MAHLSTEIN, Irina
Federal Office of Meteorology and
Climatology MeteoSwiss
Switzerland

MAHOWALD, Natalie
Cornell University
USA

MAKI, Takashi
Meteorological Research Institute
Japan

MANN, Michael
Pennsylvania State University
USA

MANNING, Martin
Victoria University of Wellington
New Zealand

MANZINI, Elisa
Max Planck Institute for Meteorology
Germany

MARAUN, Douglas
GEOMAR Helmholtz Centre
for Ocean Research
Germany

MARBAIX, Philippe
Université catholique de Louvain
Belgium

MARENGO, José
National Institute for Space Research
Brazil

MARINOVA, Dora
Curtin University
Australia

MARIOTTI, Annarita
National Oceanic and Atmospheric
Administration, Climate Program Office
USA

MAROTZKE, Jochem
Max Planck Institute for Meteorology
Germany

MARSH, Robert
University of Southampton
UK

MARTIN, Eric
Météo-France
France

MARTIN, Gill
Met Office Hadley Centre
UK

MARTÍN MÍGUEZ, Belén
Centro Tecnológico del Mar
Spain

MARTIN-VIDE, Javier
Universitat de Barcelona
Spain

MARTY, Christoph
WSL Institute for Snow and
Avalanche Research SLF
Switzerland

MASSONNET, François
Université catholique de Louvain
Belgium

MATEI, Daniela
Max Planck Institute for Meteorology
Germany

MATSUNO, Taroh
Japan Agency for Marine-Earth
Science and Technology
Japan

MATSUOKA, Kenichi
Norwegian Polar Institute
Norway

MATTHEWS, Paul
University of Nottingham
UK

MAURITSEN, Thorsten
Max Planck Institute for Meteorology
Germany

MAY, Wilhelm
Danish Meteorological Institute
Denmark

MCELROY, Charles Thomas
York University
Canada

MCINNES, Kathleen
CSIRO Marine and Atmospheric Research
Australia

MCKAY, Nicholas
University of Arizona
USA

MCKITRICK, Ross
University of Guelph
Canada

MCLEAN, John
James Cook University
Australia

MEEHL, Gerald
National Center for Atmospheric Research
USA

MEIER, Walter
National Snow and Ice Data Center
USA

MEIYAPPAN, Prasanth
University of Illinois
USA

MELSOM, Arne
Norwegian Meteorological Institute
Norway

MÉNDEZ, Carlos
Instituto Venezolano de
Investigaciones Científicas
Venezuela

MENGE, Duncan
Princeton University
USA

MENZEL, W. Paul
University of Wisconsin
USA

MERCHANT, Christopher
University of Edinburgh
UK

MEREDITH, Michael
British Antarctic Survey
UK

MERLIS, Timothy
Princeton University and National
Oceanic and Atmospheric Administration,
Geophysical Fluid Dynamics Laboratory
USA

MERRYFIELD, William
Environment Canada
Canada

MESINGER, Fedor
University of Maryland
USA

METCALFE, Daniel
Swedish University of Agricultural Sciences
Sweden

METELKA, Ladislav
Czech Hydrometeorological Institute
Czech Republic

MEYSSIGNAC, Benoit
Laboratoire d'Etudes en Géophysique
et Océanographie Spatiales
France

MICKLEY, Loretta
Harvard University
USA

MIELIKÄINEN, Kari
Finnish Forest Research Institute
Finland

MILLER, Benjamin R.
Cooperative Institute for Research
in Environmental Sciences
USA

MIMS, Forrest
Geronimo Creek Observatory
USA

MIN, Seung-Ki
CSIRO Marine and Atmospheric Research
Australia

MING, Jing
National Climate Center, China
Meteorological Administration
China

MING, Yi
National Oceanic and Atmospheric
Administration, Geophysical
Fluid Dynamics Laboratory
USA

MINSCHWANER, Kenneth
New Mexico Institute of
Mining and Technology
USA

MITCHELL, John
Met Office Hadley Centre
UK

MOBERG, Anders
Stockholm University
Sweden

MÖHLER, Ottmar
Karlsruhe Institute of Technology
Germany

MOLINIÉ, Gilles
Laboratoire de Glaciologie et Géophysique
de l'Environnement, Université Joseph Fourier
France

MONAHAN, Adam
University of Victoria
Canada

MONCKTON OF BRENCHLEY, Christopher
Science and Public Policy Institute
UK

MONTZKA, Stephen
National Oceanic and Atmospheric Administration, Earth System Research Laboratory
USA

MOORTHY, K. Krishna
Indian Space Research Organisation
India

MOOSDORF, Nils
University of Hamburg
Germany

MORGENSTERN, Olaf
National Institute of Water and Atmospheric Research
New Zealand

MORI, Nobuhito
Kyoto University
Japan

MORICE, Colin
Met Office Hadley Centre
UK

MORRISON, Hugh
National Center for Atmospheric Research
USA

MOTE, Philip
Oregon State University
USA

MSADEK, Rym
National Oceanic and Atmospheric Administration, Geophysical Fluid Dynamics Laboratory
USA

MUDELSEE, Manfred
Alfred Wegener Institute for Polar and Marine Research
Germany

MUELLER, Christoph
Justus Leibig University Giessen
Germany

MÜLLER, Rolf
Forschungszentrum Jülich
Germany

MÜLLER, Wolfgang
Max Planck Institute for Meteorology
Germany

MULLER, Christian
Belgian Institute for Space Aeronomy
Belgium

MURATA, Akihiko
Japan Agency for Marine-Earth Science and Technology
Japan

MURPHY, Brad
Bureau of Meteorology
Australia

MURPHY, Daniel
National Oceanic and Atmospheric Administration, Earth System Research Laboratory
USA

MUSCHELER, Raimund
Lund University
Sweden

MUTHALAGU, Ravichandran
Indian National Centre for Ocean Information Services
India

MYHRE, Gunnar
Center for International Climate and Environmental Research Oslo
Norway

NABBEFELD, Birgit
DLR German Aerospace Center
Germany

NAIK, Vaishali
National Oceanic and Atmospheric Administration, Geophysical Fluid Dynamics Laboratory
USA

NAKAEGAWA, Tosiyuki
Meteorological Research Institute
Japan

NAKAJIMA, Teruyuki
University of Tokyo
Japan

NASSAR, Ray
Environment Canada
Canada

NAUELS, Alexander
IPCC WGI TSU, University of Bern
Switzerland

NEELIN, J. David
University of California Los Angeles
USA

NESJE, Atle
University of Bergen and Bjerknes Centre for Climate Research
Norway

NEU, Urs
Swiss Academy of Sciences
Switzerland

NEVISON, Cynthia
University of Colorado Boulder
USA

NEWBERY, David
University of Bern
Switzerland

NEWBURY, Thomas Dunning
Amercian Association for the Advancement of Science and U.S. Department of the Interior (retired)
USA

NICHOLLS, Robert
University of Southampton
UK

NITSCHE, Helga
Deutscher Wetterdienst
Germany

NODA, Akira
Japan Agency for Marine-Earth Science and Technology
Japan

OBBARD, Jeffrey
National University of Singapore
Singapore

OBROCHTA, Stephen
University of Tokyo
Japan

O'CONNOR, Fiona
Met Office Hadley Centre
UK

OGREN, John
National Oceanic and Atmospheric Administration, Earth System Research Laboratory
USA

OGURA, Tomoo
National Institute for Environmental Studies
Japan

OHBA, Masamichi
Central Research Institute of Electric Power Industry
Japan

OHMURA, Atsumu
ETH Zurich
Switzerland

OHNEISER, Christian
Shell International B.V.
Netherlands

O'ISHI, Ryouta
University of Tokyo
Japan

OLIVIÉ, Dirk
University of Oslo
Norway

OLIVIER, Jos
Netherlands Environmental Assessment Agency
Netherlands

OOUCHI, Kazuyoshi
Japan Agency for Marine-Earth Science and Technology
Japan

OPPENHEIMER, Michael
Princeton University
USA

OREOPOULOS, Lazaros
National Aeronautics and Space Administration, Goddard Space Flight Center
USA

ORLIC, Mirko
University of Zagreb
Croatia

ORLOWSKY, Boris
ETH Zurich
Switzerland

OSBORN, Timothy
University of East Anglia
UK

OSTROM, Nathaniel
Michigan State University
USA

OVERPECK, Jonathan
University of Arizona
USA

OWENS, John
3M Company
USA

PABÓN-CAICEDO, José Daniel
Universidad Nacional de Colombia
Colombia

PADMAN, Laurence
Earth & Space Research
USA

PALMER, Matthew
Met Office Hadley Centre
UK

PAN, Genxing
Nanjing Agricultural University
China

PANDEY, Dhananjai Kumar
National Centre for Antarctic and Ocean Research
India

PARKER, Albert
University of Ballarat
Australia

PARKER, David
Met Office Hadley Centre
UK

PARRISH, David
National Oceanic and Atmospheric Administration, Earth System Research Laboratory
USA

PASSCHIER, Sandra
Montclair State University
USA

PATTYN, Frank
Université libre de Bruxelles
Belgium

PAUL, Frank
University of Zurich
Switzerland

PAVAN, Valentina
Environmental Agency of Emilia-Romagna
Italy

PAVELSKY, Tamlin
University of North Carolina
USA

PAYNE, Antony
University of Bristol
UK

PAYNTER, David
National Oceanic and Atmospheric Administration, Geophysical Fluid Dynamics Laboratory
USA

PEARSON, David
Met Office Hadley Centre
UK

PEDERSEN, Jens Olaf Pepke
Technical University of Denmark
Denmark

PELEJERO, Carles
Institució Catalana de Recerca i Estudis Avançats and Institut de Ciències del Mar
Spain

PELLIKKA, Hilkka
Finnish Meteorological Institute
Finland

PERLWITZ, Judith
Cooperative Institute for Research in Environmental Sciences
USA

PEROVICH, Donald
Cold Regions Research and Engineering Laboratory
USA

PETERS, Glen
Center for International Climate and Environmental Research Oslo
Norway

PETERS, Karsten
Monash University
Australia

PETIT, Michel
Conseil général de l'Economie,de l'Industrie, de l'Energie et des Technologies
France

PFEFFER, W. Tad
University of Colorado Boulder
USA

PHILIPONA, Rolf
Federal Office of Meteorology and Climatology MeteoSwiss
Switzerland

PIACENTINI, Rubén D.
Universidad Nacional de Rosario
Argentina

PINCUS, Robert
University of Colorado Boulder
USA

PLANTON, Serge
Météo-France
France

PLATTNER, Gian-Kasper
IPCC WGI TSU, University of Bern
Switzerland

PLUMMER, David
Environment Canada
Canada

POERTNER, Hans
Alfred Wegener Institute for Polar and Marine Research
Germany

POHLMANN, Holger
Max Planck Institute for Meteorology
Germany

POITOU, Jean
Laboratoire des Sciences du Climat et de l'Environnement, Institut Pierre Simon Laplace and Société Française de Physique
France

POKHREL, Samir
Indian Institute of Tropical Meteorology
India

POLLACK, Henry
University of Michigan
USA

POLONSKY, Alexander
Marine Hydrophysical Institute
Ukraine

PONGRATZ, Julia
Max Planck Institute for Meteorology
Germany

PORTMANN, Robert
National Oceanic and Atmospheric Administration, Earth System Research Laboratory
USA

POULTER, Benjamin
Laboratoire des Sciences du Climat et de l'Environnement, Institut Pierre Simon Laplace
France

POWER, Scott
Bureau of Meteorology
Australia

PRATHER, Michael
University of California Irvine
USA

PRENTICE, Iain Colin
Macquarie University and Imperial College
Australia

PRINN, Ronald
Massachusetts Institute of Technology
USA

PUEYO, Salvador
Institut Català de Ciències del Clima
Spain

QIAO, Bing
China Waterborne Transport Research Institute
China

QUAAS, Johannes
University of Leipzig
Germany

QUINN, Patricia
National Oceanic and Atmospheric Administration, Pacific Marine Environmental Laboratory
USA

RABATEL, Antoine
Laboratoire de Glaciologie et Géophysique de l'Environnement, Université Joseph Fourier
France

RADIĆ, Valentina
University of British Columbia
Canada

RADUNSKY, Klaus
Umweltbundesamt
Austria

RAHAMAN, Hasibur
Indian National Centre for Ocean Information Services
India

RAHIMZADEH, Fatemeh
Islamic Republic of Iran Meteorological Organization
Iran

RAHMSTORF, Stefan
Potsdam Institute for Climate Impact Research
Germany

RAIBLE, Christoph
University of Bern
Switzerland

RÄISÄNEN, Jouni
University of Helsinki
Finland

RÄISÄNEN, Petri
Finnish Meteorological Institute
Finland

RAJEEVAN, Madhavan Nair
Government of India, Ministry of Earth Sciences
India

RAMASWAMY, Venkatachalam
National Oceanic and Atmospheric Administration, Geophysical Fluid Dynamics Laboratory
USA

RAMSTEIN, Gilles
Laboratoire des Sciences du Climat et de l'Environnement, Institut Pierre Simon Laplace
France

RANDALL, David
Colorado State University
USA

RAPER, Sarah
Manchester Metropolitan University
UK

RASCH, Philip
Pacific Northwest National Laboratory
USA

RAUPACH, Michael
CSIRO Marine and Atmospheric Research
Australia

RAVISHANKARA, A.R.
National Oceanic and Atmospheric Administration, Earth System Research Laboratory
USA

RAWLS, Alec
USA

RAYNER, Nick
Met Office Hadley Centre
UK

RAYNER, Peter
University of Melbourne
Australia

REAY, David
University of Edinburgh
UK

REIS, Stefan
Centre for Ecology & Hydrology
UK

REISINGER, Andy
New Zealand Agricultural GHG Research Centre
New Zealand

REISMAN, John P.
OSS Foundation
USA

REMEDIOS, John
University of Leicester
UK

REMER, Lorraine
National Aeronautics and Space Administration, Goddard Space Flight Center
USA

REN, Guoyu
National Climate Center, China Meteorological Administration
China

RENWICK, James
Victoria University of Wellington
New Zealand

REUTEN, Christian
RWDI AIR Inc.
Canada

RIAHI, Keywan
International Institute for Applied Systems Analysis
Austria

RIBES, Aurélien
Météo-France
France

RIDDICK, Stuart
Cornell University
USA

RIDLEY, Jeff
Met Office Hadley Centre
UK

RIGNOT, Eric
University of California Irvine
USA

RIGOR, Ignatius
University of Washington
USA

RINGEVAL, Bruno
Utrecht University
Netherlands

RITZ, Christoph
Swiss Academy of Sciences
Switzerland

RIVERA, Andrés
Centro de Estudios Científicos
Chile

ROBAA, S.M.
Cairo University
Egypt

ROBERTS, Chris
Met Office Hadley Centre
UK

ROBERTSON, Iain
Swansea University
UK

ROBOCK, Alan
Rutgers University
USA

ROBSON, Jonathan
University of Reading
UK

RODHE, Henning
Stockholm University
Sweden

ROGELJ, Joeri
ETH Zurich
Switzerland

ROHLING, Eelco Johan
National Oceanography Centre
UK

ROJAS, Maisa
Universidad de Chile
Chile

ROMANOVSKY, Vladimir
University of Alaska Fairbanks
USA

RONCHAIL, Josyane
Laboratoire d'Océanographie et du Climat, Institut Pierre Simon Laplace
France

ROSEN, Sergiu Dov
Israel Oceanographic and Limnological Research
Israel

ROSENFELD, Daniel
Hebrew University of Jerusalem
Israel

ROSENLOF, Karen
National Oceanic and Atmospheric Administration, Earth System Research Laboratory
USA

ROTSTAYN, Leon
CSIRO Marine and Atmospheric Research
Australia

ROTT, Helmut
University of Innsbruck
Austria

ROWELL, David
Met Office Hadley Centre
UK

ROY, Indrani
University of Exeter
UK

ROY, Shouraseni
University of Miami
USA

RUMMUKAINEN, Markku
Swedish Meteorological and Hydrological Institute and Lund University
Sweden

RUPP, David
Oregon State University
USA

RUSSELL, Andrew
Brunel University
UK

RUTI, Paolo Michele
Italian National Agency for New Technologies, Energy and Sustainable Economic Development
Italy

SAHU, Lokesh Kumar
Physical Research Laboratory
India

SAKAGUCHI, Koichi
University of Arizona
USA

SALAS Y MELIA, David
Météo-France
France

SALZMANN, Nadine
University of Zurich and University of Fribourg
Switzerland

SAMANTA, Arindam
Atmospheric and Environmental Research
USA

SANCHEZ GOÑI, Maria Fernanda
Université Bordeaux 1
France

SANDERSON, Benjamin
National Center for Atmospheric Research
USA

SANYAL, Swarnali
University of Illinois
USA

SAROFIM, Marcus
U.S. Environmental Protection Agency
USA

SATHEESH, S.K.
Indian Institute of Science
India

SATOH, Masaki
University of Tokyo
Japan

SAUCHYN, David
University of Regina
Canada

SAULO, Celeste
Universidad de Buenos Aires
Argentina

SAUNDERS, Roger
Met Office Hadley Centre
UK

SAUSEN, Robert
DLR German Aerospace Center
Germany

SAVOLAINEN, Ilkka
VTT Technical Research Centre of Finland
Finland

SCAIFE, Adam
Met Office Hadley Centre
UK

SCHMID, Beat
Pacific Northwest National Laboratory
USA

SCHMIDT, Gavin
National Aeronautics and
Space Administration, Goddard
Institute for Space Studies
USA

SCHNEEBELI, Martin
WSL Institute for Snow and
Avalanche Research SLF
Switzerland

SCHNEIDER, Johannes
Max Planck Institute for Chemistry
Germany

SCHOENWIESE, Christian-D.
Goethe University
Germany

SCHOLES, Robert
Council for Scientific and Industrial Research
South Africa

SCHRAMA, Ernst
Delft University of Technology
Netherlands

SCHULZ, Michael
Norwegian Meteorological Institute
Norway

SCHUMANN, Ulrich
DLR German Aerospace Center
Germany

SCHUSTER, Gregory
National Aeronautics and Space
Administration, Langley Research Center
USA

SCHUUR, Edward
University of Florida
USA

SCHWARTZ, Stephen E.
Brookhaven National Laboratory
USA

SCHWARZKOPF, M. Daniel
National Oceanic and Atmospheric
Administration, Geophysical
Fluid Dynamics Laboratory
USA

SCHWEIGER, Axel
University of Washington
USA

SEDLÁČEK, Jan
ETH Zurich
Switzerland

SEHAT KASHANI, Saviz
Islamic Azad University
Iran

SEIBERT, Petra
University of Vienna
Austria

SEIDEL, Dian
National Oceanic and Atmospheric
Administration, Air Resources Laboratory
USA

SELVARAJ, Kandasamy
Xiamen University
China

SEN, Omer L.
Istanbul Technical University
Turkey

SENEVIRATNE, Sonia
ETH Zurich
Switzerland

SENSOY, Serhat
Turkish State Meteorological Service
Turkey

SENTMAN, Lori
National Oceanic and Atmospheric
Administration, Geophysical
Fluid Dynamics Laboratory
USA

SERPIL, Yaäžan
Turkish State Meteorological Service
Turkey

SETH, Anji
University of Connecticut
USA

SEXTON, David
Met Office Hadley Centre
UK

SHAO, Andrew
University of Washington
USA

SHAO, XueMei
Institute of Geographic Sciences
and Natural Resources Research,
Chinese Academy of Sciences
China

SHELL, Karen
Oregon State University
USA

SHERWIN, Toby
Scottish Association for Marine Science
UK

SHERWOOD, Steven
University of New South Wales
Australia

SHEVLIAKOVA, Elena
Princeton University
USA

SHI, Zongbo
University of Birmingham
UK

SHIBATA, Kiyotaka
Meteorological Research Institute
Japan

SHINDELL, Drew
National Aeronautics and
Space Administration, Goddard
Institute for Space Studies
USA

SHINE, Keith
University of Reading
UK

SHIOGAMA, Hideo
National Institute for Environmental Studies
Japan

SHKOLNIK, Igor
Voeikov Main Geophysical Observatory
Russian Federation

SHMAKIN, Andrey
Russian Academy of Sciences
Russian Federation

SHUMAN, Bryan
University of Wyoming
USA

SICRE, Marie-Alexandrine
Laboratoire des Sciences du Climat et de
l'Environnement, Institut Pierre Simon Laplace
France

SIDDALL, Mark
University of Bristol
UK

SIEGLE, Eduardo
Universidade de São Paulo
Brazil

SIEVERING, Herman
National Oceanic and Atmospheric Administration, Earth System Research Laboratory and University of Colorado Boulder
USA

SILLMANN, Jana
Environment Canada
Canada

SIMMONDS, Ian
University of Melbourne
Australia

SIMMONS, Adrian
European Centre for Medium-Range Weather Forecasts
UK

SINGER, S. Fred
University of Virginia
USA

SLANGEN, Aimée
Utrecht University
Netherlands

SMEDSRUD, Lars Henrik
Bjerknes Centre for Climate Research
Norway

SMITH, Doug
Met Office Hadley Centre
UK

SMITH, Ian
CSIRO Marine and Atmospheric Research
Australia

SMITH, Leonard
London School of Economics and Political Science
UK

SMITH, Sharon
Natural Resources Canada
Canada

SMITH, Stephen
Committee on Climate Change
UK

SMITH, Stephen G.G.
UK

SMITH, Steven
Pacific Northwest National Laboratory
USA

SNIDERMAN, Kale
University of Melbourne
Australia

SOLOMINA, Olga
Russian Academy of Sciences
Russian Federation

SOLOMON, Susan
Massachusetts Institute of Technology
USA

SOMERVILLE, Richard
Scripps Institution of Oceanography
USA

SONG, Shaojie
Massachusetts Institute of Technology
USA

SPAHNI, Renato
University of Bern
Switzerland

SPARRENBOM, Charlotte
Lund University
Sweden

SPARROW, Michael
Scientific Committee on Antarctic Research
UK

SPORYSHEV, Petr
Voeikov Main Geophysical Observatory
Russian Federation

SRIKANTHAN, Ramachandran
Physical Research Laboratory
India

SRIVER, Ryan
University of Illinois
USA

SROKOSZ, Meric
National Oceanography Centre
UK

STAGER, Jay Curt
Paul Smith's College
USA

STAHLE, David
University of Arkansas
USA

STAINFORTH, David
London School of Economics and Political Science
UK

STEBLER, Oliver
ETH Zurich
Switzerland

STEIG, Eric
University of Washington
USA

STEINFELDT, Reiner
University of Bremen
Germany

STENDEL, Martin
Danish Meteorological Institute
Denmark

STEPEK, Andrew
Royal Netherlands Meteorological Institute
Netherlands

STEPHENS, Graeme
National Aeronautics and Space Administration, Jet Propulsion Laboratory
USA

STEPHENSON, David
University of Exeter
UK

STERL, Andreas
Royal Netherlands Meteorological Institute
Netherlands

STERN, Harry
University of Washington
USA

STEVENSON, David
University of Edinburgh
UK

STEWART, Ronald
University of Manitoba
Canada

STIER, Philip
University of Oxford
UK

STÖBER, Uwe
University of Bremen
Germany

STOCKDALE, Timothy
European Centre for Medium-Range Weather Forecasts
UK

STOCKER, Benjamin
University of Bern
Switzerland

STOCKER, Thomas F.
Co-Chair IPCC WGI, University of Bern
Switzerland

STONE, Dáithí
Lawrence Berkeley National Laboratory
USA

STONE, Reynold
University of the West Indies
Trinidad and Tobago

STOTT, Peter
Met Office Hadley Centre
UK

STOUFFER, Ronald
National Oceanic and Atmospheric
Administration, Geophysical
Fluid Dynamics Laboratory
USA

STOY, Paul
Montana State University
USA

STRAUSS, Benjamin
Climate Central
USA

STUBENRAUCH, Claudia
Laboratoire de Météorologie Dynamique,
Institut Pierre Simon Laplace
France

STUMM, Dorothea
International Centre for Integrated
Mountain Development
Nepal

SU, Hui
National Aeronautics and Space
Administration, Jet Propulsion Laboratory
USA

SUBRAMANIAN, Aneesh
Scripps Institution of Oceanography
USA

SUGI, Masato
Japan Agency for Marine-Earth
Science and Technology
Japan

SUGIYAMA, Masahiro
Central Research Institute of
Electric Power Industry
Japan

SUN, Jianqi
Institute of Atmospheric Physics,
Chinese Academy of Sciences
China

SUN, Junying
Chinese Academy of Meteorological Sciences,
China Meteorological Administration
China

SUNDQUIST, Eric
U.S. Geological Survey
USA

SUTTON, Rowan
University of Reading
UK

SVENSSON, Gunilla
Stockholm University
Sweden

SWEENEY, Conor
University College Dublin
Ireland

SWIETLICKI, Erik
Lund University
Sweden

SWINGEDOUW, Didier
Laboratoire des Sciences du
Climat et de l'Environnement,
Institut Pierre Simon Laplace
France

TACHIIRI, Kaoru
Japan Agency for Marine-Earth
Science and Technology
Japan

TAKAHASHI, Ken
Instituto Geofísico del Perú
Peru

TAKAHASHI, Kiyoshi
National Institute for Environmental Studies
Japan

TAKAYABU, Izuru
Meteorological Research Institute
Japan

TAKAYABU, Yukari
University of Tokyo
Japan

TAKEMURA, Toshihiko
Kyushu University
Japan

TALARICO, Franco
University of Siena
Italy

TALLAKSEN, Lena M.
University of Oslo
Norway

TAMISIEA, Mark
National Oceanography Centre
UK

TANAKA, Hiroshi
University of Tsukuba
Japan

TANAKA, Katsumasa
ETH Zurich
Switzerland

TANG, Qi
Cornell University
USA

TAPIADOR, Francisco J.
Universidad de Castilla-La Mancha
Spain

TARASOV, Lev
Memorial University of Newfoundland
Canada

TAYLOR, Jeffrey
National Ecological Observatory Network
USA

TELFORD, Richard
University of Bergen
Norway

TERRAY, Laurent
Centre Européen de Recherche et de
Formation Avancée en Calcul Scientifique
France

TETT, Simon
University of Edinburgh
UK

THIELEN, Dirk
Instituto Venezolano de
Investigaciones Científicas
Venezuela

THOMAS, Robert
SIGMA Space
USA

THOMASON, Larry
National Aeronautics and Space
Administration, Langley Research Center
USA

THOMPSON, Erica
London School of Economics
and Political Science
UK

THOMPSON, Rona
Norwegian Institute for Air Research
Norway

THORNE, Peter
National Oceanic and Atmospheric
Administration, National
Climatic Data Center
USA

TIAN, Jian
University of Illinois
USA

TIGNOR, Melinda
IPCC WGI TSU, University of Bern
Switzerland

TILYA, Faustine Fidelis
Tanzania Meteorological Agency
United Republic Of Tanzania

TITUS, James G.
U.S. Environmental Protection Agency
USA

TKALICH, Pavel
National University of Singapore
Singapore

TOKINAGA, Hiroki
University of Hawaii
USA

TOMASEK, Bradley
University of Illinois
USA

TOMOZEIU, Rodica
Environmental Agency of Emilia-Romagna
Italy

TONITTO, Christina
Cornell University
USA

TOTTERDELL, Ian
Met Office Hadley Centre
UK

TRAINER, Michael
National Oceanic and Atmospheric Administration, Earth System Research Laboratory
USA

TRANVIK, Lars
Uppsala University
Sweden

TRENBERTH, Kevin
National Center for Atmospheric Research
USA

TROUET, Valerie
University of Arizona
USA

TSUSHIMA, Yoko
Met Office Hadley Centre
UK

TSUTSUI, Junichi
Central Research Institute of Electric Power Industry
Japan

TURCQ, Bruno
Institut de Recherche pour le Développement
France

TURNER, Andrew
University of Reading
UK

TZEDAKIS, Chronis
University College London
UK

UNNINAYAR, Sushel
National Aeronautics and Space Administration, Goddard Space Flight Center
USA

URREGO, Dunia H.
Université Bordeaux 1
France

VAN DEN HURK, Bart
Royal Netherlands Meteorological Institute
Netherlands

VAN DER LINDEN, Paul
Met Office Hadley Centre
UK

VAN DER WERF, Guido
VU University Amsterdam
Netherlands

VAN HUISSTEDEN, Ko
VU University Amsterdam
Netherlands

VAN KESTEREN, Line
IPCC Synthesis Report TSU
Netherlands

VAN NOIJE, Twan
Royal Netherlands Meteorological Institute
Netherlands

VAN OLDENBORGH, Geert Jan
Royal Netherlands Meteorological Institute
Netherlands

VAN OMMEN, Tasman
Australian Antarctic Division
Australia

VAN VELTHOVEN, Peter
Royal Netherlands Meteorological Institute
Netherlands

VAN WEELE, Michiel
Royal Netherlands Meteorological Institute
Netherlands

VAN YPERSELE, Jean-Pascal
Université catholique de Louvain
Belgium

VANAGS, Andrejs
The Space Exploration Society
USA

VAQUERO, José Manuel
Universidad de Extremadura
Spain

VAUGHAN, David
British Antarctic Survey
UK

VAUGHAN, Naomi
University of East Anglia
UK

VELDERS, Guus
National Institute for Public Health and the Environment
Netherlands

VERHEGGEN, Bart
ECN Energy Research Institute of the Netherlands
Netherlands

VERHOEF, Anne
University of Reading
UK

VERLEYEN, Elie
Ghent University
Belgium

VIDAL, Jean-Philippe
Institut National de Recherche en Sciences et Technologies pour l'Environnement et l'Agriculture
France

VIGNATI, Elisabetta
European Commission Joint Research Centre
Italy

VINITNANTHARAT, Soydoa
King Mongkut's University of Technology Thonburi
Thailand

VISSER, Hans
PBL Netherlands Environmental Assessment Agency
Netherlands

VOIGT, Thomas
Federal Environment Agency
Germany

VOLLMER, Martin
Swiss Federal Laboratories for Materials Science and Technology EMPA
Switzerland

VON SCHUCKMANN, Karina
Institut Français de Recherche pour l'Exploitation de la Mer
France

WAGNON, Patrick
Laboratoire de Glaciologie et Géophysique de l'Environnement, Université Joseph Fourier
France

WAHL, Eugene
National Oceanic and Atmospheric Administration, National Climatic Data Center
USA

WAHL, Terje
Norwegian Space Centre
Norway

WAHL, Thomas
University of Siegen
Germany

WALISER, Duane
National Aeronautics and Space Administration, Jet Propulsion Laboratory
USA

WALLINGTON, Timothy
Ford Motor Company
USA

WALTER, Andreas
Deutscher Wetterdienst
Germany

WANG, Bin
Institute of Atmospheric Physics, Chinese Academy of Sciences and Tsinghua University
China

WANG, Chien
Massachusetts Institute of Technology
USA

WANG, Dongxiao
South China Sea Institute of Oceanology, Chinese Academy of Sciences
China

WANG, Hailong
Pacific Northwest National Laboratory
USA

WANG, Junye
Rothamsted Research
UK

WANG, Kaicun
Beijing Normal University
China

WANG, Minghuai
Pacific Northwest National Laboratory
USA

WANG, Pinxian
Tongji University
China

WANG, Shaowu
Peking University
China

WANG, Tijian
Nanjing University
China

WANG, Ting
Lehigh University
USA

WANG, Xiaolan
Environment Canada
Canada

WANG, Xuemei
Sun Yat-sen University
China

WANG, Yingping
CSIRO Marine and Atmospheric Research
Australia

WANG, Yongguang
National Climate Center, China Meteorological Administration
China

WANG, Zhaomin
Nanjing University of Information Science and Technology
China

WANLISS, James
Presbyterian College
USA

WANNER, Heinz
University of Bern
Switzerland

WATERLAND, Robert
E. I. du Pont de Nemours & Co. Inc.
USA

WATSON, Phil
NSW Government Office of Environment and Heritage
Australia

WATSON, Thomas
Australia

WATTERSON, Ian
CSIRO Marine and Atmospheric Research
Australia

WEBB, David
National Oceanography Centre
UK

WEBB, Mark
Met Office Hadley Centre
UK

WEBB, Robert
National Oceanic and Atmospheric Administration, Earth System Research Laboratory
USA

WEEDON, Graham
Met Office Hadley Centre
UK

WEISHEIMER, Antje
European Centre for Medium-Range Weather Forecasts
UK

WEISS, Jérôme
Laboratoire de Glaciologie et Géophysique de l'Environnement, Université Joseph Fourier
France

WEISSE, Ralf
Helmholtz-Zentrum Geesthacht
Germany

WENDISCH, Manfred
University of Leipzig
Germany

WESTRA, Seth
University of Adelaide
Australia

WEYHENMEYER, Gesa
Uppsala University
Sweden

WHETTON, Penny
CSIRO Marine and Atmospheric Research
Australia

WHITE, Neil
CSIRO Marine and Atmospheric Research
Australia

WIELICKI, Bruce
National Aeronautics and Space Administration, Langley Research Center
USA

WILD, Oliver
Lancaster University
UK

WILLETT, Kate
Met Office Hadley Centre
UK

WILLIAMS, Keith
Met Office Hadley Centre
UK

WILLIAMS, Paul
University of Reading
UK

WILLIAMS, Richard G.
Liverpool University
UK

WILLIAMS, S. Jeffress
U.S. Geological Survey
USA

WILSON, Rob
University of St Andrews
UK

WITTENBERG, Andrew
National Oceanic and Atmospheric Administration, Geophysical Fluid Dynamics Laboratory
USA

WOLFF, Eric
British Antarctic Survey
UK

WOOD, Richard
Met Office Hadley Centre
UK

WOOD, Robert
University of Washington
USA

WOODS, Thomas
University of Colorado Boulder
USA

WOODWORTH, Philip
National Oceanography Centre
UK

WORDEN, Helen
National Center for Atmospheric Research
USA

WRATT, David
National Institute of Water and Atmospheric Research
New Zealand

WU, Tonghua
Cold and Arid Regions Environmental and Engineering Research Institute, Chinese Academy of Sciences
China

WURZLER, Sabine
Landesamt für Natur, Umwelt und Verbraucherschutz NRW
Germany

XIA, Chaozong
State Forestry Administration
China

XIA, Yu
IPCC WGI TSU, University of Bern
Switzerland

XIE, Shang-Ping
Scripps Institution of Oceanography
USA

XU, Chong-Yu
University of Oslo
Norway

XU, Kuan-Man
National Aeronautics and Space Administration, Langley Research Center
USA

XU, Xiaobin
Chinese Academy of Meteorological Sciences, China Meteorological Administration
China

XU, Ying
National Climate Center, China Meteorological Administration
China

XU, Yongfu
Institute of Atmospheric Physics, Chinese Academy of Sciences
China

YABI, Ibouraïma
Université d'Abomey Calavi
Benin

YASUNARI, Tetsuzo
Nagoya University
Japan

YDE, Jacob Clement
Sogn og Fjordane University College
Norway

YOKOUCHI, Yoko
National Institute for Environmental Studies
Japan

YOSHIMORI, Masakazu
University of Tokyo
Japan

YU, Rucong
China Meteorological Administration
China

YU, Zicheng
Lehigh University
USA

YUKIMOTO, Seiji
Meteorological Research Institute
Japan

ZAEHLE, Sönke
Max Planck Institute for Biogeochemistry
Germany

ZAHN, Matthias
University of Reading
UK

ZAPPA, Giuseppe
University of Reading
UK

ZEMP, Michael
University of Zurich
Switzerland

ZENG, Xubin
University of Arizona
USA

ZHANG, Chengyi
National Climate Center, China Meteorological Administration
China

ZHANG, De-er
National Climate Center, China Meteorological Administration
China

ZHANG, Gan
University of Illinois
USA

ZHANG, Guang
Scripps Institution of Oceanography
USA

ZHANG, Hua
National Climate Center, China Meteorological Administration
China

ZHANG, Rong
National Oceanic and Atmospheric Administration, Geophysical Fluid Dynamics Laboratory
USA

ZHANG, Tianyu
National Marine Environmental Forecasting Center
China

ZHANG, Xiangdong
University of Alaska Fairbanks
USA

ZHANG, Xuebin
CSIRO Marine and Atmospheric Research
Australia

ZHANG, Xuebin
Environment Canada
Canada

ZHAO, Xuepeng (Tom)
National Oceanic and Atmospheric Administration, National Climatic Data Center
USA

ZHAO, Zong-Ci
National Climate Center, China Meteorological Administration
China

ZHENG, Jingyun
Institute of Geographic Sciences and Natural Resources Research, Chinese Academy of Sciences
China

ZHOU, Guangsheng
Chinese Academy of Meteorological Sciences, China Meteorological Administration
China

ZHOU, Limin
East China Normal University
China

ZHOU, Tianjun
Institute of Atmospheric Physics, Chinese Academy of Sciences
China

ZHU, Bin
Nanjing University of Information Science and Technology
China

ZICKFELD, Kirsten
Simon Fraser University
Canada

ZORITA, Eduardo
Helmholtz-Zentrum Geesthacht
Germany

ZUIDEMA, Paquita
University of Miami
USA

ZUO, Juncheng
HoHai University
China

ZWEIFEL, Roman
Swiss Federal Institute for Forest, Snow and Landscape Research WSL
Switzerland

ZWIERS, Francis
University of Victoria
Canada

Index

This index should be cited as:
IPCC, 2013: Index. In: *Climate Change 2013: The Physical Science Basis. Contribution of Working Group I to the Fifth Assessment Report of the Intergovernmental Panel on Climate Change* [Stocker, T.F., D. Qin, G.-K. Plattner, M. Tignor, S.K. Allen, J. Boschung, A. Nauels, Y. Xia, V. Bex and P.M. Midgley (eds.)]. Cambridge University Press, Cambridge, United Kingdom and New York, NY, USA.

*Note: * indicates the term also appears in the Glossary (Annex III). Bold page numbers indicate page spans for entire chapters. Italicized page numbers denote tables, figures and boxed material.*

A

Index

D

Index

G

H

I

K

L

M

N

Q

R

Index

S

U

V

W